Mosquitoes of the World

Mosquitoes of the World

VOLUME 1

Richard C. Wilkerson, Yvonne-Marie Linton, and Daniel Strickman

Johns Hopkins University Press
BALTIMORE

Printed in Hong Kong on acid-free paper
9 8 7 6 5 4 3 2 1

Johns Hopkins University Press
2715 North Charles Street
Baltimore, Maryland 21218-4363
www.press.jhu.edu

Library of Congress Cataloging-in-Publication Data

Names: Wilkerson, Richard C., 1946– author. | Linton, Yvonne-Marie, 1972– author. | Strickman, Daniel, author.
Title: Mosquitoes of the world / Richard C. Wilkerson, Yvonne-Marie Linton, and Daniel Strickman.
Description: Baltimore, Maryland : Johns Hopkins University Press, 2020. | Includes bibliographical references and index.
Identifiers: LCCN 2019041641 | ISBN 9781421438146 (hardcover) | ISBN 9781421438153 (ebook)
Subjects: LCSH: Mosquitoes.
Classification: LCC QL536 .S798 2020 | DDC 595.77/2—dc23
LC record available at https://lccn.loc.gov/2019041641

A catalog record for this book is available from the British Library.

Special discounts are available for bulk purchases of this book. For more information, please contact Special Sales at specialsales@press.jhu.edu.

Johns Hopkins University Press uses environmentally friendly book materials, including recycled text paper that is composed of at least 30 percent post-consumer waste, whenever possible.

To my wife, Roxie Ann, with love, for her understanding, tolerance, and support, especially during preparation of this book

Richard C. Wilkerson

For my ever-patient husband, Luke, and our resilient little warriors—Messiah, Serenity, and Pierre. I chose each and every one of you, and love you completely

Yvonne-Marie Linton

With gratitude to the thousands of scientists who came before, sympathy for the millions who have died because of mosquito bites, and love for my spouse and children, who supported me during a career's worth of absences

Daniel Strickman

Contents

Preface

Within recent years, so much has been written about mosquitoes, and scattered through such a vast number of books and periodicals, that the average student now sorely feels the need of a single work, containing, in a condensed form, the essential facts so far made known in regard to the different phases of this important and highly interesting subject. It was to meet the demands of the constantly increasing army of both professional and lay workers that the present volume was designed.

Mitchell and Dupree 1907

Mitchell and Dupree would be amazed by the progress in knowledge and technology achieved since they wrote these prescient words in their introduction to *Mosquito Life* over a century ago; and today our rationale for writing *Mosquitoes of the World* is exactly the same—to facilitate the future fight against mosquitoes through a holistic understanding of their diversity, systematics, and associated biology. In many ways, this book is a celebration of the mosquito as a fascinating product of evolution. Of course, there can be no doubt that certain mosquitoes are responsible for much human suffering. Almost half a million people die from malaria and millions more suffer dengue annually, and unanticipated outbreaks of diseases like chikungunya and Zika cause fear and pain. Partnerships between science, industry, and public health agencies relentlessly strive to decrease the burden of mosquito-transmitted diseases, and we hope never to return to the days when yellow fever and malaria were major demographic forces in human ecology. The authors have spent the past eight years preparing this extensive summary of the biology, taxonomy, systematics, and global distribution of mosquitoes. Reviewing every known species certainly gives one an appreciation for the volume of scientific works that can occur during a couple of centuries, and especially for the variety of mosquitoes currently living on our planet. We were also impressed with the complexity of mosquito biology, sometimes feeling as though we were wandering through a national park full of wonders.

Volume 1 is divided in two distinct sections. Part 1 comprises a thorough walk-through of mosquito biology over 10 chapters, starting with crosscutting subjects including evolution, nomenclature, and distribution. The biology section continues through the mosquito life cycle, with focus on development, dormancy, movement, feeding, excretion, mating, and oviposition. Mosquito control was not tackled for a number of reasons, not least the availability of many other reviews, and the rapid pace of change in this field.

The second part of Volume 1 comprises a series of illustrated pages summarizing the diagnostic morphological characters, systematics, distributions, bionomics, associated pathogens and exemplar DNA sequences for each of the 41 recognized mosquito genera, and 128 carefully chosen, globally important species. These genera and species pages include published illustrations of larval characters and male genitalia, and high-resolution photographs of adult female characters. In addition to showcasing mosquito biodiversity, we intend that these pages will furnish researchers with practical morphological and molecular tools to aid future identification of these important taxa. Chapter 12 presents as an extensive Glossary of all taxonomic terms used in the second part of Volume 1.

Volume 2 is presented as a stand-alone taxonomic catalog of all described mosquito species, including fossils and synonyms. This volume comprises the first printed catalog since Knight and Stone published their seminal volume in 1977. The progression of taxonomic knowledge in the 20th century was impressive, fueled by data from newly discovered species. While Edwards (1932) listed 1,556 species, Stone, Knight, and Starcke treated 2,609 species and subspecies, and Knight and Stone reported 3,133. Our catalog (Volume 2) documents 3,700 taxa comprising 3,570 valid species and 130 subspecies, representing a further increase of 16%. Given the current global taxonomic impediment, we felt compelled to produce this printed catalog to fix and stabilize mosquito taxonomy both for the convenience of the user and, but most importantly, to serve as a baseline for future discoveries as they occur. This updated catalog provides a systematic listing of genus, subgenus, species, subspecies, and current synonyms for all valid Culicid taxa (including fossils and *nomina dubia*), fully referenced with important studies on the taxonomy, systematics, bionomics, and country-level distributions for each entity. Significant efforts were undertaken to include the etymology of each valid scientific name, and these are included in the catalog, along with a proposed informal name, for the first time. Informal species and subspecies names were proposed based on a formula of genus common names combined with type locality and a species characteristic. It is hoped that informal names will stabilize nomenclature, involve a wider audience in the biology of species that do not affect health, and stimulate productive discussion of the names themselves.

We are all thankful to the US Department of Defense Medical Research and Materiel Command, the US Army, the Walter Reed Army Institute of Research (WRAIR), and the Smithsonian Institution National Museum of Natural History (SI-NMNH) for their steadfast support of mosquito taxonomy and research. The Walter Reed Biosystematics Unit (WRBU) is housed within the SI-NMNH, where it has conducted mosquito research and curated the US National Mosquito Collection for almost 60 years, funded monetarily and in kind by both institutions. A long succession of dedicated civilian and military scientists has contributed to this patient mission, training taxonomists, advancing the field of mosquito biology, and accumulating this important reference collection to its 1.7 million specimens today. All three authors have had the privilege of leading the WRBU. Many of the photographs, illustrations, current distributions, and the taxonomic background for this book were generated from materials and products developed and maintained by the WRBU under their stewardship. We are grateful for the institutional support of WRBU, WRAIR, SI-NMNH, the US Department of Agriculture's Agricultural Research Service (USDA), and the Bill & Melinda Gates Foundation (BMGF), and the financial support of the US Army for this publication. Opinions expressed herein are solely those of the authors, and do not in any way reflect the opinions of the US Department of Defense or the US Department of the Army.

The authors have over 100 years of collective experience with mosquitoes. During this time, we have enjoyed instruction, debate, encouragement, and inspirational interactions with a long list of great mentors, colleagues and friends, including John Belkin, Mustapha Debboun, Graham Bell Fairchild, Dina Fonseca, Thomas Gaffigan, Ralph Harbach, Bruce Harrison, William Horsfall, Yiau-Min Huang, Rudy Meiswinkel, Robert Metcalf, James Pecor, EL Peyton, Ronald Rosenberg, Leopoldo Rueda, Jeff Ryan, Maria Anice Sallum, Harry Savage, Cathy Walton, Ronald Ward, and Graham White, to name but a few.

As with other large projects, there are many contributors whose efforts significantly impacted the quality of the final product. We are indebted to our skillful illustrators, Taina Litwak and Judith Stoffer, who patiently went through many iterations of the illustrations and plates, tirelessly rephotographing, redrawing, and relabeling images to realize our vision. We thank Lawrence Reeves, mosquito ecologist and nature photographer, for kindly providing the excellent mosquito images for the book covers. Vincent Burke has been a patient supporter from beginning to finish, and he deserves much credit. Rose Strickman was an exemplar of information management, doing much of the research for the review of biology, keeping references for that section straight, and drafting informal names for all the mosquitoes. Members of the WRBU team, including David Pecor and Luke Mitchell, and a steady stream of interns, were stellar in updating species distributions, sourcing references, wrangling huge data sets, and producing species distribution maps. We are beholden to them all.

Finally, and most sincerely, we thank each of our families, who put up with long absences and distracted spouses until the job was done.

PART ONE

Biology of Mosquitoes

1. Evolution

Man lives in many kinds of ecological communities and in each one of these communities man and a particular assemblage of mosquito species form an intimately related part of the ecosystem.

Ross 1964

Mosquitoes have an evolutionary history in reality and a historical evolution in science. This chapter begins with an account of the science that has led to current conclusions on how mosquitoes came to be. It then examines some of the ways in which the mechanisms of evolution affect the family. Phylogenetic studies of mosquitoes are reviewed next, with an emphasis on what those studies imply about their history. The next section discusses how current distribution of mosquitoes has biogeographical implications for our understanding of evolution. The chapter concludes with a review of fossil mosquitoes, which, though scarce, provide a stimulating view of this ancient family of insects, the Culicidae.

Historical Outlines of Evolutionary Theory

It was not until the 18th century that the body of scientific opinion began to break away from the Aristotelian (Mayr 1969) concept of static species to one that saw flora and fauna as a dynamic assembly of organisms that change over time. An appreciation that fossils represented life forms no longer in existence began as early as the 17th century with the work of Robert Hooke (Drake 1996) and Nils Steensen, or Steno (Kermit 2002), but William Smith's descriptions and explanations of geological stratigraphy in 1793 (Press & Siever 1997) must have made a great change in how people viewed the earth's history. Smith's idea was that layers of sedimentary rock were laid down in a sequence that could be compared from place to place by the kinds of fossils they contained. Therefore, patterns in one area that matched the patterns in another area were the result of the same ancient geological processes. Imparting a realistic sense of time to Earth's history was a departure from previous thinking and led directly to systematic examination of fossils, increasing the pressure to produce explanations for them.

Geological science continued to build on these concepts, erecting the systems of dating based on the fossil record that we use to this day (Walker et al. 2013). The system of geological epochs is based on the concept of uniformitarianism developed by Charles Lyell from 1830 to 1833 (McPhee 1998), which was the idea that geological structures, particularly sedimentary formations, were formed by processes no different from those observed now. The parallel between this well-developed geological concept and evolutionary theory was at a functional and conceptual level—functional because uniformitarianism was a necessary assumption before the systematic examination of the fossil record could produce the organization of geological epochs. Those epochs were determined by occurrence of particular fossils in particular strata, with the definition of each epoch based on dominance of certain organisms. The system begun by Charles Lyell was well developed by the end of the 19th century, right in the midst of early discussions of how life on Earth originated. Uniformitarianism had a conceptual parallel to evolutionary theory because it invoked natural law rather than deity to explain the formation of observed structures. Instead, uniformitarianism, like evolutionary theory, applies the principles that are currently observable and testable to the past, when they could not have been directly observed. Although the idea that physical processes have been the same throughout Earth's history may seem obvious, it was a sharp departure from European attitudes about the world in the 17th and early 18th centuries (Singer 1950).

Carl von Linné (Linnaeus) is widely credited with the establishment of binomial nomenclature, enshrined by the International Code of Zoological Nomenclature by recognition of the 10th edition of *Systema Naturae* as the first legitimate source of species names and January 1, 1758, as the starting point of recognized names (Mayr 1969). Linnaeus was a dedicated biologist with his own ideas about the origins of life, but a large part of his motivation for systematizing nomenclature was practical. Advances in technology and increases in the sheer volume of transportation across oceans had created a market for exotic plants in Europe, whether for medical, nutritional, or ornamental purposes (Wulf 2009). With utter lack of concern for invasive species, plants were moved across the globe, and it was necessary to be able to identify what kind of plant was being requested. The potential number of species was as vast then as now so that accurate designation was a challenge.[1] Linnaeus' system of binomial nomenclature was much more convenient and potentially more accurate than common names or the short descriptions used previously. Mosquitoes were part of this history, both because of a vigorous trade in cinchona bark for prevention and treatment of malaria and because of the two species of mosquitoes, *Aedes aegypti* and *Culex pipiens*, that Linnaeus described.

Although the biological theorists prior to Charles Darwin and Alfred Russel Wallace can be viewed as misguided, compromised by religious conviction, or missing the truth in one way or another, their contributions to the discussion created a framework for modern biology (Singer 1950). Prominent names include Georges-Louis Leclerc, the Comte de Buffon, who broke away from biblical explanations and discussed how spontaneously generated species could change over time. Lamarck (Jean-Baptiste Pierre Antoine de Monet, Chevalier de Lamarck) created a comprehensive theory of evolution but based it on individual development of heritable traits, an idea he shared with Charles Darwin's grandfather Erasmus Darwin. Georges Cuvier emphasized the importance of extinctions and made major contributions to paleontology, but he gave prominence to the concept that species do not change and that new species do not develop. Although such ideas are bizarre in a modern context, they stimulated others to make significant advances, like Louis Agassiz's work on the importance of glaciation to biogeography.

Today we look on the work of Darwin and Wallace as the great turning point in the lively 19th-century debate on the development of life. They knew not only how to systematize and translate large numbers of observations and scholarship into theory but also how to communicate their ideas through publications, letters, talks, and personal networks. Although we know much more about biological evolution, particularly its mechanisms, now than in the mid 1800s, the outline described back then still seems to be generally applicable. In the briefest summary, Darwin and Wallace each proposed that the current world of plants and animals developed gradually as a result of the very characteristics that distinguish biological organisms from the inanimate. They saw the importance of individual organisms in the context of communities. The four-step process leading to development of new forms was described as overproduction, struggle, selection, and inheritance. Expressed this way, each element implies a value judgment, from a human standpoint. What may be seen as a major transgression against the principle of objectivity probably made abundant sense in the 19th century when the designation of relative value was integrated into most decisions (Wimsatt & Schank 1988). To be fair to our predecessors, the current debate over the concept of progress in evolution shows that the original implication of values was not entirely irrelevant.

Mechanisms of Evolution from a Culicid Perspective

Potential Reproduction

Overpopulation implies that a female mosquito is somehow self-indulgent or wasteful in producing the usual complement of hundreds of eggs. This is a very human perspective on a process of reproduction in which the many offspring that do not live to maturity represent a bank of genetic features perhaps not currently expressed in the population, as well as a source of rapid population expansion should conditions allow. Therefore, the existence of a stable mosquito population necessarily means that on average, over long periods, each female produces only one reproducing female and one male. Such stability implies that hundreds of potential offspring per female do not reproduce because of the many sources of mortality throughout the life cycle. For most populations of mosquitoes, the majority of females probably produce no reproducing offspring, whereas certain females produce a large number. Over a short time frame on the order of months or years, the size of a population of a species may change in a given geographic area and expand or contract in spatial extent but tends to return to some sort of average for that particular environment. Perhaps rather than "overproduction," a phrase like "potential reproduction" or "dynamic reproduction" would be more accurate and less emotionally charged.

The intrinsic rate of population increase, *r*, is relatively high in the conspicuous species of mosquitoes that commonly interact with humans. In general, *r* reproductive strategists have a short generation time relative to the duration of the microhabitat in which they develop. Geological periods of rapid speciation by insects have been characterized by rapid changes in climate, an observation taken as support that *r* strategists are more likely to speciate as they generate larger numbers of individuals with lower survival exposed to a variety of environmental conditions (Southwood 1976). This concept also applies to the adults of some floodwater species. Perhaps the clearest examples are the salt marsh mosquitoes that include *Aedes taeniorhynchus* (Ritchie & Montague 1995), *Aedes sollicitans* (Zimmerman & Turner 1982), and *Aedes squamiger* (Bohart & Washino 1978) in North America; *Aedes caspius* in Europe

[1] Unfortunately, the number of species was greater—approximately 31 species of plants have gone extinct since the time of Linnaeus. International Union for Conservation of Nature, *IUCN Red List of Threatened Species*, Version 2011.2, 2011, accessed June 1, 2012, www.iucnredlist.org.

(Becker et al. 2003); and *Aedes vigilax* (Dale et al. 2008) in Asia and Australia. These species can hatch in huge numbers when conditions of tide and rain coincide to flood upper locations in salt marshes. A bank of eggs from the previous generation or accumulated from several generations hatch all at once. Abundant nutrients in the habitats fuel development of the larvae, the pupae of which eclose more or less simultaneously to release hordes of adults that often travel many kilometers in search of blood meals and then return to the salt marsh to oviposit. Although some individual females may live for weeks, the majority of the population is gone within 10–15 days. Those adults form a large and temporary population typical of *r* strategists, even though the adult habitat is of long duration.

At the opposite end of the population spectrum are *K* strategists, where *K* is the carrying capacity of a habitat. Typically, *K* strategists have short generation times relative to duration of their habitat, a larger individual size, a longer potential life span, more opportunity for intraspecific competition, higher survival rates, greater care for offspring, and a lower intrinsic reproductive rate (Southwood 1976). The concept of *K* strategy among mosquitoes is much less clear than for the *r* strategy. Although the total generation time of many species at higher latitudes is usually a year, much of that year is spent in a quiescent state as an adult (e.g., *Anopheles earlei* [Hudson 1978]), larva (e.g., *Orthopodomyia signifera* [Copeland & Craig 1990]), or egg (e.g., *Aedes impiger* and *nigripes* [Corbet & Danks 1975; Danks 2004]). Groups with predatory larvae, like the genus *Toxorhynchites* (Yasuno & Tonn 1970; Steffan et al. 1980) or *Lutzia* (formerly a subgenus of *Culex* [Bram 1967b; Berlin & Pandian 1973b]) produce larger individuals, though their time to development from egg to adult is not particularly long. Maternal care of eggs rarely occurs among mosquitoes—notably, in the egg-guarding behavior of *Trichoprosopon digitatum* (Lounibos 1983) and egg carrying by species in the subgenus *Leicesteria* of the genus *Armigeres* (Bailey et al. 1975), but certainly, no one who has experienced outbreaks of these species would think of them as typical *K* strategists. Some mosquitoes have longer larval developmental times, either because of temperature (e.g., *Aedes squamiger* [Bohart & Washino 1978]) or physiology (e.g., *Wyeomyia smithii* [Bradshaw & Holzapfel 1989]; *Anopheles cruzii* [Wilkerson & Peyton 1992]), but the time of active development is seldom more than one to several months, the larval density is not particularly low, and adults are not particularly large or long-lived. Taken as a whole, the family Culicidae appears to have few examples of life histories that approximate a *K* strategy. It is perhaps an overgeneralization, but mosquitoes typically produce many more offspring than can survive in many variations of an *r* strategy of population growth.

Reproductive Mortality

The second mechanism of biological evolution is sometimes called "struggle," the idea being that individuals within a species are subject to many sources of mortality or other stress that prevent successful reproduction by those individuals. Although struggle certainly evokes a colorful image, it also leads to the false impression that each organism has a will to live, which would be difficult to prove objectively. The most important aspect of this mechanism in evolution is that events often stop an individual from reproducing; therefore, "barriers to reproduction" might be a more objective term for this second mechanism of evolution.

Under ideal conditions, laboratory rearing of mosquitoes generally results in a high rate of survival and reproduction (Gerberg et al. 1994; Strickman 2006). Of course, achieving those ideal conditions can require a great deal of adjustment, with no real certainty that conditions are actually optimal, other than the survival rate being measured. Another source of bias in such laboratory observations is the species selected. Relatively few species have ever been colonized, and those were usually selected based on either their ease of rearing or their economic importance. Experience indicates that when all other conditions are optimized and when dealing with species that mate readily in cages or that can be induced to copulate, the major source of reproductive failure for the individual is in the egg. Relatively few studies have carefully examined the hatching rate, but it is usually well below 100% (e.g., *Aedes mediovittatus* 32%–71% hatch rate [Freier & Rosen 1988]). The reasons that eggs do not hatch are as varied as excessive temperature (Mulla & Chaudhury 1968), desiccation (Sota & Mogi 1992), predation (Focks et al. 1993a), and inhibition from other larvae (Livdahl & Edgerly 1987). Presumably, some of the same sources of hatching failure occur in the field.

Assuming that laboratory experience is representative of the high inherent probability of reproduction by individual mosquitoes, then the generally low rate of individual reproduction (e.g., Gunasekaran 1994; Harrington et al. 2001a; Suman et al. 2011) is mainly the result of the many sources of mortality. Some of those sources of mortality are usually related to density-dependent biotic factors.

Nutrition is critical for successful larval development. As noted above, most species develop from egg to pupa quickly and therefore require abundant sources of food. Larvae feed on different kinds of food. For example, *Aedes triseriatus* feed mainly on spherical bacteria, *Anopheles quadrimaculatus* on rod-shaped bacteria, algae, and protozoa, and *Coquillettidia perturbans* on spiral-shaped bacteria and algae (Walker et al. 1988). The structure of the mouthparts influences the flow of water into the brushes that filter food and varies between species (Dahl et al. 1988). Predatory larvae, like *Lutzia tigripes* (Appawu et al. 2000), have an entirely different sort of diet. Those individuals that are unable to consume adequate food die if they fail to reach a critical weight (Chambers & Klowden 1990). Larval nutrition can have a big influence on the subsequent reproductive success of the adult (Akoh et al. 1992). Some species produce smaller adult individuals when food is limited (Sumanochitrapon et al. 1998). Some field studies have shown that field populations can be under general nutritional stress (Strickman & Kittayapong 2003). Despite the ability of females of at least some species to avoid crowded oviposition sites (Zahiri et al. 1998), larval nutritional stress is likely common among natural populations of mosquitoes.

The nutrition of the adult mosquito is clearly divided between reproductive and energy requirements. Energy requirements are usually provided by sugar sources, such as floral or extrafloral nectar, fruit exudates, honeydew, or phloem (Schlein & Müller 1995; Burkett et al. 1999). Rarely, it has been demonstrated that blood is metabolized to energy-producing compounds—most notably, in the important vector *Aedes aegypti* (Harrington et al. 2001b). Of course, male mosquitoes are completely dependent on sugars and whatever reserves they acquired from larval feeding since they do not feed on blood. Nutritional requirements for production of sperm are presumably much less than for eggs, though larger male *Aedes aegypti* resulting from better larval nutrition transfer more sperm than smaller males (Ponlawat & Harrington 2009). On the other hand, a single gonotrophic cycle of *Aedes aegypti* producing 100 eggs, each weighing 4 µg at oviposition (Kliewer 1961), would represent 174% of a female weighing 0.23 mg at emergence (Focks et al. 2000). The protein, polysaccharides, lipids, and other structural elements of the eggs typically come from blood imbibed by the adult. Reserves accumulated as larvae can also be used for reproduction, as is routinely the case for species in the closely related family Chaoboridae (Strickman 1980a). Those reserves come from predation in the larval stage for all species in *Toxorhynchites*, which never take blood as adults (Collins & Blackwell 2000), a pattern also present in some species of *Topomyia* (Ramalingam 1983). Other predatory species (e.g., *Lutzia* species [Pal & Ramalingam 1981]; *Aedes* (*Mucidus*) species [Mattingly 1961; Tyson 1970c]; and *Psorophora* (*Psorophora*) species [Campos et al. 2004]) may carry over larval nutrition to the adult, but the females also take blood meals. *Wyeomyia smithii* (northern populations) is an example of a nonpredatory species that produces eggs without blood meals for several gonotrophic cycles (O'Meara et al. 1981; Lillie and Nakasone 1982). A more common pattern is autogeny, in which the female uses nutritional reserves accumulated as a larva to produce a single batch of eggs prior to a blood meal (Spielman 1971). Subsequent batches of eggs are produced normally. Facultatively autogenous species could be more widespread than suspected because of the care required to observe them. Whether an adult cannot function due to a lack of food for energy or cannot produce eggs due to a lack of food for oogenesis, the final result is failure of that individual to reproduce.

Competition is another cause of reproductive mortality. Competition for food among larvae was mentioned above, but the exact effect of the interaction between two species can be complex. Black et al. (1989) showed that mixing *Aedes albopictus* and *Aedes aegypti* increased developmental time but otherwise did not affect survival or subsequent reproduction; however, an increase in developmental time might increase the chances of mortality from predation in the field. The same study showed an increase in mating aggressiveness by *Aedes aegypti* males when both genders of both species were mixed. Male *Aedes aegypti* matings with female *Aedes albopictus* never result in offspring, whereas, male *Aedes albopictus* mated to female *Aedes aegypti* often result in female offspring (Downs 1949). In this sense, the two species may compete for mates, resulting in less reproductive success overall.

Defensive behavior by hosts approached by blood-seeking mosquitoes can interfere with feeding (Edman & Kale 1971), and a greater density of mosquitoes causes more defensive behavior (Edman et al. 1972, 1974).

Predation is an important source of mortality for both larvae and adults. Although it is beyond the scope of this chapter to list them all, many species of fish and aquatic arthropods avidly feed on mosquito larvae (Shaalan & Canyon 2009). When larvae live in habitats that are too polluted, temporary, or inaccessible, predation may not be important. More suitable habitats support a wide variety of predators, as in the example of organically grown rice fields in Korea (Lee 1998). Adults face a similarly wide array of vertebrate (Griffin et al. 1960; Becker 2006) and invertebrate predators (Linthicum et al. 1985; Strickman et al. 1997) and are sometimes eaten by the hosts from which they are probably seeking blood (Barr 1996). In a poignant example, one of the authors (D.S.) saw a malnourished child in coastal Honduras slap the abundant *Aedes taeniorhynchus* that came to feed and then carefully lick the dead mosquitoes from his hand.

Parasites have a rich interaction with mosquitoes that is only partially understood. Viruses, bacteria, fungi, protists, and nematodes include pathogens that can either kill the mosquito host or slow reproduction (Lacey & Goettel 1995; Becker 2006; Abdul-Ghani et al. 2012). In addition, mites include species that specialize as ectoparasites of adult mosquitoes; the aquatic ones climb onto them as they emerge from the pupa (Mullen 1975; Milne et al. 2009). Mosquitoes transmit many species of viruses, protists, and nematodes to vertebrates (Mullen & Durden 2002). The human pathogens are best studied, especially dengue virus (Flaviviridae), yellow fever virus (Flaviviridae), *Plasmodium* (Phylum: Apicocomplexa; Order: Haemosporida; Plasmodiidae) malaria-causing protists [*Plasmodium vivax* (Grassi & Feletti 1890), *malariae* (Feletti & Grassi, 1889), *ovale* Stephens 1922, *knowlesi* Sinton and Mulligan 1933, and *falciparum* Welch 1897], and filariae [*Wuchereria bancrofti* (Cobbold) (Phylum: Nematoda; Order: Spiruida; Onchocercidae) and *Brugia malayi* Brug (Phylum: Nematoda; Order: Spiruida; Onchocercidae)]. Filariae develop, but do not multiply, in the mosquito vector. The movement of the infective third-stage larval nematode within the mosquito and its exit from the labellum cause gross mechanical damage (Apperson et al. 1989; Erickson et al. 2009). The other pathogens are also closely associated with their vectors, but their effects on the mosquito are more subtle. Microscopic views of dengue-infected (Linthicum et al. 1996) or *Plasmodium*-infected (Golenda et al. 1990) mosquitoes show a large number of microorganisms in mosquito tissue, which is quite remarkable considering the almost normal behavior of the infected insect. Nonetheless, at least some *Anopheles* vectors mount an active physiological defense against *Plasmodium*, balanced against protection of commensal bacteria in the gut (Kumar et al. 2010b). In at least some cases, genetic refractoriness to infection is associated with an inherently lower reproductive rate, suggesting that mosquito defense mechanisms against these human pathogens can have fitness costs (Yan et al. 1997). The behavior of *Aedes aegypti* is affected by dengue virus infection, increasing the tendency to probe and increasing blood-feeding time (Platt et al. 1997). *Aedes aegypti* becomes less susceptible to repellents when it is infected with Sindbis virus (Qualls et al. 2011). Such changes in behavior may not affect reproduction or may improve reproductive potential under certain conditions that favor pathogen transmission.

Mosquitoes support a rich symbiotic bacterial flora. Studies of gut bacteria have shown a variety of species (Chapter 7), their relationship to the mosquito remaining largely unknown. In the reproductive system, one particular bacterium, *Wolbachia pipientis* Hertig 1936 (Phylum: Proteobacteria; Order: Rickettsiales; Anaplasmataceae), affects populations directly by preventing development of eggs if mates harbor incompatible strains or if one is infected and the other is not. Depending on the pair mating, *Wolbachia* can have serious consequences on reproduction, effectively creating barriers between populations of some species. In fact, the earliest indication of *Wolbachia*'s effects was in a complicated description of nonfertile populations of *Culex pipiens* in Europe (Laven 1951). Much later the incompatibility was found to be caused by *Wolbachia* rather than genetics (Yen & Barr 1973).

Humans have a significant impact on mosquitoes, particularly those species that commonly bite people. In considering sources of mortality, it is important to recognize that mosquito abatement often makes a great impact on local populations. Habitat modification can favor mosquitoes by disrupting drainage patterns, or it can be carefully and intentionally designed to eliminate larval habitats. Large-scale modifications such as the destruction of forests, pollution of rivers, and so on almost certainly reduce habitats for some species of mosquitoes, though reductions in culicid fauna go largely unnoticed unless the species involved are harmful to humans. In contrast, the effects of insecticides on mosquitoes have been carefully studied in thousands of scientific publications. Insecticides can affect reproductive survival in four ways. First, insecticides kill many mosquitoes as larvae (larvicides) or adults (adulticides), exactly as the products intend. Second, sublethal effects can temporarily or permanently lower reproductive potential of individuals. Third, physiological resistance to insecticides can result in a loss of reproductive potential, presumably because of energy costs associated with production of normally unnecessary biochemical pathways (Kumar et al. 2009). Finally, nontarget effects on insects and other arthropod predators can decrease biotic mortality of mosquitoes. Nontarget effects could conceivably reduce mosquito reproductive survival, for example, by disproportionately favoring an aquatic competitor.

Abiotic components of the environment have strong effects on mosquito populations. Basic sources of mortality like overheating, freezing, and desiccation affect species differently. For example, some groups with a majority of species in the tropics have one or a few representatives able to survive cold temperatures (Ross 1964). High temperatures experienced locally by tropical species can reduce overall reproductive potential (Strickman 2006). More subtly, high temperatures can alter reproductive fitness of northern species, possibly accounting for their failure to colonize more southern locations. Availability of water is also a primary determinant of mosquito survival. Larvae fail to develop when sites dry out too quickly, and floodwater mosquito eggs eventually die if not flooded. Even adult mosquitoes are dependent on finding microhabitats with appropriate water availability. Taken together, temperature and water are major determinants of mosquito survival (Hopp & Foley 2001).

Selection

The third mechanism of biological evolution, selection, is perhaps the most complicated. Selection operates on the axiom that reproductive mortality or success does not occur randomly for individuals within a species and that an individual's characteristics will influence whether or not it is able to reproduce and how many of its offspring will survive given the circumstances in which the individual lives. Hypothetically, identical individuals might have very different reproductive rates under different conditions. Those conditions reflect the intersection of any of the reproductive mortality factors just reviewed with the space and time occupied by the individual, imparting an aspect of randomness on survival at a small scale. On a large scale involving populations, the process of selection is not as random. One could argue that the randomness of selection decreases as population size, variety of locations, and length of time increase. As a result of the variation of selection with scale, much of selection is not easy to observe experimentally. Those examples that are clearly demonstrable probably represent the mechanisms accurately but only account for a fraction of the selection actually affecting a species.

The laboratory observation that many mosquito eggs fail to hatch suggests that selective pressure on the egg is severe. Although it is difficult to conceive the adaptive significance of some forms of ornamentation or shape in well-studied aedine eggs (Horsfall et al. 1952), their adaptations to resist drying during long periods in the soil or on the sides of containers (Savage & Strickman 2004) are clear. Over time, fewer and fewer aedine eggs hatch even under carefully controlled conditions, suggesting a mechanism for selection.

Larval nutrition is an easily observed source of selection. Some species, like *Culex pipiens* and *Trichoprosopon digitatum*, live in environments with an abundance of nutrients so that survival is not dependent on nutrient availability. Other species, like *Aedes aegypti*, live in clearer water, often with fewer nutrients, and not every larva survives under these conditions. Laboratory rearing can clearly show how some individual larvae survive while others die when nutrients are not adequate. Furthermore, some individuals grow to a larger size, with greater potential reproductive capacity as an adult (Zayed 2004).

Mammalian defensive behavior is another easily observed source of selection on mosquitoes. Every mosquito squashed or eaten is not going to convert its blood meal into eggs. From a human perspective, we find some mosquitoes much easier to kill than others. For example, *Aedes aegypti* and *Aedes albopictus* approach low to the ground, are easily disturbed by movement, and touch down on the skin gently. Many of them avoid detection in this way. Over time people develop an immunological tolerance to their bites (Peng et al. 2004b), and they become even harder to notice. Mosquitoes that bite at night probably avoid some defensive behavior if their host is sleeping. Pathogens like those that cause malaria may improve mosquito blood-feeding success by suppressing defensive reactions (Day & Edman 1983; Rossignol & Shieh 1993; McCall & Kelly 2002), suggesting there may be selective pressures on mosquitoes toward tolerance for pathogens in contradistinction to physiological defenses against infection (Luckhart & Rosenberg 1999).

Predation on larvae is another source of mortality that has apparently resulted in selection. Larval characteristics that reduce the impact of predators are probably widespread, but the number of documented adaptations is few. One clearly documented example is *Wyeomyia smithii*, which escapes predation from carnivorous chironomids by having long setae over its body (Petersen et al. 2000). The observation that many container-breeding mosquitoes have well-developed setae (e.g., *Aedes* (*Howardina*) [Berlin 1969a]) may be the result of similar selective pressure. Maternal attention to eggs by *Trichoprosopon digitatum* and *Armigeres* (*Leicesteria*) species (Macdonald 1960) may also be a response to predation that would otherwise take place (Lounibos 1983). The droplet exuded by *Culex pipiens* egg rafts is repellent to ants, protecting the eggs from predation (Hinton 1968b).

Insecticide resistance is the most carefully studied selective process in mosquitoes (Hancock et al. 2018). Mosquito control has been compromised by insecticide resistance for decades, leading to many practical studies of its occurrence, mechanisms, and management. In addition to the practical experience from mosquito control, studies of specific biochemical resistance mechanisms have made the convincing case that insecticide resistance is a snapshot of evolutionary processes in action. The link to use of insecticides in mosquito control programs (Yébakima et al. 2004) or other pest control (Eritja & Chevillon 1999) is always going to be correlative, but the increase in esterase activity and resistant forms of acetylcholinesterase was clearly the result of selection by exposing the population to organophosphate and carbamate insecticides. Changes in gene frequency can be regional, with strong genetic structure between populations, presumably in response to migration and local selection (Ben Cheikh et al. 2009). Certain genes in *Culex pipiens* appear to have altered frequency in response to selection worldwide (Ben Cheikh et al. 2008), a startling example of unintentional domestication of a wild species by human activity. Exposure to different classes of insecticides may create unpredicted selection, sometimes resulting in populations that are more resistant despite the conventional belief that alteration of modes of action or sequential use of different insecticides will tend to reduce resistance in a population (Corbel et al. 2003).

The process of mating and consequent gene transfer is, of course, a necessary part of the reproductive process. Cause-and-effect evidence is weak, but presumably many aspects of this process are the target of selective forces. Sperm competition, swarming behavior, variation in female genitalia, and no doubt a myriad of other anatomical, behavioral, and physiological characteristics of the mating process are the result of selective processes.

One aspect of mating that attracts speculation on the association of form with function and selection is the male genitalia. The incredible and highly consistent variation in male genitalia has fascinated mosquito taxonomists for generations. The male genitalia continue to be the gold standard for identification of many groups (e.g., *Culex*) and an important additional source of information for others (e.g., *Aedes*, *Anopheles*). Arnqvist (1998) suggested that variation in dipteran male genitalia is the result of selection for those forms that prevent hybridization and a consequent reduction in reproductive survival. In an examination of a group of *Drosophila* (Order: Diptera; Drosophilidae), the females of which mate multiple times (i.e., polyandrous), compared to a group of monandrous *Anopheles*, morphometric measurement of genitalic structure showed that the polyandrous group was more diverse than the monandrous group. The implication is that the higher probability of hybridization created by multiple female matings exerts selective pressure to prevent mating between species by increasing the physical barrier to cross-mating created by lock-and-key genitalic systems. Presumably, it would be difficult to demonstrate selective pressures on such complex anatomies, but the comparison of two groups correlated with morphology make the case that complex male genitalia are the result of evolutionary selection.

Inheritance

The potential reproduction, reproductive mortality, and selection of a species population would not result in evolutionary biology without genetic inheritance. Inheritance is an obvious consequence of sexual reproduction, as observed by anyone who carefully monitors the behavior, physiology, or morphology of a mosquito colony. It is possible to select for family lines of mosquitoes with a certain host preference, insecticide resistance, or body marking. Modern biochemistry, molecular biology, and population statistics have told us a lot more about the mechanics of genetic inheritance, with the implication that we have much more to learn (Ashburner 2005).

One aspect of inheritance that operates in a manner outside the well-known transmission of genes between generations is horizontal gene transfer. Transposable elements (Kidwell 2005) are segments of DNA or RNA-generated DNA that have the ability to insert themselves in the genome. Known since 1948 (in corn), transposable elements are now recognized to occur widely in genetic systems and are responsible for inactivation of genes, transfer of genes, chromosome rearrangement, and some forms of mutation. From an inheritance standpoint, transposons can affect the genome transmitted to offspring in at least three general ways. First, the process of transposon replication and insertion can create long sequences of noncoded DNA. This process may cause the dense bands of repetitive insertion by transposons and could be the cause of breaking, and subsequent inversion, of pieces of chromosome. The second way in which transposons can affect inheritance is when they behave like viruses (retroviruses are considered a special type of transposon) and spread between individuals, carrying host genes with them and causing horizontal gene transfer. Finally, transposons can be used artificially to insert genes, creating transgenic individuals that pass on new genetic traits to their offspring.

Transposons of mosquitoes are an active area of research. Kidwell (2005) summarized hundreds of endogenous transposon types in *Anopheles*, *Aedes*, and *Culex*, as well as successful transformation of five species of mosquitoes using transposons from other species. One interesting example of transformation inserted the enhanced green fluorescent protein from bacteria using the housefly [*Musca domestica* L. (Order: Diptera; Muscidae)] transposon, *Hermes*, as a vehicle (Allen et al. 2001). The transposase gene in *Hermes* was initiated by heating the embryos, which stimulated a promoter sequence provided by *Drosophila melanogaster* Meigen. The use of genetic sequences from three other species to create a functioning aspect of the genome in a mosquito shows a close relationship of mechanism and suggests that interspecific gene transfer is a possibility. Natural horizontal gene transfer was observed in four species of Thai *Anopheles* through the *Copia* retrotransposon (Rongnoparut et al. 1998), one form of which may have originated from a plant.

Hybridization between what have been considered to be separate species or separate strains is another source of the inheritance of novel genotypes. The occurrence of hybridization has sometimes been assumed based on distribution of morphological characters among species. For example, Belkin and Hogue (1959) speculated that 3 of 11 species in the genus *Deinocerites* were the result of hybridization, creating an unusual phylogenetic diagram. Crossing experiments that have indicated some level of interfertility show that genetic flow between some species may be common, or at least possible. Among the species in the clade that includes *Aedes zoosophus*, *triseriatus*, *hendersoni*, and *brelandi* in the United States, only crosses between *zoosophus* and *triseriatus* produced fertile females. All other progeny from all other crosses were sterile (Taylor 1990). Male *Aedes aegypti* that mate with female *Aedes albopictus* usually produce fertile, *aegypti*-like offspring (Downs 1949). All forms of *Aedes atropalpus*, considered at the time to be subspecies, were interfertile (O'Meara & Craig 1970), though taxonomic revision later retained two of the forms (*Aedes atropalpus* and *Aedes epactius*) as separate species, each with its own distribution and characteristics (Zavortink 1972). Similarly, species in the western part of the range of the Scutellaris Group (*Aedes hebrideus*, *alcasidi*, *katherinensis*, and *malayensis*) show various patterns of interfertility in the laboratory (Woodhill 1950; Dev & Rai 1985). It seems likely that where the ranges of these species overlap, there would be the opportunity for at least limited gene flow between them. Based on extensive hybridization, morphological, and allozyme studies, Pashley et al. (1985) suggested that the speciation events within the Scutellaris Group were explained by island formation and separation during the last 5 million years, sequentially isolating populations that later speciated.

Culex pipiens and related species may represent the best-studied example of hybridization in mosquitoes. As designated now, the group consists of four species (*Culex pipiens*, *quinquefasciatus*, *australicus*, and *globocoxitus*) separated mainly by morphological characteristics of the male genitalia supported by genetic studies. *Culex pipiens* is further divided into two subspecies (*Culex pipiens pipiens*, in Europe and Africa and *Culex pipiens pallens*, in eastern Asia). *Culex pipiens* has an autogenous, nondiapausing form that has been designated in various ways, but is here called Form Molestus (Farajollahi et al. 2011). The degree of separation between these taxonomic entities is still the subject of much discussion, with some considering that *Culex pipiens pipiens*, *Culex pipiens* Form Molestus, and *Culex quinquefasciatus* are distinct, monophyletic clades, each deserving species status (Becker et al. 2012a). The complicated pattern of interbreeding and noninterbreeding populations of *Culex pipiens* in Europe and North Africa led Laven (1959) to speculate on an inherited nonchromosomal genetic factor that caused divisions in gene flow between populations considered to be the same species. He recognized that similar isolation of populations occurred in other mosquitoes. *Wolbachia*-induced cytoplasmic incompatibility (see below; Werren 1997) was eventually recognized as the inherited cytoplasmic factor responsible for the complicated population pattern Laven had observed. Barr (1957b) recognized that *Culex pipiens* and *Culex quinquefasciatus*, both introduced species in North America, probably hybridized extensively in the middle latitudes (36°–39°N) of the United States, creating forms with male genitalia quantitatively intermediate between the two species. Actual hybridization between the species has been observed many times and elegantly demonstrated genetically (Savage & Kothera 2012). Single nucleotide polymorphisms (SNPs) marking differences in ribosomal DNA indicated a gradient of hybridization between about 30° and 40°N, leading to speculation that *Culex pipiens* was a single species introduced into North America with subsequent branching off of *Culex quinquefasciatus* (Huang et al. 2011). The observation that *Culex pipiens* distribution extends continuously from the Mediterranean through East Africa to South Africa and that *Culex quinquefasciatus* in South Africa does not hybridize with *Culex pipiens* (Cornel et al. 2003) has been attributed to *Wolbachia*-induced incompatibility rather than genetic barriers (Farajollahi et al. 2011). Reflecting some of Laven's original observations about the three forms and Rozeboom's (1951b) observations on hybrid populations of California, Strickman and Fonseca (2012) showed through analyses of DNA microsatellites and combinations of phenotypes that *Culex quinquefasciatus*, *Culex pipiens*, and *Culex pipiens* Form Molestus in coastal Central California formed a biotype of the group with unique characteristics. Cornel et al. (2012) showed the presence of forms combining the genetics of *Culex pipiens* and *Culex pipiens* Form Molestus in Central and Northern California. Regardless of the taxonomy of this species group, it represents a situation that may be more common in Culicidae in which important phenotypic traits are variously shared through hybridization. The resulting variation in phenotype would presumably make the species more adaptable to changes in distribution and to environmental changes over time.

Symbiotic bacteria are an important and quite pervasive influence on inheritance in arthropods. Primary endosymbionts of aphids, for example, appear to have developed a close, mutually beneficial association with that group early in its evolutionary history. The strains of bacteria *Buchnera aphidicola* Munson et al. 1991 (Phylum: Proteobacteria; Order: Rickettsiales; Anaplasmataceae) have developed in congruence with at least one aphid (Order: Hemiptera; Aphididae) genus *Uroleucon*, providing a clear case of cospeciation (Clark et al. 2000). Secondary or facultative endosymbionts do not necessarily inhabit specialized cells and vary in a spectrum from mutualistic relationships to pathogenic ones (Moran & Baumann 2000). Among these secondary endosymbionts are a number of different bacteria [*Wolbachia pipientis*, *Cardinium hertigii* Zchori-Fein et al. 2004 (Phylum: Bacteroidetes; Class: Sphingobacteriia; Order: Sphingobacteriales; Flexibacteraceae), *Arsenophonus nasoniae* Gherna et al. 1991 (Phylum: Proteobacteria; Order: Enterobacterales; Morganellaceae), *Spiroplasma* spp. (Phylum: Tenericutes; Order: Entomoplasmatales; Spiroplasmataceae), and *Rickettsia* spp. (Phylum: Proteobacteria; Order: Rickettsiales; Rickettsiaceae).] that affect reproduction in ways that often tend to spread and perpetuate infections. Species designation of these bacteria is difficult because of their small genomes and rapid base-pair changes and the challenges in culturing them. Although the extent to which such bacteria affect speciation has only recently been appreciated, the apparently close associations between acarines and rickettsiae (Kelly et al. 2002), particularly *Orientia tsutsugamushi*'s (Hayashi 1920) (Ogata 1929) Tamura et al. 1995 (Phylum: Proteobacteria; Order: Rickettsiales; Rickettsiaceae) effect on trombiculid mite reproduction (Oaks et al. 1983), were early indicators of a relationship that could influence evolutionary development (Dasch et al. 1996). *Wolbachia pipientis* bacteria were first identified in *Culex pipiens* mosquitoes in the 1920s, but it was not until much later (Yen & Barr 1973) that the connection was made between Laven's (1959) cytoplasmic isolation and the effect of these bacteria. Now it is recognized as a common phenomenon in mosquitoes, driven by unidirectional cytoplasmic incompatibility (infected male and uninfected female), bidirectional incompatibility (male and female with different strains of *Wolbachia*), parthenogenesis induction, conversion of genetic males to females, and selective mortality of males (Werren 1997). Infection not only affects reproductive patterns of the arthropod hosts but can also benefit the hosts' immune systems through generation of reactive oxygen compounds and antioxidant enzymes, change susceptibility to heat stress, and mediate horizontal gene transfer (Saridaki & Bourtzis 2010). Although *Wolbachia* are the most extensively studied group of bacteria that influence mosquito reproduction, there may be a complicated interaction among at least several kinds of symbionts, including *Cardinium* (Saridaki & Bourtzis 2010) and the extracellular bacteria in the genus *Asaia* (Phylum: Proteobacteria; Order: Rhodospirillales; Acetobacteraceae) (Damiani et al. 2010). The exact roles of symbiotic bacteria on mosquito evolution are not certain yet, but it is clear that they contribute to isolation of some populations, horizontal transfer of genes, and modification of genetic expression. *Wolbachia* and other endosymbionts are profoundly influencing the reproductive processes of many species of mosquitoes.

Horizontal gene transfer and nongenetic influences on the genome are more recently discovered influences on inheritance. Better-known mechanisms of chromosomal recombination and genetic sequence changes are other sources of variation inherited by subsequent generations of mosquitoes. Some changes of base pairs are nonfunctional, in that they do not result in a change of coding for amino acids. Changes between pyrimidine and purine bases are less frequent than changes within each base type (i.e., adenosine to/from guanine and cytosine to/from thymine transitions are most frequent). All mosquitoes have three pairs of chromosomes that are either metacentric or subcentric except *Chagasia bathana*, which, like the corethrellid midge *Corethra*, has four pairs of chromosomes (Kreutzer 1978). Comparisons of species show extensive translocations and inversions, though genes coding for certain enzyme systems are linked. This sort of linkage group conservation, or synteny, is typical of Culicidae and higher Diptera. Some of these linkages imply genetic homologies between *Drosophila* and Culicidae that go back at least 250 million years (Matthews & Munstermann 1994). Male anopheline, but not culicine, mosquitoes are heterogametic, meaning that the single-locus (*M*) sex-determining factor is linked to a Y chromosome that is dissimilar to its corresponding X chromosome. Female mosquitoes have similar X chromosomes (Ashburner 2005). This genetic sexing system leads the female X chromosomes in other organisms to become more susceptible to recombination, though this has not been documented in mosquitoes. The system also leads to male sterility in some hybrid pairings of species (Davies & Roderick 2005). Chromosomal morphology can be affected by selection. One striking example was demonstrated by visualizing through in situ hybridization the 32-fold replication of an esterase gene in *Culex pipiens* Form Molestus. The amplification of the gene was a direct result of selection by organophosphate insecticide. These changes so altered chromosome 2 that crosses with normal mosquitoes caused asynapsis in that chromosome (Tomita et al. 1996).

A comparison of *Anopheles gambiae* and *Anopheles funestus* gives some perspective on chromosomal recombination and sequence change in mosquitoes (Sharakhov et al. 2002). Those two species are in the same subgenus (*Cellia*) and probably shared a common ancestor 5 million years ago (MYA). In spite of the taxonomic relatedness of the two species, 70 fixed chromosome inversions were found, greatly reordering the sequences of genes. Base-pair sequences of 40 genes varied by an average of 13%, which was associated with a difference of 7.5%

in amino acid sequences. Another way of expressing this variation was that there was a 61% chance of base-pair change per site if that site did not signify an amino acid change and a 4% chance if the site caused an amino acid change. This showed not only that base-pair changes are inhibited by selection but also that these species of mosquitoes have a much higher rate of base-pair change (61 changes per site per billion years) than *Drosophila* (16 changes per site per billion years).

Mosquito Evolution and Evolutionary Theory

The four mechanisms of biological evolution—potential reproduction, reproductive mortality, selection, and inheritance—interact with genetic mechanisms during both long and short timescales. The timescale for evolutionary change, or at least the appearance of new species, has environmental as well as genetic inputs (Hubbell 2001). The genetic architecture that leads to a rate of variation from generation to generation combines with the species distribution area and its abundance to result in a wide variation in the possible rate of new species. The tropical center of mosquito distribution (Chapter 3) would tend to increase the rate of speciation according to these criteria. The result of these genetic, spatial, and temporal factors is the family Culicidae we observe today. Further generalizations about biological evolution lead to a series of concepts, each apparently with its advocates and detractors.

Perhaps the clearest conceptual disagreement is between gradualism and punctuated equilibrium (Gould & Eldredge 1993). Gradual evolution would have relatively constant, slow changes in species that eventually lead to generic, familial, and ordinal differences (Dobzhansky 1937). Major changes in biological architecture in a gradualist context would arise in a seamless, step-by-step transition. Examples relevant to this discussion would include the transition from a member of the mecopteran complex of orders to the first dipteran (Whiting 2005) or the easier-to-conceive offshoot of Dixidae to the Corethrellidae-Chaoboridae-Culicidae group of families (Wiegmann et al. 2011). Punctuated equilibrium, in contrast, proposes a series of relatively sudden (on the order of 50,000 to 100,000 years) changes interspersed with long periods of nearly static evolution. Of interest from the standpoint of the recently discovered sources of genetic change in mosquitoes is the concept that the stability of species is an active process in evolution, interrupted by events that cause a disruption of that equilibrium. The history of mosquitoes as we know it (Table 1.1) does not give strong support to either view. Among the few relevant objective facts is that the family was substantially similar to current morphological forms (Borkent & Grimaldi 2004; Poinar et al. 2019) prior to the K-T boundary separating the Cretaceous from the Tertiary, a geological boundary that corresponded to the occurrence of major worldwide extinction events. That observation might be an argument for gradual evolutionary change, especially if the fossil record were more complete. On the other hand, the species structure of at least two groups is linked to changes brought about over a short geological time frame by human activity (Farajollahi et al. 2011; Coluzzi et al. 2002). It is not difficult to speculate that as human activity dominates the entire biome, there is the possibility of other changes among Culicidae, possibly rising to the significance of punctuated change following what will turn out to have been a period of equilibrium.

Another key concept that arises out of evolutionary mechanisms is exaptation (Gould & Vrba 1982). Exaptation occurs over geological time when a phenological characteristic of an organism comes to serve a new function in the species lineage (de Queiroz 2007). An example is the development of wings in insects from evaginations of the thoracic wall that did not have this function originally. Exaptation has been proposed in contradistinction to adaptation in which selection of modifications in phenotype lead to functions adapted to alleviate the selective pressure. One carefully described example of exaptation in mosquitoes is the episystem (vector-host-pathogen relationship) of West Nile virus (WNV) transmission by a subset of *Culex* (*Culex*) mosquitoes (Tabachnick 2010). *Culex pipiens*, *quinquefasciatus*, *tarsalis*, *nigripalpus*, *univittatus*, and *vishnui* are the key vectors of WNV in various parts of the world. Other than being in the same subgenus, these species have little else in common phylogenetically, and many other species in the subgenus are not important vectors. From the frame of reference of the virus, the coincidence of physiological competence, longevity, and host range is an exaptation that makes these species powerful vectors. The "exaptive pool" has been invoked to explain major trends in evolution, a concept leading to the conclusion that interaction of features of separate species leads to mosquito-wide qualities important in making the subsequent selective environment (Gould 1988). Although the noncompetitive assemblage of ecosystem selective pressure in the environment leads to the controversial concept of group selection, the WNV episystem might be a specific example. If WNV infection imparted a selective advantage on some mosquitoes—for example, by making them less susceptible to repellents (Robert et al. 1991 [*Anopheles stephensi* and *Plasmodium berghei* or *falciparum*]; Qualls et al. 2011 [*Aedes aegypti* and Sindbis virus]), by making it easier to orient toward a febrile host (Lacroix et al. 2005 [*Anopheles gambiae* and *Plasmodium falciparum*]), or by increasing longevity under conditions of nutritional stress (Dohm et al. 1991 [*Culex pipiens* and Rift Valley fever virus]), then those species that were competent vectors might have a group selective advantage. In fact, mosquito infection by viruses or parasites associated with human disease often harms the mosquito (Lambrechts & Scott 2009; Ferguson & Read 2002) by decreasing longevity, fecundity, or both in what may often be a balance, from the pathogen's standpoint, between survival of its vector host and achievement of a transmission threshold through virulence (Ciota et al. 2011).

Although this discussion cannot comprehensively cover the consequences of biological evolution, mosquitoes offer some insights into what is possibly the most controversial aspect of this area: evolutionary progress. The subject has attracted wonderful quotations such as "Life itself needs no explicit rationale"

Table 1.1. Time line of events in culicid history

Estimated years ago	Event	Reference
260,000,000	LCA *Anopheles-Drosophila*	Krzywinski et al. 2006
260,000,000	LCA *Drosophila*-Culicidae	Arensburger et al. 2010
260,000,000	LCA Muscomorpha-Culicidae	Neafsey et al. 2015
255,500,000	ECA Diptera	Yeates and Wiegmann 2005
218,300,000	ECA Diptera major lineages	Yeates and Wiegmann 2005
217,000,000	LCA Anophelinae-Culicinae	Reidenbach et al. 2009
215,000,000	ECA Culicoidea	Krzywinski et al. 2001b
205,000,000	LCA Culicinae-*Culiseta*/*Culex*/*Uranotaenia*	Reidenbach et al. 2009
191,000,000	LCA *Culiseta-Culex*/*Uranotaenia*	Reidenbach et al. 2009
187,000,000	OF Chaoboridae	Borkent and Grimaldi 2004
187,000,000	OF Dixidae	Borkent and Grimaldi 2004
172,500,000	LCA *Anopheles-Aedes*	Krzywinski et al. 2006
172,000,000	LCA *Anopheles gambiae-Culex quinquefasciatus*/*Aedes aegypti* ancestor	Arensburger et al. 2010
168,500,000	ECA angiosperms	Wikström et al. 2001
165,000,000	LCA *Aedes*/*Armigeres*/*Coquillettidia*/*Eretmapodites*/*Haemagogus*/ *Orthopodomyia*/*Psorophora-Aedeomyia*/*Limatus*/*Malaya*/*Maorigoeldia*/ *Mimomyia*/*Sabethes*/*Shannoniana*/*Toxorhynchites*/*Trichoprosopon*/*Wyeomyia*	Reidenbach et al. 2009
158,000,000	LCA *Coquillettidia*/*Orthopodomyia-Aedes*/*Armigeres*/*Eretmapodites*/*Haemagogus*/ *Opifex*/*Psorophora*	Reidenbach et al. 2009
153,000,000	LCA *Culex-Uranotaenia*	Reidenbach et al. 2009
149,000,000	LCA Sabethines-Aedeomyia/*Mimomyia*/*Toxorhynchites*	Reidenbach et al. 2009
139,000,000	ECA Angiosperms modern lineages	Wikström et al. 2001
139,000,000	LCA *Coquillettidia-Orthopodomyia*	Reidenbach et al. 2009
138,000,000	LCA *Mimomyia-Aedeomyia*/*Toxorhynchites*	Reidenbach et al. 2009
126,000,000	LCA *Aedeomyia-Toxorhynchites*	Reidenbach et al. 2009
123,000,000	LCA *Psorophora-Aedes*/*Armigeres*/*Eretmapodites*/*Haemagogus*/*Opifex*	Reidenbach et al. 2009
121,000,000	OF Corethrellidae	Borkent and Grimaldi 2004
107,000,000	LCA Malaya-*Limatus*/*Maorigoeldia*/*Sabethes*/*Shannoniana*/*Trichoprosopon*/ *Tripteroides*/*Wyeomyia*	Reidenbach et al. 2009
100,000,000	LCA *Anopheles* (*Cellia* + *Anopheles*)-*Anopheles* (*Nyssorhynchus*)	Neafsey et al. 2015
98,000,000	LCA *Anopheles* (*Cellia*)-*Anopheles* (*Anopheles*)	Krzywinski et al. 2006
98,000,000	F *Primoculex burmanicus*	Poinar et al. 2019
95,000,000	ECA Anophelinae and subgenera	Krzywinski et al. 2001b
95,000,000	F *Burmaculex antiquus*	Borkent and Grimaldi 2004
95,000,000	LCA *Anopheles* (*Cellia*)-*Anopheles* (*Anopheles*)	White, Collins, and Besansky 2011
94,000,000	LCA New World Sabethines-*Maorigoeldia*/*Tripteroides*	Reidenbach et al. 2009
92,000,000	LCA *Aedes*/*Eretmapodites-Armigeres*/*Haemagogus*/*Aedes* (*Ochlerotatus*)/*Opifex*	Reidenbach et al. 2009
79,000,000	F *Paleoculicis minutus*	Poinar et al. 2000
77,500,000	ECA *Anopheles* modern lineages	Krzywinski et al. 2006
72,000,000	LCA *Trichoprosopon-Limatus*/*Sabethes*/*Shannoniana*/*Wyeomyia*	Reidenbach et al. 2009
66,000,000	LCA *Shannoniana-Limatus*/*Sabethes*/*Wyeomyia*	Reidenbach et al. 2009
65,000,000	LCA *Armigeres*/*Opifex-Haemagogus*/*Aedes* (*Ochlerotatus*)	Reidenbach et al. 2009
64,000,000	LCA *Aedes-Eretmapodites*	Reidenbach et al. 2009
63,600,000	LCA Anophelinae-*Bironella*	Krzywinski et al. 2001b
57,000,000	LCA *Tripteroides-Maorigoeldia*	Reidenbach et al. 2009
54,000,000	LCA *Anopheles-Bironella*	Reidenbach et al. 2009
53,000,000	LCA *Aedes* (*Ochlerotatus*)-*Haemagogus*	Reidenbach et al. 2009

Estimated years ago	Event	Reference
53,000,000	LCA *Aedes aegypti-Culex quinquefasciatus*	Arensburger et al. 2010
51,000,000	LCA *Limatus-Sabethes/Wyeomyia*	Reidenbach et al. 2009
46,000,000	F *Culiseta kishenehn* and *lemniscata*	Harbach and Greenwalt 2012
46,000,000	LCA *Armigeres-Opifex*	Reidenbach et al. 2009
43,000,000	LCA *Anopheles gambiae-Anopheles atroparvus*	Reidenbach et al. 2009
43,000,000	LCA *Sabethes-Wyeomyia*	Reidenbach et al. 2009
35,550,000	R *Anopheles* (*Cellia*)	Krzywinski et al. 2006
35,550,000	R *Anopheles pseudopunctipennis* clade	Krzywinski et al. 2001b; Krzywinski and Besansky 2003
32,500,000	GE *Anopheles* (*Anopheles*) Cyclolepteron Series NA to SA	Krzywinski et al. 2001b
32,500,000	GE *Anopheles* (*Anopheles*) Arribalzagia Series NA to SA	Krzywinski et al. 2001b
30,000,000	OF *Anopheles* (*Nyssorhynchus*)	Krzywinski and Besansky 2003
30,000,000	ECA *Anopheles* Pyretophorus Group (*gambiae, christyi, epiroticus*)	Neafsey et al. 2015
22,500,000	LCA *Anopheles rivulorum-Anopheles minimus/funestus*	Garros et al. 2005b
20,000,000	LCA *Anopheles epiroticus-Anopheles christyi*	Fontaine et al. 2015
10,000,000	E *Anopheles vagus* new habitats created by ungulates and grassland	Loaiza et al. 2012
10,000,000	LCA *Anopheles* (*Cellia*)-Neocellia Series	Loaiza et al. 2012
8,500,000	ECA *Aedes* Simpsoni Group	Bennett et al. 2018
7,000,000	LCA Micronesia-Indonesia-Melanesia *Aedes* Scutellaris Group-Polynesian *Aedes* Scutellaris Group	Pashley et al. 1985
5,330,000	LCA *Anopheles mariae-Anopheles zammittii/phoeniciae*	Coluzzi et al. 1974; García-Castellanos et al. 2009
5,000,000	LCA Melanesian *Aedes* Scutellaris Group-Micronesian *Aedes* Scutellaris Group	Pashley et al. 1985
4,350,000	R *Anopheles* Annularis Group	Morgan et al. 2009
4,000,000	ECA *Anopheles funestus*	Loaiza et al. 2012
3,950,000	R *Anopheles* Funestus Clade	Garros et al. 2005b
3,500,000	LCA *Anopheles gambiae-Anopheles coluzzii*	Lehmann et al. 2003
3,100,000	LCA *Anopheles* Jamesii Group-*Anopheles* Maculatus Group	Morgan et al. 2009
3,000,000	R *Aedes* Scutellaris Group	Pashley et al. 1985
2,600,000	ECA *Anopheles vagus*	Zarowiecki et al. 2011
2,500,000	LCA *Anopheles leesoni-Anopheles minimus*	Garros et al. 2005b
2,000,000	LCA *Aedes tongae-Aedes cooki-Aedes kesseli*	Pashley et al. 1985
1,850,000	R *Anopheles* Gambiae Complex	Fontaine et al. 2015
1,500,000	LCA *Anopheles* (*Anopheles*)-*Anopheles* (*Cellia*)	Loaiza et al. 2012
1,470,000	LCA *Anopheles melas-Anopheles arabiensis/quadriannulatus*	Fontaine et al. 2015
1,280,000	LCA *Anopheles quadriannulatus-Anopheles arabiensis*	Fontaine et al. 2015
1,250,000	LCA *Anopheles minimus* A-*Anopheles minimus* C	Garros et al. 2005b
1,000,000	E *Anopheles sundaicus* island dynamics due to tectonic and sea-level changes	Loaiza et al. 2012
900,000	E *Anopheles darlingi* new habitats with forest fragmentation	Loaiza et al. 2012
880,075	LCA *Anopheles* Group clades	Morgan et al. 2011
880,075	LCA *Anopheles* Minimus clades	Morgan et al. 2011
880,075	LCA *Anopheles* Philippinensis Group clades	Morgan et al. 2011
880,075	LCA *Anopheles* Splendidus Group clades	Morgan et al. 2011
540,000	LCA *Anopheles gambiae-Anopheles coluzzii*	Fontaine et al. 2015
370,000	LCA *Anopheles vagus* west-*Anopheles vagus* east	Zarowiecki et al. 2011
357,000	E *Anopheles minimus* B forest fragmentation created new habitats	Loaiza et al. 2012

continued

Table 1.1. *continued*

Estimated years ago	Event	Reference
234,443	E *Anopheles baimaii* forest fragmentation created new habitats	Loaiza et al. 2012
208,000	E *Anopheles minimus* A forest fragmentation created new habitats	Loaiza et al. 2012
172,202	E *Anopheles dirus* forest fragmentation created new habitats	Loaiza et al. 2012
160,000	E *Anopheles vagus* east to west mixing	Zarowiecki et al. 2011
100,000	E *Anopheles marajoara* to new savannah habitats	Loaiza et al. 2012
47,000	E *Wolbachia* sweep through *Culex quinquefasciatus* in Africa	Rasgon et al. 2006
21,000	GE *Wyeomyia smithii* further north with glacial retreat	Armbruster et al. 1998
16,500	E *Anopheles barbirostris* associated with agriculture	Loaiza et al. 2012
10,000	E human influence on *Anopheles gambiae*	Donnelly et al. 2002
10,000	GE *Aedes aegypti formosus* from sub-Sahara to humid Sahara	Tabachnick 1991
10,000	LCA *Culex pipiens* var. *molestus-Culex pipiens*	Fonseca et al. 2004
7,500	E *Anopheles gambiae* associated with agriculture	Loaiza et al. 2012
7,000	LCA *Anopheles gambiae-Anopheles quadriannulatus*	White, Collins, and Besansky 2011
4,000	E domestication of *Aedes aegypti*	Tabachnick 1991
4,000	LCA *Aedes aegypti formosus-Aedes aegypti aegypti*	Tabachnick 1991
3,250	LCA *Anopheles gambiae-Anopheles quadriannulatus*	Coluzzi et al. 2002
2,500	E *Anopheles gambiae* expanded with iron tools in agriculture	White, Collins, and Besansky 2011
1,200	GE *Aedes aegypti aegypti* from Sahara to West Africa	Tabachnick 1991
523	E first yellow fever in Haiti	Tabachnick 1991
517	E yellow fever virus Africa-NA	Tabachnick 1991
500	GE *Aedes aegypti* from West Africa to Spain and Portugal	Tabachnick 1991
400	GE *Aedes aegypti* to Western Hemisphere	Tabachnick 1991
370	E first yellow fever in Yucatan	Tabachnick 1991
160	GE *Aedes aegypti* to eastern Asia	Tabachnick 1991
39	GE *Aedes albopictus* Asia-Albania	Urbanelli et al. 2000
28	GE *Aedes albopictus* to Italy	Urbanelli et al. 2000

The nature of the events are as follows: LCA = last common ancestor (i.e., point in time when two taxa emerged from a single ancestor species); ECA = earliest common ancestor (i.e., point in time when a group first emerged); OF = oldest fossil found to date; F = significant fossil; R = radiation of a group into more species; GE = geographic expansion of range; E = general event.

(Gould 2001), "Given the same conditions, a replay of evolution would not produce the same product as we observe today" (Sepkoski 2001), "Man might be considered the reason for the existence of the entire organization of life on our planet" (Bergson 1907, as cited in Nitecki 1988), and that natural selection "achieves the aspect of purpose without the implementation of a purposer" (Simpson 1947, as cited in Nitecki 1988), as well as "Simply finding 'time's arrow' is not enough" (Raup 1988). These quotations raise ideas that can be the sources of endless debate. The early days of evolutionary theory seemed to assume that progress toward a better product was a logical necessity of natural systems (Wimsatt & Schank 1988), with the implication that humans are not just a recent product of evolution but also the most superior. Such judgments are almost outside the frame of reference of modern biologists, yet most individual scientists would look on the biological history of their areas of specialization and see what appears to be an increase in complexity and variety. Perhaps the most confounding challenge to an objective appraisal of evolutionary progress is summarized by Raup (1988) in a review of the mathematical inevitability of changes that resemble progress in systems where each step influences the next, and reversal of such changes is not possible. Assuming that at some point the ancestor species of all mosquitoes arose from its chaoborid-like antecedents, the variety of Culicidae we see today would argue for an increase in complexity and diversity. If that is the direction of mosquito evolution, then it must be shared with many other families that are equally or even more varied. Other speculative trends in evolution of the family—like increasing adaptation to human-dominated habitats—seem to be more about the response to current conditions than progress toward a goal. To quote Stephen Jay Gould (1988), "We can preserve the deeper (and essential) theme of direction in history while abandoning the intractable notion of progress."

History of Culicidae

The evolutionary history of Culicidae, like the history of almost any other group, is commonly expressed as a cladogram—the familiar diagrams in which bifurcating branches represent lineages (usually of species or populations), and their lengths are proportional to genetic distance and/or time since separation. These diagrams are so powerfully graphic that it is easy to forget that they can represent widely different kinds of data. Sometimes they are intended to represent the author's summary opinion of evolutionary history, first represented for mosquitoes by Ross (1951) and famously represented for all life by Charles Darwin and Ernst Haeckel. More recently, cladograms are usually a visual representation of quantitative analysis. These cladograms are not a representation of history but a summary of data on the measured relationships of taxa (e.g., Vesgueiro et al. 2011). Almost coming full circle to the original purpose of cladograms, recent work has integrated many sources of data to produce a best estimate of the evolutionary history of a group, beautifully represented by Wiegmann et al. (2011) for Diptera.

Willi Hennig (1965) developed the principles of quantitative cladograms constructed for the purpose of representing phylogeny, coining the term "phylogenetic systematics." He created the field as an activity that would be independent from the kinds of data inputs. As a result, the basic principles have been remarkably accommodating to new technology as it provides other kinds of data. The goal of phylogenetic systematics is to group taxa into monophyletic relationships (common evolutionary ancestor) detected by quantification of synapomorphic features (similar derived features shared by descendent taxa). Monophyly supported by synapomorphic characters must be distinguished from polyphyletic relationships (similar in observed characters but derived from separate lineages) and paraphyletic relationships (based on features not shared by all members of the actual monophyletic group). Evidence that appears synapomorphic may actually be convergent (characters known as homoplasies that arose separately in unrelated lineages) or symplesiomorphic (shared derived characters that only appear sporadically in the lineage).

The review of mosquito evolutionary history presented here is based mainly on the work of authors who adhered to the basic principles of phylogenetic systematics. Those principles are not without controversy and alternatives. For example, evolutionary systematics considers paraphyletic relationships as useful indications of "grade," in contrast to phylogenetic systematics' close adherence to "clade" (Hull 1988). Also, the core concepts of phylogenetic systematics seem circular or relative if examined closely. A polyphyletic group might be monophyletic if considered from a larger scale; groups rejected because they are paraphyletic might simply need more study; symplesiomorphy might be synapomorphy disguised by variation in expression. Finally, the two-dimensional representation and strict use of binary branching seem to limit our conception of biological systems that are undoubtedly more complicated. All that said, cladograms based on phylogenetic principles have been a successful tool for the summary of a rapidly expanding data set. Their interpretation as evolutionary history, however, must be taken with caution and constantly checked against other kinds of evidence.

One of the great strengths of phylogenetic analysis is that the method can use many kinds of data if those data can be expressed in a series of discrete choices. DNA or RNA sequence data are well suited, as are morphological data that can be expressed as presence or absence. The lack of suitability of quantitative data—for example, lengths and ratios of measurements—is a limitation that has been addressed by categorizing

Willi Hennig and Phylogenetic Systematics

Rudolf Meier (2005) describes the life and work of Willi Hennig, the originator of phylogenetic systematics. This theoretical framework, based on simple key principles of binary divergence and strict interpretation of data, has become the basis of such biological effort because of the availability of molecular and digital tools. Our current view of evolution is strongly influenced by the cladograms that are now ubiquitous in the literature. They summarize qualitative estimates of relationships and histories that were formerly portrayed as "trees," which in Darwin's *Origin of Species* appeared literally as a tree used as an analogy of evolutionary history. Much as Darwin's theory can be summarized in the simple list of overproduction, variation, selection according to fitness, and fidelity of inherited traits, Hennig's work can be summarized. The essential elements of phylogenetic systematics are (1) phylogenetic relationship, (2) monophyly, (3) paraphyly, (4) apomorphy and plesiomorphy, (5) cladism or cladistics, (6) parsimony, and (7) outgroup comparison or outgroup rooting.

The life of Willi Hennig is a significant reminder of variation in scientific path. Starting out as a bright young student who graduated from studies of extinct flying reptiles to studies of flies, he was caught up on the German side of World War II, injured on the Russian front, and eventually imprisoned by the Americans as he served in an anti-malaria unit in Italy. During about six years in prison, he wrote the first version of his phylogenetic systematics book in a notebook. He spent his entire career in Germany, leaving a position in East Germany when the Soviets shut the border in Berlin. He gradually developed his ideas to the structure of phylogenetics that is the foundation of such a wide variety of biological studies today. Tragically, he was only 63 years old when he died in 1973.

quantitative differences into a series of choices, sometimes through complicated morphometric analysis (Zelditch et al. 2000). Probably partly because of the inherent structure of the data and partly because of its abundance, morphology and genetic sequence have dominated phylogenetic analyses of the evolutionary history of mosquitoes. Other sources of relevant information exist, especially biogeography and the fossil record, but these have not been used as extensively.

Morphology and genetic sequence data often seem to be set up as rival methodologies, perhaps because each requires scientific skills that can absorb all the effort of individuals. Of course, the two sources of information are complementary rather than competitive, a point that mosquito studies have clearly shown. Hillis and Wiens (2000) explained contrasting characteristics of morphological and genetic sequence data (Table 1.2), concluding that a thorough use of either method usually results in similar phylogenetic conclusions. Historically, taxonomy started with morphological criteria, and functionally, any analysis of mosquitoes starts with collection of physical specimens. A physical examination of a specimen potentially reveals all of the results of many genes in an information-dense form, and that revelation may occur during a brief microscopic view. The conveyance of information in a description of the morphological result may require a trained person to receive it, but the information is a direct representation of the insect as observed in nature. Potentially, a much wider audience can use a morphological description rather than a genetic sequence for the purposes of identification and comparison and to simply develop a personal concept of the species. On the other hand, sequence data provide almost unlimited scope for comparisons. In theory, sequence data are not variable, or are variable within only narrow confines for a particular species. Such relatively low ambiguity is in great contrast to morphological data, which for a given specimen may depart considerably from the defined type. Ironically, identification with molecular tools often requires much less training than morphological identification, even though the techniques of molecular biology represent some of biology's most recent developments.

Concepts of dipteran evolutionary history change at a much greater rate than the flies themselves. The order Diptera is certainly an old group, probably originating in the late Permian (260–251 MYA) and settling into major lineages by the late Triassic (235–202 MYA; Yeates and Wiegmann 2005). Diptera and Strepsiptera are currently considered sister orders, the two most closely related to the mecopteran orders that include fleas (Whiting 2005). The concept of dipteran lineages changed from the old system of Nematocera (lower flies), Brachycera, and Cyclorapha (higher flies) to a series of infraorders and superfamilies (Bertone et al. 2008) that include a number of groups among lower flies and that groups other flies together in the infraorder Brachycera (Yeates & Wiegmann 1999, 2005). The infraorder that includes mosquitoes is Culicomorpha, which in turn is divided into either three superfamilies (Culicoidea, Chironomoidea, and Simulioidea; Bertone et al. 2008) or two superfamilies (Culicoidea and Chironomoidea; Yeates & Wiegmann 1999, 2005). There is strong molecular support that the families of Culicoidea are monophyletic, consisting of Chaoboridae, Corethrellidae, Culicidae, and Dixidae (Miller et al. 1997; Wiegmann et al. 2011; Fig. 1.1). The use of genetic sequence differences between taxa to calculate the years since their separation is tempting, but it must be used cautiously because different parts of the genome can change at different rates (Arbogast et al. 2002). Error terms and comparison to other methods can help maintain realistic estimates.

The close relationship of Diptera to sugar feeding as supported by special feeding, digestive, and sensory systems (Downes & Dahlem 1987) has led to speculation that diversification of flies could not have occurred prior to that of flowering plants (angiosperms). Were this the case, then the major lineages of flies could not have developed until the origin of angiosperms in the early to middle Jurassic (179–158 MYA) or even until their diversification into modern lineages in the Cretaceous (147–131 MYA; Wikström et al. 2001). However, plants that preceded angiosperms (e.g., *Gnetum*, *Ephedra*, *Cycas*, and *Zamia*, as well as some extinct groups) produced sugary exudates probably associated with attracting insects for pollination services (Labandeira 1998, 2005).

Within the family Culicidae, relationships between subfamilies, tribes, genera, and species rest on the taxonomic foundation described by Edwards (1932a). More recent efforts using quantitative treatment of both molecular and morphological characters have tended to validate, rather than repudiate, the concepts held at Edwards' time, a subject further discussed in Part 2. The relative stability of Edwards' classifi-

Table 1.2. Comparison of attributes of morphological and genetic sequence data for phylogenetic studies (Hillis and Wiens 2000)

Genetic sequence	Morphology
Objective selection of characters	Characters often selected by tradition or convenience
Gene phylogeny may not be the same as	Phylogeny is for entire organism for species
Limited or no fossil data	Fossil data available
Requires specially preserved material	Wide variety of preservation methods
Usually based on morphological first	First identification usually morphological assumptions
Nearly unlimited number of characters	Characters limited to those that are observable
Characters more likely to be independent	Characters are less likely to be independent

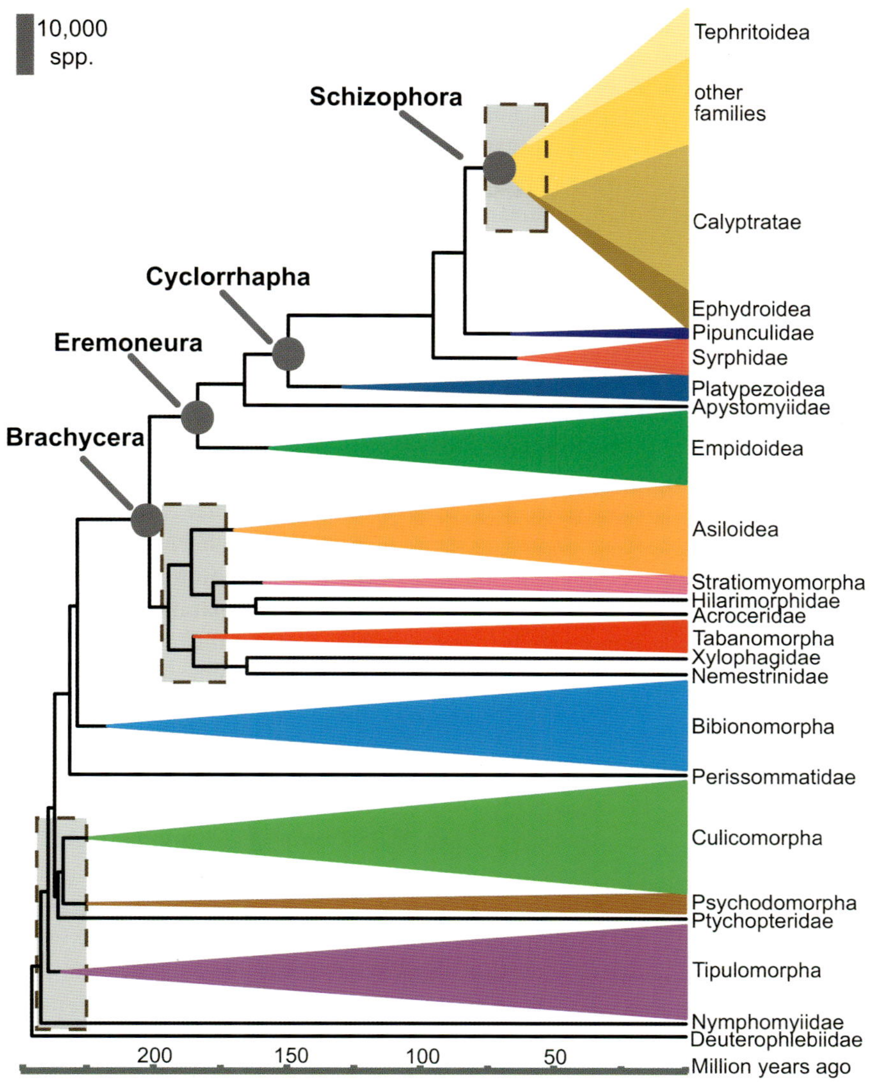

Figure 1.1. Dipteran phylogeny with estimated dates of major origins. Rapid radiation is in shaded boxes. The height of the triangles is proportional to the number of species, in accordance with the scale bar. Wiegmann et al. 2011

cation is remarkable in that he worked with only 1,400 species compared to the current number of 3,527, an increase of 152%. One of the principle differences between Edwards' general view of the family and modern concepts was that he considered the mosquitoes themselves, our current Culicidae, to be a subfamily of a larger family including Chaoboridae and Dixidae. Such a structure is phylogenetically justifiable, as the three groups (Chaoboridae has been split into Chaoboridae and Corethrellidae) are now considered to have had a common ancestor (Wiegmann et al. 2011). In fact, this structure was supported as late as Belkin's (1962) treatise on mosquitoes of the South Pacific and was not officially changed until Knight and Stone (1977).

Edwards' division above genus within the modern concept of Culicidae was probably the largest departure from current concepts of the systematics of the group. He divided mosquitoes into three tribes that would be equivalent to subfamilies today. Edwards' subfamilies would have corresponded exactly to the modern division of Anophelinae, Toxorhynchitinae, and Culicinae; however, recent 18S rDNA sequence analysis (Miller et al. 1997), morphological phylogenetics (Harbach & Kitching 1998), and mtDNA (COI and COII) and tRNA leucine sequence comparisons have led to the conclusion that *Toxorhynchites* is a one-genus tribe within Culicinae (Harbach 2007).

Edwards divided the currently defined Culicinae into five "groups." Quoting directly from his treatise:

1. The *Sabethes* group, including the genera *Harpagomyia, Topomyia, Tripteroides, Goeldia, Trichoposopon, Wyeomyia, Limatus, Sabethoides* and *Sabethes.*
2. The *Uranotaenia* group, including the three genera *Zeugnomyia, Hodgesia* and *Uranotaenia.*
3. The *Theobaldia-Mansonia* group, including the five genera *Theobaldia, Orthopodomyia, Ficalbia, Mansonia* and *Aëdeomyia.*
4. The *Aëdes* group, including the genera *Aëdes* (with its many subgenera), *Psorophora, Haemagogus, Heizmannia, Eretmopodites, Armigeres* and *Opifex.*
5. The *Culex* group, including the large genus *Culex* (with many subgenera) and the somewhat nearly related genus *Deinocerites.*

Only three of the groups (Sabethini, Aedini, and Culicini) corresponded to modern concepts. The other two groups, Theobaldia (*Theobaldia* (= *Culiseta*), *Orthopodomyia*, *Ficalbia*, *Mansonia*, *Aedomyia* (= *Aedeomyia*)) and Uranotaenia (*Zeugnomyia*, *Hodgesia*, and *Uranotaenia*) contained genera that are no longer considered to be related to each other.

Edwards described 30 genera, compared to the current system with 41 genera (Harbach 2007; Wilkerson et al. 2015). One of the old genera, *Sabethoides*, was later reduced to a subgenus (of *Sabethes*)—all the others still exist in a form similar to their original concepts, have been emended (*Aedomyia* to *Aedeomyia* and *Eretmopodites* to *Eretmapodites*), have changed in name but not concept (*Megarhinus* to *Toxorhynchites*, *Harpagomyia* to *Malaya*, and *Theobaldia* to *Culiseta*), or have split into multiple genera (*Goeldia* to *Johnbelkinia*, *Shannoniana*, *Runchomyia*, and *Isostomyia*). Of the 13 genera developed since Edwards' revision, seven are resurrections of subgenera (*Verrallina*, *Isostomyia*, *Maorigoeldia*, *Runchomyia*, *Lutzia*, *Mimomyia*, and *Coquillettidia*) with Edwards' concepts for these groups very close to current concepts. Four of the new genera (*Johnbelkinia*, *Kimia*, *Onirion*, and *Shannoniana*) represent novel concepts, though they include some species known in 1930. Only one of the modern genera (*Galindomyia*) includes a species completely unknown in 1930. Despite reorganization of the family and changes in generic status, it is remarkable that some 2,000 species described since 1930 fit so well into the taxonomic structure described by Edwards (1932).

Ross (1951) presented the first estimate of phylogenetic relationships among subfamilies and genera of mosquitoes. Based on morphology and intuition, his analysis matched some of the modern interpretations and departed from others. It took another 47 years before anyone attempted to examine the entire family's relationships at the genus level. Harbach and Kitching (1998) analyzed 73 adult, larval, and pupal morphological characters coded for 38 genera. Their study not only provided a starting point and structure for subsequent testing but also stimulated more phylogenetic research and advanced the study of mosquito morphology. The discussion that follows, attempts to present the conclusions about relationships that are documented by data analysis, pointing out some of the current disagreements and knowledge gaps.

Starting most generally, the monophyly of Culicidae is supported, placing it as a sister group to Chaoboridae and Corethrellidae (Wiegmann et al. 2011 [molecular]; Miller et al. 1997 [18S rDNA]; Harbach and Kitching 1998 [morphology]). Although morphological definitions of the family can go on for many pages, with documentation of the variation for each morphological character (Belkin 1962), the succinct diagnostic description by Edwards (1932a) has largely been repeated by other treatments of the family (e.g., Carpenter & LaCasse 1955; Harbach 2007). The group of families including Dixidae, Chaoboridae, Corethrellidae, and Culicidae is characterized by unique wing venation, small first antennal segment, enlarged second antennal segment, divided pronotum with posterior position forming parts of the lateral part of the thorax (pleuron), lack of a suture between the prescutum and the scutum, presence of an anepisternal cleft on the pleuron, and long legs. The larvae are legless and aquatic, with complete head capsules, a single pair of spiracles on the dorsal side of the eighth tergite (except *Chaoborus*, which lacks spiracles), part of a spiracular apparatus formed from elements of the eighth and ninth segments, and anal papillae projecting from the tenth abdominal segment. The most distinctive characteristic of mosquitoes is the elongated proboscis of the adults. All members of the family have scales on wing veins and other parts of the body, though this characteristic is shared with chaoborids and corethrellids. The larvae share the characteristics of an unsegmented thorax and lack of a distinct ninth abdominal segment with Chaoboridae and Corethrellidae but do not have the prehensile antennae that those two families use to capture prey. The spiracular plate of Anophelinae is not easily distinguished from those of the chaoborids and corethrellids, but the spiracular siphon of Culicinae is completely distinctive.

Another firm phylogenetic conclusion is that Anophelinae is monophyletic and basal to the remainder of the family (Trebatoski & Haynes 1969 [starch gel electrophoresis]; Miller et al. 1997; Harbach & Kitching 1998; Sallum et al. 2000b [morphology]; Sallum et al. 2002a [mtDNA COI, COII; 18ssr nuclear DNA, D2 region 28S rDNA]; Krzywinski et al. 2001b [mtDNA ND5; nuclear DNA G6PD and *white* gene, D2 region of 28S rDNA]; Harbach 2007 [review]; Reidenbach et al. 2009 [six protein-coding nuclear genes and morphology]). The monophyly of Culicinae (in the sense of including *Toxorhynchites* within the tribe Toxorhynchitini [Belkin 1962; Miller et al. 1997; Harbach & Kitching 1998; Mitchell et al. 2002 {mtDNA COI, COII, tRNA leucine}; Harbach 2007]) is not so explicitly shown because Culicidae and the only other subfamily, Anophelinae, are monophyletic; the monophyly of Culicinae is certainly implied.

The relationships of the tribes within Culicinae have been interpreted in various ways. For the most part, it is difficult to draw firm conclusions where analyses differ. The relationships of the tribes Aedeomyiini (*Aedeomyia*), Culisetini (*Culiseta*), Ficalbiini (*Ficalbia*, *Mimomyia*), Hodgesiini (*Hodgesia*), Orthopodomyiini (*Orthopodomyia*), Toxorhynchitini (*Toxorhynchites*), and Uranotaeniini (*Uranotaenia*) are uncertain (Harbach & Kitching 1998; Harbach 2007), though the summary phenogram of Reidenbach et al. (2009) suggested that *Culiseta* is basal to *Culex*, which in turn is basal to *Uranotaenia*. Furthermore, their analysis showed *Coquillettidia* as sister to *Orthopodomyia*, the pair of which were sister to Aedini. They also suggested a relationship between the three genera *Aedeomyia*, *Mimomyia*, and *Toxorhynchites* with Sabethini. Clearly, more data and interpretation will be necessary to sort out the probable phylogenetic relationships of these groups in Culicinae.

Other conclusions are firmer about groups within Culicinae. First, the tribes Aedini, Culicini, and Sabethini are each considered to be monophyletic (Judd 1996; Harbach & Kitching 1998; Harbach 2007; Reidenbach et al. 2009). Harbach and Kitching (1998) concluded that Aedini is the basal sister to Mansoniini and that Culicini is the basal sister to Sabethini, though other work has neither supported nor questioned those relationships. Their phylogenetic analysis also strongly supported the sister relationship of Aedini + Mansoniini and Culicini + Sabethini. Some authors have seen an affinity between *Toxorhynchites* and Sabethini (Belkin 1962; Mattingly 1971b; Mitchell et al. 2002 [reanalysis of *white* gene sequence]). Only the phylogenetic relationships of the two subfamilies and 5 of the 11 tribes are understood with confidence.

Judd (1996) reported on an extensive morphological study of the tribe Sabethini. The conclusions most strongly supported by the data were that Sabethini is monophyletic, that the aggregate group of sabethine genera in the Western Hemisphere (*Isotomyia*, *Johnbelkinia*, *Limatus*, *Phoniomyia*, *Runchomyia*, *Sabethes*, *Shannoniana*, *Trichoprosopon*, and *Wyeomyia*) are monophyletic, and that *Maorigoeldia* (monotypic from New Zealand) is the sister to the New World sabethines. Judd also pointed out where the current classification is not phylogenetically correct—specifically, that *Runchomyia*, *Wyeomyia*, and *Tripteroides* are not monophyletic and that Old World sabethines are paraphyletic. Reidenbach et al. (2009) indicated that the genus *Malaya* (Ethiopian and Oriental regions, 12 species) is basal to other sabethines and that *Maorigoeldia* is sister to *Tripteroides*. Harbach (2007) reviewed the addition of the genus *Kimia* (five species) from the Oriental region and *Onirion* (seven species) from the Neotropical region, challenging the concept that New World sabethines are monophyletic. This highly diverse, mainly tropical tribe will continue to be a target for further refinement, but the resolution of the monophyly of the New World genera would be an especially welcome development. An understanding of the history of a group that appears to have originated in Asia but diversified extensively in the Neotropics would have general implications for biogeography and evolution.

Genus-level phylogenetic relationships are similarly difficult to make with confidence, a subject further discussed in Part 2 of this book. *Chagasia* is considered basal within Anophelinae (Harbach 2007; Sallum et al. 2000b; Krzywinski et al. 2001a, b). It has been suggested that *Bironella* be made a part of *Anopheles* based on morphological evidence (Sallum et al. 2000; Harbach 2007), but molecular studies have suggested enough difference to maintain *Bironella* as a genus for the time being (Krzywinski et al. 2001b; Sallum et al. 2002c; Krzywinski & Besansky 2003 [review of molecular data]; Reidenbach et al. 2009). Although *Lutzia* is considered a separate genus within Culicini (Tanaka 2003c), an analysis of the secondary structure of ITS2 rDNA showed that *Lutzia bigoti* was not as different from subgenus *Culex* as other subgenera, suggesting that *Lutzia* might more accurately be considered a subgenus of *Culex*. Genus-level nomenclature of Aedini was in a state of controversy following morphological analyses by Reinert and colleagues (Reinert 2000a; Reinert, Harbach, & Kitching 2004, 2006, 2008, 2009), reflecting uncertainty about phylogenetic relationships within this important tribe. Wilkerson et al. (2015) restored the status of the genus *Aedes* to its pre-2000 status and revised subgeneric and species groups to reflect the phylogenetic progress made by Reinert and colleagues. Therefore, *Verrallina* (Reinert 1999c) was retained, while many other genera, including *Ayurakitia* (Reinert 2000h) and *Borichinda* (Harbach, Vu, & Kitching 2007), were redesignated at levels below genus. Supraspecific classification within Culicidae is widely discussed in the literature, especially as molecular evidence is joined with quantitative morphological evidence. Harbach's (2007) review of mosquito phylogeny presents in great detail the various opinions and the data that support them. His discussions of possible rearrangement of the genera in Anophelinae is a good illustration of how natural classifications based on phylogenetics and supported by cladistic analysis can sharply conflict with phenetic classification created to support identification and categorization. Whether the natural classification of Anophelinae results in a minor modification by placing *Bironella* in the subgenus *Anopheles* (Sallum et al. 2000b) or a major change by placing all species of Anophelinae in genera *Chagasia*, *Kerteszia*, *Nyssorhynchus*, and *Cellia* (Harbach 2007) or the suggestion to establish five new genera in Anophelinae based on mitochondrial protein sequences (Foster et al. 2017), natural classification schemes are based on dynamic evidence that changes depending on data and analysis. The stability of natural or phylogenetic classifications will presumably improve over time, but they are perhaps best treated as hypotheses to be debated and examined until they reach a certain level of confidence independent from their authors. Researchers performing resource-intensive phylogenetic studies have all used subsets of the taxa examined, selecting those species that were available for testing and considered to be representative. Such selection is necessary, but it creates an inherent bias that makes general conclusions difficult. Shepard et al. (2006) provided a particularly clear example of an attempt to draw general conclusions from a small and regional selection of species. Using only 39 species from the northeastern United States, the authors used 18S rDNA sequences to hypothesize about generic relationships and the separation of *Aedes* and *Ochlerotatus* despite relatively low levels of variation. On the other hand, traditional taxonomists have been able to perform revisions of taxa in which every species was examined, largely because morphological examination, especially when it is descriptive rather than quantitative, is relatively inexpensive in terms of time and resources. The conclusions presented in Figure 1.2 are probably worthy of consensus but leave many gaps in our understanding of mosquito evolution at levels that should be of great biological interest. Regardless of the lively nomenclatorial debate, typified by what have become multiple systems for aedine genera, the data on actual phylogenetic relationships and their implications for evolutionary history continue to accumulate.

Studies of phylogenetic relationships among small groups of species are scattered in the family and are sometimes motivated by importance of a group or by opportunity. These studies have an inherent advantage over attempts to determine relationships among genera and tribes in that they usually include all of the species of interest rather than a subsample. Sometimes

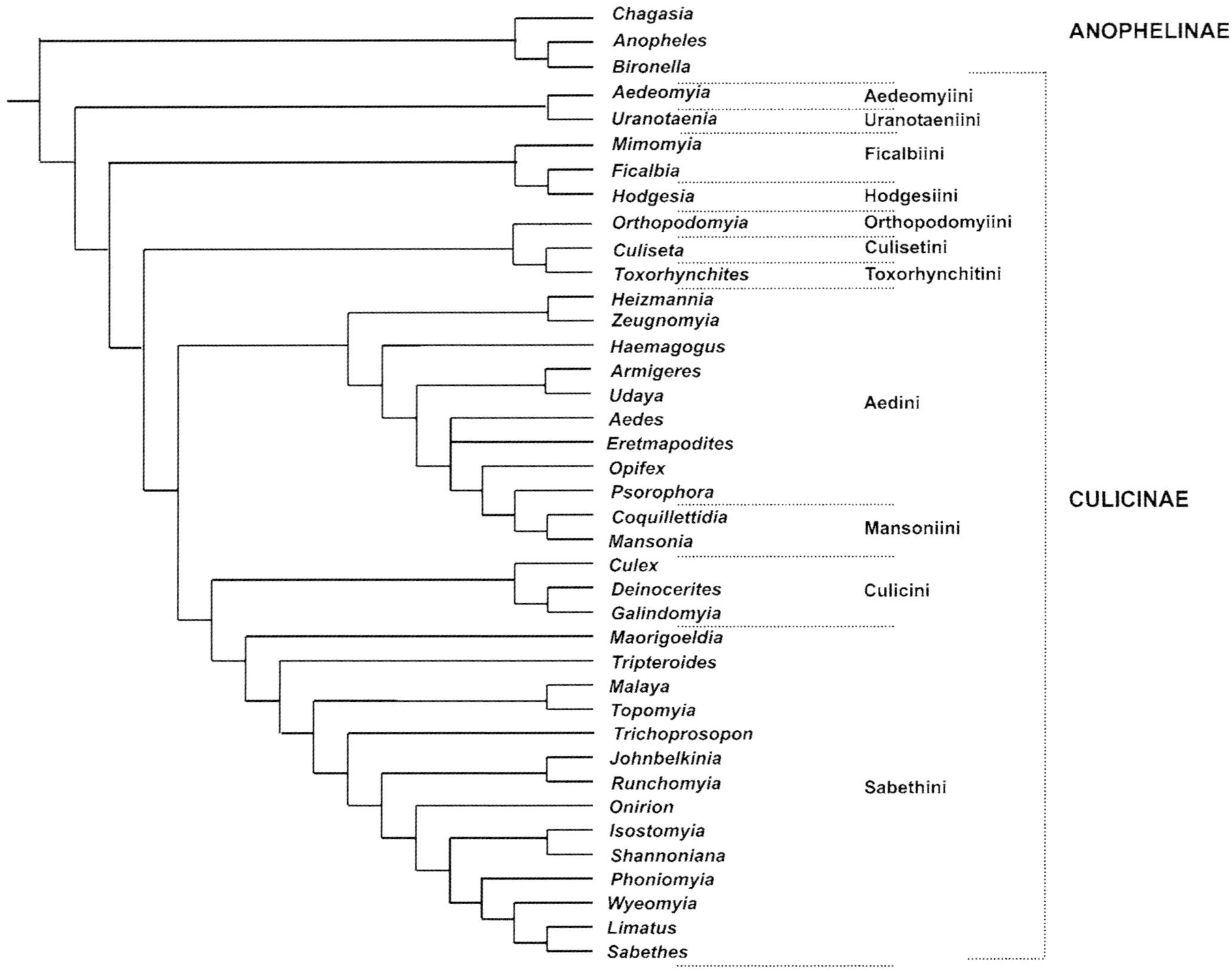

Figure 1.2. Morphologically based phylogeny of 38 mosquito genera based on 73 characters. Harbach 2007

a study with a different purpose uncovers sister relationships between pairs of species (Vesgueiro et al. 2011: *Culex nigripalpus* and *chidesteri*, *bidens*, and *declarator*). Other studies find that old designations of species actually represent more than one species, implying that the new species group is phylogenetically related (e.g., Cockburn et al. 1988; Reinert et al. 1997: *Anopheles quadrimaculatus* divided into *diluvialis*, *inundatus*, *maverlius*, *smaragdinus*, and *quadrimaculatus*; Strickman & Pratt 1989: *Culex corniger* divided into *lactator* and *corniger*). To the extent that nomenclature reflects natural classifications, any subgeneric grouping (e.g., Harbach 2004) represents phylogenetic relationships; however, just as in the case of most of the culicid tribes, the relationships among the subgeneric groups are often obscure. Therefore, the understanding of mosquito phylogeny is strongest at the highest level (related families and subfamilies) and the lowest level (multiple species groups) but weak in between (tribes, genera, subgenera).

Four groups of species are presented here as examples of detailed phylogenetic studies, none of which may have definitive results but all of which have been developed to the point of strong hypotheses. The first example is *Aedes aegypti*, one of the most important species in the world as a vector of yellow fever virus, dengue viruses, Zika virus, and chikungunya virus. A thorough review of many quantitative genetic studies (Tabachnick 1991) presented *Aedes aegypti* as two subspecies: a sylvan form, *Aedes aegypti formosus*, and a domestic form, *Aedes aegypti aegypti*. *Aedes aegypti formosus* was characterized as the original zoophilic form from the African humid tropics. It is described as darker, with white scales on the first abdominal tergite, and ovipositing in natural containers like tree holes. Using mainly electrophoretic enzyme characterization of specimens from more than 80 locations around the world, Tabachnick proposed that the genetic difference between sylvan and domestic forms was real but that West African populations could not be separated that way. The data also showed a significant difference between populations in the southeastern United States (which were similar to Asia) and those in western Texas and Mexico. The Caribbean populations and East African domestic populations were quite heterogeneous. To explain these observations, Tabachnick proposed the following evolutionary history. About 10,000 years ago, the original sylvan form moved north as the Sahara became more humid. When the area became drier about 4,000 years ago, natural larval sites disappeared, and the isolated population started using domestic sites instead. In this way, the *Aedes aegypti aegypti* subspecies was formed and later moved with the salt and gold trade to West Africa sometime in the ninth century. From there, it was transported to the Iberian Peninsula in the 15th century. *Aedes aegypti aegypti* probably moved multiple times from West Africa to the New World during the slave trade in the 15th through 17th centuries. The lack of heterogeneity in Asia supports the suggestion that the species was not introduced there until about 1850, when urban dengue first appeared in the region. The similarity between the populations in the southeastern United States and in Asia could have been due to reinvasion from Asia in the 1970s following the dramatic decline in abundance after World War II. Heterogeneity in the Caribbean may have been caused by repeated eradication attempts in the 1960s and 1970s, resulting in a series of recolonizations with different genetic stocks. West Africa's lack of distinction between sylvan and domestic forms may have been due to hybridization between the two subspecies, made more likely by the early colonization of West Africa by *Aedes aegypti aegypti*.

A more recent study (Brown et al. 2011) used microsatellite markers to examine 24 populations from 5 continents and 13 countries. This study confirmed the African origin of the species based on the greater genetic variety there. It found that the major genetic division was between African and non-African material, for which the authors maintained the subspecies designations of *Aedes aegypti formosus* and *Aedes aegypti aegypti*, respectively. They noted the presence of a non-African type in Kenya sympatric with the African type. They proposed that the West African population's domestic and sylvan habits were the result of hybridization and that domestication by *Aedes aegypti formosus* had occurred many times. Although the details of the evolutionary history of *Aedes aegypti* vary between the two studies, they both support the basic view that African *Aedes aegypti* originated as a sylvan form that has produced populations capable of close association with humans. The importance of hybridization and reintroduction adds further dimension to the dynamic nature of this species' genetic history.

The *Culex pipiens* group of species (designated "Pipiens Assemblage" by Harbach [2012]) presumably represents descendants of a common ancestor, with four extant species recognized by some systematists: *Culex pipiens*, *quinquefasciatus*, *globocoxitus*, and *australicus* (Farajollahi et al. 2011). Disagreement about nomenclature was startlingly obvious in a review of the group composed of articles by experts in the field. Authors variously followed the catalog designations used here (Harbach 2012); *pipiens*, Form Molestus, and *quinquefasciatus* as subspecies (Savage & Kothera 2012; Reisen 2012; Cornel et al. 2012); *pipiens* and *quinquefasciatus* as subspecies but Form Molestus as a biotype of *Culex pipiens pipiens* (Becker et al. 2012a); and *pipiens*, *pallens*, *quinquefasciatus*, and Form Molestus as species (Russell 2012). Harbach (2011) took the position that the use of subspecies was not helpful in this case or for other mosquitoes. *Culex pipiens* and *quinquefasciatus* are widely distributed, with *pipiens* divided into subspecies (*Culex pipiens pipiens* and *Culex pipiens pallens*). *Culex pipiens pipiens* is further divided into Form Molestus and Form Pipiens, corresponding to autogenous/stenogamous and nonautogenous/eurygamous types, respectively. Form Molestus hybridizes with Form Pipiens, at least in Portugal and the United States (Gomes et al. 2012b; Fonseca et al. 2004). In addition, *Culex pipiens* and *Culex quinquefasciatus* hybridize extensively in North America, Argentina, Madagascar, and eastern Asia. Harbach (2012) called them "pipquin populations." The origin of *Culex pipiens pipiens* is considered to be Africa and that of *Culex quinquefasciatus* to be Southeast Asia, though both places of origin are disputed. Both species have spread significantly from these sources, supposedly through movement in containers on human conveyances. The hybridization between the forms of *Culex pipiens pipiens* and between *Culex quinquefasciatus* and *Culex pipiens* (both forms) over large areas has created a web of traits that make this group of species highly adaptable to many conditions (Huang et al. 2011; Strickman & Fonseca 2012). Extensive hybridization has limits, as in southern Africa, where sympatric *Culex pipiens* and *Culex quinquefasciatus* do not hybridize, apparently due to *Wolbachia*-induced incompatibility (Cornel et al. 2003). The influence of hybridization in the group of species was elegantly indicated by Fonseca et al. (2009) in a series of specimens from eastern Asia. They showed that *Culex pipiens pallens* is a hybrid between *Culex quinquefasciatus* and *Culex pipiens pipiens* but in a special way that completely excludes expression of the *Culex pipiens pipiens* type. Males appear to have sex-linked *quinquefasciatus*-type acetylcholinesterase-2 genes (ACE) on the male portion of the sex determination site and the *pallens* type on the female portion, whereas females are homozygous for the *pallens* type. This represents introgression and fixation of the *quinquefasciatus* ACE into a nuclear gene, an example of how hybridization can create a new form. *Culex pipiens pipiens* Form Molestus appears in parts of the region, probably introduced during World War II into Japan and from Russia into China in the 1980s (Mogi 2012), and occasionally appears in the microsatellite signature of *Culex pipiens pallens*. Mogi (2012) considered *pallens* to be a separate species, possibly paralleling the adaptation of southern forms to northern climates in eastern Asia.

The Leucosphyrus Group of *Anopheles* subgenus *Cellia* consists of 20 species in Southeast Asia and India, six of which are important vectors of human malaria parasites (Sallum et al. 2005b, 2007). Recognized by a distinctive white band spanning the distal femur and proximal femur of the hind leg and by the wavy shape of Cu_1 and M veins of the wing, they were acknowledged as a group in the 1950s by D.H. Colless. EL Peyton initiated a long-term study of the group in 1977 with the support of the US Army. Notable entomologists, including B.A. Harrison, Visut Baimai, Chris Green, and Sylvie Manguin, applied a variety of morphological, cytological, and genetic methods to refine the taxonomy and systematics

demonstrated clear premating barriers, genome-wide divergences, and a pattern of SNPs that indicated a widening gap in evolutionary direction.

These species were identified by laborious examination of polytene chromosomes from larval salivary glands or from ovarian nurse cells, with the exception of *coluzzii* and *amharicus*. Significantly, karyotyping was able to distinguish all species except the former *Anopheles quadriannulatus* A and B forms, now separate species. Fixed differences in rDNA provided a more convenient method to identify all species but not to distinguish *Anopheles quadriannulatus* A and B or *Anopheles melas* from *merus*. RNA signatures for the latter two species are identical but unique within the complex, and the RNA can still be used to reliably identify both species because they are geographically disparate. Separation of the latter two species is not a problem since they are never found in the same location. The recognition of *amharicus* as distinct from *quadriannulatus* despite similarities in karyotype and some molecular markers illustrates how a deeper analysis of species is sometimes necessary to make good taxonomic decisions.

Some of the species pairs hybridize to produce fertile females but not males: *gambiae* × *arabiensis*, *gambiae* × *melas*, and *arabiensis* × *quadriannulatus*. This pattern of hybridization results from X chromosome incompatibility, affecting only males. Hybridization occurs at a low level in the field—for example, wild hybrid *gambiae* × *arabiensis* females were found at a frequency of 0.02% to 0.76% where the species occurred together in Tanzania (Ng'habi et al. 2011). Such low-level hybridization probably explains genetic introgression observed between the species.

Three phylogenies for the Gambiae Complex were proposed by White et al. (2011a). One, based on studies of X chromosome banding inversions, including of hybrids, showed *arabiensis* as basal, *gambiae* and *merus* as sisters, and *quadriannulatus* as sister to the related pair of *bwambae* and *melas*. The other phylogeny was based on mtDNA and autosomal sequences. That phylogeny showed *quadriannulatus* in the same relationship with *bwambae* and *melas*. It differed from the chromosomal phylogeny in that *merus* was indicated as sister to the related pair of *gambiae* and *arabiensis*. It is surprising that *quadriannulatus* did not appear to be clearly basal as it was hypothesized to be the species from which *gambiae* arose, with genetic input from *arabiensis*, as humans developed agriculture in the area (Donnelly et al. 2002; Coluzzi et al. 2002; White, Collins, & Besansky 2011). The doubtful relationship of *gambiae* and *arabiensis* leaves a lot of room for further study, as it is the relationship of greatest interest from a public health standpoint. The most recent attempt to establish the relationships among these species may have settled much of the question, based on analysis of full genome sequences (Fontaine et al. 2015). The autosomal chromosomes tended to group *Anopheles gambiae*, *coluzzii*, and *arabiensis*, whereas the X chromosome grouped *arabiensis* and *quadriannulatus*. Much of this pattern was explained by historical introgression dating back hundreds of thousands of years. The phylogeny that emerged from this study was that the Gambiae Complex arose from *Anopheles christyi* about 1.85 MYA. *Anopheles coluzzii* and *gambiae* form a tight sister-species pair, and a separate group includes *arabiensis*, *quadriannulatus*, and *melas*. *Anopheles merus* is relatively more distant from the other members of the complex.

The differentiation of *Anopheles gambiae* into two species has attracted special interest because of the importance of anything having to do with either controlling the species or assessing the risk from the species. Considerable work was done on the two species before they were recognized, distinguishing them as the M and S forms. As mentioned above, the five forms of the mosquito based on chromosome morphology seemed to be correlated with both ecological type and geographic location despite questionable phylogenetic significance (Gentile et al. 2002). *Anopheles coluzzii*, identified as the M form, was shown to occur in western Africa only, and *Anopheles gambiae*, identified as the S form, was revealed as present throughout the range of both species. Among the phenotypic differences recognized before the species were identified is the fact that *Anopheles coluzzii* tends to inhabit large, long-lasting larval sites associated with agriculture and the urban environment. *Anopheles gambiae*, however, usually occurs in small ground pools. Probably because of the larval habitat, *gambiae* is seasonal in the rainy season, whereas *coluzzii* continues development all year. *Anopheles gambiae* develops faster than *coluzzii* in small sites. Behavioral differences include predator escape mechanisms (holding still) by larvae of *coluzzii* but not of *gambiae* and variances in mating behavior (Gimonneau et al. 2010). Microarray scans of the genome showed that gene-poor regions near the centromeres of the chromosomes varied by 3% between the two species, leading to various speculations on how selection could occur in the absence of much expression (Lawniczak et al. 2010). Later, full genome sequencing of laboratory colonies, supported by examination of field populations, showed many genetic differences between the two forms. The sequencing also revealed that the original full genome sequence by Holt et al. (2002) inadvertently used genetic material from both species. Over 2 million SNPs and 150,000 fixed differences were distributed throughout the genome, with the highest concentration of differences on the X chromosome and more differences on chromosome 2 than 3. One area with many differences was a region with seven odor receptor genes and G protein-coupled receptors associated with neurohormones, leading to speculation that the genome at these locations reflected behavioral differences. The two species hybridize at a low level, observed as 1% in Mali and over 20% in westernmost Africa (White et al. 2011a). Another difference between the two species regards resistance to *Plasmodium* parasites. White et al. (2011b) showed that the form of a thioester-containing protein (*TEP1*) that triggers melanization of the parasite was more effective in *coluzzii* than *gambiae*.

The work performed to date on the Gambiae Complex probably provides the most accurate picture currently available of speciation in a group of mosquitoes. The process follows the principles of evolution in many specific ways, responding to new selective pressures and opportunities, sometimes in a relatively short amount of time. The ability of *Anopheles* to produce species complexes may be an inherent part of the genus' genetic architecture, described as selection for plasticity by Gimonneau et al. (2010) and suggested in a conversation with Maureen Coetzee (February 2012). Perhaps the least intuitive conclusion from the data set is the importance of horizontal gene transfer between what are considered separate species. Scarce hybridization that only results in fertile females apparently allows enough genetic contact to create meaningful gene flow between species in the complex. Both ecological and geographic separation within species causes chances for genetic isolation and separation. One result is that humans are faced with vectors finely tuned by evolution to the conditions for their own survival and, unfortunately, for transmission of a serious pathogen. The data set also gives the impression that there is no end result and that this important group of species will continue to change.

Biogeography

The difference between biogeography and distribution (Chapter 3) as used in this book is that biogeography implies logical connections between evolutionary history and location of taxa, whereas distribution only describes the temporal and physical location at a given geologic time. Most revisionary, taxonomy-based treatments about mosquitoes provide comprehensive discussions of distribution but not biogeography. Three recently developed fields should greatly improve the ability to generate biogeographic hypotheses about mosquitoes. They are global information systems that offer scalable views of spatial data with embedded analysis of correlates, ever more accurate descriptions of paleogeography and paleoclimatology, and construction of evidence-based phylogenies. Mosquito biologists have only started to apply these advantages to a better understanding of evolutionary history.

One of the first attempts at taking a larger view of mosquito biogeography was that by Belkin (1962) in the two-volume *Mosquitoes of the South Pacific*. Although limited to that region, the discussion provided a balanced view of how the various groups and species came to be where they are today in Southeast Asia, Australia, and the western Pacific Islands. Belkin recognized the existence of shared, distinctive, not-human-transported groups between Tasmania, New Zealand, and southern South America (Dixidae: *Nothodixa*) and between New Zealand and the Nearctic region (*Culiseta*, subgenus *Climacura* and *Aedes* subgenus *Ochlerotatus*). He divided the region into eight roughly sequential dispersal patterns, with particular emphasis on the significance of the New Zealand endemic species *Maorigoeldia argyropus* and *Opifex fuscus*, considered to be representatives of the oldest forms of the tribes Sabethini and Aedini, respectively. Belkin's geological framework did not include movement of continents, so the discussion frustratingly speculated on how island chains, orogenic belts, and sea level changes could account for dispersal patterns. Belkin came close to erecting his own theory of plate tectonics when he noticed in detail how some of his major dispersal patterns corresponded with geological features separated by deep submarine basins. Among Belkin's ideas worth reconsideration was his emphasis on the importance of archipelagos in the generation of new forms. He considered the Southeast Asia-Australasia region and the Caribbean (his Middle America or American Mediterranean) to be where major groups of mosquitoes originated in response to rapid environmental changes and physical distributional interruption.

Ross (1964) used detailed distribution data of contemporary mosquito taxa superimposed on intuitive phylogenies. Like Belkin, Ross did not have the advantage of modern plate tectonic theory, so he assumed movement of groups along current geographic features. Ross generally divided patterns of distribution as representing speciation originating in northern temperate conditions or in the tropics. He concluded that most groups of mosquitoes originated in the tropics with a few species that have since adapted to temperate climates. For example, the cosmopolitan northern species *Aedes vexans* was the sole temperate-adapted species in the large tropical (Africa, southern Asia, Australasia) subgenus *Aedimorphus*. This pattern of distribution, with just a few temperate or subtropical species within a larger tropical group, was observed in a wide variety of taxonomic groups. Ross cited examples of temperate species from tropical groups in *Aedes* subgenus *Aedes* (*cinereus*), *Psorophora* (*horrida*, *longipalpis*, *pygmaea*, *signipennis*), *Deinocerites* (*cancer*), *Coquillettidia* subgenus *Coquillettidia* (*perturbans*, *richiardii*), *Mansonia* (*indubitans*, *titillans*, *uniformis*), *Orthopodomyia* (*alba*), *Wyeomyia* subgenus *Wyeomyia* (*mitchellii*, *smithii*, *vanduzeei*), *Toxorhynchites* subgenus *Lynchiella* (*rutilus*), and *Uranotaenia* (*anhydor*, *lowii*, *sapphirina*, *syntheta*). Other groups were said to have originated in northern, temperate climates, including *Aedes* subgenus *Ochlerotatus*, *Culex* subgenus *Neoculex*, and *Culiseta*. Comments on *Anopheles* were limited to specific geographic associations, like the apparent links between European *maculipennis* groups and the North American *occidentalis* and *quadrimaculatus* groups and between the container-inhabiting *plumbeus* (Europe), *barianensis* (southern Asia), and *barberi* (North America).

Foley et al. (2007a) made the most recent effort to analyze patterns of mosquito distribution. Using published distributions at a country level for all mosquito species, they documented clearly that lower latitudes had more species and that higher latitudes had fewer species. Somewhat counter to Belkin's (1962) hypothesis that islands are the incubators of species, they showed a strong correlation between island size and number of species, a pattern that was not as strong for continental size. On the other hand, there was a clear relationship between species diversification and abundance of islands in the Philippines for haplotype-based populations of *Anopheles flavirostris* (Foley & Torres 2006).

Cranston's (2005) review of biogeography of flies warned that analyses of geography in relation to evolutionary history are dependent on the accuracy of phylogeny. As discussed earlier in this chapter, the current state of phylogeny seems to offer the most certainty for a few of the largest groupings and for more of the species-level groupings. It is no coincidence that many of the seemingly sensible conclusions from distributional-phylogenetic (i.e., vicariance) analyses are at a recent species level. For example, Foley et al. (2014) found that the eight species of the *Anopheles albitarsis* Group in South America were genetically divided into a northern and a southern clade, with individual species corresponding to different ecological conditions. The generalization that most of the major groups are most highly diversified in the tropics appears to be true. It is also clear that some of these tropical groups have been the genetic source of species that have either expanded their ranges into nontropical areas or have speciated into new forms outside the tropics. On the other hand, groups with widespread nontropical distributions, such as *Anopheles* (*Anopheles*), *Aedes* (*Ochlerotatus*), *Culex* (*Neoculex*), and *Culiseta*, probably reflect a different historical pattern, with the origins of distribution we see today originating many millions of years ago. Some tropical distributions can probably also be traced to extremely old geographic history.

Forming the links between geological events, presumed phylogenetic events, and current formal distribution is not as scientifically robust as a planned experiment. The analyses are selective because only a limited set of facts are available, and some major events have no corresponding link. The lack of an association between events is not informative because of incomplete knowledge, a well-based assumption that there are multiple causes for both phylogeny and distribution, and the impossibility of replication when each taxon is unique. As a result of these limitations, an attempt to assemble a geological and biogeographic history of mosquitoes is going to be speculative, however much the level of interest in phylogenetic history.

Between phylogenetic timing derived from molecular and fossil evidence (Reidenbach et al. 2009) and geological history (Cranston 2005; Lomolino et al. 2010) are fascinating correlations that may explain some aspects of modern mosquito distribution. One of the earliest indications of the family Culicidae was the split between Anophelinae and Culicinae 229–192 MYA. The earliest culicines appeared as a separate group 226–172 MYA and eventually led to the modern genus *Culiseta* (219–155 MYA). Most land masses were connected during this early period in mosquito evolution. The combined continent of South America and Africa did not separate from other southern (Gondwana) land masses until 180 MYA, suggesting that the wide distribution of anophelines, culicines, and *Culiseta* originated on Pangaea. Edwards (1941) was a leading proponent of the idea that early forms of mosquitoes, represented by what he considered to be "primitive" genera, were distributed across the New and Old Worlds because they originated before the separation of South America and Africa in the Cretaceous. Although global land masses split and moved, there was still a long period until 120 MYA when at least peninsular connections existed between most of them. Other currently widespread groups originated before that time, including Aedini (194–132 MYA), *Culex* (192–111 MYA), *Uranotaenia* (192–111 MYA), Sabethini (179–116 MYA), *Coquillettidia* (171–104 MYA), *Orthopodomyia* (171–104 MYA), *Toxorhynchites* (158–93 MYA), and *Aedeomyia* (158–93 MYA). Zavortink and Poinar (2008) also used this explanation for the wide distribution of *Toxorhynchites*, with the refinement that subsequent development of subgenera occurred in part through distribution across the northern bridges between Europe and North America. Whether because of miscalculation or a different historical sequence of events, two genera with currently restricted distributions, *Mimomyia* (Afrotropical, Australian, Oriental; 169–104 MYA) and *Psorophora* (Neotropical, Nearctic; 156–90 MYA) may have originated before the complete separation of the land masses on which they occur. This controversial area of mosquito taxonomy requires a great deal more phylogenetic work, both to verify the conclusions of Reidenbach et al. (2009) and to achieve more precision within the large genus *Aedes*. Estimates of time of origin, as well as phylogenetic pattern, can lead to hypotheses about biogeographical origins that form a separate line of support for systematics. Testing those hypotheses is difficult, but the only way to start is to pose the questions.

The implied biogeographical history of Aedini may support the notion that *Aedes* currently includes what should be a number of phylogenetically defined genera. According to the phylogeny presented by Reidenbach et al. (2009), Aedini split from a common ancestor of Sabethini, *Aedeomyia*, *Mimomyia*, and *Toxorhynchites* 194–132 MYA, when Gondwana was just beginning to break up, but peninsular connections still existed between all continents. In this phylogeny, a common ancestor of *Eretmapodites* and *Aedes* split off from other Aedini 124–61 MYA, about the time Gondwana separated completely from Laurasia and possibly before South America separated from Africa. *Aedes* became distinct from *Eretmapodites* 95–39 MYA, implying an African origin for *Aedes*. If this sequence of events is true, it seems unlikely that an African common ancestor of *Aedes* eventually gave rise to the worldwide genus we see today. It is more likely that subgenera with an African center of distribution, like *Aedimorphus*, *Neomelanoconion*, and *Stegomyia*, share a common ancestor but that those groups of *Aedes* with non-African centers of distribution, like subgenera *Ochlerotatus*, *Finlaya*, and *Protomacleaya*, separately arose from an Aedini ancestor either before or after the origination of *Aedes*. Within Africa, the cyclic separation and connection of forest habitats through drier and wetter periods, corresponding to glaciation in higher latitudes, may have caused the isolation of populations during the Pliocene and Pleistocene, resulting in groups like the Simpsoni Complex of *Aedes* (*Stegomyia*) (Bennett et al. 2018).

Correlations between geological history and distribution of Sabethini generally appear to explain important aspects of distribution. Accepting Judd's (1996) hypothesis that Old World and New World sabethines each form monophyletic clades, the formation of the tribe 179–116 MYA could have placed early forms on all the connected land masses. The estimated time of separation of Old World and New World sabethines was 124–66 MYA, corresponding to the period when South America separated from Africa and other land masses, whether through movement across the Panthalassa Ocean or, in the expanding globe model, formation of the Pacific Ocean (McCarthy 2003; Cranston 2005). The latter explanation, though controversial, provides a more direct explanation for the occurrence of most sabethines in eastern Asia and South America. Another interesting association is the correspondence of the dipteran utilization of endophytic habitats in the late Cretaceous (Labandeira 2005) and the probable radiation of sabethines at the same time.

As discussed earlier, molecular and morphological evidence has placed Anophelinae as the earliest group of Culicidae, a conclusion supported by the intuitive sense that anopheline larvae are similar to corethrellid larvae. *Chagasia* appears to have split off from the subfamily at an early stage. The subsequent biogeographical history of Anophelinae was discussed in some detail by Krzywinski et al. (2001b), reviewing molecular, distributional, and morphological evidence. These authors reached the major conclusion that the worldwide subgenus *Anopheles* originated before the breakup of Gondwana with an *Anopheles pseudopunctipennis*-like ancestor. The subgenus *Cellia* originated after South America separated from Africa and other continents about 95 MYA, accounting its absence in the Western Hemisphere. Possibly due to the occurrence of temperate-adapted species, the subgenus *Anopheles* maintained a genetic flow through northern connections between Europe and North America—connections that were not severed until the late Eocene (40 MYA). *Bironella* may have arisen as a separate group when Australia became isolated from other land masses 80–65 MYA. The ancestors of the Cyclolepteron-Arribalzagia group of subgenus *Anopheles* may have dispersed from North America to South America following either the exposure of the Antilles land bridge 49 MYA or the formation of the Isthmus of Panama 10 MYA. Although all of these speciation events associated with plate tectonics are speculative, they illustrate how geological history may have influenced the phylogenetic history of mosquitoes.

Nearer to the present are the historical biogeographical effects of two forces: glaciation and humans. It seems clear that glaciation in the Northern Hemisphere must have influenced distribution, speciation, and probably extinction. A general examination of species currently located in northern areas that were formerly covered with ice sheets shows three groups. The first contains species or species groups that are largely restricted to these areas, suggesting they originated from either relict populations that existed in gaps between ice sheets or from pioneer species that colonized land freed from receding glaciers. Examples include the *Aedes* (*Ochlerotatus*) *stimulans* group, *Anopheles earlei*, and *Anopheles messeae*. The second includes species that commonly occur outside of the northern areas and also much further south, possibly representing species that have been able to colonize northern areas. Examples in this second group include *Aedes* (*Aedimorphus*) *vexans*, *Culiseta alaskaensis* and *particeps*, and *Anopheles punctipennis*. *Wyeomyia smithii* is a particularly well-documented example, in that an analysis of restriction site-associated DNA tags (RAD tags) and associated SNPs throughout the genome showed that this species dispersed north and west from refugia in the Appalachian Mountains following recession of the Laurentide Ice Sheet 22,000–19,000 years ago (Emerson et al. 2011). Finally, a number of species have a circumpolar distribution, with or without occurrence further south. These species may have occupied northern areas before glaciation started 2 MYA, or they may have developed their distributions much more recently, following the last glacial recession approximately 15,000 years ago. *Aedes* (*Aedes*) *cinereus* and *Aedes* (*Ochlerotatus*) *sticticus* are examples. The taxa involved in these interesting distributions may have older origins on these continental land masses, but those associations have been obscured by more recent forces on speciation from glaciation (Cranston 2005). These distributions also raise many questions on genetic relationships related to selection, phylogeny, and phenotype, but there is less intensive study of these groups because of their relatively minor medical importance.

Human influences on evolution of mosquitoes have been mentioned in relation to the *Anopheles gambiae* and *Culex pipiens* groups. Invasion biology and adaptation to human activity is, of course, very recent and closely related to the subject of distribution discussed in Chapter 3. Generally, humans are most concerned with those mosquitoes that cause annoyance, transmit pathogens, or threaten animal agriculture. Ironically, those species are often unintentionally selected by human activity from among the very large pool of species that developed in the absence of humans. In one case at least, *Anopheles gambiae* is suspected to have arisen as a species because of adaptation to human agricultural activity. The domestic form of *Aedes aegypti* seems also to have arisen since people created domestic sites for what had been a forest mosquito. The movement of mosquitoes by human conveyance has provided opportunities for new genetic combinations, as seen for the *Culex pipiens* group. These hypotheses lead to a more speculative idea that we will see many more human-adapted mosquito species develop over the next 100,000 years.

The Fossil Record

The fossil record of mosquitoes is exceedingly incomplete and of limited value in understanding mosquito evolution (Belkin 1962; Harbach 2007). Nonetheless, the number of legitimate fossils continues to accumulate, and Reidenbach et al. (2009) used fossil evidence to help anchor their estimates of the age of major phylogenetic events. More recent publications have drawn some significant conclusions from fossil evidence. These studies are significant because, unlike cladistic analyses of molecular and morphological evidence, they are based on physical evidence and age graded by well-established methods.

Diptera is an old group represented by its earliest fossil, *Permotipula patricia* (Willmann 1989), dating back to the late

Permian (270–260 MYA). No entomologist could look at the clearly dipteran wing without a feeling of having made a connection between the diverse order we know today and insects that flew before modern vertebrates and flowering plants even existed. Of course, the occurrence of fossils is dependent on preservation and concentration of paleontological effort, as well as the former existence of the fauna. As a result, mosquito fossils have been from a few places representing narrow slices of the geological record (Table 1.4; see also Vol. 2, Part 3 for a list of all names associated with fossil records).

The oldest mosquito fossils found so far are from northern Burma. The deposit was Cretaceous, the amber probably formed from *Metasequoia* and *Araucaria* trees, and the original latitude was 12°N (current latitude 26°N; Grimaldi et al. 2002). One of the fossils was a small (wing, 1.9 mm long) female specimen, carefully described by Borkent and Grimaldi (2004), who estimated the age of the fossil as 100–90 million years (MY; Fig. 1.3, illustrated in Grimaldi et al. 2002). Close examination showed considerable degradation, but the fossil included most of the insect and generally resembles a modern mosquito. The most conspicuous plesiomorphic feature is an elongated but shortened proboscis only one-third the length of the antenna and also shorter than the palpi. Unlike most modern mosquitoes, the specimen has several setae on the clypeus, and the palpus lacks scales. The wing also differs from most modern forms in that scales are restricted to the hind margin and alula. Based on these characteristics, the authors judged that *Burmaculex* represented a group of culicine mosquitoes now extinct and sister to all modern culicines. They also speculated that the oldest mosquitoes, potentially dating back to the Jurassic, would be difficult to distinguish from fossilized chaoborids, a family that is more common in the fossil record. The other fossil from Myanmar amber (Poinar et al. 2019) has anopheline characteristics and an appearance very similar to a modern mosquito. Named *Priscoculex burmanicus*, the female specimen has setae as well as scales on the wing, an absence of scales on the abdomen, lack of pulvilli on the tarsi, and a smooth posterior margin of the scutellum. The 98 MY age of the formation was determined by analysis of uranium, lead, and zircon isotopes. The clearly anopheline characteristics of *Priscoculex burmanicus* suggest that older estimates of the origin of mosquitoes (Reidenbach et al. 2009) may be correct, in that the subfamily Anophelinae was already starting to diversify in the lower Cretaceous.

Another Cretaceous fossil of a mosquito is a male described as *Paleoculicis minutus* (Poinar et al. 2000), which was found in *Araucaria* amber in Alberta, Canada. The age of the fossil was estimated as 79 MY. This specimen was considered to be

Table 1.4. Summary of mosquito fossils in chronological order[1]

Name	Source	Location	Epoch (age MYA)	Reference
Primoculex burmanicus	Amber	Myanmar	Cretaceous (98)	1
Burmaculex antiquus	Amber	Myanar	Cretaceous (100–90)	2
Paleoculicis minutus	Amber	Alberta, Canada	Cretaceous (79)	3
Aetheapnomyia hoffeinsorum[2]	Amber	Baltic	Eocene (56–34)	4
Eoaedes damzeni[3]	Amber	Baltic	Eocene (56–34)	4
Aedes serafini	Amber	Baltic	Eocene (56–34)	4
Culex erikae	Amber	Poland	Eocene (56–34)	2, 5
Culex damnatorum	Compression	Wyoming, US	Eocene (56–34)	2, 5
Culex winchesteri	Compression	Colorado, US	Eocene (56–34)	2, 5
Culiseta gedanica	Amber	Baltic	Eocene	10
Culiseta kishenehn	Compression	Montana, US	Eocene (48–31)	6
Culiseta lemniscata	Compression	Montana, US	Eocene (48–31)	6
Anopheles dominicanus	Amber	Dominican Republic	Eocene-Miocene (45–15)	7
Culex malariager	Amber	Dominican Republic	Eocene-Miocene (45–15)	8
Aedes protolepis	Compression	Isle of Wight, UK	Eocene-Oligocene (37–28)	2, 5
Culex vectensis	Compression	Isle of Wight, UK	Eocene-Oligocene (37–28)	2, 5
Culex (?) *protorhinus*	Compression	Isle of Wight, UK	Eocene-Oligocene (37–28)	2, 5
Mansonia cockerelli	Compression	Isle of Wight, UK	Eocene-Oligocene (37–28)	2, 5
Anopheles (?) *rottensis*	Compression	Rott, Germany	Oligocene (28–23)	2, 5
Toxorhynchites mexicanus	Amber	Dominican Republic	Eocene-Miocene (45–15)	9

Sources: [1]Poinar et al. 2019; [2]Borkent and Grimaldi 2004; [3]Poinar et al. 2000; [4]Szadziewski 1998; [5]Evenhuis 1994; [6]Harbach and Greenwalt 2012; [7]Zavortink and Poinar 2000; [8]Poinar 2005a; [9]Zavortink and Poinar 2008; [10]Szadziewski and Gilka 2011

[1]Three species of *Neoculicites* from France and Germany are not included based on very poor preservation and uncertain status (Szadziewski 1998).
[2]Genus changed to *Aetheapnomyia* (Harbach and Greenwalt 2012).
[3]Genus changed to *Eoaedes* (Harbach and Greenwalt 2012).

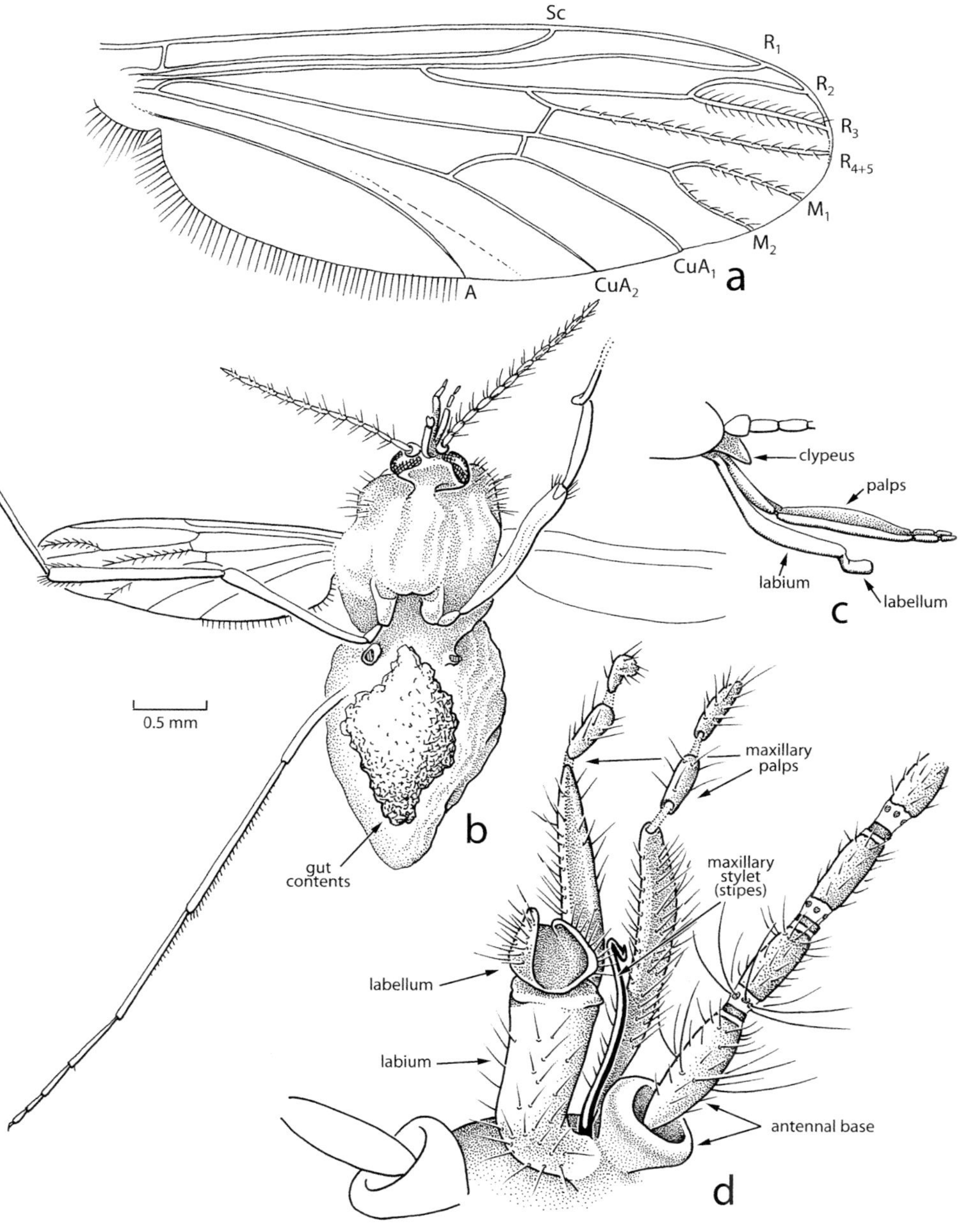

Figure 1.3. Drawing of fossil *Burmaculex antiquus* showing wing (a), habitus (b), mouthparts (c), and front of the head (d). Published in Grimaldi et al. 2002 before the full description of the species by Borkent and Grimaldi 2004

culicine, though it has the rounded scutellum and scaleless abdomen of the modern anophelines, suggesting that these two characters are plesiomorphic. *Paleoculicis* had several features not seen in modern mosquitoes, including large facets of the compound eyes, gonocoxites with strong spicules, and spines on the ninth tergite. As for *Burmaculex*, *Paleoculicis minutus* generally looked like a mosquito and, as its name implies, was relatively small.

Three fossils of modern genera have been described in great detail from amber. *Toxorhynchites mexicanus* (Zavortink & Poinar 2008) was found in Chiapas, Mexico, in Upper Oligocene deposits that implied a minimum age of 26–22.5 MY. In addition to being the only fossil representative of the genus, this specimen also appeared to be in the subgenus *Toxorhynchites*. Modern members of that subgenus are only found in the Old World, suggesting that the subgenus or its progenitor had previously been distributed more widely, when the Gondwanan continents were still connected. *Anopheles dominicanus* (Zavortink & Poinar 2000) is the only confirmed fossil in the genus. The fossil, in amber (probably from the legume *Hymenaea protera*) is from the Dominican Republic, with inclusions that place it at either 45–30 MYA or 20–15 MYA. Clearly visible spots on the wing indicate that the specimen is closest to the Myzorhynchella section of the subgenus *Nyssorhynchus*. The female apparently exuded an egg when it became stuck in the resin, the egg having a typical anopheline appearance. *Culex malariager* (Poinar 2005a) is another remarkably preserved fossil from the same formation as *Anopheles dominicanus*. Although its characters clearly identify it as a *Culex*, it has several characters not seen in modern forms. Those characters not seen in modern *Culex* included: (a) a proboscis shorter than the antennae; (b) a mixture of broad and narrow scales on the wing; a remigium (base of vein R_1) with two sets of dorsal setae; (c) a clypeus as long as broad; (d) an empodium on the foretarsus; (e) short erect setae on vein 2A; and (f), a postgenital lobe covering the base of the cerci. The piece of amber containing the mosquito also included a triatomine and a soft tick, suggesting that the original site of fossilization was near a nest. This observation was significant because the interior of the mosquito contained oocysts, sporozoites, an ookinete, and possible microgametes of *Plasmodium dominicana* (Poinar 2005b), a newly designated fossil parasite similar to the avian pathogen *P. juxtanucleare*.

Other mosquitoes from amber and compression fossils are either less well preserved or less well described. Szadziewski's (1998) descriptions of Eocene *Aetheapnomyia hoffeinsorum*, *damzeni*, and *serafini* are convincingly *Aedes* and probably in subgenus *Finlaya*, but they do not give the detail of the later work. Other kinds of fossils give some indication of how much more investigation is possible. For example, Lutz (1985) described larvae in Lower Oligocene (34–38 MYA) compression fossils from Céreste, France. Abundant arthropod fossils have been prepared using an acid treatment on Middle Miocene (16–12 MYA) calcareous deposits that originally formed in drying lakes of the Mojave Desert, California (Palmer & Bassett 1954; Palmer 1957; Park & Downing 2001). These detailed, unusual fossils are still under study and include larval mosquitoes (L.E. Park Boush, personal communication, May 17, 2011).

Conclusion

The variety that we see today in the family Culicidae is a snapshot of a group of insects that has a long history. Former species must have been adapted to many different kinds of hosts, some of which do not exist now. Their associated pathogens and symbionts have also probably changed over time. We have no way of knowing when mosquitoes began to approach the current variety of genera and species or whether an even greater variety existed previously, but most of the current genera may have been present in the Eocene or Oligocene. Evolutionary differentiation of genera and subgenera appears to have been highly influenced by the creation of separate populations following separation of the continents, possibly indicating that such changes are not likely to occur on a human timescale. Detailed examination of a few species has shown that the process of speciation continues, possibly more rapidly than many would have assumed. Therefore, we may be at a point in mosquito evolution when selection of existing species, and differentiation of new ones, occurs within the broader, stable framework of the family. Unfortunately, we know little about the extinction of rare species or species that are highly dependent on fragile habitats, as those species seldom attract much research attention. More work on phylogeny based on DNA sequences, population genetics, and the fossil record will make our knowledge of mosquito evolution increasingly precise and provide an increasingly accurate representation of insect evolution in general.

2. Nomenclature, Classification, and Identification

Microscopic work on the morphology of small insects demands, above several other things, a great deal of psychological stamina.

White 1977

Motivations for Naming and Classification

Probably no aspect of mosquito biology arouses as many strong, contrasting opinions as nomenclature. Driven by conflicting intellectual, personal, and practical motivations, the discovery of ever more diversity in the family Culicidae has resulted in the creation, destruction, re-creation, and new creation of categories and species. Unfortunately, most biologists have an incomplete knowledge of the process whereby tribes, genera, subgenera, species, and subspecies are named. The level of knowledge may be decreasing as traditional taxonomy is less and less considered a core part of the biological curriculum, a danger that has been recognized for decades (Isely 1972). This chapter describes some of the rules, issues, and current trends in mosquito nomenclature.

Why do we need names in the first place? Some of the reasons are independent from scientific efforts; people have always wanted to name the organisms around them. In what may be the most basic human urge in natural history, languages divide living creatures into categories that are useful in daily life and distinguish one from the other. Most languages recognize 250–500 "kinds" of organisms in nonscientific cultures (Winston 1999). In modern culture, even the least interested person will know a number of names of organisms, like "robin," "bass," "maple," and more. That person may not be able to identify the organisms that match the word, but he or she will still know and use the word. Applying a name under these nontechnical circumstances accomplishes a number of functions for the individual. First, and most obviously, the name allows that person to communicate what was seen. Less obviously, it also allows that person to create a symbolic memory of the experience. That memory might be used for future practical reference or near-term orientation or simply to complete the personal story of that encounter. Based on experience with the public, who often face unfamiliar pests, the name also seems to impart a sense of control or proportion to the individual, enabling that person to move on to more important aspects of life. Finally, many people find that an understanding of commonly encountered organisms based on names imparts an aesthetic appreciation of their surroundings.

The scientist is also a human being and susceptible to all the motivations for naming described above. Ideally, a scientist who classifies and names—that is, a taxonomist—puts ego aside and attempts to arrive at a system that reaches a reasonable solution to the sharply conflicting aims of the field. These aims are conflicting because a single name is supposed to communicate the identity of the taxon to a worldwide audience for a long time, as well as maximize information content about the taxon's relationship to other taxa. The basic problem is that as we learn more about those relationships and refine our ideas about their nature, the names must change to reflect the new perspectives. As a result, names shift, creating challenges for current communication as well as reviews of past work. The history of mosquito taxonomy is heavily punctuated with such changes, sometimes widely accepted and sometimes widely disputed. As it happens, mosquito biologists have departed from an extended period of relative stability in genus-level nomenclature since Edwards' review in 1932. During the last 15 years, the very large genera *Aedes* (over 900 species, including *Ochlerotatus*; Harbach 2007) and *Culex* (over 750 species; Harbach 2007) have been divided into more genera based on quantitative phylogenetic analyses of many characters. Division of *Anopheles* into a total of six genera has been suggested based on a sampling of mitochondrial protein gene sequences (Foster et al. 2017). Most recently, the extensive changes to generic designations of aedine species were reversed, in the sense that genera were returned to the status of the year 2000 (Wilkerson et al. 2015). This had the effect of synonymizing many genera back into the genus *Aedes*; however, the phylogenetic progress achieved by those who had created the many new genera (Reinert, Harbach, & Kitching 2009) was preserved by reorganizing and creating subgenera and informal species groups (Wilkerson et al. 2015; Wilkerson & Linton 2015).

The importance of names for identification was eloquently expressed by Ernst Mayr (1969). He described the tasks of taxonomy as identification, on the one hand, and classification, on the other. Whereas classification was considered an inductive process (using the characteristics of individual taxa to form the characteristics of higher-level taxa), identification was considered to be deductive (a sequential process starting with general characteristics and working down to more specific ones). Mayr stressed the importance of stability of names to identification, and he criticized the entire field of taxonomy for too-frequent name changes. The issue of nomenclatural stability is particularly important for mosquitoes. Unlike more obscure groups of insects, mosquitoes are a common target for control efforts. Sophisticated mosquito control worldwide depends heavily on knowing the identity of the species in an area. Those identifications are essential to achieve the most efficient use of resources by targeting species-specific larval sites, locations, and times of adult activity. The best control programs also monitor their efforts to assure the quality of their services and to determine how to adjust their practices in response to real-world conditions. Unlike agricultural practice, where only a few species from a field might be identified, mosquito control programs often identify all of the culicids captured as larvae and adults, reasoning that one species may indicate a habitat's suitability to other species. The general intention is to provide an early warning that can facilitate a preventive response. This routine dependence on accurate mosquito identification means that hundreds of nontaxonomists must be able to identify all of the species in their area. Those identifications are communicated widely for both operational and archival purposes. Some mosquito taxonomists have tried to be responsive to these needs by being conservative about name changes that would complicate training, identification, and record-keeping (e.g., Belkin 1962; Savage & Strickman 2004; Wilkerson et al. 2015).

Identification

The process of morphological mosquito identification commonly starts with a dichotomous key. The keys are often divided by genus (e.g., Darsie & Ward 2005), although sometimes they are dedicated to a single genus or species group (e.g., Wilkerson & Strickman 1990; Huang 1997). The vast majority of identifications can be made based on microscopic examination; however, the species of some groups can only be identified by other means (e.g., Carlson & Service 1980; Fanello et al. 2002) or are more conveniently identified by other means (e.g., Harrington & Poulson 2008). The sources of keys include taxonomic reviews (e.g., Harrison & Scanlon 1975), identification guides (e.g., Bohart & Washino 1978), regional treatises (e.g., Becker et al. 2003), medically oriented guides (e.g., Huang & Ward 1981), and more. Keys not only provide a step-by-step means of identification for the nonspecialist but also form a litmus test for the taxonomist regarding whether proposed species really can be distinguished. Additional confidence in an identification is gained by comparing a specimen to a description, illustration, or reliably identified specimens. Computer-based resources for identification have been developed for some groups using "polythetic" (Pankhurst 1978) identification that allows the process to proceed from any of several characters. Also known as multiaccess or polyclone keys (Winston 1999), they use the same principle as certain kinds of molecular identification that might examine several regions of the genome in no particular sequence. Field guides that attempt to point out necessary and sufficient characters for identification are less common for mosquitoes, but some useful attempts have been made, including charts for important species (Foote & Cook 1959), illustrations of all North American species (Carpenter & LaCasse 1955), and a compendium of illustrations of male genitalia of the subgenus *Melanoconion* of *Culex* (Pecor et al. 1992). Volume 2 of this book includes references on mosquito identification organized by taxon and geographic location.

Classification

Classification of organisms is exactly what the word implies—placing them into classes. The complication in the process arises from defining the classes. The definition of a class controls how it is used, giving rise to a great deal of biological debate. The core criterion for classifications at any level is monophyly, which is the evidence leading to the conclusion that species within a class shared a common ancestral species at some point in the past. Despite this straightforward criterion for classification, in practice we ask classification to accomplish several, sometimes conflicting, purposes.

First, classification simply provides folders into which we put smaller groups. In other words, instead of having a catalog with over 3,500 species of mosquitoes listed in alphabetical order, we arrange them in order of increasingly larger formal groups: subspecies, species, subgenera, genera, tribes, and subfamilies.

Second, classification groups mosquitoes phenetically according to demonstrable characteristics, usually morphological. One clear example is the subfamily Anophelinae, each member of which can be identified by certain common morphological features that are not shared with members of the other subfamily, Culicinae. Numerical taxonomy based on quantifiable characters and anticipating robust computational capabilities in the 1960s openly embraced phenetic classification (Sneath 1995). Now, we would probably only consider a phenetic classification as valuable if it supported classification based on monophyly.

The third type of classification is phylogenetic, based on principles described by Willi Hennig (1965). Phylogenetic classification emphasizes the widely accepted view that new species arise from existing species in a way that can be modeled as a tree with bifurcating branches. The practice of constructing such trees has been refined considerably, to the point that phylogenetic analysis is routinely used to analyze differences and similarities among populations, infective pathogens, viruses as they evolve in a single infected individual, and mosquitoes. Although it is assumed that phylogenetic analysis reflects the genealogical history of the subject group, the actual methods used to achieve that genealogy are indirect and consist of statistical comparisons of character states. Nonetheless, the stated purpose of phylogenetic classification is to create classes that are monophyletic.

The final form of classification discussed here, described by Mayr (1969), is evolutionary. As the name implies, evolutionary classification attempts to form groups based on Darwinian evolutionary history. The basis of evolutionary classification is not so different from phylogenetic in that both aim to establish monophyletic groups with biological significance based on the idea that the history of a lineage will provide the most informative classification. The two methods contrast with each other in that phylogenetic classification tends to emphasize branching without considering the degree of diversification. The criteria for establishing taxa in evolutionary classification should take into account the rate of evolution for that group compared to related groups, but such evaluations are necessarily based to a great extent on the judgment of the taxonomist.

Mosquito taxonomy as it exists today has elements that can be attributed to all four types of classification, even though current efforts are solidly based on the principle of monophyly. The continuing debate and robust change in mosquito classification clearly originate from a general intention to have a sound, biologically based taxonomy of the family. For now, the current nomenclatural instability remains a challenge for those whose needs are more about identification than classification.

Systematics has been defined as the science surrounding the study of biodiversity (Mayr 1969). Systematics is experimental science that leads to more accurate classification that in turn leads to biologically meaningful nomenclature. Taxonomy can therefore be considered a subset activity of systematics. The normal progression for study of a taxon goes from simple descriptions of species (alpha taxonomy), to relationships between species (beta taxonomy), to causal interpretations of relationships (gamma taxonomy). The potential methods for studying systematics are as varied as biological science itself, given that any aspect of an organism is particular to that kind of organism.

Current trends show a preference for quantification over intuition, history over phenetics, and evolution as the underpinning theoretical framework. The ability to sequence DNA has, of course, revolutionized systematics. Molecular biology offers a wide variety of tools for making comparisons at any level of classification. More robust genetic techniques and analysis will greatly improve taxonomists' ability to make objective classifications. To this point, morphological taxonomy has had the advantage of providing data on the entire individual mosquito in all its life stages. Most molecular studies have been limited to sequences of portions of the genome (e.g., Cywinska et al. 2006), to representative images of parts of the genome (e.g., Ruiz et al. 2005), or to biochemical measurement that closely reflects a portion of the genome (e.g., Knight & Nayar 2004 [isoenzymes]). Economical whole-genome sequencing and corresponding large-scale phylogenetic analysis will facilitate robust evolutionary histories, will stabilize nomenclature within Culicidae, and will initiate studies to gain a better understanding of the genetic basis of bionomic traits.

Rules and Procedures of Scientific Nomenclature

The centerpiece of scientific animal nomenclature is the fourth edition of the *International Code of Zoological Nomenclature* (International Commission on Zoological Nomenclature 1999). The *Code*, as it designates itself, is a comprehensive discussion and treatment of nomenclature, complete with introduction, 90 legalistic articles, and appendices, including a constitution, code of ethics, and glossary. The current *Code* is based on a long history of such guidance from entities that drew authority from a wide base of active biologists, giving the *Code* a strong foundation in the science of Western Europe. It started with the Strickland Code in 1842, commissioned by the British Association for the Advancement of Science. The International Commission on Zoological Nomenclature (or the "commission") was founded in 1898 with authority from the International Congress of Zoology. It published the first set of nomenclatural rules in 1905 as the *Règles Internationales de la Nomenclature Zoologique*. The first edition of the *Code* was published in 1961. Authority for the commission was transferred in 1972 to the International Union of Biological Sciences, which continues to support the commission, its official publication of the *Bulletin of Zoological Nomenclature*, and official lists and indices of names. The current, fourth edition of the *Code* is available online, as are other official publications of the commission. Perhaps no other effort in zoology is as broadly applicable, commonly accepted, highly organized, and likely to endure as the *Code* and activities of the commission.

The third edition of the *Code* was discussed article by article in Mayr's (1969) *Principles of Systematic Zoology*. That book is still quite a useful starting place for anyone working to understand taxonomy. Winston (1999) reviewed the fourth edition of the *Code* in what was openly intended as a guide for the nontaxonomist to describing taxa. Her intention to provide easy access to the intellectual tools required for naming new species, genera, and families is probably even more important now as the number of practicing taxonomists continues to dwindle. One of the unfortunate side effects of the reduction in number of practicing mosquito taxonomists is that large complex studies, like that of Reinert, Harbach, & Kitching (2009), are evaluated by only a few dedicated and informed individuals; other biologists who depend on the work are left making important nomenclatural changes based largely on faith rather than on scientific consensus.

The *Code* itself is an accessible document that attempts to achieve precise, unambiguous definitions in plain language. Changes made in the fourth edition were summarized in its Introduction and were more about details than major change. Significantly, electronic-only publications of taxa have some restrictions not shared with printed publications. Electronic publications must have appeared after 2011, be entered in the Official Register of Zoological Nomenclature, and state that registration was achieved. Also of general importance was the decision to adopt junior synonyms that have been commonly used. Common use was defined as being published by at least 10 different authors in at least 25 publications during a period of at least 50 years, in preference to a senior synonym that has not been used since 1899. The fourth edition requires designation of a type specimen for new species, but this has been standard practice in mosquito taxonomy for at least 200 years.

The principle provisions of the *Code* can be summarized simply, especially if the discussion is restricted to mosquitoes. The starting point of all nomenclature is Linnaeus' 10th edition of *Systema Naturae*, with publication considered to be January 1, 1758. This is important because any name published before that date is not considered to be available for use, and any name published after that date must be considered in determining seniority. The format of a species name is, as Linnaeus first established the practice, a binomial consisting of the genus and species names. Both are in italicized roman letters without diacritical marks. The first letter of the genus name, but not of the species, is capitalized. The subgenus can be inserted in italics and capitalized in parentheses between genus and species names. There may also be an italicized subspecies designation after the species name, forming a trinomial.

All scientific names are considered to be in Latin or Latinized ancient Greek unless originally designated as not in either of those languages. The gender of the species name must agree with Latin gender (masculine, feminine, or neuter) if the species is in the genitive declination—in other words, if the species name represents a modification of the genus name. As an example, acceptance of *Stegomyia* to the rank of genus would change *Aedes* (*Stegomyia*) *albopictus* to *Stegomyia albopicta*, altering both parts of the binomial. The correct interpretation of Latin grammar for constructing scientific names is usually unfamiliar to biologists. It is necessary to consult the *Code*, an authoritative text (Brown 1956), or an expert.

The names of the original author(s) of the species (usually the authors of the journal article in which the first description was published) may follow the scientific name of the species, written in roman letters. The author remains the same even if the species is transferred to another genus; however, the author's name is placed in parentheses if such a change has occurred. The *Code* encourages the use of the date of publication after the author's name, with or without a comma separating them. The author of the change of the species from one genus to another may be cited after the parenthetical species author(s). As an example, the subgenus *Neomacleaya* of *Aedes* was moved to *Verrallina* by Reinert (1999c), a genus newly elevated in rank. The species *Aedes* (*Neomacleaya*) *nubicolus* Laffoon 1946 became *Verrallina* (*Neomacleaya*) *nubicola* (Laffoon 1946) Reinert 1999. It should be of interest that the *Code* is the source of what may sometimes seem like trivial rules on formatting scientific names. These format requirements are important to keep the designation of a species informative and unambiguous.

The following principles guide the assignment of many necessary name changes: homonymy, synonymy, and seniority. Homonyms, which are the use of the same binomial name for two different species, are of two kinds. Primary homonyms occur when a new species is mistakenly given the same name as another species in the genus. Secondary homonyms occur when identical species names originally used in different genera come to be in the same genus through reclassification. This occurrence results in a name change for one of the species and consequent redesignation of the name on the type specimen. Synonymy is the converse of homonymy—the use of two different names for the same species. This has occurred frequently in the mosquito literature when differences originally thought to separate two species are later considered as not indicating a difference. Both homonyms and synonyms are corrected by assigning priority according to seniority—that is, giving the older name preference. As indicated before, one of the big changes in the fourth edition of the *Code* is that a senior synonym may be ignored if certain conditions are met, and it has not been used since 1899. A senior synonym can be ignored under other conditions if approved by the commission following application by a taxonomist. One of the conspicuous examples was *Aedes aegypti*, which was mistakenly associated with a Linnaean-type specimen, *Culex aegypti*. Although this species went through many names in the 19th century (e.g., Walter Reed [Reed & Carroll 1901] referred to it as *Stegomyia Fasciata* [*sic*] Theobald), its subsequent designation as *Aedes aegypti* became so prominent that the commission voted to accept a new type specimen and type locality, firmly connecting it to the name (Mattingly 1967).

Type Specimens

Probably no part of the mechanics of taxonomy is as obscure to the nontaxonomist as the type specimen. The very name seems to harken back to an Aristotelian concept of biology in which there is an ideal form for each living thing. In fact, the modern use of types did not begin until about 1850 (Mayr 1969). Prior to that time, the type might consist of more than one specimen and be replaced regularly in order to achieve a representation of the ideal state of the species. That sort of true Aristotelian species concept was replaced in 1901 with the functional designation of a reference specimen never to be replaced or have its name changed unless sanctioned by the commission (for example, in the case of a secondary homonym). It was not until

Table 2.1. Examples of terms coined for the description of different kinds of type specimens, selected for their unusual interest (Evenhuis 2008)

Abruptotype: A taxon hastily described to meet a project, grant, or publication deadline.

Artotype: A type specimen of a new species with clearly distinctive color patterns ultimately shown to be paint spots.

Boobootype: A holotype that should not have been described.

Concoctotype: A complete holotype made from several incomplete specimens.

Crickotype: A holotype and only known specimen of a taxon that has been completely ground up for DNA analysis.

Dermestotype: A specimen of a type series so badly eaten by dermestids that its identity cannot be confirmed.

Diplomatotype: A type specimen of a taxon described solely for political purposes, such as to access government lands otherwise inaccessible to biologists.

Ephemerotype: A type specimen that disintegrates completely after only a day and must be replaced by a neoephemerotype on a daily basis.

Fungotype: A type specimen that has been damaged by being subjected to high humidity or moisture and is covered with fungal hyphae to the point that it resembles a fuzzy white ball.

Hemitype: A type represented by only half of its original body.

Incognitotype: A type for which all labels have been lost.

Kleptotype: A type stolen from its original type depository.

Miragotype: A type of taxon in which the distinguishing characters can only be seen at certain times and under certain conditions.

Neglectotype: An uncurated type specimen found decades after an office is cleaned out.

Numerotype: A type specimen only labeled with a collection number, as the field data associated with the number are lost or destroyed.

Pluviotype: A type that has been damaged in a flood and is now encrusted with silt.

Remorsotype: A type of taxon the author regrets having described.

Scoopotype: A type specimen of a new species described more quickly than that of a colleague, who has been describing the same new species.

Tyrannotype: A type designated by the International Commission on Zoological Nomenclature.

Vampirotype: A type specimen (usually insect) impaled with a hugely oversized pin.

Xerotype: A type normally stored in fluid that has dried up.

the fourth edition of the *Code* that designation of a type was required as a part of a species description. Reference museums sometimes evaluate themselves on the number of types they possess, and some nations have laws about retention of types of species discovered in their country.

Even taxonomists have noted the irony of the type concept in a lexicon to provide "a historical context to the hobby of coining type terms" (Evenhuis 2008), which nonetheless expresses many of the technical difficulties experienced by taxonomists (Table 2.1). The contrast between the true importance of a type and the seemingly strange authority exerted by a bit of chitin mounted on a pin was expressed well by Tom Gaffigan, formerly a museum specialist at the US Army–sponsored mosquito taxonomy organization called the Walter Reed Biosystematics Unit. Tom liked to evoke the image of a military entomologist pinned down by enemy fire on the front lines desperately and self-righteously demanding to see a type specimen in order to confirm an identification. This has never happened.[1]

So what exactly is a type specimen? The type is a single example of a species selected either by the first describer of the species or by someone replacing a missing type. It retains its original name regardless of any subsequent name changes for the species, and it is usually marked with a red label clearly stating "TYPE." The collection location of the type specimen is important because that is the area to start looking for additional material should the need arise in the future. The type is not intended to represent an ideal or average for the species, though it is convenient if it is in good condition and carefully mounted. For mosquitoes, a good type specimen might consist of the larval and pupal exuviae, as well as the individual male that emerged from them, its genitalia mounted on a slide in standard format. Associated specimens can be designated as identical to the type by the author, who logically would be considered the best expert on that species at the time. Those alternate specimens might or might not be truly identical to the type itself; therefore, the type retains its prime importance as the only real reference point for the species. As the classification of mosquitoes proceeds, the type becomes essential for making taxonomic decisions. Its direct examination should be restricted to those doing revisionary work.

A type performs the following functions:

- provides firmer identifications for decisions on seniority by making comparisons to the physical type specimen
  - in case of splitting a species into two species, decides which retains the old name
  - determines whether two similar species are actually one by comparing their types
  - assigns the senior name to a species that has experienced a long series of synonymous descriptions
- provides additional descriptive details in case the original description lacked them, especially if an old description contained too few details to distinguish new species
- serves as the ultimate reference for the definition of a species

[1] To our knowledge.

The stabilization of the name *Aedes aegypti* by selecting a new type to replace Linnaeus' unreliable specimen was reviewed above. A couple of other examples illustrate the complications that can result from mosquito taxonomists' efforts to stabilize nomenclature and to be sure that names used throughout history in the scientific literature correspond to the actual organisms included. Harbach, Dahl, and White (1985) reviewed the complicated history of the taxonomy of *Culex pipiens*. The type labeled *Culex pipiens* by Linnaeus himself still exists at the Linnaean Society of London, but the specimen is actually an *Aedes* in the subgenus *Ochlerotatus*. Reviewing 18th-century literature, the authors established that Linnaeus' concept was actually that of an insect carefully illustrated by Réaumur in 1738. That illustration clearly matches our modern concept of *Culex pipiens*, and in a bold taxonomic move, Harbach et al. (1985) designated the illustration as the lectotype. They then nominated a modern specimen, collected by the authors in Sweden, as a neotype. Although the authors justified all of their actions according to the third edition of the *Code*, they certainly pushed the system by using a pre-Linnaean description as a lectotype and by designating a neotype from another location in order to get a type locality closer to that presumably intended by Linnaeus. The authors also dealt with *Culex bifurcatus*, a Linnaean type that was actually a specimen of *Anopheles* and a species synonymized under *Culex pipiens* by Martini in 1922. In this case they designated Réaumur's illustration of the male *Culex pipiens* as the lectotype and stated that designation of a neotype was unnecessary. Although complicated and legalistic, the work by Harbach et al. (1985) accomplished at least two important and positive purposes. First, it was a fascinating and thorough account of the history of a species' nomenclature dating back prior to the starting point of the Linnaean system in 1758. Second, it provided a description and type designation that stabilizes the name of *Culex pipiens*, guarding against capricious changes based on Linnaeus' still-existing types.

Another complicated example of the importance of type specimens concerned a common species in western North America and Central America, *Culex stigmatosoma* (Strickman 1988a). Described in 1907 by Dyar from specimens collected in Pasadena, California, the name was synonymized by Stone in 1958 under *Culex peus*. The name *Culex peus* was assigned in 1908 by Speiser as a replacement name for a type specimen named *Culex affinis* by Adams in 1903. Speiser took this action because the name *affinis* was preoccupied by a type named by Stephens in 1825. Alerted by the fact that the *Culex affinis* = *peus* type specimen had been collected in Oak Creek Canyon, Arizona, outside the range of the currently conceived species, Strickman examined the type carefully, as well as the types of *Culex stigmatosoma*, its synonym *Culex eumimetes*, and *Culex thriambus*. Performing a quantitative assessment of many specimens from the United States, Strickman concluded that the original Adams type was a different species, known up until then as *Culex thriambus* (originally described by Dyar in 1921). This action resurrected *Culex stigmatosoma* and changed the name of *Culex thriambus* to *Culex peus*. Eldridge and Harbach (1989) recognized that Strickman's action would create confusion in the literature because of the frequency with which the name *Culex peus* had been used between 1958 and 1988. They therefore successfully petitioned the commission to suppress the name *Culex peus*, which preserved the use of *Culex thriambus* (because it no longer had a senior synonym) and stabilized the use of *Culex stigmatosoma*, which had already been well accepted by major journals.

Alternative Codes of Nomenclature

The *International Code of Zoological Nomenclature* is one of three major nomenclatural codes (Greuter et al. 2011) based on rank and binomial nomenclature. The other two codes are the *International Code of Nomenclature of Bacteria* and the *International Code of Botanical Nomenclature*. Viruses are named in a different way, described in the *International Code of Virus Classification and Nomenclature*. Special requirements for naming human-generated plant types can be found in the *International Code of Nomenclature for Cultivated Plants* and for bacterial phytopathogens, in the *International Standards for Naming Pathogens of Phytopathogenic Bacteria*. In practice these are the guides to scientific naming used by biologists; however, there has been a recent attempt to harmonize the rules under a new code, the *BioCode* (Hawksworth 2011; Greuter et al. 2011). The *BioCode* limits its purposes to nomenclature and not ranks, though it proposes the customary rank-based system of naming. It proposes familiar systems of binomial nomenclature, use of types, priority, Latin as the linguistic base, and changes in name only if there is sufficient reason. In the absence of a rule, the *BioCode* advocates following custom. Perhaps the biggest difference proposed is a requirement for registration prior to establishment of a name. Mosquito taxonomy would not be affected much by use of the *BioCode*, but from a wider biological perspective, the rules learned for culicids would be more widely applicable to plants, bacteria, and other organisms.

The *PhyloCode* (Cantino & de Queiroz 2010; de Queiroz 2012) has a different purpose from the other codes. It distinguishes itself as a system for naming phylogenetic clades rather than classification ranks. In some ways the *PhyloCode* is the ultimate biologically based naming system and addresses Mayr's (1969) strong criticism of how current scientific nomenclature inevitably leads to instability as it attempts to name and classify with the same system. It makes a point of overcoming the artificiality of ranks and the practical difficulty of making superspecific ranks (e.g., genus) equivalent from one taxon to another. The *PhyloCode* began with some intention to replace traditional systems but in the current form attempts to superimpose clade names on rank-based systems. The *PhyloCode* uses many of the same rules as the *International Code of Zoological Nomenclature*, including precedence, priority, synonymy, and designation of types (called "specifiers" in the *PhyloCode*). On the other hand, the *PhyloCode* rejects the biological significance of ranks, even going to the extent of naming the first part of the scientific name "prenomen" and the second part "uninomen." This system increases the flexibility of grouping species into clades, defined as a monophyletic group of species or organisms. Current rank-based systems cause a cascade of name changes whenever an attempt is made to make a classification reflect a hypothesis of phylogeny, the creation of dozens of new genera of mosquitoes by Reinter, Harbach, & Kitching (2009) being an extreme example. The *PhyloCode* provides a mechanism to name each clade without creating such a cascade of name changes, even as phylogenetic study changes the relationships of clades.

Although the principles behind the *PhyloCode* are very familiar to taxonomists concerned with phylogeny, the *PhyloCode* does present a number of new ideas on the mechanics of taxonomic process and notation. First, like the *BioCode*, the *PhyloCode* creates a system for registering clade names and specifiers. The concept of specifier is similar to that of type, but it is designed to form a reference for a clade rather than a rank. As such, a specifier can be one or more specimens, one or more species, or even a particular apomorphy (shared derived character). The kinds of clades in these categories are defined in Table 2.2, where a node is a lineage at the point where it splits into two sublineages, and a branch is the lineage between two splitting events. Crown clades are based on definitions that designate descendants from a node, whereas total clades designate descendants from the earlier point on a branch that distinguishes that clade from others. Crown clades are easier to conceive since they start with a specific real or hypothetical ancestor, but total clades are defined in such a way that subsequent discoveries of species along that branch can be included in the originally named clade.

The *PhyloCode* intends for subsequent progress in taxonomy to be smoothly integrated into the existing system of clades, only modifying the concepts of the members of the clades. This nomenclatural concept would achieve much greater stability of names because they would not change as ranks were adjusted. This system would also allow taxonomists to define and name a clade in which they have confidence while leaving other clades unnamed until further study clarified their status. For example, Reinert (2000h) might have named *Aedes* and *Ochlerotatus* clades based on female genitalia and some associated characters rather than erecting two genera subsequently modified to 36 genera (Reinter, Harbach, & Kitching 2004) and finally to more than 65 genera (Reinter, Harbach, & Kitching 2009). The members of the original *Aedes* and *Ochlerotatus* clades could have been changed and other clades defined as the authors proceeded with their phylogenetic studies—all without the disruption of renaming most of the more than 900 species formerly in *Aedes*.

According to the *PhyloCode*, the traditional (= nominal) authors and names would be treated according to the *International Code of Zoological Nomenclature* with the exception of a recommended *R* in superscript or brackets after the name to designate that it is rank based. A *P* would be used to designate a clade-based (= definitional) name. Clade authors would be indicated in brackets.

Genera

The concept and naming of genera are particularly important for mosquitoes because of the intensive use of names outside of a taxonomic context. The common use of genera in mosquito biology is indicated by active designation and updates of abbreviations (e.g., Reinert 2001e and also a system described in Volume 2 of this book). Researchers who are more concerned with damage caused to human and animal health than with the insects responsible probably want to be sure that the names they use are technically correct and that other researchers using their work have an unambiguous concept of the species that appeared in their studies. The phylogenetic relationships of those study specimens to other mosquitoes are probably of less importance than a clear identification to such researchers. The other groups who use mosquito names are professionals and technicians involved in mosquito abatement and vector control. That community traps and monitors mosquitoes for the purposes of finding where problems are, what kinds of problems they have, and whether their efforts have been successful. Although the vast majority of specimens are usually one of a few locally significant species, control strategies

Table 2.2. Definitions and notations of the five kinds of clades described in the *Phylocode* (Cantino and de Queiroz 2010; de Queiroz 2012). Notations from the *Phylocode* are as follows: > = the most inclusive clade containing; < = the least inclusive clade including; & = and; v = or; ~ = but not; () = of or synapomorphic with that in; ∇ = crown clade.

Clade	Definition	Notation
Node based	Clade originating with the most recent ancestor of member included	< A & B
Branch based	Clade consisting of subject organism and all organisms or species that share a more recent common ancestor with subject organism	> A ~ Z
Apomorphy based	The clade originating with the first organism or species to possess subject apomorphy (M is apomorphy)	> M(A)
Branch-modified node-based	Clade originating with the most recent common ancestor of subject organism and all extant organisms that share a more recent common ancestor with subject than organisms not in the clade; the most inclusive crown clade containing subject organism but not nonsubject organisms	∇ A ~ Z

generally call for identification of all specimens. Several reasons exist for this, including curiosity, but probably most noteworthy is that an uncommon species may be important as an indication of habitat availability. A good technician in the field can usually sight identify common species and can reliably distinguish genera. That technician may arrange the priority of daily duties around the identifications, generalizing from a generic determination to the practical implications of the specimen in hand. Even back at the laboratory, where specialized technicians may reliably identify all species, genera are important for ordering reports and for training new personnel.

The biological concept of the genus has received relatively little attention. The lack of theory behind the genus could be the result of an inherently difficult biological concept maintained by flawed human requirements for filing names. On the other hand, efforts in systematics may have been drawn away from the genus concept by its difficulty and by the phylogenetic tools that are efficient at grouping suprageneric and species-level groups. The genus appears to fall inconveniently between those firmer definitions, even though genera have great practical importance for those who work with mosquitoes. Definitions of genus are not particularly useful,[2] though most might agree that an ideal genus should be monophyletic or paraphyletic (Dubois 1988), distinguishable by one or more character states (though possibly with exceptions; Winston 1999), and equivalent to other genera across a family or higher taxonomic category (Reinert, Harbach, & Kitching [2006b] and Harbach [2007] "principle of equivalent rank"). The concept of equivalent rank was not really defined and not always used, but it has intuitive value and probably deserves more attention. Equivalence is important in order to work toward biologically meaningful generic divisions, but basing equivalence on number of species tends to split taxa into more genera, whereas basing equivalence on shared characteristics tends to create a wide disparity in the number of species per genus.

The nature and existence of a biological basis for the genus is controversial. Linnaeus thought that a genus must be defined before describing the species within, and he considered the genus to be natural in the sense that it was created (Cain 1956). Modern methods start with a group of species and attempt to look for phylogenetic gaps that might indicate genera, sometimes leading to limited concepts of what constitutes a particular genus and its biological significance. The *Code* itself seems to endorse the idea that a genus is simply a convenient category in which to place member species. The *Code* trivializes the selection of a genus type species, with recommendations to use species that are prominently common or important, or species with names suggesting a typical status (e.g., *communis*).

The opposite extreme—that genera have evolutionary significance—raises its own controversial aspects. Namely, it requires a group selection mechanism to explain generic evolution within a context of more widely accepted species-oriented mechanisms. Nonetheless, some authors have attempted to explain genera in an evolutionary framework, even coining the term "geniation" in contradistinction to speciation (Dubois 1988). Recognizing the extremes of phylogenetic and phenetic generic concepts, Lemen and Freeman (1984) discussed three models for generic selection. They concluded that the best model for the origin of a genus involved nonrandom changes in shape and function, resulting in adaptation to ecological niches (called "adaptive zones" by the authors) that had not been used by the originating genus.

Dubois (1988) also proposed an evolutionary model for genera, considering the potential for hybridization to be a clear indication that two species were in the same genus. Similar to Lemen and Freeman, Dubois suggested that the origin of a genus begins with a clear ecological departure from the originating genus. He summarized his views on geniation as follows:

1. Geniation is not simple or frequent.
2. Geniation occurs under exceptional conditions.
3. Not all genetic revolution is geniation.
4. Geniation results in occupation of a new adaptive zone.
5. Geniation is not characterized by a change in size.

Although Dubois (1988) and Lemen and Freeman (1984) based their opinions on vertebrate groups, their conclusions shed some light on systematics of mosquito genera. Like Mayr (1969) and Winston (1999), they implied that what defines a genus is independent of the number of species in it. Their criteria were based on an evolutionary process that is reflected by phylogenetics but not totally described by phylogenies. Those criteria have a saltational aspect in which the genetic distance is not as important as the qualitative difference in adaptation. The degree of difference required to declare a separate genus again fell on subjective judgment, just as in Mayr's definition but with an important distinction. Morphological and genetic criteria for genera relied on those data as evidence for an evolutionary process that happened in the distant past. Attention to adaptive zones, on the other hand, gets much closer to the selective pressures that caused the original species to become distinctly different from its congeners. Recent revisions based on morphological and genetic phylogenies have split formerly familiar groups of mosquitoes into dozens of new, unfamiliar genera. Although the phylogenies themselves are valid presentations of testable hypotheses, the changes in nomenclature should probably be passed through the filter established by Dubois, Lemen, and Freeman: Do each of those new genera represent a distinctly different adaptive zone? Mosquito taxonomists should also consider Mayr's (1969, p. 344) words: "A generic name no longer tells much to a zoologist except in a few popular groups of animals; it has become a tool of the specialist. The degrading of the usefulness of the generic name

[2] Mayr's (1969, p. 92) definition of genus: "A category for a taxon including one species or a group of species, presumably of common phylogenetic origin, which is separated from related similar units (genera) by a decided gap, the size of the gap being in inverse ratio to the size of the unit (genus)."

has been made worse by the extreme generic splitting so prevalent in many areas of systematic zoology."

Subgeneric Taxa

Taxonomic divisions between genus and species have been used extensively in mosquito taxonomy. Most species have been assigned to a subgenus as well as a genus. Divisions within a subgenus further categorize species into sections, series, and groups. Those divisions are by convention, not by code (Harbach 2011), but they have seen a lot of use in larger subgenera. Some of those designations, particularly in *Anopheles*, are familiar even to the nonspecialist, who must work with many species in a particular tropical region. Ideally, all of those categories follow phylogenetic reality; however, they are also useful simply for filing purposes in a collection or review. In the future, use of the *PhyloCode* may provide objective criteria for formal names of these various clades, regardless of the rank assigned.

The naming of mosquito species is relatively less controversial than the naming of genera. Although groups with little medical or economic importance receive less attention, the common worldwide need to identify mosquitoes has led to revisionary work touching most species in the family. One of the great benefits of this close examination has been the discovery and study of species complexes, particularly among *Anopheles*. The previous chapter described four examples of how study of closely (phylogenetically) related and sometimes morphologically indistinguishable species sheds light on evolutionary processes. Complex groups represented by *Culex pipiens*, *Anopheles gambiae*, *Anopheles dirus*, and *Aedes aegypti* reveal some of the intricacies of population genetics, speciation, and distribution that challenge structural species concepts. The presence and importance of genetic exchange between species is particularly enlightening. Whether called introgression, hybridization, or reticulate phylogenetics, the idea that species are not so isolated from each other is in contrast to the common definition of species as reproductively isolated.

One point of general consensus is that species should be treated as the "basic rank of zoological classification" (International Commission on Zoological Nomenclature 1999). From Darwin's *On the Origin of Species* (1859) to Winston's (1999) *Describing Species*, there exists a belief that the species' taxonomic rank should correspond to a real natural entity on which the processes of evolution work. Those who use species names expect consistency of the organisms' characteristics that are of most concern, whether morphological, behavioral, physiological, ecological, or genetic. As a result, splitting an existing species into two or more may be welcomed by the community of mosquito biologists. More precise designation of species leads to better correspondence between specimens in hand and characteristics of concern.

Binomial nomenclature would have the species name as a modifier of the genus name. That traditional Linnaean system raises the importance of the genus and makes the species relatively less independent as a taxon. The argument has been made that classification systems should move toward a uninominal for species in order to achieve nomenclatural stability (Mayr 1969) and phylogenetic reality (Cantino & de Queiroz 2010). The recent turbulence in mosquito generic classification seems to have moved some practitioners in that direction. One discussion that supported the original division of *Aedes* into *Aedes* and *Ochlerotatus* (Black 2004) posed the question of whether generic designations mattered much since public health workers and evolutionary biologists are more concerned with species. From a biological standpoint, de Queiroz (2007) solves many of these problems by using a definition of species that is inclusive of many concepts, providing the flexibility to be useful for a range of biological situations. Firmly rooted in a foundation of evolutionary history, de Queiroz said that a species is "a separately evolving metapopulation lineage or a segment of such a lineage," acknowledging that species are not genetically static. He also acknowledged that there is a continuum between population-level differences and species-level differences, with the line drawn according to the judgment of the taxonomist. Higher-level groupings, as discussed above, should be based on phylogenetic clade, rather than rank.

The species names themselves (Winston 1999) are generally descriptive (e.g., *Uranotaenia sapphirina*, referring to lines of iridescent blue scales), geographic (e.g., *Culex yojoae*, named after Lake Yojoa in Honduras), or commemorative (=patronymic) (e.g., *Psorophora stonei*, named after the famous mosquito taxonomist Alan Stone). The origins of these names are briefly explained for most of the species and subspecies in Volume 2 of this book. Although the comprehensive and highly technical guide to forming scientific names is Roland W. Brown's *Composition of Scientific Words* (1956), Kitzmiller's (1982) delightful book on the origins of anopheline names provides a short and easily understood summary of why names have particular endings (Table 2.3). It also explains in detail the origins of anopheline names, serving as a history of mosquito taxonomy through the lens of those who described species and those for whom the species were named.

Unlike genus names, species names do not have to be unique within the animal kingdom. On the one hand, this gives taxonomists more flexibility in choosing names, but on the other hand, it creates nomenclatural instability when two species with the same name move from separate genera to the same one (secondary homonym). This last issue could become a common challenge in mosquito taxonomy as the sudden multiplication of generic names resettles into fewer or different designations.

A subspecies is a formally recognized rank in the *Code*, which defines it as "the species-group rank below species." More helpfully, Mayr (1969) defined a subspecies as "a geographically defined aggregate of local populations which differs taxonomically from other such subdivisions of the species." Anyone who has used an identification guide to fish, birds, or mammals will recognize that subspecies are used extensively for vertebrates. Generally, and as Mayr defined the rank, subspecies have been used for a population of a species

Table 2.3. Summary of the Latin grammar of species name endings (from the introduction to Kitzmiller 1982)

Type of name	Latin grammatical class	Ending	Example among *Anopheles*
Patronym	Masculine simple genitive singular	*-i*	*konderi*
	Feminine simple genitive singular	*-ae*	*ludlowae*
	Masculine genitive singular*	*-ii*	*cruzii*
	Commemorative suffix	*-ia*	*Howardia*
	Masculine diminutive honorific	*-ellus*	*barberellus*
	Feminine diminutive honorific	*-ella*	*Bironella*
Geographic	Locative suffix	*-ensis*	*freetownensis*
	Locative suffix	*-anus*	*georgianus*
	Locative suffix	*-icus*	*koreicus*
Descriptive	Masculine or feminine to agree with gender of genus	Various	*oiketorakras*

*Legitimacy of this form is disputed by Latin scholars. Others claim it is the masculine genitive singular of the dedicative Latin suffix *-ius*.

that is morphologically distinct and geographically separate. Seen that way, not every specimen needs to fit within a subspecies, and not every species needs to be divided into subspecies. Relatively few subspecies (130) have been designated among mosquitoes, as taxonomists have a clear preference for elevating distinct forms to full species status. For example, introgressing populations of *Anopheles gambiae* M and S forms were raised to species, not subspecies, status (Coetzee et al. 2013). *Culex pipiens* and *Culex quinquefasciatus* remain separate species despite being interfertile and originally allopatric. On the other hand, *Culex pipiens pallens* remains a legitimate designation in Asia, and *Aedes aegypti* forest (*Aedes aegypti formosus*) and domestic (*Aedes aegypti aegypti*) forms are considered subspecies.

A recent review of a group of snakes in western North America provided an objective example of the use of subspecies that may have implications for the concept's application in mosquito taxonomy. Using a combination of morphology, mitochondrial DNA sequences, and geographic distribution, Mulcahy (2008) examined 178 specimens of what had previously been a single species with 17 subspecies. The analysis resulted in designation of six species, two of which were divided into subspecies. The author recognized the subspecies based on wide geographic distribution of the molecularly defined root species and geographically cohesive distribution and morphological distinctiveness of the proposed subspecies. Part of the decision to recognize the subspecies was based on its probable utility for protecting a distinctive population. Phylogenetically, the subspecies were seen as incipient divergence within a lineage that clearly had a narrower geographic origin. This view of a subspecies required a large and carefully considered data set, but it appeared to have both phylogenetic and practical value. Although subspecies of mosquitoes have not been reviewed in this way, such a review might be a good goal, rather than eliminating all subspecies from the family, as some taxonomists have informally suggested (Harbach 2012).

Common Names

The word meaning "mosquito" appears to have changed in many languages over time, possibly in response to the wide recognition of their medical importance about 1900 (Christophers 1960). In English, they were usually called "gnats" before the 20th century. French now uses *moustique*, but Réaumur (1738) titled his review *Histoire des Cousins*. Other older French names included *moucheron* and *maringouin*. The current German name is *Stechmücken*, but mosquitoes were formerly called *Mücke*, *Schnacke*, or *Gelse*. Italian went from *zanzare* to *muskito* and Spanish from *cinife*, *cenzalo*, and *zancuda* to the familiar *mosquito* of today. The Spanish *mosquito* may have originated in the 16th century, when descriptions of the Americas started using the term *musketa*. Other older names include *myg* (Dano-Norwegian), *mygga* (Swedish), *comaro* (Russian), and *konopus* (Greek).

Popular language generally distinguishes mosquitoes only at the family level (Service 1993). Therefore, languages have not created many words like "sparrow," "bass," or "badger" for mosquitoes that identify species or genera with distinctive characteristics. Although other examples probably exist around the world, the use of "gallinipper" for *Psorophora ciliata* (a particularly large biting mosquito in the American South) and *ñatïü akua* for *Anopheles* in Paraguayan Guaraní is an unusual use of common language to provide identifications more specific than the family level. Rao (1984) reviewed an ancient Sanskrit classification by Susutra of five kinds of mosquitoes (= *mashakah*) called *samudrah* (marine or coastal), *parimandalah* (very small), *hasti* (elephant), *krishnah* (black or dark), and *pāwatiyah* (hill). It is somewhat surprising that there are not as many common names for mosquitoes as for agricultural pests. Possibly, the small size of mosquitoes and the generally similar nature of their bites explains the lack of motivation to distinguish them. As Rao pointed

Table 2.4. Comparison of guidelines for common names in English for reptiles/amphibians (Frank and Ramus 1995), birds (Gill and Donsker 2012), and insects/invertebrates (Entomological Society of America 2012)

Criterion	Amphibians/reptiles	Birds	Insects/invertebrates
Names per taxon	One	One	One
Unique	Subspecies often consist of base name with modifier	Do not include full name of one taxon in another	No explicit rule
Non-English and scientific names	Allowed	Allowed if in common use	Scientific names discouraged unless in common use
Usage in preference to accuracy	Reliance on scientific classification	Use of name (e.g., robin) in different taxa or inaccurate geography allowed	No explicit rule
Local vernacular	Not used	Formal established names preferred over vernacular	No explicit rule
Patronyms	Commonly used	No bias for or against	Sometimes used
Number of words	2–4	1–4	3 preferred, 4 allowed
Use of "island"	No explicit rule	Avoid use	No explicit rule
Capitalization	Each word	Each word	Only proper names
Scope	All species	All species	Prominent species

out, only four species of *Anopheles* had been described in India at the time that Ross showed their importance as vectors of malaria in 1898. One of the only early attempts to assign common English names to mosquitoes was included in Felt's (1904) treatment of the mosquitoes of New York State. He named only 12 of the species he described, based on morphology (*Psorophora ciliata* = giant mosquito), habitat (*Aedes canadensis* = woodland pool mosquito), or both (*Aedes sollicitans* = white-banded salt marsh mosquito).

Official lists of common names for reptiles, birds, and insects provide some insights into the motivations and rules associated with such names. An ambitious list of 12,000 amphibians and reptiles by Frank and Ramus (1995) combed over 200 references and field guides in an attempt to reflect taxonomic classification. The authors stated that "in the ever-changing world of taxonomy, it is apparent that common (or English vernacular) names are often more stable than scientific ones." Holloway (2003) wrote a compendium of over 900 common names of birds from the United States, including Hawaii. This list was annotated with explanations of the origins of both scientific and common names. More authoritatively, the International Ornithologists' Union maintains a database of bird common names officially sanctioned by a committee, formerly designated the International Ornithologists' Committee (Gill & Donsker 2012). The list is introduced by 10 rules that form a sort of miniature nomenclatural code for common names. The Entomological Society of America maintains a list of North American (United States, Canada, their possessions, and foreign locations under special circumstances) insect (and invertebrates of common concern to entomologists) common names, a function that dates back to a committee of the American Association of Economic Entomologists in 1903. The current committee has nine members, new names are approved by the entire membership of the society, and the list is kept updated on a website (Entomological Society of America 2012).

Table 2.5. Common English names of mosquitoes officially recognized by the Entomological Society of America (2012)

Scientific name	Common name
Aedes aboriginis	northwest coast mosquito
Aedes aegypti	yellowfever mosquito
Aedes albopictus	Asian tiger mosquito
Aedes cantator	brown saltmarsh mosquito
Aedes sierrensis	western treehole mosquito
Aedes sollicitans	saltmarsh mosquito
Aedes squamiger	California saltmarsh mosquito
Aedes sticticus	floodwater mosquito
Aedes vexans	vexans mosquito
Anopheles quadrimaculatus	common malaria mosquito
Culex pipiens	northern house mosquito
Culex quinquefasciatus	southern house mosquito
Deinocerites cancer	crabhole mosquito
Wyeomyia smithii	pitcherplant mosquito

Table 2.4 compares the guidelines of the reptile/amphibian, bird, and insect/invertebrate lists of common English names.

The current list of names from the Entomological Society of America includes 14 species of mosquitoes (Table 2.5). Although a small sample, these names are mostly based on an ecological or epidemiological characteristic, sometimes modified geographically. Exceptions are the Asian tiger mosquito, which is based on common usage originating from morphology ("tiger" because it is striped) and geography ("Asian" because of its origin), the northwest coast mosquito based only on geography, and the vexans mosquito based on its scientific name. Significantly, all end with the word "mosquito," and none attempt to designate genus, therefore forming true uninomials.

Table 2.6. Common names for genera of mosquitoes

Genus	Common name	Etymology
Anopheles	Nail Mosquito	Named after Guarani common name for its biting habits
Bironella	Biró Mosquito	Named after the Hungarian collector H. Biró, who captured the type specimen December 31, 1900
Chagasia	Basal Mosquito	Based on basal location in phylogeny of mosquitoes
Aedeomyia	Scaly Mosquito	Named for the generally scaly appearance of these species caused by larger erect scales and by the tuft of erect scales at the distal end of the femur of all species
Aedes	Pointy Mosquito	Refers to the relatively pointed shape of the end of the abdomen of the majority of species
Armigeres	Cord Mosquito	Refers to laterally flattened and curved proboscis
Eretmapodites	African Mosquito	This is the only genus of mosquito restricted to Africa
Haemagogus	Shoulder Mosquito	Named for the large anterior pronotal lobes of the adults
Heizmannia	Metallic Mosquito	Reference to broad iridescent or metallic-appearing scales on the scutum
Opifex	Waterwalking Mosquito	Named for the adaptations on the tarsi of males, which mate with females before they emerge from the pupal case
Psorophora	Gang Mosquito	Refers to aggressive biting habits and sudden appearance of many adults following inundation of oviposition sites
Udaya	Compressed Mosquito	Refers to thorax that is laterally compressed compared to other mosquitoes
Verrallina	Broadtail Mosquito	Named for the slightly expanded distal end of abdomen in some species and to the large female genitalia
Zeugnomyia	Silver Mosquito	Reference to large patch of silvery white scales on pleuron of adults
Culex	Typical Mosquito	Named because the Latin means "mosquito," and this was the first mosquito genus, making it the type genus for the family
Deinocerites	Crabhole Mosquito	Based on the larval habitats of these species
Galindomyia	Galindo Mosquito	Named for Pedro Galindo, who was an important Panamanian mosquito biologist who helped gather and analyze the first specimens of this one-species genus
Lutzia	Predatory Mosquito	Named because all larvae are predatory
Culiseta	Cool Mosquito	Reference to the seasonal and geographic distribution of members of this genus, which are generally either in an area with a cool climate, a high elevation in a tropical area, or the cool season of a warmer location
Ficalbia	Small Mosquito	Reference to the generally small size of adults
Mimomyia	Indistinct Mosquito	Refers to similarity of adults to various other genera, despite distinct morphological characteristics of this genus
Hodgesia	Tiny Mosquito	Refers to very small size of these mosquitoes
Coquillettidia	Underwater Mosquito	Named for the larva's and pupa's habit of attaching by siphon or respiratory tubes to stems of underwater plants in order to get oxygen
Mansonia	Submarine Mosquito	Named for the larva's and pupa's habit of attaching by siphon or respiratory tubes to stems of underwater plants in order to get oxygen
Orthopodomyia	Red Mosquito	Refers to reddish or purplish pigmentation of epidermis in the larvae in many species, making them appear pink, red, or purple
Isostomyia	Grasping Mosquito	Refers to one of the major distinguishing features of this genus, a well-developed maxillary claw of the larva in conjunction with an apical tooth, forming an organ for grasping prey; these species are apparently facultative predators
Johnbelkinia	Brushy Mosquito	Refers to the line of multibranched setae along midventer of the larval siphon to the setose male genitalia
Kimia	Ridged Mosquito	Named for the series of ridges formed by the abdominal wall on the venter of the larvae, on which it rests while feeding on the bottom
Limatus	Royal Mosquito	Refers to metallic gold and purple scales on scutum
Malaya	Ant Mosquito	Reference to unique bionomic association whereby adults are known to stimulate and then feed on regurgitation from ants
Maorigoeldia	Maori Mosquito	Named as subgenus by Edwards (1930) because of similarity to former genus *Goeldia* in the Americas, but occurrence in New Zealand

Genus	Common name	Etymology
Onirion	Dream Mosquito	Derived from the origin of the genus scientific name, Greek *oneiros*, or dream
Runchomyia	Filament Mosquito	Refers to its filamentous pecten
Sabethes	Canopy Mosquito	Refers to association of adults and larvae of this tree hole mosquito with the canopies of tropical forests
Shannoniana	Toothy Mosquito	Refers to the modified maxilla of the larvae and to pre-apical teeth on the aedeagus of the male genitalia
Topomyia	Omnivore Mosquito	Based on the currently known habits of the larvae, which feed on other mosquitoes and filter particles for nutrition
Trichoprosopon	Hairy-Lipped Mosquito	Refers to the unusual feature of setae on the clypeus in some representative species
Tripteroides	Pestle Mosquito	Derived from the Latin name meaning "like a pestle"
Wyeomyia	Bromeliad Mosquito	Refers to the common usage of leaf axils of bromeliads as immature habitats
Toxorhynchites	Carnivorous Mosquito	Named for the nutritional habit of the larvae, which are voracious predators of mosquito larvae and other aquatic arthropods
Uranotaenia	Heaven Mosquito	Derived from the Greek for heavenly (*urano*), referring to the beautiful azure blue markings on many members of the genus, though not all share this character

The catalog portion of this book proposes English common names for all species and subspecies of mosquitoes. This effort has no official sanction. Although an immediate motivation for assigning common names is the current instability of generic scientific names, a longer-term goal is to achieve stable, easily used uninomials. By including all species rather than a selection of those considered most important, it is hoped that more attention will be paid to seemingly obscure species that are actually of ecological or phylogenetic interest. As new species are described, perhaps their authors will include common names that are easily translated into other languages and follow the guidelines outlined here.

The guidelines for common names of mosquitoes followed in this book are based on those described in Table 2.4, with a few alterations. They include the following:

1. Utility of the name as a uninominal. Each name is unique within the family Culicidae, and each species has a single name.
2. Common names apply to species. Common names are for species; no attempt is made to designate genera or higher ranks.
3. Language. These guidelines are intended for English-language common names; however, it is desirable to have names that are easily translated into other languages. When a vernacular name is in common use, it may be transliterated into the language being used for the common name.
4. Offensive names. Offensive names of any kind must be avoided.
5. Capitalization. Each word is capitalized in order to distinguish the common name from other text.
6. Hyphenation. Hyphens should be used grammatically, and hyphenated words count as a single word.
7. Construction of names. Although variation is allowed, most names have four words consisting of (a) a word derived from or related to the scientific species (i.e., trivial) name, (b) the country in which the type locality is located, (c) a word related to the genus (Table 2.6), and (d) the word "Mosquito."
8. Patronyms. Patronyms may only be used if the scientific species name is a patronym, in which case the common name's patronym must match the scientific name's patronym. Only the nominative form may be used (e.g., Dyar Mosquito, not Dyar's Mosquito).
9. Abbreviation. Common names may be differentiated by use of acronyms consisting of the first letter of each word.
10. Declaration of common name. Scientific descriptions of new species are encouraged to declare a common name.

Conclusion

Nomenclature is as much a part of biology as any other aspect of the science. In some ways, it is even more important because individual scientists are at a tremendous disadvantage if they cannot identify species or do not understand how they are designated. The naming of mosquitoes and other organisms is a human activity, and it is particularly subject to egotistical motivations, which has been part of the challenge of erecting objective systems. Nomenclature of mosquitoes has evolved tremendously since the original discoveries showing the medical importance of the family, even within established systems. Alternative, thoughtful systems for nomenclature have been proposed in recent years so that the evolution of this aspect of mosquito biology may accelerate even further.

3. Distribution

Few patterns in ecology, evolution, conservation biology—and for that matter, most studies of biological diversity—make sense unless viewed in an explicit geographical context.

Lomolino et al. 2010

One of the first questions anyone asks about a species is its location. Where does it occur? Where is it most abundant? Where does it develop? Of course, the questions about "where" are matched by questions about "when." When is the species fulfilling each aspect of its life cycle? When is the population increasing or decreasing? When did the species become distinct evolutionarily? Biogeography encompasses these questions but adds another: Why?

The subject of mosquito distribution includes all species in Culicidae, deals with a range of scale from the individual to the family, and is influenced by the full complexity of interactions between physical and biological aspects of the environment. In some ways, distribution is not so much a result of evolution as an extension of it. This chapter will touch on much of the subject matter of the remainder of the book because most of mosquito biology is intricately related to the distribution of mosquitoes.

As a practical matter, where and when mosquitoes occur is essential for their management. In the first place, identification generally depends on limiting the problem to a geographic region that includes a more limited number of species. Knowing the location and predicting the presence of vector species can be an important part of disease risk assessment. Mosquito abatement depends on targeting efforts where important mosquito species reach population levels that cause a disease or nuisance problem.

Mosquito distribution is discussed here in the context of the present, past, and future. Simply describing the locations of species as they currently occur involves sophisticated theory, statistics, and technology—as well as the use of shoe leather that has always been a part of collection efforts. The current location of a mosquito species depends very much on where it was before. As a result, a discussion of past distribution includes not only paleontological considerations of evolution but more recent events like the development of agriculture and large-scale transport. Much of the interest in past and current distribution is motivated by a need to predict future distributions. Whether those predictions are short or long term, they depend on an understanding of what happened in the past and models of what will happen in the future. These themes are repeated throughout biology, and Culicidae provides a particularly rich framework for their study.

Present Distribution of Mosquitoes

Historical Efforts in Biogeography

Many of the basic principles of biogeography were developed over 200 years ago at the end of the 18th century (Lomolino et al. 2010). Work by naturalist Joseph Banks during James Cook's first expedition (1768–1771) and by Johann Reinhold Forster during his second expedition (1729–1798) showed such familiar principles as smaller numbers of species on islands, greater diversity toward the equator, and the presence of distinct species living in much the same ways but in different locations. Another key figure was Alexander von Humboldt, who traveled the world and outlined the basic patterns of plant distribution. In 1819 he described an important mosquito from Ecuador, *Sabethes chloropterus* (as *Culex chloropterus*), based on collections made during voyages from 1799 to 1804. As pointed out in Chapter 1, James Hutton established realistic views of geological stratigraphy, leading to the firm concept of uniformitarianism by Charles Lyell in 1830. Uniformitarianism was key to understanding distribution because it eliminated historical explanations that were not based on current mechanisms. Ironically, the sound principle of uniformitarianism led to centuries of confusion over major distributional patterns because the locations of continents were assumed to be fixed. Until the theory of plate tectonics was widely accepted in the 1960s, even such luminaries of mosquito and dipteran biogeography as John Belkin (1962) and Willi Hennig (1966b) resorted to speculation on land bridges and convenient sea-level changes to explain the location of species in the southern Pacific region.

The voyages of biological discovery and consequent classification of collections continued into the 19th century, perhaps most famously by Charles Darwin as a naturalist on the *Beagle* (1831–1835; Darwin 1860). Alfred Russel Wallace is often called the "father of biogeography" based on his work in Southeast Asia and South America (Wallace 1823–1913). He used his own and others' observations to synthesize and clearly explain many principles of biogeography still used today. Among them was division of the world into biogeographic realms in 1876, originally proposed by P.L. Sclater in 1858 based on what he contended were centers of origination of bird species (Lomolino et al. 2010).

During the period of biological discovery that began with Linnaeus in the 1700s, the description of mosquitoes was consistent but almost never a subject of taxonomic specialization (Pape & Thompson 2013). Up until 1850, dipterists, entomologists, and faunal catalogers in their individual lifetimes never described more than a few dozen species of mosquitoes. One of the early entomologists, J.C. Fabricius, described 16 species of mosquitoes between 1775 and 1805, examining specimens from the Americas and Europe. The only publication dedicated to mosquitoes was by Jean-Baptiste Robineau-Desvoidy, titled *Essai sur la tribu des Culicides* (1827), in which he described 17 species from Europe and the Americas. Others described more species, such as the "father of dipterology," Johann Wilhelm Meigen (34 species between 1804 and 1838), C.R.W. Wiedemann (29 species between 1819 and 1828 from Asia and the Americas), and P.J.M. Macquart (24 species between 1834 and 1850 from the Americas, Australia, Africa, and Europe). All this work was done either as a part of a larger review of Diptera or as part of a regional faunal survey. Except for von Humboldt, the early mosquito taxonomists stayed close to home and let the specimens come to them from museums and collections.

Mosquito collection and description was pursued with much more energy as the germ theory of disease led to the search for the causes of two of the greatest sources of death in the late 19th century, malaria and yellow fever (Knight & Pugh 1973). Within decades large compendia of mosquito taxonomy and distribution were produced (e.g., Theobald 1901a, 1901c, 1907, 1910c; Howard et al. 1913, 1915, 1917), forming the basis for Edward's (1932) revision that is still the foundation of mosquito taxonomy. Most mosquito distribution data were presented as country lists or simple maps until the 1960s. Belkin's introduction to *Mosquitoes of the South Pacific* (1962) made one of the early attempts to explain the current regional fauna based on geography, dispersal, and evolution. Ross (1964) pioneered what would now be called a vicariance analysis, superimposing phylogeny on distribution of mosquitoes with emphasis on those species that occur in North America.

Qualitative Analyses of Global Mosquito Distribution

Ross' phylogenies were based on intuitive groupings of taxa. These groupings gave only a rough idea of genetic distance, but they accurately reflected taxonomic divisions of the time. The analysis led to some interesting generalizations and some insights on particular groups. First, he concluded that most groups of mosquitoes reach their highest diversity in the tropics, specifically citing *Deinocerites*, *Wyeomyia*, *Culex* (*Culex*), *Culex* (*Melanoconion*), *Anopheles* (*Nyssorhynchus*), and *Aedes* (except the subgenus *Ochlerotatus*). Ross showed that within some of the tropical groups, individual species had developed the capability to live under subtropical or temperate conditions (Table 3.1), some of which were illustrated in interesting figures that combined phylogeny, morphology, climatic adaptation, distribution, and history (Fig. 3.1). Only *Aedes* (*Ochlerotatus*) and *Culiseta* were considered to be primarily temperate.

Gutsevich et al. (1974) shared some generalizations about world mosquito distribution as part of a review of mosquitoes of the Soviet Union. Like Ross (1964), they noted the greater variety of mosquitoes in the tropics, especially in the Neotropics, citing *Chagasia*, *Anopheles* (*Stethomyia*), *Anopheles* (*Kerteszia*), *Anopheles* (*Lophopodomyia*), *Anopheles* (*Nyssorhynchus*), *Psorophora*, *Deinocerites*, and *Culex* (*Melanoconion*) as examples. The biogeographic affinities of particular taxonomic groups were not considered as strong in the Ethiopian and Oriental regions and were almost absent in Australasia, with the only endemic genus being *Opifex*. The temperate North also has some distinctive patterns of mosquito distribution. The Nearctic, especially, is characterized by the absence of groups and by the great radiation at a species level of *Aedes* (*Ochlerotatus*), including some 20 species that are circumpolar, occurring in both the Palearctic and Nearctic. Gutsevich et al. (1974) pointed out the phenomenon of a few species in tropical groups of eastern Asia that have been able to expand distribution into temperate northern climates (*Aedes albopictus*, *Armigeres subalbatus*, and *Tripteroides bambusa*). Another example is *Uranotaenia novobscura*, a container-developing member of the subgenus *Pseudoficalbia* (Peyton 1977). This subgenus is principally tropical, but this species occurs as far north as Tokyo, Japan. Mogi (2012) attributed the distribution of a few species from tropical groups to the lack of significant physical barriers between the tropical and temperate areas of eastern Asia.

Foley et al. (2007b) analyzed the country distributions of all mosquitoes to document areas of greater or less diversity. These worldwide quantitative analyses documented trends that give a broad description of mosquito distribution. The greater number of species in the tropics was a strong pattern (Fig. 3.2). There was a very regular decrease in biodiversity toward the poles, with nearly 88% of all species occurring between 40°N and 40°S. To account for variations in land area, Foley et al. (2007b) also analyzed species richness per country in proportion to each country's size. Although the analysis

Table 3.1. Ross' (1964) compilation of species in predominantly tropical groups that occur in subtropical or temperate locations[1]

Taxonomic group	Species	Subtropical	Temperate
Anopheles (*Nyssorhynchus*)	*albimanus*	X	
Anopheles (*Anopheles*)			
Pseudopunctipennis Gp.	*pseudopunctipennis*	X	
Aedes (*Aedes*)	*cinereus*	X	
Aedes (*Aedimorphus*)	*vexans*	X	X
Psorophora (*Grabhamia*)	*pygmaea*	X	
	discolor	X	X
	columbiae	X	X
	signipennis	X	
Psorophora (*Janthinosoma*)	*cyanescens*	X	X
	longipalpus	X	
	varipes	X	X
	ferox	X	X
	horrida	X	
Psorophora (*Psorophora*)	*ciliata*	X	X
	howardii	X	X
Culex (*Melanoconion*)	*atratus*	X	
	opisthopus (= *taeniopus*)	X	
	abominator	X	
	peccator	X	X
	anips	X	
	mulrennani	X	
	erraticus	X	X
	pilosus	X	X
	iolambdis	X	
Deinocerites	*cancer*	X	
Coquillettidia	*perturbans*	X	
	richiardii	X	
Mansonia (*Mansonia*)	*indubitans*	X	
	titillans	X	
Mansonia (*Mansonioides*)	*uniformis*	X	
Orthopodomyia	*alba*	X	
	californica (= *signifera*)	X	
	signifera	X	X
Wyeomyia	*mitchellii*	X	
	smithii	X	X
	vanduzeei	X	
Toxorhynchites	*rutilus*	X	X
Uranotaenia	*anhydor*	X	
	lowii	X	
	sapphirina	X	X
	anhydor syntheta	X	

[1]Ross (1964) did not include two important examples that occur in subtropics and temperate areas: *Armigeres subalbatus* and *Tripteroides bambusae*

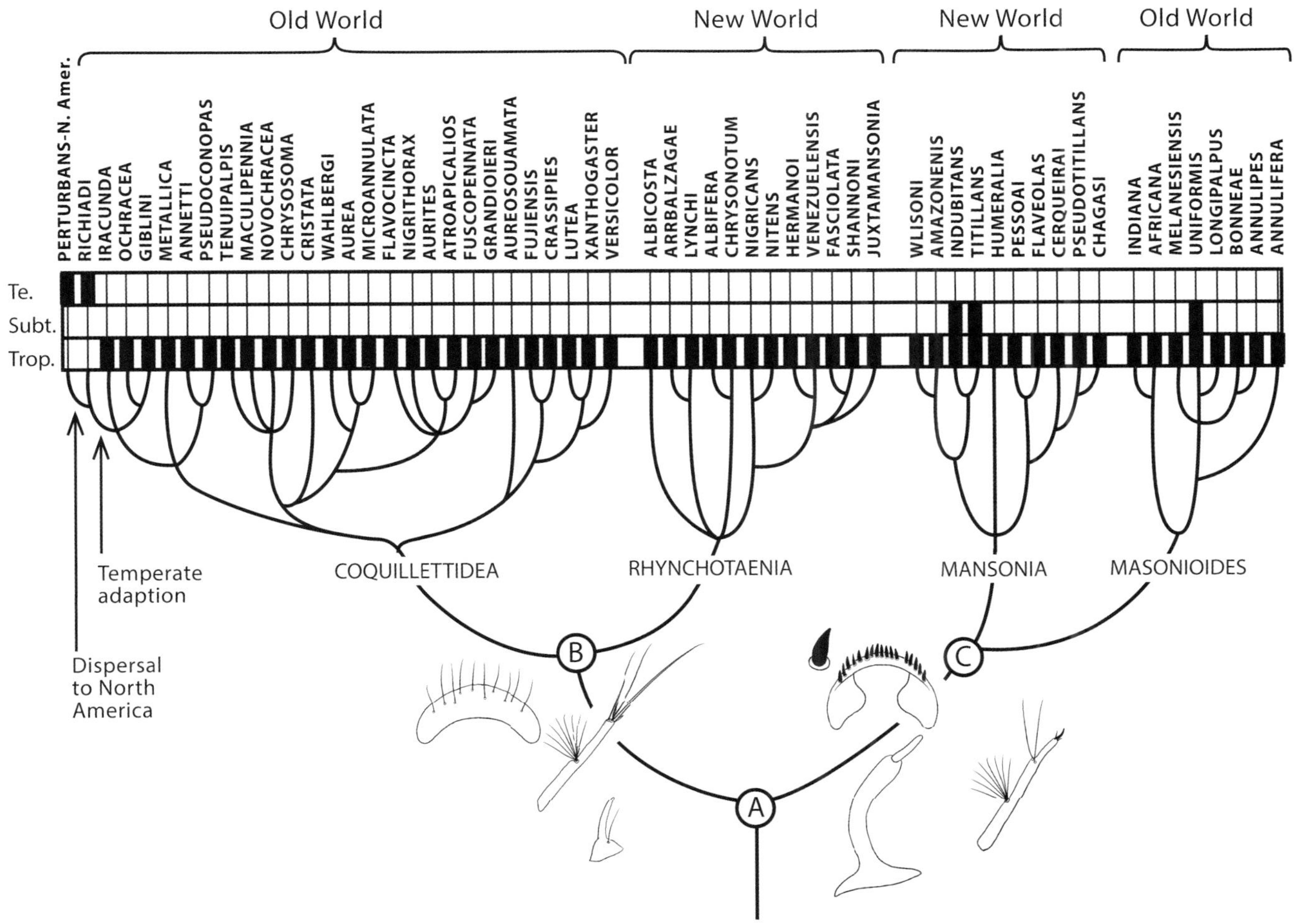

was course grained and arbitrarily categorized by political divisions, it did show more species than expected where climatic conditions were most favorable. In general, the tropical and subtropical belts had more species than expected based on land area, except for the deserts of southern Africa and the entire complex of deserts that includes the Sahara, the Middle East, and Southwest Asia. Endemic species (by country) followed the same general pattern. The generally depauperate mosquito fauna of Europe followed the commonly observed lack of diversity in other organisms, possibly as a result of the severity of previous glaciations. In contrast, the American tropics had high mosquito diversity (Foley et al. 2008b), as did Southeast Asia. Foley et al. (2007b) also performed species-area relationship analyses, finding typical log-log correlation between area and number of species. Less expected, and counter to the pattern observed in some other plants and animals, islands had greater species richness per area (slope of 0.4 if Greenland and Singapore are excluded as outliers) than did continental areas (slope of 0.33). The mosquito biodiversity on islands may be more a matter of dispersal from mainland areas by phoresis than evolution of endemic species—for example, on Malta (Gatt 1996) and Hawaii (Leong & Grace 2009).

Looking at a finer taxonomic scale but a broader geographic one, Rueda (2008) compared the mosquito species richness of the standard Sclater/Wallace biogeographical regimes (Table 3.2). Avoiding the complications of unstable generic taxonomy, an examination at the subfamily and tribe levels showed regional differences in distribution. At the coarsest level, there were about six species of Culicinae for each species of Anophelinae. That number was considerably lower, a little over four species of Culicinae per species of Anophelinae, in the Palearctic and Afrotropical regions, whereas the number was higher, over seven, in the Nearctic, Neotropical, and Australasian regions. These variations may have reflected some disparities in taxonomic effort, but they also probably resulted from the many different influences on species radiation during recent geological history. Aedini generally outnumbered Culicini, the ratio varying between 1.95 and 2.74 in every region but the Neotropics, where the ratio was only 0.72. Other smaller tribes had regional affinities, the most extreme being Ficalbiini and Hodgesiini, which do not occur in the Western Hemisphere. Hodgesiini and Aedeomyiini were the only tribes with strictly tropical distributions.

Figure 3.1. Family tree of subgenera and better-known species of *Mansonia* and *Coquillettidia* (as a subgenus of *Mansonia*). The distribution of each species is indicated by black bars at the various levels. Subt. = subtropical; Te. = temperate; Trop. = tropical. Ross 1964

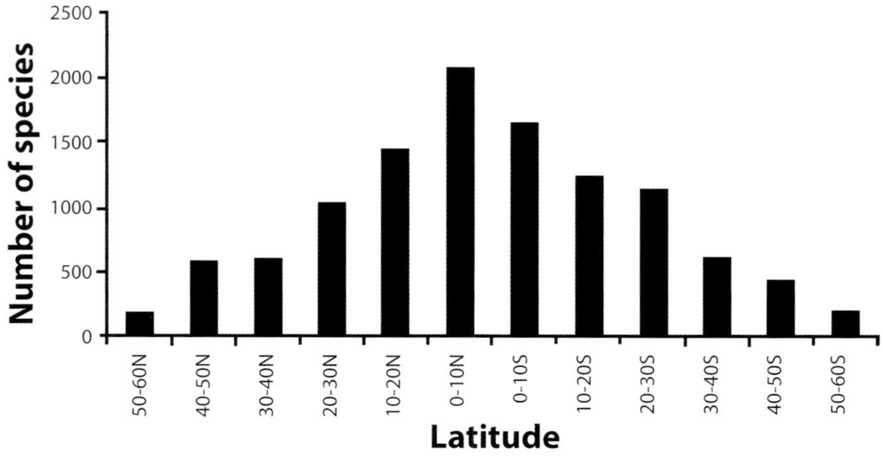

Figure 3.2. Numbers of mosquito species by intervals of 10° latitude calculated from country records in the Systematic Catalog of the Culicidae (http://www.mosquitocatalog.org). Foley et al. 2007a

Table 3.2. Number of species by subfamily and tribe in each biogeographical region (Rueda et al. 2008)

	Total number of species globally[1]						
Subfamily	**PA**	**NA**	**NT**	**AT**	**OL**	**AU**	**World**
Anophelinae	94	21	125	148	168	95	489
Culicinae	398	157	944	647	893	669	3003
Tribe (Culicinae)							
Aedeomyiini	0	0	1	3	1	2	6
Aedini	217	96	258	300	407	302	1262
Culicini	96	35	360	169	185	155	813
Culisetini	16	8	8	7	4	15	40
Ficalbiini	8	0	0	34	15	13	53
Hodgesiini	0	0	0	4	6	5	11
Mansoniini	7	2	27	26	18	26	81
Orthopodomyiini	3	3	7	7	7	1	25
Sabethini	19	4	223	6	135	89	409
Toxorhynchitini	10	5	24	22	35	15	88
Uranotaeniini	22	4	36	69	80	46	215
Total (all Culicidae)	492	178	1069	795	1061	764	3492

[1]PA = Palearctic; NA = Nearctic; NT = Neotropical; AT = Afrotropical; OL = Oriental; AU = Australasian

Fitting mosquito distribution to the classic biogeographical regions is informative, but a phylogenetic vicariance analysis of mosquitoes, similar to that done for terrestrial vertebrates by Holt et al. (2013), might provide more reasonable geographic regions for Culicidae. Various attempts to categorize mosquito distribution have been made on a limited basis (e.g., Service [1993] on *Anopheles* vectors; Rubio-Palis and Zimmerman [1997] on *Anopheles* vectors in the Neotropics), but data at a much finer scale than currently available would be needed to perform an analysis similar to Holt et al. (2013).

Mapping and Spatial Models

Principles and Methods for Spatial Modeling

Dozens of publications have included maps intended to inform the reader about the distribution of mosquito species treated in revisions or identification manuals (e.g., Belkin 1962; Darsie & Ward 2005). Those maps are usually based on a combination of collection records (from previous publications or from specimens in a collection) and expert estimation of the borders of the range. Such maps are useful to gain a quick appreciation of where the species could be collected and to eliminate some species from consideration for identification. Point source and expert estimation of range neither capture a dynamically changing distribution nor give a good estimation of the potential range of a species. Early attempts to gather fine-scale data for comparable statistics were directed at particular vector control projects, such as the World Health Organization's (1972) standardization of indices of *Aedes aegypti* abundance. Recent work on mosquito distribution has relied heavily on the availability of computer-driven geographic information systems and new applications of statistics to spatial analysis. Although they vary in the specifics of their uses, these methods are generally known as species distribution models or environmental niche models. They depend on accurate records of location, including good documentation of the technique for recording the location (Foley et al. 2009).

Species distribution models attempt to use points where the species is known to occur (true presence), points where it is known to be absent (true absence), and geographic barriers as input. These elements are correlated to various environmental layers in order to estimate the distribution of a species. Ecological niche models, in contrast, attempt to estimate where a species could occur based on ecological and physiological requirements. Therefore, species distribution models estimate the correlated conditions rather than the species itself, whereas ecological niche models estimate the distribution of a species based on its requirements expressed as a multidimensional hypervolume (Pearson 2010; Peterson et al. 2011).

One conceptual model of distribution links three biological factors with three geographic ones, producing a logical system for categorizing actual and potential species range (Peterson et al. 2011). Density-independent population growth is linked to physical (scenopoetic) requirements. Biotic interactions like competition or predation are linked to those areas where these interactions are permissive for survival. Finally, dispersal capability (whether autonomous or phoretic) is linked to the places where the species actually occurs or where it is introduced.

In practice, the environmental niche model or species distribution model estimates the potential distribution from correlated data chosen for their relevance to biological requirements (Pearson 2010). Such data layers are abundantly available, creating the danger of overfitting models with so much data that they merely present a separate way to express a known distribution with little predictive power. Another source of inaccuracy is the dynamic nature of distributions over time as the most reliable basic foci supply individuals that colonize expansion foci temporarily. Useful though such models are, judgment by knowledgeable experts is an essential part of biogeographical studies. As Peterson et al. (2011) put it, "Quite simply, no substitute exists for a good understanding of the natural history and distributional ecology of the species in question, the theoretical strengths and weaknesses of different methods, and the realities of the data available for the problem at hand, which may or may not match the assumptions of the application and/or the algorithm."

Statistical correlation techniques attempt in many ways to provide accurate models, handling combinations of discrete and continuous variables (Table 3.3). One of the techniques shown to be reliable in comparisons of training and test data sets is maximum entropy, commonly applied by the software Maxent (Larson et al. 2010). Maxent does not require absence data, which is a considerable convenience in distributional studies. It can use both categorical and continuous environmental data and produces a continuous spatial estimation of the suitability of locations. The mathematical basis of the technique is to maximize entropy or to provide the most uniform estimated distribution given the constraints imposed by the data. The "genetic algorithm for rule-set prediction," or GARP, has also been successfully used, as has boosted regression tree models (Wiebe et al. 2017). It depends on the same sort of techniques used in phylogenetic analysis, grouping data by relatedness and distance to classify locations for suitability based on multiple data layers.

The accuracy of spatial models is commonly based on the relationship of the proportion of occurrences correctly predicted (sensitivity) and the proportion of absences correctly predicted (specificity; Pearson 2010). The response operating characteristic (ROC) curve is the graph consisting of sensitivity on the vertical axis and one-minus-specificity on the horizontal axis. The worst case, where the model predicts presence and absence randomly, results in a slope of 1:1. Improved accuracy increases the curvature so that high sensitivity begins when one-minus specificity is low, and the area under the curve is maximized. The area under the curve (AUC) is often used as an overall measure of the accuracy of the model (e.g., Wiebe et al. 2017). Measurements of sensitivity and specificity require that data be divided into training and test sets.

Table 3.3. Summary of statistical techniques that correlate environmental data layers to species distribution (Pearson 2007; Khatchikian et al. 2011; Eisen and Eisen 2011)

Method	Model or software name	Species data type
Gower metric (multivariate distance metric algorithm)	DOMAIN	Presence only
Ecological niche factor analysis	BIOMAPPER	Presence and background
Artificial neural network	SPECIES	Presence and absence
Regression[1]	Statistical software	Presence and absence
Stochastic	Skeeter Buster	Presence and absence
Dynamic simulation models	CLIMEX	Presence and background
	Maxent	Presence and background
	GARP	Presence and background
Space-time risk	SaT Scan	Presence and temporal
Envelope algorithm	BIOCLIM	Presence and background

[1]Generalized linear model, generalized additive model, boosted regression trees, multivariate adaptive regression splines, logistic regression

Construction, analysis, and use of species distribution or ecological niche models have motivated the establishment of computer-based resources designed for individuals to use for studies. These tools are a big contrast to the early days of computer databases in which individual collection records were painfully checked for taxonomic accuracy, entered into nonrelational databases, and then held on individual machines for the use of only a few researchers. The Mosquito Information Project was an example, mainly directed at entering the data from John Belkin's extensive Latin American collections (e.g., Belkin & Heinemann 1975a) funded by the US Department of Defense in the 1960s and 1970s. Two individuals spent years entering and checking over 20,000 records on magnetic tape in a noncommercial database program. Although the body of information had been used to produce important taxonomic revisions (e.g., Belkin et al. 1970) and the computerized collection records were examined from time to time, a serious attempt to analyze the data was not made until 2001, when the original tapes were sent to a specialist in New York City for translation to modern systems. Although the company was located adjacent to the World Trade Center, the tapes survived the attacks on September 11, 2001, and eventually returned to the Smithsonian Institution. Finally, Foley et al. (2008) made a thorough analysis of the data with important biogeographical conclusions. Current systems take advantage of standardized headings (Hallisey et al. 2012) and publicly accessible databases that provide raw data in a standard format (Foley et al. 2010b) on species distribution (e.g., Disease Vector Database) or even customizable ecological niche models connected to extensive distribution databases (e.g., VectorMap; Moffett et al. 2009). These tools will probably become increasingly interactive and change as analytical techniques change, but their availability has greatly improved the presentation and analytical power of distributional studies.

Spatial modeling and the resulting conclusions about distribution have many uses. For example, strategic vector control to prevent dengue transmission can depend heavily on mapping, modeling concentrations of vectors over time, and making correlations with a population's immune status (Eisen & Lozano-Fuentes 2009). Species, invasion, climate change, and other examples will be discussed in more detail later in this chapter. Most of the modeling effort has concentrated on species associated with the most important mosquito-borne pathogens, ironically creating a clustered distribution of the studies themselves.

Examples of Spatial Modeling of Mosquito Distributions

Aedes aegypti is the principle vector of dengue, and *Aedes albopictus* is important regionally. Hopp and Foley (2001) used a mechanistic model to estimate global distribution of *Aedes aegypti*. They applied CIMSim, a detailed mechanistic model based on larval development (Focks et al. 1993a, 1993b), on a grid basis to the entire planet using climate data layers including water vapor deficit. They were able to conclude that *Aedes aegypti* could expand its distribution beyond its current extent into the Mediterranean Basin, in eastern Asia, and in eastern North America. Peterson et al. (2005) used a completely different approach, developing a correlative ecological niche model for Mexico based on climate layers and a remotely sensed vegetative index. Correlations were created using GARP, which produces a different model with each run. The procedure produced a predicted distribution for each month of the year, judging the most likely presence of *Aedes aegypti* when 90% or more of the models made the prediction for an individual pixel. Monthly predictions corresponded closely to the occurrence of dengue, and a representation for the year provided an informative summary of those areas of Mexico suitable for the species. That distribution of the model showed significant high elevation and dry areas of the country where *Aedes aegypti* could not survive. It also showed areas near the border with the United States where the species could thrive. Some of those areas are currently infested (southern Texas), and others were not at the time of the study (southern California). Medley (2010) also made large-scale niche models but applied them to *Aedes albopictus* and used maximum entropy rather than GARP. The models involved 18 environmental data layers, including land use, insolation, wind temperature, and humidity. Two hundred points with data on presence of *Aedes albopictus* were selected randomly from 1,052 points in North America, 5,238 in South America, and 236 in Southeast Asia. Only 73 data points were available for Europe. Models for each of these regions were excellent fits, with 98.5%–99.8% correct prediction as calculated by the area under the ROC curve. Medlock et al. (2006) used a model to show that much of southern Britain was susceptible to infestation by *Aedes albopictus*.

Some *Culex* species have attracted attention from modelers because of associated viral pathogens. Masuoka et al. (2010) used the maximum entropy method to predict Japanese encephalitis in the Republic of Korea based on the distribution of *Culex tritaeniorhynchus*. They found that the highest likelihood of finding the species was correlated to lower elevation, a moderate vegetation index related to agricultural use, and high average minimum temperatures in July. Their analysis indicated three parts of the country that were most suitable for *Culex tritaeniorhynchus*, the largest in the Northwest, surrounding Seoul. Larson et al. (2010) found that the maximum entropy method worked better than GARP for modeling the distributions of West Nile virus vectors in Iowa, United States, including *Culex pipiens*, *Culex tarsalis*, and *Aedes vexans*. The results clearly showed that *Culex pipiens* was more likely to be found in urban areas and that *Culex tarsalis* was more frequently predicted to be present along major river systems. Jacob et al. (2009) took a completely different approach to model the distribution of *Culex pipiens* in a single population center in Illinois. They successfully correlated distribution of the species to land use/land cover data produced by Landsat satellite remote sensing. They found that the key categorizations of land use were urban, nonurban, and water (e.g., ponds and lakes). These three studies show how different approaches for modeling distribution are more useful for certain purposes. At a small scale (Jacob et al. 2009), climate and rainfall variables are much less useful than land use data. In Korea (Masuoka et al. 2010), elevation was a useful predictive factor for *Culex tritaeniorhynchus* even though elevation is a surrogate for several different topographic and climatic elements of the environment. The study in Iowa (Larson et al. 2010) used ecological niche modeling, but the conclusions were largely qualitative (i.e., river systems or urban centers).

Anopheles vectors of malarial pathogens have been the subject of large- and small-scale geographic modeling efforts, motivated by interest in controlling disease. Closely related species (e.g., the Maculipennis Complex; Bates 1940) in the genus add an additional dimension of interest. The detailed differences in distribution of species within a complex are an indication of the selective pressure that led to speciation. For example, Simard et al. (2009) showed that they could distinguish locations most suitable for the closely related *Anopheles gambiae* and *Anopheles coluzzii* (formerly S and M molecular varieties; Coetzee et al. 2013) in Cameroon (Fig. 3.3). *Anopheles* has also attracted more attention than other groups because of the relatively large number of important vector species compared to other mosquito-disease systems. As a result, applied entomologists have had to consider broader questions of biogeography to design regional strategies against malaria.

Africa has a diverse fauna of anophelines and a severe problem with malaria. Gillies and de Meillon (1968) reviewed all *Anopheles* of sub-Saharan Africa, generating a map for each species. Those maps indicated the points where the species was collected and featured a dotted line showing the authors' expert opinion of the range limits, sometimes based on historical literature. Although subsequent studies reviewed below performed much more quantitative modeling of distribution, Gillies and de Meillon's treatment of all species in the genus gave insight into the geographic variety of 93 species. Of these species, 14 were surprisingly widespread across the entire sub-Saharan region except the arid zone at the southwest corner of the continent. Those widespread species were *Anopheles coustani, ziemanni, nili, rhodesiensis, funestus, leesoni, rivulorum, marshallii, gambiae, maculipalpis, pretoriensis, rufipes, pharoensis,* and *squamosus*. The authors found that more restricted distributions were similar to faunal zones established previously for birds based on climate and plant associations. The humid zone of western Africa has a characteristic fauna of *Anopheles*, including *obscurus, concolor, implexus, smithii, cinctus, vinckei, barberellus, moucheti, hancocki, hargreavesi,* and *freetownensis*. Within this region, the Cameroon Highlands share many species with the East African Highlands, a pattern also observed for birds. The distributions in eastern Africa are more complicated, with groups of species that share members across geography in a north–south axis. For example, the very northernmost part of the region that includes southern Sudan has only seven species (*Anopheles rhodesiensis, dthali, sergentii, gambiae, cinereus, rufipes,* and *salbaii*), some of which occur further south. The fauna becomes relatively more diverse toward the south. Among the 93 species reviewed by Gillies and de Meillon (1968), 37 have extremely restricted distributions (*Anopheles caliginosus, symesi, concolor, caroni, faini, lovettae, wilsoni, rickenbachi, vanhoofi, vinckei, maliensis, buxtoni, machardyi, vernus, dureni, azaniae, rodhaini, ruarinus, cameroni, fontinalis, culicifacies adenensis, tchekedii, walravensi, aruni, parensis, austenii, njombiensis, mousinhoi, mortiauxi, berghei, erepens, distinctus, keniensis, carteri, dancalicus, salbaii,* and *swahilicus*). The species with restricted distribution attract much less attention because they are only locally collected and of limited medical importance. Nonetheless, 37 of 93

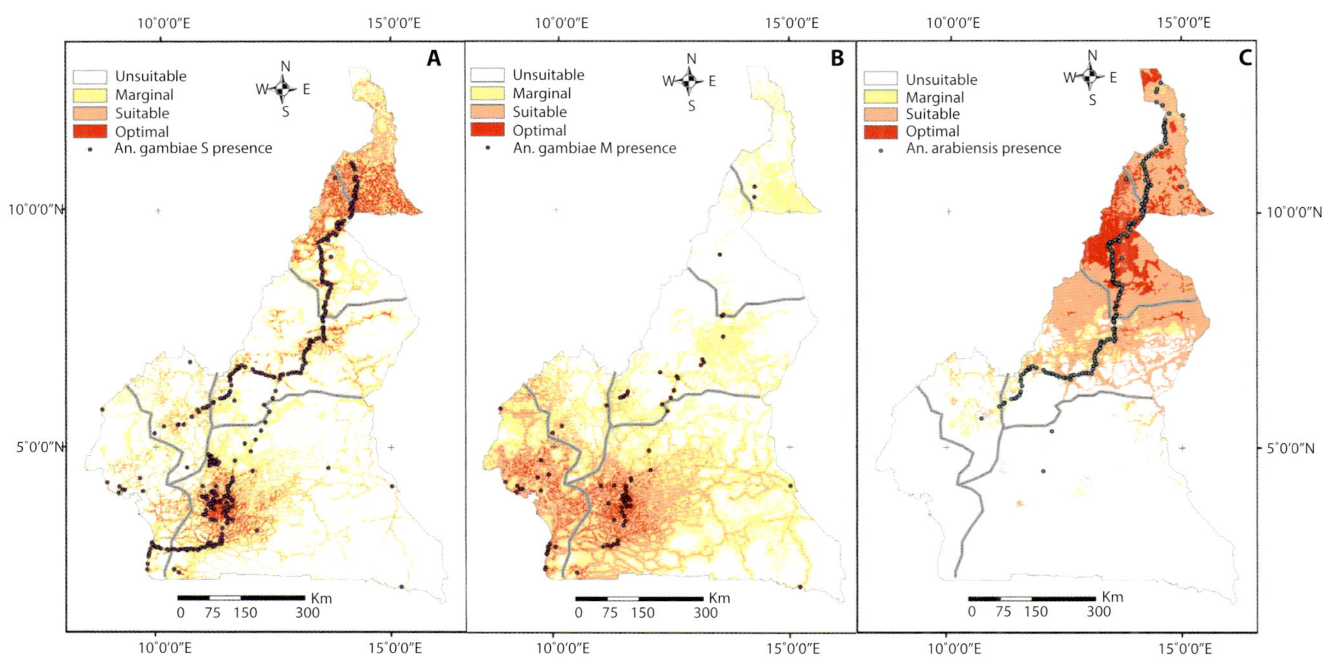

Figure 3.3. Habitat suitability in Cameroon for *Anopheles gambiae* (A), *Anopheles coluzzii* (B), and *Anopheles arabiensis* (C). Red is optimal, orange is suitable, yellow is marginal, and white is unsuitable. Simard et al. 2009

species represent a large proportion of the genetic diversity within the genus in sub-Saharan Africa, raising the question of whether any of them could become more widespread under other circumstances or with artificial dispersal. In contrast, the 14 widespread species show adaptability to a wide variety of habitats and, presumably, a significant history of dispersal.

The Malaria Atlas Project (Hay et al. 2010) is an ambitious effort to use reliable, published distribution records from 1984 onward and new documented distribution points submitted to the project. Those records were fitted by a boosted regression tree to data layers on climatic factors (decomposed to eliminate seasonal effects), topography, vegetation index, land use, and expert knowledge. The model was run for each species until the machine-learning rate started to decrease, to avoid overfitting. The project generated ecological niche model–based maps for each species, with 41,518 collection records from 22,249 locations superimposed as dots on a graded scale of the likelihood of occurrence. The project also published summaries of larval and adult characteristics for each species. The project identified 52 *Anopheles* species that were considered "dominant vector species." Some of the species were complexes, though the project had plans to treat Dirus, Albitarsis, and Culicifacies Complexes by their individual sibling species sometime in the future. The Malaria Atlas Project divided its efforts regionally: Africa, Europe, and the Middle East (Sinka et al. 2010a); the Americas (Sinka et al. 2010b); and the Asia-Pacific region (Sinka et al. 2011; Table 3.4). The maps reflected some of the general geographic patterns identified in Africa by Gillies and de Meillon (1968), but without consideration of all the *Anopheles* species, the main information imparted was about individual species. The Malaria Atlas Project maps made quantitative estimates of each vector species' occurrence, but they missed some historical distributional extensions (e.g., *Anopheles gambiae* in Egypt, *Anopheles arabiensis* in Brazil). Also, the treatment of species groups or newly subdivided species as single data sets probably distorted the distributional model. Despite these limitations, the high-resolution maps gave an excellent overview of the geographic ranges of the most important *Anopheles*. They also gave what is no doubt an accurate sense that the boundaries of distributions are "fuzzy" and that certain areas within a general range are probably unsuitable.

Table 3.4. Dominant malaria vector species mapped and summarized by the Malaria Atlas Project (Hay et al. 2010)

Region	Species of *Anopheles*
Africa	*arabiensis*
	funestus[1]
	gambiae[2]
	melas
	merus
	moucheti
	nili[1]
Europe and Middle East	*atroparvus*
	labranchiae
	messeae
	sacharovi
	sergentii
	superpictus[1]
Asia-Pacific Region	*aconitus*
	annularis
	balabacensis
	barbirostris[1]
	culicifacies[1]
	dirus[1]
	farauti[1]
	flavirostris
	fluviatilis
	koliensis
	lesteri
	leucosphyrus + *latens*
	Maculatus Group
	minimus[1]
	punctulatus[1]
	sinensis[1]
	stephensi
	subpictus[1]
	sundaicus[1]
Americas	*albimanus*
	albitarsis[1]
	aquasalis
	darlingi
	freeborni
	marajoara
	nuneztovari[1]
	pseudopunctipennis[1]
	quadrimaculatus[1]

[1]Species complex
[2]Prior to separation of *gambiae* (S molecular form) and *coluzzii* (M molecular form)

Moffett et al. (2007) performed a large-scale analysis of *Anopheles* distribution in Africa similar to that of the Malaria Atlas Project, though based on maximum entropy associations to climatic factors, elevation, and land use. Their purpose was to calculate malaria risk, and they added *Anopheles coustani* and *paludis* to the list of modeled species. Concentrating on the Gambiae Complex, Levine et al. (2004b) used GARP analysis of 12 environmental data layers to show that *Anopheles arabiensis* may have a wider distribution in central and northeastern Africa than currently documented. Their model also showed a discontinuity in suitability along the Rift Valley, reflecting the block to gene flow associated with the recent separation of *Anopheles coluzzii* (former M molecular biotype) from *Anopheles gambiae* and the separation of *Anopheles amharicus* from *Anopheles quadriannulatus* (Coetzee et al. 2013). Wiebe et al. (2017) produced what they claimed was the most accurate estimate of the distributions of each member of the Gambiae Complex based on boosted regression tree models with almost 6,000 points correlated to environmental data layers.

Modeling studies of *Anopheles* distribution in Africa have also been applied to smaller geographic areas. As might be expected, more detailed examination of smaller areas indicates the importance of distinct data layers. For example, Shililu et al. (2003) saw areas of Eritrea with more total anophelines, though they did not find a significant correlation to vegetation index or rainfall. Two different studies provided particularly detailed information about *Anopheles* distribution in northern Tanzania. Kulkarni et al. (2010) successfully modeled malaria risk on vector distribution using climate, human population, elevation, and land cover data layers correlated to *Anopheles gambiae*, *arabiensis*, and *funestus* distribution by the maximum entropy method (Fig. 3.4). They categorized land cover from 50 statistically defined vegetation types using a maximum likelihood classification, resulting in eight categories (e.g., broad-leafed evergreen, rice or irrigated, water). A separate study examined published collection records from 157 locations in northern Tanzania (Kabula et al. 2011), showing that *Anopheles gambiae* and *funestus* groups made up over 99% of *Anopheles*, despite the presence of other species. Such an uneven abundance of species in contrast to more equal distribution of presence may be typical of many mosquito groups. In West Africa, Simard et al. (2009) concentrated on *Anopheles gambiae* (as the S molecular form), *Anopheles coluzzii* (as the M molecular form), and *Anopheles arabiensis*. They clearly showed how *Anopheles arabiensis* was most likely to be present in the northern and eastern parts of Cameroon, *Anopheles gambiae* in the west and east, and *Anopheles coluzzii* in the southwest. *Anopheles moucheti* and *nili* were also more likely to be present in the southwestern portion of the country (Ayala et al. 2009).

Southeast Asia is also an important region for malaria and home to an extremely complex anopheline fauna, as indicated by the number of species that were mapped by the Malaria Atlas Project as complexes (Sinka et al. 2011). Manguin et al. (2008) reviewed the distribution of all *Anopheles* in the region, concluding that a species focus would not be practical due to the diversity of the fauna and the widespread influence of human activity. The reasons behind this diversity, at least in part, may be explained by recent geological influences (Morgan et al. 2011) discussed in this chapter. Further north, in the Republic of Korea, the anopheline fauna consists of only eight species. Foley et al. (2010a) used a maximum entropy niche model based on 104 sites and enhanced for sensitivity (i.e., statistically chose predicted presence over predicted absence). They were able to get high correlations for five of the species (*Anopheles sinensis*, *belenrae*, *pullus*, *sineroides*, and *kleini*) but not for the other three (*Anopheles lesteri*, *koreicus*, and *lindesayi japonicus*). The lack of correlation for *Anopheles lesteri* and *koreicus* may have been due to the scarcity of their collections. On the other hand, *Anopheles lindesayi japonicus* was collected frequently, suggesting that the data layers in the niche model did not include key environmental factors that influence this species' distribution.

Sweeney et al. (2006, 2007) modeled the distribution of the Farauti Complex (*Anopheles*) in northern Australia using GARP analysis and working with a set of 41 environmental data layers. Some of the data sets they found to be significant included mean weekly diurnal temperature range, lowest weekly minimum temperature, and either dew point or relative humidity at 15:00h in July. Such correlates are far from intuitive, but they successfully predicted the range of *Anopheles farauti* s.s. and *Anopheles farauti* 3, though somewhat less well. The model was not successful at predicting the presence of *Anopheles farauti* 2, possibly because it was the least abundant species. The study showed the importance of atmospheric moisture in this situation, which seemed to limit the inland occurrence of *Anopheles farauti* s.s.

An analysis of the distribution of the *Anopheles quadrimaculatus* group in the United States not only gave more precise estimations of the ranges of those recently described species but also provided a good example of how a study could

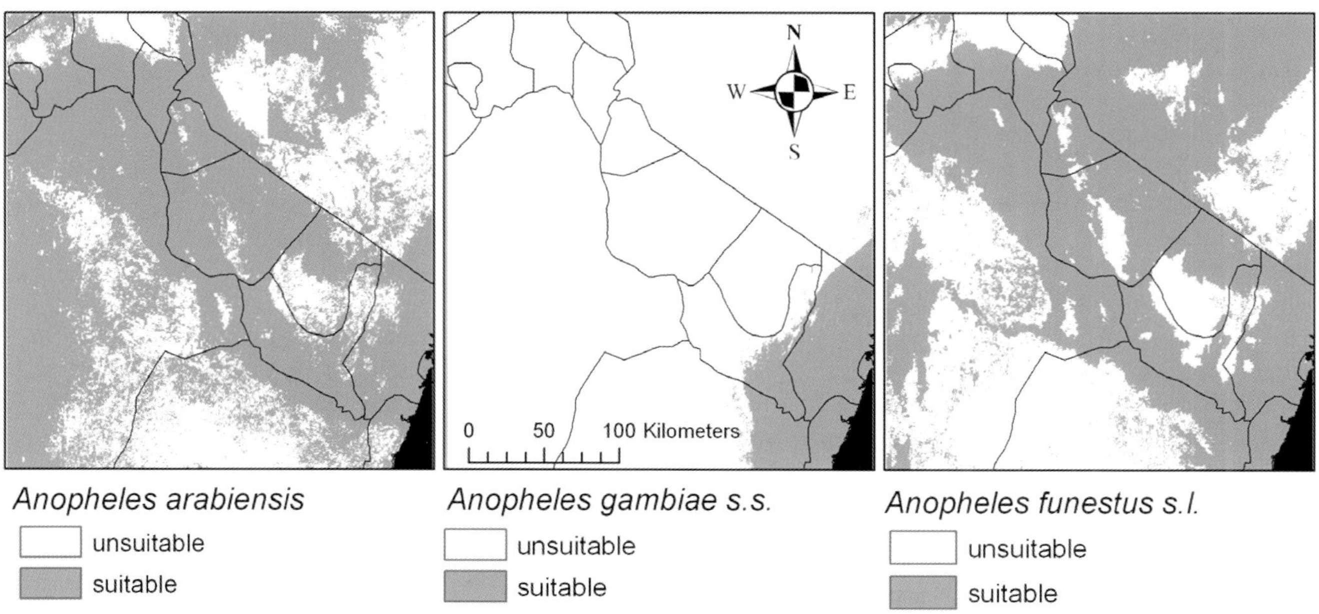

Figure 3.4. Niche models of the distribution of principal malaria vectors in northeastern Tanzania. Kulkarni et al. 2010

balance uneven collection intensity (Levine et al. 2004b). Large numbers of samples from the southeastern United States skewed results so that a procedure was developed to select no more than 10 locations chosen randomly in each state. Using GARP analysis of the relatively few collection points resulted in good correlation for all five species. These distributions provided a hypothetical picture of the ranges of these closely related species that would have been impossible by traditional methods because of limited identification during routine surveillance. The model suggests that *Anopheles quadrimaculatus*, *smaragdinus*, and *maverlius* are widespread throughout the Southeast and Middle Atlantic regions of the United States, whereas *diluvialis* and *inundatus* are more coastal.

Influences on Distribution

The discussion of modeling above stressed distribution—basically, where on a map of the earth that a species occurs. Significant data layers (e.g., average maximum temperature) might indicate their importance in the survival of populations. A closer examination of both correlative and experimental studies can move us toward an understanding of environmental and biological influences on distribution.

Landscape

"Landscape" is the term generally used for the description of visible features of a small or large area. As descriptions, landscapes are useful correlates to mosquito distribution. Landscapes may also be viewed as more than a description when considering the landscape as a real, natural entity with qualities inherent to its classification. For a landscape as a natural entity, the mosquito fauna is one of its inherent qualities. The connection between a landscape and the species of mosquitoes within it is usually a clear product of those species' biological requirements and the overall geographic ranges of the species.

The life of a mosquito is closely tied to the aquatic larval habitat. Oviposition almost always occurs close to where larvae develop, though flowing water may carry floating *Culex* eggs, or riverine floodwater *Aedes* and *Psorophora*, far from the location where they were deposited. Although adult mosquitoes are highly mobile, they generally fly no more than a few kilometers and usually much less so that the larger-scale landscape for larva and adult are the same. The salt marsh *Aedes* species are an exception, often flying significant distances inland from larval sites for blood meals and then returning to the salt marsh to oviposit. Some individual adult mosquitoes may move very far from their larval sources when blown by air currents at higher elevations (Huestis et al. 2019). In the Sahelian region of Mali, balloon-suspended sticky traps collected thousands of mosquitoes at 40 to 290 m above the ground. Using calculations from wind speed, these mosquitoes would have moved an average of 300 km during nine hours of the night.

Typical distributional studies at a landscape scale either examine where adults were captured or where larvae were collected. Besides the biological interest of these associations, each kind of study has practical implications. An understanding of adult distribution can be an indication of disease risk (e.g., Chevalier et al. 2004) and can also target control measures directed at adults. Larval distribution patterns can aid in the inspection and treatment of aquatic stages, technicians often developing detailed knowledge of where larvae are likely to occur in a local area. Modern mosquito collection for taxonomic purposes usually starts with capturing larvae. As a result, there is almost always a record of larval habitat for even the most obscure species.

Landscape can be described at various scales. For example, Rubio-Palis and Zimmerman (1997) divided the entire Neotropics into just five regions and nine subregions based on distribution of *Anopheles* vectors and geographic factors (Table 3.5). Many species occurred in a few ecological regions, but each region was represented as having a characteristic fauna. This sort of categorization lacked the quantitative correlation of a modeling effort; however, it was a useful estimate of what species would be encountered in an easily recognizable landscape. A similar categorization of ecological regions, but at a much smaller scale, was carried out by Rosa-Freitas et al. (2007) in the Brazilian state of Roraima. Also concentrating on *Anopheles* vectors, they divided forest and savannah from the previous study into eight landscapes: montane forest, submontane forest, plateau forest, lowland forest, alluvial forest, steppe, savannah, and wetland. These ecological regions were based on principle component analysis of geomorphology, vegetation, climate, rivers, land use, and vector distribution. Species under consideration were *Anopheles* (*Nyssorhynchus*) species, including *darlingi*, *janconnae* (as *albitarsis*), *triannulatus*, *nuneztovari*, *oswaldoi*, and *peryassui*. Correlations of species to landscape type were made by ecological niche modeling based on 19 collection locations and GARP analysis. Ayala et al. (2009) and Simard et al. (2009) took a similar approach in Cameroon. Another approach to regional studies has been to choose what are considered characteristic landscapes and then sample mosquitoes that occur in them. A study in Sweden, for example, quantified the increased diversity of the mosquito fauna from north to south and pointed out the parallel change in species of plants (Schäfer & Lundström 2001).

Large regional, correlative studies of landscape are useful, but they are designed to document current understanding rather than to find new associations. Other studies of natural landscapes have taken a more experimental approach by examining an abundance of species quantitatively. For example, trapping adult mosquitoes from three locations in Iowa, United States, during 111 trap nights collected 32,688 mosquitoes, half of which were *Aedes vexans* and the other half, *Aedes trivittatus*. Correlation of numbers of mosquitoes to landscape descriptions within 1,000 m showed differences between the species. *Aedes vexans* was associated with alluvial habitats, deciduous forest, and grasslands, whereas *Aedes trivittatus* was not associated with agriculture or grasslands. Those results reflected the traditional views that *Aedes vexans* larvae inhabit forested floodplains as well as open areas and that the adults fly significant distances (Horsfall et al.

Table 3.5. Ecological regions of the Neotropics and associated *Anopheles* vectors of malaria (Rubio-Palis and Zimmerman 1997)

Ecological region	Subregion	Species of *Anopheles*
Coastal		*aquasalis, albimanus, darlingi, calderoni, pseudopunctipennis,* (*Kerteszia*) spp., *punctimacula, albitarsis* s.l.
Piedmont	Western Andean	*albimanus, darlingi, nuneztovari, punctimacula, calderoni, pseudopunctipennis*
	Eastern Andean	*nuneztovari, pseudopunctipennis, darlingi, oswaldoi, trinkae, rangeli,* (*Kerteszia*) spp., *albitarsis* s.l.
	Mexico-Central American	*pseudopunctipennis, albimanus, darlingi, vestitipennis, punctimacula*
Savannah	Llanos	*darlingi, nuneztovari, albitarsis* s.l.
	Grasslands	*darlingi, nuneztovari, braziliensis, albitarsis* s.l.
	Cerrado	*darlingi, albitarsis* s.l.
	Caatinga	*darlingi, albitarsis* s.l., *aquasalis*
Interior lowland forest	Deciduous	*darlingi, albitarsis, nuneztovari, deaneorum, triannulatus*
	Evergreen	*braziliensis, oswaldoi, strode*
High valley		*pseudopunctipennis*

1973). In contrast, *Aedes trivittatus* larvae are generally restricted to sites with light to moderate tree cover, and the adults do not travel far (Carpenter & LaCasse 1955). Working with a completely distinct set of species, bird-baited traps were operated near Syracuse, New York, for a total of 641 trap nights. The majority of the collections were *Culiseta melanura*, *Culiseta morsitans*, *Coquillettidia perturbans*, and *Culex pipiens* + *restuans* (not distinguished) in open fields, in contrast to *Aedes canadensis*. That species was most abundant in wet forest (i.e., swamps). The sampling method can strongly influence perceptions of where adult mosquitoes are most concentrated. Beier et al. (1982) sampled adult mosquitoes in Indiana, United States, using power aspirators, a method that should be unbiased compared to traps with attractants. They found that *Aedes vexans* (60% of the captures), *Culex* spp. (24% of the captures), and *Aedes triseriatus* (8% of the captures) were more abundant in a part of the woodlot with dense undergrowth. Sampling with ovitraps for *Aedes triseriatus* showed no such pattern.

Consideration of more species in a greater variety of habitats shows that a mosquito species' distribution is not usually characterized by a single landscape type. One exception was reported from West Central Florida, United States, where extensive light trapping captured mostly *Aedes taeniorhynchus* and *Anopheles atropos* in salt marsh or mangrove habitats, clearly reflecting the tolerance of those species for salt water (Yan & Zhong 2005). Other common species (*Culex quinquefasciatus*, *Psorophora columbiae*, and *Anopheles crucians*) were common in a wider variety of habitats (swamps, pine flatlands, and freshwater marshes). Less common species were also found in diverse kinds of landscapes. Using a diversity index and an aggregation index, Yan and Zhong (2005) showed that salt marshes had the least diversity of mosquito species and that pine flatwoods had the greatest diversity.

Elevation is another aspect of landscape that is important to mosquito distribution. As pointed out earlier, elevation is correlated to climatic factors that affect mosquito development directly. In a given area, some species are adaptable to a wide range of elevations, whereas others are restricted to lower or higher altitudes (Table 3.6). There are good examples of northern species occupying higher elevations toward the southern ranges of their distribution. Examples include *Aedes impiger* in Colorado, United States (Baker 1961), *Culex restuans* in Central America (Strickman & Darsie 1988), and *Aedes epactius* in Mexico (Lozano-Fuentes et al. 2012b). There are also examples of many high-latitude species that do not extend their ranges into the tropics at higher elevations (e.g., *Aedes melanimon*, *Culex territans*, *Anopheles punctipennis*; Darsie & Ward 2005) and montane tropical species that do not occur at lower elevations in higher latitudes (*Aedes quadrivittatus* [Berlin 1969a], *Culex pseudostigmatosoma* [Strickman 1989b], and *Anopheles eiseni* [Wilkerson & Strickman 1990]). These examples are from the Western Hemisphere, partly because of more extensive distributional studies and probably also because the latitudinal gradient of these continents corresponds to the elevational gradients (i.e., mountain chains run north–south). Other continents have mountains that do not span as great a range of latitudes (e.g., Australia, Africa) or that are oriented roughly east–west (e.g., the Hindu Kush and the Himalayas) or form geographically restricted "islands" of mountains (e.g., the Alps).

More locally, the combination of physical conditions associated with elevation can have a profound effect on the distribution of important mosquito species. *Aedes aegypti* became progressively less common at elevations above 1,700 m in Mexico and was completely absent above 2,130 m despite the presence of suitable larval sites (Lozano-Fuentes et al. 2012a). Shililu et al. (2003) found that elevation was the dominant correlate of *Anopheles* distribution in Eritrea, where *Anopheles arabiensis* were most abundant at lower elevations (500–900 m) and *Anopheles cinereus* at higher elevations (>2,000 m). *Anopheles farauti* was restricted to the lowest elevations along the northern

Table 3.6. Distribution of mosquito species in elevation zones in the Rocky Mountains near Boulder, Colorado, United States (Baker 1961). X indicates presence of species following extensive larval sampling.

Elevation zone	Plains	Foothills	Montane	Alpine
Altitude range (m)	1,500–1,800	1,800–2,400	2,400–3,000	3,000–3,700
Species				
Aedes campestris	X			
Aedes dorsalis	X			
Aedes melanimon	X			
Aedes trivittatus	X			
Aedes vexans	X			
Aedes schizopinax		X		
Aedes cataphylla			X	X
Aedes communis			X	X
Aedes fitchii			X	X
Aedes hexodontus			X	X
Aedes implicatus			X	
Aedes pionips			X	X
Aedes pullatus			X	X
Aedes punctor			X	X
Aedes impiger				X
Culex pipiens	X			
Culex tarsalis	X	X		
Culex territans	X			
Culiseta inornata	X	X	X	
Culiseta impatiens		X	X	X
Culiseta incidens			X	X

Australian coast despite climatic conditions that were apparently suitable at slightly higher altitudes inland (Sweeney et al. 2006). Such a lack of correlation between elevation-associated climatic factors and distribution appears to be exceptional. Ahumada et al. (2004) showed that temperature explained altitudinal distribution of *Culex quinquefasciatus* on the island of Hawaii. Minakawa et al. (2002) proposed that highlands, with their lower maximum temperatures, were a barrier to *Anopheles arabiensis* distribution. In contrast, dry conditions in the low elevations of the Rift Valley interrupted the distribution of *Anopheles gambiae*.

Humans have also affected the distribution of mosquitoes—probably for thousands of years. This is particularly true of those species most closely associated with people, like *Culex pipiens*, *Aedes aegypti*, and *Anopheles gambiae*. Ross (1964) described this relationship well: "Man lives in many kinds of ecological communities and in each one of these communities man and a particular assembly of mosquito species form an intimately related part of the ecosystem." As humans affect a larger and larger portion of Earth's surface, it is logical to suppose that the distribution of more and more mosquito species will be dominated by human activity. Changes to the landscape that favor certain species include forest fragmentation and agriculture (*Culex quinquefasciatus* in Hawaii; Reiter & LaPointe 2007), irrigation and rural population concentrations (*Culex tarsalis* in the northern Great Plains of the United States; Wimberly et al. 2008), rice fields and deforestation (*Anopheles marajoara* in Brazil; Conn et al. 2002; Laporta et al. 2011), and use of water-holding containers (*Aedes aegypti* in Tucson, Arizona, United States [Walker et al. 2011]; *Aedes aegypti* and *albopictus* in Thailand [e.g., Kittayapong & Strickman 1993]; *Aedes albopictus* in Bermuda [Khatchikian et al. 2011]).

Temperature and Rainfall

Temperature is, of course, affected by factors other than elevation and latitude. The difference between a maritime temperature regime in Europe and a continental temperature regime in Asia led Mogi (2012) to conclude that *Aedes albopictus* was limited by low winter temperatures in Asia and by low summer temperatures in Europe. They extended the commonly used concept of day degrees above a calculated minimum temperature required for successful larval development to a more complicated "Thermal Suitability Index" influenced by the degree of seasonal variation. That index consisted of the ratio of day degrees above a threshold ("warmth index") to day degrees below a threshold ("cold index"). Another approach to include variation in temperature as well as thresholds used autoregressive integrated moving-average models (ARIMA) to determine the relationship of temperature and moisture to the suitability of a location for *Aedes vexans*, *Culex pipiens*, and *Culex restuans* (Trawinski & MacKay 2008). Temperature is, of course, one of the principle causes of seasonality, influencing developmental cycles directly by providing suitable conditions for current populations and indirectly by entraining genetically determined seasonal cycles. The discussion of modeling provided examples of correlations between temperature and abundance. The relationship between mosquito seasonality and temperature may be heavily dependent on off-season cool conditions (*Culex salinarius* in Maryland, United States) and rainfall (*Aedes sollicitans* in Maryland; Walsh et al. 2008).

Rainfall can be another important climatic determinant of distribution. Total annual rainfall may be a general indication of the likelihood of finding certain kinds of larval habitats (e.g., Levine et al. 2004b), but at a finer scale, the relationship may be much more complicated. One basic relationship is evaporation potential in relation to rainfall, which determines how quickly aquatic sites will dry up and how much surface water eventually reaches surface and subsurface drainage systems (Minakawa et al. 2002; Trawinski & MacKay 2008). Another factor is the timing of rainfall in relation to the seasonality of mosquitoes. For example, *Culex annulirostris* in southwestern Australia was favored by a dry spring followed by a wet summer or winter (Woodruff et al. 2002). Container species like *Aedes africanus* and *Aedes furcifer* in Ivory Coast were more abundant following any significant rain that filled their typical larval sites (Schaeffer et al. 2008). In Panama, a group of rain forest species (*Haemagogus equinus*, *lucifer*, *spegazzinii*, and *falco*) became abundant after the wet season began in May. Numbers of *Sabethes chloropterus* also increased in the wet season, and large populations persisted through the dry season (Galindo et al. 1956). *Anopheles gambiae*'s distribution was affected by the amount of precipitation in the cool season (Kulkarni et al. 2010).

The timing of rainfall can interact with local topography and the short-term history of rainfall to create conditions favorable for some species but not others. Shaman et al. (2002, 2006) calculated an index of local wetness based on a time series of water table depth to show a positive correlation with outbreaks of *Aedes vexans* and *Anopheles walkeri* but a negative correlation with numbers of *Culex pipiens* in southern Virginia, United States. Salt marsh species like *Aedes sollicitans* hatched in greater numbers when rain coincided with high tides (Shone et al. 2006). Rain can also have negative effects on populations by washing eggs, larvae, or pupae out of aquatic habitats (Koenraadt & Harrington 2008).

Behavioral Effects on Distribution

The behavior of adult mosquitoes also affects where they are likely to be. Whether seeking a host, an oviposition site, or a resting place, adult mosquitoes are mobile and respond to their physical environment in ways characteristic of their species. Studies of this kind of self-powered distribution have often concentrated on diel distribution, height above the ground, or descriptions of resting sites. The behavior of mosquitoes is the real cause of the distribution observed, and descriptions of that distribution are correlates of that behavior.

Diel distribution refers to relative abundance during a 24 h period. Mosquito biologists are usually interested in time of biting, but any aspect of behavior could be considered this way. Species in a single area may seek hosts at various times. An evening crepuscular pattern is common (e.g., *Culex erraticus*, Alabama, United States; Gray et al. 2011). In at least one tropical situation in Uganda, seven canopy species each sought most hosts at a slightly different time between 15 and 35 minutes after sunset (Haddow et al. 1968). *Anopheles sinensis* s.l. (Strickman et al. 2000a) was most abundant in the first, warmer part of the evening when temperatures were cool early in the season. This pattern changed to early morning when temperatures were warmer later in the season. Although the early-morning pattern predominated when temperatures were warm, diel periodicity varied by location for unknown reasons. The moon was one of the significant influences on abundance of biting mosquitoes; numbers were reduced when there was more moonlight. *Aedes vexans* showed no such variation in the same locations, consistently seeking most hosts shortly after sunset. Almost any other diel distribution of biting mosquitoes is present somewhere and includes day biters (e.g., *Wyeomyia* spp., *Limatus flavisetosus*, *Eretmapodites chrysogaster*, *Psorophora ferox*, and many *Aedes* species), night biters (most *Anopheles*, *Mansonia*, and *Culex*), crepuscular biters (e.g., *Aedes ingrami*), or all day and night biters (e.g. *Aedes cumminsii* and *Aedes domesticus*; Service 1993).

Vertical stratification is another common quality of mosquitoes and a general one for arthropods that distribute themselves at various heights according to resource availability, sunlight, forest structure, and more (Ulyshen 2011). Figure 3.5 summarizes the results of a common type of study in which a method of adult mosquito collection is deployed at various elevations. These studies sometimes have as their object to document those mosquitoes that stay near the ground and those that stay in the canopy of forested areas. An examination of the chart shows considerable variation in preferred vertical distribution by individual species within *Aedes*, *Anopheles*, and *Culex*. *Sabethes* and *Haemagogus* include many species that were trapped at greater heights (Pinto et al. 2009). A forest situation with many potential levels for specialization can show stratification by species according to height (Galindo et al. 1955). Within the same genus in Panama, *Haemagogus argyromeris* tended to be low in the forest, *Haemagogus equinus* and *lucifer* tended to be at a middle level, and *Haemagogus spegazzinii falco* was common in the canopy. Several studies showed that *Culex pipiens* inhabits higher elevations (Drummond et al. 2006; Lee et al. 2006; Russell & Hunter 2010; L'Ambert et al. 2012), with the exception of gravid females seeking an oviposition site (Anderson et al. 2006). Such movement to ground or subterranean sites by gravid females may account for why a lower proportion of parous female *Culex pipiens* were found at greater heights (Drummond et al. 2006;

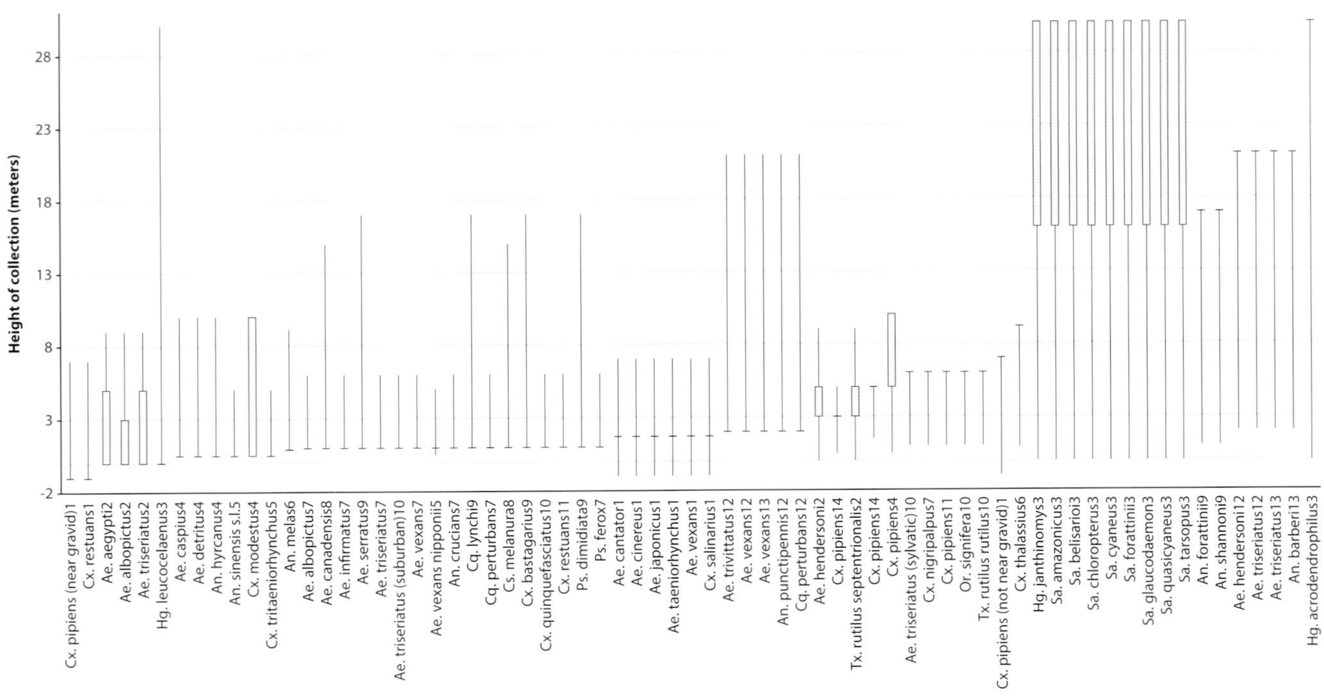

Figure 3.5. Range of heights at which representative species of mosquitoes have been collected using various methods. Vertical line is range of heights sampled; horizontal bars and columns are heights at which the species was most abundant. Numbers after species names are references: (1) Anderson et al. 2006, Connecticut, United States, carbon dioxide (CO_2)-baited traps; (2) Schreiber et al. 1988, Louisiana, United States, ovitraps; (3) Pinto et al. 2009, Pará, Brazil, netting from people; (4) L'Ambert et al. 2012, Camargue, France, CO_2 or baited traps; (5) Lee et al. 2006, Republic of Korea, CO_2 and ultraviolet traps; (6) Snow and Wilkes 1977, the Gambia, suction traps; (7) Obenauer et al. 2009a, Florida, United States, CO_2 and chemical lure; (8) Howard et al. 1983, New York, United States, bird-baited traps; (9) Julião et al. 2010, Rondônia, Brazil, landing collections; (10) Obenauer et al. 2009b, Florida, United States, ovitraps; (11) Drummond et al. 2006, New York, United States, CO_2 traps; (12) Novak et al. 1981, Indiana, United States, landing collections; (13) Novak et al. 1981, Indiana, United States, suction traps; (14) Russell and Hunter 2010, Ontario, Canada, animal-baited traps.

Anderson et al. 2006). *Anopheles melas* and *Culex thalassius* showed the opposite pattern, with the older part of the population at greater heights in a nonforested habitat of Uganda (Snow & Wilkes 1977). Individual mosquitoes may change their height preference or move vertically during the course of a day. Service (1993) commented that *Aedes africanus* and *Coquillettidia fuscopennata* performed daily vertical migration. Novak et al. (1981) remarked that *Aedes triseriatus* moves up or down between ground and canopy levels. Their observation of a distinct peak in time of biting after sunset at 21 m but not at ground level may have been due to a nightly vertical migration. Although they collected fewer *Coquillettidia perturbans* at 21 m elevation, those that were captured were found later than those at ground level, again suggesting vertical movement by some members of the population. Vertical preference can also vary seasonally, as for *Haemagogus janthinomys*, which occurs at increasing heights as the rainy season progresses in Brazil (Pinto et al. 2009). If trees are not present, mosquitoes that would occupy the forest canopy may stay close to the ground (Howard et al. 1983).

Most trapping methods capture a portion of the mosquito population engaged in a specific orientation. For example, carbon dioxide-baited traps and landing collections sample host-seeking females and sometimes males of species that mate near the host. Even suction traps capture actively flying mosquitoes, though without the influence of an attractant. However, mosquitoes spend much of their time between major activities like mating, host seeking, and oviposition. The location occupied between recognizable activities is commonly called a resting site, though the insects are probably often active even though they are not flying. For example, they might be cleaning dust from themselves or actively monitoring their environment for signals that would initiate the next activity. Although "resting" probably constitutes most of an adult mosquito's time, we know little of what happens during those periods. Numerous studies have found where some species of mosquitoes are distributed, including tree cavities, animal burrows, root masses, dense vegetation, caves, and human-made structures like barns, culverts, and basements (Panella et al. 2011). Local faunal inventories show distribution patterns characteristic of a species in that situation. Captures of *Culex quinquefasciatus* in houses in Goa, India, found 73% females and 27% males, over 60% of which were on objects hanging from the ceiling and over 20% on walls (Kaliwal et al. 2010). The majority (65%) of females had visible blood meals, suggesting that most resting mosquitoes were in the process of digesting blood prior to oviposition. A survey with electric aspirators and sweep nets in Alabama, United States, found each species with a definite preference for certain kinds of resting sites (e.g., *Aedes vexans* in vegetation, *Anopheles quadrimaculatus* and *Culex territans* in large tree cavities, Burkett-Cadena et al. 2008). Some of the species (*Anopheles quadrimaculatus*, *Culex erraticus*, *Culex peccator*, and *Culex territans*) were collected more abundantly from resting sites than with carbon dioxide-baited traps. Newly emerged *Culex pipiens* in Israel may have rested near the larval habitat, feeding on sugar (Schlein & Müller 2012).

Larval Distribution

The distribution of larvae is closely related to the distribution of adults because the female selects the oviposition site. Observations on where mosquitoes oviposit generally take special effort, whereas a record of the larval habitat is easily made simply by collecting the larvae. Taxonomic revisions generally include a discussion of larval habitat for each species because specimens for the description often originate from larval collections reared to adulthood. Broader reviews also include detailed descriptions of larval habitats (e.g., Carpenter & LaCasse 1955; Horsfall 1955). Horsfall's (1955) review was particularly thorough and in its introduction emphasized the tremendous variety of aquatic habitats used by mosquito larvae, as well as the impossibility of predicting the species of mosquitoes present based on the characteristics of the habitat.

Individual mosquito collectors have formed systems for categorizing larval habitats with several purposes in mind. Those responsible for mosquito abatement want to be able to document where larvae were found and to communicate the kind of location to separate teams charged with treatment. From a practical standpoint, such documentation can be an important guide to future efforts by decreasing emphasis on sources of mosquitoes that are less productive and increasing emphasis on sources that are more productive (Bartlett-Healy et al. 2012). Mosquito biologists have used larval habitat categorizations in individual studies, some of which are detailed below. General categorizations of larval habitats have been erected to document collections and also to summarize mosquito biology. As pointed out in 1955 by Horsfall, these attempts have not had precise predictive power, possibly because they classify habitats according to criteria observed by humans rather than by mosquitoes. The variety of broad categories is perhaps a sign of this difficulty. Laird (1988), in a volume dedicated entirely to the subject of larval habitats, considered "above-ground water" and "subterranean waters" to be the most useful division of larval habitats. He defined 11 kinds of habitats within those broad categories, further subdividing them according to observation of many species of mosquitoes. Rueda (2008) designated 10 kinds of common habitats that were like those of Laird (1988) but without the distinction between above-ground and subterranean sites. Foley et al. (2009) gave a list of habitats as a suggestion for categorization of collection records.

Service (1993) classified habitats at four levels (Table 3.7), making a list that would describe most kinds of larval sites. His broadest division between ground and container sites would probably meet the approval of most mosquito biologists. Some of the categories within the ground sites correspond to clear biological preferences (polluted waters, saltwater sites, temporary sites, and flowing waters), whereas others seem to be more for the convenience of the human observer (e.g., marshes, forest ponds, ditches). Container habitats are associated with a wide variety of mosquitoes that preferentially colonize those sorts of larval sites, but the division between natural and human-made is again a convenience for categorization and the targeting of mosquito control measures. Bamboo, bromeliads, pitcher plants (including both Sarraceniaceae in the Western Hemisphere and the unrelated family Nepenthaceae in the Eastern Hemisphere), snail shells, and crab holes are among the container habitats often associated with certain species of mosquitoes that specialize in using those sites. Service's (1993) categorization of larval habitats was a useful outline and clearly demonstrated the variety of aquatic habitats that mosquitoes employ, but a new categorization based entirely on culicid requirements might provide even more insights into the biology of this family.

Consideration of the kinds of larval sites leads to several generalizations about the distribution of mosquito species. First, the lack of appropriate larval sites can limit the distribution of species. The lack of sites might be the simple result of low rainfall that fails to fill sites that are structurally present. Sites may be scarce or absent because of the peculiarities of species with specialized habits that tolerate a narrow range of conditions met only by certain kinds of plants (e.g., bamboo) or other conditions (e.g., salt water).

A second generalization is that containers constitute a distinct category of habitats for mosquitoes. Containers formed by plants are particularly important to the biodiversity of tropical mosquitoes. For example, 20% of Neotropical species occur in bromeliads, and 153 species only occur in bromeliads (Frank & Curtis 1981). Container specialists (e.g., Sabethini) seldom colonize ground habitats. In contrast, groups dominated by species in ground habitats (e.g., *Aedes*, *Anopheles*, *Culex*) have species groups that use containers either exclusively or facultatively.

Finally, human activity influences mosquito populations by affecting larval sites. Disruption of drainage patterns, addition of organic or mineral pollutants, creation of widespread containers, importation of invasive plants, cultivation of aquatic crops, and civil works for water management can all create larval mosquito sites on a large scale, often close to human populations. Presumably, most of the species colonizing such sites did not evolve in association with people and are only finding suitable sites that happen to have been created by humans.

Table 3.7. Classification of mosquito larval habitats designed by Service (1993)

	Habitat	Genera
1. Ground habitats		
	1.1. Still waters	
	1.1.1. Permanent or semipermanent	
	(a) Marshes (b) Swamps (c) Exposed ponds, borrow pits (d) Forest ponds (e) Ditches	*Aedes, Anopheles, Aedeomyia, Coquillettidia, Culex, Culiseta, Hodgesia, Ficalbia, Mansonia, Mimomyia, Uranotaenia, Lutzia*
	(f) Rice fields	*Aedes, Anopheles, Culex, Psorophora*
	(g) Polluted waters	*Culex, Culiseta, Armigeres*
	(h) Wells	*Aedes, Anopheles, Culex*
	(i) Saltwater marshes	*Aedes, Anopheles, Culex*
	(j) Saltwater pools	*Opifex*
	(k) Subterranean	*Anopheles, Aedes, Culex*
	1.1.2. Temporary	
	(l) Exposed pools and puddles	*Aedes Anopheles, Psorophora*
	(m) Forest pools	*Aedes, Anopheles, Culex, Psorophora*
	(n) Hoofprints	*Aedes, Anopheles*
	(o) Saltwater pools	*Aedes*
	1.2. Flowing waters	
	(p) Exposed streams, ditches, irrigation channels (q) Forest streams (r) Gravel streambeds	*Anopheles, Culex*
2. Container habitats		
	2.1. Natural	
	(s) Tree holes	*Aedes, Anopheles, Culex, Haemagogus Heizmannia, Sabethes, Topomyia, Tripteroides, Orthopodomyia, Toxorhynchites*
	(t) Bamboo	*Aedes, Anopheles, Armigeres, Eretmapodites Trichoprosopon, Uranotaenia, Tripteroides*
	(u) Leaf axils, bracts, bromeliads	*Aedes, Anopheles, Culex, Malaya, Mimomyia Topomyia, Trichoprosopon, Wyeomyia, Toxorhynchites*
	(v) Pitcher plants	*Aedes, Anopheles, Armigeres, Culex, Mimomyia Toxorhynchites, Trichoprosopon, Tripteroides, Wyeomyia*
	(w) Fallen fruit husks, spathes	*Aedes, Armigeres, Culex, Eretmapodites Toxorhynchites, Trichoproso-pon, Uranotaenia*
	(x) Fallen leaves	*Aedes, Culex, Eretmapodites*
	(y) Empty snail shells	*Aedes, Eretmapodites*
	(z) Rock pools	*Aedes, Anopheles, Culex, Opifex, Toxorhynchites Uranotaenia*
	(aa) Crab holes	*Aedes, Culex, Culiseta, Deinocerites, Uranotaenia*
	2.2. Human made	
	(bb) Water tanks, cisterns	*Aedes, Anopheles, Culex, Culiseta*
	(cc) Latrines, septic tanks	*Armigeres, Culex*
	(dd) Domestic water storage pots (ee) Discarded tin cans, bottles, tires (ff) Miscellaneous	*Aedes, Culex, Toxorhynchites*

Categorizing the larval habitats of thousands of mosquito species misses some unusual examples of how larvae live. Although many other interesting larval sites exist, the few presented here extracted from reviews (Horsfall 1955; Laird 1988; Service 1993) introduce some of the unusual adaptations of mosquitoes to the aquatic environment. Mosquito larvae are commonly observed at the edges of streams or rivers because of a preference for ovipositing in that habitat or because they have been washed into such areas from other places. Larvae of *Culex starckeae*, however, live in rapidly flowing water in Australasia hooked to algal mats by modified siphonal hairs. A different kind of siphonal adaptation enables all species of *Mansonia* and *Coquillettidia*, as well as *Ficalbia ludlowae* (Philippines), to pierce air cells of submerged plants and remain submerged for extended periods. Some mosquitoes are capable of routinely completing larval development in very small accumulations of water in fallen leaves (*Eretmapodites grahami*, East Africa), cup fungi (*Aedes dobodurus*, Papua New Guinea), and, smallest of all, the leaf axil of a sedge (*Aedes woodi*, East Africa). Certain kinds of sites are simply unusual. For example, *Aedes aegypti* has been found in giant clam shells in Tuvalu, *Armigeres kuchingensis* routinely occurs in jars of feces in Singapore, *Eretmapodites penicillatus* occupies the putrid remains in dead giant African snail shells, and *Malaya* species habitually colonize water accumulated in abandoned arboreal ant nests. Probably every mosquito collector has a story about an unusual larval site. The authors of this volume would like to mention finding *Culex coronator* in an ancient Mayan sacrificial bowl in southern Mexico and *Anopheles bambusicolus* practically sealed within a bamboo internode pierced by a small beetle hole in Brazil.

Larval distribution can sometimes be a practical predictor of adult mosquito distribution. An abundance of *Anopheles sinensis* s.l. larvae dipped from rice fields in the Republic of Korea was a significant predictor of female landing collections a week later, probably because the rice fields were the only larval habitat in the area (Strickman et al. 2000a). A quantitative model of larval spatial distribution in Australia was less successful (de Little et al. 2009). Ecological niche models of larval habitats based on elevation, water presence, and vegetation only explained 24% of the variation for *Culex annulirostris* and 30% for *Aedes vigilax*. The authors attributed the weak correlation to unsystematic larval collections, which demonstrated how it is relatively more difficult to get spatially representative samples of larvae than of the more mobile adults. Working backward from historical museum collections, Dekoninck et al. (2013) attributed an increase in container-inhabiting *Anopheles plumbeus* in Belgium to the species' expansion into new larval habitats provided by abandoned pig manure tanks. Similarly, Olson et al. (2010) were able to show that adult populations of *Anopheles darlingi* increased in western Brazil when deforestation and aquaculture increased the number of larval sites.

Intraspecific and interspecific interactions have a considerable influence on larval habitats. The density of even the same species can affect survival, size, and subsequent adult fecundity. Two species, *Aedes flavopictus miyarai* and *Malaya genurostris*, in the Ryukyus Islands, Japan, were sometimes found in tiny accumulations of water in leaf axils (Mogi 1984). In that study the habitats averaged just 0.36 ml/larva, with a maximum of 43 larvae/ml. *Aedes flavopictus miyarai*, a facultative inhabitant of leaf axils, was smaller in more crowded conditions. In contrast, *Malaya genurostris*, an obligate inhabitant of axils, was the same size regardless of larval density. *Aedes sierrensis* larvae produced smaller pupae and experienced lower pupal survival at high densities in simulated tree hole habitats (Broadie & Bradshaw 2008) due to mutual interference by food-seeking larvae and unidentified substances secreted into the water. Crowded *Aedes aegypti* larvae produced smaller adults (Sumanochitrapon et al. 1998), and larval conditions with fewer nutrients were associated with smaller adults in the field (Strickman & Kittayapong 2003; Unlu et al. 2013). Evidently, competition between individuals of the same species can influence the quality of an individual larval habitat, exerting adaptive pressure on distribution of a species.

Competition between species of mosquitoes, or even with other aquatic organisms (Mokany & Shine 2003a), in the larval habitat also influences subsequent adult survival and quality. Barrera (2008) showed that starved *Aedes aegypti* larvae survived for a shorter time when other species (*Aedes triseriatus* and *albopictus*) were present. *Aedes albopictus* survived dry regimes (the water of the colony was allowed to dry completely and left two weeks before refilling) much better when *Aedes aegypti* was not present, in contrast to *Aedes aegypti*, which was not affected by the presence of the other species (Costanzo et al. 2005). When dry and constantly wet conditions fluctuated, *Aedes albopictus* colonies were not as badly affected as *Aedes aegypti*, and mixed colonies were more deleterious for *Aedes aegypti*. In the field in Florida, *Aedes albopictus* was more susceptible than *Aedes aegypti* to warmer, drier conditions (Juliano et al. 2002). *Aedes albopictus* infected with a parasite from its native Asian range, *Ascogregarina taiwanensis* (Lien & Levine) (Phylum: Apicocomplexa; Order: Eugregarinorida; Lecudinidae), lost its competitive advantage against indigenous *Aedes triseriatus* in Illinois (Aliabadi & Juliano 2002).

These sorts of negative interactions that involved invasive species may reflect an absence of long-term coadaptation that would eventually result in niche partitioning. Among bromeliad inhabitants in Costa Rica, *Anopheles neivai*, *Culex rejector*, *Wyeomyia melanopus*, and *Wyeomyia circumcincta* showed clear differences in depth occupied within a single bromeliad and in distribution between different bromeliads based on size of the plants. Tree-hole-inhabiting species in Florida lived in 115 separate holes without competition, appearing and disappearing in a nonequilibrium dynamic relationship (Bradshaw & Holzapfel 1983). *Toxorhynchites rutilus* and *Orthopodomyia signifera* maintained a constant biomass during the year, but *Culex restuans*, *Aedes triseriatus*, and *Anopheles barberi* colonized the sites only in the spring. Although niche partitioning is probably important for many different groups of coinhabiting mosquito larvae, it cannot be assumed to be a universal relationship for all species assemblages, as exceptions have been observed among other insects (Condon et al. 2008).

The presence of larval predators can make an otherwise suitable habitat unsuitable. The larvae of *Corethrella appendiculata* (Grabham) (Diptera: Corethrellidae) eat moving mosquito larvae in tree holes in Florida (Kesavaraju & Juliano 2009). No similar predation exists in Southeast Asian tree holes, so *Aedes albopictus* did not evolve defense behaviors against this kind of predator. *Aedes triseriatus* coevolved with tree-hole *Corethrella* and escapes predation by remaining motionless when a predator is near. As a result, the presence of *Corethrella appendiculata* makes a tree hole unsuitable for *Aedes albopictus* and eliminates competition for *Aedes triseriatus.* Prior to the introduction of *Aedes albopictus,* Bradshaw and Holzapfel (1983) described how *Corethrella appendiculata, Anopheles barberi, Culex restuans, Aedes triseriatus,* and *Orthopodomyia signifera* coexisted through temporal niche partitioning in Florida tree holes, held below carrying capacity by predation from *Toxorhynchites rutilus.* One implication of the study was that these species would have had a different distribution without the presence of the predator. Other examples of the effect of predation on larval distribution include elimination of *Culiseta* from stock tanks in California, United States, by *Notonecta hoffmani* Hungerford (Coleoptera: Notonectidae) (Murdoch et al. 1984) and elimination of *Aedes vigilax* in tidal saltwater pools when atherinid fish enter them (Kokkinn et al. 2009).

Larval habitats seem to be colonized by different assemblages of mosquito species in distinct parts of the world. Parallel evolution of larval ecology is a logical result of regional speciation and occurrence of similar habitats worldwide. In some cases the parallels are striking, even between species that are phylogenetically distant. One of the best examples is a special category of phytotelmata—fruit husks (Lounibos 1983). Species of one sabethine genus of the Western Hemisphere (*Trichoprosopon*), an African aedine genus (*Eretmapodites*), and an Asian aedine genus (*Armigeres*) share physiological and behavioral adaptations that allow them to inhabit viscous, highly organic media more typical of muscoid Diptera than of Culicidae. Salt marsh habitats also present physiological challenges that a series of species, mainly in the genera *Aedes* and *Anopheles,* have overcome. The *Aedes* species deposit resistant eggs in areas that will eventually flood during the highest tides or when tides and rainfall combine to flood pools in upper marshes. Those locations avoid flushing from subsequent tides and also predation from fish that follow tidal cycles. Salt marsh mosquito habitats are often created by humans due to disruption of natural drainage patterns—for example, by constructing a railroad embankment across a marsh. Rice fields are totally artificial habitats that have resulted in distinct assemblages of species usually consisting of *Anopheles, Culex,* and floodwater *Aedes.*

Past Distribution of Mosquitoes

Chapter 1 reviewed the evolution of mosquitoes, including summarizing what is known about the appearance of major groups of mosquitoes. The distributions of those earlier forms were influenced by the same factors that exist today, with the exception of human activity and transport. Distributions were affected as environmental conditions changed over geological time, influencing isolation of populations and speciation. The distributions we observe now are not only a result of current conditions but also of historical events. Some of those events were systematic changes, like glaciations, that affected the entire fauna equally. Other events were more random, such as chance dispersal or the coincidence of a species' characteristics with an environmental trend in a location. Like explanations of the weather or geology itself, we are presented with current reality and assign ourselves the task of explaining how it came to be that way. Denied the possibility of experimentation, those explanations require some skepticism, as any exceptions to systematic influences are justified based on stochastic events.

The origin of *Anopheles* may have taken place 170 million years ago somewhere on the land mass now identified as South America (Krzywinski et al. 2001b). One prediction for such an origin might be greater species richness at that location now. Within the Neotropics, Foley et al. (2008b) found the greatest species richness in northern South America in Trinidad and Tobago, Panama, and French Guiana. It is totally speculative, but this part of the world might have been where all mosquitoes originated.

Major geological events affect biological survival and distribution. The refilling of the Mediterranean Sea may be an example.[1] The Mediterranean Basin dried completely 5.6 million years ago but refilled 300,000 years later in an event called the Zanclean flood, when waters in the Atlantic found their way across the shallow entrance at the Strait of Gibraltar and cut a channel for rapid flooding (García-Castellanos et al. 2009). A complex of salt marsh *Aedes* (Coluzzi et al. 1974) may have originated at that time. Consisting of *Aedes mariae* in the western Mediterranean to Sicily and the west coast of Italy, *Aedes zammitii* from Italy to the middle of Turkey, and *Aedes phoeniciae* in the eastern Mediterranean, the speciation of this complex may have been a direct result of isolation of populations of an ancestral species.

More recent geological events are known in more detail, and their influence on mosquitoes' distributional patterns is less speculative. The most recent geological period, the Quaternary, began about 2.6 million years ago at the outset of one of only five or six glacial ages ever experienced in Earth history (Pielou 1991; Erickson 1996; Alley 2000). The Quaternary has experienced 19–20 cycles of glacial advances of about 100,000 years each. A cycle has typically lasted 60,000 to 90,000 years, during which world temperatures have been lower, overall humidity has been reduced, and ocean levels have been up to 180 m lower than they are today. These cold periods have been separated by interglacials of 10,000–40,000 years, when warmer, wetter conditions have returned. The last glacial maximum occurred approximately 22,000 years ago and receded to its current extent

[1] Thanks go to Graham White for providing this example.

about 16,000 years ago. During the last glacial maximum, the Cordilleran and Wisconsin ice sheets reached as far south in the United States as Washington State and Illinois. The Fennoscandian ice sheet covered northern Europe as far south as London and into Poland and Germany. Ice sheets also covered much of Siberia. The Southern Hemisphere did not have major ice sheets (with the exception of the Antarctic ice cap) but shared in the spread of montane glaciers even at tropical latitudes. The glacial age that started 2.6 million years ago and continues now is the first one since the appearance of mosquitoes and has affected climate and geography in many ways. Within the 100,000-year cycles are subcycles of 19,000 years, 23,000 years, and 41,000 years, caused by astronomical relationships of the earth and the sun. Other cyclical events occur at approximately 1,500-year intervals, all resulting in temperatures that have varied widely during the last 100,000 years. The last 15,000 years have been much more stable, though a period between 12,800 and 11,500 years ago, known as the Younger Dryas, was the last of a series of interruptions in the overall warming trend. From a culicid point of view, there have been strong climatic and geographical influences on distribution during the last 3 million years.

The ice sheets affected mosquitoes directly by preventing them from occupying areas that were covered. As a result, mosquitoes have lived in those formerly glaciated areas for less than 20,000 years. *Wyeomyia smithii* is an autogenous mosquito that develops in pitcher plants and is currently distributed from southern Florida to northern Canada. Allozyme genetic analysis of 34 populations from Florida (30°N) to Manitoba (54°N) showed that populations north of the maximum extent of the ice sheet had a lower rate of heterozygosity (Armbruster et al. 1998). The northern populations also had a different juvenile developmental time and a different photoperiodic response. *Wyeomyia smithii* populations had apparently migrated north, following the redistribution of pitcher plants. The lower genetic variation may have been a direct result of their recent colonization of the area. Physiological changes may have been adaptations to living at higher latitudes.

Glaciation may have had a less direct effect on mosquitoes in the tropics. Examination of the mitochondrial COII gene of *Anopheles* species in the Indo-Burma faunal zone indicated that drier conditions and lower sea levels of glacial intervals had a profound effect on genetic structure (Morgan et al. 2009, 2011; Table 3.8). Genetic evidence showed that the variation within each of nine species coalesced at the same time, probably 900,000 years ago and corresponding to when Pleistocene glacial cycles occurred more rapidly and with greater changes. The most likely evolutionary model involved two refugia at middle elevations in northeastern and northwestern portions of the region where conditions during glaciations would have remained warm and moist enough. Some species remained isolated in the refugia and developed divergent genotypes (*Anopheles annularis*, *philippinensis*, *splendidus*, and *minimus*) or even separate species (*Anopheles baimaii* and *dirus*). Others developed a more complicated genetic structure, based

Table 3.8. Characteristics and history (approximately 1 million years) of Southeast Asian *Anopheles* (*Cellia*) species that have resulted in their current distribution and genetic structure (Morgan et al. 2011)

Taxonomic grouping and species	Typical habitat[1]	Genetic structure[2]	Pleistocene refugium[3]	Altitude range[4]	Pre-expansion migration[5]
Neomyzomyia Series					
baimaii	Forest	*dirus* sister	Western	Narrow	Medium
Myzomyia Series					
Funestus Group					
minimus	Forest	Divergent	Two	Narrow	Low
aconitus	Mixed	Structured	Two	Wide	Medium
jeyporiensis	North hills	Structured	Northern	Narrow	Low
Neocellia Series					
Maculatus Group					
maculatus	Forest	Simple	Two	Wide	
sawadwongporni	Forest	Simple	Two	Wide	Medium
Annularis Group					
annularis	Mixed	Divergent	Two	Narrow	Medium
philippinensis	Mixed	Divergent	Two	Narrow	Low
Jamesii Group					
splendidus	Forest	Divergent	Two		

[1]Primarily in forest; occurs in both forested and open habitats (mixed) or in hilly regions of southern China (north hills)
[2]Two distinct clades based on mitochondrial CO_2 gene sequence (divergent), complicated set of clades (structured), or single dominant clade (simple)
[3]Indication from F_{ST} data that population was restricted to middle-altitude refugia during glaciated periods
[4]Tolerance to range of altitudes
[5]Relative rate of genetic dispersal before last glacial maximum

on periods of gene flow punctuated by periods of isolation (*Anopheles aconitus* and *jeyporiensis*). Two species remained limited in distribution and only recently expanded their ranges, creating a simple genetic structure (*Anopheles sawadwongporni* and *maculatus*). A separate study showed evidence that the same climatic changes may have caused the formation of distinct populations of *Anopheles vagus*, with the significant effect that this species is a vector of human malaria in India and Sri Lanka but nearly completely zoophilic in Southeast Asia (Zarowiecki et al. 2011). These kinds of events would probably not have occurred elsewhere, since the drier conditions of Southeast Asia were exacerbated by the exposure of much more dry land as sea level decreased. In South America, for example, there was no evidence of the existence of Pleistocene refugia.

Regional geographic variation of other mosquitoes, such as the genetically distinct populations of *Culex tarsalis* in Southern California, the Nevada-California deserts, and the remainder of California, may have been influenced by glaciations (Gimnig et al. 1999). The mountains that separate those populations are too old to account for intraspecific variation. It is possible that montane glaciers provided the real barriers, though that evidence has not been developed. The emergence of the domestic form of *Aedes aegypti* (Tabachnick 1991) and the adaptation of *Anopheles gambiae* to humans (White et al. 2011a) have been attributed to human agricultural activity, which was apparently facilitated by the retreat of the ice sheets.

Future Distribution of Mosquitoes

Predicting the future is notoriously difficult, but much of the study of mosquito biology is aimed at that goal. Risk assessment is a look into the near future for the practical purpose of targeting control efforts. New strategies for vector control, like gene drive mechanisms coupled with pathogen resistance, must make predictions about the effects of the technology. Those concerned with invasive species have the responsibility to prevent new invasions as well as manage existing ones, another task that involves making estimates of the future. Most topical of all is the prediction of the effects of climate change on mosquito distribution. Discovery of physiological mechanisms, development of new control techniques, evaluation of risk, and construction of species distribution models are all relevant to building a better understanding of climate change's effects on mosquito distribution. Predictions about the distribution of mosquitoes in the more distant future are an interesting but speculative projection of lessons we have learned from what mosquitoes have done in the past and what they are doing in the present.

Genetic Change and Invasion

For the most part, mosquito adaptation does not occur without genetic change. Although mosquitoes exhibit limited behavioral flexibility, the assumption is that most changes in behavioral or physiological capabilities are linked to genetics. Proving that assumption and showing how genetic differences result in adaptation to a location is not a small effort. Mitochondrial DNA (mtDNA) is often used as a marker for population and species differences. The connection between cytochrome c oxidase II (COII) sequences and specific phenological traits was only recently shown in *Drosophila* (Ballard et al. 2007), and there is much work left to make the connections between individual sequences and phenotypes in mosquitoes. The difference between the M and S molecular forms of *Anopheles gambiae* (now recognized as *Anopheles coluzzii* and *Anopheles gambiae* s.s., respectively; Coetzee et al. 2013) was linked to partial hybrid sterility (Costantini et al. 2009). That reproductive barrier resulted from an inversion on chromosome 2R, near the centromere. Because of the location on the chromosome, the inversion became fixed and provided the opportunity for accumulation of allelic differences. Those differences were associated with antipredation behaviors, allowing larval *Anopheles coluzzii* to live in rice fields and other larger larval habitats in which *Anopheles gambiae* s.s. experienced high mortality from aquatic predators. Others have observed that hybrid sterility was often the first indication of incipient speciation in *Anopheles* (Fritz et al. 1994).

The subject of landscape genetics is the difference in genetic structure associated with geographic variation and the degree to which populations of one species or closely related species introgress. Suitable habitats may be physically isolated from each other, creating a mosaic of populations either completely or partially cut off from each other genetically (Holderegger & Wagner 2008). Genetic differences may accumulate by random changes within isolated populations, as observed in enzyme studies of indigenous *Aedes albopictus* in Malaysia and Borneo (Black et al. 1988). The Rift Valley in East Africa was a significant genetic barrier to *Anopheles gambiae* in Kenya, based on ribosomal DNA ITS2 sequences and microsatellites (Lehmann et al. 1999). In contrast, *Anopheles punctipennis* in Vermont, United States, showed more genetic variation in mitochondrial COI genes within locations than between them (Fairley et al. 2000).

Differences between populations are far from predictable for several reasons. First, there may be significant gene flow across incomplete barriers formed by partially suitable habitats (Holderegger & Wagner 2008). Second, variation between sequences of particular genes may be influenced by the physiology of that taxon. For example, species of *Anopheles* typically have much more variation in ribosomal DNA than species of *Aedes*, *Psorophora*, or *Haemagogus* (Fritz et al. 1994). Finally, fixation of genes, even with good fitness, may take hundreds or thousands of generations if discontinuous habitats interrupt gene flow (Kiszewski & Spielman 1998).

The genetic variation observed between populations of a species may have great significance for that species if it expands its range. This concept is well documented for invasive plants, a topic of interest not only because invasive plants have undergone detailed genetic analysis but also because invasive plants can directly affect mosquito populations (Mack & Smith 2011; Stone et al. 2018). When comparing the ancestral or indigenous population to the invasive population, gene flow indicated by neutral markers (F_{ST}) may be equal or unequal to gene flow indicated by phenotypic markers (Q_{ST}). Equality between F_{ST} and Q_{ST} theoretically indicates a lack of selection during establishment of a species in its new range, but inequality suggests the selection of traits within the genetic variation that is introduced. An example is lippia (*Phyla canescens* (Kunth) Greene) introduced from Argentina to France (enhanced vegetative reproduction and growth) and to Australia (enhanced seed production; Xu et al. 2010). Another concept developed by those who study invasive plants is "evolution of increased competitive ability" (EICA), which expresses the general tendency for invasive plants to experience selection for greater ability to displace native species. The case of lippia cited above may be one example, and another is greater production of protective chemicals in invasive paperbark trees (*Melaleuca quinquenervia* (Cav.) S.F. Blake; Franks et al. 2011). The favorable changes in an invasive population's genetics may not necessarily be the result of selection from a wide range of phenotypes. Despite genetic bottlenecks, rare alleles coding for favorable phenotypes in the new environment may be rapidly fixed in the absence of competition from a larger variety of genotypes (Dlugosch & Parker 2008). Such a process would be similar mechanistically to some hypotheses on how new thoughts originate, as unique neural combinations propagate in parts of the brain where they are not swamped by existing propagation patterns (Calvin 1996).

Hybridization can also facilitate the establishment and spread of an invasive species. Invasive plants provide three clearly documented examples of how intraspecific hybridization preceded the development of invasiveness. In these cases, hybridization resulted in improved fitness in the new environment, and the species formed novel, stabilized, coalescent populations (Schierenbeck & Ellstrand 2009). Each example involved multiple intentional introductions that eventually transformed a useful imported plant into an invasive weed. An ornamental Chinese pear, *Pyrus calleryana* Decne., was selected for importation to the Unites States because it rarely produced seed and was not self-fertile. Hybridization among subsequently introduced cultivars was tracked with microsatellites to show that those introductions were the sources of trees that produced fertile seed and that became invasive in American forests (Culley & Hardiman 2009). Reed canary grass (*Phyla arundinacea* L.), an ornamental introduced from Europe to the United States, experienced genetic bottlenecks that were overcome by multiple introductions. Currently, the species is more diverse in its introduced range and has acquired characteristics that have converted it from a stable climax species (*K* strategist) to a colonizing weed (*r* strategist; Lavergne & Molofsky 2007). Intraspecific hybridization has had a similar effect on the shade tree from South America, Brazilian peppertree (*Schinus terebinthifolius* Raddi), in Florida (Geiger et al. 2011; Mukherjee et al. 2012). Human intervention may accelerate opportunities for hybridization, but hybridization also occurs naturally. Such natural hybridization was observed in Central Europe, where populations originating in glacial refugia of the Iberian Peninsula, Italy, and the Balkans have recolonized formerly glaciated areas (Keller & Taylor 2010). The occurrence of hybridization among introduced mosquitoes has been observed, particularly in the Pipiens Group of the genus *Culex*. The real importance of hybridization to mosquitoes' adaptation is a subject that requires much more study.

Invasive Mosquito Species

Mosquitoes provide many examples of invasive species (Table 3.9). Although not every mosquito that has expanded its range presents an obvious threat, others fit extremely well into the general opinion that invasive species are second only to habitat loss as a threat to biodiversity (Lowe et al. 2004). The invasion process can be subdivided into transport, establishment, spread, and impact. Often, the currently invaded range is only a portion of the potential range expansion (Jiménez-Valverde et al. 2011). Ultimately, the distribution might be influenced by continued transport, introduction of additional genetic variety, hybridization, or changes in the environment. Mosquitoes, unlike plants, are not usually intentionally shipped by humans, but accidental transport by humans has been the source of almost all nonindigenous species. Unfortunately, these modes of accidental transport tend to get repeated so that sufficient genetic material is introduced for establishment. Humans also have a role in creating suitable larval habitats for many of the introduced species. Establishment of introduced mosquitoes is not a certainty. Records of intercepted insects showed that of 22 species observed to have entered the United States, only eight (36%) eventually became established. In Hawaii the same sort of data showed that only 5 of 33 (15%) became established and in Australia, only 4 of 16 (25%) (Lounibos 2002).

The total number of nonindigenous mosquitoes is small compared to the number of all species. *Aedes aegypti*, *Aedes albopictus*, *Culex pipiens*, and *Culex quinquefasciatus* are the clearest examples of species that, through human intervention, have established worldwide distribution. *Anopheles arabiensis* (assuming this species accounted for the various range expansions of *Anopheles gambiae* s.l.) has been able to expand its distribution dramatically, but usually temporarily, thanks to vigorous eradication efforts. Five other species may be in the process of becoming important based on recent expansion and potential to infest common habitats. These are *Aedes atropalpus*, *Aedes japonicus japonicus*, *Aedes koreicus*, *Anopheles stephensi*, and *Culex coronator*. *Aedes notoscriptus* seems poised to expand its range, as it has diversified its larval habitats from rock pools and tree holes to many kinds of human-made containers (Russell 1986b). This species was introduced to and eliminated from New Zealand (Derraik 2004) but has since been introduced into California (Peterson & Campbell 2015). All the other 27 species (29 if the intentionally introduced *Toxorhynchites amboinensis* and *brevipalpis* are included) appear to have had the potential to expand to only limited geographic areas. It is within this group that a few genera other than *Aedes*, *Anopheles*, and *Culex* appear. Even at

Table 3.9. Records of mosquitoes introduced from one region to another

Species	From	To	First introduction	Conveyance	Comment	Reference
Aedeomyia catasticta	Southeast Asia and Australasia	Guam	1958			Ward 1984a
Aedes aegypti	Africa	Asia	after 1850	Ships		Lounibos 2002
Aedes aegypti	Mexico?	Arizona	1994	Ground transport?	Previously present 1931–1946 and 1951 but established in Tucson since 1994	Engelthaler et al. 1997
Aedes aegypti		Australia	shortly after 1788	Ship		Lounibos 2002
Aedes aegypti		California	2013		In two counties with current efforts to eradicate	Mulligan 2013
Aedes aegypti		Guam	early 1900s		Eradicated 1948	Ward 1984a
Aedes aegypti		Hawaii	1892	Ships	Widespread by 1913, eradicated by 1949, reappeared in 2012 probably by aircraft	Lounibos 2002; Anonymous 2012; Larish and Savage 2005; Leong and Grace 2009
Aedes aegypti		Madeira Island	2004–2005			Almeida et al. 2007
Aedes aegypti	Florida	Netherlands	2010	Tires	Imported aircraft tires	Enserink 2010; Scholte, Braks, et al. 2010b
Aedes aegypti		Polynesian islands	1924–1986	Aircraft		Manguin and Boëte 2011
Aedes aegypti		Republic of Georgia	before 2008		Near Black Sea	Medlock et al. 2012
Aedes aegypti		Russia	before 2008		Near Black Sea	Medlock et al. 2012
Aedes aegypti		Spain	before 1701		Common in 1953, absent in 1965	Eritja et al. 2005
Aedes aegypti	Europe or Africa	Western Hemisphere	1490s	Ships	Suspected in Haiti 1495; documented in Yucatan 1648	Lounibos 2002
Aedes albopictus	China?	Albania	1979	Aircraft? Tires?		Mouchet et al. 1995; Eritja et al. 2005; Adhami and Murati 1987
Aedes albopictus		Algeria	2010			Manguin and Boëte 2011
Aedes albopictus		Argentina	1998			Eritja et al. 2005
Aedes albopictus		Australia	1988		Successive introductions and eradications	Eritja et al. 2005
Aedes albopictus		Barbados	1993			Benedict et al. 2007
Aedes albopictus		Belgium	2000	Tires	Did not establish until 2004	Eritja et al. 2005; Medlock et al. 2012; Benedict et al. 2007
Aedes albopictus		Bolivia	1997			Benedict et al. 2007
Aedes albopictus		Bosnia/Herzegovina	2005	Ground transport		Medlock et al. 2012; Schaffner and van Bortel 2010
Aedes albopictus		Brazil	1985			Manguin and Boëte 2011
Aedes albopictus		Bulgaria	2011			Medlock et al. 2012

Aedes albopictus	Asia	California	2001	Lucky bamboo	First introductions seemed to disappear but reemerged in 2011 with same genetic signature in Los Angeles County	Zhong et al. 2013
Aedes albopictus		Cameroon	2000			Manguin and Boëte 2011
Aedes albopictus		Cayman Islands	1997			Benedict et al. 2007
Aedes albopictus		Central African Republic	2010			Manguin and Boëte 2011
Aedes albopictus		Colombia	1998			Benedict et al. 2007
Aedes albopictus		Corsica	2002	Ground transport		Eritja et al. 2005; Medlock et al. 2012
Aedes albopictus		Croatia	2004	Ground transport		Medlock et al. 2012; Schaffner and van Bortel 2010
Aedes albopictus		Cuba	1995			Benedict et al. 2007
Aedes albopictus		Dominican Republic	1993			Benedict et al. 2007
Aedes albopictus		Eastern Mediterranean	2010			Manguin and Boëte 2011
Aedes albopictus		El Salvador	1995			Benedict et al. 2007
Aedes albopictus		Equatorial Guinea	2001			Benedict et al. 2007
Aedes albopictus		Fiji	1988			Eritja et al. 2005
Aedes albopictus		France	1999		Original infestations in tire dumps were eradicated; infestations from ground transportation became established and started spreading in 2004	Manguin and Boëte 2011; Eritja et al. 2005; Medlock et al. 2012
Aedes albopictus	US	Geneva, Switzerland	1990	Tires		Mouchet et al. 1995
Aedes albopictus		Greece	2003	Ground transport		Medlock et al. 2012; Schaffner and van Bortel 2010
Aedes albopictus	Asia	Guam	1944–1945		Associated with WWII activities	Ward 1984a
Aedes albopictus		Guatemala	1995			Benedict et al. 2007
Aedes albopictus		Guatemala	1995			Eritja et al. 2005
Aedes albopictus	Asia	Hawaii	1896	Ship		Lounibos 2002; Larish and Savage 2005
Aedes albopictus		Honduras	1995			Benedict et al. 2007
Aedes albopictus		Israel	2003			Benedict et al. 2007
Aedes albopictus	eastern US	Italy	1990	Aircraft tires, ground transport		Manguin and Boëte 2011; Eritja et al. 2005; Medlock et al. 2012; Zhong et al. 2013

continued

Table 3.9. *continued*

Species	From	To	First introduction	Conveyance	Comment	Reference
Aedes albopictus	Java, Indonesia	Madagascar	2,000 years ago			Mouchet et al. 1995
Aedes albopictus		Malta	2009			Medlock et al. 2012; Schaffner and van Bortel 2010
Aedes albopictus	US	Mexico	1988			Manguin and Boëte 2011
Aedes albopictus		Montenegro	2001	Ground transport		Eritja et al. 2005; Medlock et al. 2012; Schaffner and van Bortel 2010
Aedes albopictus	China and Florida	Netherlands	2005	Lucky bamboo and tires		Scholte et al. 2009, Medlock et al. 2012
Aedes albopictus	Japan	New Zealand	1993			Mouchet et al. 1995
Aedes albopictus		Nicaragua	2003			Eritja et al. 2005
Aedes albopictus		Nigeria	1991			Manguin and Boëte 2011
Aedes albopictus		Panama	2002			Benedict et al. 2007
Aedes albopictus		Paraguay	1998			Benedict et al. 2007
Aedes albopictus		Reunion	1994			Eritja et al. 2005
Aedes albopictus	Singapore	Rio de Janeiro, Brazil	1986			Mouchet et al. 1995
Aedes albopictus		Slovenia	2005	Ground transport		Medlock et al. 2012; Schaffner and van Bortel 2010
Aedes albopictus		South Africa	1990			Eritja et al. 2005
Aedes albopictus		Spain	2003	Ground transport		Manguin and Boëte 2011; Medlock et al. 2012; Schaffner and van Bortel 2010
Aedes albopictus		Switzerland	2003	Ground transport		Medlock et al. 2012; Schaffner and van Bortel 2010
Aedes albopictus	Japan	Texas, US	1985	Tires	Previously introduced to US but not established	Lounibos 2002
Aedes albopictus	Japan	Texas, US	1985			Mouchet et al. 1995
Aedes albopictus		Trinidad	1983			Benedict et al. 2007
Aedes albopictus	Greece?	Turkey	2011	Ground transport?	Found near Greece	Medlock et al. 2012; Oter et al. 2013
Aedes atropalpus	eastern US	France	2003, 2005	Tires	Later eradicated	Scholte et al. 2009
Aedes atropalpus	eastern US	Italy	1996	Tires		Lounibos 2002; Medlock et al. 2012
Aedes atropalpus	eastern US	midwestern US	after 1972	Tires	Species expanded suitable larval sites from rock pools to include tires	Lounibos 2002

Aedes atropalpus	eastern US	Netherlands	2008, 2010	Tires		Scholte et al. 2009; Medlock et al. 2012
Aedes atropalpus	eastern US	Switzerland	1996	Tires	Later eradicated	Scholte et al. 2009
Aedes australis	Australia	New Zealand	1962	Ships	Otago Peninsula, South Island; displaced *Opifex fuscus*	Derraik 2004; Snell et al. 2010
Aedes bahamensis	Bahamas	southern Florida	1986	Tires	Present in only two counties	Lounibos 2002
Aedes camptorhynchus	Australia	New Zealand	1998	Ship		Derraik 2004
Aedes japonicus japonicus		Austria			Established	Medlock et al. 2012
Aedes japonicus japonicus	US	Belgium	2002	Tires		Schaffner et al. 2009
Aedes japonicus japonicus		France	2001		Eradicated	Eritja et al. 2005; Medlock et al. 2012
Aedes japonicus japonicus		Germany	2008	Ground transport	Established	Medlock et al. 2012; Schaffner and van Bortel 2010
Aedes japonicus japonicus		Hawaii	2003	Aircraft?		Larish and Savage 2005
Aedes japonicus japonicus	US	Ontario, Canada	2001	Ground transport	Established	Thielman and Hunter 2006
Aedes japonicus japonicus		Slovenia			Established	Medlock et al. 2012
Aedes japonicus japonicus		Switzerland	2008		Established	Medlock et al. 2012; Schaffner and van Bortel 2010
Aedes japonicus japonicus	Japan	US	1998	Tires	Japan probably source of original two introductions	Lounibos 2002; Fonseca et al. 2001
Aedes koreicus	Republic of Korea	Belgium	2008	Tires?	Established in small area	Medlock et al. 2012; Versteirt, Pecor, et al. 2012
Aedes koreicus	Korea	Italy	approximately 2011		Established in small area	Medlock et al. 2012
Aedes neopandani	Saipan, Tinian	Guam	1976			Ward 1984a
Aedes notoscriptus	Australia	New Zealand	no later than 1920	Ships	Abundant on North Island	Derraik 2004
Aedes notoscriptus	Australia	California	2015	Unknown	In three counties as of 2018	Peterson and Campbell 2015; website San Mateo County Mosquito and Vector Control District 2018
Aedes togoi	Asia	British Columbia and Washington	1960s	Ships and tires		Lounibos 2002

continued

Table 3.9. *continued*

Species	From	To	First introduction	Conveyance	Comment	Reference
Aedes triseriatus		France	2004		Eradicated	Medlock et al. 2012
Aedes vexans		Guam	early 1900s			Ward 1984a
Aedes vexans nocturnus	Asia or SW Pacific	Hawaii	1962	Aircraft?		Joyce and Nakagawa 1963
Anopheles albimanus		Barbados	1927	Boats	Gone by 1943	Mouchet et al. 1995
Anopheles arabiensis	East Africa	Mauritius	1866–1867	Ships	Eradicated in 19 months	Lounibos 2002
Anopheles arabiensis	Senegal	NE Brazil	1930	Ships	Eradicated 1939	Lounibos 2002; Soper 1966
Anopheles barbirostris	Vietnam	Guam	1975	Aircraft		Ward 1984a
Anopheles darlingi	Brazil	Iquitos, Peru	after 1991	Boats	Possibly associated with trade and culture of tropical fish	Lounibos 2002
Anopheles funestus		Mauritius	1865	Ships	Eventually eradicated	Mouchet et al. 1995
Anopheles gambiae s.l.	Sudan	Egypt	1942	Boats	Eradicated 1945	Mouchet et al. 1995
Anopheles gambiae s.l.	Madagascar	Mauritius	1865	Ships	Made possible by change to rapid steamship service	Mouchet et al. 1995
Anopheles gambiae s.l.	Mauritius	Reunion	1868	Wind		Mouchet et al. 1995
Anopheles indefinitus	Southeast Asia	Guam	1948		Probably introduced during WWII	Ward 1984a
Anopheles litoralis	Philippines	Guam	1975	Aircraft		Ward 1984a
Anopheles sacharovi	Turkey?	Israel	1300s		Colonized Lake Huleh due to agricultural development	MacDonald 1957
Anopheles stephensi	Arabian peninsula	Djibouti	2012	Somali migrants from Oman?	Associated with major outbreaks of malaria, expanding into Ethiopia and Somalia; potential for widespread distribution in Africa	Faulde et al. 2014
Anopheles stephensi	India	Mannar Island, Sri Lanka	2018	Boat traffic?	First record despite close distance to source; rigorous control response	Surendran et al. 2018
Anopheles subpictus	Asia	Guam	1970–1971			Ward 1984a
Anopheles vagus	Asia	Guam	1970–1971			Ward 1984a
Armigeres subalbatus	Asia	Guam	1969			Ward 1984a

Culex australicus	Australia	New Zealand	1926–1956			Fonseca et al. 2006
Culex biscaynensis	Caribbean?	Florida		Bromeliads	Possibly an indigenous species present in only one location	Lounibos 2002
Culex coronator	southwest US	Georgia, US	2006		Prior to 2003, only in south Central, southwestern, and north Central Texas, southern New Mexico, and southeastern Arizona; 2003 found in Oklahoma, Louisiana, Mississippi; 2005 in Florida	Kelly et al. 2008
Culex coronator	northern South America	southern Argentina	2006		Furthest south extent for this species	Rossi and Vezzani 2011
Culex fuscocephala	Southeast Asia	Guam	1969			Ward 1984a
Culex gelidus	Indonesia or Papua New Guinea	Australia	1996			Lounibos 2002
Culex modestus	southern Europe?	southeastern England, UK	2010	Ships?		Golding et al. 2012
Culex pipiens pipiens	Europe?	North America	1500s	Ships	Species considered originally to be from Africa	Farajollahi et al. 2011
Culex pipiens var. *molestus*	Asia	Australia	1940s		Introduced with US forces in WWII; restricted to south until late 1960s; moved north to Sydney in 1970s	Lounibos 2002; Russell 2012
Culex pipiens var. *molestus*	North America	Japan	1943	Ships?	Spread throughout Japan by 1970s	Mogi 2012
Culex pipiens var. *molestus*	Russia?	Korea, China, Taiwan	after 1989	Ground transport?		Mogi 2012
Culex pipiens var. *molestus*	Europe?	Russia	before 1950	Ground transport?	Spread east to Siberia in 1980s	Mogi 2012
Culex quinquefasciatus		Australia	1788 or 1831	Ships?	Probably introduced either with first colonists or later by American whalers	Russell 2012; Fonseca et al. 2006
Culex quinquefasciatus	Ecuador	Galapagos Islands	mid 1980s	Aircraft		Bataille et al. 2009
Culex quinquefasciatus		Guam	early 1900s			Ward 1984a
Culex quinquefasciatus	Mexico	Hawaii	1826	Ship	Original introduction traced to single ship's water barrels; subsequent introductions from southern Pacific region	Leong and Grace 2009; Fonseca et al. 2006

continued

Table 3.9. *continued*

Species	From	To	First introduction	Conveyance	Comment	Reference
Culex quinquefasciatus		New Zealand	1848	Ships	On whaling ships. Since mid 1970s has extended range south to Christchurch	Derraik 2004; Tompkins and Gleeson 2006
Culex quinquefasciatus	North America	Nigeria	after 1942			Fonseca et al. 2006
Culex quinquefasciatus	Asia?	Western Hemisphere	before 1800	Ships	Uncertainty about origin and timing; traditionally thought to be African but now thought that African populations from America. Possibly from Southeast Asia.	Farajollahi et al. 2011
Culex sitiens	Other islands	Guam	1945			Ward 1984a
Culex tritaeniorhynchus	Asia	Guam	1962			Ward 1984a
Lutzia fuscana	Asia	Guam	1968			Ward 1984a
Mansonia uniformis	Southeast Asia and Australasia	Guam	1962	Roots of aquatic plants on seaplanes		Ward 1984a
Toxorhynchites amboinensis	western Pacific region	Hawaii	1953		Intentionally introduced for biological control of *Aedes albopictus*	Goff 1980; Shroyer 1981
Toxorhynchites brevipalpis	East Africa	Hawaii	1950		Intentionally introduced for biological control of *Aedes albopictus*	Shroyer 1981
Wyeomyia mitchellii	Florida or West Indies	Hawaii	1981	Bromeliads		Shroyer 1981

that, the variety is limited to *Aedeomyia*, *Armigeres*, *Mansonia*, *Lutzia*, *and Wyeomyia*.

Although invasiveness often seems to be associated with container larval habitats, container-inhabiting groups like Sabethini, *Haemagogus*, *Armigeres*, and *Orthopodomyia* are poorly represented or unrepresented among invasive species. This lack of variety is probably the result of restrictive requirements for range expansion. These requirements include an association with humans that supports infestation and survival during transport, sufficient genetic variety among the small numbers of individuals transported, and sufficient competitive ability in the invaded range.

Despite the challenges to mosquitoes, the pace of range expansion appears to be accelerating. The 6 species that are expanding rapidly and the 28 species that have become established in a limited geographic range may be a sign that among the wide variety of mosquitoes in the world fauna, many more can expand their ranges as environmental conditions change and as opportunities for dispersal expand.

Islands appear to be more susceptible than mainland sites to establishment of nonindigenous mosquitoes. The long list of new mosquitoes on Guam, including many that have not expanded to other locations, may have been the result of the large volume of uncontrolled transport in the 1940s and a limited aquatic fauna that presented little competition (Ward 1984a). Thanks to the American military presence, Guam has been studied in much more detail than many other islands, raising the possibility that similar situations exist undetected elsewhere.

New Zealand is another well-studied location. Those islands have 4 introduced species and a limited endemic mosquito fauna of only 12 species. It is possible that the relatively small number of nonindigenous species is the result of a vigorous effort to prevent new introductions (Derraik 2004). With the exception of *Culex quinquefasciatus*, all of the introduced species are from Australia, perhaps because multiple introductions from a relatively nearby location were necessary for establishment.

The Hawaiian Islands had no mosquitoes until people introduced them. Normal shipping introduced *Aedes aegypti*, *Aedes albopictus*, and *Culex quinquefasciatus* before 1900. Subsequent introductions were by aircraft (*Aedes vexans nocturnus* [Joyce & Nakagawa 1963]; possibly, *Aedes japonicus japonicus* [Larish & Savage 2005]; possibly, reintroduction of *Aedes aegypti*; Hawaii State Department of Health [2012], retrieved from health.hawaii.gov/news/files/2013/05/12-010.pdf) in the ornamental plant trade (*Wyeomyia mitchellii*; Shroyer 1981; Yang et al. 2003) and intentionally for biological control of other mosquitoes (*Toxorhynchites amboinensis* and *brevipalpis*; Shroyer 1981). The small number of introduced species on the Hawaiian Islands is an indication of the challenges to mosquito establishment in new locations.

Studies of several mosquitoes offer more details about invasion and establishment of species outside of their indigenous ranges. Although originally identified as *Anopheles gambiae*, *Anopheles arabiensis* was probably the species (Parmakelis et al. 2008; Manguin & Boëte 2011) that suddenly expanded beyond its range several times with disastrous consequences for public health. Each introduction was by human activity: steamships to Mauritius in 1864 (Manguin & Boëte 2011, Macdonald 1957), fast warships to Natal, Brazil, in 1930 (Soper 1966), and boats to Upper Egypt in 1942 (Shousha 1948). Although it is suspected that *Anopheles arabiensis* was transported from Mauritius to Réunion by a cyclone prior to a malaria outbreak in 1868 (Manguin & Boëte 2011), that infestation would not have occurred without the original transport to Mauritius. The story of the Brazilian invasion was recorded in great detail (Causey et al. 1943a; Soper & Wilson 1943).

Anopheles arabiensis may be an exception to the general principle that establishment requires a broader genetic pool usually associated with multiple introductions. In Brazil the species became established almost as soon as rapid sea transport was initiated and caused an intense localized malaria outbreak in Natal. During the next 10 years, it expanded through this arid region, taking advantage of agricultural practices based on water storage. Eradication was finally completed in 1939–1940, following a much larger and deadly malaria epidemic that stimulated an organized campaign. Similarly, and with the trenchant advice of Fred Soper (1966), the Egyptian infestation was eliminated by 1944 following epidemics so intense that the associated disruption of transport caused famine. Ironically, niche modeling in Brazil (Levine et al. 2004b) indicated that the habitat was more suitable for *Anopheles gambiae* than for *arabiensis*. Perhaps the somewhat unfavorable environment supported the eradication effort. The series of *Anopheles arabiensis* infestations raises several questions: Is there something about this species that favors expansion of populations from limited genetic material? Why have so few other *Anopheles* species expanded their ranges significantly? And, of course, will it happen again?

The introduction of *Anopheles stephensi* into Sri Lanka by 2018 (Surendran et al. 2018), into Ethiopia in 2017 (Carter et al. 2018), and into Djibouti in 2012 (Faulde et al. 2014) is a reminder that anopheline changes in distribution can continue to happen. Despite a narrow 30 km distance between the peninsula-like Mannar Island of Sri Lanka and the Indian mainland where *Anopheles stephensi* is abundant, expansion of the species had never been recorded. The Sri Lankan government was alarmed by this development because *Anopheles stephensi* is a competent vector of malaria, and it commonly develops in urban container habitats, raising concern about urban outbreaks of malaria in a country that eliminated transmission in 2013. The situation is worse in Djibouti. That country had been almost free of malaria, and the city of Djibouti had detected few *Anopheles* of any kind. Major malaria outbreaks began in the city in 2013, almost as soon as the mosquito was introduced. Current unpublished figures show over 10,000 cases per year and indicate that the mosquito has expanded into the adjacent countries of Ethiopia and Somalia. In both countries, specimens were confirmed morphologically and by ITS2 and COI sequences.

Aedes japonicus has only recently started to expand its range into North America, Europe, and Hawaii. There was some indication of its invasiveness in the early 1990s, when it was intercepted in tires shipped to New Zealand (Andreadis et al. 2001). The first specimens appeared in the United States in 1998, and by 2000 the species was widespread. It appeared in Ontario, Canada, in 2001 (Thielman & Hunter 2006) and subsequently in Europe. Early in the American infestation, genetic evidence indicated two separate introductions, probably from Japan, in central Pennsylvania/Maryland and in New York/Connecticut/New Jersey. By 2000 there were indications of new genetic combinations in the United States (Fonseca et al. 2001). A more detailed genetic study undertaken on specimens captured between 1999 and 2005 showed that more and more new genetic combinations appeared, tending to obscure the original genetic bottlenecks (Fonseca et al. 2010). Perhaps as in the case of *Aedes atropalpus*, *Aedes japonicus* expanded from its original sylvan larval habitat that was mainly in rock pools and natural containers. Adaptation to artificial containers in its indigenous range may have led to its exportation in tires. The next step in its expansion may have depended on stochastic factors, bringing together sufficient exportation events to establish genetically viable populations. Once those populations became established, they may have produced new genetic combinations that were better adapted to the invasion and establishment cycle, as has been documented for fire ants (Ascunce et al. 2011). Subsequent invasions might have been expected to occur more rapidly as further exportation spread the preadapted invasive population.

The expansion of *Aedes albopictus*' range is the most thoroughly studied mosquito invasion (Table 3.10). Human transport of used tires initiated the infestations in the late 1970s, supplemented later by large-scale trade in lucky bamboo, a *Dracaena* species of ornamental plant that was shipped at the time in water. *Aedes albopictus* rapidly appeared across the globe. Impressively, the species colonized the entire country of Brazil between 1986 and 2004, as well as much of the eastern United States. Models of detailed climate and topographic features have indicated that even more of the world is at risk of infestation (Benedict et al. 2007), with the implication of associated disease risk (Charrel et al. 2007). Models of potential distribution have been based on temperature (e.g., ≥0°C mean temperature in January, ≥11°C mean annual temperature, and minimum of 1,350 day degrees from a base of 11°C in Italy), showing a variable distribution at a fine scale because of altitude variation (Neteler et al. 2011). Small changes in climate can make a big difference in local distribution at the margins of the potential *Aedes albopictus* range. When the first discovery of the species was reported in the Netherlands (52°N latitude) in imported lucky bamboo in 2005, it was predicted that the mosquito could not become established there based on a model requiring at least 186 days at greater than or equal to 11°C (Scholte et al. 2007). Two years later it was clear that *Aedes albopictus* had become established in the Netherlands and that changes in mean January temperatures of up to 2°C made a big difference in the extent of the country potentially infested (Takumi et al. 2009).

Table 3.10. Progression of global invasion by *Aedes albopictus*, 1979–2004 (Benedict et al. 2007)

First year infested	Country
1979	Albania
1983	Trinidad
1985	United States
1986	Brazil
1988	Mexico
1990	Italy
1991	Nigeria
1993	Barbados, Dominican Republic
1995	Cuba, Guatemala, Honduras, El Salvador
1997	Bolivia, Cayman Islands
1998	Argentina, Colombia, Paraguay
1999	France
2000	Cameroon
2001	Equatorial Guinea
2002	Panama
2003	Nicaragua, Greece, Israel, Switzerland
2004	Belgium, Spain

Detailed study of *Aedes albopictus* during its global invasion has revealed aspects of genetic adjustments that are probably representative of other mosquitoes during the years of establishment in new environments. Multiple introductions in Italy followed by genetic drift resulted in 18 populations distinguishable by allozyme patterns just 10 years after the first introduction (Urbanelli et al. 2000). Lounibos et al. (2003) found up to 10% occurrence of diapausing ability in *Aedes albopictus* from equatorial Brazil, suggesting that even fully tropical populations retain individuals with this genetically based capability. In the United States, there was a strong correlation between latitude and diapausing ability, showing that this invasive population had adjusted to local conditions on a continental scale within about 15 years. The best-fitting ecological niche models based on the invaded ranges in North America and Europe are different from the model in the native range, also suggesting that rapid genetic adaptational change has occurred (Medley 2010). Other changes involving body size, fecundity, and survivorship have also been suggested, though both the patterns and evolutionary explanations are less clear (Armbruster & Conn 2006; Leisnham et al. 2008). Even closer study would probably reveal additional patterns of phenological traits sorted by new ranges of environmental conditions. The idea that these changes occur within one or two decades is remarkable in comparison to the long-term evolutionary changes in mosquitoes that take place over thousands of years.

Aedes albopictus also provides examples of how an invasive species may interact with existing species in its new range. Competition with existing container species in Louisiana may have been reduced by *Aedes albopictus'* preference for sites close to the ground, whereas *Aedes triseriatus* and *Aedes hendersoni* seek oviposition sites up to 5 m above the ground (Schreiber et al. 1988). *Aedes albopictus* in both the United States and Brazil has tended to outcompete the previously introduced, invasive container species *Aedes aegypti*. Characteristics of *Aedes albopictus* proposed to explain how it has replaced *aegypti* in much of the Western Hemisphere include more efficient feeding, aggressive mating with *Aedes aegypti* females, and resistance to a gregarine parasite that harms *Aedes aegypti* (Juliano & Lounibos 2005). Under some conditions, *Aedes albopictus* have caused a disadvantage to *Aedes aegypti* and *Aedes triseriatus* by eating bacteria off their eggs, inhibiting hatching (Edgerly et al. 1993). Low-nutrient conditions harmed mixed larval cultures of *Aedes albopictus*, *Aedes aegypti*, and *Aedes triseriatus* more than when the species were not mixed (Barrera 2008). A different kind of stress, serial drying conditions, was also less tolerated by mixed colonies of *Aedes aegypti* and *Aedes albopictus*, though in this case *Aedes albopictus* was more severely affected (Costanzo et al. 2005). *Aedes albopictus* eggs were more susceptible to drying than those of *Aedes aegypti* (Lounibos et al. 2010), which may explain persistence of *aegypti* in southern Florida and in Texas, as well as why the species coexist in Southeast Asia (Kittayapong & Strickman 1993).

Extinction

Introduction, establishment, and genetic adaptation are the future for invasive mosquito species. The opposite process is extinction from existing ranges. The literature on mosquito extinction is limited because there are not many cases to discuss. Just as people have been responsible for most mosquito introductions, humans have also been responsible for most known cases of local extinction. As discussed previously, *Aedes aegypti* was eradicated from Brazil in the 1930s (Soper et al. 1943) and in at least 15 other Latin American countries by 1961 (Soper 1963). *Anopheles arabiensis* (reported as *Anopheles gambiae*; Parmakelis et al. 2008) was eradicated from Brazil in 1940 (Soper & Wilson 1943), following its introduction in 1930 (Shannon 1932). The same species was eradicated from Upper Egypt in 1945, following its introduction in 1942 (Shousha 1948). Less well-documented cases of eradication or disappearance from unknown causes included *Anopheles funestus* from Mauritius sometime after 1868, *Anopheles albimanus* from Barbados between 1927 and 1943, *Anopheles gambiae* from Indian Ocean islands in the 1970s, and *Aedes aegypti* from the Mediterranean Basin in 1960 (Mouchet et al. 1995). Human activity may reduce mosquito populations without eliminating them, sometimes intentionally, as is common in mosquito abatement, and sometimes unintentionally, through landscape modification. This was probably the reason for the near disappearance of *Anopheles atroparvus* in the Netherlands as saltwater habitats were eliminated in flood control efforts from 1935 to 1999 (Takken et al. 2002).

Several observations suggest that local extinction is actually common as mosquitoes enter new ranges. For example, there are records of unintentional introductions that failed to establish, like *Anopheles punctipennis* in Hawaii (Leong & Grace 2009), and intentional introductions for biological control that failed to persist (e.g., *Toxorhynchites inornatus*, *theobaldi*, and *brevipalpis* in Hawaii; Goff 1980). Local extinction has also been observed under natural conditions on a small geographic scale. *Tripteroides bambusa* disappeared from some sites during a four-year study of its occurrence in a series of 113 bamboo groves in northern Kyushu, Japan (Sunahara & Mogi 1998). Local extinction was more likely in smaller groves, presumably because drier periods eliminated all potential larval sites. A similar dynamic relationship between colonization and extinction of mosquitoes was found in tree holes in Florida modified by predation (Bradshaw & Holzapfel 1983; Ellis et al. 2006). Most mosquitoes probably expand their ranges through normal dispersal but are then excluded from some or all of the new range as conditions in a marginally suitable habitat change for the worse.

Entomologists are often asked whether mosquitoes serve an ecological "purpose" and what would happen if there were no mosquitoes. Considering the greater than 100-million-year history of Culicidae, it seems unlikely that humans will outlast mosquitoes. At least one author has speculated that the ecological consequences of mosquito elimination would be minor and geographically limited (Fang 2010). Range expansion and invasion of new geographic areas seem much more likely than extinction. Human activity has tended to support species of mosquitoes adapted to those conditions, as seen for geographical expansion of many species (Scheffer 2005). Mosquito biologists have often remarked on the greater likelihood of invasion by container-developing species and by aedine species with environmentally resistant eggs, yet many species with those qualities have not expanded beyond their indigenous ranges. A practical quantitative evaluation as performed for plants (Schmidt et al. 2012) would be a sensible precaution in light of what appears to be an acceleration of invasion. Medical entomologists, in particular, should be sensitive to the concept that nature is not always benign and that human activity can create unpredicted harm to us (Dunn 2010).

Climate Change

Climate change (Table 3.11) will certainly alter the distribution of mosquitoes. Risk estimates of mosquito-borne pathogens are of obvious concern to the medical and veterinary communities, but a key element of those estimates is the future distribution of vector species. Even a passing familiarity with insects makes obvious the connections between weather, climate, and insects' distribution. The summary result of many individual effects (Table 3.12) is likely to be extension of species' ranges into areas and during seasons that were previously too cool for development. No doubt, many changes in distribution will follow this simple pattern, but the interaction of mosquitoes with other species and geographic effects such as coastal flooding, as well as unexpected adaptational effects, will probably provide many surprises for mosquito biologists.

The realization that human-induced atmospheric changes have been modifying the natural, astronomically driven

Table 3.11. Summary of climate change effects quoted from the Intergovernmental Panel on Climate Change (Stocker et al. 2013)

Observed changes in the climate system
Warming of the climate system is unequivocal, and since the 1950s, many of the observed changes are unprecedented over decades to millennia. The atmosphere and ocean have warmed, the amounts of snow and ice have diminished, sea level has risen, and the concentrations of greenhouse gases have increased.

Atmosphere
Each of the last three decades has been successively warmer at the Earth's surface than any preceding decade since 1850. . . . In the Northern Hemisphere, 1983–2012 was *likely* the warmest 30-year period of the last 1400 years (*medium confidence*).

Ocean
Ocean warming dominates the increase in energy stored in the climate system, accounting for more than 90% of the energy accumulated between 1971 and 2010 (*high confidence*). It is *virtually certain* that the upper ocean (0 – 700 m) warmed from 1971 to 2010 . . . and it *likely* warmed between the 1870s and 1971.

Cryosphere
Over the last two decades, the Greenland and Antarctic ice sheets have been losing mass, glaciers have continued to shrink almost worldwide, and Arctic sea ice and Northern Hemisphere spring snow cover have continued to decrease in extent (*high confidence*).

Sea level
The rate of sea level rise since the mid-19th century has been larger than the mean rate during the previous two millennia (*high confidence*). Over the period 1901 to 2010, global mean sea level rose by 0.19 [0.17 to 0.21] m.

Carbon and other biogeochemical cycles
The atmospheric concentrations of carbon dioxide, methane, and nitrous oxide have increased to levels unprecedented in at least the last 800,000 years. Carbon dioxide concentrations have increased by 40% since pre-industrial times, primarily from fossil fuel emissions and secondarily from net land use change emissions. The ocean has absorbed about 30% of the emitted anthropogenic carbon dioxide, causing ocean acidification.

Drivers of climate change
Total radiative forcing is positive, and has led to an uptake of energy by the climate system. The largest contribution to total radiative forcing is caused by the increase in the atmospheric concentration of CO_2 since 1750.

Understanding the climate system and its recent changes
Human influence on the climate system is clear. This is evident from the increasing greenhouse gas concentrations in the atmosphere, positive radiative forcing, observed warming, and understanding of the climate system.

Evaluation of climate models
Climate models have improved since the AR4. Models reproduce observed continental-scale surface temperature patterns and trends over many decades, including the more rapid warming since the mid-20th century and the cooling immediately following large volcanic eruptions (*very high confidence*).

Quantification of climate system responses
Observational and model studies of temperature change, climate feedbacks and changes in the Earth's energy budget together provide confidence in the magnitude of global warming in response to past and future forcing.

Detection and attribution of climate change
Human influence has been detected in warming of the atmosphere and the ocean, in changes in the global water cycle, in reductions in snow and ice, in global mean sea level rise, and in changes in some climate extremes. . . . This evidence for human influence has grown since AR4. It is *extremely likely* that human influence has been the dominant cause of the observed warming since the mid-20th century.

Future global and regional climate change
Continued emissions of greenhouse gases will cause further warming and changes in all components of the climate system. Limiting climate change will require substantial and sustained reductions of greenhouse gas emissions.

Atmosphere: temperature
Global surface temperature change for the end of the 21st century is *likely* to exceed 1.5°C relative to 1850 to 1900 for all RCP scenarios except RCP2.6. It is *likely* to exceed 2°C for RCP6.0 and RCP8.5, and *more likely than not* to exceed 2°C for RCP4.5. Warming will continue beyond 2100 under all RCP scenarios except RCP2.6. Warming will continue to exhibit interannual-to-decadal variability and will not be regionally uniform.

Atmosphere: water cycle
Changes in the global water cycle in response to the warming over the 21st century will not be uniform. The contrast in precipitation between wet and dry regions and between wet and dry seasons will increase, although there may be regional exceptions.

Ocean
The global ocean will continue to warm during the 21st century. Heat will penetrate from the surface to the deep ocean and affect ocean circulation.

Cryosphere
It is *very likely* that the Arctic sea ice cover will continue to shrink and thin and that Northern Hemisphere spring snow cover will decrease during the 21st century as global mean surface temperature rises. Global glacier volume will further decrease.

Sea level
Global mean sea level will continue to rise during the 21st century. . . . Under all RCP [Representative Concentration Pathway] scenarios, the rate of sea level rise will *very likely* exceed that observed during 1971 to 2010 due to increased ocean warming and increased loss of mass from glaciers and ice sheets.

Carbon and other biogeochemical cycles
Climate change will affect carbon cycle processes in a way that will exacerbate the increase of CO_2 in the atmosphere (*high confidence*). Further uptake of carbon by the ocean will increase ocean acidification.

Climate stabilization, climate change commitment, and irreversibility
Cumulative emissions of CO_2 largely determine global mean surface warming by the late 21st century and beyond. . . . Most aspects of climate change will persist for many centuries even if emissions of CO_2 are stopped. This represents a substantial multi-century climate change commitment created by past, present and future emissions of CO_2.

Table 3.12. Potential influences of climate change on mosquitoes (Epstein et al. 1998; Dukes and Mooney 1999; Gubler et al. 2001; Parmesan 2006; de la Rocque et al. 2008; Gage et al. 2008; Holderegger and Wagner 2008; Lafferty 2009; Hoffmann and Sgrò 2011; Peterson and Lieberman 2012; Chen et al. 2013)

Effects directly on the mosquito
Developmental
Direct physiological effects
Developmental period
Longevity and survival
Symbiotic
Susceptibility to infection
External incubation period
Behavioral
Host selection
Blood feeding rate
Effects on groups of individuals and populations
Genetic
Population genetic structure
Genetic adaptation
Hybridization
Phenological
Seasonality
Generations per year
Timing of life cycle events
Influence of weather events
Populational
Population growth
Mass outbreaks
Indirect effects on mosquitoes
Geographic
Barriers formed by areas of unsuitable habitat
Population relationships to El Niño-Southern Oscillation cycle
Habitat structure
Landscape
Destruction of high-altitude habitats
Habitat fragmentation
Ecosystem shifts caused by changes in flora
Ecological
Distribution of blood meal hosts
Biodiversity
Conditions favoring colonizing species (*r* strategists)
Conditions that favor most adaptable pathogens and symbionts

climate cycle has spawned extensive study, from the most basic (Molod et al. 2012) to rough-scale summaries intended to incentivize (Portier et al. 2010). Governments have invested heavily in predictive, mechanistic models known as general circulation models (GCMs). These are based on advanced atmospheric physics and involve extensive use of the fastest, largest computers (Vijayavenkataraman et al. 2012). Efforts to test and coordinate these models are highly organized under large, intergovernmental frameworks (Taylor et al. 2012). One of the important categorizations for mosquito biologists has been representative concentration pathways (RCPs) that designate specific atmospheric scenarios based on anticipated extra energy on the surface of the earth caused by increases in greenhouse gases. Four scenarios have been chosen as most useful (Stocker et al. 2013): RCP 2.6 (2.6 extra watts per square meter, 421 ppm carbon dioxide (CO_2) by 2100, peaks before 2100 and then declines), RCP 4.5 (4.5 W m^{-2}, 538 ppm CO_2, continues increasing past 2100), RCP 6.0 (6.0 W m^{-2}, 670 ppm CO_2, continues increasing past 2100), and RCP 8.5 (8.5 W m^{-2}, 936 ppm CO_2, continues increasing past 2100). Various efforts to predict future mosquito distributions use RCPs as the basis for estimates of future physical conditions.

Just as Pleistocene glaciations were the big drivers of distribution in the past (Peterson & Lieberman 2012), will climate change be the big driver in the future? According to some authors, climate change-induced evolutionary and range changes have already taken place (Epstein et al. 1998; Parmesan 2006; Chen et al. 2013). Changes in distribution in response to temperature and precipitation are obvious consequences of climate change. The most disruptive effects may involve more complicated interactions. A changing biological world is likely to result in the fragmentation of landscapes, disrupting current patterns of adaptation and species interaction (Holderegger & Wagner 2008; De la Rocque et al. 2008). Species restricted to areas that do not support their effective population size (N_e, the theoretical minimum population size required to maintain the current distribution of alleles) may experience extensive genetic drift or local extinction (Hoffmann & Sgrò 2011). *Aedes japonicus* may present an example of how connected patches of distribution can result in rapid adaptation to new climate conditions, as this species exchanged genetic characteristics between two invaded locations (Virginia and Hawaii) during a period of just seven years (Egizi et al. 2015). Disruptions of landscapes may create large areas that are essentially disturbed habitats, changed by physical forces of climate change and by biological forces of ecological change. That situation could create a world in which invasion by *r* strategist species is much more common and where selective pressure tends to shift species from lower reproductive rates and specialized habitats toward higher reproduction and widely colonizing dispersal (Dukes & Mooney 1999). Human activity could be a big part of those trends, as economic disruption brought about by climate change would tend to disorganize communities, further disturbing habitats (Gubler et al. 2001).

Several studies have modeled the effects of projected climate change on mosquitoes and associated pathogens. Although not entirely about mosquitoes, these predictions of disease risk nonetheless imply changing levels of exposure to vectors of dengue (Hales et al. 2002). More extensive work has been done to model the extent and distribution of malaria risk in Africa—probably because malaria prevention there is a major global health effort. Mouchet et al. (1998) examined the redistribution of malaria during the previous 20 years, during which average temperatures increased 0.5°C, and concluded that temperature significantly changed only the altitude at which malaria occurred and that the major changes were due to rainfall and human management of the disease. Rogers and Randolph (2000) also found little predicted change based on the assumption of 1% annual increases in greenhouse gases through 2050. Ebi et al. (2005) performed a more detailed study of Zimbabwe using four different GCMs, two of which supported an indication of large increases in malaria by 2100 in highland areas if greenhouse gases reach 750 ppm. A different analysis of Botswana (Thomson et al. 2006) used the European GCM and interpretive software to show that precipitation was more influential than temperature. Parham and Michael (2010) used a model of mosquito and parasite development to show that risk was greatest at moderate temperature and rainfall. Although these examples of efforts to predict future malaria risk do not completely agree with each other, they show that risk of this disease is dependent on factors that change over relatively small distances, including the abundance of vector mosquito species. Although climate change would probably result in an overall restriction of major malaria vector species in Africa, malaria risk might increase because of shifts of distribution toward areas with more people (Peterson 2009).

Mosquitoes that develop in containers are also of special concern because of their vector status and because of their recent worldwide dispersal. In a simple historical analysis of mosquitoes in New Jersey from 2002 through 2010, Rochlin et al. (2012) concluded that wetter summers had favored *Aedes japonicus* and warmer winters had favored *Aedes albopictus*, reducing numbers of the indigenous, container-developing species *Aedes triseriatus* by 81% through competition. Predictions of *Aedes aegypti* distribution based on mechanistic developmental models for that species showed higher altitude expansion in Latin America and northward expansion in eastern China, the United States, and, potentially, in Europe (Hopp & Foley 2001). An evaluation of Australia showed how consideration of microhabitat conditions in containers influenced potential distribution by analyzing climate change effects on egg survivorship of *Aedes aegypti* (Kearney et al. 2009). Combining thermal effects on egg mortality with a mechanistic developmental model, these authors predicted a possible northward expansion where rainfall would increase and a southward expansion where temperature would increase.

Thanks to relatively abundant data and intense interest, analyses of potential shifts in the range of *Aedes albopictus* have been particularly well developed. Fischer et al. (2011) developed a Maxent-based species distribution model based on temperature and precipitation, as well as a model based on expert opinion of suitable conditions for this species. These models were then applied to conditions predicted in the European GCM, showing that suitable, uninfested habitats exist even in the present climate. Considerable expansion could occur in Europe with climate change, though the Mediterranean region might become too dry for *Aedes albopictus*. Rochlin et al. (2013) predicted similar changes in the northeastern United States, nearly doubling the number of people exposed to the mosquito.

Concern over West Nile virus has stimulated predictions of *Culex* distribution. Morin and Comrie (2010) used a mechanistic developmental model of *Culex quinquefasciatus* to compare the effects of climate change in Florida and Southern California. An earlier, warmer spring in the desert environment of California combined with hotter, drier summers would shift populations to an earlier seasonality. In contrast, the study showed few changes in Florida. These results may be indicative of the subtler effects of climate change to be expected at a distance from the margins of current distributions. In contrast, the distribution of *Culex tarsalis* in Canada could expand dramatically northward, especially if warmer, drier conditions extended the species' preferred grassland habitats (Chen et al. 2013). Similarly, *Culex pipiens* might expand its range in Canada over twofold by 2100 under a low greenhouse gas scenario and over fivefold under a high greenhouse gas scenario (Hongoh et al. 2012).

Conclusion

The distribution of mosquitoes has never been stable, is not currently stable, and will not be stable in the future. The family Culicidae has evolved to occupy a wide range of aquatic habitats as larvae, and most species remain remarkably generalist predators as adults, with the capability of both self-powered and phoretic dispersal. As a result, some species in most parts of the world are able to tolerate or even benefit from changes brought about by climate change, human-induced modifications of landscape, or other environmental shifts. Epidemics of mosquito-borne viruses (West Nile, yellow fever, dengue, Zika, and chikungunya) have been attributed to major shifts in distribution of people to urban centers, as well as to increases in human movement (Marcondes et al. 2017). Although much of the current distribution of mosquitoes is dominated by historical events associated with glaciations, it is worth remembering that the family was present in substantially the same form well before the current glacial age even began. Extinction is an important force in the development of any group of organisms, but current trends seem to point toward diversification, redistribution, and competitive readjustments of abundance. Thanks to considerable effort and scientific progress during the last 20 years, our ability to quantify and extrapolate current distribution and to predict future distribution has improved tremendously. Further application of the techniques will continue to provide a more accurate picture of what the world of mosquitoes will be in 100 years. However, genetic adaptation and events leading to invasion are largely unpredictable and powerful determinants of distribution. Beyond 100 years, mosquitoes will continue to occupy a changing array of habitats generated by human and natural forces. The few documented examples of rapid genetic adaptation may be a more general indication that mosquitoes will be able to adjust to conditions that seem disastrously disruptive to the human species.

4. Development

The insect larva has no idea of the meaning of its life or of what is to become of it. . . . Yet, physiologically, the larva is loaded with responsibilities. Not only must it maintain its own existence, but at the same time it must provide for the future nutritional needs of the pupa and for its transformation to the adult.

Snodgrass 1959

This charming quote by the old master of insect morphology R.E. Snodgrass uses a teleological analogy to point out the total dependence of mosquito survival on the intricate, successful development of each stage in preparation for the next stage. Such development is essentially a feat of engineering carried out at minute scales and at a level of detail that must equal or exceed the largest engineering projects designed by people. Loosely categorized as "development," the building of each individual mosquito is elegant, as it produces a series of stages with amazing capabilities. Development is also incredibly wasteful, in that a stable mosquito population is the result of the destruction of more than 99% of those carefully constructed bits of chitin, protein, and lipid.

The study of mosquito development started early, thanks to the obvious changes in morphology between egg, larva, pupa, and adult. Illustrations by Réaumur (1738) accurately depicted each stage and represented a widespread recognition of metamorphosis by naturalists of the time. Author after author has contributed to the literature on morphology and development of mosquitoes, culminating in some reviews used extensively in this chapter. Snodgrass (1959) turned his attention to mosquitoes, documenting many insights into how their development compared to other dipterans and to insects in general. Harbach and Knight (1980, 1981) provided an exhaustive illustrated (both drawings and scanning electron micrographs) review of mosquito morphology that emphasized terminology and the history of the terminology. Finally, Clements (2000) updated his previous work (Clements 1963) in a thorough review of development from a physiological viewpoint.

The term "development" encompasses a wide segment of mosquito biology. First, there is the descriptive aspect of the field, documenting life cycles and morphology in many different species. Using a broad scientific basis in morphology, taxonomists have provided the most thorough descriptions of the full range of species in Culicidae through two centuries. Observations on the reproductive rate, time required for development of each instar, and rates of survival—often included in the general category of "bionomics"—have also been studied for a relatively large number of species. Those kinds of studies may include experimental tests of hypotheses, but the results have been mainly descriptive. Studies of the developmental functions of behavior, biochemistry, morphology, and genetics have involved much experimentation and inventive use of technique. Of course, recent years have seen a strong emphasis on genetics, resulting in an explosion of insights on structure and function of the genome.

Research on development of mosquitoes currently emphasizes genetics, which has created a large community of scientific practice and a rapid increase in knowledge. Similar to biochemical and physiological work that began with Wigglesworth's (1930) use of *Aedes aegypti* as a model organism, modern genetic studies tend to emphasize relatively few species selected for their medical importance. One of the challenges for the future is to expand this understanding to many more species, much as has been done in purely descriptive studies of morphology associated with taxonomy. A more subtle challenge, and a more speculative one, is whether the mosquito research effort can keep up with the changes occurring in the mosquitoes themselves. As discussed in Chapter 1, careful study of a few species has shown what appears to be rapid change in response to new selective pressures. Finally, and not so much a challenge as a natural result of the accumulation of knowledge, mosquito biology would be well served by more detailed information on the lives of mosquitoes. Current common categories of larval development, pupal development, host seeking, oviposition, and more are at a rough scale that can probably come to be understood as the sum of many finer-scale aspects of mosquitoes' lives. Such an understanding would provide useful context for all of mosquito biology.

Descriptions of Development

Developmental processes vary between species of mosquitoes, making it difficult to appreciate the range of differences. As might be expected, the most detail is available on those species that are easily manipulated in the laboratory, especially *Aedes aegypti*. The accounts of development given below also rely heavily on this species (Christophers 1960; Clements 1963; Clements 2000), mainly with notes on some of the variation observed in other common species. The variety in patterns and details of development probably reflects the large spectrum of lifestyles among species of mosquitoes.

Spermatogenesis

The path to development of sperm begins in the first instar as primordial germ cells. These diploid cells multiply throughout larval development, forming the primary and secondary spermatogonia. Spermatogonia mature into diploid spermatocytes, which undergo meiosis to form haploid secondary spermatocytes. A second meiotic event forms the spermatids. The spermatids are organized into cysts, with about 20 cysts per testis.

Spermatids appear as generalized haploid cells. In a process called spermiogenesis, they take on a very different internal and external architecture to become the sperm cells themselves, or spermatozoa. Spermatid mitochondria fuse to make a mass called the nebenkern, which in turn divides into ribbon-like elements associated with the developing axoneme. The head of the spermatozoa includes the haploid nucleus and an acrosomal vesicle that contains the enzymes necessary to eventually penetrate the cellular membrane of the oocyte. The flagellum of mosquito spermatozoa is unusual among insects in that the central core of the axoneme is a single solid fiber instead of the usual pair. Another strange feature of the mosquito sperm's flagellum is that it produces two superimposed waves along its length. Flagellar action can propel the spermatozoa at 25 μ/s, which translates to 1.5 mm/min, or about 60 millionths of a mile per hour.

Spermiogenesis begins in the pupal stage of *Aedes aegypti* but in the fourth instar of *Anopheles* and *Culex*. Each testis contains approximately 14,000 spermatozoa. Some mosquitoes continue spermatogenesis at a slow pace after emergence (*Aedes*, *Culex*) but others do not (*Armigeres*).

Oogenesis

The mosquito egg is a highly specialized structure, adapted to deposition on water, in container habitats above water, on plants above water or in the water, or on soil subject to flooding. The distinct structure is especially developed in aedine genera that experience delays in hatching and that may spend a significant period of time separated from water (Savage & Strickman 2004). Of course, any insect is completely dependent on the successful development of the egg. The mosquito egg has a number of different forms, depending on the lifestyle of the species.

The account of oogenesis that follows is taken from Clements (2000) except where otherwise noted. Like the spermatozoa, eggs start as primordial germ cells in the first instar. Chromosomal replication starts in the primary ovarian follicles two to three days before the adult female emerges from the pupa. At emergence, homologous chromosomes are tightly associated (pachytene chromosomes), and it is at this time that crossing over between chromosomal elements can occur. At 40–60 hours after the female's emergence, homologous chromosomes are tightly wrapped around each other as diplotene chromosomes. It is at this stage of ovarian development that some species (e.g., *Culex pipiens*) of mosquitoes enter diapause as female adults.

A blood meal initiates a rapid change in the oocytes through a series of hormonally controlled developmental gates (Table 4.1). The nucleus grows to about 400 μm long and partially invaginates. The anterior portion of the nucleus includes the chromosomes, forming the karyosphere or germinal nucleus. The remainder of the nucleus separates into the nucleolus, which is a site for RNA synthesis. The nucleolus breaks up into fragments, and these become distributed throughout the cytoplasm of the oocyte about 13–20 hours after a blood meal. Unlike other meroistic insect ovarioles (i.e., those with

Table 4.1. General stages and developmental gates of mosquito oocyte development (Clements 2000)

Phase	Christophers' stage	Developmental gate	Stimulus to next stage
Previtellogenic	G	Germarial gate	20-hydroxyecdysone
	1a		
	Ib	Stage 1 gate	Juvenile hormone
	IIa, IIb	Previtellogenic gate	Uncertain
Initiation	IIIa	Stage III gate	Ovarian ecdysteroidogenic hormone and 20-hydroxyecdysone
Trophic	IIIb–IVb		
Posttrophic	V	Maturation gate	Syngamy

Table 4.2. Morphological descriptions of Christophers' stages of mosquito oocyte development (Clements 2000)

Stage	Oocyte description	Nurse cell description
G	Not entirely surrounded by epithelium	
Ia	Surrounded by epithelium	Not different from oocyte
Ib	No lipid droplets	Distinguished from oocyte
IIa	Small lipid droplets	
IIb	Nuclear membrane and nucleolus visible, more lipid droplets	
IIIa	50% of length of follicle, numerous large lipid droplets	
IIIb	Larger	
IVa	Follicle egg shaped, 90% of space occupied by oocyte	10% of space occupied by nurse cells
IVb	Same shape as mature oocyte, chorion appears	Nurse cells disappear
V	Full length, follicular epithelium gone, chorion structure fully formed	

nurse cells), mosquito nurse cells contribute ribosomal RNA to the oocytes—at an increasing pace following the blood meal. At the end of the rapid development of the oocytes, proteins (vitellogenin) synthesized in the fat body from amino acids of the blood meal contribute to formation of the yolk. At this point, the karyosphere has broken down, and the cell is at metaphase I. The endochorion solidifies into a continuous layer between the oocyte and epithelial cells at this time. The oocyte appears morphologically to be at Christopher's Stage IV (Table 4.2), where it may remain for hours or days.

The posttrophic phase begins once the endochorion plaques fuse to form a single layer. The exochorion is deposited in the form typical of the species, secreted by the follicular epithelium and complete 60–72 hours after the blood meal. The oocyte can continue to absorb nutrients in this stage. Interestingly, the micropylar apparatus (pore for entry of sperm) is secreted separately from the rest of the exochorion.

Much happens inside the oocyte as it is oviposited and for about two hours after oviposition. First, one or two sperm cells are injected into the egg from the female's bursa copulatrix as it passes down the oviduct. Horsfall et al. (1973) described the process for *Aedes vexans*. Immediately after oviposition the oocyte appears to be pearly white, gradually darkening to black in four to five hours. Right after oviposition the endochorion surrounds the yolk matrix, and the cytoplasm is concentrated at the poles of the oocyte. At this point the micropyle is plugged by nuclei, and one or two tailless sperm are present in the cytoplasm. The first meiotic division occurs 30 minutes after oviposition, resulting in the oocyte's haploid pronucleus and a polar body. Within an hour, the second meiosis occurs, producing two polar bodies. One of the sperm fuses with the egg pronucleus about two hours after oviposition, and the oocyte is transformed into an egg, according to physiological definition. It is ironic that the transformation of the oocyte into an egg occurs after oviposition. Strictly speaking, the mosquito has a system of external fertilization even though the sperm are deposited internally into the female.

Sex determination of the developing embryo is dependent on the genetics of the sperm and the oocyte. *Anopheles* mosquitoes have a situation similar to humans, in which distinct X and Y chromosomes determine sex, and an XY pairing is male. The *Yob* gene controlling maleness in *Anopheles gambiae* was sequenced and shown to be necessary for male survival but fatal to females (Krzywinska et al. 2016; Sinkins 2016). In other genera, the chromosomal pairs appear identical, but there is a locus for sex determination on chromosome I. In *Aedes aegypti*, the region is called the M-locus and contains a male-determining gene called *Nix* (Hall et al 2015). A similar gene was found in *Aedes albopictus* but not in *Culex quinquefasciatus* or *Anopheles gambiae*. Sex ratios are usually close to 1:1 (e.g., *Culex apicinus*; Loetti et al. 2007), but important exceptions exist. For example, *Aedes aegypti* reared at 30°C have a sex ratio of four males to three females (Tun-Lin et al. 2000). Inherited factors in the same species can reduce the proportion of females to as little as 0% (Hickey & Craig 1966) through meiotic drive (unequal transmission of alleles; Graham et al. 2004). The mechanism for female bias in *Aedes triseriatus* resulted in up to 76% females and involved the sex locus as well as a series of quantitative trait loci. Sex distortion resulting in part male, part female individuals (gynandromorphs) can result from double zygotes or from male-linked intersex genes activated by high developmental temperatures (Clements 2000). Genetic males are heterozygous for the sex-determining locus, which causes the developing male larva to have primordial cells forming both testicular and ovarian organs. The incipient ovaries disappear from males in the fourth larval instar. This is the origin of feminization of genetic males by high developmental temperature (northern *Aedes* spp.) or fungal infection. Although such false females can be female in every detail, they are unable to oviposit due to testicular remnants blocking the oviduct.

Embryogenesis

Embryonic development does not begin until after oviposition. Although the external appearance of mosquito eggs varies considerably (e.g., Horsfall et al. 1952; Reinert 2005b), their basic structure is similar throughout the family. Christophers (1960) and Horsfall et al. (1973) gave detailed descriptions of the eggs of *Aedes aegypti* and *Aedes vexans*, respectively. The egg is generally fusiform in shape, with a blunter end that includes the micropyle corresponding to the anterior of the developing embryo. The flatter side of the egg starts out as the dorsal side of the embryo, but as the embryo develops, this becomes the ventral side. The eggs are usually deposited with the morphologically dorsal (flatter) side uppermost and the posterior extruded first, following the law of Hallez that states that the three axes of the egg correspond to those of its mother. Right at oviposition, the only membranes around the egg are the thin exochorion and the thicker endochorion. It is the endochorion that darkens and that both protects the egg from desiccation and provides a surface across which there is gas exchange.

The newly deposited egg quickly absorbs water from its surroundings, more than doubling its length and decreasing its density. Aedine eggs not deposited on the water nonetheless require moist conditions for the first 24–72 hours, developing the ability to resist desiccation at 24–48 hours.

The following summary of embryogenesis is taken primarily from the detailed study of *Aedes vexans* (Horsfall et al. 1973; Table 4.3), with some details extracted from Clements (2000). Embryogenesis is similar in at least four genera (*Anopheles*, *Culex*, *Culiseta*, and *Aedes*). The time required to reach each embryonic event varies widely between species and at different temperatures. For example, *Culex theileri* requires 411 hours for embryogenesis at 12°C but only 27 hours at 36°C.

Embryogenesis begins following the two meiotic divisions and fusion of the sperm with the female pronucleus within two hours after oviposition. The germinal nucleus or karyosphere is equatorial in *Culex pipiens* and *Aedes vexans* but anterior in *Anopheles maculipennis* and *Aedes aegypti*. Two mitotic divisions of the germinal nucleus occur during the next hour in *Aedes vexans*, resulting in four nuclei. Divisions increase much more rapidly in the next hour, with approximately 95 nuclei present four hours after oviposition. During the next two hours, divisions continue without forming distinct cells. At 10 hours after oviposition, the many nuclei have migrated to the periphery of the common cytoplasm, forming the incipient blastoderm. The periplasm is a distinct, dark-staining layer just inside the blastoderm. Divisions between blastoderm cells start to form, and nucleoli appear in the next hour. The cuboidal blastoderm cells are complete 14 hours after oviposition.

Table 4.3. Summary of embryogenesis in *Aedes vexans* (Horsfall et al. 1973)

Hours after oviposition	Event or structure	Germ layer origin
1	Polar body formation	
2	Fertilization (embryogenesis begins)	
2–3	First cleavage	
5	Pole cells, primary yolk	
6	Peripheral migration of cleavage nuclei	
13	Secondary yolk cells	
15	Germ band formation	
18.5	Amnion, serosa, germ band elongation, gastrulation	
21	Segmentation	
26	Proctodeal and stomodeal invagination	
31	Antennae, mandibles, maxillae	
36–46	Germ band contraction, blastokinesis, vitelline membrane, labrum	
46	Midgut	Mesoderm
46	Pharynx, brain, Malpighian tubules, anal papillae, respiratory siphon	Ectoderm
51	Gonads	Pole cells
51	Hatching spine	Ectoderm
56	Labral musculature	Mesoderm
56	Oenocytes	Ectoderm
66	Frontal ganglion, corpora cardiaca, corpora allata	Ectoderm
71	Trachea in body, salivary glands	Ectoderm
71	Fat body	Mesoderm
81	Gastric caecae, thoracic and abdominal muscles	Mesoderm
91	Cuticle on body	Ectoderm

Other embryonic events going on about this time include production of cleavage nuclei from primordial germ cells, which develop from the eighth nuclear division. These energid cells migrate to the posterior of the egg outside of the blastoderm about five hours after oviposition, forming 12–30 pole cells. At 15 hours these pole cells become firmer and more spherical, then coalesce into a single polar body. Yolk cells (vitellophages) also originate from the eighth division about this time. They are ameboid cells scattered amid lipoprotein granules and lipid droplets that make up the yolk granules. Some of the yolk cells are produced by the blastoderm (secondary vitellophages), and some yolk cells become incorporated into the blastoderm. Eventually, the yolk cells transform so that the nucleus makes up most of the mass, and they migrate to the periphery of the yolk, forming a loose covering that is not a complete yolk sac.

Differentiation of cells within the blastoderm begins about 15 hours after oviposition. A dorsal band of columnar cells forms (the germ band), which will become the embryo. Upper lateral cells of the blastoderm are flatter and will become the amnion. The lower lateral and ventral cells become the serosa. At 16 hours after oviposition, the posterior end of the blastoderm invaginates, the pole cells occupying a dorsal position with the invagination. At 18.5 hours the germ band extends from the anterior of the egg along the dorsum, bends around the posterior of the egg, and ends about halfway to the anterior end of the egg on the ventral side. The cells differentiate into a central cord of mesoderm and lateral ectoderm, the mesoderm flattening into a two-cell layer. At this point, the pole cells are at the posterior tip of the germ band. The amnion now encloses the ventral portion of the germ band, forming the amniotic cavity. Highly attenuated cells of the serosa enclose the contents of the entire egg and later secrete the serosal cuticle.

At 21 hours, the ventral position of the germ band extends three-quarters of the way toward the anterior of the egg. Initial segmentation is visible, forming 19 segments: a pregnathal segment that will form the labrum, antennae, and cephalon; three gnathal segments (mandibular, maxillary, and labial); three thoracic segments; and 10 abdominal segments.

Gastrulation occurs at 31 hours. Strangely, gastrulation forming the digestive tract does not involve invagination of the mesoderm. Instead, mesodermal cells simply migrate toward the venter.

A process known as blastokinesis occurs from 36 to 40 hours after oviposition. The germ band contracts, rotates 180°, and migrates to the ventral side of the egg. Morphological changes occur that begin to make the embryo more recognizable as a larva. The anterolateral and lateral cephalic pouches evaginate to form preoral and cephalic lobes. The hind gut and foregut form through invagination of ectoderm at either end of the germ band. Segments 17–19 fuse to form the eighth abdominal segment, with segment 18 forming the siphon and segment 19 forming the anal segment. The three gnathal segments migrate anteriorly to be positioned under the cephalic region. Also about this time, the serosa secretes the noncellular vitelline membrane (serosal cuticle) lining the interior of the endochorion.

Up to this point, the germ band is separate from the yolk material. During the next 30 hours (46–76 hours after oviposition), the germ band successively grows dorsally to surround the yolk material, progressing from anterior to posterior. The mesodermal midgut then surrounds the yolk material. It is about this time that muscles start to form. During this process, the serosal and amniotic cells clump along the dorsum of the thorax and the anterior abdominal segments, forming a structure called the dorsal organ that disappears by 71 hours after oviposition.

Many of the elements of the mouthparts and gut have been formed by 76 hours. These include the mouth brushes, maxillary palpi, proventriculus, Malpighian tubules, and anal papillae. Much of the nervous system has also formed by this time, and the protocerebrum and deutocerebrum are distinguishable. All the ganglia (three thoracic and eight abdominal) are present, as well as the neurosecretory organs (corpus cardiacum and corpora allata). By the end of this period, the eyes begin to form from ectodermal invaginations, the musculature is largely complete, the fat body is present, and the heart is formed.

The cuticle, tracheal system, and gonads are complete in the last stages of embryogenesis. Cuticular secretion from epidermal cells starts at either end of the embryo as early as 56 hours; however, cuticle is not completed until 96 hours. Similarly, the tracheae begin formation in the siphon region at 46 hours, but tracheal tubes are not differentiated until 91 hours. The pole cells form embryonic gonads in association with mesodermal cells starting at 51 hours. At 96 hours there are 10 primordial cells in each gonad.

Larval Development

As a technicality, the true first instar might be considered as starting with secretion of the serosal cuticle while the embryo is still developing. For the purposes of this discussion, we will consider larval development to start at hatching. Much of this account of larval development is taken from Christophers' (1960) detailed description of *Aedes aegypti* and from Clements' (2000) review. Of historical interest, Réaumur (1738) described the larval and pupal stages of *Culex pipiens* in the first half of the 18th century.

Unlike some related families of flies (e.g., Chaoboridae [Strickman 1980a]), mosquito eggs hatch as a distinct cap separates from the anterior end of the chorion. Although a line of weakness or a seam is not visible, the cap breaks open in the same place and with distinct, sharp demarcation. The break is initiated by a single puncture from the egg-breaker organ of the first instar. This organ is located between the ocelli and consists of a sclerotized, pointed plate surrounded by a membrane. The first instar is considerably compressed before it emerges from the egg, with the three thoracic segments and seven of the abdominal segments occupying only one-third of the length. The larva expands rapidly as it swallows water during hatching. Remarkably, the tracheal system contains no air until later. Within one or two hours, the head swells around the mouthparts.

The larva is not purely a larva even at the earliest stage of development. Starting with the pole cells that form gonadal tissue, the larva is a mixture of larval structures that will not be present in the adult (e.g., musculature, antennae), incipient adult structures in predestined aggregations of cells, and larval structures that will survive the transition to adulthood (Malpighian tubules, nervous system, and fat body). Predestined groups of cells may form unstructured aggregations, called histoblasts, or organized aggregations, called imaginal discs, that have preliminary structures resembling the organs they will be in the adult. These cells often have multiple copies of chromosomes (polytene or endopolyploid), allowing them to develop into many cells during somatic reduction in the pupal stage. Imaginal structures become conspicuous in the third and fourth larval instars when columnar groups of imaginal cells evert into pouch-like structures, from which adult structures eventually form in a process almost like embryogenesis within the larva. For example, the adult legs appear as pear-shaped sacs at the bases of the pleural setae in the third instar. The adult legs evert from these sacs by the end of the fourth instar, as also happens with the wings, halteres, adult antennae, and adult mouthparts. Later in the fourth instar, the larval cuticle separates progressively from the adult structures, forming what has been referred to as a "prepupa." The proportion of time spent in each instar varies, with the fourth instar and pupa dominating (Table 4.4; Slater & Pritchard 1979; Rueda et al. 1990).

Ecdysis, or molting, is the process of changing from one larval instar to the next. Most of the larval structure is flexible, but the head, siphon, and setae cannot expand appreciably in size, so they appear relatively smaller in comparison to the rest of the larva as it grows during an instar. Molting allows these kinds of hard structures to expand, as they are briefly soft and expandable before new cuticle sclerotizes. Ecdysone is the steroidal hormone that stimulates ecdysis. The bioactive form of the hormone is 20-hydroxyecdysone (20E). The source of ecdysone may vary between different species of mosquitoes (Whisenton & Bollenbacher 1986; Jenkins et al. 1992; Clements 2000). Other insects produce ecdysone in the prothoracic gland cells surrounding the corpora allata, stimulated by the cerebral peptide prothoracicotropic hormone (PTTH). Mosquito larvae and pupae produce it from the abdominal and thoracic wall (Riddiford et al. 2000). The 20E hormone primes the central nervous system to react to two different peptide hormones (ecdysis triggering hormone, or ETH, and pre-ecdysis triggering hormone, or PETH) produced by Inka cells associated with the junctions of major tracheal trunks (Dai & Adams 2009). Alpha-ecdysone directly stimulates cell division and differentiation of dermal cells. As 20E diminishes in concentration, ecdysis changes from dissolution of the old cuticle to synthesis of the new one. The final stages of ecdysis are mediated by another set of genes, adding to the potential control points and intricacy of the molting process (De Loof 2008).

Table 4.4. Calculation of the proportion of time spent in each aquatic instar for *Culex quinquefasciatus*, *Aedes aegypti*, and *Aedes vexans* based on rearing larvae at a range of temperatures from 15°C to 34°C (Rueda et al. 1990) and for *Aedes vexans* at 15°C to 25°C (Slater and Pritchard 1979)

Stage	*Culex quinquefasciatus*	*Aedes aegypti*	*Aedes vexans*
First instar	14%	19%	18%
Second instar	13%	14%	13%
Third instar	17%	17%	16%
Fourth instar	33%	27%	26%
Pupa	23%	23%	27%

One of the early signs of ecdysis is the separation of the cellular epidermis from the noncellular cuticle (apolysis). Fluids fill this space and break down the endocuticle. These materials are eventually used to build the new, larger cuticle. The larva emerges from its older cuticle quickly, starting from the head end. At this point, the larva is pharate—that is, the next instar resides in the cuticle of the previous instar. The cast-off cuticle, known as the exuvium, retains many of the external features of the larva so that the fourth instar exuvium is often mounted on a slide for morphological examination by taxonomists. Within a cohort of eggs, the order in which the larvae hatch influences the time required for larval development in *Aedes aegypti* (Hotchkin 1985). Eggs in the laboratory that hatched following a second stimulus (reduction in oxygen under vacuum) developed about a day quicker than those hatched following the first stimulus. Also, larvae from three-week-old eggs developed more quickly than those at other ages from 1 to 10 weeks.

The pupal stage marks the final transition of tissues to the adult form. Up until this time, each molt has been determined to lead to another larval stage by the presence of juvenile hormone, produced by the corpora allata in response to allatotropins in the brain and midgut (Riehle et al. 2002). Allatostatins downregulate juvenile hormone production, and specialized esterases catalyze its chemical breakdown (Bai et al. 2007), resulting in the transitions to the pupal and adult stages. This process starts during the last half of the fourth instar. The fourth instar stops feeding about halfway through this stage, becoming the prepupa (Danks 2000). Much as in the pupal stage itself, the late fourth instar is motile, but it ceases to glide on currents from its mouth brushes. The pupal trumpets, compound eyes, and adult legs are clearly visible through the cuticle of the fourth instar. On emergence from the fourth instar exuvium, the pupa is light in color until it darkens in about two hours. The pupa is much more like the adult than the larva. Sheaths for the proboscis, antennae, wings, legs, and genitalia are part of the exterior shape of the cuticle. The anterior portions of the exterior pupal cuticle form an air space that makes the pupa more buoyant than the larva. Apolysis and initiation of adult cuticle formation starts early in the pupal stage, at about 17 hours in *Aedes aegypti*. Muscles formed from myoblasts (neomorphic) are large and fan-like. Muscles that persist from the larval stage (protomorphic) are long and thread-like, in bundles surrounded by a membrane.

Anyone can easily observe the process of adult emergence from the pupa, which is one of the miracles of the insect world. The pupa stretches out close to the surface and, within a few minutes, the adult comes out, rests on the surface a minute or two, and then flies off. That flight is amazing, considering that the adult at that stage has yet to secrete an endocuticle, resorb larval abdominal muscle, and excrete the remnants of the larval gut (meconium). The genitalia of males rotate 180° within about 24 hours of emergence, before which they are incapable of mating.

The cuticle undergoes a hardening process following each molt. Known as tanning, this process darkens (melanization) and strengthens (sclerotization) the cuticle through the formation of protein cross-linkages. A hormone called bursicon initiates the tanning process in conjunction with the "crustacean cardioactive peptide" in a process shared by insects and crustaceans (Van Loy et al. 2007). Bursicon was first discovered in 1965 but was not identified as a heterodimer of two cysteine knot proteins (bursicon alpha and bursicon beta) until much later (Dewey et al. 2004; Luo et al. 2004; Honegger et al. 2008). Unusual among insects, the component genes for the proteins are located on different chromosomes (Robertson et al. 2007). The two component proteins of bursicon are produced in abdominal ganglia, increasing in concentration through development (Honegger et al. 2002, 2011). In addition to stimulation of melanization and sclerotization, bursicon also controls wing maturation in the adult stage (An et al. 2012). Production ceases after adult emergence, and the neurosecretory cells that produced it are destroyed through apoptosis.

The biochemistry of tanning involves dopamine, a derivative of the amino acid tyrosine. Sclerotization occurs when dopamine is acetylated to N-acetyldopamine (NADA), and melanization occurs in the presence of N-beta-alanyldopamine (NBAD). Melanin itself is another derivative of dopamine. These are secreted into the cuticle from epidermal cells. Enzymatic oxidation produces quinone residues that cross-link to proteins, to NADA and NBAD, and to chitin.

Environmental Influences on Development

Temperature is a key environmental influence on development. It affects the speed at which larvae and pupae develop, the size of larvae and pupae (smaller at warmer temperatures), and the size of subsequent adults (Rueda et al. 1990). There are many ways to express the relationship between water temperature and larval-pupal developmental times. One method, common throughout applied entomology, is the day degree. Basically, day degrees are calculated by multiplying each day by its average temperature above a threshold considered to be the minimum temperature for development (Pruess 1983). It is usually possible to correlate day degrees to completion of developmental stages, forming a convenient way to express temperature relationships and a practical method to predict development based on easily measured parameters. Table 4.5 provides a summary of minimum developmental temperatures and day degrees. One of the striking differences between species is the range of minimum temperatures required for development from egg to adult (1.1°C–15°C), which presumably reflects the large geographic range of Culicidae. Studies that examined the minimum temperatures by stage showed some variation between stages (6°C–10.5°C for *Aedes vexans* and 7.5°C–10.7°C for *Culex quinquefasciatus*) and some variation between different studies of the same species. The number of day degrees required for each species to develop from egg to adult varied from 98 for *Aedes campestris* to 455 for *Anopheles messeae*. There was a significant exponential correlation between day degrees and minimum threshold temperature ($y=235.3e^{-0.054x}$, where y=day degrees, and x=minimum threshold temperature; $R^2=0.34$). This relationship implies that species typical of warmer aquatic habitats develop more quickly and possibly with better energetic efficiency. One of the quickest is *Aedes vittatus*, which completes larval development in 53 hours and emerges as an adult in 74 hours under high-temperature (36°C) conditions in Ugandan rock pools (McClelland & Green 1970). Females take longer to reach pupation than males for many species of mosquitoes. The mosquito strain, as well as the species, can make a big difference in developmental time. For example, an Australian strain of *Aedes aegypti* developed 19 days faster at 15°C than the same species from Louisiana (Tun-Lin et al. 2000). The fluctuating temperatures expected under natural conditions can alter the relationship between development and day degrees (Bates 1947; Bradshaw 1980; Joshi 1996; Clements 2000).

The thermal death point (Table 4.5) has been variously reported as a temperature that kills the larva during a brief period, such as 30 minutes (Christophers 1960), or, more commonly, as the lowest temperature maximum at which development is not possible. Salt content can affect this temperature. For example, *Aedes dorsalis* fails to complete development at lower salt concentrations when the temperature is relatively high (Parker 1979). For eggs and aquatic stages, the thermal limit is remarkably low, generally between 30°C and 40°C. Such temperatures are easily attained in sunlit water, possibly accounting for the relatively few species (e.g., *Anopheles albimanus*, *Anopheles gambiae*) that tolerate such conditions. It is also important to remember that larvae in the field often have a choice of temperatures in their microhabitats and probably move in response to the need to find an optimal temperature. *Aedes aegypti* in Thailand, for example, inhabits ceramic water jars of about 35 l capacity, often with loose-fitting metal lids that nonetheless allow mosquitoes to enter and oviposit (Strickman & Kittayapong 1993). When these water jars are located in open sunlight, a distinct thermocline forms during the afternoon in the water column, with the top layer well above the thermal death point. During this period, larvae spend most of their time in the lower, cooler part of the jar, only briefly surfacing through the hot layer. Similarly, *Aedes communis* in the Arctic (Churchill, Manitoba, Canada) seeks a temperature of about 16°C near the bottom of a pool

Table 4.5. Examples of how temperature affects mosquito development

Species	Stages[1]	Threshold (°C)[2]	Day-degrees to development	Lethal limit (°C)	Reference[3]
Aedes aegypti	E			40	1
Aedes aegypti	E–A	8.3	181.2		2
Aedes aegypti	4	16	100 (17 base)	42 (30 m)	3
				39 (5 d)	3
Aedes aegypti	P	18		44 (30 m)	3
				36 (to pupate)	3
Aedes albopictus	E			40	1
Aedes campestris	E–A	15	98[4]		4
Aedes campestris	E–F	3.3	177		5
Aedes cinereus	E–F	3.3	144		5
Aedes communis	E–M	3.3	151		6
Aedes communis	E–F	3.6	177		6
Aedes dorsalis	E–A	12		31	7
Aedes excrucians	E–F	4.4	244		6
Aedes flavescens	E–F	4.4	244	30	5
Aedes hexodontus	E–F	7.2	122		6
Aedes impiger	E–F	1.1	167		6
Aedes implicatus	1	4.4	44		8
Aedes implicatus	2	4.4	33		8
Aedes implicatus	3	4.4	31		8
Aedes implicatus	4	4.4	67		8
Aedes implicatus	P	4.4	47		8
Aedes implicatus	E–A	4.4	222		8
Aedes krombeini	E–A	14			9
Aedes krombeini	E–A	14		35	10
Aedes nigripes	E–F	1.1	167		6
Aedes pionips	E–F	3.9	233		6
Aedes rempeli	E–A	3.7[5]			9
Aedes vexans	1–2	6			11
Aedes vexans	3–4	9			11
Aedes vexans	P	10.5			11
Aedes vexans	E–A	10.8	109		12
Anopheles gambiae	E–A	16.5–18		41.5	13, 20
Anopheles messeae	E–A	1.9	455		14
Culex pipiens[6]	E–A	7.1	200		15
Culex restuans[7]	E–A	9.4	132		16
Culex restuans[8]	E–A	8.8	165		16
Culex quinquefasciatus	1	10.7	19		17
Culex quinquefasciatus	2	10.4	18		17
Culex quinquefasciatus	3	9	26		17
Culex quinquefasciatus	4	9	49		17
Culex quinquefasciatus	P	7.5	37		17
Culex quinquefasciatus	E–A	9.9	137		17
Haemagogus spegazzinii falco				35	18
Toxorhynchites rutilus septentrionalis	E–A		13	31	19

[1]E = egg; 1 = first instar; 2 = second instar; 3 = third instar; 4 = fourth instar; P = pupa; A = adult; M = male adult; F = female adult

[2]Minimum temperature at which development occurs

[3]1. Lounibos et al. 2010; 2. Tun-Lin et al. 2000; 3. Christophers 1960; 4. Tauthong and Brust 1977; 5. Trpis and Shemanchuk 1969; 6. Haufe and Burgess 1956; 7. Parker 1979; 8. Stewart 1974; 9. Smith and Brust 1970; 10. Joshi 1996; 11. Fouque and Baumgärtner 1996; 12. Slater and Pritchard 1979; 13. Hoshen and Morse 2004; 14. Becker 2008; 15. Strickman and Fonseca 2012; 16. Madder et al. 1983; 17. Ahumada et al. 2004; 18. Bates 1947; 19. Trimble and Smith 1978; 20. Holstein 1954

[4]Average of two populations reared at 23°C

[5]Mean temperature in Arctic, ice-bound rock pools where species develops

[6]Form molestus

[7]Ontario, Canada, population

[8]Maryland, United States, population

when the surface temperature becomes too warm (Haufe & Burgess 1956).

Temperature is not the only environmental factor that influences the duration and success of larval development (Clements 2000). Salt-tolerant species have an upper salt concentration above which they can no longer survive. Selection can increase that limit in *Aedes dorsalis* (Parker 1979). Nutrition is a common limitation on development. Some species are able to slow development in response to nutritional stresses, whereas others are not. In *Aedes aegypti*, at least, the fourth instar must achieve a critical mass in order to pupate, though the maximum mass is greater, and both are affected by temperature (Table 4.6). Addition of food to larval sites under natural conditions increased survival (from 67% to 87%) and decreased developmental time (from 24 to 12 days) of *Aedes aegypti* (Wijeyaratne et al. 1974).

The abundance of larvae interacts with nutrition (Hard et al. 1989). A European tree hole species, *Aedes geniculatus*, and a North American one, *Aedes triseriatus*, were both affected by this interaction but at different levels. Both species produced the highest proportion of females at the optimum level of food and lowest density of larvae. The success and size of pupation were greatest for *Aedes triseriatus* under similar conditions, but *Aedes geniculatus* survived better when density was moderate—suggesting some level of positive interaction between larvae. The lowest density and highest food level resulted in the shortest developmental time for females of both species but only for males of *Aedes geniculatus*. In contrast, *Aedes triseriatus* males developed fastest at lower food levels. Ikeshoji and Mulla (1970, 1974) showed that chemicals produced by extremely crowded larval populations of *Culex quinquefasciatus* (more than one-fourth instar per 9 mm^2) were toxic to first instars at higher concentrations and deleterious to development of earlier instars. The compounds (identified as 17–19 carbon hydrocarbons and 2-ethyl long-chain fatty acids) also had bacteriostatic properties. *Culex quinquefasciatus* is apparently able to regulate its own environment to prevent overcrowding.

Environmental Influences on Adults

Adult mosquitoes emerge from the pupa looking like complete males or females, but a number of developmental steps continue in the first hours or days (Clements 2000). It is remarkable that adults are able to fly within a few minutes after emergence since the indirect flight muscles and associated apodemes are still maturing. Salivary glands and the midgut also require time before becoming functional. Some other changes include voiding air from the gut, sclerotization of the cuticle, formation of muscular fibrillae in the male antennae, and 180° rotation of the male genitalia. The eyes also continue to develop for 24 hours after emergence (Snodgrass 1959).

Table 4.6. Minimum critical mass of the fourth instar of *Aedes aegypti* and maximum mass achieved under ideal nutritional conditions (Clements 2000)

Sex	Temperature	Minimum critical mass (mg)	Maximum mass (mg)
Female	22°C	2.63	4.80
Female	32°C	1.99	3.93
Male	22°C	1.91	2.94
Male	32°C	1.66	2.62

Most studies of environmental influence on adult development concern longevity and female fecundity. Any life table (Carey & Liedo 1999) or population-modeling effort (Schaeffer et al. 2008) must quantify those effects. Longevity can be easily expressed as daily survival (or its reciprocal, daily mortality). Therefore, the oft-asked question "How long does a mosquito live?" is more usefully phrased as "How many mosquitoes die each day?"

Fecundity can be viewed in a number of ways, including the number of eggs deposited per day per female for a cohort of mosquitoes. This measure is artificial in that individual mosquitoes generally do not oviposit every day. Another way to express fecundity is number of eggs per gonotrophic cycle, which is easy to count for mosquitoes that deposit eggs as a batch in rafts but more difficult for those that may put a few eggs in one place and then more in another (Colton et al. 2003). Various statistics in population biology combine longevity and fecundity into measures of reproductive rate (e.g., *r*, the net reproductive rate).

The daily survival rate changes according to many factors, but if a single value were chosen to generalize for many female mosquitoes, it might be 80% daily survival. For example, Sheppard et al. (1969) found a survival rate of 81% for *Aedes aegypti* in Bangkok, Thailand; Beier et al. (1982) found a rate of 80% for female *Aedes triseriatus* in Indiana, United States; and Chandra et al. (2010) reviewed work showing 84% for *Anopheles subpictus*. Kiszewski et al. (2004) reviewed the daily survivorship of 33 species of *Anopheles* that transmit *Plasmodium*, resulting in an average of 84.1%, a median of 85.5%, and a range from 68.2% to 98%. Midega et al. (2007) measured daily survival of *Anopheles funestus* as 83% in one Kenyan location but observed much higher rates of survival (95%) for this species in a different location. *Anopheles gambiae* in both locations had high survival rates (95%–96%). Field populations of *Culex pipiens* in Maryland, United States, had a high daily survival rate of 90.4% (Jones et al. 2012), possibly because of the proximity of resources in the area and the lack of a need to move far. Reiter (2007) made the case that longevity of mosquito populations can be greater than those observed in mark-release-recapture studies. Considering gonotrophic cycles of more than a week in field populations of *Aedes aegypti*, *Aedes africanus*, and *Haemagogus janthinomys*, the daily survival rates of these species based on observed parity rates would be greater than 90%. Some species of mosquitoes are naturally long-lived. In the laboratory, *Sabethes cyaneus* females live between 33 and 171 days, with a mean of 111 days (South & Arnqvist 2008). There is also some question of whether the daily mortality rate remains constant during the lifetime of a female mosquito (Styer et al. 2007). This is significant because infection models that represent the likelihood of transmission from vectors are dependent on the daily mortality rate raised to a power of the number of days in the extrinsic incubation period. As a result, the model is quite sensitive to the longevity of mosquitoes.

Males may live a considerably shorter time than females, as found by Tsuda et al. (2001) for *Aedes aegypti* on Hainan Island, Vietnam, where the daily survival rate for males was only 52%. Muir and Kay (1998) measured *Aedes aegypti* male daily survival of 57%–70% in Queensland, Australia, and Trpis and Hausermann (1986) measured a rate of 53% in coastal Kenya. In the Kenyan study, this meant that males lived an average of only 5.8 days and a maximum of 18 days.

Figure 4.1 illustrates how daily survivorship affects population longevity. Even at 80% daily survival rate, few mosquitoes (3.5% of the entire population) are alive 15 days after emergence. If the survival rate is only 65%, then less than 0.2% of the population would be alive at 15 days. In data not shown in the figure, less than 1% of the population would be alive at 45 days for a survival rate of 90%, at 30 days for 85%, at 22 days for 80%, and at 18 days for 75%.

Temperature has a great effect on adult survival. Fluctuating daily temperatures can increase longevity of adults compared to exposure to a constant temperature at the mean of the fluctuating regime (Joshi 1996; *Aedes krombeini*). Adult survival is generally greatest at a temperature above a minimum and below a maximum. Hence, Yang et al. (2009) noted that *Aedes aegypti* females lived longer at 27°C than at 10°C or at 31°C. Strickman (2006) observed seasonal differences in survival of caged *Aedes aegypti* exposed to ambient conditions in Thailand (Table 4.7). The high survival rates observed in cages compared to other studies of *Aedes aegypti* in the field (Table 4.8) show that other factors (e.g., predation) are probably very important to survival.

Based on longer (calculated as at least eight days based on rubidium-marking studies) gonotrophic cycles observed in the field in Puerto Rico, Reiter (2007) made the case that humidity is important for survival. *Anopheles subpictus* survives longest in the laboratory at 70% relative humidity (RH; 50% survival at 13 days at 25°C, compared to 11 days at 30% RH and 9 days at 90% RH; Mehta 1934). In the field, *Anopheles arabiensis* survives less time in the dry season (Charlwood et al. 2000). Morales-Vargas et al. (2010) found a surprising correlation between humidity experienced by the previous adult generation and size of the current generation among *Aedes aegypti* in Thailand. Rainfall can have a severe effect on adult survival, as observed for *Culex pipiens* in Maryland, where 1 cm of rain reduced daily survival by 34% (Jones et al. 2012).

Nutrition of larvae and adults can affect longevity (Joy et al. 2010). *Aedes aegypti* in the laboratory survived longer when larvae were fed less and when adults received fewer blood meals. The expected survival of 50% of the population was 11 days for females fed multiple blood meals, compared to 20 days for those fed a single blood meal or 28 days for those not blood fed. The 50% survival of female adults emerging from larvae fed a maximum diet was 36 days compared to 45 days for larvae fed a quarter of the standard diet.

The fecundity of mosquitoes ranges between 50 and 500 eggs per individual female (Clements 2000), though Holstein (1954) cites examples of *Anopheles* depositing about 1,000 eggs each during their lifetimes under ideal conditions. Variation between species of *Anopheles* was documented by Pascini et al. (2013), citing *bellator* and *homunculus* as depositing 57–62 eggs per gonotrophic cycle; *gambiae*, a mean of 64 eggs; *stephensi*, 97 eggs; and *aquasalis*, 50–118 eggs. The total number of eggs depends greatly on survival of the female as it continues to deposit eggs following gonotrophic cycles initiated by one or more blood meals. The length of each gonotrophic cycle varies according to species and temperature, but a generalization under warm conditions is about three days, as represented by *Anopheles subpictus* with a mean time of 96 hours and a range of 88 to 102 hours seasonally (Chandra et al. 2010). As discussed above, relatively few mosquitoes live long enough to go through more than two gonotrophic cycles, and care should be taken in assuming the gonotrophic cycle is the same in the field as in the laboratory. Even in the laboratory, only 14.5% of *Anopheles subpictus* deposited three or more batches of eggs. Caged *Aedes aegypti* deposited 49 to 68 eggs per gonotrophic cycle with no clear seasonal relationship (Strickman 2006), though average fecundity throughout the life of a caged

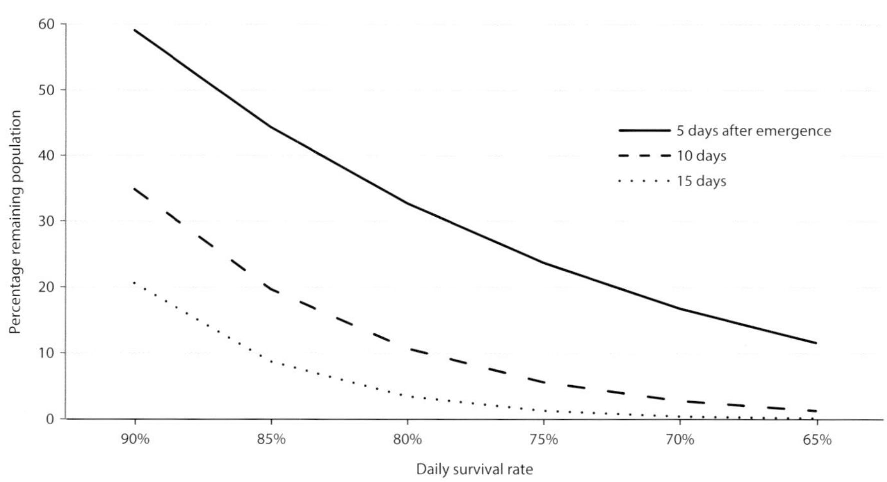

Figure 4.1. Percentage of individuals surviving 5, 10, or 15 days in relation to the population's daily survival rate.

Table 4.7. Daily survival of caged *Aedes aegypti* during 30 days exposure to ambient conditions in a house in Thailand (Strickman 2006)

Months	Range of mean daily temperatures (°C)	Range of extreme temperatures (°C)	Daily survival
Oct–Nov	25–33	22–35	99.9%
Nov–Dec	24–34	21–36	99.8%
Jan–Feb	24–34	22–36	99.3%
Mar–Apr	27–38	26–40	96.0%
May–Jun	29–37	27–41	95.0%

Table 4.8. Comparison of calculated daily survival rates of female *Aedes aegypti* in studies conducted under ambient field conditions

Daily survival rate	Location	Reference
95–99%	Chachoengsao, Thailand	Strickman 2006[1]
67–84%	Chachoengsao, Thailand	Harrington et al. 2001a[1]
81%	Bangkok, Thailand	Sheppard et al. 1969
63–88%	Bangkok, Thailand	Dye 1984
76%	Hainon Island, Vietnam	Tsuda et al. 2001
66%	Near Dar-es-Salaam, Tanzania	Conway et al. 1974
89%	Near Mombasa, Kenya	McDonald 1977
85%	Near Mombasa, Kenya	Trpis and Hausermann 1986
86–90%	Pentland, Australia	Muir and Kay 1998

[1]Strickman's study was conducted in cages exposed to ambient conditions; Harrington et al.'s study in the same district used mark-release-recapture in the field.

female (basic reproductive rate) was highest at a moderate temperature of 29°C (Yang et al. 2009). Other studies recorded averages of 58 (Dye 1984) to 86 eggs (Christophers 1960) per cycle. Temperature and season influence fecundity of other species. For example, *Culex pipiens* and *Culex restuans* deposit fewer eggs per raft in early spring in Canada (Madder et al. 1983), and *Wyeomyia smithii* deposits more eggs under conditions of fluctuating temperature (Bradshaw 1980). *Opifex fuscus* that deposited an autogenous batch of eggs lived significantly longer (mean survival of 36 days) compared to those that did not (mean survival of 18 days; Snell et al. 2010).

Genetics

The genetics of mosquitoes is arguably one of the most rapidly developing areas of culicid biology. The physical basis of genetics has been a topic of study for a long time (Clements 2000). The numbering of chromosomes was originally based on their sizes and then to linkage maps when those data became available (Sharakhova & Sharakhov 2013). Chromosomes of culicine mosquitoes consist of three pairs. Chromosomes I are the shortest and correspond to the sex chromosome. In culicines, they are similar in size, but in *Anopheles* the X chromosome is much smaller than the Y chromosome. Chromosomes II are the largest, and chromosomes III are intermediate in length. *Chagasia*, does not follow the pattern of either culicines or *Anopheles*. *Chagasia* has four chromosomes, a condition corresponding to chaoborids and presumably reflecting a basal phylogenetic position (Kreutzer 1978).

Unlike most other organisms, some mosquitoes have polytene chromosomes in particular tissues. Polytene chromosomes have been useful for species determination and physical mapping of genes (Clements 2000). The polytene chromosomes tend to occur in glandular tissue, including larval salivary glands, midgut epithelium, larval Malpighian tubules, and ovarian nurse cells. A single polytene chromosome consists of a fusion of a homologous pair of chromosomes with up to 662 copies of the genotype. A series of dark and light bands are formed by noncoding heterochromatin and coding euchromatin. Puffs appear in the heterochromatin when synthetic activity occurs at that site. Stable inversions (i.e., where the pattern is reversed between two species) were the original basis for declaring the species status of cryptic members of the *Anopheles gambiae* complex as well as other anopheline species complexes. Sometimes inversions are not characteristic of differences between species, as when the inversions occur and reverse or when inversions are polymorphic within a species. The banding pattern of polytene chromosomes can also be identical between different species (e.g., *Anopheles atroparvus* and *Anopheles labranchiae*).

Various measurements of sequence differences between genomes have been used to characterize phenotypes (e.g., insecticide resistance; Grigoraki et al. 2015) and taxonomic differences (e.g., among Turkish *Culex*; Gunay et al. 2015). Targeted approaches to detecting sequence difference include amplifying specific segments to compare size, cutting the genome or parts of it with segment-specific enzymes, probing DNA that matches short sequences, and sequencing specific portions of the genome (Collins & Paskewitz 1996). Tremendous progress has been made during the last 15 years in constructing complete genome sequences (Table 4.9). In what must be one of the most iconic papers of entomology, Holt et al. (2002) published the complete genome sequence of *Anopheles gambiae*. Among the conclusions from this study were selective introgression for certain gene regions, an X chromosome with less heterozygosity than other areas, a high concentration of transposable elements in euchromatin (16%), and a large proportion of heterochromatin (60%). Expression studies showed a high expression of enzymes associated with innate immunity and upregulation of protein and lipid metabolism after blood meals. The complete genome of *Aedes aegypti* was described in 2007 (Nene et al. 2007). This genome was larger than that of *Anopheles gambiae* due to the occurrence of transposable elements making up to 50% of the sequence. The genome of South African *Culex quinquefasciatus* contained more protein sequences

(Arensburger et al. 2010), including more genes dedicated to gustatory and olfactory reception, salivary gland secretions, and xenobiotic detoxification. The number of protein sequences in *Culex quinquefasciatus* (19,793) was 4% more than in *Aedes aegypti* and 43% more than in *Anopheles gambiae* (adjusted to counts in VectorBase in 2018; National Institute of Allergy and Infectious Diseases 2015). Five thousand single-copy genes were shared among the three genera. According to Arensburger et al. (2010), 29% of the *Culex quinquefasciatus* genome consists of transposable elements, compared to 42%–47% in *Aedes aegypti* and 11%–16% in *Anopheles gambiae.*

New, more efficient methods of sequencing have resulted in more detailed studies that treat more species. Neafsey et al. (2015) sequenced 16 *Anopheles* species, mapped and characterized genes, and calculated rates of genetic change. The results were compared to *Culex quinquefasciatus* and *Aedes aegypti*, as well as several species of *Drosophila*. They found that the X chromosomes of the *Anopheles* species were particularly dynamic, with high rates of rearrangement and gene transfer compared to *Drosophila* species and to other parts of the anopheline genome. The genome as a whole also showed a higher rate of gene gains or losses (3.12×10^{-3} gene per million years) than *Drosophila* (5.90×10^{-4} gene per million years). The highest rates of evolution were among genes assigned to the following functions: immune modulation, male accessory gland proteins, salivary peptides, odorant binding, odorant receptors, and gustatory receptors. Most genes were from widely conserved sequences, but a significant proportion were unique to Anophelinae and to each species. Overall, this study suggested that *Anopheles* has a relatively high rate of evolution, especially with respect to host selection, mating physiology, and symbiotic or pathogenic organisms.

Table 4.9. Genomes of mosquitoes in VectorBase (www.vectorbase.org), October 2015. VectorBase is a US National Institutes of Allergy and Infectious Diseases Bioinformatics Resource Center.

Species	Number of genes	Genome size (base pairs)
Aedes aegypti	17,479	1,311,011,677
Anopheles albimanus	12,335	170,508,315
Anopheles arabiensis	13,678	246,567,867
Anopheles atroparvus	14,230	224,290,125
Anopheles christyi	11,156	172,658,580
Anopheles coluzzii	14,703	224,455,335
Anopheles culicifacies A	14,882	202,998,806
Anopheles darlingi	10,948	136,935,538
Anopheles dirus A	13,239	216,307,690
Anopheles epiroticus	12,602	223,486,714
Anopheles farauti	13,703	180,984,331
Anopheles funestus	13,756	225,223,604
Anopheles gambiae	13,624	273,109,044
Anopheles maculatus B	15,046	141,984,015
Anopheles melas	16,666	227,407,517
Anopheles merus	14,415	251,805,912
Anopheles minimus A	13,128	201,793,324
Anopheles quadriannulatus A[1]	13,857	283,828,998
Anopheles sinensis[2] China strain	19,815	220,777,669
Anopheles sinensis[2] SINENSIS strain	13,331	375,763,635
Anopheles stephensi SDA-500 strain	13,764	225,369,006
Anopheles stephensi Indian strain	12,350	221,324,304
Culex quinquefasciatus	19,363	579,057,705

[1]From South Africa
[2]Apparently, with consideration of cryptic species

Figure 4.2. Diagram of the structure of cuticular layers showing finer structure of the epicuticle. Ecdysial fluid is secreted by the epidermal cells into the space between the epidermis and endocuticle, digesting the endocuticle. The exocuticle and epicuticle are shed during metamorphosis as the exuviae. Christophers 1960

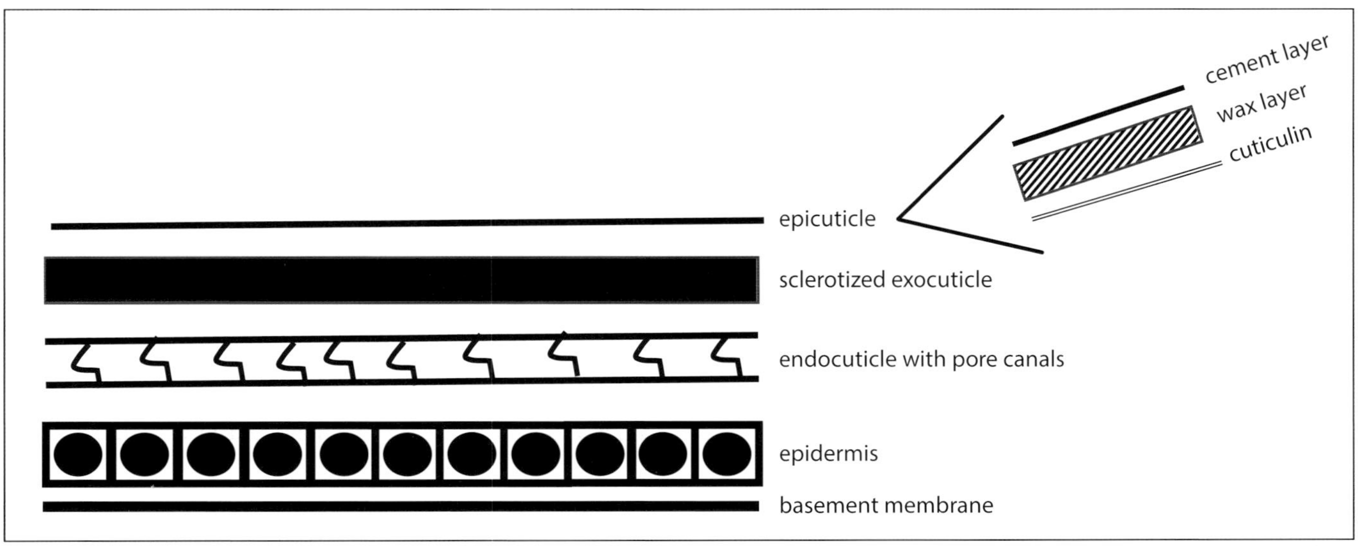

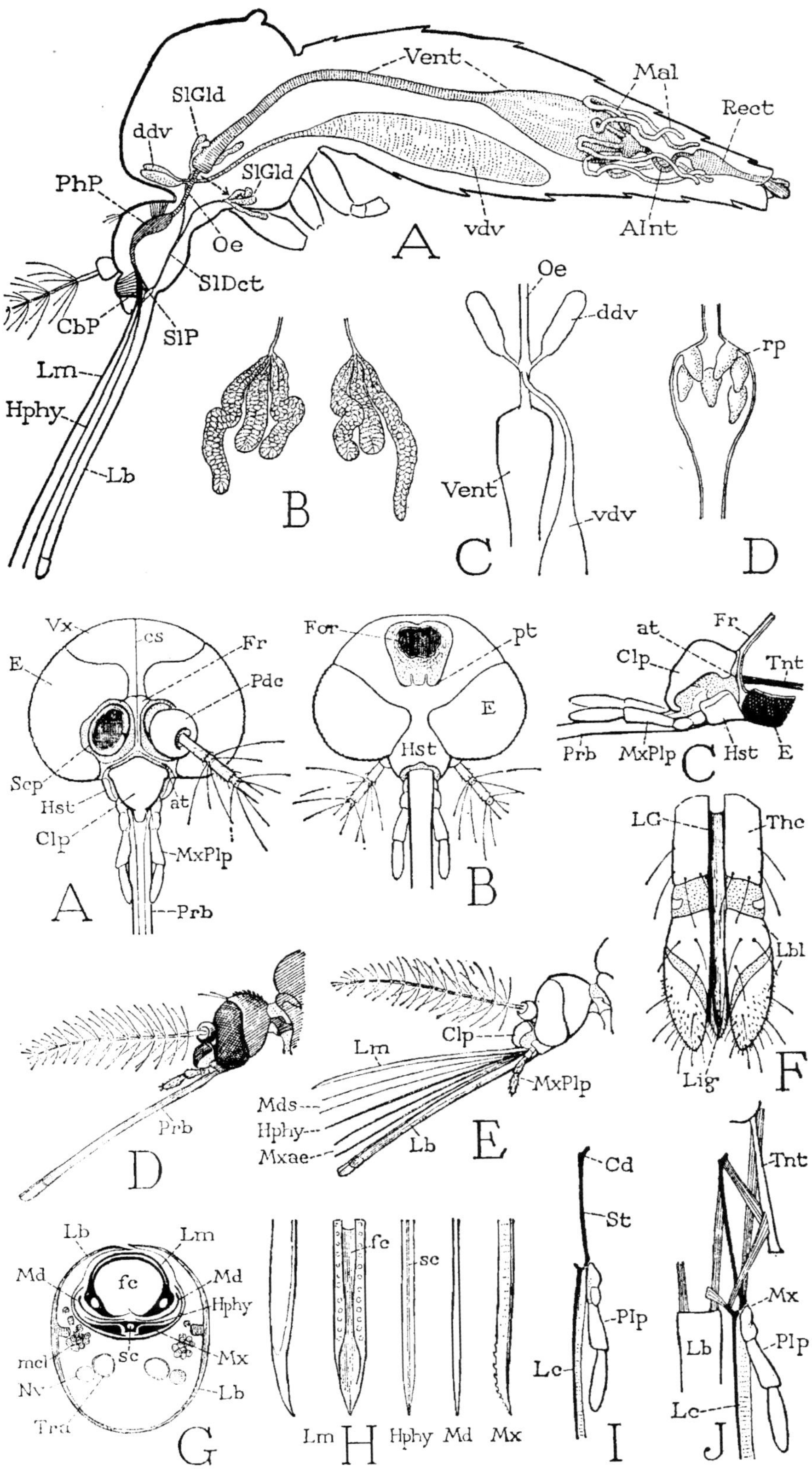

Figure 4.3. Anatomical structures of adult mosquitoes. Upper illustrations A, cross section of adult mosquito (*clockwise from left*): Lb = labium; Hphy = hypopharynx; Lm = labrum; CbP = cibarial pump; SlP = salivary pump; SlDct = salivary duct; PhP = pharyngeal pump; Oe = esophagus; ddv = dorsal diverticula; SlGld = salivary gland; vdv = ventral diverticulum or crop; Vent = ventriculus or midgut; Mal = Malpighian tubules; AInt = anterior intestine; Rect = rectum. B, salivary glands. C, detail of diverticula connections to foregut. D, interior of rectum showing rectal papillae (rp). Lower illustrations A, anterior view of the head: Prb = proboscis; MxPlp = maxillary palpus; Clp = clypeus; hst = hypostome; at = anterior tentorial pit; Scp = antennal scape; Pdc = antennal pedicel; Fr = frons; cs = median cranial sulcus; E = eye; Vx = vertex. B, posterior view of head: For = foramen; pt = posterior tentorial pit. C, lateral view of upper mouthparts: Tnt = anterior tentorial arm (visible with eye cut away). D, lateral view of head and proboscis. E, lateral view of head and proboscis with stylets separated: Mds = mandibles; Mxae = maxillae. F, detail of distal proboscis: LG = labial gutter; Thc = theca; Lbl = labellum; Lig = ligula. G, cross section of proboscis: fc = food channel; sc = salivary channel; Md = mandible; Mx = maxilla; Tra = trachea; Nv = nerve; mcl = muscle. H, distal ends of stylets. I, distal portion of maxilla: Cd = cardo; St = stipes; Plp = palp; Lc = lacinia. J, musculature of base of maxilla and labium. Snodgrass 1959

Anatomy

The external morphology of mosquitoes is reviewed later in this book as a part of the genus and species pages. Older literature explored the structure and variation of the mosquito exterior in great detail (Snodgrass 1959) and was beautifully summarized in a standard reference by Harbach and Knight (1980).

Formed by many layers that protect the interior of the insect (Fig. 4.2), the cuticular surface of an adult mosquito is a complicated structure serving many purposes. Christophers (1960) reviewed various aspects of that surface, describing sensory organs, wax channels penetrating the outer cuticle, and vestiture. The cuticle is largely covered with tiny (less than 1 μm wide, approximately 5–10 μm long) cuticular spines called microtrichia. They vary in shape and distribution depending on their position on the body. Their function is probably mechanical, as they have no obvious cellular, secretory, or sensory connections. Much larger than the microtrichia and almost covering the surface of some mosquitoes are scales of several types. Across all Culicidae, many different shapes of these chitinous structures originate as specialized epithelial cells. In *Aedes aegypti* they can be classified into several types. Flat scales cover much of the mosquito and consist of a narrow, short stalk (petiole) that expands into a striated, flat blade. Color patterns on the mosquito result from dark pigmentation of these scales, appearance of dark cuticle below them, or structural elements that create white reflectance (in *Aedes aegypti*). Narrow, curved scales taper gradually to a narrow petiole and are also striated. Their color can be dark or light. Finally, upright, forked scales end distally in a notch. The cross section of forked scales shows a flattened tube studded with striations around the outside. Considering the vigorous movement of mosquitoes during flight, mating, oviposition, and feeding, it seems clear that this scale covering has a large effect on mosquitoes' interactions with their surroundings by shedding liquid, protecting the cuticle beneath, or creating a favorable aerodynamic environment. Evidence that magnetic dust adheres to external surfaces of mosquitoes and affects orientation (Strickman et al. 2000) adds yet another dimension to the possible functions of cuticular vestiture.

The internal anatomy of mosquitoes was conveniently summarized by Snodgrass (1959), some of whose illustrations are reproduced here (Fig. 4.3). It is beyond the scope of this volume to review all aspects of the structure and function of organ systems in mosquitoes. Excellent treatments of this subject are contained in Clements (1999, 2000) and, for all insects, in Nation (2008). In general, the larval stage is the most varied between genera and species. Presumably, this variation in form reflects the variation in habitat and diet. Like the adult, larvae have obvious adaptations for sensory input (chemical, visual, vibrational) and for movement. Considering the rapid processing of food by larvae and adults, the digestive system is remarkably generalized in structure. The entry of food is the exception, as larvae are variously adapted to graze, filter, or predate, and adults have a specialized apparatus for blood or sugar feeding.

Conclusion: What Is a Mosquito?

Development is the process that results in the physical entity we call a mosquito. That development is controlled by genetics and the environment. The existence of mosquitoes is the result of hundreds of millions of years of evolution within Diptera. During the entire history of the order, each group and species had to undergo a similar process of development. Mosquito development may be just a variation on a process as old as Diptera and, to some extent, as old as insects, but the result is a group of insects that are clearly recognizable and obviously important to human health and comfort.

5. Dormancy

The study of insect dormancies is a study of misstatements and misunderstandings.

Jungreis 1978, quoted in Košťál 2006

Conditions are not always suitable for the survival of insect populations. Some of the challenges include temperature extremes, lack of water, scarce food, outbreaks of pathogens, and abundance of predators and parasitoids. These challenges are all related to time series that present species with good conditions at certain times and harsh conditions at others. We usually think of annual changes associated with climatic patterns of temperature and rainfall, but the same sorts of adaptations can give insect populations an advantage in the face of longer-term trends (e.g., the El Niño-Southern Oscillation). Although insects can escape some threats by migration, escape mechanisms, or seeking shelter, the subject of this chapter is how mosquitoes undergo various forms of decreased activity in response to time-related changes in their surroundings.

Environmental hazards may be a threat to insect species, but the entire subject of dormancy has been a considerable challenge for entomologists. The usual academic strategy of dividing phenomena into discrete activities does not work well in this area because processes are continuous, and dormancy is a process rather than a status (Košťál 2006). The responses of insects to threatening conditions necessarily involve many aspects of an insect's life; therefore, those studying such responses may find profound changes in some familiar facet of biology without considering the many other changes going on. Dormancy is an essential part of the life cycle of many species of mosquitoes.

One of the difficulties in understanding culicid dormancy is its highly variable nature. Depending on the group, species, or even individual, the manner in which mosquitoes escape unsuitable conditions ranges from complex physiological systems to simple timing. Some characteristic aspects of the family seem to have provided potential mechanisms for dormancy, like the aquatic habitat of larvae, the resistant eggs of the aedine genera, the mobility imparted by the ability to fly, and the remarkable physiology, shared with many insects, of the adults to enter a state of inactive longevity. It would be difficult to claim that the life cycles of mosquitoes include some of the most extreme examples of insect dormancy; however, the distribution of certain species in far northern locations shows that some mosquitoes are particularly well adapted to surviving extreme cold and dryness. Dormancy has allowed this essentially tropical group of insects to occupy higher latitudes and, as discussed below, plays a surprisingly large role in the life cycles of some tropical species. Considering that the long evolutionary history of mosquitoes has spanned many climatic regimes, it is perhaps not surprising that selection for dormancy is so prominent in Culicidae we observe today.

Framework for Consideration of Dormancy

Košťál's (2006) definitions of the processes of dormancy were based on considerations of insects in general, but they are very relevant to mosquitoes. It is worth quoting these definitions verbatim here, as the use of these terms has led to confusion in the past.

> Dormancy: Is a generic term covering any state of suppressed development (developmental arrest), which is adaptive (that is ecologically or evolutionarily meaningful and not just artificially induced), and usually accompanied with metabolic suppression.

Koštál divided dormancy into two categories that are very helpful in categorizing the responses of mosquitoes to adverse conditions:

> Quiescence: An immediate response (without central regulation) to a decline of any limiting environmental factor(s) below the physiological thresholds with immediate resumption of the processes if the factor(s) rise above them

and

> Diapause: A more profound, endogenously and centrally mediated interruption that routes the developmental programme away from direct morphogenesis into an alternative diapause programme of succession of physiological events; the start of diapause usually precedes the advent of adverse conditions and the end of diapause need not coincide with the end of adversity.

Denlinger and Armbruster (2014) added another aspect to the definition of diapause, stating that it requires a fixed period of latency before the resumption of development.

Koštál discussed stages of diapause as an ecophysiological process. These categories may not fit every situation, but they provide an outline for consideration of each species that undergoes diapause. The first category of processes is prediapause, consisting of the following:

> Induction phase: Occurs during genotype specific ontogenetic stage(s) (sensitive period) when cues from the environment are perceived and transduced into switching the ontogenetic pathway from direct development to diapause when the token stimuli reach some critical level (the response may be modified by other environmental factors).
>
> Preparation phase: Occurs where the phases of diapause induction and initiation are separated by a period of direct development during which the individual is covertly programmed for later expression of diapause. Behavioural and physiological preparations for diapause may take place.

The second category of processes is diapause itself:

> Initiation phase: Direct development (morphogenesis) ceases, which is usually followed by regulated metabolic suppression. Mobile diapause stages may continue accepting food, building of energy reserves and seeking suitable microhabitat. Physiological preparations for the period of adversity may take place and intensity of diapause may increase.
>
> Maintenance phase: Endogenous developmental arrest persists while the environmental conditions are favourable for direct development. Specific token stimuli may help to maintain diapause (prevent its termination). Metabolic rate is relatively low and constant. Unknown physiological process(es) lead to more or less gradual decrease of diapause intensity and increase of sensitivity to diapause terminating conditions.

Diapause ends following various terminating conditions and may be followed by a period of quiescence (as defined above), when the external environment is not yet conducive for normal development.

Masaki (1980) subdivided the life cycles of insects into five schemes according to when a species or population undergoes dormancy. In higher latitudes, dormancy often occurs during the winter (Type A, or "hibernal diapause" of Denlinger & Armbruster [2014]). If an insect uses a limited resource in the spring, or it occurs as a single generation per year (= univoltine), the population might be dormant most of the year except when the resource is available, typically in the spring (Type B). Rarely, winter dormancy is followed by summer dormancy, with the active portion of the life cycle in the fall (Type C). Another variation is activity in both spring and fall, with dormancy during both winter and summer (Type D). Finally, some insects are active in the winter, with summer dormancy (Type E). Denlinger and Armbruster (2014) refer to summer dormancy as estivation, whether diapause or quiescence. Mosquitoes have a great variety of life-cycle patterns, and individual species could probably be matched to any of the types. In fact, although this system of classifying seasonality is useful, there is such variation among mosquitoes that classification is largely a matter of making divisions across a spectrum of possibilities.

The Nature of Diapause

The definition of diapause above refers to endogenously controlled operations that initiate and maintain a state in which certain developmental processes are suspended. Denlinger and Armbruster (2014) provided a summary of which of these processes are known and which are unknown. Mosquitoes can be in diapause in any stage except the pupal, and larval diapause is usually in the fourth instar (Table 5.1). In contrast to some other groups of Diptera, mosquitoes never diapause as pupae. The reasons for this are unknown, though it might be that the culicid pupa is particularly exposed to environmental and biological challenges since it usually resides at the water surface. The physiology of diapause as an egg, which is so important in the genus *Aedes*, is not yet well understood. Unlike some other insects, mosquito eggs in diapause are always at the stage of a pharate first instar (i.e., fully embryonated). One study demonstrated elevated expression of ecdysteroid genes in diapause eggs of *Aedes albopictus*. Another study showed that the powerful juvenile hormone mimic pyriproxyfen could reverse diapause in *Aedes albopictus* eggs when absorbed from a damp substrate through the chorion (Suman et al. 2015).

Larval and adult diapause are better understood. The hormonal regulation and physiological characteristics of these kinds of diapause probably have elements in common for both hibernal and estival diapause. For larval diapause, the brain shuts down production of prothoracicotropic hormone, which in turn prevents production of ecdysone. In adult female dia-

Table 5.1. Examples of life stages that undergo diapause among mosquitoes (Denlinger and Armbruster 2014)

	Species		
Genus	**Egg**	**Larva**	**Adult**
Aedes	*albopictus, atropalpus, campestris, canadensis, dorsalis, fitchii, hexodontus, impiger, japonicus, mariae, nigripes, nigromaculis, sollicitans, squamiger, stimulans, taeniorhynchus*	*caspius, geniculatus, hendersoni, sierrensis, togoi, triseriatus*	
Anopheles	*walkeri*	*barberi, plumbeus, pulcherrimus*	*atroparvus, earlei, freeborni, superpictus, gambiae (coluzzii), hyrcanus, maculipennis, messeae, punctipennis, sacharovi*
Armigeres		*subalbatus*	
Culex			*apicalis, bitaeniorhynchus, modestus, pipiens, restuans, tarsalis, territans, tritaeniorhynchus*
Culiseta		*melanura*	*alaskaensis*
			impatiens, inornata
Orthopodomyia		*alba, pulcripalpis, signifera*	
Psorophora	*ferox*		
Toxorhynchites		*rutilus*	
Wyeomyia		*smithii*	

pause, conditioning events prevent production of juvenile hormone from the corpora allata, which in turn stops development of the ovaries. More recent work has also shown involvement of insulin receptors, with an insulin-like peptide downregulated during diapause, releasing the Forkhead transcription factor (FOXO). FOXO increases longevity, stress tolerance, fat accumulation, and resistance to oxidative stress. Transcription studies have shown a complicated genetic basis of diapause that involves multiple genes on all three chromosomes. Inhibition of the production of insulin-like proteins by RNAi shuts down ovarian development in *Culex pipiens* (Sim & Denlinger 2009), and the accumulation of FOXO induces diapause (Sim & Denlinger 2013). The diapausing females of this species accumulate cuticular hydrocarbons and have a lower rate of water loss than active insects (Benoit & Denlinger 2007).

The mechanisms for perception of day-length changes have also been described but mainly with respect to mosquitoes at higher latitudes. Tropical mosquitoes are also sensitive to circadian rhythms, with day-night cycles influencing 4.5% of genome expression in *Anopheles gambiae* (Meuti et al. 2015) and time of day influencing permethrin susceptibility of *Aedes aegypti* (Yang et al. 2010). Trehalose, a storage carbohydrate, may play a role in diapause of *Anopheles gambiae* (Liu et al. 2013). The fat body has a transporter activated during conditions of heat and low humidity that induces a higher concentration of trehalose in the hemolymph, increasing tolerance to heat stress.

Dormancy in Higher Latitudes

Observations on mosquito diapause began in the 1920s. Roubaud (1923) referred to some earlier documentation in the field of "hibernation" of *Anopheles* and, probably for the first time, induced diapause in the laboratory (*Anopheles maculipennis*). The accumulation of fat body in this species was attributed to blood meals taken by mated females late in the season. He recognized that diapause in *Culex pipiens* was different, depending on sugar sources for accumulation of fat. He also recognized that a minimum cold period was necessary to terminate diapause. Roubaud even speculated that the cause of hibernation was the accumulation of toxins due to the lack of Malpighian tubule function, terminated in late winter when the mosquito excreted.

Culex pipiens has been the convenient model for adult female diapause. The anautogenous form (i.e., not the Form Molestus) induces diapause in the larval stage in Boston (Massachusetts, United States, 42°N) when day length decreases below 14 hours and 15 minutes. Adults that emerge from diapause-conditioned larvae accumulate fat body from sugar sources and do not take blood meals (Spielman & Wong 1973). Ovarian development is minimal, with the most advanced follicles 50 μm long or about as long as the germaria. Ovarian development can be reinitiated by topical or oral administration of juvenile hormone and stopped in nondiapausing mosquitoes by removing the corpora allata (glandular sources of juvenile hormone; Spielman 1974). Secretion by the corpora allata is mediated by peptides called allatostatins that inhibit juvenile hormone secretion and other peptides called allatotropins that stimulate juvenile hormone production. Apparently, it is the lack of allatotropins that reduces levels of juvenile hormone and reduces the size of the corpora allata in diapausing *Culex pipiens* (Kang et al. 2014). Laboratory studies have shown that *Culex pipiens* in the preparation phase metabolized only one-third as much glycogen and lipid as nondiapausing mosquitoes, accumulating 1.7 times as much fat and 2.3 times as many carbohydrates (Zhou & Miesfeld 2009).

Gene expression studies of the induction phase of diapause in *Culex pipiens* have shown a complicated series of changes in the mosquito regulated by the insulin receptor-FOXO pathway and by tumor growth factor (TGF-β) and Wnt (the integrated 1 and Wingless family of peptides) signaling pathways (Sim et al. 2015; Hickner et al. 2015). These pathways have direct effects on gene expression that result in the following phenotypic characteristics: females that mate, accumulation of energy reserves, behavioral change to seek shelter, increase in cold hardiness, absence of host-seeking behavior, increase in sugar feeding, cessation of ovarian development, and elevation of antioxidant activity. Some of the genes affected and their characteristics (Sim et al. 2015) are listed in Table 5.2. Hickner et al. (2015) examined expression of 16,313 genes from initiation through induction, finding 1,130 of them significantly upregulated or downregulated. Using standardized terminology, gene ontology (GO) terms, this study helped show the complicated changes associated with diapause in *Culex pipiens* (Table 5.3).

Circadian rhythm plays an essential part in diapause of *Culex pipiens* (Meuti et al. 2015). The biochemistry of circadian clocks involves an alternating positive and negative feedback loop regulated by the interaction of light with cryptochrome 1 and cryptochrome 2. In the presence of light, these substances degrade the transcription of two gene products, period and timeless. Period and timeless create a heterodimer in the cytosol of the cell that inhibits two other proteins, clock and cycle, from entering the nucleus. Therefore, when period and timeless are not present, clock and cycle enter the nucleus, where they stimulate the E-box promoter region. This stimulates expression of more period and timeless proteins, which then form heterodimers and stop synthesis by preventing clock and timeless from entering the nucleus. Knocking out synthesis of period, timeless, cryptochrome 1, or cryptochrome 2 with dsRNA increases follicular length and stops lipid accumulation in females that are otherwise primed for diapause.

In areas where the winter temperature is moderate, as in California, United States, *Culex pipiens* may follow a surprising variation in patterns of dormancy. Apparently, diapausing females can return to normal activity as soon as the end of December if conditions are correct, as observed during a warm January in the Central Valley of California (Reisen et al.

Table 5.2. Some genes regulated in response to the Forkhead box O (FOXO) signaling pathways during the induction phase of diapause in *Culex pipiens* (Sim et al. 2015)

Gene name	General effect of gene
Oxidoreductin-1-like	Antioxidant
Major royal jelly protein 1	Antibacterial
PDZ and LIM domain protein 2	Strengthens cytoskeleton
Glycogen synthase	Energy storage
ATP-binding cassette G	Fat metabolism
LDLR chaperone boca	Fat metabolism
N-methyltransferase	Life span extension by chromatin silencing
PAX interacting protein	Localizes chromatin-silencing proteins
Homeobox protein	Prevents cell cycling
Mitotic checkpoint	Prevents cell cycling
Takeout protein	Signal between photoperiod detection and juvenile hormone pathway

Table 5.3. Top five enriched gene ontology (GO) terms for the series of genes affected by diapause induction in *Culex pipiens* (Hickner et al. 2015) and the ratio of genes upregulated compared to the total number of functional genes

GO category[1]	GO terms	Ratio[2]
BP	Organ morphogenesis	32/481
BP	Sensory organ development	23/310
CC	Chromatin-remodeling complex	8/68
MF	Calcium ion binding	10/129
MF	DNA-dependent ATPase activity	4/26
BP	Instar larval or pupal morphogenesis	17/343
BP	Epidermal cell differentiation	6/38
CC	Integral to plasma membrane	9/131
MF	Fatty-acyl CoA reductase	2/4
MF	Serine-type endopeptidase activity	5/61
BP	mRNA splicing via spliceosome	13/169
BP	Gamete generation	19/433
CC	Sin3 complex	2/2
MF	snRNA binding	2/5
CC	NuRD complex	2/5
CC	Structural constituent of cuticle	11/51
MF	Retinal binding	2/3
BP	Chitin metabolic process	3/41
BP	Chitin-based cuticle development	3/49
MF	Integral to plasma membrane	4/131
BP	Organonitrogen biosynthesis	11/187
BP	Response to alkaloid	3/11
MF	Alpha-amino acid metabolic process	5/62
MF	Fatty acid biosynthetic process	3/19
CC	ATPase complex	2/14

[1]BP = biological process; MF = molecular function; CC = cellular component
[2]Ratio, the number of module genes in category/the total number of genes in category

2010). At a similar latitude but closer to the coast in Santa Clara County, Strickman and Fonseca (2012) observed a range of winter behaviors among autogenous *Culex pipiens* that were mainly a genetic mix between the Pipiens and Molestus Forms. They observed populations that were active all winter in underground sites, typical of the Molestus Form. They also observed above-ground, autogenous populations under outdoor conditions. These mosquitoes would accumulate fat body and survive up to five months. Some individuals deposited autogenous egg rafts during this time. Females that survived the winter took blood meals in the spring. Although it is still uncertain whether this population included autogenous individuals that underwent true, hormonally regulated diapause, the phenology of that portion of the population resembled typical diapausing *Culex pipiens*. Apparently, hybridization between populations of this species resulted in a variety of dormancy traits. *Culex tarsalis* continues to take sugar meals during diapause under similar mild winter conditions (Stone & Foster 2013; Nyasembe & Torto 2014), which was probably also true of the *Culex pipiens* described above.

Anopheles punctipennis is another species that lives through the winter in a variety of ways in California (Washino & Bailey 1970). Near the coast it continues to develop slowly, but further inland all larval development ceases. Not all field-caught females brought into the laboratory take blood meals but those that do, develop eggs normally. This was in contrast to *Anopheles franciscanus* and *freeborni* from the same area, the majority of which experienced gonotrophic dissociation (i.e., blood meals did not result in ovarian development). Gonotrophic dissociation may be controlled through insulin-like peptides produced in the nervous system and found through examination of the *Anopheles gambiae* genome (Riehle et al. 2002). The peptides, like other neuropeptides, act through G protein-coupled receptors (heterotrimeric GTP-binding protein-coupled receptors). The only insulin receptor is on the ovaries, and it is expressed during the initial period of reproductive arrest following emergence and at the beginning of each gonotrophic period. The insulin and receptor interaction may be the switch for gonotrophic dissociation. Ovary ecdysteroidogenic hormone (OEH) starts up ovarian development again.

Larval dormancy is important for some species at higher latitudes. In some cases, like *Wyeomyia smithii* (Bradshaw 1980), *Aedes sierrensis* (Hawley 1985), and *Orthopodomyia pulcripalpis* (Snow & Medlock 2008), larval survival through freezing winter temperatures (Heard 1994) appears to have many parallels to adult diapause. In other cases, larval dormancy seems to be a form of quiescence, with the population possibly including active adults at the same time (*Anopheles plumbeus* and *claviger*; Roubaud & Treillard 1940), or some eggs hatching during the winter if temperatures are warm enough (*Aedes togoi*; Shestakov 1961).

Winter diapause in the egg as pharate first instars is a distinctive adaptation by many species of *Aedes* and *Psorophora* mosquitoes (Horsfall 1963). The distinctive, resistant structure of the chorion, larger food reserves, physiological adaptations to resist desiccation, and the ability to turn off the hatching response give these eggs a seed-like ability to remain in place for the winter. Remarkably, *Aedes nigripes* and *impiger* resist a long, severe winter at latitudes at least as high as 81°N (Ellesmere Island; Corbet & Danks 1975), and *Aedes impiger* achieves very high adult populations during the short summer (Thule, Greenland; Reeves et al. 2013).

The induction of egg diapause can occur in the larvae of the parental generation. *Aedes atropalpus* deposits eggs destined for diapause when the larvae of the maternal generation experience nights 9 hours long (Beach & Craig 1977); therefore, the critical photoperiod for this species is 15 hours. Some species, including *Aedes atropalpus*, *geniculatus*, *togoi*, and *triseriatus*, have a critical photoperiod that shortens northward in apparent adaptation to low temperatures that occur earlier in the fall (Focks et al. 1994).

Winter, or embryonic, diapause has been described in some detail for two species of *Aedes* with very different life histories. One is *Aedes vexans* (Horsfall et al. 1973), a species that as presently defined occurs around the world. It is typical among floodwater mosquitoes, depositing eggs in the upper few millimeters of the soil in locations likely to flood. As with most aedine eggs, they are considered to be "Type II," meaning they will hatch when exposed to the proper stimulus, in this case flooding accompanied by a decrease in dissolved oxygen concentrations (Judson 1960). The number of generations of this species depends on weather conditions, with multiple generations possible if the timing of flood events is favorable. Most eggs will survive two years if conditions for hatching are not favorable, but there are reliable records of some eggs surviving as long as five years. Eggs are resistant to hatching when deposited by a maternal generation exposed to decreasing day length at northern latitudes. These eggs are "Type III," which require a period of "conditioning" at a lower temperature before they will hatch. Following that period they are Type II again and hatch on exposure to the right conditions—usually, the following spring. Populations of *Aedes vexans* from more southern locations were not as resistant to cold, with only 33% of eggs from Alabama (United States, 35°N) surviving a cold winter, in comparison to 80% of eggs from Manitoba (Canada, 55°N). So-called hibernation by *Aedes vexans* was recognized early in the 20th century (Felt 1904).

Another species with egg diapause is *Aedes albopictus*. *Aedes albopictus* is among a small group of mosquitoes with an essentially tropical distribution that nonetheless extend their ranges northward thanks to the occurrence of overwintering mechanisms in some of their populations. The introduction of *Aedes albopictus* across the globe in the 1980s provided an opportunity to observe how diapausing and nondiapausing populations sorted themselves out following introduction to new environments. Early in the infestation of North America, the critical photoperiod that induced diapause in the maternal generation was the same throughout a wide latitudinal range (26°N through 42°N) at about 13.4 hours. This was in contrast to the situation in the native Asian range, where the critical photoperiod was progressively longer toward the north, ranging from 12.6 hours at 26°N to 13.9 hours at 38°N (Focks et al. 1994). Later, Lounibos et al. (2003) showed that during 15 years of residence in North America the species' tendency to enter diapause when exposed to short days (critical photoperiod 10 hours) was progressively greater from the south (at 26°N, 62% did not hatch) to the north (at 38°N, 99.4% did not hatch). Leisnham et al. (2008) found other aspects of populations that varied south to north, suggesting that the most northern populations were more prone to local extinction. Selection of oviposition sites was also affected in a northern location when females were induced to produce diapausing eggs in the fall (Fonseca et al. 2015). Those females were more selective, concentrating eggs in fewer, more favorable sites than summer females, which tended to distribute their eggs more widely. One of the physiological bases of diapausing *Aedes albopictus* may be the abundance of hydrocarbons on the surface of the chorion (Urbanski et al. 2010). The gene *fatty-acyl CoA elongase* is upregulated in oocytes destined for diapause, and that enzyme is probably at least partly responsible for the 33% increase in surface 19 to 51-carbon hydrocarbons (35 and 37-carbon hydrocarbons the most abundant) and the 50% decrease in water loss compared to nondiapause eggs.

Dormancy in the Subtropics

The subtropics are difficult to define precisely. For this section they can be considered those places where temperatures regularly descend below common developmental limits but only occasionally descend to the death point from cold. Such conditions usually present longer developmental seasons with a short season too cool for development. Killing frosts are rare or never occur, so sensitive plant species (e.g., orange trees, *Citrus sinensis* (L.) Osbeck) can be present, at least in favorable microclimates. From a mosquito standpoint, many species from higher latitudes occur in the subtropics, some tropical species survive in the subtropics, and a few species extend all the way from the tropics to the higher latitudes. In addition, some species appear to occur only under subtropical temperature conditions, though whether their distributions are primarily due to temperature regimes or other factors is unclear and probably depends on the adaptations and history of the individual species.

Subtropical temperature regimes present some opportunities to mosquitoes, in that abundant nutritional resources and conducive temperature conditions occur during a larger part of the year than at higher latitudes. On the other hand, species must be capable of dealing with temperatures below developmental thresholds. Either by specific adaptations or by recolonization, they must also be able to survive occasional severe freezing conditions in some locations in the subtropics. Detailed examples of overwintering in the subtropics show a variety of mechanisms for survival of mosquito populations.

Culex quinquefasciatus is the form of the Pipiens Group that is most restricted to tropical conditions, as it is not capable of diapause or other mechanisms that increase its resistance to freezing and desiccation. As described in Chapter 3, *Culex quinquefasciatus* occurs as far north into the middle latitudes as its physiology allows, hybridizing extensively with *Culex pipiens* in North America. At lower latitudes near the tropics, such as in San Antonio, Texas (United States, 29°N), populations are purely *Culex quinquefasciatus*. These populations must survive annual excursions for weeks at a time to temperatures below developmental thresholds (average January minimum 4.3°C) and occasionally experience temperatures typical of much more northern locations. Three years of records of oviposition by this species (Strickman 1998) and records of adult and larval activity in an underground storm drain (Strickman & Lang 1986) gave an indication of how this species overwintered. In a normal year, oviposition continued year-round, decreasing markedly in the fall and occurring only sporadically in the winter, when temperatures warmed sufficiently. Adult males and females were found underground, apparently moving there when temperatures at the surface were too low, and the underground habitat offered more moderate conditions. Oviposition also occurred underground, as indicated by the presence of larvae. In December 1983, record temperatures in that region (lowest temperature −10.6°C on December 25) froze all surface water and apparently resulted in a long delay in the resumption of oviposition until the following April. The population rebounded remarkably, reaching the highest levels of oviposition recorded during three years in May and June 1984. Apparently, *Culex quinquefasciatus* under these conditions slowed its activities in cool weather without ever stopping development, resuming oviposition as soon as temperatures allowed. Following an unusual weather event that must have reduced the population dramatically, some females were probably able to survive in refugia and reproduce when conditions permitted in the spring, reaching very high abundance, possibly in response to fewer predators.

Another mainly tropical species that extends its distribution into the subtropics is *Topomyia yanbarensis*, carefully studied by Okazawa et al. (1986a). This species normally has slow development (25–44 days from egg to adult), but when exposed to short day length (12 hours) and lower-than-normal temperature (24°C), it enters larval diapause in which the fourth instar may persist for up to 292 days. At 30°N with mean January temperatures of 10.5°C–11.5°C, the mosquito fauna of Tanegashima and Yakushima Islands overwinter in a variety of ways (Mogi 1996). Some reproduce all winter (*Aedes japonicus, Aedes togoi, Anopheles lindesayi japonicus, Anopheles lesteri, Culex mimeticus*, and *Culex sasai*), others are dormant adults (*Anopheles sinensis, Anopheles sineroides, Culex vagans, Culex tritaeniorhynchus, Culex pseudovishnui, Culex bitaeniorhynchus, Culex hayashii, Culex pallidothorax, Culex kyotoensis*, and *Lutzia halifaxii*), others survive as eggs (*Aedes nipponicus, Aedes riversi, Aedes albopictus, Aedes flavopictus*, and *Aedes vexans nipponii*), and others get through the winter as fourth instars (*Aedes okinawensis, Armigeres subalbatus, Orthopodomyia anopheloides, Uranotaenia novobscura*, and *Toxorhynchites towadensis*). *Tripteroides bambusae* could be found in the winter as either larvae or eggs.

Other species in the subtropics simultaneously use a number of strategies. *Orthopodomyia signifera* (as *californica*), which typically enters diapause as a fourth instar, made it through the mild, cool season of the Sacramento River Valley (California, 39°N) as second, third, or fourth instars; as adults that survived up to three months; and as eggs that delayed hatching up to 40 days (Baerg 1968). Some *Anopheles freeborni* in the Sacramento River Valley were active throughout the winter, whereas others developed large fat bodies and appeared to enter a true diapause (Rosenstiel 1947). In northern Florida (United States, 31°N), *Orthopodomyia signifera* and *Anopheles barberi* went through the winter in a variety of larval stages, *Aedes triseriatus* survived in its typical embryonic, overwintering form or as fourth instars, and *Toxorhynchites rutilus* persisted through the winter as fourth instars (Bradshaw & Holzapfel 1984). Some individuals of *Culex pipiens pallens* at 33°N in Nagasaki, Japan, continue to reproduce normally until late in the year, when temperatures become too cold for their survival. Others delay follicular development until January, enter sheltered areas, and emerge to reproduce in March (Oda 1968).

Dormancy in the Tropics

It is tempting to think of the tropics as locations where weather conditions never change and remain generally conducive to insect development. Such places exist, but they are exceptional. In fact, tropical seasonality is driven by cycles of wet and dry throughout the tropics and by cycles of temperature and day length at the margins of the tropics. Even in moderate parts of the tropics, many species of insects are dormant in the summer, corresponding to selection pressure from food availability, natural enemies, or moisture availability (Masaki 1980). Warm weather dormancy can be an adaptation to the unpredictable onset of the wet season (Tanaka 2000), to chaotic disruption created by repeated flooding events (Adis & Junk 2002), or to predators (Godfray & Hassell 1987). The positive effect on the distribution of a species can be to assure greater occupation of habitats (Masaki 1980), to enable a wider latitudinal range (Topp 1990), or to synchronize generations (Denlinger 1986).

Tropical dormancy (Denlinger 1986) is a challenge for insects because they must maintain a lower level of activity despite warm conditions, flooding, continued predation, and desiccation. Nonetheless, a wide variety of insects experience tropical diapause as eggs, larvae, pupae, or adults. Diapausing individuals in the tropics often have many of the same characteristics as those in temperate areas, including heavier bodies, more lipid content, lower metabolism, cuticle modification, greater tolerance to desiccation, and/or refuge seeking. The signals that induce diapause in the tropics are more often biological than in temperate areas, including food availability and quality as well as kairomones from plants. Physical signals such as temperature and, occasionally, humidity or rainfall are also important. Surprisingly, small variations in photoperiod can act as signals to induce diapause in the tropics at latitudes higher than 5° north or south. In general, the patterns of a species' dormancy are more variable than in temperate areas, changing geographically and with altitude.

Seasonality of the most common mosquito species collected as adults in Darwin, Australia (12°S), over 15 years was marked (Yang et al. 2008a, 2008b). The abundance of each species varied not only seasonally but also from year to year. In the case of *Aedes vigilax*, that variation was very marked. The causes of the variation were ascribed to both endogenous factors that are usually density dependent and to exogenous factors that are usually density independent. Among the exogenous factors was a change in the intensity of rainfall events. This study provided a rare glimpse of an abundance of species sharing the same geographic location over a significant time period. This data set shows the challenge of disentangling exogenous elements of the environment like the effects of weather, climate change, and human-made change from the effects of endogenous elements like dormancy.

Diapause and other forms of dormancy have not been studied much in tropical mosquitoes. Relatively few species capable of dormancy in higher latitudes with freezing winters also live in the tropics (Ross 1964; e.g., *Aedes albopictus, atropalpus, togoi, vexans*; *Anopheles crucians, lesteri, sinensis*; *Armigeres subalbatus*; *Culex erraticus, pilosus, peccator, pipiens, restuans, stigmatosoma, tritaeniorhynchus*; *Culiseta particeps*; *Psorophora ciliata, columbiae, cyanescens, ferox, howardii, varipes*; *Tripteroides bambusae*). These species do not necessarily express diapause in the tropics, though this would be difficult to document without special studies. *Aedes atropalpus* developed diapausing eggs when the maternal larval generation in El Salvador (14°N) was exposed to nine or more short day intervals, and the temperature was less than 22°C (Beach 1978). Lounibos et al. (2003) noted that 10% of *Aedes albopictus* in tropical Brazil possessed the genotype typical of diapausing populations. The tropical species *Malaya genurostris* was dormant as third and fourth instars on Ishigaki Island (Ryukyu Islands, Japan, 24°N) during winter when the mean water temperature was 18°C (Mogi 1984). Of course, slower development at lower temperatures (e.g., *Aedes aegypti*; Focks et al. 1993b) is common and probably universal among tropical species.

Dormancy among *Anopheles* mosquitoes has a special significance. This genus includes the only vectors of the pathogens responsible for human malaria and are therefore responsible for one of the major causes of disease. Knowledge about dormancy of these mosquitoes might contribute toward the means to disrupt transmission, and it might also explain how malaria can reappear following vigorous attempts to eliminate it. Tropical *Anopheles* vectors of malaria parasites could conceivably maintain the parasites during dormancy, creating a reservoir of pathogens ready to reinfect people in areas where robust control measures have already eliminated malaria from humans.

As discussed above, gonotrophic dissociation is a common mechanism for temperate *Anopheles* to survive the winter, sometimes maintaining an infection with malaria parasites. Rao (1947) documented the same phenomenon in Bhadrak, Orissa (India, 21°N) in *Anopheles annularis* and *culicifacies*. This corresponded to a 7.9% increase in fat accumulation in *Anopheles annularis* and a 2.8% increase in *Anopheles culicifacies*. Like other *Anopheles* in this condition, these Indian mosquitoes continued to take blood meals for energy requirements. This habit led these zoophagous species to take more human blood meals since homes provided the kind of shelter these mosquitoes needed in dormancy.

In Africa, early observations of dormancy of *Anopheles funestus* and *gambiae* took place in what is now Zimbabwe (Leeson 1931). Subsequent studies in Sudan (Omer & Cloudsley-Thompson 1968, 1970), with what was probably *Anopheles arabiensis*, showed other indications of dormant populations during the nine-month dry season. In small villages west of Omdurman (16°N), females (and no males) were found primarily (87.7%) in occupied huts, especially within thatch roofs. Smaller numbers of females were found in cracks in the walls of dry wells, in disused or ruined houses, and in animal burrows. These mosquitoes were all nulliparous, and 77% ($n = 213$) were engorged with blood, 91% of which was human. Ovarian development was delayed despite the blood meals but progressed slowly during the dry season so that most females were nearly ready to deposit eggs at the beginning of the wet season in July. Laboratory studies on the same population showed that when females were kept under dry, cool conditions (20°C, 10%–20% relative humidity), ovarian development ceased despite repeated blood meals, and 10% of females survived 206 days. Holstein (1954) made a number of observations of what were probably *Anopheles arabiensis* in Burkina Faso (12°N), reporting torpid, blood-feeding mosquitoes with mature ovaries that survived five to six months during the cool, dry season in dwellings, uninhabited structures, soil cracks, tree holes, dry wells, and animal pens. Within dwellings, most females were in thatch, with fewer in cracks in walls, under beds, or on surfaces. In the laboratory under ambient conditions, females given repeated blood meals survived up to 156 days and did not oviposit until humidity increased. In Mali (13°N), 6,931 *Anopheles coluzzii* were marked at the end of the rainy season, and a single female was captured 290 days later only 570 m from the site of its release (Lehmann et al. 2010). Females of this species were almost immediately abundant after rains began, which was not true of other species in the area (*Anopheles funestus*, *gambiae*, *rufipes*). One would conclude from these limited studies that part of the populations of some species of tropical *Anopheles* in Africa, including important malaria vectors, probably undergo a form of true diapause during dry seasons in which they continue to take blood meals, cease ovarian development, and alter their metabolism to attain greater longevity.

Diapause associated with gonotrophic dissociation is certainly not the only mechanism available to African *Anopheles* for survival during adverse, dry conditions. Population genetics (Taylor et al. 1993) showed that *Anopheles arabiensis* and *coluzzii* do not experience genetic bottlenecks during the dry season in parts of Mali, Nigeria, and Burkina Faso, where 90% of the annual rains (600–1,000 mm) fall in July through September. Similarly, *Anopheles gambiae* in western Kenya showed no signs of genetic bottlenecks despite low apparent populations in the dry season (Lehmann et al. 1998). Such results imply that the population does not decrease as much as casual observations would suggest. Although diapause may be an important part of the survival strategy for some species in some locations, other areas harbor reproducing populations in refugia that offer favorable conditions. *Anopheles arabiensis* was present as larvae, males, and parous females in Khartoum throughout the dry season, even though the isolated village sites near Omdurman appeared to have diapausing individuals of the same species (Omer & Cloudsley-Thompson 1970). Larvae of *Anopheles coluzzii* were collected in the middle of the dry season in Mali, where some adults were surviving through the long Sahelian dry season (Lehmann et al. 2010). In Marigat (Kenya, 0.45°N), where there is only 600 mm annual rainfall, larval sites were available all year for *Anopheles arabiensis*, *funestus*, *pharoensis*, and *coustani*. Caged *Anopheles arabiensis* continued to oviposit throughout the dry season when exposed to ambient conditions, and 76% of field-caught females were parous (Mala et al. 2014). A little further west, at Mbita Point (Kenya, 0.43°S), a survey of 128 potential larval sites during the hot, dry season found only four with water, and a separate survey found 280 *Anopheles gambiae* and *arabiensis* larvae in a few ground pools near Lake Victoria (Minakawa et al. 2001). Studies in the Kilombera Valley (Tanzania, 8°S) also showed no signs of gonotrophic dissociation among *Anopheles arabiensis*, *funestus*, and *gambiae* (Charlwood et al. 1995b, 2000). During the warm, dry season (September–December), there were larval refugia of *Anopheles arabiensis* in the field and refugia of *Anopheles gambiae* and *funestus* associated with human activity. Such refugia were not present during the hot, dry season (February–March), and the population of *Anopheles arabiensis* diminished throughout that period.

The eggs of tropical *Anopheles* can often survive longer than the few days normally required for embryonation and hatching. Although there is little evidence that tropical anopheline eggs survive for months, as in the case of the temperate *Anopheles walker* (Hurlbut 1938b), successive delays of weeks might make the difference between survival and extinction of individual populations. *Anopheles albimanus* in Panama (9°N) deposited larger eggs at the beginning of the dry season that hatched in 7–14 days rather than the usual 1–2 days during the rainy season. Field observations may have detected much longer survival of eggs (Stone & Reynolds 1939). In eastern Malaysia, eggs of *Anopheles balabacensis* were commonly stranded up to five days at the edges of pools in the field, subsequently hatching successfully. In the laboratory, 13% of eggs hatched following 14 days of storage on damp filter paper, and a few eggs (3 of 500, or 0.6%) hatched after 27 days (Rajapaksa 1971). *Anopheles baimaii* (as *dirus*) in Bangladesh commonly deposited eggs above the waterline, which remained viable for two weeks and allowed synchronous hatching of a larger cohort of mosquitoes (Rosenberg 1982). *Anopheles melas* is another mosquito that prefers to oviposit on damp surfaces rather than on open water, in this case as an adaptation to its tidal habitats in western Africa (Giglioli 1965) and timed to biweekly inundations (Muirhead-Thomson 1946). In experiments with *Anopheles gambiae*, between half and three-quarters of the eggs placed near water survived for three days in one study (Koenraadt et al. 2003), and 20% hatched after 10 days of storage on a damp surface in another study (Beier et al. 1990). Soil samples from dried larval habitats (margins of permanent pools, streambeds, temporary pools, hoofprints) collected in the dry season at Kissian (Kenya, 0°N) produced larvae of both *Anopheles gambiae* and *funestus* when flooded in the laboratory (Beier et al. 1990). Ironically, the most thorough laboratory study of delayed hatching by *Anopheles arabiensis* was conducted in Cumbe, Brazil (6°S), in association with the invasion of that country by this species in the 1930s (Chapter 3). Deane and Causey (1943) examined the survival of eggs deposited by wild females on filter paper and stored between layers of damp sand, finding that 27.4% survived 10 days, and 0.3% survived 18 days. An experiment comparing gradual drying of eggs deposited by laboratory-reared mosquitoes and stored between layers of either mud or sand showed much better survival in sand (72% survival at 8 days, 26.5% survival at 12 days) than in mud (10% survival at 8 days, 1.5% survival at 12 days). These authors also noticed a difference in morphology between eggs deposited by females reared under summer conditions (20°C–30°C) compared to those reared under winter conditions (10°C–13°C). Holstein (1954) also observed significant survival of what was probably *Anopheles arabiensis* in West Africa when eggs were stored on damp clay (33.3% survival for 10 days, 1.9% survival for 16 days).

Conclusion

Insects have very specific requirements for survival and development; however, when these conditions are not available, they must escape or delay development in order for the population to survive. Mosquitoes possess a variety of mechanisms to survive adverse conditions. These mechanisms range from highly programmed, hormonally controlled diapause in high latitudes that may result in development of a single generation per year to mixed mechanisms used by various parts of a single population in the subtropics and tropics. Dormancy can occur in any stage except the pupa, and all egg dormancy is as a pharate first instar. The longest dormancy is in some species of *Aedes*, which may survive for years as eggs under the right conditions. Adult longevity may be much extended also, usually as mated females. Dormancy also commonly occurs in the tropics, though studies to date have only begun to describe it.

Despite the great importance of dormancy to mosquitoes' lives, survival, potential to transmit pathogens, and distributions, this subject remains one of the least studied aspects of their biology. This may be because it is often difficult to find dormant mosquitoes, or it may be because people tend to study those aspects of biology in which there is a perception of progress rather than stasis. Hopefully, future efforts will concentrate on mosquito dormancy in order to develop a better understanding of how this phenomenon affects mosquito biology and human health.

6. Mosquito Movement

There must be for every living unit a certain distance which that unit may possibly cover if it continues to move all its life, with such capacity for movement as nature has given it, always in the same direction.

Ross 1905

This quote is from a seminal paper by the person credited with the discovery of malaria parasite transmission by mosquitoes. This work pointed out that effective control of mosquitoes is inextricably linked to the distance those mosquitoes can move. It is common for mosquito biologists to lament the lack of knowledge and understanding about the movement of adult mosquitoes, despite over 100 years of effort in this field. Of course, following the flight paths of individual mosquitoes is a challenging proposition even for modern technology, and we really would like to know where each mosquito is located, where it is going, and why it is going there. These questions are asked less often for larvae, but they are just as relevant to an understanding of mosquitoes and how to control them. Perhaps the single biggest generalization about mosquito movement, whether larvae or adults, is that survival of each species would be impossible without the ability to move and, in some cases, to move considerable distances. A moment of consideration of the morphological and behavioral adaptations for flight and larval swimming shows how much of what we consider to be "mosquito" is associated with mobility. The following reviews some aspects of movement for each mosquito life stage, including mechanics, variation between species, and purpose.

Egg Movement

Mosquito eggs have no means of moving other than by external physical forces. Those forces include interaction with surface tension on the water's surface, water currents, wind, and animals. Any of those forces requires proximity to the eggs; therefore, placement of the eggs by the ovipositing female is going to have a great effect on what happens to the eggs afterward. If an egg ends up in the wrong location, the larva may not be able to hatch, or the egg might be more vulnerable to predation. It would seem that any movement of the egg would be a net negative for survival since the female presumably places it in a suitable location. On the other hand, dispersion of the eggs by external forces could result in colonization of new areas or coincidental placement in a location where survival of the larva is more likely.

Floating eggs, whether in rafts or as individual eggs, are highly subject to the shape of the water surface and to interaction with each other. *Anopheles* eggs that float on the surface often self-assemble into rosettes held together by a combination of their mutual hydrophobicity and the shape of the surface tension (Burgess & Cowan 1993). Single floating eggs (e.g., *Mimomyia hybrida*) may slide toward the depression in the surface tension created by aquatic plants or the edge of a tree hole (e.g., *Toxorhynchites* spp.; Clements 1999). Egg rafts (eggs oriented vertically in a group) produced by various mosquitoes commonly move toward emergent vegetation or the edges of other structures for the same reason. In the case of *Trichoprosopon digitatum*, the female protects the raft from the disruption of surface tension caused by rain, therefore preventing its dislocation (Lounibos & Machado-Allison 1983; Clements 1999).

Although not well documented, flowing water can redistribute floating eggs deposited in still water but later dislodged into the moving portion of a stream or river. Unpublished work in Illinois, United States, showed that *Culex pipiens* egg rafts were commonly captured in screens placed in the middle of a flowing, urban stream. These rafts had probably been deposited in small embayments created by corrugated ironwork or masonry riprap but were later washed out by the temporary increase in water level common to many urban streams immediately following rains and stormwater input. Captive females in the laboratory readily deposited eggs in the slowly rotating water of a cylindrical container, and larvae successfully developed under the same conditions. The same kind of redistribution may occur for other species that deposit floating eggs at the edges of streams or rivers, like *Anopheles darlingi* or *Anopheles minimus*. Distribution of *Anopheles melas* eggs by tidal flow was observed in Sierra Leone (Muirhead-Thomson 1946). One can imagine that such a mechanism could redistribute mosquitoes for considerable distances.

Wind can also move floating eggs. The adaptation of the larvae of some *Anopheles* species to move across the surface of damp mud to water in a puddle (see "Larval Movement") may be associated with the tendency of their eggs to be blown to the edge of the water where they can become stranded. Miller et al. (2007) showed that 79% of *Anopheles gambiae* eggs became stranded in this way under simulated conditions in the laboratory.

Phoretic movement of mosquitoes on human transport may involve eggs as well as other stages. On a long trip, as on the old sailing ships, complete generations of mosquitoes were probably completed so that eggs were part of the overall movement of the species to new locations. Presumably, the original infestation started sometimes as eggs in containers. Currently, the presence of persistent aedine eggs in tires is a challenge to those attempting to prevent introduction of new species or the addition of new genetic material to an existing infestation.

Eggs deposited in soil by floodwater mosquitoes often move within the soil profile. Horsfall et al. (1973) presented a detailed review of how eggs deposited on the surface by *Aedes vexans* can be relocated to deeper levels within the soil profile. Movement of eggs up to 50 mm below the surface during prolonged periods of dryness that prevented hatching was observed. Siltation on flood plains could bury eggs deep enough to prevent successful emergence of first instars. Burial by animals was also seen, especially from the activity of earthworms.

Perhaps the strangest example of mosquito egg movement is caused by the oviposition behavior of some sabethines. Female *Sabethes chloropterus* in Panama hovered in front of small holes (2.5 cm) on the lateral sides of bamboo in the laboratory, tossing eggs individually up to 60 mm through the holes into the liquid contained within the bamboo section (Galindo 1958). The females deposited only one or two eggs at a time, ejecting the eggs rapidly and accurately. Females of *Topomyia yanbarensis* have a similar behavior (Okazawa et al. 1986a) in the Tokara Archipelago, a chain of islands in southern Japan.

Larval Movement

Mechanics

Attempts to systematize and analyze larval movement have been limited. Beyond simple description of the shapes of swimming motions and the general observation that mosquito larvae always swim in a posterior direction, there is some disagreement (Clements 1999) between the model proposed by Nachtigall (1965) and that of Strickman (1989a). Nachtigall contrasted *Aedes aegypti* with a few other aquatic dipteran larvae to describe mosquito larval movement as the result of a paddling motion of the anal fan (setae on segment 10A) combined with setal fans on the siphon of some mosquitoes. This was in contrast to the undulatory movement of other larvae, as carefully described for a ceratopogonid biting midge larva by Linley (1986). Such undulatory movement depends on the forces exerted sequentially as a sinusoidal wave moves down the length of a linear body (Gray 1953). Based on an analysis of single movie frames photographed at 1/64th of a second, Strickman made the case that typical larval swimming motions produce undulatory forces similar to those occurring in snakes or ceratopogonids. In contrast to snakes, where at least one complete sinusoidal wave is present at a given moment, mosquito larvae were said to create only half of a wave at a time (Fig. 6.1). In fact, both mechanisms probably create significant propulsive forces. Much more careful biomechanical work would be necessary to actually measure where these forces are coming from and whether the anal fan is more involved in steering than thrust, as suggested by Snodgrass (1959).

The appearance of swimming motions has been classified into basic patterns based on observations of 46 species of Central American mosquitoes (Strickman 1989a). Each type of movement is readily distinguished by the eye, but in fact they are simply variations on the sequential formation of half-sinusoidal waves based on factors such as the depth of each curve, the position of the anterior curve compared to the posterior curve, and the speed of movement (Fig. 6.2).

These observed patterns of larval movement include the following:

1. *Sustained irregular movement*: Irregular movement characterized by a smaller anterior curvature than posterior and variation in speed that gives an irregular impression. This type of movement is typical of *Culex pipiens* and has been described as having the appearance of a series of C shapes. Long-duration, sustained, irregular movement is characteristic of some species and results in a straighter line of travel, whereas short-duration, sporadic, irregular movement is characteristic of others and causes a more tumbling path (examples: *Aedeomyia squamipennis*, *Aedes* (*Howardina*) *guerrero*, *Culex* (*Carrollia*) *secundus*, *Culex* (*Culex*) *coronator*, *Culex* (*Culex*) *declarator*, *Culex* (*Culex*) *nigripalpus*, *Culex* (*Culex*) *quinquefasciatus*, *Culex* (*Culex*) *thriambus*, *Culex* (*Culex*) *pseudostigmatosoma*, *Culex* (*Melanoconion*) *erraticus*, *Culex* (*Melanoconion*) *garcesi*, *Culex* (*Melanoconion*) *pilosus*, *Culex* (*Melanoconion*) *imitator*, *Uranotaenia* (*Uranotaenia*) *lowii*).

2. *Fast or slow sinuous movement*: Sinuous movement appears more regular, as the larva forms anterior and posterior curves of about the same size. However, the posterior power stroke is faster than the anterior, detracting from the appearance of a constant speed. This leaves the impression of a series of S shapes, characteristic of *Aedes aegypti*. Some species have fast, sinuous movement that is more difficult to follow with the eye, and others have slow, sinuous movement that looks very much like sinusoidal movement (examples: *Aedes*

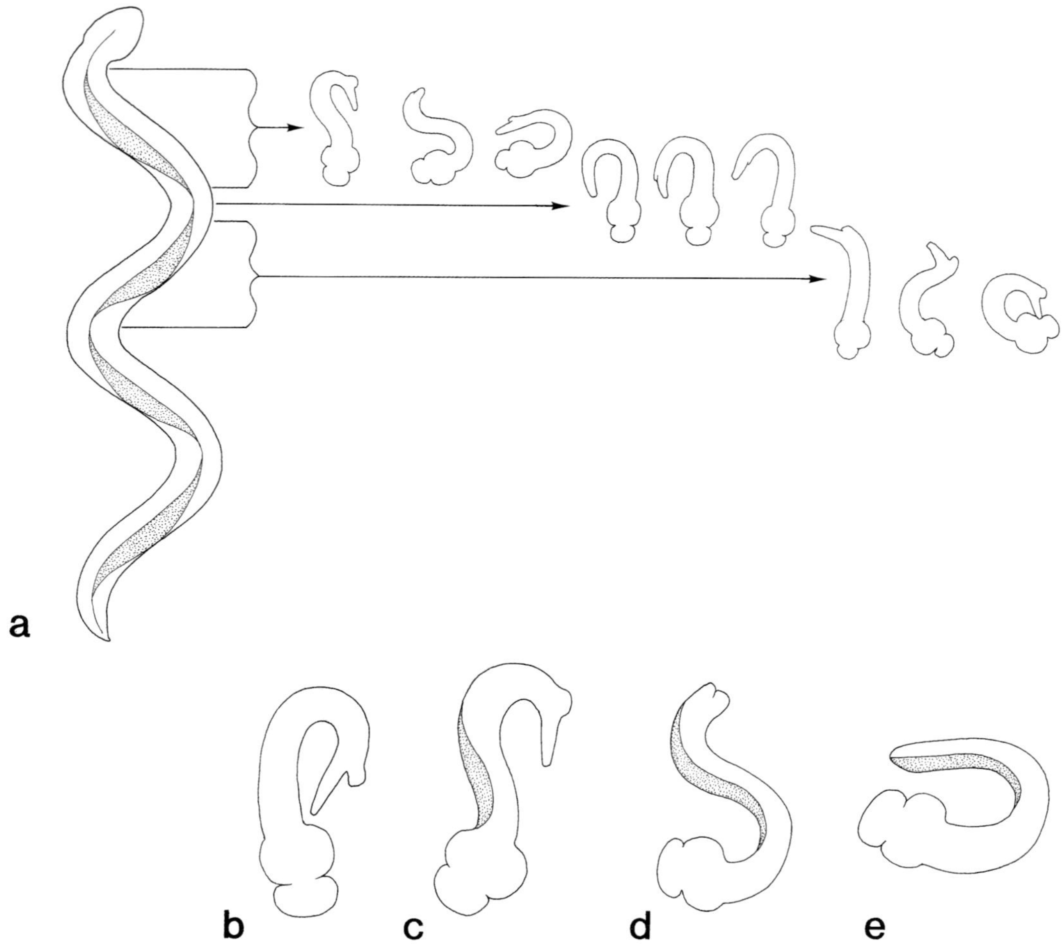

Figure 6.1. The sinusoidal nature of propulsion in mosquito larvae. a, with force qualitatively indicated by shaded areas, comparison of force exerted by one wavelength of movement by a snake to the succession of postures of a swimming mosquito larva. b–e, successive application of force by a swimming mosquito larva, showing how the summation of forces is similar to a sinusoidal wave. Strickman 1989a

(*Aedimorphus*) *vexans*, *Aedes* (*Georgecraigius*) *epactius*, *Aedes* (*Ochlerotatus*) *taeniorhynchus*, *Culex* (*Phenacomyia*) *corniger*, *Culiseta* (*Culiseta*) *particeps*, *Deinocerites pseudes*, *Haemagogus* (*Haemagogus*) *anastasionis*, *Haemagogus* (*Haemagogus*) *equinus*, *Haemagogus* (*Haemagogus*) *mesodentatus*, *Psorophora* (*Grabhamia*) *confinnis*, *Psorophora* (*Janthinosoma*) *ferox*, *Toxorhynchites* (*Lynchiella*) *theobaldi*, *Wyeomyia* (*Wyeomyia*) *pertinans*).

3. *Semisymmetric movement*: This is the most graceful in appearance, consisting of a series of deep curves alternating direction in a steady pace. The anterior flexion is almost absent, and the overall impression is that of a graceful, sinusoidal track (examples: *Aedes* (*Howardina*) *quadrivittatus*, *Culex* (*Carrollia*) *bihaicolus*, *Deinocerites cancer*, *Limatus durhamii*, *Mansonia* (*Mansonia*) *titillans*, *Orthopodomyia kummi*, *Sabethes* (*Sabethes*) *cyaneus*, *Sabethes* (*Sabethinus*) *undosus*, *Sabethes* (*Sabethoides*) *chloropterus*, *Wyeomyia* (*Wyeomyia*) *arthrostigma*, *Wyeomyia* (*Wyeomyia*) *mitchellii*).

4. *Anopheline movement*: *Anopheles* larvae move in a characteristic "kick-and-glide" motion in which flexion and deflexion are rapid (combined, usually less than 1/64th second), with the anterior curve much smaller than the posterior curve. The larva typically pauses between flexions, skidding just below the surface from the force of the movement.

Other observations describe movement in different ways and under special circumstances. Peyton (1977) said that movement by *Uranotaenia* (*Pseudoficalbia*) species was mostly aedine, which may be interpreted as being some form of sinuous movement. Under special circumstances, larvae may propel themselves by entirely different mechanisms than swimming motions and in an anterior direction. *Eretmapodites dracaenae* lives in plant axils or accumulations of water in leaves and goes in pursuit of prey with a "peculiar, rapid, gliding motion" using its mouth brushes (Hopkins 1952). A number of other larvae also glide along on substrates using their mouth brushes—for example, *Aedes vexans*, *Aedes epactius*, *Aedes taeniorhynchus*, *Culex declarator*, *Culex quinquefasciatus*, *Culex pseudostigmatosoma*, *Deinocerites cancer*, *Psorophora confinnis*, *Trichoprosopon digitatum*, *Wyeomyia arthrostigma*, and *Wyeomyia pertinans* (Strickman 1989a).

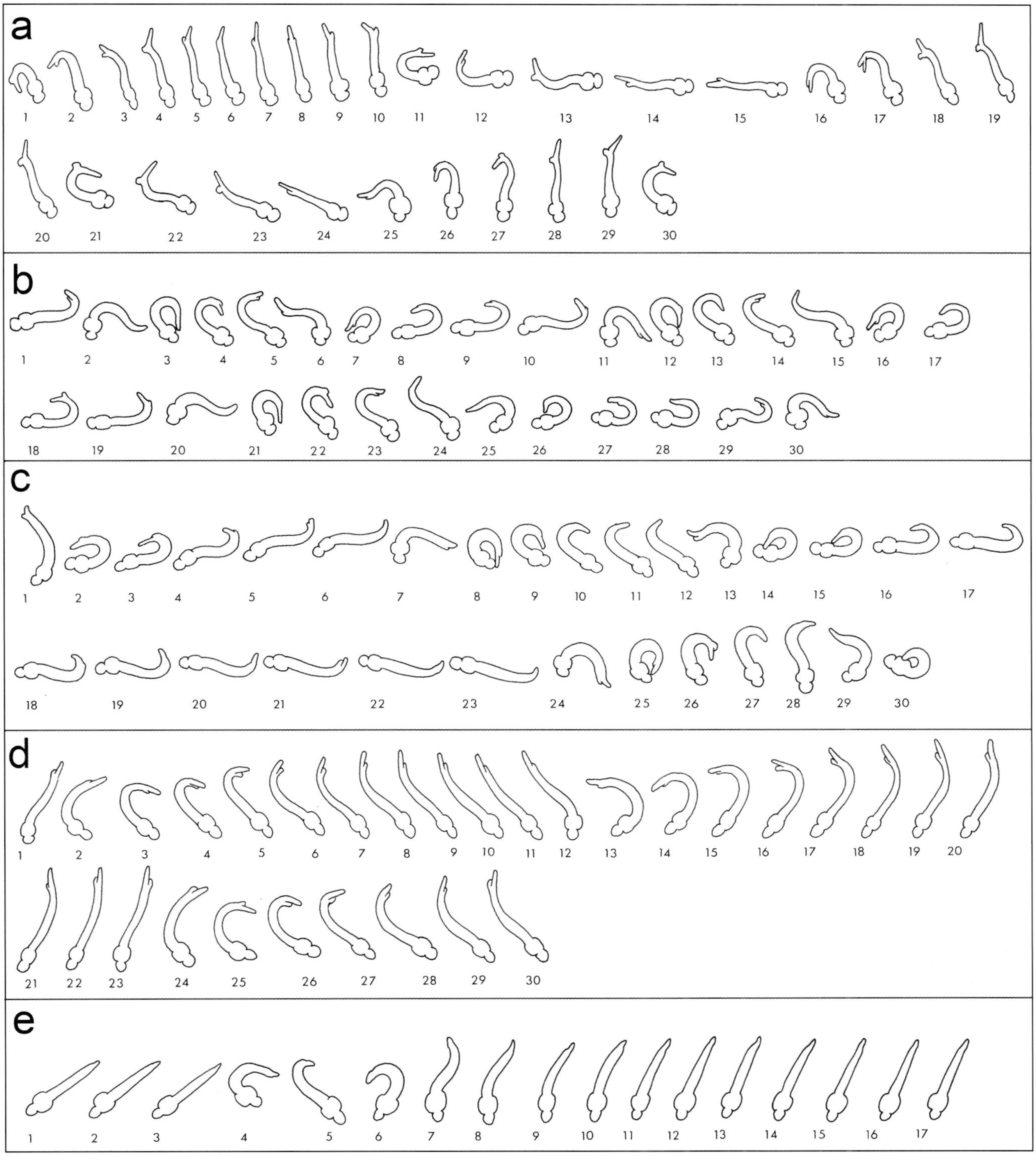

Figure 6.2. Five categories of larval movement illustrated by outlines of larvae filmed at 1/64th-second intervals. a, sustained irregular (*Culex nigripalpus*); b, fast sinuous (*Aedes taeniorhynchus*); c, slow sinuous (*Haemagogus anastasionis*); d, semisymmetric (*Trichoprosopon digitatum*); e, anopheline (*Anopheles hectoris*). Strickman 1989a

The same author also observed *Culex erraticus* and *Culex garcesi* moving under the surface tension layer by pushing themselves along with their siphons and *Shannoniana moralesi* moving on substrates by pushing itself forward with nodding motions of its head. Larvae that live in the viscous media within fruit husks, such as *Trichoprosopon digitatum*, *Eretmapodites quinquevittatus*, and *Armigeres theobaldi*, can change to a crawling motion (Lounibos 1983) interpreted as being sinusoidal movement that propels the larvae in the anterior direction.

Circumstances can lead to situations in which larval movement is essential for reaching pools of water. For example, *Aedes sollicitans* can move to more persistent water in the upper salt marsh when small pools of water start to dry up (Shone et al. 2006). *Aedes kochi* has been reported to move across damp surfaces to reach water in leaf axils and jump up to a centimeter by flexing its body (Horsfall 1955). At least some species of *Anopheles* commonly move over mud to pools of water, as the eggs are either stranded or intentionally deposited on a damp surface. Rosenberg (1982) described how *Anopheles baimaii* (cited as *Anopheles dirus* and corrected by Sallum et al. 2005b) generally deposits its eggs on damp surfaces above the waterline, and first instars crawl, moving forward, up to 53 cm to pools of water. *Anopheles gambiae* has a similar behavior but apparently at a much lower efficiency, with only 4% of larvae able to move 2 cm and 2% able to move 10 cm across damp soil to a pool of water in the laboratory (Koenraadt et al. 2003). Miller et al. (2007) worked with *Anopheles gambiae* s.l. in western Kenya and showed somewhat different results. They observed oviposition on mud adjacent to pools of water, much as observed for *Anopheles baimaii* by Rosenberg (1982), and also stranding of eggs on soil at the edge of pools either by air currents or receding water levels. First and fourth instars up to 1 cm away reached the water in one of three ways: (1) by normal undulatory motion in films of water that were greater than one-third the height of the larvae, (2) by a caterpillar-like movement in an anterior direction if the water film was less than one-third the height of the larva, or (3) by passively being washed into the pool by rain. The differences from the observations of Koenraadt et al. (2003) might have been due to working with *Anopheles arabiensis* rather than *Anopheles gambiae* s.s. In addition, Miller et al. (2007) stated that *Anopheles maculipennis* in Russia had been observed to move over damp soil.

Passive Movement

Passive movement is much more likely in fluid environments than terrestrial ones. Mosquito larvae are often in containers or small pools where water flow is minor, but even in these habitats, rain or human intervention may result in displacement. Some of that phoretic movement has involved larvae as well as eggs—for example, *Aedes aegypti* in ships and *Aedes albopictus* in tires. Other recently imported container-developing *Aedes* may or may not have been transported as larvae, as their resistant eggs might have been sufficient for the introduction. Early introductions of *Culex pipiens* and *Culex quinquefasciatus* via sailing ships must have involved all life stages, including larvae. Intentional transport of ornamental bromeliads was responsible for introducing *Culex biscaynensis* to Florida and *Wyeomyia mitchellii* to Hawaii, both species probably as larvae. *Anopheles darlingi* larvae were apparently brought into Iquitos, Peru, with tanks of tropical fish. Perhaps most unusual is the speculation that *Mansonia uniformis* was taken into Guam as larvae attached to aquatic plants on the floats of sea planes.

Many species of mosquitoes deposit eggs either directly in the water at the margins of streams or rivers (e.g., *Anopheles darlingi*, *Anopheles minimus*) or in rock pools near streams (e.g., *Aedes epactius*, *Aedes atropalpus*), or, as floodwater mosquitoes do, in the soil of floodplains (e.g., *Aedes trivittatus*, *Aedes vexans*). Larvae hatching from these eggs are subject to movement in flowing water as water levels rise or as larvae wander into the flowing portion of the water. The movement of *Culex pipiens* egg rafts in urban streams described above applied equally to the larvae that hatched from those eggs. Larvae of *Culex quinquefasciatus* in San Antonio, Texas, United States, flowed distances within underground storm drains, sometimes reaching the San Antonio River (Strickman & Lang 1986). Majambere et al. (2010) speculated that the difficulty of using larval control along the Gambia River in Africa was at least in part caused by the movement of larvae of *Anopheles gambiae*, *Anopheles arabiensis*, and *Anopheles melas* in the tidal flow.

Behavior and Posture

Movement of mosquito larvae is a big part of their behavior. One aspect of that behavior is simply their position in the water column, some species preferring to spend most of their time at the surface, middle, or bottom (Strickman 1989a). *Anopheles* are usually directly below the water surface, apparently clinging to the surface tension with specialized setae. Under artificial conditions in the laboratory, *Uranotaenia lowii* and many species of *Aedes*, *Culex*, *Culiseta*, *Deinocerites*, *Haemagogus*, *Orthopodomyia*, and *Psorophora* occupied the middle of the water column most of the time. Among the *Culex*, three species in the subgenus *Melanoconion* (*garcesi*, *imitator*, and *pilosus*) spent more time on the bottom of a container. When not attached to a plant, *Mansonia titillans* also tended to stay on the bottom, as did *Toxorhynchites theobaldi*. All of the sabethine genera observed (*Limatus*, *Sabethes*, *Trichoprosopon*, and *Wyeomyia*) spent most of their time on the bottom of a container in the laboratory.

The posture of mosquito larvae is another aspect of their behavior with respect to the water surface or features of the substrate (Fig. 6.3). Larvae hang at different angles from the surface. For example, *Wyeomyia arthrostigma* was almost perpendicular, whereas *Uranotaenia lowii*, *Culex nigripalpus*, and *Psorophora confinnis* were at a 45° angle, and the *Aedes* species were oriented to the surface at a less acute angle. *Mansonia titillans* attached to a plant held itself horizontally from the root. Larvae resting on the bottom might be oriented venter downward (e.g., *Trichoprosopon digitatum*, *Culex quinquefasciatus*) or venter upward (*Culex imitator*, *Wyeomyia arthrostigma*). The same position of lying on the bottom while filtering water has been observed in Africa for *Aedes argenteopunctatus*, *Culex poicilipes*, and *Culex sinaiticus*. This inverted habit can be associated with specialized adaptations of the siphon and development of "cephalic gills," which are bladder-like organs on the underside of the head connected to the tracheolar system. Cephalic gills may be modifications of the tracheolar system at the bases of the antennae and in *Aedes argenteopunctatus*, at least, are held open by internal air pressure. Cephalic gills have been observed in *Aedes argenteopunctatus*, *Culex argenteopunctatus*, *Culex decens*, *Culex perfuscus*, *Culex poicilipes*, *Culex sinaiticus*, and *Culex weschei* (Lewis 1949; Hopkins 1952). *Malaya genurostris* has specialized adaptations of its siphonal setae and lateral prothoracic

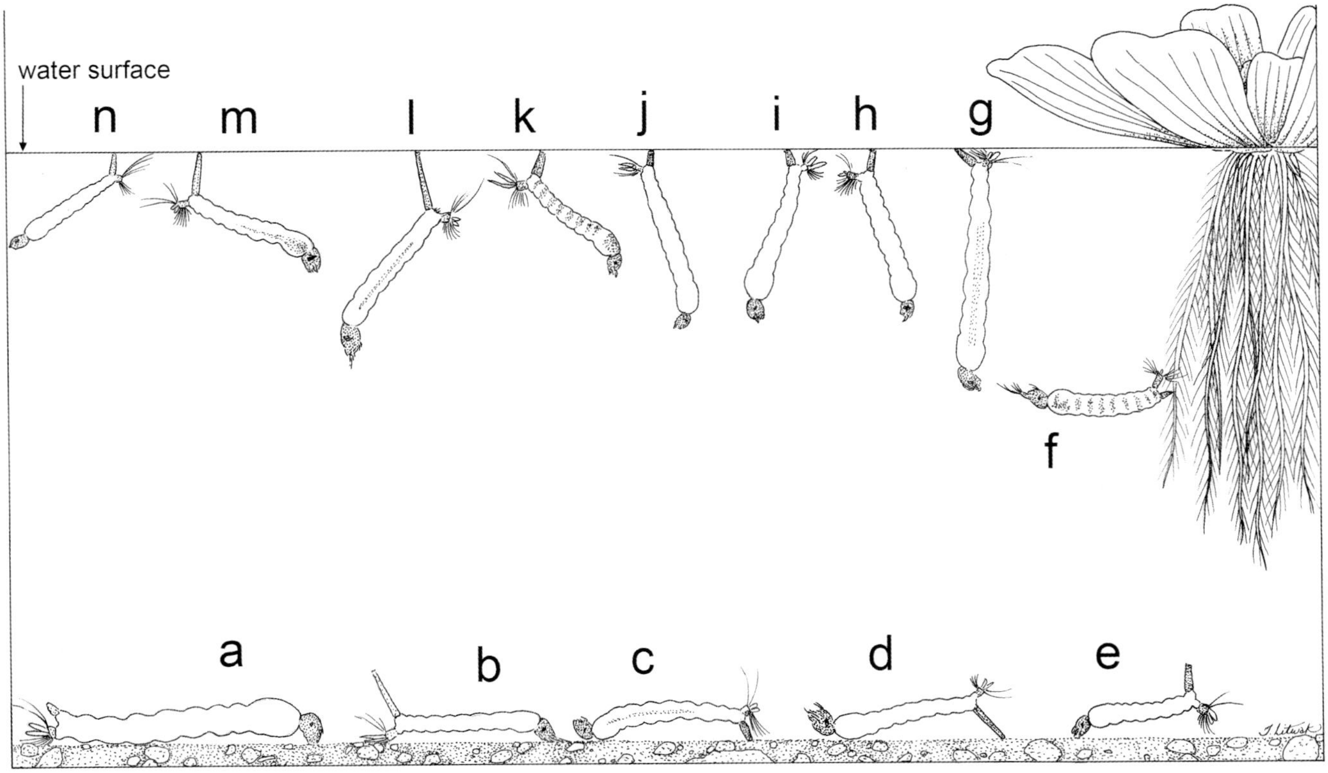

Figure 6.3. Typical postures of Central American mosquito larvae. a, *Trichoprosopon digitatum*; b, *Culex* (*Culex*) sp.; c, *Wyeomyia arthrostigma*; d, *Culex imitator*; e, *Culex quinquefasciatus*; f, *Mansonia titillans*; g, *Wyeomyia arthrostigma*; h, *Aedes taeniorhynchus*; i, *Aedes epactius*; j, *Aedes guerrero*; k, *Psorophora confinnis*; l, *Culex coronator*; m, *Culex nigripalpus*; n, *Uranotaenia lowii*.

setae to support itself while resting on the bottom (Iyengar & Menon 1948).

Detailed study of a single species can help systematize what seem like many different actions by larvae. Walker and Merritt (1991) observed, categorized, and quantified the behavior elements of *Aedes triseriatus* in the laboratory, identifying 14 individual behaviors. The majority of time was dedicated to a variety of feeding actions, including surface suspension feeding, suspension feeding in the middle of the water column, and browsing at or below the water surface. The remainder of the time was spent going from place to place by swimming or drifting and sitting still underwater. Each behavior was related to others with respect to sequence. One cluster of related behaviors at the surface often led to a dive toward the bottom and a separate cluster of behaviors.

Diving can be an escape response from predators—a response affected by the presence of predators and the age and species of larvae (Futami et al. 2008). *Aedes triseriatus* is said to dive less frequently in the presence of predatory *Toxorhynchites* larvae, and *Culex pipiens* moves less in the presence of notonectid bugs, presumably to avoid alerting the predators to their locations. For *Anopheles gambiae*, the duration of a dive was inversely proportional to the instar's stage and relatively less effective for second instars trying to escape predation from an aquatic lycosid spider when water was shallower. Pesticide exposure can also influence movement (Tomé et al. 2014). *Aedes aegypti* swam a greater distance as a fourth instar following exposure to imidacloprid as a third instar but a shorter distance when similarly exposed to spinosad. Deltamethrin slowed swimming speed, but azadirachtin had little effect on swimming. The tendency to dive varies between species, even closely related species in the same habitat (Workman & Walton 2003). *Culex erythrothorax*, compared to *Culex quinquefasciatus*, *Culex tarsalis*, and *Culex stigmatosoma*, is underrepresented in dipping samples from reconstructed wetlands because it tends not to cluster and dives more readily with and without a stimulus.

The frequency, speed, and direction of larval movement by four species was the subject of a series of studies by Rasnitsyn and Lebedeva (1998, 2000, 2002). They found that *Aedes aegypti* moved more frequently than *Anopheles sacharovi*, which moved more frequently than *Culex pipiens*. The effect of food on frequency of movement was positive in the case of *Anopheles sacharovi* but negative in the case of *Culex pipiens*. Magnetic fields may have influenced the direction of movement of *Aedes aegypti* and *Culex pipiens* since over 60% of the motions of each were in directions between NNW and NNE. *Anopheles sacharovi* had no such orientation, but its movement was characterized by changes in direction of approximately 90°. Speed of movement decreased from the second to the third instar in *Culex pipiens* (from 25 to 10 cm/s) but increased in *Aedes aegypti* (from 29 to 44 cm/s). Distances traveled per 15 minutes were remarkably long, exceeding 2 m in *Aedes aegypti* (Table 6.1). Exposure to food, as yeast paste, decreased turns and movement of *Anopheles gambiae* larvae, in contrast to presence of

Table 6.1. Distance traveled in 15 minutes by larvae in the laboratory (Rasnitsyn and Lebedeva 2002)

Species	Mean distance (cm) traveled in 15 minutes		
	Second instar	Third instar	Fourth instar
Aedes aegypti	134	255	278
Anopheles sacharovi	19	44	29
Culex pipiens	175	80	36

the common repellent active ingredient DEET, which increased turns and movement. Separate receptor systems were involved in this larval response, the food response guided by ionotropic receptor protein systems and the repellent response guided by odorant-binding protein systems (Liu et al. 2010).

Pupal Movement

Culicine pupae move actively even though they do not feed. Pupae require nearly constant contact with the surface through the trumpets, which are tubes leading to the tracheal system; therefore, when pupae dive in response to predators or other stimuli, they must soon return to the surface to maintain respiration. The mechanism for movement is quite different from that of larvae and largely powered by the expanded surfaces of the paddles. The paddles are appendages of the ninth abdominal segment and appear to arise from the previous segment (Harbach & Knight 1980). Abdominal muscles span the length of the abdomen either anteriorly or posteriorly. Along the surface, pupae typically move in an anterior direction by kicking backward; however, diving pupae maintain their dorsally oriented posture by moving in a posterior direction. Counter to previous assumptions, abdominal and thoracic setae have no role in maintaining contact with the surface of the water (Romoser & Oviedo 2011). The return to the surface is typically dependent on buoyancy (Snodgrass 1959). At the bottom of a dive, the pupa orients itself with the pupal trumpets upward in a process of self-righting (Awasthi et al. 2012). Pupae can regulate their buoyancy by compressing the ventral air space, even achieving a negative buoyancy to stay on the bottom (Romoser & Oviedo 2011; Awasthi et al. 2012).

Presumably, most movement by pupae is motivated by avoidance of predators or other stimuli. They may also dive to avoid being washed out of a habitat. Diving behavior of *Culex quinquefasciatus* was influenced by the presence of a predatory guppy fish, *Poecilia reticulata* Peters (Awasthi et al. 2012). Classifying dives as either shallow (5–20 mm), medium (20–50 mm), or deep (>50 mm), the authors found that over 15 minutes pupae only moved along the surface (39% of pupae) or made shallow dives (61% of pupae) when fish were absent. Introduction of a single fish to a container with 20 pupae resulted in a reduction of surface movement to only 14% of pupae, a reduction in shallow dives to 42% of pupae, and an increase in medium dives to 32% of pupae and in deep dives to 11% of pupae. The fish was unable to catch pupae during a dive, and pupae escaped 58% of the attacks; nonetheless, an average of 82% of the pupae were eaten in 15 minutes. The speed of pupal movement during a dive increased from an average of 70 mm/s in the absence of a fish to 99 mm/s in the presence of a fish. Surface movement increased from 57 mm/s to 94 mm/s and the distance during surface movement from 19 mm to 65 mm. Pupae were able to perceive the direction from which the fish was coming, possibly aided by mechanoreception of the first abdominal setae (Romoser & Oviedo 2011). Similar to larvae, pupae were affected by exposure to insecticides, and the behavioral results of the exposure depended on the kind of insecticide (Tomé et al. 2014).

Adult Movement

One look at a mosquito and a little thought makes it obvious that much of the visible morphology represents adaptations to movement, either by walking or flying. The long legs provide effective propulsion on surfaces while accommodating the ability to hold the body high enough to position the long proboscis for feeding on blood-meal hosts or sugar sources. Those legs end in claws, or ungues, that are often morphologically distinct between species and may or may not have a pad, the empodium, between them. The claws and empodium are almost certainly involved in grasping surfaces or providing traction or enabling the mosquito to stand on water. A single pair of wings emerging from the mesothorax are the organs that make mosquitoes one of the more powerful insect fliers. The highly organized and remarkably consistent pattern of veins on the wings is involved not only in the circulation of hemolymph but also in the formation of a highly dynamic flight surface that varies in shape during each wing stroke and according to the flight path of the mosquito. As in all Diptera, the small, club-shaped halters (alternative spelling: halteres) are posterior to the wings and are actually the highly modified second pair of wings (metathoracic wings) present in most other insects. These are said to function as a gyroscope, providing nervous feedback to the mosquito on its orientation and direction of acceleration as the forces on the rapidly beating halters alter their shape. More subtly, the prominent eyes and antennae, as well as less apparent receptors on the entire body surface, are important to the mosquito for orientation during movement.

Grooming

Grooming by female *Aedes triseriatus* was described in detail by Walker and Archer (1988). Through observation of 1,162 grooming events in the laboratory, they categorized grooming behaviors as follows (directly quoted from the article):

1. *Proboscis Groom*: The mosquito grasped the proboscis near its base with the apices of both foretibiae and repeatedly scraped the proboscis in an anterior direction, with both legs working synchronously.
2. *Antenna Groom*: The mosquito grasped an antenna with the apices of both fore-tibiae and repeatedly scraped the an-

tenna in an anterior direction, with both legs working synchronously. Both left and right antennae were groomed in this manner.
3. *Foreleg Groom*: The mosquito bent one foreleg under her body so that the leg pointed posteriorly. Then she grasped the tibia of that leg with the apices of the hind-tibiae and repeatedly scraped the foreleg in a posterior direction with both hindlegs working synchronously. Both right and left forelegs were groomed.
4. *Midleg Groom*: The mosquito bent a midleg underneath her body so that it pointed posteriorly. Then she rasped the tibia of that leg with the apices of the hind-tibiae and repeatedly scraped the midleg in a posterior direction, with both hindlegs working synchronously. Both right and left midlegs were groomed.
5. *Hindleg Groom*: The mosquito scraped in a posterior direction, the hind-tibia and tarsus of one hindleg with the apex of the contralateral hind-tibia, switching scraped and scraping legs with each stroke.
6. *Wing Groom*: In dorsal wing grooming, the mosquito deflected her wings down to the sides of her body and scraped the dorsal costa of each wing simultaneously with the apex of the ipsilateral hind-tibiae. In ventral wing grooming, the mosquito scraped the ventral costa of each wing simultaneously with the apex of the ipsilateral hind-tibiae.
7. *Abdominal Tip Groom*: The mosquito scraped, in a posterior direction, the apex of either hind-tibia against the side of the tip of the abdomen. Right and left tips were not groomed synchronously.

These behaviors were highly stereotyped. For example, the ventral surface of a wing was always done before the dorsal surface. The order of grooming followed standard patterns most of the time and most often ended with hind leg grooming. This may have had the function of ejecting any particles off the body of the insect, as the net effect of the other grooming behaviors would have concentrated particles on the hindlegs. The two pairs of legs that groomed had specialized groups of setae on the tibiae related to this function, and the authors proposed that the stout tibial spurs helped guide forelegs and hind legs during grooming. Grooming probably cleans, flattens, and orients the scales, setae, and hairs that make up the vestiture of the mosquito. Similar behavior has been observed in males and several other species, including *Aedes aegypti*, *Aedes hendersoni*, *Anopheles quadrimaculatus*, *Sabethes cyaneus*, and *Toxorhynchites brevipalpis*. Grooming is probably very important in the life of a mosquito. Some dust on the surface of a mosquito may help to detect magnetic fields (Strickman et al. 2000b), but one can well imagine that a continual accumulation of particles would soon interfere with the function of the many structures of the cuticular surface.

Walking

Each of the six legs of a mosquito follow the basic structural plan of any insect (Snodgrass 1959). The first short segment is the coxa, and the joints on either side allow movement through a range of directions. The next segment is the short trochanter, followed by the long femur, the long tibia, the five-segmented (each segment is called a tarsomere) tarsus, and, finally, the pretarsus that bears a pair of claws (ungues). All the joints from the trochanter-femur to the pretarsus are structurally restricted to movement perpendicular to the axis of the body. Therefore, all anterior-posterior movement occurs near the body, and the rest of the leg follows along.

The process of walking has been described in some detail for some insects, but not for mosquitoes. Presumably, the process of simple walking is similar because the structure is similar. Working with the stick insect (Phasmatidae), *Carausius morosus* Sinety, Cruse et al. (2007) pointed out that with 18 joints among the six legs, there are 18 degrees of freedom during insect walking. Since the insect body has only six degrees of freedom (up, down, forward, backward, left, right), that leaves 12 degrees of freedom to adjust for effective movement. They proposed six rules that describe walking movement, summarizing how the stick insect establishes a coordinated interaction between its body and the substrate while moving each leg independently:

1. Avoid liftoff to avoid static instability.
2. Facilitate early protraction to favor temporal coherence (front-to-back sequence of steps).
3. Enforce late protraction to maintain temporal coherence (subsequent steps are taken).
4. Aim touchdown location to exploit prior "knowledge" about foothold (hind leg steps on footprint of anterior leg).
5. Distribute propulsive force (all legs perform steps).
6. Enforce correction step to avoid stumbling (adjust to conditions).

Mosquitoes emerge onto the water's surface and therefore must stand for at least a few minutes on the water. Some species also stand on the water during oviposition or other activities. The extent to which mosquitoes walk on water has not been described, but it is not uncommon to see mosquitoes make their way from the water's surface to emergent vegetation. The tarsus of *Culex pipiens pallens* has adaptations for staying on the water's surface (Wu et al. 2007; Kong et al. 2015). The microscopic striations on the scales add to the hydrophobicity of the tarsal segments. The hydrophobicity of the tarsus keeps the leg from penetrating the surface tension, and the weight of the mosquito creates a dimple in the surface that exerts a force proportional to the volume of water displaced. At a maximum, that force equals 23 times the mosquito's weight on a single leg, but the mosquito can control the interaction with the water by adjusting the angle of the tarsus to the water. The ability of the tarsus to support the mosquito on water is dependent on the flexibility of the tarsus, suggesting an additional function for the joints between the five tarsal segments.

The physics of insect water walking has been described, though not for mosquitoes (Bush et al. 2008). The authors detailed a complicated relationship between structure and function in which increasingly small scales of roughness coupled with hydrophobic surfaces create the highest contact angles between water and body surface in the animal kingdom. By trapping a thin layer of air among tiny hairs on the cuticle, the water surface does not actually contact the cuticle at all. This condition, known as a Cassie-Baxter state, has been exploited in synthetic materials to make highly water-repellent surfaces. In some aquatic Hemiptera, the claws have hydrophilic surfaces that anchor the insects to the surface tension. It is not known whether this is the case for mosquito ungues, but it would be an interesting addition to the presumed function of claws in gripping features of solid substrates. Anchoring on the water might be important for mosquitoes that oviposit egg rafts on the surface and therefore must maintain a steady position as they deposit each egg.

Flight

Mechanics

Snodgrass (1959) pointed out that the structure of a mosquito wing is quite unlike that of most other insects because the base lacks plates of cuticle. Instead, a mosquito's wing has a membranous connection to the thorax, and the bulk of the wing arises from a narrow connection to those membranes. Three axillary sclerites are at the base of the wing (Harbach & Knight 1980), each with a pair of direct flight muscles that influence the angle of the wing (Clements 1999). The bulk of the thorax is taken up with the middle segment that powers the wings through the indirect mechanism common to all insects except Odonata and Ephemeroptera. The dorsoventral muscles of the middle thoracic segment cause the wing to go upward by pulling downward on the cuticular roof of the segment (the notum), leveraging the base of the wing against a cuticular fulcrum. The powerful downstroke is caused by longitudinal muscles that pull the thorax into a shorter form and pop the notum upward to leverage the wing downward. The relative strength required to move the wings is reflected in the weight of the flight muscles, which may reach over 18% of the entire body mass.

Studies of other Diptera have shown that lift and maneuver are not as straightforward as seen in aircraft. Lift depends greatly on rotation of the wing about its longitudinal axis during each up- and downstroke. Because of an insect's small size, the relative viscosity of air is much greater to an insect than to humans and our machines. This relationship is summarized by the dimensionless variable called the Reynolds number, which is the ratio of the momentum force to the force of viscosity. The Reynolds number has been measured as 170 in *Aedes aegypti* and 480 in *Culiseta annulata*, compared to 15 for a chalcid wasp, 120 for a housefly, 1,000 for a honey bee, and 2 billion for a Boeing 747 aircraft (Clements 1999).

Yan et al. (2001) performed experiments with large model *Drosophila* wings in oil to mimic the conditions and the Reynolds number experienced by an actual small mosquito wing. Their work showed the importance of rotation and that lift came about from three aerodynamic phenomena: delayed stall, wake capture, and rotation circulation. As explained by Dickinson et al. (1999), the lift from delayed stall is caused by a vortex formed along the rotating portion of the leading edge of the wing as it flaps through a high angle of attack, the wing stroke gaining extra lift by working against the stabilized vortex. Wake capture occurs as the wing passes through the disturbances in the air created by the previous stroke, increasing the effective fluid velocity. Rotational forces occur as the wing pronates before the downstroke and supinates before the upstroke, accounting for up to 35% of the lift in *Drosophila*. Rotational forces enable the insect to gain lift on both the downstroke and the upstroke, as well as provide the means for great maneuverability. Ennos (1989) described a "torsion wave" along the length of the wing as the point of rotation moved from the tip to the base. The variety of wing movements within each stroke impart a dynamic, nonsteady state to insect flight.

Using oscilloscopes, careful measurements, and photography, a few researchers have analyzed a variety of mosquito characteristics related to flight (Table 6.2). It would be interest-

Table 6.2. Anatomical and aerodynamic measurements of mosquito flight in 2-day-old mosquitoes (*Culex quinquefasciatus* from Murty et al. 1993; *Anopheles stephensi* from Hasan et al. 1984)

		Culex quinquefasciatus		*Anopheles stephensi*	
Parameter	**Unit**	**Female**	**Male**	**Female**	**Male**
Body mass	mg	1.84	1.14		
Wing length	mm	2.83	2.83	3.9	2.9
Wingspan	mm	0.666	0.629		
Wing area	mm^2	2.5	2.2		
Wing mass	mg	0.041	0.032		
Wing swept volume	mm^3	7.17	6.49	11.51	6.27
Arc length	mm	2.96	2.96		
Angular velocity	rad/s	1,320	2,415	1,689	1,821
Kinetic energy	erg	0.88	2.33		
Wing loading	mg/mm^2	0.75	0.54	0.52	0.93
Frequency	Hz	405	741	475	520

ing to know these numbers for more species, as the two species presented differ from each other. For example, wing loading (the effective weight supported by a given area of the wing) was greater in female than male *Culex quinquefasciatus*, but the relationship was just the opposite in *Anopheles stephensi*. Other statistics suggested that males of *Anopheles stephensi* were less efficient fliers than *Culex quinquefasciatus* males. These measurements were taken during hovering flight to observe a steady state. Even less is known about the parameters of flight while a mosquito is moving. Another aspect of the flight dynamics is the harmonics of sound produced by wing beats. Clements (1999) reviewed the complicated assortment of wavelengths generated by wings, stating that the frequencies of male wingbeats were always different from females for a given species. Hasan et al. (1983) identified the third harmonic of *Anopheles stephensi* as the one with the greatest amplitude.

Mosquitoes' activities can influence the nature of flight. Blood feeding has a large effect on flight (Roitberg et al. 2003). Working with *Anopheles gambiae*, these authors examined the ability of blood-fed and non-blood-fed females to escape a jumping spider (Araneae: Salticidae: *Salticus scenicus* (Clerck)). They found that both the velocity and turning ability of blood-fed mosquitoes were significantly less than in those that had not taken blood. The decrease in agility was probably related to the twofold increase in weight.

The speed of flight, as well as the shape of the flight path, can be affected by feeding as well. Angarita-Jaimes et al. (2016) used a system of infrared illumination and large Fresnel lenses to visualize the flight paths of up to 25 mosquitoes around a bed net with or without a person underneath. They found that *Culex quinquefasciatus* moved faster than *Anopheles gambiae* during swooping behavior (362 vs. 342 mm/s) but slower during bouncing (162 vs. 180 mm/s) and visiting (268 vs. 286 mm/s) behaviors. *Culex quinquefasciatus* moved faster in the presence of a human host (350 vs. 290 mm/s). Using the same apparatus, Parker et al. (2015) showed that *Anopheles gambiae* was much more active around a bed net covering a human host (120 minutes of activity for 25 mosquitoes exposed over 60 minutes) than an empty bed net (19 minutes of activity). They also revealed that mosquitoes perceived the net as a barrier and slowed about 26–41 mm away before approaching it. The studies using visual tracking of mosquitoes (Parker et al. 2015 [*Anopheles gambiae*]; Angarita-Jaimes et al. 2016 [*Culex quinquefasciatus* and *Anopheles gambiae*]) and a study that used sticky squares to capture mosquitoes (Sutcliffe & Yin 2014; *Anopheles gambiae* and *albimanus*) all demonstrated that mosquitoes tended to concentrate above the torso of the person under the bed net. Other experiments have shown that mosquitoes adjust their flight paths in response to visual stimuli as far away as 20 m (*Psorophora columbiae*) or 7.5 m (*Uranotaenia sapphirina*, *Culex quinquefasciatus*; Clements 1999).

Appentential Flight

Finding a blood meal is an essential part of the life cycle of many species of mosquitoes. The complicated set of signals involved in host seeking includes visual perception, high-volume attractants like carbon dioxide, heat, and a complex set of organic chemicals that influences choice. Such movement is "appentential" in the sense that it is initiated by a behavioral objective, rather than by chance dispersal. The initiation of movement is followed by reception of positive stimuli and, finally, attraction flight to the objective (Bidlingmayer 1994). In tracking the direction of flight, models have often started with the assumption that mosquitoes behave as particles diffusing passively through space (Barmak et al. 2011). In fact, what was probably the first model of mosquito movement described by Ronald Ross (1905) assumed this. The argument has been made that because mosquitoes might go in one direction as readily as another in pursuit of an odor plume, net movement tends to be random, and few mosquitoes travel a significant difference (Eze et al. 2011). Of course, movement is not random, and the diffusion model really is just a starting point. In general, mosquitoes fly upwind toward chemical attractants (Clements 1999). At least some species modify their paths after initiation in response to visual stimuli, as is the case for *Aedes taeniorhynchus*, which can distinguish large objects (2.4 × 1.5 m silhouette) when they are 15 m apart (Bidlingmayer 1994). Although carbon dioxide is generally attractive to host-seeking mosquitoes, the chemicals that operate at a shorter distance seem to be more specific. For example, in Kenya (Tchouassi et al. 2013) a blend of octanal, nonanal, decanal, and heptanal (aldehydes in a series imitating skin secretions of cows, donkeys, sheep, and goats) increased the catch of *Aedes mcintoshi* and *ochraceus* by about twofold but did not affect the capture of *Culex pipiens*, *univittatus*, *poicilipes*, *bitaeniorhynchus*, or *Mansonia uniformis* or *africana*. Attraction of the two *Aedes* species was increased by threefold when the proportions of the individual chemicals were optimized. Similar host specificity was observed in the distribution of *Anopheles gambiae* in Kenya (Minakawa et al. 2002). This study found a correlation between the location of humans and the location of larval sites of this species and a correlation between the number of female adults captured and the distance to larval sites. In contrast, the presence of cattle increased the abundance of zoophilic members of the *Anopheles gambiae* complex in Tanzania (Killeen et al. 2001). The movement of *Aedes aegypti* out of a house depended a great deal on whether a human was inside (Suwonkerd et al. 2006). Only 5% of mosquitoes left a house with a person inside, compared to 47% when no host was present. The time of day mosquitoes entered the house also changed, being earlier in the morning when a human was present compared to later in the morning when no one was there.

The attraction of mosquitoes to plants has been studied less than attraction to hosts, but recent work has revealed a similar set of complicated relationships. *Anopheles gambiae* showed a strong preference for the flowers of some plants, including *Ricinus communis* L. (Nyasembe et al. 2012). This attraction may have been based on plant volatile compounds. *Parthenium hysterophorus* L. was the source of a series of terpenes (D-limonene, beta-pinene, (E)-beta-ocimene, (E)-linalool oxide, (E)-beta-farnesene) and an aldehyde (hexanal), each of which was attractive to *Anopheles gambiae* and even more attractive as a blend of these chemicals in the correct proportions. Gunter Müller (personal communication, 2016) has described a series of experiments showing the importance of plants to the longevity and reproductive potential of *Anopheles gambiae* and other mosquitoes. Mosquitoes actively choose preferred plants for sugar feeding and move about during the day within plants, seeking favorable conditions of temperature and humidity.

Mating is another activity that requires flight in most species of mosquitoes. Modeling has been conducted to determine the probability of successful control of *Aedes aegypti* by the release of sterile males. This technique would put enough sterile males into a wild population to reduce the number of fertile eggs deposited by females. The models used estimates of male movement toward females (Oléron Evans & Bishop 2014), estimates of environmental effects on males' movement (Dufourd & Dumont 2013), and estimates of successful mating given random or uniform distribution of the sterile males (Ferreira et al. 2008). All the models concluded that the timing and spacing of male releases were important considerations for success, reflecting the need for males and females to move into proximity to each other. Butail et al. (2011, 2012) developed a technique to follow the tracks of individual swarming *Anopheles gambiae* males and females in the field in Mali. Using a stereo video system and statistical analysis of the tracks, they created three-dimensional representations of mating events. Over a space of about a meter and for approximately four seconds, females moved over a slightly larger area than males, culminating in a sudden coupling of the pair. The female's final approach to the male was made on an approximately horizontal plane, though with some adjustments of height.

Oviposition behavior also involves movement. *Aedes aegypti* in Puerto Rico was recaptured at over twice the rate in homes with oviposition sites (13% recapture) compared to homes without them (6% recapture), indicating detection of and movement toward those sites (Edman et al. 1998). Experiments with sticky oviposition traps showed that female *Aedes aegypti* stayed below 1.75 m in height and concentrated on the leeward side of houses in Australia during the dry season (Williams et al. 2006). Movement of *Aedes aegypti* during oviposition is increased by its habit of "skip oviposition," in which it tends to deposit eggs from a single gonotrophic cycle at multiple sites (Colton et al. 2003). In contrast, *Aedes triseriatus* showed no sign of skip oviposition in Massachusetts, United States (Edgerly et al. 1998) and responded to environmental cues to find oviposition sites (Ellis 2008). Movement requires both time and energy; therefore, greater dispersion of oviposition sites has a cost to mosquitoes, as modeled for *Anopheles gambiae* (Le Menach et al. 2005; Gu and Novak 2009). Access to oviposition sites can also influence movement of mosquitoes to them. A narrow opening decreased, but did not eliminate, oviposition by *Aedes aegypti* into containers (Strickman & Kittayapong 1993). Some *Culex quinquefasciatus* traveled at least 270 cm through horizontal tubes leading to oviposition sites in the laboratory, much preferring a horizontally oriented tube to a vertical one. In the field, this species traveled 24.4 m through drainpipes to oviposition sites (Harbison et al. 2008).

Environmental Influences on Flight

It is difficult to separate the influence of various aspects of the environment from the stimuli associated with appentential flight. For example, as mentioned above, the direction of the wind generally influences the orientation of mosquitoes as they seek a host. Wind also affects the direction of movement over longer distances (Gillies 1961; Service 1980). Classic experiments that observed the flight of *Aedes aegypti* exposed to combinations of moving stripes and air movement in the laboratory (Kennedy 1939) showed that mosquitoes tended to move upwind if they perceived they were not being blown backward and if their forward movement was not perceived to be greater than 150 cm/s. The perception of stripes was only dorsal or ventral and not lateral. Apparently, these mosquitoes could compare their visual position to their flight direction to achieve oriented, forward movement.

There is a complicated relationship between where mosquitoes go and weather conditions, diel periodicity, and landscape. *Aedes taeniorhynchus* females in the laboratory engaged in spontaneous flight activity for 40 minutes following lights off and 20 minutes at lights on, but the males' flight was much less regular (Nayar & Sauerman 1971). Suwannachote et al. (2009) saw differences in the time of day that *Aedes aegypti* entered or exited experimental huts in Kanchanaburi, Thailand. Entry into huts was greatest a little earlier (peak 0700h) in the warmer, wetter summer season than in the cooler, drier winter season (peak 0900h). Mosquitoes exited the huts much later (peak 1600h) during winter than during summer (1100h). In studies repeated in Belize and Peru simultaneously comparing populations from each location, *Anopheles darlingi* from Belize entered experimental huts in a bimodal daily pattern (1900–2000h and 0500–0600h), whereas the same species from Peru entered from 2200 to 2300h. Those studies therefore suggested a genetically controlled aspect of the daily pattern of movement. In the laboratory, *Culex annulirostris* also showed differences in patterns of activity depending on the geographic origin of the colonies (Williams & Kokkinn 2005). The season can influence the relative preference of mosquitoes for habitats. The relative numbers of *Anopheles minimus* s.l., *Anopheles maculatus*, *Anopheles aconitus*, and Hyrcanus Group in agricultural compared to forest habitats of northern Thailand changed with the seasons. In general, all the species except *Anopheles aconitus* clearly preferred the forest habitats but were relatively more abundant in agricultural landscapes when conditions were cooler and wetter in the rainy season (May–September; Overgaard et al. 2003). Tracking individual mosquitoes in the laboratory, researchers found that *Anopheles gambiae* females flew very little following a blood meal for two days, then went into a burst of flight activity for two hours coinciding with oviposition beginning at the crepuscular period of the third night (Gary & Foster 2006). Males had a similar pattern of nightly flight activity that coincided with sugar feeding.

Other influences on mosquito flight contribute to orientation and path. Some mosquito control chemicals stimulate flight (Cooperband et al. 2010). Magnetic fields may be an important part of mosquito navigation, as about two-thirds of *Anopheles freeborni*, *gambiae*, and *stephensi* line up along straight lines of magnetic force (Strickman et al. 2000b). The role of memory may be important as well (McCall & Kelly 2002). For example, *Anopheles farauti* in Papua New Guinea traveled farther when released far from the village where individuals took their first blood meals, and *Anopheles arabiensis* tended to return to its house of origin when released from two other houses. *Anopheles balabacensis* and *Culex vishnui* tended to return to the same kind of host from which they got their first blood meals.

Distance

Mike Service (1997) wrote a review of the distance that mosquitoes fly with the clever title "Mosquito Dispersal: The Long and Short of It." Following a long discussion of many studies, Service concluded that it was difficult to separate kinds of movement based on purpose, preferring to call all mosquito flight a form of dispersal. The title makes a good point that sometimes mosquitoes fly short distances and other times they fly far, even by human standards (Table 6.3). The appetential aspect of flight is also difficult to separate from flight that seems to be purely about wider distribution because some long-distance flight starts as a search for resources like resting sites, mates, nectar, blood meals, or oviposition sites. The difficulty of generalizing flight patterns by distance was illustrated by a large metadata analysis of flight distance that failed to show any taxonomic trends and had to rely on broadly defined average and maximum distances (Verdonschot & Besse-Lototskaya 2014).

The distance that mosquitoes disperse has been a constant concern due to the implications for disease-control efforts. Wider distribution of uninfected vector species can create a risk of transmission in other localities, and wider distribution of infectious vectors can spread the pathogen itself. From the mosquito species' perspective, dispersion influences species distribution, genetic flow, and the effective size of a reproducing population.

The flight distances of *Aedes aegypti* have been examined in multiple studies using various forms of mark-release-recapture. In general, this species does not fly as far as most

Table 6.3. Documented distances flown by mosquitoes reviewed by Service (1997)

Species	Location	Distance	Method
Aedes aegypti	Kenya	154 m/24 h[1]	Mark-release-recapture
Aedes serratus	Brazil	10.8 km	Mark-release-recapture
Aedes sollicitans	eastern US	177 km	Captured distance from coast
Aedes squamiger	California, US	61 km	Mark-release-recapture
Aedes taeniorhynchus	Louisiana, US	106 km	Captured distance from coast
Aedes vexans	US	32 km	Captured distance from coast
Aedes vexans	US	740 km	Distance from larval sites
Aedes vigilax	Australia	100 km	Captured distance from coast
	Australia	96.5 km	Captured distance inland from larval sites
Anopheles albimanus	Panama	19–32 km	Captured distance from larval sites
Anopheles culicifacies	Sri Lanka	2.8 km	Mark-release-recapture
Anopheles freeborni	California, US	42 km	Distance to overwintering sites
Anopheles melas	western Africa	1.6–4 km	Captured distance from larval sites
Anopheles pharoensis	Egypt	29–45 km	Captured distance from larval sites
	Israel	280 km	Captured distance from larval sites
Anopheles pulcherrimus	Saudi Arabia	25 km	Captured distance from coast
Anopheles quadrimaculatus s.l.	US	3 km	Captured distance from larval sites
Anopheles sacharovi	Israel	14 km	Distance to overwintering sites
Anopheles stephensi	Pakistan	mean 165 m	Mark-release-recapture
Culex nigripalpus	Florida, US	1.2 km[2]	Mark-release-recapture
Culex tarsalis	California, US	18 km	Distance to overwintering sites
Culex tritaeniorhynchus	Shanghai, China	200 km	Distance from larval sites
Culex quinquefasciatus	Myanmar	55 m[3]	Mark-release-recapture
Psorophora ferox	Brazil	11.5 km	Mark-release-recapture

[1]Maximum recorded; 60% of captures moved 11–50 m/24 h
[2]Maximum recorded; 82% within 400 m
[3]Distance traveled by >80% of individuals

other species of public health concern, but that distance varies depending on circumstance and location. Measurement of distance traveled by a local population of mosquitoes is complicated by considerations of rate of movement (distance per day), direction of movement (potential for movement in a different direction), and maximum movement (given all the potential reversals in direction during multiple days). The intrinsic vagility of the species might best be estimated by daily movement, but the maximum movement during the mosquito's lifetime is probably most relevant to public health.

Some of the work on *Aedes aegypti* has shown a flight range that is probably shorter than most other species of mosquitoes. Muir and Kay (1998) stated that studies up to that date had recorded maximum flight distances of 27–1,150 m but that a more average distance was less than 80 m. In their own study in northern Queensland, Australia, they observed limited movement of females (mean 56 m, 50% traveling ≤33 m, 90% traveling ≤108 m, maximum 160 m) and males (mean 35 m, 50% ≤25 m, 90% ≤82 m, maximum 160 m). They speculated that the limited range of *Aedes aegypti* in their area was due to the low humidity that inhibited longer-distance flights. Mosquitoes released from houses adjacent to a forest in Rio de Janeiro, Brazil, rarely moved the 10–100 m to the forest edge (Maciel-de-Freitas et al. 2006). Tsuda et al. (2001) showed that the geometry of release influenced the distance traveled in a rubber plantation with 60 homes located on Hainan Island, Vietnam. When they released mosquitoes at the edge of the plantation, they observed maximum daily female dispersion of 60–75 m, compared to 30–45 m when the mosquitoes were released at the center of the plantation. Most males and females went no farther than 30 m during the first day.

Other work on *Aedes aegypti* has recorded longer flight distances, though probably still shorter than most other mosquitoes. *Aedes aegypti* in coastal Kenya flew a maximum daily distance of 154 m (mean 57 m) for females and 113 m (mean 44 m) for males (Trpis & Hausermann 1986). These distances meant that over 85% of females visited no more than two houses. Similar distances were recorded in Cairns, Queensland, Australia, where mosquitoes moved preferentially toward shady, tree-lined streets to the northwest of the release point (Russell et al. 2005). The maximum distance traveled by females in this study was 200 m (mean 78 m), and 23% of the mosquitoes had gone more than 100 m by the time of the final collection on day 15. A review of 16 studies in Puerto Rico and Thailand (Harrington et al. 2005) showed no difference between males and females and somewhat longer distances in general. The range of minimum distances was 3–17 m, and the range of maximum distances was 84–566 m, including movement between villages in Thailand. *Aedes aegypti* moved further under some circumstances in Rio de Janeiro (Maciel-de-Freitas et al. 2006). In a low-density urban area, the mean distance traveled was 288 m and the maximum, 690 m; 50% were greater than 350 m. Separately, the authors saw much shorter flight ranges in high-density parts of the city, an observation related to subsequent studies of the effect of urban structure on movement of *Aedes aegypti* (David et al. 2009). Studies that detected the presence of females by genetically or chemically marked eggs resulted in longer flight ranges, possibly because this technique enabled more thorough sampling. Releases in Mississippi, United States, of genetically distinct, morphologically identifiable mosquitoes (Bond et al. 1970) and subsequent collection of eggs (Hausermann et al. 1971) showed that a female might fly 274 m just 1.5 hours after release and 488 m after four days. Thirty percent of the mosquitoes flew at least 183 m. Otero et al. (2008) reviewed the maximum distance flown by *Aedes aegypti* in a number of studies and found 2,500 m as the longest record, flown by mosquitoes released in the Israeli desert (Wolfinsohn & Galun 1953). Reiter et al. (1995) marked females with rubidium salts, so their eggs could be detected in an array of traps in San Juan, Puerto Rico. They recorded a maximum flight distance of 840 m and a mean distance of 279 m after six days. Using genetic detection of individual females, the authors calculated that a single female might fly 3,600 m during its lifetime to deposit eggs in the small batches characteristic of *Aedes aegypti*. When suspended from a fine wire in a merry-go-round arrangement known as a flight mill, this species will fly up to 18 km in a single night (Briegel et al. 2001). That observation is consistent with the calculation of relatively long total flight distances between oviposition sites, despite a much shorter total range.

Culex pipiens and *quinquefasciatus* generally fly short to moderate distances in search of specific resources. A study in an urban area of Maryland, United States (Jones et al. 2012), observed that mosquitoes reared from field-collected egg rafts only flew about 9 m per day, and the greatest distance recorded was 200 m. The authors concluded that under these circumstances this species did not have to travel great distances to find all the resources necessary for its survival. Invasive populations of *Culex quinquefasciatus* on Mauna Loa, Hawaii, United States, flew 0.81–1.89 km (mean 1.34 km) over 10 days. The maximum distance, observed only once, was 3 km. These mosquitoes dispersed along roads in the forest. The majority (70%) of the same species in the Central Valley of California, United States, also flew less than 1 km (mean 1.25 km), but the maximum distance recorded was 10.4 km (Reisen et al. 1992). In the same study, the majority (75%) of *Culex tarsalis* also flew less than 1 km (mean 0.72 km) and a maximum distance of 12.6 km. Further south, in the Coachella Valley, California, Reisen and Lothrop (1995) showed a strong ($r^2 = 0.82$) negative correlation between the log of the recapture rate of marked *Culex tarsalis* and distance from the release point. In this location they found that *Culex tarsalis* sometimes flew 5.7 km upwind and moved farthest under conditions when the population was increasing. Over the course of a season, relatively short movements by individuals can possibly accumulate to spread viruses over a large range. Venkatesan and Rasgon (2010) made the case that the spread of West Nile virus in the Great Plains of the western United States resulted from such movement within the eastern population clade of *Culex tarsalis*.

Interest in the movement of *Anopheles* goes back to the early days of concern over malaria transmission—for example, by

Ronald Ross in 1905, cited above. The copper-mining industry in what is now Kitwe, Zambia, stimulated measurement of flight distances to estimate the radius of larval control necessary to protect a community of laborers (De Meillon 1937). Maximum downwind flight varied between 1.6 and 4.5 km for *Anopheles funestus* and between 800 m and 3.2 km for *Anopheles gambiae* s.l. Upwind flight was shorter, not exceeding 1.6 km for either species. Adams (1940) recorded longer flight distances in the same location—a maximum of 2.4 km upwind and 7.2 km downwind for the two species—in one case *Anopheles funestus* covering this distance in 18 h. Gillies (1961) made more detailed observations on *Anopheles gambiae* s.l. in Tanzania. He found that mosquitoes did not move far during the first one or two days of life but that they quickly distributed themselves throughout the maximum range afterward. The range of males (mean 837 m) was a little less than that of females (mean 1.0 km), but about the same percentage of males (7%) as females (8%) flew the maximum distance of 2 km. More recently, Midega et al. (2007) saw that both *Anopheles gambiae* and *funestus* along the Kenyan coast usually flew only 200–400 m from a release point, with the maximum distance recaptured at 661 m, in contrast to a study in Tanzania that recorded a maximum range of *Anopheles funestus* of 3.2 km (Charlwood et al. 2000). In an unusual study, a female of *Anopheles coluzzii* was recaptured 570 m from its release point 155 days later during the dry season in Mali (Lehmann et al. 2010).

Longer-distance dispersal can be the result of adaptive behaviors (Service 1980, 1997). *Aedes cataphylla*, *Aedes taeniorhynchus*, and *Culex tarsalis* fly up about 6–12 m shortly after emergence, apparently to catch wind currents. *Culex tritaeniorhynchus* was reported to go up 380 m prior to northward movement near Shanghai. Reynolds et al. (1996) suspended a net from an aerodynamically shaped balloon in West Bengal, India, finding that males and females of many species (*Anopheles vagus*, *hyrcanus*; *Culex fuscocephala*, *gelidus*, *pseudovishnui*, *tritaeniorhynchus*, *vishnui*; *Mansonia annulifera*, *indiana*) were captured at this level, probably having reached this altitude through intentional flight. They calculated that mosquitoes flying for six hours under these conditions would have been transported 70–110 km. Similar work with balloons in the Sahelian region of Mali also demonstrated that a significant number of *Anopheles* and other mosquitoes reach altitudes hundreds of meters above the ground, with the implication that wind can carry them hundreds of kilometers during a single night (Huestis et al. 2019). An epidemic of malaria on the island of Réunion in 1868 was attributed to the incursion of *Anopheles gambiae* and *funestus* with a tropical cyclone (Mouchet et al. 1995). The examples of capturing mosquitoes at significant distances from the shore (Table 6.3) or long distances following emergence (Horsfall et al. 1973; *Aedes vexans*) probably also result from movement of mosquitoes in air currents, rather than self-propulsion. On the other hand, it is not unreasonable to assume that individual mosquitoes blown by the wind maintain stability and orientation through active flight. Some species travel significant distances in apparent searches for various goals, including *Aedes vexans* and *Anopheles crucians*, which rest in wooded areas and forage in the open (Bidlingmayer 1994), or *Aedes cantans*, which shifts between woods with oviposition sites and woods with hosts (Service 1997). Other species go far to reach overwintering sites (Rosenstiel 1947 and Table 6.3: *Anopheles freeborni*—42 km; *Anopheles sacharovi*—14 km; *Culex tarsalis*—18 km; Rosenstiel 1947).

Some species of *Aedes* colonize developmental sites in tidal salt marshes, where combinations of seasonal rainfall and tides create periodic conditions conducive to egg hatching and larval development. Many, if not all, of these species tend to fly long distances between oviposition sites and blood-meal sources. Among the important species are *Aedes vigilax* in Southeast Asia, the western Pacific, and Australia; *Aedes caspius* in western Asia and Europe; *Aedes sollicitans* and *taeniorhynchus* in North and Central America; and *Aedes squamiger* in California. *Aedes caspius* flies about 10 km (Becker et al. 2003), but *Aedes sollicitans* often flies 40–90 km on emigration flights (Shone et al. 2006).

Larval conditions may influence the tendency to fly distances (Nayar & Sauerman 1968). In the laboratory, crowding larvae resulted in aggregation behavior and adults with larger thoraces in relation to wing length. These adults emerged more synchronously and exhibited greater levels of activity in early adult life. Horsfall et al. (1973) described long-distance movement by a non–salt marsh species, *Aedes vexans*, associating such movement with an altered form of the mosquito (personal communication, [1999]).

The full range of movement contributes to the distribution of mosquitoes, both seasonally and long term. Expansion of mosquito distribution into new habitats involves their initial movement to those habitats, as was the case for *Culex tarsalis*' expansion into urban areas by colonizing swimming pools as larval habitats in California (Reisen 2012). Seasonal expansion and contraction of distribution is another common feature of mosquitoes, especially tropical species that experience dry-wet weather cycles. *Anopheles gambiae* and *coluzzii* maintained reproducing populations all year in the floodplain of the Niger River, Mali, and expanded to drier agricultural areas only during the wet season (Baber et al. 2010). A similar situation was observed in the Kilombero Valley, Tanzania, for *Anopheles arabiensis*. After a wide distribution during the wet season, its reproducing populations receded to small refugia near the river during the dry season (Charlwood et al. 2000). *Culex quinquefasciatus* moved into new areas less frequently at cooler, high elevations on the island of Hawaii than at lower elevations, as documented by variations in genetic signature (Keyghobadi et al. 2006).

Anthropogenic Movement

Human activity has often been the means of moving mosquitoes, sometimes over great distances. The current global spread of *Aedes albopictus* has been described as the third historical wave of anthropogenic mosquito invasions, following

Aedes aegypti in the 16th and 18th centuries and *Culex quinquefasciatus* in the 19th and 20th (Leisnham et al. 2008). Any life stage can be involved, depending on biology of the mosquito species and circumstances. Boats fitted for long journeys have transported reproducing populations of *Culex quinquefasciatus* or *Aedes aegypti* (Mouchet et al. 1995), and the importance of ships in transporting mosquitoes has been recognized for a long time (Felt 1904). The realization that *Aedes aegypti* could fly to shore from ships infected with yellow fever led the International Sanitary Convention of American Republics in 1909 to establish 200 m as the minimum distance those ships had to stay from shore (Otero et al. 2008). Smaller fishing boats that probably had fresh water in their bilges contained larvae of *Anopheles gambiae*, *funestus*, *pharoensis*, and *rufipes* in western Africa (Holstein 1954). Ships continue to carry mosquitoes between continents. Detailed examination of 734 ships from all over the world docking at Qinhuangdao port in China from 2000 to 2003 captured 3,150 adult mosquitoes (Nie et al. 2004). A large proportion of the ships (57%) contained mosquitoes, with *Aedes dorsalis*, *Culex modestus*, and *Culex tritaeniorhynchus* the most common species.

Despite efforts to treat aircraft, they probably remain an important means of transporting mosquitoes. Fifty cases of malaria in Europe between 1978 and 1995 were probably the direct result of infected *Anopheles* carried there in aircraft (Mouchet et al. 1995). A careful calculation of air traffic routes from sub-Saharan Africa based on data from 2000 showed that 19 of 1,032 routes landed in areas susceptible to local transmission (Tatem et al. 2006). The relative rarity of routes landing in climates conducive to transmission may have accounted for the relative rarity of malaria transmitted by mosquitoes originating in Africa. As insecticidal treatment of aircraft becomes less common and direct flights from Africa expand into locations like Bangkok and Hong Kong, there may be more local transmission of malaria caused by importation of infected *Anopheles* on aircraft. Examination of 38 aircraft originating in Africa and landing at Schiphol Airport between 2010 and 2011 in Amsterdam, the Netherlands, showed that 26% contained mosquitoes, including *Aedes mcintoshi*, *Culex quinquefasciatus*, and *Culex antennatus* (Scholte et al. 2010a, 2014). Considering that in 2008 Schiphol Airport received 7,605 flights from Africa, the potential volume of mosquitoes entering this one airport is great (Scholte et al. 2010b). Although *Aedes japonicus* was probably imported into the United States in tires shipped by sea, that species has also been found in aircraft (Moberly et al. 2005). Reisen (2013) suggested that West Nile virus may have been imported into the United States in *Culex pipiens* adults riding on aircraft from Israel.

The effect of human transport on the biology of mosquitoes may be much greater than just the introduction of new species. Medley et al. (2015) examined the genetic variation of *Aedes albopictus* at 739 locations in the United States. They found that gene flow was greatest where there were highways, implying that human transport of this species increased the availability of a wide variety of genotypes in a location. With a greater variety of genetic architecture available, *Aedes albopictus* could presumably adapt to a greater variety of habitats and climates. Other studies have documented the importance of roads (Moore & Mitchell 1997; Dutta et al. 1998). As a general issue, increasing globalization of trade and travel, coupled with climate change, is likely to increase the risks to humans from new types of vector mosquitoes and the pathogens they transmit (Knols & Takken 2015).

Modeling Movement

Modeling mosquito movement has clear practical implications for the effectiveness of vector control interventions. For example, the community effect of insecticide-treated bed nets, used so successfully in sub-Saharan Africa since 2000, is dependent on the distance and rate of movement of vector populations (Killeen et al. 2003). In other words, if the area with good coverage of bed nets is too small or if the vector population moves too far, then the community effect that reduces the ability of the vector population to transmit will be weakened, and people will only be protected when they are under bed nets. Conversely, the intentional distribution of sterile male mosquitoes to accomplish population suppression is dependent on the mosquitoes dispersing enough to overcome the ability of wild males to inseminate females effectively (Evans & Bishop 2014; Dufourd & Dumont 2012).

The first models of mosquito movement assumed random motion that could be expressed with the same diffusion equations used for dilution of gases (Ross 1905). More recently, various levels of sophistication have been brought to the use of agent-based models in which each mosquito is considered as a unit that follows certain rules as it encounters a series of spaces arranged in a hypothetical grid. The rule for movement within those grids can be as simple as diffusion, though still effective for drawing conclusions about release rates of sterile males (Evans & Bishop 2014). Another way to consider movement is to use what the mosquito was doing before to predict what it will do next (Markov chain) and superimpose survival projections on each individual as it moves through the grid (Otero et al. 2008). Other models have used "advection," or measurements of movement toward (or away from) stimuli, either with (Lutambi et al. 2013; De Almeida et al. 2010) or without (Dufourd & Dumont 2012) life table estimates superimposed on the population. These kinds of models have been used to estimate the effects of bed nets and the sterile insect technique cited above, as well as traps (De Almeida et al. 2010) and spatial repellents (Lutambi et al. 2013). Each of these efforts trained their models against field or large-cage data, in some cases comparing the hypothetical introduction of an intervention to real data without the intervention.

Where mosquitoes go and why have great biological interest, as well as practical importance. A conclusion from various modeling efforts is that the distribution of resources (hosts, larval sites, etc.) is as important as the distribution of the mosquitoes themselves. Lutambi et al. (2013) made the case that a mosquito population could not survive if resources were sufficiently dispersed. Conversely, Ritchie and Montague (1995) and Otero et al. (2008) pointed out that if mosquitoes were unable to move sufficiently, the population would eventually go extinct, as each subpopulation has been unable to find sufficient resources at one time or another. The need to move from where a population is to where it is not expresses the general biological pattern of maintenance foci in which a species is usually present and expansion foci periodically colonized by the species.

Conclusion

Our ability to visualize and study movement is limited by the large areas inhabited by mosquitoes, by their small size, and by their abundance. Studies find examples of how subpopulations have moved, examine the mechanisms of movement, discover stimuli that influence the direction of movement, and model the significance of movement to populations. Perhaps most impressive is how nearly every aspect of the structure, physiology, and bionomics of mosquitoes is tied to movement in one way or another. Mosquitoes are not unique in this way, but their well-developed means of movement are essential for their survival.

7. Feeding and Nutrition

Mosquitoes are by far the most important bloodsucking arthropods world-wide, causing much annoyance and disease in humans, other mammals, and birds.

Goddard 2013

People being people, they think first about the bite of bloodsucking females when they think about mosquito feeding (Fig. 7.1). That bite is a significant annoyance for those who have not been desensitized to the antigens of that species, as their local immune system reacts vigorously to the proteins in the injected saliva. The reaction to those salivary proteins makes a long-term change in immune response and therefore has biological significance. Of course, the impact of those bites is multiplied by the pathogens transmitted and the resulting disease. The biological relationship between those pathogens and the vector mosquitoes is usually an intricate one that involves development and multiplication of the pathogen as it journeys from the blood meal to the salivary glands and, finally, to a new vertebrate host. Considering the cumulative immune responses and the many pathogenic infections resulting from mosquito bites, the family Culicidae has considerable impact on vertebrates.

From a mosquito frame of reference, there are three purposes to eating. The first is growth. The mosquito larva must take on a huge nutritional load relative to its size as it develops, even providing a nutritional reserve for the pupa and resultant adult. The female usually continues to process large quantities of food in the form of blood for the development of eggs. Egg development is usually rapid, often completed in fewer than three days. Many species live in stable habitats that support a more leisurely pace, but the fragility of most aquatic habitats has exerted considerable selective pressure for speedy development.

Nutrition is important for the function of the mosquito at all stages. It is a source of energy. Energy is needed for many functions in all life stages of mosquitoes. Movement, and the metabolism to support it, is highly developed at all stages except the egg. Physiological processes like digestion, excretion, internal synthesis and secretion, and gamete development all require energy that originally comes from nutritional reserves eaten by larvae or adults. Nutrients are also processed for storage of energy, lipid, or protein reserves, even though the life of a mosquito seems short on a human timescale. Less well described is the role of nutrition for maintenance and repair of the mosquito. Wound healing seems weak or nonexistent, but encapsulation of parasites and replacement of midgut epithelial cells might be considered as maintenance.

Gamete production is a separate category of nutrient use and, of course, is absolutely essential for any species of mosquito. Sperm packets and eggs represent the largest bodily outputs from mosquitoes, other than excreta. The reproductive cells start as germinal cells in the larva but do not mature until the adult stage is reached. Mosquitoes continue to produce new sperm and eggs during adult life, requiring new nutrients or nutritional reserves.

It is difficult to separate feeding and nutrition from other life functions. Much of movement is appetentially directed at feeding. Excretion is the natural consequence of feeding, and much of secretion occurs in association with feeding, digestion, or regulation of nutrition. Discussion of evolution, movement, development, and quiescence have already touched on aspects of feeding, as will chapters on excretion, copulation, and oviposition. This chapter is organized by larval and then adult feeding and nutrition, beginning with the morphology and mechanical function of the mouthparts and digestive system and then proceeding to physiology and behavior.

Larval Feeding and Nutrition

Morphology

The pharate instar that hatches from the egg develops into its final form while absorbing the nutrition provided by the yolk,

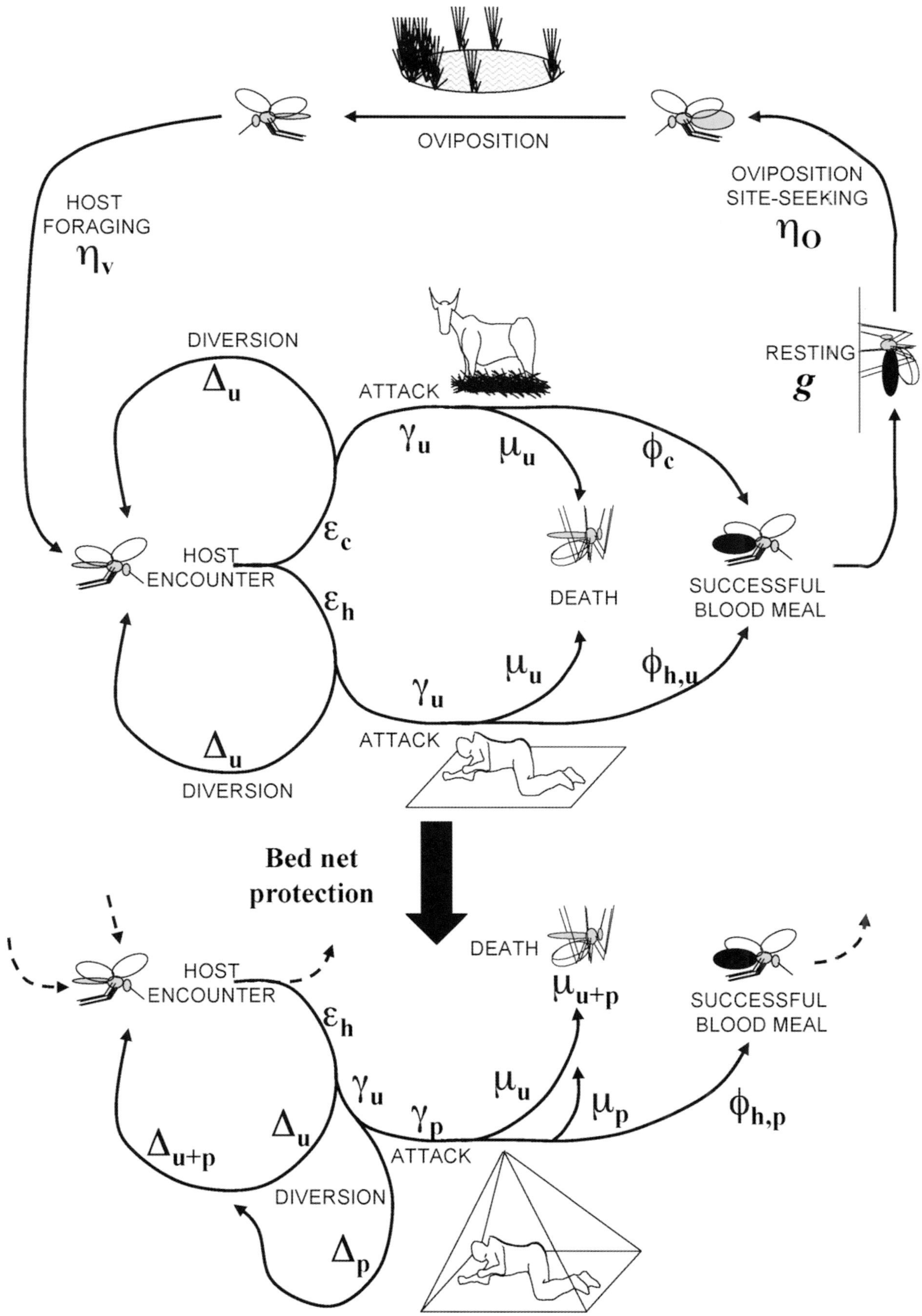

Figure 7.1. Parameterization of female mosquito feeding and other activities during the life of an adult mosquito. Killeen and Smith 2007

which is eventually surrounded by the embryo. Once the larva hatches as a first instar, it must find its nutrition from the environment. Mosquito taxonomists sometimes look on larvae with delight because there is so much morphological variation and, even more important, consistency. This means that once distinguishing features of a larva are found, these features will be similar in all individuals without the variation so often associated with the markings on adults. A bit speculatively, the marked differences in larval morphology between species and the consistency within species suggest that evolutionary, selective forces on the functions of the larvae have been strong. In a way, the larval morphology of each species has a story to tell. That story is largely about nutrition because the typical larva consumes a huge amount of material relative to its size to achieve development from first instar to pupa in less than a week under favorable conditions.

The basic plan of the larval mouthparts (Fig. 7.2) was reviewed by Clements (2000) and has also been described by Snodgrass (1959), Laffoon and Knight (1973), and Harbach and Knight (1980). For the purposes of this volume, the mouthparts and alimentary tract will be described based on functions detailed later. The references cited and others present much more information and a history of debate about the relationships of various features, as well as their functions.

The larval head is the entry point for all food. The mouthparts, easily related to the basic pattern of other insects, are formed from a combination of extensions of the cranium and three pairs of highly modified appendages. The heads of mosquito larvae are "prognathous" in that the mouthparts are not under the head but at its front; therefore, the order of the mouthparts is spatially dorsal to ventral even though they are morphologically anterior to posterior. The parts of the feeding apparatus are the same as in a generalized insect. The labrum encloses the top of the oral cavity and is hinged to the clypeus (front of the head). Next come the paired mandibles, which move in a transverse direction. Below the mandibles are the maxillae, which are constructed on the typical insect pattern of basal cardo, apical stipes formed from the ectal (outer) galea and ental (inner) lacinia and a prominent maxillary palpus adjacent to (and sometimes including part of) the galea. The floor of the preoral cavity is formed by the fused appendages forming the chitinous labium (prementum) and the fleshy hypopharynx, which together constitute the labiohypopharynx.

Typical mosquito mouthpart movements work to either filter or scrape particles into the mouth located just above the base of the labiohypopharynx. One of the mechanisms for gathering small particles of food from the water is the generation of water currents to draw particles toward the larva. Two groups of long setae located on either side of the inner surface of the apical portion of the labrum (the labropalatum) generate these currents. These labropalatal brushes are actuated through a series of rods and lever-like sclerites to rapidly (5 Hz) expand and contract like pages in a hardcover book. The labropalatum also has a shorter median brush that works to push particles into the preoral cavity. The shapes of the setae vary from lateral to mesal position and from species to species. Currents go toward the mouth and then downward for *Anopheles* larvae that are oriented horizontally just below the surface. Culicine larvae that hang downward draw water from above and below the head, then eject it horizontally and posteriorly to generate a torus of flow that enhances the overall flow toward the mouthparts. Presumably, this flow can be strong enough to propel larvae when they are moving parallel to the surface and maintaining contact with the surface at the end of the siphon, as commonly observed for many species. The mouthbrushes can vary depending on the diet in *Opifex fuscus* (Marks 1958b; McGregor 1963; Mattingly 1971b). Fed fish food in the previous instar, individuals had pectinate mouthbrush hairs and more heavily chitinized mouthparts, whereas those fed dried serum had simple hairs.

The mandibles have several functions. The apical teeth of the mandible can bite off particles, and median surfaces crush

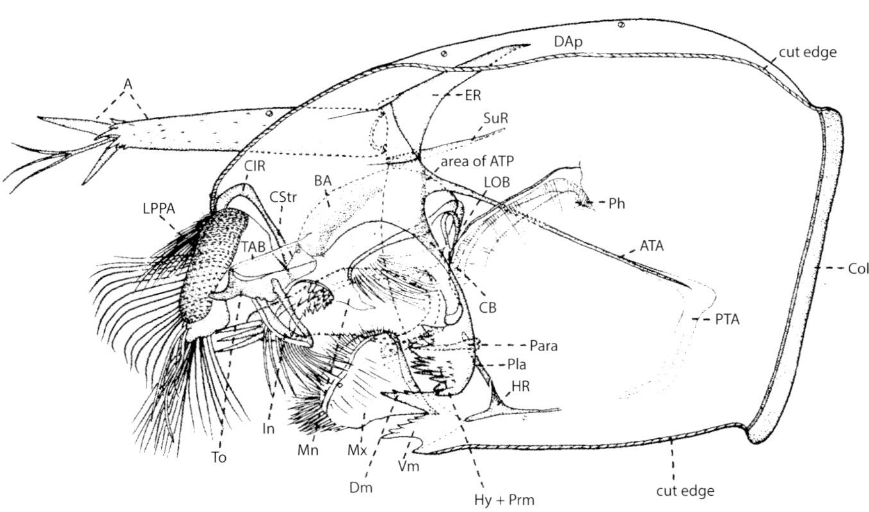

Figure 7.2. Diagram of the head of *Anopheles quadrimaculatus* larva showing a lateral view of the mouthparts and cuticular connections. A = antenna; LPPA = lateral palatal penicular area; To = torma; TAB = tormal apodemal arm; CIR = clypeolabral suture; CStr = clypeolabral strap; BA = black-spot area; In = intertorma; Mn = mandible; Mx = maxilla; Dm = dorsomentum; Vm = ventromentum; Hy + Prm = labiohypopharynx; HR = hypostomal ridge; Pla = paracoila; Para = parartis; CB = cibarial bar; LOB = lateral oral bar; ATP = anterior tentorial pit; SuR = subantennal ridge; ER = epistomal ridge; DAp = dorsal apotome; Ph = pharynx; ATA = anterior tentorial arm; PTA = posterior tentorial arm; Col = collar. Laffoon and Knight 1973

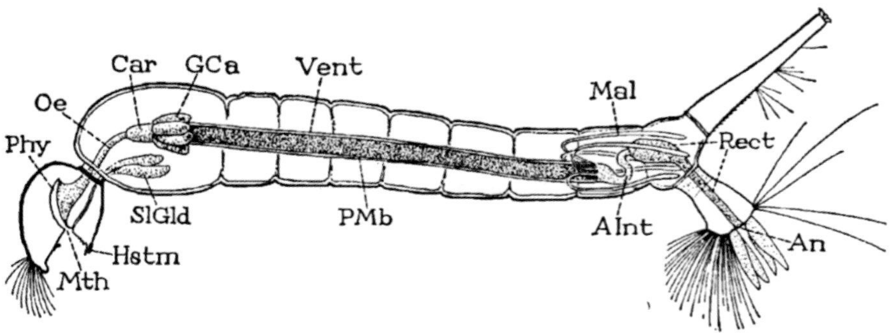

Figure 7.3. Section of *Culex* larva showing alimentary canal. Mth = mouth; Hstm = hypostome; Phy = pharynx; Oe = esophagus; SlGld = salivary gland; Car = cardia; GCa: gastric caecae; Vent = ventriculus or midgut; PMb = peritrophic membrane; Mal = Malpighian tubules; AInt = anterior intestine; Rect = rectum; An = anus. Snodgrass 1959

particles against the teeth of the labium that project between the mandibles. There are setae that catch food particles to draw them into the preoral cavity, setae that clean off the maxillae, setae that scrape food particles from the labropalatal brushes, and setae that sweep posteriorly from the lower surface of the labrum. In addition, the mandibles are equipped with chemosensory and mechanosensory organs. Movement of the mandibles is closely coordinated with that of the labropalatal brushes to keep food in motion.

Maxillae are equally as intricate as the mandibles, but their functions are less well described. Like the mandibles, the maxillae include sensory organs. Their location between the mandibles and the labiohypopharynx suggests that they serve to contain food within the preoral cavity and to manipulate it toward the mouth.

The labiohypopharynx serves as the floor of the preoral cavity. The mandibles crush food against the surface of a chitinous, toothed mound. As mentioned above, the labium represents fused appendages, and there is evidence of the original palpi fused near the base of the organ. The larval salivary duct empties into the preoral cavity at the distal end of the junction between labium and hypopharynx.

The basic morphology described above varies considerably between species. Predatory larvae in *Toxorhynchites*, *Lutzia*, *Psorophora* (*Psorophora*), and *Aedes* (*Mucidus*) have various modifications of the mouthparts that allow them to effectively grasp prey. *Toxorhynchites* have lateral palatal brushes that are much stouter than other mosquitoes and equipped to grasp prey (Clements 1999). *Lutzia* also have stout brushes for grasping prey, as well as large mandibular teeth for holding prey (Kitching et al. 2015). The organization of the maxilla also varies between species, with various levels of fusion between the palpi, galeae, laciniae, and cardo. These differences were shown to have phylogenetic significance in sabethine genera (Harbach & Peyton 1993). *Mimomyia plumosa* in Africa has reduced mouthbrushes and mentum as adaptations for feeding on soft substrates (Mattingly 1971b). Variation in mouthpart structure can even occur within species, as cited above for *Opifex fuscus* fed on different diets.

Food goes through the mouth, which is the opening to the pharynx, into the simple larval alimentary canal (Fig. 7.3; Snodgrass 1959). Larvae have a pair of salivary glands that empty into the labiopharynx through a common duct. The glands vary in different species, but a common pattern is to have a smaller anterior portion consisting of larger cells and a larger posterior portion consisting of smaller cells. As for all insects, the foregut (stomodeum) and hindgut (proctodeum) are of ectodermal origin, and the midgut (mesenteron) is of endodermal origin. The foregut consists of the pharynx in the head and the esophagus that extends through the neck and into the thorax. The esophagus joins the midgut in an organ called the cardia, which consists of an infolding of the esophagus within the midgut and is the site of secretion of the peritrophic membrane. The remainder of the midgut has been called the ventriculus. It includes eight gastric ceca near its anterior. The midgut joins the hindgut in the seventh abdominal segment. Five Malpighian tubules arise from the anterior of the hindgut, extending into the sixth abdominal segment and then looping posteriorly into the eighth abdominal segment. The hindgut itself swells slightly just posterior to the midgut, makes an S-shaped bend, and then swells into the rectum. Finally, a narrow tube extends from the rectum to the anal segment and the anus. The functions of the hindgut and Malpighian tubules will be discussed in more detail in Chapter 8.

Feeding Methods

Clements (1999, 2000) reviewed classifications of feeding mechanisms and concluded that a system based on the processes larvae use to transfer food into their preoral cavities was the most sensible method to categorize the ways in which the basic mouthpart mechanism is adapted to feeding. The one exception is predation, which is based more on what a larva eats than how it eats. Even including predation, the same larva may use more than one mechanism during its development. A few minutes of observation of a species like *Culex pipiens* will see the larvae spend a great deal of time hanging from the surface using their mouthbrushes to draw in and filter suspended particles, then occasionally descend to the substrate to nibble at food sources there. Clements' classification is useful for understanding some of the general differences between genera and for categorizing behavior:

Collecting-filtering: The common mechanism of removing suspended particles from the water column or the water surface. The mouthbrushes create currents leading to the mouthparts from the water column in culicine larvae using this method, and the particles either flow into the preoral cavity, propelled and manipulated by the mandibles and maxillae, or are scraped from the lateral palatal brushes by the mandibles. Anophelines turn their heads 180° so that the spatially ventral (morphological posterior) part of the head rests just below the surface tension layer. This layer lies just below what is usually an organic-water interface with the air in environmental waters and within a narrow (on the order of 1 μ) layer that is rich in bacteria and dissolved nutrients, known as the bacterioneuston. The anopheline larva draws this layer toward itself at remarkably high velocities, in the case of *Anopheles quadrimaculatus* reaching rates of 12 mm/s near the front of the larva and slowing progressively below and at a distance from that point. Because of the small size of the larva, the water behaves in a less turbulent and more laminar manner; this is the same physical phenomenon as for mosquito flight, where a low Reynolds number makes the medium behave in ways that would require much more viscosity at a human scale. Collecting-filtering is an important feeding mechanism for many genera, including *Anopheles*, *Culex*, *Culiseta*, *Mansonia*, and *Coquillettidia*. It is less common in the genus *Aedes*.

Collecting-gathering: The larva uses the same current-driven and filtering mechanisms as in collecting-filtering, but the food materials are resuspended by the larva from surfaces on which the particles have loosely settled. Presumably, such particles are not adherent to the surfaces, and the currents generated by the mouthbrushes, as well as perhaps direct physical contact by them, are sufficient to put the particles into the material on its way to the preoral cavity. Larvae engaged in this kind of feeding are seen to be working over objects on the substrate or gliding along the substrate, apparently powered by the mouthpart-generated water currents. This is the typical feeding mode of *Aedes*, *Haemagogus*, *Psorophora*, and *Wyeomyia*. Some species of *Anopheles* (e.g., *Anopheles farauti* and *stephensi*) also use collecting-gathering.

Scraping: Materials such as bacterial layers, algae, and protists adhering to surfaces are literally scraped off to be manipulated into the preoral cavity. Most of the scraping is probably accomplished by the tips of the mandibles, which also pulverize these materials against the prementum of the labiohypopharynx. Larvae seen to be working with their mouthparts on a single spot of the substrate are probably performing this kind of feeding. Considering the robust form of the mandibles of larvae that typically filter food, it is likely that this form of feeding is widespread among mosquitoes. Even some *Anopheles* larvae (e.g., *Anopheles albimanus*) will descend to the substrate and appear to bite at surfaces. In the laboratory, *Anopheles* will clearly take in larger particles on the bottom of a container and convert them to fine particles in the gut. *Aedes triseriatus* combines this method with collecting-filtering and probably collecting-gathering, spending 52.5% of its time at the surface (including brushing along the walls of a container) and 37.5% of its time feeding below the surface (Walker & Merritt 1991).

Shredding: In a sense, this form of feeding is a more vigorous form of scraping. Pieces of leaves, sections of filamentous algae, or parts of dead invertebrates are ingested in this way. As for scraping, the mandibles are probably the mouthparts most involved in taking the bites. Examples include *Tripteroides* (subgenus *Rachisoura*) that eat dead invertebrates, as well as *Armigeres subalbatus* and *Trichoprosopon digitatum*, which have been observed to eat dead larvae in the laboratory. *Culex bitaeniorhynchus* and *Anopheles pseudopunctipennis* bite off sections of filamentous algae. Pieces of submerged, dead leaves are eaten by *Culiseta inornata* and *Aedes triseriatus*.

Predation: The mechanisms and habits of predation vary among mosquito larvae, but the mouthparts of both obligatory and facultative predators involve modifications of the lateral palatal brushes, the mandibles, or both. Typically, the lateral palatal brushes may be stouter, sharper, or serrated—all adaptations for grabbing and killing prey. Mandibles have more pronounced apical processes, also developed for piercing and grasping. Those that consume the entire prey (e.g., *Aedes* (*Mucidus*), *Lutzia*, *Psorophora* (*Psorophora*), *Toxorhynchites*) seem to stuff the item into the preoral cavity, where the mandibles can dismember it sufficiently for entry into the mouth. *Toxorhynchites rutilus* attacks prey in three ways: entirely consuming the prey, eating pieces from it, or killing the prey without eating anything (Trimble & Smith 1978). *Anopheles baimaii* (identified as *Anopheles dirus*) commonly eats first and second instar anophelines, especially of other species, in the laboratory under crowded conditions, grabbing and macerating the larvae (Rosenberg 1982). *Eretmapodites chrysogaster* also avoids eating larvae of its own species, whereas *Trichoprosopon digitatum* commonly eats first instars in its cacao pod habitats, with the effect of establishing a cohort of larvae that pupate and emerge together (Lounibos 1983). Similar adaptive value may be a part of cannibalism observed in other species (*Toxorhynchites rutilus* and *Anopheles barberi* [Bradshaw & Holzapfel 1983]; *Culiseta longiareolata* [Spencer et al. 2002]; *Anopheles gambiae* [Futami et al. 2008]). Although described as a facultative predator, *Topomyia yanbarensis* eats larger larvae regularly (Okazawa et al. 1986a). Similarly, *Eretmopodites dracaenae* is an active predator that seeks larger prey (Haddow 1946). Some facultative predators seem to modify this process by biting their prey and then sucking out the contents (*Sabethes chloropterus* [Galindo 1958]; *Isostomyia paranensis* [Campos & Zavortink 2010]). Another form of predation is common among container-inhabiting and some other mosquitoes, in which small first instars are swept into the stream of food ingested by larger larvae.

Larvae are often surrounded by food, and navigation toward it would not seem to be important. In some situations, this is probably the case, as for *Culex pipiens* in highly eutrophic waters or *Anopheles pseudopunctipennis* surrounded by filamentous algae. As will be discussed in more detail, situations with such abundant food may be the exception. In places where nutritional resources are limited, the ability to find food becomes essential to successful development. Highly adapted predators like *Toxorhynchites* species clearly need to find large food items and respond to them quickly, with mechanoreception of movement clearly an important part of the process. In places where locations of food are aggregated, larvae that use collecting, scraping, and shredding need to find those sources of food. Larvae have chemosensory ability (Liu et al. 2010), though it is more limited than that of adult mosquitoes. Larvae express 13 odorant receptor proteins, compared to 73 in adults, with six of the proteins unique to larvae and seven shared by adults. There are only 12 olfactory sensory neurons in larval *Anopheles gambiae*, and those that affect responses to food may be located only on the antennae. Blunting the expression (with inhibitory RNA, or RNAi) of the universal odorant coreceptor (AgOR7, called ORCO in other studies) stopped responses to DEET but not to yeast. The positive yeast response was stopped by inhibition of a different odorant receptor (AgOR40) that represented an ionotropic reception path unique to larvae.

Physical surroundings can make it difficult for larvae to feed. In the laboratory, only shallow containers were suitable for several species (Briegel 2003). Only 55% of *Aedes aegypti* survived in 14 cm deep containers, fewer than 5% of *Culex pipiens* survived in containers 10 cm deep, and *Anopheles albimanus, gambiae*, and *stephensi* did best in depths less than 5 cm. These observations contrasted with the often-observed presence of *Culex pipiens* larvae in very deep water such as nitrification tanks in sewage treatment plants, abandoned swimming pools, and various vaults associated with urban structure. Such observations did not quantify survival, so it is possible that water depth did play a role in survival and the effectiveness of feeding. Field sites may also include floating objects or edges that improve the feeding situation for larvae.

Workman and Walton (2003) showed differences in the behavior of four sympatric *Culex* species in California, United States, that were affected by the amount of food present (Table 7.1). Low food conditions increased both the frequency and duration of diving but to widely different extents in the four species. *Culex erythrothorax* was the least affected and had the least contrast in total times spent diving in high food compared to low food conditions. Under favorable conditions, *Culex tarsalis* appeared to get the majority of its food from filtering at the surface, whereas *Culex erythrothorax* and *Culex stigmatosoma* spent more time feeding on substrates.

Table 7.1. Frequency and duration of fourth instar larval diving by four species of California, United States, *Culex* in the laboratory under high-food and low-food conditions (Workman and Walton 2003)

	Dives/15 min			Duration (seconds)		
Species	Low food	High food	Ratio (lo:hi)	Low food	High food	Ratio (lo:hi)
quinquefasciatus	65	8	8.1	34	5	6.8
erythrothorax	28	12	2.3	27	9	3.0
tarsalis	80	2	40.0	14	6	2.3
stigmatosoma	110	9	12.2	12	11	1.1

An account of all the behaviors associated with larval feeding are well beyond the scope of this volume. A couple of examples show how feeding can be quite specialized for some species. *Kimia decorabilis* is the topotypical species of a recently described sabethine genus (Harbach et al. 2007b). This unusual larva has pseudopod-like extensions on the ventral side of the abdominal wall of the first through seventh abdominal segments. The larva rests on these protrusions on the substrate, stabilized by long, stout hairs on segments one through six. Slowly twisting, it uses its mandibular brushes to lift material into the preoral cavity. Mogi (2010) described highly specialized species that use the liquid-filled organs of the carnivorous plant *Nepenthes mirabilis* (family: Nepenthaceae) (Lour.) Rafarin as their larval habitat. Escaping digestion by the plant, *Aedes dybasi* and *maehleri* must develop quickly before the liquid medium disappears. *Aedes dybasi* ate dead ants and bees in the laboratory, suggesting that insects trapped by the plant may be an important part of its diet.

Effects of Quantity of Larval Food

One of the effects of larval crowding and interspecies cohabitation is competition for food, resulting in less food available for each larva. Most mosquito species compensate for suboptimal conditions by producing smaller individuals on emergence, although it is common to observe pupal death when there is interference with normal feeding. For example, nutritional stress caused by a lack of food or conditions that interfere with feeding reduces survival and size of *Culex restuans* (Reiskind et al. 2004). Some species have adaptations that result in adjustments to the larval density, such as the cannibalism reported above and the oviposition repellents discussed in Chapter 10. Competition between different species has been studied explicitly to understand the relationship of newly introduced species to existing fauna and to understand this basic aspect of mosquito biology. One study (Barrera 2008) examined this topic by measuring the time to death for starved larvae: *Aedes taeniorhynchus*, a floodwater mosquito, survived 10 days in contrast to three container-developing species (*Aedes aegypti, albopictus*, and *triseriatus*) that lived a maximum of 25–30 days. Mixing species together reduced the time they could endure starvation. Interestingly, starvation slowed the respiration of the mosquitoes.

Container-developing mosquitoes appear to be more tolerant of competition. One example is the pair of species that inhabit taro (*Alocasia odora* (G. Lodd.) Spach) leaf axils on Ishigaki Island, Japan (Mogi 1984). *Aedes flavopictus miyarai* and *Malaya genurostris* manage to survive well at high densities, averaging 2.75 larvae per milliliter. The former species was smaller at high densities, but the latter was not. In Australia, the native species *Aedes notoscriptus* expanded its range of habitats from rock pools and tree holes to domestic containers starting in the 1930s (Russell 1986b). That brought this species into competition with *Aedes aegypti*, which is an introduced species common in domestic containers in parts of

Australia. Experiments in the laboratory showed that the times to pupation and survival of the two species were similar under various conditions of temperature and availability of food. When the species were reared together at various ratios, both suffered lower survival and longer time to pupation. Also, the time between 50% and 90% pupation tended to get longer. Both species did less well when they were in the minority (25% of the larvae in the container) than when there were equal numbers or when they were in the majority (75% of the larvae in the container).

The abundance of food also affects populations that consist of a single species. *Aedes aegypti* was reared at densities of 50, 100, or 200 larvae per liter, corresponding to food per larva of 3, 2.5, or 1.8 mg. The time required to pupate was proportionately shorter with more food per larva, averaging 6–8, 7–9, and 8–10 days at the corresponding levels of food per larva. The larvae with less food were also smaller in size (female wing length 2.97, 2.71, and 2.56 mm, respectively) as food was restricted (Sumanochitrapon et al. 1998). Pupae of *Aedes aegypti* from field collections in Thailand (Strickman & Kittayapong 2003) produced females that varied in wing length corresponding to the probable nutrient content of the larval habitat. For example, the average wing length from chicken-watering containers was 2.69 mm compared to 2.45 mm for those originating from clean drinking water, and covered water jars produced smaller females (2.39 mm wing length) compared to uncovered water jars (2.55 mm). It is interesting to note that despite the same general geographic origin, the *Aedes aegypti* produced in the laboratory study were larger than those collected in the field, suggesting a general nutritional deficit. Focks et al. (1993a, 1993b) created a model of productivity of container-inhabiting mosquitoes, using *Aedes aegypti* as a test species. There was close correlation between their model of the effect of food on days to pupation and survival of larvae and pupae.

Aedes sierrensis has been used as a model for larval crowding and intraspecific competition. This species lives in tree holes near the West Coast of North America from Southern California to British Columbia, Canada. Like several other mosquitoes in this region of winter rains, summer droughts, and microhabitats, there is one generation per year, and larvae develop during the winter when sites are flooded. These circumstances result in dense larval populations that develop slowly in cool temperatures. Perhaps because of the slow development or because of adaptations, the larvae can develop successfully to pupae at incredibly high densities of 10,000 larvae per liter, with pupal size inversely proportional to the log of density (Hawley 1985). A careful laboratory study showed effects on survival attributed not so much to nutritional stress as to the constant physical encounters between larvae that interrupted feeding activity (Broadie & Bradshaw 2008). The source of nutrition in the tree holes was predominately leaf litter, supplemented by benthic detritus and material in stemflow water. Bacterial action on these nutrients degraded the nutritional sources for *Aedes sierrensis* (Maciá & Bradshaw 2000).

The presence of predators can have an effect on larval feeding efficiency (Beketov & Liess 2007). *Culex pipiens* Form Molestus takes longer to develop and produces smaller adults when reared in the presence of water in which *Notonecta glauca*, a predatory backswimmer bug, had previously fed on larvae. Reared in normal water under low food conditions, larvae required 36 days to develop and produced females with a mean wing length of 3.6 mm. Larvae reared in water in which *Notonecta glauca* had fed on mosquito larvae took 46 days to develop and produced females with a wing length of 3.3 mm.

Nutritional and crowding conditions can affect more than size and survival. The pattern of light-colored scales can change on *Orthopodomyia signifera* in response to nutritional changes (Zavortink 1968). In *Aedes taeniorhynchus*, larval crowding and less nutrition brings about a series of changes associated with dispersal (Nayar & Sauerman 1968, 1971). Like some other floodwater mosquitoes, larvae under crowded conditions form aggregations that appear as roughly spherical balls with constant circulation of individuals within them. Larvae reared under those conditions produce adults with a higher wing volume to body volume ratio and females that tend to take spontaneous flight more easily. In the field, these are presumably the females that fly up and are dispersed widely either by their own flight or passively in air currents. Adult males reared from larvae under crowded conditions lived half as long or less than those reared under less crowded conditions, whether the adults were fed only water (3.6 vs. 6.4 days) or 10% sucrose solution (17.3 vs. 45.2 days). Females reared from crowded conditions lived longer than those reared under less crowded conditions when fed blood and sucrose (47.0 vs. 41.2 days) but lived a shorter time without a carbohydrate source. That result probably reflected the lower lipid and glycogen reserves in females reared from crowded larvae.

Larval Digestive Processes

Food enters the alimentary canal through the morphological mouth behind the oral cavity. It then enters the pharynx and esophagus through processes that are not well defined. As it passes the cardia, it is surrounded by a continuously secreted tube of chitinous material that forms the larval peritrophic membrane (Shao et al. 2001). Among insects, this system of producing a peritrophic membrane is designated Type 2, in contrast to Type 1 in the adult female that secretes the material from cells along the entire length of the midgut. The larval peritrophic membrane includes protein cross-links that strengthen the chitin sheet. One protein that has been identified in *Anopheles gambiae* (Ag-Lper1) is similar to involucrin in the outer layer of human skin, the function of which is also cross-linking proteins. Water flows forward outside the peritrophic membrane and is absorbed by the gastric ceca (Wigglesworth 1932).

The movement of ions across cellular membranes of the midgut is an important part of digestion. Two kinds of adenosine triphosphate enzymes (ATPases) regulate ion transport that influences water balance in the hemocoel of the larva (Patrick et al. 2006). The P-type ATPases move positive sodium (Na^+) and potassium (K^+) ions. V-type ATPases move hydronium ions (H^+). In *Aedes aegypti* these ATPases are in the gastric ceca, Malpighian tubules, hindgut, and anal papillae, which are all involved in osmoregulation. The P-type ATPases are basal in the gastric ceca, and the V-type ATPases are apical in the gastric ceca. These ionic pumping mechanisms also maintain the alkalinity of the larval gut in *Aedes aegypti* (Boudko et al. 2001). The gastric ceca have a nearly neutral pH between 7 and 8, but the adjacent, anterior portion of the midgut is highly alkaline, with a pH of 11. The posterior midgut returns to a neutral pH of 7. The V-type ATPases, combined with the activity of carbonic anhydrase, are probably responsible for maintaining alkalinity in the midgut because specific inhibitors of this enzyme tend to neutralize the pH. Midgut epithelial cells have higher carbonate, bicarbonate, and chlorine ions than the hemocoel. The carbonate and bicarbonate concentrations responsible for the alkalinity of the anterior midgut reach even higher concentrations in the lumen. The high pH in that part of the gut may dissociate protein that would otherwise be unavailable due to binding with tannin. This allows absorption of the proteins in the posterior midgut. A neuropeptide signaling pathway that involves the brain in *Aedes aegypti* regulates ion transport across the anterior midgut wall and influences the rate of peristalsis (Onken et al. 2004). Neuropeptides activate G protein-coupled receptors specific for each kind of neuropeptide (Belmont et al. 2006). The head peptides I and II, short neuropeptide F, and neuropeptide F all affect the gut wall by increasing the transepithelial voltage potential that controls movement of HCO_3^- into the gut, raising pH. This mechanism can get the pH up to 8.5, and further alkalinization to pH 11 occurs as hydronium ions are pumped out of the gut.

As the larva gets larger, the feeding stage of the fourth instar accomplishes 80%–90% of the total growth and biosynthesis of reserves that may be used by the teneral adult (Briegel 2003). Foods of animal origin are important to supply essential polyunsaturated fats (Dadd & Kleinjan 1979), sterols, vitamins, nucleotides, and amino acids (Damiens et al. 2012). Some other insects (e.g., moths and butterflies) use linoleic (18 carbons and two double bonds, designated C18:2) and linolenic acids (C18:3), which are commonly found in plants. *Culex pipiens* has requirements for small amounts (0.05 mg/100 ml of larval diet) of arachidonic acid (C20:4) in its diet late during larval development and is unable to synthesize this kind of fatty acid from plant-based foods. This is a vitamin-like need, as the requirement for vitamin B is larger (0.1 mg/100 ml of larval diet). Similar observations were made in *Anopheles stephensi*, which required both arachidonic and eicosapentaenoic (C20:5) acids. Such a requirement may not be universal among mosquitoes, as *Anopheles arabiensis* synthesized more fatty acids, including arachidonic acid (Hood-Nowotny et al. 2012). The development of a diet for *Anopheles stephensi* showed how a mixture of bovine liver powder with various plant components produced more reliable results than the plant components alone, though some larvae did complete development successfully without the liver powder (Khan et al. 2013). Survival and adult size of *Anopheles arabiensis* was affected by the origin of animal-based foods, leading to selection of bovine liver power, tuna meal, and vitamins as an optimal diet (Damiens et al. 2012).

Microflora of the Larval Alimentary Canal and Salivary Glands

The "holobiome" has been described as an organism and its associated microbes forming a unit of selection different from any of the individual organisms involved (Minard et al. 2013; Hegde et al. 2015). Mosquito larvae are no exception (Table 7.2), with dozens of species of bacteria colonizing them. Few of the species are consistently present in all larvae, but those might be considered primary symbionts if they serve a necessary role in the life of the larvae (Weiss & Aksoy 2011). Two examples are extracellular *Asaia* and intracellular *Wolbachia*, both of which are consistently present in some species. On the other hand, larvae are exposed to many different bacteria in the environment as they put almost any particulate matter through their guts, resulting in long lists of bacteria and variation between individuals and species depending on where the larvae have developed. Physiological interactions between mosquitoes and bacteria have mainly been performed in adults, though some of these interactions probably apply to larvae as well.

Experiments in which bacteria were cleared from larvae by antibiotics and then maintained in sterile conditions showed that whatever the interactions are, they are important to successful development (Coon et al. 2014). Larvae of *Aedes aegypti*, *Aedes atropalpus*, and *Anopheles gambiae* died under sterile conditions. Larvae of each species could be rescued by inoculating the medium with *Escherichia coli*, *Acinetobacter*, or *Aeromonas*; however, *Microbacteria* did not support normal development of those mosquitoes. Although definite associations between bacteria and digestive processes have not been defined, they are likely to occur given the biochemical reactions performed by some of the bacteria (Minard et al. 2013).

The majority of bacteria in the gut seem to do no harm, and sometimes benefit, the larva. That conclusion is implied by the finding of many species of bacteria during apparently normal development and the poor development of larvae in the absence of bacteria. Some naturally occurring bacteria are definitely harmful, such as *Bacillus thuringiensis israelensis* (Goldberg & Margalit 1977), the toxins of which destroy the midgut epithelial cells (Zalunin et al. 2001). In *Anopheles stephensi*, the gut microflora reduces the effects of those toxins (Patil et al. 2013). The larva may regulate the abundance of bacteria by secretion of lysozymes into the gut lumen (Li, B. et al. 2005). Two of the lysozymes identified from *Anopheles gambiae* (Lys c-3 and Lys c-8) are unlikely to have a digestive role because of the alkaline conditions of the gut. Bacteria play an active role in modulating *Plasmodium* infection of adult mosquitoes, and some of those bacterial species may have larval origins. Tchioffo et al. (2013) found that the following bacteria that decreased *Plasmodium falciparum* infection of adult *Anopheles gambiae* in Cameroon were also found in the larval gut: *Escherichia coli*, *Serratia marcescens*, *Pseudomonas*

Table 7.2. Classification of bacteria cited in text

Phylum	Class	Order	Family	Genus
Actinobacteria	Actinobacteria	Actinomycetales	Microbacteriaceae	*Microbacterium*
Actinobacteria	Actinobacteria	Actinomycetales	Micrococcaceae	*Kocuria*
Actinobacteria	Actinobacteria	Actinomycetales	Mycobacteriaceae	*Mycobacterium*
Actinobacteria	Actinobacteria	Micrococcales	Intrasporangiaceae	*Janibacter*
Actinobacteria	Actinobacteria	Propionibacteriales	Propionibacteriaceae	
Ascomycota	Saccharomycetes	Saccharomycetales	Saccharomycetaceae	*Candida*
Ascomycota	Saccharomycetes	Saccharomycetales	Saccharomycetaceae	*Hanseniaspora*
Ascomycota	Saccharomycetes	Saccharomycetales	Saccharomycetaceae	*Pichia*
Ascomycota	Saccharomycetes	Saccharomycetales	Saccharomycetaceae	*Wickerhamomyces*
Bacteroidetes	Flavobacteriia	Flavobacterales	Blattabacteriaceae	*Blattabacterium*
Bacteroidetes	Flavobacteriia	Flavobacterales	Flavobacteriaceae	*Chryseobacterium*
Bacteroidetes	Flavobacteriia	Flavobacterales	Flavobacteriaceae	*Elizabethkingia*
Bacteroidetes	Flavobacteriia	Flavobacterales	Flavobacteriaceae	*Myroides*
Basidiomycota	Wallemiomycetes	Wallemiales	Wallemiaceae	*Wallemia*
Cyanobacteria				
Firmicutes	Bacilli	Bacillales	Bacillaceae	*Bacillus*
Firmicutes	Bacilli	Bacillales	Bacillaceae	*Exiguobacterium*
Firmicutes	Bacilli	Bacillales	Listeriaceae	*Listeria*
Firmicutes	Bacilli	Bacillales	Paenibacillaceae	
Firmicutes	Bacilli	Bacillales	Staphylococcaceae	*Staphylococcus*
Firmicutes	Bacilli	Lactobacillales	Enterococcaceae	*Enterococcus*
Firmicutes	Bacilli	Lactobacillales	Leuconostocaceae	
Firmicutes	Bacilli	Lactobacillales	Streptococcaceae	*Lactococcus*
Fusobacteria				
Proteobacteria	Alpha-proteobacteria	Rhadospirillales	Acetobacteriaceae	*Asaia*
Proteobacteria	Alpha-proteobacteria	Rickettsiales	Rickettsiaceae	*Wolbachia*
Proteobacteria	Alpha-proteobacteria	Sphingomonadales	Sphingomonadaceae	*Sphingomonas*
Proteobacteria	Beta-proteobacteria	Burkholderiales	Alcadigenaceae	*Alcaligenes*
Proteobacteria	Beta-proteobacteria	Burkholderiales	Alcadigenaceae	*Bordetella*
Proteobacteria	Beta-proteobacteria	Burkholderiales	Comamonadaceae	*Comamonas*
Proteobacteria	Beta-proteobacteria	Burkholderiales	Comamonadaceae	*Delftia*
Proteobacteria	Beta-proteobacteria	Neisseriadales	Neisseriaceae	*Chromobacterium*
Proteobacteria	Beta-proteobacteria	Neisseriadales	Neisseriaceae	*Neisseria*
Proteobacteria	Beta-proteobacteria	Procabacteriales	Procabacteriaceae	
Proteobacteria	Gamma-proteobacteria	Aeromonadales	Aeromonadaceae	*Aeromonas*
Proteobacteria	Gamma-proteobacteria	Alteromonadales	Shewanellaceae	*Shewanella*
Proteobacteria	Gamma-proteobacteria	Enterobacteriales	Enterobacteriaceae	*Enterobacter*
Proteobacteria	Gamma-proteobacteria	Enterobacteriales	Enterobacteriaceae	*Escherichia*
Proteobacteria	Gamma-proteobacteria	Enterobacteriales	Enterobacteriaceae	*Klebsiella*
Proteobacteria	Gamma-proteobacteria	Enterobacteriales	Enterobacteriaceae	*Pantoea*
Proteobacteria	Gamma-proteobacteria	Enterobacteriales	Enterobacteriaceae	*Pelagibacter*
Proteobacteria	Gamma-proteobacteria	Enterobacteriales	Enterobacteriaceae	*Rahnella*
Proteobacteria	Gamma-proteobacteria	Enterobacteriales	Enterobacteriaceae	*Raoultella*
Proteobacteria	Gamma-proteobacteria	Enterobacteriales	Enterobacteriaceae	*Serratia*
Proteobacteria	Gamma-proteobacteria	Enterobacteriales	Enterobacteriaceae	*Shigella*
Proteobacteria	Gamma-proteobacteria	Enterobacteriales	Thorselliaceae	*Thorsellia*
Proteobacteria	Gamma-proteobacteria	Oceanospirillales	Halomonadaceae	*Zymobacter*
Proteobacteria	Gamma-proteobacteria	Pseudomonadales	Moraxellaceae	*Acinetobacter*
Proteobacteria	Gamma-proteobacteria	Pseudomonadales	Moraxellaceae	*Moraxella*
Proteobacteria	Gamma-proteobacteria	Pseudomonadales	Pseudomonadaceae	*Pseudomonas*
Proteobacteria	Gamma-proteobacteria	Xanthomonadales	Xanthomonadaceae	*Stenotrophomonas*

stutzeri, *Enterobacter*, and *Comamonas*. *Acinetobacter septicus* did not have this effect.

Larvae picked up many of the bacteria present in rearing water (Coon et al. 2014). *Aedes aegypti* contained 66 of the 123 species in their water, with Flavobacteriaceae accounting for 43% and Microbacteriaceae for 54% of the total bacteria in the larvae. *Aedes atropalpus* contained 91 of the 144 bacterial species in the rearing water, and Betaproteobacteria (Procabacteriaceae, Neisseriaceae, and Comamonadaceae) accounted for 63% of the bacteria. *Anopheles gambiae* larvae contained each of the 161 bacteria in the rearing water, and Flavobacteriaceae accounted for 86% of the bacteria. The microflora of larvae collected in the field has been compared to larvae in the laboratory, with the not surprising result that there is greater variety in field-collected specimens (Minard et al. 2013). Of the bacteria from field-collected larvae and pupae of *Anopheles gambiae* in Kenya, 40% were photosynthetic Cyanobacteria species (Wang et al. 2011). Field collections of *Anopheles stephensi* in southern Iran (Chavshin et al. 2012) and Haryana, India (Rani et al. 2009), documented some of the differences between lab and field-caught larvae. In India, the bacteria in larvae from the field were dominated by *Serratia marcescens*. Interestingly, they also found the extremophile *Deinococcus xinjiangensis* in Indian larvae. Larvae from Iran harbored 40 species of bacteria. Examining a coastal and an inland location, 8 of 14 bacterial genera were present in only one site or the other. *Pseudomonas*, *Aeromonas*, *Myroides*, and *Exiguobacterium* were common to both sites. This study found genera never before observed in *Anopheles* larvae: *Kocuria*, *Myroides*, and *Shewanella*.

Exposure of larvae to many kinds of bacteria in water raises the question of whether any of these bacteria are passed to the pupa and adult stages. The adult midgut contains few bacteria after emergence, at least in *Anopheles punctipennis*, *Culex pipiens*, and *Aedes aegypti* (Moll et al. 2001). The midgut microflora decreases after the final larval female's defecation but increases in the pupa. One mechanism for reducing the number of bacteria at the end of larval development and during pupal development is for the peritrophic membrane to surround the food remnants as the meconium and then kill bacteria with ecdysial fluid. This process effectively scrubs the gut of the larva and then the pupa during one meconial defecation in the former and two meconial defecations in the latter (Wang et al. 2011).

As discussed above, *Asaia* and *Wolbachia* have specific association with mosquitoes and pass from one generation to the next. Extracellular *Asaia* concentrates on the periphery of the oocytes, leading to passage to the egg as it develops prior to fertilization and oviposition (Damiani et al. 2010). Intracellular *Wolbachia* infects the egg itself, also leading to transgenerational transmission (Johnson 2015; Zouache et al. 2009).

Other bacteria are also passed transgenerationally. On Réunion Island in eastern Africa, freshly eclosed adult *Aedes albopictus* had *Acinetobacter*, *Comamonas*, *Delftia*, and *Pseudomonas* (Zouache et al. 2009). Between *Anopheles gambiae* pupa and adult, the Cyanobacteria microflora that was most abundant gives way to members of Enterobacteriaceae and Propionibacteriaceae, with *Thorsellia anopheles* the most abundant (Wang et al. 2011). The bacteria that originate in larvae may reside in the crop of the teneral adult of *Aedes aegypti* (Gusmão et al. 2007). The crop of newly emerged, unfed adults reared from larvae fed sterilized food contained *Bacillus subtilis*, *Serratia*, and the yeast *Pichia caribbica*. Another yeast, *Wickerhamomyces anomalus*, is also transferred across generations (Ricci et al. 2011a). Yeast including *Candida*, *Pichia*, *Hanseniospora*, and *Wickerhamomyces* have been found in *Anopheles gambiae*, *Anopheles stephensi*, and *Aedes aegypti*. These yeasts may also be transmitted across generations (Ricci et al. 2012).

Adult Feeding and Nutrition

Morphology

The mouthparts of adult mosquitoes are highly modified for piercing and sucking (Fig. 7.4). The basic plan is a series of five stylets surrounded by a sheath, the sheath being the labium (Snodgrass 1959). Starting at the anterior, the stylets are as follows in *Aedes aegypti*:

> Labrum: This is the thickest stylet that enters the skin or plant tissue. It is recurved to form the roof of the food canal. Muscles in the clypeus, to which it is hinged, move it up and down.
>
> Mandibles: The mandibles are two stylets, the thinnest in the mouthparts. They are each connected to a small sclerite attached to the cranium. They can each be withdrawn by muscles in the head attached to the internal skeletal structure (tentorium). The mandibles are pushed forward by the elasticity of the connection to the head.
>
> Median hypopharynx: This is a single stylet that contains the salivary canal. It does not move. The base of the hypopharynx forms the floor of the cibarial pump inside the head.
>
> Maxillary stylets: These are the stylets that actively draw the mouthparts into materials. They are tipped with recurved teeth that can grip as muscles pull on each stylet in turn. Muscles can also actively push on these stylets, using a lever arrangement within the head formed by the cardo and stipes of the maxilla. Some muscles can move the stylets laterally.

The labium does not enter the wound, folding backward along a single joint as the stylets are inserted. The tip of the labium has two small hinged portions that are the labellae, containing sensory organs and holding the stylets in place. Maxillary palpi project separately from the base of the mouthparts and are the site of sensory organs.

The mouthparts of all mosquitoes follow the basic plan described above, but modifications occur in apparent selective response to feeding needs. Males have weaker mandibles and maxillary stylets. The hypopharynx of males is integrated into the labium so that a single feeding tube is formed by the la-

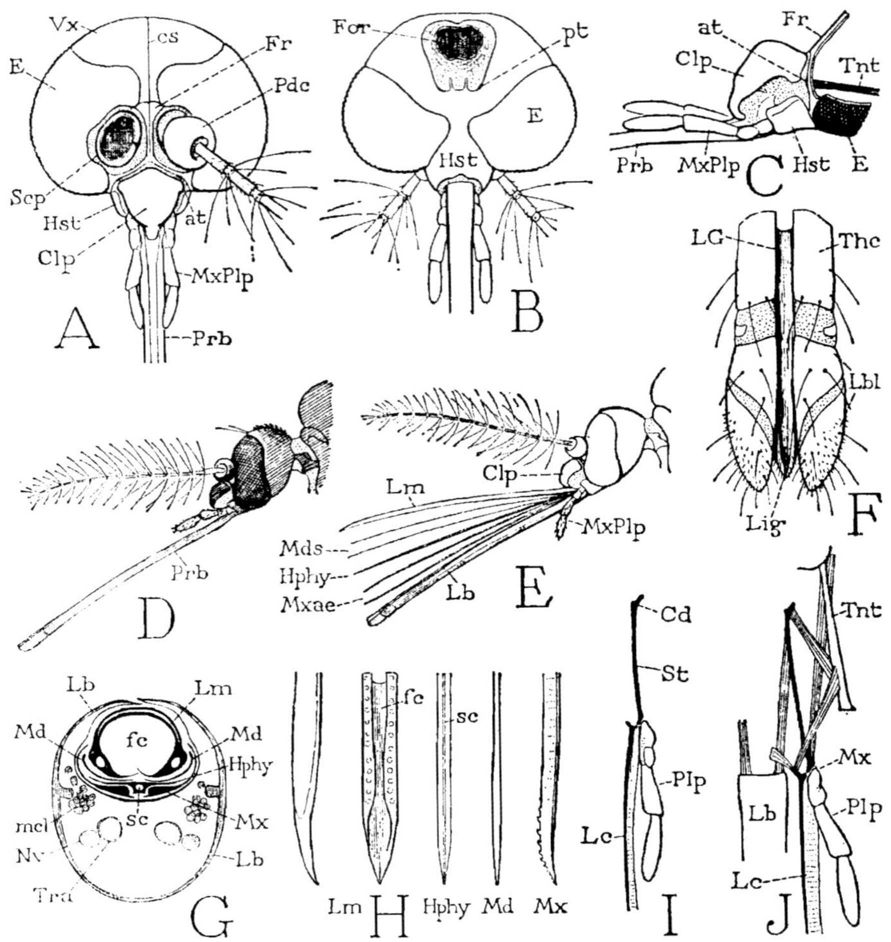

Figure 7.4. Mouthparts of *Aedes aegypti* and morphological relationships to head. A, view of head from anterior: Vx = vertex; cs = median cranial sulcus; E = eye; Fr = frons; Pdc = antennal pedicel; Scp = antennal scape; Hst = hypostome; Clp = clypeus; at = anterior tentorial pit; MxPlp = maxillary palpus; Prb = proboscis. B, view of head from posterior: For = foramen. C, lateral view of base of front of head: Clp = clypeus; Tnt = tentorium. D, lateral view of head. E, lateral view of head with mouthparts separated: Lm = labrum; Mds = mandibular stylet; Hphy = hypopharynx; Mxae = maxillary stylet; Lb = labium. F, dorsal view of tip of proboscis: LG = ligula; Thc = theca; Lbl = labella; Lig = ligula. G, cross section of proboscis: fc = food channel; sc = salivary channel; mcl = muscle; Nv = nerve; Tra = trachea. H, tips of the stylets. I, parts of the maxilla: Cd = cardo; St = stipes; Plp = palpus; Lc = lacinia. J, muscular connections to maxilla and labrum. Snodgrass 1959

brum and labiohypopharynx. *Toxorhynchites* species, none of which take blood meals, have similar mouthparts in males and females, lack mandibular stylets, and have weak maxillary stylets. *Malaya*, which feed by inserting their mouthparts into those of ants and forcing them to regurgitate, lack both mandibles and maxillary stylets.

Just as in the larva, the mouthparts are outside the alimentary canal, and the feeding tube forms the equivalent of the preoral cavity. A pair of large salivary glands extend into the thorax, each consisting of three lobes, with the middle lobe larger than the lateral ones. A small salivary pumping mechanism is formed by a muscled expansion of the common salivary duct near the base of the hypopharynx. The order of structures in the alimentary canal (Fig. 7.5) are the cibarial pump, the pharyngeal pump, the esophagus, the esophageal diverticula, the proventriculus, the midgut (also called the stomach or ventriculus), and the hindgut (Snodgrass 1959). There are three esophageal diverticula: two small sacs dorsally and a large one ventrally called the crop. The midgut has a narrower anterior portion and a broader posterior portion. The hindgut is differentiated into several regions, starting with the pyloric area. This part is narrow, contains backward-directed spines in *Aedes aegypti*, and is the site from which five long Malpighian tubules emerge. Posterior to the pyloric region, the hindgut makes an S-shaped bend and then expands to form the rectum. The rectum contains six papillae in *Aedes aegypti*. Finally, the alimentary canal ends in the anus.

Function

Feeding and the Pumping Mechanism

The female mosquito sucks up twice its weight in blood over the course of a few minutes. Considering the tiny tubes of the mouthparts, the ability of the gut and abdominal integument to stretch, and the defense mounted by a mammalian host's hemostatic system, blood feeding by mosquitoes is one of the wonders of biology. Individual species seem to have adapted to different kinds of hosts—as shown, for example, by *Culex quinquefasciatus*' relatively faster probing times on chickens compared to probing times on humans or mice (Ribeiro 2000). The variety of feeding mechanisms and adaptations is great among adult mosquitoes, but all species of both genders probably use the basic dipteran two-stage pumping mechanism of anterior cibarial pump and posterior pharyngeal pump (Fig. 7.6). This pump has a power density of 200 MW/cm^3, compared to much lower values in modern mechanical pumps (Kikuchi et al. 2008). Although it is impressive that this biological pump has a greater power-to-volume ratio than many human-made pumps, pump designers warn that higher power densities usually lead to shorter pump life. For example, the main engine turbopump of the space shuttle had a power den-

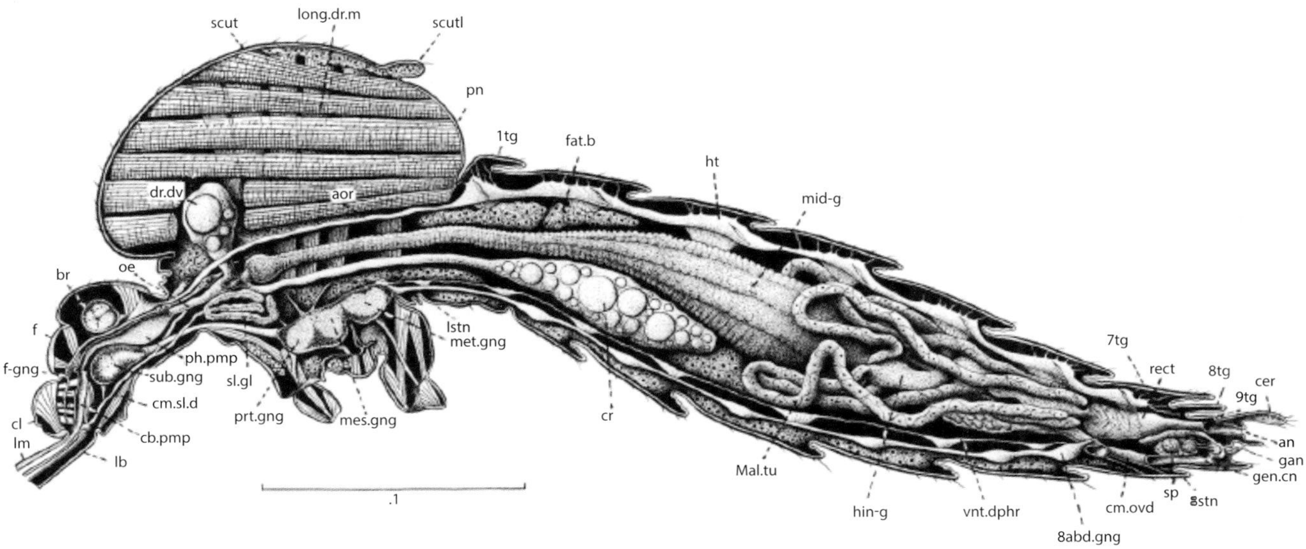

Figure 7.5. Sagittal section of female *Aedes aegypti* showing alimentary canal, nervous system, and other organs. lm = labrum; lb = labium; cl = clypeus; f-gng = frontal ganglion; cb.pmp = cibarial pump; f = frons; cm.sl.d = centromedial salivary duct; sub. gng = suboesophageal ganglion; ph.pmp = pharyngeal pump; br = brain; oe = esophagus; dr.dv = dorsal diverticulum; aor = aorta; sl.gl = salivary gland; prt.gng = prothoracic ganglion; mes. gng = mesothoracic ganglion; met.gng = metathoracic ganglion; lstn = first abdominal sternite; scut = scutum; long.dr.m = longitudinal dorsal muscle; scutl = scutellum; pn = postnotum; ltg = first abdominal tergite; fat.b = fat body; ht = heart; mid-g = midgut; cr = crop or ventral diverticulum; Mal.tu = Malpighian tubule; hin-g = hindgut; vnt.dphr = ventral diaphragm; 8abd.gng = eighth abdominal ganglion; cm.ovd = common oviduct; sp = spermatheca; 8stn = eighth abdominal sternite; gen.ch = genital chamber; gon = gonotreme; an = anus; 7tg = seventh abdominal tergite; rect = rectum; 8tg = eighth abdominal tergite; 9tg = ninth abdominal tergite; cer = cercus. Jobling 1987

sity of 36,781 MW/cm^3 but was only to be used a limited number of times. The situation in the mosquito may not be so different in that a female takes a limited number of blood meals during a relatively short life span.

Recent work has contributed greatly to the understanding of the mechanics and biophysics of this pumping mechanism. Using microparticle image velocimetry and synchrotron x-ray microimaging, researchers in Korea (Lee et al. 2009; Kim et al. 2011) used *Aedes togoi* to visualize the function of the two pumps in action and to measure physical parameters of fluid flow during feeding. During observation of live mosquitoes feeding on iodine-spiked human blood, they saw that the two pumps work in sequence, with the cibarial pump expanding first, followed by the pharyngeal pump. The cycle is timed so that the pharyngeal pump is beginning to expand before the cibarial pump has fully relaxed, thus exerting negative pressure to minimize backflow. Valves behind each pump keep blood flowing in the direction of the midgut (Ha et al. 2015). Expansion takes longer for the cibarial than for the pharyngeal pump (2.6-fold in the female and 4.2-fold in the male), giving the cibarial pump the ability to control the overall rate of flow. When the cibarial pump initially expands to its full volume of 9.43×10^{-4} µl, the negative pressure collapses the food canal, which is quickly expanded by dilator muscles. Despite these obviously effective mechanisms for keeping flow going in the direction of the midgut, the lower pressures generated by the pharyngeal pump still result in about 3% backflow toward the cibarial pump. The pumping cycle repeats itself about six times per second (6.1 Hz) on average, though the rate is slower at the beginning of feeding than at the end. This rate of pumping results in a flow rate of 0.006 µl/s, which would mean that the mosquito could take on a microliter of blood in about 2.8 minutes. In order to achieve this rate of flow through the small food canal, the fluid moves at an average of 0.416 cm/s (0.7 cm/s in the middle of the tube and 0.14 cm/s at the edges). An indication of the laminar nature of the flow (i.e., no turbulence within the feeding tube) is the low Womersley number, which is a dimensionless measure of pulsatile frequency and viscosity. The Womersley number during blood feeding is 0.117 compared to 0.00443 in a human capillary and 2.21 in a human artery.

Study of *Culex pipiens* Form Molestus (Kikuchi & Mochizuki 2011) reached the similar conclusion that blood flow in the mosquito was laminar based on the Reynolds number (dimensionless measure of the ratio of flow rate to viscosity). Under these miniature conditions, the Reynolds number was a very small 0.8 (compare to Reynolds number of 170 for mosquito flight), indicating that it was Hagen-Poiseuille flow (laminar). The main difference from *Aedes togoi* that they observed was a slower pumping cycle of 1–1.5 Hz and a suggestion that the pharyngeal pump produced more pressure than the cibarial pump.

Blood feeding on a warm-blooded host heats the mosquito (Lahondère & Lazzari 2012). In the case of *Anopheles stephensi* feeding on a mouse, the head is up to 3.3°C warmer than the abdomen during feeding, and the thorax is an intermediate

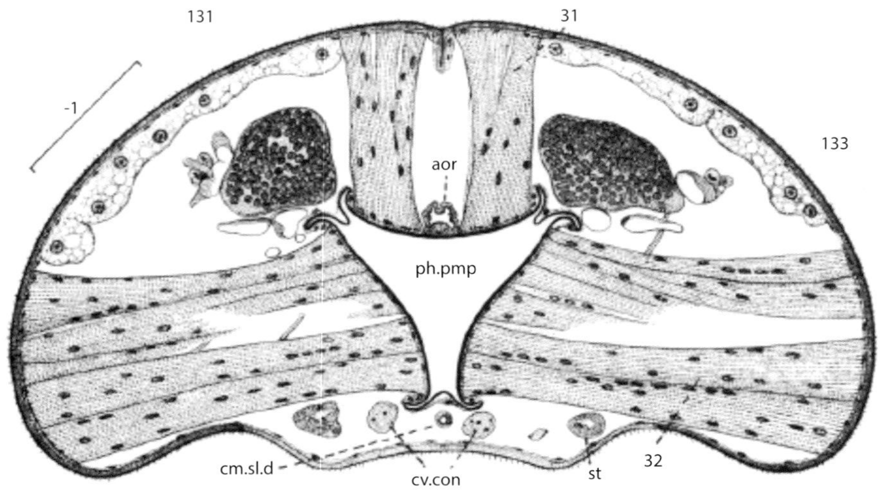

Figure 7.6. Cross section of *Aedes aegypti* female head posterior to brain, showing dorsal and lateral dilating muscles that expand the triangular pharyngeal pump. Jobling 1987

temperature. The droplet of urine excreted during feeding provides evaporative cooling to the abdomen, immediately reducing the temperature to ambient levels. *Aedes aegypti* has a warmer abdomen during feeding because the abdomen is held close to the surface of the host. The warming effect is from the blood meal itself, not from musculature action or proximity to the host, as neither sugar feeding nor resting on a host causes heating of the mosquito.

Control of the Crop

The crop is an organ for storing sugar meals before they are pumped back into the alimentary canal and digested in the anterior part of the midgut. The dipteran crop is much more than a sac, having innervation, musculature, and valve structures (Stoffolano & Haselton 2013). It is present in almost all flies. Mechanical control of entrance into the crop is provided by the cardia acting as a sphincter, innervated by the hypocerebral ganglion and influenced (demonstrated in *Culiseta inornata*) by chemosensory cues. The crop consists of epithelial cells that secrete a cuticular lining, two nerve bundles, and muscles around the duct and body of the organ. Much like the rest of the alimentary canal, the crop can control flow in and out of itself with a series of valves, musculature contraction, and the influence of hemolymph pressure. Phagostimulants like pyranose and furanose, temperature, and osmolarity influence whether a liquid meal is directed to the crop. This process is at least partly controlled by the presence of specific microRNAs, the inhibition of which prevents filling of the crop. Components of the salivary glands and substances secreted within the crop have antimicrobial activity. The crops of teneral adult mosquitoes of both sexes are filled with air bubbles pushed there from the midgut that extend the crop into the abdomen. The crops of other kinds of flies have additional functions, including production of pheromones, storage of blood (Glossinidae and Muscidae: *Stomoxys*), and inoculation of food with bacteria (Tephritidae).

Midgut Epithelium and the Peritrophic Membrane

The cardial sphincter at the posterior end of the foregut marks the beginning of the midgut and is a means of preventing backflow of food content. Starting from the interior side of the midgut (luminal or apical side) and going toward the hemolymph (basal direction), the midgut consists of a roughly single layer of cuboidal epithelial cells, a proteinaceous and noncellular basal lamina, circular and longitudinal muscles, and tracheae. Midgut epithelial cells in the anterior, sugar-absorbing portion of the organ have prominent microvilli on the luminal side, smooth endoplasmic reticula, and basal labyrinth formed by infoldings of the membrane. Posterior midgut epithelial cells have rough endoplasmic reticula and abundant mitochondria (Franz et al. 2015).

Studies of the fine structure of the midgut have shown a number of interesting features. Enterocytes distinct from the epithelial cells are scattered along the midgut adjacent to the basal lamina, serving endocrine and regenerative functions (Okuda et al. 2007). Apical microvilli are absent from a minority of the midgut epithelial cells in *Aedes aegypti* (Zieler et al. 2000). These "bare" cells have infoldings of the apical membrane rather than the bunches of hairlike microvilli appearing like rounded bushes on each cell. The microvilli have a distinct structure reinforced by nonchitinous strands. Muscles surrounding the posterior midgut are arranged in two layers in *Anopheles gambiae*, with 37 longitudinal strands overlapping to form a tube and 57 circular muscles surrounding them (Park & Shahabuddin 2000). In contrast, the 168 longitudinal muscle fibers of *Aedes aegypti* are shared between bundles, and there are 77 circular muscles.

Blood feeding causes many changes in the structure of the midgut. In *Anopheles darlingi* (Okuda et al. 2005), stretching of the midgut causes the cuboidal epithelial cells to flatten and modifies the interior structure. Death of epithelial cells has been shown in *Culex quinquefasciatus*, caused by toxic components in blood, including hemoglobin (Okuda et al. 2007). Heme is toxic because of its peroxidase activity, which produces reactive oxygen species (Pascoa et al. 2002). The mosquito removes some of the toxic components of blood by forming insoluble aggregates of heme in the lumen of the midgut of *Anopheles darlingi* or sequestering heme and ferritin in the peritrophic membrane of *Aedes aegypti* for 48 hours following a blood meal (Pascoa et al. 2002). Nonetheless, some enterocytes die following a blood meal in a process of programmed death, or apoptosis, that affects 2% of the total cells in the midgut. These dead cells are ejected through a "purse string" process, powered by actin fibers, that squeezes them into the lumen of the gut without compromising the integrity of the feeding tube. Regenerative cells replace epithelial cells killed by toxic components of the blood, but the gut eventually exhausts its supply of extra cells by the end of the third or fourth blood meal. Subsequent degradation of the gut may be an important component of the aging process.

The peritrophic membrane of the adult is different from that of the larva. In contrast to continuous secretion from the cardia (Type 2), the adult peritrophic membrane is secreted around the blood meal simultaneously by individual midgut epithelial cells (Shao et al. 2001). The membrane consists of chitin (polymerized N-acetylglucosamine) cross-linked by a number of different proteins in *Aedes aegypti* and *Anopheles gambiae*. Cross-linking (as indicated by carboxypeptidase secretion) of the peritrophic membrane peaks 4 to 6 hours after a blood meal in the latter species and 24 hours after the blood meal in the former (Abraham et al. 2005). Blood feeding also upregulates chitin production from fructose-6-phosphate in midgut epithelial cells of *Aedes aegypti* and, in addition to cross-linked proteins, chitin fibrils reinforce the membrane (Kato et al. 2006). The physiological function of the peritrophic membrane is uncertain, as its removal by addition of antibodies or chitinase speeds digestion and does not affect fecundity (Shao et al. 2001; Villalon et al. 2003). Chitinase secreted by the midgut epithelial cells may be a way in which the mosquito regulates pores in the peritrophic membrane. The peritrophic membrane may protect the gut from mechanical damage, infection, and toxic components of the blood.

Salivation

Salivation is an essential part of the blood-feeding process, facilitating penetration of the stylets and performing many biochemical roles to counteract the host's defenses. Saliva also contributes to digestion and to immunological responses. There is a certain logic in the prominent role of saliva for an insect that only imbibes liquids in that the saliva is easily mixed with the food. The size of the salivary glands, their abundant production of saliva, and the complexity of salivation certainly reflect the long process of evolution that has shaped modern mosquitoes.

The greatest challenge to blood feeding is the host's intricate system for keeping blood in blood vessels through various mechanisms of hemostasis (Ribeiro & Francischetti 2003). Approximately 500 genera and 19,000 species of blood-feeding insects have managed to overcome this problem through a wide variety of physical and biochemical adaptations. The elements of hemostasis are clotting, platelet aggregation, and vasoconstriction (narrowing of the blood vessels). Mosquitoes have overcome each of those defenses with a remarkable variety of peptides and proteins in saliva, a variety not shared between all species. Table 7.3 describes some of these substances. It is apparent that there is a similarity of function

Table 7.3. Summary of substances in mosquito saliva that affect host responses

Species	Substance name	Effect	Reference
Aedes aegypti	Adenosine deaminase	Antihemostasis	Valenzuela et al. 2002
Aedes aegypti	Alpha-glucosidase	Carbohydrate digestion	Peng and Simmons 2004
Aedes aegypti	Amylase	Carbohydrate digestion	Valenzuela et al. 2002
Aedes aegypti	Angiopoietin	Antibacterial, fibrinogen binding	Ribeiro et al. 2007; Valenzuela et al. 2002
Aedes aegypti	Antimicrobial polypeptides	Antibacterial	Ribeiro et al. 2007
Aedes aegypti	Calreticulin	Anticlotting?	Valenzuela et al. 2002
Aedes aegypti	Carboxylesterase	Digestion?	Valenzuela et al. 2002
Aedes aegypti	Chymotrypsin	Digestion	Valenzuela et al. 2002
Aedes aegypti	D7 salivary proteins	Bind biogenic amines, cysteinyl leukotrienes, and eicosanoids produced by platelets, mast cells, and neurons to prevent vasoconstriction, platelet activation, itching, swelling, and pain	Calvo et al. 2009
Aedes aegypti	DNA-binding protein	Immunity?	Valenzuela et al. 2002
Aedes aegypti	Gram-negative binding protein	Antibacterial	Valenzuela et al. 2002
Aedes aegypti	Lectin	Antibacterial	Ribeiro et al. 2007
Aedes aegypti	Mucins	Lubrication of mouthparts	Ribeiro et al. 2007

continued

Table 7.3. *continued*

Species	Substance name	Effect	Reference
Aedes aegypti	Nucleotidase	Digestion or immunity	Ribeiro et al. 2007
Aedes aegypti	Platelet-activating factor acetylhydrolase	Antiplatelet aggregation	Valenzuela et al. 2002
Aedes aegypti	Protease inhibitor	Digestion?	Ribeiro et al. 2007
Aedes aegypti	Purine hydrolase	Hydrolyzes inosin, inhibiting mast cells	Ribeiro and Francischetti 2003; Valenzuela et al. 2002
Aedes aegypti	Serine protease	Digestion	Ribeiro et al. 2007; Valenzuela et al. 2002
Aedes aegypti	Sialokinin	Suppresses vasoconstriction	Ribeiro and Francischetti 2003; Ribeiro et al. 2007
Aedes aegypti	Sphingomyelin phosphodiesterase	Digestion?	Valenzuela et al. 2002
Aedes aegypti	Sugar hydrolases	Carbohydrate digestion	Ribeiro et al. 2007
Aedes aegypti	Xa	Anticlotting factor related to serpin	Ribeiro and Francischetti 2003
Aedes aegypti, Anopheles albimanus, Culex quinquefasciatus	Adenosine diphosphate	Inhibition of platelet aggregation	Ribeiro 2000
Aedes aegypti, Anopheles albimanus, Culex quinquefasciatus	Collagen	Inhibition of platelet aggregation	Ribeiro 2000
Aedes aegypti, Anopheles albimanus, Culex quinquefasciatus	Thrombin	Inhibition of platelet aggregation	Ribeiro 2000
Anopheles gambiae	Ag5 antigens	Related to bee venom	Arcà et al. 2005
Anopheles gambiae	Amylase	Carbohydrate digestion	Arcà et al. 2005
Anopheles gambiae	Antithrombin	Anticlotting factor, smaller than that of *Aedes aegypti*	Ribeiro and Francischetti 2003
Anopheles gambiae	D7 salivary proteins	Similar to *Aedes aegypti* but none of the proteins have double ligands	Calvo et al. 2009
Anopheles gambiae	Mucins	Lubrication of mouthparts	Arcà et al. 2005
Anopheles gambiae	Protease	Digestion	Arcà et al. 2005
Anopheles gambiae	Vasodilatory peroxidase	Destroys norepinephrine and serotonin vasodilators	Ribeiro and Francischetti 2003
Culex quinquefasciatus	C-type lectin	Antibacterial	Ribeiro et al. 2004
Culex quinquefasciatus	Cystein-rich peptides	Anticlotting	Ribeiro et al. 2004
Culex quinquefasciatus	Endonuclease	Digestion or immunity	Ribeiro et al. 2004
Culex quinquefasciatus	Gambicin peptide	Antibacterial	Ribeiro et al. 2004
Culex quinquefasciatus	Hyaluronidase	Related to bee venom	Ribeiro et al. 2004
Culex quinquefasciatus	Mucins	Lubrication of mouthparts	Ribeiro et al. 2004
Toxorhynchites amboinensis	Ag5 antigens	Related to bee venom	Calvo et al. 2008
Toxorhynchites amboinensis	Amylase	Carbohydrate digestion	Calvo et al. 2008
Toxorhynchites amboinensis	Endonuclease	Digestion or immunity	Calvo et al. 2008
Toxorhynchites amboinensis	Gram-negative binding protein	Antibacterial	Calvo et al. 2008
Toxorhynchites amboinensis	Mucins	Lubrication of mouthparts	Calvo et al. 2008
Toxorhynchites amboinensis	Serine protease	Digestion	Calvo et al. 2008
Many mosquitoes	Apyrase	Suppresses inflammatory response by hydrolyzing ATP	Ribeiro and Francischetti 2003; Peng and Simmons 2004; Valenzuela et al. 2002
Many mosquitoes	Lysozymes	Antibacterial	Peng and Simmons 2004; Calvo et al. 2008; Valenzuela et al. 2002

between the few species that have been examined, but the absence of antihemostatic components in *Toxorhynchites amboinensis*, a species in a genus characterized by a complete absence of blood feeding, shows that salivary function matches feeding habits.

The details of differences between blood-feeding species might reflect subtle adaptations in accordance with host preferences, but those relationships are not obvious. For example, *Culex quinquefasciatus*, *Anopheles albimanus*, and *Aedes aegypti* have different values for anticlotting, adenosine diphosphate, vasodilation, and total protein (Ribeiro 2000). The number of allergenic proteins can be measured by quantifying immunoglobulin E (IgE) response in humans, and the total number varies between 3 in *Culiseta inornata* to 16 in *Aedes albopictus*, among eight species (Peng & Simons 2004).

Species-specific reaction to saliva can have human health implications. Detailed testing of hyperallergenic individuals in Europe showed varied reaction to different species, suggesting a specific response; however, some individuals reacted to species to which they had never been exposed, showing a more general allergic response (Peng et al. 2004a). Children naturally exposed to *Aedes vexans* gradually decreased their IgE responses between 5 and 18 years of age, in a process of desensitization (Peng et al. 2004b). Recombinant salivary components have been prepared specifically for diagnosis of allergenic reaction in humans (Simons & Peng 2001).

Examination of gene expression in salivary glands also shows the variation between mosquito species and the complexity of both function and production of the glands. Such studies have created the term "sialome" to describe gene expression specifically in salivary glands. *Aedes aegypti* (Valenzuela et al. 2002; Ribeiro et al. 2007) had a total of 614 proteins in the female salivary glands, only 41 of which had been known previously. Of those proteins, 55 were secreted into the saliva, but the functions of most of them remain unknown. Many of the genes have housekeeping functions. Surprisingly, four of the proteins appear to be prokaryotic and may represent examples of gene transfer. The sialome of *Anopheles gambiae* (Arcà et al. 2005) included 24 classes of housekeeping genes (Table 7.4), two transcription factors, 20 extracellular matrix components, and 71 secreted proteins; 27 of the proteins were only found in the salivary gland, possibly reflecting its immune functions. As in *Aedes aegypti*, antibacterial and glycosidases are a part of the sugar-feeding process. The *Anopheles gambiae* sialome also included transposable elements and *Wolbachia*-like proteins, again suggesting horizontal gene transfer. *Culex quinquefasciatus* had a more limited genome expression in the salivary glands, with only 54 transcripts detected. Many of the same functions were represented, and there were 14 new proteins (Ribeiro et al. 2004). *Toxorhynchites amboinensis* had identical salivary gland gene expression in males and females (Calvo et al. 2008). This species expressed 492 genes, including 33 secreted proteins mostly associated with carbohydrate digestion and immunity.

Table 7.4. List of functional categories of housekeeping genes found in the transcriptome of 1- to 2-day-old female *Anopheles gambiae* (Arcà et al. 2005)

Protein synthesis machinery
Signal transduction
Conserved proteins of unknown function
Transporters
Protein modification machinery
Metabolism, energy
Protein export machinery
Transcription factors
Nuclear metabolism and regulation
Cytoskeletal
Transcription machinery
Proteasome machinery
Metabolism, lipid
Metabolism, carbohydrate
Membrane proteins of unknown function
Cell adhesion
Extracellular matrix
Nitrogen excretion metabolism
Cofactor synthesis
Metabolism, amino acid
Metabolism, nucleic acid and nucleotides
Metabolism, oxidant
Metabolism, heme
Protease inhibitor

Nutrition and Regulation

Carbohydrates

In general, male and female mosquitoes ingest carbohydrates from a variety of sources that include pycnidial (fungal) nectar, honeydew of insect origin, extrafloral nectaries, and flowers (Stoffolano & Haselton 2013). The principal sugars in those sources are sucrose, fructose, and glucose (Briegel et al. 2001). *Aedes aegypti* requires a solution of sucrose between 0.5% and 50% to get any benefit. An increase in sugar concentration increases viscosity and slows the rate of feeding, extends the phase shift between the cibarial and pharyngeal pumps, and reduces backflow (Kim et al. 2013). *Aedes aegypti* (Briegel et al. 2001) can use solid sources of sugar, presumably by salivating to dissolve it, for maintenance but not to build up reserves.

Energy reserves are mainly in the form of glycogen, a small carbohydrate. Glycogen is the principal energy store for flight, as lipid cannot be used for this purpose. A certain amount of glycogen accumulated in the larval stage is present in the newly emerged adult (Vrzal et al. 2010). Glycogen reserves can fuel flight for remarkably long periods. Females of *Aedes aegypti* will fly two to nine hours during a night on a flight mill using 33%–44% of their maximum glycogen reserves. Smaller females, teneral females with low reserves, or suboptimally fed females flew less distance. Male *Anopheles coluzzii* and *gambiae* used 50% of their glycogen reserves during 25 minutes of swarming (Maïga et al. 2014). *Aedes sollicitans* and *taeniorhynchus* used up energy reserves four times more quickly in flight compared to being at rest (Van Handel 1984).

Mobilization of glycogen and lipids to simpler sugars is mediated at least in part by peptide adipokinetic hormones (Gäde 2004). This family of 40 isoforms in insects has many functions (Kaufmann & Brown 2008). Each hormone is formed from a fusion of 8–10 individual peptides. One of these hormones in *Anopheles gambiae*, Anoga-AKH-I, was characterized and found to increase trehalose levels in sugar-fed females but did not mobilize sugars from lipids. Functionally, the hormone increased flight capability. Anoga-AKH-I is produced in the corpora cardiaca, the brain, and some of the ganglia.

Mosquitoes have preferences for some sugars over others, though they can metabolize a large variety. *Culiseta inornata* (Schmidt & Friend 1991) preferred sucrose over all others. Other disaccharides readily imbibed included furanose, trehalose, and maltose; however, isomaltose, melibiose, and cellobiose required slightly higher concentrations. The disaccharides lactose, lactulose, gentiobiose, beta-glucose, and methyl beta-D-glucopyranoside did not attract feeding at all. These mosquitoes also took the trisaccharides melezitose and raffinose at low concentrations. Monosaccharides were only imbibed at higher concentrations, including fructose and alpha-glucose. Among these sugars, raffinose, furanose, trehalose, fructose, pyranoside, melibiose, and cellobiose go directly to the midgut rather than the crop.

Glucosidases are the workhorse enzymes of sugar digestion. A study of *Anopheles aquasalis* (Souza-Neto et al. 2007) identified three isoforms, designated alpha-Glu 1, alpha-Glu 2, and alpha-Glu 3, that were synthesized in the midgut. Glucosidases have also been found in saliva of *Aedes aegypti*, *Aedes albopictus*, and *Anopheles darlingi*, suggesting that some digestion takes place in the crop following ingestion of a sugar meal. In *Anopheles aquasalis*, most (70%) of the digestion and absorption took place in the posterior part of the midgut. Sugar gets metabolized to storage molecules, mostly glycogen and trehalose (Van Handel 1969). One of the metabolic pathways involves sorbitol that dehydrogenates to fructose, and in the opposite direction, fructose gets reduced to sorbitol in *Aedes sollicitans*. Other conversions of sugars do not take place—for example, sorbitol to glucose or glucose to sorbitol. Taking in sugar upregulates genes (Dana et al. 2006) associated with production of mucins that have mechanical roles in lubrication and the formation of linings of the alimentary canal. The upregulation of high-molecular-weight kininogens involved in vasodilation and of protein-digesting chymotrypsin is more difficult to explain in the context of sugar feeding.

Accumulation of fat is also associated with sugar feeding in female mosquitoes, with the fat body being the site of triglyceride synthesis from sugars (Van Handel 1984). The fat body in mature *Aedes aegypti* females is composed of carbohydrate-rich trophocytes and lipid/protein-rich oenocytes attached to the abdominal wall (Martins et al. 2011). There are morphological changes in these two cell types depending on age and diet of the mosquitoes. At emergence, the trophocytes contain more lipid and protein than later. In the unnatural situation in which females were fed exclusively on sugar, the trophocytes continued to contain lipid, and the cellular organelles (mitochondria and ribosomes) were less abundant. The more natural situation of frequent blood feeding resulted in a reduction of lipid and protein at each blood feeding, abundant cellular components, and accumulation of glycogen. The oenocytes of blood-fed females cycled between deep infoldings at blood meals and small ones between meals.

The amount of lipid in the fat body increases markedly with sugar feeding. At emergence it contains 100 μg of lipid in *Aedes aegypti*, increasing to 300 μg following seven days of sugar feeding (Ziegler & Ibrahim 2001). The amount of lipid in *Culex nigripalpus* at emergence is dependent on the abundance of the larval diet. A low-level diet results in 42.6 and 33.84 μg lipid per milligram of mosquito tissue in males and females, respectively, compared to 113.3 in males and 108.4 μg/mg in females reared from larvae given an abundant diet (Vrzal et al. 2010). *Aedes vexans* emerges with low lipid reserves, but these increase tenfold the first week of sugar feeding and 20-fold the second week (Briegel 2003).

Addition of sugar to the female diet generally increases longevity but not necessarily fecundity. Experiments in the laboratory have given the impression that sugar can have a negative effect on reproductive potential, but field studies show that sugar-starved populations can have much shorter population longevity. The relationship is not the same in all species. For example, *Aedes aegypti* survival increases linearly in relation to the log of the sugar concentration up to 10%, whereas *Anopheles gambiae* survival increases exponentially (Briegel 2003). Studies that simulated natural conditions in large cages in Kenya showed that female *Anopheles* lived only 14 days with blood alone, 29 days with glucose, and 33 days with blood and glucose (Okech et al. 2003). When denied sugar, *Aedes aegypti*, *Aedes albopictus*, *Anopheles gambiae*, and *Anopheles funestus* all have a lower age-specific survival and blood feed more frequently (Braks et al. 2006). As a result, they lay more eggs earlier in their lives, possibly increasing reproductive success under conditions of high daily mortality. *Aedes aegypti* deposited the maximum number of eggs in its lifetime when fed on human blood and no sugar sources (400 eggs per female), despite the relatively low concentration of the essential amino acid isoleucine in human blood (Harrington et al. 2001b). Mouse blood contains isoleucine but nonetheless resulted in lower reproductive potential (60 eggs per female). Artificial addition of isoleucine had no effect on fecundity unless sugar was added to the diet. At maximum life span in the laboratory (about 60 days), sugar feeding did not increase or decrease total fecundity of this species. Other experiments with *Anopheles gambiae* have shown a similar effect but with the result that females deposited more eggs in the absence of sugar (mean 379 eggs per female) than in the presence of it (mean 349 eggs per female; Gary & Foster 2001).

Larval diet also affects longevity of females fed sugar only (Joy et al. 2010). Ironically, low larval diets resulted in longer-lived *Aedes aegypti* than high larval diets. In the field, the plant source of nectar is related to survival of male and female *Anopheles gambiae*, presumably because of the quality of the sugar meals (Manda et al. 2007a).

Blood

Blood feeding in Diptera has probably evolved five times in 12 different families (Yuval 2006). The highly specialized mechanism for blood feeding in mosquitoes suggests that the family originated from a common blood-feeding ancestor. Blood meals go directly to the midgut of mosquitoes. Most of the nutritionally utilized blood, constituting 20% of all the blood material, is used for egg production in *Anopheles gambiae* (Dana et al. 2005). Amino acids in the blood signal retention of the meal and lead to initial development of the ovarian follicles from Christopher's Stage I to II (Arsic & Guerin 2008). The blood meal activates production of yolk protein precursors in the fat body to produce vitellogenin for deposition in the oocytes by upregulating vitellogenic carboxypeptidase, vitellogenic cathepsin B, and lipophorin (Attardo et al. 2005). Lipid is also transported from the blood to the fat body and subsequently to the ovaries in *Aedes aegypti* (Ziegler & Ibrahim 2001), though older females tend to retain more lipid in the fat body and contribute less to the ovaries (Briegel 2003). Poor larval nutrition can influence utilization of blood in autogenous species (i.e., those that do not take a blood meal to produce their first batch of eggs) like *Aedes atropalpus*, leading to earlier blood feeding (Telang & Wells 2004).

Blood is not just a source of nutrients to build eggs; it can also provide energy for the female adult. Some species, like *Anopheles albimanus*, divert the first meal from egg production to metabolism and concentrate a volume of 10–12 μl of blood into the 2–3 μl volume capacity of the midgut (Briegel 2003). Use of blood for energy would create dangerous levels of nitrogenous waste as deamination of the amino acids is necessary for the other components of those molecules to enter the metabolic cycle (Pennington et al. 2003). In *Aedes aegypti*, the nitrogen is sequestered in the amino acid proline. This provides a sink for the nitrogen as it is more gradually excreted in the forms of urea, uric acid, and ammonia. The deaminated proline residue can be used as energy.

The essential components of blood for a mosquito have not been completely defined. The importance of cholesterol for *Aedes aegypti* was shown during development of artificial blood meals (Talyuli et al. 2015). Small females, indicating poor larval nutrition, used 50% of their cholesterol reserves in order to deposit 25% fewer eggs than large females. Supplementary cholesterol in an artificial blood meal eliminated this difference. The importance of dietary cholesterol from larval reserves and in blood is a consequence of mosquitoes' inability to synthesize cholesterol from lipids (Kumar et al. 2010a). It has been possible to make artificial blood for *Culex quinquefasciatus* that is nearly as good as real blood in terms of fecundity (Griffith & Turner 1996). The diet consisted of salt, sodium bicarbonate, soy milk infant formula, ovalbumin (powdered egg white), gamma globulin, and adenosine triphosphate in water.

Blood feeding has a large effect on genomic expression. The original description of the *Anopheles gambiae* genome (Holt et al. 2002) also looked at gene expression before and after a blood meal. They found 97 transcripts upregulated and 71 downregulated (Table 7.5). The clear theme of the change was an increase in digestion, immunity, and synthesis and a decrease in metabolism. Time series examinations of gene expression have added further definition to the physiological changes brought about by blood feeding in *Anopheles gambiae*. Changes that affected 50% of gene expression began within three hours of a blood meal and were most apparent one, two, and three days after (Marinotti et al. 2006). At 24 hours post blood meal (pbm), gene expression was reduced in the midgut (22% of total expressed genes upregulated, 47% downregulated) and fat body (10% upregulated, 54% downregulated) but increased in the ovaries (56% upregulated, 5% downregulated) as functions shifted from digestion to egg formation. Dana et al. (2005) found a similar shift in function indicated by gene expression during the first 48 hours after a blood meal. The impressive volume of genes involved gives some idea of the complex physiological activity taking place during digestion and oogenesis. At 24 hours after a blood meal, out of a total of 14,900 genes, *Anopheles gambiae* upregulated 2,388 genes and downregulated 2,536, for most of which the function was unknown (Marinotti et al. 2005). A comparison of the expression of genomic copies of five genes showed differences 30 hours after sugar or blood meals (Dana et al. 2006). Blood-fed females expressed relatively more cathepsin-B-like protease and vitellogenin and relatively less mucin, high-molecular-weight kininogen, and chymotrypsin.

Blood feeding affects a much smaller proportion of gene expression in *Aedes aegypti* (Sanders et al. 2003). Of 1,778 genes expressed, only 333 changed during blood feeding. The percentage of genes upregulated or downregulated following a blood meal was 2.6% at three hours, 12.7% at 12 hours, 14.1% at 24 hours, and down to 1.2% at 72 hours. The upregulated genes were associated with digestion and synthesis, while the

Table 7.5. Gene functions upregulated and downregulated following a blood meal by *Anopheles gambiae* (Holt et al. 2002)

Upregulated	Downregulated
Cellular and nuclear signaling	Muscle processes
Digestion	Glycolysis
Ammonia excretion	Sugar digestion
Lipid synthesis and transport	Vision
Gene translation	Circadian rhythm
Egg development	Structural protein
Fat metabolism	
Lysosomal enzymes	
Melanization	

downregulated genes were associated with maintenance and sensory functions. The contrast in gene expression between *Anopheles gambiae* and *Aedes aegypti* is interesting, considering that the two species often feed on the same kind of host, use blood for metabolism as well as egg production, and often feed multiple times in one gonotrophic cycle. This contrast raises the question of whether the differences originated from selective forces that are not easy to define or from more random differences that have occurred since Anophelinae and Culicidae split tens of millions of years ago.

The grossly swollen gut of a mosquito following a full blood meal is a visible reminder that the midgut is the location of intense activity. It has also been the target of intense study because it is a principal component of the cycle of events leading to human disease from mosquito-borne pathogens. Regulation of the midgut happens from a combination of neural and endocrine systems. In *Anopheles gambiae*, a neuropeptide common to all insects, myosuppressin, downregulates gut muscular activity by interacting with the first neuropeptide receptor described in a mosquito (Schöller et al. 2005). Based on work with *Aedes aegypti*, mosquito midguts are typically innervated by four neurons of two types (Moffett & Moffett 2005). The two types are characterized by their reactivity to antibodies that recognize the neuropeptide FMRFamide (two neurons designated RF-I) and molluscan small cardioactive peptide b (two neurons designated SCP-I). Neuropeptide F is analogous to neuropeptide Y in vertebrates. Neuropeptide Y controls feeding, circadian rhythms, and other functions (Stanek et al. 2002). It is mostly expressed in the head and midgut. It reaches its highest concentration in the hemolymph before a blood meal and 24 hours after a blood meal. RF-I neurons originate in the region of the esophagus and anastomose along the entire midgut. SCP-I neurons have branches along the midgut itself. These nerves control operation of sphincters, including the pyloric valve as well as the musculature surrounding the gut. Enteroendocrine cells in the midgut interact with the nerves for this function. The four types of enteroendocrine cells are not evenly distributed along the gut and are more abundant in females than in males. One microRNA (miR-1174) localized in the posterior portion of the midgut represents a different system of control (Liu et al. 2014). This mRNA controls serine hydroxymethyltransferase expression, influencing sugar absorption, fluid excretion, and blood intake. Silencing miR-1174 production with inhibitory RNA has profound effects on *Aedes aegypti* and *Anopheles gambiae*, decreasing survival, causing the crop to bloat following a sugar meal, and preventing blood feeding.

The pH of the adult gut is controlled by the secretion of carbonic anhydrase, expressed by at least 14 different genetic transcripts in *Anopheles gambiae* (Del Pilar Corena et al. 2005). This enzyme raises pH by increasing the concentration of bicarbonate and carbonate ions. Among *Aedes aegypti*, *Aedes albopictus*, *Aedes taeniorhynchus*, *Anopheles quadrimaculatus*, *Culex nigripalpus*, and *Culex quinquefasciatus*, the enzyme was present in some of the species' anterior midguts of females, but never males. Carbonic anhydrase was present in the posterior midgut of females of all these species but in males of only *Aedes aegypti*, *Aedes albopictus*, and *Culex nigripalpus*. The pH of the posterior midguts of these species varied between 8.0 and 9.5—alkaline, as is typical of many insects (Harrison 2001), but not as alkaline as the anterior midgut of mosquito larvae.

Digestion and absorption of nutrients from the midgut involve enzymes that break molecules into pieces. Enzymes are secreted by midgut endocytes (Okuda et al. 2002) along the entire surface of the blood meal bolus (Rascón 2010). Some of the digestive products are secreted as zymogens that must be activated by other enzymes (Rascón 2010), particularly trypsin (Vizioli et al. 2001). The principal enzymes (considering *Aedes aegypti* as representative) cleave amino acids off proteins by making cuts in the interior of the protein molecule (endoproteases; Brackney et al. 2010). They are serine-type proteases, so called because serine amino acid is located at the active site. This family of enzymes includes trypsins (cleave lysine from arginine), chymotrypsins (cleave phenylalanine, tyrosine, and tryptophan), elastases (break fibers, including those of bacterial cell walls, at glycine, alanine, and valine residues), and collagenases (cleave collagen proteins). Many different serine proteases have been expressed in the genome, but relatively few of these have been found in the midgut. Chymotrypsin-like serine proteases are expressed in the midgut in response to blood meals following activation of a gene by juvenile hormone shortly after female emergence (Bian et al. 2008). Some enzymes are present constantly, and others are expressed at particular stages of blood meal digestion. Other digestive enzymes include carboxypeptidases (cleave amino acids from the C-terminal end of the protein or peptide), aminopeptidases (cleave amino acids from the N-terminal end of the protein or peptide), and chitinases (cleave glycosidic bonds in chitin, a polymer of N-acetylglucosamine; Zhang et al. 2011).

The orchestration of enzyme secretion following the blood meal varies somewhat between species. In *Aedes aegypti* trypsins are expressed in two peaks at 1–3 hours pbm and 8–36 hours pbm, stimulated by the brain's production of insulin-like protein ILP-3 (Gulia-Nuss et al. 2011). The first trypsins are different from the second and less important for egg production (Brackney et al. 2010; Rascón 2010). Aminopeptidase is secreted 4–12 hours pbm (Briegel 2003), and carboxypeptidases peak at 24 hours pbm (Moreira et al. 2000). In *Anopheles darlingi* aminopeptidases are secreted in two peaks, at 36 and 48 hours pbm, trypsins at 16–36 hours pbm, and chymotrypsins at 12–48 hours pbm (Okuda et al. 2005). *Culex quinquefasciatus* absorbs water, salts, and small molecules from the blood meal in the anterior portion of the midgut for the first 6–18 hours pbm, then secretes digestive enzymes in the posterior midgut, with peak levels of chymotrypsin, elastase, and aminopeptidase at 36 hours pbm and continuing until 72 hours pbm (Okuda et al. 2002).

Aedes aegypti terminates trypsin synthesis with trypsin-modulating oostatic factor (TMOF; Borovsky 2003). It is a decapeptide secreted by the follicular epithelium of the ovary starting 18 hours after a blood meal, peaking at 33 hours, and stopping at 48 hours. It is quite stable and can be absorbed orally by adults or larvae; however, it kills larvae. The lethal larval effect depends on only four amino acids (tyrosine, aspartic acid, proline, and alanine) and has this affect against *Aedes aegypti*, *Aedes taeniorhynchus*, *Anopheles quadrimaculatus*, *Culex nigripalpus*, and *Culex quinquefasciatus*.

Cholesterol is a key nutrient for mosquitoes because they are metabolically incapable of synthesizing it from simple lipids (Kumar et al. 2010a). Adults start their lives with a reserve of cholesterol stored by larvae as they ingested phytosterol and converted it. Nutritionally, sterols are essential for construction of cellular membranes and to make certain hormones, like ecdysteroids. Sterol carrier protein-2 (SCP-2) is a protein that binds to sterols and fatty acids to move them across membranes and within cells. In *Aedes aegypti*, silencing its form of SCP-2 with RNAi lowers egg production and eventually leads to the death of the insect. Some natural plant substances, like alpha mangostin, are toxic to mosquitoes because they bind to and inactivate SCP-2.

The digestive process involves more than the alimentary canal as nutrients influence other organ systems, particularly the fat body and ovaries (Fig. 7.7). The system of chemical communication within the mosquito to control these processes is complex, possibly with the adaptive advantage of providing many points of adjustment in response to the availability of nutrient resources. Initiation and repression of blood feeding is regulated by the 4E-binding protein (4E-BP), which is phosphorylated by the target of rapamycin (TOR) complex in response to insulin-like growth factors secreted in response to the availability of amino acids following a blood meal (Arsic & Guerin 2008; Shiao et al. 2008; Roy & Raikhel 2012). The 4E-BP system is general throughout eukaryotes (Teleman et al. 2005). The binding protein has many phosphorylation sites, and when most of them are filled, the protein is unable to bind to the eIF4E protein (initiation factor). That initiation factor connects messenger RNA to the ribosomal complex so that when it is repressed by 4E-BP, genetic translation in general is repressed. *Aedes aegypti*, but not *Anopheles gambiae*, has another regulator of feeding called the head peptide (AaegHP) that suppresses blood feeding (Riehle et al. 2002). AaegHP cycles in abundance in the brain, terminal ganglia, and midgut (Stracker et al. 2002). The gene for the head peptide, *Aea-HP-1*, codes for three peptides that are processed posttranslation. They originate from neurosecretory cells and midgut endocrine cells.

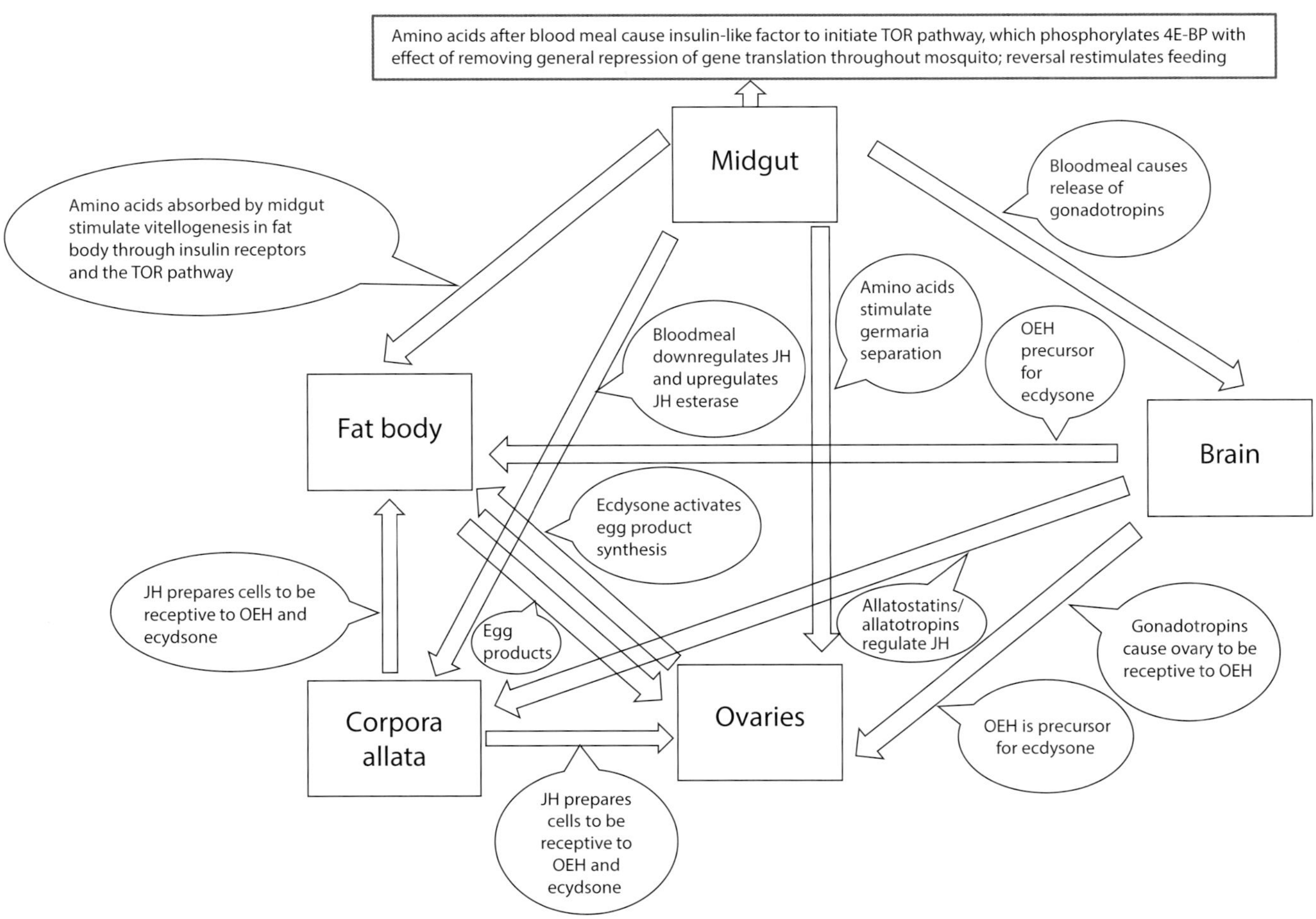

Figure 7.7. Interorgan regulatory pathways associated with blood feeding by mosquitoes. These relationships are summarized from studies mainly done on *Anopheles gambiae* and *Aedes aegypti* by Brown and Cao 2001; Raikhel et al. 2002; Attardo et al. 2005; Dana et al. 2005; Teleman et al. 2005; Arsic and Guerin 2008; Shiao et al. 2008; Gulia-Nuss et al. 2011; and Roy and Raikhel 2012.

One of the main effects of blood feeding is provision of materials for egg development. The ovarian follicles enter a previtellogenic state following emergence (i.e., they go from Christopher's Stage I to II, meaning that the ovarian follicles separate from the germaria), stimulated by juvenile hormone (Dana et al. 2005). The secretion of juvenile hormone from the corpora allata is regulated by neuropeptides from the brain that either stimulate secretion (allatotropins) or suppress it (allatostatins; Riehle et al. 2002). Juvenile hormone also prepares the fat body cells and the ovaries to be receptive to ovarian ecdysteroidogenic hormone (OEH) and ecdysone. The blood meal initiates a process that downregulates juvenile hormone production, both by slowing production from the corpora allata and by destroying juvenile hormone with esterases (Attardo et al. 2005). Following the blood meal, the brain secretes gonadotropins that also prepare the ovaries, as well as OEH from neurosecretory cells (Riehle et al. 2002). OEH from the brain goes through neurons to the midgut endocrine cells, the perivisceral organs (recipients of neurosecretory peptides located near the ganglia), and the ovaries (Brown & Cao 2001). OEH causes the ovaries, through the insulin signaling pathway, to secrete ecdysone that is converted to 20-hydroxyecdysone in the fat body and activates expression of genes that produce vitellogenin, egg yolk proteins, and other egg products that go to the ovaries. Then 20-hydroxyecdysone binds to a specific receptor (EcR-USP), directly stimulating the genes (Raikhel et al. 2002). Further regulation is provided in *Aedes aegypti* by microRNA, miR-275 (Bryant et al. 2010). Production of miR-275 is regulated by 20-hydroxyecdysone and the TOR pathway. Although its exact function is unknown, inhibition of this microRNA disrupts messenger RNA regulation and causes severe disruption to blood feeding and digestion, including the rerouting of blood meals to the crop.

Effects of Parasites and Pathogens on Digestion

The first intimate contact of the female adult mosquito with parasites and pathogens is usually through the alimentary canal. Subsequent growth of these microorganisms affects the physiology of the mosquito to a greater or lesser extent. It is difficult to separate the discussion of what are usually considered commensal bacteria from parasites and pathogens, and some species that might be considered commensal or incidental could be pathogenic. A complete list of all the known relationships between parasitic or pathogenic organisms and mosquitoes is beyond the scope of this volume, but this section presents some significant examples spanning viruses to filariae.

Arboviruses transmitted by mosquitoes are important pathogens of human and animal health. With few exceptions, transmission involves replication of the viruses in the mosquito, exposing the virus to the complex physiological processes of digestion. If there is a pattern to arbovirus infection, it is that the mosquito mounts various defenses to modulate the infection, and the virus has adapted closely to cellular and extracellular components to maintain an infection. Dengue virus provides some examples of extracellular effects on a virus. Dengue virus in *Aedes aegypti* is affected by serine proteases secreted in the early and late stages of blood digestion. Spiking blood meals with serine protease inhibitors or RNAi silencing of late trypsins increased infectivity of dengue serotype 2 (Brackney et al. 2008). In contrast, early trypsin secretion not only increased infection with this virus but appeared to be necessary (Molina-Cruz et al. 2005). Trypsin inhibitors added to the blood meal caused up to 97% reduction in midgut infection, and in contrast, the addition of trypsin to the blood meal increased infection. Apparently, the correct trypsins at the correct time prepare the virus for midgut epithelial infection. Extracellular structures can be pathways or barriers for virus infection (Romoser et al. 2005; Franz et al. 2015). Rift Valley fever virus has been observed in the tracheae and tracheoles of *Culex pipiens* that penetrate the midgut epithelium, suggesting that the virus particles use these channels to get through the midgut epithelium and the midgut basal lamina. Similar observations have been made on Venezuelan equine encephalitis virus in *Aedes taeniorhynchus* (Romoser et al. 2004). The salivary gland basal lamina may be the major barrier that prevents *Anopheles* from being important vectors of arboviruses.

Studies on several viruses have shown some of the intracellular processes that affect infection. Sindbis virus causes observable cellular damage in the midgut epithelium and in the muscles surrounding the alimentary canal of *Aedes albopictus* (Vo et al. 2010). Enveloped viruses like Sindbis enter cells through two routes, either by being surrounded by folds of the membrane (endocytosis) followed by fusion with the endosomal membrane under low pH conditions or by direct fusion at the plasma membrane through specific protein-receptor interaction and fusion with lipid elements of the membrane (Hernandez et al. 2001, Rawat et al. 2003). Sometimes few cells are initially infected in the gut (Smith et al. 2007, 2008). Venezuelan equine encephalitis virus can initiate a systemic infection of *Aedes taeniorhynchus* from an initial infection of just one to five midgut epithelial cells, usually in the posterior portion of the midgut. Infection is initiated by recognition of envelope glycoprotein spikes on the virus particle. Intracellular immune responses are also important. The immune system of *Aedes aegypti* actively responds to dengue virus infection through RNAi silencing and the Toll pathway, possibly costimulated by the midgut microflora (Xi et al. 2008). The amount of virus that makes its way to the salivary glands depends on the species of mosquito involved, probably accounting for the relative transmission potential of species (Colton et al. 2005).

Wolbachia can infect many cells of mosquitoes, including fat tissue around the midgut, salivary glands, and the cardia (Moreira et al. 2009). Natural infections have little effect on longevity, but *Wolbachia* from other species of flies or mosquitoes can have a fitness cost and shorten life span. For example, introduction of a *Drosophila* strain, *w*MelPop, into *Aedes aegypti* not only shortens life span but also interferes with blood feeding (Turley et al. 2009). Infected mosquitoes took progressively smaller blood meals, from 3.0 mg at 5 days to only 1.6 mg at 35 days. This contrasted with uninfected mosquitoes that took 3.8 mg of blood at 5 days and 3.2 mg at 35 days. At the older age, infected females fed aberrantly, the stylets not emerging from the labium. *Wolbachia* infection can also have a profound effect on infection of a mosquito with viruses or *Plasmodium* (*Plasmodium gallinaceum* in *Aedes aegypti* [Moreira et al. 2009]; *Plasmodium falciparum* in *Anopheles coluzzii* [Shaw et al. 2016]), at least some of these effects occurring in the midgut epithelium. Inhibition of virus infection is related to the abundance of *Wolbachia* in the mosquito, which in turn tends to be highest when mosquitoes are infected by a strain that does not naturally occur in that species (Rainey et al. 2014). Not every virus-mosquito system is affected—for example, *Wolbachia* does not protect *Armigeres subalbatus* from Japanese encephalitis virus infection, *Aedes polynesiensis* or *albopictus* from dengue virus, or *Aedes albopictus* from chikungunya virus. The mechanisms for protection include competition for essential nutrients like cholesterol (Caragata et al. 2013; Rainey et al. 2014) and upregulation of various components of the mosquito's immune system (Moreira et al. 2009; Rainey et al. 2014). Among those immune mechanisms are the Toll and IMD pathways leading to increased antimicrobial peptides (cecropin, defensin, thioester-containing proteins, and C-type lectins), the JAK-STAT (Janus kinases and signal transducer and activator transcription proteins) pathway that stimulates an antiviral immune response, and the phenoloxidase melanization pathway.

The effect of fungal infection on blood feeding is not at all subtle. *Anopheles gambiae* infected with *Metarhizium anisopliae*, a common pathogen of insects, blood fed less often, took less blood at each feeding, and produced fewer eggs (Scholte et al. 2006). Only half of the infected mosquitoes took a second blood meal. The effect was proportional to the size of the initial inoculum. Uninfected mosquitoes fed an average of 4.39 times when offered 8 blood meals, lightly infected females took 3.4 blood meals, and heavily infected females took 2.07 blood meals. Eggs deposited per gonotrophic cycle were also reduced, from 65.6 eggs from uninfected females to 54.8 eggs from lightly infected females and 45.5 eggs from heavily infected ones.

The interactions between *Plasmodium* and mosquito digestive processes are extensive and have been studied in some detail because of interest in malaria. The presence of *Plasmodium berghei* gametocytes in rat blood imbibed by *Anopheles gambiae* was associated with changes in genetic expression that have been adaptive for the parasites in some cases or immunogenic against the parasites in others (Dana et al. 2006). Decreases in mucin and chymotrypsin gene expression might work to protect the parasites while they are still in the gut lumen. An increase in high-molecular-weight kininogen might retard clotting of the blood meal to allow more time for the parasites to escape to the midgut epithelium. On the other hand, an increase in cathepsin B-like protein might represent a greater immune response by the mosquito against the parasite. Artificially, the normal regulation of the early digestive process has been linked to a variety of foreign effector genes to kill parasites and reduce the chances of mosquito infection (Kim et al. 2004; Abraham et al. 2005).

Natural immunity by mosquitoes to *Plasmodium* limits damage caused by the parasites as they cross the midgut epithelium. Gametocytes are the haploid sex cells of *Plasmodium* produced by the secondary mammalian host. The following account of the events leading to the formation of a productive *Plasmodium* oocyst 16 days pbm and infective salivary glands 18 days pbm is taken from an *Anopheles gambiae* and *Plasmodium berghei* model at 20.5°C (Whitten et al. 2006). Ingested by mosquitoes, the gametocytes quickly lose their flagella in the lumen of the midgut and merge to form the zygote that transforms to the ookinete at about one hour pbm. As its name implies, the diploid ookinete is motile, and it is the stage that penetrates the midgut epithelium to eventually form an oocyst on the apical, or hemocoel, side of the midgut. Midgut invasion by the ookinete starts about 20 hours pbm, and the ookinete reaches the midgut basal lamina about 4 hours later. The peritrophic membrane is completed about 28 hours pbm, somewhat delayed by the presence of *Plasmodium*. Ookinetes of *Plasmodium falciparum* are equipped with chitinases to penetrate the peritrophic membrane before protein cross-links solidify it, whereas *Plasmodium vivax* invades the midgut epithelium before peritrophic membrane formation.

The process by which ookinetes (Beerntsen et al. 2000) traverse the midgut epithelium is central to the survival of parasites, and pathogenicity to the mosquito has resulted in selection for immune mechanisms. While the mosquito actively targets the parasites, ookinetes move from midgut cell to midgut cell, or between them (disputed by some; Baton & Ranford-Cartwright 2005), killing each cell they pass through (Han et al. 2000). The ookinetes secrete serine proteases to penetrate the cells, and they are capable of squeezing through small openings. Dying midgut epithelial cells secrete nitric oxide, which may kill parasites remaining in the cell, as well as increase the toxicity of the lumen environment for the parasites. In a model using *Aedes aegypti* and *Plasmodium gallinaceum*, the luminal side of the midgut discourages ookinete invasion with lectins while the ookinetes target those cells that lack villi (bare cells with folds of the luminal membrane; Zieler et al. 2000). The ookinetes may target areas of the midgut for invasion by going to junctions of susceptible cells (Baton & Ranford-Cartwright 2005). As discussed above, cells killed by *Plasmodium* parasites are squeezed into the lumen by a purse-string mechanism and replaced by regenerative cells in a process that maintains the integrity of the midgut (Han et al. 2000; Baton & Ranford-Cartwright 2005). Those few ookinetes that escape the midgut defenses rely on substances that interact with the basal side of the midgut for adhesion of the oocyst (Templeton et al. 2000).

The complex relationship between mosquitoes and *Plasmodium* results in the differences that lead to vector competence or incompetence, sometimes at the level of pairings or mispairings between strains of the parasite and the species of vector (Beerntsen et al. 2000). The mosquito mounts an impressive array of defenses, including rejecting cells by apoptosis, secreting antimicrobial peptides (defensins) and nitric oxide into the midgut by the midgut epithelial cells, and lining the gut wall with mucins (Abraham & Jacobs-Lorena 2004). Two other immune mechanisms are the C-type lectin proteins that bind various carbohydrates and leucine-rich repeat immune genes (LRIM; Osta et al. 2004). Of the 22 lectins expressed by *Anopheles gambiae*, one protected parasites from melanization in a probable case of the parasite using the host's immunity for its own protection. Some of the immune responses are stimulated by memory of an earlier infection (Ramirez et al. 2015), as new ookinetes trigger immune pathways primed by increased levels of evokin, a carrier of lipid signal compounds (Ramirez et al. 2012).

The negative effects of infection have demonstrated the importance of immunity to the mosquito. In an *Anopheles gambiae* and *Plasmodium berghei* model, mosquitoes genetically engineered to express an effective antiplasmodial peptide had improved longevity and fecundity (Marrelli et al. 2007). In the field, infected *Anopheles gambiae* were less likely to feed successfully and more likely to die trying, possibly because of disruption to apyrase production in the saliva (Anderson et al. 2000; Beerntsen et al. 2000).

Filarial infection is also affected by the feeding process. Some mosquitoes physically disrupt the microfilariae with the cibarial armature, which are teeth in the cibarial pump formed from the wall of the hypopharynx that is the floor of the pump (Beerntsen et al. 2000). In a model system using a filarial parasite of rodents, bats, and marsupials (*Litomosoides chagasfilhoi*) and *Culex quinquefasciatus*, the microfilariae invaded the midgut within two to three hours (Santos et al. 2006). This mosquito develops a peritrophic membrane quickly, so the filariae had to penetrate it. Each microfilaria lost its outer sheath as it went through the peritrophic membrane. By three hours pbm, the peritrophic membrane was too dense to penetrate. The invasion process damaged midgut epithelial cells, sometimes causing them to detach. The microfilariae entered a midgut cell by lying flat among the microvilli, then entering perpendicularly.

Bacterial Flora in the Alimentary Canal and Salivary Glands

Diptera have many special relationships with bacteria. Bacteria influence location of food, identification of oviposition sites, defense against pathogens, modulation of the immune response, nutritional interactions (Engel & Moran 2013), and possibly even insecticide resistance (Nkya et al. 2013). The Culicidae are also heavily affected by bacteria, including events in the alimentary canal and salivary glands. The concept that an organism does not live independently from its microbial associates has been called the holobiome, and this concept certainly applies to adult mosquitoes (Minard et al. 2013; Hegde et al. 2015). The diversity of bacteria is lower than in some animals, and there is great variability between species and even between individuals. Some extracellular genera of bacteria occur widely within Culicidae, such as *Acinetobacter, Plantoea, Pseudomonas, Asaia, Serratia,* and *Enterobacter.* Proteobacteria dominates the microflora throughout the family, though other phyla are commonly present. Most extracellular bacteria are in the alimentary canal, salivary glands, and gonads, whereas intracellular *Wolbachia* is more widely distributed.

Inventories of bacteria (Table 7.2) in the alimentary canal performed by culture, metagenomics, or a combination of both methods have revealed a lot of variety and a few patterns. Some species of bacteria appear to be unique to mosquitoes, such as *Janibacter anophelis* and *Thorsellia anopheles* in *Anopheles arabiensis* and *Pantoea stewartii* and *Elizabethkingia anopheles* in *Anopheles gambiae* (Ricci et al. 2012). *Asaia* is a genus of bacteria that specializes in colonization of insect guts, and Culicidae is no exception. Another indication of coadaptation of bacteria is strain specificity of *Pantoea stewartii* in *Aedes aegypti* and generally more successful colonization of the midgut by strains originating from mosquitoes compared to strains from environmental samples (Terenius et al. 2012).

Laboratory-reared adult mosquitoes generally have lower bacterial diversity, but this is variable over time and between colonies (Azambuja et al. 2005). It is likely that the bacteria found in one colony have limited relationships with other colonies. Nonetheless, it is worth considering the bacteria that have been observed. *Enterobacter cloacae* and *Pseudomonas aeruginosa* were the dominant gut bacteria in *Anopheles albimanus* and *stephensi* (Azambuja et al. 2005). Dong et al. (2009) also recorded variation cage to cage and generation to generation. The dominant species in *Anopheles gambiae* were *Enterobacter asburiae, Microbacterium, Sphingomonas, Serratia,* and *Chryseobacterium meningosepticum.* A colony of *Culex pipiens* in Egypt (Zayed & Bream 2003) had the following cultured gut bacteria: *Bacillus sphaericus, Bacillus subtilis, Enterococcus durans, Enterococcus faecalis, Listeria murrayai, Microbacterium imperiale, Moraxella caviae, Staphylococcus aureus, Staphylococcus saccharolyticus,* and *Staphylococcus saprophyticus. Acinetobacter calcoaceticus, Neisseria mucosa,* and *Staphylococcus haemolyticus* only appeared after a blood meal. The presence of so many pathogens raises the question of whether those who maintained the colony were the source of some of the bacteria. Colonized *Culex quinquefasciatus* in India (Mourya et al. 2001b) had many of the same genera as in the Egyptian colony: *Pseudomonas, Acinetobacter junii, Staphylococcus epidermis,* and *Aeromonas culicicola.*

Variability of bacteria in field populations across eight species of mosquitoes was studied by deep sequencing following sterile dissection of non-blood-fed females captured in coastal Kenya (Osei-Poku et al. 2012). A total of 798 operational taxonomic units (OTUs—each roughly representing a species of bacteria) were found, but the median per mosquito was only 42 OTUs and the maximum, 71 OTUs. In general, one OTU dominated the microflora, and the most common four OTUs accounted for 90% of the bacteria. Overall, Gammaproteobacteria were the most common (62.3%), followed by Alphaproteobacteria (18.3%) and Flavobacteria (11.6%). The most common genera were *Aeromonas* (38.7%), *Asaia* (13.2%), *Chryseobacterium* (9.1%), and *Zymobacter* (6%). Individuals within species of mosquitoes were not similar to each other. In order of increasing uniformity, they were *Anopheles gambiae, Anopheles funestus, Mansonia africana, Aedes bromeliae,*

Mansonia uniformis, *Culex quinquefasciatus*, and *Aedes aegypti*. Considering the ecology of these species, it is difficult to assign any reason for these differences or even to guess whether the measurements would be replicated in another location. The hypothesis left to be tested is whether conditions within some mosquito species tend to expand or restrict the variety of bacteria that can colonize their alimentary canals.

Much of the work describing the microflora of mosquitoes has concentrated on *Anopheles* vectors of *Plasmodium*. *Anopheles stephensi* in India demonstrated the difference between field and lab evaluations of adults (Rani et al. 2009). This species had 53 genera of bacteria cultivated from the midguts of field-collected specimens in contrast to only 15 from lab colonies. Genetic evaluations of bacteria were similar, with 171 OTUs in field-collected specimens compared to 44 from colonies. The dominant kinds of bacteria were also different. From the field, Paenibacillaceae dominated in males and *Serratia marcescens* in females, whereas in the colonies *Chryseobacterium meningosepticum* was one of the dominant species. A similar contrast between field and colony mosquitoes was observed in *Anopheles gambiae* in Cameroon (Boissière et al. 2012). The number of genera was less than for *Anopheles stephensi* in India, with only 28 genera of bacteria greater than 1% of the microflora of *Anopheles gambiae*. Proteobacteria dominated in the field, and Flavobacteria dominated in the colony. Other phyla present were Bacteroides, Actinobacteria, Firmicutes, and Fusobacteria.

The origin of bacteria in adult mosquitoes is not clear. Some species, like *Asaia* and *Wolbachia*, are transmitted either transgenerationally or by mating, but the majority of bacteria in the alimentary canal are also common outside of mosquitoes. Cameroonian *Anopheles gambiae* collected in the field had 11 of 24 OTUs common to both immature stages and adults (Tchioffo et al. 2013), implying that the bacteria somehow transferred from the larvae through the pupae to the adults. Ten bacteria were only in the immature forms, and four were only in the adults (*Klebsiella* and *Serratia* species). In southern Iran, *Anopheles stephensi* adults harbored 48 species of bacteria, with an indication that some of them (*Pseudomonas*, *Aeromonas*, *Myroides*) were transferred from larva to adult (Chavshin et al. 2012). *Alcaligenes* and *Bordatella* were only in adults. The influence of environmental factors on the presence of a bacteria was clearly shown by the presence of 10 bacterial genera in mosquitoes from one field site or the other, compared to 4 genera in common between the two sites.

The bacterial flora of *Anopheles* changes significantly during the life of the adult female. In *Anopheles gambiae* in semifield cages in Kenya (Wang et al. 2011), Enterobacteriaceae and Propionobacteria predominated on emergence, with *Thorsellia anophelis* most abundant. The bacteria are replaced by Flavobacteriaceae (*Bacteroidetes*) as the mosquitoes sugar feed. Blood feeding changes things again as the total number of bacteria increases, but the diversity of species decreases, and Proteobacteria comes to dominate the gut. Diversity increases two days pbm, and by seven days pbm, the microflora again switched to Flavobacteriaceae (*Elizabethkingia*).

Bacteria in other field-caught mosquitoes show some variation from *Anopheles*. *Culex quinquefasciatus* from Pune, India, had abundant *Acinetobacter* (41% of bacteria) and Lactococcus (13%; Pidiyar et al. 2004). *Aedes aegypti* (whole body) contained predominantly Firmicutes in Madagascar, and *Aedes albopictus* had more Proteobacteria (Zouache et al. 2011). The latter species also showed significant association with locations that were different for males and females. Abundant genera included *Bacillus*, *Acinetobacter*, *Agrobacterium*, and *Enterobacter*. Enterobacteriaceae was also the predominant bacterial group in guts of *Aedes albopictus* from Virginia, United States (though many other groups were found as well; Joyce et al. 2011). In contrast, the most abundant bacteria in *Aedes aegypti* in India was *Aeromonas culicicola* in one study (Mourya et al. 2002) and *Serratia odorifera* in another (Apte-Deshpande et al. 2012).

The number of bacteria in the midgut increases greatly following the blood meal. Sugar-fed *Anopheles gambiae* had 10^4 colony-forming units per midgut compared to 10^6 in blood-fed females, a 100-fold increase (Dong et al. 2009). Azambuja et al. (2005) measured a much greater increase in some individuals of *Anopheles albimanus* and *stephensi*, up to 16,000-fold. In *Anopheles gambiae*, the diversity of bacterial species decreases even as the number of bacteria goes up (Wang et al. 2011). Enterobacteriaceae (*Serratia*, *Raoultella*, *Enterobacter*, and *Klebsiella*) come to dominate the microflora. By four days pbm, the bacteria of the midgut are back to their pre–blood meal state. The growth of bacteria following a blood meal has been ascribed to the increase in temperature and downregulation of oxidative stress and immune responses (Minard et al. 2013). There is some indication that the residual bacteria left in the midgut following completion of blood digestion are important in recolonizing the next blood meal. Addition of penicillin and streptomycin to the first three blood meals had a lifetime effect on the bacterial microflora, reducing subsequent increases following blood meals and changing the bacterial species composition (Gendrin et al. 2015).

Bacteria in the gut can be commensal in the sense that they provide some physiological services to mosquitoes. *Serratia* and *Enterobacter* secrete enzymes that could affect blood digestion. *Asaia* may provide vitamins, and *Acinetobacter* in *Aedes albopictus* appears to be involved in sugar and blood digestion (Minard et al. 2013). *Acinetobacter* is commonly associated with other insects that use nectar sources of sugar (Engel & Moran 2013). Bacteria may also contribute to maintenance of a chemically reducing environment in the gut (Wang et al. 2011). Studies of *Aedes aegypti* have shown that administration of any of a number of antibiotics (carbenicillin, tetracycline, gentamycin, or kanamycin—but not spectinomycin) interfered with lysis of red blood cells and resulted in poor digestion, leading to lower fertility (De O Gaio et al. 2011), implying that bacteria are important to this species for blood digestion. Another possible benefit from the bacterial microflora is its inhibitory effect on parasites, as described below.

Under other circumstances, the midgut flora interferes with the mosquito, reducing longevity and fertility. In the laboratory,

Anopheles gambiae were more fecund (53% increase in eggs) and long-lived (35% increase in survival, to 10 days) when fed enough penicillin and streptomycin to reduce diversity of bacteria and increase the proportion of Acinetobacteriaceae compared to Enterobacteriaceae (Gendrin et al. 2015). Mosquitoes mount natural immune responses, further indicating that uncontrolled bacterial growth would be detrimental. Injection of bacteria (*Salmonella typhimurium* or *Staphylococcus aureus*) into the hemocoel of *Anopheles gambiae* upregulates the immune response, including production of lectins, fibrinogen proteins, and gram-negative bacteria-binding proteins (Aguilar et al. 2005). Lysozymes that would suppress bacteria are produced in the midgut, salivary glands, and Malpighian tubules of *Anopheles gambiae* (Li, B. et al. 2005). Other mechanisms that may protect the midgut epithelium from bacterial infection include digestive enzymes, antimicrobial peptides, nitric oxide secretion, the prophenoloxidase complex involved in encapsulation, and even production of antimicrobials by some of the bacteria themselves (Azambuja et al. 2005). Commensal bacteria also form a protective layer against the luminal side of the peritrophic membrane (Cirimotich et al. 2011). The production of nitric oxide and of other reactive oxygen species that can kill bacteria is induced by the blood meal and concentrated in the endoperitrophic space between the peritrophic membrane and midgut epithelium, which is further protected from infection by a layer of mucins (Wang et al. 2011).

Asaia and *Wolbachia*

Asaia and *Wolbachia* are examples of bacteria that have close relationships with mosquitoes, the former extracellular and the latter intracellular. As discussed above, other species, like *Pantoea stewartii*, seem to be closely related to mosquitoes, but much less is known about them. Although the widespread infection of many mosquitoes and other arthropods with *Wolbachia* (Kittayapong et al. 2000) has attracted much attention, other intracellular bacteria commonly occur in arthropods, such as *Rickettsia* in ticks and *Orientia* in chiggers. Competition and possibly other interactions between mosquito-associated bacteria may be quite extensive, as occurs between *Asaia*, *Wolbachia*, and *Acinetobacter* (Hegde et al. 2015). *Asaia* is not known to affect pathogen transmission by mosquitoes (Cirimotich et al. 2011).

The effects of *Wolbachia* can be favorable or unfavorable to mosquitoes. Whole groups of mosquitoes are not infected by *Wolbachia* (Kittayapong et al. 2000), and until recently no natural infections of *Anopheles* were known. Shaw et al. (2016) found that 46% of *Anopheles coluzzii* in Burkina Faso were infected and 32% of *Anopheles arabiensis*. Infection of these species with naturally occurring strains of *Wolbachia* made no difference to the mosquitoes' survival, fecundity, or sex ratio, but it did shorten the gonotrophic cycle by a mean of 0.73 days. In a small sample, infected *Anopheles coluzzii* was less likely to be infected with *Plasmodium falciparum*. On the other hand, artificial infection of *Anopheles gambiae* with an *Aedes albopictus* strain of *Wolbachia* (*w*AlbB) increased *Plasmodium* infection of the mosquito (Ricci et al. 2012).

Wolbachia often inhibits viral infection, protecting the mosquito as well as limiting transmission of pathogenic viruses. The magnitude of this effect is influenced by the strain of *Wolbachia* and species of mosquito, and inhibition is generally proportional to the abundance of the bacteria (Johnson 2015). Some examples of virus-mosquito systems unaffected by natural *Wolbachia* infection include Japanese encephalitis virus in *Armigeres subalbatus*, dengue virus in *Aedes polynesiensis* and *albopictus*, chikungunya virus in *Aedes albopictus*, and West Nile virus in *Aedes aegypti*. Competition for cholesterol may be one of the principal mechanisms by which *Wolbachia* inhibits viral development, and also involved are enhancement of immune mechanisms and microRNA-mediated responses (Caragata et al. 2014; Rainey et al. 2014; Hegde et al. 2015; Johnson 2015).

Although the competition for cholesterol may be protective against viral infection, such competition for nutrients also represents a limitation to development of the mosquito. Infection of *Aedes aegypti* with a *Wolbachia* originating from *Drosophila melanogaster* (*w*MelPop) efficiently reduced dengue infection but also reduced the number of eggs and those eggs' viability. Supplementation of the mosquito diet with amino acids restored viability (Caragata et al. 2014). *Wolbachia* also has riboflavin requirements (Fallon et al. 2014) that may compete with nutritional requirements of the mosquito itself.

The clearest example of disruption by *Wolbachia* concerns reproduction. Cytoplasmic incompatibility was recognized first in *Culex pipiens*, though *Wolbachia* was not known to be the cause (Laven 1959). Populations of this species could not interbreed with other populations in Europe, leading to the false conclusion that genetic differences between populations were the cause. In fact, different strains of *Wolbachia* prevented reproduction. This pattern has been observed in other species, in which an infected male mating with an uninfected female fails to produce viable offspring, whereas an infected female can produce offspring following mating with an uninfected male (Saridaki & Bourtzis 2010; Rainey et al. 2014). This has the net effect of driving the infection through a population and maintaining the infection once it is in place. The mechanism for *Wolbachia*-induced cytoplasmic incompatibility may be production of an enzyme that blocks the activity of ubiquitin, a polypeptide involved in posttranslational modification of proteins (Beckmann et al. 2017). Regulation of the activity of this enzyme by two different genes may explain how the system is rescued when either the female or both sexes are infected. Carriage of these genes by *Wolbachia* prophage viruses into the mosquito genome can enhance the effect (LePage et al. 2017).

Interaction of the Midgut Microflora with *Plasmodium*

The gut microflora tends to inhibit invasion of the midgut epithelium by *Plasmodium* ookinetes, reducing overall infection with these parasites. In one experiment, feeding newly emerged females with bacteria cultured from pupal midguts reduced the number of *Plasmodium falciparum* oocysts in *Anopheles gambiae* by 40%–55% and the number of infected mosquitoes by 25%–40% (Tchioffo et al. 2013). *Serratia marcescens* was observed to decrease *Plasmodium vivax* infection in *Anopheles albimanus* (Azambuja et al. 2005), and *Enterobacter* decreased *Plasmodium falciparum* infection of *Anopheles gambiae* (Boissière et al. 2012). Experiments that decreased the number of bacteria in the midgut had the opposite effect, increasing *Plasmodium* infection. Examples included *Plasmodium falciparum* infection of *Anopheles gambiae* and *arabiensis* fed methylparaben (Benedict et al. 2009), aseptic *Anopheles gambiae* infected with *Plasmodium falciparum* (Dong et al. 2009), and gentamycin-treated *Anopheles funestus* challenged with *Plasmodium falciparum* (Azambuja et al. 2005). Experiments using clinical concentrations of antibiotics imply that antibiotic use may raise the risk of malaria by increasing the chances of mosquito infection with *Plasmodium* (Gendrin et al. 2015; Hegde et al. 2015).

Microbes in the Salivary Glands

Surprisingly, the salivary gland can have a rich assortment of bacteria. Using a laboratory colony of sugar-fed *Anopheles culicifacies*, Sharma et al. (2014) showed that the diversity of bacteria in laboratory-reared mosquitoes was greater in the salivary glands than in the midgut. As in the midgut, the majority were Proteobacteria, with Bacteroides, Firmicutes, Actinobacteria, and Acidobacteria also represented. Very small numbers of other bacteria were detected, including Archaea, Cyanobacteria, and the extremophile genus *Deinococcus*. This study revealed that 11% of species were common to the midgut and salivary glands, but 54 were only found in the latter. Mosquitoes generally discharge 50%–60% of their saliva during feeding, and 20% is sucked back in (Schneider & Higgs 2008). This might account for the presence of some of the bacteria.

Bacteria in the salivary glands are presumably commensal or accidental. In contrast, transmitted pathogens find their way into the salivary glands after having passed the midgut in a process that uses the mosquito's physiology quite specifically. Once in the salivary glands, such pathogens are transmitted to a host in the saliva. *Plasmodium* blood-stage parasites recognize sulfated polysaccharides in the saliva of *Anopheles gambiae* (Arcà et al. 2005), providing a mechanism for possible tropism of parasites to the mosquito during feeding. At the opposite end of the mosquito infection process, parasites find receptor proteins on the distal ends of the lateral lobes of the salivary glands (King et al. 2011). The proteins involved are important immunogenic components of the saliva, and they are also found in the proximal ducts of the lateral lobes. It can be difficult to show the relationship between saliva and transmission of the *Plasmodium* parasite to a mammalian host (Kebaier et al. 2010). Viral infection was of longer duration, earlier onset, greater pathogenicity, or higher titer when transmitted by a mosquito rather than inoculated by needle for the following systems: *Aedes aegypti*-mouse-dengue 2 (Conway et al. 2014), *Aedes aegypti/Aedes triseriatus/Culex quinquefasciatus*-mouse-Cache Valley virus, *Aedes triseriatus*-rodents-LaCrosse virus, *Aedes triseriatus*-mouse-vesicular stomatitis virus (Limesand et al. 2000), *Aedes aegypti*-mouse-West Nile virus, *Culex tarsalis*-mouse-West Nile virus (Styer et al. 2010), and *Culex pipiens*-chicken-West Nile virus. In contrast, transmission of Saint Louis encephalitis and western equine encephalitis viruses was the same whether introduced into birds by a needle or the bite of *Culex tarsalis*.

At least one of the reasons for salivary enhancement of virus infection is the suppression of the host's immune system by the saliva. Some of the same pharmacokinetics that counter hemostasis also suppress the antiviral response of the host (Schneider & Higgs 2008). Immunosuppressive qualities of the saliva may also mitigate the immune response to the feeding process itself. The net effect is a suppression of the T_H1 immune response, decreasing T cells, tumor necrosis factor alpha, interleukin-2, and interferon gamma. Typical of a T_H1 to T_H2 shift, interferon gamma causes an increase in interleukin-10, resulting in inhibition of the ability of the host to recognize viral antigens. Interference with clotting mechanisms also decreases antimicrobial factors in the blood. Some of these immunological effects last for over a week, raising the possibility that mosquitoes gain a community benefit from the pharmacoactive effects of their saliva. The mechanism for enhancement of dengue transmission cited above was serine protease digestion of extracellular matrix proteins (Conway et al. 2014).

Adult Feeding Behavior

Blood-Feeding Behavior

One of the first questions asked about a species of mosquito is the host preference of the females. It is a complicated question because most blood-feeding species can bite a variety of different animals. It is also a significant question because those species that bite humans by preference or circumstance are more likely to transmit pathogens. If they take mostly human blood, they are better suited to maintain pathogens like *Plasmodium falciparum* (malaria) that only infect people. On the other hand, species that commonly bite animals and humans are more likely to transmit zoonotic pathogens. For example, West Nile virus is most common in birds, but mosquitoes that are usually avian feeders, like *Culex pipiens*, can be a vector of the virus to people when they occasionally bite humans. Host preference can vary geographically, seasonally, by microhabitat, and by learning to bite certain kinds of hosts (Lyimo & Ferguson 2009).

Some species and groups of mosquitoes have females with distinctly specialized feeding habits, and unusual feeding patterns occur through proximity of mosquitoes to unusual hosts. Perhaps the most extreme example are males and females of species of *Malaya* that feed on regurgitated food from ants in the genus *Crematogaster* (Clements 1999; Downes 1958). Female mosquitoes stand in the way of a column of ants and stimulate one by touching it with its forelegs and antennae. The modified mouthparts suck up the droplet of regurgitated food while the mosquito vibrates its wings and sways back and forth. *Uranotaenia sapphirina* in Florida, United States, feeds almost exclusively on annelids, mostly earthworms (the majority on *Sparganophilus tennesseensis* Reynolds), and some on leeches (*Macrobdella ditetra* Moore and

Philobdella floridana (Verrill); Reeves et al. 2018). Observations have also been made of aberrant feeding on other blood-feeding mosquitoes by *Aedes trivittatus* (Strickman 1977) and possibly more regular feeding on caterpillars. A variety of mosquitoes feed opportunistically on fish-like mudskippers that can be exposed in mud. Some mosquitoes normally feed on reptiles and amphibians, such as *Deinocerites*, *Culex* (*Neoculex*), *Mimomyia*, and some species of *Uranotaenia* (Peyton 1977; Clements 1999), whereas a wide variety of other species will occasionally feed on reptiles. Among the examples of mosquitoes that feed mainly on birds are *Aedeomyia squamipennis* (Clements 1999), *Culex stigmatosoma*, and *Culex thriambus* (Reisen 2012).

Studies of host selection have approached the problem in various ways: direct observation (Corbet & Downe 1966), baited traps (Russell & Hunter 2010), choice tests, or analyses of gut contents immunologically or genetically (Chaves et al. 2010; Takken & Verhulst 2013). Studies in which blood-fed mosquitoes are collected from the field and then analyzed for hosts would seem to offer the best hope of understanding the comparative preferences of species that are exposed to the same populations of hosts. The main bias in these studies is the collection method, which could preferentially pick up mosquitoes feeding on one host or another. For example, *Culex quinquefasciatus* commonly feeds on humans indoors but not outdoors. *Anopheles subpictus* in India feeds more on humans indoors (as high as 73% of the mosquitoes) compared to outdoors (0%–25%; Chandra et al. 2010). Reviews have indicated that although a mosquito species in each area may have an overall preference for a certain host, that species usually also feeds on other hosts as circumstances permit. Some species will switch host preference in response to selection pressure, as observed for *Anopheles gambiae* changing from predominantly human feeding to cattle feeding in Burkina Faso following widespread use of insecticide-treated bed nets. St. Laurent et al. (2016) performed collections on cattle and found *Anopheles hyrcanus, barbirostris, nivipes, and peditaeniatus*, infected with human *Plasmodium*, showing that these mosquitoes were feeding on both hosts. Some species regularly change their host preference seasonally, such as *Culex pipiens* and *tarsalis* shifting from birds to mammals in the fall in the western United States (Kilpatrick et al. 2006; Reisen 2012). The flexible nature of mosquito host preference has led to the suggestion of intentionally selecting for animal feeding in species that transmit human pathogens (Egeth & Kurzban 2012). Table 7.6 shows how different host preferences can be observed in different studies of the same species.

Analysis of data from such studies can involve a number of statistics that compare the biting rate on each host to the abundance of that host in the range of biting mosquitoes. Among those statistics (Chaves et al. 2010) are the following:

Host Utilization Rate: Proportion of Blood Meals per Host Species

Forage ratio: Host utilization rate divided by the proportion of hosts among other hosts.

Feeding index: Proportion of blood meals on one host compared to others divided by expected proportion based on abundance or size of host.

Null model testing: Statistic *C* that tests randomness of the association of all mosquito species with a host, where a small *C* indicates aggregation by the mosquito population on some hosts, and a large *C* shows segregation and specialization by individual mosquito species.

Meta-analysis of blood-feeding studies showed that feeding among mosquito species was aggregated (shared at least one host species) wherever multiple sites in a region were analyzed. The conclusion from the meta-analysis was that host utilization is dominated by host availability rather than by any physiological advantage or adaptation to particular hosts.

Few studies have been able to integrate host abundance and mosquito blood feeding. The results can show considerable preference for certain hosts. In Brazil, near São Paulo, Gomes et al. (2003) used a feeding index to compare human feeding to other common animals in the area (Table 7.7). It was clear that *Aedes albopictus* and *Culex quinquefasciatus* had strong preferences for feeding on humans, especially compared to chickens for the former and to cattle for the latter. *Aedes scapularis* preferred cattle and dogs over humans, and *Culex nigripalpus* preferred horses. In the eastern United States, *Culiseta melanura* and *Culex pipiens* fed preferentially on certain birds in different locations, but those birds were not the most abundant species (Apperson et al. 2004). The same study showed a high level of aggregation of mammal-feeding species in New Jersey, where the majority of *Anopheles bradleyi*, *Anopheles crucians*, *Anopheles quadrimaculatus*, *Culex salinarius*, *Aedes japonicus*, *Aedes trivittatus*, and *Aedes vexans* fed on deer.

Direct observation of mosquitoes can give important insights into host selection, as already pointed out for the strange feeding behavior of *Malaya* on ants. In Arctic Canada at Lake Hazen on Ellesmere Island (81°N), Corbet and Downe (1966) found that *Aedes nigripes* and *Aedes impiger* fed on king eiders (Anatidae: *Somateria spectabilis* (Linnaeus)) and musk oxen (Bovidae: *Ovibus moschatus* Blainville). Other species avoided bites in various ways. Long-tailed jaegers (Stercorar-

Table 7.6. Compilation of numbers of studies in which prominent mosquito species showed a preference for either humans or animals (Takken and Verhulst 2013)

	Number of studies (number of studies indoors)	
Species	**>50% of blood meals on humans**	**>50% of blood meals on animals**
Aedes aegypti	7 (3)	1 (0)
Anopheles arabiensis	11 (11)	9 (4)
Anopheles funestus	7 (6)	4 (0)
Anopheles gambiae	10 (9)	2 (0)
Culex quinquefasciatus	6 (5)	11 (2)

Table 7.7. Feeding indices of mosquitoes near Sâo Paulo, Brazil, comparing the rate of feeding on humans to other hosts based on the abundance of hosts. Each number is the ratio of human feeding compared to the expected feeding rate if the host selection was random (Gomes et al. 2003).

Human feeding index compared to:	Feeding index			
	Aedes albopictus	*Aedes scapularis*	*Culex nigripalpus*	*Culex quinquefasciatus*
Chickens	42.00	5.00	1.35	3.57
Cattle	1.45	0.24	0.80	13.70
Dogs	9.64	0.11	1.00	1.31
Horses	3.86	2.37	0.11	4.60

iidae: *Stercorarius longicaudus* Vieillot) were avoided, possibly because of naturally occurring repellent compounds. Feathers or fur appeared to protect Arctic hares (Leporidae: *Lepus arcticus* Ross) and red-throated loons (Gaviidae: *Gavia stellata* (Pontoppidan)), and ermine (Mustelidae: *Mustela erminea* Linnaeus) had effective defensive behaviors.

Many studies have collected blood-fed females and evaluated the identity of hosts immunologically or genetically. Blood-fed mosquitoes are not always easy to collect in numbers, and replication can be a problem in these studies. For example, Chang et al. (2008) collected only 101 blood-fed females in Taiwan. They showed that the majority of 10 *Anopheles minimus* fed on cattle and that the majority of 44 *Anopheles sinensis* s.l. and of 27 *Culex tritaeniorhynchus* fed on pigs. Larger samples in southern Zambia (Fornadel et al. 2011), analyzed genetically, resulted in the surprising conclusion that predominant malaria vectors were *Anopheles coustani* and *squamosus*, 86% ($n=59$) and 65% ($n=34$) of which fed on humans, respectively. Gingrich and Williams (2005) presented a comprehensive picture of host selection in Delaware, United States, by using four different collection methods in four habitats for two years to get 2,558 blood-fed specimens. Using immunologically based ELISAs, they had enough specimens to provide meaningful results for eight species. Considering 10% or more of blood meals as a cutoff, only two species concentrated their feeding on two hosts (*Anopheles quadrimaculatus* and *Coquillettidia perturbans* on dogs and horses). The remainder had three or four hosts, with most feeding on birds (*Culex pipiens*), on dogs (*Aedes albopictus*), and on rabbits (*Aedes sollicitans*, *Aedes vexans*, *Culex restuans*, and *Culex salinarius*).

Interest in the biting habits of *Anopheles* has been intense over the years because of the association of this genus with transmission of the parasites that cause human malaria. Table 7.8 summarizes four different papers that assembled the results of dozens of field studies that examined human feeding by malaria vectors. Despite doubts about these comparisons based on species identification and comparison of dissimilar methods, there are several generalizations from these summaries. First, a comparison of studies that found predominantly human feeding (anthropophily) versus studies that found predominantly animal feeding (zoophily) usually reflected mean calculations of the human feeding index. Closer examination of these numbers raises the point that even the most anthropophilic of species sometimes feed mainly on animals (e.g., *Anopheles gambiae* and *funestus*). A second point is that species clearly more likely to bite animals, like *Anopheles albimanus*, can nonetheless be important vectors of malaria to humans. Third is the impressive number of species in the tropics that feed significantly on humans, a contrast from Gingrich and Williams' (2005) study in Delaware that did not find a large proportion of any mosquito species taking human blood meals. Finally, and reflecting back to the discussion of invasive mosquito species in Chapter 3, major shifts could occur in the transmission potential of local anopheline fauna if some of these species expand their ranges into new areas.

Human-associated species may be in close proximity to humans, but their feeding habits cover a range of specificity. Perhaps the extreme example is *Aedes aegypti*, which in domestic situations in Puerto Rico and Thailand feeds almost exclusively on people (Scott et al. 2000). Deprived of human blood meal sources, *Aedes aegypti* will feed on other animals, and the species also increases its sugar feeding (Scott & Takken 2012). *Anopheles gambiae* is another highly anthropophilic species, feeding preferentially on adults, pregnant women, and people with malaria (Takken & Verhulst 2013). The preference for human blood meals was shown experimentally in Tanzania using a choice of rooms containing a calf or a person, with the result that about 70% fed on the human (Gillies 1964a). Of course, that means that about 30% fed on the calf, so even this species was flexible in host choice. In Kenya, 88% of the mosquitoes collected inside houses at Mbita Point were *Anopheles gambiae*, and the availability of human hosts influenced larval distribution of the species (Minakawa et al. 2002). *Aedes albopictus* commonly feeds on humans but typically uses birds and other mammals as blood sources, too, as observed in Macao, where 44% of this species fed on humans and 23% on chickens. A study using bird-baited traps, carbon dioxide-baited traps, and human landing collections during a single year on the Camargue Plain of southern France showed considerable human feeding, with over 40% of *Anopheles hyrcanus* and *Culex modestus* collected from people (L'Ambert et al. 2012).

The *Culex pipiens* group of species offers an interesting example of host range that has been studied in some detail because of the association of these species with transmission of West Nile and Saint Louis encephalitis viruses. As described

Table 7.8. Calculation of the proportion of malaria vectors that fed on humans, compiled from multiple studies

	Human blood index (number of studies)		Number of studies anthropophilic/zoophilic
Anopheles species	Kiszewski et al. 2004	Moffett et al. 2007	Sinka, Bangs, et al. 2010; Sinka, Rubio-Palis, et al. 2010; Sinka et al. 2011
aconitus			1/3
albimanus	0.102 (16)		2/2
albitarsis			2/2
annularis			0/8
aquasalis	0.109 (3)		1/1
arabiensis	0.871 (32)	0.526 (21)	11/14
atroparvus	0.245 (8)		1/5
balabacensis			4/0
barbirostris	0.127 (9)		
coustani		0.157 (2)	
culicifacies	0.052 (55)		0/9
darlingi	0.458 (2)		12/0
dirus	0.355 (18)		12/1
farauti	0.658 (19)		5/0
flavirostris	0.300 (9)		2/4
fluviatilis	0.034 (27)		5/6
freeborni	0.019 (8)		1/1
funestus	0.980 (30)	0.844 (12)	19/6
gambiae	0.939 (36)	0.815 (16)	12/4
koliensis			3/0
labranchiae	0.151 (17)		2/3
lesteri	0.010 (1)		2/2
maculatus	0.155 (10)		1/9
marajoara			2/2
melas	0.690 (6)	0.576 (3)	5/3
merus		1.0 (1)	3/2
messeae	0.172 (14)		1/2
minimus	0.425 (12)		6/14
moucheti		0.931 (3)	5/0
multicolor	0.008 (15)		
nili		0.949 (3)	6/0
nuneztovari	0.222 (11)		2/4
pharoensis	0.520 (17)		
pseudopunctipennis	0.477 (13)		3/2
pulcherrimus	0.062 (12)		
punctulatus	0.855 (7)		2/0
quadriannulatus		0.011 (1)	
sacharovi	0.087 (47)		
sergentii	0.100 (18)		1/6
sinensis	0.018 (21)		1/9
stephensi	0.023 (37)		2/4
subpictus			1/9
superpictus	0.093 (18)		
sundaicus	0.611 (17)		1/1

in Chapter 1, this group is divided into several species, some of which freely hybridize to create local populations that may recombine various aspects of phenotype (Strickman & Fonseca 2012). The generally accepted pattern of host preference is that northern populations of *Culex pipiens* feed on birds, and southern populations of *Culex quinquefasciatus* have more of a tendency to feed on mammals. Form Molestus of *Culex pipiens* also tends to feed on mammals. Studies like those of Gingrich and Williams (2005) and baited trap comparisons in Ontario, Canada (Russell & Hunter 2010), and New Jersey (Faraji & Gaugler 2015) support the idea that northern populations of American *Culex pipiens* feed primarily on birds. However, a detailed comparison of northern *Culex pipiens*, hybrid populations with *Culex quinquefasciatus* in middle latitudes, and *Culex quinquefasciatus* in the southern United States failed to show a clear pattern of increasing mammalian feeding (Savage & Kothera 2012). Personal observations of the ease with which locally collected colonies of *Culex pipiens* can be selected for mammalian feeding also support the idea that this group of species is highly adaptable and opportunistic about selection of hosts.

The frequency of blood feeding is probably as variable as the selection of hosts. Many *Anopheles* take a blood meal soon after emergence to replenish nutritional reserves and not to develop eggs (Beier 1996; Scott & Takken 2012). Subsequent meals in these species contribute to egg development—sometimes a single blood meal (*Anopheles tessellatus*) and sometimes several meals to support a gonotrophic cycle (*Anopheles caroni*, *culicifacies*, *gambiae*, *funestus*). Nutritional reserves may be replenished by a blood meal late in egg development (*Anopheles culicifacies*, *stephensi*, *subpictus*), and *Plasmodium* infection can increase feeding frequency (Takken & Verhulst 2013). *Aedes aegypti* regularly takes multiple blood meals during each gonotrophic cycle (Trpis & Hausermann 1986; Reiter et al. 1995; Scott et al. 2000). On the other hand, *Aedes sierrensis* takes blood meals at a rather long interval of eight days (Hawley 1985).

Highly adapted to blood feeding though mosquitoes are, there are nevertheless significant costs of this activity for at least some species. The further a female must search for a host, the less likely that individual will survive long enough to complete incubation of *Plasmodium* parasites, according to models of *Anopheles gambiae* and *arabiensis* (Killeen & Smith 2007). Experimental evidence of how mosquitoes will expend more energy in response to barriers to reaching hosts has been observed in populations of *Anopheles funestus* in Benin, which shifted their time of biting in response to the widespread use of insecticide-treated bed nets (Moiroux et al. 2012). The concentration of *Anopheles gambiae* near its favored human hosts (Minakawa et al. 2002) is an indication that this species minimizes the distance required to get from hosts to oviposition sites. Hosts can also exert significant defensive behavior, either increasing the effort required to successfully obtain a blood meal or killing the mosquito that attempts to bite. Experiments with restrained and unrestrained hosts in the laboratory showed that *Aedes aegypti* was over 100 times more successful at feeding on restrained hosts (mice, deer mice, chickens) and that 56%–90% of the mosquitoes in a cage were killed when these hosts were unrestrained (Day & Edman 1984). Field observations have shown that the behavior of the host species can influence the success of blood-feeding mosquitoes. The black-crowned night heron (*Nycticorax nycticorax* (Linnaeus)) and the great blue heron (*Ardea herodias* Linnaeus), which must stand still while waiting to prey on fish, suffered approximately 10 times more bites from *Aedes taeniorhynchus* than the cattle egret (*Bubulcus ibis* Bonaparte) or the little blue heron (*Egretta caerulea* (Linnaeus), which were able to actively defend against bites (Edman et al. 1984). Corbet and Downe (1966) recorded observations of how ermine avoided the bites of *Aedes impiger* and *nigripes* by head shaking. On the other hand, the density of *Anopheles arabiensis* was independent of feeding success in Tanzania, leading to the conclusion that human defensive behavior was not important (Charlwood et al. 1995b). The decrease in flying ability following a full blood meal increases the susceptibility of *Anopheles gambiae* to predation by jumping spiders (Roitberg et al. 2003).

The means by which mosquitoes find their hosts are complicated compared to insects that are more specialized in their feeding preferences. The mechanisms for host determination and detection closely fit the common patterns of blood feeding described, in which species will feed on a variety of hosts depending on availability and broad host preferences. Considering that the same individuals also seek specific sugar sources, the sensing mechanisms of mosquitoes have to be well developed. Mosquitoes commonly use visual signals from a distance, heat and carbon dioxide up close, and many kinds of chemical signals from moderate to close distances (Enserink 2002; Bernier et al. 2003). The goals of these abilities are to detect the signals, put them in the correct behavioral order, and use them to successfully blood feed. There is also considerable redundancy and flexibility in the process, as observed when mosquitoes will feed on droplets of blood or blood substitutes in the absence of many of the cues that come from a real host. Not surprisingly, the neural responses to odors depend on the physiological state, at least in female *Anopheles gambiae* mosquitoes that cease to respond to odors associated with hosts after a blood meal (Takken et al. 2001).

The precise nature of chemical cues and these cues' detection by mosquitoes have been the subjects of study for many decades (Debboun et al. 2007). It became clear with the work of Bernier and others (2007a, 2007b) that attraction to a particular host was mediated by detection of many chemicals, some of which were attractive and others that were repellent. Bacterial decomposition of organic compounds on the skin, especially in the sebum (Luntz 2003), contributes to chemicals affecting mosquito behavior (Enserink 2002). Detecting all of these compounds simultaneously informs a mosquito about the relative attractiveness of a host, or its repellence. Some animals have exploited chemical detection systems by either secreting repellent chemicals (e.g., the gaur, *Bos gaurus* Smith) or anointing themselves with particular plants or ants (Weldon & Carroll 2007). *Plasmodium falciparum* infection increased human hosts' attractiveness to *Anopheles gambiae* (LaCroix et al. 2005). There is the possibility that the highly flexible detection systems in mosquitoes have evolved in response to hosts' development of chemical defenses, much as plant defensive compounds can become attractants to some herbivorous insects and their parasitoids or predators (Jones et al. 2011b). Given the individual compounds and mosquito feeding behavior, White (2007) coined the term "phagomone" to include all chemicals that affect feeding behavior.

Sequencing and interpretation of the *Anopheles gambiae* genome (Holt et al. 2002) opened the door to a much deeper understanding of both olfactory and gustatory chemical detection. The complete picture of the physiological mechanisms is not complete, but a summary of the system answers many questions. Odorant reception begins when a compound binds to one of many odorant-binding proteins (Zwiebel & Takken 2004). The proteins are water soluble and cysteine rich. They are associated with odorant receptors located at the distal end of odorant receptor neurons in structures on the antennae, palps, and other organs. Sensitivity of the sensilla trichodea on the antennae varies in relation to time after a blood meal in *Aedes aegypti* (Siju et al. 2010). Appropriately stimulated, the odorant receptors undergo a cascade of proteins that ultimately cause the sensory neuron to fire. The cascade starts with the release of G proteins on the interior of the receptor cell, which in turn release another protein causing production of effector enzymes in the odorant receptor neuron. Secondary messengers include cyclic adenosine monophosphate and inositol 1,4,5-triphosphate. This complex system, which results in a nervous signal in response to a class of environmental compounds, allows for modification by other chemicals and desensitization in the event of overexposure. Arrestin proteins are involved in both desensitization and resensitization. Odorant receptors operate in conjunction with a single coreceptor called "Orco" (Bohbot & Dickens 2012). The signal chain can also be affected by modulating chemicals that interact with either the odorant receptor or Orco at a different location on the protein (allosteric modulator). These modulating chemicals can be semiochemicals that operate at extremely low concentrations or artificially inhibiting chemicals, like repellents, that operate at much higher concentrations. The odorant receptors are a highly conserved system classed with G protein-coupled receptors (Fox et al. 2001). There are only a few kinds of these receptors, and one of them has only been found in females in studies of *Anopheles gambiae*. The process of odor reception in *Aedes aegypti* is regulated in some ways by a complex of 28 neuropeptides that act directly on the olfactory neurons in the antenna (Siju et al. 2014).

Taste has a separate system of detection using gustatory receptors (Pitts et al. 2017). These also work with highly conserved coreceptors. Ionotropic receptors are important for taste, detecting amines and carboxylic acids. There are differences in the amino acid sequences of gustatory receptors of Anophelinae and Culicinae, but both sugar and carbon dioxide gustatory receptors are conserved (Sparks et al. 2013). Gustatory receptors of *Aedes aegypti* were similar between males and females, with 35 of them commonly shared, 6 unique to males, and 3 unique to females.

The locations of odorant and gustatory receptors can be characterized through genetic expression studies or by single-neuron electrophysiological recording. Most host detection organs are on the antennae and maxillary palpi (Zwiebel & Takken 2004), with carbon dioxide detection on the latter. Grooved peg sensillae on the antennae are sensitive to (L)-lactic acid (*Aedes aegypti*) and ammonia (*Anopheles gambiae*). Sensillae trichodea are sensitive to a range of carboxylic acids, geranyl acetone, 3-methyl-1-butanol, 6-methyl-5-heptene-2-one, and indole (*Anopheles gambiae*, *arabiensis*, *quadriannulatus*, and *atroparvus*). *Aedes aegypti* antennal trichoid sensillae are sensitive to 6–10 carbon aldehydes (Ghaninia et al. 2008).

Most gustatory receptors are in sensory hairs with porous tips on the labella, labrum, inner mouthparts, wing margins, genitalia, and tarsi (Sparks et al. 2013). The labella of *Aedes aegypti* is the site of one gustatory receptor that responds to repellent compounds (DEET, citronellal, IR3535, and picaridin) and two that respond to salt and sugar (Sanford et al. 2013). However, ionotropic gustatory receptors are mostly on the antennae (Pitts et al. 2017). In *Culiseta inornata*, the labral gustatory sense organs detect salt, sugar, water, and nucleotides. The cibarium also contains a series of sense organs, including gustatory ones (Ignell et al. 2005).

The maxillary palpus of *Aedes aegypti* is a good example of a multipurpose sensory organ (Bohbot et al. 2014). It detects mechanical and heat signals, as well as gustatory and odorant chemical signals. Noninnervated structures on the palpi include microtrichiae (tiny hairs), scales, and basiconica. Nonporous sensilla chaetica are mechanoreceptors on the dorsolateral sides of the fourth segments. Porous sensilla chaetica on the same location detect octenol, carbon dioxide, and host odors. Campaniform sensillae are also mechanoreceptors. Typically, the maxillary palpi extend outward along the axis of the body during feeding, and the fifth segment vibrates.

Among the phagomones, carbon dioxide has a special place by both activating searching behavior and orienting the mosquito toward a host. Carbon dioxide receptor proteins are heterodimers in the maxillary palpi of males and females of *Aedes aegypti* and *Culex quinquefasciatus* (Erdelyan et al. 2012). Activation of *Aedes aegypti* by carbon dioxide can be enhanced by adding short-chain aliphatic acids (Bosch et al. 2000). Chemicals (butanal, pentanal, hexanol, butanedione) that activate carbon dioxide receptors can disrupt host seeking by blocking normal attraction (Turner et al. 2011). Carbon dioxide in the breath also orients mosquitoes toward a host, usually observed as catches of more mosquitoes in traps or choice tests.

Carbon dioxide attraction can be enhanced by addition of various chemicals at the proper concentrations. Based on exhalations from animals, some of the early examples were 1-octen-3-ol, acetone, lactic acid, butanone, and 3-methyl-2-hexenoic acid (Luntz 2003). Addition of chemical blends (6–10 carbon aldehydes) to carbon dioxide increased the number of *Aedes mcintoshi* and *ochraceus* threefold in

Kenya (Tchouassi et al. 2013) and addition of (L)-lactic acid increased captures of *Anopheles funestus* by 44% (Murphy et al. 2001). Eccrine human sweat contains (L)-lactic acid attractive to *Anopheles gambiae*, but incubated sweat was more attractive even after removal of lactic acid enzymatically (Braks et al. 2001).

Attractive blends of chemicals based on host odors have been developed both as research tools and as practical tools for mosquito collection. The first report of a blend as attractive as a human hand to *Aedes aegypti* without addition of carbon dioxide included the following compounds: acetone, dimethyl disulfide, and dichloromethane (Bernier et al. 2003). Unlike *Aedes aegypti*, *Anopheles gambiae* was not attracted to lactic acid alone (Smallegange et al. 2005). The two species were also different in the concentration of ammonia that was attractive, the former to 2-17 ppb and the latter to 136 ppb. *Anopheles gambiae* was most attracted to a blend of lactic acid, ammonia, and a mixture of 12 saturated carboxylic acids. Another successful blend used in the field to attract *Anopheles gambiae* and *funestus* in Kenya included ammonia, (S)-lactic acid, tetradecanoic acid, and 3-methyl-1-butanol (Mukabana et al. 2012). A basic blend of attractants for *Anopheles gambiae* (ammonia, lactic acid, tetradecanoic acid, and carbon dioxide) was enhanced by addition of 3-heptanol at 0.025 ppm, by 2-methyl propanal at 0.05 ppm, or by 4,5-dimethylthiazole at 0.73 ppm (Smallegange et al. 2012).

Repellents are another kind of phagomone. Synthetic repellents like DEET have a complicated effect on mosquitoes, being specifically detected by odorant receptor neurons of the antenna (*Culex quinquefasciatus*; Syed & Leal 2008) and gustatory receptors of the maxillary palpi (*Aedes aegypti*; Sanford et al. 2013) but also having a general repellent effect against male and female mosquitoes regardless of their activity (Frances 2007; Obermayr 2014). A wide variety of repellent compounds exist (Debboun et al. 2007; Paluch et al. 2010), including many naturally occurring plant chemicals (Strickman et al. 2009). The influence of these common botanical chemicals on mosquito behavior in the field is unknown, though studies on sugar feeding to be described may indicate that they are involved in selection of nectar sources.

Chemicals that are secreted by feeding mosquitoes and that attract other mosquitoes are a remarkable, but little studied, kind of phagomone. Early Russian work indicated that *Aedes communis* feeds more on hosts where other mosquitoes of the same species are feeding, describing this as an "invitation effect" (Alekseev et al. 1976). When a person stays in one place, bites seem to come in groups, suggesting that the mosquitoes somehow attract each other. Another anecdotal indication of an attractive phagomone produced by feeding mosquitoes is the apparent resistance to repellents when biting densities are high. The author (DS) observed this personally in Paraguay when *Aedes scapularis* was biting at rates of hundreds per minute, sometimes feeding directly through a droplet of DEET repellent. Controlled experiments in the laboratory have shown that 73% of *Aedes sierrensis* chose a guinea pig with open, screened cups containing other feeding mosquitoes compared to feeding on a host without other mosquitoes (Ahmadi & McClelland 1985). The effect was almost as strong (70% of feeding mosquitoes) when *Aedes aegypti* were used as the attracting species. *Aedes aegypti* was more attracted to chicken skin membranes over blood that already had feeding mosquitoes (Cavanagh & Townson 1986). *Aedes cantans* behaved similarly, with 70% feeding on a leg that had feeding mosquitoes compared to the other leg (Charlwood et al. 1995a). The same level of preference was not observed among Tanzanian *Anopheles*, though four species (*Anopheles coustani*, *funestus*, *squamosus*, and *pharoensis*) were more numerous on a leg with other feeding *Anopheles*. An invitation effect was also invoked to possibly explain a feeding preference by *Aedes albopictus* in Cambodia for the head, arms, posterior trunk, and feet, as well as a concentration of bites where other bites had recently occurred (Charlwood et al. 2014). These experiments all suggest attraction to volatile chemicals, but they do not produce clear evidence of whether the attractant was coming from the host (kairomone) or from other mosquitoes (pheromone). The experiments with *Aedes sierrensis* suggest the latter because mosquitoes feeding in closed cups provided no extra attraction to the host.

Host attractiveness to mosquitoes can be modified by the host's own response to pathogens transmitted by mosquitoes. *Plasmodium falciparum* gametocytes secrete a chemical ((E)-4-hydroxy-3-methyl-but-2-enyl pyrophosphate) that causes human red blood cells to release carbon dioxide, aldehydes, and monoterpenes attractive to blood-seeking mosquitoes. The chemical also directly stimulates feeding and increases the susceptibility of *Anopheles gambiae* to parasite infection (Emami et al. 2017).

Associative learning is one of the behaviors that influences both blood and sugar feeding. Learning results in individual mosquitoes returning to the same species of host (Takken & Verhulst 2013), the same sugar source, or the same location. First suggested in 1934 to explain behavior by *Anopheles maculipennis*, several examples of learning have been documented through field observations (McCall & Kelly 2002). *Anopheles farauti* in Papua New Guinea flies farther if released in a different village than when released in the village from which it came (Charlwood et al. 1988). Similarly, 68% of *Anopheles arabiensis* returned to the house from which they were originally captured when released from a different house. *Anopheles balabacensis* and *Culex vishnui* in Thailand tended to take their second blood meals from the same host species from which they took their first blood meal. Laboratory experiments on *Culex quinquefasciatus* showed that females could be trained on artificial scents associated with defibrinated bovine blood or sugar and that males could be trained on sugar (Tomberlin et al. 2006). The experiments consisted of exposing mosquitoes to capillary tubes filled with the nutrient and coated with either artificial vanilla or strawberry scent. Mosquitoes experienced three brief (10 second) exposures and two minutes later were given the choice between empty pipettes coated with either the associated scent or the other scent. About 90% of the mosquitoes went to the empty pipette with the scent associated with sugar, and about 60% of females went to the pipette associated with blood. The effect was slightly stronger with strawberry scent, suggesting that some odors are either more easily detected or more easily associated with a nutrient source. *Anopheles gambiae* can be trained to associate either visual or olfactory signals with a more suitable membrane (sausage skin compared to parafilm) or a more suitable blood source (unsalted compared to salted blood; Chilaka et al. 2012). Mosquitoes learned the associations after only a few exposures and retained the association for three, but not four, days. Male and female *Culex pipiens* Form Molestus retained conditioning from exposure to a blend of flower volatile compounds for at least 61 hours, the unconditioned mosquitoes going to a sugar source with the volatile blend 50% of the time and the conditioned mosquitoes going to it 81% of the time (Jhumur et al. 2006). Associative ability may give individuals the advantage of avoiding false sugar sources (Wright & Schiestl 2009) or increase the chances of successfully completing the next blood meal.

Sugar-Feeding Behavior

As discussed previously, sugar feeding is as much a part of the mosquito life cycle as blood feeding. This was recognized early based on field observations of Arctic mosquitoes (*Aedes communis* and *hexodontus* [Downes 1958]). Mosquitoes' survival can depend on getting a sugar meal, probably always for males and usually for females (Gary & Foster 2004). Of course, non-blood-feeding groups, like *Toxorhynchites*, rely entirely on larval reserves and sugar feeding for adult nutrition (Stone & Foster 2013). Some species, like *Aedes aegypti*, *Anopheles culicifacies*, *Anopheles gambiae*, *Anopheles stephensi*, and *Anopheles subpictus*, have been shown able to use blood in place of sugar to replenish energy reserves, though conversion from blood components to glycogen reserves requires about 24 hours (Foster & Takken 2004; Scott & Takken 2012). The rate of mobilization can be better in some species (e.g., *Anopheles gambiae*) than in others (e.g., *Anopheles atroparvus*; Stone & Foster 2013).

Evaluations of the importance of sugar feeding under natural conditions show dramatically large effects. Mark-release-recapture studies of *Anopheles sergentii* in an Israeli oasis was one such evaluation (Gu et al. 2011). In the presence of natural *Acacia raddiana* (plant taxonomic designations, common names, and native ranges are listed in Table 7.9) sugar sources, the daily survival rate was 0.93, and the gonotrophic cycle was 2.36 days, compared to a survival rate of 0.72 and a gonotrophic cycle of 3.33 days in the absence of floral nectar sources (Gu et al. 2011). *Anopheles gambiae* lived longer in Kenyan mesocosms when many sugar sources were available, males surviving 12.5 days longer and females surviving 7.4 days longer (Stone et al. 2012a). Removal of flowers from invasive mesquite trees (*Prosopis*) during the dry season in Mali was associated with a twofold drop in the number of female *Anopheles* and a ninefold drop in the number of males. This decline in the population was associated with decreases in the rate of sugar feeding by about 70% and a shift in species composition from a balance between *gambiae*, *arabiensis*, and *coluzzii* to mostly *coluzzii* (Müller et al. 2017). *Opifex fuscus* required sugar meals in the laboratory to be able to deposit autogenous eggs in the first gonotrophic cycle (Snell et al. 2010).

In some habitats a large proportion of mosquitoes have taken sugar. For example, in an Israeli oasis with only *Acacia raddiana* blooming, 80%–90% of *Anopheles sergentii* and *Aedes caspius* contained sugar meals. Under the same circumstances, applying an insecticide-treated bait to the plants reduced female populations by 93.5% and male populations by 89% (Müller & Schlein 2006; Müller et al. 2010). *Culex pipiens* populations in Israel also fed intensively on sugar sources, as indicated by a three-week 90% reduction in this species following application of sugar and spinosad insecticide on flowering plants (Schlein & Müller 2008) or a more attractive blend of juices, wine, sugar, and spinosad (Müller et al. 2010).

Like blood feeding, sugar feeding occurs more often during certain times of the day. Based on trapping *Anopheles gambiae* s.l. (97% *Anopheles gambiae*, 3% *Anopheles arabiensis*) on flowering *Acacia macrostachya* in Mali, most feeding took place early (1930–2200h) or late (0400–0500h) during hours of darkness (Müller et al. 2010). Kenyan *Anopheles gambiae* males in cages fed most actively during the dusk-to-dawn transition and often fed more than once per night (mean 2.02 sugar meals per day; Gary & Foster 2006). Females fed on sugar at the same times they sought blood meals, but their sugar-feeding activity decreased during the two days following a blood meal and resumed after oviposition. Female *Culex annulirostris* observed in the laboratory took sugar meals most often during darkness, with a mean of 1.32 sugar meals during scotophase compared to 0.57 meals per photophase (Williams & Kokkinn 2005). Diel periodicity of sugar feeding has not received as much attention as blood feeding, but it seems reasonable to assume that across many species of mosquitoes, the pattern would vary according to the availability of sugars secreted by plants and the microclimatic conditions tolerated by individual mosquitoes.

One of the surprising aspects of mosquitoes' sugar feeding is their ability to pierce plant tissue to extract juices directly from stems and leaves. Starch-digesting amylase in saliva is one of the indications that mosquitoes are adapted to this kind of feeding (Stone & Foster 2013). *Anopheles sergentii* prefers to feed on nectar sources in Israeli desert habitats, but 40%–45% of males and females feed on plant-tissue juices when flowers are unavailable in the dry season (Junnila et al. 2010). Feeding on plant tissue was indicated in the mosquitoes by the presence of fructose in the absence of flowers or extrafloral nectaries, by the presence of cellulose, and by the presence of plant DNA. Of 24 plant species that were present in the dry season, only three (*Suaeda asphaltica*, *Malva nicaeensis*, and *Conyza dioscoridis*) were utilized by the mosquitoes for tissue feeding, and those plants represented less than 1% of the total plant biomass. In the laboratory, *Anopheles sergentii* tended to feed on tissue from *Ochradenus baccatus* and nectar from *Tamarix nilotica*, even though both were flowering. Other studies in the same area (Müller & Schlein 2005) showed that

Table 7.9. Plant taxonomic designations, common names, and native ranges for species mentioned in the text

Species	Family	Common name	Native range
Acacia macrostachya	Fabaceae	Sama	Sahel and western Africa
Acacia raddiana Savi	Fabaceae	Twisted Acacia	Mediterranean region
Acacia saligna (Labill.) H.L. Wendl.	Fabaceae	Cujong	Australia
Acacia senegal (L.) Willd.	Fabaceae	Senegal Gum	Africa
Asclepias syriaca L.	Apocynaceae	Common Milkweed	Northern and eastern North America
Bauhinia reticulata DC.	Fabaceae	Camel's Foot	Sahelian Africa
Boscia angustifolia A. Rich.	Capparaceae	Basari	Sahelian Africa, Arabia
Buddleja davidii Franch.	Scrophulariaceae	Butterfly Bush	China, Japan
Cascabela thevetia (L.) Lippold	Apocynaceae	Yellow Oleander	Mexico, Central America
Cassia sieberiana DC.	Fabaceae	Drumstick Tree	Africa
Ceratonia siliqua L.	Fabaceae	Carob Tree	Mediterranean, western Asia, Macaronesia
Citrus × aurantium L.	Rutaceae	Bitter Orange	Southeast Asia
Conyza dioscoridis (L.) Desf.	Asteraceae	Conyza	Middle East, northeastern Africa
Crotalaria L.	Fabaceae	Rattlepods	Africa
Cucumis melo L.	Cucurbitaceae	Muskmelon	Iran, Anatolia, Caucasus
Delonix regia (Boj. Ex Hook) Raf.	Fabaceae	Royal Poinciana	Madagascar
Eriobotyra japonica (Thunb.) Lindl.	Rosaceae	Loquat	China
Faidherbia albida (Delile) A. Chev.	Fabaceae	Apple-Ring Acacia	Africa, Middle East
Fallopia baldschuanica (Regel) Holub	Polygonaceae	Mile-a-Minute	Asia
Ficus carica L.	Moraceae	Common Fig	Middle East, western Asia
Ficus thonningii Blume	Moraceae	Fig	Africa
Guiera senegalensis J.F. Gmel.	Combretaceae	Moshi Medicine	Sahelian Africa
Hamelia patens Jacq.	Rubiaceae	Firebush	Subtropical and tropical Americas
Hyptis suaveolens (L.) Poit.	Lamiaceae	Pignut	Tropical Americas
Ipomoea batatas (L.) Lam.	Convolvulaceae	Sweet Potato	Central and South America
Lantana camara L.	Verbenaceae	Big-Sage	Central and South America
Leptadenia pyrotechnica (Forssk.) Decne.	Apocynaceae	Khimp	Africa to India
Malva nicaeensis All.	Malvaceae	French Mallow	Eurasia, North Africa
Mangifera indica L.	Anacardiaceae	Mango	South Asia
Manihot esculenta Crantz	Euphorbiaceae	Cassava	Brazil
Ochradenus baccatus Delile	Resedaceae	Taily Weed	Middle East
Opuntia ficus-indica (L.) Mill.	Cactaceae	Prickly Pear	Mexico
Parthenium hysterophorus L.	Asteraceae	Whitetop Weed	Tropical Americas
Phragmites australis Adans.	Poaceae	Common Reed	Cosmopolitan
Plantanthera obtusata (Banks ex Pursch) Lindl.	Orchidaceae	Small Northern Bog Orchid	Northern America, Europe, Asia
Polygonum equisetiforme	Polygonaceae	Horsetail Knotweed	Middle East, North Africa
Prosopis farcta	Fabaceae	Syrian Mesquite	Western and southern Asia, North Africa
Psidium guajava L.	Myrtaceae	Guava	Tropical Americas
Punica granatum L.	Lythraceae	Pomegranate	Iran
Ricinus communis L.	Euphorbiaceae	Castor Bean	Southeast Mediterranean, eastern Africa, India
Rubus ulmifolius sanctus (Schreb.) Sudre	Rosaceae	Holy Bramble	Europe, Asia
Saccharum officinarum L.	Poaceae	Sugarcane	Southeast Asia
Senna didymobotrya (Fresen.) Irwin & Barneby	Fabaceae	African Senna	Africa
Senna siamia (Lam.) Irwin et Barneby	Fabaceae	Kassod Tree	South and Southeast Asia
Silene otites (L.) Wibel ex Sm.	Caryophyllaceae	Spanish Catchfly	Europe
Spiraea latifolia (Aiton) Borkh.	Rosaceae	Broadleaf Meadowsweet	Eastern North America
Suaeda asphaltica	Amaranthaceae	Dead Sea Sea-Blite	Middle East, North Africa

continued

Table 7.9. *continued*

Species	Family	Common name	Native range
Tamarix chinensis Lour.	Tamaricaceae	Salt Cedar	China, Korea
Tamarix jordanus Boiss.	Tamaricaceae	Jordan Tamarisk	Middle East, western Asia
Tamarix nilotica (Ehrenb.) Bunge	Tamaricaceae	Nile Tamarisk	North Africa, Middle East
Tecoma stans (L.) Juss. ex Kunth	Bignoniaceae	Yellow Trumpetbush	Americas
Trillium grandiflorum (Michx.) Salisb.	Melanthiaceae	White Trillium	Northern and eastern North America
Vachellia nilotica (L.) P.J.H. Hurter & Mabb.	Fabaceae	Gum Arabic Tree	Africa to India
Viola labradorica Schrank	Violaceae	American Dog Violet	Eastern North America, Greenland
Vitex agnuscastus L.	Verbenaceae	Chaste Tree	Mediterranean
Ziziphus mauritiana Lam.	Rhamnaceae	Chinese Date	Southeast Asia
Ziziphus spinus-christi L. (Desf.)	Rhamnaceae	Christ's Thorn Jujube	Africa, southern and western Asia

Anopheles sergentii, *Anopheles claviger*, *Aedes caspius*, and *Culex pipiens* all fed on plant tissue when other sugar sources were scarce. Tissue feeding by *Culex pipiens* differed from blood feeding, in that the labium bent only a little, and the stylets were apparently inserted less deeply. Mosquitoes that only had access to plant tissue as a sugar source were stressed, surviving less time, depositing fewer eggs, and experiencing longer gonotrophic cycles. Stem and leaf feeding has also been observed in Kenya, where *Anopheles gambiae* fed on *Parthenium hysterophorus*, *Lantana camara*, and *Ricinus communis* (Manda et al. 2007a), as well as on *Ipomoea batatas* (Impoinvil et al. 2004).

The observations on tissue feeding, though limited to a couple of ecological situations, are another indication of the importance of sugar in the diet and the intricate adaptations of mosquitoes to get it. Studies on nectar and fruit feeding provide more detail on selectivity of plant hosts and the importance of those hosts to survival and reproduction. Early studies with P_{32}-labeled plants in northern Ontario, Canada, showed that local mosquitoes (*Aedes implacabilis*, *diantaeus*, and *punctor*), as well as *Aedes aegypti*, only fed on 3 plants (*Viola labradorica*, *Trillium grandiflorum*, and *Spiraea latifolia*) out of 24 native plants that were blooming at the time (Abdel-Malek & Baldwin 1961). Feeding only occurred when leaves were present on the plant samples, suggesting that volatiles from the leaves were essential to recognition of the nectar sources.

Although mosquitoes choose sugar sources selectively, they are also adaptable in that choice. Studies in Israel showed that a native species of mosquito, *Culex pipiens*, and an imported one, *Aedes albopictus*, fed on a variety of native and imported plants (Schlein & Müller 2008; Müller et al. 2011). Male and female *Aedes albopictus* fed selectively on flowers of four of six ornamental plants (*Tamarix chinensis*, *Vitex agnuscastus*, *Polygonum baldchuanicum*, and *Buddleja davidii*) and four native species (*Ziziphus spinus-christi*, *Polygonum equisetiforme*, *Prosopis farcta*, and *Ceratonia siliqua*). It also fed on the seedpod of *Ceratonia siliqua* and on the rotting fruits of a number of imported (*Opuntia ficus-indica* and *Eriobotyra japonica*) and native (*Ficus carica*, *Rubus ulmifolius sanctus*, and *Punica granatum*) plants. Out of 26 plants tested, *Culex pipiens* males and females fed primarily on flowers of the native species *Tamarix jordanus*, as well as on *Polygonum equisetiforme* and *Acacia saligna*. In the field, about 10% also fed on *Phragmites australis*, dominated by younger mosquitoes that were freshly emerged.

Anopheles gambiae and related species followed similar patterns of sugar feeding. Males and females had similar preferences, mosquitoes were highly selective among the plant species available, mosquitoes were adaptable enough to feed on a variety of plants, and mosquitoes fed on a variety of sugar-bearing plant parts. Field tests of 26 plant species in Mali (Müller et al. 2010) showed female preferences for five species of fruit or seedpods (*Psidium guajava*, *Cucumis melo*, *Bauhinia reticulata* [formerly *Piliostigma reticulatum*], *Ficus thonningii*, and *Saccharum officinarum*). Males preferred the first four species listed and the seedpods of *Faidherbia* (formerly *Acacia*) *albida*. Both males and females had the strongest preference for *Psidium guajava*, the fruits of which attracted almost six times as many mosquitoes as did a sugar solution. Flowers from nine species of the plants were attractive to females (*Acacia macrostachya*, *Faidherbia albida*, *Boscia angustifolia*, *Ziziphus mauritiana*, *Guiera senegalensis*, *Vachellia* [formerly *Acacia*] *senegal*, *Acacia nilotica*, *Leptaolenia pyrotechnica*, and *Crotalaria* sp.), and the first eight on the list were attractive to males. One type of flower (*Hyptis suaveolens*), also noted by Conti et al. (2012), and one fruit (*Citrus* × *aurantium*) were repellent, based on being less attractive than a sugar solution. In Kenya (Manda et al. 2007b), *Anopheles gambiae*, given a choice of 13 plants found near homes, chose two imported species (*Parthenium hysterophorus* [stems and leaves] and *Tecoma stans*) and two native species (*Ricinus communis* [extrafloral nectaries] and *Senna didymobotrya*) over all others. *Hamelia patens* was not among the plants tested, but it was commonly fed upon. In Burkina Faso, 69%–74% of male *Anopheles coluzzii* had fed on sugar, preferring *Mangifera indica*, *Delonix regia*, *Cascabela thevetia*, *Senna siamea*, and *Cassia sieberiana* (Gouagna et al. 2010). In three-hour no-choice tests, the largest proportion of males (60%) fed on *Mangifera indica* and *Delonix regia*. The selection of a plant for a sugar source can make a big difference to the success of a mosquito. *Anopheles gambiae* in the laboratory in Kenya survived longer on *Manihot esculenta* (19.2 days for females, 26.3 days for males) than on *Ricinus communis* (7.8 days for females, 12.7 days for males) or on *Lantana camara* (2.9 days for females, 3.5 days for males; Gary & Foster 2004). Impoinvil et al. (2004) observed a similar difference for female *Anopheles gambiae* between *Ricinus communis* (mean survival 7.0 days) and *Lantana camara* (2.5 days) or *Ipomoea batatas* (2.3 days).

The fact that mosquitoes readily feed on sucrose solutions in the laboratory shows that they are able to detect it in the absence of any plant volatiles that might be associated with a sugar meal. Nonetheless, plants and their flowers or other parts produce phagomones associated with sugar feeding, including phenols, aldehydes, alcohols, ketones, and terpenes (Nyasembe & Torto 2014). These compounds are found in various plants and, as described above, mosquitoes are able to recognize many different plants as food sources. This suggests that mosquitoes detect blends of compounds to identify sugar sources, much as occurs during host seeking for blood meals. Careful experimentation with plant volatiles collected from *Parthenium hysterophorus* used electroantennograms of *Anopheles gambiae* coupled to gas chromatography-mass spectrometry to first identify active volatile compounds (Nyasembe et al. 2012). A blend of the following compounds in the right proportions was most attractive: D-limonene, hexanal, beta-pinene, (E)-beta-ocimene, (E)-linalool oxide, and (E)-beta-farnesene. Culicines were not attracted to the hexanal component, and higher concentrations of these compounds were repellent.

Plant phagomones that affect *Culex pipiens* were studied in a dual port olfactometer based on compounds from *Asclepias syriaca* (Otienoburu et al. 2012). Each of three compounds (benzaldehyde (E)-beta-ocimene, phenylacetaldehyde benzyl alcohol nonanal, and (E)-2-nonenal) were as attractive as a blend of them, which was as attractive as a pentane extraction of the entire flower. Similar work using a *Culex pipiens* Form Molestus and *Silene otites* model (Jhumur et al. 2006, 2007, 2008) identified phenyl acetaldehyde, vertrole, 2-methoxyphenol, phenyl ethyl alcohol, acetophenone, linalool oxide, linalool, and (Z)-3-hexenyl acetate as attractive. (Z)-3-hexenyl acetate was only attractive to females and is associated with green leaf odor. As with *Anopheles gambiae*, the pattern seems to be identification of combinations of compounds to identify sugar sources.

Sugar feeding is affected by and affects other behaviors. *Anopheles gambiae* females in an olfactometer preferred honey odors to host-like odors (dirty socks) early after emergence but switched to a preference for host odors following sugar-saturation and at five days post emergence (Foster & Takken 2004). In a more natural environment, females took human blood meals in preference to sugar post emergence and readily took sugar when a host was not available (Stone et al. 2012b). *Plasmodium falciparum* infection caused female *Anopheles gambiae* to take more sugar meals during oocyst formation, a stimulation that stopped after sporozoites were in the salivary glands (Nyasembe et al. 2014). This implies some remarkable coadaptation in which the parasite extends longevity and delays blood feeding during the extrinsic incubation period, then releases this influence when the parasite has reached an infectious stage and requires introduction to the human host in order to complete its life cycle. Grooming, at least by male *Aedes albopictus* in Réunion, increases following sugar feeding, presumably to remove sticky traces of nectar (Jacquet et al. 2012). The amino acids in nectar may have little influence on nutrition, as addition of amino acids that mimicked the nectar of *Lantana camara* did not affect longevity of male *Culex quinquefasciatus* and increased longevity of females by only 5% (Vrzal et al. 2010). Floral feeding by mosquitoes has only been shown to result in pollination for a few plants, including *Silene otites* by *Culex pipiens* and *Culiseta annulata* and *Platathera* (formerly *Habenaria*) *obtusata* by *Aedes* spp. (Thien 1969; Jhumur et al. 2006).

Conclusion

Many of the adaptations we observe in larval and adult mosquitoes are related to feeding. This general observation applies to most insects and is the result of a world in which they must find, compete for, and utilize nutritional resources that are often limited. Those limitations include the abundance of nutritional resources over space and time, with implications for movement, visual and chemical orientation mechanisms, and selection of food. Mosquitoes are among those insects that are generally adaptable in their selection of food but have clear preferences for certain larval foods, blood-meal hosts, and sugar sources. Several aspects of feeding are surprising, including the role of associative learning in selection, the highly varied and essential role of commensal bacteria throughout the alimentary canal and the salivary glands, and the importance of sugar feeding, which includes nectar and plant-tissue sources. Of course, the main reason that mosquitoes have received so much attention from entomologists is because of their blood-feeding habits. Blood feeding brings these insects into intimate contact with humans, sometimes resulting in transmission of diseases that have helped shape the demography of our species.

8. Excretion

Nature deserves praise and admiration for making machines so small.

Marcello Malpighi, cited from *Opera Posthuma* (1697) in Adelmann (1966)

As solids and liquids make their way from anterior to posterior in the alimentary canal, it is difficult to distinguish the point where absorption ends and excretion begins. Those larvae that filter feed are selective about where they feed but much less selective about what gets swept into their mouths. As a result, much of the material they ingest goes through the guts unchanged, and the process of excretion might be said to begin at the same time as ingestion. Adult female mosquitoes taking a blood meal must deal with a huge bulk of material in the midgut that contains a toxic quantity of nitrogenous compounds, with the challenges of detoxification starting soon after ingestion. At the other end of the process, the kidney-like Malpighian tubules and the rectum of the hindgut both absorb as well as excrete. Therefore, for convenience as much as for logic, we emphasize the morphology and functions of the hindgut, with acknowledgment that other parts of the gut contribute to excretion.

Morphology of Excretory Organs

Larva

The gastric ceca that evaginate from the anterior of the midgut contribute to osmotic balance (Clements 2000). Mosquito larvae have eight of them, and they point posteriorly. There are four cell types, the distributions of which vary among species. One odd feature of the gastric ceca is that they secrete a chitinous membrane across their opening or deeper inside. This membrane provides a barrier against the contents of the gut outside the peritrophic membrane.

Posterior to the midgut is the hindgut, or proctodeum. The five Malpighian tubules enter the hindgut at its most anterior end in the pyloric region. Each tubule consists of some 60 cells. A little less than 20% of the cells are smaller stellate cells, and the remainder are called the principal cells. Tubules extend anteriorly from the hindgut, then make a sharp bend to extend most of their lengths posteriorly. The posterior portion of the tubules becomes opaque in the fourth instar due to the deposition of granules in the cells. The pylorus is immediately behind the openings of the Malpighian tubules. It extends a ring of cuticular spines into the lumen and is surrounded by a sphincter-like arrangement of muscles. The next portion of the hindgut is the anterior intestine, or ileum. It is narrow, with deep infoldings of the cells on the lumenal side. The anterior intestine has a sharp bend, or even a loop, before it joins the bulbous rectum. Rectal papillae are lacking in the larvae, but there are deep infoldings. The rectum is usually uniform in structure along its length, but it is differentiated into two sections in saline-water species of *Aedes*. The short anal canal is at the posterior end of the rectum.

The hindgut, like the foregut, is ectodermal in embryonic origin and has a cuticular lining (except the Malpighian tubules). Two chitin synthase genes have been found in *Anopheles gambiae*, which are similar in sequence to those from *Aedes aegypti* (Zhang et al. 2012). Although these genes are expressed in all life stages, they should have the most functional significance following the molting process at each instar.

Adult

The tissues involved in excretion are in many places within the adult mosquito. The sources of neuropeptides that regulate water balance and excretion are widespread in the body. They are produced by neurons or endocrine cells in a preliminary form and then processed into smaller sizes and stored in the nervous system, midgut endocrine cells, neurosecretory cells of the central nervous system, and peripheral neurosecretory cells. The corpora cardiaca are the main storage organs for neuropeptides (Gäde 2004).

The Malpighian tubules and hindgut are the organs directly involved in excretion (Beyenbach 2003). Five long tubules open into the anterior of the hindgut, in the region of the pyloric valve. Each tubule is a blind tube, uniform in appearance along its length and well supplied with respiratory tracheoles. The Malpighian tubules of males are shorter than those of females. The tubules consist of principal and stellate cells. The principal cells are more numerous (about 80% of the cells) and larger than the stellate cells (Pietrantonio et al. 2000). The principal cells have a smooth basal membrane facing the hemolymph and an apical brush border membrane facing the lumen of the tubule. Principal cells of females are unusual in that they contain granules of calcium, magnesium, or potassium salts. Another unusual feature is the presence of elongated mitochondria that move in and out of the brush-border microvilli in accordance with metabolic needs of the hydronium ion pump.

Just as in the larva, the first long part of the hindgut is the anterior intestine. It is behind the pyloric region of the hindgut. In adults, the intestine forms an S-shaped bend, or even a loop. It leads into the bulbous rectum, which has six small papillae directed into the lumen of the anterior part (Snodgrass 1959). The pear-shaped rectal papillae are made up of about 10 cells each. They have a lumen that opens into the hemolymph in structures called rectal pads, reflecting their function in reabsorption for osmoregulation. The second, larger part of the hindgut is called the rectum and empties through a sphincter-controlled anus to the exterior of the mosquito. The circular muscles of the hindgut are more prominent than those of the midgut in *Aedes albopictus* (Vo et al. 2010). The cell wall of the rectum consists of columnar cells interspersed with cone-shaped endocrine cells (Kersch & Pietrantonio 2011).

Excretory Regulation

Water Regulation in Larval Mosquitoes

Like any aquatic organism, larval mosquitoes must actively manage the balance between the osmolarity of their interior hemolymph and that of the exterior water medium. The anal gills, more correctly called the anal papillae, project from the 10th larval segment around the anus as four soft and hollow fingers (Wigglesworth 1932). Length varies greatly according to species and osmotic conditions of the medium. Wigglesworth regarded these as prolapsed rectal papillae, whereas Harbach and Knight (1980) suggested they were derived from ancestral cerci of the now lost 11th segment. In either case they are the principal organ for water absorption in the larva, though their removal does not kill the larva. Water is also reabsorbed in the hindgut and possibly the gastric ceca, which are exposed to the fluid trapped between the peritrophic membrane and midgut epithelium. The gastric ceca and the anal papillae contain ATPases associated with active transport of sodium and potassium ions, a mechanism used in the gastric ceca, midgut, Malpighian tubules, and rectum to move water (Patrick et al. 2006; Tsujimoto et al. 2013). The integument of the larva is not waterproof, with greater permeability toward the posterior. The Malpighian tubules perform a similar function to that in the adult (see below), ridding the larva of excess water and waste solutes (Beyenbach 2003). In media with high osmolarity (e.g., salt water), the larva may increase levels of trehalose and proline (Pennington et al. 2003) in the hemolymph to decrease the pressure differential.

Water moves by diffusion into and out of cells through selective channels formed by aquaporin proteins (see the box below). Aquaporin expression is low in the gastric ceca, midgut, and hindgut, possibly to limit water intake from the gut contents into the hemolymph (Marusalin et al. 2012). Nonetheless, aquaporin 1 (AQP1) is present in the gastric ceca of *Anopheles gambiae* (Tsujimoto et al. 2013). Most aquaporin expression is in the Malpighian tubules (with four different kinds of aquaporin proteins) and the anal papillae (with two different kinds). This distribution corresponds to the functions of the organs, in that Malpighian tubules get water out of the larva, and the anal papillae take water in. Characteristically for aquaporins, mercury ions (Hg^{2+}) inhibit function and stop the anal papillae from absorbing water.

Acid balance between the environment and the internal tissues of the larva is another physiological challenge for mosquitoes related to water balance (Harrison 2001). Hormonally regulated secretion in the Malpighian tubules is the principal mechanism for achieving a suitable pH in the hemocoel. In acid water, expelling chloride ions electrically balances intake of bicarbonate to raise internal pH. In alkaline water, secretion of sodium ions balances intake of hydronium ions that lower pH. As in vertebrates, many of the solutes in cells and in intercellular spaces have a buffering effect that helps keep pH constant. Among these are proteins, phosphorus compounds, and organic acids. The rectum can have an important role also. *Aedes dorsalis* lives in alkaline, salty water and maintains an internal pH of 7.6 against environmental conditions of pH 10.5 by secreting bicarbonate ions in exchange for chloride ions in the rectum.

Other signaling systems that are less well described affect water transport in larvae. In *Aedes aegypti* and *Anopheles gambiae*, the medial neurosecretory cells of the brain produce the ovary ecdysteroidogenic hormone type I (OEH I; Brown & Cao 2001). The hormone is distributed in axons through the neurohemal organ associated with the aorta. The hormone does not affect ecdysteroid secretion in the larva, but it does affect the hindgut pyloric valve and the midgut cardiac valve and is probably involved in food and water transport.

Defecation in Larval Mosquitoes

Larval mosquitoes usually develop quickly and therefore must process a great deal of food for rapid growth. That food may come in discrete packets for predatory species or as a more or less constant stream of particles when larvae filter feed. Even

Aquaporins

Aquaporins are a class of major intrinsic proteins discovered in the 1990s following a search for a molecule that allowed water to pass through cellular membranes (Spring et al. 2009; Cohen 2012). Their discovery was a major one in cell biology, resulting in a Nobel Prize for Peter Agre, who later turned his attention to mosquitoes and malaria. The basic structure of the 28 kDa protein consists of six loops within the membrane separated by five interior loops. The aquaporins occur in sets of four within the membrane, and a pair of the interior loops of each join to form a selective channel through which water and other small molecules can travel. Charged amino acids at the entrance and middle of the channel keep out charged ions. Of the 200 aquaporins identified (Agre et al. 2002), a variety of types exist. Some are "strict," in that they only allow water molecules to pass, each one oriented to the selective shape and electrostatic pattern of the channel that blocks hydronium ($H^+ = H_3O^+$) and other ions. These strict aquaporins can allow a tremendous flow of water through the membrane. For example, human Aqp1 can handle 3 billion molecules of water per second. Other aquaporins (aquaglyceroporins) allow a larger variety of molecules to pass, including glycerol and other polyols, urea, and gases. All aquaporins are nonenergetic, depending on osmotic gradients to power the flow of molecules, and they are all poisoned by mercury ions (Hg^{2+}).

Among mosquitoes, seven aquaporins have been found in *Anopheles gambiae* (Liu et al. 2010) and six in *Aedes aegypti* (Drake et al. 2010). One or several of them are expressed in virtually all tissues. The expression of aquaporins varies from tissue to tissue and within a tissue according to physiological state. At least one aquaporin gene in *Anopheles gambiae* can be expressed in two different forms by splice variation. One of them, *Ag*AQP1A, is only expressed in the adult ovaries, whereas the other, *Ag*AQP1B, is in the midgut and Malpighian tubules of all life stages and in Johnston's organ at the base of the female adult's antennae.

Some of the characteristics of all six aquaporins of *Aedes aegypti* have been described (Drake et al. 2010; Drake et al. 2015):

AQP1: This is the homolog of *Drosophila melanogaster*'s *DRIP* gene. It is highly selective for water. This protein occurs in all tissues, but it is highest in the Malpighian tubules. It is upregulated in those organs beginning three hours after a blood meal.

AQP2: Also upregulated beginning three hours after a blood meal, this protein is present in all tissues except the fat body. It is most abundant in the Malpighian tubules, foregut, midgut, hindgut, and ovaries. The protein can transport glycerol, urea, and sugars as well as water.

AQP3: This is a homolog of *Drosophila melanogaster*'s *BIB*. It is weakly expressed in the Malpighian tubules, midgut, and ovaries. After a blood meal, it is upregulated in the ovaries and fat body.

AQP4: This protein is most highly expressed in the midgut and Malpighian tubules. It is downregulated in the former and upregulated in the latter after a blood meal.

AQP5: Only the ovaries have low expression of this aquaporin. It is upregulated following a blood meal.

AQP6: The thorax is the main location for this protein, and the highest expression is in the foregut.

Our understanding of the functions of mosquito aquaporins is far from complete, but they clearly interact with other regulatory systems to maintain water balance, transport small molecules in and out of cells, and accomplish urinary excretion. Variation between *Anopheles gambiae* and *Aedes aegypti* suggests that aquaporins reflect ancient selective forces and stochastic genetic events, raising the question of how other mosquitoes use these interesting proteins.

for the predatory species, the volume of food is normally so great that the midgut and hindgut are filled with material at any one time. Uric acid may be one of the principal nitrogenous wastes of mosquito larvae, though other forms of waste nitrogen are less well known (Clements 2000). Uric acid and toxic zinc, manganese, and phosphorus compounds are not only excreted but are also sequestered in various tissues.

Anopheles punctipennis, *Culex pipiens*, and *Aedes aegypti* eliminate their complete gut contents three times during development (Moll et al. 2001), implying that this pattern is general for mosquitoes. The first expulsion of gut contents is by the fourth instar during what is sometimes called the prepupal stage, when the larva has stopped feeding. The second time is by the pupa immediately after pupation. And the third time is by the pupa just before adult emergence, when the pupa is really a pharate adult. During these defecations, the gut contents are surrounded by the peritrophic membrane in a packet called the meconium, a term borrowed from the designation for fetal feces in mammals. These meconial defecations appear to have the function of literally cleaning the gut before further development, as they result in a major reduction of gut bacteria. Other insects typically have only one meconial defecation after adult emergence, suggesting that mosquitoes have sensitivity to the bacteria from a previous stage of development.

Urination and Diuresis in Adult Mosquitoes

A mosquito takes in 10 times its own hemolymph volume during a blood meal (Kersch & Pietrantonio 2011), creating a tremendous physiological challenge to expel excess water and ions, as well as concentrate the nutritional elements of the blood. The biochemical orchestration of this feat has been studied extensively in the past 20 years, providing an example of the complex systems that regulate processes in mosquitoes. Unlike mammals that use blood pressure across renal tubules to filter water and ions, mosquitoes and other insects have no significant hydrostatic pressure and must rely on osmotic gradients created by secretion of ions (Pietrantonio et al. 2005). The mechanisms for excretion are amazingly effective. Following a blood meal, the female can excrete fluid at a rate of 200 times its total extracellular fluid per day, or 60 nl/min, compared to only 0.09 nl/min during normal excretion in the male mosquito or a human renal turnover rate of 12.8 times per day (Beyenbach 2003). Much of the physiology has been worked out using *Aedes aegypti* as a model organism; therefore, in the discussion below, such processes can be attributed to that species if another is not mentioned.

Insects use the Malpighian tubules as their principal organ of diuresis (O'Donnell & Spring 2000), though in mosquitoes the salivary glands, midgut, and hindgut are also involved (Beyenbach 2003). The hemolymph, nervous system, and endocrine cells carry hormonal signaling substances, many of which are neuropeptides (see the box below). The tubules are regulated entirely through such signaling substances, lacking any connection to the nervous system. Although the structure of the tubules appears relatively similar along their length, their ends absorb solutes and fluid from the hemolymph, and the portions closer to the alimentary canal perform some selective reabsorption. Regulation of diuresis by hormones was first recognized from decapitation experiments in both larvae and adults (Wheelock et al. 1988). In other insects, these hormones may be synergistic, cooperative, or antagonistic. Most of what has been found in mosquitoes supports the cooperative category of regulation, in which cation and anion secretion are regulated separately and in coordination.

Diuretic peptides fall into two general categories (Beyenbach 2003). The leukokinin-like peptides (LLPs or kinins or leukokinins) are small (6–15 amino acids) peptides with conserved C-terminal pentapeptide sequences. The corticotropin-releasing factor-like diuretic peptides (CRF-DPs; Gäde 2004) are larger (30–46 amino acids), more variable in sequence, and often species specific in their activity. The common neurotransmitter serotonin (5-hydroxy tryptamine, or 5-HT) also has a role in diuresis among insects.

Mosquito LLPs are homologues of cockroach myotropic leukokinins (Pietrantonio et al. 2005). The mosquito LLPs (Kersch & Pietrantonio 2011) consist of three different peptides released from the abdominal ganglia and coded by the

Neuropeptides

Neuropeptides are among the many substances used by animals to regulate physiological processes (Predel et al. 2010). Like other hormones, these compounds travel through the hemolymph at low concentrations to a target tissue as a messenger for a process. The initial secretion of the hormone may be caused by an environmental factor or other condition, such as taking a blood meal. Once at the target, the neuropeptide binds to a receptor, which is often specific for that peptide. Receptors are either G protein-coupled receptors (GPCRs) or tyrosine kinase receptors. Binding results in release of other messengers with more local effect, such as cyclic adenosine monophosphate (cAMP), cyclic guanidine monophosphate (cGMP), or direct effects on ion channels. One or several more steps may be involved to achieve the physiological response. The importance of GPCRs is indicated by their abundant variation in the genome, *Anopheles gambiae* having 276 putative sequences for them (Lee & Pietrantonio 2003).

Genomic analyses have yielded an estimate of 35 neuropeptide genes in *Anopheles gambiae* and 22 in *Aedes aegypti*. However, a biochemical analysis using a technique that can sequence a peptide on its target tissue (matrix-assisted laser adsorption/desorption time-of-flight mass spectroscopy [MALDI-TOF MS]) identified a total of 73 different putative neuropeptides on nervous and midgut tissues. This complete list of neuropeptides in larval and adult *Aedes aegypti* was categorized according to probable function, as shown in the table below:

Functional category	Number of neuropeptides detected
Adipokinetic hormone	1
Allatostatin-A	7
Allatostatin-C	1
Allatotropin	1
Capability (capa)	6
Corazonin	1
Diuretic hormone	1
Ecdysis-triggering hormone	2
Extended FMRFamides (phenylalanine-methionine-arginine-phenylalanine amide)	11
SIFamide peptide (serine-isoleucine-phenylalanine amide)	2
Kinins	3
Myoinhibitory enzyme	5
Myosuppressin	1
Short neuropeptide F	6
Neuropeptide-like precursor 1	9
Pigment-dispensing hormone	1
Pyrokinins	5
Sulfakinins	2
Tachykinin-related peptides	5
Orcokinins	3

same gene (Lu et al. 2011). *Culex* mosquitoes have variations on these LLPs, including culekinin and the *Culex salinarius* diuresin (Pietrantonio et al. 2001). In *Aedes aegypti*, each bind to the same multiligand receptor protein, *Ae*KR. It is about 65.2 kDa in size and is one of the kinds of G protein-coupled receptors (GPCRs) in mosquitoes, at least 25 of which are known from *Anopheles gambiae* (Olsen et al. 2007). The receptor can be modified after it is formed, which changes its function.

*Ae*KR has been found in the head, stellate cells of the Malpighian tubules, posterior midgut, hindgut, and ovaries. In the hindgut, *Ae*KR is distributed in the endocrine cells and in a collar-like, torus-shaped arrangement around the opening of each rectal papilla into the hemolymph. The abundance of *Ae*KR corresponds to peak diuresis for 15 minutes after a blood meal and to post–peak diuresis for another 45 minutes, an hour during which 40% of post–blood meal urine is excreted. Knocking out *Ae*KR with dsRNA reduces diuresis by 50%, indicating roles of the LLPs and *Ae*KR in regulation. LLPs also cause hindgut contractions (Kersch & Pietrantonio 2011).

In the Malpighian tubules, LLPs increase diuresis and fluid secretion by stimulating the movement of chloride ions ($2Cl^-$) into the lumen (apical direction) through septate junctions between cells (paracellular transport) and through apical membranes of stellate cells (transcellular transport; O'Connor & Beyenbach 2001). The mechanism is a calcium cascade as the *Ae*KR activation by LLP releases phospholipase C, mediating secretion of inositol (1,4,5)-triphosphate (IP-3) and diglycerol, which in turn cause the release of calcium ions (Ca^{2+}) from intracellular stores and activation of calcium channels (Beyenbach 2003; Lu et al. 2011) through reorganization of cytoskeleton of the septate junction (Beyenbach 2012). Chloride permeability increases as a result and corresponds to a negative transepithelial voltage potential in the stellate cells (O'Donnell & Spring 2000), which can be neutralized by dinitrophenol or cyanide (Masia et al. 2000). Although the associated transepithelial voltage potential is maintained by an energy-utilizing process, the flow of chloride ions themselves is passive across that electrical potential (Pietrantonio et al. 2005). The action of LLPs on chloride transport is rapid but rapidly reversible, possibly through the action of metalloproteases (Gäde 2004). It works in conjunction with positive ion (potassium, sodium, and magnesium) active transport to maintain the proper electrical balance between the cytoplasm of the Malpighian tubule cells and the lumen of the tubule.

The proton pump that is the mechanism for active transport has received much study, in part thanks to the ability to study Malpighian tubules in vitro. One of the early indications that there was a proton pump came from experiments in which ethylisopropyl amiloride (EIPA), a specific inhibitor of Na^+/H^+ exchange channels, resulted in a rapid decrease in intracellular pH of principal cells and a 78% reduction in urine flow (Petzel et al. 2000). Stopping cellular respiration with dinitrophenol or cyanide also decreased sodium ion secretion and subsequent urine flow (Beyenbach & Masia 2002). The pump is powered with the help of V-type H^+-ATPase, known also from other kinds of insects (Weng et al. 2003), and by P-type Na^+/K^+ ATPase (Patrick et al. 2006). This enzyme is concentrated on the apical side of the principal cells in the small microvilli of the brush border. This is also the location of potassium cotransporter proteins (Piermarini et al. 2011). The distribution of the two kinds of ATPases is not uniform throughout the various tissues that secrete potassium ions. The hindgut only contains the P-type. In the Malpighian tubules, the relative concentrations of the two enzymes vary along the length of the organ (Patrick et al. 2006).

Although the apical membrane is the site of energy release by ATPases and of cotransporter proteins, the basolateral membrane (hemolymph side) has both energy-consuming and passive routes of entry for monovalent cations (Fig. 8.1). Abundant potassium channels there allow ions to reach equilibrium quickly (Scott et al. 2004). Potassium accounts for 64% of the ion transport across the basolateral membranes of the Malpighian tubule's principal cells. From genome sequencing, the potassium ion channels in the Malpighian tubules appear to be closely related to those in humans (Rouhier et al. 2014). They are called Kir channels (K^+ inward rectifier), and they are affected by the same toxins that affect Kir channels in humans, as well as by newly discovered molecules (Raphemot et al. 2013). The function of these channels is essential, and the toxins that inhibit potassium channels are mortal for the mosquito. The hindgut also has Kir channels—two kinds that are not the same as those in Malpighian tubules and one kind that is shared with the midgut. Variant splicing of gene products adds more variety to the kinds of Kir channel proteins, unlike vertebrates (Rouhier & Piermarini 2014). The apical membrane is also the site of cation proton antiporter proteins, common to bacteria and eukaryotes (Kang'ethe et al. 2007). Designated Na^+/K^+ exchanger proteins (NHE), they play a role, for example, in the conversion of yeast from salt intolerant to tolerant. Although located in the basolateral membrane, NHE proteins are powered by the V-type H^+-ATPases of the apical membrane.

The CRF-DPs regulate secretion of cations into the lumen of the Malpighian tubules. Unlike chloride ions, potassium (K^+) and sodium (Na^+) ions are actively transported across the principal cells (Beyenbach 2003), as described. Sodium predominates during and immediately after the blood meal, then potassium increases as the red blood cells are digested. The first indications of these hormones were from experiments with head tissues, identifying two natridiuretic hormones, interpreted as coupling blood feeding to diuresis (Petzel 1987). It was also recognized that one of the downstream effects of these hormones was an increase in concentrations of cyclic adenosine monophosphate (cAMP). Whether induced by hormones or drugs, increases in cAMP stimulated NaCl secretion in the urine and Na^+ absorption through the basolateral membranes (hemolymph side) of the principal cells. cAMP also causes a redistribution of beta-actin molecular fibrils within the principal cells, probably associated with redistribution of mitochondria and transporter molecules in preparation for urination (Karas et al. 2005).

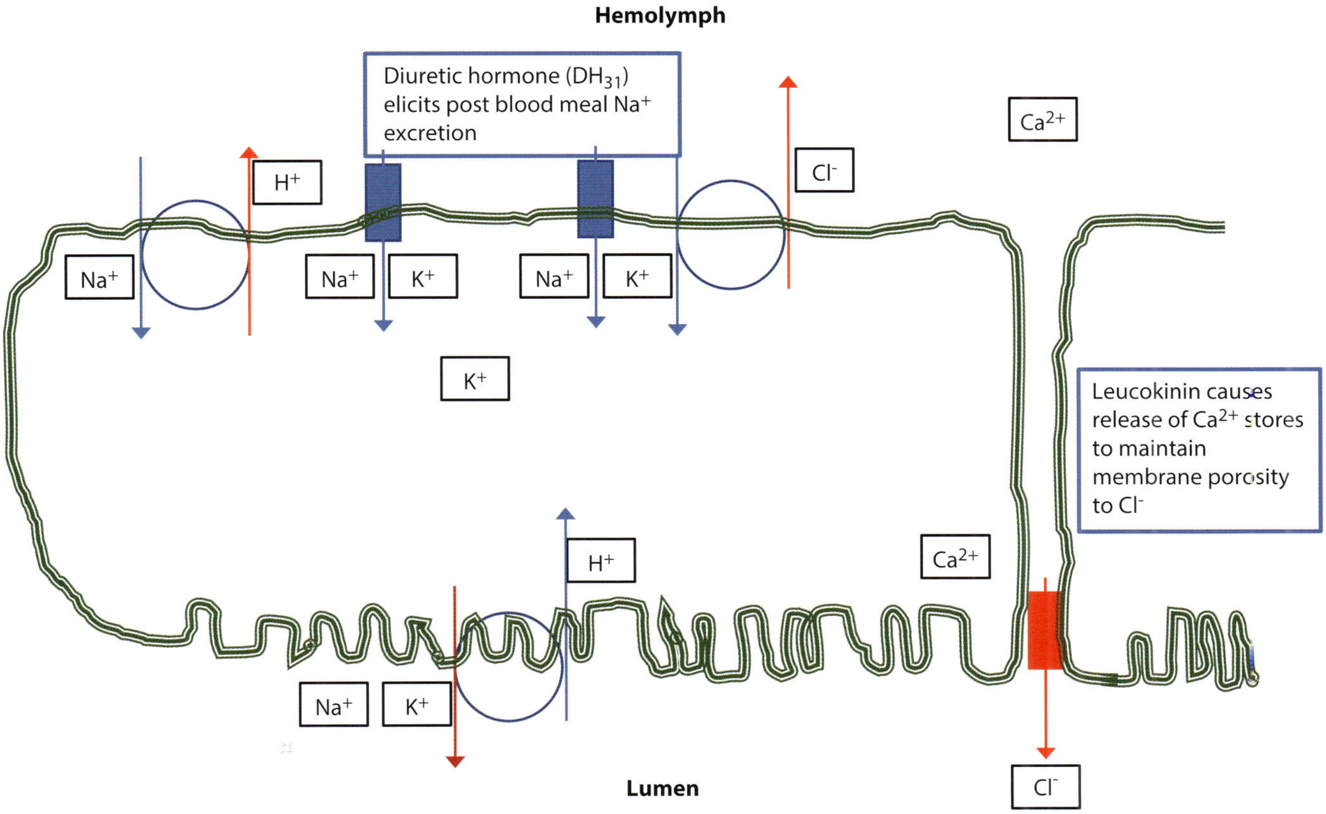

Figure 8.1. Diagrammatic summary of the mechanisms for ion flow in principal cells of the Malpighian tubules. *Blue*, flow in the intracellular direction; *red*, flow out of the cell; *blue circles*, ATP-energized pumps; *rectangles*, ion-specific pores in the cell wall or between cells. Sodium is pumped into the cell from the hemolymph in an exchange with hydronium ions, thus maintaining electrical balance. Sodium and potassium are also pumped into the cell in a chlorine exchange pump, and they flow into the cell through passive channels. On the apical side of the tubule epithelium (toward the lumen), the brush border is the site of a proton pump that pushes sodium and potassium out of the cell. Chlorine ions also flow by diffusion into the lumen through intercellular interstices. Redrawn from Scott et al. 2004, with modifications from Beyenbach 2003; Kwon et al. 2012

Two peptide hormones are involved in cation transport within the Malpighian tubules: the calcitonin-like DH_{44} and the corticotropin-releasing factor-like DH_{31} (Coast et al. 2005). DH_{31} has the same effect as cAMP, stimulating secretion of sodium-rich urine and depolarizing the basolateral membrane, showing that it is the mosquito natriuretic protein. DH_{31} also causes contraction of hindgut muscles by interacting with a receptor protein (GPCRCAL1) expressed on them (Kwon & Pietrantonio 2013). The secretion of DH_{31} is in proportion to the blood meal size and is further regulated by degradation in the hemolymph by GPCR kinases and by desensitization of the receptors by beta-arrestins (Kwon et al. 2012). Remarkably, the principal cells of the Malpighian tubules do not uniformly express the GPCRCAL1 receptor, being more abundant basally (toward the tip of the tubule, most distant from the alimentary canal). The receptor is concentrated on the basolateral membrane and occurs in patches of cells. There is also a gradient of expression of the NHE proteins, with NHE8, a general cation/proton exchanger, more abundant distally and NHE3, a specific sodium ion/proton exchanger, more abundant basally. As a result, it would be expected that DH_{31} would have relatively more effect on the basal portion of the tubule. Such specialization along the length of the tubule would not be expected, given its morphological uniformity.

DH_{44} has a more general effect on cation transport, facilitating both potassium and sodium transport (Jagge & Pietrantonio 2008). The receptor in the Malpighian tubules, GPRDIH1, is expressed in parallel with the urination cycle. Secretion of DH_{44} and expression of GPRDIH1 increase after eclosion and after blood meals. The temporal variation of the receptor, as well as the hormone, creates two different ways to regulate this urinary mechanism. DH_{44} has a wide variety of functions during the life cycle, working with LLPs in the immature stages to regulate pre-ecdysis behavior.

Serotonin (5-hydroxytryptamine, or 5-HT) is a common neurotransmitter, affecting feeding, neuroendocrine activity, development, learning, the midgut, and salivation (Lee & Pietrantonio 2003). In mosquito excretion, it influences the rhythm of hindgut contractions of the female adult and stimulates Malpighian tubule secretion in larvae and adults (Cady & Hagedorn 1999; Pietrantonio et al. 2001). The serotonin receptor in *Aedes aegypti* (*Aedes* 5-HT7 or pC5-HT_7) was the first neurohormone receptor cloned from any mosquito (Pietrantonio et al. 2001; Lee & Pietrantonio 2003). The receptor is much less sensitive to serotonin than *Ae*KR is to

LLPs; therefore, serotonin is not operating at the low concentrations typical of hormones. Despite the relatively high concentrations of serotonin required to stimulate the receptor in both *Aedes aegypti* and *Culex salinarius*, it is not a general receptor for neurotransmitters, as it is not stimulated by dopamine, octopamine, or tyramine (Lee & Pietrantonio 2003). *Aedes* 5-HT7 is located on the tracheoles of the Malpighian tubules and in the hindgut. Although serotonin stimulates tubule secretion, its effect may be through respiration rather than directly on ion transport.

The *Aedes* 5-HT7 receptor affects expression of aquaporins. These proteins form channels in cell membranes that, depending on the kind of aquaporin, allow only water or water and various small molecules to pass through. Within the Malpighian tubules, aquaporins are expressed in the tracheolar cells of *Aedes aegypti* (Duchesne et al. 2003; Spring et al. 2009). One of the proteins (AQP1, equivalent to DRIP in *Drosophila melanogaster*) is a "strict" aquaporin that only allows water molecules to pass. It is 26.11 kDa in size and similar to AQP4 in humans (Pietrantonio et al. 2000; Drake et al. 2010). Expression of these proteins causes fluid levels to increase along the lengths of the tracheoles, increasing air to other tissues. There are three other aquaporins expressed in the Malpighian tubules (AQP2, AQP4, and AQP5; Drake et al. 2015). All are upregulated for 24 hours following a blood meal, probably in preparation for the massive urination required, especially for paracellular water flow (Benoit et al. 2014). Aquaporins are present throughout the Malpighian tubules of *Anopheles gambiae*, more abundantly in the principal cells basally and more abundantly in the stellate cells distally (Tsujimoto et al. 2013), though other studies have found aquaporins only in the principal cells of this species (Liu et al. 2010). This species has a total of seven aquaporins, in comparison to six in *Aedes aegypti* (Drake et al. 2010; Liu et al. 2011). Inhibition of the four aquaporins in the Malpighian tubules of *Aedes aegypti* causes an 82% reduction in urination.

Other, less well-defined, processes also influence diuresis. A microRNA designated miR-1174, localized in the posterior midgut, is essential for several functions in *Aedes aegypti* and *Anopheles gambiae*, including sugar absorption, blood intake, and fluid excretion. This mRNA regulates expression of serine hydroxymethyl transferase enzyme. Without miR-1174, mosquitoes cannot efficiently secrete fluid from sugar or blood feeding, leading to death (Liu et al. 2014). CAPA (cardioacceleratory peptide) receptors with GPCR function and two of the six corresponding neuropeptides may also have a role in control of diuresis (Olsen et al. 2007). Another signaling system using the theme of the cardioexcitatory tetrapeptide, FMRFamide (phenylalanine-methionine-arginine-phenylalanine), has a corresponding receptor in *Anopheles gambiae* (Duttlinger et al. 2003).

Regulation of hemolymph circulation is probably closely related to diuresis in terms of transporting solutes to the Malpighian tubules and away from the midgut and hindgut. The hindgut of males and females has abundant receptors of a different GPCR-mediated hormone (Paluzzi et al. 2014). The hormone that activates the receptor is a glycoprotein cysteine knot-forming heterodimer of a type common in mammals. The receptors are also distributed in the midgut and Malpighian tubules, and they are sparsely distributed in juvenile stages. In the adult hindgut, activation of the receptors inhibits sodium and enhances potassium ion excretion. This function is important in the latter stages of blood digestion, when breakdown of the red blood cells causes an increase in potassium after sodium has already been excreted in the urine.

After a female mosquito expels excess water from a blood meal, it is faced with the opposite problem of avoiding water loss and replacing water as needed (Benoit & Denlinger 2010; Benoit et al. 2010a). Of course, this is the usual situation for males. Water is lost directly through the cuticle during respiration through the spiracles, as a part of defecation, and as a result of the secretion of urine. Mosquitoes lose water more easily than most blood-feeding arthropods. Of *Culex pipiens*, *Aedes aegypti*, and *Anopheles gambiae*, all consist of 66%–68% water when fully hydrated, but their water loss per hour and tolerance to dehydration varies. At the extremes, *Aedes aegypti* loses 2.6% of its water per hour and can tolerate a 36% loss. *Anopheles gambiae* is significantly more susceptible, losing 5.1% water per hour and tolerating only 29% loss. Dehydration and rehydration can be a costly process for the mosquito. *Culex pipiens* females produced fewer eggs and lost weight when exposed to repeated bouts of dehydration (25% loss of water) and rehydration (Benoit et al. 2010b). This effect was reversed if sugar was added to the diet, suggesting that the dehydration/rehydration process is energetic.

Defenses against water loss (Benoit & Denlinger 2010; Benoit et al. 2010a) include the development of a thicker layer of cuticular hydrocarbons following the first blood meal, as the stretching process of the abdomen increases permeability of the cuticle. Mosquitoes can also drink water directly, and it is possible that some of the plant tissue feeding by mosquitoes is more about replacing water than sugar (see Chapter 7). If a mosquito faces drying conditions, water stores in the hemolymph decrease more rapidly than water in tissues. Other potential mechanisms to survive longer under drying conditions include the production of heat-shock proteins to protect other proteins, production of antioxidants, bracing cells with cytoskeletal proteins, increases in trehalose concentration in the hemolymph, and sequestration of ions in the fat body. One impressive example of the effectiveness of these measures was demonstrated by shutting down the expression of a single heat-shock protein (Hsp) with RNAi in *Aedes aegypti* and observing a decrease in tolerance of water loss from 36% to either 28% (Hsp 70) or 26% (Hsp 90).

Not all mosquitoes use deposition of solutes in the hemolymph as a mechanism to protect against dehydration. *Culex pipiens* can only tolerate a 40% loss of water whether in or out of diapause, and it can only replace water by drinking (Benoit & Denlinger 2007). During diapause, females do not accumulate trehalose, sorbitol, or glycerol; instead, they deposit approximately twice as much cuticular hydrocarbon to achieve greater waterproofing of their integument. Even so, water loss rates at 25°C were impressive, with diapausing females losing 2.5%–2.75% of water per hour and nondiapausing females losing 3.5%–4% per hour. There is also evidence that mosquitoes can regulate the expression of aquaporins to reduce water loss under stress conditions, like diapause (Benoit et al. 2014). Expression of AQP1 was suppressed by *Anopheles gambiae* in a drying environment, and it survived longer when AQP1 expression was suppressed by RNAi (Liu et al. 2010).

Defecation in Adult Mosquitoes

As in most organisms, mosquitoes excrete solid and liquid waste through the same orifice. The physiological need of mosquitoes to eliminate large volumes of liquid following blood and, to some extent, sugar, meals tends to separate the process of urination from defecation. Mosquitoes defecate in small, sticky spots that are much less liquid than the urine following a blood meal. Defecation is controlled by the terminal abdominal ganglion providing nervous signals to the pyloric and anal sphincters, as well as the circular and longitudinal muscles of the hindgut (Caroci & Noriega 2003). Some level of endocrine control is also involved, as allatostatins cause relaxation of the pyloric valve (Pietrantonio et al. 2001). The presence of free amino acids stimulates longer retention of the hindgut contents, probably signaled through the midgut. If the formation of free amino acids is inhibited by addition of a trypsin inhibitor, the protein contents of the blood meal are rapidly excreted. Endocrine control is less well defined, but allatostatins (hormones that reduce secretion by the corpora allata) secreted by midgut endocrine cells may influence hindgut activity. That activity includes considerable absorption 18 hours after a blood meal, when the midgut contents are sealed off by the peritrophic membrane in *Culex quinquefasciatus* (Okuda et al. 2002).

The content of feces following a blood meal are rich in nitrogen, as the mosquito does not store nitrogen separate from egg-making nutritional requirements (Briegel 1986). The amino acid proline may serve this function as it builds up to higher concentrations if protein digestion is inhibited by enzymatic inhibitors or a lack of sugar to power metabolic processes (Pennington et al. 2003). Most of the feces consist of water, uric acid, urea, ammonium urate, free and bound amino acids (with histidine predominating), and hematin. The balance between urea and uric acid is affected by access to water and by the host species. For example, human blood has more indigestible nitrogenous components compared to rodent blood, so the proportion of fecal uric acid is greater following a blood meal on a human by *Aedes aegypti*. Uric acid is produced with the help of xanthine dehydrogenase, an enzyme containing molybdenum that can oxidize a variety of polycyclic compounds (von Dungern & Briegel 2001). Most uric acid is excreted soon after the blood meal and hematin follows later.

Other components of the blood meal also affect the contents of the feces. For example, DNA from *Brugia malayi* in *Aedes aegypti* or from *Plasmodium vivax* in *Anopheles stephensi* is detectable in the feces (Pilotte et al. 2016). *Anopheles stephensi* and *Aedes triseriatus* in the laboratory defecate infectious hepatitis B virus particles for three days following infected blood meals, even though none of the mosquito tissues become infected (Blow et al. 2002).

Mosquitoes are naturally exposed to many toxins in the environment, which was once referred to as the "xenobiotic assault" (Strode et al. 2006). Mosquitoes respond to this assault with metabolic processes to break the toxic molecules into nontoxic, hydrophilic components that are presumably excreted in the feces. Study of this process has been stimulated by attempts to understand insecticide resistance. The primary enzymes involved are cytochrome P450s, glutathione transferases, carboxylesterases, and cholinesterases, with many variants expressed differentially by life stages, males, and females. *Aedes aegypti* has a more rapidly evolving and varied set of detoxification enzymes compared to *Anopheles gambiae* (Strode et al. 2008). Induction of these metabolic processes by one compound can influence the mosquito's tolerance to other compounds, which is an interesting parallel to the way bacteria can increase the immune responses to parasites (Chapter 7). One example was an increase in resistance to insecticides (imidacloprid and permethrin) when larvae of *Aedes aegypti* were exposed to glyphosate herbicide or benzpyrene polyaromatic hydrocarbon (Riaz et al. 2009). The pre-exposure induced increases in P450s and esterases. In another study, fluoranthene polyaromatic hydrocarbon induced both P450s, and glutathione S-transferases and copper induced P450s, resulting in greater resistance to permethrin and temephos insecticides (Poupardin et al. 2008).

Conclusion

Mosquitoes' process of excretion is one of those subjects that has greatly benefited from proteomics and genomics. Proteomics have provided the means to sequence amino acids on receptor sites, definitively identifying regulatory messengers. Genomics has led to identification of receptors, studies of gene expression, and comparison of processes with other organisms. Using such techniques, we now understand the functioning of the Malpighian tubules in great depth, introducing a fascinating aspect of the lives of mosquitoes that involves elegant biochemical regulation in which both messengers and receptors are expressed according to functions at a place and time within the mosquitoes' bodies. Perhaps these processes have parallels in other insects and, in many cases, with most other organisms of any kind. Mosquitoes face a challenge following a blood meal, when huge quantities of liquid and excess ions must be excreted quickly by individual females to increase chances of survival. These feats of excretion are impressive anywhere in the animal kingdom, though shared by other blood-feeding arthropods. There is much more to learn about excretion in mosquitoes, as most of the work has concentrated on diuresis in adult females of only two species.

9. Copulation and Insemination

They swarm through our windows at nightfall, filling our barracks with an evil humming, to celebrate their bacchanalian orgies.

Emilio A. Goeldi, quoted by Mitchell and Dupree (1907)

Morphology

The morphology of sex organs in mosquitoes has been said to follow the pattern of other insects, in which variation in structure is driven by an evolutionary process that selects for differences to keep species separate and maximize the chances of successful insemination (Arnqvist 1998). Generally, monandrous species have a tendency toward "lock-and-key" systems wherein the genitalia between different species simply do not match well mechanically. In contrast, polyandrous species often have additional specialization to assure that a given copulation event results in successful insemination. Corresponding to this contrast, insects that mate more than once have a faster rate of divergence in organs related to copulation than in other organs. Mosquitoes appear somewhere in the middle of this spectrum, having groups with distinctive, species-specific differences in male (Khalin 2009; Motoki et al. 2009b; Yadav et al. 2014) and female (Reinert 2000a, 2000h, 2010k) genitalia that presumably contribute to the prevention of hybridization. Possibly because male mosquitoes mate multiple times and females usually mate once, there is selection for successful insemination on the part of the females (South & Catteruccia 2016), as well as selection to prevent wasted energy and semen on hybridization by the males. It is possible that the association of genitalia diversity with strong selection is false, in that taxonomists have often started with the assumption that a difference in genitalic structure is equivalent to a species difference (Huber 2003).

The following discussion of internal and external male and female morphology presents a background for understanding the mating process. It also points out representative variation across the family.

Internal Male

Illustrated in Figure 9.1 and proceeding from the anterior to the posterior, the internal organs of male reproduction (Snodgrass 1959) begin with the paired testes located in the seventh abdominal segment. Each testicle is a bundle of sheathed tubes, in which spermatozoa are formed and mature. The testes are pyriform in shape and often coated in fat cells (Sinclair et al. 2007). The sperm go through three general phases of development in the testes, progressing from growth in the anterior, to maturation in the middle, and to transformation into spermatozoa at the posterior end (Pitts et al. 2014).

The spermatozoa drain into a narrow testicular duct, sometimes called the vas deferens in imitation of vertebrate anatomy. These tubes are thin walled anteriorly and surrounded by circular musculature posteriorly. In culicines, the testicular ducts either expand at their posterior ends to form a pair of seminal vesicles, or these vesicles are united into a single sac. The posterior end of the seminal vesicle(s) joins the short ejaculatory duct, which ends in the genital pore (gonopore) within the intromission organ of the mosquito, the aedeagus. In anophelines, the sperm are stored in the testes rather than in seminal vesicles (Klowden 2007). In *Anopheles gambiae*, the testis is organized into distinct zones from anterior to posterior representing growth, maturation, transformation, and storage: germ cells, spermatogonia, spermatocytes, spermatids, and spermatozoa (Pitts et al. 2014). The number of sporocysts decrease and the portion of the testes dedicated to storage increases with age in *Anopheles gambiae* and *stephensi* (Scott et al. 2008). The large male accessory glands empty into the ejaculatory duct at its anterior end. These glands are surrounded by transverse muscle bands and at least sometimes differentiated into anterior and posterior zones. They vary in

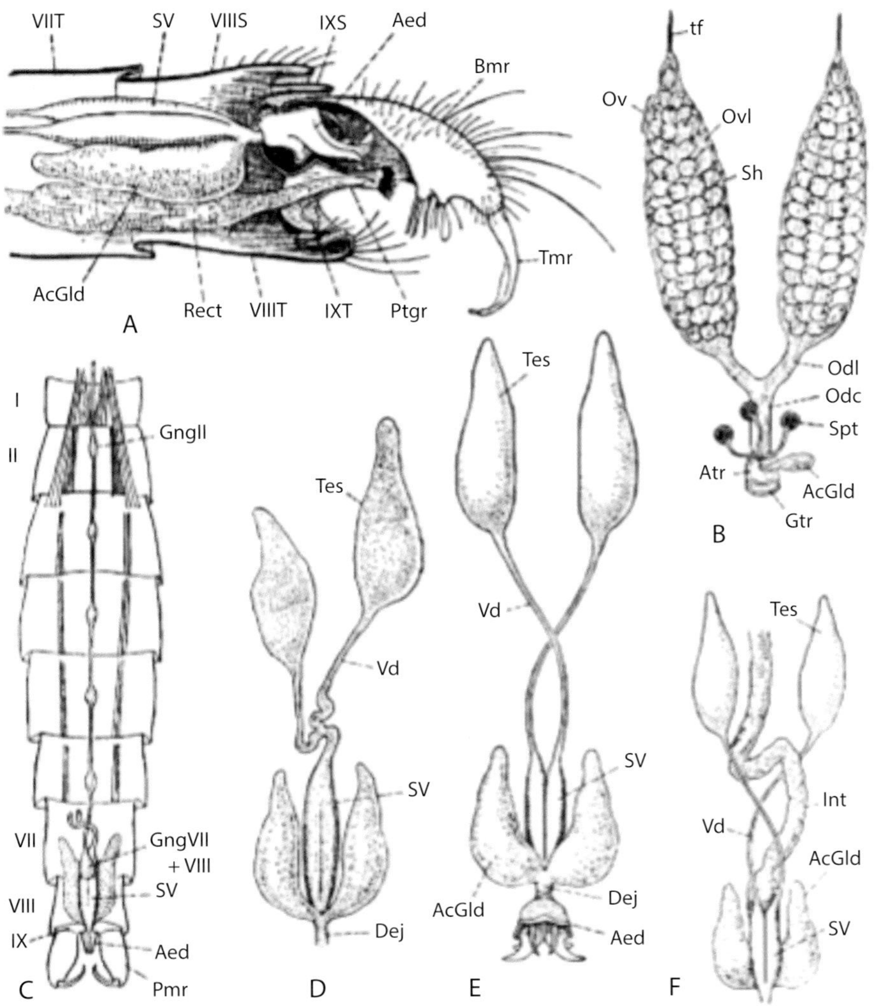

Figure 9.1. Internal mosquito reproductive system. A, sagittal section of terminal segments of male *Culex quinquefasciatus*: VIIT = seventh tergum; SV = seminal vesicle; VIIIS = eighth sternite; IXS = ninth sternite; Aed = aedeagus; Bmr = basimere or gonocoxite; Tmr = telomere or gonostylus; Ptgr = proctiger; IXT = ninth tergite; VIIIT = eighth tergite; Rect = rectum; AcGld = accessory gland. B, female reproductive system: Atr = atrium; Ov = ovary; tf = terminal filament; Ovl = ovariole; Sh = sheath of ovary; Odl = lateral oviduct; Odc = common oviduct; Spt = spermatheca; AcGld = accessory gland. C, male *Aedes aegypti* reproductive system and nerves: IX = ninth segment; VIII = eighth segment; VII = seventh segment; II = second segment; I = first segment; GngII = second abdominal ganglion; GngVII + VIII = terminal abdominal ganglion. D, *Aedes aegypti* male ventral view of reproductive organs: Dej = ejaculatory duct; Vd = vas deferens; Tes = testis. E, male reproductive organs of *Culex quinquefasciatus* after genital rotation. F, male reproductive organs of *Culex quinquefasciatus* in relation to hindgut. Snodgrass 1959

shape in different species, being globular, tubular, or pyriform. The ejaculatory duct itself is composed of three layers: an inner cuticular lining, a thin epithelial cellular layer, and a layer of transverse muscle fibers (Spielman 1964).

One of the peculiarities of mosquito male genitalia is that they rotate sometime after emergence and before mating, an event that occurs in all Culicidae and a few other dipteran families (Wood 1991). The rotation allows mosquitoes to mate when the position is venter to venter or end-to-end, facing in opposite directions (Oliva et al. 2014). Rotation may also protect the morphologically ventral parts of the genitalia. Rotation of the genitalia at the junction of the seventh and eighth abdominal segments causes the testicular ducts to cross over each other (e.g., in *Culex pipiens*) or twists the male reproductive tract at their junction (e.g., in *Aedes aegypti*). The testes remain ventral to the alimentary canal, and the remaining organs to the posterior assume a position dorsal to the alimentary canal following rotation. The rotation is accomplished through muscular action and can occur in a clockwise or counterclockwise direction (Oliva et al. 2011). The time of rotation postemergence varies a great deal between species, with some examples as follow: *Aedes albopictus*, 11–25 hours; *Verrallina iriomotensis*, 12 hours; *Aedes aegypti*, 18–24 hours; *Aedes taeniorhynchus*, 30 hours; *Aedes atriisimilis*, 40 hours; *Aedes provocans*, 96 hours; *Anopheles arabiensis*, *gambiae*, and *stephensi*, greater than 24 hours; *Culex tritaeniorhynchus* and *quinquefasciatus*, 19 hours; and *Culiseta inornata*, 6–12 hours. Temperature affects the timing of rotation, requiring 30 hours at 29°C and 48 hours at 18°C for *Anopheles funestus* (Dahan & Koekemoer 2014). The final rotation varies for this species between 135° and 180°. The time from emergence to male genitalia rotation may be much more plastic than usually assumed. A 125-generation-old laboratory colony of *Anopheles arabiensis* self-selected for a much shorter time to rotation than wild individuals from Sudan (Oliva et al. 2011). The laboratory strain rotated genitalia as early as 3.5 hours after emergence, and 96% of individuals had rotated by 17 hours, when none of the wild males had rotated yet. Selection for early rotation in the laboratory seems logical because the earlier a male is ready to mate, the more likely it will mate with monandrous females. On the other hand, it is not clear why the same selection pressure does not occur in the field.

A mosquito sperm consists of a head and a flagellum (Degner & Harrington 2016). The head is the same width as the flagellum, 0.5–0.6 μm in *Aedes aegypti*. Internally, the flagellum contains two mitochondrial elements and the axoneme.

The axoneme has tubules surrounding a central single fiber. In cross section, there are two concentric circles of tubules, each with nine tubules. Length of the flagellum is variable between species and sometimes within species. On the short side, *Anopheles albimanus* sperm are about 100 μm long, compared to 570 μm for *Culiseta inornata*.

External Male

Terminology of dipteran male genitalia can be frustratingly variable. The Glossary and Section II of this volume present a standardized system largely based on Harbach and Knight (1980). That reference includes detailed comparisons of terminology for other Diptera and for Culicidae. Despite the variability of terms, a consideration of the general dipteran pattern of male genitalia helps understand what is common between mosquitoes and what is different from related Diptera. The ground plan (hypothesis describing the most probable ancestral condition) for male genitalia can be divided into six parts (Wood 1991). These parts are the epandrium (morphological tergum of the ninth abdominal segment), the hypandrium (morphological sternum of the ninth abdominal segment), paired appendage-like gonocoxopodites on either side of and posterior to the intromission organ, a median tubular aedeagus (the intromission organ), parameres that extend the aedeagus through a lever action, and the proctiger, which consists of the remainder of the genitalia posterior to the ninth segment and consisting of elements of segments X and XI. The gonocoxopodites have a basal, fingerlike gonocoxite and a subterminal, clawlike gonostylus. The aedeagus is formed from paired parts that surround a sclerotized extension (phallus) of the internal ejaculatory duct. Most Diptera have a sperm pump between the phallus and the ejaculatory duct, but this typically dipteran adaptation is absent in mosquitoes and other members of the infraorder Culicomorpha, and the phallus is not sclerotized (Sinclair et al. 2007). The membranous intromission organ of mosquitoes may not be homologous to the aedeagus of other Diptera but is instead formed by elements of the parameres. Mosquitoes are also different in that the epandrium and hypandrium are fused to form a ring, a feature that is evident when preparing culicid male genitalia for morphological examination.

Mosquito taxonomists have used male genitalia extensively for species determination because in some groups the characters are consistent and distinctive. There are standard methods for preparation and examination of male genitalia, which is useful for comparisons but tends to obscure functional features of these complex organs. Standard mounts usually use potassium hydroxide-cleared specimens to see cuticular features more clearly, but this eliminates most of the soft tissues that are important in the living insect. Standard mounts also tend to obscure the three-dimensional structure of the male genitalia (Kirti 2014a, 2014b). Of course, the variety seen between species suggests variety in function, but even without a careful comparison of different species, the details of one should help understand the family.

Spielman performed what must be considered a classic study of mating by *Aedes aegypti* (Spielman 1964). Observing hundreds of matings, flash freezing pairs while mating (in copula) at different stages of the rapid mating process, and carefully examining the morphology, he produced a detailed account of the morphology and function of male and female mosquitoes during copulation. There are three basic parts to the genitalia, each bilateral and each supported by a separate apodeme (ingrowth of cuticle forming a tendon-like muscular attachment).

Starting from the most lateral position, the first part of the genitalia is the gonocoxopodite (= clasper), which is the same as the gonopod of the dipteran ground plan. It is the large, fingerlike appendage that is clearly oriented for clasping toward the center. In modern terminology (Harbach & Knight 1980) the appendage consists of (from distal to basal ends) a small gonostylar claw set into a socket within the large, clawlike gonostylus. The gonostylus articulates with the roughly cone-shaped gonocoxite. The mesal surface of the gonocoxite has characteristic and sometimes very distinctive, large accessory setae innervated from the terminal abdominal ganglion. The lateral surface usually has long setae directed posteriorly. The base of the dorsal surface of the gonocoxite extends into an apodeme that has a complex three-dimensional structure and that twists to extend horizontally. At the base of this apodeme on the dorsal side is a ridge that articulates with the apodeme of the aedeagus.

Moving medially, the next cuticular part of the male genitalia is the paraproct (= claspette). It consists of a stemlike basal portion, an apodeme, capped by a roughly triangular or leaf-shaped plate. One corner of the plate extends posteriorly, forming a hook visible beyond the membranous structures in the intact male. The corner posterior and medial to the hook forms a lobe that articulates with the end of the aedeagus (= paramere). The third corner of the plate hinges with the apodeme. Three-dimensional examination of a range of *Aedes* species has shown considerable variation in the shape of the paraproct plate (Khalin 2006, 2009).

The functional aedeagus is in the middle of the genitalia complex. It is the intromission organ of the male during mating. Viewed from the ventral side, it consists of two crescent-shaped lobes tipped with a row of teeth on the posterior end of each. The ventral edges of the lobes are convex and not connected to each other. In contrast, the dorsal edges are joined by a membrane associated with the ejaculatory duct. The actual genital opening is near the base of the aedeagus and is not visible from the exterior. The apodeme of the aedeagus emerges from each lobe's dorsal side and extends at an angle (viewed from the ventral side). It is about as long as the aedeagus itself. The apodeme articulates with a lobe of the gonocoxopodite apodeme.

The sternum and tergum of the ninth abdominal segment (IX-S and IX-Te) form a ring at the base of the genital complex. This ring is important for mechanical stabilization of the genitalia, and the lobes of the ninth tergum (IX-TL) (which faces ventrally because of the genital rotation) cover the anal opening. At least in *Aedes albopictus*, the shape of IX-TL can vary between populations (Kaur 2014). Two membranous structures emerge from the ninth segment. The proctiger represents segments X, XI, and the telson. It is continuous with the ventral margin of the aedeagus and the dorsal margin of the anus. Seen from a ventral view, the proctiger obscures the bases of the paraprocts and the entire aedeagus. The second membranous structure arising from the ninth segment is the aedeagal pouch, which is a pocket in which the aedeagus rests.

Seven muscles actuate each side of the male genitalia. The gonostylus has an extensor and a retractor in the body of the gonocoxite. A retractor of the gonocoxite extends from IX-Te to the gonocoxopodite apodeme, capable of moving the gonocoxite ventrally and balanced by a muscle from IX-S that can move the gonocoxite dorsally. Another muscle attached to IX-Te goes to the distal end of the aedeagal apodeme; therefore, its retraction tends to extend the aedeagus through a levering action and pulls the aedeagal apodeme against the gonocoxopodite's apodeme, exerting lateral pressure against the base of the gonocoxopodite. A muscle that retracts the lateral sclerite of the paraproct causes the sclerite to rub against the gonocoxopodite apodeme. Finally, there is a muscle between the midpoint of the aedeagal apodeme and the gonocoxopodite apodeme.

Internal Female

The portions of the female reproductive tract involved in copulation and insemination (Snodgrass 1959) are posterior to the common oviduct where it ends in the gonopore. The gonopore, located in the eighth abdominal segment, leads into the atrium, or genital chamber. The atrium is an invagination of the body wall and is not developmentally part of the reproductive tract. The opening of the atrium outside the mosquito is called the gonotreme. The single female accessory gland empties into the atrium near the place where the spermathecal ducts are attached. The function of the female accessory gland is unknown (Shaw et al. 2015).

In culicines, represented by *Aedes aegypti* (Pascini et al. 2012), there are usually two spermathecal ducts, one leading to a single larger (100 μm diameter) spermatheca and the other branching into two smaller (75 μm diameter) spermathecae, for a total of three spermathecae. Sphincters between each duct and its spermatheca are sometimes referred to as spermathecal pumps. Glandular cells concentrated on the distal portion of the spermatheca and on the duct form a mass almost as large as the spermatheca itself. These glandular cells penetrate the spermatheca through cuticular tubules, distributing substances necessary for the survival of the sperm. The spermathecae are cuticular, forming a continuous layer with the ducts. Invaginations in the cuticle before mating may represent how the spermathecae expand. A layer of flat nucleated epithelial cells coats the spermatheca, and columnar epithelial cells surround the ducts. The number of spermathecae is variable in culicines. Among those with three spermathecae are *Aedes polynesiensis* and *albopictus* (Oliva et al. 2011). *Uranotaenia* always has a single spermatheca (Peyton 1977). *Mimomyia* has either one or three (Belkin 1962), and *Tripteroides* also usually has either one (most in the Oriental biogeographic region) or three (most in the Australasian region), but one group can have two (Mattingly 1980).

Anophelines, represented by *Anopheles aquasalis* (Pascini et al. 2013) and *Anopheles gambiae* (Shaw et al. 2014), have a single spermatheca and therefore a single duct opening into the atrium. The spermatheca has two layers, consisting of a thick cuticular layer in contact with the sperm and an outer thin epithelial layer below glandular and fat body cells. The posterior third of a spermatheca has pores 2–3 μm in diameter that contain projections of both the epithelial and glandular cells. The fat body cells have large nuclei and rough endoplasmic reticulum and are filled with glycogen granules and lipid droplets, presumably to provide nutrition for the sperm. Epithelial cells probably secrete the cuticle of the spermatheca. The spermathecal volume capacity is limited compared to *Aedes aegypti* since the diameter of the single spermatheca of *Anopheles aquasalis* is 82 μm. Spermathecal ducts have a star-shaped cross section and are thicker and shorter than in *Aedes aegypti* and muscled by helical fibers. The size of spermathecae is said to be related to the length of the sperm in that species (Klowden & Chambers 2004).

A major difference between anophelines and at least some culicines is the structure of the atrium itself (Degner & Harrington 2016). In anophelines, the atrium is expanded into the organ that receives the full load of semen from the male. Culicines have a more complicated structure in which the atrium is a narrow tube leading to a large bursa copulatrix that receives the full load of semen. The bursa copulatrix empties through a thickened pad of muscular and connective tissue, known as the spermathecal eminence (Spielman 1964), into an expanded portion of the common oviduct called the spermathecal vestibule. There is a group of hairs at the opening called the ventral tuft. The spermathecal ducts empty into the vestibule anterior to the duct leading to the female accessory gland.

External Female

The external genitalia of female mosquitoes have not received the same attention from taxonomists as the male genitalia. John Reinert dedicated much of his career to understanding variation in female genitalia in aedines (e.g., Reinert 2000a) and based some of his taxonomic decisions on that morphology (e.g., Reinert 2000h). The account here is based mainly on Reinert's (2000e) and Spielman's (1964) description of *Aedes aegypti*, as well as interpretations of morphology described in Harbach and Knight (1980).

Female genitalia do not rotate following emergence, so morphological dorsal and ventral orientations are the same as those observed. The most conspicuous feature of aedine female genitalia are the cerci, which are the modified appendages of the eleventh segment. They appear as single segmented, scaled lobes on the dorsal side of the genitalia. Although not unique to aedines, the cerci and the abdominal segment XI protrude from the abdomen when the mosquito is at rest, giving aedines the "pointed abdomen" that separates them so conveniently from *Culex*. Viewed from a ventral aspect, there is a single lobe of cuticle extending from the eleventh segment between the cerci. This is the postgenital lobe. The genital opening lies immediately below the postgenital lobe, and the anus is immediately dorsal to it.

The genital opening is surrounded by the vaginal lips, which are a collar formed by dorsal and ventral straps, hinged laterally. The outer edges of these structures are folded over to form a liplike entrance. They are not part of one of the abdominal segments but part of the genital tract. At rest, the dorsal and ventral parts press together, closing the entrance. The vaginal lips are hidden by terminal abdominal terga and sterna. They are continuous with the bursa copulatrix and other structures described above. With proper preparation, the spermathecal eminence, accessory gland duct base, and sclerites associated with the bursa are visible and can be used to separate some taxa. The insula is another morphological feature, consisting of a short tongue projecting posteriorly from the lower vaginal lip.

Anatomical Mechanics of Copulation

In other insects, the presence of multiple spermathecae is associated with separation of sperm from different males, but mosquitoes usually mate only once (Yuval 2006). In those culicines with three spermathecae, it is rare that all of them are filled. In *Aedes aegypti*, *albopictus*, and *polynesiensis*, usually the largest and one of the smaller spermathecae are filled (Oliva et al. 2014), though more individuals have all three spermathecae filled in the laboratory, where multiple matings may be common. Only about two-thirds of the sperm injected by the male make it to the spermathecae. As described above, *Anopheles* have only one spermatheca, as do the families most closely related to mosquitoes, Chaoboridae and Corethrellidae. The majority of mosquitoes mate on the wing, or at least initiate sexual contact while flying. This habit appears to be associated with rapid mating and is aided by the relatively large diameter and musculature of the ejaculatory duct (Downes 1969).

During mating, retraction of the muscle attached to the gonocoxopodite apodeme straightens out the apodeme and extends the gonocoxopodite caudally and dorsally (Spielman 1964). That pulls the aedeagal apodeme, extending the aedeagus posteriorly and everting the aedeagal pouch. Retraction of the muscle between the ninth segment and the tip of the aedeagal apodeme holds it against the gonocoxopodite and assures that the aedeagus remains in an extended position with the two lobes separated and the long axis of the organ now perpendicular to the long axis of the male. The aedeagus pushes against the lateral sclerite of the paraproct, pushing the hooklike corner further out. The lateral sclerite of the paraproct is pressed against the gonocoxopodite apodeme by the retractor muscle of the paraproct, stabilizing the paraproct and pressing the anal cone more tightly against the ventral side of the erect aedeagus. As the muscles attached to the ninth segment retract, the entire genital complex swings to a more dorsal position. Unlike some insects, the aedeagus does not extend itself deeply into the female. The membranous anal cone and everted aedeagal pouch seal the aedeagus to the female genital opening while the gonocoxopodites grip the female cerci tightly. The paraprocts extend caudally and hold the anal cone against the female's genital tract. During semen transfer to the bursa copulatrix, the teeth at the end of the aedeagus are in contact with the brush of setae on the spermathecal eminence. The female atrium expands to receive the copulatory apparatus and quickly contracts again about 30 seconds after mating (*Aedes aegypti*; Oliva et al. 2014).

Male Accessory Gland and Seminal Fluid Protein Substances

Substances in Semen

The functions and contents of the male accessory glands (MAGs) have received considerable attention. They produce most of the nonsperm components of the semen, including dozens of different proteins. Other nonsperm components of the semen are contributed by seminal vesicles (in culicines), the ejaculatory duct, and the testes themselves (Avila et al. 2011). Combined with the MAG substances in semen, these other components have been collectively called seminal fluid proteins (SFPs). Many of the proteins and functions have been shown for mosquitoes following detailed studies on *Drosophila melanogaster* in the 1980s (Chapman 2009). Seminal fluids affect the mated female's cell structure, circadian rhythm, sperm storage, flight, response to hosts, blood digestion, mobilization of nutritional resources, ovarian development, antimicrobial defensive substances integrated into the eggs, fertility, oviposition, and mating refractoriness (Gillott 2003; Avila et al. 2011; Klowden 2007; Scott et al. 2008). Studies in which male accessory glands were transplanted into the female's hemocoel were conducted in the 1960s and 1970s, but the ability to extract, amplify, and sequence either the messenger RNA or the proteins (by mass spectrographic identification of amino acids) themselves has greatly advanced this field of study. Function has been discovered by genomic comparison, knockout mutants, posttranscriptional silencing (inhibitory RNA), or overexpression. In general, mated females have increased egg development, higher oviposition rates, higher blood digestion rates, larger blood meal size and less frequent feeding, reduced tendency to mate again, and reduced tendency to take flight (Sirot et al. 2011).

Multiple studies have inventoried the MAG proteins by amplification of messenger RNA in the glands. Comparison of these studies is challenging because the number of proteins detected depends on technique and probably the exact stage at which the insect was sampled. Almost all these studies have been conducted in *Anopheles gambiae* or *Aedes aegypti*. In *Anopheles gambiae*, 46 proteins were identified, 25% of which were unique to that species and 40% of which were shared with *Drosophila melanogaster* (Dottorini et al. 2007). These proteins included a sperm chemoattractant, a protein that reduces mating by females, and a series of functional proteins involved in lipid transport and metabolism, serine protease inhibition, antimicrobial action, carboxylesterases, chaperonins, and redox proteins. The protein sequences showed a great rate of evolutionary change, possibly greater than in any other mosquito tissue (Chapman 2009). A more recent study (Baldini et al. 2012) found 121 MAG proteins, 26% of which were unique to the glands (Table 9.1).

Aedes aegypti MAGs produced 198 proteins, 36% of which were like other proteins in the mosquito; 31% were similar in function to proteins in *Culex quinquefasciatus*; 30%, similar to proteins in *Anopheles gambiae*; and 21%, similar to proteins in human semen (Boes et al. 2014), though the sequences of the proteins were quite different between the species (Sirot et al. 2008). Although 62% of the proteins could not be assigned to a function, the others were associated with cytoskeletal elements, the extracellular matrix, oxidation, metabolism, immunity, hormones, protein regulation, signal transduction, and synthesis. Another study found 139 protein sequences, significantly including lectins and odorant-binding proteins (Baldini et al. 2012). One of the odorant-binding proteins, OBP 22, is sensitive to ligands with two aromatic rings connected by a rigid chain and is found on external sense organs of the mosquito as well as in the semen (Li et al. 2008). The total number of proteins in the MAGs and seminal vesicles has been estimated at 250, 63 of which were secretory and only 53 of which were found among the 93 seminal fluid proteins (SFPs; Sirot et al. 2008, 2011). That study identified major functional compounds as proteases (25% of the proteins), protease inhibitors (11% of the proteins), chaperonins, cysteine-rich secreted proteins (CRISPs), defensins, ecdysteroidogenesis compounds, lectins, lipases, oxireductases, RNAases, and sperm utilization proteins. In addition, there were small peptides. It also showed that some of the MAG proteins have an apocrine (pinched off from cells) origin. According to other authors (South & Catteruccia 2016), there are 200 *Aedes aegypti* SFPs, compared to only 100 in *Aedes albopictus*, the contrast between closely related species again showing the large evolutionary pressure on this aspect of mosquito physiology.

Table 9.1. Functional categories of proteins in male accessory gland substances determined from gene expression studies (Baldini et al. 2012)

	Percentage of proteins	
Functional category of protein	***Anopheles gambiae* (*n* = 121)**	***Aedes aegypti* (*n* = 139)**
Specific function of male accessory gland	26.45	28.06
Redox	11.57	10.79
Proteases	11.57	17.27
Nucleic acid binding	8.26	
Hydrolases	4.96	5.04
Protein folding	4.96	2.88
Transmembrane proteins	3.31	5.76
Immune peptides	2.48	0.72
Serpins	1.65	
Nucleic acid metabolism	1.65	0.72
Protein metabolism	1.65	
Lipases	1.65	1.44
Kinases	0.83	1.44
Cysteine-rich secretory proteins	0.83	0.72
Lectins		2.88
Odorant binding		2.16
Other	9.09	10.79

The MAGs continue to mature following eclosion (Alfonso-Parra et al. 2014; Takken et al. 2006). Based on a marker protein, the quantity increased for three days posteclosion. The contents were sufficient to complete five matings and were replenished in 48 hours. Mating increased the rate of replenishment.

The semen of *Aedes aegypti* also includes nonprotein substances that affect the female and fertilization. Remarkably, male messenger RNA in the semen is transcribed in the female reproductive tract (Alfonso-Parra et al. 2016). Inserted in the female during mating, juvenile hormone III (JH III) decreases the rate of previtellogenic resorption and increases the amounts of stored ovarian lipids (Clifton et al. 2014). Sugar nutrition of the males influences the quantity of JH III transferred to the female and consequent fecundity of the female. *Anopheles gambiae* transfers 20-hydroxyecdysone (20E) in the semen but not juvenile hormone (Shaw et al. 2015). The 20E has a similar effect on female reproduction in *Anopheles gambiae* as JH III in *Aedes aegypti* (Baldini et al. 2013). Other substances in the semen include free and protein-bound carbohydrates, free amino acids, and amines (Gillott 2003).

Symbionts in Semen

Semen contains many microorganisms. The first example of venereal transfer of endosymbiotic bacteria was *Asaia* among *Anopheles stephensi* (Damiani et al. 2008). Evidence for transfer of Zika virus among *Aedes aegypti* included transfer from intrathoracically infected males to females and a low rate of transmission from orally infected females to males (Campos et al. 2017). Metagenomic examination of the reproductive tracts of mating pairs taken from swarms of *Anopheles gambiae* and *coluzzii* in Burkina Faso showed the presence of a wide variety of bacteria (Segata et al. 2016). Thirty samples contained 240 to 763 different bacterial species, as indicated by operational taxonomic units (OTUs). *Acinetobacter lwoffii* was the most abundant bacteria. Ninety percent of the bacteria were in 54 OTUs. The 12 most abundant OTUs were in the groups *Pseudomonas*, *Acinetobacter*, *Geobacillus*, *Corynebacterium*, Comamonadaceae, *Staphylococcus*, and *Micrococcus*. Among the endosymbionts detected were *Asaia*, *Rickettsia*, *Spiroplasma*, *Thorsellia anophelis*, and *Wolbachia*. The most abundant groups were different in males and females. The lower reproductive tracts of females were enriched with Gammaproteobacteria and *Acinetobacter* with occasional *Delftia* and the ovaries with Deinococcales, a group of extremophiles.

MAGs had abundant *Ruminococcus* and *Trabulsiella*, and the testes had more *Neisseria* than other tissues.

Wolbachia occurs in sperm and often causes cytoplasmic incompatibility (expressed as failure to fertilize the female pronucleus in the egg) in the next generation if the female has a different strain or lacks *Wolbachia* (Sanogo et al. 2007). Transmission between generations is highly efficient, observed as 98.6% in the field in California, United States, among *Culex pipiens* and *quinquefasciatus* (Rasgon & Scott 2003). One of the possible mechanisms for this phenomenon is the presence of *Wolbachia* proteins that prevent pronuclei derived from male sperm to associate properly with female histones in the egg (Beckmann et al. 2013). These proteins are coded from genes (*w*Pip-282 and *w*Pip283) located within a transposon-like element or associated with mobile elements derived from *Wolbachia* phage virus. Ankyrin protein genes are also located in these genomic regions of the *Wolbachia* and are good candidates for disruption of fertilization, as this class of proteins usually affects interactions of proteins and transcription (Sinkins et al. 2005). Phages may be able to move these compatibility factors between mosquitoes, preventing complete genetic separation of populations with incompatible *Wolbachia*. Such horizontal transfer and rescue from cytoplasmic incompatibility may be unusual, considering extensive field observations on the association of *Wolbachia* infection with limitation of mitochondrial haplotypes in *Culex pipiens* and *quinquefasciatus* in southern Africa (Rasgon et al. 2006). The common presence of phage mobile elements and transposons in *Wolbachia* have resulted in the accumulation of many pseudogenes.

Among the *Culex pipiens* group of species, there are five *Wolbachia* types, designated *w*Pip I–V (Dumas et al. 2013). These strains are not taxon-specific between *Culex quinquefasciatus*, *pipiens*, *pipiens* variety molestus, and *pipiens pallens*. Individual specimens usually (75%) have a single strain. Some of the strains are more abundant according to geographic locations (Type I in sub-Saharan Africa, South America, and Southeast Asia; Type II in North America), but in Europe, for example, all five strains are present. Some of the variety in Europe may be caused by the greater susceptibility of aboveground populations to high temperature events that challenge *Wolbachia* in the mosquitoes, compared to underground populations of the molestus variety that are almost always infected (Mahilum 2003). *Culex pipiens* in California only have a single strain of *Wolbachia* (Rasgon & Scott 2003). Since inheritance of the *Wolbachia* type is matrilineal, it is not surprising that the type is correlated to mitochondrial DNA haplotypes.

The presence of *Wolbachia* is greater in *Culex pipiens* with the organophosphate insecticide resistance genes *Ester* (metabolic resistance) or *Ace-1* (target site; Duron et al. 2006). These genes have a considerable fitness cost, including an increase in immature-stage mortality. For example, under conditions of great selection pressure, *Ester* expression may account for 12% of all soluble proteins in a mosquito. It is possible that compromising the immune system of the resistant mosquitoes makes them more susceptible to *Wolbachia*.

Mating Plug

The semen of some mosquitoes forms a coagulated plug in the atrium of the female, which is not like the spermatophores of other arthropods that form a packet containing the sperm before deposition in or acquisition by the female (Sinclair et al. 2007). The mating plug is formed to various degrees in anophelines and probably not at all in culicines (Shaw et al. 2015). Species that form a well-coagulated plug are *Anopheles gambiae*, *arabiensis*, *funestus*, and *stephensi*. An intermediately coagulated plug forms in *Anopheles atroparvus*, *dirus*, *farauti*, and *sinensis*. *Anopheles albimanus* and other species in the subgenus *Nyssorhynchus* do not form mating plugs at all. Despite the often-repeated statement that *Aedes* have no mating plug, Oliva et al. (2014) claim that granular material in the semen clusters and coagulates in the bursa inseminalis, inhibiting further mating 40 minutes after copulation.

The mating plug of *Anopheles gambiae* consists of 27 proteins (Rogers et al. 2009). The male contributes 15 of the proteins, the female contributes six, and six come from both male and female. One glutamine-rich protein that comes from the anterior of the MAG, called "Plugin," is probably responsible for the coagulation of the plug. Cross-links are created when a transglutaminase from the posterior of the MAG interacts with semen in the atrium (Sirot et al. 2011). This is one of three transglutaminases in *Anopheles* (*Aedes aegypti* only has two; Baldini et al. 2012). This kind of transglutaminase is unique to *Anopheles* that form a mating plug (Avila et al. 2011) and the only one produced in the MAG. It is absent from species that do not form a plug (Chapman 2009). When production of this enzyme is inhibited by RNAi, a plug does not form, and most (75%) sperm are unable to reach the spermatheca. Proteases secreted by the female into the plug break it down in 24–36 hours. The presence of a plug does not seem to inhibit deposition of sperm in the spermatheca from second matings, which was one of the original suggestions for the plug. The major function of the mating plug seems to be facilitation of sperm movement into the spermatheca, but it may have a role in transfer of other substances to action sites within the female.

Male Mating Capabilities

Influences on Mating Success

Males have a limited capacity for successful mating. Their age postemergence, nutritional state, recent mating history, environmental conditions, and genetics all have an influence. *Aedes albopictus* from Réunion Island was examined in the laboratory by exposing single males to groups of females (Boyer et al. 2011). Males were most active during the first six days postemergence and did not mate at all if more than 14 days old. A male was able to inseminate a mean of 9.5 females in a week, filling a mean of 15.5 spermathecae during that time. There was individual variation among the males, some achieving 14 inseminations in a week and occasionally filling all three spermathecae of a single female. *Aedes aegypti* males can mate a mean of 3–5.8 times per day in the laboratory, but they generally mate only 8–9 times in their lifetimes (Bargielowski et al. 2011). Mating shortens the males' life span, from 37 days for virgin males to 12 days for males exposed to four females per day. A typical *Anopheles gambiae* male mates zero to three times in its lifetime (Stone & Foster 2013), and only 7% of a mixed population of *Anopheles coluzzii* and *gambiae* in Burkina Faso had mated (Maïga et al. 2014). Even in the laboratory, only 33%–38% of males mate per day during the peak in reproductive activity four to eight days postemergence (Sawadogo et al. 2013). In Tanzania, only 62% of pregravid *Anopheles funestus* had been inseminated (Charlwood et al. 2000). *Anopheles culicifacies* generally mated no more than three times in their lifetimes and became less sexually active a week after emergence (Howell & Knols 2009). *Anopheles stephensi* had somewhat higher mating rates, with 60% of males mating during a 50-minute swarm in the field (Downes 1969) but only 42% of males mating in the laboratory (Howell & Knols 2009).

The size of males can make a difference in their mating success. For *Anopheles coluzzii* and *gambiae* in Burkina Faso, the larger half of the male population was more likely to have mated; however, the largest males were less likely to have mated than those of a more moderate size (Maïga et al. 2014). The mean wing length of males in a swarm was 2.76 mm compared to 2.82 mm for those in copula (Sawadogo et al. 2013). In the laboratory, *Anopheles gambiae* males reared to fit into three size categories were most successful at mating when they were medium in size (53% copulated, and mated males had an average wing length of 2.75 mm), followed by small males (31% and 2.6 mm) and large males (7.5% and 3.05 mm; Ng'habi et al. 2008). These results suggest that males less than the maximum size are most successful at mating, even though their nutritional reserves and longevity are less than the largest males. Under natural conditions the maximum size is probably seldom achieved so that the relatively larger males are those most likely to mate.

The successes of mating by *Opifex fuscus* (Yuval 2006) and *Aedes aegypti* (South & Catteruccia 2016) are also related to male size, with small *Aedes aegypti* males (wing length 1.86 mm) capable of three complete inseminations in a row compared to larger males (wing length 2.33 mm) able to mate four times (Helinski & Harrington 2011). Larger males also transfer more sperm, at least in some species, including *Aedes aegypti*, *Anopheles freeborni*, *Anopheles gambiae*, and *Wyeomyia smithii* (Cator 2011).

Potentially related to size, male nutrition also influences mating success. Swarming requires a great deal of energy, making it necessary for *Anopheles gambiae* males to feed on sugar following each evening they swarm (Stone & Foster 2013), their performance depending on access to sugar the night before swarming (Lees et al. 2014). In mesocosms in the laboratory, 50% of the males of this species mated if given sugar, but only 11% mated without. *Aedes albopictus* males fed sugar were able to mate more successfully and lived longer than males without sugar when offered three new virgin females each day (Villiard & Gaugler 2015). The unfed males filled more spermathecae per mating, suggesting a functional premonition of unlikely subsequent mating.

Not surprisingly, weather affects mating. Males of *Aedes aegypti* in Australia were able to copulate more at 25°C–35°C than at cooler or warmer temperatures (Bader & Williams 2012). These temperature differences affected not only the insemination rate but also the number of sperm per insemination (3,000 at 18°C, 4,000 at 25°C, and 5,000 at 29°C and 35°C).

Individual males may be more fertile because of seemingly unrelated traits of a genetic origin. Insecticide resistance can affect the mating ability of males, with organophosphate-susceptible males performing 76% of matings when they competed with resistant males (Berticat et al. 2002). Inbred colonies of *Anopheles coluzzii* seemed to prefer to mate with their congeners in a cage but were much more fertile when forced to mate with hybrid males (Ekechukwu et al. 2015). Hybrid male matings with each laboratory strain resulted in more eggs (271% or 71%) per female and subsequently more larvae.

Number of Sperm

The total number of sperm in males is affected by radiation treatment in sterile-insect technique programs. Normal males of *Anopheles arabiensis* continued to produce sperm during six days of observation, increasing the total from 3,232 two days postemergence to 4,396 at six days (Damiens et al. 2013). Another study counted 5,022 sperm per male at three days postemergence and 8,214 sperm at 12 days (Helinski & Knols 2009). Males irradiated with 70 Gy as pupae had a decrease in sperm from 2,104 to 1,098. Strangely, dieldrin exposure to this resistant strain was protective, with radiation-treated males having 2,175 sperm at two days postemergence and 3,492 sperm at six days.

Quantitative assay of Y chromosome DNA in *Anopheles coluzzii* indicated a large range in the number of sperm transferred to a female (Ekechukwu et al. 2015). In individual matings, the number of sperm transferred varied between 1,372 to 3,171 per female. The number was much lower in cages with multiple individuals, ranging from 344 to 670. Dehydration of males (hydric stress) increased the number of sperm transferred, possibly as an adaptation to increase reproductive potential when there were indications of early death (Ekechukwu 2015).

Aedes aegypti males continued to produce more sperm for about 10 days, and the total number of sperm started to decline 20 days postemergence (Ponlawat & Harrington 2007). The number of sperm per male was influenced by size. The number of sperm adjusted for male size varied between strains, with a wild Thai strain having 6,500 sperm at 15 days postemergence compared to 5,200 in the standard Rockefeller strain and 4,500 in a Thai lab strain. The number of sperm transferred per mating followed similar trends (Ponlawat & Harrington 2009). Field mosquito males that were one day postemergence transferred 1,152 sperm, compared to 1,892 sperm from 10-day old males. Larger males of the Thai lab strain (wing length 2.15–2.31 mm) transferred more sperm than smaller males (wing length 1.77–2.07 mm), but this difference was most distinct at 1 and 5 days postemergence and less distinct at 10 days.

Response of the Female to Mating

Mating causes many changes in the female, ranging from structural changes of its cells to egg development to modification of subsequent behavior. More details on some of these changes are discussed below, but all are influenced by gene expression following successful mating (Alfonso-Parra et al. 2016). There is a progression of down- and upregulation for 24 hours following mating in *Aedes aegypti*. Immediately after mating, 14 sequences are downregulated and 11 are upregulated. At 6 hours after mating, the numbers of sequences are 15 and 34, respectively, and at 24 hours they are 110 and 86. There are a total of 150 distinct sequences upregulated and 130 downregulated. Among the effects are a general increase in proteases and in the immune system. Clearly, mating has a much greater effect on the female than simply transfer of sperm.

Mating Refractoriness

Many female Diptera tend to mate only a single time in response to inhibitory signals (moncoitic substances) associated with the first copulation (Gillott 2003). For mosquitoes, the usual pattern is that the female mates once and stores enough sperm to inseminate all the eggs deposited during its lifetime. The tendency toward monandry is an indication of intense evolutionary pressure to develop adaptations contributing to successful insemination from a single mating (South & Catteruccia 2016).

Mating refractoriness of *Aedes aegypti*, *Anopheles quadrimaculatus*, and *Culex* can be caused by injection or transplantation of MAG substances into the hemolymph of virgin females (Baldini et al. 2012). Those kinds of experiments did not inhibit mating by *Anopheles gambiae*, even though females that mated normally are subsequently refractory. In contrast, other experiments (Shutt et al. 2010) showed increases in mating refractoriness of 57%–86% after injection of MAG substances into virgin female *Anopheles gambiae*, as well as *Anopheles coluzzii*, *Anopheles stephensi*, and *Aedes aegypti*. The variation might be explained by absorption rates within the female, potentially enhanced under some circumstances by proteases in the atrium. A competing idea that sperm in the spermatheca cause mating refractoriness in *Anopheles* was based on a failure of MAG substances to inhibit mating in *Anopheles gambiae* and *albimanus* (Klowden 2001). The evidence for the involvement of sperm and the spermatheca was based on observations that *Anopheles gambiae*, unlike *Aedes aegypti*, ceases to mate when the testes are removed and that mated females do not oviposit if the spermatheca is removed (Klowden 2006). These observations were contradicted by the experiments of Shutt et al. (2010) cited previously and by experiments in which sperm production was halted by either RNAi or hybridization (*gambiae* × *melas*; Thailayil et al. 2011). Such sperm-sterile males still caused mating refractoriness in *Anopheles gambiae*. A nonprotein component of MAG substances, 20E, is also said to contribute to mating refractoriness in *Anopheles gambiae* (Shaw et al. 2015).

Unlike other SFPs, the monocoitic substances in mosquitoes can be similar between species. Matrone was identified as a partially purified mixture of peptides and proteins (Sirot et al. 2008) extracted from the MAGs in *Aedes aegypti* and *Culex tarsalis* that caused mating refractoriness when injected into the female hemocoel (Takken et al. 2006). The active component is assumed to be a hormone-like peptide (Shaw et al. 2015). One of the peptides is the sex peptide (Acp70A), identified in *Aedes aegypti* based on sequence similarity with this peptide in *Drosophila melanogaster* (Baldini et al. 2012). This peptide binds to G protein–coupled receptors in the female reproductive tract and may act on the main nerve centers for the copulatory apparatus in the terminal abdominal ganglion. The inhibitory effect of SFPs is extremely powerful, as shown by mating refractoriness induced in virgin female *Aedes aegypti* for 34 days following injection with about 3% of the dose that a female would receive from a normal mating (Helinski et al. 2012a). Induction of refractoriness may be delayed after exposure of the female to monocoitic substances, which may explain the polyandry discussed below.

Polyandry: Multiple Mating by Females

Although the typical pattern for mosquitoes is monoandry, in which females mate only once in their lifetimes, there is evidence that polyandry occurs regularly. The difficult part of judging the extent of polyandry is that mosquitoes concentrated in cages tend to do this more than mosquitoes in the wild (*Anopheles gambiae*; Baldini et al. 2012). About 25% of female *Anopheles arabiensis* in small cages mated multiple times when introduced to mixed groups of normal and radiation-sterilized males (Helinski et al. 2008). Using genetic markers in the laboratory to detect multiple matings, 18.2% of *Anopheles stephensi* females mated more than once in cages (Jayaprakash & Karthikeyan 2014). In a review of field rates of *Anopheles* multiple matings, Scott et al. (2008) estimated the following rates of multiple matings by *Anopheles* females: *dirus*, 0%, *freeborni*, 4%, *gambiae*, 2.5%, *maculipennis*, 0.13%, *messeae*, 0%, and *nuneztovari*, 15%.

The rate of polyandry for *Aedes aegypti* was much higher (Scott et al. 2008; Oliva et al. 2014). Up to 30% of females were mated more than once, as indicated by genotypes of sperm in the spermathecae, and 14% were twice-mated in semifield conditions indicated by stable isotope markers (Helinski et al. 2007, 2012b). There is an indication that females mate later in their lives as they experience more gonotrophic cycles, with 6% remating after four gonotrophic cycles, 22% after five cycles, 38% after six cycles, and 48% after seven cycles. Multiple matings in the laboratory tend to fill more of the spermathecae, with 92% having all three filled in cage matings. Another potential source of multiple mating is during the 40-minute period following copulation, when the female remains receptive to other males. Males contribute somewhat to monoandry in this species, as they partially select virgin females for mating attempts (only 33% of mating attempts in the field are on parous females). Males can also withhold sperm even though they copulated. *Aedes albopictus* females can also apparently mate multiple times, as 26% of females in the field deposited eggs from different male partners.

Effect on Blood Feeding

Aedes aegypti MAGs are the source of a decapeptide, Aea-HP-1, originally found in the head and shown to inhibit blood feeding (Naccarati et al. 2012). It is produced in the anterior portions of the glands, separate from proteases produced in the posterior portions. Aea-HP-1 is labile in the female, lasting only about two hours. It does not bind to the sex protein receptor but appears to have its effect in the brain. The adaptive benefit may be to lower the risk to females following mating or possibly to cause them to distance themselves from the host and males that would attempt further mating. Aea-HP-1 seems to be unique to *Aedes aegypti*, suggesting rapid evolution of this control system. A similar protein is produced by the midgut 1 hour and 30 hours after blood feeding, probably to achieve the right timing of blood meals.

Effect on Egg Development

Egg development in *Anopheles gambiae* only occurs following both mating and blood feeding. The major mechanism for this interaction involves 20E transferred by the male, which causes the female reproductive tract to secrete "mating-induced stimulator of oogenesis," or MISO (Shaw et al. 2015). MISO is part of a signaling cascade that directs nutrients from a blood meal to egg development via ecdysone receptors in the female reproductive tract (Baldini et al. 2013). Stimulation of those receptors by MISO causes production of vitellogenic lipid transporter. The 20E may also act directly on the female's fat body to increase fecundity. Mating also improves oogenesis in *Anopheles stephensi* (Clifton et al. 2014). Injection of amino acids, feeding amino acids with sugar, or blood feeding induced ovarian development in this species more often in mated than unmated females; however, a proportion of the unmated, blood fed (73%) and amino-acid fed (3%) females developed eggs (Uchida et al. 2003).

Matrone and juvenile hormone stimulate vitellogenesis in *Aedes aegypti* and *Aedes taeniorhynchus* (Gillott 2003; South & Catteruccia 2016), the juvenile hormone acting directly on the female fat body.

Effect on Oviposition

Neither *Aedes aegypti* nor *Anopheles gambiae* oviposit if they are unmated (Klowden 2006). Removal of the spermatheca in the latter species, but not in the former, stops oviposition in mated females (Klowden 2001). Semen fluids stimulate oviposition in *Aedes aegypti* and *albopictus* (Helinski et al. 2012a), increasing the likelihood of oviposition by autogenous *Aedes taeniorhynchus* (Clifton et al. 2014). In *Anopheles gambiae* this effect was shown to be independent from deposition of sperm (Thailayil et al. 2011; Baldini et al. 2012), suggesting that inhibition from removal of the spermatheca is related to a secretion of the spermatheca itself. Not all mosquitoes require mating to oviposit. Virgin female *Aedes vexans*, *Aedes trivittatus*, *Psorophora columbiae*, and *Psorophora ferox*, for example, all deposit sterile eggs if they have a blood meal. Unmated *Sabethes chloropterus* will also deposit sterile eggs following a blood meal (Galindo 1958).

A portion of *Aedes aegypti* matrone stimulates oviposition. This portion is a midsized protein with an alpha fraction of 60 kDa and a beta fraction of 30 kDa (Gillott 2003), the alpha portion probably responsible for oviposition (Gillott 2003). This is too big to pass through the atrial wall, suggesting that the female processes matrone into smaller components, or the matrone causes the female to secrete other signaling compounds. In contrast, 20E transferred by male *Anopheles gambiae* is necessary and sufficient to assure oviposition by mated females (Shaw et al. 2015).

Sperm Physiology in the Female

As described above, the distribution of sperm within the female is an active process. For example, the formation of the mating plug through the action of semen-originated transglutaminase is essential to the sperms' ability to reach the spermatheca of *Anopheles gambiae* (Shaw et al. 2015). Some semen components help protect the sperm in their journey from the male to the female spermatheca (Avila et al. 2011). The fat body cells around the spermatheca provide nutrition to the sperm, and nutritional reserves of the female are mobilized by mating, at least in part to provide for the sperm (Klowden 2007; Shaw et al. 2014). In *Anopheles gambiae*, 20E transferred in the semen induces the female to upregulate expression of the *HPX15* gene, causing secretion of heme peroxidase into the spermatheca so that sperm are protected from oxidative chemicals, especially those that are abundant following a blood meal. Upregulation of the female genome also increases the concentration of cytochrome P450 enzymes in the sperm environment. Thioredoxin in the MAG substances of *Aedes aegypti* may serve the same purpose (Sirot et al. 2008).

MAG substances influence movement of sperm in the female. Mitogen-activated protein kinase (MAPK) and influx of extracellular calcium ions in *Culex quinquefasciatus* influence the form of flagellar movement and its speed (Thaler et al. 2013). The slowest form of movement is low-amplitude, long-wavelength movement of the 200 μm flagellum, powering the sperm forward by 5 μm/s. Two superimposed helical waveforms enable the sperm to move at 45 μm/s. The highest speed of 75 μm/s is achieved with a single rapid helical movement of the flagellum. Sperm can also move backward when the calcium ion concentration is low. In *Aedes aegypti*, sperm do not start moving until they are mixed with MAG substances in the female bursa (Degner & Harrington 2016). Sperm movement is also influence by the size of the flagellum (Klowden & Chambers 2004; Helinski & Knols 2009). Some *Anopheles* males contain a large range of sizes (*gambiae*, 26–500 μm, *quadriannulatus*, 40–450 μm, *arabiensis*, 100–400 μm), and larger sperm are more abundant in the spermathecae.

As in human sperm, mosquito sperm express odorant-binding proteins (Pitts et al. 2014). Three of the proteins in the sperm are the same as in antennae, and seven are more abundant in the adult maxillary palps, the proboscis, and the larval antennae. Odorant-binding proteins are matched by the cofactor Orco, which is concentrated in the flagellum of the sperm. Addition of artificial Orco agonists activates the flagella, suggesting this system may be related to internal directional systems in the sperm, or it may be part of the system for activating movement. Further indications of the physiology of sperm orientation include the presence in MAG substances of chemoattractants for sperm, specifically a protein called Cats (CRISP/Antigen5/PR-1; Dottorini et al. 2007) and odorant-binding proteins (Li et al. 2008).

Mating Behavior

Yuval (2006) emphasized the tremendous evolutionary selection of mating systems, writing: "Every species of sexually reproducing organism manifests a unique mating system. Each such system is a species-specific work in progress molded by sexual selection acting on the differential mating success of individuals in the population." The variety of mating habits among mosquitoes is a reminder of this selective process. After all, mating and insemination by pairs of the same species are essential to reproduction.

Mating as a Mechanism for Genetic Separation

Some examples from species complexes show how mating can be part of the isolating mechanism between species. At one end of the spectrum, species in the *Culex pipiens* group have much hybridization in some areas (Fonseca et al. 2009). Forms that can be distinguished based on morphology, behavior, and physiology nonetheless hybridize, resulting in stable populations like *Culex pipiens pallens* in Asia and the extensive population of mixed types in North America. *Anopheles gambiae* and *coluzzii* hybridize rarely, but regularly (Tripet et al. 2001, 2003). In Mali, 2 of 195 female *coluzzii* contained *gambiae* sperm (1.03%), and 1 of 55 female *gambiae* contained *coluzzii* sperm (1.82%). A single hybrid female was captured among these collections in the field. Hybrids between *Anopheles gambiae* and *arabiensis* were even more rare, reported at a rate of 3 per 17,705 females. Hybrids between members of the Gambiae Complex are rarely fertile; in fact, the original identification of most of the species was through crossing studies that resulted in sterile offspring from matings between sibling species (White et al. 2011a). Even within *Anopheles gambiae*, assortative mating can result in subpopulations of this species (Scott et al. 2008). In spite of the mating barriers between members of the Gambiae Complex, uncommon hybridization can have important effects on the genomes of its member species (Barrón et al. 2019). In the *Anopheles farauti* complex in Papua Province, Indonesia, at least eight species are almost completely isolated from each other through mating barriers and geographic isolation (Bangs et al. 2015).

Aedes in the subgenus *Stegomyia* often have weak mating barriers that result in sterile offspring. Mixed in cages, *Aedes polynesiensis* females mated preferentially with conspecific males, but about a third of them mated with male *albopictus* (Ali & Rozeboom 1973). Single male *Aedes polynesiensis* and *albopictus* always chose conspecific females when kept in cages, with a choice of three females of the same species and three females of *aegypti*, *mascarensis*, or *polynesiensis* (Nijhout & Craig 1971). In contrast, male *Aedes aegypti*, *simpsoni*, and *mascarensis* mated with females of the other species 15%–40% of the time.

Weak mating barriers have affected *Aedes aegypti* and *albopictus* in North America, where both species are invasive. In a phenomenon called satyrization (Tripet et al. 2011), male *Aedes albopictus* induced mating refractoriness in female *Aedes aegypti*, even though the offspring failed to hatch. If a male *Aedes aegypti* mates with a female *Aedes albopictus*, the female retains the ability to mate and is not induced to lay eggs; therefore, *Aedes albopictus* tends to replace *Aedes aegypti* where satyrization is common enough. Measurements of the presence of *albopictus* sperm in *aegypti* females may underestimate the phenomenon because cross species matings often result in transfer of MAG substances without sperm (Carrasquilla & Lounibos 2015). Those MAG substances suppress further mating by about 80% in *aegypti* females but not at all in the opposite direction (*aegypti* male MAG substances in *albopictus* females). *Aedes aegypti* females are also behaviorally more likely to mate with *albopictus* than vice versa, though over time there is selection for *aegypti* females to be more recalcitrant (Bargielowski et al. 2015), at a cost of less conspecific mating (Bargielowski et al. 2016).

Swarming

The males of many mosquito species form aerial groups, nuptial assemblies, into which females enter preparatory to mating (Seidahmed 2015). Females can also form loose swarms separate from the males, as in *Culex quinquefasciatus* and *Aedes hexodontus* (Downes 1969). Swarms have been compared to a mammalian lek, characterized by male clustering at a stable location with much male-to-male interaction, lack of male territory but apparent competition for position, no male offering of resources to the females, no male parental care of offspring, and female selection of mates (Cabrera & Jaffe 2007; Diabaté et al. 2011; South & Catteruccia 2016). Some swarms have been called "hot spot" leks, on the assumption that females are attracted to the site, and males gather there in consequence. It is uncertain whether females choose males or vice versa. Some authors assume that females can make a choice, and others claim that males are competing for females (Shishika et al. 2014). The evidence as described below seems to support both views.

Swarming has been proposed as the ancestral mating system in Diptera (Yuval 2006). The first swarms may have formed near emergence sites, and mate selection was only exercised by males. Under those circumstances, the tendency was toward female monogamy, which is generally the case in modern Culicidae. Swarming mating systems with no choice of mates by females tend to evolve toward mating without swarms on substrates, in which females choose mates and are polyandrous. The uncertainty of culicid phylogeny makes this sort of interpretation difficult, and any sense of the direction of the evolution of mosquito mating systems is unclear.

Swarming in *Anopheles*

Attention to *Anopheles* vectors of malaria has advanced the knowledge of swarming tremendously. *Anopheles gambiae*, *funestus*, *culicifacies*, *maculipennis*, *messeae*, *stephensi*, *subpictus*, *franciscanus*, *pseudopunctipennis*, and *freeborni* all swarm during crepuscular periods (Roubaud & Treillard 1940; Yuval 2006). The exact light level at which swarming occurs varies between species. For example, *Anopheles culicifacies* swarms when light wanes from 592 to 18 lux, whereas *Anopheles freeborni* swarms between 350 and 0.5 lux (Howell & Knols 2009). Members of the subgenus *Nyssorhynchus* do not swarm at all according to older accounts, mating at resting sites near larval habitats or near a blood meal host. More recent work involving close observation and acoustic recording shows that *Anopheles albimanus* forms swarms in cages (Pantoja-Sánchez et al. 2019). *Anopheles atroparvus* is stenogamous and does not swarm.

The locations of swarms vary between species, contributing to their separation (Howell & Knols 2009; South and Catteruccia 2016). *Anopheles gambiae* has been observed to form swarms at 0.5–2 m, 2–3 m, or 1–4 m in various studies. The location of the swarms is associated with places that are open to the sky. *Anopheles funestus* swarms at 2–4 m and orients on markers appearing on the horizon. *Anopheles hyrcanus* and *philippinensis* swarm at a lower elevation, 0.3–0.6 m over shrubs. *Anopheles culicifacies* swarms over mounds. There can be as many as 5,000 males in a swarm (*Anopheles franciscanus*) or as few as 15 (*Anopheles culicifacies*). Swarming is an energetic activity, males generally using 50% of their caloric reserves in a single evening. The cause of the females' initial attraction to a swarm is not known with certainty (Takken et al. 2006), but there are usually many fewer females than males in a swarm (between 0.1% and 20%; Howell & Knols 2009).

The careful descriptions of *Anopheles coluzzii* and *gambiae* swarms by Diabaté and colleagues in Burkina Faso (Diabaté et al. 2011; Diabaté 2012; Diabaté & Tripet 2015) form a detailed foundation for consideration of other species. The males of *Anopheles coluzzii* initially formed a swarm when individuals assembled 1–3 m above a visible marker, entering the location in zigzag flight. Markers were objects on the ground providing a contrasting pattern to their surroundings, usually in an area open to air circulation and exposed to sunlight. This started 5 minutes after sunset, and the swarm reached its maximum size in 9–15 minutes, ending just 25–30 minutes after the first males arrived. The swarms consisted of males and, consistently (97.4% of 190 swarms), only *Anopheles coluzzii*. Other species collected in the swarms were *Anopheles rufipes*, *Anopheles pharoensis*, and a species of *Culex*. It is possible that conditions at the time of swarming influence interspecific matings, as Segata et al. (2016) captured 13.3% (4 of 30) mixed couples from swarms in Burkina Faso. Females, usually previously blood fed (Scott & Takken 2012), entered briefly prior to pairing with a male, flying off together while in copula. Mating success increased with the size of the swarm. Strangely, the number of males per swarm occurred in a series of modes (100, 180, 360, 420, or 690 males) rather than being normally distributed. In a rice-growing location with an abundant mosquito population, there were 190 swarm locations in a village with an area of only 760 m × 340 m. The most successful males were in the middle of the range of sizes, suggesting either female selection or better male capability to locate a female.

Under other circumstances, *Anopheles gambiae* and *coluzzii* shared swarms (Dabiré et al. 2013). The males of each species were affected by the habitat, possibly because of the relative abundance of larval development sites for each species. In a wooded savannah, 21 of 28 swarms consisted of *Anopheles gambiae* only, and 7 swarms were mixed between the two species. In contrast, a rice-growing area had 38 of 62 swarms containing only *Anopheles coluzzii* until the conclusion of the rainy season, when most swarms were mixed. Overall, 60% of the swarms were monotypic. The mix of species in 40% of the swarms did not result in the same proportion of mixed pairs—only 4 of 33 pairs consisted of a male or female of the other species. None of those four pairs resulted in deposition of sperm from the opposite species, indicating a high degree of separation between the two.

Males of *Anopheles gambiae* s.l. flew surprisingly long distances in Tanzania, up to 3.6 km, with 45% of a population flying further than 624 m and 7% further than 1.6 km (Gillies 1961). The implication of this movement is the potential for species to mix in swarms. The tendency to have a single species in a swarm is an adaptation, possibly as a first behavioral layer against hybridization. The lack of mating between species even in mixed swarms suggests that there are mechanisms at an individual level to prevent mating and insemination between species.

Detailed movement of individual mosquitoes in natural swarms of *Anopheles gambiae* and *coluzzii* in Mali gave an indication of some of the specific behaviors leading to copulation during the few seconds in which a female encounters a male (Butail et al. 2011, 2012). Using computer analysis of images, the authors were able to track the hundreds of individuals in a swarm and the individual females that entered. They saw that the movement of both males and females mostly occurred in a horizontal plane, although significant vertical movement was noted as well. During a period of just two or three seconds, the female flew into the swarm with more changes of movement and velocity than the males in what might be interpreted as a search pattern. A male quickly found the female in a series of tight turns and the two mated. More detailed analysis of 450,000 data points (Shishika et al. 2014) showed that males made a turn about three times per second. There was no pause when a male encountered another male, but there was a 0.04 s pause on encountering a female. Interactions were like the physics of a "damped spring," in that movement of individuals in the swarm tended to return them toward the center as they interacted with each other, and those interactions were perceived throughout the swarm. There was structure within the swarm, as subgroups of 2 to 12 individuals formed, and these tended to interact with isolated individuals. Males were generally more densely distributed in the middle of the swarm (Howell & Knols 2009).

Anopheles arabiensis generally follows the same pattern of swarming (Seidahmed 2015). It rarely mates with any of its sibling species in the field during swarms that occur during the evening crepuscular period for about 50 minutes. Males leave the swarm when they find a mate and may return later. In the lab, some males remain apart from the swarm with their antennal fibrillae unextended. Fibrillar extension is a requirement for mating and occurs about an hour before swarming. By 48 hours after emergence, this species is ready to mate, but there is low insemination up to 72 hours (Howell & Knols 2009), and males are most effective at 5 to 7 days after emergence (Takken et al. 2006).

Swarming in Nonanophelines

Aedines and culicines that swarm usually do so near the emergence site (Yuval 2006), though there are exceptions like *Psorophora confinnis* and *Aedes cataphylla* (Downes 1969). The mating habits of most species are not documented, so it is hard to generalize about these groups. Certainly, some species in *Aedes* (*Ochlerotatus*), *Psorophora*, and *Culex* mate in swarms; however, some species do not swarm (e.g., *Aedes rempeli* [Smith & Brust 1970], *Culex pipiens* form molestus [Reisen 2012]). Some species mate near the blood meal host, including *Aedes* in subgenera *Stegomyia* (*aegypti*, *albopictus*, *polynesiensis*, and *scutellaris*), *Fredwardsius* (*vittatus*), *Ochlerotatus* (*diantaeus*, *sierrensis*, and *spencerii*), and *Aedes* (*varipalpus*) (Downes 1969). *Mansonia uniformis* and *africana* also form mating swarms near the host. In some cases, at least, swarming near the host may be a result of the attraction of males to concentrations of females. In other cases, it is known that the males are attracted to host odors. Possibly, other odors can be involved, as there is some indication that *Culex pipiens* form molestus mates at locations based on associated floral odors (Jhumur et al. 2006).

Visual markers and prevailing wind often influence the location, size, and shape of swarms (Downes 1969). When mosquitoes are abundant, the number of visual markers for swarms appears to be a limiting factor, as swarms get larger when markers are less abundant. Some species have been observed to share markers but partition the space vertically (*Aedes hexodontus*, *flavescens*, and *excrucians* or *Uranotaenia alboabdominalis* and *Mansonia fuscopennata*). Very large swarms of species, like *Psorophora confinnis* and *Aedes hexodontus*, that extend over 30 m in height may show a plume of mosquitoes at the top where the mosquitoes are flying into gentle wind. When wind is stronger, the swarms get lower to avoid the stronger air movement. Individual mosquitoes tend to face into the wind.

Swarming behavior is not necessarily stereotyped in each species. Although *Aedes polynesiensis* is known to swarm near a host, in American Samoa males formed swarms at the bases of trees. Swarms formed rather early, between 1645h and 1750h. During the first half of this period, the males were smaller and had more nectar in their crops (Tuten et al. 2013). The groups of males were small (fewer than 10 individuals) and usually located on the northern and eastern sides of the trees. The males were selective about the species of tree and preferred trunks with a larger diameter.

Aedes aegypti is also variable in its swarming behavior, sometimes mating individually and sometimes forming spherical swarms of fewer than 30 males and about 30 cm across near hosts at certain times of the day (Cabrera & Jaffe 2007; Oliva et al. 2014) or near oviposition sites outdoors (Chadee & Gilles 2014). The initiation of photophase stimulated mating swarms, as did the presence of a rat blood-meal host. In the field in Trinidad, half of the copulations took place in the first two or the last two hours of daylight, whether mosquitoes were indoors or outdoors. More copulation took place in larger swarms, and insemination was more successful indoors (11.5% of copulations failed to inseminate) than outdoors (38.6% of copulations failed to inseminate). Females of both *Aedes aegypti* and *albopictus* commonly reject males by kicking with the hind legs, flying away, or refusing copulation after genital contact (Helinski & Harrington 2012; Benelli 2015). Successful mating lasts an average of 63 seconds in *Aedes albopictus*, with 39 seconds of preliminaries.

Acoustic Signaling

People strongly associate the presence of mosquitoes with the whining buzz created by the insects' wingbeats. That association may originate from the annoying habit of some species to hover over the ear at night or simply the common experience of hearing the buzz followed by feeling the bite. Perhaps for these reasons, biologists started examining the acoustic abilities of mosquitoes early in the history of culicid biology. Johnston (1855) used *Culex* (probably *pipiens*) mosquitoes as the model for his assertion that the antennae are the hearing organs of insects. He recognized the expansion of the base of the antennae, now called Johnston's organ, as the likely site where vibrations from the fibrillae were communicated to the nervous system. He also recognized that males locate females for mating through perception of sound. These ideas were applied to practice, with the first suggestion to use sounds for trapping mosquitoes in 1878 (Diabaté & Tripet 2015). Recording and analysis of mosquito sounds have gone on for decades, but it is only recently that some intricate and significant interactions have been described.

Johnston's organ in male *Aedes aegypti* contains over 15,000 sensory cells, forming a sensitive detector of fibrillar deflection (Aldersley 2015). The mosquito can detect the movement of a fibrilla by 7 nm, or 0.0001 degree of angle. The sensory cells are arranged radially, implying that each antenna is capable of directional detection of sound. The antenna can respond to the amplitude of sounds by becoming stiffer in loud environments. Females are much less sensitive to sound and have less than half as many sensory cells in their Johnston's organs. Each antennal fibrilla is innervated, but the detection of sound is restricted to the Johnston's organ (Downes 1969). The fibrillae of some species are erected perpendicular to the antenna prior to swarming, using a hydrostatic mechanism.

As far as is known, the males of all mosquitoes that mate in flight orient on the females' sounds produced by the wings beating (Yuval 2006). Initial attraction to those sounds is not necessarily specific (Downes 1969). For example, *Aedes aegypti* males are initially attracted by wingbeats of *Anopheles gambiae*, *Culex pipiens*, or *Aedes vexans*. Those that do not mate in flight are not sensitive to sound. The nature of sound production in mosquitoes can be observed at a rough level by seeing the effect of frequencies on behavior. Exposing males to wingbeat frequencies produced by an oscillator or tuning fork showed responses by *Aedes aegypti* at 400–550 Hz (Nijhout & Craig 1971). Females respond to 700 Hz (Cator 2011). Mating was disrupted in field populations of *Culex tarsalis* and *Culex tritaeniorhynchus* exposed to the appropriate wingbeat frequencies (Diabaté & Tripet 2015).

The specificity of response to frequencies at the final stages before copulation is paralleled by the differences in frequencies produced by species. Even closely related species may have different wingbeat frequencies, such as *Anopheles gambiae* (male, 682 Hz, female, 460 Hz) and *Anopheles coluzzii* (male, 704 Hz, female, 462 Hz; Pennetier et al. 2010). *Toxorhynchites brevipalpis* males and females have a similar frequency of 450–500 Hz, but they vary this frequency as they approach each other in apparent gender identification (Aldersley 2015). When opposite sexes of *Aedes aegypti* approach each other, the males have a mean wingbeat frequency of 673 Hz and females, 465 Hz.

Larger individuals had a higher frequency than smaller ones, though the effect was weaker or absent in females (Cator 2011). For example, male *Anopheles gambiae* with a wing length of 3.14 mm produced a fundamental frequency of 696 Hz, compared to those with a wing length of 2.71 mm that produced a frequency of 639 Hz. In male *Aedes aegypti*, large males with a wing length of 2.36 mm produced tones at 735 Hz, and those with a wing length of 1.77 mm produced tones at 676 Hz. Temperature may also influence frequency, as *Aedes aegypti* in the field in Thailand had higher frequencies than mosquitoes in the laboratory (Cator et al. 2011).

Wingbeat frequencies can be characteristic of species. Some have taken advantage of the consistency of flight tones to detect species of mosquitoes up to 100 mm away from a microphone, even suggesting the use of mobile phones for this purpose (Mukundarajan et al. 2017).

Acoustic detection by mating mosquitoes is one of the more surprising aspects of their biology. The fundamental frequency that corresponds to the number of times per second that wings beat is the loudest component of the mosquito sound. The sound is more complex, consisting of a series of harmonics that represent integer multiples of the fundamental frequency. As described above, male mosquitoes will orient on the fundamental frequency of females, but much more goes on acoustically between the male and the female in the final moments before copulation. In *Aedes aegypti*, the two sexes actively matched the male's second harmonic and the female's third at about 1,200 Hz as they approached each other (Cator et al. 2009), maintaining the convergence for 1–10 seconds. More detailed studies showed that this sort of convergence occurred in 92% of matings and varied during those few seconds, ranging on and off or between different harmonics (male to female harmonic ratio of 5:3 or 3:2) in 70% of the couples (Aldersley 2015). Evidently, males and females can distinguish the frequency spectrum produced by a partner and their own sounds to achieve such a complicated "mating song." Couples that produce frequency matching are more likely to have successful matings and to produce offspring capable of successful mating (Cator & Harrington 2011). The direct perception of sound goes up to only 2,000 Hz, but some of the harmonic matching goes above that frequency. The mechanism to detect matching at those high frequencies may be detection of interference between harmonics and its cessation when they match (Pennetier et al. 2010).

Males also match frequencies about half the time as they approach each other—but only the fundamental frequencies and not the harmonics (Aldersley 2015). Male-male interaction involves a constant frequency, unlike the ranging up-and-down characteristic of male-female interaction. Male perception and matching of frequencies of other males is probably an important part of maintaining the structure of a swarm, possibly accounting for the "damped spring" dynamic described in the section on swarming.

Harmonic frequency matching is not restricted to *Aedes aegypti*, also having been observed in *Culex* and *Anopheles* (Pennetier et al. 2010). Closely related species may distinguish each other through this mechanism, as mixed pairs of *Anopheles gambiae* and *coluzzii* do not match frequencies, whereas monotypic pairs do. In *Anopheles gambiae*, at least, flight tone matching just before mating is like that in *Aedes aegypti*, matching the male second harmonic and the female third at a frequency of about 1,300 Hz for small individuals and 1,400 Hz for large ones.

Influence of Pheromones

Both contact and volatile pheromones have been implicated in the response of male and female mosquitoes to each other. It has been suggested that the lack of a functional Johnston's organ at the bases of *Aedes aegypti* female antennae implies they are unable to orient on males through acoustic signals (Cabrera & Jaffe 2007). Males of this species produce a volatile pheromone that stimulates flight by females, and cuticular hydrocarbons on the surface of females are attractive to males. The relative proportions of female cuticular hydrocarbons (n-heptadecane, n-petacosane, and n-hexacosane) change after mating (Polerstock et al. 2002), possibly making another mechanism for monogamy by enabling males to recognize nonvirgin females. Cuticular hydrocarbons may also be involved in attraction of females to males in swarming *Anopheles arabiensis* (Lees et al. 2014). During swarming, the males produce more octanal, nonanal, 2-hydroxy-2-butanone, decanal, sulcatone, and 2,3-butanediol. Although males are attracted to female wingbeat frequencies, they will mate with females experimentally deprived of their wings (Nijhout & Craig 1971). Experiments have shown that *Aedes aegypti* of opposite genders are attracted to each other in an olfactometer and also to certain volatiles emitted by the mosquitoes (2,6,6-trimethylcyclohex-2-ene-1,4-dione; 2,2,6-trimethylcyclohexane-1,4-dione; and 1-(4-ethylphenyl)ethanone; Fawaz et al. 2014). Not all species in the subgenus *Stegomyia* behave similarly. *Aedes mascarensis* and *albopictus* would mate with wingless females, but *simpsoni* and *scutellaris* would not. *Aedes albopictus* would attempt to mate with a recently (less than 1 hour, but not 10 hours) dead female, perceiving pheromones through fore- and midtarsi.

Chemical cues may be a secondary means for species to identify conspecific mates when swarming time and place overlap, as in some *Anopheles* with similar wingbeat harmonics (Yuval 2006). Species that do not mate in flight are more likely to depend on chemical signals for location of mates and for species recognition. One of the best-described examples is *Culiseta inornata* (Lang 1977). Males of this species will attempt to mate with dead females (fewer than 36 days old) or clay balls embedded with female legs. The effect is only through contact. Presumably, a contact pheromone is also involved for species like *Deinocerites cancer* and *Opifex fuscus* that manage to mate with females before they have completely emerged from the pupal case (Nijhout & Craig 1971; Cabrera & Jaffe 2007).

Forced or Adapted Copulation in the Laboratory

Mating and insemination can be a limiting factor for colonization of some mosquito species. For small colonies, it is often possible to artificially induce insemination by anesthetizing the female and bringing her genitalia in proper contact with a decapitated male's genitalia. This technique was first developed for the floodwater *Aedes stimulans* and *vexans* (McDaniel & Horsfall 1957). It was then adapted to *Anopheles punctipennis*, *quadrimaculatus*, *freeborni*, *albimanus*, and *earlei*, some of these even being induced to mate with different species (Baker et al. 1962). Decapitation removes the inhibition from the subesophageal ganglion to the terminal abdominal ganglion that controls the genitalia and mating. Males of *Aedes stimulans* and *vexans* could be induced to mate artificially three times in succession, but each time with less chance of insemination. In the case of *Anopheles pseudopunctipennis*, males can mate two or three times (Ekechukwu et al. 2015). *Anopheles* (*Nyssorhynchus*) *trinkae*, *argyritarsis*, *darlingi*, *rangeli*, and *triannulatus* would not mate under these conditions.

Some mosquitoes can also change their behavior in the laboratory to form mating strains. One example is *Anopheles pseudopunctipennis* (Ekechukwu et al. 2015). Normally requiring outdoor mating conditions, a colony in Bolivia was exposed to stroboscopic (140/min) blue (470 nm) light for 20 minutes each night (Lardeux et al. 2007). This procedure induced enough mosquitoes to mate in a $30 \times 30 \times 30$ cm cage that it was possible to select for a cage-mating strain.

Specific Mating Behaviors

Considering the similarity in basic structure of mosquito male and female genitalia, it is remarkable that mating behavior varies so greatly. Selection for those behaviors may have been associated with determination of partner fitness, partner species identification, the effectiveness of the insemination process, the spatial concentration of individuals, or other aspects of the environmental context for a species. There is an element of sexual conflict in mosquito mating systems, as the polyandrous males experience selection for multiple matings and the monogamous females do not (Hosken et al. 2009). The few examples described below are illustrative of the variety reported so far, but the range of mating behaviors probably extends much further.

Variations on Mating in Flight

ANOPHELES CULICIFACIES

Probably typical of swarming *Anopheles*, the antennal fibrillae of the males become erect about an hour before swarming and fold again after mating (Howell & Knols 2009). On encountering a female in the swarm, a male grasps the partner with the tarsal claws of the forelegs. The male then swings its abdomen up to grasp the female genitalia. Once interlocked, the male releases its grip with the foretarsi, and the pair fly out of the swarm to complete mating. The female can reject a male, but there is no apparent male selection of females. After mating, the male must sugar feed, usually before dawn.

ARMIGERES DIGITATUS

This species develops in fresh bamboo internodes in Sarawak, Malaysia. It was colonized, providing an opportunity to observe mating (Okazawa et al. 1991). Males began mating activities three days and females four days after emergence. There was no indication of swarming in the description, the males flying to females in flight, capturing the females by the legs, and then assuming a venter-to-venter position. The males repeatedly touched the tip of the female abdomen until copulation was initiated. Copulation was usually completed in flight over a period of just a few seconds.

ERETMAPODITES CHRYSOGASTER

Eretmapodites is an aedine genus restricted to Africa and Madagascar. Their larvae are predatory in leaf axils, collections of water in leaves, and flooded snail shells. Females can be avid blood feeders. The mating behavior of *Eretmapodites chrysogaster* was described by Haddow (1946) as "peculiar and violent." The male finds the female in flight, the genitalia are engaged, and the pair settle with the male suspended by its genitalia from the female. The pair stay in this position for over an hour.

CULEX PIPIENS, TARSALIS, QUINQUEFASCIATUS, AND *STIGMATOSOMA*

These species, the most abundant members of the subgenus *Culex* in California, are responsible for a proportion of both human and zoonotic arbovirus transmission (Reisen 2012). They typically mate at dusk from large male swarms one or two days after emergence. *Culex tarsalis* will use bushes as markers for large swarms and other kinds of markers for small ones. These markers may be used by multiple generations of mosquitoes. Pairs engage their genitalia in flight and quickly settle to a surface, where they complete copulation in an end-to-end position. In the field, *Culex pipiens* used upright markers, like fence posts. These species are remarkably adaptable to mating in cages in the laboratory. It is possible to select for stenogamy in *Culex tarsalis*, and the other species mate readily in smaller spaces, as does *Culex pipiens* in Japan (Oda 1968).

Species That Mate with Females at Emergence from the Pupa

DEINOCERITES CANCER

Larvae of this species develop in flooded, coastal land crab holes in the Caribbean basin. These sites are unusual, as the level of the salty water varies with the tide, and this water is heavily contaminated with organic material associated with the crabs. Males will seek females on the wing but more typically seek female pupae before emergence, or actively during emergence (Downes 1966; Conner & Itagaki 1984; Yuval 2006). Males spend much of their time on the water in their search for pupae. Evidently operating on chemical cues, the male examines the water surface with its nonplumose, long antennae. The males examine male pupae but for much briefer intervals than older female pupae. From 14 hours before emergence, the attraction of males to female pupae, but not to male pupae, increases exponentially as measured by the amount of time they spend with the pupae. On encountering a female pupa, the antennal tips vibrate and form figure eights around the respiratory trumpets. The male then shuffles its foretarsi over the surface of the water above the pupa and finally engages the foretarsal claws around the respiratory trumpets. The male may move the pupa to another position, and it actively defends the pupa from other males. That defense takes the form of warding off a male with one of its forelegs or even shoving another male away with its entire body. As the female begins to emerge from the pupal case, the male fans its wings and seeks to mate. The male engages the female genitalia with its elongated gonocoxopodites before the female can emerge from the pupal case, the male abdomen twisting around under the female's. Once the female emerges, the pair shift to an end-to-end position, and the male tows the female to a solid substrate. The couple remain in copula for up to 70 minutes (average 40–50 minutes) in an end-to-end position, typically filling all three spermathecae. Other males continue to be attracted to the empty pupal exuviae for about three minutes. Surprisingly, males are also attracted to pupal exuviae of *Aedes aegypti* and *Culex pipiens*.

OPIFEX FUSCUS

This endemic species from New Zealand has several unusual characteristics, presumably associated with the long isolation of those islands. Larvae occupy fresh or saline pools in rocks by the windy coastline. Like *Deinocerites cancer*, males seek female pupae before emergence (Marks 1958b; Slooten & Lambert 1983). The males spend most of their time on the surface of the water while they seek pupae, forming what might be considered a two-dimensional swarm. The male mosquitoes are attracted to older pupae and, unlike *Deinocerites cancer*, do not distinguish pupal gender. Attraction seems to be based on visual and vibration signals. In a laboratory colony, an excess of males may drown female pupae in their attempts to mate with them (Snell et al. 2010). On encountering a pupa, the male faces in the same direction and grasps the respiratory trumpets with its foretarsal claws, reaching below the water surface with its forelegs, head, and anterior body to do so. Sometimes at this stage, which lasts any time up to two minutes, the male may be dragged underwater by the pupa. At the surface, the male then places its abdomen along the notum of the pupa and grasps the pupa with the gonocoxites of its genitalia along a groove in the pupal notum, disengaging its forelegs from the respiratory trumpets. The male may rest this way on the pupa for 10–20 minutes, waiting for the emergent adult to split the pupal case. As soon as the pupal case opens, the male adult makes genital contact with the emergent adult for three to five minutes before attempting to copulate. If the emergent adult is a male, contact is broken, but if it is a female, copulation begins. As the female emerges completely from the pupal case, they change position to being end-to-end and mating is completed.

This unusual mating habit is supported by several morphological specializations (Marks 1958b; Slooten & Lambert 1983). The male's antennae are not plumose, and Johnston's organ is reduced, associated with a lack of the use of sound for mate recognition. The foretarsal claws are longer than usual and covered with hydrofuge hairs, as are the eyes. The gonocoxites are stout and fit into the saddle-shaped grooves on the pupal notum. At any one time, there is an average of 260 males available for one emerging female, indicating a high operational sex ratio. Visual indications of competition between males are unusual, though, rarely, several males may try to mate with one female. Nonetheless, there appears to be some competition for mates, as the males in copula (mean wing length 3.8 mm) are slightly larger than randomly collected males (mean wing length 3.75 mm; Slooten & Lambert 1984).

CULISETA INORNATA

Although not typical of this species, males sometimes engage females as they emerge from the pupal exuviae in a manner like *Deinocerites cancer* (Downes 1966). More typically, the males patrol emergence sites and approach females right after emergence (Yuval 2006). As described above, a contact pheromone is involved in stimulating the males to mate, and mating usually lasts over an hour.

AEDES MAEHLERI

Aedes maehleri is the only member of the subgenus *Stegomyia* in which the males mate with the females as they emerge from the pupal case (Mogi 2010). This species occurs in the western Pacific and develops in the pitcher-like containers of *Nepenthes mirabilis* (Lour.) Rafarin.

Sabethines

Sabethines all develop in containers, and most are diurnally active. There is clearly a visual element to their interactions (Downes 1969) in that they are usually ornamented with colors, projections of scales, or iridescence. One possible function of such ornamentation is in courtship (South & Arnqvist 2008) that results in more successful mating and fewer instances of hybridization. Most sabethines mate on surfaces following location of females by patrolling males (Yuval 2006). This suggests the possibility of female selection of mates. There are some sabethine species that mate in flight, including *Tripteroides aranoides* and *bambusae* (Okazawa et al. 1986a).

SABETHES CHLOROPTERUS

Males of this arboreal, diurnal species respond to flying females and their flight tones, but mating occurs on a vertical surface (Zsemlye et al. 2005). The male may locate the female during flight and follow her to a substrate, or the male may patrol vertical surfaces until it finds a resting female. Both sexes are highly ornamented with iridescent green, silver, gold, and purple, but the male has extra white patches on the sides of the midfemora and ventral side of the proboscis. The male antennae are not more plumose than those of the female, but each fibrilla is longer. On approach, the male midleg taps the female hind leg, possibly to detect semiochemicals. The male moves from side to side and finally grasps the female's wing with its midleg, then twists its abdomen below the female's while balancing itself on its forelegs. The pair then shift to an end-to-end position and complete insemination for six seconds. The entire mating process takes an average of 3.7 minutes but can last up to 34 minutes. The female can reject the male and sometimes does so by kicking with the hind legs.

SABETHES CYANEUS

This tropical Central American species has striking "paddles" of dark blue and white scales on the midlegs, extending from the distal half of the tibiae and proximal portion of the tarsi. Removal of the paddles from the female, but not from the male, interfered with successful mating (South and Arnqvist 2008). The mating process was like that of *Sabethes chloropterus*, courtship lasting seven minutes (South et al. 2009) and mating, only 50 seconds. Courtship consisted of the male hovering anterior to the female, waving its legs, shifting its position laterally back and forth, and making superficial genital contact (Philips et al. 1996). Females only mated once, even if the first mating was with a sterile male. They rejected males 79% of the time by kicking them or by lowering their (females') abdomens. Males spent almost a quarter of their time in courtship, with a severe fitness cost. Males without access to females lived 69 days, whereas males that mated throughout their lives lived only 48 days.

TOPOMYIA YANBARENSIS

This autogenous, larval-predatory sabethine species occurs on southern Japanese islands in the subtropics. Larvae live in bamboo internodes, the access to which is restricted to small holes created by beetles. Mating activity started two days after emergence and continued intensively for two weeks (Okazawa et al. 1986a). Males perform a vertical up-and-down flight in front of bamboo while seeking mates. Locating a female, the pair mate on the surface of the bamboo if the female does not reject the male. Males often attempt to mate with each other.

WYEOMYIA ARTHROSTIGMA

This species can be common in Central America, developing in leaf axils of bromeliads, as is typical for the genus. Males patrol vertical surfaces about two to six centimeters away, hovering near females perched on the substrate. Males make a series of darting movements or looping flights oriented toward females. The male grasps the costal margin of the female's wing with midtarsal claws, causing the male to twist its venter toward the lateral side of the female. The male plants its forelegs on the surface between the female's forelegs. The male hind legs arch upward, the male holds its wings out horizontally, and, as the hind legs wave, the genitalia are placed in repeated superficial contact with the female's genitalia. Finally, the male fully engages the female genitalia, insemination occurs, and the female kicks the male away. The whole process takes about 2.3 minutes, with a minute for the preliminaries, a minute for copulation, and 30 seconds for superficial genital contact.

WYEOMYIA SMITHII

These mosquitoes develop in pitcher plants in North America, ranging from Florida to southern Canada. They are completely autogenous, never taking a blood meal and depositing multiple batches of eggs. Males approach females that are resting on vertical surfaces shortly after emergence during daylight (O'Meara & Lounibos 1981). In the laboratory, 55% of males from Massachusetts and 35% of males from Florida were able to mate 30 minutes after emergence with females only 1 hour old.

Practical Applications of Mating Biology

Population-level manipulation of insects is a goal of modern insect pest management, the idea being to reduce or change an insect population so that it no longer exceeds a threshold at which it causes damage (Metcalf & Luckmann 1975). Chemical mosquito control can be highly effective, but it is challenged by the need to treat areas near people and to treat areas that take a lot of work to find. Methods that treat mosquito populations on an area-wide basis and that do not use insecticides are likely to have appeal. The sterile-insect technique (SIT) is just such a method. Originally conceived by E.F. Knipling (1955) in the 1930s, it saw its first application against the North American screwworm fly (*Cochliomyia hominivo-*

rax (Coquerel)) in the early 1950s, eliminating this devastating pest from the island of Curaçao. The basic concept of SIT is that release of enough sterile males can compete with wild fertile males and result in a large proportion of the female population depositing nonviable eggs. As the natural population declines, repeated releases of sterile males become increasingly effective to the point of eliminating the pest population. As difficult as this process sounds in both practice and theory, it worked dramatically well against the screwworm. By 2006, this species had been eliminated from its former range in North and Central America. Release of radiation-sterilized males of screwworms and tephritid fruit flies are still routine, based on programs managed by the US Department of Agriculture.

Mosquitoes have been the target of SIT programs (Benedict & Robinson 2003; Seidahmed 2015). Chemosterilization, cytoplasmic incompatibility, irradiation, hybrid incompatibility, incompatible genetic translocation, and other genetic incompatibility were among the methods used to sterilize male mosquitoes. Of 28 trials between 1959 and 1979 (against *Aedes aegypti*, *Aedes albopictus*, *Anopheles albimanus*, *Anopheles culicifacies*, *Anopheles gambiae*, *Anopheles quadrimaculatus*, *Culex pipiens*, *Culex quinquefasciatus*, and *Culex tritaeniorhynchus*), four have been considered the most successful. The first success was in Myanmar in 1967 against *Culex quinquefasciatus* using naturally occurring cytoplasmic incompatibility induced by *Wolbachia* infection. Release of chemosterilized males of the same species in 1970 on an island off the Florida coast was also considered effective. *Culex pipiens* was controlled in an area near Montpelier, France, in 1971 by release of males with a sex-linked translocation that resulted in sterile matings with wild-type females. Chemosterilized males were used in El Salvador in 1974 against *Anopheles albimanus*, taking advantage of a strain in which the males were resistant to dieldrin insecticide. Added to the larval medium, this powerful insecticide killed all of the females so that mechanical sorting to assure male-only releases was unnecessary. This promising trial was terminated by civil unrest, but it resulted in significant reduction of mosquito populations. Other trials that showed a reduction in wild populations of the target species were directed against *Aedes aegypti* in India, *Culex quinquefasciatus* in India and California, *Anopheles culicifacies* in Pakistan, and *Anopheles quadrimaculatus* in Florida.

Failures of SIT trials were due to a variety of factors, including failure of reared males to mate sufficiently with wild-type females, difficulty rearing sufficient numbers, inefficient systems for separating male from female mosquitoes, excessive immigration of wild-type males into the trial area, and social opposition to mosquito release. Reduced fitness of laboratory-reared males has been considered a continuing challenge, and such adaptation can arise within a few generations of colonization (Ferguson et al. 2005). When radiation is used to sterilize males, the dosage must be carefully calibrated to achieve maximum sterility with minimum fitness cost. A program with *Aedes albopictus* was one example of such calibration. Careful dose ranging identified 35 Gy exposure of pupae as optimal, in that 93% of males were sterile, and they were fit enough to support a release rate of five-to-one sterile to wild males (Oliva et al. 2012; Madakacherry et al. 2014). Spatially explicit mathematical models have indicated that mosquito movement in response to environmental stimuli, spacing of release sites, and frequency of releases are all critical components of a successful program (Ferreira et al. 2008; Dufourd & Dumont 2013; Oléron Evans and Bishop 2014).

Recently, interest in SIT for mosquito control has increased (Lees et al. 2014). Many of the recent articles on mating behavior and physiology state that the use of SIT, or a variation of it, was an important practical motivation for the research. The renewed interest in mass release methods for mosquito control is partly the result of new technology and partly a response to current limitations on chemically based control techniques. Although a matter of opinion, there is also a feeling in the professional community that the battle against disease vectors is being lost despite considerable effort to overcome them. The success of SIT against screwworm and various area-wide successes to control agricultural pests are examples of how mosquito control might be accomplished without the need to perform carefully targeted, geographically fine-scaled application of toxicants.

Current efforts use a variety of methods. The International Atomic Energy Agency of the United Nations has provided technical support for an ongoing program of radiation-sterilized male *Anopheles arabiensis* in northern Sudan and a planned program in South Africa (Azrag et al. 2016; Munhenga et al. 2016). They take advantage of a Y chromosome-linked dieldrin resistance gene that has been introgressed through mating with local strains. Using this trait, the program can rear large colonies, then select males in the generation for release by putting dieldrin in the rearing medium. Males have been sterilized as pupae by exposing them to 70 Gy of X-radiation (Damiens et al. 2013). It has been calculated that because of fitness costs of the laboratory-reared mosquitoes, 10 sterilized males must be released for every wild male.

Another effort takes a commercial approach based on release of males infected by a strain of *Wolbachia* that is incompatible with the wild population (Dobson et al. 2016; Mains et al. 2016). This firm has taken the interesting approach of registering these males as biopesticides against *Aedes polynesiensis*, *albopictus*, and *aegypti*. The first males were infected by a series of hybridizations with a naturally infected species (O'Connor et al. 2012), but subsequently, methods have been developed for infection of embryos with the *Wolbachia* of choice. Releases have been made in American Samoa and Kentucky, New York, and California, United States. An ongoing experimental project in Fresno, California, combines advanced, automated rearing and male separation to achieve efficient rearing. This project is in support of controlling the recent invasion of *Aedes aegypti* in California.

Wolbachia are also the basis of a completely different approach to stopping viral transmission by *Aedes aegypti* (Ye et al. 2015; Caragata et al. 2016). Unnatural combinations of

Wolbachia strains can have a strong negative effect on the ability of the infected mosquito to transmit pathogens, especially viruses by *Aedes aegypti*. The use of a *Wolbachia* strain from *Drosophila melanogaster* has resulted in *Aedes aegypti* that are nearly as fit as wild types and that reduce transmission of dengue, chikungunya, and Zika viruses. Introduced to a community in males and females, the *Wolbachia* tends to establish itself, as noninfected females produce sterile eggs due to cytoplasmic incompatibility. The effectiveness of this method is still being established at the time of this writing, but its appeal resulted in widespread approval and pilot implementation in Latin America, Australia, Southeast Asia, and the western Pacific.

Genetic modification is the basis for other mosquito control schemes. A commercial enterprise has had good success suppressing populations of *Aedes aegypti* in Latin America by releasing males that carry a sex-linked gene that kills only females in the next generation (Carvalho et al. 2015). The surviving males continue to transmit the gene, extending the effect of the release for a few weeks. The lethal gene uses a mechanism counteracted by any tetracycline drug, so colonies can be raised normally when drugs are added to the larval medium. What is more, the generation reared for release without tetracycline produces only males, making other forms of sex sorting unnecessary. The genetically modified male mosquito is moderately less capable of mating than wild types, living four days less and mating with about half as many females in its lifetime (Bargielowski et al. 2011). As with any release of genetically modified organisms, the main obstacle to more widespread use of this technique has been social rather than technical.

More recently, "gene drive" has been used to alter mosquitoes in ways intended to either kill the next generation or modify it so that it cannot transmit pathogens (Burt et al. 2018). There has been a history of looking for a way to drive genes into a wild population using selfish genes like *MEDEA* (Enserink 2007), translocations, transposons, meiotic drive, and symbionts (Christophides 2005). Using the CRISPR/Cas9 system (clustered regularly interspaced short palindromic repeats and CRISPR-associated protein 9), it is possible to insert any gene in a target area of the genome (Dong et al. 2018). Repair mechanisms then place the same gene in the matching chromosome, resulting in homozygous insertion of genes and greatly reducing the time and effort required to create genetically modified organisms. Although far from practical yet, there is much optimism that male and female mosquitoes could be released that spread the target genes throughout the population as each generation is converted to the homozygous condition (Eckhoff et al. 2017). In the most hopeful scenarios, relatively few releases would be necessary to get the effect, though directional gene flow might make the distribution of releases more complicated (Scott et al. 2002). Of course, such an uncontrollable change to wild populations of any organism has gotten a lot of attention from those concerned with unintended consequences. Regulation and social acceptance may be the biggest challenge to development (Caplan et al. 2015). An evenhanded treatment of the risks and mitigations of such a program was produced by the US National Academy of Sciences (Committee on Gene Drive Research in Non-Human Organisms 2016).

Conclusion

Those familiar with the general appearance and biology of mosquitoes have reason to be mystified by the mating process in these insects. Why are there so many males when each one mates more than once, and females mate only once? Why is sexual dimorphism so pronounced? What does swarming accomplish for a species? Why do the male terminalia rotate? How do mates find each other? The literature addressing these questions has been enriched considerably by research accomplished in the last two decades, thanks in large part to advances in genetic, protein, and computational techniques. As might be expected, the additional information raises many other questions and, as in all of mosquito biology, begs the question of what happens across the family rather than in just a few species.

Behavioral and physiological processes involved in mosquito mating are highly selected, resulting in intricate adaptations to get sperm from males to females and to keep those sperm viable during storage in the spermathecae. The relative importance of these adaptations in the appearance of new species is impossible to assess, but variety of the process among mosquitoes would suggest that changes in mating can be one of the first ways in which two populations begin to separate genetically. On the other hand, the complicated morphological, physiological, and biochemical aspects of mating must put real restrictions on the amount of variation that can occur without a fitness cost. Mosquito mating will certainly remain a subject of fascination for entomologists, whether their interest is biological or practical.

10. Egg Development and Oviposition

Because of the great diversity of oviposition sites many different sampling techniques would be required if the eggs of the different species were to be adequately sampled. However, relatively little attention has been devoted to sampling egg populations; consequently few methods have been developed. It is disappointing that there has been so little effort to study the biology and ecology of the eggs, because much valuable information can be obtained from the egg population.

Service 1976b

Oviposition is in many ways the culmination of the life of a mosquito. It is at this point that the female brings all the evolutionary history, development, nutrition, behavior, and insemination to bear on the singularly most important event for the species—reproduction. The female's accomplishment of placing eggs in locations where they can develop is the success of another generation in the long history of the family. Mosquitoes are *r* strategists ecologically, for the most part depositing many eggs to assure survival of a small percentage of offspring; however, a review of oviposition leads to no doubt that evolutionary forces on this process have been intense. Mosquito biologists would do well to follow the advice from Michael Service in the quotation referenced above.

Egg Development and Structure

Chapter 4 discussed the general pattern of egg development and embryogenesis. Here, more detail is presented on regulation of the process leading to completion of the egg oviposited by a female. Egg development may be divided into two sequential processes. The first starts right after emergence, when the ovaries consist of a series of ovarioles, each with a distal (i.e., toward the tip of the ovary) germarium (primordial cells) and proximal follicle (the oocyte and seven nurse cells surrounded by the follicular epithelium). At this stage the follicle is connected to the lateral oviduct by a solid rod, the pedicel, and a hollow tube, the calyx (Clements 2000). The germarium, follicle, and pedicel are surrounded by an extracellular matrix called the intima. The ovariolar sheath surrounds the intima and forms a connective tissue envelope around it with branching muscle fibers that continue to the calyx. Muscles activating the calyx are only longitudinal. There is an additional outer layer of circular muscles beginning at the lateral oviduct within the ovary. These lead from the two ovaries to a common oviduct separated from the vagina by a sphincter-equipped gonopore.

Stereospecific juvenile hormone III (JHIII) secreted by the corpora allata stimulates maturation (Mayoral et al. 2009; Fig. 10.1). Allatotropin from the brain stimulates the corpora allata to produce JHIII, which is less abundant in females that emerge with low teneral reserves (Hernández-Martinez et al. 2007). As the concentration of JH increases, gene expression in the fat body goes through early, middle, and late stages in anticipation of processing blood-meal nutritive components into yolk proteins (Zhu et al. 2003; Fig. 10.2). In *Aedes aegypti*, secretion of JH is highest 36 hours after emergence (Riddiford 2013).

The mosquito regulates the level of JH as maturation proceeds, actively reducing both the current level of the hormone and slowing further production. A specific esterase cleaves JH in *Culex quinquefasciatus* as a part of homeostatic control (Kamita et al. 2011). PISCF (=type C)-allatostatin from the brain slows production of JH in *Aedes aegypti* and *Anopheles gambiae*, reaching its highest level at three days postemergence when JH production is minimal (Li et al. 2004, 2006).

The early set of gene expression peaks 6 hours posteclosion, the middle set at 24 hours, and the late set at 66 hours. During this series of processes, the functions of the fat body shift from mainly lipid and carbohydrate metabolism to protein synthesis. At the end of this preliminary maturation process, the maturing follicle includes nurse cells and an oocyte. The oocyte is the cell that will become the female contribution to the embryo (Mahilum 2003). The oocyte is surrounded by a clear layer that is the location for yolk protein accumulation. JH controls gene expression through the activity of a combination (heterodimer) of two proteins: methoprene-tolerant

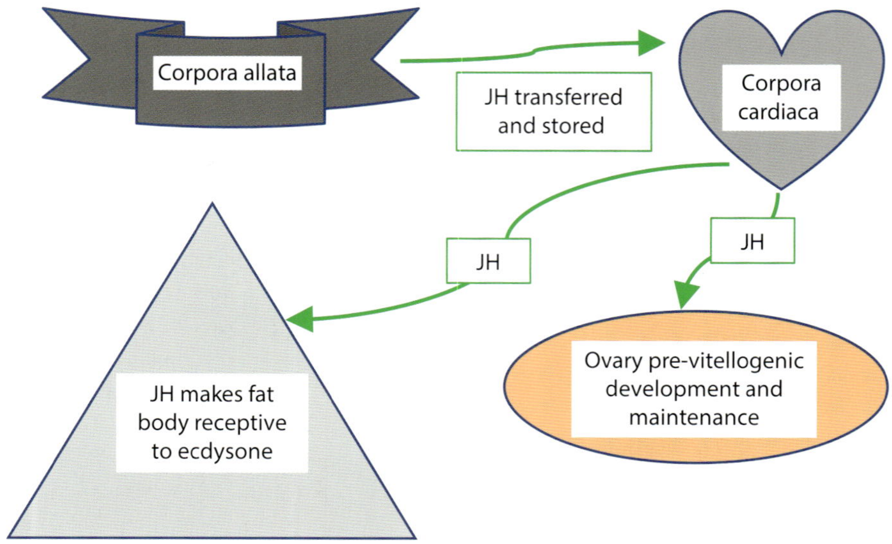

Figure 10.1. Regulation of ovary for about two days after eclosion. Green arrows indicate a chemical messenger. JH = juvenile hormone. Shaw et al. 2015

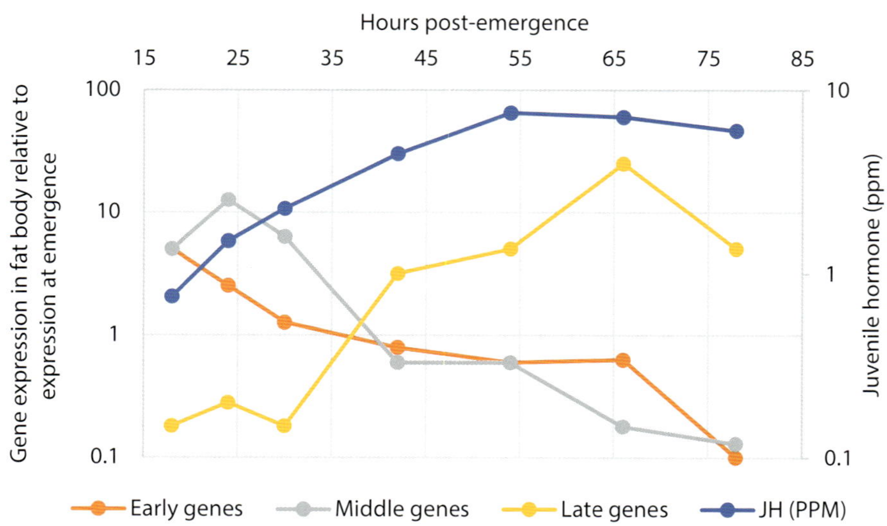

Figure 10.2. Gene expression in the fat body of female *Culex quinquefasciatus* in relation to juvenile hormone (JH) levels during the first few days following emergence. Redrawn from Zhu et al. 2003

(MET) and cycle (CYC; Shin et al. 2012). An alternative view of the JH receptor involves a homo- or heterodimer of MET plus a steroid reception coactivator called FISC (Riddiford 2013). This heterodimer binds to target genes in the presence of JH (Li et al. 2011). During previtellogenic ovarian development, a protein designated AHR38 inhibits formation of the receptor heterodimer involved in stimulation of vitellogenesis following a blood meal (Zhu et al. 2000), thus keeping the process at the correct stage at the correct time. This first stage of ovarian development ends with completion of previtellogenic maturation of the ovaries, in which the developing follicles separate from the germarium (Zhu et al. 2010), and the fat body becomes capable of producing yolk precursor proteins (YPP) in response to the genetic cascade initiated by 20-hydroxyecdysone (20E; Raikhel et al. 2002).

The next stage of egg development begins following a blood meal and is also related to mating (Fig. 10.3), the sequence of which varies between mosquito species and sometimes within species (Scott & Takken 2012). JH in male accessory gland substances increases fecundity in *Aedes aegypti* by directing nutrient sources to production, enlarging ovarian follicles, and preventing follicle resorption (Shaw et al. 2015). In *Anopheles gambiae*, 20E performs a similar function. The female in this species produces a protein that stimulates oogenesis following mating, known as the mating-induced stimulator of oogenesis (MISO) (Baldini et al. 2013). Produced in the female atrium in response to 20E binding to the ecdysone receptor (EcR), MISO stimulates production of a vitellogenic lipid transporter. Mating also influences egg development in blood-fed *Anopheles stephensi*, increasing egg development from 73% in unmated females to 100% in mated females (Uchida et al. 2003).

Blood feeding increases the concentration of amino acids in the hemolymph of *Anopheles stephensi*, peaking at about 75 nmol/μl (1.03% weight to volume, considering average amino acid molecular weight as 137) 24 hours after feeding (Uchida et al. 2003). Experimentally, a much higher concentration of amino acids (10%) is necessary to induce egg development in 78% of mated females, suggesting that other factors in the blood meal are involved in stimulation of the ovaries. In *Aedes aegypti*, larval nutrition affects the sensitivity of the fat body to the signals initiated by the increase in concentration of amino acids in the hemolymph (Shiao et al. 2008). The balance of individual amino acids in blood can affect fecundity (Braks et al. 2006). Neuropeptide F from the brain is part of the signaling pathway that indicates a blood meal (Garczynski et al. 2005, 2007). In *Aedes aegypti*, the cationic and branched amino acids (leucine, tryptophan, methionine, valine, histidine, lysine, phenylalanine, arginine, asparagine, and threonine) derived from the blood meal stimulate insulin-like peptides and ovary ecdysteroidogenic hormone (OEH) from the brain to cause ovarian production of ecdysone (Attardo et al. 2006; Riddiford 2013). The OEH is produced by medial neurosecretory cells of the brain and is distributed through axons that extend out through the neurohemal organ associated with the aorta. The axons of these nerves branch extensively along the midgut (Brown & Cao 2001). The blood meal also induces release of OEH from abdominal metameric perivisceral organs that are connected to abdominal ganglia by neurosecretory cells.

The insulin-like peptides initiate phosphorylation of ribosomal S6 kinase, a part of the target-of-rapamycin (TOR) pathway (Roy et al. 2007). More specifically, TOR kinase is stimulated by cationic amino acid transporters (CATs; Shiao et al. 2008). The kinase activates S6 kinase through phosphorylation, leading to GATA transcription factor (named for affinity to GATA bases) activating *Vg* expression. Regulation of this pathway is part of the control over vitellogenesis, with phosphatase and tensin homolog (PTEN) a factor that down-regulates the pathway (Arik et al. 2009).

The ecdysone is converted to 20E in the fat body, working with the insulin-like peptide to activate vitellogenic synthesis by the *YPP* gene. That gene has three upstream regulatory regions, requiring a combination of activations to function (Raikhel et al. 2002). The 20E binds at the EcR receptor coupled with mandatory partner proteins ultraspiracle (USP) A (during early and late vitellogenesis) or USP B during the middle phase (Wang et al. 2000) and a responsive element of one of the regulatory regions of the *YPP* gene called VgEcRE1 (Martin et al. 2001). The response of *YPP* to 20E binding is modulated by transcription protein factors produced by the early genes *E75* (Zhu et al. 2000), *Broad* (Chen et al. 2004), and *E74A* (activator)/*E74B* (repressor) in *Aedes aegypti*. The EcR also occurs in two forms, EcR A being less sensitive and peaking at the time 20E is most abundant, 16–20 hours post-blood meal (pbm; Wang et al. 2002). The E74 proteins work in conjunction with 20E, E74B being most abundant 24 hours pbm and E74A most abundant at the end of vitellogenesis (30–32 hours pbm; Sun et al. 2002). E75 also occurs in various forms,

Figure 10.3. Regulation of ovary following mating and a blood meal. The influence of ecdysone secreted by the male in the seminal fluid is described from *Anopheles gambiae*, and the overall pattern of regulation is taken from *Aedes aegypti*. Blue arrows indicate a composite process, green arrows indicate a chemical messenger, and black arrows indicate transfer of materials. 20E = 20 hydroxyecdysone; EcR/USP = receptor complex formed from ecdysone receptor and ultraspiracle receptor; IIS = insulin or insulin-like growth factor I signaling; MISO = mating-induced stimulator of oogenesis protein; OEH = ovarian ecdysteroidogenic (= ecdysiotropic) hormone; PTEN = phosphatase and tensin homolog (downregulates IIS); TOR = target of rapamycin kinase signal transduction pathway; YPP = yolk protein precursor. Brown and Cao 2001; Chen et al. 2004; Attardo et al. 2006; Arik et al. 2009; Baldini et al. 2013; Shaw et al. 2015

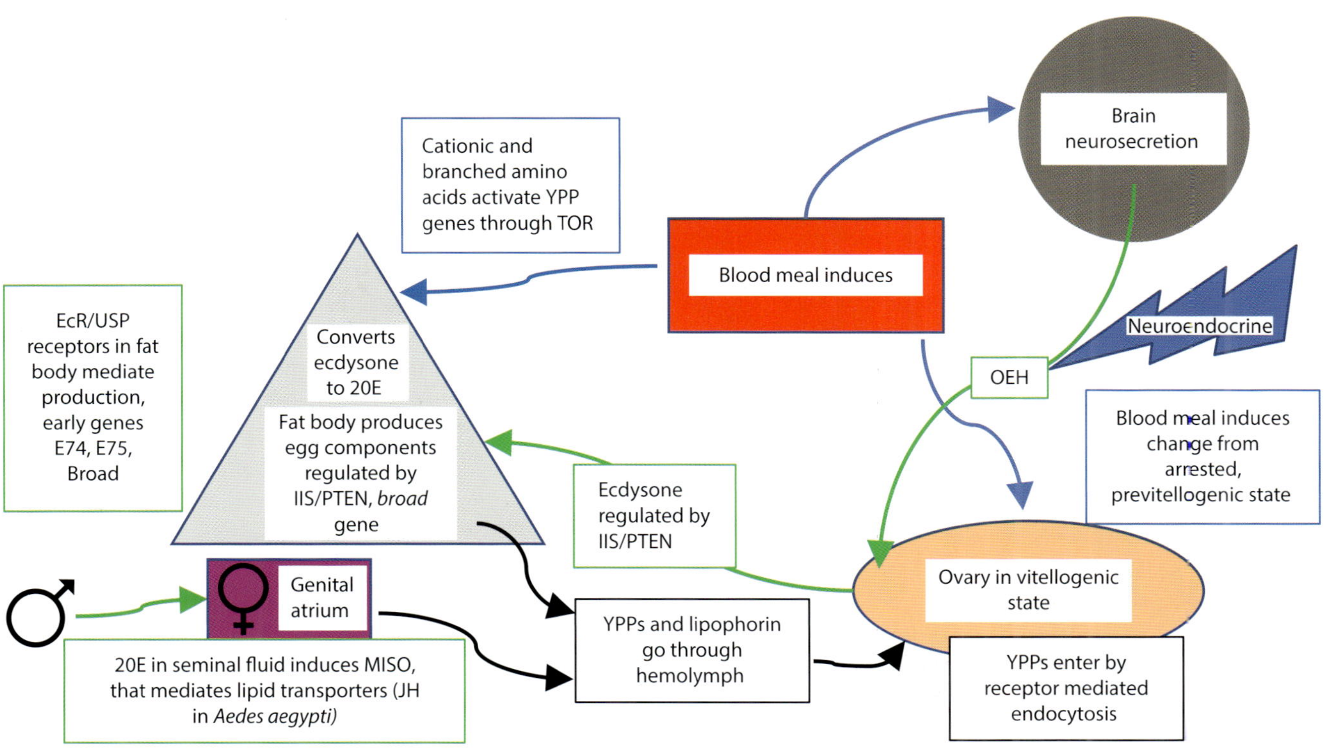

designated A, B, and C (Cruz et al. 2012). They interact with the EcR A receptor at different times following the blood meal.

The effect of ecdysone on the female adult is completely different from that on a molting larva because of the presence of 20E-induced proteins (βFTZ-F1, AHR3) that control the genetic response in the fat body and ovaries (Li et al. 2000; Kapitskaya et al. 2000; Zhu et al. 2003). Yolk protein precursors consisting of vitellogenin, vitellogenic carboxypeptidase (*VCP* gene with similar requirement for a combination of activations), lipophorin, and vitellogenic cathepsin B reach maximum production in fat body trophocytes 24 hours pbm, and production ceases at 30–32 hours (Zhu et al. 2000, 2010). The enzymatic components are proenzymes later activated during embryogenesis, which follows oviposition and fertilization. During the 72-hour period in which eggs are formed following a blood meal, approximately 7,500 genetic transcripts are expressed in the fat body (Zou et al. 2013; Roy et al. 2015). Amino acids and JH upregulate the first wave of genetic transcription 3–12 hours pbm, followed by ovarian-origin 20E 12–36 hours pbm, by an orphan nuclear receptor (HR3) at 36–48 hours pbm, and, finally, by JH and MET 48–72 hours pbm. The fat body shifts from lipid and carbohydrate storage to protein production until the eggs are completed, when the organ resorbs synthetic cells by autophagy and returns to a storage function. Earlier work with limited genome arrays showed how rapidly gene expression shifts following a blood meal (Marinotti et al. 2006). Just three hours after a blood meal, gene expression starts to shift in *Anopheles gambiae*. During the three days of egg development, 50% of all genes measured had changed their levels of expression.

The blood meal also initiates changes in the structure of the ovaries (Clements 2000). The calyces expand within 24 hours pbm, and the lateral oviducts widen at 72 hours. Near the end of oogenesis, the pedicel cells break down, leaving an outer membranous layer forming a tube from the developing egg to the calyx. At the final stage of egg development, the follicular epithelium secretes the chorion around the follicle, detaching from the posterior (distal) end first.

Fecundity can be influenced by temperature, probably because of size variation, in *Aedes aegypti* (Bader & Williams 2012). The mean number of ovarioles was 100 at 18°C, 85 at 25°C or 29°C, and 65 at 35°C. Aging reduces fecundity in *Anopheles gambiae*, caused at least in part by an accumulation of reactive oxygen species. Administration of antioxidants can reverse decreases in fecundity (DeJong et al. 2007). *Plasmodium falciparum* was more successful in mosquito strains selected for lower levels of reactive oxygen species and higher fecundity, suggesting that there is a natural selection for more susceptible strains of *Anopheles gambiae*. Estivation also reduces fecundity of *Anopheles arabiensis*, estivating individuals in the laboratory in Burkina Faso depositing a maximum of 480 eggs in four gonotrophic cycles over a period of up to five months, compared to nonestivating individuals depositing 763 eggs during five cycles (Holstein 1954).

Parasites and symbionts can also affect fecundity. For example, *Brugia malayi* Brug filarial infection delays egg development in *Armigeres subalbatus* (Santa-Ana et al. 2007). Artificial infection of *Aedes aegypti* with a strain of *Wolbachia* (*w*MelPop) almost eliminates the ability of this mosquito to produce viable eggs from chicken, mouse, or Guinea pig blood, despite normal fecundity when fed human blood (McMeniman et al. 2011). The oocyte itself can be infected with symbionts or pathogens, as mentioned in Chapter 7 in a discussion of the microbial flora of mosquitoes. The real importance of transovarial transmission of pathogens remains a point of discussion among medical entomologists, but abundant examples from a variety of mosquito-borne viruses indicate there is some role in the ecology of those diseases. On the one hand, there are indications that the rate of transovarial transmission could be underestimated by sampling eggs too early. For example, dengue-2 virus was detected in only 6%–9% of *Aedes aegypti* eggs immediately following oviposition and four days after intrathoracic inoculation, whereas 50%–63% of eggs were infected 60 days after oviposition (Mourya et al. 2001a).

Development

Autogeny

Autogeny in mosquitoes is the deposition of one or more batches of eggs without a blood meal. There are groups of mosquitoes that never take a blood meal and produce multiple batches of eggs (e.g., all *Toxorhynchites* species, *Topomyia* species, and *Wyeomyia smithii*). Some mosquitoes have just one autogenous gonotrophic cycle before starting the typical cycle of blood meals and oviposition. The nutritive contributions to the autogenous eggs are entirely dependent on lipids and proteins stored from larval feeding so that the success of such autogenous oviposition depends on larval nutrition (Attardo et al. 2005). Some of these species always deposit a batch of autogenous eggs (e.g., *Aedes atropalpus*, *Deinocerites cancer* [Downes 1966]), whereas others are influenced by photoperiod, temperature, or nutrition (e.g., *Culex tarsalis*, *Culex pipiens* Form Molestus). *Aedes taeniorhynchus* goes long distances from salt-marsh oviposition sites in southern Florida, United States, to inland sites for blood feeding, but females usually deposit a small (approximately 25 eggs) autogenous brood before embarking on the long flight for blood meals (Ritchie & Montague 1995). Autogeny may be a more widespread capability among mosquitoes than expected in that species usually considered anautogenous have rare autogenous strains (*Aedes aegypti*, *albopictus*).

Sugar feeding by females was necessary for *Opifex fuscus* to deposit eggs autogenously, despite abundant larval nutritional reserves (Snell et al. 2010). A mean of 30 (maximum 100) autogenous eggs were deposited as early as 7 days and an average of 12 days after emergence. In contrast, the occurrence of autogeny in *Aedes atropalpus* is not affected by larval nutrition or adult sugar feeding, though the number of eggs varied between a mean of 159 when larval nutrition was high to 71 when larval nutrition was low (Telang & Wells 2004). Autogenous oviposition by *Culex tarsalis* varies seasonally in Southern California, United States, being greatest in September and least in February (Reisen & Lothrop 1995).

Individual Egg

The detailed external morphology of eggs was described by Harbach and Knight (1978, 1980), concentrating on *Aedes aegypti*, *Anopheles aconitus*, and *Toxorhynchites brevipalpis*. The eggs of many other species have been described. More generally, the egg is usually some variation of oblong in shape with rounded ends. In *Aedes* and *Anopheles*, at least, the blunter end with the micropyle (where sperm enter) is anterior, the sharper end is posterior, and the more convex side is dorsal (Clements 2000). The maternally secreted chorion forms the exterior "shell" and is divided into two layers, the endochorion and the exochorion. The endochorion is the layer that darkens and hardens within hours of oviposition through a process of protein cross-linking accomplished by enzymes of maternal origin. The exochorion forms enduring floats, frills, and decks in some genera (e.g., *Orthopodomyia*, *Anopheles*, *Chagasia*) or a friable layer that adds contrast to underlying reticulations (e.g., *Aedes*, *Psorophora*).

External egg morphology has been used for classification of some *Anopheles* (White 1977), including identification of sibling species in the *maculipennis* complex. Interest in this genus that includes vectors of human *Plasmodium* combined with the intricate patterns of the exochorion have made eggs a target of study for decades. The species of *Aedes* and *Psorophora* that deposit eggs in soil in anticipation of subsequent flooding (i.e., floodwater mosquitoes) have also been identified from egg morphology. In this case, eggs have not been important in the taxonomy of species designation but have proven to be surprisingly practical for field identification (Horsfall et al. 1952; Horsfall & Craig 1956). Extraction of eggs from soil (Horsfall 1956) followed by microscopic examination of shape, color, and reticulation is an efficient method to survey for floodwater mosquitoes in North America. Submerged against a matte black background, the diagnostic characteristics of the eggs are easily seen in a single plane so that they can be quickly sorted by species under a dissecting microscope.

The common characteristics of aedine eggs have been used as one of the reasons uniting the concept of *Aedes* as a single large genus (Savage & Strickman 2004). All known eggs from the genus have an exochorion that is friable and often degrades within weeks of oviposition. The asymmetric profile (dorsal side more arched) and the presence of an egg cap indicated by a line of separation in the serosal cuticle are also characteristic of the genus. *Aedes* species do not deposit their eggs on the water and usually have Type II (require hatching stimulus after embryonation) or Type III (require conditioning prior to hatching stimulus) eggs. The resistance of *Aedes* eggs to desiccation varies, even among species with similar ecology. For instance, eggs of *Aedes albopictus* are more resistant than those of *polynesiensis*, despite their close phylogenetic and ecological relationship (Gubler 1970). Some other mosquito eggs have some resistance to drying, including *Opifex fuscus*; *Orthopodomyia signifera*, *kummi*, and *alba*; *Eretmapodites*; *Anopheles arabiensis*, *dirus*, and *gambiae*; *Malaya*; *Wyeomyia*; and *Tripteroides bambusae* (Horsfall 1955; Clements 1999).

Groups of Eggs

Many species deposit eggs in clusters. The most commonly observed kind of cluster is a raft, in which each egg stands vertically and adheres to the sides of its neighbors. The genera that deposit rafts are *Culex*, *Lutzia*, *Culiseta*, *Coquillettidia*, *Uranotaenia*, and *Armigeres* (Clements 2000). Deposition of egg rafts probably arose multiple times evolutionarily within the subfamily, considering the wide variety of culicines that deposit eggs as rafts. Another indication of the adaptive, rather than phylogenetic, origins of raft formation is the presence of species in predominantly raft-forming genera that deposit single eggs (e.g., *Culiseta victoriensis*, *hilli*, and *frenchii*; *Culex* (*Microculex*) [Harbach 2011]; *Culex* (*Melanoconion*) *alogistus*, *ocossa*, *panocossa*, *pilosus*, and *vexillifer*) and raft-forming species in groups that usually deposit single eggs (e.g., *Trichoprosopon digitatum* and *Uranotaenia* (*Pseudoficalbia*) [Peyton 1977]). In *Culex*, the eggs adhere lightly to each other laterally with interlocking tubercles. They tend to stand vertically on the water because the anterior end faces the water and has a hydrophobic center and raised hydrophilic collar (the corolla). *Trichoprosopon digitatum* eggs are hydrophilic anteriorly and hydrophobic posteriorly, presumably achieving the same vertical orientation in water as for *Culex*. *Trichoprosopon* eggs also adhere to each other with interlocking tubercles.

Eggs form other kinds of groups as well. *Anopheles* eggs on the water often form "rosettes" because the ends of the eggs are attracted to each other. *Anopheles melas*, unusually for the genus, rests in one place while ovipositing on damp surfaces and thus can create a pile of eggs that stick to each other as they harden over a period of two to three hours (Giglioli 1965). *Mansonia* eggs are deposited underwater, attached to the leaves of aquatic plants. The eggs are secured by their posterior ends to form a different kind of rosette and encased in a bubble with hydrophilic anterior ends protruding into the water. Other genera with species that cement eggs to plants are *Mansonia* and *Ficalbia* and *Culex* (*Melanoconion*), (*Neoculex*), and (*Lophoceraomyia*) (Horsfall 1955; Clements 1999). *Armigeres* (*Leicesteria*) *digitatum* deposits eggs that rest vertically in the water with the spiny, anterior end protruding above the surface (Okazawa et al. 1991). The spines hold the eggs together in a group. Perhaps the strangest grouping of eggs occurs in *Verrallina carmenti*, which deposits its eggs above the waterline in a ribbonlike gelatinous matrix (Belkin 1962), reminiscent of the egg clusters of chaoborids (Strickman 1980a). Another species using a gelatinous matrix for its eggs is *Culex* (*Microculex*) *gaudeator*, though the mass is deposited on the water (Clements 1999).

Ovulation and Oviposition

The path of the oocyte from the follicle to the ovipositor is powered by the muscle layers described above and controlled by the terminal abdominal ganglion (Clements 1999). Neither the calyces nor the lateral oviducts are open until after the first ovulation. The opening of the various tubes leading to the ovipositor starts with the pedicel cells breaking down near the end of oogenesis, leaving the outer membranous layers as the first duct through which the oocyte will pass. Changes in the calyces start about 24 hours pbm, expanding and changing cellular structure. By 72 hours pbm, the lateral oviducts are open. After the tubes are open, the oocyte's movement into the vagina is controlled by a sphincter in the internal gonopore. The oocyte proceeds to the final opening in the female genital tract, the gonotreme, posterior first so that the micropyle faces anteriorly relative to the female. Sperm are allocated to the oocyte in the vagina as the oocyte is partially extruded through the gonotreme, and the sperm enter through the micropyle. Technically, the female extrudes an oocyte and not an egg because fusion of the sperm and oocyte cell does not take place until after oviposition. Seminal fluid proteins and hormones stimulate oviposition in *Anopheles gambiae*, *Aedes aegypti*, and *albopictus* (Helinski et al. 2012a; Shaw et al. 2015).

Adaptations for Maintaining Position of Eggs Following Oviposition

Some of the ways in which groups of eggs are formed into rafts or clusters were described previously. Individual eggs can also have adaptations to maintain their positions (Savage & Strickman 2004). *Anopheles*, *Toxorhynchites*, *Trichoprosopon*, *Wyeomyia*, *Orthopodomyia*, *Mimomyia*, and *Tripteroides* usually deposit eggs singly on the water surface, and the eggs, presumably, maintain position at the surface by having a specific gravity less than that of water. Some of these mosquitoes have obvious chorionic adaptations to trap gas to keep the eggs afloat. *Mansonia*, *Ficalbia*, some species in the *Culex* subgenera *Melanoconion*, *Neoculex*, and *Lophoceraomyia* cement eggs to the leaves of aquatic plants. Some mosquitoes secure their eggs by fibers (*Haemagogus* and a few *Aedes*) or adhesive (*Orthopodomyia signifera*, *kummi*, and *alba*) above the waterline. *Aedes aegypti* forms a gelatinous chorionic pad on the dorsal side, cementing eggs in place with the ventral side facing away from the substrate (Christophers 1960).

Process of Egg Deposition

Mechanics

Relatively few detailed accounts exist on the process of oviposition. The observations that have been made, even at a general level, show that the family Culicidae brings much of the adaptive variety to egg deposition that it brings to larval feeding, mating, dormancy, and other life processes. Mosquitoes oviposit during flight or while in contact with a substrate, such as a plant, a vertical container surface, or the soil. Oviposition during flight requires precise control as the mosquito places eggs in the kinds of places on the water most suitable for them or even tosses the eggs into narrow holes. Oviposition from a surface takes advantage of the derrick-like structure of the legs to allow the positioning of the end of the abdomen. The abdomen itself can direct eggs by bending at the terminal three segments, forming a short functional ovipositor sometimes used to place eggs among particles of soil, within textured surfaces, under the surface of an aquatic plant, or sequentially in a highly ordered egg raft. The descriptions of oviposition below will give some idea of what a mosquito does when it deposits an egg.

Anopheles melas, a member of the Gambiae Complex, were observed in *Avicennia* black mangrove swamps of the Gambia (Giglioli 1965) but not in *Rhizopora* red mangrove (Muirhead-Thomson 1946). Females performed a preoviposition behavior for about 30 minutes, during which they flew back and forth to the mud surface, patting it with their hind tarsi. The ovipositing female stood on the mud, on the aerial root of a mangrove, or on another plant, holding its abdomen with the tip upward at a 45° angle. Eggs were ejected forcefully from the abdomen to the mud surface, forming a pile. Each egg appeared posterior first from the genital pore and remained only partially extruded while it was inseminated through the micropyle, which was still inside the female atrium. There was a short pause after each egg was ejected, creating a 10–12 second interval for deposition of each egg. Eggs are mostly deposited during the two highest tides of each month, creating cohorts of mosquitoes every two weeks during seasons corresponding to highest freshwater inputs to *Avicennia* groves.

In the high Arctic on Ellesmere Island (81°N), the floodwater mosquito *Aedes nigripes* searches for the warmest, sunniest, and best wind-sheltered spots around receding snowmelt pools in the tundra (Corbet & Danks 1975). They place their eggs about 2.5 cm above the level of the standing water, which sometimes puts the eggs in moss or at the base of sedge stems meters away from the water. Probably because of the harsh temperature conditions, tiny differences in microtopography can affect the distribution of eggs.

One of the stranger oviposition habits is shared by six species in the aedine *Armigeres* subgenus *Leicesteria* (*flavus*, *annulitarsis*, *balteatus*, *inchoatus*, *magnus*, and *traubi*; Lounibos 1983). These mosquitoes deposit their eggs on their own hind legs, carrying the eggs until just before eclosion. At that time, the mosquitoes go to a suitable open fruit husk that holds water and dip the eggs in the water, where they hatch immediately. At least one of these species, *Armigeres flavus*, holds the raft between the hind tarsi and tibiae, then releases the entire raft when it finds a suitable site, and the eggs are ready to hatch (Miyagi et al. 2005). The only other form of maternal behavior by mosquitoes has been described for the sabethine *Trichoprosopon digitatum* (Lounibos & Machado-Allison 1983). These mosquitoes deposit a raft of eggs on the surface of foul water in cut cacao pods or bamboo. Females stand on the surface of the water, holding the rafts between their legs until the larvae hatch.

Not all species in the subgenus *Leiscesteria* exhibit maternal behavior. *Armigeres digitatus* deposits eggs during the day in newly cut bamboo internodes in Sarawak, Malaysia (Okazawa et al. 1991). The female flies up and down in the bamboo container, finally landing on the water. It shakes its abdomen, and the undarkened posterior end of the egg appears partially extruded. The egg remains in this position for 20–30 seconds, then the female bends its abdomen down to the water and inserts the egg below the surface. Once released, the egg floats vertically, and the female places subsequent eggs adjacent to the first ones so that they form a group connected by their anterior, spiny ends that project above the surface.

Topomyia yanbarensis is a sabethine mosquito that lives in the Tokara Island chain of southern Japan (Okazawa et al. 1986a). This species never takes blood and deposits only about two eggs per day. Females about to oviposit rest on bamboo stems, examining small holes with antennae and the proboscis. During oviposition, they hover above a hole leading to water in the internode. With the midlegs contacting the bamboo's surface, the mosquito accurately propels a single egg into the hole.

Sabethes chloropterus, a sabethine in the Western Hemisphere, also propels its eggs through holes into bamboo (Galindo 1958). This mosquito prefers vertically oriented stems that are covered on top and a hole that is 2.5 cm wide. Females approach the stem slowly, circle it two or three times, and then hover in front of the hole with their hind legs and forelegs flexed over the dorsum and almost touching. The midlegs extend out ventrally. Very quickly, the mosquito flexes its head and thorax backward, thrusts its abdomen forward, and tosses one or two eggs into the hole. This may be repeated in the same or a different hole up to 18 times, either with or without a resting period between oviposition events.

Behavior

SKIP OVIPOSITION

Container-developing species that deposit eggs singly may engage in "skip oviposition," in which females use more than one site for distribution of eggs from a single gonotrophic cycle (Mogi & Mokry 1980). Skip oviposition was suggested as the standard term for this phenomenon, based on habits of *Aedes albopictus* in Japan and *Wyeomyia smithii* in Newfoundland, Canada. In each case the distribution of fewer eggs among more sites was considered an adaptation to the limitation on nutritional resources within a small container habitat.

Although the movement of female *Aedes aegypti* is usually limited to a radius of about 200 m, the species has been shown to fly much farther in quest of oviposition sites (Reiter 2007). Separate studies in Puerto Rico, using rubidium markers, and in Trinidad, using genetic signatures, showed that many females deposited eggs from a single gonotrophic cycle in more than one site (Reiter et al. 1995; Colton et al. 2003). *Aedes albopictus* distributes its eggs more widely in the summer months in New Jersey, United States (Fonseca et al. 2015). The generation of females depositing eggs that overwinter concentrate their eggs in open habitats with more organic matter. Skip oviposition tends to blur oviposition site preferences, as the most preferable sites will not necessarily get more eggs than other sites (Corbet & Chadee 1993).

Opifex fuscus is among the container-developing species that use skip oviposition, in this case the container being a rock pool (Snell et al. 2010). During the first autogenous gonotrophic cycle, this species deposited 58% of its eggs in two or more (maximum of eight) different sites. *Eretmapodites dracaenae* also distributes its eggs among oviposition sites, as observed in plant axils in Uganda (Haddow 1946).

Oviposition Habitat Selection

Oviposition habitat selection has been defined as "a pattern of oviposition selective for a measurable attribute of the habitat" (Spencer et al. 2002). In general, female mosquitoes choose where the larvae will develop based on where they put their eggs (Afify & Galizia 2015). Using olfactory cues at a greater distance and visual, tactile, and gustatory cues at a shorter distance, females execute a wide variety of behaviors to place eggs (Navarro-Silva et al. 2009). Direct selection by the successful or unsuccessful development of the eggs to the adult stage combined with adaptation of the immature stages to various kinds of sites has resulted in coadaptation of females and immature stages. In some sense, this coadaptation is like that observed between two separate organisms and has been the source of much of the niche development in the family Culicidae. Although mosquitoes are more similar to *r* strategists characterized by aggressive colonization of new sites, high reproductive rates, and high mortality rates, they nonetheless compete for resources in the aquatic habitat as larvae, and niche separation may be important for many, if not most, species.

Searching for oviposition sites may be an important cost to mosquitoes during their life cycles. The skip oviposition habit described above causes the female to fly greater distances and extends the gonotrophic period of *Aedes aegypti* from three or four days in the laboratory to at least eight days in the field (Reiter 2007). The abundance of *Aedes aegypti* near oviposition sites (Pates & Curtis 2005) may reflect the need of this species to minimize the already considerable time females spend finding places to deposit eggs. Females seeking oviposition sites were also influenced by a geographic barrier (large road) and by attractive habitat (shady area with trees) in Queensland, Australia (Russell et al. 2005), or by a leeward orientation during the dry season (Williams et al. 2006).

Extension of the gonotrophic period by delays in finding oviposition sites was also supported by a model of *Anopheles* survival and reproductive rate (Gu et al. 2006; Gu & Novak 2009). The relative abundance of oviposition sites and the amount of time required to find them could extend the gonotrophic cycle considerably, in theory reducing the basic reproductive rate. For example, if 2.5% of the landscape is suitable for oviposition and a mosquito flies 250 m per day, it will take 13.6 days for a female to find a place to deposit eggs. If 12.5% of the landscape is suitable, then the gonotrophic cycle is only 4.1 days. Of course, mosquitoes cut search time through orientation toward chemical or other characteristics of appropriate oviposition sites. If there are no places to oviposit, then females must wait longer. Many different circumstances might lead to such a delay, but in one case *Aedes vexans* females in Central Illinois, United States, were observed to cease oviposition for 11 days when suitable sites were flooded (Strickman 1980b). Salt marsh species, like *Aedes sollicitans*, that fly inland

40–90 km for blood meals usually return to the salt marsh to deposit eggs, expending a great deal of energy on finding oviposition sites (Shone et al. 2006). *Anopheles melas*, a salt-marsh species on the western coast of Africa, also travels inland for blood meals and then back to the salt marsh to oviposit (Giglioli 1965). The search for oviposition sites may benefit a population beyond putting the eggs at the most suitable sites by effectively distributing individuals over a larger geographic range (Edman et al. 1998; Le Menach et al. 2005).

Characteristic Kinds of Oviposition Sites

Chapter 3 includes detailed discussions of what kinds of mosquitoes are found where and at what times. Larval habitats discussed in the chapter and summarized in Table 3.7 could be used as a categorization of oviposition sites as well because the larvae are for the most part distributed where females put their eggs. The descriptions below summarize observations about influences on oviposition at those sites.

Anopheles species usually deposit their eggs directly on the water, though some species prefer to place their eggs on damp surfaces. *Anopheles gambiae* s.l. deposits 77% of its eggs on mud adjacent to pools in western Kenya (Miller et al. 2007), preferring fresh, sunlit, transient, turbid water devoid of vegetation (Sumba et al. 2004). Oviposition was greatest when the soil was saturated with moisture, regardless of particle size (Huang et al. 2006). Even more eggs come to rest on the mud either because the size of the pool decreases or because wind blows them there. The first instars can crawl over the mud to the water. *Anopheles baimaii* in Bangladesh (reported as *Anopheles dirus*) behaves similarly, depositing its eggs above the waterline (Rosenberg 1982). The eggs remain viable for two weeks until either hatched by flooding or the first instars crawl across damp margins of the ground pool to water. The observation of a single *Anopheles gambiae* larva hatching from a soil sample that had been dry for 1.5 months in western Kenya calls into question the conclusion from laboratory studies that eggs of this species can only live about two weeks before hatching (Minakawa et al. 2001).

Anopheles arabiensis deposits its eggs in more sunny and exposed locations than *Anopheles funestus* in the Kilombera Valley, Tanzania, resulting in more pronounced seasonality of the former than the latter (Charlwood et al. 1995b). In the same location, an interesting trapping technique that floated clear sheets of acetate with glue on the water showed that gravid *Anopheles*, *Culex*, and *Mansonia* came to oviposit (Harris et al. 2011). The status of rice fields can influence oviposition by *Anopheles arabiensis* (Mutero et al. 2000). Mosquitoes preferred shallow depths of less than 4 cm and fields that were flooded before rice was transplanted to the field.

Anopheles pseudopunctipennis in Central America prefers sites that contain filamentous green algae (Bond et al. 2005). The algae are clearly important for nutrition, as they constitute almost half of the gut contents of larvae in Chiapas, Mexico. In the laboratory, pure bacteria-free cultures of the alga *Spirogyra majuscula* Kützing were sufficient food for developing larvae. Larvae were unable to avoid predation by fish in the absence of algae, indicating another selective value for females to choose oviposition sites with algae. Oviposition habitat selection may be the cause of the seasonality of this species in much of its range. It is generally more abundant in the dry season when the flow in rivers and streams slows or stops, giving a chance for filamentous algae to grow.

Culex quinquefasciatus oviposits in streams when sewage outflow enriches the nutrient content (Benzon & Apperson 1988). In Florida, enrichment, season (represented by temperature), presence of conspecific larvae, and both current and previous rainfall were important components of a predictive model of oviposition. The observation of oviposition in moving water is interesting, especially since this species colonizes such a wide variety of habitats. In Illinois, the closely related *Culex pipiens* was observed by the author (D.S.) to oviposit in polluted urban streams, the egg rafts sometimes floating down the middle of the channel and becoming caught in a screen placed across it. In the laboratory, females oviposited in moving water when they had no choice, and the larvae hatching from those egg rafts successfully completed development in the moving water.

Some *Culex* species deposit eggs above the waterline (Clements 1999). *Culex territans* sometimes puts its raft above the water. *Culex abominator* typically oviposits on the tops of duckweed (*Lemna*), and both *Culex erraticus* and *theobaldi* place eggs on plants above the water. *Culex alogistus*, *ocossa*, *panocossa*, *pilosus*, and *vexillifer* deposit single eggs on plants or most surfaces above the waterline. In these cases the first instar must crawl to the water when it hatches, though *alogistus*, *pilosus*, and *vexillifer* may wait up to a month before hatching. *Orthopodomyia signifera*, *kummi*, and *alba* also deposit eggs above the waterline, gluing them in place (Horsfall 1955; Clements 1999).

Many mosquitoes place their eggs on soil in locations that are likely to flood at some time in the future. These floodwater mosquitoes (Horsfall 1963) are all in the genera *Aedes* and *Psorophora*. Selection of oviposition sites is influenced by several factors—most notably, elevation above the existing water level of a ground pool or shoreline. The elevation probably reflects moisture level at the time of oviposition, as seen in the laboratory with *Aedes vexans* (Strickman 1980d). The amount of moisture may be quite important during the first critical hours following deposition when the egg must absorb water, and the chorion is less resistant to drying. The embryo is also more susceptible to drying, probably until the formation of the serosal cuticle toward the end of development. Females also choose locations based on proximity to objects, possibly because of the objects' effect on surface moisture near them and because of shading (Strickman 1980c). *Aedes vexans*, at least, has a strong preference for soils that have flooded before (Strickman 1980b). It is not known whether the mosquito distinguishes such soil by physical or chemical signals, but this behavior reduces the chances that eggs would be wasted on rain-soaked soil or adjacent to bodies of water that do not change level.

Containers are important oviposition sites for many mosquitoes. Such sites are not in contact with soil and can be categorized as plant based (phytotelmata), human made (artificial), or geologic (rock pools). Rarely, containers of animal origin (zootelmata) are important, such as *Eretmapodites* in snail shells or *Deinocerites* in crab holes. Mosquitoes that use these oviposition sites are often referred to as "container breeding," using breeding in the husbandry sense of conducting an operation in which animals are produced from mating to slaughter. Except for intentional production of mosquitoes for research or control purposes, container breeding is a misnomer for mosquito biologists. A more accurate designation would be "container developing," the term we have tried to use consistently.

Container-developing species sometimes show specialization in selection of habitats. For example, *Culex pinarocampa* only oviposits in rainwater accumulated in palm spathes in Central America. *Aedes dybasi* and *maehleri* of Palau and Yap, respectively, put their eggs in the pitchers of *Nepenthes*, avoiding the digestive properties of these plant parts (Mogi 2010). *Uranotaenia ascidiicola*, *xanthomelaena*, and *gigantea* also occur in *Nepenthes* (Peyton 1977). *Wyeomyia smithii* develops exclusively in *Sarracenia purpurea* L. pitcher plants. *Topomyia yanbarensis*, mentioned above, is just one example of a large series of species that oviposit in water accumulated in bamboo stems, sometimes further specializing into preferences for age, position, or condition of the stem.

Other container-developing species are opportunistic or have expanded their repertory of containers to include artificial containers. The primary examples of such generalists are *Aedes aegypti* and *albopictus*. These two species have expanded their distributions globally through phoretic movement of containers by humans. The variety of container habitats used by *Aedes aegypti* have been described in various settings as a part of efforts to eliminate sources of the mosquitoes (e.g., Kittayapong & Strickman 1993; Wong et al. 2011), showing greater abundance in certain kinds of common sites. Water jars commonly used in rural Thailand for water use and storage represented more than half of the infested sites. Interestingly, covered water jars were infested more often than uncovered jars. Lab experiments showed that females would enter the water jars through the small crevices left between the cover and the jar (Strickman & Kittayapong 1993), though the lids reduced the number of eggs in the laboratory and in the field in Peru (Wong et al. 2011).

Some members of the subgenus *Culex* are facultative about their use of containers for oviposition. This is particularly noticeable for *Culex pipiens* and *Culex quinquefasciatus*. These species are commonly found in a wide variety of containers, including cemetery vases, buckets (Strickman 1988b; Strickman & Fonseca 2012), and storm drains (Strickman & Lang 1986). They are also found in ponds and pools containing water with high organic content (Becker et al. 2003). Their common colonization of underground sites and containers with narrow access led to studies of their ability to travel up pipes to seek oviposition sites (Harbison et al. 2008). *Culex quinquefasciatus* females were more likely to travel along horizontal than vertical pipes. They went as far as 24 m, though proportionately fewer traveled a greater distance. Narrower diameter pipes (1.3 cm) discouraged ovipositing females, though a few went up to 2.7 m through such narrow tubes.

The optical qualities of water can have a big influence on oviposition habitat selection (Bernáth et al. 2012). *Culex restuans* is more attracted to black-lined pools, and gravid females of both *Aedes aegypti* and *atropalpus* will hover over surfaces with the proper reflectance, such as glass on top of black paper or a black-and-white contrasting surface. *Anopheles gambiae* orients toward the movement of larvae on the surface of water, and *Anopheles minimus* turns toward the sight of water. In the absence of other cues, ovipositing *Aedes aegypti* orient orthogonally to the polarization of light from the water's surface.

Chemical Signals for Oviposition

Dozens of organic chemicals affect mosquito oviposition (Afify & Galizia 2015). The origins of those chemicals include plants, bacteria, aquatic predators, and other mosquitoes. Some animals may produce chemicals that attract oviposition coincidentally, as the presence of these animals would be unlikely in larval habitats (*Aedes albopictus* attraction to water with carpet shells [*Paphia undulata* Bom] or giant tiger prawns [*Penaeus monodon* Fabricius]; Thavara et al. 2004). Inorganic chemicals, like NH_4^+ and PO_4^{-3} ions, can influence oviposition, though attraction to them may be the result of their effect on the bacterial flora (Nguyen et al. 2012). In addition, synthetic chemicals unrelated to any known natural product affect oviposition. The effects of these chemicals can be categorized as those that influence the rate of oviposition (deterrents and stimulants) or orientation to the oviposition site (repellents and attractants).

INFLUENCE OF PLANTS

Table 10.1 (Afify & Galizia 2015) summarizes the effects of plants on oviposition by mosquitoes. The dominant effects are positive, either attraction or stimulation, but this could be an artifact of the species of plants and mosquitoes selected for testing. Grass infusions were about equally attractive to *Culex quinquefasciatus* as the egg-raft pheromone (see below) in Tanzania, but together the infusion and the pheromone were synergistic (Mboera et al. 2000). Plant extracts were often found to be deterrent, but this could be an artifact of looking for substances to discourage oviposition. Chemicals associated with plants have both negative and positive effects (Table 10.2), usually dependent on concentration. The chemicals are in a variety of categories but terpenes predominate. Some of the chemicals are those that affect adult mosquito orientation toward sugar sources, including phenols, aldehydes, alcohols, ketones, terpenes, linalool, pinene, and nonanal.

Some specific associations have been shown (Navarro-Silva et al. 2009). Attractive compounds include p-cresol (4-methylphenol, extracted from birchwood) for *Aedes triseriatus*, grass-derived chemicals (phenol, 4-methylphenol, 4-ethylphenol, indole, 3-methylindole or skatole) for *Culex quinquefasciatus*, and skatole for *Culex tarsalis* and *stigmatosoma*. Combinations of chemicals can be synergistically

Table 10.1. Plant extracts and infusions that affect mosquito oviposition (Afify and Galizia 2015)

Plant species	Genera tested	Effects
Family: Acanthaceae		
Andrographis lineata	*Anopheles*	Deter
Andrographis paniculate	*Anopheles*	Deter
Family: Apiaceae		
Apium graveolens (celery)	*Aedes*	Deter
Family: Apocynaceae		
Solenostemma argel	*Culex*	Deter
Family: Asteraceae		
Ageratum houstonianum	*Aedes, Anopheles, Culex*	Deter
Artemisia annua (wormwood)	*Aedes, Anopheles, Culex*	Deter
Solidago canadensis (goldenrod)	*Aedes, Anopheles, Culex*	Deter
Spilanthes mauritiana	*Aedes, Anopheles, Culex*	Deter
Family: Cannabaceae		
Celtis occidentalis (hackberry)	*Aedes*	Attract
Family: Ceratophyllaceae		
Ceratophyllum demersum (hornwort)	*Anopheles*	Stimulate, deter
Family: Cyperaceae		
Fimbristylis spadicea	*Anopheles*	Stimulate, deter
Family: Fabaceae		
Acacia schaffneri (acacia)	*Culex*	Attract
Cassia obtusifolia	*Anopheles*	Deter
Medicago sativa (alfalfa)	*Aedes, Culex*	Attract, stimulate
Family: Fagaceae		
Quercus alba (white oak)	*Aedes*	Attract, repel, stimulate, deter
Quercus nigra (water oak)	*Aedes*	Attract, stimulate
Quercus virginiana (live oak)	*Culex*	Attract
Family: Lamiaceae		
Mentha piperita (peppermint)	*Aedes*	Deter
Ocimum basilicum (basil)	*Aedes*	Deter
Rosemarinus officinalis (rosemary)	*Aedes*	Deter
Family: Menispermaceae		
Cocculus hirsutus (broom creeper)	*Anopheles*	Deter
Family: Moringaceae		
Moringa oleifera (benzoil tree)	*Aedes*	Stimulate
Family: Myrtaceae		
Syzygium cumini (Java plum)	*Aedes, Anopheles, Culex*	Deter
Family: Pinaceae		
Pinus palustris (longleaf pine)	*Aedes*	Repel, stimulate
Family: Poaceae		
Arundinaria gigantea (bamboo)	*Aedes*	Attract, stimulate
Bambusa sp. (bamboo)	*Aedes*	Stimulate
Brachiaria mutica (para grass)	*Anopheles*	Stimulate, deter

Plant species	Genera tested	Effects
Cymbopogon nardus	*Aedes*	Deter
Cynodon dactylon (Bermuda grass)	*Aedes, Culex*	Attract, repel, stimulate, deter
Digitaria sp.	*Culex*	Attract, stimulate
Eleusine indica (goose grass)	*Culex*	Stimulate
Jouvea straminea	*Anopheles*	Stimulate, deter
Megathyrsus maximus (Guinea grass)	*Aedes*	Attract, stimulate
Pennisetum purpureum (Napier grass)	*Aedes*	Stimulate
Stenotaphrum secundatum	*Aedes*	Attract, stimulate
Family: Rutaceae		
Aegle marmelos (bael)	*Anopheles*	Deter
Family: Solanaceae		
Solanum tribolatum	*Anopheles*	Deter

Table 10.2. Summary of the effect of plant-derived chemicals on mosquito oviposition (Afify and Galizia 2015)

Chemical	Class	Genera tested	Effect on oviposition
Nonanal	Aldehyde	*Culex*	Attract, stimulate
4-methylcyclohexanol	Alicyclic alcohol	*Aedes, Toxorhynchites*	Attract, stimulate
2,3-benzopyrrole (indole)	Aromatic heterocyclic	*Aedes, Culex*	Attract, repel, stimulate, deter
3-methylindole (skatole)	Aromatic heterocyclic	*Aedes, Culex, Toxorhynchites*	Attract, repel, stimulate, deter
Naphthalene	Aromatic polycyclic	*Culex*	Stimulate
2-tridecanone	Ketone	*Culex*	Attract
Dimethyltrisulfide	Organic trisulfide	*Culex*	Attract, repel
2,3-dimethylphenol	Phenolic	*Aedes*	Stimulate
2,4-dimethylphenol	Phenolic	*Aedes*	Attract, stimulate
2-methylphenol (o-cresol)	Phenolic	*Aedes, Toxorhynchites*	Attract, stimulate
3-methylphenol (*m*-cresol)	Phenolic	*Aedes, Toxorhynchites*	Attract, stimulate, deter
4-ethylphenol	Phenolic	*Aedes*	Stimulate, deter
4-methylphenol (*p*-cresol)	Phenolic	*Aedes, Culex, Toxorhynchites*	Attract, repel, stimulate, deter
Phenol	Phenolic	*Aedes*	Attract, stimulate, deter
Eugenol	Phenylpropanoid	*Aedes*	Deter
Trans anethole	Phenylpropanoid	*Aedes*	Deter
Beta-pinene	Terpene	*Aedes*	Stimulate
Borneol	Terpene	*Aedes*	Stimulate
Borneol acetate	Terpene	*Aedes*	Stimulate
Camphor	Terpene	*Aedes*	Stimulate
Cineol	Terpene	*Aedes*	Deter
Citronellal	Terpene	*Aedes*	Deter
Confertifolin	Terpene	*Aedes*	Deter
Linalool	Terpene	*Aedes*	Deter
p-cymene	Terpene	*Aedes*	Deter
Pulegone	Terpene	*Aedes*	Deter
Thymol	Terpene	*Aedes*	Deter
Trimethylamine	Tertiary amine	*Culex*	Attract

attractive, as is the case for a combination of skatole and mosquito oviposition pheromone (MOP) for *Culex quinquefasciatus*. Isobutyric, isovaleric, and hexanoic acids are repellent to ovipositing *Aedes aegypti*, *Culex quinquefasciatus*, and *Culex tarsalis*. *Aedes aegypti* and *albopictus* are repelled by C_{21} fatty acid esters, including hexadecyl pentanoate, tetradecyl heptanoate, and tridecyl octanoate; however, propyl octadecanoate is attractive. Female *Anopheles stephensi* are also sensitive to C_{21} fatty acid esters in a concentration-dependent way, depositing fewer eggs at some concentrations and more eggs at others (Sharma et al. 2009). *Anopheles albimanus* are repelled from ovipositing in water with the plant-derived chemicals guaiacol, phenol, isoeugenol, longifolene, caryophyllene, phenylethyl alcohol, or p-cresol.

The odorant-binding proteins (OBPs) of *Culex quinquefasciatus* sensitive to 4-methylphenol and 4-ethylphenol were discovered in a survey of OBPs identified in the *Anopheles gambiae* genome (Zhu et al. 2013). Injection of RNAi against expression of the genes *OBP37* or *OBP99* into pupae stopped detection of 4-methylphenol and 4-ethylphenol by emerging females. Expressing these genes in frog oocytes together with the universal coreceptor gene *ORCO* caused the cell membranes of the oocytes to respond to low concentrations of the two oviposition attractants.

Oviposition attractants represent another example of learning in mosquitoes. *Culex quinquefasciatus* reared in water containing a concentration of skatole normally repellent for egg deposition subsequently chose skatole-containing water for oviposition, in preference to water with the normal attractant p-cresol (McCall & Eaton 2001). Similarly, it was possible to alter the preference to oviposit in water with hay infusion or guinea pig feces depending on rearing conditions of the females. The next generation had normal oviposition preferences, suggesting that the change in preference was due to conditioning rather than genetic selection. Both *Culex quinquefasciatus* and *Aedes aegypti* can be trained to prefer oviposition sites with artificial odors (Chilaka et al. 2012). *Anopheles stephensi* shows a similar learning behavior (McCall & Kelly 2002). When this species is reared under crowded conditions, females avoid oviposition in sites that have conspecific larvae. On the other hand, when the larvae are reared in uncrowded conditions, females do not avoid sites with larvae.

Kairomones from Predators. Predatory species can deter some species of mosquitoes from ovipositing (Afify & Galizia 2015). Among the predators that affect mosquito oviposition are fish (*Gambusia affinis* Baird and Girard, *Betta splendens* Regan), insects (Hemiptera: Notonectidae: *Notonecta maculata* Fabricius, *Anisops debilis* Gerstaeker; Odonata: Aeshnidae: *Anax imperator* Leach; Coleoptera: Dytiscidae: *Eretes griseus* (Fabricius)), and a copepod (*Mesocyclops longisetus* (Thièbaud)). A meta-analysis of 32 studies showed that the presence of other predatory species resulted in 47% fewer mosquito eggs deposited, suggesting that predators reduce survival not only by consuming larvae but also by causing females to search farther for fewer oviposition sites (Vonesh & Blaustein 2010).

Paradoxically, predators can also be attractive to ovipositing females. The copepod predator *Mesocyclops longisetus* produces chemicals, probably terpenes, that attract ovipositing *Aedes aegypti*, possibly representing selection for copepods that attract their prey (Torres-Estrada et al. 2001). The presence of predatory *Toxorhynchites theobaldi* larvae combined with *Aedes aegypti* larvae is attractive to ovipositing females of the latter species (Albeny-Simões et al. 2014). The mechanism for this attraction was shown to be enrichment of bacteria in the water from decomposition of damaged prey larvae, and the same effect could be achieved by mechanically killing larvae. It is speculative, but the feeding behavior of *Toxorhynchites theobaldi* that leaves pieces of larvae unconsumed may be an adaptation to increase the attractiveness of the site to ovipositing prey species.

The effect of the backswimmer bug *Notonecta maculata* has been studied in Israel (Eitam & Blaustein 2004), where avoidance of this predator by ovipositing *Culiseta longiareolata* increases the equilibrium population size (Spencer et al. 2002). Even a single backswimmer in a pool exposed to natural populations of mosquitoes discouraged oviposition by *Culiseta longiareolata* and *Culex laticinctus*. These species deposited less than half as many egg rafts in bug-infested water. The bugs also discouraged oviposition by *Anopheles gambiae* in the laboratory (Munga et al. 2006; Silberbush et al. 2010). In pairwise comparisons, a greater number of bugs had a greater repellent effect on *Culiseta longiareolata* (Silberbush & Blaustein 2011). The presence of a volatile kairomone from the bugs was suggested by a repellent effect in an oviposition site surrounded by a screened channel holding water in which the bugs had swum (Silberbush & Blaustein 2008). Chemicals that deterred oviposition were identified from analysis of water in which *Notonecta maculata* had swum (Silberbush et al. 2010). All samples contained n-heneicosane and n-tricosane, each of which was repellent to ovipositing *Culiseta longiareolata*, though not as repellent as the bug water.

A different backswimmer, *Anisops sardea* Herrich-Schaeffer, discouraged oviposition by *Culiseta longiareolata*, though not *Culex laticinctus*. The reduction in oviposition by the former species was observed whether the bugs were free-swimming or caged. In Missouri, United States, water in which *Notonecta irrorata* Uhler had swum discouraged oviposition by *Culex pipiens* and *restuans* for up to two days after removing the bugs (Blaustein et al. 2005).

Ovipositing mosquitoes can also be sensitive to chemicals produced by vertebrate predators. Some tadpoles can discourage oviposition. *Anopheles gambiae* deposited 31% fewer eggs in water conditioned with *Xenopus* tadpoles (Munga et al. 2006). In Australia, *Aedes australis* deposited 80% fewer eggs in water containing a *Crinia signifera* Girard tadpole and 63% fewer in water that had held a tadpole (Mokany & Shine 2003b). There are probably many examples of how the presence of fish discourages oviposition since the selection against larval survival can be severe. The mosquito fish, *Gambusia af-*

finis (Baird & Girard), leaves exudates in the water that reduce oviposition by *Culex pipiens*, *quinquefasciatus*, and *tarsalis* but not by *Aedes aegypti* (Angelon & Petranka 2002; Van Dam & Walton 2008).

INFLUENCE OF MOSQUITO EGGS AND JUVENILE STAGES

The presence of mosquitoes in an aquatic stage might favor selection of that site for further oviposition based on signals that represent a suitable habitat. On the other hand, mosquito larvae already present in the aquatic site might signal competition and a reduced chance of survival. The Allee effect has been coined as a term for how the presence of conspecifics can be beneficial up to a certain level (Williams et al. 2008). *Aedes aegypti*, for example, was observed to deposit three times as many eggs in sites containing 11–38 eggs, compared to sites with either no eggs or with more than 39 eggs. Called "superoviposition," females in the field will regularly deposit eggs where there are already some eggs, but they may prefer to avoid their own eggs (Chadee et al. 1990). In the laboratory, females oviposit regardless of the density of previously deposited eggs if they have no choice (Williams et al. 2008). Work on the chemicals involved in attraction of this species to its eggs has shown that dodecanoic acid and (Z)-9-hexadecanoic acid extracted from the eggs is attractive to ovipositing females (Ganeson et al. 2006). The methyl esters of these organic acids were repellent or deterrent.

The Allee effect also applies to the presence of *Aedes aegypti* larvae (Benzon & Apperson 1988). The presence of larvae was attractive to ovipositing females at densities of 0.5, 1, or 2 larvae/ml but distinctly less attractive at densities of 0.25 or 4 larvae/ml. In the field in Iquitos, Peru, oviposition was mostly influenced by the presence of larvae but also by the source of water (tap water, intentionally collected rainwater, or accidentally collected rainwater), by the absence of a lid, and by exposure to the sun (Wong et al. 2011). Again, the greatest effect was from the presence of larvae overall (correlation coefficient 1.66) compared to high densities of larvae (>100 larvae, correlation coefficient 0.784) or low densities of larvae (1–10 larvae, correlation coefficient 0.037). Heneicosane, a C_{21} alkane, is apparently the principal attractive component produced by larvae (Mendki et al. 2000). This chemical is present in the larval cuticle and is distributed sufficiently in the water to be attractive to ovipositing females.

The presence of conspecifics can encourage or discourage oviposition, depending on the species (Edgerly et al. 1998; Afify & Galizia 2015). Eggs of *Aedes triseriatus* deter oviposition by this species, and eggs of *Culex quinquefasciatus* can repel *Anopheles gambiae*. Conspecific eggs have a positive effect on oviposition by *Culex quinquefasciatus* or *Aedes albopictus*. The presence of conspecific larvae stimulates oviposition by *Aedes aegypti*, *Aedes atropalpus*, *Aedes australis*, *Aedes togoi*, *Aedes triseriatus*, and *Culex quinquefasciatus*. *Anopheles gambiae* deposited about 40% fewer eggs in the presence of conspecific larvae (Munga et al. 2006). Stimulation of oviposition by the presence of offspring would be an example of the "P-P hypothesis," which stands for "oviposition-preference/offspring-performance" (Yoshioka et al. 2012). The ideal free distribution theory of oviposition, in which females sense the potential quality of the habitat regardless of the presence of offspring, could explain some of these distributions, though association of oviposition with the pheromones described below suggests otherwise.

Chemicals mediate these responses, including a pheromone secreted by *Culex quinquefasciatus* eggs (erythro-6-acetoxy-5-hexadecanolide), a variation of the MOP from *Culex tarsalis* (erythro-5,6-dihydroxyhexadecanoic acid), and a pheromone from *Aedes aegypti* larvae (n-heneicosane; Navarro-Silva et al. 2009). Stereochemistry (Gao & Hall 2003; Dhotare et al. 2005; Prasad & Anbarasan 2007) affects the activity of the *Culex quinquefasciatus* MOP, with the (5*R*, 6*S*) stereoisomer more attractive. For *Culex quinquefasciatus*, both grass infusions (Mboera et al. 2000) and skatole (Olagbemiro et al. 2004) acted synergistically with MOP as oviposition attractants. Other compounds associated with immature stages that affect oviposition include (Z)-9-hexadecanoic acid, methyl dodecanoate, methyl tetradecanoate, methyl (Z)-9-hexadecenoate, methyl hexadecanoate, methyl (Z)-9-octadecenoate, methyl octadecanoate, octadecenoic acid, and hexadecanoic acid. Immature stages infected with various mosquito pathogens can deter oviposition.

The gene for the OBP for the *Culex quinquefasciatus* MOP has been identified (Pelletier et al. 2010) among 130 OBP recognized in this species (Zhu et al. 2013). The protein from this gene is expressed in the antennae. Experiments that knocked out production of the protein using RNAi showed a reduced response to MOP but no change in the response to blood-host odors, including skatole, indole, and nonanal. The implication is that some sensory organs on the antennae of this species are adapted specifically for detection of MOP. CquiOBP1 protein has the same six-helix structure of other OBPs, including a molecular pocket that binds a lactone and hydrophobic moiety matching the MOP (Mao et al. 2010). The OBP made a bond to MOP, octanal, nonanal, decanal, and geranylacetone. OBP1 from *Aedes aegypti* and *Anopheles gambiae* also bound MOP but to a lesser extent than for *Culex quinquefasciatus*. Sensilla trichodea on the antennae of *Aedes aegypti* have increased sensitivity at 72 hours after a blood meal to undecanone, acetic acid, and propionic acid, possibly as these organs switch to sensitivity for oviposition sites (Siju et al. 2010).

BACTERIAL INFLUENCE

Leaf infusions known to stimulate oviposition by *Aedes aegypti* contain many different bacteria, which in the aggregate produce chemicals affecting oviposition (Ponnusamy et al. 2008). Bacteria identified molecularly with a certainty over 95% included some of the genera found in the mosquito alimentary tract, including *Bacillus*, *Enterobacter*, *Klebsiella*, *Lactococcus*, *Pseudomonas*, and *Sphingomonas*. Bacteria from North American bamboo (*Arudinaria gigantea*) leaf infusions were in phylum Firmicutes, class Bacilli (*Bacillus thuringiensis* and *Lactococcus lactis*), phylum Proteobacteria, class Alphaproteobacteria (*Rhizobium huautlense* and *Brevundimonas vesicularis*), and class Gammaproteobacteria (*Citrobacter freundii*; *Enterobacter asburiae*, *cancerogenus*, *gergoviae*, and *ludwigii*; *Klebsiella oxytoca*, *granulomatis*; *Pseudomonas fulva*, *plecoglossicida*). Bacteria from white oak (*Quercus alba*) leaf infusions included some of the same species (*Bacillus thuringiensis*, *Brevundimonas vesicularis*, *Enterobacter asburiae*, *Klebsiella granulomatis*, and *Rhizobium huautlense*) as well as additional species in class Alphaproteobacteria (*Agrobacterium tumefaciens*, *Azorhizobium caulinodans*, *Caulobacter fusiformis*, *Porphyrobacter* sp., *Sphingomonas aromaticivorans*), class Betaproteobacteria (*Curvibacter gracilis*, *Variovorax koreensis*), and Gammaproteobacteria (*Pseudomonas lanceolata*, *syringae*). Of the bacteria isolated from *Arundinaria* leaf and white oak infusions, over 90% of the species were attractive to ovipositing *Aedes aegypti*. Chemical extraction of carboxylic acids and esters from lyophilized bacteria produced four chemicals with an effect on oviposition on surfaces wetted by solutions or by plain water. Mosquitoes deposited significantly more eggs in solutions of nonanoic acid, tetradecanoic acid, or the methyl ester of tetradecanoic acid. Solutions with the methyl ester of hexadecanoic acid received fewer eggs than surfaces wetted with plain water. Not surprisingly, these effects were highly dependent on concentration of the chemicals.

Most other work on the effect of bacteria has shown them to be stimulants for oviposition by *Aedes aegypti*, *Aedes albopictus*, and *Culex quinquefasciatus* (Afify & Galizia 2015). Bacteria from alfalfa infusions or natural larval habitats were attractants for *Culex quinquefasciatus* and *Anopheles gambiae*, respectively. Of two bacteria associated with rearing *Aedes aegypti*, cultures of one (*Acinetobacter calcoaceticus*) were significantly attractive to ovipositing females, and the other, *Enterobacter cloacae*, was not as attractive (Benzon & Apperson 1988). High concentrations of bacteria or the chemicals they produce were oviposition deterrents for *Anopheles gambiae*, *Aedes aegypti*, and *Culex quinquefasciatus* under various circumstances.

Table 10.3. Synthetic chemicals that affect mosquito oviposition (Afify and Galizia 2015)

Compound	Genera tested	Effects
Heptadecyl butanoate	*Aedes, Anopheles*	Deter
Pentadecyl hexanoate	*Aedes, Anopheles*	Deter
Tetradecyl heptanoate	*Aedes, Anopheles*	Deter
Tridecyl octanoate	*Aedes, Anopheles*	Deter
Butyl heptadecanoate	*Aedes, Anopheles*	Deter
Octyl tridecanoate	*Aedes*	Deter
Undecyl decanoate	*Aedes, Anopheles*	Stimulate, deter
Decyl undecanoate	*Aedes, Anopheles*	Stimulate, deter
Nonyl dodecanoate	*Aedes, Anopheles*	Stimulate, deter
Dodecyl nonanoate	*Aedes, Anopheles*	Stimulate, deter
Octadecyl propanoate	*Aedes, Anopheles*	Stimulate, deter
Hexyl pentadecanoate	*Aedes, Anopheles*	Stimulate, deter
Heptyl tetradecanoate	*Aedes*	Deter
Propyl octadecanoate	*Aedes, Anopheles*	Attract, stimulate, deter
Pentyl hexadecanoate	*Anopheles*	Stimulate
Hexadecyl pentanoate	*Aedes, Anopheles*	Repel, deter
Ethyl 2-((4-nitrophenyl)diazenyl)-3-oxobutanoate	*Aedes*	Attract, stimulate
Ethyl 2-((4-fluorophenyl)diazenyl-3-oxobutanoate	*Aedes*	Repel, deter
Isobutyl 2-((4-flurophenyl)diazenyl)-3-oxobutanoate	*Aedes*	Repel, deter
Methyl 2-((3-chlorophenyl)diazenyl-3-oxobutanoate	*Aedes*	Repel, deter
Ethyl 2-((1-hydroxynaphthalen-2-yl)diazenyl)-3-oxobutanoate	*Aedes*	Attract, stimulate
Diethyl *m*-toluamide (DEET)	*Aedes, Culex*	Stimulate, deter
Diethyl phenyl acetamide (DEPA)	*Aedes, Culex*	Stimulate, deter
Diethyl benzamide (DEB)	*Aedes, Culex*	Stimulate, deter
3-(3-cyclohexen-1-ylcarbonyl)-piperidine	*Aedes*	Deter
1-(3-cyclohexen-1-ylcarbonyl)-2-methylpiperidine	*Aedes*	Deter
Ethyl anthranilate		
Butyl anthranilate	*Aedes*	Deter
Acetic acid	*Culex*	Stimulate

SYNTHETIC CHEMICALS

Chemicals have been used to manipulate mosquito behavior in various ways, with the objective of protecting people and domestic animals from bites and the possibility of pathogen transmission (Strickman 2018). Motivated by a search for compounds that could disrupt oviposition, esters and repellent compounds have been tested for their influence on oviposition. Concentration-dependent deterrence has been the most common effect, but in some cases there are concentrations in which attraction and stimulation of oviposition have been observed (Afify & Galizia 2015; Table 10.3). The importance of proportions and quantity of attractant chemicals was observed in Brazil using a blend of 83% tetradecanoic acid, 16% nonanoic acid, and 1% tetradecanoic acid methyl ester at a carefully controlled 6 ng/μl (Barbosa et al. 2010). This blend increased oviposition by *Aedes aegypti* by over 100% in large cages and in homes.

Conclusion

There is an impressive parallel between feeding and oviposition in female adult mosquitoes. In both cases, and unlike many other insects, mosquitoes use an extensive suite of chemical and physical cues to find the targets of their attention. The orientation toward a host, sugar source, or oviposition site uses multiple olfactory and gustatory sense organs, which are tuned for the detection of certain chemicals by a system of proteins, each with a specific range of sensitivities. Experience with the odors or tastes can lead to conditional learning that influences subsequent behavior, or even the behavior of an adult based on the experience of the larva from which it emerged. Just as in the case for feeding, variation in oviposition contributes to a wide variety of behaviors. This enables the family Culicidae to exploit many kinds of habitats, even as each species performs the same functions for survival and reproduction.

The physiological basis of egg development is also complex, with multiple points of regulation from both the nervous and endocrine systems. It is not clear whether the system is conservative and represents those feedback systems that are necessary or if the system is the result of the adjustment of multiple mechanisms accumulated over evolutionary time. Another aspect of egg development worth some attention is its rapidity. Many species repeat gonotrophic cycles every few days, cycling through the complicated series of physiological changes repeatedly, on a scale of hours. The contrast between the rapid changes for the individual and the long evolutionary history that has led up to it is nothing short of startling.

Oviposition is a fitting place to conclude this review of mosquito biology. It is not only the culmination of the inputs to the life cycle, it is also representative of many aspects of the family. As pointed out, oviposition accomplishes the same objective for all species but in a wide variety of ways. That variety uses the entire toolbox of mosquito capabilities, from larval efficiency to excellent flight and from exquisite chemical reception to precise endocrine control. From a purely human standpoint, oviposition is also a reminder of the entropy inherent in evolution. Incredible organization of the individual organism nonetheless leads to incredible waste in the population, where so few of those individuals successfully reproduce. During the past 200 million years, that cycle of selection has resulted in a family of insects with a wide variety of genetic capabilities, prepared to adapt to changes likely to occur in their environment.

PART TWO

Mosquito Diversity, Systematics, and Medical Importance

11. Genera and Medically Important Species Pages

Subfamily *Culicinae*

Tribe Aedini

Tribe Culicini

Mosquitoes of the World is an ambitious title that brings to mind a book that will tell you everything you want to know about all mosquitoes. Because of the large number of mosquitoes, and their importance to human and animal health, we knew we could not conquer the subject completely—but we did get your attention, no? To live up to the implied promise and our respective background strengths, we developed a holistic strategy to introduce the glorious diversity and the devastating disease transmission capacity of mosquitoes while investigating their form, function, distribution, and vectorial and taxonomic status in concise two-page spreads. Adopting the mantra that a picture speaks a thousand words, we provide 168 beautifully illustrated pages covering all 41 mosquito genera and detailing bionomic, genetic, and morphological characters of 128 mosquito species of pest, invasive, or medically important status.

Limited by publishing costs and space constraints, an initial short list of 170 species (about 5% of named species) was proposed. This was later whittled back to 128 using a highly scientific "expert opinion" method that was purely our best judgment. A species may have been important enough to make our initial target list, but where we lacked access to good specimens to photograph or high-quality immature stage illustrations were not available in the literature, we had to forego it. This is why, in this section, you may not find included a favorite or expected species. For images, we relied heavily on the more than 1.7 million reference specimens housed in the US National Mosquito Collection, curated by the staff of the Walter Reed Biosystematics Unit (WRBU) on behalf of the Smithsonian Institution (National Museum of Natural History, or NMNH). Two species (*Aedes taylori* Edwards, 1936, and *Ae. furcifer* (Edwards, 1913)) were so similar that they are treated on the same species page (see p. 408). Many species, especially in *Anopheles*, comprise species complexes; initially, component taxa of medical importance had separate species pages, but these were later combined. Thus, the 128 species formally presented here represent more than 200 species in reality.

The presentation sequence of the genus and species pages reflects the overall systematics of the family Culicidae where genera are listed alphabetically by subfamily and tribe (where appropriate), and species are listed alphabetically within subgenera (see Table 11.1). For both genera and species, morphologically defining characters are given for adult females and larvae, as these are the most commonly encountered stages. Within the genera pages, exemplar male genitalia illustrations are also given, but these are not illustrated for each species. With the exception of the *Anopheles maculipennis* complex (see species pages, p. 312), the first cryptic species to be discovered largely through egg differences, we do not treat the egg stages for any species. Pupae can provide highly diagnostic characters, particularly in closely related species groups—e.g., *Anopheles* (*Ano.*) *mediopunctatus* (Lutz, 1903)—and serious students of mosquito systematics should include all life stages in their studies.

The "Diagnostic Characters" sections accompanying each genus and species page are a novel and (we hope) useful method of capturing and illustrating key taxonomic characters of each taxon treated. The final characters chosen for illustration were extensively researched and derived from the most recent keys, original descriptions, redescriptions, and taxonomic revisions for adult females and larvae. Those chosen were usually consensus characters narrowly defined for a given taxon, or for it and closely related taxa. In some cases, group and genus characters were also used to enhance the overall picture of what the species or genus looks like. Choosing definitive diagnostic characters was complicated by the common occurrence of sibling (cryptic) species, especially in *Anopheles*. In those species complexes, as with the genera, we aimed for a "gestalt" picture of a subset of the most important characters. The sources for the characters can be found in the "General References" section at the beginning of the taxonomic catalog (Vol. 2), in the "Important references" at the start of the genus entries, and in the "Refs." section of each species. The catalog species entries (Vol. 2) contain the most-utilized sources of information for the species pages. If one desires to use the genus and species pages for teaching or identification purposes, it is highly recommended that source literature also be consulted. Following the genus and species pages, a Glossary of Diagnostic Morphological Terms is provided (see Chapter 12, p. 548), which includes annotated illustrations of the mosquito habitus (see Fig. 12.1, p. 549), *Anopheles* wing spots (see Fig. 12.2, p. 550), and both *Anopheles* (see Fig. 12.3, p. 551) and *Culex* larvae (see Fig. 12.4, p. 552) to assist the nonspecialist audience.

Most everyone would like an easy way to identify mosquitoes. DNA identification tools based on DNA barcoding or whole genome assemblies could one day make that a reality, but taxonomically verified voucher samples are needed to underpin gold standard reference DNA sequences. There is a valid place for DNA-based taxonomy—especially while determining

Table 11.1. Culicid genera and biomedically important species showcased in Chapter 11. Placement order follows the accepted current classification.

Family	Subfamily	Tribe	Genus	Page	Subgenus	Species	Page
Culicidae	Anophelinae		***Anopheles***	***212***	*Anopheles*	*aztecus*	**294**
						bancroftii	**296**
						barbirostris	**298**
						campestris	**300**
						claviger	**302**
						donaldi	**304**
						freeborni	**306**
						lesteri	**308**
						letifer	**310**
						maculipennis	**312**
						nigerrimus	**314**
						pseudopunctipennis	**316**
						punctimacula	**318**
						punctipennis	**320**
						quadrimaculatus	**322**
						sinensis	**324**
					Cellia	*aconitus*	**326**
						annularis	**328**
						annulipes	**330**
						balabacensis	**332**
						culicifacies	**334**
						dirus	**336**
						farauti	**338**
						flavirostris	**340**
						fluviatilis	**342**
						funestus	**344**
						gambiae	**346**
						jeyporiensis	**348**
						koliensis	**350**
						latens	**352**
						maculatus	**354**
						minimus	**356**
						moucheti	**358**
						multicolor	**360**
						nili	**362**
						pharoensis	**364**
						pulcherrimus	**366**
						sergentii	**368**
						stephensi	**370**
						subpictus	**372**
						sulawesi	**374**
						sundaicus	**376**
						superpictus	**378**
						tessellatus	**380**
					Kerteszia	*cruzii*	**382**
						neivai	**384**
						pholidotus	**386**
					Nyssorhynchus	*albimanus*	**388**
						albitarsis	**390**
						aquasalis	**392**
						darlingi	**394**
						nuneztovari	**396**
						triannulatus	**398**
			Bironella	***214***			
			Chagasia	***216***			
	Culicinae	Aedeomyiini	***Aedeomyia***	***218***			
		Aedini	***Aedes***	***220***	*Aedimorphus*	*cumminsii*	**400**
						ochraceus	**402**
						vexans	**404**

Family	Subfamily	Tribe	Genus	Page	Subgenus	Species	Page
					Dahliana	*geniculatus*	**406**
					Diceromyia	*furcifer*	**408**
						taylori	**408**
					Downsiomyia	*niveus*	**410**
					Finlaya	*fijiensis*	**412**
						kochi	**414**
						poicilius	**416**
					Georgecraigius	*atropalpus*	**418**
					Hulecoeteomyia	*japonicus*	**420**
						koreicus	**422**
					Neomelaniconion	*mcintoshi*	**424**
					Ochlerotatus	*angustivittatus*	**426**
						atlanticus	**428**
						canadensis	**430**
						caspius	**432**
						dorsalis	**434**
						infirmatus	**436**
						melanimon	**438**
						scapularis	**440**
						sollicitans	**442**
						squamiger	**444**
						taeniorhynchus	**446**
						trivittatus	**448**
						vigilax	**450**
					Protomacleaya	*triseriatus*	**452**
					Rampamyia	*notoscriptus*	**454**
					Stegomyia	*aegypti*	**456**
						africanus	**458**
						albopictus	**460**
						bromeliae	**462**
						luteocephalus	**464**
						polynesiensis	**466**
						scutellaris	**468**
					Tanakaius	*togoi*	**470**
			Armigeres	***222***	*Armigeres*	*subalbatus*	**472**
			Eretmapodites	***224***			
			Haemagogus	***226***	*Haemagogus*	*janthinomys*	**474**
						mesodentatus	**476**
						spegazzinii	**478**
			Heizmannia	***228***			
			Opifex	***230***			
			Psorophora	***232***	*Grabhamia*	*columbiae*	**480**
					Janthinsoma	*ferox*	**482**
					Psorophora	*ciliata*	**484**
			Udaya	***234***			
			Verrallina	***236***			
			Zeugnomyia	***238***			
		Culicini	***Culex***	***240***	*Culex*	*annulirostris*	**486**
						antennatus	**488**
						fuscocephala	**490**
						gelidus	**492**
						nigripalpus	**494**
						perexiguus	**496**
						pipiens	**498**
						quinquefasciatus	**500**
						restuans	**502**
						salinarius	**504**
						sitiens	**506**
						tarsalis	**508**
						theileri	**510**
						tritaeniorhynchus	**512**

continued

Table 11.1. *continiued*

Family	Subfamily	Tribe	Genus	Page	Subgenus	Species	Page
						univittatus	514
						vishnui	516
					Melanoconion	*erraticus*	518
						ocossa	520
						spissipes	522
						taeniopus	524
						vomerifer	526
					Oculeomyia	*bitaeniorhynchus*	528
						poicilipes	530
			Deinocerites	*242*			
			Galindomyia	*244*			
			Lutzia	*246*			
		Culisetini	***Culiseta***	*248*	*Climacura*	*melanura*	532
					Culiseta	*inornata*	534
		Ficalbiini	***Ficalbia***	*250*			
			Mimomyia	*252*			
		Hodgesiini	***Hodgesia***	*254*			
		Mansoniini	***Coquillettidia***	*256*	*Coquillettidia*	*perturbans*	536
			Mansonia	*258*	*Mansonia*	*titillans*	538
					Mansonioides	*indiana*	540
						uniformis	542
		Orthopodomyiini	***Orthopodomyia***	*260*			
		Sabethini	***Isostomyia***	*262*			
			Johnbelkinia	*264*			
			Kimia	*266*			
			Limatus	*268*			
			Malaya	*270*			
			Maorigoeldia	*272*			
			Onirion	*274*			
			Runchomyia	*276*			
			Sabethes	*278*	*Sabethoides*	*chloropterus*	544
			Shannoniana	*280*			
			Topomyia	*282*			
			Trichoprosopon	*284*		*digitatum*	546
			Tripteroides	*286*			
			Wyeomyia	*288*			
		Toxorhynchitini	***Toxorhynchites***	*290*			
		Uranotaeniini	***Uranotaenia***	*292*			

the taxonomic status of incipient or cryptic species—but ultimately, taxonomy relies on a unique name being formally assigned to the species to ensure that, globally, we all speak the same language. Without being formally described, the molecular "species" has no status and is neither included in valid species lists nor is it reflected in our updated taxonomic catalog (see Vol. 2). It remains a huge impediment in the fight against mosquito-borne disease that only an estimated half of extant taxa have been named. There are many species we have not yet seen, and many more unnamed species that suffer in silence while the last steadfast classical taxonomists retire, or worse. Here, for both the genus and species pages, we discuss the taxonomic status and systematic placement and give exemplar mitochondrial COI "DNA barcode" sequences to facilitate identification of the taxonomic unit under treatment. As proponents of DNA barcoding as a highly useful method for species recognition, we include GenBank accession numbers for sequences obtained for morphologically verified mosquito species derived from efforts of the Mosquito Barcoding Initiative (MBI) and researchers worldwide. Where the differentiation of the group primarily uses another marker (e.g., ITS2), these GenBank accession numbers were given instead. In some cases, neither COI nor ITS2 were available, and exemplar sequences were given based on what was available in GenBank at the time of writing. Somewhat incredibly, at the time of writing, there were no DNA sequences available for some of the most important disease vectors (e.g., *Anopheles donaldi*).

Reflected in both the species pages presented here and in the taxonomic catalog (Vol. 2) are the updated distributions for each species, as illustrated by maps. Depending on the complexity of the species group treated, these maps often reflect the distribution of the nominotypical member of the respective complex as "sensu lato," and where this is the case, it is explicitly stated, or otherwise corresponds to the species in question, as "sensu stricto," (= s.s.). or Species distributions were care-

Table 11.2. List of viruses detected in various mosquito genera. Data includes virus reports from insect pools and laboratory studies and should not be misconstrued to infer vector competency of these viruses in nature.

Abbreviation	Virus	Associated Culicid vector genera
AACV	*Anopheles* associated C virus	*Anopheles*
AAOV	*Anopheles annulipes* orbivirus	*Anopheles*
AATV	Australian *Anopheles* totivirus	*Anopheles*
ABRV	Abras virus	*Culex*
ACAV	Acara virus	*Anopheles, Culex*
ACDV	Acado virus	*Culex*
AgDNV	*Anopheles gambiae* densovirus	*Anopheles*
AHOV	*Anopheles hinesorum* virus	*Anopheles*
AINOV	Aino virus	*Culex*
AKAV	Akabane virus	*Aedes, Culex*
ALFV	Alfuy virus	*Aedeomyia*
ALFV	Alfuy virus	*Aedes*
AMBV	Anhembi virus	*Trichoprosopon*
AMIV	*Anopheles minimus* virus	*Anopheles*
AMTV	Arumowot virus	*Culex*
ANAV	*Anopheles* A virus	*Anopheles*
ANBV	*Anopheles* B virus	*Anopheles*
AnCPV	*Anopheles* cypovirus	*Anopheles*
AnCV	*Anopheles* C virus	*Anopheles*
ANDV	Andasibe virus	*Anopheles*
AnFV	*Anopheles* flavivirus	*Anopheles*
AngFV	*Anopheles gambiae* flavivirus	*Anopheles*
AnsFV	*Anopheles squamosus* flavivirus	*Anopheles*
ANTV	Antequera virus	*Aedeomyia, Culex*
ANUV	Ananindeua virus	*Coquillettidia*
APEUV	Apeu virus	*Aedes, Anopheles*
ARKV	Arkonam virus	*Anopheles, Culex*
ARUV	Aruac virus	*Culex, Psorophora, Sabethes, Trichoprosopon, Wyeomyia*
ARV	Adelaide River virus	*Culex*
AsTV	*Armigeres subalbatus* totivirus	*Armigeres*
AToV	*Anopheles* totivirus	*Anopheles*
AURAV	Aura virus	*Aedes, Culex*
BABV	Babahoyo virus	*Culex*
BAGV	Bagaza virus	*Culex, Mimomyia*
BAKV	Bakau virus	*Culex*
BAMV	Bamaga virus	*Aedes*
BANV	Banzi virus	*Anopheles, Culex, Mansonia*
BARV	Barur virus	*Anopheles, Culex, Mansonia*
BATV	Batai virus	*Aedes, Anopheles, Culex*
BBKV	Babanki virus	*Aedes, Anopheles, Culex, Eretmapodites, Lutzia, Mansonia, Mimomyia*
BEBV	Bebaru virus	*Aedes, Culex, Verrallina*
BEFV	Bovine ephemeral fever virus	*Aedes, Anopheles, Culex*
BELV	Belmont virus	*Culex*
BFV	Barmah Forest virus	*Aedes, Coquillettidia, Culex, Opifex*
BGIV	Bangui virus	*Anopheles*
BGNV	Bangoran virus	*Culex*

continued

Table 11.2. *continued*

Abbreviation	Virus	Associated Culicid vector genera
BIAV	Bobia virus	*Lutzia*
BIMV	Bimiti virus	*Culex*
BIRV	Birao virus	*Anopheles*
BKNV	Batken virus	*Aedes*
BOAV	Bolahum virus	*Anopheles*
BORV	Boraceia virus	*Anopheles*
BOTV	Botambi virus	*Culex*
BOUV	Bouboui virus	*Aedes, Anopheles*
BOZOV	Bozo virus	*Aedes, Culex, Anopheles*
BQSV	Barranqueras virus	*Culex*
BSQV	Bussuquara virus	*Coquillettidia, Culex, Mansonia, Trichoprosopon*
BTKV	Boteke virus	*Mansonia*
BUNV	Bunyamwera virus	*Aedes, Anopheles, Mansonia*
BVSV	Bovine vesicular stomatitis virus	*Culex*
BWAV	Bwamba virus	*Aedes, Anopheles, Mansonia*
CABV	Cabassou virus	*Culex, Coquillettidia, Limatus, Mansonia, Wyeomyia*
CALV	California serogroup viruses	*Coquillettidia*
CAPV	Capim virus	*Culex*
CARV	Caraparu virus	*Anopheles, Culex, Limatus, Wyeomyia*
CATUV	Catu virus	*Anopheles, Culex, Coquillettidia*
CBV	Coot Bay virus	*Anopheles*
CEV	California encephalitis virus	*Aedes, Anopheles, Culex, Culiseta, Psorophora*
CHGV	Chagres virus	*Sabethes*
CHIKV	Chikungunya virus	*Aedes, Anopheles, Coquillettidia, Culex, Mansonia, Opifex*
CNAV	Cananeia virus	*Culex*
COCV	Cocal virus	*Aedes, Anopheles, Culex, Trichoprosopon*
CORV	Corriparta virus	*Culex*
COTV	Cotia virus	*Aedes, Coquillettidia, Culex, Limatus, Psorophora*
CQIV	Calchaqui virus	*Aedeomyia*
CVOV	Calovo virus	*Aedes, Anopheles, Coquillettidia*
CVV	Cache Valley virus	*Aedes, Anopheles, Culiseta, Psorophora*
CWV	Cape Wrath virus	*Anopheles*
DABV	Dabakala virus	*Culex*
DENV	Dengue virus	*Aedes, Anopheles, Armigeres, Culex, Haemagogus, Trichoprosopon*
DSwV	Dairy Swamp virus	*Anopheles*
DUGV	Dugbe virus	*Aedes*
EEEV	Eastern equine encephalomyelitis virus	*Aedes, Anopheles, Coquillettidia, Culiseta, Culex, Psorophora, Uranotaenia*
EHDV	Epizootic hemorrhagic disease of deer virus	*Anopheles*
EHV	Edge Hill virus	*Aedes, Anopheles, Culex*
EILV	Eliat virus	*Anopheles*
ENSV	Enseada virus	*Culex*
ENTV	Entebbe bat virus	*Culex*
EUBV	Eubenangee virus	*Anopheles*
EVEV	Everglades virus	*Aedes, Anopheles, Culex*
FLAV	Flanders virus	*Culex, Culiseta*
GAMV	Gamboa virus	*Aedeomyia, Anopheles, Coquillettidia, Culex, Limatus, Mansonia, Uranotaenia*
GERV	Germiston virus	*Aedes, Anopheles, Culex*

Abbreviation	Virus	Associated Culicid vector genera
GETV	Getah virus	*Aedes, Armigeres, Culex, Tripteroides, Verrallina*
GGV	Gan Gan virus	*Aedes, Culex*
GJAV	Guajara virus	*Culex, Limatus*
GLOV	Gray Lodge virus	*Culex*
GLV	Gumbo Limbo virus	*Aedes, Culex*
GMAV	Guama virus	*Anopheles, Coquillettidia, Mansonia*
GOMV	Gomoka virus	*Anopheles, Culex*
GROV	Guaroa virus	*Anopheles, Psorophora*
GTRV	Guaratuba virus	*Anopheles*
HaCV	Haslams Creek virus	*Anopheles*
HJV	Highlands J virus	*Aedes, Culiseta*
HPV	Hart Park virus	*Culex, Culiseta*
IACOV	Iaco virus	*Wyeomyia*
IERIV	Ieri virus	*Psorophora*
ILEV	Ilesha virus	*Anopheles, Mansonia*
ILHV	Ilheus virus	*Aedes, Anopheles, Coquillettidia, Culex, Haemogogus, Psorophora, Trichoprosopon, Wyeomyia*
INGV	Ingwavuma virus	*Aedes, Culex*
INKV	Inkoo virus	*Aedes*
ISFV	Isfahan virus	*Aedes*
ITPV	Itaporanga virus	*Culex*
ITQV	Itaqui virus	*Culex*
ITUV	Itupiranga virus	*Aedes, Psorophora*
ITV	Israel turkey meningoencephalitis virus	*Culex*
JACV	Jacareacanga virus	*Culex*
JBEV	Japanese encephalitis virus	*Aedes, Anopheles, Armigeres, Culex, Mansonia, Verrallina*
JCV	Jamestown Canyon virus	*Aedes, Culiseta*
JSV	Jerry Slough virus	*Culiseta*
JUGV	Jugra virus	*Aedes, Uranotaenia*
JURV	Jurona virus	*Haemagogus*
KAIV	Kaikalur virus	*Culex*
KAMV	Kamese virus	*Culex, Lutzia*
KASV	Kasba virus	*Culex*
KAVV	New genotype of Torque teno virus (TTV), designated KAV according to the patient's initials	*Mansonia*
KDV	Kadipiro virus	*Armigeres*
KEDV	Kedougou virus	*Aedes*
KETV	Ketapang virus	*Culex*
KEYV	Keystone virus	*Aedes, Anopheles, Culex*
KIMV	Kimberley virus	*Culex*
KINV	Kindia virus	*Aedes*
KNAV	Kununurra virus	*Aedeomyia*
KOKV	Kokobera virus	*Aedes, Culex*
KOOV	Koongol virus	*Anopheles, Culex*
KOTV	Kotonkan virus	*Anopheles*
KOWV	Kowanyama virus	*Anopheles*
KPKV	Kampung Karu virus	*Anopheles*
KRBV	Karumba virus	*Anopheles*
KRIV	Kairi virus	*Mansonia, Psorophora*
KUNV	Kunjin virus	*Culex*

continued

Table 11.2. *continued*

Abbreviation	Virus	Associated Culicid vector genera
KWAV	Kwatta virus	*Culex*
KYZV	Kyzylagach virus	*Culex*
LACV	La Crosse virus	*Aedes, Culex*
LATV	Latino virus	*Culex*
LEAV	Leanyer virus	*Anopheles*
LEBV	Lebombo virus	*Aedes, Mansonia*
LEDV	Lednice virus	*Culex*
LJV	La Joya virus	*Culex*
LLSV	Llano Seco virus	*Culex*
LMV	Las Maloyas virus	*Anopheles, Culex, Mansonia, Psorophora*
LOKV	Lokern virus	*Culex, Psorophora*
LPKV	Long Pine Key virus	*Anopheles*
LUKV	Lukini virus	*Aedes, Anopheles*
LUMV	Lumbo virus	*Aedes*
MADV	Madrid virus	*Culex*
MAGV	Maguari virus	*Mansonia, Psorophora*
MALV	Malakal virus	*Mansonia*
MAPV	Mapputta virus	*Anopheles*
MAYV	Mayaro virus	*Anopheles, Haemagogus, Coquillettidia, Mansonia, Psorophora*
MBAV	Manitoba virus	*Culex*
MBGV	Marburg virus	*Anopheles*
MBOV	Mboke virus	*Aedes*
MCAV	Macaua virus	*Sabethes*
MCOV	Marco virus	*Anopheles*
McPV	Mac Peak virus	*Anopheles*
MDCV	Mojui dos Campos virus	*Psorophora, Wyeomyia*
MDOV	Monte Dourado virus	*Anopheles*
MDV	Main Drain virus	*Anopheles*
MELV	Melao virus	*Aedes, Psorophora*
MERV	Mermet virus	*Culex*
MEV	Mink enteritis virus	*Anopheles*
MHV	Mahogany Hammock virus	*Culex*
MIDV	Middelburg virus	*Aedes, Mansonia*
MINV	Minnal virus	*Culex*
MIRV	Mirim virus	*Aedes, Culex, Psorophora*
MOJUV	Moju virus	*Mansonia, Culex, Coquillettidia*
MORV	Moriche virus	*Culex*
MOSV	Mossuril virus	*Culex*
MPKV	Maprik virus	*Aedes, Mansonia, Ficalbia*
MPOV	M'Poko virus	*Culex*
MQOV	Mosqueiro virus	*Culex, Mansonia, Wyeomyia*
MTBV	Marituba virus	*Culex*
MUCV	Mucambo virus	*Aedes, Culex, Haemagogus,* 'Sabethini,' *Wyeomyia*
MURV	Murutucu virus	*Anopheles, Coquillettidia, Culex*
MVEV	Murray Valley encephalitis virus	*Aedes, Culex*
MYXV	Myxoma virus	*Anopheles*
NDOV	Nyando virus	*Aedes, Anopheles*
NDUV	Ndumu virus	*Aedes, Culex, Mansonia*

Abbreviation	Virus	Associated Culicid vector genera
NEGV	Negishi virus	*Aedes*
NEPV	Nepuyo virus	*Culex*
NGOV	Ngoupe virus	*Aedes*
NJLV	Naranjal virus	*Culex*
NKOV	Nkolbisson virus	*Aedes, Culex, Eretmapodites*
NODV	Nodamura virus	*Culex*
NOLAV	Nola virus	*Culex*
NORV	Northway virus	*Aedes, Culiseta*
NRIV	Ngari virus	*Aedes, Anopheles*
NTAV	Ntaya virus	*Anopheles*
ODRV	Odrenisrou virus	*Culex*
OKOV	Okola virus	*Eretmapodites*
OLIV	Olifantsvlei virus	*Culex, Mansonia*
ONNV	O'nyong-nyong virus	*Anopheles, Mansonia*
ORIV	Oriboca virus	*Aedes, Anopheles, Culex, Mansonia, Psorophora*
OROV	Oropouche virus	*Aedes, Anopheles, Coquillettidia, Culex, Psorophora*
ORUV	Orungo virus	*Aedes, Anopheles*
OSSAV	Ossa virus	*Culex*
OUBIV	Oubi virus	*Culex*
OUBV	Oubangui virus	*Culex*
PAHV	Pahayokee virus	*Culex*
PALV	Palyam virus	*Culex*
PARAV	Para virus	*Culex*
PATAV	Pata virus	*Aedes*
PATV	Patois virus	*Culex*
PCAV	Pacora virus	*Culex*
PERV	Perinet virus	*Anopheles, Culex, Mansonia*
PGAV	Pongola virus	*Aedes, Mansonia*
PIAV	Picola virus	*Culex*
PIXV	Pixuna virus	*Anopheles, Trichoprosopon*
PLAV	Playas virus	*Aedes*
PLSV	Palestina virus	*Culex*
POTV	Potosi virus	*Psorophora*
POWV	Powassan virus	*Aedes, Anopheles*
PRV	Paroo River virus	*Culex*
PUCV	Puchong virus	*Mansonia*
PURV	Purus virus	*Psorophora*
PVV	Pueblo Viejo virus	*Aedeomyia*
RBUV	Rochambeau virus	*Coquillettidia*
RBV	Rio Bravo virus	*Anopheles*
RESV	Restan virus	*Culex*
ROCV	Rocio virus	*Psorophora*
RRV	Ross River virus	*Aedes, Anopheles, Culex, Mansonia, Opifex*
RTAV	Resistencia virus	*Culex*
RVFV	Rift Valley fever virus	*Aedes, Anopheles, Culex, Coquillettidia, Eretmapodites, Mansonia*
SAGV	Sagiyama virus	*Aedes, Culex*
SANV	Sango virus	*Mansonia*
SARV	Santa Rosa virus	*Anopheles*

continued

Table 11.2. *continued*

Abbreviation	Virus	Associated Culicid vector genera
SATV	Sathuperi virus	*Culex*
SAV	San Angelo virus	*Anopheles, Psorophora*
SDNV	Serra do Navio virus	*Aedes*
SEPV	Sepik virus	*Armigeres, Mansonia*
SFV	Semliki Forest virus	*Aedes, Anopheles, Culex, Eretmapodites*
SHOV	Shokwe virus	*Aedes, Anopheles, Mansonia*
SHUV	Shuni virus	*Culex*
SIMV	Simbu virus	*Anopheles, Coquillettidia*
SINV	Sindbis virus	*Aedes, Anopheles, Culex, Culiseta, Coquillettidia, Mansonia, Opifex*
SJV	San Juan virus	*Aedeomyia*
SLEV	St. Louis encephalitis virus	*Aedes, Anopheles, Culex, Deinocerites, Mansonia, Psorophora, Trichoprosopon, Wyeomyia*
SORV	Sororoca virus	'Sabethini'
SPOV	Spondweni virus	*Aedes, Culex, Eretmapodites, Mansonia*
SRV	Shark River virus	*Anopheles, Culex*
SSHV	Snowshoe hare virus	*Anopheles, Culiseta*
STRV	Stratford virus	*Anopheles*
TAHV	Tahyna virus	*Aedes, Anopheles, Culex, Culiseta*
TAIV	Tai virus	*Aedes, Culex*
TANV	Tanga virus	*Anopheles*
TATV	Tataguine virus	*Anopheles, Coquillettidia, Culex*
TCMV	Tacaiuma virus	*Anopheles, Haemagogus*
TENV	Tensaw virus	*Aedes, Anopheles, Coquillettidia, Culex, Mansonia, Psorophora*
TERV	Termeil virus	*Aedes, Culex*
TETEV	Tete virus	*Aedes*
TETV	Tettnang virus	*Aedes*
TIBOV	Tibet orbivirus	*Anopheles*
TILV	Tilligerry virus	*Anopheles*
TLAV	Tlacotalpan virus	*Anopheles, Mansonia*
TMEV	Tembe virus	*Anopheles*
TMUV	Tembusu virus	*Aedes, Anopheles, Culex*
TNTV	Triniti virus	*Trichoprosopon*
TONV	Tonate virus	*Coquillettidia, Mansonia, Wyeomyia*
TRUV	Trubanaman virus	*Anopheles*
TsDNV	*Toxorhynchites splendidus* densovirus	*Toxorhynchites*
TURV	Turlock virus	*Culex*
TVTV	Trivittatus virus	*Aedes, Culex, Mansonia*
UGSV	Uganda S virus	*Aedes*
UMAV	Umatilla virus	*Culex*
UMBV	Umbre virus	*Culex*
UNAV	Una virus	*Aedes, Coquillettidia, Psorophora*
USUV	Usutu virus	*Aedes, Anopheles, Coquillettidia, Culex, Mansonia*
VEEV	Venezuelan equine encephalitis virus	*Aedes, Anopheles, Culex, Deinocerites, Limatus, Mansonia, Psorophora, Wyeomyia*
VELV	Vellore virus	*Culex*
VINV	Vinces virus	*Culex*
VKTV	Venkatapuram virus	*Culex*
VRV	Virgin River virus	*Anopheles, Culex*

Abbreviation	Virus	Associated Culicid vector genera
VSIV	Vesicular stomatitis (Indiana serotype) virus	*Aedes*
VSNJV	Vesicular stomatitis (New Jersey serotype) virus	*Culex, Mansonia*
WANV	Wanowrie virus	*Culex*
WARV	Warrego virus	*Anopheles*
WEEV	Western equine encephalomyelitis virus	*Anopheles, Culex, Culiseta, Psorophora*
WGRV	Wongorr virus	*Aedes*
WHAV	Whataroa virus	*Culex, Culiseta, Opifex*
WITV	Witwatersrand virus	*Culex*
WNV	West Nile virus	*Aedes, Anopheles, Culex, Psorophora*
WONV	Wongal virus	*Culex, Coquillettidia*
WSLV	Wesselsbron virus	*Aedes, Anopheles, Culex, Mansonia*
WYOV	Wyeomyia virus	*Aedes, Anopheles, Coquillettidia, Culex, Culiseta, Haemagogus, Johnbelkinia, Limatus, Mimomyia, Psorophora, Trichoprosopon, Uranotaenia, Wyeomyia*
XIBV	Xiburema virus	*Sabethes*
YACV	Yacaaba virus	*Aedes*
YAOV	Yaounde virus	*Culex*
YATAV	Yata virus	*Mansonia*
YFV	Yellow fever virus	*Aedes, Anopheles, Haemagogus, Mansonia, Sabethes, Runchomyia*
ZEGV	Zegla virus	*Anopheles*
ZIKAV	Zika virus	*Aedes, Anopheles, Eretmapodites, Mansonia*
	Breu Branco virus	*Anopheles, Culicines*
	Chittoor virus	*Anopheles*
	Cypvirus (unnamed)	*Anopheles*
	Dianke virus	*Culex*
	Nanay virus	*Culex*
	Tyagiri virus	*Anopheles*

fully reviewed in the additive database maintained in the WRBU's online mosquito catalog (www.wrbu.org), as well as corrected, added to, or deleted country records, according to published country lists, and documented new species records. Our large source of new country-level distribution data was developed by comparing our database with the 1.7 million specimens housed in the US National Mosquito Collection (NMNH) and also the vouchered specimens in the MBI (see BOLD: Barcode of Life Database, at www.boldsystems.org). These were augmented by additional passes of biologically and spatially curated name/country-level records of more than 500,000 data points in the WRBU's online curated mosquito distribution portal VectorMap (www.vectormap.org), as well as the 98,657 online Culicid records for the Arthropodes d'Intérêt Médical (ARIM) archive collection of the Institut de Recherche pour le Développement (IRD), France (https://arim.ird.fr/arim/Conservation_etats/etatCollection). Additionally, 459,980 Culicid records from the Global Biodiversity Information Facility (GBIF; https://www.gbif.org/species/3346) were curated and included where appropriate. Large compendiums of regional mosquito distributions (e.g., Kyalo et al. 2017, Robert et al. 2019, Irish et al. 2020), especially those that included molecular identifications, were incredibly helpful in this process, as were revised country-level checklists that updated or discarded doubtful records.

Extensive literature searches were conducted to establish the associated pathogens for all species and genera treated here. The Centers for Disease Control and Prevention (CDC) online ArboCat (https://wwwn.cdc.gov/arbocat/) was extremely helpful as a baseline to build this reference database and was used to guide official virus abbreviations. A comprehensive list of mosquito-associated viruses is provided for each species page and added to each genus page. Virus names, abbreviations, and a summary of associated genera are given in Table 11.2. Associated pathogens include pathogenic bacteria, viruses, and parasites, including malaria. We carefully chose the wording "Associated Pathogens" instead of the more typically listed "Medical Importance." This subtle but important distinction serves as a caveat, particularly for the virus associations. Viral biosurveillance studies typically screen multiple whole mosquitoes in pools, meaning that "virus-positive" sources can invariably include host blood meals. Some mosquitoes are effective laboratory vectors of certain viruses but would never come in contact with them in nature. Here we do not differentiate between proven virus vectors and viruses recovered from wild-caught mosquito species.

Genus *Anopheles* Meigen, 1818
Subfamily Anophelinae
"Nail Mosquito"

Systematics and Distribution

Together with the small and geographically restricted genera *Bironella* and *Chagasia*, *Anopheles* mosquitoes make up the distinctive subfamily Anophelinae. The genus, represented by *An. maculipennis* Meigen, is subdivided into eight subgenera: *Anopheles*, *Baimaia*, *Cellia*, *Christya*, *Kerteszia*, *Lophopodomyia*, *Nyssorhynchus*, and *Stethomyia* (see Vol. 2, p. 1080, for internal relationships). Although only around 70 *Anopheles* species worldwide are responsible for the transmission of the five *Plasmodium* species that cause human malaria, an increasing number are incriminated in the transmission of arboviruses (see Associated Pathogens), including the *Anopheles*-only viruses O'nyong-nyong spread by *An. gambiae* Giles and *An. funestus* Giles in Africa, and Guaroa virus by *An. neivai* Howard, Dyar & Knab in the Neotropics. *Anopheles* are found globally, except at the polar ice caps and on some isolated island groups in the Pacific and Atlantic Oceans, including Fiji and New Zealand.

Bionomics

Typically *Anopheles* immatures are raised in clean, fresh, still, or slow-moving waters with plentiful floating vegetation, although there are many exceptions. Immature *An. litoralis* King are found in highly saline waters (≥2.5% to 3%) in Philippine salt flats, whereas *An. cucphuongensis* Vu et al. immatures individually occupy egg-cup-sized pools in limestone outcrops in the Cuc Phuong forest in Vietnam.

Our recent helium balloon captures in Mali showed that African *Anopheles*, including *An. coluzzii* Coetzee & Wilkerson, *An. gambiae* s.s., and *An. coustani* Laveran, escape the harsh arid conditions of the Sahara Desert as fully gravid females, migrating long distances at altitudes of over 240 m. This indicates that *Anopheles* mosquitoes can land in new zones potentially already infected with *Plasmodium* or other arboviruses, and may explain ongoing malaria in sub-Saharan African countries with active malaria eradication programs.

Associated Pathogens

AACV, AAOV, AATV, ACAV, AgDNV, AHOV, AMIV, ANAV, ANBV, AnCV, AnCPV, ANDV, AnFV, AngFV, AnsFV, APEUV, ARKV, AToV, BANV, BARV, BATV, BBKV, BEFV, BGIV, BIRV, BOAV, BORV, BOUV, BOZOV, Breu Branco virus, BUNV, BWAV, CARV, CATUV, CBV, CEV, CHIKV, Chittoor virus, COCV, CVOV, CVV, CWV, unnamed Cyprivirus, DENV, DSwV, EEEV, EHDV, EHV, EILV, EUBV, EVEV, GAMV, GERV, GMAV, GOMV, GROV, GTRV, HaCV, ILEV, ILHV, JBEV, KEYV, KOOV, KOTV, KOWV, KPKV, KRBV, LEAV, LMV, LPKV, LUKV, MAPV, MAYV, MBGV, MCOV, McPV, MDOV, MDV, MEV, MURV, MYXV, NDOV, NRIV, NTAV, ONNV, ORIV, OROV, ORUV, PERV, PIXV, POWV, RBV, RRV, RVFV, SARV, SAV, SFV, SHOV, SIMV, SINV, SLEV, SRV, SSHV, STRV, TAHV, TANV, TATV, TCMV, TENV, TIBOV, TILV, TLAV, TMEV, TMUV, TRUV, Tyagiri virus, USUV, VEEV, VRV, WARV, WEEV, WNV, WSLV, WYOV, YFV, ZEGV, ZIKAV; *Brugia malayi*, *Plasmodium falciparum*, *P. knowlesi*, *P. malariae*, *P. ovale*, *P. vivax*, *Wuchereria bancrofti*.

Diagnostic Characters

(a) Habitus (♀). (b) Head (♂): Palpus (MPlp) about as long as proboscis (P), apically clubbed. **(c) Head (♀):** Palpus (MPlp) about as long as proboscis (P). **(d) Thorax (dorsal):** Posterior margin of scutellum (Stm) evenly rounded, with evenly distributed setae. **(e) Abdomen:** Largely devoid of scales or with only small scale patches; setae present. **(f) Wing:** Usually with regular pattern of pale and dark spots; wing scales all dark in a few species; veins M_{1+2} and M_{3+4} straight or curved but not wavy. **(g) Larva (thorax):** Seta 1-M not palmate. **(h) Larva (terminal segments):** Siphon (S) absent. **(i) Larva (abdomen):** Seta 1 usually palmate, at least on abdominal segments III–VI. Image source: WRBU.

Exemplar whole genome sequences: *Anopheles* (*Anopheles*): *atroparvus* (GCA_000473505); ***Anopheles* (*Cellia*):** *funestus* (GCA_003951495.1), *gambiae* (GCF_000005575), *nili* (GCA_000439205.1); ***Anopheles* (*Nyssorhynchus*):** *albimanus* (GCA_000349125), *darlingi* (GCA_000211455).

Catalog entry: Vol. 2, p. 625.

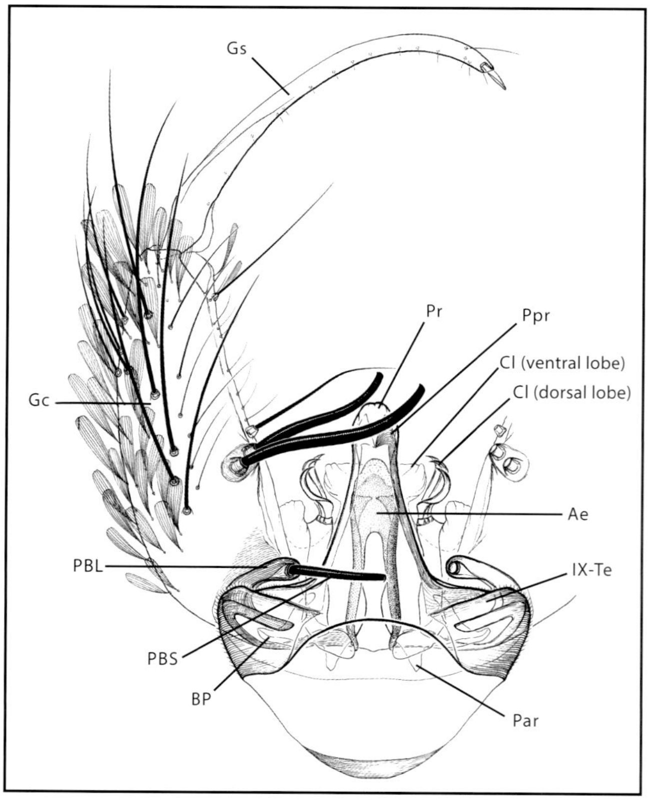

***Anopheles* (*Nyssorhynchus*) *strodei* Root, 1926 (♂ genitalia)**

Ae: Aedeagus; BP: Basal Piece; Cl: Claspette; Gc: Gonocoxite; Gs: Gonostylus; Par: Paramere; PBL: Parabasal Lobe; PBS: Parabasal Seta; Ppr: Paraproct; Pr: Proctiger; IX-Te: Tergum IX. (Modified after Faran, 1980).

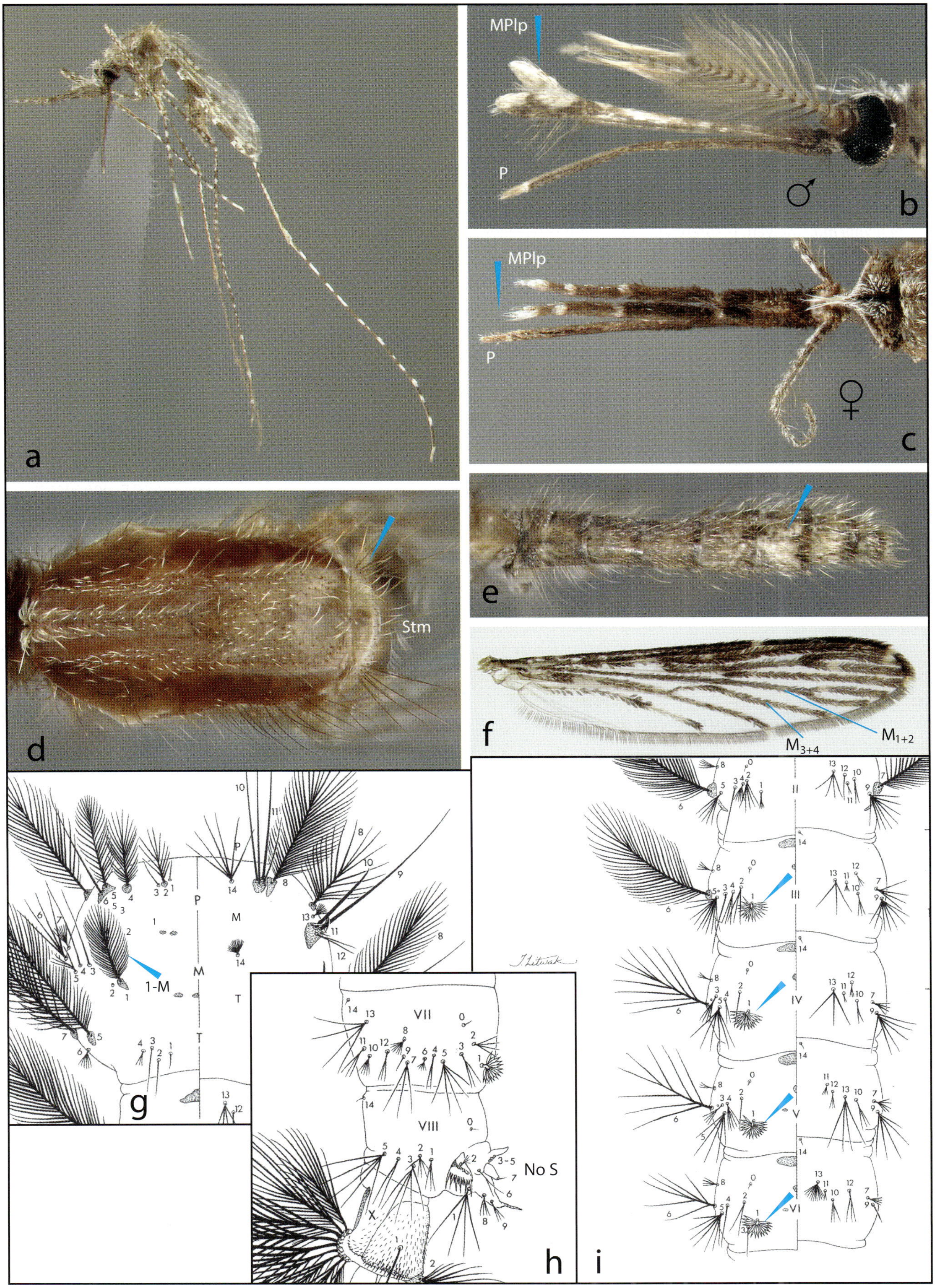

a
MPlp
P
♂
b
MPlp
P
♀
c
d
Stm
e
f
M_{3+4}
M_{1+2}
P
M
T
1-M
g
VII
VIII
X
No S
h
II
III
IV
V
VI
i

Genus *Bironella* Theobald, 1905
Subfamily Anophelinae
"Biró Mosquito"

Systematics and Distribution

The delicate members of the genus *Bironella* are restricted to primary forests of the Australasian Region, and are rarely encountered. The genus comprises only eight species, further split into three subgenera—*Bironella*, *Brugella*, and *Neobironella*. The type species *Bironella gracilis* Theobald is by far the most widespread of all *Bironella* species, being reported from the Moluccas to the Bismarck Archipelago. The subgenus *Bironella* comprises two species: *Bi. gracilis* and *Bi. simmondsi* Tenorio, which has only been reported from Queensland, Australia. Species assigned to the subgenus *Brugella* include *Bi. hollandi* Taylor (Bismarck Archipelago and Solomon Islands), *Bi. obscura* (New Guinea), and *Bi. travestita* (Moluccas [Ceram] and New Guinea). The remaining three described taxa are placed in the subgenus *Neobironella*. These include *Bi. confusa* Bonne-Wepster, found on New Guinea and the islands of the Bismarck Archipelago, *Bi. derooki* Soesilo & van Slooten (New Guinea), and *Bi. papuae* (Swellengrebel & Swellengrebel de Graaf), reported from the Moluccas, New Guinea, and northern Queensland, Australia. The overall placement of *Bironella* within the Anophelinae remains unresolved. Phylogenetic studies to date have placed it either as a sister taxa to *Anopheles* or embedded within the genus *Anopheles* itself. Diagnostic morphological characters separating *Bironella* from species in the two other genera (*Anopheles* and *Chagasia*) in the subfamily Anophelinae are illustrated (see Diagnostic Characters).

Bionomics

Bironella immatures are most commonly found among floating vegetation and debris associated with heavily shaded freshwater sites. *Bironella* (*Bironella*) *gracilis* and *Bi.* (*Bir.*) *simmondsi* are typically associated with deep-forest water bodies, such as swamps, pools, and rivers, but the larvae of *Bi. gracilis* have also been harvested from flower bracts. The immature stages of species in the subgenus *Brugella* include larger habitats (lagoons, stream margins, swamps, and ground pools), as well as animal wallows, flooded ruts in the road, and, occasionally, artificial containers. *Bironella* (*Neobironella*) *confusa* immatures have been collected in container habitats, including a tree hole and discarded tin can, but more common habitats for subgenus *Neobironella* species are the margins of ponds and streams, stream pools, and rock pools.

Little is known on the adult behaviors of *Bironella* species. Adults of *Bi.* (*Brugella*) *obscura*, *Bi.* (*Bru.*) *travestita*, and *Bi.* (*Nbi.*) *confusa* have been collected in malaise traps. *Bironella* (*Bru.*) *travestita* and *Bi.* (*Bir.*) *gracilis* are known to bite humans, with *Bi. gracilis* also reported to enter human dwellings.

Associated Pathogens

None.

Diagnostic Characters

(a) Habitus (♀). (b) Thorax (dorsal): Posterior margin of scutellum (Stm) evenly rounded, with evenly distributed setae. **(c) Abdomen:** Covered with setae but nearly devoid of scales. **(d) Wing:** Veins M_{1+2} and M_{3+4} distinctly sinuate (wavy). **(e) Larva (thorax):** Seta 1-M and 3-T palmate; seta 4-P closer to 1-3-P than 5-7-P. **(f) Larva (head).** Images sources: WRBU; Belkin (1962).

Exemplar DNA barcodes: *Bironella gracilis COI* (AF417725), *Bi.* (*Bru.*) *hollandi* (whole mitochondrial genome MF381612 and NC_037796).

Catalog entry: Vol. 2, p. 694.

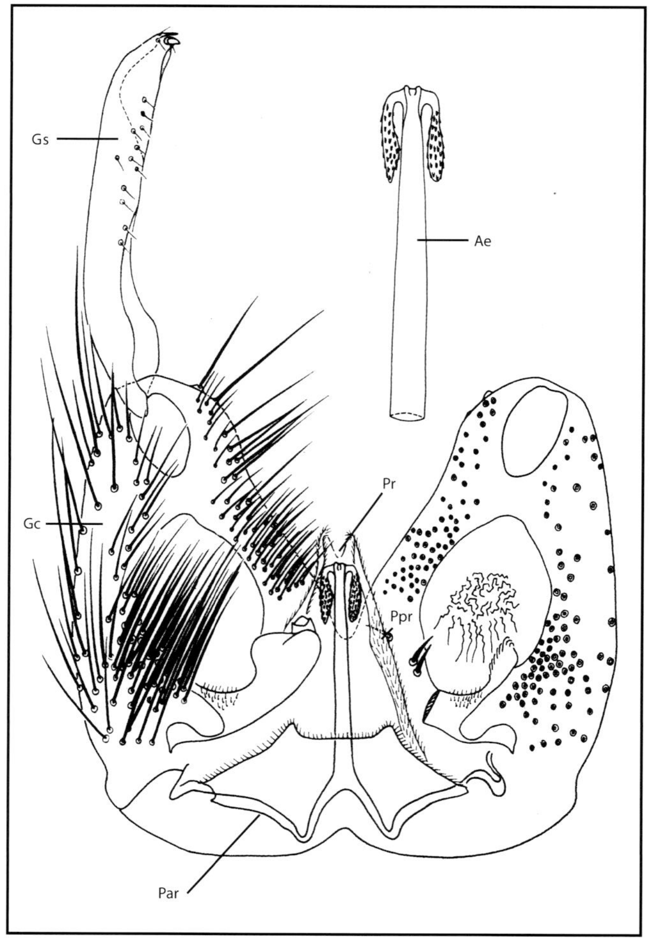

***Bironella* (*Bironella*) *gracilis* Theobald (♂ genitalia)**

Ae: Aedeagus; Gc: Gonocoxite; Gs: Gonostylus; Par: Paramere; Ppr: Paraproct; Pr: Proctiger. (Modified after Tenorio, 1977).

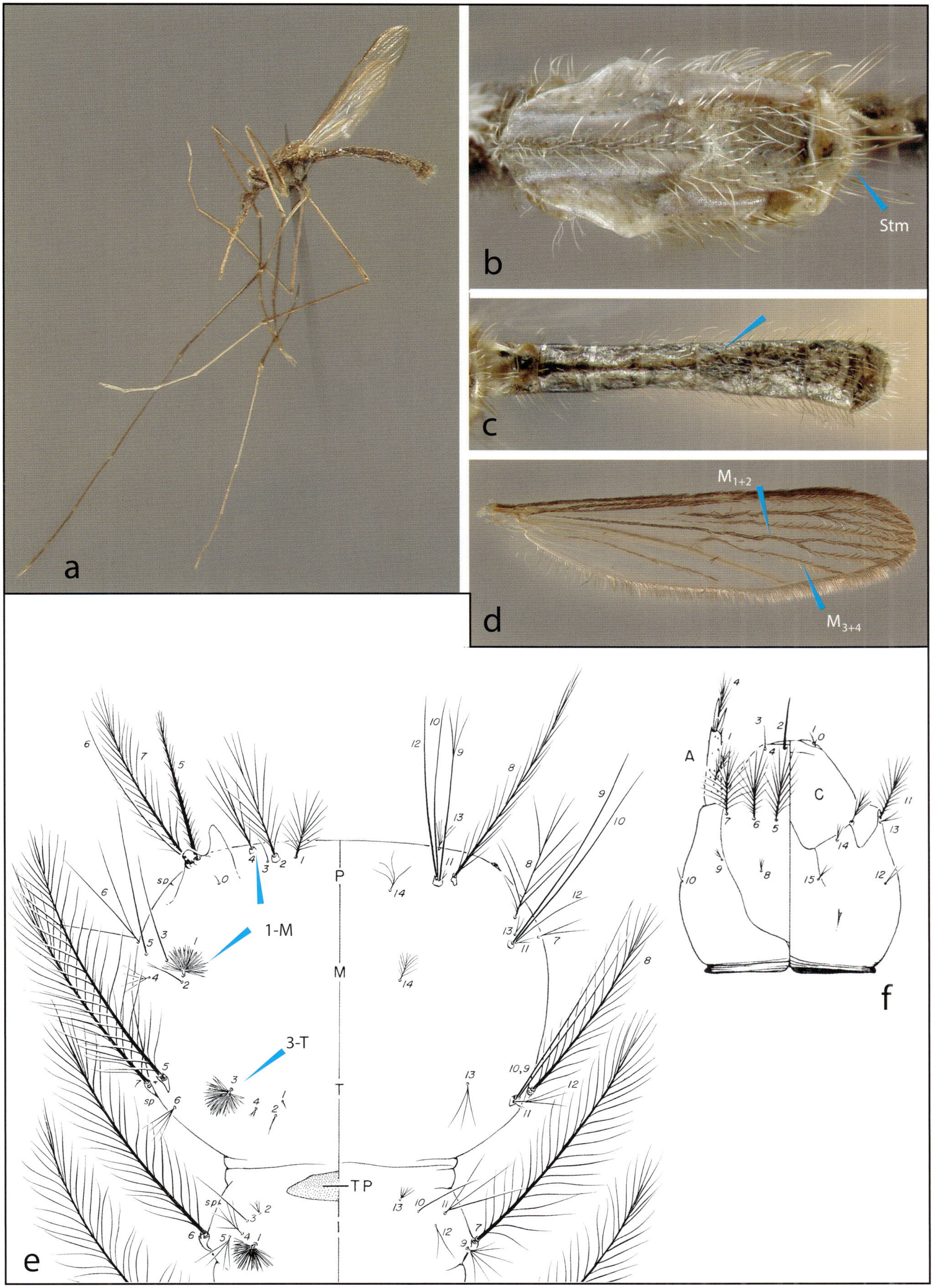
a
b
Stm
c
d
M_{1+2}
M_{3+4}
e
1-M
3-T
P
M
T
TP
f
A
C

Genus *Chagasia* Cruz, 1906
Subfamily Anophelinae
"Basal Mosquito"

Systematics and Distribution

Together with the genera *Anopheles* and *Bironella*, the Neotropical genus *Chagasia* comprises one of three genera in the subfamily Anophelinae. The genus currently comprises only five described species: *Ch. ablusa* Harbach, *Ch. bathana* (Dyar), *Ch. bonneae* Root, *Ch. fajardi* (Lutz), and *Ch. rozeboomi* Causey, Deane & Deane. The original type species of the genus is *Ch. neivae* Cruz, which is currently in synonymy with *Ch. fajardi*. The range of *Ch. bathana* extends from Ecuador, Colombia, and Venezuela through Central America into southern Mexico (Belize, Colombia, Costa Rica, Ecuador, Honduras, French Guiana, Guatemala, Mexico, Nicaragua, Panama, and Venezuela), whereas all four other *Chagasia* species are restricted to South America. *Chagasia ablusa* is found in Colombia, Peru, and Venezuela; *Ch. bonneae* in Bolivia, Brazil, Colombia, Ecuador, French Guiana, Guyana, Peru, Suriname, and Venezuela; *Ch. fajardi* in Argentina, Bolivia, Brazil, Colombia, and Guyana; and *Ch. rozeboomi* in Brazil. Recently, a new species (*Chagasia* nr. *fajardi)* was discovered by our team in collections made in the Amazonian forest of eastern Ecuador.

Mosquitoes of the genus *Chagasia* exhibit interesting morphological and genetic characteristics that are shared mostly by the subfamily Anophelinae, but some characters echo those thought to be diagnostic for the subfamily Culicinae (e.g., trilobed scutellum with setae in three groups, parallel resting stance, postpronotal setae, arched scutum). The most notable difference between *Chagasia* and all other Anophelinae and Culicinae mosquitoes examined to date is the number of chromosomes—*Ch. bathana* has three pairs of autosomes and a pair of sex chromosomes ($2n = 8$), in comparison to the $2n = 6$ reported in all other mosquitoes. Midges of the family Chaoboridae, once regarded as a subfamily within Culicidae, also share the same chromosomal arrangement ($2n = 8$). These attributes, along with evidence from multiple phylogenetic treatments, suggest that the genus *Chagasia* is a monophyletic lineage basal to other Anophelinae and that it may be both the basal ancestral lineage within Culicidae and the possible New World origin for all mosquitoes.

Bionomics

Immatures of the genus *Chagasia* are normally associated with shady streams, where they can be found hiding in the grassy margins, on dead leaves or floating debris of stream margins, in the slow-flowing waters caught around tree roots, or in clear rock pools in stream basins. Adult *Chagasia* often rest in the vegetation close to the stream margins or in the shade of the nearby humid forest canopy. True forest species, female *Chagasia* bite during the day and night in heavily shaded forest environs where they seldom encounter humans. Although females will feed on humans under laboratory conditions, *Chagasia* are essentially zoophilic.

Associated Pathogens

None.

Diagnostic Characters

(a) Habitus (♀). (b) Thorax (dorsal): Scutellum (Stm) slightly trilobed with setae in three groups. **(c) Abdomen:** With plentiful setae but almost devoid of scales. **(d) Head (dorsal). (e) Wing:** Dark scaled, or with dark and pale scales, not forming distinct spots. **(f) Larva (head):** Setae 2,4,6-C with very long stems ending in multiple long branches. **(g) Larva (abdomen):** Seta 1-III–V palmate with distinctive spatulate leaflets. **(h) Larva (terminal segments):** Anterior spiracular lobe with elongate median process (AMPc); spiracular median plate with a fringe of long spicules laterally. Images sources: WRBU; Mattingly (1971); Harbach & Howard (2009).

Exemplar DNA barcodes: *Chagasia bathana* (KX420410–14), *Ch. bonneae* (KF671005–11, KX420415), *Ch. sp.* nr. *fajardi* (KF671012–13).

Catalog entry: Vol. 2, p. 696.

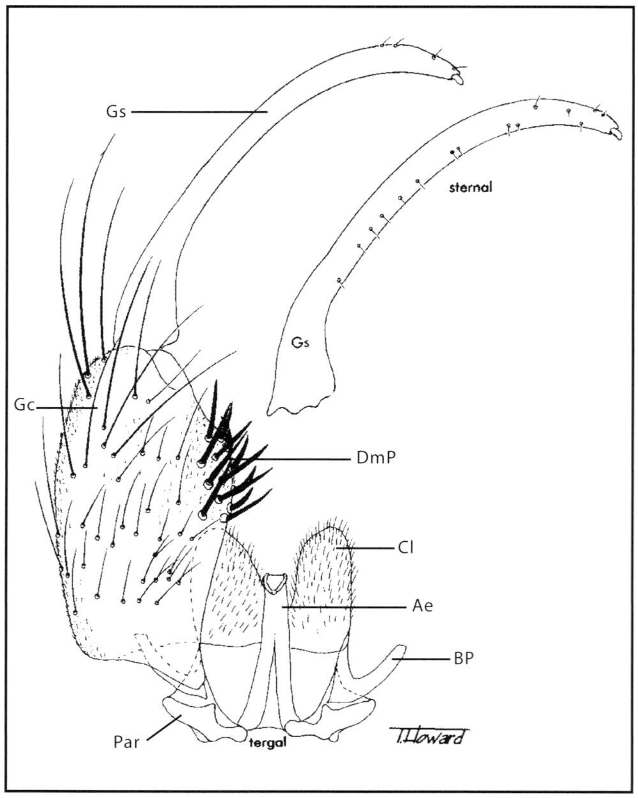

***Chagasia bonneae* Root (♂ genitalia)**

Ae: Aedeagus; Gc: Gonocoxite; Gs: Gonostylus; Par: Paramere; Pr: Proctiger. (Modified after Harbach and Howard, 2009).

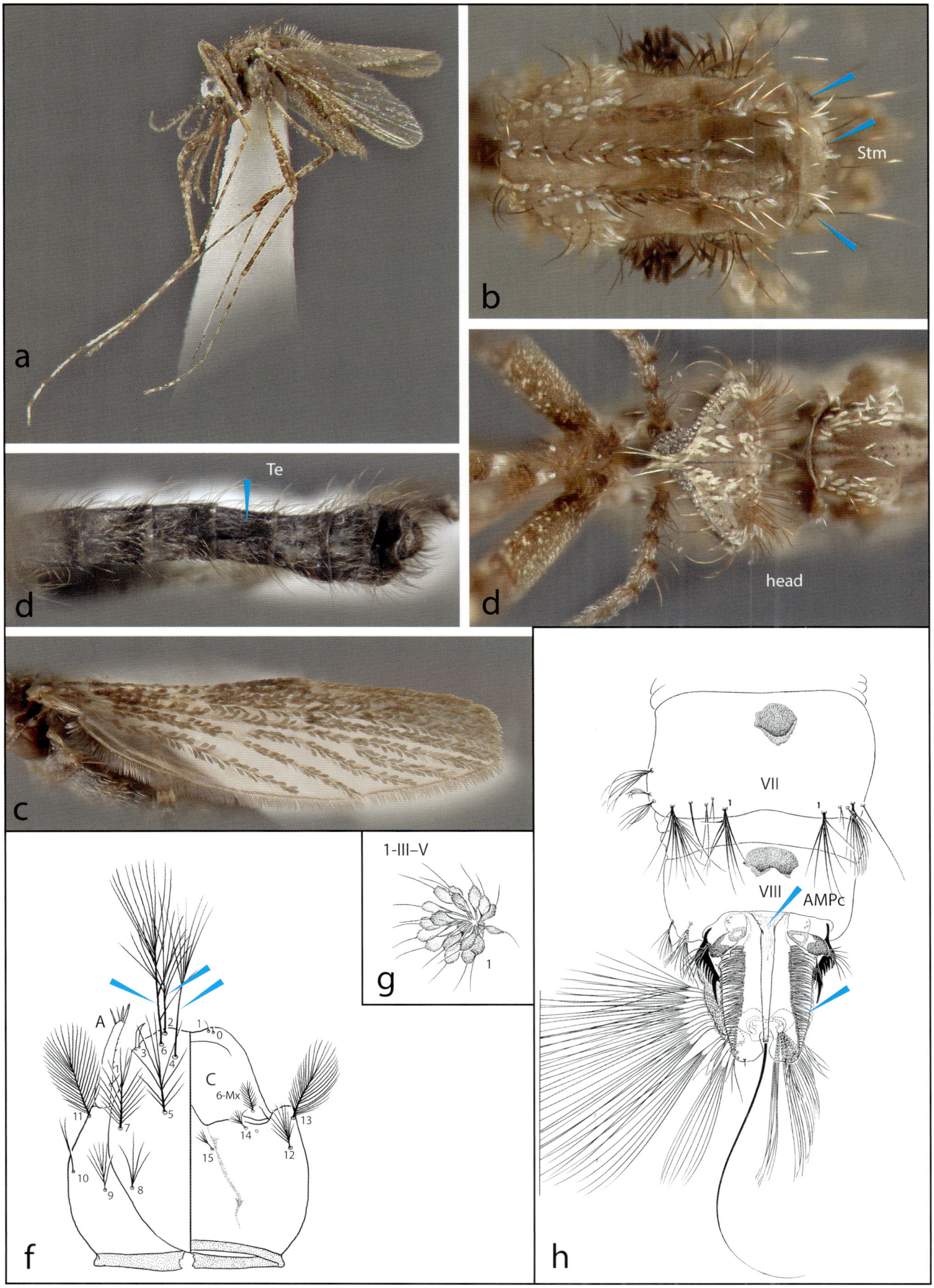
a
b
Stm
Te
d
head
d
c
VII
VIII
AMPc
h
1-III–V
g
A
C
6-Mx
f

Genus *Aedeomyia* Theobald, 1901
Subfamily Culicinae, Tribe Aedeomyiini
"Scaly Mosquito"

Systematics and Distribution

The genus *Aedeomyia*, the sole representative of the tribe Aedeomyiini, is distributed in the Afrotropical, Australasian, Neotropical, and Oriental Regions. The genus *Aedeomyia* comprises seven species subdivided into two subgenera: six in the subgenus *Aedeomyia* (*Ad. africana* Neveu-Lemaire, *Ad. catasticta* Knab, *Ad. madagascarica* Brunhes, Boussès & da Cunha Ramos, *Ad. pauliani* Grjebine, *Ad. squamipennis* (Lynch Arribálzaga), and *Ad. venustipes* (Skuse)) and one in *Lepiothauma* (*Ad. furfurea* (Enderlein)), found across sub-Saharan Africa and on Madagascar. *Aedeomyia squamipennis* currently serves as the type species for the genus. Species of the genus *Aedeomyia* are covered with broad scales and superficially resemble mosquitoes of the tribe Aedini, except that all female flagellomeres and the apical two flagellomeres of the males are uniquely stout. This character is also notable in the larva, where the antennae are swollen, strongly curved, and longer than the head capsule.

Bionomics

Larvae of *Ad.* (*Lep.*) *furfurea* are usually collected in larger permanent water bodies (swamps or ponds) that maintain high levels of aquatic vegetation, especially the water lettuce (*Pistia* spp.). Immature stages of *Ad. furfurea* have also been collected in densely vegetated rice fields in Madagascar. When under threat of predation, *Aedeomyia* larvae stay submerged for long periods, possibly using their enlarged antennae to facilitate cuticular respiration.

Adult *Aedeomyia* are most active for a few hours after sunset. Although *Aedeomyia* species are primarily ornithophilic canopy feeders, *Ad.* (*Lep.*) *furfurea* and *Ad.* (*Aed.*) *africana* females will opportunistically attack humans at ground level and are highly attracted to people positioned high up in the canopy. In Thailand, *Ad.* (*Aed.*) *catasticta* often bites humans. Several arboviruses have been recovered from wild-caught *Ad. catasticta* and *Ad. squamipennis*.

Associated Pathogens

ALFV, ANTV, CQIV, GAMV, KNAV, PVV, SJV; bird *Plasmodium* spp.

Diagnostic Characters

(a) Habitus (♀). (b) Head (♀): Broad scales on pedicel (Pe), Flm_1, and clypeus (Clp). **(c) Head (♂):** $Flm_{12,13}$ elongate in ♂♂. **(d) Thorax (side):** Lower mesepimeral setae (MeSL) present; prespiracular (PsS) and postspiracular (PS) setae absent. **(e) Leg (III):** Fe-II, III with large apical scale tufts. **(f) Leg (III):** $Ta\text{-}III_1$ longer than $Ta\text{-}III_{2-5}$ combined. **(g) Wing:** Dense covering of broad white, yellow, and brown scales. **(h) Larva (thorax):** Setae 9,10-P,M,T usually quite long. **(i) Larva (head):** Antennae (A) swollen, strongly curved, longer than the head. **(j) Larva (terminal segments):** Pecten (Pt) absent; siphon (S) uniformly pubescent; pair of curved hooks and a pair of many-branched setae at apex; siphonal trachea reduced; comb scales (CS) in a single row on small sclerotized plate (CP). Images sources: WRBU.

Exemplar DNA barcodes: *Aedeomyia africana* (KU186980–82, KU186985), *Ad. catasticta* (AY729969), *Ad. furfurea* (KU186979, KU186983–84, KU186986), *Ad. squamipennis* (KT766383–90).

Catalog entry: Vol. 2, p. 697.

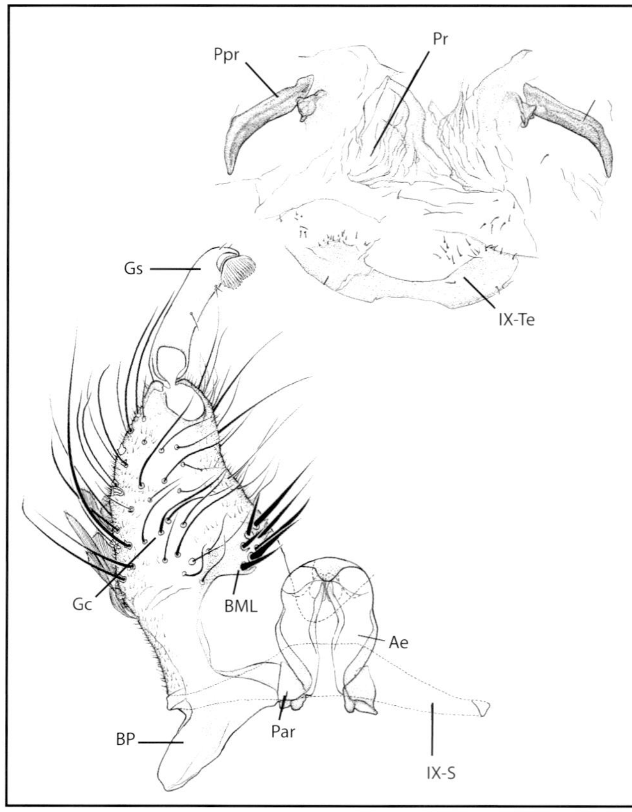

***Aedeomyia* (*Aedeomyia*) *catasticta* Knab (♂ genitalia)**

Ae: Aedeagus; BML: Basal Mesal Lobe; BP: Basal Piece; Gc: Gonocoxite; Gs: Gonostylus; Par: Paramere; Ppr: Paraproct; Pr: Proctiger; IX-Te: Tergum IX; IX-S: Sternum IX. (Modified after Tyson, 1970).

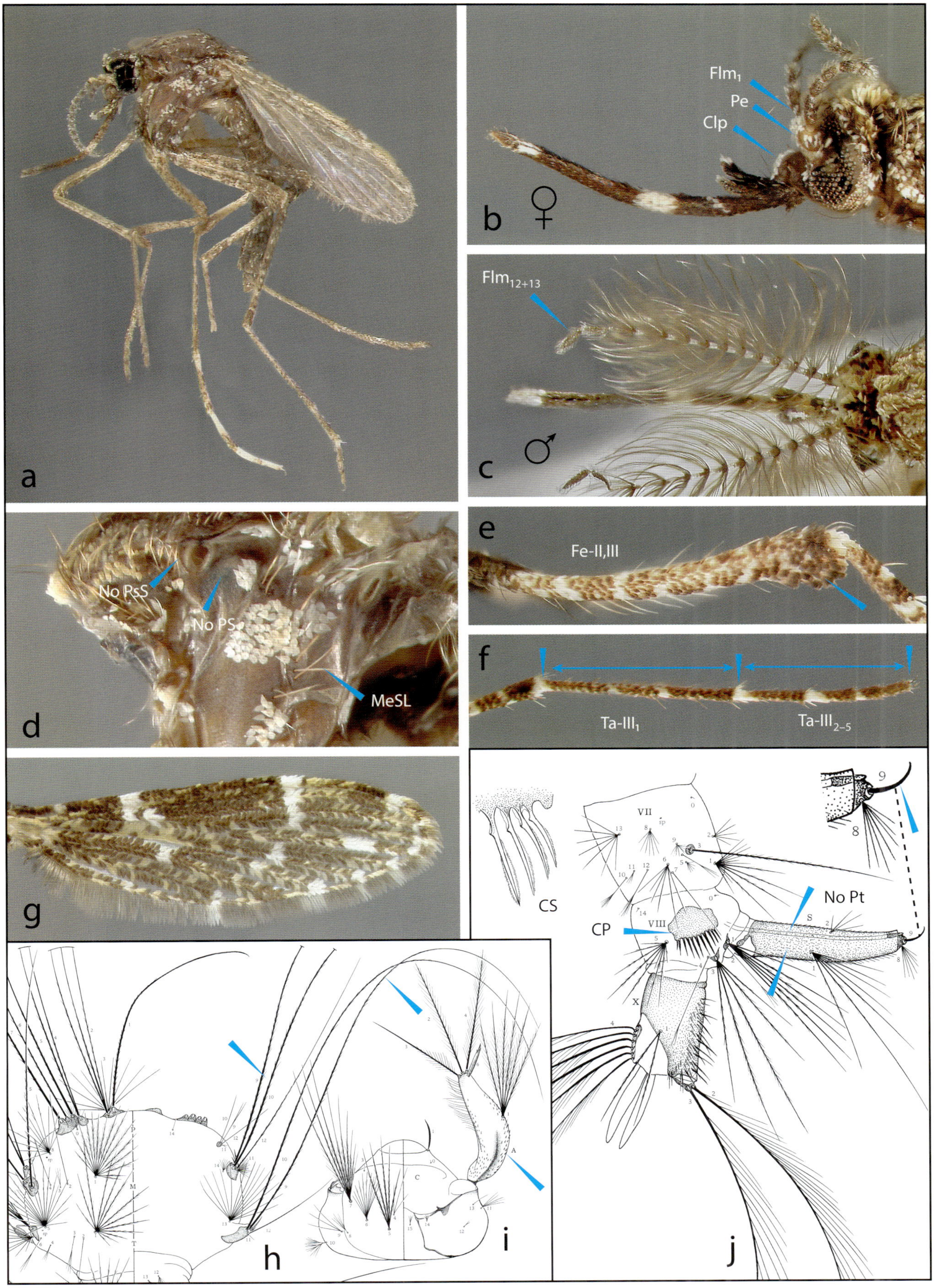
a
Flm1
Pe
Clp
b
♀
Flm12+13
c
♂
No RsS
No PS
MeSL
d
e
Fe-II,III
f
Ta-III1
Ta-III2–5
g
CS
VII
CP
VIII
No Pt
S
X
h
i
j

Genus *Aedes* Meigen, 1818
Subfamily Culicinae, Tribe Aedini
"Pointy Mosquito"

Systematics and Distribution

The genus *Aedes* is the largest in the tribe Aedini, comprising over 900 species, including many vectors of debilitating viral diseases to humans. A series of morphological phylogenies and resultant taxonomic actions by Reinert, Harbach, and Kitching since 2000 removed and elevated many species, species groups, and subgenera from the genus to new generic-level ranks, reducing the genus to only 12 species. These new genera and subgenera were largely undiagnosable, resulting in widespread taxonomic instability and confusion. In 2015 *Aedes* was stabilized, and the genus now comprises the original 12 genera and 74 subgenera. Genome-level molecular analyses are underway in our laboratory to resolve apparent paraphyletic groupings in tribe Aedini and genus *Aedes*.

Aedes mosquitoes are distributed in all regions of the world except Antarctica and include some of the mosquito worlds' most infamous travelers—*Ae.* (*Stegomyia*) *aegypti* (Linnaeus), *Ae.* (*Stg.*) *albopictus* (Skuse), *Ae.* (*Hulecoeteomyia*) *japonicus* (Theobald), and *Ae.* (*Hul.*) *koreicus* (Edwards).

Bionomics

Aedes females bite during the day and are known to feast on a wide variety of mammalian, bird, amphibian, and reptilian hosts, often taking blood from more than one host per egg cycle. *Aedes* (*Pat.*) *dobodurus* King & Hoogstraal breeds exclusively in small amounts of rainwater in cup fungi in Papua New Guinea. Many temperate *Aedes* are typical floodwater mosquitoes, with egg hatching stimulated by the water inundation from snowmelt, resulting in high-density, early season populations. Other tropical forest species, including *Aedes* (*Stegomyia*), that typically lay eggs in small natural cavities, have rapidly domesticated to artificial containers in urban and semiurban environments, where they act as vectors of zoonotic spillover diseases.

Arboviral infections in *Aedes* mosquitoes appear to facilitate transmission. When *Aedes aegypti* is infected with dengue virus, it probes more and takes longer to feed; when infected with Sindbis virus, it becomes less susceptible to repellents. Several arboviruses are transmitted transovarially by *Aedes* mosquitoes. Balloon captures at heights up to 240 m above the Sahara Desert showed the active long-distance migration of gravid *Aedes* females, potentially transporting already infected offspring to new zones.

Associated Pathogens

AKAV, ALFV, APEUV, AURAV, BATV, BBKV, BEBV, BEFV, BFV, BKNV, BOUV, BOZOV, BUNV, BWAV, CEV, CHIKV, COCV, COTV, CVOV, CVV, DENV, DUGV, EEEV, EHV, EVEV, GERV, GETV, GGV, GLV, HJV, ILHV, INGV, INKV, ISFV, ITUV, JBEV, JCV, JUGV, KEDV, KEYV, KINV, KOKV, LACV, LEBV, LUKV, LUMV, MBOV, MELV, MIDV, MIRV, MPKV, MUCV, MVEV, NDOV, NDUV, NEGV, NGOV, NKOV, NORV, NRIV, ORIV, OROV, ORUV, PATAV, PGAV, PLAV, POWV, RRV, RVFV, SAGV, SDNV, SFV, SHOV, SINV, SLEV, SPOV, TAHV, TAIV, TENV, TERV, TETEV, TETV, TMUV, TVTV, UGSV, UNAV, USUV, VEEV, VSIV, WGRV, WNV, WSLV, WYOV, YACV, YFV, ZIKAV.

Diagnostic Characters

(a) Habitus (♀). (b) Thorax (side): Prespiracular setae (PsS) absent; postspiracular setae (PS) nearly always present; mesopostnotal setae (MpnS) usually absent; base of hind coxa (C-III) usually below base of mesomeron (Msm). **(c) Abdomen:** Apex usually (not always) more or less pointed in ♀♀. **(d) Wing:** Remigial setae (ReS) absent. **(e) Larva (thorax):** Seta 9–12-T sometimes moderately reduced; longest seta ≥0.5 times length of 9–12-M. **(f) Larva (terminal segments):** Single pair of seta 1-S, inserted near or beyond middle of siphon (S); pecten (Pt) always present (may be reduced to one or two spines). Image sources: WRBU; Belkin et al. (1970); Tanaka et al. (1979).

Exemplar whole genome sequences: *Ae.* (*Stg.*) *aegypti* (GCF_002204515), *Ae.* (*Stg.*) *albopictus* (GCF_001876365).

Catalog entry: Vol. 2, p. 699.

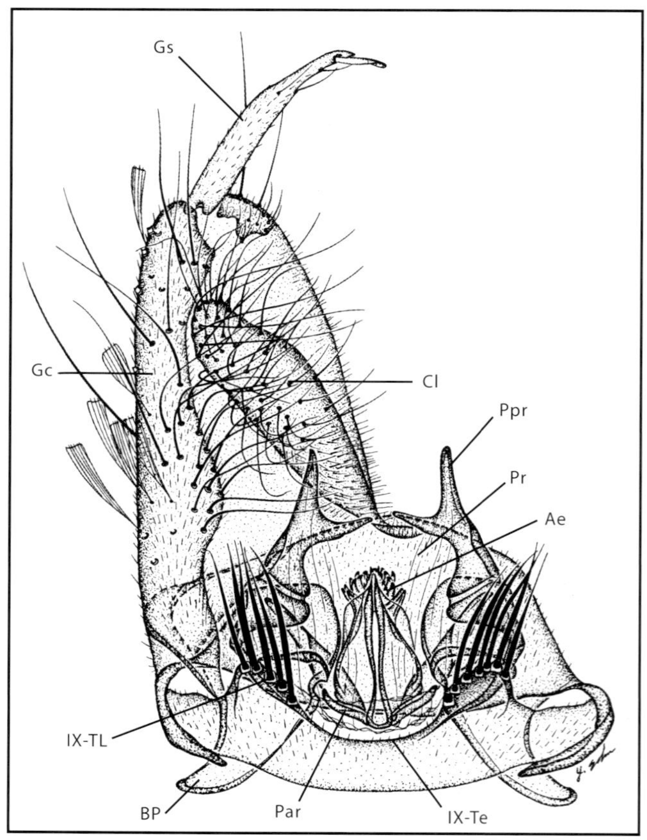

***Aedes* (*Stegomyia*) *bromeliae* (Theobald) (♂ genitalia)**

Ae: Aedeagus; BP: Basal Piece; Cl: Claspette; Gc: Gonocoxite; Gs: Gonostylus; Par: Paramere; Ppr: Paraproct; Pr: Proctiger; IX-Te: Tergum IX; IX-TL: Tergum IX Lobe. (Modified after Huang, 2004).

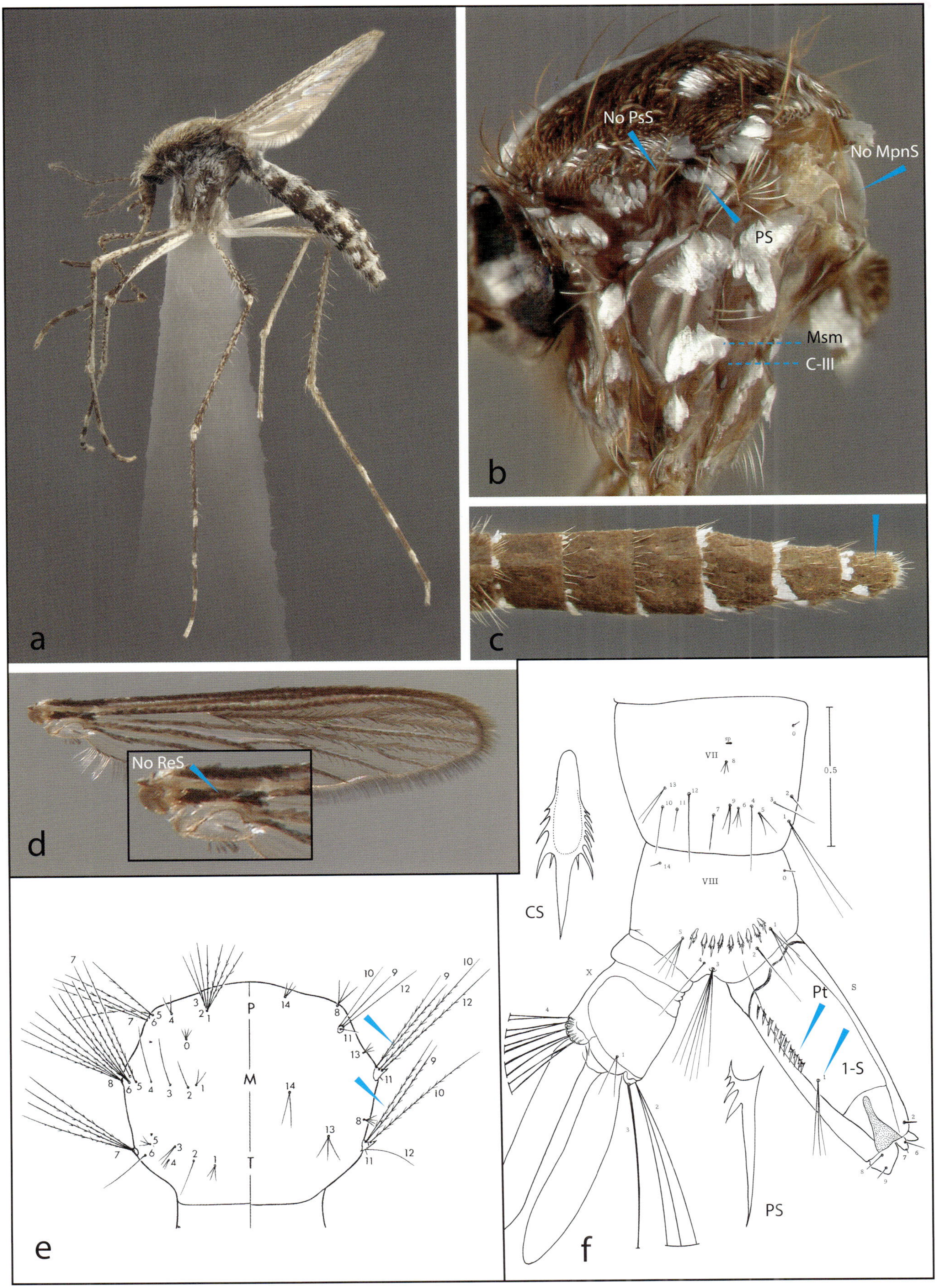
a
No PsS
No MpnS
PS
Msm
C-III
b
c
No ReS
d
P
M
T
e
CS
VII
VIII
0.5
X
Pt
S
1-S
PS
f

Genus *Armigeres* Theobald, 1901
Subfamily Culicinae, Tribe Aedini
"Cord Mosquito"

Systematics and Distribution

Mosquitoes of the genus *Armigeres* are among the largest in the tribe Aedini. They are easily recognized by their downturned and laterally flattened proboscis and two long rows of scales that separate their eyes. As larvae, the absence of a pecten differentiates *Armigeres* species from all other aedines. The genus contains 58 species subdivided into two subgenera: *Armigeres* (40 spp.) and *Leicesteria* (18 spp.). *Armigeres obturbans* (Walker, 1859) serves as the type species of the genus. *Armigeres* are restricted to the Oriental Region, except *Ar.* (*Leicesteria*) *annulipalpis* (Theobald) that reaches south to Ceram Island (Australian Region), and *Ar.* (*Armigeres*) *subalbatus* (Coquillett) that ranges north into the Palearctic Region (China, Korea, and Japan).

Bionomics

Armigeres species are mainly associated with jungle and jungle-edge plantation environs, although some species now thrive in urban and semiurban areas, particularly in those regions associated with poor sanitation. *Armigeres* (*Leicesteria*) species are found in lowland and dipterocarp forests but never in primary forest environments. Many species of *Armigeres* readily attack humans. They are often reported to bite during the day but are most active at dawn and dusk. One author recalls the ferocity of *Ar. subalbatus* (see species pages, p. 472) in the Cuc Phuong National Forest in Vietnam. Host-seeking females emerged in huge densities for an intense 30–60 minutes as the sun set, and their numbers, persistence, and painful bites often forced even the most hardened field collectors to shelter indoors during that period.

Armigeres eggs are laid singly. Post ovipositional care is noted in some species of the subgenus *Leicesteria*, including *Ar. annulitarsis* (Leicester), *Ar. balteatus* Macdonald, *Ar. flavus* (Leicester), *Ar. inchoatus* Barraud, *Ar. magnus* (Theobald), and *Ar. traubi* Macdonald. Females gather all their eggs and attach them to their hind legs before releasing them or incompletely hatched larvae onto preferred water surfaces or into tiny protected habitats, such as bamboo insect rot holes (*Ar. flavus*). Egg batches still attached to the hind legs of these *Armigeres* species have been noted on museum specimens and have been reported on actively host-seeking females in the wild.

Immatures are most commonly found in small container habitats, with foul waters or those with high organic content. Common larval habitats include tree holes, hollow logs, tree, banana, and palm stumps, split and broken bamboo, pitcher plants, leaf axils, fallen leaves/spathes, and fallen fruit husks, as well as a number of artificial containers. In urban areas, they are common in septic tanks, cesspools, and cement tanks. The modified larval mouthparts of some *Armigeres* (*Armigeres*) suggest that certain species—for example, *Ar. obturbans* and *Ar. subalbatus*—are predatory as larvae or rasp at fruit husks.

Associated Pathogens

AsTV, DENV, GETV, JBEV, Kadipiro virus, SEPV; *Brugia pahangi*, *Dirofilaria immitis*, *Wuchereria bancrofti*.

Diagnostic Characters

(a) Habitus (♀). (b) Wing: Alula (Al) and upper calypter (UC) with marginal hairlike scales. **(c) Thorax (side):** Prespiracular setae (PsS) absent; postspiracular area (PA) with scales, seta present or absent; base of small mesomeron (Msm) more or less even with base of C-III; head scales primarily broad and flat dorsally. **(d) Thorax (dorsal):** Acrostichal (AcS) and dorsocentral (DS) setae absent. **(e) Head (ventral):** Eyes separated ventrally by two long rows of scales. **(f) Head (♀):** Proboscis (P) laterally compressed and slightly downturned at tip. **(g) Larva (terminal segments):** Pecten (Pt) absent. **(h) Larva (head). (i) Head (♂):** Palpus (MPlp) about as long as proboscis (P); MPlp long, slender, upturned, and three segmented, with the last two segments nearly bare. Image sources: WRBU; Tanaka et al. (1979); Rattanarithikul et al. (2010).

Exemplar DNA barcodes: *Ar.* (*Arm.*) *kesseli* (MG709137), *Ar.* (*Arm.*) *malayi* (MG242529), *Ar.* (*Arm.*) *subalbatus* (KJ768155, KJ768109, KC970284–91), *Ar.* (*Lei.*) *flavus* (KU530125).

Catalog entry: Vol. 2, p. 812.

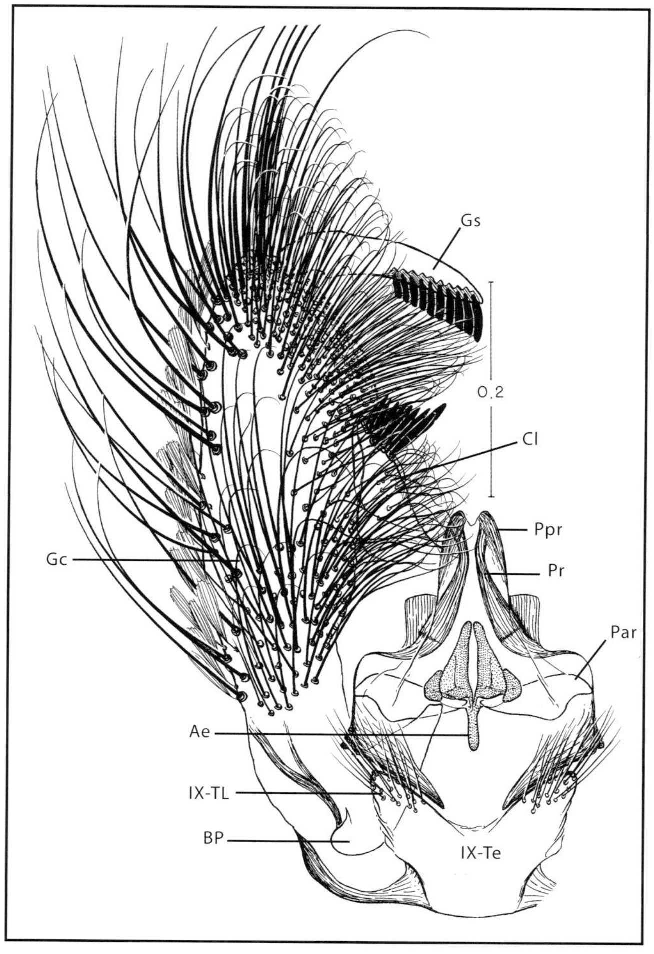

***Armigeres* (*Armigeres*) *breinli* (Taylor) (♂ genitalia)**

Ae: Aedeagus; BP: Basal Piece; Cl: Claspette; Gc: Gonocoxite; Gs: Gonostylus; Par: Paramere; Ppr: Paraproct; Pr: Proctiger; IX-Te: Tergum IX; IX-TL: Tergum IX Lobe. (Modified after Belkin, 1962).

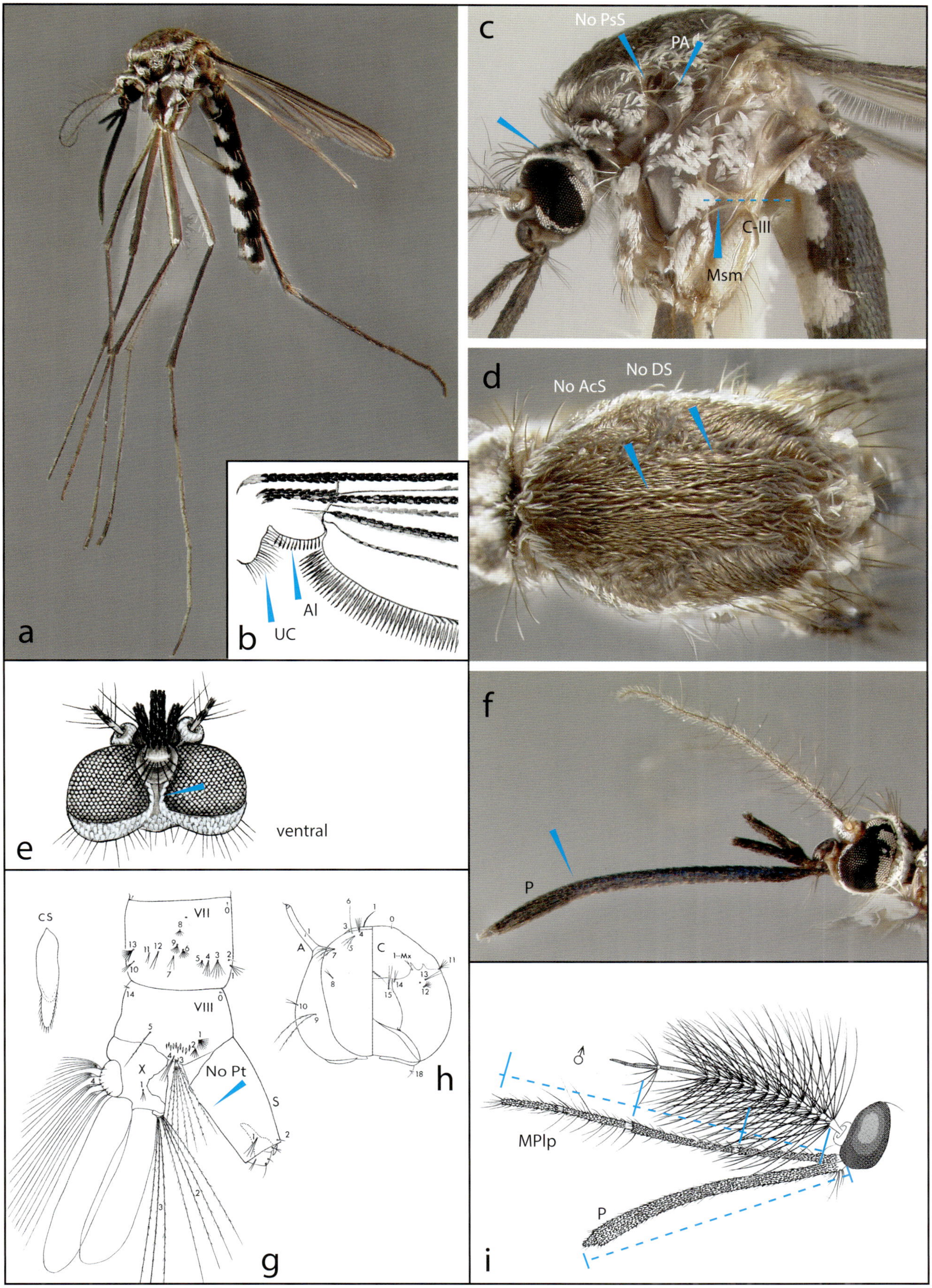
a
b
UC
Al
c
No PsS
PA
C-III
Msm
d
No AcS
No DS
e
ventral
f
P
g
CS
VII
VIII
X
No Pt
S
h
A
C
1-Mx
i
♂
MPlp
P

Genus *Eretmapodites* Theobald, 1901
Subfamily Culicinae, Tribe Aedini
"African Mosquito"

Systematics and Distribution

Mosquitoes of the Afrotropical genus *Eretmapodites* are yellowish brown and sport a distinctive broad patch of flat silver scales separating the eyes posterior to the antennae (see *e*). Unlike other African aedines, the paratergite of *Eretmapodites* species is bare and distinctly broad (see *b*). The Afrotropical genus comprises some 48 described species, which are not further subdivided. Three of these, *Er. plioleucus* Edwards, *Er. silvestris* Ingram & de Meillon, and *Er. wansoni* Edwards, comprise two subspecies. The genus has no formal subgeneric divisions, and *Er. quinquevittatus* Theobald, originally described from specimens in Freetown, Sierra Leone, serves as the type species. The genus *Eretmapodites* is the only mosquito genus entirely restricted to Africa.

Bionomics

Eretmapodites species are common inhabitants of African forests. Much like species of the Neotropical forest-dwelling genus *Trichoprosopon*, *Eretmapodites* immatures are found in almost all water collections in natural forest containers on the forest floor, including fruit husks, snail shells, fallen leaves, leaf spathes, and split bamboo. They are also found in small water collections in discarded trash and even in the small collections of water held by certain species of fungi. *Eretmapodites* immatures have been collected in the axils of banana plants—for example, *Er. silvestris*, which is common in *Dracaena* leaf axils in South Africa, and *Er. silvestris conchobius*, found in *Sansevieria* leaf axils in Sudan. *Eretmapodites* larvae are secondarily predacious, feeding on other larvae when resources are scarce.

In Uganda, *Eretmapodites* females are diurnal, biting at ground level in heavily shaded environs of the forest. Although they probably feed on forest birds and mammals, some will feed on humans. Female *Er. silvestris conchobius* have been described as "fierce and fearless biters," attacking humans through their clothes in bright sunlight more than 20 m from their heavily shaded resting sites. Several *Eretmapodites* species have been circumstantially incriminated in the transmission of Rift Valley fever virus, and *Er. chrysogaster* is a proven vector of yellow fever. *Eretmapodites chrysogaster* has been successfully colonized in the laboratory.

Associated Pathogens

BEBV, JBEV, NKOV, OKOV, RVFV, SFV, SPOV, ZIKAV.

Diagnostic Characters

(a) Habitus (♀): Proboscis (P) long, slender, and slightly downturned apically. **(b) Thorax (side):** Postspiracular setae (PsS) present; paratergite (Pa) broad, without scales or setae. **(c) Abdomen (side):** Terga (Te) black with silvery patches. **(d) Abdomen (ventral):** Abdomen laterally compressed, golden yellow ventrally (not evident here except on terminal segments). **(e) Head (dorsal):** Interocular space (IS) and margins of eyes with broad, appressed, silvery scales. **(f) Larva (terminal segments):** Pecten (Pt) with only 1–4 pecten spines (PS); seta 4-X single or double, plumose, inserted on a small grid (G); comb scales (CS) in a patch. **(g) Larva (head):** Antenna (A) short and cylindrical. **(h) Larva (abdomen):** Setae 6-I–VI often borne on distinct projections. Image sources: WRBU; Mattingly (1971).

Exemplar DNA barcodes: *Er. intermedius* (GQ165770, GQ165779, GQ165790).

Catalog entry: Vol. 2, p. 818.

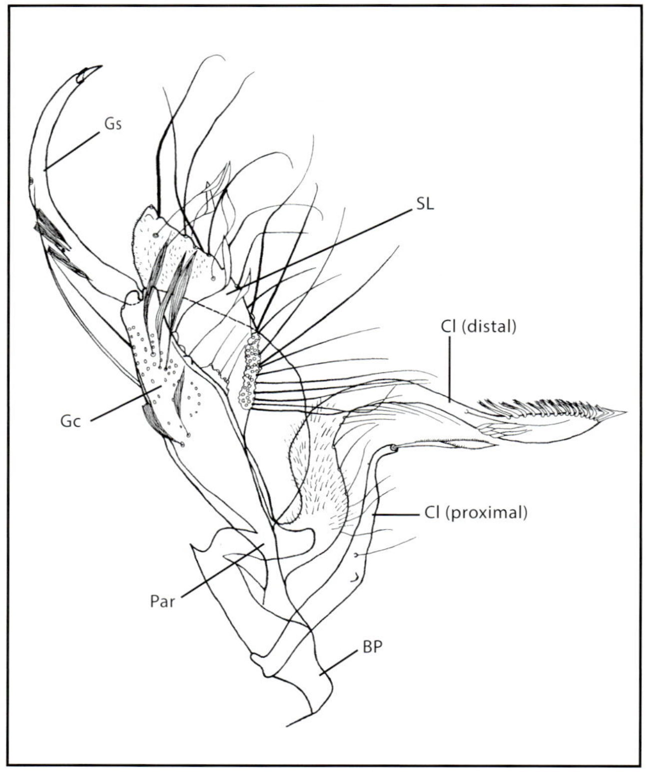

***Eretmapodites rickenbachi* Ferrara & Eouzan, 1974 (♂ genitalia)**

BP: Basal Piece; Cl: Claspette; Gc: Gonocoxite; Gs: Gonostylus; Par: Paramere; SL: Subapical Lobe. (Modified after Ferrara and Eouzan, 1974).

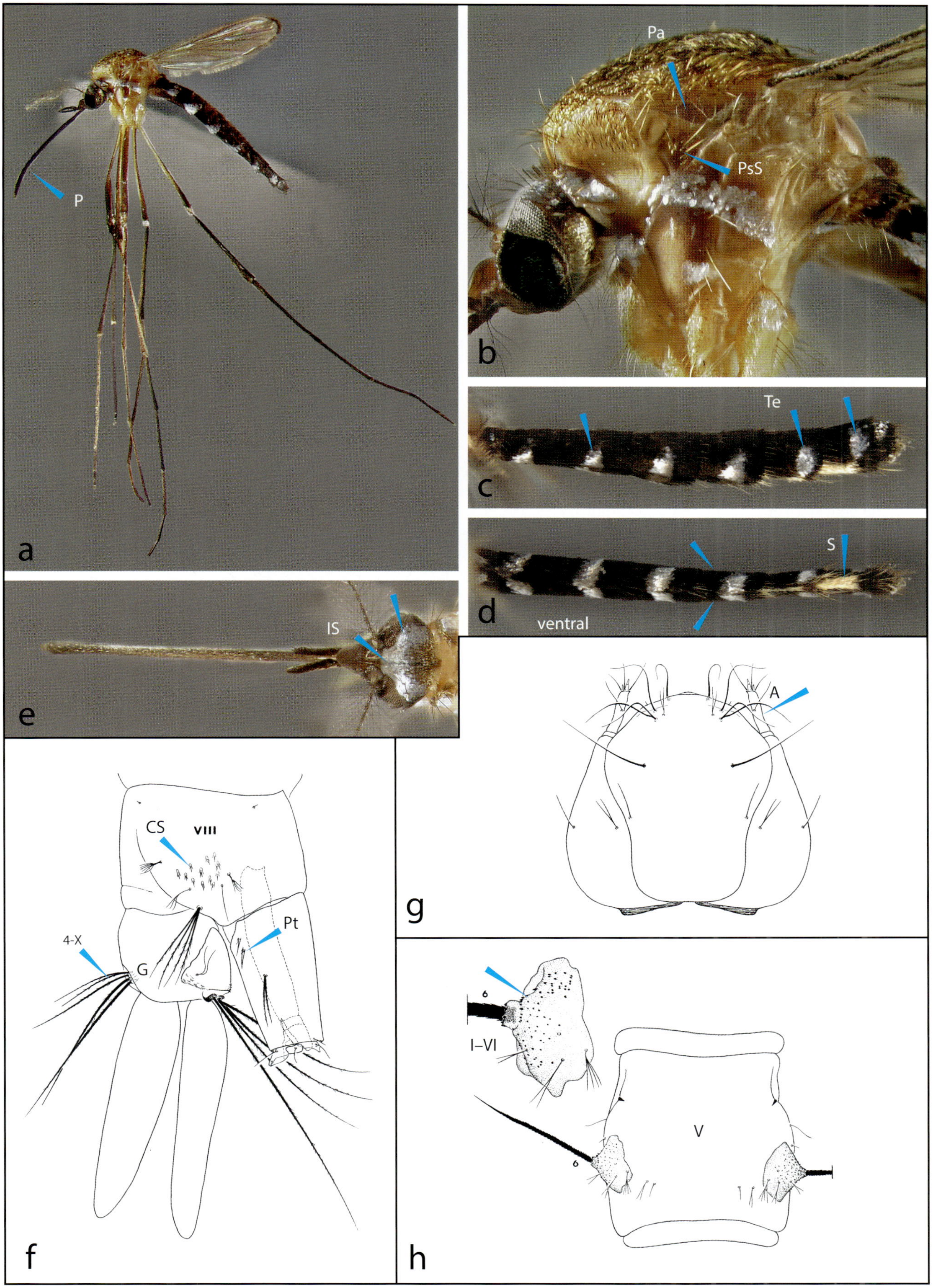
Pa
PsS
P
Te
S
ventral
IS
A
CS
VIII
Pt
4-X
G
6
I–VI
V
6
a
b
c
d
e
f
g
h

Genus *Haemagogus* Williston, 1896
Subfamily Culicinae, Tribe Aedini
"Shoulder Mosquito"

Systematics and Distribution

Haemagogus species are attractive mosquitoes, often with a metallic sheen. The aedine genus *Haemagogus* comprises some 28 valid species, subdivided into two subgenera: *Conopostegus* (4 spp.) and *Haemagogus* (24 spp.). The palpi of *Hg. equinus* Theobald are as long as the proboscis, a trait that is more commonly associated with anopheline mosquitoes. *Haemagogus* (*Haemagogus*) mosquitoes are found in Neotropical South and Central America, with some species creeping into the southern United States. Species of the small subgenus *Conopostegus* occur from Honduras in Central America down into South America as far as northern Argentina and southern Brazil.

Bionomics

Haemagogus species are primarily associated with dense primary and gallery Neotropical forest habitats and coastal mangroves, Some are found in remnant forest patches in cities. *Haemagogus* immatures mainly occupy natural forest containers, including tree holes, bamboo, and bromeliad axils, but some raise in rock holes, ground pools, and artificial containers. Eggs of *Haemagogus* species can survive desiccation for several months. Many *Haemagogus* species will readily attack humans, particularly in forest edges, clearings, and mangroves. Females are diurnal and feed on birds, primates, and forest-dwelling and domestic mammals. Some species, like *Hg. leucocelaenus* Dyar & Shannon, are sylvatic, spend almost their entire life cycle high in the canopy, and only rarely descend to ground level. Oviposition occurs in natural cavities at varying heights throughout the canopy, including ovitraps when set at 7 m and 9 m. The acrodendrophilic *Hg. janthinomys* Dyar is routinely captured at human bait in the Amazon forest canopy of Peru and ventures to ground level only at certain seasons in the forests of French Guiana. As *Haemagogus* mosquitoes feed both on humans and primates and have long life spans, they have increased capacity for arbovirus contact, amplification, and zoonotic disease transfer. *Haemagogus* species, including those in the subgenus *Conopostegus* (e.g., *Hg. leucocelaenus*) and the subgenus *Haemagogus* (e.g., *Hg. capricornii* Lutz, *Hg. equinus*, *Hg. janthinomys*, *Hg. lucifer* (Howard, Dyar & Knab), *Hg. mesodentatus* (Komp & Kumm), *Hg. spegazzinii* Brèthes), are important vectors of yellow fever virus in the Americas.

Associated Pathogens

DENV, ILHV, JURV, MAYV, MUCV, TCMV, UNAV, WYOV, YFV.

Diagnostic Characters

(a) Habitus (♀): Head, thorax, and abdomen covered with broad, distinct, reflective metallic scales; pleuron with a broad vertical band of silver scales or with three arcs of silver-scaled bands. **(b) Thorax (side):** Antepronotum (Ap) large, closely approximated dorsally; prespiracular setae (PsS) absent; C-III larger than C-II; dorsal margin of C-III slightly below or level with upper border of mesomeron (Msm). **(c) Wing:** Entirely dark. **(d) Thorax (dorsal):** Acrostichal (AcS), dorsocentral (DS), and prescutellar (PrsS) setae mostly absent. **(e) Larva (head):** Pale brownish head with darker collar. **(f) Larva (thorax):** Setae 5–7-P on small, isolated tubercles. **(g) Larva (abdomen):** Seta 12-I present and/or seta 7-III–V distinctly longer than corresponding seta 9-III–V. **(h) Larva (terminal segments):** Seta 3-VII well developed and usually single; seta 5-VIII inserted well below dorsal margin of segment X; seta 4-X made up of five or six pairs of setae; saddle (Sa) extends around c. 0.5 of segment X; single pair of seta 1-S inserted beyond pecten (Pt); comb (C) with comb scales (CS) and pecten (Pt) present. Image sources: WRBU; Arnell (1973).

Exemplar DNA barcodes: *Hg. janthinomys* (KY859893–96, *Hg. lucifer* (KT766493–517), *Hg. leucocelaenus* (KX758543, MG242532), *Hg. equinus* (JX260698).

Catalog entry: Vol. 2, p. 822.

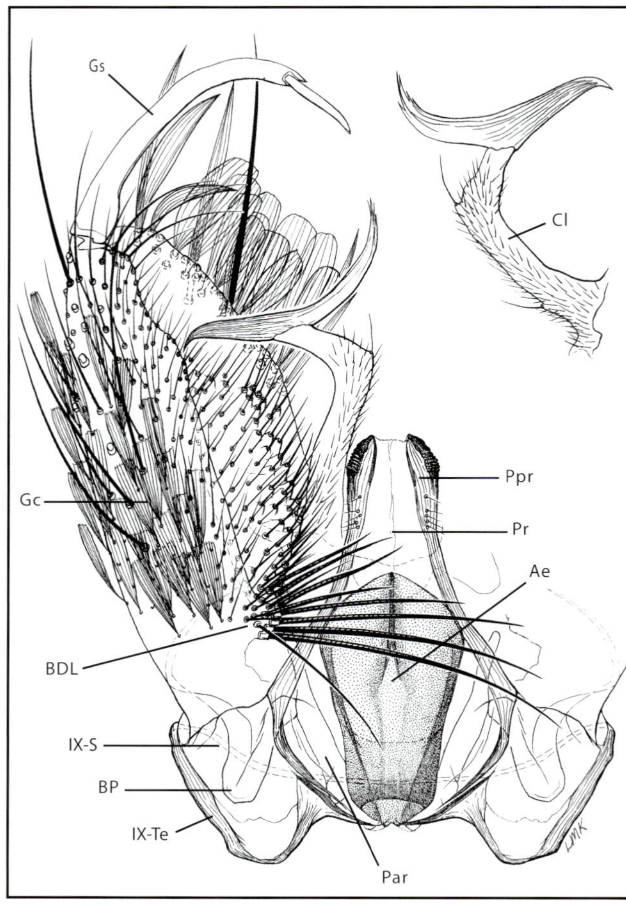

***Haemagogus* (*Haemagogus*) *splendens* (Williston) (♂ genitalia)**

Ae: Aedeagus; BDL: Basal Dorsomeral Lobe; BP: Basal Piece; Gc: Gonocoxite; Gs: Gonostylus; Par: Paramere; Ppr: Paraproct; Pr: Proctiger; IX-Te: Tergum IX; IX-S: Sternum IX. (Modified after Arnell, 1973).

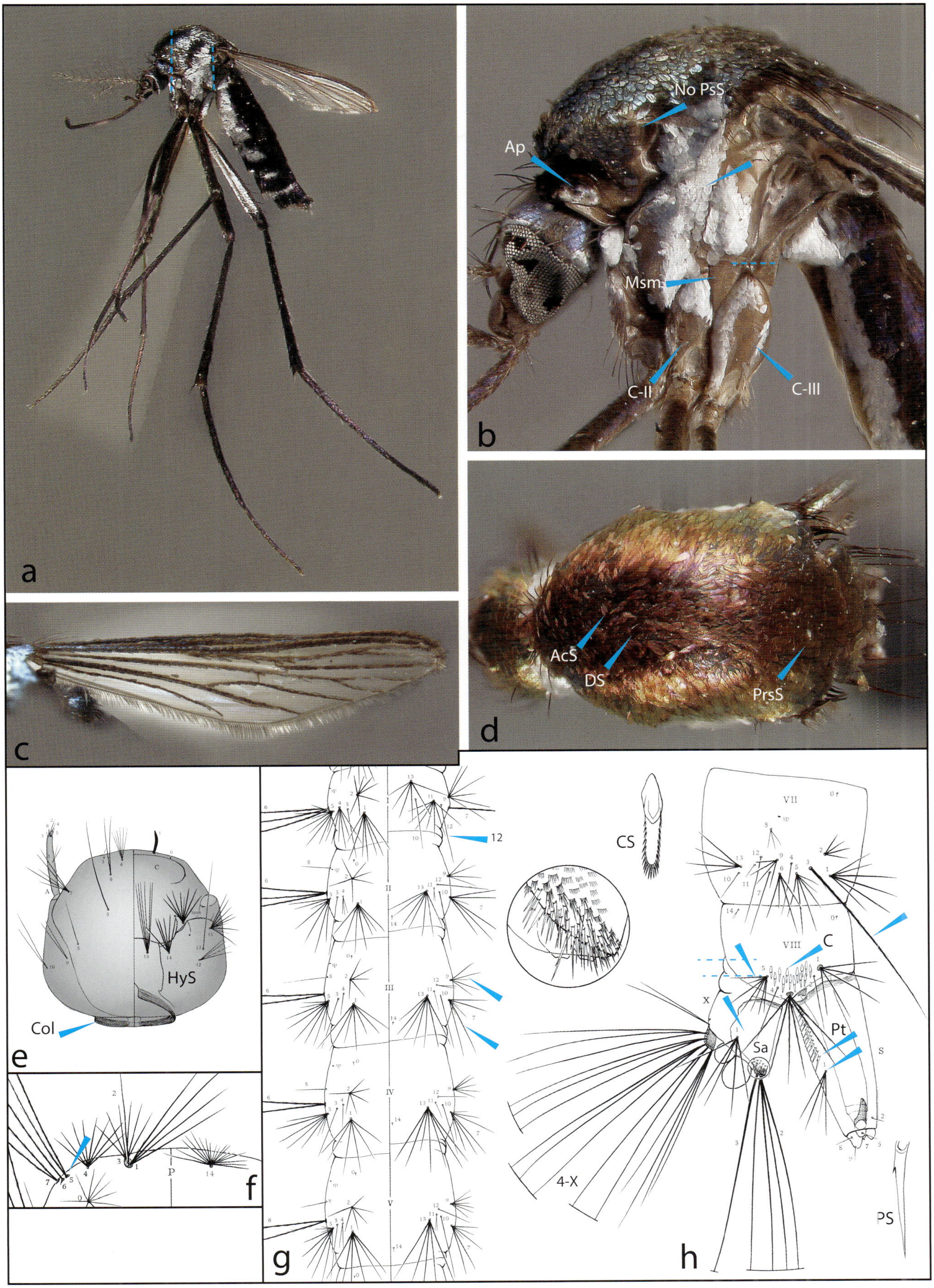
a
No PsS
Ap
Msm
C-II
C-III
b
c
AcS
DS
PrsS
d
HyS
Col
e
f
12
g
CS
VII
VIII
C
X
Pt
Sa
S
4-X
PS
h

Genus *Heizmannia* Ludlow, 1905
Subfamily Culicinae, Tribe Aedini
"Metallic Mosquito"

Systematics and Distribution

Species of the aedine genus *Heizmannia* are vividly decorated with colored, iridescent scales more commonly associated with the Neotropical sabethines (see plate). The genus *Heizmannia* comprises 40 species in two subgenera: *Heizmannia* (34 spp.) and *Mattinglyia* (6 spp.). The type species of the genus is *Heizmannia scintillans* Ludlow. Species of *Heizmannia* are mainly restricted to the Oriental Region, although some occur in the southeastern Palearctic. *Heizmannia aurea*—the only Australasian representative—is endemic to Halmahera Island in the Moluccas.

Bionomics

All *Heizmannia* species are forest associated. Adults are reported to bite humans during the day, particularly in rubber plantations, although little is known about other host preferences. Females commonly lay their eggs in elevated natural arboreal containers, including tree holes and split, cut, and insect-bored standing bamboo. However, immatures have also been recovered from habitats at ground level, including fallen or split bamboo, small ground pools, crab holes, discarded coconut shells, and artificial containers associated with rubber tapping. In India, *Hz. chandi* immatures are commonly found in tree holes in association with immature *Ae. albopictus* (Skuse) (see species page, p. 460), *Ar. subalbatus* (Coquillett) (see species page, p. 472), and several *Toxorhynchites* species.

Associated Pathogens

None.

Diagnostic Characters

(a) Habitus (♀): Legs mostly dark. **(b) Thorax (side):** Erect scales low on occiput (Occ), occasionally absent; prespiracular setae (PsS) absent; postspiracular setae (PS) mostly absent; upper mesokatepisternal setae (MkSU) absent; lower mesepimeral setae (MeSL) present; subspiracular (SSc), postspiracular (PoSc), and paratergal (PaSc) scales present; scales present on laterotergite (not shown). **(c) Wing:** Vein 1A terminates beyond mediocubital crossvein (mcu). **(d) Wing:** Alula (Al) dorsal edge with broad scales; upper calypter (UC) fringed with setae (not shown). **(e) Thorax (dorsal):** Scutum (Scu) with broad scales; acrostichal (AA) and dorsocentral (DA) areas without setae; head vertex (V) with dark, broad appressed scales and patches of broad white or silver scales laterally and between eyes. **(f) Abdomen (side):** Abdominal terga with distinct white or silver triangular basolateral markings. **(g) Larva (head):** Maxilla with a brush (MxB) or a bundle of filaments. **(h) Larva (terminal segments):** Comb (C) with ≤20 scales (CS); pecten (Pt) with c. 15 spines; seta 1-S long, usually 1–3 branched, inserted at or beyond midpoint of siphon (S); saddle (Sa) incomplete. Image sources: WRBU; Mattingly (1970).

Exemplar DNA barcodes: *Hz.* (*Hez.*) *chandi* (AY917208), *Hz.* (*Mat.*) *discrepans* (AY834242, EU259296).

Catalog entry: Vol. 2, p. 826.

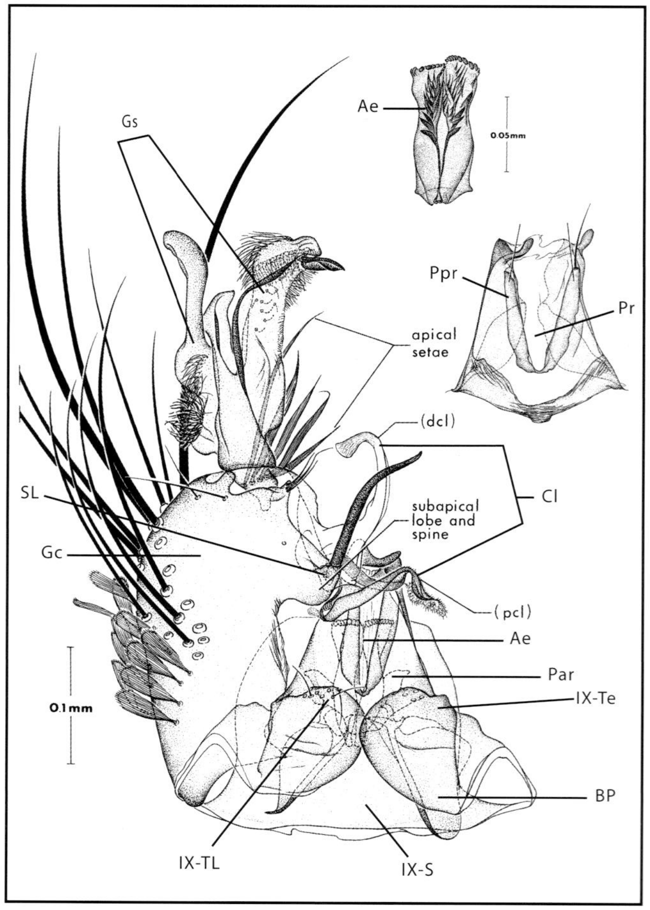

***Heizmannia* (*Heizmannia*) *complex* (Theobald) (♂ genitalia)**

Ae: Aedeagus; BP: Basal Piece; Cl: Claspette (distal [dcl], proximal [pcl]); Gc: Gonocoxite; Gs: Gonostylus; Par: Paramere; Ppr: Paraproct; Pr: Proctiger; SL: Subapical Lobe; IX-Te: Tergum IX; IX-TL: Tergum IX Lobe; IX-S: Sternum IX. (Modified after Mattingly, 1970).

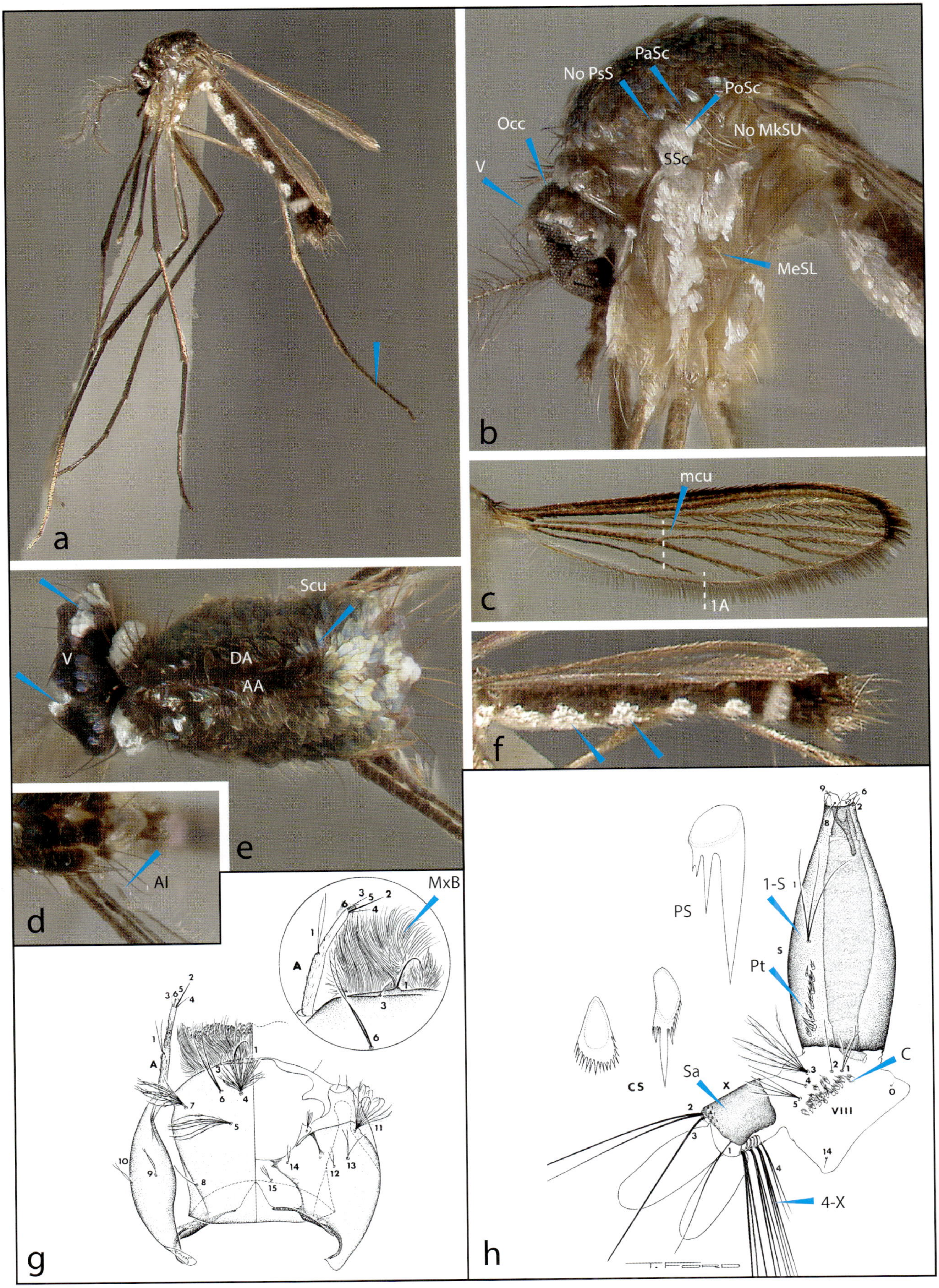
a
PaSc
No PsS
PoSc
Occ
No MkSU
SSc
V
MeSL
b
mcu
1A
c
Scu
V
DA
AA
e
Al
d
f
MxB
A
g
PS
1-S
S
Pt
CS
Sa
C
X
VIII
4-X
h
T. FORD

Genus *Opifex* Hutton, 1902
Subfamily Culicinae, Tribe Aedini
"Waterwalking Mosquito"

Systematics and Distribution

The aedine genus *Opifex* comprises only two species, split in two monotypic subgenera: *Nothoskusea* and *Opifex*. *Opifex* have evolved a highly specialized mating strategy—mating with females before they are fully emerged from the pupal case—and as such, several morphological specializations associated with auditory mate recognition in swarms (e.g., plumosity of male antennae and large Johnston's organs) are markedly reduced. The genus is endemic to coastal regions of New Zealand. The type species *Op. fuscus* Hutton is found on the North and South Islands, the Snares Islands, and the Kermadec Islands. It was the sole member of the genus until recently, when *Op. chathamicus* was described and appointed its own subgenus, *Nothoskusea*. *Opifex chathamicus* is endemic to the Chatham Islands and is by far the largest mosquito in New Zealand.

Bionomics

Opifex fuscus primarily feeds on seabirds, and sports the most remarkable "eyelashes"—large setae around the eyes, presumably for protection from spiky feathers. It is also a well-known nuisance species to persons conducting outdoor coastal pursuits, delivering a painful bite. The females are autogenous and oviposit in brackish to supersaline coastal rock pools, with a preference for sites that are occupied by conspecific immatures, perhaps related to mating. Male *Op. fuscus* rest on the aqueous surface of the immature habitats, ready to pounce and mate with newly emerged females, often before eclosion is fully completed. Upon locating a mature pupa, the males grasp the respiratory trumpets with their extensive foretarsal claws (see *c*), waiting for the pupal case to split. If the emergent adult is a male, contact is broken, but if it is a female, copulation begins (see Chapter 9). Egg deposition is staggered 3–12 days after mating. Experimentally, *Op. fuscus* is a competent vector for several arboviruses, but its naturally short life span decreases the likelihood of transovarial transmission.

Opifex (*Nothoskusea*) *chathamicus* is easily recognized by its abdominal sterna, dark scaled with basal lateral pale patches and a pale apical median patch on some segments. *Opifex chathamicus* is rare, and immatures have been recovered from rockpools at or above the high tide line and filled with fresh to brackish water at only three discrete sites on the Chatham Islands. Only eight adults of *Op. chathamicus* have ever been caught in the wild—two females collected at human bait on South East Island, one female captured indoors 2 km from the coast at Wharekauri, and five specimens collected indoors at 2200h southeast of Waitangi, some 5 km from the shore. These sightings indicate that *Op. chathamicus* are either strong fliers, or that they occupy a wider range of immature habitats than has been currently described.

Associated Pathogens

BFV, CHIKV, RRV, SINV, WHAV.

Diagnostic Characters

(a) Habitus (♀). **(b) Wing:** Subcostal vein (Sc) short, with numerous basal setae. **(c) Leg:** Unguis (U-I) long (*Opifex fuscus*). **(d) Head (side):** Antenna with a fully developed scape (Sc) behind pedicel (Pe); vertex (V) and occiput (Occ) with short setae instead of erect scales. **(e) Head (dorsal):** Eyes widely separated. **(f) Larva (abdomen):** Seta 6-I–V with ≥3 branches. **(g) Larva (thorax/abdomen):** Seta 9–12-T reduced, single; seta 7-I with ≥2 branches. **(h) Larva (terminal segments):** Saddle (Sa) small; pecten spines (PS) few to absent. Image sources: WRBU; Belkin (1962); Mattingly (1970).

Exemplar DNA sequences: *Opifex fuscus* (*arginine kinase* gene GQ906818, *CAD* gene GQ906843, *catalase* gene GQ906867, *enolase* gene GQ906891, *hunchback* gene GQ906915, *white* gene GQ906934).

Catalog entry: Vol. 2, p. 830.

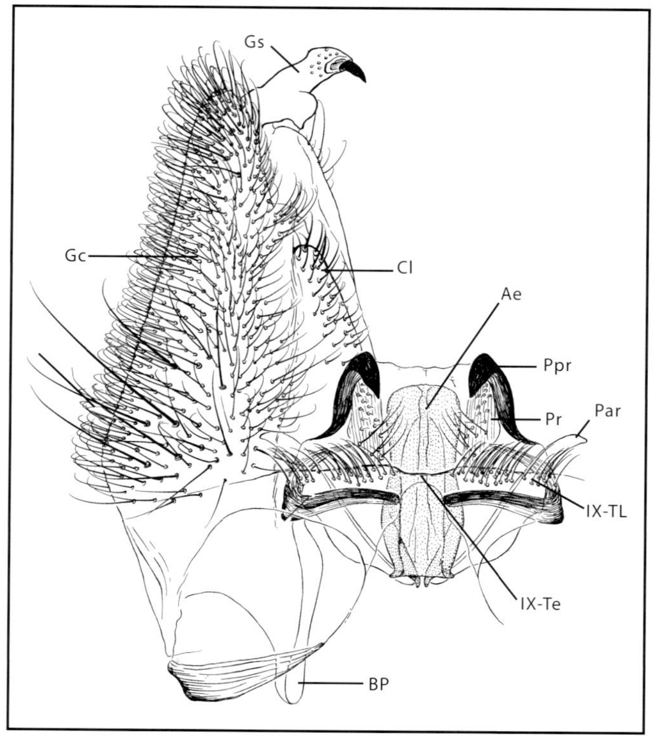

***Opifex* (*Opifex*) *fuscus* Hutton (♂ genitalia)**

Ae: Aedeagus; BP: Basal Piece; Cl: Claspette; Gc: Gonocoxite; Gs: Gonostylus; Par: Paramere; Ppr: Paraproct; Pr: Proctiger; IX-Te: Tergum IX; IX-TL: Tergum IX Lobe. (Modified after Belkin, 1962).

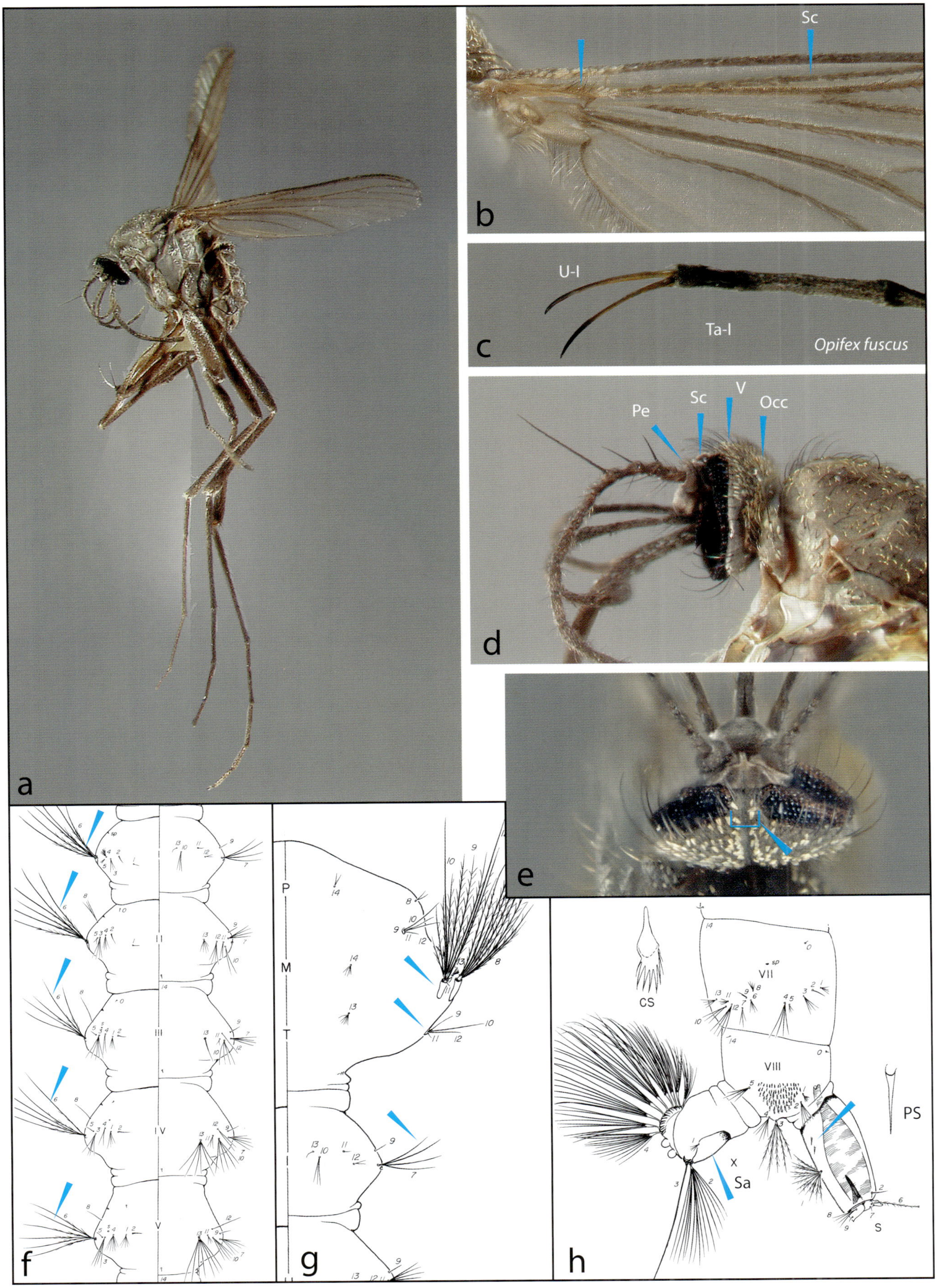
Sc
b
U-I
Ta-I
Opifex fuscus
c
Pe
Sc
V
Occ
d
e
a
f
P
M
T
g
CS
VII
VIII
PS
Sa
X
S
h

Genus *Psorophora* Robineau-Desoidy, 1827 Subfamily Culicinae, Tribe Aedini "Gang Mosquito"

Systematics and Distribution

Species of the aedine genus *Psorophora* are large and can deliver a painful bite. The genus comprises 49 species in three subgenera: *Grabhamia* (16 spp.), *Janthinsoma* (23 spp.) and *Psorophora* (10 spp.). The type species of the genus is *Psorophora ciliata* (as *Culex ciliata* Fabricius). *Psorophora* species are restricted to the warmer tropical and temperate regions of the New World.

Bionomics

In common with many other aedines, *Psorophora* are floodwater mosquitoes. Eggs are deposited in damp or dry soils in areas prone to flooding and can tolerate many months of desiccation. The immature stages of species of subgenus *Janthinosoma*, such as *Ps. ferox* (von Humboldt), are found in heavily shaded woodland pools, whereas species of the subgenus *Grabhamia* and *Psorophora* are found in open, sunlit temporary brackish and freshwater ground pools, including irrigated pastures, rice fields, and the flooded margins of swamps. All larvae of the species in the subgenus *Psorophora* are predacious. The females of many species avidly bite domestic animals and will readily attack humans, and several species are medically important arbovirus vectors. *Psorophora* (*Pso.*) *ciliata* (Fabricius) and *Ps.* (*Pso.*) *columbiae* (Fabricius) are strong fliers and are commonly detected 5–10 miles from their larval habitats.

Associated Pathogens

ARUV, CEV, COTV, CVV, EEEV, GROV, IERIV, ILHV, ITUV, KRIV, LMV, LOKV, MAGV, MAYV, MDCV, MELV, MIRV, ORIV, OROV, Potosi virus, PURV, ROCV, SAV, SLEV, TENV, UNAV, VEEV, WEEV, WNV, WYOV.

Diagnostic Characters

(a) Habitus (♀). (b) Thorax (side): Postspiracular setae (PS) and prespiracular setae (PsS) present. **(c) Abdomen (dorsal):** Apex of abdomen pointed; mesopostnotum (Mpn) bare. **(d) Larva (head). (e) Larva (terminal segments):** Comb (C) with comb scales (CS) in a single row, usually on a sclerotized plate (CP); siphon (S) much expanded toward base; saddle (Sa) complete; seta 4-X well developed, extending to end of segment X. Image sources: WRBU; Belkin et al. (1970).

Exemplar DNA barcodes: *Psorophora* (*Gra.*) *cingulata* (KM592989), *Ps.* (*Gra.*) *confinnis* (KY859921–22), *Ps.* (*Gra.*) *columbiae* (JX260111–12), *Ps.* (*Gra.*) *discolor* (KM592989), *Ps.* (*Gra.*) *insularius* (JX260113, JX260717), *Ps.* (*Gra.*) *pygmaea* (JX260115–16), *Ps.* (*Gra.*) *signipennis* (JX260719–20), *Ps.* (*Jan.*) *cyanescens* (HE600019, MG242535), *Ps.* (*Jan.*) *ferox* (M593037–38), *Ps.* (*Jan.*) *horrida* (MG242537–38), *Ps.* (*Jan.*) *longipalpus* (JX260718, JX260114), *Ps.* (*Pso.*) *ciliata* (MG242533), *Ps.* (*Pso.*) *howardii* (JX260715–16).

Catalog entry: Vol. 2, p. 831.

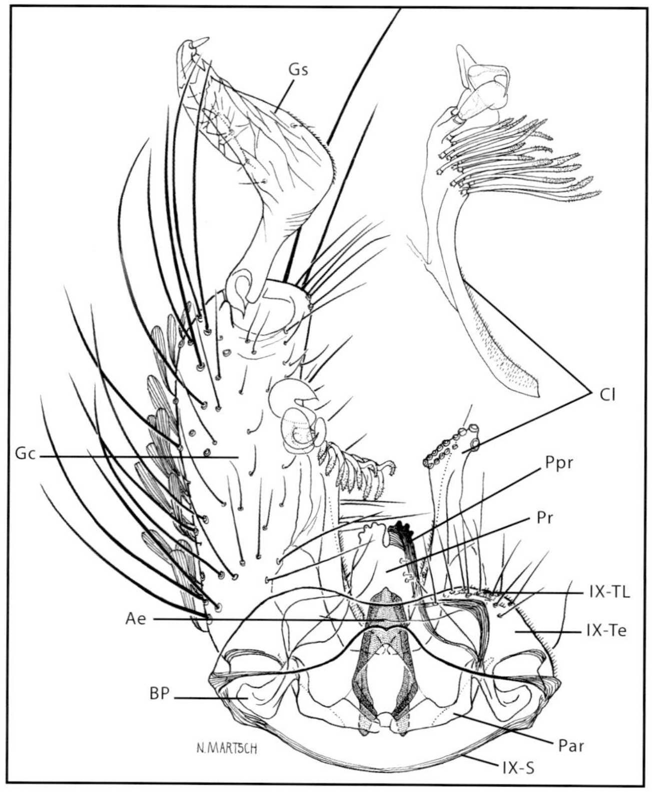

***Psorophora* (*Janthinosoma*) *johnstonii* (Grabham) (♂ genitalia)**

Ae: Aedeagus; BP: Basal Piece; Cl: Claspette; Gc: Gonocoxite; Gs: Gonostylus; Par: Paramere; Ppr: Paraproct; Pr: Proctiger; IX-Te: Tergum IX; IX-TL: Tergum IX Lobe; IX-S: Sternum IX. (Modified after Belkin, Heinemann, & Page, 1970).

a
PS
PsS
b
Mpn
c
C
A
d
4-X
PcrS
Sa
X
VIII
VII
CP
PS
S
CS
e

Genus *Udaya* Thurman, 1954
Subfamily Culicinae, Tribe Aedini
"Compressed Mosquito"

Systematics and Distribution

Mosquitoes of the Oriental aedine genus *Udaya* are small, highly ornate species, spattered with distinctive dense patches of silvery scales on their head, pleura, and abdominal terga (see *a*, *b*). The eyes of both sexes are separated by a distinctive spoon-shaped patch of silver scales (see *d*), and the wing bears broad, flat scales on the alula (see *f*). The genus shares highest morphological similarities with *Aedes*, *Heizmannia*, and *Zeugnomyia* species in the tribe Aedini. *Udaya* comprises only three described species: The type species *Udaya argyrurus* (Edwards) (as *Paraedes argyrurus* Edwards) was first described from a female collected in Nongoh, Assam, India, but has also been reported from Malaysia, Thailand, and Singapore. *Udaya lucaris* Macdonald & Mattingly was described in all life stages from the Ulu Gombak Forest Reserve, Selangor, Malaysia. *Udaya subsimilis* (Barraud) was described from a single male collected as an immature in bamboo in Sukuna, Bengal, India. As *Udaya* species are rarely encountered, the true species distributions and taxon composition in the genus are most probably underrepresented.

Bionomics

Little is known of the biology of any of the *Udaya* species, except that immatures of *Ud. argyrurus* and *Ud. subsimilis* have been collected in bamboo. To protect their offspring from predation, females shoot single eggs into the small holes resulting from insect damage in standing bamboo. The effect of this oviposition strategy is apparent in the morphology of the adult flies, whose bodies are small and elongated to facilitate extraction from the tiny bamboo holes (see *c*, *d*). The feeding preferences of *Udaya* species have not been studied, but *Ud. argyrurus* females have been collected in traps alongside other biting flies, suggesting that they are at least host seeking.

Associated Pathogens

None.

Diagnostic Characters

(a) Habitus (♀). (b) Thorax (side): Antepronotal setae (ApS) in two groups separated by patch of broad silver antepronotal (ApSc) scales; paratergite with a distinct patch of broad silver scales (PaSc); postpronotal scales (PpSc) and postspiracular scales (PoSc) absent. **(c) Thorax (dorsal):** Acrostichal area (AA) without setae. **(d) Head (dorsal):** Distinctive interocular patch of silver scales extending to antennal pedicels (Pe). **(e) Wing. (f) Wing (enlargement):** Alula (Al) with broad marginal scales. **(g) Larva (head):** Setae 4–6-C close to anterior head margin; seta 7-C inserted well medial of the dorsal apotome lateral margin. **(h) Larva (terminal segments):** Comb (C) with a single row of ≤10 scales (CS); posterior margin of saddle (Sa) without strong spines. Image sources: WRBU; Mattingly (1971).

Exemplar DNA sequence: *Ud. subsimilis* (mtDNA *Cyt b* KF687465).

Catalog entry: Vol. 2, p. 839.

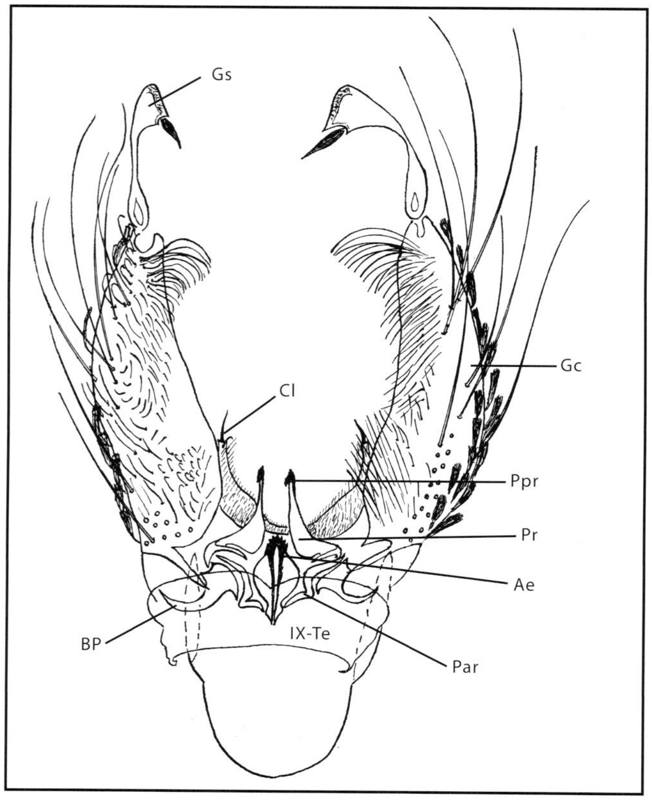

***Udaya argyrurus* (Edwards) (♂ genitalia)**

Ae: Aedeagus; BP: Basal Piece; Cl: Claspette; Gc: Gonocoxite; Gs: Gonostylus; Par: Paramere; Ppr: Paraproct; Pr: Proctiger; IX-Te: Tergum IX. (Modified after Thurman, 1954).

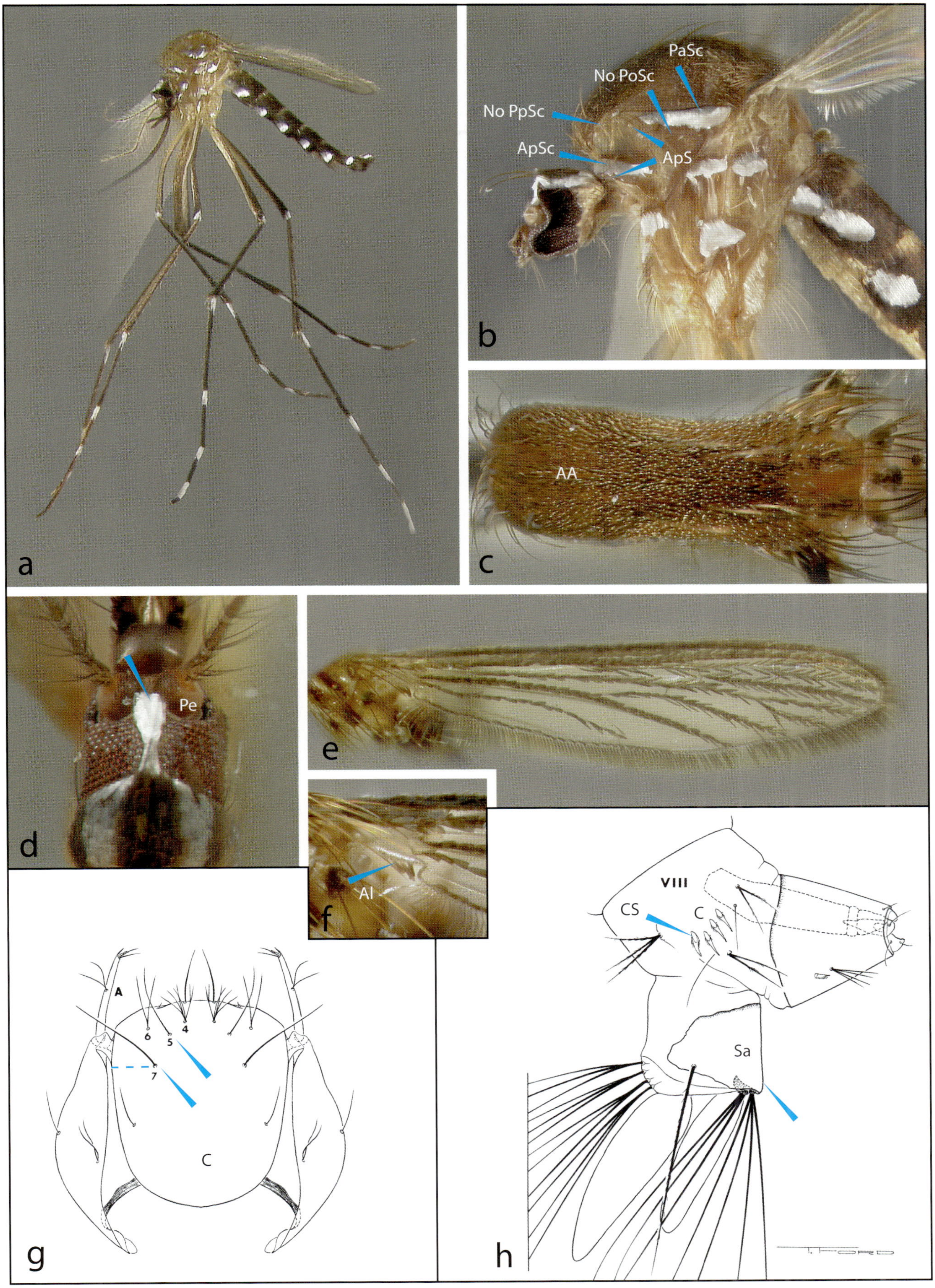
a
PaSc
No PoSc
No PpSc
ApSc
ApS
b
AA
c
Pe
d
e
Al
f
A
6
5
4
7
C
g
VIII
CS
C
Sa
h
T. FORD

Genus *Verrallina* Theobald, 1903
Subfamily Culicinae, Tribe Aedini
"Broadtail Mosquito"

Systematics and Distribution

The highly distinctive features of both male and female genitalia distinguish the genus *Verrallina* from all other genera of the tribe Aedini. The genus *Verrallina* comprises 95 valid species, subdivided into three subgenera: *Harbachius* (13 spp.), *Neomaclaya* (52 spp.), and *Verrallina* (30 spp.) (see Vol. 2, p. 1103, for internal relationships). *Verrallina* species occur from the Indian Subcontinent north to Russia and Japan, south through the Oriental Region, and southeast into the Papuan Subregion and northeastern Australia. The type species *Verrallina butleri* (Theobald) (as *Aedes butleri* Theobald) is restricted to the Oriental Region.

Bionomics

Overall, little is known about the bionomics of *Verrallina* species, except that they are mainly jungle creatures that also reportedly inhabit coastal mangrove swamps and bamboo groves. Females of several species readily attack humans during the day in shaded areas of the forest, causing issues for loggers, and on palm oil plantations. *Verrallina carmenti* (Edwards) and *Ve. lineata* (Taylor) are nuisance pests in the South Pacific. Some species (e.g., *Ve. carmenti*, *Ve. lineata*, *Ve. reesi* (King & Hoogstraal)) even enter human dwellings to feed. Habitats of immatures generally include temporary freshwater ground pools in forests or plantations, stream pools, flood pools, road ruts, puddles, or animal hoofprints, as well as coastal brackish water pools and swamps (e.g., *Ve. butleri* (Theobald)). Some species of the subgenus *Neomacleaya* inhabit natural containers, including crab holes (e.g., *Ve. clavata* (Barraud), *Ve. parasimilis* (King & Hoogstraal)), plant axils (e.g., *Ve. uncus* (Theobald)), and artificial container habitats, including wells (e.g., *Ve. vallistris* (Barraud)) and tin cans (e.g., *Ve. campylostylus* (Laffoon)). Species of the subgenus *Verrallina* lay their eggs in areas prone to flooding. Uniquely among mosquitoes, *Verrallina* eggs are protected by gelatinous strings, as with toads (e.g., family Bufonidae), to prevent egg desiccation.

Associated Pathogens

BEBV, GETV, JBEV; *Brugia malayi*.

Diagnostic Characters

(a) Habitus (♀): Mesopostnotum usually bare (not shown). **(b) Thorax (side):** Paratergite (Pa) without scales or setae; postpronotum (Ppn) bare or with all curved and narrow scales; prespiracular area (PsA) bare; postspiracular setae (PS) present. **(c) Wing:** Alula (Al) with fairly broad marginal scales. **(d) Thorax (dorsal):** Scutum (Scu) with dorsocentral setae (DS) present; acrostichal setae (AcS) sparse to numerous; all scutal and scutellar scales curved and narrow. **(e) Head (dorsal):** Decumbent scales of vertex (V) broad. **(f) Leg:** Ta-I–III dark scaled. **(g) Larva (head):** Antenna (A) with numerous spicules (inset); characteristic position and form of setae 4–7-C, short and multibranched; 5–7-C stout and about the same length; seta 11-C short and multibranched. **(h) Larva (abdomen):** Seta 7-I long, 1–3 branched; seta 7-II short, branched. **(i) Larva (thorax):** Seta 8-P long and thickened. **(j) Larva (terminal segments):** Saddle (Sa) incomplete; seta 4-X usually with 10 setae on a grid and 2–3 precratal setae (PcrS); pecten (Pt) with 6–20 pecten spines (PS), with the distal-most 5 more widely spaced. Image sources: WRBU; Reinert (1974).

Exemplar DNA barcodes: *Verrallina* (*Har.*) *nobukonis* (LC054379), *Ve.* (*Nma.*) *indica* (AY834245), *Ve.* (*Nma.*) *pseudomediofasciata* (EU259310), *Ve.* (*Ver.*) *butleri* (KF564665–69), *Ve.* (*Ver.*) *funerea* (MG712565–66), *Ve.* (*Ver.*) *lugubris* (DQ424950, DQ424955–56, EU259290–91).

Catalog entry: Vol. 2, p. 840.

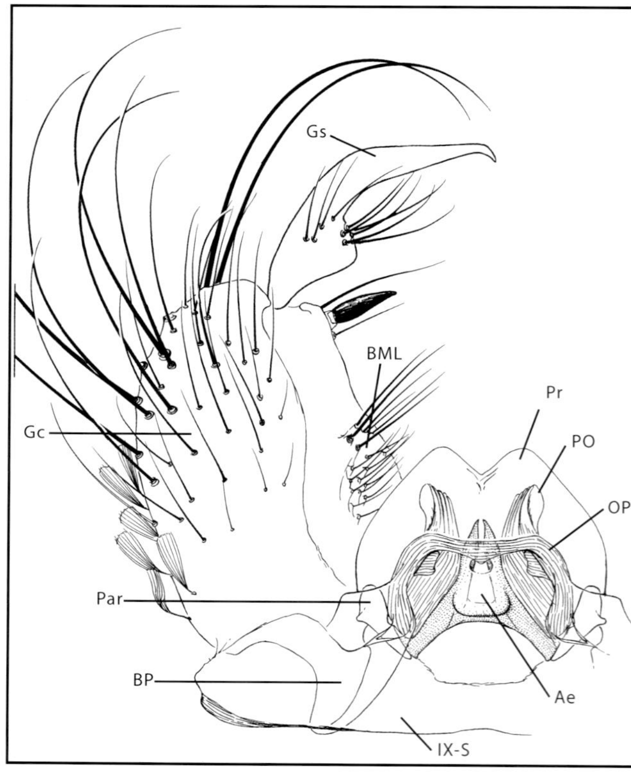

***Verrallina* (*Verrallina*) *carmenti* (Edwards) (♂ genitalia)**

Ae: Aedeagus; BML: Basal Mesal Lobe; BP: Basal Piece; Gc: Gonocoxite; Gs: Gonostylus; OP: Opisthophallus; Par: Paramere; PO: Prosophallus; Pr: Proctiger; IX-S: Sternum IX. (Modified after Belkin, 1962).

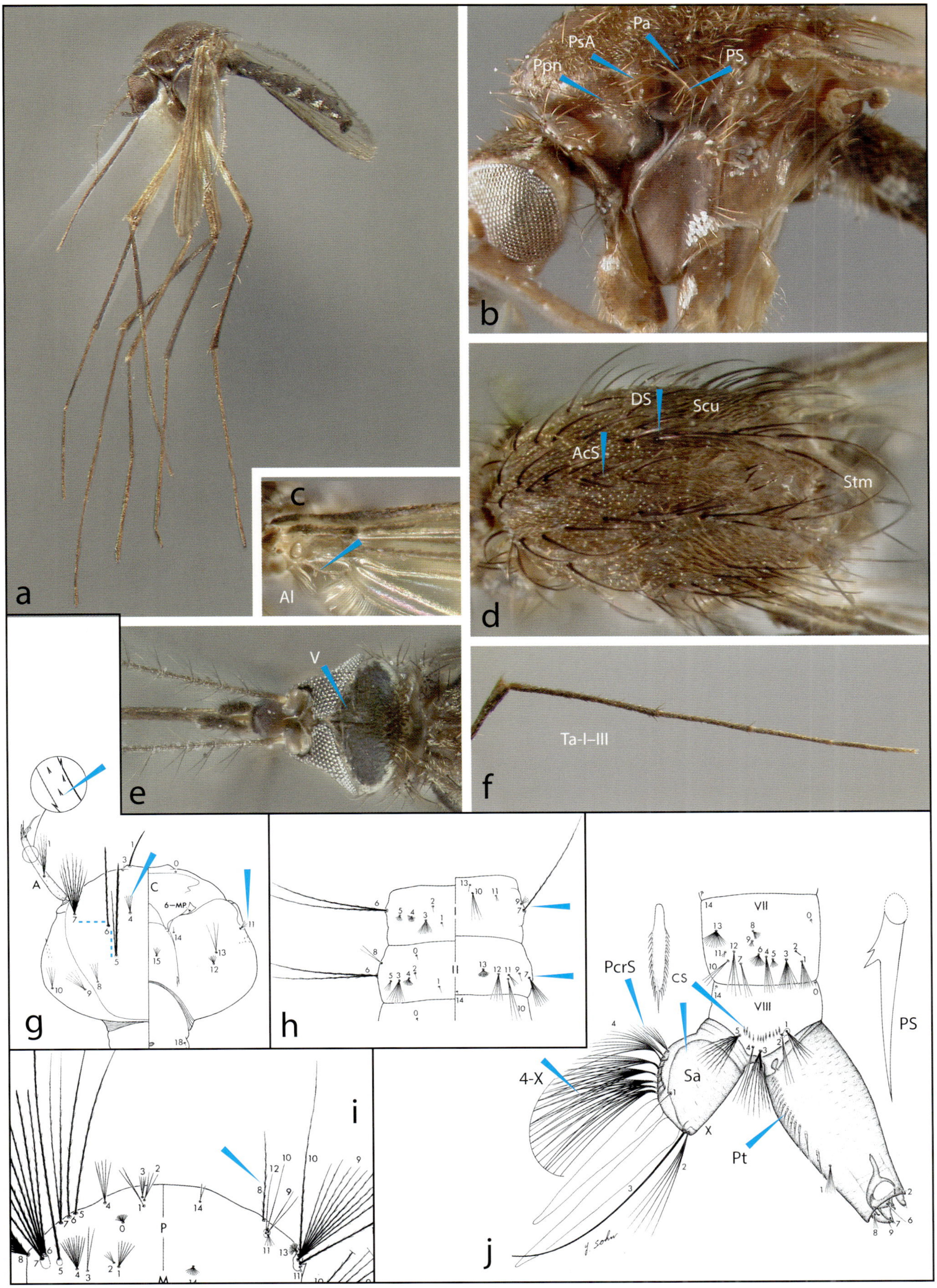
a
Pa
PsA
Ppn
PS
b
c
Al
DS
Scu
AcS
Stm
d
V
e
Ta-I–III
f
A
C
6–MP
g
I
II
h
i
P
M
VII
VIII
PcrS
CS
4-X
Sa
X
Pt
PS
j
Y. Sohn

Genus *Zeugnomyia* Leicester, 1908
Subfamily Culicinae, Tribe Aedini
"Silver Mosquito"

Systematics and Distribution

Adult *Zeugnomyia* are typically small mosquitoes, with a disproportionately large proboscis. Adults of both sexes are characterized by broad vertical stripes of brilliant silver scales on the sides of the thorax and distinctly flat patches of silver scales adorning the head, scutum, and abdomen. The genus *Zeugnomyia* is not subdivided into subgenera and comprises only four described species: *Ze. aguilari* Baisas & Feliciano, *Ze. fajardoi* Baisas & Feliciano, *Ze. gracilis* Leicester, and *Ze. lawtoni* Baisas. *Zeugnomyia aguilari*, *Ze. fajardoi*, and *Ze. lawtoni* were described on the basis of male genitalia and are inseparable as adult females or as larvae. The type species *Zeugnomyia gracilis* females differ from the other *Zeugnomyia* species by the dark scales on all scutellar lobes, and in males the gonostylus lacks the spine prominent in all other species. *Zeugnomyia* larvae differ from other aedines through a combination of characters of seta 7-C, the comb scales, and the saddle (see *f*, *h*). All species appear to be geographically restricted to the Philippines, except *Ze. gracilis*, which has also been reported from Singapore, Malaysia, and Thailand.

Bionomics

Little is known about the biology of *Zeugnomyia* species. Adults are generally associated with shaded forest environs. In Malaysia, female *Ze. gracilis* feed readily on humans, but the feeding habits of other species remain unknown. Larval habitats typically include small pools of rainwater in natural containers, including fallen leaves (abaca and coconut), anahaw palm leaf axils (*Saribus rotundifolius*), tree holes, and cut bamboo. Larvae were once recovered from a discarded tin can. The larvae of *Ze. gracilis* are predacious, feeding readily on other species sharing their larval habitat.

Associated Pathogens

None.

Diagnostic Characters

(a) Habitus (♀). (b) Thorax (dorsal): Dorsocentral setae (DS) absent. **(c) Wing:** Alula (Al) bare; upper calypter (UC) fringed with setae; vein 1A terminates only slightly beyond level of mediocubital crossvein (mcu); cell R_2 distinctly shorter than vein R_{2+3}. **(d) Thorax (side):** Pleuron with a vertical stripe of silver scales; lower mesepimeral setae (MeSL) present; prespiracular (PsS) and postspiracular setae (PS) absent; patches of flat silver scales on head. **(e) Abdomen (dorsal):** Patches of flat silver scales. **(f) Larva (head):** Seta 7-C slender, 1–2 branched. **(g) Larva (head):** Hypostomal suture (HyS) reaches posterior tentorial pit (PTP). **(h) Larva (terminal segments):** Saddle (Sa) incomplete, large spines on posterior margin; pecten (Pt) present; comb (C) with single row of comb scales (CS). Image sources: WRBU; Edwards & Given (1928).

Exemplar DNA barcodes: *Zeugnomyia gracilis* (KF564776, KF564778).

Catalog entry: Vol. 2, p. 850.

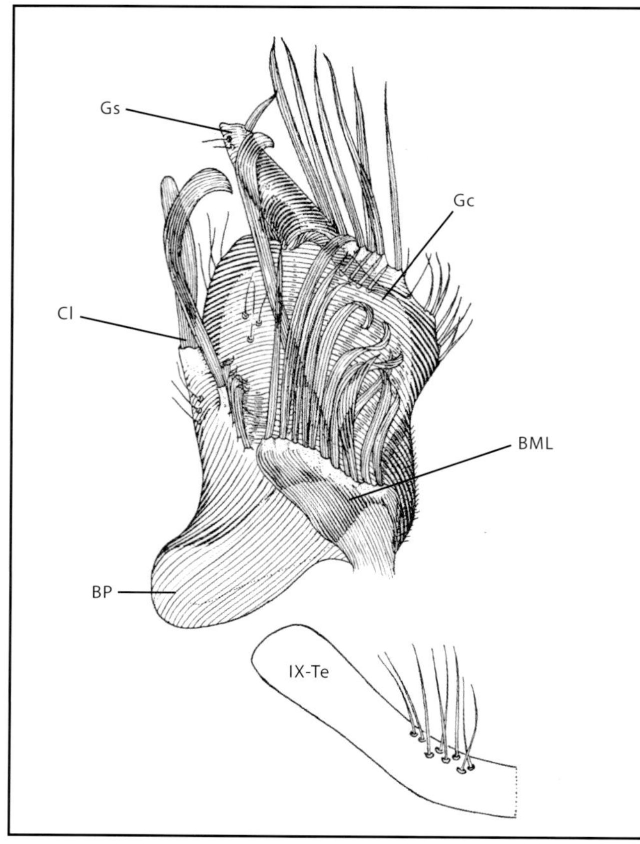

***Zeugnomyia aguilari* Baisas & Feliciano (♂ genitalia)**

BML: Basal Mesal Lobe; BP: Basal Piece; Cl: Claspette; Gc: Gonocoxite; Gs: Gonostylus; IX-Te: Tergum IX. (Modified after Baisas and Feliciano, 1953).

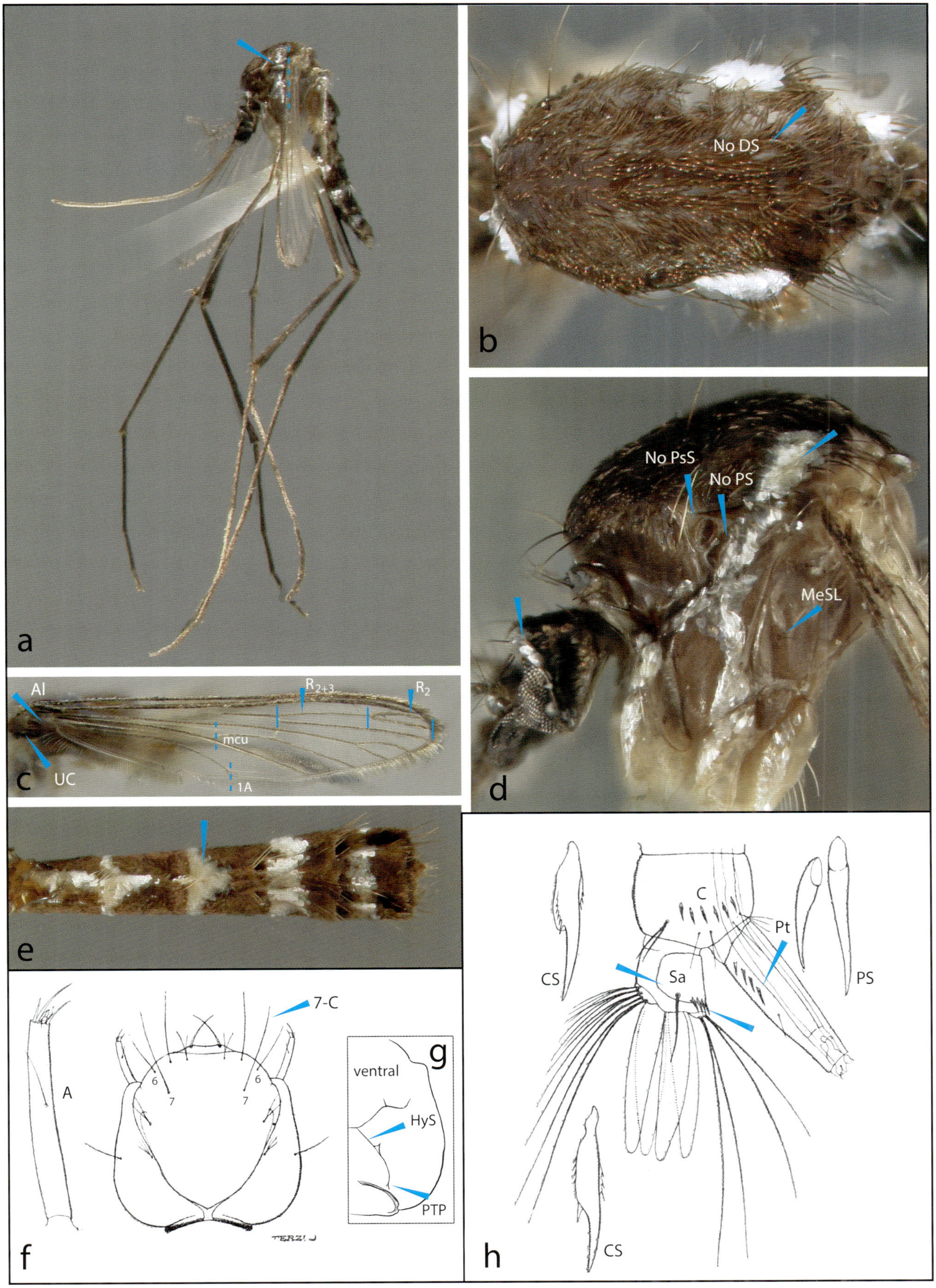
a
No DS
b
No PsS
No PS
MeSL
d
Al
R_{2+3}
R_2
mcu
UC
1A
c
e
7-C
A
6
6
7
7
ventral
g
HyS
PTP
f
C
Pt
CS
Sa
PS
CS
h

Genus *Culex* Linnaeus, 1758
Subfamily Culicinae, Tribe Culicini
"Typical Mosquito"

Systematics and Distribution

The biomedically important genus *Culex* has 769 species in 26 formally recognized subgenera, including *Acalleomyia* (1 sp., Indo-Malaya), *Acallyntrum* (8 spp., Australasia), *Aedinus* (4 spp., Neotropics), *Afroculex* (1 spp., South Africa), *Allimanta* (1 sp., Argentina), *Anoedioporpa* (12 spp., Neotropics), *Barraudius* (4 spp., Palaearctic and Afrotropics), *Belkinomyia* (1 sp., Colombia), *Carrollia* (18 spp., Neotropics), *Culex* (200 spp., global), *Culiciomyia* (54 spp., Afrotropics, Orient, and Australasia), *Eumelanomyia* (77 spp., Afrotropics, Orient, and Australasia), *Kitzmilleria* (1 sp., Africa), *Lasiosiphon* (1 sp., Africa, Southwest Asia), *Lophoceraomyia* (112 spp., Orient and Australasia), *Maillotia* (9 spp., Africa, Southwest Asia), *Melanoconion* (160 spp., Nearctic, Neotropics), *Micraedes* (8 spp., Caribbean), *Microculex* (33 spp., Neotropics), *Neoculex* (26 spp., Nearctic and Old World), *Nicaromyia* (1 sp., Cuba), *Oculeomyia* (19 spp., Afrotropics, Palearctic, Orient, Australasia), *Phenacomyia* (3 spp., Neotropics), *Phytotelmatomyia* (4 spp., Neotropics), *Sirivanakarnius* (1 sp., Bonin Islands), and *Tinolestes* (3 spp., New World), and seven species remain unplaced (see Vol. 2, p. 1087, for internal relationships).

Bionomics

Adults feed on birds, reptiles, amphibians, and mammals, including humans. Immature stages typically occupy groundwater habitats that can be fresh, brackish, clean, or heavily polluted. Some species occupy natural containers, including phytotelmata (e.g., tree holes, rock holes, leaf axils, flower bracts, bamboo, pitcher plants, fruit husks, and fallen spathes or leaves). Immatures of the subgenera *Belkinomyia*, *Nicaromyia*, and *Tinolestes* rear exclusively in crab holes. Eggs are usually laid in rafts, but some *Microculex* lay gel-covered individual eggs to prevent desiccation. In laboratory colony settings, *Cx. pipiens* can exhibit cannibalism. Immatures of the Australasian *Cx.* (*Ocu.*) *starckeae* Stone & Knight live in rapid-flowing waters, hooking onto algal mats using their modified siphonal hairs.

Associated Pathogens

ABRV, ACAV, ACDV, AINOV, ANUV, AMTV, ANTV, APEUV, ARKV, ARV, AURV, AURAV, BABV, BAGV, BANV, BAKV, BARV, BATV, BBKV, BEBV, BEFV, BELV, BFV, BGNV, BIMV, BOTV, BOZOV, BQSV, BSBV, BSQV, BVSV, CABV, CAPV, CARV, CATUV, CEV, CHIKV, CHPV, CNAV, COCV, CORV, COTV, DABV, DENV, Dianke, EEEV, EHV, ENSV, ENTV, EVEV, FLAV, GAMV, GERV, GETV, GGV, GJAV, GLOV, GLV, GOMV, HPV, ILHV, INGV, ITPV, ITQV, ITV, JACV, JBEV, KAIV, KAMV, KASV, KETV, KEYV, KIMV, KOKV, KOOV, KUNV, KWAV, KYZV, LACV, LATV, LEDV, LJV, LLSV, LMV, LOKV, MADV, MBAV, MERV, MHV, MINV, MIRV, MOJUV, MORV, MOSV, MPOV, MQOV, MUCV, MURV, MVEV, Nanay, NDUV, NEPV, NJLV, NKOV, NODV, NOLAV, ODRV, OLIV, ORIV, OROV, OSSAV, OUBIV, OUBV, PAHV, PALV, PARAV, PATV, PCAV, PERV, PIAV, PLSV, PRV, RESV, RRV, RTAV, RVFV, SAGV, SATV, SFV, SHUV, SINV, SLEV, SPOV, SRV, TAHV, TAIV, TATV, TENV, TERV, TMUV, TURV, TVTV, UMAV, UMBV, USUV, VEEV, VELV, VINV, VKTV, VRV, VSNJV, WANV, WEEV, WHAV, WITV, WNV, WONV, WSLV, WYOV, YAOV; *Wuchereria bancrofti*, bird malarias, including *Plasmodium relictum*.

Diagnostic Characters

(a) Habitus (♀). (b) Thorax (side): Prespiracular (PsS) and postspiracular setae absent. **(c) Wing:** Alula (Al) with narrow fringe of scales. **(d) Larva (head):** Seta 2-C usually absent; seta 3-C located dorsally. **(e) Larva (thorax/abdomen):** Seta 12-I always present. **(f) Leg:** Well-developed pulvillus (Pv) present. **(g) Larva (terminal segments):** Pecten (Pt) and comb (C) present on siphon (S); seta 1-S with three or more pairs of prominent setae; saddle (Sa) usually complete, sometimes incomplete and greatly reduced, but never divided into dorsal and ventral sclerites. Image sources: WRBU; Strickman & Darsie (1988).

Exemplar DNA barcodes: *Cx.* (*Barraudius*) *modestus* (KJ012102), *Cx.* (*Cux.*) *pipiens* (KJ012110–148), *Culex* (*Cux.*) *quinquefasciatus* (KJ012162–173, whole genome GCF_000209185), *Cx.* (*Cux.*) *theileri* (KJ012174–234), *Cx.* (*Cux.*) *torrentium* (KJ012235–242), *Cx.* (*Cux.*) *tritaeniorhynchus* (KJ012243–250); *Cx.* (*Maillotia*) *hortensis* (KJ012068–080).

Catalog entry: Vol. 2, p. 851.

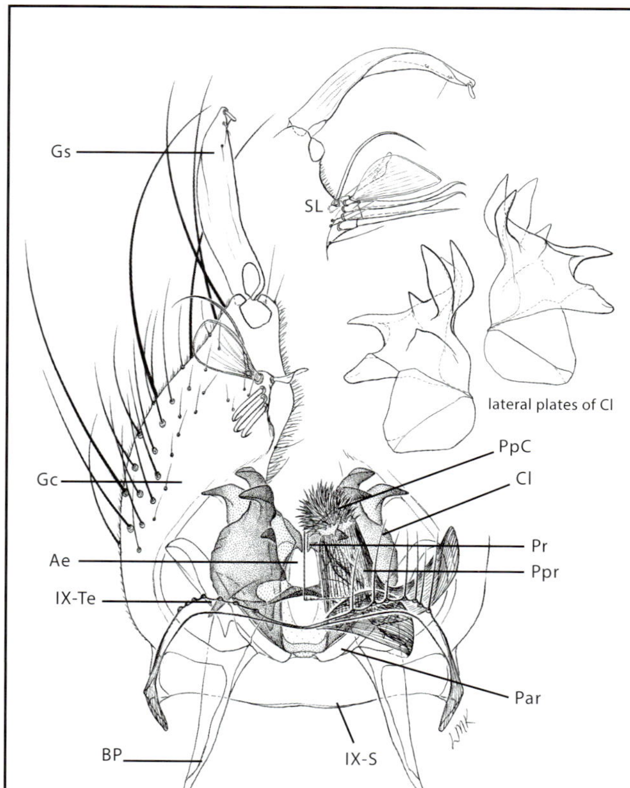

***Culex* (*Culex*) *declarator* Dyar & Knab (♂ genitalia)**

Ae: Aedeagus; BP: Basal Piece; Cl: Claspette; Gc: Gonocoxite; Gs: Gonostylus; Par: Paramere; PpC: Paraproct Crown; Ppr: Paraproct; Pr: Proctiger; SL: Subapical Lobe; IX-Te: Tergum IX; IX-S: Sternum IX.

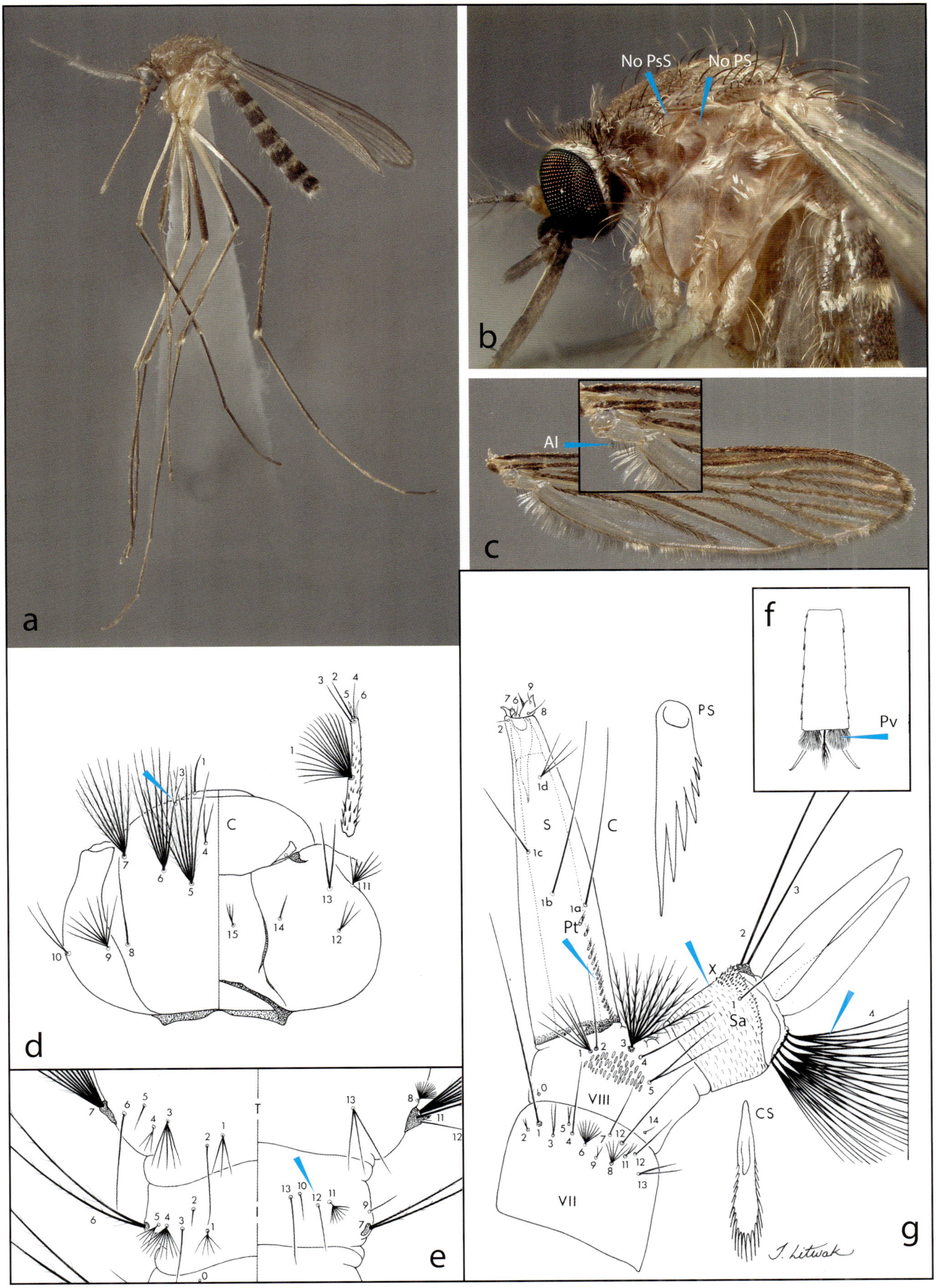
a
No PsS
No PS
b
Al
c
d
C
e
T
f
Pv
g
PS
S
C
1a
1b
1c
1d
Pt
X
Sa
VIII
VII
CS
J. Litwak

Genus *Deinocerites* Theobald, 1901
Subfamily Culicinae, Tribe Culicini
"Crabhole Mosquito"

Systematics and Distribution

The culicine genus *Deinocerites* comprises some 18 described species, which are not further subdivided into subgenera. *Deinocerites* most closely resembles *Culex*; however, in *Deinocerites* species, the maxillary palpus is short (see *c*), the antenna is much longer than the proboscis (see *c*), and the first flagellomere is usually long (see *a*). *Deinocerites* species are restricted to the Neotropics, although the ranges of *De. cancer*, *De. mathesoni* Belkin & Hogue, and *De. pseudes* Dyar & Knab extend into the southern United States. *Deinocerites cancer* Theobald is the type species of the genus.

Bionomics

Known as crabhole mosquitoes, immatures of *Deinocerites* species are almost always recovered from these sites, although some species have also been collected in artificial containers and small rock holes and tree holes. *Deinocerites pseudes* has been observed individually dotting eggs onto the sides of the crab holes, just above the waterline.

Female *Deinocerites* are believed to be active at night and spend the day resting in the humid crab holes. In Colombia, host-seeking females *De. atlanticus* Adames were collected in Shannon traps set close to rural forest patches, farms, and mangroves in the early evening (1800–2100h). Overall, the host preferences of *Deinocerites* species have not been well studied. In a detailed study in Panama, *De. pseudes*, *De. epitedeus* (Knab), and *De. cancer* were shown to be highly opportunistic, feeding on a wide variety of hosts including mammals, humans, and birds, as well as reptiles and amphibians, whereas *D. melanophylum* Dyar & Knab and *D. dyari* Belkin & Hogue preferred reptiles. *Deinocerites pseudes* is an incriminated vector of St. Louis and Venezuelan equine encephalitis viruses, but no other *Deinocerites* species is believed to be of medical significance.

Associated Pathogens

SLEV, VEEV.

Diagnostic Characters

(a) Habitus (♀). **(b) Head (dorsal):** Flm_1 usually very long. **(c) Head (side):** Antenna (Ant) considerably longer than proboscis (P); palpus (MPlp) very short. **(d) Larva (head):** Head with distinct lateral expansion, at least as wide as long, widest at insertion of antennae (A); base of mandible with distinct, highly spiculose lobe. **(e) Larva (terminal segments):** Pecten (Pt) with 2–3 branched pecten spines (PS); saddle (Sa) with ill-defined separate dorsal and ventral plates. Image sources: WRBU; Adames (1971).

Exemplar DNA barcodes: *Deinocerites atlanticus* (MF179287–306), *De. magnus* (MH376751–52).

Catalog entry: Vol. 2, p. 944.

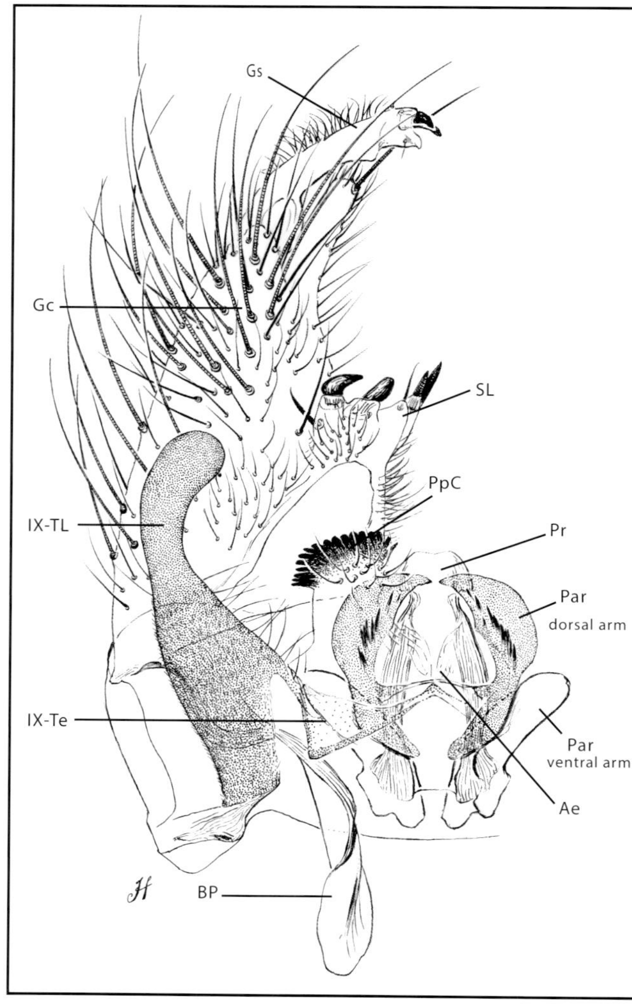

***Deinocerites nicoyae* Adames & Hogue (♂ genitalia)**

Ae: Aedeagus; BP: Basal Piece; Gc: Gonocoxite; Gs: Gonostylus; Par: Paramere; PpC: Paraproct Crown; Pr: Proctiger; SL: Subapical Lobe; IX-Te: Tergum IX; IX-TL: Tergum IX Lobe. (Modified after Adames, 1971).

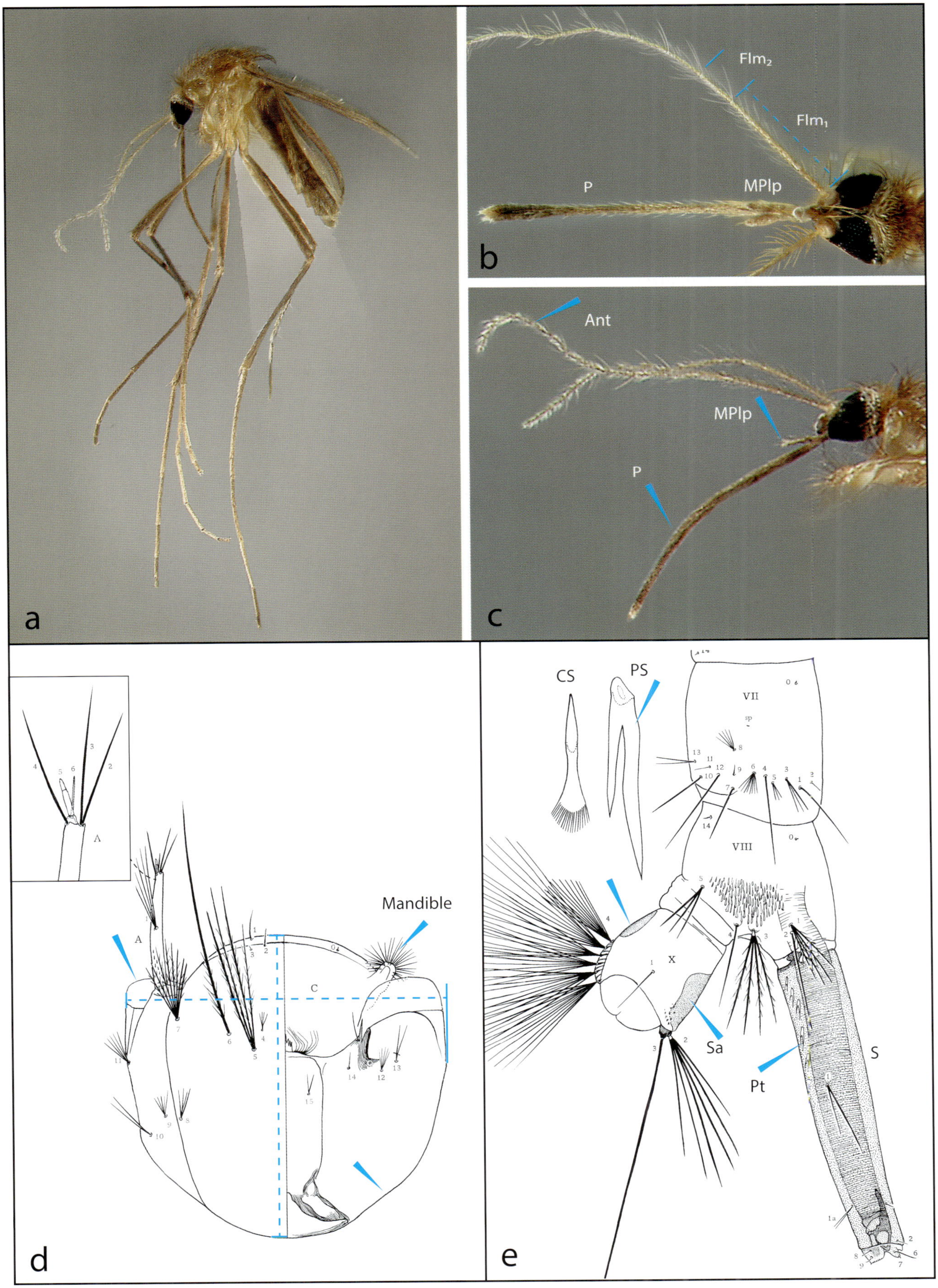
a
b
Flm₂
Flm₁
P
MPlp
c
Ant
MPlp
P
d
A
Mandible
C
A
e
CS
PS
VII
VIII
X
Sa
Pt
S

Genus *Galindomyia* Stone & Barreto, 1969 Subfamily Culicinae, Tribe Culicini "Galindo Mosquito"

Systematics and Distribution

The tribe Culicini comprises four genera: *Culex*, *Deinocerites*, *Lutzia*, and *Galindomyia*. *Galindomyia leei* Stone & Barreto, named in honor of Vernon Harold Lee, is the sole representative of this Neotropical genus. *Galindomyia leei* shares morphological similarities with both *Culex* and *Deinocerites* species in both the adult and larval stages. In both *Deinocerites* and *Galindomyia*, the palpus is short, and the antenna is much longer than the proboscis (see *a*), distinguishing species in these two genera from those in the genus *Culex*. In *Galindomyia*, the apex of the proboscis is swollen in both males and females (see *e*), differentiating it from members of *Deinocerites*.

Bionomics

Galindomyia leei was first described based on adult specimens collected resting in crab-holes in mangrove swamps where the Río Raposo meets the Pacific Coast of Colombia. Other Culicini mosquitoes occupying the same crab-holes included *Cx.* (*Belkinomyia*) *eldridgei* Adames & Galindo and *Cx.* (*Tinolestes*) *latisquama* (Coquillett). To facilitate the formal description of the immature stages, gravid females collected resting in crab-holes in Curico, Colombia, were induced to oviposit in the laboratory in cylinders of plaster designed to simulate crab-holes. As with *Deinocerites pseudes* Dyar & Knab, females of *Galindomyia leei* were observed to lay their eggs individually on the walls of the imitation crab-holes, just above the waterline. Unlike floodwater mosquitoes, no fluvial inundation was necessary for egg eclosion, and the neonates simply dropped or slid down the walls of the protected habitat to complete their development. Male *Galindomyia leei* have also been collected in light traps in Ecuador, expanding the known geographic range of the genus.

Associated Pathogens

None.

Diagnostic Characters

(a) Habitus (♀): Antenna (Ant) longer than proboscis (P); pulvilli (Pv) present (not shown). **(b) Thorax (side):** Prespiracular (PsS) and postspiracular (PS) setae absent; acrostichal setae (AcS) absent (not shown); paratergite (Pa) bare; mesepimeron with a patch of upper setae (MeSU) and a single lower seta (MeSL). **(c) Wing:** Vein 1A ends beyond mediocubital crossvein (mcu); vein R_2 distinctly longer than vein R_{2+3}; membrane with distinct microtrichiae (not shown). **(d) Head (dorsal):** Vertex (V) with broad decumbent scales, with a few erect head scales on occiput (Occ); eyes contiguous. **(e) Head (side):** Flagellomeres Flm_1 and Flm_{13} quite long, about equal length. **(f) Larva (head):** Occipital foramen (OF) circular, collar incompletely developed (Col); setae 2,3-C absent. **(g) Larva (abdomen):** Seta 1-VI medial of 2-VI. **(h) Larva (terminal segments):** Comb (C) with comb scales (CS) in a triangular patch; pecten (Pt) with pecten spines (PS) extending to midsiphon (S); saddle (Sa) not well demarcated, with poorly sclerotized dorsal and ventral plates; seta 1-S with three pairs of setae, two large and one small; seta 4-X with five pairs of setae on a grid (G). Image sources: WRBU; Adames & Galindo (1975).

Exemplar DNA sequences: None.

Catalog entry: Vol. 2, p. 946.

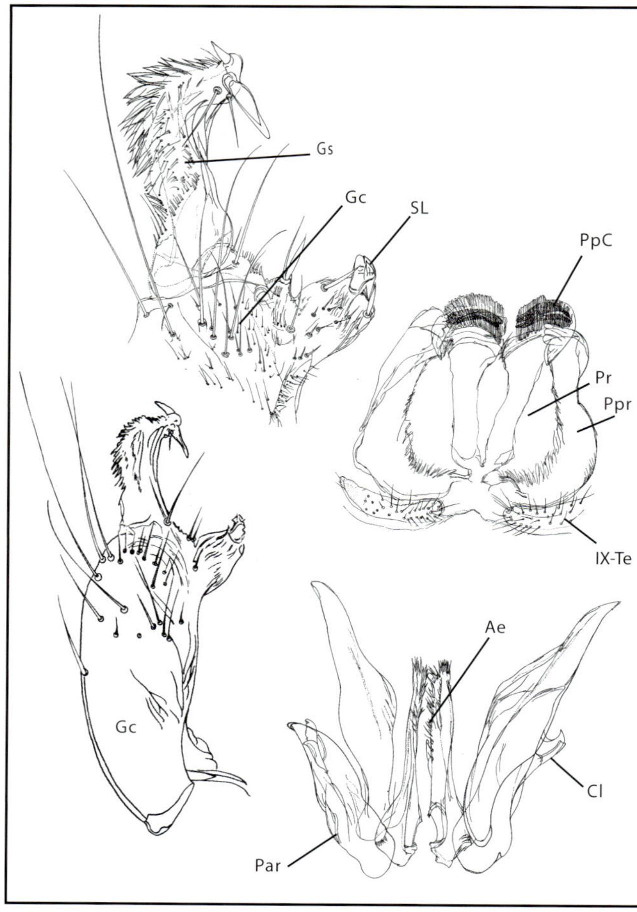

Galindomyia leei **Stone & Barreto (♂ genitalia)**

Ae: Aedeagus; Cl: Claspette; Gc: Gonocoxite; Gs: Gonostylus; Par: Paramere; PpC: Paraproct Crown; Ppr: Paraproct; Pr: Proctiger; SL: Subapical Lobe; IX-Te: Tergum IX. (Modified after Stone and Barreto, 1969).

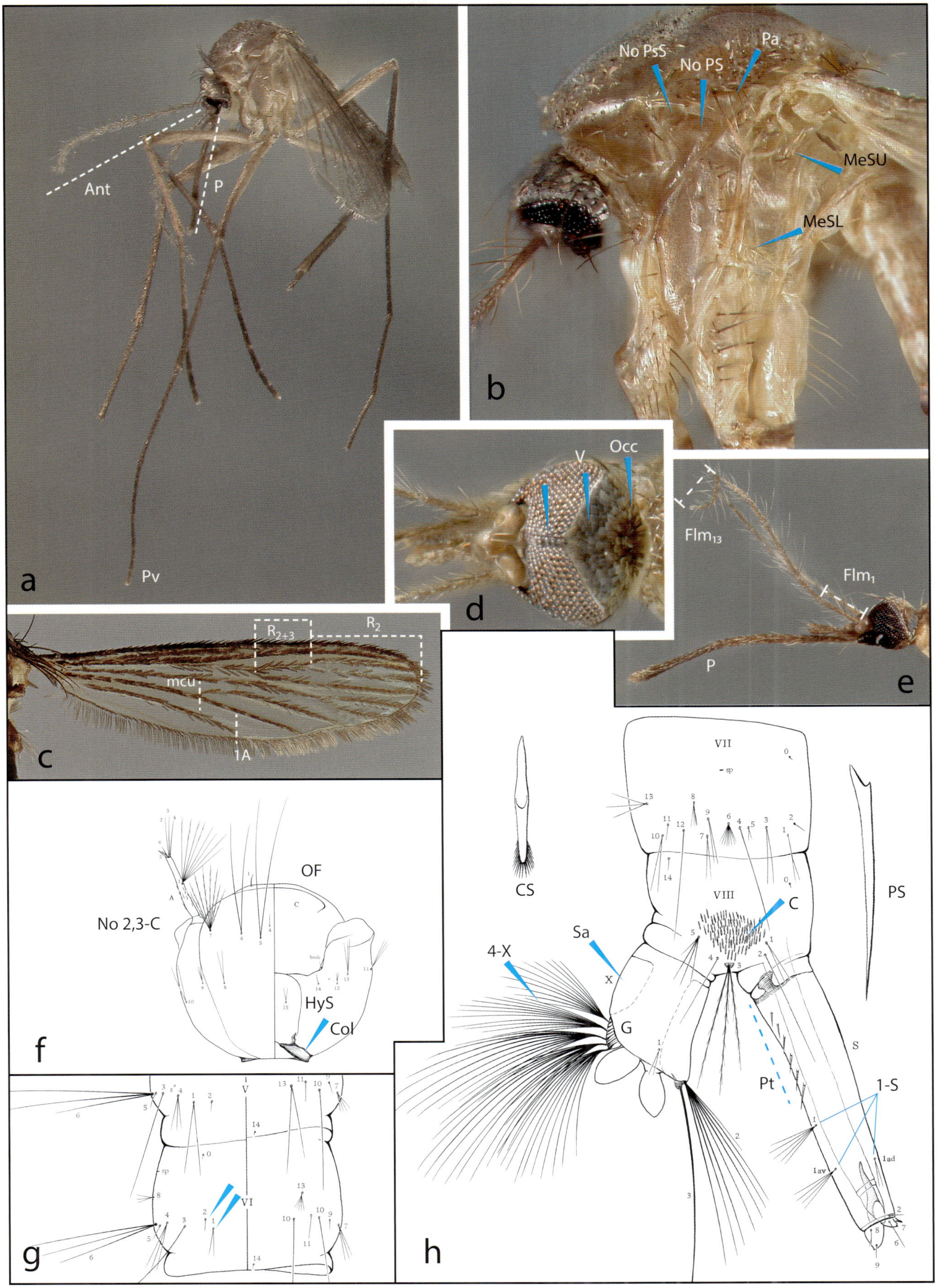

a
Ant
P
Pv
b
No PsS
No PS
Pa
MeSU
MeSL
c
R_{2+3}
R_2
mcu
1A
d
V
Occ
e
Flm_{13}
Flm_1
P
f
No 2,3-C
OF
HyS
Col
g
V
VI
h
VII
VIII
CS
PS
C
Sa
4-X
X
G
S
Pt
1-S

Genus *Lutzia* Theobald, 1903
Subfamily Culicinae, Tribe Culicini
"Predatory Mosquito"

Systematics and Distribution

Mosquitoes belonging to genus *Lutzia* are large, drab-colored mosquitoes that superficially resemble those in the genus *Culex*; some still consider *Lutzia* to be a subgenus of genus *Culex*. The genus comprises nine species, arranged in three subgenera. The monotypic subgenus *Insulalutzia* is represented by *Lutzia shinonagai* (Tanaka, Mizusawa & Saugstad) from Japan; subgenus *Lutzia* comprises two Neotropical species, *Lt. allostigma* Howard, Dyar & Knab and *Lt. bigoti* (Bellardi); and subgenus *Metalutzia* includes six species. The subgenus *Metalutzia* includes the Indian *Lt. agraensis* Singh & Prakash; Afrotropical *Lt. tigripes* (de Grandpre & de Charmoy); Oriental *Lt. vorax* Edwards and *Lt. chiangmaiensis* Somboon & Harbach; *Lt. fuscana* (Wiedemann), present across the Orient to Timor in the Australasian Region; and *Lt. halifaxii* (Theobald), found in the Australasian Region through the Papuan Subregion and Oriental Region to Pakistan. *Lutzia bigoti* is the type species of the genus.

Bionomics

Little is known of the bionomics of the monotypic genus *Insulalutzia*, except that immatures of *Lt. shinonagai* were collected in a tree hole. Larvae of all *Lutzia* species are ferocious predators and obligate cannibals, and thus all necessary energy stores required for oogenesis are absorbed in the immature stages. At least *Lt.* (*Lut.*) *allostigma* and *Lt.* (*Lut.*) *bigoti* are known not to require blood meals for oogenesis. Immature habitats of the subgenus *Metalutzia* appear to be broad, consisting of any temporary or permanent water bodies and natural or artificial containers, with the only defined characteristic being sufficient prey species for larval nutrition.

Associated Pathogens

BBKV, BIAV, KAMV.

Diagnostic Characters

(a) Habitus (♀). (b) Thorax (side): Antepronotum (Ap) with narrow scales; 4–8 long mesepimeral setae (MeS) (alveoli evident in photo). **(c) Thorax (dorsal):** Scutellum (Stm) with narrow scales. **(d) Leg:** Fe-I, II and Ti-I, II with anterior pale scale patches. **(e) Wing:** Crossvein rm nearly at same level as base of vein M_{3+4}. **(f) Leg:** Pulvillus (Pv) large. **(g) Larva (head):** Mouthparts modified for predation; mandible (Mn) with strong teeth; maxillary brush (MxB) with many strong pectinate hairs. **(h) Larva (terminal segments):** Siphon (S) short, with continuous midposterior row of setae; pecten (Pt) usually extends along entire length of siphon; saddle (Sa) complete, extending much farther dorsally than ventrally. Image sources: WRBU; Belkin (1962).

Exemplar DNA barcodes: *Lt. allostigma* (MF172338), *Lt. tigripes* (KU187026–30), *Lt. halifaxii* (KM609458), *Lt. fuscana* (KF407917–18, KF564755–56, HQ398896–97).

Catalog entry: Vol. 2, p. 947.

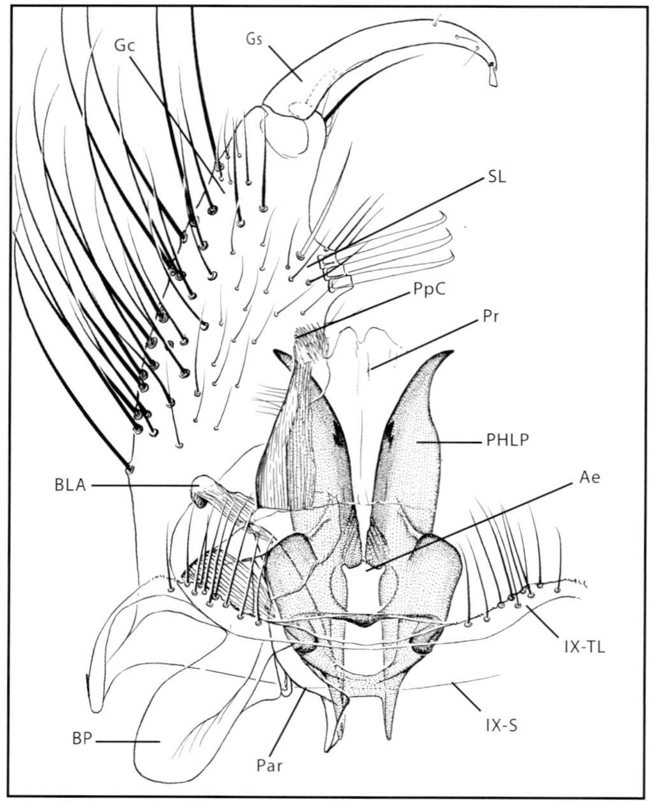

***Lutzia* (*Metalutzia*) *halifaxii* Theobald (♂ genitalia)**

Ae: Aedeagus; BLA: Basal Lateral Arm; BP: Basal Piece; Gc: Gonocoxite; Gs: Gonostylus; Par: Paramere; PHLP: Phallosome Lateral Plate; PpC: Paraproct Crown; Pr: Proctiger; SL: Subapical Lobe; IX-TL: Tergum IX Lobe; IX-S: Sternum IX.

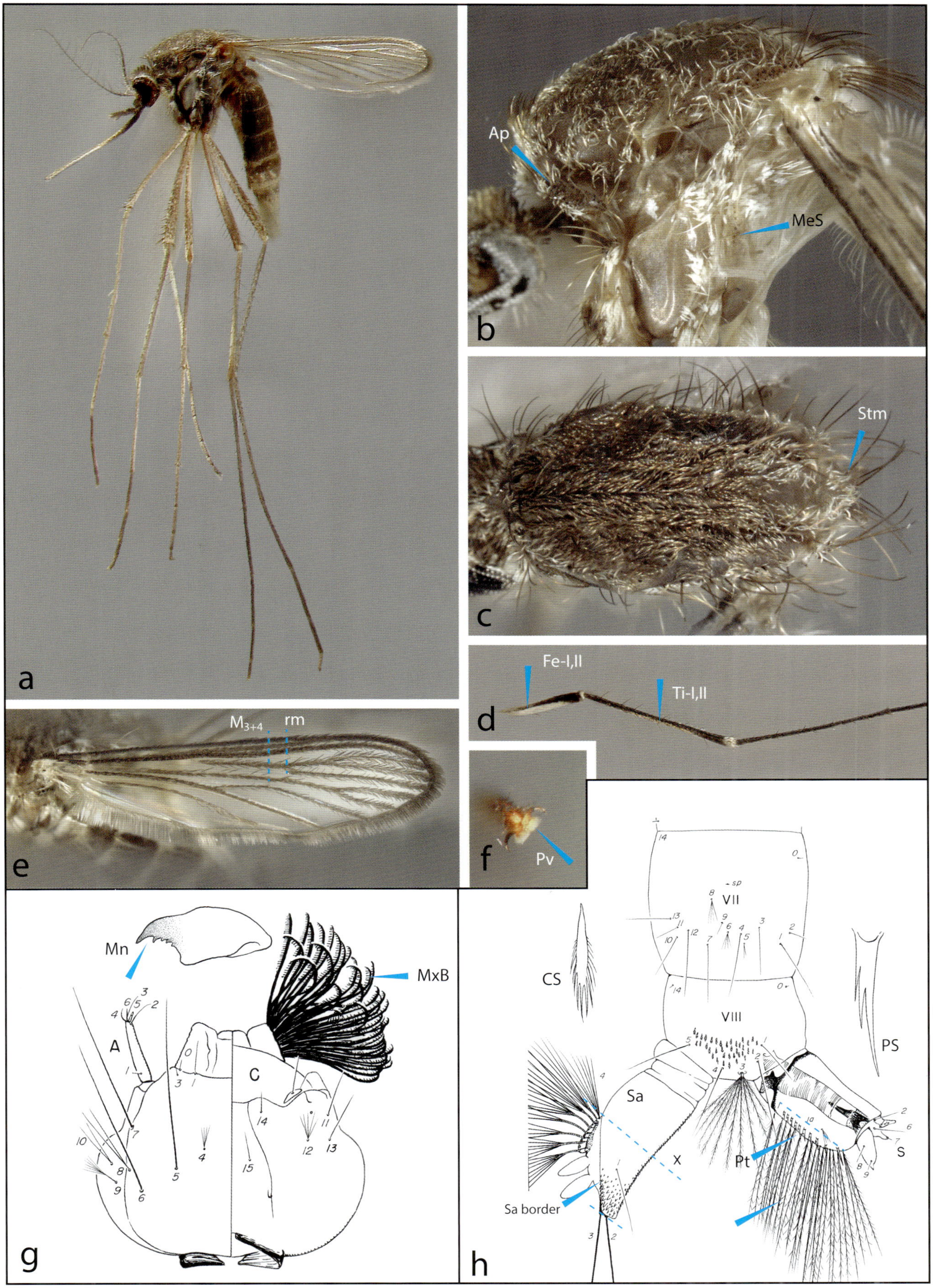
a
Ap
MeS
b
Stm
c
Fe-I,II
Ti-I,II
d
M3+4
rm
e
Pv
f
Mn
MxB
A
C
g
CS
VII
VIII
PS
Sa
X
Pt
S
Sa border
h

Genus *Culiseta* Felt, 1904
Subfamily Culicinae, Tribe Culisetini
"Cool Mosquito"

Systematics and Distribution

Mosquitoes of the genus *Culiseta* are large and superficially resemble those of the genus *Culex*. The genus *Culiseta* comprises the sole genus of the tribe Culisetini, and the species *Culiseta impatiens* (Walker) represents the genus. *Culiseta* comprises 39 species: 2 (*Cs. arenivaga Marks* and *Cs. wui* Lin, Tseng & Lien) are unplaced, and the other 37 are subdivided across seven subgenera: *Allotheobaldia* (1 sp.), *Austrotheobaldia* (1 sp.), *Climacura* (6 spp.), *Culicella* (14 spp.), *Culiseta* (12 spp.), *Neotheobaldia* (2 spp.) and *Theomyia* (1 sp.) (see Vol. 2). In contrast with most groups of mosquitoes, the genus *Culiseta*, informally dubbed the "Cool Mosquito," has its highest species-level biodiversity in northern temperate regions. *Culiseta* species are most common in the Palearctic, Nearctic, and Australasian Regions, although there are some representatives in the Afrotropical and Oriental Regions and in the Neotropics.

Bionomics

Adult *Culiseta* species feed on birds and mammals, including domestic animals and humans, with some species feeding on reptiles. Several species are of medical importance, particularly the Nearctic species *Cs. morsitans* (Coquillett), *Cs. inornata* (Williston) (see species pages, p. 534), *Cs. melanura* (Coquillett) (see species pages, p. 532), and the Palearctic *Cs. annulata* (Schrank, 1776). Although most *Culiseta* females oviposit in groundwater habitats, including ponds, marshes, stream edges, and ditches, the immatures of some species are also found in wells and rock pools. The unusual Australasian forest-associated subgenus *Neotheobaldia* comprises two species—*Cs. frenchii* (Theobald, 1901) and *Cs. hilli* (Edwards, 1926)—whose immatures inhabit the fresh water that collects in underground crayfish tunnels. Immatures pass the winter underground as larvae. *Culiseta fraseri* (Edwards, 1914) of the monotypic subgenus *Theomyia* is found in sub-Saharan forests, where tree holes are utilized as larval habitats.

Associated Pathogens

CEV, CVV, EEEV, FLAV, HJV, HPV, JCV, JSV, NORV, SINV, SSHV, TAHV, WEEV, WHAV, WNV; *Plasmodium circumflexum*.

Diagnostic Characters

(a) Habitus (♀). (b) Thorax (side): Prespiracular setae (PsS) present; postspiracular area (PA) usually without setae or scales; lower mesepimeral setae (MeSL) present. **(c) Leg:** Pulvillus (Pv) absent. **(d) Wing:** Ventral setae present at base of subcosta (Sc) (except *Theomyia*). **(e) Wing. (f) Larva (head):** Hypostomal suture (HyS) complete. **(g) Larva (terminal segments):** Seta 5-VIII below dorsal margin of segment X; seta 1-S inserted at base of siphon; seta 4-X usually with ≥3 pairs of setae. Image sources: WRBU; Dobrotworsky (1971).

Exemplar DNA barcodes: *Culiseta annulata* (KX675388–89), *Cs. morsitans* (KU748496, KU748460), *Cs. inornata* (MG087325, MG087024), *Cs. subochrea* (MG962351–52).

Catalog entry: Vol. 2, p. 950.

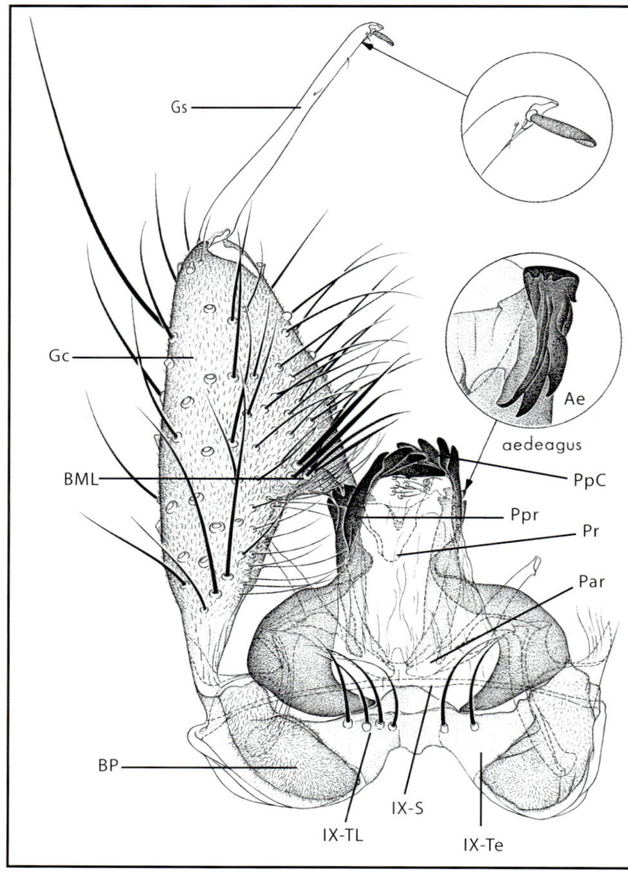

***Culiseta* (*Climacura*) *marchettei* Garcia, Jeffery & Rudnick (♂ genitalia)**

Ae: Aedeagus; BML: Basal Mesal Lobe; BP: Basal Piece; Gc: Gonocoxite; Gs: Gonostylus; Par: Paramere; PpC: Paraproct Crown; Ppr: Paraproct; Pr: Proctiger; IX-Te: Tergum IX; IX-TL: Tergum IX Lobe; IX-S: Sternum IX. (Modified after Dobrotworsky, 1971).

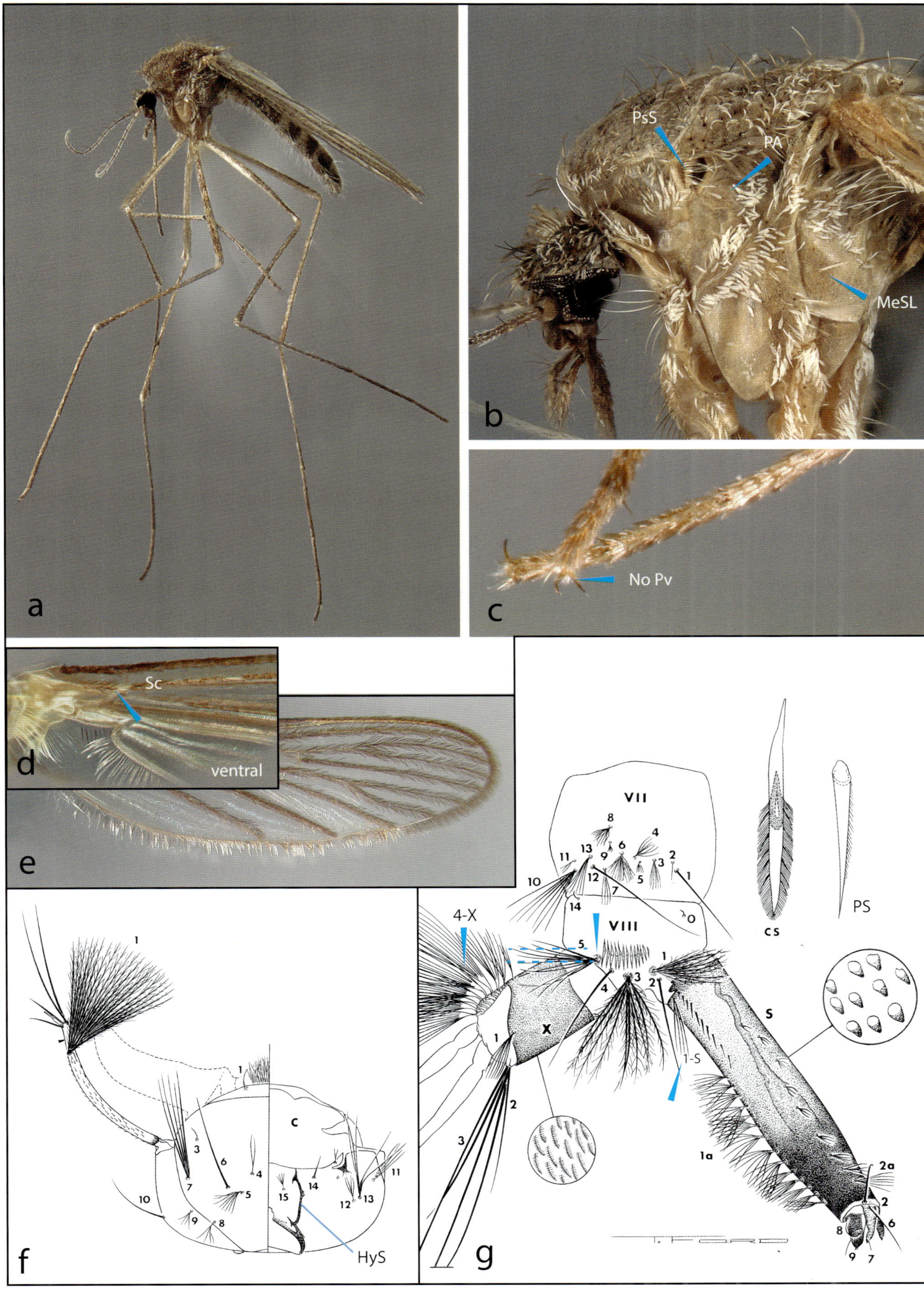
a
PsS
PA
MeSL
b
No Pv
c
Sc
d
ventral
e
f
c
HyS
VII
VIII
4-X
X
1-S
S
1a
2a
CS
PS
T. FORD
g

Genus *Ficalbia* Theobald, 1903
Subfamily Culicidae, Tribe Ficalbiini
"Small Mosquito"

Systematics and Distribution

As their informal name suggests, species of the genus *Ficalbia* are some of the smallest-known mosquitoes. As a result, they are often overlooked in trap collections or misidentified as ceratopogonids, and only eight species are described. These include *Ficalbia circumtestacea* (Theobald), *Fi. ichiromiyagii* Toma & Higa, *Fi. jacksoni* Mattingly, *Fi. ludlowae* Brunetti, *Fi. malfeyti* (Newstead), *Fi. minima* (Theobald), *Fi. nigra* (Theobald), and *Fi. uniformis* (Theobald). *Ficalbia minima* serves as the type species for the genus. Although *Ficalbia* and *Mimomyia* are currently both placed in the tribe Ficalbiini, cladistic analysis of morphological traits indicated that these are not sister taxa. Revisions of both the internal relationships among *Ficalbia* species and the phylogenetic status of the genus within the family Culicidae are overdue.

Ficalbia species are mainly distributed in the Afrotropical and Oriental Regions. *Ficalbia ichiromiyagii* was recently described from, and is so far restricted to, the Iriomote Island of the Japanese Ryukus Archipelago, extending the overall distribution of *Ficalbia* species into the southeastern Palearctic. The range of *Fi. minima* extends from the mainland and islands of the Oriental Region and Pacific islands across to Australia. *Ficalbia jacksoni* is endemic to China. *Ficalbia circumtestacea*, *Fi. malfeyti*, *Fi. minima*, *Fi. nigra*, and *Fi. uniformis* are Afrotropical species, with *Fi. circumtestacea* and *Fi. uniformis* also found on Madagascar. *Ficalbia ludlowae* occurs on the islands of Indonesia and the Philippines.

Bionomics

Little is known of the bionomics of adult or immature *Ficalbia* species. Immature *Fi. uniformis* reportedly occupies flooded meadows, canals, and marshes with dense vegetation, as well as deep, clear water sites in Madagascar. The immature stages of *Fi. ichiromiyagii* are found in densely vegetated jungle swamps in the undisturbed Otomi swamp forest regions of Iriomote Island, Japan. Resting adults have been collected in both the canopy and lower vegetation at night, but nothing is known on the biting preferences of *Ficalbia* species. *Ficalbia modesta* and *Fi. flavens* pools have tested positive for Maprik virus (MPKV).

Due to the lack of comprehensive exemplar illustrations, there was no available image of the male genitalia of *Ficalbia* to include in this genus.

Associated Pathogens

MPKV.

Diagnostic Characters

(a) Habitus (♀): Proboscis (P) slightly swollen at tip (♀). **(b) Head (dorsal):** Erect head scales only on occiput (Occ). **(c) Thorax (side):** Prespiracular seta (PsS) absent; postspiracular area (PA) bare; prealar scales (PrSc) absent; lower mesepimeral setae (MeSL) absent; flagellomere Flm_1 2–6 times longer than Flm_2. **(d) Wing:** Vein 1A terminates beyond base of mediocubital crossvein (mcu). **(e) Thorax (dorsal):** Scutellum (Stm) with narrow scales; acrostichal (AcS) and dorsocentral (DS) setae present. **(f) Wing:** Alula (Al) and upper calypter (not shown) with fringe of narrow scales. **(g) Larva (abdomen):** Abdominal seta 12-I present. **(h) Head (♂):** Proboscis (P) greatly swollen at tip. **(i) Larva (head):** Distal and basal portions of antenna (A) fused, not articulated; hypostomal suture (HyS) absent or incomplete. **(j) Larva (terminal segments):** Single row of comb scales (CS); seta 1-S inserted near base of siphon (S); pecten (Pt) reduced, with 1–2 spines; saddle (Sa) complete; seta 4-X with four pairs of setae, and sometimes 1–2 minute extra setae. Image sources: WRBU; Mattingly (1971).

Exemplar DNA barcodes: *Ficalbia minima* (KF564761–63, KF564767).

Catalog entry: Vol. 2, p. 957.

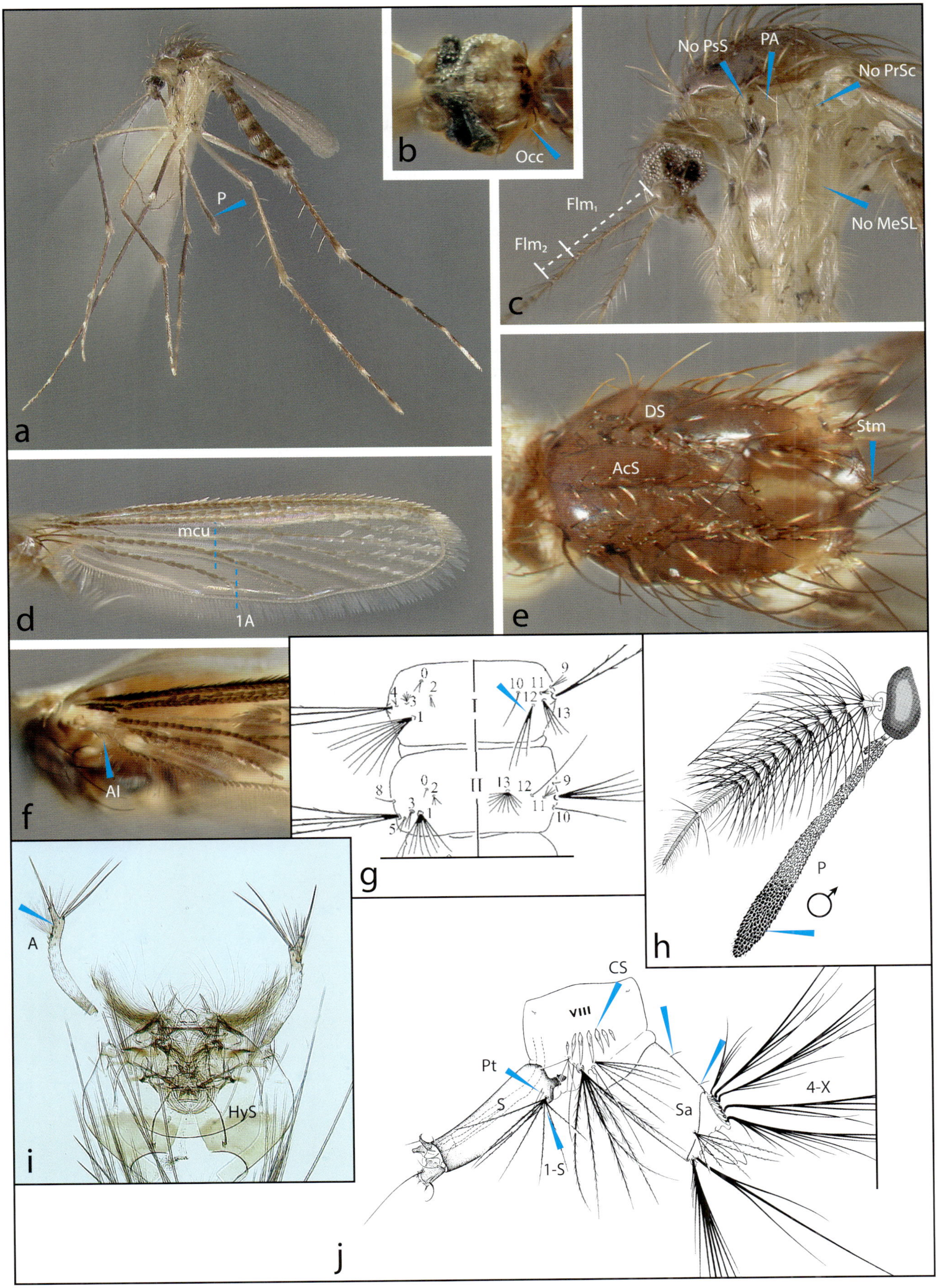
a
P
b
Occ
No PsS
PA
No PrSc
Flm₁
Flm₂
No MeSL
c
mcu
1A
d
DS
AcS
Stm
e
Al
f
I
II
g
P
♂
h
A
HyS
i
CS
VIII
Pt
S
1-S
Sa
4-X
j

Genus *Mimomyia* Theobald, 1903
Subfamily Culicinae, Tribe Ficalbiini
"Indistinct Mosquito"

Systematics and Distribution

Mimomyia species are tiny mosquitoes, most closely related to those in the genus *Ficalbia*, and together these two genera comprise the tribe Ficalbiini. The 45 described species are subdivided into three subgenera (see Vol. 2, p. 1096, for internal relationships) and are represented by the type species *Mimomyia splendens* Theobald. The seven species of subgenus *Etorleptiomyia* are found in the Australasian, Oriental, and Afrotropical Regions. The subgenus *Ingramia* (21 sp.) is reported from Madagascar and surrounding islands and various regions of the Orient. The subgenus *Mimomyia* is composed of 17 species, including *Mi.* (*Mim.*) *chamberlaini* Ludlow, which has two valid subspecies: *clavipalpus* (Theobald), and *metallica* (Leicester). Species of the subgenus *Mimomyia* are found in the Afrotropical, Australasian, and Oriental Regions.

Bionomics

Immatures of the subgenera *Etorleptiomyia* and *Mimomyia* typically occur in ground pools, swamps, and marshy depressions filled with dense vegetation. In common with mosquitoes of the tribe Mansoniini (*Mansonia* and *Coquillettidia*), the larval siphon of some *Mimomyia* species are modified for piercing, suggesting these taxa can also attach to, and respire through, aquatic vegetation in their larval habitats. Immatures of subgenus *Ingramia* are found only in phytotelmata—tree holes, cut bamboo, pitcher plants, and the leaf axils of *Colocasia*, *Pandanus*, *Ravenala*, and *Typhonodorum*. Amphibians are the preferred hosts of many *Mimomyia* species. Female *Mi. chamberlaini* feeds on the frog *Rana tigrine* in India, and *Mi.* (*Eto.*) *elegans* (Taylor) feeds on the introduced toad *Bufo marinus* in Australia. In Kenya, both amphibian and human blood was detected in *Mi.* (*Eto.*) *mediolineata* (Theobald) and *Mi.* (*Mim.*) *splendens* Theobald, and *Mi.* (*Mim.*) *hispida* (Theobald) was shown to have fed on other mammals, including cattle. Although several species are known to bite humans, *Mimomyia* species have typically not been regarded to be of medical importance. However, recent biosurveillance studies of pooled samples of wild-caught *Mi. hispida* in Senegal detected West Nile, Bagaza, and Babanki viruses, suggesting that their medical impact, at least locally, could be underestimated.

Associated Pathogens

BAGV, BBKV, WNV.

Diagnostic Characters

(a) Habitus (♀). (b) Thorax (dorsal): Acrostichal (AcS) and dorsocentral (DS) setae present. **(c) Wing:** Alula (Al) bare or with broad decumbent scales dorsally; upper calypter (UC) with setae or slender, hair-like scales. **(d) Wing:** Vein 1A terminates beyond level of mediocubital crossvein (mcu). **(e) Leg:** Pulvillus (Pv) absent or indistinct. **(f) Head (dorsal):** Erect scales on vertex (V) and occiput (Occ), or only on occiput. **(g) Larva (thorax):** Setae 9–12-M,T often on prominent tubercles; seta 12-I present. **(h) Larva (head):** Distal part of antenna (A) articulated. **(i) Larva (terminal segments):** Siphon (S) usually sharply narrowed distally, sometimes modified for piercing plants; siphon with a single pair of subventral setae (1-S); seta 4-X with 4–6 setal pairs; pecten (Pt) absent or reduced to ≤4 spines. Image sources: WRBU; Mattingly (1971); Tanaka et al. (1979).

Exemplar DNA barcodes: *Mi. chamberlaini* (AY729979, KY352274), *Mi. splendens* (KU187093–94, KU380454, KU380447).

Catalog entry: Vol. 2, p. 959.

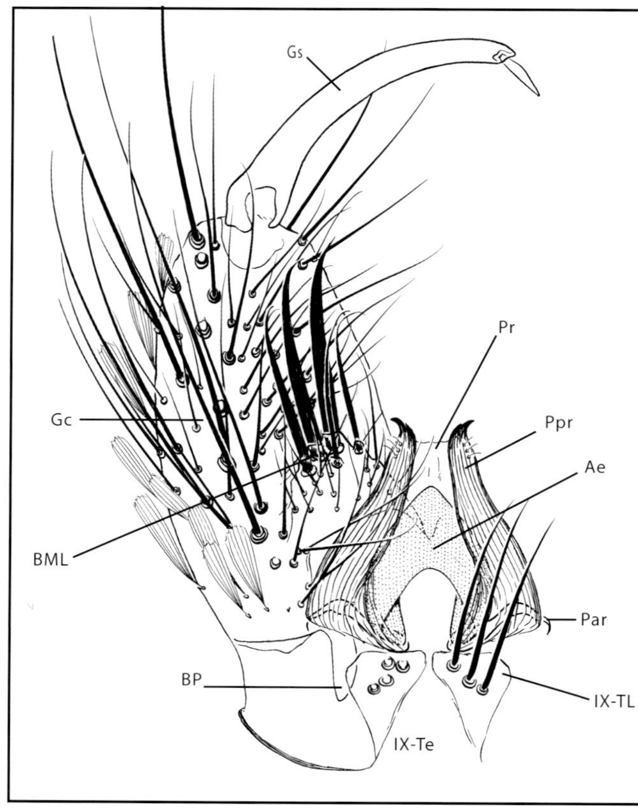

***Mimomyia* (*Mimomyia*) *gurneyi* (Belkin) (♂ genitalia)**

Ae: Aedeagus; BML: Basal Mesal Lobe; BP: Basal Piece; Gc: Gonocoxite; Gs: Gonostylus; Par: Paramere; Ppr: Paraproct; Pr: Proctiger; IX-Te: Tergum IX; IX-TL: Tergum IX Lobe. (Modified after Belkin, 1962).

a
b
DS
AcS
Scu
c
UC
Al
d
mcu
1A
e
No Pv
f
V
Occ
g
P
M
T
I
h
A
C
i
VII
VIII
X
4-X
CS
S

Genus *Hodgesia* Theobald, 1904
Subfamily Culicinae, Tribe Hodgesiini
"Tiny Mosquito"

Systematics and Distribution

Hodgesia mosquitoes are instantly recognizable by their tiny size and the unique, elongated, fork-tipped scales on the distal half of their wings. Species of the genus *Hodgesia* are thought to be sister taxa to *Ficalbia* (Ficalbiini, Culicinae) and, like *Culiseta*, share the unusual basal positioning of seta 1-S on the siphon (see *g*). The fusion of the larval cardo and maxillary palps and the complete hypostomal suture distinguish *Hodgesia* larvae from those belonging to the genus *Ficalbia*. The eleven described species in genus *Hodgesia* comprise the entirety of the culicine tribe Hodgesiini. There are no assigned subgenera at this time. Species include *Ho. bailyi* Barraud, *Ho. cairnsensis* Taylor, *Ho. cyptopus* Theobald, *Ho. lampangensis* Thurman, *Ho. malayi* Leicester, *Ho. nigeriae* Edwards, *Ho. psectropus* Edwards, *Ho. quasisanguinae* Leicester, *Ho. sanguinae* Theobald, *Ho. solomonis* Belkin, and *Ho. spoliata* Edwards. *Hodgesia sanguinae* is the type species of the genus.

Hodgesia species are distributed in the Afrotropical, Australasian, and Oriental Regions. *Hodgesia quasisanguinae* and *Ho. solomonis* span both Oriental and Australasian Regions. However, the geographic distribution of the genus is most likely to be underreported, as *Hodgesia* species are so tiny that they are often mistaken for *Culicoides* biting midges and may escape through the mesh of light trap catches designed for larger Culicid species.

Bionomics

Little is known about the biological traits of *Hodgesia* species. Immatures are most commonly associated with jungle environments, and immature habitats include permanent freshwater jungle ground pools and swamps teaming with dense vegetation. *Hodgesia* immatures often share similar habitats with those of the genus *Mimomyia*, and these taxa are often collected in sympatry. Several *Hodgesia* species have been collected biting humans during the day near their larval habitats, including the Afrotropical *Ho. sanguinae* and the Oriental *Ho. malayi* and *Ho. quasisanguinae*. *Hodgesia solomonis* is renowned for its vicious attacks on humans in jungle areas of the Solomon Islands. Biting behaviors of other species have not been described.

Associated Pathogens

None.

Diagnostic Characters

(a) Habitus (♀). **(b) Wing:** Vein 1A terminates level with base of mediocubital crossvein (mcu). **(c) Wing:** Membrane with distinct microtrichia; scales on distal half long and notched at tips. **(d) Thorax (side):** Pleuron (p) with a patch of silvery scales and much-reduced setae; acrostichal setae (AcS) absent, dorsocentral (DS) setae prominent. **(e) Head (dorsal):** Head scales flat and broad, silvery scale patch toward vertex (V); scutum usually without markings, except broad silvery scales on the antepronotum (Ap). **(f) Larva (head):** Seta 1-A large, inserted near apex of antenna (A). **(g) Larva (terminal segments):** Siphon (S) short, with seta 1-S attached at base; pecten (Pt) well developed; saddle (Sa) complete. Image sources: WRBU; Mattingly (1971).

Exemplar DNA barcodes: *Hodgesia bailyi* (EU259292).

Catalog entry: Vol. 2, p. 964.

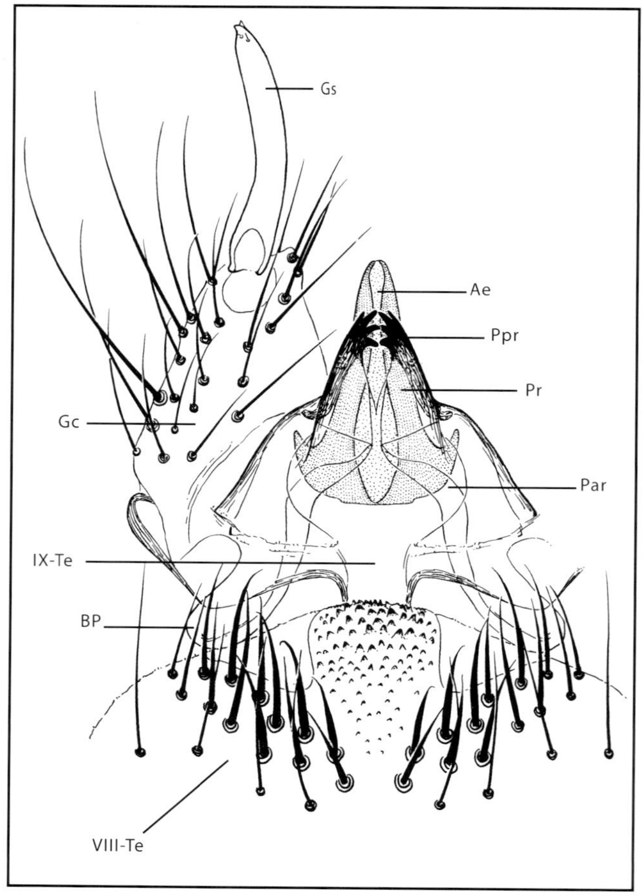

***Hodgesia solomonis* Belkin** (♂ **genitalia)**

Ae: Aedeagus; BP: Basal Piece; Gc: Gonocoxite; Gs: Gonostylus; Par: Paramere; Ppr: Paraproct; Pr: Proctiger; VIII-Te: Tergum VIII; IX-Te: Tergum IX. (Modified after Belkin, 1962).

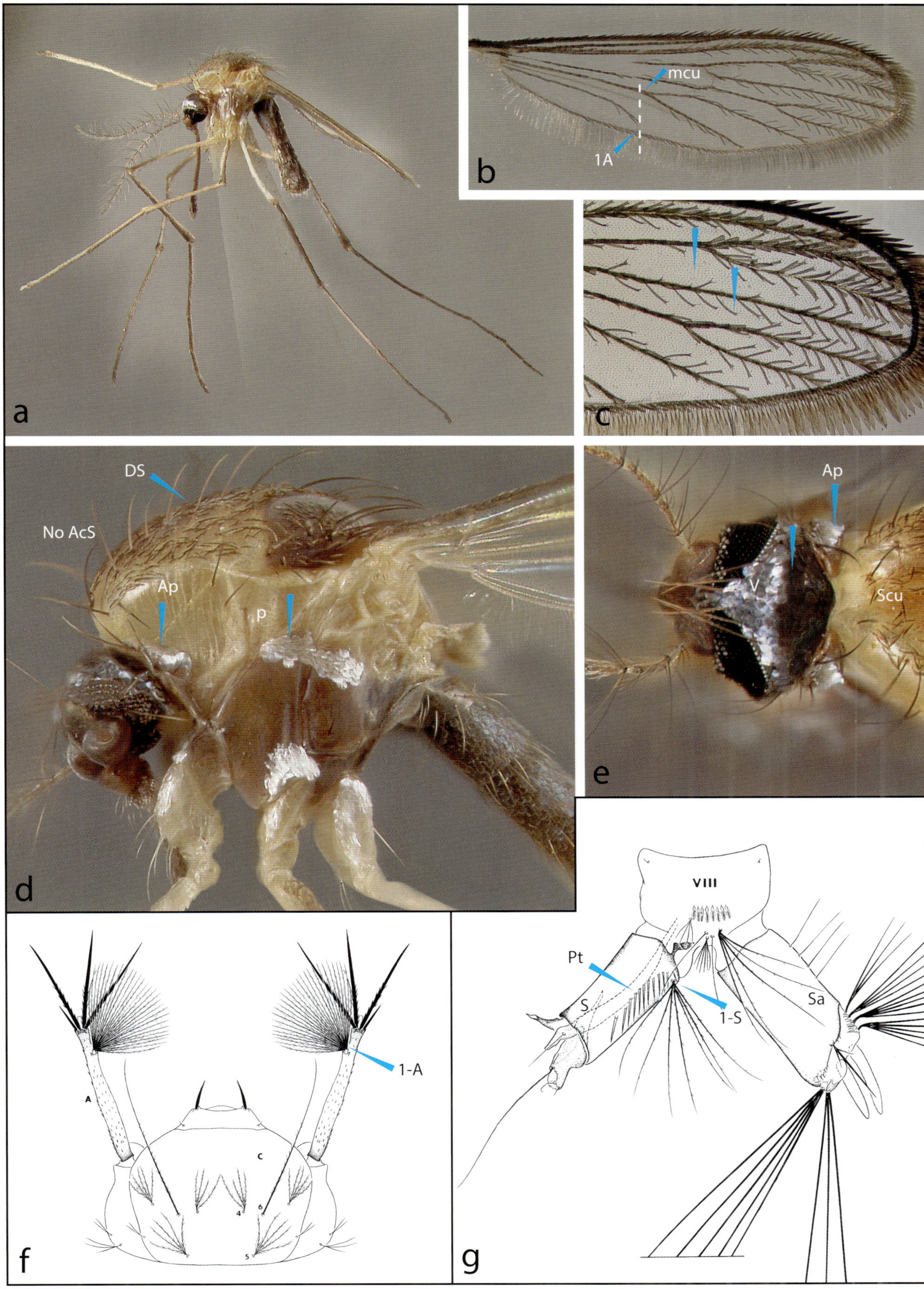
mcu
1A
b
c
a
DS
No AcS
Ap
p
d
Ap
V
Scu
e
VIII
Pt
S
1-S
Sa
1-A
A
C
4
6
5
f
g

Genus *Coquillettidia* Dyar, 1905 Subfamily Culicidae, Tribe Mansoniini "Underwater Mosquito"

Systematics and Distribution

Mosquitoes of the genus *Coquillettidia* are distinctive, large, and yellowy orange (see *a*). The genus contains 58 species divided into three subgenera: *Austromansonia* (1 sp.), *Coquillettidia* (44 spp.), and *Rhynchotaenia* (13 spp.). The monotypic *Austromansonia* is represented by *Cq. tenuipalpis* (Edwards), which is found only in New Zealand. The largest subgenus *Coquillettidia* has representatives in the Afrotropical, Palearctic, Oriental, and Australasian Regions, plus the type species *Cq. perturbans* (Walker), which is present in the Nearctic. Species of the subgenus *Rhynchotaenia* are confined to the Neotropics. Together, mosquitoes of the genera *Coquillettidia* and *Mansonia* comprise the tribe Mansoniini.

Bionomics

Immature sites for *Coquillettidia* are much conscribed by the availability of a variety of essential aquatic plants, including grasses and reeds, through which they respire. In common with *Mansonia*, *Coquillettidia* larvae attach themselves to the roots and stems of aquatic plants using their highly modified siphons (see *i*) to obtain oxygen for respiration. This predation-evasion strategy allows the larvae to remain almost motionless and undetected as they develop, but the larvae can readily detach from their host plants and will rapidly disperse if threatened. In some regions, *Coquillettidia* species feed primarily on birds, particularly waterfowl. However, in Europe, Africa, and the Nearctic, several species are ferocious pest species and biters of humans and domestic livestock and contribute to the transmission of zoonotic arboviruses (e.g., *Cq.* (*Rhy.*) *arribalzagae* (Theobald), *Cq.* (*Coq.*) *fuscopennata* (Theobald), *Cq.* (*Coq.*) *metallica* (Theobald), *Cq.* (*Coq.*) *perturbans* (Walker) (see species pages, p. 536), *Cq.* (*Rhy.*) *venezuelensis* (Theobald)), filariasis (e.g., *Cq.* (*Coq.*) *crassipes* (van der Wulp)), and several avian malarias (e.g., *Cq.* (*Coq.*) *aurites* (Theobald), *Cq.* (*Coq.*) *pseudoconopas* (Theobald), *Cq.* (*Coq.*) *metallica*). Both diurnal and night biters have been reported.

Associated Pathogens

ANUV, BFV, BSQV, CABV, CALV, CATUV, CHIKV, COTV, CVOV, EEEV, GMAV, ILHV, MAYV, MOJUV, MURV, OROV, RBUV, RVFV, SIMV, SINV, TATV, TENV, TONV, UNAV, USUV, WONV, WNV, WYOV; *Brugia malayi*, avian *Plasmodium* sp.

Diagnostic Characters

(a) Habitus (♀). **(b) Head (dorsal):** Clypeus (Clp) bare, longer than wide. **(c) Legs:** Fe-I–III dark. **(d) Abdomen:** Apex blunt in females; VIII-Te unmodified, with only basal half sclerotized. **(e) Wing:** Wing scales somewhat broad and usually symmetrical, mixed dark and pale. **(f) Genitalia** (♀): Cerci (Ce) spatulate, articulated with IX-Te; IX-Te bi-lobed, with few long setae; three equal-sized spermathecae (not shown). **(g) Legs:** Unguis U-I–III simple; pulvillus (Pv) absent. **(h) Larva (head):** Antenna (A) very long, articulated, with distal portion elongate, usually as long as the basal portion. **(i) Larva (terminal segments):** Siphon modified for piercing plant tissues (SAp); seta 1-X multibranched or absent. Image sources: WRBU; Belkin (1962); Belkin (1968).

Exemplar DNA barcodes: *Coquillettidia aurites* (GQ165776), *Cq. metallica* (KU187112–13), *Cq. nigricans* (KY859871–72), *Cq. perturbans* (MG086163), *Cq. richiardii* (JX040513).

Catalog entry: Vol. 2, p. 966.

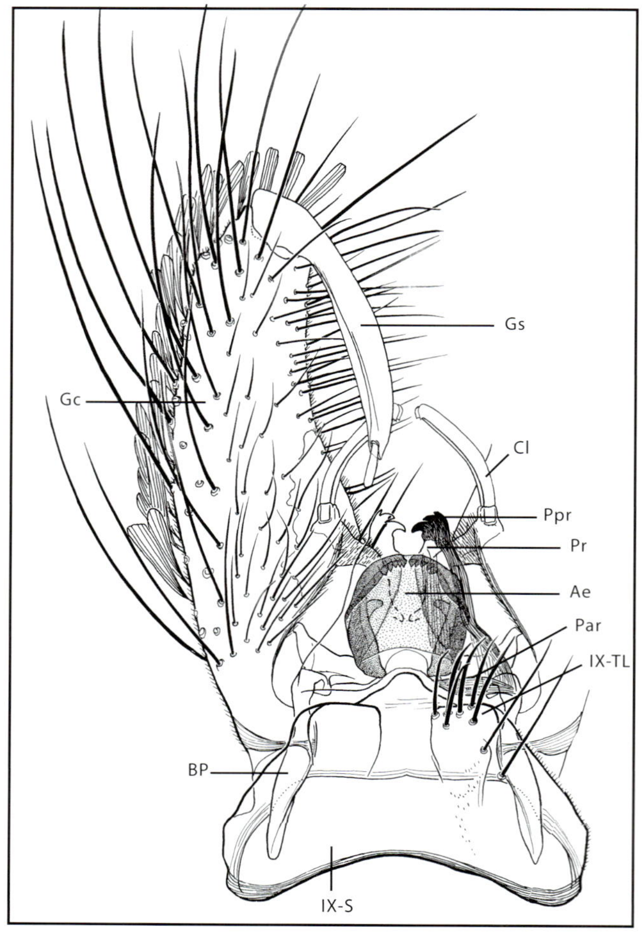

***Coquillettidia* (*Rhynchotaenia*) *nigricans* (Coquillett) (♂ genitalia)**

Ae: Aedeagus; BP: Basal Piece; Cl: Claspette; Gc: Gonocoxite; Gs: Gonostylus; Par: Paramere; Ppr: Paraproct; Pr: Proctiger; IX-TL: Tergum IX Lobe; IX-S: Sternum IX. (Modified after Belkin, Heinemann and Page, 1970).

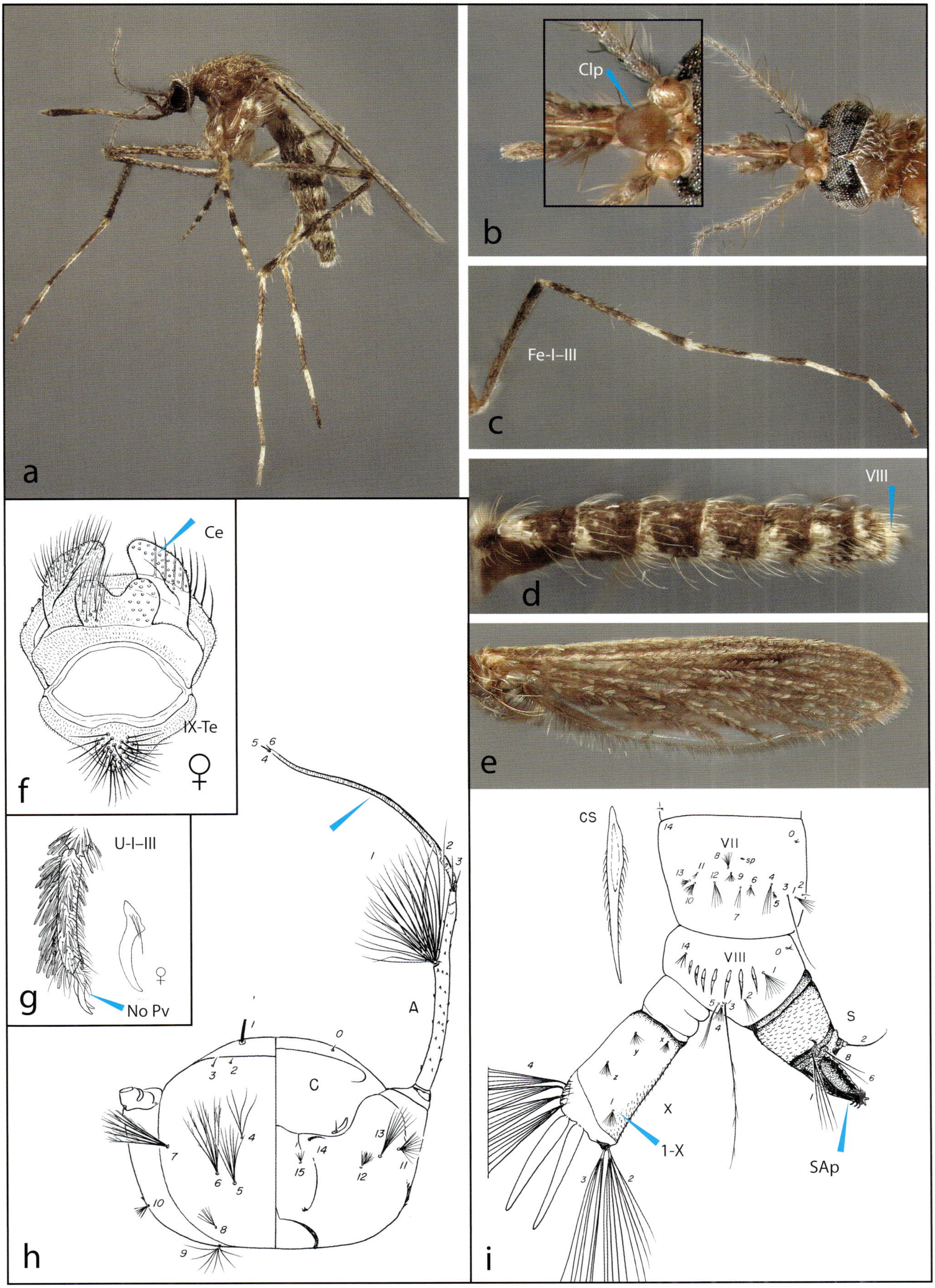
a
Clp
b
Fe-I–III
c
VIII
d
e
Ce
IX-Te
♀
f
U-I–III
♀
No Pv
g
A
C
h
CS
VII
sp
VIII
S
X
1-X
SAp
i

Genus *Mansonia* Blanchard, 1901
Subfamily Culicinae, Tribe Mansoniini
"Submarine Mosquito"

Systematics and Distribution

Mansonia species are usually large mosquitoes, distinguished from all other mosquito genera by the presence of broad, mixed dark and pale asymmetrical scales on the wings (see *f*). They most commonly occur in tropical regions worldwide, but some species exhibit cold tolerance, surviving in more northerly climes. The genus comprises 25 valid species subdivided into two geographically distinct subgenera: *Mansonia* (15 spp.) restricted to the Old World and the New World subgenus *Mansonioides* (10 spp.). Most New World species are Neotropical, but *Ma. dyari* Heinemann & Page and *Ma. titillans* (Walker) are also present in central and southern United States. *Mansonia titillans* (as *Culex titillans* Walker) is the type species of the genus.

Bionomics

Mansonia species avidly bite man and are often considered nuisance pests. *Mansonia* species also feed on a wide range of birds, mammals, and reptiles. Alligator blood was recovered from both *Ma. dyari* and *Ma. titillans* in Florida, United States. Females are predominantly crepuscular but will also bite at any time during the night and during the day when disturbed. *Mansonia* are strong fliers and can be found many miles from the larval habitats. *Mansonia* species oviposit in large permanent water bodies such as ponds and lakes, especially those supporting water hyacinth (*Eichornia crassipes*) and water lettuce (*Pistia stratiotes*). The females lay their eggs in star-shaped clusters on the underside of floating leaves, facilitated by specially adapted, hook-like abdominal setae (see *g*). Immatures of some species burrow into the muddy substrate, whereas others cling to floating root masses. However, larvae and pupae of all species attach to and respire through the roots and stems of aquatic plants using their modified siphons (see *i*). Herbicides are widely used to control *Mansonia* populations by removing the host plants upon which the immatures rely.

Associated Pathogens

BANV, BARV, BBKV, BSQV, BTKV, BUNV, BWAV, CABV, CHIKV, GMAV, ILEV, JBEV, KAVV, KRIV, LEBV, LMV, MAGV, MALV, MAYV, MIDV, MOJUV, MPKV, MQOV, NDUV, OLIV, ONNV, ORIV, PERV, PGAV, PUCV, RRV, RVFV, SANV, SEPV, SHOV, SINV, SLEV, SPOV, TENV, TLAV, TONV, TVTV, USUV, VEEV, VSNJV, WSLV, YATAV, YFV, ZIKAV; *Brugia malayi, Wuchereria bancrofti*.

Diagnostic Characters

(a) Habitus (♀): All Fe and Ti dark, speckled with pale scales. **(b) Leg:** Unguis (U-I–III) simple; pulvilli (Pv) absent. **(c) Thorax (side):** Postspiracular setae (PS) present. **(d) Head (dorsal):** Clypeus (Clp) bare, nearly as broad as long. **(e) Abdomen (side):** IX-Te poorly developed and lightly sclerotized; apex blunt in females; cercus (Ce) well developed, curved upward; two large spermathecae, with third absent or very small (not shown). **(f) Wing:** Scales broad and asymmetrical, dark and pale scales intermixed. **(g) Abdomen:** VIII-Te distinctly modified with short, robust, hook-like setae. **(h) Larva (head):** Antenna (A) about as long as width of head. **(i) Larva (terminal segments):** Spiracular apparatus (SAp) modified for piercing plant tissues; segment X with multiple sharply pointed, stout setae. Image sources: WRBU; Belkin et al. (1970).

Exemplar DNA barcodes: *Mansonia (Mansonia) flaveola* (JX260705, JX260065), *Ma. (Man.) indubitans* (MH118158), *Ma. (Man.) titillans* (KY859898–902), *Mansonia (Mansonioides) africana* (KU380478–79), *Ma. (Mnd.) annulata* (HQ341635–42), *Ma. (Mnd.) annulifera* (KX816503–11), *Ma. (Mnd.) bonneae* (KX816475–85), *Ma. (Mnd.) dives* (KX816465–74), *Ma. (Mnd.) indiana* (KX816496–502), *Ma. (Mnd.) uniformis* (KX816486–96).

Catalog entry: Vol. 2, p. 973.

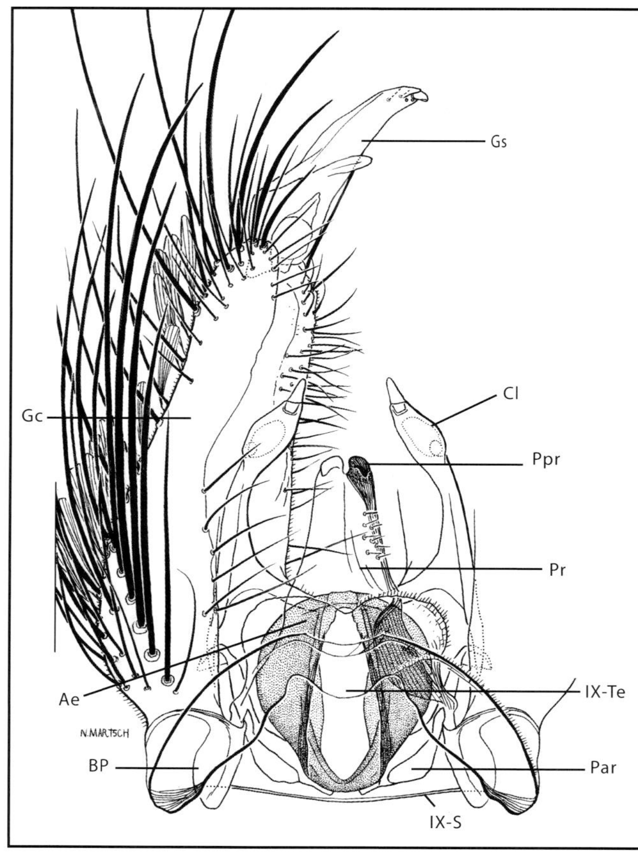

***Mansonia (Mansonia) titillans* Walker (♂ genitalia)**

Ae: Aedeagus; BP: Basal Piece; Cl: Claspette; Gc: Gonocoxite; Gs: Gonostylus; Par: Paramere; Ppr: Paraproct; Pr: Proctiger; IX-Te: Tergum IX; IX-S: Sternum IX. (Modified after Belkin, Heinemann and Page, 1970).

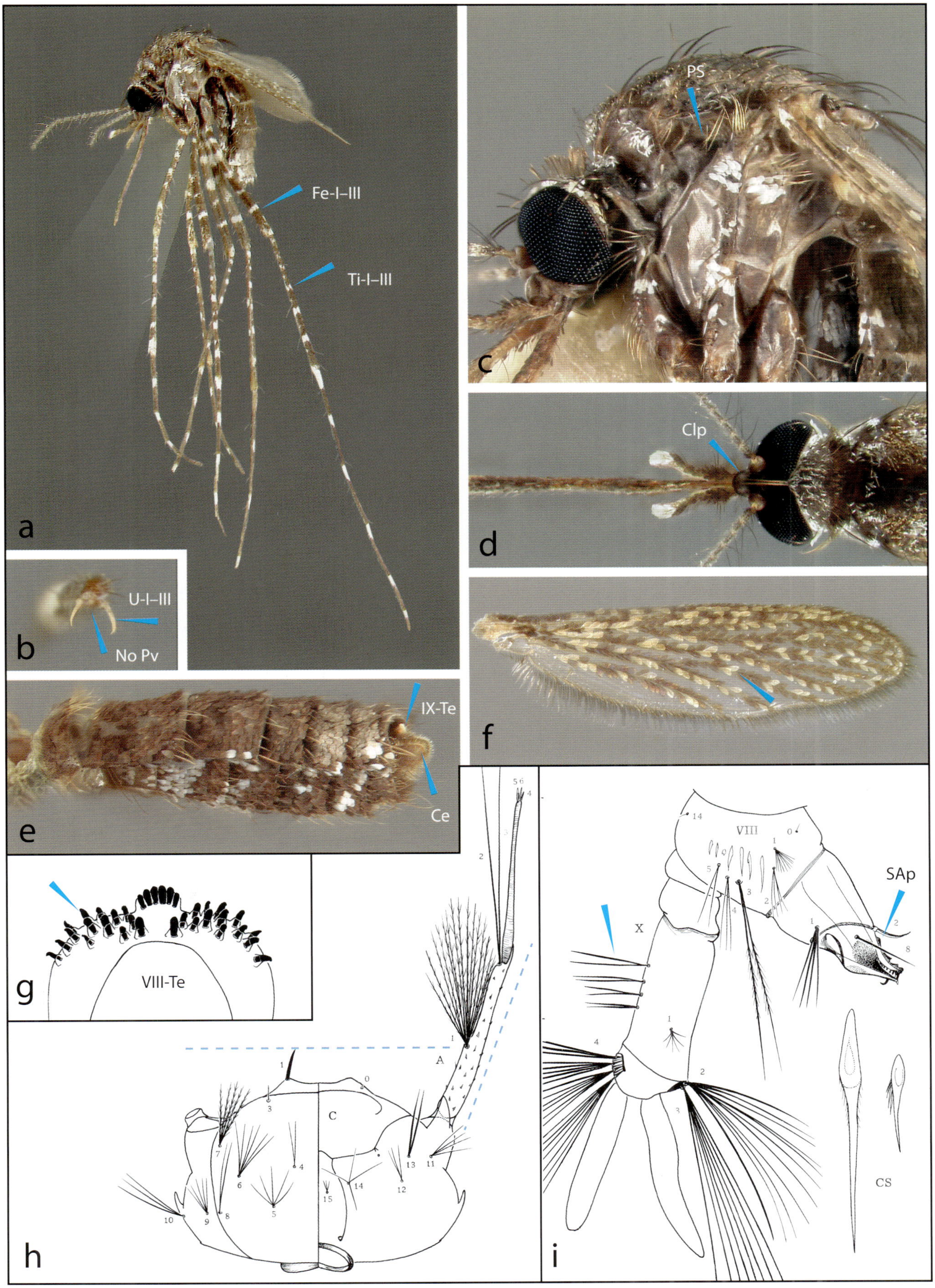

Fe-I–III
Ti-I–III
a
U-I–III
No Pv
b
PS
c
Clp
d
IX-Te
Ce
e
f
VIII-Te
g
A
C
h
VIII
SAp
X
CS
i

Genus *Orthopodomyia* Theobald, 1904 Subfamily Culicinae, Tribe Orthopodomyiini "Red Mosquito"

Systematics and Distribution

Adult *Orthopodomyia* are somewhat reddish in color and strikingly marked with bands and lines of silvery-white scales (see *b*, *c*). They are differentiated from all other mosquito genera by the curious character of an incredibly short fourth tarsal segment ($Ta_4 < Ta_5$; see *e*), with Ta_1 longer than Ta_{2-5} combined (see *d*) on legs I and II. Members of the genus *Orthopodomyia* comprise the tribe Orthopodomyiini. No formal subgenera have been adopted for the genus, but the 36 species are subdivided into eight informal groups (see Vol. 2, p. 1097, for internal relationships). *Orthopodomyia albipes* Leicester is the type species of the genus. *Orthopodomyia* species (n = 36) are primarily distributed in the Oriental and Neotropical Regions, although *Or. alba* Baker, *Or. kummi* Edwards, and *Or. signifera* (Coquillett) extend up into the Nearctic. In the Afrotropical Region, only three occur on the mainland, while nine species are endemic to Madagascar. *Orthopodomyia pulcripalpis* (Rondani) is the only species to occur in the Palearctic Region.

Bionomics

Orthopodomyia species are closely associated with forests or woods. Immature stages are found primarily in tree holes but also in rainwater collections in bamboo and *Heliconia* flower spathes, as well as numerous artificial container habitats, including tires. In Illinois, United States, *Orthopodomyia alba* and *Or. signifera* were extremely common in residential areas, and immatures were most commonly ound in deep rot holes in trees that had extended to the heartwood and were filled from the trees' vascular systems. *Orthopodomyia signifera* lays its eggs on, or slightly above, the waterline in these tree cavities. Although birds appear to be the primary hosts of most *Orthopodomyia* species, the Oriental species *Or. albipes* and *Or. andamanensis* Barraud are known to bite humans in the forest during the night. In northern areas of North America, the New World *Or. alba* overwinters as larvae, and there are reports of second and third instar larvae surviving freezing in solid ice for a week. In the warmer climes of Texas, United States, the species overwinters as adults.

Associated Pathogens

None.

Diagnostic Characters

(a) Habitus (♀). **(b) Thorax (dorsal):** Strikingly marked, usually with stripes of white, silver, or gold scales, some species with distinctive mottling. **(c) Thorax (side):** Postspiracular setae (PS) absent. **(d) Leg:** Ta-I, II_1 longer than combined Ta-I, II_{2-5}. **(e) Leg:** Ta-I, II_4 shorter than Ta-I, II_5. **(f) Larva (head):** Seta 1-A inserted on basal half of antenna (A); setae 4-7-C well developed, multibranched. **(g) Larva (terminal segments):** Segment VIII with a distinct dorsal plate (except *Or. alba*); comb plate absen; comb scales (CS) usually in two rows, with posterior row scales (CSP) usually markedly longer than anterior (CSA); siphon (S) with a single pair of seta 1-S; saddle (Sa) usually complete; seta 4-X with ≥6 pairs of setae; pecten (Pt) absent. Image sources: WRBU; Zavortink (1971).

Exemplar DNA barcodes: *Orthopodomyia alba* (GU908120), *Or. anopheloides* (AY917200, LC054512–16), *Or. fascipes* (MF172345–46), *Or. pulcripalpis* (KY608735).

Catalog entry: Vol. 2, p. 977.

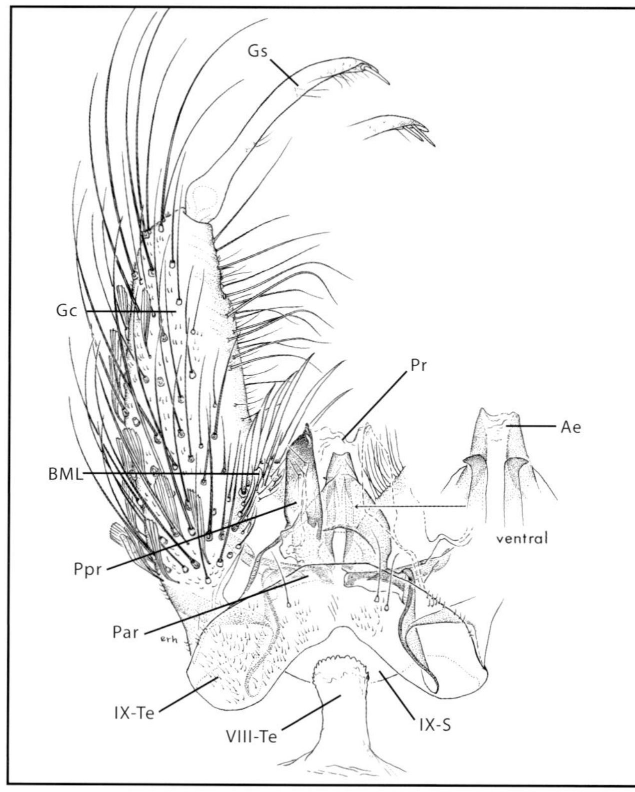

***Orthopodomyia albipes* Leicester (♂ genitalia)**

Ae: Aedeagus; BML: Basal Mesal Lobe; Gc: Gonocoxite; Gs: Gonostylus; Par: Paramere; Ppr: Paraproct; Pr: Proctiger; VIII-Te: Tergum VIII; IX-Te: Tergum IX; IX-S: Sternum IX. (Modified after Zavortink, 1971).

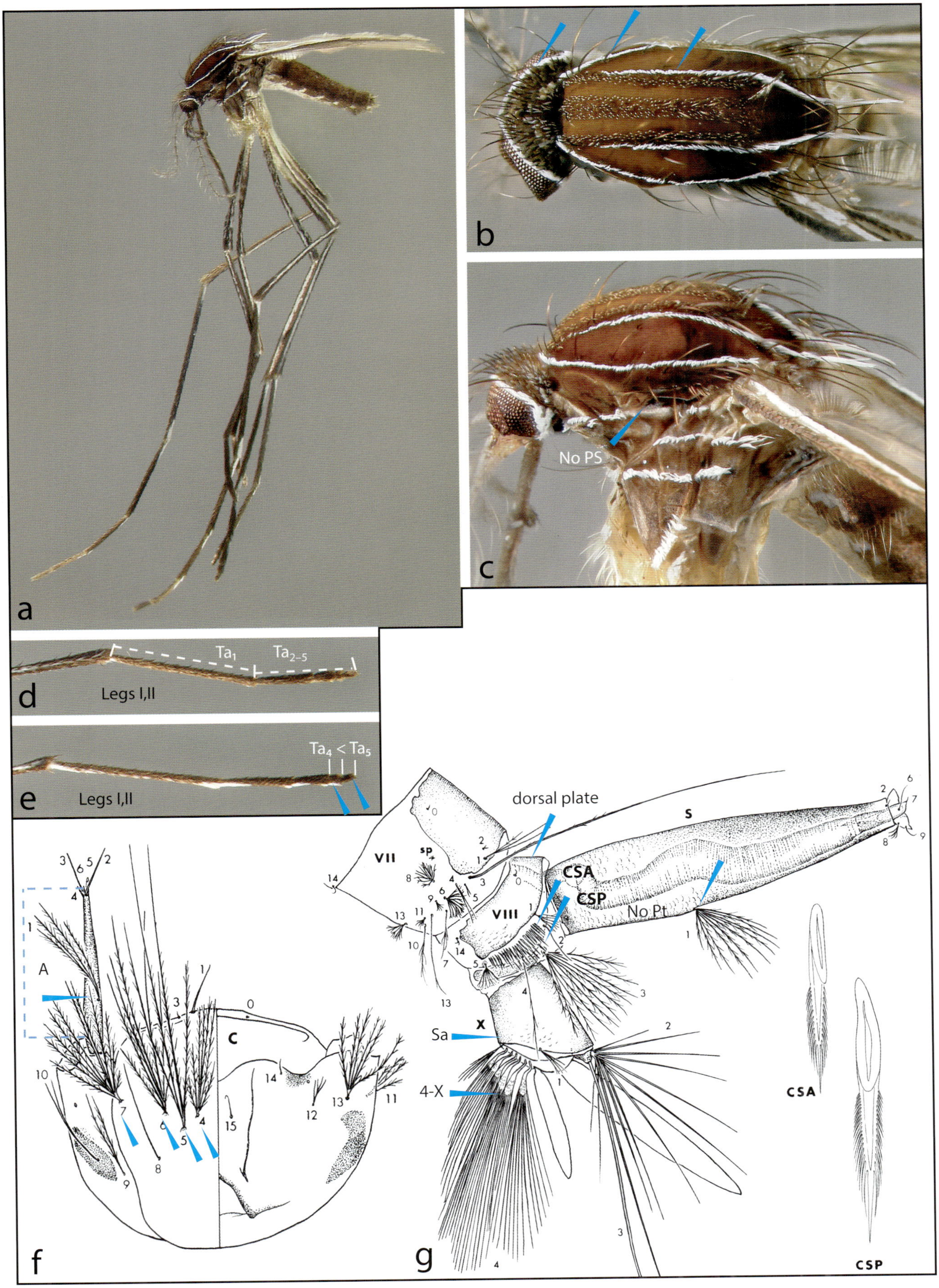
a
b
c
No PS
d
Ta1
Ta2–5
Legs I,II
e
Ta4 < Ta5
Legs I,II
f
A
C
g
dorsal plate
S
VII
VIII
X
sp
CSA
CSP
No Pt
Sa
4-X
CSA
CSP

Genus *Isostomyia* Coquillett, 1906
Subfamily Culicinae, Tribe Sabethini
"Grasping Mosquito"

Systematics and Distribution

Compared with the colorful plumage donning other sabethine genera, species comprising the genus *Isostomyia* are somewhat dull except for their muted greenish-blue dorsal head scales (see *b*). *Isostomyia* comprises only four valid species: *Isostomyia espini* (Martini), *Is. lunata* (Theobald), *Is. paranensis* (Brèthes), and *Is. perturbans* (Williston). *Isostomyia perturbans* (original combination: *Aedes perturbans* Williston) serves as the type species for the genus. All four species were pulled from the genus *Trichoprosopon* and placed as a subgenus of *Runchomyia* by Zavortink in 1979 before being elevated to generic status by Harbach and Peyton in 1993. Prior phylogenetic treatments suggest close relationships between *Isostomyia*, *Shannonia*, and *Runchomyia* (*Ctenogoeldia*) based on morphological characters. In *Is. paranensis*, seta 13-T sits on a common plate with 9–12-T, an unusual character only otherwise present in sabethine species belonging to the genera *Kimia* and *Trichoprosopon*. All *Isostomyia* species are Neotropical.

Bionomics

Immature *Is. espini* occupy aroid leaf axils in forest and agricultural areas, whereas *Is. paranensis* is found year-round in marshy wetlands near Buenos Aires, Argentina, that are dominated by the large rhizomatous plant *Scirpus giganteus*, which grows up to 1.5 m high. Interestingly, fourth instar larvae of *Is. paranensis* killed each other in the laboratory, but only algae were recovered from the digestive tracts of wild-caught larva, suggesting that this predatory behavior might have been stress induced. Predatory larvae commonly accrue enough nutritional reserves to sustain oogenesis in adulthood. In all *Isostomyia* larvae, the elements of the maxillary brush have fused to form a strong claw (see *e*), strongly indicative of predacious activity in this life stage. This supposition is further supported by a yearlong study of seasonal adult age structure of mosquitoes in Punta Lara and La Plata in the Argentinian province of Buenos Aires. In this study, adult females were dissected and age-grouped according to ovarian tracheation and ovariolar stages. All field-collected *Is. paranensis* females were found to have laid eggs (parous), irrespective of season, suggesting that the species does not require a blood meal (autogenous). Later studies showed that although *Is. paranensis* females readily fed on humans outdoors, lab-reared females were autogenous, with egg batches similar in size in both blood-induced and autogenous oogenesis. *Isostomyia espini* females are attracted to Centers for Disease Control and Prevention (CDC) light traps and will readily feed on humans. Adults have been collected outdoors, and females are known to enter houses to bite, especially in rural agricultural villages close to aroid (*Philodendron*) plantations.

Associated Pathogens

None.

Diagnostic Characters

(a) Habitus (♀): Fe-I shorter than proboscis (P). **(b) Thorax (dorsal):** Dorsal head scales with muted green to blue reflective colors; scutum (Scu) with broad, flat scales. **(c) Thorax (side):** Posterior postpronotal setae (PpS) absent; dorsal margin of hind coxa (C-III) about level with base of mesepimeron (Mm). **(d) Legs:** Ta-I–III$_{1-5}$ all dark. **(e) Larva (maxilla):** Maxillary brush (MxB) elements coalesced into a claw. **(f) Larva (head).** **(g) Larva (terminal segments):** Seta 6-S strongly developed, recurved at tip; pecten scales (PS) fine and hair-like; seta 3-X branched. Image sources: Harbach & Peyton (1993); Campos & Zavortink (2010).

Exemplar DNA sequences: None.

Catalog entry: Vol. 2, p. 981.

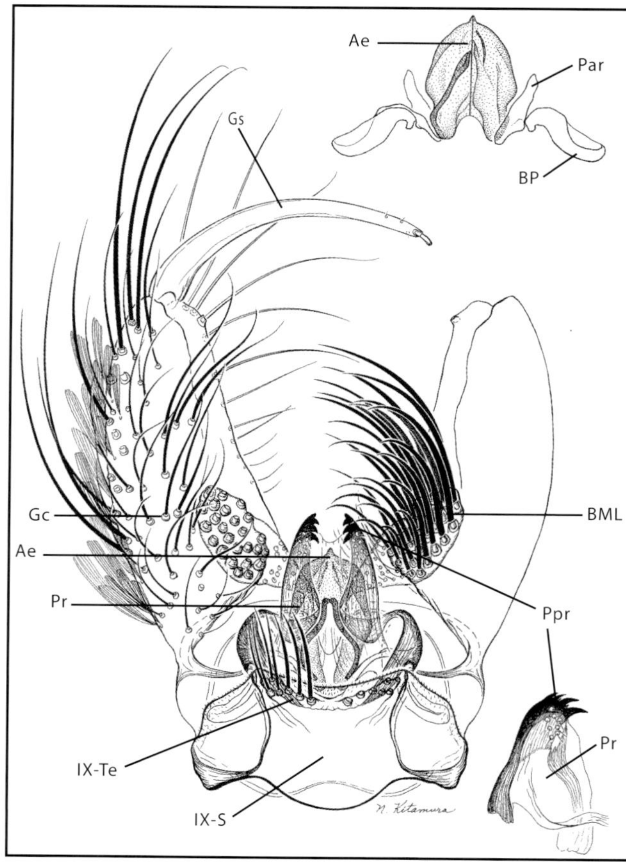

***Isostomyia paranensis* (Brèthes) (♂ genitalia)**

Ae: Aedeagus; BML: Basal Mesal Lobe; BP: Basal Piece; Gc: Gonocoxite; Gs: Gonostylus; Par: Paramere; Ppr: Paraproct; Pr: Proctiger; IX-Te: Tergum IX; IX-S: Sternum IX. (Modified after Campos and Zavortink, 2010).

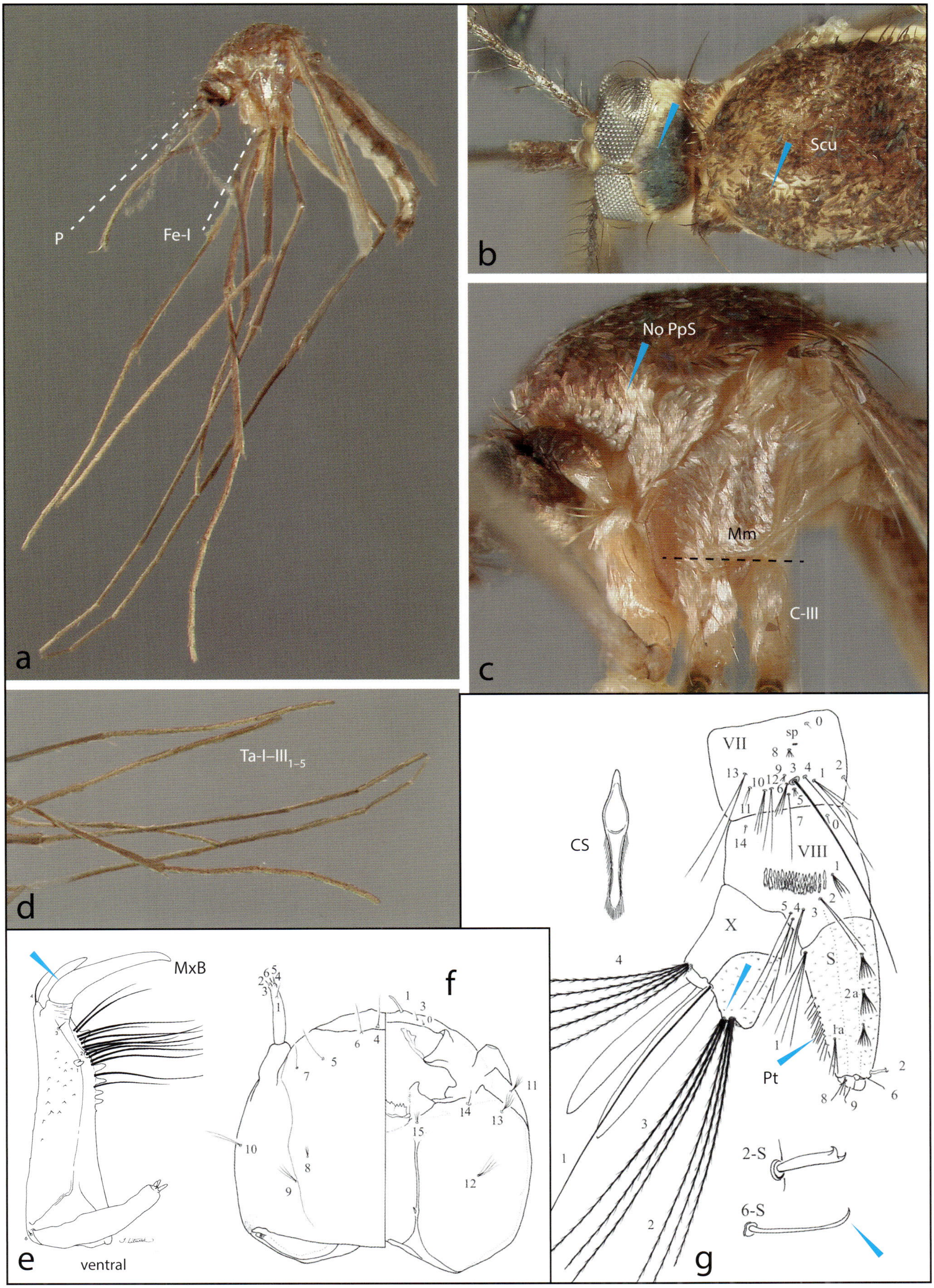
a
P
Fe-I
b
Scu
c
No PpS
Mm
C-III
d
$Ta\text{-}I\text{–}III_{1\text{–}5}$
e
MxB
ventral
f
g
CS
VII
sp
VIII
X
S
Pt
2-S
6-S

Genus *Johnbelkinia* Zavortink, 1979
Subfamily Culicinae, Tribe Sabethini
"Brushy Mosquito"

Systematics and Distribution

Adults of the sabethine genus *Johnbelkinia* sport brightly iridescent silver or blue scales on the vertex and occiput (see *c*), and their proboscis is distinctly longer than their forefemur (Fe-I). *Johnbelkinia* comprises only three species: *Johnbelkinia leucopus* (Dyar & Knab), *Jb. longipes* (Fabricius), and *Jb. ulopus* (Dyar & Knab). The genus is distributed in the Neotropics, with *Jb. leucopus* reported from Brazil, Costa Rica, Ecuador, Nicaragua, and Panama, and *Jb. longipes* in Brazil, Bolivia, Colombia, Costa Rica, Ecuador, French Guiana, Guatemala, Guyana, Honduras, Mexico, Nicaragua, Panama, Paraguay, Peru, Suriname, Trinidad and Tobago, and Venezuela. *Johnbelkinia longipes* (original combination: *Culex longipes* Fabricius), the type species of the genus, is found in Bolivia, Brazil, Colombia, Costa Rica, Ecuador, French Guiana, Guatemala, Guyana, Nicaragua, Panama, Paraguay, Peru, Suriname, and Venezuela.

Bionomics

Mosquitoes of the genus *Johnbelkinia* are associated with forest environments. *Johnbelkinia ulopus* was shown to be diurnal in Venezuela and is known to feed on a variety of mammals, including humans. Females oviposit in waters collected in the leaf axils and flower bracts of plants, including species of *Alocasia*, *Colocasia*, *Heliconius*, and *Xanthosoma*. Immatures are also occasionally found in wet tree holes, and in water collected in fallen palm spathes and fruit husks on the ground. In Colombia, *Jn. ulopus* immatures were reported from bamboo stumps and in CDC light traps set in bamboo groves at 848 m above sea level and in the leaf axils of *Xanthosoma* aroids in forest environs as high as 1,751 m above sea level. Larvae are thought to be predacious and have been reported feeding on *Wyeomyia* larvae. Adult *Johnbelkinia* have been collected transporting the eggs of *Dermatobia hominis*, the human botfly.

Associated Pathogens

WYOV.

Diagnostic Characters

(a) Habitus (♀): Ta-II, III_{1-5} partially pale scaled. **(b) Thorax (side):** Scutum (Scu) with broad, flat scales, dull to moderately iridescent; mesokatepisternal (MkSL) setae usually below lower margin of mesepimeron (Mm); dorsal margin of C-III about level with base of mesepimeron (Mm); 1–2 posterior postpronotal setae (PpS) present (indistinct in photo). **(c) Head (dorsal):** Vertex (V) and occipital (Occ) scales with bright silver or blue reflections. **(d) Head/legs:** Proboscis (P) long and slender, 1.2–1.4 times longer than Fe-I. **(e) Larva (head):** Maxilla (Mx) modified for grasping. **(f) Larva (head). (g) Larva (abdomen):** Seta 6-I–III long, multibranched. **(h) Larva (terminal segments):** Comb (C) with comb scales (CS) in large patch; siphon (S) with multibranched accessory setae (1a-S) present midventrally; siphon without pecten. Image sources: WRBU; Zavortink (1979); Harbach & Peyton (1993).

Exemplar DNA barcodes: *Jb. longipes* (MF172322–27), *Jb. ulopus* (MF172328–29).

Catalog entry: Vol. 2, p. 982.

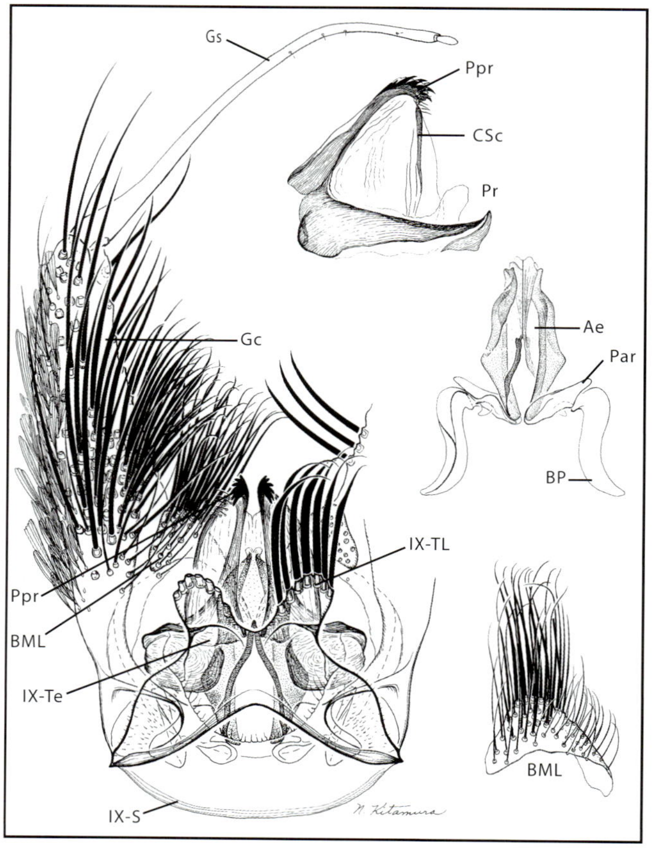

***Johnbelkinia ulopus* (Dyar & Knab) (♂ genitalia)**

Ae: Aedeagus; BML: Basal Mesal Lobe; BP: Basal Piece; CSc: Cercal Sclerite; Gc: Gonocoxite; Gs: Gonostylus; Par: Paramere; Ppr: Paraproct; Pr: Proctiger; IX-Te: Tergum IX; IX-TL: Tergum IX Lobe; IX-S: Sternum IX. (Modified after Zavortink, 1979).

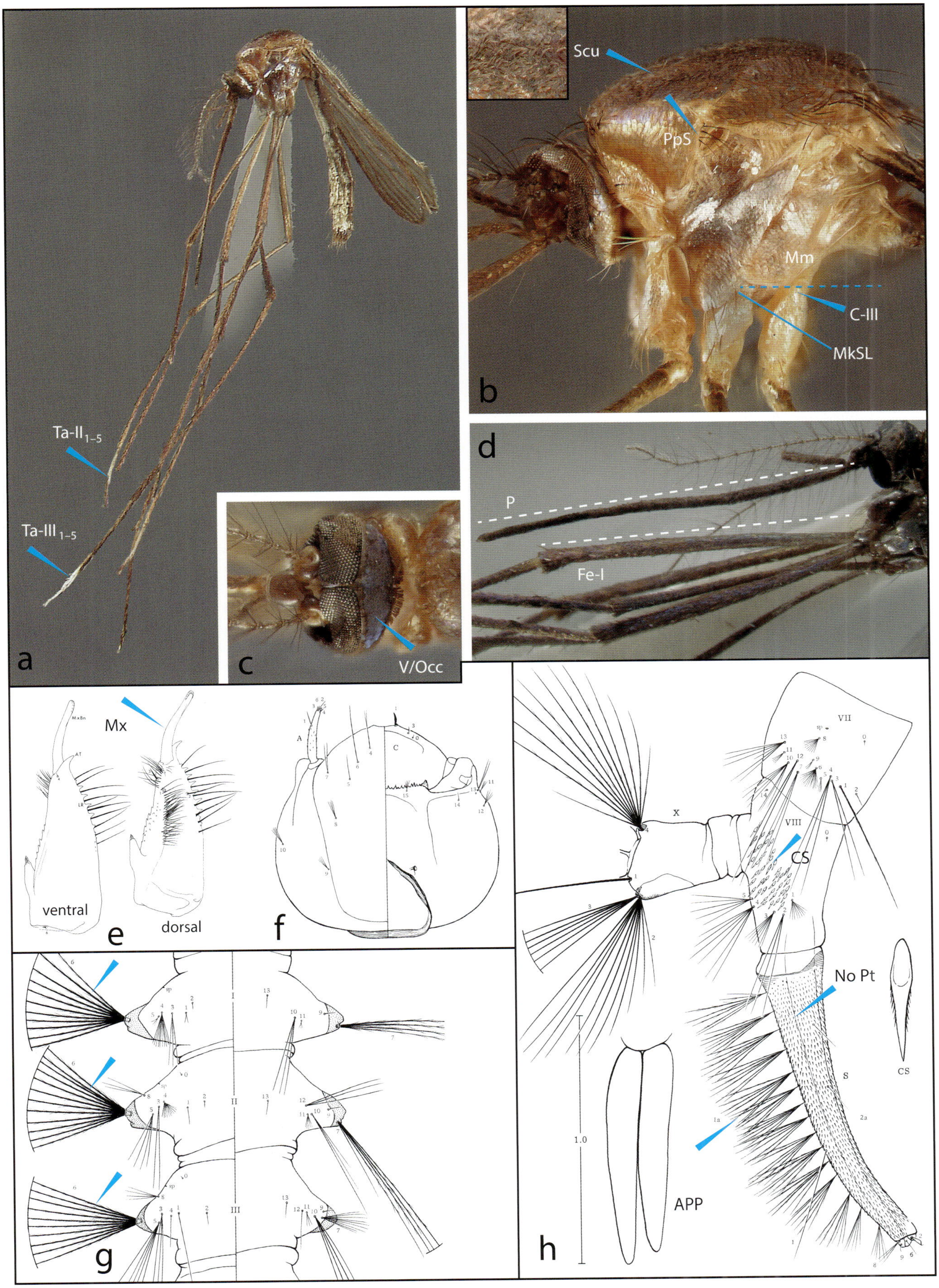
Ta-II1–5
Ta-III 1–5
a
Scu
PpS
Mm
C-III
MkSL
b
V/Occ
c
d
P
Fe-I
Mx
ventral
dorsal
e
f
g
CS
No Pt
APP
h

Genus *Kimia* Vu Duc Huong & Harbach, 2007 Subfamily Culicinae, Tribe Sabethini "Ridged Mosquito"

Systematics and Distribution

Species of the genus *Kimia* bear distinctive broad silver scale patches on the antepronotum, pleuron, and sternum. In common with species of the genera *Malaya* and *Topomyia*, these small mosquitoes share the same distinctive median stripe of silvery-white scales down the center of the scutum (see *c*). The genus *Kimia* comprises only five species: *Km. decorabilis* (Leicester), *Km. imitata* (Baisas, 1946), *Km. miyagii* (Toma & Mogi, 2003), *Km. nemorosa* (Gong), and *Km. suchariti* (Miyagi & Toma). *Kimia decorabilis* (Leicester) serves as the type species of the genus. All five species are restricted to the eastern Oriental Region.

Bionomics

Little is known about the bionomics of any *Kimia* species. Adult *Kimia* have never been collected in the wild and are only known from reared larvae. Immatures are associated with bamboo, being taken from water collection in insect bore holes in standing bamboo internodes and rainwater collections in upright bamboo stumps. The maxillae of *Kimia* larvae are modified for grasping. During the laboratory rearing of the holotype of *Km. suchariti*, the larvae were observed to be predacious, and observed capturing and eating several *Armigeres* and *Tripteroides* larvae. That *Kimia* adults have never been captured suggests that the females are not attracted to light or animal lures and, like species of the closely related genera *Malaya* and *Topomyia*, adult *Kimia* most probably do not blood feed. Given the predacious nature of the larvae, it is likely that enough reserves are acquired at the immature stages to forego the need for extra nutrients for oogenesis.

Associated Pathogens

None.

Diagnostic Characters

(a) Habitus (♀): Ornamentation distinctive, with flat, broad silvery scales. **(b) Thorax (side):** Silvery scales on posterior antepronotum (ApSc), pleuron (p), C-I–III, and abdominal sternum (S). **(c) Thorax (dorsal):** Silvery, flat, broad scales in two rows along midline of scutum (Scu); scutellar lobes with broad, dark scales (not shown). **(d) Abdomen:** Terga (Te) dark scaled; sterna (S) with silvery scales, IX-S without setae and scales (not shown). **(e) Head (dorsal):** Vertex (V) with silvery scales. **(f) Wing:** Upper calypter (UC) lacking setae. **(g) Larva (terminal segments):** Comb (C) without comb plate; comb scales (CS) in a single irregular row; siphon without pecten (Pt), with dorsal and ventral double rows of setae along entire length; seta 4-X multibranched, single. **(h) Larva (head):** Maxillary brush modified into an articulated rigid bundle (MxBn); maxillary apical tooth (AT) stout. **(i) Larva (head).** Image sources: WRBU; Harbach et al. (2007).

Exemplar DNA sequences: None.

Catalog entry: Vol. 2, p. 983.

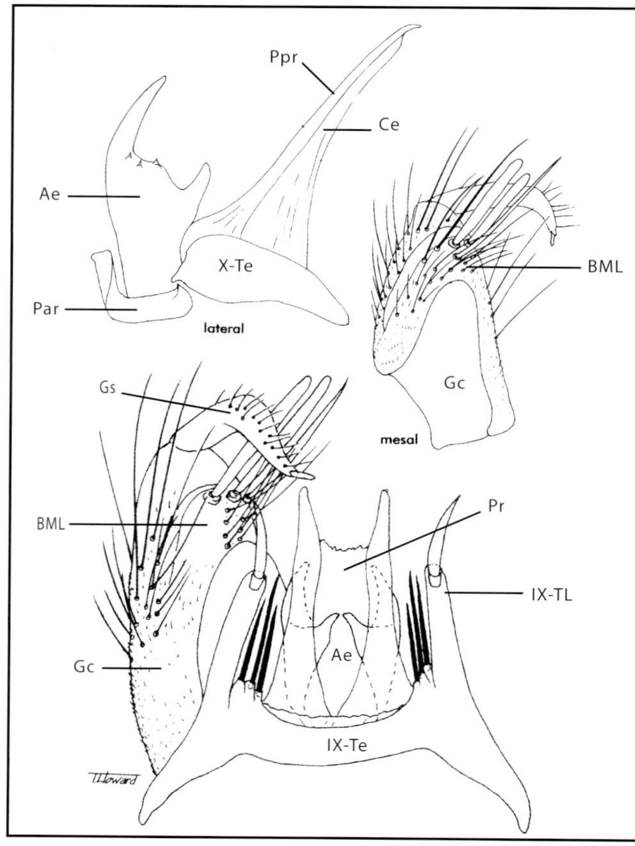

***Kimia decorabilis* (Leicester) (♂ genitalia)**

Ae: Aedeagus; BML: Basal Mesal Lobe; Ce: Cercus; Gc: Gonocoxite; Gs: Gonostylus; Par: Paramere; Ppr: Paraproct; Pr: Proctiger; IX-Te: Tergum IX; IX-TL: Tergum IX Lobe; X-Te: Tergum X. (Modified after Harbach, Huong and Kitching, 2007).

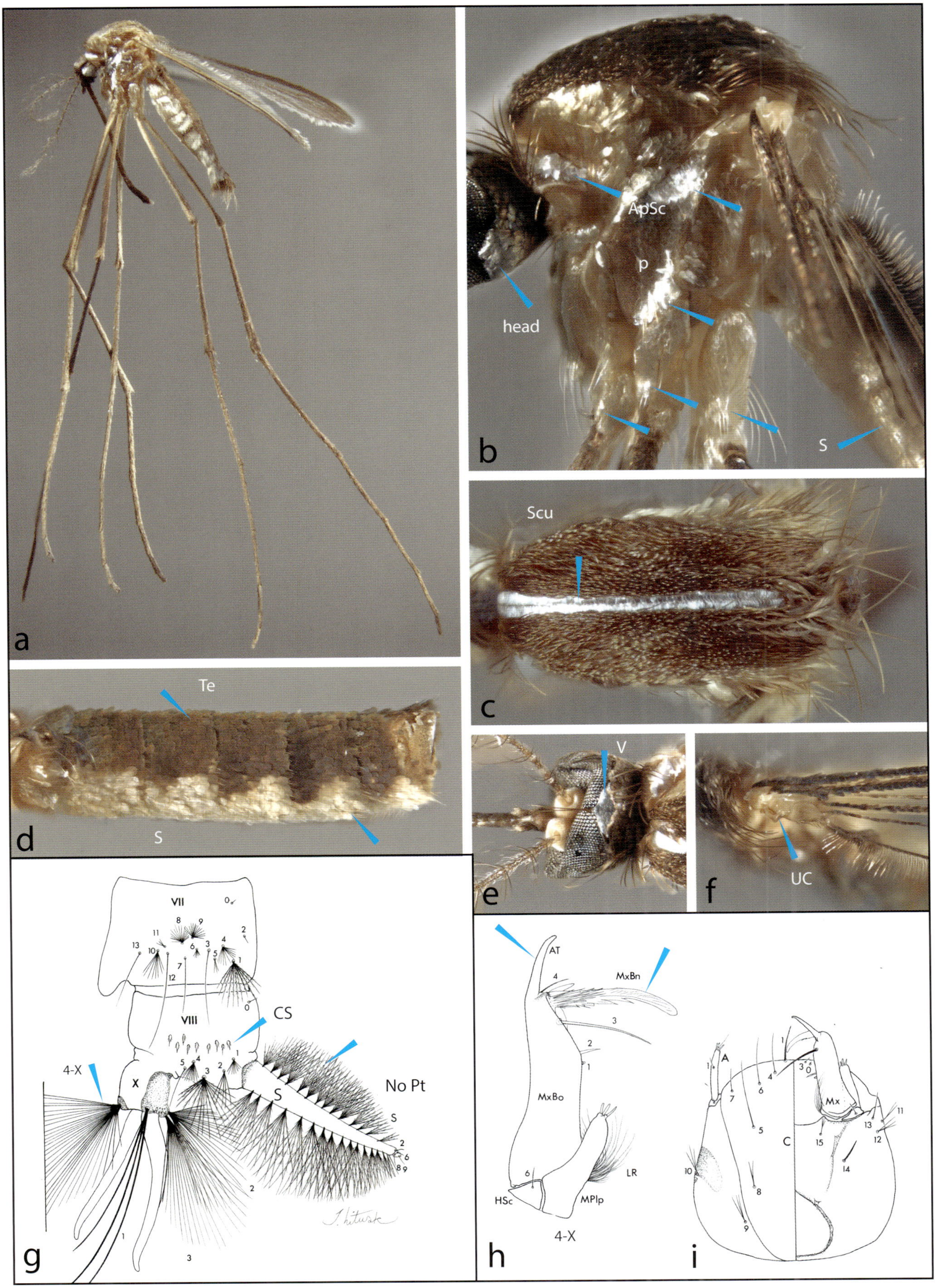
a
ApSc
p
head
S
b
Scu
c
Te
S
d
V
e
UC
f
VII
VIII
CS
4-X
X
S
No Pt
g
AT
MxBn
MxBo
LR
HSc
MPlp
4-X
h
A
C
Mx
i

Genus *Limatus* Theobald, 1901
Subfamily Culicinae, Tribe Sabethini
"Royal Mosquito"

Systematics and Distribution

Unique among mosquitoes, adults of the sabethine genus *Limatus* bear a single unguis on their hind legs (U-III; see *d*). This Neotropical genus comprises nine species as follows: *Limatus andinus* Levi-Castillo, *Li. asulleptus* (Theobald), *Li. durhamii* Theobald, *Li. flavisetosus* de Oliveira Castro, *Li. guayasi* Levi-Castillo, *Li. hoffmani* Root, *Li. martiali* Senevet & Abonnenc, *Li. paraensis* (Theobald) and *Li. pseudomethysticus* (Bonne-Wepster & Bonne). The genus is represented by the type species *Li. durhamii*. *Limatus* species are found on Trinidad and Tobago in the West Indies and from Mexico to Argentina on mainland Central and South America.

Bionomics

Limatus are forest mosquitoes. Larvae have been found in cavities in broken bamboo canes, tree holes, coconut husks, cacao pods, fallen palm spathes, snail shells, rock holes along streams, and a variety of small artificial containers (e.g., tires, buckets). In Argentina, the number of *Limatus* larvae found in each *Megalobulimus* snail shell averaged 11.6 (range 1–27).

Adult *Limatus* are diurnal, feed readily on humans, and are considered nuisance pests in forest regions. Adults of *Li. durhamii* have been captured during landing-biting collections and their eggs recovered from ovitraps at ground level and on towers in the forest in Ecuador. *Limatus asulleptus* and *Li. durhamii* have been reported from sea level to 1,800 m in Costa Rica. *Limatus durhamii* has been collected sharing larval habitats in urban forests with the invasive *Aedes albopictus* (see species page, p. 460) in Rio de Janeiro, Brazil, but the species are inversely correlated in the Brazilian Atlantic Forest. *Limatus durhamii, Li. flavisetosus*, and *Li. pseudomethysticus* are involved or implicated in the transmission of the arboviruses listed below.

Associated Pathogens

CABV, CARV, COTV, GAMV, GJAV, VEEV, WYOV.

Diagnostic Characters

(a) Habitus (♀). (b) Thorax (dorsal): Scutum (Scu) covered with gold, blue, and violet scales. **(c) Thorax (side):** Prespiracular scales (PsSc) present, but no setae (PsS). **(d) Leg:** Unguis (U-III) single. **(e) Larva (head):** Occipital foramen (OF) with short and broad caudolateral slits; setae 1-C closely approximated. **(f) Larva (head):** Maxilla (Mx) without an apical tooth (AT). **(g) Larva (terminal segments).** Image sources: WRBU; Mattingly (1971); Harbach & Peyton (1993).

Exemplar DNA barcodes: *Limatus durhamii* (MF172330–34), *Li. flavisetosus* (KF671027, MF172335–37).

Catalog entry: Vol. 2, p. 984

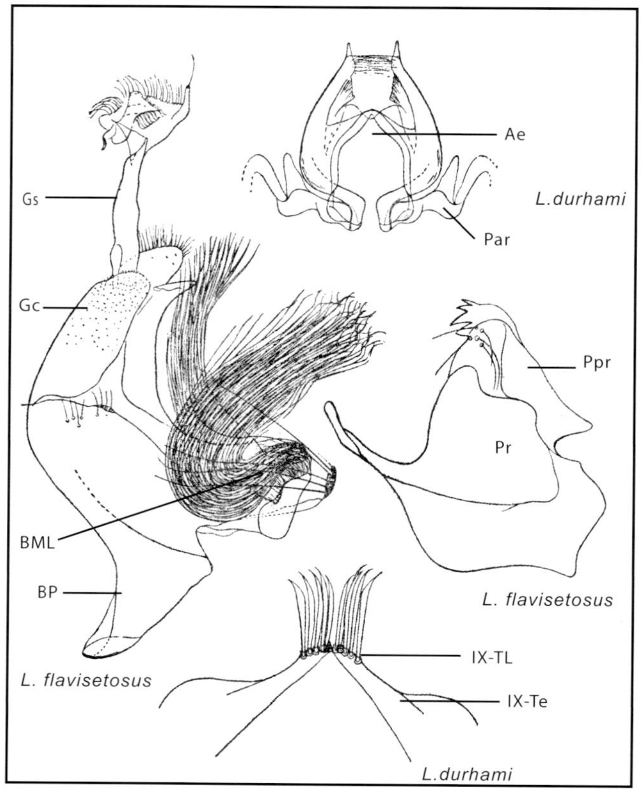

***Limatus durhamii* Theobald and *Limatus flavisetosus* Castro (♂ genitalia)**

Ae: Aedeagus; BML: Basal Mesal Lobe; BP: Basal Piece; Gc: Gonocoxite; Gs: Gonostylus; Par: Paramere; Ppr: Paraproct; Pr: Proctiger; IX-Te: Tergum IX; IX-TL: Tergum IX Lobe. (Modified after Lane and Cerqueira, 1942).

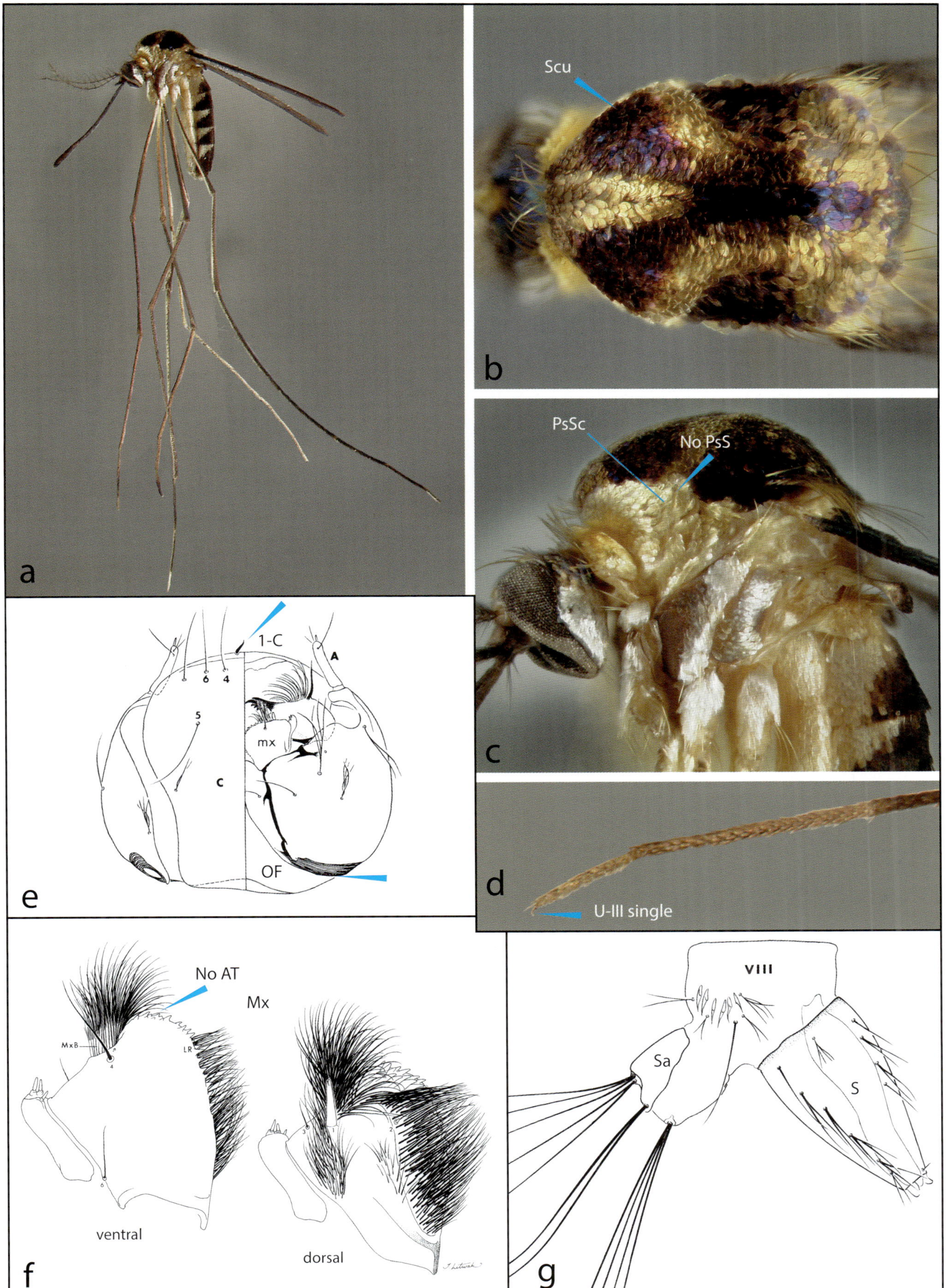
a
Scu
b
PsSc
No PsS
c
d
U-III single
1-C
6
4
A
5
mx
C
OF
e
No AT
Mx
MxB
LR
ventral
dorsal
f
VIII
Sa
S
g

Genus *Malaya* Leicester, 1908
Subfamily Culicinae, Tribe Sabethini
"Ant Mosquito"

Systematics and Distribution

Adults of the genus *Malaya* are small, distinctive mosquitoes. Like *Kimia* and *Topomyia* species, they bear a line of silvery or white scales down the center of the scutum (see *c*) but are distinguished from these and all other genera by their distinctive hairy, clubbed proboscis (see *b*).

The sabethine genus *Malaya* comprises the following 12 species: *Ml. farquharsoni* (Edwards), *Ml. fraseri* (Edwards), *Ml. genurostris* Leicester, *Ml. incomptas* Ramalingam & Pillai, *Ml. jacobsoni* (Edwards), *Ml. leei* (Wharton), *Ml. marceli* (Mattingly), *Ml. moucheti* (Hamon & Adam), *Ml. solomonis* (Wharton), *Ml. splendens* (de Meijere), *Ml. taeniarostris* (Theobald), and *Ml. trichorostris* (Theobald). *Malaya genurostris* is the type species of the genus. All species of the genus *Malaya* occur in the Afrotropical, Oriental, and Australasian Regions.

Bionomics

The feeding behavior of *Malaya* species is unique in the family Culicidae. Females are incapable of piercing and taking blood meals due to the extensive modification of their mouthparts and have evolved to feed entirely on the regurgitation of ants. To feed, the adult *Malaya* seeks and mounts ants in the daytime, drumming on their maxillary palps with its club-like proboscis. This induces the ant to regurgitate a drop of sugary nectar, which is rapidly sucked up by the mosquito before it flies off to find another willing victim. Unlike ant relationships with other insects, this is not a symbiotic association, but the ants apparently are not harmed in the process.

Larvae are commonly found in water-filled plant cavities, including tree holes, pitcher plants, and the leaf axils of domestic and wild banana (*Musa textilis*), *Alocasia*, *Colocasia*, and *Pandanus* and in water collected in arboreal ant nests. Immatures are also found closer to the ground, in the water collected in pineapple or taro leaf axils and, less frequently, in bamboo stumps.

Associated Pathogens

None.

Diagnostic Characters

(a) Habitus (♀). (b) Head (♀): Modified proboscis (P) with many erect setae, upturned and apically swollen. **(c) Thorax (dorsal):** Scutum (Scu) usually with median longitudinal white or silver stripe. **(d) Wing:** Upper calypter (UC) bare, alula fringed (not shown); vein 1A turns and terminates level with base of mediocubital crossvein (mcu). **(e) Larva (thorax):** Seta 13-P absent. **(f) Larva (head):** Maxillary palpus (MPlp) fused with maxillary body (MxBo). **(g) Larva (head):** Occipital foramen (OF) slit-like. **(h) Larva (terminal segments):** Siphon (S) with two or more subdorsal setae and assorted ventral or subventral setae; seta 4-X single; comb (C) with comb scales (CS) fringed and in an irregular patch. Image sources: WRBU; Belkin (1962); Harbach & Peyton (1993).

Exemplar DNA barcodes: *Malaya genurostris* (AY917204, EU259293).

Catalog entry: Vol. 2, p. 986.

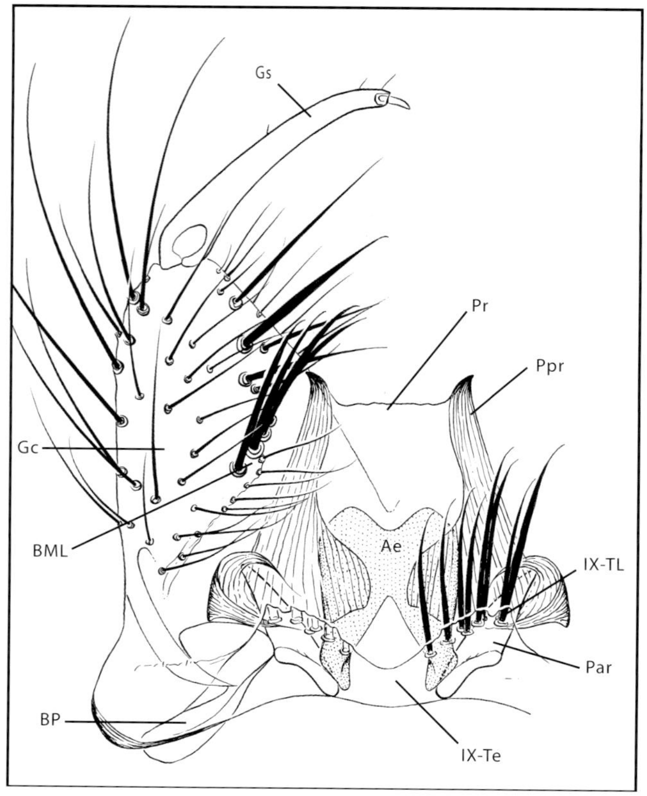

***Malaya solomonis* (Wharton) (♂ genitalia)**

Ae: Aedeagus; BML: Basal Mesal Lobe; BP: Basal Piece; Gc: Gonocoxite; Gs: Gonostylus; Par: Paramere; Ppr: Paraproct; Pr: Proctiger; IX-Te: Tergum IX; IX-TL: Tergum IX Lobe. (Modified after Belkin, 1962).

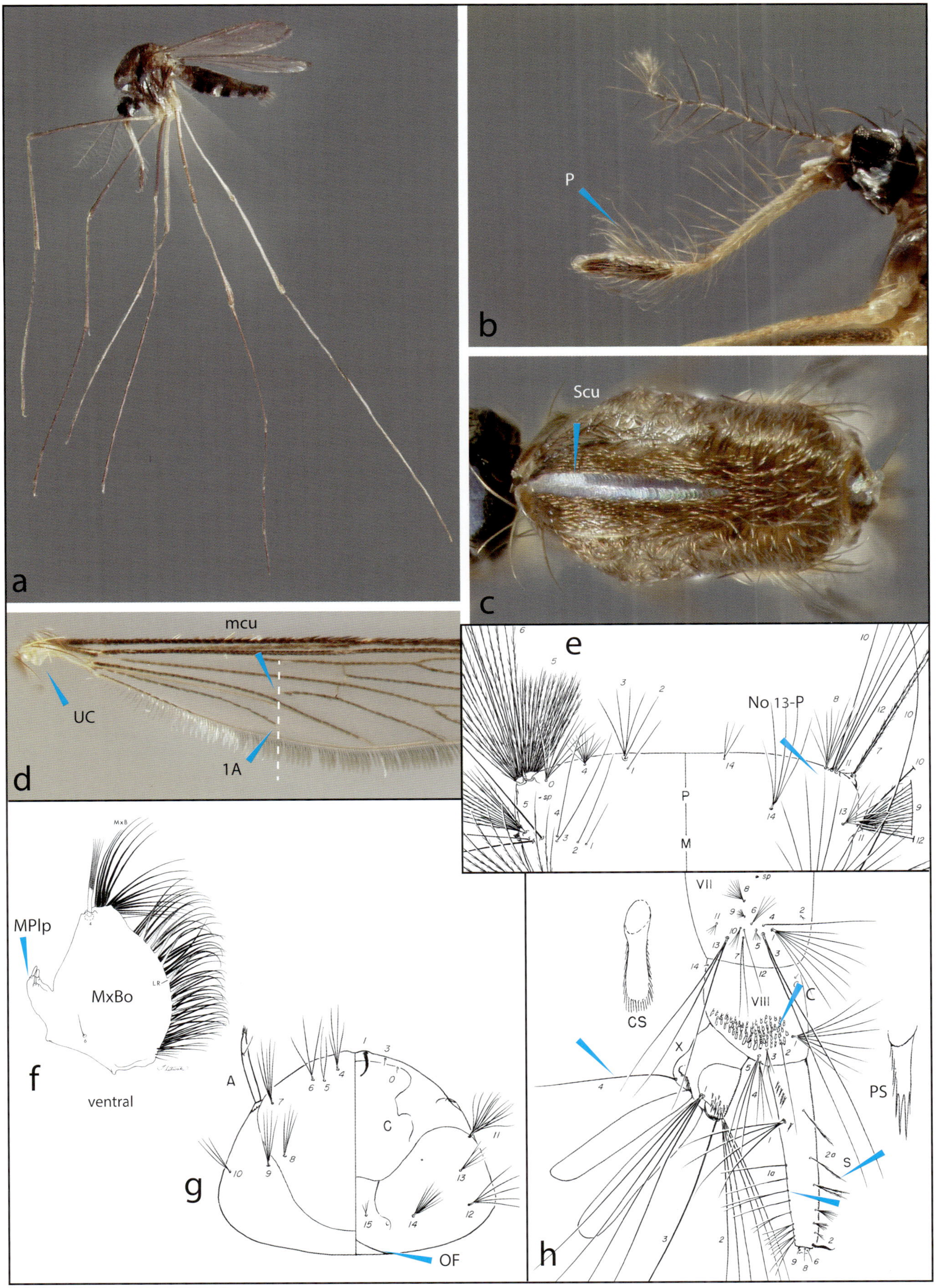
a
P
b
Scu
c
mcu
UC
1A
d
e
No 13-P
P
M
MxB
MPlp
MxBo
LR
f
ventral
A
C
g
OF
VII
VIII
CS
C
X
PS
S
h

Genus *Maorigoeldia* Edwards, 1930
Subfamily Culicinae, Tribe Sabethini
"Maori Mosquito"

Systematics and Distribution

The sabethine genus *Maorigoeldia* was named after the Maori people of New Zealand and Emilio Augusto Goeldi. The genus is monotypic and represented by the species *Maorigoeldia argyropus* (Walker). It was originally described as *Culex argyropus* and is informally named the "Silver-footed New Zealander Maori Mosquito." The adult *Mg. argyropus* is a large and extremely attractive mosquito, with highly distinctive, thick bluish-silver stripes across its head and thorax. It bears unique primitive morphological characters, including well-developed acrostichal setae (see *d*), suggesting it may be a basal lineage within Sabethini. However, its evolutionary relationships with regard to other sabethine taxa are still not well defined.

This arboreal species is endemic to New Zealand and has been collected both on the North and South Islands and on Little Barrier Island. Its current distribution is notably patchy and wholly restricted to remaining large fragments of near-pristine native lowland forests, suggesting that the survival of this species and genus may be under threat.

Bionomics

Immature *Mg. argyropus* is most commonly found in phytotelmata in lowland forests, including tree holes and rot holes of *Nothofagus* species, fallen *Rhopalostylis sapida* palm fronds, and in water collections in stumps of the silver tree fern, *Cyathea dealbata*. The species appears to have also adapted to natural containers and has been collected in rain barrels and discarded tins, jars, and tires. *Maorigoeldia argyropus* has been collected in artificial containers in sympatry with the invasive *Aedes notoscriptus* (Skuse) (see species pages, p. 454). The large *Mg. argyropus* larvae typically rest upside down on the bottom of the rearing site, similar to genus *Tripteroides*, returning to the surface infrequently to respire. In the laboratory, adult *Mg. argyropus* exhibited three cycles of oogenesis with sustenance from sugar water, suggesting that the species is adapted to autogenous egg production. Indeed, *Mg. argyropus* is not attracted to CO_2-baited light traps and has never been reported landing or feeding on humans.

Associated Pathogens

None.

Diagnostic Characters

(a) Head (♀): Erect head scales restricted to occiput (Occ); prespiracular setae (PsS) present. **(b) Genitalia (♀) (lateral):** Three rounded spermathecae. **(c) Head (♂). (d) Thorax (dorsal):** Scutum (Scu) with strong acrostichal (AcS), dorsocentral (DS), prescutellar (PrsS), and supralar (SaS) setae; scutum with pale bluish and silver scales (not shown). **(e) Wing:** Vein 1A terminates beyond base of mediocubital crossvein (mcu); wing upper calypter (UC) with marginal setae. **(f) Larva (head):** Maxilla. **(g) Larva (head):** Occipital foramen (OF) circular with distinct collar (Col): seta 3-C ventral; seta 13-P present (not shown). **(h) Larva (terminal segments):** Comb (C) with comb scales (CS) in a large patch of ≥100; seta 4-X with a single pair of setae. Image sources: WRBU; Belkin (1962); Belkin (1968); Harbach & Peyton (1993).

Exemplar DNA sequences: *Mg. argyropus—arginine kinase* gene (GQ906815), *CAD* gene (GQ906840), *catalase* gene (GQ906864), *endolase* gene (GQ906888), *hunchback* gene (GQ906912), *white* gene (GQ906932).

Catalog entry: Vol. 2, p. 988.

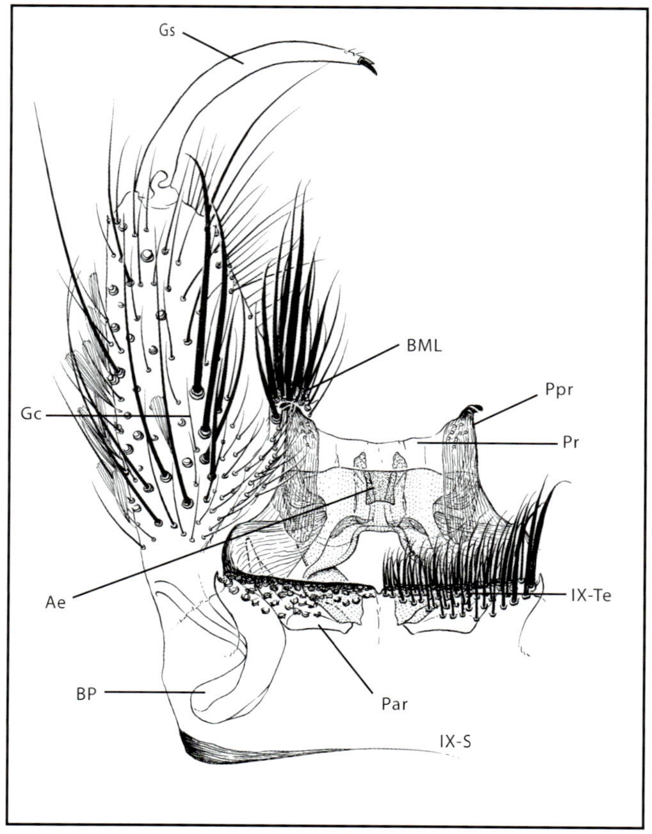

***Maorigoeldia argyropus* (Walker, 1848) (♂ genitalia)**

Ae: Aedeagus; BML: Basal Mesal Lobe; BP: Basal Piece; Gc: Gonocoxite; Gs: Gonostylus; Par: Paramere; Ppr: Paraproct; Pr: Proctiger; IX-Te: Tergum IX; IX-S: Sternum IX. (Modified after Belkin, 1962).

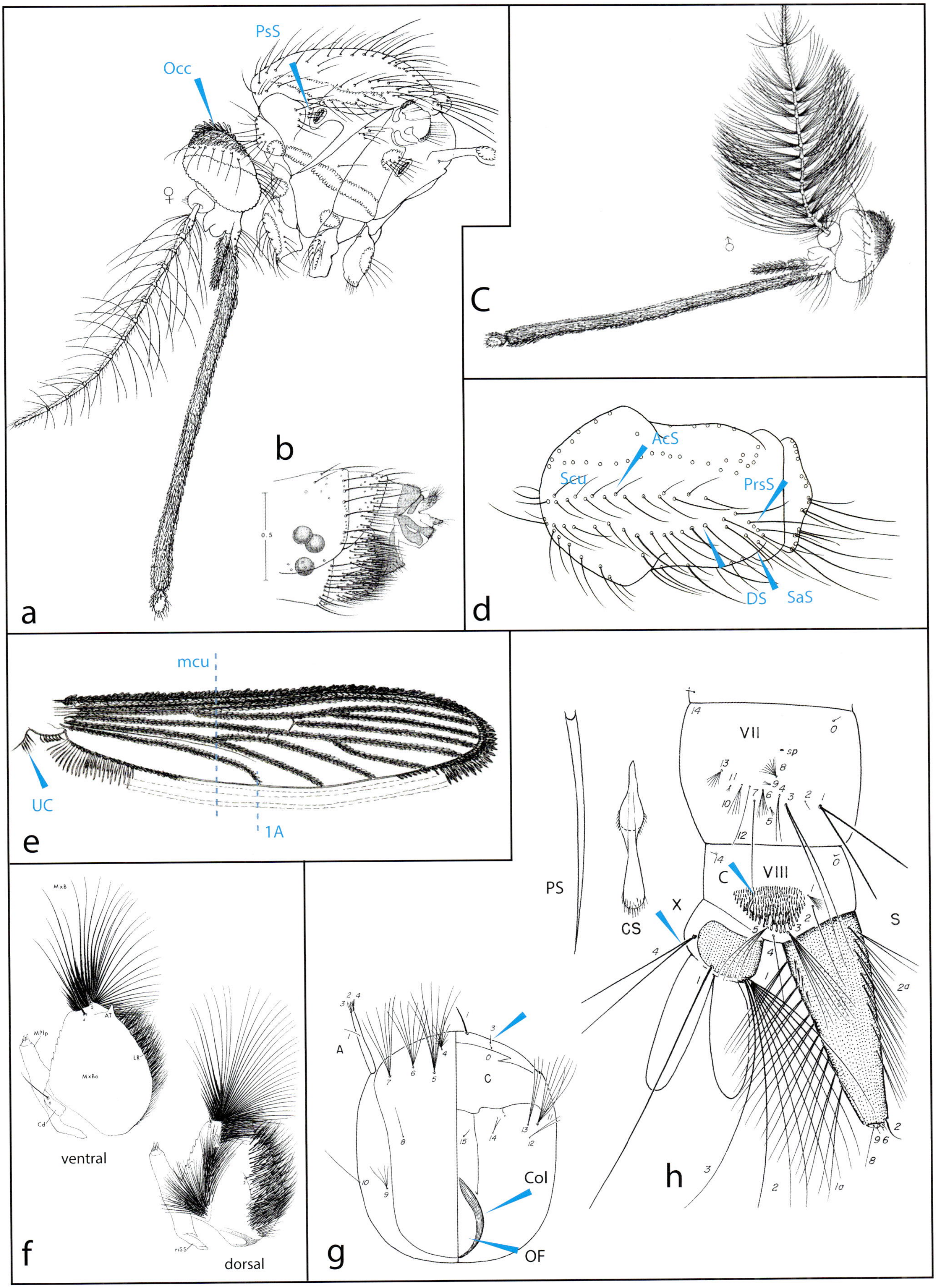
PsS
Occ
a
b
c
d
AcS
Scu
PrsS
DS
SaS
mcu
UC
1A
e
f
ventral
dorsal
g
A
C
Col
OF
h
PS
CS
VII
VIII
X
S

Genus *Onirion* Peyton & Harbach, 2000
Subfamily Culicinae, Tribe Sabethini
"Dream Mosquito"

Systematics and Distribution

Adults of the sabethine genus *Onirion* most closely resemble *Wyeomyia* species. The long setae arising from the male proctiger (see male genitalia illustration), also known as cercal setae, are characteristic of this genus, along with a combination of adult female (see *a–e*) and larval characters (see *f, g*). This Neotropical genus comprises the following seven species: *Onirion aenigma* Harbach, *On. brucei* (del Ponte & Cerqueira), *On. celatum* Peyton & Harbach, *On. imparis* Peyton & Harbach, *On. personatum* (Lutz), *On. regale* Peyton & Harbach, and *On. sirivanakarni* (Duret). *Onirion personatum* is the type species of the genus.

Bionomics

All species of the genus *Onirion* are closely associated with tropical forest habitats. Few studies have captured bionomic information on *Onirion* species, so the biology of many species remains elusive. Adults are diurnal, ovipositing and host seeking during the day. Females of several species are known to feed on humans in the forest. *Onirion personatum* was frequently collected in CDC CO_2 traps and CDC traps baited with both CO_2 and 1-octen-3-ol in the southeastern Atlantic Forest, Brazil, suggesting this species is attracted to large mammals. Immature *Onirion* are largely thought to be associated with water collections in bamboo and ground bromeliads. *Onirion imparis* immatures were collected in bamboo internodes in the Yasuní National Park in eastern Ecuador. Although *Onirion* species have not yet been implicated in arbovirus transmission, this is probably mostly due to insufficient study. *Onirion personatum* has been collected laden with eggs of the human botfly, *Dermatobia hominis*, in São Paulo, Brazil.

Associated Pathogens

None.

Diagnostic Characters

(a) Habitus (♀): Ta-II, III with white scales on at least one tarsomere. **(b) Head (dorsal):** Head scales broad and flat. **(c) Leg. (d) Thorax (side):** Silvery-white scales on lower part of head, antepronotum (Ap), postpronotum (Ppn), pleuron (p), and C-I–III. **(e) Abdomen (side):** Abdominal scales broad and flat; dark scales sometimes with slight blue iridescence; terga (Te) dark; sterna (S) with distinct white scaling in a well-demarcated line. **(f) Larva (thorax):** Setae 4-P and 7-T with distinct aciculae on one side; seta 11-T short with distinct sharp branches. **(g) Larva (terminal segments):** Pecten (Pt) filamentous; numerous setae along length of siphon. Image sources: WRBU; Harbach & Peyton (2000).

Exemplar DNA barcodes: *Onirion* sp. stA (MF172340–44).

Catalog entry: Vol. 2, p. 989.

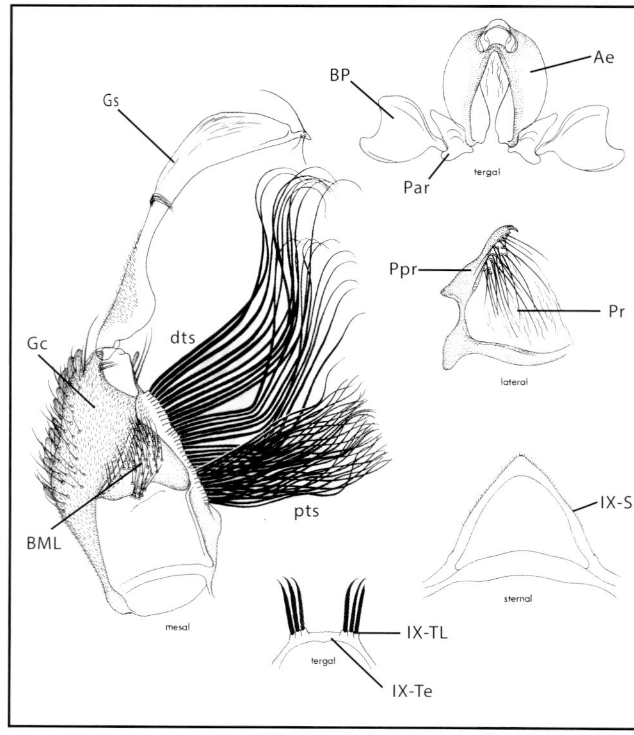

***Onirion personatum* (Lutz) (♂ genitalia)**

Ae: Aedeagus; BML: Basal Mesal Lobe; BP: Basal Piece; dts: Distal Tergal Setae; Gc: Gonocoxite; Gs: Gonostylus; Par: Paramere; Ppr: Paraproct; Pr: Proctiger; pts: Proximal Terminal Setae; IX-Te: Tergum IX; IX-TL: Tergum IX Lobe; IX-S: Sternum IX. (Modified after Harbach and Peyton, 2000).

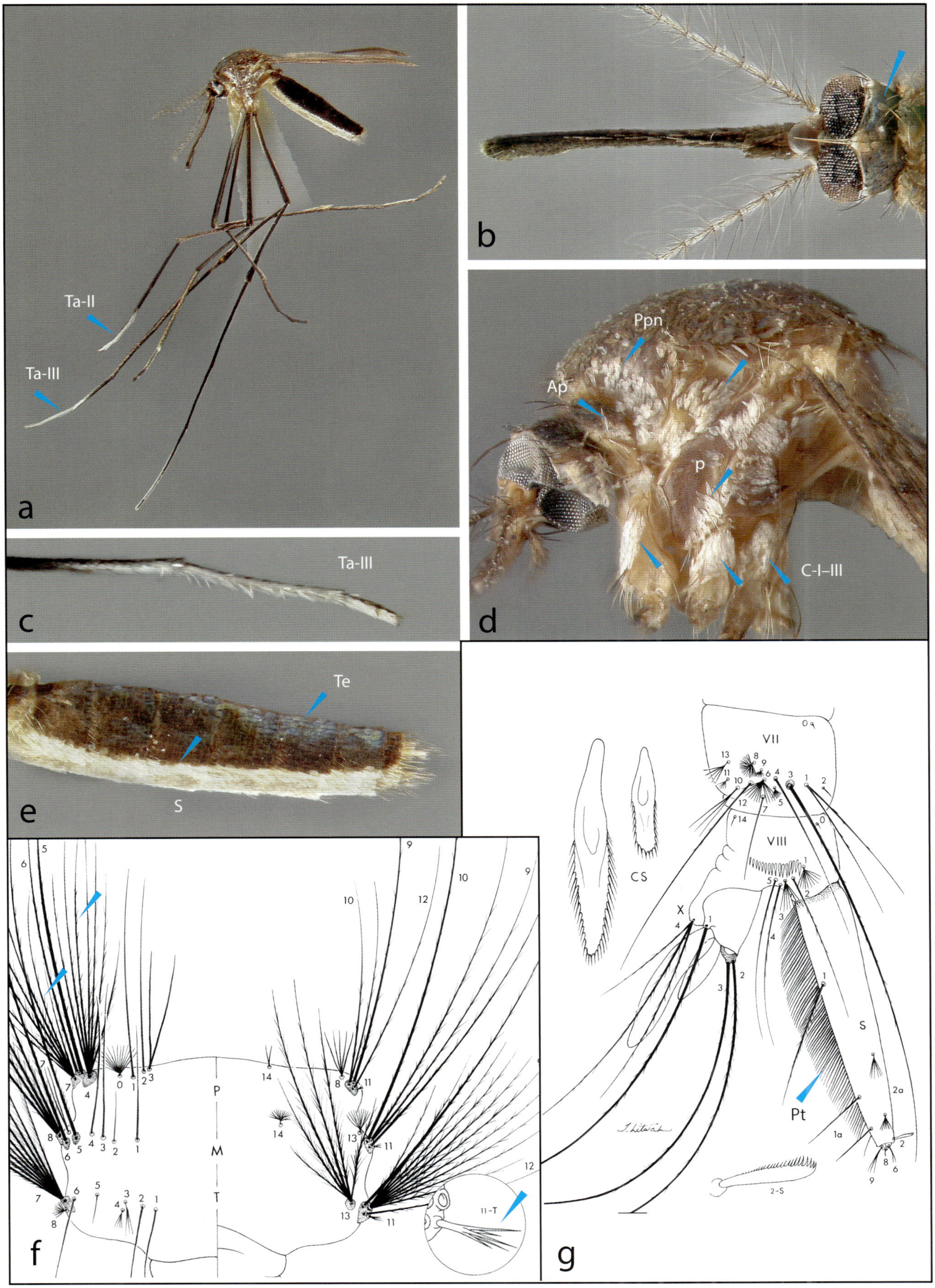

a
Ta-II
Ta-III
b
Ppn
Ap
p
C-I–III
d
c
Ta-III
e
Te
S
f
P
M
T
11-T
g
VII
VIII
CS
X
S
Pt
2-S
1a
2a

Genus *Runchomyia* Theobald, 1903
Subfamily Culicinae, Tribe Sabethini
"Filament Mosquito"

Systematics and Distribution

Species of the Neotropical genus *Runchomyia* are typically attractive sabethines, with dark legs contrasting with their bright blue and silver head scales (see *b*). The genus includes eight species divided into two subgenera: *Ctenogoeldia* (two species—*Runchomyia magna* (Theobald) and *Ru. walcotti* (Lane & Cerqueira)) and *Runchomyia* (six species—*Ru. cerqueirai* (Stone), *Ru. frontosa* Theobald, *Ru. humboldti* (Lane & Cerqueira), *Ru. rapax* (Dyar & Knab), *Ru. reversa* (Lane & Cerqueira), and *Ru. theobaldi* (Lane & Cerqueira)). The type species of the genus—*Ru.* (*Run.*) *frontosa*—is a proven field vector of yellow fever, which can cause paralysis and death in humans. Little is known about the vector capacity of other *Runchomyia* species.

Runchomyia magna occurs widely across the Neotropical regions of Central and South America, whereas *Ru. walcotti* is only recorded from Brazil. Species of the subgenus *Runchomyia* are largely restricted to South America, but *Ru. cerqueirai* and *Ru. frontosa* are also reported from Panama, and *Ru. frontosa*, *Ru. rapax*, and *Ru. theobaldi* are found on Trinidad and Tobago. Five of the six species in the subgenus *Runchomyia* (*Ru. cerqueirai*, *Ru. frontosa*, *Ru. humboldti*, *Ru. reversa*, and *Ru. theobaldi*) are found in sympatry in the Unidade de Conservação Ambiental Desterro forest on Santa Catarina Island, Brazil.

Bionomics

Species of the genus *Runchomyia* are forest associated. *Runchomyia* larvae are typically collected in small phytocontainers, including the flower bracts and leaf axils of *Calathea*, *Heliconia*, and *Xanthosoma* plants and in both arboreal and terrestrial bromeliads. However, *Ru. reversa* immatures occupy the internodes of *Guadua chacoensis* and *Guadua trinii* bamboos, and one *Ru. magna* pupa was found in ground seepage water in open sunlight at the forest edge in Yasuní National Park, Ecuador. Interestingly, *Ru. frontosa* immatures are commonly collected in the juices of the carnivorous bromeliad *Brocchinia reducta*. Like most sabethines, *Runchomyia* species are diurnal, although there are rare reports of some species also being active at night. In Rio de Janeiro, *Ru. frontosa* and *Ru. reversa* were collected in CDC light traps in the Atlantic Forest and Guapiaçu Ecological Reserve, respectively. *Runchomyia frontosa*, *Ru. magna*, and *Ru. theobaldi* are attracted to and feed on humans in forest environments. *Runchomyia magna* females were collected host seeking human bait in the early evening (1700–1800h).

Associated Pathogens

YFV.

Diagnostic Characters

(a) Habitus (♀). (b) Head/leg: Head scales bright blue and silver dorsally; proboscis (P) approximately 1.2–1.4 times length of Fe-I. **(c) Thorax (side):** Lower edge of mesepimeron (Mm) above lower mesokatepisternal setae (MkSL); at least one postpronotal setae (PpS) present. **(d) Legs:** Ta-I–III all dark. **(e) Thorax (dorsal):** Scutal (Scu) scales narrow and curved. **(f) Larva (head):** Maxillary brush modified to form bundle (MxBn). **(g) Larva (head):** Occipital foramen (OF) slit-like. **(h) Larva (terminal segments):** Pecten (Pt) filamentous; seta 3-X single; seta 6-S small, not hooked at tip. Image sources: WRBU; Lane & Cerqueira (1942); Harbach & Peyton (1993).

Exemplar DNA barcodes: *Runchomyia magna* (MF172350–54).

Catalog entry: Vol. 2, p. 990.

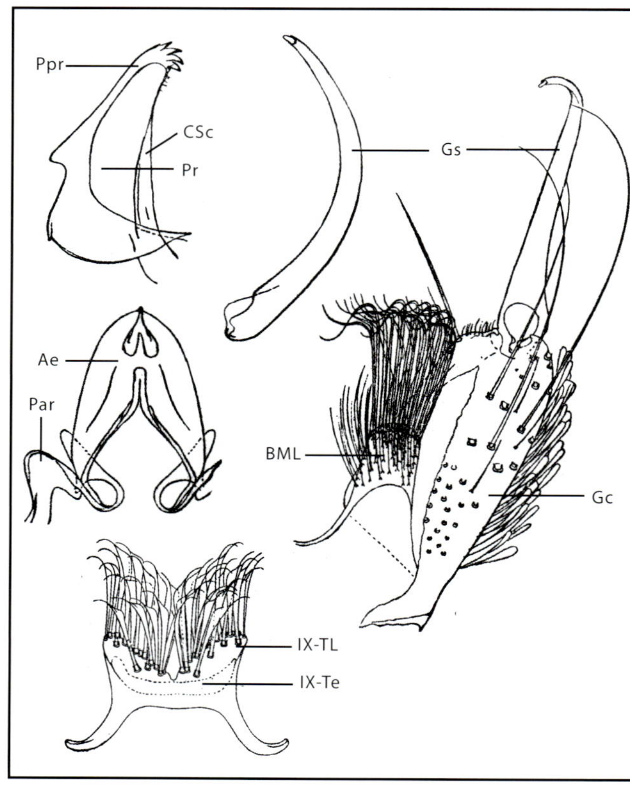

***Runchomyia* (*Runchomyia*) *frontosa* Theobald (♂ genitalia)**

Ae: Aedeagus; BML: Basal Mesal Lobe; CSc: Cercal Sclerite; Gc: Gonocoxite; Gs: Gonostylus; Par: Paramere; Ppr: Paraproct; Pr: Proctiger; IX-Te: Tergum IX; IX-TL: Tergum IX Lobe. (Modified after Lane and Cerqueira, 1942).

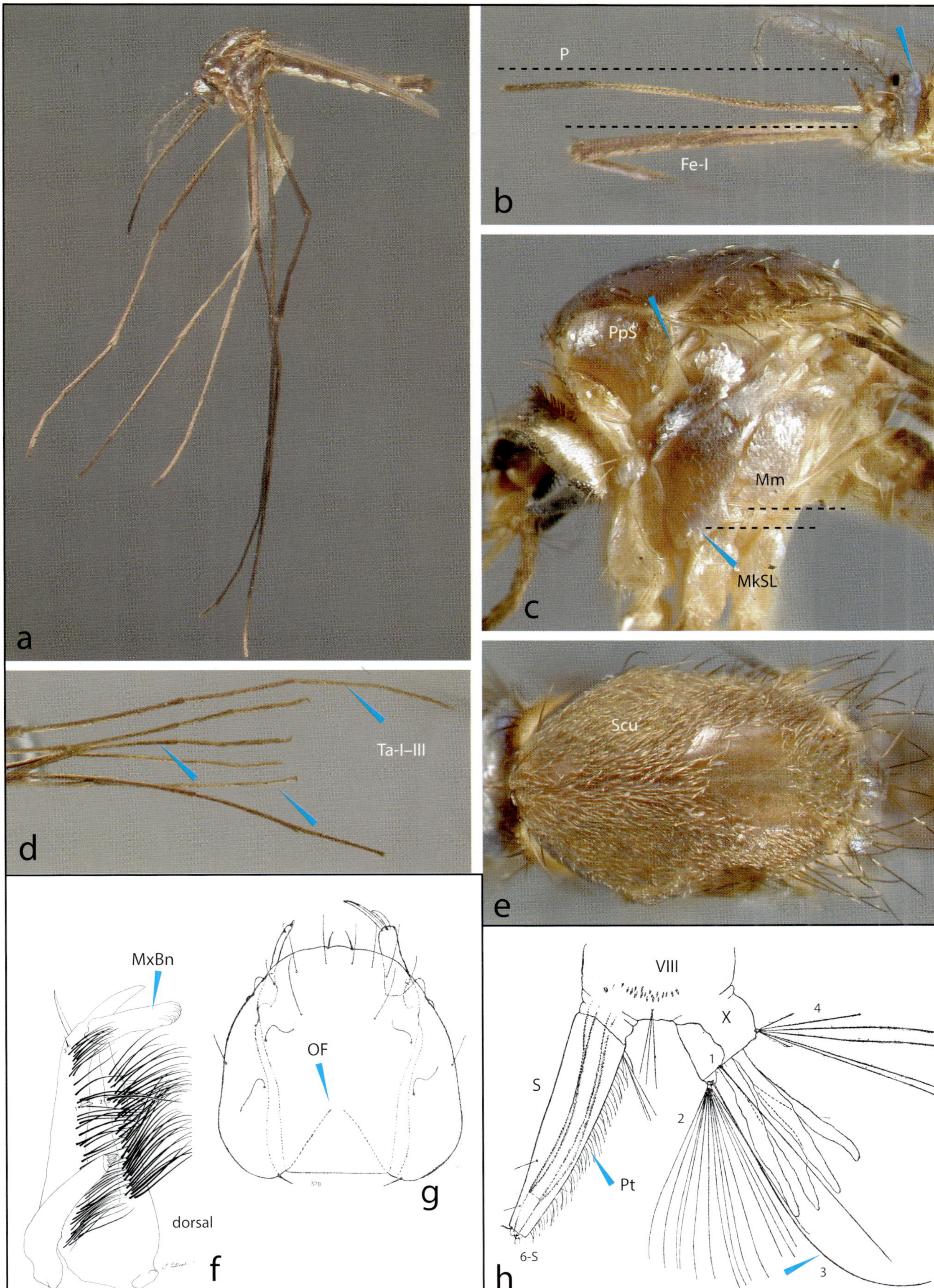
P
Fe-I
PpS
Mm
MkSL
Ta-I–III
Scu
MxBn
dorsal
OF
VIII
X
S
Pt
6-S
a
b
c
d
e
f
g
h
1
2
3
4

Genus *Sabethes* Robineau-Desvoidy, 1827
Subfamily Culicinae, Tribe Sabethini
"Canopy Mosquito"

Systematics and Distribution

Mosquitoes belonging to the genus *Sabethes* are the Broadway glitz of the mosquito world. Densely clothed with large, broad colorful scales on the pleura (see *b*) and bright metallic scales on the scutum (see *d*), *Sabethes* mosquitoes truly are the most spectacular of all mosquitoes. The genus *Sabethes* comprises 41 species, subdivided into five subgenera: *Davismyia* (1 sp.), *Peytonulus* (12 spp.), *Sabethes* (18 spp.), *Sabethinus* (6 spp.), and *Sabethoides* (4 spp.). *Sabethes locuples* Robineau-Desvoidy serves as the type species of the genus. *Sabethes petrocchiae* (Shannon & del Ponte) comprises the monotypic subgenus *Davismyia* and is unique in that it is the only sabethine outside of the genus *Wyeomyia* to bear prealar setae. *Sabethes* mosquitoes are primarily associated with the Neotropical forests of South America, but two new species of *Sabethes* (*Sabethinus*)—*Sa. andreae* Harbach and *Sa. theresae* Harbach—have recently been described from Costa Rica, expanding the known range of the genus into Central America.

Bionomics

Sabethes immatures develop in a wide range of natural containers available in tropical forest environs, including bamboo, leaf axils and aroids, flower bracts, palm and tree stumps, tree holes, and fallen spathes. Some species (e.g., *Sa.* (*Sab.*) *gymnothorax* Harbach & Petersen and *Sa.* (*Sab.*) *bipartipes* Dyar & Knab) are predacious as larvae.

Sabethes females slowly dance their way through heavily shaded forests in the day, seeking suitable hosts, including humans. *Sabethes* (*Dav.*) *petrocchiae* was collected biting humans during the day in gallery forests in Bolivia, and many species of the subgenus *Sabethes* (e.g., *Sa. amazonicus* Gordon & Evans, *Sa. cyaneus* (Fabricius), and *Sa. quasicyaneus* Peryassú) have been captured at human bait, both in the canopy and at ground level in Peru, increasing their potential to vector zoonotic pathogens. Species in the subgenus *Sabethoides* are primarily arboreal, becoming active in the late afternoon, when most mammalian hosts rest. *Sabethes* (*Sbo.*) *chloropterus* (von Humboldt) is a highly competent vector of yellow fever. Some species of *Sabethes*, including *Sa. cyaneus* and *Sa. chloropterus*, partake in elaborate courtship displays (see *Sabethes chloropterus* species, p. 544). The ornamental leg paddles (see *b*) of both sexes of species of the subgenus *Sabethes* appear to serve a role in mate attraction in males. When female leg paddles were removed, males rarely approached to initiate mating, whereas when males with leg paddles removed approached females, courtship behaviors continued unhindered, suggesting female leg paddles are important cues for mate finding in males.

Associated Pathogens

ARUV, CHGV, MCAV, SLEV, XIBV, YFV.

Diagnostic Characters

(a) Habitus (♀): Covered with distinctive multicolored metallic scales; paddles of outstanding scales on some legs (subgenus *Sabethes*). **(b) Thorax (side):** Mesopostnotum (Mpn) well developed; prealar (PaS) setae absent; pleuron (p) densely covered with large scales and a few setae. **(c) Abdomen (side):** Abdomen truncate, with a shaggy appearance on the posterior segments. **(d) Thorax (dorsal):** Antepronotum (Ap) large, nearly meeting dorsally; scutum (Scu) with overlapping, broad, bright metallic colored scales. **(e) Thorax (dorsal):** Setae restricted to the anterior margin and posterior to wing base (latter shown). **(f) Larva (head):** Maxilla (Mx) with prominent apical tooth (AT); maxillary brush (MxB) reduced and made up of individual fibers; maxillary palpus (MPlp) and cardo fused with the body of the maxilla (MxBo). **(g) Legs:** Fe-I, II always longer than Fe-III; Ta-III_{1-5} longer than Fe-III plus Ti-III combined. Image sources: WRBU; Harbach & Petersen (1992).

Exemplar DNA barcodes: *Sabethes belisarioi* (MH248002), *Sa. intermedius* (GU992952–53), *Sa. paradoxus* (MF172362–63), *Sa. soperi* (MF172364–66), *Sa. undosus* (MF172373–77).

Catalog entry: Vol. 2, p. 992.

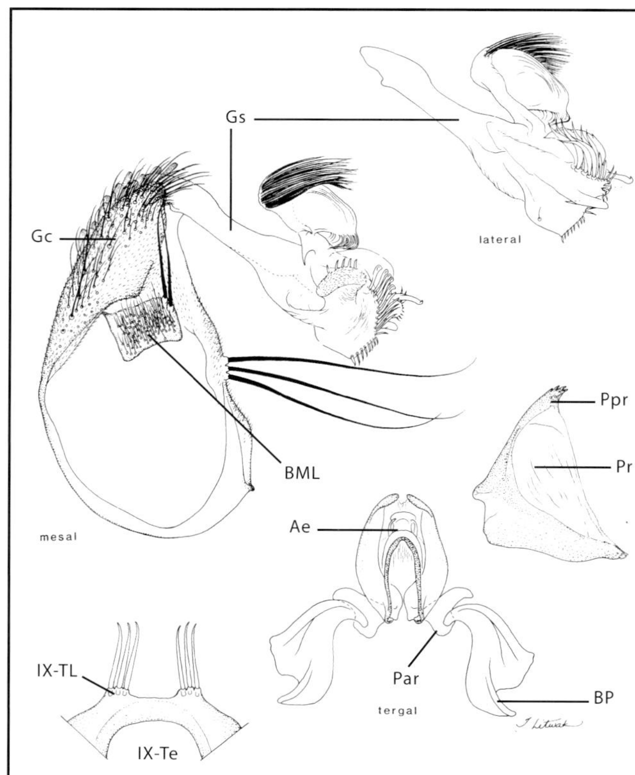

***Sabethes* (*Sabethes*) *tarsopus* Dyar & Knab (♂ genitalia)**

Ae: Aedeagus; BML: Basal Mesal Lobe; BP: Basal Piece; Gc: Gonocoxite; Gs: Gonostylus; Par: Paramere; Ppr: Paraproct; Pr: Proctiger; IX-Te: Tergum IX; IX-TL: Tergum IX Lobe. (Modified after Harbach and Petersen, 1992).

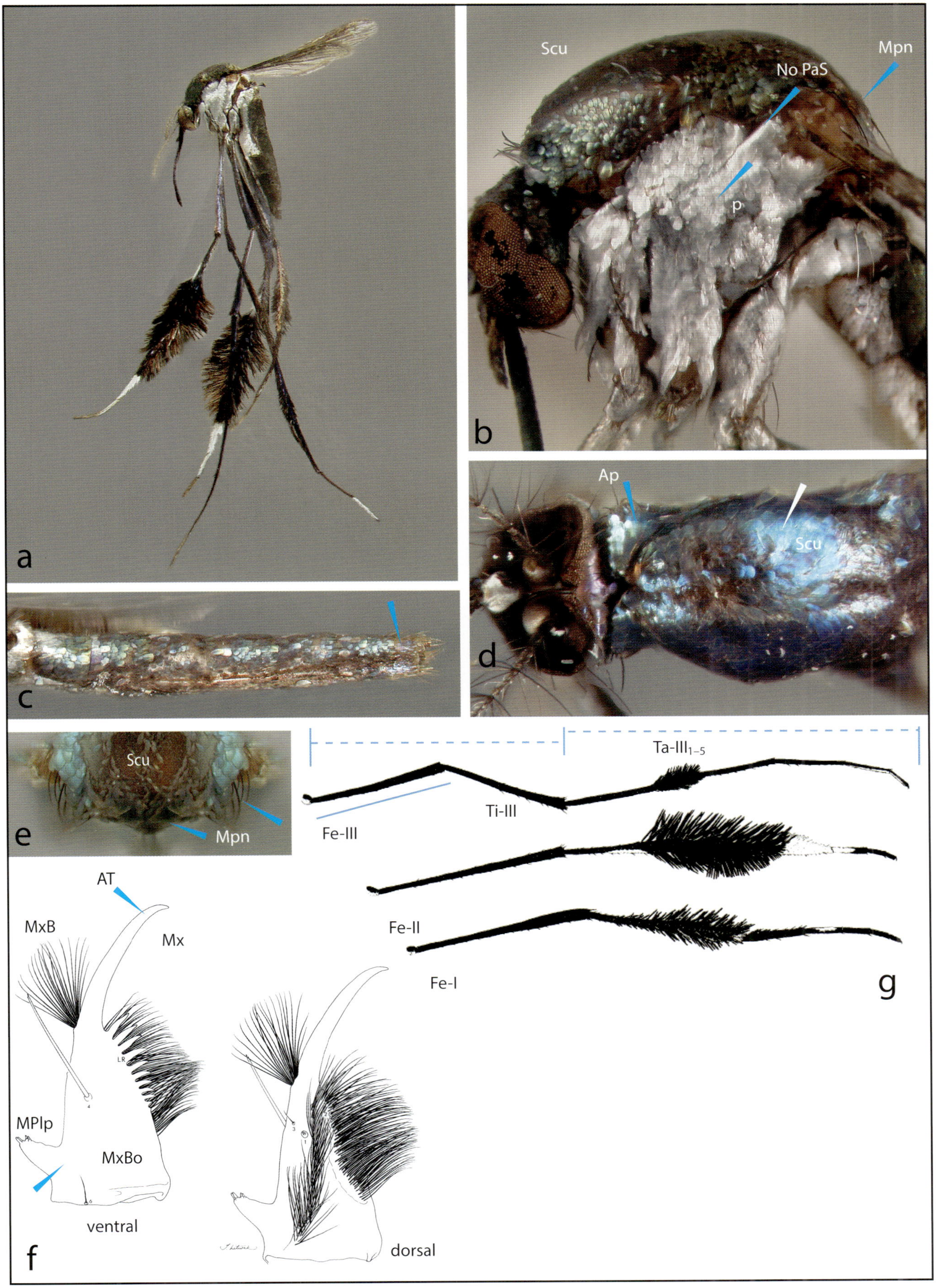
a
Scu
No PaS
Mpn
p
b
Ap
Scu
c
d
Scu
Mpn
e
Ta-III1–5
Fe-III
Ti-III
Fe-II
Fe-I
g
AT
MxB
Mx
MPlp
MxBo
ventral
dorsal
f

Genus *Shannoniana* Lane & Cerqueira, 1942
Subfamily Culicinae, Tribe Sabethini
"Toothy Mosquito"

Systematics and Distribution

The small genus *Shannoniana* comprises only three formally described species—*Shannoniana fluviatilis* (Theobald), *Sh. moralesi* (Dyar & Knab), and *Sh. schedocyclia* (Dyar & Knab)—but other new, as yet undescribed species have been suggested. *Shannoniana fluviatilis* is the type species of the genus. The genus has been informally dubbed the "Toothy Mosquito" in recognition of one of its most distinctive characteristics. *Shannoniana* larvae share the presence of both a prominent tooth-like process of the maxillary body and maxillary claw only with sabethine species of the New World *Isostomyia* and Old World *Tripteroides*. Species of *Shannoniana* are found only in Central and South America (Mexico to Bolivia and Brazil).

Bionomics

Like many other sabethine genera, *Shannoniana* species are forest associated, and are active in the heavily shaded forest environs during the day. *Shannoniana fluviatilis* adults have been collected while biting and landing on horses, while seeking human bait in a national forest in Rio de Janeiro, Brazil, and while using a hand net in a premontane forest in Antioquia, Colombia. In nature, immature *Shannoniana* have been found in water collections in standing, broken bamboo and in the leaf axils of Araceae. In the laboratory, *Shannoniana* larvae have been observed to be predacious and cannibalistic, accounting for their modified larval maxillae. *Shannoniana moralesi* was noted to move in an unusual way as larvae, attaching to vertical and horizontal substrates using its head, then pulling its body forward to continue forward propulsion.

Associated Pathogens

None.

Diagnostic Characters

(a) Habitus (♀). (b) Thorax (side): Lower mesokatepisternal setae (MkSL) extend above the level of edge of mesepimeron (Mm); postprocoxal scales (PSc) present on postprocoxal membrane (PM); scutellum (Stm) without silvery scales. **(c) Head (♀, dorsal):** Palpus (MPlp) ≥0.3 times length of proboscis. **(d) Abdomen (dorsal):** I-Te densely scaled laterally. **(e) Head/legs:** Fe-I about as long as proboscis (P); Ti-III with a broad median pale band. **(f) Larva (terminal segments):** Accessory midventral setae of siphon (S) absent. **(g) Larva (head):** Occipital foramen (OF) circular with a distinct collar (Col). **(h) Larva (head):** Maxilla (Mx) modified for grasping; maxillary brush (MxB) forming a claw; maxilla (Mx) strongly articulated with head, ventrad of the palpus (MPlp). Image sources: WRBU; Harbach & Peyton (1993).

Exemplar DNA barcodes: *Shannonia fluviatilis* (MF172380–84), *Sh. schedocyclia* (MF172378–79).

Catalog entry: Vol. 2, p. 997.

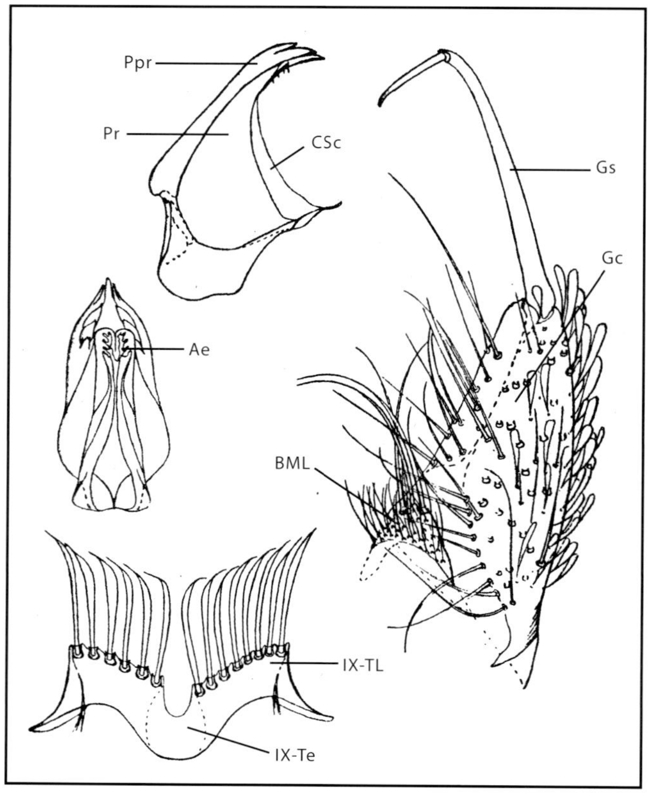

Shannoniana fluviatilis **(Theobald) (♂ genitalia)**

Ae: Aedeagus; BML: Basal Mesal Lobe; CSc: Cercal Sclerite; Gc: Gonocoxite; Gs: Gonostylus; Ppr: Paraproct; Pr: Proctiger; IX-Te: Tergum IX; IX-TL: Tergum IX Lobe. (Modified after Lane and Cerqueira, 1942).

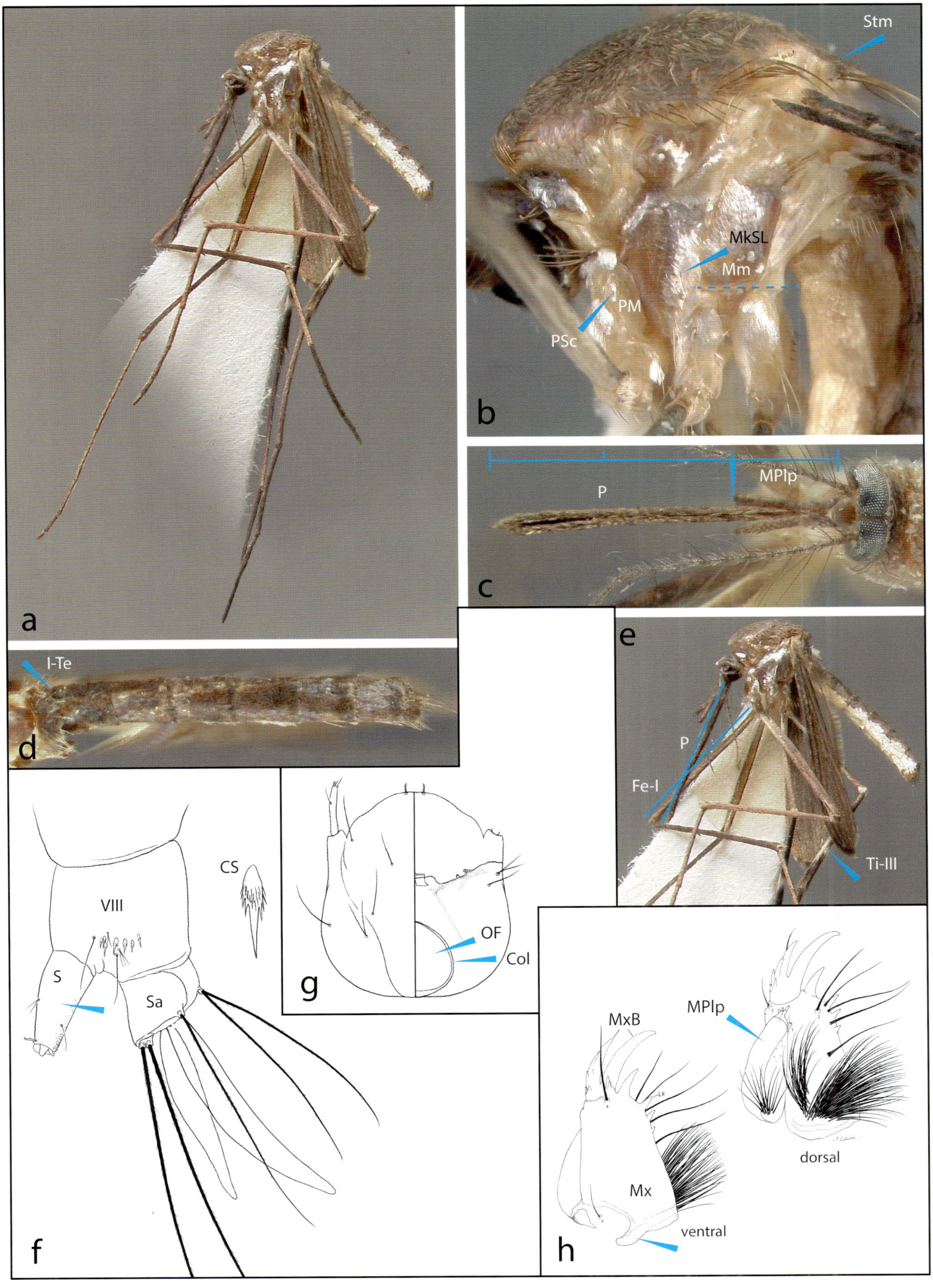

a
Stm
MkSL
Mm
PM
PSc
b
P
MPlp
c
I-Te
d
e
P
Fe-I
Ti-III
VIII
CS
S
Sa
f
OF
Col
g
MxB
MPlp
Mx
ventral
dorsal
h

Genus *Topomyia* Leicester, 1908
Subfamily Culicinae, Tribe Sabethini
"Omnivore Mosquito"

Systematics and Distribution

Species of the sabethine genus *Topomyia* are small mosquitoes that are morphologically most similar to those of *Malaya*, with long central lines of silver or white scales along the scutum (see *e*). However, unlike the bent, hairy club proboscis of *Malaya* mosquitoes, the proboscis of *Topomyia* species are straight and unmodified (see *b*). The genus *Topomyia* is subdivided into three subgenera: the monotypic *Miyagiella*, comprising only *Topomyia discors* Harbach, *Suaymyia* (22 spp.), and *Topomyia* (42 spp.). The type species is *Topomyia minor* Leicester. The immature stages of many species are still unknown, and the male genitalia are by far the most reliable characters for species recognition. *Topomyia* species are most common in the Oriental Region, although some are found as far north as Japan and south to the islands of the Australasian Region.

Bionomics

Topomyia females have been collected in heavy shade, resting on trees in humid jungle settings or along the banks of jungle streams. It is supposed that *Topomyia* females do not blood feed. They are not attracted to animal-baited traps, and they have never been reported landing on, or attempting to bite, humans. *Topomyia* (*Suaymyia*) *yanbarensis* has been colonized in the laboratory. Males were noted to mate with resting females. Once mated, females oviposited single eggs into small bamboo bore holes (3–4 mm), laying several eggs daily and more than 100 eggs in a single ovicycle. Immatures of the subgenus *Topomyia* are mainly associated with rainwater collections in a wide variety of leaf axils and in pitcher plants, whereas immatures of the subgenus *Suaymyia* are mainly recovered from bamboo internodes and stumps. Most *Suaymyia* larvae are predacious, whereas larvae of most species of the subgenus *Topomyia* are browsers, gaining nutrition by rasping the plant tissues of their immature habitats. The male holotype of *To.* (*Miyagiella*) *discors* was captured in a light trap in Sabah, Malaysian Borneo, but the characteristics of its immature habitats remain unknown.

Associated Pathogens

None.

Diagnostic Characters

(a) Habitus (♀). (b) Head (side): Proboscis (P) unmodified. **(c) Abdomen:** Terga (Te) usually with pale metallic markings. **(d) Wing:** Upper calypter (UC) bare; vein 1A turns abruptly and terminates more or less level of base of mediocubital crossvein (mcu). **(e) Thorax (dorsal):** Scutum (Scu) with a median narrow strip of broad white, silvery, or brown scales. **(f) Larva (head):** Maxillary palpus (MPlp) articulated, attached to the maxillary body (MxBo). **(g) Larva (head):** Occipital foramen (OF) slit-like. **(h) Larva (thorax):** Seta 8-M present; seta 13-P absent; seta 13-T not inserted on sclerotized area with 9–12-T; setae 6-M and 7-T not modified as stout spines. **(i) Larva (terminal segments):** Pecten (Pt) present; seta 4-X with a single pair of branched setae; siphonal setae present and variable in number. Image sources: WRBU; Mattingly (1971); Tanaka et al. (1979); Harbach & Peyton (1993).

Exemplar DNA sequences: *Topomyia houghtoni—28S* rDNA (KF687531, KF687599), mtDNA *COII* gene (KF687370), mtDNA *Cyt b* gene (KF687441).

Catalog entry: Vol. 2, p. 998.

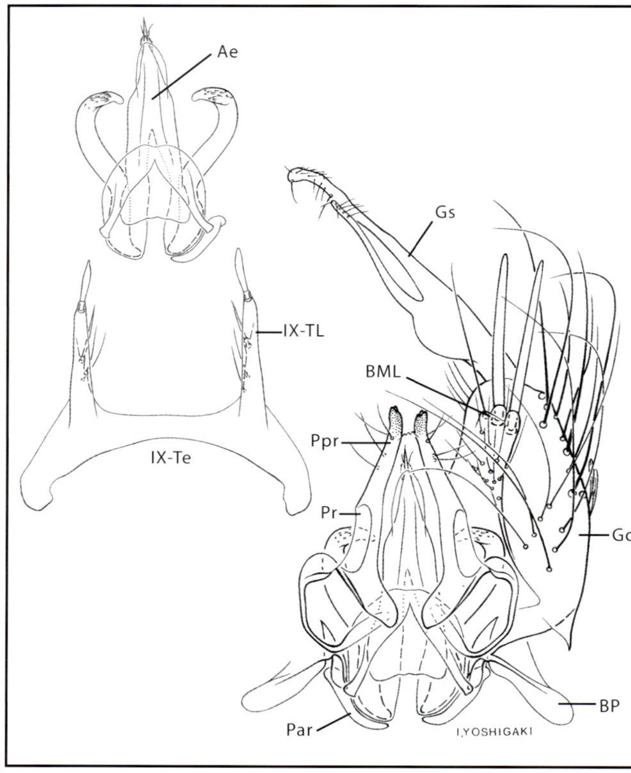

***Topomyia* (*Suaymyia*) *yanbarensis* Miyagi (♂ genitalia)**

Ae: Aedeagus; BML: Basal Mesal Lobe; BP: Basal Piece; Gc: Gonocoxite; Gs: Gonostylus; Par: Paramere; Ppr: Paraproct; Pr: Proctiger; IX-Te: Tergum IX; IX-TL: Tergum IX Lobe. (Modified after Thurman, 1959).

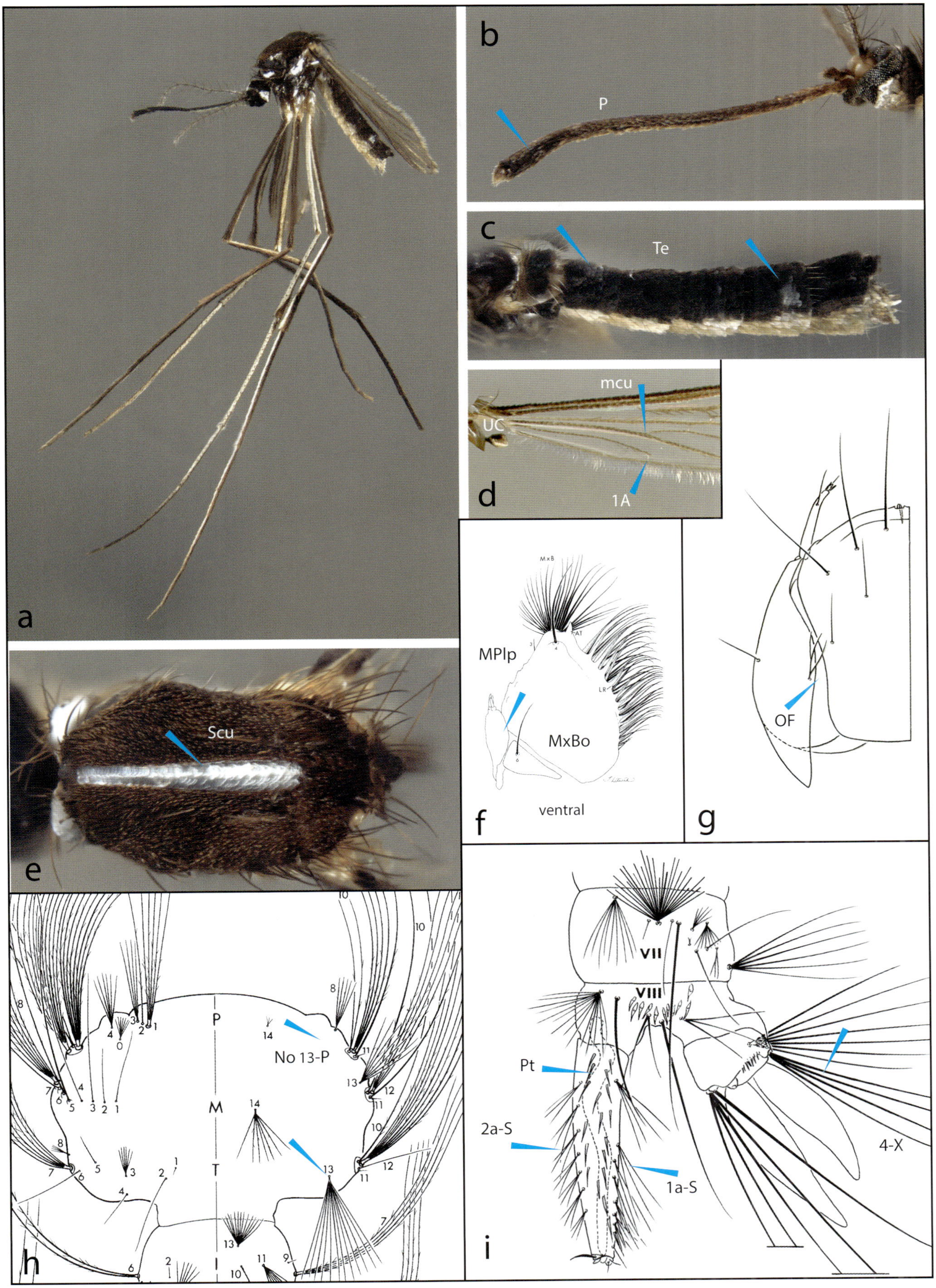
a
b
P
c
Te
d
mcu
UC
1A
e
Scu
f
MPlp
MxBo
ventral
g
OF
h
P
M
T
No 13-P
i
VII
VIII
Pt
2a-S
1a-S
4-X

Genus *Trichoprosopon* Theobald, 1901 Subfamily Culicinae, Tribe Sabethini "Hairy-lipped Mosquito"

Systematics and Distribution

Trichoprosopon are large mosquitoes with attractive markings and coloration and a characteristically hairy clypeus. The genus comprises 13 formally described species and two subspecies: *Tr. andinum* Levi-Castillo, *Tr. brevipes* (da Costa Lima), *Tr. castroi* Lane & Cerqueira, *Tr. compressum* Lutz (and ssp. *mogilasium* (Dyar & Knab)), *Tr. digitatum* (Rondani) (and ssp. *townsendi* Stone), *Tr. evansae* Antunes (in Lane & Cerqueira), *Tr. lampropus* (Howard, Dyar & Knab), *Tr. lanei* (Antunes), *Tr. obscurum* Lane & Cerqueira, *Tr. pallidiventer* (Lutz), *Tr. simile* Lane & Cerqueira, *Tr. soaresi* Lane & Cerqueira, and *Tr. vonplesseni* (Dyar & Knab). The species *Tr. digitatum* currently represents the genus. *Trichoprosopon* species are native to the Central and South American Neotropics.

Bionomics

Trichoprosopon species can become locally abundant. In a study in a fragmented patch of Atlantic Forest in western Santa Catarina, southern Brazil, *Tr. pallidiventer* was the most commonly collected mosquito around sunset, comprising over 59% of total catch. The immature stages of all *Trichoprosopon* species are restricted to phytotelmata, especially bamboo internodes and fallen leaves, spathes, and fruit husks. The larvae of some *Trichoprosopon* species, including *Tr. compressum*, *Tr. digitatum*, *Tr. pallidiventer*, and *Tr. soaresi*, are known obligate or facultative predators. *Trichoprosopon compressum* and *Tr. pallidiventer* immatures occupy bamboo internodes with larger exit holes and foul waters caused by decaying organic matter, whereas *Tr. soaresi* tends to utilize internodes with smaller (2 mm) holes and clean, clear waters without organic contamination. *Trichoprosopon digitatum* (see species pages, p. 546) is by far the most commonly reported *Trichoprosopon* species. It appears to be ecologically diverse, host seeking and biting humans and other mammals and birds up to 34 m in the canopy and at ground level. It oviposits in a wide range of natural and artificial containers both on the ground and high into the canopy. It is the only species common in plant habitats created by humans, rendering it a pest species in forest plantations. Female *Tr. digitatum* lay their eggs in loose rafts and guard them from habitat flooding until they hatch, up to 48 hours later.

Associated Pathogens

AMBV, ARUV, BSQV, COCV, DENV, ILHV, PIXV, SLEV, TNTV, WYOV.

Diagnostic Characters

(a) Habitus (♀): I-Te with sparse scales distally (not shown). **(b) Thorax (dorsal):** Antepronotum (Ap) well separated, small. **(c) Head/leg:** Proboscis (P) not long; about same size as Fe-I. **(d) Thorax (side):** Long row of setae on the lower mesokatepisternum (MkSL); lower margin of mesepimeron (Mm) above level of lower mesokatepisternal setae (MkSL); mesopostnotal setae present (MpnS); postprocoxal scales (PSc) absent; prespiracular setae (PsS) present. **(e) Head (side):** Clypeus (Clp) usually with lateral setae. **(f) Larva (terminal segments):** Pecten (Pt) absent; siphon (S) without midventral filaments. **(g) Larva (head):** Maxilla (Mx) with a maxillary brush (MxB); cardo (Cd) fused with the base of the maxillary palpus. **(h) Larva (head):** Hypostomal suture (HyS) complete; occipital foramen (OF) circular with a distinct collar (Col). **(i) Larva (head):** Larval mandible. **Other characters (not shown):** Larval metathoracic seta 8-M absent. Image sources: WRBU; Mattingly (1971); Harbach & Peyton (1993).

Exemplar DNA barcodes: *Trichoprosopon compressum* (MF172405), *Tr. digitatum* (KF671032–33, MF172406–08), *Tr. evansae* (KM593039), *Tr. pallidiventer* (MF172409–11).

Catalog entry: Vol. 2, p. 1003.

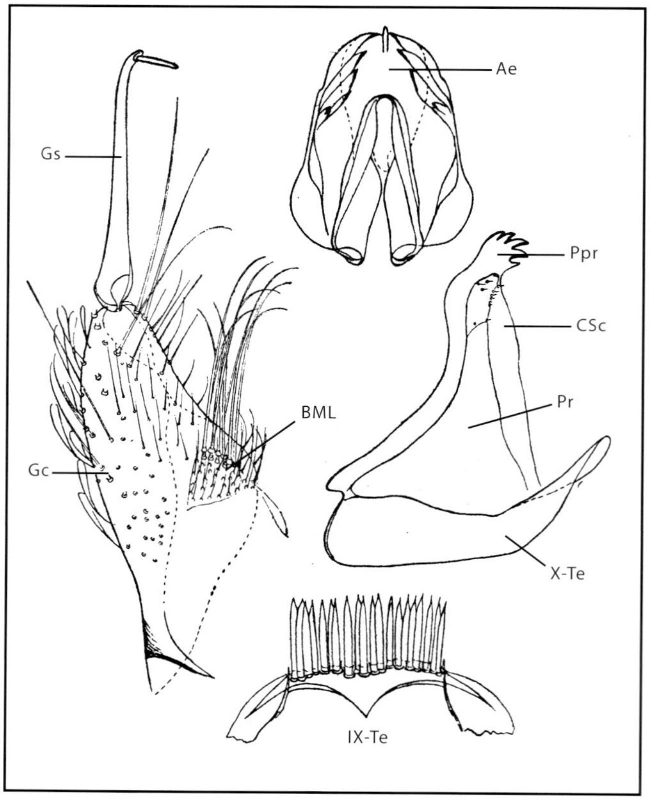

***Trichoprosopon digitatum* (Rondoni) (♂ genitalia)**

Ae: Aedeagus; BML: Basal Mesal Lobe; CSc: Cercal Sclerite; Gc: Gonocoxite; Gs: Gonostylus; Ppr: Paraproct; Pr: Proctiger; IX-Te: Tergum IX; X-Te: Tergum X. (Modified after Lane and Cerqueira, 1942).

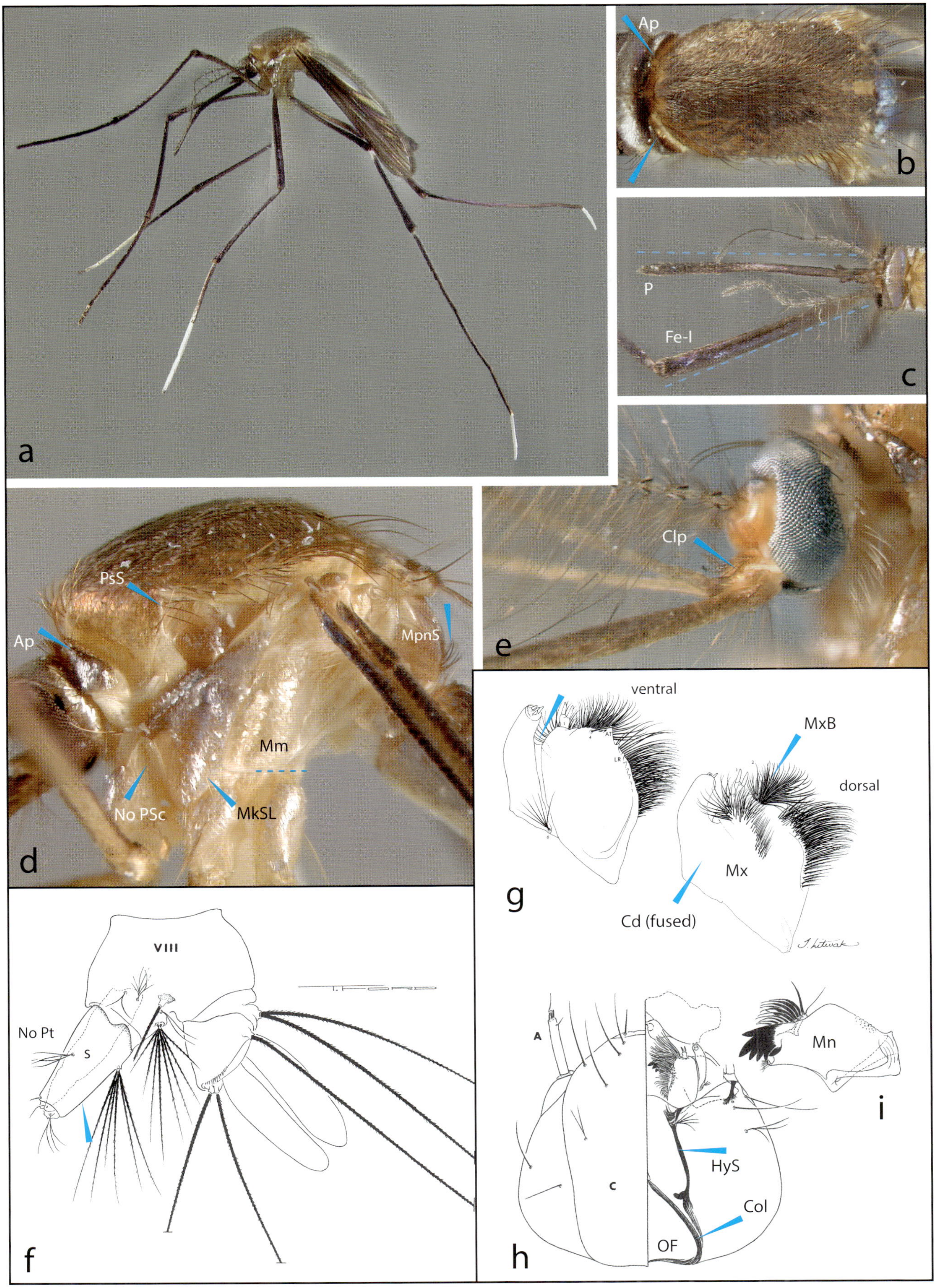

a
Ap
b
P
Fe-I
c
Clp
e
PsS
MpnS
Ap
Mm
No PSc
MkSL
d
ventral
MxB
dorsal
Mx
Cd (fused)
g
VIII
No Pt
S
f
A
C
Mn
HyS
Col
OF
h
i

Genus *Tripteroides* Giles, 1904
Subfamily Culicinae, Tribe Sabethini
"Pestle Mosquito"

Systematics and Distribution

The genus *Tripteroides* is a large, complex assemblage of 122 species, subdivided into five subgenera—*Polylepidomyia* (19 spp.), *Rachionotomyia* (14 spp.), *Rachisoura* (28 spp.), *Tricholeptomyia* (9 spp.), and *Tripteroides* (52 spp.). The genus includes both jungle-dependent species and those adapted to domestic environments. The species comprising the genus are highly diverse, and the internal relationships of species within the genus and its position within the Culicinae are in need of revision. The relationships among species in the genera *Tripteroides* and *Trichoprosopon* remain unresolved as many characters overlap. The type species of the genus is *Runchomyia philippinesis* Giles, which is currently a subjective synonym of *Tp. nitidoventer* (Giles). *Tripteroides* species are distributed across the Oriental and Australian Regions, from India to Japan and south to Australia, including the islands of Southeast Asia and the South Pacific.

Bionomics

Immature *Tripteroides* are commonly collected in pockets of rainwater found in standing, fallen, and discarded phytotelma habitats (e.g., bamboo and tree stumps, leaf axils, including *Alocasia*, *Colocasia*, *Nipa*, and *Pandanus*, ginger flower bracts, tree holes, snail shells, coconut shells and husks, *Nepenthes* pitcher plants, and fallen leaves). Some species including *Tp. aranoides* (Theobald) are also collected in artificial containers, and immatures of the Australian endemic *Tp. tasmaniensis* (Strickland) solely inhabit rock holes. Similar to *Maorigoeldia argyropus* (Walker), the larvae of *Tripteroides* rest upside down at the base of the container, looking up through the water in search of food. While larvae of the subgenus *Rachisoura*, with their enlarged, spiny maxillae, are predacious, other *Tripteroides* species feed primarily on decaying insect matter. *Tripteroides aranoides* has been successfully colonized and its laboratory bionomics studied in detail. With plentiful food resources, egg to adult eclosion takes around three weeks at 28°C.

In the laboratory, mating was observed to be initiated in flight in *Tp. aranoides*, with copulation occurring on the vertical surfaces of the cage walls. Female *Tp. aranoides* are autogenous, producing their first batch of eggs without needing to feed; however, subsequent oogenesis requires blood meals. Oviposition occurs on the wing, with females "shooting" out eggs one at a time into bamboo cups and through small bore holes (11 × 4 mm) in bamboo internodes by flicking their abdomens. Little is known of the host preferences of most *Tripteroides* species, as few species appear to bite humans. In the Australian Region, several species (e.g., *Tp. solomonis* Edwards) are regarded as pests, even entering houses to bite, whereas in the Oriental Region, only *Tp. aranoides* attacks humans. The arboviral competencies of *Tripteroides* species are largely unstudied, but *Tp. bambusa* (Yamada) is susceptible to infection with Getah virus.

Associated Pathogens

GETV.

Diagnostic Characters

(a) Habitus (♀). (b) Head (dorsal): Erect head scales restricted to occiput (Occ). **(c) Thorax (dorsal):** Acrostichal setae (AcS) absent. **(d) Thorax (side):** Prespiracular setae (PsS) present. **(e) Wing:** Vein 1A terminates beyond level of the mediocubital crossvein (mcu); upper calypter (UC) with marginal setae (not shown). **(f) Larva (head):** Seta 3-C ventral; occipital foramen (OF) circular, with a distinct collar (Col). **(g) Larva (thorax):** Seta 13-P present; seta 6-M and/or seta 7-T often as a stout spine (not shown). **(h) Larva (terminal segments):** Comb (C) with comb scales (CS) in a single row; one multibranched seta 4-X. Image sources: WRBU; Tanaka et al. (1979); Mattingly (1981).

Exemplar DNA barcodes: *Tripteroides affinis* (EU259298), *Tp. aranoides* (AY917210), *Tp. atripes* (KU495085–86, KU495088), *Tp. bambusa* (KT358398–400, LC054517–21, EF647970), *Tp. tasmaniensis* (KU495089).

Catalog entry: Vol. 2, p. 1005.

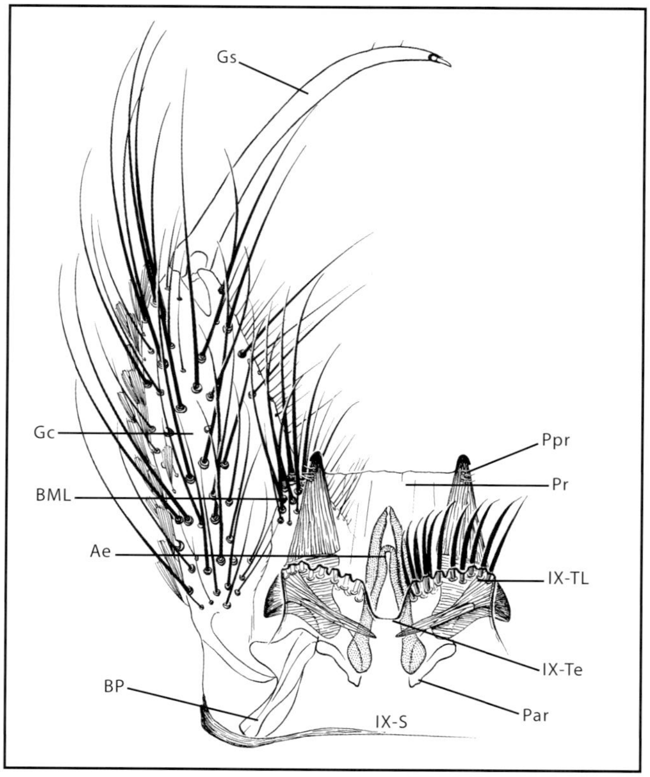

***Tripteroides* (*Tripteroides*) *bambusa* (Yamada) (♂ genitalia)**

Ae: Aedeagus; BML: Basal Mesal Lobe; BP: Basal Piece; Gc: Gonocoxite; Gs: Gonostylus; Par: Paramere; Ppr: Paraproct; Pr: Proctiger; IX-Te: Tergum IX; IX-TL: Tergum IX Lobe; IX-S: Sternum IX. (Modified after Tanaka, Mizusawa and Saugstad, 1979).

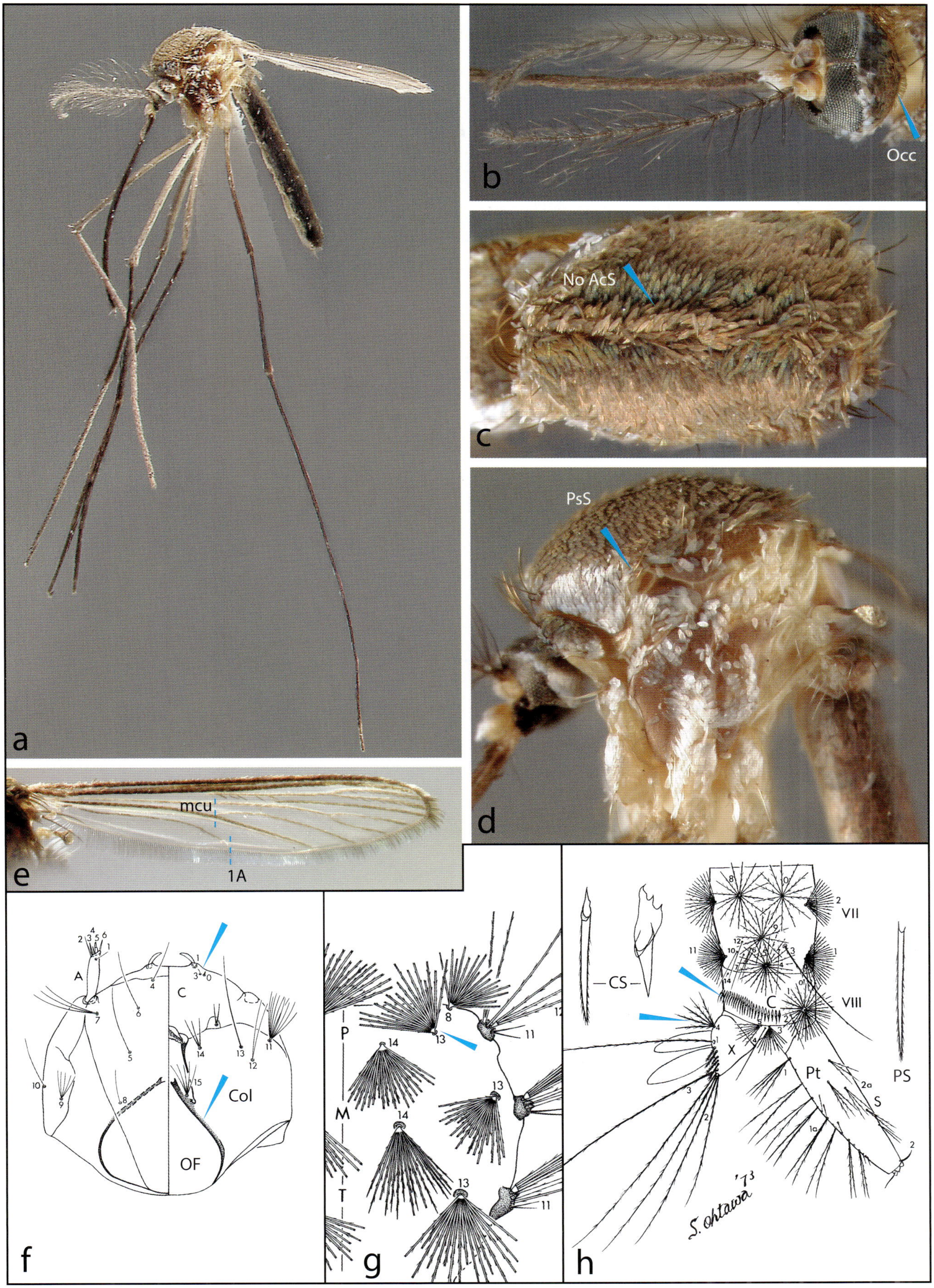
a
b
Occ
c
No AcS
d
PsS
e
mcu
1A
f
A
C
Col
OF
g
P
M
T
h
CS
VII
VIII
C
X
Pt
S
PS
S. Ohtawa '73

Genus *Wyeomyia* Theobald, 1901
Subfamily Culicinae, Tribe Sabethini
"Bromeliad Mosquito"

Systematics and Distribution

With 139 species, *Wyeomyia* comprises the largest sabethine genus. Morphologically, species of the genus *Wyeomyia* are most similar to the New World sabethine genera *Limatus* and *Sabethes*. *Wyeomyia* species are differentiated from all other sabethine mosquitoes (except *Sabethes* (*Davismyia*) *petrocchiae* (Shannon & del Ponte) by the presence of prealar setae (see *b*). Following recent assessment of the internal relationships among component taxa using morphological and allozyme data, the species were divided into the following 17 subgenera: *Antunesmyia* (3 spp.), *Caenomyiella* (1 sp.), *Cruzmyia* (4 spp.), *Decamyia* (3 spp.), *Dendromyia* (6 spp.), *Dodecamyia* (1 sp.), *Exallomyia* (3 spp.), *Hystatomyia* (9 spp.), *Menolepis* (1 sp.), *Miamyia* (7 spp.), *Nunezia* (3 spp.), *Phoniomyia* (22 spp.), *Prosopolepis* (1 sp.), *Spilonympha* (7 spp.), *Triamyia* (2 spp.), *Wyeomyia* (35 spp.), and *Zinzala* (2 spp.). Twenty-nine species remain unplaced. *Wyeomyia grayii* Theobald is the type species of the genus. Species of the genus *Wyeomyia* are found across the Neotropics, the Caribbean islands, and into the eastern United States.

Bionomics

Like many sabethines, *Wyeomyia* species are most commonly encountered in Neotropical forests. Adults oviposit in waters gathered in natural containers, including bamboo (broken, internodes, and stumps), tree holes, pitcher plants, bromeliad and aroid leaf axils, flower bracts, and the occasional artificial container. Immatures of *Wy.* (*Wyo.*) *smithii* (Coquillett) develop in the fluids gathered in the northern pitcher plant (*Sarracenia purpurea gibbosa*) and the southern pitcher plant (*Sarracenia purpurea venosa*), gaining protein from the decomposing insects in the pitchers. This autogenous species is restricted to acidic sphagnum bogs in the eastern United States where these host plants are found. As larvae, *Wy. smithii* protects itself from predation by coexisting carnivorous chironomid species by having long setae all over its body. This most northern of *Wyeomyia* species overwinters as larvae and can tolerate freezing.

Adult *Wyeomyia* are active during the day, rarely wandering far from their larval habitats. They occupy all elevations throughout the forest, depending on host preferences. Typical of all sabethine mosquitoes is the distinctive resting pose, with hind legs bent over the head, but peculiarly, the tiny *Wy. smithii* females maintain this posture even during flight. Many *Wyeomyia* females readily feed on humans, as well as many mammals and birds, and some are implicated in zoonotic viral transmission. Few studies have focused on arboviral surveillance of *Wyeomyia* species, possibly due to difficulties in species identification, and their medical importance is likely underestimated as a result. Several strains of Ilheus and Venezuelan encephalitis viruses have been isolated from *Wy.* (*Wyo.*) *medioalbipes* Lutz in Trinidad alone.

Associated Pathogens

ARUV, CABV, CARV, IACOV, ILHV, MDCV, MQOV, MUCV, SLEV, TONV, VEEV, WYOV.

Diagnostic Characters

(a) Habitus (♀). (b) Thorax (side): Prespiracular setae (PsS) present; prealar setae (PaS) present (indistinct in photo). **(c) Thorax (dorsal):** Scutum (Scu) with dull, bronzy metallic scales through brightly golden. **(d) Head (dorsal):** Clypeus (Clp) without seta; head without erect scales. **(e) Legs:** Ta-I–III dark. **(f) Larva (head):** Maxillary palpus (MPlp) articulated, attached to maxillary body (MxBo); maxillary brush (MxB) unmodified, usually present. **(g) Larva (head):** Setae 1-C not closely approximated. **(h) Larva (abdomen):** Seta 8-M present. Image sources: WRBU; Porter (2009); Porter (2014).

Exemplar DNA sequences: *Wyeomyia occulta* (MF172468–73), *Wy. pertinans* (MF172474–84), *Wy. pseudopecten* (MF172485–94), *Wy. robusta* (MF17249599), *Wy. smithii* (MF822694, MF827000), *Wy. splendida* (MF172500–04), *Wy. surinamensis* (MF172510–14), *Wy. testei* (MF172515–24), *Wy. ypsipola* (MF172522–24).

Catalog entry: Vol. 2, p. 1016.

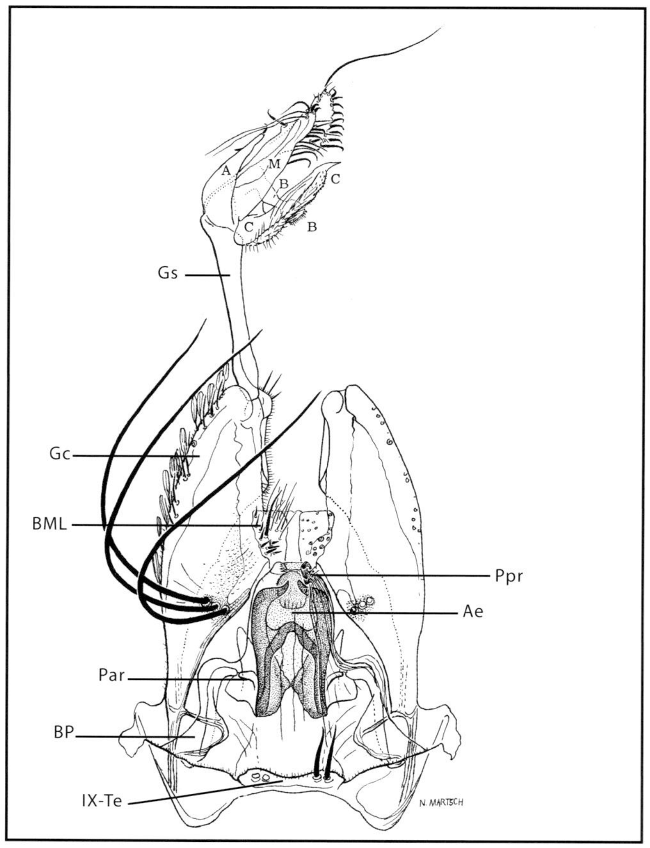

***Wyeomyia* (*Wyeomyia*) *mitchellii* (Theobald) (♂ genitalia)**

Ae: Aedeagus; BML: Basal Mesal Lobe; BP: Basal Piece; Gc: Gonocoxite; Gs: Gonostylus; Par: Paramere; Ppr: Paraproct; IX-Te: Tergum IX. (Modified after Belkin, Heinemann and Page, 1970).

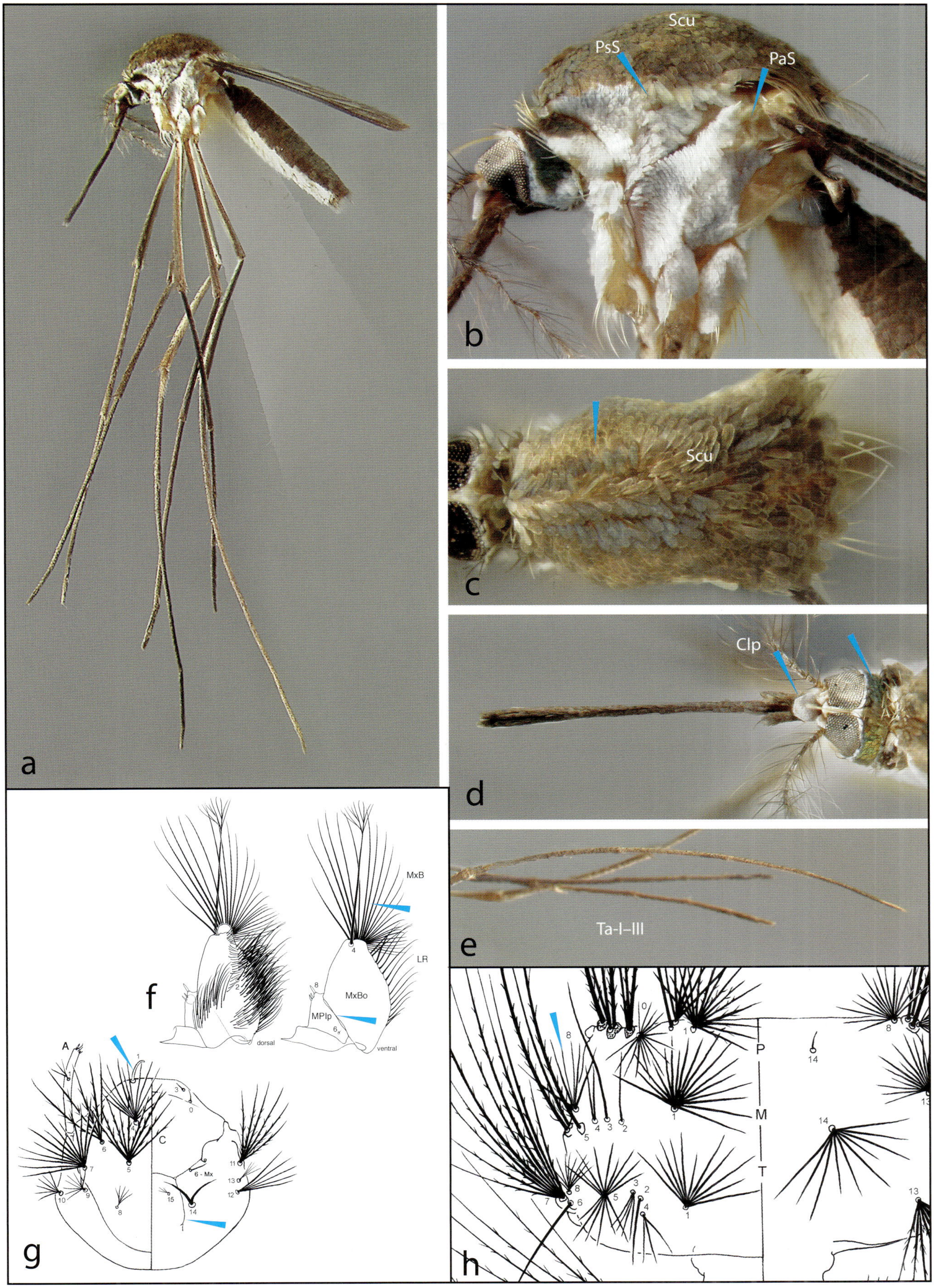
a
b
Scu
PsS
PaS
c
Scu
d
Clp
e
Ta-I–III
f
MxB
LR
MxBo
MPlp
dorsal
ventral
g
A
C
6 - Mx
h
P
M
T

Genus *Toxorhynchites* Theobald, 1901 Subfamily Culicinae, Tribe Toxorhynchitini Lahille, 1904 "Carnivorous Mosquito"

Systematics and Distribution

Forest-associated, *Toxorhynchites* are often highly ornate as adults, with colorful, iridescent scales and accessory scale tufts on the posterior abdominal segments. Unique within Culicinae, the scutellum of *Toxorhynchites* species lacks the median scutellar lobe and appears evenly rounded—a character most commonly associated with anopheline mosquitoes (except *Chagasia*). As adults, *Toxorhynchites* species are the gentle giants of the mosquito world, with nectar-feeding adults measuring up to 18 mm in length and boasting impressive wingspans of greater than 24 mm. In contrast, their aggressive, carnivorous larvae are stout, with heavily chitinized mouthparts for grasping and consuming aquatic prey.

The genus *Toxorhynchites* comprises 90 recognized species in four subgenera: *Afrorhynchus* (19 spp., Africa), *Ankylorhynchus* (4 spp., South America), *Lynchiella* (16 spp., 15 confined to South America, with *Tx. rutilus* extending from South America through the eastern United States into Northeast Canada), and *Toxorhynchites* (51 spp., Old World). The genus is represented by *Tx. brevipalpus* Theobald and is the sole member of the tribe Toxorhynchitini.

Bionomics

Toxorhynchites females lay their eggs in natural container habitats in forest environs (e.g., pitcher plants, tree holes, bamboo stumps). Their larvae are carnivorous and highly predatory, with a ferocious appetite for consuming other mosquito larvae—a characteristic that has been exploited for biocontrol in high-density larval sites of malaria vector mosquitoes, such as in rice fields, and in the container breeder habitats of *Aedes aegypti* and *Ae. albopictus* (see species pages, p. 456 and p. 460, respectively). Having already accumulated all the necessary stores to facilitate egg production as larvae, adult females do not blood feed and are sustained instead on carbohydrate-rich honeydew, nectar, fruits, and decaying organic refuse. *Toxorhynchites* adults are active during the day.

Associated Pathogens

TsDNV.

Diagnostic Characters

(a) Habitus (♀): Very large mosquito, often covered with intensely colored iridescent scales; long downturned proboscis. **(b) Thorax (side):** Posterolateral abdominal scale tufts present. **(c) Wing:** Alula (Al) and upper calypter (UC) bare; distinct sclerotization in cell M_4 in front of vein Cu; wing margin distinctly emarginate near the terminus of vein CuA. **(d) Thorax (dorsal):** Scutellum (Stm) evenly rounded, not lobed; no setae on acrostichal (AA), dorsocentral (DA), and prescutellar (PsA) areas. **(e) Larva (terminal segments). (f) Larva (head):** Mouthparts modified for predation; lateral palatal brush (LPB) of ≤10 flattened filaments. **(g) Head (side):** Proboscis (P) tapered and bent distinctively downward, originates from a deep cavity on the capsule. **(h) Head (dorsal):** Clypeus (Clp) short. Image sources: WRBU; Mattingly (1971).

Exemplar DNA barcodes: *Tx.* (*Lynchiella*) *guadeloupensis* (MF172385–88), *Tx.* (*Lyn.*) *haemorrhoidalis haemorrhoidalis* (MF172389–93), *Tx.* (*Lyn.*) *haemorrhoidalis superbus* (MF172394–403), *Tx.* (*Lyn.*) *moctezuma* (MF172404), *Tx.* (*Lyn.*) *theobaldi* (KY782648–55), *Tx.* (*Toxorhynchites*) *splendens* (KY411577–89), *Tx.* (*Tox.*) *towadensis* (LC054522–26).

Catalog entry: Vol. 2, p. 1032.

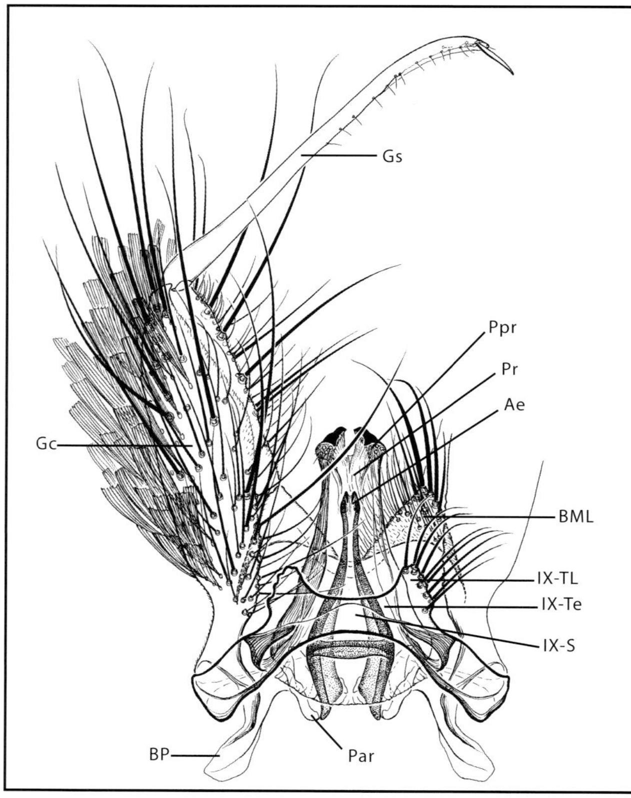

***Toxorhynchites* (*Lynchiella*) *haemorrhoidalis superbus* (Dyar & Knab) (♂ genitalia)**

Ae: Aedeagus; BML: Basal Mesal Lobe; BP: Basal Piece; Gc: Gonocoxite; Gs: Gonostylus; Par: Paramere; Ppr: Paraproct; Pr: Proctiger; IX-Te: Tergum IX; IX-TL: Tergum IX Lobe; IX-S: Sternum IX. (Modified after Belkin, 1962).

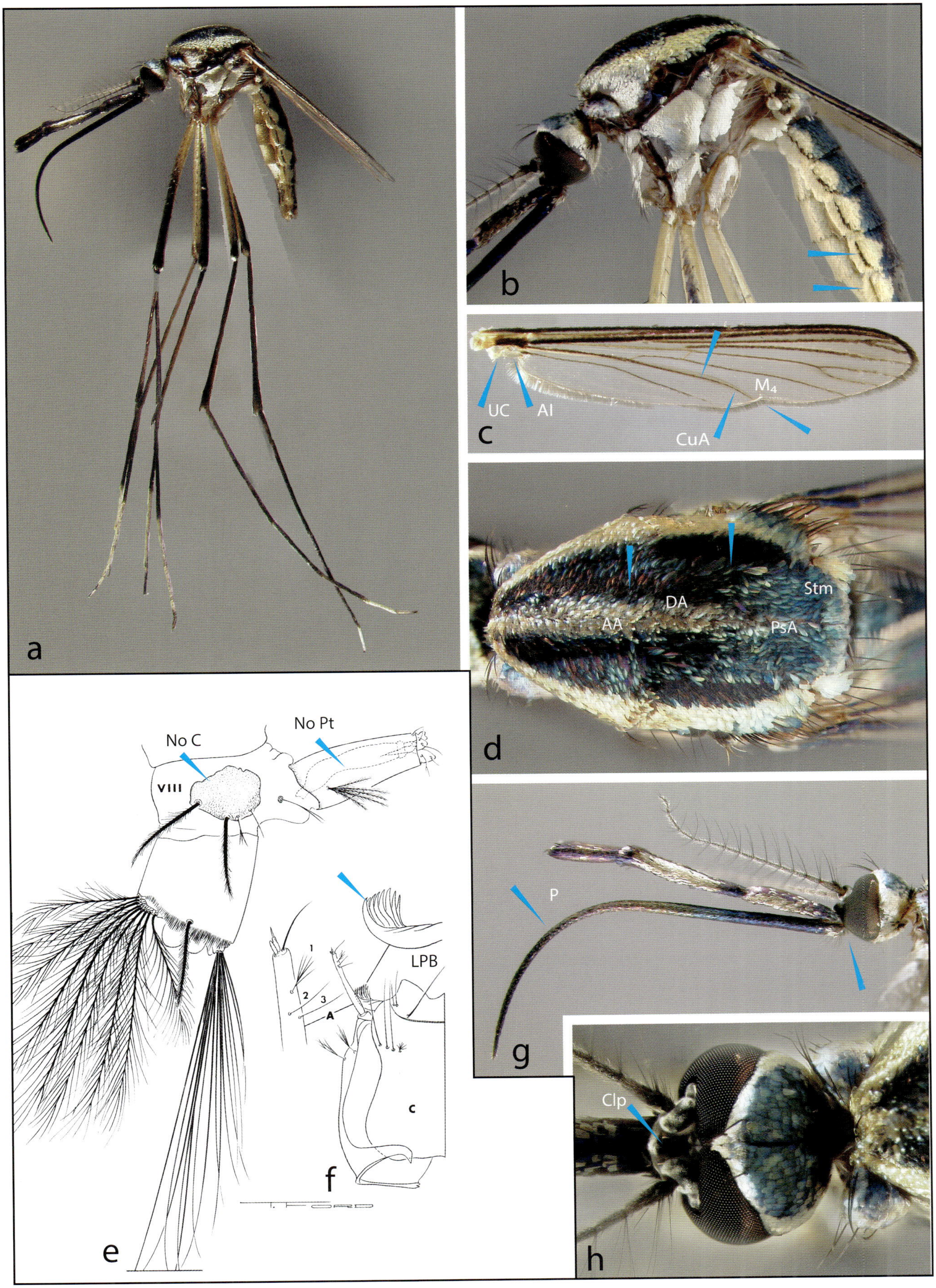

a
b
UC
AI
M4
CuA
c
DA
AA
Stm
PsA
d
No C
No Pt
VIII
1
2
3
A
LPB
C
f
T. FORD
e
P
g
Clp
h

Genus *Uranotaenia* Lynch Arribálzaga, 1891
Subfamily Culicinae, Tribe Uranotaeniini "Heaven Mosquito"

Systematics and Distribution

Uranotaenia species are small, brightly ornamented mosquitoes. The large genus *Uranotaenia* comprises 271 species and 6 subspecies, and is most speciose in Afrotropical and Oriental Regions. It is the sole representative of the tribe Uranotaeniini. *Uranotaenia* is represented by the type species *Ur. pulcherrima* Lynch Arribálzaga and is subdivided into two subgenera (see Vol. 2, p. 1099, for internal relationships). The subgenus *Pseudoficalbia* comprises 150 species distributed through the Afrotropics, Indian Subcontinent, Oriental Region, and Australasia, with a single species and subspecies found in the New World. The 121 species in the subgenus *Uranotaenia* are found in the tropical and subtropical regions of the Afrotropical, Middle East, Australasian, Neotropical, and Oriental Regions. The New World EEV and WNV vector *Ur. sapphirina* (Osten Sacken) occurs from Mexico north to Canada.

Bionomics

Female *Uranotaenia* do not commonly feed on humans, instead preferring mammals, birds, amphibians, and reptiles. In Florida, female *Ur. sapphirina* feeds exclusively on annelids, including earthworms and leeches, whereas in Tennessee, blood meals included deer, humans, and eastern box turtles, a finding that could be explained by the presence of cryptic taxa. *Uranotaenia lowii* Theobald feeds only on anurans (frogs and toads), and host-seeking females of this species are attracted to their hosts by the vibrations of their songs. Immature stages of most *Uranotaenia* are found in permanent groundwaters, including swamps and marshes with floating and emergent vegetation. Others are found in temporary waters collected in natural containers, including leaf axils, tree holes, bamboo, flower bracts, and fallen plant spathes or leaves, as well as in rock holes, crab holes, and other artificial containers. In temperate regions, females overwinter as mated females and are multivoltine.

Associated Pathogens

EEEV, JUGV, GAMV, WNV, SVV.

Diagnostic Characters

(a) Habitus (♀). (b) Thorax (side). (c) Wing: Cell R_2 always shorter than vein R_{2+3}. **(d) Wing:** Vein 1A sharply curved downward, ending before base of mediocubital crossvein (mcu); wing membrane microtrichiae small, not visible using ordinary magnification (not shown). **(e) Larva (head):** Seta 1-C on a distinct prominence on anterior margin of the head; seta 14-C at the anterior margin of the head; hypostomal suture (HyS) absent. **(f) Larva (terminal segments):** Single row of posterior comb scales (CS) present on comb plate (CP); pecten (Pt) with few to many pecten spines (PS); seta 1-S inserted far from base of siphon. Image sources: WRBU; Belkin et al. (1970).

Exemplar DNA barcodes: *Uranotaenia balfouri* (KM097030), *Ur. bicolor* (AY917205), *Ur. calosomata* (KF671034), *Ur. lowii* (AF417728, AF417764, HE600029), *Ur. sapphirina* (GU908125–30), *Ur. socialis* (JX260253).

Catalog entry: Vol. 2, p. 1042.

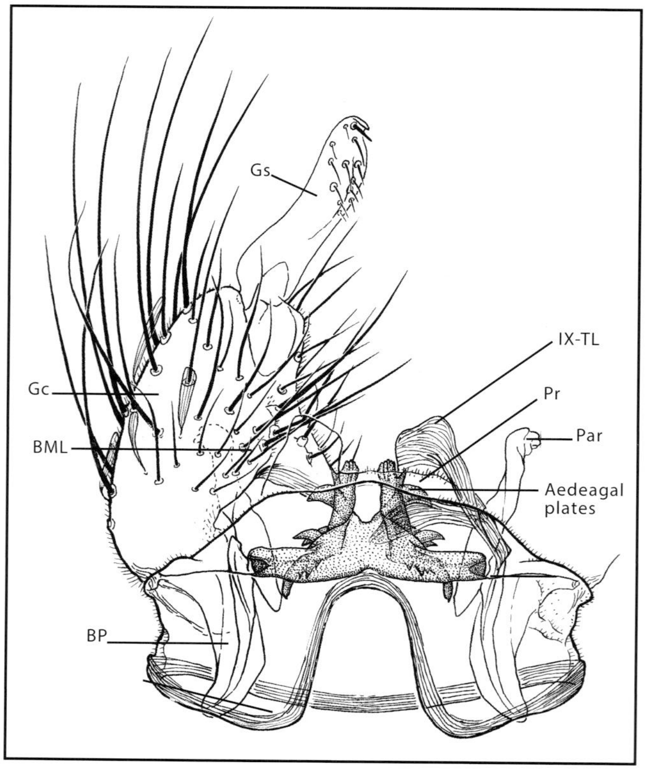

***Uranotaenia* (*Uranotaenia*) *sapphirina* (Osten Sacken) (♂ genitalia)**

BML: Basal Mesal Lobe; BP: Basal Piece; Gc: Gonocoxite; Gs: Gonostylus; Par: Paramere; Pr: Proctiger; IX-TL: Tergum IX Lobe. (Modified after Belkin, Heinemann and Page, 1970).

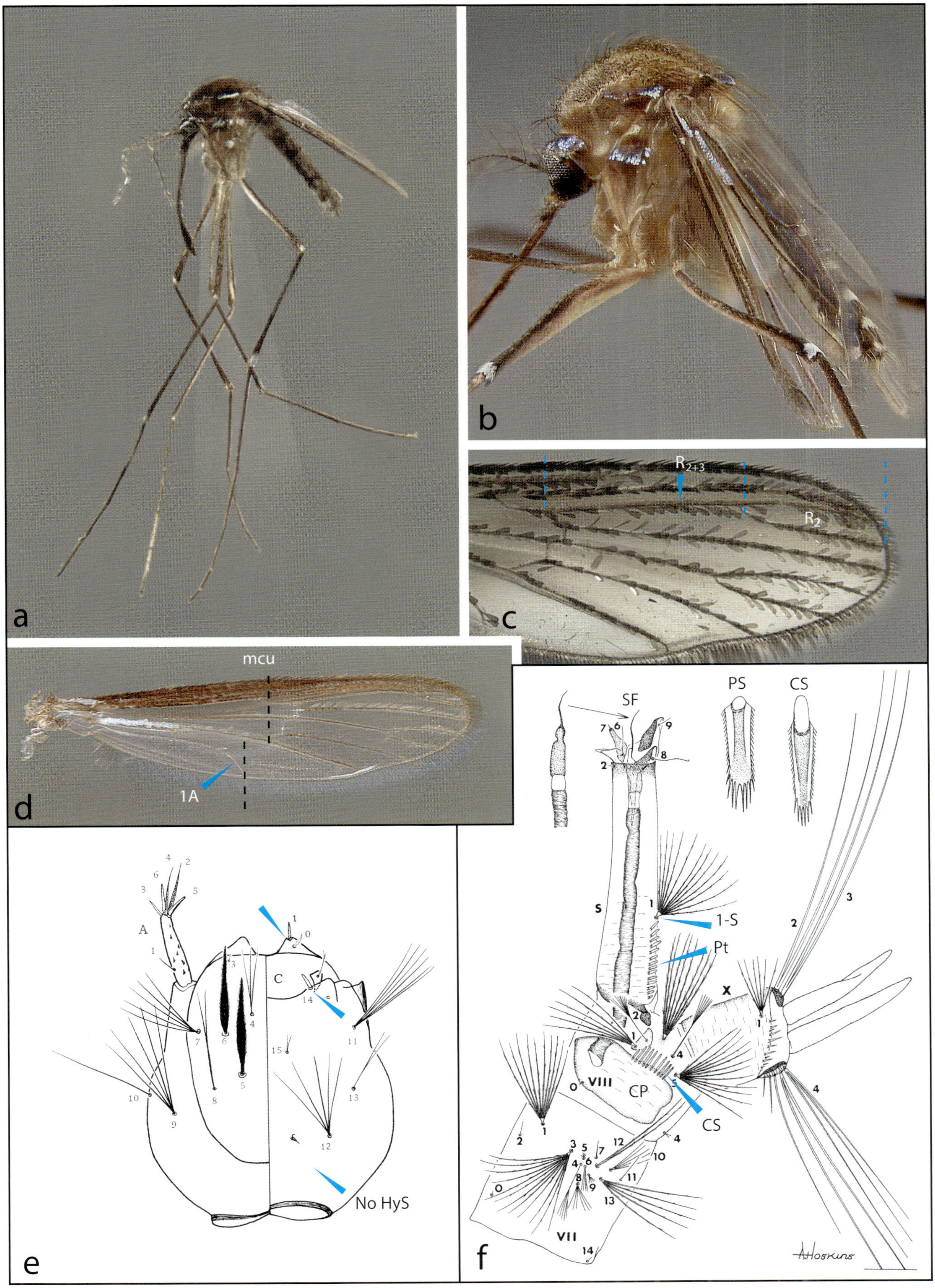
a
b
R2+3
R2
c
mcu
1A
d
A
C
No HyS
e
SF
PS
CS
S
1-S
Pt
X
VIII
CP
CS
VII
A Hoskins
f

Anopheles (*Anopheles*) *aztecus* Hoffman, 1935 Subfamily Anophelinae "Aztec Mexican Nail Mosquito"

Systematics and Distribution

Anopheles aztecus was originally described as a subspecies of *Anopheles maculipennis* Meigen (see species page, p. 312), from specimens collected in and around Mexico City. It is one of 19 species in the Holarctic Maculipennis Group (see Vol. 2, p. 1081, for internal relationships) and is the most southerly member of the Nearctic Maculipennis Group, occurring from the very southwestern United States only and down into Mexico. Note that the highly limited US distribution is not indicated on this map to avoid misrepresentation. In common with its southernmost Palearctic cousin, *An. sacharovi* Favre, the species is considerably paler than any of its northerly relatives. Wing scales are denser and/or darker at cross veins, giving the wings a spotted look characteristic of all members of the Maculipennis Group.

Anopheles aztecus remains unplaced within the Maculipennis Group. Chromosomal differences were noted between *An. aztecus* and other Maculipennis Group species in the Americas; however, DNA barcoding of four taxonomically verified *An. aztecus* archive specimens housed at the US National Collection returned 100% DNA match with sequences of *An. quadrimaculatus* Say (see species page, p. 322), leaving these authors to doubt the taxonomic uniqueness of the species within the Maculipennis Group. Further investigation is necessary to confirm its validity.

Bionomics

Surprisingly, little is documented on the biology of *Anopheles aztecus*. In common with other species in the Nearctic Maculipennis Group, immatures are found in surface freshwater habitats, including ditches, lakes, ponds, and canals. *Anopheles aztecus* has also been collected in gutters, and in pools of clear water. In the Xochimilco region of the Valley of Mexico, *An. aztecus* is known to be sympatric with but always more abundant than *An. pseudopunctipennis* Theobald (see species page, p. 316), which is also a highly competent malaria vector.

Anopheles aztecus avidly bites humans and is the primary vector of *P. vivax* malaria in Mexico. The species is readily colonized and has been used in many laboratory *Plasmodium* infection studies. Experimental infections with *P. berghei* have resulted in midgut infection rates of up to 70%.

Associated Pathogens

Plasmodium berghei, *P. falciparum*, *P. gonderi*, *P. vivax*.

Diagnostic Characters

(a) Habitus (♀): Legs without contrasting bands. **(b) Thorax (side). (c) Wing:** With only dark scales (including fringe at apex); scales darker and/or denser at crossveins and furcations, forming faint dark spots. **(d) Head (dorsal):** Scales and setae of interocular space (IS) and usually pale scales of vertex (V), golden yellow. **(e) Head (side):** Palpus (MPlp) unicolorous, without contrasting pale markings. **(f) Larva (head):** Setae 2,3-A pointed; seta 2-C thick, separated by a space >2 times diameter of one alveolus at base; seta 3-C dendritic, highly branched. **(g) Larva (thorax):** Setae 9–12-P single. **(h) Larva (abdominal segments):** Seta 0-IV,V minute. Image source: WRBU.

Exemplar DNA sequences: *An. aztecus 28S rDNA*: L78466.

Catalog entry: Vol. 2, p. 629.

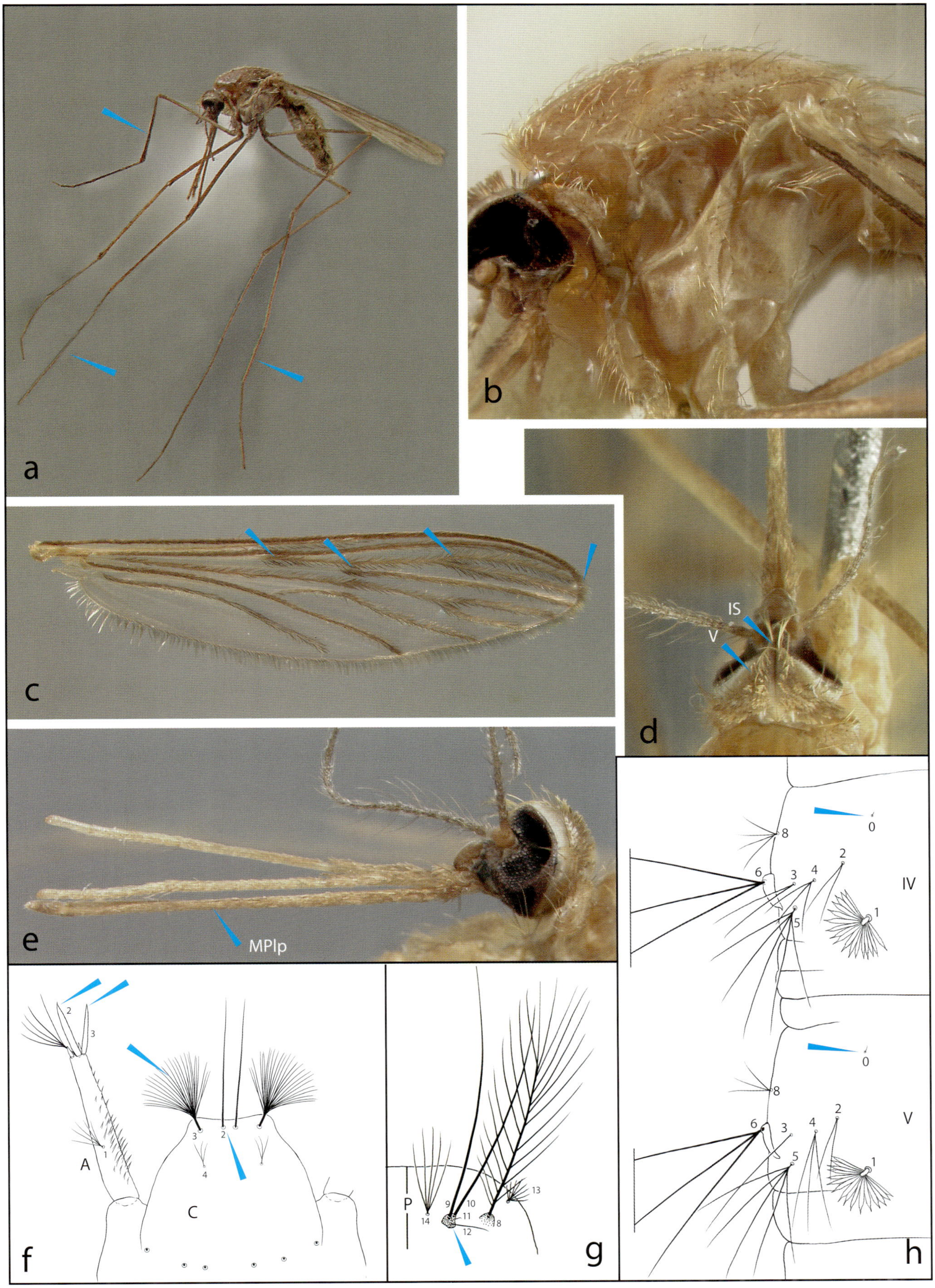
a
b
c
IS
V
d
MPlp
e
2
3
1
A
3
2
4
C
f
P
14
9
10
11
12
8
13
g
0
8
2
6
3
4
IV
5
1
0
8
6
3
4
2
V
5
1
h

Anopheles (Anopheles) bancroftii Giles, 1902
Subfamily Anophelinae
"Bancroft Australian Nail Mosquito"

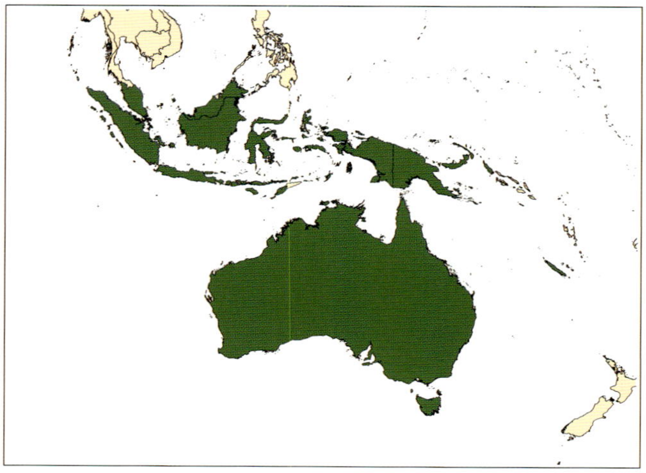

Systematics and Distribution

Anopheles bancroftii is recognized by its dark shaggy palps (see *f*), fully dark hind legs except for narrow apical bands on Ta-III$_{1-4}$, and its distinctive wing spot characters (see *g*). Until it was recognized as a valid species, *Anopheles bancroftii* was commonly mistaken for the primary malaria vector *An. barbirostris* van der Wulp. *Anopheles bancroftii*, originally described from specimens collected in Queensland, Australia, is now known to comprise two valid subspecies: *An. bancroftii bancroftii* and *An. bancroftii barbiventris* Brug, with the latter described from the Indonesian island of Celebes. Molecular studies have revealed the presence of at least six molecularly identifiable forms of *An. bancroftii* (A1, A2, B, C1, C2, and D) based on nuclear rDNA *ITS*-2 restriction profiles. *Anopheles pseudobarbirostris* Ludlow, described from the Philippines, is morphologically quite similar to *An. bancroftii* and varies only in the patterning and size of some pale patches on the wing fringe, leading some taxonomists to query the validity of these as separate taxa. However, these two taxa are reported in sympatry in Papua New Guinea and were shown to exhibit unique *ITS*-2 restriction profiles with respect to each other and the other *An. bancroftii* genotypes. Hybrids between these forms have been observed in nature and in laboratory conditions. Further work is needed to exactly define the taxonomic status and distribution of *An. bancroftii bancroftii*, *An. bancroftii barbiventris*, and *An. pseudobarbirostris* and to formally describe any additional cryptic taxa within the Bancroftii Group. *Anopheles bancroftii* s.l. has been reported from Australia, Borneo, Celebes, Indonesia, the Moluccas, Malaysia, and Papua New Guinea, and recent molecular studies have confirmed the introduction of *An. bancroftii* (genotype A1) in New Caledonia.

Bionomics

As expected for a cryptic species complex, population-level bionomic distinctions occur in *An. bancroftii* s.l. Preferred immature habitats for *An. bancroftii* s.s. in Queensland, and s.l. in the Northern Territories of Australia include dark swamps with a rich and diverse plant community, such as vegetated freshwater swamps (especially with water hyacinth), and among weeds, grass, reeds, or roots in water holes or stream margins. In Papua New Guinea, the immature habitats are limited to jungle ponds and cutoff pools in the Digoel River, where aquatic vegetation is composed mainly of *Azolla*, algae, and reeds.

In Queensland (genotype A2) and the Northern Territories (genotype A1), female *An. bancroftii* can be extremely aggressive biters of humans and other vertebrates during the night and have even been reported to bite during the day among shady vegetation. In Merauke and Tanah Merah, Papua New Guinea (genotype A2), *Plasmodium* infection rates of 4.3% have been reported, and precipitin tests confirmed 90% human blood meals, even among those mosquitoes collected resting in cattle sheds. In contrast, other populations (e.g., Cairns, Australia) appear totally indifferent to people, even at high densities.

Associated Pathogens

BEFV, DSwV, EUBV, MEV, RRV; *Plasmodium* spp., *Wuchereria bancrofti*.

Diagnostic Characters

(a) Habitus (♀): Fe-I–III with many scattered pale scales ventrally (not evident in photo). **(b) Thorax (side):** Mesepimeral scales (MeSc) sparse. **(c) Leg (III):** Ta-III$_5$ dark, Ta-III$_{1-4}$ with narrow apical pale bands. **(d) Abdomen (side):** II–VIII-S with basal tufts; VII-S with laterally visible scale tuft. **(e) Leg (I):** Fe-I distinctly swollen basally. **(f) Head (side):** Palpus (MPlp) entirely dark and shaggy in appearance. **(g) Wing:** Subcostal pale spot (SCP) present; wing fringe with pale patches at ends of veins M$_1$, M$_2$, M$_{3+4}$, and CuA and without a pale patch at termination of vein R$_{4+5}$. **(h) Larva (abdominal segments):** Seta 1 well-developed palmate, at least on some segments. **(i) Larva (spiracular apparatus):** Anterior median process (AMPc) present on spiracular apparatus (SAp). **(j) Larva (head):** Antennal shaft swollen basally, narrowing beyond seta 1-A; seta 1-A large, multi-branched; seta 2-C indistinctly aciculate; seta 3-C 20–40 branched. Image source: WRBU.

Exemplar DNA sequences: *An. bancroftii* s.l. *ITS*2: *An. bancroftii* genotype A1 (AF203381), genotype A2 (AF203378), genotype B (AF203379), genotype C1 (AF203382), genotype C2 (AF203380), genotype D (AF203383).

Catalog entry: Vol. 2, p. 630.

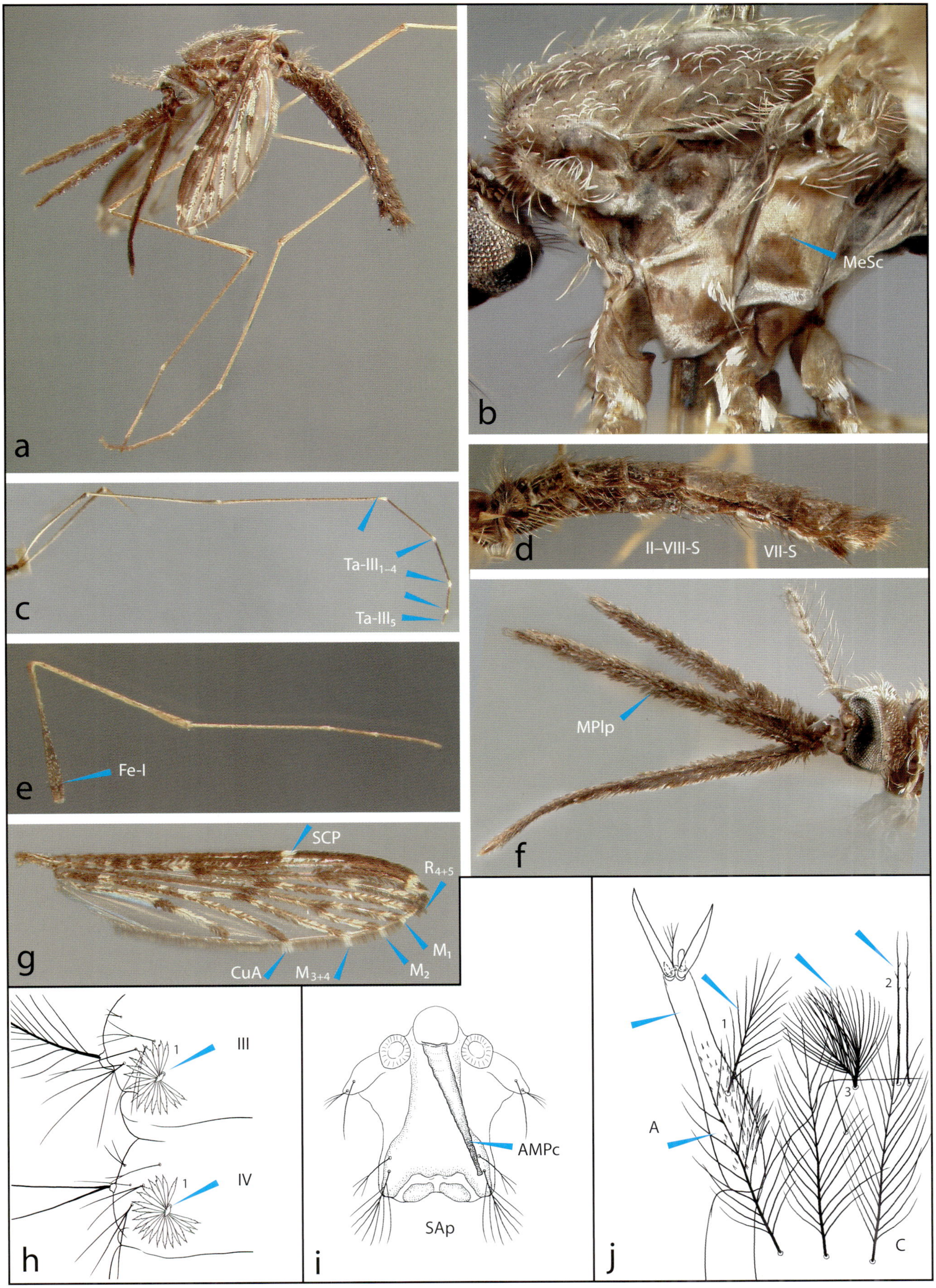
a
b
MeSc
c
Ta-III$_{1-4}$
Ta-III$_5$
d
II–VIII-S
VII-S
e
Fe-I
f
MPlp
g
SCP
R_{4+5}
M_1
M_2
M_{3+4}
CuA
h
1
III
1
IV
i
AMPc
SAp
j
1
2
3
A
C

Anopheles (Anopheles) barbirostris van der Wulp, 1884
Subfamily Anophelinae
"Beard-beaked Indonesian Nail Mosquito"

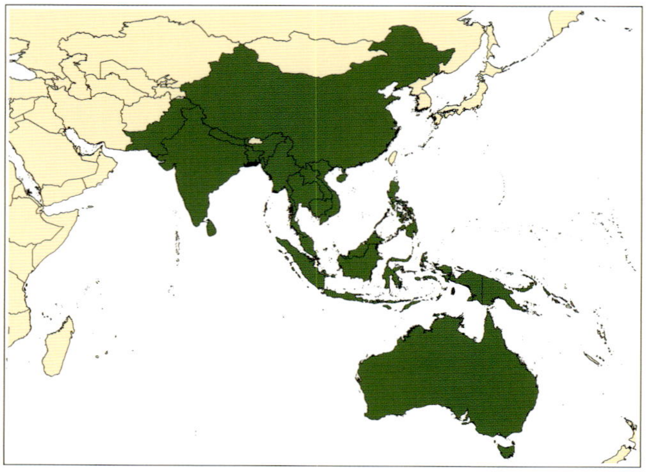

Systematics and Distribution

Anopheles barbirostris takes its name from the Latin for "bearded beak," in reference to its proboscis with erect scales that, together with its extremely shaggy-scaled long palps, gives the species a distinctive top-heavy appearance (see *a*). *Anopheles barbirostris* is the nominotypical member of the Barbirostris Complex, Barbirostris Subgroup, and Barbirostris Group of the Myzorhynchus Series (*Anopheles* subgenus *Anopheles*) (see Vol. 2, p. 1080, for internal relationships). The group has long been recognized for its taxonomic complexity, but a recent series of integrated molecular and morphological studies has formally recognized six taxa in the Barbirostris Complex, stabilizing the internal systematics of the complex. This taxonomic assessment was further complicated by the discovery of two isomorphic *An. barbirostris* s.l. species at the type locality in Mount Ardjoeno in Java. Along with *An. barbirostris* s.s., members of the Barbirostris Complex include *An. campestris* Reid (see p. 300), *An. dissidens* Taai & Harbach, *An. saeungae* Taai & Harbach, *An. vanderwulpi* Townson & Harbach, and *An. wejchoochotei* Taai & Harbach (see p. 300). Another cryptic taxon (*An. barbirostris* species A3) is reported in Kanchanaburi, Thailand, and at least one more is suspected in Sulawesi. Nuclear *ITS2* rDNA and mtDNA *COI* sequences generated from *An. barbirostris* s.l. specimens collected from Kalimantan by the authors do not match any other Barbirostris Group species to date. Two species remain in synonymy with *An. barbirostris*—*martini* Laveran, described from specimens from Cambodia, and *innominata* Stoker & Waktoedi, described from Sulawesi.

Anopheles barbirostris s.l. is reported from the Indian subcontinent eastward across to Vietnam and south through the Southeast Asian peninsula as far as Indonesia and Borneo. Given the recent taxonomic clarification of members of the Barbirostris Complex, *An. barbirostris* s.s. has only been molecularly confirmed in Borneo (Kalimantan), Java, Thailand, and Vietnam. The distribution illustrated is that of *An. barbirostris* s.l. (see map). Further molecular studies are needed to confirm the distributions of all component taxa.

Bionomics

Anopheles barbirostris s.s. is often found in hill and foothill environments, particularly those associated with rice cultivation. In this environment, *An. barbirostris* s.s. is found in almost all stagnant groundwater pools, wells, drainage ditches, and animal footprints, as well as at the margins of flowing rivers and streams. Typically, sites have some vegetation and are in open sunlight or partial shade. The jury is still out as to the malaria vector status of *An. barbirostris* s.s. The species mainly feeds on large mammals, including cows and horses, but will feed on people where its preferred hosts are scarce. *Anopheles barbirostris* s.s. has been collected resting inside human dwellings and in animal shelters. Associated pathogens listed below refer to *An. barbirostris* s.l. and will need to be modified as integrated molecular vector and pathogen confirmation becomes the norm.

Associated Pathogens

Chittoor virus, JBEV; *Plasmodium falciparum*, *P. vivax*, *Wuchereria bancrofti*.

Diagnostic Characters

(a) Habitus (♀): Fe-III mostly dark, without preapical pale band; Ta-II_{1-5} not banded. **(b) Thorax (side):** Antepronotal scales (ApSc) present (not distinct in photo); pleuron with white scale patches. **(c) Abdomen (dorsal):** Sterna (S) with median patches of pale scales and rows of pale scales on lateral margins; VII-S with a tuft of dark scales. **(d) Head (dorsal):** Proboscis (P) entirely dark scaled; palpus (MPlp) with dark, numerous, and erect scales; pedicel (Pe) with dorsal and lateral scales; clypeus (Clp) without scales. **(e) Wing:** Costa (C) lacking a presector pale spot (PSP); wing with three large dark spots on costa (C) and veins R–R_1; apex with two small fringe spots. **(f) Larva (thorax/abdominal segments):** Seta 1-P well developed; seta 1-I well-developed palmate; seta 1 well-developed palmate on ≥5 segments. **(g) Larva (head):** Seta 1-A well developed, long; seta 3-C usually with ≥40 stiff branches. **(h) Larva (abdominal segments):** Palmate seta 1-V. Image sources: WRBU; Harrison & Scanlon (1975).

Exemplar DNA sequences: *An. barbirostris ITS2* (Thailand AB971283, EU812764), (Kalimantan EU812759), *COI* (Thailand AB971312); *An. barbirostris* species A3 *ITS2* (U60410, AB362232–34), *COI* (Thailand AB32238–40); *An. dissidens ITS2* (Thailand AB971284–96, EU812769, EU81278), *COI* (Indonesia AB971313–25); *An. saeungae ITS2* (Thailand AB971297–305, EU812795, EU812791), *COI* (Thailand AB971326–34); *An. vanderwulpi ITS2* (Sumatra EU812766–68). (For *An. campestris* and *An. wejchoochotei*, see *An. campestris* species p. 300).

Catalog entry: Vol. 2, p. 630.

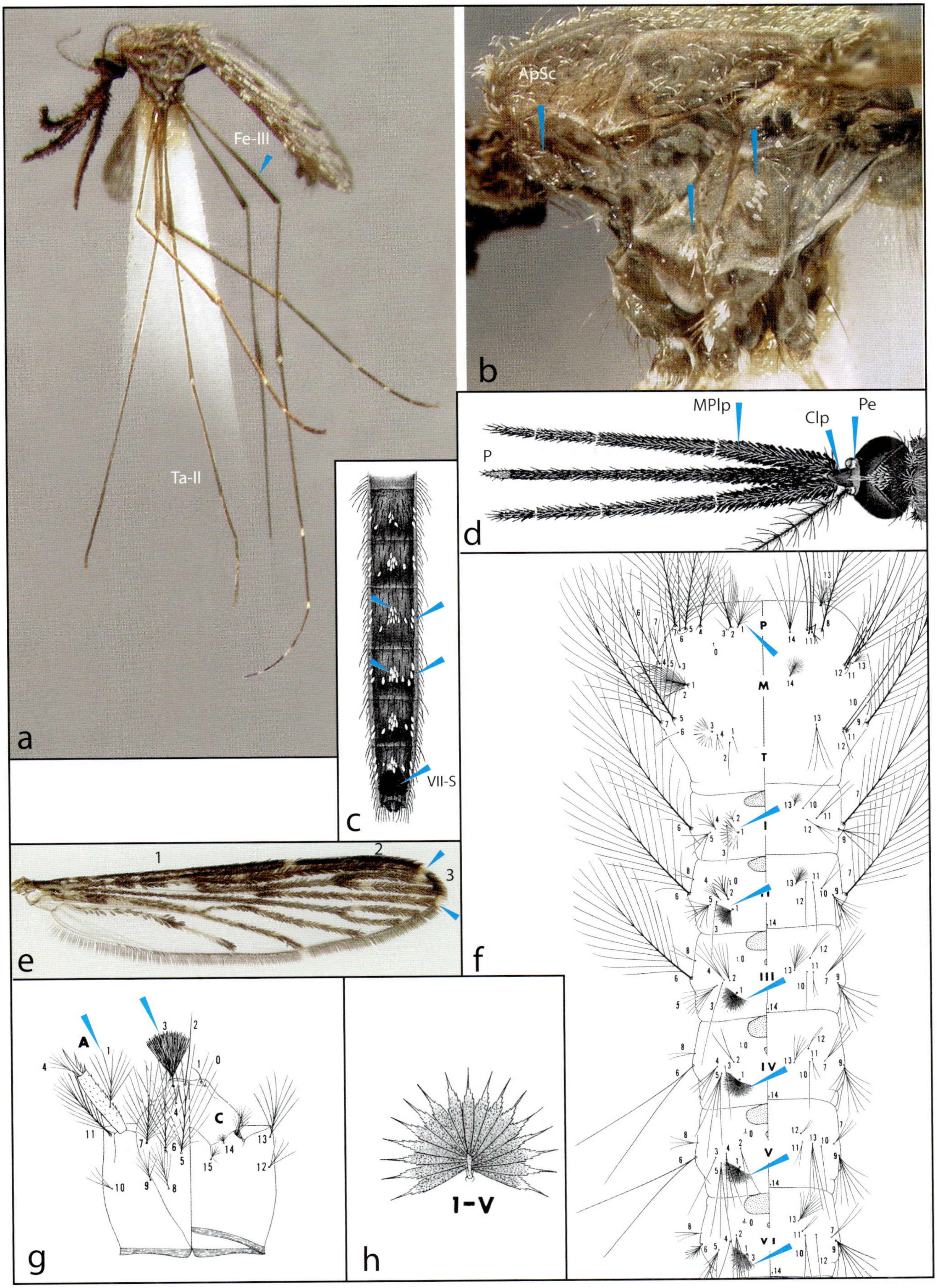
Fe-III
Ta-II
a
ApSc
b
MPlp
Clp
Pe
P
d
VII-S
c
1
2
3
e
P
M
T
I
II
III
IV
V
VI
f
A
C
g
I-V
h

Anopheles (Anopheles) campestris Reid, 1962
Subfamily Anophelinae
"Alluvial Malaysian Nail Mosquito"

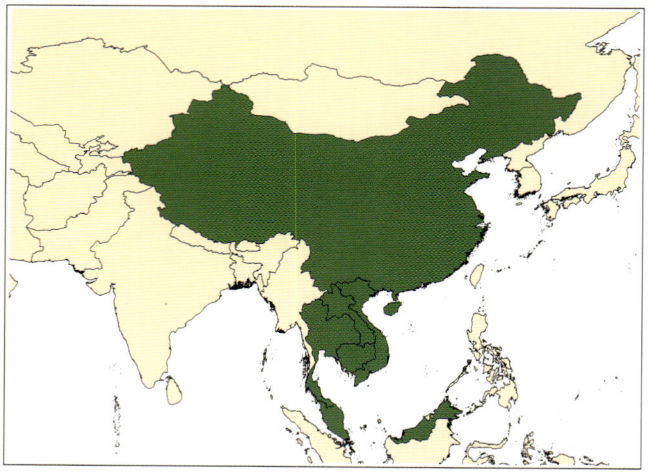

Systematics and Distribution

Anopheles campestris is one of six formally recognized species in the Barbirostris Complex (Myzorhynchus Series, Barbirostris Group, Barbirostris Subgroup) (see Vol. 2, p. 1082, for internal relationships). Other species in the Barbirostris Complex include *An. barbirostris* van der Wulp, *An. dissidens* Taai & Harbach, *An. saeungae* Taai & Harbach, *An. vanderwulpi* Townson & Harbach, and *An. wejchoochotei* Taai & Harbach. *Anopheles wejchoochotei* represents the taxon previously dubbed "*campestris*-like," first separated from *An. campestris* s.s. on the basis of cross-mating studies and divergent mitotic karyotypes. The mitotic karyotype of *An. campestris* has a telecentric Y chromosome, unique among other members of the Barbirostris Complex. Three karyotypic forms have been noted in *An. wejchoochotei.*

The type locality of *An. campestris* is Rantau Panjang, Klang, Selangor, Malaysia, and *An. wejchoochotei* was described from Chiang Mai, Thailand. The current distribution of *An. campestris* s.l. includes mainland Southeast Asia (Cambodia, Malaysia, Thailand, Vietnam), doubtful records from China, and the invasive establishment on Guam (see map). However, since the discovery of *An. wejchoochotei, An. campestris* has only been molecularly confirmed in peninsular Malaysia and in Thailand. *Anopheles wejchoochotei* is widely distributed in Thailand, and the species is likely to occur elsewhere. Further molecular studies are needed to fully determine the distribution of these component taxa.

Bionomics

Named from the Latin "of fields," *An. campestris* is closely associated with low-lying open habitats and broad alluvial plains at elevations of no more than 200 m. *Anopheles campestris* females are reportedly the most anthropophilic species of the Barbirostris Complex and readily enter homes to feed on people. They are regarded as primary malaria vectors, especially in coastal areas of Malaysia. Little is known about the biting habits of *An. wejchoochotei*, but the mother of the type series was collected at human bait, indicating that the species is attracted to humans.

Immatures of *An. campestris* are found in shaded, still, or slow-moving groundwater, with at least some floating, emergent, or submerged vegetation. Typical habitats include ponds, swamps, pools, rice fields, marshes, drainage ditches, canals, wells, and deep animal footprints in pasture lands. Immatures of *An. wejchoochotei* have been collected in rice fields at elevations of 310 m in Chiang Mai, Thailand.

Associated Pathogens

BANV, SIMV; *Plasmodium* spp., *Brugia malayi.*

Diagnostic Characters

(a) Habitus (♀): Fe-III without distal broad preapical white bands. **(b) Thorax (side):** Antepronotal scales (ApSc) present. **(c) Wing:** With three dark marks (denoted as 1, 2, 3) on costa (C) and veins R–R_1; costa without presector pale spot (PSP); apex with two small pale spots (denoted as i, ii), without pale spot at end of vein R_2. **(d) Head (side):** Clypeus (Clp) without patch of scales; palpus (MPlp) shaggy in appearance. **(e) Abdomen (dorsal):** Sterna (S) with many white scales scattered between median patch and lateral rows; VII-S with tuft of black scales. **(f) Larva (thorax):** Seta 1-P with ≥4 branches. **(g) Larva (head):** Seta 1-A with very long branches; seta 3-C multibranched. **(h) Larva (abdominal segments):** Seta 1 palmate on segments I–VII, well developed on segments II–VII. Image sources: WRBU; Harrison & Scanlon (1975).

Exemplar DNA sequences: *An. campestris* (AB331582-88, AB436102-129), *An. wejchoochotei ITS2* (AB971306–311, EU812808–809), *An. wejchoochotei COI* (AB971335–40).

Catalog entry: Vol. 2, p. 632.

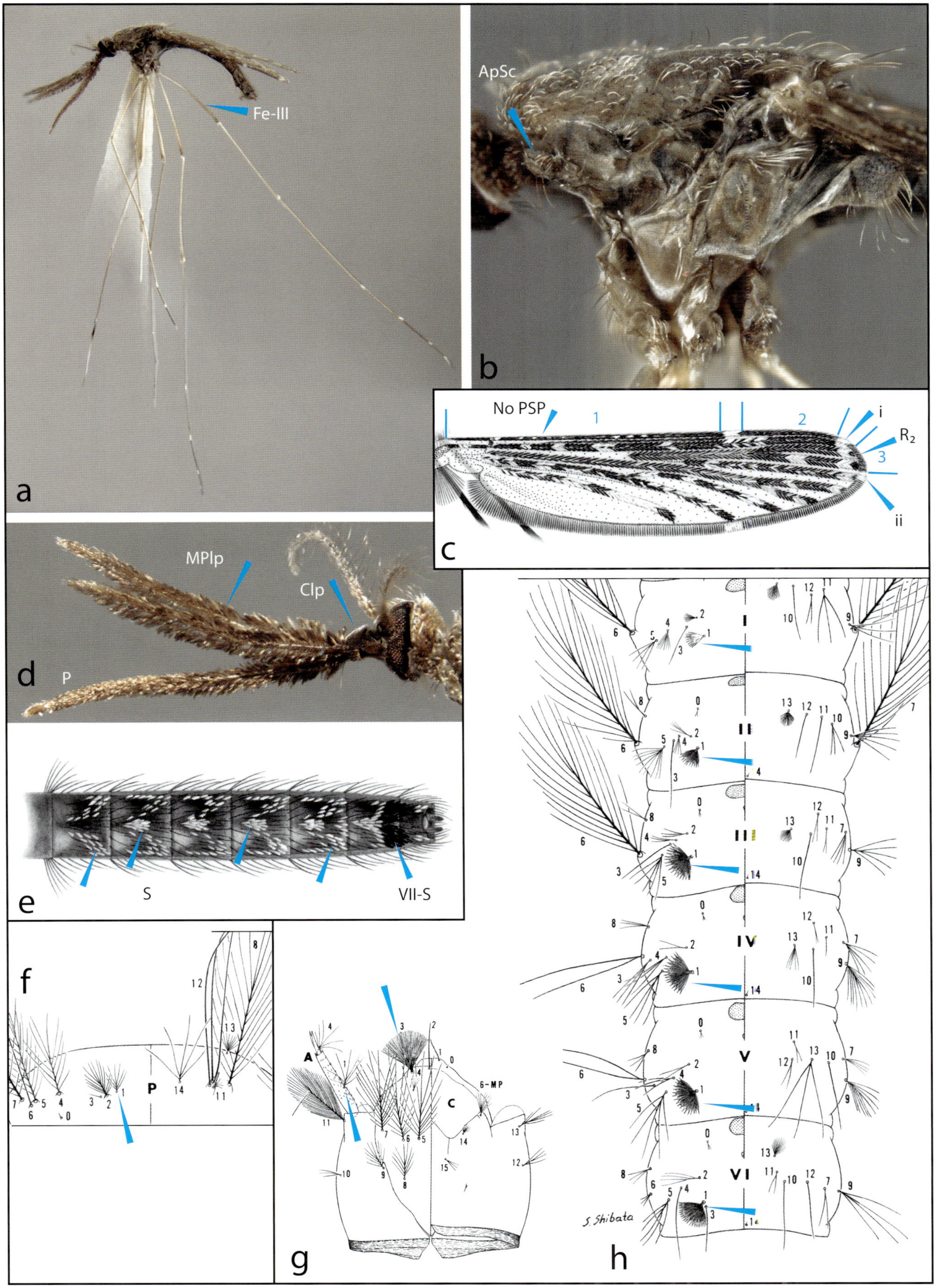
a
Fe-III
b
ApSc
c
No PSP
1
2
3
i
ii
R2
d
MPlp
Clp
P
e
S
VII-S
f
P
g
A
C
6-MP
h
I
II
III
IV
V
VI
S. Shibata

Anopheles (*Anopheles*) *claviger* (Meigen, 1804)
Subfamily Anophelinae
"Club-lipped German Nail Mosquito"

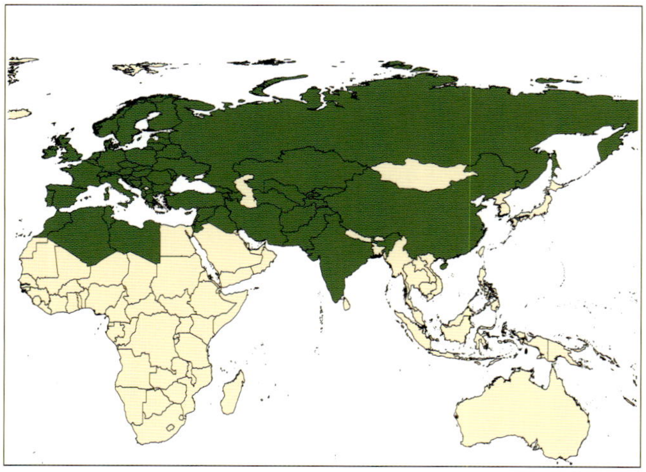

Systematics and Distribution

First described from specimens collected in Germany, *Anopheles claviger* is a large, drab species with entirely dark wings and legs. The species takes its name from the Latin for "club bearing," referring to the shape of the male palps. *Anopheles claviger* represents the Claviger Complex of *Anopheles* subgenus *Anopheles* (see Vol. 2, p. 1080, for internal relationships), which also includes *An. petragnani* del Vecchio. These two taxa are purportedly distinguishable by subtle larval characters and most reliably identified using molecular techniques. During collections made in the United Kingdom, one of the authors determined another novel cryptic taxon, collected and link reared from a woodland pool in sympatry with *An. claviger* s.s. The new species was recovered on the basis of mtDNA *COI* and nuclear *ITS2* sequence differences and is currently pending formal description. Seven synonyms exist for *An. claviger* s.s., including *villosus* Robineau-Desvoidy, *grisescens* Stephens, *turkestani* Shingarev, *amaurus* Martini, *habibi* Mulligan & Puri, *missiroli* Del Vecchio, and *pollutus* Cañamares. It remains unclear if the new taxon conforms to any of these existing valid names. *Anopheles claviger* s.l. occurs throughout the western Palearctic, from the United Kingdom across Europe as far as the Middle East and south to North Africa. While the nominotypic species is found all over the above-mentioned geographic range, *An. petragnani* is largely restricted to the Mediterranean Basin and is sometimes found in sympatry with *An. claviger*.

Bionomics

Adult *An. claviger* s.s. is primarily exophilic, feeding on both humans and large mammals after dusk, and can often be collected resting in animal shelters. The species is multivoltine (producing multiple generations per year), with northern populations peaking in May and September. Females are relatively long-lived, averaging four to five egg cycles per generation, although egg batches are notably smaller in the late summer populations. Unlike most *Anopheles* that oviposit directly on the water surface, *An. claviger* s.s. lays its eggs in the damp substrate slightly above the waterline.

Immature habitats for *An. claviger* s.s. are often in cool, clean waters, such as shaded woodland pools, springs, or wells, and they are commonly collected in cisterns in the Mediterranean. Once common in Tajikistan, noticeably lower densities of *An. claviger* s.s. were correlated to the disappearance of spring-water sources in the 1960s. In colder northerly climates, *An. claviger* s.s. overwinters as larvae and is able to develop and pupate in winter. In warmer southern parts of its distribution, the species is active year-round. *Anopheles claviger* s.s. is a competent malaria vector and indeed was one of the first mosquitoes used to demonstrate experimental transmission of *Plasmodium falciparum* to humans. By contrast, *An. petragnani* tolerates hotter larval habitat waters, appears solely zoophilic, and has never been implicated in human disease transmission.

Associated Pathogens

RVFV; *Plasmodium falciparum*, *P. vivax*.

Diagnostic Characters

(a) Habitus (♀). (b) Thorax (side): Lower proepisternal setae (PeSL) present. **(c) Wing:** Scales uniformly distributed, without concentrations of dark scales. **(d) Leg (I):** Ta-I_1 long, about equal in length to Ta-I_{2-5} combined. **(e) Head (side):** Vertex (V) with distinct tuft of broad white scales, lateral and posterior scales dark brown; labellum (La) and proboscis (P) dark, concolorous; palpomere $MPlp_5$ short, not longer than half the length of $MPlp_4$. **(f) Thorax (dorsal):** Scutum (Scu) with narrow pale scales on broad median area. **(g) Larva (head):** Antenna (A) spiculate; seta 1-A basally branched; seta 2-C inserted close together; setae 2–3-C single or with 2–3 apical branches; seta 4-C shorter than 3-C, 2–5 branched; setae 5–7-C branched, long and plumose. **(h) Larva (abdomen):** Leaflets of palmate setae. **(i) Larva (thorax/abdomen):** Seta 1-I not palmate; seta 1-II palmate with 10–15 leaflets; setae 2-IV,V 3–5 branched; setae 6-IV–VI branched but not plumose; tergal plates (TP) narrow, with height of plate ≤3 times the width. Image sources: WRBU; Ross & Roberts (1943).

Exampler DNA sequences: *Anopheles claviger* s.s. *COI* (KM258237–44; KP942734–36; JX255719–JX255720). *Anopheles petragnani ITS2* (AY129232-33).

Catalog entry: Vol. 2, p. 632.

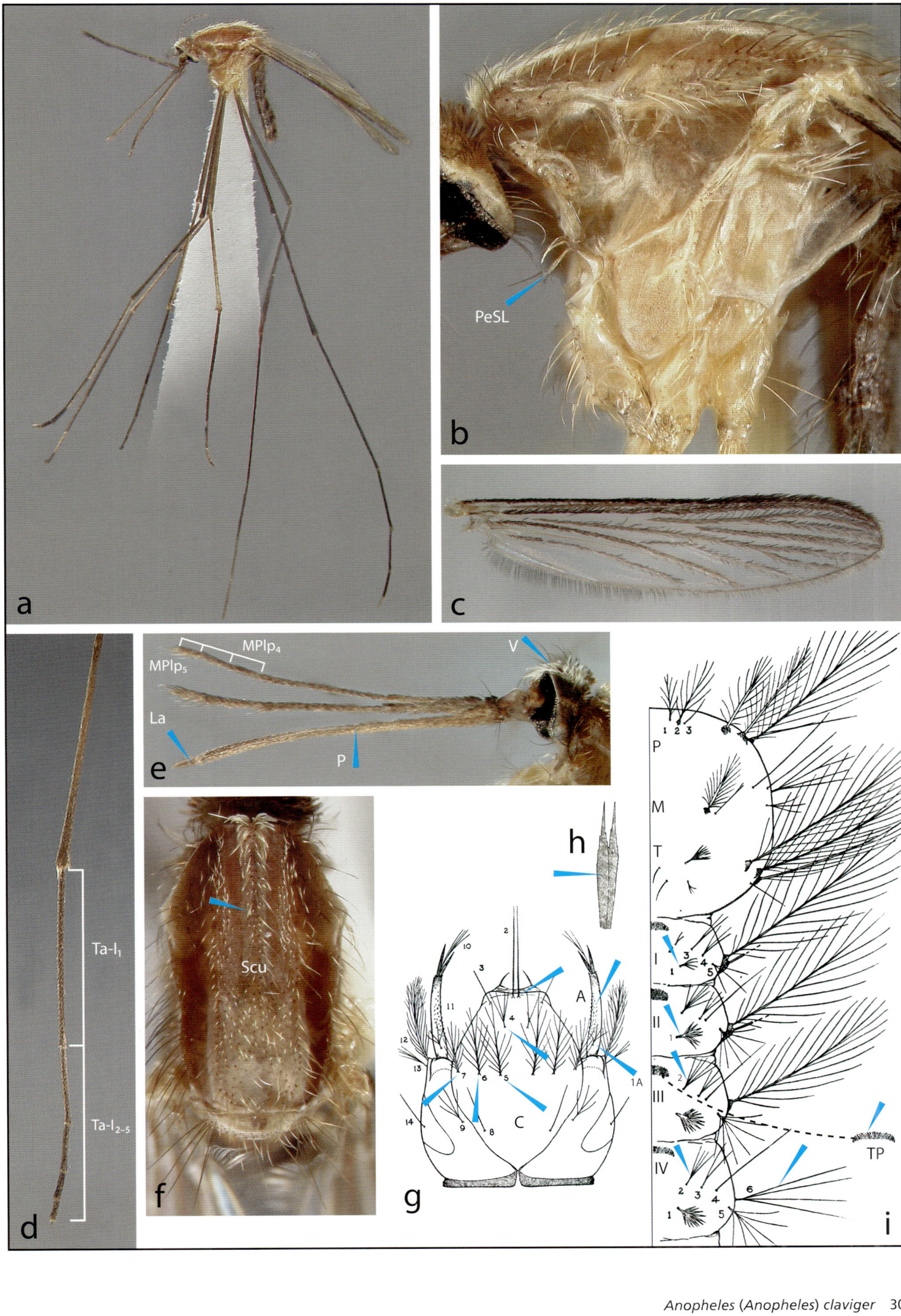
a
PeSL
b
c
MPlp$_4$
MPlp$_5$
V
La
P
e
Ta-I$_1$
Ta-I$_{2-5}$
d
Scu
f
h
A
1A
C
g
P
M
T
I
II
III
IV
TP
i

Anopheles (Anopheles) donaldi Reid, 1962
Subfamily Anophelinae
"Donald Malaysian Nail Mosquito"

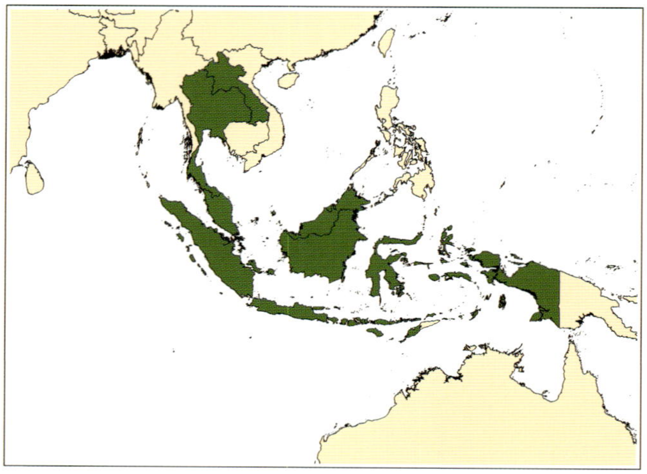

Systematics and Distribution

Anopheles donaldi appears bushy headed, with a plethora of dark, erect scales clothing its palps and proboscis. It is most easily distinguished from other members of the Barbirostris Group at the egg stage, or by the following diagnostic characters in the adults: pale accessory spot on wing fringe at R_2 (see *f*), narrow pale bands on Ta-II and broader pale bands on Ta-I (see *a*, *b*), and numerous pale scales on the abdominal sterna (see *e*). Named in honor of Donald Colless, *Anopheles donaldi* is one of 17 species belonging to the diverse Oriental Barbirostris Group (*Anopheles* subgenus *Anopheles*) (see Vol. 2, p. 1082, for internal relationships). *Anopheles donaldi* is placed within the Barbirostris Subgroup, along with *An. franciscoi* Reid, *An. hodgkini* Reid, *An. pollicaris* Reid, and the six members of the Barbirostris Complex (*An. barbirostris* van der Wulp, *An. campestris* Reid, *An. dissidens* Taai & Harbach, *An. saeungae* Taai & Harbach, *An. vanderwulpi* Townson & Harbach, and *An. wejchoochotei* Taai & Harbach).

Originally described from central Malaysia, the known distribution of *An. donaldi* includes Indonesia, Laos, Malaysia (including Sarawak, Borneo), and Thailand (see map).

Bionomics

Anopheles donaldi is an important localized vector of human malaria. Adults have been associated with jungle-fringe habitats in hilly areas and swamps with dense tree cover in lowland sites. *Anopheles donaldi* is regarded as a primary vector in the Malaysian state of Sarawak in Northeast Borneo, where sporozoite rates of 0.23% have been reported. *Anopheles donaldi* has replaced *An. balabacensis* as the main malaria vector in Kinabatangan, Sabah, due to prolific deforestation in the region. Resting *An. donaldi* take refuge in the dense, cool jungle vegetation, and will readily attack people during the day. Female *An. donaldi* can be found biting during the day in darker places, such as shaded areas among vegetation. The species is anthropophilic but most likely also feeds on domestic animals and wildlife. It is endophilic and will invade houses in search of blood meals at night. *Anopheles donaldi* accounted for 22.8% of all *Anopheles* found in traditional Dayak longhouses in Sarawak and was the dominant *Anopheles* species (49.9%) captured in farm settings. By contrast in deep forest sites, very few specimens were collected at human- or monkey-baited traps. One *An. donaldi* collected in a *Plasmodium knowlesi* incrimination study was reported to be positive for *Plasmodium* oocysts, but it was not explicitly stated if it was *P. knowlesi*.

Immature *Anopheles donaldi* is found in shaded or heavily vegetated freshwater ground habitats, known to include rice fields and river swamps, jungle pools, swamp forest, sedge swamps, and overgrown drainage canals.

Associated Pathogens

Brugia malayi, *Plasmodium* spp., *P. knowlesi* (?)

Diagnostic Characters

(a) Habitus (♀): Fe-III without a preapical pale band; Ta-II_{1-5} with narrow apical pale bands. **(b) Leg (I):** First pale band on Ta-I about as long as Ta-I_5. **(c) Head (side):** Flm_1 only with pale scales. **(d) Head (dorsal):** Clypeus (Clp) without dark scale patch. **(e) Abdomen (side):** I–VI-S each with ≥20 median pale scales; VII-S with tuft of dark scales. **(f) Wing:** Costa (C) without presector pale spot (PSP) or preapical pale spot (PP); wing apex with three small pale spots, center spot near end of vein R_2. **(g) Larva (head):** Seta 1-A with long branches; seta 2-C closely approximated; seta 3-C many branched, broom-like; seta 9-C 5–11 branched. **(h) Larva (terminal segments). (i) Larva (thorax/abdominal segments):** Seta 1 palmate, fully developed on most segments; seta 1-II palmate, usually pigmented; seta 1-P with ≥4 branches. Image sources: WRBU; Harrison & Scanlon (1975).

Exemplar DNA sequences: None.

Catalog entry: Vol. 2, p. 634.

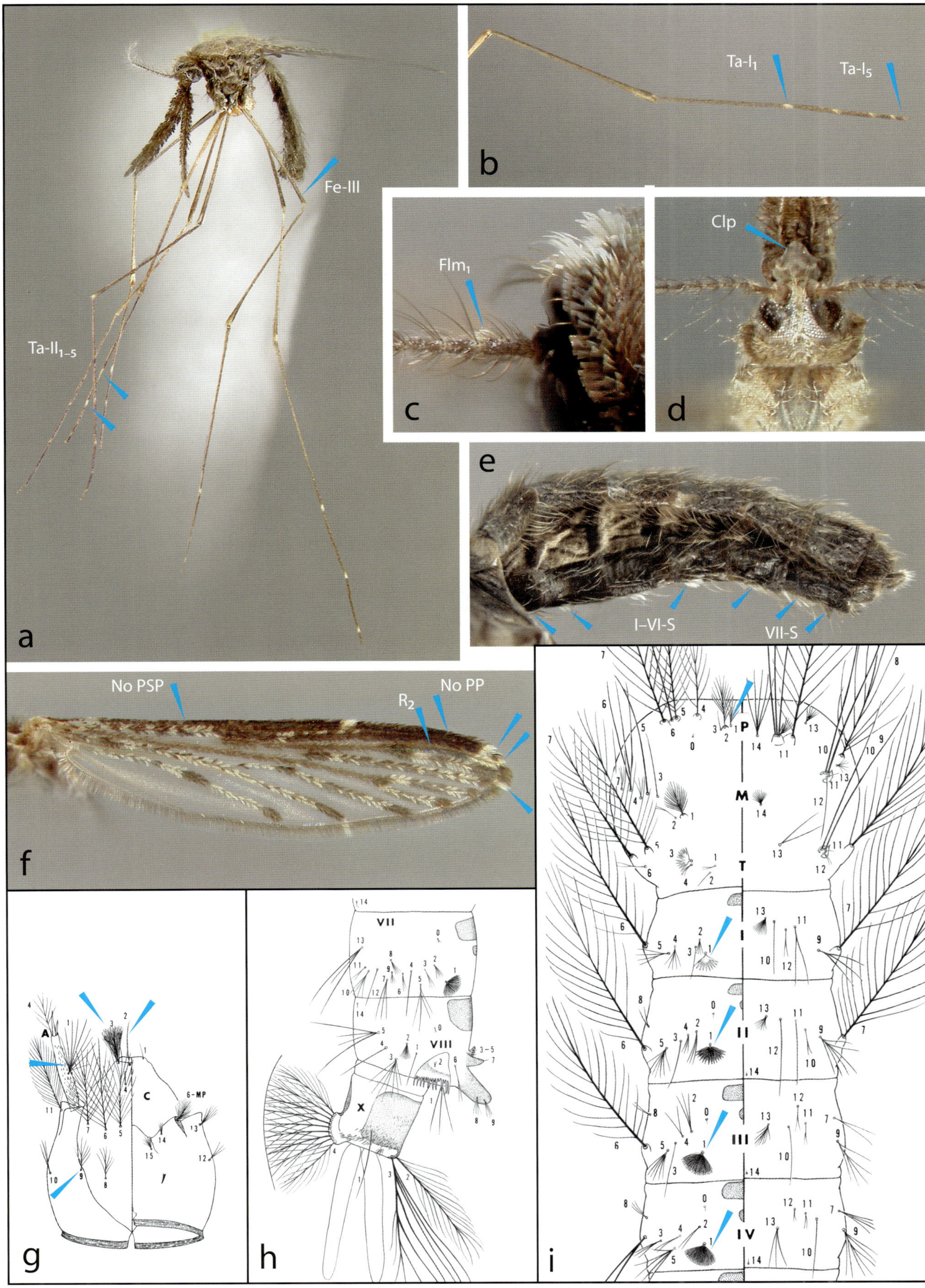

Fe-III
Ta-II$_{1-5}$
a
Ta-I1
Ta-I5
b
Flm$_1$
c
Clp
d
e
I–VI-S
VII-S
No PSP
R$_2$
No PP
f
A
C
6-MP
g
VII
VIII
X
h
P
M
T
I
II
III
IV
i

Anopheles (Anopheles) freeborni Aitken, 1939 Subfamily Anophelinae "Freeborn American Nail Mosquito"

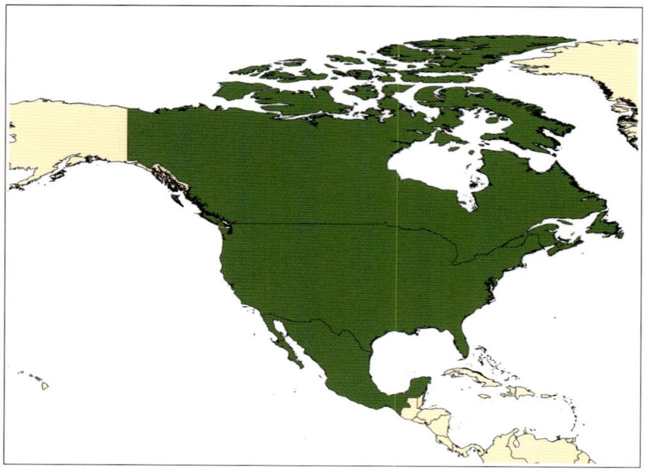

Systematics and Distribution

Named in deference to Stanley Barron Freeborn, *Anopheles freeborni* is a large mosquito with dark legs and palps and attractive spotted wings. Together with *An. earlei* Vargas, *An. hermsi* Barr & Guptavanij, and *An. occidentalis* Dyar & Knab, *An. freeborni* belongs to the Freeborni Subgroup of the Holarctic Maculipennis Group (*Anopheles* Series; Angusticorn Section) of *Anopheles* subgenus *Anopheles* (see Vol. 2, p. 1080, for internal relationships). *Anopheles freeborni* was originally described as a subspecies of *An. maculipennis* Meigen, based on specimens from Davis, California, United States, and was designated full species status two years later. The closely related species *An. hermsi* is restricted to the southwestern United States, where it is often found in sympatry with *An. freeborni*. The species are morphologically indistinguishable, and the presence of *An. hermsi* was only confirmed following crossing experiments, chromosome analysis, and DNA studies. The species can be reliably differentiated using the internal (*ITS*) and external (*ETS*) transcribed spacer regions of the nuclear rDNA.

Anopheles freeborni is found in the western arid and semiarid regions of the United States and Mexico, and up into Canada up to latitudes of under 48°.

Bionomics

Immature *An. freeborni* is often found in rice fields, with highest densities at the field margins where the waters are warmer, and algal growth is plentiful. The immature stages are also found in other shallow ground pools of clear seepage water (e.g., roadside pools) in open sunlight and, more rarely, in heavy shade.

Anopheles freeborni females are known to aggressively attack humans, and will enter houses to feed at dusk and through the night. On cloudy days, or in heavy shade, they also can bite during the day. One study of *An. freeborni* in a rice agriculture system in Northern California showed that over 99% of blood meals were mammalian and, of these, only 3% were from humans, suggesting that *An. freeborni* feeds freely on animals in the absence of human hosts. Rabbits accounted for over 45% of blood meals in the Sacramento Valley. *Anopheles freeborni* is a highly effective vector of *Plasmodium* species, both in the wild and in laboratory settings. *Anopheles freeborni* adults mate freely in confined spaces and are easily colonized, so the species has become a valuable asset in controlled human malaria challenges to trial candidate vaccines. In northern parts of their range, adults will overwinter in diapause.

Associated Pathogens

BUTV, MDV, VEEV, VRV; *Plasmodium berghei*, *P. falciparum*, *P. malariae*, *P. vivax*.

Diagnostic Characters

(a) Habitus (♀). (b) Wing (inset shows wing scales): Wing scales narrow, short, and sparse; wing membrane readily visible on apical half of wing; wing entirely dark scaled (including fringe at apex), with scales darker and/or denser at cross veins and bifurcations, forming faint dark spots. **(c) Head (side):** Palpus (MPlp) unicolorous. **(d) Legs:** Entirely dark. **(e) Head (dorsal):** Scales and setae of interocular space (IS) and pale scales of vertex (V) yellowish white. **(f) Larva (head):** Seta 1-A large, plumose; seta 2-C single, closely approximated; seta 3-C strongly branched (>25 branches); setae 5–7-C large, plumose. **(g) Larva (thorax/abdominal segments):** Seta 1-I,II very small or greatly reduced, not palmate; seta 2-IV,V usually 2–3 branched; seta 6-I–III plumose. **(h) Larva (abdomen):** Leaflets of palmate seta 1-III–VII serrated. Image sources: WRBU; Carpenter & LaCasse (1955).

Exemplar DNA barcodes: *An. freeborni* (AF417717).

Catalog entry: Vol. 2, p. 636.

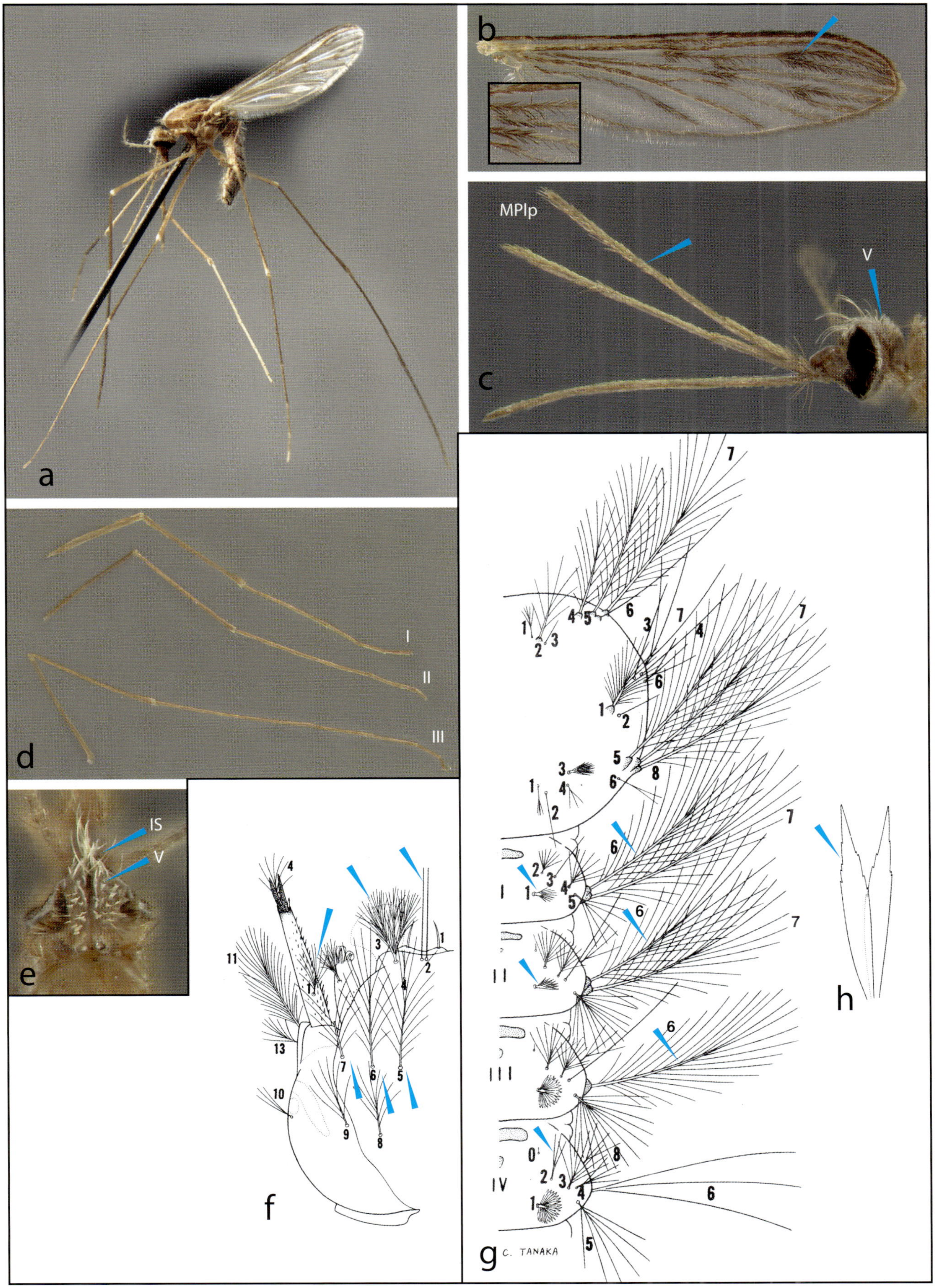
a
b
MPlp
V
c
d
I
II
III
e
IS
V
f
g
C. TANAKA
h

Anopheles (*Anopheles*) *lesteri* Baisas & Hu, 1936
Subfamily Anophelinae
"Leste Philippine Nail Mosquito"

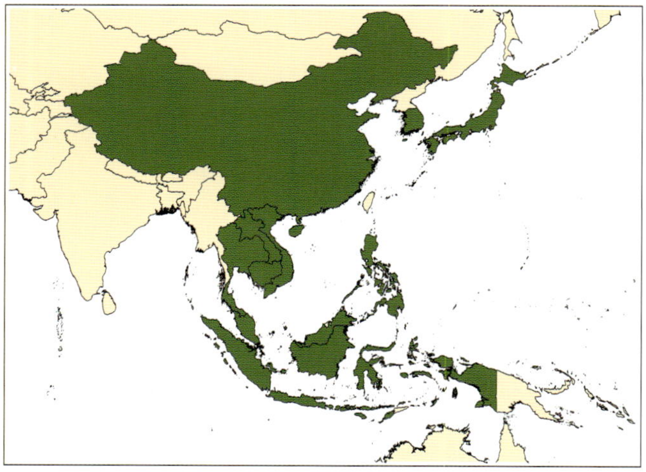

Systematics and Distribution

Anopheles lesteri is a dark species with short apical bands on the legs, and a predominantly dark leading edge to the wing, which also has noticeably absent or reduced presector and humeral pale spots. *Anopheles lesteri* was named in deference to Henry Leste. It is placed in the Lesteri Subgroup of the Hyrcanus Group (Myzorhynchus Series, *Anopheles* subgenus *Anopheles*) (see Vol. 2, p. 1080, for internal relationships), along with *An. crawfordi* Reid, *An. peditaeniatus* (Leicester), and *An. vietnamensis* Nguyen, Tran & Nguyen.

The taxonomic history of *An. lesteri* has been complex, but recent molecular studies have resolved many long-standing queries on the taxonomic status of the species, resulting in four synonyms. These include *paraliae* Sandosham, described from the Malaysian mainland and Borneo, and *kiangsuensis* Xu & Feng, *anthropophagus* Xu & Feng, and *dazhaius* Ma, all described from China. *Anopheles paraliae* was originally regarded to be a subspecies of *An. lesteri* and was later elevated to full species status on the basis of differing wing characters (apical fringe spot is narrower in *paraliae*), and perceived differences in larval habitat. However, DNA analyses of the nuclear *ITS2* rDNA and mtDNA *COI* and *COII* regions showed no species-level genetic differences, and *An. paraliae* was recently reduced to a junior synonym once more. Molecular comparison of the *ITS2* regions of *An. lesteri* and *An. anthropophagous* from China also confirmed the conspecificity of these taxa, and the latter was placed in synonymy with *An. lesteri*.

Anopheles lesteri was first described from Santa Mesa of Manila, Luzon, Philippines, but the type locality is now fully urbanized and lacks typical habitats. The current distribution of *An. lesteri* includes Brunei, Cambodia, China (including Hong Kong), Guam, Indonesia (including Kalimantan [Borneo] and Sulawesi), Japan, Malaysia, the Philippines, Singapore, South Korea, Thailand, and Vietnam.

Bionomics

Anopheles lesteri is a highly competent malaria vector. Females are endophilic and endophagic, readily entering homes to feed and rest. In northwest Malaysia, daily survival rates of *An. lesteri* were estimated at 0.68, and mark-recapture studies examining temporal distributions estimated 12-day gonotrophic cycles. Somewhat unusually in *Anopheles*, *An. lesteri* can survive the winter both as egg or adult stages. *Anopheles lesteri* immatures can be found in shaded, cool impounded water, like that found in rice fields, marshes, ground pools, or ponds. Late in the rice-growing season, when plants shade the deep, cooler water, rice fields become highly prolific sites for *An. lesteri* immatures.

Associated Pathogens

Plasmodium falciparum, *P. vivax*, *Brugia malayi*.

Diagnostic Characters

(a) Habitus (♀). (b) Thorax (side): Antepronotal scales (ApSc) present; C-II with distinct upper patch of pale scales. **(c) Head (side):** Clypeus (Clp) with patches of dark scales laterally; basal 4–8 flagellomeres of flagellum (Fl) with pale scales; palpomere $MPlp_3$ with pale basal band at most as wide as other pale bands. **(d) Wing:** Costa (C) usually without humeral pale spot (HP); pale fringe spot at termination of vein CuA usually present; presector pale spot (PSP) vaguely defined or missing; vein 1A with two dark spots. **(e) Leg (III):** Fe-III without distal broad preapical white band; $Ta\text{-}III_{1-5}$ with apical pale bands only. **(f) Larva (terminal segments). (g) Larva (head):** Antenna (A) spiculate; seta 1-A with branches reaching beyond mid-point on antenna; seta 2-C single, closely approximated; seta 3-C with ≥10 branches. **(h) Larva (thorax/abdominal segments):** Seta 1 well-developed, palmate on ≥5 segments; seta 1-P without branches on basal 0.5, single or with short branches on distal 0.5 (not distinct in illustration). **(i) Larva (abdomen):** Leaflets of palmate seta 1-III–VII serrated.

Exemplar DNA sequences: *An. lesteri* COI (KC855663–64, EU699001–055); *ITS2* (Philippines AY375469, AY375468; South Korea AB733020–22, AY375467; China, as *An. anthropophagus*—AY803792; Thailand, as *An. paraliae*—AB733023–27), *COI* (KC855663–64, EU699001–055). Image sources: WRBU; Tanaka et al. (1979).

Catalog entry: Vol. 2, p. 639.

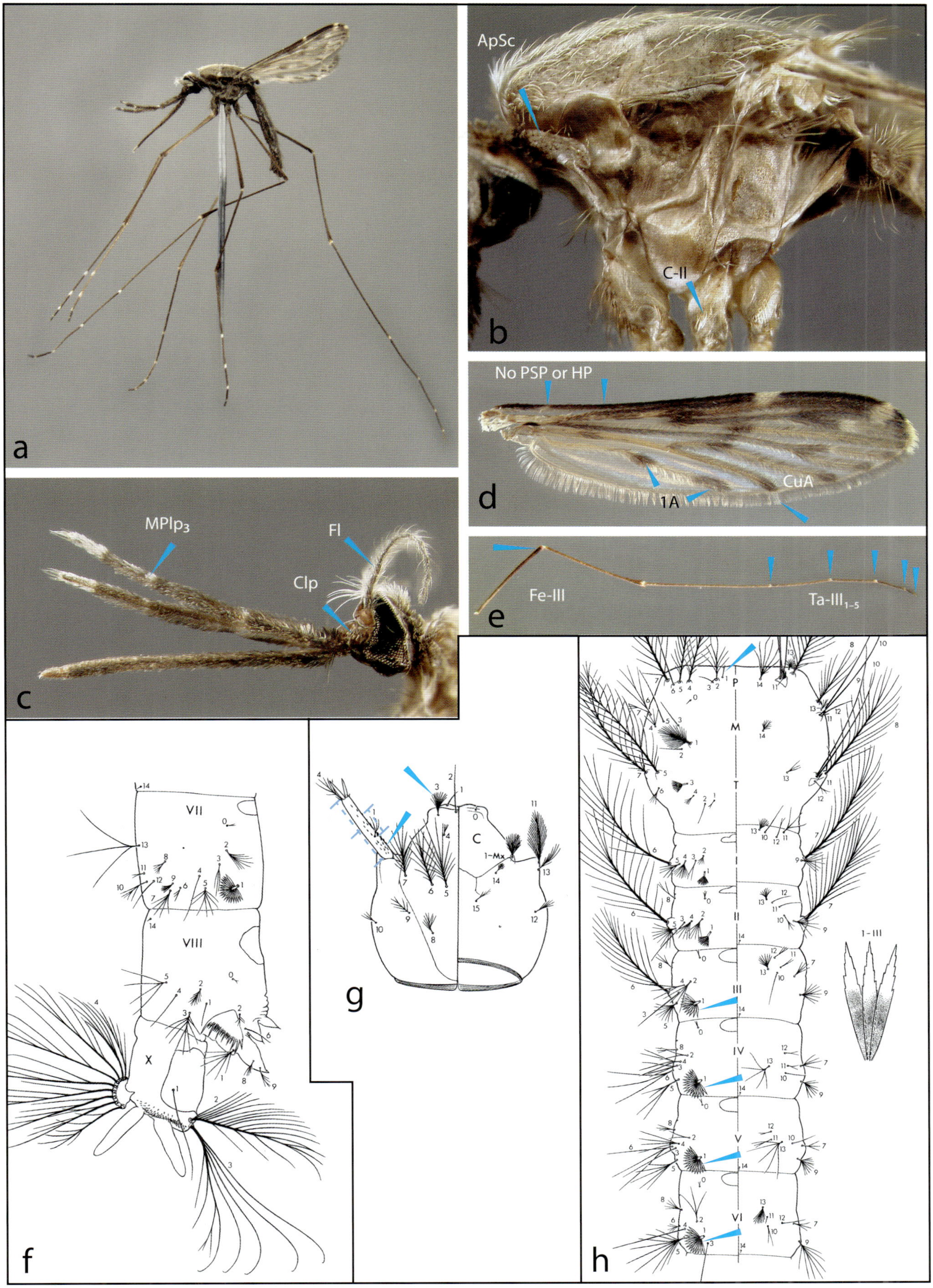
a
ApSc
C-II
b
No PSP or HP
CuA
1A
d
MPlp₃
Fl
Clp
c
Fe-III
Ta-III₁₋₅
e
VII
VIII
X
f
C
1-Mx
g
P
M
T
I
II
III
IV
V
VI
1-III
h

Anopheles (Anopheles) letifer Sandosham, 1944 Subfamily Anophelinae "Deadly Malaysian Nail Mosquito"

Systematics and Distribution

Anopheles letifer is somewhat drab, with unicolorous palps and proboscis, no distinct pale bands on the foretarsomeres, and only small basal pale bands on the hind tarsomeres (see *c*). It is the nominal species of the Letifer Subgroup (Umbrosus Group, Myzorhynchus Series of *Anopheles* subgenus *Anopheles*), which also includes *An. collessi* Mattingly, *An. roperi* Reid, and *An. whartoni* Reid. As adults, *Anopheles letifer* can be distinguished from all other members of the Umbrosus Group in that the base of wing vein CuA is entirely dark, and the base of vein 1A is either entirely pale or has one to five dark scales at the origin (see *d*). In the pupal stage, the T-shaped tragus of *An. letifer* is entirely unique but can be difficult to see (not shown). Many early records of *An. letifer* are recorded under the name *An. umbrosus* (Theobald).

Anopheles letifer was first described in 1944, based on specimens collected in the village of Sijangkang in Selangor, mainland Malaysia. Today the species is reported from flat coastal forested areas of mainland Southeast Asia (Cambodia, Malaysia, Singapore, Thailand, and Vietnam), in Indonesia, and the Malaysian state of Sarawak in northwestern Borneo.

Bionomics

Anopheles letifer is aptly named after the Latin word for "deadly." Although females will feed on monkeys, cattle, and chickens, this species is highly anthrophophilic (up to 90% human blood meals reported) and serves as the primary vector of human malaria in coastal regions of mainland Malaysia and Borneo. It is primarily exophilic, though readily enters houses in forest environments to bite, and biting activity occurs through the night, from dawn to dusk. In heavily shaded sites, females can be induced to bite during the day. Adults rest outdoors during the day and have been collected on tree trunks and on vegetation surrounding cultivated areas. Abundance of adult *An. letifer* peaks during the rainy season, as the availability of suitable immature habitats increases. *Anopheles letifer* and *An. nigerrimus* Giles (see p. XX) are often found in sympatry.

Anopheles letifer is closely associated with peat forests but is never found in virgin swamp forest, which typically marks the start of *An. umbrosus* (Theobald) habitat. The immatures of *An. letifer* are found in the dark-brown, peaty (acidic; pH 5.8 or less) groundwaters in plantations of coconut, pineapple, rubber, and coffee and are commonly associated with the broken forest habitats caused by palm oil plantations. Although immature *An. letifer* have been recovered from brackish water pools in coastal locations in Borneo, females will never oviposit in these environments unless no freshwater pools are available.

Associated Pathogens

Plasmodium falciparum, *P. traugli* (deer malaria), *P. vivax*.

Diagnostic Characters

(a) Habitus (♀): Legs without tuft of dark scales on Fe-III, Ta-II_{1-5}, or Ta-III_{1-5}. **(b) Thorax (side):** Antepronotal scales (ApSc) present; upper proepisternal setae (PeSU) usually absent. **(c) Leg (III):** Fe-III without distal broad preapical white band; Ti-III dark scaled or with small pale spot basally; Ta-III_5 dark scaled. **(d) Wing:** Costa (C) without presector pale spot (PSP); base of veins R and CuA with dark scales; wing apex with two pale spots, without pale spot at vein R_2; vein 1A pale scaled proximal to median dark spot, infrequently with two or three dark scales near base. **(e) Abdomen (dorsal):** VII-S without tuft of black scales. **(f) Head (side):** Palpus (MPlp) unicolorous; clypeus (Clp) without lateral patches of dark scales; flagellomere Flm_1 only with pale scales (indistinct in photo). **(g) Larva (head):** Seta 1-A with very long branches (usually ≤20), reaching well beyond midpoint on antenna; seta 2-C single or with ≤4 branches near tip; seta 4-A with 1–2 stout, slightly hooked branches. **(h) Larva (terminal segments)**. **(i) Larva (abdominal segments):** Seta 0-III usually with <5 branches, at least on one side; seta 1 usually not palmate, at most on two or three segments; seta 6-IV with <7 branches. Image sources: WRBU; Harrison & Scanlon (1975).

Exemplar DNA barcodes: *An. letifer* (KF564694–95).

Catalog entry: Vol. 2, p. 639.

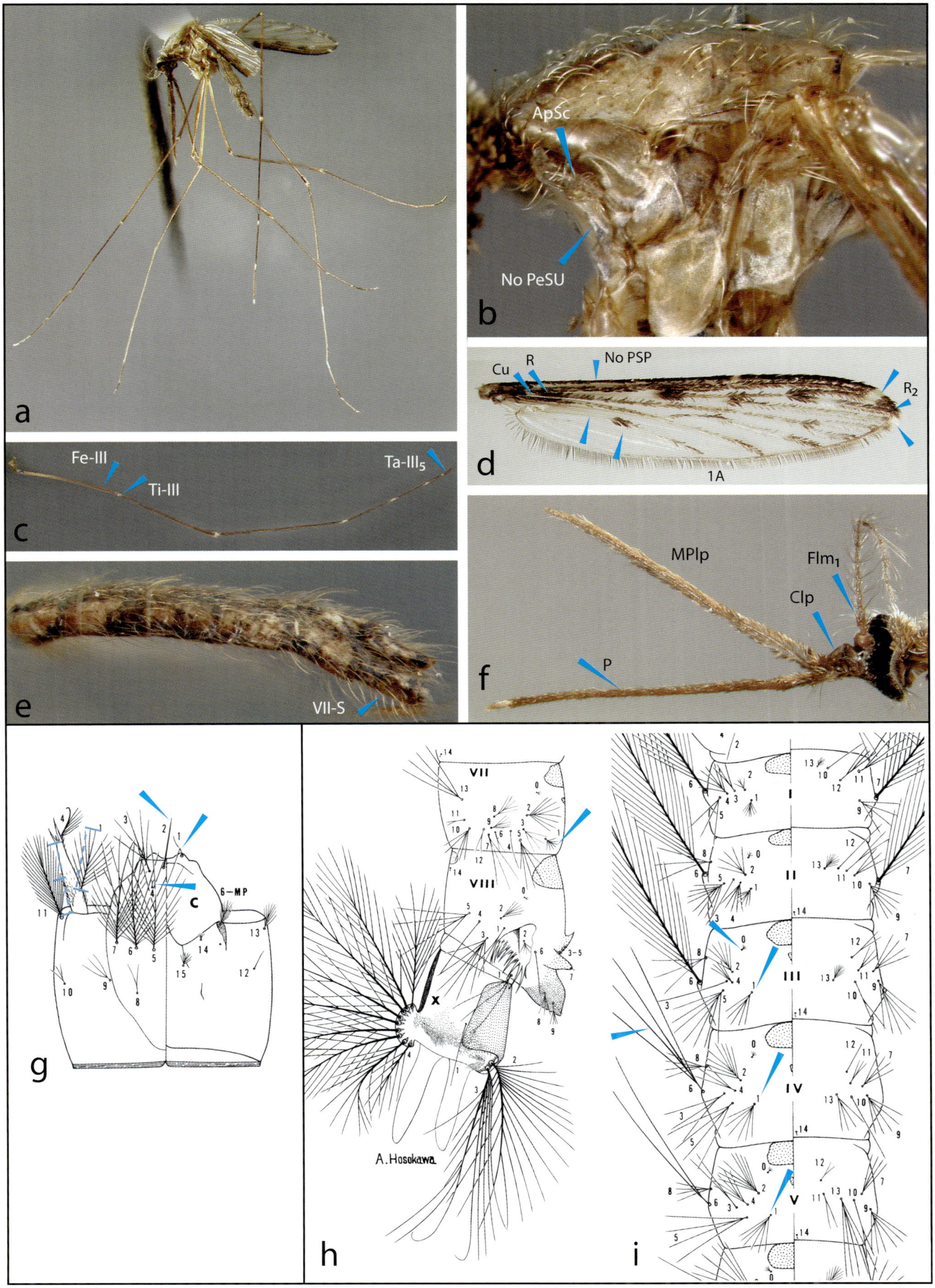
a
ApSc
No PeSU
b
Cu
R
No PSP
R2
1A
d
Fe-III
Ti-III
Ta-III5
c
MPlp
Flm1
Clp
P
f
VII-S
e
6-MP
c
g
VII
VIII
x
A. Hosokawa
h
I
II
III
IV
V
i

Anopheles (Anopheles) maculipennis Meigen, 1818
Subfamily Anophelinae
"Spot-winged German Nail Mosquito"

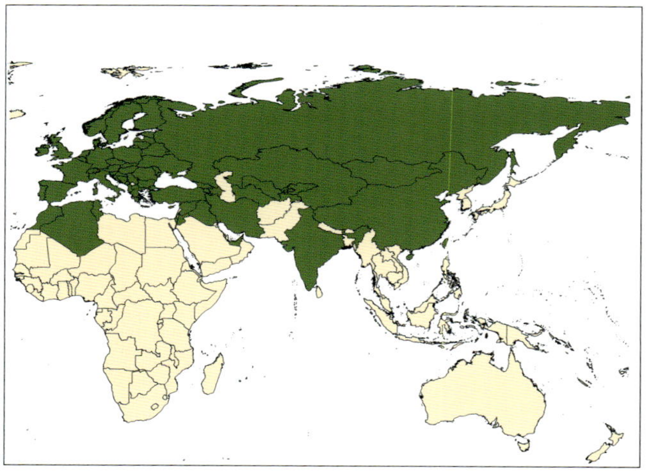

Systematics and Distribution

Anopheles maculipennis is the largest Palearctic *Anopheles*, easily recognized by its charismatic spotted wings—thus, its Latin-derived name. *Anopheles maculipennis* s.s. has three synonyms: *alexandraeschingarevi* Shingarev, *basilii* Falleroni, and *typicus* Hackett & Missiroli.

Anopheles maculipennis is the type species of the genus *Anopheles* and was the first *Anopheles* mosquito to be exposed as a species complex following the discovery of the phenomenon of "Anophelism without malaria" in parts of Europe. The Holarctic Maculipennis Group includes 24 species (see Vol. 2, p. 1081, for composite taxa), separated into the Maculipennis (Palearctic), Quadrimaculatus (Holarctic), and Freeborni (Nearctic) Subgroups, with four Nearctic taxa still unplaced (see other species pages—*An. aztecus* p. 294, *An. freeborni* p. 306, *An. quadrimaculatus* p. 322). The Palearctic Maculipennis Subgroup comprises *An. maculipennis* s.s. and the following nine species: *An. artemievi* Gordeyev, Zvantsov, Goryacheva, Shaikevich & Yezhov, *An. atroparvus* van Thiel, *An. daciae* Linton, Nicolescu & Harbach, *An. labranchiae* Falleroni, *An. martinius* Shinagarev, *An. melanoon* Hackett, *An. messeae* Falleroni, *An. persiensis* Linton, Sedaghat & Harbach, and *An. sacharovi* Favre. Egg morphology remains a valuable tool in the recognition of the component species (see *g*), but most studies now define the species using diagnostic *ITS2* sequences.

Anopheles maculipennis s.s. was originally described from Germany. It is widely distributed in Europe, the Caucasus Region, and the Middle East, and was once intercepted as far away as New Zealand. Once common in Cyprus, the species is believed to have been eradicated from the island following insecticide campaigns. The map indicates the distribution of *An. maculipennis* s.l.

Bionomics

Anopheles maculipennis s.s. is predominantly zoophilic. Although a minor vector when compared to *An. sacharovi*, *An. maculipennis* s.s. can be locally important in malaria transmission—for example, in the Biga Plains of Turkey—and it is implicated in the transmission of a significant number of viruses. Immatures of *Anopheles maculipennis* s.s. are found in highest densities in ponds, lakes, large permanent ground pools, and at stream margins or drainage canals in direct sunlight at altitudes up to 2,300 m. They can also be found in rice fields when the waters are deep. Preferred sites are those with clear fresh water with floating debris, filamentous algal mats, and moderate to abundant emergent vegetation; however, one author collected the species, together with *Cx. hortensis* Ficalbi and *An. superpictus* Grassi (see species page, p. 378), in a foul-smelling, stagnant pool in a drying, rocky riverbed in northern Greece. Immatures are often found in association with other Maculipennis Complex species, including *An. messeae*, *An. daciae*, *An. melanoon*, and sometimes *An. sacharovi*.

Associated Pathogens

AACV, BATV, BTKV, CVOV, MBGV, MYXV, RVFV, WEEV, WNV; *Plasmodium vivax*.

Diagnostic Characters

(a) Habitus (♀): Relatively large, dark, or pale brown mosquitoes. **(b) Thorax (dorsal):** Scutum (Scu) laterally darkened, with distinct anterior tuft of pale scales. **(c) Wing:** Predominantly dark scaled, with more and/or darker scales at cross veins and bifurcations forming dark spots. **(d) Head (side). (e) Larva (thorax/abdominal segments):** Abdominal palmate setae well developed, but less so on first two segments. **(f) Larva (head):** Seta 2-C closely approximated, single or weakly branched apically; seta 3-C multibranched. **(g) Eggs: i.** *An. sacharovi*, **ii.** *An. melanoon* (dark form), **iii.** *An. atroparvus*, **iv.** *An. melanoon* (light form, syn. *subalpinus*), **v.** *An. labranchiae*, **vi.** *An. messeae*, **vii.** *An. maculipennis* s.s., **viii.** *An. beklemishevi*. Image sources: WRBU; Linton et al. (2003).

Exemplar DNA sequences: *An. maculipennis* s.s. *COI* (AF491682–736), *ITS2* (AF342716–22).

Catalog entry: Vol. 2, p. 640.

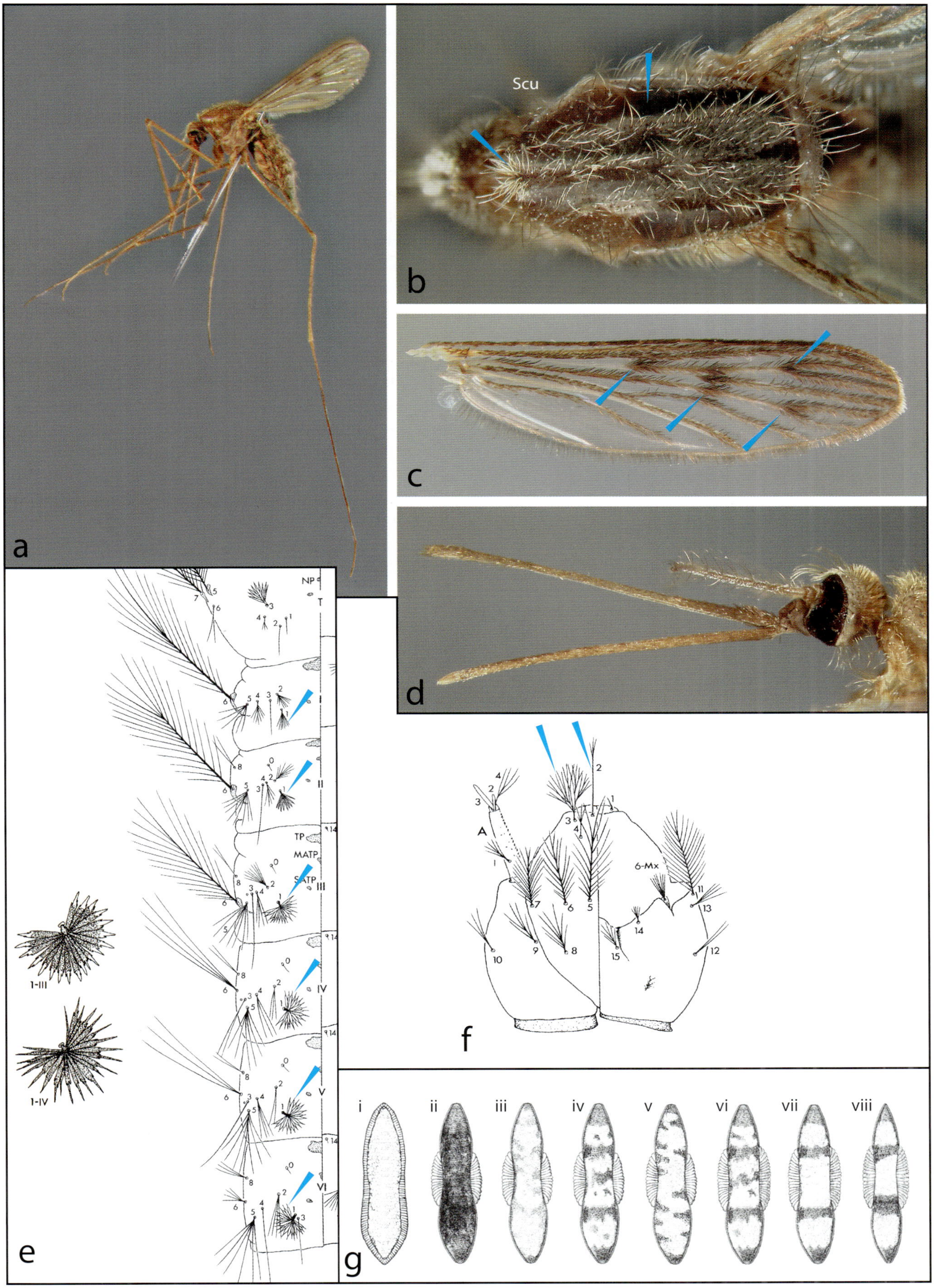
a
b
Scu
c
d
e
NP
T
TP
MATP
SATP
I
II
III
IV
V
VI
1-III
1-IV
f
A
6-Mx
g
i
ii
iii
iv
v
vi
vii
viii

Anopheles (Anopheles) nigerrimus Giles, 1900
Subfamily Anophelinae
"Sable Indian Nail Mosquito"

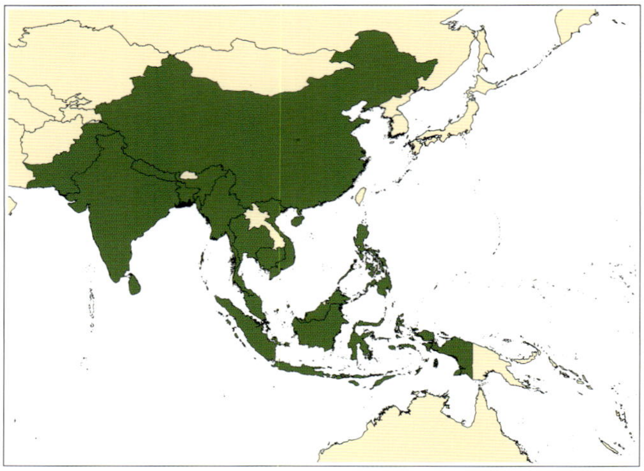

Systematics and Distribution

Anopheles nigerrimus was first described from specimens collected in Calcutta, India, and it takes its name for the Latin for "blackest." It is a large, dark mosquito with a cluster of dark scales on the clypeus and pale scales interspersed on the palps and on the basal 4–8 antennal flagellomeres (see *d*). *Anopheles nigerrimus* belongs to the large Hyrcanus Group (Myzorhynchus Series) (see Vol. 2, p. 1082, for internal relationships), an assemblage of 25 species, many of which are major malaria vectors. Within the Hyrcanus Group, *An. nigerrimus* gives its name to the Nigerrimus Subgroup, which also includes *An. nitidus* Harrison, Scanlon & Reid, *An. pseudosinensis* Baisas, and *An. pursati* Laveran. Five synonyms of *An. nigerrimus* exist: *indiensis* Theobald, *bentleyi* Bentley, *minutus* Theobald, *williamsoni* Baisas & Hu, and *venhuisi* Bonne-Webster.

Anopheles nigerrimus is distributed across the Oriental Region, from Pakistan across the Indian subcontinent, including Sri Lanka, and throughout the Southeast Asian mainland. It has also been reported from the islands of Borneo, Indonesia, and the Philippines.

Bionomics

Anopheles nigerrimus is common in lowlands and river valleys. It is readily collected at human and cattle bait and in light traps, and is often collected in sympatry with the primary malaria vector *An. sinensis* Wiedemann (see species page, p. 324), which is also in the Hyrcanus Group. Adult females are most active just after dusk (1800–2000h) and are attracted to both cattle and humans in Pakistan. The species has been reported to be a competent malaria vector in Celebes, Java, Malaysia, and in the rice areas of Southern Thailand, but has never been found naturally infected in India. In Bangladesh, *An. nigerrimus* was found naturally infected with *Plasmodium falciparum* and *P. vivax* (VK-210). Due to its association with intensive rice agriculture, evidence for insecticide resistance has been noted in this species.

Immatures occupy deep, cooler freshwater sites shaded with dense aquatic vegetation. Typical immature habitats for *An. nigerrimus* include large open marshes, large stream pools, and rice fields. In Sri Lanka, larval sites also included agricultural and earth wells, borrow pits, rock pools, and tank margins. In Malaysia, large densities of *An. nigerrimus* immatures are commonly found in fish ponds with plentiful water lettuce (*Pistia stratiotes*), whereas in Java and Celebes, the species prefers swampy rice fields and deep swamps with copious aquatic vegetation, including *Azolla*, *Eichhornia*, *Jussiaea*, *Hydrilla*, *Pistia*, and *Spirodella*.

Associated Pathogens

Plasmodium falciparum, *P. vivax*, *Wuchereria bancrofti*.

Diagnostic Characters

(a) Habitus (♀). (b) Thorax (side): Antepronotal scales (ApSc) present. **(c) Head (side):** Clypeus (Clp) with lateral patch of dark scales. **(d) Head (dorsal):** Basal 4–8 flagellomeres of flagellum (Fl) with pale scales. **(e) Legs:** Legs without tuft of dark scales; Fe-III without distal broad preapical white band; Ta-III_{1-5} with basal and apical pale bands. **(f) Larva (abdominal segments):** Seta 1 well-developed palmate on ≥5 segments. **(g) Wing:** Costa (C) without presector pale spot (PSP). **(h) Larva (thorax):** Seta 1-P without branches on basal 0.5, single or with short branches on distal 0.5 (indistinct in illustration). **(i) Larva (head):** Seta 3-C 40-plus branches, branched from near base. **(j) Larva (terminal segments).** Image sources: WRBU; Harrison & Scanlon (1975).

Exemplar DNA barcodes: *An. nigerrimus* (AB781757–61, AB778787–99, LC33324).

Catalog entry: Vol. 2, p. 643.

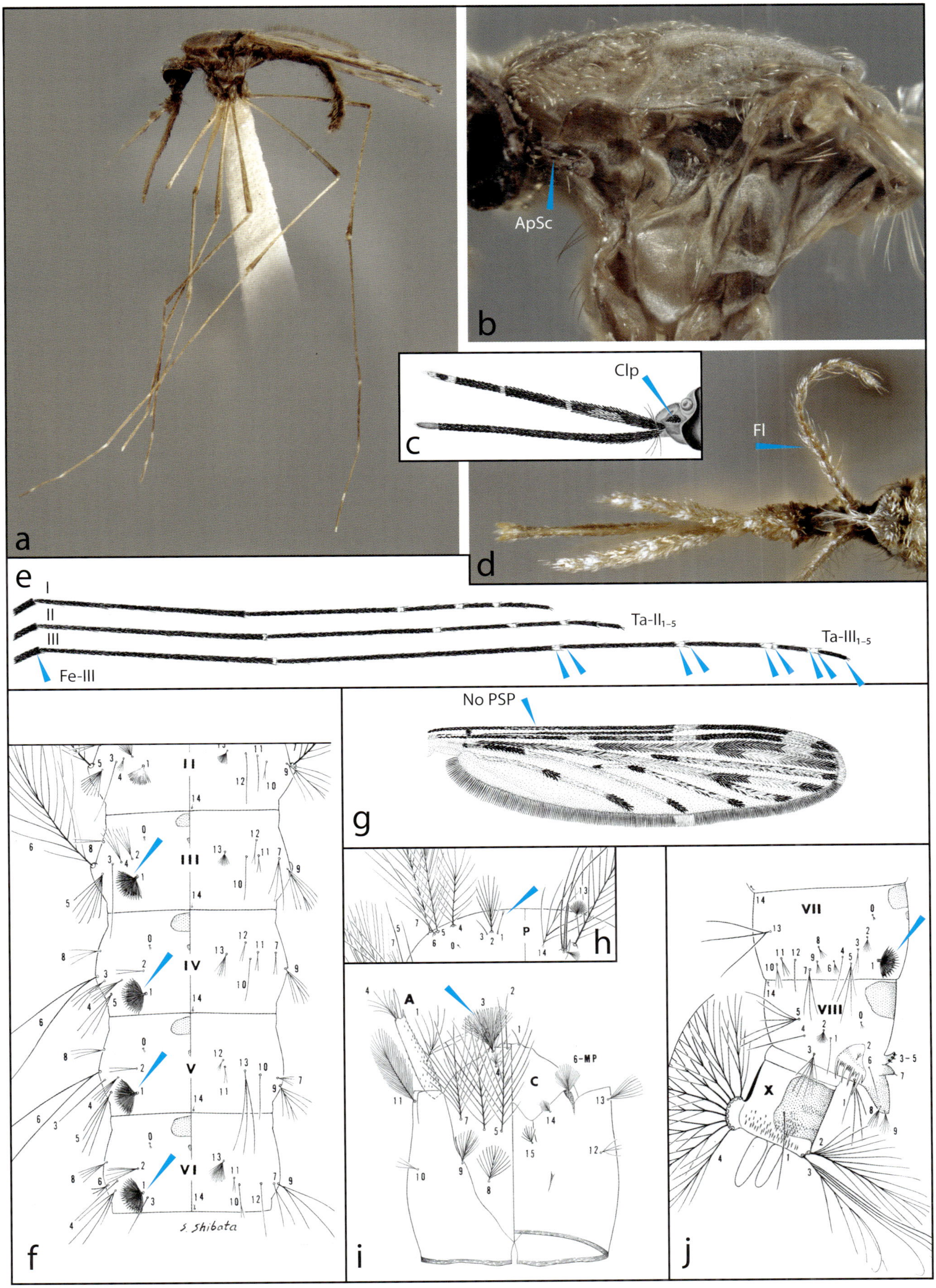
a
b
ApSc
c
Clp
d
Fl
e
I
II
III
Ta-II1–5
Ta-III1–5
Fe-III
No PSP
g
f
II
III
IV
V
VI
S. Shibata
h
P
i
A
C
6-MP
j
VII
VIII
X

Anopheles (*Anopheles*) *pseudopunctipennis* Theobald, 1901
Subfamily Anophelinae
"False-winged Grenada Nail Mosquito"

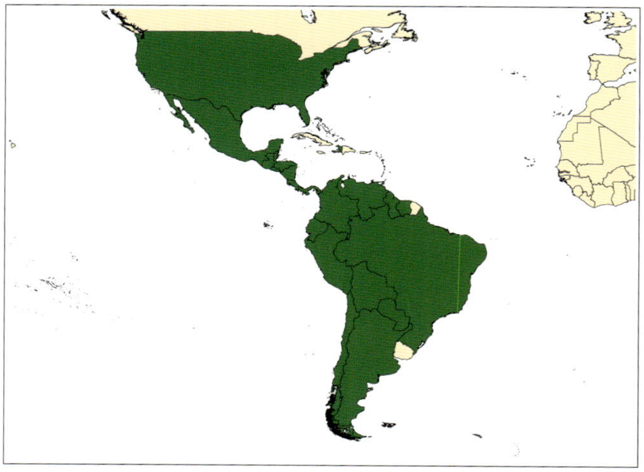

Systematics and Distribution

Anopheles pseudopunctipennis has a shock of erect white scales on its head, stretching forward as a frontal tuft (see *b*), with pale scales and setae continuing over the scutum and scutellum. Its legs are unicolorous, without bands or spots (see *a*). It serves as the nominotypical member of the Pseudopunctipennis Group of the Angusticorn Section (*Anopheles* subgenus *Anopheles*). Other members of the Pseudopunctipennis Group include *An. chiriquiensis* Komp, *An. eiseni* Coquillett, *An. franciscanus* McCracken, *An. hectoris* Giaquinto-Mira, *An. parapunctipennis* Martini, and *An. tibiamaculatus* (Neiva). *Anopheles pseudopunctipennis* is differentiated from other members of the group in that the terminal palpomere (MPlp$_5$) is pale (see *b*), the base of wing vein R$_s$ is white scaled, and the scales on wing vein M are predominately pale (see *d*).

Anopheles pseudopunctipennis is possibly the most highly diverse, taxonomically unresolved *Anopheles* of medical importance. The species currently has five recognized subspecies (*An. levicastilloi* Leví-Castillo, *An. neghmei* Mann, *An. noei* Mann, *An. patersoni* Alvarado & Heredia, *An. rivadeneirai* Leví-Castillo), described from Argentina, Chile, Colombia, and Ecuador. There are three synonyms—*peruvianus* Tamayo, *argentinus*, and *tucumanus* Lahille. Differences in wing pattern have been noted and are illustrated (see *a*). Besides the above, cytogenetics and crossing experiments showed that *An. pseudopunctipennis* likely comprises many more independently evolving lineages. Efforts are clearly needed to resolve the taxonomic status of the species.

Anopheles pseudopunctipennis is a widely distributed malaria vector, found from Argentina and Chile in South America north to the southern United States and on several islands in the Caribbean (Antigua and Barbuda, Curacao, Grenada, Haiti, and Trinidad and Tobago).

Bionomics

Anopheles pseudopunctipennis is associated with valleys and foothills in mountainous areas, where it is important in dry-season malaria transmission. It has been found actively transmitting malaria at elevations up to 2,800 m in Bolivia. In Argentina, California, Guatemala, and Mexico, the species is highly anthropophilic, entering houses to feed and flying considerable distances to find blood meals. In the West Indies and Central America, the species is regarded as exophilic, and in Costa Rica, Grenada, and the Canal Zone in Panama, the species is never found biting indoors.

Immature habitats typically include open, sunlit freshwater stream pools or margins, floodplain ground pools, and drainage ditches, usually with abundant filamentous green algae (e.g., *Spirogyra*), upon which the larvae rely for food and protection from predators. Since heavy rainfall washes away the larvae and reduces available suitable habitat, *An. pseudopunctipennis* populations peak in the dry season.

Associated Pathogens

SAV, SARV, VEEV; *Plasmodium* spp.

Diagnostic Characters

(a) Habitus (♀): Legs mostly dark scaled, without distinct rings or spots; noted differences in wing tip pattern illustrated. **(b) Head (side):** Palpomere MPlp$_5$ pale scaled; vertex (V) and occiput (Occ) with numerous erect scales. **(c) Thorax (dorsal). (d) Wing:** Base of wing vein R$_s$ white scaled; costa (C) with apical pale (AP) and subcostal pale (SCP) spots, otherwise dark scaled; cubitus (Cu) predominantly pale scaled with dark spots at each end; vein 1A with apical 0.5 dark scaled, distal 0.5 pale scaled, and with small pale spot at apex; vein M predominantly pale scaled. **(e) Larva (thorax):** Seta 9-M,T about twice length of 10-M,T. **(f) Larva (head):** Seta 3-C single. **(g) Spiracular apparatus:** Spiracular apparatus (SAp) with two posterolateral spiracular lobe plates (PSLP), each with elongate, slender, sclerotized projection or "tail" from inner caudal margin. **(h) Larva (pecten plate). (i) Larva (terminal segments). (j) Larva (abdominal segments):** Seta 1-III–VII palmate, leaflets with slightly serrated margins, notched on apical 0.34; seta 2-IV single. Image sources: WRBU; Rueda et al. (2004).

Exemplar DNA barcodes: *An. pseudopunctipennis* (AF417721, HM022407, JN412834–38, KF698878).

Catalog entry: Vol. 2, p. 646.

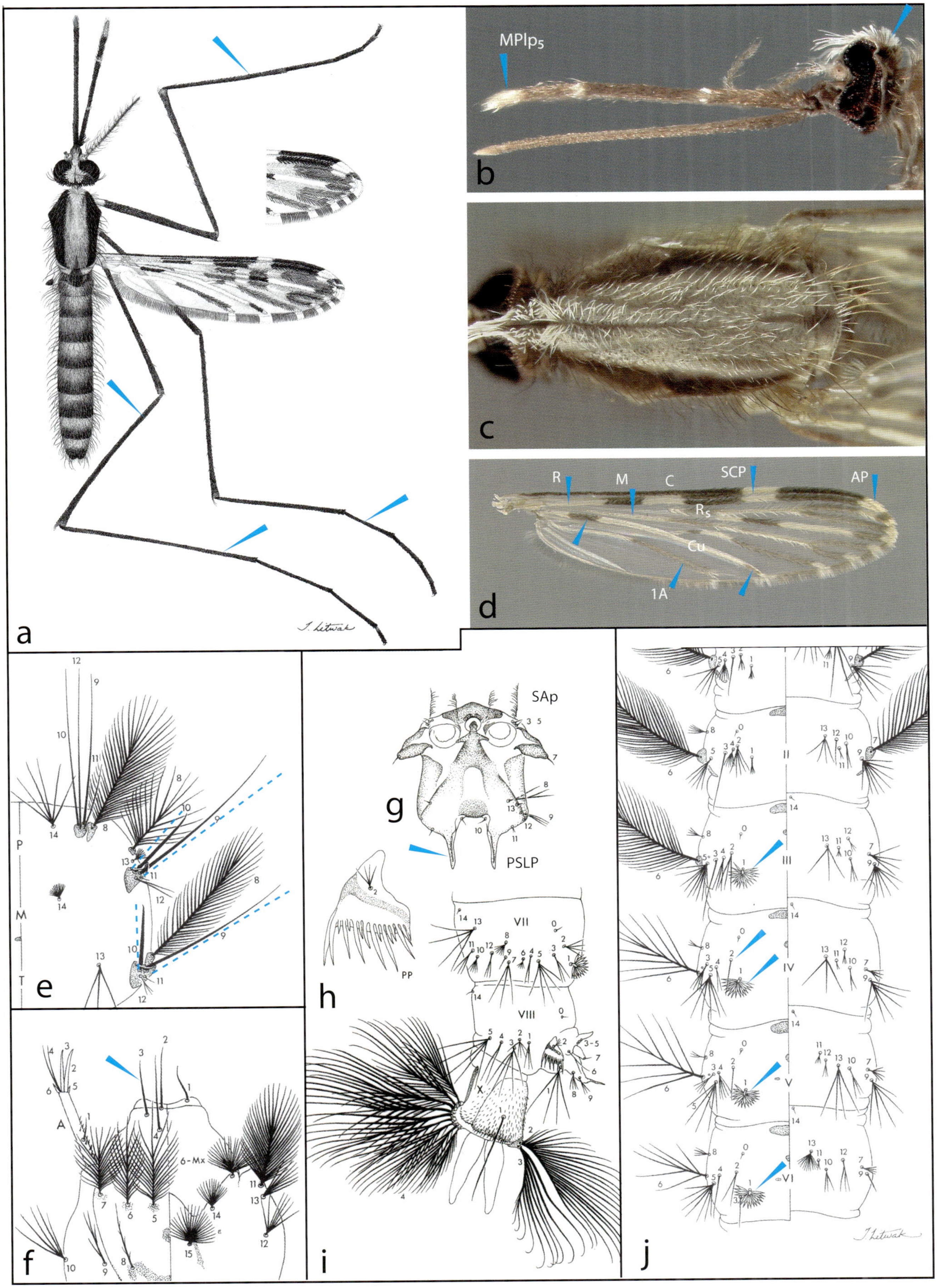
a
b
MPlp5
c
d
R
M
C
SCP
AP
Rs
Cu
1A
e
P
M
T
f
A
6-Mx
g
SAp
PSLP
h
PP
i
VII
VIII
X
j
II
III
IV
V
VI

Anopheles (Anopheles) punctimacula Dyar & Knab, 1906 Subfamily Anophelinae "Dotted Panamanian Nail Mosquito"

Systematics and Distribution

Anopheles punctimacula is an attractive mosquito notable for its spotted wings, after which it was most likely named. *Anopheles punctimacula* is one of 24 unplaced species within the Arribalzagia Series, Laticorn Section, of the *Anopheles* subgenus *Anopheles* (see Vol. 2, p. 1082). The species has two current synonyms—*strigimacula* Dyar & Knab, described from Mexico, and *venezuelae* Evans, from Venezuela. The morphologically similar taxa *An. punctimacula*, *An. calderoni* Wilkerson, *An. malefactor* Dyar & Knab, and *An. guarao* Anduze & Capdevielle are commonly misidentified. Molecular studies using nuclear rDNA and mtDNA *COI* sequences strongly support six independently evolving lineages in *An. punctimacula* populations collected in Panama and Costa Rica. *Anopheles malefactor* was recently elevated from synonymy with *An. punctimacula*, and additional cryptic taxa have since been discovered closely related to both it and to *An. calderoni*, suggesting that the taxonomy of these Neotropical Arribalzagia species are far from resolved.

Anopheles punctimacula is a common Neotropical anopheline found from Mexico through to Argentina and Uruguay in South America. There are no records of the species from any of the islands in the Caribbean.

Bionomics

Anopheles punctimacula s.l. females are predominantly anthropophilic and are found in high numbers feeding and resting in human dwellings. Females are known to feed on livestock, are readily attracted to both human and animal bait, and are easily collected in light traps. In Panama the species is associated with jungle habitats and reportedly flies long distances in search of suitable hosts. *Anopheles punctimacula* s.l. is regarded as a primary malaria vector at high altitudes in Ecuador, Peru, and southwestern Colombia. Although reported naturally infected with *Plasmodium* parasites in Panama, it is not considered a primary vector there.

Anopheles punctimacula s.l. immatures are often found in partially or completely shaded habitats in streams of clear water among vegetation along the margins. They can also be found developing along the margins of swamps, or even in the depths of large swamps. In Ecuador, *An. punctimacula* s.l. immatures were negatively associated with floating algae and positively associated with high levels of dissolved oxygen in the larval habitats.

Associated Pathogens

VEEV; *Plasmodium falciparum*, *P. vivax*.

Diagnostic Characters

(a) Habitus (♀). **(b) Thorax (side):** Upper mesepimeron without pale scales (UMSc). **(c) Abdomen (dorsal):** Abdominal segments with erect or semierect posterolateral scale tufts. **(d) Wing:** Costal vein with a small subcostal dark spot (SCD) (spot at junction with subcostal vein (Sc)); postsubcostal pale spot (POSCP) on the costal vein contiguous with the corresponding spot on vein R_1; sector pale spot (SP) interrupted by accessory sector dark spot (ASD). **(e) Larva (head):** Seta 2-A truncate; seta 3-C 3–8 branched; seta 4-A long with lateral branches. **(f) Palmate seta:** Seta 1 palmate with leaflets notched or distinctly serrate, more developed in segments III–VII. **(g) Larva (thorax):** Seta 1-P branched; setae 9–12-P single. **(h) Larva (terminal segments):** Seta 1-X not inserted on the saddle, sometimes in a notch on the margin of the saddle; tergal plates (TP) three times wider than long. **(i) Legs:** Fe and Ti speckled with abundant large pale spots; Ta-III_5 usually with a dark band or spot. Image sources: WRBU; Wilkerson (1990).

Exemplar DNA barcodes: *An. calderoni COI* (Colombia HQ642964–70, Ecuador HQ642971–74), *ITS2* (HQ622618–HQ622622); *An. punctimacula COI* (Panama HQ622626; Brazil MH844166–69), *An. malefactor COI* (Panama HQ622625); *An. guarao* (none).

Catalog entry: Vol. 2, p. 647.

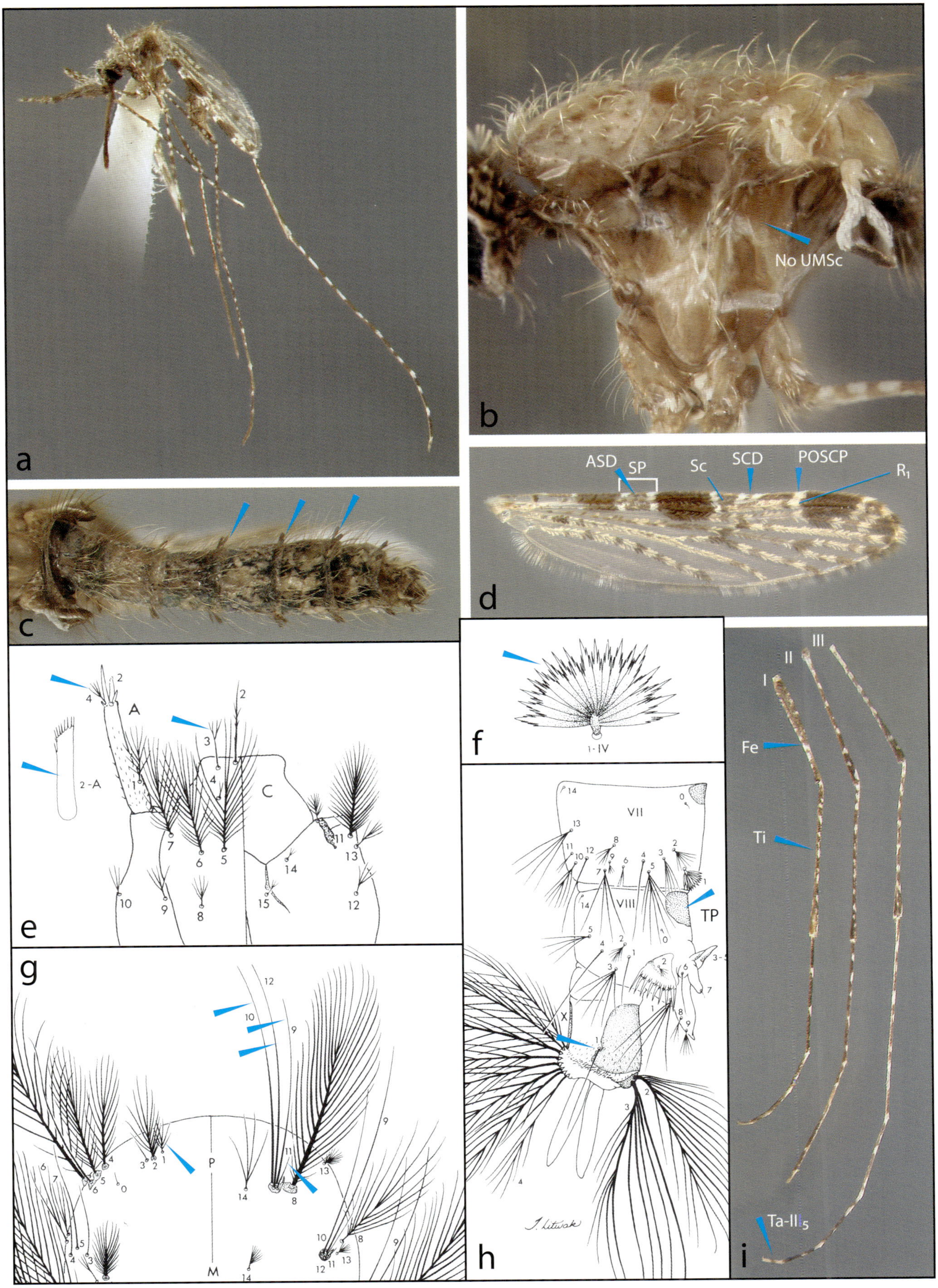
a
No UMSc
b
c
ASD
SP
Sc
SCD
POSCP
R_1
d
e
2-A
A
C
f
1-IV
g
P
M
h
VII
VIII
TP
X
I
II
III
Fe
Ti
Ta-III5
i

Anopheles (Anopheles) punctipennis (Say, 1823)
Subfamily Anophelinae
"Spot-winged American Nail Mosquito"

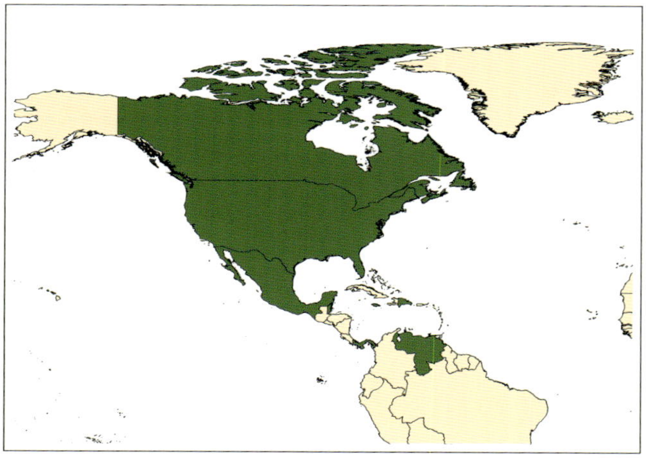

Systematics and Distribution

Anopheles punctipennis has completely dark shaggy palps (see *b*) and distinctly crisp spotted wings (see *a*, *c*). Diagnostic *An. punctipennis* wing markings include vein 1A with both apical and basal dark spots and a yellowish subcostal pale spot (variable in size) on the costa and vein R_1 (see *c*). *Anopheles punctipennis* is the nominotypical member of the Punctipennis Group of the Augusticorn Section (*Anopheles* subgenus *Anopheles*) (see Vol. 2, p. 1081, for internal relationships). *Anopheles punctipennis* has two synonyms—*hyemalis* Fitch, described from the eastern United States, and *stonei* Vargas, described from Mexico. *Anopheles perplexens* is the only other currently described member of the Punctipennis Group. The two are reliably separated on egg and larval characters. Wing characters have been proposed to separate the taxa as adults, but these appear to be unreliable. Wing markings illustrated for *An. punctipennis* (see *c*) also pertain to *An. perplexens*.

Cross-mating studies, nuclear rDNA *28S*, and DNA barcoding studies suggest the presence of at least three distinct taxa within *An. punctipennis* s.l. Unidirectional sterility was noted between Mexican females and South American males, distinct *28S* rDNA signatures were noted between eastern and western populations in the United States, and our DNA barcoding studies have shown the presence of three species-level lineages in *An. punctipennis* s.l. across its range. Efforts are needed to elucidate the component taxa within *An. punctipennis* s.l.

Anopheles punctipennis was originally described from Chestertown in Kent County, Maryland, United States. Its current patchy distribution (as s.l.) stretches from Canada to Venezuela and includes the Caribbean islands of the Dominican Republic and Jamaica.

Bionomics

Anopheles punctipennis s.l. is a highly competent laboratory vector of malaria, but it plays a secondary role in transmission in the wild. Adults are common in hilly areas of dense woodland. Peak biting occurs outdoors after dusk, but they will occasionally venture inside dwellings to feed and can be enticed to bite during the day in shaded woodland environs. Other hosts include deer. The species appears bivoltine, with population peaks in early spring and late fall. Females overwinter as adults, often in human habitations or outhouses where they will occasionally bite.

Immature *An. punctipennis* s.l. is found in a wide array of freshwater habitats, both permanent and temporary. Reported immature habitats include ponds, stream pools, ground pools, springs, animal wallows, roadside puddles, and wheel ruts. The immatures are also collected in artificial containers, including rainwater barrels.

Associated Pathogens

TENV; *Dirofilaria immitis*, *Plasmodium falciparum*, *P. vivax*.

Diagnostic Characters

(a) Habitus (♀). (b) Head (side): Palpus (MPlp) completely dark scaled. **(c) Wing:** Subcostal pale spot (SCP) on costa (C) and vein R_1 of variable size, yellowish; vein 1A with basal and apical dark spots. **(d) Larva (head):** Seta 1-A branched, inserted on basal 0.25–0.50 of antenna; seta 2-C single, without fine aciculae, set close together (< width of one alveolus); seta 3-C with 15–40 or more branches; setae 2,3-C inserted anterior to seta 4-C. **(e) Larva (abdomen):** Leaflets of palmate seta. **(f) Larva (thorax/abdominal segments):** Segment II with very small accessory tergal plate (ATP) just posterior to tergal plate (TP); seta 0-IV,V minute; seta 1-II usually rudimentary, not well-developed palmate; seta 1-III–VII palmate, large, well developed; seta 6-VI very small (not shown). Image sources: WRBU; Carpenter & LaCasse (1955).

Exemplar DNA barcodes: *An. punctipennis* s.l. (KT111921, KR653634, KR669447, KR666470, KR665669, KR670272).

Catalog entry: Vol. 2, p. 648.

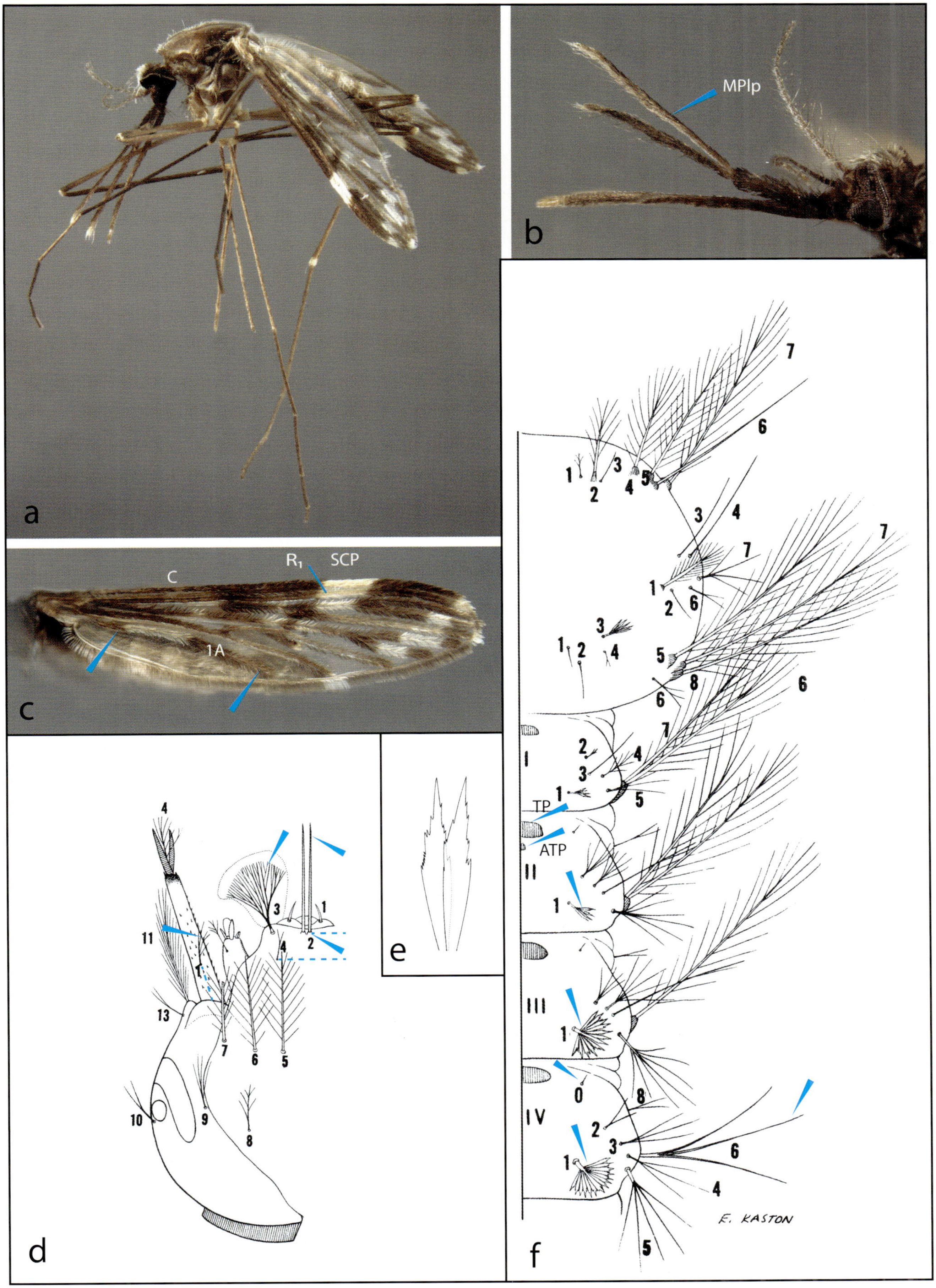
a
b
MPlp
c
C
R_1
SCP
1A
d
4
11
13
1
3
1
2
4
7
6
5
10
9
8
e
f
7
6
1
2
3
4
5
3
4
7
1
2
6
3
1
2
4
5
8
6
6
I
2
3
4
7
1
5
TP
ATP
II
1
III
1
0
8
IV
2
3
1
6
4
5
E. KASTON

Anopheles (*Anopheles*) *quadrimaculatus* Say, 1824
Subfamily Anophelinae
"Four-spotted American Nail Mosquito"

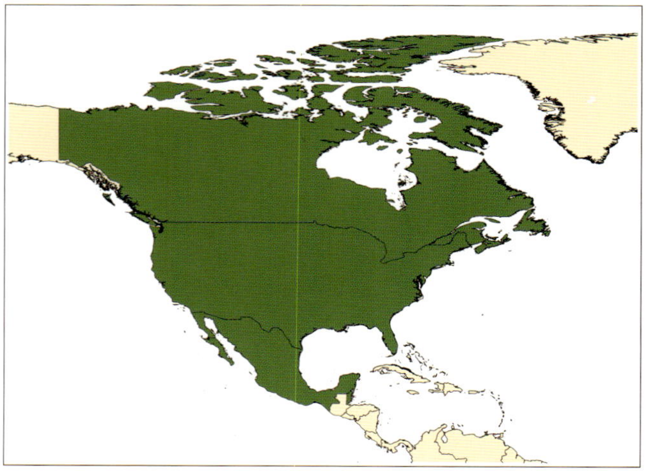

Systematics and Distribution

Anopheles quadrimaculatus is a large mosquito named after the distinctive four dark spots on its wings. The species lends its name to the Quadrimaculatus Subgroup of the Holarctic Maculipennis Group (Anopheles Series, Angusticorn Section) of *Anopheles* subgenus *Anopheles*, where it is placed with four other closely related taxa—*An. diluvialis* Reinert, *An. inundatus* Reinert, *An. maverlius* Reinert, and *An. smaragdinus* Reinert). The latter four species are essentially isomorphic with respect to each other and *An. quadrimaculatus* and were described following a series of genetic crossing, isozyme, and cytological studies. DNA-based assays are routinely used to differentiate these taxa in biosurveillance studies. Also placed within the Quadrimaculatus Subgroup is the Palearctic species *An. beklemishevi* Stegnii & Kabanova, which appears basal to all Nearctic taxa in phylogenetic treatments. *Anopheles quadrimaculatus* has one valid synonym—*annulimanus* van der Wulp.

The type locality of *Anopheles quadrimaculatus* s.s. is Wabasha, Minnesota, United States, but the species has been reported from southeastern Canada and the eastern United States, on the Caribbean islands of Costa Rica, Cuba, the Dominican Republic, and Puerto Rico, and down through Central America into Belize. The distributions of *An. diluvialis*, *An. inundatus*, *An. maverlius*, and *An. smaragdinus* are limited to the southeastern United States.

Bionomics

Anopheles quadrimaculatus often occurs in incredibly high numbers, especially in the Gulf of Mexico. Traps set in the Mississippi Delta often collect over 7,000 mosquitoes per night, 99% of which are *An. quadrimaculatus*. These extraordinary densities, coupled with a ferocious appetite for human blood, secured the species major malaria vector status. It played a dominant role in malaria endemics in the pre-1950s in the southeastern United States. Fertilized adult females gather in high densities in basements, sheds, and tree hollows to overwinter, even in the southern parts of their range where larvae are still present.

Typical immature habitats include slightly alkaline fresh water in open sunlit ponds, swamps, bayous, and borrow pits with abundant floating and emergent vegetation, but *An. quadrimaculatus* larvae have also been reported in many "odd" sites—tin cans, barrels, cornfield furrows, and even in heavily polluted waters. The larvae feed on algae, rod-shaped bacteria, and protozoa.

Associated Pathogens

ACAV, APEUV, BUNV, CARV, CATUV, CBV, CVV, CWV, EHDV, GMAV, GROV, KOTV, MAYV, MCOV, MODV, MURV, NTAV, ORIV, OROV, RBV, SFV, TENV, WNV, VEEV, ZEGV; *Dirofilaria immitis*, *Plasmodium berghei*, *P. falciparum*, *P. vivax*.

Diagnostic Characters

(a) Habitus (♀): Palpus (MPlp) all dark scaled. **(b) Thorax (dorsal):** Length of dorsal dark scutal setae <0.5 times width of scutum (Scu). **(c) Head/thorax (front):** Vertex (V) and frontal tuft (FT) with pale scales and setae; C-I without black scales. **(d) Thorax (halteres):** Capitellum (Ca) of halter (Hl) dark scaled. **(e) Wing:** Wing without contrasting pale markings and with four distinct concentrations of dark scales; apex of wing all brown scaled; vein 1A dark scaled. **(f) Larva (head):** Seta 2-C usually single or finely aciculate, separated by width of one alveolus or more. **(g) Larva (head):** Seta 3-C with 15–40-plus branches. **(h) Larva (head):** Seta 1-A branched, inserted on basal 0.25–0.50 of antenna; seta 4-C inserted posterior to 2,3-C. **(i) Larva (thorax/abdominal segments):** Seta 1-II palmate, usually well developed; seta 6-VI minute, branched. Image sources: WRBU; Reinert et al. (1990).

Exemplar DNA barcodes: *An. quadrimaculatus* s.s. (GU908030–41, KT707126, KR663155).

Catalog entry: Vol. 2, p. 648.

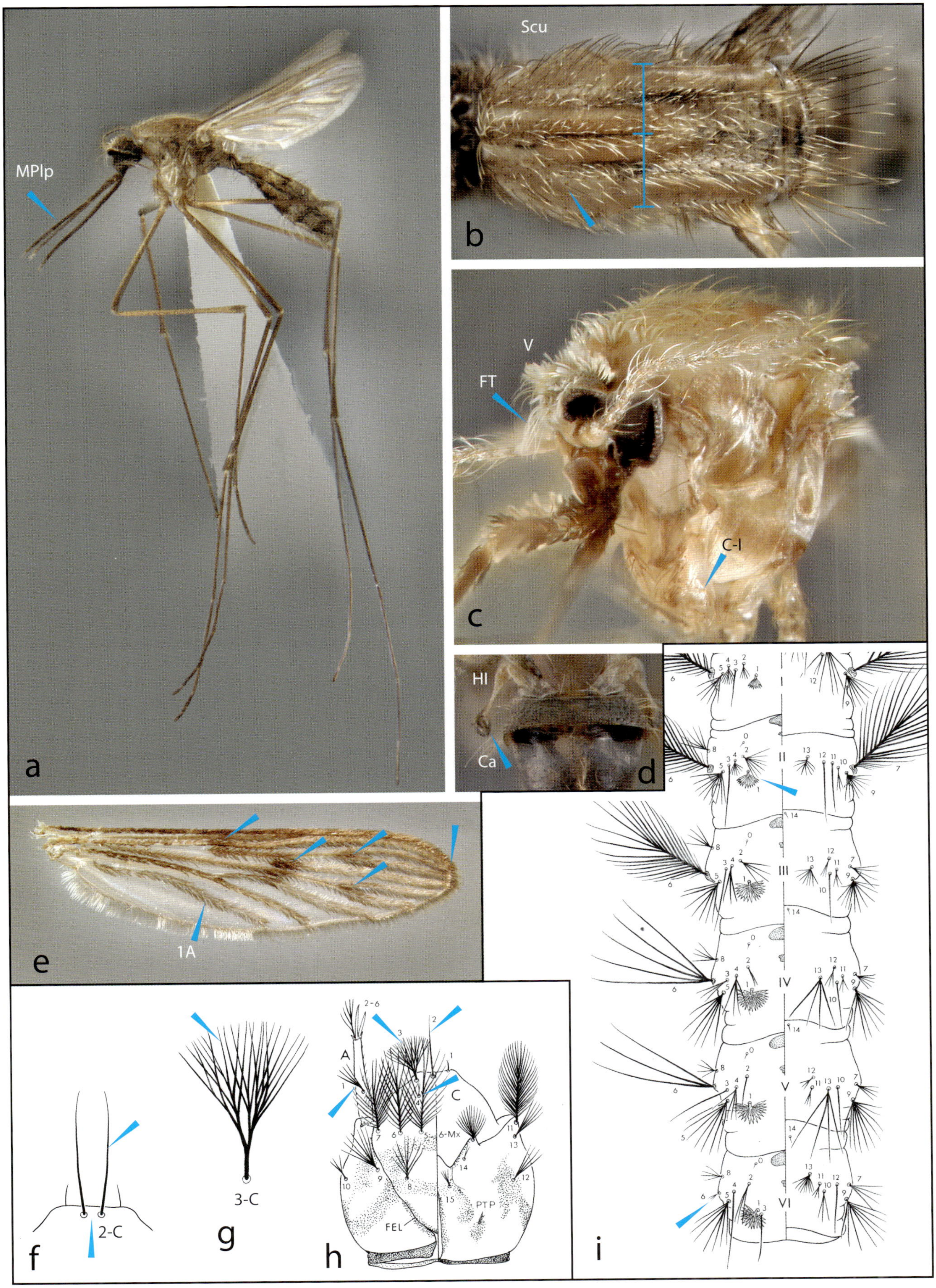
MPlp
a
Scu
b
V
FT
C-I
c
Hl
Ca
d
1A
e
2-C
f
3-C
g
2-6
3
2
1
A
C
1
7
6
5
6-Mx
11
13
10
9
8
14
15
12
PTP
FEL
h
I
II
III
IV
V
VI
i

Anopheles (Anopheles) sinensis Wiedemann, 1828
Subfamily Anophelinae
"Han Chinese Nail Mosquito"

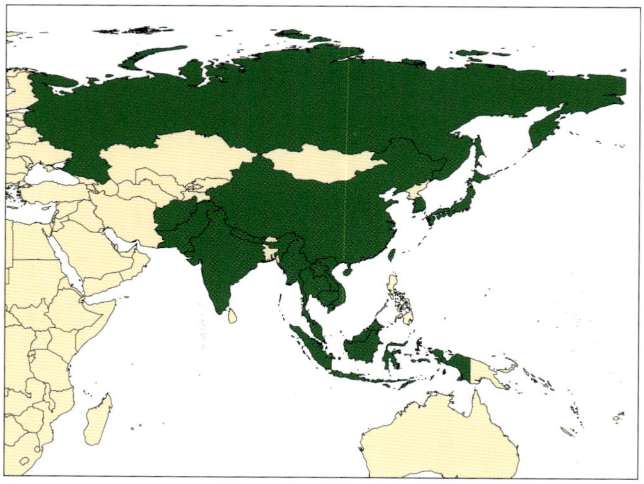

Systematics and Distribution

Anopheles sinensis is a large mosquito, with pale scales on its basal 4–8 flagellomeres, a lateral patch of dark scales on the clypeus (see *d*), and wings with no humeral or presector pale spots and two dark spots on vein 1A (see *c*). *Anopheles sinensis* is one of 16 members of the Hyrcanus Group of the Myzorhynchus Series (*Anopheles* subgenus *Anopheles*) (see Vol. 2, p. 1082, for related taxa), which also includes five species in the Lesteri Subgroup and four in the Nigerrimus Subgroup. *Anopheles sinensis* has three synonyms: *plumiger* Dönitz, *jesoensis* Tsuzuki, and *changfus* Ma.

Anopheles sinensis was originally described from China and has a wide distribution. The species is reported from Afghanistan through China and the Indian subcontinent and to Southeast Asia. It is present across the countries of mainland Southeast Asia, from Taiwan southeast through to the Indonesian island of Timor (see map).

Bionomics

Because there are so many new species in the Hyrcanus Group, previously hidden under and reported as *An. sinensis*, it is difficult to tease out the bionomic attributes of each of the component taxa. Female *An. sinensis* mosquitoes are essentially zoophilic, biting people only when other mammalian hosts are absent. Cattle and buffalo are preferred hosts in South Korea (89.8%), but in Taiwan, 86.4% of bloodmeals were porcine, with only 9.1% of bloodmeals bovine. The species is dominantly exophilic and exophagic and rarely found indoors. *Anopheles sinensis* is routinely implicated in malaria transmission, but it is nowhere near as effective a vector as other Hyrcanus group members such as *An. nigerrimus* Giles (see p. 314), *An. lesteri* Baisas & Hu (see p. 308), *An. pullus* Yamada, or *An. kleini* Rueda. In northern parts of their range, adult females overwinter as mated nullipars, typically entering diapause when temperatures drop to 13°C–15°C.

Anopheles sinensis immatures are found in a wide variety of open freshwater habitats with emergent and/or floating vegetation. They are common in shallow waters in rice fields, grassy flood pools, seepage waters, marshes, swamps, and ponds and are often reported in smaller habitats, including streams and rock pools, and wheel ruts.

Associated Pathogens

JEV; *Plasmodium falciparum*, *P. vivax*, *Wuchereria bancrofti*.

Diagnostic Characters

(a) Habitus (♀). (b) Thorax (side): Antepronotal scales (ApSc) present (not distinct in photo); C-II with distinct upper patch of pale scales. **(c) Wing:** Costa (C) usually without humeral pale spot (HP); CuA usually present; presector pale spot (PSP) ill-defined or absent; vein 1A with two dark spots; wing apex with narrow pale fringe spot, not extending beyond veins R_1 to R_3; pale fringe spot usually present at termination of vein CuA. **(d) Head (dorsal):** Flagellum (Fl) with pale scales on basal 4–8 flagellomeres; clypeus (Clp) with patch of dark scales laterally; $MPlp_3$ with pale basal band at most as wide as other pale bands. **(e) Leg (III):** Fe-III without distal broad preapical white band; Ta-III_{1-5} with apical pale bands only. **(f) Thorax (halteres):** Capitellum (Ca) of haltere (Hl) dark. **(g) Larva (thorax):** Seta 1-P without branches on basal 0.5, single (shown) or with short branches on distal 0.5. **(h) Larva (head):** Antenna (A) spiculate; seta 1-A with long branches reaching beyond midpoint of antenna; seta 2-C single, closely approximated; seta 3-C with ≥10 branches. **(i) Larva (terminal segments). (j) Larva (abdominal segments):** Seta 1 fully developed palmate on at least five segments. Image sources: WRBU; Harrison & Scanlon (1975).

Exemplar DNA barcodes: *An. sinensis COI* (AB781782–84), whole genomes (GCA_000441895, GCA_000472065).

Catalog entry: Vol. 2, p. 649.

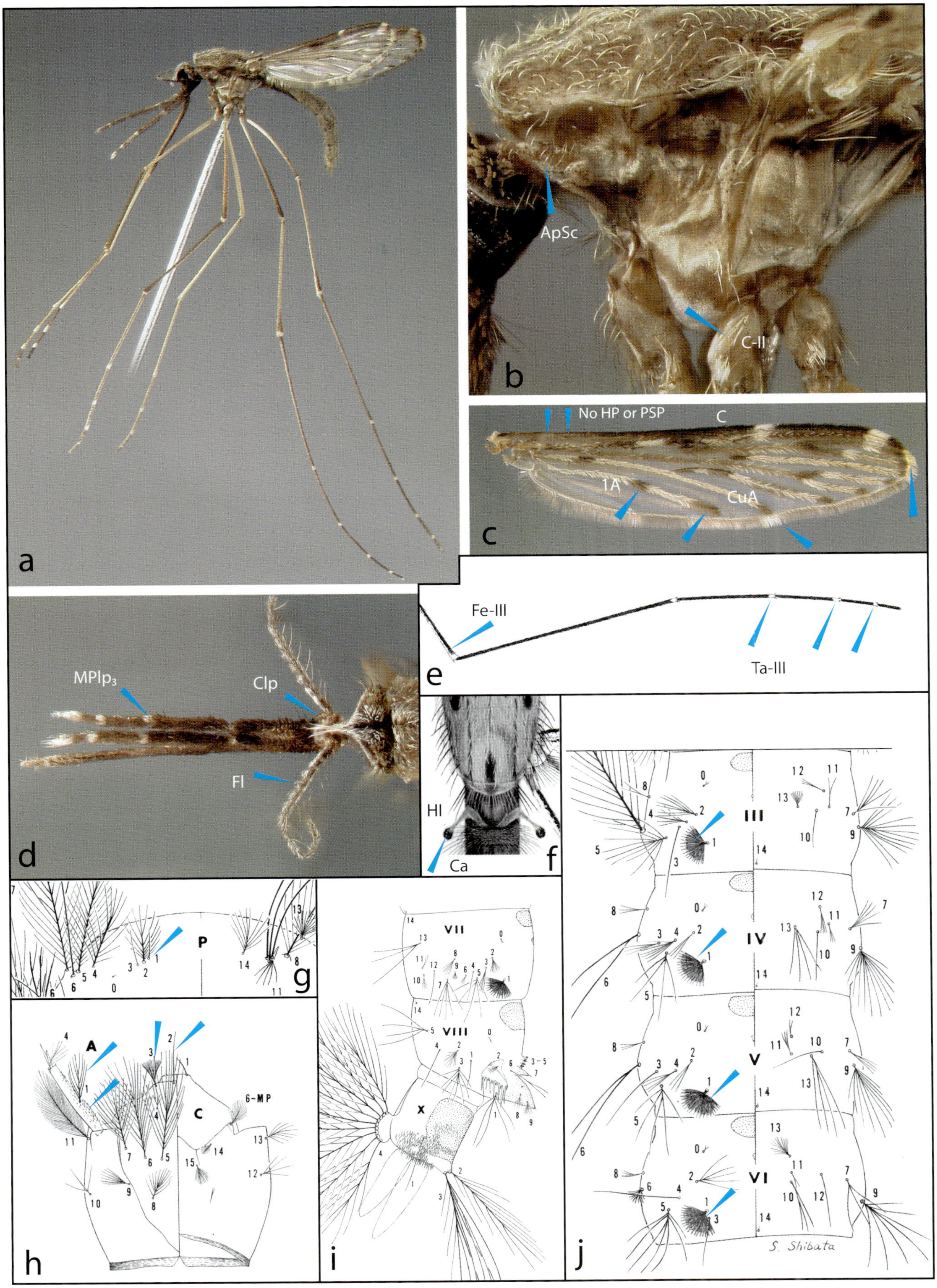
a
b
ApSc
C-II
c
No HP or PSP
C
1A
CuA
d
MPlp3
Clp
Fl
e
Fe-III
Ta-III
f
Hl
Ca
g
P
h
A
C
6-MP
i
VII
VIII
X
j
III
IV
V
VI
S. Shibata

Anopheles (*Cellia*) *aconitus* Dönitz, 1902
Subfamily Anophelinae
"Clean Indonesian Nail Mosquito"

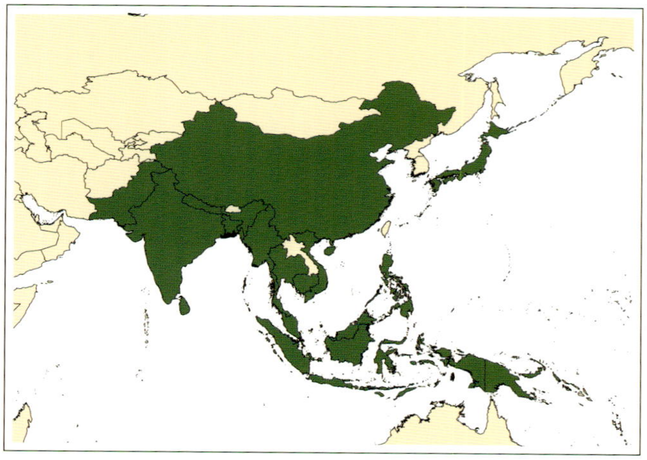

Systematics and Distribution

Anopheles aconitus is a small, commonly encountered mosquito. It is recognized by a long pale-tipped band on the proboscis separated only by a tiny dark band, a shock of erect white scales on its vertex (see *d*), and the unique characters of its ornately decorated wings (see *e*). The species gives its name to the informally recognized Aconitus Subgroup of the Funestus Group of *Anopheles* (*Cellia*) (see Vol. 2, p. 883). Other species in the Aconitus Subgroup include *An. filipinae* Manalang, *An. mangyanus* (Banks), *An. pampanai* Büttiker & Beales, and *An. varuna* Iyengar. *Anopheles aconitus* has two recognized synonyms: *albirostris* Theobald and *brahmacharii* Christophers. Four karyotypic forms of *An. aconitus* have been reported—form A (X_1, X_2, Y_1), form B (X_1, X_2, Y_2), and form C (X_1, X_2, Y_3)—all found in sympatry in Maetang District, Chiang Mai Province, Northern Thailand, and form D (X_3, X_4, Y_4) from Java, Indonesia. A scanning electron microscopy study on egg morphology together with morphometric analyses concluded that karyotypic form B and form C from Thailand were not distinct species, but instead comprise cytological races of the same species. Specimens collected in both Kajoe Tanam, Sumatra, and Willem Island, Soekaboemi, Java (Indonesia) were used in the original description of *An. aconitus*. The species is widely distributed across the mainland of the Indo-Oriental Region, from Sri Lanka, India, and Pakistan northeast through Nepal to Hainan Island, China, and south down to the islands of Indonesia before stretching southeastward into Papua New Guinea in the Australasian Region.

Bionomics

Anopheles aconitus s.l. can be found throughout the year in most open, rural agricultural sites and in forest edge agricultural settings. The species is reported at altitudes ranging from sea level to 800 m, although occasional records of up to 1,000 m have been reported. However, peak adult abundance occurs with rainy seasons and rice harvests (March–April, August–September), when plants are close to maturity and around 1.5 m high. As well as in rice fields, immature *An. aconitus* s.l. is found in clear freshwater sites with little or no water movement, including ponds, marshes, swamps, riverbed pools, ditches, and irrigation channels. Reports from smaller habitats, including wells, wheel ruts, and hoofprints, are documented but rare. *Anopheles aconitus* s.l. is primarily zoophilic, feeding from dusk to around midnight both indoors and out, and will readily feed on humans in areas where cattle are scarce.

Although considered a secondary malaria vector across most of its range, the medical importance of *An. aconitus* increases substantially where it sustains high population density. In infectivity studies, form B/C was shown to be susceptible to infection with both *Plasmodium falciparum* and *P. vivax* parasites. The vector status of forms A and D is not explicitly determined, but Indonesian populations of *An. aconitus* (presumably form D) are considered primary malaria vectors.

Associated Pathogens

Plasmodium falciparum, *P. vivax*.

Diagnostic Characters

(a) Habitus (♀): Legs entirely dark or with narrow apical pale bands or dorsal patches on some tarsomeres (Ta). **(b) Thorax (side):** Antepronotal scales (ApSc) absent. **(c) Thorax (dorsal):** Anterior promontory (AnP) with long, white, slender scales; scutum (Scu) center appearing nearly bare except for setae, or with slender seta-like pale scales reaching to scutellum (Stm). **(d) Head (side):** Proboscis (P) pale distally; palpus (MPlp) with three pale bands; preapical dark palpal band variable from as long as the two adjoining pale bands to completely absent (short in photo); vertex (V) with white erect scales. **(e) Wing:** Costal area with ≥4 dark spots; hind margin of wing usually with pale fringe spot at vein 1A; vein 1A with two dark spots on distal 0.5; vein R_2 with median pale spot. **(f) Abdomen (side):** Abdominal segments VII and VIII and ♀ cercus (Ce) without scales. **(g) Larva (thorax):** Only one of setae 9,10-T single. **(h) Larva (head):** Seta 2-C aciculate; seta 4-C branched. **(i) Larva (abdominal segments):** Tergal plates (TP) on segments IV–VII large, enclosing small median accessory tergal plates (ATP). Image sources: WRBU; Harrison (1980).

Exemplar DNA barcodes: *An. aconitus* s.l. (MN319655-72, AY423055, DQ000253–64).

Catalog entry: Vol. 2, p. 653.

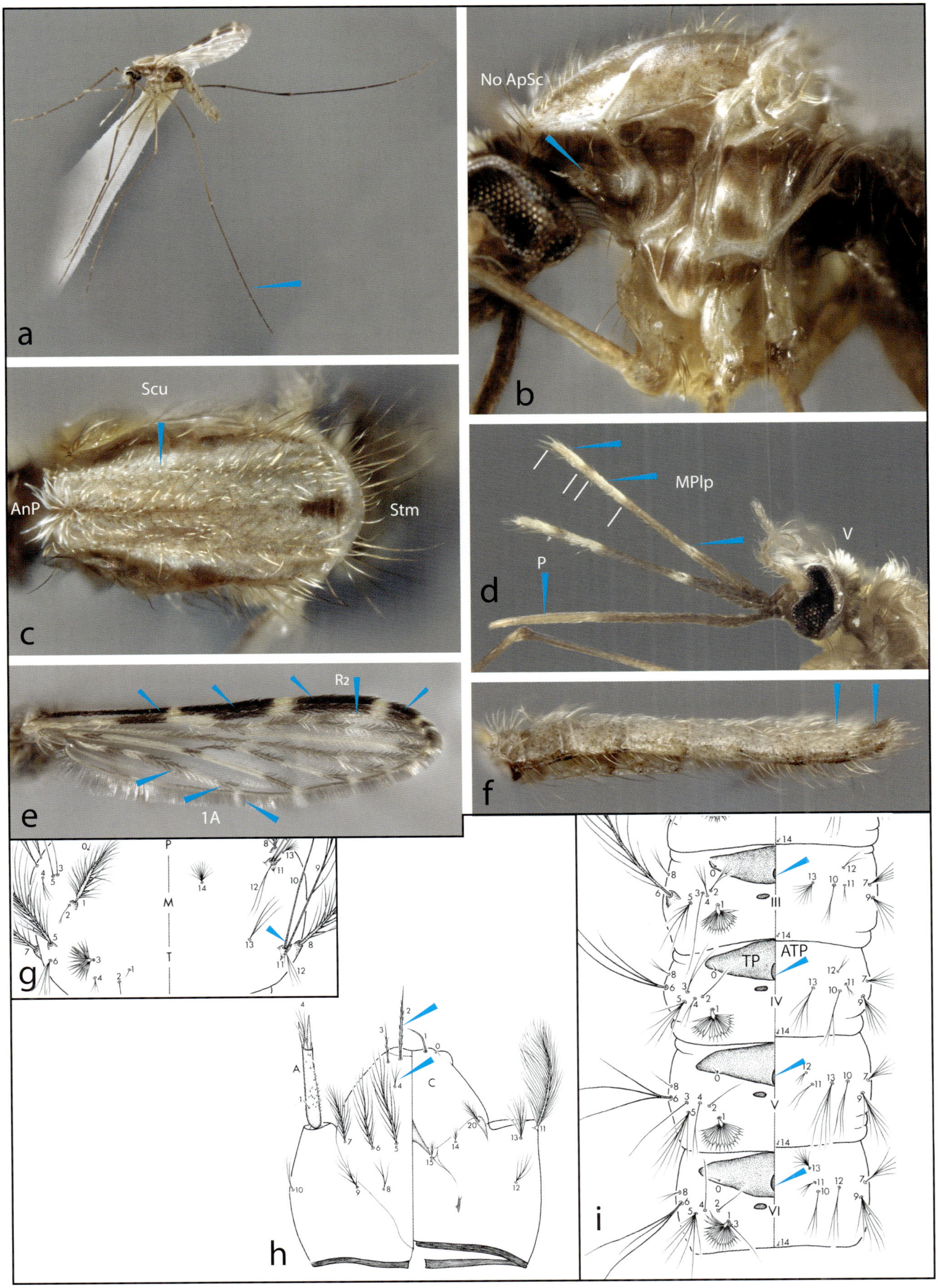
a
No ApSc
b
Scu
AnP
Stm
c
MPlp
V
P
d
R2
1A
e
f
P
M
T
g
A
C
h
TP
ATP
III
IV
V
VI
i

Anopheles (Cellia) annularis van der Wulp, 1884
Subfamily Anophelinae
"Ringed Indonesian Nail Mosquito"

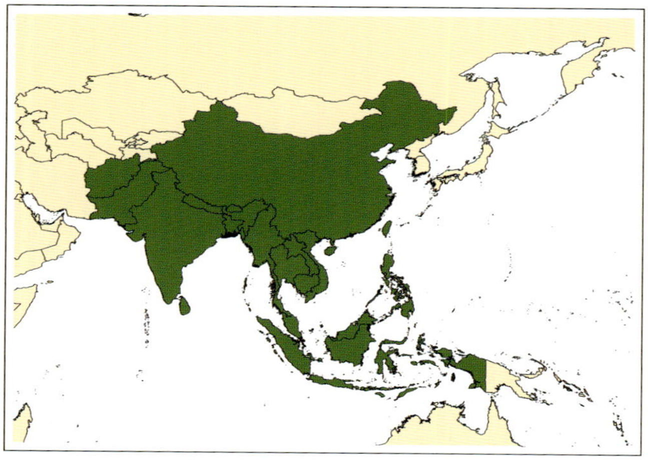

Systematics and Distribution

Anopheles annularis is an attractive common mosquito with three crisp white palpal bands, wide white tarsal stripes on its forelegs, and distinctive white "socks" on its hind legs. The wing is composed of interspersed dark and cream scales that turn to bright white on the wing fringes and costa. *Anopheles annularis* is the nominotypical species of the informally designated Annularis Group and Annularis Complex in the *Anopheles* subgenus *Cellia*. The Annularis Group comprises five species split into two species complexes: Annularis Complex (composed of *An. annularis*, *An. pallidus* Theobald, *An. philippinensis* Ludlow, and *An. schueffneri* Stanton) and the Nivipes Complex (containing only *An. nivipes* (Theobald), which is known to comprise two cytogenetic forms).

Cytogenetic studies conducted in India showed support for two species (A and B) in *An. annularis*, characterized by $+j^1$ and j^1 arrangements in chromosome arm 2, respectively. Nuclear rDNA *ITS*-2 markers were correlated with these forms in India and in Sri Lanka, and we recently detected both forms in the foothills of Nepal. Six formerly described taxa are currently in synonymy with *An. annularis*: *fuliginosus* Giles, *leucopus* Dönitz, *jamesii* Liston, *nagpori* James & Liston, *lineata* Ludlow, and *adiei* James & Liston, the latter of which was commonly thought to be a seasonal form of *An. annularis*. It remains unknown how species A and B correspond to topotypic populations of *An. annularis* s.s., and which, if any, of these synonyms correspond to the other emergent taxa in the Annularis Complex.

Anopheles annularis was originally described from Mount Ardjoeno in Java, Indonesia. It is found (as s.l.) across the mainland and islands of Southeast Asia, in the Philippines, and north to Taiwan and Nepal and through to the Indian subcontinent as far west as Afghanistan and Pakistan. Species A has been molecularly verified in China, India, Myanmar, Nepal, the Philippines, Timor, Thailand, and Sri Lanka. Species B has been detected in Sri Lanka, India, and Nepal, and has been matched to one Cambodian sample in GenBank.

Bionomics

Key bionomic traits governing vector competency in adult *An. annularis* s.l. are highly variable depending on region. In Yunnan, China, *An. annularis* s.l. is a formidable pest, feeding on humans in high numbers throughout the night. In Bengal, India, the species is anthropophilic and zoophilic and found resting in houses and stables in roughly equal numbers. In Malaysia, *An. annularis* s.l. is regarded as zoophilic, rarely biting humans. The species has been found naturally infected in low numbers with *Plasmodium* spp. in China, Malaysia, Myanmar, the Philippines, and the Thai-Cambodian border region but is thought to be of low medical importance across the rest of its range. In India, only species A has been implicated in malaria transmission.

Female *An. annularis* s.l. mosquitoes are reportedly strong fliers and disperse a considerable distance from larval sites. The immatures stages of *An. annularis* s.l. are most commonly collected in large freshwater sites with ample aquatic vegetation, including lakes, ponds, rice fields, ditches, and swamps. They have also been recovered from large artificial water tanks. In northern India, the species is thought to overwinter both as adults and as larvae.

Associated Pathogens

JBEV, KOWV, TILV, TRUV; *Wuchereria bancrofti*, *Plasmodium falciparum*, *P. vivax*.

Diagnostic Characters

(a) Habitus (♀). (b) Thorax (side): Anterior promontory (AnP) with erect white, slender scales; upper proepisternal setae (PeSU) absent. **(c) Head (dorsal):** Frontal tuft (FT) with long, white slender scales; palpus (MPlp) with three pale bands. **(d) Wing:** Vein CuA with more dark than pale scaling and dark spot at fork with crossvein mcu. **(e) Leg (III):** Fe-III, Ti-III, and $Ta\text{-}III_1$ without pale speckling; $Ta\text{-}III_{3-5}$ entirely white. **(f) Larva (terminal segments). (g) Larva (head):** Seta 3-C multibranched, brushlike; seta 8-C single or double. **(h) Larva (thorax/abdominal segments):** Abdominal seta 1-I palmate, moderately developed. Image sources: WRBU.

Exemplar DNA sequences: *An. annularis* s.l. species A—*ITS2* (India DQ351853, Sri Lanka FJ526607, Myanmar FJ526609, Philippines FJ526613, China EU919720, Thailand FJ526615, Timor GU062187); *An. annularis* s.l. species B *ITS2* (India DQ351854).

Catalog entry: Vol. 2, p. 654.

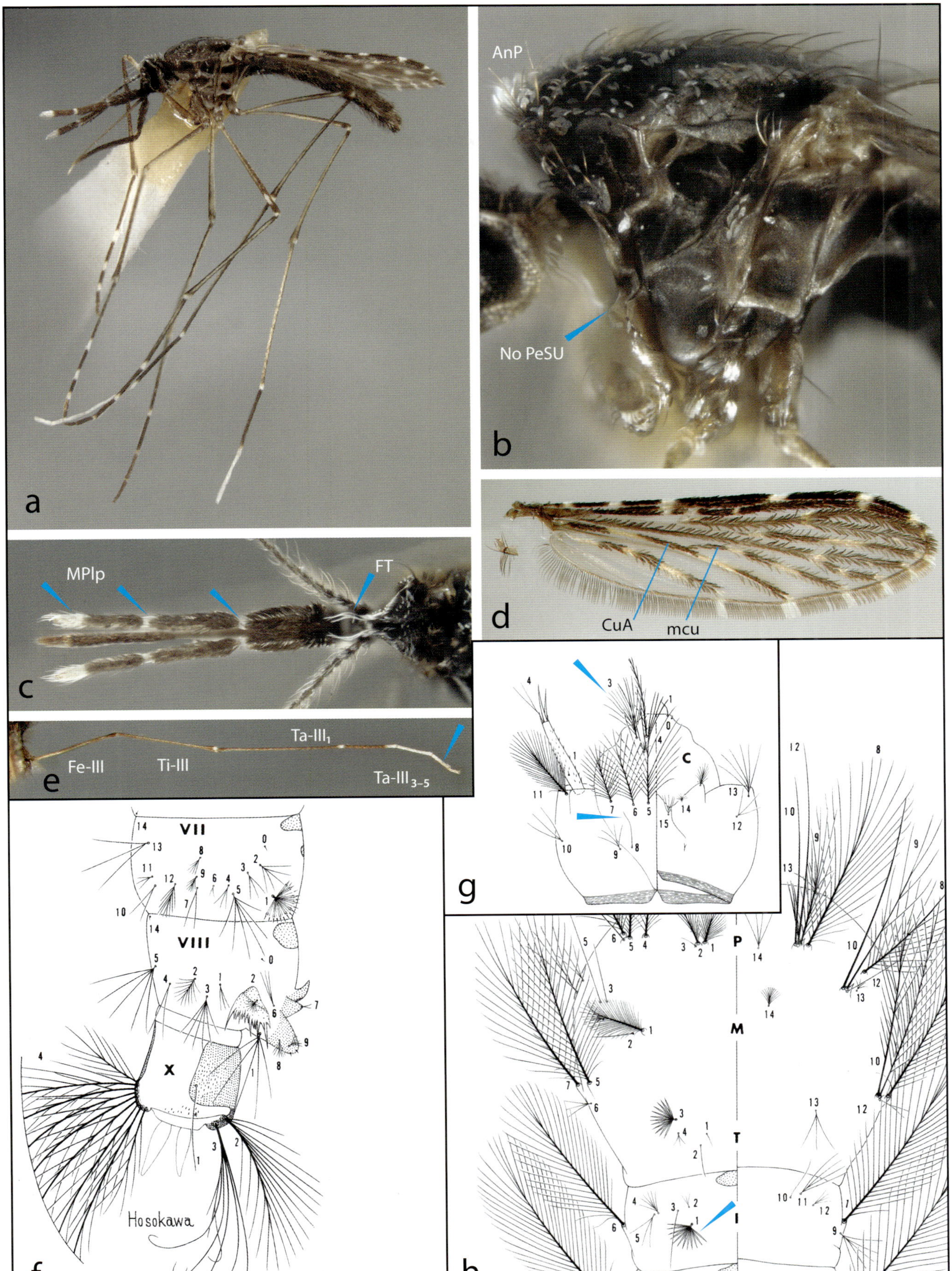
a
AnP
No PeSU
b
MPlp
FT
c
CuA
mcu
d
Ta-III1
Fe-III
Ti-III
Ta-III3–5
e
VII
VIII
X
Hosokawa
f
C
g
P
M
T
I
h

Anopheles (*Cellia*) *annulipes* Walker, 1856
Subfamily Anophelinae
"Ring-footed Australian Nail Mosquito"

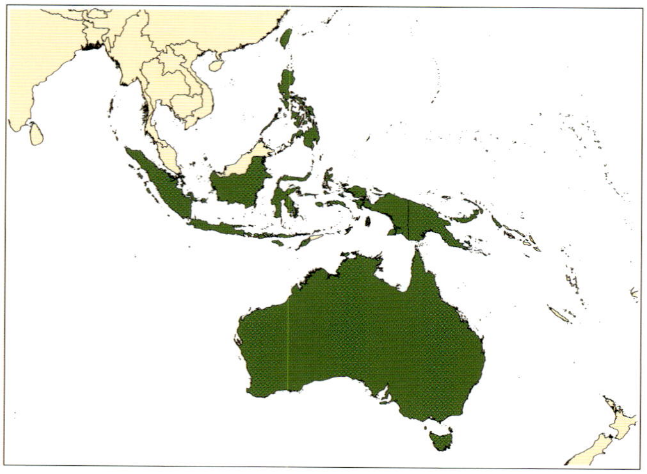

Systematics and Distribution

Anopheles annulipes takes its name from the Latin for "ringed feet," and indeed these, along with its dominantly white palps, are highly distinctive. The white bands on the legs appear to be more widely spread the farther they are away from the body. *Anopheles annulipes* most closely resembles *An. punctulatus* Dönitz. Molecular analysis of *An. annulipes* s.l. specimens across its range using mitochondrial markers, including *COI* barcodes, indicates the presence of 17 unique species-level lineages (*An. annulipes* A–Q), making this the most speciose cryptic species group exposed within mosquitoes to date. The putative species belong to two major clades (north and south), suggesting that climate was a driver of species radiation. For now, the Annulipes Complex is unplaced within the Neomyzomyia Series of the *Anopheles* subgenus *Cellia* (see Vol. 2, p. 1084, for internal relationships). There are five synonyms of *An. annulipes*—*musivus* Skuse, and *mastersi* Skuse described from New South Wales, Australia, and *perplexus* Taylor, *persimilis* Taylor, and *derricki* Taylor from Queensland, Australia. Efforts are needed to formally describe the composite taxa within this important group of malaria vectors.

Anopheles annulipes was originally described from specimens collected at Van Diemen's Land, Tasmania, Australia. Since then, the species (as s.l.) has been described from Indonesia, Papua New Guinea, the Philippines, and Taiwan (see map).

Bionomics

The original description of *An. annulipes* indicates it was found in dry inland sites up to 5,000 ft in Tasmania, Australia, and at sea level in coastal areas of New Guinea. *Anopheles annulipes* s.l. immatures are typically found in sunlit, shallow freshwater habitats, including rock pools, grassy ground pools, and at the edges of marshes and slow-flowing rivers. *Anopheles annulipes* s.l. is anthropophilic and readily enters houses to feed. The species is found resting in rabbit burrows and is implicated in the natural transmission of myxomatosis in Australia. *Anopheles annulipes* A appears to be a strong flier, with maximum recorded flights of 5 km in capture-mark-recapture experiments, compared with 1.5 km for *An. annulipes* G.

Associated Pathogens

AAOV, AATV, HaCV, MYXV, STRV, TILV; *Plasmodium* spp., *P. berghei*.

Diagnostic Characters

(a) Habitus (♀): Capitellum (Ca) of haltere black scaled dorsally; Ta-I_{2-5} and Ta-II_{2-5} banded or spotted. **(b) Thorax (dorsal):** Scutum (Scu) with scales on lateral areas. **(c) Head (side):** Antenna (Ant) not variably colored; $MPlp_3$ never divided apically by a black band; $MPlp_5$ dark basally. **(d) Wing:** Many dark spots on wing, with at least four and sometimes more on CuA and 1A. **(e) Leg (II). (f) Larval (head):** Seta 1-A very small, inconspicuous; seta 2-C smooth or weakly aciculate; seta 3-C at most 0.75 times length of seta 2-C ≥6, sometimes highly branched; seta 4-C about 0.25 times length of seta 2-C; seta 4-C always strongly branched from near the base; setae 5,6-C plumose. **(g) Larva (thorax/abdominal segments):** Seta 1,2-P on separate or fused tubercles; setae 9–12-P,M,T single (not shown); palmate setae well developed on segments II to VII; tergal plates (TP) not unusually developed, present on segments I to VIII, with small median accessory plates (ATP) on III to VII. Image sources: WRBU; Ross & Roberts (1943).

Exemplar DNA sequences: *An. annulipes s.l. COI* (not barcode region): Species A (DQ420429–31); B (DQ420432–34); C (DQ420435–36; DQ420449–50; DQ420473–74; DQ420481–83); D (DQ420437–39); E (DQ420440–42); F (DQ420443–45); G (DQ420446–48); H (DQ420451–53); I (DQ420454–55); J (DQ420456–58); K (DQ420459–61); L (DQ420462–64); M (DQ420465–67); N (DQ420468–69, DQ420476–77); O (DQ420470–72); P (DQ420478–79); Q (DQ420484–85).

Catalog entry: Vol. 2, p. 654.

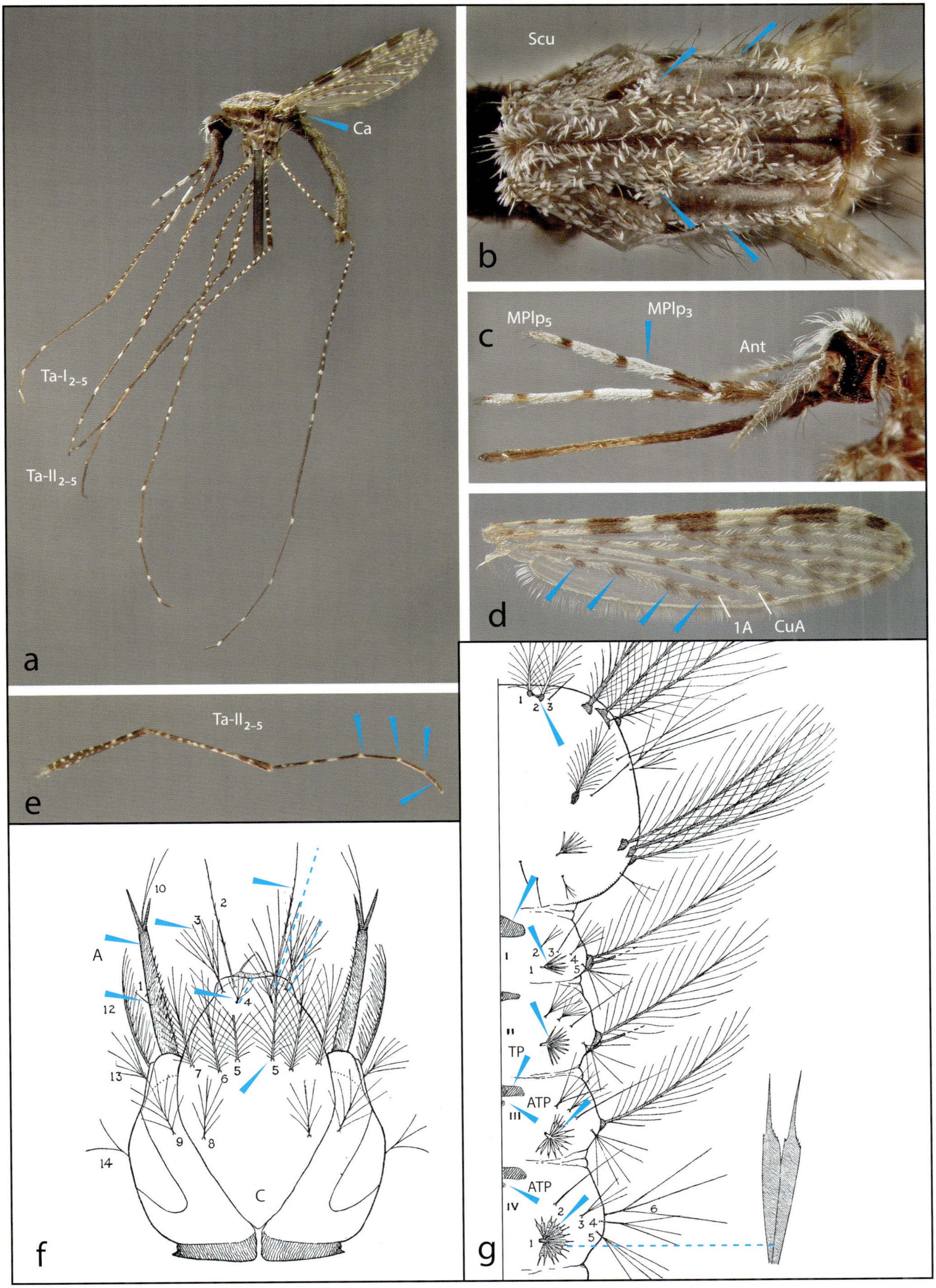
Ca
Ta-I$_{2-5}$
Ta-II$_{2-5}$
a
Scu
b
MPlp$_5$
MPlp$_3$
Ant
c
1A
CuA
d
Ta-II$_{2-5}$
e
10
2
3
A
1
12
4
13
7
6
5
5
9
8
14
C
f
1
2
3
I
II
TP
ATP
III
ATP
IV
4
5
6
g

Anopheles (*Cellia*) *balabacensis* Baisas, 1936 Subfamily Anophelinae "Balabac Philippine Nail Mosquito"

Systematics and Distribution

Anopheles balabacensis is a highly ornate species, with alternately pale and dark scales on its palps (see *b*), legs (see *d*), and wings (see *c*) and white scale patches on the abdomen, giving a decidedly spotted appearance to the whole fly (see *a*). It sports the classic Leucosphyrus Group white hind knees (see *d*). Formerly regarded as a member of the Dirus Complex, a taxonomic revision of the Leucosphyrus Group now places *An. balabacensis* in the Leucosphyrus Complex (Leucosphyrus Subgroup, Leucosphyrus Group) of the subgenus *Anopheles* (*Cellia*) (see Vol. 2, p. 1085). Other members of the Leucosphyrus Complex include *An. baisasi* Colless, *An. introlatus* Colless, *An. latens* Sallum & Peyton (see species page, p. 352), and *An. leucosphyrus* Dönitz. All except *An. introlatus* are vectors of human malaria.

Anopheles balabacensis takes its name from Balabac Island in the Philippines, where it was first described. Previous literature reports document the species from Pakistan and India across mainland Southeast Asia as far north as Taiwan and south to the islands of the Philippines, Borneo, and Indonesia. However, following an integrated systematics revision, these records were determined to belong to several newly defined taxa in the Dirus and Leucosphyrus Complexes. The current verified distribution of *An. balabacensis* is limited to forested areas of the southern Philippine islands of Balabac, Culion, and Palawan; Borneo (including Brunei, South and East Kalimantan, and eastern Sarawak and Sabah); and the Indonesian islands of Java, Lombok, Sumba, and Sumbawa.

Bionomics

Anopheles balabacensis is a forest-dependent species typically associated with higher elevations (foothills and mountainous areas). In deforested areas of Sabah, this species is rare. The species is attracted both to monkeys and humans, and in Sabah, the species is primarily exophilic; it will enter houses to feed but returns outside to rest. In Sabah, *P. falciparum* infection rates of 1.3% were reported in *An. balabacensis*, and it is considered the primary vector in northeast Borneo, Central Java, and Lombok Island. As well as human malarias, 1.61% of wild-caught *An. balabacensis* ($n = 1{,}437$) from the Malaysian state of Sabah (eastern Borneo) tested positive for simian malaria parasites, including *Plasmodium coatneyi*, *P. inui*, *P. fieldi*, *P. cynomolgi*, and *P. knowlesi*, the latter of which is also known to cause disease in humans.

Immature *An. balabacensis* is typically collected in shaded, temporary freshwater pools, including rock pools, wheel ruts, ditches, puddles, animal wallows, and animal footprints. Less frequently, larvae have been collected from rice fields and stream edges and in containers of natural (e.g., fruit husks) and artificial (e.g., plastic buckets) origin. Somewhat unusually for *Anopheles*, eggs of *An. balabacensis* exhibit desiccation tolerance. Maintained in humid conditions, eggs collected from drying mud in breeding pools in Sabah were found to be viable up to 27 days after removal from water.

Associated Pathogens

Plasmodium coatneyi, *P. inui*, *P. falciparum*, *P. fieldi*, *P. cynomolgi*, *P. knowlesi*, *P. pitheci*, *P. silvaticum*, *Brugia malayi*, *Wuchereria bancroftii*, *W. kalimantani*.

Diagnostic Characters

(a) Habitus (♀): Proboscis (P) as long as or slightly longer than Fe-I. **(b) Head (dorsal):** Apical pale band on $MPlp_5$ distinctly white or whitish, not strongly contrasting with pale bands on $MPlp_2$ and $MPlp_3$. **(c) Wing:** With ≥4 dark spots in the costal area; veins M and M_{3+4} noticeably wavy; pale scales of wing cream colored to golden to yellowish; accessory sector pale spot (ASP) usually on veins C, Sc, and R (not on C in photo); split of vein R_{2+3} basal to split of vein M_{1+2}. **(d) Leg (III):** Apex of Ti-III and base of $Ta\text{-}III_1$ with large white band; $Ta\text{-}III_4$ with small to indistinct basal pale band; $Ta\text{-}III_5$ without basal pale band. **(e) Larva (thorax/abdominal segments):** Seta 1-M basal sclerotized tubercle not produced into distinct tooth; seta 1-P basal sclerotized tubercle with prominent tooth or spine (not distinct in illustration); seta 1-II moderately developed palmate but distinctly smaller than 1-III; seta 2-IV,V 3,4 branched. **(f) Larva (head):** Seta 1-A small, usually single; seta 2-C widely separated; setae 3,4-C short; seta 3-C single, extending to or only slightly beyond anterior margin of head; seta 4-C usually reaching base of seta 2-C; seta 5-C conspicuously longer than antenna (A); setae 5–7-C long and branched. **(g) Larva (terminal segments).** Image sources: WRBU; Sallum et al. (2005).

Exemplar DNA sequences: *An. balabacensis COI* (MH032606–676), *ITS2* (KY883194–201).

Catalog entry: Vol. 2, p. 656.

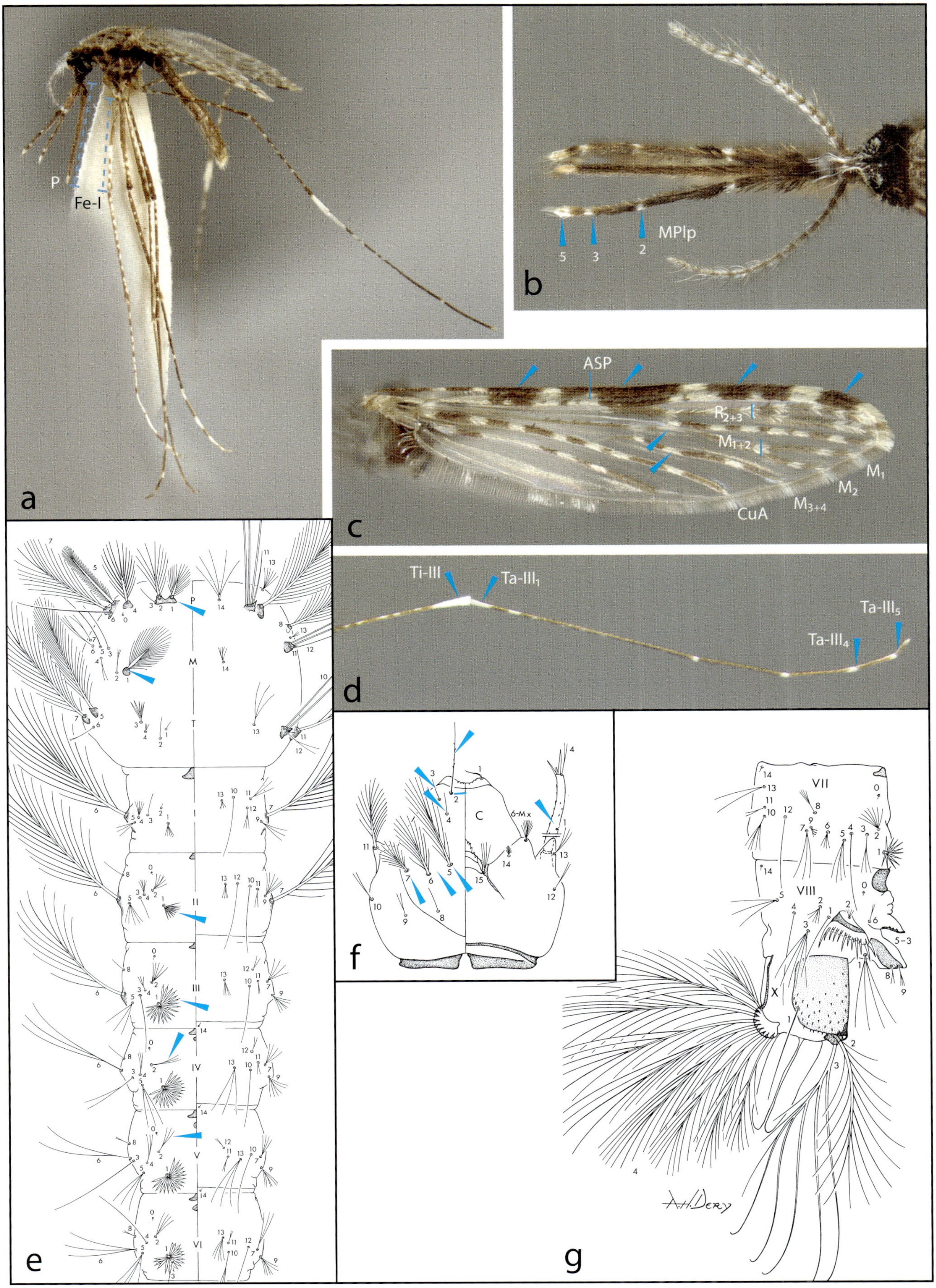
P
Fe-I
a
MPlp
5
3
2
b
ASP
R2+3
M1+2
M1
M2
M3+4
CuA
c
Ti-III
Ta-III1
Ta-III5
Ta-III4
d
P
M
T
I
II
III
IV
V
VI
e
C
6-Mx
f
VII
VIII
X
A.H.Dery
g

Anopheles (*Cellia*) *culicifacies* Giles, 1901
Subfamily Anophelinae
"Gnat-faced Indian Nail Mosquito"

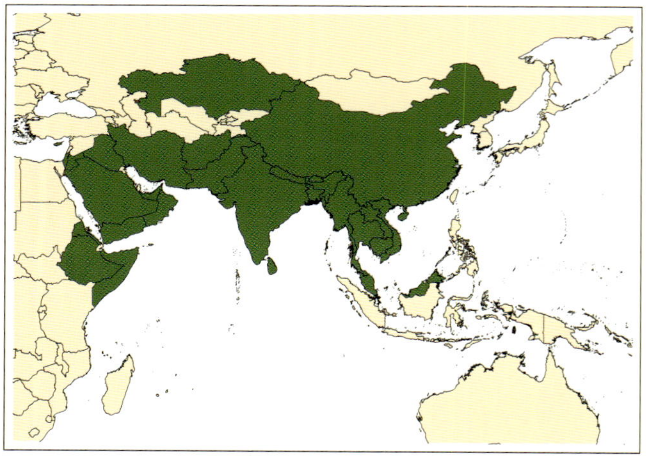

Systematics and Distribution

Anopheles culicifacies is a pale-bodied species that lacks antepronotal scales (see *b*) and has no, or slender, scales on the central scutum (see *c*). *Anopheles culicifacies* is placed in the Culicifacies Subgroup of the Funestus Group (Myzomyia Series, *Anopheles* subgenus *Cellia*) (see Vol. 2, p. 1083, for internal relationships). Early records of *An. culicifacies* were reported under *listoni* Giles, which is now in synonymy with *An. culicifacies*, along with *indica* Theobald, *punjabensis* James, and *adenensis* Christophers. Five distinct species-like lineages of *An. culicifacies* have been recognized on the basis of cytology, referred to as species A, B, C, D, and E. Species A, B, C, and D are reported from India, and B and E from Sri Lanka, but the taxonomic status of other geographical populations remain undetermined. Biological differences exist in the behavior, vectorial capacity, distribution, and seasonal prevalence of the putative species, and it remains uncertain whether any of these conform to any of the above synonyms.

Originally described from Hoshangabad in the Central Provinces of India, *An. culicifacies* s.l. is now widely reported from the Indian subcontinent, eastward through China, Nepal, and Bangladesh, down into the countries of the Southeast Asian mainland. It also occurs in the Horn of Africa, the Arabian Peninsula, the Middle East, and the eastern boundaries of the Mediterranean (including Cyprus) as far north as Kazakhstan (see map, as *An. culicifacies* s.l.).

Bionomics

Anopheles culicifacies is commonly associated with alluvial plains but has been reported at elevations up to 6,500 ft in the Himalayas. After feeding, females rest in huge numbers in cool, dark houses or cattle sheds during the day to mature their eggs. *Anopheles culicifacies* s.l. is predominantly anthropophilic, although some populations are reportedly zoophilic, preferring instead to feed on cattle and birds. Species A, C, and D are competent malaria vectors but have low human blood indices in wild populations, suggesting they prefer to feed on other mammals. Species E is a highly competent vector in southern India and Sri Lanka. Species B is highly zoonotic and is not implicated in human pathogen transmission. In northern India, or at high elevations, the species pass the winter as diapausing larvae.

Clean, open, sunlit freshwater bodies with grassy edges attract ovipositing *An. culicifacies* s.l. Typical habitats include slow-moving streams, drainage ditches, fallow rice fields, rain pools, canals, pools in riverbeds, shallow tanks, pits, wells, and, occasionally, brackish water sites. Immatures have been collected between 35 and 960 m in Thailand, only above 960 m in Vietnam, and as high as 2,286 m in Pakistan.

Associated Pathogens

Plasmodium falciparum, *P. vivax*.

Diagnostic Characters

(a) Habitus (♀): Legs all dark or with some pale patches or pale bands on apices of tarsomeres (Ta). **(b) Thorax (side):** Antepronotal scales (ApSc) absent. **(c) Thorax (dorsal):** Center of scutum (Scu) lacking scales or with slender white scales. **(d) Wing:** Vein R_1 without a pale interruption at preapical dark spot (PD); vein R_{4+5} dark, except at base. **(e) Wing:** Remigium (Re) dark scaled. **(f) Head (side):** Palpus (MPlp) with three pale bands, apical pale band much shorter than preapical dark band. **(g) Larva (head):** Seta 1-A much reduced; seta 2-C widely separated; setae 5–7-C long, branched. **(h) Larva (abdominal segments). (i) Larva (terminal segments):** Segments IV–VII with small tergal plate (TP), not enclosing median accessory tergal plate (ATP). Image sources: WRBU; Harrison (1980).

Exemplar DNA sequences: *Anopheles culicifacies* s.l. *ITS2*: species A (AF402297, AJ534246, EF462897); species B (AY167747, EF462896, AJ534247); species C (AJ534643); species D (AJ534644); species E (AY168883).

Catalog entry: Vol. 2, p. 658.

a
b
No ApSc
c
Scu
d
R_{4+5}
PD
R_1
e
Re
f
MPlp
g
A
C
h
IV
V
VI
i
TP
ATP
VII
VIII
X

Anopheles (*Cellia*) *dirus* Peyton & Harrison, 1979
Subfamily Anophelinae
"Dire Thai Nail Mosquito"

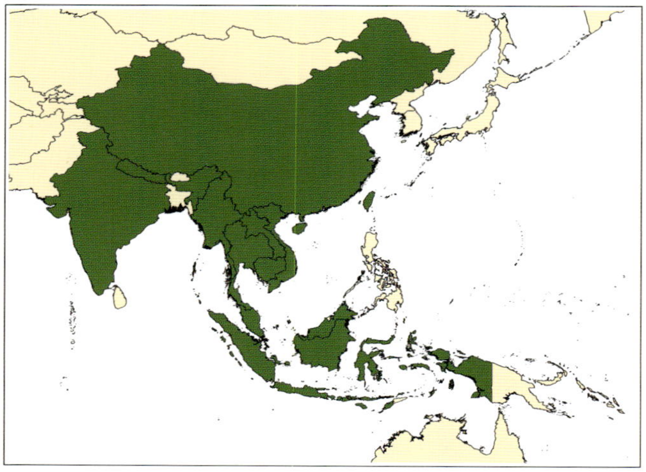

Systematics and Distribution

Anopheles dirus is a very attractive species, with both wings and legs (femora and tibiae) generously patterned with a series of contrasting short patches, or spots, of dark and pale scales. It has bright, shiny white scales on the presector and sector pale spots and the halters, contrasting with the duller white and yellowish scales elsewhere on the wing. It took several years and multiple crossing, cytogenetics, allozyme, and molecular studies to tease out the component members of the Dirus Complex, which has been complicated due to their close affinity with members of the Leucosphyrus Complex (see *An. latens*, p. 352) (see Vol. 2, p. 1085, for internal relationships). Currently, these include *Anopheles dirus* (formerly *dirus* A) and six other formally named species: *An. cracens* Sallum & Peyton (formerly B), *An. scanloni* Sallum & Peyton (formerly C), *An. baimaii* Sallum & Peyton (formerly D), *An. elegans* (James) (formerly E), *An. nemophilous* Peyton & Ramalingam (formerly F), *An. takasagoensis* Morishita, and another recognized form, "aff. *takasagoensis*."

The type locality of *Anopheles dirus* s.s. is Ban Bu Phram, Prachin Buri, Thailand, and it is the most widespread of all the component taxa. Members of the Oriental Dirus Complex (as s.l.) are reported from eastern India, Bangladesh, and Myanmar and down throughout the mainland of Southeast Asia to Malaysia. The complex is most diverse in Myanmar and Thailand, where many species can be found in sympatry.

Bionomics

Members of the Dirus Complex are forest-dependent species in mountainous areas and foothills across their range. They are also present in disturbed forest habitats, including rubber plantations, logging areas, and agricultural settings or villages on the forest fringe. The component taxa vary in malaria transmission, with the most important being *An. baimaii*, *An. dirus*, and *An. scanloni* due to their highly anthropophilic behavior. Whereas these species readily feed on people both indoors and out, *An. nemophilous* and *An. takasagoensis* are staunchly zoophilic, and the feeding preferences of *An. cracens* and *An. elegans* remain unclear.

Immatures of *An. dirus* s.l. occupy shaded, shallow, often temporal freshwater habitats, and therefore densities of these species increase exponentially in the rainy seasons. Typical immature habitats include puddles, ground pools, animal wallows, large animal footprints, and wheel ruts. Recently, *An. baimaii* was reported breeding in wells in Myanmar, indicating a worrying encroachment of this effective malaria species into human settlements.

Associated Pathogens

Plasmodium falciparum, *P. vivax*, *Wuchereria bancrofti*.

Diagnostic Characters

(a) Habitus (♀): Proboscis (P) slightly longer than forefemur. **(b) Thorax (dorsal). (c) Leg (III):** Apex of Ti-III and base of Ta-III_1 with large white band; Ti-III apical pale band without dark interruption; Ta-III_4 always with conspicuous basal pale patch on dorsal surface. **(d) Head (dorsal):** Proboscis (P) with apex not pale; palpus (MPlp) with ≥4 pale bands. **(e) Thorax (side):** Antepronotal scales (ApSc) present; presector dark spot (PSD) of vein R with 1–4 pale interruptions, most frequently two. **(f) Wing:** Bright, shiny white scales on presector pale (PSP) and sector pale spots (SP), nearly as pure white scales on the halter (Hl) (not shown), distinctly contrasting with yellowish scales of the subcostal pale (SCP) and preapical pale spots (PP); accessory sector pale spot (ASP) always absent on vein C; vein 1A, at c. level of presector dark spot (PSD), with pale spot c. equal to others on that vein. **(g) Larva (thorax):** Seta 1-P basal sclerotized tubercle with prominent tooth. **(h) Abdomen (side):** Sternum VI (VI-S) without posteromedian patch of dark scales. **(i) Palmate setae 1-II and 1-VII. (j) Larva (head):** Seta 3-C single; seta 4-C usually not reaching base of 2-C; seta 5-C distinctly longer than antenna. **(k) Larva (abdominal segments):** Seta 1-II weakly developed palmate; setae 1-III–VI well-developed palmate, with narrow lanceolate or filiform leaflets. **Larva (terminal segments):** Seta 1-X inserted on or at edge of saddle (Sa), never in marginal notch. Image sources: WRBU; Sallum et al. (2005).

Examplar DNA barcodes: *An. baimai* (DQ897952–54), *An. cracens* (DQ897947–48), *An. dirus* (DQ897943–46), *An. elegans* (EU259300, DQ897957–58), *An. nemophilous* (DQ897959–61), *An. scanloni* (AM180766–874), *An. takasagoensis* (DQ897962–64), *An.* aff. *takasagoensis* (AB518534–38).

Catalog entry: Vol. 2, p. 659.

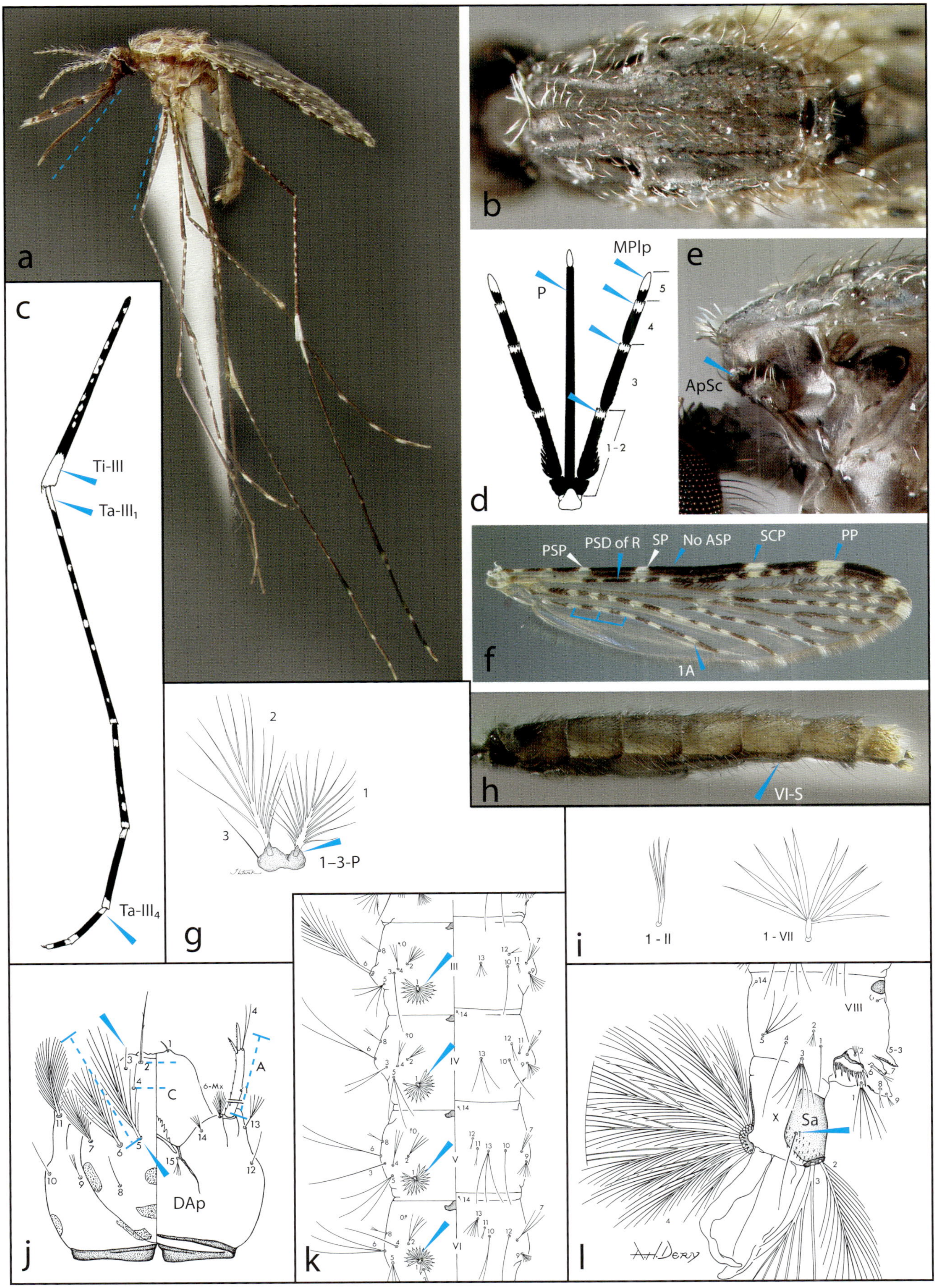
a
b
c
Ti-III
Ta-III1
Ta-III4
d
P
MPlp
5
4
3
1–2
e
ApSc
f
PSP
PSD of R
SP
No ASP
SCP
PP
1A
g
1
2
3
1–3-P
h
VI-S
i
1 - II
1 - VII
j
C
A
6-Mx
DAp
k
III
IV
V
VI
l
VIII
X
Sa
5-3

Anopheles (*Cellia*) *farauti* Laveran, 1902
Subfamily Anophelinae
"Faraut Ni-Vanuatu Nail Mosquito"

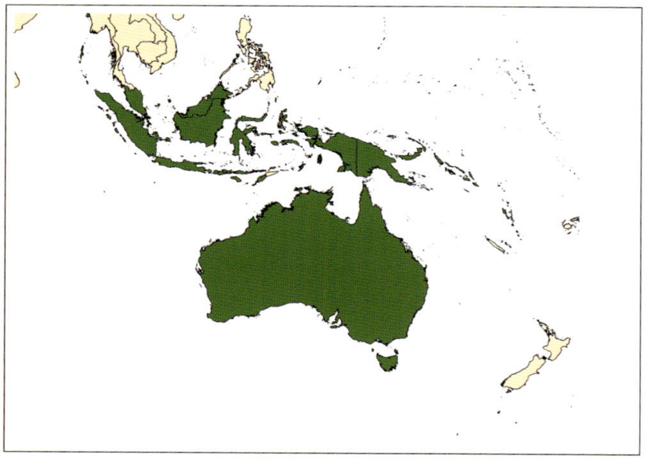

Systematics and Distribution

Anopheles farauti is a handsomely ornate species crowned with a shock of white erect scales on its head and plentiful small pale spots over its legs and wings. *Anopheles farauti* was described from specimens collected in Faureville in the New Hebrides and was named in honor of Gaudens Charles Faraut. The species has one synonym—*moluccensis* Swellengrebel & Swellengrebel de Graaf—which was described from various localities in Papua New Guinea and the Moluccas. *Anopheles farauti* is the nominotypical member of the Farauti Complex of the Punctulatus Group (Neomyzomyia Series, *Anopheles* subgenus *Cellia*) (see Vol. 2, p. 1085, for internal relationships). The Farauti Complex comprises *An. farauti* s.s., *An. hinesorum* Schmidt (=*An. farauti* 2), *An. irenicus* Schmidt (=*An. farauti* 7), *An. oreios* Bangs & Harbach (=*An. farauti* 6), and *An. torresiensis* Schmidt (=*An. farauti* 3). Three others, *An. farauti* 4, 5, and 8, are recognized pending formal description, although phylogenetic treatments suggest *An. farauti* 4 is more closely related to *An. punctulatus* Dönitz than to any other taxa within the Punctulatus Group. Molecular differentiation using the nuclear rDNA *ITS2* is proving the most reliable method for species recognition within the Farauti Complex. The newly discovered *An. farauti* 8 displays the same *ITS2* sequences as *An. farauti* s.s. These taxa can be differentiated by nuclear DNA *ITS1* sequences and by a species-diagnostic *ITS1* PCR-RFLP using the restriction enzyme *Msp* I. *Anopheles hinesorum* comprises multiple *ITS2* genotypes, two of which are zoophilic, suggesting even further hidden diversity in the Farauti Complex.

The known distribution of *An. farauti* s.l. includes Australia, Indonesia (including Borneo, Sulawesi, and the Moluccas), Malaysia, Papua New Guinea, the Solomon Islands, and Vanuatu (see map).

Bionomics

Anopheles farauti s.s. is a widespread, common coastal species usually found within a kilometer of the coast. *Anopheles farauti* 4 is common inland but can also be collected in low numbers in coastal sites, *Anopheles farauti* 8 occurs only inland, and *An. farauti* 6 is a highland species. *Anopheles farauti* s.s. and *An. farauti* 4 are regarded as primary malaria vectors, and *An. farauti* 8 is a proven field vector. The largest of all the Farauti Complex species, *An. farauti* 6, has been circumstantially incriminated.

Immatures of *An. farauti* s.s. typically occupy saline or brackish water habitats, including lagoons, coastal mangrove swamps, and brackish pools. The immatures are usually found in heavily shaded habitats with emergent, floating, and submerged vegetation or algae. There are reports of *An. farauti* being collected in artificial containers, but it remains unclear whether these larvae are indeed *An. farauti* s.s.

Associated Pathogens

EUBV, KOOV, McPV; *Plasmodium falciparum*, *P. malariae*, *P. vivax* 210/247, *Wuchereria bancrofti*.

Diagnostic Characters

(a) Habitus (♀). (b) Head (side): Proboscis (P) dark scaled except for indistinct pale band at apex; $MPlp_3$ with a broad apical band, a narrow dark subbasal band, and a narrow basal white band; $MPlp_5$ dark basally. **(c) Thorax (side):** Halter (Hl) capitellum (Ca) black scaled dorsally; sterna (S) mostly bare; terga (Te) without dense scales. **(d) Thorax (dorsal):** Scutum (Scu) with broad decumbent pale scales throughout. **(e) Wing:** Numerous dark and pale spots on wing; veins R–R_1 with 9–17 dark spots or patches. **(f) Legs (I,II):** Ta-I_{2-5} and Ta-II_{2-5} black-and-white spotted. **(g) Larva (thorax/abdominal segments):** Setae 9–12-P long, single: basal tubercles of setae 1–3-P usually fused; seta 1-M plumose; seta 1-I small, distinctly palmate with broad leaflets. **(h) Larva (head):** Seta 2-C widely separated. **(i) Larva (terminal segments).** Image sources: WRBU; Belkin (1962).

Exemplar DNA sequences: *An. farauti* whole genome (GCA_000473445, GCA_000956265).

Catalog entry: Vol. 2, p. 661.

a
b
MPlp$_3$
MPlp$_5$
P
c
Ca
Te
Hl
S
Scu
d
R
R$_1$
e
Ta-II
Ta-I
f
P
M
T
g
A
C
h
VIII
S
X
i

Anopheles (*Cellia*) *flavirostris* (Ludlow, 1914)
Subfamily Anophelinae
"Yellow-beaked Philippine Nail Mosquito"

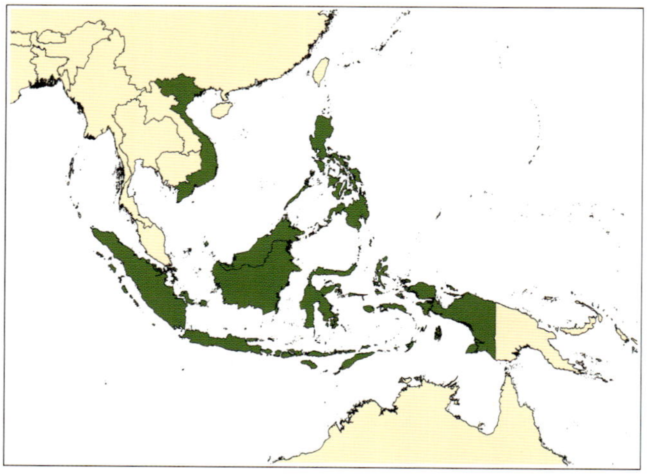

Systematics and Distribution

Anopheles flavirostris takes its name from the Latin for "yellow beak"—referring to the distinctly yellow tip to its proboscis (see *b*). This tiny species has three distinct pale bands on the proboscis (see *b*) and three large pale spots on the leading edge of its wing (see *e*). *Anopheles flavirostris*, together with *An. leesoni* Evans and *An. longipalpis* (Theobald), are members of the Minimus Subgroup within the Funestus Group (Cellia Series, *Anopheles* subgenus) (see Vol. 2, p. 1083, for internal relationships). Morphologically, *An. flavirostris* is similar to—and was long regarded as a subspecies of—*An. minimus* (Theobald). The two species have been routinely misidentified, leading to confusion on bionomics, distribution, and vector competency. Hybridization studies and noted differences of the cibarial armature, along with DNA approaches, support the separate status of *An. flavirostris*, and all historical records of *An. minimus* from the Philippines, Indonesia and eastern Borneo are now attributed to *An. flavirostris*. Larval illustrations are not available for *An. flavirostris*, thus here *An. flavirostris* larval characters are illustrated on drawings of *An. minimus* (see *f–i*).

Bionomics

Anopheles flavirostris is typically regarded as a lowland foothill species, but specimens have been collected from sea level up to 1,500 m. *Anopheles flavirostris* feeds readily on large mammals—buffalo, cattle, and humans. Overall, *Anopheles flavirostris* is considered a primary vector across much of the Philippines. Typical detected infection rates are low in wild-caught specimens (even at elevations over 1,000 m in Luzon) but are consistent enough to maintain low-level transmission cycles. In eastern Borneo (Sabah and East Kalimantan), *An. flavirostris* is highly anthropophilic and will enter houses to feed, whereas in other islands of the Indonesian archipelago, it is rarely collected at human bait, is predominantly exophilic, and is only regarded as a local vector. Host seeking occurs throughout the night, peaking toward midnight.

Anopheles flavirostris immatures are typically found in shaded portions of clear, slow-flowing streams, particularly those sites with emergent vegetation close to human habitation. Other freshwater immature habitats include slow-moving waters in canals, drainage ditches, and seepage pools. Immature *An. flavirostris* has a low tolerance for saline and stagnant waters. This author clearly remembers collecting immature *An. flavirostris* at the edges of a fast-flowing stream in a village in Mindanao, Philippines. The edge of the stream was choked with tall reeds, and here, among the heavily shaded calm waters, the immatures of this malaria species were plentiful. Sustainable vector control strategies involved the routine removal of the reeds from the stream edge; this significantly reduced the number of *An. flavirostris* in subsequent collections and significantly reduced malaria burden in localized areas.

Associated Pathogens

Plasmodium falciparum, *P. vivax*, *Wuchereria bancrofti*.

Diagnostic Characters

(a) Habitus (♀): Legs entirely dark scaled. **(b) Head (dorsal):** Proboscis (P) with ventral pale scale patch (indistinct in photo); palpus (MPlp) with three pale bands; $MPlp_5$ pale basally. **(c) Thorax (side):** Antepronotum (Ap) without scales; proepisternum with 1–4 setae (PeS). **(d) Thorax (dorsal):** Center of scutum (Scu) pale, setose, mostly lacking scales. **(e) Wing:** With relatively few dark spots; ≤3 dark spots on CuA and vein 1A; fringe without pale spot at end of vein 1A; base of vein R with white or yellow-white scales. **(f) Larva (head):** Seta 1-A single, small; seta 2-C single, inserted ≥ the distance between 2-C and 3-C on one side. **(g) Larva (thorax):** Seta 3-T palmate, leaflets without long fine points. **(h) Larva (abdominal segments):** Seta 0-IV–VI arising just posterolateral to tergal plate (TP) or just on or at edge of plate; seta 0-IV–VII small, 1–2 branched; tergal plates (TP) on III–VII very large, ≥0.5 width of segment, enclosing small median posterior accessory plate (ATP); tergal plates rounded posteriorly. **(i) Larva (terminal segments).** Image sources: WRBU; Harrison (1980).

Exemplar DNA sequences: *An. flavirostris COI* (not barcode region): (DQ059585, AY943650-65).

Catalog entry: Vol. 2, p. 661.

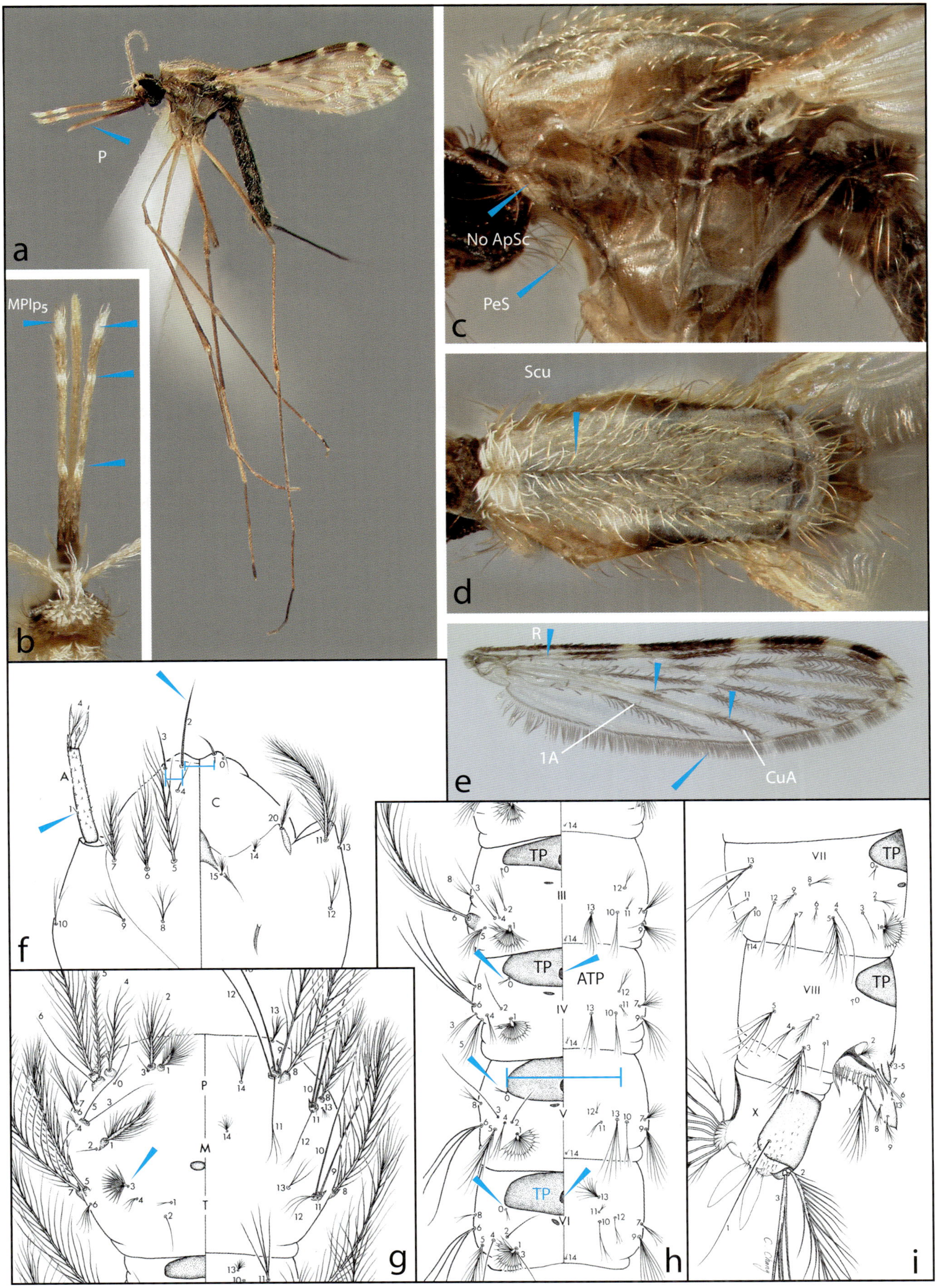
a
P
b
MPlp5
c
No ApSc
PeS
d
Scu
e
R
1A
CuA
f
A
C
g
P
M
T
h
TP
ATP
III
IV
V
VI
i
VII
VIII
X
C. Chang

Anopheles (*Cellia*) *fluviatilis* James, 1902 Subfamily Anophelinae "Riverine Indian Nail Mosquito"

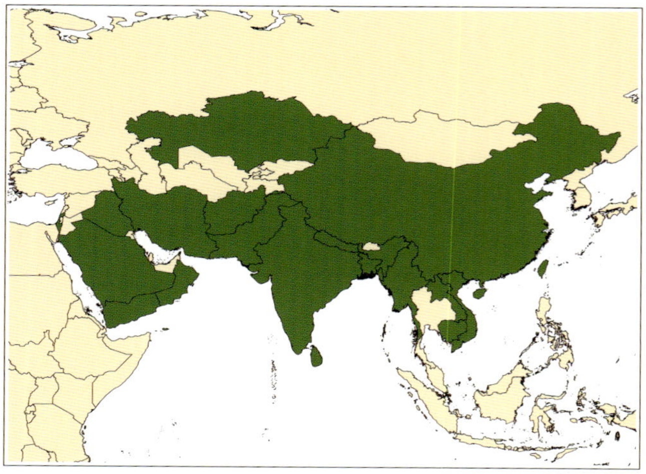

Systematics and Distribution

Anopheles fluviatilis is a small, almost entirely dark-legged mosquito with pale spotty wings and a shock of white erect scales on its head extending forward to a frontal tuft. *Anopheles fluviatilis* is known to comprise a species complex of three chromosomally identified forms (S, T, and U), and these plus two additional forms (V and X) have been recovered by DNA sequences. Form S is common in India and Nepal. Form T, found in India, Iran, Nepal, and Pakistan, exhibits three *ITS2* haplotypes (T1, T2, and Y). Form U, closely related to form T, is found in Uttar Pradesh, India, and in southeastern Iran and can reportedly hybridize. Forms V and X have recently been detected in Iran and India, respectively. Members of the Fluviatilis Complex are closely related to those in the Oriental Minimus Complex and share diagnostic characters of very large tergal abdominal plates in the larval stages. Recent molecular studies concluded that *An. fluviatilis* S is conspecific with *An. minimus* C. Three synonyms of *An. fluviatilis* exist: *listonii* Liston, *leptomeres* Theobald, and *arabica*, Christophers & Chand, but it remains unknown which, if any, correlate to the genetic forms identified. Earlier references often refer to *An. fluviatilis* as its junior synonym *listonii* Liston or as *christophersi* Theobald (=junior synonym of *An. minimus*).

The current distribution of *An. fluviatilis* s.l. includes the Middle and Far East, throughout the Indian subcontinent, including Sri Lanka, into the northern Oriental Region of Nepal and China and into Taiwan. *Anopheles fluviatilis* was originally described from specimens collected in Duars, Jeypore Hill Tracts, and Nagpur in Northeast, East, and Central India, respectively; however, the whereabouts of the type specimen are unknown. A molecularly typed neotype needs to be designated to allow the formal description of the known genetic forms.

Bionomics

Anopheles fluviatilis s.l. is associated with rocky hill or foothill regions at higher elevations. Immatures are widely dispersed in areas of suitable habitats and have never been found concentrated in high numbers, complicating their effective control. Typical immature habitats for *An. fluviatilis* s.l. include slow-flowing, heavily vegetated sites, such as stream pools, irrigation ditches, lake margins, ponds, and swamp edges; occasionally, larvae have been recovered from rice fields and in water tanks. Immatures were collected at 7,300 ft in the Yemen, in pools within a ditch heavily vegetated with filamentous green algae and stonewort rushes. In Southern India, adult *An. fluviatilis* is found at low elevations (<1,000 ft) but have been reported as high as 6,000 ft in Kashmir and 7,500 ft in Muree (Pakistan). Adults appear to be strong fliers, with females captured over 2,600 ft from their oviposition sites. Form S (=*An. minimus* C) is strongly anthropophilic, feeding avidly on humans, whereas forms T and U are exclusively or primarily zoophilic, preferring cattle. Form S (=*An. minimus* C) starts host seeking around 1.5 hours after dusk, and biting continues until midnight. It is regarded as a primary malaria vector in India.

Associated Pathogens

Plasmodium falciparum, *P. vivax*.

Diagnostic Characters

(a) Habitus (♀). (b) Thorax (side): Upper proepisternal setae (PeSU) present; scutal fossa (SF) without scales. **(c) Thorax (dorsal):** Center of scutum (Scu) mostly pale with narrow pale scales on median area; sides of scutum markedly darker than median area. **(d) Wing:** Costa (C) entirely dark scaled in the basal third, with no pale spots basal to sector pale spot (SP); radius (R) with only white or yellowish scales just distal to humeral crossvein (h); vein R_3 mostly dark; vein R_{4+5} mostly pale; wing fringe usually with at least four pale spots on posterior margin. **(e) Head (side):** Proboscis (P) all dark; palpus with one broad and two narrower bands; $MPlp_4$ with a narrow basal pale band, around half size of pale band on $MPlp_3$; $MPlp_5$ all pale; erect head scales broad, white on vertex (V) and frontal tuft (FT) and dark brown laterally and posteriorly. **(f) Larva (head):** Seta 1-A small, single; seta 2-C simple, widely spaced. **(g) Larva (thorax/abdominal segments):** Anterior tergal plates (TP) on segments III–VII very large, >0.5 width of segment, enclosing accessory tergal plate (ATP); seta 0-IV–VI arising posterolateral to primary tergal plate (TP), or at or on edge of plate; seta 0-IV–VII relatively large, 2–6 branched; seta 4-M with sum of branches of both sides 8–11 (indistinct in illustration); one of setae 9,10-T unbranched. Note: In lieu of available whole *An. fluviatilis* larval drawings, the diagnostic larval characters are illustrated here using *An. minimus*. Image sources: WRBU; Ross & Roberts (1943).

Exemplar DNA sequences: *Anopheles fluviatilis* s.l. *COI* (EU256365–67, GQ906976–87).

Catalog entry: Vol. 2, p. 662.

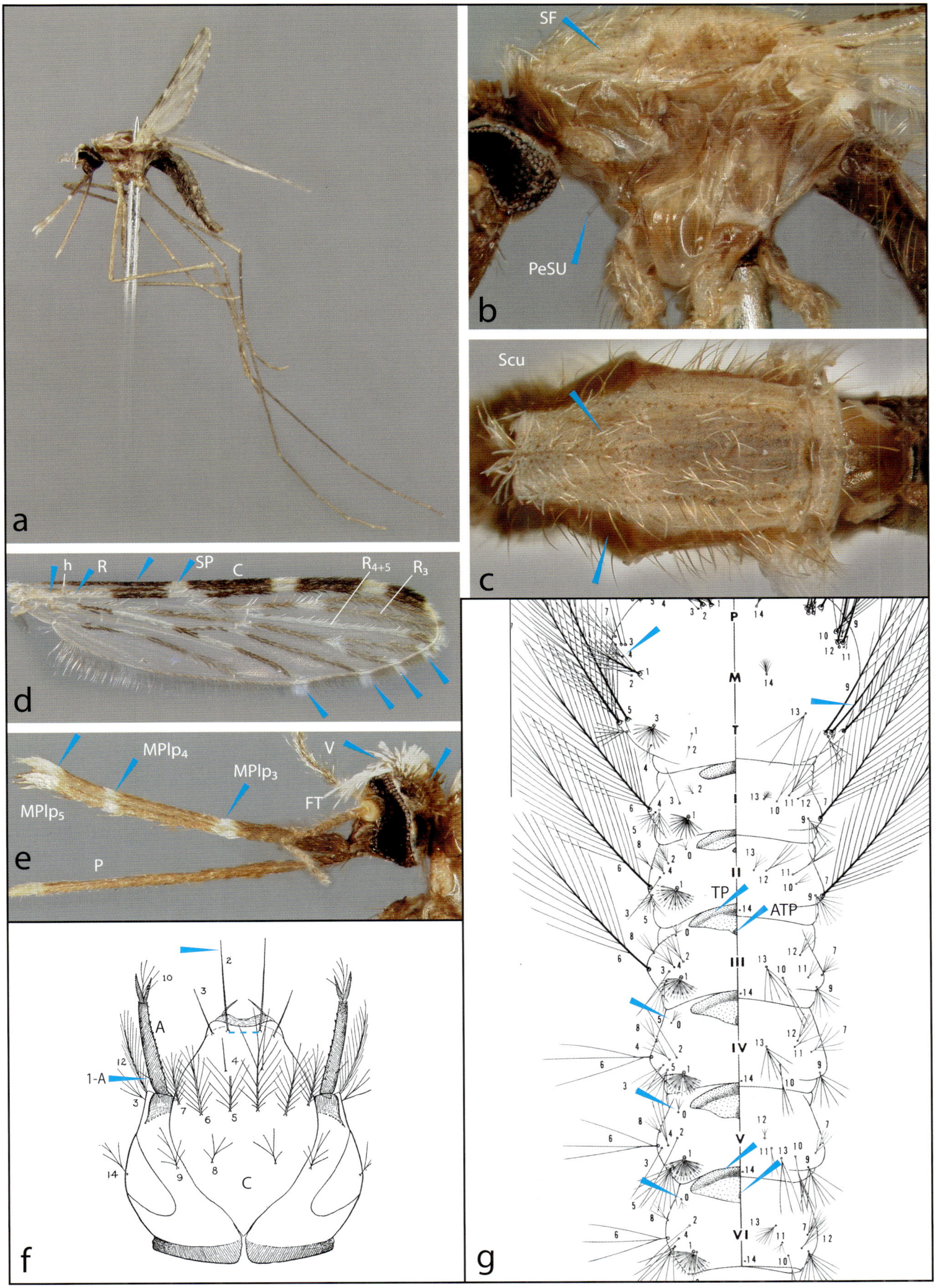
a
SF
PeSU
b
Scu
c
h
R
SP
C
R_{4+5}
R_3
d
$MPlp_4$
$MPlp_3$
$MPlp_5$
V
FT
P
e
A
1-A
C
f
P
M
T
I
II
TP
ATP
III
IV
V
VI
g

Anopheles (*Cellia*) *funestus* Giles, 1900 Subfamily Anophelinae "Deadly Sierra Leonean Nail Mosquito"

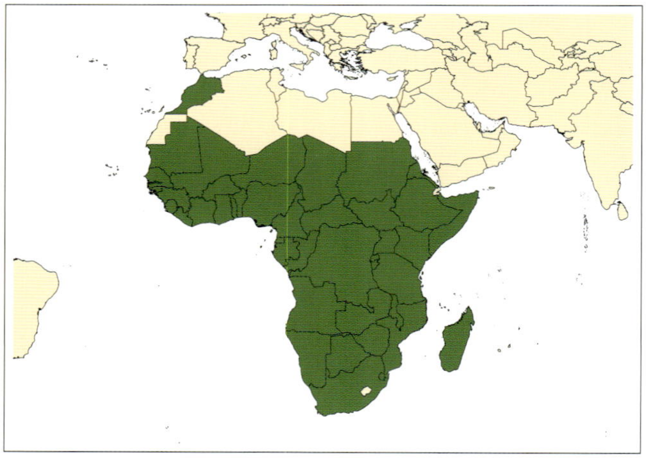

Systematics and Distribution

Anopheles funestus is a small but deadly creature. This super effective malaria vector is somewhat nondescript, bearing a three-banded proboscis with the apical and subapical ones narrower than the intervening black band (see *d*) and legs with only the smallest of apical bands on the hind tarsi (see *e*). However, several species-diagnostic characters are apparent in wing spot patterns (see *c*). *Anopheles funestus* is the nominotypical member of both the Funestus Group and Funestus Subgroup of the *Anopheles* subgenus *Cellia* (see Vol. 2, p. 1083, for internal relationships). The Funestus Group comprises 23 formally described Afrotropical and Oriental species and a number of genetically defined taxa, including "*funestus*-like." The group is further subdivided into the Afrotropical Funestus and Rivulorum Subgroups, and Oriental Aconitus and Culicifacies Subgroups, with both Afrotropical and Oriental species placed within the Minimus Subgroup. The Oriental species *An. jeyporiensis* James (see p. 348) is unplaced. Species in the Funestus Subgroup include *An. aruni* Sobti, *An. confusus* Evans & Leeson, *An. funestus*, *An. funestus*-like, *An. longipalpis* (Theobald), *An. parensis* Gillies, and *An. vaneedeni* Gillies & Coetzee. DNA-based identification, including species-diagnostic polymerase chain reaction (PCR) assays based on the nuclear *ITS2* rDNA region, are commonly used to differentiate species of the Funestus Group. There are currently seven synonyms for *An. funestus* s.s.—*kumasii* Chalmers, *hebes* Dönitz, *umbrosa* Theobald, *subumbrosa* Theobald, *anisochloros* Theobald, *bisignata* Grünberg, and *imerinensis* Monier & Treillard.

Anopheles funestus s.s. was first described from Freetown, Sierra Leone. It is widely distributed throughout Africa, including on the islands of Madagascar and Zanzibar, where its range overlaps significantly with areas of highest malaria transmission.

Bionomics

Anopheles funestus s.s. is the most competent malaria vector in the Funestus Group. *Anopheles funestus* is commonly collected in biosurveillance surveys as this species is highly anthropophilic and strongly endophilic, and likes to rest in human dwellings after feeding. *Anopheles funestus* s.s. is commonly collected in light traps, at human bait, and in indoor resting collections or pyrethroid spray catches. The other Afrotropical Funestus Group species—*An. rivulorum* Leeson, *An. parensis*, and *An. leesoni* Edwards—are also collected resting indoors. There is some evidence that implementation of indoor residual spray campaigns can result in behavioral modifications in *An. funestus* s.l. In Toloki, Benin, the proportion of outdoor biting *An. funestus* s.l. changed from 45% to 68.1% before and after one year of initiation of a deltamethrin indoor residual spray campaign, and the peak biting times switched from 0300h to 0500h.

Immature *An. funestus* s.s. is typically collected in large, deep bodies of clear water, including lakes, ponds, streams, rivers, and ditches, especially where emergent vegetation is present. *Anopheles funestus* s.s. is also collected in high densities in older, maturing rice fields.

Associated Pathogens

BOZOV, BUNV, BWAV, GERV, NDOV, NRIV, ONNV, ORUV, SFV, TANV, TATV; *Plasmodium falciparum*, *P. vivax*, *Wuchereria bancrofti*.

Diagnostic Characters

(a) Habitus (♀). **(b) Larva (head):** Setae 2-4-C single. **(c) Wing:** No pale spots on the costa (C) basal to sector pale spot (SP); accessory sector pale spot (ASP) continuous to costa; preapical dark spot (PD) without pale interruption, longer than preapical pale spot (PP); vein R all pale basally; no pale fringe at end of vein 1A; vein M_{3+4} with a pale spot. **(d) Head (dorsal):** Palpus (MPlp) with three pale bands, including one apically. **(e) Leg (III):** Legs not speckled; Ta-III_{1-5} with very narrow apical pale bands. **(f) Larva (thorax/abdominal segments):** Tergal plate (TP) on segments IV and V wider than distance between palmate setae; TP large, c. two times wider than deep; tergal accessory plates usually absent; setae 9,10-M long and single (not shown); sides of thorax and abdomen not spiculate. Image sources: WRBU; Ross & Roberts (1943).

Exemplar DNA sequences: *An. funestus* s.s. *COI* (JQ424633–85, MG190073–76); whole genome (GCA_003951495, GCA_000349085).

Catalog entry: Vol. 2, p. 662.

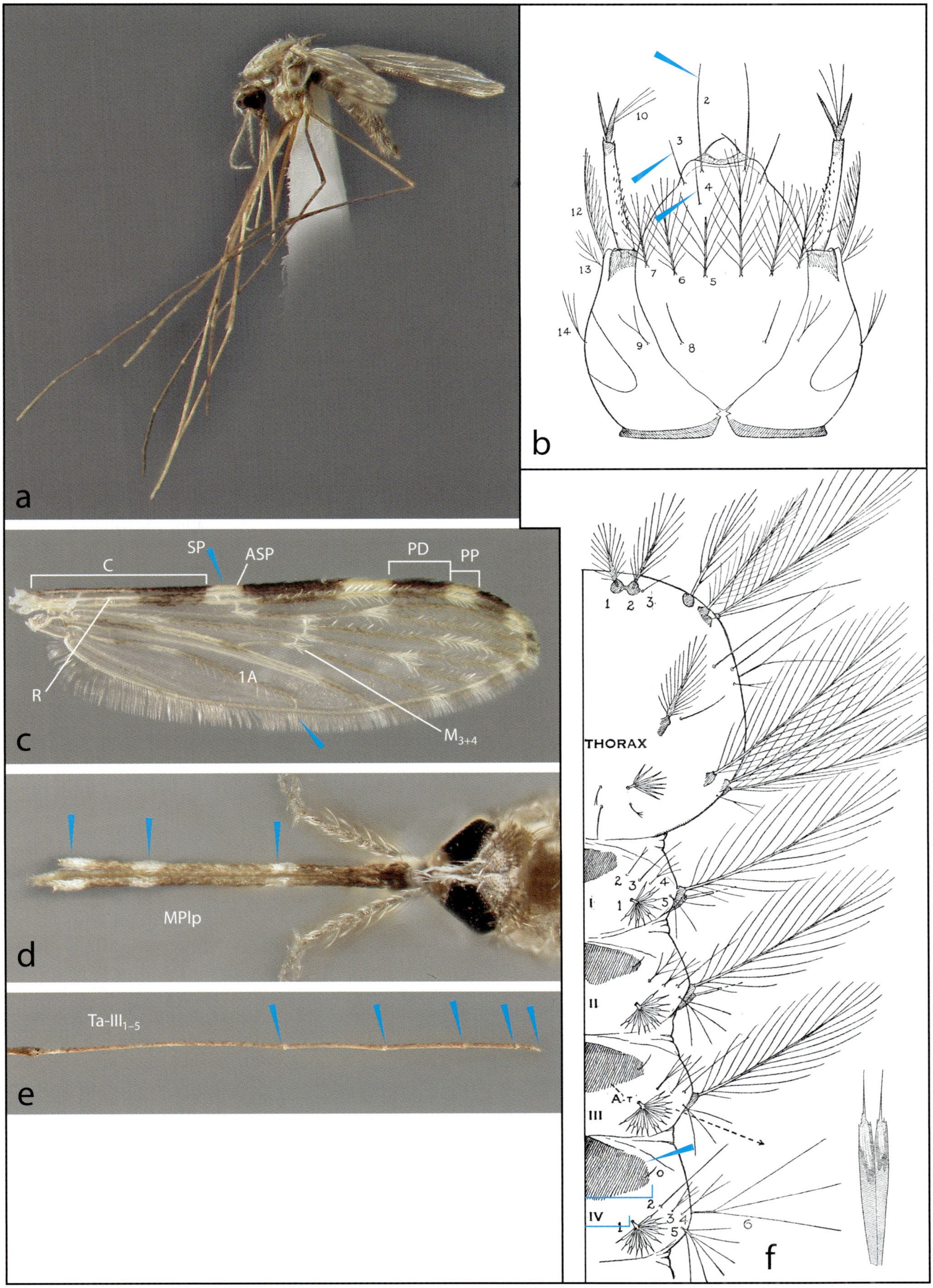

a
b
2
10
3
4
12
13
7
6
5
14
9
8
c
C
SP
ASP
PD
PP
R
1A
M_{3+4}
d
MPlp
e
Ta-III_{1-5}
f
THORAX
1 2 3
I
II
III
IV
A-T
6

Anopheles (*Cellia*) *gambiae* Giles, 1902
Subfamily Anophelinae
"Malaria Gambian Nail Mosquito"

Systematics and Distribution

Anopheles gambiae is a common Afrotropical mosquito, commonly recognized by the patterns on its wing. These include four or more dark spots on the costa and veins $R–R_1$ and pale fringe spots up to CuA or 1A (see *e*). As the dominant Afrotropical vector of cerebral malaria, *Anopheles gambiae* is quite literally the world's most dangerous animal known to humans. *Anopheles gambiae* is the nominotypical member of the infamous Gambiae Complex (Pyretophorus Series, *Anopheles* subgenus *Cellia*) (see Vol. 2, p. 1086). The taxonomy of two chromosomally distinct but essentially isomorphic forms of *An. gambiae*—Mopti (M) and Savanna (S)—were recently clarified with the formal description of *Anopheles coluzzii* Coetzee & Wilkerson (= M form), named in deference to Mario Coluzzi, who worked tirelessly on this complex. The name *An. gambiae* was retained for the most common form (S). There are noted differences in the eggs of these species but not in the larvae or adults, and thus characters and images shown here pertain to both species and to others in the Gambiae Complex. The taxonomic status of three other reported chromosomal forms of *An. gambiae* (Bamako, Forest, and Bissau) and their relation (if any) to any of the four valid species synonyms—*costalis* Giles, *gracilis* Dönitz, *minor* Holstein, and *typicus* Holstein—have still to be resolved.

Formally described species in the Gambiae Complex now include *An. amharicus* Hunt, Wilkerson & Coetzee (formerly *An. quadriannulatus* B), *An. arabiensis* Patton, *An. bwambae* White, *An. coluzzii*, *An. fontenillei* Barrón, Paupy, Rahola, Akone-Ella, Ngangue, Wilson-Bahun, Pombi, Kengne, Costantini, Simard, González & Ayala, *An. gambiae*, *An. melas* Theobald, *An. merus* Dönitz, and *An. quadriannulatus* (Theobald). *Anopheles gambiae*, described from the Gambia, is widespread across sub-Saharan Africa and has encroached into the Yemen and Saudi Arabia on the Arabian Peninsula. Its range overlaps the areas of most intensive *Plasmodium falciparum* malaria transmission zones in Africa.

Bionomics

The most important larval habitats of *An. gambiae* and *An. coluzzii* are man-made; thus, both species are commonly associated with human activities and dwellings, especially in areas affected by deforestation, agriculture, road making, and veld burning. Immature *An. gambiae* s.s. is associated with temporary, rain-dependent habitats, including ground pools, quarries, puddles, wheel ruts, and hoofprints, whereas *An. coluzzii* tends to occupy slightly larger water bodies associated with agriculture, including rice fields, ditches, and reservoirs.

Anopheles gambiae and *An. coluzzii* are both highly anthropophilic, feeding only on other mammals when people are not present, and are extremely effective malaria vectors. *Anopheles gambiae* s.s. and *An. coluzzii* are sympatric in Mali, and both species were collected on sticky traps posted at high altitudes (40 m–240 m) over the Sahel Region, suggesting they utilize long-distance migration survival strategies.

Associated Pathogens

AnCV, AnCPV, AgDNV, AngFV, AToV, BARV, BWAV, CVOV, ILEV, NRIV, ONNV, ORUV, TAHV, TATV, WSLV, ZIKV; *Plasmodium falciparum*, *P. gallinaceum* (chicken), *P. malariae*, *P. ovale*.

Diagnostic Characters

(a) Habitus (♀). (b) Abdomen (dorsal): Abdomen without posterolateral scale tufts; sparse or no dorsal abdominal scales. **(c) Head (dorsal):** Palpus (MPlp) usually with three pale bands. **(d) Leg (III):** $Ta\text{-}III_{1-5}$ not entirely pale; legs sparsely speckled. **(e) Wing:** Costa (C) and veins $R–R_1$ with ≥4 dark spots; preapical dark spot (PD) on vein R_1 with or without pale interruption; wing with pale fringe spots up to CuA or 1A. **(f) Larva (head):** Seta 1-A very small; seta 2-C single or aciculate, aciculae can be long; seta 2-C widely separated > than distance between 2-C and 3-C; seta 3-C single (shown) or ≤8 branched. **(g) Larva (thorax/abdominal segments):** Seta 1-P not well developed, basal tubercle absent or small; setae 9,10-M usually single; spine at base of setae 9–12-M distinct, curved, and sharply pointed; seta 1 palmate on all or most segments; segment V with tergal plate (TP) small, <0.67 times distance between palmate setae; no spicules on sides of larval thorax and abdomen. Image sources: WRBU; White (1985) (characters illustrated on *An. bwambae* larvae).

Exemplar DNA sequences: Whole genomes: *An. arabiensis* (GCA_000349185); *An. bwambae* (GCA_007018985); *An. coluzzii* (GCA_000150765, GCA_004136515); *An. gambiae* s.s. (GCF_000005575, GCA_001542645, GCA_001014525, GCA_00015025); *An. melas* (GCA_000473525); *An. merus* (GCA_000473525); *An. quadriannulatus* (GCA_000349065). *An. amharicus* ITS2 (GQ870316).

Catalog entry: Vol. 2, p. 663.

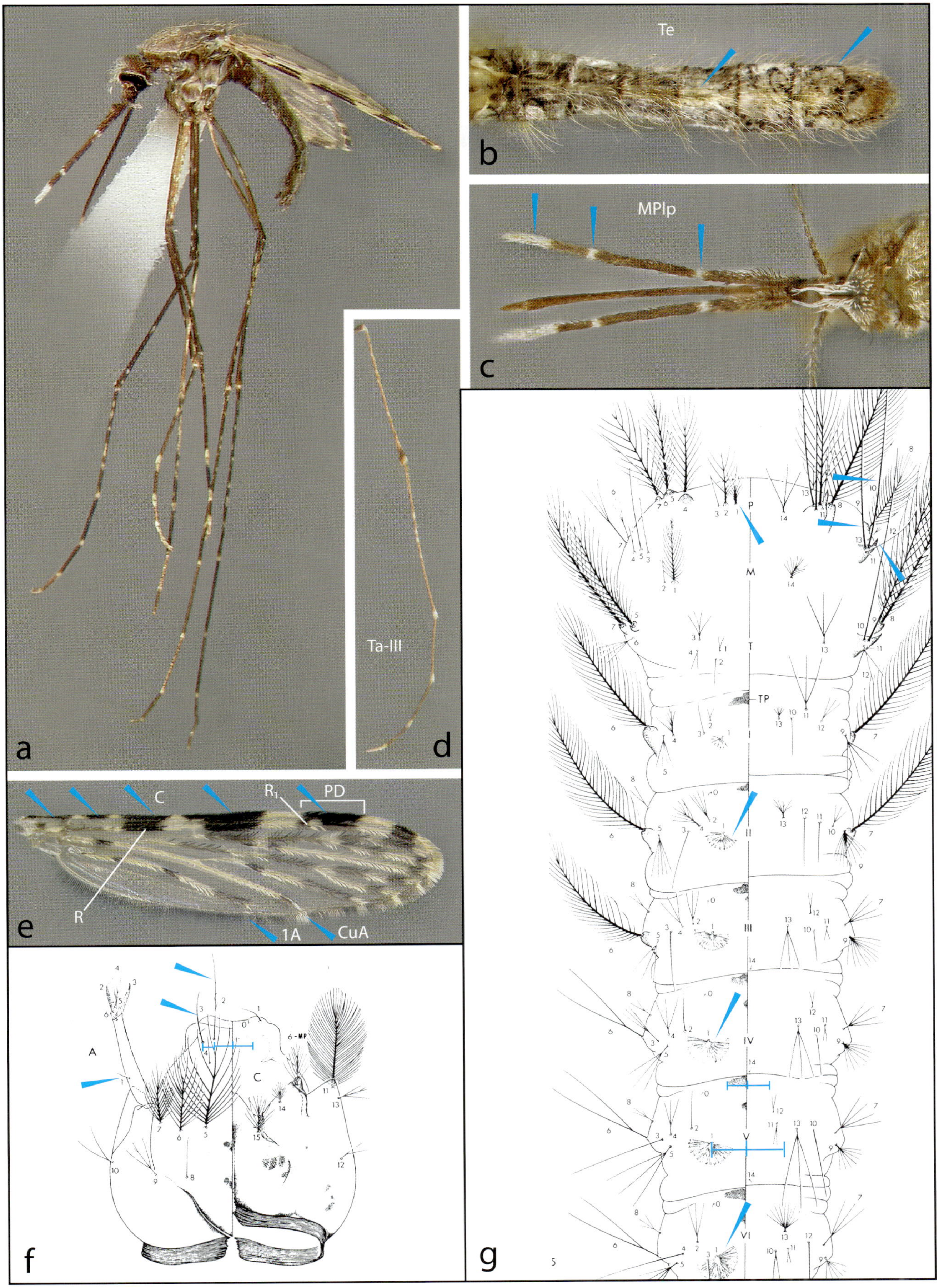

a
Te
b
MPlp
c
Ta-III
d
C
R1
PD
R
1A
CuA
e
A
6-MP
C
f
P
M
T
TP
I
II
III
IV
V
VI
g

Anopheles (*Cellia*) *jeyporiensis* James, 1902 Subfamily Anophelinae "Jeypur Indian Nail Mosquito"

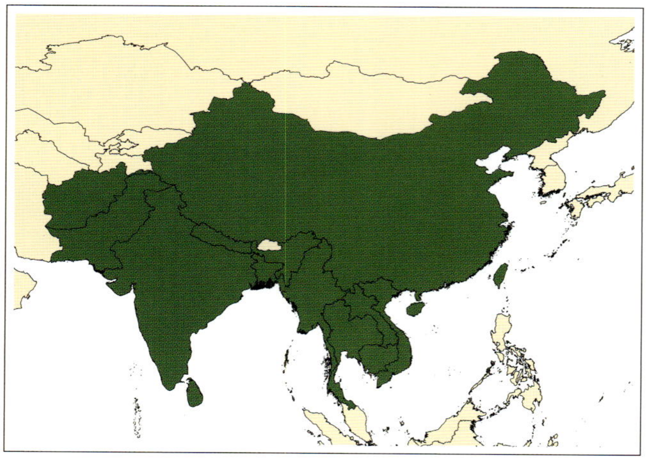

Systematics and Distribution

Anopheles jeyporiensis looks handsomely snow dusted, with a unique covering of short, oblong white scales on the scutum extending all the way back to the scutellum. *Anopheles jeyporiensis* is the only unplaced member of the Funestus Group (*Cellia Series*, *Anopheles* subgenus *Cellia*) (see Vol. 2, p. 1083, for internal relationships). The species is highly variable, especially in morphological characters of the wing and palps, and four chromosomal forms (A, B, C, D) have been reported from Thailand. Three taxa are currently regarded as synonyms: *jeyporensis* Theobald, described from India, *candidiensis* Koidzumi (Taiwan), and *tonkinensis* Toumanoff (Vietnam).

Anopheles jeyporiensis was first described from Nagpur, Jeypur [Jeypore State, in the Central Provinces], India. Its current distribution (as s.l.) centers on the Indian subcontinent (including Sri Lanka) westward to Pakistan and eastward through the northern countries of mainland Southeast Asia to Taiwan.

Bionomics

Anopheles jeyporiensis is regarded as a primary malaria vector, and reported infection rates for *Plasmodium* range from 2.5% to 4.5%. The species is associated with hilly, mountainous, or high plateau regions and has been reported at elevations up to 1,829 m. It is most common following the monsoon and in the cooler periods (November to March) of the year in India. In Laos and Vietnam, *An. jeyporiensis* is predominantly endophilic, entering houses to feed on humans, although it is also captured biting cattle outdoors. In Laos, indoor biting times peaked at 2200h, and the biting rates were sustained through the early hours of the morning. Exophilic bites peaked at midnight.

Anopheles jeyporiensis immatures are found in seepage pools, ponds, and swamps with abundant emerging vegetation where the stagnant or slow-moving fresh waters are cool (23°C–33°C) and clear. Immatures have also been collected in water tanks and from terraced rice fields, but only when water replacement was occurring. In Thailand, the four chromosomal forms (A, B, C, D) are reportedly associated with specimens collected from marshy depressions with plentiful submerged and emergent vegetation.

Associated Pathogens

Brugia malayi, *Plasmodium falciparum*, *P. vivax*, *Wuchereria bancrofti*.

Diagnostic Characters

(a) Habitus (♀). (b) Abdomen (dorsal): Terminal abdominal segments without scales. **(c) Thorax (dorsal):** Center of scutum (Scu) covered with short white scales and golden setae. **(d) Thorax (side):** Antepronotal scales (ApSc) absent. **(e) Leg (I):** Ta-I_1 with short apical pale band nearly two times width of tarsomere. **(f) Leg (III):** Ta-III pale bands narrow; Ta-III_5 not all white. **(g) Head (side):** Palpus (MPlp) with three pale bands; frontal tuft (FT) with distinctive, long white scales. **(h) Wing:** Vein R with accessory spot on preapical dark (PD) area. **(i) Larva (head):** Setae 2,3-C with numerous short aciculae; seta 4-C 2–5 branched. Image sources: WRBU; Harrison (1980).

Exemplar DNA barcodes: *An. jeyporiensis* (DQ154157, DQ154159, DQ317591–93, GQ259195, HQ877379).

Catalog entry: Vol. 2, p. 665.

a
b
Scu
c
No ApSc
d
Ta-I1
I
e
III
f
Ta-III5
MPlp
FT
g
R
PD
h
A
C
i

Anopheles (*Cellia*) *koliensis* Owen, 1945
Subfamily Anophelinae
"Koli Solomon Islander Nail Mosquito"

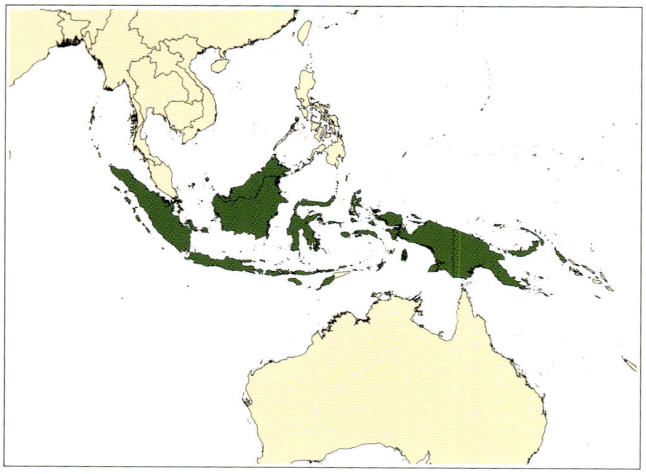

Systematics and Distribution

Anopheles koliensis is the Dalmatian of the mosquito world, with heavily spotted wings and a plethora of short white rings and spots on all its legs. *Anopheles koliensis* is an unplaced member of the Punctulatus Group (Neomyzomyia Series, *Anopheles* subgenus *Cellia*) (see Vol. 2, p. 1085, for internal relationships). Molecular differentiation based on the nuclear rDNA *ITS2* is proving the most reliable method for species recognition within the Punctulatus Group. Studies of *An. koliensis* in Papua New Guinea revealed three unique *ITS2* PCR-RFLP genotypes, suggesting the species may comprise additional cryptic taxa. The three variants were designated as M (Madang), W (Wosera), and MW (Madang-Wosera) based on collection site, with the latter corresponding to the classic type and with some evidence of interbreeding indicated between M and MW forms. Within the Punctulatus Group, the evolutionary relationships between *An. koliensis* and the remaining species are still unresolved. Using mtDNA *cytochrome c oxidase* II gene (*COI*), *An. koliensis* sits basal to all other taxa, is grouped with *An. punctulatus* and *An. farauti* 4 using whole mtDNA genomes, and sits between *An. farauti* Laveran and *An. punctulatus* Dönitz in studies using nuclear rDNA *ITS2* sequences. Whole genome markers are needed to fully resolve the phylogenetic relationships and better understand the evolutionary relationships between *An. koliensis* and other species in the Punctulatus Group.

Anopheles koliensis takes its name from the type locality of Koli Area, Guadalcanal, in the Solomon Islands. The known distribution of the species (as s.l.) includes Indonesia (including Borneo and Sulawesi), New Guinea, including Papua New Guinea, and the Solomon Islands.

Bionomics

Anopheles koliensis is regarded as one of the top five malaria vectors in the Southwest Pacific. It is endophilic, endophagous, and highly anthropophilic. Due to its endophilic behaviors, vector control efforts, including insecticide-treated bed nets and indoor residual spraying programs, have been highly successful against *An. koliensis* and have pushed the species to apparent extinction in the Solomon Islands and on the islands of Buka and Bougainville in the Solomon Archipelago. Like *An. farauti*, *An. farauti* 4, *An. hinesorum*, and *An. punctulatus*, *An. koliensis* s.l. are associated with lowland, inland environments in close proximity to humans. Only one specimen has ever been reported over 1,000 m. In Papua New Guinea, *ITS* genotype *An. koliensis* M was found along the Madang coastline, fewer than 5 km from the sea, and *An. koliensis* W was found only in Wosera, more than 30 km from the coast. The typical form of *An. koliensis* is most often between 5 km–15 km from the sea but it has occasionally been detected closer to the coast.

In West Papua, immature habitats for *An. koliensis* are numerous in peridomestic sites, including freshwater collections in drainage ditches, natural ground pools, and agricultural plots of water spinach (*Ipomea reptans*). Studies here suggest that short periods of drought lengthen the gonotrophic cycle of *An. koliensis*, leading to a reduction in biting activity by parous females and consequently lowering malaria transmission. Upon rainfall and availability of suitable immature habitats, the gonotrophic cycle returns to normal and malaria transmission reignites.

Associated Pathogens

Plasmodium falciparum, *P. malariae*, *P. vivax*, *Wuchereria bancrofti*.

Diagnostic Characters

(a) Habitus (♀): Terga (Te) without dense scales. **(b) Thorax (side):** Mesepimeron without a central patch of flat pale scales (MeSc); halter (Hl) capitellum (Ca) dark scaled dorsally. **(c) Thorax (dorsal):** Scutum (Scu) with scales on lateral areas. **(d) Wing:** Numerous dark spots; base of vein R with one or two black spots. **(e) Legs (I,II):** Ta-I and Ta-II ringed or black and white spotted. **(f) Head (side):** Proboscis (P) with pale scales on apical 0.33 to 0.50, in addition to apical pale band; $MPlp_3$ with apical white portion divided by a black band; $MPlp_5$ dark ventrally. **(g) Larva (thorax/abdomen):** Seta 1-M plumose; seta 1-P with stout expanded stem; setae 1–3-P on common tubercle; seta 9–12-P long, single; seta 1-I not palmate, with slender or narrow hairlike branches. **(h) Larva (head):** Ventromentum (Vm) relatively narrow and long. **(i) Larva (terminal segments):** Tergal plate (TP) of segment VIII small; TP on IV,V approx. two times wider than deep (not shown); tergal accessory plates usually absent. Image sources: WRBU; Belkin (1962).

Exemplar DNA sequences: *An. koliensis* s.l. *ITS2* (KU139113–15), *COI* (KU662130–306), partial mitochondrial genome (JX219742–43).

Catalog entry: Vol. 2, p. 667.

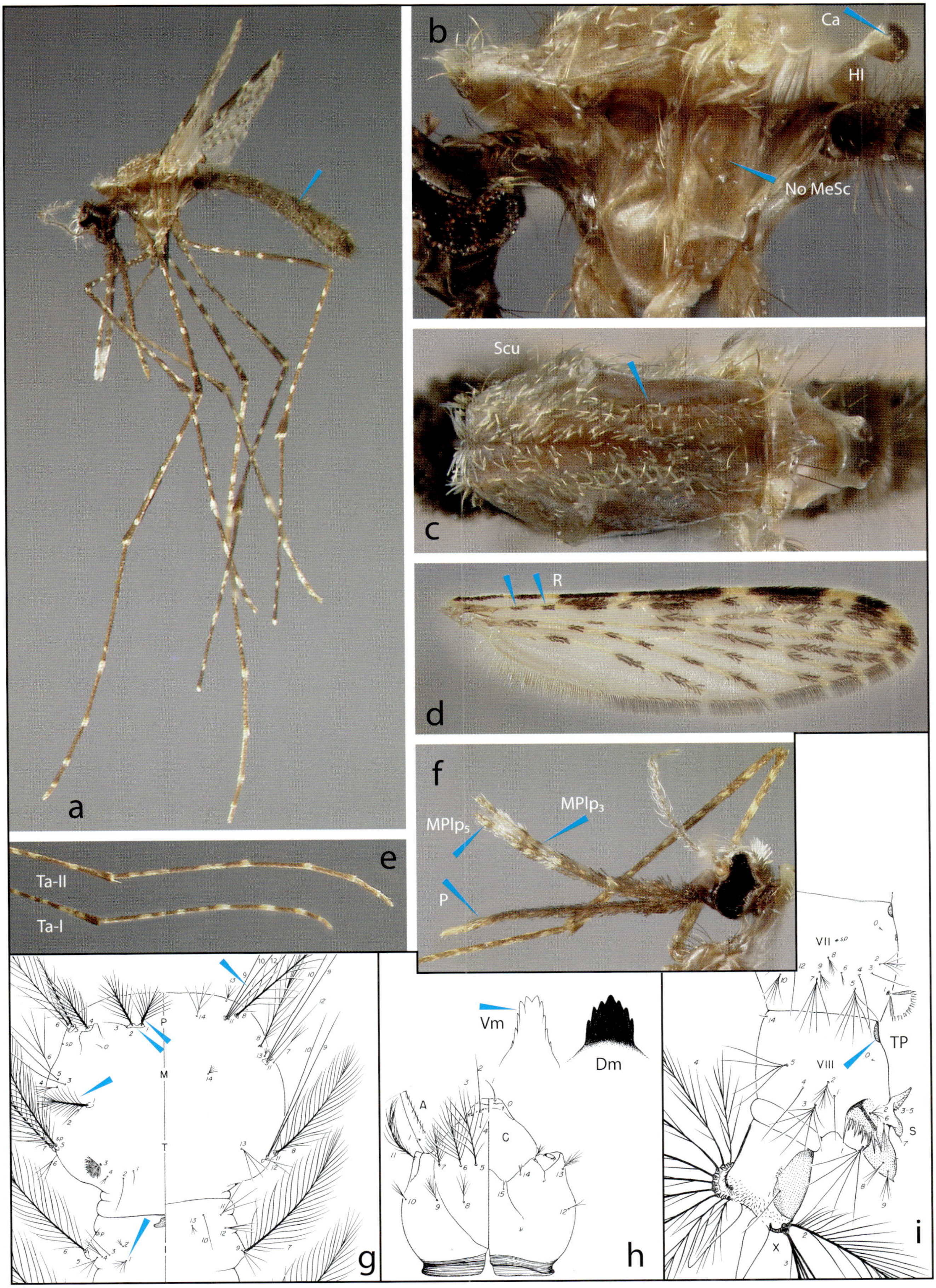
a
b
Ca
Hl
No MeSc
c
Scu
d
R
e
Ta-II
Ta-I
f
MPlp5
MPlp3
P
g
P
M
T
h
Vm
Dm
A
C
i
VII
TP
VIII
S
X

Anopheles (*Cellia*) *latens* Sallum & Peyton, 2005
Subfamily Anophelinae
"Secret Thai Nail Mosquito"

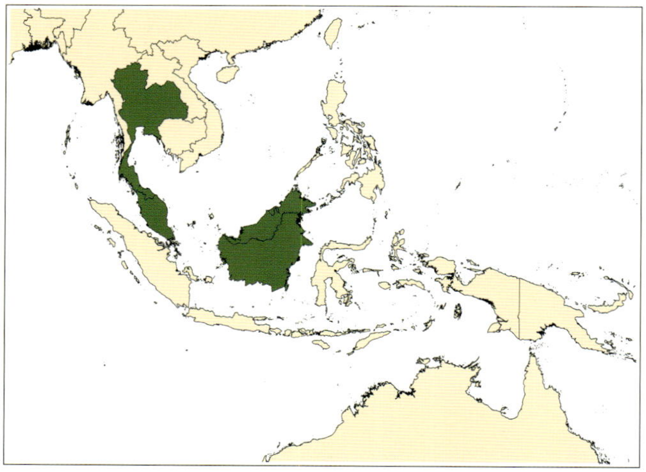

Systematics and Distribution

Anopheles latens is a striking mosquito, beautifully pebble dashed with small pale scale patches on the legs (see *a*), white hind "knees," and an intricate dark-light mosaic pattern on the wings (see *c*). *Anopheles latens* takes its name from the Latin for "concealed" as the species was long hidden under *An. leucosphyrus* Dönitz until chromosomal studies indicated the presence of two cytotypes. *Anopheles leucosphyrus* and *An. latens* (formerly *An. leucosphyrus* A) are morphologically identical as adults, pupae, and larvae. Thus, this species page serves to identify both species, which are regarded as sister taxa in the Leucosphyrus Complex (Leucosphyrus Subgroup, Leucosphyrus Group) of the subgenus *Anopheles* (*Cellia*) (see Vol. 2, p. 1085, for internal relationships). Along with *An. latens* and *An. leucosphyrus*, the Leucosphyrus Complex comprises a further three species: *An. baisasi* Colless, *An. balabacensis* Baisas, and *An. introlatus* Colless.

Anopheles latens and *An. leucosphyrus* are allopatric. Following their taxonomic revision, it appears that *Anopheles leucosphyrus*, originally described from Kajoe Tanam in Sumatra, is endemic only to forested parts of that island, whereas the established range of *An. latens* stretches from the forests of Southern Thailand through peninsular Malaysia and into the northern and east parts of Borneo (see map). The type locality of *An. latens* is Phangnga, Ban Bang Kaeo, Thailand.

Bionomics

Within the Leucosphyrus Complex, *An. baisasi*, *An. balabacensis*, and *An. latens* are all highly effective human malaria vectors. *Anopheles introlatus* is highly zoophilic, and the natural vector competency of *An. leucosphyrus* s.s. in Sumatra needs to be evaluated. *Anopheles latens* is a primary vector of human malaria (*Plasmodium falciparum*, *P. vivax*) in the Malaysian state of Sarawak, northeast Borneo. *Anopheles latens* is a forest mosquito known to feed on both monkeys and humans outdoors; it also readily enters houses late at night to feed on humans. This plasticity in feeding behavior is of concern in forested areas, where it is a proven vector in the novel foci of *P. knowlesi* (a simian malaria) to humans in Sarawak.

Ovipositing *An. latens* are attracted to shaded, temporary ground pools and natural containers of clear or turbid water in forested areas. Immature *An. latens* have been found in small, shaded forest pools, forest streams, and tree-covered swamps. Little is known on the true bionomics of *An. leucosphyrus* s.s.

Medical Importance

Plasmodium knowlesi, *P. falciparum*, *P. vivax*, *Wuchereria bancrofti*.

Diagnostic Characters

(a) Habitus (♀): Proboscis (P) slightly longer than forefemur. **(b) Head (dorsal):** $MPlp_5$ apical pale band distinctly light-cream colored, contrasting with silvery-white bands on $MPlp_{2,3}$. **(c) Wing:** Presector dark spot (PSD) of vein R entirely dark or with 1–5 pale interruptions; accessory sector pale spot (ASP) usually extending onto vein C on at least one wing; veins M_{3+4} and M noticeably wavy; fork of cell R_2 in line or basal to fork of cell M_{1+2}; fringe with pale spot between veins 1A and CuA. **(d) Thorax (side). (e) Leg (III):** Apex of Ti-III and base of Ta-III_1 with large white band; Ta-III_4 without basal pale band; Ta-III_5 with apical pale band. **(f) Larva (thorax):** Basal sclerotized tubercle of seta 1-P without prominent tooth or spine arising from posterodorsal margin, conspicuously smaller than that of seta 2-P; setae 1,2-P basal sclerotized tubercles usually separate, distance between bases wide, equal to or greater than basal width of tubercle of seta 1-P. **(g) Larva (head):** Seta 3-C single; seta 5-C conspicuously longer than antenna (A). **(h) Larva (abdominal segments):** Seta 1-II not fully developed, noticeably smaller than seta 1-III–VI. Image sources: WRBU; Sallum et al. (2005).

Exemplar DNA sequences: *An. latens ITS2* (MG008596–612), *COI* (DQ897936–37); *An. leucosphyrus COI* (DQ897939), *NADH6* (DQ899798–99).

Catalog entry: Vol. 2, p. 667.

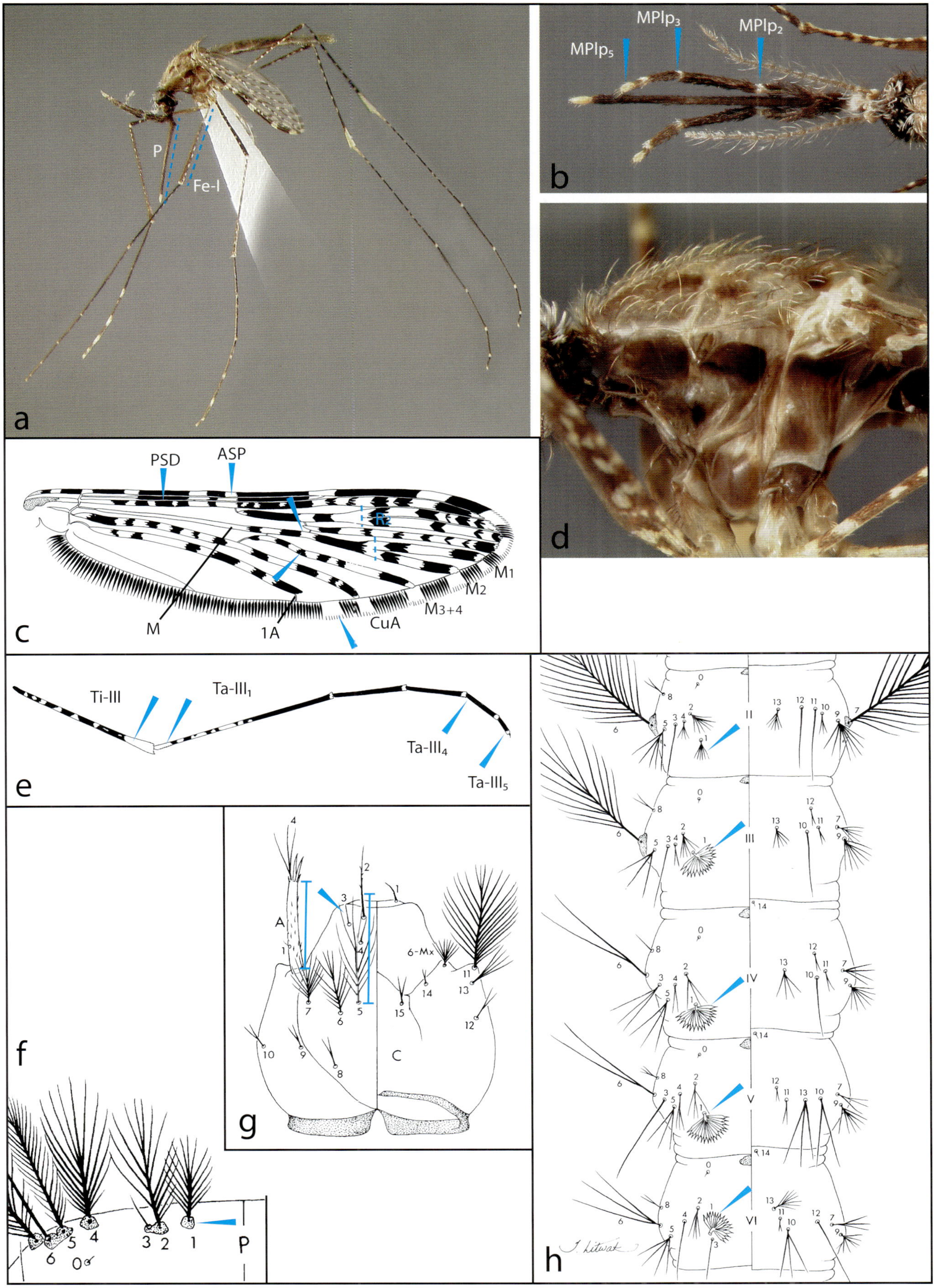
a
P
Fe-I
b
MPlp5
MPlp3
MPlp2
d
c
PSD
ASP
R2
M1
M2
M3+4
CuA
M
1A
e
Ti-III
Ta-III1
Ta-III4
Ta-III5
f
P
g
A
C
6-Mx
h
II
III
IV
V
VI

Anopheles (*Cellia*) *maculatus* Theobald, 1901 Subfamily Anophelinae "Spotted Chinese Nail Mosquito"

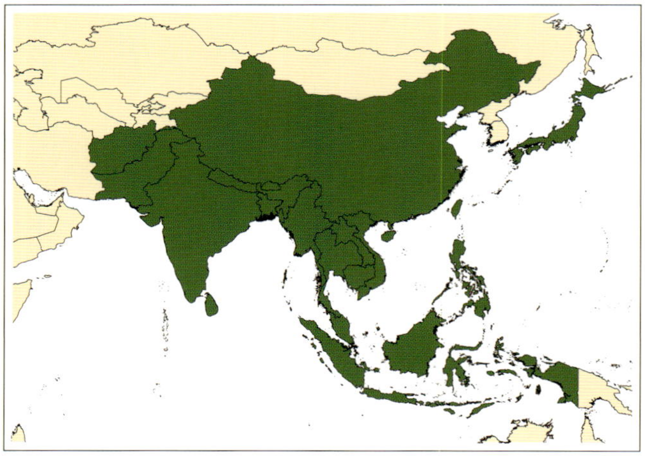

Systematics and Distribution

Anopheles maculatus is a small mosquito, with many tiny white spots and bands along its legs (see *a*, *c*). It has four distinct dark spots on the leading edge of the wing, with the remainder of the wing predominantly pale scaled (see *d*). Cryptic taxa were first proposed on the basis of chromosomal differences, and nine formally described species are now recognized within the Maculatus Group (Neocellia Series) (see Vol. 2, p. 1084). These include *Anopheles maculatus* s.s. (= form B), *An. sawadwongporni* Rattanarithikul & Green (= form A), *An. dravidicus* Christophers (= form C), *An. greeni* Rattanarithikul & Harbach (= form D), *An. notanandai* Rattanarithikul & Green (= form G), *An. willmori* (James) (= form H), *An. pseudowillmori* (Theobald) (= form I), *An. dispar* Rattanarithikul & Harbach (= form J), and *An. rampae* Harbach & Somboon (= form K). Within the Maculatus Group, *An. maculatus* s.s. and its sister taxa *An. dravidicus* are placed in the Maculatus Subgroup, and *An. notanandai*, *An. rampae*, and *An. sawadwongporni* make up the Sawadwongporni Subgroup. All others remain unplaced. *Anopheles maculatus* s.s. has one synonym—*hanabusai* Yamada, described from Taiwan.

Anopheles maculatus s.s. was originally described from specimens collected in Hong Kong and it has been molecularly confirmed as far away as India, suggesting the species is probably widely distributed in the Oriental Region. *Anopheles maculatus* s.l. is reported from the Indian subcontinent and throughout the Southeast Asian mainland and islands as far south and east as Timor (see map).

Bionomics

Species of the Maculatus Group are typically found in hilly, forested areas close to human settlements. Immatures are found in a wide range of permanent and temporal habitats. *Anopheles maculatus* s.s. immatures are collected in sunlit rock pools near waterfalls or in drying stream pools, whereas *An. pseudowillmori* tends to be collected in larger and more permanent habitats, including ponds, lakes, and rice fields. *Anopheles maculatus* and *An. sawadwongporni* are cited as the most anthropophilic of the group, biting both indoors and outdoors from dusk until around 2100h. *Anopheles maculatus* s.s. is a primary vector in India, Java, Malaysia, and Thailand, as is *An. sawadwongporni* in Thailand, *An. willmori* in Nepal, and *An. pseudowillmori* at the Myanmar-Thai border. In the Philippines, *An. dispar* and *An. greeni* are predominantly zoophilic and secondary vectors at best. *Anopheles dravidicus*, *An. rampae*, and *An. notanandai* are not implicated in malaria transmission.

Medical Importance

TIBOV; *Plasmodium falciparum*, *P. vivax*, *Wuchereria bancrofti*.

Diagnostic Characters

(a) Habitus (♀). (b) Thorax (side): Upper proepisternal setae (PeSU) absent. **(c) Leg (III):** Fe-III, Ti-III, and Ta-III_1 speckled; Ta-III_5 pale scaled. **(d) Wing:** Veins C and R–R_1 with ≥4 dark spots. **(e) Thorax (dorsal). (f) Head (side). (g) Larva (thorax):** Seta 1-P with ≥15 branches; seta 11-P slender with fine branches. **(h) Larva (head):** Distance between bases of seta 2-C > distance between bases of setae 2-C and 3-C on one side; seta 1-A small, single, rarely 2,3 branched; setae 2,3-C aciculate; setae 5–7-C long, branched. **(i) Larva (abdominal segments):** Seta 6-III usually with ≥20 branches; seta 6-V,VI 3–6 branched. Image source: WRBU.

Exemplar DNA sequences: *Anopheles maculatus* s.s. whole genome (GCA_002091835, GCA_000473185); *An. dispar ITS2* (AF234778); *An. dravidicus ITS2* (JQ446414–16); *An. greeni ITS2* (AF234779); *An. notanandai* (None); *An. rampae ITS2* (DQ518626–27, KY039086); *An. pseudowillmori ITS2* (DQ518628–29, JQ446421–27); *An. sawadwongporni ITS2* (JQ446417–20); *An. willmori ITS2* (JQ446411–12, AF512552).

Catalog entry: Vol. 2, p. 668.

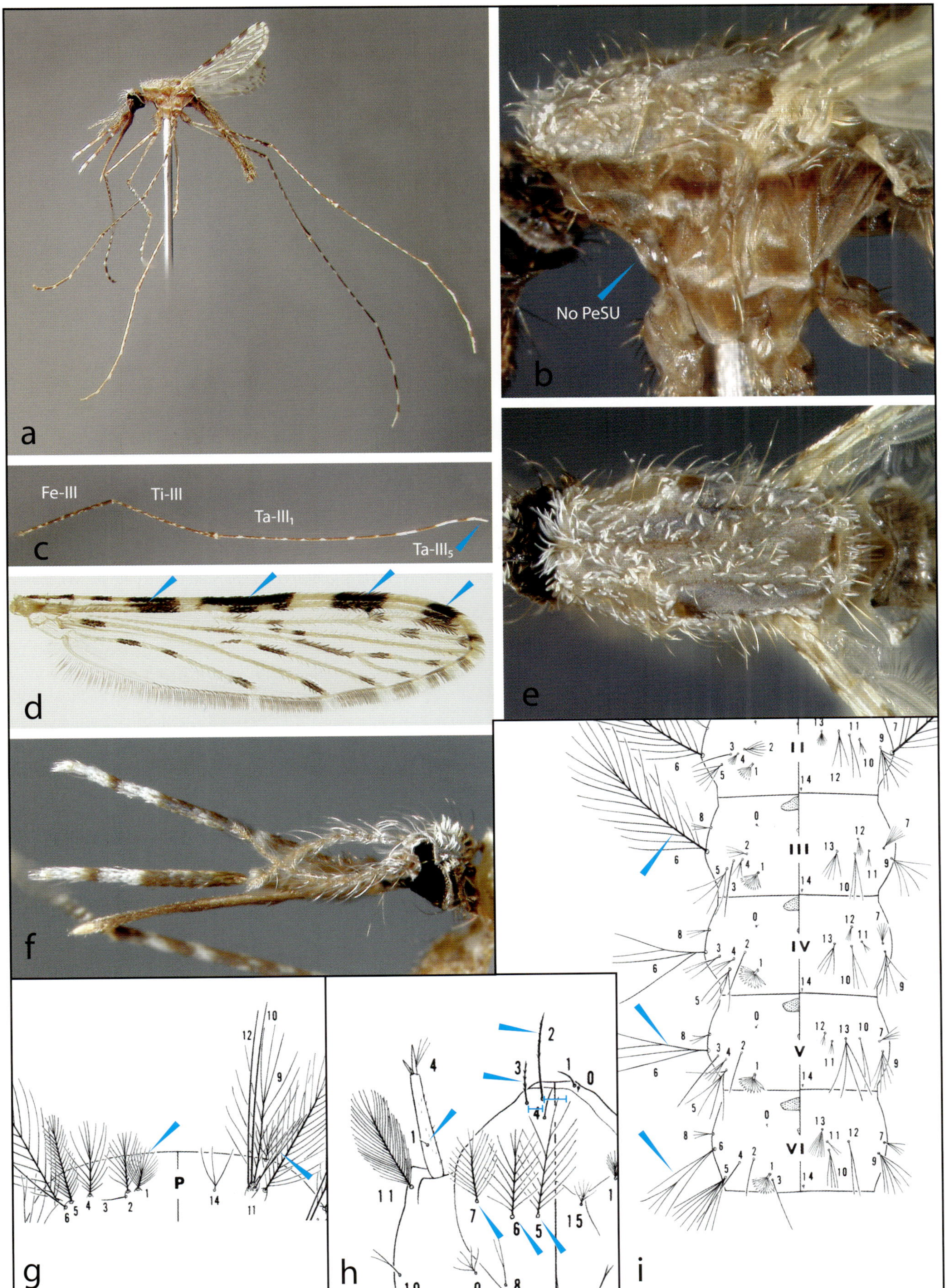
a
No PeSU
b
Fe-III
Ti-III
Ta-III$_1$
Ta-III$_5$
c
d
e
f
P
g
h
II
III
IV
V
VI
i

Anopheles (Cellia) minimus Theobald, 1901
Subfamily Anophelinae
"Tiny Chinese Nail Mosquito"

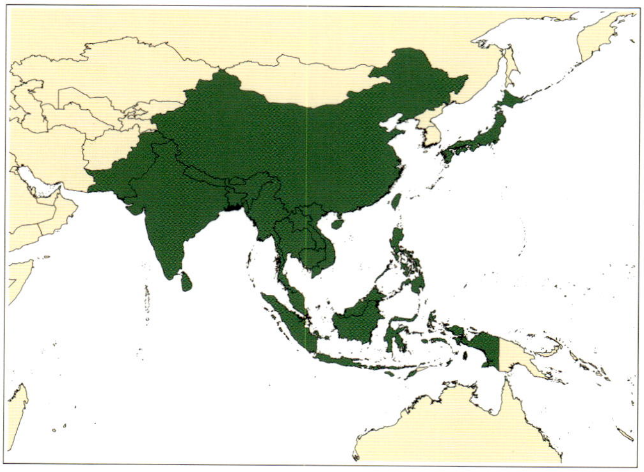

Systematics and Distribution

Anopheles minimus is a tiny pale mosquito, often separated by eye in trap catches purely by the virtue of its size. *Anopheles minimus* is the nominotypical member of the Minimus Complex (Minimus Subgroup, Funestus Group, Myzomyia Series) (see Vol. 2, p. 1083, for internal relationships). *Anopheles minimus* s.l. comprises three undisputed taxa: *An. minimus* Theobald (= species A), *An. harrisoni* Harbach & Manguin (= species C), and *An. yaeyamaensis* Somboon & Harbach (= species E). There are no morphological characters to reliably distinguish the adults or immatures, and thus molecular methods are mostly used. Two other putative entities—species D (chromosomal variant of A) and another represented by a single specimen (D3 sequence variant)—have been reported but their taxonomic status remains unresolved. Most recently, 28S sequence comparison of *An. harrisoni* from Thailand and Vietnam with *An. fluviatilis* S (Fluvialitis Complex, Funestus Group) from India confirmed the conspecificity of these taxa, adding to the taxonomic complexity of the Funestus Group. There are five synonyms of *An. minimus*: *alboapicalis* Theobald, *christophersi* Theobald, *cohaesa* Dönitz, *formosaensis* I Tsuzuki, and *vincenti* Laveran. It remains to be seen whether any of these valid names correspond to the above molecular forms.

The type locality of *An. minimus* s.s. is Pokfulam, Hong Kong. Following a series of in-depth studies on the taxonomy and distributions of the Funestus Group as a whole, many early records attributed to *An. minimus* were found to be erroneous. The clarified distribution of *An. minimus* s.l. includes Bangladesh, Cambodia, southern China, India, Japan, Laos, Malaysia, Myanmar, Nepal, Taiwan, Thailand, and Vietnam (see map for s.l.). *Anopheles minimus* s.s. and *An. harrisoni* are reported from southern China, Laos, Thailand, and Vietnam. *Anopheles minimus* s.s. is also found in Cambodia and Taiwan, and *An. yaeyamaensis* is known only from Ishigaki Island in the Japanese Ryukyu Archipelago.

Associated Pathogens

AMIV; *Plasmodium falciparum*, *P. vivax*.

Diagnostic Characters

(a) Habitus (♀): Legs almost all dark, with indistinct pale marking on apices of Ti; Ta-I–III$_{1-4}$ with faint dorsoapical pale spots and Ta-III$_5$ dark scaled (not distinct in photo). **(b) Abdomen (dorsal):** Segments VII, VIII, and ♀ cercus (Ce) without scales. **(c) Thorax (side):** No antepronotal scales (ApSc); upper proepisternum with 1–4 brown to golden setae (not distinct in photo). **(d) Thorax (dorsal):** Scutum (Scu) and scutellum (Stm) with broad median pale pruinose area contrasting with darker lateral areas; scutum clothed with setae but not scales. **(e) Head (side):** Proboscis (P) dark or with some pale scales ventrally; palpus (MPlp) with three pale bands, preapical dark band approximately equal to or shorter than apical pale band. **(f) Wing:** Remigium (Re) and base of vein R white scaled; costa (C) usually with presector pale spot (PSP) but without humeral pale (HP) spot; base of R with only white or yellow-white scales; vein R$_2$ dark except at base and apex; vein 1A with one long dark spot on distal half; hind margin of wing without pale fringe spot at vein 1A; vein M$_{3+4}$ with two dark spots distal to mcu fork. **(g) Larva (head):** Seta 1-A small and single; seta 2-C simple, widely spaced. **(h) Larva (thorax):** Seta 4-M with sum of branches on both sides 8–11; setae 9,10-T with only one seta simple. **(i) Larva (abdominal segments):** Seta 0-IV–VI arising posterolateral to primary tergal plate (TP), or at or on edge of plate. **(j) Larva (abdominal segments):** Seta 0-IV–VII relatively large, 2–6 branched; anterior tergal plates (TP) on segments III–VII very large, >0.5 width of segment, enclosing median accessory tergal plate (ATP). Image sources: WRBU; Harrison (1980).

Exemplar DNA barcodes: *An. minimus* A: *28S* (AF114018, AF416782, AJ512720), *ITS2* (AB088379–80); *An. minimus* C: *28S* (DQ665847, AJ512719, AF425594, AF114017, AF437880, as *fluviatilis* S, India), *ITS2* (AF194506, DQ238493, AF416784); *An. minimus* E: *28S* (AB088039–40), *ITS2* (AB088376–77).

Catalog entry: Vol. 2, p. 670.

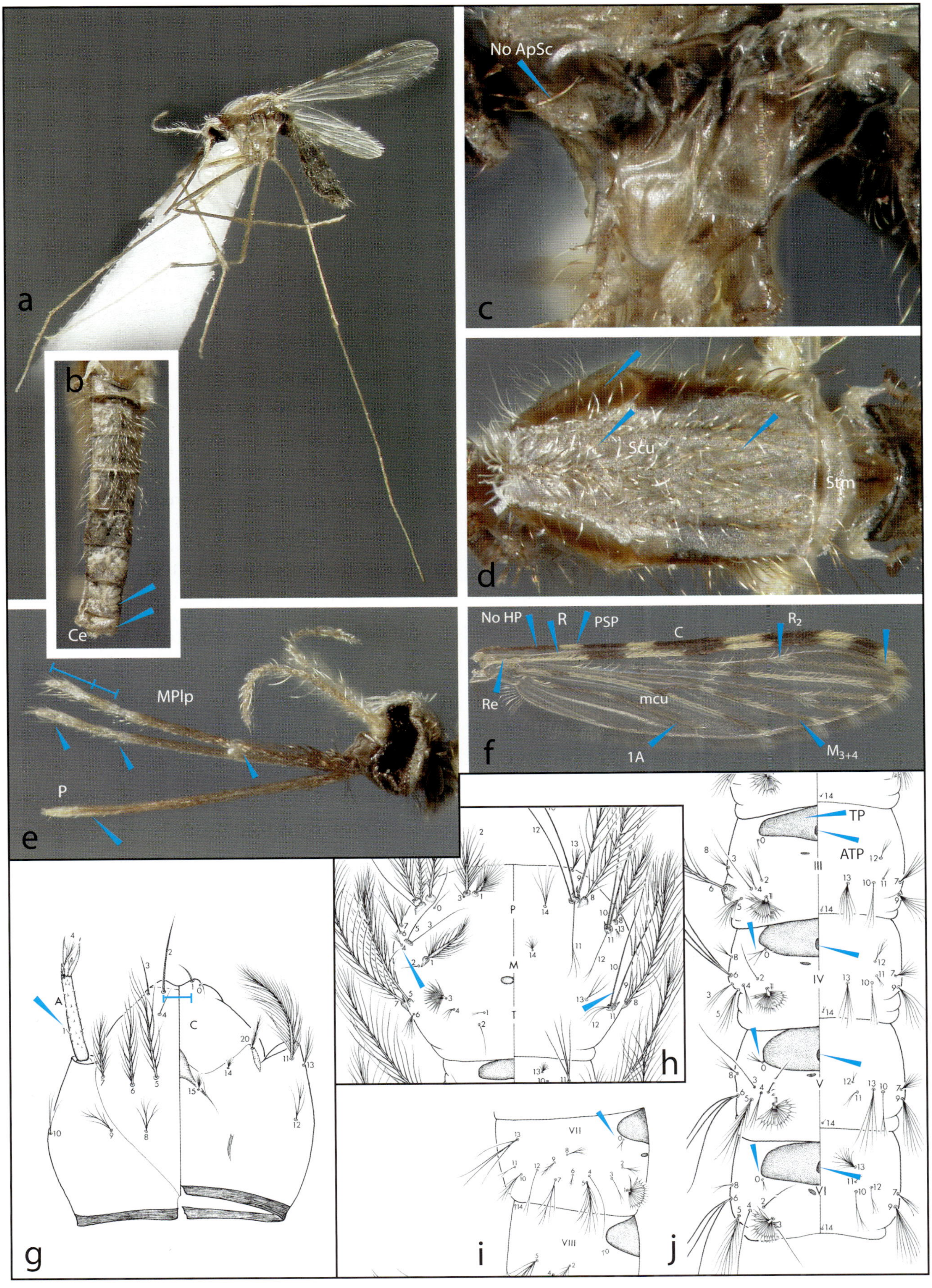
a
No ApSc
c
b
Ce
Scu
Stm
d
No HP
R
PSP
C
R_2
Re
mcu
1A
M_{3+4}
f
MPlp
P
e
A
C
g
P
M
T
h
TP
ATP
III
IV
V
VI
VII
VIII
i
j

Anopheles (*Cellia*) *moucheti* Evans, 1923 Subfamily Anophelinae "Mouchet Congolese Typical Mosquito"

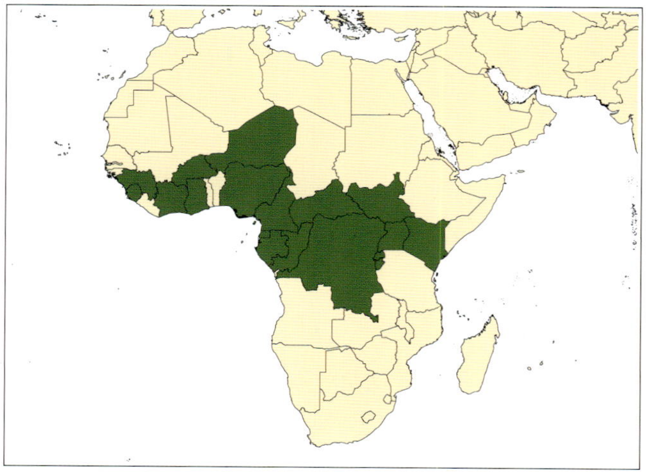

Systematics and Distribution

Anopheles moucheti is a tiny Afrotropical mosquito named after René Libert Joseph Mouchet. The species was first described as a variety of *An. marshallii* Theobald, and these species can easily be confused. The taxonomic history of *An. moucheti* is a little complex. Currently, three entities are defined within *An. moucheti* s.l.—*An. moucheti* Evans, *An. moucheti nigeriensis* Evans, and *An. bervoetsi* D'Haenans (recently elevated to full species status), all of which are unplaced within the Myzomyia Series of the *Anopheles* subgenus *Cellia* (see Vol. 2, p. 1083). Morphological and genetic differences have been noted between *An. moucheti* and topotypic populations of *An. moucheti nigeriensis* and *An. bervoetsi*, and a diagnostic PCR assay has been developed to identify all three based on DNA sequence differences in the nuclear rDNA *ITS*1 region.

Anopheles moucheti was first described from Stanleyville, Belgian Congo (= Kisangani, Democratic Republic of the Congo), and is reported from most countries across the west and central equatorial belt of the African Continent, including Madagascar and the Comoros Islands. *Anopheles moucheti nigeriensis* has only been reported from near its type locality near Lagos in Nigeria, and *An. bervoetsi* is known only from the Tsakalakuku, Democratic Republic of the Congo.

Bionomics

Anopheles moucheti s.l. is associated with equatorial forest-edge habitats, caused by rivers and road building, or in cultivated forest clearings and rural villages where slow-flowing rivers cut through equatorial forest or forest remnants. It is an important malaria vector wherever it occurs in high densities. The immature stages are most often found in turbid waters at the margins of slow-flowing streams or rivers, especially where water lettuce (*Pistia* spp.) is present. *Anopheles moucheti* s.l. immatures are often collected in association with *An. nili* s.s. (Theobald) and *An. carnevalei* Brunhes, Le Goff & Geoffroy.

It is highly anthropophilic and endophilic and can often be found resting inside human habitations in Cameroon. Females bite through the night, with highest intensity from midnight to dawn. *Anopheles moucheti* s.s. is thought to also feed on other forest mammals. Little has been documented on sylvan host choices, but *Anopheles moucheti* s.s. has been implicated in the potential zoonotic crossover of *Plasmodium* spp. (and potentially arboviral agents) between great apes and humans. While *An. moucheti* s.s. is fully susceptible to DDT, permethrin, and deltamethrin, its association with rural environs poses significant practical challenges to control.

Associated Pathogens

Plasmodium falciparum, *P. praefalciparum*.

Diagnostic Characters

(a) Habitus (♀): Legs not speckled; Ta-I_4 dark or indistinctly pale at apex. **(b) Thorax (dorsal):** Scutal scales (Scu) on posterior third sparse, narrow, and yellowish brown. **(c) Wing:** Preapical dark spot (PD) spot on R_1 not interrupted by a pale spot; PD much larger then preapical pale spot (PP); vein M_{3+4} with three pale spots, one at apex; vein 1A usually without a pale fringe spot. **(d) Head (dorsal):** Palpus (MPlp) with three pale bands; preapical dark band distinctly shorter than apical pale and preapical pale bands; $MPlp_5$ all pale. **(e) Leg (III):** Apices of Ta-III_{1-3} or Ta-III_{1-4} pale. **(f) Abdomen (dorsal):** Abdomen without posterolateral scale tufts. **(g) Larva (head):** Seta 2-C single; seta 3-C much shorter than 2-C, with ≥3 short, stiff branches. **(h) Larva (thorax):** Setae 1–3-P well developed, arising from distinct basal tubercles. **(i) Larva (abdomen):** Leaflets of abdominal palmate seta. **Other larval characters (not illustrated):** Abdominal segment V without median accessory tergal plates (ATP); width of tergal plate (TP) on segment V 0.67 or less distance between palmate setae; seta 1-X usually double. Image sources: WRBU; Evans (1938).

Exemplar DNA sequences: *An. moucheti* s.s. *ITS*1 (AM232661); *An. m. nigeriensis ITS*1 (AM232662); *An. bervoetsi ITS*1 (AM232663).

Catalog entry: Vol. 2, p. 671.

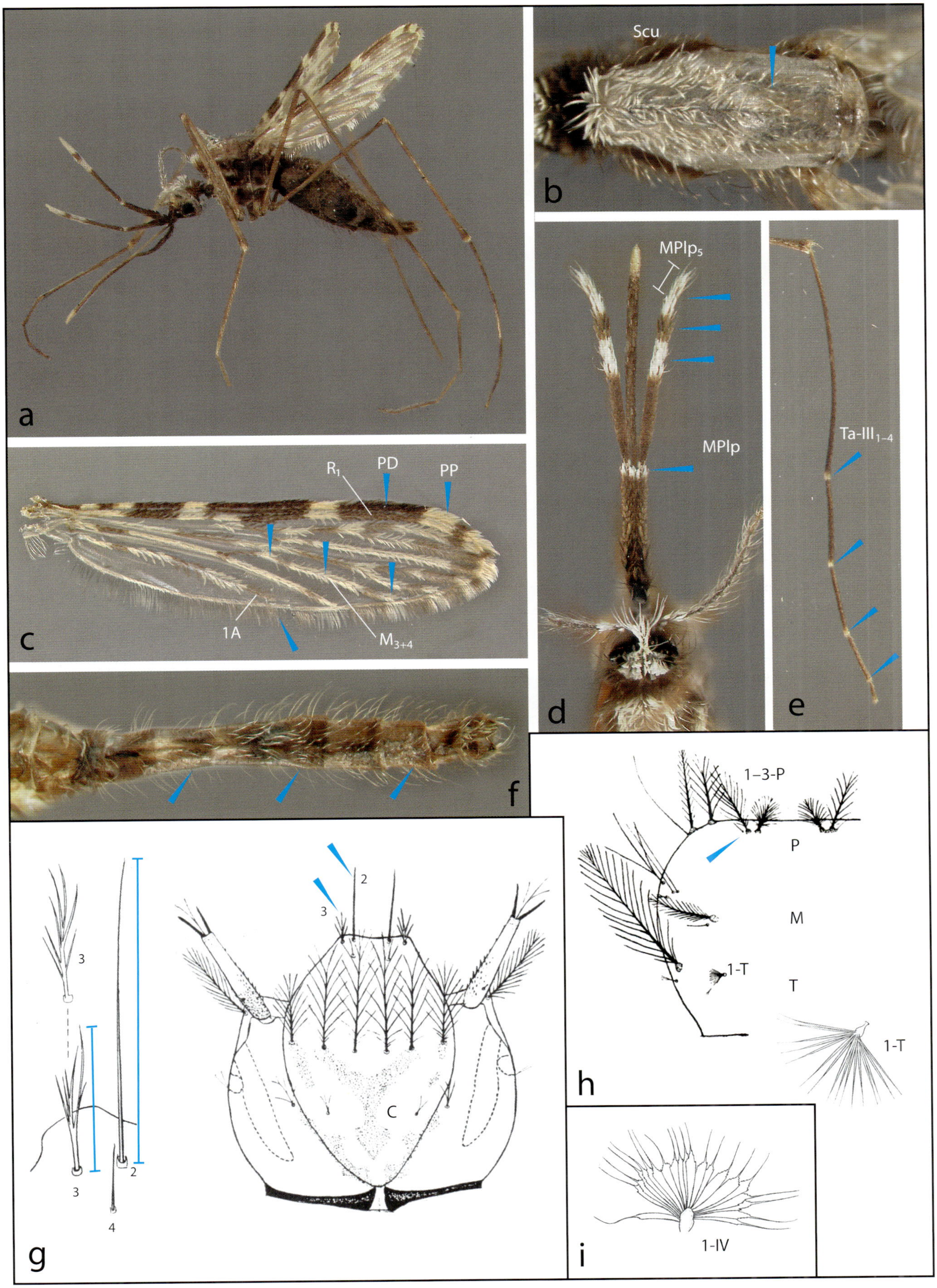
Scu
b
MPlp5
MPlp
d
Ta-III1–4
e
a
R1
PD
PP
1A
M3+4
c
f
1–3-P
P
M
1-T
T
1-T
h
2
3
3
3
2
4
C
g
1-IV
i

Anopheles (*Cellia*) *multicolor* Cambouliu, 1902 Subfamily Anophelinae "Multicolored Egyptian Nail Mosquito"

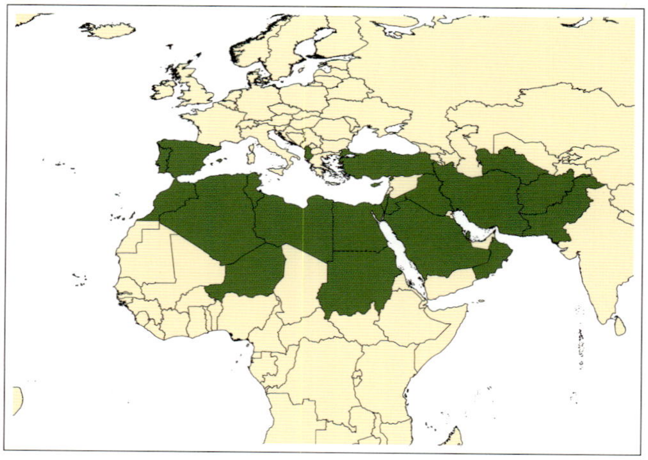

Systematics and Distribution

Anopheles multicolor is a pale species, with light-colored scales on the scutum (see *b*) and head vertex (see *c*). The palps of *An. multicolor* display three distinct pale bands with the terminal palpal segment ($MPlp_5$) entirely dark (see *c*), and the wings have a distinctive broad, pale spot at the base of the costa (see *d*). *Anopheles multicolor* was first described from specimens collected in Egypt—adults collected from the Suez Canal, and larvae from nearby Ismailia. Together with *An. listeri* De Meillon and *An. seretsei* Abdulla-Khan, Coetzee & Hunt, *An. multicolor* comprises the Listeri Group of the Paramyzomyia Series of the *Anopheles* subgenus *Cellia* (see Vol. 2, p. 1085). Three synonyms exist: *chaudoyei* Theobald (Algeria), *impunctus* Dönitz (Egypt), and *nigrifasciatus* Theobald (Pakistan). Unlike most *Anopheles* vectors, there is no evidence to suggest that *An. multicolor* comprises a species complex.

Anopheles multicolor is much regarded as a desert species. Its range extends from North Africa (including the Canary Islands) through the eastern Mediterranean Basin and Cyprus into Southwest Asia and eastward to Pakistan. In addition, it has been reported from Murcia in mainland Spain.

Bionomics

The immatures of *An. multicolor* are found in inland or coastal sites in semiarid or desert regions. Although the larvae have been found in freshwater oases, the species prefers brackish water. Some populations of *An. multicolor* have been found tolerating extremely high salinities of up to 5.96% in small water collections in salt pans. One report details the recovery of *An. multicolor* larvae from a small salt-encrusted pool where the salinity was twice that of seawater. Other reported habitats include cesspools, old shallow wells, stagnant and flowing drains, and ground pools, with or without vegetation. In northern Africa, larval densities of *An. multicolor* peak between April and July and again in November to December. *Anopheles multicolor* is often recovered in association with *Aedes caspius* (Pallas) (see p. 432), *Culiseta longiareolata* (Macquart, 1838), and *An. cinereus hispaniola* (Theobald).

Adult *An. multicolor* are strong fliers, and host-seeking females have been collected up to 13 km from suitable immature habitats. Surprisingly, little is known about the bionomics of adults. Along with *An. pharoensis* Theobald (see p. XX), *An. multicolor* is considered a major malaria vector in Egypt, where it readily enters houses to feed on humans. However, other populations are considered primarily exophilic and of little medical importance, even at high densities. Integrated systematics studies are needed on populations of *An. multicolor* from Egypt and across its range to determine if hidden cryptic taxa exist that account for the disparate bionomic characterization in both larva and adult.

Associated Pathogens

RVFV; *Plasmodium falciparum*.

Diagnostic Characters

(a) Habitus (♀). (b) Thorax (dorsal): Scutum (Scu) with numerous pale scales and setae; scutal fossa (SF) with scattered pale scales. **(c) Head (side):** Erect head scales broad, white on vertex (V), dark brown posteriorly and laterally; palpus (MPlp) with three distinct pale bands; $MPlp_5$ dark apically. **(d) Wing:** Costa (C) with at least four separate dark areas; base of costa with broad pale spot (basal pale + prehumeral pale, basal dark missing). **(e) Larva (thorax/abdominal segments):** Seta 1-P with very small unsclerotized base; one of setae 9–12-M single (not shown); setae 1-T rudimentary, not palmate; tergal plates (TP) narrower than distance between palmate setae. **(f) Larva (head):** Setae 2,3-C single; seta 2-C distinctly longer than seta 4-C; setae 5–7-C long, plumose. **(g) Abdomen (dorsal):** Tergum (Te) without posterolateral scale tufts. Image sources: WRBU; Ross & Roberts (1943).

Exemplar DNA sequences: *An. multicolor ITS2* (AY564228–29), *VGSC* gene intron (AY902754).

Catalog entry: Vol. 2, p. 671.

a
SF
Scu
b
MPlp5
MPlp
V
c
C
d
P
M
T
TP
I
II
III
IV
e
f
Te
g

Anopheles (*Cellia*) *nili* (Theobald, 1904) Subfamily Anophelinae "Nile Sudanese Nail Mosquito"

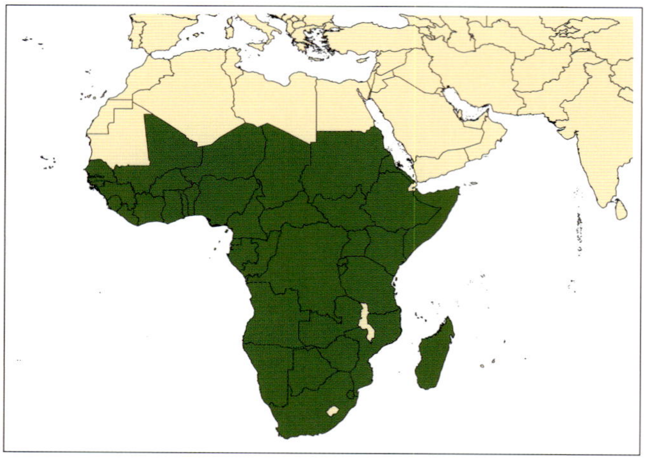

Systematics and Distribution

Anopheles nili is an alluring species, with striking pale spots on the leading edge of its wings together with pale-tipped palps and a shock of pale scales on its crown. *Anopheles nili* is the nominotypical member of the Nili Complex of the Neomyzomyia Series of the *Anopheles* subgenus *Cellia*. There are two species currently placed in synonymy with *An. nili*: *unicolor* Grünberg, described from Ghana, and *nigra* Christophers, described from Sudan. Other closely isomorphic members of the complex include *An. carnevalei* Brunhes, Le Goff & Geoffroy, *An. ovengensis* Awono-Ambene, Kengne, Simard, Antonio-Nkondjio & Fontenille, and *An. somalicus* Rivola & Holstein. The species can be differentiated on subtle morphological characters of the adults and larvae, but are most easily recognized using molecular identification techniques. Recent microsatellite analyses suggest that *An. nili* from Cameroon may comprise additional cryptic diversity.

The distribution of *An. nili* s.s. includes almost all forest and humid savannah zones in sub-Saharan Africa, including Madagascar. *Anopheles somalicus* has been reported from West Africa (Burkina Faso, Cameroon, Côte d'Ivoire, Guinea, Mali, Sierra Leone, and Somalia). *Anopheles carnevalei* appears restricted to West Africa (Cameroon, Côte d'Ivoire, Equatorial Guinea), and *An. ovengensis* has only been recorded in Cameroon.

Bionomics

Anopheles nili, *An. carnevalei*, and *An. ovengensis* are all highly anthropophilic and are involved in the transmission of human malarias. By contrast, *An. somalicus* is intensely zoophilic. As a forest-savannah transition species, *An. nili* is found in river valleys, degraded forest, agricultural clearings, villages, and even city margins, increasing its impact in human malaria transmission. Infection rates of 1%–3% are typical for *An. nili*. Forest *An. nili* is endophilic and anthropophilic, whereas the savannah populations appear exophilic. *Anopheles carnevalei* and *An. ovengensis* are found deeper in the forest. *Anopheles carnevalei* feeds mainly on animals throughout the night but can be found biting humans outdoors, particularly at riverine bathing sites near dusk. Biting rates of 50 to 300 bites per human per night have been recorded for *An. ovengensis* along river habitats in Equatorial Guinea, and *Plasmodium* infection rates of 0.4%–4.4% have been reported in Cameroon and Equatorial Guinea.

Immatures of *An. nili* are associated with vegetation and debris collections at the edge of sunlit, fast-flowing rivers. In contrast, immatures of the deep forest species *An. carnevalei* and *An. ovengensis* are typically found among water lettuce (*Pistia* spp.) in the cooler waters of heavily shaded, slow-moving rivers.

Associated Pathogens

Plasmodium spp.

Diagnostic Characters

(a) Habitus (♀). (b) Head (side): MPlp dark except for pale apex. **(c) Wing comparison:** (*nili* and *somalicus*, *carnevalei*, *ovengensis*): No pale fringe spot at end of vein M_{3+4} (*nili* and *somalicus*) or pale fringe present at end of M_{3+4} (*ovengensis* and *carnevalei*). Other possible wing differences noted on illustrations. **(d) Wing (s.l.):** Preapical dark spot (PD) on vein R_1 without pale interruption; vein M_{3+4} with a single pale spot; preapical pale spot (PP) subequal to length of apical dark spot (AD). **(e) Palmate seta:** 1-III: All palmate setae with long filaments, extending to an acute point. **(f) Larva (thorax):** Seta 1-P flattened, inserted on distinct basal tubercle; setae 9,10-M single. **(g) Larva (head):** Seta 2-C strongly branched apically, branches mostly mesal; seta 3-C 0.5 or more length of 2-C. Image sources: WRBU; Evans (1938); Manguin et al. (2013).

Exemplar DNA sequences: *Anopheles nili ITS2* (AJ429048), COI (KR014837); *An. carnevalei ITS2* (AJ429050); *An. ovengensis ITS2* (AJ429049); *An. somalicus ITS2* (AJ438689).

Catalog entry: Vol. 2, p. 672.

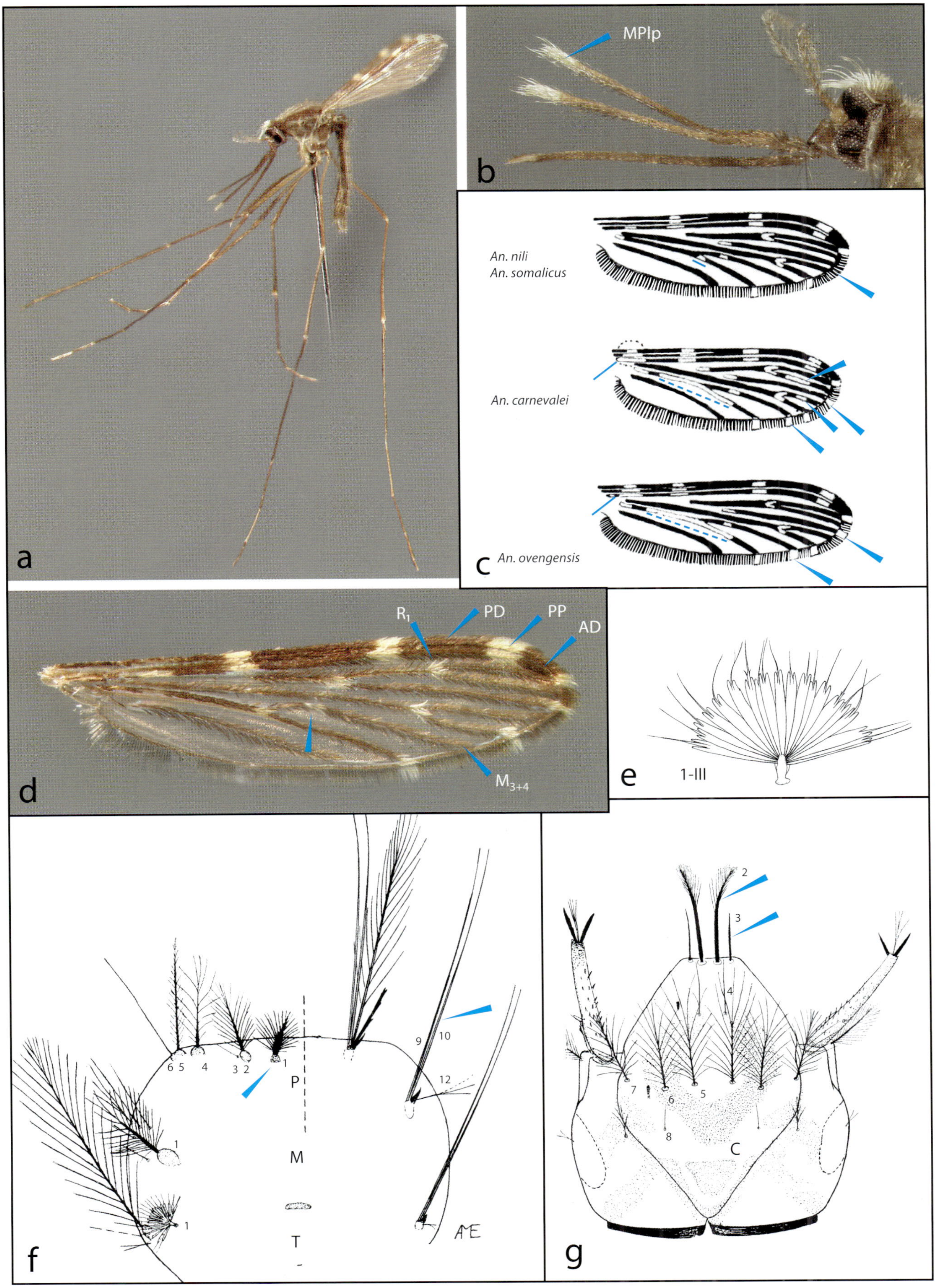
a
MPlp
b
An. nili
An. somalicus
An. carnevalei
c
An. ovengensis
R1
PD
PP
AD
M3+4
d
1-III
e
6 5 4 3 2 1
P
M
T
9 10
12
1
1
AE
f
2
3
4
7
6
5
8
C
g

Anopheles (*Cellia*) *pharoensis* Theobald, 1901
Subfamily Anophelinae
"Pharaoh Egyptian Nail Mosquito"

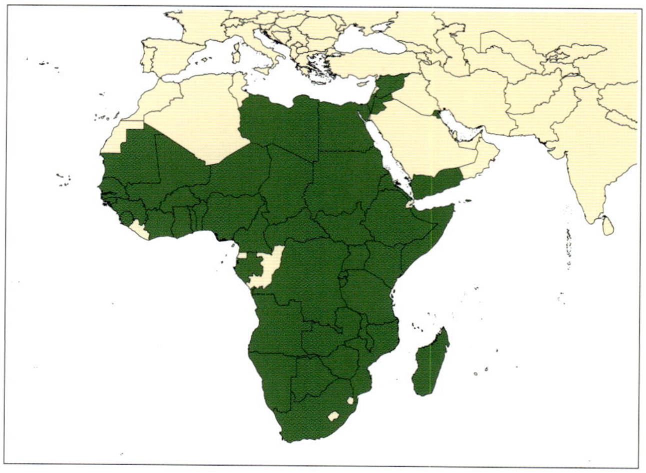

Systematics and Distribution

Anopheles pharoensis is a grayish mosquito with a pale-scaled abdomen sporting dark, laterally projecting scale tufts (see *e*). This species was originally described from Cairo, Egypt. It serves as the type species of the subgenus *Cellia*, where it is informally placed within the Cellia Series (see Vol. 2, p. 1083, for internal relationships), along with *An. squamosus* Theobald, among others. *Anopheles pharoensis* and *An. squamosus* are the only two African *Cellia* species in which larval seta 3-C is strongly branched or dendroid (see *i*). *Anopheles pharoensis* is commonly reported across the African Continent, from the tropical zones to South Africa (including Madagascar) and the Middle East, with a seemingly patchy distribution from Yemen to as far north as Syria. Four species are currently in synonymy with *An. pharoensis*—*albofimbriatus* Giles, *albus* Ventrillon, *bozasi* Neveu-Lemaire, and *maculicosta* Becker. Chromosomal inversion studies indicated that populations from Zululand (KwaZulu-Natal, South Africa) exhibited different X-chromosome arrangements and mating incompatibility when compared with other African and topotypic Egyptian populations. This study, together with reports of two distinctly variable morphological character states of the larval palmate setae in East Africa (see *h*) and variable vector status across its range, strongly implies that the species may comprise a species complex. Further integrated systematics studies are needed to elucidate the taxonomic status of *An. pharoensis* s.l. across its range.

Bionomics

Immature *An. pharoensis* are commonly found in freshwater ground pools densely populated with reeds, papyrus, or grasses, as well as muddy pools in rice fields. The species is a highly effective vector of human malaria in Egypt, Cameroon, and Senegal, but it is not regarded as an important vector elsewhere in Africa. Although reported to enter houses and to bite humans in Africa, human blood indices of Kenyan *An. pharoensis* were as low as 0.16, with the majority of blood feeds attributed to cattle. In contrast, *An. pharoensis* is highly anthropophilic in cultivated areas of the Nile Delta, where its population peaks during the warmer summer months. *Anopheles pharoensis* adults were collected on sticky nets suspended from helium balloons at heights of 40 to 240 m above Sahelian villages in Mali, suggesting that the species undertakes long-distance migrations at high altitudes to evade desiccation in the inhospitable hot, dry season. Females were collected fully gravid, meaning they had taken at least one blood meal before migrating, increasing the risk of importing malarial and arboviral pathogens into new regions upon descent.

Associated Pathogens

BGIV, BIRV, RVFV, SINV; *Brugia* spp., *Plasmodium falciparum*, *Plasmodium vivax*.

Diagnostic Characters

(a) Habitus (♀). (b) Leg (I): Ta-I,II$_5$ dark (II not shown). **(c) Leg (III):** Ta-III$_{3,4}$ with dark and pale rings, Ta-III$_{3,4}$ pale over apical half; Ta-III$_5$ entirely pale. **(d) Head (side):** Palpus (MPlp) with four pale bands. **(e) Abdomen (dorsal):** Terga (Te) densely covered with broad pale scales; II–VII-Te with posterolateral dark scale tufts. **(f) Wing:** Costa (C) with at least four dark areas. **(g) Larva (thorax/abdominal segments):** Pecten spines (PS) with well-developed spinules (not shown). **(h) Leaflets of abdominal palmate setae:** Leaflets either broad, markedly notched with very short filaments (typical form), or much narrower ending in a single long point. **(i) Larva (head):** Seta 3-C dendroid, with single stem divided into ≥8 main branches. Image sources: WRBU; Evans (1938); Ross & Roberts (1943).

Exemplar DNA barcodes: *An. pharoensis* s.l. (KU380430, KU380435, KU380470, KR014844).

Catalog entry: Vol. 2, p. 674.

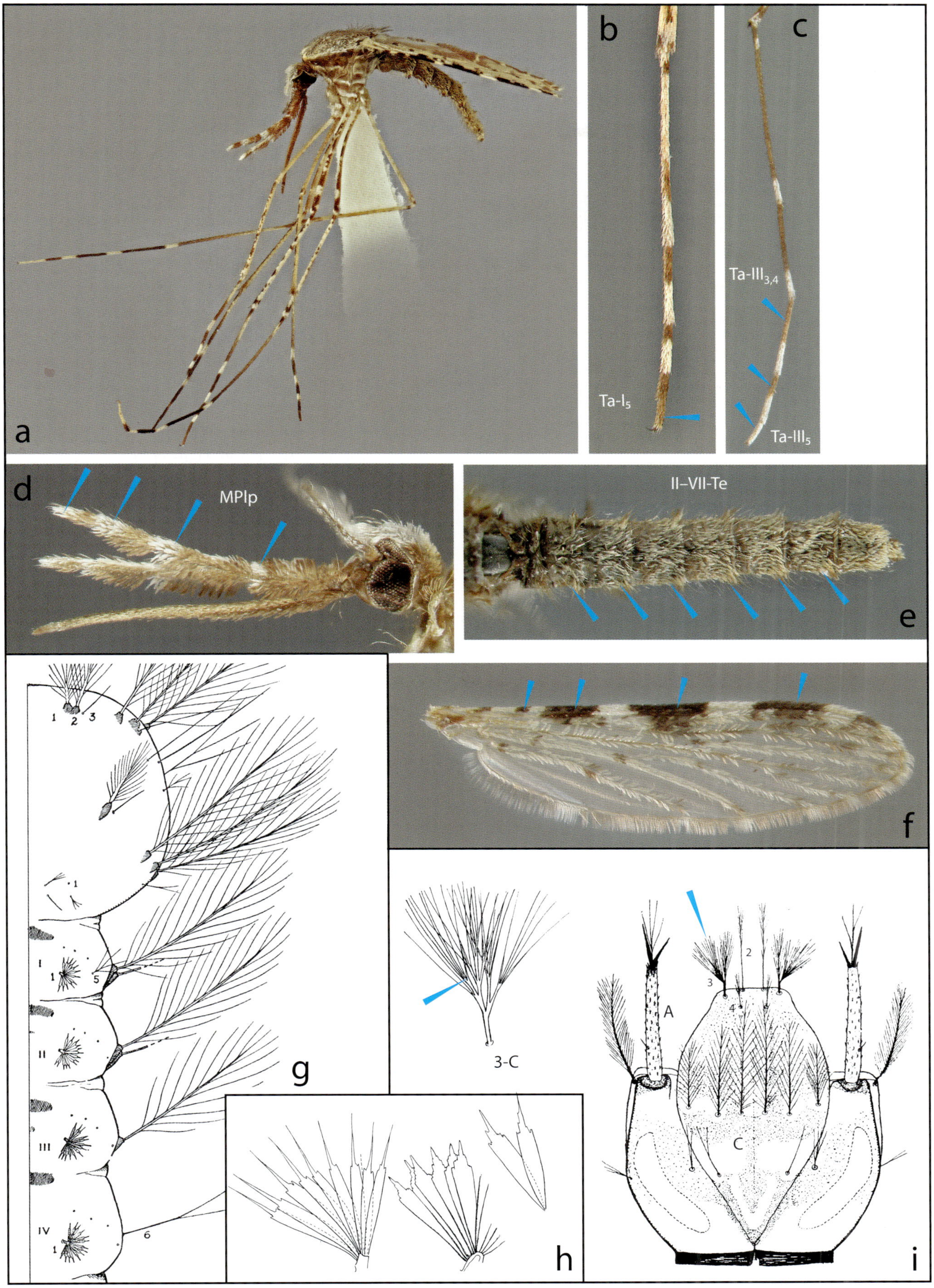
a
b
c
Ta-I5
Ta-III3,4
Ta-III5
d
MPlp
II–VII-Te
e
f
g
1 2 3
1
I
1 5
II
III
IV
1 6
h
3-C
A
2
3
4
C
i

Anopheles (*Cellia*) *pulcherrimus* Theobald, 1902 Subfamily Anophelinae "Beautiful Pakistani Nail Mosquito"

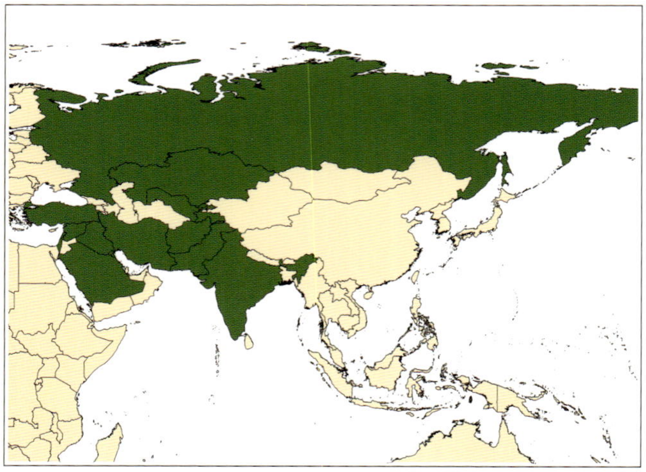

Systematics and Distribution

Anopheles pulcherrimus is a beautiful light-colored mosquito, heavily dusted with white and golden scales, most notably on the scutum, head, and antennae (see *a*). It is an unplaced member of the informally designated Neocellia Series of *Anopheles* subgenus *Cellia* (see Vol. 2, p. 1084, for component taxa). The species was first described from Lahore, Pakistan, and has one synonym, *atropotenae* Lindtrop. *Anopheles pulcherrimus* has been reported from Central Asia (Afghanistan, Kazakhstan, Tajikistan, Uzbekistan), the Middle East (Bahrain, Iran, Iraq, Israel, Lebanon, Saudi Arabia, Syria, Turkey), India, Pakistan, and Russia.

Bionomics

Anopheles pulcherrimus is found in lowland river valleys in Uzbekistan and Tajikistan at elevations ranging to 1,000 m above sea level. Immature *An. pulcherrimus* are restricted to warm, sunlit, stagnant habitats with abundant submerged vegetation, including pools in riverbeds and flood land lakes. Where water temperatures are sufficiently warm, young rice fields are also productive larval habitats for this species.

Anopheles pulcherrimus females are highly anthropophilic but will feed on cattle. Females are readily attracted to light traps and have been collected resting during the day in cattle sheds and in pit shelters. The species is mainly exophilic, but specimens have been collected indoors. Extremely high densities of *An. pulcherrimus* were described in Tajikistan in the 1970s, with average human biting rates of 110–120 every five minutes in the evenings and at night. Malaria resurgence after the cessation of intensive insecticide spraying was attributed to *An. pulcherrimus* in Afghanistan and Iraq. During the summer months, many people sleep outside to evade the stifling heat, and *An. pulcherrimus* is known to readily feed on humans outdoors. Females are active between 12°C and 34°C, with maximum biting activity noted between 24°C and 27°C. In Herat, Afghanistan, *An. pulcherrimus* populations peak in the latter half of July and throughout August. However, two distinct seasonal peaks are noted in *An. pulcherrimus* populations from Sistan and Baluchistan Provinces in Iran (May–June and October–November), but no consistent genetic differences were noted between these populations. Bivoltine populations echoing those found in Iran have also been noted in Iraq.

Associated Pathogens

Tyagin virus, RVFV; *Plasmodium falciparum*, *P. vivax*.

Diagnostic Characters

(a) Habitus (♀). **(b) Wing:** Leading edge of wing with four distinct dark areas. **(c) Legs (I,II,III):** Ti-I–III and Fe-I–III with sparse pale scales; Fe-II,III with longitudinal white lines terminating in an oval spot (not distinct in photo); Ta-III_2 with at least apical 0.5 pale; Ta-III_{3-5} all pale. **(d) Head (dorsal):** Palpus (MPlp) with four pale bands. **(e) Abdomen (dorsal):** Terga densely covered with broad pale scales; II–VII-Te with prominent posterolateral dark scale tufts. **(f) Larva (various parts illustrated):** Seta 3-C with 4–12 branches; seta 4-C at most 0.5 as long as 2,3-C; seta 1-P with many lateral branches arising from stout main stem; tergal plates (TP) on segments III–VII somewhat reduced, with concave posterior border not subsuming small rounded posterior accessory tergal plate (ATP) (not shown). Image sources: WRBU; Gutsevich et al. (1974).

Exemplar DNA barcodes: *An. pulcherrimus* (JX255709–12).

Catalog entry: Vol. 2, p. 675.

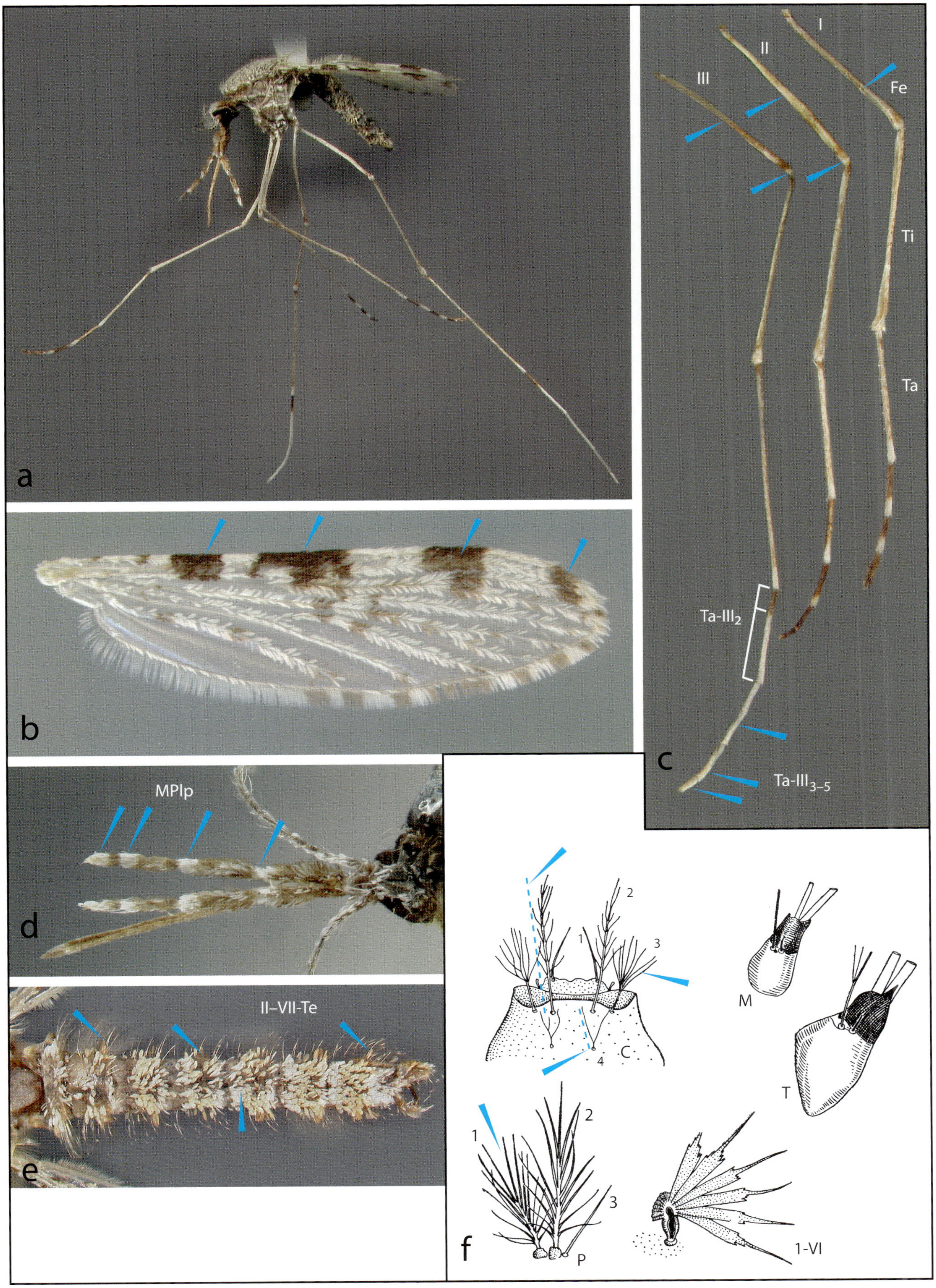
a
III
II
I
Fe
Ti
Ta
Ta-III$_2$
Ta-III$_{3-5}$
b
c
MPlp
d
II–VII-Te
e
1
2
3
4
C
M
T
P
1-VI
f

Anopheles (*Cellia*) *sergentii* (Theobald, 1907) Subfamily Anophelinae "Sergent Algerian Nail Mosquito"

Systematics and Distribution

Anopheles sergentii is a light-colored species with distinctive pale bands on the palps. It was named in deference to Edmond Sergent. *Anopheles sergentii* comprises two valid subspecies—*An. sergentii sergentii* and *An. sergentii macmahoni* Evans, described from Kenya. The subspecies *macmahoni* has one synonym—*barkhuusi* Giaquinto-Mira, described from Ethiopia. These taxa are placed in the Demeilloni Group (Myzomyia Series), along with *An. carteri* Evans & de Meillon, *An. demeilloni* Evans, *An. freetownensis* Evans, *An. garnhami* Edwards, *An. keniensis* Evans, and *An. lloreti* Gil Collado.

Anopheles sergentii is a major malaria vector in arid regions of northern Africa, the Mediterranean Basin, and the Middle East and is often referred to as the "desert malaria vector." *Anopheles sergentii* was described from Algeria, and its range extends through Africa as far south as Kenya and Cameroon, northeast to Italy, and southeast to Pakistan. The subspecies *An. sergentii macmahoni* appears limited to countries in the sub-Saharan belt of Africa.

Bionomics

Anopheles sergentii s.l. is a highly efficient malaria vector, despite its propensity for zoophily. Female *An. sergentii* s.l. feed avidly on humans and cattle, but also on sheep, goats, horses, and birds. In dual human/animal housing, the presence of animals is thought to act as zooprophylaxis, diverting host seeking *An. sergentii* s.l. away from people to instead feed on livestock. After blood feeding, female *An. sergentii* s.l. rests indoors in houses or animal sheds to mature their eggs.

An. sergentii s.l. females utilize desert oases and a range of water collections in irrigated areas for oviposition. The water in larval habitats can be shaded or unshaded, clean or semipolluted, fresh or slightly brackish, slow moving or stagnant, but is always with vegetation and/or algae. Under laboratory conditions (27°C+/−2°C, 70%–80% relative humidity), the mean life expectancy at emergence for *An. sergentii* s.l. was 23.3 days. *Anopheles sergentii* s.l. can overwinter either as larvae or as adult females. Integrated systematics studies are needed to ascertain the taxonomic status of the two subspecies and better define their respective roles in malaria transmission.

Associated Pathogens

RVFV; *Plasmodium falciparum*, *P. ovale*, *P. vivax*.

Diagnostic Characters

(a) Habitus (♀). (b) Head (side view): Palpus (MPlp) with three pale bands; $MPlp_5$ entirely pale; vertex (V) with broad pale scales. **(c) Wing (base):** Remigium (Re) pale scaled. **(d) Thorax (side view):** Upper proepisternal setae (PeSU) present; vertex (V) with broad pale scales, lateral and posterior scales dark brown. **(e) Wing:** Presector pale spot (PSP) present; prehumeral pale (PHP) and humeral pale spots (HP) absent; fringe spots well marked at all veins except anal vein (1A); vein R without basal dark spot (BD) distal to humeral crossvein; vein R_{4+5} dark scaled except for small pale spots at base and apex. **(f) Thorax (dorsal view):** Scutal fossa (SF) without scales; scutum (Scu) with broad silvery pruinose stripe with obvious pale scales in addition to setae, some erect pale scales anteriorly. **(g) Leg (III):** Ta-III_5 dark. **(h) Larva (abdomen):** Seta 1-II–VII well-developed palmate, filaments > half leaflet length; tergal plates (TP) on segments III–VII not exceptionally large, with concave posterior border not encompassing small posterior accessory plates (ATP); segment VII tergal plate (TP) wider than distance between palmate setae. **(i) Larva (thorax):** Seta 3-T palmate, well developed. **(j) Larva (thorax):** Seta 9-M,T plumose and 10-M,T single. **(k) Abdomen (dorsal):** II–VII-Te without distinct scale tufts. **(l) Larva (head):** Seta 1-A single; setae 2,3-C usually single or with inconspicuous aciculae; seta 4-C <0.67 length of 2-C; seta 5-C slightly longer than 6-C; dorsal apotome (DAp) with transverse pigmented band below bases of setae 5-7-C. Image source: WRBU.

Exemplar DNA barcodes: *An. sergentii COI* (MF631763–68), *ITS2* (KT160246–47).

Catalog entry: Vol. 2, p. 678.

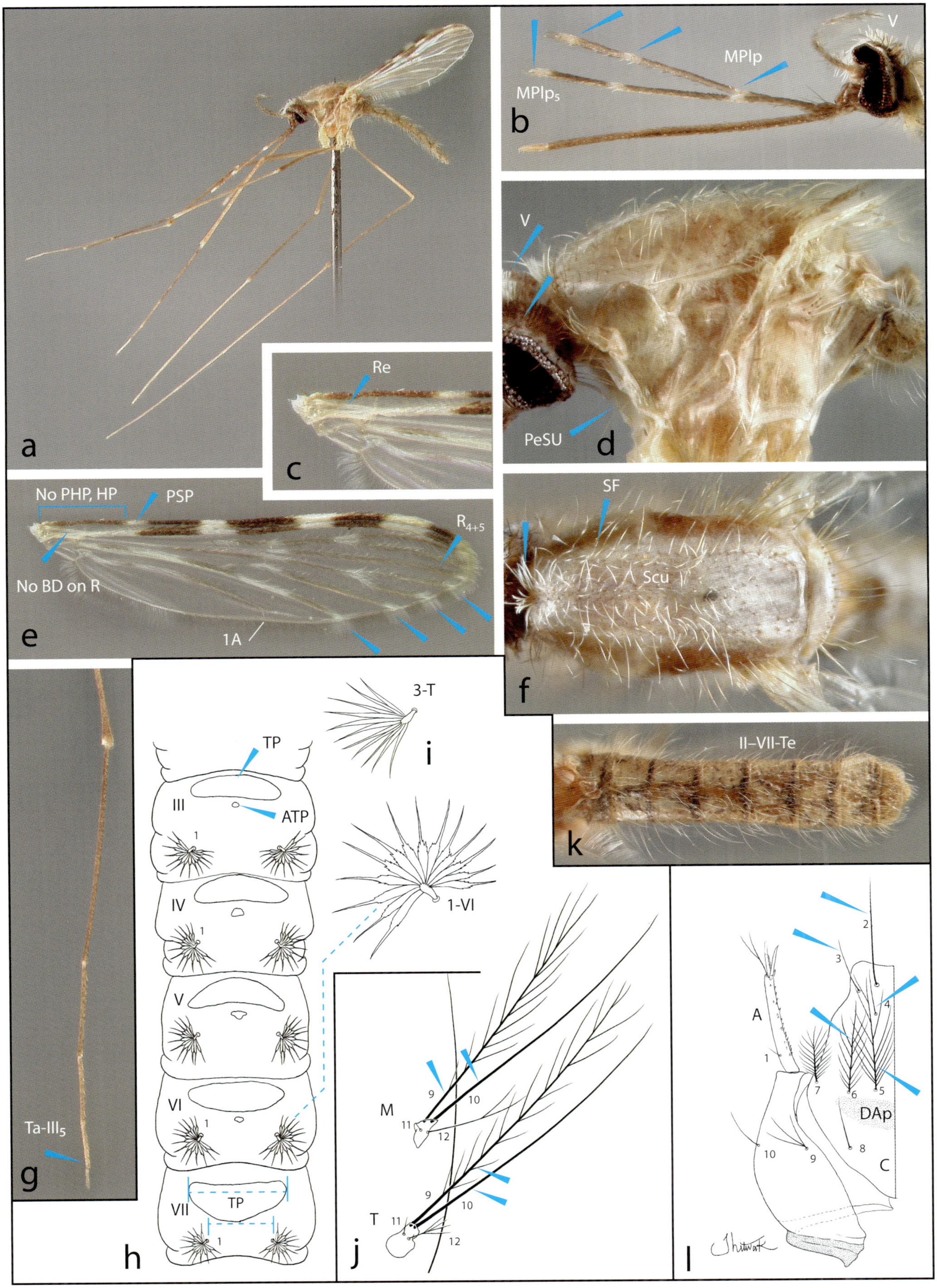
V
MPlp
MPlp₅
b
a
Re
c
V
PeSU
d
No PHP, HP
PSP
R4+5
No BD on R
1A
e
SF
Scu
f
3-T
i
TP
III
ATP
1
IV
1
1-VI
V
VI
1
VII
TP
1
h
Ta-III₅
g
II–VII-Te
k
2
3
4
A
1
7
6
5
DAp
M
9
10
11
12
9
10
T
11
12
j
10
9
8
C
l

Anopheles (*Cellia*) *stephensi* Liston, 1901 Subfamily Anophelinae "Stephens Indian Nail Mosquito"

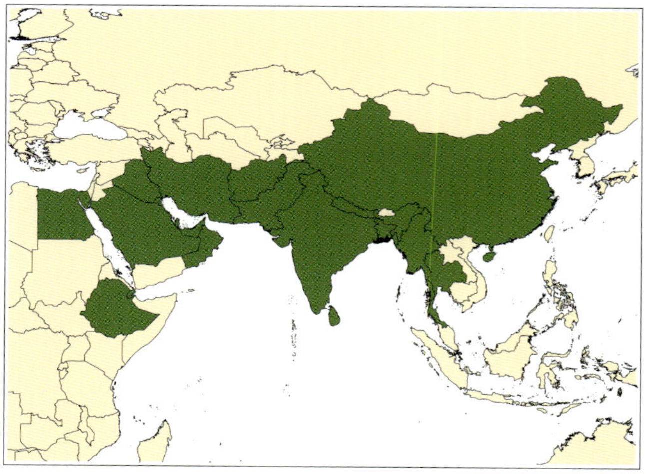

Systematics and Distribution

Anopheles stephensi is an attractive species, with ornate, heavily spotted upper legs. It was first described from Ellichpur, India, and named in deference to John William Watson Stephens. *Anopheles stephensi* is unplaced within the Neocellia Series of the *Anopheles* subgenus *Cellia* (see Vol. 2, p. 1084, for internal relationships). Four species are currently in synonymy with *An. stephensi*: *metaboles* Theobald, *intermedia* Rothwell, *folquei* de Mello, and *mysorensis* Sweet & Rao. Three egg forms, differing in size and number of floats, have been recognized and referred to as type, intermediate, and *mysorensis* forms. Genetic analysis using mtDNA *COI* barcoding clearly indicates at least two separate taxa: the first in India, Pakistan, Djibouti, Ethiopia, and derived colony strains and another distinct species-level grouping in Saudi Arabia. Clearly, the systematics of *An. stephensi* are ripe for further taxonomic studies.

The distribution of *An. stephensi* centers on the Indian subcontinent, stretching west to the Middle East and Arabian Peninsula and east to Myanmar, Cambodia, and Thailand, but the species appears to be on the move. In 2014, *An. stephensi* (type form) was molecularly confirmed from Djibouti, marking its first incursion onto the African Continent. Subsequently, the species has been reported from several localities in Ethiopia. A new invasion of molecularly confirmed type form was recently reported in Mannar Island, Sri Lanka, as well.

Bionomics

There appear to be distinctive biological traits associated with the urban (type) and rural (*mysorensis*) forms of *An. stephensi*. In urban areas the species is highly anthropophilic and endophilic, and immatures are found in high densities in larger artificial containers, including wells, cisterns, fountains, and water barrels. In rural areas the species is uncommon and strongly zoophilic and exophilic; immatures are found in stream pools, stream margins, irrigation canals, and springs. Of note, the invasive African populations conform to the type form genetically and are closely associated with urban populations. Immatures were detected in water cisterns in Djibouti City and Camp Lemonnier, Djibouti, and in various urban/periurban container habitats, including abandoned tires, in Ethiopia. This close association with humans is highly troubling, and experts fear that the presence of *An. stephensi* in high-density cities could instigate urban malaria outbreaks in Africa for the first time. To further complicate the story, laboratory studies showed that colony *An. stephensi* (= type form) mosquitoes infected with *Plasmodium berghei* or *P. falciparum* rendered them less susceptible to repellents.

Associated Pathogens

AHSV, CHIKV, CHPV, TAHV, VEEV, unnamed Cypovirus; *Plasmodium berghei*, *P. falciparum*, *P. vivax*.

Diagnostic Characters

(a) Habitus (♀). (b) Head (side): Maxillary palpus (MPlp) with pale bands; $MPlp_5$ entirely pale; erect head scales broad, white on vertex (V) and dark brown laterally and posteriorly. **(c) Thorax (dorsal):** Scutum (Scu) with obvious pale scales in addition to setae; scutal fossa (SF) with scattered pale scales. **(d) Abdomen (side):** V–VIII-S usually with pale scales. **(e) Wing:** Wing with pale spots present on nearly all veins; vein 1A with three dark spots. **(f) Abdomen (dorsal):** II–VII-Te without dark scale tufts. **(g) Larva (head):** Seta 1-A small, unbranched; setae 2,3-C smooth, not aciculate; distance between seta 2-C > distance between setae 2-C and 3-C on one side; setae 5–7-C long and branched. **(h) Larva (thorax/abdominal segments):** Setae 1,2-P inserted on dark, sclerotized tubercles; setae 9,10-P and 9,10-M not all unbranched (not shown); seta 9-M plumose, branched from base (not shown); setae 10–12-M unbranched (not shown); setae 9,10-T branched (not shown); tergal plates (TP) on segments IV–VII small, not enclosing small median accessory plates (ATP). **(i) Leg (III):** All Fe and Ti spotted with pale scales. Image sources: WRBU; Ross & Roberts (1943).

Exemplar DNA barcodes: *An. stephensi* whole genome (GCA_003448975, GCA_003448955, GCA_00349045, GCA_000300775), *COI* (KR817728, MG571266, AY729980 (India), KF933378 (Djibouti, Ethiopia), KJ528887–95 (Saudi Arabia)).

Catalog entry: Vol. 2, p. 679.

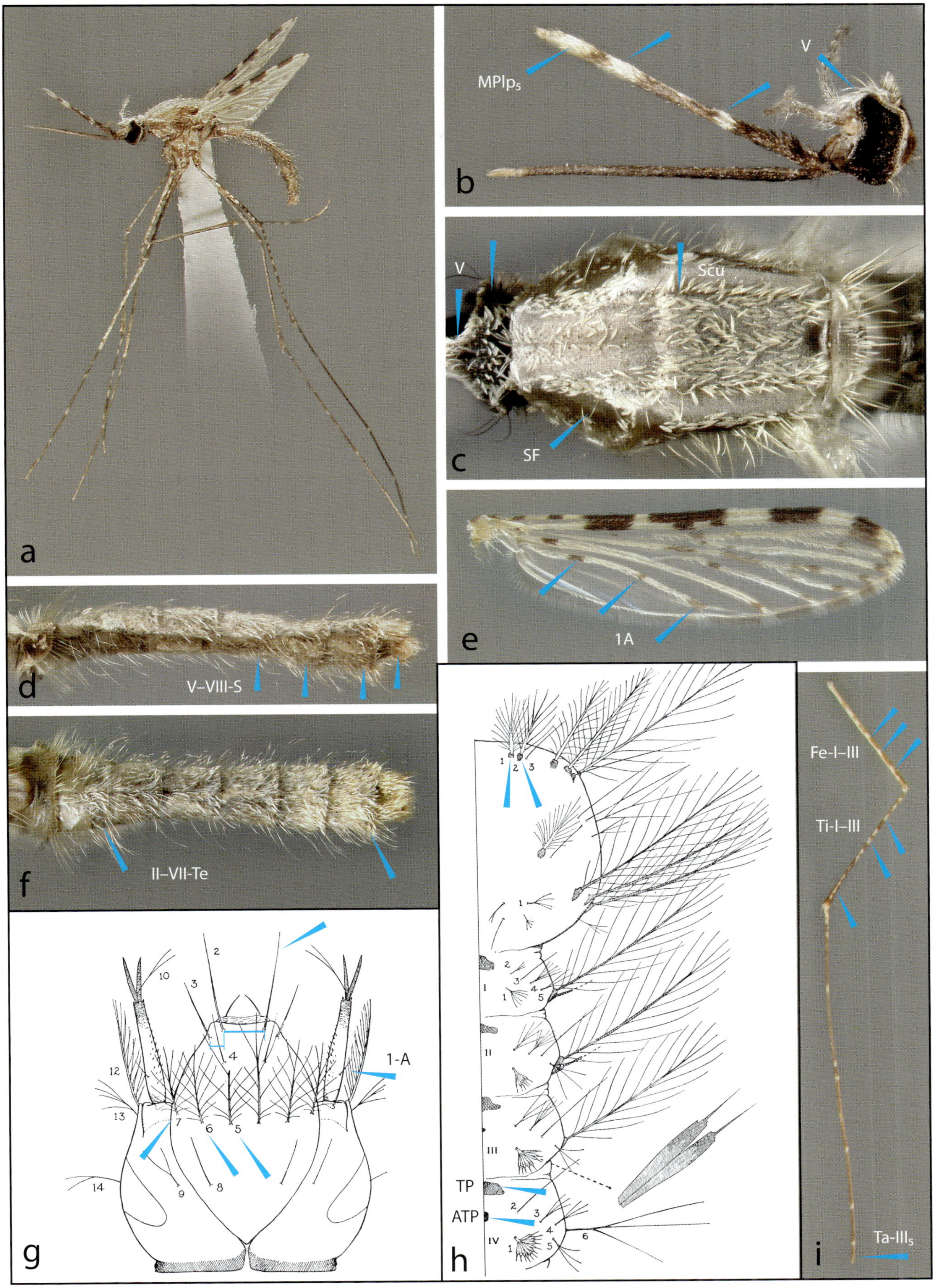

MPlp5
V
b
V
Scu
SF
c
a
1A
e
V–VIII-S
d
II–VII-Te
f
2
3
10
4
12
1-A
13
7
6
5
14
9
8
g
1
2
3
1
I
2
3
4
5
1
II
III
TP
ATP
IV
2
3
4
5
6
1
h
Fe-I–III
Ti-I–III
Ta-III5
i

Anopheles (*Cellia*) *subpictus* Grassi, 1899 Subfamily Anophelinae "Small-painted Indian Nail Mosquito"

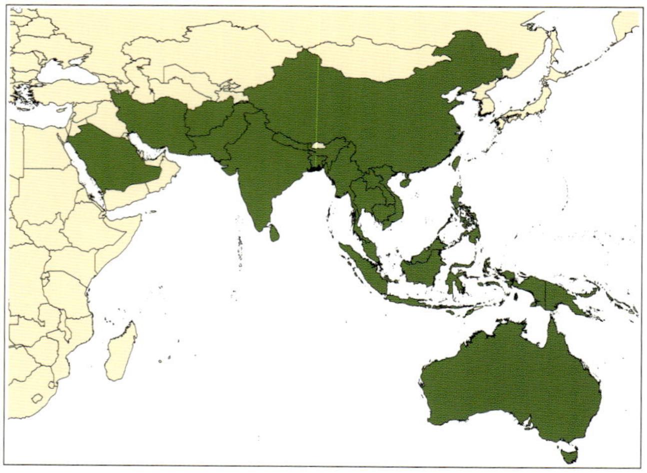

Systematics and Distribution

Anopheles subpictus is recognized by the broad pale bands on its tibia and lack of speckles on its femora and tibiae (see *a*, *d*); the preapical dark band of the palps is at least half as wide as the apical pale band (see *e*). The taxonomic status of *Anopheles subpictus* has long been under scrutiny. Three subspecies of *An. subpictus* were reported until morphological studies concluded these taxa comprised only two distinct species—*An. subpictus* and *An. indefinitus* (Ludlow), with the third purported subspecies (*An. subpictus malayensis* Hacker) sunk as a junior synonym of the latter. *Anopheles indefinitus* from the Philippines was regarded as the freshwater form, with *An. subpictus* referring to the brackish-water breeder. Far from solving the conundrum, additional efforts to understand the biology of these taxa across their known ranges resulted in finding additional freshwater populations of varying vector competencies, leading to the documentation of four additional biological and chromosomal forms within *An. subpictus*—*An. subpictus* A, B, C, and D. There are two synonyms of *An. subpictus*—*rossi* Giles and *error* Theobald, which, as for *An. subpictus* s.s., were all originally described from Calcutta in India. It is not clear if any of the forms and available names correspond. *Anopheles subpictus* taxa are placed within the Subpictus Complex of the Pyretophorus Series of the *Anopheles* subgenus *Cellia* (see Vol. 2, p. 1086, for internal relationships).

Anopheles subpictus s.l. has been reported across the Indian subcontinent westward to Afghanistan and Iran and eastward into the Southeast Asia mainland. *Anopheles subpictus* is limited to the coastal regions of the Southeast Asian peninsula and the islands as far east as Papua New Guinea and the Solomon Islands (see map).

Bionomics

Anopheles subpictus s.l. is reported as predominantly zoophilic and exophagic, with populations biting from dusk throughout the night. However, populations of the brackish-water breeder *An. subpictus* B are anthropophilic and will enter houses to feed. It is the only species thought to be involved in malaria transmission. Wild-caught *An. subpictus* B have been found positive for *Plasmodium* species in Southeast India, Sri Lanka, Malaysia, and Indonesia, where it is regarded to be of minor importance as a malaria vector. In Sulawesi, however, *An. subpictus* is regarded as the primary malaria vector.

Anopheles subpictus has been reported as immatures in waters ranging from 0.56% to 5.36% salinity. *Anopheles subpictus* B is the only species restricted to brackish water coastal localities, including lagoons and salt marshes. *Anopheles subpictus* A, C, and D are typically found in inland freshwater sites, including rice fields, muddy river pools, animal wallows, and artificial containers. They are often found in close association with humans. Although forms A, C, and D do show some tolerance to brackish water, they are never found there in high numbers.

Associated Pathogens

ARKV, JBEV, WNV; *Plasmodium* spp., *Wuchereria bancrofti*.

Diagnostic Characters

(a) Habitus (♀). (b) Thorax (side): Antepronotal scales (ApSc) absent; upper proepisternal setae (PeSU) present. **(c) Abdomen (dorsal):** Segments VII and VIII and cercus (Ce) with at least a few scales. **(d) Legs (II/III):** Legs not speckled; some articulations of Ta comprising both apical and basal pale bands. **(e) Head (side):** Palpus (MPlp) with three pale bands, apical pale band about two times length of preapical dark band and subapical pale band narrow, ≤0.33 length of preapical dark band. **(f) Larva (thorax/abdominal segment):** Seta 1-P usually with 12 or fewer branches; setae 1,2-P with bases not heavily sclerotized; seta 4-M usually 1–2 branched. **(g) Larva (abdominal segments):** Seta 1-I reduced palmate, most often with six or fewer smooth lanceolate leaflets; segments IV–VII with small tergal plates (TP) not encompassing small accessory plates (ATP). **(h) Larva (head):** Setae 2,3-C single; seta 4-C inserted well posterior of seta 2-C, bases widely separated. Image source: WRBU.

Exemplar DNA barcodes: *An. subpictus COI* (KF406709–747), *Anopheles subpictus* s.l. NPK-2007 *COI* (EU143303–08).

Catalog entry: Vol. 2, p. 679.

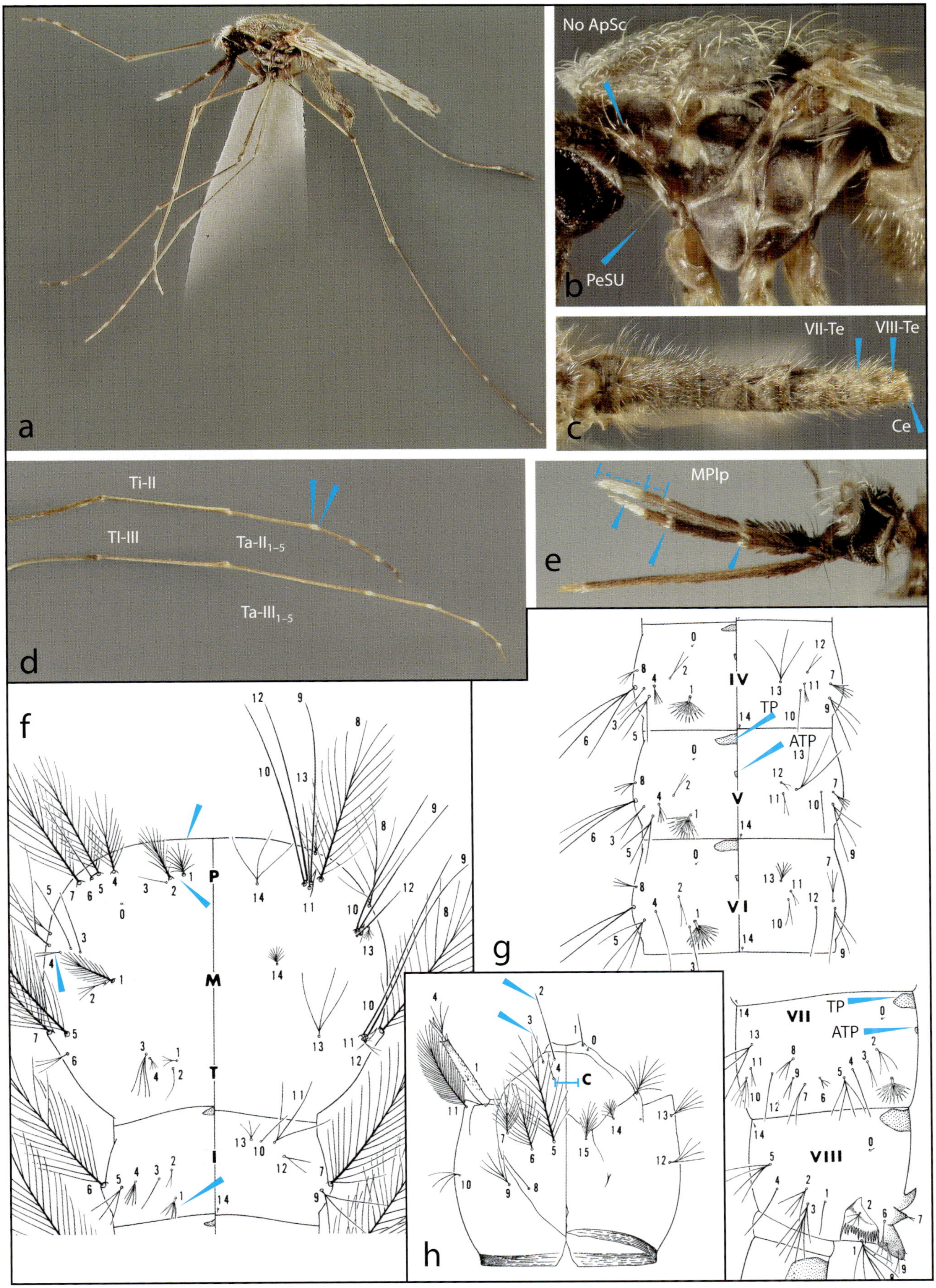

a
No ApSc
PeSU
b
VII-Te
VIII-Te
Ce
c
Ti-II
TI-III
Ta-II1–5
Ta-III1–5
d
MPlp
e
f
P
M
T
I
g
IV
V
VI
TP
ATP
h
C
VII
VIII

Anopheles (Cellia) sulawesi Waktoedi Koesoemawinangoen, 1954
Subfamily Anophelinae
"Sulawesi Indonesian Nail Mosquito"

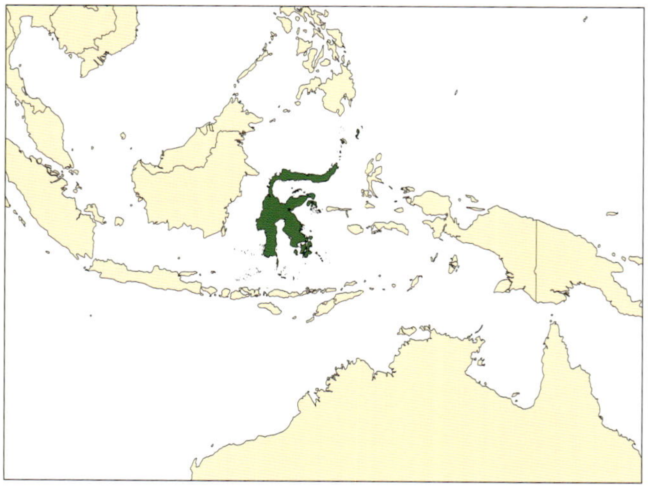

Systematics and Distribution

Anopheles sulawesi is a highly attractive anopheline, with long, delicate legs, heavily decorated with white bands and spots (see *a*, *e*), and wings with distinct spots of alternating dark and pale scales (see *f*). *Anopheles sulawesi* is a member of the Leucosphyrus Subgroup of the Neomyzomyia Series of the *Anopheles* subgenus *Cellia* (see Vol. 2, p. 1085), where it is most closely related to *An. elegans* (James), *An. hackeri* Edwards, and *An. pujutensis* Colless. *Anopheles elegans* is restricted to India and Sri Lanka, whereas the three others are Oriental species. The long proboscis (c. 1.25×forefemur) and shorter palps (0.85–1.15×forefemur) of *An. sulawesi* (see *c*) are shared with *An. elegans*, *An. hackeri*, and *An. pujutensis* and serve to differentiate these three Oriental species from all other anophelines in the region. *Anopheles sulawesi* can easily be recognized by the presence of one to four spots in the PSD spot on vein R (see *f*); these are absent in *An. hackeri* and *An. pujutensis*. *Anopheles sulawesi* larvae are distinctive in the length and placement of seta 4-C (see *g*).

Anopheles sulawesi was first described based on adult specimens collected in a forested area of the Bone-Dumoga Forest Reserve (00°35'S, 123°54'E) in Toraut, on the Indonesian Island of Sulawesi. The species is regarded as a Sulawesian endemic, being found only on Sulawesi and some small adjacent islands, including Boeton, Kaboena, Sangihe, and Talaud.

Bionomics

Immatures have been collected from shaded rock pools, from marginal pools overhung by vegetation, and from drying riverbeds. Immature habitats usually have a high organic content and are typically lined with decaying leaf matter or thick black muddy soil substrates. In common with several other species in the Leucosphyrus Group, *An. sulawesi* is a competent transmitter of human malaria and is regarded as a primary vector in Sulawesi. Two archive specimens in the US National Collection are blood fed and bear labels indicating human feeding behavior; another is labeled "Indoor-resting." However, a study reported no human landing captures of the species between 1800h and 2300h over three nights, despite collectors sitting 3–5 m from the adult emergence sites. This suggests either that species bite later in the night or that there is host seeking farther away from the larval sites. That the species has been found indoors naturally infected with human malaria parasites on several occasions clearly indicates anthropophily.

Associated Pathogens

Plasmodium spp.

Diagnostic Characters

(a) Habitus (♀). (b) Leg (III): Ti-III apical pale band with narrow, dark longitudinal extension on ventral surface in addition to separate dark spot. **(c) Head (side):** Palpus (MPlp) with apical pale band c. two times length of preapical dark band; $MPlp_5$ mainly dark scaled; proboscis (P) mainly dark scaled with band of pale scales at apex, basal to labella (La) (not distinct in photo). **(d) Abdomen (dorsal):** VIII-Te with posterior golden scales. **(e) Leg (III):** $Ta\text{-}III_{2-5}$ without basal pale scales. **(f) Wing:** Both M and M_{3+4} veins noticeably wavy; accessory sector pale spot (ASP) absent from costa (C), present on vein R; presector dark (PSD) spot on vein R with 1–4 pale interruptions; preapical dark spot (PD) of vein R_1 with 1–4 pale spots; 6–10 pale interruptions from PSD to PD area of vein R_1; sector dark spot (SD) of vein R_1 with 2–5 pale spots. **(g) Larva (head):** Seta 3-C single, extending beyond head margin; seta 4-C long, extending well beyond base of 2-C; seta 5-C conspicuously longer than antenna. **(h) Thorax (dorsal). (i) Larva (thorax):** Seta 1-M basal sclerotized tubercle without a distinct pointed or angular tooth; seta 1-P basal sclerotized tubercle with prominent tooth on posterodorsal margin. **(j) Larva (abdominal segments):** Seta 1-II not fully developed palmate; individual leaflets of palmate setae with long and slender apical filaments; tergal plate on segments IV,V about two times wider than deep, accessory plates usually absent. Image sources: WRBU; Sallum et al. (2005).

Exemplar DNA barcodes: *An. sulawesi COI* (DQ897967).

Catalog entry: Vol. 2, p. 680.

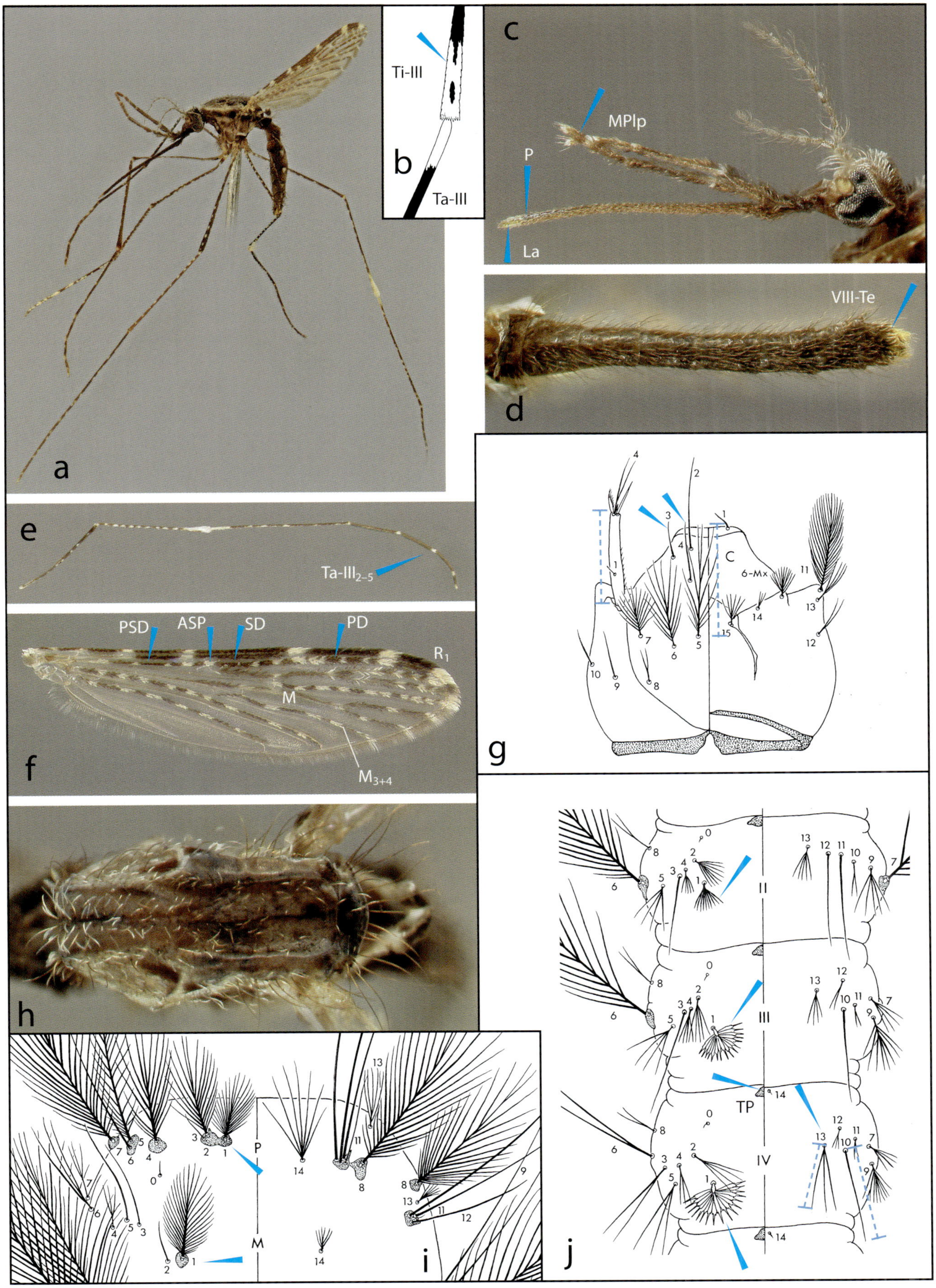
a
b
Ti-III
Ta-III
c
MPlp
P
La
d
VIII-Te
e
Ta-III$_{2-5}$
f
PSD
ASP
SD
PD
R_1
M
M_{3+4}
g
C
6-Mx
h
i
P
M
j
II
III
IV
TP

Anopheles (*Cellia*) *sundaicus* (Rodenwaldt, 1925)
Subfamily Anophelinae
"Sunda Indonesian Nail Mosquito"

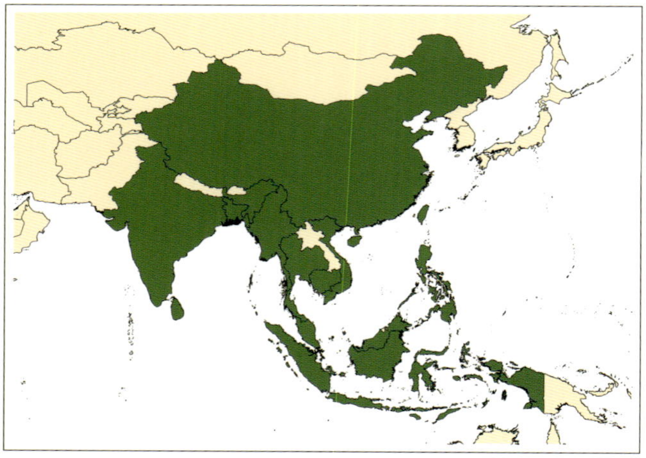

Systematics and Distribution

Anopheles sundaicus s.l. has been reported from the islands and mainland of Southeast Asia through to India and Sri Lanka. Chromosomal and isoenzyme studies first exposed *An. sundaicus* as a species complex containing three distinct forms—the brackish-water tolerant forms A and C and freshwater inland form B. Form A, detected in coastal brackish-water populations from Trat Province in Thailand, was later described as *An. epiroticus* Linton & Harbach. A neotype of *An. sundaicus* s.s. was described from Pandan Beach, Kuching, Sarawak, but its chromosomal typing as form A was at odds with genetic sequence data. All three chromosomal forms were reportedly in sympatry at the costal locality of Asahan in North Sumatra, but only one form was recovered in targeted sampling at this site following the catastrophic tsunami that hit this region on Boxing Day in 2004. Samples collected in South Tapanuli, Java (inland form B), were also not genetically different from any other Indonesian population. Indonesian populations (collectively referred to as *An. sundaicus* species D, not forms A, B, or C) were genetically somewhat distinct from both *An. sundaicus* s.s. and *An. epiroticus* and are reportedly most similar to Sri Lankan and Indian populations of *An. sundaicus* s.l. Most recently, *An. sundaicus* species E, distinguished from all others on the basis of *ITS2* sequences, was reported on the Nicobar Islands. *Anopheles sundaicus* and associated taxa belong to the Pyretophorus Series of the *Anopheles* subgenus *Cellia*.

Bionomics

Anopheles sundaicus s.l. females are avid biters of cattle and readily attack humans outdoors, peaking from 1800h to 2100h. Indoor collections have also been reported. *Anopheles sundaicus* s.s. was collected from rock pools along a sandy shoreline in Pandan Beach, Sarawak, Malaysian Borneo. The immature habitats were sunlit and contained submerged, decaying leaves. The water was somewhat brackish, consisting mainly of rainwater with saltwater spray from nearby crashing waves. Females were collected biting cattle and humans outdoors in villages along the coast and at an inland site, Miri, Sarawak. In Miri, the *An. sundaicus* populations were phenomenally high, along with *An. barbirostris* s.l., and specimens were picked like cherries from the vegetation on which they rested. *Anopheles epiroticus* immatures are common all along the coastal regions of the mainland of Southeast Asia to India, predominating in brackish-water fish or shrimp ponds and lagoons with filamentous green algae. In Indonesia, *An. sundaicus* D has been shown to tolerate a high range of salinity, ranging from 0% to 11%.

Associated Pathogens

Plasmodium falciparum, *P. vivax*.

Diagnostic Characters

(a) Habitus (♀). (b) Thorax (side): Antepronotal scales (ApSc) absent; upper proepisternal setae (PeSU) usually absent (present here). **(c) Head (side):** Proboscis (P) entirely dark scaled; palpus (MPlp) with three pale rings, subapical dark band ≥0.5 times apical pale band and subapical pale band much narrower than apical band and equal to basal band. **(d) Wing:** Costa (C) basal to humeral crossvein (h) dark scaled; pale fringe spot absent between cubital (CuA) and anal (1A) veins; vein 1A with two dark spots. **(e) Legs:** Legs speckled with basal and apical pale bands on some tarsomeres; Ta-I with broad pale bands; apex of Ti-III and base of Ta-III_1 without distinct contiguous white marking. **(f) Larva (thorax):** Seta 3-M not palmate; seta 4-M 3,4 branched from near base (not distinct in illustration); seta 1-P usually with ≥8 branches; setae 9,10,12-P,M single; setae 9,10-T branched or plumose. **(g) Larva (head):** Seta 1-A single; seta 3-C ≥0.5 times length of seta 2-C. **(h) Larva (abdominal segments):** Seta 1-I palmate but less developed than on following segments; tergal plates (TP) on segments III–VI not very large, with concave posterior border not encompassing median posterior accessory tergal plate (ATP); distance between bases of abdominal palmate setae > width of anterior tergal plates; seta 6-IV–VI branched at base. Image sources: WRBU; Linton et al. (2001).

Exemplar DNA barcodes: *An. sundaicus* s.s. *COI* (AY789709–26).

Catalog entry: Vol. 2, p. 680.

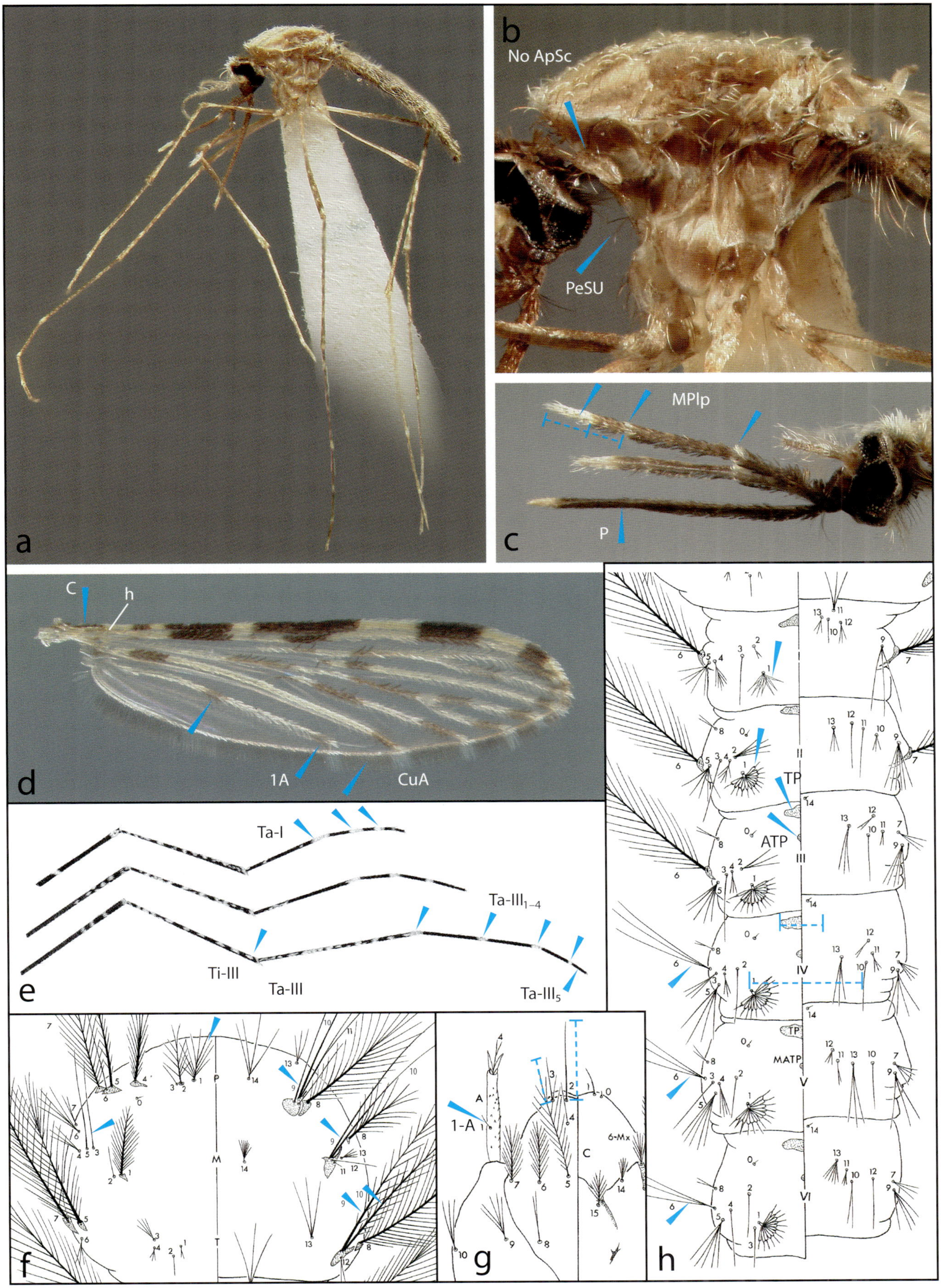
a
b
No ApSc
PeSU
c
MPlp
P
d
C
h
1A
CuA
e
Ta-I
Ta-III1–4
Ti-III
Ta-III
Ta-III5
f
P
M
T
g
1-A
A
C
6-Mx
h
I
II
III
IV
V
VI
TP
ATP
MATP

Anopheles (*Cellia*) *superpictus* Grassi, 1899
Subfamily Anophelinae
"Super-painted Italian Nail Mosquito"

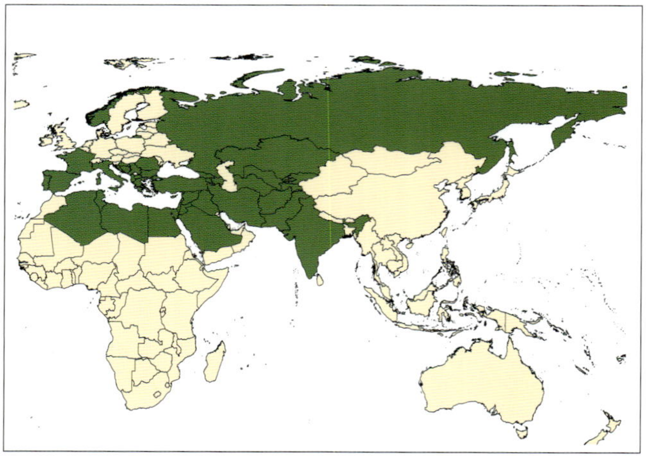

Systematics and Distribution

Anopheles superpictus is a pale-bodied mosquito clothed with pale scales on the scutum (see *d*) and head vertex and flecked throughout the flagellomeres (see *c*); unusually, the cuticle of the abdomen and abdominal setae are also pale (see *h*). *Anopheles superpictus* belongs to the Neocellia Series of the *Anopheles* subgenus *Cellia* (see Vol. 2, p. 1084, for internal relationships), which includes 14 other species including the malaria vectors *An. pulcherrimus* Theobald (see p. 366), and *An. stephensi* Liston (see p. 382). Recent studies have elucidated two forms of *An. superpictus* (A and B) in Iran, based on morphological differences in the larval and adult stages, mitochondrial DNA differences, and perceived differences in bionomics (differences in climatic preferences and malaria competence). Three nuclear *ITS2* genotypes (X, Y, Z) were recovered, suggesting further incipient speciation. Eight synonyms are recognized for *An. superpictus*: *palestinensis* Theobald, *nursei* Theobald, *cardamatisi* Newstead & Carter, *vassilievii* Portschinsky, *macedoniensis* Cot & Hovasse, *berestnevi* Shingarev, *atheniensis* Cardamatis, and *hellenicus* Peus. A detailed integrated study is needed to assess the taxonomic status of the species using samples from the type locality and across its range.

Originally described from Italy, *An. superpictus* s.l. is found along the south and eastern Mediterranean Basin (including northern Africa and the Middle East), in eastern Europe, and into the Caucasus Region of the Western Palearctic (see map).

Bionomics

Anopheles superpictus is an effective malaria vector across its range. It is common in mountain valleys and foothills across its range and has been reported at elevations of up to 2,800 m in Tajikistan. It is rarely collected in lowland or dominantly agricultural areas. Females are typically endophilic and readily enter homes to feed on humans. In colder climes the females reportedly enter diapause en masse by the end of October but have been noted to come out of hibernation and feed in warm houses during the winter period.

Immatures of *An. superpictus* s.l. are associated with shallow stream pools of stagnant or slow-flowing water in rocky riverbeds—waters typified by high calcium salt concentrations. Occasional reports exist of immatures collected in drying rice fields at elevation. Optimal temperatures for larval development occurs between 30°C and 35°C.

Associated Pathogens

RVFV; *Plasmodium* spp.

Diagnostic Characters

(a) Habitus (♀). (b) Thorax (side): Upper proepisternal setae (PeSU) absent. **(c) Head (side):** Palpus (MPlp) with three distinct pale bands; $MPlp_5$ all pale; vertex (V) with broad white scales, posterior and lateral scales dark brown. **(d) Thorax (dorsal):** Scutal fossa (SF) without scales; scutum (Scu) with many narrow pale scales. **(e) Wing:** One or two pale spots at base of costa (C), humeral pale (HP) and presector pale (PSP); presector pale spot (PSP) present; accessory sector dark spot (ASD) absent due to fusing of sector pale and accessory sector pale spots; pale fringe spot absent between cubital vein (CuA) and vein 1A; vein 1A with two dark spots, distal spot much longer; most of vein R_{4+5} pale scaled, with small basal and apical dark spots. **(f) Larva (thorax/abdominal segments):** Seta 1-P with weakly developed tubercle; setae 1,2-P basal tubercles not fused; seta 9-M branched and 10-M single (not shown); seta 3-T forming weakly palmate seta with 3–9 branches; setae 9,10-T branched (not shown); abdominal segments III–VII with tergal plates (TP) not encompassing posterior accessory plates (ATP); width of tergal plate (TP) on segment V usually about 0.67 distance between palmate setae (not shown); seta 1-I not palmate; seta 1-II–VII well developed, palmate; leaflets of seta 1-III,IV with long apical filaments; seta 6-IV usually five branched. **(g) Larva (head):** Seta 2-C widely separated; setae 2,3-C usually single, slightly aciculate; seta 4-C <0.67 length of seta 2-C. **(h) Abdomen (dorsal):** Terga (Te) not pale scaled, without distinct dark posterolateral scale tufts. **(i) Legs (I–III):** Femora and tibia without speckling; Ta-III dark, some tarsomeres may be slightly pale apically; Fe-II,III without longitudinal white stripe. Image sources: WRBU; Ross & Roberts (1943).

Exemplar DNA barcodes: *An. superpictus COI* (AY900634–50, JF966745, JX255703–08).

Catalog entry: Vol. 2, p. 680.

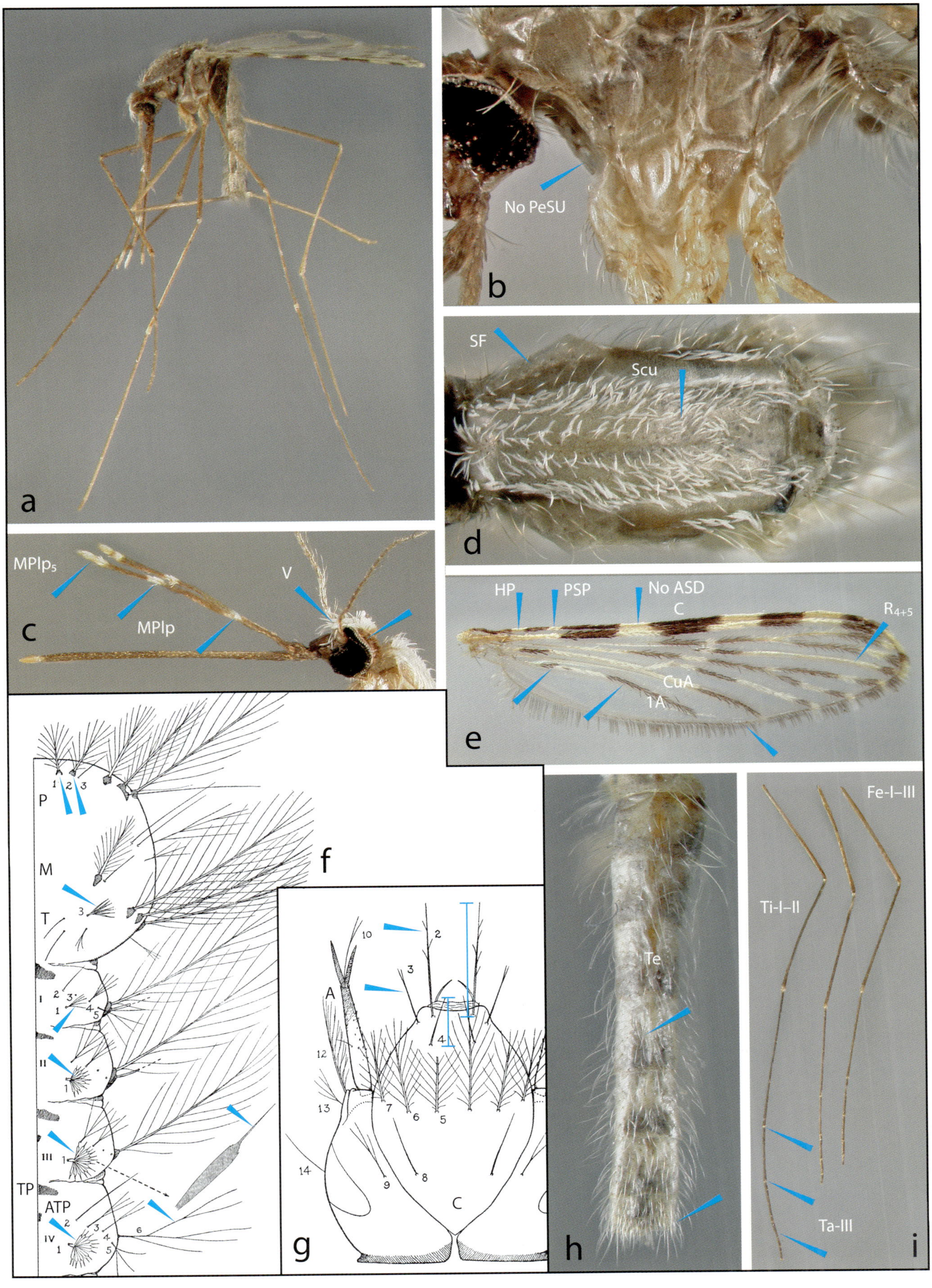

a
No PeSU
b
SF
Scu
d
MPlp5
V
c
MPlp
HP
PSP
No ASD
C
R4+5
CuA
1A
e
P
M
T
I
II
III
TP
ATP
IV
f
10
2
3
A
4
12
13
7
6
5
14
9
8
C
g
Te
h
Fe-I–III
Ti-I–II
Ta-III
i

Anopheles (*Cellia*) *tessellatus* Theobald, 1901 Subfamily Anophelinae "Tessellated Malaysian Nail Mosquito"

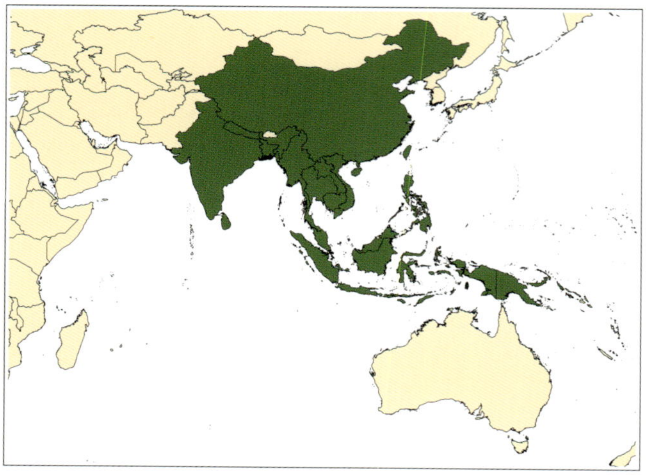

Systematics and Distribution

Excessive banding on its legs and palps and the ornate patterning of its wings give *Anopheles tessellatus* a distinctly spotty look. Originally described from Perak in western peninsular Malaysia, this species is widespread from Sri Lanka, India, and Nepal through mainland Southeast Asia to the islands of Indonesia, the Philippines, New Guinea, and the Solomon Islands. *Anopheles tessellatus* is the only representative of the Tessellatus Group (Neomyzomyia Series). Two subspecies are recognized: *An. tessellatus kaladana* Stoker & Waktoedi and *An. tessellatus orientalis* (Swellengrebel & Swellengrebel de Graaf). Our examination of material from across its range, including topotypic material through DNA barcoding, suggests the species comprises a complex of at least four distinct taxa across its range. Although we lack exemplars from the type locality in mainland Malaysia, our preliminary data suggests that the representative "*tessellatus*" found in China, Japan, Laos, and Vietnam comprise the most widespread form, with at least two additional forms in the Philippines and Borneo, and another in Sri Lanka. Efforts are underway to publish these data and formally elevate at least two of the available synonyms.

Bionomics

Anopheles tessellatus s.l. is predominantly zoophilic, with up to 20 times more individuals collected on cattle than on humans. Females will enter houses to bite humans and take several blood meals in a gonotrophic cycle. Indian populations of *An. tessellatus* s.l. raised under laboratory colony conditions took three to four, seven to nine, and one to two days, respectively, for the egg, larval, and pupal stages to mature. A period of five to seven days is needed for the newly emerged females to mate, blood feed, and lay their first batch of eggs. In the wild it is rare to find *An. tessellatus* s.l. larvae in high densities, suggesting that females spread their eggs across multiple locations rather than ovipositing all eggs in the same habitat. Somewhat unusually for *Anopheles* mosquitoes, females choose dirty, stagnant water for oviposition, either in open or shaded habitats. Sometimes these pools include muddy, organic wallows frequented by water buffalo. Although highly competent to transmit human malarias, *An. tessellatus* s.l. becomes locally significant only where populations are high or where cattle, buffalo, sheep, and goats are scarce, such as in the Maldives, where it also acts as the vector of Bancroftian filariasis.

Associated Pathogens

Chittoor virus, JBEV, KPKV; *Plasmodium falciparum*, *P. vivax*, *Wuchereria bancrofti*.

Diagnostic Characters

(a) Habitus (♀): Fe-I–III and Ti-I–III speckled; abdomen without ventral median tufts of dark scales (not shown). **(b) Head (side):** Palpomeres ($MPlp_{3-5}$) with broad apical pale bands; proboscis (P) with pale-yellowish scales on apical half. **(c) Thorax (side):** Antepronotal scales (ApSc) present (not evident in photo); upper proepisternal setae (PeSU) present. **(d) Thorax (haltere):** Halter (Hl) capitellum (Ca) white scaled dorsally. **(e) Wing:** Pale fringe spots at end of all wing veins. **(f) Leg (III):** Apex of Ti-III with a narrow pale band; $Ta\text{-}III_{1-4}$ with small apical pale bands. **(g) Larva (thorax):** Seta 1-P usually 2–5 branched, inserted on lightly pigmented basal tubercle (not evident in illustration); setae 9,10-P,M,T single. **(h) Larva (head):** Seta 1-A small, single. **(i) Larva (abdominal/terminal segments):** Segments IV–VII with small tergal plates (TP) not enclosing small median accessory tergal plates (ATP); segments IV–VII with tergal plates smaller than distance between bases of each pair of palmate setae; seta 1-I,II much reduced palmate; seta 1-III–VII palmate with apical filaments usually <0.5 as long as basal portion. Image sources: WRBU; Tanaka et al. (1979).

Exemplar DNA barcodes: *An. tessellatus* s.l. *COI* (MT256975–MT257041).

Catalog entry: Vol. 2, p. 681.

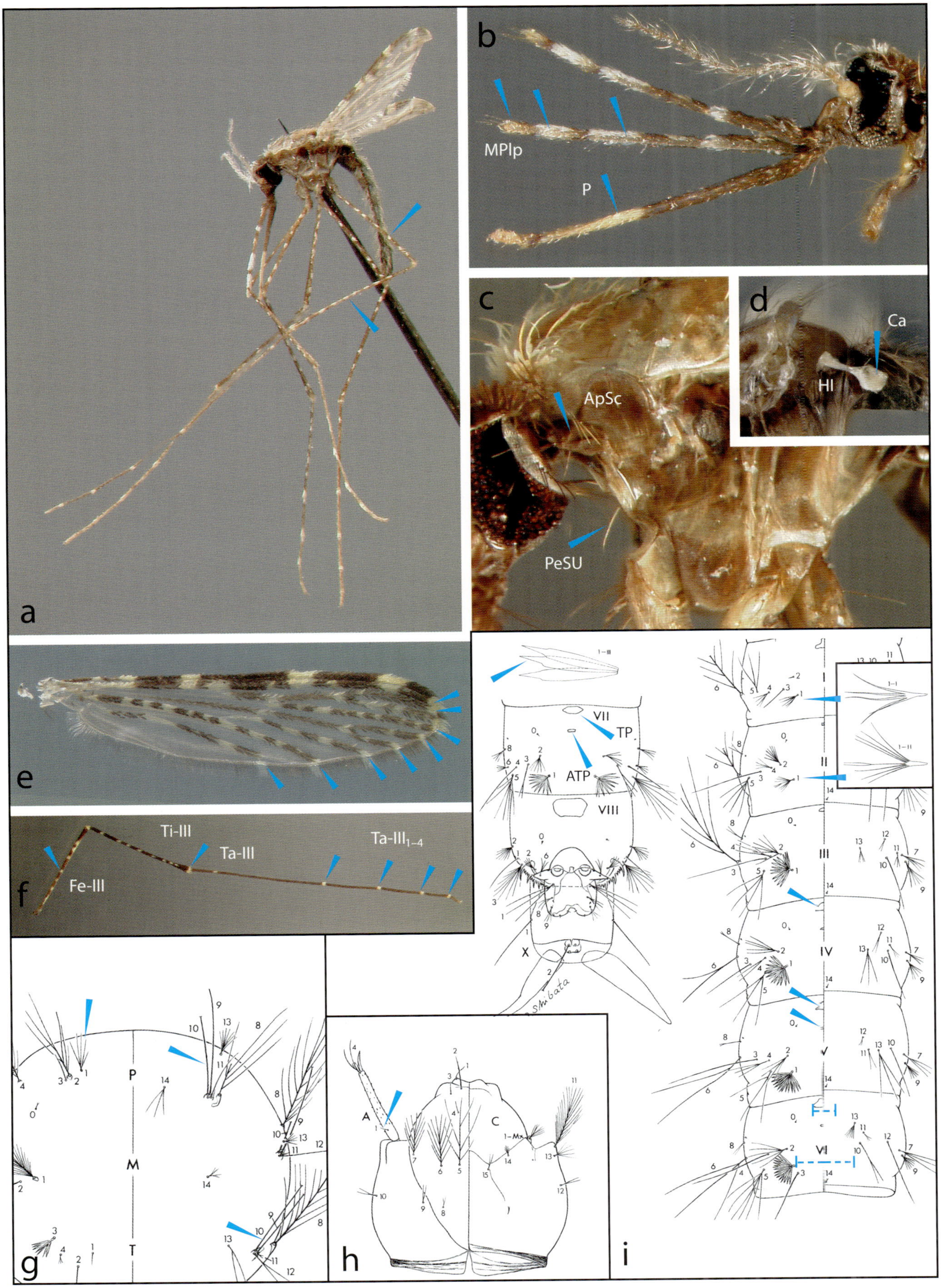
a
b
MPlp
P
c
ApSc
PeSU
d
Ca
Hl
e
f
Fe-III
Ti-III
Ta-III
Ta-III1–4
g
P
M
T
h
A
C
1-Mx
i
VII
TP
ATP
VIII
X
I
II
III
IV
V
VI

Anopheles (*Kerteszia*) *cruzii* Dyar & Knab, 1908
Subfamily Anophelinae
"Cruz Brazilian Nail Mosquito"

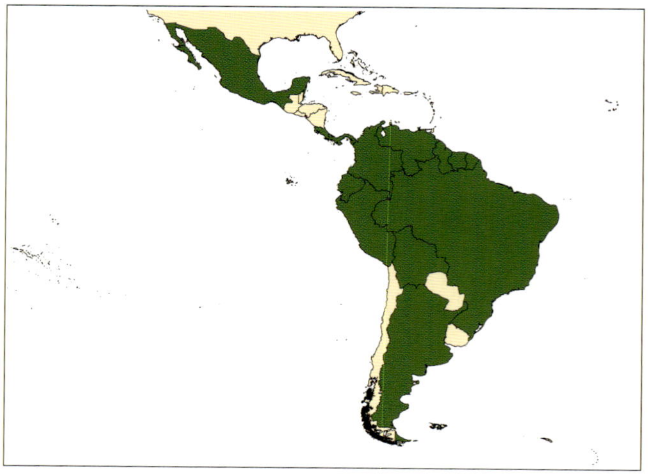

Systematics and Distribution

This taxon was first erroneously described as *Anopheles lutzii* Theobald, 1901—a name that was found to be preoccupied by *An.* (*Nyssorhynchus*) *lutzii* Cruz, 1901. Subsequently, it was given the replacement name *cruzii* by Dyar and Knab in 1908. Immatures of *An. cruzii* are most similar to those of *An.* (*Ker.*) *homunculus* Komp and *An.* (*Ker.*) *bellator* Dyar & Knab. *An. cruzii* adults are similar to *An. homunculus*, and differentiation of the adult females can be tricky. Both chromosome studies and DNA analyses have shown that *An. cruzii* is a cryptic species complex comprising at least three distinct genetic entities, and that the Itatiaia population from Rio de Janeiro State, Brazil, is in the process of incipient speciation. There are currently three synonyms—*adolphoi* Neiva, *lutzii* Theobald, and *montemor* Correa—and it remains to be determined which of these, if any, correspond to the genetically identified taxa. Originally described from the Guanabara Region (= Rio de Janeiro municipality) of Brazil, *An. cruzii* s.l. is widely distributed across mainland Central and South America from Costa Rica to Argentina.

Bionomics

Immature *An. cruzii* s.l. is almost always found in the leaf axils of bromeliads, from ground level to 15 m up into the canopy, although there has been the odd report of *An. cruzii* s.l. immatures collected from other freshwater sites, including rainwater pools and ditches. Immatures develop incredibly slowly—35 days under laboratory conditions and up to four months in nature. From the third larval stage and as pupae, the immatures have a unique characteristic reddish pigmentation in their cuticle that is not exhibited in any other *An.* (*Kerteszia*) species.

In their tropical environs, wherever the species is present, *An. cruzii* s.l. is often the dominant *Anopheles* collected. *Anopheles cruzii* s.l. mosquitoes are long-lived—up to 56 days in the wild—increasing their opportunity to be infected with, and transmit, arboviral and pathogen agents. They serve as a primary malaria vector in littoral south and southeastern Brazil, where they occur in a variety of habitats from the coastal flats up into the mountains. Females are highly aggressive biters of humans, with bicrepuscular biting habits peaking for around 90 minutes at dusk, with a smaller peak at dawn. They have also been reported biting people during the day in shaded forested areas. *Anopheles cruzii* s.l. has been reported biting monkeys and humans both at the ground level and high in the canopy and, less commonly, out in the open at the forest edge. Females readily enter homes to bite people but do not rest indoors, evading residual spraying control efforts. Natural infection of the pathogenic flagellate *Herpetomonas pessoi*, which affects egg maturation, was detected in *An. cruzii* s.l. in Santa Catarina, Brazil, highlighting the potential utility of this natural interaction for biocontrol.

Associated Pathogens

BORV, GTRV, ICOV, TCMV; *Plasmodium brasilianum*, *P. falciparum*, *P. simium*, *P. vivax*.

Diagnostic Characters

Habitus (♀). (b) Head (side): Scales on $MPlp_{3,4}$ predominately decumbent, those on base of 3 may be slightly erect. **(c) Thorax (dorsal):** Scutal (Scu) integument silvery pruinose with four dark longitudinal stripes; scutum (Scu) without pale scales on acrostichal (AA), dorsocentral (DA), and prescutellar (PrA) areas. **(d) Thorax (side):** Mesepimeron (Mm) with two small scale patches (upper and middle). **(e) Wing:** Accessory sector pale spot (ASP) near middle of costa (C), nearly equidistant between sector pale (SP) and subcostal pale spots (SCP); vein M with dark scales basal to level of mcu. **(f) Leg (III):** $Ta\text{-}III_{2\text{-}5}$ with broad apical pale bands 0.4–0.7 times length of each tarsomere. **(g) Larva (abdomen):** Palmate seta 1-I–VII with narrow pointed leaflets (1-IV illustrated). **(h) Larva (head):** Setae 2-C, 3-C, or both, single; seta 3-C single or with a few aciculae; setae 5–7-C single, double, or branched apically, never plumose; seta 8-C not extending past the base of 6-C. **(i) Larva (abdominal/terminal segments):** Seta 1-I palmate, less developed than on following segments; seta 6-VI long, aciculate; saddle (Sa) lightly sclerotized; seta 1-S single.

Exemplar DNA barcodes: *An. cruzii* s.l. *COI* (JF923692, KC992738–791, KM507293–314). Image sources: WRBU; Wilkerson & Peyton (1991).

Catalog entry: Vol. 2, p. 684.

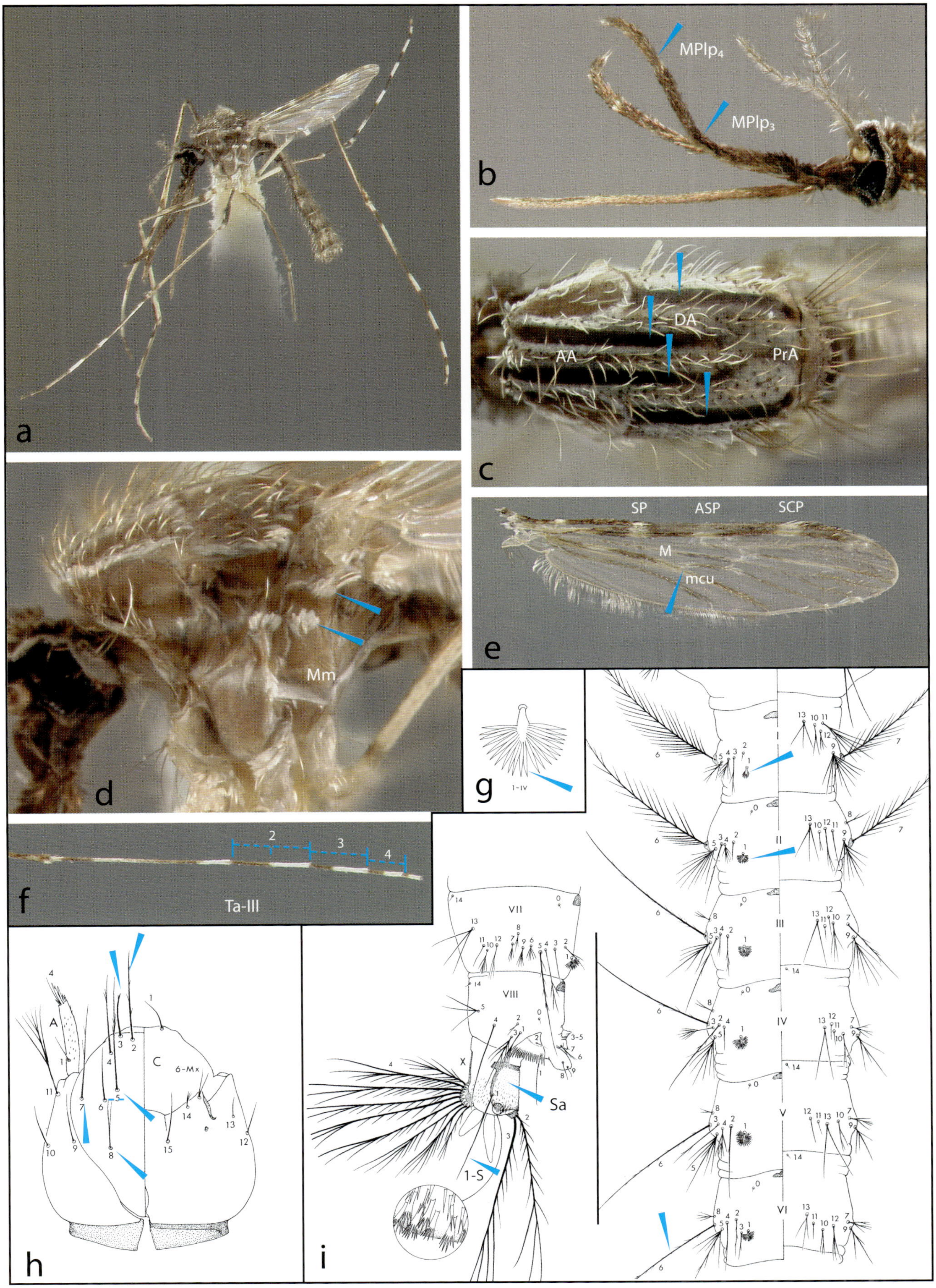
a
b
$MPlp_4$
$MPlp_3$
c
DA
AA
PrA
d
Mm
e
SP
ASP
SCP
M
mcu
f
2
3
4
Ta-III
g
1-IV
h
A
C
6-Mx
i
VII
VIII
X
Sa
1-S
I
II
III
IV
V
VI

Anopheles (*Kerteszia*) *neivai* Howard, Dyar & Knab, 1913
Subfamily Anophelinae
"Malaria Panamanian Nail Mosquito"

Systematics and Distribution

Anopheles neivai is perhaps the most distinctive of all species in the subgenus *Kerteszia*. It has white wing tips (see *d*) and heavy dustings of white scales covering the coxa and long parallel lines of white scales running down the femur and tibia in contrast to the broad white stripes on the tarsi (see *a*). The species was first described from specimens collected at Fort San Felipe, in the Porto Bello bay region of Panama, and named in honor of the Brazilian medical entomologist Artur Neiva (1880–1943). However, morphological identification of *An. neivai* is not straightforward in the adult females, and doubtful occurrence records persist and are perpetuated, particularly those reported at higher altitudes. DNA barcoding approaches in our laboratory found at least four cryptic species within samples morphologically identified as *An. neivai*: *An. neivai* s.s. from Panama, *Anopheles neivai* 2 from Colombia and Venezuela, *Anopheles neivai* 3 from Venezuela and Ecuador, and *An. neivai* 4 from Ecuador. Much work remains to formally describe each of these taxa and determine the corresponding distributions of all the component taxa in the Neivai Complex. *Anopheles neivai* s.l. has an extensive distribution stretching from Chiapas in southern Mexico through Central America to Peru, Bolivia, and Brazil in South America (see map, as s.l.).

Bionomics

Immatures of *An. neivai* s.l. develop in rainwater collections formed in the leaf axils of terrestrial and epiphytic bromeliads, some of which can be many meters high in the canopy. In the tropical rainforests in Panama, *An. neivai* s.l. immatures are commonly found high up in arboreal bromeliads, whereas in deciduous forests, they are more likely to be found in bromeliads at ground level. Occasionally, *An. neivai* s.l. larvae have been recovered from axils of aroids and in tree holes. Populations of *An. neivai* s.l. increase following periods of heavy rain, because oviposition sites become plentiful.

Female *An. neivai* s.l. readily feed on humans. The biting behaviors of *Anopheles neivai* s.l. also change according to habitat. Adults tend to be more aggressive in sylvatic areas, in comparison to intra- and peridomestic areas, and are the least aggressive when found inside houses. In the mangrove swamps along the Pacific coast of Colombia, *An. neivai* (presumably *An. neivai* 2) is highly anthropophilic and regarded as a primary malaria vector. In Buenaventura, Valle del Cauca, Colombia, *Plasmodium* infection rates of 4.76% were reported in *An. neivai*, and 1.5% of females dissected showed evidence of at least 10 oviposits, indicating extremely high longevity and multiple blood feeds—bionomic factors that exponentially increase vector competence for both parasites and viruses. In the same region, biting differentials were noted in relation to population age structure, with older females feeding at dawn (0530h–0630h), and nulliparous females seeking blood meals in the evening, after dusk (1800h–1900h).

Associated Pathogens

Guaroa virus, ILHV, VEEV, YFV; *Plasmodium falciparum*, *P. vivax*.

Diagnostic Characters

(a) Habitus (♀). (b) Thorax (side): Mesepimeron (Mm) with small patch of scales (UMSc) inserted near upper mesepimeral setae (MeSU). **(c) Thorax (dorsal):** Scutal (Scu) integument silvery pruinose with four dark longitudinal stripes; scutum (Scu) without pale scales on acrostichal (AA), dorsocentral (DA), and prescutellar (PrA) areas. **(d) Wing:** Accessory sector pale spot (ASP) near middle of costa, nearly equidistant between sector pale (SP) and subcostal pale spots (SCP); large pale apical fringe spot, rarely divided into two small pale spots. **(e) Abdomen (dorsal):** Terga (Te) and sterna (S) without scales, occasionally with a few scales on VII–VIII-Te and cercus. **(f) Leg (III):** Ta-III$_{2-4}$ with broad white bands on distal 0.5–0.7 of each tarsomere; Ta-III$_5$ with base dark, apical 0.35–0.60 pale. **(g) Larva (head):** Setae 5,6-C usually single; seta 11-C single or with 2,3 apical branches. **(h) Larva (abdominal/terminal segments):** Seta 1-I single, not palmate; seta 1-II–VII leaflets smooth sided; seta 1-VII palmate; seta 6-VI long, plumose, similar to 6-III–V; seta 9-II–VII branched; seta 13-III–V usually three branched but shorter than respective segments. Image sources: WRBU; Zavortink (1973).

Exemplar DNA barcodes: *Anopheles neivai* s.l. *COI* (KM234385, KM234400, KM234415, KM234433).

Catalog entry: Vol. 2, p. 685.

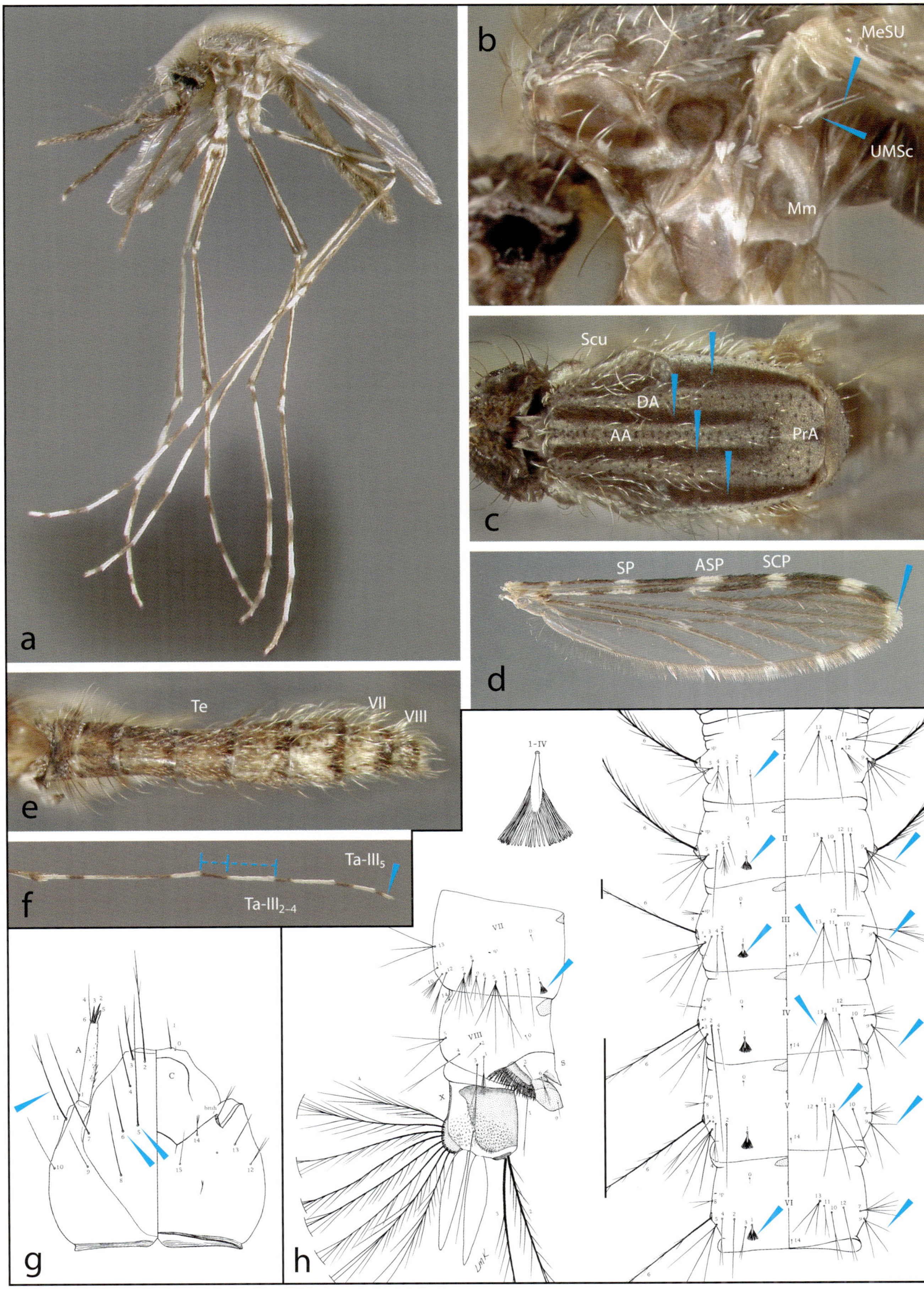
b
MeSU
UMSc
Mm
c
Scu
DA
AA
PrA
d
SP
ASP
SCP
e
Te
VII
VIII
f
Ta-III$_5$
Ta-III$_{2-4}$
g
h
1-IV

Anopheles (*Kerteszia*) *pholidotus* Zavortink, 1973
Subfamily Anophelinae
"Armored Panamanian Nail Mosquito"

Systematics and Distribution

Female *Anopheles pholidotus* and *An. lepidotus* Zavortink share the unique feature of a C-shaped row of long projecting white scales on the mesepimeron (see *c*). Additionally, *An. pholidotus* has a pale wing fringe spot adjacent to the tip of vein M_1 (see *e*) and distal transverse rows of long erect dark scales on the upper surface of abdominal segments V–VII (see *g*). *Anopheles pholidotus* was described from a male holotype, female allotype, and paratypes—including a male, a female, larval and pupal exuviae, and whole larvae—collected in Bocas del Toro, Panama. Concurrently, the same author described another closely related species, *An. lepidotus* Zavortink, and inadvertently, the descriptions and illustrations of the female—the most commonly collected stage in incrimination studies—were switched, leading to several years of taxonomical confusion and the incorrect incrimination of *An. lepidotus* as a primary malaria vector in Latin America. Laboratory rearing of both species, combined with molecular systematics approaches, have clarified the situation and resulted in the corrected assignment of *An. pholidotus* as the vector taxon. Recently, reexamination of archive specimens and DNA studies have confirmed the distribution of *An. pholidotus* in Bolivia, Colombia, Costa Rica, Ecuador, Panama, Peru, and Venezuela.

Bionomics

Because of the long-standing confusion surrounding the taxonomy of *An. pholidotus*, the species has often been misreported as *An. lepidotus* or, prior to the description of these two species, as *An. boliviensis* Theobald. The taxonomic confusion also clouds the bionomics reported, and thus truly little is known about *An. pholidotus* with regard to its biology and life cycle. As for most other *Kerteszia* species, the preferred immature habitat is in the rainwater collections within the bracts of bromeliads. In Ecuador, immature stages of *An. lepidotus* were collected in arboreal bromeliads.

In the malaria focal area of Cunday-Villarrica, Tolima, Colombia, molecularly confirmed *An. pholidotus* was collected in forest environs and in human landing collections conducted both inside and around houses, which showed high population densities of *An. pholidotus*. This study verified *An. pholidotus*, confirming the misidentification and erroneous incrimination of *An. lepidotus* as a primary malaria vector. *Anopheles pholidotus* is the malaria vector, whereas *An. lepidotus* is rare and of no medical importance. *Anopheles pholidotus* is also a mechanical vector of the eggs of the human botfly, *Dermatobia hominis*.

Medical Importance

Dermatobia hominis; *Plasmodium* spp.

Diagnostic Characters

(a) Habitus (♀). (b) Head (dorsal): Palpomeres $MPlp_{2-5}$ with apical white bands. **(c) Thorax (side):** Mesepimeron (Mm) with one long ventrally projecting C-shaped row of long white scales. **(d) Thorax (dorsal):** Scutal (Scu) integument silvery pruinose with four dark longitudinal stripes. **(e) Wing:** Accessory sector pale spot (ASP) near middle of costa, nearly equidistant between sector pale (SP) and subcostal pale spots (SCP); vein M_1 with pale fringe spot adjacent to tip. **(f) Leg (III):** $Ta\text{-}III_{1,2}$ with narrow apical white bands from dorsal view. **(g) Abdomen (dorsal):** Segments V–VII-Te with distal transverse rows of long erect dark scales. **(h) Larva (head):** Seta 3-A long and tapered, much longer than 2-A; seta 3-C moderately long, slightly stouter than 2-C, with long attenuated tip. **(i) Larva (thorax):** Seta 11-P moderately developed, long. **(j) Larva (terminal segments):** Pecten (Pt) with two rows of spinules on each spine (PS). **(k) Larva (abdominal segments):** Seta 1-I long, single; seta 1-II–VI palmate, brushlike; seta 6-VI with several strong basal branches; seta 9-IV–VI single. Image sources: WRBU; Zavortink (1973).

Exemplar DNA sequences: *An. pholidotus COI* (JQ041287–88, MH843473), *ITS2* (JN967768–70).

Catalog entry: Vol. 2, p. 685.

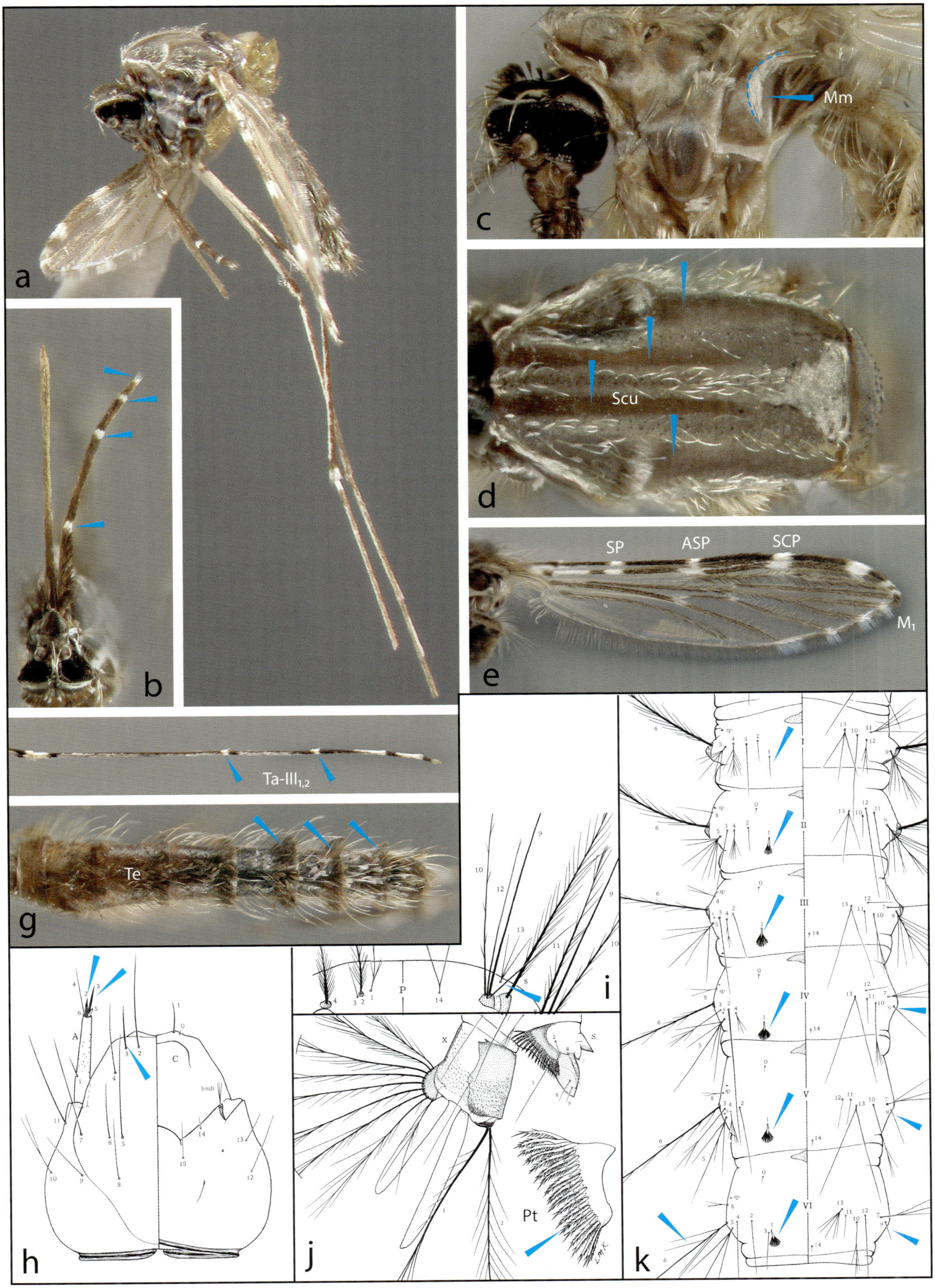
a
b
Mm
c
Scu
d
SP
ASP
SCP
M1
e
Ta-III1,2
Te
g
A
C
h
P
i
X
S
Pt
j
k

Anopheles (*Nyssorhynchus*) *albimanus* Wiedemann, 1821
Subfamily Anophelinae
"White-handed Dominican Nail Mosquito"

Systematics and Distribution

Anopheles albimanus is a common mosquito easily recognized by its ornately spotted wings (see *e*) and long white hind tarsi ($Ta\text{-}III_{3-5}$), broken only by a narrow, dark basal band on $Ta\text{-}III_5$ (see *a*). Within the Neotropical subgenus *Nyssorhynchus*, *An. albimanus* lends its name to the Albimanus Section and is the sole representative of the Albimanus Series. Due to its wide distribution and primary malaria vector status, *Anopheles albimanus* has been closely studied, with no record of cryptic taxa. However, a novel species—dubbed *An. albimanus* B pending formal description—was detected in our laboratory using DNA approaches. *Anopheles albimanus* B was collected in Guayas, Ecuador, and later detected in archive samples collected in Pichincha and Esmeraldas, Ecuador, and again in biosurveillance studies in Tumbes, Peru, suggesting this taxon is fairly widespread in the southern parts of its range. Only *A. albimanus* B has been detected in these regions to date. Six species are currently placed in synonymy with *An. albimanus*, including *cubensis* Agramonte, *albipes* Theobald, *tarsimaculata* Goeldi, *gorgasi* Dyar & Knab, *bisignatus* Hoffman, and *trisignatus* Hoffman. However, *An. albimanus* B is geographically quite removed from these taxa and is therefore unlikely to correspond to any current synonym. Populations from the type locality in the Dominican Republic need to be genetically assessed to define *An. albimanus* s.s. before the true composition of the Albimanus Complex can be defined. Currently, the distribution of *An. albimanus* s.l. reaches from the southern United States through Central America and some islands of the Caribbean and then south to Peru and Brazil.

Bionomics

Adult *An. albimanus* s.l. is present year-round, with populations generally highest in the dry season. It is primarily associated with humid lowlands at altitudes under 500 m above sea level. *Anopheles albimanus* s.l. mosquitoes are strong fliers and have been detected up to 3 km from release sites. Biting patterns vary across their range. Females are strongly exophilic in Haiti, yet in Mexico, females rest in houses in high numbers. In other regions of Latin America, females reportedly enter houses to feed, yet they rarely rest there, instead returning to the humid sanctuary of the forest before sunrise. Host-seeking females are readily attracted to human-, horse-, and donkey-baited traps, with the highest numbers collected between 1800h and 2400h, while the reported biting peak is later in Colombia, between 2200h and 2400h. Horse-baited traps are the standard method for assessing adult densities of *An. albimanus* s.l. in Latin America. *Anopheles albimanus* B was found resting in horse stables and was relentlessly biting humans during the day in Ecuador, suggesting this taxon could also be a formidable malaria vector.

Immature *An. albimanus* s.l. is usually found in permanent, open, sunlit ground pools with muddy substrata, abundant algal vegetation—particularly blue-green algae—and clear water. Larval habitats can include fresh- or brackish-water ground pools, lakes and swamps, drainage ditches, rice fields, and water-filled cattle hoofprints in the rainy season.

Associated Pathogens

CHIKV, SINV, SFV, TLAV; *Plasmodium berghei*, *P. falciparum*, *P. vivax*.

Diagnostic Characters

(a) Habitus (♀): $Ta\text{-}III_{3,4}$ entirely white; $Ta\text{-}III_5$ with a basal dark band. **(b) Thorax (dorsal). (c) Head (side):** Palpomere $MPlp_4$ with dark or yellow to golden-brown scales on mediolateral surface, scales never white or cream colored. **(d) Abdomen (dorsal):** II-Te without dark posterolateral scale tufts. **(e) Wing:** Accessory sector pale spot (ASP) about 0.4 from wing base, conspicuously closer to sector pale (SP) than to subcostal pale spot (SCP). **(f) Leg (I):** $Ta\text{-}I_5$ usually all dark. **(g) Larva (head):** Seta 2-C aciculate; seta 4-C moderately long to long, usually extending to anterior margin of head. **(h) Larva (thorax):** Seta 1-P plumose, with filiform branches; setae 1,2-P on large common tubercle. **(i) Larva (abdominal segments VII–VIII):** Seta 1 palmate leaflets smooth sided. **(j) Larva (abdominal segments I–VI):** Seta 13-I–IV moderately large to large. Images sources: WRBU; Faran (1980).

Exemplar DNA barcodes: *An. albimanus* whole genome (GCA_000349125), *COI* (MN319673–76, FJ516463–553); *An. albimanus* B (MN319677–91).

Catalog entry: Vol. 2, p. 687.

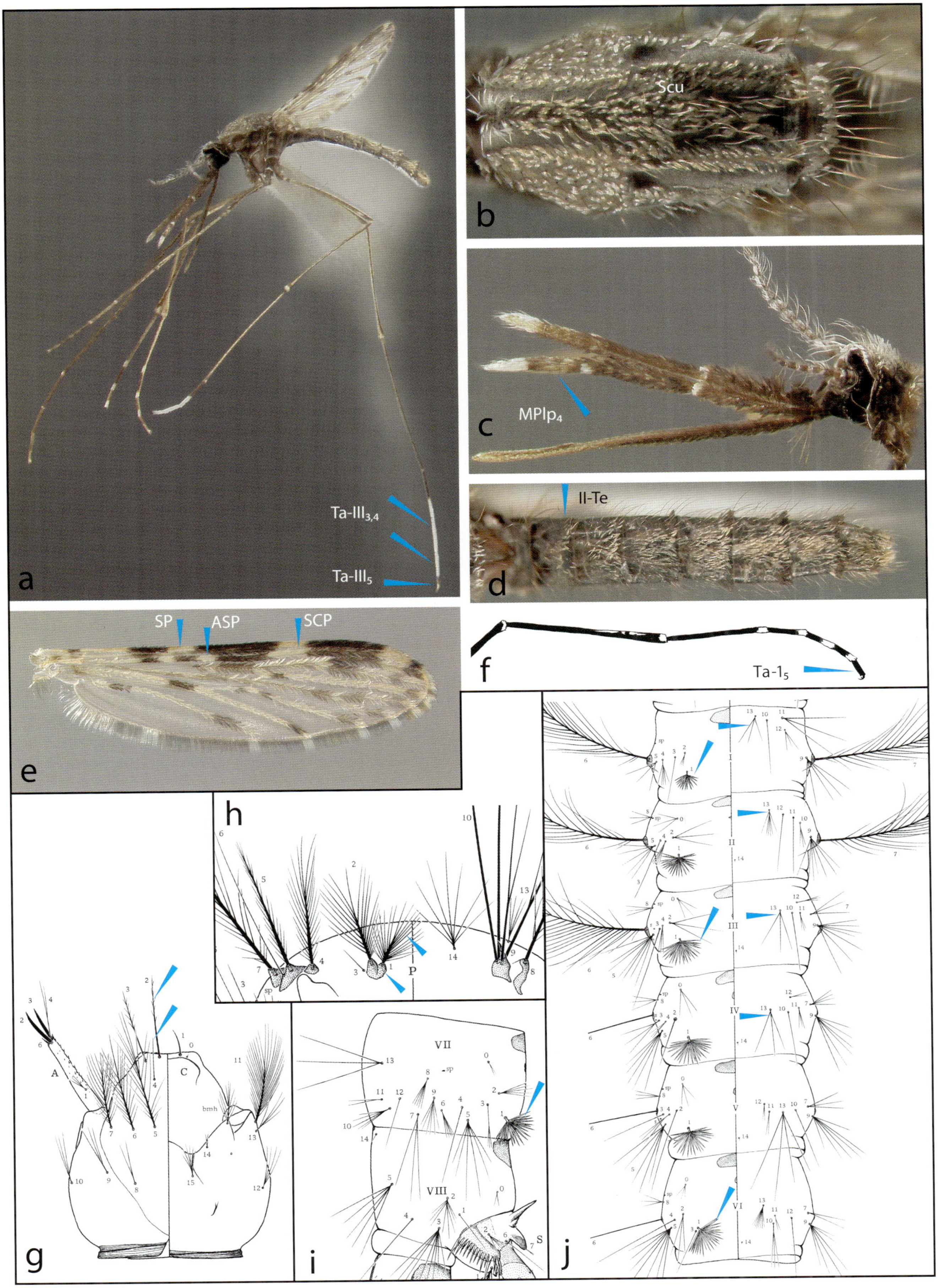
a
Ta-III3,4
Ta-III5
b
Scu
c
MPlp4
d
II-Te
e
SP
ASP
SCP
f
Ta-I5
g
A
C
bmh
h
P
sp
i
VII
VIII
S
j
I
II
III
IV
V
VI

Anopheles (Nyssorhynchus) albitarsis Lynch Arribálzaga, 1878 Subfamily Anophelinae "White-footed Argentinian Nail Mosquito"

Systematics and Distribution

Anopheles albitarsis has long white "socks" on its hind legs (see *a, d*) and scales on the abdominal surface, at least on segments II-VII (see *b*). *Anopheles albitarsis* is the name-bearing species for the Albitarsis Complex, Albitarsis Group, and Albitarsis Series of the Argyritarsis Section of the *Anopheles* subgenus *Nyssorhynchus* (see Vol. 2, p. 1087, for internal relationships). Together with *An. braziliensis*, members of the Albitarsis Complex are easily recognized by a unique pair of white scale stripes on sternum I (see *c*). The Albitarsis Complex comprises five formerly described taxa (molecularly confirmed distributions in parentheses): *An. albitarsis* (northern Argentina, southern Brazil, and Paraguay), *An. deaneorum* Rosa-Freitas (southwestern Brazil), *An. janconnae* Wilkerson & Sallum (northern Brazil), *An. marajoara* Galvão & Damasceno (central Brazil), and *An. oryzalimnetes* Wilkerson & Motoki (Brazil, Paraguay). In addition, several new taxa—*An. albitarsis* species F (Colombia, Trinidad, Venezuela), species G (Brazil), lineage H (Brazil), species I (Colombia, Venezuela), and species J (Brazil)—have been proposed in our laboratory, and additional novel genetic lineages are recognized in *An. marajoara* (lineages 1 and 2) and *An.* sp. nr. *janconnae*, suggesting more work is needed to elucidate all component taxa across its range.

All taxa within the Albitarsis Complex are isomorphic, except *An. deaneorum*, which can be differentiated on immature characters (see *g*). *Anopheles marajoara* is commonly reported in the literature as having a wide distribution across South America, but it appears to have been a catchall for all the cryptic species within *An. albitarsis* s.l. *Anopheles janconnae*, *An. albitarsis* F, and *An. albitarsis* I occur north of the Amazon, whereas all other taxa appear to be restricted to south of the Amazon basin. As DNA approaches are needed to reliably identify all member taxa, the illustrated distribution reflects the distribution of *An. albitarsis* s.l. *Anopheles albitarsis* s.s. is found only in northern Argentina, Paraguay, and southern Brazil.

Bionomics

The Albitarsis Complex is of great epidemiological importance, as three of the five formally described species (*An. deaneorum*, *An. janconnae*, and *An. marajoara*) are highly anthropophilic and endophagic and are proven regional malaria vectors in Brazil. *Anopheles albitarsis* s.s. and *An. oryzalimnetes* are predominantly zoophilic and not implicated in malaria transmission. A spatial ecological and evolutionary variance analysis (SEEVA) showed a high correlation between the distributions of *Anopheles albitarsis* G and H and areas of intensive malaria transmission, suggesting these species may also be competent malaria vectors, at least locally. Given the number of new taxa discovered, the bionomics and vector status of all taxa need to be revisited.

Medical Importance

LMV, SLEV, WEEV; *Plasmodium falciparum*, *P. vivax*.

Diagnostic Characters

(a) Habitus (♀): Palpomere $MPlp_4$ with at least some white or cream-colored scales on mediolateral surface (not shown). **(b) Abdomen (dorsal):** Abdominal terga (Te), at least segments II–VII-Te, covered with scales; posterolateral scale tufts absent from II-Te and often III-Te and IV-Te. **(c) Abdomen (ventral):** Sternum I (I-S) with two longitudinal lines of white scales. **(d) Leg (III):** $Ta\text{-}III_5$ entirely white scaled. **(e) Wing:** Costa (C) without sector pale spot (SP); accessory sector pale spot (ASP) c. 0.4 from wing base, conspicuously closer to sector pale (SP) than to subcostal pale spot (SCP). **(f) Larva (thorax):** Setae 1–3-P usually inserted on common tubercle. **(g) Larva (head):** Seta 3-C with short aciculae (all species except *An. deaneorum*), branched distally in *An. deaneorum*) (not shown); setae 5–7-C plumose. **(h) Larva (abdominal segments III-VI):** Seta 1 palmate leaflets smooth sided; seta 6-IV,V single. Image sources: WRBU; Linthicum (1988).

Exemplar DNA barcodes: *An. albitarsis* s.s. (KJ492407–432), *An. deaneorum* (KJ492739–771), *An. janconnae* (JQ615347–441), *An. marajoara* (KJ011940–999), *An. oryzalimnetes* (KJ492830–94), *An. albitarsis* F (JQ614998–5041), *An. albitarsis* G (JQ615042–146), *An. albitarsis* H (KJ492406–677), *An. albitarsis* I (JQ615189–200).

Catalog entry: Vol. 2, p. 687.

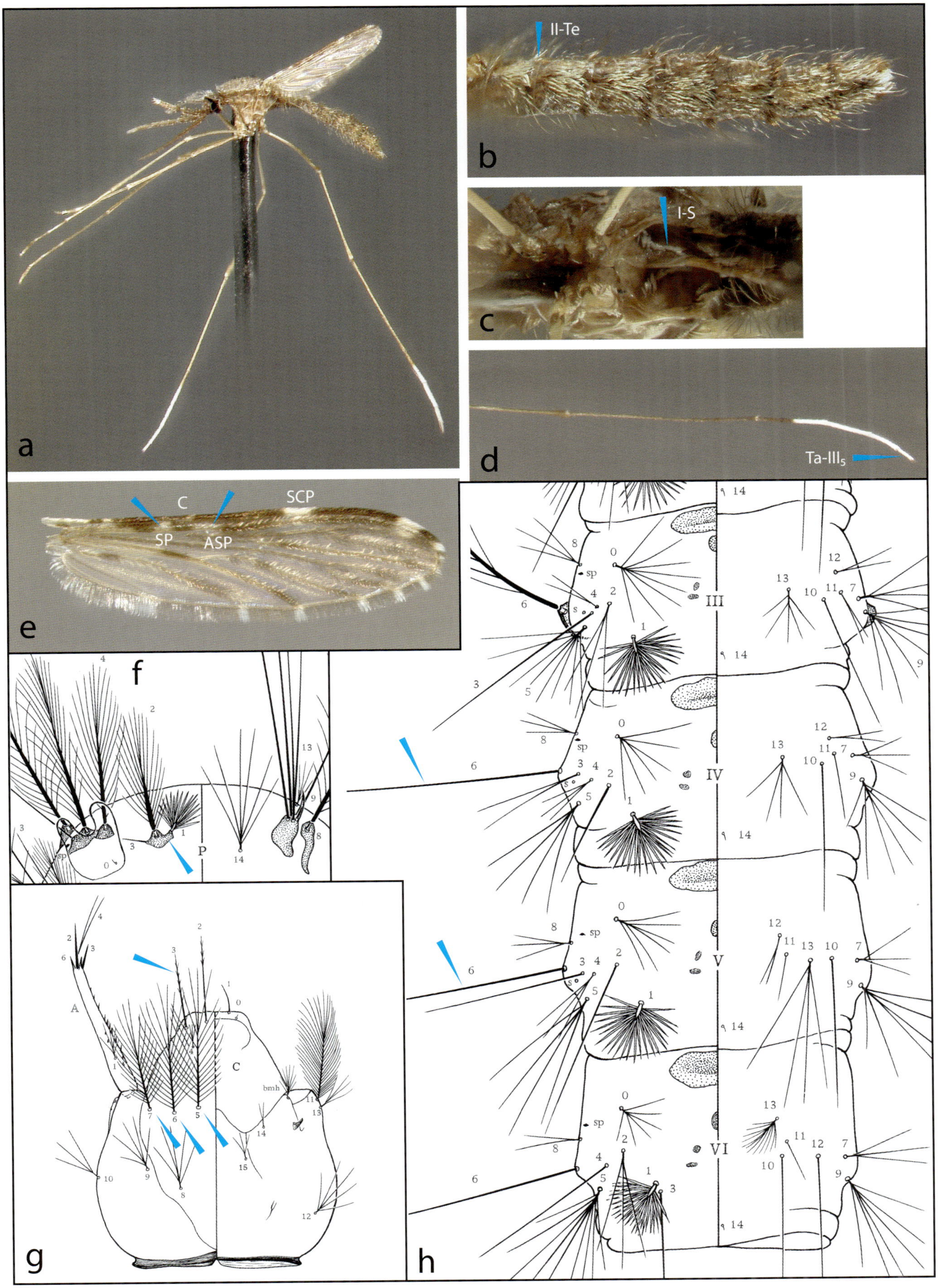
a
b
II-Te
c
I-S
d
Ta-III5
e
C
SCP
SP
ASP
f
P
sp
g
A
C
bmh
h
III
IV
V
VI
sp

Anopheles (Nyssorhynchus) aquasalis Curry, 1932
Subfamily Anophelinae
"Saltwater Panamanian Nail Mosquito"

Systematics and Distribution

Anopheles aquasalis mosquitoes have a somewhat golden glow, as all the scales on the wings and costa are yellowish and never white. *Anopheles aquasalis* is placed in the Oswaldoi Subgroup (Oswadoi Group, Oswaldoi Series, Albimanus Section) of the *Anopheles* genus *Nyssorhynchus* (see Vol. 2, p. 1086, for internal relationships), along with *An. anomalophyllus* Komp, *An. evansae* (Brèthes), *An. galvaoi* Causey, and *An. ininii* Senevet & Abonnenc. There are currently five synonyms of *An. aquasalis*—*emilianus* Komp, *guarujaensis* Ramos, *delta* Anduze, *guarauno* Anduze, and *deltaorinoquensis* Cova García, Pulido & Amarista.

Anopheles aquasalis was originally described from specimens collected in the Canal Zone of Panama and is now known to be widely distributed in the coastal areas of Central and South America (from Mexico to Paraguay), including many islands in the Caribbean.

Bionomics

The name *aquasalis* means "from salt water" and aptly describes the brackish water habitats exploited by ovipositing females and developing immatures. Immature habitats of *An. aquasalis* include coastal brackish-water ground pools and mangrove swamps. However, the species has also been collected from freshwater habitats several kilometers inland, and in the laboratory, *An. aquasalis* females will oviposit in fresh water. Following forced mating of the F_1 generation of *An. aquasalis* collected in Belem, Brazil, the species began to mate freely in cages and has now been successfully colonized. This colony rears successfully in well water (0.04% salinity), whereas another *An. aquasalis* colony established in Rio de Janeiro has been successfully maintained using 10% salt concentrations for larval rearing.

Anopheles aquasalis females are highly anthropophilic and will enter houses to feed and rest. The species plays a significant role in malaria transmission in coastal areas from Venezuela to southeastern Brazil, especially when present in high densities. It is regarded as the primary vector of malaria in Venezuela and the Caribbean islands of Trinidad and Tobago and the Lesser Antilles.

Medical Importance

VEEV; *Plasmodium vivax* VK210 and VK247.

Diagnostic Characters

(a) Habitus (♀): Ta-III_2 with basal dark band 0.3–0.6 times length of segment; Ta-$III_{3,4}$ entirely white; Ta-III_5 with a basal dark band. **(b) Thorax (side):** Postpronotal setae (PeSU) and scales (PScU) absent; pale scales on C-I-III yellowish, not white. **(c) Abdomen (dorsal):** II-Te with dark posterolateral scale tufts. **(d) Thorax (dorsal). (e) Wing:** Pale scales on wing yellowish, not white; humeral pale spot (HP) large, >1.3 times length of prehumeral dark spot (PHD); accessory sector pale spot (ASP) about 0.4 from wing base, conspicuously closer to sector pale (SP) than to subcostal pale spot (SCP). **(f) Head (side):** $MPlp_4$ with at least some white or cream-colored scales on mediolateral surface. **(g) Larva (head):** Seta 1-A short, <2.0 times width of antenna; setae 2,3-C branched, not dendritic; seta 3-C, and usually 2-C, with apical 0.5 plumose, with moderately long to long branches. **(h) Larva (thorax):** Seta 1-P palmate, leaflets smooth sided. **(i) Larva (terminal segments):** Seta 1-X longer than saddle (Sa); seta 1-X at or near ventral margin of Sa. Image sources: WRBU; Faran (1980).

Exemplar DNA sequences: *An. aquasalis* whole genome (GCA_002846955), *COI* (AF548888–902, KT766410–14); whole genome (GCA_002846955).

Catalog entry: Vol. 2, p. 688.

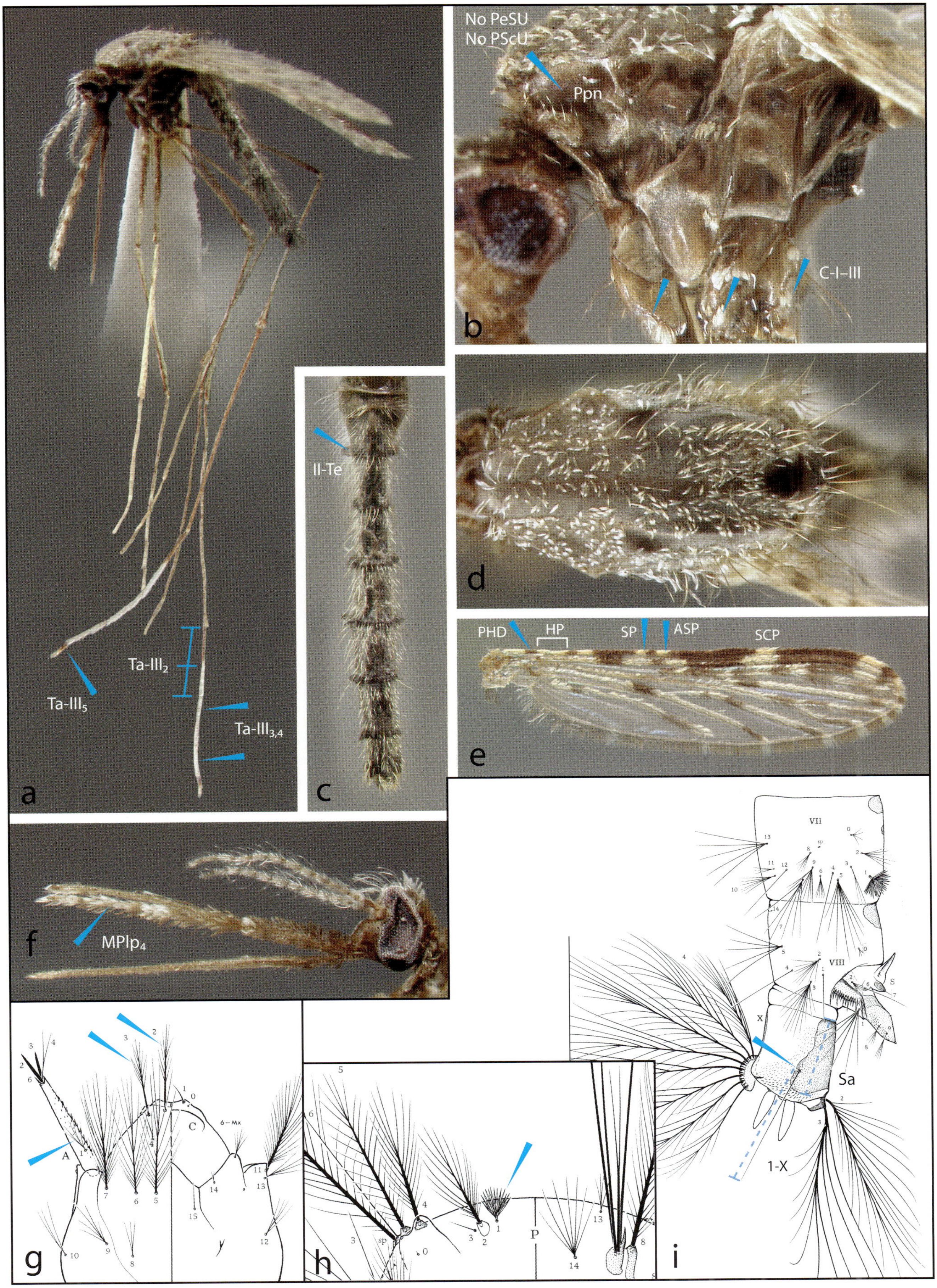
a
Ta-III$_5$
Ta-III$_2$
Ta-III$_{3,4}$
b
No PeSU
No PScU
Ppn
C-I–III
c
II-Te
d
e
PHD
HP
SP
ASP
SCP
f
MPlp$_4$
g
A
C
6-Mx
h
P
i
VII
VIII
X
S
Sa
1-X

Anopheles (*Nyssorhynchus*) *darlingi* Root, 1926 Subfamily Anophelinae "Darling Brazilian Nail Mosquito"

Systematics and Distribution

Anopheles darlingi is one of the best-known Neotropical *Anopheles*, recognized by its distinctive hind leg markings consisting of dark femora, tibia scattered with a few pale scales, and entirely white Ta-$III_{4,5}$ (see *f*). *Anopheles darlingi*, named after Samuel T. Darling, is a primary vector of human malaria in Central and South America and is regarded as the most anthropophilic of all Neotropical *Anopheles*. *Anopheles darlingi* is the monotypic member of the Darlingi Group of the Argyritarsis Series of the *Anopheles* subgenus *Nyssorhynchus* (see Vol. 2, p. 1087, for internal relationships). It is the only *Nyssorhynchus* malaria vector species that has not yet been formally exposed as a species complex, although significantly different nuclear rDNA *ITS2* sequences were noted in a population from Dourado, São Paulo State, in southeastern Brazil. Perhaps not coincidentally, the only valid synonym of *An. darlingi*—*paulistensis* Galvão, Lane & Correa— was originally described from Novo Oriente (near Lussanvira), Bacia do Tietê, São Paulo, Brazil. Its range stretches from southern Mexico to northeast Argentina, with the exception of Nicaragua, Costa Rica, and Panama (see map).

Bionomics

Immature *An. darlingi* is typically found in partially shaded, clear freshwater habitats with emergent or floating vegetation. Representative habitats include debris-dense stream margins, ponds with muddy bottoms, grassy ground pools, and swamps, typically within plantations or cultivated fields, although immatures have also been reported in abandoned gold mines in Venezuela and in brackish water in Belize.

Adult *An. darlingi* mosquitoes are associated with humid lowland, riverine forest environments and are particularly numerous in areas affected by deforestation, human settlements, and agriculture. *Anopheles darlingi* is highly anthropophilic, with higher numbers collected in houses even when bait animals are used. The species is regarded as endophilic, although exophagy has been noted in populations where indoor insecticidal spraying is implemented. The anomalous Dourado population was noted to be exophilic. Irrespective of where their blood meals are taken, *An. darlingi* mosquitoes rest outdoors. Efforts are needed to clarify the taxonomic status of the Dourado population of *An. darlingi*.

Associated Pathogens

Plasmodium falciparum, *P. malariae*, *P. vivax* VK210 and VK247, *Wuchereria bancrofti*.

Diagnostic Characters

(a) Habitus (♀). **(b) Thorax (side):** Anterior mesepimeron with small patch of appressed scales (AMSc); upper mesepimeron without pale scales (UMSc). **(c) Head (side):** Palpomere $MPlp_4$ with at least some white or cream-colored scales on mediolateral surface; $MPlp_5$ all pale. **(d) Abdomen (dorsal):** Terga (Te) with posterolateral scale tufts, covered with scales; I-S bare, rarely with few, sparse scales (not shown). **(e) Wing:** Accessory sector pale spot (ASP) about 0.4 from wing base, conspicuously closer to sector pale (SP) than to subcostal pale spot (SCP); prehumeral dark spot (PHD) well developed, 3–4 times length of the humeral pale spot (HP); vein 1A with fewer than five pale and five dark alternating spots. **(f) Leg (III):** Fe-III and Ti-III dark with sparse pale scales, not speckled; Ta-$III_{4,5}$ entirely white scaled. **(g) Larva (thorax):** Seta 1-P plumose with slender branches. **(h) Larva (head):** Setae 5–7-C plumose. **(i) Larva (terminal segments):** Seta 13-S well developed, long, c. 2.2–2.5 times length of saddle (Sa). **(j) Larva (abdominal segments):** Seta 1 palmate leaflets smooth sided; seta 6-IV–V single. Image sources: WRBU; Linthicum (1988).

Exemplar DNA sequences: *An. darlingi COI* (DQ298209–39, EU918350–62, GQ285644–55, JQ898319–27), whole genome (GCA_000211455).

Catalog entry: Vol. 2, p. 689.

a
No UMSc
AMSc
b
MPlp5
MPlp4
c
PHD
HP
SP
SCP
ASP
1A
e
Te
d
f
Ta-III4,5
Ti-III
Fe-III
g
h
i
j
VII
VIII
Sa
s
x
II
III
IV
V
VI
N. Kitamura

Anopheles (Nyssorhynchus) nuneztovari Gabaldón, 1940 Subfamily Anophelinae "Nuñez-Tovar Venezuelan Nail Mosquito"

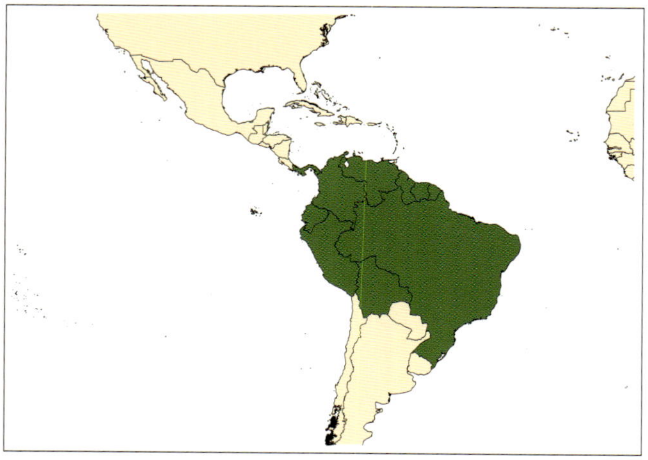

Systematics and Distribution

Anopheles nuneztovari is noted by its long white-ended palps and hind tarsi, broken only by the narrowest of dark bands (see *a*, *c*, *f*), and bushy lateral scale tufts on segments II–VII of the abdomen (see *d*). *Anopheles nuneztovari*, originally described from San Carlos, Cojedes, Venezuela, is the nominotypical member of the Nuneztovari Complex (Oswaldoi Subgroup, Oswaldoi Group, Oswaldoi Series) of the *Anopheles* subgenus *Nyssorhynchus* (see Vol. 2, p. 1086). Female *Nyssorhynchus* species are often difficult to distinguish, with many overlapping taxonomic characters causing difficulty in correctly assigning species identification in host-seeking females—the most common stage collected in biosurveillance surveys. Indeed, the wing morphology of *An. nuneztovari* is highly variable even within a single progeny. As with many other *Nyssorhynchus* malaria vectors, *An. nuneztovari* has been exposed as a species complex on the basis of cytogenetics, isozyme analysis, and molecular systematics approaches. Today the Nuneztovari Complex comprises three formally named species—*An. dunhami* Causey, *An. goeldii* Rozeboom & Gabaldón, and *An. nuneztovari* s.s. and an additional five independently evolving lineages, dubbed *An. nuneztovari* lineages 1–5. There are three known cytotypes—*nuneztovari* A from Brazil and B/C (which represent a single species) from Colombia and Venezuela.

Anopheles nuneztovari s.l. has been reported from Bolivia, Brazil, Colombia, Ecuador, French Guiana, Guyana, Panama, Peru, Suriname, and Venezuela (see map, as s.l.). Molecularly confirmed *An. nuneztovari* s.s. has been reported from Bolivia, Colombia, and Venezuela. Molecularly verified *An. dunhami* and *An. goeldii* are found in Brazil, Colombia, and Peru, and Brazil and Suriname, respectively.

Bionomics

As in other emergent species complexes, there are inherent difficulties associated with correctly assigning distributions and bionomic information, vector competence, biting behaviors, and distribution of component taxa. *Anopheles nuneztovari* s.s. (cytotype B/C) is a competent malaria vector in Bolivia, Colombia, and Venezuela and bites late in the evening, peaking between 2200h and 0200h and through the night. *Anopheles nuneztovari* A is thought to be primarily zoophilic and is not incriminated in human malaria transmission. Its peak biting activity is between 1800h and 2000h. Although *An. nuneztovari* s.s. bites indoors, other populations of *An. nuneztovari* (as s.l.) have been noted to rest outdoors.

The immature habitats of *An. nuneztovari* s.l. include lakes, lagoons, rivers, flooded gold mines, fish ponds, ground pools, tire ruts, and animal hoofprints. *Anopheles nuneztovari* s.l. immatures have been recovered in still and flowing aquatic habitats with fresh, clear water with plentiful floating and/or emergent vegetation.

Associated Pathogens

Plasmodium falciparum, *P. vivax*

Diagnostic Characters

(a) Habitus (♀). (b) Thorax (dorsal): Prescutellar (PrA) area with distinct small dark spot, never entirely covering prescutellar area. **(c) Head (side):** Palpomere $MPlp_4$ with white or cream-colored mediolateral scales but no yellow scales. **(d) Abdomen (dorsal):** At least II–VII-Te covered with scales and with posterolateral dark scale tufts. **(e) Wing:** Pale wing scales yellowish, at least on anterior veins; accessory sector pale spot (ASP) about 0.4 from wing base, conspicuously closer to sector pale spot (SP) than to subcostal pale spot (SCP); SCP on costa <0.5 times length of sector dark spot (SD); vein M stem predominantly pale scaled. **(f) Leg (III):** Fe-III and Ti-III dark with sparse pale scales, not speckled; $Ta\text{-}III_2$ with basal dark band ≥0.25 length of segment; $Ta\text{-}III_3$ all white; $Ta\text{-}III_5$ with basal dark band. **(g) Larva (thorax):** Seta 1-P with narrow lanceolate leaflets; setae 1–3-P not inserted on common tubercle. **(h) Larva (head):** Setae 2,3-C aciculate, sometimes apically branched; seta 4-C 0.3–0.6 times length of seta 3-C; seta 4-C usually single or double, moderately long, usually extending beyond the base of 2-C; setae 5–7-C plumose. **(i) Larva (terminal segments). (j) Larva (abdominal segments I-VI):** Seta 0-II 5–8 branched, subequal or longer than length of leaflets of seta 1-II; seta 1 palmate leaflets smooth sided; seta 6-IV,V single; seta 11-I very long, 2–4 branched; seta 13-I small, usually >3 branches. Images sources: WRBU; Faran (1980).

Exemplar DNA sequences: *An. nuneztovari* s.l. *ITS2* (AY028121, AY028109, AY028128).

Catalog entry: Vol. 2, p. 691.

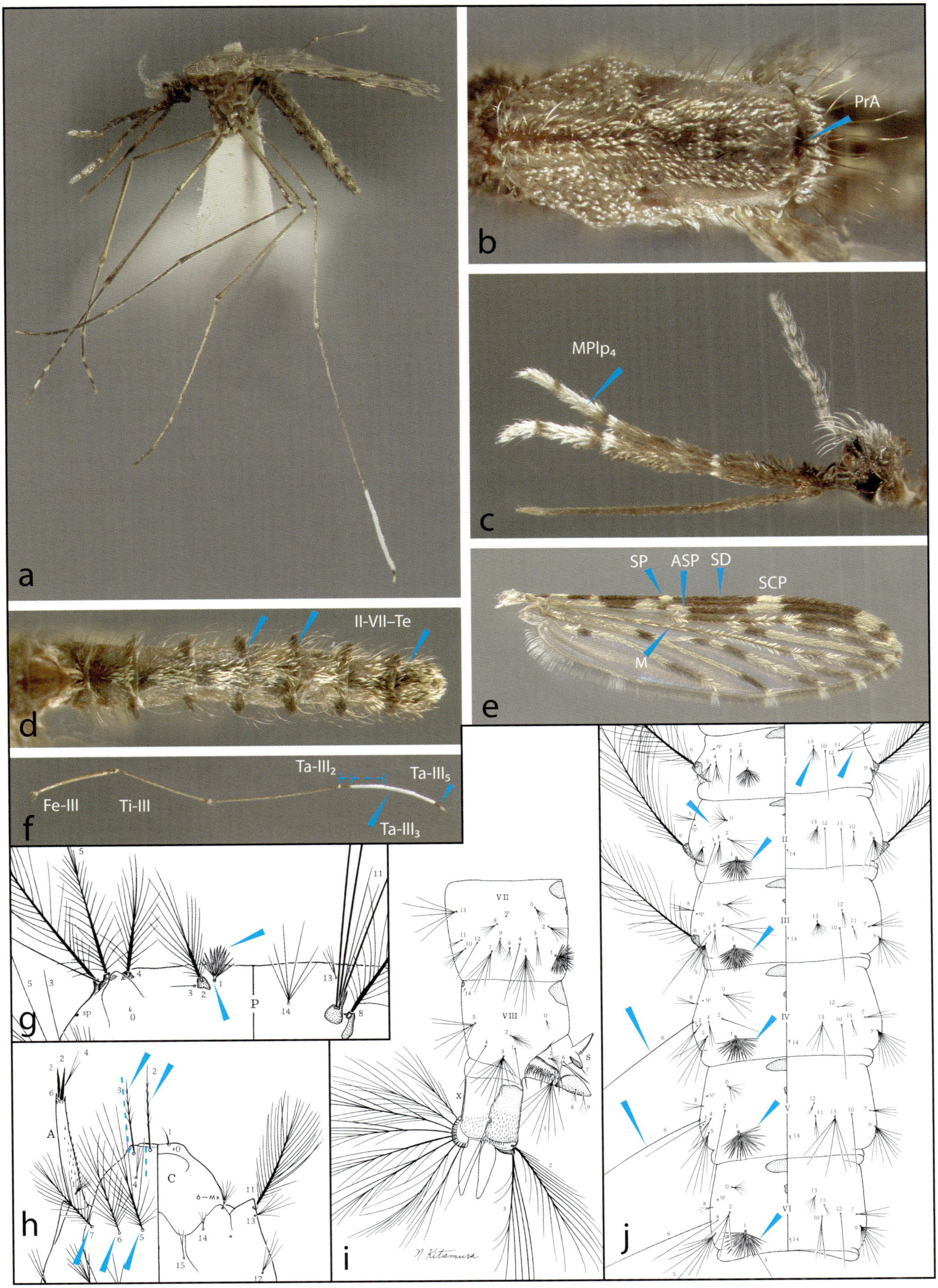
a
b
PrA
c
MPlp$_4$
d
II-VII–Te
e
SP
ASP
SD
SCP
M
f
Fe-III
Ti-III
Ta-III$_2$
Ta-III$_5$
Ta-III$_3$
g
h
i
N. Kitamura
j

Anopheles (*Nyssorhynchus*) *triannulatus* (Neiva & Pinto, 1922)
Subfamily Anophelinae
"Three-ringed Brazilian Nail Mosquito"

Systematics and Distribution

Anopheles triannulatus is a small mosquito, with three distinctive palpal bands and long white "socks" on its hind legs, broken only by a dark basal band on Ta-III_5. *Anopheles triannulatus* is the nominotypical species of the Triannulatus Group (Oswaldoi Series, Albimanus Section) of the *Anopheles* subgenus *Nyssorhynchus* (see Vol. 2, p. 1087, for internal relationships). Based on allozyme and RAPD markers of sympatric *An. triannulatus* s.l. from Salobra, Mato Grosso do Sul State in Brazil, the Triannulatus Group now comprises *An. triannulatus*, *An. halophylus* Silva-do-Nascimento & Lourenço-de-Oliveira, and *An. triannulatus* C. Females of all three taxa are essentially isomorphic, but reliable egg, larva, and male genitalia characters exist to differentiate *An. triannulatus* and *An. halophylus*. *Anopheles triannulatus* C appears to share intermediate, sharing morphological characters of both species. Five synonyms exist for *An. triannulatus*: *bachmanni* Petrocchi, *chagasi* Galvão, *cuyabensis* Neiva & Pinto, *davisi* Paterson & Shannon, and *perezi* Shannon & Del Ponte. *Anopheles triannulatus* C is more closely related to *An. halophylus* and does not appear to correspond morphologically to any of the current synonyms. Two distinct lineages of *An. triannulatus* s.l. have been reported from Northwest and Southeast Colombia using mtDNA *COI* and nuclear rDNA *ITS*2 sequences. Form I is found in both locations, whereas form II is apparently restricted to southeastern Colombia. It is unclear how these forms correspond to the currently defined taxa in the Triannulatus Group.

Anopheles triannulatus s.s. was first described from specimens collected at Rio Cuiabá, Fazenda Sao João, Mato Grosso, Brazil. It is a common Neotropical forest mosquito, with a wide distribution from Argentina and Paraguay in South America north to Nicaragua, and is reported from Trinidad and Tobago and Puerto Rico in the Caribbean. The distribution of *An. triannulatus* s.l. is illustrated (see map).

Bionomics

Adult *An. triannulatus* s.l. mosquitoes are closely associated with forest environments and bite early in the night (1800h–2100h). In Gamboa, a rain forest settlement in the Canal Zone, Panama, one author recalls overwhelming densities of savage *An. triannulatus* s.l. biting us at the edge of a large lake with plentiful water lettuce, as soon as the sun went down. In Trinidad, females are readily collected at human bait. In Amapá, Brazil, the species appears zoophilic and is rarely captured biting humans. Blood-meal analysis of 154 *An. triannulatus* from the Matapí River showed high zoophilic feeding preferences for pigs and cattle. Only 2% tested positive for human blood. Mixed blood feeds were common (15.6%).

Immature *An. triannulatus* s.s. is found in freshwater lakes, ponds, or large ground pools, especially in or around the crowns of water lettuce (*Pistia stratiotes* Linnaeus). In Upper Orinoco, Venezuela, *An. triannulatus* s.l. immatures were most abundant in lagoons with submerged aquatic plants and sparse emergent grasses. *Anopheles halophylus* breeds exclusively in brackish water habitats in Salobra, Brazil.

Associated Pathogens

Breu Branco virus, LMV; *Plasmodium falciparum*, *P. vivax*.

Diagnostic Characters

(a) Habitus (♀). (b) Thorax (side): Anterior mesepimeron with a patch of small white scales (AMSc). **(c) Head (side):** Palpomere $MPlp_4$ with at least some white or cream-colored mediolateral scales but no yellow scales. **(d) Wing:** Accessory sector pale spot (ASP) about 0.4 from wing base, conspicuously closer to sector pale (SP) than to subcostal pale spot (SCP); prehumeral dark spot (PHD) relatively long, extending to the humeral crossvein (h); subcostal pale spot (SCP) small; vein 1A with fewer than five pale and five dark alternating spots. **(e) Abdomen (dorsal):** Terga (Te) covered with scales; II-Te with posterolateral scale tufts. **(f) Leg (III):** Fe-III and Ti-III dark with sparse pale scales, not speckled; Ta-III_{3-4} white; Ta-III_5 with basal dark band. **(g) Larva (thorax):** Seta 1-P with narrow lanceolate leaflets; setae 1–3-P not inserted on common tubercle. **(h) Larva (head):** Setae 2-C well-separated, clypeal index <2.5; setae 2,3-C single or aciculate, sometimes apically branched; setae 5–7-C plumose. **(i) Larva (terminal segments):** Median plate (MdP) with long lateral arms. **(j) Larva (abdominal segments):** Seta 1 palmate, leaflets smooth sided; seta 1-I well developed palmate; seta 11-I long, 3–7 branched; seta 13-I long, 2–4 branched; seta 6-IV,V single. Image sources: WRBU; Faran (1980).

Exemplar DNA sequences: *An. triannulatus* s.s. *COI* (JN085964–6138, KR909593–665); *An. halophylus COI* (JN086076, JN086121), *ITS*2 (JF973007–08); *An. triannulatus* C *COI* (JN086075, JN086077).

Catalog entry: Vol. 2, p. 692.

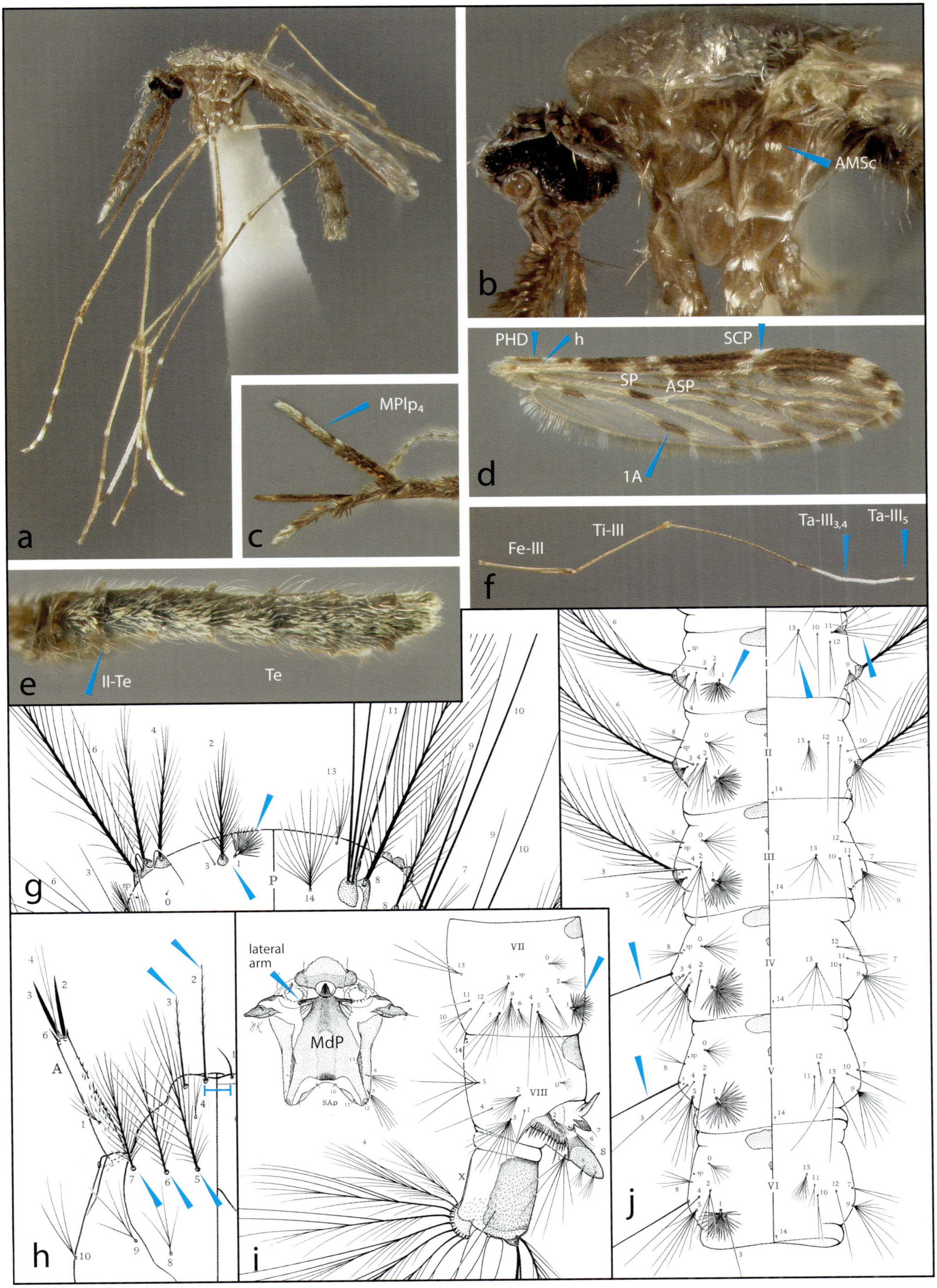
a
b
AMSc
c
MPlp4
d
PHD
h
SCP
SP
ASP
1A
e
II-Te
Te
f
Fe-III
Ti-III
Ta-III3,4
Ta-III5
g
P
h
A
i
lateral arm
MdP
SAp
VII
VIII
X
S
j
I
II
III
IV
V
VI

Aedes (Aedimorphus) cumminsii (Theobald, 1903) Subfamily Culicinae, Tribe Aedini "Cummin Sudanese Pointy Mosquito"

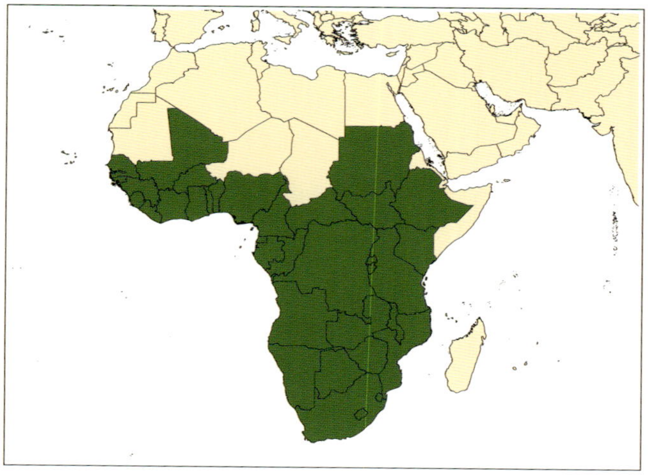

Systematics and Distribution

Aedes cumminsii is widely spread across the African Continent south of the Sahara Desert. It is one of seven of the 22 species in the subgenus *Aedimorphus* in Africa. *Aedes cumminsii*, named after Stephenson Lyle Cummin, was originally described from the Central African country now recognized as South Sudan. *Aedes cumminsii* has one valid subspecies—ssp. *mesostictus* Harbach—which was originally described from Ashanti in Ghana. This taxon has been placed into synonomy with and elevated from synonomy with *Ae. cumminsii* several times, suggesting additional studies are needed to establish its taxonomic status. *Aedes cumminsii* belongs to the informal Dentatus Group, which consists of five species, of which *Ae. cumminsii* is the most abundant in South Africa.

Bionomics

Given its extensive distribution and medical importance, relatively little information is available on the biology of this species. The females feed preferentially on cattle, but also attack humans. Like other members of the subgenus, *Ae. cumminsii* is a floodwater mosquito, depositing its eggs in the soil of depressions above the current waterline in anticipation of future flooding. The eggs are deposited in floodplain depressions as well as poorly drained depressions in grasslands called "dambos," and larvae have been collected in temporary turbid water pools at the edges of marshes. Rain that floods the site tends to hatch an entire cohort of eggs, so this species occurs in larger numbers after heavy rains, and adults may be absent between flooding events.

Associated Pathogens

MIDV, NKOV, RVFV, SHOV, SINV, SPOV.

Diagnostic Characters

(a) Habitus (♀): Ti-III with apical yellow spot. **(b) Thorax (side):** Postspiracular (PoSc) and subspiracular (SSc) scale patches present. **(c) Head:** Proboscis (P) entirely dark scaled. **(d) Thorax (dorsal):** Scutum (Scu) indistinctly marked; scutellum (Stm) with narrow yellow scales. **(e) Abdomen (dorsal). (f) Larva (head):** Integument spiculate. **(g) Larva (terminal segments):** Comb with spinelike comb scales (CS); seta 1-S small, inserted basal to distal-most of c. 14 pecten (Pt); pecten spines (PS) closely spaced basally, three distal-most widely spaced; siphon index ≥4; c. six precratal setae (PcrS); tiny chitinous plates on integument of larval thorax and abdomen. Image sources: WRBU; Hopkins (1952).

Exemplar DNA barcodes: *Ae. cumminsii COI* (MK300225, MG232546, MG232440, KU187001).

Catalog entry: Vol. 2, p. 703.

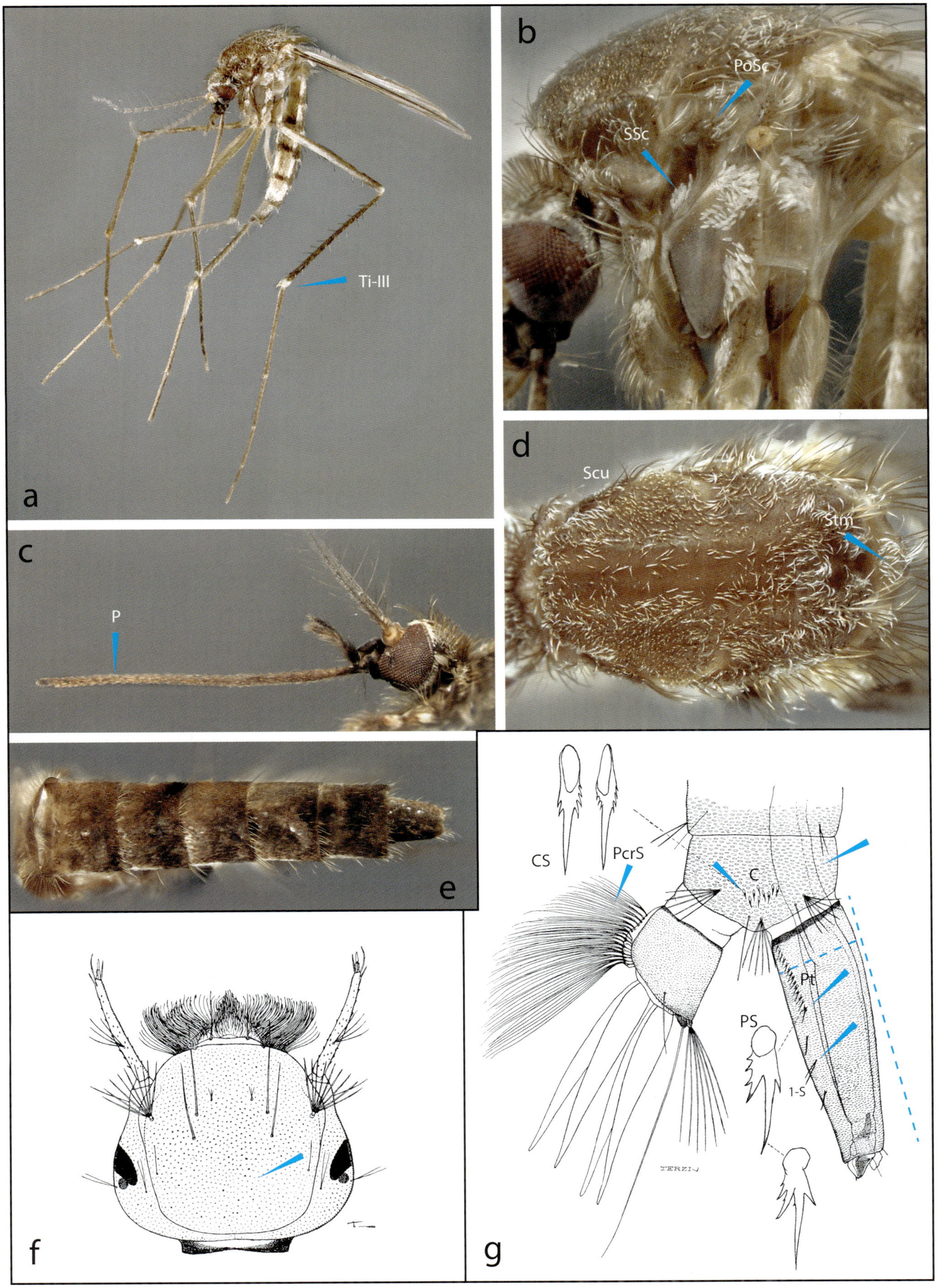
a
Ti-III
b
PoSc
SSc
c
P
d
Scu
Stm
e
f
g
CS
PcrS
C
Pt
PS
1-S

Aedes (Aedimorphus) ochraceus (Theobald, 1901) Subfamily Culicinae, Tribe Aedini "Yellow Zimbabwean Pointy Mosquito"

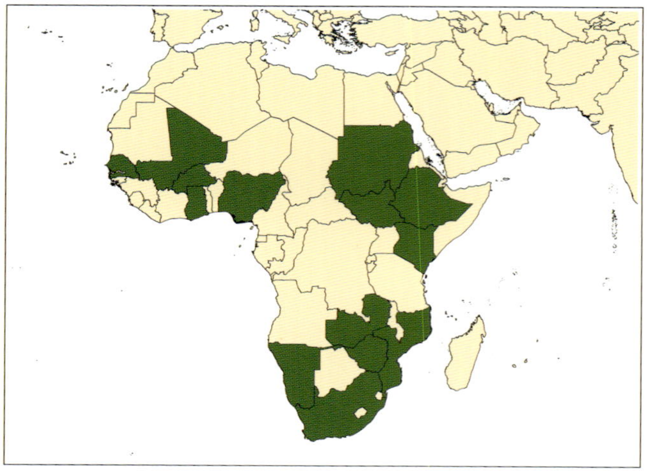

Systematics and Distribution

Originally described from what is now Harare in Zimbabwe, *Aedes ochraceus* is a morphologically unique species that was previously classified as the monotypic member of the Ochraceus Group, which occurs in African countries south of the Sahelian ecological zone to South Africa. It is present in both western and eastern Africa but appears much more abundant in the east of the continent. Population genetics studies indicated that the species was the result of a relatively new, single introduction into Kenya, sharing close genetic identity with a Rift Valley fever virus (RVFV)-competent population in Senegal. This mosquito is considered highly recognizable, with extremely distinctive adult characteristics and an elongated siphon in the larval stage.

Bionomics

Aedes ochraceus is a floodwater mosquito, and seeks out wooded savannah habitats where it directly oviposits its eggs into the surface layer of the soil, in places likely to flood in the future. Transovarian viral infections of RVFV have been reported in *Ae. ochraceus*. During the 2006–2007 RVFV outbreak in Garissa, Kenya, *Ae. ochraceus* was by far the dominant species in all catches (76%). Females were shown to feed preferentially on goats (37.6%), cattle (16.4%), donkeys (10.7%), and sheep, but they will also feed on humans (5.3%). At high densities, this species can become a biting nuisance to people. Studies undertaken in coastal Kenya found both male and female *Ae. ochraceus* heavily infected with CHIKV, suggestive of transovarial or possible sexual transmission of the virus. The species was also field collected positive for dengue virus (DENV) strains 1–3, indicating that this species is of high medical importance.

Associated Pathogens

CHIK, DENV(1–3), RVFV, WSLV.

Diagnostic Characters

(a) Habitus (♀). (b) Thorax (dorsal): Scutum (Scu) with stripes of yellow and brown scales. **(c) Wing:** Costa (C) yellow scaled. **(d) Legs:** Tarsomeres lack distinct pale bands; all tibia and femura with longitudinal yellow stripes bordered by dark stripes. **(e) Larva (head):** Setae 5-7-C plumose, each c. eight branched. **(f) Larva (terminal segments):** Comb (C) with comb scales (CS) basally fringed, in irregular patch of 8–11; seta 1-S small, inserted c. 0.67 from base; pecten (Pt) with 13–18 spines (PS), widely spaced distally; siphon(S) long, index 6.0–8.0. Image sources: WRBU; Hopkins (1952).

Exemplar DNA barcodes: *Ae. ochraceus COI* (KU187000–01).

Catalog entry: Vol. 2, p. 706.

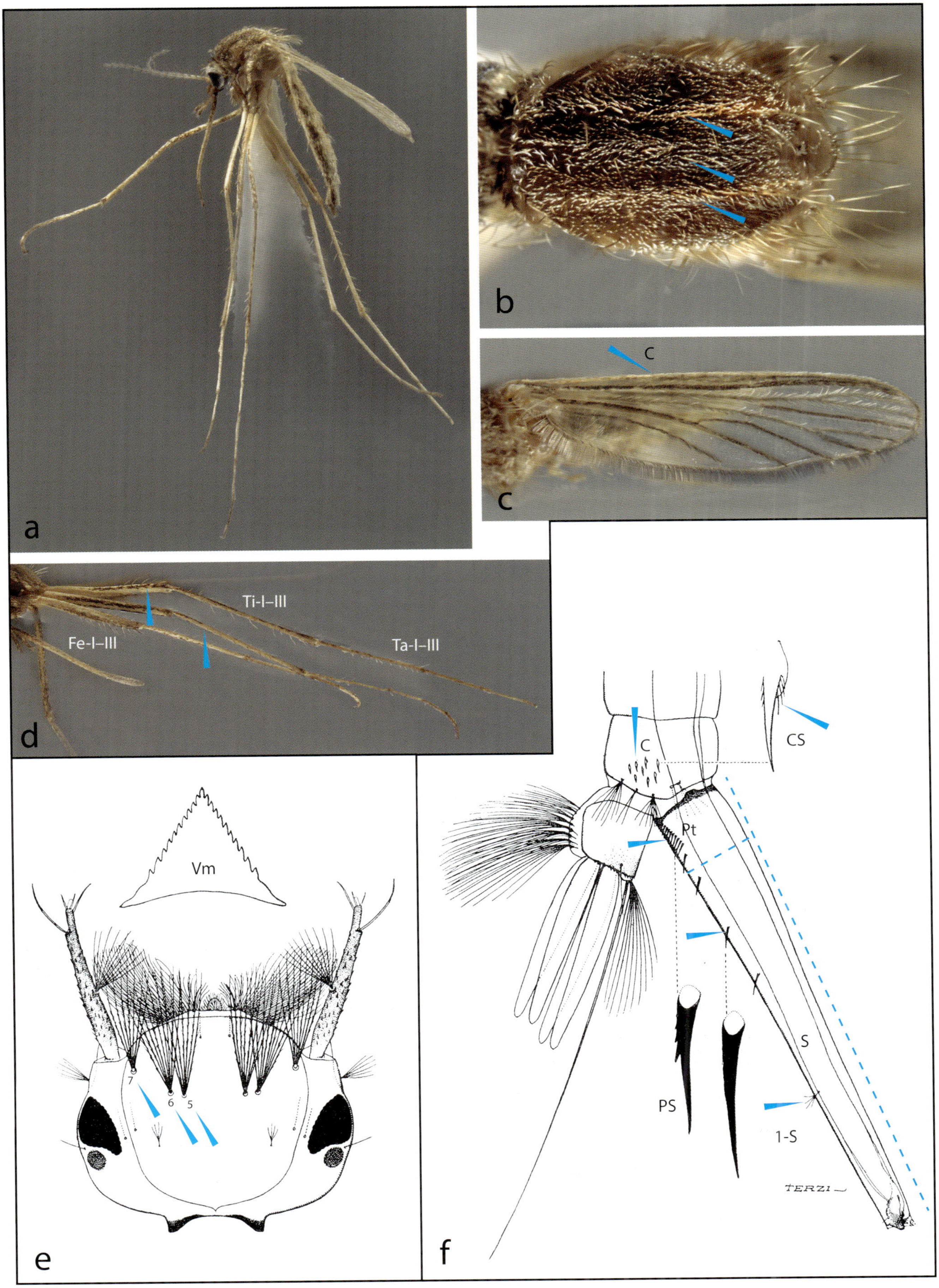
a
b
C
c
Ti-I–III
Fe-I–III
Ta-I–III
d
Vm
7
6
5
e
C
CS
Pt
S
PS
1-S
TERZI
f

Aedes (Aedimorphus) vexans (Meigen, 1830)
Subfamily Culicinae, Tribe Aedini
"Vexing German Pointy Mosquito"

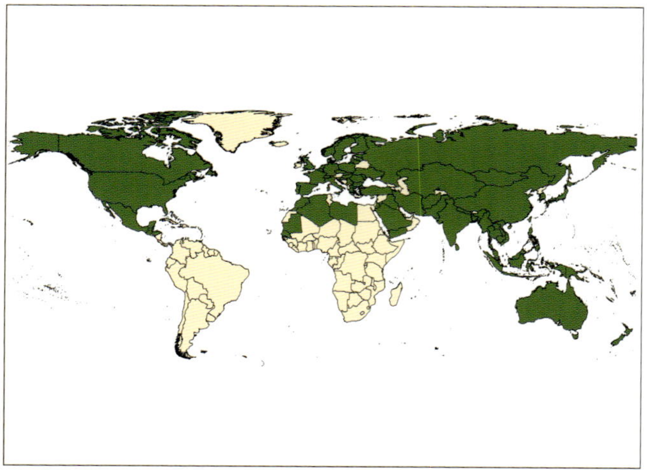

Systematics and Distribution

Aedes vexans is a common, medium-sized floodwater mosquito. It is one of the most widely distributed species in the world and the only member of the subgenus *Aedimorphus* found in the Nearctic and Neotropics. Morphological studies revealed three geographically structured subspecies—ssp. *arabiensis* (Patton) in Africa, ssp. *nipponii* (Theobald) in Southeast Asia, and ssp. *nocturnus* (Theobald) in the Australasian region—suggesting that *Ae. vexans* comprises cryptic taxa across its range. DNA barcoding efforts indicate the separate species status of the topotypic *Aedes vexans* from Germany from those in North America and Korea (=ssp. *nipponii*). This is of significance, as different populations are incriminated as vectors of different arboviruses in different geographical parts of the species' range.

Bionomics

Aedes vexans overwinters as eggs in temperate climates. The species is multivoltine, with full development from egg to adult lasting 8–10 days. Long-living adults at least 113 days old have been detected; thus, populations can become quite large under optimal climatic conditions. Extended longevity plays a critical role in the acquisition and transmission of pathogens. Mating occurs in swarms, and copulation occurs on the wing. In Canadian populations, egg batches were large, with 108–182 eggs. *Aedes vexans* eggs are deposited in the soil, and eggs can survive desiccation and weather extremes for up to three years. Larvae can be found in mainly sunlit habitats, including flood pools on prairies and partially wooded floodplains, ditches, swamps, rice fields, storm water facilities, dredge spoil sites, ditches, drying bogs, woodland pools, elephant footprints, and, rarely, in containers. Habitats usually have little aquatic vegetation or algae.

Blood feeding begins the second day after emergence in *Ae. vexans*, and these ready biters are nondiscriminatory, feeding on any available hosts. *Aedes vexans* disperses considerable distances (5–15 miles) from its immature habitats in search of hosts, and thus nuisance from this species can be quite widespread. Distinctive for a freshwater species, *Ae. vexans* females sometimes undertake migratory flights that, aided by weather systems, can distribute individuals hundreds of miles from their eclosion sites. As its name suggests, *Ae. vexans* commonly reaches levels of abundance that create significant biting annoyance and harassment for humans and cattle, its primary hosts. It is the dominant biting mosquito in much of the northern plains of the United States and Canada, as well in the Rhine valley of Germany. *Aedes vexans* females are readily attracted to and collected in light traps.

Given the illumination of distinct species within the *Ae. vexans* species complex, efforts are now needed to tease out the true biology and vector competency of each component taxon.

Associated Pathogens

BAV, BATV, CHAOV, EEEV, GETV, JBEV, JCV, KEYV, LCV, NEGV, POTV, SFV, SLEV, TAHV, TVTV, VRV, VEEV, WEEV, WNV, ZIKV; *Dirofilaria immitis*.

Diagnostic Characters

(a) Habitus (♀). (b) Head (side): Proboscis (P) brown with numerous white scales ventrally; vertex (V) with median narrow scales. **(c) Thorax (dorsal):** Scutellum (Stm) with pale, narrow scales. **(d) Legs:** Anterior surfaces of Fe-II,III speckled with pale scales. **(e) Legs:** Ta-II,III with basal white bands. **(f) Wing:** No pale fringe scales along posterior margin; wing scales sparse, dark, and pale (indistinct in photo). **(g) Larva (head):** Dorsal apotome (DAp) with median spiculose area; seta 4-C anterior to 5-C; seta 6-C single or double; seta 8-C single. **(h) Larva (terminal segments):** Comb (C) with spinelike comb scales (CS), basally fringed and in irregular row. Image sources: WRBU; Reinert (1973).

Exemplar DNA barcodes: *Ae. vexans COI* (AF253041–47, KT358460–62, KY694972–77).

Catalog entry: Vol. 2, p. 708.

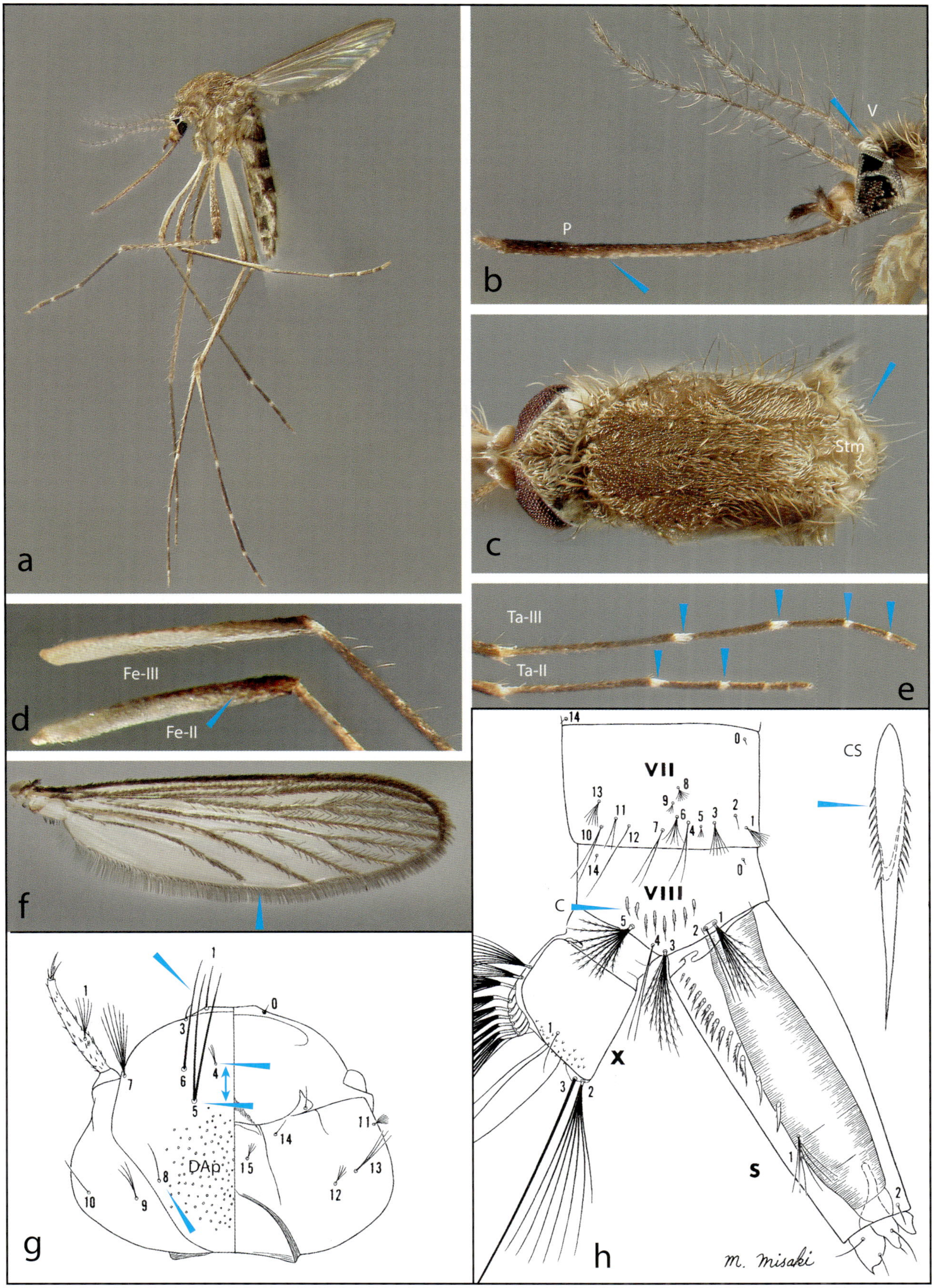
a
V
P
b
Stm
c
Fe-III
Fe-II
d
Ta-III
Ta-II
e
f
DAp
g
VII
VIII
CS
C
x
s
h
M. Misaki

Aedes (Dahliana) geniculatus (Olivier, 1791)
Subfamily Culicinae, Tribe Aedini
"White-kneed French Pointy Mosquito"

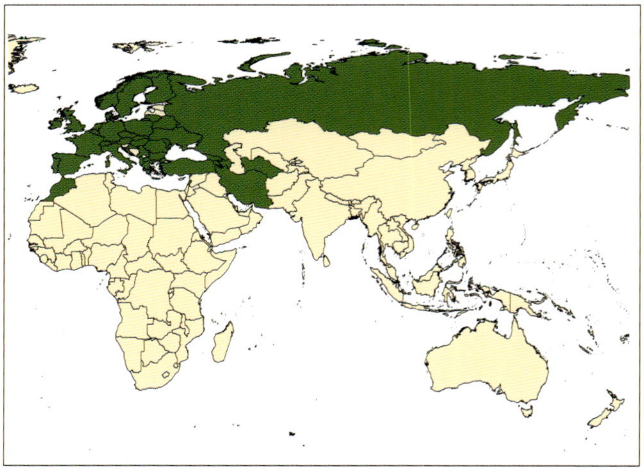

Systematics and Distribution

In nature, *Aedes geniculatus* is easily recognized as strikingly ornamented, with golden-yellow, jet-black, and white markings (Note: the markings in the species plate are notably dulled, being from an archive specimen). Originally described from Paris, France, the species is dendrolimnic and thus associated with deciduous forests across Europe (from England to Turkey), the Caucasus, and North Africa. The ecology of the species, but not its phylogeny, is similar to that of *Ae. triseriatus* (Say) in North America. The greater abundance of tree-hole-developing aedines in the Americas compared to Europe suggests that disruptions from the most recent glaciation were more severe in Europe, possibly due to the narrower area of refugia south of the ice sheet. It is closely related to two other Palearctic species: *Ae. echinus* (Edwards) and *Ae. gilcolladoi* (Sánchez-Covisa, Rodríguez Rodríguez & Guillén Llera).

Bionomics

Aedes geniculatus bites during the day. When abundant, as in southeastern Europe, it can be an annoying pest of humans. *Aedes geniculatus* lays batches of about 70 eggs in the mid-afternoon in tree holes of various kinds and heights. It prefers tree holes up to 4 m high in broadleaf trees like birch, alder, and nonnative maple, and is also found in oaks. Larvae have occasionally been recovered from small ground pools. In England, adult emergence occurs from May to September, with highest biting densities, and seasonal oviposition peaks during the warmest months of July and August. Few eggs hatch in the same year they are laid, with most eggs surviving the winter to hatch the following spring, suggesting that the northerly populations of this species are univoltine. In warmer climes, *Ae. geniculatus* will overwinter both as eggs and as larvae.

Associated Pathogens

CHIK, EEEV, WEEV, WNV, YFV; *Ascogregarina geniculati*, *Dirofilaria immitis*, *Dirofilaria repens*.

Diagnostic Characters

(a) Habitus (♀): Proboscis (P) distinctly longer than Fe-I; Ti-I–III and Ta-I–III all dark; Fe-I–III with apical pale spots. **(b) Genitalia (♀):** Cercus (Ce) short, blunt. **(c) Head (side):** Proboscis (P) and palpus (MPlp) black scaled. **(d) Thorax (dorsal):** Scutum (Scu) with distinctive white and blackish pattern dominated by two dorsocentral blackish stripes; scutellar (Stm) scales narrow, yellowish. **(e) Thorax (side):** Pleuron with patches of broad, whitish scales. **(f) Abdomen (dorsal):** Distinct lateral triangular white patches dorsally; dark scales with purplish sheen (indistinct in photo). **(g) Larva (head):** Antenna (A) without spicules, 0.5 times length of head; seta 1-A single. **(h) Larva (abdominal segments):** Numerous stellate setae on thorax and abdomen; branches of stellate setae c. same length as segment. **(i) Larva (terminal segments):** Elongate comb scales (CS) in a single row of 8–15; CS with strong apical spinules and small lateral spinules; pecten (Pt) with 15–19 long spines (PS); seta 1-S 4,5 branched; seta 4-X with 7–10 setae on grid and 1–2 precratal setae (PcrS); siphon index 2.3–3.2; acus (SA) present on base of siphon. Image sources: WRBU; Seguy (1924); Marshall (1938); Reinert (1973); Reinert (2002); Becker (2010).

Exemplar DNA barcodes: *Ae. geniculatus COI* (KM457576–78, KP942765–68, KP942753–55, MH392201–05).

Catalog entry: Vol. 2, p. 719.

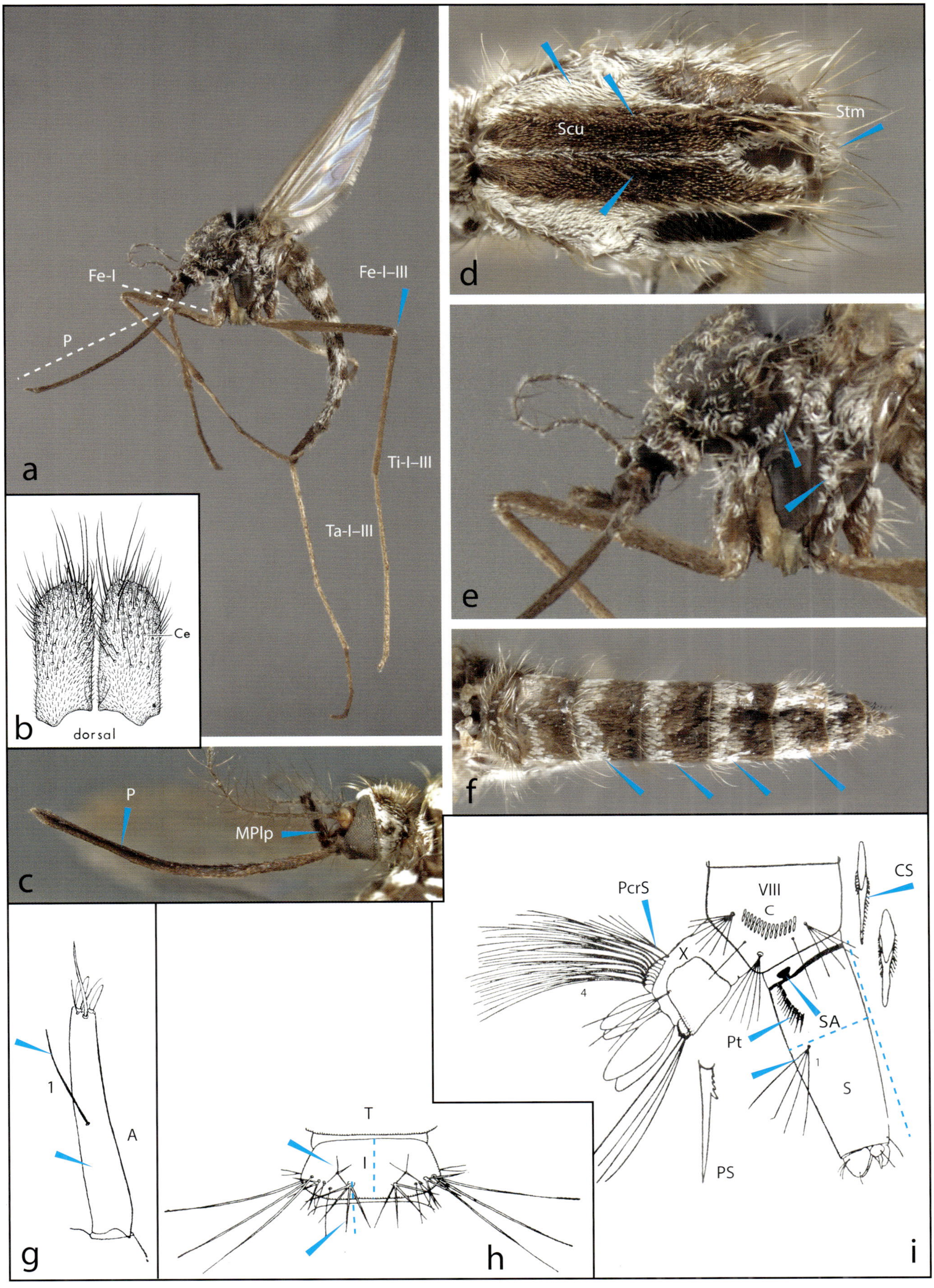

Fe-I
Fe-I–III
P
a
Ti-I–III
Ta-I–III
Ce
b
dorsal
Scu
Stm
d
e
f
P
MPlp
c
1
A
g
T
I
h
PcrS
VIII
C
CS
X
4
SA
Pt
1
S
PS
i

Aedes (Diceromyia) furcifer (Edwards, 1913) and *Ae. (Dic.) taylori* Edwards, 1936
Subfamily Culicinae, Tribe Aedini
Ae. furcifer–"Taylor Nigerian Pointy Mosquito" *Ae. taylori*–"Forked Sudanese Pointy Mosquito"

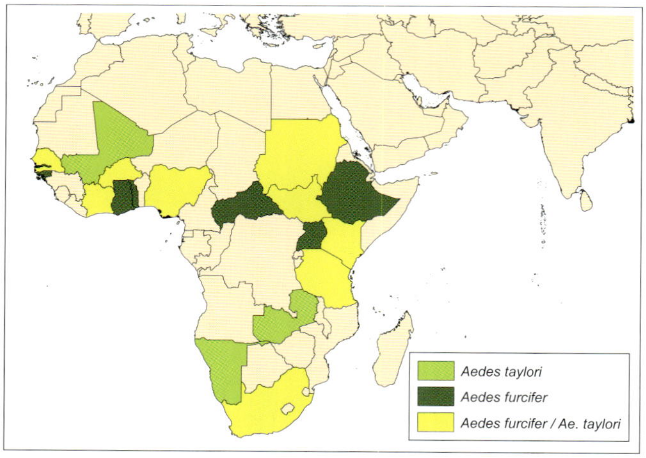

Systematics and Distribution

Aedes furcifer is most closely related to *Ae. taylori*, and *Ae. cordellieri* Huang. These three species are most reliably differentiated on characters of the male genitalia, although, purportedly diagnostic differences for adult female *Ae. furcifer* and *Ae. taylori* are noted in the abundance and patterning of the abdominal scales (see *e*, *f*). Populations previously referred to as *Ae. taylori* in Kenya, Natal Province (South Africa), Tanzania, and Uganda were subsequently recognized as *Ae. cordellieri*. The true identity of the "*Ae. taylori*" populations reportedly responsible for arbovirus transmission in Sudan and Zimbabwe still needs to be verified. *Aedes furcifer* and *Ae. taylori* occur across sub-Saharan Africa, including the Sahelian region (see map).

Bionomics

Due to their close morphological similarity, *Ae. furcifer* and *Ae. taylori* were often not differentiated in early ecological and biosurveillance studies, and thus some aspects of their true distributions, ecology, and vector capacities remain intertwined. Most reports cite "*furcifer-taylori*," making no effort to differentiate the species. *Aedes furcifer* and *Ae. taylori* are closely associated with Afrotropical forest and forest edge habitats, where they feed on monkeys in the canopy and readily go down to ground level to feed on passing persons or baboons, making them important zoonotic disease vectors. Immatures of *Ae. taylori* have been collected from water collections in tree holes, from pineapple axils, in steps cut into coconut palms, in rock pools, and in artificial containers. *Aedes furcifer* oviposits in tree holes, preferring rot holes.

Associated Pathogens

Ae. furcifer—BOUV, BUNV, CHIKV, DENV, RVFV, YFV, ZIKV; *Ae. taylori*—BOUV, CHIKV, DENV-2, YFV.

Diagnostic Characters

(a) Habitus (♀) **(*furcifer, taylori*):** Proboscis (P) with broad median pale band; Fe-I-III, Ti-I-III, and Ta-I-III$_1$ speckled with pale scales. **(b) Wing (*furcifer, taylori*):** Alula (Al) with broad scales; wing scales broad on all veins, dark mixed with white; tertiary fringe scales (TFS) mix of pale and dark. **(c) Thorax (side) (*furcifer, taylori*):** Erect forked scales numerous, not restricted to occiput (Occ) (indistinct in photo); paratergite with scales (PaSc); upper prealar area with broad scales (UPrSc); mesepimeral scales (MeSc) present; lower mesepimeral setae (MeSL) present. **(d) Thorax (dorsal) (*furcifer, taylori*):** Scutellar (Stm) lobes all with broad scales; acrostichal (AcS) and dorsocentral (DS) setae well developed; scutum (Scu) anteriorly uniformly pale, usually golden with dark spots at scutal angles (ScA) (*taylori*); scutum (Scu) mottled dark anteriorly (*furcifer*, not shown); prescutellar setae (PrsS) well developed. **(e) Abdomen (dorsal):** Terga (Te) and sterna (S) with no speckling (*taylori*). **(f) Abdomen (dorsal):** Terga (Te) and sterna (S) with pale speckling (*furcifer*). **(g) Genitalia** (♂): *Ae. taylori*. **(h) Genitalia** (♂): Claspettes (Cl) of *Ae. taylori* and *Ae. furcifer*. Image sources: WRBU; Huang (1986).

Exemplar DNA barcodes: *Ae. furcifer COI* (AY645210–25, KU187185, MG242495); *Ae. taylori COI* (AY645226–33, MG242519).

Catalog entries: *Ae. furcifer*: Vol. 2, p. 721. *Ae. taylori*: Vol. 2, p. 722.

a
furcifer-taylori
P
Fe-I–III
Ti-I–III
Ta-I–III
c
Occ
PaSc
UPrSc
MeSC
MeSL
d
Scu
ScA
Stm
DS
AcS
PrsS
taylori
b
Al
TFS
furcifer-taylori
e
taylori
f
Te
furcifer
g
gonostylus
gonocoxite
claspette
paraproct
aedeagus
IX tergum
taylori
h
(1)
(2)
(3)
furcifer

Aedes (*Downsiomyia*) *niveus* (Ludlow, 1903)
Subfamily Culicinae; Tribe Aedini
"Snowy Philippine Pointy Mosquito"

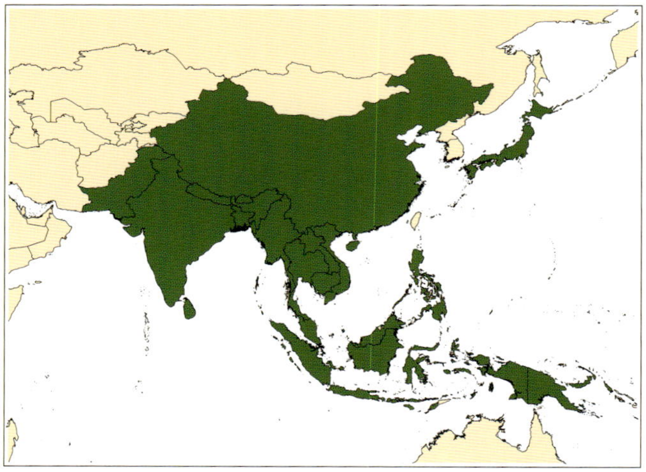

Systematics and Distribution

The tropical species *Aedes niveus* was named after the Latin name for snow, in reference to the heavy dusting of white scales adorning the top of the scutum (see *a*, *d*). Originally described based on specimens from Oras, Eastern Samar, in the Philippines, *Aedes niveus* s.s. is known only to occur from the islands of the Philippine archipelago. It is the nominotypical member of the Niveus Group of mosquitoes in the *Aedes* subgenus *Downsiomyia* (see Vol. 2, p. 1077, for internal relationships). Dozens of species make up the Niveus Group, which is widely distributed from Pakistan to New Guinea and from China to Singapore. For the most part, species of the Niveus Group are distinguishable only by characters of the male genitalia. In general, these species are understudied, and integrated systematics studies of the group throughout its range are needed before the true array of component taxa can be determined.

Bionomics

Aedes niveus s.s. is a day biter that deposits its eggs in tree holes. The species is an important vector of periodic *Wuchereria bancrofti* filariasis in the Philippines.

Members of the Niveus Group are generally regarded as zoophilic, but some bite humans and are of medical importance. At least one member of the group is a canopy feeder and is implicated as a sylvatic vector of dengue (DENV) among primates in Malaysia, substantiated by the presence of all four serotypes of the virus in Niveus Group species in this habitat. They are also probably the vector of a fifth serotype of DENV identified in Malaysia. Laboratory tests in Thailand showed that another Niveus Group member *Ae. albolateralis* is a competent vector of DENV, being almost as susceptible to DEN-2 as *Ae. albopictus* (Skuse, 1895). *Aedes albolateralis* immatures have been recovered from tree holes and bamboo stumps in Thailand. Another important vector species in the Niveus Group is *Aedes harinasutai* Knight, 1978. *Aedes harinasutai* readily bites humans and has been implicated as the principal vector of subperiodic *Wuchereria bancrofti* in Western Thailand. Its immature bionomics are unknown.

Associated Pathogens

DENV; *Wuchereria bancrofti*.

Diagnostic Characters

(a) Habitus (♀). (b) Thorax (side): Paratergite (Pa) without scales; postpronotum (Ppn) without broad white scales. **(c) Wing. (d) Thorax (dorsal):** Scutum (Scu) with anterior half conspicuously pale scaled. **(e) Larva (terminal segments):** Seta 4-X with five pairs of setae; precratal setae (PcrS) present; comb (C) with comb scales (CS) not on sclerotized plate. **(f) Legs:** Ta-I–III all dark; Fe-II without silvery spot in middle of anterior surface; Fe-III with apical dark area for ≥0.25 segment. Image source: WRBU.

Exemplar DNA barcodes: *Aedes niveus* s.l. (KY817567).

Catalog entry: Vol. 2, p. 724.

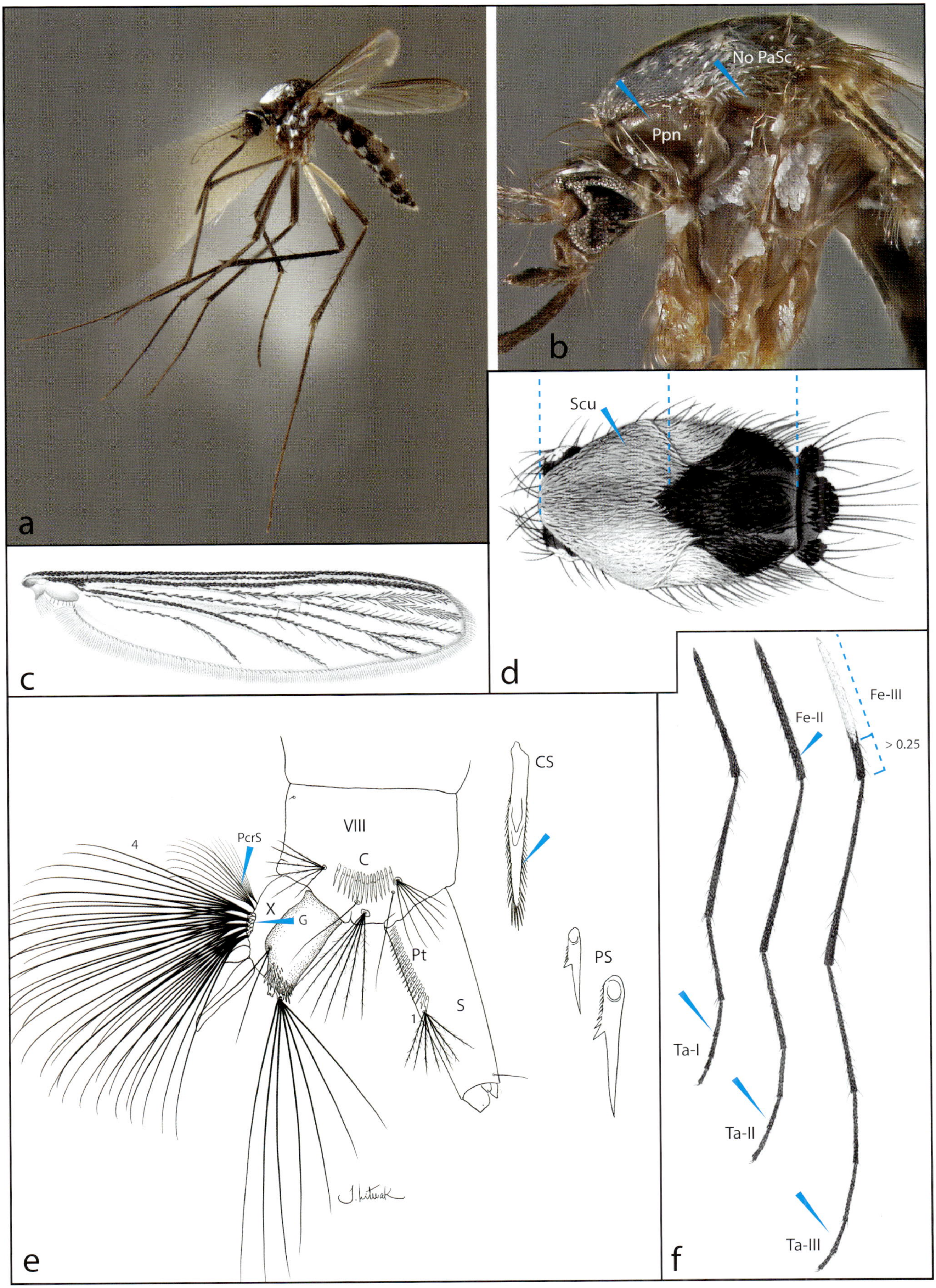

No PaSc
Ppn
b
a
Scu
c
d
VIII
C
4
PcrS
X
G
Pt
S
1
CS
PS
J. Litwak
e
Fe-III
Fe-II
> 0.25
Ta-I
Ta-II
Ta-III
f

Aedes (Finlaya) fijiensis Marks, 1947
Subfamily Culicinae, Tribe Aedini
"Viti Fijian Underwater Mosquito"

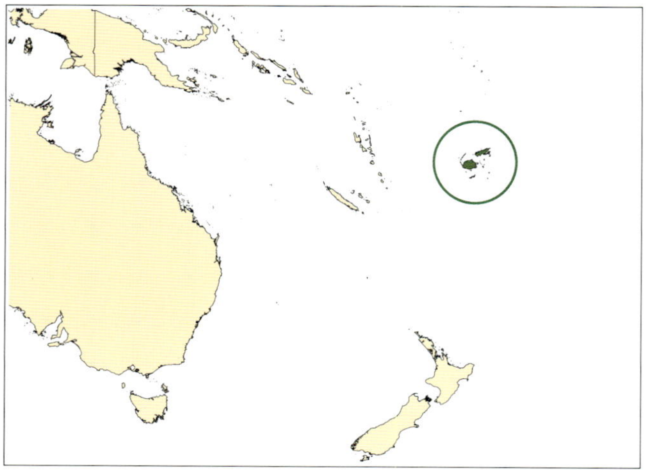

Systematics and Distribution

Aedes fijiensis is a distinctive mosquito, with ornate multiple white leg bands of varying thicknesses and highly decorated wings and abdomen with plentiful clusters of white scales, giving the illusion of spots. The species is a Fijian endemic, originally described from Samabula, Viti Levu, Fiji. *Aedes fijiensis* is a member of the informal Kochi Group (see *Aedes kochi* species page, p. 414), which are distinguished from all other *Aedes* species as larvae by multiple stellate setae on their thorax and abdomen, and a comb bearing 50 or more comb scales. Fiji is the most easterly distribution of any member of the Kochi Group, suggesting that this species is the farthest migration from areas to the west where the group is more speciose.

Bionomics

Aedes fijiensis is a nocturnal pest species in Fiji that aggressively attacks humans. *Aedes fijiensis* oviposits in water collections in natural leaf axil containers, especially those of *Pandanus* (screw palm) and *Colocasia* (taro or elephant ears). Screw palms are farmed for fiber, and taro is a valuable food source in the tropics, placing the developmental sites for this species close to people. Despite its localized distribution, the species is the primary vector of filariasis in Fiji.

Associated Pathogens

Wuchereria bancrofti.

Diagnostic Characters

(a) Habitus (♀). (b) Thorax (dorsal): Scutal (Scu) pale scales bright white; scutellar lateral lobe scales (LSSc) pale, dark scales present on midlobe (MSSc); halter (Hl) largely dark scaled. **(c) Head (dorsal):** Proboscis (P) with broad median pale band. **(d) Abdomen (side):** VI,VII-S with apical tufts of erect dark scales. **(e) Wing:** Regular spots of pale and dark scales; prehumeral pale spot (PHP) distinct; preapical pale spot (PP) not developed on vein C. **(f) Legs:** Fe-I–III and Ti-I–III with many pale spots and rings; Fe-I–III with prominent apical scale tufts; Ta-III$_{2,3}$ with basal ≥0.5 dark; Ta-III$_4$ all dark. **(g) Larva (head):** Seta 4-C usually 2–5 branched, without enlarged base, posterior to (or level with) seta 6-C; seta 14-C large stellate, multibranched. **(h) Larva (terminal segments):** Comb (C) with ≥50 comb scales (CS) in triangular patch; distal row of CS with 2–4 lateral spinules and fine spinules (fringe) apically; seta 1-X usually 2–3 branched; seta 4-X with five pairs of setae, each with long basal stalk, all arising from basal boss, without distinct bars; siphon (S) with uniform, dense short spicules. **(i) Larva (thorax and abdominal segments):** Thorax and abdomen with conspicuous stellate setae; seta 1-III–V stellate with numerous branches of same size or slightly longer as setae 2 and 3 on corresponding segments; seta 6-III–V, usually 2–3 branched; seta 12-I present; seta 13-II–V usually ≥4 branched. Image sources: WRBU; Belkin (1962).

Exemplar DNA sequences: None.

Catalog entry: Vol. 2, p. 727.

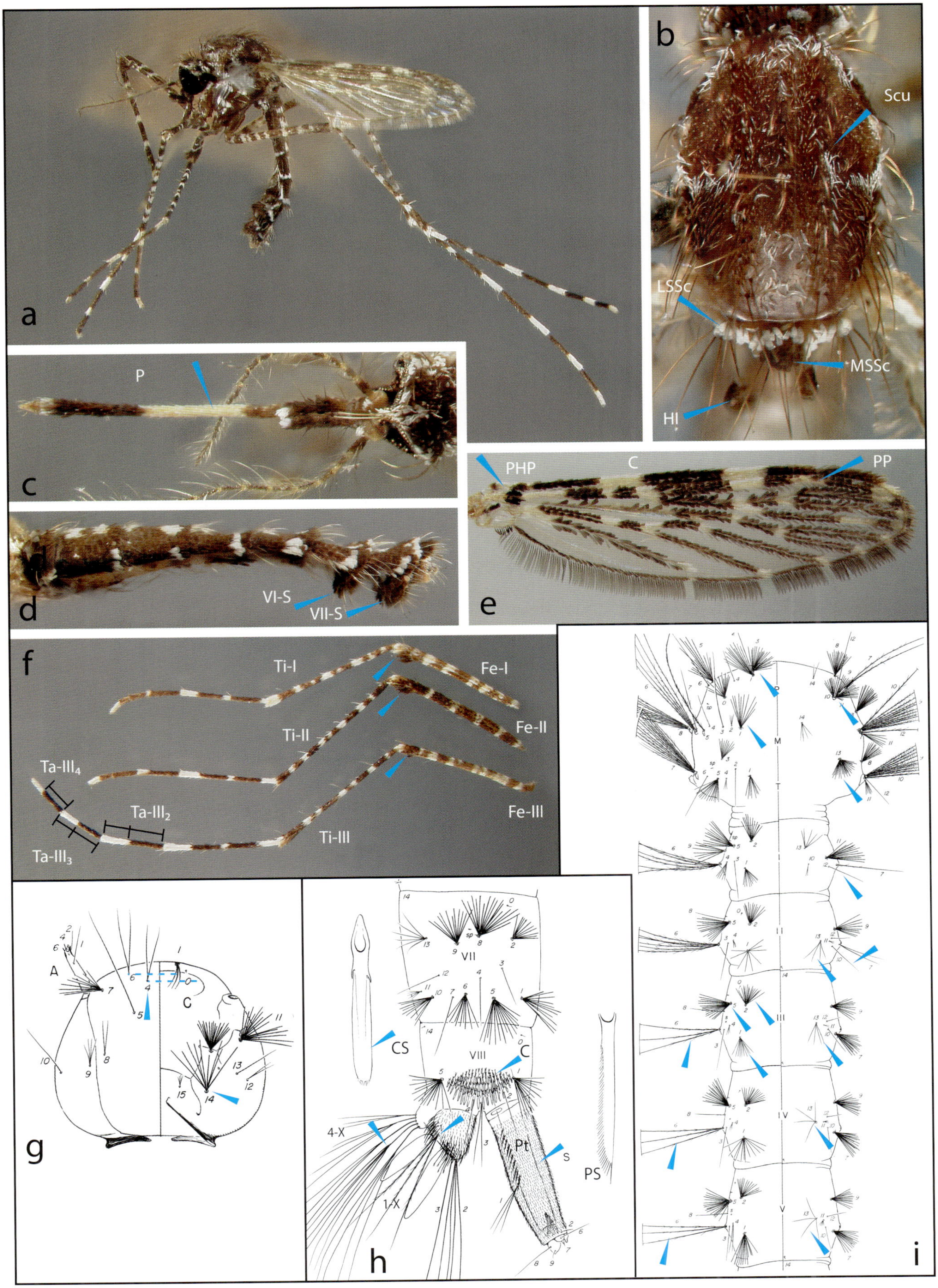
a
b
Scu
LSSc
MSSc
Hl
c
P
d
VI-S
VII-S
e
PHP
C
PP
f
Ti-I
Fe-I
Ti-II
Fe-II
Ta-III$_4$
Ta-III$_2$
Ta-III$_3$
Ti-III
Fe-III
g
A
C
h
VII
VIII
CS
C
4-X
1-X
Pt
S
PS
i
P
M
T
I
II
III
IV
V

Aedes (*Finlaya*) *kochi* (Dönitz, 1901)
Subfamily Culicinae, Tribe Aedini
"Koch Papuan Pointy Mosquito"

Systematics and Distribution

Aedes kochi is an extremely ornate species, brown and covered with white scales in distinct patches, or "spots," and multiple leg bands of various sizes, giving a paint-spattered appearance. The pale and dark wing spots are analogous to those in genus *Anopheles*. *Aedes kochi* is distributed on the islands of Australasia and the Pacific Region, including Australia, Indonesia, Fiji, New Guinea, the Philippines, Samoa, the Solomon Islands, and Tonga. Adult females of the Kochi Group can be difficult to distinguish from each other (see *Aedes fijiensis* species page, p. 412).

Bionomics

While all workers agree on the pest status of *Ae. kochi*, reports of the biting behaviors of *Ae. kochi* are somewhat conflicting. Females readily bite day and night in New Guinea, whereas they are reported only to bite at night on New Britain. In Australia, biting activity starts at 1500h but peaks from dusk until 2300h. The species has been reported biting both outdoors and indoors, and they are reported to tenaciously squeeze their way through mesh netting with a wide enough weave. In addition to humans, the species feeds on horses but was not collected at chicken-baited traps. *Aedes kochi* are readily collected in light traps.

Immatures are found in the leaf axils of a wide variety of plants, including *Alocasia*, banana, *Crinum*, pineapple, *Pandanus*, sago palm, and taro. These habitats create a situation in which the species is associated with human activity and habitat disturbance. Eggs are deposited just above the larval site on the petioles of leaves. Larvae can move across moist surfaces and have the unusual behavior of flexing their bodies to jump a centimeter or more.

The widely differing bionomic reports suggest that *Ae. kochi* may comprise a species complex across its range. Recent DNA barcoding studies in our laboratory revealed two distinct, sympatric species of *Ae. kochi* in Queensland, Australia, although how either of these relate to the nominotypical species described from New Guinea is unknown.

Associated Pathogens

EUBV, RRV; *Wuchereria bancrofti*.

Diagnostic Characters

(a) Habitus (♀): Fe-I–III and Ti-I–III with many pale spots and rings; Fe-I–III with prominent apical scale tufts. **(b) Leg (III):** Ta-$III_{2,3}$ with basal ≥0.5 dark; Ta-III_4 all dark. **(c) Head (side):** Proboscis (P) with broad median pale band. **(d) Thorax (dorsal):** Scutal (Scu) pale scales bright white; lateral scutellar lobes scales (LSSc) pale, dark scales present on midlobe (MSSc). **(e) Abdomen (dorsal):** VIII-Te with medial basal pale patch. **(f) Wing:** Contrasting pattern of dark and pale scales; prehumeral pale spot (PHP) distinct; preapical pale spot (PP) on vein R, not vein C. **(g) Abdomen (side):** Halter (Hl) with pale stem and dark-scaled capitellum (Ca); VI,VII-S with apical tufts of erect dark scales. **(h) Larva (head):** Seta 5-C single; seta 6-C 2,3 branched. **(i) Larva (terminal segments):** Comb (C) with ≥50 comb scales (CS) in several rows making a triangular patch; distal row of CS with 2–4 lateral spinules and spinelike apex; siphon (S) without spicules; pecten spines (PS) illustrated. Larval thorax and abdomen with conspicuous stellate setae (not shown). Image sources: WRBU; Marks (1947).

Exemplar DNA sequences: None.

Catalog entry: Vol. 2, p. 728.

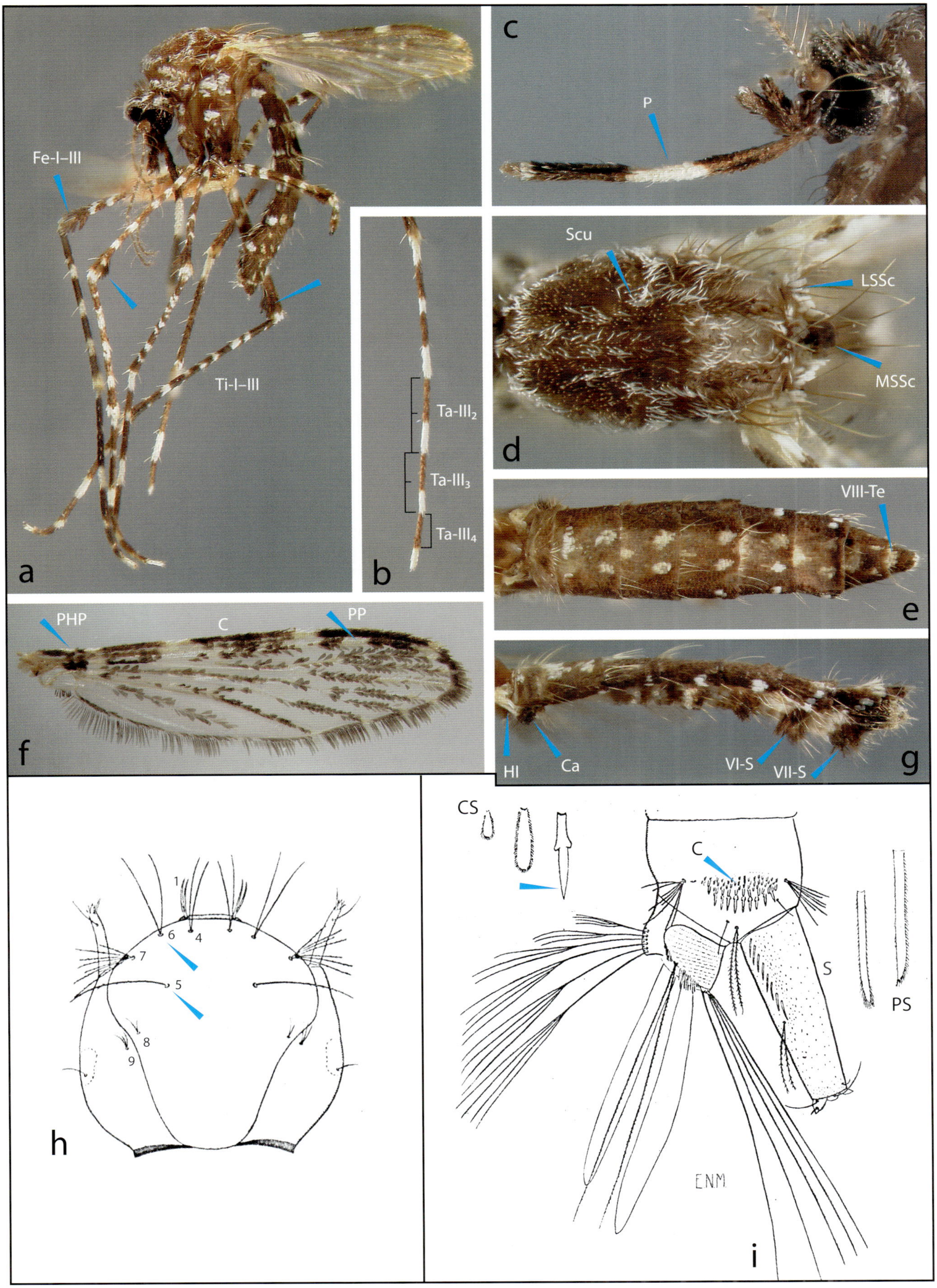

a
Fe-I–III
Ti-I–III
b
Ta-III2
Ta-III3
Ta-III4
c
P
d
Scu
LSSc
MSSc
e
VIII-Te
f
PHP
C
PP
g
Hl
Ca
VI-S
VII-S
h
1
4
5
6
7
8
9
i
CS
C
S
PS
ENM

Aedes (*Finlaya*) *poicilius* (Theobald, 1903)
Subfamily Culicinae, Tribe Aedini
"Variegated Malaysian Pointy Mosquito"

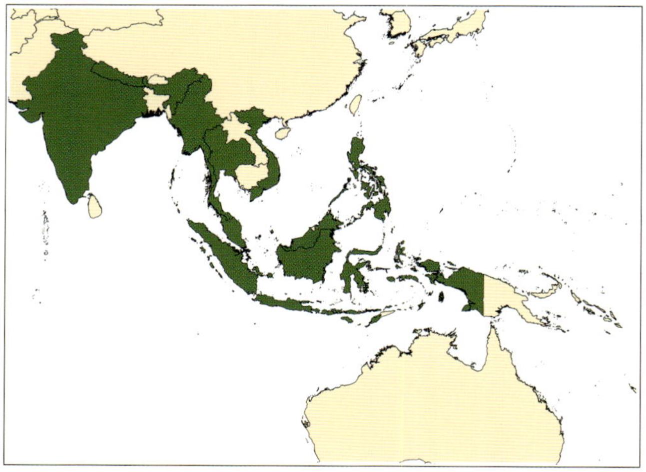

Systematics and Distribution

Aedes poicilius is quite attractive, with a distinctly spotted appearance. All legs sport a plethora of contrasting white bands of different sizes, its wings are heavily spotted, and the proboscis bears a broad median band and a white tip. Originally described from Pulau Jerejak in Penang, mainland Malaysia, the species occurs from India and Nepal, through the Southeast Asian mainland, and across to the islands of Borneo, Indonesia, and the Philippines.

Bionomics

Aedes poicilius females can become nuisance pests near plantations as they bite humans aggressively, day and night, indoors and out. In parts of the Philippines, *Ae. poicilius* is the most abundant human-feeding mosquito and acts as the principal vector of filariasis in rural areas. Comparative human-landing catches with cow-baited traps clearly also support zoophily in this species, with females also feeding on domestic water buffalo.

Aedes poicilius immatures develop in a variety of plant leaf axils, including aroids, abaca (a species of banana used for the fibers of its stem), banana, and pineapple. Abundance of *Ae. poicilius* increases with the presence of agricultural plantations of the above crop plants, especially the abaca. In the Philippines, *Ae. poicilius* is commonly found in wild banana axils in association with *Ae. flavipennis* Giles and *Malaya genurostris* Leicester. Larvae and pupae survive the dry season in the small amounts of water and damp detritus held in the tighter upper-leaf axils of the wild banana, thought to retain water longer and replenish small amounts of water through dew condensation. This dry-season survival strategy is echoed by members of the *Ae.* (*Stegomyia*) *simpsoni* complex occupying banana leaf axils in Africa.

Associated Pathogens

Setaria labiotapapillosa, *Wuchereria bancrofti*.

Diagnostic Characters

(a) Habitus (♀): Fe-I–III with apical ventrolateral tufts; Ta-I–III_{1-5} mostly with apical pale bands. **(b) Thorax (dorsal):** Scutellar (Stm) scales broad and flat. **(c) Head (dorsal):** Vertex (V) with broad, flat scales; proboscis (P) with median pale band. **(d) Thorax (side):** Postpronotal scales (PpSc) broad, flat. **(e) Wing:** Regularly spotted. **(f) Larva (abdomen):** Seta 12-I present. **(g) Larva (thorax):** Seta 0-P large, stellate. **(h) Larva (terminal segments):** Comb scales (CS) numerous, arranged in triangular patch; pecten (Pt) spines (PS) long and narrow; seta 1-S branched; saddle (Sa) with long marginal spicules; siphon (S) long; siphon index ≥2.0. Image sources: WRBU; Mattingly (1971a).

Exemplar DNA sequences: None.

Catalog entry: Vol. 2, p. 728.

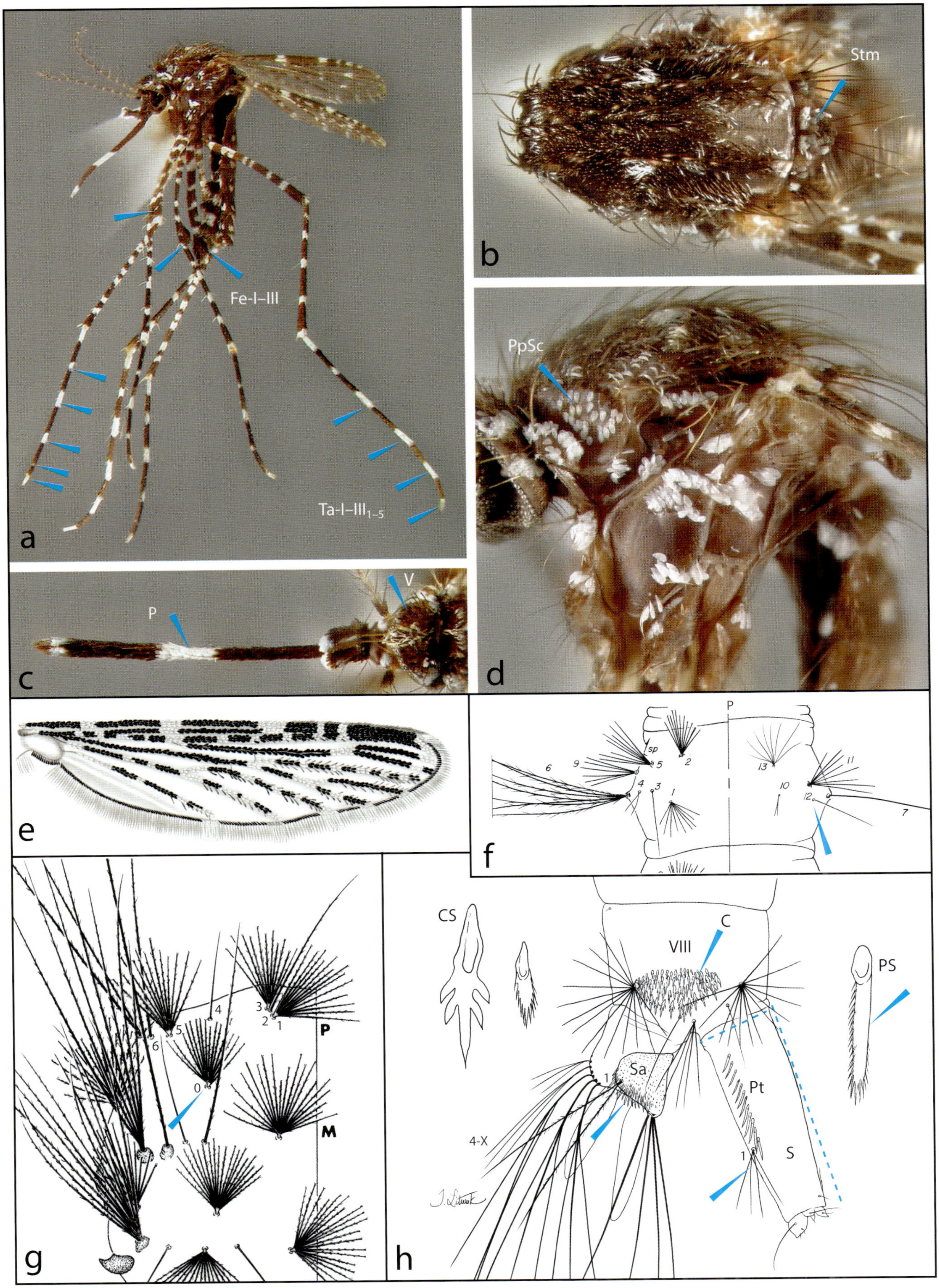
a
Fe-I–III
Ta-I–III1–5
b
Stm
c
P
V
d
PpSc
e
f
P
sp
1
2
3
4
5
6
7
9
10
11
12
13
g
0
1
2
3
4
5
6
P
M
h
CS
C
VIII
PS
Sa
Pt
S
1
4-X

Aedes (Georgecraigius) atropalpus (Coquillett, 1902)
Subfamily Culicinae, Tribe Aedini
"Black-lipped American Pointy Mosquito"

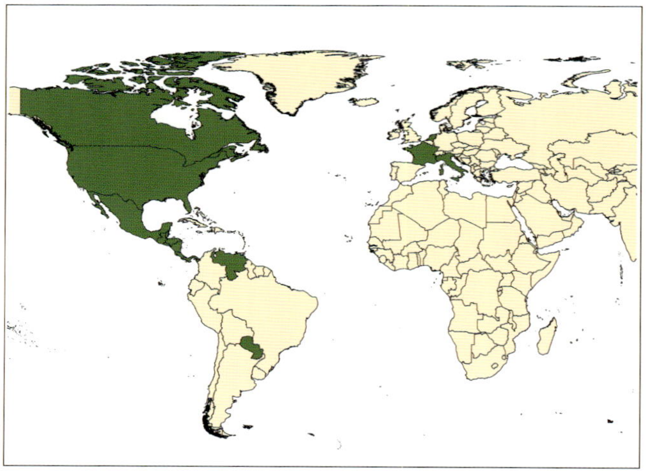

Systematics and Distribution

Aedes atropalpus is named after its dark palps. Its native range is in the eastern United States and Canada. It expanded its range within North America and invaded Europe, where it became established in Italy in 1996 and was detected and eliminated from France in 2003 and 2005, Switzerland in 2008, and the Netherlands in 2008 and 2010. The subgenus *Georgecraigius* comprises only three species—*Ae. atropalpus* and *Ae. epaticus* Dyar & Knab, and *Ae. fluviatilis* (Lutz). It has been proposed that *Ae. atropalpus* may hybridize with *Ae. epaticus* in the southern part of the former species' range, but robust evidence for this is lacking.

Bionomics

Day biters, *Ae. atropalpus* avidly feeds on humans, especially close to its immature habitats. This species' preferred immature habitats are rock holes or rock pools associated with streams and rivers, although it has also been reported occupying tree holes, small ground pools, and *Agave* plant axils. It can be a common and conspicuous mosquito along rocky streams and rivers like the Potomac, just to the west of Washington, DC, United States, much to the annoyance of those who fish there. Sometime after 1972, the species adapted to immature development in tires, causing a rapid expansion of range and apparent abundance. In contrast to the closely related *Ae. epaticus*, females of *Ae. atropalpus* are autogenous, depositing their first batch of eggs without requiring a blood meal, and can produce several egg batches per year in warmer climates. Eggs of *Ae. atropalpus* are deposited on the walls of containers and hatch when flooded. *Aedes atropalpus* overwinters as eggs, and the combination of the above bionomic characteristics predisposes the species to be an effective invasive species, capable of readily establishing populations outside its natural range. Saint Louis encephalitis (SLEV) and La Crosse encephalitis (LACV) can be transmitted transovarially in this species in the laboratory.

Associated Pathogens

EEEV, JEV, LACV, MVEV, SLEV, VEEV, WEEV, WNV.

Diagnostic Characters

(a) Habitus (♀). (b) Thorax (side): Paratergite (Pa) bare; postspiracular scales (PoSc) present; C-III with scales on lateral surface. **(c) Thorax (dorsal):** Scutum (Scu) with posterior dorsocentral setae (PDS) and broad lateral pale-scaled area reaching dorsocentral area, scales pale yellowish to pale golden with background scales mostly dark, bronze colored; scutellum (Stm) dark scaled, without broad white scales on lobes; posterior scutal fossa (SF) without setae. **(d) Head (dorsal):** Clypeus (Clp) without scales. **(e) Leg (III):** Ta-III_{1-4} with basal and apical rings; Ta-III_5 white scaled. **(f) Larva (terminal segments):** Base of siphon (S) irregularly sclerotized with an extensive basal membranous-appearing area ventrally between rows of pecten spines; comb scales (CS) numerous, usually 34–62, rounded with distinct fringe; pecten (Pt) spines (PS) widely spaced distally; seta 4-X on grid (G), without precratal setae (PcrS). **(g) Larva (abdominal segments):** Seta 6-III,IV usually 4,5 branched. **(h) Larva (head):** Anterior portion of head and antenna distinctly more pigmented than posteriorly. Image sources: WRBU; Zavortink (1972).

Exemplar DNA barcodes: *Ae. atropalpus* (JF868951, MG242470, KM457565–67 as *Ochlerotatus atropalpus*).

Catalog entry: Vol. 2, p. 730.

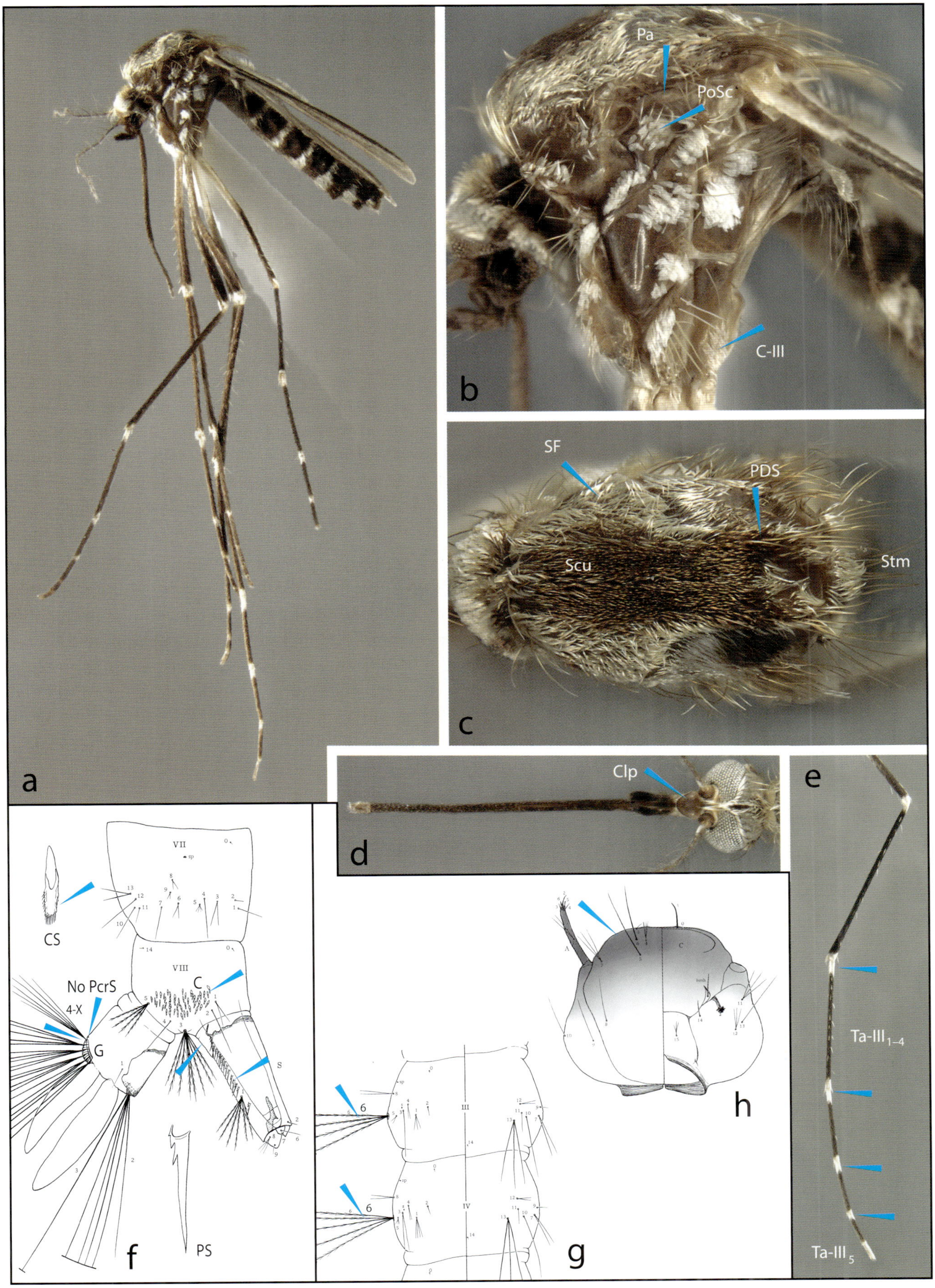
a
Pa
PoSc
C-III
b
SF
PDS
Scu
Stm
c
Clp
d
e
Ta-III$_{1-4}$
Ta-III$_5$
VII
CS
VIII
No PcrS
4-X
C
G
S
PS
f
III
IV
6
6
g
h

Aedes (*Hulecoeteomyia*) *japonicus* Theobald, 1901
Subfamily Culicinae, Tribe Aedini
"Tokyo Japanese Pointy Mosquito"

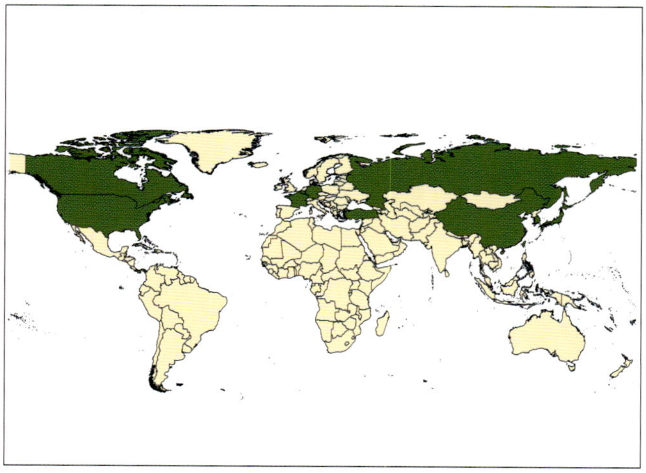

Systematics and Distribution

Aedes japonicus adults have a distinctive bronze lyre-shaped pattern on the scutum. The nominotypical form, described from Tokyo, Japan, is sometimes referred to as *Ae. japonicus japonicus* to differentiate it from its three valid and morphologically distinct subspecies. These include *Ae. japonicus shintienensis* Tsai & Lien, described from Taiwan, and *Ae. japonicus amamiensis* Tanaka, Mizusawa & Saugstad, and *Ae. japonicus yaeyamensis* Tanaka, Mizusawa & Saugstad, both described from different parts of the Ryukus Islands, Japan. Given recent molecular studies, it seems likely that these subspecies will soon be elevated to species in their own right

Of these three subspecies, only *Ae. japonicus japonicus* is considered highly invasive, expanding from its native range of Korea and Japan through a series of sequential invasions starting in the United States (Pennsylvania) in 1998. By 2012, it was established in other regions of the United States (Hawaii, Washington, Oregon, and the eastern states) and Canada (Quebec and Ontario). It has now also become established in much of central Europe (Austria, Belgium, Germany, Slovenia, Switzerland), and has been intercepted and eradicated in France and New Zealand. The similarly invasive Southeast Asian species *Ae. koreicus* (Edwards) (see species page, p. 422) is so closely related that it is often considered a sibling species to *Ae. japonicus*.

Bionomics

In its natural environment, *Ae. japonicus* s.l. inhabits forested areas and is active during the day. Females bite humans and large animals, especially horses, and readily feed on smaller mammals including mice and chicks in the laboratory. *Aedes japonicus japonicus* immatures occupy a wide variety of natural and artificial containers, usually preferring shaded places and water containing rich organic matter. Rock holes appear to be the most favored immature habitat for this subspecies. *Aedes japonicus japonicus* females oviposit in a variety of shaded container habitats containing fresh water rich in organic matter. They place their eggs on the walls of the container or just above the waterline on floating objects. They prefer stone or cement containers with moderate levels of organic matter, often with turbid or colored water, though larvae can also be found in tree holes, bamboo, artificial containers, and even ground pools. Larval development proceeds successfully at cooler temperatures above 10°C. The species overwinters as eggs in cooler portions of its range and as larvae in warmer areas. Its successful invasion into Europe and North America has been credited to its larval habitat plasticity and adult cold tolerance that allows an expanded seasonal activity range in temperate climates.

Associated Pathogens

DENV, CHIKV, CVV, EEEV, JEV, LVC, ORUV, RVFV, SLEV, WEEV, WNV.

Diagnostic Characters

(a) Habitus (♀). (b) Thorax (dorsal): Scutellar (Stm) lobes with long, narrow scales; scutum (Scu) with golden stripes; distinctive lyre-shaped stripes and two submedian and a median stripe. **(c) Head (side):** Proboscis (P) dark scaled, without pale band. **(d) Leg (III):** Ta-III_{1-3} with broad pale basal bands; Ta-III_4 all dark or with a few pale scales at base; Ta-III_5 all dark. **(e) Larva (terminal segments):** Seta 1-S inserted within the pecten (Pt); one or more distal pecten spines (PS) widely spaced; saddle (Sa) incomplete; seta 1-X close to posterior margin of saddle; seta 4-X inserted on grid (G); precratal setae (PcrS) absent. **(f) Larva (head/thorax):** Setae 5,6-C multibranched and inserted near anterior edge of head, 5-C medial to 6-C. Image sources: WRBU; Knight (1968).

Exemplar DNA barcodes: *Ae. japonicus* s.l. (JX259639–46, KF211480–505, KF874592–303, KM457579–98).

Catalog entry: Vol. 2, p. 737.

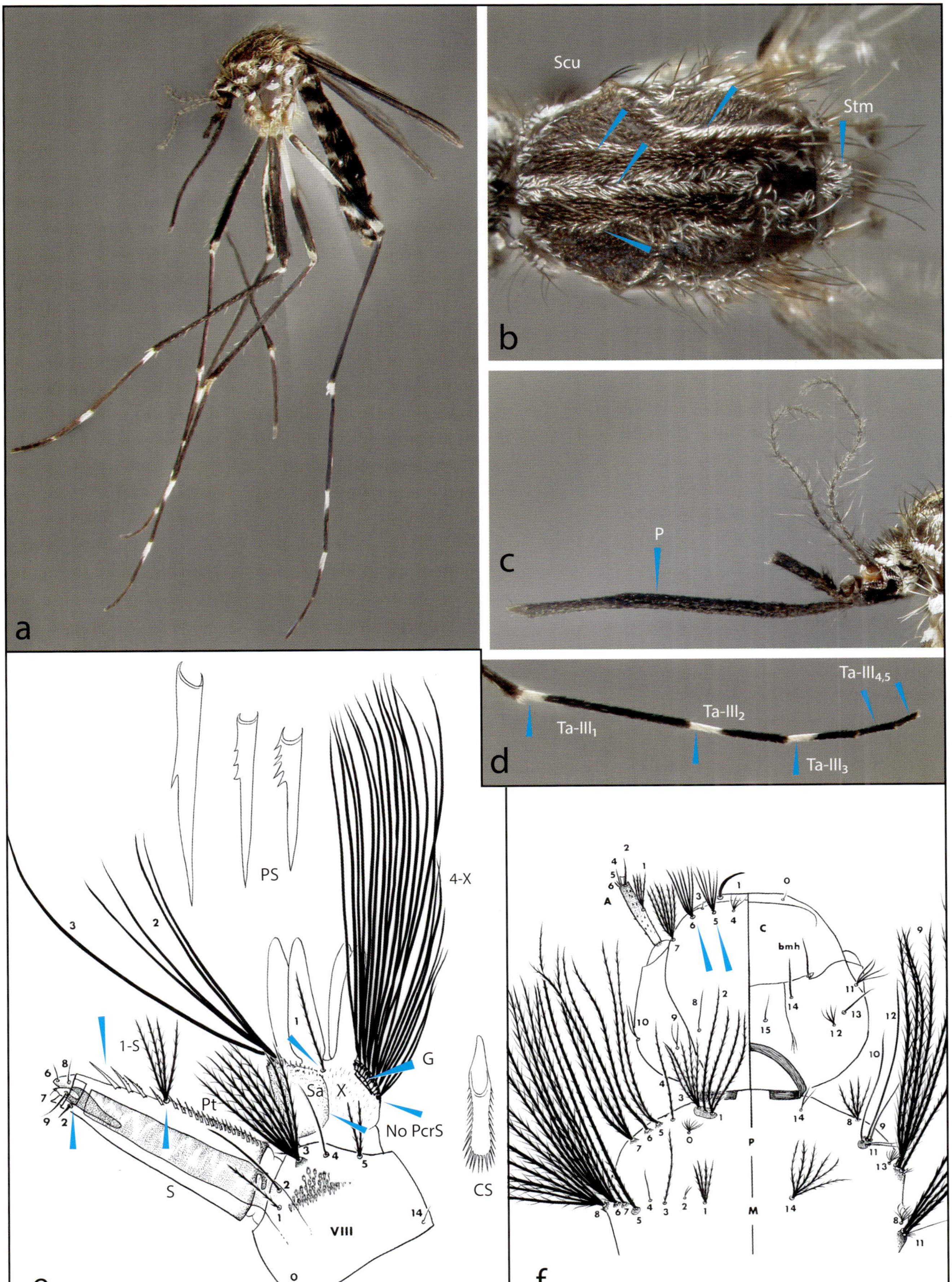
a
Scu
Stm
b
P
c
Ta-III1
Ta-III2
Ta-III3
Ta-III4,5
d
PS
4-X
3
2
1-S
1
G
Sa
X
Pt
No PcrS
S
VIII
CS
e
A
C
bmh
P
M
f

Aedes (Hulecoeteomyia) koreicus (Edwards, 1917)
Subfamily Culicinae, Tribe Aedini
"Joseon Korean Pointy Mosquito"

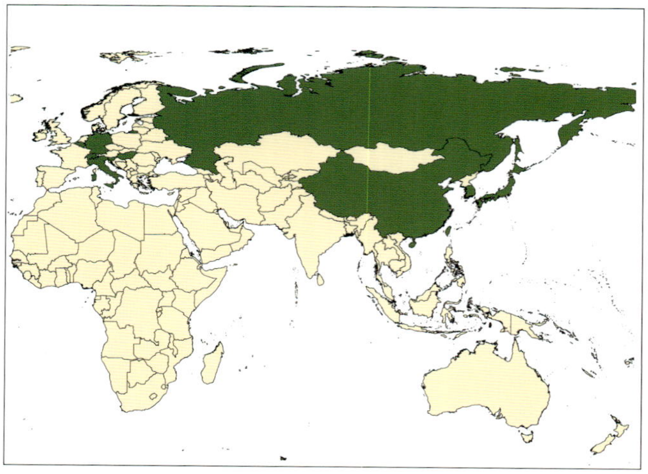

Systematics and Distribution

Aedes koreicus is an attractive mosquito, with golden scutal stripes that contrast with the bright silvery-white scale spots of the pleuron. It belongs to the recently reestablished *Aedes* subgenus *Hulecoeteomyia* (originally described as a genus and previously recognized as the Chrysolineatus Subgroup of genus *Aedes*), which includes *Ae. bhutanensis* Somboon & Harbach, *Ae. cherrapunjiensis* (Natarajan, Rajavel & Jambulingam), *Ae. chrysolineatus* (Theobald), *Ae. formosensis* Yamada, *Ae. harveyi* (Barraud), *Ae. japonicus* (Theobald), *Ae. jugraensis* (Leicester), *Ae. nigrorhynchus* Brug, *Ae. pallirostris* Brug, 1931, *Ae. reinerti* Rattanarithikul & Harrison, *Ae. rizali* (Banks), *Ae. saxicola* Edwards, *Ae. sherki* Knight, and *Ae. yunnanensis* (Gaschen), all of which (except *Ae. koreicus*) are distributed across Southeast Asia. Species within the subgenus are best identified on a combination of larval and adult characters, as male genitalia and pupal stages lack species-diagnostic characters.

Aedes koreicus was first described from specimens collected in South Korea and is native to Northeast China, Russia, the Korean Peninsula, and Japan. In 2008 *Ae. koreicus* was discovered out of its range as an established population in an industrial area of Belgium, but the mechanism of the introduction was unclear. It was also later detected in Italy in 2011 and then in Switzerland, and predictive models show that most of northern Europe provides suitable climatic and environmental characteristics to allow establishment of this species. *Aedes koreicus* has no synonyms or subspecies, but a morphological variant is reported from the southern Korean island of Jeju-do, which bears a distinctive basal hind band on the Ta-III_5. It is this Jeju-do morphotype that has invaded Europe.

Bionomics

In their native range, larvae of *Ae. koreicus* are found in rock pools, tree holes, plant axils, and sometimes artificial containers, including garden ponds and water barrels. Preferred oviposition sites are small, still freshwater sites with decaying tree leaves. The invasive population in Europe has mainly been found in artificial containers like tires and unused construction equipment, as well as in temporary ground pools at times. Females bite humans. The species overwinters as eggs. Unlike its sister taxa *Ae. japonicus* (see species page, p. 420), which is more rurally associated, *Ae. koreicus* is found in urban areas in close proximity to humans, heightening concern regarding its role in arthropod-borne disease transmission.

Associated Pathogens

JBEV; *Brugia malayi*, *Dirofilaria immitis*.

Diagnostic Characters

(a) Habitus (♀). (b) Thorax (side): Postpronum (Ppn) usually with a few broad dark scales; subspiracular area (SA) usually with patch of broad white scales. **(c) Head (side):** Proboscis (P) entirely dark scaled; pedicel (Pe) usually with more pale scales than dark, often all scales pale. **(d) Thorax (dorsal):** Scutum (Scu) yellow striped with median longitudinal pale stripe, forked posteriorly at prescutellar area (PrA); lateral scutellar lobes (LSSc) with narrow scales. **(e) Leg (II, III):** Fe-II with anterior and dorsal surfaces white preapically; Ta-III_{1-4} white banded basally; all legs with basal tarsal bands. **(f) Larva (head):** Seta 5-C 3–7 branched. **(g) Larva (thorax):** Seta 7-T with greatly elongate tapered branches, generally similar to seta 6-M. **(h) Larva (terminal segments):** Seta 1-S relatively short, none of the branches reaching apex of siphon (S), usually at end or beyond pecten (Pt); saddle (Sa) spiculate on posterior margin; seta 1-X single, on saddle; comb (C) with comb scales (CS) in patch, CS paddle shaped and fringed. Image sources: WRBU; Tanaka et al. 1979.

Exemplar DNA barcodes: *Ae. koreicus* (JF430393, KM258298–99, KM457599–600, KT358407, KT962063).

Catalog entry: Vol. 2, p. 738.

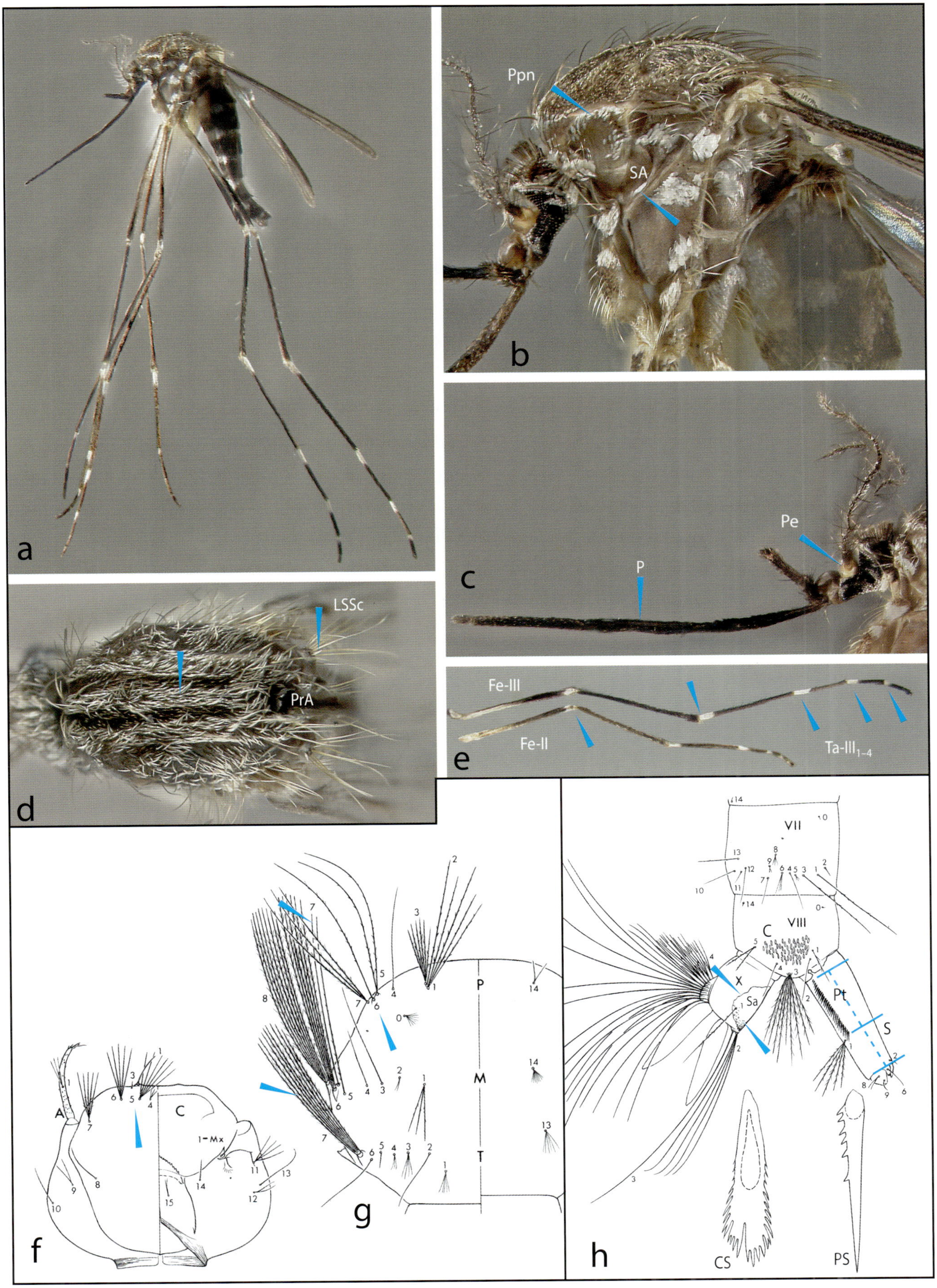
a
b
Ppn
SA
c
P
Pe
d
LSSc
PrA
e
Fe-III
Fe-II
$Ta\text{-}III_{1-4}$
f
A
C
1-Mx
g
P
M
T
h
VII
VIII
C
X
Sa
Pt
S
CS
PS

Aedes (*Neomelaniconion*) *mcintoshi* Huang, 1985
Subfamily Culicinae, Tribe Aedini
"McIntosh South African Pointy Mosquito"

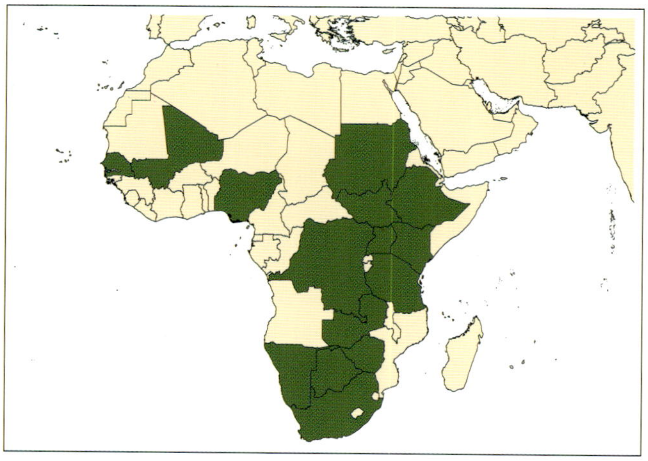

Systematics and Distribution

Aedes mcintoshi was first described from Onderstepoort in South Africa in 1985 and named after Bruce M. McIntosh, in recognition of his contributions to the knowledge of South African mosquitoes. For many years it was misidentified as the morphologically similar Southeast Asian species *Ae. lineatopennis* (Ludlow), originally described from the Philippines. Earlier records of *Ae. lineatopennis* in Africa are considered to refer to this species. *Aedes mcintoshi* is widely distributed, extending south from sub-Sahelian Africa on both the west and east sides of the continent.

Along with *Ae. mcintoshi* and *Ae. lineatopennis*, *Ae. circumluteolus* (Theobald), *Ae. luridus* McIntosh, *Ae. luteolateralis* (Theobald), and *Ae. unidentatus* McIntosh comprised the previously recognized the informal Lineatopennis Group of the subgenus *Neomelanoconion*, which are morphologically difficult to distinguish in the adult stages. These taxa are quite unique and share the following distinctive morphological characteristics: abdominal terga with basal bands, scutum with wide golden yellow scales on either side and on wing veins 1A, R_{4+5} and costa all with dark scales, and radial vein (R) with all white scales (at least from base to the base of Rs).

Bionomics

Aedes mcintoshi is a major vector of RVFV in Africa. Females bite at night, preferring cattle but commonly feeding on humans outdoors. As floodwater mosquitoes, females deposit their eggs in the upper layer of soil in grassy sites most likely to flood. Egg desiccation tolerance in floodwater mosquitoes is often associated with transovarian transmission of viruses between successive generations. In *Ae. mcintoshi*, eggs may remain in a dormant state for years, hatching only following sustained submersion. Larvae have been found in ground pools and residual pools left by receding streams. *Aedes mcintoshi* has been collected occupying the same immature habitats as *Ae. unidentatus* in South Africa and *Ae. circumluteolus* in Kenya. Landscape genetics approaches supported the single-species status of this species but revealed high genetic variation in subpopulations, which was found to be correlated with soil clay content and recent precipitation.

Associated Pathogens

BBKV, BUNV, NDUV, NRIV, MIDV, PGAV, RVFV, WESV, WSLV; *Plasmodium* spp.

Diagnostic Characters

(a) Habitus (♀): Fe-I–III without pale bands. **(b) Thorax (side):** Paratergite (Pa) bare; mesepimeron with a lower anterior seta (MeSL). **(c) Head (dorsal):** Vertex (V) with narrow decumbent scales. **(d) Thorax (dorsal):** Scutum (Scu) bordered with lateral band of yellow scales from anterior promontory (AnP) to scutellum (Stm); scutellar scales all narrow; acrostichal (AcS) and dorsocentral (DS) setae present. **(e) Head (side):** Erect forked scales on head numerous, not restricted to occiput (Occ). **(f) Abdomen (dorsal):** Terga (Te) with basal pale bands. **(g) Wing:** Veins C, R_{4+5} and 1A all dark scaled; veins R_s, R_{2+3}, and M with white plume scales. **(h) Larva (head):** Antenna (A) spiculate; seta 1-A small, with few branches; seta 6-C slightly posterior to a small seta 4-C; seta 7-C positioned anterior to setae 4–6-C. **(i) Larva (terminal segments):** Pecten spines (PS) extend ≥0.5 times length of siphon (S); seta 1-S usually at end of, or beyond, pecten (Pt); precratal setae (PcrS) present; stout comb scales (CS) in a single row. Image sources: WRBU; Knight & Hull (1953).

Exemplar DNA barcodes: *Ae. mcintoshi* (KJ940551–692, KJ940755–59, KU187018).

Catalog entry: Vol. 2, p. 749.

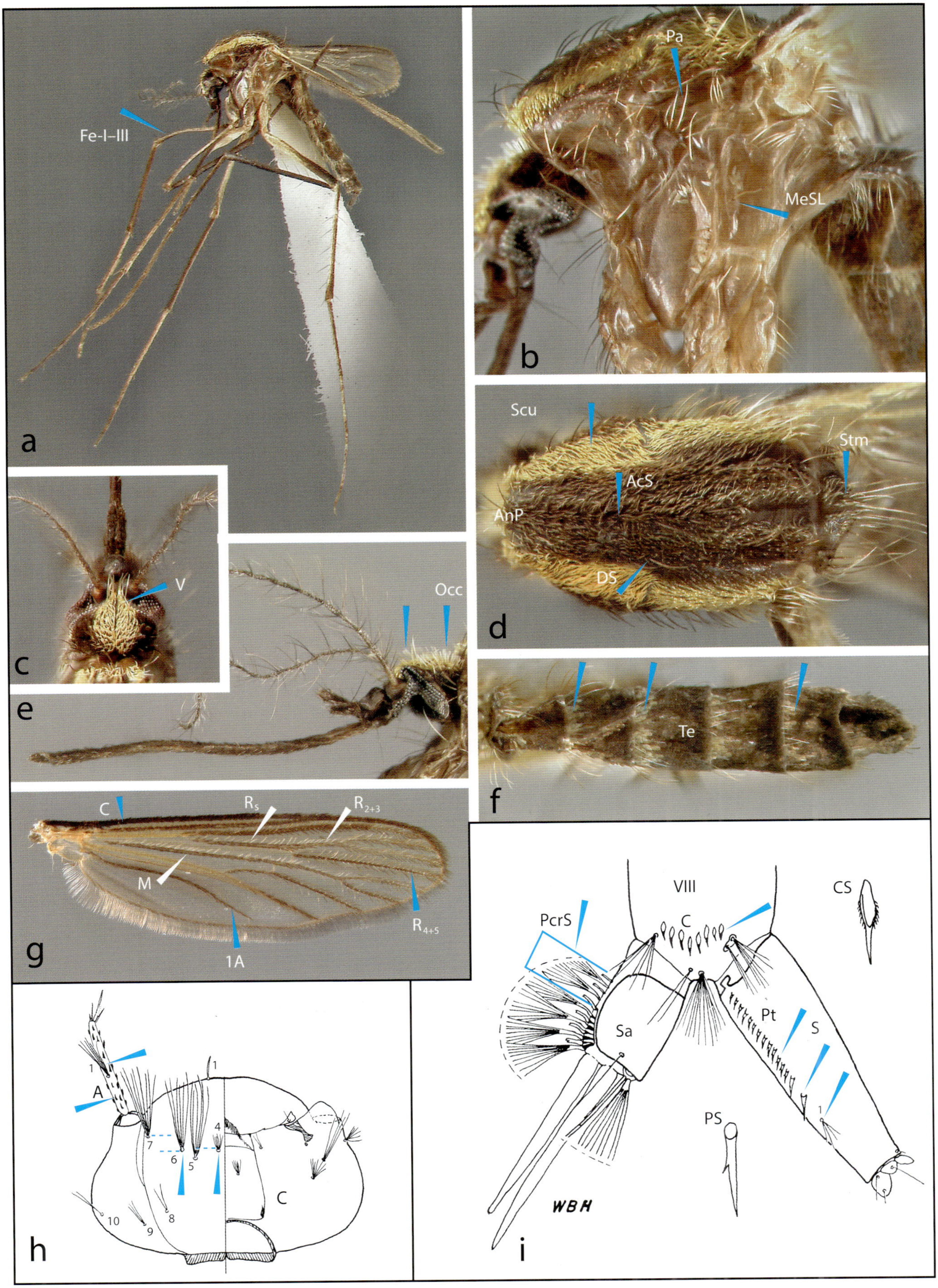
Fe-I–III
a
Pa
MeSL
b
c
V
Occ
e
Scu
AcS
Stm
AnP
DS
d
Te
f
C
Rs
R2+3
M
R4+5
1A
g
A
1
1
7
6
5
4
10
9
8
C
h
VIII
CS
PcrS
C
Sa
Pt
S
PS
1
WBH
i

Aedes (Ochlerotatus) angustivittatus Dyar & Knab, 1907
Subfamily Culicinae, Tribe Aedini
"Narrow-banded Costa Rican Pointy Mosquito"

Systematics and Distribution

First described from Costa Rica, *Aedes angustivittatus* ranges from Mexico to Argentina, and it is one of the most common *Aedes* in Central America. It takes its name from the characteristic narrow stripes on its thorax. It is currently placed in the Ochlerotatus Group of the subgenus *Ochlerotatus*, where it is most closely related to *Ae. atactavittatus* Arnell, *Ae. meprai* Martinez & Prosen, and *Ae. trivittatus* (Coquillett). It is suggested that hybridization may occur between *Ae. angustivittatus* and *Ae. trivittatus* in parts of their overlapping ranges.

Bionomics

In common with all floodwater mosquitos, *Ae. angustivittatus* deposits eggs in the upper layer of the soil of a wide variety of sites that are likely to flood. Immatures are usually reported in temporary ground pools formed by rains and stream overflow, pools in drying or intermittent streams, animal hoofprints in marshy pastures, and margins of freshwater swamps and ponds, usually in wooded areas. It is primarily a lowland species, although collections of *Ae. angustivittatus* have been reported up to 2,000 m. These mosquitoes generally prefer feeding on mammals during crepuscular periods but readily feed on humans during the daytime in shaded areas. *Aedes angustivittatus* is utilized by the botfly *Dermatobia hominis* (Diptera: Oestridae) as a phoretic host of its eggs.

Associated Pathogens

ILHV, SARV, VEEV.

Diagnostic Characters

(a) Habitus (♀): Ta-I-III without pale bands. **(b) Thorax (side):** Lower mesepimeral setae (MeSL) absent. **(c) Thorax (dorsal):** Scutum (Scu) with broad lateral pale-scaled areas; scutum with narrow, curved scales. **(d) Legs:** Fe-I–III without apical white spots; Ti-III with anterior stripe of white scales (indistinct in photo). **(e) Larva (thorax):** Integumental spicules long. **(f) Larva (head):** Setae 5,6-C single. **(g) Larva (abdomen):** Seta 1-IV,V usually double. **(h) Larva (terminal segments):** Comb scales (CS) with many subequal spinules. Image sources: WRBU; Arnell (1976).

Exemplar DNA barcodes: *Ae. angustivittatus* (KM592987, KM593061, KY859903–918).

Catalog entry: Vol. 2, p. 752.

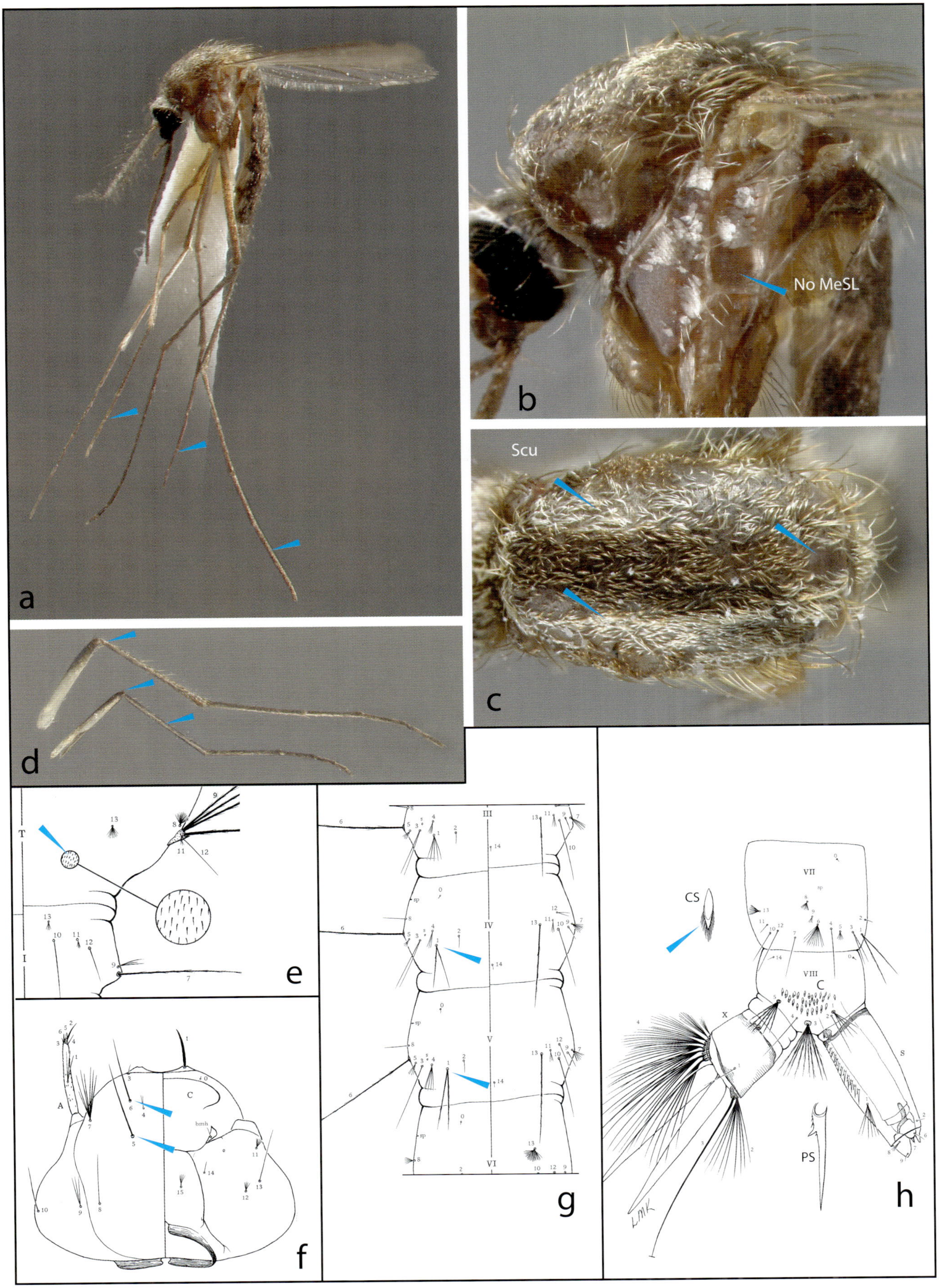
a
No MeSL
b
Scu
c
d
e
T
I
f
A
C
bmh
g
III
IV
V
VI
h
CS
VII
VIII
C
X
S
PS
LMK

Aedes (Ochlerotatus) atlanticus Dyar & Knab, 1906
Subfamily Culicinae, Tribe Aedini
"Atlantic American Pointy Mosquito"

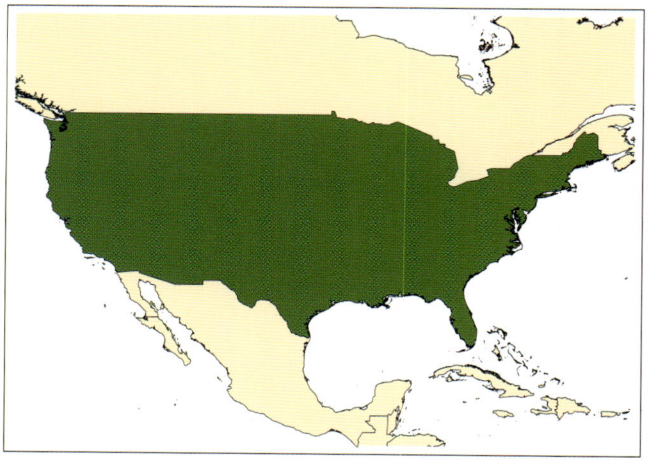

Systematics and Distribution

Aedes atlanticus mosquitoes are generally dark, but they sport a broad median stripe of white scales on the head that stretches forward between the eyes and a narrow stripe of silvery-white to pale-yellow scales along the whole length of the scutum. The wing scales are narrow and dark. *Aedes atlanticus* is one of 10 species assigned to the Protoculex Group of the *Aedes* subgenus *Ochlerotatus* (see Vol. 2, p. 1078, for internal relationships). First described from Florida, United States, *Ae. atlanticus* is a common pest species in the southeastern United States from Texas to the Atlantic coast and as far north as Maryland. Although not reported from the Caribbean, it does occur on the Florida Keys. Its range extends south to Guatemala.

Bionomics

Aedes atlanticus deposits its eggs in a wide variety of floodwater sites, both in open and wooded areas. In coastal areas it may colonize brackish water sites. Larvae have been detected in temporary pools in open fields and woodlands, often in association with *Ae. infirmatus*. In Virginia, larvae have been reported from March to November. Females are ready biters in shaded areas but will feed in bright sunlight. Hosts include humans and a wide range of mammals, birds, and reptiles. Adults are closely associated with the immature habitats, with short flight ranges noted of up to half a mile.

Associated Pathogens

EEEV, EVEV, KEYV, TENV, TVTV, WNV.

Diagnostic Characters

(a) Habitus (♀). **(b) Thorax (side):** Antepronotum (Ap) and postspiracular area (PA) without pale scales; paratergite scales (PaSc) absent; postprocoxal membrane (PM) bare; subspiracular scales (SSc) absent. **(c) Thorax (dorsal):** Median longitudinal white stripe along entire length of scutum (Scu), narrower than lateral brownish areas. **(d) Leg:** Ta-III without pale bands. **(e) Larva (terminal segments):** Comb scales (CS) 4–6, in single row; siphon (S) width about equal to width of segment X; siphon index c. 2.0–2.5; seta 1-S inserted at or beyond last pecten spine (PS); seta 2-X distinctly shorter than seta 3-X; anal papillae (APP) ≤5 times width of segment X. **(f) Larva (head):** Setae 5,6-C gradually tapering from base; seta 8-C long, reaching forward to base of seta 5-C. **(g) Head (dorsal):** Lateral black scale patches extend forward to eye margin. Image sources: WRBU; Mattingly (1979).

Exemplar DNA barcodes: *Ae. atlanticus* (JX259518–19, MG242469).

Catalog entry: Vol. 2, p. 753.

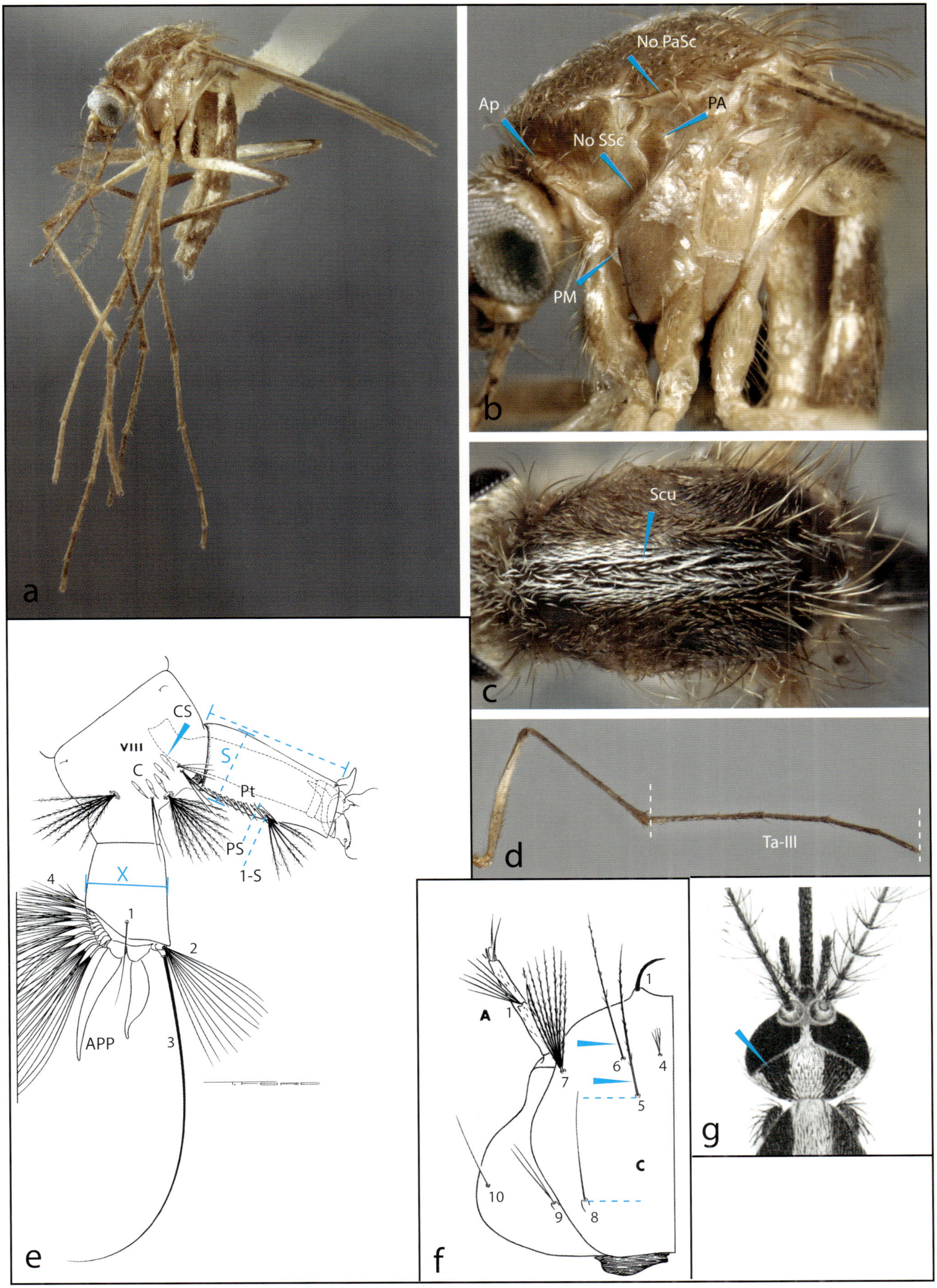

a
No PaSc
Ap
PA
No SSc
PM
b
Scu
c
Ta-III
d
CS
VIII
S
C
Pt
PS
1-S
X
4
1
2
APP
3
e
A
1
6
4
7
5
C
10
9
8
f
g

Aedes (*Ochlerotatus*) *canadensis* (Theobald, 1901)
Subfamily Culicinae, Tribe Aedini
"Iroquois Canadian Pointy Mosquito"

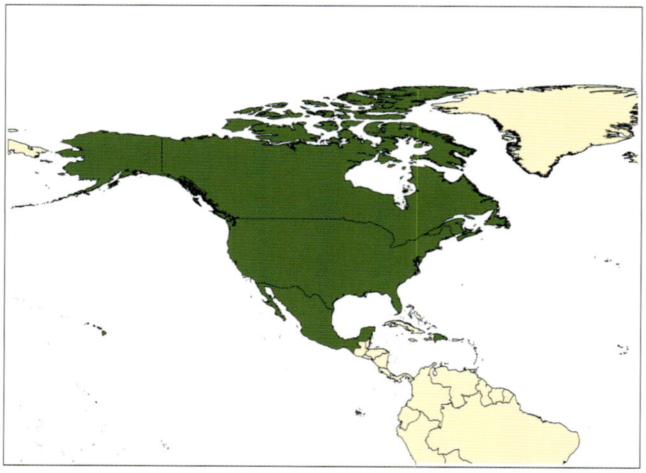

Systematics and Distribution

Aedes canadensis is a dark mosquito with a dusting of white scale patches on the pleuron and distinctly banded hind legs. It is distributed from Alaska and Canada to the southern United States, Dominican Republic, and Mexico. In North America, it is present along the Eastern Seaboard, from northern Florida to Labrador in Canada. It is commonly found east of the Mississippi but is absent from the arid Southwest and the Pacific coast. The northern population of *Ae. canadensis* is morphologically distinct from its darker subspecies *Ae. canadensis mathesoni* Middlekauf. The two entities are reportedly found in sympatry in the southeastern United States. This variation in coloration has been attributed to environmental influence, but after further genetic scrutiny, it may well represent a species-level difference. *Aedes canadensis* and its ssp. *mathesoni* are the sole members of the Culicada Group of the subgenus *Ochlerotatus*.

Bionomics

Aedes canadensis deposits its eggs at the margins of woodland pools, and eggs do not hatch until they have been conditioned through the winter. In the northern part of its range, eggs hatch in the early spring in pools formed from melting snow, though some eggs delay hatching until later in the summer or during the fall. Typically, the species has one generation per year (univoltine), and adults are long-lived, often being present until late summer. Due to their longevity, older adults collected in traps can be almost devoid of scales and are difficult to identify. In atypical years, heavy rainfalls associated with hurricanes can cause a large adult surge in the fall. Details of the life cycle in Mexico and the Dominican Republic are uncertain.

Adult females are weak fliers and can be nuisance biters, particularly in shaded habitats close to their larval sites. Somewhat unusually, *Ae. canadensis* is highly attracted to turtles and can be seen in clouds around egg-laying females in the spring. It has been reported feeding on the painted turtle, *Chrysemys picta* (Schneider), and the box turtle, *Terrapine carolina* (Linnaeus). Other hosts include humans, large and small mammals, birds, and other reptiles. Mixed-sex gynandromorphs have been reported.

Associated Pathogens

BUNV, EEEV, HJV, JCV, KEYV, RVFV, SSHV, VEEV, WNV; *Dirofilaria immitis*.

Diagnostic Characters

(a) Habitus (♀). **(b) Thorax (dorsal):** Scutum (Scu) covered with brown scales. **(c) Head (dorsal):** Proboscis (P) without a pale band; palpus (MPlp) with white-scaled apex, remainder with scattered pale scales. **(d) Thorax (side):** Hypostigmal area (HyA) and postprocoxal membrane (PM) bare. **(e) Wing:** Costa (C) entirely dark basally. **(f) Leg:** Ta-III$_5$ all white. **(g) Larva (head/thorax):** Antenna (A) spiculate and short, c. 0.5 head length; seta 1-A multibranched; 5-C 4–10 branched; 6-C 3–8 branched; 5–7-C not in straight line, 6-C far forward of 5-C; 1-M relatively short, about as long as 3,4-M, not reaching beyond bases of prothoracic setae. **(h) Larva (terminal segments):** Comb scales (CS) in patch, numerous; CS without apical frill of minute spinules; siphon (S) short, index 2.5–4.0; pecten (Pt) with evenly spaced spines (PS); anal papillae (APP) ≥ length of saddle (Sa); saddle incomplete. Image sources: WRBU; Carpenter & LaCasse (1955).

Exemplar DNA barcodes: *Ae. canadensis* s.l. (KM638737, KM937806, KR427986, KR518784, KR715197, KR730530, KR967937, KR969949, MF825957, MG087709, MG242476).

Catalog entry: Vol. 2, p. 755.

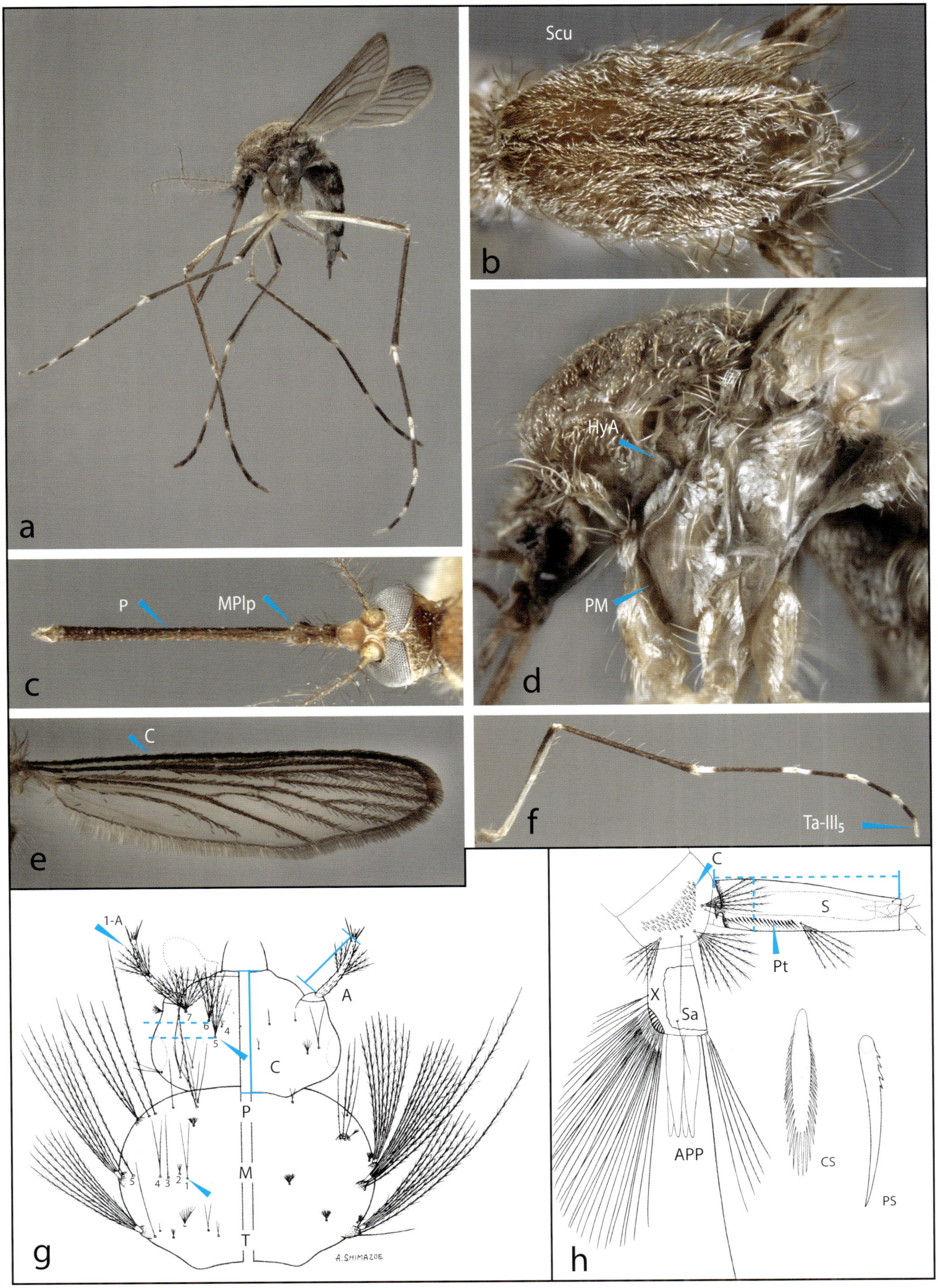
Scu
b
a
HyA
PM
d
P
MPlp
c
C
e
f
Ta-III5
1-A
A
C
7
6
4
5
P
M
T
5
4
3
2
1
A. SHIMAZOE
g
C
S
Pt
X
Sa
APP
CS
PS
h

Aedes (Ochlerotatus) caspius (Pallas, 1771) Subfamily Culicinae, Tribe Aedini "Caspian Old World Pointy Mosquito"

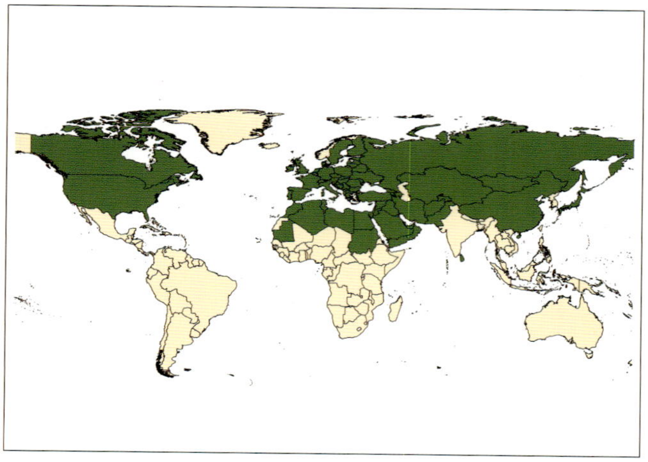

Systematics and Distribution

Named after the Caspian Sea from where it was originally described, *Aedes caspius* is one of the most commonly collected species in many areas of Europe and the Middle East. *Aedes caspius* occurs across the Palearctic, from Ireland to Mongolia and south into North Africa, the Arabian Peninsula, and Iran. The species is also reported from the Nearctic Region, in the United States and Canada. It is most closely related to the common Holarctic vector species *Ae. dorsalis* (Meigen). Overall, *Ae. caspius* tends to be found in hotter, drier areas than *Aedes dorsalis*.

Bionomics

Aedes caspius is common in coastal areas of the United Kingdom and in saline lakes and pools around the Mediterranean, where it is often numerous. In the remainder of its range, it is found in both fresh and saline marshes. Females deposit eggs in a wide variety of shaded or unshaded floodwater sites, with multiple generations per year. Some populations are autogenous, depositing one batch of eggs prior to a blood meal. Eggs are the dormant stage of the mosquito in areas with colder winters. Larvae can develop in fresh or salty water, and the species is often abundant in salt marshes. Like other salt marsh mosquitoes, females often make long-distance flights shortly after emergence. *Aedes caspius* can be a significant pest when it is abundant, biting day and night, indoors and out. At lower population levels, biting usually occurs outdoors during crepuscular periods. It is a competent vector of arboviruses and microsporidia and one of the only mosquitoes proven to transmit *Francisella tularensis*, the causative agent of tularemia.

Associated Pathogens

IKV, ISFV, RVFV, TAHV, WNV; *Cristulospora aedis*, *Francisella tularensis*, *Spiroplasma sabaudiense*.

Diagnostic Characters

(a) Habitus (♀). (b) Thorax (dorsal): Scutum (Scu) with golden scales and narrow dorsocentral lines of white scales. **(c) Abdomen (dorsal):** Terga (Te) with median pale stripe or all pale scaled. **(d) Wing:** Speckled with dark and pale scales; costa (C) predominantly dark scaled. **(e) Leg:** Ta-III_{1-4} with basal and apical bands. **(f) Larva (head):** Antenna (A) usually uniformly spiculate; seta 1-A 3–12 branched. **(g) Larva (terminal segments):** Comb scales (CS) 18–28, in triangular patch; siphon (S) with 1-S multibranched; pecten (Pt) with c. 20 spines (PS), reaching approximately to the middle of the siphon (S). Image sources: WRBU; Hopkins (1952).

Exemplar DNA barcodes: *Aedes caspius* (AF253024–25, KM452948, KM457568–69, KP942727, KP942758, KT361854–55, KT634315, KT876464, KT876468, MH559350, MH634424–25).

Catalog entry: Vol. 2, p. 755.

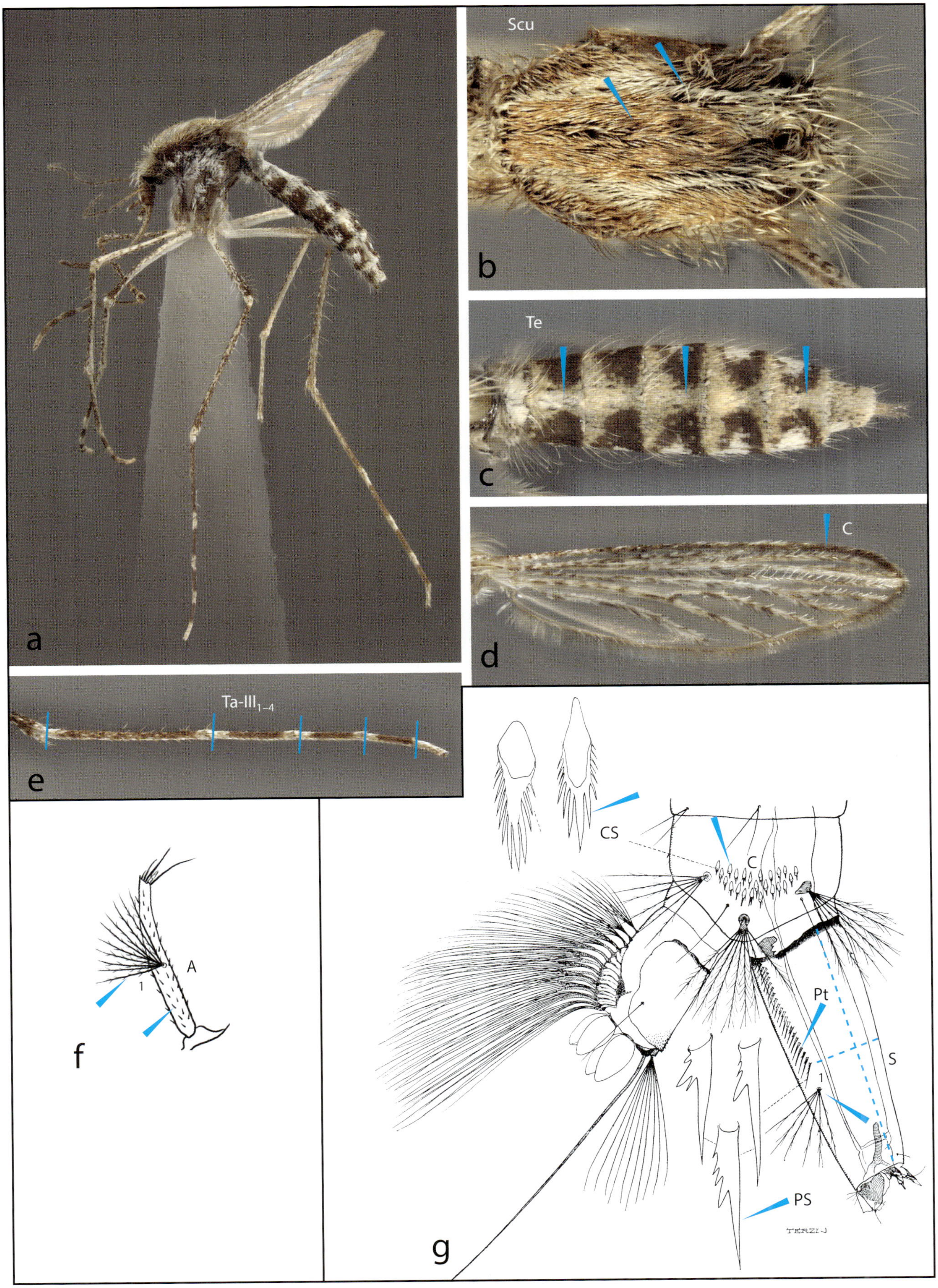
a
Scu
b
Te
c
C
d
Ta-III1–4
e
A
1
f
CS
C
Pt
S
1
PS
TERZI-J
g

Aedes (Ochlerotatus) dorsalis (Meigen, 1830)
Subfamily Culicinae, Tribe Aedini
"Ridged German Pointy Mosquito"

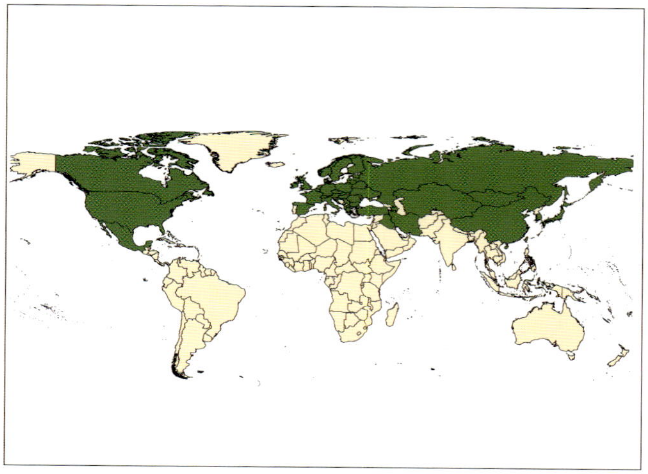

Systematics and Distribution

Aedes dorsalis is a floodwater mosquito, typified by its characteristic patterned scutum with two dark stripes. The species occurs in both the Nearctic (Canada south to Mexico) and the Palearctic, where it has been reported throughout Europe, reaching east into China, Mongolia, and Siberia. It is sympatric with its sister taxon *Ae. caspius* (see species page, p. 432) in certain regions of Europe and along the shores of the Black and Caspian Seas.

Bionomics

Aedes dorsalis immatures are salt tolerant. Larvae can be found in coastal and inland saline ponds, lakes, and marshes. They have been collected developing in up to a 12% concentration of salt in waters adjacent to the Great Salt Lake, Utah, United States, but in Europe, Mongolia, and Siberia, immatures are only reported from freshwater lakes. The adults start to emerge in early spring, and the population dwindles by late August. Males emerge before the females and form mating swarms 1.5–2 m above the ground.

Notoriously strong fliers, they have been observed to disperse up to 30 km from their larval habitats after mating. Females deposit their eggs deep (c. 2 cm) into the mud of drying water bodies in unshaded, grassy sites that are subject to flooding. Numbers of generations produced each year are climatically controlled, with more generations produced in favorable hot, wet summers. Facultative diapause occurs when temperatures drop below 15.5°C. Overwintering occurs as eggs, with habitat flooding stimulating egg eclosion the following spring. Female *Ae. dorsalis* uses a variety of hosts, exhibiting crepuscular-nocturnal activity. Females can be important nuisance pests to humans, biting incessantly, especially on cloudy days.

Associated Pathogens

Banna, CEV, CVV, GLOV, HPV, Liaoning, LLSV, SLEV, VRV, WEE, WNV.

Diagnostic Characters

(a) Habitus (♀). (b) Head (dorsal): Proboscis (P) dark, speckled with pale scales basally. **(c) Leg:** Ta-III_{1-3} with white rings basally and apically; Ta-III_4 with basal white ring and a few white scales at apex; Ta-III_5 almost entirely white. **(d) Thorax (side):** White hypostigmal scales (HySc) and postprocoxal scales (PSc) present. **(e) Abdomen (dorsal):** Nearly covered with white scales. **(f) Wing:** With distinct white areas, nonwhite areas speckled with black scales. **(g) Larva (head):** Seta 1-A inserted near middle of antenna, branched, long but not reaching tip of antenna; setae 5,6-C 1–4 branched. **(h) Larva (terminal segments):** Comb scales (CS) rounded apically, in irregular double row; saddle (Sa) incomplete, not encircling segment X; seta 1-X shorter than length of saddle; siphon (S) without setae except for 1-S and 2-S; siphon index c. 4.0; pecten (Pt) with spines (PS) evenly spaced. Image sources: WRBU; Tanaka et al. (1979).

Exemplar DNA barcodes: *Ae. dorsalis* (JF868943, JF868954, JX259585–92, JX260433–34, KC913569, KM571446, KP942725–26, KT358408–10, MG242488).

Catalog entry: Vol. 2, p. 759.

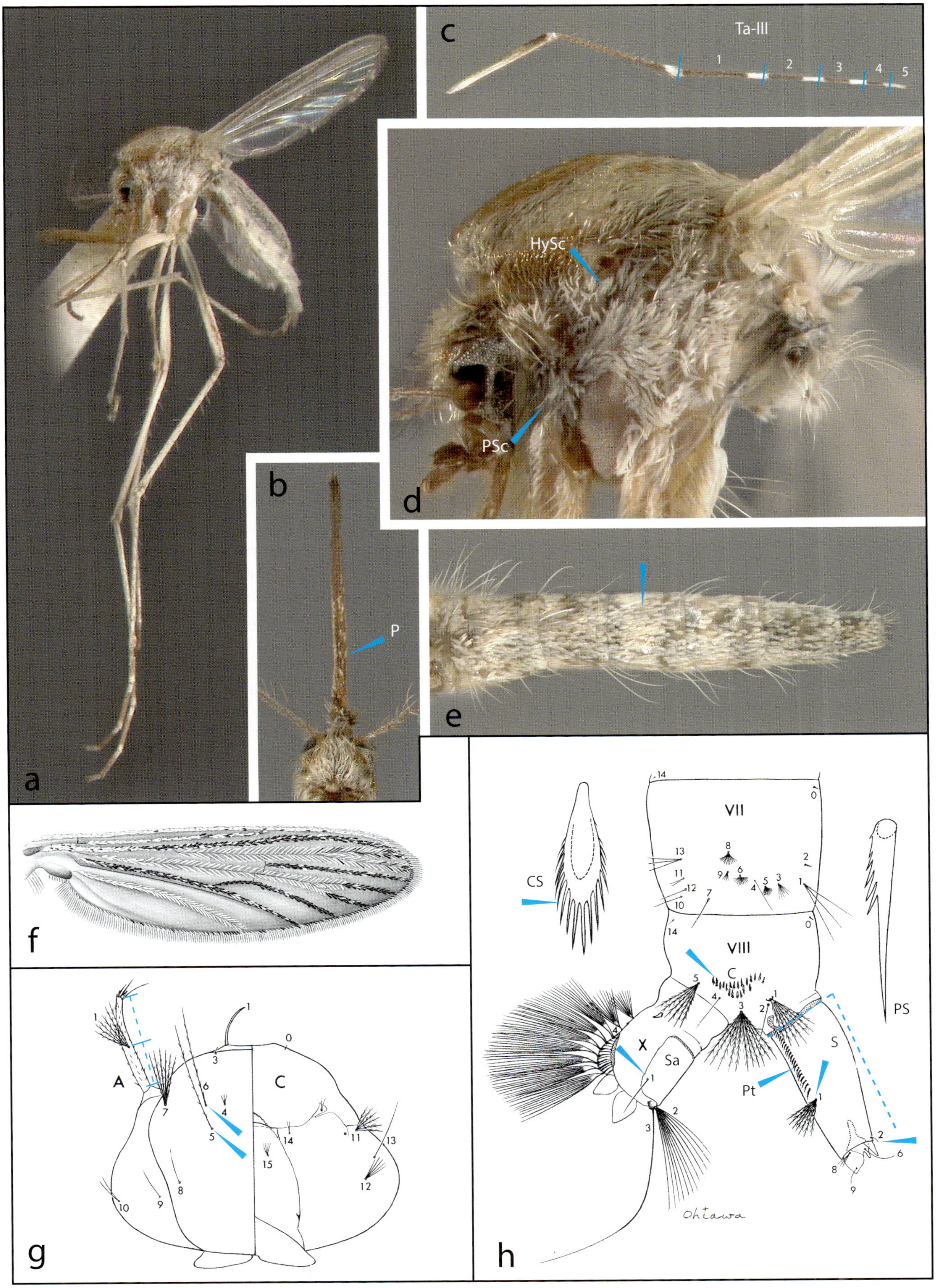
a
b
P
c
Ta-III
1
2
3
4
5
d
HySc
PSc
e
f
g
A
C
h
VII
VIII
CS
PS
C
X
Sa
S
Pt
Ohtawa

Aedes (Ochlerotatus) infirmatus Dyar & Knab, 1906
Subfamily Culicinae, Tribe Aedini
"Infirm American Pointy Mosquito"

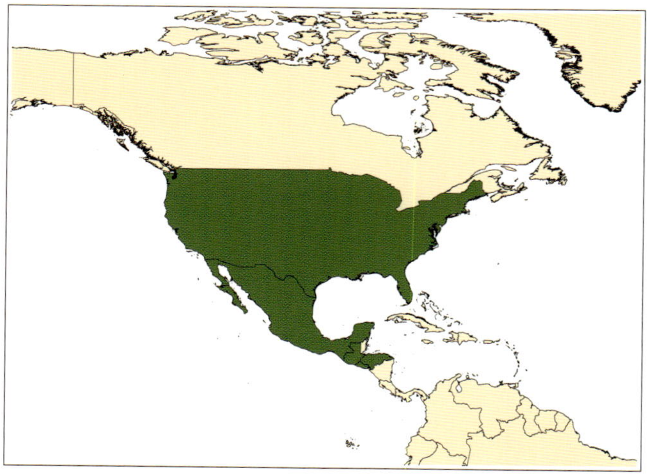

Systematics and Distribution

Aedes infirmatus is a medium-sized, dark-brown mosquito with a broad patch of silvery scales on its scutum. The species is one of 23 assigned to the Ochlerotatus Group of the subgenus *Ochlerotatus* (see Vol. 2, p. 1078, for other component taxa). Originally described from specimens collected in Baton Rouge, Louisiana, United States, the distribution of *Ae. infirmatus* is limited to woodland environments in Mexico and the southern United States.

Bionomics

Aedes infirmatus is closely associated with woodlands. *Ae. infirmatus* females are opportunistic and persistent biters and can become nuisance pests when numbers are high. The species will readily feed on passing hosts during the day in cool, shaded areas in or near the woods. Although host-seeking females have been collected near houses at night, they rarely enter the buildings to feed unless population sizes are high. Females lay their eggs in low-lying grassy areas liable to flooding, and hatching is stimulated by water inundation. The immatures rear through to adults quickly in these temporary floodwater pools and can be found, often in association with *Ae. atlanticus* Dyar & Knab, 1906 (see species page, p. 428), from early spring to late fall. The species overwinters as eggs.

Associated Pathogens

EEEV, KEYV, TENV, TVTV, WNV.

Diagnostic Characters

(a) Habitus (♀): Ti-III all dark; Fe-I–III dark apically. **(b) Abdomen (dorsal):** Terga (Te) with small to large pale basolateral scale patches. **(c) Thorax (dorsal):** Scutum (Scu) with broad anterior silvery-white or yellow area; scutal scales narrow and curved. **(d) Thorax (side):** Paratergite (PaSc) without scales; antepronotum (Ap) and postspiracular area (PA) without pale scales; subspiracular scales (SSc) pale; lower mesepimeral setae (MeSL) absent. **(e) Head (side):** Proboscis (P) entirely dark scaled. **(f) Larva (head). (g) Larva (terminal segments):** Comb scale (CS) with spinelike central spinule, 2.5–3.0 times length of lateral spinules; pecten (Pt) ending before or near seta 1-S; anal papillae (APP) <5 times length of saddle (Sa). **(h) Larva (abdominal segments):** Seta 6-I,II 2–3 branched; seta 6-III–VI usually single. Image sources: WRBU; Arnell (1976).

Exemplar DNA barcodes: *Ae. infirmatus* (JX259637–38).

Catalog entry: Vol. 2, p. 764.

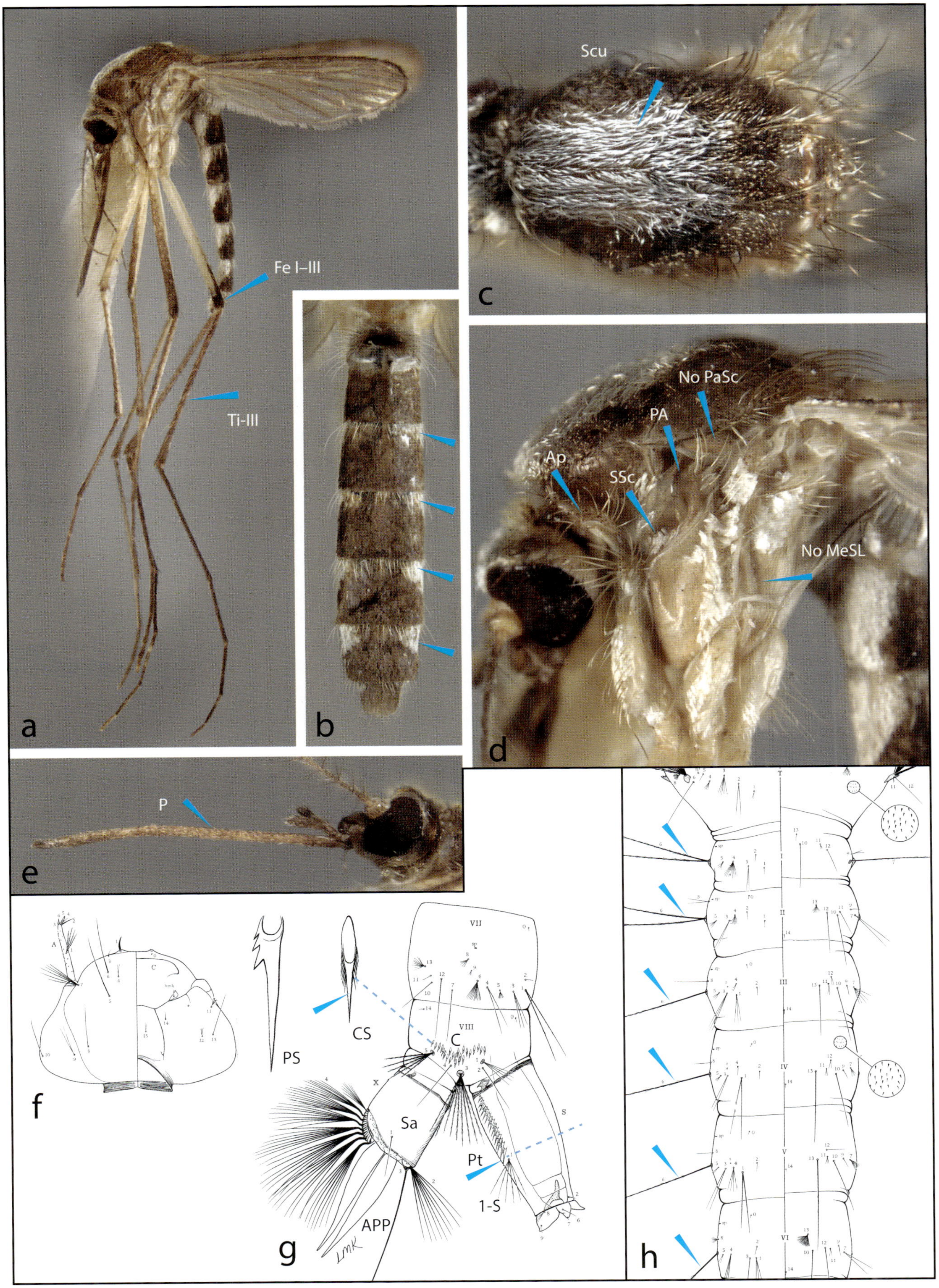

Scu
Fe I–III
Ti-III
No PaSc
PA
Ap
SSc
No MeSL
P
PS
CS
C
Sa
Pt
1-S
APP
a
b
c
d
e
f
g
h

Aedes (Ochlerotatus) melanimon Dyar, 1924
Subfamily Culicinae, Tribe Aedini
"Encephalitis American Pointy Mosquito"

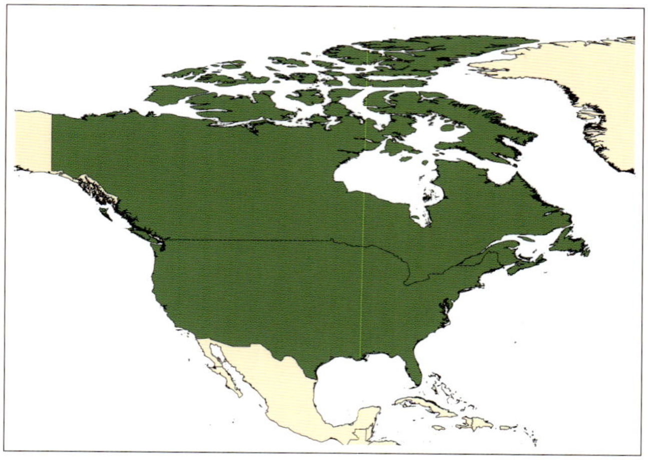

Systematics and Distribution

Aedes melanimon, first described from specimens collected in Bakersfield, California, is distributed throughout the western United States and Canada. Salt-and-pepper in appearance, its close morphological similarity to the Palearctic species *Ae. dorsalis* (Meigen) was noted by Dyar in his original description. It most probably takes its name from its primarily dark wing scales, which differentiate it from *Ae. dorsalis*, along with characteristics of the male genitalia. The species is one of four closely related North American species previously placed (including *Ae. canadensis* (Theobald), *Ae. campestris* Dyar & Knab, and *Ae. dorsalis*). With the exception of *Ae. canadensis*, these species lack clear diagnostic morphological characters and can only be reliably diagnosed using molecular techniques, leading to some confusion in distribution and bionomic attributions.

Bionomics

Aedes melanimon, *Ae. dorsalis*, and *Ae. nigromaculis* (Ludlow) form an ecological gradation of floodwater species occurring in the flooded fields of California. *Aedes nigromaculis* is the least salt tolerant and is dependent on flooded pastures with emergent vegetation. *Aedes dorsalis* is the most salt tolerant and occurs as a salt marsh mosquito, as well as in inland sites with alkaline water. *Aedes melanimon* tolerates brackish water but occurs in a variety of freshwater sites, including rice fields. It is locally abundant in areas of Central California, where agricultural practices create favorable larval habitats of temporary pools filled by occasional rain in irrigated pastures. Although it occurs in coastal areas, it is most abundant inland. The adults rest in long grasses close to the larval sites and can become significant pests, biting day or night.

Aedes melanimon feeds primarily on humans and large mammals but has also been noted to take blood meals from birds. The number of generations per year depends on the elevation at which it develops, and the species can be univoltine in mountain sites above 2,000 meters. Southern populations overwinter as eggs or larvae, whereas northerly populations estivate only as eggs. *Aedes melanimon* is a major vector of western equine and California group encephalitis viruses and is capable of transovarian transmission of CEV.

Associated Pathogens

CEV, CVV, LOKV, JSV, MDV, SLEV, WEEV, WNV.

Diagnostic Characters

(a) Habitus (♀). **(b) Thorax (side):** Postprocoxal scales (PSc) present. **(c) Leg:** Ta-III usually with pale basal and apical bands covering articulations. **(d) Wing:** Dark and pale scales intermixed; vein 1A and costa (C) with considerably more black than pale scales. **(e) Abdomen (dorsal):** VII-Te usually with more dark than pale scales. **(f) Larva (head):** Total branches of setae 5-C + 6-C on both sides ≤ 6; Seta 1-M much shorter than antenna and much smaller than 3-M (not shown). **(g) Larva (terminal segments):** Comb scales (CS) usually with prominent pointed central spinule; pecten spines (PS) in regular row reaching to middle of siphon (S); seta 1-S inserted at or beyond middle of siphon; saddle (Sa) incomplete, not encircling segment X; seta 1-X weak, single, and no longer than saddle. Image sources: WRBU; Bohart & Washino (1978).

Exemplar DNA barcodes: *Aedes melanimon* (JX259647, KM571866).

Catalog entry: Vol. 2, p. 766.

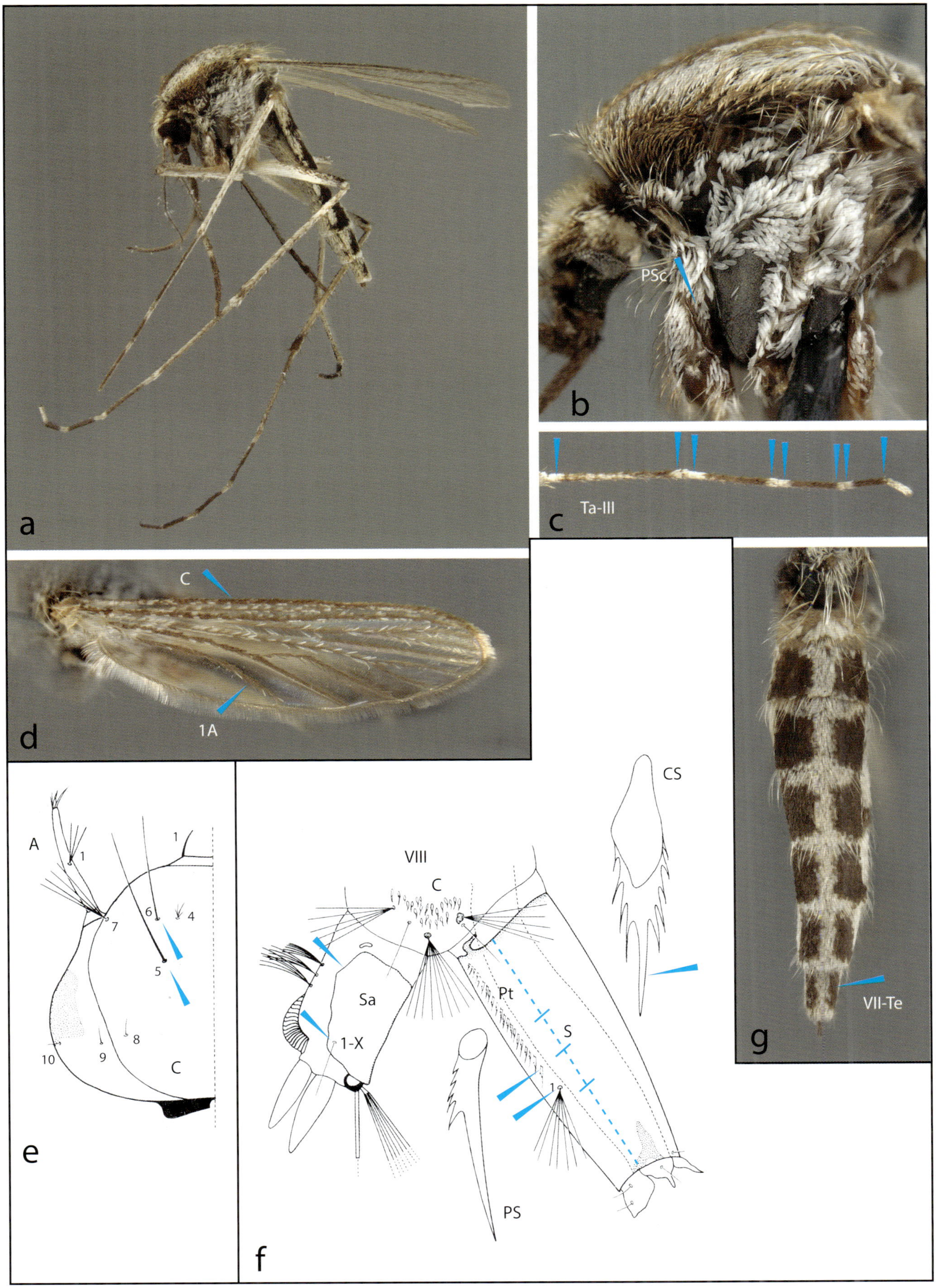
a
PSc
b
Ta-III
c
C
1A
d
VII-Te
g
A
1
7
6
4
5
10
9
8
C
e
VIII
C
Sa
1-X
Pt
S
1
PS
CS
f

Aedes (Ochlerotatus) scapularis (Rondani, 1848)
Subfamily Culicinae, Tribe Aedini
"Scapular Brazilian Pointy Mosquito"

Systematics and Distribution

Aedes scapularis is named after the distinctive hexagonal patch of yellowy-white narrow scales on its "shoulders" (see *c*). It is the type species of both the subgenus *Ochlerotatus* and the Ochlerotatus Group, the largest informal group in the subgenus (see Vol. 2, p. 1078, for a list of the 23 component taxa). *Aedes scapularis* is closely related to *Ae. phaeonotus* and *Ae. comitatus* Arnell (endemic to Grenada and Colombia, respectively), and all bear a long line of white scales on the anterior side of the hind tibia (Ti-III) that is unique within the group (see *b*). *Aedes scapularis* is a widespread species, occurring throughout most of tropical and subtropical America from the southern United States through Central and South America to Argentina, including the islands of the Caribbean. It occurs at low to moderate elevations and is one of the most common and widely distributed mosquitoes in the New World.

Bionomics

Eggs of *Ae. scapularis* are tolerant to desiccation, and females oviposit in grasses prone to flooding during periods of high rainfall. Although most common in temporary freshwater pools with soil bases, immatures have also been collected in ponds, swamp margins, rock holes and crabholes, and a variety of artificial containers. They can be especially abundant at the vegetated margins of large low-elevation lakes. Under those circumstances, large banks of eggs may be flooded at once to produce mosquito densities abundant enough to kill cattle. *Aedes scapularis* appears well adapted to man-made environments and is present in great numbers in large rural and urban environments, especially urban parks. It readily enters homes to feed. Host-seeking females in Brazil are reportedly most active during the evening crepuscular period, but *Ae. scapularis* populations in Suriname and French Guyana are reportedly diurnal, with biting peaks between 1200h and 1500h. DNA barcoding analysis has shown high genetic diversity in *Ae. scapularis* populations in Brazil, with genetically similar clusters correlated with diagnosable wing geometric patterns. All these bionomic and morphological differences warrant further study to elucidate component cryptic taxa within *Ae. scapularis* s.l. across its range.

Associated Pathogens

IHLV, KRIV, LUKV, MAGV, MAYV, MELV, OROV, PLAV, ROCV, SLEV, VEEV, WYOV, YFV; *Dirofilaria immitis*, *Wuchereria bancrofti*.

Diagnostic Characters

(a) Habitus (♀). (b) Leg: Ti-III anteriorly and Ta-III basally with white scales in long stripes. **(c) Thorax (dorsal):** Scutum (Scu) with broad area of yellowish to white scales extending from anterior margin to a little beyond middle; scutal (Scu) scales narrow and curved. **(d) Thorax (side):** Lower mesepimeral setae (MeSL) absent. **(e) Larva (terminal segments):** Comb scales (CS) in 2–3 rows, evenly fringed; pecten spines (PS) evenly spaced ending just beyond middle of siphon (S). **(f) Larva (head):** Seta 5-C single. **(g) Larva (thorax/abdominal segments):** Integumental spicules long and conspicuous; seta 3-P with ≥3 branches; seta 1-III usually three branched. Image sources: WRBU; Arnell (1976).

Exemplar DNA barcodes: *Ae. scapularis* (KT766398–404, MF172265–67).

Catalog entry: Vol. 2, p. 771.

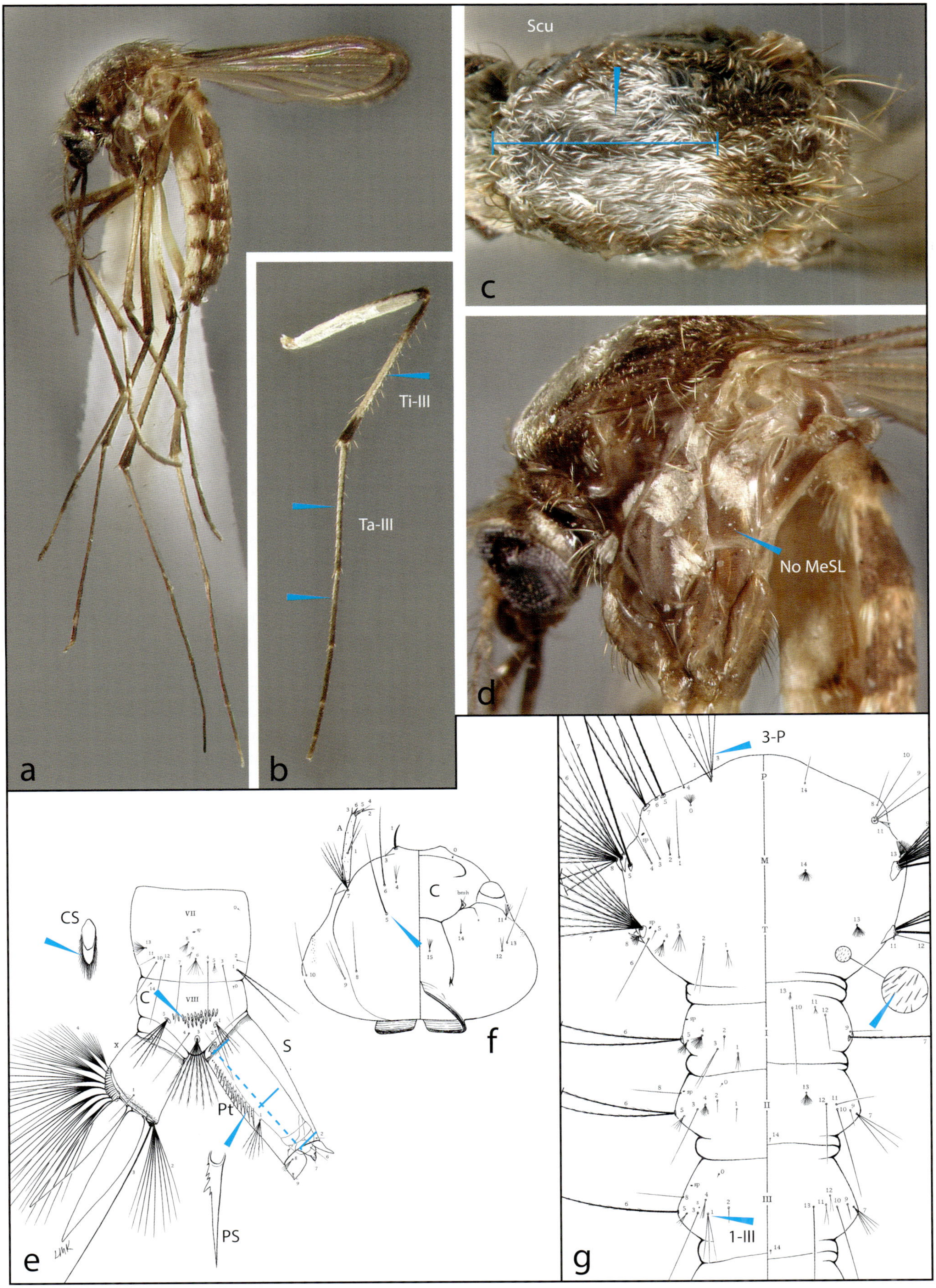
Scu
Ti-III
Ta-III
No MeSL
3-P
P
M
T
I
II
III
1-III
CS
VII
VIII
C
x
S
Pt
PS
A
C
a
b
c
d
e
f
g

Aedes (Ochlerotatus) sollicitans (Walker, 1856)
Subfamily Culicinae, Tribe Aedini
"Disturbing American Pointy Mosquito"

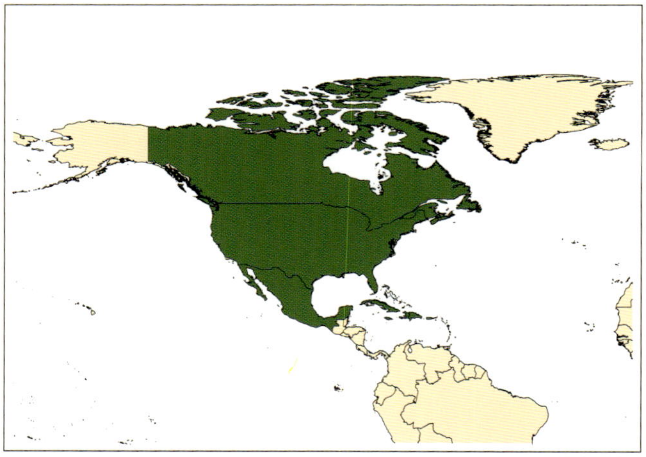

Systematics and Distribution

Aedes sollicitans sports a thick stripe of white scales across its thorax, contrasting with the golden-brown scales of the scutum. Its leg joints are white, and all tarsi are banded with thick white stripes with a distinctive median yellow stripe. *Aedes sollicitans* is a salt marsh floodwater mosquito, originally described from Charleston, South Carolina, United States. It dominates many coastal areas of the eastern and southern United States, goes as far north as northeastern Canada, and extends south into Mexico and the Caribbean. Populations of *Ae. sollicitans* have also been found in inland brackish water swamps in the United States, particularly in oil fields. *Aedes sollicitans* is closely related to *Ae. taeniorhynchus* (Wiedemann) (see species page, p. 446), which, along with *Ae. mitchellae* (Dyar) and *Ae. nigromaculis* (Ludlow), make up the Culicelsa Group of the subgenus *Ochlerotatus* (see Vol. 2, p. 1078).

Bionomics

Females often undertake long-distance flights inland or across bodies of water to seek suitable hosts, finding their way back to the salt marsh for oviposition. *Aedes sollicitans* most commonly deposits its eggs in areas of salt marshes that flood as a result of high tides, causing fresh water from storms to flood the upper areas of the marshes. These areas tend to be less subject to the flushing action of tides and also to the presence of predatory fish. Because of the extensive nature of these kinds of sites in coastal locales, *Ae. sollicitans* can occur in extremely high numbers. They are exceptionally strong fliers, and adults often migrate in large numbers to human settlements far from their breeding sites (up to 100 km). They are incredibly aggressive, biting both day and night, even in bright sunlight, and their bites leave a painful, distinct sting, adding to their pest status. Blood is taken from a wide variety of vertebrate hosts, and the species is an important pest of humans. *Aedes sollicitans* overwinters as eggs along the North Atlantic Seaboard, with populations active from April or May until October. In the far south of their distribution, populations are ever present.

Associated Pathogens

CVV, EEEV, KEYV, RVFV, VEEV, WNV.

Diagnostic Characters

(a) Habitus (♀): Fe-I–III and Ti-I–III speckled with pale scales. **(b) Abdomen (dorsal):** Terga (Te) with white to yellow dorsomedian longitudinal stripe; lateral pale scales white, basal and medial scales pale yellow. **(c) Thorax (side):** Postprocoxal scales (PSc) white. **(d) Head (dorsal):** Proboscis (P) with broad median pale band. **(e) Wing:** Speckled with dark and pale scales. **(f) Leg:** Ta-III$_1$ with median yellowish band; Ta-III$_{1-4}$ with broad basal pale bands; Ta-III$_5$ entirely white. **(g) Larva (head):** Seta 1-A long, reaching tip of antenna; setae 5,6-C single, gradually tapering. **(h) Larva (terminal segments):** Comb (C) with sharply pointed scales (CS) with long median spinule; saddle (Sa) complete, posterior margin without spicules; siphon index c. 2.0–2.5; seta 1-VIII ≤0.5 times siphon length; seta 1-S inserted beyond pecten (Pt); seta 2-X with ≥6 branches, distinctly shorter than seta 3-X. Image sources: WRBU; Bohart & Washino (1978).

Exemplar DNA barcodes: *Ae. sollicitans* (HQ978779, GU907937-41, MG242514, MN520650 as *Ochlerotatus sollicitans*).

Catalog entry: Vol. 2, p. 772.

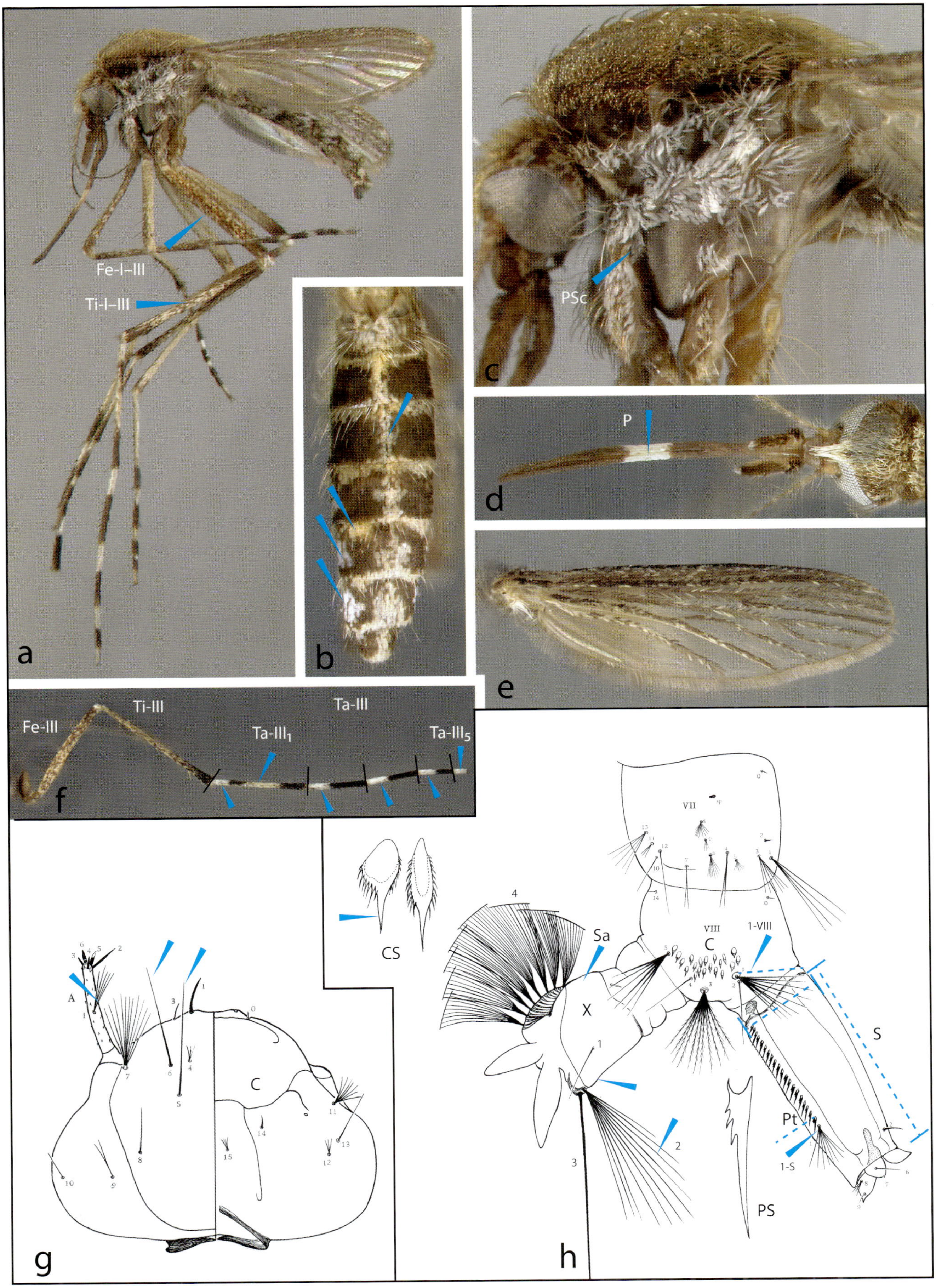
Fe-I–III
Ti-I–III
a
b
PSc
c
P
d
e
Fe-III
Ti-III
Ta-III
Ta-III1
Ta-III5
f
CS
A
C
g
VII
VIII
1-VIII
Sa
C
X
S
Pt
1-S
PS
h

Aedes (Ochlerotatus) squamiger (Coquillett, 1902)
Subfamily Culicinae, Tribe Aedini
"Scaly American Pointy Mosquito"

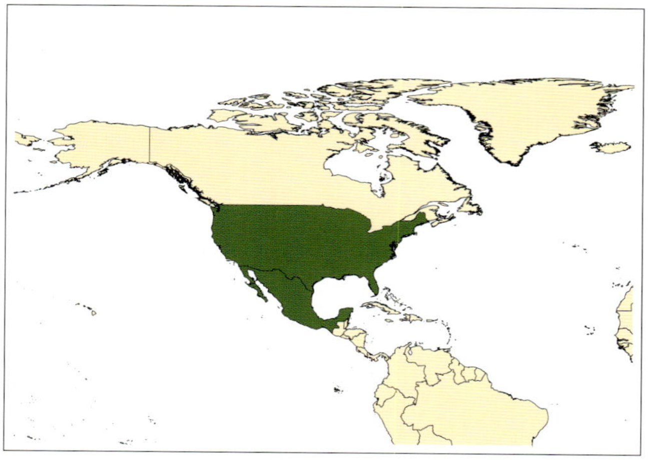

Systematics and Distribution

Aedes squamiger is a coastal mosquito with a restricted distribution, being endemic to the West Coast of the United States and Baja California, Mexico. It occurs as far north as Bodega Bay in Sonoma County, California. *Aedes squamiger* was originally described based on specimens collected in Palo Alto, California. The species remains unplaced with regard to informal groups within the subgenus *Ochlerotatus*, but it is thought to be most closely related to the high mountain snowmelt mosquito, *Ae. hexodontus* Dyar. It has one synonym, *deniedmannii* Ludlow, described from Benicia in California.

Bionomics

Aedes squamiger is a salt marsh, floodwater mosquito with an unusual winter life cycle. Its annual generation begins with females depositing drought-resistant eggs in portions of the salt marsh protected from strong tidal flow but flooded by the combination of high tides and rain that occur in this region in the winter. Human construction sites, such as housing developments, roads, and railroads, tend to create suitable habitat, whereas appropriate larval sites are relatively scarce in natural salt marshes. In the northern part of its range where it is most abundant, neonates emerge in early winter when water temperatures go below about 10°C, and larval development continues slowly through the winter in the cold, highly oxygenated water. Adults emerge in huge cohorts, mate, and then females travel dozens of kilometers inland in search of blood meals. Diapause is interrupted in April, following the temperature rise and early spring rains. As a result of these attributes, the species is a significant human pest in California. Low numbers of *Ae. squamiger* have been collected in CO_2-baited encephalitis-virus-surveillance (EVS) traps.

Associated Pathogens

WNV.

Diagnostic Characters

(a) **Habitus (♀):** Proboscis (P) with many dark and pale scales intermixed, without median pale band; $Ta\text{-}III_{1-5}$ with broad, only basal pale bands; Pale band on $Ta\text{-}III_2 \geq 0.3$ of segment. **(b) Abdomen (dorsal):** Terga (Te) with basal pale bands. **(c) Thorax (dorsal):** Scutum (Scu) with mixed brown and pale scales laterally. **(d) Thorax (side):** Postpronotum (Ppn) without contrastingly colored scale patches. **(e) Wing:** With broad triangular-shaped, dark, and pale scales evenly intermixed; mixed black-and-white tertiary fringe scales (not distinct in photo). **(f) Larva (head):** Antenna (A) spiculate; seta 1-A multibranched; seta 5-C 1–4 branched; seta 6-C usually single or double. **(g) Larva (terminal segments):** Comb scales (CS) fringed with subequal spinules; seta 1-X as long or longer than Sa; saddle (Sa) incomplete; siphon (S) with one pair of setae, excluding seta 2-S. Image sources: WRBU; Carpenter & LaCasse (1955).

Exemplar DNA barcodes: *Ae. squamiger* (JX259673–75).

Catalog entry: Vol. 2, p. 773.

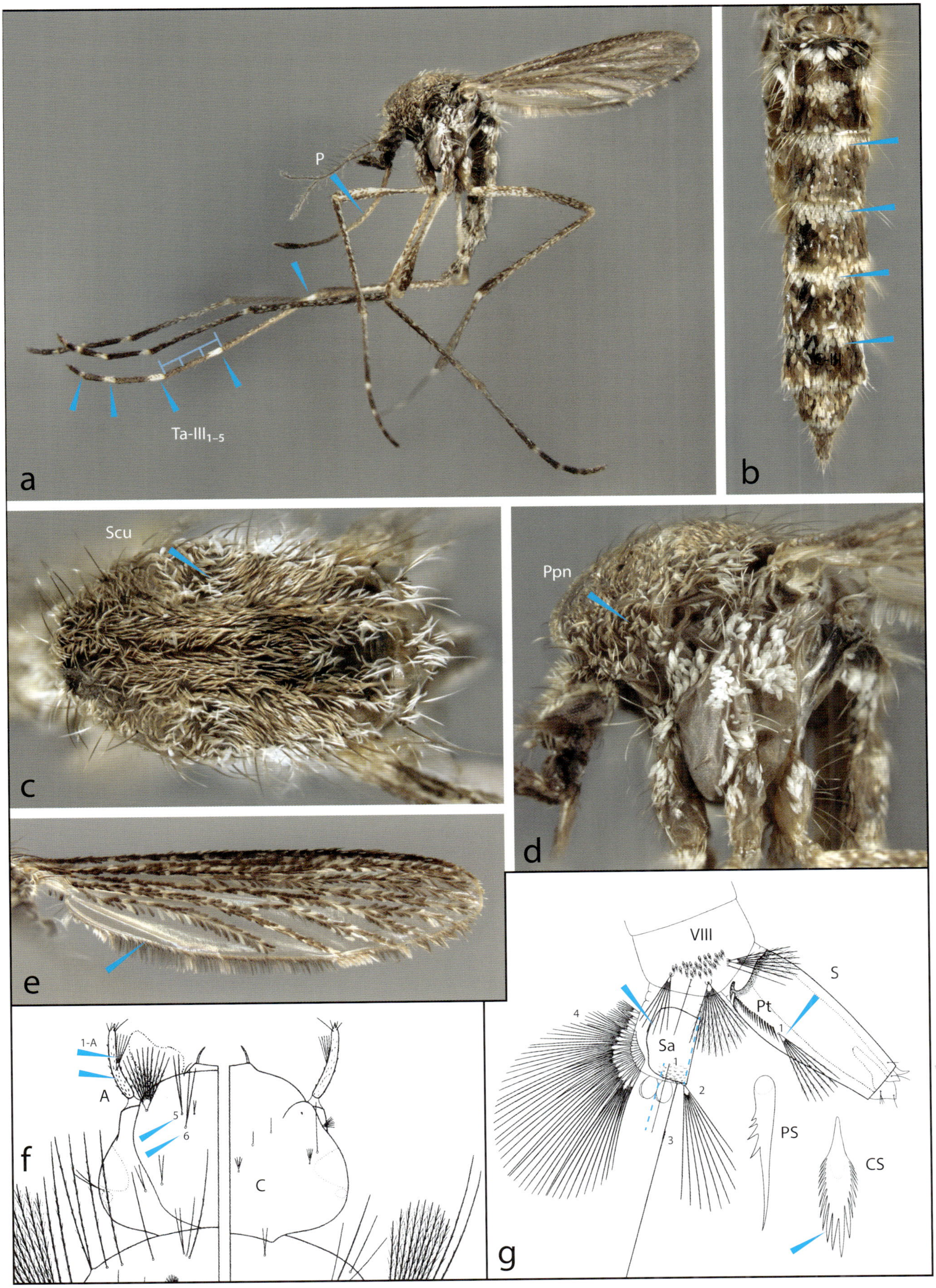

P
Ta-III1–5
a
b
Scu
c
Ppn
d
e
1-A
A
5
6
f
C
VIII
S
4
Pt
Sa
1
2
3
PS
CS
g

Aedes (*Ochlerotatus*) *taeniorhynchus* (Wiedemann, 1821)
Subfamily Culicinae, Tribe Aedini
"Band-beaked Mexican Pointy Mosquito"

Systematics and Distribution

Aedes taeniorhynchus is a medium to small dark mosquito with white-tipped palps and banding on the proboscis, abdomen, and tarsi. *Aedes taeniorhynchus* is predominantly a salt marsh mosquito, occurring along the Atlantic coast of the United States from New England southward (including most Caribbean islands) to Brazil, and on the Pacific coast of the United States from Southern California down through Central America to Peru.

Together with *Ae. mitchellae* (Dyar), *Ae. nigromaculis* (Ludlow), and *Ae. sollicitans* (Walker) (see species page, p. 442), *Ae. taeniorhynchus* belongs to the Culicelsa Group of the subgenus *Ochlerotatus*. The species has five synonyms: *damnosus* Say from the United States, *niger* Giles from Antigua, *portoricensis* Ludlow from Puerto Rico, *epinolus* Dyar & Knab from Peru, and *pix* Martini from Belize. Although most specimens of *Ae. taeniorhynchus* sport a broad median white band on the proboscis, these bands were notably absent in around 10% of samples collected in Virginia and North Carolina. Specimens from Jamaica and other Antillean islands are also considerably darker than typical exemplars, and these morphological differences could denote the presence of hidden biodiversity.

Bionomics

Aedes taeniorhynchus females feed predominantly on large mammals, including cattle and humans. Although peak biting times are crepuscular, this species is highly aggressive and will feed at any time of the day—even in bright sunlight. Females are strong fliers, even under windy conditions, undertaking long flights of dozens of kilometers to search for suitable hosts. *Aedes taeniorhynchus* often migrate in large numbers to brackish-water inland sites, including oil fields, many miles from the coast. The species' high density and aggressive biting render it a nuisance pest in many parts of its range, particularly in the southern United States and in coastal regions of the Caribbean islands.

Mating occurs in swarms, close to the emergence sites, and females need a blood meal to produce their first egg batch. Females deposit their eggs in the open at the muddy bases of vegetation in salt marshes and mangroves. Immatures are commonly found in transient pools that can be fresh but are more normally at least slightly brackish. In Jamaica, *Ae. taeniorhynchus* was reported from large crabholes and in irrigation channels. Population explosions are tightly linked to periods of heavy rainfall and high tides in the mid- and late summer. Adults and immatures are found year-round in the far south of its range, but the species overwinters as eggs in the north.

Associated Pathogens

CVV, EEEV, EVEV, EYEV, GLV, KEYV, ORIV, PLAV, RVFV, TENV, TVTV, VEEV, WNV, WYOV.

Diagnostic Characters

(a) Habitus (♀): Ta-III$_1$ without median white band; Ta-III$_{1-5}$ with basal white bands. **(b) Wing:** All dark scaled. **(c) Thorax (side):** Postprocoxal membrane (PM) bare. **(d) Head (side):** Proboscis (P) with broad, median pale band. **(e) Leg:** Fe-I and Ti-I all dark. **(f) Abdomen (dorsal):** Terga (Te) with basal pale bands, without central spots or median stripe. **(g) Larva (head):** Setae 5,6-C single, gradually tapering. **(h) Larva (terminal segments):** Comb scales (CS) rounded apically, 9–20 in irregular patch; seta 1-VIII at least 0.5 as long as siphon (S); siphon about as wide as segment X; seta 1-S inserted beyond pecten (Pt); saddle (Sa) complete with many posterior spicules; anal papillae (APP) shorter than segment X, without tracheae; seta 2-X with ≥6 strong branches, distinctly shorter than 3-X. Image sources: WRBU; Belkin et al. (1970).

Exemplar DNA barcodes: *Ae. taeniorhynchus* (JX259676, KT766538–47, MG242518, KY859919–20).

Catalog entry: Vol. 2, p. 775.

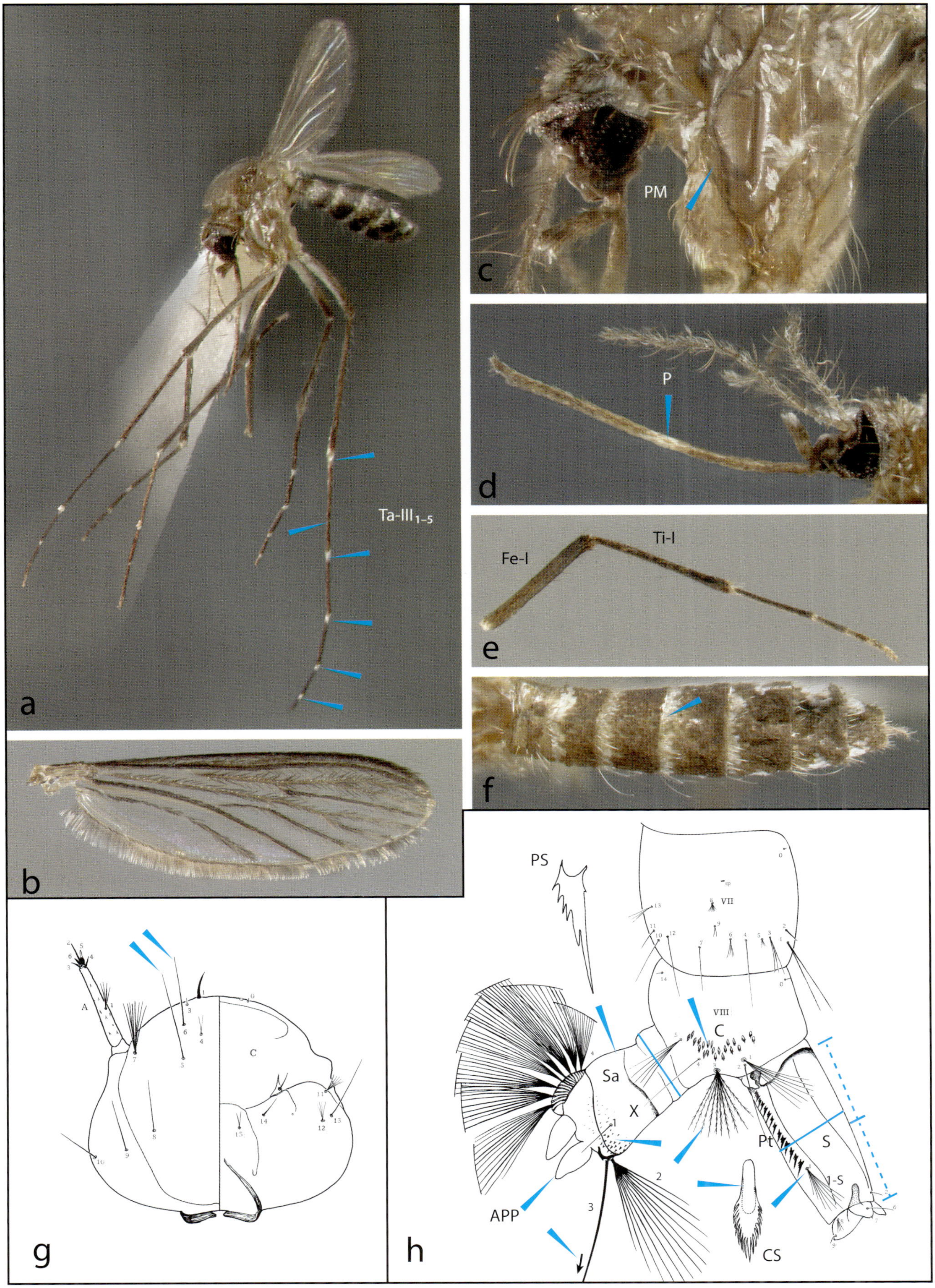
a
Ta-III1–5
b
c
PM
d
P
e
Fe-I
Ti-I
f
g
A
C
h
PS
VII
VIII
C
Sa
X
Pt
S
1-S
APP
CS

Aedes (Ochlerotatus) trivittatus (Coquillett, 1902)
Subfamily Culicinae, Tribe Aedini
"Three-banded American Pointy Mosquito"

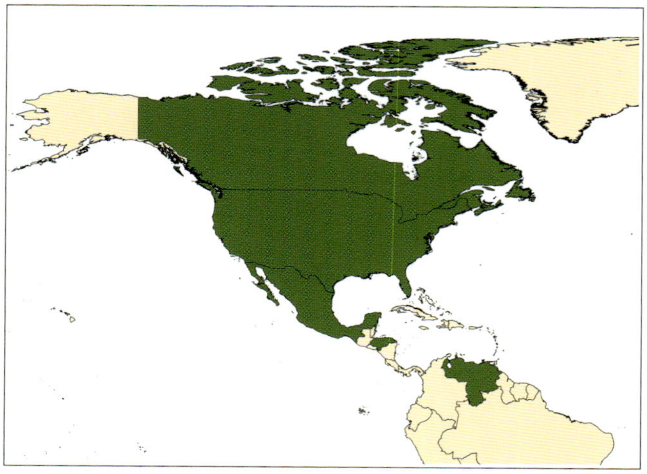

Systematics and Distribution

Aedes trivittatus is named after the three characteristic stripes of black scales that decorate its scutum. It is one of four closely related floodwater mosquito species in the Ochlerotatus Group, which also includes *Ae. atactavittatus* Arnell, *Ae. angustivittatus* Dyar & Knab, and *Ae. meprai* Martinez & Prosen. Of these, *Ae. trivittatus* and *Ae. angustivittatus* are the only two species implicated in pathogen transmission. *Aedes trivittatus* is common in the eastern and central United States, and its distribution reaches north into southern Canada and south into Venezuela. It is one of the few Nearctic representatives of the Neotropical Scapularis Group. There are indications that *Ae. trivittatus* may hybridize with *Ae. angustivittatus* in the southern part of its range.

Bionomics

Aedes trivittatus females are multivoltine, with adults surviving some five to six weeks. Females deposit their eggs (1–100 per batch, mean = 55) at the grassy margins of freshwater ponds, streams, or temporary pools. Characteristically, this species only places its eggs in locations with shade, even if provided by a single tree. *Aedes trivittatus* overwinters as eggs, emerging in early June, with adults active until September. Larvae can be found throughout the summer in flooded meadows and woodland pools. As with other members of the Scapularis Group, the larval thoracic and abdominal integument is spiculose, giving the larvae a mottled appearance. Minimum time from egg to adult emergence is eight days.

Biting females are usually also concentrated in forested areas or in the ecotone between woods and open fields. Although more active at dusk, the species is active day and night, which may explain its association with shaded areas. The females will feed even in bright sunshine. *Aedes trivittatus* is at once catholic yet highly plastic in its feeding strategy, with an overwhelming preference for cottontail rabbits. Other occasional hosts include humans, birds, raccoons, and cattle, with infrequent meals taken from a multitude of other mammals (horses, squirrels, opossums, cats) and amphibians and reptiles. Where populations are high following heavy rains, *Ae. trivittatus* becomes a troublesome pest of humans.

Associated Pathogens

EEEV, LACV, TVTV, WEEV, WNV; *Dirofilaria immitis*.

Diagnostic Characters

(a) Habitus (♀): All tarsi without pale rings. **(b) Thorax (side):** Lower mesepimeral setae (MeSL) absent. **(c) Leg. (d) Thorax (dorsal):** Scutum (Scu) with dorsocentral line of white or occasionally yellowish-white scales broadened at and beyond supraalar area (SaA); scutal (Scu) scales narrow and curved. **(e) Larva (head):** Setae 5,6-C single. **(f) Larva (terminal segments):** Spinules of comb scales (CS) subequal, with longer moderately differentiated median spinule; pecten (Pt) ends at, or near, level of seta 1-S. **(g) Larva (abdominal segments):** Seta 6-I,II double (rarely single); seta 6-III–VI single (rarely double); seta 13-III usually single, subequal in length to seta 13-IV,V; thoracic and abdominal integument with numerous conspicuous spicules. Image sources: WRBU; Arnell (1976).

Exemplar DNA barcodes: *Ae. trivittatus* (JX259689–97, KR681661).

Catalog entry: Vol. 2, p. 776.

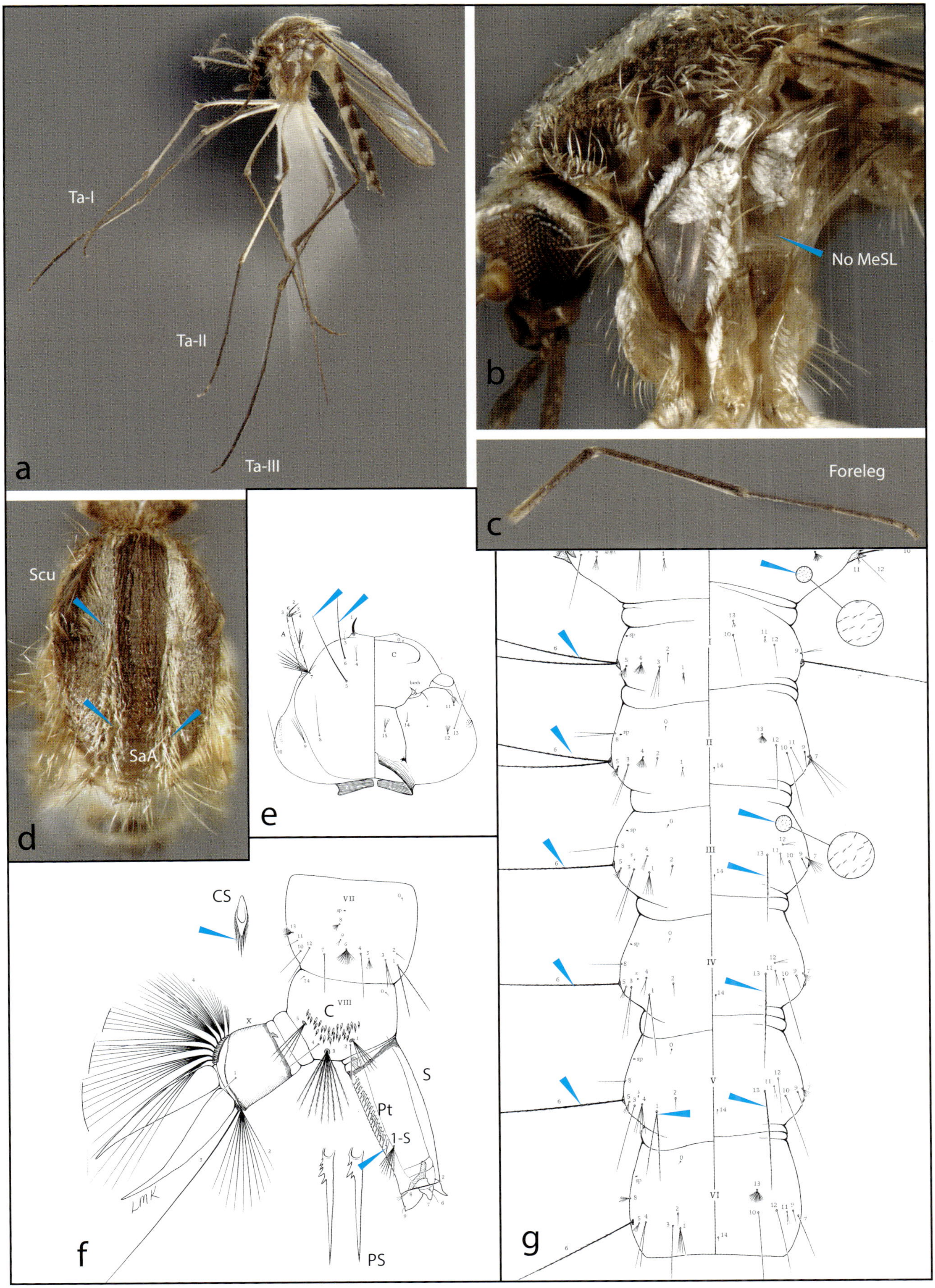

a
Ta-I
Ta-II
Ta-III
b
No MeSL
c
Foreleg
d
Scu
SaA
e
f
CS
C
S
Pt
1-S
PS
g

Aedes (Ochlerotatus) vigilax (Skuse, 1889)
Subfamily Culicinae, Tribe Aedini
"Watchful Australian Pointy Mosquito"

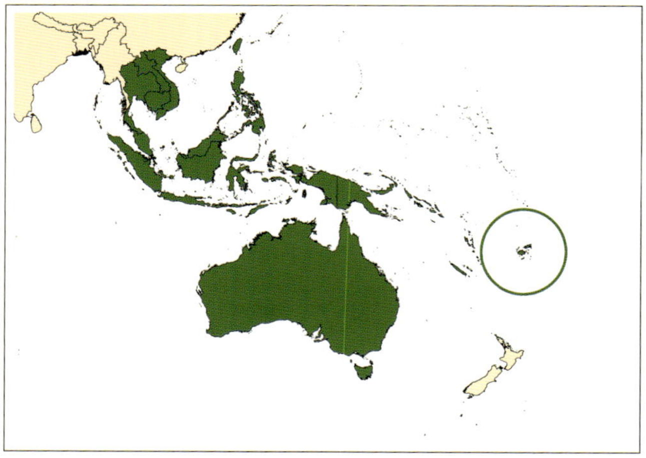

Systematics and Distribution

The body of *Aedes vigilax* has a drab salt-and-pepper appearance, in sharp contrast to the broad white stripes donning its abdomen and tarsi. Originally described from populations in New South Wales and Queensland, Australia, *Ae. vigilax* is widely distributed in salt marshes and mangrove swamps across the Australasian Region. The species has five synonyms and two valid subspecies: *Ae. vigilax ludlowae* (Blanchard) from the Philippines, Thailand, and Taiwan, and *Ae. vigilax vansomerenae* Mattingly from the Seychelles. *Aedes vigilax* and its two subspecies comprise the informal Empihals Group of the *Aedes* subgenus *Ochlerotatus*.

Bionomics

Eggs of *Ae. vigilax* appear to be less drought-resistant than other *Aedes*, and oviposition choices minimizing desiccation appear to be a key survival strategy in these mosquitoes. Although eggs have been recovered from samphire and the pneumatophores of the grey mangrove, female *Ae. vigilax* preferentially seeks out marine couch (*Sporobolus* spp.) plants and their sods, attaching the eggs between 0 cm and 4 cm from the ground, perhaps using exochorionic exudates and the rough texture of the cigar-shaped egg to minimize egg detachment from these humid spots. After flooding of the salt marshes occurs, the eggs undergo installment hatching, and newly hatched larvae are dispersed across the marshes one to two hours after tidal inundation. First and second instar larvae actively move to the lower terrain, where water is liable to gather in brackish water ground pools large enough to facilitate their successful eclosion.

Aedes vigilax is a powerful flier and can travel considerable distances from coastal breeding sites to find its hosts. Documented hosts of *Ae. vigilax* include dogs, birds, humans, brushtail possums, cats, and flying foxes. Females are aggressive biters and will bite any time of the day or night, peaking about sunset. Somewhat unsurprisingly, given its opportunistic feeding habits, *Ae. vigilax* is involved in the transmission of many arboviruses, and the species is recognized as the principal vector of nonperiodic filariasis in New Caledonia and the Loyalty Islands.

Associated Pathogens

BAMV, BEFV, BFV, EHV, GETV, GGV, JFV, KOKV, MVEV, RRV, SINV, TERV, YACV; *Dirofiliaria immitis*, *Wuchereria bancrofti*.

Diagnostic Characters

(a) Habitus (♀): Terga (Te) with white basal bands of uniform width; Ta-I–III with distinct white basal bands. **(b) Head (dorsal):** Decumbent scales of vertex (V) mostly narrow; erect scales of head cover most of dorsal surface. **(c) Thorax (side):** Antepronotum (ApSc) with scales; proepisternum (PeSc) with scales; subspiracular area without scales (SSc); postspiracular patch of broad whitish scales (PoSc); paratergite (PaSc) with scales; prealar area with patch of broad white scales (PrSc); lower mesepimeral setae (MeSL) absent; scutum (Scu) without distinctive markings; scutum (Scu) just anterior to scutellum with anterior pale scale patch entirely white, sometimes with pale brown scales along acrostichal line (not shown). **(d) Leg:** Ta-III with basal pale bands ≥0.25 length of tarsomeres. **(e) Head (side):** Basal 0.67 of proboscis (P) with mid-ventral pale scaling. **(f) Wing:** With small areas of white scaling. **(g) Larva (thorax):** Seta 5-P single or double. **(h) Larva (abdominal segments):** Most abdominal setae comparatively short; seta 1-III–V short, branched; seta 6-III branched. **(i) Larva (thorax):** Lateral spine of plate bearing setae 9–12-M,T small and lightly pigmented. **(j) Larva (head):** Seta 4-C short, nearer level of 6-C than 5-C; setae 11,12-C short. **(k) Larva (terminal segments):** Comb scales (CS) in 1–3 irregular rows; saddle (Sa) incomplete; anal papillae (APP) shorter than 0.5 times length of saddle; seta 3-X single; siphon index <2.0; pecten spines (PS) composing pecten (Pt) evenly spaced, none widely spaced beyond seta 1-S. Image sources: WRBU; Belkin (1972).

Exemplar DNA barcodes: *Ae. vigilax* (GQ143720–32, MG242526).

Catalog entry: Vol. 2, p. 776.

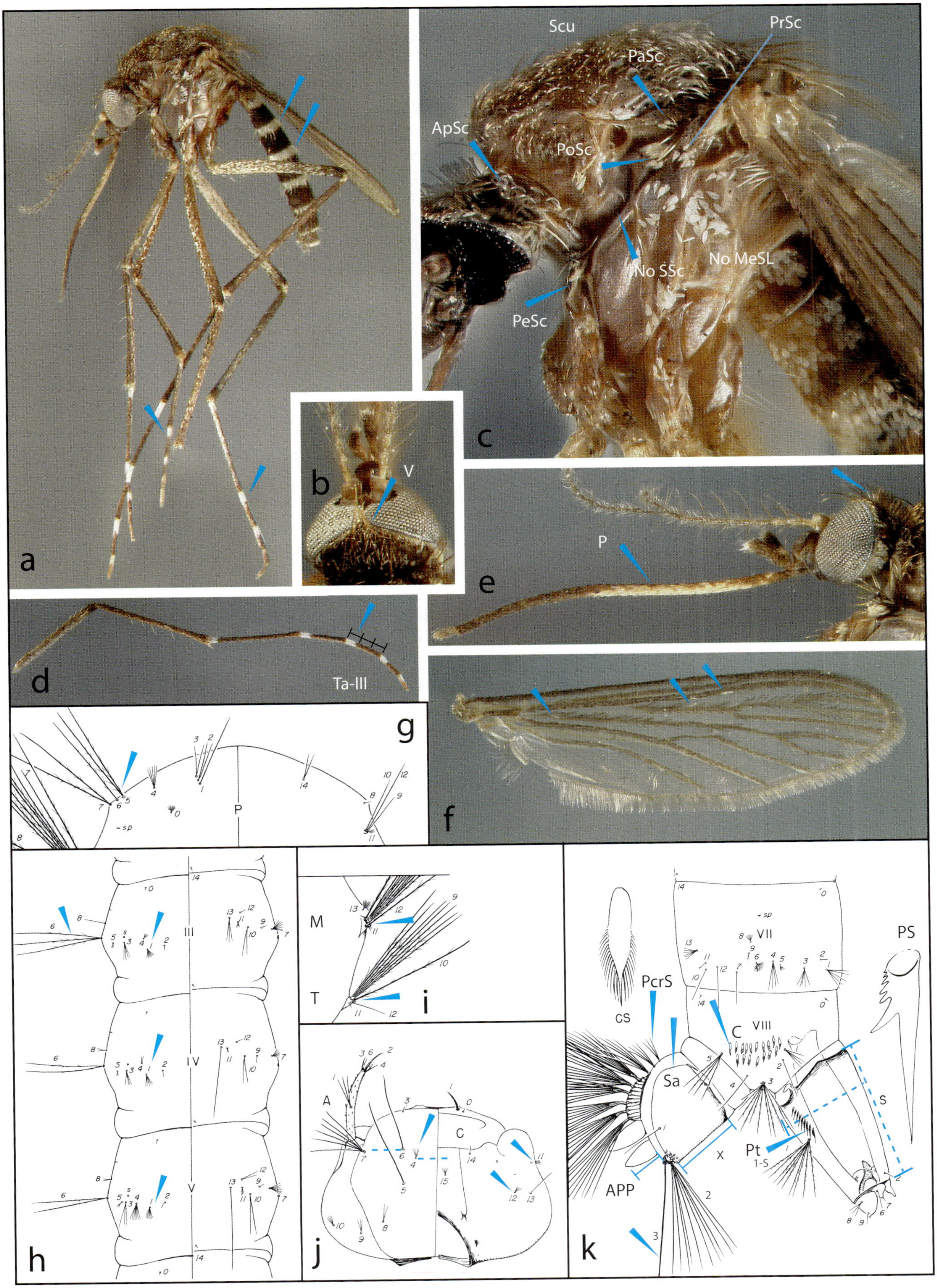
a
b
V
c
Scu
PrSc
PaSc
ApSc
PoSc
No SSc
No MeSL
PeSc
d
Ta-III
e
P
f
g
P
h
III
IV
V
i
M
T
j
A
C
k
VII
VIII
PS
PcrS
CS
C
Sa
S
X
Pt
1-S
APP

Aedes (Protomacleaya) triseriatus (Say, 1823)
Subfamily Culicinae, Tribe Aedini
"Three-spotted American Pointy Mosquito"

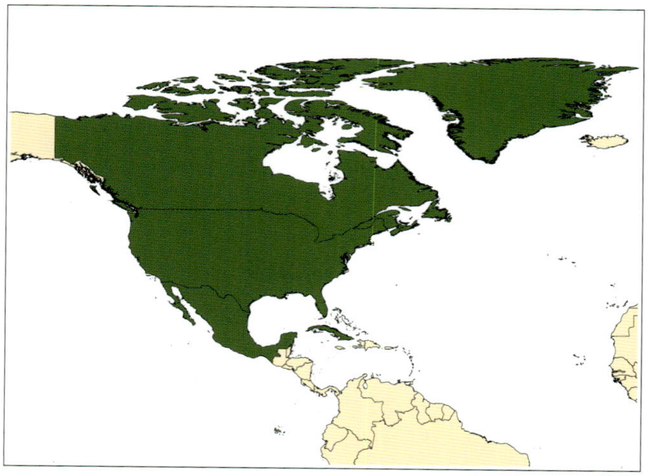

Systematics and Distribution

Aedes triseriatus is a striking mosquito, with dark legs, contrasting silvery-white markings across its body segments, and three white spots on its abdominal segments, for which it was named. Originally described from Pennsylvania, United States, it is found in hardwood forests in eastern and central North America, in Mexico, and in Cuba. In the United States, it is found as far west as Texas, and is present on the East Coast from the Florida Keys as far north as Canada and Greenland. *Aedes triseriatus* is not considered invasive *per se*, but its bionomics could support establishment out of its natural range. In France in 2004, *Ae. triseriatus* was intercepted in a batch of used tires imported from Louisiana, but it was eradicated before the species could become established. Where larvae are found in sympatry with *Ae. albopictus* (Skuse) (see species page, p. 460), *Ae. triseriatus* will often be outcompeted.

Bionomics

Adult *Ae. triseriatus* is common in forested areas, where it feeds opportunistically on a wide range of woodland mammals, including deer, squirrels, opossums, rats, rabbits, and chipmunks. Birds, amphibians, and reptiles can act as hosts where mammals are rare, as can people who are active in the forest environs. The species rests in heavily shaded areas close to the larval habitats but will fly into open areas to feed. Immatures preferentially rear in tree holes up to 27 m high into the canopy but can also be found in a range of shaded artificial container habitats in forest environs. Occasionally, tires in open sunlight have proven productive habitats for *Ae. triseriatus*. Larvae have also been collected in pools, ponds, streams, and marshes, but these occurrences are rare. Except as fourth instar larvae, *Ae. triseriatus* is almost indistinguishable from the closely related *Ae. hendersoni* Cockerell. Conflicting bionomic traits reported in the literature can be a result of misidentifications of these two species.

Aedes triseriatus is univoltine. This life trait is sometimes incorrectly reported as multivoltine, where in fact the species exhibits staggered egg eclosion. Larval stages, which feed mainly on spherical bacteria in the water, are present from March to October, peaking in early July, with several subsequent peaks until late August, when the population starts to decline. *Aedes triseriatus* overwinters predominantly as diapausing eggs in northern populations. Neonates hatch following the thaw of tree hole waters. Overwintering as larvae is reported in more temperate areas where the aquatic habitats do not fully freeze. Proven transovarian transmission of LACV is deemed responsible for the overwintering of this virus in the United States. In the laboratory, females infected with the virus took smaller blood meals, refeeding several times during one egg development cycle. Venereal transmission of LACV by the males has also been noted. If these mechanisms occur in nature, these could be of high epidemiological significance.

Associated Pathogens

CVV, EEEV, JCV, LACV, POTV, RVFV, TVT, WNV.

Diagnostic Characters

(a) Habitus (♀): Striking mosquito with contrasting silvery-white markings, anteriorly and laterally on the scutum and pleuron and basoventrally and basolaterally on the abdomen; all tarsi entirely dark scaled. **(b) Thorax (dorsal):** Scutum (Scu) with lateral silvery-white stripes; prescutellar area (PrA) dark, bordered by a line of slender white scales. **(c) Head (side):** Proboscis (P) and palpus (MPlp) dark scaled; clypeus (Clp) without scales; vertex (V) with numerous erect and narrow curved scales. **(d) Thorax (side):** Pleuron with silvery-white scale patches on paratergite (Pa), mesepimeron (Mm), and postpronotum (Ppn); C-III lateral surface with scales. **(e) Abdomen (ventral):** Sternum (S) with basal silvery-white bands; tergum (Te) with basolateral silvery-white scale patches. **(f) Wing:** Base of costa (C) usually dark scaled. **(g) Larva (abdominal segments):** Seta 12-I present. **(h) Larva (head):** Seta 1-C thickened, distinct. **(i) Larva (terminal segments):** Comb scales (CS) fringed, 10–13, in irregular rows; seta 4-X with four pairs of setae, and two pairs of precratal setae (PcrS). Image sources: WRBU; Zavortink (1972).

Exemplar DNA barcodes: *Ae. triseriatus* (AF417730, MG242523, JX259685–88, JX260473).

Catalog entry: Vol. 2, p. 785.

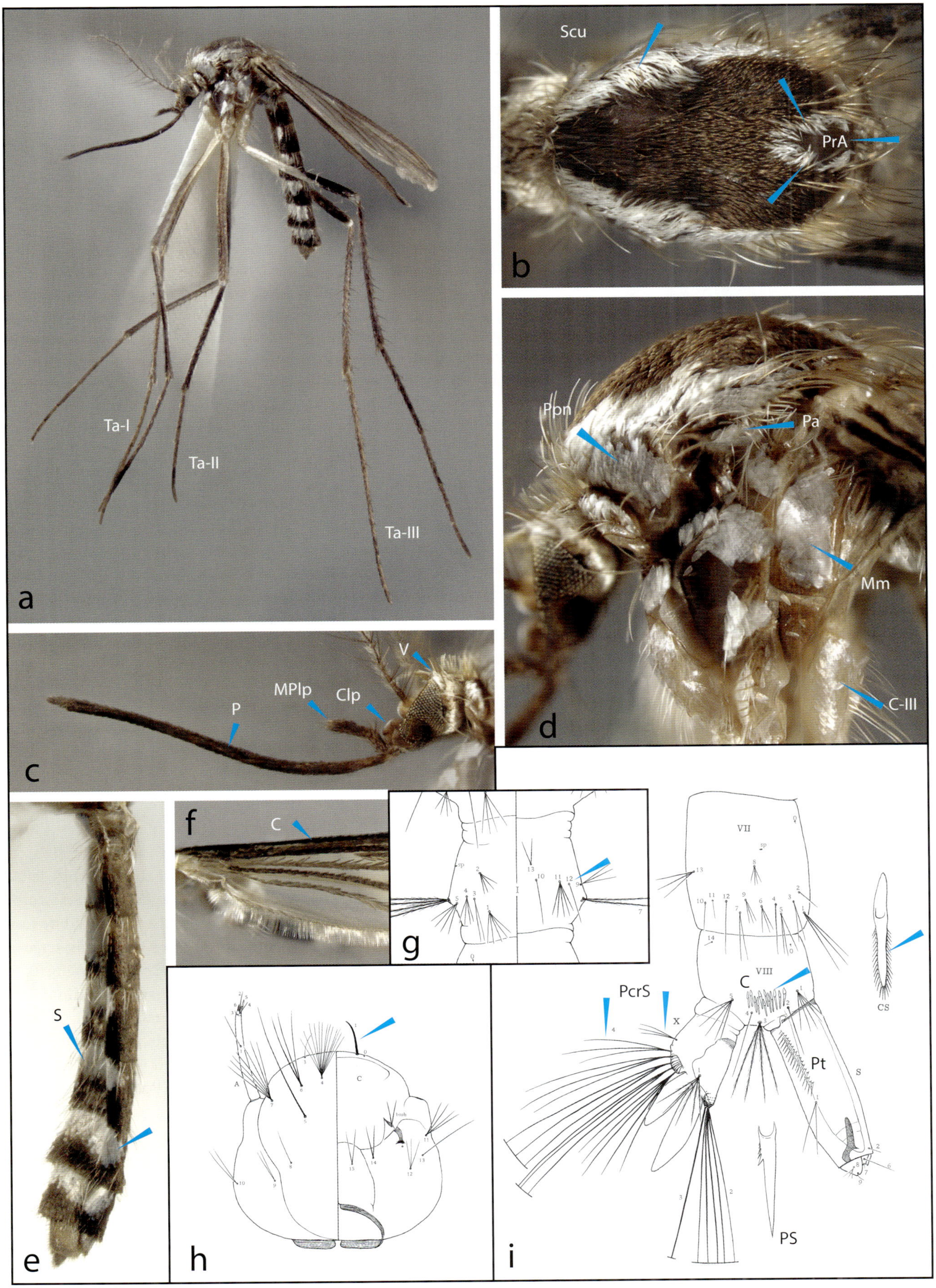

a
Ta-I
Ta-II
Ta-III
b
Scu
PrA
c
P
MPlp
Clp
V
d
Ppn
Pa
Mm
C-III
e
S
f
C
g
h
i
VII
VIII
PcrS
X
C
Pt
S
PS
CS

Aedes (*Rampamyia*) *notoscriptus* (Skuse, 1889)
Subfamily Culicinae, Tribe Aedini
"Scriptwriting Australian Pointy Mosquito"

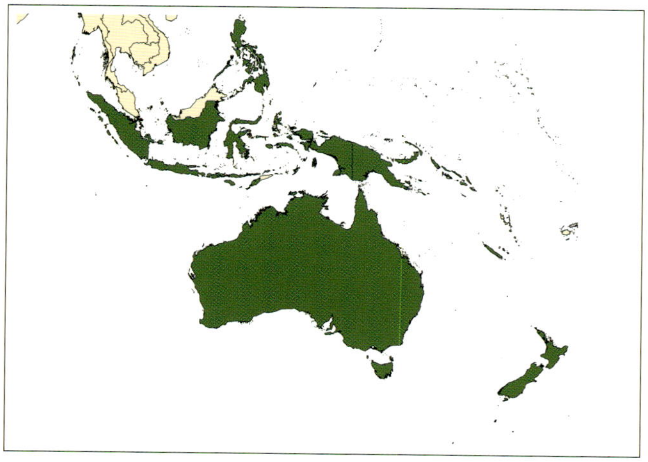

Systematics and Distribution

Aedes notoscriptus is an attractively marked mosquito, with clusters of bright-white scales on the ends of the palps and across the lateral thorax, rendering it a distinctly spotted look. *Aedes notoscriptus*, along with two other Australasian taxa—*Ae. albilabris* Edwards and *Ae. quinquelineatus* Edwards—compose the subgenus *Rampamyia*. *Aedes notoscriptus* currently has one recognized subspecies—*Ae. notoscriptus montanus* Brug, originally described from Lembang, Java. Native to Australia, the invasive *Ae. notoscriptus* became established in New Zealand sometime before 1920 and is now abundant on the North Island. It also occurs in Indonesia, New Guinea, the Solomon Islands, and New Caledonia. Most recently, the species has invaded and become established in three counties of California, United States.

In his treatment of mosquitoes of the South Pacific, John Belkin noted that *Ae. notoscriptus* was "one of the most variable mosquitoes I have ever seen," with differences noted in the adult ornamentation and caudal segments of the larva. Cryptic genetic lineages have been noted in populations of *Ae. notoscriptus* in Australia and New Zealand, which correlate with wing centroid size, suggesting this species comprises up to four cryptic taxa. Vector competency varies geographically, and allozyme data indicated at least five distinct genetic signatures in over 17 populations sampled—four in New South Wales and another in Darwin. Whole mitochondrial genomes have been sequenced for two distinct species within the Notoscriptus Complex, and geographical differences in egg morphology have also been reported; however, there have been no efforts to formally describe the component taxa. Given its invasive nature and role in arboviral transmission, comprehensive integrated systematic studies of the taxa across their range are needed to clarify the taxonomic status and medical importance of its component members.

Bionomics

Given the above, the reported biological attributes of *Ae. notoscriptus* undoubtedly reflect more than one species. Immatures of the invasive container-breeder *Ae. notoscriptus* s.l. rear in a wide variety of natural containers, including leaf axils, rock pools, tree holes, bamboo stumps, coconut husks and shells, palm fronds, and a variety of large and small artificial containers made of cement, wood, and metal. *Aedes notoscriptus* s.l. has recently adapted to peridomestic environs, where it is extremely common in roof gutters and has been shown to outcompete *Ae. aegypti* (see species page, p. 456) in New South Wales. It has become a common urban pest throughout Australia, biting people in shaded areas during the day. Other proven hosts include dogs, brushtail possums, birds, cats, and flying foxes. In New Zealand, *Ae. notoscriptus* s.l. is a ferocious biter and frequently enters houses to feed, but biting behaviors are known to vary. In native forests of New Zealand, *Ae. notoscriptus* s.l. was detected at both ground level and 10 m up in the canopy. A source population from Brisbane, Australia, was successfully colonized. Under laboratory conditions, the average time from egg eclosion to adult emergence was 11 days under typical Brisbane summer temperatures of 20.5°C–28.9°C and 20 days under typical winter temperatures of 10.1°C–21.2°C.

Associated Pathogens

BFV, CHIKV, DENV(1–4), RRV, WHAV; *Dirofilaria immitis*.

Diagnostic Characters

(a) Habitus (♀). (b) Thorax (dorsal): Scutum (Scu) with lyre-shaped lateral silvery lines, narrow median silvery lines, and anterior sublateral golden lines of scales. **(c) Wing:** Scales all dark dorsally. **(d) Head (dorsal):** Proboscis (P) dark scaled with median pale ring. **(e) Thorax (side):** Postpronotum (Ppn) with small distinct lower patch of silvery scales; antealar area (AnA) with distinct silvery scale patch. **(f) Leg:** Fe-II and Ti-III with anterior lines of pale scales along nearly whole length. **(g) Larva (head):** Seta 1-C robust, sharply hooked; seta 5-C single. **(h) Larva (terminal segments):** Comb scales (CS) in irregular triangular patch; siphon (S) not pilose; siphonal acus (SA) large and broadly joined to siphon (S); seta 1-S short, <0.5 times length of siphon (S). **(i) Larva (thorax/abdominal segments):** Without conspicuous stellate setae. Image sources: WRBU; Belkin (1972).

Exemplar DNA barcodes: *Aedes notoscriptus* s.l. (KU495076–80, MG242508).

Catalog entry: Vol. 2, p. 787.

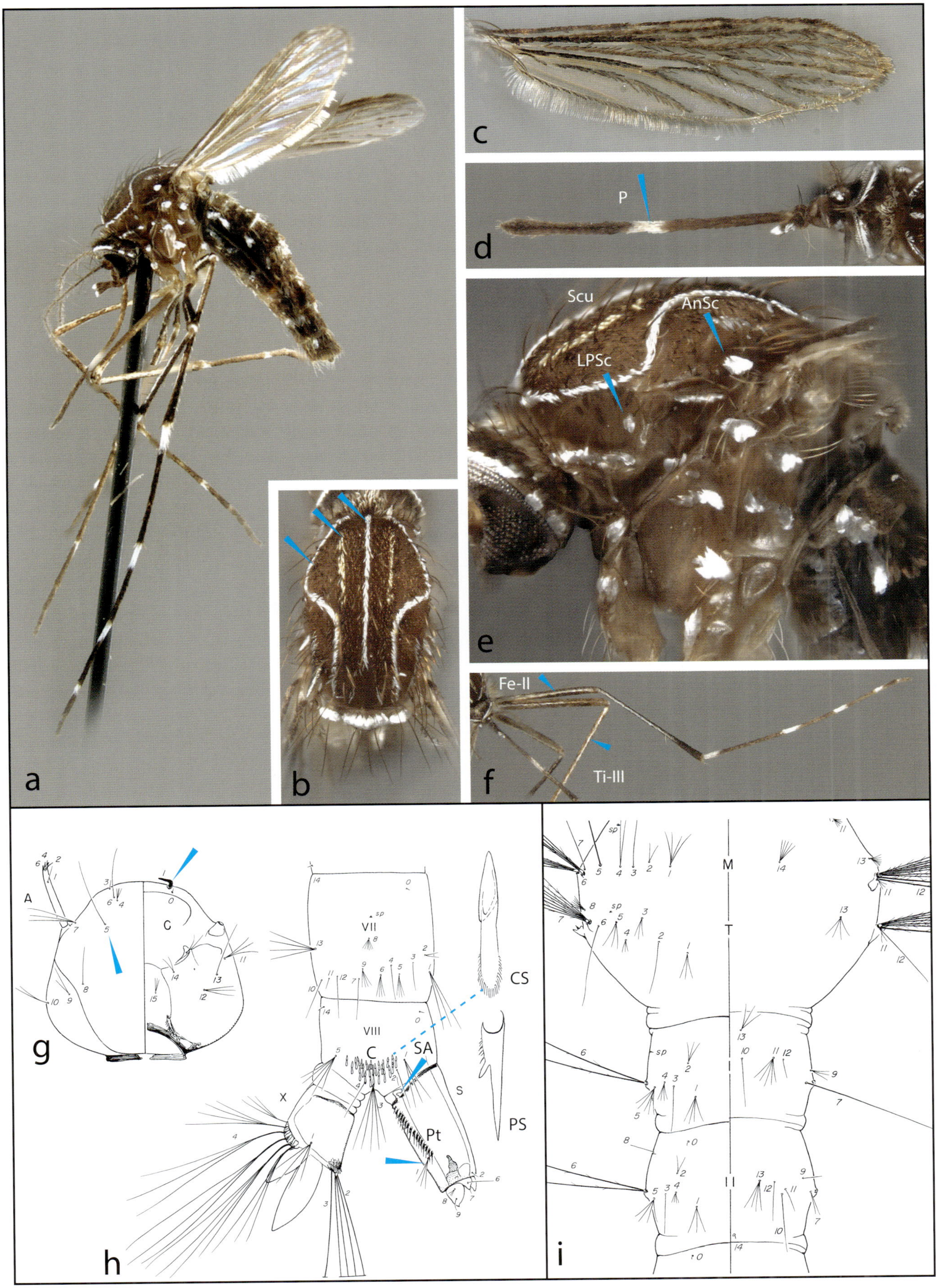
a
b
c
P
d
Scu
AnSc
LPSc
e
Fe-II
Ti-III
f
A
C
g
VII
VIII
C
SA
S
X
Pt
CS
PS
h
M
T
I
II
i

Aedes (Stegomyia) aegypti (Linnaeus, 1762)
Subfamily Culicinae, Tribe Aedini
"Yellow Fever Malaysian Pointy Mosquito"

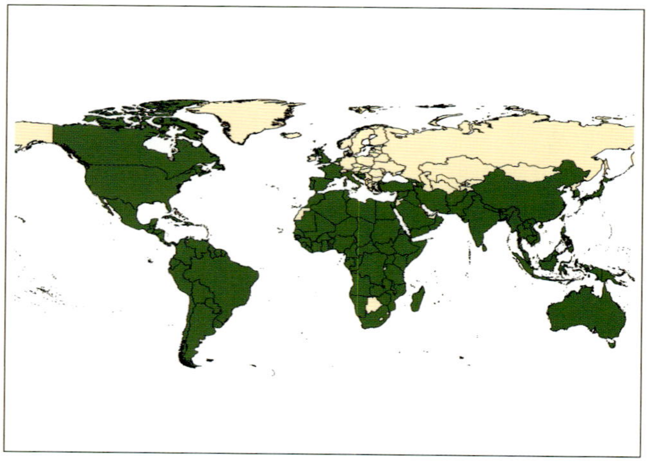

Systematics and Distribution

Aedes aegypti is characterized by its contrasting black-and-white coloration and the distinctive lyre-shaped pattern of silvery-white scales on its scutum. *Aedes aegypti* is by far and away the best studied of all mosquito species, boosted into notoriety by Walter Reed's discovery that it was responsible for the transmission of yellow fever virus (YFV) in Cuba. *Aedes aegypti* has 27 synonyms and one valid subspecies, *Ae. aegypti formosus* (Walker). This sylvatic, presumed ancestral, entity was described from Sierra Leone, but it has been reported in forested regions in eastern and southern sub-Saharan Africa. The distribution of *Ae. aegypti* is essentially panglobal in tropical regions.

Current theory based on genetics and history suggests that the sylvatic form followed humans as agriculture was developed in Africa. Inadvertent transportation on goods within ships gradually expanded the range almost globally, throughout the subtropics and tropics. Eastern Asia may have been the last major geography invaded, the species not being widespread in Southeast Asia until the middle of the 20th century. *Aedes aegypti* has recently expanded its range into the western United States, has reinvaded southern Europe, including Turkey, and has adapted to urban conditions in Australia.

Bionomics

Aedes aegypti is commonly found in urban and semiurban environments in close association to humans, where it will occupy all natural and man-made containers as immatures. In rural environments, the species will oviposit in tree holes and rotten tree stumps. Its wide distribution and tendency to form genetically isolated populations has resulted in rapid adaptation to many conditions, though it seems to be sensitive to competition with other container-developing species, including *Ae. albopictus* (Skuse) (see species page, p. 460). *Aedes aegypti* oviposits only in fresh water, and females can distinguish saline waters containing 5, 10, 17.5, and 35 salt parts per 1,000 from fresh water. Eggs are laid in artificial containers, where they can survive months before hatching. In these barren environments, immatures are often stressed for adequate nutrition, and the resultant adults appear somewhat weakened. Diurnal adults mate near the blood-meal host, and tend not to fly farther than necessary to find resources. Females readily enter houses to feed. Several studies have shown a high preference for feeding on humans, even when other animals are available; however, some populations probably feed primarily on other animals, including birds.

Associated Pathogens

AINOV, AHSV, BOZOV, BSQV, BUNV, CATUV, CHIKV, CHPV, CPV, CVV, DENV, EEEV, EHDV, GROV, HPV, ILHV, ITV, IRIV, JAPV, JEV, JOIV, KETV, KUNV, LACV, MAYV, MBGV, MCOV, MEBV, MELV, MTBV, MUCV, MVEV, NAVV, NEPV, NOLAV, NTAV, ORIV, ORUV, RESV, RVFV, SFV, SINV, TAHV, TSUV, TYUV, VEEV, VSIV, WARV, WNV, WSLV, YAOV, YFV, ZEGV, ZIKV; *Plasmodium gallinaceum*, *P. lophurae*.

Diagnostic Characters

(a) Habitus (♀): Distinctively marked with silvery or white scale patches on scutum and legs; Ta-III$_{1-5}$ with pale basal bands, sometimes with faint apical bands; Ta-III$_5$ all or mostly white. **(b) Thorax (side):** Lower proepisternal (PScL) and mesepimeral (MeSc) scales present; postpronotal scales (PpSc) present; postspiracular area (PoSc) without scales; paratergite (PaSc) with broad white scales; subspiracular area with broad white scales (SSc). **(c) Thorax (dorsal):** Scutum (Scu) with lyre-shaped white markings; scutellum (Stm) with broad white scales on all lobes. **(d) Larva (head):** Seta 4-C prominent, many branched; seta 5-C single, at same level as 7-C; setae 4,6-C distinctly anterior to 7-C. **(e) Head (dorsal):** Palpus (MPlp) with apical white scales; proboscis (P) entirely dark scaled; clypeus (Clp) with patch of pale scales; pedicel (Pe) with pale scales on lateral surfaces. **(f) Larva (terminal segments):** Comb scales (CS) in single row, each with distinct lateral spinules; seta 4-X with five pairs of setae. **(g) Larva (thorax):** Support plates of setae 9–12-M,T with stout hooked spines. Image sources: WRBU; Belkin (1972).

Exemplar DNA barcodes: *Aedes aegypti* (AY645248–71, JQ926676–704, KT339679–83, KT766391–97, MG004689–714, whole genome: GCF_002204515.2).

Catalog entry: Vol. 2, p. 792.

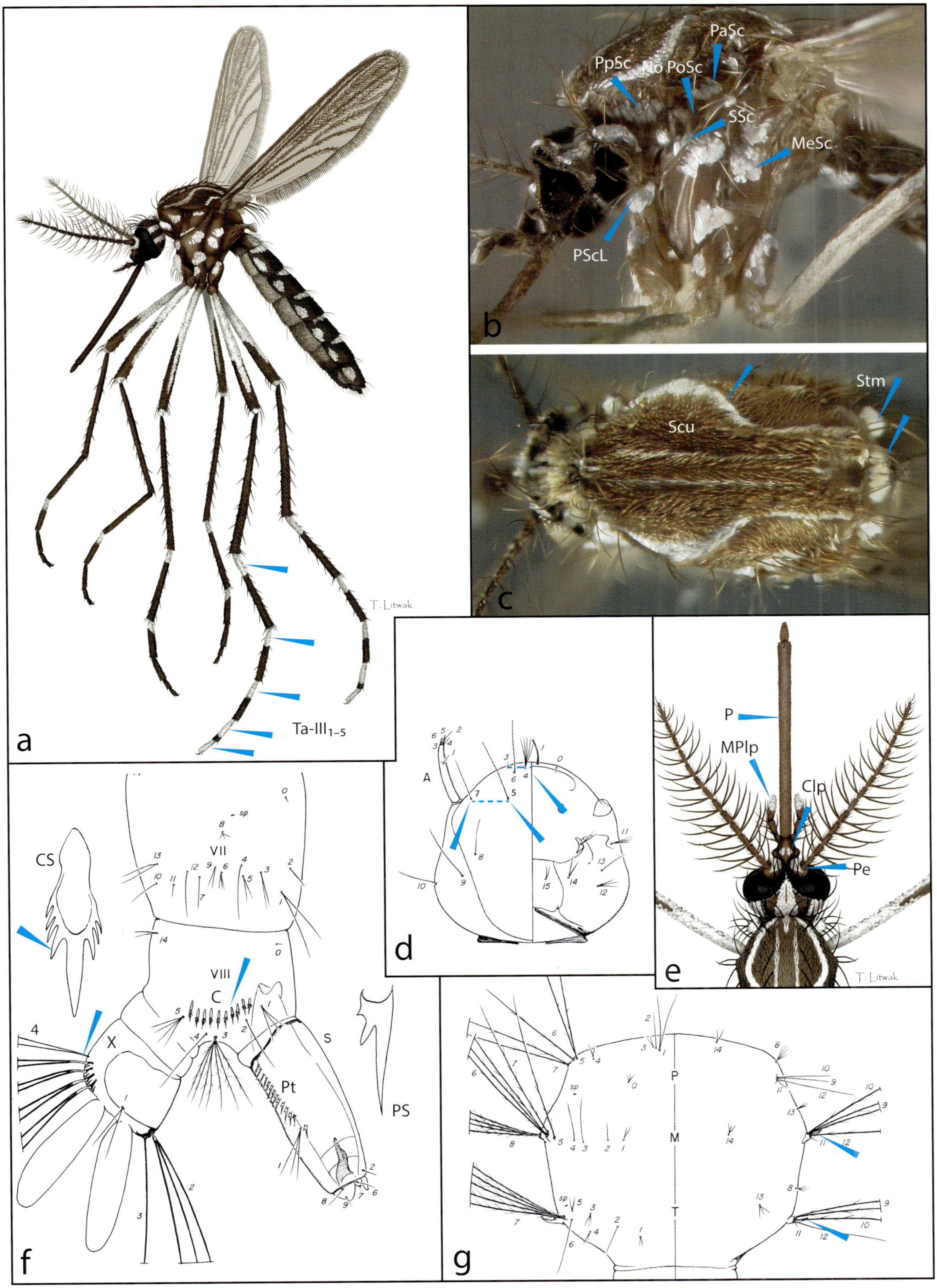
PaSc
PpSc
No
PoSc
SSc
MeSc
PScL
b
Stm
Scu
c
T. Litwak
Ta-III1–5
a
A
d
P
MPlp
Clp
Pe
T. Litwak
e
CS
VII
VIII
C
X
S
Pt
PS
f
P
M
T
g

Aedes (Stegomyia) africanus (Theobald, 1901)
Subfamily Culicinae, Tribe Aedini
"Local Nigerian Pointy Mosquito"

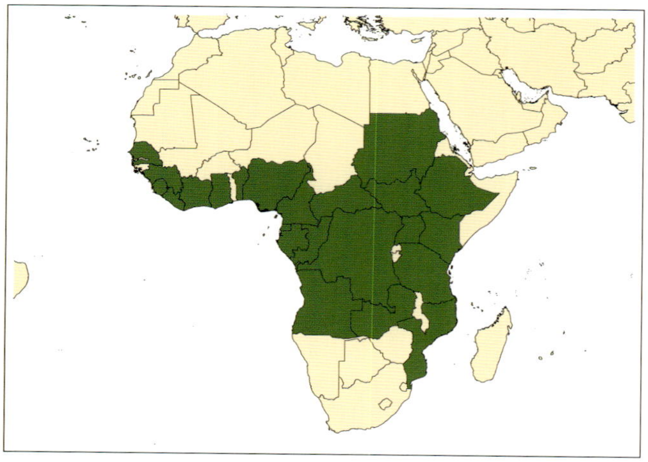

Systematics and Distribution

Aedes africanus is recognized as one of the most important Afrotropical arbovirus vectors, especially for its role in the yellow fever cycle. Like most mosquitoes in the subgenus *Stegomyia*, it is highly ornate. It was originally described from specimens collected in Duke Town and Bonny in Nigeria, and it is widely distributed throughout sub-Saharan Africa except for Madagascar. It is the nominotypical member of the Africanus Group (see Vol. 2, p. 1079, for internal relationships), which includes eight species, seven of which are known to feed on humans, including the medically important *Ae. luteocephalus* (Newstead), with which it is often confused (see species page, p. 464).

Bionomics

Aedes africanus is most closely associated with forested environments and has been reported in arboreal environs from sea level to altitudes of at least 1,800 m. Crepuscular blood feeding occurs at both ground and canopy level. *Aedes africanus* primarily feeds on primates, but this species is known to bite humans outdoors and will readily enter houses to feed. It serves as an effective vector of zoonotic pathogens between nonhuman primates and humans, especially yellow fever. Females preferentially deposit their eggs in plant-based containers close to the ground, but immatures are also found in bamboo stumps, cut bamboo, tree holes, branch forks, rock holes, artificial containers, including discarded metal cans and tires, and suitable habitats indoors. In at least some locations, females are present year-round. Eggs withstand desiccation and seasonality is suppressed by installment hatching, even when flooding of the oviposition sites occurs.

Associated Pathogens

BBKV, BOUV, BOZOV, CHIKV, DENV-2, RVFV, YFV, ZIKV.

Diagnostic Characters

(a) Habitus (♀): White "knee" spots absent from all femora. **(b) Thorax (side):** Paratergite with broad white scales (PaSc); postspiracular scales absent (PoSc); subspiracular (SSc) and postpronotal scales (PpSc) broad, white. **(c) Head (side):** Proboscis (P) entirely dark scaled; palpus (MPlp) with apical white scales; pedicel (Pe) with lateral white scales; vertex (V) with small number of erect forked scales; narrow decumbent scales absent from occiput (Occ) and vertex (V); patch of broad white supraalar scales (SaSc) present. **(d) Thorax (dorsal):** Anterior promontory (AnP) with median stripe of broad white scales; scutal fossa (SF) and lateral lobes of scutellum with broad white scales (LSSc). **(e) Leg (mid):** Fe-II with three large white patches on anterior surface. **(f) Leg (III):** Ti-III with a white stripe on ventral surface in basal 0.2–0.3; Ta-III_{1-4} with basal white band; Ta-III_4 with basal 0.20–0.33 white; Ta-III_5 all dark. **(g) Larva (head):** Antenna (A) mostly smooth, not spiculate; seta 1-A very small; seta 4-C well developed, branched; setae 5–7-C usually single; setae 5-C and 7-C inserted approximately at same level. **(h) Larva (terminal segments):** Comb (C) with spatulate comb scales (CS) in a single row; siphon short, index about 2.0; pecten spines (PS) c. six times as long as wide, with well-developed denticle, sometimes with one or two smaller basal denticles; no precratal setae (PcrS); four pairs of seta 4-X. Image sources: WRBU; Huang (1990).

Exemplar DNA barcodes: *Aedes africanus* (GQ165782, MG242464).

Catalog entry: Vol. 2, p. 794.

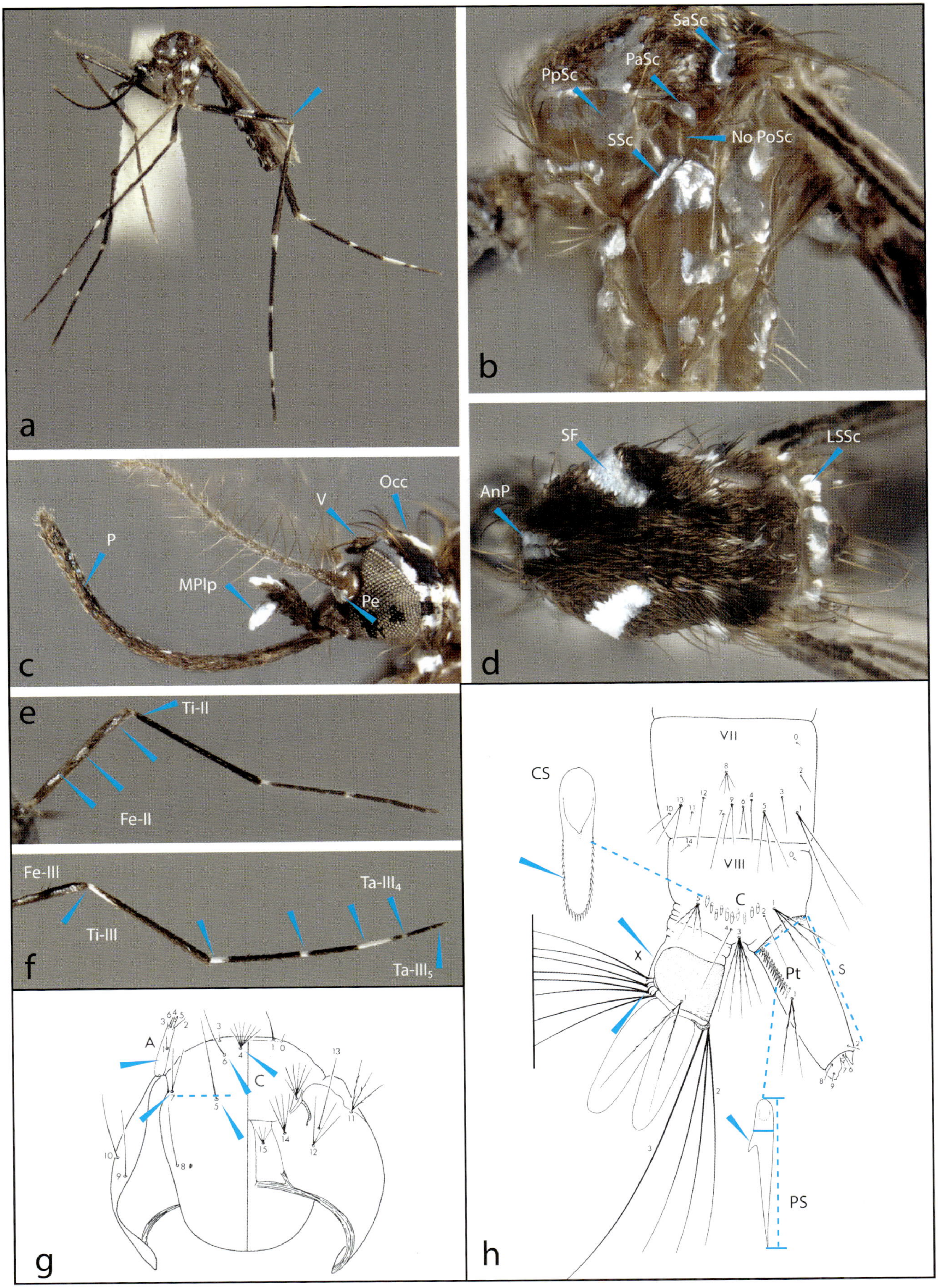

a
SaSc
PaSc
PpSc
SSc
No PoSc
b
SF
LSSc
AnP
d
V
Occ
P
MPlp
Pe
c
e
Ti-II
Fe-II
Fe-III
Ta-III4
Ti-III
Ta-III5
f
A
C
g
VII
VIII
CS
C
X
Pt
S
PS
h

Aedes (*Stegomyia*) *albopictus* (Skuse, 1895)
Subfamily Culicinae, Tribe Aedini
"Asian Tiger Mosquito"

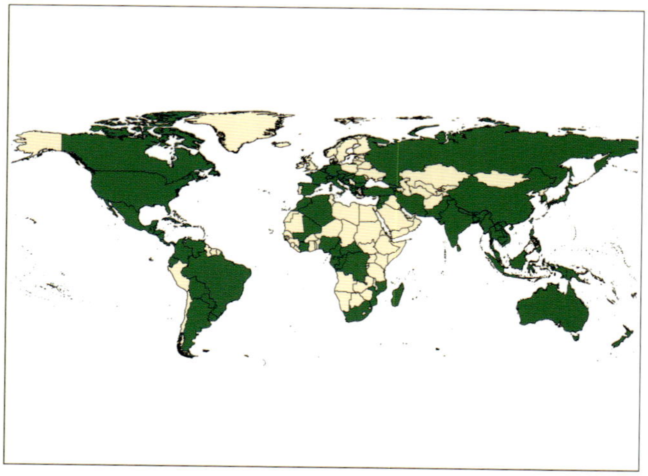

Systematics and Distribution

Aedes albopictus is one of the most commonly recognized black-and-white striped mosquitoes, with a close association with human habitats. The species is the nominotypical member of the Albopictus Subgroup (Scutellaris Group), which comprises 11 species. The species has three synonyms—*samarensis, nigritia,* and *quasinigrita,* all described by Clara Southmayd Ludlow from the Philippines. Originally described from Calcutta in India, this species has attracted much attention due to its rapid global invasion, and it is now present on all continents. Present across Southeast and eastern Asia, the species reaches as far north as the Primorye region of Russia. Although it became established in Hawai'i at the end of the 19th century and in Guam in the 1940s, its major invasive expansion did not begin until 1979 with its introduction into Albania through the used tire trade. The species became abundant in the United States during the 1980s and has since spread to most parts of that country with suitable habitat, as well as much of the rest of the Western Hemisphere. During the 2000s, *Ae. albopictus* became established in much of Europe and the Middle East. The origins of tropical populations in India and Africa are uncertain.

Bionomics

In Southeast Asia, *Ae. albopictus* females oviposit primarily in plant-based containers and receptacles containing parts of plants. Even in its native range, the species also colonizes artificial containers, sometimes cohabiting the same site with *Ae. aegypti* (Linnaeus) (see species page, p. 456). Where *Ae. aegypti* or *Ae. guamensis* Farner & Bohart occur in sympatry with *Ae. albopictus, Ae. albopictus* usually outcompetes them. Northern populations of *Ae. albopictus* deposit diapausing eggs in autumn that are resistant to temperatures moderately below freezing. In North America, *Ae. albopictus* occurs in forested sites in natural containers in the southeastern United States, but it is most abundant as an urban mosquito occupying artificial containers, especially cemetery vases, gutters, water collections in garden environments, and water pools in discarded trash. Eggs survive desiccation, enabling the inadvertent introduction of the species through global trade, primarily of tires and plants such as the so-called lucky bamboo (*Dracaena sanderiana* Mast.: Asparagaceae), which is not a true bamboo. *Aedes albopictus* feeds during the day, taking blood meals from many kinds of hosts, including mammals and birds, with a ferocious appetite for humans. *Aedes albopictus* bites multiple hosts per egg cycle, intensifying its ability to transmit pathogens. Due to its close association with human dwellings and its day-biting habits, it can become an intense pest in built-up areas.

Associated Pathogens

AMTV, ARKV, CHIKV, CHPV, CVV, DENV, EEEV, ITPV, JBEV, KASV, KUNV, LACV, SFV, TAHV, USUV, VEEV, VSAV, WEEV, WNV, YFV, ZIKV; *Dirofilaria immitis, Plasmodium lophurae, P. galliceum, P. fallax.*

Diagnostic Characters

(a) Habitus (♀): Distinctively marked with silvery or white scale patches on scutum and legs; Ta-I–III$_{1-5}$ with only basal bands; tergal (Te) scales basal, often not connected with lateral pale scales; I-Te without median patch of white scales. **(b) Thorax (side):** Antealar area with patch of broad pale scales (AnS); mesepimeron with lower scales (LMSc); paratergite with scales (PaSc); postpronotal scales (PpSc) present; postspiracular scales absent (PoSc); proepisternal scales (PeSc) present; scutal angles (ScA) without pale scales; subspiracular area with broad white scales (SSc). **(c) Thorax (dorsal):** Scutum (Scu) with median longitudinal stripe. **(d) Head (dorsal):** Proboscis (P) entirely dark scaled; palpus (MPlp) with white scales at apex; pedicel (Pe) with scales on lateral surfaces. **(e) Larva (abdomen):** Seta 4-V ≤0.5 times length of seta 3-V; seta 2-IV–VI usually with 1–2 branches; seta 13-III–V with unequal branches, with one branch distinctly longer than others. **(f) Larva (terminal segments):** Seta 2-VII usually with 1–2 branches; comb (C) with comb scales (CS) in single row; seta 4-X with four pairs of setae on grid; seta 4a-X longer than 0.75 times 4b-X (not shown). **(g) Larva (head):** Seta 1-A very small, weakly developed; seta 4-C prominent, many branched; seta 5-C single, at same level as 7-C; setae 4-C and 6-C distinctly anterior to 7-C. Image sources: WRBU; Belkin (1962).

Exemplar DNA barcodes: *Aedes albopictus* (AB738121–22, KP843392–401, KP877563–75, KP896550–75, KR061437–56, whole genome: GCF_006496715).

Catalog entry: Vol. 2, p. 794.

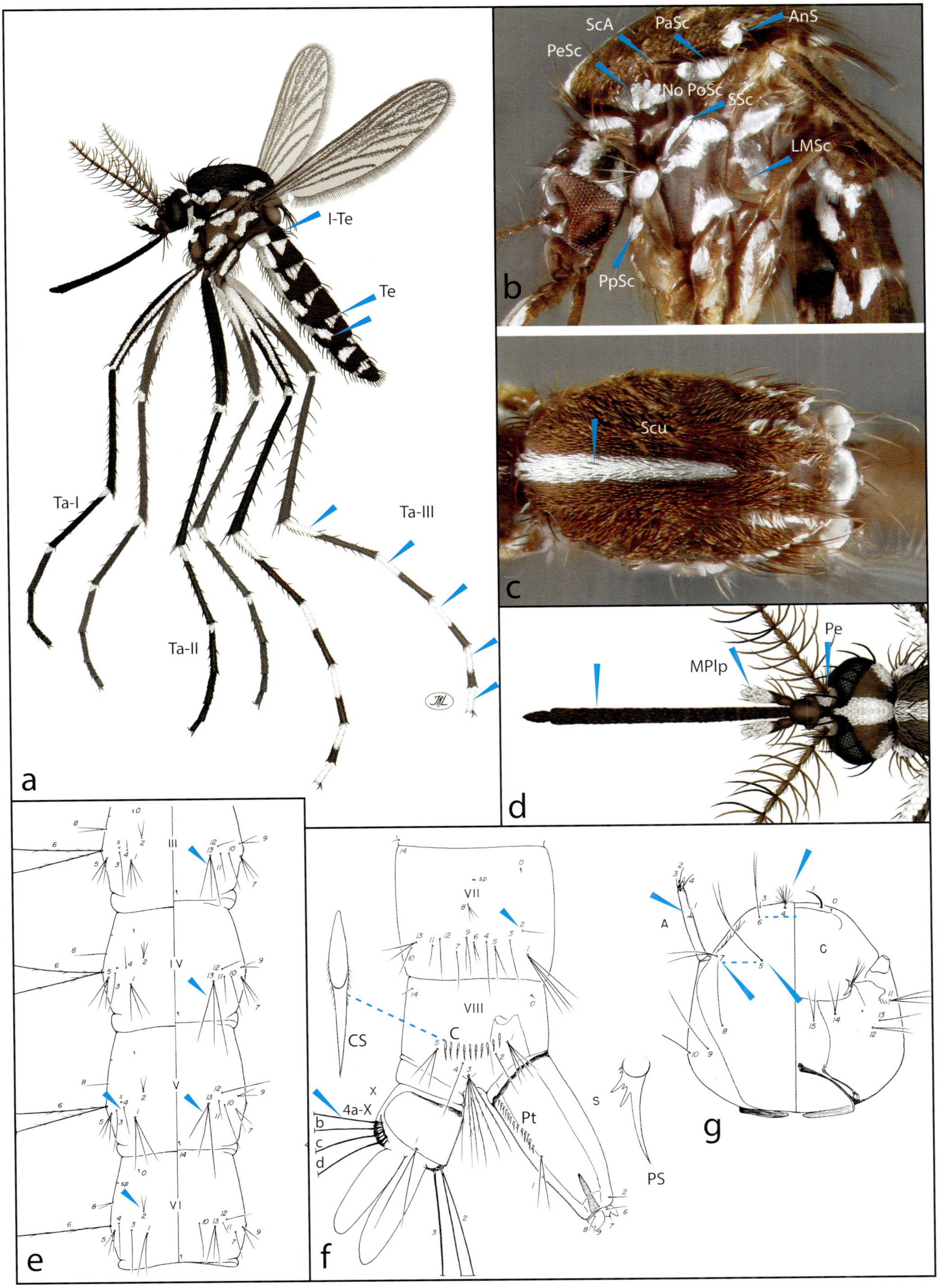
I-Te
Te
Ta-I
Ta-II
Ta-III
a
ScA
PaSc
AnS
PeSc
No PoSc
SSc
LMSc
PpSc
b
Scu
c
MPlp
Pe
d
III
IV
V
VI
e
VII
VIII
CS
C
X
4a-X
b
c
d
Pt
S
PS
f
A
C
g

Aedes (Stegomyia) bromeliae Theobald, 1911
Subfamily Culicinae, Tribe Aedini
"Bromeliad Ugandan Pointy Mosquito"

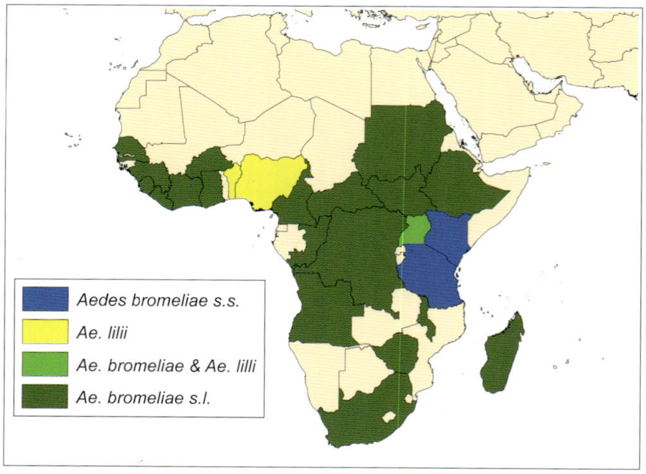

Systematics and Distribution

The highly ornamented *Aedes bromeliae* was originally described from the Kampala swamp in Uganda. It is one of ten species in the Simpsoni Group, and shares closest similarity with *Ae. simpsoni* (Theobald) and *Ae. lilii* (Theobald). Genetic techniques are needed to reliably differentiate these species. Recent molecular studies have confirmed that *Ae. simpsoni*, the nominotypical member of the complex, is restricted to southern Africa, while *Ae. bromeliae* is the most widespread and abundant across most of central and eastern sub-Saharan Africa. *Aedes lilii* is more common in western Africa and is the only member of the group found in the Republic of Benin. Despite intensive surveys, only *Ae. bromeliae* was found in Tanzania, but in neighboring Uganda, *Ae. bromeliae* and *Ae. lilii* are found in sympatry.

Aedes bromeliae is a highly efficient YFV vector and "the principal vector in the greatest known African yellow fever epidemic" in Ethiopia in the 1960s. It is epidemiologically important to differentiate *Ae. bromeliae* and *Ae. lilii*, as the latter is zoophilic and thought to be of no medical importance. However, introgression was noted in Uganda where the taxa are sympatric, as this could result in a higher propensity for hybrid *Ae. lilii* to feed on humans and become involved in the human YFV cycle.

Bionomics

Aedes bromeliae females are peridomestic and highly associated with agriculture. They deposit their eggs in a wide variety of plant-based containers, including leaf axils of a number of food plants (pineapple, taro, banana), palm spathes, and tree holes. Oviposition sites can include water collections in tree spathes on the forest floor, but the immatures are readily collected in banana axils up to 2 m above ground level. *Aedes bromeliae* is not thought to seek hosts in the forest canopy but instead readily bites humans and monkeys that enter the plantations.

Medical Importance

YFV.

Diagnostic Characters

(a) Habitus (♀): Distinctively marked with silvery or white scale patches on scutum and legs; Ta-I–III with only basal bands. **(b) Thorax (side):** Lower proepisternal (PScL) and mesepimeral scales (MeSc) present; paratergite with broad white scales (PaSc); postspiracular area without scales (PoSc); subspiracular area with broad white scales (SSc). **(c) Head (side):** Proboscis (P) entirely dark scaled; palpus (MPlp) with white scales at apex; pedicel (Pe) with lateral scales. **(d) Thorax (dorsal):** Scutum (Scu) with submedian stripes; scutal fossa (SF) with patch of broad white scales; scutellum (Stm) with broad white scales on lobes. **(e) Leg (III):** Fe-III with apical white "knee" spot; Ta-III_4 variable in color, where white band is present, always basal (all dark in photo); Ta-III_5 all or mostly white. **(f) Leg (II):** Fe-II with apical "knee" spot and white spot on anterior surface; Ta-II_2 with at least basal 0.5 white. **(g) Larva (head):** Seta 1-A very small, weakly developed; seta 4-C prominent, branched; seta 5-C single, at same level as 7-C; setae 4,6-C distinctly anterior to 7-C. **(h) Larva (terminal segments):** Comb scales (CS) in single row; four pairs of seta 4-X; precratal setae (PcrS) absent. Image source: WRBU.

Exemplar DNA barcodes: *Ae. bromeliae* (KT998409–28, MF183226–58, KF135496–97), *Ae. lilii* (KT998389–408).

Catalog entry: Vol. 2, p. 796.

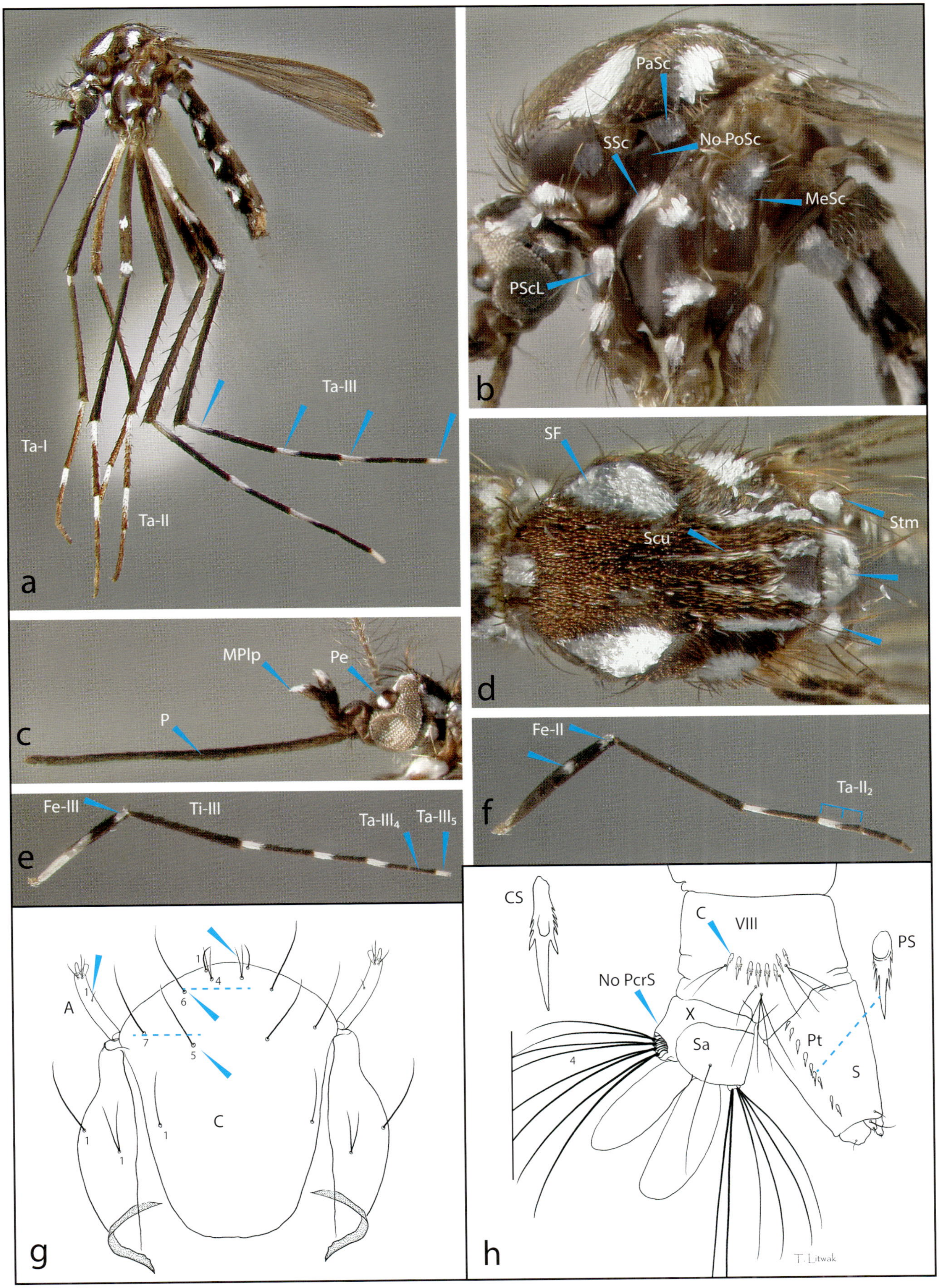

a
Ta-I
Ta-II
Ta-III
b
PaSc
SSc
No PoSc
MeSc
PScL
c
MPlp
Pe
P
d
SF
Scu
Stm
e
Fe-III
Ti-III
$Ta-III_4$
$Ta-III_5$
f
Fe-II
$Ta-II_2$
g
A
C
1
4
5
6
7
h
CS
C
VIII
PS
No PcrS
X
Sa
Pt
S
4
T. Litwak

Aedes (Stegomyia) luteocephalus Newstead, 1907
Subfamily Culicinae, Tribe Aedini
"Yellow-faced Congolese Pointy Mosquito"

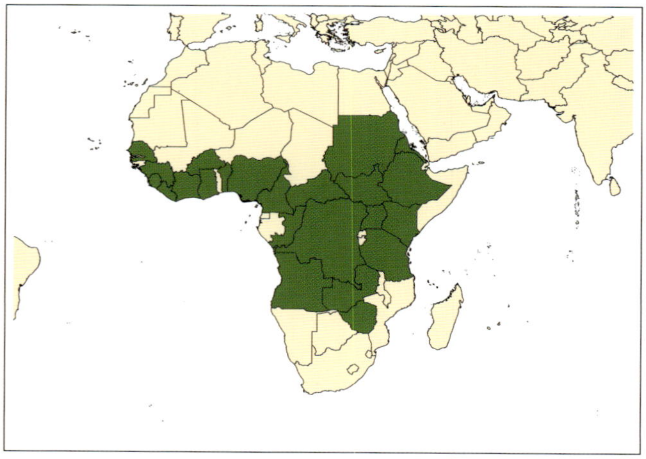

Systematics and Distribution

Aedes luteocephalus was originally described from the Belgian Congo (= Democratic Republic of the Congo), and the species occurs widely across sub-Saharan Africa, north of South Africa. The species is one of 17 major global vectors responsible for the transmission of the "big four" *Aedes*-borne fever viruses–Chikungunya virus (CHIKV), Dengue (DENV), Yellow fever (YFV), and Zika virus (ZIKV). *Aedes luteocephalus* is one of eight species comprising the Africanus Group of the subgenus *Stegomyia* (see Vol. 2, p. 1079, for internal relationships) and is easily confused with the nominotypical member of that group, *Ae. africanus* (Theobald) (see species page, p. 458), both as adults (males and females) and larvae. As adults, *Ae. luteocephalus* is most readily distinguished from *Ae. africanus* by the hind femur, which has a large pale band at the base and two large white patches on the anterior median and apical areas, whereas the hind femur of *Ae. africanus* bears three large white patches on the anterior basal, median, and apical areas. The larvae of *Ae. luteocephalus* and *Ae. africanus* can be differentiated using characters of the comb scales and pecten spines.

Bionomics

Originally a forest mosquito, *Ae. luteocephalus* is now commonly found in village and urban sites, highlighting its potential for global expansion and establishment in nonnative localities. Experimental biting rates were greater in the forest canopy than at ground level, supporting the status of this species as an important bridge virus vector between primates and humans. It readily bites at ground level in villages and in the forest and is most active during the evening crepuscular period. In the forest, eggs are laid in water collections in tree holes, branch forks, and rot holes, occasionally utilizing broken or cut bamboo stems. Artificial containers are opportunistically utilized in domestic and peridomestic environs. Eggs have been collected in plastic bottles some 9 m up in the canopy. Eggs are laid above the waterline and are stimulated to emerge on flooding of the larval habitat. *Aedes luteocephalus* eggs undergo installment hatching, contributing to the long season that this species is present.

Associated Pathogens

CHIK, DENV, YFV, ZIKV.

Diagnostic Characters

(a) Habitus (♀): Distinctively marked with silvery or white scale patches on scutum and legs; terga (Te) with basal bands and basolateral white spots (not shown); Ta-I–III with only basal bands. **(b) Thorax (side):** Postpronotal scales (PpSc) present; proepisternal scales (PeSc) white; subspiracular area with broad white scales (SSc); mesepimeron (Mm) covered with white scales. **(c) Head (dorsal):** Proboscis (P) entirely dark scaled; palpus (MPlp) with white scales at apex; pedicel (Pe) with lateral and mesal scales. **(d) Leg (II):** Fe-II with three large white anterior spots. **(e) Thorax (dorsal):** Scutum (Scu) with distinct median line of narrow golden scales; lateral scutellar lobe scales (LSSc) all broad. **(f) Leg (III):** Fe-III anterior surface with basal pale band and white spots on median and apical areas; Ta-III_3 about 0.5–0.8 white; Ta-III_4 dark. **(g) Larva (head):** Seta 1-A very small, weakly developed; seta 4-C prominent, many branched; seta 5-C single, about same level as 7-C; setae 4,6-C distinctly anterior to 7-C. **(h) Larva (terminal segments):** Comb scales (CS) spatulate, apex rounded but slightly narrowed apically, surrounded by short fringe; CS in single row; four pairs of seta 4-X; pecten spines (PS) with two ventral and one or two small dorsal denticles. Image sources: WRBU; Huang (1990).

Exemplar DNA barcodes: *Ae. luteocephalus* (AY645234–40, KU380354, KU380387, KU380427, KU380469, KU380472, MG242504).

Catalog entry: Vol. 2, p. 801.

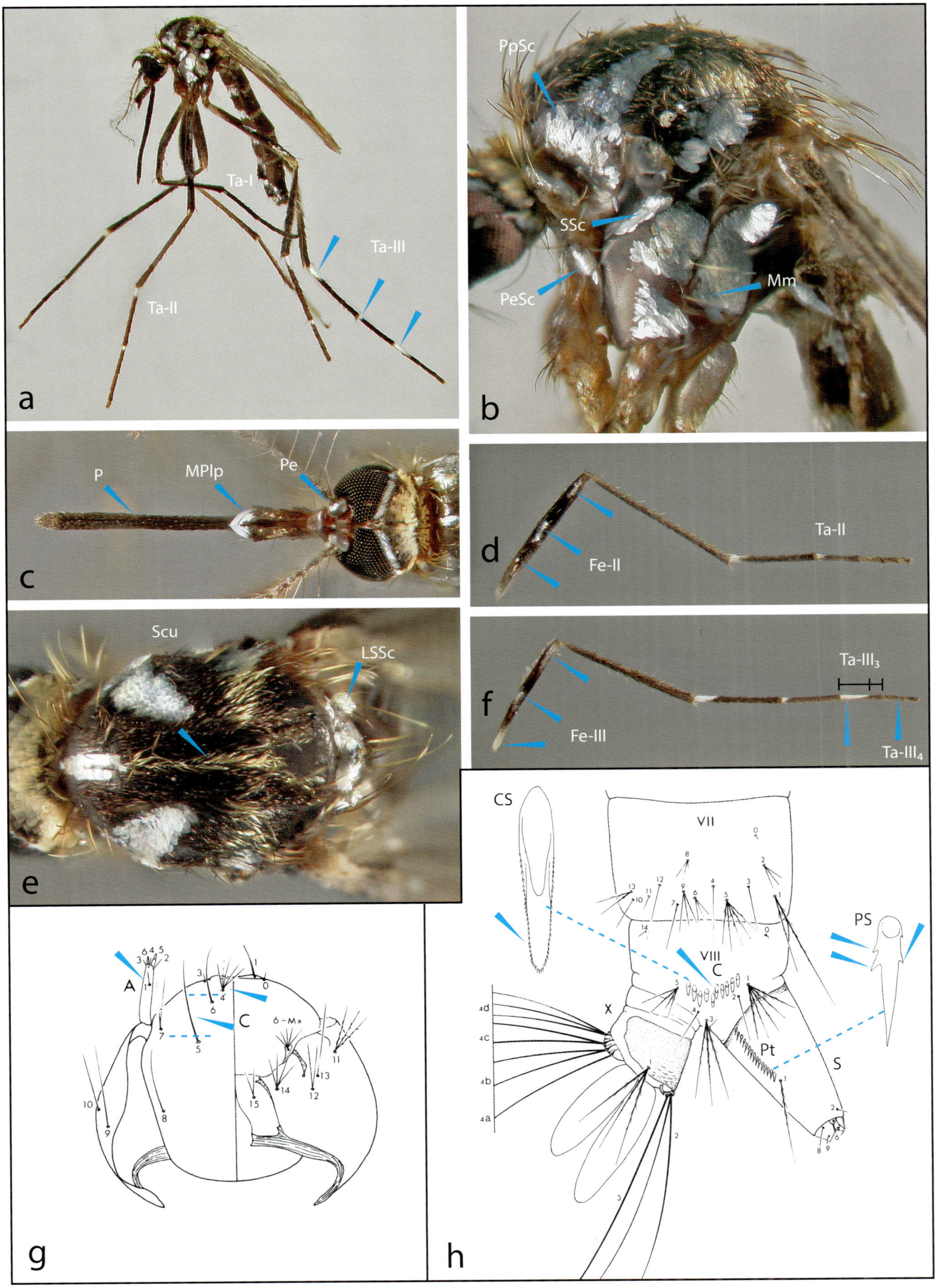

a
Ta-I
Ta-II
Ta-III
b
PpSc
SSc
PeSc
Mm
c
P
MPlp
Pe
d
Fe-II
Ta-II
e
Scu
LSSc
f
Fe-III
Ta-III$_3$
Ta-III$_4$
g
A
C
6-Mx
h
CS
VII
VIII
C
X
Pt
S
PS

Aedes (*Stegomyia*) *polynesiensis* Marks, 1951
Subfamily Culicinae, Tribe Aedini
"Polynesia Fijian Pointy Mosquito"

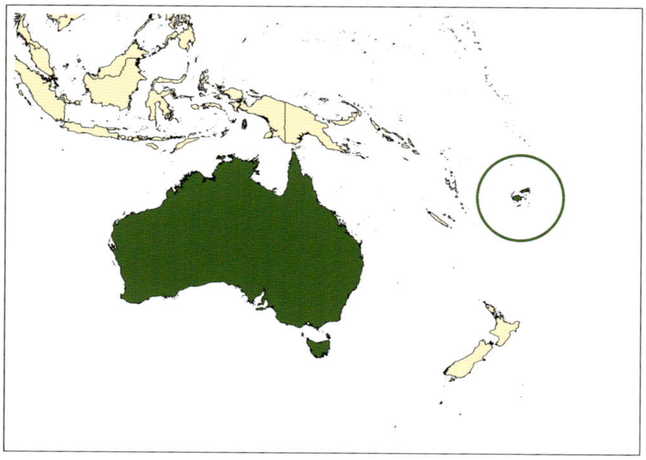

Systematics and Distribution

Aedes polynesiensis is an ornate species bearing a single wide midscutal stripe, two distinct lines of silvery-white scales across the lateral thorax, and large white bands on its hind tarsi. The species was first described from Taveuni, Fiji Islands, and has a wide island distribution from the South China Sea in the north, south to Australia, and east to Pitcairn Island. It is common throughout Polynesia, Fiji, and the islands of the western Pacific. *Aedes polynesiensis* is a member of the large and complex Scutellaris Group of the subgenus *Stegomyia*, which comprises 33 species (see Vol. 2, p. 1079, for internal relationships). It is an important vector of filariasis and a secondary dengue vector across its range.

Bionomics

As with many other *Stegomyia*, *Ae. polynesiensis* immatures develop in a wide variety of containers, including tree holes, banana stumps, leaf axils, coconut shells and husks, cacao pods, rock holes, and many other artificial containers such as tires, bottles, and canoes. Unusually within the subgenus, female *Ae. polynesiensis* will also deposit eggs in restricted accumulations of water on the ground, including holes made by the land crab *Cardisoma carnifex* (Herbst, 1794) (Gecarcinidae).

Its readiness to colonize artificial containers and egg desiccation tolerance has probably led to its widespread distribution through the islands by the Polynesians. Genetic studies suggest that the species became established in the islands with the arrival of people some 1,500–3,000 years ago, and interisland population studies support patterns usually associated with human-assisted mosquito movement.

Females readily bite humans diurnally, with biting peaks in the late afternoon and early morning. As with many other *Stegomyia*, mating usually takes place near the host. Females also readily mate with male *Ae. albopictus* (Skuse, 1895) (see species page, p. 460), which has led to some interesting attempts at intentional control.

Associated Pathogens

DENV, MRV, RRV, ZIKV; *Wuchereria bancrofti*.

Diagnostic Characters

(a) Habitus (♀): Distinctively marked with silvery or white scale patches on scutum and legs; Fe-II without anterior median white line; Fe-III with anterior broad white longitudinal stripe broadened on basal 0.3–0.5; Ta-III_5 white. **(b) Thorax (dorsal):** Scutum (Scu) with median silvery stripe; scutellar midlobe with broad white scales and dark apical scales (MSSc). **(c) Thorax (side):** Scales on antepronotum (ApSc), postpronotum (PpSc) and paratergite (PaSc); lower mesepimeral scales white (LMSc); proepisternal white scale patch (PeSc); subspiracular area without scales (SSc); supraalar scales (SaSc) as complete white line; antealar scales broad (AnSc); costa (C) with distinct silvery spot at base. **(d) Head (dorsal):** Proboscis (P) entirely dark scaled; palpus (MPlp) with white scales at apex; pedicel (Pe) with lateral and mesal scales. **(e) Abdomen (dorsal). (f) Larva (head):** Seta 1-A very small, weakly developed; seta 4-C prominent, many branched; seta 5-C single, at same level as 7-C; setae 4,6-C distinctly anterior to 7-C. **(g) Larva (terminal segments):** Comb scales (CS) in a single row, with fine fringe at base and spinelike apex; saddle (Sa) complete; seta 2-X usually 3 branched; four pairs of seta 4-X. Image sources: WRBU; Belkin (1972).

Exemplar DNA sequences: *Ae. polynesiensis ITS*2 (AY822661–62).

Catalog entry: Vol. 2, p. 803.

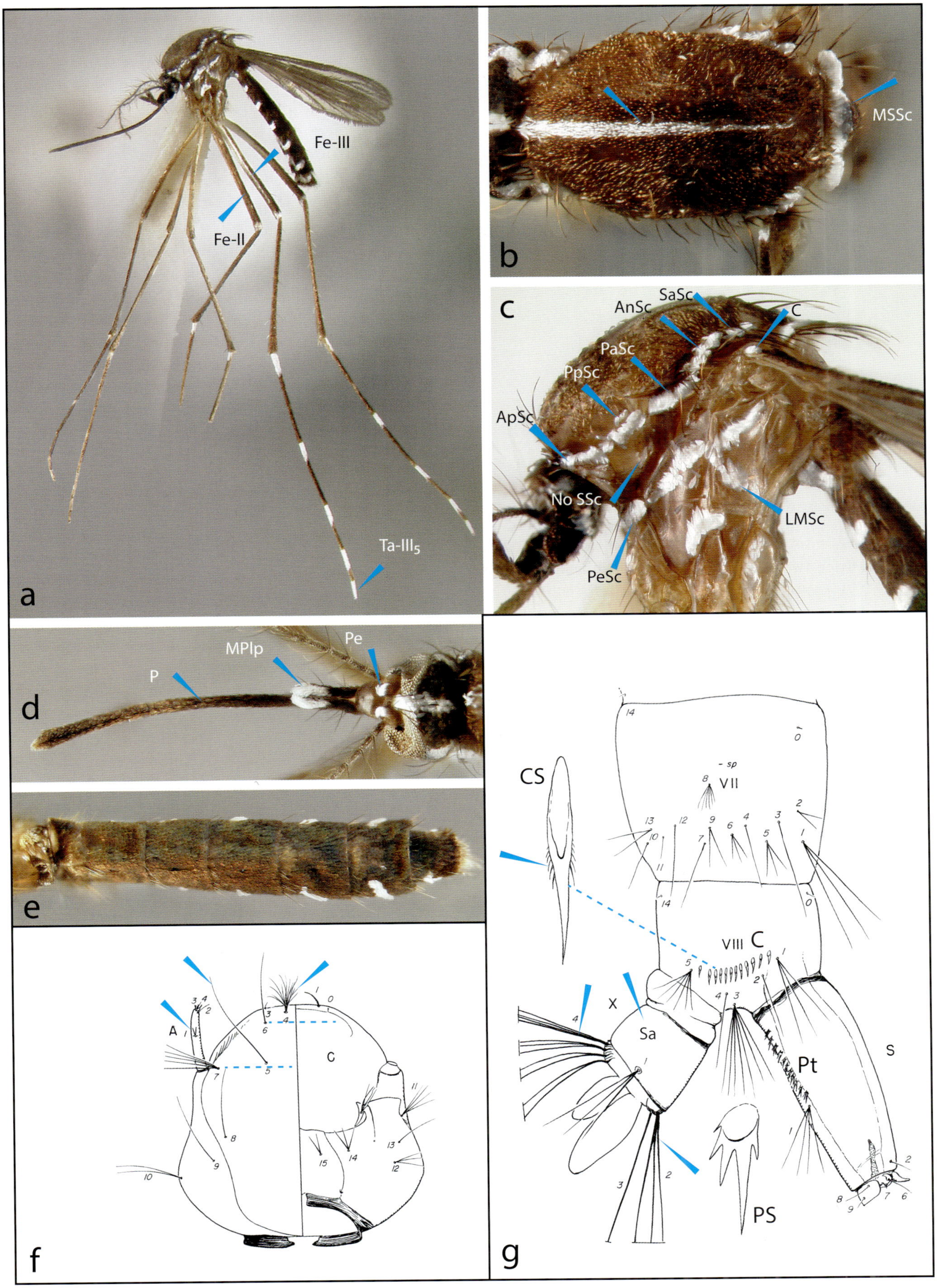
a
Fe-III
Fe-II
Ta-III5
b
MSSc
c
SaSc
AnSc
C
PaSc
PpSc
ApSc
No SSc
LMSc
PeSc
d
P
MPlp
Pe
e
f
A
C
g
CS
VII
VIII C
X
Sa
Pt
S
PS

Aedes (*Stegomyia*) *scutellaris* (Walker, 1860)
Subfamily Culicidae, Tribe Aedini
"Plated Papuan Pointy Mosquito"

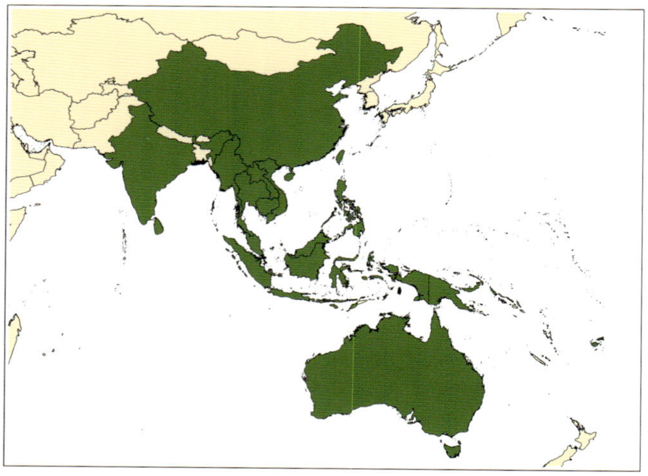

Systematics and Distribution

Aedes scutellaris is a highly ornate dark mosquito, with legs, abdomen, and thorax decorated with thick rings and lines of silvery-white scales, giving it a decidedly striped appearance. The species was first described from specimens found on the Aroe Islands (New Guinea), and it occurs in the western Pacific, including Fiji and the Philippines, in Southeast Asia, in the Australasian islands, and in Australia. It is the nominotypical member of the Scutellaris Group (see Vol. 2, p. 1079, for internal relationships), the most speciose of all *Aedes* groups, containing 33 species including the effective invasive vector *Ae. albopictus* (Skuse) (see species page, p. 460). Members of the Scutellaris Group have probably speciated during the last 5 million years as ancestral stem species experienced successive isolation events with the rise and fall of sea level. Hybridization between members of the Scutellaris Group is common, even apparently between those most distant phylogenetically. The Scutellaris Group contains many important vectors, with 11 of the 12 species tested determined capable of vertical transmission of DENV-1. The Scutellaris Group is further subdivided into the Albopictus and Scutellaris Subgroups. Members of the Scutellaris Subgroup all sport well-developed supraalar white lines, with broad flat scales extending over the wing root toward the scutellum.

Bionomics

Laboratory studies have revealed that male *Ae. scutellaris* will not mate with wingless females, suggesting that copulation takes place during flight. Mating occurs near the host. The peak biting time for *Ae. scutellaris* females is in the evening, but they are readily stimulated to bite opportunistically during the day in shaded areas. Females deposit their eggs in the crevices of coconut shells, tree holes, palm fronds, *Pandanus* leaf axils, bamboo, and artificial containers such as empty cans. Occasionally, larvae are seen in small ground pools and wells. Studies in New Guinea showed that oviposition choices in *Ae. scutellaris* were not driven by differences in organic pollution, salinity, water flow, temperature, pH, specific gravity, or vegetation, but instead, physical properties of the sites were critical to oviposition. These included visual and tactile stimuli, with females preferring a combination of rough-textured sites with low levels of reflected light. Although the visual stimulus was initially important in attracting the ovipositing female, the tactile stimulus governed the proportion of eggs laid. Larvae develop quickly, completing their full developmental cycle from egg to imago in about seven days for males and about nine days for females at ambient temperatures.

Associated Pathogens

DENV; *Wuchereria bancrofti*.

Diagnostic Characters

(a) Habitus (♀): Distinctively marked with silvery or white scale patches on scutum and legs; Fe-II without anterior median white line; Ta-I–VIII with basal pale bands. **(b) Abdomen (dorsal):** Terga (Te) with complete or nearly complete pale bands. **(c) Thorax (side):** Lower proepisternal (PScL) and lower mesepimeral (LMSc) scales white; paratergite with white scales (PaSc); postpronotal scales (PpSc) present; subspiracular area with white scales (SSc); supraalar area (SaSc) with line of well-developed, broad white scales. **(d) Thorax (dorsal):** Scutum (Scu) with median silvery stripe. **(e) Head (side):** Proboscis (P) entirely dark scaled; palpus (MPlp) with white scales at apex; pedicel (Pe) with mesal and lateral scales. **(f) Larva (head):** Seta 1-A very small, weakly developed; seta 4-C prominent, many branched; seta 5-C single, about same level as 7-C; setae 4,6-C distinctly anterior to 7-C. **(g) Larva (terminal segments):** Comb scales (CS) in single row; CS spatulate, apex rounded but slightly narrowed apically, surrounded by short fringe; pecten (Pt) with 10–16 spines (PS); PS with 1–3 ventral and one or two small dorsal denticles; four pairs of seta 4-X; seta 1-S inserted at or before middle of siphon (S). **(h) Leg (III):** Ta-III_3 with basal 0.5 white; Ta-III_4 with basal 0.75 white; Ta-III_5 white scaled. Image sources: WRBU; Huang (1972).

Exemplar DNA barcodes: *Ae. scutellaris* (DQ397899, KP843372–401).

Catalog entry: Vol. 2, p. 805.

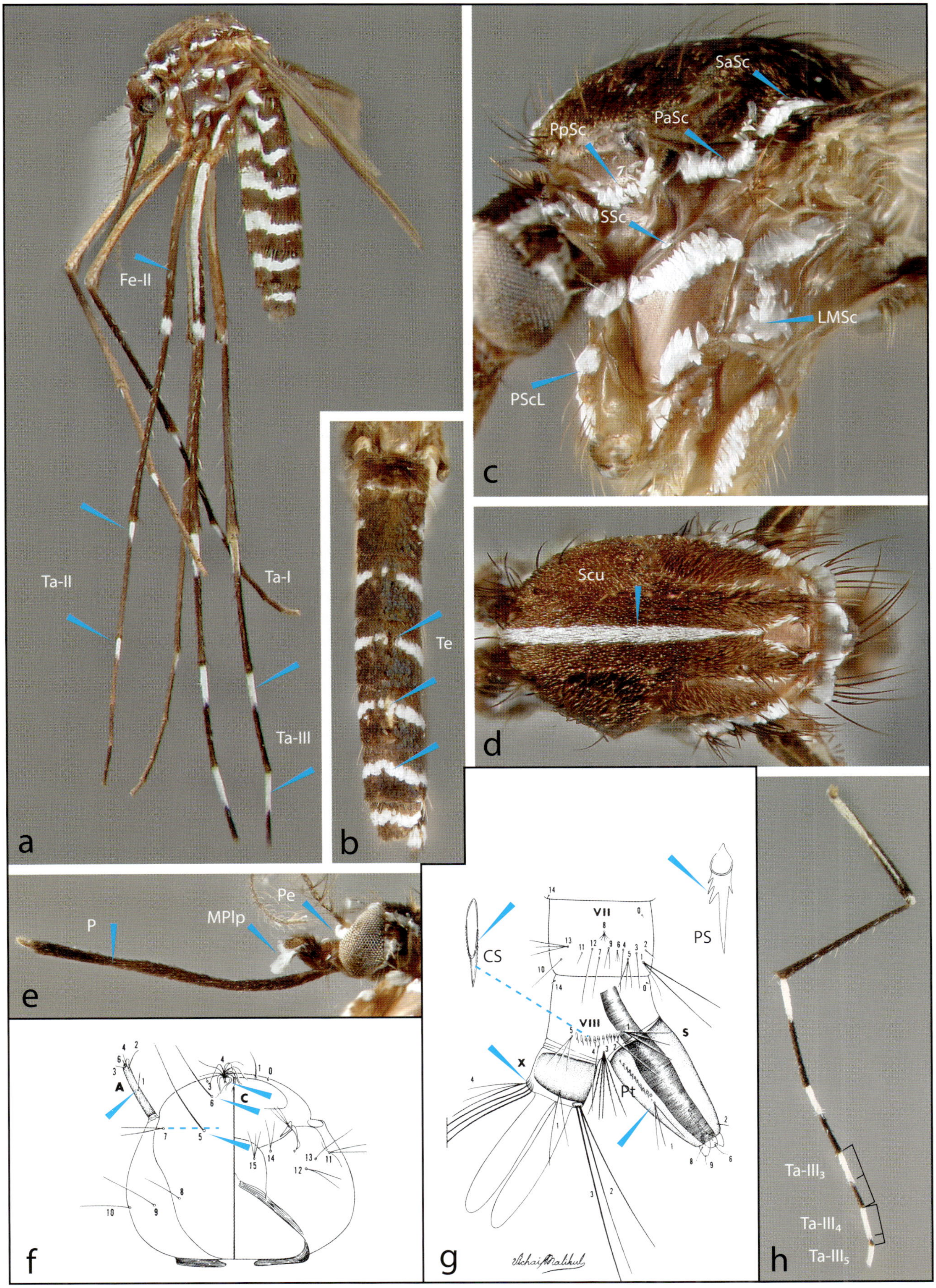
Fe-II
Ta-II
Ta-I
Ta-III
a
Te
b
SaSc
PaSc
PpSc
SSc
LMSc
PScL
c
Scu
d
P
Pe
MPlp
e
A
C
f
VII
VIII
CS
PS
S
X
Pt
g
Ta-III$_3$
Ta-III$_4$
Ta-III$_5$
h

Aedes (Tanakaius) togoi (Theobald, 1907)
Subfamily Culicinae, Tribe Aedini
"Togo Japanese Pointy Mosquito"

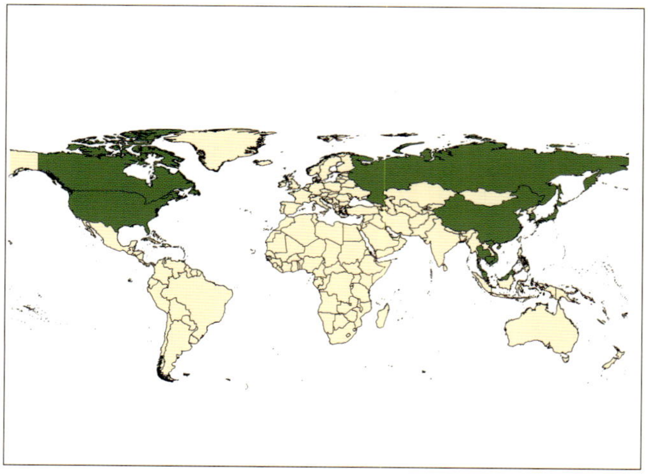

Systematics and Distribution

Aedes togoi was first described from Osaka, Japan. It is native to eastern Asia, from the Southeast Asian mainland (Cambodia, Malaysia, Thailand) up north to the Primorye region of Russia, Korea, Taiwan, and Japan, but due to its tolerance of egg desiccation, the species has become invasive. It became established in the northwestern United States and British Columbia, Canada, in the 1960s—most probably brought to the area in used tires. *Aedes togoi* and the Japanese endemic *Ae. savoryi* Bohart are the only two species comprising the newly defined *Aedes* subgenus *Tanakaius*. As adults, mosquitoes in the subgenus *Tanakaius* most closely resemble those in the subgenus *Collessius*. Subgenus *Tanakaius* species differ in that they lack the characteristic narrow yellowish stripe of scales on the anterior surface of Fe-II and bear a lower mesepimeral seta (MeSL; see figure *b*).

Bionomics

Aedes togoi females deposit their eggs in tidal pools, rock pools, and cement containers (including water tanks) near coastal areas. Other kinds of containers are also utilized, especially those made from hard nonmetallic substances such as glass, clay, and stone. Natural sites including ground pools, ditches, and gutters have also been reported as immature habitats. Ovipositing females have been shown to be attracted by the presence of conspecific larvae, presumably as a survival strategy from predation.

Some aspects of the reported bionomics of *Ae. togoi* conflict and may be indicative of cryptic speciation. *Aedes togoi* is most commonly associated with coastal regions, but it has been known to encroach inland. Females of some populations readily bite humans any time during the day and are common enough to cause a pest nuisance in seaside towns. Other populations are reluctant to feed on humans at all, preferring other mammals and birds. Some populations of this species are autogenous, depositing one batch of eggs without a blood meal, while others require a blood meal to stimulate oogenesis. Oviposition sites are usually in brackish or salt water, although immatures have been reported from freshwater sites. Different populations of *Ae. togoi* overwinter in considerable numbers in artificial containers as larvae in Japan, whereas farther south in Asia, the species is reportedly active all year-round.

Associated Pathogens

JBEV, POWV, SFV; *Brugia malayi*, *Dirofilaria immitis*, *Wuchereria bancrofti*.

Diagnostic Characters

(a) Habitus (♀): Fe-I–III without pale bands; VIII-Te wide, only semiretracted into segment VII. **(b) Thorax (side):** Occiput (Occ) covered with erect scales; lower mesepimeral seta (MeSL) present. **(c) Thorax (dorsal):** Acrostichal (AA) and dorsocentral areas (DA) with distinct narrow lines of pale scales. **(d) Abdomen (dorsal):** Cercus (Ce) short, broad; VII-Te compressed laterally (not evident in photo). **(e) Head (dorsal):** Palpus (MPlp) with white scales at apex; vertex (V) covered with erect scales, decumbent scales mostly narrow. **(f) Legs:** Fe-II with narrow anterior stripe of yellowish scales; Ta-III_{1-5} with basal and apical pale bands. **(g) Wing:** Upper calypter (UC) with fringe of long setae; alula (Al) with narrow marginal scales. **(h) Larva (terminal segments):** Comb scales (CS) rounded and evenly fringed; seta 1-S inserted on distal 0.25 of siphon (S); siphon short, at most as long as segment X; pecten (Pt) with evenly spaced spines (PS), inserted at or below base of seta 1-S; anal papillae (APP) shorter than segment X; seta 1-X inserted outside saddle (Sa); seta 2-X with ≥5 branches; seta 4-X on grid, with grid shorter than length of segment X. **(i) Larva (head):** Setae 4–6-C inserted well anterior to seta 7-C; setae 5,6-C with multiple branches, arranged in transverse row. **(j) Genitalia (♀) (ventral).** Image sources: WRBU; Tanaka et al. (1979).

Exemplar DNA sequences: *Ae. togoi* (AB738293, JQ728038–39, KC510142, KM497421–23, KT358401–03, LC025658–728).

Catalog entry: Vol. 2, p. 807.

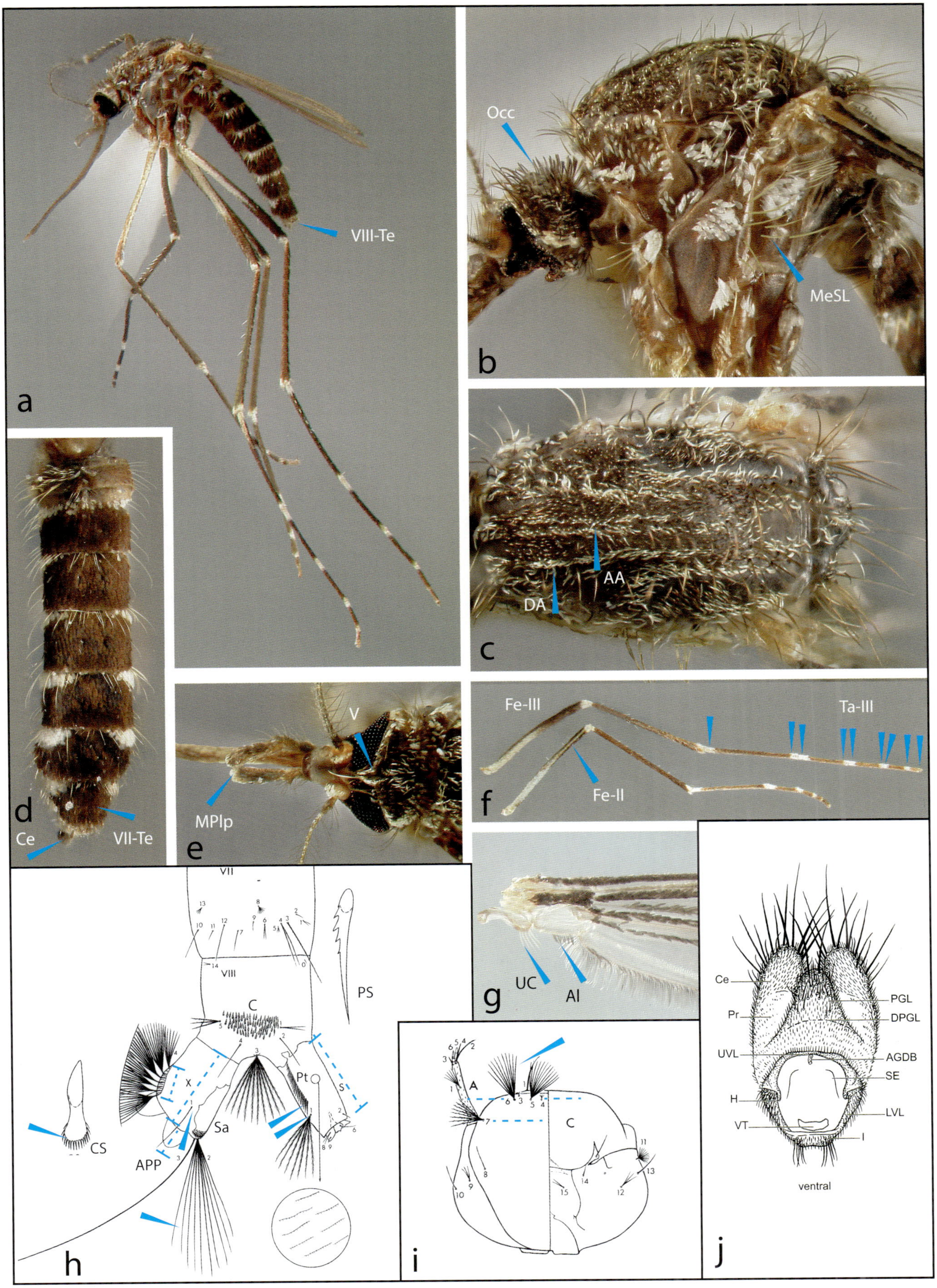
a
VIII-Te
b
Occ
MeSL
c
AA
DA
d
Ce
VII-Te
e
V
MPlp
f
Fe-III
Ta-III
Fe-II
g
UC
Al
h
VII
VIII
C
PS
X
Pt
S
Sa
APP
CS
i
A
C
j
Ce
Pr
UVL
H
VT
PGL
DPGL
AGDB
SE
LVL
I
ventral

Armigeres (Armigeres) subalbatus (Coquillett, 1898)
Subfamily Culicinae, Tribe Aedini
"White-bellied Japanese Cord Mosquito"

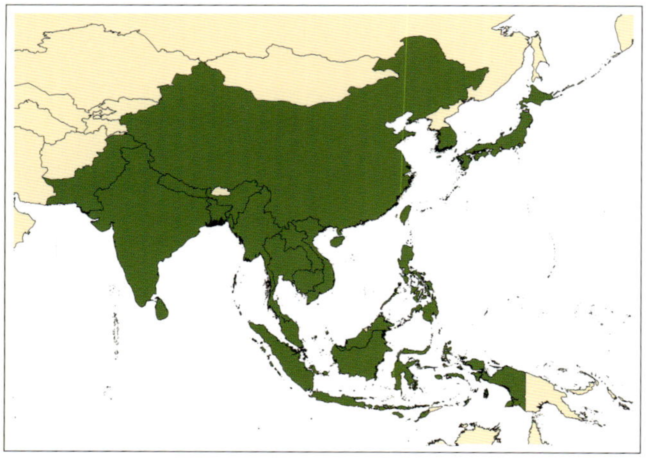

Systematics and Distribution

Armigeres subalbatus is a large mosquito, generously ornamented with distinctive patches of white scales. Originally forest associated, the species now thrives in rural and suburban areas and is closely associated with human settlements with poor sanitation. *Armigeres subalbatus* occurs in lower-altitude areas across the Indo-Oriental Region and has become established on Guam. Reported from China, Japan, and Korea, it is the only member of the subgenus *Armigeres* known to occur in the Palearctic Region. In common with all subgenus *Armigeres* species, *Ar. subalbatus* lacks a pecten on the larval siphon, easily distinguishing it from all other immature aedines within its range.

Bionomics

Populations of *Ar. subalbatus* peak following heavy rainfall, causing a pest nuisance in some areas. This suggests that eggs of *Ar. subalbatus* may tolerate desiccation, and embryonic larvae are stimulated from dormancy by flooding. *Armigeres subalbatus* is highly plastic in oviposition site choice, favoring contained sites with stagnant, strongly polluted (foul) waters or those with high organic content. Documented larval habitats include both natural containers, including *Pandanus* axils, sago palm and banana stumps, fallen leaves and spathes, flower bracts, pitcher plants, bamboo, fruit shells and husks, hollow logs, rock holes, and tree holes, and various artificial containers, including septic tanks. Morphological adaptations of the larval mandible and maxilla suggest that the species may be carnivorous, at least in part.

Armigeres subalbatus follows distinct circadian rhythms: egg hatching and eclosion occur during the day, while oviposition, larval molts, and pupation occur at night. Females are highly anthropophilic but are also known to feed on birds and other mammals (cows, rabbits). Females exhibit crepuscular biting cycles, with highest activity at dusk (1700h–1800h), with a lesser peak at dawn. Elevated host temperatures positively influence host attractiveness. Strongly moderated by light/dark cycles, female *Ar. subalbatus* has been reported biting throughout the day in deep, dark forest, and host-seeking female density fluctuates with the phases of the moon, heightening during the full moon.

Associated Pathogens

JEV; *Brugia pahangi*, *Dirofilaria immitis*, *Wuchereria bancrofti*.

Diagnostic Characters

(a) Habitus (♀). (b) Head (dorsal): Eyes ventrally with two rows of scales. **(c) Head (side):** Proboscis (P) curved downward and laterally compressed. **(d) Thorax (side):** Postspiracular setae (PS) present; lower anterior mesepimeral seta (MeSL) present. **(e) Abdomen (side):** Terga (Te) with lateral pale patches not extending dorsally; III–VI-S with apical dark bands. **(f) Larva (terminal segments):** Comb scales (CS) narrow with rounded apex, evenly fringed apically and laterally with fine spicules; siphon (S) without pecten (Pt). **(g) Thorax (dorsal):** Scutum (Scu) without pair of submedian golden stripes, lateral margins with complete line of pale scales; prescutellar setae (PrsS) present. **(h) Leg (III):** Fe-III posteroventral surface with broad white band extending to or nearly to apex. Image sources: WRBU; Tanaka et al. (1979); Rattanarithikul et al. (2010).

Exemplar DNA barcodes: *Ar. subalbatus* (HQ398903–08, KC970285, KC970288–89, KM497419, KT358440–42, MG242530).

Catalog entry: Vol. 2, p. 815.

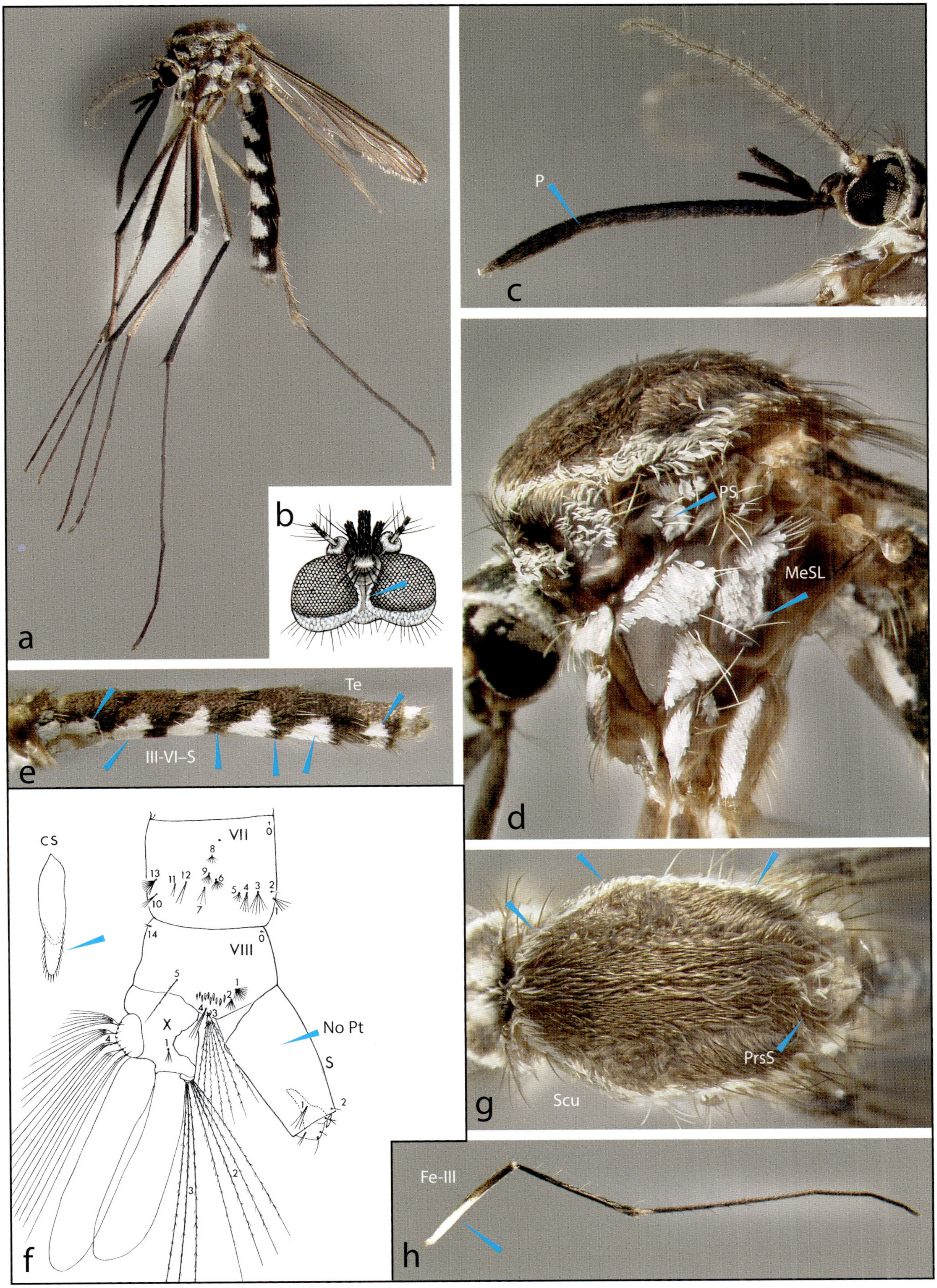
a
b
P
c
PS
MeSL
d
Te
III-VI–S
e
CS
VII
VIII
X
No Pt
S
f
PrsS
Scu
g
Fe-III
h

Haemagogus (*Haemagogus*) *janthinomys* Dyar, 1921
Subfamily Culicinae, Tribe Aedini
"Violet Trinbagonian Shoulder Mosquito"

Systematics and Distribution

Haemagogus janthinomys is a handsome mosquito, with a thorax clothed in broad silver, golden, and greeny-blue scales and vivid violet scales adorning the upper surface of the abdominal segments. The range of *Hg. janthinomys* includes Central and South America, including Trinidad and Tobago.

Haemagogus janthinomys belongs to the Albomaculatus Section of the subgenus *Haemagogus* and is readily distinguishable from all other species in this group except *Hg. capricornii*. Male genitalia characters—in particular the spicules, paraproct, and apical process—are the most reliable diagnostic characters to differentiate these two species. These difficulties may have confused bionomic and distributional assignments for these two species.

Bionomics

Haemagogus janthinomys is a true arboreal mosquito, active during the day (0600h–1800h) and prevalent when relative humidity reaches 70% or above. It is found exclusively in primary tropical rain forest throughout most of its range, reaching its highest densities at heights over 100 m up into the forest canopy. The species is multivoltine, with some females laying as many as seven egg batches. Population densities are more than five times higher in the wet season than in the dry season. *Haemagogus janthinomys* utilizes natural container habitats for breeding, primarily tree holes. It seldom ventures to ground level to seek blood meals and then, only when the forest is disturbed. In French Guiana, ground host-seeking activity by *Hg. janthinomys* was shown to be seasonal. *Haemagogus janthinomys* is extremely plastic in host choice, feeding on primates, birds, rodents, opossums, cattle, horses, dogs, and humans, with several individuals noted to take blood from up to three different host species in a single ovicycle. This eclectic feeding strategy renders *Hg. janthinomys* an important zoonotic vector for viruses, particularly yellow fever, across its range. Another unusual bionomic aspect promoting viral transmission by this species is that, on average, seven eggs were retained by the females per oviposition cycle. This carryover of "old eggs" may promote transovarial transmission in infected females and, most importantly, indicates that blood-feeding behaviors are not inhibited by egg production.

Associated Pathogens

ILHV, YFV.

Diagnostic Characters

(a) Habitus (♀): Fe-II,III without apical pale spots; Fe-III with silver scales anteriorly extending to apex. **(b) Thorax (side):** Scutum (Scu) with contrasting silver scales in antealar area (AnA); pleuron with single broad vertical band of silver scales; lower mesokatepisternum (Mkm) with a well-developed seta (not evident in photo); postpronotum (Ppn) densely scaled; mesopostnotum (Mpn) bare (not shown); C-I,II with patches of dark scales. **(c) Wing:** Vein R_{2+3} usually greater than 0.55 of R_2. **(d) Thorax (dorsal):** Scutal scales of various metallic colors (green, blue, copper, bronze). **(e) Larva (head):** Seta 6-C single. **(f) Abdomen (dorsal):** IV–VIII-Te with dark bluish-green scales on distal margins, contrasting with predominantly purple tergal scales. **(g) Larva (thorax/abdominal segments):** Setae 5–7-P on common tubercle; seta 12-I absent. **(h) Larva (terminal segments):** Comb scales (CS) attached to sclerotized plate (CP), 4–11 CS in a single row, with single pointed, minutely fringed spine; saddle (Sa) with well-developed spines on caudal margin; boss (B) strongly sclerotized; pecten (Pt) with spines (PS) 10–14, in about basal 0.5 of siphon (S); seta 1-S 2,3 branched, inserted in about basal 0.55 of siphon. Image sources: WRBU; Arnell (1973).

Exemplar DNA barcodes: *Hg. janthinomys* (KM593034–35, KY859893–97, MF172319–21, MG242531).

Catalog entry: Vol. 2, p. 824.

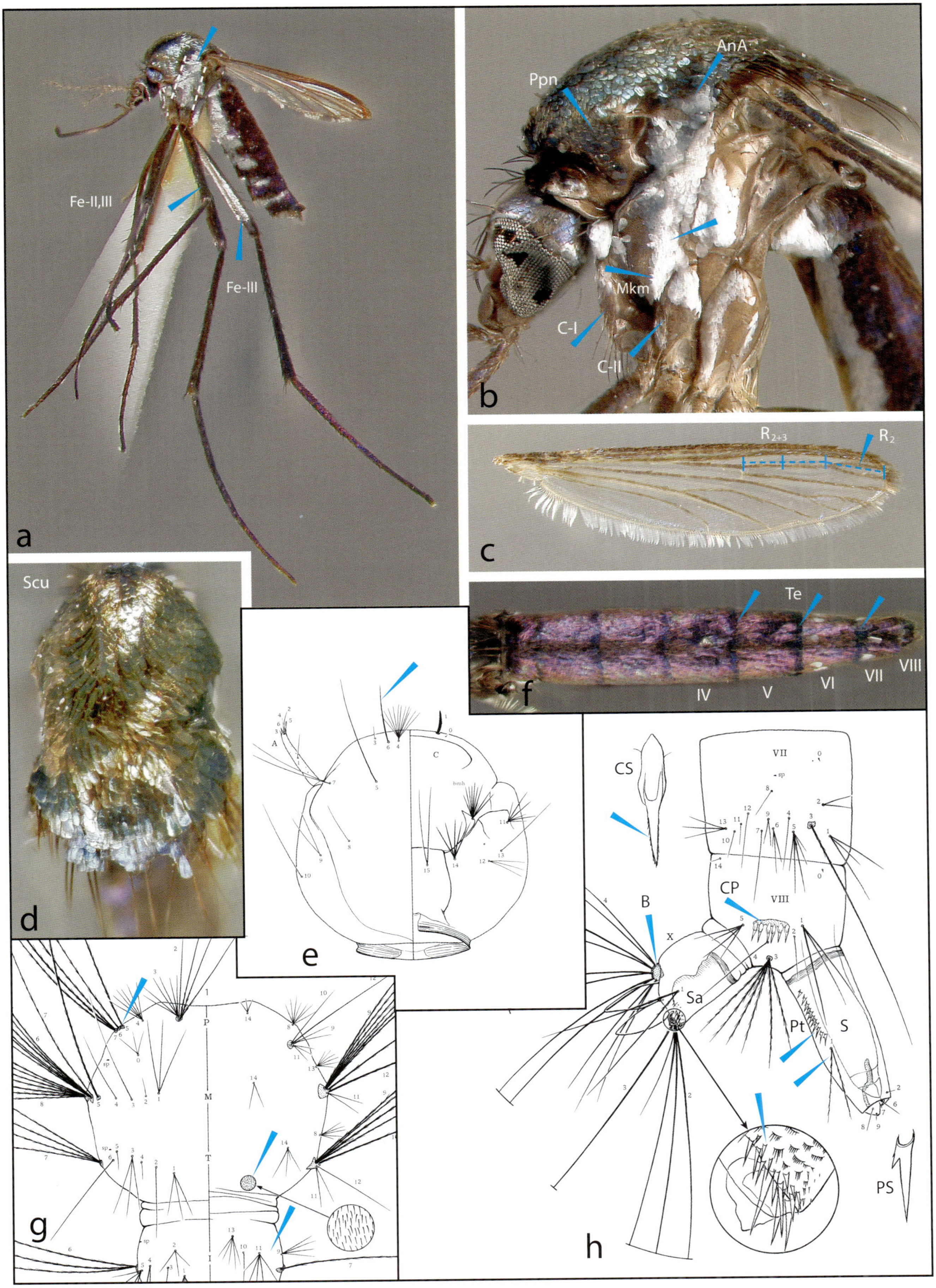
a
Fe-II,III
Fe-III
b
Ppn
AnA
Mkm
C-I
C-II
c
R_{2+3}
R_2
d
Scu
e
f
Te
IV
V
VI
VII
VIII
g
h
CS
CP
B
X
Sa
Pt
S
PS

Haemagogus (Haemagogus) mesodentatus Komp & Kumm, 1938
Subfamily Culicinae, Tribe Aedini
"Toothy Costa Rican Shoulder Mosquito"

Systematics and Distribution

Haemagogus mesodentatus is stunningly decorated, with blue, green, copper, and bronze iridescent scales on its head and scutum and along the leading edges of its wings, contrasting with its dense silver pleura and violet-sheened legs. The name *mesodentatus* is in reference to its toothed aedeagus. The species closely resembles *Hg. janthinomys* in the larval stage (see p. 474). However, the comb scales of *Hg. mesodentatus* larvae are not positioned on a comb plate, as in *Hg. janthinomys*; instead, 4–20 comb scales attach directly to the eighth abdominal segment in a perfectly straight row (see *f*). Originally described from Costa Rica, *Hg. mesodentatus* is reported throughout Central America, from Mexico to Panama.

Bionomics

Little is known about the specific biology of this species, with many studies referring to it as "the same as *Hg. janthinomys*." Adult females are diurnal, and like other *Haemagogus* mosquitoes, they seldom leave the forest. Females are known to attack humans in the forest, especially those involved with jungle-clearing activities, such as logging and agricultural practices. Immatures of *Hg. mesodentatus* are found in tree holes and in cut bamboo. The species is a highly competent vector of yellow fever, and it has been held responsible for the catastrophic collapse of the Classic lowland Mayan civilization of the Petén Region of Guatemala.

Associated Pathogens

CHIK, ONNV, VEEV, WNV, YFV.

Diagnostic Characters

(a) Habitus (♀): Pleuron with single broad vertical band of silver scales; wing vein R_{2+3} usually greater than 0.55 of R_2 (not shown, see *Hg. janthinomys*, p. 475). **(b) Thorax (side):** Scutum (Scu) with contrasting silver scales in antealar area (AnA); postpronotum (Ppn) densely scaled; mesopostnotum (Mpn) bare. **(c) Leg (II):** Ta-$II_{2,3}$ with white or gray scales on outer surface. **(d) Thorax (dorsal):** Scutal (Scu) scales vary from coppery to gold to metallic green and blue. **(e) Larva (thorax/abdominal segments):** Seta 12-I present. **(f) Larva (terminal segments):** Comb scales (CS) 4–20 in single row, not attached to sclerotized plate; seta 4-X with five pairs of setae (rarely six). Image sources: WRBU; Arnell (1973).

Exemplar DNA sequences: *Hg. mesodentatus ITS2* (M95128).

Catalog entry: Vol. 2, p. 824.

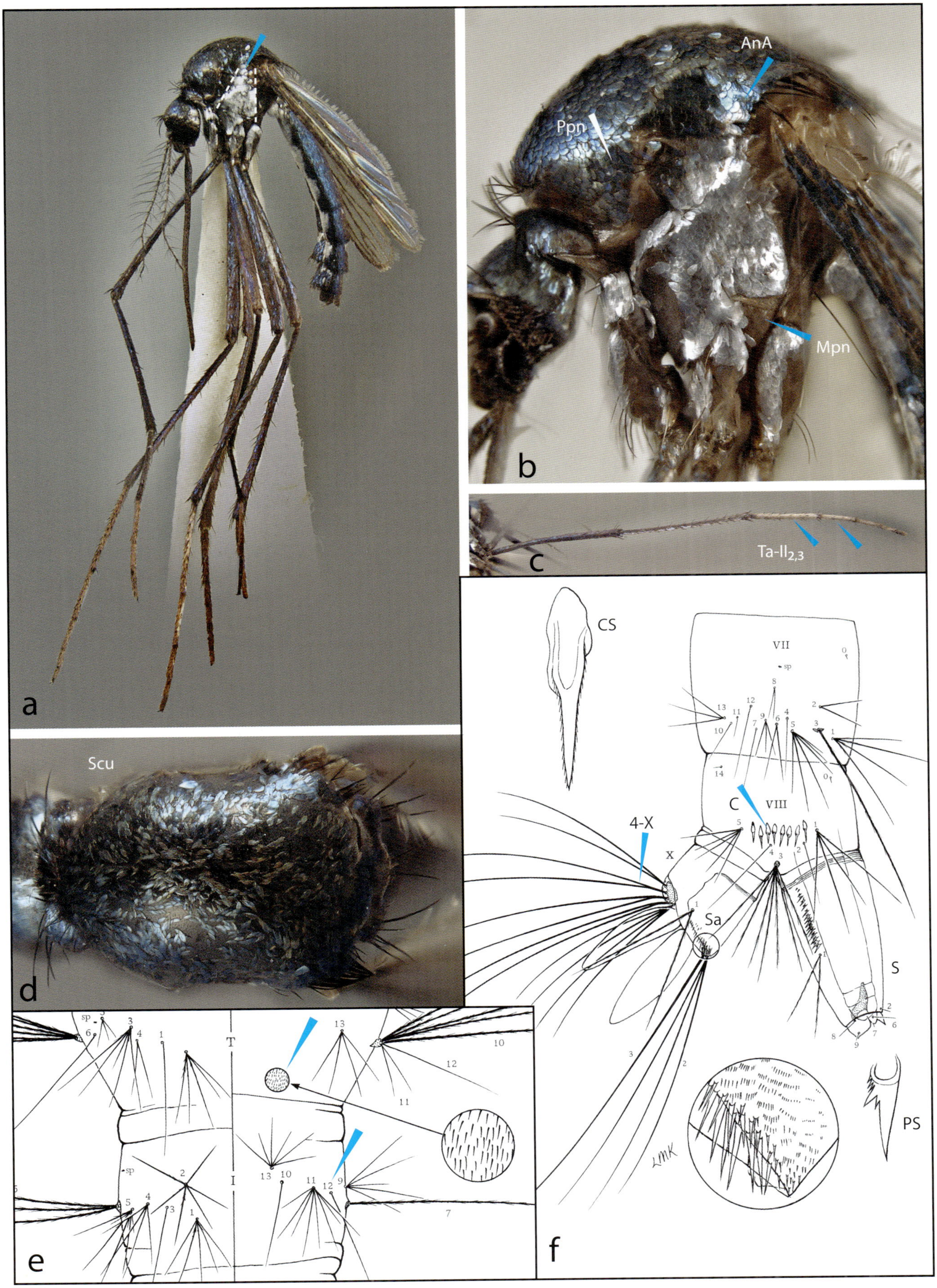
a
AnA
Ppn
Mpn
b
Ta-II2,3
c
Scu
d
CS
VII
sp
VIII
C
4-X
X
Sa
S
PS
LMK
f
T
I
e

Haemagogus (*Haemagogus*) spegazzinii Brèthes, 1912
Subfamily Culicinae, Tribe Aedini
"Spegazzini Argentinian Shoulder Mosquito"

Systematics and Distribution

Haemagogus spegazzinii is distributed from Colombia to Argentina and to the islands of Trinidad and Tobago. The species is one of two in the subgenus *Haemagogus* (n = 24). The history of the species is taxonomically complex. *Haemagogus spegazzinii*, along with *Hg. capricornii* Lutz, was synonymized under *Hg. equinus* Theobald. This was difficult to agree with and most definitely the result of lack of access to male specimens. As with most Culicidae, the palps of *Hg. spegazzinii* and *Hg. capricornii* are about one-eighth of the length of the proboscis, whereas palps of *Hg. equinus* are unique, being as long as the proboscis—a trait more common with mosquitoes in the subfamily Anophelinae.

Bionomics

This species is most commonly collected as larvae, which rear in tree holes and bamboo internodes. Given the large population sizes at certain times of the year, it is supposed that this species must occupy other niches, but these have never been located, possibly due to their inaccessible locations in the canopy. Mating occurs in swarms, also high in the canopy. Females are rare in light traps, and males are almost never sampled in nature unless link reared from larval collections. *Haemagogus spegazzinii* can become locally dominant at certain times of the year, accounting for more than 80% of mosquitoes captured. It is the most common species of *Haemagogus* in Colombia and one of only two *Haemagogus* species reported from east of the Andes. The subspecies *Hg. spegazzinii falco* is associated with low-lying areas under 500 m above sea level.

Associated Pathogens

ILHV, YFV.

Diagnostic Characters

(a) Habitus (♀): Pleuron with single broad vertical band of silver scales; Fe-II,III lacking apical pale spots; Ta-II dark scaled. **(b) Thorax (side):** Antepronotal scales (ApSc) nearly all silvery; mesopostnotum bare (not shown); postpronotal scales (PpSc) dense, coppery to golden; lower mesokatepisternum seta (MkSL) absent or underdeveloped; antealar area (AnA) with silver scales; coxae silver scaled. **(c) Abdomen (side):** Terga (Te) without silver scales—instead blue, purple, green, or golden. **(d) Thorax (dorsal):** Scutal scales metallic colored, mostly coppery to light bronze. **(e) Head (dorsal):** Vertex (V) decumbent scales green to gold. **(f) Larva (thorax/abdominal segments):** Setae 5,6-P on common tubercle; seta 12-I absent. **(g) Larva (head):** Setae 5,6-C single. **(h) Larva (terminal segments):** Comb scales (CS) sharply pointed with minute fringe, 8–10 in single row; pecten spines (PS) 20–24; siphon index 3.2–3.4; saddle (Sa) with well-developed spines on caudal margin; boss (B) strongly sclerotized; seta 1-S 3–4 branched, inserted about 0.60 distance from base of siphon (S). Image sources: WRBU; Arnell (1973).

Exemplar DNA barcodes: *Hg. spegazzinii* (MH118155).

Catalog entry: Vol. 2, p. 825.

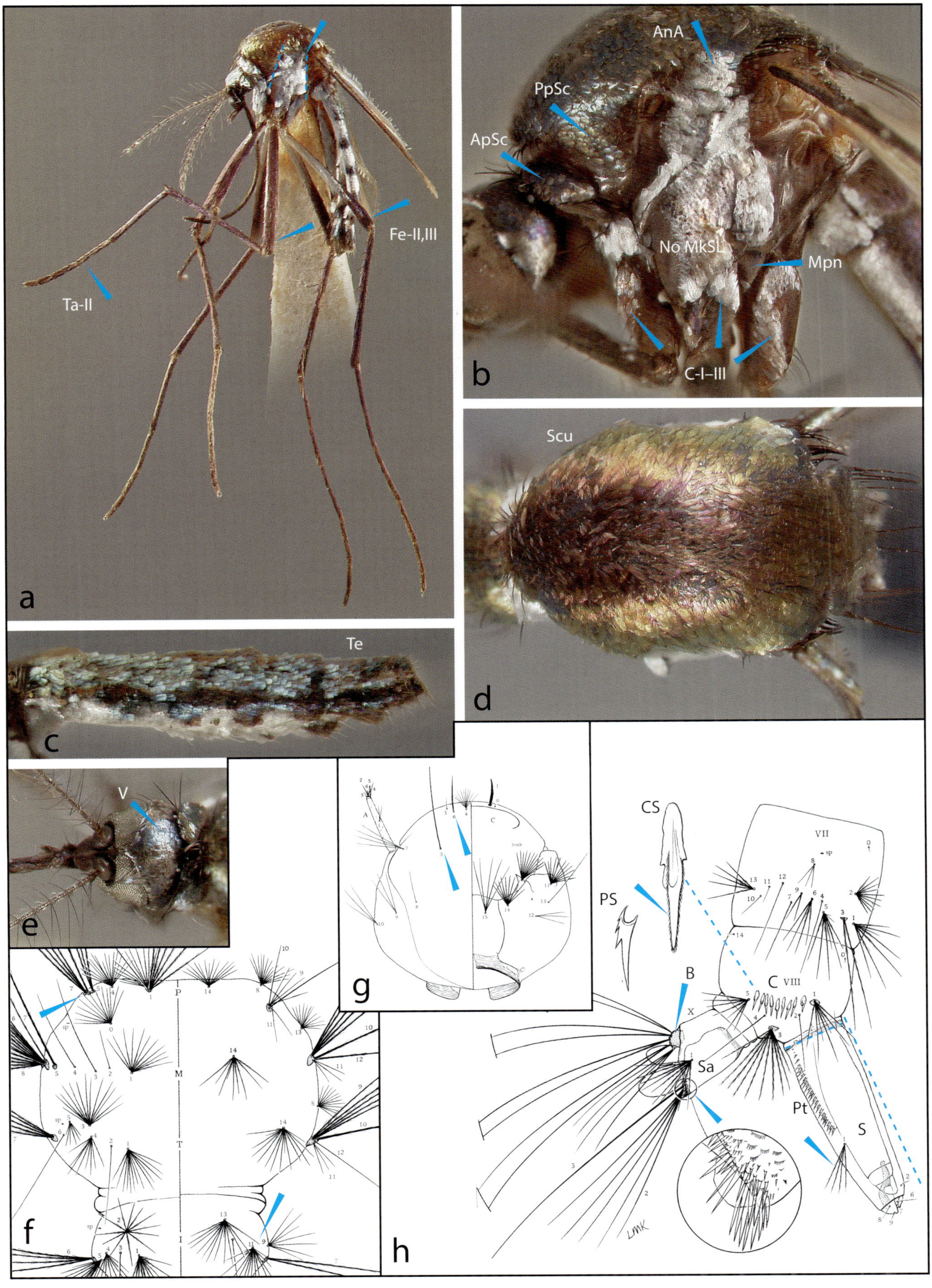
Ta-II
Fe-II,III
a
AnA
PpSc
ApSc
No MkSL
Mpn
C-I–III
b
Te
c
Scu
d
V
e
P
M
T
I
f
A
C
g
CS
PS
B
C
VIII
VII
X
Sa
Pt
S
LMK
h

Psorophora (Grabhamia) columbiae (Dyar & Knab, 1906)
Subfamily Culicinae, Tribe Aedini
"Columbia American Gang Mosquito"

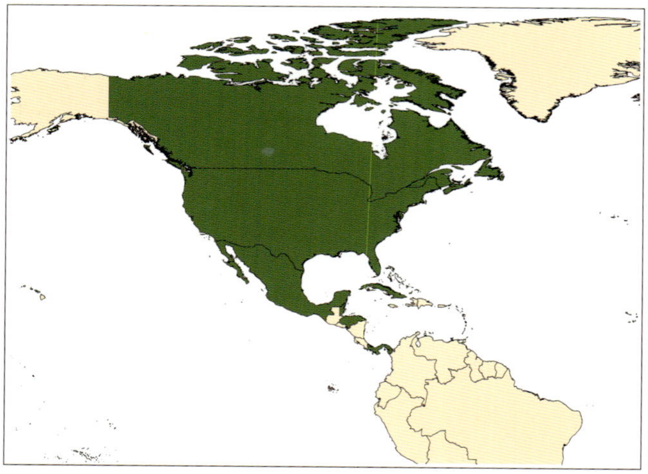

Systematics and Distribution

Psorophora columbiae is a medium-sized dark mosquito with banded legs and a spattering of broad white appressed scales adorning the pleura and abdominal segments that give it an overall salt-and-pepper look. The species is present in southern Canada, the eastern United States, and across the south-central desert belt, into Mexico. It has also been reported on islands in the Caribbean, including the Bahamas and Cayman Islands. *Psorophora columbiae* is a member of the *Ps. confinnis* species complex.

Bionomics

The females bite incessantly throughout the day and night. *Psorophora columbiae* feeds entirely opportunistically, attacking all mammals in its vicinity, but it does show a marked preference for cattle. Reported hosts also include humans, rabbits, deer, and armadillos. Populations peak from mid-July to late August. At high densities, *Ps. columbiae* is known to cause deaths of adult cows through exsanguination and of calves by suffocation.

Psorophora columbiae is a floodwater mosquito. Eggs withstand desiccation and *Ps. columbiae* passes the winter in this stage. Immatures are found in vegetation-rich temporary pools of accumulated rainwater (sometimes brackish), with submerged and emergent grasses. Rice fields provide an ideal habitat for this species. The larval stages are short, usually between 4 and 10 days, ensuring multiple generations per individual and enabling rapid population increases in suitable conditions. Males emerge several hours before the females and wait for the females to emerge, mating almost immediately. Although mosquitoes of the genus *Psorophora* are extremely strong fliers, *Ps. columbiae* has a reportedly small range (<1 mile). Females are attracted to light and are readily captured in Centers for Disease Control and Prevention (CDC) and New Jersey light traps.

Associated Pathogens

POTV, RVFV, VEEV, WNV.

Diagnostic Characters

(a) Habitus (♀): Very large species; Fe-III pale basally without preapical white band, with apical black scale tufts; U-I–III with distinct tooth basally (not shown). **(b) Thorax (side):** Proepisternum (Ps) without pale scales; lower mesepimeron with 5–11 setae (MeSL) (not distinct in photo); postprocoxal scales (PSc) white. **(c) Head (side):** Proboscis (P) with golden scales on distal half, apex dark; occiput (Occ) with slender erect scales with notched tips (not distinct in photo). **(d) Thorax (dorsal):** Scutum (Scu) with distinct median stripe of golden-brown scales along entire length. **(e) Wing:** Wing scales narrow, brownish with inconspicuous pale scales intermixed on costal (C) and subcostal (Sc) veins and vein R_1. **(f) Larva (terminal segments):** Spines (PS) of pecten (Pt) long, ≥15; seta 1-X short, 3,4 branched; comb scales (CS) illustrated. **(g) Larva (head):** Head truncated anteriorly; antenna (A) short, not or barely reaching, front of head. Image source: WRBU.

Exemplar DNA barcodes: *Ps. columbiae* (JX260111–12).

Catalog entry: Vol. 2, p. 831.

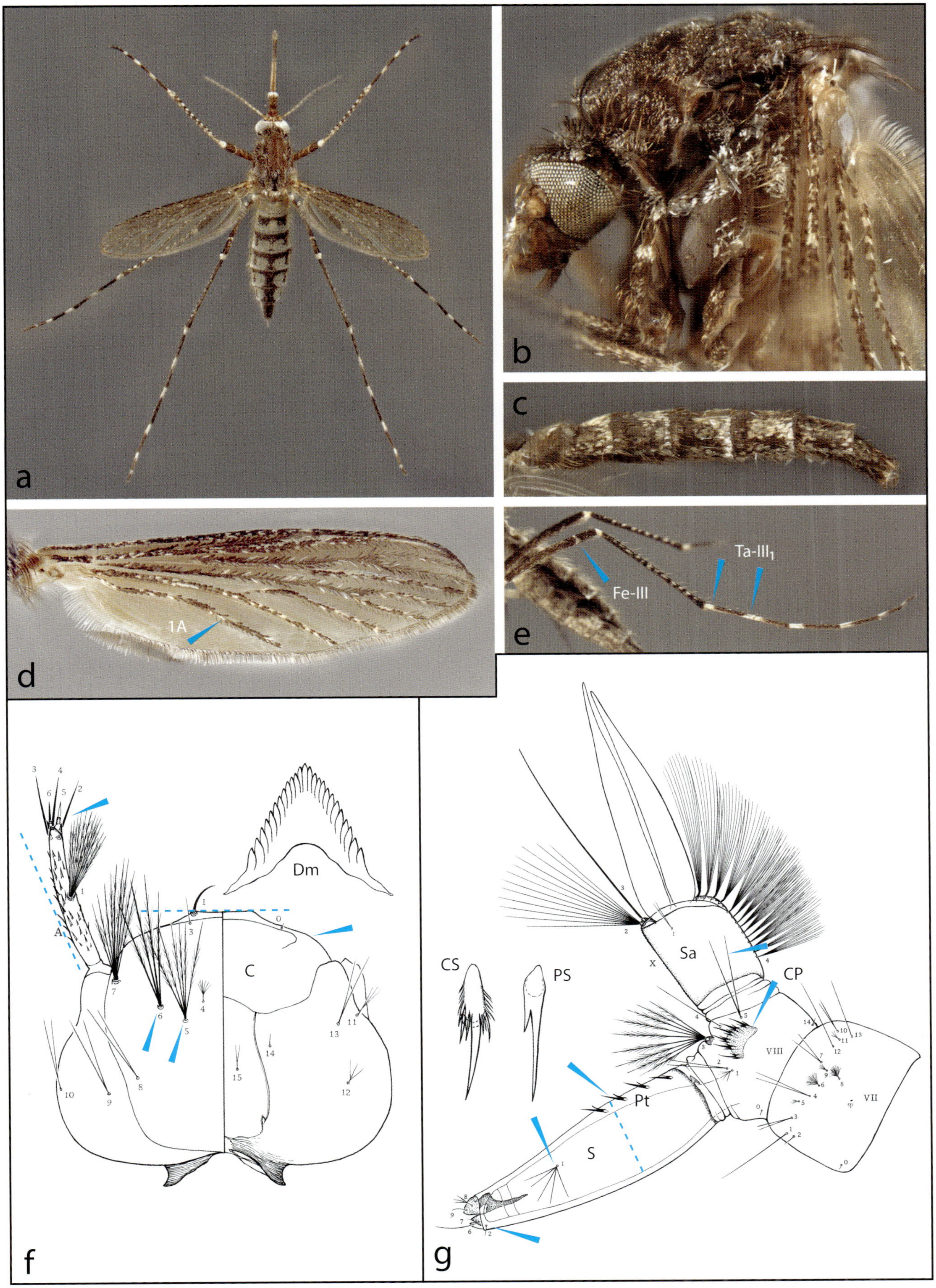
a
b
c
d
1A
e
Fe-III
Ta-III$_1$
f
A
C
Dm
g
Sa
CP
CS
PS
Pt
S
VIII
VII

Psorophora (Janthinosoma) ferox (von Humboldt, 1819)
Subfamily Culicinae, Tribe Aedini
"Wild Ecuadorian Gang Mosquito"

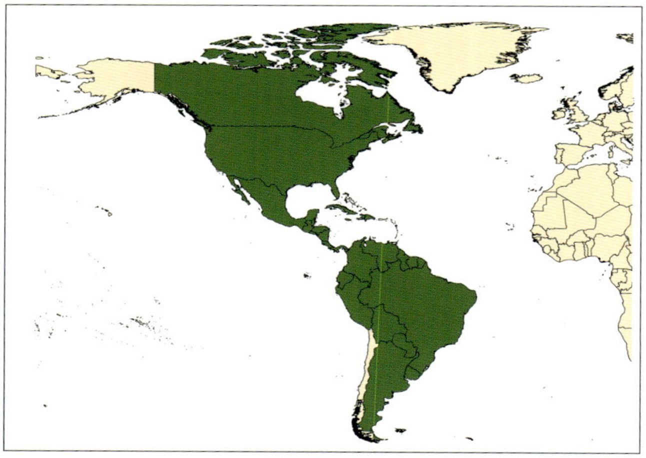

Systematics and Distribution

Psorophora ferox is a common medium-sized floodwater mosquito, with a long dark proboscis and bright-white hind tarsi. The species is native to much of the Americas, from Argentina to southeastern Canada. In the United States, its distribution is limited to the eastern half. *Psorophora ferox* is one of 23 described species in the subgenus *Janthinosoma*.

Bionomics

Psorophora ferox is closely associated with wet woodland habitats, and larvae are found in temporary rain-filled pools and in overflow stream pools, especially those heavily shaded by bushes. Mating swarms occur in cloudy/overcast conditions in open woodlands. It is a highly aggressive and painful biter, even attacking humans in the open on cloudy days. The species feeds opportunistically throughout the day and night. Females feed mainly on mammals but also on birds. Up to 80 eggs per brood are laid, and like other floodwater species, eggs of *Ps. ferox* are tolerant to desiccation and are laid in regions prone to flooding. The species overwinters as eggs. Larvae develop during the summer in the temperate climes of North America. *Psorophora ferox* larvae have distinctive antennae that are as long or longer than the head capsule.

In Panama and Colombia, this species is commonly infested by eggs of the human botfly (*Dermatobia hominis*), which use the mosquitoes as transportation to their hosts. The *Dermatobia* eggs are stimulated to hatch by the warmth of the mammalian host upon which the mosquito is feeding, and then the first instar larvae drop off the mosquito and enter the host through the puncture wound caused by the mosquito bite.

Associated Pathogens

GROV, IERIV, ILHV, KRIV, MAGV, MAYV, ORIV, OROV, ROCV, UNAV, VEEV, WNV, WYOV.

Diagnostic Characters

(a) Habitus (♀): Fe-III pale basally, without preapical white band; Fe-III without distinct erect dark scales apically; Ta-$III_{4,5}$ and often apex of Ta-III_3 white scaled; U-I–III with distinct tooth basally (not shown). **(b) Thorax (dorsal):** Scutum (Scu) without a distinct pattern of pale and dark scales. **(c) Abdomen (dorsal):** I-Te with patch of purple scales medially; II–VI-Te dark, purple, with apicolateral patches of white scales. **(d) Thorax (side):** Proepisternum with pale scales (PeSc); subspiracular patch of white scales (SSc). **(e) Head (side):** Occiput (Occ) erect scales broad, distinctly forked apically. **(f) Wing:** Wing with narrow dark scales. **(g) Larva (head):** Head rounded anteriorly; antenna (A) long, distinctly longer than head length; setae 5,6-C usually double and longer than 7-C; seta 7-C ≥5 branched. **(h) Larva (terminal segments):** Seta 1-VIII small and multibranched; seta 2-VIII much longer than 1-VIII; 6–8 comb scales (CS), often on posterior border of weakly sclerotized comb plate (CP); siphon (S) expanded medially; seta 1-S very small, ≥3 branched; seta 2-S small, weak, and straight; pecten (Pt) with ≤8 small spines (PS). **(i) Larva (abdominal segments):** Seta 6-IV–VI longer than length of respective segments. Image sources: WRBU; Belkin et al. (1970).

Exemplar DNA barcodes: *Ps. ferox* (KM593037–38, KM593062, KY859923–32, MF172347–49, MG242536).

Catalog entry: Vol. 2, p. 834.

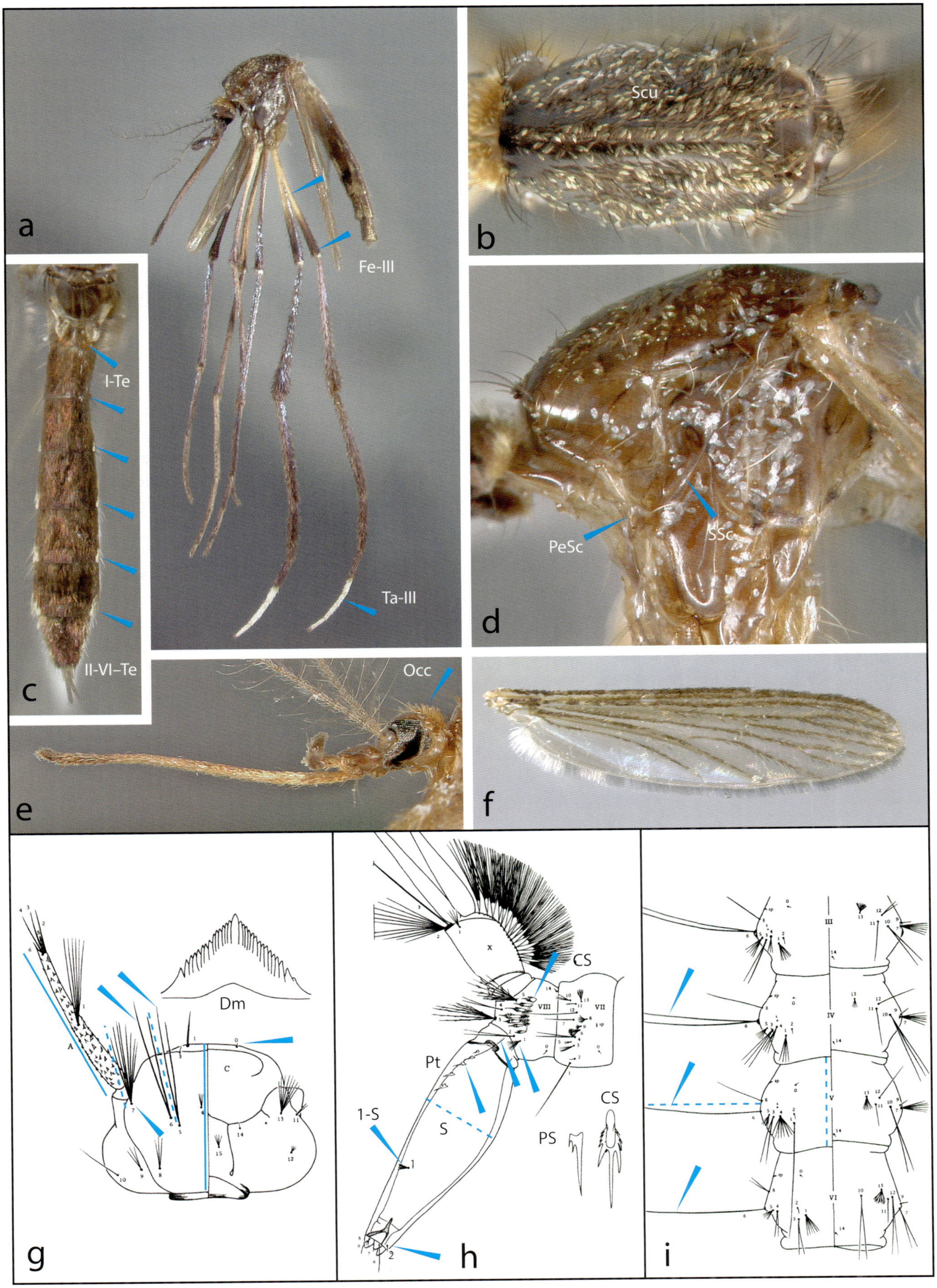
a
Fe-III
Ta-III
b
Scu
c
I-Te
II-VI–Te
d
PeSc
SSc
e
Occ
f
g
Dm
A
C
h
CS
Pt
S
1-S
PS
CS
i
III
IV
V
VI

Psorophora (Psorophora) ciliata (Fabricius, 1794) Subfamily Culicidae, Tribe Aedini "Eyelid American Gang Mosquito"

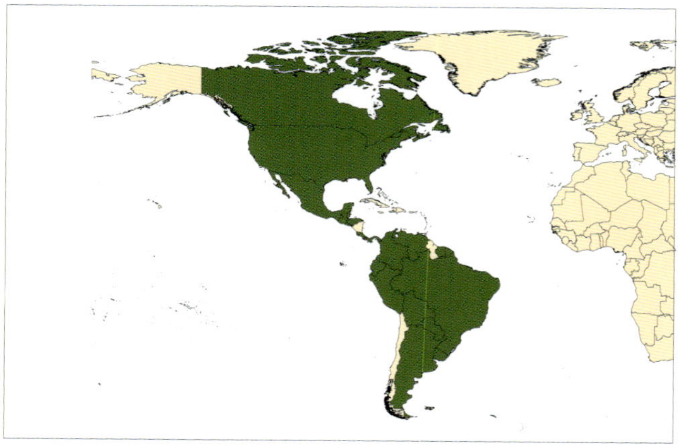

Systematics and Distribution

Psorophora ciliata is a giant, shaggy-legged mosquito renowned for its ferocity and painful bites and the fact that it is the largest mosquito in North America. Flanery (1897) referred to *Ps. ciliata* as "the little zebra-legged thing—the shyest, slyest, meanest and most venomous of them all." Originally described from South Carolina in the United States, *Ps. ciliata* is native to much of Central and South America and has been reported from Cuba, the Cayman Islands, and Jamaica in the Caribbean.

Psorophora ciliata is the nominotypical species of the genus *Psorophora*. Specimens from the United States, Mexico and Jamaica correspond to the type, while specimens from Venezuela show notable morphological differences in leg and thorax markings, suggesting that *Ps. ciliata* comprises at least two species across its range. The species illustrated here represents "true" *Ps. ciliata*. Three synonyms currently exist: *cyanopennis* (Humboldt) from Colombia, *tibialis* (Robineau-Desvoidy) from Brazil, and *lynchi* Brèthes from Argentina. Given the biomedical importance of this species, an integrated systematics study to ascertain the taxonomic status of the species across its range is long overdue.

Bionomics

Psorophora ciliata is a typical floodwater mosquito. Females lay their elongate ovoid eggs in areas subject to flooding in heavy rains. Eggs, carefully tucked into unshaded cracks in the soil, potholes, or small ground depressions, aptly survive desiccation and are believed to estivate until they have overwintered, in spite of flooding stimulus. The large, square-headed larvae have modified mouthparts designed to grip and immobilize prey. As second through fourth instars, the larvae of *Ps. ciliata* are predacious and, somewhat unusually among mosquitoes, feed primarily on other mosquito larvae. However, there are scattered reports of cannibalism and feeding on tadpoles. They develop quickly (four to six days) in these temporary environments, with the pupal stage lasting only two days. The species is found as larvae from May to September in the southern United States. It is found in highest numbers in rice fields, although the species is relatively rare overall. *Psorophora ciliata* mosquitoes are painful, persistent biters, attacking at any time of day. In Florida, *Ps. ciliata* is primarily mammalophilic, with hosts shown to include humans, ruminants, ovids, rabbits, armadillos, and raccoons.

Associated Pathogens

SLEV, TENV, VEEV, WEEV, WNV.

Diagnostic Characters

(a) Habitus (♀). (b) Thorax (side). (c) Abdomen (side): Dark and white scales randomly scattered. **(d) Wing:** Vein 1A speckled black and white along entire length. **(e) Leg (III):** Fe-III with narrow preapical white band; Ta-III_1 with basal and median pale bands; U-III long, single, without a basal tooth (not shown). **(f) Larva (head):** Head rounded anteriorly; antenna (A) long, reaching well beyond front of head; setae 5,6-C ≥4 branched. **(g) Larva (terminal segments):** Seta 5-VIII 2,3 branched, 5–6 comb scales (CS) usually on edge of weakly sclerotized plate (CP); <8 small pecten spines (PS); siphon (S) centrally swollen; seta 1-S small or tiny with ≥3 branches; seta 2-S small, weak, and straight. Image sources: WRBU; Belkin et al. (1970).

Exemplar DNA barcodes: *Ps. ciliata* (MG242533).

Catalog entry: Vol. 2, p. 836.

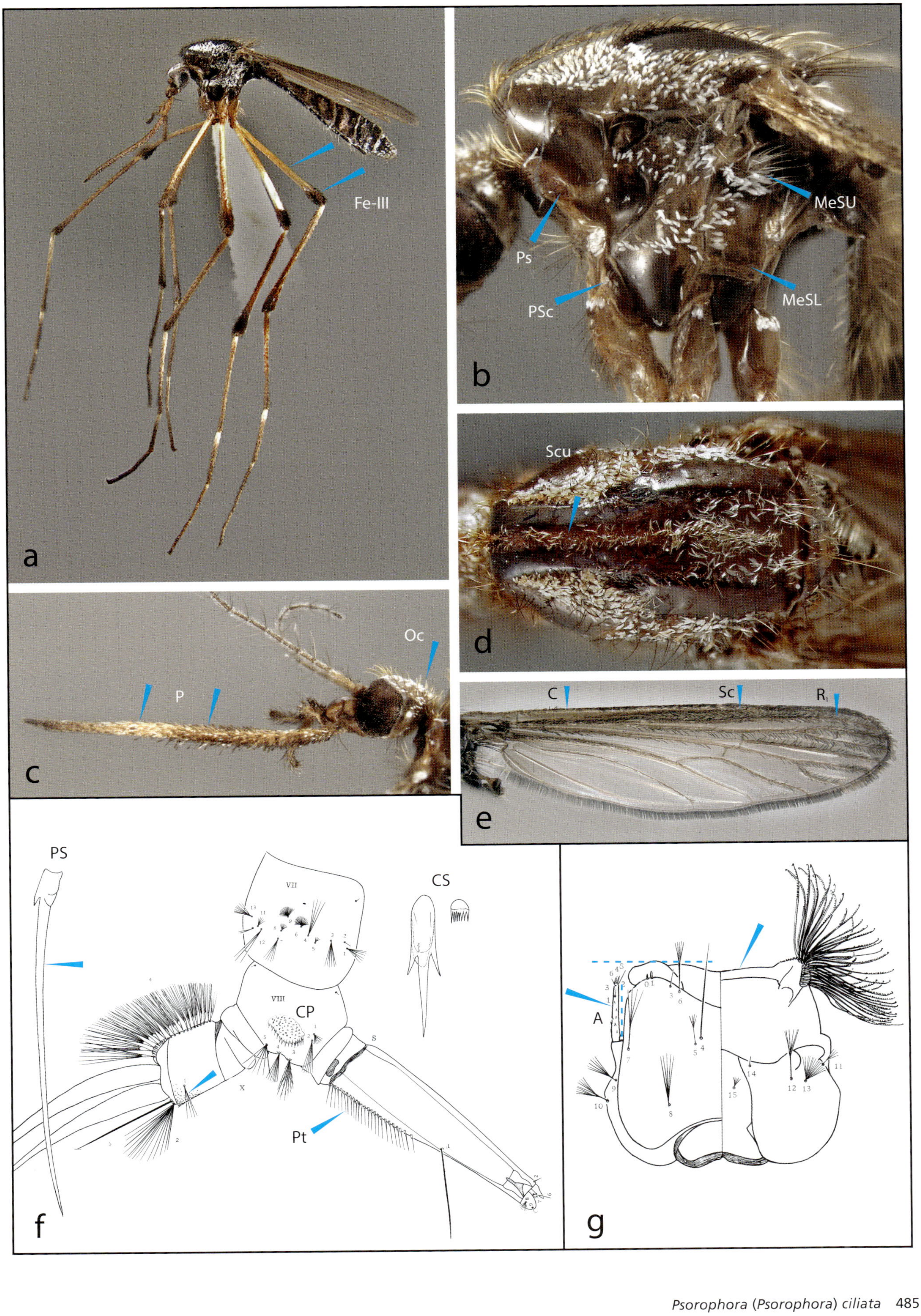

Fe-III
a
MeSU
Ps
PSc
MeSL
b
Scu
d
Oc
P
c
C
Sc
R1
e
PS
VII
CS
VIII
CP
X
S
Pt
f
A
g

Culex (Culex) annulirostris Skuse, 1889
Subfamily Culicinae, Tribe Culicini
"Ring-beaked Australian Typical Mosquito"

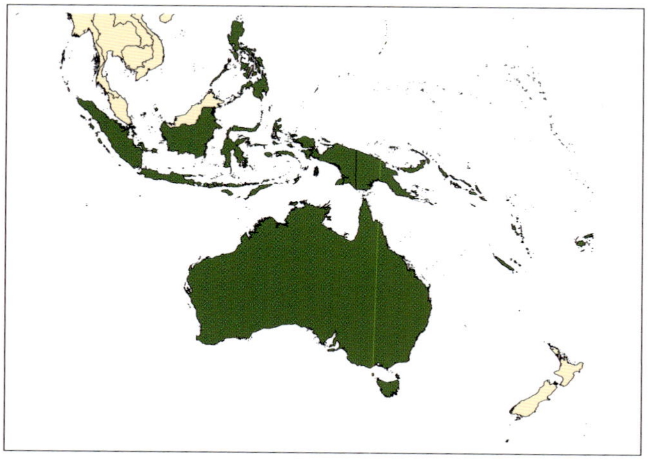

Systematics and Distribution

Culex annulirostris is a medium-sized brown to dark-brown mosquito with a distinct median pale band on the proboscis and banding on its legs. It belongs to the Sitiens Subgroup and Sitiens Group of the subgenus *Culex*, along with *alis* Theobald, *litoralis* Bohart, *palpalis* Taylor, sitiens Wiedemann, and *whitmorei* (Giles). It closely resembles *Cx. sitiens* (see species page, p. 506), with which it is sympatric in parts of its range, both of which belong to the Sitiens Group and Subgroup within the *Culex* subgenus *Culex*. The species takes its name from the Latin, with *annulus* meaning "ring" and *rostrum* meaning "bill or beak." It is found from Indonesia east to the southern Pacific islands and from Australia north to the Philippines.

Bionomics

Ovipositing *Cx. annulirostris* has been observed laying eggs in shallow pools of water with emergent grass immediately following rainfall but generally seeks out established ground pools with well-developed vegetation, such as rice fields and swamps. Across its range, however, immatures have been reported in natural container habitats, including cacao shells, bamboo stumps, and artificial containers. Immatures are found in both urban and rural environments, and oviposition sites chosen range from full sunlight to heavily shaded, clean or polluted, with fresh or brackish water. Under laboratory conditions, ovipositing females prefer black-colored containers with water that is 10 cm deep with conspecific larvae present, indicating that this species is using a combination of visual and chemical cues to choose oviposition sites.

Culex annulirostris females are active from spring to late autumn and are multivoltine, allowing population levels to get high. They feed primarily outdoors in the evening but have been reported feeding indoors and feeding during the day. They exhibit high plasticity in host choices, literally feeding on whatever warm-blooded animal is readily available. Confirmed hosts include dogs, cats, brushtail possums, macropods, horses, humans, birds, cats, and fruit bats. Population-specific host preferences are reported, with some populations preferentially feeding on mammals, while others clearly prefer birds. In Australia, *Cx. annulirostris* is active throughout the year, but highest numbers have been recorded between October and May, when average daily temperatures rise above 17.5°C. Populations reach their peak when mean temperatures reach 25°C.

The ecological plasticity of this species suggests that closer inspection is needed to better establish the taxonomic status of this competent vector.

Associated Pathogens

BFV, BEFV, BELV, CORV, EHV, EUBV, GGV, GETV, JBEV, KIMV, KOKV, KOOV, KUNV, MVEV, MYXV, PIAV, PRV, RRV, WONV; *Wuchereria bancrofti*.

Diagnostic Characters

(a) Habitus (♀): Abdominal terga pale basally. **(b) Thorax (side):** Postspiracular area (PA) bare; lower mesepimeral setae (MeSL) absent; abdominal terga pale basally. **(c) Wing:** Wing all dark scaled. **(d) Head (side):** Proboscis (P) with median pale band 0.25–0.30 of total length. **(e) Leg (III):** Fe-III usually with some pale scales intermixed with dark. **(f) Leg (II):** Fe-II usually intermixed with pale scales but not forming definite spot. **(g) Larva (thorax/ abdomen):** Setae 1-M,T small; seta 6-VI double (not shown); seta 7-I double. **(h) Larva (terminal segments):** Comb (C) with scales (CS) apically rounded with even fringe of fine spinules; siphon index <6.0; seta 1-S tufts well developed, 6–12 branched; pecten (Pt) with spines (PS) bearing 8–9 lateral denticles. **(i) Larva (head):** Dorsal apotome (DAp) distinctly separated from median labral plate (MLP); seta 1-C short, stout, apically blunt. Image sources: WRBU; Sirivanakarn (1976).

Exemplar DNA barcodes: *Cx. annulirostris* (KU494989–95).

Catalog entry: Vol. 2, p. 858.

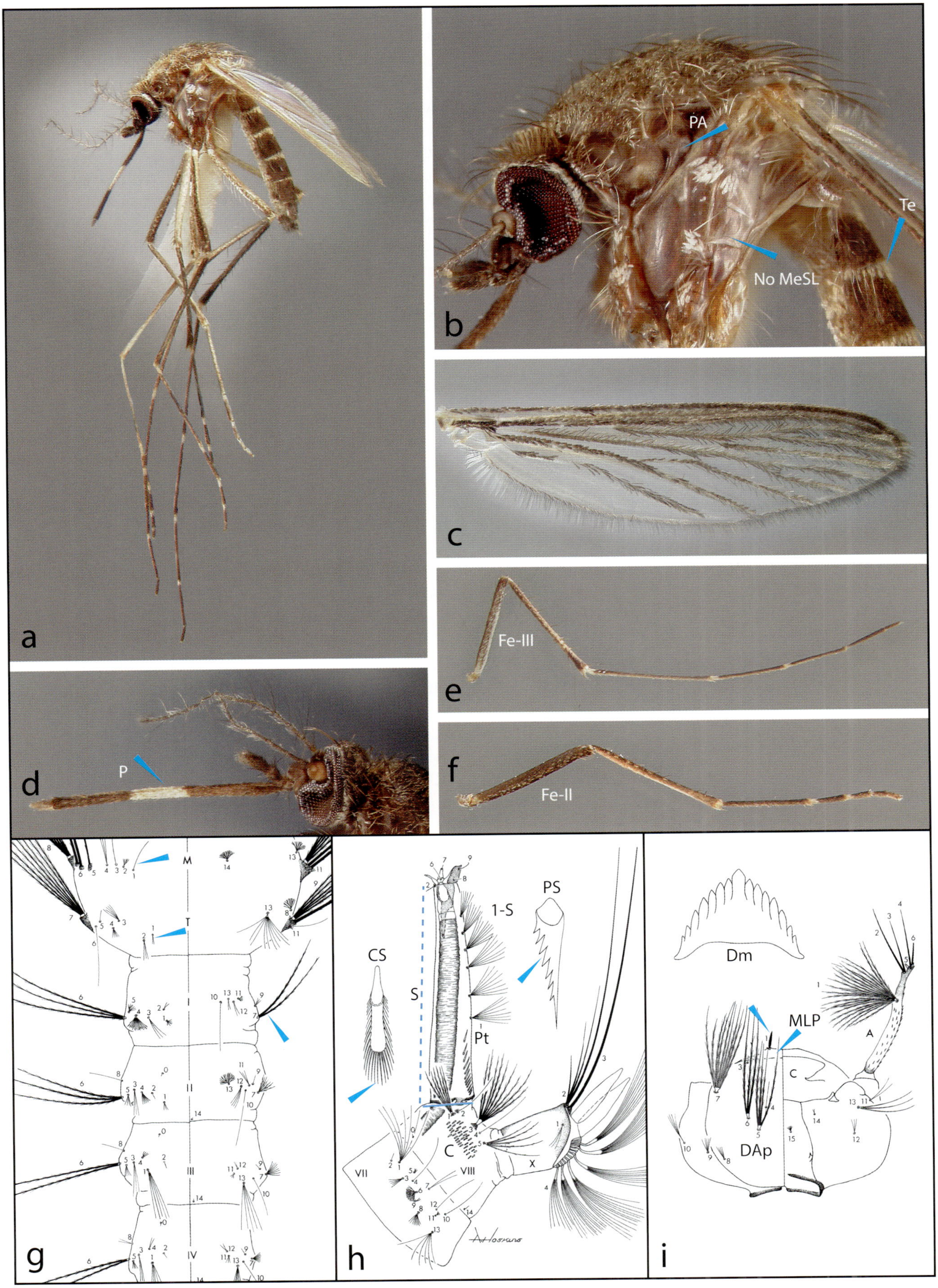
a
PA
Te
No MeSL
b
c
Fe-III
e
P
d
Fe-II
f
M
T
I
II
III
IV
g
1-S
PS
CS
S
Pt
C
VII
VIII
X
h
Dm
MLP
A
C
DAp
i

Culex (*Culex*) *antennatus* (Becker, 1903)
Subfamily Culicinae, Tribe Culicini
"Antenna Egyptian Typical Mosquito"

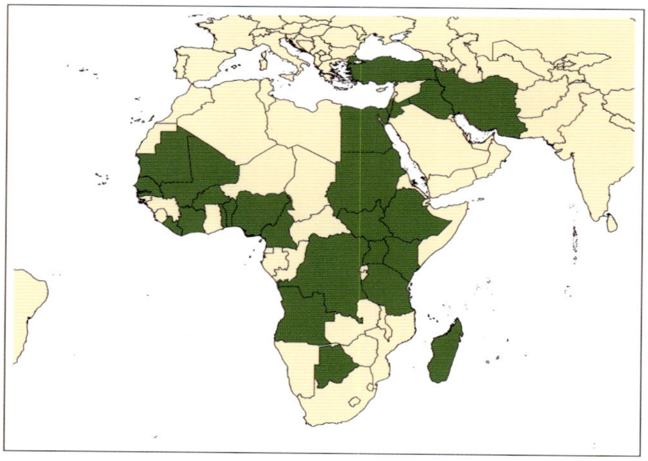

Systematics and Distribution

Originally described from Cairo, Egypt, *Culex antennatus* is a typical *Culex* (*Culex*) species, with notably drab coloration and muted ornamentation. *Culex antennatus* belongs to the Decens Subgroup and Pipiens Group of the subgenus *Culex* (see Vol. 2, p. 1089, for internal relationships). Other members of the Decens Subgroup include *Cx. decens* Theobald, *Cx. invidiosus* Theobald, with its two valid subspecies—*vicinalis* de Meillon & Laviopierre and *vexillatus* Edwards—and *Cx. litwakae* Harbach. It ranges from Turkey and the Middle East through the Rift Valley into Africa as far south as Mozambique and Madagascar, stretching across to the west coast of the African continent.

Bionomics

Larvae of *Cx. antennatus* are found in groundwater habitats, including swamps, pools, rice fields, animal footprints, and seepage drains with dense vegetation cover. Larvae are found predominantly in fresh water but have occasionally been observed in brackish-water habitats as well. Adult *Cx. antennatus* feeds primarily on cattle and humans but will feed on other domestic livestock, including goats and sheep, if cattle densities are low. Females generally feed in the evening. *Culex antennatus* and *Cx. pipiens* Linnaeus (see species page, p. 498) typically dominate cattle shed and indoor collections and will readily enter houses to feed on humans if other hosts are absent. *Culex antennatus* is readily caught in light traps, comprising 92%–98% of all catches during the 2003 RVFV outbreak in Egypt. Blood-meal analyses reveal that *Cx. antennatus* collected indoors fed mainly on humans, while those collected outdoors fed mainly on domesticated livestock.

Associated Pathogens

ACDV, AMTV, PERV, RVFV, WNV.

Diagnostic Characters

(a) Habitus (♀). **(b) Thorax (side):** Postspiracular scales (PoSc) absent; prealar scales (PrSc) absent; lower mesepimeron with at least one seta (MeSL). **(c) Thorax (dorsal). (d) Head (side):** Proboscis (P) without pale bands. **(e) Abdomen (side):** Pale stripes of cream scales on VI,VII-Te and small pale spots on the lateral borders of II–V-Te (not visible in dorsal aspect). **(f) Legs (I–III):** Ti-I–III and Fe-I,II with dark-scaled anterior surfaces; tarsomeres without pale bands. **(g) Larva (head):** Seta 1-C relatively long, slender; seta 5-C usually 2,3 branched. **(h) Larva (terminal segments):** Distal spines (PS) of pecten (Pt) with 2–5 basal denticles; siphon index <8.5; six pairs of seta 1-S (1a–f), three of them lateral; seta 1-S tufts relatively short, not longer than diameter of siphon (S). **(i) Larva (abdominal segments IV–VI):** Seta 1-IV,V often >3 branched, relatively short, not reaching margin of following segment; seta 6-VI usually single. Image sources: WRBU; Harbach (1988).

Exemplar DNA barcodes: *Cx. antennatus* (KU187048, KU187050, KU187037–38, MK033248).

Catalog entry: Vol. 2, p. 859.

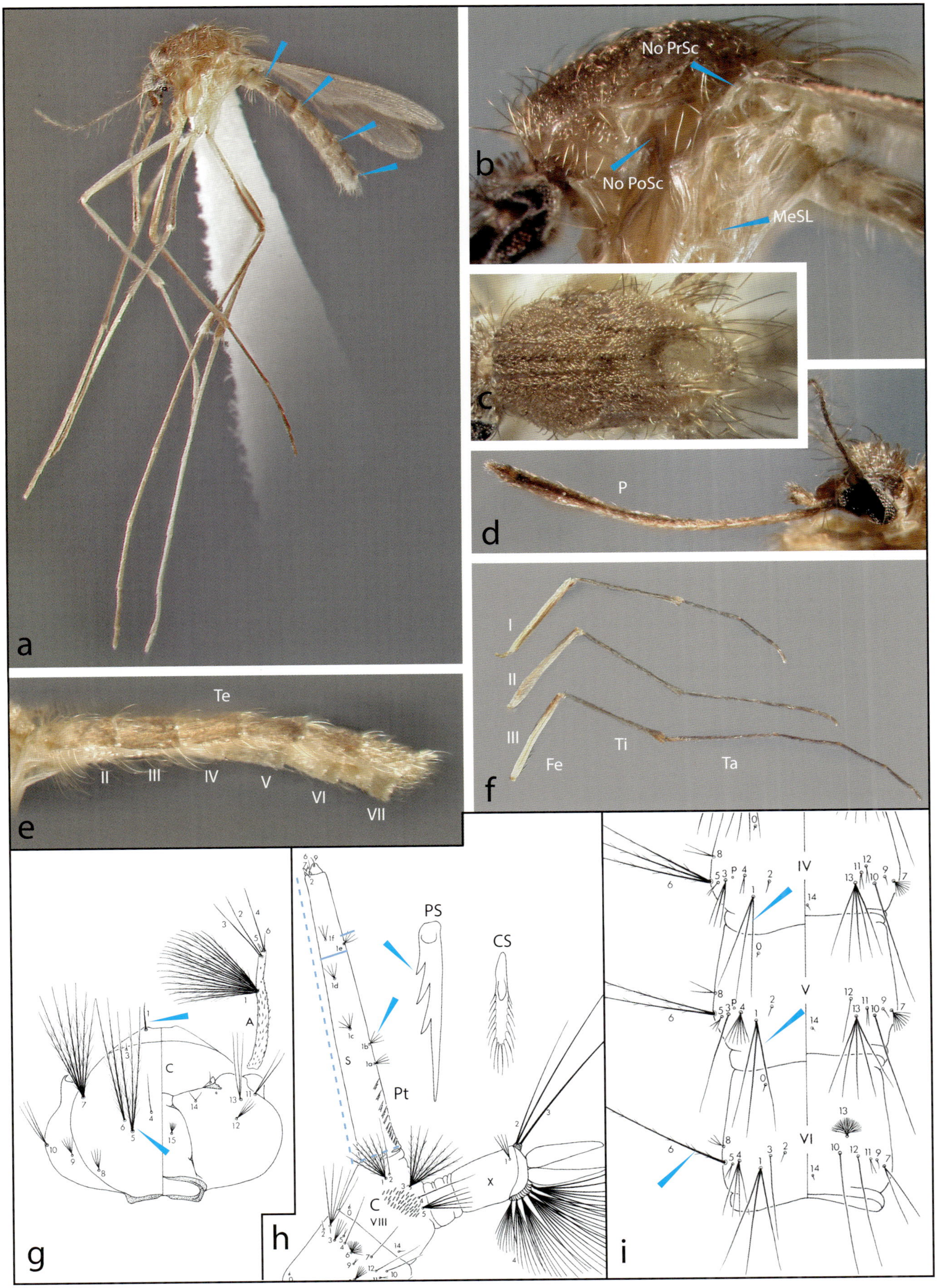
a
No PrSc
No PoSc
MeSL
b
c
P
d
Te
II
III
IV
V
VI
VII
e
I
II
III
Fe
Ti
Ta
f
A
C
g
PS
CS
S
Pt
C
VIII
X
h
IV
V
VI
i

Culex (*Culex*) *fuscocephala* Theobald, 1907
Subfamily Culicinae, Tribe Culicini
"Dark-headed Sri Lankan Typical Mosquito"

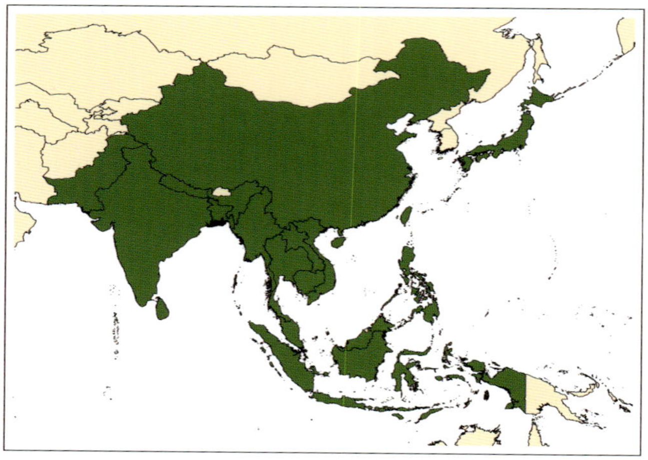

Systematics and Distribution

Culex fuscocephala is differentiated from other *Culex* species by a distinctly dark-colored head and rows of acrostichal setae on the scutum. This species has a broad geographic range and is found from Pakistan to Sri Lanka, throughout Indochina, Japan, the Philippines, and all Southeast Asian islands, and from Indonesia through to New Guinea. *Culex fuscocephala* belongs to the medically important Univittatus Subgroup and Pipiens Group of the subgenus *Culex*, which also includes *Cx. neavei* Theobald, *Cx. perexiguus* Theobald (see species page, p. 496), and *Cx. univittatus* Theobald (see species page, p. 514).

Bionomics

Culex fuscocephala immatures are found primarily in rice fields, ground pools, animal burrow pits, and temporary puddles, but they will occasionally utilize artificial containers such as clay pots and coconut shells. Adult *Cx. fuscocephala* feeds outdoors on cattle and other domesticated livestock and will only feed on humans if no other preferred host is available. Due to its mostly tropical range, this species is active virtually all year-round. In Malaysia, population peaks of *Cx. fuscocephala* are tied to the monsoon rains, meaning the timing of population ebbs and flows are difficult to predict. The species is a highly effective vector of Japanese encephalitis, with the virus multiplying to high titers within 10 days of ingesting low-titer infected blood meals. Viremia was maintained for at least four weeks, indicating that this multivoltine species plays a critical role in Japanese encephalitis virus (JBEV) transmission alongside the primary regional vector *Cx. tritaeniorhynchus* Giles (see species page, p. 512).

Associated Pathogens

JBEV; *Brugia malayi, Wuchereria bancrofti.*

Diagnostic Characters

(a) Habitus (♀): Ta-I–III all dark. **(b) Thorax (dorsal):** Distinct double row of acrostichal setae (AcS) present. **(c) Abdomen (dorsal):** Terga all dark scaled. **(d) Larva (head):** Dorsal apotome (DAp) distinctly separated from median labral plate (MLP); seta 1-C slender; setae 5,6-C double. **(e) Head (dorsal):** Proboscis (P) entirely dark scaled. **(f) Thorax (side):** Mesepimeron with one or two setae (MeS); pleuron (p) with broad, dark integumental stripes and distinct scale patches. **(g) Larva (terminal segments):** Siphon (S) tapers gradually. Image sources: WRBU; Tanaka et al.(1979).

Exemplar DNA barcodes: *Cx. fuscocephala* (AB738104–05, HQ398886–88, HQ398890, JQ728237–39, JQ728354, JQ728383, KF406796–800).

Catalog entry: Vol. 2, p. 864.

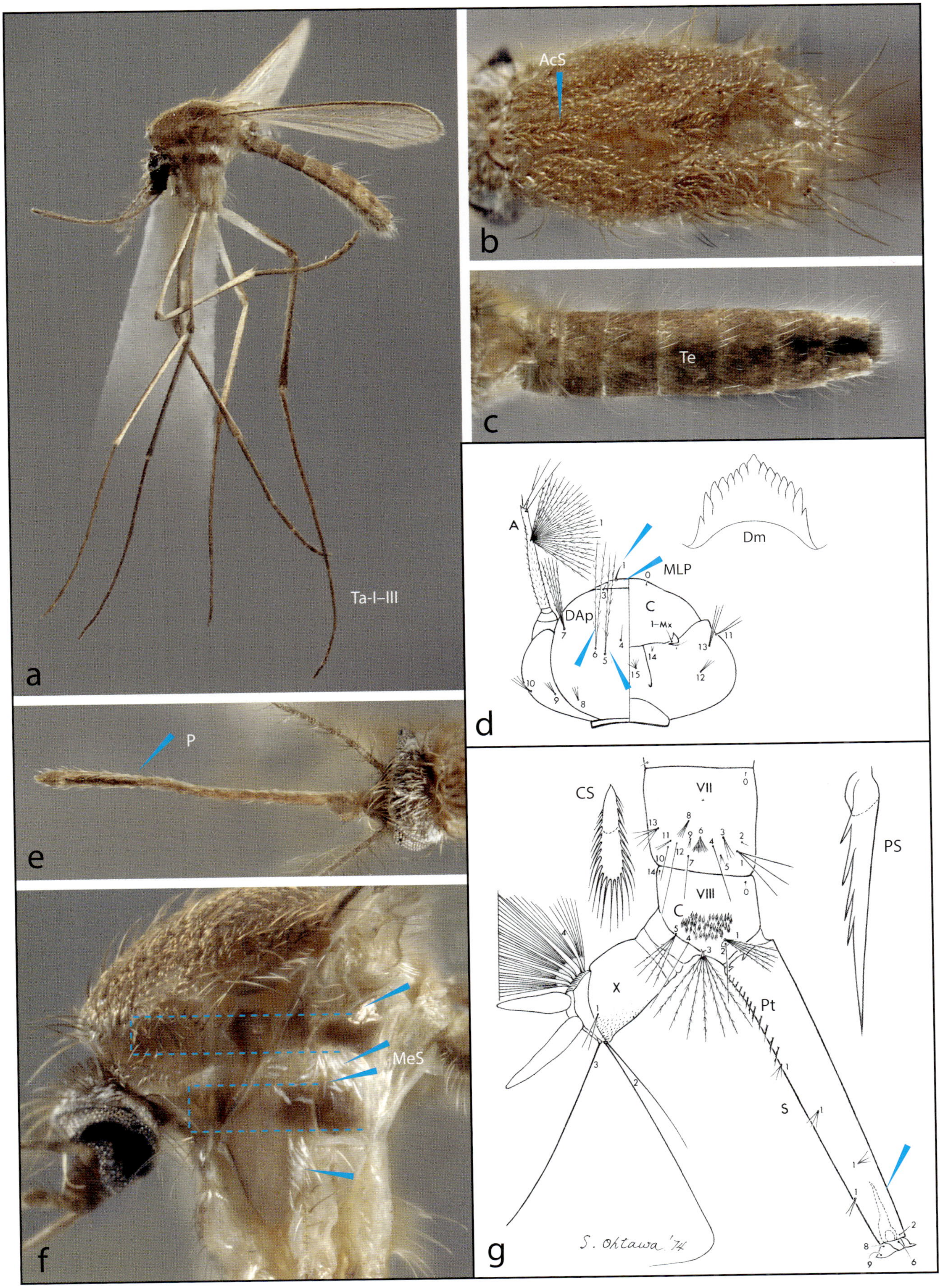
a
Ta-I–III
AcS
b
Te
c
A
MLP
Dm
C
DAp
1–Mx
d
P
e
MeS
f
CS
VII
VIII
C
PS
X
Pt
S
S. Ohtawa '74
g

Culex (Culex) gelidus Theobald, 1901
Subfamily Culicinae, Tribe Culicini
"Icy Malaysian Typical Mosquito"

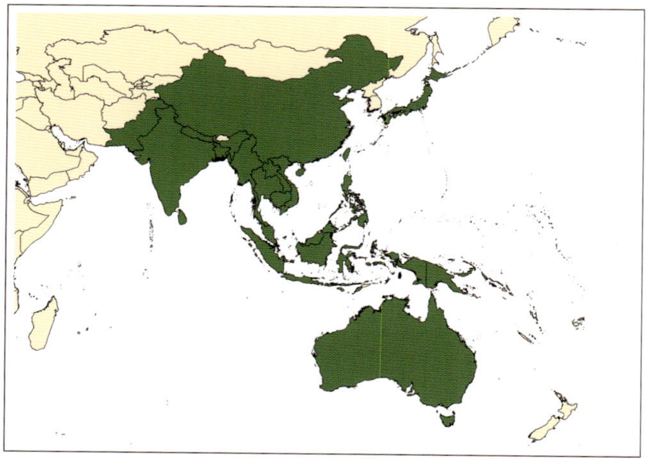

Systematics and Distribution

Adult *Culex gelidus* are medium-sized mosquitoes, with a distinct white-scaled scutum and broad pale banding on the proboscis. *Culex gelidus* is the monotypic member of the Gelidus Subgroup of the Pipiens Group of the subgenus *Culex*. Originally described from mainland Malaysia, its native range is considered to be mainland Southeast Asia. However, the species has become a highly successful invasive that is widely distributed from Pakistan across the Indian subcontinent, to China and Japan in the North, and south through mainland Southeast Asia through the islands of Southeast Asia and the South Pacific to Australia. In recent years, populations of this efficient vector species have notably increased in several countries, particularly in urban and suburban areas, giving cause for concern.

Bionomics

Culex gelidus utilizes a wide variety of oviposition sites and demonstrates an ability to quickly adapt to new environments. Immatures can be found in temporary and semipermanent water sources, such as rice fields, ground pools, stream pools, small streams, and freshwater swamps. Observations indicate that this species prefers natural habitats in its native range yet utilizes both natural and polluted water in areas where it is considered invasive. Water with high levels of organic matter, such as wastewater treatment plant holding tanks, agriculture seepage pools, and artificial containers (water barrels and tanks), are used by this species opportunistically. *Culex gelidus* larvae have also been found in tires.

Peak biting times of female *Cx. gelidus* are in the early morning before dawn (0300h–0600h). Adults are vicious biters, preferring to feed on large domestic animals such as cows, horses, and pigs. In the absence of domesticated livestock, this species will readily feed on humans and has been documented taking blood meals from wild birds, chickens, goats, and deer. The plasticity of *Cx. gelidus* with regard to its oviposition sites and broad host range has facilitated the successful expansion of its range. Increasing pressures for rice cultivation have increased suitable habitats for this species, sustaining high population densities. The association of *Cx. gelidus* with multiple hosts, including pigs, has placed it of high potential importance in zoonotic pathogen transmission, especially in the transmission of Japanese encephalitis to humans.

Associated Pathogens

BATV (Chandipura, Chittoor), CHIKV, JBEV, KUNV, GETV, MVEV, RRV, SINV, TMUV, WNV.

Diagnostic Characters

(a) Habitus (♀): Ta-I–III with pale bands at articulations. **(b) Thorax (dorsal):** Distinct double row of acrostichal setae (AcS) present; scutum (Scu) white-scaled anterior to prescutellar area (PrA). **(c) Thorax (side):** Pleuron (p) with distinct scale patches; lower mesepimeral setae (MeSL) absent. **(d) Head (side):** Proboscis (P) with broad median pale band; vertex (V) with erect white scales. **(e) Abdomen (dorsal):** Terga (Te) with broad basal pale bands. **(f) Legs (I–III):** Fe-II,III without pale speckling on anterior surfaces. **(g) Larva (head):** Seta 1-C distinct, pointed; dorsal apotome (DAp) distinctly separated from median labral plate (MLP). **(h) Larva (terminal segments):** Siphon (S) short, expanded medially; dorsal surface of siphon smooth, without a strong subapical spine dorsally; anal papillae (APP) longer than saddle (Sa). Image sources: WRBU; Sirivanakarn (1976).

Exemplar DNA barcodes: *Cx. gelidus* (HQ398892, HQ398894–95).

Catalog entry: Vol. 2, p. 865.

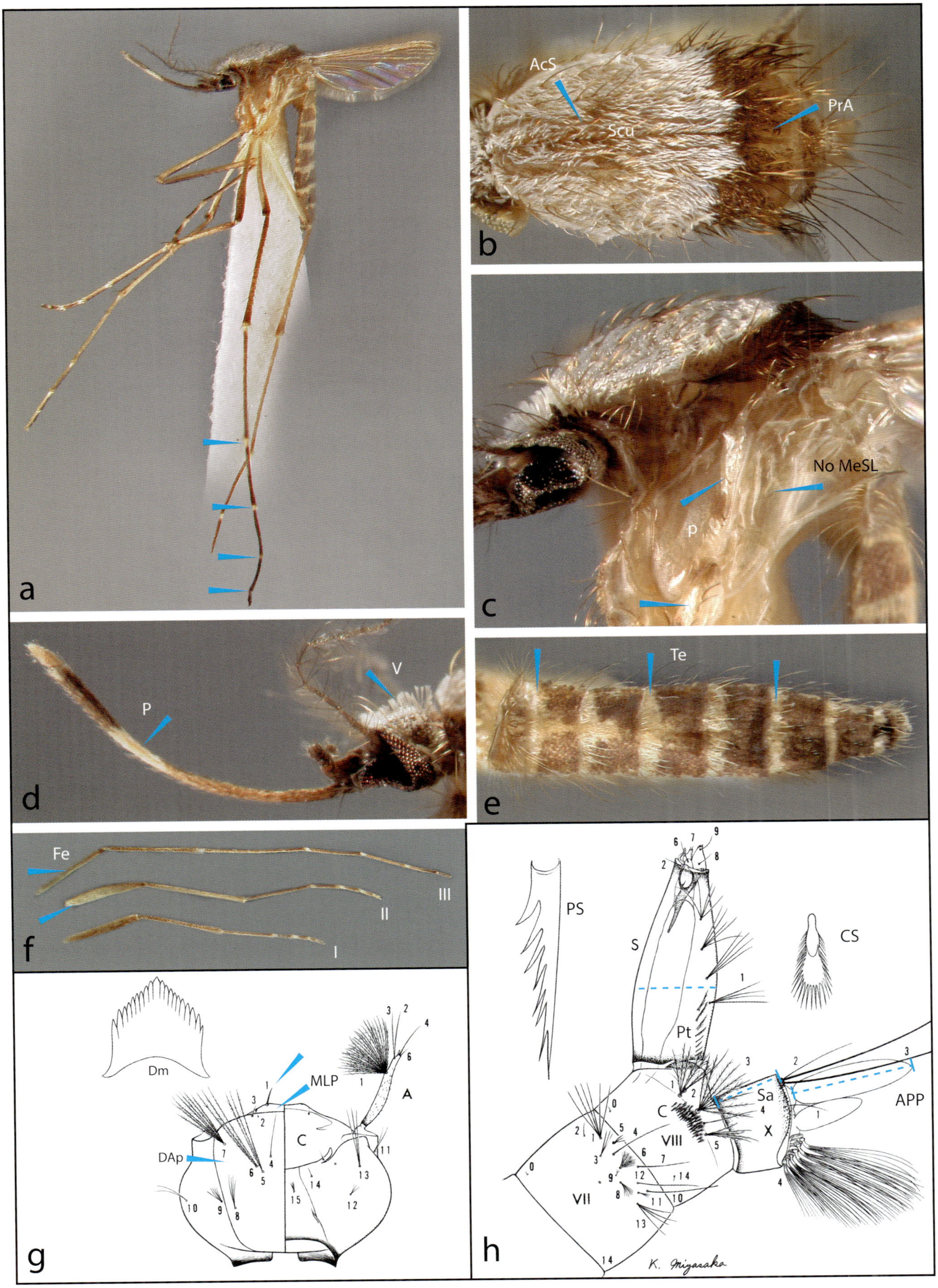
a
AcS
Scu
PrA
b
No MeSL
p
c
V
P
d
Te
e
Fe
III
II
I
f
Dm
MLP
A
C
DAp
g
PS
S
CS
Pt
Sa
APP
VIII
X
VII
h
K. Miyasaka

Culex (Culex) nigripalpus Theobald, 1901
Subfamily Culicinae, Tribe Culicini
"Black-lipped Saint Lucian Typical Mosquito"

Systematics and Distribution

Culex nigripalpus, a medium-sized mosquito named after its dark palpus, was first described from the Caribbean island of Saint Lucia. It is unusually colorful for a *Culex* species, sporting black scales and bronze to metallic blue-green reflections on the abdomen, with dark, reflective bronze to metallic blue-green scaled legs. It is placed in the Apicinus Subgroup of the Pipiens Group (subgenus *Culex*) and is widespread across the Caribbean and Central and South America, reaching as far north as the United States. Mixed-sex gynandromorphs have been reported in this species.

Bionomics

Although *Cx. nigripalpus* eggs have been found in a wide variety of habitats, the species is generally described as a floodwater mosquito. Thus, abundance of this species is closely linked to rainfall wherever it occurs. Females primarily lay eggs in freshly flooded ditches and agricultural seepage pits. Ovipositing females target those sites that flood infrequently yet, once flooded, remain so for at least 10–14 days. Eggs are laid in rafts of 90–210 eggs on the water surface and hatch in 24–36 hours. In the northern temperate zones, *Cx. nigripalpus* larvae are found in freshwater grassy pools and seepage pits. In the southern Atlantic and Gulf Coast regions, immatures can be found in crabholes filled with brackish water, and adults often rest on the walls of these humid tubes. In the tropical areas of Central and South America, larvae are found in polluted water pools, as well as in bromeliad axils. *Culex nigripalpus* larvae have occasionally been found in wheel ruts and artificial containers as well. The timing of larval development is temperature dependent and is hastened in midsummer, even when water temperatures exceed 100°F. Each larval stage lasts approximately 24 hours at summer temperatures, taking double that time during winter. Once the fourth instar larvae molt to pupae, the adult emerges in 36–48 hours.

During the winter and spring months, *Cx. nigripalpus* generally feeds on birds. However, during the summer and fall, females shift host preference to mammals, including domesticated livestock and humans. This biphasic host preference is common among other zoonotic vectors. Females feed primarily at night and are regarded as mostly exophagic. However, when in large numbers, the species has been known to enter buildings to seek out blood meals. Females are active from early evening to late into the night, searching for hosts and oviposition sites. In various studies, light traps set with green light, followed by blue light, have proven the most attractive to *Cx. nigripalpus*.

Associated Pathogens

CABV, EEEV, EVEV, HPV, KEYV, MAGV, ROCV, SLEV, TENV, VEEV, VSNJV, WNV, WYOV.

Diagnostic Characters

(a) Habitus (♀). (b) Abdomen (side): VII-Te dark, with large lateral patches of white scales. **(c) Head (side):** Proboscis (P) dark scaled, paler ventrally on basal half. **(d) Thorax (side):** Head with narrow curved scales dorsally and long erect forked scales; mesepimeron without median patch of white scales (MeSc). **(e) Leg (III):** Ta-I–III all dark scaled. **(f) Abdomen (dorsal):** II,III-Te with indistinct or very narrow basal pale bands; VII-Te mostly dark, with large lateral patches of white scales; VII,VIII-Te sometimes also with narrow basal white bands. **(g) Thorax (dorsal):** Acrostichal setae (AcS) present; scutellar midlobe with basal patch of fine dark-brown scales (MSSc). **(h) Larva (thorax):** Integument with tiny spicules; seta 3-P ≥0.67 times length of 1-P. **(i) Larva (thorax):** Seta 1-M very small, equal in length to 2-M. **(j) Larva (terminal segments):** Seta 1-S tufts inserted ventrally and laterally, not in even row; seta 1-S basal-most tuft usually single, length nearly twice width of siphon (S) at insertion; seta 2-X 2,3 branched, shorter than 3-X; siphon (S) without small spines near apex; siphon index 8.0. **(k) Larva (head):** Seta 1-A large, inserted on apical 0.34 of antenna; setae 5-7-C with ≥3 aciculate branches; seta 6-C 3,4 branched. Image sources: WRBU; Belkin et al. (1970).

Exemplar DNA barcodes: *Cx. nigripalpus* (KM592991–92, KM593008, KM593058, KP211381–93, KY859885–92).

Catalog entry: Vol. 2, p. 870.

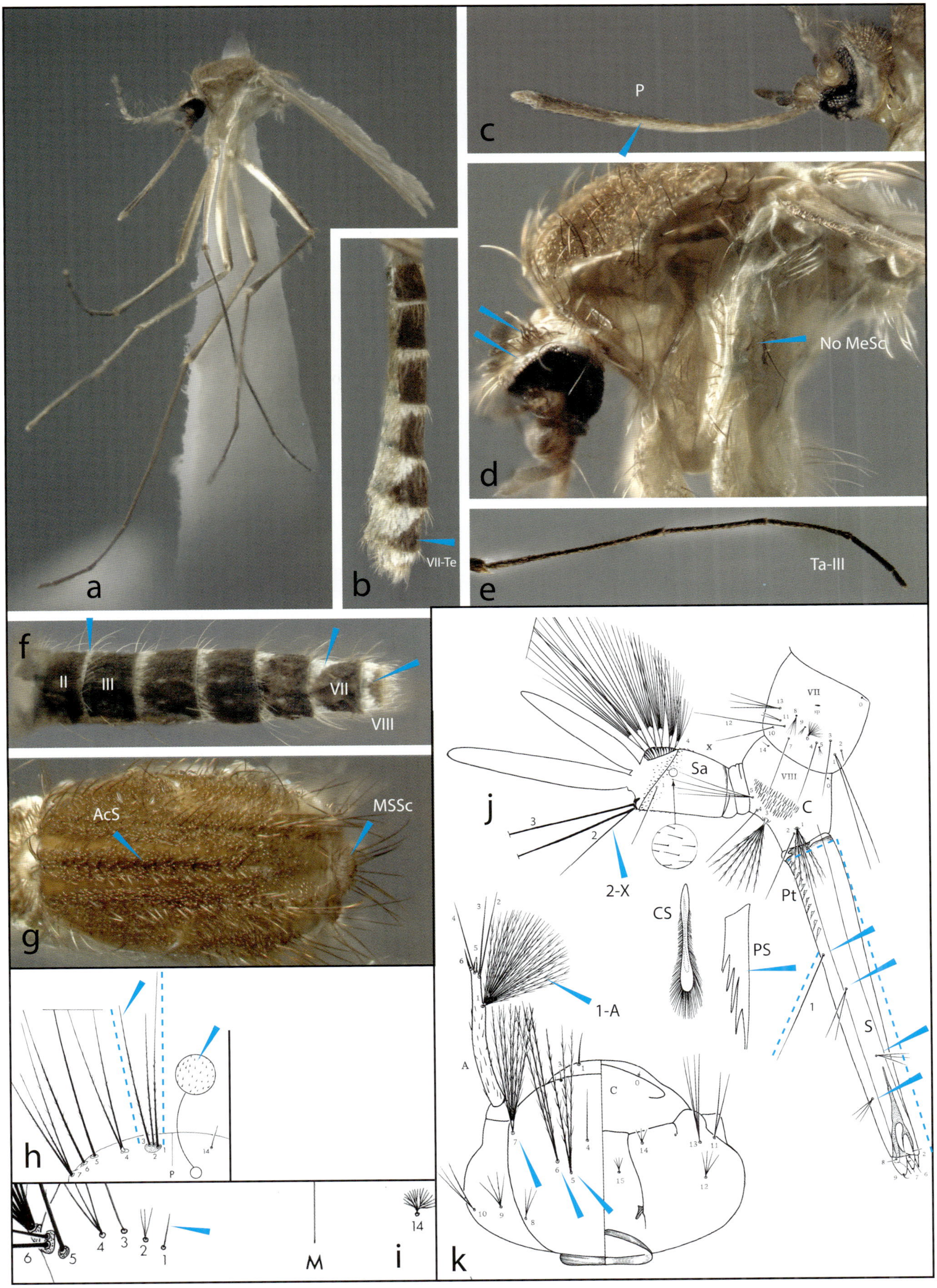
a
b
VII-Te
P
c
No MeSc
d
Ta-III
e
f
II
III
VII
VIII
AcS
MSSc
g
h
i
M
14
j
Sa
2-X
3
2
VII
VIII
C
Pt
CS
PS
S
1-A
A
k
C

Culex (Culex) perexiguus Theobald, 1903
Subfamily Culicinae, Tribe Culicini
"Tiny Lebanese Typical Mosquito"

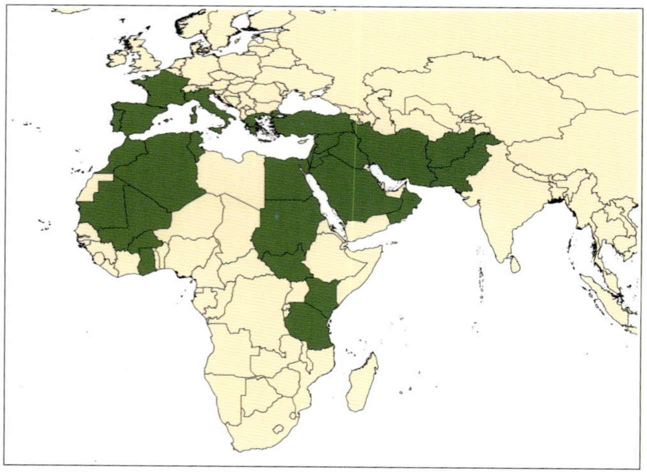

Systematics and Distribution

Culex perexiguus is a medium-sized mosquito that closely resembles other *Culex* (*Culex*) species, including *Cx. quinquefasciatus* Say (see species page, p. 500) and *Cx. univittatus* Theobald (see species page, p. 514). Distinguishing characteristics include the lack of a complete anterior pale stripe on the midfemur and no scales on wing vein 2A, as well as the fact that in males the ventral arm of the phallosome is short. *Culex perexiguus* was previously synonymized under *Cx. univittatus*, causing some confusion on the true distributions of these taxa. Although European records of *Cx. univittatus* are now understood to comprise *Cx. perexiguus*, the taxonomic status of these species is not entirely clear across their range. Given the medical importance of this species, it is imperative to determine the geographical distribution of the component members of the Univittatus Subgroup, which currently comprises only four taxa—*Cx. fuscocephala* Theobald (see species page, p. 490), *Cx. neavei* Theobald, *Cx. perexiguus* and *Cx. univittatus*.

Bionomics

Culex perexiguus is a common mosquito throughout its range and remains active from spring to autumn. *Culex perexiguus* larvae are found in a variety of standing-water bodies like swamps, stream pools, and ponds, generally away from human habitation. Larvae require clean water with little or no pollution and emergent vegetation. Immatures are generally found in fresh water but are able to tolerate some salinity. They readily adapt to human-influenced environments, with immatures occasionally found in artificial containers, including wells. Like other zoonotic vectors, *Cx. perexiguus* mosquitoes feed on a variety of warm-blooded hosts. They feed exclusively on birds early in the year, switching to larger mammals, including domesticated livestock in the summer and autumn. In the absence of birds or livestock, and when large numbers are present, *Cx. perexiguus* will readily feed on humans both indoors and outdoors. *Culex perexiguus* was the dominant species passively collected on sticky nets suspended from balloons at high altitude over the Sahel, suggesting the species uses wind currents to facilitate long-distance migration away from harsh climatic conditions. Given its role in the transmission of RVFV, this life strategy is epidemiologically highly significant.

Associated Pathogens

Barkedji virus, RVFV, SINV, USUV, WNV.

Diagnostic Characters

(a) Habitus (♀): Proboscis (P) all dark; Ta-I–III without pale bands. **(b) Thorax (side):** Postspiracular (PoSc) and prealar (PrSc) scales present; lower mesokatepisternal scales (MScL) distinctly separate from upper scale patch (MScU); lower mesepimeron with at least one seta (MeSL). **(c) Legs (II,III):** Fe-II usually lacks a complete anterior pale stripe; Ti-III with broad anterior pale stripe. **(d) Abdomen (dorsal):** Terga (Te) basal pale bands without median expansion. **(e) Wing:** Costa (C) with a few pale scales at base (not distinct in photo); vein 2A present, bare or with only a few scales. **(f) Larva (head):** Seta 1-C slender, about as thick as setae 5,6-C; seta 5-C usually 2,3 branched. **(g) Larva (terminal segments):** Comb scales (CS) evenly fringed at apex and sides; Pecten (Pt) with distal pecten spines (PS) with 2–5 ventral denticles arising proximally; siphon index 5.5–7.9; seta 1-S ≤ diameter of siphon (S); seta 1-S with 3–6 pairs of tufts, 1–3 pairs laterally. **(h) Larva (abdominal segments III–VI):** Seta 1-IV,V usually with a few branches and less than length of two abdominal segments combined; seta 6-VI usually single. Image sources: WRBU; Harbach (1988).

Exemplar DNA barcodes: *Cx. perexiguus* (KF406802, KJ012103–09, KU380348, KU380382, KU380423, KU380445, KU380476).

Catalog entry: Vol. 2, p. 871.

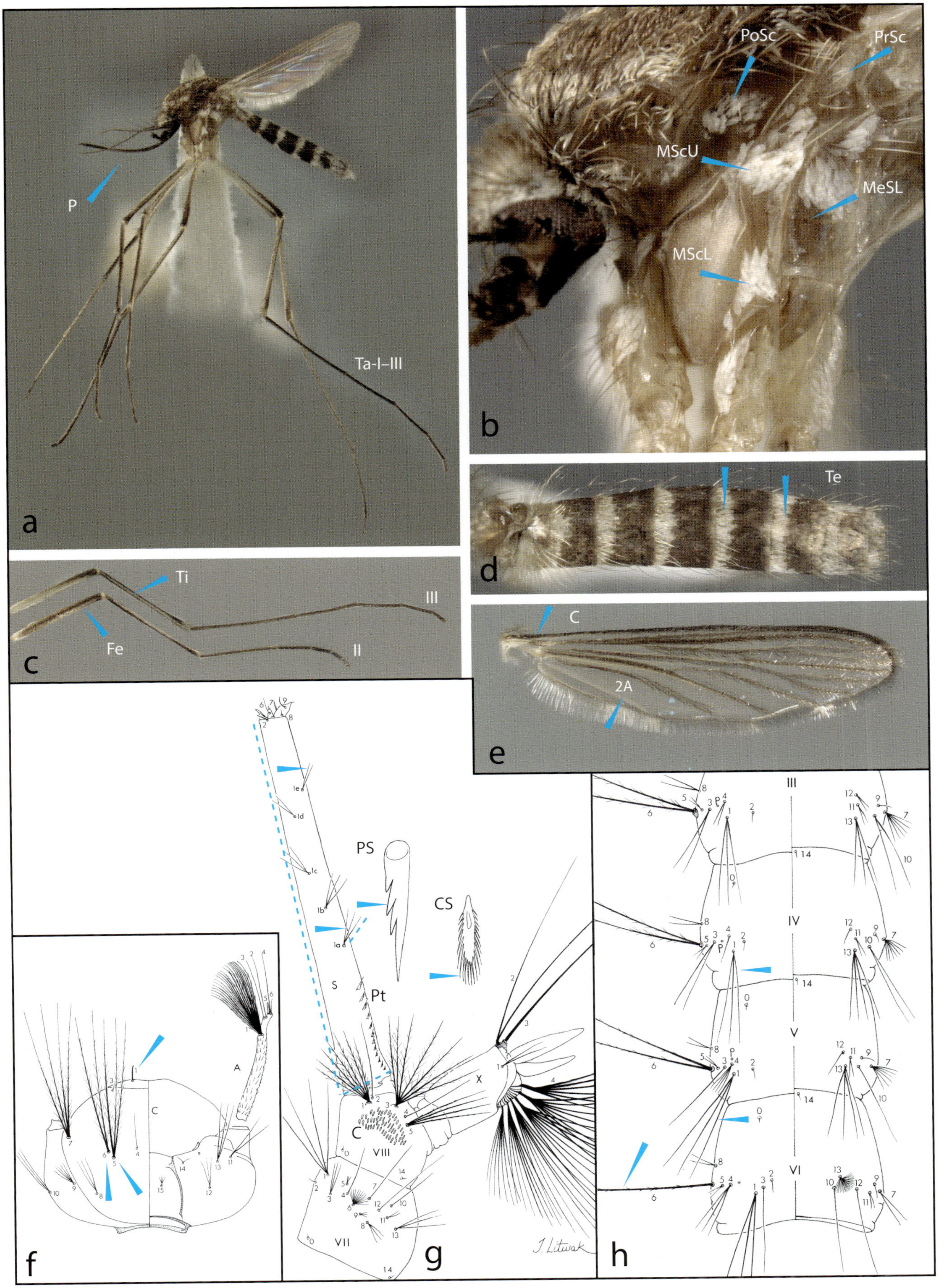
P
Ta-I–III
a
PoSc
PrSc
MScU
MeSL
MScL
b
Te
d
Ti
III
Fe
II
c
C
2A
e
PS
CS
S
Pt
C
VIII
X
VII
g
A
C
f
III
IV
V
VI
h
J. Litwak

Culex (*Culex*) *pipiens* Linnaeus, 1758
Subfamily Culicinae, Tribe Culicini
"Piping Swedish Typical Mosquito"

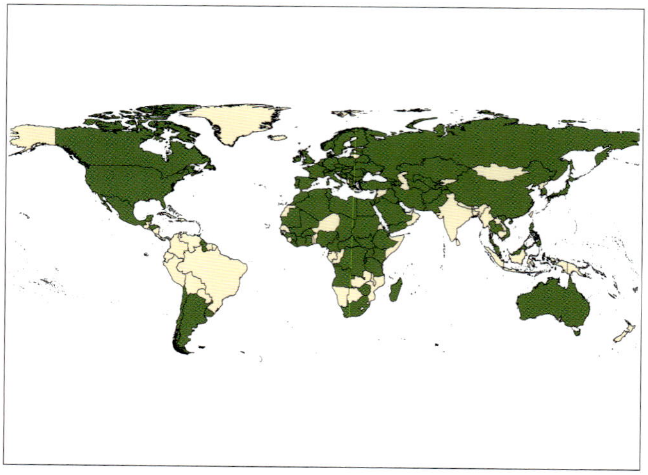

Systematics and Distribution

Culex pipiens is a medium-sized, light-brown mosquito with few distinguishing characteristics. The species has a panmictic, essentially global distribution, and populations exhibit morphological, physiological, and bionomic differences across its range. The taxonomic history of this species is complicated, with mosquito workers disagreeing as to whether *Cx. pipiens* comprises a single polytypic species or a species complex. *Culex pipiens* is often confused with other closely related taxa within the Pipiens Group (see Vol. 2, p. 1089, for component taxa), particularly *Cx. quinquefasciatus* (see species page, p. 500), with which is it known to extensively hybridize where the two species meet in North America. These hybrids have been highlighted as potential bridge vectors of West Nile virus in the New World. In northern Europe, microsatellite studies reported that *Cx. pipiens* s.s. and the "molestus" ecotype display unique fingerprints with no evidence for gene flow between them and implied this is indicative of separate species status. Opinions on the true taxonomic status of these ecotypes have been widely debated, but at present they are regarded as subspecific forms, with no formal taxonomic status. Mixed-sex gynandromorphs have been reported in *Culex pipiens*, *Cx. pipiens* "molestus," and *Cx. quinquefasciatus*.

Bionomics

Given the confusion in taxonomic status, the bionomics of *Culex pipiens* is also somewhat unclear. In Europe, *Cx. pipiens* is represented by two ecotypes—*pipiens* and "molestus," which can be genetically differentiated using DNA barcoding and other markers. It is widely accepted that "molestus" is autogenous (can lay its eggs without requiring a blood meal), bites humans and other large domestic animals, and occupies underground larval habitats, including flooded basements, sewers, wells, and septic tanks, whereas the typical *pipiens* ecotype feeds almost exclusively on birds, is found in overground habitats, and needs a blood meal before producing eggs. Overground populations of immature *Cx. pipiens* are found in highly polluted, human-made water sources and artificial containers as well as natural, clean water pools in both heavy shade and full sun. *Culex pipiens* egg rafts exude a substance that is repellent to ants, protecting the eggs from predation. Larval morphology is affected by environmental conditions—larvae collected in more polluted water sources tend to have shorter siphons than those collected in cleaner water. *Culex pipiens* feeds at night and has been collected both indoors and out.

Associated Pathogens

AHSV, AINOV, FMV, GROV, ITV, JEV, LACV, LMV, OCKV, ROCV, RVFV, SINV, SFV, TAHV, TURV, TVTV, TYUV, VEEV, WNV, USUV; *Wuchereria bancrofti*.

Diagnostic Characters

(a) Habitus (♀): Ta-I–III entirely dark. **(b) Abdomen (dorsal):** Terga (Te) with pale-yellowish basal bands; bands on Te darker than white basolateral spots (not distinct in photo); sterna not banded, some dark scales on midline and posterolateral corners (not visible in photo). **(c) Thorax (side):** Postspiracular scales (PoSc) absent; lower mesepimeron with one seta (MeSL); prealar scales (PrSc) absent; C-I with some dark scales. **(d) Leg:** Anterior surface of Fe-I,II and Ti-I–III without median longitudinal pale stripes. **(e) Head (side):** Proboscis (P) midventral area usually pale. **(f) Wing:** Wing entirely dark scaled; subcostal vein (Sc) inserted on costa at or beyond level of branching of R_{2+3}; vein R_{2+3} short, <0.25 length of cell R_2. **(g) Thorax (dorsal):** Scutum (Scu) with golden reddish-brown scales; posterolateral scales and integument darker than rest of scutum (Scu) giving the impression of a pair of oval spots. **(h) Larva (thorax):** Seta 1-M usually single. **(i) Larva (head):** Seta 1-C relatively long, slender; seta 5-C usually 4–8 branched. **(j) Larva (terminal segment):** Comb (C) with comb scales (CS) evenly fringed from sides to apex; pecten (Pt) with 10–18 pecten spines (PS); distal pecten spines with 2–5 basal denticles; siphon/saddle index >3.5; seta 1-S relatively long, longer than diameter of siphon (S) at point of insertion; seta 1-S usually four pairs, 1a-S near distal pecten spine, 1c-S not in line with other setae; seta 1-X usually single. **(k) Larva (abdominal segments III–VI):** Sum of branches of seta 1-III–V on one side <10, usually <6; seta 1-III,IV usually double, relatively short, not reaching margin of following segment; seta 6-VI usually double. Image sources: WRBU; Harbach (1988).

Exemplar DNA barcodes: *Cx. pipiens* s.l. (KJ012113, KJ012119, KJ012127, KJ012129, KJ012139–45).

Catalog entry: Vol. 2, p. 872.

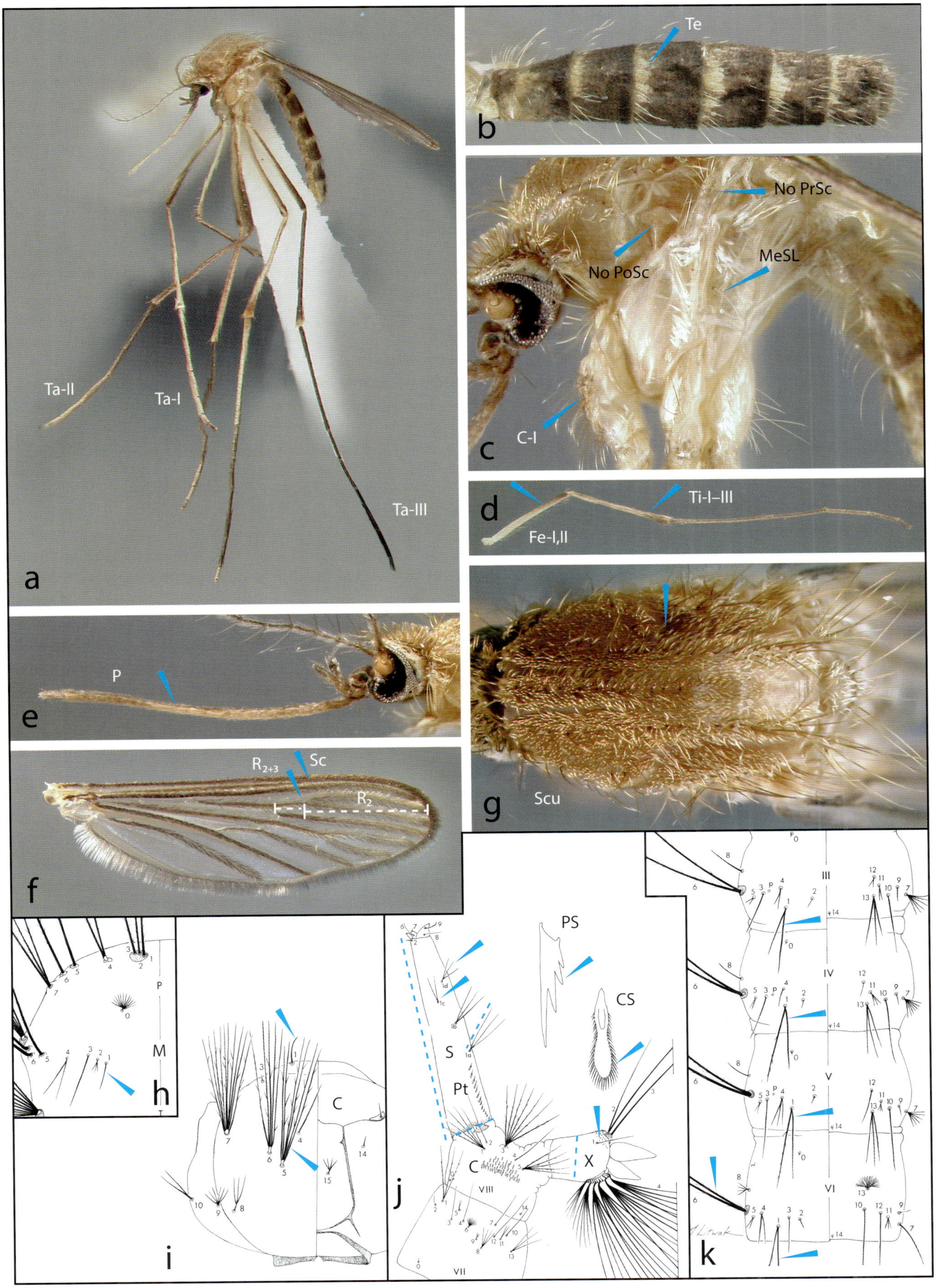

a
Ta-II
Ta-I
Ta-III
b
Te
c
No PrSc
No PoSc
MeSL
C-I
d
Fe-I,II
Ti-I–III
e
P
f
R_{2+3}
Sc
R_2
g
Scu
h
P
M
i
C
j
PS
CS
S
Pt
C
X
VIII
VII
k
III
IV
V
VI

Culex (*Culex*) *quinquefasciatus* Say, 1823
Subfamily Culicinae, Tribe Culicini
"Five-banded American Typical Mosquito"

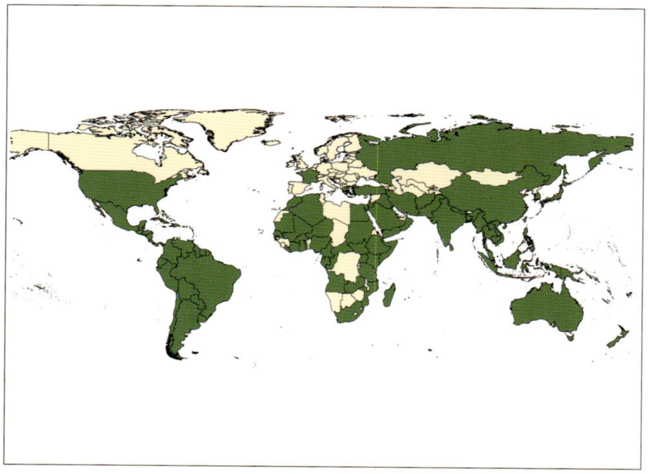

Systematics and Distribution

Culex quinquefasciatus is a medium-sized, light golden brown mosquito, with a characteristically banded abdomen. It belongs to the Pipiens Complex (see p. 498) and is commonly found in tropical to subtropical climates. The reports from temperate zones may be of *Cx. pipiens* hybrids. Hybrid offspring often display characteristics of both species, and hybridization is frequent where the two species ranges overlap, typically in subtropical to temperate zones. Male genitalia characters are the only reliable way to differentiate these two species by morphology. *Culex pipiens* s.l. involved in West Nile transmission in Turkey was recently confirmed as *Cx. quinquefasciatus* through DNA barcoding, delineating its distribution in Eastern Europe. *Culex quinquefaciatus* is placed within the Pipiens Subgroup, along with *Cx. australicus* Dobrotworsky & Drummond, *Cx. globocoxitus* Dobrotworsky, and *Cx. pipiens* Linnaeus. The species has 40 valid synonyms (see Vol. 2).

Bionomics

Culex quinquefasciatus eggs cannot withstand desiccation and will rapidly die if the water source is lost. Females lay eggs in loosely cemented oval rafts of about 100 eggs, which will normally hatch in 24 to 30 hours. Larval heads are short and stout, with darker pigmentation toward the base. Neonates take five to eight days to complete their development. Larvae are found in both domestic and peridomestic habitats, in both clean and polluted water. Common larval sites are organically rich, including sewers, ground pools, ditches, agriculture seepage pits, and sewage treatment plant holding tanks. *Cx. quinquefasciatus* will also readily lay eggs in large containers, such as trash cans. On emergence, both males and females take sugar meals from plants. *Cx. quinquefasciatus* is autogenous and can lay its first batch of eggs without taking a blood meal. In contrast to *Cx. pipiens*, which is primarily ornithophilic, *Cx. quinquefasciatus* is predominantly anthropophilic but will feed on other warm-blooded vertebrates opportunistically, including birds, domestic animals, and small mammals. Females feed mostly at night both indoors and outdoors and have been found resting both inside and outside buildings. This species is found in much higher numbers in urban areas as compared to rural populations. In Brazil, the survival rate of *Cx. quinquefasciatus* varies widely across different populations, indicating that vector capacity may also vary regionally. Mixed-sex gynandromorphs have been reported in *Cx. quinquefasciatus*.

Associated Pathogens

AMTV, APEUV, APV, BEFV, BUNV, CHIKV, CHPV, CWV, EEEV, EHDV, GFV, INGV, JEV, KOTV, KOWV, KRIV, KUNV, MAGV, MVEV, NEPV, NTAV, ORIV, OROV, PARAV, ROCV, RVFV, SFSV, SFV, SLEV, STRV, TURV, USUV, VEEV, VSAV, VSIV, VSNJV, WANV, WNV, ZEGV; *Plasmodium relictum*, *Wuchereria bancrofti*.

Diagnostic Characters

(a) ♀ (habitus): Terga (Te) with basal pale bands; wing entirely dark scaled; Ta-I–III entirely dark. **(b) Thorax (side):** Postspiracular scales (PoSc) absent; prealar scales (PrSc) absent; lower mesepimeron with one seta (MeSL); C-I with a few dark scales. **(c) Thorax (dorsal):** Scutal (Scu) integument yellowish or pale brown. **(d) Legs (II,III):** Anterior surface of Fe-I (not shown), Fe-II and Ti-I–III without median longitudinal pale stripes. **(e) Head (side):** Proboscis (P) without distinct median pale band. **(f) Abdomen (dorsal):** Terga (Te) bands basal, yellowish. **(g) Larva (head):** Seta 1-C relatively long, slender; seta 5-C usually 4–8 branched; setae 5,6-C 4–6 branched. **(h) Larva (terminal segments):** Comb (C) with comb scales (CS) evenly fringed from sides to apex; siphon/saddle index <3.5; siphon wider medially; pecten (Pt) with 10–18 pecten spines (PS); distal PS with 2–5 basal denticles; seta 1-S relatively long, longer than diameter of siphon (S) at point of insertion; seta 1-S usually with four pairs, 1a-S near distal pecten spine, 1c-S not in line with other setae; seta 1-X usually single. Siphonal variation shown. **Other characters not illustrated** (see *Cx. pipiens* page, p. 498): **Adult:** Wing vein R_{2+3} short, <0.25 length of cell R_2; abdominal sterna (S) not banded. **Larva:** Seta 1-M usually single; sum of branches of seta 1-III–IV on one side <10; seta 1-III,IV usually double, relatively short, not reaching margin of following segment; seta 6-VI usually double. Image sources: WRBU; Harbach (1988).

Exemplar DNA barcodes: *Cx. quinquefasciatus COI* (FJ210901, FJ210909–10, KJ012162, KJ012164–73); whole genome GCF_000209185.

Catalog entry: Vol. 2, p. 875.

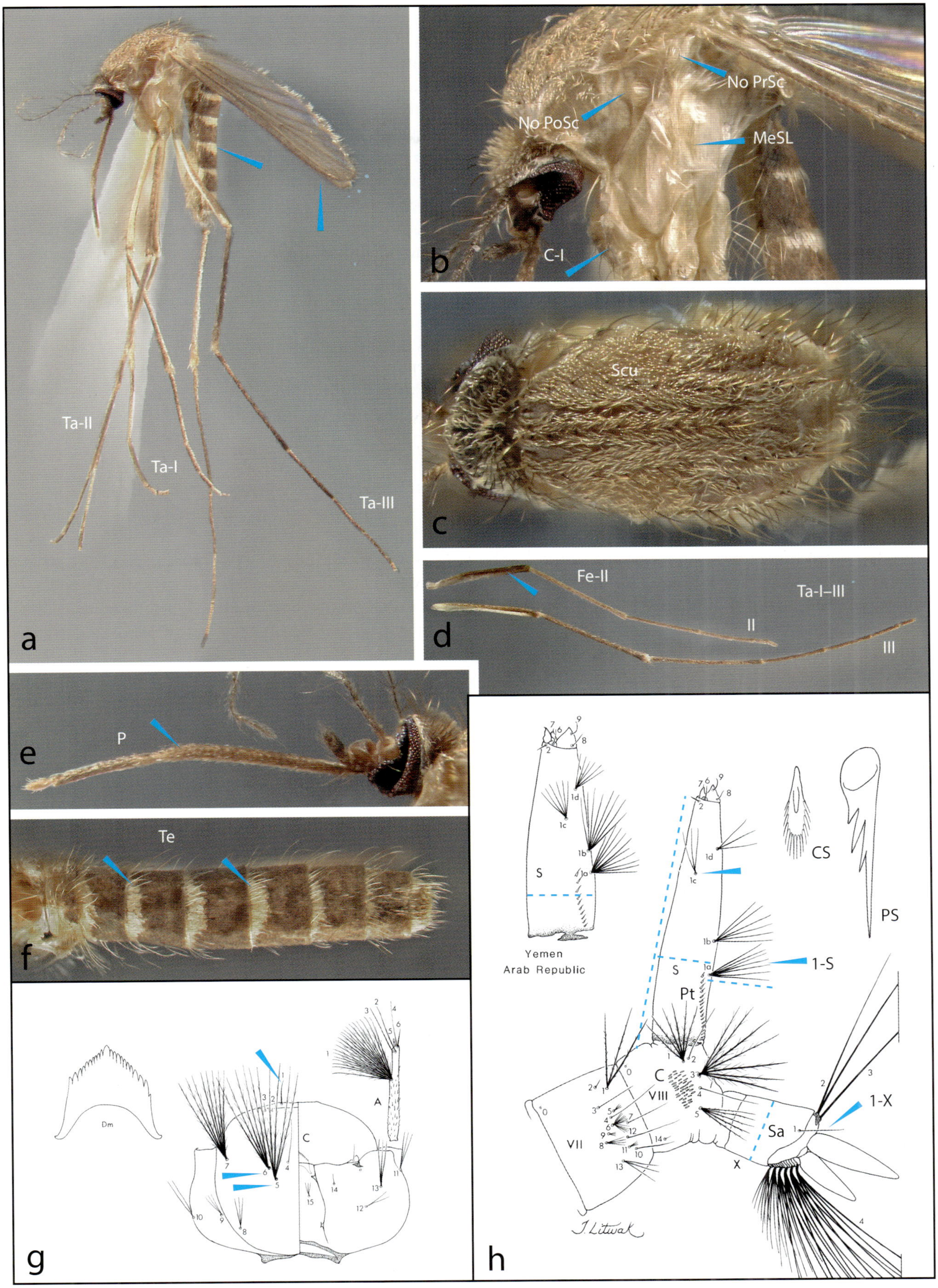

a
Ta-II
Ta-I
Ta-III
b
No PrSc
No PoSc
MeSL
C-I
c
Scu
d
Fe-II
Ta-I–III
II
III
e
P
f
Te
g
Dm
A
C
h
S
Yemen
Arab Republic
S
Pt
1-S
CS
PS
C
VIII
VII
Sa
X
1-X
J. Litwak

Culex (*Culex*) *restuans* Theobald, 1901
Subfamily Culicinae, Tribe Culicini
"Watchful Canadian Typical Mosquito"

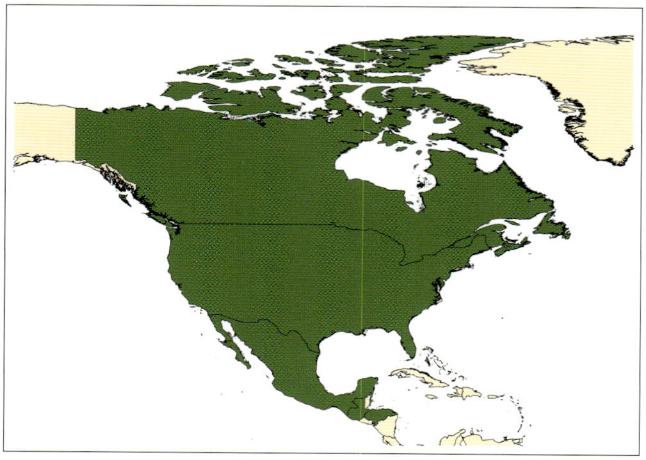

Systematics and Distribution

Culex restuans is a medium-sized Nearctic mosquito with golden-brown scales. The species is found from Central America to Alaska. *Culex restuans* is the nominotypical species of Restuans Complex, which also includes *Cx. acharistus* Root, *Cx. brethesi* Dyar, and *Cx. laticlasper* Galindo & Blanton. The species has one valid synonym, *brehmei* Knab. Identification of *Cx. restuans* based solely on adult female morphology is problematic. *Culex restuans* is often mistaken for *Cx. pipiens* or *Cx. quinquefasciatus*, which looks somewhat similar and occupies similar biological niches. This character fusion causes frequent misidentification and misleading biosurveillance reports. Frequently, "pools" contain two or more taxa, leading to incorrect associations of pathogens and bionomic traits. Some studies indicate that *Cx. restuans* may even be more common than *Cx. pipiens* in North America, and imply that misidentifications may overestimate *Cx. pipiens* density.

Bionomics

Oviposition sites are usually shaded and located under dense tree canopies, and immatures are found in a variety of habitats, including ditches, stream pools, and artificial containers. Females have been observed laying more than 10 times the number of egg clutches in containers with added nutrients than in control containers with only water. Decreased numbers are found in containers with conspecific larvae. In addition, it has been observed that greater habitat abundance in a given environment decreases egg clutch density. Interestingly, it appears that *Cx. restuans* larvae learn to respond to cues from predators in a threat-sensitive manner, based on a single conditioning event. Populations peak in spring and usually decline by late summer and early fall. Females are primarily ornithophilic but will feed on other vertebrate hosts opportunistically. Host-seeking females will readily feed on humans if other hosts are not available and will enter houses to bite humans. *Culex restuans* overwinters as adults in the eastern United States, and remaining eggs and immatures do not survive the cold season.

Associated Pathogens

EEEV, HPV, MERV, SLEV, WNV.

Diagnostic Characters

(a) Habitus (♀): Ta-III$_{1-5}$ dark scaled. **(b) Head (dorsal):** Vertex (V) with narrow curved and erect forked scales on median and lateral areas, dark scales on occiput (Occ) (see also *e*). **(c) Thorax (dorsal):** Scutum (Scu) with acrostichal setae (AcS), fine dark-brown scales and usually two small round white spots; scutellar midlobe with basal patch of white or cream-colored scales (MSSc). **(d) Abdomen (dorsal):** Terga (Te) with distinct basal broad bands II–VII-Te. **(e) Head (side):** Proboscis (P) entirely dark scaled. **(f) Thorax (side):** Pleuron (p) with small patches of broad pale scales. **(g) Larva (thorax):** Seta 3-P at least 0.67 length of seta 1-P. **(h) Larva (head):** Seta 1-A inserted 0.3–0.5 from base of antenna; seta 4-C usually 3,4 branched (2 illustrated); seta 5–7-C with ≥3 branches, all with small aciculae. **(i) Larva (terminal segments):** Pecten (Pt) with 12–20 spines (PS); siphon index 4.0–4.5; siphon (S) without small spines near apex; siphon dorsally slightly convex basally and slightly concave apically; seta 2-X single, equal in length to 3-X. **(j) Larva (abdominal segments I–VI):** Seta 6-I,II double, seta 6-III–VI long and single. Image sources: WRBU; Strickman & Darsie (1988).

Exemplar DNA barcodes: *Cx. restuans* (JX259911–15, GU908085–95).

Catalog entry: Vol. 2, p. 877.

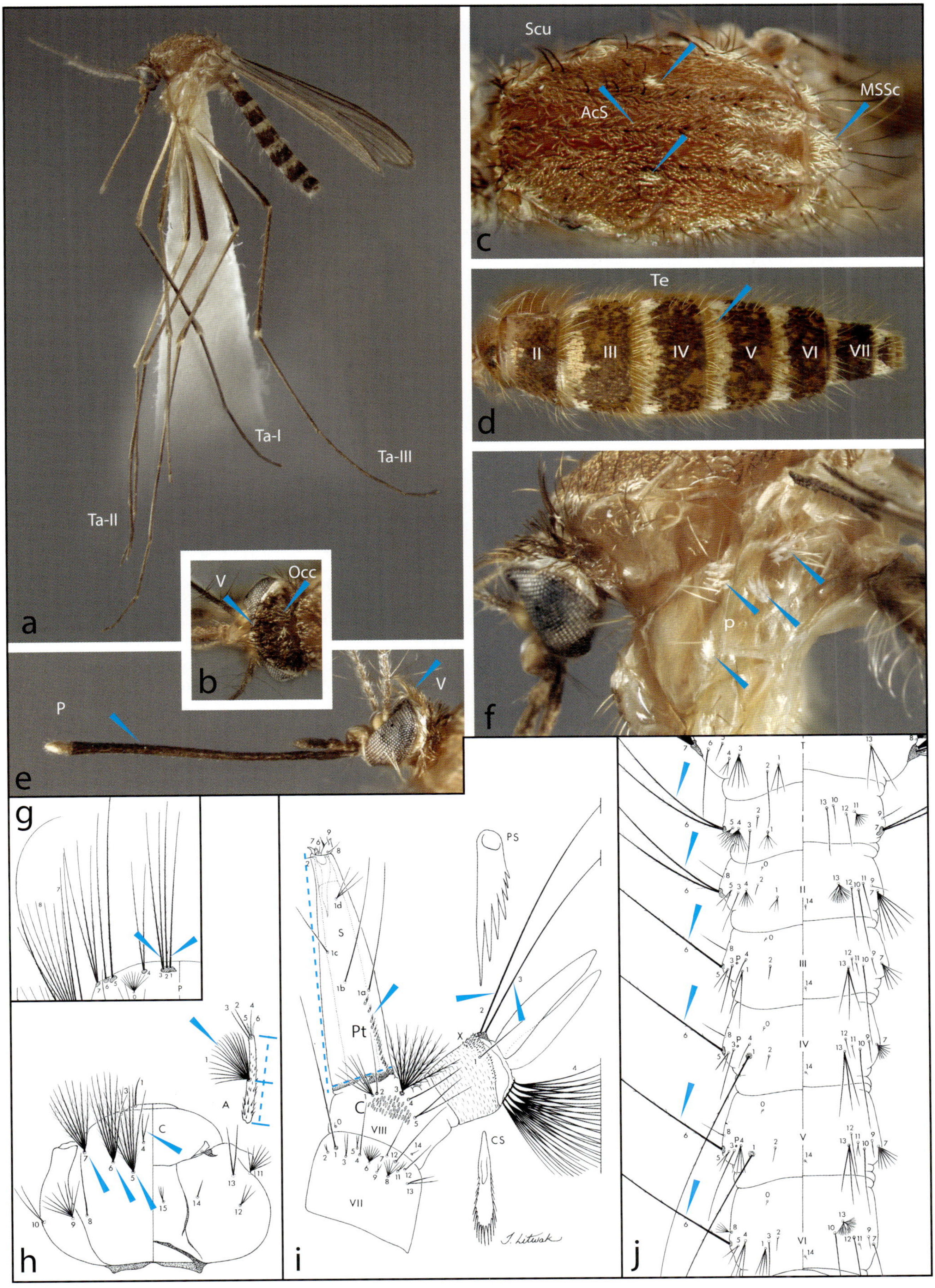

a
Ta-I
Ta-III
Ta-II
b
V
Occ
c
Scu
AcS
MSSc
d
Te
II
III
IV
V
VI
VII
e
P
V
f
P
g
h
A
C
i
S
Pt
C
VIII
VII
PS
CS
X
J. Litwak
j
T
I
II
III
IV
V
VI

Culex (*Culex*) *salinarius* Coquillett, 1904
Subfamily Culicinae, Tribe Culicini
"Saline American Typical Mosquito"

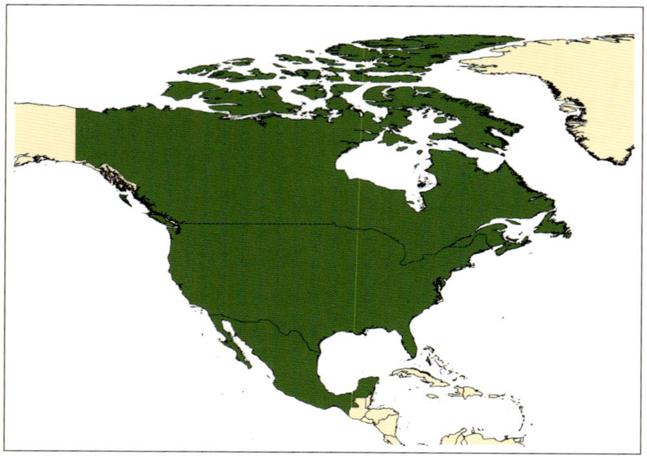

Systematics and Distribution

Culex salinarius is a Nearctic species found from Mexico to Alaska. The dark-scaled proboscis of *Cx. salinarius* differentiates it from all other drab *Culex* (*Culex*) in the Nearctic. This species can be found in great numbers in coastal areas but, despite its misleading name, *Cx. salinarius* immatures are found mostly in fresh water and do not tolerate salinity well. Ironically, draining marshes and installing freshwater impoundment to control *Aedes sollicitans* (Walker) creates excellent immature habitats for *Cx. salinarius*. *Culex salinarius* is the nominotypical member of the Salinarius Complex of the Pipiens Group (subgenus *Culex*), which also includes *Cx. alani* Forattini, *Cx. archegus* Dyar, *Cx. bilineatus* Theobald, Cx. dolosus (Lynch Arribálzaga), and Cx. spinosus Lutz.

Bionomics

Culex salinarius females are aggressive nocturnal biters. Large swarms form at twilight, when the species emerges from its daytime resting sites in cattails and other marshland vegetation. Adult *Cx. salinarius* mosquitoes are strong fliers and have been observed flying over 12 km from their larval habitats seeking blood meals and oviposition sites. During daylight hours, adults can be found resting on walls of buildings or on shaded trees, and at night they readily enter houses to feed. They can be an extreme nuisance to humans when populations peak, but they are true opportunistic feeders, with host preference closely tied to host density. That said, maximal eggs per batch are produced following bird feeds ($n = 50$), with fewer than half this number of eggs generated after imbibing a suboptimal human or mammal feed.

In the North, *Cx. salinarius* populations build gradually from late spring through summer, and there is usually a late-season population peak in autumn, which persists until the winter weather forces the adults into diapause. Unusual among North American *Culex* species, *Cx. salinarius* overwinters in natural shelters such as muskrat huts and other animal burrows. Ovarian dissection studies revealed that nearly all of the overwintering northern *Cx. salinarius* sampled in the fall were nulliparous, though over 90% of the specimens collected in the spring were parous. These data suggest that northern *Cx. salinarius* take a blood meal late in the season and complete a gonotrophic cycle before leaving diapause. In the southern parts of its range, immatures and adults are active year-round.

Females lay eggs in rafts in both fresh and polluted water sources, including in the fresher water portions of salt marshes, grassy pools, ditches, and ponds and, occasionally, in rain barrels, animal footprints, and even tree holes. Famously, this species was found in the *Phragmites* marshes of Staten Island, following the West Nile outbreak in New York.

Associated Pathogens

EEEV, SFV, SLEV, TENV, UGSV, WNV.

Diagnostic Characters

(a) Habitus (♀). **(b) Leg (III):** Ta-III$_{1-5}$ dark scaled. **(c) Thorax (dorsal):** Scutum (Scu) with acrostichal setae (AcS); scutellar midlobe with basal patch of fine dark-brown scales (MSSc) (not distinct in photo). **(d) Thorax (side):** Head with narrow curved scales and long erect forked scales dorsally; pleuron (p) with several groups of broad scales; mesepimeron with distinct median patch of white scales (MeSc). **(e) Head (side):** Proboscis (P) dark scaled, slightly paler ventrally. **(f) Abdomen (dorsal):** II,III-Te with indistinct or very narrow basal pale bands; VII,VIII-Te mostly covered by coppery scales (not distinct in photo). **(g) Larva (thorax):** Seta 1-M long, 2,3 branched, much longer than 2-M. **(h) Larva (head):** Seta 6-C 3,4 branched. **(i) Larva (terminal segments:** Siphon (S) without small spines near apex; siphon index 5.7–8.5; seta 1-S basal-most tuft 2,3 branched, just slightly longer than width of siphon (S) at insertion; seta 2-X 3-branched. Image sources: WRBU; Darsie & Day (2006).

Exemplar DNA sequences: *Cx. salinarius* (JX260663–67, KP281746–51, MG086044, MH801135–37).

Catalog entry: Vol. 2, p. 878.

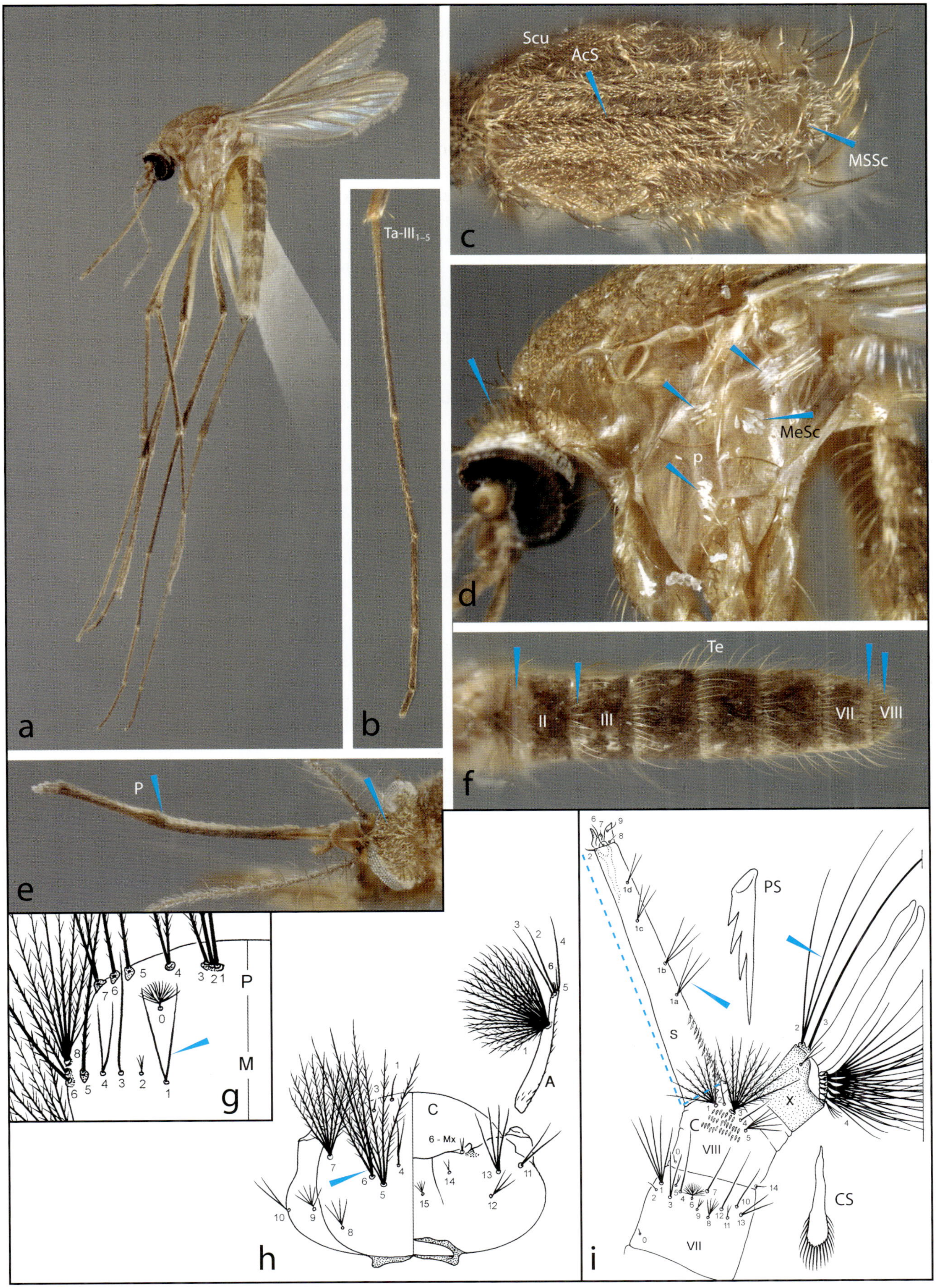
a
b
Ta-III1–5
c
Scu
AcS
MSSc
d
MeSc
p
e
P
f
Te
II
III
VII
VIII
g
P
M
h
C
A
6 - Mx
i
PS
CS
S
C
X
VIII
VII
1a
1b
1c
1d

Culex (*Culex*) *sitiens* Wiedemann, 1828
Subfamily Culicinae, Tribe Culicini
"Thirsty Indonesian Typical Mosquito"

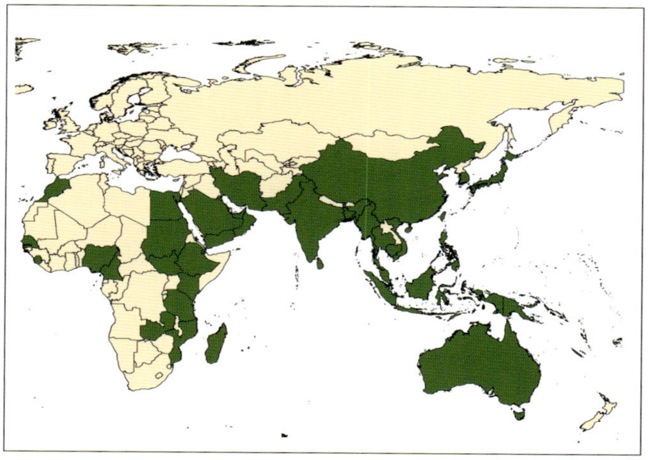

Systematics and Distribution

Culex sitiens is primarily dark brown in color, with a large median white band on the proboscis and pale stripes across the abdominal segments and leg joints, giving it an almost wasp-like appearance. Originally described from the Indonesian island of Sumatra, *Cx. sitiens* has a complex taxonomic history and 14 valid synonyms. *Culex sitiens* is one of six recognized species in the Sitiens Subgroup, which also includes *Cx. alis* Theobald, *Cx. annulirostris* Skuse (see p. 486), *Cx. litoralis* Bohart, *Cx. palpalis* Taylor, and *Cx. whitmorei* (Giles). *Culex sitiens* has 14 valid synonyms (see Catalog entry).

Bionomics

Culex sitiens females bite throughout the night, with a single peak between 1900h and 2000h. Although the species reportedly feeds on birds and pigs, human biting rates of 135 bites/person/hour have been reported in Thailand, indicating that it is a significant pest nuisance. In the South Pacific, *Cx. sitiens* is a ferocious human biter. Adults are readily caught in light traps, in cow-baited traps, and at human bait.

Females deposit large egg batches (mean 158, range 103–223), and neonates develop to adults in 8–12 days, allowing population densities to rapidly increase. Both males and females are long-lived, and the species is multivoltine. Autogeny was noted in 1% of females fed on high-sucrose diets (25% sucrose) in the laboratory, but this was deemed unlikely in nature. Immatures of *Cx. sitiens* occur in sunlit fresh, brackish, and seawater habitats in coastal mangrove and nipa swamps and tidal marshes, especially where water bodies are polluted with trash. These can include puddles, ditches, rock pools, ponds, crabholes, water collections in boats, and some smaller trash containers (e.g., cans, jars) close to populated coastal areas. The anal gills of the larvae are short and bulbous, which is believed to be an adaptation to salinity.

Associated Pathogens

EUBV, JEBV, MOSV, RRV, TMUV; *Brugia malayi*, *Plasmodium juxtanucleare*.

Diagnostic Characters

(a) Habitus (♀): Terga (Te) with basal pale bands only. **(b) Thorax (dorsal):** Scutum with dark-brown scales and small pale scale patches on margins and on prescutellar area (PrA); acrostichal setae (AcS) present. **(c) Head (side):** Proboscis (P) with broad median pale area; vertex (V) all brown scaled. **(d) Thorax (side):** Pleuron (p) with distinct patches of scales; lower mesepimeral setae (MeSL) absent. **(e) Wing:** Wing dark scaled; vein R_{2+3} shorter than vein R_2. **(f) Legs (I,II):** Fe-I,II speckled anteriorly; Ta-I–III with pale bands at articulations. **(g) Larva (terminal segments):** Pecten (Pt) on basal 0.2–0.34 of siphon (S); pecten spines (PS) with c. 9 fine lateral barbs; saddle (Sa) longer than anal papillae (APP), complete, encircling segment X. **(h) Larva (head):** Seta 1-C evident and flattened; setae 2,3-A subapical on antennal shaft. Image sources: WRBU; Harbach (1988).

Exemplar DNA barcodes: *Cx. sitiens COI* (AB738201–02, AB738205–06, AB738271, DQ154161, DQ154163, DQ310144, JQ728396–98, JQ728400, JQ728402, MK300239, MK300241–44).

Catalog entry: Vol. 2, p. 879.

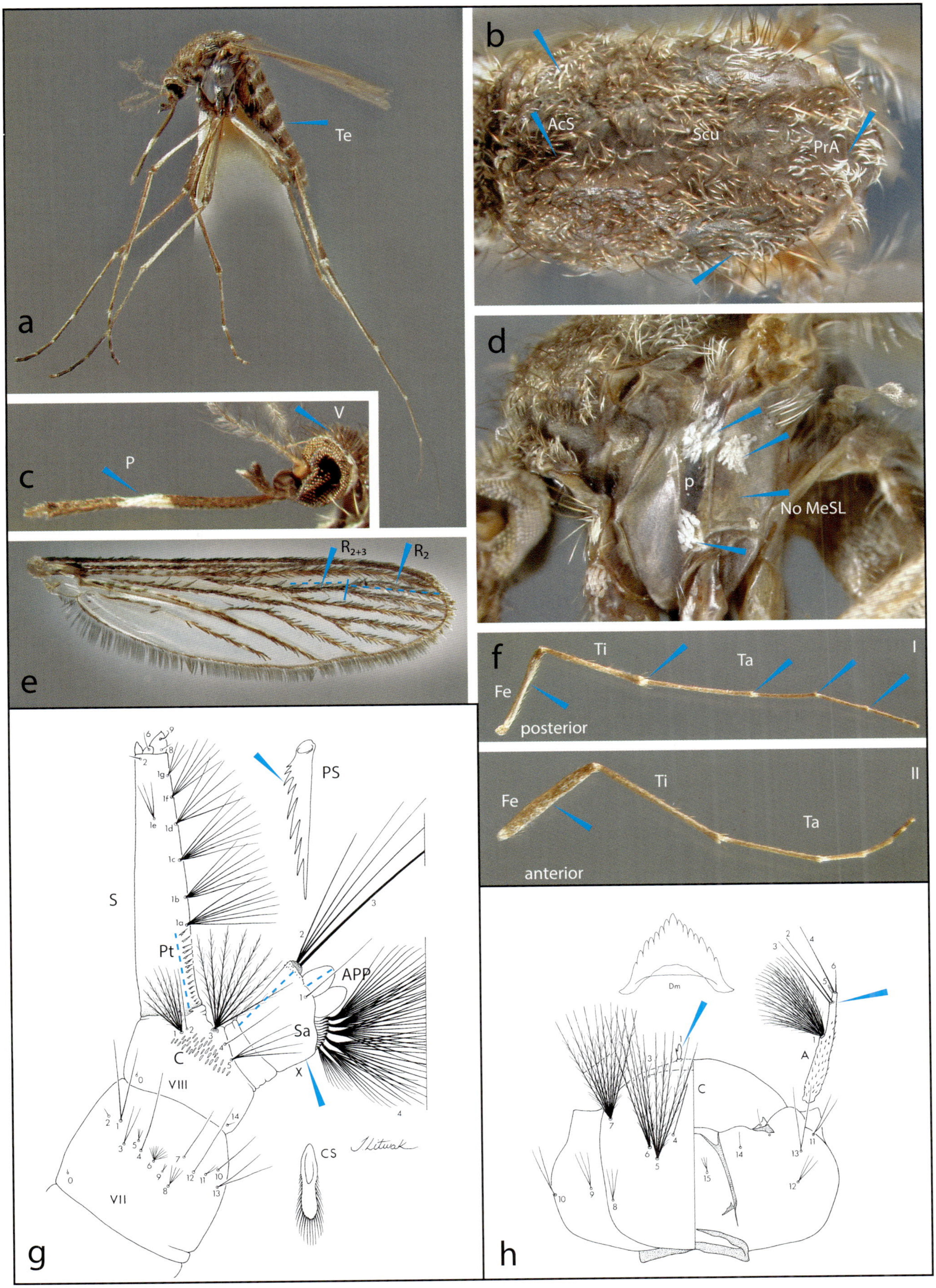
a
Te
b
AcS
Scu
PrA
c
V
P
d
p
No MeSL
e
R_{2+3}
R_2
f
Ti
Ta
I
Fe
posterior
II
anterior
g
PS
S
Pt
APP
Sa
C
VIII
X
VII
CS
h
Dm
A
C

Culex (*Culex*) *tarsalis* Coquillett, 1896
Subfamily Culicinae, Tribe Culicini
"Leggy American Typical Mosquito"

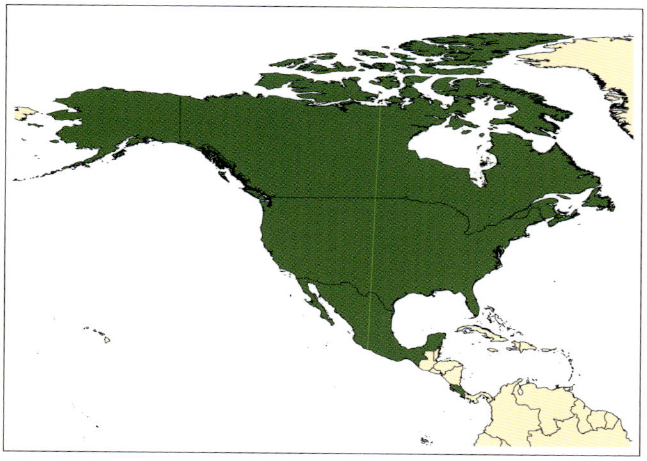

Systematics and Distribution

Culex tarsalis is readily recognizable, as the proboscis sports a prominent bright-white median band, and its legs have white bands that span both sides of the joints. *Culex tarsalis* gives its name to the Tarsalis Subgroup, which comprises 25 species. The species has the following two valid synonyms: *kelloggii* Theobald and *willistoni* Giles. This species extends from Alaska to Central America but appears best suited to the Great Plains and arid western areas of North America. *Culex tarsalis* has the dubious honor of delivering one of the most painful bites of all mosquitoes.

Bionomics

Unique within North American *Culex* species, *Cx. tarsalis* eggs, laid in rafts, can survive desiccation for short periods of time. Larvae are found in a wide variety of both shaded and open sunlit habitats, including ditches, irrigation systems, ground pools, marshes, pools in streambeds, rain barrels, and animal footprints. Foul water habitats, such as those created by animal corrals and abattoirs, appear to be favorite larval habitats in many localities. Somewhat unexpectedly, immatures of *Cx. tarsalis* have also been reported at altitudes up to 7,000 ft in the Sierra Nevada of California and up to 9,000 ft in the Rocky Mountains, Utah. At these high altitudes, larval habitats typically comprise clear, clean waters, potentially suggestive of cryptic speciation.

Culex tarsalis is an aggressive, persistent, diurnal biter that will readily enter buildings seeking blood meals. In one study, host-seeking females were observed traveling 4 km upwind and males traveling 1.6 km upwind, suggesting that adults are strong fliers. When females are searching for a blood meal, host preference appears to be heavily dependent on host availability. In rural, undeveloped areas, *Cx. tarsalis* feeds almost exclusively on wild and domestic birds; however, in more developed habitats, females will feed on cattle, horses, and humans, although these are regarded as incidental hosts. These habits facilitate zoonotic transmission of pathogens, particularly viral encephalitis, to humans.

The species is long-lived and multivoltine. In temperate zones, *Cx. tarsalis* is active from late spring to early autumn, but in subtropical areas of its distribution (southern United States and Mexico), it breeds throughout the year. Favored adult resting sites include small man-made structures such as chicken coops, and natural shelters such as wood rat burrows. In the colder months, females can be found overwintering as adults in the protection of basements, sheds, cellars, and caves. The species has been successfully colonized.

Associated Pathogens

CEV, FLAV, GLOV, HPV, JCV, JSV, KEDV, LLSV, LOKV, MDV, MRV, NTAV, PGAV, ROCV, SFV, SHOV, TURV, VEEV, WEEV, WNV, WSLV; *Plasmodium* spp.

Diagnostic Characters

(a) Habitus (♀): Ta-I–III with distinct white bands, most prominent on Ta-III. **(b) Head (side):** Proboscis (P) with broad median pale band; palpus (MPlp) with pale scales. **(c) Abdomen (side):** Terga (Te) with broad basal bands of white or yellowish-white scales; sterna (S) with inverted V-shaped dark markings on each segment. **(d) Wing:** Scattered pale scales on costal (C) and subcostal (Sc) veins (scattering indistinct in photo). **(e) Larva (head):** Seta 1-A large, multi-branched, inserted at a constriction on apical 0.34 of antenna (A); setae 5–7-C with ≥3 aciculate branches. **(f) Larva (terminal segments):** Siphon index 4.5–5.5; seta 1-S usually with five pairs of ventral tufts inserted in straight line; seta 2-X 2,3 branched, shorter than 3-X. Image source: WRBU.

Exemplar DNA barcodes: *Cx. tarsalis* (MG087219, MG079988, MG079101, KT117301, KR681424, KR695532).

Catalog entry: Vol. 2, p. 881.

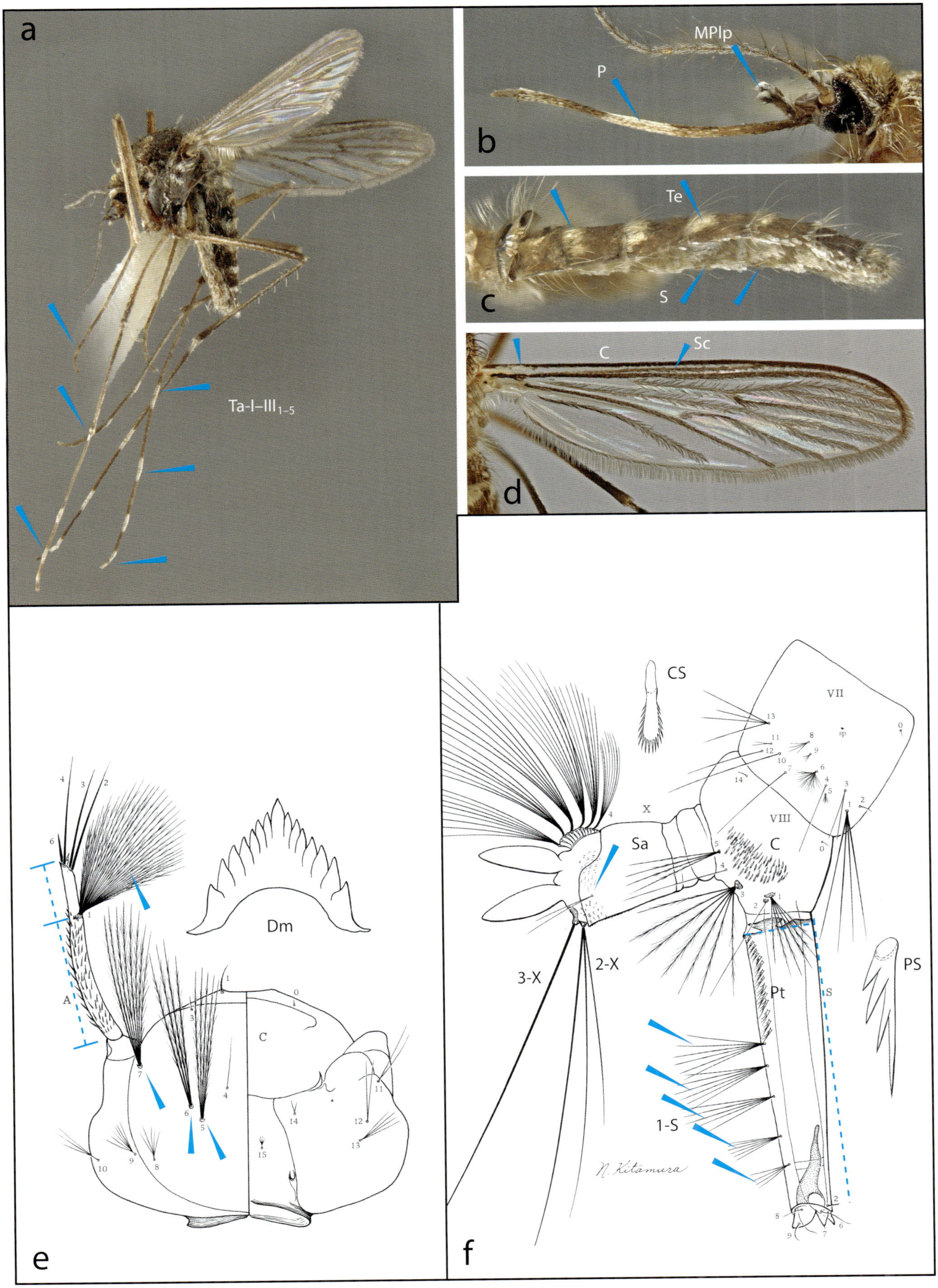
a
Ta-I–III1–5
b
MPlp
P
c
Te
S
d
C
Sc
e
A
Dm
C
f
CS
VII
VIII
X
Sa
C
3-X
2-X
Pt
S
PS
1-S
N. Kitamura

Culex (Culex) theileri Theobald, 1903
Subfamily Culicinae, Tribe Culicini
"Theiler South African Typical Mosquito"

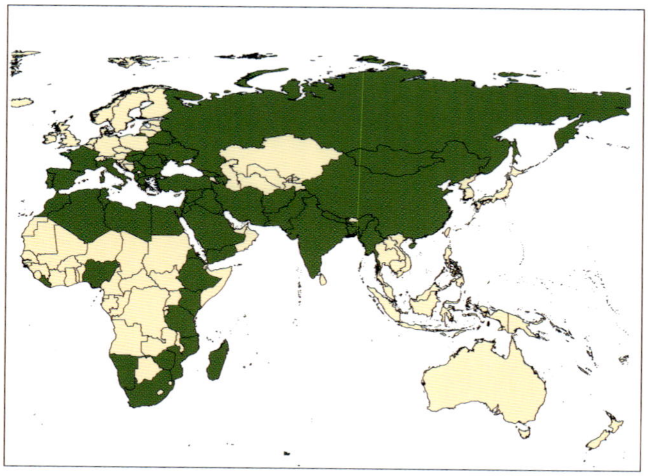

Systematics and Distribution

Culex theileri is a large distinct mosquito bearing pale bands on the abdomen, widened medially to form a yellowish triangular pattern; postspiracular and prealar scales; and with pale stripes on the fore- and midfemora and all tibiae. *Culex theileri* is the nominotypical member of the Theileri Subgroup, which comprises only three species, including *Cx. laticinctus* Edwards and *Cx. mattinglyi* Knight. *Culex theileri* has five valid synonyms: *annulata* Theobald, *alpha* Séguy, *creticus* Theobald, *onderstepoortensis* Theobald, and *pettigrewii* Theobald. Originally described from Pretoria, South Africa, *Cx. theileri* is widely distributed in the western Oriental, southern Palearctic, and Afrotropical Regions, and is particularly common in the temperate highlands of eastern Africa (see map).

Bionomics

Culex theileri is a highly adaptable species that is also extremely focal, appearing in large numbers in some places but apparently quite rare in others. Oviposition sites appear highly plastic, with immatures found in temporary and permanent water bodies that can contain fresh, brackish, salt, or foul water in open sunlight or shaded locations. Larvae are found in large ground pools, slow-moving streams and stagnant stream pools, mangrove and nipa palm swamps, cement tanks, seepage pits, and cesspools. Artificial containers are also utilized, including cement tanks, canoes, jars, and cans. The immatures can be sympatric with *Cx. antennatus* (Becker) (see species page, p. 488), *Cx. pipiens* Linnaeus (see species page, p. 498), or *Cx. perexiguus* Theobald (see species page, p. 496) but are often found alone. In the Indomalayan Region, *Cx. theileri* populations peak in late spring to early summer. In Turkey, immatures appear from May to October and are generally absent in the winter period, indicating they overwinter as adults.

Host preferences appear similarly variable. In Egypt, females feed mainly on pigs and birds, but are known to enter houses to bite humans. In South Africa, host-seeking female *Cx. theileri* preferentially feed on cattle, and opportunistically on birds and humans. On Gran Canaria in the Spanish Canary Islands, *Cx. theileri* fed primarily on goats, dogs, and sheep, while a smaller number were observed feeding on cats, cattle, and humans. The species has been reported taking blood meals from multiple hosts for egg production, increasing the opportunity for the uptake and spread of zoonotic diseases.

In the Himalayan Region, *Cx. theileri* is common at high elevations (1,000 to 3,000 m), but elsewhere in its range, it is found at lower elevations. Wing size and shape geometry indicate a positive correlation between size and altitude. This trait appears to be adaptive for survival, as larger individuals possess smaller surface-to-volume ratios, conducive to heat conservation in cold climates.

Associated Pathogens

CTFV, GERV, OCKV, RVFV, SHUV, SINV, WEEV, WNV; *Dirofilaria immitis*.

Diagnostic Characters

(a) Habitus (♀): Ta-I–III dark. **(b) Thorax (side):** Postspiracular (PoSc) and prealar scales (PrSc) present; lower mesepimeron with at least one seta (MeSL); scales of upper (MScU) and lower mesokatepisternum (MScL) and posterior mesanepisternum (prealar area) (PrSc) merged. **(c) Head (side):** Proboscis (P) without pale bands. **(d) Abdomen (dorsal):** Tergal (Te) pale bands widened medially. **(e) Larva (thorax/abdominal segment I):** Seta 14-P single (not shown); seta 7-I much shorter than seta 6-I. **(f) Larva (head):** Seta 1-C slender; seta 5-C 3,4 branched. **(g) Larva (terminal segments):** Comb scales (CS) pointed, with proximal fringe; siphon with most elements of seta 1-S mostly close to posterior midline, multibranched (11–14). Image sources: WRBU; Harbach (1988).

Exemplar DNA barcodes: *Cx. tarsalis* (FJ210898–900, HE610458–60, JN051388–90, KF407831, KJ012174–234).

Catalog entry: Vol. 2, p. 882.

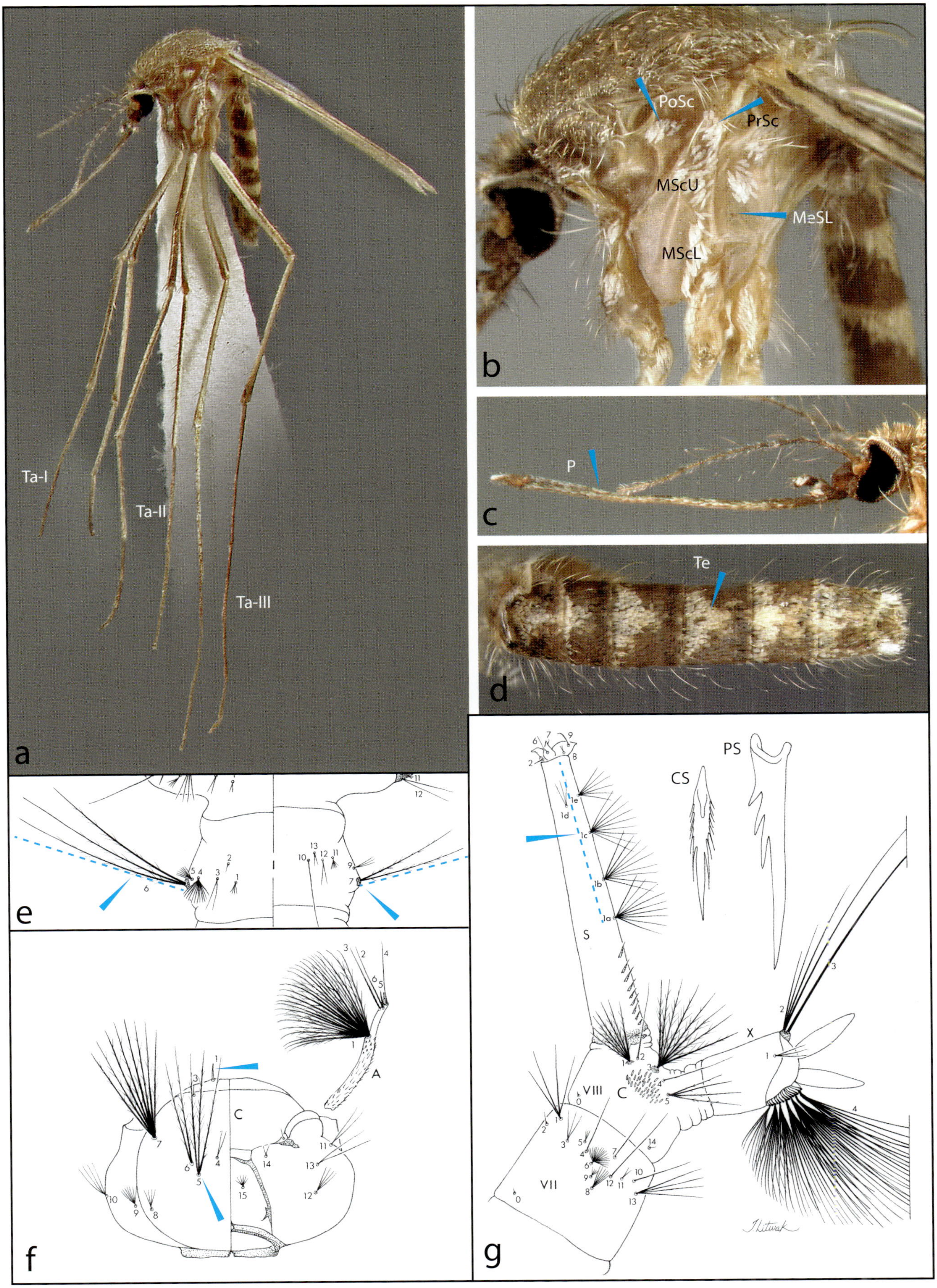

a
Ta-I
Ta-II
Ta-III
b
PoSc
PrSc
MScU
MeSL
MScL
c
P
d
Te
e
f
C
A
g
PS
CS
S
VIII
VII
C
X

Culex (Culex) tritaeniorhynchus Giles, 1901
Subfamily Culicinae, Tribe Culicini
"Three-banded Indian Typical Mosquito"

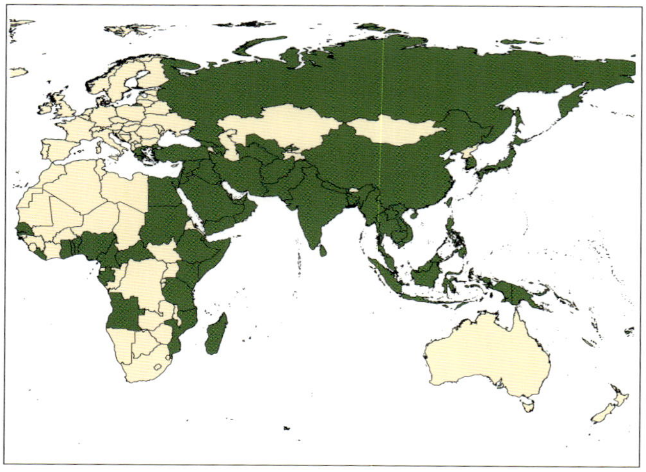

Systematics and Distribution

Culex tritaeniorhynchus is a small drab mosquito belonging to the Vishnui Subgroup of *Culex* (*Culex*), which also includes *Cx. alienus* Colless, *Cx. annulus* Theobald, *Cx. incognitus* Baisas, *Cx. perplexus* Leicester, *Cx. philippinensis* Sirivanakarn, *Cx. pseudovishnui* Colless, *Cx. vishnui* Theobald (see species page, p. 516), and *Cx. whitei* Barraud. Of all the Vishnui Subgroup, *Cx. tritaeniorhynchus* has the widest distribution, stretching from eastern Africa through the Middle East and the Indian subcontinent and across the mainland and islands of Southeast Asia, including the Philippines. *Culex tritaeniorhynchus* has three valid synonyms: *biroi* Theobald, *summorosus* Dyar, and *siamensis* Barraud & Christophers.

Bionomics

Culex tritaeniorhynchus populations can explode where suitable habitats prevail. Its numbers increase in response to rice cultivation, with highest larval densities occurring when the rice is high and dense, creating the optimal shaded habitat for immature *Cx. tritaeniorhynchus*. Like the closely related *Cx. vishnui*, *Cx. tritaeniorhynchus* is highly attracted to cows and pigs and only occasionally feeds on people. *Culex tritaeniorhynchus* is strongly exophilic, and its anthropophagy increases when people sleep outdoors during the rice harvest period or the hot seasons. These feeding traits are not shared by the essentially isomorphic species *Cx. pseudovishnui*, which feeds almost exclusively on cows and water buffalo (98.7%), only feeding on pigs and humans when cows are not available. In India, *Cx. tritaeniorhynchus* is the dominant species collected at cattle sheds and pigsties, especially at dusk, when biting is at its most intense.

Due to its affiliation with pigs, it is an important vector of Japanese encephalitis virus, and vertical transmission has been noted, with F_1 progeny capable of transmitting the virus. The species has been successfully colonized, indicating stenogamy—mating in a restricted space and requiring no nuptial flight. When life table characteristics were compared under laboratory conditions in four Asian strains (Japan, Taiwan, Pakistan, and Bangladesh), female life expectancy ranged from 18.2–28.5 days, with males significantly lower at 11.8–17.2 days. *Culex tritaeniorhynchus* from Japan was distinctly larger, produced fewer egg batches, and exhibited longer generation times than strains from Pakistan, Bangladesh, and Taiwan. Japanese populations of *Cx. tritaeniorhynchus* are known to estivate in cold conditions, overwintering as adult females.

Associated Pathogens

AINOV, AKAV, ARKV, BAGV, EEV, DENV-3, JBEV, KAIV, GETV, RVFV, SAGV, SINV, TMUV, WNV, YOUV; *Brugia malayi, Spiroplasma taiwanense, Wuchereria bancrofti.*

Diagnostic Characters

(a) Habitus (♀): Fe-I,II dark anteriorly; Ta-I–III with pale bands. **(b) Thorax (dorsal):** Scutum (Scu) with unicolorous dark scales; acrostichal setae (AcS) present. **(c) Abdomen (side):** Terga (Te) only with basal pale bands. **(d) Thorax (dorsal):** Pleuron (p) with distinct scale patches; postspiracular scales (PoSc) absent; mesepimeral (MeSL) setae absent; lower mesokatepisternal scales (MScL) present. **(e) Head (side):** Proboscis (P) with broad median pale band; vertex (V) with erect brownish scales. **(f) Wing:** Wing dark scaled; vein R_{2+3} shorter than R_2. **(g) Larva (head):** Dorsal apotome (DAp) distinctly separated from median labral plate (MLP); seta 1-C sharp, tapered; seta 6-C double. **(h) Larva (thorax/abdominal segment I):** Seta 4-P usually double; seta 7-I double. **(i) Larva (terminal segments):** Comb scales (CS) evenly fringed; siphon (S) long; pecten (Pt) not reaching apex of siphon (S); saddle (Sa) shorter than anal papillae (APP). Image sources: WRBU; Harbach (1988).

Exemplar DNA barcodes: *Cx. tritaeniorhynchus* (KJ012243–50, KM350641–69, KM362833–50).

Catalog entry: Vol. 2, p. 883.

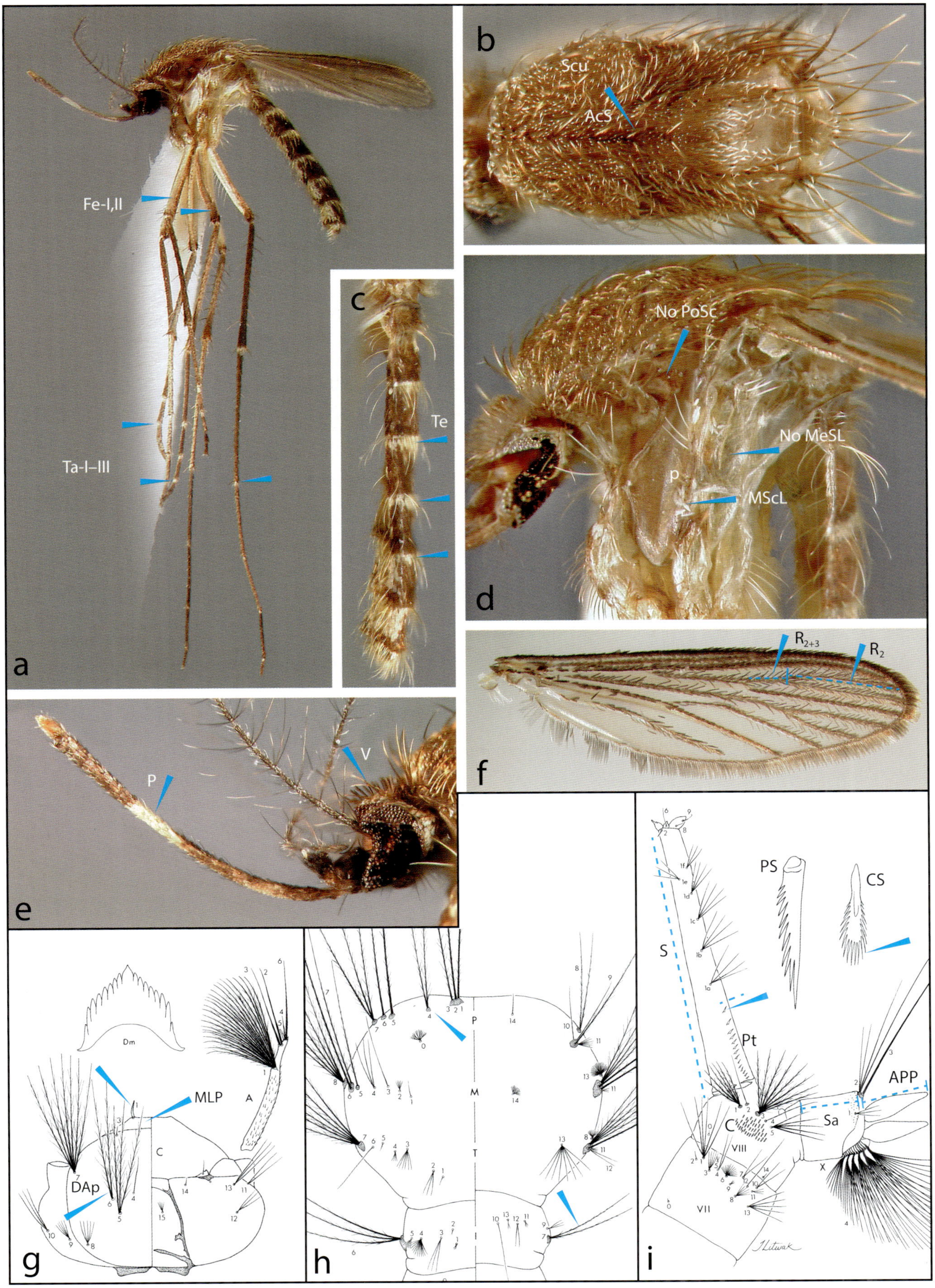
a
Fe-I,II
Ta-I–III
b
Scu
AcS
c
Te
d
No PoSc
No MeSL
p
MScL
f
R_{2+3}
R_2
e
P
V
g
Dm
MLP
A
C
DAp
h
P
M
T
i
S
PS
CS
Pt
C
Sa
APP
VIII
VII
X

Culex (Culex) univittatus Theobald, 1901
Subfamily Culicinae, Tribe Culicini
"One-banded Zimbabwean Typical Mosquito"

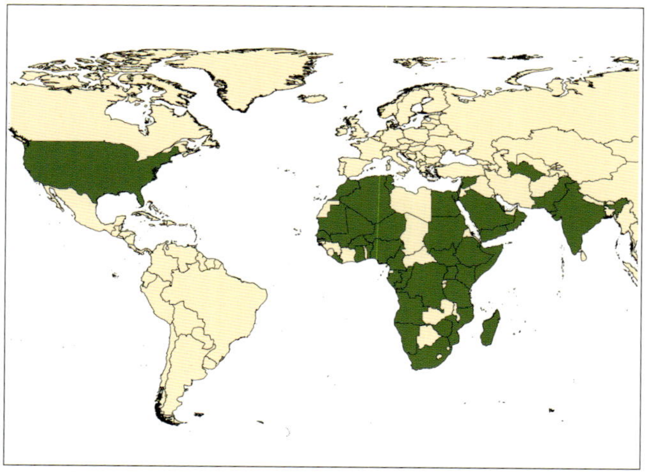

Systematics and Distribution

Culex univittatus is a small- to medium-sized species, with a highly distinctive line of scales marking the position of wing vein 2A. This character is not reported in any other mosquito species and could represent a primitive character of the family Culicidae. There are four currently recognized synonyms of *Cx. univittatus*: *ataeniatus* Theobald, *goughii* Theobald, *montforti* Ventrillon, and *simplex* Theobald. *Culex univittatus* is the nominotypical member of the Univittatus Subgroup, which comprises four species, including *Cx. fuscocephala* Theobald (see species page, p. 490), *Cx. neavei* Theobald, and *Cx. perexiguus* Theobald (see species page, p. 496).

This distribution of this species needs to be reverified in some parts of its range. Previous reports of *Cx. univittatus* in Europe are now recognized as pertaining to *Cx. perexiguus*, which is also reported widely in Africa. The authors collected both species in sympatry in Tanzania. *Culex univittatus* and *Cx. neavei* are allopatric in South Africa, with *Cx. univittatus* in the temperate Transvaal Highveld and *Cx. neavei* occurring in the subtropical coastal lowlands of Natal. *Culex univittatus* appears to be most common in more temperate areas of the Afrotropical Region, but it has been reported in nearly every country in Africa, into the Middle East and Indian subcontinent.

Bionomics

Given the complex taxonomic history of *Cx. univittatus*, *Cx. neivai*, and *Cx. perexiguus* and overlapping distributions in Africa, there still remains some doubt over their true distributions, and the correct assignment of bionomic and pathogen associations with each species. *Culex univittatus* immatures can be found in freshwater ground pools, marshes, ditches, swamp and stream margins, and animal footprints. *Culex univittatus* is a typical ground pool breeder, with fewer numbers found in larger water bodies like rice fields. Larvae are found in partially shaded to fully sunlit habitats with filamentous algae present. Larvae are highly mobile and have been noted shooting more than 3 m across the habitat substrate following potential predatory disturbances. Adult females are primarily ornithophilic but demonstrate highly diverse feeding habitats, including many mammalian species, and readily enter houses to feed on humans. *Culex univittatus* populations in South Africa apparently do not feed on humans. Females become active at sunset and continuing to sunrise, peaking at about midnight.

Associated Pathogens

BAGV, OLIV, SINV, SPOV, USUV, WEEV, WSLV, WNV; *Wuchereria bancrofti*.

Diagnostic Characters

(a) Habitus (♀): Fe-II with complete anterior pale stripe; Ta-I–III not banded. **(b) Thorax (side):** Postspiracular (PoSc) and prealar scales (PrSc) present; upper (MScU) and lower (MScL) mesokatepisternal scale patches separate; lower mesepimeron with one or two setae (MeSL). **(c) Abdomen (dorsal):** Terga (Te) with basal bands narrow, of even width; abdominal basal pale bands slightly expanded mesally. **(d) Wing:** Base of costa (C) with line of pale scales; vein 2A present. **(e) Head (side):** Proboscis (P) entirely dark scaled. **(f) Larva (head):** Seta 1-C stout. **(g) Larval (terminal segments):** Comb (C) with average of 43 scales; comb scales (CS) apically rounded with even fringe of fine spinules; siphon (S) relatively long, index <8.5; seta 1-S shorter than width of siphon, with five or six pairs of tufts, two lateral; seta 1a-S usually situated in the pecten; seta 1-X with 2–4 branches; pecten (Pt) with pecten spines (PS) with three lateral denticles; seta 6-VI usually single (not shown). Image sources: WRBU; Harbach (1988).

Exemplar DNA barcodes: *Cx. univittatus COI* (KU187056–60, KU380425, LC088986–99, LC100115).

Catalog entry: Vol. 2, p. 884.

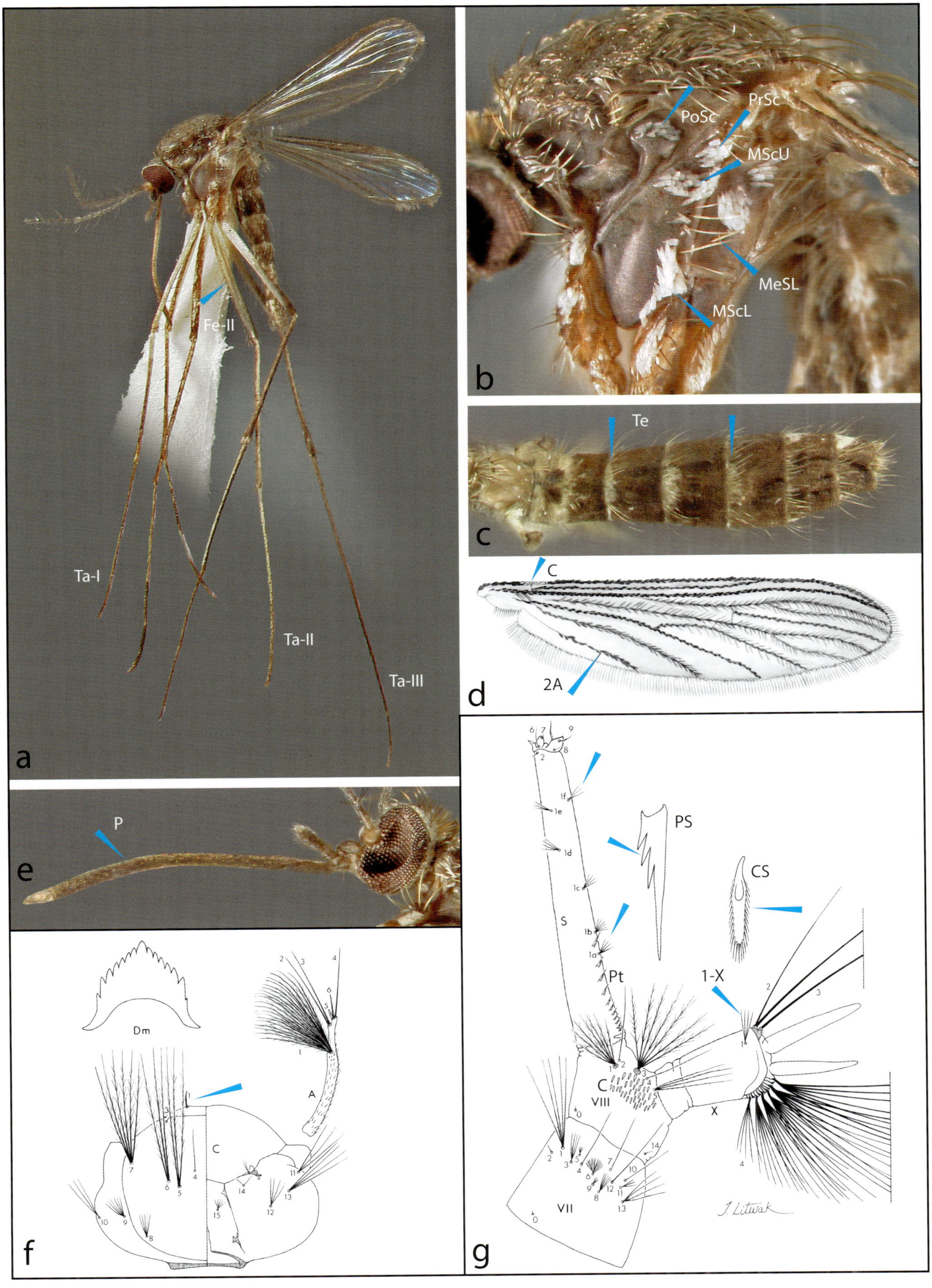
a
Fe-II
Ta-I
Ta-II
Ta-III
b
PoSc
PrSc
MScU
MeSL
MScL
c
Te
d
C
2A
e
P
f
Dm
A
C
g
PS
CS
S
Pt
1-X
C
VIII
X
VII
J. Litwak

Culex (*Culex*) *vishnui* Theobald, 1901
Subfamily Culicinae, Tribe Culicini
"Vishnu Indian Typical Mosquito"

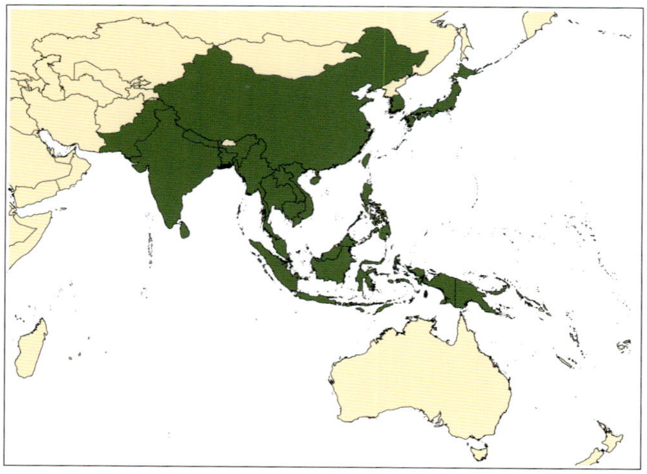

Systematics and Distribution

Culex vishnui is the nominotypical member of the Vishnui Subgroup which comprises eight species, including three important vectors of Japanese encephalitis virus, *Cx. tritaeniorhynchus* Giles (see species page, p. 512), *Cx. pseudovishnui* Colless, and *Cx. vishnui* Theobald (see species page, p. 516). Other members include *Cx. alienus* Colless, *Cx. annulus* Theobald, *Cx. incognitus* Baisas, *Cx. perplexus* Leicester, *Cx. philippinensis* Sirivanakarn, and *Cx. whitei* Barraud. All members of the Vishnui Subgroup are difficult to differentiate as adults, but their larvae possess clearly defined species-diagnostic characters. As adults, *Cx. vishnui* is most commonly sympatric with *Cx. tritaeniorhynchus* and is separated by possessing a distinct median pale band on the proboscis. *Culex vishnui* has a wide distribution. Its range extends from India in the West through all of mainland Asia and Southeast Asia, from Japan and the Korean Peninsula south to include the Indonesian islands, Borneo, and the Philippines east to Papua New Guinea.

Bionomics

Host-seeking females of *Cx. vishnui* are collected in high densities in both light traps and animal-baited traps across their range. In Vietnam, cattle are the preferred hosts, and *Cx. vishnui* density is positively correlated with cattle density. In southern India, blood meal studies were revealed to be 64.3% bovine and 35.5% porcine, with the remainder from humans, goats, and birds. Feeding occurs mainly at dusk. The proportion of human feeding increases markedly in the hot season in India, when most people sleep outdoors to escape the unbearable heat within their dwellings. Although *Cx. vishnui* is primarily exophilic and zoophilic, this species has been reported entering homes to bite humans, especially when populations are high. Gestating females are commonly found resting in animal shelters or in cultivated fields with dense vegetation, such as sugarcane.

Population densities of *Cx. vishnui* are closely linked to the rice cultivation cycle. Immatures are dominant in rice fields early in the season, but as the rice vegetation grows taller, they are typically replaced by *Cx. tritaeniorhynchus*. Immatures have been found in stream and ground pools, drainage ditches, road ruts, and animal footprints.

Associated Pathogens

GANV, GETV, JBEV, INGV, PALV, SATV, UMBV, VKTV, WNV.

Diagnostic Characters

(a) Habitus (♀)**:** Ta-I–III pale banded. **(b) Thorax (dorsal):** Scutum (Scu) pale brown scaled; acrostichal setae (AcS) present. **(c) Head (side):** Proboscis (P) with distinct median band not extending basally onto ventral surface; vertex (V) pale brown scaled. **(d) Thorax (side):** Pleuron (p) with distinct scale patches; postspiracular scales (PoSc) absent; lower mesepimeron setae (MeSL) absent; terga (Te) with only basal bands. **(e) Wing:** Vein R_{2+3} shorter than vein R_2. **(f) Legs (I–III):** Fe-I,II without anterior speckling; Fe-III with apical dark band not contrasting with pale scaling. **(g) Larva (mouthparts). (h) Larva (thorax):** Integument with numerous spicules. **(i) Larva (head):** Dorsal apotome (DAp) distinctly separated from median labral plate (MLP); seta 1-C stout. **(j) Larva (terminal segments):** Comb scales (CS) fringed with long spinules, apex pointed, spine-like, >10 in three or four rows; seta 1-S tufts with ≥5 branches; anal papillae (APP) > saddle (Sa) length; siphon (S) long, index >5.0. Image sources: WRBU; LaCasse & Yamaguti (1948); Sirivanakarn (1976).

Exemplar DNA barcodes: *Cx. vishnui COI* (AB690844, AB738095, AB738149, AB738195–96, AY729973, HM769285–86, KC876678–79, MF179237–40).

Catalog entry: Vol. 2, p. 885.

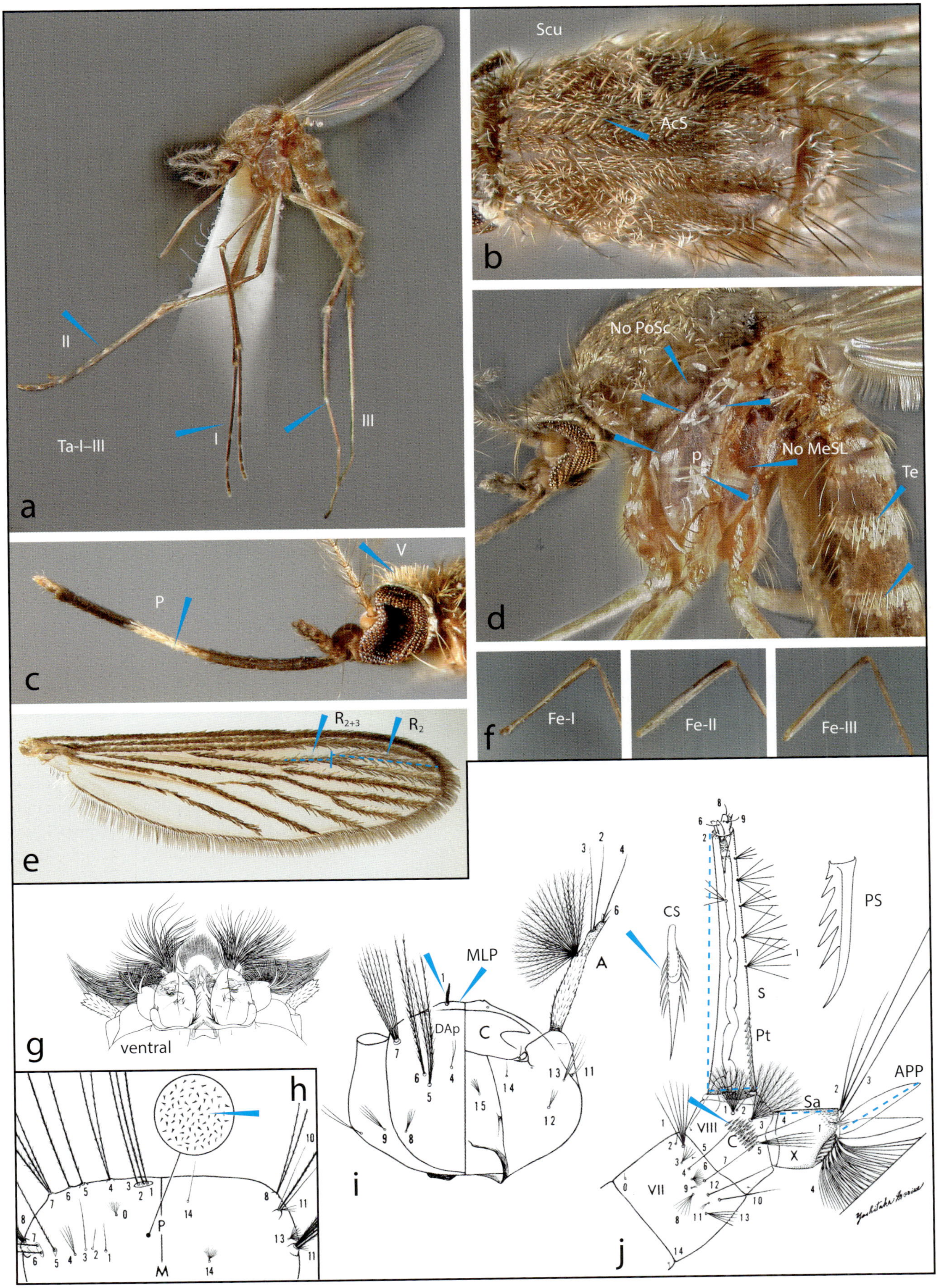
a
II
I
III
Ta-I–III
b
Scu
AcS
c
V
P
d
No PoSc
P
No MeSL
Te
e
R_{2+3}
R_2
f
Fe-I
Fe-II
Fe-III
g
ventral
h
P
M
i
MLP
DAp
C
A
j
CS
PS
S
Pt
Sa
APP
VIII
C
X
VII

Culex (Melanoconion) erraticus (Dyar & Knab, 1906)
Subfamily Culicinae, Tribe Culicini
"Erratic American Typical Mosquito"

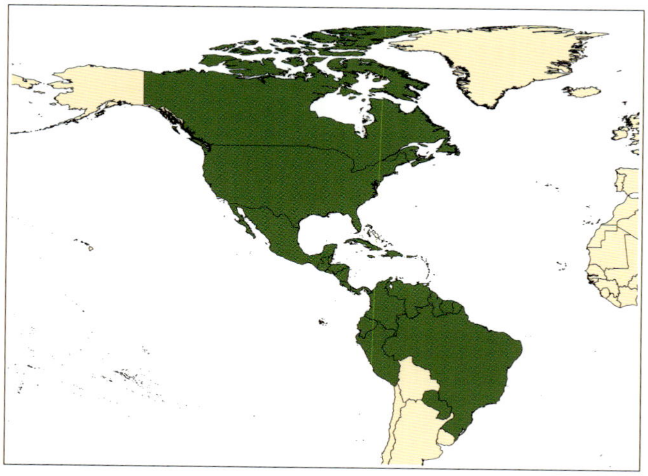

Systematics and Distribution

Culex erraticus is a medium-sized mosquito, with a slender proboscis slightly enlarged at the tip. The mesonotum is brown, with evenly distributed fine golden scales, and the abdomen is black, typically without banding. *Culex erraticus* is the nominotypical member of both the Erraticus Group and Erraticus Subgroup of the subgenus *Melanoconion*. The Erraticus Subgroup comprises three species: *Cx. erraticus*, *Cx. aureonotatus* Duret & Barreto, and *Cx. invocator* Pazos. *Culex erraticus* has 10 valid synonyms: *borinqueni* Root, *degustator* Dyar, *egberti* Dyar & Knab, *homoeopas* Dyar & Ludlow, *leprincei* Dyar & Knab, *moorei* Dyar, *peribleptus* Dyar & Knab, *pose* Dyar & Knab, *tovari* Evans, and *trachycampa* Dyar & Knab. This common New World species has been reported from Canada in the North through Central and South America to Brazil and is found on the islands of the Caribbean.

Bionomics

Culex erraticus females deposit their eggs as small masses glued to the surfaces of floating leaves. Immature habitats typically include permanent water sources—large swamps and stream pools—with heavily overgrown surface vegetation. In Panama, *Cx. erraticus* immatures are reportedly common along the grassy edges of slow-moving streams, freshwater dams, and lakes, especially those shaded by tree canopies. In Venezuela, immature *Cx. erraticus* are associated with unshaded, warm, vegetated waters in flooded open pastures and swamps. Areas with heavy duckweed (genus *Lemna*) growth are distinctly unfavorable to this species.

Culex erraticus is a competent vector of Eastern Equine Encephalitis and West Nile viruses, and a vector of the lizard malaria *Plasmodium floridense*. Host-seeking females can travel up to 3 km in search of a blood meal. Host preference in *Cx. erraticus* appears strongly correlated to avian breeding seasons. In spring, *Cx. erraticus* females feed almost exclusively on nesting birds and hatchlings. As the summer progresses and hatchlings leave the nests, *Cx. erraticus* switches to feeding primarily on mammals, including dogs, white-tailed deer, cats, horses, and people, but will also take blood from reptiles, amphibians, and wading birds, including the great blue and yellow-crowned night herons. The timing of this shift varies greatly from year to year, but colder winters seem to delay the host shift. Throughout the year, females can commonly be found resting in hollow trees.

Associated Pathogens

EEE, WNV; *Plasmodium floridense*.

Diagnostic Characters

(a) Habitus (♂): Ta-I–III all dark (both sexes). **(b) Thorax (dorsal):** Acrostichal setae (AcS) absent. **(c) Head (dorsal):** Vertex (V) with narrow decumbent scales centrally. **(d) Wing:** Veins R_2, R_3 with broad plume scales. **(e) Thorax (side):** Mesepimeron with patch of large scales (MeSc); eyes nearly encircled by broad scales. **(f) Larva (terminal segments):** Comb scales (CS) long, with pointed, unfringed apical section; saddle (Sa) with sharp spicules on margin. Image sources: WRBU; Belkin, Heinemann, & Page (1970).

Exemplar DNA barcodes: *Cx. erraticus* (KM593023–24, KM593029, KM593031–32, KM593041, KM593045, KT766417–31, KX389310, KY859873–84).

Catalog entry: Vol. 2, p. 917.

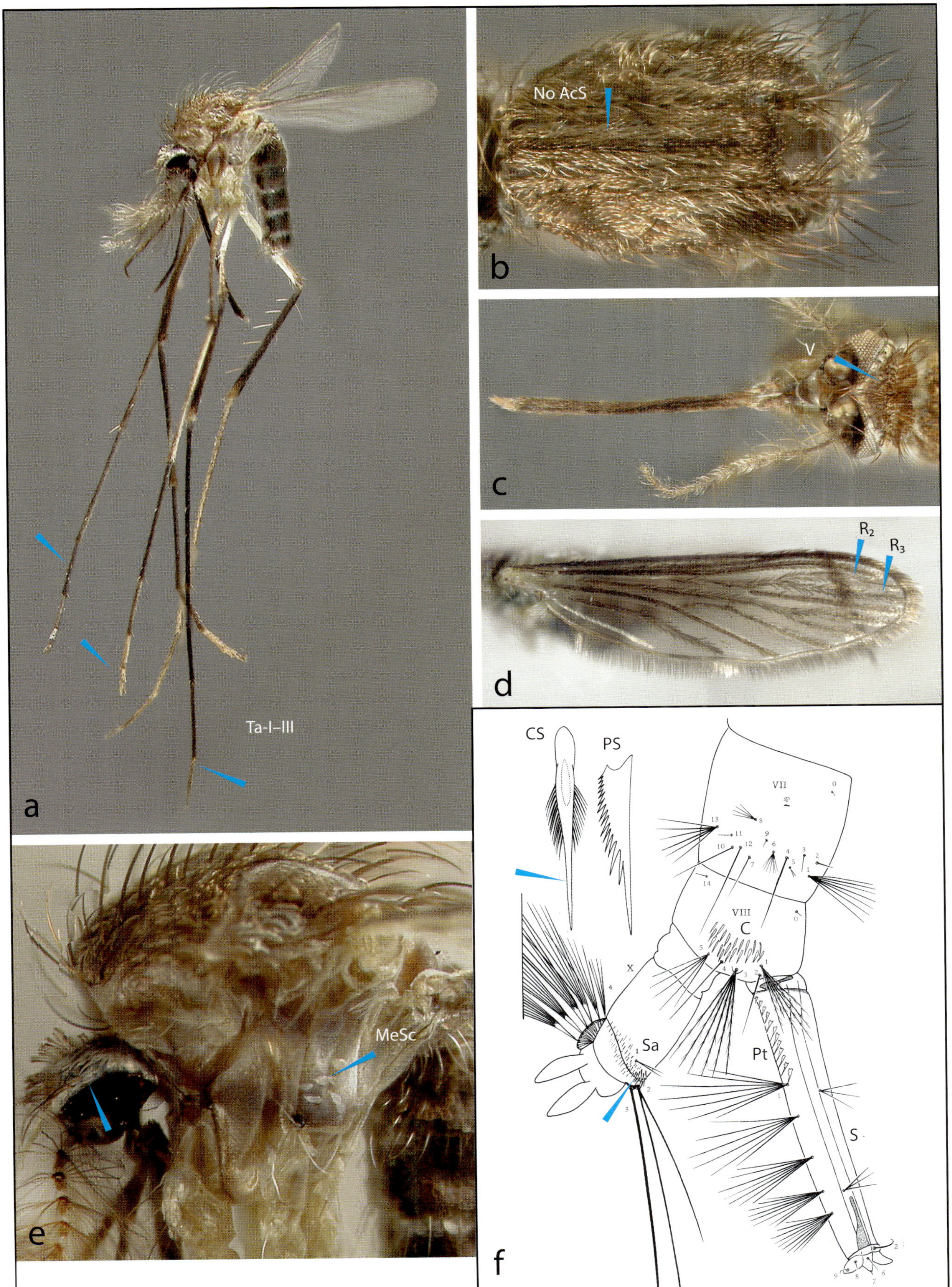
No AcS
b
V
c
R_2
R_3
d
Ta-I–III
a
MeSc
e
CS
PS
VII
VIII
C
X
Sa
Pt
S
f

Culex (Melanoconion) ocossa Dyar & Knab, 1919
Subfamily Culicinae, Tribe Culicini
"Bone Guyanese Typical Mosquito"

Systematics and Distribution

Culex ocossa is distributed across most of Central and South America. It is a small dark mosquito, with dark forked scales on its head. *Culex ocossa* is the nominotypical member of the Ocossa Group within the Spissipes Section of the subgenus *Culex*, which comprises only *Cx. ocossa* and *Cx. panocossa* Dyar & Knab. *Culex ocossa* is essentially isomorphic with *Cx. panocossa*, but adult females can be differentiated by the number of cibarial teeth (character not shown). *Culex ocossa* typically has 31 or 32 teeth, while *Cx. panocossa* has 24 or 25. *Culex ocossa* is distributed across most of Central and South America. Until 1970, *Cx. ocossa* was considered a synonym of *Cx. aikenii* Aiken & Rowland, the main vector of Venezuelan equine encephalitis (VEE) and other arboviruses, including Apeu virus (APEUV) and Marituba virus (MTBV) in the Neotropics, and it seems likely that the vectorial role of *Cx. ocossa* may still be underestimated as a result.

Bionomics

Immature *Cx. ocossa* is typically found in permanent slow-moving or stagnant freshwater bodies with muddy bottoms and abundant floating vegetation, such as *Pistia*, *Eichhornia*, *Salvinia*, and *Azolla*. Sites include sunlit or partially shaded river, lake, and swamp margins in forested areas, artificial lakes in urban gardens, and roadside ditches near cacao plantations. Larvae have also been collected in cement-lined pools.

In southern Brazil, host-seeking female *Cx. ocossa* were collected associated with human dwellings and feeding on humans. Adults are active at night, and biting peaks at midnight, tapering off until dawn.

Associated Pathogens

BABV, Nanay virus, PARAV, VEEV, WEEV.

Diagnostic Characters

(a) Habitus (♀): Pleural integument yellowish with or without a dark spot. **(b) Thorax (dorsal):** Scutum (Scu) with yellowish integument, with several dark spots (indistinct in photo). **(c) Abdomen (dorsal):** Terga (Te) with basolateral patches of white scales. **(d) Head (dorsal):** Vertex (V) narrow scales dark, whiteish laterally. **(e) Head (side):** Erect forked scales on vertex (V) all dark. **(f) Leg (III):** Ta-III$_{1-5}$ all dark. **(g) Larva (head):** Area between the two seta 1-C concave; seta 2-C present, medial to 1-C; seta 5-C long, with 5–7 aciculate branches. **(h) Larva (terminal segments):** Comb (C) with 13–25 short comb scales (CS) interspersed with 4–6 large spinelike scales; seta 1-S with six posterior and two anterior pairs of tufts. Image sources: WRBU; Belkin, Heinemann, & Page (1970).

Exemplar DNA sequences: None.

Catalog entry: Vol. 2, p. 923.

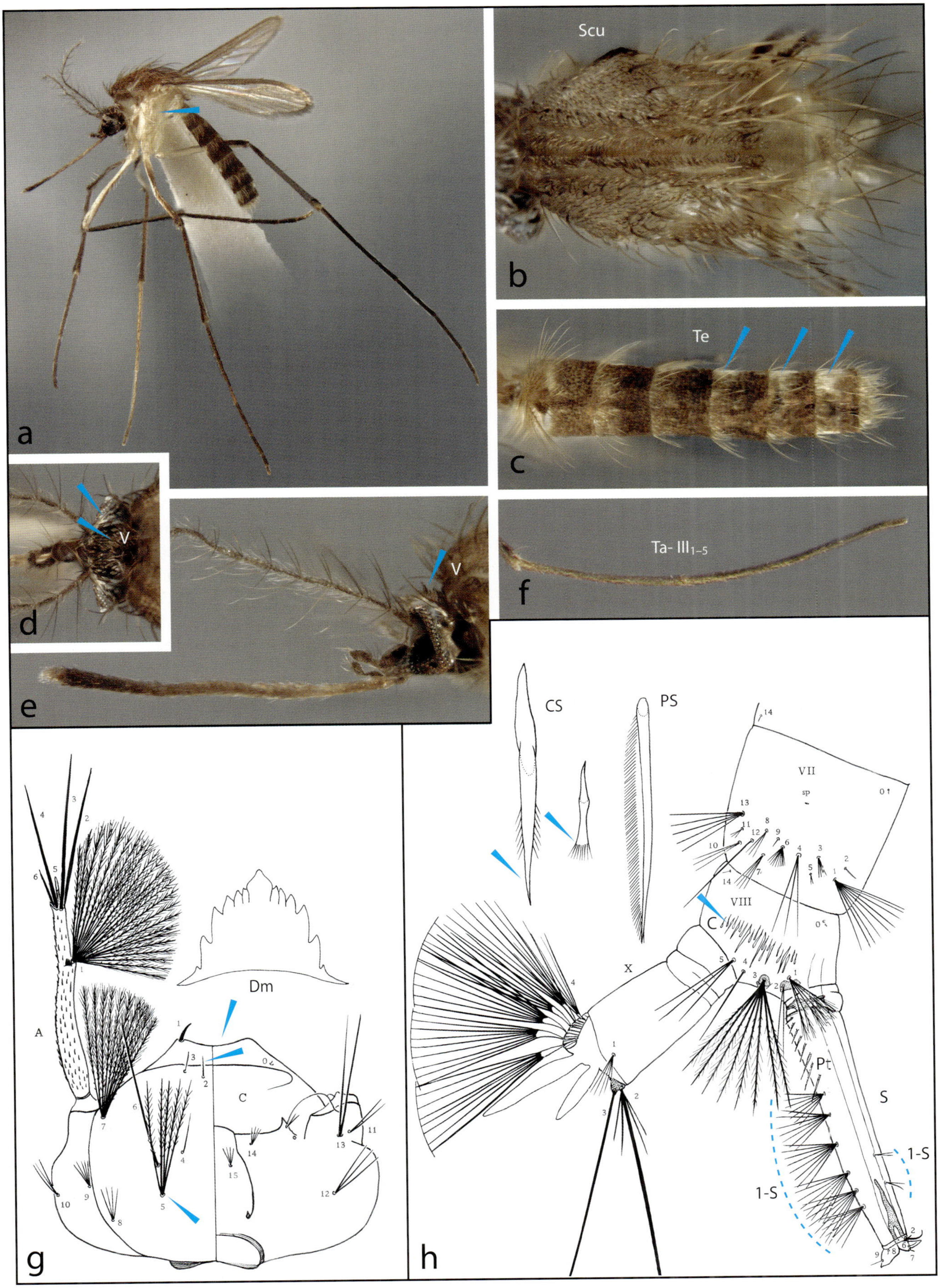
a
b
Scu
c
Te
d
V
e
V
f
Ta- III1-5
g
A
Dm
C
h
CS
PS
VII
sp
VIII
C
X
Pt
S
1-S
1-S

Culex (*Melanoconion*) *spissipes* (Theobald, 1903)
Subfamily Culicinae, Tribe Culicini
"Thick-footed Trinbagonian Typical Mosquito"

Systematics and Distribution

Culex spissipes is a medium-sized mosquito with distinctive all-dark legs. It is found throughout mainland Central and South America from Mexico to southern Brazil and Paraguay. Although it was first described from Trinidad and Tobago, it has not been reported from other Caribbean islands. The species is the only member of the Spissipes Group and nominotypical member of the Spissipes Section of the Neotropical *Culex* subgenus *Melanoconion* which comprises 23 species. *Culex spissipes* has four synonyms: *alvarezi* Sutil Oramas, Pulido Florenzano & Amarista Meneses, *fur* Dyar & Knab, *menytes* Dyar, and *haynei* Komp & Curry.

Bionomics

Biting activity in female *Cx. spissipes* peaks at midnight, with lesser peaks noted at dawn and dusk. Host-seeking females have been collected using Shannon traps at human bait in forested areas and at the margins of lakes and using CDC light traps supplemented with dry ice. Although detailed host-preference data for *Cx. spissipes* are not available, the species has been collected during human-landing catches and in mouse- and chicken-baited traps at night.

Culex spissipes females deposit their egg rafts directly on the water surface or on wet leaves. Preferred oviposition sites include heavily or partially shaded margins of permanent bodies of fresh water, including ground pools, lakes, and swamps. Sites typically have abundant emergent and floating vegetation or submerged and floating fallen leaves. Immature life-stage durations are related to ambient temperature, which appears optimal between 20°C and 31°C. Within this range, neonates emerge in 2 days, larvae take 13–15 days to fully develop, and adults emerge from the pupal stage in about 3 days. Due to lack of available larval illustrations, diagnostic characters of the male genitalia are provided here instead.

Associated Pathogens

BIMV, BVSV, CARV, ITQV, ORIV, SLEV, TONV, VEEV.

Diagnostic Characters

(a) Habitus (♀). (b) Thorax (dorsal): Acrostichal setae (AcS) present. **(c) Thorax (side):** Mesokatepisternum without a patch of scales in upper corner (MScU). **(d) Head (dorsal):** Narrow decumbent scales of vertex (V) pale golden posteriorly, bronzy anteriorly, and erect forked scales all dark. **(e) Leg (III):** $Ta\text{-}III_{1\text{-}5}$ all dark. **(f) Genitalia (♂):** Gonocoxite (mesal). **(g) Genitalia (♂):** Gonostylus. **(h) Genitalia (♂):** Foliform leaflet of subapical lobe. **(i) Genitalia (♂):** Lateral plate of aedeagus. **(j) Genitalia (♂):** Subapical lobe of gonocoxite. **(k) Genitalia (♂):** Aedeagus. **(l) Genitalia (♀):** 9th tergum. **(m) Genitalia (♂):** 9th tergal lobe. Image sources: WRBU; Sallum & Forattini (1996).

Exemplar DNA barcodes: *Cx. spissipes* (KX779868–70).

Catalog entry: Vol. 2, p. 928.

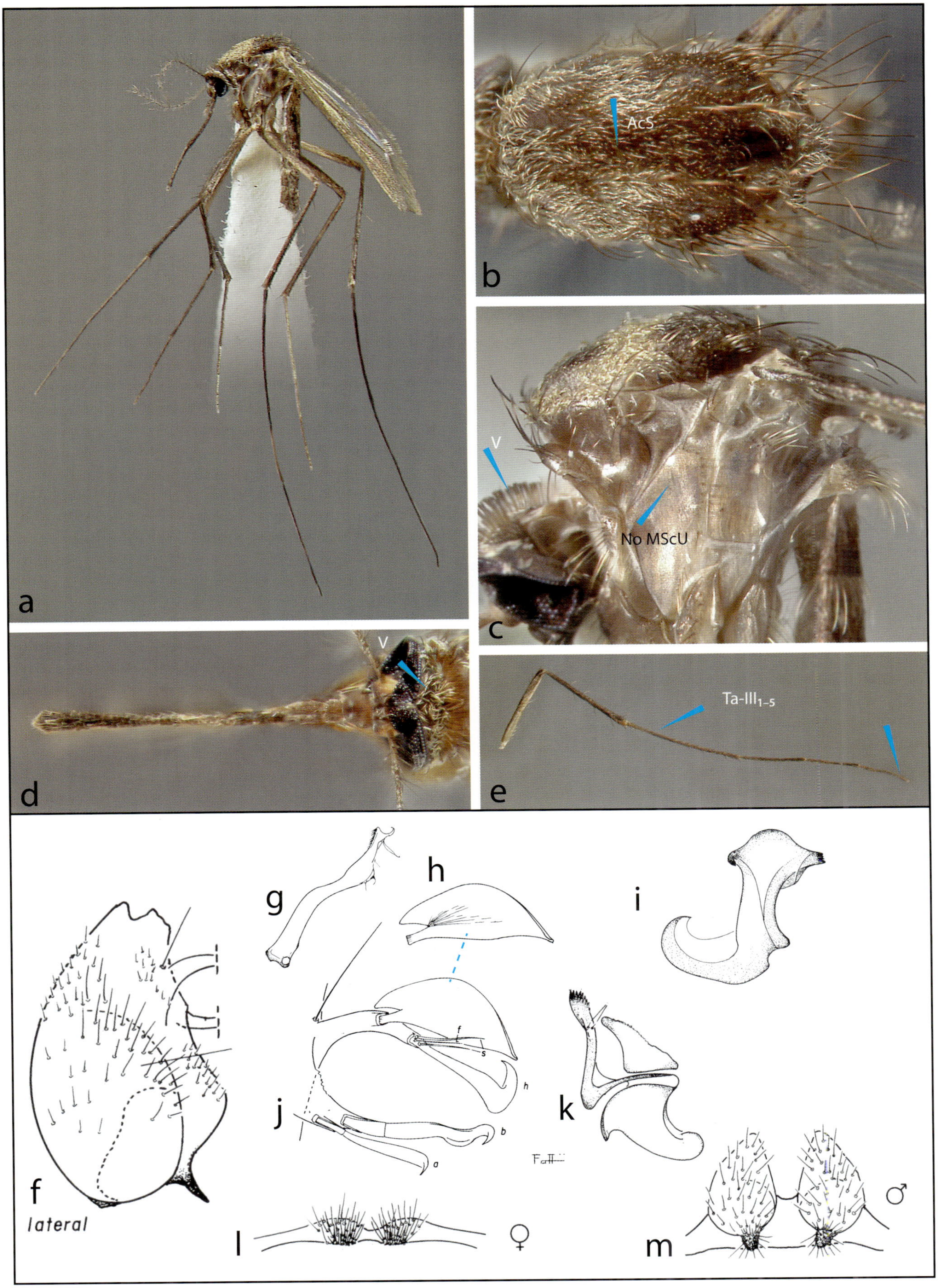

a
b
AcS
c
V
No MScU
d
V
e
Ta-III1–5
f
lateral
g
h
i
j
k
l
♀
m
♂

Culex (Melanoconion) taeniopus Dyar & Knab, 1907
Subfamily Culicinae, Tribe Culicini
"Ribbon-footed Nicaraguan Typical Mosquito"

Systematics and Distribution

Culex taeniopus is the nominotypical member of the Taeniopus Group of the Spissipes Section of the subgenus *Melanoconion* (see Vol. 2, p. 1094, for internal relationships), which also includes *Cx. akritos* Forattini & Sallum, *Cx. cedecei* Stone & Hair, and *Cx. ikelos* Forattini & Sallum. *Culex taeniopus* is a medium-sized brown mosquito with a somewhat shaggy scutum lightly covered with long setae, and an almost nude pleuron bearing only the lightest spattering of white scales. The hind tarsi have conspicuous basal and apical rings of white scales, and Ta-III_5 is entirely white, distinguishing it from *Cx. cedecei.* Its dark capitellum and lack of broad, spatulate, white postspiracular scales differentiates it from *Cx. akritos* and *Cx. ikelos. Culex taeniopus* was originally described from material collected in Bluefields, Nicaragua, and the species is found all the way from the southern United States throughout Central and South America.

Bionomics

Culex taeniopus adults have been collected resting in dense fully or partially shaded vegetation, in fissures in coralline limestone scarps, and in crabholes. The preferred hosts are rodents, but the females will also feed on both humans and birds. Adult females are rarely collected indoors but will readily feed on humans outdoors. They are collected in large numbers at human bait at and after sundown, and in CDC light traps. In blood-meal assessment, *Cx. taeniopus* appeared to show host preference toward the opossum (*Didelphis*), along with species of other bare-tailed marsupials (*Marmosa* and *Caluromys*) and squirrels. In Trinidad and Tobago, female biting activity has two distinct peaks, the first just after sunset (1800h–2000h) and the second late at night, between 0000h and 0200h. Elsewhere, only the first of these biting peaks has been reported. The species is most common near the larval habitats, but adults have been shown to migrate up to 7 km during periods of heavy rains in Venezuela, presumably to find suitable oviposition sites.

Immatures of *Cx. taeniopus* are typically found at the margins of swamps and rivers but have also been reported from stagnant water ground pools. In Venezuela, the species is most common in marshy lowland areas in association with the coquillo (*Manicaria saccifera*) and silica (*Raphia taedigera*) palms. Life table studies carried out in Jamaica (as *opisthopus*) and in Florida, showed that immatures develop slowly and that immature survival rates are negatively correlated with temperature.

Associated Pathogens

ANUV, BIMV, BSQV, CAPM, ENSV, EEEV, GMAV, GTRV, MAGV, MIRV, OSSAV, SLEV, VEEV.

Diagnostic Characters

(a) Habitus (♀): Ta-III_{1-4} with basal and apical white bands. **(b) Thorax (dorsal):** Scutal (Scu) scales mainly dark, some golden scales on prescutellar area. **(c) Head (dorsal):** Decumbent scales of vertex (V) and occiput (Occ) narrow, dark medially, gray to white laterally. **(d) Head (side):** Proboscis (P) entirely dark scaled. (e) **Thorax (side):** Postspiracular area without broad white scales (PoSc); upper mesokatepisternal scales (MScU) broad and white; capitellum (Ca) of halter (Hl) dark. **(f) Leg (III):** Ta-III_5 white. **(g) Larva (head):** Seta 1-C positioned at tip of projecting lobe of labrum. **(h) Larva (terminal segments):** Comb (C) with comb scales (CS) small with even fringe at apex; siphon (S) long, index about 17; seta 1-S with distal subdorsal tuft inserted near apex of siphon (S); seta 3-P <0.5 length of 1-P. **(i) Larva (thorax)**: Seta 3-P less than 0.5 size of 1-P. Image sources: WRBU; Belkin, Heinemann, & Page (1970, as *opisthopus*).

Exemplar DNA barcodes: *Cx. taeniopus* (JX259916).

Catalog entry: Vol. 2, p. 928.

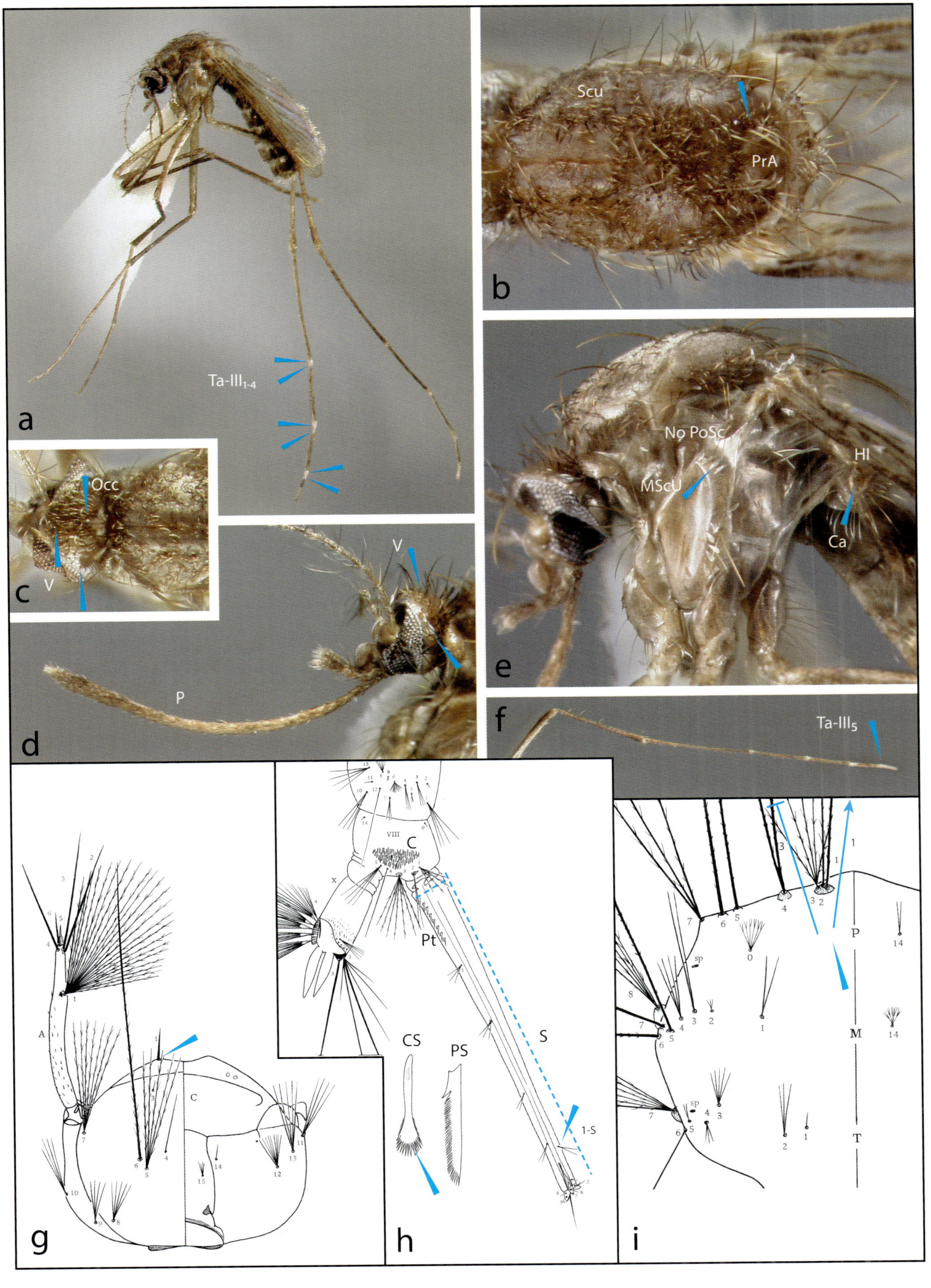
a
Ta-III1-4
b
Scu
PrA
c
Occ
V
d
V
P
e
No PoSc
MScU
Hl
Ca
f
Ta-III5
g
A
C
h
VIII
C
X
Pt
S
CS
PS
1-S
i
P
M
T

Culex (Melanoconion) vomerifer Komp, 1932

Subfamily Culicinae, Tribe Culicini

"Plowman Panamanian Typical Mosquito"

Systematics and Distribution

Culex vomerifer is a small dark-brown species with dark legs and four conspicuous dark spots on its light-colored pleura. The species is the nominotypical member of the Vomerifer Group (Sallum & Forattini 1996) in the Spissipes Section of the subgenus *Melanoconion* (see Vol. 2, p. 1094, for internal relationships). The Vomerifer Group also includes *Cx. gnomatos* Sallum, Hutchings, Leila & Ferreira, *Cx. portesi* Senevet & Abonnenc, and *Cx. sacchettae* Sirivanakarn & Jakob. *Culex vomerifer* and *Cx. gnomatos* are essentially isomorphic as adult females, with the only reliable distinguishing character being the absence of a patch of tiny, light-golden setae in the midmesepimeron in *Cx. vomerifer*. These species can more easily be distinguished by the characters of the male genitalia, which is plow shaped in *Cx. vomerifer*, giving this species its name. Given the difficulty in reliably separating *Cx. vomerifer* and *Cx. gnomatos*, these species are often lumped together in biosurveillance studies, likely confusing the medical significance of both taxa. *Culex gnomatos* has only been reported in Brazil and Peru, whereas the distribution of *Cx. vomerifer* includes Brazil, Colombia, Ecuador, French Guiana, Panama, Peru, Trinidad and Tobago, and Venezuela.

Bionomics

Culex vomerifer adults are common in forest edge habitats, including secondary growth vegetation in partially cleared forest areas or near mangroves and swamps. They have been collected both in the canopy and at ground level, suggesting that *Cx. vomerifer* actively utilizes both strata for host location. The species is readily attracted to CDC light traps and Disney traps, especially when baited with CO_2 or small rodents (mice, hamsters). The bionomic traits of *Cx. vomerifer* are seemingly at odds in the adult and immature stages. Despite being one of the commonest *Melanoconion* species collected as host-seeking females in human biting collections, its larvae are extremely difficult to locate in the field. As a result, no high-quality larval illustrations are available, thus characters of the male genitalia are used here to illustrate species diagnostic characters (see *e-g*). Immatures can be collected from shallow pools and in buttress root caves in lowland swamp margins in deep shade. Sites are normally highly organic, with colored waters from submerged leaves and woody debris. *Culex vomerifer* males were collected parasitized with *Arrenurus* and *Durenia* mites in the Atlantic Forest region of southern Brazil.

Associated Pathogens

ANUV, BSQV, CARV, GMAV, ITQV, MADV, MOJUV, Murutuca virus, OSSAV, PATV, VEEV, VINV.

Diagnostic Characters

(a) Habitus (♀). (b) Thorax (side): Vertex (V) with numerous narrow falcate scales; mesepimeron (Mm) with patch of small pale golden setae on median portion; pleural integument yellowish or whitish, contrasting with pale-brown to dark-brown scutal integument. **(c) Leg (III):** Ta-III$_{1-5}$ all dark. **(d) Thorax (dorsal):** Acrostichal setae (AcS) present on anterior promontory (AnP) and on posterior portion of acrostichal area (AA). **(e) Genitalia (♂). (f) Genitalia (♂): Lateral plate of aedeagus. (g) Genitalia (♂):** 9th tergal lobe. Image sources: WRBU; Sirivanakarn & Jacob (1982 (1981)).

Exemplar DNA barcodes: *Cx. vomerifer* (KX779893–95).

Catalog entry: Vol. 2, p. 930.

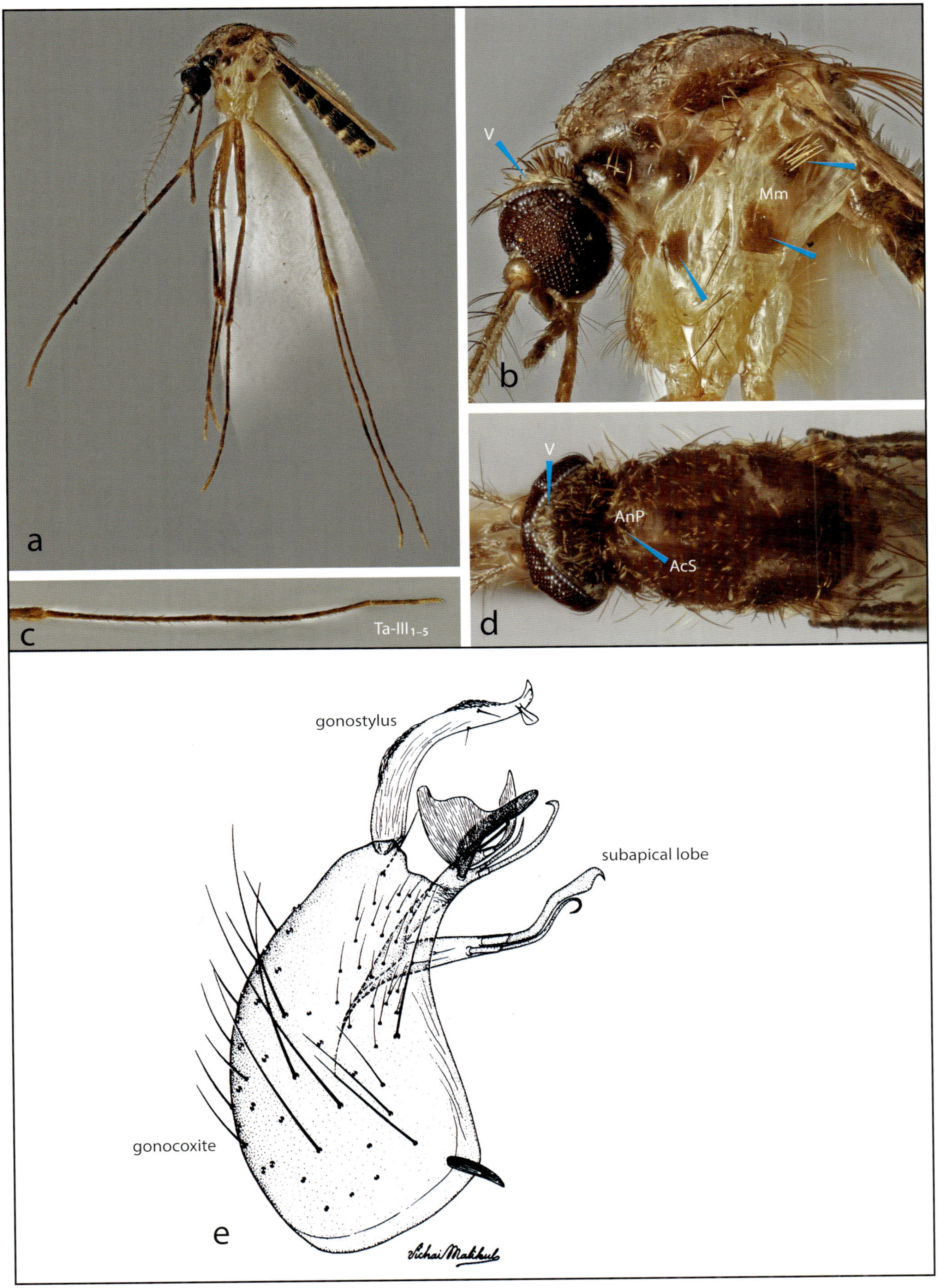
a
V
Mm
b
Ta-III1–5
c
V
AnP
AcS
d
gonostylus
subapical lobe
gonocoxite
e
Vichai Malikul

Culex (Oculeomyia) bitaeniorhynchus Giles, 1901
Subfamily Culicinae, Tribe Culicini
"Double-banded Indian Typical Mosquito"

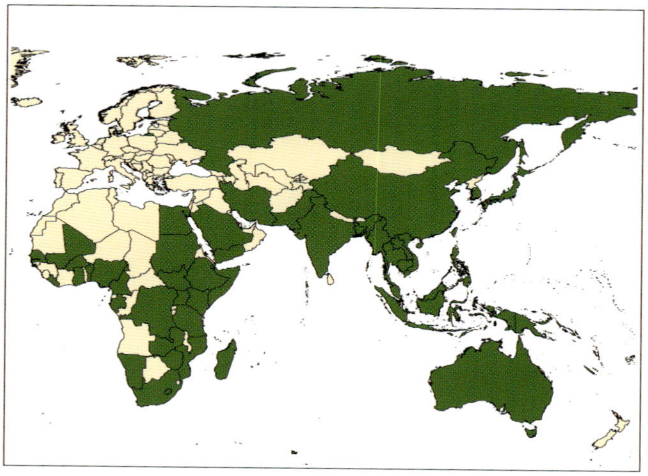

Systematics and Distribution

Culex bitaeniorhynchus is a medium-sized mosquito, with distinct apical bands of pale-yellow scales on the abdomen. The proboscis has a median pale band and two dorsolateral pale spots on the labellum. The wings and legs of this species are speckled with pale and dark scales. *Culex bitaeniorhynchus* is the nominotypical member of the Bitaeniorhynchus Complex of the subgenus *Oculeomyia*, which also includes *Cx. infula* Theobald, *Cx. longicornis* Sirivanakarn, *Cx. luzonensis* Sirivanakarn, *Cx. pseudosinensis* Colless, and *Cx. selangorensis* Sirivanakarn. *Culex bitaeniorhynchus* has the following four valid synonyms: *abdominalis* Taylor, *ager* Giles, *ethiopicus* Edwards, and *karatsuensis* Mochizuki. *Culex bitaeniorhynchus* has a wide-ranging distribution and is present in tropical and subtropical zones of the Afrotropical, southern Palearctic, and Indomalayan Regions and the mainland and islands of the Southeast Asian and Australasian Regions.

Bionomics

The biting habits of *Cx. bitaeniorhynchus* vary markedly by region but are viewed as primarily ornithophilic, including feeding on both wild and domestic birds in Thailand and elsewhere. The species also feeds on a wide variety of non-avian hosts, including cows, pigs, and dogs. In Malaysia, New Guinea, and Vietnam, females have been observed ferociously attacking humans; however, populations in Africa, southwestern Asia, and India appear not to be attracted to people at all. Adults begin host seeking just after sunset and are active throughout the night. In colony, females started to blood feed just two days after emergence.

Culex bitaeniorhynchus immatures are found in permanent or semipermanent large groundwater habitats in partial or full sunlight, including swamps, rice fields, marshy depressions, irrigation ditches, and ponds. Immatures have been collected at sites in or near forests, on plains, and in mountainous areas at a wide range of elevations. The key factor influencing oviposition is the presence of *Spirogyra* (water silk) algal mats, which serve as protection from predation and form the primary food source for the larvae. It has been shown that females that consume algal chlorophyll as larvae lay green eggs, whereas those that feed on decaying algae as larvae lay cream-colored eggs. Neonates emerge by forcing off a lower circular egg cap, often leaving the entire upper surface of the egg raft seemingly intact.

Associated Pathogens

AINOV, AKAV, BATV, DENV-3, GETBV, JBEV, KAIV, MVEV, RVFV, SAGV, SINV, TMTV, UMBV; *Brugia malayi*, *Plasmodium relictum*, *Wuchereria bancrofti*.

Diagnostic Characters

(a) Habitus (♀): Lower mesepimeral setae (MeSL) absent; Ta-I–III with basal pale bands. **(b) Leg:** Fe-I–III and Ti-I–III without rows of pale spots. **(c) Thorax (dorsal):** Acrostichal setae (AcS) present. **(d) Abdomen (dorsal):** Abdomen with distinct apical bands of pale-yellow scales. **(e) Head (side):** Proboscis (P) with median pale band and two dorsolateral pale spots at labellum (La). **(f) Wing:** With intermixed pale and dark scales. **(g) Larva (head):** Setae 2,3-A inserted apically on antenna (A); antennal spicules absent or sparse; seta 1-C slender; no demarcation between dorsal apotome (DAp) and median labral plate (MLP); dorsomentum (Dm) with fine serrations. **(h) Larva (terminal segments):** Comb scales (CS) 4–7; pecten spines (PS) <7; seta 1-S with four pairs of 1–4 branched tufts, positioned laterally or posterolaterally. Image sources: WRBU; Harbach (1988).

Exemplar DNA barcodes: *Cx. bitaeniorhynchus* (AB690839, AB738178, AB738227–28, DQ154162, HQ398898–99, JQ728034, JQ728200, JQ728349, KF406792–95, KT358430–32).

Catalog entry: Vol. 2, p. 938.

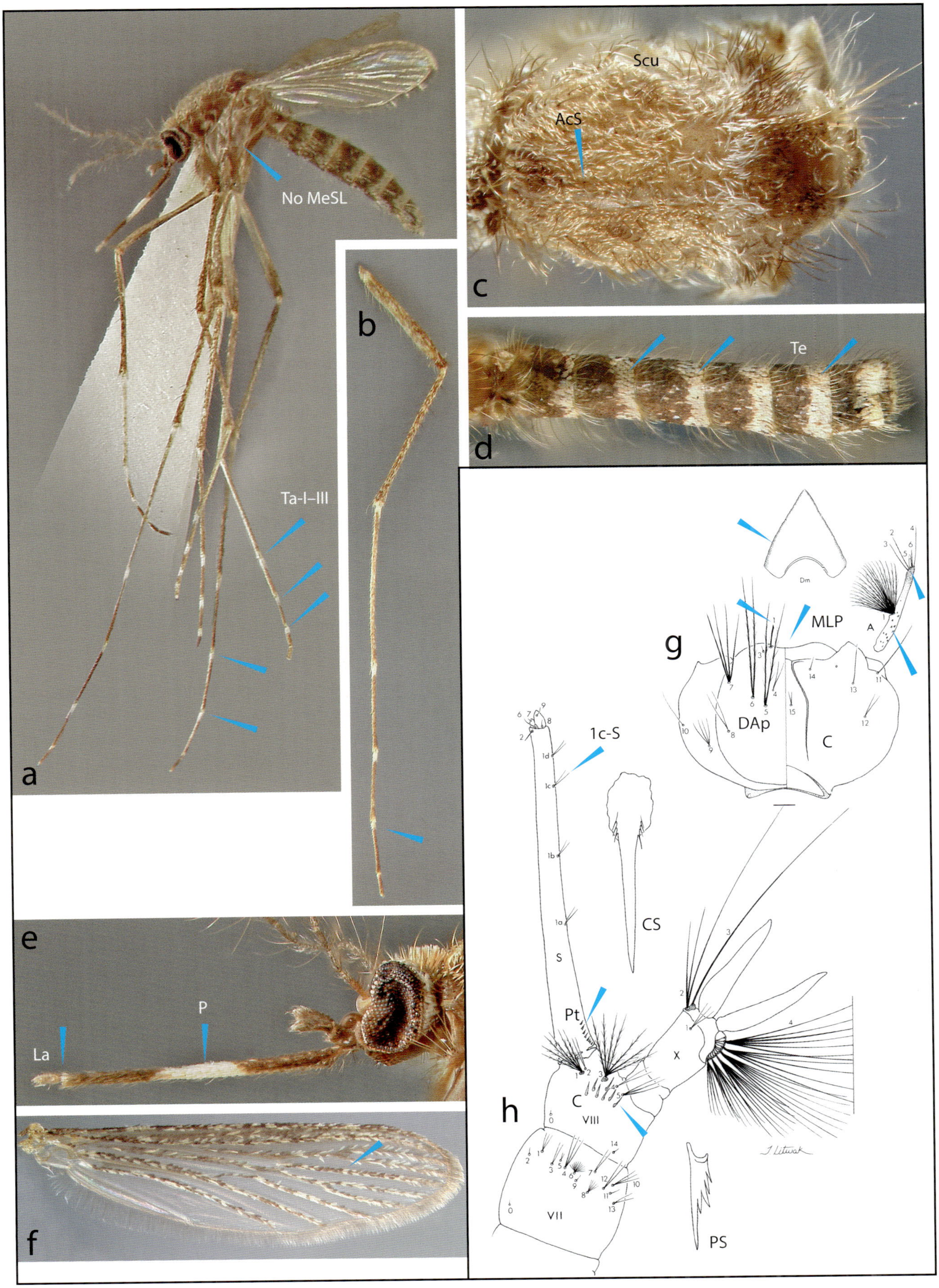
No MeSL
Ta-I–III
a
b
Scu
AcS
c
Te
d
e
La
P
f
Dm
MLP
A
g
DAp
C
1c-S
CS
S
Pt
X
C
VIII
h
VII
PS
J. Litwak

Culex (Oculeomyia) poicilipes (Theobald, 1903)
Subfamily Culicinae, Tribe Culicini
"Sock Nigerian Typical Mosquito"

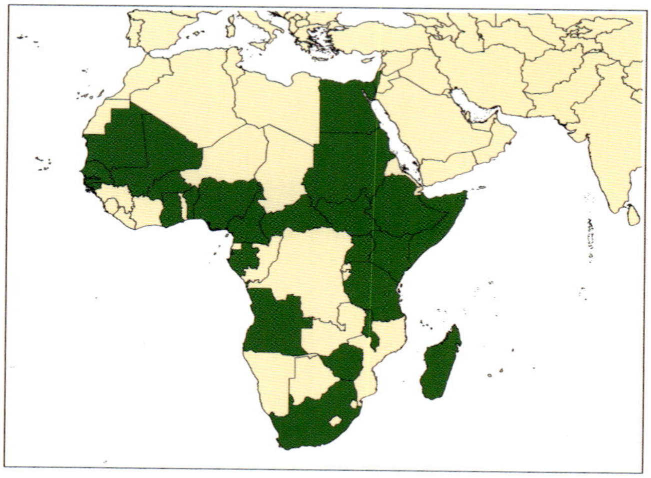

Systematics and Distribution

Culex poicilipes is a large dark species, readily recognized from all other species in the subgenus *Oculeomyia* by the attractive rows of pale spots that adorn its femora and tibiae. The proboscis is dark scaled, with a distinct median pale band. *Culex poicilipes* was originally described as *Lasioconops pocilipes*, and has seven valid synonyms, including: *auritaenia* Enderlein, *maculipes* Theobald, *madagascariensis* Ventrillion, *par* Newstead, *punctipes* Theobald, *quasigelidus* Theobald, and *taeniorhynchoides* Giles. *Culex poicilipes* is found throughout the tropical and subtropical areas of Africa and Madagascar, and has also been reported from Israel.

Bionomics

Immature habitats of *Cx. poicilipes* are primarily clean freshwater sites with abundant emergent or floating vegetation, including reeds, *Pistia*, and *Ceratophyllum* plants. These commonly include lakes, swamps, marshy depressions near rivers, and mature rice fields but can also include small ground habitats, such as ground pools and hoofprints, heavily shaded with grass and green algae. In Sahelian Regions, the species is commonly found in temporary groundwater pools. Rarely, larvae have also been collected in foul and turbid water. Larvae appear to rarely go to the surface for air, relying instead on air bubbles 60 to 120 cm beneath the surface. In Kenya, larvae have been found in greater numbers in rice fields with taller rice plants; however, numbers seem to decline when the rice reaches 100 cm tall. Populations tend to peak toward the end of the rainy season when ground pools are mostly covered by vegetation. Immature *Cx. poicilipes* is commonly found in association with *An. pharoensis* Theobald (see species page, p. 364), *Cx. antennatus* (Becker) (see species page, p. 488), and *Cx. perexiguus* Theobald (see species page, p. 496).

As adults, *Cx. poicilipes* is often the most abundant species in light traps set close to water bodies. Females of *Cx. poicilipes* readily bite humans across their range and will enter houses to feed in Egypt. They have also been collected biting cattle, sheep, goats, and chickens in Senegal, while in Kenya they are thought to be primarily ornithophilic. One blood meal sequenced from *Cx. poicilipes* in Kenya was identified as rat (*Rattus norvegicus*; KP858491). In KwaZulu-Natal, South Africa, this species was dominant in human-biting catches conducted on a 30-foot tree platform, suggesting it could also feed on arboreal fauna. Adult females bite mainly outdoors after sunset and at night. *Culex poicilipes* is the most common species collected in light traps set close to water in the Senegal River Basin, and arboviral screening revealed 36 strains of RVFV in wild *Cx. poicilipes* collected there, evidencing its role as a major RVFV vector in the Afrotropics.

Associated Pathogens

BBKV, BAGV, Dianke virus, OLIV, RVFV, Sanar virus, WNV.

Diagnostic Characters

(a) Habitus (♀). **(b) Thorax (side):** Lower proepisternal scales (PScL) present; mesepimeral setae (MeS) absent. **(c) Legs (I–III):** Fe-I–III and Ti-I–III with rows of pale spots anteriorly; Ti-II without pale stripe; Ta-I–III with narrow basal pale rings. **(d) Head (dorsal):** Proboscis (P) with median pale band. **(e) Larva (abdominal segment I):** Seta 7-I about as long as 6-I, usually single (double shown). **(f) Wing:** Dark scaled. **(g) Larva (head):** Dorsal apotome (DAp) distinctly separated from median labral plate (MLP); seta 1-C stout, thicker than branches of setae 5,6-C; seta 14-C single. **(h) Larva (terminal segments):** Comb (C) with ≤10 comb scales (CS); CS spinelike, with pointed apex and proximal fringe; siphon (S) curved in lateral view; pecten (Pt) with pecten spines (PS) on basal 0.3 of siphon; seta 1-S 2.5–3.0 times width of siphon. Image sources: WRBU; Harbach (1988).

Exemplar DNA sequences: None.

Catalog entry: Vol. 2, p. 939.

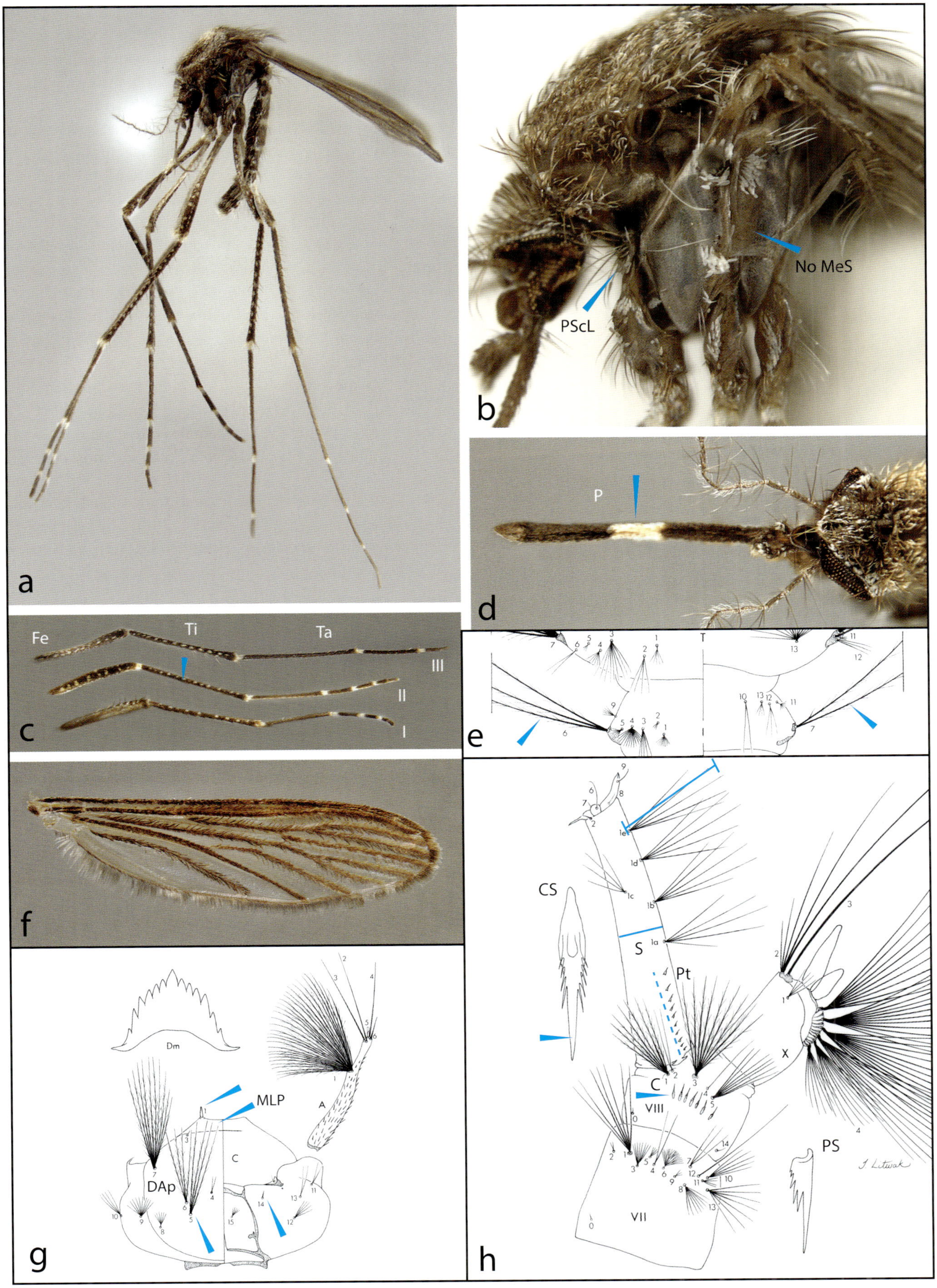

a
b
PScL
No MeS
c
Fe
Ti
Ta
III
II
I
d
P
e
f
g
Dm
MLP
A
C
DAp
h
CS
S
Pt
C
VIII
X
PS
VII

Culiseta (Climacura) melanura (Coquillett, 1902)
Subfamily Culicinae, Tribe Culisetini
"Black-tailed American Cool Mosquito"

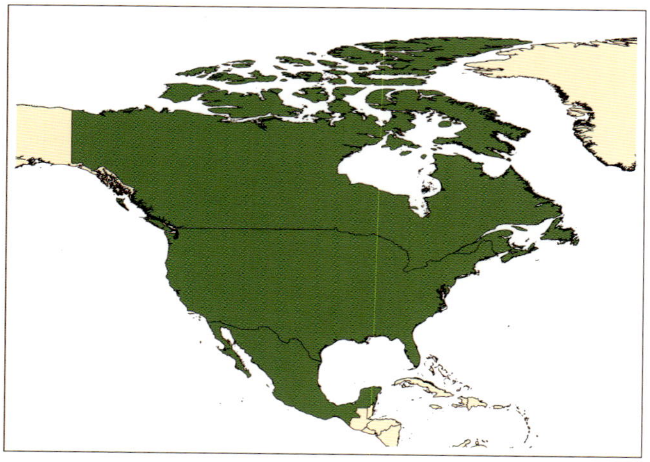

Systematics and Distribution

Adults of *Culiseta melanura* are chocolate-brown, medium-sized mosquitoes with little in the way of obvious distinctive markings. The species is one of six in the subgenus *Climacura*, which also includes *Cs. antipodea* Dobrotworksky (Australia), *Cs. marchettei* Garcia, Jeffery, & Rudnick (Cambodia, Malaysia), *Cs. novaezealandiae* Pillai (New Zealand), *Cs. taiwanica* Lien, Lin, & Weng (Taiwan), and *Cs. tonnoiri* (Edwards) (New Zealand). It is the sole representative of this subgenus in North America (Canada, United States, and Mexico).

Bionomics

Detailed knowledge of the adult biology of this species is woefully lacking. The species feeds preferentially on songbirds and plays a significant maintenance role in the bird cycle of Eastern equine encephalitis virus (EEEV). The species occasionally feeds on reptiles and other forest-associated mammals but rarely attacks people unless they are near to the larval habitats. Adults are often represented in light trap collections placed in forested areas and in artificially placed "resting boxes" but have never been reported inside human dwellings. Reported life strategies for the species in different regions in northern America strongly indicate the presence of an as yet undetermined cryptic species within this taxon. The species is multivoltine in all zones, but *Cs. melanura* populations in the North persist through essential female diapause, which is not reported in southern populations. In New Jersey, diapause has been noted both for females and for larvae, whereas *Cs. melanura* populations in Rhode Island survive in the winter as larvae, and female diapause has not been reported. In spring, summer, and fall, immatures of *Cs. melanura* are readily collected in groundwater habitats, including swamps, floodplains, ground pools, and other small, permanent water bodies. Females oviposit in cool, acidic waters (c. pH 5.0).

Associated Pathogens

EEEV, FLAV, FMV, HPV, SLEV, WEEV, WNV.

Diagnostic Characters

(a) Habitus (♀): Proboscis (P) entirely dark scaled. **(b) Abdomen (side):** Sternal scales cream colored or yellowish, sometimes with dark scales intermixed. **(c) Leg (III):** Ta-III dark. **(d) Thorax (side):** Antepronotum (Ap) without scales or setae; prespiracular area with 1–5 dark setae (PsS); lower paratergite (Pa) with 1–6 slender setae; paratergite scales (PaSc) absent. **(e) Wing:** Costa (C) entirely dark scaled. **(f) Larva (head/thorax):** Antennae at least as long as head; seta 1-A, multiple, aciculate, inserted in notch at outer 0.25 of shaft and extending much beyond tip. **(g) Larva (terminal segments):** Comb (C) with about 25 large comb scales (CS) arranged in distinct bar-like row; seta 1-S with 8–16 tufts in midventral line, starting within pecten (Pt) and ending near apex of siphon (S); siphon with small dorsolateral setae; siphon index 6–7; a few precratal setae (PcrS) present. Image sources: WRBU; Carpenter & LaCasse (1955).

Exemplar DNA barcodes: *Cs. melanura* (JX259942–81, JX260675).

Catalog entry: Vol. 2, p. 951.

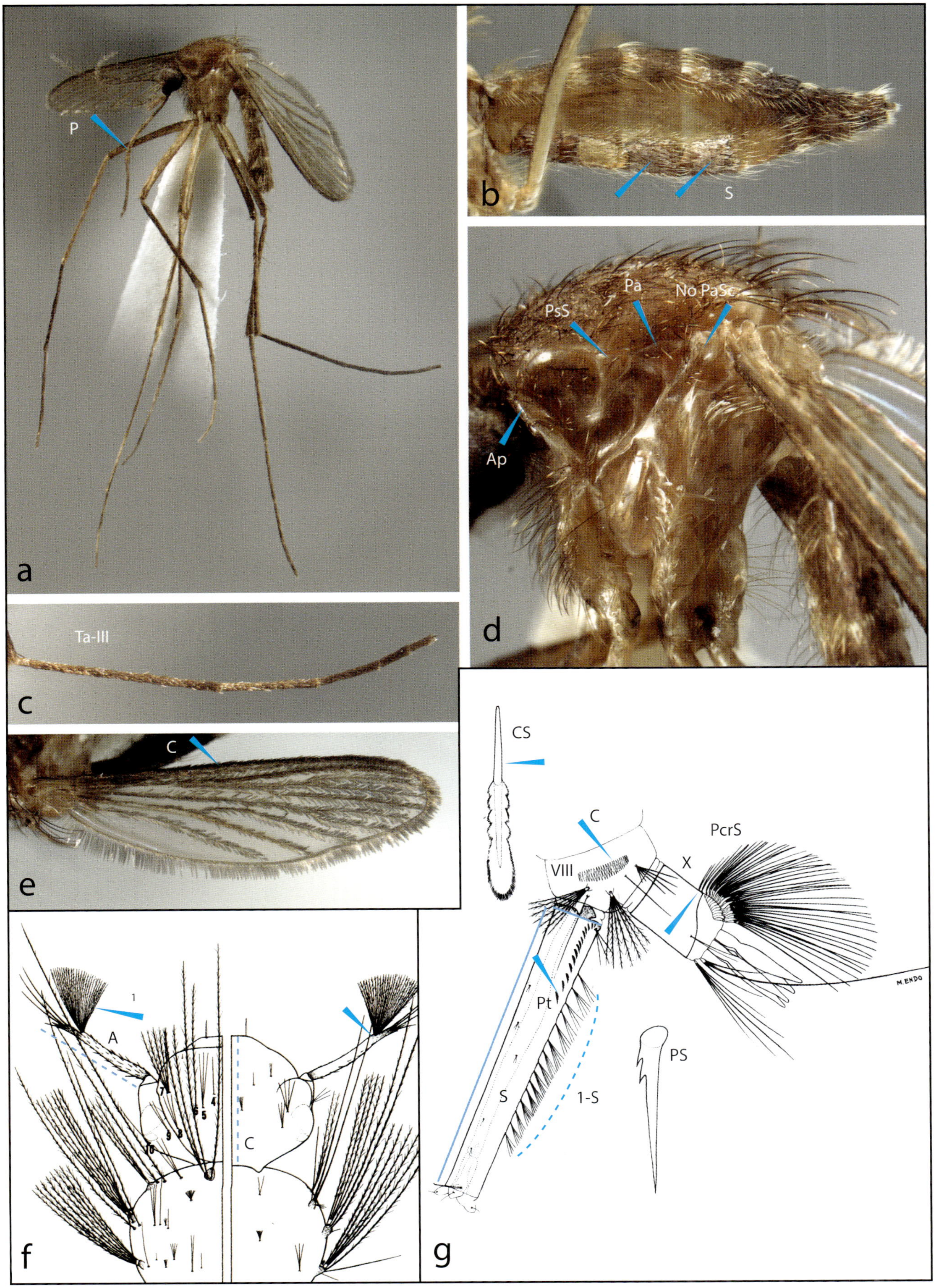
P
a
b
S
Pa
PsS
No PaSc
Ap
d
Ta-III
c
C
e
1
A
7
6 5 4
9 8
10
C
f
CS
C
PcrS
VIII
X
Pt
S
1-S
PS
M. ENDO
g

Culiseta (*Culiseta*) *inornata* (Williston, 1893)
Subfamily Culicinae, Tribe Culisetini
"Unadorned American Cool Mosquito"

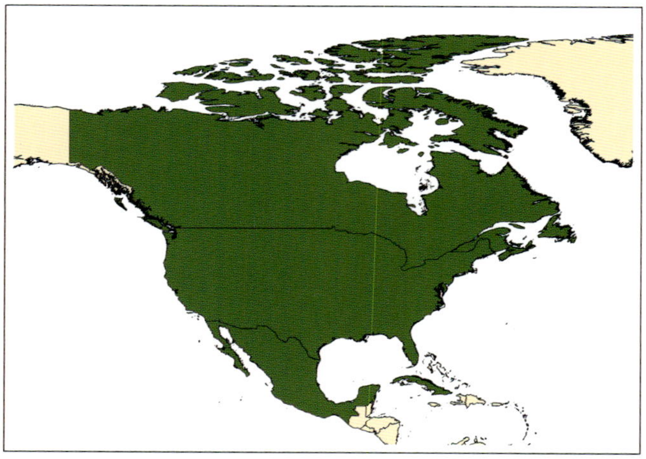

Systematics and Distribution

Culiseta inornata is a large mosquito found throughout the mainland United States and into western Canada, and southward to Mexico and Cuba. As its name suggests, it is a somewhat drab mosquito, with no distinct banding patterns or otherwise obvious diagnostic characters. *Culiseta inornata* is one of 13 species placed in the subgenus *Culiseta*, and has one valid synonym, *magnipennis* Felt.

Bionomics

Immature *Cs. inornata* is found in ephemeral pools of temporary water, including puddles, ground pools, and ditches. Occasionally, immatures are also found occupying artificial water containers, including neglected swimming pools. Water in the chosen habitats is often grossly polluted. Larvae have also been collected in coastal brackish water marshes in California, and they are capable of completing their immature development in dilutions of seawater ranging from 50–700 mos mol/kg.

Culiseta inornata adults are active in the cooler months of the year (September–June), producing multiple generations per year (multivoltine). In the southern United States, females are active all through the winter and disappear through the hot summer months. Females estivate as adults in cellars and barns, but males do not survive the summer. Estivating populations include either females that have laid at least one batch of eggs (parous) or those that are fully gravid. Larvae are extremely cold-hardy and have been recovered from ice-covered ground pools in northern parts of the species' range.

Female *Cs. inornata* mosquitoes feed almost exclusively on horses and ruminants, such as deer and cows, rarely feeding on people, birds, or pigs. They are fierce biters and can be troublesome to livestock in wooded areas in early spring. Adults are active at night and are readily collected in CDC light traps. *Culiseta inornata* is stenogamous and has been observed mating tip-to-tip on fallen trees close to lakes.

Associated Pathogens

CEV, CVV, FMV, JCV, JEV, JSV, MDV, SSHV, WEEV, WNV.

Diagnostic Characters

(a) Habitus (♀). (b) Thorax (side): Pale scales present on paratergite (PaSc) and postspiracular area (PoSc). **(c) Legs:** Fe-I–III, Ti-I–III, and Ta-I–III$_{1,2}$ distinctly speckled with cream-colored scales; Ta-III$_5$ entirely dark. **(d) Wing:** Wing with speckled brown-and-white scales on three anterior veins. **(e) Larva (head):** Antenna (A) shorter than head length; seta 1-A inserted near midpoint of antenna; seta 4-C 3–5 branched; seta 6-C distinctly longer than 5,7-C. **(f) Larva (abdominal segments):** Seta 1-VI,VII as long as seta 1-V. **(g) Larva (terminal segments):** Siphon (S) index about 4; seta 1-S with even row of long setae distal to pecten (Pt); saddle (Sa) smooth, without large spicules at apex; seta 1-X 1–5 branched, as long or longer than saddle (Sa). Image source: WRBU.

Exemplar DNA barcodes: *Cs. inornata* (HQ569463, JX259935–41, JX260672–74).

Catalog entry: Vol. 2, p. 955.

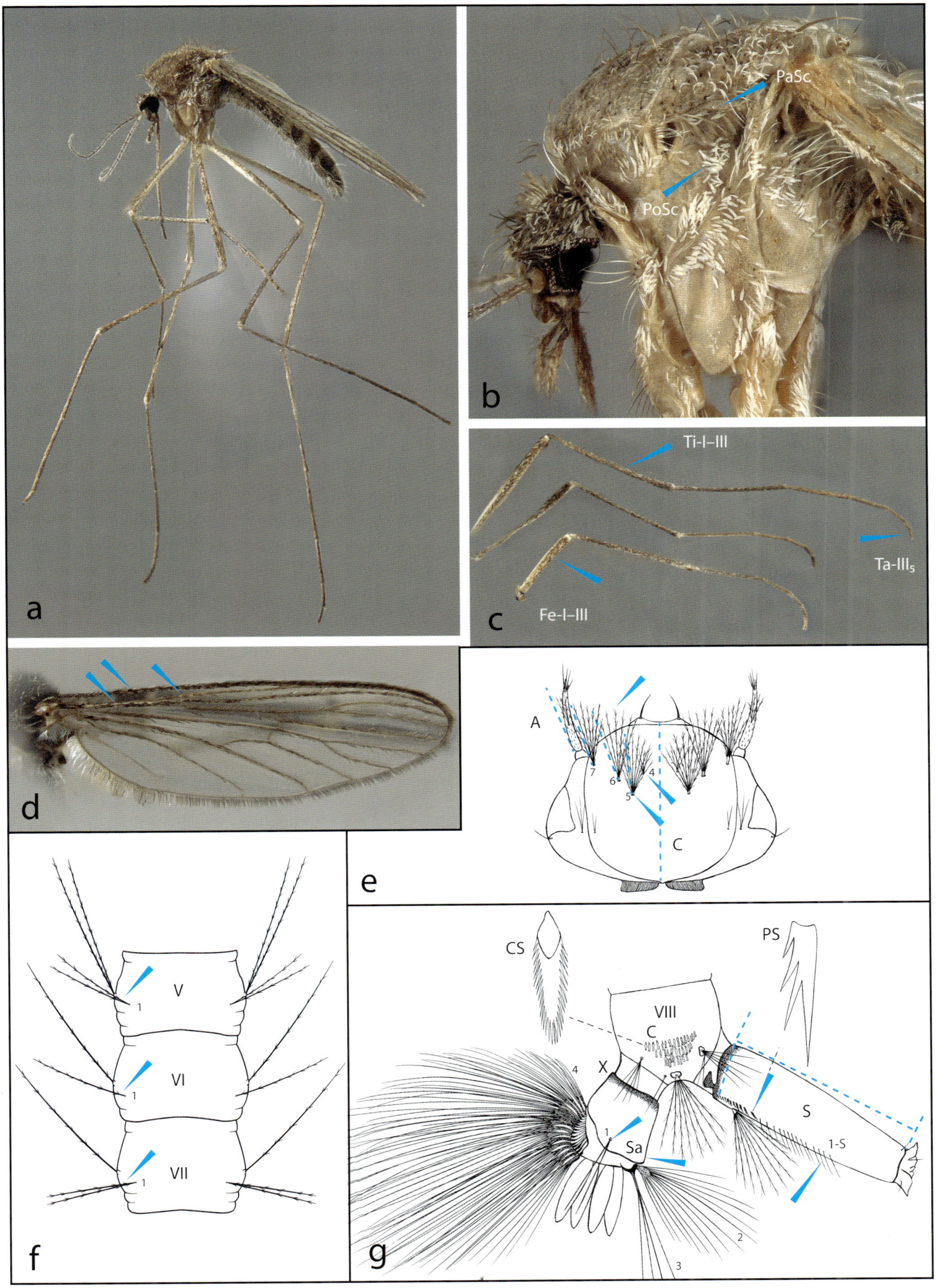
a
PaSc
PoSc
b
Ti-I–III
Ta-III$_5$
Fe-I–III
c
d
A
7
6
5
4
C
e
V
VI
VII
1
f
CS
PS
VIII
C
X
4
1
Sa
S
1-S
2
3
g

Coquillettidia (*Coquillettidia*) *perturbans* (Walker, 1856)
Subfamily Culicinae, Tribe Mansoniini
"Agitated American Underwater Mosquito"

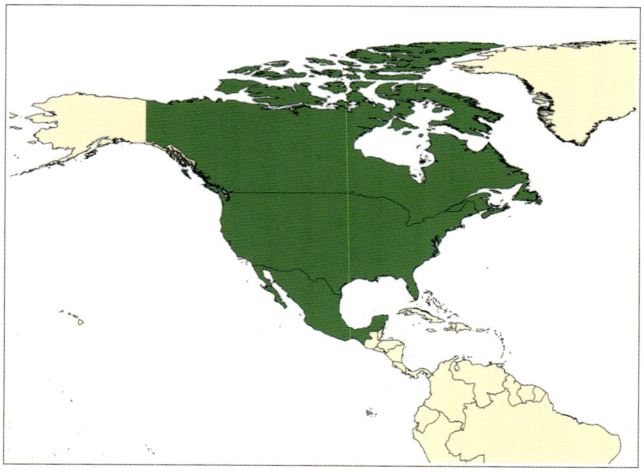

Systematics and Distribution

Coquillettidia perturbans is an attractive mosquito, dusted with a mix of broad brown, white, and black scales. *Coquillettidia perturbans* serves as the type species of the genus (as *Culex perturbans* Walker), and has two valid synonyms—*ochropus* Dyar & Knab, and *testaceus* van der Wulp. This is the only species of the genus *Coquillettidia* known to occur in North America. *Coquillettidia perturbans*, along with all members of the subgenus *Coquillettidia* and *Austromansonia*, lacks postspiracular setae, a character that easily differentiates it from *Mansonia* species and other aedines in its range. Common in forest and marshland habitats in the eastern United States and southern Canada, *Cq. perturbans* has a patchier distribution in the western continental United States and Mexico.

Bionomics

Persistent and painful biters, with a strong proboscis capable of penetrating clothing, the females can be a severe nuisance to both humans and animals where abundant. Although *Cq. perturbans* is strongly anthropophilic, females feed opportunistically on a wide variety of wild and domestic birds (e.g., chickens, quail), wild mammals (e.g., armadillos, rabbits, raccoons, opossums), and domestic mammals, including cattle. Females actively host seek in the early evening but are known to bite during the day in shady spots. Adults are phototropic and are readily captured in light traps. Adults are strong fliers, able to travel up to five miles from their immature habitats. Obligatory diapause occurs in the larval stage, which can last up to nine months in northern parts of their range, and after which synchronized emergence of the adults occurs in late spring. Climatic conditions also have an impact on yearly generations of *Cq. pertubans*. Populations are univoltine in Canada, whereas two to three overlapping generations are noted in the warmer climes of Florida, with densities peaking from June through September.

Females deposit high numbers of eggs per oviposition ($n = 150–350$), which are laid in a single raft on or near the leaves of emergent aquatic plants rooted in thick, water-logged, humus-rich soils. Associated aquatic plants include *Calla* spp. (water arum), *Carex* spp. (sedges), *Juncus* spp. (rushes), *Nymphaea* spp. (water lily), *Phragmites* spp. (reeds), *Pontederia* spp. (pickerelweed), *Sagittaria* spp. (arrowhead), and *Typha* spp. (cattails). Larval stages of *Cq. perturbans* feed primarily on algae and spiral-shaped bacteria. Neonates submerge and pierce the submerged roots or stems of these plants with their highly sclerotized, sawlike siphons for respiration. Larvae are difficult to collect, as they quickly detach from the roots and burrow into the soil when disturbed. Pupae attach to the plants using modified respiratory trumpets, where they remain until the imago is ready to emerge.

Associated Pathogens

EEEV, TENV, TVTV, WNV.

Diagnostic Characters

(a) Habitus (♀). (b) Abdomen: VIII-Te without chitinized hooks. **(c) Thorax (side):** Postspiracular setae (PS) absent. **(d) Leg (III):** Ta-III$_1$ with pale median ring. **(e) Abdomen (dorsal):** Abdomen blunt. **(f) Wing:** Scales medium to broad, symmetrical. **(g) Genitalia (♀) (external):** Cercus (Ce) unmodified; IX-Te bilobed. **(h) Larva (terminal segments):** Saddle (Sa) ventrally with or without 1–2 very small setae; seta 1-VIII long and multibranched; seta 1-X long and multibranched; spiracular apparatus (SAp) modified for piercing plants. **(i) Larva (head/thorax):** Antenna (A) very long, at least twice length of head; two short setae distal to seta 1-A. Image sources: WRBU; Carpenter & LaCasse (1955); Ronderos & Bachmann (1963 (1962)).

Exemplar DNA barcodes: *Cq. pertubans* (JF867703–04, JF867742–43, JF867810, JF867850, JF868118, JN302425, JX259890–97).

Catalog entry: Vol. 2, p. 969.

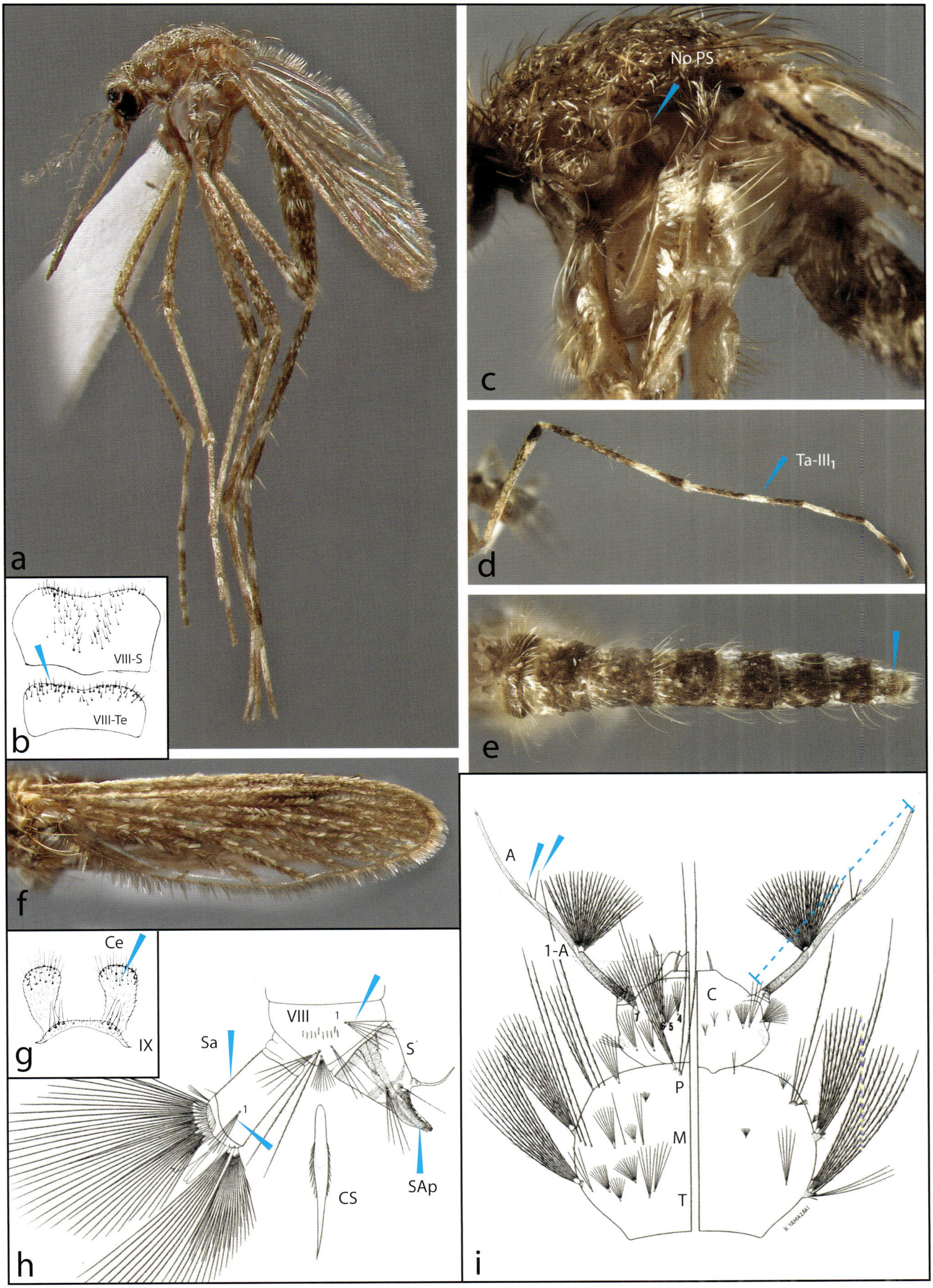
a
No PS
c
Ta-III1
d
VIII-S
VIII-Te
b
e
f
Ce
IX
g
VIII
1
Sa
S
1
CS
SAp
h
A
1-A
C
7
6 5 4
P
M
T
K YAMAZAKI
i

Mansonia (Mansonia) titillans (Walker, 1848)
Subfamily Culicinae, Tribe Mansoniini
"Ticklish Brazilian Submarine Mosquito"

Systematics and Distribution

Mansonia titillans, the type species of the genus (as *Culex titillans* Walker), is a medium-sized, dark mosquito that can be a severe biting nuisance in some parts of its distribution. The species is reported in Central and South America, Trinidad and Tobago, the Antilles and Caribbean islands, and into the southeastern corner of the United States, where its distribution is somewhat scattered. Dense populations flourish in Florida, where they peak in late fall.

Bionomics

Mansonia mosquitoes are strong fliers, and *Ma. titillans* has been collected several miles from suitable immature habitats. Both males and females of this species are attracted to light and are often captured in CDC light traps. Female *Ma. titillans* feeds preferentially on mammals, including humans, and is easily attracted to animal-baited traps. It also opportunistically feeds on a wide range of animals, including birds and even alligators in Florida, facilitating the cross-species transfer of zoonotic pathogens. In Guatemala, 90% of all blood meals were from equines and cattle. In the United States, the species geographically overlaps with and is often confused with the morphologically and biologically similar species, *Ma. indubitans* Dyar & Shannon and *Ma. dyari* Belkin, Heinemann & Page.

Mansonia titillans is closely associated with marshes, lakes, or large freshwater ponds that support high densities of aquatic vegetation, the stems and roots to which *Mansonia* larvae and pupae attach to facilitate respiration. Water lettuce (*Pistia* sp.) is the principal plant host, although the species has been collected associated with water hyacinth (*Eichhornia crassipes*). In Panama, *Pistia* plants are rare, but *Ma. titillans* populations are plentiful; thus, other, as yet unidentified, plant hosts must be suitable for this species. The females oviposit underwater, using the spines on the VIII-Te (see *i*) to rasp the underside of the submerged/floating leaves for egg adhesion. The rosette-like egg masses are enveloped in an air bubble and transported under the leaves by the female abdomen.

Associated Pathogens

BSQV, CABV, KRIV, SLEV, TLAV, TONV, VEEV, WNV; *Dirofilaria immitis*.

Diagnostic Characters

(a) Habitus (♀). (b) Thorax (side): Postspiracular setae (PS) present; supraalar area (SaA) with elongate scales with simple apices. **(c) Head (side):** Proboscis (P) with a distinct postmedian white band; Flm_1 with patch of broad dark scales medially (not distinct in photo). **(d) Wing:** With distinct white speckling, scales broad and asymmetrical. **(e) Legs (I–III):** Fe-I–III speckled, without distinct preapical white band anteriorly. **(f) Abdomen (♀):** VII-Te with row of short spiniform setae. **(g) Leg (III):** $Ta\text{-}III_1$ without a median pale ring. **(h) Larva (terminal segments):** Comb scales (CS) with a strong primary spine and a few small spinules; seta 1-VIII short and multibranched; modified spiracular apparatus (SAp) for piercing plants; seta 1-X short, multibranched; seta 4-X with four precratal tufts (PcrS). **(i) Abdomen (♀):** VIII-Te with group of robust spines posteriorly. **(j) Larva (head):** Antenna (A) apically articulated, long with two setae as long as basal part of antenna. Image sources: WRBU; Belkin, Heinemann, & Page (1970).

Exemplar DNA barcodes: *Ma. titillans* (KT766518–37, KY859898–902).

Catalog entry: Vol. 2, p. 974.

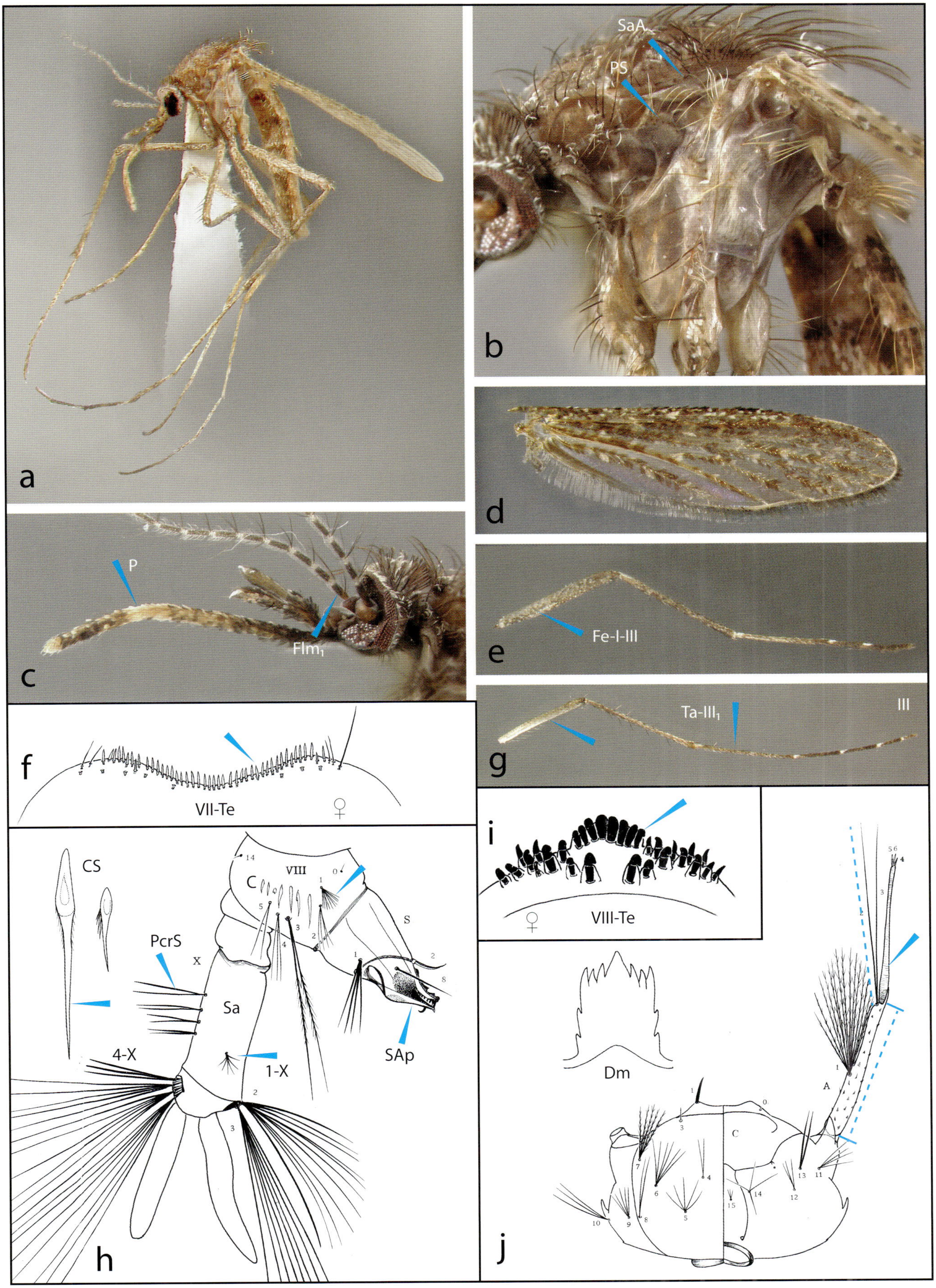
a
SaA
PS
b
d
P
Flm$_1$
c
Fe-I-III
e
Ta-III$_1$
III
g
f
VII-Te
♀
CS
PcrS
C
VIII
S
X
Sa
4-X
1-X
SAp
h
i
♀
VIII-Te
Dm
A
C
j

Mansonia (Mansonioides) indiana Edwards, 1930
Subfamily Culicinae, Tribe Mansoniini
"Yawadvipa Indonesian Submarine Mosquito"

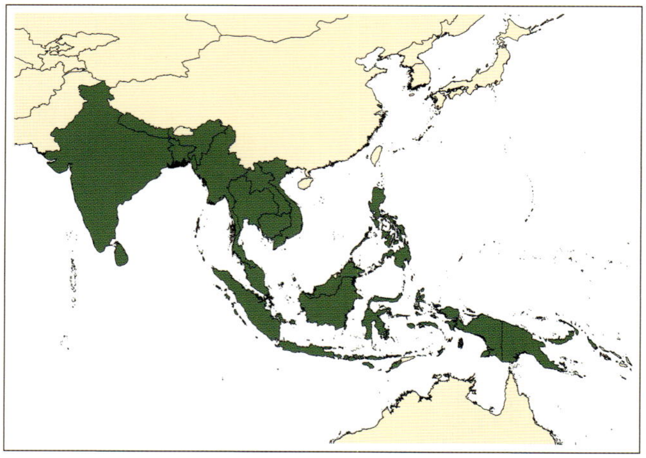

Systematics and Distribution

Mansonia indiana has a salt-and-pepper appearance caused by a dusting of broad white scales on its body and wings. The species was originally described from the Indonesian island of Java, and its distribution stretches from India and Nepal in the West through the mainland countries and islands of Southeast Asia to New Guinea. It is one of 10 described species in the subgenus *Mansonoides*.

Bionomics

Immature *Ma. indiana* is most commonly associated with coastal rice fields but is also found in open swamps. Like other *Mansonia*, this species uses specially modified siphons as larvae and modified trumpets as pupae to pierce stems and roots of aquatic vegetation to obtain oxygen for respiration Immature life cycles are reportedly similar to *Ma. uniformis* (see species page, p. 542), with which this species is sympatric.

Little is known about the adult bionomics of *Ma. indiana*. Females are crepuscular, biting at dawn and dusk, most commonly outdoors. In Thailand, the species will enter houses but avoids biting indoors. A study on host preference in Kerala, India, showed that *Ma. indiana* feeds preferentially on people, but blood meals were also detected from a range of domestic animals, including sheep/goats, cattle, and pigs. Mixed ovid and bovine blood meals were reported.

Associated Pathogens

JBEV; *Wuchereria malayi*.

Diagnostic Characters

(a) Habitus (♀). (b) Thorax (dorsal): Scutum (Scu) without well-defined round spots or greenish stripes. **(c) Thorax (side):** Pleuron not contrasting with scutum (Scu) or coxae (C-I–III); postpronotum (Ppn) with appressed white scales. **(d) Wing:** Wing scales mixed dark and pale, broad and asymmetrical. **(e) Leg (III):** Fe-III with c. five bands of pale scales; pale bands incomplete on all tarsomeres. **(f) Head (dorsal):** Palpus (MPlp) with apical white scales. **(g) Larva (terminal segments):** Seta 3-VIII double; comb (C) with two robust blunt comb scales; modified spiracular apparatus (SAp) for piercing plants; saddle (Sa) two times longer than wide; precratal setae (PcrS) present. **(h) Larva (head). Other characters (not shown):** III–VII-S mostly pale scaled; VIII-Te with strong chitinized hooks (see *Mansonia titillans* species page, p. 538); larval antenna with dark rings at base and at origin of antennal shaft; seta 6-III,IV single. Image sources: WRBU; Barraud (1934).

Exemplar DNA barcodes: *Ma. indiana* (KX816496–502).

Catalog entry: Vol. 2, p. 975.

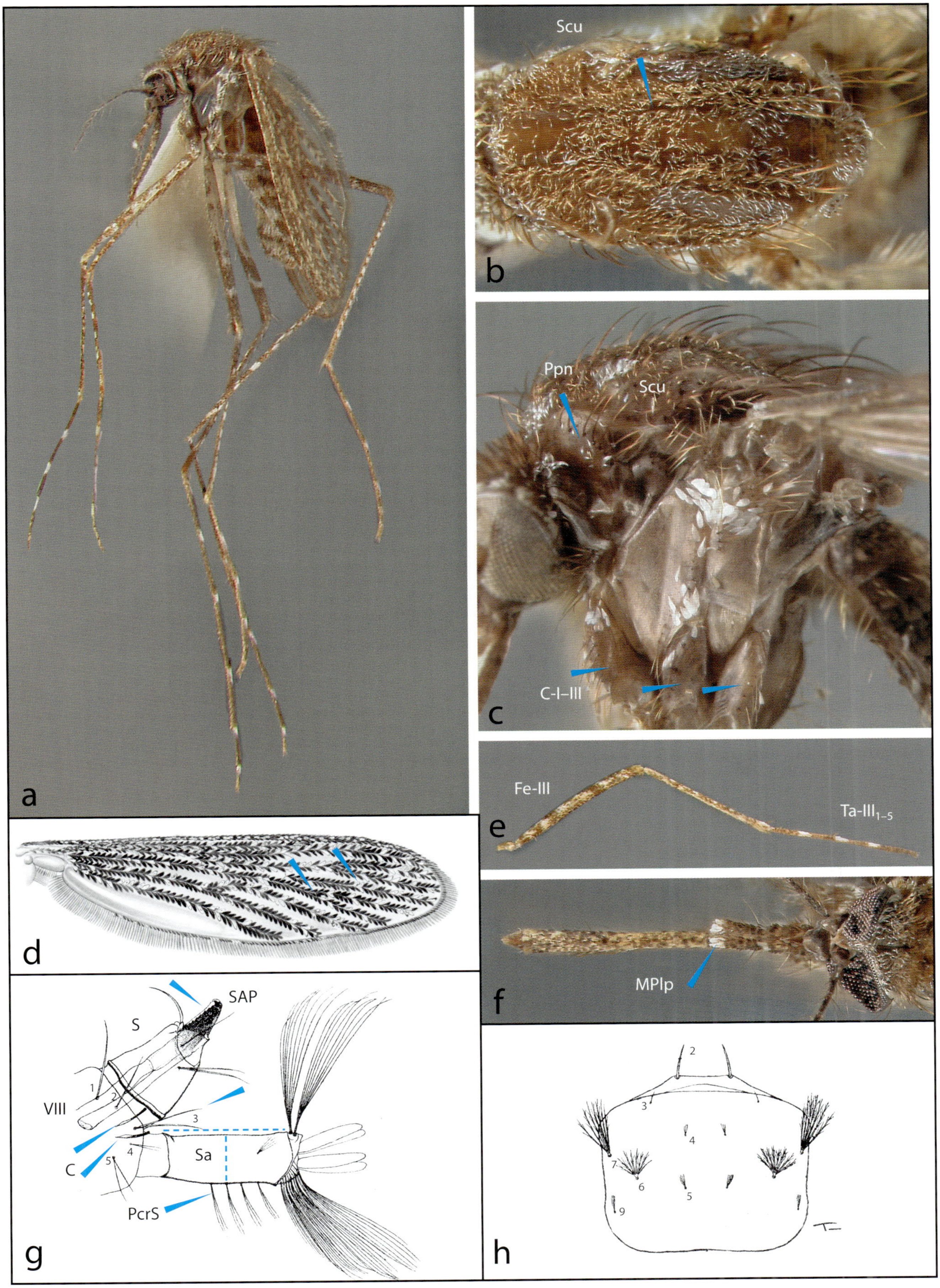
a
Scu
b
Ppn
Scu
C-I–III
c
d
Fe-III
Ta-III$_{1-5}$
e
MPlp
f
SAP
S
VIII
1
2
3
4
5
C
Sa
PcrS
g
2
3
4
7
6
5
9
h

Mansonia (Mansonioides) uniformis (Theobald, 1901)
Subfamily Culicinae, Tribe Mansoniini
"Uniform Indian Submarine Mosquito"

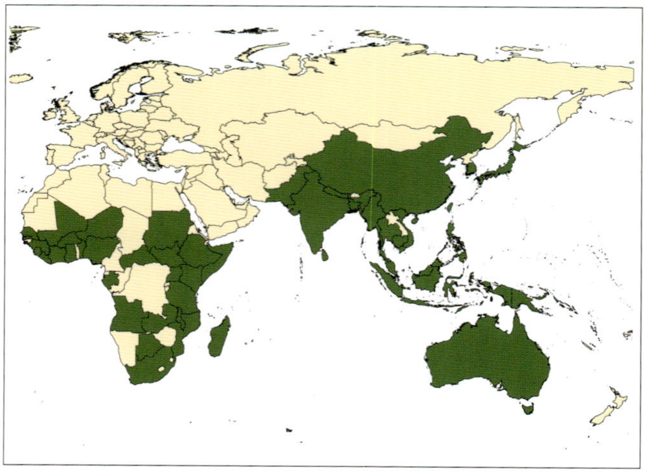

Systematics and Distribution

Mansonia uniformis is an extremely scaly, medium-sized mosquito that has a vast distribution, from sub-Saharan Africa through Pakistan, India, the mainland and islands of Southeast Asia, the Pacific islands, and across to northern Australia. Originally described from India, this species is exceptionally common and widely distributed in sub-Saharan Africa, where it is commonly collected with *Ma. africana*. *Mansonia africana* is restricted to the continent of Africa. Mating, without insemination, has been noted between these species in the laboratory. Crosses of African *Ma. uniformis* with populations from Malaysia and Sri Lanka are viable, but crosses between the African strains and populations from farther afield showed marked reduction in egg viability. These findings, along with our as yet unpublished DNA barcoding results, support cryptic speciation within *Ma. uniformis* across its range. *Mansonia uniformis* has three recognized synonyms: *australiensis* Giles, *marquesensis* Dyar, and *reversus* Theobald.

Bionomics

Mansonia uniformis is strongly exophagic, feeding primarily on cattle outdoors. Peak biting occurs between 1900h and 2000h, just following sunset. The species frequently bites people and a swathe of other mammals, including primates. In Ganibeia, Illubabor Province, Ethiopia, *Ma. uniformis* was the most common species in human bait collections, with peak populations coinciding with periods of high rainfall. *Mansonia uniformis* is stenogamous and mates readily in small cages. The females look for their first blood meal within 24 hours of emergence and oviposit within three to five days, immediately seeking another blood meal. The species is multivoltine, and population densities can increase rapidly. The species is considered a notable vector of several pathogens and viruses, and its importance is heightened by its high densities and close association with people.

Mansonia immatures use their modified siphons and trumpets to pierce the roots and stems of aquatic vegetation for respiration. In nature, immature *Ma. uniformis* is found closely associated with large unshaded swamps, although the species has readily adapted to occupy man-made rice field habitats, where it is often the dominant taxa collected. In Kenya, *Ma. uniformis* was shown to breed throughout the rice cycle, seemingly unaffected by changes in water depth. The species is a strong flier, with flights detected up to 3.5 km (average, 1.5 km) from the breeding sites.

Associated Pathogens

BUNV, JBEV, KAVV, MALV, NDUV, ONNV, PGAV, PUCV, RRV, SPOV, YATAV, YFV; *Brugia patei*, *B. malayi*, *B. pahangi*, *Dirofilaria repens*, *Plasmodium cynomolgi bastianellii*, *Wuchereria bancrofti*.

Diagnostic Characters

(a) Habitus (♀): Palpus (MPlp) with apical yellowish scales. **(b) Thorax (dorsal):** Scutum (Scu) with pair of longitudinal greenish stripes, without well-defined round spots. **(c) Abdomen:** VIII-Te with strong chitinized hooks. **(d) Thorax (side):** Postpronotum (Ppn) with narrow, curved scales; C-I–III not contrasting with pleuron (p) or scutum (Scu). **(e) Leg (III):** Fe-III with approximately five bands of pale scales; Ta-III_{1-5} pale bands complete posteriorly. **(f) Wing:** Wing scales broad and asymmetrical, mixed dark and pale. **(g) Abdomen (side):** III–VII-S mostly pale scaled. **(h) Larva (head/thorax).** **(i) Larva (terminal segments):** Seta 3-VIII double; two robust blunt comb scales (CS); saddle (Sa) <2 times as long as wide; modified spiracular apparatus (SAp) for piercing plants. **Other characters not shown:** Larva: Antenna with dark rings at base and at origin of antennal shaft; seta 6-III,IV single. Image sources: WRBU; Tanaka, Mizusawa, & Saugstad (1979).

Exemplar DNA barcodes: *Ma. uniformis* (KX816486–95, MF179284–85, MG712562–64, MK033254).

Catalog entry: Vol. 2, p. 976.

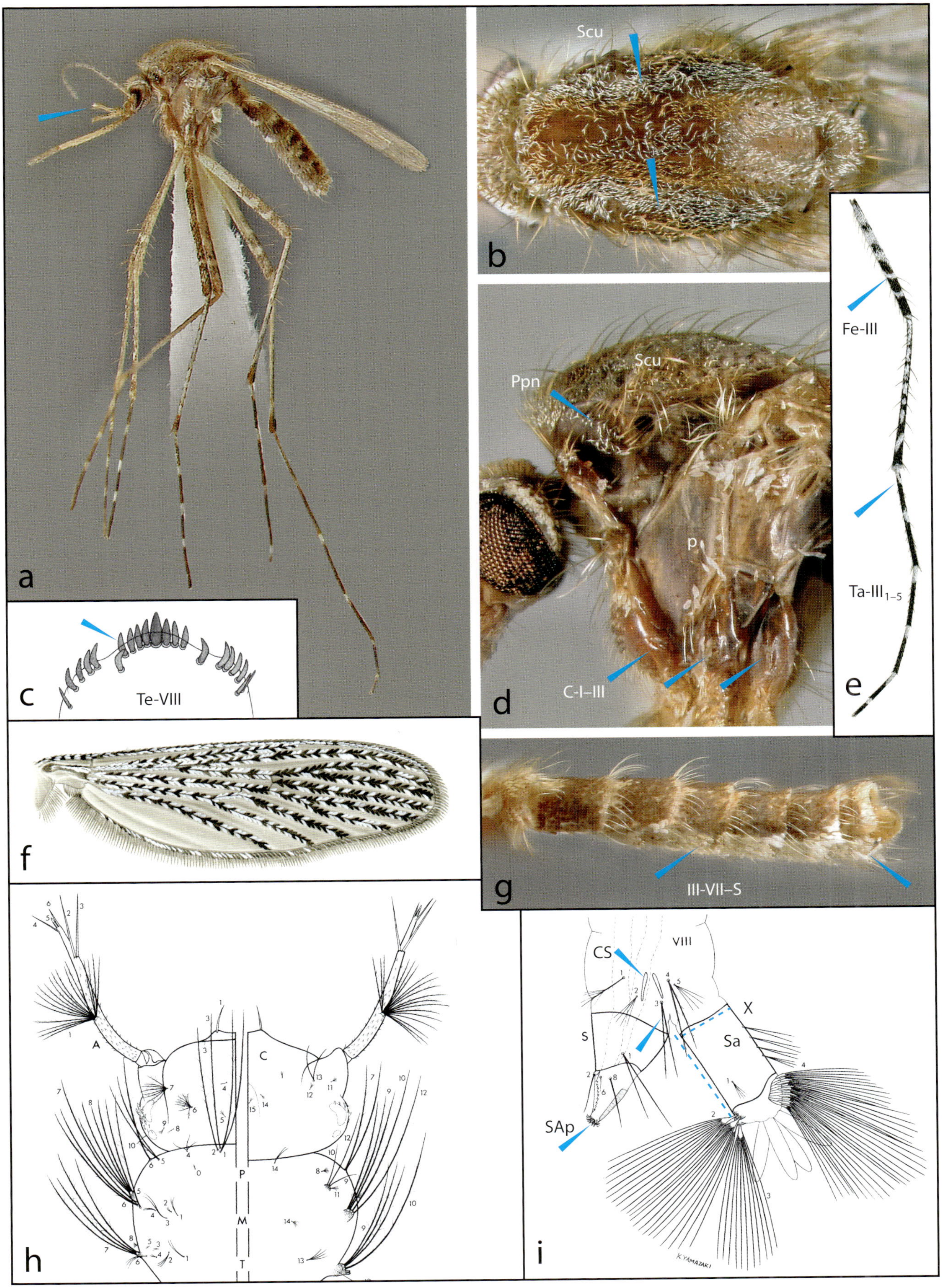
a
Scu
b
Fe-III
Ppn
Scu
p
Ta-III$_{1-5}$
C-I–III
d
e
c
Te-VIII
f
g
III-VII–S
h
A
C
P
M
T
i
CS
VIII
X
S
Sa
SAp
K.YAMASAKI

Sabethes (Sabethoides) chloropterus (von Humboldt, 1819)
Subfamily Culicinae, Tribe Sabethini
"Green-winged Ecuadorian Canopy Mosquito"

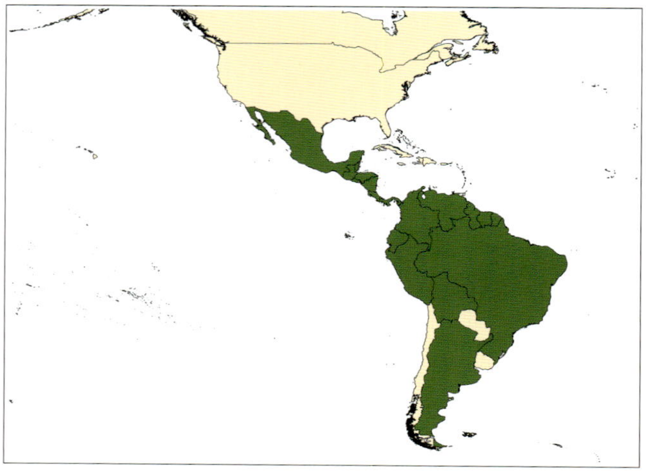

Systematics and Distribution

Sabethes chloropterus is a stunning mosquito, clothed with a copious covering of gold, green, and white scales and featuring green wings, from which it takes its name. The proboscis is exceptionally long and, in common with other taxa in the subgenus *Sabethoides*, its leg setae are not modified into paddles. The mesopostnotum is well developed, giving the mosquito a distinctly "humped" appearance. The species, one of only four in the subgenus *Sabethoides*, has three valid synonyms—*confusus* Theobald, *imperfectus* Bonne-Webster & Bonne, and *rangeli* Surcouf & Gonzales-Rincones. *Sabethes chloropterus*, an acrodendrophile, is found year-round in treetop habitats in dense, humid forest environs across Central and Latin America, from Mexico to northern Argentina and southern Brazil, including Trinidad and Tobago.

Bionomics

Adults are exclusively diurnal, with highest biting activity occurring in the early afternoon (1400h–1500h). Adults feed on mammals, including monkeys high in the canopy, but also feed readily on people at ground level. Females have occasionally been collected in CDC light traps and at human bait. *Sabethes chloropterus* always bites high on the body of humans, favoring the nose and lips. The species is stenogamous, with females perching on vertical surfaces waiting to be noticed by patrolling males searching for potential mates. Males perform elaborate courtship rituals lasting up to 34 minutes and involving highly regimented oscillations of the midleg before copulation occurs. Without this courtship, copulation will not take place.

Sabethes chloropterus females lay rhomboid eggs and strongly favor natural containers, particularly those with small openings and large water-holding capacity, such as tree rot holes and insect-bored (not cut) standing bamboo. Females hover in front of the holes, shooting one or two eggs into the highly protected habitat. Larvae are both predaceous and cannibalistic, sucking the body contents of their prey through puncture holes they create in the cuticle.

Associated Pathogens

CHGV, ILHV, SLEV, YFV.

Diagnostic Characters

(a) Habitus (♀): Covered with multicolored metallic scales; leg setae not modified into "paddles"; abdomen much longer than Fe-III; Fe-III, and Ti-III combined shorter than length of all tarsomeres; Fe-II shorter than Fe-I; Ta-II largely white scaled ventrally. **(b) Habitus (♀) (side, close):** Proboscis (P) ≤ length of Fe-I; upper proepisternal setae (PeSU) present; lower mesokatepisternal seta (MkSL) present; prealar setae (PaS) absent; lower mesokatepisternal seta (MkSL) present; upper mesepimeral setae (MeSU) short, not reaching to middle of mesopostnotum; mesopostnotum large; I–VIII-Te scales metallic greenish blue, violet basally and pale yellow or silvery white basolaterally; I–VIII-S pale. **(c) Head/thorax (dorsal):** Vertex (V) with violet or bluish scales, not green; antepronotum (Ap) large, nearly contiguous dorsally; antepronotal and scutal scales metallic green or bluish, appearing golden in some lights. **(d) Larva (head):** Seta 4-C mesal to 1-C. **(e) Larva (thorax/abdominal segments):** Seta 1-I mesal to seta 2-I. **(f) Larva (terminal segments):** Comb scales (CS) small and close together with fringe of minute spicules, not on a sclerotized plate; siphon (S) long and narrow; seta 1-S 2.5–3.0 times width of siphon; seta 1-X double; seta 4-X 3,4 branched. Image sources: WRBU; Harbach (1991); Harbach & Peyton (1993).

Exemplar DNA barcodes: *Sa. chloropterus* whole mitochondrial genome (MF957172).

Catalog entry: Vol. 2, p. 995.

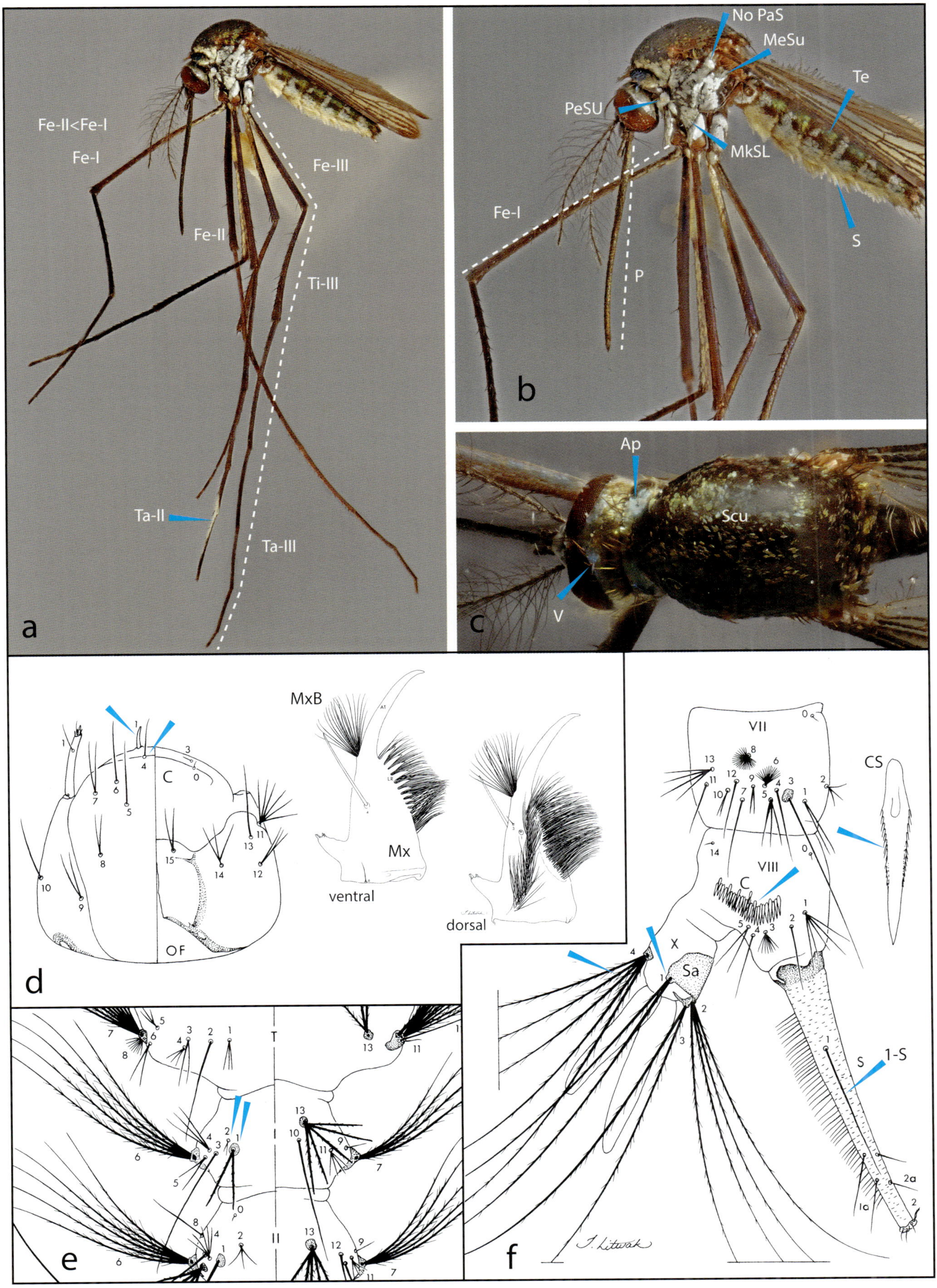
Fe-II<Fe-I
Fe-I
Fe-II
Fe-III
Ti-III
Ta-II
Ta-III
a
No PaS
MeSu
Te
PeSU
MkSL
Fe-I
P
S
b
Ap
Scu
V
c
MxB
Mx
ventral
dorsal
C
OF
d
T
I
II
e
VII
CS
VIII
C
X
Sa
S
1-S
f

Trichoprosopon digitatum (Rondani, 1848)
Subfamily Culicinae, Tribe Sabethini
"Digital Brazilian Hairy-lipped Mosquito"

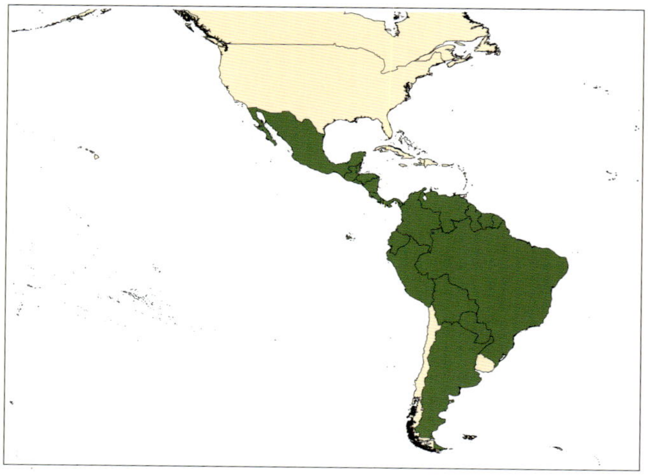

Systematics and Distribution

The instantly recognizable *Trichoprosopon digitatum*, with its purple sheen and white hindleg "bobby socks," is extremely common in forest and forest edge habitats of South and Central America. The species, one of 13 in the New World sabethine genus *Trichoprosopon*, has one valid subspecies—*townsendi* Stone—and four valid synonyms: *nivipes* Theobald, *splendens* Lutz, *subsplendens* Martini, and *wilsoni* Ludlow.

Bionomics

Unique within the typically diurnal Sabethini, *Tr. digitatum* instead seeks its hosts under the cover of darkness and is more active during darker periods in the lunar cycle. *Trichoprosopon digitatum* adults are readily collected in Shannon traps but are always more numerous in traps augmented with light or gas lanterns. They are found resting both outdoors and within human habitations. Adults can be trapped both at ground level and in the canopy, suggesting that the species moves freely in forested environs, seeking hosts and suitable oviposition sites. Under field conditions, diel oviposition was noted throughout the day (0500h–1800h).

Trichoprosopon digitatum is one of only a handful of mosquitoes that are known to guard their eggs. Females lay 60–80 eggs in a loose raft and stand over the raft, guiding it under their bodies with their legs for up to 24 hours until the eggs hatch. This is thought to serve as a strategy to prevent "washout" of the eggs from the pluvial downpours common in the tropics. The species uses a wide vertical range of the forest for oviposition, and *Tr. digitatum* immatures have been collected in "ovipots" placed at ground level, and at 5 m, 8 m, 25 m, and 34 m up into the canopy. The species is extremely plastic in oviposition choice, utilizing a wide variety of natural and artificial container habitats, including young floral bracts of *Heliconia aurea*, fallen fruit husks, internodes in water collected in dead bamboo, bamboo stumps, fallen palm bracts, tree holes in standing and fallen trees, rainwater pockets in a folded tarpaulin on the forest floor, and discarded water receptacles. Cacao husks support high densities of *Tr. digitatum*, and larvae have been noted to rasp the fruit endocarp with their mandibles. Although *Tr. digitatum* prefers to oviposit in receptacles containing a high level of organic matter, females will also lay their eggs in natural and man-made receptacles that contain little food. The larvae can exhibit stage-specific cannibalism, preying only on younger conspecifics, and finding uniform-aged cohorts of larvae in larval habitats is common. Females show active preference for oviposition in sites where other female *Tr. digitatum* are already guarding their eggs, perhaps in an effort to decrease losses to conspecific predation by larger instars. In *Heliconius*, larvae have been observed to "crawl" from one bract to another to evade predation. *Trichoprosopon digitatum* is protandrous (i.e., males emerge significantly earlier than females). As people inadvertently create many suitable immature habitats, especially through agriculture, *Tr. digitatum* is more commonly found in domestic or peridomestic areas. As with every other aspect of its life history, feeding choices are plastic, and this species is known to feed on all available mammals and birds, albeit with a distinct preference for humans. Intersex exemplars (gynandromorphs) of *Tr. digitatum* have been observed.

Associated Pathogens

ARUV, BSQV, COCV, DENV, ILHV, PIXV, SLEV, TNTV, WYOV.

Diagnostic Characters

(a) Habitus (♀). b, (c) Head: Clypeus (Clp) with dense patches of setae laterally. **(d) Thorax (side):** Mesepimeral (Mm) integument paler than rest of pleuron. **(e) Wing. (f) Thorax (dorsal):** Upper calypter (UC) with setae. **(g) Leg (III):** $Ta\text{-}III_1$ with small basal scale patch. **(h) Larva (thorax):** Seta 0-P 1,2 branched; seta 7-P 1,2 branched. **(i) Larva (head). (j) Larva (abdominal segment):** Seta 12-II single, borne on sclerotized tubercle. **(k) Larva (abdominal segment)**: Seta 6-VI long. Image sources: WRBU; Mattingly (1971a); Zavortink (1983).

Exemplar DNA barcodes: *Tr. digitatum* (KF671032–33, MF172406–MF172408).

Catalog entry: Vol. 2, p. 1003.

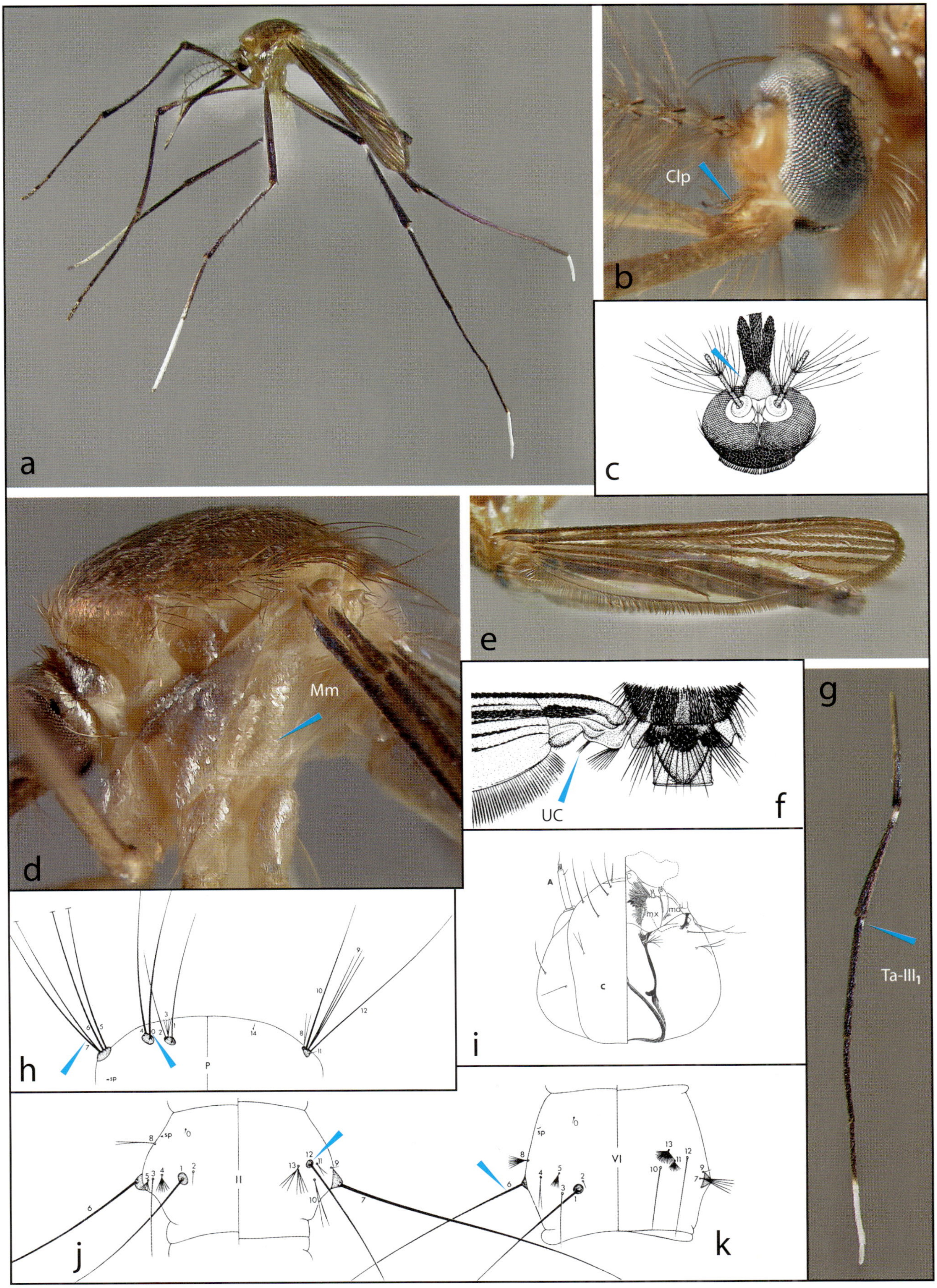
a
Clp
b
c
Mm
d
e
UC
f
g
Ta-III$_1$
h
P
i
II
j
VI
k

12. Glossary of Diagnostic Morphological Terms

This morphological glossary is included to enhance the reader's understanding of the diagnostic taxonomic characters used in the genus and species pages. Chapters 1–10 contain references to or explanations of terms used in Part 1. The glossary was compiled from several sources, including Harrison et al. (2016), the Mosquito Taxonomic Inventory, Museum of Natural History, London (http://mosquito-taxonomic-inventory.info/), Harbach and Knight (1980, 1981), Cumming and Wood (2009), and McAlpine (1981). We primarily followed the format of Harrison et al. (2016), with the word "plural" abbreviated as "pl." and the plural forms of structures either following the singular in parentheses or with a parenthetical "(s)" following the singular form. Prominent abbreviations are indicated in parentheses, with the applicable life stage (L = larva, P = pupa) and/or sex (♂ = male, ♀ = female) indicated in brackets (e.g., [♀, ♂, L]). For the purpose of space on the illustrations and in the text, we modified the naming of the coxa, trochanter, fore-, mid- and hindlegs, to coxa (C-I to C-III), trochanter (Tr-I to Tr-III), femur (Fe-I to Fe-III), tibia (Ti-I to Ti-III), and tarsus (Ta-I to Ta-III). Thus, C-II is the midcoxa, and Ta-III$_5$ refers to tarsomere 5 of the hindleg. Much of the glossary phrasing is taken verbatim from the respective sources, with changes for clarity or to combine information. Labeled illustrations of an example adult female (Fig. 12.1), *Anopheles* wing spots (Fig. 12.2), and exemplar larval stages of *Anopheles* (Fig. 12.3) and *Culex* (Fig. 12.4) precede the glossary.

There is some variability in the literature concerning how terms such as "mesal" and "lateral" are used. McAlpine (1981) recommends: "Terms such as mesal for medial, and mesad, laterad, and distad for medially, laterally, and distally, are often found in entomological works, but as stated by Mackerras (1970) there is not good reason for using them." Medial, lateral and distal are therefore used in this book.

Figure 12.1. Annotated line drawings of generalized external characteristics of the adult female mosquito and wing (modified from Wilkerson & Strickman 1990 and Rattanarithikul et al. 2005a).

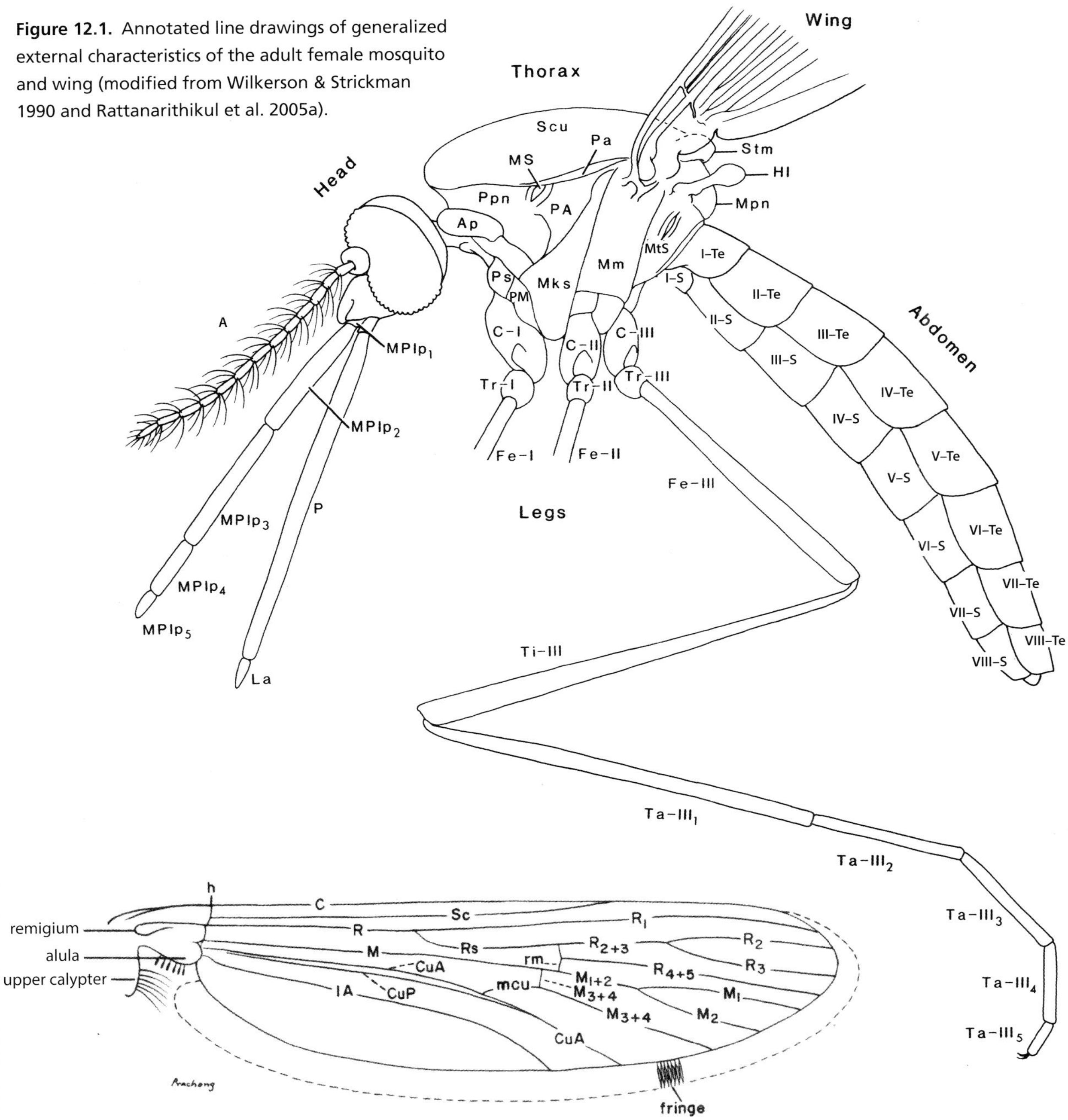

Figure 12.2. Annotated illustrations of the wings of four *Anopheles* species, indicating standardized nomenclature of wing spots (from Wilkerson & Peyton 1990).

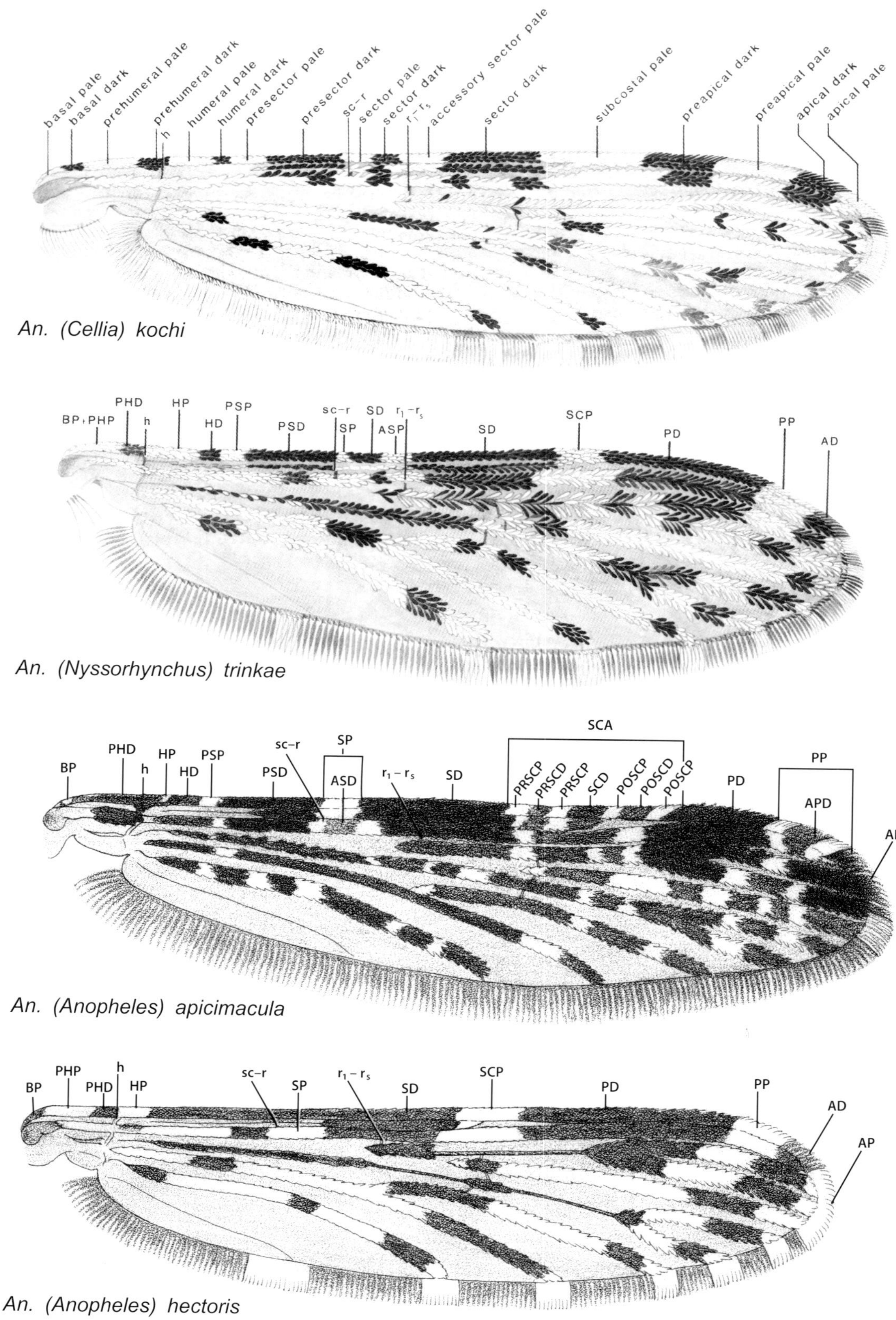

Figure 12.3. Example larval stage drawing indicating standardized nomeclature of *Anopheles* larval characters and numbering of setae (*Anopheles shannoni*, from Wilkerson et al. 1997).

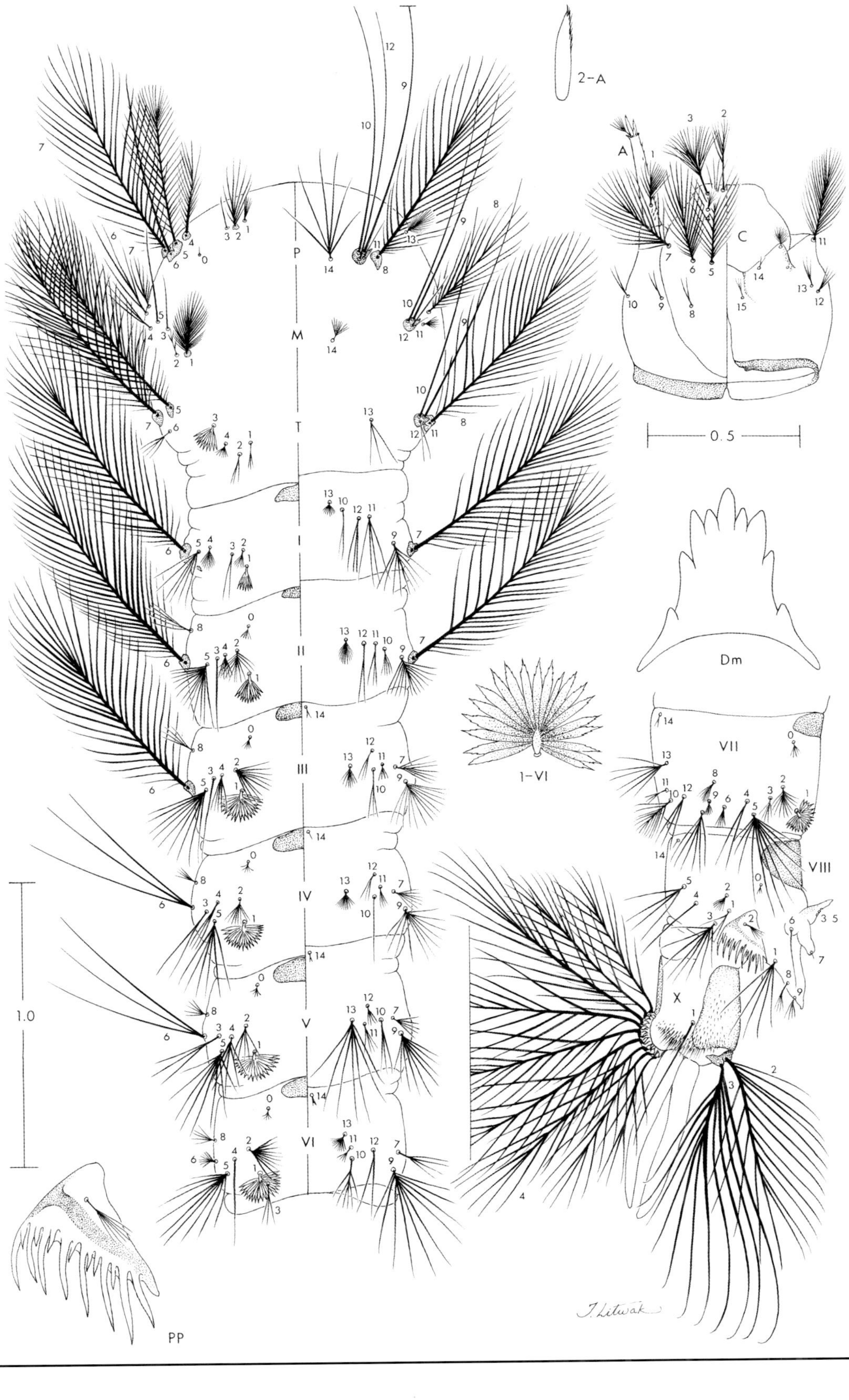

Figure 12.4. Example larval stage drawing indicating standardized nomeclature of Culicine larval characters and numbering of setae (*Culex litwakae*, from Harbach 1985).

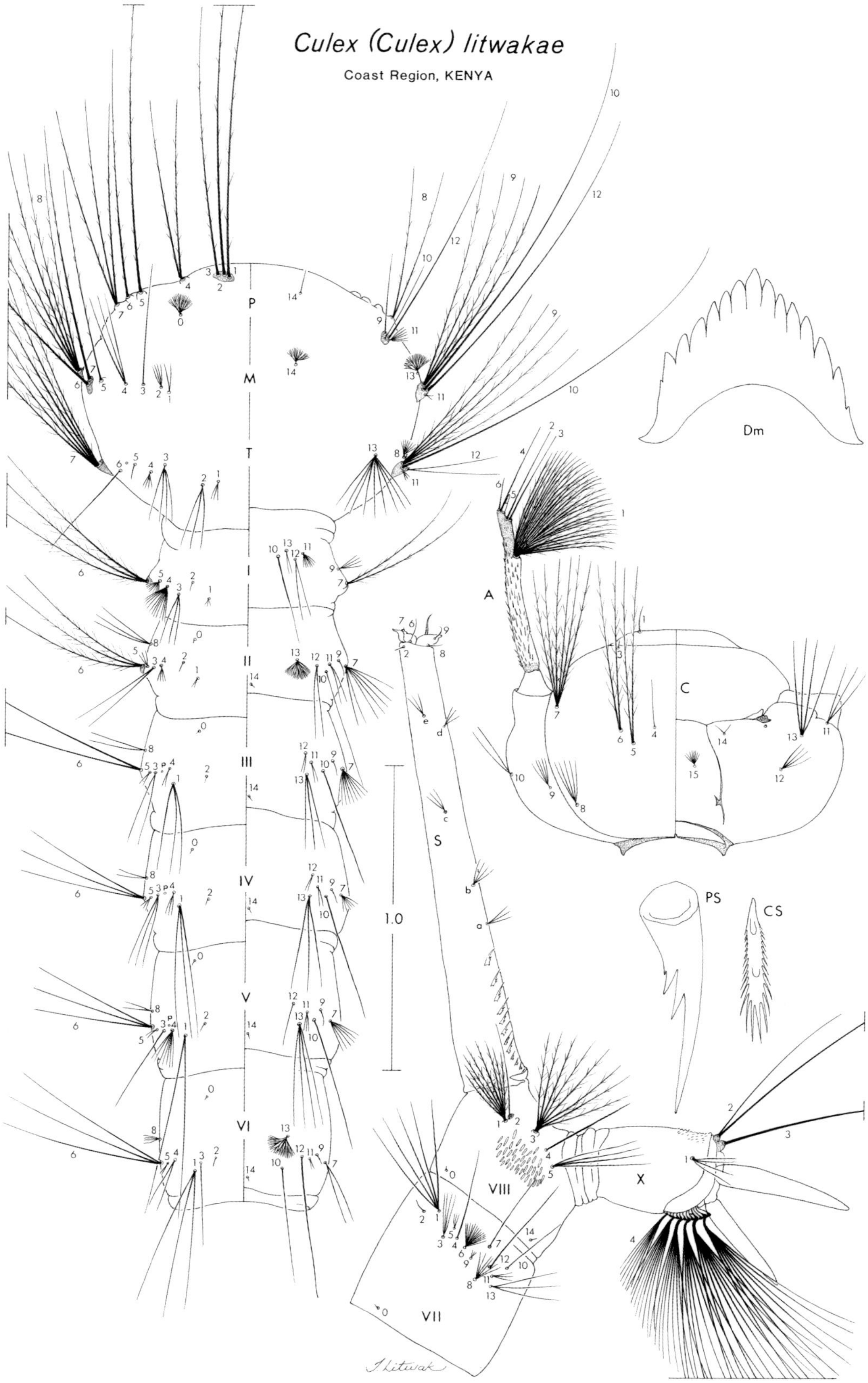

A

abdomen—[♀, ♂, L]. The third or most posterior major division (tagma) of the body. Ten segments are recognized in adults and nine in larvae (segments I–VIII and X; IX is not apparent externally). The segments are typically numbered with Roman numerals.

abdominal segment—[♀, ♂, L]. One of the subdivisions of the insect abdomen. These segments are rounded or oval, with dorsal and ventral sclerotized plates and membranes separating those plates on each segment. On the adults, the plates have setae and/or scales of taxonomic importance, but generally, features on the dorsal plates do not match those on the ventral plates. The integument of the larval segments is primarily membranous, possesses various types of setae, and may have small sclerotized plates.

accessory tergal plate (ATP)—[L]. One or more small median or submedian plates behind the dorsomedian tergal plate on certain (Te-I–VII) abdominal segments of some anophelines.

acicula (pl. aciculae)—[L]. A small, slender, rigid, needle- or thorn-like spicule (see **spicule**); often seen in a fringe on the posterior margin of the larval saddle, called a saddle fringe. Those on individual setae may appear more slender and frail. A seta without lateral processes, such as aciculae, is described being "simple."

acrostichal area (AA)—[♀, ♂]. The median longitudinal area straddling the midline of the scutum from the anterior promontory to the prescutellar area, usually bearing bilaterally paired acrostichal setae and/or scales. Certain genera and subgenera can be identified by the absence of acrostichal setae or scales.

acrostichal scale (AcSc)—[♀, ♂]. One of the scales usually occurring in one or two lines on the acrostichal area of the scutum; when two rows are present, they are borne on either side of the acrostichal setae (unless absent), sometimes merging posteriorly with the prescutellar scales.

acrostichal seta (AcS)—[♀, ♂]. One of the setae occurring in a median longitudinal row on the acrostichal area of the scutum; absent in some genera and species.

acrotergite (Ac)—[♀, ♂]. The part of a tergum anterior to the antecostal suture. In the thorax of winged insects, the acrotergite of the metanotum has become the mesopostnotum, while that of the mesonotum has remained as a narrow flange bearing the first phragma. Because the acrotergite of the metanotum is functionally assigned to the mesothorax as the mesopostnotum, the mesonotum of mosquitoes can be said to have two acrotergites.

aculea (pl. aculeae)—[♀, ♂, L]. Tiny, sharp pointed spicules that create a covering (see **tormentum**) over the cuticula. On the wing membrane, these structures are called microtrichia. Not common on larvae.

adult—[♀, ♂]. A fully developed sexually mature female or male; the final stage in the arthropod life cycle. Synonymous with imago.

aedeagal sclerite (AeS)—[♂]. In males of the Dixidae and certain Culicidae—for example, *Aedes* (*Verrallina*)—a supporting sclerite on either side of the phallus; in most male mosquitoes, one of a pair of lateral sclerotizations of the aedeagus.

aedeagus (Ae)—[♂]. In most male mosquitoes, the central body of the phallosome serving as the intromittent organ; in certain species—for example, *Aedes* (*Verrallina*)—the distal part of the phallus.

alula (Al)—[♀, ♂]. In many Diptera, a lobe of the posterior margin of the wing bounded proximally by the upper calypter, distally by the axillary incision, and anteriorly by the base of the anal vein; weakly differentiated in mosquitoes.

alveolus (pl. alveoli)—[♀, ♂, L]. A cuplike sensory structure in the integument that serves as a socket for an associated seta (of various types); this structure can be used to confirm the past presence of a seta where a specimen is damaged and the seta has been rubbed off.

anal papilla (pl. papillae) (APP)—[L]. One of the soft, clear, or translucent and elongate processes extending posteriorly from abdominal segment X and surrounding the anus. There are usually two pairs of these, a dorsal pair and a ventral pair.

anal vein (A)—[♀, ♂]. From a dorsal view, the most posterior vein on the wing that also extends distally to join the hind margin of the wing. Scale colors and spots on this wing vein are important in identifying *Anopheles*. The majority of mosquitoes have a single anal vein designated 1A; however, *Culex* (*Culex*) *univittatus* has more than one.

antealar area (AnA)—[♀, ♂]. The lateral area of the scutum at the level of the paratergite bearing the antealar scales (AnSc) and setae (AnS).

antecosta (pl. antecostae) (An)—[♀, ♂]. The anterior, internal transverse ridge of a tergum or sternum marked externally by the antecostal suture, serving for the attachment of longitudinal muscles; forms an acrotergite or acrosternite. Best known in mosquitoes in the thorax, where antecostae bear the phragmata.

antecostal suture (AtS)—[♀, ♂]. The external groove of the first and second phragmata. That of the first phragma is not readily visible in mosquitoes.

antenna (pl. antennae) (Ant for adult, A for larva)—[♀, ♂, L]. One of the bilateral paired appendicular organs projecting from the head, originating between the eyes on females and anterior to the eyes on larvae. On adult mosquitoes, the antenna consists of a narrow, ringlike basal segment (scape) that is underneath the second segment (pedicel), which is globular and tire-shaped, and a third segment (flagellum) comprising a series of 13–14 flagellomeres bearing whorls of sensory setae. On the head of larval mosquitoes, inserted anterolaterally and consisting of a narrow, ringlike scape that is united with a large distal tubular part, the fused pedicel and flagellum.

antennal seta (1-A)—[L]. Seta on the shaft of the larval antenna.

anteprocoxal membrane (AM)—[♀, ♂]. The membrane anterior to the forecoxa, between it and the proepisternum, sometimes with anteprocoxal scales.

antepronotal scale (ApSc)—[♀, ♂]. One of the scales occurring in a group on the antepronotum, sometimes forming a tuft.

antepronotum (Ap)—[♀, ♂]. One of the paired anterior divisions of the pronotum, the first part of the thorax; visible as an anterolateral lobe beneath the anterior promontory area on the scutum, bearing setae and occasionally scales and laterally situated just above the dorsal extension of the proepisternum.

anterior. A directional clue pointing to some point near or on the head. Examples: the head is anterior to the abdomen; the anterior end of an abdominal segment is actually the basal (attachment-wise) part of that segment, and the posterior end of that segment is most distal from the head; segments on the legs have anterior aspects.

anterior dorsocentral area (ADA)—[♀, ♂]. A pair of submedian longitudinal rows of setae and/or scales that are near but lateral to the acrostichal (median) row on the scutum; divided into two regions, the anterior dorsocentral area and the posterior dorsocentral area, with setae and/or scales that occur laterally and just anterior to the beginning of the prescutellar area.

anterior median process (AMPc)—[L]. In anopheline larvae, an elevated dome- or knob-like membranous area situated posteriorly on the midline of the anterior spiracular lobe of the spiracular apparatus; in some species, produced into an elongate process—for example, in species of *Chagasia*.

anterior mesanepisternum (AMas)—[♀, ♂]. The somewhat triangular, anterior part of the mesanepisternum bearing the mesothoracic spiracle and separated from the posterior mesanepisternum and the mesokatepisternum by the anepisternal cleft. In mosquitoes, it is divided into four areas: the prespiracular area anterior to the spiracle, the hypostigmal area immediately below the spiracle, the subspiracular area ventral to the hypostigmal area, and the postspiracular area posterior to the hypostigmal and subspiracular areas.

anterior mesepimeral scale (AMSc)—[♀, ♂]. One of the mesepimeral scales occurring in a group on the cephalodorsal quarter of the mesanepimeron.

anterior promontory (AnP)—[♀, ♂]. The broad median area of the mesonotum at the anterior end of the acrostichal area that usually projects over the cervix (= neck).

apex. The furthest or most distal point on a structure from the base.

apical. A directional relationship to the apex of a structure; typically, something beyond the midpoint on a lengthy structure—for example, like a preapical pale band on a leg segment, or apical pale bands on the abdominal segments, or apical bands on hind tarsomeres.

apical tooth (AT)—[L]. An apical process of the maxillary body, typical of sabethine larvae. Usually, a small bump or protuberance at the distal end of the maxilla, but in some it has developed into a prominent tooth- or claw-like process.

apicodorsal lobe (ADL)—[♂]. The most distal area of the dorsal (prerotation sense) surface of the gonocoxite; an apical dorsomesal lobe.

apodeme—[♀, ♂, L]. Infoldings of the integument (see **antecostae** and **phragmata**); usually marked externally by a suture or pit.

apodome of gonocoxite (AG)—[♂]. An internal continuation of the proximal mesal dorsal (prerotation sense) corner of the gonocoxite; fused or articulated with the basal piece.

arculus—[♀, ♂]. A crossvein near the wing base between the radius and cubitus (see **remigium**).

attenuated. The gradual loss of diameter (width) toward the apex of a structure (for example, a sharp spine).

B

basal. A directional relationship to something closer to the base than the midpoint on a lengthy structure; typically used in describing structures—for example, basal pale bands on the hind tarsomeres, basal transverse stripes on the abdominal segments, or a basal row of pecten spines on the venter of a siphon (= proximal).

basalare (Ba)—[♀, ♂]. The episternal epipleurite giving insertion to the anterior pleural wing muscles; in the mesothorax of mosquitoes, the lobe borne at the apex of the posterior mesanepisternum anterior to the pleural wing process and in the metathorax of mosquitoes, the small lobe at the apex of the metepisternum anterior to the pleural wing process.

basal dorsomesal lobe (BDL)—[♂]. In male mosquitoes, a basal lobe on the mesal margin of the dorsal (prerotation sense) surface of the gonocoxite.

basal lateral arm (BLA)—[♂]. In *Culex* mosquitoes, a laterally directed basal process of the paraproct; may bend in caudal and/or ventral directions. Same as basal sternal process.

basal mesal lobe (BML)—[♂]. A small lobe or sclerite located basomesally on the gonocoxite; usually connected ventrally with the lobe on the other side; possibly homologous with the claspette.

basal piece (BP)—[♂]. One of the pair of basal supporting sclerotizations of the phallosome, lying within abdominal segment IX and often extending anteriorly into VIII, articulated with the paramere, sometimes articulated with the base of the proctiger, and often connected with the apodeme of the gonocoxite; probably primitively an apodeme of the mesal margin of the gonocoxite. Same as the gonocoxal apodeme.

basal tooth—[♀, ♂]. A small basoventral sharp projection on the ungues (claws) of some mosquitoes; its length and shape in relation to the ungues can be important in species identification.

boss (B)—[L]. In certain mosquito larvae, a more or less sclerotized elevated area without a grid at the base of the ventral brush (see **grid**).

C

capitellum (Ca)—[♀, ♂]. The apical part of the halter, which is knob-like and has scales.

cardo (Cd)—[♀, ♂, L]. The basal subdivision of the maxilla. In adults, it is embedded in the membrane near the posterior tentorial pit, fused anteriorly with the rod-like stipes to form the stipitocardinal rod, and distinguishable from the stipes by muscle insertions. In larvae, the structure is close to the basic insect pattern, the cardo forming the base of the maxilla, basal to the stipes. These structures vary between species according to feeding specialization and are generally weaker in males.

central spine of a comb scale—[L]. The median spine and typically the largest spine on a larval comb scale. Morphologically, this structure is a spicule, not a scale.

cercal sclerite (CSc)—[♀, ♂]. Any sclerite forming a part of the cercus. In males, the more or less distinctly sclerotized area between the paraprocts on the dorsal (prerotation sense) surface of the proctiger or one of a pair of sclerites in the same area.

cercus (pl. cerci) (Ce)—[♀]. A pair of conspicuous lobes representing appendages of segment XI (that itself is reduced and not distinguishable) at the apex of the abdomen beyond tergum IX. The cerci are involved in both mating and oviposition. The shape varies between genera. The cerci are inconspicuous in males, forming part of the sclerotization of the proctiger/paraproct complex.

cibarium (Clb)—[♀, ♂, L]. The preoral food cavity between a floor formed by the hypopharynx and a roof formed by the clypeopalatum. In adults, the cibarium is just below the clypeus, which contains muscles activating the cibarial pump, the first stage of the sucking mechanism (the second being the pharyngeal pump). In larvae, the cibarium is not formed into a pump.

claspette (Cl)—[♂]. A descriptive term applied to a variably shaped lobe arising mesally at the base of the gonocoxite.

clypeus (Clp)—[♀, ♂, L]. A large sclerotized structure on the head below and anterior to the antennal pedicels and above the bases of the maxillary palpi. The clypeus, prominent in adults, is delimited by the epistomal suture dorsally and the clypeolabral suture ventrally. The same structure occurs in larvae but is less prominent.

comb (C)—[L]. A patch or row(s) of spicules on the posterolateral part of abdominal segment VIII, sometimes arising from a sclerite (the comb plate).

comb plate (CP)—[L]. A small or large lateral sclerite on abdominal segment VIII bearing the comb spicules, which occur on certain culicine larvae; sometimes the paired lateral sclerites are joined dorsally.

comb scale (CS)—[L]. A generalized name for spicules that occur in the comb on abdominal segment VIII. Each spicule is nonarticulated and joined continuously with the integumental membrane or a comb plate. Individual spicules may have lateral and apical spinules or denticles. Those with apical spinules often have a central spine designated the lateral spine. Some comb scales lack spinules or denticles and appear lanceolate or sharply pointed and claw-like. Comb scales are absent on *Toxorhynchites* and only occur on *Anopheles* during the first larval instar.

compound eye(s)—[♀, ♂, L, P]. Light perception organs consisting of an aggregation of optic elements (ommatidia) generally located on each side of the head of adult mosquitoes; they are convex, kidney-shaped, and occupy much of the dorsal and lateral surface of the head. The elements of the compound eyes appear in the late fourth instar and pupa.

costa (C)—[♀, ♂]. The anterior-most (= front) longitudinal vein on the wing, normally extending from the humeral plate at the base of the wing to the wing apex.

coxa (pl. coxae) (C-I, C-II, C-III)—[♀, ♂]. The basal-most sclerotized segment of each leg, articulated by a membrane with the thorax. C-I, C-II, C-III are abbreviations for the procoxa, mesocoxa, and metacoxa, respectively.

cratal seta (CrS)—[L]. One of the ventral brush setae borne on the grid and corresponding to the series of setae that are 4-X (= fourth seta on the tenth abdominal segment).

cubitus (Cu)—[♀, ♂]. The fifth longitudinal vein on the wing, with two main branches: the cubitus anterior (CuA) and the cubitus posterior (CuP). The CuA is located dorsally on the wing and has scales, while the CuP is posterior to the CuA, is projected ventrally and is thus most easily seen on the venter of the wing, and is long, slender, and without scales.

cubitus anterior (CuA)—[♀, ♂]. The anterior branch of the cubitus. Unbranched in mosquitoes.

cubitus posterior (CuP)—[♀, ♂]. The posterior branch of the cubitus. In mosquitoes, it is unbranched and weakly developed, and closely parallels the cubitus anterior.

D

decumbent scales—[♀, ♂]. Scales that lie flat or appressed on the surface of origin—for example, on top of the head, abdomen, legs, or wing veins.

distal. Located away from the center of the body or center of a structure or point of attachment. Often used in terms like "most-distal" or beyond the midpoint of a structure toward the apex—for example, a preapical band on the tibia is distal to the midpoint (lengthwise) on the tibia. On larvae, it is most often used in discussing forks occurring on setae or distal aciculate branches on a seta (= apical, therefore a relative term).

distal division of subapical lobe (dSL)—[♂]. On the mesal or dorsomesal portion of the gonocoxite in *Culex* (see **subapical lobe**).

distally (adv). Toward the apex, away from the center of the insect body.

distal tergal setae (dts)—[♂]. Descriptive term for distal tergal setae of gonocoxite.

dorsal. Relating to the dorsum of an organism or body structure—for example, the dorsum (or dorsal aspect) of a leg segment, wing, proboscis, thorax, head, abdomen, etcetera.

dorsal apotome (DAp)—[L]. The cranial area bounded laterally by the frontal ecdysial lines and apically by an imaginary line between the most apical parts of the frontal ecdysial lines.

dorsocentral area (DA)—[♀, ♂]. A longitudinal area of the scutum on each side of the acrostichal area extending from the anterior promontory to the prescutellar area; it is indefinitely divided into anterior and posterior dorsocentral areas at the level of the antealar area and bears the dorsocentral scales and setae.

dorsocentral seta (DS)—[♀, ♂]. On the scutum, seta occurring on the dorsocentral area; divided into anterior and posterior rows near the median posterior corner of the scutal fossa at about the level of the antealar area.

dorsomentum (Dm)—[L]. The upper and more distal of the two subdivisions produced when the mentum is completely divided by a transverse inflection of membrane.

dorsum. Entire dorsal area of a segment (= tergum (Te), called tergite by many).

E

epimeron—[♀, ♂]. A main division of the wing-bearing thoracic segment found posterior to the pleural suture. The upper part is the anepimeron, and the lower part is the katepimeron. These are separated by an anapleural suture (see **mesepimeron**).

episternum—[♀, ♂]. A main division of a wing-bearing thoracic segment found anterior to the pleural suture. The upper part is the anepisternum, and the lower part is the katepisternum. These are separated by an anapleural suture (see **epimeron**, **mesepisternum**).

erect forked scale—[♀]. Typically long, narrow, stalked scales that are erect and stand vertically from the top of the head. They have a slightly widened and blunt apex with a median notch (depth varying per species). These occur on the vertex and occiput of adult mosquitoes.

exuviae (singular and plural spelling)—[L]. The entire outer covering (= skin) of a larva (also a pupa), including the membranous and sclerotized parts. Typically, exuviae are shed between the four larval instars and while the adult is eclosing from the pupa. These "skins" are very important to taxonomists identifying adults, especially when sympatric species are extremely similar or identical.

F

femur (Fe)—[♀, ♂]. The third segment of the leg. The femur is usually the largest leg segment in mosquitoes. The femora of the legs from anterior to posterior are abbreviated Fe-I, Fe-II, and Fe-III, corresponding to the forefemur, midfemur, and hindfemur, respectively.

flagellomere (Flm)—[♀, ♂]. An individual unit of the antennal flagellum. There is usually a whorl of long, thin setae around the base of each flagellomere.

flagellum (Fl)—[♀, ♂]. The distal and third segment of the antenna. In adult mosquitoes, it is composed of 13–14 smaller flagellomeres.

frontal tuft (FT)—[♀, ♂]. A grouping of erect, elongate, simple setae and/or fusiform scales arising from the interocular space and the immediately adjacent portion of the vertex of the head.

G

galea (pl. galeae) (Gl)—[L]. The outer endite lobe of the maxilla. In mosquito larvae, either united with part of the stipes to form the galeastipes or united with the mesal part of the stipes and the lacinia to form the maxillary body; in anopheline larvae, a small triangular area partly separated from the main part of the maxillary body by the galeastipital fissure.

galeastipes (GSt)—[L]. Recognizable in many mosquito larvae, mostly culicines, as the lateral half of the maxillary body. It can be separated from the laciniastipes by dorsal and ventral maxillary sutures. The galeastipes represents the fused stipes and galea, which are partly separated in anophelines. The galeastipes usually bears the maxillary brush and setae 1-, 3-, 4- and 5-Mx. The maxillary palpus is articulated laterally at its base.

gonocoxal apodeme—[♂]. Also commonly called the basal piece, it consists of a process that projects anteriorly from each gonocoxite or the gonocoxal region of the composite hypandrial sclerite. The gonocoxal apodemes (as a pair) indicate the position of the medially attached parameres.

gonocoxite (Gc)—[♂]. The largest unit of the genitalia bearing the gonostylus apically; fused together midventrally at their base.

gonocoxopodite (Gp)—[♂]. Originating as the morphological coxa of abdominal segment IX, the gonocoxopodite consists of the gonocoxite, stylus, and gonostylus.

gonostylus (Gs)—[♂]. The one-segmented stylus of the gonocoxopodite; movable and attached at or near the apex of the gonocoxite.

grid (G)—[L]. The ventral area of sclerotized ridges on abdominal segment X that bear the cratal setae (pairs of seta 4-X) of the ventral brush. Smaller, more anterior paired (precratal) setae of the ventral brush that are attached ventrally to the saddle or membrane of segment X may occur on larvae of certain species. The network of sclerotized ridges that bear the cratal setae have transverse grid bars at the base of individual setae, sometimes with lateral grid bars that may be confluent with the ventroposterior margin of the saddle (see **boss**).

gular area (GuA)—[L]. The midventral rear edge of the cranium and a large area of the gular (i.e., neck) membrane posterior to the head.

H

habitus. The characteristic form and appearance of an organism; here we use it to refer to renderings of entire adult specimens on the genus and species pages.

halter(s) (Hl)—[♀, ♂]. The paired drumstick-shaped structures on the metathorax below and behind the wing, consisting of a thick base (scabellum), a slender stemlike portion (pedicel), and an expanded distal part (capitellum). The halters are homologous with the second pair of insect wings.

head—[♀, ♂, L]. The most anterior major section (tagma) of the body, bearing eyes, antennae, and mouthparts. It is separated from the thorax by the cervix in adults and larvae.

hypandrium—[♂]. The sternum of abdominal segment IX (prerotation sense), fused laterally with the tergum (epandrium) to form a ring. In mosquitoes the hypandrium/epandrium are separate from the gonocoxites.

hypopharynx (Hy)—[♀, ♂, L]. In the larva, the fleshy hypopharynx is fused to the chitinous labium to form the labiohypopharynx, which is the floor of the preoral cavity. There is often a chitinous mound on the labiohypopharynx, against which the mandibles crush food particles. The larval salivary duct empties into the preoral cavity at the junction of the labium and the hypopharynx. In adults, the hypopharynx is a single stylet containing the salivary channel and forming the floor of the food channel. The base of the hypopharynx forms the floor of the cibarium.

hypostigmal area (Hya)—[♀, ♂]. A weakly sclerotized or membranous area immediately below the mesothoracic spiracle and just above the subspiracular area. The hypostigmal area is part of the anterior mesanepisternum.

hypostigmal scale (HySc)—[♀, ♂]. One of the scales occurring in a small cluster on the hypostigmal area of the anterior mesanepisternum.

I

instar(s)—[L]. A term used to denote a particular larval developmental period. Mosquito larvae go through four of these periods, shedding the exuviae between each period. The periods are usually denoted as first, second, third, and fourth instars. After the fourth period is completed, the larva changes to a pupa.

integument. The outer surface of the body, comprised of membranes and sclerites.

interocular space (IS)—[♀, ♂]. The narrow part of the vertex on the head, between the compound eyes and above the margins of the antennal pedicels (demarcated by the postfrontal sutures).

intersegmental membranes (IM)—[♀, ♂, L]. The membrane connecting the segments of the body and of the appendages.

L

labellum (pl. labellae) (La)—[♀]. One lobe of a partially bilobed structure situated at the apex of the labium.

labial palpus (pl. labial palpi) (LP)—[L]. Vestigial labial palpi are represented by a membranous area on each side of the ligula or prementoligular teeth, usually bearing four or five peg-like setae.

labium (Lb)—[♀, ♂, L]. In adults, the elongate sheath-like covering of the proboscis, which covers and protects the internal mouthparts (stylets). It consists of the prementum forming the sheath, the labellae, and the ligula. In larvae, the labium is the base of the labiohypopharynx, with posterior limit (submentum) in the gular area and anterior limit (prementum) at the exit of the salivary duct.

lacinia (Lc)—[L]. The inner endite lobe of the maxilla, either united with part of the stipes to form the laciniastipes or united with the mesal part of the stipes and the galea to form the maxillary body.

lateral. A directional term used in referring to the "side" (in respect to the midline) of an organism or structure.

laterally (adv). On the side.

lateral palatal brush (LPB)—[L]. Part of the feeding mechanism formed by a series of hairs forming a brush on each side of the labropalatum, distal to the labral plate and torma; often stouter and strengthened in predatory species.

lateral plate (LP)—[♂]. In *Culex*, one of the pair of variously developed lateral sclerotizations of the phallosome located between the aedeagus and proctiger; often consisting of inner and outer divisions that may also be subdivided.

lateral scutellar lobe—[♀]. In culicines and *Chagasia*, one of two lateral lobes on the posterior of the scutellum.

lateral scutellar scale (LSSc)—[♀, ♂]. One of the scales occurring in a small group on each lateral scutellar lobe.

lateral scutellar seta (LSS)—[♀, ♂]. One of the setae occurring in a small group on each lateral scutellar lobe.

lateral spine—(L). The spinule extending posteriorly from the tip of a comb scale.

laterotergite (Lt)—[♀, ♂]. The lateral part of abdominal tergum I when more or less distinctly separated from the median dorsal part; usually not visible from above. Each laterotergite extends down to the posterior thoracic spiracle and is sometimes divided into a dorsal anetergite and a ventral katatergite.

leg(s)—[♀, ♂]. One of six paired appendages originating from ventrolateral positions on the thorax. The legs are composed of segments called (from proximal to distal): coxa (C-I to C-III), trochanter (Tr-I to Tr-III), femur (Fe-I to Fe-III), tibia (Ti-I to Ti-III), and tarsus (Ta-I to Ta-III). The three pairs of legs and each segment are denoted by the fore-, mid-, and hind-, as appropriate. The forelegs are articulated with the prothorax; midlegs, with the mesothorax; and hindlegs, with the metathorax.

ligula (Lg)—[♀, ♂, L]. In adults, a structure forming the floor of the labium between the two labellae. The ligula is formed from the fused glossae and paraglossae of the labium. In larvae, the glossae and paraglossae are fused or separate and may form an elevated, cusped, tongue-like structure between the labial palpi. In other larvae, they are indistinguishably fused into a complex structure called the prementoligular teeth.

lower mesepimeral scale (LMSc)—[♀, ♂]. One of the scales occurring in a group on the ventroanterior quarter of the mesanepimeron.

lower mesepimeral seta (MeSL)—[♀, ♂]. One of the setae occurring in groups on the anterior, middle, and/or posterior area of the mesanepimeron below the level of the metathoracic spiracle.

lower mesokatepisternal scale (MScL)—[♀, ♂]. One of the scales occurring in a group located immediately in front of the lower mesokatepisternal setae.

lower mesokatepisternal seta (MkSL)—[♀, ♂]. One of the setae that occur in a roughly vertical line along the posterior margin of the mesokatepisternum, approximately at the level of the top of the mesomeron (= mesotrochantin); sometimes continuous with the upper mesokatepisternal setae.

lower proepisternal scale (PScL)—[♀, ♂]. One of the proepisternal scales occurring in a group basal to the forecoxa and below the upper proepisternal scales; sometimes continuous with the upper proepistemal scales dorsolaterally.

lower proepisternal seta (PeSL)—[♀, ♂]. One of the proepisternal setae occurring in a group mesal to the forecoxa and below the upper proepisternal setae.

M

mandible (Mn)—[♀, ♂, L]. One of the first pair of gnathal appendages. In females, occurring just ventral to the labrum in the premental gutter; reduced or absent in males. In larvae, borne on the underside of the head, where each is implanted obliquely in the membranous area that extends anteriorly from the margin of the lateralia to the cibarial bar; they are typically flattened lobes with their mesal ends developed into strongly sclerotized teeth and a spiculate lobe.

maxilla (Mx)—[♀, ♂, L]. The pair of maxillae are the second set of mouthparts, located ventral to the hypopharynx. In adults, the maxilla consists of a stylet-like lacinia fitted with barbs near the tip in females and reduced in size in males. The prominent palpus emerges from the base, where there is an internal rod composed of the stipes and cardo. In larvae, borne on the transverse margin of the lateralia lateral to the mentum and posterior to each mandible; typically shaped as a flattened lobe consisting of the variously developed and fused parts of the generalized insect maxilla.

maxillary brush (MxB)—[L]. A collection of variously developed spicules arising distally from the maxillary body or the galeastipes when distinguishable.

maxillary bundle (MxBn)—[L]. A form of the maxillary brush in some sabethines in which the spicules are sheathed or fused.

maxillary palpus (pl. maxillary palpi) (MPlp; $MPlp_{1-5}$)—[♀, ♂, L]. Paired sensory structures inserted on the head immediately below the clypeus and lateral to the base of the proboscis; vary widely in length and form according to genus and sex and primitively consist of five segments called palpomeres. In larvae, the structure is a laterally located cylindrical or fusiform lobe fused basally with the maxillary body.

media (M)—[♀, ♂]. The fourth principal longitudinal vein on the wing, with two areas of branching; beyond the wing crossveins, M changes to M_{1+2} and more distally splits into two separate veins, M_1 and M_2, that extend to the wing margin just posterior to the wing apex; also at the crossveins, a more posterior branch, M_{3+4}, extends to the hind margin of the wing before vein CuA.

medial. Near median line/area (= mesal of other authors).

median accessory tergal plate (MATP)—[L]. In some anopheline larvae, a small dorsal sclerite located immediately posterior to the tergal plate on the midline of abdominal segments I–VII.

median area. Area down the middle.

median labral plate (MLP)—[L]. The dorsal sclerite of the labrum; separated from the dorsal apotome by the clypeolabral suture.

median line. Down the middle.

median plate (MdP)—[L]. In anopheline larvae, a large sclerite located centrally on the dorsal surface of the spiracular apparatus. The median plate receives the attachment of the muscles that close the spiracular apparatus; it is homologous with the spiracular apodeme, posterolateral lobe plates I, and the posterior median plate found in culicine larvae.

median scutellar lobe—[♀, ♂]. The posterior lobe on the midline of the scutellum. Present on culicine mosquitoes except *Toxorhynchites* and absent in all anophelines except *Chagasia*.

median scutellar scale (MSSc)—[♀, ♂]. A scale on the median scutellar lobe.

media-one (M_1)—[♀, ♂]. The anterior branch of media-one-plus-two (M_{1+2}) that reaches the margin of the wing near the apex.

media-one-plus-two (M_{1+2})—[♀, ♂]. The anterior branch of the media (media posterior) that splits into media-one (M_1) and media-two (M_2) near the wing apex.

media-three-plus-four (M_{3+4})—[♀, ♂]. The posterior branch of the media (media posterior), with a short basal crossvein-like segment that traditionally has been considered the mediocubital crossvein; the longer distal part has been considered to be cubitus anterior (CuA).

media-two (M_2)—[♀, ♂]. The posterior branch of media-one-plus-two (M_{1+2}).

mediocubital crossvein (mcu)—[♀, ♂]. The crossvein extending between the media and cubitus, connecting media-three-plus-four (M_{3+4}) and the cubitus anterior (CuA).

mentum (Mt)—[L]. A sclerite at the base of the labium formed by the morphological postmentum and located distal to the submentomental suture and proximal to the submentum. The mentum is often shaped into a proximal fold (ventromentum) and a distal, triangular part (dorsomentum).

meron (m)—[♀, ♂]. The lateral postarticular area of the coxal base proximal to the basicostal suture on the mesothorax (mesomeron) and metathorax (metameron). It is more associated with the pleuron than the coxa.

mesal. See medial.

mesanapleural suture (MAnS)—[♀, ♂]. The suture occurring between the mesanepisternum and the mesokatepisternum.

mesanepimeral ridge (MAeR)—[♀, ♂]. An internal ridge along the ventral margin of the mesanepimeron serving as an attachment for wing muscles.

mesanepimeron (Mam)—[♀, ♂]. The large upper area of the mesepimeron bearing scales and setae, referred to as mesepimeral elements.

mesanepisternum (Mas)—[♀, ♂]. The upper area of the mesepisternum separated from the mesokatepisternum by the mesanapleural suture.

mesepimeral scale (MeSc)—[♀, ♂]. Scale on the mesepimeron.

mesepimeral seta (MeS)—[♀, ♂]. Seta on the mesepimeron.

mesepimeron (Mm)—[♀, ♂]. The large lateral thoracic sclerite immediately posterior to the mesokatepisternum, below the wing base, and above the mesomeron; divided into a large upper portion, the mesanepimeron, and a narrow horizontal meskatepimeron just below the upper part.

mesepisternum (Ms)—[♀, ♂]. The area of the mesopleuron anterior to the mesopleural suture. Usually used for the fused mesanepisternum (upper) and mesokatepisternum (lower).

mesokatepimeron (Mkm)—[♀, ♂]. The lower division of the mesepimeron represented by a small strip of cuticle projecting over the dorsal edge of the mesomeron.

mesokatepisternum (Mks)—[♀, ♂]. A large, diagonally projecting lateral thoracic sclerite between the postspiracular area and the mesepimeron; the lower part is broad and projects downward between the fore- and midcoxae. The upper part joins with the small posterior mesanepisternum sclerite that narrows to a rounded point called the prealar knob just in front of the wing base. On the ventral side, the two bilateral mesokatepisterna join between the pro- and mesosterna.

mesomeron (Msm)—[♀, ♂]. The meron of the mesothorax, which is a lateral thoracic sclerite below the mesepimeron that projects downward between the bases of the mid- and hindcoxae.

mesonotum (Mn)—[♀, ♂]. The dominant tergal (dorsal) element of the thorax, derived from the notum of the mesothorax and in mosquitoes divided into the anterior narrow acrotergite, the large scutum, and the posterior scutellum.

mesopleuron (Mp)—[♀, ♂]. The pleuron of the mesothorax consisting of the basalare, subalare, mesepisternum, mesepimeron, and mesotrochantin.

mesopostnotum (Mpn)—[♀, ♂]. A large dorsomesal lobe of the mesothorax that projects posteriorly from under the scutellar lobes to join with the dorsum of the abdomen. Internally, this is the location of the second phragma. The large median portion of the mesopostnotum is the mediotergite, and the smaller lateral areas are the pleurotergites.

mesothoracic spiracle (MS)—[♀, ♂]. A large opening on the side of the mesothorax forming a connection between the atmosphere and the tracheal system. The spiracle is located just below the scutum, behind the postpronotum and prespiracular area, and in front of the large posterior portion of the postspiracular sclerite.

mesothorax (M)—[♀, ♂, L]. The second or middle portion of the thorax. In adults, this segment bears the wings and midlegs; it is the largest portion of the thorax. In larvae the segments of the thorax are fused, with the middle part bearing 14 pairs of setae numbered in a standard fashion from the dorsal to the ventral midlines so that numbered setae are homologous between species.

metameron (Mem)—[♀, ♂]. The meron of the metathorax, which is a small, narrow sclerite located above the base of the hindcoxa, posterior to the mesepimeron and mesomeron, and beneath the metathorax.

metanotum (Mtn)—[♀, ♂]. The tergal element or notum of the metathorax, applied only to the primary sclerotization of the dorsum lying between the meso- and metapostnota. The acrotergite of the metanotum is functionally assigned to the mesothorax as the mesopostnotum.

metapleural suture (mts)—[♀, ♂]. The external groove (pleural suture) of the metapleural ridge separating the metepisternum from the metepimeron. It has the appearance of an irregular groove extending downward from near the base of the halter; it is sometimes interrupted by the apophyseal pit.

metapleuron (Mtp)—[♀, ♂]. The pleuron of the metathorax consisting only of a reduced metepisternum and metepimeron.

metapostnotum (Mtpn)—[♀, ♂]. The postnotum of the metathorax, consisting of a narrow plate behind the intersegmental cleft and extending ventrally to the metameron of each side. It is reflected upon and united with the anterior part of abdominal segment I.

metathoracic spiracle (MtS)—[♀, ♂]. An opening on the metathorax that forms a connection between the atmosphere and the tracheal system. It is smaller than the mesothoracic spiracle and is located between the upper part of the mesepimeron and the halter, ventral to the pleurotergite.

metathorax (in adult, Mt; in larva, T)—[♀, ♂, L]. The third and most posterior segment of the thorax. In adults, it is located below the wing and just behind mesepimeron and mesomeron, bearing the metathoracic spiracles, hindlegs, and halters. In larvae, it is the most posterior third (T) of the fused thorax, bearing 13 pairs of numbered setae. The setae are numbered in a standardized system starting near the dorsal midline and continuing laterally and ventrally to near the ventral midline.

metepimeron (Mtm)—[♀, ♂]. A small triangular part of the metapleuron posterior to the metapleural suture. It is sometimes poorly separated from the metepisternum and appears to be continuous dorsally with the metanotum.

metepisternum (Mts)—[♀, ♂]. The area of the metapleuron anterior to the metapleural suture. It is immediately behind and below the metathoracic spiracle.

N

notum (N)—[♀, ♂]. The tergum of a thoracic segment formed from the primary segmental sclerotization and the acrotergite, which is the intersegmental sclerotization from the preceding segment. The acrotergite of the mesothorax is a narrow anterior strip of the mesonotum, whereas the acrotergite of the metathorax is the postnotum and is functionally assigned to the mesonotum.

O

occipital foramen (OF)—[♀, ♂, L]. The posterior opening of the head into the cervix (= neck).

occiput (Occ)—[♀, ♂]. The extreme posterior portion of the top of the head adjacent to the postocciput and separated by the postoccipital suture. It is not clearly differentiated from the vertex (dorsally) and the postgena (laterally).

opisthophallus (OP)—[♂]. A transverse lobiform element of the phallosome located between the phallus and proctiger. It is supported on either side by the opisthophallic sclerite. The opisthophallus is strongly developed in *Maorigoeldia* and *Verrallina*.

P

palmate seta (pl. palmate setae)—[L]. A seta with flattened, movable, and usually horizontal branches that radiate from a common point on a short basal stem; when spread out, it resembles a fully extended collapsible hand fan. Palmate setae are commonly found on *Anopheles* larvae—for example, seta 1—on a variable number of abdominal segments.

parabasal lobe (PBL)—[♂]. A variably developed lobe located dorsobasally (prerotation sense) on the gonocoxite of anophelines. It bears one or more strongly developed parabasal setae.

parabasal seta (PBS)—[♂]. Any distinctive seta on the parabasal lobe.

paramere (Par)—[♂]. A lateral sclerite of the phallosome interposed between, and articulating with, the basal piece and the aedeagal sclerite.

paranotal suture (PnS)—[♀, ♂]. The strongly developed suture separating the paratergite from the scutum; often appearing to be continuous posteriorly with the supraalar groove.

paraproct (Ppr)—[♂]. A paired sclerite located laterally on the proctiger and basally articulated with tergum X.

paraproct crown (PpC)—[♂]. A group of spicules located at the apex of the paraproct in some species of *Culex*.

paratergal scale (PaSc)—[♀, ♂]. Any scale on the paratergite.

paratergite (Pa)—[♀, ♂]. A narrow lateral sclerite of the mesonotum that is separated by a suture from the scutum; positioned ventral to the scutum and anterior to the wing root.

pecten (Pt)—[L]. Spicules arranged in a comb-like row and located posterolaterally on both sides of the siphon (culicines) or on the posterior margin of the pecten plate (anophelines).

pecten plate (PP)—[L]. A plate on both sides of *Anopheles* larvae located ventrolaterally from the spiracular apparatus on abdominal segment VIII. The pair of sclerites may be connected posteriorly in a U-shaped band.

pecten spine(s) (PS)—[L]. An individual spicule of the pecten.

pedicel (Pe)—[♀, ♂]. The second or subbasal segment of the antenna, usually enlarged and appearing functionally as the basal segment of the antenna (see **antenna**).

phallosome (PH)—[♂]. The complex of structures surrounding the gonopore between the proctiger, gonocoxites, and sternum IX. It consists of various components, the homologies of which are unknown, and includes the basal pieces, parameres, prosophallus, phallus, and opisthophallus.

phallosome lateral plate (PHLP)—[♂]. One of the pair of variously developed lateral sclerotizations of the phallosome of *Culex*. It is located between the aedeagus and proctiger; often with various subdivisions.

phallus (Ph)—[♂]. The median lobe of the phallosome located between the prosophallus and opisthophallus. It is present only in some mosquitoes and is supported by the aedeagal sclerite.

pleural coxal process (PCP)—[♀, ♂]. The process of the pleuron that articulates with the coxa, which is well developed only on the proepisternum.

pleural membrane (PMe)—[♀, ♂]. The membrane between the tergum and sternum of a body segment.

pleural suture (pls)—[♀, ♂]. The external groove of the internal pleural ridge separating the episternum and epimeron of each segment.

pleural wing process (pwp)—[♀, ♂]. A process at the upper end of the mesopleural or metapleural suture. The mesopleural wing process is posterior to the basalare and at the apex of the posterior mesanepisternum, serving as a fulcrum for the wing. A metapleural wing process is in a similar position in relation to the basalare and mesanepisternum of the metathorax.

pleuron (p)—[♀, ♂]. The sclerotization of the pleural (=lateral) area of a body segment; on the thorax, termed pro-, meso- or metapleuron as appropriate.

pleurotergite (Pt)—[♀, ♂]. Upper and lower pleurotergites are the lateral areas of the mesopostnotum, located above the metathoracic spiracle. Each upper and lower pleurotergite is the site of an internal apodeme.

plumose seta (pl. plumose setae)—[L]. Referring to a seta that is long with many regularly arranged, or alternating, branches on each side of the main stem, similar to a plume or feather. The branching plume can be widespread or narrow.

posterior. A direction opposite to the direction of the anterior end, which is the head on mosquitoes. However, leg segments and many structures can have posterior sides or margins, as well as anterior, dorsal, and ventral sides/margins.

posterior dorsocentral seta (PDS)—[♀, ♂]. Setae in a row on the scutum in the posterior dorsocentral area, the setae forming a continuous group with the prescutellar setae.

posterior mesanepisternum (PMas)—[♀, ♂]. A narrow vertically oriented sclerite extending from a ventral position at the upper portion of the mesokatepisternum to a vertical position at the prealar knob. It is part of the mesanepisternum, separated from the remainder of the sclerite by the weakly developed mesanapleural suture.

posterolateral spiracular lobe plate I (PSLP′)—[L]. In culicines, the inner sclerite (that nearest the spiracular openings) of one of the posterolateral spiracular lobes of the spiracular apparatus. In anophelines the homologous sclerite may be part of the median spiracular plate.

posterolateral spiracular lobe plate II (PSLP″)—[L]. The often poorly delimited outer sclerite of one of the posterolateral spiracular lobes of the spiracular apparatus. The plate is at the distal margin of culicine siphons and at the part of the spiracular apparatus closest to abdominal segment VIII in anophelines.

postgenital lobe (PGL)—[♀]. The sclerotized median lobe ventral to the anus and dorsal to the upper vaginal lip.

postlabium (PLb) or postmentum (Pmt)—[♀, ♂, L]. In adults, the basal part of the labium proximal to the prelabium. The postlabium is either a single sclerite corresponding to the postmentum or divided by the submentomental suture into a distal mentum and proximal submentum. In larvae the postlabium is a basal portion of the mouthpart, extending anteriorly to the prementum, laterally to the hypostomal suture, posteriorly to the level of the tentorial pits, and continuously with the gula.

postnotum (Pn)—[♀, ♂]. The phragma-bearing plate in the dorsum of a wing-bearing thoracic segment; derived from the anterior part (acrotergite) of the following notum.

postprocoxal membrane (PM)—[♀, ♂]. The membrane between the forecoxa and the mesokatepisternum on the side of the thorax.

postpronotal scale (PpSc)—[♀, ♂]. Any scale borne on the postpronotum, occurring in groups of upper and lower postpronotal scales.

postpronotum (Ppn)—[♀, ♂]. A large lateral thoracic sclerite ventral to the scutum, anterior to the scutal angle and mesothoracic spiracle, and immediately posterior to the antepronotum. This sclerite has a narrow ventral extension that meets the upper part of the proepisternum and is not clearly separated from the proepimeron.

postspiracular area (PA)—[♀, ♂]. A small sclerotized area ventral and posterior to the mesothoracic spiracle, formed from a portion of the anterior anepisternum. The postspiracular area is connected or continuous with the subspiracular area.

postspiracular scales (PoSc)—[♀, ♂]. Scales on the postspiracular sclerite. Groups of scales may extend posteriorly onto the membrane of the anepisternal cleft.

postspiracular seta (PS)—[♀, ♂]. One of the setae occurring in a group on the postspiracular area.

prealar knob (PK)—[♀, ♂]. A portion of the posterior mesanepisternum that is formed into a dorsally projecting knob anterior to the wing base and ventral to the paratergite.

prealar scale (PrSc)—[♀, ♂]. Any scale borne anterior to the wing on the posterior mesanepisternum; an upper group occurs among and above the prealar setae, and a lower group sometimes occurs below them.

prealar seta (PaS)—[♀, ♂]. One of the setae occurring in a group on the prealar knob.

preapical. A directional term used for a scale band or seta that is found near but not at the apex of a structure.

precratal seta (pl. precratal setae) (PcrS)—[L]. Paired setae on the venter of a complete saddle or membranous area on segment X that are just anterior to the ventral brush and anterior to the sclerotized grid. Both the ventral brush and the precratal setae are morphologically setae 4-X.

prelabium (pLb)—[L]. The distal part of the labium formed from the prementum, ligula, and labial palpi. It is separated from the hypopharynx by the salivary meatus but united with it laterally by the cibarial bars to form the labiohypopharynx.

prementum (Prmt)—[♀, ♂]. The stipital region of the labium, formed into the sheath that contains the feeding stylets fitted into the median dorsal groove. It also bears the ligula and labella distally.

prescutellar area (PrA)—[♀, ♂]. The median posterior area of the scutum between the posterior acrostichal area and the scutellum.

prescutellar seta (PrsS)—[♀, ♂]. Any seta on the anterior or lateral margins of the prescutellar area. These setae are usually in several rows and are often difficult to distinguish anteriorly from the dorsocentral setae.

prespiracular area (PsA)—[♀, ♂]. The small triangular area of the anterior mesanepisternum just anterior to the mesothoracic spiracle; set off from the postpronotum by a strong ridge.

prespiracular scale (PsSc)—[♀, ♂]. One of the scales occurring in a small group on the prespiracular area.

prespiracular seta (PsS)—[♀, ♂]. Any seta occurring on the prespiracular area.

proboscis (P)—[♀, ♂]. A generalized word for the greatly elongated sheath-like labium and the enclosed mouthparts.

proctiger (Pr)—[♂]. The entire complex of structure formed by the postgenital abdominal segments X and XI.

proepimeron (Pm)—[♀, ♂]. The area of the propleuron posterior to the propleural suture. The proepimeron is not usually well demarcated from the lower margin of the postpronotum.

proepisternal scale (PeSc)—[♀, ♂]. Any scale on the proepisternum occurring in upper and lower groups or as one large group.

proepisternum (Ps)—[♀, ♂]. An anterolateral sclerite anterior to the propleural suture on each side of the thorax that extends from the base of the antepronotum down to the base of the forecoxa and further anteriorly around the coxa under the cervix to join together on the midline.

pronotum (PrN)—[♀, ♂]. The notum of the prothorax, which is much reduced medially but well developed laterally into the antepronotum and postpronotum.

prosophallus (PO)—[♂]. A ventral (prerotation sense), lobe-like element of the phallosome located between the bases of the gonocoxites known only from *Verrallina*.

prothorax (P)—[♀, ♂, L]. The anterior segment of the thorax. In adults, parts include the antepronotum, postpronotum, proepisternum, postprocoxal membrane, and foreleg. In larvae it is the most anterior part of the fused thorax, bearing 14 pairs of setae numbered in standardized manner starting from the dorsal midline.

proximal. More basal, toward the insect body.

proximal division of subapical lobe (pSL)—[♂]. The proximal portion of the subapical lobe in *Culex*.

proximal terminal setae (pts)—[♂]. Proximal tuft of gonocoxal setae in *Onirion*.

pulvillus (pl. pulvilli) (Pv)—[♀, ♂]. One of two pad-like or divided lobes on the fifth tarsomere, one arising below the base of each unguis.

R

radial sector (R_s)—[♀, ♂]. The stem of radius-one-plus-two (R_{1+2}) and radius-three-plus-four (R_{3+4}); splitting from the stem of the radius (R) with radius-one (R_1).

radiomedial crossvein (rm)—[♀, ♂]. The crossvein extending between the radius-four-plus-five (R_{4+5}) and the medial vein-one-plus two (M_{1+2}).

radius (R)—[♀, ♂]. The third principal longitudinal vein of the wing, dividing into four branches. Radius one (R_1) extends from the base of the wing to the distal wing margin, anterior to the apex. Radius two-plus-three (R_{2+3}) emerges posteriorly from R_1 at about the middle of the length of that vein. In the apical third of the wing, R_{2+3} branches into radius two (R_2) and radius three (R_3), both reaching the wing margin near the apex. Radius four-plus-five (R_{4+5}) emerges posteriorly from R_{2+3} apical to the midpoint of the wing, extending to the apex. There is a short crossvein-like segment between the point where R_{4+5} emerges from R_{2+3} and the radiomedial crossvein.

remigial seta (ReS)—[♀, ♂]. Any seta occurring on the posterior margin of the radius proximal to the arculus.

remigium (Re)—[♀, ♂]. The part of the radius basal to the arculus.

S

saddle (Sa)—[L]. The large dorsal sclerite covering most of the dorsal and lateral surfaces of abdominal segment X. It can completely circle the segment. The segment supports seta 1-X and, often, many tiny nonarticulating spicules.

saw (SAW)—[L]. In *Mansonia* and *Coquillettidia*, the anterior serrated plate of the spiracular apparatus used for insertion and attachment of the siphon into plants; at least partly homologous with the anterior spiracular lobe of other mosquito larvae.

scutal angle (ScA)—[♀, ♂]. The acute angular projection on the lateral margins of the scutum just anterior and dorsal to the prespiracular area or just anterior to the prescutal suture when present.

scutal fossa (SF)—[♀, ♂]. A slightly concave anterolateral area on the scutum extending posteriorly to the level of the scutal angle and from the dorsocentral row to the edge of the scutum.

scutellum (Stm)—[♀, ♂]. A transverse area on the dorsum of the thorax immediately posterior to the scutum and separated from it by the scutoscutellar suture. The mosquito scutellum is morphologically the mesoscutellum. The hind margin is trilobed in culicines except *Toxorhynchites* and evenly rounded in anophelines except *Chagasia*.

scutum (Scu)—[♀, ♂]. The principal dorsal area of the thorax belonging to the mesonotum. It extends posteriorly from the anterior promontory to the scutellum. The scutum combines the prescutum and scutum because the prescutal suture is only present laterally.

siphon (S)—[L]. In culicines, a dorsal tube originating on abdominal segment VIII. The spiracular apparatus (that contains elements of abdominal segment IX) is at its apex, and it contains the posterior ends of the dorsal tracheal trunks. The shapes of siphons vary, as do their vestiture with setae and spicules.

siphon acus (SA)—[L]. In some culicines, a small sclerite attached posterolaterally to the base of the siphon. The acus forms a complete ring around the siphon in *Mansonia*.

spicule(s)—Minute, nonarticulating spinelike structures that can occur in dense patches on both sclerotized and membranous parts of adults and larvae.

spine(s)—[L]. A large, sturdy, immovable spicule, with a sharp or narrowly rounded tip.

spinule(s)—[L]. A minute, spinelike spicule; always stiff; commonly occurring in short rows on the saddle and siphon.

spiracular apparatus (Sap)—[L]. A five-lobed valvular structure at the tip of the siphon in culicines and on abdominal segment VIII in anophelines.

sternum (pl. sterna) (S)—[♀, ♂, L]. The ventral sclerotized part of the exoskeleton, usually used for the abdominal segments, also used in reference to individual segments.

sternum. Entire venter of a segment (called a sternite by many).

sternum IX (IX-S)—[♂]. Abdominal sternum IX is a prominent and mechanically important part of the male genitalia of most mosquitoes. It extends laterally and merges with abdominal tergum IX, forming a ring.

stipes (Stp)—[♀, ♂, L]. In adults, a rod-like structure located inside the head ventral to the tentorial arms; posteriorly fused with the cardo to form the stipitocardinal rod; anteriorly articulated with the lacinia and providing support to the maxillary palpus. In larvae, often merged with the galea to form the galeastipes that constitutes much of the lateral side of the maxilla and supports the maxillary brushes and the maxillary palpus.

stipitocardinal rod (SCR)—[♀, ♂]. The fused cardo and stipes.

subapical lobe (SL)—[♂]. In *Culex*, a mesal or dorsomesal (prerotation sense) lobe located near or distal to the middle of the gonocoxite. The subapical lobe is divided into distal (dSL) and proximal (pSL) portions.

subapical lobe setae—[♂]. Setae on the subapical lobe of *Culex* have distinct, characteristic forms. The designation of these setae has been standardized as follows:

a = seta *a* of pSL
b = seta *b* of pSL
f = flat seta of dSL
h = hooked seta of dSL
l = foliform seta of dSL
s = saber-like seta of dSL

subcosta (= subcostal vein) (Sc)—[♀, ♂]. The second primary longitudinal vein on the wing; a single vein immediately behind the costa that extends lengthwise toward the apex of the wing anterior to radius one (R_1). The subcosta joins the costa proximal to the wing apex.

submedian accessory tergal plate (SATP)—[L]. In some anopheline larvae, one of usually a pair of small dorsal sclerites borne on each side of the midline immediately posterior to the median accessory tergal plates of abdominal segments I–VII.

submentum (Smt)—[♀, ♂, L]. The basal region of the labium. In adults, the submentum is present when there is a submentomental suture dividing the postmentum into the distal mentum and proximal submentum. In larvae, the submentum is the base of the labium adjacent posteriorly to the gula and together forming the labiogula.

subspiracular area (SA)—[♀, ♂]. An area just above the dorsoanterior extension of the mesokatepisternum and ventral to the hypostigmal area and mesothoracic spiracle. It is usually continuous with or connected to the postspiracular area. The subspiracular area is part of the anterior mesanepisternum.

subspiracular scale (SSc)—[♀, ♂]. One of the scales occurring in a group on the subspiracular area.

supraalar area (SaA)—[♀, ♂]. A small lateral area of the scutum just above and in front of the wing.

supraalar scale (SaSc)—[♀, ♂]. One of the scales occurring on the supraalar area of the scutum; sometimes difficult to distinguish from the antealar scales.

supraalar seta (SaS)—[♀, ♂]. One of the setae occurring in a cluster on the supraalar area of the scutum; a row of supraalar setae may extend anteriorly above the antealar area or be more or less continuous with the posterior end of a row of antealar setae.

T

tarsal claw (U, U-I, U-II, U-III)—[♀, ♂]. Commonly used terminology for the unguis (pl. ungues) on the fifth and most distal tarsomere of each leg—fore- (U-I), mid- (U-II) and hind- (U-III) ungues.

tarsomere (Ta-I$_{1-5}$, Ta-II$_{1-5}$, Ta-III$_{1-5}$)—[♀, ♂]. One of the five individual subsegments of a tarsus, indicated by a subscript in order from proximal to distal. The fifth tarsomere bears the two ungues and, in some species, the pulvilli.

tarsus (pl. tarsi) (Ta-I, Ta-II, Ta-III)—[♀, ♂]. The fifth leg segment (designated fore-, mid-, or hindtarsus) located distal to the tibia. The tarsus consists of five tarsomeres.

tergal plate (TP)—[L]. A small to large anteromesal sclerite occurring on the dorsum of each of the abdominal segments I–VIII of *Anopheles* larvae. In both anophelines and culicines, abdominal segment X is typically covered by a tergal plate called the saddle. The saddle extends laterally, sometimes completely around the segment.

tergite—[♀]. A subdivision of a tergum—typically, small paired plates associated with tergum X of females; often incorrectly applied to an entire tergum.

tergum (pl. terga) (Te)—[♀, ♂, L]. The dorsal sclerotization of a body segment; called the notum on the thorax but usually used in referring to the dorsal sclerites on the abdominal segments. Such plates are much less common on larvae.

tergum lobe (IX-TL)—[♂]. One of a pair of lobes on the ninth tergum of the male genitalia.

tergum IX (IX-Te)—[♀, ♂]. On females, a small plate between tergum VIII and the cerci. On males, the ventral (postrotation sense) portion of a sclerite forming the base of the genitalia; usually bilobed.

tergum X (X-Te)—[♀, ♂]. In male mosquitoes, usually developed as a pair of tergites laterally at the base of the proctiger, often articulating with tergum IX and the basal piece and sometimes connected to the basal part of the paraproct sclerotization. In female mosquitoes, a small sclerite situated dorsally between tergum IX and the cerci in some species, particularly anophelines.

tertiary fringe scales (TFC)—[♀, ♂]. Two of the four rows of scales forming a fringe around the trailing edge of the wing. The tertiary fringe scales are the smallest.

thorax (pl. thoraces)—[♀, ♂, L]. The second or intermediate division (tagma) bearing legs and wings and comprising pro-, meso-, and metathoracic segments.

tibia (pl. tibiae) (Ti-I, Ti-II, Ti-III)—[♀, ♂]. The fourth segment of the leg, just distal to the femur and referred to as fore-, mid- and hindtibia.

trachea (pl. tracheae)—[♀, ♂, L]. Gas-filled, sclerotized tubes forming the respiratory system and connecting to the atmosphere, the primary structures of which are the paired dorsal tracheal trunks. The tracheae lead to smaller, noncuticular, liquid-filled tubes (tracheoles) that convey gases in dissolved form to tissues.

transnotal suture (TnS)—[♀, ♂]. The transverse suture of the notum separating the antepronotum and the postpronotum.

U

unguis (pl. ungues) (U)—[♀, ♂]. The tarsal claw inserted as a pair on the fifth tarsomere (post-tarsus). Referred to as fore- (U-I), mid- (U-II), and hind ungues (U-III).

upper calypter (UC)—[♀, ♂]. A lobe of the wing located at the base of the trailing edge.

upper mesepimeral scale (UMSc)—[♀, ♂]. One of the scales occurring in a group on the dorsoposterior quarter of the mesanepimeron. Mixed with the upper mesepimeral setae.

upper mesepimeral seta (McSU)—[♀, ♂]. One of the setae occurring in a group on the upper posterior area of the mesanepimeron, anterior to the metathoracic spiracle.

upper mesokatepisternal scale (MScU)—[♀, ♂]. One of the scales occurring in a horizontal line on the upper part of the mesokatepisternum.

upper mesokatepisternal seta (MkSU)—[♀, ♂]. One of the setae occurring in a horizontal line on the upper part of the mesokatepisternum.

upper proepisternal seta (PeSU)—[♀, ♂]. Any of the setae on the upper proepisternum.

V

vein 1A—[♀, ♂]. See **anal vein**.

vein C—[♀, ♂]. See **costa**.

vein CuA and vein CuP—[♀, ♂]. See **cubitus**.

vein M, vein M_1, vein M_2, and vein M_{3+4}—[♀, ♂]. See **media**.

vein R_1, vein R_2, vein R_3, and vein R_{4+5}—[♀, ♂]. See **radius**.

vein Sc—[♀, ♂]. See **subcosta**.

venter. Entire ventral area of a segment (= sternum, called sternite by many).

ventral brush (seta 4-X)—[L]. Technically known as seta 4-X and projecting as a linear series of irregularly or regularly paired setae borne posteroventrally on the ventral midline of abdominal segment X; can appear as one to two pairs of setae in certain species. The longer, more posterior setae in a brush usually originate on a grid or boss, and anterior preboss (= precratal) setae, if present, are usually weaker and shorter.

ventromentum (Vm)—[L]. The portion of the mentum proximal to the submentomental suture. In anophelines, it is a well-developed, apically dentate structure. In culicines, it can be difficult to distinguish as a lightly sclerotized spiculate area.

vertex (V)—[♀, ♂]. The area on the dorsal and lateral surfaces of the head just posterior to the compound eyes and interocular space and anterior to the occiput.

W

wing fringe—[♀, ♂]. The scales on the apex and trailing edge of the wing consisting of four rows of scales: one of long fringe scales, one of medium-length secondary scales, and two rows of short tertiary scales.

wing spots—[♀, ♂]. Dark or light-colored scales that form a pattern of contrasting spots on various veins of the wing. The nomenclature of these spots is standardized and represents homologies across genera. They are as follows:

accessory preapical dark (APD)
accessory sector dark (ASD)
accessory sector pale (ASP)
apical dark (AD)
apical pale (AP)
basal dark (BD)
basal pale (BP)
humeral dark (HD)
humeral pale (HP)
postsubcostal dark (POSCD)
postsubcostal pale (POSCP)
preapical dark (PD)
preapical pale (PP)
prehumeral dark (PHD)
prehumeral pale (PHP)
presector dark (PSD)
presector pale (PSP)
presubcostal pale (PRSCP)
sector dark (SD)
sector pale (SP)
subapical dark (preapical dark + apical dark) (SAD)
subbasal dark (humeral dark + presector dark) (SBD)
subcostal area (SCA)
subcostal dark (SCD)
subcostal pale (SCP)

Index

Page numbers followed by f and t indicate figures and tables. Page numbers in *italic* indicate glossary definitions, and page numbers in **bold** indicate the main discussion.

Mosquitoes of the World

Mosquitoes of the World

VOLUME 2

Richard C. Wilkerson, Yvonne-Marie Linton, and Daniel Strickman

Johns Hopkins University Press
BALTIMORE

Printed in Hong Kong on acid-free paper
9 8 7 6 5 4 3 2 1

Johns Hopkins University Press
2715 North Charles Street
Baltimore, Maryland 21218-4363
www.press.jhu.edu

Library of Congress Cataloging-in-Publication Data

Names: Wilkerson, Richard C., 1946– author. | Linton, Yvonne-Marie, 1972– author | Strickman, Daniel, author.
Title: Mosquitoes of the world / Richard C. Wilkerson, Yvonne-Marie Linton, and Daniel Strickman.
Description: Baltimore, Maryland : Johns Hopkins University Press, 2020. | Includes bibliographical references and index.
Identifiers: LCCN 2019041641 | ISBN 9781421438146 (hardcover) | ISBN 9781421438153 (ebook)
Subjects: LCSH: Mosquitoes.
Classification: LCC QL536 .S798 2020 | DDC 595.77/2—dc23
LC record available at https://lccn.loc.gov/2019041641

A catalog record for this book is available from the British Library.

Special discounts are available for bulk purchases of this book. For more information, please contact Special Sales at specialsales@press.jhu.edu.

Johns Hopkins University Press uses environmentally friendly book materials, including recycled text paper that is composed of at least 30 percent post-consumer waste, whenever possible.

Contents

Expanded Contents

PART THREE

Taxonomic Catalog of Culicidae

Genus Color Key

Anopheles
Bironella
Chagasia
Aedomyia
Aedes
Armigeres
Eretmapodites
Haemogogus
Heizmannia
Opifex
Psoraphora
Udaya
Verrallina
Zeugnomyia
Culex
Deinocerites
Galindomyia
Lutzia
Culiseta
Ficalbia
Mimomyia
Hodgesia
Coquillettidia
Mansonia
Orthopodomyia
Isostomyia
Johnbelkinia
Kimia
Limatus
Malaya
Maorigoeldia
Onirion
Runchomyia
Sabethes
Shannoniana
Topomyia
Trichoprosopon
Tripteroides
Wyeomyia
Toxorhynchites
Uranotaenia

PART THREE

Taxonomic Catalog of Culicidae

Introduction

How to Become an Entomologist

Oh, to name a species
Now that spring is here.
And who would name a species
Would leave his name for 'ere.

Proclaim the tarsi "longer"!
Confuse the setal hair.
Declare the palpi "darker."
A muddled key, prepare.

So, go you! Name your species.
Describe in verbiage prim,
From cerci to antennae
This holotype so trim.

Go, you, name the species,
If taxonomy's your whim.
However it is published, it's
Another synonym.

—Albro T. Gaul

Bulletin of the Brooklyn Entomological Society
1946, Vol. XLI, p. 83

The system for naming organisms, including mosquitoes, had its official start with the 10th edition of *Systema Naturae* by Carl Linnaeus in 1758. Linnaeus himself is credited with naming a number of mosquito species, such as *Aedes aegypti* (the yellow fever mosquito, a.k.a. the Yellow Fever Malaysian Pointy Mosquito) and *Culex pipiens* (the northern house mosquito, a.k.a. the Piping Swedish Typical Mosquito) (shown are the current scientific names and informal names designated herein and in the official list of common names from the Entomological Society of America [ESA]). Between 1758 and 1800, not much describing and classifying was underway in the new discipline codified by Linnaeus, even though classification of plants, animals, and things has been carried out by humans ever since they, and probably other human congeners, have been in existence. By 1800 there were 14 valid mosquito species, named by 11 authors, all placed in genus *Culex* (see Knight 1972 for a review of 18th-century mosquito classification). Two more genera, *Anopheles* and *Aedes*, were added by Meigen early in the 19th century, and currently recognizable genera *Haemagogus*, *Psorophora*, *Sabethes*, and *Uranotaenia* were in use by 1900. See Harbach (2018c), a comprehensive resource for mosquito-related names and classification, for an overview of the history of mosquito classification.

By 1900 only 183 names had been proposed for mosquito species. However, the pace of study and descriptions soon picked up, presumably because in 1897 Ronald Ross proved that *Anopheles* mosquitoes are intermediate hosts for human malaria parasites and because Carlos Finlay's hypothesis that mosquitoes transmit yellow fever virus was proven correct. See Figure 1 for a graphic showing the number of species (all names, including current synonyms) described by year and cumulatively over time. It suggests, in addition to the initial burst of species descriptions, that more taxonomic activity took place during times of global conflict and the US Vietnam War (see the first three arrows, Figure 1), but much less (our inference) occurred after the development of DNA methods for species identification and phylogenetic inference. The increase indicated by the fourth arrow was mostly a result of excellent work done in Madagascar and Africa by a few authors (Brunhes, da Cunha Ramos, Ribeiro, & Le Goff). The flattening of the curve in the figure suggests there are either fewer and fewer new species to describe or fewer taxonomists to

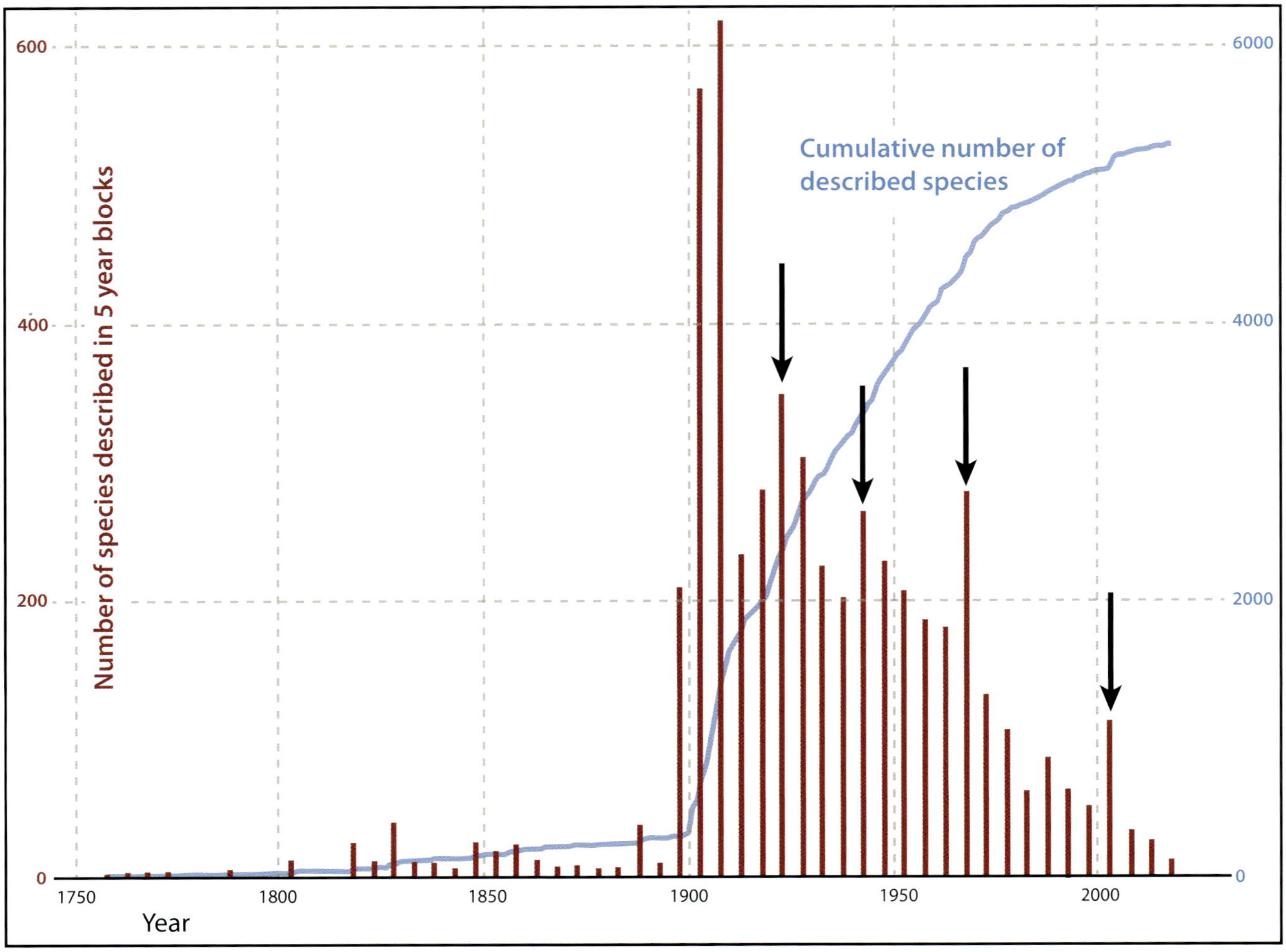

Figure 1. Mosquito new species descriptions in five-year increments (red bars), and cumulatively (blue line). Arrows indicate assumed correlation of increased taxonomic activity in wartime (see text for further comments).

Table 1. Relative efficiency of species description for a few select authors

Author	Total names	Valid species	% valid
Dyar	177	88	50
Edwards	425	357	84
Theobald	584	318	54.5
Ludlow	68	33	48.5
Taylor	84	36	43
Leicester	112	82	73
Belkin	69	68	98.5

describe them. We prefer the latter explanation, based on informal museum collection inventories and the backlog of unnamed species known to researchers from around the world. The use of DNA, especially *COI* barcoding methods, has resulted in a flood of previously unrecognized taxa, but not their naming. A conservative estimate is that at minimum only half the extant species have been described (Unpublished data: Richard Wilkerson).

The number of names proposed in a given time period gives one a sense of taxonomic activity but not necessarily accuracy. Table 1 gives examples of what could be termed "noise" produced when authors are unaware, for whatever reason, that a species they are describing has already been named. In total about 30% of species described have been subsequently determined to be synonyms.

Today, and in the catalog presented here, we have a total of 5,403 species names (including fossils). Included in the above total are 3,570 valid species; 130 valid subspecies; 59 *nomina dubia*; 28 fossil species; 81 *nomina nuda* and rejected names; and 1,537 synonyms—that is, species described at later dates for species described earlier, priority being paramount in determining valid species (International Commission on Zoological Nomenclature 1999).

Family Culicidae is a monophyletic yet diverse and "old" family (see Vol. 1, Chapter 1). Until recently, Culicidae was considered to have three subfamilies: Culicinae, Anophelinae, and Toxorhynchitinae (until the 1950s, the families Dixidae and Chaoboridae were also included in Culicidae). The latter (consisting of genus *Toxorhynchites*) was reduced to a tribe in subfamily Culicinae by Harbach & Kitching (1998) and Mitchell

Table 2. Number of taxa in respective categories (Knight & Stone 1977; Harbach 2018c)

	Edwards 1932	Stone, Knight & Starke 1959	Knight & Stone 1977	Harbach 2018c
Genera	30	31	34	41
Subgenera	89	95	120	162
Generic group synonyms	211	228	220	232
Species	1,400	2,401	2,960	3,559
Subspecies and varieties	146	208	173	1301[1]
Species group synonyms	1,286	1,423	1,478	1,537[2]
Nomina dubia		19	35	59
Nomina nuda		37	46	76[3]
Fossil species			16	25
Rejected names			3	3
Unavailable names				1

[1]Varietal names no longer valid according to the International Commission on Zoological Nomenclature (1999); implemented for Culicidae by Harbach & Howard (2007).
[2]Includes two fossil species.
[3]Includes one fossil name.

et al. (2002). Presently, there are 11 tribes in Culicidae. A tribe is a suprageneric category of convenience that implies relatedness of constituent taxa. Some have single genera, but others are quite large. Belkin (1962) established 10 tribes based on groupings mostly articulated by Edwards (1932). And, as mentioned above, the 11th tribe, Toxorhynchitini, was added later.

The catalog presented here is built on the work of hundreds of taxonomists. A trivia number for sure, but here are 899 single or multiple author descriptions written by 1,110 individuals. In large part, the current classification follows Edwards (1932). Subsequently, Stone et al. (1959) produced a comprehensive taxonomic catalog to the Culicidae, which was then followed by Knight & Stone's (1977) catalog, with supplements by Knight (1978b), Ward (1984b), and Ward (1992). The present catalog is built on these catalogs and supplements, and we mostly follow the format—and sometimes use verbatim information—from those publications. Table 2 is a continuation of a table in the introduction to Knight & Stone's (1977) catalog, showing numbers of taxonomic groups over time.

Some Current Issues in Mosquito Systematics

Tribe Aedini. Tribe Aedini has been the focus of intense morphological phylogenetic study for the last 20–30 years. The following, from Wilkerson et al. 2015 and Wilkerson & Linton (2015), is reproduced here, mostly verbatim, to summarize the various studies and classification changes of interest in current aedine systematics.

The tribe Aedini comprises about a third of all recognized mosquito species and includes many vectors of viral diseases debilitating to humans, such as dengue and chikungunya. Within the tribe, the genus *Aedes*, in the traditional sense, is the largest, with 970 species and subspecies. Other aedine genera are *Armigeres*, *Eretmapodites*, *Haemagogus*, *Heizmannia*, *Opifex*, *Psorophora*, *Udaya*, *Verrallina*, and *Zeugnomyia*. During the past 11 years, based on a series of morphological phylogenetic studies by Reinert, Harbach, & Kitching (RH&K) (2004, 2006, 2008, 2009) and the taxonomic actions resulting from those studies, the original genus *Aedes* was split into 74 genera, reducing the genus *Aedes* from over 900 species (Knight & Stone 1977; Knight 1978; Ward 1984b; Ward 1992) to 12. Chief among the reasons given by RH&K to elevate so many genera was the authors' claim of an unreferenced "principle of equivalent rank." This implied that if traditionally accepted genera were phylogenetically coequal with other clusters of species in their analyses, the newly recognized groups should be given similar taxonomic status. These taxonomic actions were highly controversial (Savage & Strickman 2004; Savage 2005) and resulted in widespread confusion about which names to apply to most vectors of disease organisms in genus *Aedes* (see Table 1 in Wilkerson et al. 2015). For example, during this period *Aedes japonicus* (Theobald), an invasive species and proven vector of West Nile virus and Cache Valley virus, was known variously as *Aedes (Finlaya) japonicus* (Knight & Stone 1977), *Ochlerotatus (Finlaya) japonicus* (Reinert 2000h), '*Ochlerotatus*' ('*Finlaya*') *japonicus* (Reinert, Harbach, & Kitching 2004), and *Hulecoeteomyia japonica* (Reinert, Harbach, & Kitching 2006).

Close scrutiny of the RH&K phylogenetic results and a reanalysis of their data set led Wilkerson et al. (2015) to conclude that based on the evidence provided by RH&K, the classification changes they promoted, which had resulted in the split of the well-known genus *Aedes* into so many genera, were not warranted. *Aedes*, in a more traditional sense, was therefore reinstated (Wilkerson et al. 2015; Wilkerson & Linton 2015). But to preserve the phylogenetic hypotheses of RH&K, their genera were reduced in rank to subgenera of *Aedes*. Any subgenera in the fragmented RH&K system were reduced to putative informal group status. Rationalization for reinstatement of genus *Aedes* to include all "traditionally" accepted species was based on opinions promoting a conservative approach to classification change based on new phylogenetic analyses (Wiley 1979, 1981; Wiley & Liebermann 2011). These opinions were comprehensively solidified by Vences et al. (2013), who, in detail, discussed the relationship between nomenclatorial utility and phylogenetic accuracy. As a guide to determine the suitability for classification changes, they proposed a number of taxon-naming criteria (TNCs). Appropriate TNCs were cited to reinstate the "traditional" species in genus *Aedes* (Wilkerson et al. 2015; Wilkerson & Linton 2015). Central to these arguments reinstating genus *Aedes* while retaining other traditional aedine genera were TNC 2, Clade Stability; TNC 3,

Phenotypic Diagnosability; TNC 8, Manageability; TNC 10, Nomenclatural Stability; and TNC 11, Community Consensus. Since, to these authors (Wilkerson et al. 2015), no compelling evidence warranted changing the classification of traditional diagnosable genera, such genera in tribe Aedini should be retained until strong, multiple lines of evidence are produced showing the contrary. Harbach (2018c) partly follows this reasoning but, justifiably, instead cites Edwards (1932) as the source for economy of change and restraint from fragmentation of taxa ("splitting"). Harbach (2018c) also supports the many-genera approach for Aedini, terming it a "phylogenetic" classification and calling the classification of Wilkerson et al. a "modified traditional polyphyletic classification." In fact, the two aedine classifications are essentially the same (except for some subgenera that were reduced to informal groupings), but the ranks of the numerous genera were reduced to subgenera. Except for relatively minor differences, Wilkerson et al.'s (2015) analysis found no alternative overall phylogeny, which was not surprising since they used the same data set but different analytical methods. It is worth noting that in addition to the Reinert et al. papers, others (e.g., Soghigian et al. 2017) have reinforced findings that genus *Aedes* is paraphyletic and is internally composed of other aedine genera. So far, however, no adequate way forward has been posited to reflect these as yet not-well-supported findings into a new workable classification.

New replacement name. A replacement name, *Aedes* (*Paulianius*) *brucei* Wilkerson, is proposed here for *Aedes* (*Paulianius*) *hirsutus* (Brunhes & Boussès 2017), original combination *Paulianius hirsutus* Brunhes & Boussès. This species became a junior secondary homonym of *Aedes* (*Aedimorphus*) *hirsutus* (Theobald 1901a), original combination *Culex hirsutum*, when Wilkerson et al. 2015 designated genus *Paulianius* Brunhes & Boussès as a subgenus in genus *Aedes* Meigen, making it the senior secondary homonym.

Genus *Anopheles*. A similar question regarding rankings of genus group taxa is a proposal by Foster et al. (2017) that, based on whole mitochondria sequence of mostly New World species, elevates the subgenera *Kerteszia*, *Nyssorhynchus*, *Lophopodomyia*, and *Stethomyia* to generic rank. Further, their genus *Nyssorhynchus* was divided into two subgenera, *Nyssorhynchus* and *Myzorhynchella*. Refer also to Foster et al. (2017) for a number of species realignments and a proposal to describe a new genus in Anophelinae for *An. parvus*. We do not incorporate their classification here because it was published after we finalized our catalog. Additionally, we prefer to first assess general community acceptance to the utility of their classification in reference to criteria in Vences et al. (2013), as described above for tribe Aedini. See Harbach (2018a) for further discussion.

Informal species groups. Informal species groups (see p. 1077) have been proposed in some genera for similar-appearing species. Listed here are genera *Anopheles*, *Aedes*, *Culex*, *Uranotaenia* Old World/New World, *Mimomyia*, *Orthopodomyia*, *Tripteroides*, and *Toxorhynchites*. Since they are not covered by the *International Code of Zoological Nomenclature*, a wide range of category names have been used. In some cases there have been successful attempts (*Anopheles*, *Culex*) to standardize the names and their equivalency. Usually, there is no proven phylogenetic basis for these groups, but they serve as potential hypotheses of relatedness. As a convention for presenting these group names, we capitalize both names (e.g., Leucosphyrus Subgroup) and do not italicize the group name, even if it is a scientific name (i.e., in this case it is derived from *Anopheles* (*Cellia*) *leucosphyrus*). This distinction is important to avoid confusion from mixing names covered by the *Code* with those that are not. Though other conventions are used (e.g., *leucosphyrus* subgroup), we do not recommend it. When in doubt, to express that a species group going by a specific name is being discussed, we recommend using, for example, *Anopheles leucosphyrus sensu lato* (*s.l.*) (in a broad sense). We think this better serves communication, and in distinction use *sensu stricto* (*s.s.*) if the identity of a species is not in doubt.

The groups presented here have been proposed by others, and attribution is given with each entry. Genera *Culex* and *Anopheles* species groups were taken mostly verbatim from reviews done by Ralph Harbach (see references in those sections).

Generic and subgeneric abbreviations. As a convenience and to improve communication, we suggest the use of two- and three-letter abbreviations for genera and subgenera, respectively (Wilkerson et al. 2015; Wilkerson & Linton 2015). Table 3 contains a modification of a number of such lists proposed by Reinert over the years.

Terms used in the taxonomic catalog. As in Knight & Stone (1977), the gender of generic-group names is summarized here in Table 3, which also gives the number of species currently recognized in each genus or subgenus: f (feminine), m (masculine), n (neuter). This is important because gender agreement of genus-group names and adjectival and participial species-group names is required by the International Commission on Zoological Nomenclature (Article 32.2). See Harbach (2018c, p. 142) for a complete discussion since we do not pursue how to name species, just catalog them. In addition to *sensu lato* (*s.l.*) and *sensu stricto* (*s.s.*) (see above) a few other Latin terms can be found in the catalog: *auctorum* (of authors) and *sensu auctorum* (a particular author's concept of a taxon); *nomen dubium*, plural *nomina dubia* (a name of unknown or doubtful application); *nomen nudum*, plural *nomina nuda* (name(s) without status, not available); and, *nomen oblitum* (a forgotten name that can be designated as no longer available).

Genus-group entries. Each valid genus is accompanied by author-date, type species, and a name derivation, if known. Genus-group synonyms are also given. For most valid genera, there is a section with what we judge to be important references pertaining to the genus in general. This is straightforward if no subgenera are listed under a given genus. If there are subgenera, we might provide literature that applies to all the subgenera, which will be found there. Otherwise, these references will appear under the nominotypical subgenus—for example, for *Anopheles*, under *Anopheles* subgenus *Anopheles*.

Format for genus-group taxonomic entries usually follows Knight & Stone (1977) and supplements (Knight 1978b; Ward 1984b; Gaffigan & Ward 1985; Ward 1992). A departure from Knight & Stone and the rest is the use of the term "type spe-

Table 3. Gender of valid genus-group names used herein, number of valid species, subspecies, and total in each, and suggested two-letter genus (in bold) and three-letter subgenus abbreviations (*nomina dubia* and fossil species not included)

Subfamily/tribe	Genus	Subgenus	Gender	No. of sp.	No. of ssp.	Total	Abbreviation
Anophelinae Grassi 1900	***Anopheles*** Meigen 1818			**476**	**33**	**509**	***An.***
		Anopheles Meigen 1818	m	185	21	206	*Ano.*
		Baimaia Harbach, Rattanarithikul & Harrison 2005	f	1		1	*Bmi.*
		Cellia Theobald 1902	f	225	12	237	*Cel.*
		Christya Theobald 1903	f	2		2	*Chs.*
		Kerteszia Theobald 1905	f	12		12	*Ker.*
		Lophopodomyia Antunes 1937	f	6		6	*Lph.*
		Nyssorhynchus Blanchard 1902	m	40		40	*Nys.*
		Stethomyia Theobald 1902	f	5		5	*Ste.*
	Bironella Theobald 1905f			**8**		**8**	***Bi.***
		Bironella Theobald 1905	f	2		2	*Bir.*
		Brugella Edwards 1930	f	3		3	*Bru.*
		Neobironella Tenorio 1977	f	3		3	*Nbi.*
	Chagasia Cruz 1906	Monotypic	f	**5**		**5**	***Ch.***
Culicinae Meigen 1818							
Aedeomyiini Theobald 1901	***Aedeomyia*** Theobald 1901b			**7**		**7**	***Ad.***
		Aedeomyia Theobald 1901b	f	6		6	*Ady.*
		Lepiothauma Enderlein 1923	f	1		1	*Lpi.*
Aedini Neveu-Lemaire 1902	***Aedes*** Meigen 1818			**935**	**35**	**970**	***Ae.***
		Abraedes Zavortink 1970	m	1		1	*Abr.*
		Acartomyia Theobald 1903a	f	3		3	*Act.*
		Aedes Meigen 1818	m	12		12	*Aed.*
		Aedimorphus Theobald 1903a	m	65	8	73	*Adm.*
		Alanstonea Mattingly 1960	f	2		2	*Ala.*
		Albuginosus Reinert 1987	m	9		9	*Alb.*
		Ayurakitia Thurman 1954	f	2		2	*Ayr.*
		Aztecaedes Zavortink 1972	m	1		1	*Azt.*
		Belkinius Reinert 1982	m	1		1	*Blk.*
		Bifidistylus Reinert, Harbach, & Kitching 2009	m	2	1	3	*Bif.*
		Borichinda Harbach & Rattanarithikul 2007	m	1		1	*Bor.*
		Bothaella Reinert 1973	f	6		6	*Bot.*
		Bruceharrisonius Reinert 2003	m	9		9	*Brh.*
		Cancraedes Edwards 1929	m	10		10	*Can.*
		Catageiomyia Theobald 1903	f	29		29	*Cat.*
		Catatassomyia Dyar & Shannon 1925	f	1		1	*Cts.*

continued

Table 3. ***continued***

Subfamily/tribe	Genus	Subgenus	Gender	No. of sp.	No. of ssp.	Total	Abbreviation
Aedini Neveu-Lemaire 1902 (cont.)	***Aedes*** Meigen 1818 (cont.)	*Christophersiomyia* Barraud 1923	f	5		5	*Chr.*
		Coetzeemyia Huang, Mathis, & Wilkerson 2010	f	1		1	*Coe.*
		Collessius Reinert, Harbach, & Kitching 2006	m	9	1	10	*Col.*
		Cornetius Huang 2005	m	1		1	*Cor.*
		Dahliana Reinert, Harbach, & Kitching 2006	f	3		3	*Dah.*
		Danielsia Theobald 1904	f	3		3	*Dan.*
		Dendroskusea Edwards 1929	f	5		5	*Dsk.*
		Diceromyia Theobald 1911	f	8		8	*Dic.*
		Dobrotworskyius Reinert, Harbach, & Kitching 2006	m	7		7	*Dob.*
		Downsiomyia Vargas 1950	f	30		30	*Dow.*
		Edwardsaedes Belkin 1962	m	3		3	*Edw.*
		Elpeytonius Reinert, Harbach, & Kitching 2009	m	2		2	*Elp.*
		Finlaya Theobald 1903	f	36		36	*Fin.*
		Fredwardsius Reinert 2000	m	1		1	*Fre.*
		Georgecraigius Reinert, Harbach, & Kitching 2006	m	3		3	*Grg.*
		Geoskusea Edwards 1929	f	10		10	*Geo.*
		Gilesius Reinert, Harbach, & Kitching 2006	m	2		2	*Gil.*
		Gymnometopa Coquillett 1906	f	1		1	*Gym.*
		Halaedes Belkin 1962	m	3		3	*Hal.*
		Himalaius Reinert, Harbach, & Kitching 2006	m	2		2	*Him.*
		Hopkinsius Reinert, Harbach & Kitching 2008	m	7		7	*Hop.*
		Howardina Theobald 1903	f	34		34	*How.*
		Huaedes Huang 1968	m	3		3	*Hua.*
		Hulecoeteomyia Theobald 1904	f	15	3	18	*Hul.*
		Indusius Edwards 1934	m	1		1	*Ind.*
		Isoaedes Reinert 1979	m	1		1	*Isa.*
		Jarnellius Reinert, Harbach, & Kitching 2006	m	5		5	*Jar.*
		Jihlienius Reinert, Harbach, & Kitching 2006	m	3		3	*Jih.*
		Kenknightia Reinert 1990	f	12		12	*Ken.*
		Kompia Aitken 1941	f	1		1	*Kom.*
		Leptosomatomyia Theobald 1905	f	1		1	*Lep.*
		Levua Stone & Bohart 1944	f	1		1	*Lev.*
		Lewnielsenius Reinert, Harbach & Kitching 2006	m	1		1	*Lew.*

Subfamily/tribe	Genus	Subgenus	Gender	No. of sp.	No. of ssp.	Total	Abbreviation
		Lorrainea Belkin 1962	f	5		5	*Lor.*
		Luius Reinert, Harbach, & Kitching 2008	m	1		1	*Lui.*
		Macleaya Theobald 1903	f	11		11	*Mac.*
		Molpemyia Theobald 1910	f	3		3	*Mol.*
		Mucidus Theobald 1901	m	14	1	15	*Muc.*
		Neomelaniconion Newstead 1907	n	28	1	29	*Neo.*
		Nyctomyia Harbach 2013	f	2		2	*Nyc.*
		Ochlerotatus Lynch Arribálzaga 1891	m	186	10	196	*Och.*
		Paraedes Edwards 1934	m	9		9	*Par.*
		Patmarksia Reinert, Harbach & Kitching 2006	f	13		13	*Pat.*
		Paulianius Brunhes & Boussès 2017	m	9		9	*Pln.*
		Petermattinglyius Reinert, Harbach, & Kitching 2009	m	5		5	*Pet.*
		Phagomyia Theobald 1905	f	16	1	17	*Phg.*
		Polyleptiomyia Theobald 1905	f	2		2	*Pmt.*
		Pseudalbuginosus Huang & Rueda 2015	m	1		1	*Pal.*
		Protomacleaya Theobald 1907	f	40		40	*Pro.*
		Pseudarmigeres Stone & Knight 1956	m	5	2	7	*Psa.*
		Pseudoskusea Theobald 1907	f	4		4	*Psk.*
		Rampamyia Reinert, Harbach, & Kitching 2006	f	3	1	4	*Ram.*
		Rhinoskusea Edwards 1929	f	4		4	*Rhi.*
		Rusticoidus Shevchenko & Prudkina 1973	m	9	1	10	*Rus.*
		Sallumia Reinert, Harbach, & Kitching 2008	f	2		2	*Sal.*
		Scutomyia Theobald 1904	f	9		9	*Sct.*
		Skusea Theobald 1903	f	4		4	*Sku.*
		Stegomyia Theobald 1901	f	128	6	134	*Stg.*
		Tanakaius Reinert, Harbach, & Kitching 2004	m	2		2	*Tan.*
		Tewarius Reinert 2006	m	4		4	*Tew.*
		Vansomerenis Reinert, Harbach & Kitching 2006	f	3		3	*Van.*
		Zavortinkius Reinert 1999	m	11		11	*Zav.*
		Subgenus uncertain *Aedes*		29		29	
	Armigeres Theobald 1901			**58**	**1**	**59**	***Ar.***
		Armigeres Theobald 1901	m	40	1	41	*Arm.*
		Leicesteria Theobald 1904	f	18		18	*Lei.*
	Eretmapodites Theobald 1901	Monotypic	m	**48**	**3**	**51**	***Er.***

continued

Table 3. *continued*

Subfamily/tribe	Genus	Subgenus	Gender	No. of sp.	No. of ssp.	Total	Abbreviation
Aedini Neveu-Lemaire 1902 (cont.)	***Haemagogus*** Williston 1896			**28**		**28**	***Hg.***
		Conopostegus Dyar 1925	m	4		4	*Con.*
		Haemagogus Williston 1896	m	24		24	*Hag.*
	Heizmannia Ludlow 1905			**40**		**40**	***Hz.***
		Heizmannia Ludlow 1905	f	34		34	*Hez.*
		Mattinglyia Lien 1968	f	6		6	*Mat.*
	Opifex Hutton 1902			**2**		**2**	***Op.***
		Nothoskusea Dumbleton 1962	f	1		1	*Not.*
		Opifex Hutton 1902	m	1		1	*Opi.*
	PsorophoraRobineau-Desvoidy 1827			**49**		**49**	***Ps.***
		Grabhamia Theobald 1903	f	16		16	*Gra.*
		Janthinosoma Lynch Arribálzaga 1891	n	23		23	*Jan.*
		Psorophora Robineau-Desvoidy 1827	f	10		10	*Pso.*
	Udaya Thurman 1954	Monotypic	m	**3**		**3**	***Ud.***
	Verrallina Theobald 1903			**95**		**95**	***Ve.***
		Harbachius Reinert 1999	m	13		13	*Har.*
		Neomacleaya Theobald 1907	f	52		52	*Nma.*
		Verrallina Theobald 1903	f	30		30	*Ver.*
	Zeugnomyia Leicester 1908	Monotypic	f	**4**		**4**	***Ze.***
Culicini Meigen 1818	***Culex*** Linnaeus 1758			**769**	**30**	**799**	***Cx.***
		Acalleomyia Leicester 1908	f	1		1	*Aca.*
		Acallyntrum Stone & Penn 1948	n	8		8	*Acl.*
		Aedinus Lutz 1904	m	4		4	*Ads.*
		Afroculex Danilov 1989	m	1		1	*Afc.*
		Allimanta Casal & García 1968	f	1		1	*Alm.*
		Anoedioporpa Dyar 1923	f	12		12	*And.*
		Barraudius Edwards 1921	m	4		4	*Bar.*
		Belkinomyia Adames & Galindo 1973	f	1		1	*Bel.*
		Carrollia Lutz 1905	f	18		18	*Car.*
		Culex Linnaeus 1758	m	200	18	218	*Cux.*
		Culiciomyia Theobald 1907	f	54	1	55	*Cui.*
		Eumelanomyia Theobald 1909	f	77	2	79	*Eum.*
		Kitzmilleria Danilov 1989	f	1		1	*Kit.*
		Lasiosiphon Kirkpatrick 1925	m	1		1	*Las.*
		Lophoceraomyia Theobald 1905	f	112	1	113	*Lop.*

Subfamily/tribe	Genus	Subgenus	Gender	No. of sp.	No. of ssp.	Total	Abbreviation
		Maillotia Theobald 1907	f	9	3	12	*Mai.*
		Melanoconion Theobald 1903	n	160		160	*Mel.*
		Micraedes Coquillett, 1906	m	8		8	*Mca.*
		Microculex Theobald 1907	m	33	2	35	*Mcx.*
		Neoculex Dyar 1905	m	26		26	*Ncx.*
		Nicaromyia Broche & Rodríguez Rodríguez 2001	f	1		1	*Nic.*
		Oculeomyia Theobald 1907	f	19	3	22	*Ocu.*
		Phenacomyia Harbach & Peyton 1992	f	3		3	*Phc.*
		Phytotelmatomyia Rossi & Harbach 2008	f	4		4	*Phy.*
		Sirivanakarnius Tanaka 2004	m	1		1	*Sir.*
		Tinolestes Coquillett 1906	m	3		3	*Tin.*
		Subgenus uncertain *Culex*		7		7	
	Deinocerites Theobald 1901	Monotypic	m	**18**		**18**	***De.***
	Galindomyia Stone & Barreto 1969	Monotypic	f	**1**		**1**	***Ga.***
	Lutzia Theobald 1903			**9**		**9**	***Lt.***
		Insulalutzia Tanaka 2003	f	1		1	*Ilt.*
		Lutzia Theobald 1903	f	2		2	*Lut.*
		Metalutzia Tanaka 2003	f	6		6	*Mlt.*
Culisetini Belkin 1962	***Culiseta*** Felt 1904			**39**	**2**	**41**	***Cs.***
		Allotheobaldia Broelemann 1919	f	1		1	*All.*
		Austrotheobaldia Dobrot-worsky 1954	f	1		1	*Aut.*
		Climacura Howard, Dyar, & Knab 1915	f	6		6	*Cli.*
		Culicella Felt 1904	f	14		14	*Cuc.*
		Culiseta Felt 1904	f	12	1	13	*Cus.*
		Neotheobaldia Dobrot-worsky 1958	f	2	1	3	*Net.*
		Theomyia Edwards 1930	f	1		1	*Thm.*
		Subgenus uncertain *Culiseta*		2		2	
Ficalbiini Belkin 1962	***Ficalbia*** Theobald 1903	Monotypic	f	**8**		**8**	***Fi.***
	Mimomyia Theobald 1903			**45**	**2**	**47**	***Mi.***
		Etorleptiomyia Theobald 1904	f	7		7	*Eto.*
		Ingramia Edwards 1912	f	21		21	*Ing.*
		Mimomyia Theobald 1903	f	17	2	19	*Mim.*
Hodgesiini Belkin 1962	***Hodgesia*** Theobald 1904	Monotypic	f	**11**		**11**	***Ho.***
Mansoniini Belkin 1962	***Coquillettidia*** Dyar 1905			**58**		**58**	***Cq.***

continued

Table 3. ***continued***

Subfamily/tribe	Genus	Subgenus	Gender	No. of sp.	No. of ssp.	Total	Abbreviation
Mansoniini Belkin 1962 (cont.)	***Coquillettidia*** Dyar 1905 (cont.)	*Austromansonia* Belkin 1968	f	1		1	*Aus.*
		Coquillettidia Dyar 1905	f	44		44	*Coq.*
		Rhynchotaenia Brèthes 1910	f	13		13	*Rhy.*
	Mansonia Blanchard 1901			**25**	**1**	**26**	***Ma.***
		Mansonia Blanchard 1901	f	15		15	*Man.*
		Mansonioides Theobald 1907	m	10	1	11	*Mnd.*
Orthopodomyiini Belkin, Heinemann & Page 1970	***Orthopodomyia*** Theobald 1904	Monotypic	f	**36**		**36**	***Or.***
Sabethini Blanchard 1905	***Isostomyia*** Coquillett 1906	Monotypic	f	**4**		**4**	***Is.***
	Johnbelkinia Zavortink 1979	Monotypic	f	**3**		**3**	***Jb.***
	Kimia Vu Duc Huong & Harbach 2007	Monotypic	f	**5**		**5**	***Km.***
	Limatus Theobald 1901	Monotypic	m	**9**		**9**	***Li.***
	Malaya Leicester 1908	Monotypic	f	**12**		**12**	***Ml.***
	Maorigoeldia Edwards 1930	Monotypic	f	**1**		**1**	***Mg.***
	Onirion Peyton & Harbach 2000	Monotypic	n	**7**		**7**	***On.***
	Runchomyia Theobald 1903			**8**		**8**	***Ru.***
		Ctenogoeldia Edwards 1930	f	2		2	*Cte.*
		Runchomyia Theobald 1903	f	6		6	*Run.*
	Sabethes Robineau-Desvoidy 1827			**41**		**41**	***Sa.***
		Davismyia Lane & Cerqueira 1942	f	1		1	*Dav.*
		Peytonulus Harbach, 1991	m	12		12	*Pey.*
		Sabethes Robineau-Desvoidy 1827	m	18		18	*Sab.*
		Sabethinus Lutz 1904	m	6		6	*Sbn.*
		Sabethoides Theobald 1903	m	4		4	*Sbo.*
	Shannoniana Lane & Cerqueira 1942	Monotypic	f	**3**		**3**	***Sh.***
	Topomyia Leicester 1908			**65**		**65**	***To.***
		Miyagiella Harbach 2014	f	1		1	*Myg.*
		Suaymyia Thurman 1959	f	22		22	*Sua.*
		Topomyia Leicester 1908	f	42		42	*Top.*
	Trichoprosopon Theobald 1901	Monotypic	n	**13**	**2**	**15**	***Tr.***
	Tripteroides Giles 1904			**122**	**4**	**126**	***Tp.***

Subfamily/tribe	Genus	Subgenus	Gender	No. of sp.	No. of ssp.	Total	Abbreviation
		Polylepidomyia Theobald 1905	f	19		19	*Pol.*
		Rachionotomyia Theobald 1905	f	14		14	*Rah.*
		Rachisoura Theobald 1910	f	28		28	*Rac.*
		Tricholeptomyia Dyar & Shannon 1925	f	9		9	*Tri.*
		Tripteroides Giles 1904	m	52	4	56	*Trp.*
	Wyeomyia Theobald 1901			**139**	**1**	**140**	***Wy.***
		Antunesmyia Lane & Cerqueira 1942	f	3		3	*Ant.*
		Caenomyiella Harbach & Peyton 1990	f	1		1	*Cae.*
		Cruzmyia Lane & Cerqueira 1942	f	4		4	*Cru.*
		Decamyia Dyar 1919	f	3		3	*Dec.*
		Dendromyia Theobald 1903	f	6		6	*Den.*
		Dodecamyia Dyar, 1918	f	1		1	*Dod.*
		Exallomyia Harbach & Peyton 1992	f	3		3	*Exm.*
		Hystatomyia Dyar 1919	f	9		9	*Hys.*
		Menolepis Lutz 1905	f	1		1	*Men.*
		Miamyia Dyar 1919	f	7	1	8	*Mia.*
		Nunezia Dyar 1928	f	3		3	*Nuz.*
		Phoniomyia Theobald 1903	f	22		22	*Pho.*
		Prosopolepis Lutz 1905	m	1		1	*Prl.*
		Spilonympha Motta & Lourenço-de-Oliveira 2005	f	7		7	*Spi.*
		Triamyia Dyar 1919	f	2		2	*Tra.*
		Wyeomyia Theobald 1901	f	35		35	*Wyo.*
		Zinzala Zavortink 1986	f	2		2	*Zin.*
		Subgenus uncertain		29		29	
Toxorhynchitini Lahille 1904	***Toxorhynchites*** Theobald 1901			**90**	**10**	**100**	***Tx.***
		Afrorhynchus Ribeiro 1992	m	19	1	20	*Afr.*
		Ankylorhynchus Lutz 1904	m	4		4	*Ank.*
		Lynchiella Lahille 1904	f	16	3	19	*Lyn.*
		Toxorhynchites Theobald 1901	m	51	6	57	*Tox.*
Uranotaeniini Lahille 1904	***Uranotaenia*** Lynch Arribálzaga 1891			**271**	**6**	**277**	***Ur.***
		Pseudoficalbia Theobald 1912	f	150	3	153	*Pfc.*
		Uranotaenia Lynch Arribálzaga 1891	f	121	3	124	*Ura.*
Species, subspecies, total				**3570**	**130**	**3700**	

cies" of a genus instead of the more specific terms haplotype, orthotype, logotype, diatype, or tautotype, which can be derived from the literature if needed. Subsequent genus- and species-group name and literature changes and additions were updated mostly by one of us (RCW) in the online database mosquitocatalog.org (static since 2016) and in RCW's personal catalog database spreadsheet. Our taxonomic catalog is the first comprehensive print version of this information since Knight & Stone (1977).

Elements of Species-Group Entries (See Print and Database History)

Species-group names. Valid species are in **bold** and arranged alphabetically under each genus. If applicable, species are followed by synonyms (in italics) and subspecies (also in **bold**) in chronological order, as was done in the Knight & Stone (1977) catalog. Subspecies synonyms follow subspecies regardless of publication date.

Original description. An original description is a citation in which the species was named (see below for annotations).

Type loc. = type locality. Efforts were made by Knight and Stone (1977) and others to include verbatim, or close to verbatim, information from the original description, as do we. Sometimes this information was absent or sketchy, and authors (us included) interpreted where the species was first collected. Herein we indicated what we added, interpreted, or updated by using square brackets. An abbreviation indicating where type specimens were deposited is provided in parentheses following the type locality. See the type depository section for an explanation of the acronyms shown.

Distr. = distribution. The countries in these lists came from various sources and are sometimes not documented with certainty, as was the case with the original Knight and Stone (1977) catalog. We have tried to improve this documentation, however recording reliable attributed distributions is inherently difficult as not all species identifications presented in the literature or in archive collections are validated by taxonomic experts. However, we did our best to geographically and biologically curate the additional country records prior to inclusion (see Vol. 1, Part 2). Readers should treat our lists, except for those also represented as species pages in Vol. 1, as a starting point, and readers should do their own verifications before drawing conclusions. One must remember, too, that hidden in these lists of countries are cryptic species that may have, or probably have, different or overlapping distributions. Where complex taxa remain to be defined, distribution entries for the nominotypical member of the complex will in many cases still include occurences of other cryptic taxa. The countries used were chosen mostly to accommodate where species have actually been collected and not necessarily to accommodate political boundaries. See www.vectormap.si.edu for a more formal accounting of recognized countries.

Refs. = literature citations. The literature chosen for the catalog was usually only a subset of that available. In general, an effort was made, by us and the authors of the Stone et al. (1959) and Knight & Stone (1977) catalogs, to include information relating to taxonomy or classification, especially taxonomic acts. In addition to the obvious—descriptions, redescriptions, revisions, and comprehensive keys—we included pertinent reviews of distributions and ecology and more. This grew to include quite important molecular citations as well. This proved to be not a totally objective task, and references that are only marginally relevant found their way into the species entries while others, possibly (probably), have been overlooked. All are good reason for further library research and updates in the form of supplements, as has been done for previous catalogs.

For the original descriptions and other citations related to the species, there is usually a brief interpretation of the components of the reference (life stages, in order of presentation): M = male (could be further annotated with G, or gen., for genitalia); F = female (possibly F gen., as for the male); P = pupa; L = larva; E = egg. If any part of a life stage is illustrated, however minor, the abbreviation is accompanied by an asterisk (e.g., P*). Other annotations then follow if available: bion. = any biological information; syn. = synonymy; distr. = distribution; ssp. = subspecies; desig. = designation—of a lectotype, for instance; tax. = taxonomic information of any sort; class. = classification information of any sort; phyl. = phylogenetic information of any sort, often modified by mol. = molecular.

Ety. = etymology. Often termed "name derivations." Name derivations are provided for species and subspecies, not synonyms. It is highly recommended that authors state what their species names mean, but especially for older names, often nothing is stated, and Latin and Greek dictionaries can be of limited help to nonclassical scholars. Derivations are annotated so the reader knows where to start in order to pursue those that are doubtful. If (usually) a direct statement is not made about the species name, the convention "n.s." (not stated) is used; if the derivation was still a mystery after researching it, "n.k" (not known) is used. Of special use was the online resource Whitaker's Words Online, Williams (2015), and Brown (1956). Square brackets are liberally used to indicate our interpretations when explanations are not explicitly stated. Sometimes the answer is found by reading the description (most of which were read by R.W., with some help). Harbach (2018c) published a compendium of all names associated with culicids, to include derivations of select valid species and genera. He articulated and separated the various themes authors used to form names, such as patronyms, geography, et cetera. This is in the spirit of Jozwiak et al.'s (2015) "Taxonomic Etymology—in Search of Inspiration." Since we worked in parallel with Harbach without knowing what the other was doing, a comparison of results might prove interesting and, we would hope, complementary. His derivations were not used for what is presented here, except when he is cited. For insight into how scientific names are formed, see Harbach (2018c), McDowall (1991), International Commission on Zoological Nomenclature (1999), Brown (1956), and Winston (1999).

Inf. name = informal name. An innovation we are providing is an informal name for all valid species and subspecies. A rationale for formation of these names is given in Vol. 1. We use the term "informal name" to describe our names in order to draw a distinction between, and avoid conflict with, our names and the list of common names maintained by the ESA. There are only 15 ESA names, which are cited with those species. The informal names have elements taken from the name derivation and type locality along with a unique name for each genus (e.g., the nail mosquito is genus *Anopheles*). Some informal names were quite problematical to form, so we sometimes adopted imaginative names, which have rationales given in square brackets.

Synonyms. Synonym entries are treated differently from valid names and do not have distributions since they would be the same as the senior synonym and, as stated above, no name derivation (too time consuming) or informal names (they would be synonyms too). Most literature associated with synonyms is associated with the senior synonym unless the literature has relevance to the synonymy itself.

Abbreviations Used for Type Depositories

A	Personal collection of species author
AM	Division of Entomology, Zoological Museum [The Zoological Museum], Amsterdam University, Amsterdam, Netherlands
AMC	Institute of Microbiology and Epidemiology, Academy of Military Medical Science, PLA, Beijing, People's Republic of China
AMNH	American Museum of Natural History, New York, New York, United States
ANIC	Australian National Insect Collection, Canberra, Australia
ANS	Academy of Natural Sciences of Philadelphia, Philadelphia, Pennsylvania, United States [now Academy of Natural Sciences of Drexel University]
B	Naturhistorisches Museum, Basel, Switzerland
BA	Museo Nacional de Ciencias Naturales Bernardino Rivadavia, Buenos Aires, Argentina [now Museo Argentino de Ciencias Naturales; includes types formerly in MLP and MEPRA]
BBM	Bernice P. Bishop Museum, Honolulu, Hawai'i, United States
BC	Bigot Collection, in collection of J.E. Collin, Newmarket, United Kingdom [now in the Oxford University Museum of Natural History (?)]
BEH	Bureau of Entomology, Hangchow, People's Republic of China
BM	British Museum (Natural History), London, United Kingdom [see current designation, NHMUK]
BMC	Department of Parasitology, Bethune Medical College, Jilin, People's Republic of China [now Norman Bethune College of Medicine]
BNITM	Bernhard-Nocht-Instituts für Schiffs- und Tropenkrankheiten, Hamburg, Germany
CAS	California Academy of Sciences, San Francisco, California, United States
CEIE	Centro Ecuatoriano de Investigaciones Entomológicas, Guayaquil, Ecuador [nonextant]
CLM	Central Laboratory, Redent, Mauritius
CM	Field Museum of Natural History, Chicago, Illinois, United States [formerly Chicago Natural History Museum]
CMC	Department of Parasitology, Chung-Shan Medical University, Canton, People's Republic of China
CMNZ	Canterbury Museum, Christchurch, New Zealand
CMPHM	Centro de Medicina Preventiva e Higiene Militar, Buenos Aires, Argentina [All types now in USNM]
CMT	Le Musée royal de l'Afrique centrale, Tervuren, Brussels, Belgium
CNC	Canadian National Collection, Ottawa, Ontario, Canada
CPRR	Centro de Pesquisas René Rachou, Brazil [present name of INER and DNER]
CRME	Centre for Research in Medical Entomology, Madurai, Tamil Nadu, India
CSIR	Council for Scientific and Industrial Research Museum, Canberra, Australia [now in ANIC]
CUMC	Chinese Union Medical College, Peking, People's Republic of China
DAC	Department of Agriculture, Cairo, Egypt
DEI	Deutsches Entomologisches Institut, Berlin-Friedrichshagen, Germany [now IP, see below]
DERM	Laboratorio de Entomología de la División de Endemias Rurales, Maracay, Venezuela [replaces MDM]
DMB	División de Malariología, Bogotá, Colombia [if a type still exists, it would be in Instituto Nacional para Programas Especiales de Salud]
DMN	Durban Natural Science Museum, Kwa-Zulu Natal, Durban, South Africa
DMNZ	Dominion Museum, Wellington, New Zealand
DNER	Departamento Nacional de Endemias Rurais, Rio de Janeiro, Brazil [now CPRR]
DPFA	División de Profilaxis de la Fiebre Amarilla, Caracas, Venezuela [types probably in DERM]
DPHU	Department of Parasitology, Hebrew University, Jerusalem, Israel
DSIR	Entomology Division, Department of Scientific and Industrial Research, Christchurch, New Zealand
EIHU	Entomological Institute, Hokkaido University, Hokkaido, Japan
EMD	Laboratoire de Parasitologie de l'École de Médecine de Dakar, Senegal
EMH	l'École de Médecine (laboratory of parasitology), Hanoi, Vietnam [extant?]
EMIP	Entomological Museum of the Institute of Preventive Medicine, National Defense Medical Center, Taipei, Taiwan
ESE	Entomological Society of Egypt, Cairo, Egypt
ESPG	Entomoteca de Sanidad Pública de Guatemala, Guatemala City, Guatemala [nonextant]
EUO	Sección de Entomología, Departamento de Parasitología y Microbiología, Escuela de Medicina, Universidad de Oriente, Ciudad Bolívar, Venezuela

FH	Faculdade de Higiene e Saúde Pública de Universidade de São Paulo, São Paulo, Brazil [now Faculdade de Saúde Pública, see FSP]
FM	Museo Zoologico, Universita degli Studi, Firenze
FMI	Fan Memorial Institute of Biology, Peking, People's Republic of China
FMP	Faculté de Médicine, University of Paris, Paris, France [Laboratoire de Parasitologie et Mycologie]
FMRP	Parasitologia-Departamento de Biologia Celular e Molecular e Bioagentes Patogênicos, Faculdade de Medicina de Ribeirão Preto, São Paulo, Brazil
FMSP	Departamento de Parasitologia da Faculdade de Medicina de Universidade de São Paulo, São Paulo, Brazil
FSP	Faculdade de Saúde Pública, Universidade de São Paulo, São Paulo, Brazil [= FH]
GDZ	Department of Zoology, University of Glasgow, Glasgow, Scotland, United Kingdom
GLB	Geneeskundig Laboratorium, Jakarta, Indonesia
GMC	Biology Department, Guiyang Medical College, Guizhou Province, People's Republic of China
GML	Gorgas Memorial Laboratory, Panama City, Panama [all types now in USNM]
GPIUH	Geological-Palaeontological Institute and Museum, University of Hamburg, Hamburg, Germany
HIS	Hygiene Institute, Skoplji, Serbia [= Institute of Public Health of Serbia]
HM	Kansallismuseo, Helsinki, Finland
HNM	Magyar Nemzeti Múzeum, Budapest, Hungary
HUJ	Entomological Institute, Hokkaido University, Hokkaido, Japan
IB	Instituto Butantan, São Paulo, Brazil
IBBA	Instituto Bacteriológico del Departamento Nacional de Higiene, Buenos Aires, Argentina [= FNM; collection held at the Departamento de Entomología Sanitaria, Instituto Nacional de Microbiología, Av. Velez Sarsfield 563, Buenos Aires, Argentina] [Relevant types not known to us]
IBSP	Instituto Biológico, São Paulo, S.P., Brazil [no types now]
IEE	Instituto Español de Entomología, Madrid, Spain
IERT	Institut d'enseignement et de recherches tropicales, Bondy, Seine, France [now Centre de Faunistique de l'ORSTOM a Bondy, 93-France; see SSC]
IHC	Instituto de Higiene, Caracas, Venezuela [probably no types; a few now in FH]
IHMT	Department of Entomology, Instituto de Higiene e Medicina Tropical, Lisbon, Portugal [= Universidade Nova de Lisboa Instituto de Higiene e Medicina Tropical]
IID	Medical Zoology Laboratory, Institute for Infectious Diseases, University of Tokyo, Tokyo, Japan
IM	Zoological Survey of India, Indian Museum, Calcutta, India
IMEM	Istituto di Malariologia "Ettore Marchiafava," Rome, Italy
IMP	E.I. Martsinovskiy Institute of Medical Parasitology and Tropical Medicine, Moscow, Russia [= MTI]
INER	Instituto Nacional de Endemias Rurais, Belo Horizonte, Minas Gerais, Brazil [now CPRR]
INM	Instituto Nacional de Microbiología, Buenos Aires, Argentina
INPA	Instituto Nacional de Pesquisas da Amazônia, Belém, Pará, Brazil [no types now]
IOC	Instituto Oswaldo Cruz, Rio de Janeiro, Brazil [FIOCRUZ]
IP	Institut für Pflanzenschutzforschung, Zweigstelle Eberswalde, Abteilung, Taxonomie der Insekten [formerly DEI, Deutsches Entomologisches Institut, Eberswalde, Germany]
IPD	Institute of Parasitic Disease, Kweichow, People's Republic of China
IPS	Institut de Parasitologie, Faculté de Médecine, Université de Strasbourg, Strasbourg, France
IRSM	Institut Recherche Scientifique de Madagascar, Tananarive, Madagascar
ISET	Laboratorio de Entomología, Instituto de Salubridad y Enfermedades Tropicales, Mexico City, D.F., Mexico
ISN	Institut Royal des Sciences Naturelles de Belgique, Brussels, Belgium
ISS	Istituto Superiore de Sanità, Rome, Italy
ITH	Instituut voor Tropische Hygiëne en Geographische Pathologie, Amsterdam, the Netherlands
IZ	Institute of Zoology, Academica Sinica, Beijing, People's Republic of China
IZAA	Institute of Zoology, Kazakhstan Academy of Sciences Almaty, Kazakhstan
IZC	Instituto de Biología "Juan Noé," Santiago, Chile
K	Christian-Albrechts Universität, Kiel, Germany
KCB	Kastenholz Collection, Bonn, Germany
KIM	Kunming Institute of Medical Research, Kunming, People's Republic of China
KLPH	Kyoto Laboratory of Public Health, Kyoto, Japan
KMC	Biology Department, Kweiyang Medical College, Kweiyang, People's Republic of China
KU	Kyushu University, Kyushu, Japan
KZI	Kazakh Institute of Zoology, Almaty, Kazakhstan

LACM	Los Angeles County Museum of Natural History, Los Angeles, California, United States [now The Natural History Museum of Los Angeles County]
LGUL	Laboratoire de Géologie, Université de Lyon, Lyon, France
LM	State Museum of Natural History, Leiden, the Netherlands
LPFM	Laboratoire de Parasitologie de la Faculté de Médecine d'Alger, Université d'Alger, Algeria [no longer exists; all remaining types in FMP]
LS	University of Lund, Lund, Sweden
LSL	Linnean Society of London, London, England, United Kingdom
LSTM	Liverpool School of Tropical Medicine, Liverpool, England, United Kingdom
LU	Location unknown
MCZ	Museum of Comparative Zoology, Harvard University, Cambridge, Massachusetts, United States
MDM	Museo de División de Malariología, Caracas, Venezuela [now DERM]
MEPRA	Instituto de Microbiología y Parasitología, Facultad de Medicina, Buenos Aires, Argentina [all types now in BA]
MHNC	Museo Nacional de Historia Natural, Santiago, Chile [types probably lost]
MIT	Malaria Institute, Amani, Tanzania
MLP	Museo de la Plata, Argentina [all types now in BA]
MM	Macleay Museum of Natural History, University of Sydney, Sydney, Australia
MMC	Department of Parasitology, Seventh Military Medical College, People's Republic of China
MNB	Museum für Naturkunde, Paläotologisches Museum, Berlin, Germany
MNHP	Muséum National d'Histoire Naturelle, Paris, France
MNRJ	Museu Nacional, Rio de Janeiro, R.J., Brazil [types assumed lost in fire of 2019]
MRI	Department of Entomology, Medical Research Institute, Colombo, Sri Lanka
MSI	Malaria Institute of India, Delhi, India [collections transferred to the National Institute of Communicable Diseases, Delhi, India (Reinert 1973e: 32); see NICD]
MTI	Moscow Tropical Institute, Moscow, USSR [= IMP]
MU	Museum, Madrid University School of Agriculture, Madrid, Spain
NB	Malaria Research Headquarters, Sabah, Malaysian Borneo
ND	None designated
NE	Type nonexistent
NELS	Same as NSIR
NHMUK	Natural History Museum, London, England, United Kingdom
NICD	National Institute of Communicable Diseases, Delhi, India
NIV	National Institute of Virology, Sandringham, Johannesburg, Republic of South Africa
NMC	Naturwissenschaftlichen Museum der Coburger Landestiftung, Coburg, Germany
NML	Museum of the, Escola Nacional de Saúde Pública e de Medicina Tropical, Lisboa, Portugal
NMM	National Museum of Victoria, Melbourne, Australia
NMNH	National Museum of Natural History. See USNM
NMST	National Science Museum, Tokyo, Japan
NMW	Naturhistorisches Museum, Vienna, Austria
NSIR	Entomology Division, Department of Scientific and Industrial Research, Nelson, New Zealand [now in Auckland]
NTU	Institute of Tropical Medicine, National Taiwan University, Taipei, Taiwan
NY	New York State Museum, Albany, New York, United States
O	Hope Collection, Oxford University, Oxford, England, United Kingdom
PIA	Institut Pasteur d'Algérie, Algiers, Algeria [no types; now all in FMP]
PIG	Institut Pasteur, French Guiana
PIH	Institut Pasteur, Hanoi, Vietnam
PIP	Institut Pasteur, Paris, France
QM	Queensland Museum, South Brisbane, Queensland, Australia [contains specimens from SPHTM, and UQIC]
PNM	Philippine National Museum, Manila, Philippines
RIE	Department of Medical Zoology, Research Institute of Endemics, Nagasaki University, Japan
RTMP	Royal Tyrell Museum of Paleontology, Drumheller, Alberta, Canada
RUM	Università degli Studi, Rome, Italy
SA	South African Museum, Cape Town, Republic of South Africa
SAIM	South African Institute for Medical Research, Johannesburg, Republic of South Africa
SAM	South Australian Museum, Adelaide, South Australia, Australia

SBC	Sarawak Biodiversity Council, Kuching, Sarawak, Malaysian Borneo
SDEI	Senckenberg Deutsches Entomologisches Institut, Munich, Germany
SIP	Sichuan Institute of Parasitic Disease Control, Chengdu, People's Republic of China
SM	Naturhistoriska riksmuseet, Stockholm, Sweden
SMC	Department of Parasitology, Szechwan Medical College, Chengtu, People's Republic of China
SMMU	Department of Parasitology, Second Military Medical University, People's Republic of China
SMNS	Staatliches Museum für Naturkunde Stuttgart, Zweigstelle, Ludwigsburg, Germany
SMT	Staatliches Museum für Tierkunde, Dresden, Germany [= The State Museum of Zoology]
SNG	Senckenberg Forschungsinstitut und Naturmuseum Frankfurt, Germany
SPM	Serviço de Profilaxia da Malaria do Estado de São Paulo, São Paulo, S.P., Brazil [all types transferred to FH]
SSC	Centre de Faunistique de 1'ORSTOM, France [formerly designated IERT]
STMPR	School of Tropical Medicine, San Juan, Puerto Rico
SU	Stanford University, Palo Alto, California, United States [all types in LACM]
TMDU	Department of Medical Zoology, Tokyo Medical Dental University, Tokyo, Japan
TMMC	Department of Parasitology, Third Military Medical College, Chongqing, People's Republic of China
TMP	Transvaal Museum, Pretoria, Republic of South Africa [now = Ditsong National Museum of Natural History]
TMRI	Taiwan Provincial Malaria Research Institute, Nankang, Taipei County, Taiwan
TMS	Tomsk Malaria Station, Tomsk, Siberia, Russia
TORINO	Instituto e Museu di Zoologia, Università degli Studi, Torino, Italy
UM	Laboratorio de Entomología, Instituto de Biología, Universidad de México, Ciudad de México, D.F., México
UQIC	University of Queensland Insect Collection, Brisbane, Queensland, Australia [closed, specimens sent to QM]
US	School of Public Health and Tropical Medicine, University of Sydney, Sydney, New South Wales, Australia
USNM	United States National Museum, Washington, DC, United States
VBDK	Division of Vector-Borne Diseases, Ministry of Health, Nairobi, Kenya
VMS	Vladimir Malaria Station, Vladimir, Ivanovo, Russia
VRL	Agricultural Research Council-Veterinary Research Laboratory Onderstepoort, Transvaal, Republic of South Africa [now = Onderstepoort Veterinary Research Institute]
WAM	Museum of Western Australia, Perth, Western Australia
YIE	Yunan Institute of Epidemiology, Xiaguan, People's Republic of China
YIM	Yunan Institute of Malaria Control, Simao, People's Republic of China
ZIL	Zoological Institute, Russian Academy of Sciences, Leningrad, Russia
ZM	Zoologisches Museum der Humboldt Universität, Berlin, Germany
ZMC	Universitetets Zoologiske Museum, Copenhagen, Denmark
ZMM	State Central Museum of Zoology, Moscow, Russia
ZMVU	Vilnius University Zoological Museum, Vilnius, Lithuania

General Publications, Family Culicidae

(In alphabetical order, with the exception of recent catalogs)

Species Monographs

Christophers 1960 (bionomics, morphology; *Aedes aegypti*)
Hawley 1988 (bionomics; *Aedes albopictus*)
Horsfall et al. 1973 (bionomics, embryology; *Aedes vexans*)
Mian & Zavortink 2012 (bionomics, taxonomy; *Culex pipiens* group)
Vinogradova 2000 (bionomics, taxonomy; *Culex pipiens*)

Biology

Bates 1949 (general natural history)
Becker et al. 2010 (bionomics, control)
Benedict & Seawright 1987 (pigmentation and environment)
Bright & Hogue 1972: 22–43 (bionomics)
Clements 1963, 1999, 2000, 2011 (bionomics, physiology, behavior, associated pathogens)
Horsfall 1955 (general ecology, behavior, physiology)
Mattingly 1969a (disease transmission)
Mattingly 1971b (ecological aspects of mosquito evolution)
Mullen & Durden 2019 (bionomics, associated pathogens)
Nielsen & Haeger 1960 (swarming and mating)
Service 1976b (collection methods)
Sinka et al. 2010a, 2010b, 2011 (bionomics, distribution; *Anopheles*)
Spielman 1964 (copulation)
Spielman 2001 (bionomics, associated pathogens)
Strickman et al. 2009 (personal protection)

Significant Internet Sites

Arbovirus Catalog—CDC Division of Vector-Borne Diseases (DVBD): https://wwwn.cdc.gov/Arbocat
Barcode of Life Database (BOLD; DNA barcodes): https://www. Boldsystems.org
Companion Vector-Borne Diseases (commercial with wide connection to general information): https://www.cvbd.org
GenBank (DNA sequences): https://www.ncbi.nlm.nih.gov/genbank/
Global Biodiversity Information Facility (GBIF; dist.): https://www.gbif.org
Malaria Atlas Project (MAP; bion., dist.; *Anopheles*): https://map.ox.ac.uk
Mosquito Taxonomic Inventory (MTI; taxonomy, morphology): http://mosquito-taxonomic-inventory.info
VectorBase (genomics): https://www.vectorbase.org
VectorMap (bionomics, distribution): http://www.vectormap.si.edu
Walter Reed Biosystematics Unit (WRBU; bion., tax.): http://www.wrbu.org

Associations

American Mosquito Control Association (http://www.mosquito.org)
European Mosquito Control Association (http://www.emca-online.eu)
Pan-African Mosquito Control Association (http://www.pamca.org)
Society for Vector Ecology (http://www.sove.org)

Taxonomic Anatomy

Belkin 1952 (larval and pupal chaetotaxy)
Belkin 1953b (larval and pupal chaetotaxy)
Belkin 1954b (pupal "setal ring")
Belkin 1960 (pupal and larval setal innervations)
Belkin 1962: 547–563 (terminology)
Cumming & Wood 2009 (Diptera adult morphology and terminology)
Hara 1957 (female terminalia)
Harbach & Knight 1980, 1981 (morphology, anatomy)
Knight 1971 (larval mandibles)
Knight & Chamberlain 1948 (pupal chaetotaxy)
Mohrig 1967 (female terminalia; Aedini)
Reid 1963b (pupal trumpet)
Reinert 1976a (cervical sclerite)
Senevet et al. 1957: 83–89 (larval preclypeal spines), 90–98 (dorsal prothoracic hairs of *Culex*)
Snodgrass 1959 (mosquito morphology)

Recent Taxonomic Catalogs (Chronological)

Stone, Knight & Starcke 1959 (world catalog)
Knight & Stone 1977 (world catalog)
Knight 1978b (first supplement to world catalog)
White 1980 (Africa)
Ward 1984b (second supplement to world catalog)
Gaffigan & Ward 1985 (index to second supplement to world catalog)
Pecor et al. 1992 (*Culex* subgenus *Melanoconion*)
Ward 1992 (third supplement to world catalog)
Torres-Guteirrez & Sallum 2015 (*Culex* subgenus *Melanoconion*; continental South America)
Tanaka 2018 (Japan)

General Taxonomy

Belkin 1950b (L; chaetotaxy)
Belkin 1962 (South Pacific)
Bohart 1946a (China)
Bonne-Wepster 1954 (Indo-Australian Region)
Bonne-Wepster & Swellengrebel 1953 (Anophelinae, Indo-Australian Region)
Boyd 1949 (Anophelinae, malaria)
Crovello 1969: 63–67 (numerical taxonomy)
Dodge 1966 (first-stage larvae)
Edwards 1932a (taxonomy summary, world mosquitoes)

Foote & Cook 1959 (pictorial keys to medically important species)
Higa et al. 2006 (types in Hokkaido, Japan)
Horsfall 1972 (review, bionomics)
Kiknadze 1967 (review of chromosome studies)
Kitzmiller 1963 (chromosomes, review)
Kitzmiller 1967 (chromosomes, review, North America)
Kitzmiller 1976 (chromosomes, review)
Lee & Craig 1983 (F*; cibarial comparisons)
Lee et al. 1980, 1982, 1984, 1987a, 1987b, 1988a, 1988b, 1988c, 1989a, 1989b, 1989c (Australasian Region)
Marchon-Silva et al. 1996 (types in FIOCRUZ [IOC])
Mattingly 1971a (keys to genera)
Mattingly 1973d (keys to genera in French)
Reinert 1974b (dissection methods, morphological terms)
Reinert (numerous publications on mosquito female genitalia)
Reinert, Harbach, & Kitching 2009: 782 (generic abbreviations for tribe Aedini as interpreted by these authors)
Reinert 2010l (E; summary)
Reinert, Harbach, & Kitching 2004, 2006, 2008, 2009 (phylogenetics, tribe Aedini)
Russell et al. 1943 (Anophelinae)
Senevet 1958 (Anophelinae)
Snow 2003 (European mosquito names: see series, Parts 1–11)
Stone & Knight 1955, 1956a, 1956b, 1957a, 1957b, 1957c (types in USNM)
Vargas 1940b (adult generic key for Americas)
Watt 1979 (collection abbreviations)
Wiley 1979 (phylogeny, classification, taxonomy)
Wiley 1981 (phylogeny, classification, taxonomy)
Wiley & Liebermann 2011 (phylogeny, classification, taxonomy)
Wilkerson & Peyton 1990 (*Anopheles* wing spots)

Australian Region

Bartholomew 1913 (history)
Belkin 1956 (expedition to Southwest Pacific)
Belkin 1962 (South Pacific)
Belkin 1968a (New Zealand)
Belkin, Knight, & Rozeboom 1945: 241–265 (Anophelinae; Solomon Islands and New Hebrides)
Bohart 1957 (Micronesia)
Bohart & Ingram 1946b (Okinawa and Central Pacific)
Dobrotworsky 1965 (Victoria)
Dumbleton 1968 (larvae; New Zealand)
Edwards 1924b (taxonomy, adult stage)
Iyengar 1960 (South Pacific)
Knight et al. 1944 (keys, bionomics; Australasia)
Lee 1944a (culicine larvae)
Lee & Woodhill 1944 (Anophelinae)
Lee et al. 1980–1989 (see General Taxonomy)
Maffi 1973b (Rennell & Bellona, British Solomon Islands)
Marks 1969a (Queensland's coral cays)
Marks 1973 (Queensland)
Paine 1943 (Fiji)
Penn 1949b (pupae; New Guinea)
Peters 1962 (anopheline pupal key)
Reisen et al. 1972 (Guam)
Slooff 1972 (Solomon Islands)
Snell 2005 (keys; New Zealand)
Steffan 1966 (checklist; Papuan subregion)
Taylor 1973 (Rennell & Bellona, British Solomon Islands)
van den Assem & Bonne-Wepster 1964 (New Guinea)

Ethiopian Region

Bartholomew 1913 (history)
Coetzee 2017 (taxonomy, keys, bionomics; Afrotropics)
Edwards 1941 (culicine adults)
Gillies & De Meillon 1968 (Anophelinae)
Hopkins 1952 (culicine larvae)
Mekuria 1968 (catalog, culicines; Ethiopia)
Mekuria et al. 1978 (checklist; Ethiopia)
Rioux 1960 (North Chad; keys)

Africa

Benabadji & Larrouy 1968 (checklist; Algeria)
Brunhes et al. 2000 (keys, summary; North Africa)
Edwards 1911b (*Culex*, taxonomy; Africa)
Edwards 1915a (checklist; Natal)
Kyalo et al. 2017 (African *Anopheles*)

Nearctic Region

Barr 1958 (keys; Minnesota)
Belkin et al. 1968 (type information)
Breeland 1982 (bibliography: Florida, United States)
Breeland et al. 1961 (Tennessee Valley)
Carpenter & LaCasse 1955 (keys, taxonomy; North America)
Darsie 1951 (culicine pupae)
Darsie 2005 (Pupal key; Florida, United States)
Darsie & Ward 1981, 2005 (keys, distribution; North America)
Dodge 1963 (larvae; eastern North America)
Dyar 1922i (summary; United States)
Frohne 1955 (larval saddle spines)
Gjullin & Eddy 1972 (northwestern United States)
Gjullin et al. 1961 (Alaska, United States)
Harmston 1949 (checklist; Colorado, United States)
Harrison et al 2016 (keys, mid-Atlantic United States)
King et al. 1960 (keys; southeastern United States)
Nielsen & Rees 1961 (Utah)
Steward & McWade 1961 (Ontario)
Stojanovich 1960 (southeastern United States)
Stojanovich 1961 (northeastern United States)
Ward & Darsie 1982 (taxonomy; North America, additions)

Neotropical Region

Barreto-Reyes 1955 (checklist; Colombia)
Belkin & Heinemann 1972 (Hispaniola)

Belkin et al. 1967 (project description; English/Spanish)
Belkin et al. 1968 (type information)
Belkin, Heinemann & Page 1970 (Jamaica)
Belkin, Schick & Heinemann 1971 (species originally described from Brazil)
Bertram 1971 (British Honduras [Belize])
Calderón Falero et al. 1994 (female stage key; Peru)
Chaverri 2009 (key; Central America)
Consoli & Lourenço-de-Oliveira 1994 (keys; Brazil)
Cova García 1961 (Anophelinae; Venezuela)
Cova García, Oramas & Rausseo 1966 (Toxorhynchitinae and Culicinae; Venezuela)
Darsie et al. 2010 (checklist; Bahamas)
Del Ponte, Castro, & García 1951 (*Aedes* and *Psorophora*; Argentina)
Fauran 1961b (catalog; French Guiana)
Forattini 1962 (Anophelinae)
García & Ronderos 1962 (Anophelini; Argentina)
González & Carrejo 2007 (*Anopheles* keys; Colombia)
Komp 1942 (Caribbean anophelines)
Kumm et al. 1940 (Costa Rica)
Lane 1939 (catalog)
Lane 1953 (2 vol. revision; Neotropics)
Lane & Cerqueira 1942 (Sabethini)
Linthicum 1988 (revision; *Nyssorhynchus*)
Martini 1935 (Mexico)
Morales-Ayala 1971 (Peru)
Perez Vigueras 1956 (Cuba)
Shannon 1931 (Brazil)
Suarez 1988 (*Anopheles* key, Colombia)

Oriental Region

Amerasinghe 1990 (adults; Sri Lanka)
Amerasinghe 1991 (larvae; Sri Lanka)
Aslamkhan 1972 (Pakistan)
Baisas & Dowell 1965 (Philippines; *Anopheles*)
Barras de Carvalho & de Mello-Patiu 2008 (forensics)
Barraud 1934 (Culicinae; India)
Basio 1971b (Philippines)
Bohart 1945 (Philippines)
Borel 1930 (Cochinchina and South Annam)
Delfinado 1966a (Culicini; Philippines)
Khin-Maung-Kyi 1971b, 1971c: 281–296, 297–305 (Anophelinae; Myanmar)
Khin-Maung-Kyi 1971a, 1971d: 473–483, 485–493 (Myanmar)
Lambrecht & van Someren 1971 (Chagos Archipelago, Indian Ocean)
Lu & Chen 1983 (P; key)
Morgan et al. 2009 (mol. phyl; Neocellia Series)
Namgay et al. 2018 (*Anopheles*, distribution, bionomics; Bhutan)
Mamedniyazov & Yerokhin 2005 (*Anopheles* keys; Middle Asia)
Rao et al. 1973 (western Himalayas)
Rattanarithikul et al. 2005 a,b; 2006 a,b; 2007; 2010 (keys, taxonomy, bionomics; Thailand)
Ree 2003 (taxonomy review, keys; Korea)
Stoker & Waktoedi Koesoemawinangoen 1949 (Anophelinae; Indonesia)
Stone & Delfinado 1973 (catalog)
Tanaka 1971 (Ryukyu Islands)
Thurman 1959b (Northern Thailand)
Toma & Miyagi 1986 (pupae and larvae, keys and illustrations; Ryuku Archipelago)
Waktoedi Koesoemawinangoen 1954 (Anophelinae; Indonesia)
Yin Yin May 1972 (Myanmar)

Palearctic Region

Abdel-Malek 1960 (Syria; larval key)
Abul-Hab 1967, 1969 (culicine larvae; Iraq)
Aitken 1954 (Sardinia and Corsica)
Aslamkhan 1971b (Pakistan)
Cranston et al. 1987 (review, British mosquitoes)
Dubitzky 1970b (Kazakhstan)
Gad & Salit 1972 (Red Sea area, Egypt)
Gutsevich 1973a (former USSR; generic and subgeneric key to slide-mounted females)
Gutsevich 1973b (former USSR; key to females)
Gutsevich et al. 1971, 1974 (USSR)
Kramar 1958 (Czechoslovakia)
La Casse & Yamaguti 1950 (Japan)
Lee & Lien 1970 (Korea)
Margalit & Tahori 1974 (Sinai)
Marshall 1938 (British)
Martini 1929, 1930, 1931a, 1931b, 1931c (reviews)
Mattingly & Knight 1956 (Arabia)
Mihályi 1955b (Hungary)
Minar 1971 (Mongolia)
Mohrig 1969 (Germany)
Monchadskii 1951 (larvae; USSR)
Natvig 1948 (Danish and Fennoscandian)
Polyakova 1970b (northern Siberia)
Rageau et al. 1972 (distribution, France)
Ribeiro et al. 1980a (distribution; Cabo Verde)
Rioux 1958 (southern France)
Rioux & Arnold 1955 (France)
Robert et al. 2019 (Euro-Mediterranean)
Senevet & Andarelli 1959 (Mediterranean)
Shalaby 1972 (Libya)
Sicart & Ruffie 1960 (key to pupae)
Skierska 1971 (Poland)
Snow et al. 1997 (UK mosquito bibliography)
Snow 2002–2003 (European mosquito names: see series, Parts 1–11)
Stackelberg 1937 (former USSR; adults)
Ward 1972: 93–97 (Afghanistan)

Taxonomic Catalog

Family Culicidae

Subfamily Anophelinae

Genus *Anopheles* Meigen

(Vol. 1, p. 212)

Important References: Christophers 1915 (M gen.*); Senevet 1934 (P*; key, review); Evans 1938 (M*, F*, P*, L*, E*; keys to F and L; Ethiopian Region); Causey, Deane, & Deane 1946 (M gen.; key, Amazon Anophelini); De Meillon 1947b (M*, F*, P*, L*; keys; Ethiopian Region); Jepson et al. 1947 (bion., keys; Mauritius); Colless 1948 (keys, distr.; Borneo); Chow 1949b (keys, distr.; China); Baba 1950: 1 (keys, distr.; China); Doucet 1951b (keys; Madagascar); Romeo Viamonte & Castro 1951 (F*; buccopharyngeal armature); Bonne-Wepster & Swellengrebel 1953: 1 (distr.; Indo-Australian Region); Lane 1953 (revision, keys; Neotropical Region); Gelfand 1954 (M, F; keys; Liberia); Pringle 1954 (keys; Iraq); Weyer 1954 (key; Palearctic Region); Senevet & Andarelli 1955a (keys; Mediterranean Region); Senevet & Andarelli 1955c (antepalmate hairs); Mattingly & Knight 1956 (Arabia); Vargas & Martínez-Palacios 1956 (keys; Mexico); Guy 1959 (keys; Morocco); Hara 1959 (M gen.; keys; Japan); Vargas 1959 (keys; New World); Otsuru & Ohmori 1960 (keys; Japan); Cova García 1961 (Venezuela); Forattini 1961: 169 (M, F, L; keys; Neotropical Region); Belkin 1962 (tax., distr.; South Pacific); Forattini 1962 (review, keys; Neotropical Region); Service 1962 (keys; W. Africa); Do-Van-Quy 1963 (keys; Vietnam); various authors in Russell et al. 1963 (keys); DuBose & Curtin 1965 (keys; Mediterranean area); Grjebine 1966 (Madagascar); Peyton & Scanlon 1966 (F; key; Thailand); Gillies & De Meillon 1968 (F*, L*; keys, review; Ethiopian Region); Hinton 1968a, 1968b (E*; bion.); Reid 1968 (M*, F*, P*, L*; keys, revision; Malaysia and Borneo); Scanlon et al. 1968 (checklist; Thailand); Mattingly 1969c (E*; tax.); Belkin et al. 1970 (Jamaica); Cagampang-Ramos & Darsie 1970 (F, L; keys, distr.; Philippines); Zavortink 1970b (New World tree hole spp.); Gorham et al. 1973 (keys; western South America); Postiglione et al. 1973 (Turkey); Rattanarithikul & Harrison 1973 (L; keys; Thailand); Zavortink 1973 (M*, F*, P*, L*; keys; subg. *Kerteszia*); Gutsevich et al. 1974 (former USSR); Coluzzi & Kitzmiller 1975: 285 (genetics); Cova García & Sutil Oramas 1975a: 201 (F*; key; Venezuela); Harrison & Klein 1975 (Cambodia); Harrison & Scanlon 1975 (subg. *Anopheles*; Thailand); Ribeiro & da Cunha Ramos 1975 (F, L; keys; Angola); Vargas 1975 (F; key; Costa Rica); Cova García & Sutil Oramas 1976a: 13 (M gen.*; key; Venezuela); Cova García & Sutil Oramas 1977 (M gen.*, F*, L*, E*; keys; Venezuela); Kitzmiller 1977 (chrom.); Klein 1977b (review; Cambodia); White 1977 (tax., morph.; *Anopheles* sp. complexes); Tanaka et al. 1979 (revision; Japan); Wood et al. 1979 (keys, review; Canada); Faran 1980 (Albimanus Section, subg. *Nyssorhynchus*); Harrison 1980 (revision, Myzomyia Series, subg. *Cellia*; Thailand); Darsie & Ward 1981, 2005 (keys; North America); Faran & Linthicum 1981 (M*, F*, P, L*; keys; subg. *Nyssorhynchus*; Amazonia); Lu & Li 1982a (China); Clark-Gil & Darsie 1983 (keys, distr.; Guatemala); Rao 1984 (India); Gillies & Coetzee 1987 (F, L; keys; Afrotropical Region); Lee et al. 1987b (review, key; Australasian Region); Linthicum 1988 (M*, F*, P*, L*; keys, revision, Argyritarsis Section, subg. *Nyssorhynchus*); Suárez et al. 1988 (keys, Colombia); Darsie & Pradhan 1990, 1991 (Nepal); Das et al. 1990 (keys; India); Wilkerson

& Peyton 1990 (*Anopheles* wing spots); Wilkerson & Strickman 1990 (illus. key; Mexico and Central America); Glick 1992 (F; key; southwestern Asia and Egypt); Peyton et al. 1992 (tax.; subg. *Kerteszia* and *Nyssorhynchus*); Wilkerson et al. 1993 (1990 Central America Spanish key translation); Harbach 1994a (rev. genus *Anopheles* intern. class.); Rattanarithikul & Panthusiri 1994 (keys, med. imp. spp.; Thailand); Lambert & Spencer 1995 (genetics, evolution); Nagpal & Sharma 1995 (India); Rubio-Palis & Zimmerman 1997 (review, bion., malaria vectors); Lu et al. 1997 (China); Harbach 2004 (class., phyl.); Sallum et al. 2000b (phyl.); Krzywinski et al. 2001a, 2001b (mol. phyl.); Krzywinski et al. 2001a, 2001b (mol. phyl.); Sallum et al. 2002c (phyl.); Krzywinski & Besansky 2003 (review); Harbach et al. 2005 (*Baimaia*, new subg.); Harbach & Kitching 2005 (phyl.); Rattanarithikul & Harrison et al. 2005 (F, L; tax., keys, bion.; Thailand); Rattanarithikul et al. 2006b (F*, L*; keys, bion., review; Thailand); González & Carrejo 2007 (keys, Colombia); Harbach 2007 (phyl.); Azari-Hamidian & Harbach 2009 (keys, distr.; Iran); Morgan et al. 2009 (mol. phyl.; Neocellia Series); Service 2010 (review, *Anopheles* names); Sinka et al. 2010a (vector distr.; Africa, Europe, Middle East); Sinka et al. 2010b (vector distr.; Americas); Sinka et al. 2011 (vector distr.; Asia-Pacific); Antonio-Nkondjio & Simard 2013 (tax., review; *nili*, *moucheti*); Beebe et al. 2013 (syst., bion.; SW Pacific); Boete & Beisel 2013 (transgen. mosq.); Choochote & Saeung 2013 (syst. methods); Conn et al. 2013 (phylogeograpy, malaria; S America); Dev & Sharma 2013 (review, malaria vectors; India); Dia et al. 2013 (tax.; Funestus Group); Drame et al. 2013 (salivary biomarkers); Duchet et al. 2013 (control issues); Durnez & Coosemans 2013 (malaria trans.); Garros & Dujardin 2013 (syst., gen., morph.); Gendrin & Christophides 2013 (mosq. microbiota); Harbach 2013b (phyl., class.); Lahondère & Lazzari 2013 (bion.); Lanzaro & Lee 2013 (tax., gen., distr.; Gambiae Complex); Manguin et al. 2013 (*Anopheles* bacterial fauna); Obsomer et al. 2013 (surveillance methods); Rejmánková et al. 2013 (L; bion.); Roberts et al. 2013 (control issues); Sinka 2013 (distr., bion., niche models); Suwonkerd et al. 2013 (bion., malaria; SE Asia); Vythilingam & Hii 2013 (monkey malaria; SE Asia); Gunathilaka et al. 2014 (L; key; Sri Lanka); Freitas et al. 2015 (mol. phyl.); Harbach & Kitching 2015 (*Anopheles* phyl., *Christya* to subg.); Neafsey et al. 2015 (mol. phyl., 16 genomes); Gunathilaka 2017 (key, adults; Sri Lanka); Foster et al. 2017 (phyl.); Kyalo et al. 2017 (distr.; sub-Saharan Africa); Bourke et al. 2018 (mtDNA phyl., Neotropical *Anopheles*); Harbach 2018a (*Anopheles* genera comment); Irish et al. 2020 (distr.; sub-Saharan Africa).

Subgenus *Anopheles* Meigen

Meigen 1818: 10. Type species: *Anopheles maculipennis* Meigen (International Commission on Zoological Nomenclature 1959, Opinion 547).

Name derivation: useless, unprofitable, hurtful (Gr).

syn. *Cycloleppteron* Theobald 1901b: 234 (July 15), 1901c: 312 (Nov. 23) (as genus). Type species: *Anopheles grabhamii* Theobald.

syn. *Cyclolepidopteron* Blanchard 1901: 1046 (Dec. 6) (as genus). Type species: *Anopheles grabhamii* Theobald.

syn. *Rossia* Theobald 1902b: 183 (not Owen 1835, not Bonaparte 1838) (as genus). Type species: *Anopheles sinensis* Wiedemann.

syn. *Myzorhynchus* Blanchard 1902: 795 (as genus). Type species: *Anopheles sinensis* Wiedemann.

syn. *Arribalzagia* Theobald 1903a: 13, 81 (as genus). Type species: *Arribalzagia maculipes* Theobald.

syn. *Lophoscelomyia* Theobald 1904c: 12 (Jan.) (as genus). Type species: *Lophoscelomyia asiatica* Leicester.

syn. *Lophomyia* Giles 1904a: 366 (as genus). Type species: *Lophomyia asiatica* Giles.

syn. *Arribalzagaia* Blanchard 1905: 624 (emend. of *Arribalzagia*).

syn. *Nototricha* Coquillett 1906d: 13 (as genus). Type species: *Cycloleppteron mediopunctatus* Theobald.

syn. *Coelodiazesis* Dyar & Knab 1906d: 177 (as genus). Type species: *Anopheles barberi* Coquillett.

syn. *Neostethopheles* James 1910b: 98 (as genus). Type species: *Anopheles aitkenii* James.

syn. *Patagiamyia* James 1910b: 98 (as genus). Type species: *Anopheles gigas* Giles.

syn. *Cyclophorus* Eysell 1912: 422 (not Agassiz 1846, not Pfeiffer 1847, not Kraatz 1880) (as genus). Type species: *Anopheles nigripes* Staeger.

syn. *Proterorhynchus* Brèthes 1912: 10, 14 (as genus). Type species: *Proterorhynchus argentinus* Brèthes.

syn. *Memnemyia* Strickland 1915: 204 (as genus). Type species: *Anopheles brevipalpis* Roper.

syn. *Anopheles* subgenus *Shannoniella* da Fonseca & da Silva Ramos 1940a (March) (1939): 384. Type species: *Shannoniella limai* da Fonseca & da Silva Ramos.

syn. *Anopheles* subgenus *Shannonesia* da Fonseca & Ramos 1940b: 966. Type species: *Shannonesia limai* da Fonseca & Ramos.

syn. *Anopheles* subgenus *Ayrozamyia* Coutinho & Farias 1942: 343 (as genus). Type species: *Ayrozamyia tibiamaculatus* Neiva.

syn. *Anopheles* subgenus *Russellia* Vargas 1943a: 60, 63, 68–70. Type species: *Anopheles xelajuensis* de Leon.

syn. Species group, *Maculipennia* Buonomini & Mariani 1946: 105 (not valid since no type was designated, not absolute tautonymy).

syn. *Anopheles* subgenus *Missirolimyia* Vargas 1950a: 80. Type species: *Missirolymyia algeriensis* Theobald.

syn. *Anopheles* subgenus *Maculipennia* Buonomini & Mariani 1953: 178 (A, L, E; tax.) Type species: *Maculipennia maculipennis* Meigen.

Important references: Reid 1950 (keys; Umbrosus Group); Reid 1953 (keys; Hyrcanus Group; SE Asia); Lane 1953 (Neotropical Region); Mattingly & Knight 1956 (Arabia);

Ohmori 1959 (P; key; Japan); Reid & Knight 1961 (class.); Cova-García 1961 (Venezuela); Reid 1962 (keys; Barbirostris Group); Forattini 1962 (Neotropical Region); Reid 1965 (key; Aitkenii Group); Kitzmiller et al. 1967 (genetics; Maculipennis Complex); Gillies & De Meillon 1968 (Afrotropical Region); Reid 1968 (Malaysia, Borneo); Belkin et al. 1970 (Jamaica); Darsie & Cagampang-Ramos 1972 (keys; Guam); Harrison 1972 (class.; Hyrcanus Group); Gutsevich et al. 1974 (former USSR); Harrison & Scanlon 1975 (revision; Thailand); Xu & Feng 1975 (Hyrcanus Group; China); Floore et al. 1976 (tax.; Crucians Subgroup); Guy et al. 1976: (Maculipennis Complex); Stegnii & Kabanova 1976: 192 (cytogen; Maculipennis Complex); Klein 1977b (Cambodia); Tanaka et al. 1979 (Japan); Wood et al. 1979 (Canada); Darsie & Ward 1981 (keys; North America); Lu & Li 1982a (China); Clark-Gil & Darsie 1983 (Guatemala); Rao 1984 (India); Lee et al. 1987b (Australasian Region); Gillies & Coetzee 1987 (Afrotropical Region); Das et al. 1990 (keys; India); Darsie & Pradhan 1990, 1991 (Nepal); Wilkerson & Strickman 1990 (keys; Central America and Mexico); Glick 1992 (keys; southwestern Asia and Egypt); Rattanarithikul & Panthusiri 1994 (keys, med. imp. spp.; Thailand); Nagpal & Sharma 1995 (India); Lu et al. 1997 (China); Sallum et al. 2000b (phyl., subg. *Anopheles* paraphyly); Sallum et al. 2002c (phyl., subg. *Anopheles* paraphyly); Linton et al. 2003 (review; Maculipennis Complex); Harbach 2004 (class.); Harbach & Kitching 2005 (phyl.; subg. *Anopheles* paraphyly); Darsie & Ward 2005 (keys; North America); Rueda et al. 2005 (tax., type info.); Rattanarithikul et al. 2006b (keys; Thailand); Collucci & Sallum 2007 (phyl.; subg. *Anopheles* monophyletic); Rueda et al. 2009 (E*; Hyrcanus Group); Beltrán-Aguilar et al. 2011 (M gen.*, F*, P*, L*; subg. diagnosis, bion., tax., distr.; Mexico); Townson et al. 2013 (Barbirostris Complex); Taai & Harbach 2015 (review; Barbirostris Complex); Harbach & Kitching 2015 (phyl.; subg. *Anopheles* paraphyly); Freitas et al. 2015 (phyl.; subg. *Anopheles* monophyletic); Kyalo et al. 2017 (distr., Sub-Saharan Africa); Irish et al. 2020 (distr., Sub-Saharan Africa).

aberrans Harrison & Scanlon
1975: 152 (M*, F*, P*, L*). **Type loc.:** Doi Suthep, Muang, Chiang Mai, Thailand (USNM). **Distr.:** Bhutan, Malaysia, People's Republic of China, Thailand. **Refs.:** Rattanarithikul et al. 2006b (F*, L*; bion., distr., keys); Namgay et al. 2018 (bion., distr.; Bhutan). **Ety.:** aberrant (L); banding pattern on adults and larvae deviates from unicolored appearance of others in the group. **Informal name:** Aberrant Thai Nail Mosquito.

acaci Baisas
1946a: 30 (M*, F, P*, L*). **Type loc.:** Malangam Hill, Llavac [Laguna], Luzon, Philippines (NE). **Distr.:** Malaysia (Borneo, Sabah), Philippines. **Refs.:** Reid 1965: 115 (L*; syn.); Reid 1968: 241 (M*, P*, L*; tax.); Basio 1971b: 33 (M*, F*; bion.). **Ety.:** Alejandro B. Ac-ac. **Informal name:** Ac-ac Philippine Nail Mosquito.

ahomi Chowdhury
1929: 986 (M, L*; *barbirostris* var.). **Type loc.:** Assam, India (LU). **Distr.:** India, Nepal. **Refs.:** Christophers 1933: 161 (tax.); Reid 1962: 37 (to sp., distr.); Reid 1968: 120 (key); Pasahan 1974 (chrom.*); Darsie & Pradhan 1990, 1991 (F, L; tax., keys, bion., distr.; Nepal). **Ety.:** Ahom tribe, Assam, India. **Informal name:** Ahom Indian Nail Mosquito.

aitkenii James
1903 (F*). In Theobald 1903a: 22. **Type loc.:** Karwar, Bombay (near Goa Frontier), India (NHMUK). **Distr.:** Andaman Isl., Bangladesh, Cambodia, India, Indonesia (West Papua), Malaysia, Myanmar, Nepal, People's Republic of China, Philippines, Sri Lanka, Taiwan, Thailand, Vietnam. **Refs.:** Christophers 1933: 103 (M*, F*, P, L, E; tax.); Crawford 1938: 36 (P*); Bonne-Wepster & Swellengrebel 1953: 86 (M*, F*, L*); Reid & Knight 1961: 477 (P*); Reid 1965: 107 (M*, F, P, L*, E; syn.); Nguyen Thuong Hien 1968 (F*, L*; keys, tax., bion., distr.; Vietnam); Reid 1968: 237 (M*, F*, P*, L*); Basio 1971b: 33 (M*, F*; bion.); Khin-Maung-Kyi 1971b: 282; Baisas 1974: 44 (M*, F*, P, L*; tax., bion., distr.; Philippines); Ahmed 1987 (distr.; Bangladesh); Lee et al. 1987b: 9 (F; key, tax., bion., distr., review); Darsie & Pradhan 1990, 1991 (F, L; tax., keys, bion., distr.; Nepal); Qu & Zhu 2008: 211 (distr.; People's Republic of China). **Ety.:** E.H. Aitken. **Informal name:** Aitken Indian Nail Mosquito.

ssp. ***britanniae*** Strickland & Chowdhury
1931: 3 (*aitkenii* var.). **Type loc.:** n.k. **Refs.:** Harbach 2018c: 25 (as ssp.). **Ety.:** n.s. (Britain). **Informal name:** British Indian Nail Mosquito.

syn. *pallida* Ludlow
1905a: 129 (not Theobald 1901a (F; *Stethomyia*)). **Type loc.:** Camp Stotsenberg [Angeles], Pampanga, Luzon, Philippines (USNM). **Refs.:** Bohart 1945: 12 (syn.).

albotaeniatus (Theobald)
1903a: 88 (F*; *Myzorhynchus*). **Type loc.:** Perak, Malaya [Malaysia] (NHMUK). **Distr.:** Indonesia, Laos, Malaysia. **Refs.:** Bonne-Wepster & Swellengrebel 1953: 164 (M*, F*, L*); Reid 1968: 143 (M*, F*, P*, L*, E*); Harrison & Scanlon 1975: 31 (distr.); Lee et al. 1987b: 11 (F; key, tax., bion., distr., review). **Ety.:** n.s. [white ribbon (L); prob. ref. to white banding on the legs]. **Informal name:** White-banded Malaysian Nail Mosquito.

algeriensis Theobald
1903a: 21 (F*). **Type loc.:** Algeria (NHMUK). **Distr.:** Afghanistan, Albania, Algeria, Austria, Bosnia and Herzegovina, Bulgaria, Croatia, Cyprus, Egypt, Estonia, France (includes Corsica), FYRO Macedonia, Georgia, Germany, Greece, Hungary, Iran, Iraq, Ireland, Israel (and Gaza Strip and West Bank), Italy (includes Sardinia and Sicily), Jordan, Kosovo, Kyrgyzstan, Lebanon, Libya, Moldova, Montenegro, Morocco, Netherlands, Portugal, Romania, Russia (Southern Dists.), Serbia, Spain

(includes Balearic Isl.), Sweden, Syria, Tajikistan, Tunisia, Turkey, Ukraine, United Kingdom (Wales). **Refs.:** Séguy 1924: 144 (E*); Marshall 1938: 130 (M*, F*, P*, L*, E*; bion.); Aitken 1953: 339 (A, P, L, E; bion.); Gutsevich et al. 1974: 82 (M*, F, L*); Glick 1992: 129 (distr.); Snow & Ramsdale 1999 (distr.; Europe); Alten et al. 2000 (distr.; Turkey); Ramsdale & Snow 2000: 2 (distr.; Europe); Becker et al. 2010: 165 (M*, F*, L*; keys, tax., distr., bion.); Osório et al. 2010 (distr.; Portugal); Bueno Mari et al. 2011 (distr.; Spain)); Habirov et al. 2012 (mol. tax., bion., distr.; Tajikistan); Robert et al. 2019 (distr.; western Palearctic). **Ety.:** Algeria. **Informal name:** Berber Algerian Nail Mosquito [the Berbers are considered the indigenous people of Algeria].

syn. *lukisii* Christophers
1916a: 120 (F). **Type loc.:** Amara [on the Tigris], Iraq (NHMUK, MSI). **Refs.:** Edwards 1932a: 37 (syn.).

alongensis Venhuis
1940: 173 (L*). **Type loc.:** Baie d'Along, Tonkin, Indochina [Hanoi province, Vietnam] (GLB). **Distr.:** Vietnam. **Refs.:** Bonne-Wepster & Swellengrebel 1953: 102 (M*, F*, L*); Nguyen Thuong Hien 1968 (F*, L*; keys, tax., bion., distr.; Vietnam). **Ety.:** Bay of Along. **Informal name:** Along Vietnamese Nail Mosquito.

syn. *tonkinensis* Galliard & Ngu
1940: 595 (M*, F*, P*, L*; not Toumanof 1931). **Type loc.:** Golfe du Tonkin, Indochina [Hanoi province, Vietnam] (EMH?). **Refs.:** Senevet 1947c: 214 (syn.).

anchietai Corrêa & Ramalho
1968: 115 (M*, F*, P*, L*). **Type loc.:** Cangaiba, São Paulo, Brazil (FH). **Distr.:** Argentina, Brazil. **Refs.:** Belkin et al. 1971: 2 (tax.). **Ety.:** José de Anchieta, S.J. **Informal name:** Anchieta Brazilian Nail Mosquito.

annandalei Prashad
1918: 123 (M*, P*, L*). **Type loc.:** Sureil (5,000 ft), Darjeeling Dist. [West Bengal], India (IM). **Distr.:** Cambodia, India, Indonesia, Nepal, Vietnam. **Refs.:** Iyengar 1922 (L*); Christophers 1933: 139 (A, L); Bonne-Wepster & Swellengrebel 1953: 151 (M*, F*, L*); Nguyen Thuong Hien 1968 (F*, L*; keys, tax., bion., distr.; Vietnam); Darsie & Pradhan 1990, 1991 (F, L; tax., keys, bion., distr.; Nepal). **Ety.:** Thomas Nelson Annandale. **Informal name:** Annandale Indian Nail Mosquito.

syn. *djajasanensis* Brug
1926: 805 (A, L; *Lophoscelomyia annandalei* var.). **Type loc.:** Djajasana, near Garoet, Java (NHMUK). **Refs.:** Puri 1929a (syn.); Bonne-Wepster & Swellengrebel 1953: 155 (tax.); Reid 1963a: 111 (syn.).

annulipalpis Lynch Arribálzaga
1878: 149 (F). **Type loc.:** Baradero, Buenos Aires, Argentina (NE). **Distr.:** Argentina, Uruguay. **Refs.:** Alvarado & Ponte 1935 (L); Del Ponte 1940: 602 (M*); Romeo Viamonte & Castro 1951: 320 (F*); Lane 1953: 179 (F*, L); García & Casal 1965a: 5 (P*); Rossi 2017 (M*, F*, P*, L*; tax., phyl.). **Ety.:** ringed feeler (L); palpus. **Informal name:** Ringed Argentinian Nail Mosquito.

apicimacula Dyar & Knab
1906a: 136 (A). **Type loc.:** Livingston, Guatemala (USNM). **Distr.:** Argentina, Belize, Colombia, Costa Rica, Ecuador, El Salvador, French Guiana, Guatemala, Guyana, Honduras, Mexico, Nicaragua, Panama, Suriname, Trinidad and Tobago, Venezuela. **Refs.:** Senevet 1931: 96 (P*); Kumm 1941: 92 (E*); Komp 1942: 55 (F*), 99 (P*, L*), 144 (M*); Pelaez 1945: 74 (F*; key); Cova García 1946: 45 (M*, F*, L*, E*); Stone & Knight 1956b: 214 (type info.); Beltrán-Aguilar et al. 2011: 638 (M*, F*, P*, L*; tax., keys, distr.). **Ety.:** spotted tip (L); apical black wing spot. **Informal name:** Spot-tipped Guatemalan Nail Mosquito.

arboricola Zavortink
1970b: 14 (M*, L*; as *arboricolus*). **Type loc.:** "Bajo Grande," near Cerro Punta, Chiriquí, Panama (USNM). **Distr.:** Panama. **Ety.:** tree inhabitant (L). **Informal name:** Tree-dwelling Panamanian Nail Mosquito.

argyropus (Swellengrebel)
1914: 334 (F*; *Myzorhynchus*). **Type loc.:** Deli [East Coast], Sumatra [Indonesia] (NE). **Distr.:** Cambodia, India, Indonesia (includes Sumatra), Malaysia, Myanmar, People's Republic of China, Thailand, Timor, Vietnam. **Refs.:** Christophers 1924a: 30 (as *hyrcanus* var.); Edwards 1932a: 41 (as *hyrcanus* "aberration"); Nguyen Thuong Hien 1968 (F*, L*; keys, tax., bion., distr.; Vietnam); Reid 1953: 36 (M*, F*, P*, L*, E*; from "aberration"); Rattanarithikul & Harrison 1973 (L; key; Thailand); Harrison & Klein 1975 (distr.); Harrison & Scanlon 1975: 50 (M*, F*, P*, L*; distr.); Lee et al. 1987b: 12 (F; key, tax., bion., distr., review); Whelan & Hapgood 2000 (bion, distr.; East Timor); Oo et al. 2004 (distr.; Myanmar); Rattanarithikul et al. 2006b (F*, L*; bion., distr., keys); Qu & Zhu 2008 (distr.; People's Republic of China); Saeung et al. 2014 (E*). **Ety.:** silver foot (Gr); Ta-III_{3-5} silvery. **Informal name:** White-footed Indonesian Nail Mosquito.

artemievi Gordeyev, Zvantsov, Goryacheva, Shaikevich & Yezhov
2005 (M, F; mol. tax., chrom.). **Type loc.:** Alga Settlement, Leileksky District, Batkensk Region, South Kyrgyzstan (LU?). **Refs.:** Habirov et al. 2012 (mol. tax., bion., distr.; Tajikistan). **Distr.:** Kyrgyzstan, Tajikistan, Turkmenistan. **Ety.:** Mikhail Artemiev. **Informal name:** Artemiev Russian Nail Mosquito.

asiaticus Leicester
1903: 291 (A). **Type loc.:** Ampang Jungle, near Kuala Lumpur [Selangor], Malaya [Malaysia] (NHMUK). **Distr.:** Malaysia, Thailand. **Refs.:** Theobald 1904c: 13 (M, F); Strickland 1914 (L*; bion.); Bonne-Wepster & Swellengrebel 1953: 156 (M*, F*, L*); Reid & Knight 1961:

477 (P*); Reid 1968: 196 (M, F*, P*, L*, E*); Rattanarithikul & Harrison 1973 (L*; key; Thailand); Harrison & Scanlon 1975: 136 (M*, F*, P*, L*); Rattanarithikul et al. 2006b (F*, L*; bion., distr., keys). **Ety.:** Asian. **Informal name:** Asia Malaysian Nail Mosquito.

atratipes Skuse
1889: 1755 (F). **Type loc.:** Berowra, New South Wales, Australia (ANIC). **Distr.:** Australia. **Refs.:** Bancroft 1908: 12 (F; as genus *Pyretophorus*); Lee & Woodhill 1944: 100 (M*, F*, L*); Reid & Knight 1961: 477 (P*); Dobrotworsky 1965: 45 (M*, F, L*); Lee et al. 1987b: 16 (F; key, tax., bion., distr., review, type info.). **Ety.:** clothed foot (L); dark legs. **Informal name:** Black-footed Australian Nail Mosquito.

atroparvus van Thiel
1927: 389 (A; *maculipennis* var.). **Type loc.:** Bolsward, Friesland, Holland (NE). **Distr.:** Austria, Belarus, Belgium, Bosnia and Herzegovina, Bulgaria, Crimean Peninsula, Croatia, Czech Republic, Denmark, France (includes Corsica), FYRO Macedonia, Germany, Greece, Hungary, Iran, Ireland, Italy, Latvia, Lithuania, Moldova, Montenegro, Netherlands, People's Republic of China, Poland, Portugal, Romania, Russia (Southern Dists.), Serbia, Slovakia, Slovenia, Spain, Sweden, Switzerland, Ukraine, United Kingdom. **Refs.:** Hackett & Missiroli 1935 (M*, L*, E*; bion., distr.); Bates 1940: 351 (to ssp.); Rioux 1958: 108 (E*; tax.); Sicart & Ruffie 1960: 635 (P); Kitzmiller et al. 1967: 158 (chrom.*); Bianchi 1968 (to sp.); Farci et al. 1973 (chrom.*); Fraccaro et al. 1976: 27 (chrom.); Mezzanotte & Ferrucci 1978 (tax.; *s.l.*); White 1978 (tax., review, keys, distr.); Ramsdale & Snow 2000 (distr., Europe); Linton et al. 2002b (bion., mol. tax., distr.; England); Linton et al. 2003 (review; Maculipennis Complex); Nicolescu et al. 2004 (mol. tax., distr.; Romania); Sinka et al. 2010a: 117 (bion., review, distr., niche model); Becker et al. 2010: 173 (E*; key, tax., distr., bion.); Danabalan et al. 2014 (bion., distr.; United Kingdom); Robert et al. 2019 (distr.; western Palearctic). **Ety.:** black, small (L). **Informal name:** Black Dutch Nail Mosquito.

syn. *fallax* Roubaud
1934: 740 (M, E*; *maculipennis* var.). **Type loc.:** Near Caen [Calvados], Normandy, France (PIP?). **Refs.:** Bates 1940: 354 (syn.).

syn. *cambournaci* Roubaud & Treillard
1936: 730 (F, L, E*; *maculipennis* biotype). **Type loc.:** Aquas de Moura Region [Setubal], Portugal (PIP?). **Refs.:** Bates 1940: 354 (syn.).

atropos Dyar & Knab
1906b: 160 (F). **Type loc.:** Florida Keys, Florida, United States (USNM). **Distr.:** Cayman Isl., Cuba, Jamaica, Mexico, United States (continental). **Refs.:** Hinman 1932: 138 (L; bion.); Roth 1944: 104 (M*), 107 (M; keys); Penn 1949a: 57 (P*); Vargas 1950a: 74 (E); Yamaguti & LaCasse 1950b: 14 (M*, F*, L*); Carpenter & LaCasse 1955: 31 (M*, F*, L*; keys); Stone & Knight 1956b: 276 (type info., lectotype desig.); Dodge 1966: 361 (first instar L*; key); Belkin 1968b: 10 (syn.); Kreutzer, Narang, & Kitzmiller 1969 (chrom.*); Belkin et al. 1970: 27 (M*, F, P*, L*); Kreutzer & Kitzmiller 1971b (chrom.*; as *bradleyi*, misident. (B. Harrison, pers. comm.)); Floore et al. 1976: 37 (A, L, P; tax.); Linley 1992: 40 (E*); Darsie & Ward 2005 (F*, L*; keys, distr.); Harrison et al. 2016 (F*, L*; keys, distr.). **Ety.:** n.s. [Atropos (Gr), a fate of Greek of mythology]. **Informal name:** Atropos American Nail Mosquito.

syn. *ferruginosus* Wiedemann
1828: 12 (F). **Type loc.:** New Orleans, Louisiana, United States (NMW). **Refs.:** Coquillett 1906d: 7 (syn.); Belkin 1968b: 9 (lectotype desig.; from syn. with *quinquefasciatus*); Belkin et al. 1970: 27 (*nomen oblatum*, according to Art. 23(b) of the International Commission on Zoological Nomenclature).

aztecus Hoffmann
1935: 5 (M*, F*, L*, E*; *maculipennis* ssp.). **Type loc.:** Valle de Mexico [Mexico City and vicinity, Mexico] (UM). **Distr.:** Mexico, United States (continental). **Refs.:** Vargas 1939: 356 (to sp.); Vargas 1941d: 113 (E*); Aitken 1945: 304 (M, F*, L, E; tax.); Pelaez 1945: 74 (F*; key); Penn 1949b: 58 (P*); Pratt 1952 (P*, L*; tax., distr.); Vargas & Martínez Palacios 1956 (M*, F, P*, L*, E; keys); Baker & Kitzmiller 1964 (chrom.*); Dodge 1966: 361 (first instar L*; key); Kitzmiller et al. 1967: 173 (chrom.*); White 1978 (tax., review, keys, distr.); White 1979: 23 (as Nearctic member of Maculipennis Complex); Wilkerson & Strickman 1990 (F*; key, distr.); Porter & Collins 1996 (mol. phy.; Nearctic); Muñoz-Cabrera et al. 2006 (M, F, L; bion., distr.). **Ety.:** Aztec area of Central Mexico. **Informal name:** Aztec Mexican Nail Mosquito. (Vol. 1, p. 294)

baezai Gater
1933: 162 (M*, F*, P, L). **Type loc.:** Pulau Langkawi [Perlis], Malaya [Malaysia] (NHMUK). **Distr.:** Cambodia, Guam, Indonesia, Malaysia, Philippines, Singapore, Thailand, Vietnam. **Refs.:** Colless 1948: 88 (M*, F*, L*; *baezai* ssp.); Bonne-Wepster & Swellengrebel 1953: 177 (M*, P*, L*); Nguyen Thuong Hien 1968 (F*, L*; keys, tax., bion., distr.; Vietnam); Reid 1968: 165 (M*, F*, P*, L*, E*); Basio 1971b: 34 (M*; bion.); Basio & Reisen 1971: 60 (L; distr.); Rattanarithikul & Harrison 1973 (L*; key; Thailand); Baisas 1974: 45 (M, F, P*, L*; tax., bion., distr.; Philippines); Harrison & Klein 1975: 10 (distr.); Harrison & Scanlon 1975: 114 (M, F*, P*, L*; lectotype desig., distr.); Klein 1977a: 110 (distr.); Lee et al. 1987b: 20 (F; key, tax., bion., distr., review); Rattanarithikul et al. 2006b (F*, L*; bion., distr., keys). **Ety.:** J.I. Baeza. **Informal name:** Baeza Malaysian Nail Mosquito.

syn. *gateri* Baisas
1936b: 78 (P*, L). **Type loc.:** Iwahig, Palawan, Philippines (NE; paratype USNM). **Refs.:** Reid 1950: 290 (syn.).

baileyi Edwards
1929c: 323 (F; *gigas* var.). **Type loc.:** Yatung, Tibet (NHMUK). **Distr.:** Bangladesh, Cambodia, India, Laos, Nepal, People's Republic of China, Taiwan, Thailand, Vietnam. **Refs.:** Christophers 1933: 137 (M*, L); Ahmed 1987 (distr.; Bangladesh); Darsie & Pradhan 1990, 1991 (F, L; tax., keys, bion., distr.; Nepal); Harrison et al. 1991: 209 (from *gigas* var.); Rattanarithikul et al. 2006b (F*, L*; bion., distr., keys). **Ety.:** F.M. Bailey. **Informal name:** Bailey Chinese Nail Mosquito.

balerensis Mendoza
1947: 173 (M*). **Type loc.:** Baler, Quezon [Luzon], Philippines (NE). **Distr.:** Philippines. **Refs.:** Basio 1971b: 34 (M*, F*; bion.). **Ety.:** Baler, Philippines. **Informal name:** Baler Philippine Nail Mosquito.

bancroftii Giles
1902: 511 (A). **Type loc.:** Burpengary, Queensland, Australia (NHMUK). **Distr.:** Australia, Indonesia (including Borneo, Sulawesi, Moluccas), Malaysia, New Caledonia, Papua New Guinea, Sri Lanka. **Refs.:** Bancroft 1908: 14 (F; in genus *Myzorhynchus*); Lee & Woodhill 1944: 111 (M*, F*, L*, E*); Bonne-Wepster & Swellengrebel 1953: 214 (M*, F*, L*); Peters 1962: 164 (P*); Lee et al. 1987b: 22 (F; key, tax., bion., distr., review); Beebe et al. 2001 (mol. tax.; genotypes A1, A2, B, C1, C2, D); Beebe et al. 2013 (Bancroftii Group); Pol et al. 2018 (distr.; New Caledonia). **Ety.:** Thomas Lane Bancroft. **Informal name:** Bancroft Australian Nail Mosquito. (Vol. 1, p. 296)

ssp. ***barbiventris*** Brug
1938 (M, F*, L). **Type loc.:** Kalawara, Palow, Res Menado, Celebes [Sulawesi] (NHMUK). **Distr.:** Indonesia (Sumatra). **Refs.:** Bonne-Wepster & Swellengrebel 1953: 222 (M*, F*, L*); Lee et al. 1987b: 38 (F; key, tax., bion., distr., review; as variety); Harbach & Howard 2007: 37. **Ety.:** n.s. (in German) [bearded venter (L); poss. ref. to scale patches]. **Informal name:** Bearded Indonesian Nail Mosquito.

barberi Coquillett
1903b: 310 (F). **Type loc.:** Plummer's Island, Maryland, United States (USNM). **Distr.:** Canada, Puerto Rico, United States (continental). **Refs.:** Vargas 1942d: 329 (E*); Roth 1944: 98 (M*), 107–110 (M gen.; keys); Pelaez 1945: 74 (F*; key); Jenkins & Carpenter 1946: 35 (bion.); Darsie 1949: 524 (P*); Yamaguti & LaCasse 1950b: 17 (M*, F*, L*); Carpenter & LaCasse 1955: 32 (M*, F*, L*; keys); Stone & Knight 1956b: 276 (type info.); Dodge 1966: 362 (first instar L*; key); Zavortink 1970b: 7 (M*, F*, P*, L*); Smith & Trimble 1973 (distr.; Canada); Darsie & Ward 2005 (F*, L*; keys, distr.); Harrison et al. 2016 (F*, L*; keys, distr.; note 15). **Ety.:** Herbert Spencer Barber. **Informal name:** Barber American Nail Mosquito.

barbirostris van der Wulp
1884: 248 (F). **Type loc.:** Mount Ardjoeno, Java [Indonesia] (LM). **Distr.:** Australia, Bangladesh, Cambodia, Guam, India (includes Andaman Isl.), Indonesia (includes Borneo, Java, Flores, Moluccas, Sulawesi, Sumatra [includes Keulauan Riouw Archipelago, Timor]), Laos, Malaysia, Myanmar, Nepal, Papua New Guinea, Pakistan, People's Republic of China (includes Hainan), Philippines, Sri Lanka, Thailand, Vietnam. **Refs.:** Davis 1926 (F*); Christophers 1933: 155 (M*, F*, P, L, E); Penn 1949b: 12 (P*); Bonne-Wepster & Swellengrebel 1953: 201 (M*, F*, L*); Reid 1962: 7 (M, F*, P*, L, E*); Nguyen Thuong Hien 1968 (F*, L*; keys, tax., bion., distr.; Vietnam); Reid 1968: 122 (M*, F*, P*, L*, E*); Chowdaiah et al. 1970 (chrom.*); Aslamkhan 1971b (distr.; Pakistan); Rattanarithikul & Harrison 1973 (L*; key, distr.; Thailand; *s.l.*); Ameen & Talukdar 1974c: 92 (P*); Harrison & Scanlon 1975: 83 (M*, F*, P*, L*; distr.); Lien et al. 1977 (tax.); Lee et al. 1987b: 39 (F; key, tax., bion., distr., review); Darsie & Pradhan 1990, 1991 (F, L; tax., keys, bion., distr.; Nepal); Whelan & Hapgood 2000 (bion., distr.; East Timor); Oo et al. 2004 (distr.; Myanmar); Rattanarithikul et al. 2006b (F*, L*; bion., distr., keys); Sinka et al. 2011: 89 (Barbirostris Complex; bion., review, distr., niche model); Taai & Harbach 2015: 245 (M*, F*, P*, L*; phyl., redescr., bion.). **Ety.:** n.s. [bearded beak (L); proboscis and palpi with many erect scales]. **Informal name:** Beard-beaked Indonesian Nail Mosquito. (Vol. 1, p. 298)

syn. *martini* Laveran
1902a: 907 (F). **Type loc.:** Near Pursat, Cambodia, Indochina (PIP). **Refs.:** Reid 1947: 89 (syn.).

syn. *innominata* Stoker & Waktoedi Koesoemawinangoen
1949: 10 (A*; *barbirostris* ssp.). **Type loc.:** Celebes [Sulawesi] (LU). **Refs.:** Bonne-Wepster & Swellengrebel 1953: 209 (as syn. of *vanus*); Reid 1968: 122 (syn.).

barbumbrosus Strickland & Chowdhury
1927a: 18 (in footnote: new name for *pallidus* Swellengrebel, not Theobald 1901a). **Distr.:** Bangladesh, Bhutan, Cambodia, India, Indonesia, Malaysia, Nepal, People's Republic of China (includes Hainan, Hong Kong), Sri Lanka, Taiwan, Thailand, Timor, Vietnam. **Refs.:** Chow 1949a: 5 (A); Bonne-Wepster & Swellengrebel 1953: 212 (M*, F*, L*); Reid 1962: 29 (M*, F*, P*, L*; distr.); Wattal et al. 1962: 63 (M*, F*; distr.); Nguyen Thuong Hien 1968 (F*, L*; keys, tax., bion., distr.; Vietnam); Reid 1968: 140 (distr.); Rattanarithikul & Harrison 1973 (L; key; Thailand); Harrison & Klein 1975 (distr.); Harrison & Scanlon 1975: 102 (M*, F*, P*, L*; distr.); Ahmed 1987 (distr.; Bangladesh); Lee et al. 1987b: 47 (F; key, tax., bion., distr., review); Darsie & Pradhan 1990, 1991 (F, L; tax., keys, bion., distr.; Nepal); Whelan & Hapgood 2000 (bion., distr.; East Timor); Rattanarithikul et al. 2006b (F*, L*; bion., distr., keys); Qu & Zhu 2008 (distr.; China); Namgay et al. 2018 (bion., distr.; Bhutan). **Ety.:** n.s. [shady or dark beard (L)]. **Informal name:** Black-bearded Indonesian Nail Mosquito.

syn. *pallidus* Swellengrebel
1919: 8 (syntype M*, L?; *Myzorhynchus barbirostris* var.). **Type loc.:** Possibly Mandailing, Sumatra, and Nusa

Kambangan Island (= Noesa Kemba) [Indonesia] (type not designated). **Refs.:** Townsend 1990: 116 (type info.).
syn. *pseudumbrosus* Haga
1930 (1929): 364 (M, F, L; in key couplets) [not listed by Edwards 1932a; Stone et al. 1959; Knight & Stone 1977; or Knight 1978b]. **Type loc.:** Bolseard, Friesland, Holland [Netherlands] (NE). **Refs.:** Ward 1979: 132 (*lapsus calami* of *barbumbrosus*; syn.).

barianensis James
1911 (A). In James & Liston 1911: 76. **Type loc.:** Barian (7,000 ft), Murree Hills, Punjab Hill States [India] (A). **Distr.:** Afghanistan, India, Pakistan, Tajikistan. **Refs.:** Christophers 1933: 117 (M*, F*, P*, L*, E; bion.); Rueda & Pecor et al. 2008: 397 (distr.). **Ety.:** Barian, Punjab, India. **Informal name:** Barian Indian Nail Mosquito.

beklemishevi Stegnii & Kabanova
1976: 6 (E*; chrom.*). **Type loc.:** (not designated). **Distr.:** Finland, Russia (northwestern, Volga), Sweden. **Refs.:** White 1978 (tax., review, keys, distr.); Ramsdale & Snow 2000: 3 (distr.; Europe); Linton et al. 2003 (review; Maculipennis Complex); Becker et al. 2010: 173 (E*; key); Robert et al. 2019 (distr.; western Palearctic). **Ety.:** Vladimir Nikolaievitch Beklemishev. **Informal name:** Beklemishev Russian Nail Mosquito.

belenrae Rueda
2005: 2 (M*, F*, P*, L*). **Type loc.:** Tongilchon, Gyeonggi-do, South Korea (USNM). **Distr.:** People's Republic of China, South Korea. **Refs.:** Li, Lee et al. 2005 (mol. tax.); Qu & Zhu 2008 (distr.; People's Republic of China). **Ety.:** Belen P. Rueda (spouse). **Informal name:** Belen Korean Nail Mosquito.

bengalensis Puri
1930: 955 (M, L; *aitkeni* var. [*sic*]). **Type loc.:** Marianbarie, Bengal Terai [near Sukna, West Bengal], India (NHMUK, MSI [NICD]). **Distr.:** Bangladesh, Bhutan, Cambodia, India, Indonesia, Japan, Malaysia, Myanmar, Nepal, People's Republic of China (includes Hong Kong), Philippines, Taiwan, Thailand, Vietnam. **Refs.:** Christophers 1933: 109 (A, L; tax.); Baisas 1938: 221 (P*, L*); Puri 1949: 483 (to ssp.); Reid 1965: 113 (M*, F, P*, L*, E*; to sp.); Kanda & Kamimura 1967: 108 (M*, F*, P*, L*; distr.); Reid 1968: 245 (M, F, P*, L*, E*); Aslamkhan 1971b (distr.; Pakistan); Khin-Maung-Kyi 1971b: 282 (as *aitkeni* [*sic*] var.); Basio 1971b: 34 (M*, F*; bion.); Rattanarithikul & Harrison 1973 (L*; key; Thailand); Baisas 1974: 47 (M, F, P, L; tax., bion., distr.; Philippines); Harrison & Scanlon 1975: 157 (M*, F*, P*, L*; distr.); Harrison & Klein 1975 (distr.; Cambodia); Tanaka et al. 1975c: 210 (distr.); Tanaka et al. 1979: 57 (M*, F*, P, L*); Lee et al. 1987b: 50 (F; key, tax., bion., distr., review); Darsie & Pradhan 1990, 1991 (F, L; tax., keys, bion., distr.; Nepal); Oo et al. 2004 (distr.; Myanmar); Rattanarithikul et al. 2010 (F*, L*; bion., distr., keys); Namgay et al. 2018 (bion., distr.; Bhutan). **Ety.:** Bengal Region, India. **Informal name:** Bengal Indian Nail Mosquito.

borneensis McArthur
1949: 49 (M*, L*; *aitkeni* var. [*sic*]). **Type loc.:** Kampong Naudu, Tambunan Plain, North Borneo, Borneo (LU). **Distr.:** Malaysia (Borneo). **Refs.:** Bonne-Wepster & Swellengrebel 1953: 94 (M*, L*); Reid 1965: 115 (syn.); Reid 1968: 245 (M*, F, P*, L*; from syn.). **Ety.:** Borneo. **Informal name:** Borneo Malaysian Nail Mosquito.

bradleyi King
1939: 468 (M*, F, P, L*; *crucians* var.). **Type loc.:** Saint Johns River, Brevard County, Florida, United States (USNM). **Distr.:** Honduras, Mexico, Nicaragua, United States (continental). **Refs.:** King & Bradley 1941: 65 (to sp.); Vargas 1941d: 112 (E); Roth 1944: 99 (M*; tax), 107 (M; keys); Pelaez 1945: 74 (F*; key); Penn 1949a: 64 (P*); Yamaguti & LaCasse 1950b: 41 (M*, F*, L*); Carpenter & LaCasse 1955: 34 (M, F, L*; keys); Kreutzer & Kitzmiller 1970: 554 (tax.); Kreutzer et al. 1970 (chrom.*); Kreutzer & Kitzmiller 1971b (misident.; see *atropos*); Floore et al. 1976: 30 (M*, F*, P*, L*; tax.); Darsie & Ward 2005 (F*, L*; keys, distr.); Beltrán-Aguilar et al. 2011: 643 (M*, F*, P*, L*; tax., keys, distr.); Harrison et al. 2016 (F*, L*; keys, distr.; notes 20, 22). **Ety.:** George Hirst Bradley. **Informal name:** Bradley American Nail Mosquito.

brevipalpis Roper
1914: 141 (M*, F*). **Type loc.:** Membakut Swamp, North Borneo, [Malaysian] Borneo (NHMUK). **Distr.:** Indonesia, Malaysia (Borneo). **Refs.:** Strickland 1915 (L, E); Reid 1950: 304 (M, F, P*, L*); Bonne-Wepster & Swellengrebel 1953: 119 (M*, F*, L*); Reid 1968: 169 (M*, F*, P*, L*, E*); Lee et al. 1987b: 51 (F; key, tax., bion., distr., review). **Ety.:** short feeler (L); short palpus. **Informal name:** Short-lipped Malaysian Nail Mosquito.

brevirostris Reid
1950: 300 (M*, F*, P*, L*). **Type loc.:** Bangi, Selangor, Malaya [Malaysia] (NHMUK). **Distr.:** Malaysia, Singapore. **Refs.:** Bonne-Wepster & Swellengrebel 1953: 187 (M*, F*, L*). **Ety.:** short beak (L); proboscis and palpus very short. **Informal name:** Short-beaked Malaysian Nail Mosquito.

bulkleyi Causey
1937a: 543 (M). **Type loc.:** Chandbury, Thailand (NE). **Distr.:** Thailand. **Refs.:** Rattanarithikul & Harrison 1973 (L*; key; Thailand); Harrison & Scanlon 1975: 134 (M); Rattanarithikul et al. 2006b (F*, L*; bion., distr., keys). **Ety.:** Lucius Constant Bulkley. **Informal name:** Bulkley Thai Nail Mosquito.

bustamentei Galvão
1955: 255 (F*). **Type loc.:** Rebeirao [Ribeirão] Pequeno, Santa Catarina, Brazil (FH). **Distr.:** Brazil. **Ety.:** Fernando Machado de Bustamente. **Informal name:** Bustamente Brazilian Nail Mosquito.

calderoni Wilkerson
1991: 25 (M*, F*, P*, L*). **Type loc.:** Salitral, Department of Piura, Peru (USNM). **Distr.:** Colombia, Ecuador, Peru, Venezuela. **Refs.:** Rubio-Palis & Moreno 2003: 159 (distr.; Venezuela); González et al. 2010 (F*, L; mol. tax., morphometrics, distr.; Colombia, Ecuador, Peru); Becker et al. 2010: 374 (F*, L*; key, tax., distr., bion.); Bourke et al. 2018 (mol. tax.; as *s.l.*). **Ety.:** Guillermo Calderón Falero. **Informal name:** Calderón Peruvian Nail Mosquito.

caliginosus De Meillon
1943: 91 (F; *coustani* ssp.). **Type loc.:** Jadotville [Katanga], Belgian Congo [Democratic Republic of the Congo] (SAIM). **Distr.:** Angola, Botswana, Democratic Republic of the Congo, Eswatini. **Refs.:** De Meillon 1947b: 47 (F*); Hopkins 1952: 195 (tax.); Gillies & De Meillon 1968: 38 (F*; to sp.); Ribeiro & da Cunha Ramos 1975: 13 (distr.); Kyalo et al. 2017 (distr.; sub-Saharan Africa). **Ety.:** n.s. [dark (L)]. **Informal name:** Black Congolese Nail Mosquito.

campestris Reid
1962: 15 (M*, F*, P, L, E). **Type loc.:** Rantau Panjang, Klang, Selangor, Malaya [Malaysia] (NHMUK). **Distr.:** Cambodia, Guam, Malaysia, People's Republic of China, Thailand, Vietnam. **Refs.:** Nguyen Thuong Hien 1968 (F*, L*; keys, tax., bion., distr.; Vietnam); Rattanarithikul & Harrison 1973 (L; key; Thailand); Harrison & Klein 1975 (distr.); Harrison & Scanlon 1975: 89 (M*, F*, P*, L*); Rattanarithikul et al. 2006b (F*, L*; bion., distr., keys); Taai & Harbach 2015: 258 (M, F, P, L; bion., genetics). **Ety.:** of fields (L); from broad alluvial plains. **Informal name:** Alluvial Malaysian Nail Mosquito. (Vol. 1, p. 300)

chiriquiensis Komp
1936b: 156 (M*, F*, L). **Type loc.:** Chiriquí Volcano, NW Panama (USNM). **Distr.:** Guatemala, Panama. **Refs.:** Dampf 1939: 281 (syn.); Hoffmann 1939 (M*, F*); Stone & Knight 1956b: 277 (type info., lectotype desig.); Wilkerson 1990a: 179 (tax.; from syn. with *parapunctipennis*). **Ety.:** Volcán de Chiriquí, Panama. **Informal name:** Chiriquí Panamanian Nail Mosquito.

chodukini Martini
1929: 141 (F). **Type loc.:** Tashkent [Uzbek, USSR] (LU). **Distr.:** Uzbekistan. **Ety.:** Nikolai Chodukin. **Informal name:** Chodukin Uzbek Nail Mosquito.

claviger (Meigen)
1804: 4 (M, F; *Culex*). **Type loc.:** [Germany], Europe (NE). **Distr.:** Afghanistan, Albania, Algeria, Armenia, Austria, Azerbaijan, Belarus, Belgium, Bosnia and Herzegovina, Bulgaria, Crimean Peninsula, Croatia, Cyprus, Czech Republic, Denmark, Estonia, Finland, France (includes Corsica), FYRO Macedonia, Georgia, Germany, Greece (includes Crete), Hungary, India, Iran, Iraq, Ireland, Israel (and Gaza Strip and West Bank), Italy (includes Sicily), Jordan, Kazakhstan, Kyrgyzstan, Latvia, Lebanon, Libya, Liechtenstein, Lithuania, Luxembourg, Moldova, Montenegro, Morocco, Netherlands, Norway, Pakistan, Palestine, People's Republic of China, Poland, Portugal, Romania, Russia (Central, northwestern, Southern Dists., Volga), Serbia, Slovakia, Slovenia, Spain, Sweden, Switzerland, Syria, Tajikistan, Tunisia, Turkey, Turkmenistan, Ukraine, United Kingdom, Uzbekistan. **Refs.:** Marshall 1938: 126 (M*, F*, P*, L*, E*; bion.); Ross & Roberts 1943b: 3 (M*, F*, L*; tax., distr., bion.); Torres Cañamares 1945 (M*, F, P*, L*; tax., bion.); Bates et al. 1949: 422 (A, L; tax.); Aitken 1954: 451 (A, P, L, E); Senevet & Andarelli 1955a: 41 (M*, F, P*, L*, E; tax.); Senevet & Andarelli 1955a: 128 (P, E*; tax.; as *missiroli*); Senevet & Andarelli 1955b: 129 (E*; as *missiroli*); Senevet & Andarelli 1955b (tax.); Service 1973 (bion.); Gutsevich et al. 1974: 86 (M*, F*, L*); Utrio 1975: 63 (distr.); Utrio 1976 (L*); Glick 1992 (F; key, distr.); Ramsdale & Snow 2000: 3 (distr.); Hadjivassilis 2000: 38 (distr.; Cyprus); Trari et al. 2002: 330 (distr.; Morocco); Beck et al. 2003: 24 (distr.; Luxembourg); Qu & Zhu 2008 (distr.; People's Republic of China); Becker et al. 2010: 166 (M*, F*, P, L*; keys, tax., bion., distr.); Habirov et al. 2012 (mol. tax., bion., distr.; Tajikistan); Robert et al. 2019 (distr.; western Palearctic). **Ety.:** club bearing (L); ref. to shape of male palpus. **Informal name:** Club-lipped German Nail Mosquito. (Vol. 1, p. 302)

syn. *villosus* Robineau-Desvoidy
1827: 411 (M, F). **Type loc.:** Paris, France (NE). **Refs.:** Edwards 1932a (syn.).

syn. *grisescens* Stephens
1828: 503 (two syntypes). **Type loc.:** Ripley, Surrey, England (NHMUK). **Refs.:** Edwards 1932a (syn.); Townsend 1990: 80 (type info.).

syn. *turkestani* Shingarev
1926: 47 (M, F; *algeriensis* var.). **Type loc.:** Turkestan (LU). **Refs.:** Edwards 1932a: 38 (syn. suggested).

syn. *amaurus* Martini
1929: 135 (M*, F). **Type loc.:** [Ben Anen, Taschkent], Turkestan (NHMUK). **Refs.:** Senevet & Andarelli 1955a: 128 (tax.); Stone et al. 1959: 16 (syn.).

syn. *habibi* Mulligan & Puri
1936: 68 (F*). **Type loc.:** Hudda village on banks of Habib Nullah, Quetta, Baluchistan [Pakistan] (NHMUK). **Refs.:** Aslamkhan 1971b (distr.; Pakistan); Glick 1992: 126 (syn.).

syn. *missiroli* Del Vecchio
1939: 34 (E*; *Anopheles claviger* (*bifurcatus*) var.). **Type loc.:** Latium, Italy (LU). **Refs.:** Bates et al. 1949: 422 (syn.).

syn. *pollutus* Torres Cañamares
1945: 232 (L*; *claviger* var.). **Type loc.:** Near Cuenca, Spain (LU). **Refs.:** M. Bates in Boyd 1949 (syn.); Senevet

& Andarelli 1955a: 128 (P, L; tax.); Stone et al. 1959: 17 (syn.).

colledgei Marks
1956: 41 (M*, F*, P*, L*). **Type loc.:** Rex Creek, Mossman, Queensland, Australia (ANIC). **Distr.:** Australia. **Refs.:** Lee et al. 1987b: 52 (F; key, tax., bion., distr., review); Townsend 1990: 62 (type info.). **Ety.:** William Robert Colledge. **Informal name:** Colledge Australian Nail Mosquito.

collessi Reid
1963a: 103 (M, F*, P, L*, E*). **Type loc.:** Ulu Gombak near Kuala Lumpur, Selangor, Malaya [Malaysia] (NHMUK). **Distr.:** Borneo, Brunei, Malaysia. **Ety.:** Donald Henry Colless. **Informal name:** Colless Malaysian Nail Mosquito.

concolor Edwards
1938 (M*, F*). In Evans 1938: 56–58. **Type loc.:** Kabila, Kwango Province [Leopoldville], Belgian Congo [Democratic Republic of the Congo] (NHMUK). **Distr.:** Angola, Cameroon, Democratic Republic of the Congo. **Refs.:** Vincke & Leleup 1949a (M*, L*, E*); Gillies & De Meillon 1968: 47 (M*, F*, P, L*); Ribeiro & da Cunha Ramos 1975: 15 (distr.); Kyalo et al. 2017 (distr.; sub-Saharan Africa). **Ety.:** uniformly colored (L); dark. **Informal name:** Brown Congolese Nail Mosquito.

corethroides Theobald
1907: 35 (M, F*). **Type loc.:** [Burpengary], Queensland, Australia (NHMUK). **Distr.:** Australia. **Refs.:** Bancroft 1908: 11 (M, F); Lee & Woodhill 1944: 81 (syn.); Marks 1956: 49 (M*, F; tax.); Lee et al. 1987b: 54 (F; key, tax., bion., distr., review). **Ety.:** similar to *Corethra* [genus of family Corethrellidae]. **Informal name:** Broom-backed Australian Nail Mosquito.

costai da Fonseca & da Silva Ramos
1940a: 385 (M*). **Type loc.:** São Vicente, São Paulo, Brazil (FH). **Distr.:** Bolivia, Brazil, Colombia, French Guiana, Peru, Suriname, Venezuela. **Refs.:** Bonne 1923c (M*; as *mediopunctatus*); Lane 1953: 221 (syn.); Belkin et al. 1971: 3 (tax.); Sallum et al. 1999: 288 (from syn.); Sallum & Flores 2004 (E*); Bourke et al. 2018 (mol. tax.; as *s.l.*). **Ety.:** Arthur Costa Filho. **Informal name:** Costa Brazilian Nail Mosquito.

syn. *bonnei* da Fonseca & da Silva Ramos
1940a: 386 (F). **Type loc.:** Suriname (NE). **Refs.:** Belkin, Schick, & Heinemann 1965: 61 (type info.); Sallum et al. 1999: 282 (syn.; transfer from *mediopunctatus*; emend. from *bonnei* to *bonneorum*; according to International Commission on Zoological Nomenclature Art. 32.5, the original *bonnei* is correct).

coustani Laveran
1900: 109 (F). **Type loc.:** Madagascar (SAIM). **Distr.:** Angola, Benin, Botswana, Burkina Faso, Burundi, Cameroon, Central African Republic, Chad, Comoros, Côte d'Ivoire, Democratic Republic of the Congo, Egypt, Eritrea, Eswatini, Ethiopia, Gabon, The Gambia, Ghana, Guinea, Guinea-Bissau, Israel, Jordan, Kenya, Liberia, Madagascar (includes Glorioso and Juan de Nova Isl.), Malawi, Mali, Mauritania, Mauritius, Mayotte, Morocco, Mozambique, Namibia, Nigeria, Oman, Republic of the Congo, Republic of South Africa, La Réunion, Rwanda, São Tomé & Príncipe, Saudi Arabia, Senegal, Sierra Leone, Somalia, South Sudan, Sudan, Tanzania (includes Zanzibar), Togo, Uganda, Yemen, Zambia, Zimbabwe. **Refs.:** Blanchard 1905 (F*); Macfie & Ingram 1922 (F*; as *mauritianus* de Grandpré & de Charmoy); Séguy 1924: 153 (L*; as *An. mauritianus*); De Meillon 1936b (E*); Evans 1938: 59 (M*, F*, P*, L*, E*); De Meillon 1947b: 43 (M*, F*, P*, L*, E*); Senevet & Andarelli 1955a: 101 (M*, F*, P, L*, E); Gillies & De Meillon 1968: 35 (type loc. info.); Duznetsov 1971; 441 (distr.; Yemen); Margalit & Tahori 1974: 80 (distr.); Service 1976a (distr.; Gabon); Coetzee 1980 (tax.; *s.l.*; Coustani Group); Coetzee 1994a (type loc., neotype desig.); Kyalo et al. 2017 (distr.; sub-Saharan Africa); Robert et al. 2019 (distr.; western Palearctic). **Ety.:** Camille César Clément Adolphe Coustan. **Informal name:** Coustan Malagasy Nail Mosquito.

syn. *mauritianus* de Grandpré & de Charmoy
1901: 8 (A). **Type loc.:** Mauritius (NE). **Refs.:** Senevet 1932b: 478 (syn.; publ. date).

syn. *similis* Theobald
1901a: 129 (M, F*; *paludis* var.). **Type loc.:** Mashonaland [Salisbury, Zimbabwe]; British Central Africa [Malawi; Gold Coast (Ghana)]; Mauritius (NHMUK).

crawfordi Reid
1953: 41 (M*, F*, P, L, E*). **Type loc.:** Kuala Lumpur, Selangor, Malaya [Malaysia] (NHMUK). **Distr.:** Bhutan, Cambodia, India, Indonesia, Malaysia, People's Republic of China, Thailand, Vietnam. **Refs.:** Peyton & Scanlon 1966 (F*; distr.); Nguyen Thuong Hien 1968 (F*, L*; keys, tax., bion., distr.; Vietnam); Reid 1968: 102 (distr.); Rattanarithikul & Harrison 1973 (L; key; Thailand); Harrison & Klein 1975 (distr.); Harrison & Scanlon 1975: 54 (M*, F*, P*, L*; distr.); Harrison et al. 1991: 223 (distr.); Rattanarithikul et al. 2006b (F*, L*; bion., distr., keys); Qu & Zhu 2008 (distr.; People's Republic of China); Ngo et al. 2013 (tax.; *dangi* [a *nomen nudum*] unnecessarily declared a syn. of *crawfordi*); Saeung et al. 2014 (E*); Namgay et al. 2018 (bion., distr.; Bhutan). **Ety.:** Robert Crawford. **Informal name:** Crawford Malaysian Nail Mosquito.

crucians Wiedemann
1828: 12 (A). **Type loc.:** New Orleans, Louisiana, United States (NMW). **Distr.:** Bahamas, Belize, Canada, Costa Rica, Cuba, Dominican Republic, El Salvador, Guatemala, Haiti, Honduras, Jamaica, Mexico, Nicaragua,

Puerto Rico, United States (continental). **Refs.:** Davis 1926 (F*); Bellamy 1939 (ref. to brackish *crucians*); Roth 1944: 99 (M*), 107 (M; keys); Pelaez 1945: 74 (F*; key); Darsie 1949: 525 (P*); Penn 1949a: 63 (P*); Bellamy & Repass 1950: 84 (E*); Yamaguti & LaCasse 1950b: 38 (M*, F*, L*); Breeland 1951: 224 (L*); Carpenter & LaCasse 1955: 35 (M*, F*, L*; keys); Horsfall 1955: 75 (review); Dodge 1966: 362 (first instar L*; key; *s.l.*; includes *bradleyi* and *georgianus*); Belkin 1968b: 9 (lectotype desig.); Belkin et al. 1970: 17 (M*, F*, P*, L*); Kreutzer et al. 1970 (chrom.*); Kreutzer & Kitzmiller 1971b (chrom.*, tax.); Vargas 1975: 77 (distr.); Floore et al. 1976: 15 (M*, F*, P*, L*; tax.); Wilkerson et al. 2004 (mol. tax.; *An. crucians* complex); Darsie & Ward 2005 (F*, L*; keys, distr.); Beltrán-Aguilar et al. 2011: 647 (M*, F*, P*, L*; tax., keys, distr.); Harrison et al. 2016 (F*, L*; keys, distr., tax.; *s.l.*; notes 8, 19); Giordano 2019 (distr.: Canada). **Ety.:** n.s. [torment, torture (L)]. **Informal name:** Excruciating American Nail Mosquito.

crypticus Coetzee
1994a: 129 (A, P, L; chrom.). **Type loc.:** Northcliff, Johannesburg, Transvaal [Republic of South Africa] (JSAIM). **Distr.:** Republic of South Africa. **Refs.:** Kyalo et al. 2017 (distr.; sub-Saharan Africa). **Ety.:** concealed, cryptic (L); similar to *coustani*. **Informal name:** Cryptic South African Nail Mosquito.

cucphuongensis Vu, Nguyen, Tran, & Nguyen
1991: 145 (M*, F*, P*, L*). **Type loc.:** Cuc Phuong National Forest, Ha Nam Ninh Province, Vietnam (USNM). **Distr.:** Vietnam. **Ety.:** Cuc-Phuong National Forest, Vietnam. **Informal name:** Cuc-Phuong Vietnamese Nail Mosquito.

culiciformis Cogill
1903: 333 (M, F, L*). **Type loc.:** Karwar, India (NHMUK). **Distr.:** India, People's Republic of China. **Ety.:** n.s. [gnat-appearing (L)]. **Informal name:** Deceptive Indian Nail Mosquito [culiciformis = "gnat-appearing" = Deceptive].

syn. *culiciformis* James & Liston
1904: 122. **Type loc.:** Karwar, India (LU). **Refs.:** Christophers 1916b: 457 (syn.).

daciae Linton, Nicolescu & Harbach
2004 (E*; mol. tax., distr.). In Nicolescu et al. 2004: 530. **Type loc.:** Budeni (44 ° 11'55"N, 26°06'43"E), Giurgiu, Muntenia, Romania (NHMUK). **Distr.:** Czech Republic, Finland, France, Germany, Kazakhstan, Poland, Romania, Serbia, Slovakia, United Kingdom. **Refs.:** Di Luca et al. 2004 (distr. Kazakhstan, as *messae*); Linton et al. 2005 (bion., mol. tax., distr.; United Kingdom); Danabalan et al. 2014 (bion., distr.; United Kingdom); Rydzanicz et al. 2017 (distr.; Poland); Robert et al. 2019 (distr., western Palearctic). **Ety.:** Dacia [former country name of modern Romania]. **Informal name:** Dacia Romanian Nail Mosquito.

diluvialis Reinert
1997 (M*, F*, P*, L*, E*). In Reinert et al. 1997: 15. **Type loc.:** Magnolia Camp, Manatee Springs State Park, Levy County, Florida, United States (USNM). **Distr.:** United States (continental). **Refs.:** Levíne et al. 2004 (mol. tax., distr., niche model); Darsie & Ward 2005 (F*, L*; keys, distr.); Harrison et al. 2016 (F*, L*; keys, distr.). **Ety.:** flood (L); ref. to larvae in flooded swamps. **Informal name:** Marshland American Nail Mosquito.

dissidens Taai & Harbach
2015: 253 (M, F, P, L; phyl., bion.). **Type loc.:** Ban Wang Jom, Chiang Dao District, Chiang Mai Province, Thailand (NHMUK). **Distr.:** Thailand. **Ety.:** different or disagreeing (L); ref. to difference from *An. vanderwulpi*. **Informal name:** Dissident Thai Nail Mosquito.

donaldi Reid
1962: 17 (M*, F*, P, L, E*). **Type loc.:** 14th mile Puchong-Klang Road, Selangor, Malaya [Malaysia] (NHMUK). **Distr.:** Borneo (Sarawak), Indonesia, Laos, Malaysia, Thailand. **Refs.:** Peyton & Scanlon 1966: 1 (F*; distr.); Reid 1968: 132 (M*, F*, P, L, E*; keys, bion.); Rattanarithikul & Harrison 1973 (L; key; Thailand); Harrison & Scanlon 1975: 94 (M*, F*, P*, L*); Reid 1980 (tax.); Rattanarithikul et al. 2006b (F*, L*; bion., distr., keys); Tan et al. 2008 (bion., simian malaria); Suwonkerd et al. 2013 (bion., med. imp.). **Ety.:** Donald Colless. **Informal name:** Donald Malaysian Nail Mosquito. (Vol. 1, p. 304)

earlei Vargas
1943b: 9 (M*). **Type loc.:** Jefferson County, Wisconsin, United States (USNM). **Distr.:** Canada, United States (Alaska, continental). **Refs.:** Darsie 1949: 518 (P*); Penn 1949a: 60 (P*); Pratt 1952: 484 (P*, L*; tax.); Rozeboom 1952: 477 (M*, L, E*; bion., tax.); Carpenter & LaCasse 1955: 37 (M*, F*, L*; keys); Stone & Knight 1956b: 277 (type info.); Kitzmiller & Baker 1965: 275 (chrom.*); Wallis & Whitman 1971 (bion.); White 1978 (tax., review, keys, distr.); Porter & Collins 1996 (mol. phy.; Nearctic); Darsie & Ward 2005 (F*, L*; keys, distr.); Harrison et al. 2016 (F*, L*; keys, distr.). **Ety.:** Walter Clarence Earle. **Informal name:** Earle American Nail Mosquito.

eiseni Coquillett
1902b: 192 (M, F). **Type loc.:** Aguna, Guatemala (USNM). **Distr.:** Belize, Bolivia, Brazil, Colombia, Costa Rica, Ecuador, El Salvador, French Guiana, Guatemala, Guyana, Honduras, Mexico, Nicaragua, Panama, Peru, Suriname, Trinidad and Tobago, Venezuela. **Refs.:** Davis 1926 (F*); Rozeboom 1941: 104 (review); Corrêa 1942: 25 (P, L*, E*); Vargas 1942e (E*); Pelaez 1945: 74 (F*; key); Deane, L.M. et al. 1946 (F*; key, bion.; NE Brazil & Amazonas); Causey et al. 1946 (M*; key; NE Brazil & Amazonas); Deane, M.P. et al. 1946 (L*; key, bion.; NE Brazil & Amazonas); Lane 1953: 168 (M*, F*, P, L*, E); Stone & Knight 1956b: 277 (type info.); Beltrán-Aguilar

et al. 2011: 654 (M*, F*, P*, L*; tax., keys, distr.); Berti et al. 2015 (distr.; Venezuela). **Ety.:** Gustavus August Eisen. **Informal name:** Eisen Guatemalan Nail Mosquito.

syn. *niveopalpis* Ludlow
1919: 166 (F). **Type loc.:** Comacho Reservoir, Empire, Canal Zone, Panama (USNM). **Refs.:** Edwards 1932a: 38 (syn.).

ssp. ***geometricus*** Corrêa
1944: 408 (M, P, L, E). **Type loc.:** Guarujá, Ilha de Santo Amaro [São Paulo], Brazil (NE). **Distr.:** Brazil. **Ety.:** geometric (L); pattern on egg chorion. **Informal name:** Geometric Brazilian Nail Mosquito.

ejercitoi Mendoza
1947: 171 (F*). **Type loc.:** Baler, Quezon [Luzon], Philippines (NE). **Distr.:** Philippines. **Refs.:** Basio 1971b: 35 (M*, F*; bion.). **Ety.:** Antonio Liza Ejercito. **Informal name:** Ejercito Philippine Nail Mosquito.

engarensis Kanda & Oguma
1978: 46 (M, F, P, L). **Type loc.:** Engaru, Hokkaido Island, Japan (NHMUK). **Distr.:** Japan, South Korea. **Ety.:** Engaru, Hokkaido Isl., Japan. **Informal name:** Engaru Japanese Nail Mosquito.

evandroi da Costa Lima
1937: 3 (F*). **Type loc.:** São Bento, Rio de Janeiro, Brazil (IOC). **Distr.:** Argentina, Brazil. **Ety.:** Evandro Chagas. **Informal name:** Evandro Brazilian Nail Mosquito.

fausti Vargas
1943a: 66 (F, L). **Type loc.:** Tamazunchale, San Luis Potosí, Mexico (ISET). **Distr.:** Mexico, Nicaragua. **Refs.:** Pelaez 1945: 74 (F*; key); Vargas & Martínez Palacios 1956: 113 (M*, F*, L*); Zavortink 1970b: 12 (M*, F*, L*; tax., distr.). **Ety.:** Ernest Carroll Faust. **Informal name:** Faust Mexican Nail Mosquito.

fluminensis Root
1927d: 602 (M*, P*). **Type loc.:** Itaperuna, Rio de Janeiro, Brazil (USNM). **Distr.:** Argentina, Bolivia, Brazil, Colombia, Ecuador, Panama, Paraguay, Peru, Venezuela. **Refs.:** Corrêa & Ramos 1942: 47 (M*); Deane, L.M. et al. 1946: 13 (F*); Corrêa 1950b: 82 (L); Lane 1953: 224 (M*, F*, P, L, E*); Lounibos et al. 1997b (E*); Bourke et al. 2018 (mol. tax.; as *s.l.*). **Ety.:** n.s. [fluvial (L); native to Fluminense Region of Rio de Janeiro State, Brazil]. **Informal name:** Fluminense Brazilian Nail Mosquito.

syn. *holmbergi* Del Ponte & Heredia
1945: 382 (M*, F, L*). **Type loc.:** Corpus, San Ignacio, Misiones, Argentina (INM). **Refs.:** Lane 1953: 224 (syn.).

forattinii Wilkerson & Sallum
1999 (M*, F*, P*, L*; bion.). **Type loc.:** Costa Marques, Rondônia, Brazil (USNM). **Distr.:** Brazil, Colombia, Ecuador, French Guiana, Peru, Venezuela. **Refs.:** Pecor et al. 2000 (tax., distr.; Peru); Hutchings & Sallum 2001 (bion.); Linton et al. 2013 (bion., mol. tax., distr.; Ecuador). **Ety.:** Oswaldo Paulo Forattini. **Informal name:** Forattini Brazilian Nail Mosquito.

fragilis (Theobald)
1903f: 257 (M; *Stethomyia*). **Type loc.:** Kuala Lumpur [Selangor], Malaya [Malaysia] (NHMUK). **Distr.:** Indonesia, Malaysia, Nepal, Philippines, Thailand. **Refs.:** Christophers 1933: 109 (= *bengalensis*?); Reid 1965: 109 (M*, F, P*, L*; syn.); Reid 1968: 232 (M*, F*, P*, L*; distr.); Basio 1971b: 35 (M*, F*; bion.); Rattanarithikul & Harrison 1973 (L*; key; Thailand); Baisas 1974: 48 (M, F, P, L*; tax., bion., distr.; Philippines); Harrison & Scanlon 1975: 162 (M*, F, P*, L*); Lee et al. 1987b: 57 (F; key, tax., bion., distr., review); Townsend 1990: 75 (type info.); Darsie & Pradhan 1990, 1991 (F, L; tax., keys, bion., distr.; Nepal); Rattanarithikul et al. 2006b (F*, L*; bion., distr., keys). **Ety.:** n.s. [fragile (looking) (L)]. **Informal name:** Dainty Malaysian Nail Mosquito.

syn. *treacherii* Leicester
1908: 19 (M, F). **Type loc.:** Malaya [Malaysia] (NHMUK). **Refs.:** Puri 1949: 483 (to ssp.); Reid 1965: 109 (syn.; lectotype desig.).

syn. *stantoni* Puri
1949: 483 (L). **Type loc.:** Malaya [Malaysia] and Netherlands East Indies (LU). **Refs.:** Reid 1965: 109 (syn.).

franciscanus McCracken
1904: 12 (M, F*, L*). **Type loc.:** Felt Lake, near Stanford University [Santa Clara County], California, United States (CAS). **Distr.:** Mexico, United States (continental). **Refs.:** Howard et al. 1917: 1023 (syn. with *pseudopunctipennis*); Aitken 1945: 327 (M, F*, P, E*; bion.; to ssp. from syn. with *pseudopunctipennis*; lectotype desig.); Freeborn 1949: 391 (tax., bion.); Penn 1949a: 66 (P*); Yamaguti & LaCasse 1950b: 35 (M*, F*, L*); Belkin et al. 1951: 23 (bion.); Carpenter & LaCasse 1955: 45 (M*, F, L*; keys); Smithson 1972: 501 (to sp.); Smithson & McClelland 1972: 80 (chrom.*; ssp.); Bohart & Washino 1978: 20 (M, F*, L*; keys, tax., bion., distr.); Arnaud 1979: 57 (type info.); Darsie & Ward 2005 (F*, L*; keys, distr.). **Ety.:** San Francisco. **Informal name:** San Francisco American Nail Mosquito.

syn. *boydi* Vargas
1939: 361 (E). **Type loc.:** Central California, United States (NE). **Refs.:** Aitken 1942: 164 (tax.); Ross & Roberts 1943a: 24 (to syn.); Russell et al. 1943: 23 (syn.); Aitken 1945: 344 (from syn. to *pseudopunctipennis franciscanus* var. *boydi*); Freeborn 1949: 392 (to syn. with *franciscanus*); Mann 1950: 40 (E*).

syn. *willardi* Vargas
1941e, 1941f: 74 (E; *pseudopunctipennis* var.). **Type loc.:** Chihuahua, Mexico (ISET). **Refs.:** Russell et al. 1943: 23 (to syn. with *franciscanus*); Ross & Roberts 1943a: 24 (to syn.); Aitken 1945: 268 (tax.); Osorno-Mesa & Muñoz-Sarmiento 1948: 107 (M*, L*, E*); Mann 1950: 40 (E*).

franciscoi Reid
1962: 27 (M, F, P*, L, E*). **Type loc.:** Iwahig, Palawan, Philippines (NHMUK). **Distr.:** Philippines. **Refs.:** Cagampang-Ramos & Darsie 1970 (F*, L*; keys); Basio 1971b: 35 (M*, F*; bion.); Baisas 1974: 48 (M, F, P, L; tax., bion., distr.; Philippines). **Ety.:** Francisco Baisas. **Informal name:** Francisco Philippine Nail Mosquito.

freeborni Aitken
1939: 192 (M, F, E; *maculipennis* ssp.). **Type loc.:** Davis, California, United States (CAS). **Distr.:** Canada, Mexico, United States (continental). **Refs.:** King & Bradley 1941: 65 (to sp.); Aitken 1945: 296 (M, F, P*, L*, E*; tax.); Pelaez 1945: 74 (F*; key); Penn 1949a: 55 (P*); Ross & Roberts 1950: 7 (M*, F*, L*; bion., distr., as *maculipennis freeborni*); Yamaguti & LaCasse 1950b: 25 (M*, F*, L*); Pratt 1952: 484 (L*; tax.); Belkin 1953a: 318 (P*, L*); Rozeboom 1953: 677 (tax.); Barr 1954: 445 (tax.); Carpenter & LaCasse 1955: 39 (M*, F, L*; keys); Horsfall 1955: 82 (review); Kitzmiller & Baker 1963: 254 (chrom.*); Dodge 1966: 362 (first instar L*; key); White 1978 (tax., review, keys, distr.); Bohart & Washino 1978: 22 (M*, F*, L*; keys, tax., bion., distr.); Porter & Collins 1996 (mol. phyl.; Nearctic); Wilkerson & Strickman 1990 (F*; key); Fritz et al. 1991 (mol. tax., chrom.); Darsie & Ward 2005 (F*, L*; keys, distr.); Sinka et al. 2010b (distr., bion.); Sinka 2013 (distr., bion.). **Ety.:** Stanley Barron Freeborn. **Informal name:** Freeborn American Nail Mosquito. (Vol. 1, p. 306)

freyi Meng
1957: 36 (F*, L*). **Type loc.:** mainland China [People's Republic of China] (LU). **Distr.:** People's Republic of China. **Refs.:** Feng 1958: 91 (F*, L*). **Ety.:** Richard Frey. **Informal name:** Frey Chinese Nail Mosquito.

fuscicolor van Someren
1947b: 43 (M*, F*, P*, L). **Type loc.:** Fianarantsoa, Madagascar (NHMUK). **Distr.:** Madagascar (includes Glorioso and Juan de Nova Isl.). **Refs.:** Kyalo et al. 2017 (distr.; sub-Saharan Africa). **Ety.:** n.s. [dark color (L)]. **Informal name:** Brown Malagasy Nail Mosquito.

gabaldoni Vargas
1941c: 389 (M*, F*). **Type loc.:** Tenosique, Tabasco, Mexico (ISET). **Distr.:** Belize, Guatemala, Mexico. **Refs.:** Pelaez 1945: 74 (F*; key); Vargas & Martínez Palacios 1946: 19 (P*, L*, E*); Vargas & Martínez Palacios 1956: 103 (M*, F*, L*); Beltrán-Aguilar et al. 2011 (M*, F*, P*, L*; keys, distr., tax.). **Ety.:** Arnaldo Gabaldón. **Informal name:** Gabaldón Mexican Nail Mosquito.

georgianus King
1939: 462 (M*, F, P, L*; *crucians* var.). **Type loc.:** Quitman, Brooks County, Georgia, United States (USNM). **Distr.:** United States (continental). **Refs.:** King & Bradley 1941: 65 (to sp.); Roth 1944: 99 (M*), 107–110 (M gen.; keys); Penn 1949a: 61 (P*); Bellamy & Repass 1950: 84 (E*); Yamaguti & LaCasse 1950b: 43 (M*, F*, L*); Carpenter & LaCasse 1955: 41 (M, F, L*; keys); Floore et al. 1976: 40 (M*, F*, P*, L*; tax.); Darsie & Ward 2005 (F*, L*; keys, distr.); Harrison et al. 2016 (F*, L*; keys, distr.; note 23). **Ety.:** Georgia [US state]. **Informal name:** Georgia American Nail Mosquito.

gigas Giles
1901b: 196 (M, F). **Type loc.:** Conoor (6,000 ft), Nilgiri Hills [Madras], India (NHMUK). **Distr.:** Bangladesh, India, Indonesia, Myanmar, Nepal, Philippines, Sri Lanka, Thailand. **Refs.:** Christophers 1923b (F*); Christophers 1933: 130 (M*, F*, L*); Mochtar & Walandouw 1934: 932 (F*, L*; *s.l.*); Bonne-Wepster & Swellengrebel 1953: 130 (tax.); Narayandas 1954: 19 (E*); Nguyen Thuong Hien 1968 (F*, L*; keys, tax., distr.; Vietnam); Reid 1968: 215 (M*, F*, L*; tax.); Aslamkhan 1971b (distr.; Pakistan); Townsend 1990: 134 (type info.); Darsie & Pradhan 1990, 1991 (F, L; tax., keys, bion., distr.; Nepal); Oo et al. 2004 (distr.; Myanmar). **Ety.:** giant (Gr); a large mosquito. **Informal name:** Giant Indian Nail Mosquito.

ssp. ***formosus*** Ludlow
1909a: 22 (F; as sp.). **Type loc.:** Camp John Hay, Benguet [Mountain, Luzon], Philippines (USNM). **Distr.:** Indonesia, Philippines. **Refs.:** Christophers 1924a: 25, 85 (to *gigas* var.); Bonne-Wepster & Swellengrebel 1953: 136 (M*, F*, L*); Stone & Knight 1956b: 277 (type info.); Basio 1971b: 35 (M*, F*; bion.); Lien et al. 1977 (distr.); Harbach & Howard 2007: 39 (from *gigas* var.). **Ety.:** n.s. [beautiful (L)]. **Informal name:** Formosa Philippine Nail Mosquito [Formosa was a mythical island long believed by Europeans to be located in Asia].

ssp. ***simlensis*** (James)
1911 (A*, L; *Patagiamyia*; as sp.) . In James & Liston 1911: 66. **Type loc.:** Mahasu, near Simla [East Punjab], India (NHMUK). **Distr.:** Bangladesh, India, Nepal, Pakistan, People's Republic of China. **Refs.:** Christophers 1916b: 467 (to syn. of *gigas*); Christophers 1924b: 12 (to *gigas* var.); Christophers 1933: 136 (M*, F*, P, L); Reid 1968: 215 (M*, F*; tax.); Aslamkhan 1971b (distr.; Pakistan); Ahmed 1987 (distr.; Bangladesh); Townsend 1990: 134 (type info.); Darsie & Pradhan 1990, 1991 (F, L; tax., keys, bion., distr.; Nepal); Glick 1992: 129 (distr.); Harbach & Howard 2007: 39 (from *gigas* var.); Qu & Zhu 2008 (distr.; People's Republic of China). **Ety.:** Simla, India. **Informal name:** Mahasu Indian Nail Mosquito.

ssp. ***refutans*** Alcock
1913: 161 (A; *gigas* var.). **Type loc.:** [Maskeliya], Ceylon [Sri Lanka] (NHMUK?). **Distr.:** Sri Lanka. **Refs.:** Christophers 1933: 138 (A, L); D'Abrera 1944: 344 (E*); Reid 1968: 215 (E*); Harbach & Howard 2007: 39 (from *gigas* var.). **Ety.:** n.s. [refute (L)]. **Informal name:** Refusal Sri Lankan Nail Mosquito.

ssp. ***danaubento*** Mochtar & Walandouw
1934: 932 (F*, L*). **Type loc.:** Danau-Bento, West Coast, Sumatra [Indonesia] (GLB?). **Distr.:** Indonesia (Sumatra).

Refs.: Bonne-Wepster & Swellengrebel 1953: 140 (M*, F*, L*); Harbach & Howard 2007: 39 (from *gigas* var.). **Ety.:** Danau Bento, Central Sumatra. **Informal name:** Danau-Bento Indonesian Nail Mosquito.

ssp. ***oedjalikalah*** Nainggolan
1939: 163 (M*, F*, L*). **Type loc.:** Oedjali Kalah, Sumatra [Indonesia] (LU). **Distr.:** Indonesia (Sumatra). **Refs.:** Bonne-Wepster & Swellengrebel 1953: 142 (M*, F*, L*; as *oedjalikalahensis*); Harbach & Howard 2007: 39 (from *gigas* var.). **Ety.:** Mt. Kerintji, Sumatra. **Informal name:** Oedjali Kalah Indonesian Nail Mosquito.

ssp. ***pantjarbatu*** Waktoedi Koesoemawinangoen
1954: 49 (L). **Type loc.:** Sumatra [Indonesia] (LU). **Distr.:** Indonesia. **Refs.:** Harbach & Howard 2007: 39 (from *gigas* var.). **Ety.:** Pancurbatu [Pantcurbatu], Indonesia. **Informal name:** Pancurbatu Indonesian Nail Mosquito.

ssp. ***crockeri*** Colless
1955a: 337 (F*, P*, L*). **Type loc.:** Saung-Saung, Sunsuran Trace, North Borneo, Borneo (NHMUK). **Distr.:** Malaysia (Borneo). **Refs.:** Reid 1968: 219 (F*, P*, L*; to ssp.). **Ety.:** n.s. [prob. Crocker Range, Sabah, Borneo, Malaysia]. **Informal name:** Crocker Bornean Nail Mosquito.

grabhamii Theobald
1901a: 205 (M*, F). **Type loc.:** Jamaica (NHMUK). **Distr.:** Cayman Isl., Cuba, Dominican Republic, Haiti, Jamaica, Puerto Rico, United States (continental), Virgin Isl. **Refs.:** Senevet 1931: 89 (P*); Rozeboom 1941: 104 (review); Pratt 1943: 1717 (L*); Lane 1953: 177 (M*, F*, L*); Reid & Knight 1961: 477 (P*); Dodge 1966: 362 (first instar L*; key); Belkin et al. 1970: 34 (M*, F*, P*, L*); Darsie et al. 2002: 430 (distr.; United States). **Ety.:** Michael Grabham. **Informal name:** Grabham Jamaican Nail Mosquito.

guarao Anduze & Capdevielle
1949: 119 (M, F, P*, L*). **Type loc.:** Caripito, Monagas, Venezuela (A). **Distr.:** Trinidad and Tobago, Venezuela. **Refs.:** Lane 1953: 198 (M*, F, P*, L*). **Ety.:** Guaraó [Carib Amerindian tribe]. **Informal name:** Guaraó Venezuelan Nail Mosquito.

hailarensis Xu & Luo
1998: 49 (M*, F*, P*, L*, E*). **Type loc.:** Hailar, Nei Mongol Autonomous Region, People's Republic of China (IPD). **Distr.:** People's Republic of China. **Ety.:** Hailar City, Mongolia. **Informal name:** Hailar Chinese Nail Mosquito.

hectoris Giaquinto-Mira
1931: 606 (M*, F*, L*). **Type loc.:** Guatemala City, Guatemala (A). **Distr.:** El Salvador, Guatemala, Honduras, Mexico. **Refs.:** Dampf 1939: 282 (M*); Rozeboom 1941: 106 (review); Vargas 1941d: 110 (E*); Vargas & Martínez Palacios 1942 (M*, F*, L*, E*); Pelaez 1945: 74 (F*; key); Vargas 1949a: 233 (P); Lane 1953: 166 (M*, F*, L*, E); Baker et al. 1966 (chrom.). **Ety.:** Héctor Giaquinto. **Informal name:** Héctor Guatemalan Nail Mosquito.

syn. *cricillium* Martini
1932b: 99 (F). **Type loc.:** San Cristóbal de las Casas, Chiapas, Mexico (LU). **Refs.:** Martini 1935: 21 (syn.).

heiheensis Ma
1981b: 66 (M*, F*, E). **Type loc.:** Aihui, Heilungjang Province, China [People's Republic of China] (IZ). **Distr.:** People's Republic of China. **Ety.:** Heihe, Heilungjang Prov., Sichuan, People's Republic of China. **Informal name:** Heihe Chinese Nail Mosquito.

hermsi Barr & Guptavanij
1989: 353 (M, F, P, L, E). **Type loc.:** Malibu Canyon at Pacific Coast Highway, Los Angeles County, California, United States (USNM). **Distr.:** United States (continental). **Refs.:** Porter & Collins 1996 (mol. phy.; Nearctic); Darsie & Ward 2005 (F*, L*; keys, distr.). **Ety.:** William Brodbeck Herms. **Informal name:** Herms American Nail Mosquito.

hodgkini Reid
1962: 20 (M, F*, P*, L*, E*). **Type loc.:** 16th mile Ulu Gombak Road, Selangor, Malaya [Malaysia] (NHMUK). **Distr.:** Cambodia, Malaysia, Thailand, Vietnam. **Refs.:** Rattanarithikul & Harrison 1973 (L; key; Thailand); Rattanarithikul et al. 2006b (F*, L*; bion., distr., keys). **Ety.:** Ernest Pease Hodgkin. **Informal name:** Hodgkin Malaysian Nail Mosquito.

hunteri (Strickland)
1916a: 263 (M*, F, L*; *Myzorhynchus*). **Type loc.:** Tebong Negri Sembilan, Malaya [Malaysia] (LU). **Distr.:** Indonesia, Malaysia. **Refs.:** Reid 1950: 310 (tax.); Bonne-Wepster & Swellengrebel 1953: 199 (M*, F*, L*); Reid 1968: 193 (M, L; tax.); Harrison & Klein 1975 (distr.); Harrison & Scanlon 1975: 97 (M*, F*, P*, L*; distr.). **Ety.:** Mr. Hunter [poss. ref. to Peter Sinclair Hunter]. **Informal name:** Hunter Malaysian Nail Mosquito.

hyrcanus (Pallas)
1771: 475 (A, *Culex*). **Type loc.:** [Caspian Sea] (LU). **Distr.:** Afghanistan, Albania, Armenia, Austria, Azerbaijan, Bosnia and Herzegovina, Bulgaria, Cambodia, Crimean Peninsula, Croatia, Cyprus, Czech Republic, France (includes Corsica), FYRO Macedonia, Georgia, Greece, Hungary, Iran, Iraq, Italy (includes Sardinia and Sicily), Japan, Jordan, Kazakhstan, Kosovo, Kyrgyzstan, Lebanon, Libya, Malaysia, Moldova, Mongolia, Montenegro, Morocco, People's Republic of China, Romania, Russia (Southern Dists.), Serbia, Slovakia, Spain, Sri Lanka, Syria, Tajikistan, Turkey, Turkmenistan, Ukraine, Uzbekistan. **Refs.:** Martini 1930: 146 (M*, F*); Christophers 1933: 152 (tax.); Ross & Roberts 1943b: 5 (M*, F*, L*; tax., distr., bion.); Bates et al. 1949 (tax.); Reid 1953: 51 (tax.); Gutsevich et al. 1974: 105 (M*, F*, L*); Margalit & Tahori 1974: 80 (distr.); Gutsevich 1976: 148 (tax.); Lee

et al. 1987b: 58 (F; key, tax., bion., distr., review); Glick 1992: 129 (distr.; tax.); Ramsdale & Snow 2000: 4 (distr.); Halgos & Benkova 2004: 68 (distr.; Slovakia); Becker et al. 2010: 169 (M*, F*, L*; keys, tax., distr., bion.); Robert et al. 2019 (distr.; western Palearctic). **Ety.:** ancient land of Hyrcania. **Informal name:** Hyrcania Old World Nail Mosquito.

syn. *pictus* Loew
1845: 4 (M). **Type loc.:** Rhodes Island [Greece] (ZM). **Refs.:** Senevet & Andarelli 1955a: 109 (M*, F*, P, L*).

syn. *flerowi* Portschinsky
1910: 605 (F*). **Type loc.:** River Amu-Darya, Uzbek [Uzbekistan] (LU).

syn. *mesopotamiae* Christophers & Chand
1915: 196 (M, F*; *sinensis* var.). **Type loc.:** Lower Iraq (MSI). **Refs.:** Christophers 1933: 152 (syn.).

syn. *marzinovski* Shingarev
1926: 48 (F; *hyrcanus* var.). **Type loc.:** Karayazy Steppe, South Caucasus [USSR] (LU). **Refs.:** Stone et al. 1959: 20 (syn.).

syn. *popovi* Shingarev
1928: 50 (M*, F). **Type loc.:** Turkestan [Turkmenistan] (LU). **Refs.:** Martini 1930: 150 (A).

syn. *mahmuti* Martini
1930: 149 (F*; as var.). **Type loc.:** Asia Minor (LU). **Refs.:** Christophers 1933: 152 (syn.).

insulaeflorum (Swellengrebel & Swellengrebel de Graaf)
1920: 2 (tax.; *Stethomyia aitkenii* var. for unclassified Larva No. 1) and 1920: 23 (L*). **Type loc.:** Noesa Kambangan, South Java [Indonesia] (NE). **Distr.:** Cambodia, India, Indonesia (includes Java), Malaysia, Myanmar, People's Republic of China, Philippines, Taiwan, Thailand, Vietnam. **Refs.:** Christophers 1933: 111 (A, L); Baisas 1938: 221 (P*); Lee & Woodhill 1944: 97 (M*, F*, L); Bonne-Wepster & Swellengrebel 1953: 97 (M*, F, L*); Reid 1965: 119 (M*, F, P*, L*); Nguyen Thuong Hien 1968 (F*, L*; keys, tax., bion., distr.; Vietnam); Basio 1971b: 36 (M*, F*; bion.); Harrison & Scanlon 1975: 167 (M*, F*, P*, L*; distr.); Lee et al. 1987b: 60 (F; key, tax., bion., distr., review); Oo et al. 2004 (distr.; Myanmar); Rattanarithikul et al. 2006b (F*, L*; bion., distr., keys); Qu & Zhu 2008 (distr.; People's Republic of China). **Ety.:** n.s. [island flower (L)]. **Informal name:** Floral Indonesian Nail Mosquito.

interruptus Puri
1929a: 387 (M, P*, L*, E*; as var.). **Type loc.:** Sukna, Darjeeling District [West Bengal], India (NHMUK, MSI, IM). **Distr.:** Borneo; Brunei, Bhutan, Cambodia, India, Malaysia, Nepal, People's Republic of China, Sri Lanka, Thailand, Vietnam. **Refs.:** Christophers 1933: 139 (M*, F*, L*, E*); Puri 1949: 483 (to sp.); Reid 1968: 203 (M*, F*, P*, L*, E*; tax.); Ramalingam 1974 (distr.); Harrison & Scanlon 1975: 140 (M*, F*, P*, L*; distr.); Darsie & Pradhan 1990, 1991 (F, L; tax., keys, bion., distr.; Nepal); Rattanarithikul et al. 2006b (F*, L*; bion., keys, distr.); Namgay et al. 2018 (bion., distr.; Bhutan). **Ety.:** n.s. [interrupted (L)]. **Informal name:** Spotted Indian Nail Mosquito.

inundatus Reinert
1997 (M*, F*, P*, L*, E*). In Reinert et al. 1997: 15. **Type loc.:** Rook's Bluff Fish Camp, Walton County, Florida, United States (USNM). **Distr.:** United States (continental). **Refs.:** Levíne et al. 2004 (mol. tax., distr., niche model); Darsie & Ward 2005 (F*, L*; keys, distr.); Harrison et al. 2016 (F*, L*; keys, distr.). **Ety.:** inundated plains (L); larval habitat. **Informal name:** Floodwater American Nail Mosquito.

judithae Zavortink
1969a: 28 (M*, F, P*, L*). **Type loc.:** Cochise Stronghold Recreation Area, Dragoon Mountains, Cochise County, Arizona, United States (USNM). **Distr.:** Mexico, United States (continental). **Refs.:** Zavortink 1970b: 10 (M*, F, P*, L*; distr., tax.); Darsie & Ward 2005 (F*, L*; keys, distr.). **Ety.:** Judith Bergland. **Informal name:** Judith American Nail Mosquito.

kleini Rueda
2005: 13 (M*, F*, P*, L*). **Type loc.:** Cheongpyeong, Gyeongsang buk-do, South Korea (USNM). **Distr.:** South Korea. **Ety.:** Terry Allen Klein. **Informal name:** Klein Korean Nail Mosquito.

koreicus Yamada & Watanabe
1918: 206 (F*). **Type loc.:** Pyeoncheing [= P'yongch'ang], Korea (IID). **Distr.:** Japan, People's Republic of China, South Korea. **Refs.:** LaCasse & Yamaguti 1950: 27 (M*, F*, P*, L*; bion., distr., tax.); Hara 1957: 47 (F*); Hara 1957 (F*); Ohmori 1959: 222 (P*); Hara 1959 (F*); Tanaka et al. 1979: 65 (M*, F*, P, L*). **Ety.:** Korea. **Informal name:** Pyeongchang Korean Nail Mosquito.

syn. *punctibasis* Edwards
1921d: 274 (M, F). **Type loc.:** Nagasaki [Kyushu], Japan (NHMUK).

syn. *edwardsi* Yamada
1924: 238 (F). **Type loc.:** Inokashira, near Tokyo [Honshu], Japan (IID). **Refs.:** LaCasse & Yamaguti 1950: 32 (M*, F*, P*, L*; bion., distr., tax.); Hara 1957: 48 (F*); Stone et al. 1959: 21 (syn.).

syn. *hisaoe* Tsuchimoto
1948: 366 (M, F*, L; *koreicus* var.). **Type loc.:** Kameyama-Mura, Hiroshima Prefecture [Honshu], Japan (LU). **Refs.:** Kamimura 1966: 8 (syn.) [cited by Tanaka et al. 1979].

kweiyangensis Yao & Wu
1944: 1 (M*, F*, P*, L*, E*). **Type loc.:** Kweiyang Municipality, Kweichow Province, China [People's Republic of China] (LU). **Distr.:** People's Republic of China. **Ety.:** After type locality. **Informal name:** Kweiyang Chinese Nail Mosquito.

labranchiae Falleroni
1926: 564 (E*; *claviger* var.). **Type loc.:** [Pontine Marshes, Rome, Latium], Italy (LU). **Distr.:** Algeria, Bosnia and

Herzegovina, Bulgaria, Croatia, Czech Republic, Egypt, France (Corsica), Greece, Iran, Italy (includes Sardina and Sicily), Libya, Morocco, Romania, Spain, Tunisia, United Kingdom. **Refs.:** Bates 1940: 354 (to sp.); Bates et al. 1949 (F, L, E*; keys); Aitken 1953: 307 (A, P, L, E; bion.); Mezzanotte & Ferrucci 1978 (tax.; *s.l.*); White 1978 (tax., review, keys, distr.); Linton et al. 2003 (review; Maculipennis Complex); Sinka et al. 2010a: 117 (bion. review, distr., niche model); Becker et al. 2010: 174 (E*; key, tax., distr., bion.); Labouti et al. 2011 (review., tax., bion., distr.); Robert et al. 2019 (distr., western Palearctic). **Ety.:** Antonio Labranca. **Informal name:** Labranca Italian Nail Mosquito.

syn. *sicaulti* Roubaud
1935a: 107 (M*, L*, E*; *labranchiae* var.). **Type loc.:** Near Rabat, French Morocco (PIP?). **Refs.:** Bates 1940: 353 (syn. suggested); Romeo Viamonte 1950 (as var. of *labranchiae*); Senevet & Andarelli 1956c (as *labranchiae* ssp.); Guy & Holstein 1968 (syn. with *labranchiae*); White 1978 (from syn. with *labranchiae*; neotype desig.); de Zulueta et al. 1983: 73 (eggs, chrom., allozymes, syn.); Labouti et al. 2011 (mol. tax. review, syn. with *labranchiae*).

syn. *pergusae* Missiroli
1935: 333 (E; *maculipennis* ssp.). **Type loc.:** Lake Pergusa, Enna Province, Sicily (LU). **Refs.:** Bates 1940: 352 (syn.).

lesteri Baisas & Hu
1936: 229 (F, P*, L*, E*; *hyrcanus* var.). **Type loc.:** Santa Mesa of Manila [Rizal, Luzon], Philippines (NE). **Distr.:** Brunei, Cambodia, Guam, Hong Kong, Indonesia (includes Kalimantan, Sulawesi), Japan, Malaysia, People's Republic of China, Philippines, Singapore, South Korea, Thailand, Vietnam. **Refs.:** Ohmori 1957 (M*, E*); Ohmori 1959: 222 (P*); Nguyen Thuong Hien 1968 (F*, L*; keys, tax., bion., distr.; Vietnam); Reid 1968: 105 (M*, F*, P*, L*, E*; as *paraliae*); Basio 1971b: 36 (M*, F*; bion.); Basio & Reisen 1971: 60 (L; distr.); Tanaka 1971: 4 (distr.); Harrison 1973b: 4 (tax.); Baisas 1974: 50 (M, F*, P, L*; tax., bion., distr.; Philippines); Harrison & Scanlon 1975: 61 (M*, F*, P*, L*; distr.; as *paraliae*); Tanaka et al. 1975c: 212 (tax., distr.); Kanda & Oguma 1976: 325 (tax.); Otsuru et al. 1976: 301 (E); Tanaka et al. 1979: 83 (M*, F*, P, L*, E); Linley et al. 1995: 59 (E*; as *paraliae*); Ma et al. 2000a (tax.); Wilkerson et al. 2003 (tax.); Rueda et al. 2005 (M*, F*, P*, L*; neotype desig.); Rattanarithikul et al. 2006b (F*, L*; bion., distr., keys; as *paraliae*); Becker et al. 2010: 339 (F*, L*; key, tax., distr., bion.); Sinka et al. 2011: 89 (bion., review, distr., niche model); Saeung et al. 2014 (E*; as *paraliae*). **Ety.:** Henry Leste. **Informal name:** Leste Philippine Nail Mosquito. (Vol. 1, p. 308)

syn. *paraliae* Sandosham
1959: 189 (F, L; *lesteri* ssp.). **Type loc.:** Malaya [Malaysia] and Borneo (NHMUK). **Refs.:** Reid 1963a: 101 (lectotype desig, distr.); Harrison et al. 1991: 208 (from ssp. of *lesteri*); Taai et al. 2013 (syn. with *lesteri*);.

syn. *kiangsuensis* Xu & Feng
1975 (F, E*). **Type loc.:** Wukiang, Kiangsu Province, China [People's Republic of China] (Chinese Medical College). **Refs.:** Qu & Xhu 2008 (syn.).

syn. *anthropophagus* Xu & Feng
1975: 81 (M*, F*, P*, L*, E*; as *lesteri* ssp.). **Type loc.:** Wukiang, Kiangsu Province, China [People's Republic of China] (Chinese Medical College). **Refs.:** Ma 1981b: 60 (from ssp. of *lesteri*); Ma et al. 2000a (tax.); Wilkerson et al. 2003 (syn.).

syn. *dazhaius* Ma
1981b: 65 (M*, F*, E*). **Type loc.:** Sichuan Province, China [People's Republic of China] (IZ). **Refs.:** Miao et al. 1988 (hybridization experiments; *dazhaius* conspecific with *lesteri*; syn. with *anthropophagus*); Qu & Zhu 2008 (review).

letifer Sandosham
1944: 129 (F*, L*). **Type loc.:** Kg. Sijangkang, Selangor, Malaya [Malaysia] (NHMUK). **Distr.:** Cambodia, Indonesia, Malaysia (includes Sarawak), Singapore, Thailand, Vietnam. **Refs.:** Reid 1950: 294 (M*, F*, P*, L*; neotype desig., tax.); Bonne-Wepster & Swellengrebel 1953: 180 (M, F*, L*); Nguyen Thuong Hien 1968 (F*, L*; keys, tax., bion., distr.; Vietnam); Reid 1968: 173 (M*, F*, P*, L*, E*); Harrison & Klein 1975 (distr.); Harrison & Scanlon 1975: 118 (M*, F*, P*, L*; distr.); Lee et al. 1987b: 65 (F; key, tax., bion., distr., review); Rahman et al. 1997 (review, bion., Malaysia); Rattanarithikul et al. 2006b (F*, L*; bion., distr., keys); Manguin 2013 (tax., bion., med. imp., control). **Ety.:** n.s. [causing death, deadly (L)]. **Informal name:** Deadly Malaysian Nail Mosquito. (Vol. 1, p. 310)

lewisi Ludlow
1920b: 74 (M, F). **Type loc.:** Selenga, Verkne, Udinsk, Siberia [USSR] (USNM). **Distr.:** Japan, Russia. **Refs.:** Aitken 1945: 308 (M, F; preoccupies *messeae*?); Stone & Knight 1956b: 278 (type info.); Stone et al. 1959: 22 (to sp.). **Ety.:** William Figures Lewis. **Informal name:** Lewis Russian Nail Mosquito.

syn. *selengensis* Ludlow
1920b: 77 (F). **Type loc.:** Selenga, Siberia [USSR] (LU). **Refs.:** Aitken 1945: 308 (syn.); Stone & Knight 1956b: 279 (type info., lectotype desig.).

liangshanensis Kang, Tan, Cao, Cheng, Yang & Huang
1984: 11 (M*, F*, P*, L*, E*). **Type loc.:** Zhaojiao County, Sichuan Province, China [People's Republic of China] (SIP). **Distr.:** People's Republic of China. **Ety.:** Liangshan Yi, Sichuan Prov., China. **Informal name:** Liangshan Chinese Nail Mosquito.

syn. *kunmingensis* Dong & Wang
1985b: 117 (M*, F*, P*, L*, E*). **Type loc.:** Long-tan Ba, Kumming Province, China [People's Republic of China] (YIM). **Refs.:** Ma et al. 2000b (syn.; mol. tax.); Ma & Xu 2005 (from syn. with *liangshanensis*); Qu & Zhu 2008: 211 (syn. with *liangshanensis*).

lindesayi Giles

1900a: 166 (A*; as *lindesaii*). **Type loc.:** Bakloh (7,000 ft) [Punjab Hill States, India] (NHMUK). **Distr.:** Afghanistan, Bhutan, India, Japan, Myanmar, Nepal, Pakistan, People's Republic of China, South Korea, Taiwan, Tajikistan, Thailand, Vietnam. **Refs.:** Christophers 1933: 123 (M*, F*, P, L*, E; tax.); Reid & Knight 1961: 477 (P*); Nguyen Thuong Hien 1968 (F*, L*; keys, tax., bion., distr.; Vietnam); Reid 1968: 208 (M*, F*, L*, E*); Aslamkhan 1971b (distr.; Pakistan); Gutsevich et al. 1974: 102 (M*, F, L*); Darsie & Pradhan 1990, 1991 (F, L; tax., keys, bion., distr.; Nepal); Glick 1992: 129 (distr.); Oo et al. 2004 (distr.; Myanmar); Rattanarithikul et al. 2006b (F*, L*; bion., distr., keys); Qu & Zhu 2008 (distr.; People's Republic of China); Namgay et al. 2018 (bion., distr.; Bhutan). **Ety.:** Victor Lindesay. **Informal name:** Lindesay Indian Nail Mosquito.

syn. *maculata* Theobald

1910a: 1 (F; *lindesayi* var., not Theobald 1901a). **Type loc.:** Kurseong, Darjeeling District [West Bengal], E Himalayas, India (IM).

ssp. ***japonicus*** Yamada

1918: 689 (F*; as sp.). **Type loc.:** [Kanayama], Hokkaido [and Mount Myogi], Japan (IID). **Distr.:** Japan, People's Republic of China, South Korea. **Refs.:** LaCasse & Yamaguti 1950: 17 (M*, F*, P*, L*; bion., distr., tax.); Tajima 1952: 90 (E*); Hara 1957: 48 (F*); Hara 1957 (F*); Ohmori 1959: 222 (P*); Hara 1959 (F*); Tanaka et al. 1979: 61 (M*, F*, P, L*). **Ety.:** Japan. **Informal name:** Myogi Japanese Nail Mosquito.

ssp. ***pleccau*** Koidzumi

1924: 97 (A, L; as sp.). **Type loc.:** Musha, Formosa [Taiwan, Republic of China] (LU; syntype in NHMUK). **Distr.:** People's Republic of China, Taiwan. **Refs.:** Puri 1949: 485 (to ssp.). **Ety.:** *pleccau*, word for mosquito in Takasago language, Japan. **Informal name:** Pleccau Taiwanese Nail Mosquito.

ssp. ***cameronensis*** Edwards

1929c: 323 (A; as var.). **Type loc.:** Cameron's Highlands, Malay Peninsula [Malaysia] (NHMUK). **Distr.:** Malaysia, Thailand. **Refs.:** Puri 1949: 485 (to ssp.); Bonne-Wepster & Swellengrebel 1953: 126 (M*, F*, L*); Reid 1968: 208 (M*, F*, P*, L*; tax.); Harrison et al. 1991: 196 (tax.; Thailand); Rattanarithikul et al. 2010 (F*, L*; bion., distr., keys). **Ety.:** Cameron's Highlands, Malaysia. **Informal name:** Cameron Malaysian Nail Mosquito.

ssp. ***benguetensis*** King

1931: 753 (M*, F*, L*; as var.). **Type loc.:** Baguio, Benguet Subprovince, Luzon, Philippines (USNM). **Distr.:** Philippines. **Refs.:** Puri 1949: 485 (to ssp.); Cagampang-Ramos & Darsie 1970 (F*, L*; keys); Wanson & Berteaux 1954: 33 (L*; as *mortiauxi*); Basio 1971b: 36 (M*, F*; bion.). **Ety.:** Baguio, Benguet Subprov., Luzon, Philippines. **Informal name:** Benguet Philippine Nail Mosquito.

maculipennis Meigen

1818: 11 (M, F*). **Type loc.:** Germany (MNHP). **Distr.:** Albania, Algeria, Armenia, Austria, Azerbaijan, Belarus, Belgium, Bosnia and Herzegovina, Brazil, Bulgaria, Crimean Peninsula, Croatia, Cyprus, Czech Republic, Denmark, Estonia, Finland, France, FYRO Macedonia, Georgia, Germany, Greece, Hungary, India, Iran, Iraq, Ireland, Israel, Italy (includes Sicily), Jordan, Kazakhstan, Kyrgyzstan, Latvia, Lithuania, Luxembourg, Moldova, Mongolia (or Outer Mongolia), Montenegro, Morocco, Netherlands, New Zealand [intercepted], Norway, People's Republic of China, Poland, Portugal, Romania, Russia (Central, northwestern, Southern Dists., Volga), Serbia, Slovakia, Slovenia, Spain, Sweden, Switzerland, Syria, Taiwan, Tajikistan, Tunisia, Turkestan, Turkey, Turkmenistan, Ukraine, United Arab Emirates, United Kingdom, Uzbekistan. **Refs.:** Séguy 1924: 150 (L*, E*); Freeborn 1926 (F*); Bates et al. 1949 (F, L, E*; keys); Buonomini & Mariani 1953: 178 (A, L, E; tax.); Aitken 1954: 453 (tax.); Horsfall 1955: 92 (review); Rioux 1958: 106 (tax.); Sicart & Ruffie 1960: 635 (P*); Polyakova 1970b: 132 (distr.); Gutsevich et al. 1974: 92 (tax.); White 1978 (F; tax., review, keys, distr.; lectotype desig.); Korvenkontio et al. 1979 (E*; mol. tax.; *s.l.*); Sichinava et al. 1983: 420 (distr.; Georgia); Lee et al. 1987b: 66 (F; key, tax., bion., distr., review); Ramsdale & Snow 2000 (distr., Europe); Linton et al. 2002c (mol. tax., distr.; Greece); Beck et al. 2003: 24 (distr.; Luxembourg); Linton et al. 2003: 41 (M*, F, P*, L*, E; tax., distr.); Gordeyev et al. 2004: 21 (distr.; Tadjikistan); Segaghat et al. 2003b (mol. tax., distr.; Iran); Nicolescu et al. 2004 (mol. tax., distr.; Romania); Linton et al. 2007 (mol. tax., bion., distr.; Greece); Becker et al. 2010: 170 (M*, F*, L*; keys, tax., bion., *s.l.*); Becker et al. 2010: 174 (E*; key, tax., distr., bion., as *s.s.*); Namazov 2014 (distr.; Azerbaijan); Robert et al. 2019 (distr., western Palearctic). **Ety.:** spotted wings (L). **Informal name:** Spot-winged German Nail Mosquito. (Vol. 1, p. 312)

syn. *alexandraeschingarevi* Shingarev

1928: 48 (M*, F). **Type loc.:** Vladimir and Penzenskaya, USSR (LU). **Refs.:** Zhelochovtsev 1937: 707 (syn.); White 1978 (tax., review, keys, distr.; corrected specific epithet; Art. 26, International Commission on Zoological Nomenclature 1964).

syn. *basilii* Falleroni

1932: 29 (E; *maculipennis* ssp.). **Type loc.:** Maccarese [Latium], Italy (LU). **Refs.:** Hackett & Missiroli 1935: 93 (syn.).

syn. *typicus* Hackett & Missiroli

1935: 45 (*Anopheles maculipennis* form). **Type loc.:** [Neither type nor type locality designated]. **Refs.:** Bates 1940: 353 (syn.).

maculipes (Theobald)

1903a: 81 (F*; *Arribalzagia*). **Type loc.:** São Paulo, Brazil (NHMUK). **Distr.:** Argentina, Bolivia, Brazil, French Guiana, Paraguay, Trinidad and Tobago, Uruguay. **Refs.:**

da Costa Lima 1929a: 286 (M*, F*, P*, L*); Mullin-Díaz 1943: 1 (F*; tax.); Causey et al. 1944: 5 (E*); Corrêa & Rabelo 1951: 167 (M*); Wilkerson 1988 (tax.). **Ety.:** spotted foot (L); legs. **Informal name:** Spot-footed Brazilian Nail Mosquito.

malefactor Dyar & Knab
1907c: 198 (M, F). **Type loc.:** Chagres River, Panama (USNM). **Distr.:** Colombia, Costa Rica, Ecuador, Panama, Venezuela. **Refs.:** Stone & Knight 1956b: 278 (type info., lectotype desig.); Wilkerson 1990b: 235 (M*, F*, P*, L*; tax.; from syn. with *punctimacula*); Berti et al. 2015 (distr.; Venezuela); Bourke et al. 2018 (mol. tax.; as *s.l.*). **Ety.:** n.s. [evildoer (L)]. **Informal name:** Evil Panamanian Nail Mosquito.

manalangi Mendoza
1940: 368 (M*, F*, L*). **Type loc.:** Jose Panganiban, Camarines Norte, Philippine Islands (LU). **Distr.:** Philippines. **Refs.:** Reid 1962: 35 (M, F, P*, L, E*); Cagampang-Ramos & Darsie 1970 (F*, L*; keys); Basio 1971b: 36 (M*; bion.); Baisas 1974: 51 (M, F*, P, L*, E*; tax., bion., distr.; Philippines). **Ety.:** Cristóbal Manalang. **Informal name:** Manalang Philippine Nail Mosquito.

marteri Senevet & Prunnelle
1927: 529 (M*, F, L*). **Type loc.:** Camp du Marechal, Camp des Chenes, and Hammam Melouane, Algeria (PIA, NHMUK). **Distr.:** Albania, Algeria, Bulgaria, Cyprus, France (Corsica), Greece, Iran, Iraq, Israel (and Gaza Strip and West Bank), Italy (Sardinia and Sicily), Jordan, Lebanon, Libya, Morocco, Portugal, Spain, Syria, Tajikistan, Tunisia, Turkey. **Refs.:** Hadjinicolaou 1938: 44 (E*); Aitken 1953: 343 (A, P, L, E; bion.); Senevet & Andarelli 1955a: 91 (M*, F*, P*, L*, E; bion., tax.); Shahgudian 1956: 71 (L*; tax.; as *sogdianus*); Postiglione et al. 1973: 132 (distr.; as *sogdianus*); Ribeiro et al. 1987 (syn.); Townsend 1990: 101 (type info.); Glick 1992: 129 (distr.); Ramsdale & Snow 2000: 6 (distr.); Hadjivassilis 2000: 38 (distr.); Becker et al. 2010: 177 (M*, F*, L*; keys, tax., distr., bion.); Robert et al. 2019 (distr.; western Palearctic). **Ety.:** Marie Louise Marter. **Informal name:** Marter Algerian Nail Mosquito.
syn. *sogdianus* Keshishian
1938: 888 (M*, F*, P*, L*; as sp.). **Type loc.:** Karatag, Tovil-dora, Varzob [Stalinabad, Tajikistan] (MTI). **Refs.:** Bates et al. 1949: 431 (to ssp.); Ribeiro et al. 1987: 108 (syn.).
syn. *conquensis* Torres Cañamares
1946: 47 (M*, F*, L*, E*; *marteri* var.). **Type loc.:** Cuenca, Spain (IEE). **Refs.:** Shahgudian 1956: 74 (syn.); Ribeiro et al. 1987 (syn.).

martinius Shingarev
1926: 47 (F). **Type loc.:** Near Tashkent, Turkestan [Kazakhstan] (MTI). **Distr.:** Afghanistan, Iran, Kazakhstan, Kyrgyzstan, Uzbekistan. **Refs.:** Zhelochovtsev 1937: 709 (syn.); De Meillon 1947b: 77 (M*, F*; ? = *neireti* Blanchard); Stegnii 1976 (chrom., diff. from *sacharovi*); White 1978: 21 (tax., review, keys, distr.; from syn. with *sacharovi*); Glick 1992: 129 (distr.); Linton et al. 2003 (review; Maculipennis Complex); Sedaghat et al. 2003b (syst. review). **Ety.:** Erich Martini. **Informal name:** Martini Kazakhstani Nail Mosquito.
syn. *relictus* Shingarev
1928: 47 (M, F*). **Type loc.:** Chikmek, Samarkand, Uzbekistan (LU). **Refs.:** Martini 1930: 146 (syn.); White 1978: 21 (syn.; transfer from *sacharovi*).
syn. *elutior* Martini
1930: 145 (M*, F; *Anopheles* (*Anopheles*) *elutus* var.). **Type loc.:** Turkestan [Kazakhstan] (LU). **Refs.:** Bates 1940: 354 (syn.); White 1978: 21 (syn.; transfer from *sacharovi*).

mattogrossensis Lutz & Neiva
1911: 297 (F). **Type loc.:** "lagôa de Manicoré," Mato Grosso, Brazil (IOC). **Distr.:** Bolivia, Brazil, Colombia, Ecuador, Peru, Venezuela. **Refs.:** Gabaldón 1939 (M*); Deane, L.M. et al. 1946 (F*; key, bion.; NE Brazil & Amazonas); Causey et al. 1946 (M*; key; NE Brazil & Amazonas); Deane, M.P. et al. 1946 (L*; key, bion.; NE Brazil & Amazonas); Lane 1953: 181 (M*, F*, L*, E*); Belkin et al. 1971: 3 (type info.); Linton et al. 2013 (bion., mol. tax., distr.; Ecuador); Berti et al. 2015 (distr.; Venezuela). **Ety.:** Mato Grosso, Brazil. **Informal name:** Mato Grosso Brazilian Nail Mosquito.
syn. *amazonicus* Christophers
1923b: 71 (F*). **Type loc.:** Manaus, Amazonas, Brazil (NHMUK). **Refs.:** Shannon 1933b: 135 (syn.); Belkin 1968b: 9 (tax.); Belkin et al. 1971: 3 (type loc. info.); Wilkerson 1988 (tax.).

maverlius Reinert
1997 (M*, F*, P*, L*, E*). In Reinert et al. 1997: 18. **Type loc.:** Choctawhatchee River, Bruce, Walton County, Florida, United States (USNM). **Distr.:** United States (continental). **Refs.:** Levine et al. 2004a (mol. tax., distr., niche model); Darsie & Ward 2005 (F*, L*; keys, distr.); Harrison et al. 2016 (F*, L*; keys, distr.). **Ety.:** acronym: Center for Medical, Agricultural and Veterinary Entomology. **Informal name:** Maverlius American Nail Mosquito.

medialis Harbach
2018c: 3 (new name for *intermedium* Chagas, not *intermedia* Rothwell 1907). **Distr.:** Argentina, Bolivia, Brazil, Colombia, Costa Rica, Ecuador, French Guiana, Guyana, Nicaragua, Paraguay, Peru, Suriname, Trinidad and Tobago, Venezuela. **Refs.:** Davis 1926 (F*); da Costa Lima 1929a: 281 (M*, F*, P*, L*, E*; tax.); Causey et al. 1944: 5 (E*); Deane, L.M. et al. 1946 (F*; key, bion.; NE Brazil & Amazonas; as *intermedius*); Causey et al. 1946 (M*; key; NE Brazil & Amazonas; as *intermedius*); Deane, M.P. et al. 1946 (L*; key, bion.; NE Brazil & Amazonas; as *intermedius*); Floch & Abonnenc 1951: 39 (M*, F*, L*, E*); Lane 1953: 216 (M*, F*, P, L, E*); Belkin et al. 1971: 3

(tax.); Forattini, Sallum, & Flores 1997 (E*; as *intermedius*); Dusfour et al. 2012 (malaria incrim.; as *intermedius*). **Ety.:** n.s. [in the middle, intermediate, of or belonging to the middle (L)]. **Informal name:** Middle Brazilian Nail Mosquito.

syn. *intermedium* Chagas
1908: 85 (F*; *Cycloleppteron*) In Perryasú 1908. **Type loc.:** Rio de Janeiro and Xerém, Rio de Janeiro, Brazil (IOC?).

mediopunctatus (Lutz)
1903 (M*, F*; *Cycloleppteron*). In Theobald 1903a: 60. **Type loc.:** Santos, São Paulo State, Brazil (NHMUK). **Distr.:** Argentina, Brazil, Colombia, French Guiana, Guyana, Peru, Suriname, Trinidad and Tobago. **Refs.:** Shannon 1933b (F*); Komp 1942: 60 (F*), 106 (P*, L*), 147 (M*); Causey et al. 1944: 5 (E*); Leví-Castillo 1945: 79 (M*, F*, P*, L*, E*); Cova García 1946: Fig. 1 (E*), 51 (L*), 101 (F*), 136 (M*); Deane, L.M. et al. 1946 (F*; key, bion.; NE Brazil & Amazonas); Causey et al. 1946 (M*; key; NE Brazil & Amazonas); Deane, M.P. et al. 1946 (L*; key, bion.; NE Brazil & Amazonas); Wilkerson 1988 (tax.); Calderón et al. 1995 (distr.; Peru); Sallum et al. 1999: 282 (M*, F*, P*, L*; authorship change, tax.); Sallum & Flores 2004 (E*). **Ety.:** spotted in the middle (L); wing. **Informal name:** Spot-waisted Brazilian Nail Mosquito.

syn. *rockefelleri* Peryassú
1923b: 68 (F*; *Cyclolepidopteron*). **Type loc.:** vicinity of Rio de Janeiro, Brazil (NHMUK). **Refs.:** Belkin et al. 1971: 3 (type loc. info.); Townsend 1990: 128 (type info.).

syn. *limai* da Fonseca & da Silva Ramos
1940a: 384 (M*). **Type loc.:** São Vicente, São Paulo, Brazil (FH). **Refs.:** Belkin et al. 1971: 3 (tax.).

syn. *costalimai* Coutinho
1944: 425 (M*, F*, P*, L*). **Refs.:** Lane 1951: 333 (syn.).

melanoon Hackett
1934: 113 (M*, L*, E*). **Type loc.:** [Viareggio, Tuscany], Italy (LU). **Distr.:** Albania, Azerbaijan, Bosnia and Herzegovina, Bulgaria, France (includes Corsica), FYRO Macedonia, Georgia, Greece, Iran, Iraq, Italy (includes Sardinia), Kosovo, Moldova, Netherlands, Portugal, Romania, Russia (Southern Dists.), Serbia, Spain, Turkey. **Refs.:** Bates et al. 1949 (F, L, E*; keys; to sp; includes *subalpinus*); Bates et al. 1949: 439 (E*); Aitken 1953: 332 (E; bion.); Rioux 1958: 106 (M*; tax.); Postiglione et al. 1970 (distr.; includes *subalpinus*); White 1979 (tax., review, keys, distr.); Sichinava et al. 1983 (distr.; Georgia; as *subalpinus*); Glick 1992: 129 (distr.; as *subalpinus*); Alten et al. 2000 (distr.; Turkey); Eritja et al. 2000: 11 (distr.; Spain); Linton et al. 2002a (mol. tax., syn. of *subalpinus*); Linton et al. 2003 (review; Maculipennis Complex); Nicolescu et al. 2004 (mol. tax., distr.; Romania); Linton et al. 2007 (mol. tax., bion., distr.; Greece); Rueda, Pecor et al. 2008 (distr., Iraq); Becker et al. 2010: 175 (E*; key, tax., distr., bion.; includes *subalpinus*); Namazov 2014 (distr.; Azerbaijan); Robert et al. 2019 (distr.; western Palearctic). **Ety.:** black egg (Gr). **Informal name:** Black-egg Italian Nail Mosquito.

syn. *subalpinus* Hackett & Lewis
1935: 378 (M, L, E*; *maculipennis* var.). **Type loc.:** Albania (LU). **Refs.:** Cianchi et al. 1987 (chromosomes, as ssp.); Ribeiro et al. 1988: 234 (to sp. from ssp. *melanoon*); Linton et al. 2002a (E, mol. tax.; syn.).

menglangensis Ma
1981a: 277 (M*, F*). **Type loc.:** Mengalang, Yunan, China [People's Republic of China] (AMMS). **Distr.:** People's Republic of China. **Refs.**: Dong et al. 1989 (P*, L*). **Ety.:** Mengla Co., China. **Informal name:** Menglang Chinese Nail Mosquito.

messeae Falleroni
1926: 564 (E*; *claviger* var.). **Type loc.:** [Pontine Marshes, Latium], Italy (LU). **Distr.:** Albania, Austria, Belarus, Belgium, Bosnia and Herzegovina, Bulgaria, Crimean Peninsula, Croatia, Czech Republic, Denmark, Estonia, Finland, France, FYRO Macedonia, Germany, Greece, Hungary, Iran, Ireland, Italy, Kazakhstan, Kosovo, Kyrgyzstan, Latvia, Lithuania, Moldova, Mongolia, Montenegro, Netherlands, Norway, People's Republic of China, Poland, Romania, Russia (Central, northwestern, Southern Dists., Volga), Serbia, Slovakia, Slovenia, Sweden, Switzerland, Turkey, Ukraine, United Kingdom. **Refs.:** Bates 1940: 353 (to sp.); Aitken 1945: 308 (? = *lewisi*); Bates et al. 1949 (F, L, E*; keys); Rioux 1958: 102 (tax.); Minar 1971: 92 (distr.); White 1978 (tax., review, keys, distr.); Ramsdale & Snow 2000: 7 (distr.); Alten et al. 2000 (distr.; Turkey); Linton et al. 2002b (bion., mol. tax., distr.; England); Linton et al. 2002c (mol. tax., distr.; Greece); Nicolescu et al. 2004 (E*; mol. tax., distr.; Romania); Linton et al. 2003 (review; Maculipennis Complex); Linton et al. 2005 (distr., mol. tax., distr.; United Kingdom, Ireland); Linton et al. 2007 (mol. tax., bion., distr.; Greece); Qu & Zhu 2008 (distr.; People's Republic of China); Sinka et al. 2010a: 117 (bion., review, distr., niche model); Becker et al. 2010: 175 (E*; key, tax., distr., bion.); Danabalan et al. 2014 (bion., distr.; United Kingdom); Robert et al. 2019 (distr., western Palearctic). **Ety.:** Alessandro di Federico Messea. **Informal name:** Messea Italian Nail Mosquito.

minor da Costa Lima
1929c: 1101 (M, F). **Type loc.:** Estrelia [Rio de Janeiro], Brazil (IOC). **Distr.:** Argentina, Brazil, French Guiana, Paraguay, Suriname. **Refs.:** Pinto 1939: 428 (M*); Deane, L.M. et al. 1946 (F*; key, bion.; NE Brazil & Amazonas); Causey et al. 1946 (M*; key; NE Brazil & Amazonas); Deane, M.P. et al. 1946 (L*; key, bion.; NE Brazil & Amazonas); Galvão 1951: 363 (L*); Lane 1953: 202 (M*, F*, P*, L*, E*); Belkin et al. 1971: 3 (type info.); Fauran & Pajot 1974a: 100 (distr.). **Ety.:** n.s. [smallest (L); poss. ref.

to small size]. **Informal name:** Minor Brazilian Nail Mosquito.

montanus Stanton & Hacker
1917: 273 (M, F*, L; *albotaeniatus* var.). **Type loc.:** Ulu Gombak, Selangor, Malaya [Malaysia] (NHMUK). **Distr.:** Indonesia, Malaysia, Thailand. **Refs.:** Bonne-Wepster & Swellengrebel 1953: 168 (M*, F*, L*); Reid 1968: 148 (M*, F*, P*, L*, E*); Harrison & Scanlon 1975: 34 (M*, F*, P*, L*); Rattanarithikul et al. 2006b (F*, L*; bion., distr., keys). **Ety.:** n.s. [mountainous (L)]. **Informal name:** Montane Malaysian Nail Mosquito.

namibiensis Coetzee
1984: 2 (M, F*, P*, L*, E). **Type loc.:** Mahongo, Kavango, Namibia (USNM). **Distr.:** Guinea-Bissau, Kenya, Mali, Mozambique, Namibia, Tanzania. **Refs.:** Kyalo et al. 2017 (distr.; sub-Saharan Africa). **Ety.:** Namibia. **Informal name:** Damara Namibian Nail Mosquito [the Damara are an indigenous people of Namibia].

neomaculipalpus Curry
1931: 643 (M, F, P*, L). **Type loc.:** Near Gatun, east of Panama City, Panama (LU). **Distr.:** Argentina, Belize, Bolivia, Brazil, Colombia, Costa Rica, Ecuador, El Salvador, Guatemala, Honduras, Mexico, Nicaragua, Panama, Paraguay, Peru, Trinidad and Tobago, Venezuela. **Refs.:** Senevet 1934: 33 (P*); Kumm 1941: 91 (E*); Komp 1942: 59 (F*), 104 (P*, L*), 145 (F*); Pelaez 1945: 74 (F*; key); Cova García 1946: 48 (F*, L*); Vargas & Martínez Palacios 1956: 105 (M*, F*, L*); Kitzmiller et al. 1966 (chrom.); Heinemann & Belkin 1977b: 414 (distr.); Beltrán-Aguilar et al. 2011: 662 (M*, F*, P*, L*; tax., keys, distr.). **Ety.:** new to the Neotropical fauna plus with spotted palpus (L). **Informal name:** Spotty-lipped Panamanian Nail Mosquito.

nigerrimus Giles
1900a: 162 (F*). **Type loc.:** Calcutta [West Bengal], India (NHMUK). **Distr.:** Bangladesh, Brunei, Cambodia, India, Indonesia (includes Kalimantan), Malaysia (includes Sabah, Sarawak), Myanmar, Nepal, Pakistan, People's Republic of China (includes Hainan, Hong Kong), Philippines, Sri Lanka, Thailand, Timor, Vietnam. **Refs.:** Baisas 1947b (first instar L; as *hyrcanus* ssp.); Reid 1953: 24 (M*, F*, P*, L*, E*; to sp.); Nguyen Thuong Hien 1968 (F*, L*; keys, tax., bion., distr.; Vietnam); Reid 1968: 98 (M*, F*, P*, L*, E*; as *indiensis*); Seetharam & Chowdaiah 1971: 429 (chrom.*); Aslamkhan 1971b (distr.; Pakistan); Rattanarithikul & Harrison 1973 (L; key, distr.; Thailand; as *indiensis*); Ameen & Talukdar 1974c: 91 (P*); Harrison & Scanlon 1975: 65 (M*, F*, P*, L*; distr.); Lee et al. 1987b: 67 (F; key, tax., bion., distr., review); Darsie & Pradhan 1990, 1991 (distr.); Glick 1992: 129 (distr.); Linley et al. 1995: 54 (E*); Whelan & Hapgood 2000 (bion., distr.; East Timor); Oo et al. 2004 (distr.; Myanmar); Rattanarithikul et al. 2006b (F*, L*; bion., distr., keys); Qu & Zhu 2008 (distr.; People's Republic of China); Songsawatkiat et al. 2013 (mol. tax.; *s.l.*); Saeung et al. 2014 (E*). **Ety.:** blackest (L). **Informal name:** Sable Indian Nail Mosquito. (Vol. 1, p. 314)

syn. *indiensis* Theobald
1901a: 145 (A*; *Anopheles* (*Anopheles*) *sinensis* ssp.). **Type loc.:** Madras, India (NE). **Refs.:** Harrison et al. 1973: 265 (to syn.); Townsend 1990: 86 (type info.).

syn. *bentleyi* Bentley
1902: 15 (F*). **Type loc.:** Tezpur, Assam, India (LU).

syn. *minutus* Theobald
1903a: 91 (F; *Myzorhynchus*, not Macquart 1834). **Type loc.:** Lahore, Punjab [Pakistan] (NHMUK). **Refs.:** Reid 1953: 51 (tax.).

syn. *williamsoni* Baisas & Hu
1936: 222 (P*, L*; *Anopheles hyrcanus* var.). **Type loc.:** Penang, Malaya [Malaysia] (NE). **Refs.:** Reid 1953: 27 (tax.; syn.?).

syn. *venhuisi* Bonne-Wepster
1951a: 284 (M*, P*, L*, E). **Type loc.:** Java [Indonesia] (LU). **Refs.:** Bonne-Wepster & Swellengrebel 1953: 242 (A, L; tax.); Reid 1953: 24 (syn.).

nilgiricus Christophers
1924b: 13 (A; as var.). **Type loc.:** [Nilgiri Hills, Madras], India (NHMUK). **Distr.:** India. **Refs.:** Christophers 1933: 129 (M*, F*, L); Puri 1949: 485 (to ssp.); Darsie & Pradhan 1990, 1991 (F, L; tax., keys, bion., distr.; Nepal); Harrison et al. 1991: 199 (from ssp. of *lindesayi*). **Ety.:** Nilgiri Hills, Madras, India. **Informal name:** Nilgiri Indian Nail Mosquito.

nimpe Nguyen, Tran & Harbach
2000 (M*, F*, P*, L*). In Nguyen et al. 2000: 189–198. **Type loc.:** Village TK97, Tan Tien Commune, Ca Mau Province, Vietnam (NHMUK). **Distr.:** Vietnam. **Ety.:** acronym: National Institute of Malariology, Parasitology and Entomology, Hanoi, Vietnam. **Informal name:** NIMPE Vietnamese Nail Mosquito.

nitidus Harrison, Scanlon & Reid
1973: 266 (A). **Type loc.:** Selangor, Malaysia (NHMUK). **Distr.:** Cambodia, India, Indonesia, Malaysia, Myanmar, People's Republic of China, Thailand, Vietnam. **Refs.:** Harrison & Scanlon 1975: 58 (M*, F*, P*, L*; distr.); Oo et al. 2004 (distr.; Myanmar); Rattanarithikul et al. 2006b (F*, L*; bion., distr., keys); Qu & Zhu 2008 (distr.; People's Republic of China); Saeung et al. 2014 (E*). **Ety.:** shining, bright (L); well-defined wing pattern. **Informal name:** Bright-winged Malaysian Nail Mosquito.

noniae Reid
1963a: 108 (M, F*, P*, L*). **Type loc.:** Ulu Langat, Selangor, Malaya [Malaysia] (NHMUK). **Distr.:** Malaysia, Philippines. **Refs.:** Reid 1968: 201 (M*, F*, P*, L*; tax.). **Ety.:** Nonie Reid. **Informal name:** Nonie Malaysian Nail Mosquito.

obscurus (Grünberg)
1905: 380 (F*; *Myzorhynchus*). **Type loc.:** Kamerun [Cameroon] (ZM). **Distr.:** Angola, Benin, Burkina Faso, Cameroon, Central African Republic, Côte d'Ivoire, Democratic Republic of the Congo, Equatorial Guinea, Ethiopia, Gabon, Ghana, Guinea, Liberia, Mali, Nigeria, Republic of the Congo, Sierra Leone, South Sudan, Tanzania (Zanzibar), Togo, Uganda. **Refs.:** Evans 1938: 79 (M*, F*, P*, L*, E*); De Meillon 1947b: 49 (M*, F*, L*, E*); Lacan 1956: 150 (distr.); Gillies & De Meillon 1968: 45 (M*, F*, L*, E*; distr.); Ribeiro & da Cunha Ramos 1975: 15 (distr.); Service 1976a (distr.; Gabon); Kyalo et al. 2017 (distr.; sub-Saharan Africa). **Ety.:** n.k. (in German) [dark, poorly known (L)]. **Informal name:** Obscure Cameroonian Nail Mosquito.
syn. *strachanii* Theobald
1907: 85 (F*; *Myzorhynchus*). **Type loc.:** Lagos [Nigeria] (NHMUK).
syn. *nowlini* Evans
1932: 98 (M, F, P, L; *Anopheles obscurus* var.). **Type loc.:** Firestone Plantation, Liberia (NHMUK). **Refs.:** White 1980: 116 (from *obscurus* var.).

occidentalis Dyar & Knab
1906b: 159 (A). **Type loc.:** Stanford University, California, United States (USNM). **Distr.:** United States (continental). **Refs.:** Dampf 1936 (F*); King & Bradley 1941: 65 (to sp.); Aitken 1945: 282 (M, F, P*, L*, E*); Penn 1949a: 54 (P*); Yamaguti & LaCasse 1950b: 22 (M*, F*, L*); Pratt 1952: 484 (P*, L*; tax.); Carpenter & LaCasse 1955: 43 (M*, F*, L*; keys); Stone & Knight 1956b: 279 (type info.); Kitzmiller et al. 1967: 176 (chrom.*); White 1978 (tax., review, keys, distr.); Bohart & Washino 1978: 25 (M, F, L*; keys, tax., bion., distr.); Porter & Collins 1996 (mol. phyl., Nearctic); Darsie & Ward 2005 (F*, L*; keys, distr.). **Ety.:** the setting sun, west (L); western United States. **Informal name:** Western American Nail Mosquito.

omorii Sakakibara
1959: 288 (M*, F*, P*, L*). **Type loc.:** Mount Tochu, Misakubo City, Iwatagun, Sizuoka Prefecture, Japan (RIE). **Distr.:** Japan. **Refs.:** Tanaka et al. 1979: 59 (M*, F*, P, L*). **Ety.:** Nanzaburo Omori. **Informal name:** Omori Japanese Nail Mosquito.

palmatus (Rodenwaldt)
1926: 794 (L*; *Stetomyia* [Lapsus for *Stethomyia aitkenii* var. *palmata*]). **Type loc.:** Sidoeagoeng (Preanger Regencies), Java (LU). **Distr.:** Indonesia (includes Java), Malaysia, People's Republic of China, Thailand. **Refs.:** Bonne-Wepster & Swellengrebel 1953: 98 (A, L*); Reid 1965: 117 (M, L*); Reid 1968: 246 (M, L*); Harrison & Scanlon 1975: 173 (M*, F*, P*, L*; distr.); Lee et al. 1987b: 70 (F; key, tax., bion., distr., review); Rattanarithikul et al. 2006b (F*, L*; bion., distr., keys); Qu & Zhu 2008 (distr.; People's Republic of China). **Ety.:** palm branches (L) [prob. ref. to large palmates]. **Informal name:** Palm Indonesian Nail Mosquito.

paludis Theobald
1900b: 75 (F). **Type loc.:** Katunga, Sierra Leone (NHMUK). **Distr.:** Angola, Benin, Burkina Faso, Burundi, Cameroon, Central African Republic, Côte d'Ivoire, Democratic Republic of the Congo, Ethiopia, Gabon, Ghana, Guinea, Kenya, Liberia, Malawi, Mali, Nigeria, Rwanda, São Tomé and Príncipe, Senegal, Sierra Leone, Somalia, South Sudan, Sudan, Tanzania (includes Zanzibar), Togo, Uganda. **Refs.:** Evans 1938: 74 (M*, F*, P, L); De Meillon 1947b: 51 (M*, F*, P*, L); Gillies & De Meillon 1968: 38 (M*, F*, P, L, E; distr.); Service 1976a (distr.; Gabon). **Ety.:** n.s. [marsh, swamp (L); ref. to larval habitat]. **Informal name:** Marshland Sierra Leonean Nail Mosquito.

papuensis Dobrotworsky
1957: 185 (M*, F, P*, L*). **Type loc.:** Minj, Papua, New Guinea (CSIR). **Distr.:** Indonesia (West Papua), Papua New Guinea. **Refs.:** Lee et al. 1987b: 72 (F; key, tax., bion., distr., review). **Ety.:** Papua New Guinea. **Informal name:** Minj Papuan Nail Mosquito.

parapunctipennis Martini
1932b: 101 (M, F). **Type loc.:** San Cristóbal de las Casas, Chiapas, Mexico (LU). **Distr.:** Guatemala, Mexico. **Refs.:** Pelaez 1945: 74 (F*; key); Vargas 1949a: 233 (P); Lane 1953: 163 (M*, F*, L*); Vargas & Martínez Palacios 1956: 80 (M*, F*, P*, L*); Wilkerson 1989 (type info.); Wilkerson 1990a: 180 (neotype desig.); Beltrán-Aguilar et al. 2011: 667 (M*, F*, P*, L*; tax., keys, distr.). **Ety.:** similar (Gr); close to *pseudopunctipennis*. **Informal name:** Chiapas Mexican Nail Mosquito.
ssp. ***guatemalensis*** de León
1938: 416 (M*, F*, P*, L*; *chiriquiensis* var.). **Type loc.:** Cumbre del Aire, Department of Totonicapán, Guatemala (USNM). **Distr.:** Guatemala, Mexico. **Refs.:** Vargas 1940b: 68 (L; tax.); Vargas 1941d: 112 (E*); Lane 1953: 166 (M, F, L); Wilkerson 1990a: 181 (neotype desig.; from var. to ssp.). **Ety.:** Guatemala. **Informal name:** Maya Guatemalan Nail Mosquito [the Maya are indigenous peoples of Central America].

peditaeniatus (Leicester)
1908: 31 (M, F; *Myzorhynchus*). **Type loc.:** Malaya [Malaysia] (NHMUK). **Distr.:** Afghanistan, Bangladesh, Bhutan, Cambodia, India, Indonesia, Iran, Malaysia, Myanmar, Nepal, Pakistan, People's Republic of China, Philippines, Sri Lanka, Thailand, Vietnam. **Refs.:** Reid 1953: 32 (M*, F*, P*, L*, E*); Nguyen Thuong Hien 1968 (F*, L*; keys, tax., bion., distr.; Vietnam); Basio 1971b: 37 (M*, F*; bion.); Harrison 1973b: 3 (distr.); Baisas 1974: 52 (M, F*, P, L*, E*; tax., bion., distr.; Philippines); Harrison & Scanlon 1975: 71 (M*, F*, P*, L*; distr.); Lee et al. 1987b: 75 (F; key, tax., bion., distr., review); Darsie & Pradhan 1990, 1991 (F, L; tax., keys, bion., distr.; Nepal); Glick 1992: 129 (distr.);

Suleman et al. 1993 (distr.; Pakistan); Linley et al. 1995: 49 (E*); Oo et al. 2004 (distr.; Myanmar); Azari-Hamidian et al. 2006 (distr.; Iran); Rattanarithikul et al. 2006b (F*, L*; bion., distr., keys); Saeung et al. 2014 (E*); Namgay et al. 2018 (bion., distr.; Bhutan). **Ety.:** foot band (L); ref. to bands on legs. **Informal name:** Band-footed Malaysian Nail Mosquito.

perplexens Ludlow
1907: 267 (F). **Type loc.:** Camp Roosevelt, Mount Gretna, Pennsylvania, United States (USNM). **Distr.:** Canada, United States (continental). **Refs.:** Bellamy 1956: 515 (M, F, P, L, E*; tax., bion.; from syn. with *punctipennis*); Mattingly 1971a: Pl. 11 (P paddle*, trumpet*); Kreutzer & Kitzmiller 1971a: 409 (chrom.*, tax.); Kreutzer et al. 1972: 191 (chrom.*); Linley & Kaiser 1994: 43 (E*); Darsie & Ward 2005 (F*, L*; keys, distr.); Harrison et al. 2016 (F*, L*; keys, distr.). **Ety.:** perplexing. **Informal name:** Perplexing American Nail Mosquito.

persiensis Linton, Sedaghat, & Harbach
2003 (E, mol. tax., syst.). In Sedaghat et al. 2003b: 527–535. **Type loc.:** Amol (36°26'N, 52°24'E), Amol County, Mazandaran Province, Iran (NHMUK). **Distr.:** Azerbaijan, Iran. **Refs.:** Linton et al. 2003 (review; Maculipennis Complex); Sedaghat et al. 2009 (M, F, P, L, E*); Robert et al. 2019 (distr., western Palearctic). **Ety.:** Persia. **Informal name:** Persia Iranian Nail Mosquito.

peryassui Dyar & Knab
1908: 53 (new name for *lutzi* Cruz, not Cruz 1901). **Distr.:** Bolivia, Brazil, Colombia, French Guiana, Guyana, Peru, Suriname, Venezuela. **Refs.:** Bonne 1923b (E*; as *mediopunctatus*); Shannon & Davis 1930: 481 (M*, F*, P*, L*; tax.); Causey et al. 1944: 4 (E*); Deane, L.M. et al. 1946 (F*; key, bion.; NE Brazil & Amazonas); Causey et al. 1946 (M*; key; NE Brazil & Amazonas); Deane, M.P. et al. 1946 (L*; key, bion.; NE Brazil & Amazonas); Lane 1953: 183 (M*, F*, P*, L*, E*); Linley & Lounibos 1994: 25 (E*); Berti et al. 2015 (distr.; Venezuela); Bourke et al. 2018 (mol. tax.; as *s.l.*). **Ety.:** Antonio Gonçalves Peryassú. **Informal name:** Peryassú Brazilian Nail Mosquito.

syn. *lutzi* Cruz
1907a (F; *Manguinhosia*). **Type loc.:** Margens do Rio Bicudo, Minas Gerais, Brazil (IOC?). **Refs.:** Belkin et al. 1971: 3 (tax.).

syn. *alagoanii* Peryassú
1925: 258 (M, F). **Type loc.:** suburbs of Maceio, Alagoas, Brazil (LU).

syn. *celidopus* Dyar & Shannon
1925b: 41 (F). **Type loc.:** Carmo, Rio Branco, Amazonas, Brazil (USNM).

petragnani Del Vecchio
1939: 30 (E*). **Type loc.:** Latium, Italy (LU). **Distr.:** Algeria, France (includes Corsica), Germany, Greece, Italy (includes Sardinia and Sicily), Morocco, Portugal, Spain (includes Balearic Isl.), Tunisia, United Kingdom. **Refs.:** Bates et al. 1949: 422 (syn.); Senevet & Andarelli 1955a: 128 (P*, E*); Senevet & Andarelli 1955b: 129 (E*); Coluzzi 1962: 1025 (from syn.); Ramsdale & Snow 2000: 7 (distr.); Eritja et al. 2000: 11 (distr.; Spain); Becker et al. 2010: 168 (M*, F*, P, L*; keys, tax., distr., bion.); Robert et al. 2019 (distr.; western Palearctic). **Ety.:** Giovanni Petragnani. **Informal name:** Petragnani Italian Nail Mosquito.

syn. *portucaliensis* Figueiredo
1904: 223 (F*, L*; *Anopheles bifurcatus* var.). **Type loc.:** Barro, Loures [Lisbon], Portugal (LU). **Refs.:** Ribeiro et al. 1988: 235 (from var. *bifurcatus* to syn.).

peytoni Kulasekera, Harrison, & Amerasinghe
1988: 303 (M*, F, P*, L*). **Type loc.:** Kanneliya, Galle District, Sri Lanka (USNM). **Distr.:** Sri Lanka. **Ety.:** EL Peyton. **Informal name:** Peyton Sri Lankan Nail Mosquito.

pilinotum Harrison & Scanlon
1974: 33 (M*, F*, P*, L*). **Type loc.:** Baong, N Ecija, Luzon, Philippines (USNM). **Distr.:** Indonesia, Philippines. **Refs.:** Lee et al. 1987b: 76 (F; key, tax., bion., distr., review). **Ety.:** hairy notum (L). **Informal name:** Hairy-back Philippine Nail Mosquito.

pinjaurensis Barraud
1932a: 353 (M*; *aitkeni* [*sic*] var.). **Type loc.:** Pinjaur, near Kalka [Punjab Hill States], India (MSI). **Distr.:** India. **Refs.:** Christophers 1933: 110 (M*; to sp.). **Ety.:** Pinjaur, Patiala State, India. **Informal name:** Pinjaur Indian Nail Mosquito.

plumbeus Stephens
1828: 503 (A). **Type loc.:** North Ireland (LU). **Distr.:** Albania, Algeria, Armenia, Austria, Azerbaijan, Belarus, Belgium, Bosnia and Herzegovina, Bulgaria, Crimean Peninsula, Croatia, Cyprus, Czech Republic, Denmark, Estonia, France (includes Corsica), FYRO Macedonia, Georgia, Germany, Greece, Hungary, Iran, Ireland, Italy (includes Sardinia and Sicily), Kosovo, Liechtenstein, Lithuania, Luxembourg, Moldova, Montenegro, Morocco, Netherlands, Poland, Portugal, Romania, Russia (Southern Dists.), Serbia, Slovakia, Slovenia, Spain, Sweden, Switzerland, Syria, Tunisia, Turkey, Ukraine, United Kingdom. **Refs.:** Séguy 1924: 145 (E*), 145, 146 (L*); Marshall 1938: 134 (M*, F*, P*, L*; bion.); Aitken 1953: 346 (A, P, L, E; bion.); Senevet & Andarelli 1955a: 56 (M*, F, P*, L*, E); Coluzzi & Cancrini 1971 (chrom.); Gutsevich et al. 1974: 84 (M*, F, L*); Glick 1992 (F*; key, distr.); Ramsdale & Snow 2000: 8 (distr.); Beck et al. 2003: 24 (distr.; Luxembourg); Becker et al. 2010: 178 (M*, F*, L*; keys, tax., distr., bion.); Osório et al. 2010 (distr.; Portugal); Namazov 2014 (dist.; Azerbaijan); Robert et al. 2019 (distr.; western Palearctic). **Ety.:** lead color (L). **Informal name:** Leaden Irish Nail Mosquito.

syn. *nigripes* Staeger
1839: 552 (A). **Type loc.:** Charlottenlund, Denmark (ZMC). **Refs.:** Martini 1930: 166 (M*, F, L*, E*).
syn. *corsicanus* Edwards
1928c: 166 (F; as var.). **Type loc.:** Porto, Corsica (NHMUK). **Refs.:** Bates et al. 1949: 423 (syn. suggested).
syn. *intermedius* Shingarev
1928: 49 (not Peryassú 1908) (M*). **Type loc.:** Moskva [USSR] (MTI). **Refs.:** Bates et al. 1949: 424 (syn. suggested).

pollicaris Reid
1962: 24 (M, F*, P, L*, E*). **Type loc.:** 16th mile Ulu Gombak Road, Selangor, Malaya [Malaysia] (NHMUK). **Distr.:** Malaysia, Thailand. **Refs.:** Peyton & Scanlon 1966: 1 (distr.); Harrison & Scanlon 1975: 100 (M*, F*, P*, L*); Rattanarithikul et al. 2006b (F*, L*; bion., distr., keys). **Ety.:** thumb-like (L); thumb-like stigmal process of larva. **Informal name:** Thumb-like Malaysian Nail Mosquito.

powderi Zavortink
1970b: 17 (F*, P*). **Type loc.:** 10 km SE of La Sierra, San José, Costa Rica (USNM). **Distr.:** Costa Rica, Panama. **Refs.:** Lee & Woodhill 1944: 86 (M*, F*, L*). **Ety.:** Anthony William Powder. **Informal name:** Powder Costa Rican Nail Mosquito.

powelli Lee
1944b: 21 (M*, F*, L*). **Type loc.:** Adelaide River, Northern Territory, Australia (CSIR). **Distr.:** Australia. **Refs.:** Marks 1956: 51 (M, L; bion.); Mattingly 1971a: Pl. 25 (L head*, thorax*, terminal segs.*); Lee et al. 1987b: 80 (F; key, tax., bion., distr., review). **Ety.:** Owen Watkins Powell. **Informal name:** Powell Australian Nail Mosquito.

prachongae Rattanarithikul & Harrison
2017 (M*, F, P*, L*; review; Gigas Complex). In Harbach, Rattanarithikul, & Harrison 2017. **Type loc.:** Loei Province, Phu Luang District, Huai Pong (17°7'N, 101°32'E), Thailand (USNM). **Distr.:** Laos, Thailand. **Refs.:** Rattanarithikul et al. 2006b (F*, L*; keys, bion.; as *Anopheles* n. sp. near *gigas*). **Ety.:** Prachong Panthusiri. **Informal name:** Prachong Thai Nail Mosquito.

pseudobarbirostris Ludlow
1902a: 129 (F). **Type loc.:** Hagonoy, Bulacan [Luzon], Philippines (USNM). **Distr.:** Australia, Indonesia, Papua New Guinea, Philippines. **Refs.:** Baisas 1935c: 309 (P*); Bonne-Wepster & Swellengrebel 1953: 220 (M*, P*, L); Basio 1971b: 37 (M*, F*; bion.); Harrison 1973a: 369 (distr.); Baisas 1974: 53 (M, F*, P*, L*, E*; tax., bion., distr.; Philippines); Lee et al. 1987b: 83 (F; key, tax., bion., distr., review). **Ety.:** false (Gr); similar to *barbirostris*. **Informal name:** Mustache Philippine Nail Mosquito [*pseudobarbirostris* = "false bearded" = Mustache].

pseudomaculipes (Chagas)
1908 (A*; *Arribalzagia*). In Peryassú 1908: 108. **Type loc.:** Xerem, Rio de Janeiro, Brazil (IOC?). **Distr.:** Brazil, Colombia, Peru, Uruguay. **Refs.:** Davis 1926 (F*); Dyar 1928: 466 (M*, F); Vargas 1942b: 73 (from syn. with *maculipes*); Mullin-Diaz 1943: 1 (F*); Stone & Knight 1956b: 279 (type info., lectotype desig.); Belkin et al. 1971: 4 (tax.). **Ety.:** false (Gr); similar to *maculipes*. **Informal name:** False-spotted Brazilian Nail Mosquito.

pseudopictus Grassi
1899 (A). In Grassi et al. 1899: 100–104. **Type loc.:** Italy (LU). **Distr.:** Afghanistan, France (includes Corsica), Iran, Iraq, Italy (Sicily, Sardinia), Romania, Spain, Turkey, Turkmenistan, Uzbekistan. **Refs.:** Dow 1953: 690 (tax.); Senevet & Andarelli 1955a: 109 (M*, F*, P, L*); Glick 1992: 125 (from syn. with *hyrcanus*; distr.); Alten et al. 2000 (distr.; Turkey); Becker et al. 2010: (M*, F*, L*; keys); Miaoulis et al. 2018 (morph. and mol. tax.; evidence for syn. with *hyrcanus*). **Ety.:** n.s. **Informal name:** False-spotted Italian Nail Mosquito.

pseudopunctipennis Theobald
1901c: 305 (M, F). **Type loc.:** Grenada (NE). **Distr.:** Argentina, Belize, Bolivia, Brazil, Chile, Colombia, Costa Rica, Ecuador, El Salvador, Guatemala, Guyana, Haiti, Honduras, Lesser Antilles (includes Antigua and Barbuda, Aruba, Curaçao, Grenada), Mexico, Nicaragua, Panama, Paraguay, Peru, Suriname, Trinidad and Tobago, United States (continental), Venezuela. **Refs.:** Rozeboom 1937 (E*); Roth 1944: 98 (M*), 107 (M; keys); Aitken 1945: 327 (M, F*, P*, L, E*; tax., bion.); Pelaez 1945: 74 (F*; key); Yamaguti & La Casse 1950b: 31 (M*, F*, L*); Lane 1953: 160 (M*, F*, P*, L*, E*); Carpenter & LaCasse 1955: 46 (M*, F*, L*; keys); Vargas & Martínez Palacios 1956: 83 (M*, F*, P*, L*); Dodge 1966: 362 (first instar L*; key); Vargas 1984 (tax.); Molez et al. 1987 (distr.; Haiti); Lee et al. 1987b: 88 (F; key, tax., bion., distr., review); Wilkerson & Strickman, 1990 (F* key; Central America); Rejmánková et al. 1991 (bion.); Estrada-Franco et al. 1992 (mol. tax.); Estrada-Franco et al. 1993a (mol. tax.); Estrada-Franco et al. 1993b (cross-mating, isoenzymes, mol. tax.; as sp. complex); Manguin et al. 1995 (gen. tax.); Manguin et al. 1996 (bion.); Coetzee et al. 1999 (tax., chrom.); Rueda et al. 2004: 13 (M*, F*, P*, L*; neotype desig.); Darsie & Ward 2005 (F*, L*; keys, distr.); Harbach & Howard 2007: 39 (exclusion of *bifoliata*); Sinka et al. 2010b: 72 (bion., review, distr., niche model); Becker et al. 2010: 374 (F*, L*; key, tax., distr., bion.); Beltrán-Aguilar et al. 2011: 671 (M*, F*, P*, L*; tax., keys, distr.); Linton et al. 2013 (dist.; Ecuador). **Ety.:** false (Gr); similar to *punctipennis*. **Informal name:** False-winged Grenada Nail Mosquito. (Vol. 1, p. 316)
syn. *peruvianus* Tamayo
1907 (M*, P*, L*). In Tamayo & García 1907: 35. **Type loc.:** Los Aguas de Huacachina, San Pedro de Lloc, and Chanchamayo Province, Peru (NE). **Refs.:** Lane 1953 (syn.).

syn. *argentinus* Brèthes
1912: 15 (M, F; *Proterorhynchus*). **Type loc.:** Tucumán, Salta, Jujuy, Santiago del Estero, Argentina (BA). **Refs.:** Belkin et al. 1968: 9 (lectotype desig. by Casal (BA)).

syn. *tucumanus* Lahille
1912: 253 (P*). **Type loc.:** Rio Sali, Argentina (NE).

ssp. ***levicastilloi*** Leví-Castillo
1944: 1 (M, E; as sp.). **Type loc.:** Guayasa, Ecuador (LU). **Distr.:** Ecuador. **Refs.:** Leví-Castillo 1945: 41 (M*, F, P, L*, E*; to ssp.); Lane 1953: 160 (? = *p. pseudopunctipennis*). **Ety.:** José Roberto Leví-Castillo. **Informal name:** Leví-Castillo Ecuadorian Nail Mosquito.

ssp. ***rivadeneirai*** Leví-Castillo
1945: 33 (M*, F*, L*, E*). **Type loc.:** Provinces of Carchi, Imbabura, Pichincha, Cotopaxi, Tungurahua, Chimborazo, Canar, Azuay, and Loja; Interandian Region, Ecuador (LU). **Distr.:** Ecuador. **Refs.:** Lane 1953: 160 (? = *p. pseudopunctipennis*). **Ety.:** Jaime Rivadeneira Davilla. **Informal name:** Rivadeneira Ecuadorian Nail Mosquito.

ssp. ***patersoni*** Alvarado & Heredia
1947: 73 (E*; as var.). **Type loc.:** Tucumán, Argentina (NE). **Distr.:** Argentina. **Refs.:** Lane 1953: 161 (to ssp.). **Ety.:** William Cleland Paterson. **Informal name:** Paterson Argentinian Nail Mosquito.

ssp. ***neghmei*** Mann
1950: 34 (M, F*, L*, E*). **Type loc.:** Ouebrada de Minemine, Tarapaca, Chile (IZC). **Distr.:** Chile. **Ety.:** Amador Neghme. **Informal name:** Neghme Chilean Nail Mosquito.

ssp. ***noei*** Mann
1950: 37 (M, F, L*, E*). **Type loc.:** Oasis de Suca, Tarapaca, Chile (IZC). **Distr.:** Chile. **Ety.:** Juan Noé. **Informal name:** Noé Chilean Nail Mosquito.

pseudosinensis Baisas
1935c: 297 (M*, F, P*, L*; *hyrcanus* var.). **Type loc.:** Calauan, Laguna, Luzon, Philippines (NE; paratypes (USNM)). **Distr.:** Philippines. **Refs.:** Reid 1953 (to sp. status); Cagampang-Ramos & Darsie 1970 (F*, L*; keys); Basio 1971b: 37 (M*, F*; bion.); Harrison 1972: 77 (tax.). **Ety.:** false (Gr); similar to *sinensis*. **Informal name:** Plastic Philippine Nail Mosquito [*pseudosinensis* = "false china" = Plastic].

pseudostigmaticus Dobrotworsky
1957: 184 (M*, F, P*, L*). **Type loc.:** Maroondah, Victoria, Australia (NMM). **Distr.:** Australia. **Refs.:** Dobrotworsky 1965: 42 (M*, F, L*); Lee et al. 1987b: 89 (F; key, tax., bion., distr., review). **Ety.:** false (Gr); similar to *stigmaticus*. **Informal name:** Fake-spotted Australian Nail Mosquito.

pullus Yamada
1937: 238 (M*, F*, P*, L*, E*). **Type loc.:** Keiki-do and South Tyuseido, Chosen, Korea (IID). **Distr.:** Australia, Japan, People's Republic of China, South Korea. **Refs.:** Reid 1953: 52 (tax.); Ohmori 1957 (M*, E*); Ohmori 1959: 222 (P*; as *yatsushiroensis*); Otsuru & Ohmori 1960: 50 (M*, F*, P, L, E*; keys; as *yatsushiroensis*); Hong & Ree 1968: 118 (M*, F; distr.; as *yatsushiroensis*); Tanaka et al. 1979: 79 (M*, F*, P, L, E); Tanaka et al. 1979: 80 (M, F*, P, L, E; as *yatsushiroensis*); Qu & Zhu 2008 (distr.; People's Republic of China). **Ety.:** dark, dusky (L) [overall color]. **Informal name:** Dusky Korean Nail Mosquito.

syn. *yatsushiroensis* Miyazaki
1951: 195 (M*, F*, E*). **Type loc.:** Yatsushiro City, Kumamoto Prefecture, Kyushu, Japan (KU). **Refs.:** Shin & Hong 2001 (syn.); Rueda et al. 2003 (to sp.); Wilkerson et al. 2003 (syn.); Hwang et al. 2004: 99 (molecular support for syn.); Qu & Zhu 2008 (syn. with *pullus*).

syn. *xiaokuanus* Ma
1981b: 67 (M*, F*, E*). **Type loc.:** Mudanjiang, Heilongjiang Province, China [People's Republic of China] (IZ). **Refs.:** Miao et al. 1988 (same as *An. yatsushiroensis* (= *pullus*) from Sichuan).

syn. *junlianensis* Lei
1996 (M*, F*, P, L*). **Type loc.:** Sichuan, China [People's Republic of China] (SIP). **Refs.:** Ma et al. 2000b (syn. with *yatsushiroensis*); Ma & Xu 2005 (as sp.); Qu & Zhu 2008: 211 (as syn. of *pullus*).

punctimacula Dyar & Knab
1906a: 136 (A). **Type loc.:** Colon, Canal Zone, Panama (USNM). **Distr.:** Argentina, Belize, Bolivia, Brazil, Colombia, Costa Rica, Ecuador, El Salvador, Guatemala, Guyana, Honduras, Mexico, Nicaragua, Panama, Peru, Uruguay, Venezuela. **Refs.:** Davis 1926 (F*); Senevet 1934: 36 (P*); Simmons 1937: 191 (F*); Kumm 1941: 91 (E*); Komp 1942 (M*, F*, P*, L*; keys); Ross & Roberts 1943a: 25 (M*, F*, L*); Pelaez 1945: 74 (F*; key); Deane, L.M. et al. 1946 (F*, key, bion.; NE Brazil & Amazonas); Causey et al. 1946 (M*; key; NE Brazil & Amazonas); Deane, M.P. et al. 1946 (L*; key, bion.; NE Brazil & Amazonas); Kreutzer et al. 1969 (chrom.*); Panday 1975a, 1975b (distr.; Suriname); Wilkerson & Strickman 1990: 7 (F*; distr.; Central America); Wilkerson 1990a (M*, F*, P*, L*; tax.); Rejmánková et al. 1998 (bion.); Rodríguez et al. 2002: 1 (E*); Rossi & Martínez 2003: 471 (distr.; Uruguay); Becker et al. 2010: 375 (F*, L*; key, tax., distr., bion.); Beltrán-Aguilar et al. 2011: 682 (M*, F*, P*, L*; tax., keys, distr.); Pinault & Hunter 2012 (bion.); Linton et al. 2013 (dist.; Ecuador); Loaiza et al. 2013 (mol. tax.); Berti et al. 2015 (distr.; Venezuela); Bourke et al. 2018 (mol. tax.; as *s.l.*). **Ety.:** n.s. [spotted (L)]. **Informal name:** Dotted Panamanian Nail Mosquito. (Vol. 1, p. 318)

syn. *strigimacula* Dyar & Knab
1906a: 136 (A). **Type loc.:** Cordoba [Veracruz], Mexico (USNM).

syn. *venezuelae* Evans
1922: 214 (F*). **Type loc.:** La Cabrera, Carabobo, Venezuela (NHMUK). **Refs.:** Belkin 1968b: 11 (tax.); Wilkerson 1988 (tax.).

punctipennis (Say)
1823: 9 (A; *Culex*). **Type loc.:** Chestertown, Kent County, Maryland, United States (NE). **Distr.:** Canada, Dominican Republic, Jamaica, Mexico, Panama, United States (continental), Venezuela. **Refs.:** Shannon & Davis 1930 (F*); Roth 1944: 103 (M*), 107 (M; keys); Aitken 1945: 345 (A, P*, L, E*); Pelaez 1945: 74 (F*; key); Darsie 1949: 514 (P*); Penn 1949a: 59 (P*); Yamaguti & LaCasse 1950b: 29 (M*, F*, L*); Breeland 1951: 224 (L*); Carpenter & LaCasse 1955: 48 (M*, F*, L*; keys); Horsfall 1955: 126 (review); Belkin et al. 1966: 2 (restriction of type loc.); Dodge 1966: 364 (first instar L*; key); Kreutzer & Kitzmiller 1971a: 409 (tax.); Kreutzer et al. 1972: 191 (chrom.*); Kreutzer 1977: 265 (relationship to Maculipennis Group); Bohart & Washino 1978: 26 (M, F*, P*, L*; keys, tax., bion., distr.); Wilkerson & Strickman 1990 (F* key; Central America); Fritz et al. 1991 (tax., morph.); Estrada-Franco & Ma et al. 1993 (cross-mating, cryptic species); Linley & Kaiser 1994: 43 (E*); Porter & Collins 1996 (mol. phyl.; as *punctipennis* complex, E in eastern United States, W in western United States); Darsie & Ward 2005 (F*, L*; keys, distr.); Beltrán-Aguilar et al. 2011: 690 (M*, F*, P*, L*; tax., keys, distr.); Harrison et al. 2016 (F*, L*; keys, distr., tax.; *s.l.*, notes 7, 21). **Ety.:** spotted wing (L). **Informal name:** Spot-winged American Nail Mosquito. (Vol. 1, p. 320)

syn. *hyemalis* Fitch
1847: 281 (M, F; *Culex*). **Type loc.:** eastern New York State, United States (LU). **Refs.:** Stone & Knight 1956b: 278 (type info., lectotype desig.).

syn. *stonei* Vargas
1941a: 179 (M*, F; *punctipennis* var.). **Type loc.:** Monterrey, Nuevo Leon, Mexico (ISET, USNM). **Refs.:** Freeborn 1949: 390 (syn.).

pursati Laveran
1902a: 907 (F). **Type loc.:** Near Pursat, Cambodia, Indochina (PIP). **Distr.:** Cambodia, Malaysia, Thailand, Vietnam. **Refs.:** Reid 1953: 27 (tax.); Reid 1963a: 98 (M*, F*, P*, L*, E*; from syn.); Harrison & Scanlon 1975: 75 (M*, F*, P*, L*); Rattanarithikul et al. 2006b (F*, L*; bion., distr., keys); Saeung et al. 2014 (E*). **Ety.:** Pursat (Pouthisat), Cambodia. **Informal name:** Pursat Cambodian Nail Mosquito.

quadrimaculatus Say
1824 (F). In Keating 1824: 356. **Type loc.:** Wabasha, Minnesota, United States (NE). **Distr.:** Belize, Canada, Costa Rica, Cuba, Dominican Republic, Mexico, Panama, Puerto Rico, United States (continental). **Refs.:** Howard et al. 1913: Fig. 695 (E*); Roth 1944: 101 (M*), 107 (M; keys); Aitken 1945: 280 (tax.); Pelaez 1945: 74 (F*; key); Ross 1947 (F*); Penn 1949a: 56 (P*); Yamaguti & LaCasse 1950b: 27 (M*, F*, L*; *s.l.*); Breeland 1951: 224 (L*); Carpenter & LaCasse 1955: 50 (M*, F*, L*; keys); Belkin et al. 1966: 3 (restriction of type loc.); Dodge 1966: 364 (first instar L*; key; *s.l.*); Wilkerson & Strickman 1990 (F* keys; Central America); Linley et al. 1993a (E*); Porter & Collins 1996 (mol. tax.; Nearctic); Reinert et al. 1997: 22 (M*, F*, P*, L*, E*; keys, bion., med. imp., neotype desig.); Reinert 1997 (bibliography); Levíne et al. 2004 (mol. tax., distr., niche model); Darsie & Ward 2005 (F*, L*; keys, distr.); Sinka et al. 2010b: 72 (Quadrimaculatus Subgroup; bion. review, distr., niche model); Beltrán-Aguilar et al. 2011: 695 (M*, F*, P*, L*; tax., keys, distr.); Harrison et al. 2016 (F*, L*; keys, distr., tax.; *s.s.* and *s.l.*, notes 9, 18). **Ety.:** four spots (L); on wing. **Informal name:** Four-spotted American Nail Mosquito. Official Entomological Society of America common name: common malaria mosquito. (Vol. 1, p. 322)

syn. *annulimanus* van der Wulp
1867: 129 (M*). **Type loc.:** Wisconsin, United States (LM). **Refs.:** Belkin et al. 1966: 3 (restriction of type loc.); Belkin 1968b: 9 (tax.).

rachoui Galvão
1952: 69 (F*). **Type loc.:** Acude São Bento, Municip. Santo Amaro, Bahia, and Paranaguá, Parana, Brazil (FH). **Distr.:** Brazil. **Refs.:** Belkin et al. 1971: 4 (tax., lectotype desig.). **Ety.:** René Guimarães Rachou. **Informal name:** Rachou Brazilian Nail Mosquito.

reidi Harrison
1973a: 365 (F*). **Type loc.:** Peradeniya Botanical Gardens, Kandy District, Central Province, Sri Lanka (USNM). **Distr.:** Sri Lanka. **Ety.:** John Alexander Reid. **Informal name:** Reid Sri Lankan Nail Mosquito.

roperi Reid
1950: 298 (M*, F*, P*, L*). **Type loc.:** Kuala Kubu Bahru, Selangor, Malaya [Malaysia] (NHMUK). **Distr.:** Cambodia, India, Indonesia, Malaysia, Thailand. **Refs.:** Bonne-Wepster & Swellengrebel 1953: 184 (M*, F*, L*); Reid 1968: 183 (M*, F*, P*, L*, E*); Harrison & Klein 1975 (distr.); Harrison & Scanlon 1975: 124 (M*, F*, P*, L*; distr.); Rattanarithikul et al. 2006b (F*, L*; bion., distr., keys). **Ety.:** Richard Snelling Roper. **Informal name:** Roper Malaysian Nail Mosquito.

sacharovi Favre
1903: 189 (A). **Type loc.:** Evlaxe, Yelizavetpol District, Transcaucasia [Elisabethpol, Caucasus, USSR] (ZMM). **Distr.:** Afghanistan, Albania, Algeria, Armenia, Austria, Azerbaijan, Bosnia and Herzegovina, Bulgaria, Crimean Peninsula, Croatia, Cyprus, France, Georgia, Greece, Iran, Iraq, Israel (and Gaza Strip and West Bank), Italy (includes Sardinia), Jordan, Lebanon, Macedonia, Moldova, Montenegro, Palestine, People's Republic of China, Portugal, Russia (Southern Dists.), Spain, Syria, Tajikistan, Turkey. **Refs.:** Martini 1929: 143 (M*, F*, L*, E*; as *elutus*); Ross & Roberts 1943b: 1 (M*, F*, L*; tax., distr., bion.); Bates et al. 1949 (F, L, E*; keys); White 1979 (tax., review, keys, distr.; lectotype desig., male); Glick 1992: 129 (distr.); Hadjivassilis 2000: 38 (distr.; Greece);

Ramsdale & Snow 2000 (distr.; Europe); Sedaghat et al. 2003a: 243 (M*, F, P*, L*, E; tax.); Sedaghat et al. 2003b (distr., Iran); Linton et al. 2003 (review; Maculipennis Complex); Qu & Zhu 2008 (distr.; People's Republic of China); Linton et al. 2007 (mol. tax., bion., distr.; Greece); Sinka et al. 2010a: 117 (bion., review, distr., niche model); Becker et al. 2010: 175 (F*, E*; keys, tax., distr., bion.); Namazov 2014 (distr.; Azerbaijan); Robert et al. 2019 (distr., western Palearctic). **Ety.:** Nikolai Alekseevich Sacharov. **Informal name:** Sacharov Russian Nail Mosquito.

syn. *elutus* Edwards
1921d: 273 (M, F, L, E*). **Type loc.:** Kishon, Palestine [Israel] (NHMUK).

saeungae Taai & Harbach
2015: 256 (M, F, P, L; bion., gen., phyl.). **Type loc.:** Ban Don Tham, Ko Kha District, Lampang Province, Thailand (NHMUK). **Distr.:** Indonesia, Thailand. **Ety.:** Atiporn Saeung. **Informal name:** Saeung Thai Nail Mosquito.

samarensis Rozeboom
1951a: 502 (M*, F*, P*, L*). **Type loc.:** Osmena, Samar, Philippines (USNM). **Distr.:** Philippines. **Refs.:** Cagampang-Ramos & Darsie 1970 (F*, L*; keys); Basio 1971b: 38 (M*, F*; bion.); Baisas 1974 (P*). **Ety.:** Samar, Philippines. **Informal name:** Samar Philippine Nail Mosquito.

saperoi Bohart & Ingram
1946a: 46 (F*, L*, E*). **Type loc.:** Chizuka, Okinawa, Ryukyu-Retto [Japan] (USNM). **Distr.:** Japan. **Refs.:** Bohart & Ingram 1946b: 53 (F*, L*, E; bion., distr., tax.); Reid 1968: 143 (keys); Tanaka et al. 1975c: 210 (tax., distr.); Tanaka et al. 1979: 68 (M*, F*, P, L*). **Ety.:** James Joseph Sapero. **Informal name:** Sapero Japanese Nail Mosquito.

ssp. ***ohamai*** Ohama
1947: 11 (M*, L*) (mimeographed). **Type loc.:** Ishigaki Island [Ryukyu-Retto, Japan] (NMST). **Distr.:** Japan. **Refs.:** Reid 1968: 143 (key); Tanaka et al. 1979: 71 (to ssp. of *saperoi*). **Ety.:** Shinken Ohama. **Informal name:** Ohama Japanese Nail Mosquito.

separatus (Leicester)
1908: 36 (F; *Myzorhynchus*). **Type loc.:** Kuala Lumpur [Selangor], Malaya [Malaysia] (NHMUK). **Distr.:** Cambodia, Indonesia, Malaysia, Singapore, Thailand, Vietnam. **Refs.:** Crawford 1938: 63 (P*); Bonne-Wepster & Swellengrebel 1953: 195 (M*, F*, L*); Nguyen Thuong Hien 1968 (F*, L*; keys, tax., bion., distr.; Vietnam); Reid 1968: 187 (M*, F*, P*, L*, E*); Harrison & Klein 1975 (distr.); Harrison & Scanlon 1975: 128 (M*, F*, P*, L*; distr.); Lee et al. 1987b: 92 (F; key, tax., bion., distr., review); Rattanarithikul et al. 2006b (F*, L*; bion., distr., keys); Jeffery et al. 2010 (distr.; Singapore). **Ety.:** separate, distinct (L); from *sinensis*. **Informal name:** Separate Malaysian Nail Mosquito.

syn. *snijdersi* Swellengrebel
1921: 114 (M; *Myzorhynchus sinensis* var.). **Type loc.:** Medan [East Coast], Sumatra [Indonesia] (NE).

shannoni Davis
1931: 345 (F*). **Type loc.:** Belém, Pará, Brazil (USNM). **Distr.:** Bolivia, Brazil, Colombia, Ecuador, Guyana, Peru, Suriname, Venezuela. **Refs.:** Shannon 1933b: 136 (M, F*, P*); Deane, L.M. et al. 1946 (F*; key, bion.; NE Brazil & Amazonas); Causey et al. 1946 (M*; key; NE Brazil & Amazonas); Deane, M.P. et al. 1946 (L*; descr., key, bion.; NE Brazil & Amazonas); Lane 1953: 204 (M, F*, P*, L*, E*); Lounibos et al. 1997b (E*); Wilkerson et al. 1997: 461 (M*, F*, P*, L*). **Ety.:** Raymond Corbett Shannon. **Informal name:** Shannon Brazilian Nail Mosquito.

similissimus Strickland & Chowdhury
1927a: 19 (new name for *similis* Strickland, not Theobald 1901a). **Distr.:** Malaysia. **Refs.:** Reid 1950: 310 (L; tax.). **Ety.:** most similar to *similis* (L). **Informal name:** Similar Malaysian Nail Mosquito.

syn. *similis* Strickland
1917: 611 (F, L; *Myzorhynchus*). **Type loc.:** Malay Peninsula [Malaysia] (LU).

sinensis Wiedemann
1828: 547 (M, F). **Type loc.:** [Canton], China [People's Republic of China] (ZMC). **Distr.:** Afghanistan, Cambodia, India, Indonesia (includes Flores, Java, Kalimantan, Sulawesi, Sumatra [includes Ketulauan Riouw Archipelago, Timor]), Japan, Laos, Macau, Malaysia, Myanmar, Nepal, Pakistan, People's Republic of China (includes Hong Kong), Russia, Singapore, South Korea, Taiwan, Thailand, Vietnam. **Refs.:** Bohart & Ingram 1946b: 51 (M*, F*, L*; bion., distr., tax.); LaCasse & Yamaguti 1950 (F*); Reid 1953: 10 (M*, F*, P*, L*, E*; tax.); Hara 1957: 47 (F*); Ohmori 1957 (M*, E*); Ohmori 1959: 221 (P*); Hara 1959 (F*); Nguyen Thuong Hien 1968 (F*, L*; keys, tax., bion., distr.; Vietnam); Harrison 1973b: 7 (lectotype desig, distr.); Lee et al. 1973: 383 (distr.); Harrison & Scanlon 1975: 45 (M*, F*, P*, L*; distr.); Tanaka et al. 1975c: 210 (tax., distr.); Kanda & Oguma 1976: 325 (A*; tax.); Otsuru et al. 1976: 301 (E); Kanda & Oguma 1977: 115 (hybridization with *sineroides*); Tanaka et al. 1979: 72 (M*, F*, P, L*, E); Beales 1984 (tax.); Lee et al. 1987b: 94 (F; key, tax., bion., distr., review); Darsie & Pradhan 1990, 1991 (F, L; tax., keys, bion., distr.; Nepal); Linley et al. 1995: 44 (E*); Gornostaeva 2000: 477 (distr.; Russia); Whelan & Hapgood 2000: 405 (bion., distr.; East Timor); Oo et al. 2004 (distr.; Myanmar); Rattanarithikul et al. 2006b (F*, L*; bion., distr., keys); Becker et al. 2010: 340 (F*, L*; key, tax., distr., bion.); Sinka et al. 2011: 89 (bion. review, distr., niche model); Saeung et al. 2014 (E*). **Ety.:** China. **Informal name:** Han Chinese Nail Mosquito [the Han are the dominant ethnic group of China].
(Vol. 1, p. 324)

syn. *plumiger* Dönitz
1901: 37 (M, F). **Type loc.:** Hong Kong [People's Republic of China] (ZM).

syn. *jesoensis* Tsuzuki
1902c: 764 (M). **Type loc.:** Hokkaido, Japan (LU). **Refs.:** Reid 1953: 51 [Variously spelled as *jezoensis* & *jezoensis*].

syn. *changfus* Ma
1981b: 65 (M*, F*, E). **Type loc.:** Emei, Sichuan Province, China [People's Republic of China] (IZ). **Refs.:** Miao et al. 1988 (syn. with *sinensis*); Qu & Zhu 2008 (review; syn. with *sinensis*).

sineroides Yamada
1924: 233 (M, F). **Type loc.:** Bibai [Hokkaido], Japan (IID). **Distr.:** Japan, People's Republic of China, South Korea. **Refs.:** LaCasse & Yamaguti 1950: 23 (M*, F*, P*, L*; keys); Asanuma & Nakagawa 1954a: 370 (P*); Hara 1957: 47 (P*); Ohmori 1957 (M*, E*); Ohmori 1959: 222 (P*); Otsuru & Ohmori 1960: 53 (M*, F*, P, L, E*; tax.); Kanda & Oguma 1977: 115 (hybridization with *sinensis*); Tanaka et al. 1979: 81 (M*, F*, P, L*, E). **Ety.:** like (Gr); resembling *sinensis*. **Informal name:** China Japanese Nail Mosquito.

sintoni Puri
1929b: 401 (M*, F, L*). **Type loc.:** Malabar, Madras, India (NHMUK, MSI). **Distr.:** India. **Refs.:** Christophers 1933: 116 (M*, F*, L*); Reid 1968: 223 (M*, F*, L*); Dong & Wang 1985a (P*, L*); Tewari & Hiriyan 1991: 170 (P*). **Ety.:** John Alexander Sinton. **Informal name:** Sinton Indian Nail Mosquito.

sintonoides Ho
1938a: 279 (M*, F, P*, L*). **Type loc.:** Hainan Island [People's Republic of China] (FMI). **Distr.:** Malaysia, People's Republic of China, Thailand, Vietnam. **Refs.:** Nguyen Thuong Hien 1968 (F*, L*; keys, tax., bion., distr.; Vietnam); Reid 1968: 223 (M*, F*, P*, L*; tax., distr.); Harrison & Scanlon 1975: 181 (M*, F*, P*, L*); Rattanarithikul et al. 2006b (F*, L*; bion., distr., keys). **Ety.:** near *sintoni* (Gr). **Informal name:** Hainan Chinese Nail Mosquito.

smaragdinus Reinert
1997 (M*, F*, P*, L*, E*). In Reinert et al. 1997: 28. **Type loc.:** North End Trail, Manatee Springs State Park, Levy County, Florida, United States (USNM). **Distr.:** United States (continental). **Refs.:** Levíne et al. 2004 (mol. id., distr., niche model); Darsie & Ward 2005 (F*, L*; keys, distr.); Harrison et al. 2016 (F*, L*; keys, distr.); Giordano 2019 (distr.; Canada). **Ety.:** emerald green (L); larva and pupa. **Informal name:** Emerald American Nail Mosquito.

stigmaticus Skuse
1889: 1758 (M, F). **Type loc.:** Blue Mountains, New South Wales, Australia (ANIC). **Distr.:** Australia. **Refs.:** Bonne-Wepster & Swellengrebel 1953: 114 (M*, F*, L*); Marks 1956: 46 (M*, F*, L; bion.); Dobrotworsky 1957: 181 (M*, F, P*, L*; tax.); Dobrotworsky 1965: 41 (M*, F*, L*); Lee et al. 1987b: 95 (F; key, tax., bion., distr., review; type depos.). **Ety.:** n.s. [mark (L); ref. to wing being "hyaline, tinged with pale yellowish-brown at the stigmatic region"]. **Informal name:** Spot-winged Australian Nail Mosquito.

stricklandi Reid
1965: 121 (M*, F*, P*, L*). **Type loc.:** Johore, Malaya [Malaysia] (NHMUK). **Distr.:** Malaysia, Thailand. **Refs.:** Rattanarithikul & Harrison 1973: 2 (L*; key, distr.); Harrison & Scanlon 1975: 171 (M*, F, P*, L*); Rattanarithikul et al. 2006b (F*, L*; bion., distr., keys). **Ety.:** Cyril Strickland. **Informal name:** Strickland Malaysian Nail Mosquito.

sumatranus Swellengrebel & Rodenwaldt
1932: 78 (F*, L*; *gigas* var.). **Type loc.:** Karoo-Hochebene & Kotaradja, Sumatra [Indonesia] (LU). **Distr.:** Indonesia (Sumatra). **Refs.:** Bonne-Wepster & Swellengrebel 1953: 144 (F, L*); Harbach & Howard 2007: 39 (from *gigas* var.); Harbach, Rattanarithikul, & Harrison 2017 (to sp.). **Ety.:** Sumatra, Indonesia. **Informal name:** Sumatra Indonesian Nail Mosquito.

symesi Edwards
1928a: 269 (F). **Type loc.:** Kisumu, Kenya (NHMUK). **Distr.:** Democratic Republic of the Congo, Kenya, South Sudan, Uganda. **Refs.:** Evans 1938: 77 (M, F*, P*, L*); Gillies & De Meillon 1968: 41 (M, F*, P, L; distr.); Kyalo et al. 2017 (distr.; sub-Saharan Africa). **Ety.:** Cecil Bernard Symes. **Informal name:** Symes Kenyan Nail Mosquito.

tasmaniensis Dobrotworsky
1966: 130 (M*, F*, P*, L*, E*). **Type loc.:** Birralee, Tasmania (CSIR). **Distr.:** Australia. **Refs.:** Lee et al. 1987b: 103 (F; key, tax., bion., distr., review);. **Ety.:** Tasmania. **Informal name:** Tasmania Australian Nail Mosquito.

tenebrosus Dönitz
1902: 53 (F*). **Type loc.:** Wadi [el] Natrun [south of Alexandria], Egypt (ZM). **Distr.:** Angola, Botswana, Burundi, Cameroon, Chad, Democratic Republic of the Congo, Egypt, Ethiopia, Gabon, Israel (and Gaza Strip and West Bank), Jordan, Kenya, Libya, Madagascar (includes Glorioso and Juan de Nova Isl.), Malawi, Mozambique, Oman, Republic of the Congo, Republic of South Africa, Rwanda, Saudi Arabia, Sierra Leone, Somalia, Tanzania (includes Zanzibar), Uganda, Zambia, Zimbabwe. **Refs.:** Kirkpatrick 1925b: 44 (M*, F, P*, L*; as *mauritianus*); de Meillon 1936b (E*); De Meillon 1947b: 47 (F, P, L, E*); Gillies & De Meillon 1968: 41 (F*, P, E*); Ribeiro & da Cunha Ramos 1975: 14 (distr.); Service 1976a (distr.; Gabon); Glick 1992: 129 (distr.); Kyalo et al. 2017 (distr.; sub-Saharan Africa); Robert et al. 2019 (distr.; western

Palearctic). **Ety.:** n.s. [dark, gloomy (L)]. **Informal name:** Gloomy Egyptian Nail Mosquito.
syn. *subtilis* Pressat
1905: 48 (A*). **Type loc.:** Ismailia [Suez Canal], Egypt (NE).

tibiamaculatus (Neiva)
1906: 288 (F; *Myzomyia*). **Type loc.:** Oliveira, Minas Gerais, Brazil (LU). **Distr.:** Argentina, Bolivia, Brazil, Peru. **Refs.:** Galvão & Barretto 1941: 67 (from syn. with *eiseni*); Coutinho & Farias 1942 (M; tax.); Lane 1953: 175 (M*; F*); Belkin et al. 1971: 4 (type info.; tax.). **Ety.:** spotted tibia (L); ref. to apical 0.25 of hind tibia with a very characteristic white spot. **Informal name:** Spot-legged Brazilian Nail Mosquito.

tigertti Scanlon & Peyton
1967: 19 (M*, F, P*, L*). **Type loc.:** Ban Bu Phram, Prachinburi Province, Thailand (USNM). **Distr.:** Thailand. **Refs.:** Rattanarithikul & Harrison 1973: 2 (L*; key); Harrison & Scanlon 1975: 164 (M*, F, P*, L*); Rattanarithikul et al. 2006b (F*, L*; bion., distr., keys). **Ety.:** William A. Tigertt. **Informal name:** Tigertt Thai Nail Mosquito.

umbrosus (Theobald)
1903a: 87 (F*; *Myzorhynchus*). **Type loc.:** [Pekan], Penhang [Pahang], Malaya [Malaysia] (NHMUK). **Distr.:** Bangladesh, Cambodia, India, Indonesia, Malaysia, Philippines, Singapore, Thailand, Vietnam. **Refs.:** Christophers 1933: 162 (M*, F*, P, L*); Crawford 1938: 67 (P*); Ross & Roberts 1943b: 7 (M*, F*, L*; tax., distr., bion.); Colless 1948: 91 (M*, F*, L*; tax.); Reid 1950: 284 (M*, F*, P*, L*); Bonne-Wepster & Swellengrebel 1953: 170 (M*, F*, L*); Nguyen Thuong Hien 1968 (F*, L*; keys, tax., bion., distr.; Vietnam); Reid 1968: 156 (M*, F*, P*, L*, E*); Aslamkhan 1971b (distr.; Pakistan); Harrison & Scanlon 1975: 110 (M*, F*, P*, L*); Lee et al. 1987b: 104 (F; key, tax., bion., distr., review); Rattanarithikul et al. 2006b (F*, L*; bion., distr., keys). **Ety.:** n.s. [shady (L); poss. ref. to darker (wings) appearance in comp. to *barbirostris*]. **Informal name:** Shaded Malaysian Nail Mosquito.
syn. *novumbrosus* Strickland
1916b: 272 (A, L; *Myzorhynchus*). **Type loc.:** Malaya [Malaysia] (LU). **Refs.:** Colless 1948: 91 (syn.).

vanderwulpi Townson & Harbach
2013 (M, F, P, L; mol. tax.). In Townson et al. 2013: 183. **Type loc.:** Krajan Hamlet, Kelurahan, Jambu Subdistrict, Semarang District, Central Java Province, Indonesia (NHMUK). **Distr.:** Indonesia. **Ety.:** Frederik Maurits van der Wulp. **Informal name:** Wulp Indonesian Nail Mosquito.

vanus Walker
1859: 91 (M). **Type loc.:** Makassar, Celebes [Sulawesi] (NHMUK). **Distr.:** Indonesia (includes Sulawesi), Malaysia, Philippines. **Refs.:** Bonne-Wepster & Swellengrebel 1953: 209 (M*, F*, L); Reid 1962: 33 (M, F, P*, L, E*); Reid 1968: 141 (M, P*, L, E*; distr.); Cagampang-Ramos & Darsie 1970 (P*, L*; distr.); Basio 1971b: 38 (M*, F*; bion.); Baisas 1974: 54 (M, F, P, L, E*; tax., bion., distr.; Philippines); Lee et al. 1987b: 105 (F; key, tax., bion., distr., review). **Ety.:** n.s. [empty, void (L)]. **Informal name:** Useless Indonesian Nail Mosquito.

veruslanei Vargas
1979: 87 (F*). **Type loc.:** Quintana Roo, Yucatan, Mexico (ISET). **Distr.:** Mexico. **Refs.:** Vargas 1979b (tax.). **Ety.:** true *lanei* (L) [for C. John Lane]. **Informal name:** Lane Mexican Nail Mosquito.

vestitipennis Dyar & Knab
1906a: 136 (A). **Type loc.:** Cacao Trece Aguas, Alta Vera Paz, Guatemala (USNM). **Distr.:** Belize, Colombia, Costa Rica, Cuba, Dominican Republic, El Salvador, Guatemala, Guyana, Haiti, Honduras, Jamaica, Mexico, Nicaragua, Panama, Paraguay, Puerto Rico. **Refs.:** Gerry 1932: 37 (F*); Vargas 1941d: 110 (E*); Pratt 1943: 1717 (L*); Stone, in Russell et al. 1943: 34 (lectotype desig.); Pelaez 1945: 74 (F*; key); Lane 1953: 172 (M*, F*, P*, L*); Stone & Knight 1956b: 280 (type info.); Dodge 1966: 364 (first instar L*; key); Belkin et al. 1970: 31 (M*, F*, P*, L*); Rejmánková et al. 1998 (bion.); Ulloa et al. 2005 (tax., bion.); Beltrán-Aguilar et al. 2011: 702 (M*, F*, P*, L*; tax., keys, distr.). **Ety.:** n.s. [clothed wing (L)]. **Informal name:** Cloth-winged Guatemalan Nail Mosquito.

vietnamensis Nguyen, Tran, & Nguyen
1993: 215 (M*, F*, P*, L*). **Type loc.:** Muong Va Commune, Song Ma District, Son La Province, Vietnam (NIMPE). **Distr.:** Vietnam. **Ety.:** Vietnam. **Informal name:** Hung Vietnamese Nail Mosquito [the Hong Bang dynasty of the Hung kings is considered the earliest Vietnamese state].

walkeri Theobald
1901a: 199 (F*). **Type loc.:** Lake Simcoe, Ontario, Canada (NHMUK). **Distr.:** Canada, Mexico, United States (continental). **Refs.:** Hurlbut 1938a: 149 (L*); Hurlbut 1938b: 521 (E*); Roth 1944: 105 (M*), 107 (M; keys); Pelaez 1945: 74 (F*; key); Penn 1949a: 52 (P*); Yamaguti & LaCasse 1950b: 20 (M*, F*, L*); Carpenter & LaCasse 1955: 53 (M*, F*, L*; keys); Dodge 1966: 364 (first instar L*; key); Belkin 1968b: 11 (lectotype desig.); Kitzmiller et al. 1974: 22 (chrom.); Porter & Collins 1996 (mol. phy.; Nearctic); Darsie & Ward 2005 (F*, L*; keys, distr.); Beltrán-Aguilar et al. 2011: 707 (M*, F*, P*, L*; tax., keys, distr.); Harrison et al. 2016 (F*, L*; keys, distr.; notes 16, 17). **Ety.:** Edmund Murton Walker. **Informal name:** Walker Canadian Nail Mosquito.

wejchoochotei Taai & Harbach
2015: 259 (M, F, P, L; bion., gen., phyl.). **Type loc.:** Ban Nong Chom, San Sai District, Chiang Mai Province,

Thailand (NHMUK). **Distr.:** Thailand. **Refs.:** Suwannamit et al. 2009 (cross mating); Thongsahuan et al. 2009 (as *campestris*-like; chrom.). **Ety.:** Wej Choochote. **Informal name:** Wej Choochote Thai Nail Mosquito.

wellingtonianus Alcock
1912: 1 (F*). **Type loc.:** Larut Hills, Malaya [Malaysia] (NHMUK). **Distr.:** Malaysia. **Refs.:** Bonne-Wepster & Swellengrebel 1953: 146 (M*, F*, L*). **Ety.:** Arthur Robarts Wellington. **Informal name:** Wellington Malaysian Nail Mosquito.

whartoni Reid
1963a: 106 (M, F*, P*, L, E). **Type loc.:** Singgora, Maran, Pahang, Malaya [Malaysia] (NHMUK). **Distr.:** Cambodia, Malaysia, Thailand. **Refs.:** Rattanarithikul & Harrison 1973: 2 (L*; distr.); Harrison & Klein 1975 (distr.); Harrison & Scanlon 1975: 122 (M*, F*, P*, L*; distr.); Rattanarithikul et al. 2006b (F*, L*; bion., distr., keys). **Ety.:** Ronald Harry Wharton. **Informal name:** Wharton Malaysian Nail Mosquito.

xelajuensis de León
1938: 421 (M*). **Type loc.:** Cerro Quemado, near Xelajú, Quetzaltenango, Guatemala (ESPG; not known, depository nonextant). **Distr.:** Guatemala, Mexico. **Refs.:** Vargas 1942c: 169 (M*, F, L*); Vargas 1943: 69 (P*, L*); Pelaez 1945: 74 (F*; key); Lane 1953: 171 (M*, F, P, L*; in part); Vargas & Martínez Palacios 1956: 116 (M*, F*, L*); Zavortink 1970b: 15 (M*, F*, P*, L*). **Ety.:** Xelajú (Quezaltenango, Guatemala). **Informal name:** Xelajú Guatemalan Nail Mosquito.

xui Dong, Zhou, Dong & Mao
2007: 37 (M*, F*, P*, L*, E*). **Type loc.:** Yunnan Province, China [People's Republic of China] (Yunnan Institute of Parasitic Diseases). **Distr.:** People's Republic of China. **Ety.:** Xu Jin-jiang. **Informal name:** Xu Chinese Nail Mosquito.

ziemanni Grünberg
1902: 550 (F). **Type loc.:** Wuri, Cameroon (ZM). **Distr.:** Angola, Benin, Botswana, Burkina Faso, Burundi, Cameroon, Central African Republic, Chad, Côte d'Ivoire, Democratic Republic of the Congo, Egypt, Ethiopia, Gabon, The Gambia, Ghana, Guinea, Guinea-Bissau, Kenya, Liberia, Libya, Malawi, Mali, Mauritania, Morocco, Mozambique, Namibia, Niger, Nigeria, Republic of the Congo, Republic of South Africa, Rwanda, Senegal, Sierra Leone, South Sudan, Sudan, Tanzania (includes Zanzibar), Togo, Tunisia, Uganda, Zambia, Zimbabwe. **Refs.:** Evans 1938: 70 (F*, P, L, E*); De Meillon 1947b: 48 (F*, P, L, E*); Fox 1958: 216 (tax.); Gillies & De Meillon 1968: 42 (F*, E*); Service 1976a (distr.; Gabon); Trari et al. 2002: 331 (distr.; Morocco); Kyalo et al. 2017 (distr.; sub-Saharan Africa); Robert et al. 2019 (distr., western Palearctic). **Ety.:** Hans Richard Paul Ziemann. **Informal name:** Ziemann Cameroonian Nail Mosquito.

Subgenus *Baimaia* Harbach, Rattanarithikul, & Harrison

2005: 752 (M*, F, P*, L; bion.) (*Anopheles* subg.). Type species: *Anopheles kyondawensis* Abraham.

Name derivation: Visut Baimai.

Important references: Harrison & Scanlon 1975: 179 (L*; key); Harbach & Kitching 2005 (phyl.); Rattanarithikul et al. 2006b (F*, L*; keys, bion., distr.; Thailand); Harbach & Kitching 2016 (phyl.).

kyondawensis Abraham
1947: (L*). **Type loc.:** Kyondaw, Burma [Myanmar] (NHMUK). **Distr.:** Myanmar, Thailand. **Refs.:** Rattanarithikul & Harrison 1973: 1 (L*); Harrison & Scanlon 1975: 179 (L*; distr.); Harbach et al. 2005 (transfer to *Baimaia*); Rattanarithikul et al. 2006b (F*, L*; bion., distr., keys). **Ety.:** Kyondaw Chaung, Burma [Myanmar]. **Informal name:** Kyondaw Burmese Nail Mosquito.

Subgenus *Cellia* Theobald

1902b: 181, 1903a: 107 (as genus). Type species: *Anopheles pharoensis* Theobald (June 16). Stone et al. 1959 (to subg.).

Name derivation: A. (Angelo) Celli of the University of Rome.

syn. *Grassia* Theobald 1902b: 181 (as genus) (not Fisch 1885). Type species: *Anopheles rossi* Giles.
syn. *Howardia* Theobald 1902b: 181 (as genus) (not Berlese & Leonardi 1896, not Dalla Torre 1897). Type species: *Anopheles costalis* Loew.
syn. *Myzomyia* Blanchard 1902: 795 (as genus). Type species: *Anopheles rossi* Giles (July 4).
syn. *Pyretophorus* Blanchard 1902: 795 (as genus). Type species: *Anopheles costalis* Loew (July 4).
syn. *Aldrichia* Theobald 1903a: 353 (as genus) (not Coquillett 1894, not Vaughan 1900). Type species: *Aldrichia error* Theobald.
syn. *Feltinella* Theobald 1907: 56 (as genus). Type species: *Feltinella pallidopalpi* Theobald.
syn. *Neocellia* Theobald 1907: 111 (as genus). Type species: *Neocellia indica* Theobald (James 1910b: 101).
syn. *Pseudomyzomyia* Theobald 1907: (errata and addenda to p. 74; error for p. 47) (as genus). Type species: *Anopheles rossi* Giles.
syn. *Calvertia* Ludlow 1908b: 10, 1909a: 22 (as genus; not Bourguignat 1880, not Warren 1908). Type species: *Chagasia lineata* Ludlow.
syn. *Calvertina* Ludlow 1909c: 234 (new name for *Calvertia* Ludlow 1908b). Type species: *Chagasia lineata* Ludlow.

syn. *Nyssomyzomyia* James 1910a: 37, 1910b: 101 (as genus). Type species: *Anopheles rossi* Giles.
syn. *Christophersia* James 1910a: 33, 1910b: 103 (as genus). Type species: *Christophersia halli* James.
syn. *Neomyzomyia* Theobald 1910c: 29 (as genus). Type species: *Myzomyia elegans* James.
syn. *Aldrichinella* Theobald 1910c: 3, 77 (as genus). Type species: *Aldrichia error* Theobald.
syn. *Dactylomyia* Newstead & Carter 1910: 377 (as genus) (not Aldrich 1893). Type species: *Dactylomyia ceylonica* Newstead & Carter.
syn. *Calviana* Strand 1928: 48 (new name for *Calvertina* Ludlow 1909c). Type species: *Chagasia lineata* Ludlow.
syn. *Anopheles*, subgenus *Myzomyia*, group *Paramyzomyia* Christophers & Barraud 1931: 168. Type species: *Anopheles turkhudi* Liston (Christophers 1933: 251).
syn. *Anopheles*, subgenus *Palaeomyzomyia* Swellengrebel & Rodenwaldt 1932: 114. Type species: *Myzomyia parangensis* Ludlow.
syn. *Anopheles*, subgenus *Myzomyia*, group *Eomyzomyia* Evans 1934: 560. Type species: *Myzomyia wilsoni* Evans.

Important references: Baisas 1936b (P*; key, *Myzomyia*; Philippines); Rozeboom & Knight 1946 (F*, L*, P*; keys; Punctulatus Complex); Mattingly & Knight 1956 (Arabia); Colless 1956 (F*, L*; keys, revision; Leucosphyrus Group); Belkin 1962 (tax.; South Pacific); DuBose & Curtin 1965 (keys; Mediterranean area); Grjebine 1966 (Madagascar); Gillies & De Meillon 1968 (F*, L*; keys, review; Ethiopian Region); Nguyen Thuong Hien 1968 (F*, L*; keys, tax., bion., distr.; Vietnam); Reid 1968 (keys, review; Malaysia and Borneo); Cagampang-Ramos & Darsie 1970 (keys; Philippines); Bryan 1974 (morph.; Punctulatus Complex); Gutsevich et al. 1974 (former USSR); Klein 1977b (Cambodia); Mattingly 1977 (tax.; Gambiae Complex); Tanaka et al. 1979 (M*, F*, L*; keys, review; Japan, Korea); Harrison 1980 (Myzomyia Series; Thailand); Lu & Li 1982a (People's Republic of China); Rao 1984 (India); Lee et al. 1987b (F; key, review; Australasian Region); Gillies & Coetzee 1987 (F, L; keys; Afrotropical Region); Das et al. 1990 (keys; India); Darsie & Pradhan 1990, 1991 (Nepal); Peyton 1990: 197 (class.; Leucosphyrus Group); Glick 1992 (F; key; southwestern Asia and Egypt); Rattanarithikul & Panthusiri 1994 (keys, med. imp. spp.; Thailand); Green 1995: 225–237 (chrom.; Series Myzomyia, Neocellia) 225–237; Nagpal & Sharma 1995 (India); Lu et al. 1997 (People's Republic of China); Sallum et al. 2000b (phyl.); Krzywinski et al. 2001a, 2001b (mol. phyl.); Krzywinski et al. 2001a, 2001b (mol. phyl.); Anthony et al. 2002 (Pyretophorus Series); Sallum et al. 2002c (mol. phyl.); Chen et al. 2003 (mol. phyl.); Harbach 2004 (phyl., class.); Harbach & Kitching 2005 (phyl.); Sallum et al. 2005a(tax., six n. sp., Leucosphyrus Group); Sallum et al. 2005b (revision; Leucosphyrus Group); Rattanarithikul et al. 2006b (F*, L*; keys, bion.; Thailand); Sallum et al. 2007 (mol. phyl.; Leucosphyrus Group); Mohanty et al. 2009 (mol. phyl.); Dixit et al. 2010 (mol. phyl.; India); Sathe & Jagtap 2012 (mol. phyl., tax.); Norris & Norris 2015 (mol. phyl.; southern Africa); Neafsey et al. 2015 (mol. phyl., 16 genomes); Harbach & Kitching 2015 (phyl., review); Freitas et al. 2015 (mol. phyl., review); Kyalo et al. 2017 (distr.; Sub-Saharan Africa); Irish et al. 2020 (distr.; Sub-Saharan Africa).

aconitus Dönitz
1902: 70 (F*). **Type loc.:** Kajoe Tanam [north of Padang, Tapanuli], Sumatra and Willem Island, Soekaboemi, Java (ZM). **Distr.:** Bangladesh, Bengal, Bhutan, Cambodia, India (includes Andaman Isl.), Indonesia (Flores, Java, Kalimantan, Moluccas, Sulawesi, Sumatra [includes Ketulauan Riouw Archipelago, Timor, West Papua]), Japan, Laos, Malaysia, Myanmar, Nepal, Pakistan, People's Republic of China (includes Hainan, Hong Kong), Philippines, Sri Lanka, Thailand, Timor, Vietnam. **Refs.:** Dönitz 1902: English translation; Christophers & Barraud 1931: 183 (E*); Christophers 1933: 216 (M*, F*, L*, E); Crawford 1938: 86 (P*); Bonne-Wepster & Swellengrebel 1953: 365 (M*, F*, L*); Nguyen Thuong Hien 1968 (F*, L*; keys, tax., bion.; Vietnam); Reid 1968: 320 (M*, F*, P*, L*, E*; distr.); Aslamkhan 1971b (distr.; Pakistan); Rattanarithikul & Harrison 1973 (L*; key; Thailand); Harrison 1980: 33 (M*, F*, P*, L*; distr.); Lee et al. 1987b: 117 (F; key, tax., bion., distr., review); Darsie & Pradhan 1990, 1991 (F, L; tax., keys, bion., distr.; Nepal); Baimai et al. 1996 (chrom.); Whelan & Hapgood 2000 (bion., distr.; East Timor); Chen et al. 2003 (mol. phyl.); Garros et al. 2005b (phyl.); Rattanarithikul et al. 2006b (F*, L*; bion., distr., keys); Sinka et al. 2011: 89 (bion., review, distr., niche model); Chen et al. 2012 (genetics); Chen et al. 2012 (pop. gen.); Morgan et al. 2011 (distr., phylogeography); Namgay et al. 2018 (bion., distr.; Bhutan). **Ety.:** n.k. [in German (third rib without a black spot toward the base?); poss. ref. to absence of dark spot on R separating sector pale spot from the accessory sector pale spot]. **Informal name:** Clean Indonesian Nail Mosquito. (Vol. 1, p. 326)
syn. *albirostris* Theobald
1903a: 24 (M, F*; *Myzomyia*). **Type loc.:** Malaya [Malaysia] (NHMUK). **Refs.:** Harrison 1980: 38 (lectotype desig.; syn.).
syn. *brahmacharii* Christophers
1912 (F*; *Myzomyia*). In Brahmachari Bahadur 1912: 186. **Type loc.:** Calcutta [West Bengal], India (MSI).

ainshamsi Gad, Harbach, & Harrison
2006 (M*, F, P*, L*). **Type loc.:** Ras Shukeir District, El-Bahr El-Ahmar (Red Sea), Egypt (USNM). **Distr.:** Egypt. **Refs.:** Morgan et al. 2009 (mol. phyl; Neocellia Series). **Ety.:** Ain Shams University. **Informal name:** Ain Shams Egyptian Nail Mosquito.

amharicus Hunt, Wilkerson, & Coetzee
2013 (M, F*, P*, L*). In Coetzee et al., 2013: 264). **Type loc.:** Babo village (7°42'4500N, 36°58'0000E) outside Jimma,

Ethiopia (USNM). **Distr.:** Ethiopia. **Refs.:** Kyalo et al. 2017 (distr.; sub-Saharan Africa). **Ety.:** Amharic (Ethiopian language). **Informal name:** Amharic Ethiopian Nail Mosquito.

amictus Edwards
1921c: 71 (M, F). **Type loc.:** Townsville, Queensland, Australia (NHMUK). **Distr.:** Australia, Indonesia. **Refs.:** Lee & Woodhill 1944: 152 (M*, F*, L*, E*); Colless 1976 (tax.); Lee et al. 1987b: 121 (F; key, tax., bion., distr., review). **Ety.:** unmixed, pure (Gr); contrast to *annulipes*. **Informal name:** Pure Australian Nail Mosquito.

annularis van der Wulp
1884: 249 (F). **Type loc.:** Mount Ardjoeno, Java [Indonesia] (LM). **Distr.:** Afghanistan, Bangladesh, Bhutan, Cambodia, India, Indonesia (Java, Flores, Kalimantan, Moluccas, Sulawesi, Sumatra [includes Ketulauan Riouw Archipelago, Timor]), Malaysia, Myanmar, Nepal, Pakistan, People's Republic of China, Philippines, Sri Lanka, Taiwan, Thailand, Timor, Vietnam. **Refs.:** Christophers 1933: 300 (M*, F*, P, L*, E); Ross & Roberts 1943b: 31 (M*, F*, L*; tax., distr., bion.); Bonne-Wepster & Swellengrebel 1953: 438 (M*, F*, L*); Horsfall 1955: 193 (review); Hara 1959: 110 (F*); Nguyen Thuong Hien 1968 (F*, L*; keys, tax., bion., distr.; Vietnam); Reid 1968: 373 (M*, F*, P*, L*, E*; keys, tax.); Basio 1971b: 38 (M*, F*; bion.); Rattanarithikul & Harrison 1973 (L*; key, distr.; Thailand); Ameen & Talukdar 1974c: 90 (P*); Baisas 1974: 56 (M, F*, P, L*; tax., bion., distr.; Philippines); Dash et al. 1983 (morph., tax.; *s.l.*); Ahmed 1987 (distr.; Bangladesh); Lee et al. 1987b: 129 (F; key, tax., bion., distr., review); Darsie & Pradhan 1990, 1991 (F, L; tax., keys, bion., distr.; Nepal); Suleman et al. 1993 (distr.; Pakistan); Atrie et al. 1999 (*s.l.*; chrom.); Whelan & Hapgood 2000 (bion., distr.; East Timor); Rattanarithikul et al. 2006b (F*, L*; bion., distr., keys); Alam et al. 2007 (*s.l.*; mol. tax); Rueda, Pecor et al. 2008 (distr.); Morgan et al. 2009 (mol. phyl; Neocellia Series); Surendran et al. 2013 (mol. tax.; distr.; Sri Lanka; as *annularis* A); Sinka et al. 2011: 89 (bion., review, distr., niche model); Morgan et al. 2011 (phylogeography, distr.); Namgay et al. 2018 (bion., distr.; Bhutan). **Ety.:** ringed (L); palpi, tarsi. **Informal name:** Ringed Indonesian Nail Mosquito. (Vol. 1, p. 320)
syn. *fuliginosus* Giles
1900a: 161 (F*). **Type loc.:** Calcutta [West Bengal], India (NHMUK). **Refs.:** Christophers & Barraud 1931: 177 (E*); Senevet 1931: 62 (P*); Townsend 1990: 76 (type info.).
syn. *leucopus* Dönitz
1901: 37 (M, F). **Type loc.:** Doerian [Palembang], Sumatra [Indonesia] (ZM).
syn. *jamesii* Liston
1901: 441 (A; not Theobald 1901a). **Type loc.:** Ellichpur [Central Provinces], India (LU).
syn. *lineata* Ludlow
1908a: 50 (A; *Chagasia*?). **Type loc.:** Camp Gregg, Pangasinan, Luzon, Philippines (USNM).
syn. *nagpori* James & Liston
1911: 90 (A; *Nyssorhynchus fuliginosus* var.). **Type loc.:** Nagpur [Central Provinces], India (LU).
syn. *adiei* James & Liston
1911: 89 (A; *Nyssorhynchus fuliginosus* var.). **Type loc.:** Ferozepore and Amritsar, Punjab [Hill States], India (LU). **Refs.:** Christophers 1933: 300 (syn.; melanic variation).

annulatus Haga
1930 (1929) (M, F, L; *longirostris* var.). **Type loc.:** Upper Digoel [Western], New Guinea (NHMUK). **Distr.:** Australia, Indonesia (West Papua). **Refs.:** Bonne-Wepster & Swellengrebel 1953: 311 (M, F, L*; to sp.); Ward 1979: 130 (authorship changed from de Rook (1930) to Haga); Lee et al. 1987b: 129 (F; key, tax., bion., distr., review, lectotype desig.); Townsend 1990: 43 (type info.). **Ety.:** ringed (L); legs. **Informal name:** Ring-legged Papuan Nail Mosquito.

annulipes Walker
1856a: 433 (A). **Type loc.:** Van Diemen's Land [Tasmania], Australia (NHMUK). **Distr.:** Australia, Indonesia (includes West Papua), Philippines, Taiwan. **Refs.:** Bancroft 1908: 14 (M, F; as genus *Nyssorhynchus*; "Spear Mosquito"); Ross & Roberts 1943b: 9 (M*, F*, L*; tax., distr., bion.); Lee & Woodhill 1944: 121 (M*, F*, L*, E*); Lee 1948: 54 (tax.); Peters 1962: 166 (P*); Dobrotworsky 1965: 47 (M*, F*, L*); Booth & Bryan 1986 (cyt., tax.; *s.l.*); Lee et al. 1987b: 136 (F; key, tax., distr., review; *s.l.*); Foley & Bryan 1991 (tax., bion.; *s.l.*); Bryan et al. 1991 (distr., bion., tax.; *s.l.*); Foley et al. 2007a (tax., phyl.); Boyd & Foley 2007 (*s.l.*; distr.); Foley et al. 2007b (mol. phyl.; *s.l.*); Foley et al. 2007c (mol. phyl.; *s.l.*). **Ety.:** ringed feet (L); tarsi. **Informal name:** Ring-footed Australian Nail Mosquito. (Vol. 1, p. 330)
syn. *musivus* Skuse
1889: 1754 (F). **Type loc.:** Elizabeth Bay, near Sidney, Mount Kembla and Illawarra, New South Wales, Australia (ANIC). **Refs.:** Lee et al. 1987b: 136 (tax., type depos., lit. review, distr., bion.).
syn. *mastersi* Skuse
1889: 1757 (M, F). **Type loc.:** Blue Mountains, New South Wales, Australia (ANIC). **Refs.:** Lee et al. 1987b: 136 (tax., lit. review, distr., bion.).
syn. *perplexus* Taylor
1943a: 153 (M*, F*). **Type loc.:** Adelaide River, Northern Territory, Berner Creek, Watten and Cairns, Queensland, Australia (ANIC). **Refs.:** Lee & Woodhill 1944: 121 (syn.); Lee et al. 1987b: 136 (tax., lectotype desig., type depos., lit. review, distr., bion.).
syn. *persimilis* Taylor
1943a: 155 (F*; *Anopheles* (*Myzomyia*) *perplexus* var.). **Type loc.:** Irvinebank, Queensland, Australia (ANIC).

Refs.: Lee & Woodhill 1944: 121 (syn.); Lee et al. 1987b: 136 (tax., type depos., lit. review, distr., bion.).

syn. *derricki* Taylor

1943a: 155 (type M*). **Type loc.:** Irvinebank, Queensland, Australia (ANIC). **Refs.:** Lee & Woodhill 1944: 121 (syn.); Lee et al. 1987b: 136 (tax., lit. review, distr., bion.).

apoci Marsh

1933: 193 (M, F*, L*). **Type loc.:** Southwest Persia [Iran] (NHMUK). **Distr.:** Iran, Iraq. **Refs.:** Pringle 1954: 60 (tax.); Pringle & Garrett-Jones 1957: 90 (tax.); Townsend 1990: 44 (type info.); Robert et al. 2019 (distr.; western Palearctic). **Ety.:** acronym: Anglo-Persian Oil Company. **Informal name:** APOC Iranian Nail Mosquito.

arabiensis Patton

1905: 625 (M*, F*, L, E*). **Type loc.:** Aden Hinterland [West Aden Protectorate = Yemen] (LU). **Distr.:** Angola, Benin, [Brazil], Burkina Faso, Burundi, Cabo Verde, Cameroon, Chad, Democratic Republic of the Congo, Djibouti, Egypt, Eritrea, Eswatini, Ethiopia, The Gambia, Ghana, Guinea, Guinea Bissau, Kenya, La Réunion, Madagascar (includes Glorioso and Juan de Nova Isl.), Malawi, Mali, Mauritania, Mauritius, Mozambique, Namibia, Niger, Nigeria, Republic of South Africa, Rwanda, Saudi Arabia, Senegal, Somalia, South Sudan, Sudan, Tanzania (includes Zanzibar), Togo, Uganda, Yemen, Zambia, Zimbabwe. **Refs.:** White 1975: 311 (distr.; from syn. with *gambiae*); Mattingly 1977: 324 (tax.); Rodhain et al. 1977 (distr.); Sinka et al. 2010a: 117 (bion., review, distr., niche model); Becker et al. 2010: 325 (distr., bion.); Kyalo et al. 2017 (distr.; sub-Saharan Africa); Robert et al. 2019 (distr.; western Palearctic). **Ety.:** Arabia. **Informal name:** Arab Yemeni Nail Mosquito.

syn. *davidsoni* Ribiero, da Cunha Ramos, Pires, & Capela

1979: 75 (M*, F*, P*, L*; *quadriannulatus* ssp.). **Type loc.:** Ribiera do Mangue, Santiago Island, Cape Verde Republic (IHMT). **Refs.:** Cambournac et al. 1982: 266 (to syn. of *arabiensis*).

ardensis (Theobald)

1905g: 17 (M, F*; *Pyretophorus*). **Type loc.:** Durban, Natal [KwaZulu-Natal, Republic of South Africa] (NHMUK). **Distr.:** Angola, Democratic Republic of the Congo, Ethiopia, Kenya, Nigeria, Republic of the Congo, Republic of South Africa, Rwanda,Tanzania, Uganda, Zimbabwe. **Refs.:** Evans 1938: 100 (M*, F*, P*, L*, E*); Gillies & De Meillon 1968: 68 (M*, F*, P, L*, E*; keys); Kyalo et al. 2017 (distr.; sub-Saharan Africa). **Ety.:** n.s. [burning, fiery hot (L)]. **Informal name:** Flaming South African Nail Mosquito.

syn. *pyretophoroides* Theobald

1907: 48 (M*, F*; *Myzomyia*). **Type loc.:** Pretoria [Transvaal, Republic of South Africa] (NHMUK).

argenteolobatus (Gough)

1910: 116 (F; *Cellia*). **Type loc.:** Onderstepoort, Transvaal [Republic of South Africa] (TMP, LSTM). **Distr.:** Angola, Botswana, Burkina Faso, Democratic Republic of the Congo, Republic of South Africa, Senegal, Tanzania, Zambia, Zimbabwe. **Refs.:** Evans 1938: 389 (M*, F*, P*, L*); De Meillon 1947b: 243 (M*, F*, L*, E*); Mattingly 1947: 243 (P*); Vincke & Leleup 1950: 1603 (bion.); Kyalo et al. 2017 (distr.; sub-Saharan Africa). **Ety.:** n.s. [silvery lobes (L); poss. ref. to apex of abdomen]. **Informal name:** Silvery South African Nail Mosquito.

syn. *pseudosquamosa* Newstead & Carter

1911b: 236 (F; *Cellia*). **Type loc.:** Chinyanta's Village, Luombwa River, NE Rhodesia [Zambia] (NHMUK). **Refs.:** Townsend 1990: 124 (type dep.).

aruni Sobti

1968: 481 (M*, F*, E). **Type loc.:** Kaburi Kikombe, Zanzibar [Tanzania], East Africa (NHMUK). **Distr.:** Tanzania (Zanzibar). **Refs.:** De Meillon et al. 1977: 657 (tax. of *aruni*?); Dia et al. 2013 (tax., bion., distr.; Africa); Kyalo et al. 2017 (distr.; sub-Saharan Africa). **Ety.:** Mr. Arun (collector). **Informal name:** Arun Tanzanian Nail Mosquito.

aurirostris (Watson)

1910: 251 (F*; *Myzomyia*). **Type loc.:** Merton Estate, Klang, Malaya [Selangor, Malaysia] (NHMUK). **Distr.:** Malaysia. **Refs.:** Reid 1968: 252 (M*, F*, P*, L*). **Ety.:** n.s. [gold beak (L); prob. ref. to apical half of proboscis golden scaled]. **Informal name:** Gold-beaked Malaysian Nail Mosquito.

austenii (Theobald)

1905b: 102 (F; *Pyretophorus*). **Type loc.:** Bihe, Angola (NHMUK). **Distr.:** Angola, Democratic Republic of the Congo. **Refs.:** De Meillon 1947b: 101 (F*); Lips & Hamon 1955: 872 (M*, F*, P*, L*); Kyalo et al. 2017 (distr.; sub-Saharan Africa). **Ety.:** Edward Ernest Austen. **Informal name:** Austen Angolan Nail Mosquito.

azaniae Bailly-Choumara

1960: 532 (M*, F*, P*, L*). **Type loc.:** British Somaliland [Somalia] (IERT). **Distr.:** Djibouti, Ethiopia, Kenya, Saudi Arabia, Somalia, Yemen. **Refs.:** Gillies & De Meillon 1968: 316 (distr.); Rodhain et al. 1977 (distr.); Ahmed et al. 2011 (distr.; Saudi Arabia); Kyalo et al. 2017 (distr.; sub-Saharan Africa). **Ety.:** Azania (ancient name of NE Africa). **Informal name:** Azania Somalian Nail Mosquito.

azevedoi Ribeiro

1969: 113 (M*, F*, P*, L*, E*). **Type loc.:** Moçâmedes, Angola (NML). **Distr.:** Angola, Namibia, Republic of South Africa. **Refs.:** Kyalo et al. 2017 (distr.; sub-Saharan Africa). **Ety.:** J. Fraga de Azevedo. **Informal name:** Azevedo Angolan Nail Mosquito.

syn. *deaconi* De Meillon & van Eeden

1976: 336 (M, F, P*, L*). **Type loc.:** Voorsorg, Upington, Cape Province, South Africa [Republic of South Africa] (SAIM). **Refs.:** White 1980: 117 (syn.).

baimaii Sallum & Peyton
2005 (M, F, P*, L*). In Sallum et al. 2005a: 175–180. **Type loc.:** Mae Sot, Ban Kariang, Thum Rua, Thailand (USNM). **Distr.:** Bangladesh, India, Myanmar, People's Republic of China, Thailand. **Refs.:** Sallum et al. 2005b: 42 (M*, F*, P*, L*); Rattanarithikul et al. 2006b (F*, L*; bion., distr., keys); Morgan et al. 2011 (phylogeography, distr.). **Ety.:** Visut Baimai. **Informal name:** Baimai Thai Nail Mosquito.

baisasi Colless
1957c: 137 (F, P, L). **Type loc.:** Los Banos [Laguna], Luzon, Philippines (NHMUK). **Distr.:** Philippines. **Refs.:** Cagampang-Ramos & Darsie 1970 (F*, L*; keys); Peyton 1990: 204 (as sp.); Sallum et al. 2005b: 24 (M, F*, P*, L*). **Ety.:** Francisco Edlagan Baisas. **Informal name:** Baisas Philippine Nail Mosquito.

balabacensis Baisas
1936b: 65 (M, F*, P*, L; *leucosphyrus* var.). **Type loc.:** Balabac, Balabac Island [Palawan], Philippines (USNM). **Distr.:** Brunei, Indonesia (Java, Kalimantan, Lombok, Sumba, Sumbawa), Malaysian Borneo (Sabah, Sarawak), Philippines (Culion, Palawan). **Refs.:** Baisas & Ubaldo-Pagayon 1956: 224 (L*, E*); Colless 1956: 43 (M, F*, P*, L*, E); Colless 1957c: 135 (F, L; to sp.); Nguyen Thuong Hien 1968 (F*, L*; keys, tax., bion., distr.; Vietnam); Reid 1968: 291 (M*, F*, P*, L*, E*; keys, tax.); Basio 1971b: 39 (M*, F*; bion.); Mattingly 1971a: Pl. 1 (A*; scutum), Pl. 8 (P paddle*), Pl. 22 (L* terminal segs.); Rajapaksa 1971 (bion.; Sabah); Sallum et al. 2005b: 20 (M, F*, P*, L*; lectotype desig.); Yang 1983 (review; tax., bion.); Sallum et al. 2007: 27 (phyl.); Maekawa et al. 2009 (distr.; Sumbawa); Sinka et al. 2011: 89 (bion., review, distr., niche model); Chua et al. 2017 (simian malaria incrim.; Sabah); Manin et al. 2018 (pop. gen.). **Ety.:** Balabac Isl., Philippines. **Informal name:** Balabac Philippine Nail Mosquito. (Vol. 1, p. 332)

barberellus Evans
1932: 94 (M*, F*, L*). **Type loc.:** Firestone Plantation, Liberia (NHMUK). **Distr.:** Angola, Côte d'Ivoire, Democratic Republic of the Congo, Guinea, Liberia, Nigeria, Republic of the Congo, Sierra Leone. **Refs.:** De Meillon 1947b: 103 (M*, F*, P, L*); Peters 1955a: 89 (E*); Gillies & De Meillon 1968: 98 (M*, F*, P, L*, E*; distr.); Kyalo et al. 2017 (distr.; sub-Saharan Africa). **Ety.:** M.A. Barber. **Informal name:** Barber Liberian Nail Mosquito.

berghei Vincke & Leleup
1949c: 248 (A*, L). **Type loc.:** Kishibi forest, Kanzenze region, Kolwezi Territory, Katanga, Belgian Congo [Democratic Republic of the Congo] (CMT). **Distr.:** Democratic Republic of the Congo. **Refs.:** Gillies & De Meillon 1968: 156 (M, F*); Kyalo et al. 2017 (distr.; sub-Saharan Africa). **Ety.:** L. Van den Berghe. **Informal name:** Berghe Congolese Nail Mosquito.

bervoetsi D'Haenans
1961: 189 (M*, F*, L*). **Type loc.:** Tsakalakuku, Feshi, Leopoldville, Congo [Kinshasa, Republic of the Congo] (CMT). **Distr.:** Central African Republic, Cameroon, Democratic Republic of the Congo, Republic of the Congo. **Refs.:** Wanson & Berteaux 1954: 33 (L*; as *mortiauxi*); Gillies & De Meillon 1968: 4,100 (F*, L*; to sp.); Brunhes et al. 1998a: 397 (to ssp.); Antonio-Nkondjio et al. 2008 (from ssp. of *moucheti*); Kyalo et al. 2017 (distr.; sub-Saharan Africa). **Ety.:** Wilfred Pierre Bervoets. **Informal name:** Bervoets Congolese Nail Mosquito.

brohieri Edwards
1929c: 322 (F; *Anopheles theileri* var.). **Type loc.:** Wa, Gold Coast [Ghana] (NHMUK). **Distr.:** Benin, Burkina Faso, Burundi, Cameroon, Central African Republic, Côte d'Ivoire, Democratic Republic of the Congo, The Gambia, Ghana, Guinea, Mali, Nigeria, Republic of the Congo, Senegal, Sierra Leone, South Sudan, Uganda. **Refs.:** de Meillon 1947b: 167 (F); Mattingly 1949a: 23 (P*?; to sp.); Hamon 1954a: 579 (F*, P*, L*); Adam et al. 1956: 748 (M*, F*; to var. of *hancocki* Edwards); Gillies & De Meillon 1968: 4, 158 (F*, L*; syn.; new); Service 1977: 234 (F*; distr.); Kyalo et al. 2017 (distr.; sub-Saharan Africa). **Ety.:** Samuel Lindsay Brohier. **Informal name:** Brohier Ghanaian Nail Mosquito.

syn. *septentrionalis* Evans
1934: 558 (M, F*, L*; *theileri* var.). **Type loc.:** Arua [Northern Region], Uganda (NHMUK). **Refs.:** Adam, Hamon et al. 1956: 758 (syn.); Stone & Knight 1957c: 199 (type info.).

syn. *masseguini* Hamon
1954a: 575 (F*, P*, L*; *Anopheles hancocki* var.). **Type loc.:** Tiera, Bobo Dioulasso, Haute Volta [Burkina Faso] (IERT). **Refs.:** Adam et al. 1956: 750 (M, F*); Harrison 1973c: 278 (type info.).

syn. *gilroyi* Service
1960c: 87 (M*, F*, P, L*; *Anopheles hancocki* var.). **Type loc.:** Itchi, near Enugu, eastern Nigeria (NHMUK).

brucei Service
1960c: 85 (M*, F*, P, L*). **Type loc.:** Lokoja, northern Nigeria (NHMUK). **Distr.:** The Gambiae, Nigeria. **Refs.:** Dia et al. 2013 (tax., bion., distr.; Africa); Kyalo et al. 2017 (distr.; sub-Saharan Africa). **Ety.:** Leonard Jan Bruce-Chwatt. **Informal name:** Bruce Nigerian Nail Mosquito.

brumpti Hamon & Rickenbach
1955a: 342 (F*). **Type loc.:** Oue, Cercle de Tougan, Haute Volta [Burkina Faso] (IERT). **Distr.:** Burkina Faso. **Ety.:** Alexandre Joseph Émile Brumpt. **Informal name:** Brumpt Burkinabe Nail Mosquito.

brunnipes (Theobald)
1910c: 64 (F*; *Nyssorhynchus*). **Type loc.:** Bihe, Angola (NHMUK). **Distr.:** Angola, Benin, Burkina Faso, Cameroon, Côte d'Ivoire, Democratic Republic of the

Congo, Equatorial Guinea, The Gambia, Ghana, Guinea, Madagascar (includes Glorioso and Juan de Nova Isl.), Mali, Mozambique, Nigeria, Republic of the Congo, Senegal, Seychelles, Sierra Leone, South Sudan, Sudan, Tanzania, Togo, Zambia, Zimbabwe. **Refs.:** Davey 1941: 5 (P*, L*); De Meillon 1947b: 105 (M*, F*, L*); Kyalo et al. 2017 (distr.; sub-Saharan Africa). **Ety.:** n.s. [brown foot (L); poss. ref. to leg color]. **Informal name:** Brown-footed Angolan Nail Mosquito.

buxtoni Service
1958: 176 (M*, P, L). **Type loc.:** Oku, 20 miles north of Kumbo, Bamenda Province, Cameroon (NHMUK). **Distr.:** Cameroon. **Refs.:** Gillies & De Meillon 1968: 70 (M*, P, L*); Townsend 1990: 55 (lectotype desig.); Kyalo et al. 2017 (distr.; sub-Saharan Africa). **Ety.:** Patrick Alfred Buxton. **Informal name:** Buxton Cameroonian Nail Mosquito.

bwambae White
1985: 506 (M*, F*, P*, L*, E*). **Type loc.:** Mongiro Semliki Forest, Bwamba County, Uganda (NHMUK). **Distr.:** Uganda. **Refs.:** Becker et al. 2010: 326 (distr., bion.); Kyalo et al. 2017 (distr.; sub-Saharan Africa). **Ety.:** Bwamba Co., Uganda. **Informal name:** Bwamba Ugandan Nail Mosquito.

cameroni De Meillon & Evans
1935: 91 (F*). **Type loc.:** Fishhoek, near Capetown, Cape Province [Republic of South Africa] (NHMUK). **Distr.:** Republic of South Africa. **Refs.:** De Meillon 1947b: 90 (F*); Coetzee & du Toit 1979 (M*, F*, L*, P*; tax.); Townsend 1990: 57 (type dep.); Kyalo et al. 2017 (distr.; sub-Saharan Africa). **Ety.:** William Alexander James Cameron. **Informal name:** Cameron South African Nail Mosquito.

carnevalei Brunhes, Le Goff & Geoffroy
1999: 552 (F*). **Type loc.:** Tiassalé, Ivory Coast (LTVO). **Distr.:** Cameroon, Côte d'Ivoire, Equatorial Guinea, Gabon. **Refs.:** Cano et al. 2003: 435 (distr.; Equatorial Guinea); Moreno et al. 2006 (L*); Antonio-Nkondjio et al. 2009 (L; bion.); Antonio-Nkondjio & Simard 2013 (bion., mol. tax., review); Kyalo et al. 2017 (distr.; sub-Saharan Africa). **Ety.:** Pierre Jean Louis Carnevale. **Informal name:** Carnevale Ivorian Nail Mosquito.

caroni Adam
1961 (F*). **Type loc.:** Matouridi (cave) "Moyen-Congo," Republic of the Congo (IERT). **Distr.:** Gabon, Republic of the Congo. **Refs.:** Pajot & Adam 1964: 626 (M*, F*, P*, L, E*); Kyalo et al. 2017 (distr.; sub-Saharan Africa). **Ety.:** René Jean Gilbert Caron. **Informal name:** Caron Congolese Nail Mosquito.

carteri Evans & De Meillon
1933: 276 (M, F*, L*; *demeilloni* var.). **Type loc.:** Johannesburg, Transvaal [Republic of South Africa] (NHMUK). **Distr.:** Democratic Republic of the Congo, Lesotho, Republic of South Africa, Zimbabwe. **Refs.:** De Meillon 1947b: 179 (M, F*, P, L*); Gillies & De Meillon 1968: 4, 185 (F*, L; to sp.); Townsend 1990: 58 (type dep.); Kyalo et al. 2017 (distr.; sub-Saharan Africa). **Ety.:** Henry Francis Carter. **Informal name:** Carter South African Nail Mosquito.

christyi (Newstead & Carter)
1911b: 238 (F; *Neocellia*?). **Type loc.:** Uganda (NHMUK). **Distr.:** Burundi, Cameroon, Chad, Democratic Republic of the Congo, Eritrea, Ethiopia, Kenya, Rwanda, Tanzania, Uganda. **Refs.:** Evans 1938: 291 (M*, F*, P, L*, E*); De Meillon 1947b: 205 (M*, F*, L*, E*); Lambrecht & Leleup 1951: 328 (M*, F*); Kyalo et al. 2017 (distr.; sub-Saharan Africa). **Ety.:** Cuthbert Christy. **Informal name:** Christy Ugandan Nail Mosquito.

cinctus (Newstead & Carter)
1910: 381 (F*; *Cellia*). **Type loc.:** Broomassie [Ashanti, Ghana] (NHMUK). **Distr.:** Angola, Burkina Faso, Cameroon, Central African Republic, Côte d'Ivoire, Democratic Republic of the Congo, Equatorial Guinea, Ethiopia, Gabon, Ghana, Guinea, Liberia, Nigeria, Republic of the Congo, Republic of South Africa. **Refs.:** Evans 1938: 126 (M, F*, P, L*); De Meillon 1947b: 63 (M*, F*, L*); Ribeiro & da Cunha Ramos 1975: 16 (distr.); Kyalo et al. 2017 (distr.; sub-Saharan Africa). **Ety.:** girded, belted (L); rings on legs. **Informal name:** Belted Ghanaian Nail Mosquito.

cinereus Theobald
1901a: 161 (F*). **Type loc.:** Salisbury, Mashonaland [Zimbabwe] (NHMUK). **Distr.:** Algeria, Chad, Egypt, Eritrea, Eswatini, Ethiopia, FYRO Macedonia, Georgia, Greece, Guinea-Bissau, Israel (and Gaza Strip and West Bank), Italy (includes Sardinia and Sicily), Jordan, Libya, Lithuania, Kenya, Madeira, Malawi, Morocco, Mozambique, Nambia, Niger, Portugal, Republic of the Congo, Republic of South Africa, Saudi Arabia, Somalia, Spain, Sudan, Tanzania, Tunisia, Uganda, Ukraine, Yemen, Zimbabwe. **Refs.:** Evans 1938: 329 (M*, F*, P*, L*, E*); De Meillon 1947b: 209 (M*, F*, L*, E*); Mattingly & Knight 1956 (tax.); Gillies & De Meillon 1968: 316 (distr.); Ramsdale & Snow 2000: 9 (distr.); Trari et al. 2002: 331 (distr.; Morocco); Ahmed et al. 2011 (distr.; Saudi Arabia); Kyalo et al. 2017 (distr.; sub-Saharan Africa); Robert et al. 2019 (distr.; western Palearctic). **Ety.:** n.s. [ash colored (L); poss. ref. to color]. **Informal name:** Ashy Zimbabwean Nail Mosquito.

syn. *jehafi* Patton
1905: 630 (M*, F*, L*, E*). **Type loc.:** [Jebel Jehaf, D'thalia] Aden Hinterland [Yemen] (NHMUK).

ssp. ***hispaniola*** (Theobald)
1903a: 49 (M, F*; *Myzomyia*). **Type loc.:** Spain and Santa Cruz, Tenerife, Canary Islands (NHMUK). **Distr.:** Algeria, Bulgaria, France, Greece, Italy, Morocco, Spain (includes Canary Isl.), Yemen. **Refs.:** Séguy 1924: 156 (L*,

E*); Senevet 1930: 312 (P*); Sergent 1937: 102 (E*); Romeo Viamonte 1950: 271 (M*, F*, L*); Aitken 1953: 348 (A, P, L, E; tax., bion.); Horsfall 1955: 237 (review); Mattingly & Knight 1956: 130 (tax.); Senevet & Rioux 1960: 530 (? = *cinereus* ssp.); Dahl & White 1978: 392 (as syn. of *cinereus*); Ribeiro et al. 1980b: 132 (to ssp.); Becker et al. 2010: 181 (M*, F*, L*; keys, tax., distr., bion.). **Ety.:** diminutive of Hispania, the Roman name for Spain. **Informal name:** Hispaniola Spanish Nail Mosquito.

ssp. syn. *myzomyifacies* Theobald
1907: 69 (F*; *Pyretophorus*). **Type loc.:** Algeria (NHMUK).

ssp. syn. *italicus* Raffaele
1928: 11 (M*, F*, L*, E*). **Type loc.:** Calbria, Italy (LU). **Refs.:** Aitken 1953: 349 (syn.).

ssp. syn. *rifenus* Baeza Cuéllar
1933: 87 (M*, F*, P*, L*; *Myzomyia*). **Type loc.:** Er Rif [Morocco] (LU). **Refs.:** Romeo Viamonte 1950: 283 (syn.).

clowi Rozeboom & Knight
1946: 114 (F*, P*, L*). **Type loc.:** Humboldt Bay, Hollandia, Western New Guinea (Papua) [Indonesia] (USNM). **Distr.:** Indonesia (West Papua), Papua New Guinea. **Refs.:** Bonne-Wepster & Swellengrebel 1953: 352 (F*, L*); Lee et al. 1987b: 161 (F; key, tax., bion., distr., review); Cooper et al. 2000: 841 (distr.; Papua New Guinea); Beebe et al. 2015 (review). **Ety.:** Richard James Clow. **Informal name:** Clow Indonesian Nail Mosquito.

coluzzii Coetzee & Wilkerson
2013 (M, F*, P*, L*). In Coetzee et al. 2013: 249. **Type loc.:** Soukourani, Niono District (270 km NE of Bamako; 14°13'1399N, 06°03'0283W), Mali (USNM). **Distr.:** Angola, Benin, Burkina Faso, Cameroon, Central African Republic, Chad, Comoros, Côte d'Ivoire, Democratic Republic of the Congo, Equatorial Guinea, Gabon, The Gambia, Ghana, Guinea, Guinea-Bissau, Liberia, Mali, Mauritania, Niger, Nigeria, Republic of the Congo, São Tomé and Príncipe, Senegal, Sierra Leone, Somalia, Togo, Zimbabwe. **Refs.:** Kyalo et al. 2017 (distr.; sub-Saharan Africa). **Ety.:** Mario Coluzzi. **Informal name:** Coluzzi Malian Nail Mosquito.

comorensis Brunhes, Le Goff, & Geoffroy
1997: 178 (F*). **Type loc.:** Sada, Mayotte Island, Comores archipelago (LTVO). **Distr.:** Comoros. **Refs.:** Kyalo et al. 2017 (distr.; sub-Saharan Africa). **Ety.:** Comoros Isl. **Informal name:** Mayotte Comoran Nail Mosquito.

confusus Evans & Leeson
1935: 40 (M, F*, P*, L*, E*). **Type loc.:** Umtali and Sinoia, Southern Rhodesia [Zimbabwe] (NHMUK). **Distr.:** Democratic Republic of the Congo, Ethiopia, Kenya, Mozambique, Republic of South Africa, Tanzania, Zimbabwe. **Refs.:** De Meillon 1947b: 123 (M, F*, P, L*, E*); Gillies 1962: 81 (new); Kyalo et al. 2017 (distr.; sub-Saharan Africa). **Ety.:** n.s. [confused (L)]. **Informal name:** Malaria Zimbabwean Nail Mosquito.

cracens Sallum & Peyton
2005 (M, F, P*, L*). In Sallum et al. 2005a: 165. **Type loc.:** Terengganu, Malaysia [West] (USNM). **Distr.:** Indonesia, Malaysia, Thailand. **Refs.:** Rattanarithikul et al. 2006b (F*, L*; bion., distr., keys); Sallum et al. 2007: 27 (phyl.). **Ety.:** n.s. [neat, graceful (L)]. **Informal name:** Graceful Malaysian Nail Mosquito.

cristatus King & Baisas
1936: 80 (M*, F*, L*). **Type loc.:** Titunod, Kolambugan, Lanao Province, Mindinao, Philippines (USNM). **Distr.:** Philippines. **Refs.:** Colless 1956: 70 (M, F, P, L*); Cagampang-Ramos & Darsie 1970 (F*, L*; keys); Basio 1971b: 39 (M*, F*; bion.); Lee et al. 1987b: 163 (F; key, tax., bion., distr., review); Sallum et al. 2005b: 78 (M*, F*, P*, L*). **Ety.:** n.s. [crested (L)]. **Informal name:** Crested Philippine Nail Mosquito.

cristipalpis Service
1977: 233 (F*). **Type loc.:** Kafanchan, Kagoro Hills of Jos Plateau, Nigeria (NHMUK). **Distr.:** Nigeria. **Refs.:** Kyalo et al. 2017 (distr.; sub-Saharan Africa). **Ety.:** crested feeler (L); palpus. **Informal name:** Plume-lipped Nigerian Nail Mosquito.

culicifacies Giles
1901b: 197 (M, F). **Type loc.:** Hoshangabad [Central Provinces], India (NHMUK). **Distr.:** Afghanistan, Bahrain, Bangladesh, Bhutan, Cambodia, Cyprus, Eritrea, Ethiopia, India, Iran, Iraq, Israel, Jordan, Kazakhstan, Laos, Malaysia, Myanmar, Nepal, Oman, Pakistan, People's Republic of China, Saudi Arabia, Somalia, Sri Lanka, Thailand, Vietnam, Yemen (Socotra). **Refs.:** Christophers 1933: 197 (M*, F*, P, L*, E; tax.); Evans 1938: 172 (M*, F*, P, L, E; as *adenensis*); Ross & Roberts 1943b: 13 (M*, F*, L*; tax., distr., bion.); De Meillon 1947b: 99 (M*, F*, P*, L*; as *adenensis*); Horsfall 1955: 199 (review); Viswanathan, Rao & Halgeri 1955: 371 (bion.); Gillies & De Meillon 1968: 104 (M*, F*, P, L*; tax.; as *adenensis*); Rattanarithikul & Harrison 1973 (L*; key; Thailand); Nguyen Thuong Hien 1968 (F*, L*; keys, tax., bion., distr.; Vietnam); Beidas & Gillies 1980: 172 (E*; as *adenensis*); Harrison 1980: 55 (M*, F*, P*, L*; lectotype desig.); Ahmed 1987 (distr.; Bangladesh); Das 1990 (F*) (as sp. A and B); Darsie & Pradhan 1990, 1991 (F, L; tax., keys, bion., distr.; Nepal); Baimai et al. 1996 (*s.l.*; chomosomes); Kar et al. 1999 (*s.l.*; tax.); Rattanarithikul et al. 2006b (F*, L*; bion., distr., keys); Manonmani et al. 2007 (*s.l.*; mol. tax.); Becker et al. 2010: 341 (F*, L*; key, tax., distr., bion.); Sinka et al. 2011: 89 (Culicifacies Complex; bion., review, distr., niche model); Dev & Sharma 2013 (tax., bion.; Culicifacies Complex.); Kirti 2014a (M*, F*); Sharma, Tyagi et al. 2014 (*s.l.*; tax., distr.); Kyalo et al. 2017 (distr.; sub-Saharan Africa); Kyalo et al. 2017 (distr.; sub-Saharan

Africa; as *adensis*); Namgay et al. 2018 (bion.; distr., Bhutan). **Ety.:** n.s. [gnat face (L); gnat resembling]. **Informal name:** Gnat-faced Indian Nail Mosquito.

syn. *indica* Theobald
1901a: 183 (F). **Type loc.:** Madras, India (NHMUK).

syn. *listoni* Giles
1901b: 197 (M, F). **Type loc.:** Ellichpur [Central Provinces], India (NHMUK).

syn. *punjabensis* James
1911 (A*; *Myzomyia*; *culicifacies* var.). In James & Liston 1911: 72. **Type loc.:** Punjab, India (NHMUK).

syn. *adenensis* Christophers
1924c: 296 (A; *culicifacies* var.). **Type loc.:** Dar al-A'mir, Aden Hinterland [Yemen] (NHMUK). **Refs.:** Mattingly & Knight 1956: 95 (to ssp.); Harrison 1980: 52 (syn.; lectotype desig.).

cydippis De Meillon
1931a: 318 (L*). **Type loc.:** [Richards Bay], Zululand, Angola (LU). **Distr.:** Angola, Botswana, Burkina Faso, Burundi, Cameroon, Central African Republic, Chad, Democratic Republic of the Congo, Ethiopia, Eswatini, Madagascar (includes Glorioso and Juan de Nova Isl.), Mozambique, Republic of the Congo, Republic of South Africa, Rwanda, Tanzania, Uganda, Zambia, Zimbabwe. **Refs.:** Evans 1938: 386 (P, L*, E*); Lips 1962: 108 (to sp.); Gillies & De Meillon 1968: 253 (M, F*, P, L*, E*); Ribeiro & da Cunha Ramos 1975: 28 (distr.); Segerman 1990 (type info.); Kyalo et al. 2017 (distr.; sub-Saharan Africa). **Ety.:** flirtatious Athenian maiden. **Informal name:** Cydippe South African Nail Mosquito.

syn. *entebbiensis* Evans
1938: 388 (L; *Anopheles* (*Myzomyia*) *squamosus* var.). **Type loc.:** Entebbe, Uganda (LSTM). **Refs.:** De Meillon 1947b: 241 (syn.).

dancalicus Corradetti
1939b: 277 (M, F, P, L). **Type loc.:** Dobi Valley, Dancalia [Ethiopia] (ISS, NHMUK). **Distr.:** Djibouti, Eritrea, Ethiopia. **Refs.:** Corradetti 1940a (M*, F*, P*, L*); Edwards 1941: 446 (M, F, P, L); De Meillon 1947b: 231 (M*, F*, L*); Townsend 1990: 65 (type info.); Kyalo et al. 2017 (distr.; sub-Saharan Africa). **Ety.:** Dancalia Region, northeast Africa. **Informal name:** Dancalia Ethiopian Nail Mosquito.

daudi Coluzzi
1958: 69 (F*). **Type loc.:** Gololei Daùd (near Villaggio Duca degli'Abruzzi), Scebeli Valley, Somalia (IMEM, NHMUK). **Distr.:** Somalia. **Refs.:** Gillies & De Meillon 1968: 199 (subg. transfer); Townsend 1990: 65 (type info.); Kyalo et al. 2017 (distr.; sub-Saharan Africa). **Ety.:** Gololei Daùd, Somalia. **Informal name:** Daùd Somalian Nail Mosquito.

deemingi Service
1970a: 647 (F*). **Type loc.:** Mambilla Plateau, near Ngel Nyaki, Nigeria (NHMUK). **Distr.:** Cameroon, Nigeria. **Refs.:** Brunhes et al. 1997: 178 (M*); Brunhes et al. 2003: 179 (M*, P*, L*; tax.); Kyalo et al. 2017 (distr.; sub-Saharan Africa). **Ety.:** John C. Deeming. **Informal name:** Deeming Nigerian Nail Mosquito.

demeilloni Evans
1933: 266 (M, F*). **Type loc.:** Tzaneen, Transvaal [Republic of South Africa] (NHMUK). **Distr.:** Angola, Botswana, Burundi, Cameroon, Côte d'Ivoire, Democratic Republic of the Congo, Eritrea, Ethiopia, Ghana, Kenya, Malawi, Mauritania, Mozambique, Namibia, Republic of the Congo, Republic of South Africa, Rwanda, Somalia, South Sudan, Sudan, Tanzania, Uganda, Yemen, Zambia, Zimbabwe. **Refs.:** Evans 1938: 263 (M*, F*, P*, L*, E*); De Meillon 1947b: 175 (M*, F*, L*, E*); Gillies & De Meillon 1968: 316 (distr.); Ribeiro & da Cunha Ramos 1975: 21 (distr.); Kyalo et al. 2017 (distr.; sub-Saharan Africa). **Ety.:** Botha De Meillon. **Informal name:** De Meillon South African Nail Mosquito.

dirus Peyton & Harrison
1979: 41 (M, F*, P*, L*). **Type loc.:** Ban Bu Phram, Prachin Buri, Thailand (USNM). **Distr.:** Cambodia, India, Indonesia, Laos, Malaysia, Myanmar, Nepal, People's Republic of China, Taiwan, Thailand, Vietnam. **Refs.:** Ahmed 1987 (distr.; Bangladesh); Baimai et al. 1987 (*dirus s.l.*); Baimai et al. 1988a: (*dirus s.l.*); Darsie & Pradhan 1990, 1991 (F, L; tax., keys, bion., distr.; Nepal); Oo et al. 2004 (distr.; Myanmar); Sallum et al. 2005a: 29 (M, F*, P*, L*; distr.); Sallum et al. 2005b (tax.; *s.l.*); Rattanarithikul et al. 2006b (F*, L*; bion., distr., keys); Sallum et al. 2007: 30 (phyl.); Becker et al. 2010: 342 (F*, L*; key, tax., distr., bion.); Sinka et al. 2011: 89 (bion., review, distr., niche model; Dirus Complex); Dev & Sharma 2013 (tax., bion.; Dirus Complex.). **Ety.:** Kitzmiller (1982) reported to a revised species description [not seen] was added "In recognition of the vector efficiency of the new Thailand species, we have chosen the Latin name *dirus* (meaning: dire, direful, fearful, dreadful, ominous) to reflect its role in the transmission of malaria." **Informal name:** Dire Thai Nail Mosquito. (Vol. 1, p. 336)

dispar Rattanarithikul & Harbach
1991: 176 (M*, F, P*, L*). **Type loc.:** Subic Bay, Pastolan, Bataan Province, Luzon, Philippines (USNM). **Distr.:** Philippines. **Ety.:** dissimilar, unequal (L); comp. to *greeni*. **Refs.:** Morgan et al. 2009 (mol. phyl; Neocellia Series). **Informal name:** Unique Philippine Nail Mosquito.

distinctus (Newstead & Carter)
1911b: 234 (A; *Pyretophorus*). **Type loc.:** Luapula River, below Chingola's Village, NE Rhodesia [Zambia] (NHMUK). **Distr.:** Angola, Botswana, Democratic Republic of the Congo, Malawi, Namibia, Tanzania, Zambia, Zimbabwe. **Refs.:** De Meillon 1947b: 111 (M*, F*, P*, L*, E*); Ribeiro & da Cunha Ramos 1975: 20 (distr.); Kyalo et al. 2017 (distr.; sub-Saharan Africa).

Ety.: n.s. [separate, distinct (L)]. **Informal name:** Distinct Zambian Nail Mosquito.

syn. *melanocosta* Newstead & Carter
1911b: 236 (A; *Pyretophorus distinctus* var.). **Type loc.:** Luapula River, below Chingola's Village, NE [Northern] Rhodesia [Zambia] (NHMUK). **Refs.:** Evans 1934: 549 (syn.).

domicolus Edwards
1916: 363 (F*). **Type loc.:** Zungeru [Northern Provinces], Nigeria (NHMUK). **Distr.:** Benin, Burkina Faso, Cameroon, Central African Republic, Côte d'Ivoire, Democratic Republic of the Congo, Ethiopia, Ghana, Guinea, Mali, Mauritania, Niger, Nigeria, Senegal, Sierra Leone, Seychelles, South Sudan, Sudan, Uganda, Zambia, Zimbabwe. **Refs.:** Evans 1938: 214 (F*, L); De Meillon 1947b: 110 (F*, L; ? = *longipalpus*); Adam et al. 1956: 100 (M*, F*, P, E*; to var.); Gillies & De Meillon 1968: 107 (F*, L*; to sp.); White 1980: 118 (unjustified emendation to *domicolus*); Zavortink 1987: 2 (distr.); Kyalo et al. 2017 (distr.; sub-Saharan Africa). **Ety.:** house inhabitant (L). **Informal name:** Domestic Nigerian Nail Mosquito.

syn. *gingeroi* Corradetti & Archetti
1947: 677 (L). **Type loc.:** Gingero, SW Ethiopia (LU).

dravidicus Christophers
1924c: 297 (A; as var.). **Type loc.:** Nilgiri Hills [Madras], India (NHMUK). **Distr.:** Bhutan,India, Laos, Nepal, People's Republic of China, Thailand, Vietnam. **Refs.:** Rattanarithikul & Green 1987: 261 (from syn. of *maculatus*; lectotype desig.); Darsie & Pradhan 1990, 1991 (F, L; tax., keys, bion., distr.; Nepal); Harrison et al. 1991: 223 (distr.); Rattanarithikul et al. 2006b (F*, L*; bion., distr., keys); Qu & Zhu 2008 (distr.; People's Republic of China); Morgan et al. 2009 (mol. phyl; Neocellia Series); Namgay et al. 2018 (bion., distr.; Bhutan). **Ety.:** Dravidians (Sanskrit for early inhabitants of India). **Informal name:** Dravidian Indian Nail Mosquito.

dthali Patton
1905: 627 (M*, F*, L*, E*). **Type loc.:** D'thala, Hardeba, Sulek, and Nobat, Aden Hinterland [Yemen] (LU). **Distr.:** Afghanistan, Algeria, Chad, Djibouti, Egypt, Eritrea, Ethiopia, India, Iran, Iraq, Israel (and Gaza Strip and West Bank), Jordan, Kenya, Lebanon, Mauritania, Mali, Morocco, Niger, Oman, Pakistan, Saudi Arabia, Somalia, Sudan, Syria, Tunisia, Yemen. **Refs.:** Christophers & Puri 1931b; Christophers 1933: 188 (M*, F*, P, L, E); De Meillon 1947b: 107 (M*, F*, L*, E*); Mattingly & Knight 1956: 95 (tax.); Gillies & De Meillon 1968: 316 (distr.); Rodhain et al. 1977 (distr.); Kyalo et al. 2017 (distr.; sub-Saharan Africa). **Ety.:** D'thala (Emirate of Dhala, Federation of South Arabia). **Informal name:** D'thala Yemeni Nail Mosquito.

syn. *wardi* Leeson & Theodor
1948: 222 (M, F*, P, L*; *Anopheles* (*Myzomyia*) var.). **Type loc.:** Socotra [Yemen] (NHMUK). **Refs.:** Mattingly & Knight 1956: 95 (syn.).

dualaensis Brunhes, Le Goff & Geoffroy
1999: 554 (L*). **Type loc.:** Douala, Cameroon (LTVO). **Distr.:** Cameroon. **Refs.:** Kyalo et al. 2017 (distr.; sub-Saharan Africa). **Ety.:** Duals, Cameroon. **Informal name:** Duals Cameroonian Nail Mosquito.

dureni Edwards
1938 (M*, F*). In Evans 1938: 123. **Type loc.:** Kibubu River, Bayaka District, Kwango Province, Belgian Congo [Democratic Republic of the Congo] (CMT). **Distr.:** Angola, Burkina Faso, Côte d'Ivoire, Democratic Republic of the Congo, Republic of the Congo, Rwanda. **Refs.:** De Meillon 1947b: 65 (M*, F*, L*, E*); Vincke & Leleup 1949b: 253 (bion.); Ribeiro & da Cunha Ramos 1975: 16 (distr.); Brunhes et al. 1997: 180 (F*; tax.); Kyalo et al. 2017 (distr.; sub-Saharan Africa). **Ety.:** Albert Nicholas Duren. **Informal name:** Duren Congolese Nail Mosquito.

elegans (James)
1903 (F*; *Myzomyia*?). In Theobald 1903a: 51. **Type loc.:** Karwar, Bombay, India (NHMUK). **Distr.:** India, Solomon Isl., Sri Lanka, Thailand. **Refs.:** Reid 1949: 46 (from syn. with *leucosphyrus*); Colless 1956: 75 (M, F*, P*, E; tax.); Mendis et al. 1983 (P, L; bion.); Townsend 1990: 70 (improper type info.); Sallum et al. 2005a: 180 (M, F, P*, L*; type info, tax.); Sallum et al. 2005b: 49 (M, F*, P*, L*); Sallum et al. 2007: 30 (phyl.). **Ety.:** n.s. [fine, elegant (L)]. **Informal name:** Elegant Indian Nail Mosquito.

eouzani Brunhes, Le Goff & Boussès
2003: 182 (F*, P). **Type loc.:** Bamboutos Mountains, West Cameroun [Cameroon] (LTV). **Distr.:** Cameroon, Gabon. **Refs.:** Kyalo et al. 2017 (distr.; sub-Saharan Africa). **Ety.:** Jean-Pierre Eouzan. **Informal name:** Eouzan Cameroonian Nail Mosquito.

epiroticus Linton & Harbach
2005 (M*, F*, P*, L*, bion., mol. tax.). In Linton et al. 2005: 332. **Type loc.:** Village Van Duc A, An Trach Commune, Gia Rai District, Bac Lieu Province, Thailand (NHMUK). **Distr.:** Cambodia, Laos, Malaysia, Myanmar, Singapore, Thailand, Vietnam. **Refs.:** Nguyen Thuong Hien 1968 (F*, L*; keys, tax., bion.; Vietnam, as *sundaicus*); Dusfour et al. 2004a,b (mol tax., as *sundaicus s.l.*); Rattanarithikul et al. 2006b (F*, L*; bion., distr., keys). **Ety.:** from the mainland (Gr); ref. to "its primary distribution in Southeast Asia." **Informal name:** Mainland Thai Nail Mosquito.

erepens Gillies
1958: 9 (M*, F*, P, L, E*). **Type loc.:** South Pare District, Kihurio, Tanganyika [Tanzania] (NHMUK). **Distr.:** Kenya, Tanzania. **Refs.:** Gillies & De Meillon 1968: 174 (M*, F*, L*, E*; tax.); Gillies & Coetzee 1987: 95 (from ssp. of *wellcomei*); Kyalo et al. 2017 (distr.; sub-Saharan Africa; as *wellcomei* ssp.). **Ety.:** crawl up (L); ref. to larva

on plants above waterline. **Informal name:** Creeping Tanzanian Nail Mosquito.

erythraeus Corradetti
1939c: 287 (L*). **Type loc.:** Ghinda, 3,000 ft, Eritrea (ISS). **Distr.:** Eritrea. **Refs.:** Kyalo et al. 2017 (distr.; sub-Saharan Africa). **Ety.:** Eritrea (L). **Informal name:** Ghinda Eritrean Nail Mosquito.

ethiopicus Gillies & Coetzee
1987: 76 (F*). **Type loc.:** Gambela, Illubabor Province, Ethiopia (NHMUK). **Distr.:** Ethiopia. **Ety.:** Ethiopia. **Informal name:** Menelik Ethiopian Nail Mosquito [Menelik I (c. 950 B.C.) is traditionally considered the first Solomonic emperor of Ethiopia].

faini Leleup
1952: 151 (M*, F*, P*, L). **Type loc.:** Yolohafiri Grotto, Mont Hoyo Kibali-Ituri [Orientale], Belgian Congo [Democratic Republic of the Congo] (CMT). **Distr.:** Democratic Republic of the Congo, Gabon. **Refs.:** Laarman 1959: 147 (M*, F*, P*, L*; as *vanthieli*); Gillies & De Meillon 1968: 56 (M*, F*, P, L*; syn.; *vanthieli*); Kyalo et al. 2017 (distr.; sub-Saharan Africa). **Ety.:** Alexandre Marie Alphonse Jean Fain. **Informal name:** Fain Congolese Nail Mosquito.

syn. *vanthieli* Laarman
1958: 18 (F). **Type loc.:** Irangi, Kivu, Belgian Congo [Democratic Republic of the Congo] (CMT). **Refs.:** Leleup & Zaghi 1960: 925 (to ssp.); Gillies & De Meillon 1968: 56 (syn.).

farauti Laveran
1902b: 908 (F). **Type loc.:** Faureville, Vate [Efate], New Hebrides Islands (PIP). **Distr.:** Australia, Indonesia (includes Borneo, Moluccas, Sulawesi), Malaysia, Papua New Guinea, Solomon Isl., Vanuatu. **Refs.:** Lee & Woodhill 1944: 140 (M, F*, L*, E*; as *moluccensis*); Knight & Farner 1944: 132 (from syn. with *punctulatus*); Rozeboom & Knight 1946: 120 (F*, P*, L*; tax.); Reid 1947: 90 (tax.); Horsfall 1955: 210 (bion.); Belkin 1962: 138 (M*, F*, P*, L*); Bryan & Coluzzi 1971 (chrom.*); Bryan 1973: 64 (tax.); Maffi 1973a: 29 (tax.); Bryan 1974: 413 (tax.); Mahon & Miethke 1982 (*s.l.*; *farauti* 3); Sweeney 1987 (bion.); Lee et al. 1987b: 165 (F; key, tax., bion., distr., review); Foley & Bryan 1993 (mol. tax.; *s.l.*); Foley et al. 1993 (mol. tax.; *s.s.*, *s.l.*); Schmidt et al. 2001: 391 (M, F, P*, L*; tax.; *s.l.*); Bower et al. 2008 (mol. tax.); Cooper et al. 2009 (bion., biomed. imp.; *s.l.*); Becker et al. 2010: 356 (F*, L*; key, tax., distr., bion.; *s.l.*); Sinka et al. 2011: 89 (Farauti Complex; bion. review, distr., niche model); Bangs et al. 2014 (*s.l.*; review, bion., tax.); Beebe et al. 2015 (review). **Ety.:** Gaudens Charles Faraut. **Informal name:** Faraut Ni-Vanuatu Nail Mosquito. (Vol. 1, p. 338)

syn. *moluccensis* Swellengrebel & Swellengrebel de Graaf
1920: 1 (L*; *Nyssorhynchus annulipes* var.), 29 (M, F; *annulipes* var.). **Type loc.:** Several localities in Moluccas and Western New Guinea (Papua) (NE). **Refs.:** Knight & Farner 1944: 132 (syn.).

filipinae Manalang
1930: 258 (F, L*; *aconitus* var.). **Type loc.:** [Bulutong stream at Tungcong Manga, Luzon], Philippines (NE). **Distr.:** Philippines. **Refs.:** Christophers & Puri 1931a: 491 (to sp.); King 1932b: 515 (M*, F*, L*; tax.); Baisas 1936a: 215 (P*); Simmons & Aitken 1942: 134 (bion.); Bohart 1945: 17 (M*, F*); Bonne-Wepster & Swellengrebel 1953: 376 (M*, F*, L*); Cagampang-Ramos & Darsie 1970 (F*, L*; keys); Basio 1971b: 40 (M*, F*; bion.); Baisas 1974: 57 (M, F*, P, L*, E*; tax., bion., distr.; Philippines). **Ety.:** inhabitant of the Philippines (L). **Informal name:** Tagalog Philippine Nail Mosquito [the Tagalog are the dominant ethnicity in the Philippines].

flavicosta Edwards
1911a: 142 (F). **Type loc.:** Baro [Northern Provinces], Nigeria (NHMUK). **Distr.:** Angola, Benin, Burkina Faso, Cameroon, Central African Republic, Côte d'Ivoire, The Gambia, Ghana, Guinea, Kenya, Madagascar (includes Glorioso and Juan de Nova Isl.), Mali, Niger, Nigeria, Republic of South Africa, Senegal, Seychelles, Sierra Leone, South Sudan, Togo. **Refs.:** De Meillon 1947b: 131 (M, F*, L*); Mattingly 1949a: 24 (P*); Hanney 1959: 169 (M*, F*, L, P*; bion.); Coz et al. 1960: 742 (M*, F*, L*; tax.); Service 1960c: 89 (M, F, L, P; tax.); Gillies & De Meillon 1968: 316 (distr.); Ribeiro & da Cunha Ramos 1975: 22 (distr.); Kyalo et al. 2017 (distr.; sub-Saharan Africa). **Ety.:** yellow (L); wing costal margin. **Informal name:** Yellow-winged Nigerian Nail Mosquito.

flavirostris (Ludlow)
1914a: 30 (F; *Myzomyia*). **Type loc.:** Camp Wilhelm, Tayabas [= Quezon (Luzon)], Philippines (USNM). **Distr.:** Indonesia (includes E. Kalimantan, Sumatra Java, Sulawesi, Bali, Lombok, Sumbawa, Sumba, Flores, West Timor), Malaysia (Borneo-Sabah), Philippines, Timor, Vietnam. **Refs.:** King 1932b: 491 (M*, F*, L*; to sp.); Baisas 1936a: 215 (P*); Baisas 1947b (first instar L; as *minimus* ssp.); Gapuz & Santiago 1949: 127 (L*); Mendoza & Abinoja 1952: 53 (bion.); Bonne-Wepster & Swellengrebel 1953: 371 (M*, F*, L*); Stone & Knight 1956b: 277 (type info., lectotype desig.); Baisas 1957: 68 (to sp.); Cagampang-Ramos & Darsie 1970 (F*, L*; keys); Basio 1971b: 43 (M*, F*; bion.; as *minimus* ssp.); Pinhão 1974 (distr.; Timor); Baisas 1974: 59 (M, F*, P, L*, E*; tax., bion., distr.; Philippines); Harrison 1980 (F, L, P; keys, tax., review); Lee et al. 1987b: 196 (F; key, tax., review); Somboon et al. 2000: 476 (support of sp.); Whelan & Hapgood 2000 (bion, distr.; East Timor); Foley et al. 2002 (bion.); Foley et al. 2003 (bion.); Torres et al. 2006 (mol. tax.); Foley & Torres 2006 (pop. gen.); Becker et al. 2010: 343 (F*, L*; key, tax., distr., bion.); Sinka et al. 2011: 89 (bion., review, distr., niche model); Sinka et al. 2011

(bion., review, niche model; Asia-Pacific Region). **Ety.:** yellow beak (L); proboscis. **Informal name:** Yellow-beaked Philippine Nail Mosquito. (Vol. 1, p. 340)

syn. *merak* (*cohesia*) Mangkoewinoto

1919: 57 (A, L*; *Myzomyia aconita* var.). **Type loc.:** Kali Madaksa in Merak (Bantam), West Java (NE). **Refs.:** Harrison 1980: 40 (syn.; transfer from *minimus*).

fluviatilis James

1902: 31 (A*, L*). **Type loc.:** Duars (Jalpaiguri District) [West Bengal], Nagpur [Central Provinces] and the Jeypore Hill Tracts, India (LU). **Distr.:** Afghanistan, Bahrain, Bangladesh, Bhutan, India, Iran, Iraq, Israel, Kazakhstan, Myanmar, Nepal, Oman, Pakistan, People's Republic of China (includes Hong Kong), Saudi Arabia, Sri Lanka, Taiwan, Vietnam, Yemen. **Refs.:** Christophers 1933: 203 (M*, F*, P, L*, E); Ross & Roberts 1943b: 17 (M*, F*, L*; tax., distr., bion.); Horsfall 1955: 214 (bion.); Bhombore et al. 1956 (bion.); Nguyen Thuong Hien 1968 (F*, L*; keys, tax., bion.; Vietnam); Maffi 1971: 449 (L*; distr.); Chowdaiah & Seetharam 1975: 495 (chrom.*); Harrison et al. 1991: 223 (distr.); Glick 1992 (F*; key; Southwest Asia); Oo et al. 2004 (distr.; Myanmar); Singh et al. 2004 (mol. tax.); Garros et al. 2005b (mol. tax., morph.; Funestus and Minimus Groups); Chen et al. 2006 (mol. tax., syst., distr.; *fluviatilis* X, Y); Becker et al. 2010: 344 (F*, L*; key, tax., distr., bion.); Sinka et al. 2011: 89 (bion., review, distr., niche model; Fluviatilis Complex); Dev & Sharma 2013 (tax., bion.; Fluviatilis Complex.); Kumar et al. 2013 (mol. tax.); Namgay et al. 2018 (bion., distr.; Bhutan). **Ety.:** n.s. [of rivers (L); larval hab. streams]. **Informal name:** Riverine Indian Nail Mosquito. (Vol. 1, p. 342)

syn. *listonii* Liston

1901: 365 (A*; not Giles 1901b). **Type loc.:** Ellichpur [Amaraoti District, Central Provinces], India (NHMUK). **Refs.:** Christophers 1911a: 66 (E*).

syn. *leptomeres* Theobald

1903a: 38 (F; *Myzomyia*). **Type loc.:** India (NHMUK).

syn. *arabica* Christophers & Chand

1915: 189 (M, F, L; *funestus* var.). **Type loc.:** Muscat [Oman] (NHMUK). **Refs.:** Mattingly & Knight 1956: 92 (syn.); Townsend 1990: 45 (type dep.).

fontenillei Barrón, Paupy, Rahola, Akone-Ella, Ngangue, Wilson-Bahun, Pombi, Kengne, Costantini, Simard, González, & Ayala

2019 (F; chromosomes, mol. phyl.). **Type loc.:** La Lopé National Park, Gabon. **Distr.:** Gabon. **Ety.:** Didier Fontenille. **Informal name:** Fontenille Gabonian Nail Mosquito.

fontinalis Gillies & De Meillon

1968: 115 (F*). **Type loc.:** Zesfontein, Kaokoveld, South-west Africa [Namibia] (NHMUK). **Distr.:** Namibia, Republic of South Africa. **Refs.:** Kyalo et al. 2017 (distr.; sub-Saharan Africa). **Ety.:** n.s. [fountain (L); type loc. a mountain spring]. **Informal name:** Fountain Namibian Nail Mosquito.

freetownensis Evans

1925c: 461 (M, F*; *marshalli* var.). **Type loc.:** Kissy Bridge, Freetown, Sierra Leone (NHMUK). **Distr.:** Burkina Faso, Cameroon, Central African Republic, Côte d'Ivoire, Ghana, Guinea, Mauritania, Nigeria, Republic of the Congo, Sierra Leone, Senegal. **Refs.:** Evans 1929b: 422 (M*; as *marshalli* var.); De Meillon 1947b: 179 (M*, F*, P, L*); Kyalo et al. 2017 (distr.; sub-Saharan Africa). **Ety.:** Freetown (Sierra Leone). **Informal name:** Freetown Sierra Leonean Nail Mosquito.

funestus Giles

1900b: 50 (M, F*). **Type loc.:** Freetown [Sierra Leone] (NHMUK). **Distr.:** Angola, Benin, Botswana, Burkina Faso, Burundi, Cameroon, Central African Republic, Chad, Comoros, Côte d'Ivoire, Democratic Republic of the Congo, Equatorial Guinea, Eritrea, Eswatini, Ethiopia, Gabon, The Gambia, Ghana, Guinea, Guinea-Bissau, Kenya, Liberia, Madagascar (includes Glorioso and Juan de Nova Isl.), Malawi, Mali, Mauritania, Mayotte, Morocco, Mozambique, Namibia, Niger, Nigeria, Republic of the Congo, Republic of South Africa, Rwanda, São Tomé and Príncipe, Senegal, Sierra Leone, Somalia, South Sudan, Sudan, Tanzania (includes Zanzibar), Togo, Uganda, Zambia, Zimbabwe. **Refs.:** Macfie & Ingram 1922 (F*); Ross & Roberts 1943b: 15 (M*, F*, L*; tax., distr., bion.); De Meillon 1947b: 114 (M*, F*, P*, L*, E*); Doucet 1951b: 68 (A; as *imerinensis*); Service 1960b: 77 (M*, F*, E*; tax.); Gillies & De Meillon 1968: 131 (distr.); De Meillon et al. 1977 (tax.); Green & Hunt 1980 (chrom.); Gillies & Coetzee 1987 (F, L; keys, bion., distr.); Garros et al. 2005a (mol. & morph. tax.; *s.l.*); Temu et al. 2007 (tax.; *s.l.*); Koekemoer et al. 2009 (mol. tax.; *s.l.*); Sinka et al. 2010a: 117 (bion., review, distr., niche model); Becker et al. 2010: 322 (F*, L*; key, tax., distr., bion.); Dia et al. 2013 (tax., bion.; Funestus Group); Vezenegho et al. 2013 (tax.; *s.l.*); Kyalo et al. 2017 (distr.; sub-Saharan Africa; as *s.l.*). **Ety.:** n.s. [fatal, causing death (L)]. **Informal name:** Deadly Sierra Leonean Nail Mosquito. (Vol. 1, p. 344)

syn. *kumasii* Chalmers

1900: 1262 (M, F). **Type loc.:** Kumasi, Ashanti, Gold Coast [Ghana] (NE).

syn. *hebes* Dönitz

1902: 84 (A*). **Type loc.:** Dar-es-Salaam [Tanzania] (ZM).

syn. *umbrosa* Theobald

1903a: 34 (A; *Myzomyia funesta* var.). **Type loc.:** Gambia (NHMUK). **Refs.:** Evans 1938: 145 (syn.); Townsend 1990: 145 (type info.).

syn. *subumbrosa* Theobald

1903a: 34 (F*; *Anopheles funestus* var.). **Type loc.:** Gambia (NHMUK). **Refs.:** Townsend 1990: 138 (type dep.).

syn. *anisochloros* Theobald
1903c: pl. VI, Fig. 4 (A*; *funestus* var.). **Type loc.:** Gambia (NHMUK). **Refs.:** Evans 1938: 145 (syn.); Townsend 1990: 43 (type info.).

syn. *bisignata* Grünberg
1905: 378 (F*; *Myzomyia*). **Type loc.:** Kete Kratje, Togo [Ghana] (ZM). **Refs.:** Evans 1938: 145 (syn.?).

syn. *imerinensis* Monier & Treillard
1935: 572 (F; *Anopheles* (*Myzomyia*) *funestus* var.). **Type loc.:** Tananarive, Madagascar (LU). **Refs.:** De Meillon 1947b: 125 (tax.); White 1980: 118 (syn., from *funestus* var.).

fuscivenosus Leeson
1930: 125 (M, F*). **Type loc.:** Shamva, Southern Rhodesia (NHMUK). **Distr.:** Zimbabwe. **Refs.:** De Meillon 1947b: 123 (M, F); Gillies & De Meillon 1968: 4, 146 (M, F*; to sp.); Ribeiro & da Cunha Ramos 1975: 18 (distr.); Dia et al. 2013 (tax., bion., distr.; Africa); Kyalo et al. 2017 (distr.; sub-Saharan Africa). **Ety.:** dark veins (L); comp. to *funestus*. **Informal name:** Dark-veined Zimbabwean Nail Mosquito.

gabonensis Rahola, Makanga & Paupy
2014 (F*). In Rahola et al. 2014: 630. **Type loc.:** La Lékédi Park (178897S, 1299311E, Alt: 606 m), Gabon (IRD). **Distr.:** Gabon. **Refs.:** Kyalo et al. 2017 (distr.; sub-Saharan Africa). **Ety.:** Gabon. **Informal name:** Orungu Gabonese Nail Mosquito [the Kingdom of Orungu (1700–1927) was a small state in present-day Gabon].

gambiae Giles
1902: 511 (F). **Type loc.:** Gambia Valley, West Africa [Gambia] (NHMUK). **Distr.:** Angola, Benin, Burkina Faso, Burundi, Cameroon, Central African Republic, Chad, Comoros, Côte d'Ivoire, Democratic Republic of the Congo, Equatorial Guinea, Eswatini, Ethiopia, Gabon, The Gambia, Ghana, Guinea, Guinea-Bissau, Kenya, Liberia, Madagascar (includes Glorioso and Juan de Nova Isl.), Malawi, Mali, Mauritius, Mayotte, Morocco, Mozambique, Namibia, Niger, Nigeria, Oman, Republic of South Africa, Republic of the Congo, Rwanda, Saudi Arabia, Senegal, Seychelles, Sierra Leone, Somalia, South Sudan, Sudan, Tanzania (includes Zanzibar), Togo, Uganda, Yemen, Zambia, Zimbabwe. **Refs.:** Theobald 1903c: 2 (as *gambiensis*, lapsus); Macfie & Ingram 1922 (F*; as *costalis* Theobald); Séguy 1924: 161 (L*), as *costalis*; Evans 1938: 299 (M*, F*, P*, L*, E*; bion.); Causey et al. 1943a (bion.; *s.l.*); Deane, L.M. et al. 1946 (F*; key, bion.; NE Brazil & Amazonas; *s.l.*); Causey et al. 1946 (M*; key; NE Brazil & Amazonas; *s.l.*); Deane, M.P. et al. 1946 (L*; key, bion.; NE Brazil & Amazonas; *s.l.*) De Meillon 1947b: 190 (M*, F*, L*, E*; bion.); Horsfall 1955: 223 (review); Coluzzi 1964: 197 (tax.); Chauvet & Dejardin 1968: 69 (L); Gillies & De Meillon 1968: 208 (M*, F*, P*, L*, E*; bion.); Chauvet et al. 1969: 51 (tax.; Gambiae Complex); Service 1970b: 131 (tax.); Green 1971: 421 (tax.); White & Muniss 1972: 793 (tax.); Davidson & Hunt 1973: 121 (Gambiae Complex); Reid 1973: 87 (tax.); White 1973: 249 (cytogenetics, tax.; sp. D); Reid 1975a (tax.); Reid 1975b: 299 (pupal diff. of spp. A and B); White 1975: 313 (tax.); Mahon et al. 1976: 25 (mol. tax.; Gambiae Complex); Gatti et al. 1977: 105 (mol. tax.; Gambiae Complex); Mattingly 1977: 323 (Gambiae Complex); Sinka et al. 2010a: 117 (bion., review, distr., niche model); Becker et al. 2010: 323 (*s.l.*; F*; key, tax., distr., bion.); Becker et al. 2010: 324 (*s.s.*; F*, L*; tax., distr., bion.); Ahmed et al. 2011 (distr.; Saudi Arabia); Lanzaro & Lee 2013 (bion., gen.; Gambiae Complex); Crawford et al. 2016 (mol. tax.; *s.l.*); Kyalo et al. 2017 (distr.; sub-Saharan Africa; as *s.l.* and *s.s.*). **Ety.:** Gambia Valley, West Africa. **Informal name:** Malaria Gambian Nail Mosquito. (Vol. 1, p. 346)

syn. *costalis* Giles
1900b: 49 (M*, F*; not Loew 1866). **Type loc.:** West Coast of Africa (NHMUK). **Refs.:** Mattingly 1977: 323 (tax.).

syn. *gracilis* Dönitz
1902: 76 (M*, F). **Type loc.:** Togo and Kamerun [Cameroon] (NE).

syn. *minor* Holstein
1949: 156 (F). **Type loc.:** Koua, Haute Volta [Burkina Faso] (NE).

syn. *typicus* Holstein
1949: 156 (F; *Anopheles gambiae* ssp.). **Type loc.:** Koua, Haute Volta [Burkina Faso] (NE).

garnhami Edwards
1930a: 292 (M, F*). **Type loc.:** Kericho, Kenya (NHMUK). **Distr.:** Burundi, Democratic Republic of the Congo, Eritrea, Ethiopia, Kenya, Republic of South Africa, Rwanda, Somalia, South Sudan, Tanzania, Uganda, Zimbabwe. **Refs.:** Evans 1938: 278 (M*, F*, P, L*); Corradetti 1940b: 141 (A*, L*, E*); De Meillon 1947b: 183 (M*, F*, L*); Kyalo et al. 2017 (distr.; sub-Saharan Africa). **Ety.:** Percy Cyril Claude Garnham. **Informal name:** Garnham Kenyan Nail Mosquito.

syn. *walshi* Evans & De Meillon
1933: 280 (M*, F*, P*, L*; *Anopheles garnhami* var.). **Type loc.:** Magoeba's Kloof, Tzaneen, Transvaal [Republic of South Africa] (NHMUK). **Refs.:** De Meillon 1947b: 185 (A*, L*; syn. suggested).

ssp. ***basilewskyi*** Leleup
1957: 69 (M, F). **Type loc.:** Mount Meru [Northeast Tanzania] (CMT). **Distr.:** Tanzania. **Refs.:** Leleup 1960: 402 (M*, F*); Gillies & De Meillon 1968: 4 (syn.); Brunhes et al. 1998a: 397 (from syn. of *garnhami*). **Ety.:** P. Basilewsky. **Informal name:** Basilewsky Tanzanian Nail Mosquito.

gibbinsi Evans
1935: 469 (M, F, P, L*, E). **Type loc.:** Fort Portal [Western Region], Uganda (NHMUK). **Distr.:** Burundi, Central African Republic, Democratic Republic of the Congo, Ethiopia, Kenya, Republic of the Congo, Rwanda, South

Sudan, Tanzania, Uganda. **Refs.:** De Meillon 1947b: 149 (M*, F*, P, L*, E*); Peters 1955b: 100 (F*); Gillies & De Meillon 1968: 4, 160 (M*, F*, P, L*, E*; to sp.); Kyalo et al. 2017 (distr.; sub-Saharan Africa; also as *gibbonsi* [*sic*]). **Ety.:** Ernest Gerald Gibbins. **Informal name:** Gibbins Ugandan Nail Mosquito.

syn. *kibena* Peters
1955b: 100 (F*). **Type loc.:** Njombe, Tanganyika [Tanzania] (NHMUK). **Refs.:** Gillies & de Meillon 1968: 160 (syn.).

grassei Grjebine
1953b: 203 (M*, F*, P*, L*). **Type loc.:** Antongil, near Maroantsetra, Tamatave, Madagascar (IRSM). **Distr.:** Madagascar (includes Glorioso and Juan de Nova Isl.). **Refs.:** Gillies & De Meillon 1968: 187 (M*, F*, P, L*, E*; syn.); Kyalo et al. 2017 (distr.; sub-Saharan Africa). **Ety.:** Pierre-Paule Grassé. **Informal name:** Grassé Malagasy Nail Mosquito.

greeni Rattanarithikul & Harbach
1991: 170 (M, F*, P*, L*). **Type loc.:** Wawa, Montalban, Rizal Province, Luzon, Philippines (USNM). **Distr.:** Philippines. **Refs.:** Morgan et al. 2009 (mol. phyl; Neocellia Series). **Ety.:** Christopher A. Green. **Informal name:** Green Philippine Nail Mosquito.

grenieri Grjebine
1964: 38 (L*). **Type loc.:** Ampandroatraka, Anosibe District, Madagascar (IRSM). **Distr.:** Madagascar (includes Glorioso and Juan de Nova Isl.). **Refs.:** Kyalo et al. 2017 (distr.; sub-Saharan Africa). **Ety.:** Paul Fernand Grenier. **Informal name:** Grenier Malagasy Nail Mosquito.

griveaudi Grjebine
1961: 939 (F*). **Type loc.:** Manjakatompy, Ambahara Forest, Madagascar (IRSM). **Distr.:** Madagascar (includes Glorioso and Juan de Nova Isl.). **Refs.:** Kyalo et al. 2017 (distr.; sub-Saharan Africa). **Ety.:** n.s. [prob. Paul Griveaud]. **Informal name:** Griveaud Malagasy Nail Mosquito.

hackeri Edwards
1921c: 70 (F; *leucosphyrus* var.). **Type loc.:** Malaya [Malaysia] (NHMUK). **Distr.:** Indonesia, Malaysia, Philippines, Thailand. **Refs.:** Bonne-Wepster & Swellengrebel 1953: 295 (to sp.); Colless 1956: 78 (M, F*, P*, L*, E); Reid 1968: 300 (M, F*, P*, L*, E; tax.); Rattanarithikul & Harrison 1973 (L*; key; Thailand); Sallum et al. 2005b: 67 (M*, F*, P*, L*; distr.); Rattanarithikul et al. 2006b (F*, L*; bion., distr., keys). **Ety.:** Henry Pollard Hacker. **Informal name:** Hacker Malaysian Nail Mosquito.

hamoni Adam
1962: 154 (M*, F*, P*, L*). **Type loc.:** Maya-N'louari Cave, 3 km SW of Maya, Kindamba, Mayama, Congo [Republic of the Congo] (IERT). **Distr.:** Republic of the Congo. **Refs.:** Gillies & De Meillon 1968: 58 (M*, F*, P, L*); Kyalo et al. 2017 (distr.; sub-Saharan Africa). **Ety.:** Jacques Pierre Jean Hamon. **Informal name:** Hamon Congolese Nail Mosquito.

hancocki Edwards
1929c: 322 (M, F; *theileri* var.). **Type loc.:** Bulambuli, Uganda (NHMUK). **Distr.:** Angola, Burkina Faso, Cameroon, Central African Republic, Congo, Côte d'Ivoire, Democratic Republic of the Congo, Gabon, The Gambia, Ghana, Guinea, Kenya, Liberia, Mali, Nigeria, Senegal, Sierra Leone, Uganda. **Refs.:** Evans 1934: 552 (M; to sp.); De Meillon 1947b: 137 (M*, F*, L*, E*); Kyalo et al. 2017 (distr.; sub-Saharan Africa). **Ety.:** George Leonard Rhys Hancock. **Informal name:** Hancock Ugandan Nail Mosquito.

hargreavesi Evans
1927: 40 (F*; *marshalli* var.). **Type loc.:** Mabang, Sierra Leone (NHMUK). **Distr.:** Benin, Cameroon, Central African Republic, Côte d'Ivoire, Democratic Republic of the Congo, Gabon, Ghana, Guinea, Guinea-Bissau, Liberia, Nigeria, Republic of the Congo, Sierra Leone, Togo, Uganda. **Refs.:** Evans 1929b: 420 (as *marshalli* var.); Evans 1931a: 131 (to sp.); De Meillon 1947b: 133 (M*, F*, P, L*, E*); Mattingly 1947: 243 (P*); Service 1976a (distr.; Gabon); Kyalo et al. 2017 (distr.; sub-Saharan Africa). **Ety.:** Ernest Hargreaves. **Informal name:** Hargreaves Sierra Leonean Nail Mosquito.

harperi Evans
1936: 533 (M, F*, P, L*). **Type loc.:** Kitale [Rift Valley], Kenya (NHMUK). **Distr.:** Djibouti, Eritrea, Kenya, Uganda. **Refs.:** De Meillon 1947b: 135 (M, F*, L*); De Meillon & Hardy 1953: 31 (F); Kyalo et al. 2017 (distr.; sub-Saharan Africa). **Ety.:** J.O. Harper (collector). **Informal name:** Harper Kenyan Nail Mosquito.

harrisoni Harbach & Manguin
2007 (A, P, L; genetics). In Harbach, Garros et al. 2007: 44. **Type loc.:** Village of Khoi, Phu Cuong Commune, Tan Lac District, Hoa Binh Province, Vietnam (NHMUK). **Distr.:** Laos, Myanmar, People's Republic of China, Thailand, Vietnam. **Ety.:** Bruce Arthur Harrison. **Informal name:** Harrison Vietnamese Nail Mosquito.

hervyi Brunhes, Le Goff & Geoffroy
1999: 554 (F*). **Type loc.:** Zinder, Niger (LTVO). **Distr.:** Niger. **Ety.:** John-Paul Hervy. **Informal name:** Hervy Nigerian Nail Mosquito.

hilli Woodhill & Lee
1944: 63 (M, F*, L*, E*). **Type loc.:** Adelaide River, Northern Territory, Australia (CSIR). **Distr.:** Australia, Papua New Guinea. **Refs.:** Bonne-Wepster & Swellengrebel 1953: 329 (M, F*, L*); Lee et al. 1987b: 199 (F; key, tax., bion., distr., review). **Ety.:** n.s. [prob. Gerald Freer Hill]. **Informal name:** Hill Australian Nail Mosquito.

hinesorum Schmidt

2001 (M, F, P*, L*; tax.). In Schmidt et al. 2001: 400. **Type loc.:** Etty Bay (17°34'S, 146°06'E), Queensland, Australia (UQIC). **Distr.:** Australia, Indonesia, Papua New Guinea, Solomon Isl. **Refs.:** Beebe et al. 2015 (review). **Ety.:** Judith and Michael Hines. **Informal name:** Hines Australian Nail Mosquito.

hughi Lambert & Coetzee

1982: 329 (M, F, P*, L*, E*). **Type loc.:** Makonde, Transvaal Province, Republic of South Africa (USNM). **Distr.:** Republic of South Africa. **Refs.:** Kyalo et al. 2017 (distr.; sub-Saharan Africa). **Ety.:** Hugh Patterson. **Informal name:** Hugh South African Nail Mosquito.

incognitus Brug

1931b: 136 (L*). **Type loc.:** Merauke, Western New Guinea (Papua) (LU). **Distr.:** Indonesia (West Papua), Papua New Guinea. **Refs.:** Bonne-Wepster & Swellengrebel 1953: 331 (L*); Lee et al. 1987b: 209 (F; key, tax., bion., distr., review). **Ety.:** n.s. [unknown (L); poss. ref. to "description of a hitherto unknown Anopheline larva" (English translation)]. **Informal name:** Unknown Papuan Nail Mosquito.

indefinitus (Ludlow)

1904c: 299 (F; *Myzomyia rossi* var.). **Type loc.:** Philippines (USNM). **Distr.:** Cambodia, Guam, Indonesia, Laos, Malaysia, Mariana Isl., Nepal, People's Republic of China, Philippines, Sri Lanka, Taiwan, Thailand, Vietnam. **Refs.:** Yamaguti & LaCasse 1950a (M*, F*, L*; keys, tax., bion., distr.; Guam); Bonne-Wepster & Swellengrebel 1953: 421 (M*, F*, L; as *malayensis*); Stone & Knight 1956b: 278 (type info., lectotype desig.); Bohart 1957: 17 (M, F, L; as *subpictus indefinitus*); Reid 1959: 101 (L; as *malayensis*); Reid 1966: 327 (to sp.; syn.); Reid 1968: 334 (F*, P, L*, E*); Pratt & Siren 1971 (distr.); Basio 1971b: 45 (M*, F*; bion.; as *subpictus* var.); Rattanarithikul & Harrison 1973 (L*; key, distr.; Thailand); Baisas 1974: 62 (M, F*, P, L*, E*; tax., bion., distr.; Philippines); Lien et al. 1977 (distr.); Darsie & Pradhan 1990, 1991 (F, L; tax., keys, bion., distr.; Nepal); Harrison et al. 1991: 223 (Thailand.); Rattanarithikul et al. 2006b (F*, L*; bion., distr., keys). **Ety.:** n.s. [uncertain (L); poss. ref. to similarity to two other species]. **Informal name:** Indefinite Philippine Nail Mosquito.

syn. *formosaensis II* Tsuzuki

1902a: 28 (A, L). **Type loc.:** Formosa [Taiwan, Republic of China] (NHMUK). **Refs.:** King 1932a: 328 (? = *subpictus indefinitus* (Ludlow)); Reid 1968: 336 (syn.); Townsend 1990: 74 (type info.).

syn. *trimaculata* Tsuzuki

1907: 549 (M, F; *formosaensis II* var.). **Type loc.:** Formosa [Taiwan, Republic of China] (LU). **Refs.:** Stone et al. 1959: 57 (syn.); Reid 1968: 336 (syn.).

syn. *malayensis* Hacker

1921: 1 (F; *Anopheles subpictus* var.). **Type loc.:** Malaya [Malaysia] (NHMUK). **Refs.:** Reid 1966: 327 (syn.).

introlatus Colless

1957c: 137 (F, P, L). **Type loc.:** 15th mile of Ulu Gombak Road, Selangor, Malaya [Malaysia] (NHMUK). **Distr.:** Indonesia, Malaysia, Thailand. **Refs.:** Reid 1968: 297 (tax.); Rattanarithikul & Harrison 1973 (L; key; Thailand); Hii et al. 1988: 43 (from ssp. of *balabacensis*); Sallum et al. 2005b: 17 (M, F*, P*, L*; distr.); Rattanarithikul et al. 2006b (F*, L*; bion., distr., keys). **Ety.:** n.s. [carried in (L)]. **Informal name:** Scattered Malaysian Nail Mosquito.

irenicus Schmidt

2003 (M, F, P*, L*; keys). In Schmidt et al. 2003: 517. **Type loc.:** Poha River (9°23'S, 159°53'E), Guadalcanal, Solomon Islands (QM). **Distr.:** Solomon Isl. **Refs.:** Beebe et al. 2015 (review). **Ety.:** aiming at peace (Gr); nonhuman biter. **Informal name:** Peaceful Solomon Islander Nail Mosquito.

jamesii Theobald

1901a: 134 (F*). **Type loc.:** Quilon, Travancore, India (NHMUK). **Distr.:** Bangladesh, Bhutan, Cambodia, India, Laos, Malaysia, Myanmar, Nepal, People's Republic of China, Sri Lanka, Thailand, Vietnam. **Refs.:** Davis 1926 (F*); Christophers 1933: 291 (M*, F*, L, E); Jayewickreme 1952: 47 (bion.); Nguyen Thuong Hien 1968 (F*, L*; keys, tax., bion, distr.; Vietnam); Reid 1968: 369 (M*, F*, P*, L*, E*); Aslamkhan 1971b (distr.; Pakistan); Darsie & Pradhan 1990, 1991 (F, L; tax., keys, bion., distr.; Nepal); Oo et al. 2004 (distr.; Myanmar); Rattanarithikul et al. 2006b (F*, L*; bion., distr., keys); Morgan et al. 2009 (mol. phyl; Neocellia Series); Namgay et al. 2018 (bion., distr.; Bhutan). **Ety.:** James [collector?]. **Informal name:** James Indian Nail Mosquito.

jebudensis Froud

1944: 73 (M*, F, P, L*). **Type loc.:** Ijebu-ode, Southern Nigeria (NHMUK). **Distr.:** Angola, Cameroon, Côte d'Ivoire, Democratic Republic of the Congo, Gabon, Nigeria, Republic of the Congo. **Refs.:** Gillies & De Meillon 1968: 60 (M*, F*, P, L*); Kyalo et al. 2017 (distr.; sub-Saharan Africa). **Ety.:** Ijebu-ode, Nigeria. **Informal name:** Ijebu-ode Nigerian Nail Mosquito.

jeyporiensis James

1902: 32 (A*, L*). **Type loc.:** Nagpur, Jeypur State [Jeypore State, Central Provinces], India (LU). **Distr.:** Afghanistan, Bangladesh, Cambodia, India, Laos, Macau, Myanmar, Nepal, Pakistan, People's Republic of China (includes Hainan Isl., Hong Kong), Sri Lanka, Taiwan, Thailand, Vietnam. **Refs.:** Christophers 1933: 220 (M*, F*, P, L*, E); Bonne-Wepster & Swellengrebel 1953: 392 (M*, F*, L; as *candidiensis*); Nguyen Thuong Hien 1968 (F*, L*; keys, tax., bion.; Vietnam); Aslamkhan 1971b (distr.; Pakistan); Harrison & Klein 1975: 10 (distr.); Harrison 1980: 69 (M*, F*, P*, L*; type info.); Darsie & Pradhan 1990, 1991 (F, L; tax., keys, bion., distr.; Nepal); Harrison et al. 1991: 223 (distr.; Thailand); Baimai et al. 1996 (chrom.; as A, B, C, D); Vythilingam et al. 2003 (bion., med. imp., distr.; Laos); Oo et al. 2004 (distr.; Myanmar); Rattanarithikul et al. 2006b (F*, L*; bion., distr., keys); Qu & Zhu 2008 (distr.; People's Republic of China); Rueda, Pecor et al. 2008: 397 (distr.); Morgan et al. 2011 (distr., phylogeography); Manguin 2013 (tax., bion., med. imp., control); Gunathilaka et al. 2015 (distr.; Sri Lanka). **Ety.:** Jeypur State, India. **Informal name:** Jeypur Indian Nail Mosquito. (Vol. 1, p. 348)

syn. *vincenti* Laveran

1901c: 993 (F). **Type loc.:** Van-Linh, Tonkin, Indochina [Hanoi province, Vietnam] (PIP). **Refs.:** Reid 1947: 88 (syn.); Harbach 2018c (senior synonym pending application to the International Commission on Zoological Nomenclature; also see Harbach, Garros et al. 2007).

syn. *jeyporensis* Theobald

1903a: 66 (M*, F; *Pyretophorus*). **Type loc.:** Jeypore Agency, India (NHMUK). **Refs.:** Christophers 1933: 220 (tax.); Harrison 1980: 65 (junior secondary homonym; lectotype desig.).

syn. *candidiensis* Koidzumi

1924: 98 (A). **Type loc.:** Lake Candidius, Formosa [Taiwan, Republic of China] (LU). **Refs.:** Christophers 1933: 225 (to var.); Russell et al. 1943: 116 (ssp.); Chow 1970: 47 (tax.); Harrison 1980: 83 (syn.).

syn. *tonkinensis* Toumanoff

1931: 576 (M, L; *Anopheles* (*Myzomyia*) *aconitus* var.). **Type loc.:** Tonkin, Indochina [Hanoi province, Vietnam] (LU). **Refs.:** Harrison 1980 (syn.).

karwari (James)

1903 (F*; *Nyssorhynchus*). In Theobald 1903a: 102. **Type loc.:** Karwar, Bombay, India (NHMUK). **Distr.:** Bangladesh, Bhutan, Cambodia, India, Indonesia, Laos, Malaysia, Myanmar, Nepal, Papua New Guinea, People's Republic of China, Philippines, Sri Lanka, Thailand, Vietnam. **Refs.:** D'Abrera 1944: 357 (E*); Colless 1948: 104 (M*, F*, L*); Penn 1949b: 14 (P*); Bonne-Wepster & Swellengrebel 1953: 460 (M*, F*, L*); Nguyen Thuong Hien 1968 (F*, L*; keys, tax., bion., distr.; Vietnam); Reid 1968: 360 (M*, F*, P*, L*, E*; tax.); Aslamkhan 1971b (distr.; Pakistan); Basio 1971b: 40 (M*, F*; bion.); Lee et al. 1987b: 213 (F; key, tax., bion., distr., review); Darsie & Pradhan 1990, 1991 (F, L; tax., keys, bion., distr.; Nepal); Oo et al. 2004 (distr.; Myanmar); Rattanarithikul et al. 2006b (F*, L*; bion., distr., keys); Namgay et al. 2018 (bion, distr.; Bhutan). **Ety.:** Karwar, India. **Informal name:** Karwar Indian Nail Mosquito.

syn. *nigrans* Stanton

1912: 7 (new name for *karwari* James 1903; in error). **Refs.:** Christophers 1916b (syn.).

keniensis Evans

1931a: 130 (M*, F*; *marshalli* var.). **Type loc.:** Taveta, Kenya (NHMUK). **Distr.:** Democratic Republic of the Congo, Kenya, Tanzania, Uganda, Zimbabwe. **Refs.:** Evans 1933: 266 (to sp.); De Meillon 1947b: 185 (M*, F*, P, L, E); Gillies 1955: 160 (E*); Kyalo et al. 2017 (distr.; sub-Saharan Africa). **Ety.:** Kenya. **Informal name:** Kikuyu Kenyan Nail Mosquito [the Kikuyu are an indigenous people of Kenya].

kingi Christophers

1923a: 1011 (F*). **Type loc.:** Wakite, Livingstone Range, Tanganyika [Tanzania] (NHMUK). **Distr.:** Democratic Republic of the Congo, Ethiopia, Kenya, Tanzania, Uganda. **Refs.:** Evans 1938: 117 (M, F*, P*, L*); De Meillon 1947b: 71 (M*, F*, L*); Kyalo et al. 2017 (distr.; sub-Saharan Africa). **Ety.:** Harold Holmes King. **Informal name:** King Tanzanian Nail Mosquito.

kochi Dönitz

1901: 36 (M, F). **Type loc.:** Padang [Tapanuli], Sumatra [Indonesia] (ZM, NHMUK). **Distr.:** Bangladesh, Cambodia, India, Indonesia, Laos, Malaysia, Myanmar, Nepal, People's Republic of China, Philippines, Singapore, Thailand, Vietnam. **Refs.:** Christophers 1933: 172 (M*, F*, L*, E); Crawford 1938: 71 (P*); Bonne-Wepster & Swellengrebel 1953: 265 (M*, F*, L); Baisas & Ubaldo-Pagayon 1956: 224 (L*, E*); Reid 1968: 269 (distr.); Aslamkhan 1971b (distr.; Pakistan); Basio 1971b: 40 (M*, F*; bion.); Townsend 1990: 91 (type info.); Darsie & Pradhan 1990, 1991 (F, L; tax., keys, bion., distr.; Nepal); Oo et al. 2004 (distr.; Myanmar); Rattanarithikul et al. 2006b (F*, L*; bion., distr., keys). **Ety.:** n.s. Heinrich Hermann Robert Koch (Harbach 2018c). **Informal name:** Koch Indonesian Nail Mosquito.

syn. *ocellatus* Theobald

1901a: 174 (F*). **Type loc.:** Taipang [*sic* = Taiping], Perak, Malaya [Malaysia] (NHMUK). **Refs.:** Townsend 1990: 114 (unavailable name, Art 11, International Commission on Zoological Nomenclature 1999).

syn. *flava* Ludlow

1908a: 32 (M, F; *Cellia*). **Type loc.:** Camp Wilhelm [Quezon, Luzon], Philippines (USNM). **Refs.:** Stone & Knight 1956b: 277 (type info., lectotype desig.).

syn. *halli* James

1910a: 33 (M, F*; *Christophersia*). **Type loc.:** Sylhet, Assam, India (IM).

kokhani Vythilingam, Jeffery & Harbach

2007 (F*). In Vythilingam et al. 2007: 464. **Type loc.:** Kapit, Sarawak, Malaysian Borneo (NHMUK). **Distr.:** Malaysia. **Ety.:** Ng. Kok Han. **Informal name:** Kok Han Malaysian Nail Mosquito.

kolambuganensis Baisas

1932: 252 (F, L; as *kulambuganensis* [*sic*]). **Type loc.:** Kolambugan [*sic*, as Kulambugan] [Lanaol, Mindanao], Philippines (NE). **Distr.:** Philippines. **Refs.:** Cagampang-Ramos & Darsie 1970 (F*, L*; key); Basio 1971b: 41 (M*, F*; bion.). **Ety.:** Kolambugan, Mindanao, Philippines. **Informal name:** Kolambugan Philippine Nail Mosquito.

koliensis Owen

1945: 236 (M, F*, L*). **Type loc.:** Koli Area, Guadalcanal, Solomon Islands (USNM). **Distr.:** Indonesia (includes Kalimantan, Sulawesi), Papua New Guinea, Solomon Isl. **Refs.:** Penn 1949b: 20 (P*); Bonne-Wepster & Swellengrebel 1953: 349 (M, F*, L*); Belkin 1962: 144 (M*, F*, P*, L*); Bryan 1973: 64 (tax.); Bryan 1974: 413 (tax.); Lee et al. 1987b: 226 (tax., lit. review, distr., bion.); Foley & Bryan 1993 (*s.l.*; mol. tax.); Foley et al. 1993 (mol. tax.); Benet et al. 2004 (mol. tax., bion., distr.; Papua New Guinea; as *s.l.*); Cooper et al. 2009 (bion., biomed. imp.); Sinka et al. 2011: 89 (bion. review, distr., niche model); Beebe et al. 2015 (phyl., control, distr., review). **Ety.:** Koli Dist., Guadalcanal. **Informal name:** Koli Solomon Islander Nail Mosquito. (Vol. 1, p. 350)

kosiensis Coetzee, Segerman, & Hunt

1987: 26 (F, P, L, E*; X chrom.*). **Type loc.:** Kosi Bay, northern Zululand, Republic of South Africa (SAIM). **Distr.:** Republic of South Africa. **Refs.:** Kyalo et al. 2017 (distr.; sub-Saharan Africa). **Ety.:** Kosi Bay, Republic of South Africa. **Informal name:** Kosi South African Nail Mosquito.

lacani Grjebine

1953a: 483 (F*, P*, L*). **Type loc.:** La Mandraka, Manjakandriana District [Tananarive], Madagascar (IRSM). **Distr.:** Madagascar (includes Glorioso and Juan de Nova Isl.). **Refs.:** Kyalo et al. 2017 (distr.; sub-Saharan Africa). **Ety.:** André Marie Maurice Lacan. **Informal name:** Lacan Malagasy Nail Mosquito.

latens Sallum & Peyton

2005. In Sallum et al. 2005: 160 (M, F, P*, L*). In Sallum et al. 2005a. **Type loc.:** Phang Nga, Ban Bang Kaeo, Thailand (USNM). **Distr.:** Indonesia, Malaysia (includes Borneo, Sabah), Thailand. **Refs.:** Rattanarithikul et al. 2006b (F*, L*; bion., distr., keys); Sallum et al. 2007: 30 (phyl.); Sinka et al. 2011: 89 (bion. review, distr., niche model). **Ety.:** concealed, hidden (L); in Leucosphyrus Complex. **Informal name:** Secret Thai Nail Mosquito. (Vol. 1, p. 352)

leesoni Evans

1931b: 546 (M, P, L; *funestus* ssp.). **Type loc.:** Southern Rhodesia [Zimbabwe] (LSTM). **Distr.:** Angola, Benin, Botswana, Burkina Faso, Cameroon, Central African Republic, Côte d'Ivoire, Democratic Republic of the Congo, Equatorial Guinea, Ethiopia, Ghana, Guinea, Kenya, Mali, Mozambique, Nigeria, Republic of the Congo, Republic of South Africa, Senegal, South Sudan, Sudan, Tanzania (includes Zanzibar), Uganda, Zambia, Zimbabwe. **Refs.:** Evans 1938: 168 (M*, F*, P*, L*, E*); De Meillon 1947b: 129 (M*, F*, P*, L*, E*); Ribeiro & da Cunha Ramos 1975: 18 (distr.); Dia et al. 2013 (tax., bion., distr.; Africa); Kyalo et al. 2017 (distr.; sub-Saharan Africa). **Ety.:** Herbert Sefton Leeson. **Informal name:** Leeson Zimbabwean Nail Mosquito.

letabensis Lambert & Coetzee

1982: 328 (M, F, P*, L*, E*). **Type loc.:** Grey Stones, Transvaal Province, Republic of South Africa (USNM). **Distr.:** Mozambique, Tanzania, Republic of South Africa. **Refs.:** Kyalo et al. 2017 (distr.; sub-Saharan Africa). **Ety.:** Letaba Dist., Province of Limpopo, Republic of South Africa. **Informal name:** Letaba South African Nail Mosquito.

leucosphyrus Dönitz

1901: 37 (F). **Type loc.:** Kajoe Janam [*sic* = Kajoe Tanam [Kajutanam] north of Padang [Tapanuli], Sumatra [Indonesia] (ZM, NHMUK). **Distr.:** Indonesia (Sumatra). **Refs.:** Crawford 1938 (P*); Bonne-Wepster & Swellengrebel 1953: 279 (M*, F*, L*); Colless 1956: 55 (M, F*, P*, L*); Colless 1957c: 133 (F, L in part; distr.; Sumatra); Reid 1968: 386 (M*, F*, P*, L*; in part, distr.; Sumatra); Aslamkhan 1971b (distr.; Pakistan); Ramalingam 1974 (distr.); Baimai et al. 1988b (*dirus s.l.*); Peyton 1990: 197 (class.; Leucosphyrus Group); Townsend 1990: 94 (type info.); Sallum et al. 2005b: 6 (M, F*, P*, L*; distr.); Sallum et al. 2007: 30 (phyl.); Sinka et al. 2011: 89 (bion., review, distr., niche model). **Ety.:** white ankle (Gr); white hind tibia and hind tarsomere 1. **Informal name:** White-legged Indonesian Nail Mosquito. (Vol. 1, see *latens* p. 352)

limosus King

1932a: 330 (M*, F*, L*; as var.). **Type loc.:** Balintawak [Rizal], Luzon, Philippines (USNM). **Distr.:** Indonesia, Philippines. **Refs.:** Baisas 1936a: 216 (P*); Colless 1948: 115 (M*, F*, L*; to ssp.); Bonne-Wepster & Swellengrebel 1953: 432 (M*, F*, L*); Cagampang-Ramos & Darsie 1970 (F*, L*; keys); Basio 1971b: 47 (M*, F*; bion.; as *vagus* ssp.); Ramalingam 1974 (as valid sp.); Baisas 1974: 74 (M, F*, P, L*, E*; tax., bion., distr.; Philippines). **Ety.:** muddy or growing in mud (L); larval habitat. **Informal name:** Muddy Philippine Nail Mosquito.

listeri De Meillon

1931b: 482 (F). **Type loc.:** M'singa, Zululand [KwaZulu-Natal, Republic of South Africa] (NHMUK). **Distr.:** Angola, Botswana, Mozambique, Namibia, Republic of South Africa, Zimbabwe. **Refs.:** De Meillon 1936a (tax.); De Meillon 1947a: 35 (P*), 217 (M*, F*, L*, E*); Kyalo et al. 2017 (distr.; sub-Saharan Africa). **Ety.:** Frederick Spencer Lister. **Informal name:** Lister South African Nail Mosquito.

litoralis King

1932a: 311 (M*, F*, L*). **Type loc.:** Paranaque, Rizal, Luzon, Philippines (USNM). **Distr.:** Malaysia, Mariana Isl., Philippines. **Refs.:** Baisas 1936a: 216 (P*); Bonne-Wepster & Swellengrebel 1953: 402 (M*, F*, L*); Cagampang-Ramos & Darsie 1970 (F*, L*; keys); Basio 1971b: 41 (M*, F*; bion.); Ramalingam 1974 (distr.); Baisas 1974: 63 (M*, F*, P, L*; tax., bion., distr.; Philippines); Ward et al. 1976: 99 (distr.). **Ety.:** beach or seashore (L). **Informal name:** Seashore Philippine Nail Mosquito.

lloreti Gil Collado
1936: 313, 318 (F*, P*, L*). **Type loc.:** Rebola, Fernando Po Island (MU). **Distr.:** Equatorial Guinea. **Refs.:** Evans 1938: 214 (F*, P*, L*); Maffi 1962: 63 (tax.); Kyalo et al. 2017 (distr.; sub-Saharan Africa). **Ety.:** Abelardo Lloret Peralt. **Informal name:** Lloret Equatoguinean Nail Mosquito.

longipalpis (Theobald)
1903a: 37 (F; *Myzomyia*). **Type loc.:** Zomba, Nyasaland [Malawi] (NHMUK). **Distr.:** Angola, Botswana, Burkina Faso, Burundi, Cameroon, Democratic Republic of the Congo, Ethiopia, Guinea, Kenya, Malawi, Mozambique, Republic of the Congo, Republic of South Africa, South Sudan, Tanzania (includes Zanzibar), Uganda, Zambia, Zimbabwe. **Refs.:** De Meillon 1947b: 141 (M*, F*, P, L*, E*); Ribeiro & da Cunha Ramos 1975: 22 (distr.); Kyalo et al. 2017 (distr.; sub-Saharan Africa). **Ety.:** long feeler [palpus] (L). **Informal name:** Long-lipped Malawian Nail Mosquito.

longirostris Brug
1928a (F*). **Type loc.:** Boven-Digoel, [Western] New Guinea (NHMUK). **Distr.:** Australia, Indonesia (West Papua), Papua New Guinea. **Refs.:** Bonne-Wepster & Swellengrebel 1953: 305 (M*, F*, L*); Peters 1962: 167 (P*); Lee et al. 1987b: 232 (F; key, tax., bion., distr., review); Beebe et al. 2013 (Longirostris Complex). **Ety.:** long snout (L); proboscis. **Informal name:** Long-beaked Indonesian Nail Mosquito.

lounibosi Gillies & Coetzee
1987: 73 (F, L*). **Type loc.:** Kombenii River, Rabai, Kilifi District, Kenya (NHMUK). **Distr.:** Kenya. **Refs.:** Kyalo et al. 2017 (distr.; sub-Sa
haran Africa). **Ety.:** Leon Philip Lounibos. **Informal name:** Lounibos Kenyan Nail Mosquito.

lovettae Evans
1934: 564 (F*). **Type loc.:** Amani, Tanganyika [Tanzania] (NHMUK). **Distr.:** Tanzania. **Refs.:** De Meillon 1947b: 73 (F*); Gillies & De Meillon 1968: 60 (M*, F*, P, L*); Kyalo et al. 2017 (distr.; sub-Saharan Africa). **Ety.:** Margaret Elizabeth Lovett. **Informal name:** Lovett Tanzanian Nail Mosquito.

ludlowae (Theobald)
1903a: 42 (F*; *Myzomyia*; as *ludlowii*). **Type loc.:** Luzon, Philippines (NHMUK). **Distr.:** India (includes Andaman Isl.), Indonesia (Borneo, Flores, Java, Moluccas, Sulawesi, Sumatra [includes Ketulauan Riouw Archipelago, Timor]), People's Republic of China (includes Hainan), Philippines, Singapore, Sri Lanka, Taiwan, Thailand. **Refs.:** Baisas 1936a: 216 (P*); Urbino 1936: 267 (E*); Bonne-Wepster & Swellengrebel 1953: 404 (M*, F*, L*); Stone 1957a: 333 (emend. to *ludlowae*); Hara 1959: 111 (F*); Baisas & Dowell 1967 (F*, L*; keys; Philippines); Mattingly 1969c (E*); Cagampang-Ramos & Darsie 1970 (F*, L*; keys); Basio 1971b: 41 (M*, F*; bion.); Basio 1971b: 42 (M*, F*; bion.; as *cabrerai*); Baisas 1974: 64 (M*, F*, P*, L*, E*; tax., bion., distr.; Philippines); Baisas 1974: 66 (M, F*, P, L; tax., bion., distr.; Philippines; as *cabrerai*); Darsie & Cagampang-Ramos 1977 (tax.; *cabrerai* to syn. with *ludlowae*); Lee et al. 1987b: 236 (F; key, tax., bion., distr., review); Whelan & Hapgood 2000. **Ety.:** Clara Ludlow. **Informal name:** Ludlow Philippine Nail Mosquito.

syn. *formosensis* Koidzumi
1917: 354 (F; as var.). **Type loc.:** Port Ari, Ako Prefecture, Formosa [Taiwan, Republic of China] (LU).

syn. *hatorii* Koidzumi
1920: 24 (F, L). **Type loc.:** Formosa [Taiwan, Republic of China] (LU). **Refs.:** King 1932a: 322 (syn.).

ssp. ***torakala*** Stoker & Waktoedi Koesoemawinangoen
1949: 58 (M*, F*). **Type loc.:** Torakala, Sulewesi [Sulawesi], Indonesia (LU). **Distr.:** Indonesia (Sulawesi). **Refs.:** Baisas 1974: 65 (M, F, E*; tax., bion., distr.; Philippines); Kitzmiller 1982: 539 (type loc. info.); Harbach & Howard 2007: 40 (to ssp.). **Ety.:** Torakala, South Sulawesi **Informal name:** Torakala Indonesian Nail Mosquito.

syn. *cabrerai* Cagampang-Ramos & Darsie
1969 (F; *ludlowae* ssp.). In Darsie & Cagampang-Ramos 1969: 34. **Type loc.:** Philippines (LU). **Refs.:** Darsie & Cagampang-Ramos 1977: 123 (syn. with *ludlowae*).

lungae Belkin & Schlosser
1944: 269 (M*, F*, L*). **Type loc.:** Tassafaronga Swamp, Guadalcanal, Solomon Islands (USNM). **Distr.:** Papua New Guinea, Solomon Isl. **Refs.:** Belkin et al. 1945: 251 (M, F, P*, L; bion.); Bonne-Wepster & Swellengrebel 1953: 313 (M, F*, L*); Belkin 1962: 147 (M*, F*, P*, L*, E; tax.); Lee et al. 1987b: 240 (F; key, tax., bion., distr., review). **Ety.:** Lunga Dist., Solomon Is. **Informal name:** Lunga Solomon Islander Nail Mosquito.

macarthuri Colless
1956: 67 (M, F*, P*, L*). **Type loc.:** Long Tebangan, Sarawak, Malaysian Borneo (NHMUK). **Distr.:** Indonesia, Malaysia, Thailand. **Refs.:** Hii et al. 1988: 43 (from ssp. *riparius*); Sallum et al. 2005b: 80 (M, F*, P*, L*); Rattanarithikul et al. 2006b (F*, L*; bion., distr., keys); Sallum et al. 2007: 30 (phyl.). **Ety.:** John MacArthur. **Informal name:** MacArthur Malaysian Nail Mosquito.

machardyi Edwards
1930b: 545 (A). **Type loc.:** Amani, Tanganyika [Tanzania] (NHMUK). **Distr.:** Tanzania. **Refs.:** Evans 1938: 107 (M*, F*, P, L*); Gillies 1955: 158 (E*); Kyalo et al. 2017 (distr.; sub-Saharan Africa). **Ety.:** John William MacHardy. **Informal name:** MacHardy Tanzanian Nail Mosquito.

maculatus Theobald
1901a: 171 (M*, F*). **Type loc.:** Hong Kong [People's Republic of China] (NHMUK). **Distr.:** Bangladesh, Cambodia, India, Indonesia (includes Flores, Java, Kalimantan, Moluccas, Sumatra [includes Ketulauan Riouw Archipelago, Sulawesi, Timor]), Japan, Laos,

Lebanon, Malaysia, Myanmar, Nepal, Pakistan, People's Republic of China (includes Hainan, Hong Kong), Philippines, Singapore, Sri Lanka, Taiwan, Thailand, Vietnam. **Refs.:** Christophers & Barraud 1931: 179 (E*); Christophers 1933: 278 (M*, F*, L*, E; tax.); Crawford 1938: 93 (P*); Ho 1938b: 402 (M*, F*, P*); Ross & Roberts 1943b: 33 (M*, F*, L*; tax., distr., bion.); Bonne-Wepster & Swellengrebel 1953: 454 (M*, F*, L*); Hara 1959: 111 (F*); Reid 1968: 353 (M*, F*, P*, L*, E*; tax.); Aslamkhan 1971b (distr.; Pakistan); Basio 1971b: 42 (M*, F*; bion.); Narang et al. 1973b: 13 (chrom.*); Baisas 1974: 68 (M, F*, P, L*; tax., bion., distr.; Philippines); Green et al. 1985a (chrom.; *maculatus s.l.*); Rattanarithikul & Green 1987: 252 (lectotype desig., distr.); Darsie & Pradhan 1990, 1991 (F, L; tax., keys, bion., distr.; Nepal); Suleman et al. 1993 (distr.; Pakistan); Whelan & Hapgood 2000 (bion., distr.; East Timor); Oo et al. 2004 (distr.; Myanmar); Rattanarithikul et al. 2006b (F*, L*; bion., distr., keys); Becker et al. 2010: 345 (F*, L*; key, tax., distr., bion.); Sinka et al. 2011: 89 (bion. review, distr., niche model; Maculatus Group); Singh et al. 2012 (mol. tax.; *s.l.*). **Ety.:** spotted (L); wings. **Informal name:** Spotted Chinese Nail Mosquito. (Vol. 1, p. 354)

syn. *hanabusai* Yamada
1925: 471 (M, F; *Myzomyia*). **Type loc.:** Kagi, Formosa [Taiwan, Republic of China] (IID).

maculipalpis Giles
1902: 297 (M*). **Type loc.:** Salisbury, Mashonaland [Southern Rhodesia] (LU). **Distr.:** Angola, Benin, Botswana, Burkina Faso, Burundi, Cameroon, Central African Republic, Comoros, Côte d'Ivoire, Democratic Republic of the Congo, Eswatini, Ethiopia, Gabon, The Gambia, Ghana, Guinea, Guinea-Bissau, Kenya, Madagascar (includes Glorioso and Juan de Nova Isl.), Malawi, Mali, Mauritius, Mayotte, Mozambique, Namibia, Niger, Nigeria, Republic of South Africa, Rwanda, Senegal, South Sudan, Sudan, Tanzania (includes Zanzibar), Togo, Uganda, Zambia, Zimbabwe. **Refs.:** Evans 1938: 365 (M*, F*, P, L*); De Meillon 1947b: 227 (M*, F*, L*); Gillies & De Meillon 1968: 316 (distr.); Lee et al. 1987b: 243 (F; key, tax., bion., distr., review); Morgan et al. 2009 (mol. Phyl.; Neocellia Series); Kyalo et al. 2017 (distr.; sub-Saharan Africa). **Ety.:** spotted feeler (L); palpus. **Informal name:** Spot-lipped Zimbabwean Nail Mosquito.

majidi Young & Majid
1928: 469 (M, F*, L*; *karwari* var.). **Type loc.:** Coorg, India (MSI, NHMUK). **Distr.:** Bangladesh, India, Myanmar, Nepal. **Refs.:** Iyengar 1929 (to sp.); Christophers 1933: 226 (M*, F*, L*); Khin-Maung-Kyi 1971d: 481 (distr., bion.); Ahmed 1987 (distr.; Bangladesh); Townsend 1990: 100 (type info.); Darsie & Pradhan 1990, 1991 (F, L; tax., keys, bion., distr.; Nepal); Oo et al. 2004 (distr.; Myanmar). **Ety.:** Sayid Abdul Majid. **Informal name:** Majid Indian Nail Mosquito.

maliensis Bailly-Choumara & Adam
1959: 3742 (M, F, L). **Type loc.:** Mali, Fouta-Djalon, Guinea (IERT). **Distr.:** Guinea, Mali. **Refs.:** Bailly-Choumara & Adam 1960: 110 (M*, F*, P*, L*); Kyalo et al. 2017 (distr.; sub-Saharan Africa). **Ety.:** Mali [Province of Mali, Republic of Guinea]. **Informal name:** Mali Guinean Nail Mosquito.

mangyanus (Banks)
1906b: 991 (F; *Myzomyia*). **Type loc.:** Rio Baco, Chicago, Mindoro, Philippines (NY). **Distr.:** Philippines. **Refs.:** King 1932b: 508 (M*, F*, L*; tax.); Baisas 1936a: 214 (P*); Gapuz & Santiago 1949: 127 (L*); Bonne-Wepster & Swellengrebel 1953: 378 (M*, F*, L*); Basio 1971b: 43 (M*, F*; bion.); Baisas 1974: 69 (M, F*, P*, L*, E*; also with first stage larva; tax., bion., distr.; Philippines); McCabe & Johnson 1980: 28 (type info.). **Ety.:** Mangyanes tribe of Mindoro, Phiippines. **Informal name:** Mangyanes Philippine Nail Mosquito.

syn. *febrifera* Banks
1914: 405 (M, F; *Myzomyia*). **Type loc.:** Canlubang, near Calamba [Laguna], Luzon, Philippines (NE). **Refs.:** King 1932b: 486, 508 (syn.).

marshallii (Theobald)
1903a: 77 (F*; *Pyretophorus*). **Type loc.:** Salisbury, Mashonaland [Southern Rhodesia] (NHMUK). **Distr.:** Angola, Botswana, Burundi, Cameroon, Central African Republic, Côte d'Ivoire, Democratic Republic of the Congo, Eswatini, Ethiopia, Gabon, Ghana, Guinea, Kenya, Liberia, Malawi, Mozambique, Namibia, Nigeria, Republic of the Congo, Republic of South Africa, Rwanda, Sierra Leone, South Sudan, Sudan, Tanzania (includes Zanzibar), Togo, Uganda, Zambia, Zimbabwe. **Refs.:** De Meillon et al. 1936 (tax., morph.); De Meillon 1947b: 145 (M*, F*, P, L*, E*); Peters 1955b: 97 (F*; tax.); Chauvet 1962: 1145 (A); Gillies & De Meillon 1968: 152 (M*, F*, P, L*, E*; syn.); White 1975: 315, 317 (*neireti* Ventrillon incorrectly assoc. with this sp. as *neireti* Blanchard, an invalid comb.; distr.); Service 1976a (distr.; Gabon); Lambert 1979 (tax.); Lambert 1981 (tax.); Kyalo et al. 2017 (distr.; sub-Saharan Africa). **Ety.:** Guy Anstruther Knox Marshall. **Informal name:** Marshall Zimbabwean Nail Mosquito.

syn. *pitchfordi* Giles
1904c: 34 (F*; *Pyretophorus*; as ssp.). **Type loc.:** 80 miles north of Eshowe, Zululand [KwaZulu-Natal, Republic of South Africa] (NHMUK). **Refs.:** De Meillon 1947b: 151 (F*, P, L*).

syn. *neireti* Blanchard
1906: 188 (A; *Myzomyia funesta* var.). Type loc.: Tanarive, Madagascar (FMP).

syn. *transvaalensis* Carter
1910a: 237 (F*; *Pyretophorus*). **Type loc.:** Leydsdorp, Transvaal [Republic of South Africa] (NHMUK). **Refs.:** Evans 1938: 180 (syn.).

syn. *pseudocostalis* Theobald
1910c: 41 (F*; *Pyretophorus*). **Type loc.:** Bihe, Angola (NHMUK).

mascarensis De Meillon
1947a: 116 (M, F). **Type loc.:** Diego Suarez area, Madagascar (SAIM). **Distr.:** Comoros, Madagascar (includes Glorioso and Juan de Nova Isl.), Mayotte. **Refs.:** Grjebine & Chauvet 1961: 119 (P*, L*); Chauvet 1962: 1145 (F*); Coetzee 1994b (E*); Kyalo et al. 2017 (distr.; sub-Saharan Africa). **Ety.:** Mascarene Isl. **Informal name:** Mascarene Malagasy Nail Mosquito.

melas (Theobald)
1903a: 76 (F; *Pyretoforus costalis* var. *melas*). **Type loc.:** [Bathurst], Gambia (NHMUK). **Distr.:** Angola, Benin, Cameroon, Côte d'Ivoire, Democratic Republic of the Congo, Equatorial Guinea, Gabon, The Gambia, Ghana, Guinea, Guinea-Bissau, Liberia, Mauritania, Nigeria, Republic of the Congo, São Tomé and Príncipe, Senegal, Sierra Leone, Togo. **Refs.:** De Meillon 1947b: 203 (tax.); Gelfand 1954: 6 (tax.); Coluzzi 1964: 204 (M*, P, L, E*); Gillies & De Meillon 1968: 220 (F*, L*, E*; bion.); Mahon et al. 1976: 25 (mol. tax.); Service 1976a (distr.; Gabon); Mattingly 1977: 326 (tax.); Sinka et al. 2010a: 117 (bion., review, distr., niche model); Becker et al. 2010: 326 (distr., bion.); Kyalo et al. 2017 (distr.; sub-Saharan Africa). **Ety.:** black (L); melanistic. **Informal name:** Black Gambian Nail Mosquito.

meraukensis Venhuis
1932: 1040 (F*, L*). **Type loc.:** Merauke, Western New Guinea (Papua) (GLB). **Distr.:** Australia, Indonesia (West Papua). **Refs.:** Bonne-Wepster & Swellengrebel 1953: 319 (M*, F*, L*); Lee et al. 1987b: 248 (F; key, tax., bion., distr., review). **Ety.:** Merauke, Papua Prov., Indonesia. **Informal name:** Merauke Indonesian Nail Mosquito.

merus Dönitz
1902: 77 (M, F*). **Type loc.:** Dar-es-Salaam, Tanganyika [Tanzania] (ZM). **Distr.:** Comoros, Eswatini, Kenya, Madagascar (includes Glorioso and Juan de Nova Isl.), Mauritius, Mayotte, Mozambique, Republic of South Africa, Somalia, Tanzania (includes Zanzibar), Zambia, Zimbabwe. **Refs.:** Coluzzi 1964: 204 (M*, P, L, E*); Gillies & De Meillon 1968: 223 (tax., bion.); Cross 1983: 155 (dist.; Republic of South Africa); Sinka et al. 2010a: 117 (bion., review, distr., niche model); Becker et al. 2010: 326 (distr., bion.); Kipyab et al. 2013 (bion.); Kyalo et al. 2017 (distr.; sub-Saharan Africa). **Ety.:** n.s. [pure, unmixed, genuine (L)]. **Informal name:** Unadulterated Tanzanian Nail Mosquito.

syn. *litoralis* Halcrow
1957: 133 (M, F, P, L, E; *gambiae* ssp.). **Type loc.:** Pointe aux Piments, Mauritius (A). **Refs.:** Stone et al. 1959: 43 (syn.?).

syn. *tangensis* Kuhlow
1962: 448 (M, F). **Type loc.:** Tanga, Tanganyika [Tanzania] (BNITM).

millecampsi Lips
1960a: 52 (F, E*). **Type loc.:** Lubumbashi, Katanga Province, Congo [Republic of the Congo] (NHMUK). **Distr.:** Republic of Congo. **Refs.:** White 1975: 314 (tax.); Brunhes et al. 1997: 180 (F*; from ssp. of *dureni*). **Ety.:** Albert Millecamps. **Informal name:** Millecamps Congolese Nail Mosquito.

milloti Grjebine & Lacan
1953 (M*, F*, P*, L*). In Grjebine 1953a: 495. **Type loc.:** La Mandraka, District Manjakandriana [Tananarive], Madagascar (IRSM). **Distr.:** Madagascar (includes Glorioso and Juan de Nova Isl.). **Refs.:** Kyalo et al. 2017 (distr.; sub-Saharan Africa). **Ety.:** Jacques Millot. **Informal name:** Millot Malagasy Nail Mosquito.

minimus Theobald
1901a: 186 (F*). **Type loc.:** Pokfulam, Hong Kong [People's Republic of China] (LU). **Distr.:** Bangladesh, Bhutan, Cambodia, China (includes Hong Kong), India, Indonesia, Japan, Laos, Malaysia, Malta, Myanmar, Nepal, Pakistan, People's Republic of China (includes Hong Kong), Philippines, Sri Lanka, Taiwan, Thailand, Vietnam. **Refs.:** Christophers 1933: 209 (M*, F*, P, L, E); Wu 1936: 269 (E*); Ross & Roberts 1943b: 19 (M*, F*, L*; tax., distr., bion.); Bonne-Wepster & Swellengrebel 1953: 369 (M*, F*, L*); Hara 1959: 110 (F*); Nguyen Thuong Hien 1968 (F*, L*; keys, tax., bion., distr.; Vietnam); Reid 1968: 314 (M*, F*, P*, L*, E*; distr.); Aslamkhan 1971b (distr.; Pakistan); Tanaka et al. 1975c: 210 (bion., distr.); Tanaka et al. 1979: 48 (M*, F*, L*); Harrison 1980: 78 (M*, F*, P*, L*; distr.); Darsie & Pradhan 1990, 1991 (F, L; tax., keys, bion., distr.; Nepal); Green et al. 1990 (chrom.; *minimus s.l.*); Sharpe et al. 2000 (phyl.; Minimus Group); Chen et al. 2002 (mol. tax.); Toma et al. 2003: 267 (distr.; Japan); Oo et al. 2004 (distr.; Myanmar); Garros et al. 2005b (tax., bion.); Somboon et al. 2005 (crossing; *minimus* C, E); Harbach et al. 2006 (M, F, P*, L*; neotype desig.); Rattanarithikul et al. 2006b (F*, L*; bion., distr., keys); Harbach, Garros et al. 2007 (tax., bion., gen.; form C = *harrisoni*); Foley et al. 2008c (niche model., distr.); Becker et al. 2010: 346 (F*, L*; key, tax., distr., bion.); Somboon et al. 2010 (syst., tax., bion., gen.; form E = *yaeyamensis*); Morgan et al. 2011 (distr., phylogeography); Chen et al. 2011 (mol. phyl.; *s.l.*); Sinka et al. 2011: 89 (bion., review, distr., niche model; Minimus Complex); Dev & Sharma 2013 (tax., bion.; Minimus Complex.); Namgay et al. 2018 (bion, distr.; Bhutan). **Ety.:** smallest (L); ref. to being an extremely small species. **Informal name:** Tiny Chinese Nail Mosquito. (Vol. 1, p. 356)

syn. *christophersi* Theobald
1902a: 378 (F*). **Type loc.:** Duars [West Bengal], India (NHMUK). **Refs.:** Harrison 1980: 83 (lectotype desig.); Harbach, Garros et al. 2007 (syn.).

syn. *formosaensis I* Tsuzuki
1902b: 288 (M, F). **Type loc.:** Formosa [Taiwan, Republic of China] (NHMUK; on slides). **Refs.:** Harrison 1980: 83 (lectotype desig.).

syn. *cohaesa* Dönitz
1903: 233 (*aconitus* var., new name for *formosaensis I* Tsuzuki Sept. 1902, not *formosaensis II* Tsuzuki Feb. 1902). **Refs.:** Harbach, Garros et al. 2007 (syn.).

syn. *alboapicalis* Theobald
1910c: 25 (F; *Myzomyia christophersi* ssp.). **Type loc.:** Meenglas, Jalpaiguri, and Duars [West Bengal], India (NHMUK).

mirans Sallum & Peyton
2005 (M, F, P*, L*). In Sallum et al. 2005a: 160. **Type loc.:** Kalatuara, Morapitiya, Sinharaja, western Sri Lanka (USNM). **Distr.:** India, Sri Lanka. **Refs.:** Sallum et al. 2007: 30 (phyl.). **Ety.:** to wonder at, look at (L). **Informal name:** Wonderful Sri Lankan Nail Mosquito.

moghulensis Christophers
1924c: 296 (A; *jeyporiensis* var.). **Type loc.:** Quetta [Baluchistan, Pakistan], Kasauli [Punjab Hill States], Bombay and Satpura Hills, India (NHMUK). **Distr.:** Afghanistan, India, Iran, Pakistan. **Refs.:** Puri 1928 (to sp.); Christophers & Barraud 1931: 176 (E*); Senevet 1931: 26 (P*); Christophers 1933: 270 (M*, F*, P, L*, E). **Ety.:** Moghul Empire. **Informal name:** Moghul Indian Nail Mosquito.

mortiauxi Edwards
1938 (M*, F*). In Evans 1938: 204. **Type loc.:** Kabila, Kwango Province [Leopoldville], Belgian Congo [Democratic Republic of the Congo] (NHMUK). **Distr.:** Democratic Republic of the Congo. **Refs.:** De Meillon 1947b: 157 (M*, F*); Kyalo et al. 2017 (distr.; sub-Saharan Africa). **Ety.:** Roger Maurice Modeste Mortiaux. **Informal name:** Mortiaux Congolese Nail Mosquito.

moucheti Evans
1925b: 211 (M, F*; *marshalli* var.). **Type loc.:** Stanleyville, Belgian Congo [Kisangani, Tshopo Province, Democratic Republic of the Congo] (CMT). **Distr.:** Burundi, Cameroon, Central African Republic, Côte d'Ivoire, Democratic Republic of the Congo, Equatorial Guinea, Gabon, Ghana, Guinea, Kenya, Nigeria, Republic of the Congo, Rwanda, Sierra Leone, South Sudan, Tanzania, Uganda. **Refs.:** Evans 1929b (M*, P*, L*; as *marshalli* var.); Evans 1938 (M*, F*, L*, E*); De Meillon 1947b: 153 (M*, F*, P, L*, E); Wanson et al. 1947: 39 (bion.); Gillies & De Meillon 1968 (M*, F*, L*, E*; keys); Service 1976a: 260 (distr.; Gabon); Gillies & Coetzee 1987 (F, L, P; keys); Brunhes et al. 1998a (tax., review); Antonio-Nkondjio et al. 2009b (L; bion.); Sinka et al. 2010a: 117 (bion., review, distr., niche model); Antonio-Nkondijo & Simard 2013 (bion., mol. tax.; *s.l.*; review); Paupy et al. 2013 (ape *Plasmodium*); Kyalo et al. 2017 (distr.; sub-Saharan Africa). **Ety.:** René Libert Joseph Mouchet. **Informal name:** Mouchet Congolese Nail Mosquito. (Vol. 1, p. 358)

ssp. ***nigeriensis*** Evans
1931a: 138 (M, P, L*, E*). **Type loc.:** Yaba [Lagos], Southern Nigeria (NHMUK). **Distr.:** Cameroon, Niger, Nigeria, Guinea. **Refs.:** De Meillon 1947b: 156 (F, L*); Gillies & De Meillon 1968: 122 (as ssp.); Brunhes et al. 1998a (syn. with moucheti); Kyalo et al. 2017 (distr.; sub-Saharan Africa). **Ety.:** Nigeria. **Informal name:** Yaba Nigerian Nail Mosquito.

mousinhoi De Meillon & Pereira
1940: 80 (M, F*, P, L*). **Type loc.:** Manjacaze, Sul do Save, Mozambique (SAIM). **Distr.:** Cameroon, Democratic Republic of the Congo, Mozambique, Republic of South Africa, Zimbabwe. **Refs.:** De Meillon 1947b: 151 (F*, P, L*); Gillies & De Meillon 1968; 4, 168 (M*, F*, P, L*; to sp.); Kyalo et al. 2017 (distr.; sub-Saharan Africa). **Ety.:** Joaquin Augusto Mousinho. **Informal name:** Mousinho Mozambican Nail Mosquito.

multicinctus Edwards
1930a: 291 (M, F*). **Type loc.:** Taveta, Kenya (NHMUK). **Distr.:** Cameroon, Central African Republic, Democratic Republic of the Congo, Kenya, Zimbabwe. **Refs.:** Evans 1934: 569 (to var.); Parent & Vincke 1945: 24 (P*); De Meillon 1947b: 79 (F*); Gillies & De Meillon 1968: 79 (syn.); White 1980: 120 (from syn. with *natalensis*; as form); Brunhes et al. 1998a: 397 (from ssp. of *natalensis*); Kyalo et al. 2017 (distr.; sub-Saharan Africa). **Ety.:** many rings (L); on legs. **Informal name:** Multi-ringed Kenyan Nail Mosquito.

syn. *puselis* De Meillon
1934: 269 (M*, P; *Anopheles natalensis* var.). **Type loc.:** Tzaneen, Transvaal [Republic of South Africa] (SAIM). **Refs.:** Evans 1938: 116 (syn.).

multicolor Cambouliu
1902: 705 (F). **Type loc.:** Isthme de Suez, Egypt (FMP). **Distr.:** Afghanistan, Albania, Algeria, Cyprus, Egypt, Iran, Iraq, Israel (and Gaza Strip and West Bank), Jordan, Libya, Morocco, Niger, Pakistan, Portugal, Qatar, Saudi Arabia, Spain (includes Canary Isl.), Sudan, Tunisia, Turkmenistan. **Refs.:** Séguy 1924: 157 (E*), 159 (L*), as *chaudoyei*; Christophers 1933: 257 (M*, F*, P, L*, E); Evans 1938: 343 (M*, F*, P, L*, E); Ross & Roberts 1943b: 25 (M*, F*, L*; tax., distr., bion.); De Meillon 1947b: 219 (M*, F*, L*, E*); Senevet et al. 1956 (L*; tax.); Pringle et al. 1960 (L*; key, bion.); Gillies & De Meillon 1968: 316 (distr.); Mattingly 1969c (E*); Aslamkhan 1971b (distr.; Pakistan); Glick 1992 (F*; key, tax., distr.); Abdullah & Merdan 1995 (distr.; Saudi Arabia); Amr et al. 1997 (distr.; Jordan); Eritja et al. 2000: 11 (distr.; Spain); Hadjivassilis 2000: 38 (distr.); Trari et al. 2002: 331 (distr.; Morocco); Becker et al. 2010: 182 (M*, F*, L*; keys, tax., distr., bion.); Ahmed et al. 2011 (distr.; Saudi Arabia); Kyalo et al. 2017 (distr.; sub-Saharan Africa); Robert et al. 2019 (distr.; western Palearctic). **Ety.:** many colors (L). **Informal name:** Multicolored Egyptian Nail Mosquito. (Vol. 1, p. 360)

syn. *impunctus* Dönitz
1902: 67 (A). **Type loc.:** Wadi [el] Natrun [south of Alexandria], Egypt (NE).
syn. *chaudoyei* Theobald
1903a: 68 (M*, F*; *Pyretophorus*). **Type loc.:** Touggourt, Algeria (NHMUK).
syn. *nigrifasciatus* Theobald
1907: 65 (F; *Pyretophorus*). **Type loc.:** Peshin [Baluchistan, Pakistan] (NHMUK).

murphyi Gillies & De Meillon
1968: 253 (M, F, P, L, E). **Type loc.:** Keneba, The Gambia (NHMUK). **Distr.:** Burkina Faso, The Gambia. **Refs.:** Kyalo et al. 2017 (distr.; sub-Saharan Africa). **Ety.:** Dennis Hugh Murphy. **Informal name:** Murphy Gambian Nail Mosquito.

natalensis (Hill & Haydon)
1907: 152 (M, F*, L*; *Myzorhynchus*). **Type loc.:** Natal [KwaZulu-Natal, Republic of South Africa] (DMN). **Distr.:** Angola, Burkina Faso, Burundi, Cameroon, Central African Republic, Côte d'Ivoire, Democratic Republic of the Congo, Ethiopia, Gabon, Kenya, Mozambique, Republic of the Congo, Republic of South Africa, Rwanda, Tanzania, Uganda, Zimbabwe. **Refs.:** Evans 1938: 110 (M*, F*, L*); De Meillon 1947b: 77 (M*, F*, L*); O'Connor 1967 (distr.); Gillies & De Meillon 1968: 79 (M*, F*, P, L*); Kyalo et al. 2017 (distr.; sub-Saharan Africa). **Ety.:** Natal, Republic of South Africa. **Informal name:** Natal South African Nail Mosquito.
syn. *aureosquamiger* Theobald
1907: 73 (F; *Pyretophorus*). **Type loc.:** Pretoria [Transvaal, Republic of South Africa] (NHMUK).

nataliae Belkin
1945: 315 (M, F*, P, L*). **Type loc.:** Natalia Springs, Poha River Valley, Guadalcanal, Solomon Islands (USNM). **Distr.:** Solomon Isl. **Refs.:** Belkin 1962: 149 (M*, F*, P*, L*); Lee et al. 1987b: 256 (F; key, tax., bion., distr., review). **Ety.:** Natalia Yantsin. **Informal name:** Natalia Solomon Islander Nail Mosquito.

nemophilous Peyton & Ramalingam
1988: 2276 (M*, F*, P*, L*). **Type loc.:** Ban Bang Ra Ko, Phangnga Province, Thailand (USNM). **Distr.:** Malaysia, Thailand. **Refs.:** Sallum et al. 2005b: 55 (M*, F*, P*, L*; distr.); Rattanarithikul et al. 2006b (F*, L*; bion., distr., keys); Sallum et al. 2007: 30 (phyl.). **Ety.:** woods loving (Gr). **Informal name:** Forest Thai Nail Mosquito.

nili (Theobald)
1904b: 66 (F*; *Myzomyia*). **Type loc.:** Jebel Akmet-Aga on the White Nile, Sudan (NHMUK). **Distr.:** Angola, Benin, Botswana, Burkina Faso, Burundi, Cameroon, Central African Republic, Chad, Côte d'Ivoire, Democratic Republic of the Congo, Eswatini, Ethiopia, Gabon, The Gambia, Ghana, Guinea, Guinea-Bissau, Kenya, Liberia, Mali, Mozambique, Namibia, Niger, Nigeria, Republic of the Congo, Republic of South Africa, Rwanda, Senegal, Seychelles, Sierra Leone, Somalia, South Sudan, Sudan, Tanzania, Togo, Uganda, Zambia, Zimbabwe. **Refs.:** Macfie & Ingram 1922 (F*); Evans 1938: 137 (M, F*, P*, L*, E*); Hamon 1954a: 583 (L*); Lewis 1956a: 477 (L); Peters 1956: 531 (tax.); Gillies & De Meillon 1968: 83 (M*, F*, P*, L*, E*); Ribeiro & da Cunha Ramos 1975: 16 (distr.); Gillies & Coetzee 1987 (F*, L*; key, bion.); Kengne et al. 2003 (mol. tax.; *s.l.*); Awono-Ambene et al. 2006 (mol. tax.; *s.l.*); Antonio-Nkondjio et al. 2009a (phyl.; as *moucheti* Nili Complex); Antonio-Nkondjio et al. 2009b (L; bion.); Sinka et al. 2010a: 117 (bion., review, distr., niche model; Nili Complex); Antonio-Nkondjio & Simard 2013 (bion., mol. tax.; *s.l.*, review); Ndo et al. 2013 (*s.l.*; mol. tax.); Antonio-Nkondjio & Simard 2013 (mol. tax.; *s.l.*); Kyalo et al. 2017 (distr.; sub-Saharan Africa). **Ety.:** Nile River, Sudan (L). **Informal name:** Nile Sudanese Nail Mosquito. (Vol. 1, p. 362)
syn. *unicolor* Grünberg
1905: 379 (F*; *Myzomyia*). **Type loc.:** Kete Kratje, Togo [Ghana] (ZM).
syn. *nigra* Christophers
1924a: 45 (A; *Myzomyia funesta* var.). **Type loc.:** Mid Sobal, Sudan (NHMUK).

nivipes (Theobald)
1903f: 258 (M; *Nyssorhynchus*). **Type loc.:** Kuala Lumpur [Selangor], Malaya [Malaysia] (NHMUK). **Distr.:** Bangladesh, Bhutan, Cambodia, India, Laos, Malaysia, Nepal, People's Republic of China, Thailand, Vietnam. **Refs.:** Reid 1967: 175 (M*, P*, L*; from syn.); Darsie & Pradhan 1990, 1991 (F, L; tax., keys, bion., distr.; Nepal); Harrison et al. 1991: 223 (distr.; Thailand.); Darsie et al. 1996 (distr.; Nepal); Subbarao et al. 2000: 71 (distr.; India); Rattanarithikul et al. 2006b (F*, L*; bion., distr., keys); Qu & Zhu 2008 (distr.; People's Republic of China); Morgan et al. 2009 (mol. Phyl.; Neocellia Series); Namgay et al. 2018 (bion, distr.; Bhutan). **Ety.:** snowy foot (L); white leg markings. **Informal name:** Snow-footed Malaysian Nail Mosquito.

njombiensis Peters
1955b: 95 (F*). **Type loc.:** Njombe, Tanganyika [Tanzania] (NHMUK). **Distr.:** Angola, Democratic Republic of the Congo, Tanzania. **Refs.:** Gillies & De Meillon 1968: 168 (M, F*, P, L*); Kyalo et al. 2017 (distr.; sub-Saharan Africa). **Ety.:** Njombe, Tanzania. **Informal name:** Njombe Tanzanian Nail Mosquito.

notanandai Rattanarithikul & Green
1987: 256 (M, F*, E*). **Type loc.:** Ban Tha Cha Lao, Tha Yang, Phetchaburi Province, Thailand (USNM). **Distr.:** Laos, Thailand. **Refs.:** Rattanarithikul et al. 2006b (F*, L*; bion., distr., keys). **Ety.:** Vimol Notananda. **Informal name:** Notananda Thai Nail Mosquito.

notleyi van Someren
1949a: 3 (M*, F, P, L*). **Type loc.:** Sakaramy, near Diego Suarez, Madagascar (NHMUK). **Distr.:** Madagascar (includes Glorioso and Juan de Nova Isl.). **Refs.:** Kyalo et al. 2017 (distr.; sub-Saharan Africa). **Ety.:** Frank Brian Notley. **Informal name:** Notley Malagasy Nail Mosquito.

novaguinensis Venhuis
1933: 203 (M*, F*, L*; *punctulatus* var.). **Type loc.:** Kakajoe and Donggeap on Merauke River, Western New Guinea (Papua) (GLB). **Distr.:** Australia, Indonesia. **Refs.:** Taylor 1943a: 155 (F; only of type series; as *derricki*; syn. by Lee & Woodhill 1944: 121); Lee & Woodhill 1944: 164 (M*, F*, L*, E*); Bonne-Wepster & Swellengrebel 1953: 334 (M*, F*, L*); Lee et al. 1987b: 258 (F; key, tax., bion., distr., review). **Ety.:** New Guinea. **Informal name:** New Guinea Indonesian Nail Mosquito.
syn. *breinli* Taylor
1943a: 156 (F*). **Type loc.:** Irvinebank, Queensland, Australia (US). **Refs.:** Lee & Woodhill 1944: 164 (syn.).

oreios Bangs & Harbach
2014 (M*, F*, P*, L*). In Bangs et al. 2014: 69. **Type loc.:** Piramid, Baliem Valley, Papua province [West Papua], Indonesia (NHMUK). **Distr.:** Indonesia (West Papua), Papua New Guinea. **Ety.:** of the mountains (Gr); ref. to mountainous areas of New Guinea. **Informal name:** Montane Indonesian Nail Mosquito.

ovengensis Awono-Ambene, Kengne, Simard, Antonio-Nkondjio, & Fontenille
2004: 563 (F*, L*). **Type loc.:** Oveng (02°10'N, 10°30'E), near Campo, Cameroon (IRD). **Distr.:** Cameroon, Equatorial Guinea. **Refs.:** Antonio-Nkondjio et al. 2009b (L; bion.); Antonio-Nkondjio & Simard 2013 (bion., mol. tax., review); Kyalo et al. 2017 (distr.; sub-Saharan Africa). **Ety.:** Oveng, Cameroon. **Informal name:** Oveng Cameroonian Nail Mosquito.

pallidus Theobald
1901a: 134 (F*; *fuliginosus* var.). **Type loc.:** Madras and Behar, India (NHMUK). **Distr.:** Bangladesh, Cambodia, India, Indonesia, Laos, Malaysia, Myanmar, Nepal, Pakistan, Sri Lanka, Thailand, Vietnam. **Refs.:** Christophers & Barraud 1931: 177 (E*); Christophers 1933: 309 (M*, F, L, E); Bonne-Wepster & Swellengrebel 1953: 448 (A, L*); Nguyen Thuong Hien 1968 (F*, L*; keys, tax., bion., distr.; Vietnam); Reid 1968: 389 (tax.); Aslamkhan 1971b (distr.; Pakistan); Darsie & Pradhan 1990, 1991 (F, L; tax., keys, bion., distr.; Nepal); Harrison et al. 1991: 223 (distr.); Oo et al. 2004 (distr.; Myanmar); Morgan et al. 2009 (mol. Phyl.; Neocellia Series); Kyalo et al. 2017 (distr.; sub-Saharan Africa). **Ety.:** pale, ashen (L); fringe scales yellow in comp. to black in *fuliginosus*. **Informal name:** Pale Indian Nail Mosquito.
syn. *fowleri* Christophers
1911a: 64 (M, F), 130 (F*); *Neocellia* [part of Christophers 1911b]. **Type loc.:** Amritsar [Punjab Hill States], India (NHMUK).

paltrinierii Shidrawi & Gillies
1988: 202 (M*, F, P, L*). **Type loc.:** Dank and Nizra Rivers, Oman (NHMUK). **Distr.:** Oman, United Arab Emirates. **Refs.:** Morgan et al. 2009 (mol. Phyl.; Neocellia Series). **Ety.:** Adriano B. Paltrinieri. **Informal name:** Paltrinieri Omani Nail Mosquito.

pampanai Büttiker & Beales
1959: 63 (M*, F*, L*; as *pampanae*). **Type loc.:** Ksim 2 (Prek Chimeang), 10 km north of Snoul, Kratie Province, Cambodia (NHMUK). **Distr.:** Cambodia, Laos, Thailand, Vietnam. **Refs.:** Peyton & Scanlon 1966: 1 (F*); Reid 1976: 111 (lectotype desig.); Harrison 1980: 99 (M*, F*, P*, L*; distr.); Rattanarithikul et al. 2006b (F*, L*; bion., distr., keys). **Ety.:** Emilio J. Pampana. **Informal name:** Pampana Cambodian Nail Mosquito.

parangensis (Ludlow)
1914b: 129 (F; *Myzomyia*). **Type loc.:** Parang [Cotabato], Mindanao, Philippines (USNM). **Distr.:** Indonesia, Malaysia, Philippines. **Refs.:** Baisas 1936a: 216 (P*); Bonne-Wepster & Swellengrebel 1953: 408 (M*, F*, L*); Stone & Knight 1956b: 279 (type info., lectotype desig.); Reid 1968: 328 (distr.); Cagampang-Ramos & Darsie 1970 (F*, L*; keys); Baisas 1974 (M*, F*, L*); Lee et al. 1987b: 262 (F; key, tax., bion., distr., review). **Ety.:** Parang, Mindanao, Philippine Isl. **Informal name:** Parang Philippine Nail Mosquito.

parensis Gillies
1962: 82 (M*, F*, P, L, E). **Type loc.:** Kihurio, South Pare District, Tanganyika [Tanzania] (NHMUK). **Distr.:** Botswana, Ethiopia, Kenya, Malawi, Mozambique, Republic of South Africa, Tanzania (includes Zanzibar), Uganda, Zambia, Zimbabwe. **Refs.:** Dia et al. 2013 (tax., bion., distr.; Africa); Kyalo et al. 2017 (distr.; sub-Saharan Africa). **Ety.:** Pare Dist., Tanzania. **Informal name:** Pare Tanzanian Nail Mosquito.

pattoni Christophers
1926: 871 (M, F). **Type loc.:** Tsi-nan-fu, Shantung, China [People's Republic of China] (NHMUK). **Distr.:** People's Republic of China. **Refs.:** Martini 1930: 179 (M, F); Feng 1931: 494 (P*, L*; tax., bion.); Russell et al. 1943 (F, L; keys, bion.); Bates et al. 1949 (F; key); Chow 1949b (F, L; keys, bion.); Yu & Zhang 1981 (F*). **Ety.:** Walter Scott Patton. **Informal name:** Patton Chinese Nail Mosquito.

pauliani Grjebine
1953a: 469 (M*, F*, P*, L*). **Type loc.:** Majunga, Madagascar (IRSM, NHMUK). **Distr.:** Madagascar (includes Glorioso and Juan de Nova Isl.). **Refs.:** Townsend 1990: 118 (type info.); Coetzee 1994b (E*); Kyalo et al. 2017

(distr.; sub-Saharan Africa). **Ety.:** Renaud Paulian. **Informal name:** Paulian Malagasy Nail Mosquito.

pharoensis Theobald
1901a: 169 (M, F*). **Type loc.:** Cairo, Egypt (NHMUK). **Distr.:** Angola, Benin, Botswana, Burkina Faso, Burundi, Cameroon, Central African Republic, Chad, Côte d'Ivoire, Democratic Republic of the Congo, Djibouti, Egypt, Eritrea, Ethiopia, Gabon, The Gambia, Ghana, Guinea, Guinea-Bissau, Israel (and Gaza Strip and West Bank), Jordan, Kenya, Kuwait, Libya, Madagascar (includes Glorioso and Juan de Nova Isl.), Malawi, Mali, Mauritania, Mozambique, Namibia, Niger, Nigeria, Republic of South Africa, Rwanda, São Tomé and Príncipe, Senegal, Sierra Leone, Somalia, South Sudan, Sudan, Syria, Tanzania, Togo, Uganda, Yemen, Zambia, Zimbabwe. **Refs.:** Macfie & Ingram 1922 (F*); Davis 1926 (F*); Evans 1938: 370 (M, F*, P*, L*, E*); Ross & Roberts 1943b: 35 (M*, F*, L*; tax., distr., bion.); De Meillon 1947b: 233 (M, F*, L*, E*); Senevet & Andarelli 1955a: 188 (M*, F*, P*, L, E); Gillies & De Meillon 1968: 253 (M, F*, L*); Miles et al. 1983 (gen.; *s.l.*); Gillies & Coetzee 1987 (F*, L*; key, bion.); Glick 1992 (F*; key, tax., distr.); Mudegu & Mukiama 1995 (mol. tax.; *s.l.*); Al-Houty 1997 (distr.; Kuwait); Becker et al. 2010: 326 (F*, L*; key, tax., distr., bion.); Kyalo et al. 2017 (distr.; sub-Saharan Africa); Robert et al. 2019 (distr.; western Palearctic). **Ety.:** pharaohs [land of the]. **Informal name:** Pharaoh Egyptian Nail Mosquito. (Vol. 1, p. 364)
syn. *albofimbriatus* Giles
1902: 302 (F*; as var.). **Type loc.:** Tel Lahmul, Palestine (NHMUK).
syn. *maculicosta* Becker
1903: 69 (F). **Type loc.:** Cairo, Egypt (ZM).
syn. *albus* Ventrillon
1904 (F). In Laveran 1904: 211. **Type loc.:** Madagascar (MNHP).
syn. *bozasi* Neveu-Lemaire
1905a: 32 (F; *Nyssorhynchus*). **Type loc.:** Doufilé, "Afrique centrale" [White Nile] (PIP). **Refs.:** Rodhain & Boutonnier 1985: 273 (type dep.).

philippinensis Ludlow
1902b: 426 (Aug. 23) (F), 1902: 128 (Sept.). **Type loc.:** San José, Abra, Luzon, Philippines (USNM). **Distr.:** Bangladesh, Bhutan, Cambodia, India, Indonesia, Laos, Malaysia, Myanmar, Nepal, People's Republic of China, Philippines, Thailand, Vietnam. **Refs.:** Crawford 1938: 96 (P*); Roy & Siddons 1939: 159 (E*); Colless 1948: 109 (M*, F*, L*; tax.); Masters 1948: 178 (L); Stone & Knight 1956b: 279 (type info.); Reid 1967: 175 (M*, P*, L*; tax.); Nguyen Thuong Hien 1968 (F*, L*; keys, tax., bion., distr.; Vietnam); Aslamkhan 1971b (distr.; Pakistan); Basio 1971b: 44 (M*, F*; bion.); Baisas 1974: 71 (M, F*, P, L*; tax., bion., distr.; Philippines); Darsie & Pradhan 1990, 1991 (F, L; tax., keys, bion., distr.; Nepal); Oo et al. 2004 (distr.; Myanmar); Rattanarithikul et al. 2006b (F*, L*; bion., distr., keys); Morgan et al. 2009 (mol. Phyl.; Neocellia Series); Morgan et al. 2011 (distr., phylogeography); Namgay et al. 2018 (bion., distr.; Bhutan). **Ety.:** Philippine Isl. **Informal name:** Barangay Philippine Nail Mosquito ["barangay" is a Philippine term for village, ward, or district].
syn. *freerae* Banks
1906b: 993 (F; *Pyretophorus*). **Type loc.:** Manila [Rizal], Luzon, Philippines (NE).
syn. *pampangensis* Brunetti
1920: 114 (new name for *philippinensis* Ludlow 1902b; in error).
syn. *hainanensis* Takei
1941: 417 (A; *Anopheles* (*Myzomyia*)). **Type loc.:** Pao-ting, Nanch'en, Nan-feng, and Tahsing, Hainan Island [People's Republic of China] (LU). **Refs.:** Reid 1968: 385 (syn.).

pretoriensis (Theobald)
1903a: 99 (F*, L*; *Nyssorhynchus*). **Type loc.:** Pretoria [Transvaal, Republic of South Africa] (NHMUK). **Distr.:** Angola, Benin, Botswana, Burkina Faso, Cameroon, Cabo Verde, Central African Republic, Comoros, Côte d'Ivoire, Democratic Republic of the Congo, Eritrea, Eswatini, Ethiopia, Gabon, Ghana, Guinea, Kenya, Liberia, Madagascar (includes Glorioso and Juan de Nova Isl.), Malawi, Mali, Mauritania, Mayotte, Mozambique, Namibia, Niger, Nigeria, Republic of the Congo, Republic of South Africa, Rwanda, Saudi Arabia, Senegal, Seychelles, Somalia, South Sudan, Sudan, Tanzania (includes Zanzibar), Togo, Uganda, Yemen, Zambia, Zimbabwe. **Refs.:** Evans 1938: 358 (M*, F*, P, L*, E*); De Meillon 1947b: 225 (M*, F*, L*, E*); Service 1970a: 648 (F*); Ahmed et al. 2011 (distr.; Saudi Arabia); Kyalo et al. 2017 (distr.; sub-Saharan Africa). **Ety.:** Pretoria, Republic of South Africa. **Informal name:** Pretoria South African Nail Mosquito.
syn. *tibani* Patton
1905: 629 (M*, F*, L*, E*). **Type loc.:** Hinterland up to Jehat, 6,800 ft [West Aden Protectorate] [Yemen] (NHMUK). **Refs.:** Mattingly & Knight 1956: 96 (tax.).

pseudojamesi Strickland & Chowdhury
1927b: 241 (M*, F*, L*). **Type loc.:** [Bengal], India (LU). **Distr.:** Bangladesh, Bhutan, Cambodia, China, India, Indonesia, Myanmar, Nepal, Thailand, Vietnam. **Refs.:** Christophers & Barraud 1931: 178 (E*; as *ramsayi*); Christophers 1933: 294 (M*, F*, L*, E; as *ramsayi*); Crawford 1938: 100 (P*; as *ramsayi*); Nguyen Thuong Hien 1968 (F*, L*; keys, tax., bion., distr.; Vietnam; as *ramsayi*); Reid 1968: 365 (distr.); Reid 1968: 365 (M*, F*, P*, L*, E*; distr.; as *ramsayi*); Aslamkhan 1971b (distr.; Pakistan; as *ramsayi*); Ahmed 1987 (distr.; Bangladesh; as *ramsayi*); Darsie & Pradhan 1990, 1991 (F, L; tax., keys, bion., distr.; Nepal); Harrison et al. 1991: 223 (distr.; Thailand); Oo et al. 2004 (distr.; Myanmar); Rattanarithikul et al. 2006b (F*, L*; bion., distr., keys); Qu & Zhu 2008 (distr.; People's

Republic of China; as *ramsayi*); Morgan et al. 2009 (mol. Phyl.; Neocellia Series). **Ety.:** false (Gr); similar to *jamesi*. **Informal name:** 007 Indian Nail Mosquito [*pseudojamesi* = "false James" = James Bond = 007, as popularized in the novels by Ian Fleming (1908–1964)].

syn. *ramsayi* Covell

1927: 1019 (M, F*). **Type loc.:** Cachar, Assam, India (NHMUK). **Refs.:** Nurul Huda & Harrison 1985: 49 (syn.).

pseudosundaicus Tyagi, Hiriyan, Teware, Ayanar, Samuel, Arunachalam, Paramasivan, Krishnamoorthy, Dhananjeyan, Leo & Rajendran

2009 (M*, F*, P*, L*). **Type loc.:** Kuzhithura, Kollam District, Kerala State, India (CRME). **Distr.:** India. **Ety.:** false (Gr); similar to *sundaicus*. **Informal name:** Kollam Indian Nail Mosquito.

pseudowillmori (Theobald)

1910c: 65 (F*; *Nyssorhynchus*). **Type loc.:** Meenglas, Duars, and Jalpaiguri [West Bengal], India (IM). **Distr.:** Bangladesh, Bhutan, India, Laos, Nepal, People's Republic of China, Thailand, Vietnam. **Refs.:** Rattanarithikul & Green 1987: 252 (from syn. with *maculatus*); Darsie & Pradhan 1990, 1991 (F, L; tax., keys, bion., distr.; Nepal); Harrison et al. 1991: 223 (distr.; Thailand.); Green et al. 1992 (chrom.; *An. willmori*, *An. pseudowillmori*); Rattanarithikul et al. 2006b (F*, L*; bion., distr., keys); Qu & Zhu 2008 (distr.; People's Republic of China); Morgan et al. 2009 (mol. Phyl.; Neocellia Series); Namgay et al. 2018 (bion., distr.; Bhutan). **Ety.:** false (Gr); similar to *willmori*. **Informal name:** Meenglas Indian Nail Mosquito.

pujutensis Colless

1948: 101 (M*, F*, L*; *leucosphyrus* ssp.). **Type loc.:** Pujut, Sarawak, [Malaysian] Borneo (CSIR). **Distr.:** Indonesia, Malaysia (includes Sarawak), Thailand. **Refs.:** Colless 1956: 72 (M, F*, P*, L*; to sp.); Reid 1968: 303 (M, F*, P*, L*; keys); Sallum et al. 2005b: 63 (M, F*, P*, L*); Rattanarithikul et al. 2006b (F*, L*; bion., distr., keys). **Ety.:** Pujut, Sarawak, Malaysia. **Informal name:** Pujut Malaysian Nail Mosquito.

pulcherrimus Theobald

1902a: 369 (F*). **Type loc.:** Lahore [Punjab, Pakistan] (NHMUK). **Distr.:** Afghanistan, Bahrain, India, Iran, Iraq, Israel (and Gaza Strip and West Bank), Kazakhstan, Kuwait, Lebanon, Pakistan, Russia, Saudi Arabia, Syria, Tajikistan, Turkey, Uzbekistan. **Refs.:** Christophers & Barraud 1931: 180 (E*); Senevet 1931: 41 (P*); Christophers 1933: 311 (M*, F*, P, L*, E); Russell et al. 1943 (F, L; keys, bion.); Bates et al. 1949: 434 (bion.); Macan 1950 (bion.); Foote & Cook 1959 (key, bion.); Chinayev 1965 (bion., distr.; Central Asia); Baker et al. 1968 (chrom.); Gokberk 1970: 20 (distr.); Gutsevich et al. 1974: 110 (M*, F, L*); Glick 1992 (F*; key, tax., distr.); Zaim et al. 1993 (bion.); Djadid et al. 2003 (mol. tax.); Ahmed et al. 2011 (distr.; Saudi Arabia); Morgan et al. 2009 (mol. phyl.; Neocellia Series); Robert et al. 2019 (distr.; western Palearctic). **Ety.:** most beautiful (L). **Informal name:** Beautiful Pakistani Nail Mosquito. (Vol. 1, p. 366)

syn. *atropotenae* Lindtrop

1924: 40 (A; *Cellia pulcherrima* var.). **Type loc.:** Azerbaydzhan, USSR [Azerbaijan] (LU).

punctulatus Dönitz

1901: 36 (M, F). **Type loc.:** Stephansort, New Guinea, and Herbertshoehe, Bismarck Archipelago (ZM, NHMUK). **Distr.:** Australia, Indonesia, Papua New Guinea, Solomon Isl. **Refs.:** Ross & Roberts 1943b: 11 (M*, F*, L*; tax., distr., bion.); Lee & Woodhill 1944: 134 (M, F*, L*); Belkin et al. 1945: 253 (M, F, P*, L); Rozeboom & Knight 1946: 95 (F*, P*, L*; tax.); Woodhill 1946: 276 (tax., bion.); Penn 1949b: 16 (P*); Belkin 1962: 136 (M*, F*, L*, first instar L*, E; keys, bion., distr.; *s.l.*, *s.s.*); Bryan 1973: 64 (tax.); Bryan 1974: 413 (tax.); Lee et al. 1987b: 265 (F; key, tax., bion., distr., review; *s.l.*); Townsend 1990: 125 (type info.); Foley & Bryan 1993 (mol. tax.; *s.l*); Foley et al. 1993 (mol. tax.); Foley et al. 1994 (mol. tax.; *s.l.*); Foley et al. 1995 (*s.l.*); Cooper et al. 2002 (bion., distr.; *s.l.*); Beebe & Cooper 2002 (bion., distr.; *s.l.*); Schmidt et al. 2003 (keys; *s.l.*); Sinka et al. 2011: 89 (bion., review, distr., niche model; Punctulatus Group, Punctulatus Complex); Beebe et al. 2013 (Punctulatus Group); Logue et al. 2013 (key, mol. tax.; *s.l.*); Harbach 2013b (phyl., tax.; *s.l.*); Sinka 2013 (bion.; *s.l.*); Beebe et al. 2013 (tax., bion., review, distr.; SW Pacific; *s.l.*); Beebe et al. 2015 (review; *s.s.* and *s.l.*). **Ety.:** spotted (wing) (L). **Informal name:** Spotted Indonesian Nail Mosquito.

quadriannulatus (Theobald)

1911b: 242 (F*; *Pyretophorus*). **Type loc.:** Onderstepoort, Transvaal [Republic of South Africa] (NHMUK). **Distr.:** Botswana, Eswatini, Malawi, Mozambique, Namibia, Republic of South Africa, Tanzania (includes Zanzibar), Uganda, Zambia, Zimbabwe. **Refs.:** White 1975: 315 (tax.); Mattingly 1977: 325 (from syn. with *gambiae* to provide name for *gambiae* C); Coetzee 1987 (M, F, P, L; tax.); Becker et al. 2010: 325 (tax., distr., bion.; *s.l.*); Kyalo et al. 2017 (distr.; sub-Saharan Africa). **Ety.:** four rings (L); on palpus. **Informal name:** Four-ringed South African Nail Mosquito.

radama De Meillon

1943: 90 (M, F*). **Type loc.:** [Diego Suarez], Madagascar (SAIM). **Distr.:** Madagascar (includes Glorioso and Juan de Nova Isl.). **Refs.:** Doucet 1951b: 58 (M*, F*, L*); Kyalo et al. 2017 (distr.; sub-Saharan Africa). **Ety.:** n.s. [poss. Radàma Isl.]. **Informal name:** Radàma Malagasy Nail Mosquito.

rageaui Mattingly & Adam

1954: 55 (F*). **Type loc.:** Oliga, Cameroon (NHMUK). **Distr.:** Cameroon, Côte d'Ivoire, Democratic Republic of the Congo, Ghana, Guinea, Republic of the Congo. **Refs.:** Adam & Mattingly 1956: 734 (M*, F*, L*, E*; to ssp.); Brunhes et al. 1999: 556 (F*, L*; from syn. with *smithi*);

Kyalo et al. 2017 (distr.; sub-Saharan Africa). **Ety.:** J. Rageau. **Informal name:** Rageau Cameroonian Nail Mosquito.

rampae Harbach & Somboon
2011 (M*, F, P*, L*, E). In Somboon et al. 2011. **Type loc.:** Nong Bua Lam Phu, Udon Thani Province, Thailand (NHMUK). **Distr.:** Cambodia, India, Thailand, Vietnam. **Ety.:** Rampa Rattanarithikul. **Informal name:** Rampa Thai Nail Mosquito.

ranci Grjebine
1953a: 488 (M*, F*, P*, L*). **Type loc.:** Ambodisoka, District Marolambo and Anosibe, District Moramanga, Madagascar (IRSM, NHMUK). **Distr.:** Madagascar (includes Glorioso and Juan de Nova Isl.). **Refs.:** Townsend 1990: 127 (type info.); Kyalo et al. 2017 (distr.; sub-Saharan Africa). **Ety.:** André Ranc. **Informal name:** Ranc Malagasy Nail Mosquito.

recens Sallum & Peyton
2005 (M, F, P*, L*). In Sallum et al. 2005a: 185. **Type loc.:** Sumatra Island, Gunong Dempo, Indonesia (USNM). **Distr.:** Indonesia. **Ety.:** n.s. ["fresh, young, recent" (L)]. **Informal name:** Recent Indonesian Nail Mosquito.

rennellensis Taylor & Maffi
1991: 196 (F, P, L). **Type loc.:** Rennell Island, Solomon Islands (NHMUK). **Distr.:** Solomon Isl. **Refs.:** Harbach & Ward 1994: 73 (lectotype desig.); Beebe et al. 2015 (review). **Ety.:** Rennell Isl., Solomon Isl. **Informal name:** Rennell Solomon Islander Nail Mosquito.

rhodesiensis Theobald
1901a: 184 (M*, F*). **Type loc.:** Mashonaland, Southern Rhodesia [Salisbury, Zimbabwe] (NHMUK). **Distr.:** Angola, Benin, Botswana, Burkina Faso, Cameroon, Central African Republic, Chad, Côte d'Ivoire, Democratic Republic of the Congo, Djibouti, Eritrea, Eswatini, Ethiopia, Gabon, Ghana, Guinea, Kenya, Malawi, Mali, Mauritania, Morocco, Mozambique, Namibia, Niger, Nigeria, Republic of the Congo, Republic of South Africa, Saudi Arabia, Senegal, Seychelles, Sierra Leone, Somalia, South Sudan, Sudan, Tanzania, Uganda, Zambia, Zimbabwe. **Refs.:** Christophers & Puri 1931b; De Meillon 1947b: 91 (M*, F*, L*); Gillies 1955: 159 (E); Ahmed et al. 2011 (distr.; Saudi Arabia; as *rupicola*); Kyalo et al. 2017 (distr.; sub-Saharan Africa). **Ety.:** Rhodesia (Zimbabwe). **Informal name:** Rhodesia Zimbabwean Nail Mosquito.

ssp. ***rupicolus*** Lewis
1937: 181 (M*, F*, P, L; as sp.). **Type loc.:** Gebel [Jebel] Moya, near Sennar, Blue Nile, Sudan (NHMUK). **Distr.:** Algeria, Chad, Egypt, Eritrea, Ethiopia, Jordan, Lebanon, Morocco, Niger, Saudi Arabia, Sudan, Syria, Yemen. **Refs.:** De Meillon 1947b: 97 (M*, F*, L*, E*; ? = *rhodesiensis*); Lewis 1956a: 478 (tax.); Mattingly & Knight 1956: 96 (to ssp.); Senevet et al. 1959: 596 (P*); Kyalo et al. 2017 (distr.; sub-Saharan Africa); Robert et al. 2019 (distr.; western Palearctic). **Ety.:** rock dweller (L); "many hundreds [larvae] in a deep cleft in the rock." **Informal name:** Rock-dwelling Sudanese Nail Mosquito.

ssp. syn. *aegypti* Salem
1938: 1 (M*, F*, P*, L*, E*). **Type loc.:** Wadi Taba, Sinai, Egypt (LU). **Refs.:** Edwards 1941: 442 (? = *rupicolus*).

ssp. syn. *dthalisimilis* Corradetti
1939a: 58 (M, F, P*, L; *rhodesiensis* var.). **Type loc.:** Semien District [Ethiopia] (LU). **Refs.:** Edwards 1941: 442 (? = *rupicolus*).

riparis King & Baisas
1936: 83 (M*, F*, L*; *leucosphyrus* var.). **Type loc.:** Titunod, Kolambugan, Lanao, Mindanao, Philippines (USNM). **Distr.:** Malaysia, Philippines. **Refs.:** Colless 1956: 65 (M, F*, L*; to sp.); Reid 1968: 305 (M, F*, P*, L*; tax., distr.); Basio 1971b: 44 (M*, F*; bion.); Sallum et al. 2005b: 76 (M, F*, P*, L*). **Ety.:** n.s. [prob. ref. to larvae at stream banks, shore (L)]. **Informal name:** Freshwater Philippine Nail Mosquito.

rivulorum Leeson
1935: 70 (M, F, P, L; *funestus* var.). **Type loc.:** Muzari River, Sinoia, Southern Rhodesia [Zimbabwe] (NHMUK). **Distr.:** Angola, Benin, Botswana, Burkina Faso, Cameroon, Côte d'Ivoire, Democratic Republic of the Congo, Eritrea, Eswatini, Ethiopia, Guinea, Kenya, Malawi, Mali, Mozambique, Namibia, Niger, Nigeria, Republic of the Congo, Republic of South Africa, Senegal, Seychelles, South Sudan, Sudan, Tanzania (includes Zanzibar), Uganda, Zambia, Zimbabwe. **Refs.:** Evans 1938: 167 (A, P, L, E*; as *garnhamellus*); De Meillon 1947b: 125 (M*, F*, P*, L*, E*); Ribeiro & da Cunha Ramos 1975: 18 (distr.); Dia et al. 2013 (tax., bion., distr.; Africa); Kyalo et al. 2017 (distr.; sub-Saharan Africa). **Ety.:** little streams (L). **Informal name:** Brookside Zimbabwean Nail Mosquito.

syn. *garnhamellus* Evans & Leeson
1937: 384 (A*, P, L; *rivulorum* var.). **Type loc.:** Kisumu, Kenya (LSTM). **Refs.:** De Meillon 1947b: 125 (syn.).

rodhaini Leleup & Lips
1950: 303 (M*, F*, L). **Type loc.:** Kakontwe, Katanga, Belgian Congo [Democratic Republic of the Congo] (CMT). **Distr.:** Democratic Republic of the Congo. **Refs.:** Leleup & Lips 1951 (tax., morph.); Kyalo et al. 2017 (distr.; sub-Saharan Africa). **Ety.:** F. Rodhain. **Informal name:** Rodhain Congolese Nail Mosquito.

roubaudi Grjebine
1953a: 476 (M*, F*, P*, L*). **Type loc.:** La Mandraka (District Manjakandriana) [Tananarive], Madagascar (IRSM). **Distr.:** Madagascar (includes Glorioso and Juan de Nova Isl.). **Refs.:** Kyalo et al. 2017 (distr.; sub-Saharan Africa). **Ety.:** Émile Roubaud. **Informal name:** Roubad Malagasy Nail Mosquito.

ruarinus Edwards
1940: 93 (M, F, P*, L*). **Type loc.:** Ruia Estate, PO M'Sonneddi, Salisbury [Mashonaland], Southern Rhodesia [Zimbabwe] (NHMUK). **Distr.:** Angola, Eswatini, Namibia, Republic of South Africa, Zimbabwe. **Refs.:** De Meillon 1947b: 95 (M, F*, L*); Kyalo et al. 2017 (distr.; sub-Saharan Africa). **Ety.:** rock pools, *ruari*. **Informal name:** Ruari Zimbabwean Nail Mosquito.

rufipes (Gough)
1910: 119 (F; *Nyssorhynchus pretoriensis* var.). **Type loc.:** Onderstepoort, Transvaal [Republic of South Africa] (LU). **Distr.:** Algeria, Angola, Benin, Botswana, Burkina Faso, Cameroon, Central African Republic, Chad, Côte d'Ivoire, Democratic Republic of the Congo, Eritrea, Eswatini, Ethiopia, Gabon, The Gambia, Ghana, Guinea, Guinea-Bissau, Kenya, Madagascar (includes Glorioso and Juan de Nova Isl.), Malawi, Mali, Mauritania, Mozambique, Namibia, Niger, Nigeria, Republic of the Congo, Republic of South Africa, Senegal, Seychelles, Sierra Leone, South Sudan, Sudan, Tanzania (includes Zanzibar), Togo, Uganda, Zambia, Zimbabwe. **Refs.:** Macfie & Ingram 1922 (F*); Evans 1938: 349 (M, F*, P, L*, E*); De Meillon 1947b: 221 (M*, F*, L*, E*); Hamon, Taufflieb, & Dyemkouma 1961: 24 (tax.); Gillies & De Meillon 1968: 244 (M*, F*, P, L*, E*; syn.); Ribeiro & da Cunha Ramos 1975: 27 (distr.); Service 1976a (distr.; Gabon); Kyalo et al. 2017 (distr.; sub-Saharan Africa). **Ety.:** reddish foot (L). **Informal name:** Red-footed South African Nail Mosquito.

syn. *watsoni* Edwards
1911a: 143 (M, F). **Type loc.:** Katagum [Northern Provinces], Nigeria (NHMUK). **Refs.:** Kyalo et al. 2017 (distr., Ghana).

ssp. ***broussesi*** Edwards
1929a: 83 (M*, F*, L*). **Type loc.:** Djanet [Fort Charlet], southern Tassili-n-Ajjer Mountains [Saharan Oasis], Algeria (NHMUK). **Distr.:** Algeria, Burkina Faso, Chad, Egypt, Ghana, Libya, Mali, Niger, Nigeria. **Refs.:** Martini 1930: 170 (M, F, L; tax.; as *brousseri*, *lapsus*); Senevet 1933: 201 (P*); De Meillon 1947b: 225 (A, P, L; as *ingrami*); Mattingly 1949a: 23 (P*; as *ingrami*); Senevet & Andarelli 1955a: 172 (A*, P*, L*); Rioux 1960: 62 (to ssp.); Gillies & De Meillon 1968: 246 (F*; syn.); Robert et al. 2019 (distr.; western Palearctic). **Ety.:** André Louis Gaston Brousses. **Informal name:** Brousses Algerian Nail Mosquito.

ssp. syn. *ingrami* Edwards
1929c: 322 (A; *Anopheles rufipes* var.). **Type loc.:** Bole, Gold Coast [Ghana] (NHMUK).

ssp. syn. *seneveti* Rioux
1959: 733 (A). **Type loc.:** Tibesti, North Chad, French Equatorial Africa [Chad] (LU).

ssp. syn. *brucechwatti* Hamon, Taufflieb, & Dyemkouma
1961: 25 (M*, F*). **Type loc.:** Largeau, Chad, French Equatorial Africa (IERT).

salbaii Maffi & Coluzzi
1958: 48 (M*, F*, P*, L*). **Type loc.:** Salbai, Further Giuba Region, Somalia (IMEM, MIT, NHMUK). **Distr.:** Djibouti, Egypt, Eritrea, Ethiopia, Kenya, Niger, Somalia. **Refs.:** Gillies & De Meillon 1968: 316 (distr.); Rodhain et al. 1977 (distr.); Townsend 1990: 130 (type info.); Kyalo et al. 2017 (distr.; sub-Saharan Africa). **Ety.:** Tuc Salbai [a small stream in Somalia]. **Informal name:** Salbai Somalian Nail Mosquito.

saungi Colless
1955a: 334 (M*, F*, P*, L*). **Type loc.:** Saung-Saung, Sunsuran Trace [Sabah], North Borneo [Malaysia Borneo] (NHMUK). **Distr.:** Malaysia. **Refs.:** Reid 1968: 278 (M*, F*, P*, L*; keys). **Ety.:** Saung-Saung, Sunsuran Trace, North Borneo. **Informal name:** Saung-Saung Malaysian Nail Mosquito.

sawadwongporni Rattanarithikul & Green
1987: 256 (M, F*, E*). **Type loc.:** Ban Thung Khua Mu, Rong Kwang, Phrae Province, Thailand (USNM). **Distr.:** India, Laos, People's Republic of China, Thailand, Vietnam. **Refs.:** Rattanarithikul et al. 2006b (F*, L*; bion., distr., keys); Qu & Zhu 2008 (distr.; People's Republic of China); Morgan et al. 2011 (distr., phylogeography). **Ety.:** Phorn Sawadwongporn. **Informal name:** Sawadwongporn Thai Nail Mosquito.

scanloni Sallum & Peyton
2005 (M, F, P*, L*). In Sallum et al. 2005: 170. **Type loc.:** Sai Yok (Office of Livestock), Kanchanaburi, Thailand (USNM). **Distr.:** Thailand, Vietnam. **Refs.:** Rattanarithikul et al. 2006b (F*, L*; bion., distr., keys); Sallum et al. 2007: 30 (phyl.). **Ety.:** John E. Scanlon. **Informal name:** Scanlon Thai Nail Mosquito.

schueffneri Stanton
1915: 373 (F*). **Type loc.:** Lampong, Sumatra (NHMUK). **Distr.:** Indonesia. **Refs.:** Bonne-Wepster & Swellengrebel 1953: 451 (M*, F*, L*); Reid 1968: 352 (F, L; tax, keys). **Ety.:** Wilhelm Augustus Paul Schüffner. **Informal name:** Schüffner Indonesian Nail Mosquito.

schwetzi Evans
1934: 555 (F*). **Type loc.:** Kalunga, Katanga, Belgian Congo [Democratic Republic of the Congo] (NHMUK). **Distr.:** Angola, Democratic Republic of the Congo, Gabon, Mali, Republic of South Africa, Tanzania, Zambia, Zimbabwe. **Refs.:** De Meillon 1947b: 159 (M*, F*, P, L*, E; as *michaeli*); Peters 1955b: 101 (F*; to var.); Gillies & De Meillon 1968: 123 (M*, F*, P, L*, E*; to sp.); Ribeiro & da Cunha Ramos 1975: 22 (distr.); Kyalo et al. 2017 (distr.; sub-Saharan Africa). **Ety.:** Jacques Schwetz. **Informal name:** Schwetz Congolese Nail Mosquito.

syn. *michaeli* De Meillon & Leeson
1940: 63 (M, F*, P*, L*). **Type loc.:** Ndola, Northern Rhodesia [Zambia] (SAIM). **Refs.:** Gillies & De Meillon 1968: 123 (syn.).

seretsei Abdulla-Khan, Coetzee & Hunt
1998 (M, F*, P*, L*, E*). **Type loc.:** Kasane, Botswana (SAIMR). **Distr.:** Botswana. **Refs.:** Kyalo et al. 2017 (distr.; sub-Saharan Africa). **Ety.:** Seretse Khama. **Informal name:** Seretse Botswanan Nail Mosquito.

sergentii (Theobald)
1907: 68 (F*; *Pyretophorus*). **Type loc.:** Algeria (NHMUK). **Distr.:** Albania, Algeria, Bulgaria, Burkina Faso, Cameroon, Chad, Côte d'Ivoire, Djibouti, Egypt, Eritrea, Greece, Iran, Iraq, Israel (and Gaza Strip and West Bank), Italy (includes Sicily), Jordan, Libya, Kenya, Morocco, Pakistan, Portugal, Saudi Arabia, Senegal, Spain (includes Canary Isl.), South Sudan, Sudan, Tunisia, Turkmenistan. **Refs.:** Séguy 1924: 161 (L*); Theodor 1925: 381 (E*); Senevet 1930: 303 (P*); Christophers 1933: 193 (M*, F*, P, L*, E); Saliternik 1955: 445 (L*, E*; bion.); Mattingly & Knight 1956: 97 (tax.); Gillies & De Meillon 1968 (F, L; keys; comp. to ssp. *macmahoni*); Glick 1992 (F*; key, tax., distr.); Abdullah & Merdan 1995 (distr.; Saudi Arabia); Amr et al. 1997 (distr.; Jordan); Trari et al. 2002: 331 (distr.; Morocco); Mamedniyazov & Yerokhin 2005 (keys, distr.; Turkmenistan); Azari-Hamidian & Harbach 2009 (F, L; keys, distr.; Iran); Sinka et al. 2010a: 117 (bion. review, distr., niche model); Becker et al. 2010: 183 (M*, F*, L*; keys, tax., distr., bion.); Kyalo et al. 2017 (distr.; sub-Saharan Africa); Robert et al. 2019 (distr.; western Palearctic). **Ety.:** Edmond Sergent. **Informal name:** Sergent Algerian Nail Mosquito. (Vol. 1, p. 368)

ssp. ***macmahoni*** Evans
1936: 538 (M, F, P, L*; as sp.). **Type loc.:** Isiolo [central], Kenya (NHMUK). **Distr.:** Algeria, Burkina Faso, Côte d'Ivoire, Djibouti, Eritrea, Ethiopia, Ghana, Guinea, Kenya, Mali, Republic of South Africa, Senegal, Somalia, South Sudan, Sudan. **Refs.:** De Meillon 1947b: 187 (M*, F*, L*); Lewis 1956a: 481 (L); Mattingly & Knight 1956: 97 (tax.; to ssp.); Rodhain et al. 1977 (distr.); Kyalo et al. 2017 (distr.; sub-Saharan Africa). **Ety.:** James Patrick MacMahon. **Informal name:** MacMahon Kenyan Nail Mosquito.

ssp. syn. *barkhuusi* Giaquinto-Mira
1950: 294 (L; *Anopheles* (*Myzomyia*) *macmahoni* var.). **Type loc.:** Valley of Becilo, Ethiopia (NE). **Refs.:** Stone et al. 1959: 54 (syn.).

seydeli Edwards
1929c: 322 (A; *theileri* var.). **Type loc.:** La Lunga, Katanga, Belgian Congo [Democratic Republic of the Congo] (NHMUK). **Distr.:** Burundi, Democratic Republic of the Congo, Ethiopia, Malawi, Mozambique, Tanzania, Zambia, Zimbabwe. **Refs.:** De Meillon 1947b: 161 (M*, F*, L*; to sp.); Gillies 1955: 160 (E*); O'Connor 1967 (distr.); Kyalo et al. 2017 (distr.; sub-Saharan Africa). **Ety.:** Charles Henri Victor Seydel. **Informal name:** Seydel Congolese Nail Mosquito.

smithii Theobald
1905b: 101 (F). **Type loc.:** Sierra Leone (NHMUK). **Distr.:** Benin, Cameroon, Côte d'Ivoire, Equatorial Guinea, Gabon, Guinea, Guinea-Bissau, Kenya, Liberia, Nigeria, Sierra Leone. **Refs.:** De Meillon 1947b: 85 (M*, F*, L*); Gillies & De Meillon 1968: 53 (M*, F*, P, L*, E*; syn.); Service 1976a (distr.; Gabon); Brunhes et al. 1999: 556 (F*, L*; tax.); Kyalo et al. 2017 (distr.; sub-Saharan Africa; as *s.s.* and syn. *cavernicolus*). **Ety.:** Frederick Smith. **Informal name:** Smith Sierra Leonean Nail Mosquito.

syn. *pallidopalpi* Theobald
1907: 57 (M*; *Feltinella*). **Type loc.:** Mount Aureol, Sierra Leone (NHMUK).

syn. *cavernicolus* Abonnenc
1954: 288, 1956: 802 (M*, F*). **Type loc.:** Dalaba, French Guinea (EMD). **Refs.:** Brunhes et al. 1999: 556 (tax.; syn. with *smithii*).

solomonis Belkin, Knight & Rozeboom
1945: 247 (M, F*, P*, L*). **Type loc.:** Poha River tributary, Guadalcanal, Solomon Islands (USNM). **Distr.:** Papua New Guinea, Solomon Isl. **Refs.:** Belkin 1962: 150 (M*, F*, P*, L*, E); Lee et al. 1987b: 286 (F; key, tax., bion., distr., review). **Ety.:** Solomon Isl. **Informal name:** Poha Solomon Islander Nail Mosquito.

somalicus Rivola & Holstein
1957: 382 (M*, P*, L*; as *nili* var.). **Type loc.:** Villabruzzi, Audegle, and Uebi Scebeli, Somalia (IERT, NHMUK). **Distr.:** Burkina Faso, Cameroon, Côte d'Ivoire, Guinea, Mali, Sierra Leone, Somalia. **Refs.:** Gillies & De Meillon 1968: 4, 87 (M*, F, P*, L*; to sp.); Townsend 1990: 135 (type info.); Antonio-Nkondjio et al. 2009b (L; bion.); Antonio-Nkondjio & Simard 2013 (bion., mol. tax., review); Kyalo et al. 2017 (distr.; sub-Saharan Africa). **Ety.:** Somalia. **Informal name:** Ajuran Somalian Nail Mosquito [the Ajuran Sultanate was a medieval Muslim empire located in present-day Somalia].

splendidus Koidzumi
1920: 23 (M, F, L). **Type loc.:** Suisha of Jitsugetsutan, Nanto Prefecture and Rivers Sobiu and Dakusui, Ako Prefecture, Formosa [Taiwan, Republic of China] (LU). **Distr.:** Afghanistan, Bangladesh, Bhutan, Cambodia, India, Laos, Myanmar, Nepal, Pakistan, People's Republic of China, Taiwan, Thailand, Vietnam. **Refs.:** Christophers 1933: 296 (M*, F*, P, L*, E); Wu 1936: 268 (E*); Bonne-Wepster & Swellengrebel 1953: 469 (M, F*, L*); Hara 1959: 111 (F*); Nguyen Thuong Hien 1968 (F*, L*; keys, tax., bion., distr.; Vietnam); Aslamkhan 1971b (distr.; Pakistan); Darsie & Pradhan 1990, 1991 (F, L; tax., keys, bion., distr.; Nepal); Oo et al. 2004 (distr.; Myanmar); Gupta & Chaudhry 2005 (E*); Rattanarithikul et al. 2006b (F*, L*; bion., distr., keys); Morgan et al. 2009 (mol. phyl.; Neocellia Series); Morgan et al. 2011 (distr., phylogeography); Namgay et al. 2018 (bion., distr.; Bhutan). **Ety.:** n.s. [clear, bright, beautiful (L)]. **Informal name:** Splendid Taiwanese Nail Mosquito.

syn. *indiensis* Theobald
1903a: 99 (not Theobald 1901a) (A; *Nyssorhynchus maculipalpis* var.). **Type loc.:** [Nagpur, Central Provinces], India (NHMUK).

squamosus Theobald
1901a: 167 (F*). **Type loc.:** British Central Africa [Malawi] (NHMUK). **Distr.:** Angola, Benin, Botswana, Burkina Faso, Burundi, Cameroon, Central African Republic, Chad, Côte d'Ivoire, Democratic Republic of the Congo, Eritrea, Eswatini, Ethiopia, Gabon, The Gambia, Ghana, Guinea, Guinea-Bissau, Kenya, Liberia, Madagascar (includes Glorioso and Juan de Nova Isl.), Malawi, Mali, Mauritania, Mozambique, Namibia, Niger, Nigeria, Republic of the Congo, Republic of South Africa, Rwanda, Senegal, Sierra Leone, Somalia, South Sudan, Sudan, Tanzania (includes Zanzibar), Togo, Uganda, Zambia, Zimbabwe. **Refs.:** Macfie & Ingram 1922 (F*); Evans 1938: 379 (M, F*, P*, L*); De Meillon 1947b: 239 (M*, F*, L*); Duznetsov 1971: 441 (distr.; Arabian Peninsula); Kyalo et al. 2017 (distr.; sub-Saharan Africa). **Ety.:** scaly (L); a very scaly species. **Informal name:** Scaly Malawian Nail Mosquito.

syn. *tananariviensis* Ventrillon
1906c: 198 (M, F; *Cellia*). **Type loc.:** Tananarive, Madagascar (MNHP, NHMUK). **Refs.:** Townsend 1990: 140 (type info.).

syn. *arnoldi* Stephens & Christophers
1908: 175 (A; *Cellia*). **Type loc.:** Transvaal [Republic of South Africa] (NHMUK).

syn. *pretoriensis* Gough
1910: 117 (A; *Cellia*). **Type loc.:** Onderstepoort, Transvaal [Republic of South Africa] (LSTM).

stephensi Liston
1901: 441 (F*). **Type loc.:** Ellichpur [Amaroati District, Central Provinces], India (NHMUK). **Distr.:** Afghanistan, Bahrain, Bangladesh, Djibouti, Egypt, Ethiopia, India, Iran, Iraq, Kuwait, Myanmar, Nepal, Oman, Pakistan, People's Republic of China, Philippines, Qatar, Saudi Arabia, Sri Lanka, Thailand, Vietnam. **Refs.:** Senevet 1930: 325 (P*); Christophers & Barraud 1931: 179 (E*); Christophers 1933: 273 (M*, F*, L*); Rao et al. 1938: 261 (E*); Ross & Roberts 1943b: 29 (M*, F*, L*; tax., distr., bion.); Pringle 1952: 779 (L*; tax.); Aslamkhan 1971b (distr.; Pakistan); Aslamkhan et al. 1972: 29 (morphological variation); Green 1982: 49–61 (chrom.; *stephensi* species group); Townsend 1990: 136 (type info.); Darsie & Pradhan 1990, 1991 (F, L; tax., keys, bion., distr.; Nepal); Glick 1992 (F*; key, distr., tax.); Oo et al. 2004 (distr.; Myanmar); Rattanarithikul et al. 2006b (F*, L*; bion., distr., keys); Morgan et al. 2009 (mol. phyl.; Neocellia Series); Becker et al. 2010: 347 (F*, L*; key, tax., distr., bion.); Sinka et al. 2011: 89 (bion., review, distr., niche model); Dev & Sharma 2013 (tax., bion.; *stephensi* and variants); Robert et al. 2019 (distr., western Palearctic). **Ety.:** John William Watson Stephens. **Informal name:** Stephens Indian Nail Mosquito. (Vol. 1, p. 370)

syn. *metaboles* Theobald
1902a: 374 (F*). **Type loc.:** Lahore [Punjab, Pakistan] (NHMUK).

syn. *intermedia* Rothwell
1907: 34 (M, F; *Neocellia*). **Type loc.:** Deesa [Bombay], India (NHMUK).

syn. *folquei* de Mello
1918: 363 (F*). **Type loc.:** Pragana, Nagar Avely, Damao [India] (IM).

syn. *mysorensis* Sweet & Rao
1937: 673 (E; *stephensi* var.). **Type loc.:** India (LU). **Refs.:** Rao et al. 1938: 262 (E*); Puri 1949: 495 (to ssp.); Rutledge et al. 1970: 1024 (syn.).

stookesi Colless
1955a: 331 (M*, F*, P*, L*). **Type loc.:** Tambunan, North Borneo [Sabah], [Malaysian] Borneo (NHMUK). **Distr.:** Malaysia (Borneo, Sabah). **Refs.:** Reid 1968: 275 (A*, P*, L*; keys). **Ety.:** Valentine Alexander Stookes. **Informal name:** Stookes Malaysian Nail Mosquito.

subpictus Grassi
1899 (A). In Grassi et al. 1899: 101. **Type loc.:** India (RUM). **Distr.:** Afghanistan, Australia, Bangladesh, Bhutan, Cambodia, India, Indonesia, Iran, Laos, Malaysia, Maldives, Mariana Isl., Myanmar, Nepal, Pakistan, Papua New Guinea, People's Republic of China, Philippines, Saudi Arabia, Solomon Isl., Sri Lanka, Taiwan, Thailand, Timor, Vietnam. **Refs.:** Davis 1926 (F*; as *rossi* Giles); Christophers & Barraud 1931: 175 (E*); Christophers 1933: 231 (M*, F*, P, L*, E); Ross & Roberts 1943b: 21 (M*, F*, L*; tax., distr., bion.); Penn 1949b: 15 (P*); Bonne-Wepster & Swellengrebel 1953: 415 (M*, F*, L*); Hara 1959: 111 (F*); Reid 1966 (tax.); Nguyen Thuong Hien 1968 (F*, L*; keys, tax., bion., distr.; Vietnam); Reid 1968 (M*, F*, P*, L*; keys, tax., bion.); Aslamkhan 1971b (distr.; Pakistan); Basio 1971b: 45 (M*, F*; bion.); Darsie & Cagampang-Ramos 1971b: 28 (distr.); Narang et al. 1973a (chrom.); Ameen & Talukdar 1974c: 89 (P*); Harrison & Klein 1975: 11 (distr.); Suguna 1982 (tax; *s.l.*); Reuben & Suguna 1983 (tax.; *s.l.*); Lee et al. 1987b: 288 (F; key, tax., review); Darsie & Pradhan 1990, 1991 (F, L; tax., keys, bion., distr.; Nepal); Suguna et al. 1994 (tax., chrom.; *s.l.*); Whelan & Hapgood 2000 (bion., distr.; East Timor); Oo et al. 2004 (distr.; Myanmar); Singh et al. 2004 (tax.; *s.l.*); Chaudhry et al. 2005 (cyto. tax.; *s.l.*); Rattanarithikul et al. 2006b (F*, L*; bion., distr., keys); Chandra et al. 2010 (tax.; *s.l.*); Sinka et al. 2011: 89 (bion., review, distr., niche model; Subpictus Complex); Chhilar & Chaudhry 2012 (mol. tax.; *s.l.*); Surendran et al. 2013 (mol. tax.; *s.l.*); Kirti 2014b (morphology); Kyalo et al. 2017 (distr.; sub-Saharan Africa); Namgay et al. 2018 (bion., distr.; Bhutan). **Ety.:** nearly (L); near (smaller than) *pictus*. **Informal name:** Small-painted Indian Nail Mosquito. (Vol. 1, p. 372)

syn. *rossii* Giles
1899: 63 (M, F). **Type loc.:** Calcutta [West Bengal], India (NHMUK).
syn. *error* Theobald
1903a: 353 (F; *Aldrichia*). **Type loc.:** [Calcutta, West Bengal], India (NHMUK).

sulawesi Waktoedi Koesoemawinangoen
1954: 63 (A; as *An.* (*Myzomyia*) *leucosphyrus* var.). **Type loc.:** Bone-Dumoga Forest Reserve (0°35'S, 123°54'E), Toraut, Sulawesi (USNM). **Distr.:** Sulawesi. **Refs.:** Colless 1956: 66 (F*; as *leucosphyrus*, Celebes form); Reid 1968: 303 (F*; to sp.); Hii et al. 1988: 43 (authorship change); Sallum et al. 2005b: 72 (M*, F*, P*, L*; neotype desig., corrected type loc.); Sallum et al. 2007: 30 (phyl.). **Ety.:** Sulawesi, Indonesia. **Informal name:** Sulawesi Indonesian Nail Mosquito. (Vol. 1, p. 374)

sundaicus (Rodenwaldt)
1925: 185 (A*; *Myzomyia ludlowi* var.). **Type loc.:** Sunda Islands [neotype: Pandan Beach near Lundu, Sarawak, Malaysia] (Neotype in Sarawak Biodiversity Council (SBC), Kuching, Sarawak (Borneo), Malaysia). **Distr.:** Malaysia (Borneo, Sarawak); (as *s.l.*, Bangladesh, India, Indonesia, People's Republic of China, Philippines, Taiwan, Timor). **Refs.:** Christophers & Barraud 1931: 175 (E*); Christophers 1933: 245 (M*, F*, L, E; to sp.); Crawford 1938: 103 (P*); Ross & Roberts 1943b: 23 (M*, F*, L*; tax., distr., bion.); Colless 1948: 111 (M*, F*, L*); Bonne-Wepster & Swellengrebel 1953: 395 (M*, F*, L*); Reid 1959: 101 (L); Foote & Cook 1959 (key); Reid 1968: 343 (M*, F*, P*, L*, E*; tax); Aslamkhan 1971b (distr.; Pakistan); Lee et al. 1987b: 296 (F; key, tax., bion., distr., review); Sukowati & Baimai 1996 (chrom.; *s.l.*); Sukowati et al. 1999 (mol. tax.; *s.l.*); Whelan & Hapgood 2000 (bion., distr.; East Timor); Linton et al. 2001a (M*, F, P*, L*; mol. tax., neotype desig.); Oo et al. 2004 (distr.; Myanmar); Dusfour et al. 2004a (bion., syst.; *s.l.*); Dusfour et al. 2004b (tax., bion.; *s.l.*); Dusfour et al. 2007 (mol. tax.; *s.l.*); Dusfour et al. 2007 (tax.; *s.l.*); Becker et al. 2010: 348 (F*, L*; key, tax., distr., bion.); Sinka et al. 2011: 89 (bion., review, distr., niche model; Sundaicus Complex); Dev & Sharma 2013 (tax., bion.; Sundaicus Complex). **Ety.:** Sunda Isl. **Informal name:** Sunda Bornean Nail Mosquito. (Vol. 1, p. 376)

superpictus Grassi
1899: 560 (A). **Type loc.:** Italy (LU). **Distr.:** Afghanistan, Albania, Algeria, Armenia, Azerbaijan, Bosnia and Herzegovina, Bulgaria, Croatia, Cyprus, Egypt, France (Corsica), FYRO Macedonia, Georgia, Greece, India, Iran, Iraq, Israel (and Gaza Strip and West Bank), Italy (includes Sicily), Jordan, Kazakhstan, Kyrgyzstan, Lebanon, Libya, Montenegro, Norway, Pakistan, Portugal (Azores), Romania, Russia (Southern Dists.), Serbia, Spain, Syria, Tajikistan, Tunisia, Turkey, Turkmenistan, Uzbekistan. **Refs.:** Theodor 1925: 381 (E*); Senevet 1930: 319 (P*); Christophers 1933: 264 (M*, F*, P, L*, E); Ross & Roberts 1943b: 27 (M*, F*, L*; tax., distr., bion.); Monchadskii 1951: 96 (L*); Senevet & Andarelli 1956c (tax.); Mattingly 1969c (E*); Gutsevich et al. 1974: 114 (M*, F*, L*; tax.); Glick 1992 (F*; key, tax., distr.); Amr et al. 1997 (distr.; Jordan); Ramsdale & Snow 2000: 10 (distr.); Hadjivassilis 2000: 38 (distr.; Cyprus); Voyadjoglou & Harbach 2001 (F; key, distr.; Greece); Oshaghi et al. 2007 (mol. tax.; *s.l.*); Azari-Hamidian & Harbach 2009 (F, L; keys, distr.; Iran); Morgan et al. 2009 (mol. phyl.; Neocellia Series); Sinka et al. 2010a: 117 (bion., review, distr., niche model); Becker et al. 2010: 185 (M*, F*, L*; keys, tax., distr., bion.); Habirov et al. 2012 (mol. tax., bion., distr.; Tajikistan); Robert et al. 2019 (distr.; western Palearctic). **Ety.:** really (*pictus*). **Informal name:** Super-painted Italian Nail Mosquito. (Vol. 1, p. 378)
syn. *palestinensis* Theobald
1903a: 71 (M*, F*; *Pyretophorus*). **Type loc.:** several localities in Palestine [= Lebanon and Syria] and Cyprus (NHMUK).
syn. *nursei* Theobald
1907: 66 (F; *Pyretophorus*). **Type loc.:** Quetta [Baluchistan, Pakistan] (NHMUK).
syn. *cardamatisi* Newstead & Carter
1910: 379 (M*, F*; *Pyretophorus*). **Type loc.:** Athens [Attica], Greece (NHMUK).
syn. *vassilievii* Portschinsky
1911: 58 (A*; *Anopheles superpictus* var.). **Type loc.:** Tashkent [Uzbekistan] and Golodnaya Steppe [Karaganda, Kazakhstan] (LU).
syn. *macedoniensis* Cot & Hovasse
1917: 890 (A*, L; *Pyretophorus superpictus* var. and as *Pyretophorus macedoniensis*). **Type loc.:** Macedonia, Greece (FMP).
syn. *berestnevi* Shingarev
1926: 48 (F; *Anopheles* (*Myzomyia*) *superpictus* var.). **Type loc.:** Near Tashkent [Uzbekistan] (LU). **Refs.:** Christophers 1933: 264 (syn.).
syn. *atheniensis* Cardamatis
1931: 129 (A*; *Pyretophorus*). **Type loc.:** Athens [Attica], Greece (LU). **Refs.:** Pandazis 1935: 4 (syn.) [an earlier record of syn. than recorded by Ward 1984b: 269].
syn. *hellenicus* Peus
1954: 76 (E*). **Type loc.:** Langadas Loutra, Thessaloniki, Macedonia, Greece (NE).

swahilicus Gillies
1964b: 67 (M*, F*, P, L*). **Type loc.:** Goshi, Molindi District, Kenya (NHMUK). **Distr.:** Kenya, Tanzania (includes Zanzibar). **Refs.:** Kyalo et al. 2017 (distr.; sub-Saharan Africa). **Ety.:** Swahili peoples of East Africa. **Informal name:** Swahili Kenyan Nail Mosquito.

takasagoensis Morishita
1946: 21, 65 (M, F; *leucosphyrus* var.). **Type loc.:** Takeng, Taiwan Valley, Pei Tung, Taichung, Taiwan [Republic of China] (USNM). **Distr.:** Cambodia, People's Republic of China, Taiwan, Thailand, Vietnam. **Refs.:** Stone 1963:

119 (Colless, in *litt.*; syn.); Peyton & Harrison 1980: 337 (from syn. with *balabacensis*); Sallum et al. 2005b: 51 (M*, F*, P*, L*; neotype desig., type loc. info.); Sallum et al. 2007: 30 (phyl.); Qu & Zhu 2008 (distr.; People's Republic of China). **Ety.:** Takasagozaku [Japanese name for Taiwanese people]. **Informal name:** Takasagozaku Taiwanese Nail Mosquito.

tchekedii De Meillon & Leeson
1940: 64 (F*). **Type loc.:** Okavango Swamp, Bechuanaland [Botswana] (NHMUK). **Distr.:** Angola, Botswana. **Refs.:** Gillies & De Meillon 1968: 125 (M*, F*, P, L*; keys); Segerman 1990: 53 (type dep.); Kyalo et al. 2017 (distr.; sub-Saharan Africa). **Ety.:** Tchekedi Khama. **Informal name:** Tchekedi Botswanan Nail Mosquito.

tessellatus Theobald
1901a: 175 (F*). **Type loc.:** Taipang [= Taiping], Perak, Malaya [Malaysia] (NHMUK). **Distr.:** Bangladesh, Bhutan, Cambodia, Guam, India, Indonesia, Laos, Malaysia, Maldives, Myanmar, Nepal, Papua New Guinea, People's Republic of China, Philippines, Solomon Isl., Sri Lanka, Taiwan, Thailand, Timor, Vietnam. **Refs.:** Christophers & Barraud 1931: 174 (E*); Christophers 1933: 182 (M*, F*, P, L*, E); Crawford 1938: 82 (P*); D'Abrera 1944: 353 (E); Colless 1948: 98 (M*, F*, L*); Bonne-Wepster & Swellengrebel 1953: 274 (M*, F*, L*; as *tesselatus*); Baisas & Ubaldo-Pagayon 1956: 224 (first instar L*, E*); Hara 1959: 112 (F gen.*); Baisas & Dowell 1967 (F*, L*; keys; Philippines); Nguyen Thuong Hien 1968 (F*, L*; keys, tax., bion., distr.; Vietnam); Reid 1968: 257 (M*, F*, P*, L*, E*); Miyagi et al. 1969: 33 (tax.); Aslamkhan 1971b (distr.; Pakistan); Basio 1971b: 46 (M*, F*; bion.); Basio & Reisen 1971: 60 (L; distr.); Rattanarithikul & Harrison 1973 (L*; key; Thailand); Narang et al. 1974 (chrom.); Baisas 1974: 72 (M, F*, P, L*; tax., bion., distr.; Philippines); Tanaka et al. 1975c: 210 (bion., distr.); Tanaka et al. 1979 (M*, F*, P, L*; keys); Lee et al. 1987b: 301, 306 (*s.l.*; F key, tax., bion., distr., review); Darsie & Pradhan 1990, 1991 (F, L; tax., keys, bion., distr.; Nepal); Amerasinghe 1995 (P*; key; Sri Lanka); Whelan & Hapgood 2000 (bion., distr.; East Timor); Oo et al. 2004 (distr.; Myanmar); Rattanarithikul et al. 2006b (F*, L*; bion., distr., keys); Oo et al. 2006 (F*, L*; keys; Myanmar); Namgay et al. 2018 (bion., distr.; Bhutan). **Ety.:** mosaic, checkered (L); ref. to thoracic pattern. **Informal name:** Tessellated Malaysian Nail Mosquito. (Vol. 1, p. 380)

syn. *formosae* Hatori
1901: 275 (A*). **Type loc.:** Taihoku, Kielun, and Tansui, Formosa [Taiwan, Republic of China] (LU).

syn. *deceptor* Dönitz
1902: 60 (F*). **Type loc.:** Sumatra [Indonesia] (NE).

syn. *thorntonii* Ludlow
1904a: 69 (F; *Myzomyia*). **Type loc.:** Cottabato [Cotabato], Mindanao, Philippines (USNM). **Refs.:** Stone & Knight 1956b: 280 (type info., lectotype desig.).

syn. *ceylonica* Newstead & Carter
1910: 377 (A*; *Dactylomyia*). **Type loc.:** Trincomalee, Ceylon [Sri Lanka] (NHMUK).

syn. *kinoshitai* Koidzumi
1917: 135 (F). **Type loc.:** Ryukokosho, Taihoku (Taipei), Formosa [Taiwan, Republic of China] (LU).

syn. *taiwanensis* Koidzumi
1917: 135 (F). **Type loc.:** Garden of Central Inst., Formosan government and Banshoryo, Ako Prefecture, Formosa [Taiwan, Republic of China] (LU).

ssp. ***orientalis*** (Swellengrebel & Swellengrebel de Graaf)
1920: 3 (L*; *Neomysomyia* punctulata var.). **Type loc.:** North Celebes [Sulawesi], Moluccas, and East Java [Indonesia] (NE). **Distr.:** Indonesia. **Refs.:** Bonne-Wepster & Swellengrebel 1953: 279 (F*, L*); Harbach & Howard 2007: 39 (to ssp.). **Ety.:** n.s. [from the East (L)]. **Informal name:** Eastern Indonesian Nail Mosquito.

ssp. ***kalawara*** Stoker & Waktoedi Koesoemawinangoen
1949: 54 (A*). **Type loc.:** Kalawara, Sulawesi Utara, Indonesia (LU). **Distr.:** Indonesia. **Refs.:** Harbach & Howard 2007: 40 (to ssp.). **Ety.:** named after type locality. **Informal name:** Kalawara Indonesian Nail Mosquito.

theileri Edwards
1912d: 247 (new name for *albipes* Theobald 1911b, not Theobald 1901a). **Distr.:** Angola, Burkina Faso, Burundi, Democratic Republic of the Congo, Ethiopia, Gabon, Iraq, Kenya, Kuwait, Mozambique, Nigeria, Republic of South Africa, Sierra Leone, Tanzania, Uganda, Zambia, Zimbabwe. **Refs.:** Evans 1938: 228 (M, F*, P*, L*, E*); De Meillon 1947b: 165 (M*, F*, L*, E*); O'Connor 1967 (distr.); Ribeiro & da Cunha Ramos 1975: 20 (distr.); Al-Houty 1997 (distr., Kuwait); Rueda et al. 2008 (distr.); Osório et al. 2010 (distr.; Portugal); Kyalo et al. 2017 (distr.; sub-Saharan Africa). **Ety.:** Arnold Theiler. **Informal name:** Theiler South African Nail Mosquito.

syn. *albipes* Theobald
1911b: 243 (F*; *Pyretophorus*). **Type loc.:** Onderstepoort, Transvaal [Republic of South Africa] (VRL, NHMUK). **Refs.:** Townsend 1990: 39 (type info.).

theobaldi Giles
1901b: 198 (F*). **Type loc.:** Ellichpur, Amraoti District, Central Provinces, India (NHMUK). **Distr.:** Bangladesh, India, Myanmar, Nepal, Pakistan. **Refs.:** Christophers & Barraud 1931: 180 (E); Christophers 1933: 287 (A, L); Bhombore et al. 1954; Aslamkhan 1971b (distr.; Pakistan); Darsie & Pradhan 1990, 1991 (F, L; tax., keys, bion., distr.; Nepal); Oo et al. 2004 (distr.; Myanmar). **Ety.:** Frederic Vincent Theobald. **Informal name:** Theobald Indian Nail Mosquito.

torresiensis Schmidt
2001 (M, F, P*, L*; tax.). In Schmidt et al. 2001: 400. **Type loc.:** Jabiru (12°40'S, 132°50'E), Northern Territory, Australia (UQIC). **Distr.:** Australia, Papua New Guinea. **Refs.:** Foley & Bryan 1993 (mol. tax.; as *farauti* 3); Beebe

et al. 2015 (review). **Ety.:** "Torresian zoogeographic province of northern Australia & southern New Guinea, the coastline of which was discovered in 1606 by Captain Luis Baez de Torres." **Informal name:** Torresian Australian Nail Mosquito.

turkhudi Liston
1901: 441 (F*). **Type loc.:** Ellichpur [Amaraoti District, Central Provinces], India (NHMUK). **Distr.:** Afghanistan, Algeria, Bangladesh, Djibouti, Egypt, Eritrea, Ethiopia, India, Iran, Iraq, Israel (and Gaza Strip and West Bank), Jordan, Morocco, Nepal, Pakistan, Saudi Arabia, Somalia, Sudan, Yemen. **Refs.:** Senevet 1931: 19 (P*); Christophers 1933: 252 (M*, F*, P, L*, E); Evans 1938: 336 (M*, F*, P, L*, E); De Meillon 1947b: 213 (M*, F*, L*, E*); Mattingly & Knight 1956: 98 (tax.); Rodhain et al. 1977 (distr.); Darsie & Pradhan 1990, 1991 (F, L; tax., keys, bion., distr.; Nepal); Kyalo et al. 2017 (distr.; sub-Saharan Africa); Robert et al. 2019 (distr.; western Palearctic). **Ety.:** Dnanesh Atmaran Turkhud. **Informal name:** Turkhud Indian Nail Mosquito.
syn. *azriki* Patton
1905: 632 (M*, F*, L*). **Type loc.:** Aziriki Spring near D'thala [West Aden Protectorate] [Yemen] (NE). **Refs.:** F.W. Edwards. In Evans 1938: 336 (syn.).
syn. *flaviceps* Edwards
1921c: 69 (M, F). **Type loc.:** Erkowit, Sudan (NHMUK). **Refs.:** Evans 1938: 336 (syn.).
syn. *persicus* Edwards
1921d: 280 (M; *Anopheles* (*Myzomyia*) *turkhudi* var.). **Type loc.:** East Iran (NHMUK). **Refs.:** Christophers 1933: 252 (tax.).
ssp. ***telamali*** Saliternik & Theodor
1942: 433 (M*, F*, L*). **Type loc.:** Tel Amal, Plain of Esdraelon, Israel (DPHU). **Distr.:** Israel (and Gaza Strip and West Bank). **Refs.:** Harbach & Howard 2007: 40 (to ssp.). **Ety.:** Tel Amal [Israel]. **Informal name:** Tel Amal Israeli Nail Mosquito.
syn. *amutis* De Burca
1943: 233 (M; L is different sp.). **Type loc.:** Seganeiti, Eritrea (NHMUK). **Refs.:** Gillies & De Meillon 1968: 228, 230 (syn.; in part); Townsend 1990: 42 (lectotype desig.).

vagus Dönitz
1902: 80 (M*, F*). **Type loc.:** Fort de Kock [West Coast], Sumatra, Banjoe-Biroe, Java, and numerous other Indonesian localities (ZM). **Distr.:** Afghanistan, Bangladesh, Bhutan, Cambodia, Guam, India, Indonesia, Laos, Malaysia, Mariana Isl., Myanmar, Nepal, People's Republic of China, Philippines, Singapore, Sri Lanka, Thailand, Vietnam. **Refs.:** Christophers & Barraud 1931: 175 (E*); Christophers 1933: 241 (M*, F*, P, L*, E); Crawford 1938: 106 (P*); Bonne-Wepster & Swellengrebel 1953: 425 (M*, F*, L*); Nguyen Thuong Hien 1968 (F*, L*; keys, tax., bion., distr.; Vietnam); Reid 1968: 242 (distr.); Aslamkhan 1971b (distr.; Pakistan); Basio 1971b: 46 (M*; bion.); Darsie & Cagampang-Ramos 1971a: 399 (distr.); Darsie & Cagampang-Ramos 1971b: 28 (distr.); Ameen & Talukdar 1974c: 88 (P*); Lee et al. 1987b: 308 (F; key, tax., bion., distr., review); Darsie & Pradhan 1990, 1991 (F, L; tax., keys, bion., distr.; Nepal); Oo et al. 2004 (distr.; Myanmar); Rattanarithikul et al. 2006b (F*, L*; bion., distr., keys); Rueda, Pecor et al. 2008: 397 (distr.); Zarowiecki et al. 2011 (distr., phylogeography); Namgay et al. 2018 (bion., distr.; Bhutan). **Ety.:** roving, wandering (L); ref. to its wide distribution. **Informal name:** Wandering Indonesian Nail Mosquito.
syn. *immaculata* James
1902: 35 (A). **Type loc.:** Ennur, Madras, India (NHMUK).
syn. *immaculatus* Theobald
1903a: 23 (F). **Type loc.:** "Evidently from Goa, India" (NHMUK; same type specimen as for *immaculata* James). **Refs.:** Christophers 1933: 241 (tax.).
syn. *flava* Swellengrebel
1917: 807 (M, F; *Myzomyia indefinita* var.; not Ludlow 1908). **Type loc.:** Soerabaia, Java [Indonesia] (NHMUK).
syn. *flavescens* Swellengrebel
1921: 47 (F; *Myzomyia*). **Type loc.:** Soerabaia, Java [Indonesia] (NE). **Refs.:** Christophers 1933: 245 (? = *sundaicus* Rodenwaldt); Reid 1968: 347 (syn.).
syn. *albino* Stoker & Waktoedi Koesoemawinangoen
1949: 26 (A*; *Anopheles* (*Myzomyia*) *vagus albino*). **Type loc.:** Indonesia (LU). **Refs.:** Reid 1968: 342 (syn.).

vaneedeni Gillies & Coetzee
1987: 81 (M*, F*, P, L). **Type loc.:** Pusela, Tzaneen, Republic of South Africa (NHMUK). **Distr.:** Botswana, Kenya, Namibia, Republic of South Africa, Zambia, Zimbabwe. **Refs.:** Dia et al. 2013 (tax., bion., distr.; Africa); Kyalo et al. 2017 (distr.; sub-Saharan Africa). **Ety.:** Gideon van Eeden. **Informal name:** Eeden South African Nail Mosquito.

vanhoofi Wanson & Lebied
1945: 118 (M*, F*, P*, L*). **Type loc.:** Grand Grotte at Thysville, Leopoldville, Belgian Congo [Kinshasa, Democratic Republic of the Congo] (CMT). **Distr.:** Democratic Republic of the Congo, Republic of the Congo. **Refs.:** Gillies & De Meillon 1968: 63 (M*, F*, P, L*); Kyalo et al. 2017 (distr.; sub-Saharan Africa). **Ety.:** Lucien van Hoof. **Informal name:** Hoof Congolese Nail Mosquito.

varuna Iyengar
1924: 24 (F*). **Type loc.:** Calcutta [West Bengal], India (LU). **Distr.:** Bangladesh, Bhutan, India, Indonesia, Laos, Myanmar, Nepal, Pakistan, People's Republic of China, Sri Lanka, Thailand, Vietnam. **Refs.:** Christophers 1933: 214 (M*, F*, L*; to sp.); Rao & Ramakrishna 1940 (L); D'Abrera 1944: 348 (E*); Reid 1968: 313 (M*, F*, L*; tax.); Aslamkhan 1971b (distr.; Pakistan); Lien et al. 1977 (distr.); Harrison 1980: 107 (M*, F*, P*, L*; distr.); Darsie & Pradhan 1990, 1991 (F, L; tax., keys, bion., distr.; Nepal); Harrison et al. 1991: 223 (distr.; Thailand.); Van Bortel et al. 2001: 730 (distr.; Vietnam and Myanmar

(as Burma)); Oo et al. 2004 (distr.; Myanmar); Rattanarithikul et al. 2006b (F*, L*; bion., distr., keys); Qu & Zhu 2008 (distr.; People's Republic of China); Namgay et al. 2018 (bion., distr.; Bhutan). **Ety.:** Varuna [Sanskrit, God of Rain]. **Informal name:** Varuna Indian Nail Mosquito.

vernus Gillies & De Meillon
1968: 89 (F*). **Type loc.:** Giant's Castle Reserve, Drakensberg Mountains, Natal [KwaZulu-Natal], [Republic of] South Africa (NHMUK). **Distr.:** Republic of South Africa. **Refs.:** Kyalo et al. 2017 (distr.; sub-Saharan Africa). **Ety.:** n.s. [of spring (L)]. **Informal name:** Springtime South African Nail Mosquito.

vinckei De Meillon
1942: 87 (M, F*, L). **Type loc.:** Congo River [Orientale], Belgian Congo [Democratic Republic of the Congo] (LU). **Distr.:** Democratic Republic of the Congo, Gabon. **Refs.:** Segerman 1990: 53 (type info.); Kyalo et al. 2017 (distr.; sub-Saharan Africa). **Ety.:** Ignace Henri Vincke. **Informal name:** Vincke Congolese Nail Mosquito.

walravensi Edwards
1930a: 290 (M, F). **Type loc.:** Elisabethville [Katanga], Belgian Congo [Democratic Republic of the Congo] (NHMUK). **Distr.:** Angola, Botswana, Democratic Republic of the Congo, Tanzania, Zambia, Zimbabwe. **Refs.:** De Meillon 1947b: 173 (M*, F*, P, L*, E*); Kyalo et al. 2017 (distr.; sub-Saharan Africa). **Ety.:** Paul Adelson François Walravens. **Informal name:** Walravens Congolese Nail Mosquito.

watsonii (Leicester)
1908: 38 (F; *Pyretophorus*). **Type loc.:** Near Klang [Selangor], Malaya [Malaysia] (NHMUK). **Distr.:** Malaysia. **Refs.:** Christophers 1923b (F*); Bonne-Wepster & Swellengrebel 1953: 262 (M*, F*, L*); Reid 1968: 270 (M*, F*, P*, L*); Kyalo et al. 2017 (distr.; sub-Saharan Africa). **Ety.:** Dr. Watson. **Informal name:** Watson Malaysian Nail Mosquito.

wellcomei Theobald
1904b: 64 (F*). **Type loc.:** Baro River, Sudan (NHMUK). **Distr.:** Angola, Benin, Botswana, Burkina Faso, Cameroon, Central African Republic, Chad, Côte d'Ivoire, Democratic Republic of the Congo, Eritrea, Ethiopia, Gabon, The Gambia, Ghana, Guinea, Kenya, Mali, Mauritania, Mozambique, Niger, Nigeria, Senegal, Seychelles, South Sudan, Sudan, Tanzania (includes Zanzibar), Togo, Uganda, Zambia, Zimbabwe. **Refs.:** De Meillon 1947b: 171 (M*, F*, P, L, E*); Kyalo et al. 2017 (distr.; sub-Saharan Africa). **Ety.:** n.s. [poss. after Henry Wellcome / Wellcome Laboratories]. **Informal name:** Wellcome Sudanese Nail Mosquito.

ssp. ***ugandae*** Evans
1934: 553 (F, P*, L*). **Type loc.:** King's Lake Area, Kampala, Uganda (NHMUK). **Distr.:** Angola, Botswana, Burundi, Democratic Republic of the Congo, Gabon, Tanzania, Uganda. **Refs.:** De Meillon 1947b: 114 (F, P, L); Gillett 1955: 36 (M; tax.); Gillies & De Meillon 1968: 176 (F*); Ribeiro & da Cunha Ramos 1975: 20 (distr.); Knight 1978b: 11 (to sp. in error); Kyalo et al. 2017 (distr.; sub-Saharan Africa). **Ety.:** Uganda. **Informal name:** Kampala Ugandan Nail Mosquito.

ssp. syn. *milesi* De Meillon & Evans
1935: 93 (F*; *Anopheles walravensi* var.). **Type loc.:** Victoria Falls, Southern Rhodesia [Zimbabwe] (NHMUK). **Refs.:** De Meillon 1947b: 175 (? = *michaeli* De Meillon & Leeson); Gillies & De Meillon 1968: 4 (syn.).

ssp. ***ungujae*** White
1975: 316 (F). **Type loc.:** Kibonde, Nzungu, Zanzibar [Tanzania] (NHMUK). **Distr.:** Tanzania (Zanzibar). **Ety.:** Unguja, Swahili name for Zanzibar. **Informal name:** Unguja Tanzanian Nail Mosquito.

willmori (James)
1903 (F*; *Nyssorhynchus*). In Theobald 1903a: 100. **Type loc.:** Kashmir [Pakistan] (NHMUK). **Distr.:** Afghanistan, Bangladesh, Bhutan, India, Myanmar, Nepal, Pakistan, People's Republic of China, Sri Lanka, Thailand, Vietnam. **Refs.:** Christophers 1933: 285 (A, P, L, E; as var.); Stone 1967: 200 (emend.); Aslamkhan 1971b (distr.; Pakistan); Rattanarithikul & Green 1987: 254 (emend. to orig. spelling; lectotype desig.); Darsie & Pradhan 1990, 1991 (F, L; tax., keys, bion., distr.; Nepal); Harrison et al. 1991: 223 (distr.; Thailand.); Green et al. 1992 (chrom.; *An. willmori*, *An. pseudowillmori*); Oo et al. 2004 (distr.; Myanmar); Rattanarithikul et al. 2006b (F*, L*; bion., distr., keys); Qu & Zhu 2008 (distr.; People's Republic of China); Rueda, Pecor et al. 2008: 397 (distr.); Morgan et al. 2009 (mol. phyl.; Neocellia Series); Namgay et al. 2018 (bion., distr.; Bhutan). **Ety.:** "Lieut. Willmore, I.M.S." **Informal name:** Willmore Pakistani Nail Mosquito.

syn. *indica* Theobald
1907: 111 (M, F*; *Neocellia*). **Type loc.:** Dehra Dun [United Provinces], India (NHMUK). **Refs.:** Christophers 1931 (syn.).

syn. *dudgeonii* Theobald
1907: 113 (M, F*; *Neocellia*). **Type loc.:** Kangra Valley [Western Himalayas; East Punjab], India (NHMUK).

syn. *maculosa* James & Liston
1911: 112 (A; *Neocellia willmori* var.). **Type loc.:** Pathankot [Punjab Hill States] and Kurseong [West Bengal], India (NHMUK).

wilsoni Evans
1934: 560 (M*, F*, P, L). **Type loc.:** Amani, Tanganyika [Tanzania] (NHMUK). **Distr.:** Kenya, Tanzania. **Refs.:** Evans 1938: 96 (M*, F*, P, L*); Gillies 1955: 158 (E*); Kyalo et al. 2017 (distr.; sub-Saharan Africa). **Ety.:** Donald Bagster Wilson. **Informal name:** Wilson Tanzanian Nail Mosquito.

yaeyamaensis Somboon & Harbach
2010 (M, F, P, L, E; mol. tax.). In Somboon et al. 2010: 44. **Type loc.:** Nishihama stream, Nosoko, Ishigaki City,

Ishigaki Island, Ryukyu Archipelago, Japan (NHMUK). **Distr.:** Japan. **Ety.:** Yaeyama Isl., Japan. **Informal name:** Yaeyama Japanese Nail Mosquito.

Subgenus *Christya* Theobald

1903b: 34 (as genus). Type species: *Christya implexa* Theobald.

Name derivation: Cuthbert Christy.

Important references: Sallum et al. 2000b (phyl.; as Christya series of subg. *Anopheles*); Harbach & Kitching 2005 (phyl.; Christya series of subg. *Anopheles*); Harbach & Kitching 2016 (phyl., class.; to subgen. from syn. with *Anopheles*, subg. *Anopheles*).

implexus (Theobald)
1903b: 34 (F; *Christya*). **Type loc.:** Uganda (NHMUK). **Distr.:** Angola, Burkina Faso, Burundi, Cameroon, Central African Republic, Côte d'Ivoire, Democratic Republic of the Congo, Ethiopia, Ghana, Guinea, Kenya, Nigeria, Republic of the Congo, Republic of South Africa, Rwanda, South Sudan, Sudan, Tanzania, Uganda, Zambia. **Refs.:** Evans 1938: 88 (M*, F*, P*, L*, E*); De Meillon 1947b: 55 (M*, F*, P, L*, E*); Lambrecht 1954: 204 (M*, F*, L*; tax.); Gillies & De Meillon 1968: 48 (M*, F*, L*, E*; distr.); Ribeiro & da Cunha Ramos 1975: 15 (distr.); Harbach & Kitching 2015 (from subg. *Anopheles* to subg. *Christya*); Kyalo et al. 2017 (distr.; sub-Saharan Africa). **Ety.:** n.s. [involved, entwined, enfolded, interwoven (L)]. **Informal name:** Entangled Ugandan Nail Mosquito.
syn. *henrardi* Wolfs
1945b: 174 (F*; *Anopheles* (*Anopheles*) *implexus* var.). **Type loc.:** Coquilhatville, Belgian Congo [Mbandaka, Democratic Republic of the Congo] (CMT). **Refs.:** De Meillon 1947b: 55 (syn.).

okuensis Brunhes, Le Goff, & Geoffroy
1997: 174 (M, F*, L*). **Type loc.:** Sabga, Cameroon (LTVO). **Distr.:** Cameroon. **Refs.:** Harbach: http://mosquito-taxonomic-inventory.info/simpletaxonomy/term/8742; Irish et al. 2020 (tax.). **Ety.:** Mt. Oku region, Cameroon. **Informal name:** Oku Cameroonian Nail Mosquito.

Subgenus *Kerteszia* Theobald

1905f: 66. Type species: *Kerteszia boliviensis* Theobald (as genus).

Name derivation: Kálmán Kertész.

syn. *Anopheles*, subgenus *Dendropaedium* Dyar & Knab 1918a: 141. Type species: *Dendropaedium cruzii* Dyar & Knab (Edwards 1932a: 43).

Important references: Komp 1937b (tax.); Coutinho 1946b: 13 (keys); Cova-García 1961 (Venezuela); Zavortink 1973 (M*, F*, L*, P*; keys, revision); Gorham et al. 1973 (keys, bion.); Peyton et al. 1992 (comp. with subg. *Nyssorhynchus*); Sallum et al. 2000b (phyl.); Sallum et al. 2002c (mol. phyl.); Collucci & Sallum 2003 (phyl.); Harbach & Kitching 2005 (phyl., review); Beltrán-Aguilar et al. 2011: 711 (M gen.*, F*, P*, L*; bion., distr., tax.; Mexico); Harrison et al. 2012 (F; key); Harbach & Kitching 2015 (phyl., review); Freitas et al. 2015 (mol. phyl.).

auyantepuiensis Harbach & Navarro
1996: (F*, P*, L*). **Type loc.:** Auyantepi, Canaima National Park, Bolivar State, [SE] Venezuela (LBVM). **Distr.:** Venezuela. **Ety.:** Auyantepui, Venezuela. **Informal name:** Auyantepui Venezuelan Nail Mosquito.

bambusicolus Komp
1937b: 515 (F, L). **Type loc.:** La Union, Meta, Colombia (USNM). **Distr.:** Argentina, Bolivia, Brazil, Colombia, Ecuador, French Guiana, Guyana, Peru, Suriname, Venezuela. **Refs.:** Coutinho 1946a (M*); Lane 1953: 281 (M*, F, L); Komp 1956: 40 (tax.); Stone & Knight 1956b: 276 (type info., lectotype desig.); Zavortink 1973: 18 (M*, F*, P*, L*); Linton et al. 2013 (bion., distr.; Ecuador). **Ety.:** bamboo inhabitant (L). **Informal name:** Bamboo Columbian Nail Mosquito.

bellator Dyar & Knab
1906b: 160 (A). **Type loc.:** Trinidad [Trinidad & Tobago] (USNM). **Distr.:** Brazil, Guyana, Suriname, Trinidad and Tobago, Venezuela. **Refs.:** Anduze 1942b: 435 (F*); Komp 1942: 75 (F*), 127 (L), 164 (M*); Rozeboom 1942a: 242 (E*); Ross & Roberts 1943a: 37 (M*, F*, L*); Pittendrigh 1950a, 1950b (bion.); Stone & Knight 1956b: 276 (type info., lectotype desig.); Zavortink 1973: 26 (M*, F*, P*, L*); Forattini & Marucci 1993 (E*). **Ety.:** n.s. [warlike, warrior (L)]. **Informal name:** Warrior Trinbagonian Nail Mosquito.
syn. *bromelicola* Dyar
1925c: 27 (F; *Anopheles* (*Dendropaedium*) *bellator* race). **Type loc.:** Manoa, Orinoco River, Venezuela (USNM). **Refs.:** Stone & Knight 1956b: 276 (type info., lectotype desig.).

boliviensis (Theobald)
1905f: 66 (F*; *Kerteszia*). **Type loc.:** Songo, Bolivia (HNM). **Distr.:** Bolivia, Brazil, Colombia, Ecuador, French Guiana, Guyana, Paraguay, Peru, Suriname, Venezuela. **Refs.:** Komp & Osorno-Mesa 1936 (M*, L*); Komp 1937b: 503 (M*); Anduze 1942a: 435 (F*); Leví-Castillo 1945: 128 (M*, F*, L*); Lane 1953: 279 (M*, F, L*, E*); Komp 1956: 40 (L); Belkin 1968b: 9 (tax.); Zavortink 1973: 30 (F*); Linton et al. 2013 (distr.; Ecuador). **Ety.:** Bolivia. **Informal name:** Aymara Bolivian Nail Mosquito [an indigenous people of the Andes, primarily based in Bolivia, Chile, and Peru].

cruzii Dyar & Knab
1908: 53 (new name for *lutzii* Theobald 1901a, not Cruz 1901). **Distr.:** Argentina, Bolivia, Brazil, Colombia, Costa

Rica, Ecuador, French Guiana, Guyana, Mexico, Panama, Peru, Suriname, Venezuela. **Refs.:** Davis 1926 (F*); Komp 1937b: 504 (M*, L; tax.; to sp.); Netto 1940: 32 (L*); Cova García 1946 (M, F*, L, E*); Barreto & Coutinho 1951: 177 (tax.); Lima 1952: 401 (L*); Lane 1953: 285 (M*, F, L); Zavortink 1973: 23 (M*, F*, L); Wilkerson & Peyton 1991 (M*, F*, P*, L*; bion.); Forattini & Marucci 1993 (E*); Ramirez & Dessen 2000a (chrom., tax.; *s.l.*); Ramirez & Dessen 2000b (chrom., tax.; *s.l.*); de Carvalho-Pinto & Lourenço-de-Oliviera 2004 (mol. tax.); Calado & Navarro-Silva 2005 (mol. tax.); Rona et al. 2009 (mol. tax.; *s.l.*); Rona et al. 2010a (tax; *s.l.*); Rona et al. 2010b (mol. tax.; *s.l.*); Harrison et al. 2012 (tax., key); Lorenz et al. 2012 (tax., morph.); Linton et al. 2013 (distr.; Ecuador); Rona et al. 2013 (mol. tax.; *s.l.*); Berti et al. 2015 (distr.; Venezuela). **Ety.:** Oswaldo Gonçalves Cruz. **Informal name:** Cruz Brazilian Nail Mosquito. (Vol. 1, p. 382)

syn. *lutzii* Theobald
1901a: 177 (F*). **Type loc.:** Rio de Janeiro, Brazil (NHMUK). **Refs.:** Belkin 1968b: 10 (lectotype desig.).

syn. *adolphoi* Neiva
1908b: 457 (new name for *lutzii* Theobald 1901a, not Cruz 1901; *Myzorhynchella*). **Type loc.:** Caraguatatuba, São Paulo, Brazil (FH). **Refs.:** Barreto & Coutinho 1951: 177 (syn.).

syn. *montemor* Corrêa
1950a: 53 (M*). **Type loc.:** Caraguatatuba, São Paulo, Brazil (FH). **Refs.:** Barretto & Coutinho 1951: 177 (syn.); Belkin et al. 1971: 7 (tax.).

gonzalezrinconesi Cova García, Pulido F. & Escalante de Ugueto
1977: 140 (M*, F*, P*, L*). **Type loc.:** Mata Mulas, Municipio Delicias, Táchira, Venezuela (DERM). **Distr.:** Venezuela. **Ety.:** Rafael Antonio González Rincones. **Informal name:** González Venezuelan Nail Mosquito.

homunculus Komp
1937b: 509 (M*, F, L). **Type loc.:** Restrepo [Meta], Colombia (USNM). **Distr.:** Bolivia, Brazil, Colombia, French Guiana, Guyana, Peru, Trinidad and Tobago, Venezuela. **Refs.:** Anduze 1942b: 435 (F*); Anduze 1942d (E*); Cova García 1946: Fig. 1 (E*), 60 (L), 106 (F*), 130 (M*); Pittendrigh 1950a, 1950b (bion.); Lima 1952: 401 (L*); Lane 1953: 287 (M*, F, L, E*); Stone & Knight 1956b: 278 (type info., lectotype desig.); Zavortink 1973: 20 (M*, F*, P*, L*; distr.); Calado et al. 2005 (mol. tax.); Cardoso et al. 2012 (bion., distr.); Sallum et al. 2009 (M*, P*, L*; mol. tax.). **Ety.:** little man, mannekin (L); mannekin-like appearance of the aedeagus. **Informal name:** Mannekin Colombian Nail Mosquito.

syn. *anoplus* Komp
1937b: 514 (M*, L). **Type loc.:** Restrepo [Meta], Colombia (USNM). **Refs.:** Lane 1953: 287 (syn.).

laneanus Corrêa & Cerqueira
1944: 112 (M*; as sp.). **Type loc.:** Campos do Jordão, Serra da Mantiqueira, São Paulo, Brazil (FH). **Distr.:** Argentina, Bolivia, Brazil, Peru. **Refs.:** Corrêa 1946: 605 (M*, E*); Martínez 1949: 13 (F*); Corrêa 1950a: 53 (syn. with *lutzii* Theobald 1901a; as *cruzii*); Barretto & Coutinho 1951: 177 (from syn.); Lane 1951 (to ssp.); Romeo Viamonte & Castro 1951: 313 (F*); Lane 1953: 286 (M*, F, L); Forattini 1962: 445 (to sp.); Zavortink 1973: 29 (M*, F*, L; distr.); Calderón et al. 1995 (distr.; Peru); Forattini et al. 1997: 368 (E*); Sallum et al. 2000a (M*, F, P*, L*; tax.). **Ety.:** John (João) Lane. **Informal name:** João Brazilian Nail Mosquito.

lepidotus Zavortink
1973: 17 (M*, F*, L). **Type loc.:** Restrepo, Meta, Colombia (USNM). **Distr.:** Bolivia, Colombia, Ecuador, Peru. **Refs.:** Harrison et al. 2012 (M*, F, P*, L*; key, mol. tax.); Linton et al. 2013 (bion., mol. tax., distr.; Ecuador). **Ety.:** scaly (Gr). **Informal name:** Scaly Colombian Nail Mosquito.

neivai Howard, Dyar, & Knab
1913 vol. 2: Pls. 41, 130 and 1917: 986 (F*, L*; keys). **Type loc.:** Fort San Felipe, Porto Bello Bay, Panama (USNM). **Distr.:** Belize, Bolivia, Brazil, Colombia, Costa Rica, Ecuador, El Salvador, French Guiana, Guatemala, Guyana, Mexico, Nicaragua, Panama, Peru, Suriname, Venezuela. **Refs.:** Komp 1937b (M*, F*, L; tax.; from syn. with *cruzii*); Senevet & Abonnenc 1938: 509 (M*, P*, L*); Leví-Castillo 1945: 118 (M*, F*, L*); Pelaez 1945: 74 (F*; key); Floch & Abonnenc 1951: 70 (M*, F*, L*); Stone & Knight 1956b: 279 (type info., lectotype desig.); Zavortink 1973: 11 (M*, F*, P*, L*; distr.); Baerg & Boreham 1974a: 629 (E*); Murillo et al. 1988 (bion.); Murillo et al. 1988 (bion.); Astaiza-V et al. 1988 (bion.); Murillo et al. 1989 (bion.); Carvajal et al. 1989 (bion., distr.); Calderón et al. 1995 (distr.; Peru); Solarte et al. 1996 (bion.); Beltrán-Aguilar et al. 2011: 711 (M*, F*, P*, L*; tax., keys, distr.); Linton et al. 2013 (distr.; Ecuador). **Ety.:** Arthur Neiva. **Informal name:** Malaria Panamanian Nail Mosquito. (Vol. 1, p. 384)

syn. *hylephilus* Dyar & Knab
1917: 38 (F). **Type loc.:** Gatún, Canal Zone, Panama (USNM). **Refs.:** Komp 1937b: 521 (syn.); Belkin, Schick, & Heinemann1965: 44 (lectotype desig.).

pholidotus Zavortink
1973: 15 (M*, F*, P*, L*). **Type loc.:** La Zorra, Bocas del Toro, Panama (USNM). **Distr.:** Bolivia, Colombia, Costa Rica, Ecuador, Panama, Peru, Venezuela. **Refs.:** Harrison et al. 2012 (M, F, P, L; key, distr.; Bolivia, Colombia, Ecuador, Venezuela); Parker et al. 2012 (bion., distr.; Peru); Escovar et al. 2014 (M*, F*, L*; mol. tax., bion., distr.; Colombia, Costa Rica). **Ety.:** clad in scales (Gr). **Informal name:** Armored Panamanian Nail Mosquito. (Vol. 1, p. 386)

rollai Cova García, Pulido F., & Escalante de Ugueto
1977: 241 (new name for *hilli* Cova García, Pulido F., & de Ugueto 1976, not *hilli* Woodhill & Lee 1944). **Distr.:** Venezuela. **Ety.:** Rolla Bennett Hill. **Informal name:** Rolla Venezuelan Nail Mosquito.

syn. *hilli* Cova García, Pulido F., & de Ugueto 1976: 344 (M*, F*, P*, L*). **Type loc.:** La Chimenea, Municipality Altamira, State of Barinas and Mata Mulas, Municipality Delicias, state of Táchira, Venezuela (ND).

Subgenus *Lophopodomyia* Antunes

1937a: 69 (as subg.). Type species: *Anopheles* (*Lophopodomyia*) *squamifemur* Antunes 1937a.

Name derivation: lopho-, tuft; -podo-, foot; -myia, little fly (Gr); ref. to distinguishing tuft of scales on the hind femur.

syn. *Anopheles*, subgenus *Arthuromyia* Galvão 1941b: 564. Type species: *Myzorhynchella gilesi* Peryassú.

Important references: Lane 1953 (review; South America); Cova-García 1961 (tax.; Venezuela); Forattini 1962 (review; South America); Wilkerson & Strickman 1990 (keys; Central America and Mexico); Sallum et al. 2000b (phyl.); Sallum et al. 2002c (mol. phyl.); Harbach & Kitching 2005 (phyl., review); Harbach & Kitching 2005 (phyl., review); Freitas et al. 2015 (mol. phyl.)

gilesi (Neiva)
1908 (M; *Myzorhynchella*). In Peryassú 1908: 103. **Type loc.:** Lassance, Brazil (LU). **Distr.:** Bolivia, Brazil, Colombia, Ecuador, Paraguay, Venezuela. **Refs.:** Galvão 1941b: 565 (M*; tax.); Deane et al. 1943: 167 (P*, L*, E*); Deane, L.M. et al. 1946 (F*; key, bion.; NE Brazil & Amazonas); Causey et al. 1946 (M*; key; NE Brazil & Amazonas); Deane, M.P. et al. 1946 (L*; key, bion.; NE Brazil & Amazonas); Lane 1953: 189 (M*, F*, P, L*, E*); Belkin et al. 1971: 7 (type info.). **Ety.:** George Michael James Giles. **Informal name:** Giles Brazilian Nail Mosquito.

gomezdelatorrei Leví-Castillo
1955d: 510 (M*, F, P*, L*). **Type loc.:** Chiltazón, Carchi, Ecuador (CEIE). **Distr.:** Ecuador. **Ety.:** José Gómez de la Torre. **Informal name:** Gómez Ecuadorian Nail Mosquito.

oiketorakras Osorno-Mesa
1947: 431 (M*, F*, P*, L*). **Type loc.:** Bogotá [Cundinamarca], Colombia (USNM). **Distr.:** Colombia. **Refs.:** Stone & Knight 1956b: 279 (type info., lectotype desig.). **Ety.:** inhabitant of a mountaintop (Gr). **Informal name:** Montane Columbian Nail Mosquito.

pseudotibiamaculatus Galvão & Barretto
1941: 65 (F*). **Type loc.:** Casa Grande (Municipio Mogi das Cruzes), São Paulo, Brazil (FMSP). **Distr.:** Brazil. **Refs.:** Lane et al. 1947 (M*, P*, L); Lane 1953: 192 (M*, F*, P*, L). **Ety.:** false (Gr); similar to *tibiamaculatus*. **Informal name:** Freckle-legged Brazilian Nail Mosquito [*pseudotibiamaculatus* = "false-spot legged" = Freckle-legged].

squamifemur Antunes
1937a: 69 (F*). **Type loc.:** Vegagrande, Restrepo, Meta, Colombia (IOC). **Distr.:** Bolivia, Brazil, Colombia, Ecuador, French Guiana, Panama, Peru, Seychelles, Venezuela. **Refs.:** Flock & Abonnenc 1946d (distr.; French Guiana); Deane et al. 1949: 497 (M*); Lane 1953: 186 (M*, F*); Boreham & Baerg 1974: 564 (M*, P*, L*, E*; tax.); Vargas 1975: 77 (distr.); Calderón et al. 1995 (distr.; Peru); Linton et al. 2013 (mol. tax., bion., dist.; Ecuador); Berti et al. 2015 (distr.; Venezuela). **Ety.:** scale, femur (L); tuft on hind femur. **Informal name:** Scale-tufted Columbian Nail Mosquito.

vargasi Gabaldón, Cova García, & Lopez
1941b: 25 (M*, F*, P*, L*, E*). **Type loc.:** Ouebrada de la Cueva del Guácharo, Monagas, Venezuela (DERM). **Distr.:** Venezuela. **Refs.:** Lane 1953: 194 (M*, F*, P, L*, E; keys). **Ety.:** José María Vargas and Luís Vargas. **Informal name:** Vargas Venezuelan Nail Mosquito.

Subgenus *Nyssorhynchus* Blanchard

1902: 795 (as genus). Type species: *Anopheles argyritarsis* Robineau-Desvoidy.

Name derivation: n.s. [nysso-, prick, pierce, puncture; -rhynchus, nose, snout, muzzle. From Blanchard 1902: a "trompe piqueuse" biting trunk (like the trunk of an elephant)].

syn. *Laverania* Theobald 1902b: 183 (as genus) (not Grassi & Feletti 1890). Type species: *Anopheles argyritarsis* Robineau-Desvoidy.
syn. *Myzorhynchella* Theobald 1907: 23, 78 (as genus). Type species: *Myzorhynchella nigra* Theobald.
syn. *Manguinhosia* Cruz 1907a: 271 (as genus). Type species: *Manguinhosia lutzii* Cruz.

Important references: Galvão 1940 (notes; Brazil); Galvão 1941b (M*, F*, P*, L*; keys, bion., tax.; *Nys.* and *Myzorhynchella*); Lane 1953 (M*, F*, P*, L*; keys, bion., tax.); Cova-García 1961 (keys, tax.; Venezuela); Forattini 1962 (M*, F*, P*, L*; tax., keys, bion.); Faran 1980 (M*, F*, P*, L*; keys, revision; Albimanus Section); Faran & Linthicum 1981 (M*, F*, L*; keys, tax., bion.; Amazon region); Steiner et al. 1982 (rev., genetics and evolution); Clark-Gil & Darsie 1983 (M, F; keys; Guatemala); Linthicum 1988 (revision, keys; Argyritarsis Section); Wilkerson & Strickman 1990 (keys; Central America and Mexico); Peyton et al. 1992 (tax., *Nys.—Ker.* comparison); Sallum et al. 2000b (phyl.); Rubio-Palis 2000 (review); Krzywinski et al. 2001a, Krzywinski et al. 2001b (mol. phyl.); Sallum et al. 2002c (mol. phyl.); Calle et al. 2002 (*Nys.* morphometrics; Colombia); Harbach & Kitching 2005 (phyl., review); Sallum et al. 2008 (tax., review); Bourke et al. 2010 (mol. phyl.); Bourke et al. 2011 (mol. phyl.; Myzorhynchella Section); Beltrán-Aguilar et al. 2011 (M*, F*, P*, L*; keys, bion., distr., tax.; Mexico); Foster et al. 2013 (mol. phyl.); Freitas et al. 2015 (mol. phyl.); Harbach & Kitching 2015 (phyl., review); Neafsey et al. 2015 (mol. phyl., 16 genomes)

albertoi Unti
1941: 9 (E*; *strodei* var.). **Type loc.:** Rio Paraiba, São Paulo, Brazil (NE). **Distr.:** Brazil. **Refs.:** Faran 1980: 123 (from *evansi* var. to syn. of *strodei*); Sallum et al. 2010 (M*, E*; phylogenetics, tax.). **Ety.:** Alberto da Silva Ramos. **Informal name:** Alberto Brazilian Nail Mosquito.

albimanus Wiedemann
1821: 10 (F). **Type loc.:** Saint Domingo Institute [Dominican Republic] (NMW). **Distr.:** Argentina, Bahamas, Belize, Brazil, Cayman Isl., Colombia, Costa Rica, Cuba, Dominican Republic, Ecuador, El Salvador, Guatemala, Guyana, Haiti, Honduras, Jamaica, Lesser Antilles (Antigua and Barbuda, Barbados, Curaçao, Guadeloupe, Montserrat, Saint Kitts and Nevis, Sint Maarten, Trinidad and Tobago, Virgin Isl.), Mexico, Nicaragua, Panama, Paraguay, Peru, Puerto Rico, Suriname, United States (continental), Uruguay, Venezuela. **Refs.:** Gerry 1932 (F*); Rozeboom 1938: 99 (E*); Rozeboom & Gabaldón 1941: 95 (tax.); Ross & Roberts 1943a: 35 (M*, F*, L*); Roth 1944: 98 (M*), 107–110 (M gen.; keys); Floch & Abonnenc 1945a (M*, F*, L*; distr.; Guadeloupe); Pelaez 1945: 74 (F*; key); Vargas 1948: 153 (M*); Penn 1949a: 68 (P*); Yamaguti & LaCasse 1950b: 11 (M*, F*, L*); Belkin 1952: 122 (L*); Carpenter & LaCasse 1955: 55 (M*, F*, L*; keys); Dodge 1966: 361 (first instar L*; key); Belkin et al. 1970: 24 (M*, F*, P*, L*); Keppler et al. 1973 (chrom.*); Faran 1980: 35 (M*, F*, P*, L*; keys, tax., bion., distr.); Breeland 1980 (bibliography); Rejmánková et al. 1991 (bion.); Rodriquez et al. 1992 (E*); Rejmánková et al. 1996 (bion.); Solarte et al. 1996 (bion.); Bejarano 2003: 517 (distr.; Colombia); Darsie & Ward 2005 (F*, L*; keys, distr.); Sinka et al. 2010b: 72 (bion., review, distr., niche model); Moreno et al. 2010 (genome); Becker et al. 2010: 376 (F*, L*; key, tax., distr., bion.); Beltrán-Aguilar et al. 2011: 718 (M*, F*, P*, L*; tax., keys, distr.); Conn et al. 2013 (bion., vector); Gomez et al. 2014 (morph., bion.; *s.l.*). **Ety.:** n.s. [white hand (L); prob. ref. to white Ta-III_5]. **Informal name:** White-handed Dominican Nail Mosquito. (Vol. 1, p. 388)

syn. *cubensis* Agramonte
1900: 460 (M, F*). **Type loc.:** Several localities in Cuba (LU). **Refs.:** Theobald 1907: 106 (syn.).

syn. *albipes* Theobald
1901a: 125 (M*, F; *Anopheles argyrotarsis* [*sic*] ssp.). **Type loc.:** Jamaica (NHMUK). **Refs.:** Belkin 1968b: 8 (lectotype desig.).

syn. *tarsimaculata* Goeldi
1905: 133 (new name for *albipes* Theobald 1901a, in error; as *Cellia argyrotarsis* var.). **Refs.:** Rozeboom & Gabaldón 1941: 96 (syn.).

syn. *gorgasi* Dyar & Knab
1907c: 198 (F). **Type loc.:** La Boca, Canal Zone, Panama (USNM). **Refs.:** Rozeboom & Gabaldón 1941: 96 (syn.); Faran 1980: 14 (from syn. with *albimanus* to *nomen dubium*).

syn. *bisignatus* Hoffmann
1938: 176 (A*; as var.). **Type loc.:** Several localities in Tamaulipas, Mexico (UM). **Refs.:** Ross & Roberts 1943a: 36 (syn.).

syn. *trisignatus* Hoffmann
1938: 177 (A*; *Anopheles albimanus* var.). **Type loc.:** Tampico, Tamaulipas, Mexico (UM). **Refs.:** Ross & Roberts 1943a: 36 (syn.).

albitarsis Lynch Arribálzaga
1878: 150 (F). **Type loc.:** Baradero, Buenos Aires, Argentina (IOC). **Distr.:** Argentina, Brazil; [as *albitarsis s.l.*, Colombia, Costa Rica, French Guiana, Guatemala, Guyana, Jamaica, Panama, Paraguay, Peru, Trinidad and Tobago, United States (continental), Uruguay, Venezuela]. **Refs.:** Senevet 1934: 45 (P*); Pinto 1939: 345 (M*, F*, P*, L*, E); Rozeboom 1942a: 238 (E*; tax.); Ross & Roberts 1943a: 31 (M*, F*, L*); Roth 1944: 97 (M*), 107–110 (M gen.; keys); Cova-García 1946: Fig. 1 (E*), 31 (L*), 83 (F*), 118 (M*); Deane, L.M. et al. 1946 (F*; key, bion.; NE Brazil & Amazonas; *s.l.*); Causey et al. 1946 (M*; key; NE Brazil & Amazonas; *s.l.*); Deane, M.P. et al. 1946 (L*; key, bion.; NE Brazil & Amazonas; *s.l.*); Romeo Vaimonte & Castro 1951: 321 (F*); García & Casal 1965a: 6 (P*); Kreutzer et al. 1976: 473 (chrom.*); Patriota Nascimento & Lourenço-de-Oliveira 1984 (tax.; *s.l.*); Rosa-Freitas & Deane 1989: 289 (neotype desig.); Rosa-Freitas et al. 1990 (mol. tax; *s.l.*); Calderón et al. 1995 (distr.; Peru); Wilkerson et al.1995a (mol. tax.); Wilkerson et al. 1995c (mol. tax.; *s.l.*); Monnerat et al. 1999 (E*; *s.l.*); Krzywinski et al. 2001a, 2001b (mtDNA, phyl.; *s.l.*); Wilkerson et al. 2005 (phyl.; *s.l.*); Li & Wilkerson 2005 (mol. tax.; *s.l.*); Lehr et al. 2005 (mol. tax., phyl.; *s.l.*); Brochero et al. 2007 (*s.l.*; *albitarsis* sp. F); Li & Wilkerson 2007 (mol. tax.; *s.l.*); Motoki et al. 2009a (M*, F, P*, L*); Sinka et al. 2010b: 72 (bion., review, distr., niche model, *s.l.*); Ruiz-Lopez et al. 2012 (mol. tax.; *s.l.*); Foley et al. 2014 (distr., bion., model; *s.l.*); Laporta et al. 2015 (bion., distr., niche model; *s.l.*). **Ety.:** n.s. [white foot (L); ref. to white tarsal markings, esp. hind tarsomeres]. **Informal name:** White-footed Argentinian Nail Mosquito. (Vol. 1, p. 390)

syn. *limai* Galvão & Lane
1937b: 227 (A, E*; *Anopheles* (*Nyssorhynchus*) var.). **Type loc.:** Suburbs of Pinheiros and Butantan, São Paulo, Brazil (NE). **Refs.:** Lane 1953: 244 (syn.); Belkin et al. 1971: 4 (tax.).

syn. *imperfectus* Corrêa & Ramos
1943: 246 (F; *albitarsis* ssp.). **Type loc.:** Vera Cruz, São Paulo, Brazil (FH). **Refs.:** Lane 1953: 244 (syn.); Belkin et al. 1971: 4 (tax.).

anomalophyllus Komp
1936d: 163 (M*, L). **Type loc.:** Wenham's Farm, near Almirante, Panama (USNM). **Distr.:** Costa Rica, Nicaragua, Panama. **Refs.:** Kumm 1941: 91 (E*); Komp

1942: 75 (F), 125 (L), 161 (M*); Faran 1980: 95 (M*, F*, P*, L*; keys, tax., bion., distr.). **Ety.:** uneven leaf (Gr); unusual aedeagal leaflets. **Informal name:** Leaved Panamanian Nail Mosquito.

antunesi Galvão & Franco do Amaral
1940: 150 (M*, F, L*, E*). **Type loc.:** Vila Emilio Ribas, Campos do Jordão, São Paulo, Brazil (FMSP). **Distr.:** Argentina, Brazil, Uruguay. **Refs.:** Galvão 1941b: 552 (M*, F*, P*, L*, E*); Belkin et al. 1971: 4 (type info.); Forattini et al. 1997: 369 (E*); Rossi & Martínez 2003: 471 (distr.; Uruguay); Nagaki et al. 2010 (M*, F*, L*, E*; redescr.). **Ety.:** Paulo César de Azevedo Antunes. **Informal name:** Antunes Brazilian Nail Mosquito.

aquasalis Curry
1932 (M*, L; *tarsimaculatus* var.). **Type loc.:** Canal Zone, Panama (LU). **Distr.:** Argentina, Brazil, Colombia, Costa Rica, Dominica, Dominican Republic, Ecuador, French Guiana, Guyana, Lesser Antilles (includes Antigua and Barbuda, Grenada, Guadeloupe, Martinique, Montserrat, Saint Kitts and Nevis, Saint Lucia, Saint Vincent and the Grenadines, Virgin Isl.), Mexico, Nicaragua, Panama, Paraguay, Suriname, Trinidad and Tobago, Uruguay, Venezuela. **Refs.:** Gabaldón et al. 1940a: 10 (P*, L*); Rozeboom & Gabaldón 1941: 91 (M*; tax.; to sp.); Rozeboom 1942a: 242 (E*; tax.); Ross & Roberts 1943a: 33 (M*, F*, L*); Floch & Abonnenc 1945a (M*, F*, L*; distr.; Guadeloupe); Cova-García 1946: Fig. 1 (E*), 35 (L*), 87 (F*), 122 (M*); Deane, L.M. et al. 1946 (F*; key, bion.; NE Brazil & Amazonas); Causey et al. 1946 (M*; key; NE Brazil & Amazonas); Deane, M.P. et al. 1946 (L*; key, bion.; NE Brazil & Amazonas); Coher 1949: 87 (F); Downs 1950 (morph.); Kitzmiller & Chow 1971 (chrom.*); Faran 1980: 75 (M*, F*, P*, L*; keys, tax., bion., distr.); Pérez & Conn 1992 (chrom.); Linley et al. 1993b (E*); Flores-Mendoza & Lourenço-de-Oliveira 1996 (bion.); Maldonado et al. 1997 (E*; tax.); Grillet et al. 1998 (bion.); Lopes & Khouri 2000 (F gen.); Fairley et al. 2002 (pop. gen.); Fairley et al. 2005 (mol. tax.); Sinka et al. 2010b: 72 (bion. review, distr., niche model); Becker et al. 2010: 377 (F*, L*; key, tax., distr., bion.); Conn et al. 2013 (bion., vect.); Pascini et al. 2013 (morph.); Martinez-Villegas et al. 2019 (mt. genome). **Ety.:** from salt water (L). **Informal name:** Saltwater Panamanian Nail Mosquito. (Vol. 1, p. 392)
syn. *emilianus* Komp
1941: 794 (M*, F, L, E*). **Type loc.:** Belém, Pará, Brazil (USNM). **Refs.:** Causey et al. 1945a: 248 (syn.).
syn. *guarujaensis* Ramos
1942: 61 (M*, L*, E*; *oswaldoi* var.). **Type loc.:** Ilha de Guaruja, São Paulo, Brazil (LU). **Refs.:** Lane 1949 (syn.).
syn. *delta* Anduze
1948: 18 (A; as var.). **Type loc.:** Pedernales, Delta Amacuro, Venezuela (NE). **Refs.:** Lane 1953: 260 (syn.).
syn. *guarauno* Anduze
1948: 18 (A; *aquasalis* var.). **Type loc.:** Pedernales, Delta Amacuro, Venezuela (NE). **Refs.:** Lane 1953: 260 (syn.).
syn. *deltaorinoquensis* Cova García, Pulido F., & Amarista 1977: 150 (M*, F*, P*, L*, E*). **Type loc.:** Sucre and Monagas states, Delta Amacuro Federal Territory, Venezuela (DERM). **Refs.:** Faran 1980: 75 (syn.).

argyritarsis Robineau-Desvoidy
1827: 411 (F). **Type loc.:** Rio de Janeiro, Guanabara, Brazil (NE; originally in MNHP). **Distr.:** Argentina, Belize, Bolivia, Brazil, Colombia, Costa Rica, Dominica, Dominican Republic, Ecuador, El Salvador, French Guiana, Guatemala, Guyana, Honduras, Lesser Antilles (Antigua and Barbuda, Guadeloupe, Grenada, Martinique, Montserrat, Saint Lucia, Saint Vincent and the Grenadines, Trinidad and Tobago), Mexico, Nicaragua, Panama, Paraguay, Peru, Uruguay, Venezuela. **Refs.:** Pinto 1939: 363 (M*, F*, P, L*, E*); Coutinho & Verano 1942 (L; clypeal setae); Komp 1942: 62 (F*), 109 (L*), 149 (M*); Floch & Abonnenc 1945a (M*, F*, L*; distr.; Guadeloupe); Pelaez 1945: 74 (F*; key); Cova-García 1946: Fig. 1 (E*), 28 (L*), 81 (F*), 115 (M*); Deane, L.M. et al. 1946 (F*; key, bion.; NE Brazil & Amazonas); Causey et al. 1946 (M*; key; NE Brazil & Amazonas); Deane, M.P. et al. 1946 (L*; key, bion.; NE Brazil & Amazonas); Senevet 1948d: 437 (P); Belkin 1968b: 9 (tax.); Belkin et al. 1971: 5 (tax., type loc. info.); Kreutzer et al. 1975 (chrom.*); Linthicum 1988: 139 (M*, F*, P*, L*); Calderón et al. 1995 (distr.; Peru); Beltrán-Aguilar et al. 2011: 734 (M*, F*, P*, L*; tax., keys, distr.); Berti et al. 2015 (distr.; Venezuela). **Ety.:** silvery tarsi (Gr). **Informal name:** Silver-footed Brazilian Nail Mosquito.
syn. *rooti* Brèthes
1926b: 106 (M; *Cellia*). **Type loc.:** Tucuman, Argentina (BA). **Refs.:** Edwards 1932a: 44 (syn.).

arthuri Unti
1941: 9 (E*; *strodei* var.). **Type loc.:** Rio Paraiba São Paulo, Brazil (NE). **Distr.:** Brazil. **Refs.:** Faran 1980: 123 (from *evansi* var. to syn. with *strodei*); Sallum et al. 2010 (M*, E*; phyl., tax.); Bourke et al. 2018 (mol. tax.; *s.l.*). **Ety.:** Arthur Costa Filho. **Informal name:** Arthur Brazilian Nail Mosquito.
syn. *artigasi* Unti
1941: 9 (E*; *strodei* var.). **Type loc.:** Rio Paraiba, São Paulo, Brazil (NE). **Refs.:** Galvão 1943: 144 (syn.); Faran 1980: 123 (from syn. with *arthuri* to syn. with *strodei*); Sallum et al. 2010 (syn. with *arthuri*).

atacamensis González & Sallum
2010 (M*, F, P*, L*; tax., mol. tax.). **Type loc.:** Los Perales, Provincia de Copiapó, III Región de Atacama, Chile (FH). **Distr.:** Chile. **Ety.:** Atacama, Chile. **Informal name:** Atacama Chilean Nail Mosquito.

benarrochi Gabaldón, Cova-García, & Lopez
1941a: 3 (M*, F, P*, L*, E*). **Type loc.:** La Ceiba, Trujillo, Venezuela (DERM). **Distr.:** Bolivia, Brazil, Colombia, Paraguay, Peru, Puerto Rico, Venezuela. **Refs.:** Cova-García 1946: Fig. 1 (E*), 41 (L*), 93 (F*), 127 (M*); Deane,

L.M. et al. 1946 (F*; key, bion.; NE Brazil & Amazonas); Causey et al. 1946 (M*; key; NE Brazil & Amazonas); Deane, M.P. et al. 1946 (L*; key, bion.; NE Brazil & Amazonas); Coher 1949: 88 (F); Faran 1980: 136 (M*, F*, P*, L*; keys, tax., bion., distr.); Lounibos et al. 1997a (E*); Quiñones et al. 2001 (tax. as *benarrochi* B; Colombia); Estrada et al. 2003 (E* as *benarrochi* B; Colombia); Ruiz et al. 2005 (mol. tax.; as *benarrochi* B). **Ety.:** E.I. Benarroch. **Informal name:** Benarroch Venezuelan Nail Mosquito.

braziliensis (Chagas)
1907: 18 (F; *Cellia*). **Type loc.:** Lassance, Brazil (IOC). **Distr.:** Argentina, Bolivia, Brazil, Colombia, French Guiana, Guyana, Peru, Suriname, Trinidad and Tobago, Venezuela. **Refs.:** Davis 1926 (F*); Townsend 1934: 494 (to sp.); Cova García 1946: Fig. 1 (E*), 32 (L*), 84 (F*), 119 (M*) [as *pessoai*]; Deane, L.M. et al. 1946 (F*; key, bion.; NE Brazil & Amazonas; as *pessoai*); Causey et al. 1946 (M*; key; NE Brazil & Amazonas; as *pessoai*); Deane, M.P. et al. 1946 (L*; key, bion.; NE Brazil & Amazonas; as *pessoai*); Senevet 1948b: 280 (M*, P*, L*; as *pessoai*); Senevet 1948d: 437 (P); Galvão 1950: 38 (L*; tax.; as *pessoai*); Lane 1953: 245 (M*, F, L*, E); Belkin et al. 1971: 5 (type loc. info.; tax.); Linthicum 1988: 213 (M*, F, P*, L*); Calderón et al. 1995 (distr.; Peru). **Ety.:** Brazil. **Informal name:** Tupi Brazilian Nail Mosquito [an indigenous people of Brazil considered the major ancestors of modern Brazilians].
syn. *pessoai* Galvão & Lane
1937a: 67 (M*, F, P, L*). **Type loc.:** Pinheiros near São Paulo, São Paulo, Brazil (FMSP). **Refs.:** Lane 1953: 245 (syn.).

darlingi Root
1926: 706 (M*, F*, P*, L*, E*). **Type loc.:** Caxiribu, near Porto das Caixas, Rio de Janeiro, Brazil (USNM). **Distr.:** Argentina, Belize, Bolivia, Brazil, Colombia, Ecuador, El Salvador, French Guiana, Guatemala, Guyana, Honduras, Mexico, Panama, Paraguay, Peru, Suriname, Uruguay, Venezuela. **Refs.:** Komp 1940 (distr.; Belize, Guatemala); Komp 1942: 66 (F*), 112 (P*, L*), 152 (M*); Rozeboom 1942a: 237 (E*; tax.); Ross & Roberts 1943a: 29 (M*, F*, L*); Pelaez 1945: 74 (F*; key); Cova García 1946: Fig. 1 (E*), 30 (L*), 82 (F*),116 (M*); Deane, L.M. et al. 1946 (F*; key, bion.; NE Brazil & Amazonas); Causey et al. 1946 (M*; key; NE Brazil & Amazonas); Deane, M.P. et al. 1946 (L*; key, bion.; NE Brazil & Amazonas); Romeo Vaimonte & Castro 1951: 320 (F*); Stone & Knight 1956b: 277 (type info.); Kreutzer et al. 1972 (chrom.*); Linthicum 1988: 170 (M*, F, P*, L*); Linley 1992: 40 (E*); Manguin et al. 1996 (L; bion.); Lounibos et al. 1998 (mol. tax.); Malafronte et al. 1999 (mol. tax; *s.l.*?); Scarpassa & Conn 2007 (mol. tax., phyl.); Loaiza et al. 2009 (distr.; Panama); Sinka et al. 2010b: 72 (bion. review, distr., niche model); Becker et al. 2010: 378 (F*, L*; key, tax., distr., bion.); Hiwat & Bretas 2011 (bion., review); Motoki et al. 2012 (tax., morph.); Conn et al. 2013 (bion., distr.); Berti et al. 2015 (distr.; Venezuela); Emerson et al. 2015 (mol. tax.; *s.l.*; distr.); Laporta et al. 2015 (bion., distr., niche model). **Ety.:** Samuel T. Darling. **Informal name:** Darling Brazilian Nail Mosquito. (Vol. 1, p. 394)
syn. *paulistensis* Galvão, Lane, & Corrêa
1937: 37 (M*, F, E*; *darlingi* var.). **Type loc.:** Novo Oriente (near Lussanvira), Bacia do Tietê, São Paulo, Brazil (LU). **Refs.:** Lane 1949 (syn.).

deaneorum Rosa-Freitas
1989a: 535 (M*, F*, P, L*, E). **Type loc.:** Guajará-Mirim, Palheta, Rondônia, Brazil (IOC). **Distr.:** Argentina, Bolivia, Brazil, Paraguay. **Refs.:** Klein et al. 1990 (tax., bion., colony); Klein et al. 1991 (tax., DNA hybridization); Motoki et al. 2009a (M*, F, P*, L*); Ruiz-Lopez et al. 2012 (mol. tax.; *s.l.*); Foley et al. 2014 (distr., bion., model; *s.l.*); Laporta et al. 2015 (bion., distr., niche model). **Ety.:** Leônidas M. Deane and Maria P. Deane. **Informal name:** Deane Brazilian Nail Mosquito.

dunhami Causey
1945: 231 (M*, F, L, E*). **Type loc.:** Tefé, Amazonas, Brazil (USNM). **Distr.:** Brazil, Colombia, Peru. **Refs.:** Lane 1953: 268 (syn. with *nuneztovari* without comment); Stone & Knight 1956b: 277 (type info.); Peyton 1993 (from syn. with *nuneztovari* to senior syn. of *trinkae*); Calderón et al. 1995 (distr.; Peru); Lounibos et al. 1997a (E*); Lounibos et al. 1998: 830 (genetics, tax.; to sp.); Calado et al. 2008 (M gen, mol. tax., distr., Brazilian Amazon); Ruiz et al. 2010 (mol. tax., distr.; Colombia); Foster et al. 2013 (mol. phyl.; to Nuneztovari Complex). **Ety.:** George Clark Dunham. **Informal name:** Dunham Brazilian Nail Mosquito.

evansae (Brèthes)
1926b: 106 (M; *Cellia*). **Type loc.:** Tucumán, Argentina (BA). **Distr.:** Argentina, Bolivia, Brazil, Colombia, Ecuador, Nicaragua, Panama, Paraguay, Peru, Uruguay, Venezuela. **Refs.:** Senevet 1932a: 252 (P); Pinto 1939: 425 (F*, L*; from syn. with *tarsimaculatus*); Belkin et al. 1968: 10 (tax.); Faran 1980: 67 (M*, F*, P*, L*; keys, tax., bion., distr.; as *noroestensis*); Faran 1981 (M*; lectotype desig.); Peyton et al. 1983: 70 (emend. to *evansae*); Calderón et al. 1995 (distr.; Peru); Sallum et al. 2002b: 11 (E*); Rossi & Martínez 2003: 471 (distr.; Uruguay). **Ety.:** Alwen Myfanwy Evans. **Informal name:** Evans Argentinian Nail Mosquito.
syn. *metcalfi* Galvão & Lane
1937b: 218 (M, E*; *Anopheles* (*Nyssorhynchus*) *oswaldoi* var.). **Type loc.:** Porto das Caixas, Rio de Janeiro, Brazil (USNM). **Refs.:** Lane 1949: 403 (syn.); Belkin et al. 1971: 6 (tax., lectotype desig.); Faran 1981: 87 (from syn. with *oswaldoi* to syn. with *noroestensis*).
syn. *noroestensis* Galvão & Lane
1937b: 220 (A, M, E*; *Anopheles* (*Nyssorhynchus*) *oswaldoi* var.). **Type loc.:** Novo Oriente [Pereira Barreto], near Lussanvira, São Paulo, Brazil (FH). **Refs.:** Galvão & Damasceno 1942a: 125 (to sp.); Rozeboom 1942a: 240 (E*); Galvão et al. 1943; Deane, L.M. et al. 1946 (F*; key, bion.;

NE Brazil & Amazonas); Causey et al. 1946 (M*; key; NE Brazil & Amazonas); Deane, M.P. et al. 1946 (L*; key, bion.; NE Brazil & Amazonas); Lane 1953: 264 (M*, F, P, L*, E*); Belkin et al. 1971: 6 (lectotype desig.); Faran 1981 (M*; syn. with *evansi* [*sic*?]; lectotype desig.).

syn. *ayrozai* Unti
1940b: 379 (F, P, L, E*; *Anopheles* (*Nyssorhynchus*) *oswaldoi* var.). **Type loc.:** Guaratingueta, São Paulo, Brazil (NE). **Refs.:** Lane 1949: 403 (syn.); Belkin et al. 1971: 6 (tax.); Faran 1981: 87 (syn.).

syn. *clarki* Komp
1943: 197 (M*, F). **Type loc.:** Monteros, Tucumán, Argentina (LU). **Refs.:** Lane 1949: 403 (syn.); Faran 1981: 88 (syn.).

galvaoi Causey, Deane, & Deane
1943b: 293 (M*, F*, E*). **Type loc.:** Rio Branco, Acre, Brazil (FMSP). **Distr.:** Argentina, Brazil, Peru. **Refs.:** Deane, L.M. et al. 1946 (F*; key, bion.; NE Brazil & Amazonas); Causey et al. 1946 (M*; key; NE Brazil & Amazonas); Deane, M.P. et al. 1946 (L*; key, bion.; NE Brazil & Amazonas); Coher 1949: 88 (F); Belkin et al. 1971: 5 (lectotype desig.); Faran 1980: 35 (M*, F*, P*, L*; keys, tax., bion., distr.); Calderón et al. 1995 (distr.; Peru); Sallum, Bergo et al. 2002 (M*, F*, P*, L*); Sallum et al. 2002b: 10 (E*). **Ety.:** Augusto Leopoldo Ayroza Galvão. **Informal name:** Galvão Brazilian Nail Mosquito.

goeldii Rozeboom & Gabaldón
1941: 89 (M*, F*, L*, E). **Type loc.:** Rio Tapajós, Boa Vista (Fordlandia), Brazil (USNM). **Distr.:** Brazil, Suriname. **Refs.:** Deane, L.M. et al. 1946 (F*; key, bion.; NE Brazil & Amazonas); Causey et al. 1946 (M*; key; NE Brazil & Amazonas); Deane, M.P. et al. 1946 (L*; key, bion.; NE Brazil & Amazonas); Floch & Abonnenc 1946c (syn. with *nuneztovari*); Lane 1953: 548 (syn.); Faran 1980 (syn.); Gabaldón 1981 (from syn.); Bergo et al. 2007 (M gen.); Calado et al. 2008 (M*; DNA sequences, phylogenetics; validation of species status); Sant'Ana et al. 2015 (M*, F*, P*, L*, E*; redescr., mol. tax.). **Ety.:** Emilio Augusto Goeldi. **Informal name:** Goeldi Brazilian Nail Mosquito.

guarani Shannon
1928 (F). In Dyar 1928: 444. **Type loc.:** Iguacu [Parana], Brazil (USNM). **Distr.:** Argentina, Brazil. **Refs.:** Nagaki et al. 2011: 375 (M*, F, P, L*; redescr.; to sp.). **Ety.:** Guaran [southern S. American indigenous tribe]. **Informal name:** Guaraní Brazilian Nail Mosquito.

halophylus Silva-do-Nascimento & Lourenço-de-Oliveira
2002: 801 (M*, F*, P*, L*, E*). **Type loc.:** Salobra (20°12'40"S, 56°29'30"W), Mato Grosso, Brazil (IOC). **Distr.:** Brazil. **Refs.:** Silva-do-Nascimento et al. 2006 (mol. tax.); Silva-do-Nascimento et al. 2011 (mol. tax.; *s.l.*). **Ety.:** salt loving (Gr); larval habitat. **Informal name:** Saline Brazilian Nail Mosquito.

ininii Senevet & Abonnenc
1938: 494 (M*, P*, L*). **Type loc.:** Mines de Saint-Elie [Inini], French Guiana [Guiana] (NE). **Distr.:** Argentina, Bolivia, Brazil, French Guiana, Guyana, Paraguay. **Refs.:** Senevet 1948a: 277 (M*; tax.); Senevet 1948c: 431 (P*); Floch & Abonnenc 1951: 50 (M*, F, L*); Belkin 1968b: 53 (type info.); Faran 1980: 89 (M*, F*, P*, L*; keys, tax., bion., distr.). **Ety.:** administrative district of Inini, French Guiana. **Informal name:** Inini French Guianese Nail Mosquito.

janconnae Wilkerson & Sallum
2009 (M*, F, P, L; morph., morphometrics). In Motoki et al. 2009a: 846. **Type loc.:** Boa Vista municipality, Roraima, Brazil (FH). **Distr.:** Brazil, Venezuela. **Refs.:** Ruiz-Lopez et al. 2012 (mol. tax.; *s.l.*); Foley et al. 2014 (distr., bion., model; *s.l.*); Laporta et al. 2015 (bion., distr., niche model). **Ety.:** Jan Evelyn Conn. **Informal name:** Conn Brazilian Nail Mosquito.

konderi Galvão & Damasceno
1942a: 115 (M*, P, L*). **Type loc.:** Solimões River at Coari, Amazonas, Brazil (IOC). **Distr.:** Bolivia, Brazil, Peru. **Refs.:** Causey et al. 1944: 5 (E*); Deane, L.M. et al. 1946 (F*; key, bion.; NE Brazil & Amazonas); Causey et al. 1946 (M*; key; NE Brazil & Amazonas); Deane, M.P. et al. 1946 (L*; key, bion.; NE Brazil & Amazonas); Lane 1953: 262 (syn.); Belkin et al. 1971: 6 (tax.); Lounibos et al. 1997a (E*); Flores-Mendoza et al. 2004: 118 (M*, F*, P*, L*, E*; neotype desig.; to sp.). **Ety.:** Valerio Augusto Regis Konder. **Informal name:** Konder Brazilian Nail Mosquito.

lanei Galvão & Franco do Amaral
1938: 8 (M*, F*, P, L*, E*). **Type loc.:** Vila Emilio Ribas, Prefeitura Sanitária de Campos do Jordão, Serra de Mantiqueira, São Paulo, Brazil (FMSP). **Distr.:** Brazil. **Refs.:** Pinto 1939: 387 (M*, F, L*); Lane 1953: 249 (M*, F, P, L*, E*); Senise & Sallum 2007 (M*, F, P*, L*; redescr.). **Ety.:** John Lane. **Informal name:** Lane Brazilian Nail Mosquito.

lutzii Cruz
1901: 423 (F*, L, E*). **Type loc.:** Lagoa Rodrigo de Freitas, Rio de Janeiro (Guanabara), Brazil (IOC). **Distr.:** Argentina, Bolivia, Brazil, Paraguay, Peru. **Refs.:** Davis 1926 (F*); Ramos 1937: 215 (M*); Galvão 1941b: 521 (M*, F*, P*, L*, E*); Romeo Viamonte & Castro 1951: 313 (F*); Calderón et al. 1995 (distr.; Peru); Forattini et al. 1998: 259 (E*); Nagaki et al. 2011: 375 (M*, F, P, L*; redescr.). **Ety.:** Adolfo Lutz. **Informal name:** Lutz Brazilian Nail Mosquito.

syn. *nigra* Theobald
1907: 78 (M, P*; *Myzorhynchella*). **Type loc.:** Cantareira (São Paulo), Brazil (NHMUK). **Refs.:** Belkin 1968b: 10 (lectotype desig.).

marajoara Galvão & Damasceno
1942b: 424 (M, F, L). **Type loc.:** Cachoeira do Arari, Brazil (FMSP). **Distr.:** Bolivia, Brazil, Colombia, French Guiana, Panama, Trinidad and Tobago, Venezuela. **Refs.:** Galvão &

Damasceno 1942c: 60–66 (repeat of orig. descr.?); Lane 1953: 244 (syn.); Belkin et al. 1971: 4, 46 (tax., type info.); Linthicum 1988: 198 (from syn. with *albitarsis*); Rubio-Palis et al. 2003; Senise et al. 2006 (type info.); Motoki et al. 2009a (M*, F, P*, L*); Sinka et al. 2010b: 72 (bion., review, distr., niche model); Ruiz-Lopez et al. 2012 (mol. tax.; *s.l.*); Foley et al. 2014 (distr., bion., model; *s.l.*); Laporta et al. 2015 (bion., distr., niche model). **Ety.:** Marajó Isl., Brazil. **Informal name:** Marajó Brazilian Nail Mosquito.

syn. *domesticus* Galvão & Damasceno 1944: 78 (M, F, E*). **Type loc.:** Cachoeira, Ilha de Marajó, Pará, Brazil (LU). **Refs.:** Corrêa et al. 1950: 280 (bion.); Belkin et al. 1971: 4 (type info.); Linthicum 1988: 198 (syn.).

nigritarsis (Chagas) 1907: 12 (F; *Myzorhynchella*). **Type loc.:** Brazil (NHMUK). **Distr.:** Argentina, Brazil. **Refs.:** Rachou & Sousa 1950: 234 (F*); Townsend 1990: 110 (type info.). **Ety.:** black tarsi (L). **Informal name:** Black-footed Brazilian Nail Mosquito.

nuneztovari Gabaldón 1940: 5 (M*). **Type loc.:** San Carlos, Cojedes, Venezuela (DERM). **Distr.:** Bolivia, Brazil, Colombia, Ecuador, French Guiana, Guyana, Panama, Peru, Suriname, Venezuela. **Refs.:** Rozeboom & Gabaldón 1941: 91 (M*); Cova García 1946: Fig. 1 (E*), 36 (L*), 88 (F*), 124 (M*); Floch & Abonnenc 1946c: 3 (M, F, P, L); Kitzmiller et al. 1973 (chrom.); Sutil Oramas 1976: 33 (M*, F*, P*, L*, E*); Faran 1980: 112 (M*, F*, P*, L*; keys, tax., bion., distr.); Steiner et al. 1980 (chrom.; *s.l.*); Savage 1986 (tax.); Conn 1990 (chrom.); Conn et al. 1993 (chrom.; *s.l.*); Hribar 1994 (M gen.; morph.); Fritz et al. 1994 (mol. tax.; *s.l.*); Hribar 1995a (F; morph.); Hribar 1995b (P; morph.); Linley et al. 1996 (E*); Lounibos et al. 1997a (E*); Lounibos et al. 1998: 830 (genetics, tax.); Sierra et al. 2004 (mol. tax.; *s.l.*); Calado et al. 2008 (mol. tax.; *s.l.*); Sinka et al. 2010b: 72 (bion., review, distr., niche model); Becker et al. 2010: 380 (F*, L*; key, tax., distr., bion.); Scarpassa & Conn 2011 (mol. tax., phyl.; *s.l.*); Dusfour et al. 2012 (malaria incrim.); Conn et al. 2013 (bion., distr.; *nuneztovari s.l.*); Berti et al. 2015 (distr.; Venezuela). **Ety.:** Manuel Nuñez Tovar. **Informal name:** Nuñez Tovar Venezuelan Nail Mosquito. (Vol. 1, p. 396)

oryzalimnetes Wilkerson & Motoki 2009 (M*, F, P*, L*). In Motoki et al. 2009a: 841. **Type loc.:** Centro de Desenvolvimento Agrícola do Vale do Ribeira (CEDAVAL), Registro municipality, São Paulo, Brazil (FH). **Distr.:** Brazil, Paraguay. **Refs.:** Ruiz-Lopez et al. 2012 (mol. tax.; *s.l.*); Foley et al. 2014 (distr., bion., model; *s.l.*); Laporta et al. 2015 (bion., distr., niche model; *s.l.*). **Ety.:** rice inhabiting (Gr); ref. to larval habitat. **Informal name:** Rice Brazilian Nail Mosquito.

oswaldoi (Peryassú) 1922b: 179 (M*, F*; *Cellia*). **Type loc.:** Valle do Rio Doce [Espirito Santo], (Baixada Fluminense), Brazil (MNRJ). **Distr.:** Argentina, Bolivia, Brazil, Colombia, Costa Rica, Ecuador, French Guiana, Guyana, Panama, Paraguay, Peru, Suriname, Trinidad and Tobago, Uruguay, Venezuela. **Refs.:** Davis 1926 (F*); Senevet 1934: 49 (P*; to sp.); Galvão & Lane 1938 (tax.); Rozeboom & Gabaldón 1941: 91 (M*, L, E); Coutinho & Verano 1942: 353 (L*); Komp 1942: 74 (F*), 124 (L), 160 (M*); Rozeboom 1942a: 239 (E*); Cova García 1946: Fig. 1 (E*), 34 (L*), 86 (F*), 121 (M*); Deane, L.M. et al. 1946 (F*; key, bion.; NE Brazil & Amazonas); Causey et al. 1946 (M*; key; NE Brazil & Amazonas); Deane, M.P. et al. 1946 (L*; key, bion.; NE Brazil & Amazonas); Belkin et al. 1971: 6 (type info.); Faran 1980: 55 (M*, F*, P*, L*; keys, tax., bion., distr.); Lounibos et al. 1997a (E*); Rossi & Martínez 2003: 471 (distr.; Uruguay); Estrada et al. 2003 (E*, = *oswaldoi* B; *oswaldoi* A; Colombia); Ruiz et al. 2005 (mol. tax. = *oswaldoi* B; Colombia); Motoki et al. 2007 (redescr.); Ruiz-Lopez et al. 2013 (mol. tax., Oswaldoi Complex). **Ety.:** Oswaldo Gonçalves Cruz. **Informal name:** Oswaldo Brazilian Nail Mosquito.

syn. *aquacaelestis* Curry 1932 (M*, L; *Anopheles* (*Nyssorhynchus*) *tarsimaculatus* var.). **Type loc.:** Canal Zone, Panama (LU). **Refs.:** Rozeboom 1941: 102 (syn.).

parvus (Chagas) 1907: 4 (M, F; *Myzorhynchella*). **Type loc.:** [Oliveira, Minas Gerais], Brazil (IOC?). **Distr.:** Argentina, Bolivia, Brazil, Chile, Colombia, Paraguay, Peru, Venezuela. **Refs.:** Galvão 1941b: 536 (M*, F*, P*, L*); Cova García 1946: Fig. 1 (E*), 42 (L*), 94 (F*), 117 (M*); Deane, L.M. et al. 1946 (F*; key, bion.; NE Brazil & Amazonas); Causey et al. 1946 (M*; key; NE Brazil & Amazonas); Deane, M.P. et al. 1946 (L*; key, bion.; NE Brazil & Amazonas); Belkin et al. 1971: 6 (type info.); Calderón et al. 1995 (distr.; Peru); Forattini et al. 1998: 258 (E*). **Ety.:** small (L) [comp. to *lutzi*]. **Informal name:** Small Brazilian Nail Mosquito.

pictipennis (Philippi) 1865: 596 (M; *Culex*). **Type loc.:** Aconcagua Province, Chile (NE). **Distr.:** Argentina, Brazil, Chile. **Refs.:** Galvão 1941b: 560 (F*; tax.); Lane 1953: 254 (M*, F, L*, E*); Lane & Neghme 1956 (M*, F, L*); Linthicum 1988: 165 (M, F, L); Rueda, Rodriguez et al. 2008 (tax., distr.); González & Sallum 2009 (M*, L*; neotype desig.). **Ety.:** n.s. [painted wing (L)]. **Informal name:** Paint-winged Chilean Nail Mosquito.

syn. *variegatus* Blanchard 1852 (A; *Culex*). In Gay 1852: 333. **Type loc.:** Arquero, Chile (MNHP). **Refs.:** Belkin 1968b: 10 (lectotype desig.; change in syn.).

syn. *bigotii* Theobald 1901a: 135 (F*). **Type loc.:** Chile (NE, originally in NHMUK). **Refs.:** Belkin 1968b: 9 (tax.).

syn. *chilensis* R. Blanchard 1905: 289 (new name for *variegatus* Blanchard, not Schrank 1781; *Culex*). **Refs.:** Edwards 1932a: 126 (tax.); Stone et al. 1959: 131 (*nomen dubium* in *Psorophora*); Belkin 1968b: 10 (syn.).

pristinus Nagaki & Sallum
2010 (M*, F*, L*, E*; tax., mol. tax.). In Nagaki et al. 2010: 82. **Type loc.:** Fazenda Saint Claire, Pico do Itapeva, border of municipalities of Pindamonhangaba/Campos do Jordão, São Paulo, Brazil (FH). **Distr.:** Brazil. **Ety.:** pristine, pure (L); ref. to larval habitat. **Informal name:** Pristine Brazilian Nail Mosquito.

rangeli Gabaldón, Cova García & Lopez
1940a: 13 (M*, P*, L*, E*). **Type loc.:** Puerto Cabello, Carabobo, Venezuela (DERM). **Distr.:** Bolivia, Brazil, Colombia, Ecuador, Peru, Venezuela. **Refs.:** Rozeboom & Gabaldón 1941: 91 (M*, L, E; tax.); Cova García 1946: Fig. 1 (E*), 37 (L*), 89 (F*), 123 (M*); Deane, L.M. et al. 1946 (F*; key, bion.; NE Brazil & Amazonas); Causey et al. 1946 (M*; key; NE Brazil & Amazonas); Deane, M.P. et al. 1946 (L*; key, bion.; NE Brazil & Amazonas); Faran 1980: 99 (M*, F*, P*, L*; keys, tax., bion., distr.); Linley & Lounibos 1993: 158 (E*); Estrada et al. 2003 (E*; Colombia); Berti et al. 2015 (distr.; Venezuela). **Ety.:** Rafael Rangel. **Informal name:** Rangel Venezuelan Nail Mosquito.

rondoni (Neiva & Pinto)
1922a: 322 (F; *Cellia*). **Type loc.:** Ladario, on Rio Paraguai, Mato Grosso [now in Matto Grosso do Sul], Brazil (IOC). **Distr.:** Argentina, Bolivia, Brazil, Paraguay, Venezuela. **Refs.:** Davis 1933a: 278 (L*; tax.); Pinto 1939: 398 (M*, F*, L*; bion.); Corrêa & Ramos 1944 (E); Deane, L.M. et al. 1946 (F*; key, bion.; NE Brazil & Amazonas); Causey et al. 1946 (M*; key; NE Brazil & Amazonas); Deane, M.P. et al. 1946 (L*; key, bion.; NE Brazil & Amazonas); Belkin et al. 1971: 6 (type loc. info., lectotype desig.); Faran 1980: 132 (M*, F*; keys, tax., bion., distr.); Wilkerson et al. 1995b: 575 (F*; tax.); Sallum & Wilkerson 1997: 365 (P*, L*); Forattini et al. 1998: 258 (E*). **Ety.:** Cândido Mariano Rondon. **Informal name:** Rondon Brazilian Nail Mosquito.

sanctielii Senevet & Abonnenc
1938: 501 (M*). **Type loc.:** Mines de Saint-Elie, Sinnamary River [Inini], French Guiana [Guiana] (NE). **Distr.:** French Guiana. **Refs.:** Floch & Abonnenc 1946e: 4 (key); Floch & Abonnenc 1951: 51 (M*); Belkin 1968b: 53 (type info.). **Ety.:** n.s. [Mines of Saint-Élie, French Guiana]. **Informal name:** Saint Élie French Guianese Nail Mosquito.

sawyeri Causey, Deane, Deane, & Sampaio
1943: 11 (M*, F*, P*, L*, E*; as sp.). **Type loc.:** Serra de Ibiapaba, near São Benedito, Ceará, Brazil (USNM). **Distr.:** Brazil. **Refs.:** Deane, L.M. et al. 1946 (F*; key, bion.; NE Brazil & Amazonas); Causey et al. 1946 (M*; key; NE Brazil & Amazonas); Deane, M.P. et al. 1946 (L*; key, bion.; NE Brazil & Amazonas); Linthicum 1988: 155 (from ssp. of *argyritarsis*). **Ety.:** Wilbur A. Sawyer. **Informal name:** Sawyer Brazilian Nail Mosquito.

striatus Sant'Ana & Sallum
2017 (M*, F, P*, L*, E*). **Type loc.:** Coronel Pacheco, road from Juiz de Fora to Uba municipality, Embrapa (21°38895S, 43°19936W), Minas Gerais, Brazil (FH). **Distr.:** Brazil. **Ety.:** wrinkled, striated (L); surface of ventral claspette. **Informal name:** Wrinkled Brazilian Nail Mosquito.

strodei Root
1926: 711 (M*, P*, L*). **Type loc.:** Agua Limpa near Juizde Fora, Minas Gerais, Brazil (USNM). **Distr.:** Argentina, Bolivia, Brazil, Colombia, Costa Rica, Ecuador, Guatemala, Guyana, Honduras, Mexico, Nicaragua, Panama, Paraguay, Peru, Suriname, Trinidad and Tobago, Venezuela. **Refs.:** Senevet 1934: 52 (P*); Galvão & Lane 1936: 275 (M, F, L*, E*); Pinto 1939: 425 (syn.); Galvão 1941b: 516 (M*, F, L*); Komp 1942: 71 (F), 120 (L*), 158 (M*); Rozeboom 1942a: 240 (E*; tax.); Galvão 1943: 143 (M*, L*, tax.); Pelaez 1945: 74 (F*; key); Cova García 1946: Fig. 1 (E*), 40 (L*), 91 (F*), 126 (M*); Deane, L.M. et al. 1946 (F*; key, bion.; NE Brazil & Amazonas); Causey et al. 1946 (M*; key; NE Brazil & Amazonas); Deane, M.P. et al. 1946 (L*; key, bion.; NE Brazil & Amazonas); Stone & Knight 1956b: 280 (type info., lectotype desig.); Faran 1980: 123 (M*, F*, P*, L*; keys, tax., bion., distr.; from syn. with *evansi*); Calderón et al. 1995 (distr.; Peru); Lounibos et al. 1997a (E*); Beltrán-Aguilar et al. 2011 (M*, F* P*, L*; keys, distr., tax.); Berti et al. 2015 (distr.; Venezuela). **Ety.:** George King Strode. **Informal name:** Strode Brazilian Nail Mosquito.

syn. *ramosi* Unti
1940a: 489 (L*; *strodei* var.). **Type loc.:** Lorena, São Paulo, Brazil (NE). **Refs.:** Faran 1980: 123 (from *evansi* var. to syn. with *strodei*); Sallum et al. 2010.

syn. *lloydi* Unti
1941: 10 (E; *strodei* var.). **Type loc.:** Panama (NE). **Refs.:** Faran 1980: 123 (from *evansi* var. to syn. with *strodei*).

triannulatus (Neiva & Pinto)
1922b: 356 (F*; *Cellia*). **Type loc.:** Rio Cuiabá, Fazenda São João, Mato Grosso, Brazil (LU). **Distr.:** Argentina, Bolivia, Brazil, Colombia, Costa Rica, Ecuador, French Guiana, Guatemala, Guyana, Nicaragua, Panama, Paraguay, Peru, Puerto Rico, Suriname, Trinidad and Tobago, Venezuela. **Refs.:** Pinto 1939: 406 (M*, F*, P*, L*, E*; tax.); Komp 1942: 69 (F*), 118 (L*), 156 (M*); Rozeboom 1942a: 241 (E*); Floch & Abonnenc 1944 (E*); Cova García 1946 (as *davisi*), Fig. 1 (E*), 38 (L*), 90 (F*), 125 (M*); Deane, L.M. et al. 1946 (F*; key, bion.; NE Brazil & Amazonas); Causey et al. 1946 (M*; key; NE Brazil & Amazonas); Deane, M.P. et al. 1946 (L*; key, bion.; NE Brazil & Amazonas); Belkin et al. 1971: 6 (tax.); Faran 1980: 141 (M*, F*, P*, L*; keys, tax., bion., distr.); Conn 1991 (chrom.); Lounibos et al. 1997a (E*); Chadee & Wilkerson 2005 (distr.; Trinidad & Tobago); Zimmerman et al. 2006 (bion.); Silva-do-Nascimento et al. 2006 (mol. tax.; *s.l.*); Silva-do-Nascimento et al. 2011 (mol. tax.;

s.l.); Rosero et al. 2012 (mol. tax.; *s.l.*, distr.; Colombia); Moreno et al. 2013 (phylogeography; *s.l.*); Berti et al. 2015 (distr.; Venezuela). **Ety.:** n.s. [three ringed (L)]. **Informal name:** Three-ringed Brazilian Nail Mosquito. (Vol. 1, p. 398)

syn. *cuyabensis* Neiva & Pinto
1923: 235 (F*; *Cellia*). **Type loc.:** Rio Cuiabá, Fazenda São João, Mato Grosso, Brazil (IOC). **Refs.:** Pinto 1939: 406 (syn.); Belkin et al. 1971: 7 (lectotype desig.).

syn. *bachmanni* Petrocchi
1925: 70 (M*, F*; author name Petrochi [*sic*]). **Type loc.:** Corrientes, Entre Rios, Formosa, Argentina (NE). **Refs.:** Pinto 1939: 406 (syn.); Cova García 1961: 39, 91, 128 (M*, F*, L*, E*; from syn. as ssp.; syn.); Faran 1980: 141 (from ssp. of *triannulatus*).

syn. *davisi* Paterson & Shannon
1927: 5 (M, F, L). **Type loc.:** Tres Pozos, Embarcación, Salta, Argentina (USNM). **Refs.:** Galvão 1940: 412 (L*, E*; to ssp.); Stone & Knight 1956b: 277 (type info.); Faran 1980: 141 (from syn. with *bachmanni* to syn. with *triannulatus*).

syn. *perezi* Shannon & Del Ponte
1928: 56 (F). **Type loc.:** Santa Barbara, Tucumán, Argentina (NE). **Refs.:** Pinto 1939: 406 (syn.); Faran 1980: 141 (from syn. with *bachmanni*).

syn. *chagasi* Galvão
1941a: 92 (F, E*; as var.). **Type loc.:** Manaus, Brazil (LU). **Refs.:** Lane 1949 (syn.); Belkin et al. 1971: 7 (type loc. info.).

trinkae Faran
1979: 27 (M*, F*, P*, L*). **Type loc.:** 15 km S of Puyo, Pastaza, Ecuador (USNM). **Distr.:** Bolivia, Colombia, Ecuador, Peru. **Refs.:** Faran 1980: 106 (M*, F*, P*, L*; keys, tax., bion., distr.); Linley & Lounibos 1993: 162 (E*); Lounibos et al. 1997a (E*); Lounibos et al. 1998: 830 (genetics, tax.). **Ety.:** Trinka Coster. **Informal name:** Trinka Ecuadorian Nail Mosquito.

Subgenus *Stethomyia* Theobald

1902b:182 (as genus). Type species: *Stethomyia nimbus* Theobald.

Name derivation: n.s. [stetho-, breast; -myia, fly (Gr); prob. ref. to "prothoracic lobes mammilated"].

Important references: Edwards 1932a (to subg. of genus *Anopheles*); Lane 1953 (M*, F, P*, L*, E*; keys, bion.; Neotropics); Cova-García 1961 (keys, tax.; Venezuela); Forattini 1962 (M*, F*, P*, L*; keys, tax., bion.); Wilkerson & Strickman 1990 (F; key; Central America and Mexico).

acanthotorynus Komp
1937a: 358 (M*). **Type loc.:** Iquitos, Loreto, Peru (USNM). **Distr.:** French Guiana, Peru. **Refs.:** Senevet & Abonnenc 1940: 442 (M*, F). **Ety.:** n.s. [thorn spoon (Gr); poss. ref. to spoon-shaped claspette ending in a terminal spine]. **Informal name:** Thorny Peruvian Nail Mosquito.

canorii Floch & Abonnenc
1945d: 1 (M*). **Type loc.:** Saut Canori (Haut Approuague), French Guiana [Guiana] (PIP). **Distr.:** French Guiana. **Refs.:** Harrison 1973c: 277 (type info.). **Ety.:** Saut Canori (Approuague), French Guiana. **Informal name:** Canori French Guianese Nail Mosquito.

kompi Edwards
1930a: 288 (M; *nimbus* var.). **Type loc.:** Almirante Panama (USNM). **Distr.:** Brazil, Colombia, Costa Rica, French Guiana, Panama, Peru, Suriname, Venezuela. **Refs.:** Senevet 1934: 39 (P*); Komp 1942: 48 (F*), 86 (L*), 134 (M*); Deane, L.M. et al. 1946 (F*; key, bion.; NE Brazil & Amazonas); Causey et al. 1946 (M*; key; NE Brazil & Amazonas); Deane, M.P. et al. 1946 (L*; key, bion.; NE Brazil & Amazonas); Lane 1953: 152 (M*, F, P*, L*, E*); Kitzmiller et al. 1976: 483 (chrom.). **Ety.:** William H. Wood Komp. **Informal name:** Komp Panamanian Nail Mosquito.

nimbus (Theobald)
1902b: 182 (F*; *Stethomyia*). **Type loc.:** Cabacaburi, Pomeroon River, British Guiana [Guyana] (NHMUK). **Distr.:** Bolivia, Brazil, Colombia, French Guiana, Guyana, Peru, Trinidad and Tobago, Venezuela. **Refs.:** da Costa Lima 1929a: 293 (M*); Cova García 1946: Fig. 1 (E*), 53 (L), 109 (F*), 133 (M*); Deane, L.M. et al. 1946 (F*; key, bion.; NE Brazil & Amazonas); Causey et al. 1946 (M*; key; NE Brazil & Amazonas); Deane, M.P. et al. 1946 (L*; key, bion.; NE Brazil & Amazonas); Floch & Abonnenc 1951: 27 (M*, L*); Romeo Viamonte & Castro 1951: 313 (F*); Lane 1953: 150 (M*, F, P, L, E*); Belkin 1968b: 10 (lectotype desig.); Calderón et al. 1995 (distr.; Peru); Berti et al. 2015 (distr.; Venezuela). **Ety.:** n.s. [black cloud (L)]. **Informal name:** Cloudy Guyanese Nail Mosquito.

syn. *lineata* Lutz
1905: 202 (M; *Rhynchomyia*). **Type loc.:** Brazil (IOC?). **Refs.:** Belkin et al. 1971: 2 (tax.).

thomasi Shannon
1933a: 58 (new name for *lewisi* Shannon 1931, not Ludlow 1920). **Distr.:** Brazil, Colombia, Peru, Venezuela. **Refs.:** Deane, L.M. et al. 1946 (F*; key, bion.; NE Brazil & Amazonas); Causey et al. 1946 (M*; key; NE Brazil & Amazonas); Deane, M.P. et al. 1946 (L*; key, bion.; NE Brazil & Amazonas); Cova García 1946: 56 (L), 109 (F*), 133 (M*). **Ety.:** H. Wolferstan Thomas. **Informal name:** Thomas Brazilian Nail Mosquito.

syn. *lewisi* Shannon
1931: 154 (M*, F; *Stethomyia*). **Type loc.:** Rio Coruripe, near Salvador City, Bahia, Brazil (USNM). **Refs.:** Shannon 1933a (syn.); Stone & Knight 1956b: 278 (type info.).

Genus *Bironella* Theobald

(Vol. 1, p. 214)

Subgenus *Bironella* Theobald

1905f: 69. Type species: *Bironella gracilis* Theobald (as genus and subg.) (International Commission on Zoological Nomenclature 1965, Opinion 737; name validation).

Name derivation: Lajos Biró (collector); biro- plus a diminutive suffix -ella (L).

Important references: Edwards 1922d: 98 (generic status confirmed); Christophers 1924a (as subg. of *Anopheles*); Bonne-Wepster & Swellengrebel 1953: 50 (M*, F*, L*; keys, bion., tax.); Belkin 1962: 129 (M*, F*, P*, L*; keys, tax.; South Pacific); Tenorio 1977 (M*, F*, P*, L*; keys, bion., distr.); Lee et al. 1988a (F; key, tax., bion., distr.); Besansky & Fahey 1997 (mol. tax.); Harbach & Kitching 1998 (phyl.); Foley et al. 1998 (mol. phyl.); Sallum et al. 2000b (phyl.); Krzywinski et al. 2001a, 2001b (mol. phyl.); Sallum et al. 2002c (mol. phyl.); Harbach & Kitching 2005 (phyl., class.).

gracilis Theobald
1905f: 69 (M*). **Type loc.:** Muina, [Northeast] New Guinea (HNM). **Distr.:** Australia, Indonesia, Papua New Guinea. **Refs.:** Brug & de Rook 1922 (M*, F*, L*; tax.); Lee & Woodhill 1944: 51 (M*, F*, L*); Bonne-Wepster & Swellengrebel 1953: 52 (M*, F*, L*; tax.); Stone 1957b: 171 (tax.); International Commission on Zoological Nomenclature 1965 (Opinion 737; name validation); Tenorio 1977: 323 (M*, F*, P*, L*; keys, bion., distr.; Papuan subregion); Lee et al. 1988a: 38 (F; key, tax., bion., distr., review). **Ety.:** slender (L); body. **Informal name:** Slender Papuan Biró Mosquito.
syn. *walchi* Soesilo
1932: 1035 (L*; *Anopheles*). **Type loc.:** Wadapi, Japen Island, New Guinea (LU). **Refs.:** Lee & Woodhill 1944: 79 (? = *gracilis*); Bonne-Wepster & Swellengrebel 1953: 76 (L*; tax.).
syn. *occulta* Bonne-Wepster
1951b: 71 (M*). **Type loc.:** Pionnierbivak, [Western] New Guinea (NE). **Refs.:** Stone 1957b: 171, 174 (syn.; tax.).

simmondsi Tenorio
1977: 331 (M*, F*, L). **Type loc.:** Dinner Creek, Innisfail, Queensland, Australia (UQ). **Distr.:** Australia, Papua New Guinea. **Refs.:** Lee et al. 1988a: 44 (F; key, tax., bion., distr., review). **Ety.:** John Howard Simmonds. **Informal name:** Simmonds Australian Biró Mosquito.

Subgenus *Brugella* Edwards

Edwards 1930a: 288 (as subg. of *Bironella*). Type species: *Anopheles travestitus* Brug.

Name derivation: Steffen Lambert Brug.

Important references: Bonne-Wepster & Swellengrebel 1953: 50 (M*, F*, L*; keys, bion., tax.); Belkin 1962: 129 (M*, F*, P*, L*; keys, tax.; South Pacific); Tenorio 1977 (M*, F*, P*, L*; keys, bion., distr.).

hollandi Taylor
1934: 229 (M, F). **Type loc.:** Kavieng, New Ireland, Bismarck Archipelago (ANIC). **Distr.:** Papua New Guinea, Solomon Isl. **Refs.:** Belkin et al. 1945: 244 (M*, F*, P*, L*, E*; tax.); Dumbleton 1946: 250 (E*); Belkin 1962: 130 (M*, F*, P*, L*, E; tax., bion.); Mattingly 1971a: Pl. 1 (wing*); Pl. 22 (L head*, thorax*); Tenorio 1977: 335 (M*, F, P*, L*); Lee et al. 1988a: 47 (F; key, tax., bion., distr., review, type dep.). **Ety.:** Edwin Archibald Holland. **Informal name:** Holland Papuan Biró Mosquito.

obscura Tenorio
1975: 734 (M*, F, P*, L*). **Type loc.:** Sepik, Vanimo, Daunda, Sepik District, NE New Guinea (BBM). **Distr.:** Indonesia, Papua New Guinea. **Refs.:** Tenorio 1977: 339 (M*, P*, L*); Lee et al. 1988a: 50 (F; key, tax., bion., distr., review). **Ety.:** black, indistinct (L); lacks basotergal lobe and setae on gonocoxite. **Informal name:** Black Papuan Biró Mosquito.

travestita (Brug)
1928b (M*, F*, L*; *Anopheles*). **Type loc.:** Piroe, Ceram, Moluccas [Indonesia] (LU). **Distr.:** Guinea, Indonesia (Moluccas), Papua New Guinea. **Refs.:** Lee & Woodhill 1944: 70 (M*, F*, L*); Bonne-Wepster & Swellengrebel 1953: 73 (M*, F*, L*); Tenorio 1977: 342 (M*, F, P*, L*; lectotype desig.); Lee et al. 1988a: 51 (F; key, tax., bion., distr., review). **Ety.:** across clothed (L); similar structure and vestiture of antenna and palpus in both sexes. **Informal name:** Matching Indonesian Biró Mosquito.

Subgenus *Neobironella* Tenorio

1977: 345 (as subg. of genus *Bironella*). Type species: *Bironella confusa* Bonne-Wepster.

Name derivation: new *Bironella*.

Important references: Tenorio 1977 (M*, F*, P*, L*; keys, bion., distr.); Lee et al. 1988a (F; key, tax., bion., distr.)

confusa Bonne-Wepster
1951b: 67 (M*). **Type loc.:** [Boemi District or Hollandia], Western New Guinea [Papua New Guinea] (NHMUK). **Distr.:** Indonesia (West Papua). **Refs.:** Slooff 1963a: 138 (M*, F, P*, L*); Tenorio 1977: 345 (M*, F, P*, L*; distr., transfer); Lee et al. 1988a: 57 (F; key, distr., review). **Ety.:** n.s. [confused (L); poss. confused with another species]. **Informal name:** Confusing Papuan Biró Mosquito.

derooki Soesilo & van Slooten
1931b: 126 (M*; as *de rooki*). **Type loc.:** Boemi District or Hollandia, [Western] New Guinea (NHMUK). **Distr.:** Indonesia (West Papua). **Refs.:** Lee & Woodhill 1944: 64 (L*; as *soesiloi*); Bonne-Wepster 1951b: 68 (syn.); Bonne-Wepster & Swellengrebel 1953: 68 (M*, F, L*; tax.; as *soesiloi*); Peters 1962: 163 (P*; as *soesiloi*); Tenorio 1977: 355 (M*, F, P*; transfer); Lee et al. 1988a: 64, (F; key, tax., bion., distr., review; as *soesiloi*); Evenhuis & Gon 1989: 194 (from syn. with *soesiloi*). **Ety.:** Hendrik De Rook. **Informal name:** De Rook Papuan Biró Mosquito.

syn. *soesiloi* Strickland & Chowdhury
1931a: 12 (L*; *Anopheles* (*Bironella*)). **Type loc.:** [Manokwari, Western New Guinea [Papua New Guinea]] (LU). **Refs.:** Evenhuis & Gon 1989: 194 (syn.).

papuae (Swellengrebel & Swellengrebel de Graaf)
1920: 3 (L*; *Stethomyia aitkenii* var.). **Type loc.:** West New Guinea (NE). **Distr.:** Australia, Indonesia (Papua), Papua New Guinea. **Refs.:** de Rook & Soesilo 1930 (M*, F, L*); Lee & Woodhill 1944: 58 (M*, F*, L*; tax.); Bonne-Wepster & Swellengrebel 1953: 62 (M*, F*, L*; tax.); Bonne-Wepster & Swellengrebel 1953: 67 (M*; tax.; as *brugi*); Tenorio 1977: 352 (M*, F; distr., transfer); Lee et al. 1988a: 60 (F; key, tax., bion., distr., review). **Ety.:** Papua New Guinea. **Informal name:** Irian Papuan Biró Mosquito [Irian is the historical name for Papua New Guinea].

syn. *brugi* Soesilo & van Slooten
1931b: 126 (M*; *papuae* var.); 1931a: 1256 (in Dutch). **Type loc.:** Boemi District or Hollandia, [Western] New Guinea (NHMUK). **Refs.:** Tenorio 1977: 352 (syn.).

Genus *Chagasia* Cruz

(Vol. 1, p. 216)

1906: 199. Type species *Chagasia neivae* Cruz (subjective synonym of *Chagasia fajardi* (Lutz 1904)).

Name derivation: Evandro Chagas.

Important references: Lane 1953: 138 (M*, F, L*, P*, E*; keys, tax., bion.); Kreutzer 1978 (chrom. number); Harbach & Howard 2009 (M*, F*, P*, L*; keys, tax., review); Beltrán-Aguilar et al. 2011 (M gen.*, F*, P*, L*; bion., distr., tax.; Mexico).

ablusa Harbach
2009 (A*, P, L). In Harbach & Howard 2009: 5. **Type loc.:** Misión Cutivireni (12S, 74W), Satipo, Junin, Peru (USNM). **Distr.:** Colombia, Peru, Venezuela. **Refs.:** Berti et al. 2015 (distr.; Venezuela). **Ety.:** n.s [different (L)]. **Informal name:** Strange Peruvian Basal Mosquito.

bathana (Dyar)
1928 (F, P, L; as *Anopheles* (*Chagasia*) *bathanus*). In Curry 1928: 244. **Type loc.:** Gatún, Canal Zone, Panama (USNM). **Distr.:** Belize, Colombia, Costa Rica, Ecuador, French Guiana, Guatemala, Guyana, Honduras, Mexico, Nicaragua, Panama, Peru, Venezuela. **Refs.:** Senevet 1934: 50 (P); Gabaldón et al. 1940b: 58 (F, P, L); Rozeboom 1941: 98 (bion., distr.); Komp 1942: 46 (M, F, P, L); Leví-Castillo 1945: 17 (M, F, P, L); Knight & Chamberlain 1948: 11 (P); Lane 1953 (F, P, L, E); Horsfall 1955: 41 (bion., distr.); Vargas & Martínez-Palacios 1956: 52 (M, F, P, L); Villanueva 1961: 217 (bion., distr.); Mattingly 1971a: Pl. 8 (P abdomen*), Pl. 22 (L terminal segs.*); Baerg & Boreham 1974b: 631 (E); Fauran & Pajot 1974a: 100 (distr.); Calderón et al. 1995: 5 (bion., distr.; Peru); Harbach & Howard 2009: 9 (A*, P, L; distr.); Beltrán-Aguilar et al. 2011: 736 (M*, F*, P*, L*; tax., keys, distr.). **Ety.:** C.H. Bath. **Informal name:** Bath Panamanian Basal Mosquito.

bonneae Root
1927b: 474 (M, F, P*, L*). **Type loc.:** Dam and Moengo, Suriname (USNM). **Distr.:** Bolivia, Brazil, Colombia, Ecuador, French Guiana, Guyana, Peru, Suriname, Venezuela. **Refs.:** Deane, L.M. et al. 1946 (F*; key, bion.; NE Brazil & Amazonas); Causey et al. 1946 (M*; key; NE Brazil & Amazonas); Deane, M.P. et al. 1946 (L*; key, bion.; NE Brazil & Amazonas); Lane 1953: 144 (M*, F, P*, L*); Calderón et al. 1995 (distr.; Peru); Harbach & Howard 2009: 13 (A*, P*, L*; distr.); Linton et al. 2013 (distr., bion., DNA; Ecuador); Berti et al. 2015 (distr.; Venezuela). **Ety.:** Johanna Bonne-Wepster. **Informal name:** Bonne Surinamese Basal Mosquito.

fajardi (Lutz)
1904 (F; *Pyretophorus*). In Bourroul 1904: 64. **Type loc.:** Coqueiros (21.0'0"S, 47.50'0"W), Jardinopolis, São Paulo State, Brazil (USNM). **Distr.:** Argentina, Bolivia, Brazil, Colombia, Ecuador, Peru. **Refs.:** Galvão & Barretto 1939: 114 (L*, E*); Romeo Viamonte & Castro 1951: 324 (F*); Lane 1953: 140 (M*, F, P*, L*, E*); Mattingly 1971a: Pl. 1 (A scutum*); Belkin et al. 1971: 2 (tax.); Calderón et al. 1995 (distr.; Peru); Harbach & Howard 2009: 15 (A*; neotype desig., distr.). **Ety.:** Jacobo Fajardo. **Informal name:** Fajardo Brazilian Basal Mosquito.

syn. *neivae* Cruz
1906: 199 (F). **Type loc.:** Juiz de Fora, Minas Gerais, Brazil (LU). **Refs.:** Theobald 1907: 123 (syn.).

syn. *maculata* Peryassú
1921b: 141 (A; as *fajardi* var.). **Type loc.:** Parque em Cambuquira, Minas Gerais, Brazil (LU). **Refs.:** Edwards 1932a: 32 (syn.); Belkin et al. 1971: 2 (type info.).

syn. *stigmopteryx* Martini
1932a: 276 (A; as *fajardi* var.). **Type loc.:** Butantan, São Paulo, Brazil (IP). **Refs.:** Belkin et al. 1971: 2 (tax.); Belkin 1971a: 31 (type info.).

rozeboomi Causey, Deane & Deane
1944: 3 (E*). **Type loc.:** Ceara, Brazil (NE). **Distr.:** Brazil. **Refs.:** Causey et al. 1945b: 343 (M*, F*, P*, L*, E*); Deane, L.M. et al. 1946 (F*; key, bion.; NE Brazil & Amazonas); Causey et al. 1946 (M*; key; NE Brazil & Amazonas); Deane, M.P. et al. 1946 (L*; key, bion.; NE Brazil & Amazonas); Stone & Knight 1956b: 280 (type info., lectotype desig.); Harbach & Howard 2009: 17 (A*, P*, L). **Ety.:** Lloyd Eugene Rozeboom. **Informal name:** Rozeboom Brazilian Basal Mosquito.

Subfamily Culicinae
Tribe Aedeomyiini

Genus *Aedeomyia* Theobald

(Vol. 1, p. 218)

Important references: Giles 1902: 475, 478 (unjustified emendation to *Aedomyia*); Edwards 1929c: 326 (as *Aedomyia*; key); Mackerras 1937 (as *Aedomyia*; tax.); Lane 1953 (Neotropics); Thurman 1959b: 60 (as *Aedeomyia* w/o comment); Belkin 1962: 273 (tax.; *Aedomyia* invalid emend; South Pacific); Tyson 1970a: 1 (F, P, L; keys, revision); da Cunha Ramos & Ribeiro 1975 (L; distr.; Angola); Lee et al. 1982: 1–15 (tax.; Australasian Region); Service 1990 (L; bion., tax., distr.; Africa); Rattanarithikul & Harrison et al. 2005 (F, L; tax., keys, bion.; Thailand); Rattanarithikul et al. 2006b (F, L; keys, distr.; Thailand); Brunhes et al. 2011 (genus and subg. descr., key).

Subgenus *Aedeomyia* Theobald

1901b: 235 (July 15); 1901a: 98; 1901c: 218 (Nov. 23): 218 (as genus and subg.). Type species: *Aedes squamipenna* Lynch Arribálzaga. Brunetti 1914: 54 (type species desig.).

Name derivation: n.s. [aedeo-, bashful, shy; -myia, fly (Gr)].

syn. *Aedomyia* Giles 1902: 478 (emend.).

africana Neveu-Lemaire
1906: 273 (M*). **Type loc.:** Near Bobo-Dioulasso, Burkina Faso (IRD). **Distr.:** Angola, Benin, Burkina Faso, Cameroon, Côte d'Ivoire, Democratic Republic of the Congo, Ethiopia, Gabon, Ghana, Kenya, Malawi, Mali, Mauritania, Mozambique, Nigeria, Senegal, Republic of the Congo, Republic of South Africa, South Sudan, Sudan, Tanzania, Uganda. **Refs.:** Edwards 1941: 65 (M*, F*), 373 (P*); Hopkins 1952: 73 (L*); Gillett 1955 (E*); Ribeiro 1966: 163 (L*); da Cunha Ramos & Ribeiro 1975: 134 (L*; tax., bion., keys, distr.; Angola); White 1980: 123 (to subg. *Lepiothauma*); Jupp 1996 (M*, F*, L*; tax., key); Brunhes et al. 2011: 110 (M*, F*, P*, L*; distr., tax., neotype desig.; as subg. *Aedeomyia*). **Ety.:** Africa. **Informal name:** Burkinabe African Scaly Mosquito.

catasticta Knab
1909: 387 (F). **Type loc.:** Samal, Bataan [Luzon], Philippines (USNM). **Distr.:** Australia, Bangladesh, Cambodia, Fiji, Guam, India, Indonesia, Malaysia, Mariana Isl., Micronesia (Wake Isl.), Nepal, Palau, Papua New Guinea, Philippines, Singapore, Solomon Isl., Thailand, Timor, Vietnam. **Refs.:** Barraud 1923h (L*); Mackerras 1937: 259 (A, L*; from syn. with *venustipes*); Knight & Chamberlain 1948: 15 (P*); Coher 1949: 107 (F*); Bonne-Wepster 1954: 55 (M*, F*, L*); Bohart 1957: 6 (distr.; Micronesia); Stone & Knight 1957c: 196 (type info., lectotype desig.); Belkin 1962: 276 (M*, F*, P*, L*); Delfinado 1966a: 72 (M, F, P*, L*); Tyson 1970a: 8 (M*, F*, P*, L*); Basio 1971b: 32 (M*, L*; bion.); Mattingly 1971a: Pl. 3 (M, F antenna*, leg*), Pl. 11 (P abdomen*), Pl. 27 (L head*, thorax*, terminal segs.*); Baisas 1974: 42 (M, F*, P*, L*; tax., bion., distr.; Philippines); Lien et al. 1977 (distr.); Lee et al. 1982: 4 (F; key, tax., bion., distr., review); Ahmed 1987 (distr.; Bangladesh); Darsie et al. 1993 (distr.; Nepal); Whelan & Hapgood 2000 (bion., distr.; East Timor); Rattanarithikul et al. 2006b (F*, L*; bion., distr., keys). **Ety.:** n.s. [clever strike? (L)]. **Informal name:** Butterfly Philippine Scaly Mosquito [*Catasticta* is a large genus of butterflies].

madagascarica Brunhes, Boussès, & da Cunha Ramos
2011: 119 (M*, F*, P, L). **Type loc.:** La Station Forestiere d'Ivoloina, Toamasina [Atsinanana Region], Madagascar (IRD). **Distr.:** Madagascar (includes Glorioso and Juan de Nova Isl.). **Ety.:** Madagascar. **Informal name:** Merina Malagasy Scaly Mosquito [the Merina are the dominant ethnic group of the Malagasy Highlands].

pauliani Grjebine
1953a: 467 (L*). **Type loc.:** Ambila-Lemaitso, Tamatave, Madagascar (IRSM). **Distr.:** Madagascar (includes Glorioso and Juan de Nova Isl.). **Refs.:** Brunhes et al. 2011: 116 (L*). **Ety.:** Renaud Paulian. **Informal name:** Paulian Malagasy Scaly Mosquito.

squamipennis (Lynch Arribálzaga)
1878: 151 (M, F; *Aedes*). **Type loc.:** Buenos Aires, Argentina (NE). **Distr.:** Argentina, Belize, Bolivia, Brazil, Colombia, Costa Rica, Cuba, Dominican Republic, Ecuador, French Guiana, Guatemala, Guyana, Honduras, Mexico, Nicaragua, Panama, Paraguay, Peru, Suriname, Trinidad and Tobago, Uruguay, United States (continental), Venezuela. **Refs.:** Senevet & Abonenc 1939d: 590 (M*, L*); Coher 1949: 106 (F*); Lane 1953: 587 (M*, F, P*, L*); Heinemann & Belkin 1977b: 414, 452 (distr.); Lee et al. 1982: 11 (misident.); Petersen & Linley 1995 (E*); Rossi & Martínez 2003: 471 (distr.; Uruguay); Pena & Chadee 2004 (distr.; Dominican Republic); Berti et al. 2015 (distr.; Venezuela); Burkett-Cadena et al. 2017 (distr.; Florida, United States). **Ety.:** scale wing (L); ref. to wings distinctly covered with blackish scales. **Informal name:** Scale-winged Argentinian Scaly Mosquito.

venustipes (Skuse)
1889: 1761 (F*; *Aedes*). **Type loc.:** Elizabeth Bay near Sydney, New South Wales, Australia (MM). **Distr.:** Australia. **Refs.:** Mackerras 1937: 261 (tax.); Douglas 1962: 262 (M*, F, P*, L*); Dobrotworsky 1965: 68 (M*, F*, L*); Tyson 1970a: 12 (M, F*, P*, L*). **Ety.:** n.s. [attractive, lovely

(venust) feet? (L)]. **Informal name:** Lovely-footed Australian Scaly Mosquito.

Subgenus *Lepiothauma* Enderlein

1923: 25. Type species: *Lepiothauma furfurea* Enderlein (Tyson 1970a: 2) (as genus).

Name derivation: n.s. [lepio-, scale (Gr); thauma-, wonder, marvel (Gr); prob. ref. to distinctive scaling].

Important references: Edwards 1925: 262 (syn. with *Aedomyia*); Edwards 1941: 64 (M, F); Hopkins 1952: 71 (L; distr.; Africa); Tyson 1970: 1 (to subg., revision).

furfurea (Enderlein)
1923: 26 (F*; *Lepiothauma*). **Type loc.:** Mao Godi, Cameroun [Cameroon] (ZM). **Distr.:** Angola, Botswana, Burkina Faso, Cameroon, Democratic Republic of the Congo, Kenya, Madagascar (includes Glorioso and Juan de Nova Isl.), Mali, Mozambique, Namibia, Oman, Republic of the Congo, Republic of South Africa, Senegal, South Sudan, Sudan, Tanzania, Uganda, Zambia, Zimbabwe. **Refs.:** Edwards 1941: 66 (M, F); De Meillon et al. 1945: 100 (P); Hopkins 1952: 75 (L*); da Cunha Ramos & Ribeiro 1975: 135 (L*; tax., bion., keys, distr.; Angola); Jupp 1996 (M*, F*; key); Brunhes et al. 2011: 101 (M*, F*, P*, L*). **Ety.:** n.s. [grainy texture (L)]. **Informal name:** Grainy Cameroonian Scaly Mosquito.

Tribe Aedini

Savage & Strickman 2004 (tax., discussion; Aedini); Savage 2005 (tax., discussion; Aedini; Reinert 2009f (genus level abbreviations with Reinert, Harbach, & Kitching Aedini interpretation); Wilkerson et al. 2015 (genus level abbreviations according to their interpretation given herein)

Genus *Aedes* Meigen

(Vol. 1, p. 220)

Important references: Dyar 1920n: 103–106; Bauer 1928 (yellow fever vectors not *aegypti*); Barraud 1934 (South Asia); Knight & Hull 1951b (keys; Philippines); Lane 1953 (Neotropical Region); Senevet & Andarelli 1954b (L*; key; North Africa); Muspratt 1956a (M*, F*, L*); keys, bion, distr., subg. *Stegomyia*, South Africa); Sazanova 1958 (F; key, forest zone; former USSR); Smith 1958 (F; key; New England, United States); Mattingly 1958a (keys, subg. *Paraedes*, *Rhinoskusea*, *Cancraedes*, Indomalayan area); Hedeen 1959 (L; key; France); Mattingly 1959:1 (keys; subg. *Skusea*, *Diceromyia*, *Geoskusea*, *Christophersiomyia*; Indomalayan area); Price 1960 (key, first stage larval; Minnesota, United States); Mattingly 1961 (keys; subg. *Mucidus*, *Ochlerotatus*, *Neomelaniconion*; Indomalayan area); Belkin 1962 (M*, F*, P*, L*; keys, tax., class.; South Pacific); Dobrotworsky 1965 (tax., bion., distr.; Australia); DuBose & Curtin 1965 (keys; Mediterranean area); Forattini 1965a (Neotropical Region); Cova-García et al. 1966 (Venezuela); Myers 1967 (key to eggs; California and Nevada, United States); Huang 1968b (key to subg.; Papuan Region); Belkin 1968a (New Zealand); Smith 1969a (L; key; New England, United States); Smith 1969b (second and third instar larval recognition); Berlin 1969a (M*, F*, P*, L*; keys, bion., revision; subg. *Howardina*); Rjazantzeva 1970 (F gen.*); Belkin et al. 1970 (M*, F*, P*, L*; keys; Jamaica); Reinert 1970a (subg. *Diceromyia*); Schick 1970a and Schick 1970b (former *Ae. terrens* group; Neotropical Region); Tyson 1970b (subg. *Mucidus*); Lunt & Nielsen 1971a (F thoracic setae); Lunt & Nielsen 1971b (F thoracic setae); McIntosh 1971 (M, F; keys; subg. *Neomelaniconion*; southern Africa); Reinert 1972c (subg. *Ayurakitia*); Huang 1972c (M*, F*, P*, L*; keys, revision; *Ae. scutellaris* group; SE Asia); Zavortink 1972 (various subg. formerly in *Ae.* (*Finlaya*); New World); Moriya, Yabe & Harada 1973 (E*; morph.); Ribeiro & da Cunha Ramos 1973:107 (F, L; keys, bion, tax., distr.; Angola); Reinert 1973b (subg. *Aedimorphus*); Reinert 1973e (subg. *Bothaella*); Gutsevich 1974 (F; Palearctic); Reinert 1974a (*Verrallina* as subg.); Gutsevich et al. 1974 (former USSR); Gutsevich 1975 (F; head morph.); Hartberg 1975 (phyl.); McIntosh 1975 (subg. *Aedimorphus*; southern Africa); Reinert 1976c (subg. *Indusius* and *Edwardsaedes*); Reinert 1976d (subg. *Rhinoskusea*); Arnell 1976 (subg. *Ochlerotatus*, *Ae. scapularis* group; New World); Wolff & Nielsen 1977 (chaetotaxy; snowpool *Aedes*; United States); Huang 1977a (subg. *Stegomyia*; Oriental Region); Abercrombie 1977 (subg. *Christophersiomyia*); Huang 1979a (subg. *Stegomyia*; Oriental Region); Tanaka et al. 1979 (Japan); Wood et al. 1979 (Canada); Reinert 1981 (subg. *Paraedes*); Darsie & Ward 1981 (F, L; keys; North America); Lee et al. 1982 (Vol. 2; Australasian Region); Lu & Li 1982a (People's Republic of China); Clark-Gil & Darsie 1983 (keys; Guatemala); Reinert 1984 (subg. *Verrallina*; Sri Lanka); Lee et al. 1984 (Vol. 3; Australasian Region); Reinert 1985 (subg. *Scutomyia*); Darsie 1985 (F, L; keys; Argentina); Reinert 1987 (subg. *Albuginosus*); Lee et al. 1987a (Vol. 4; Australasian Region); Reinert 1990 (subg. *Kenknightia*); Darsie & Pradhan 1990, 1991 (Nepal); Huang 1990 (subg. *Stegomyia*, *Ae. africanus* group; Afrotropical Region); Reinert 1993 (subg. *Molpemyia*); Rattanarithikul & Panthusiri 1994 (keys, med. imp. spp; Thailand); Jupp 1996 (keys, tax.; southern Africa); Reinert 1999a (subg. *Zavortinkius*); Reinert 2000a (F gen.; intro.; Aedini); Reinert 2000g (subg. *Fredwardsius*); Reinert 2000i (subg. *Ayurakitia* to genus); Reinert 2002e (F gen.*); Huang 2001 (key to subg. of *Aedes*; Afrotropical Region); Tanaka 2003a (P*; Japan); Reinert, Harbach, & Kitching 2004 (phyl., class.); Huang 2004 (subgenus *Stegomyia*; Afrotropical Region); Darsie & Ward 2005 (F, L; keys; North America); Rattanarithikul & Harrison et al. 2005 (F, L; tax., keys, bion.; Thailand); Reinert, Harbach, & Kitching 2006 (phyl., class.); Reinert, Harbach, & Kitching 2008 (phyl., class.); Reinert, Harbach, & Kitching 2009 (phyl., class.); Huang et al. 2010 (subg. *Levua*; phyl.); Huang et al. 2012 (subg. *Levua*; tax., morph.); Wilkerson et al. 2015 (phyl., class.); Huang & Rueda 2015 (subg. *Pseudalbuginosus*; tax., morph.); Mohamed et al. 2017 (F, L; keys, bion.; Sudan); Brunhes et al. 2017 (subg. *Paulianius* [as genus]; tax., bion., keys); Soghigian et al. 2017 (phyl.).

Subgenus *Abraedes* Zavortink

1970c (as subg. of *Aedes*). Type species: *Aedes* (*Abraedes*) *papago* Zavortink.

Name derivation: n.s. [Habr = delicate (Gr) and *Aedes* (Harbach 2018c: 238)].

Important references: Reinert 2000h: 177 (to subg. of genus *Ochlerotatus*); Reinert 2002e: 39 (F gen.*); Reinert, Harbach, &Kitching 2004 (to genus); Reinert, Harbach, & Kitching 2009 (as genus; phyl., class.); Wilkerson et al. 2015 (phyl., class.; to subg. of genus *Aedes*).

papago Zavortink
1970c: 2 (M*, F, P*, L*). **Type loc.:** Mendoza Canyon, Coyote Mountains, Pima County, Arizona, United States (USNM). **Distr.:** United States (continental). **Refs.:**

Zavortink 1972: 4 (M*, F, P*, L*; keys, distr.); Reinert 2002e: Fig. 40 (F*); Darsie & Ward 2005 (F*, L*; keys, distr.). **Ety.:** n.s. [Papago is a name used by Europeans for the indigenous Native American tribe Tohono O'odhaim, the "Desert People," whose members reside in the Sonoran Desert of Southern Arizona and Northwestern Mexico]. **Informal name:** Tohono American Pointy Mosquito.

Subgenus *Acartomyia* Theobald

1903a: 251 (as genus). Type species: *Acartomyia zammitii* Theobald.

Name derivation: n.s. [prob. ref. to "head clothed with flat *irregularly-disposed* scales all over" and "The ragged appearance of the head is very marked"].

Important references: Coluzzi & Sabatini 1968 (as Mariae Complex); Coluzzi et al. 1976 (as Mariae Complex); Reinert, Harbach, & Kitching 2008 (phyl., class.; to subg. of genus *Ochlerotatus*); Reinert, Harbach, & Kitching 2009 (phyl., class.; to genus); Becker et al. 2010 (tax.); Wilkerson et al. 2015 (phyl., class.; to subg. of genus *Aedes*).

mariae (Sergent & Sergent)
1903: 62 (M*, F*, L*; *Culex*). **Type loc.:** Alger, Algeria (LU). **Distr.:** Algeria, France (includes Corsica), Greece, Israel (Gaza Strip and West Bank), Italy (includes Sardinia and Sicily), Morocco, Portugal, Spain (includes Balearic Isl.). **Refs.:** Séguy 1924: Pl. XXV, 101 (L*), 99 (M gen.*), as *Ae. desbansi*; Aitken 1954: 477 (M, F, L; tax.); Rioux 1958: 147 (M*, F*, L*); Mohrig 1967 (F*); Coluzzi & Sabatini 1968: 68 (F, L; tax.); Gutsevich et al. 1971: 196 (M*, F*, L*; distr.); Coluzzi et al. 1974: 321 (tax., distr.); Coluzzi et al.1976: 221 (tax.; Mariae Complex); Eritja et al. 2000: 11 (distr.; Spain); Trari et al. 2002: 331 (distr.; Morocco); Becker et al. 2010: 240 (M*, F*, L*; keys, tax., distr., bion.); Martini 1930: 302 (M*, F, L*); Robert et al. 2019 (distr.; Western Palearctic). **Ety.:** Mary [granddaughter of an author]. **Informal name:** Mary Algerian Pointy Mosquito.
syn. *desbansi* Séguy
1923: 205 (M, F). **Type loc.:** Le Pradet, Var [Provence-Alpes-Côte d'Azur], France (MNHP).

phoeniciae Coluzzi & Sabatini
1968: 68 (L). **Type loc.:** Beirut, Lebanon (NHMUK). **Distr.:** Cyprus, Israel (Gaza Strip and West Bank), Lebanon, Syria, Turkey. **Refs.:** Coluzzi et al. 1970: 123 (to sp.); Coluzzi et al. 1974: 321 (tax., distr.). **Ety.:** Phoenicia was an ancient civilization composed of independent city-states located along the coast of the Mediterranean Sea and stretching through what is now Syria, Lebanon, and northern Israel. **Informal name:** Phoenicia Lebanese Pointy Mosquito.

zammitii (Theobald)
1903a: 252 (M*, F*, P*, L*; as *Acartomyia*). **Type loc.:** Malta (NHMUK). **Distr.:** Albania, Bulgaria, Croatia, Greece, Israel, Italy (includes Sicily), Lebanon, Malta, Montenegro, Romania, Turkey. **Refs.:** Clavero 1946: 9 (F*; as *zammittii*, *lapsus*); Coluzzi & Sabatini 1968: 68 (tax.); Coluzzi & Bullini 1971: 455 (to sp.); Coluzzi et al. 1974: 321 (tax.); Becker et al. 2010: 242 (M*, F*, L*; keys, tax., distr., bion.); Robert et al. 2019 (distr.; Western Palearctic). **Ety.:** Dr. Zammit. **Informal name:** Zammit Maltese Pointy Mosquito.

Subgenus *Aedes* Meigen

1818: 13. Type species: *Aedes cinereus* Wiedemann (as genus).

Name derivation: distasteful, disagreeable, unpleasant, odious, troublesome (Gr).

Important references: Belkin 1962: 412 (tax.); Sabrosky 1999: 40 (attrib. *Aedes cinereus* to Wiedemann); Gornostaeva 2003 (tax.); Reinert, Harbach, & Kitching 2004 (phyl., class.; as genus); Reinert, Harbach, & Kitching 2006 (phyl., class.; as genus); Reinert, Harbach, & Kitching 2008 (phyl., class.; as genus); Reinert, Harbach, & Kitching 2009 (phyl., class.); Becker et al. 2010 (keys, tax.; Europe); Wilkerson et al. 2015 (phyl., class.); Soghigian et al. 2017 (phyl.).

alexandrei Gornostaeva
2005: 485 (M*). **Type loc.:** United Kingdom. **Distr.:** Russia. **Ety.:** n.s [Alexander (Aleksandr) Shtakelberg]. **Informal name:** Alexandre Russian Pointy Mosquito.

cinereus Meigen
1818: 13 (M, F). **Type loc.:** Europe (MNHP). **Distr.:** Algeria, Austria, Belarus, Belgium, Bulgaria, Canada, Crimean Peninsula, Croatia, Czech Republic, Denmark, Estonia, Finland, France (includes Corsica), Georgia, Germany, Hungary, Ireland, Italy, Latvia, Kaliningrad Oblast, Lithuania, Luxembourg, Macedonia, Moldova, Mongolia, Netherlands, Norway, Poland, Romania, Russia (Central, Northwestern, Southern, and Volga regions), Serbia, Slovakia, Slovenia, Spain, Sweden, Switzerland, Turkey, Ukraine, United Kingdom, United States (Alaska, continental). **Refs.:** Brolemann 1920 (F*); Séguy 1924: 88 (E*), 71 (M gen.*); Martini 1930 (F*); Gjullin 1937 (F*); Jobling 1938 (F*); Marshall 1938 (F*); Edwards 1941 (F*); Ross 1947 (F*); Natvig 1948: 401 (M*, F, L*; bion.); Darsie 1951: 12 (P*); Yamaguti & LaCasse 1951d: 256 (M*, F*, L*); Bohart 1954: 358 (L); Carpenter & LaCasse 1955: 265 (M*, F*, L*; keys); Lvov 1956: 929 (M*, F*; tax.); Craig & Horsfall 1958: 854 (E*); Kramar 1958 (F*); Snodgrass 1959 (F*); Price 1960: 551 (first instar L*); Clements 1963 (F*); Ross & Horsfall 1965 (M*, F*, L*, E*; keys); Dodge 1966: 351 (first instar L; key); Mohrig 1967 (F*); Myers 1967 (E*); Kalpage & Brust 1968 (E*); Mohrig

1969 (F*); Peus 1970: 22 (M*, F*); Rjazantzeva 1970 (F*); Peus 1972: 182 (M*, F, L*); Horsfall & Voorhees 1972 (E*); Gutsevich et al. 1974: 334 (M*, F, L); Dahl 1975: 90 (distr.); Utrio 1976: 129, 130, 134 (L*); Minar 1976: 337 (distr.); Wolff & Nielsen 1976 (bion., distr.); Skierska 1977 (F*); Bohart & Washino 1978: 49 (M, F, L*; keys, tax., bion., distr.; as *hemiteleus*); Bickley 1980 (as ssp. *hemiteleus*); Mgeladze 1991: 329 (distr.); Dahl 1997 (E*); Alten et al. 2000 (distr.; Turkey); Darsie & Ward 2000 (F, L; keys, distr.); Reinert 2000e: Fig. 1 (F*); Spungis 2000: 9 (distr.; Latvia); Beck et al. 2003: 14 (distr.; Luxembourg); Darsie & Ward 2005 (F*, L*; keys, distr.); Becker et al. 2010: 189 (M*, F*, L*; keys, tax., distr., bion.); Harrison et al. 2016 (F*, L*; keys, distr.; note 11); Robert et al. 2019 (distr.; Western Palearctic). **Ety.:** n.s. [ash colored (L); poss. ref. to color]. **Informal name:** Ashy European Pointy Mosquito.

syn. *rufus* Gimmerthal
1845: 295 (F). **Type loc.:** Russia (LU).

syn. *nigritulus* Zetterstedt
1850: 3459 (F; *Culex*). **Type loc.:** Lapponia Lulensi at Qvickjock [= Kvikkjokk, Norrbotten], Sweden (LS).

syn. *fuscus* Osten Sacken
1877: 191 (M, F). **Type loc.:** Cambridge, Massachusetts, United States (MCZ). **Refs.:** Carpenter & LaCasse 1955: 265 (syn.).

syn. *leucopygos* Eysell
1903: 299 (M, F). **Type loc.:** Germany (LU). **Refs.:** Peus 1972: 192 (tax.).

syn. *pallidohirta* Grossbeck
1905: 359 (F; *Culex*). **Type loc.:** Orange Mountains, New Jersey, United States (USNM). **Refs.:** Theobald 1910c: 612 (as *pallidocephala*, *lapsus*); Stone & Knight 1956a: 223 (type info., lectotype desig.).

syn. *hemiteleus* Dyar
1924h: 179 (M, F; *Aedes cinereus* race). **Type loc.:** Lakes Center Camp, Plumas County, California, United States (USNM). **Refs.:** Carpenter & LaCasse 1955: 265 (syn.); Stone & Knight 1956a: 218 (type info., lectotype desig.).

dahuricus Danilov
1987b: 35 (M*, F*, L*). **Type loc.:** Kubukhai, Onan District, Chita Province, USSR (IMP). **Distr.:** Russia. **Refs.:** Tanaka 2000a (distr.). **Ety.:** n.s. **Informal name:** Daur Russian Pointy Mosquito [Dahur is a former derogatory name for the Daur people].

dmitryi Gornostaeva
2005: 485 (M*). **Type loc.:**? In Russian. **Distr.:** Russia. **Ety.:** n.s. [Dmitry Konstantinovich Lvov (Harbach 2018c)]. **Informal name:** Dmitry Russian Pointy Mosquito.

esoensis Yamada
1921: 77 (M*, F). **Type loc.:** Kanayama [Hokkaido], Japan (IID). **Distr.:** Japan, People's Republic of China, Russia, South Korea. **Refs.:** Marshall 1938 (F*); Sasa et al. 1950: 632 (M*); La Casse & Yamaguti 1950: 95 (M*, F*, P*, L; keys, tax., bion.); Asanuma & Nakagawa 1954a: 374 (P*); Lvov 1956: 929 (M*, P*; as *cinereus* ssp.); Hara 1957: 63 (F*); Hara 1958 (F); Gutsevich et al. 1974: 336 (M*, F, L); Tanaka et al. 1979: 427 (M*, F*, L*); Tanaka 2000a: 2 (P*; tax., key). **Ety.:** n.s. [prob. ref. to the Ezo region of Japan]. **Informal name:** Kanayama Japanese Pointy Mosquito ["Ezo" is a former derogatory term for the Ainu people of Hokkaido; named for type locality].

syn. *asanumai* Sasa, Kano, & Takahasi
1950: 637 (M*). **Type loc.:** Rubeshibe, Hokkaido, Japan (IID). **Refs.:** Hara 1958: 25 (M*); Tanaka et al. 1975b: 48 (syn.).

syn. *pseudoesoensis* Yamaguti & Tamaboko
1954: 418 (M*, F*, P, L*). **Type loc.:** Mount Hakusan, Ishikawa Prefecture [Honshu], Japan (LU). **Refs.:** Tanaka et al. 1975b: 49 (syn.).

geminus Peus
1970: 22 (M*, F*). **Type loc.:** Berlin-Hermsdorf, am Tegeler Fliess, Germany (A). **Distr.:** Czech Republic, Finland, France, Germany, Hungary, Ireland, Italy, Liechtenstein, Moldova, Poland, Romania, Russia (Central Federal Dist.), Slovakia, Spain, Sweden, United Kingdom. **Refs.:** Peus 1972: 177 (M*, F, P*, L*); Dahl 1975: 91 (distr.); Eritja et al. 2000: 11 (distr.; Spain); Orszagh et al. 2001: 6 (distr.; Slovakia); Melero-Alcíbar 2005a (P; tax.); Becker et al. 2010: 191 (M*, F*, L*; keys, tax., distr., bion.); Robert et al. 2019 (distr.; Western Palearctic). **Ety.:** twin of (close to) (L) *cinereus*. **Informal name:** Twin German Pointy Mosquito.

mubiensis Luh & Shih
1958: 226 (M*, F). **Type loc.:** Mubien, Guanxi Autonomous Region, China [People's Republic of China] (AMC). **Distr.:** People's Republic of China. **Refs.**: Chen & Zhao 1984 (L*, P*). **Ety.:** n.s. [Mubien, China]. **Informal name:** Mubien Chinese Pointy Mosquito.

nataliae Gornostaeva
2005: 483 (M*). **Type loc.:**? In Russian. **Distr.:** Russia. **Ety.:** n.s. [Natalia Yantsin, author's first wife (Harbach 2018c)]. **Informal name:** Natalia Russian Pointy Mosquito.

rossicus Dolbeskin, Gorickaja, & Mitrofanova
1930: 253 (M*, L*). **Type loc.:** Dnepropetrovsk (previously Ekaterinoslav), Perno, Orenburg [USSR] (LU). **Distr.:** Austria, Belarus, Croatia, Czech Republic, France, Germany, Hungary, Japan, Norway, Poland, Russia, Slovakia, Sweden, Ukraine. **Refs.:** Monchadskii 1951: 232 (L*); Hara 1958: 23 (M*, F*, L*); Skierska 1970: 611 (M*, L*; distr.); Peus 1972: 189 (M*, F, L*; to ssp. of *esoensis*); Gutsevich et al. 1974: 336 (M*, F, L*); Tanaka et al. 1975b, d: 49 (tax.); Becker et al. 2010: 192 (M*, F*, L*; keys, tax., distr., bion.). **Ety.:** n.s. ["from Russia" (Harbach 2018c)]. **Informal name:** Ross Russian Pointy Mosquito.

syn. *tarnogradskii* Martini
1930: 259 (L). **Type loc.:** Caucasus [USSR] (NHMUK). **Refs.:** Mattingly 1955c: 30 (tax.).

sasai Tanaka, Mizusawa, & Saugstad
1975b: 41 (M*, F, L*). **Type loc.:** Umegashima spa, Akaishi Mountains, Shizuoka Prefecture, Honshu, Japan (NMST). **Distr.:** Bhutan, Japan, Russia. **Refs.:** Tanaka et al. 1975d (tables 2 and 7 for Tanaka et al. 1975b); Tanaka et al. 1979: 228 (M*, F*, L*); Gornostaeva 2000 (distr.; Russia); Tanaka 2000a (P*; tax., key). **Ety.:** Manabu Sasa. **Informal name:** Sasa Japanese Pointy Mosquito.

valeryi Gornostaeva
2005: 488 (M*). **Type loc.:**? In Russian. **Distr.:** Russia. **Ety.:** n.k. [translated from Russian: Valery Nikolayevich Danilov (Harbach 2018c)]. **Informal name:** Valery Russian Pointy Mosquito.

yamadai Sasa, Kano, & Takahasi
1950: 635 (M*, F). **Type loc.:** Shimoyubetsu, Hokkaido, Japan (IID). **Distr.:** Japan, Russia. **Refs.:** Hara 1957: 63 (F*); Hara 1958 (F); Tanaka et al. 1979: 432 (M*, F*, L*); Gornostaeva 2000 (distr.; Russia); Tanaka 2000a (P*; tax., key). **Ety.:** Shinichiro Yamada. **Informal name:** Yamada Japanese Pointy Mosquito.

Subgenus *Aedimorphus* Theobald

1903a: 290 (July). Type species: *Uranotaenia domestica* Theobald 1901d (as genus).

Name derivation: n.s. [in the shape, semblance, form of *Aedes* (Gr)].

syn. *Eculex* Felt 1904: 391c (as genus). Type species: *Culex sylvestris* Theobald.
syn. *Reedomyia* Ludlow 1905a: 94 (as genus). Type species: *Reedomyia pampangensis* Ludlow.
syn. *Lepidotomyia* Theobald 1905f: 80 (as genus). Type species: *Lepidotomyia alboscutellata* Theobald.
syn. *Pecomyia* Theobald 1905g: 23 (as genus). Type species: *Pecomyia maculata* Theobald.
syn. *Pseudograbhamia* Theobald 1905h: 243 (as genus). Type species: *Pseudograbhamia maculata* Theobald.
syn. *Duttonia* Newstead 1907. In Newstead et al. 1907: 17. Type species: *Duttonia tarsalis* Newstead (Brunetti 1914: 59).
syn. *Mimeteculex* Theobald 1908c: 258 (as genus). Type species: *Mimeteculex kingii* Theobald.
syn. *Geitonomyia* Leicester 1908: 134 (as genus). Type species: *Culex caecus* Theobald.
syn. *Bathosomyia* Theobald 1909b: 9 (as genus). Type species: *Bathosomyia abnormalis* Theobald.
syn. *Leslieomyia* Christophers 1911: 68 (as genus). Type species: *Leslieomyia taeniorhynchoides* Christophers.

Important references: Hamon, Service et al. 1961 (M; key; Tarsalis Group); Pao & Knight 1970a (L mandibles*; tax.); Reinert 1973b (tax.; SE Asia); McIntosh 1975 (tax.; southern Africa); Lee et al. 1982: 23 (tax., bion.; Australasia); Tewari & Hiriyan 1994b: 105 (keys; Tarsalis Group); Reinert 2000e (F gen.*); Tanaka 2003a (P*; key, tax.; Japan); Reinert, Harbach, & Kitching 2004 (phyl., class.); Reinert, Harbach, & Kitching 2009 (phyl., class.; to genus); Rattanarithikul et al. 2010 (F*, L*; keys, bion.; Thailand); Huang & Rueda 2015 (key, tax., Afrotropical Region); Wilkerson et al. 2015 (phyl., class.; to subg. of genus *Aedes*); Soghigian et al. 2017 (mol. phyl.).

abnormalis (Theobald)
1909b: 10 (M*; *Bathosomyia*). **Type loc.:** Obuasi, Ashanti, Gold Coast [Ghana] (NHMUK). **Distr.:** Burkina Faso, Central African Republic, Côte d'Ivoire, Ghana, Nigeria, Senegal. **Refs.:** Edwards 1941: 180 (M*). **Ety.:** n.s. [abnormal (L)]. **Informal name:** Abnormal Ghanaian Pointy Mosquito.

ssp. ***kabwachensis*** Edwards
1941: 181, 395, 432 (M*, F, P, L). **Type loc.:** Kabwach Forest, Kisii, Kavirondo, Kenya (NHMUK). **Distr.:** Kenya, Uganda. **Refs.:** Mattingly 1956a: 38 (lectotype desig.). **Ety.:** Kabwach Forest, Kenya. **Informal name:** Kabwach Kenyan Pointy Mosquito.

aerarius McIntosh
1975: 281 (M*, F*, L*). **Type loc.:** Oslo Beach, South Africa (NIV). **Distr.:** South Africa. **Refs.:** Segerman 1990: 53 (holotype from SIAM to NIV); Jupp 1996 (M*, F*; key). **Ety.:** n.s. [coppery (L)]. **Informal name:** Coppery South African Pointy Mosquito.

albodorsalis Fontenille & Brunhes
1985: 151 (F*). **Type loc.:** Fenerive, Tamatave Province, Madagascar (MNHP). **Distr.:** Madagascar (includes Glorioso and Juan De Nova Isl.). **Ety.:** n.s. [white dorsum (L); prob. ref. to anterior two-thirds of scutum covered with silvery scales]. **Informal name:** White-backed Malagasy Pointy Mosquito.

alboscutellatus (Theobald)
1905f: 80 (F*; *Lepidotomyia*). **Type loc.:** Simbang, Huon Gulf, New Guinea [West Papua] (HNM). **Distr.:** Australia, Cambodia, India, Indonesia (includes West Papua), Japan, Laos, Malaysia, Myanmar, Papua New Guinea, Philippines, Solomon Isl., South Korea, Sri Lanka, Thailand, Timor, Vietnam. **Refs.:** Barraud 1934: 250 (M*, F); Penn 1949c (P*, L*); La Casse & Yamaguti 1950: 130 (M*, F*); Knight & Hull 1953: 457 (M*, F, L*); Hara 1957: 66 (F*); Belkin 1962: 465 (M*, F, P*, L*); Mohrig 1967 (F*); Pao & Knight 1970a: 105 (L*); Basio 1971b: 10 (M*; bion.); Reinert 1973b: 17 (M*, F*, P*, L*); Tanaka et al. 1979: 407 (M*, F*, L*); Lee et al. 1982: 25 (tax., review); Whelan & Hapgood 2000 (bion., distr.; East Timor); Reinert 2000e: Fig. 9 (F*); Tanaka 2003a: 12 (P*; tax., key); Rattanarithikul et al. 2010 (F*, L*; keys, bion., distr.; Thailand). **Ety.:** white scutellum [shield] (L);

"scutellum densely silvery-white scaled." **Informal name:** White-shielded Papuan Pointy Mosquito.

syn. *argentinotus* Banks
1909: 547 (M, F; *Culex*). **Type loc.:** Pinagsalaan well, Taytay, Rizal, Luzon, Philippines (NE).

syn. *omurensis* Yamada
1921: 73 (M*, F). **Type loc.:** Omura, Kyushu, Japan (IID).

alboventralis (Theobald)
1910c: 251 (F*; *Protomacleaya*). **Type loc.:** Bihe and Katema Bihe, Angola (NHMUK). **Distr.:** Angola, Congo, Côte d'Ivoire, Democratic Republic of the Congo, Gabon, Kenya, South Africa, Sudan and South Sudan, Tanzania, Uganda, Zimbabwe. **Refs.:** McIntosh 1975: 271 (M*, F, L; tax.); Service 1976a (distr.; Gabon); Jupp 1996 (M*, F*; key). **Ety.:** white venter (L); ref. to characteristic white abdominal venter. **Informal name:** White-bellied Angolan Pointy Mosquito.

argenteoscutellatus Carter & Wijesundara
1948: 139 (M*). **Type loc.:** Ceylon [Sri Lanka] (NHMUK). **Distr.:** Sri Lanka. **Ety.:** silvery tray (L); scutellum. **Informal name:** Silver-backed Sri Lankan Pointy Mosquito.

bambiotai Geoffroy
1987: 101 (M*, F*, E*). **Type loc.:** Andiopodoume Forest, Ivory Coast [Côte d'Ivoire] (USNM). **Distr.:** Central African Republic, Côte d'Ivoire. **Ety.:** Bamio Region, Central African Republic, and Tai National Park, Republic of Côte d'Ivoire. **Informal name:** Bamio Equatorian Pointy Mosquito.

bancoi Geoffroy
1987: 103 (M*, F*, E*). **Type loc.:** Banco Forest, near Abijan [Abdijan], Ivory Coast [Côte d'Ivoire] (USNM). **Distr.:** Côte d'Ivoire. **Ety.:** Banco Forest, Republic of Côte d'Ivoire. **Informal name:** Banco Ivorian Pointy Mosquito.

bevisi (Edwards)
1915b: 275 (F; *Ochlerotatus*). **Type loc.:** Umbilo, Durban, Natal [KwaZulu-Natal, South Africa] (NHMUK). **Distr.:** South Africa. **Refs.:** Edwards 1941: 194 (F); Muspratt 1950: 77 (M*, F, P, L*); Muspratt 1953b: 88 (M); McIntosh 1975: 278 (M*, F*, L); Jupp 1996 (M*, F*; key). **Ety.:** L. Bevis. **Informal name:** Bevis South African Pointy Mosquito.

caecus (Theobald)
1901a: 413 (F*; *Culex*). **Type loc.:** Klang mangrove swamps, Selangor, Malaya [Malaysia] (NHMUK). **Distr.:** Bangladesh, Cambodia, India, Indonesia, Malaysia, Mariana Isl., Myanmar, Nepal, Pakistan, People's Republic of China, Philippines, Singapore, Thailand, Vietnam. **Refs.:** Barraud 1934: 257 (M*, F, L*); Bonne-Wepster 1954: 87 (M, F*, L*); Aslamkhan 1971b (distr.; Pakistan); Reinert 1973b: 23 (M*, F*, P*, L*); Lee et al. 1982: 29 (tax., review); Darsie & Pradhan 1990, 1991 (F, L; tax., keys, bion., distr.; Nepal); Reinert 2000e: Fig. 6 (F*); Rattanarithikul et al. 2010 (F*, L*; keys, bion., distr.; Thailand). **Ety.:** n.s. [dark, gloomy (L)]. **Informal name:** Dark Malaysian Pointy Mosquito.

caliginosus (Graham)
1910: 268 (M, F; *Culex*). **Type loc.:** West Africa [Lagos, Nigeria] (NHMUK). **Distr.:** Central African Republic, Nigeria. **Refs.:** Edwards 1941: 191 (M*, F); Hopkins 1952: 195 (tax.). **Ety.:** n.s. [dark (L)]. **Informal name:** Dark Nigerian Pointy Mosquito.

centropunctatus (Theobald)
1913d: 599 (M, F; *Culicelsa*). **Type loc.:** Acholi, Lado District, and River Mewri, Mongalla District, Sudan (NHMUK). **Distr.:** Central African Republic, Ghana, Nigeria, Sudan and South Sudan. **Refs.:** Edwards 1941: 196 (M*, F), 398 (P*); Hopkins 1952: 199 (L*). **Ety.:** n.s. [central spot (L)]. **Informal name:** Center-spotted Sudanese Pointy Mosquito.

culicinus Edwards
1922a: 271 (M*, F). **Type loc.:** Delhi, India (MSI). **Distr.:** Cambodia, India, Indonesia, Pakistan, Thailand, Vietnam. **Refs.:** Barraud 1934: 252 (M*, F); Aslamkhan 1971b (distr.; Pakistan); Reinert 1973b: 29 (M*, F*, P*, L*); Reinert 2000e (F); Rattanarithikul et al. 2010 (F*, L*; keys, bion., distr.; Thailand). **Ety.:** n.s. [gnat, midge, or similar insect (L)]. **Informal name:** Midge-like Indian Pointy Mosquito.

cumminsii (Theobald)
1903a: 214 (F*; *Culex*). **Type loc.:** Bahr-el-Ghazal, Central Africa [South Sudan] (NHMUK). **Distr.:** Angola, Benin, Burkina Faso, Cameroon, Central African Republic, Côte d'Ivoire, Democratic Republic of the Congo, Ethiopia, Ghana, Kenya, Liberia, Mali, Mozambique, Nigeria, Senegal, Sierra Leone, South Africa, Sudan and South Sudan, Uganda, Zambia. **Refs.:** Edwards 1941: 193 (M*, F), 398 (P); Hopkins 1952: 195 (L*); Worth & De Meillon 1960: 242 (distr.); Pao & Knight 1970a: 115 (L*); McIntosh 1975: 279 (M*, F, L; distr.); Jupp 1996 (M*, F*; key). **Ety.:** Stephenson Lyle Cummin. **Informal name:** Cummin Sudanese Pointy Mosquito. (Vol. 1, p. 400)

syn. *fuscopalpalis* Theobald
1909b: 18 (M*; Culicada). **Type loc.:** Obuasi, Ashanti, Ghana (NHMUK).

ssp. ***mesostictus*** Harbach
2018c: 3 (new name for *mediopunctatus* Theobald 1909b, not *mediopunctatus* Theobald 1905h). **Distr.:** Ghana, Nigeria. **Refs.:** Edwards 1941: 193 (M*, F); Haddow et al. 1951: 224 (syn.); White 1975: 306 (elev. from syn. with *cumminsii*); White 1980: 125 (as form; emend. from *mediopunctatus*); Ward 1984b: 233 (*mediopunctata* Theobald 1909b as junior homonym of *mediopunctatus* Theobald 1905h); Harbach & Howard 2007: 45 (to ssp.). **Ety.:** spotted in middle (Gr). **Informal name:** Middle-spotted Ghanaian Pointy Mosquito.

ssp. syn *mediopunctata* (Theobald)
1909b: 18 (F; Culicada). **Type loc.:** Obuasi, Ashanti, Ghana (NHMUK).
ssp. syn. *daruensis* Evans
1925a: 123 (M*, F; as *Aedes* (*Aedimorphus*) *cumminsi* [*sic*] var.). **Type loc.:** Moa River, Daru, Sierra Leone (NHMUK). **Refs.:** Edwards 1941: 193 (syn. with *mediopunctata*); Harbach 2018c (syn. with *mesostictus*).

dalzieli (Theobald)
1910c: 234 (F*; *Culiciomyia*). **Type loc.:** Katagum, Northern Nigeria (NHMUK). **Distr.:** Angola, Benin, Burkina Faso, Democratic Republic of the Congo, Madagascar (includes Glorioso & Juan De Nova Isl.), Mali, Mozambique, Nigeria, Senegal, Sudan and South Sudan, Zimbabwe. **Refs.:** Edwards 1941: 183 (M*, F); Hopkins 1952: 186 (L); Ribeiro & da Cunha Ramos 1973: 128 (distr.); McIntosh 1975: 273 (M*, F, L); Jupp 1996 (M*, F*; key). **Ety.:** n.s. [poss. ref. to collector Dalziel]. **Informal name:** Dalziel Nigerian Pointy Mosquito.
syn. *rhecter* Dyar
1921d: 51 (M*). **Type loc.:** Lomagundi, Portuguese East Africa [Mozambique] (USNM). **Refs.:** Edwards 1941: 183 (syn.); Stone & Knight 1956a: 224 (type info., lectotype desig.).

davidi Basio
1971a: 51 (M*). **Type loc.:** UP College of Agriculture Campus, Los Banos, Laguna, Luzon, Philippines (A). **Distr.:** Philippines. **Refs.:** Reinert 1973b: 33 (tax.). **Ety.:** Romeo David. **Informal name:** David Philippine Pointy Mosquito.

dentatus (Theobald)
1904b: 75 (F*; *Culex*). **Type loc.:** Abyssinia [Ethiopia] (NHMUK). **Distr.:** Angola, Burkina Faso, Central African Republic, Democratic Republic of the Congo, Ethiopia, Kenya, Mozambique, Senegal, South Africa, Sudan and South Sudan, Uganda, Zimbabwe. **Refs.:** Edwards 1941: 190 (M*, F); Hopkins 1952: 194 (L); Pao & Knight 1970a: 116 (L*); McIntosh 1975: 274 (M*, F, L*; distr.); Linley & Turell 1993 (E*); Jupp 1996 (M*, F*; key). **Ety.:** n.s. [toothed (L)]. **Informal name:** Toothed Ethiopian Pointy Mosquito.
syn. *pallidopunctata* Theobald
1911b: 267 (F*; *Culex*). **Type loc.:** Onderstepoort, Transvaal [South Africa] (NHMUK).

domesticus (Theobald)
1901d: xiv (A; *Uranotaenia*). **Type loc.:** Vice-Consulate, Old Calabar [Eastern Provinces], Nigeria (NHMUK). **Distr.:** Angola, Benin, Burkina Faso, Cameroon, Central African Republic, Congo, Côte d'Ivoire, Democratic Republic of the Congo, Gabon, Ghana, Kenya, Liberia, Mozambique, Nigeria, Uganda. **Refs.:** Macfie & Ingram 1922 (F); Edwards 1941: 172 (M*, F), 393 (P); Hopkins 1952: 173 (L*); van Someren, Teesdale, & Furlong 1955: 477 (M, F); Worth & De Meillon 1960: 241 (distr.); Reinert 1972a: 60 (E*); Reinert 2000e: Fig. 2 (F*). **Ety.:** domestic (L); "a domestic species." **Informal name:** Domestic Nigerian Pointy Mosquito.

durbanensis (Theobald)
1903a: 246 (F*; *Grabhamia*). **Type loc.:** Durban [South Africa] (NHMUK). **Distr.:** Angola, Congo, Democratic Republic of the Congo, Ethiopia, Madagascar (includes Glorioso & Juan De Nova Isl.), Mozambique, South Africa, Tanzania. **Refs.:** Edwards 1941: 199 (M*, F); De Meillon et al. 1945: 97 (P*); Hopkins 1952: 204 (L*); Ribeiro 1966: 163 (M*); Ribeiro & da Cunha Ramos 1974: 93 (M*, L*); Linley & Service 1994: 60 (E*); Jupp 1996 (M*, F*; key). **Ety.:** Durban, South Africa. **Informal name:** Durban South African Pointy Mosquito.
syn. *ocellata* Theobald
1910c: 284 (M*, F*; *Grabhamia*). **Type loc.:** Lourenço Marques [Mozambique] (NHMUK).
ssp. ***angolae*** Ribeiro & da Cunha Ramos
1974: 94 (M*, F, L*). **Type loc.:** Mocamedes, Angola (IHMT). **Distr.:** Angola. **Ety.:** Angola. **Informal name:** Mocamedes Angolan Pointy Mosquito.

ebogoensis Rickenbach & Ferrara
1965: 26 (M*). **Type loc.:** Ebogo, Cameroon (IERT). **Distr.:** Cameroon, Central African Republic. **Ety.:** Ebogo, Dept. of Nyong and Sanaga, Cameroon. **Informal name:** Ebogo Cameroonian Pointy Mosquito.

eritreae Lewis
1942b: 165 (M*, F). **Type loc.:** Acchele [Akele] Guzai, Debre Axum, 20.3 km from Saganeiti [Segeneiti] along Dessie Road [Eritrea] (NHMUK?). **Distr.:** Eritrea, South Africa. **Refs.:** Mattingly 1956a: 39 (holotype in NHMUK?); McIntosh 1975: 272 (M*, F, L); Jupp 1996 (M*, F*; key). **Ety.:** Eritrea. **Informal name:** Acchele Eritrean Pointy Mosquito.

fowleri (de Charmoy)
1908: 258 (F*; *Culex*). **Type loc.:** Vacoa, Mauritius (LSTM). **Distr.:** Benin, Burkina Faso, Central African Republic, Comoros, Democratic Republic of the Congo, Ethiopia, Gabon, Ghana, Kenya, Madagascar (includes Glorioso and Juan De Nova Isl.), Malawi, Mali, Mauritius, Nigeria, Reunion, Senegal, Sierra Leone, South Africa, Sudan and South Sudan, Tanzania, Togo, Uganda, Zimbabwe. **Refs.:** Theobald 1901a: 392 (M only of type series; as *hirsutum*; syn. by Edwards 1941: 198); Edwards 1941: 198 (M*, F), 399 (P); Hopkins 1952: 202 (L*); Pao & Knight 1970a: 117 (L*); Linley & Turell 1993 (E*); Jupp 1996 (M*, F*; key). **Ety.:** George Fowler. **Informal name:** Fowler Mauritian Pointy Mosquito.
syn. *nigeriensis* Theobald
1910c: 281 (F*; *Grabhamia*). **Type loc.:** Katagum, Northern Nigeria (NHMUK). **Refs.:** Edwards 1941: 198 (syn.).

gibbinsi Edwards
1935a: 135 (M). **Type loc.:** Namwamba Valley, 8,000 ft, Ruwenzori, Uganda (NHMUK). **Distr.:** Kenya, Uganda.

Refs.: Edwards 1941: 187 (M*, F), 397 (P*); Hopkins 1952: 192 (L). **Ety.:** Ernest Gerald Gibbins. **Informal name:** Gibbins Ugandan Pointy Mosquito.

gouldi Reinert
1972e: 357 (F*). **Type loc.:** Shah Zada, Lahore [Punjab], West Pakistan [Pakistan] (USNM). **Distr.:** Pakistan. **Ety.:** Douglas J. Gould. **Informal name:** Gould Pakistani Pointy Mosquito.

hirsutus (Theobald)
1901a: 392 (M*, F; *Culex*). **Type loc.:** Salisbury, Mashonaland [Zimbabwe] (NHMUK). **Distr.:** Benin, Burkina Faso, Côte d'Ivoire, Ethiopia, Kenya, Madagascar (includes Glorioso and Juan De Nova Isl.), Mali, Mauritania, Namibia, Nigeria, Senegal, South Africa, Sudan and South Sudan, Togo, Zambia, Zimbabwe. **Refs.:** Edwards 1941: 196 (M*, F), 399 (P*); Hopkins 1952: 201 (L*); Pao & Knight 1970a: 118 (L*); Jupp 1996 (F*; key). **Ety.:** hairy, hirsute (L). **Informal name:** Hairy Zimbabwean Pointy Mosquito.
syn. *transvaalensis* Theobald
1903a: 165 (F*; *Culex*). **Type loc.:** Pretoria, Transvaal [South Africa] (NHMUK).
ssp. ***adenensis*** Edwards
1941: 198 (M, F; as *hirsutus* var.). **Type loc.:** Jebel Jehaf [West Aden Protectorate; Yemen] (NHMUK). **Distr.:** Yemen. **Refs.:** Mattingly & Knight 1956: 101 (tax.); Mattingly 1956a: 39 (lectotype desig.); White 1975: 307 (from var. of *hirsutus*). **Ety.:** Aden, Yemen. **Informal name:** Aden Yemeni Pointy Mosquito.

holocinctus Edwards
1941: 193 (M; *cumminsi* ssp.). **Type loc.:** Gede, near Mombasa, Kenya (NHMUK). **Distr.:** Ethiopia, Kenya, Uganda. **Refs.:** Ovazza et al. 1956: 161 (P*, L*; to sp.). **Ety.:** entirely girded (Gr, L); ref. to complete basal pale bands on all abdominal terga. **Informal name:** Belted Kenyan Pointy Mosquito.

jamesi (Edwards)
1914b: 77 (F; *Ochlerotatus*). **Type loc.:** Colombo, Ceylon [Sri Lanka] (NHMUK). **Distr.:** India, Sri Lanka. **Refs.:** Barraud 1934: 249 (M*, F); Carter & Wijesundara 1948: 140 (L*). **Ety.:** S.P. James. **Informal name:** James Sri Lankan Pointy Mosquito.

karooensis Muspratt
1961: 93 (M*, F, L*). **Type loc.:** Noupoort, 22 mi north of Middleburg, Cape Province [South Africa] (SAIM). **Distr.:** South Africa. **Refs.:** McIntosh 1975: 272 (M*, F, L; elev. from ssp. of *eritreae*); Jupp 1996 (M*, F*; key). **Ety.:** Karoo Region, South Africa. **Informal name:** Karoo South African Pointy Mosquito.

leesoni Edwards
1932b: 562 (M). **Type loc.:** Shamva, Southern Rhodesia [Zimbabwe] (NHMUK). **Distr.:** Angola, Botswana, Mozambique, South Africa, Sudan and South Sudan, Zimbabwe. **Refs.:** Edwards 1941: 183 (M*); Hopkins 1952: 185 (L*; as *verna*); van Someren, Teesdale, & Furlong 1955: 479 (L; as *verna*); McIntosh 1975: 269 (M*, F*, L*; distr.); Jupp 1996 (M*, F*; keys). **Ety.:** Herbert Sefton Leeson. **Informal name:** Leeson Zimbabwean Pointy Mosquito.
syn. *verna* Lewis
1944: 27 (M*, F, P, L*). **Type loc.:** Danagla Forest, Blue Nile Province, Sudan (NHMUK). **Refs.:** Mattingly 1956a: 39 (type info.); McIntosh 1975: 269 (syn.); White 1980: 126 (emend. from *verna* to *vernus*; noun in apposition according to Chris Thompson pers. comm.).

leptolabis Edwards
1936: 51 (M). **Type loc.:** Nabadzidza, Uganda (NHMUK). **Distr.:** Benin, Cameroon, Central African Republic, Congo, Côte d'Ivoire, Democratic Republic of the Congo, Gabon, Kenya, Liberia, Nigeria, Uganda, Zambia. **Refs.:** Edwards 1941: 173 (M*); van Someren, Teesdale, & Furlong 1955: 477 (M); van Someren 1956: 4 (P*, L); Service 1976a: 260 (distr.; Gabon). **Ety.:** n.s. [prob. ref. to slender gonostylus (Gr)]. **Informal name:** Slender-styled Ugandan Pointy Mosquito.

leucarthrius (Speiser)
1909: 43 (F; *Culex*). **Type loc.:** Meru [Tanzania] (SM). **Distr.:** Tanzania. **Refs.:** Edwards 1941: 188 (F). **Ety.:** n.s. [white jointed (Gr)]. **Informal name:** White-jointed Tanzanian Pointy Mosquito.

longiseta Edwards
1936: 50 (M). **Type loc.:** Stanleyville, Zaire [Kisangani, Tshopo Province, Democratic Republic of the Congo] (NHMUK). **Distr.:** Central African Republic, Côte d'Ivoire, Democratic Republic of the Congo. **Refs.:** Edwards 1941: 173 (M*). **Ety.:** long seta (L); ref. to seta on basal lobe. **Informal name:** Long-bristled Congolese Pointy Mosquito.

lowisii (Theobald)
1910c: 257 (M*, F*; *Reedomyia*). **Type loc.:** Andaman Islands (NHMUK). **Distr.:** India, Indonesia, Malaysia, Papua New Guinea, Philippines. **Refs.:** Barraud 1934: 250 (M, F); Knight & Hull 1951a: 200 (tax.); Knight & Hull 1953: 459 (tax.; as *mindoroensis*); Basio 1971b: 11 (M*; bion.; as *mindoroensis*); Reinert 1973b: 35 (M*, F*; lectotype desig.); Lee et al. 1982: 31 (tax., review). **Ety.:** Lowis [collector?]. **Informal name:** Lowis Indian Pointy Mosquito.
syn. *mindoroensis* Knight & Hull
1951a: 199 (M*, F). **Type loc.:** Calapan, Mindoro, Philippines (NHMUK). **Refs.:** Reinert 1973b: 35 (syn.).

mansouri Qutubuddin
1959: 21 (M*, F). **Type loc.:** Juba, Sudan Republic [South Sudan] (NHMUK). **Distr.:** Sudan and South Sudan. **Ety.:** Mansour Ali Haseeb. **Informal name:** Haseeb Sudanese Pointy Mosquito.

masoalensis Fontenille & Brunhes
1985: 153 (F*). **Type loc.:** Onive Valley, Masola, Diego Suarez Province, Madagascar (MNHP). **Distr.:** Madagascar (includes Glorioso and Juan De Nova Isl.). **Ety.:** Masoala, Madagascar. **Informal name:** Masaola Malagasy Pointy Mosquito.

mathioti Fontenille & Brunhes
1985: 154 (F*). **Type loc.:** Onive Valley, Masola, Diego Suarez Province, Madagascar (MNHP). **Distr.:** Madagascar (includes Glorioso and Juan De Nova Isl.). **Ety.:** M. Christian Mathiot. **Informal name:** Mathiot Malagasy Pointy Mosquito.

mattinglyi Hamon & Rickenbach
1954: 930 (M*, F). **Type loc.:** Nasso, Cercle de Bobo Dioulasso, Haute Volta [Burkina Faso] (IERT). **Distr.:** Benin, Burkina Faso, Cameroon, Central African Republic, Côte d'Ivoire, Mali. **Refs.:** Hamon & Le Berre 1961: 718 (P*, L*). **Ety.:** Peter Frederick Mattingly. **Informal name:** Mattingly Burkinabe Pointy Mosquito.

mediolineatus (Theobald)
1901c: 113 (F; *Culex*). **Type loc.:** Thayetmyo [Magwe], Upper Burma [Myanmar] (NHMUK). **Distr.:** Cambodia, Indonesia, Malaysia, Myanmar, People's Republic of China, Thailand, Vietnam. **Refs.:** Borel 1930: 268 (M*, F, L*); Barraud 1934: 263 (M*, F); Reinert 1973b: 38 (M*, F*, P*, L*, E*); Reinert 2000e: Fig. 4 (F*); Rattanarithikul et al. 2010 (F*, L*; keys, bion., distr.; Thailand). **Ety.:** middle lined (L); ref. to widely separated parallel lines on thorax. **Informal name:** Lined Burmese Pointy Mosquito.
syn. *trilineatus* Theobald
1901c: 105 (F*; *Culex*). **Type loc.:** Thayetmyo [Magwe], Upper Burma [Myanmar] (NHMUK).

natronius Edwards
1932b: 562 (M). **Type loc.:** Lake Maseche, Uganda (NHMUK). **Distr.:** Angola, Democratic Republic of the Congo, Ethiopia, Kenya, Madagascar (includes Glorioso and Juan De Nova Isl.), Uganda, Yemen. **Refs.:** Edwards 1941: 199 (M*, F); Hopkins 1952: 205 (L*); Knight 1953a: 224 (tax.); Jupp 1996 (M*, F*; key). **Ety.:** n.s. [containing sodium (Na), component of the mineral salt natron; prob. ref. to "larvae in alkaline pools"]. **Informal name:** Alkaline Ugandan Pointy Mosquito.

nigricephalus (Theobald)
1901d: iv (F*; *Stegomyia*). **Type loc.:** Bonny [Eastern Provinces], Nigeria (NHMUK). **Distr.:** Angola, Benin, Côte d'Ivoire, Democratic Republic of the Congo, Gabon, Gambia, Liberia, Nigeria, São Tomé and Príncipe. **Refs.:** Edwards 1941: 185 (M*, F), 397 (P*); Hopkins 1952: 188 (L*). **Ety.:** n.s. [black head (L); ref. to "head black"]. **Informal name:** Black-headed Nigerian Pointy Mosquito.
syn. *paludosus* Graham
1910: 270 (M, F; *Myxosquamus*). **Type loc.:** [Lagos, Nigeria], West Africa (NHMUK).

nigrostriatus (Barraud)
1927c: 549 (M*, F; *Aedimorphus*). **Type loc.:** Golaghat, Assam, India (NHMUK). **Distr.:** India, Myanmar. **Refs.:** Barraud 1934: 262 (M*, F); Reinert 1973b: 43 (M*, F*; lectotype desig.); Bhattacharyya et al. 1995 (P*, L*). **Ety.:** black line (L); "a pair of conspicuous black stripes on the mesonotum." **Informal name:** Black-striped Indian Pointy Mosquito.

oakleyi Stone
1939: 163 (M*, F). **Type loc.:** Root Farm, Guam [Mariana Islands] (USNM). **Distr.:** Guam, Mariana Isl., Zimbabwe. **Refs.:** Bohart & Ingram 1946b: 41 (M*, F, P*, L; bion., distr., tax.); Yamaguti & LaCasse 1950a: 67 (M*, F*, L*); Lee et al. 1982: 42 (tax., review). **Ety.:** R.G. Oakley. **Informal name:** Oakley Mariana Islander Pointy Mosquito.

ochraceus (Theobald)
1901c: 103 (F*; *Culex*). **Type loc.:** Salisbury, Mashonaland [Zimbabwe] (NHMUK). **Distr.:** Burkina Faso, Ethiopia, Ghana, Kenya, Mali, Mozambique, Namibia, Nigeria, Senegal, South Africa, Sudan and South Sudan, Zambia, Zimbabwe. **Refs.:** Edwards 1941: 200 (M, F*); Hopkins 1952: 207 (L*); Qutubuddin 1959: 22 (M*); Pao & Knight 1970a: 114 (L*); McIntosh 1975: 284 (M*, F*, L; distr.); Jupp 1996 (F*; key). **Ety.:** pale yellow (Gr); overall yellow color. **Informal name:** Yellow Zimbabwean Pointy Mosquito. (Vol. 1, p. 402)
syn. *kingii* Theobald
1908c: 258 (M, F*; *Mimeteculex*). **Type loc.:** Faufikia, Upper White Nile, Sudan (NHMUK).

orbitae Edwards
1922a: 260 (new name for *taeniata* Leicester, not Wiedemann 1828). **Distr.:** Malaysia, Singapore, Thailand. **Refs.:** Edwards & Given 1928: 344 (L); Edwards, in Barraud 1934: 250 (M, F); Reinert 1973b: 45 (M*, F*, P*, L*; lectotype desig.); Rattanarithikul et al. 2010 (F*, L*; bion., distr., keys) [as *orbitae*]. **Ety.:** n.s. [orbit (L); poss. ref. to nature of orbital bristles]. **Informal name:** Orbital-bristle Malaysian Pointy Mosquito.
syn. *taeniatus* Leicester
1908: 133 (M, F; *Lepidotomyia taeniata*). **Type loc.:** Kuala Lumpur, Federal Malay States [Malaysia] (NHMUK).

ovazzai Hamon & Adam
1959c: 147 (M*). **Type loc.:** Adiopodoume (Abidjan), Ivory Coast [Côte d'Ivoire], French West Africa (IERT). **Distr.:** Central African Republic, Côte d'Ivoire. **Ety.:** M. Ovazza. **Informal name:** Ovazza Ivorian Pointy Mosquito.

pachyurus Edwards
1936: 51 (M). **Type loc.:** Kaaimans Gat, Cape Province [South Africa] (NHMUK). **Distr.:** South Africa. **Refs.:** Edwards 1941: 191 (M*); Muspratt 1953b: 87 (M, F, L); McIntosh 1975: 277 (M*, F, L); Jupp 1996 (M*, F*; key).

Ety.: n.s. [thick, stout, large (Gr); poss. ref. to hypopygium]. **Informal name:** Stout South African Pointy Mosquito.

pallidostriatus (Theobald)
1907: 410 (M*, F; *Culex*). **Type loc.:** India (NHMUK). **Distr.:** Bangladesh, India, Malaysia, Nepal, Pakistan, People's Republic of China, Sri Lanka, Thailand. **Refs.:** Barraud 1934: 261 (M*, F, L*); Aslamkhan 1971b (distr.; Pakistan); Reinert 1973b: 50 (M*, F*, P*, L*; lectotype desig.); Darsie & Pradhan 1990 (F, L; tax., keys, bion., distr.; Nepal); Reinert 2000e (F); Rattanarithikul et al. 2010 (F*, L*; keys, bion., distr.; Thailand). **Ety.:** pale lines (L); ref. to lines of pale scales on the scutum. **Informal name:** Pale-striped Indian Pointy Mosquito.
syn. *parascelos* Theobald
1910a: 18 (F; *Culex*). **Type loc.:** Madras Town, India (IM). **Refs.:** Townsend 1990: 118 (syntype F, NHMUK).

pampangensis (Ludlow)
1905a: 94 (F; *Reedomyia*). **Type loc.:** Angeles, Pampanga, Luzon, Philippines (USNM) (neotype). **Distr.:** India, Indonesia, Nepal, Philippines, Thailand, Vietnam. **Refs.:** Barraud 1934: 251 (M, F; as *niveoscutella*); Bohart 1945: 63 (from syn.); Knight & Hull 1953: 454 (M*, F, L*); Stone & Knight 1956a: 223 (type info.); Basio 1971b: 11 (M*; bion.); Reinert 1973b: 54 (M*, F*, P*, L*; neotype desig.); Darsie et al. 1993 (distr.; Nepal); Reinert 2000e: Fig. 5 (F*); Rattanarithikul et al. 2010 (F*, L*; keys, bion., distr.; Thailand). **Ety.:** Angeles, Pampanga, Luzon, Philippine Isl. **Informal name:** Pampanga Philippine Pointy Mosquito.
syn. *niveoscutellum* Theobald
1905g: 22 (F*; *Reedomyia niveoscutella* [*sic*]). **Type loc.:** India (NHMUK). **Refs.:** Reinert 1973b: 54 (syn.).

pipersalatus (Giles)
1902: 372 (M*, F; *Stegomyia*). **Type loc.:** Jhansi, Gonda, [United] Provinces, India (NHMUK). **Distr.:** Bangladesh, Cambodia, India, Laos, Myanmar, Nepal, Pakistan, Sri Lanka, Thailand. **Refs.:** Barraud 1934: 258 (M*, F, L*); Aslamkhan 1971b (distr.; Pakistan); Reinert 1973b: 59 (M*, F*, P*, L*); Ahmed 1987 (distr.; Bangladesh); Darsie & Pradhan 1990, 1991 (F, L; tax., keys, bion., distr.; Nepal); Rattanarithikul et al. 2010 (F*, L*; keys, bion., distr.; Thailand). **Ety.:** n.s. [dance (L); poss. ref. to male lekking behavior]. **Informal name:** Dancing Indian Pointy Mosquito.
syn. *maculata* Theobald
1905h: 243 (M, F; *Pseudograbhamia*). **Type loc.:** Galgamuwa, Sri Lanka (NHMUK).

pubescens Edwards
1925: 264 (M). **Type loc.:** Daru, Sierra Leone (NHMUK). **Distr.:** Cameroon, Central African Republic, Côte d'Ivoire, Democratic Republic of the Congo, Ghana, Sierra Leone, Uganda. **Refs.:** Edwards 1941: 194 (M*, F). **Ety.:** n.s [downy (L); poss. ref. to short fine hairs on mesepimeron]. **Informal name:** Downy Sierra Leonean Pointy Mosquito.

punctifemoris (Ludlow)
1921: 690 (F; *Stegomyia*). **Type loc.:** Fort William McKinley, Rizal [Luzon], Philippines (USNM). **Distr.:** Bangladesh, India, Nepal, Pakistan, Philippines. **Refs.:** Bohart 1945: 64 (M, F*); Knight & Hull 1953: 459 (M*, F); Aslamkhan 1971b (distr.; Pakistan); Aslamkhan 1971a: 61 (M*, F; distr.); Reinert 1973b: 63 (M*, F*); Darsie & Pradhan 1990, 1991 (F, L; tax., keys, bion., distr.; Nepal); Reinert 2000e: Fig. 8 (F*). **Ety.:** spot, femur (L); femora dark "with a marked sprinkling of white scales on the dorsal aspect and a small white apical spot." **Informal name:** Spot-legged Philippine Pointy Mosquito.

quasiunivittatus (Theobald)
1901c: 32 (F*; *Culex*). **Type loc.:** Salisbury, Mashonaland [Zimbabwe] (NHMUK). **Distr.:** Angola, Cameroon, Democratic Republic of the Congo, Ethiopia, Kenya, Malawi, Malaysia, Mali, Mozambique, Nigeria, South Africa, Sudan and South Sudan, Zimbabwe. **Refs.:** Edwards 1941: 189 (M*, F*); Hopkins 1952: 193 (L*); Pao & Knight 1970a: 117 (L*); McIntosh 1975: 282 (M*, F*, L); Jupp 1996 (M*, F*; key). **Ety.:** nearly (L); similar to *univittatus*. **Informal name:** Single-striped Zimbabwean Pointy Mosquito.

rickenbachi Hamon & Adam
1959c: 151 (M*). **Type loc.:** Tiebissou, Ivory Coast [Côte d'Ivoire], French West Africa (IERT). **Distr.:** Cameroon, Côte d'Ivoire, Nigeria. **Refs.:** Rickenbach & Ferrara 1964: 972 (P*, L*). **Ety.:** André Rickenbach. **Informal name:** Rickenbach Ivorian Pointy Mosquito.
syn. *hamoni* Mattingly
1963: 166 (M*, F*). **Type loc.:** Akute Village, Ogun River, near Iju Waterworks, S. Nigeria (NHMUK). **Refs.:** White 1975: 308 (syn.).

semlikiensis van Someren
1950b: 40 (M*, F*). **Type loc.:** Semliki River, Bwamba, Toro District, Uganda (NHMUK). **Distr.:** Uganda. **Ety.:** Semliki River, Uganda. **Informal name:** Semliki Ugandan Pointy Mosquito.

senyavinensis Knight & Hurlbut
1949: 27 (M*, F, P*, L*). **Type loc.:** Rekisau, Not District, Ponape [Pohnpei] Island [Federated States of Micronesia] (USNM). **Distr.:** Micronesia (Wake Isl.). **Refs.:** Lee et al. 1982: 43 (tax., distr.; as *senyaviensis* [*sic*]). **Ety.:** Senyavin Isl., Federated States of Micronesia. **Informal name:** Senyavin Micronesian Pointy Mosquito.

stenoetrus (Theobald)
1907: 395 (F; *Culex*). **Type loc.:** Maskeliya, Ceylon [Sri Lanka] (NHMUK). **Distr.:** India, Sri Lanka. **Refs.:** Barraud 1934: 255 (M*, F); Reinert 1973b: 84 (M*, F*; tax.). **Ety.:** narrow abdomen (Gr); "markedly narrowed abdomen."

Informal name: Narrow-bellied Sri Lankan Pointy Mosquito.

syn. *pseudostenoetrus* Theobald

1910c: 343 (F*; *Culex*). **Type loc.:** Hakgala, Ceylon [Sri Lanka] (NHMUK).

subdentatus Edwards

1936: 51 (M). **Type loc.:** Grahamstown, Cape Province [South Africa] (NHMUK). **Distr.:** South Africa. **Refs.:** Edwards 1941: 191 (M*, F); McIntosh 1975: 277 (M*, F, L); Jupp 1996 (M*, F*; key). **Ety.:** nearly (L); near *dentatus*. **Informal name:** Chewy South African Pointy Mosquito [*subdentatus* = "beneath teeth" = Chewy].

syntheticus Barraud

1928b: 375 (new name for *fisheri* Barraud, not Dyar 1917a). **Distr.:** India. **Ety.:** n.s. [synthetic (L)]. **Informal name:** Synthetic Indian Pointy Mosquito.

syn. *fisheri* Barraud

1928a: 662 (M*, F, L*). **Type loc.:** Tavargatti, Bombay, Deccan, India (NHMUK).

taeniorhynchoides (Christophers)

1911b: 68–72 (M*, F*; *Leslieomyia*). **Type loc.:** Amritsar [Punjab Hill States, India] (LU). **Distr.:** Cambodia, India, Nepal, Pakistan, Sri Lanka, Thailand, Vietnam. **Refs.:** Barraud 1934: 260 (M*, F); Aslamkhan 1971b (distr.; Pakistan); Reinert 1973b: 85 (M*, F*; tax., distr.); Townsend 1990: 140 (type dep. from NHMUK to LU); Darsie et al. 1996 (distr.; Nepal); Reinert 2000e: Fig. 7 (F*). **Ety.:** similar (Gr); like *taeniorhynchus* [ribbon beaked]. **Informal name:** Ribbon-beaked Indian Pointy Mosquito.

syn. *maculata* Theobald

1905g: 23 (F*; *Pecomyia*; not Meigen 1804). **Type loc.:** India (NHMUK).

taufliebi Rickenbach & Ferrara

1965: 24 (M*). **Type loc.:** Mbega, Dept Dyong, and Sanaga, Cameroon (IERT). **Distr.:** Cameroon, Central African Republic, Senegal. **Ety.:** R. Taufflieb. **Informal name:** Taufflieb Cameroonian Pointy Mosquito.

tricholabis Edwards

1941: 179 (M*, F). **Type loc.:** Gede, near Mombasa [Coast], Kenya (NHMUK). **Distr.:** Benin, Burkina Faso, Central African Republic, Côte d'Ivoire, Kenya, Mali, Nigeria. **Refs.:** Mattingly 1956a: 38 (lectotype desig.). **Ety.:** n.s. [hair lip (Gr); prob. ref. to long setae on the gonostylus]. **Informal name:** Hairy-junk Kenyan Pointy Mosquito.

ssp. ***bwamba*** van Someren

1950c: 67 (M, P*, L; *Aedes* (*Aedimorphus*)). **Type loc.:** Bwamba, Toro District, Uganda (NHMUK). **Distr.:** Benin, Kenya, Nigeria, Uganda. **Refs.:** Hopkins 1952: 182 (L); van Someren, Teesdale, & Furlong 1955: 478 (L); Service 1960a: 230 (L); Pao & Knight 1970a: 106 (L*). **Ety.:** Toro Dist., Bwamba, Uganda. **Informal name:** Bwamba Ugandan Pointy Mosquito.

trimaculatus (Theobald)

1905f: 86 (F*; *Culex*). **Type loc.:** Bombay, India (HNM). **Distr.:** Guam, India. **Refs.:** Barraud 1934: 264 (M*, F). **Ety.:** three spotted (L); scutum with three spots. **Informal name:** Three-spotted Indian Pointy Mosquito.

syn. *littoralis* Barraud

1927c: 551 (M*, F; *Aedimorphus*). **Type loc.:** Trombay, Bombay Harbour [Bombay], India (NHMUK). **Refs.:** Barraud 1934: 264 (syn.).

trukensis Bohart

1957: 63 (F). **Type loc.:** Ton, Truk Islands, Caroline Islands [Federated States of Micronesia] (KU). **Distr.:** Fiji, Malaysia, Micronesia (Wake Isl.), New Caledonia, Papua New Guinea, Vanuatu. **Refs.:** Lee et al. 1982: 44 (tax., distr.). **Ety.:** Truk [Eastern Caroline Isl.]. **Informal name:** Truk Micronesian Pointy Mosquito.

vexans (Meigen)

1830b: 241 (F; *Culex*). **Type loc.:** Berlin, Germany (MNHP). **Distr.:** Afghanistan, Albania, Algeria, Armenia, Australia, Austria, Azerbaijan, Bangladesh, Belarus, Belgium, Belize, Bosnia and Herzegovina, Bulgaria, Cambodia, Canada, Corsica, Crimean Peninsula, Croatia, Czech Republic, Denmark, Estonia, Fiji, Finland, France (includes Corsica), FYRO Macedonia, Gambia, Georgia, Germany, Greece (includes Crete), Guam, Guatemala, Honduras, Hong Kong, Hungary, India, Indonesia (includes Java, Flores, Kalimantan, Timor), Iran, Iraq, Italy (includes Sardinia and Sicily), Japan, Jordan, Kazakhstan, Kiribati, Kosovo, Laos, Latvia, Liberia, Libya, Lithuania, Macedonia, Malaysia, Malta, Mariana Isl., Mauritania, Mexico, Micronesia (Wake Isl.), Moldova, Mongolia, Montenegro, Morocco, Myanmar, Nepal, Netherlands, New Caledonia, New Zealand, Norway, Pakistan, Papua New Guinea, People's Republic of China, Philippines, Poland, Portugal, Romania, Russia, Samoa (Ind. State of Samoa, American Samoa, Western Samoa), Saudi Arabia, Serbia, Singapore, Slovakia, Slovenia, Solomon Isl., South Africa, South Korea, Spain (includes Balearic Isl.), Sri Lanka, Sweden, Switzerland, Taiwan, Tajikistan, Thailand, Tonga, Turkey, Turkmenistan, Tuvalu, Ukraine, United Kingdom, United States (Alaska, continental, Hawai'i), Uzbekistan, Vanuatu, Vietnam, Yemen. **Refs.:** Mitchell 1907b (E*; as *sylvestris*); Howard et al. 1913 (E*); Howard et al. 1917 (E; as *sylvestris*); Séguy 1924: 88 (E*), 71 (M gen.*); Barraud 1928a (F*); Barraud 1934: 253 (M*, F, L); Gjullin 1937 (F*); Marshall 1938 (E*); Ross 1947 (F*); Natvig 1948: 412 (M*, F, L*); Penn 1949b: 60 (P*); Darsie 1951: 13 (P*); Yamaguti & LaCasse 1951d: 251 (M*, F*, L*); Carpenter & LaCasse 1955: 263 (M*, F*, L*; keys); Muspratt 1955: 169 (M*, F, L; tax.); Horsfall & Craig 1956: 370 (E*); Craig & Horsfall 1958 (E*); Price 1960: 560 (first instar L*); Belkin 1962 (F*); Mohrig 1967 (F*); Ross & Horsfall 1965 (M*, F*, L*, E*; keys); Hamon et al. 1966: 373 (syn.); Dodge 1966: 359 (first instar L*; key); Mohrig 1967 (F*); Myers

1967 (E*); Kalpage & Brust 1968 (E*); Mohrig 1969 (F*); Horsfall et al. 1970: 1713 (E*); Pao & Knight 1970b: 115 (L*); Rjazantzeva 1970 (F*); Aslamkhan 1971b (distr.; Pakistan); Belkin & Heinemann 1971: 27 (distr.); Matsuo, Yoshida, & Kunou 1972 (E*); Horsfall et al. 1973 (E); Reinert 1973b: 66 (M*, F*, P*, L*, E; world distr.); Bickley 1976 (distr.; Alaska, United States); Utrio 1976: 129, 135 (L*); Ramalingam 1976: 307 (tax., bion., keys., distr.; Samoa and Tonga); Skierska 1977 (F*); Bohart & Washino 1978: 120 (M, F*, L*; keys, tax., bion., distr.); Moussiegt 1979 (annotated bib.); Lee et al. 1982: 32 (distr.); Moussiegt 1982 (annotated bib.); Moussiegt 1988b (annotated bib.); Linley 1990 (E*); Mgeladze 1991 (distr.; Georgia); Darsie et al. 1996 (distr.; Nepal); Jupp 1996 (F*; key); Reinert 2000e: Fig. 3 (F*); Spungis 2000: 9 (distr.; Latvia); Whelan & Hapgood 2000: 405 (bion., distr.; East Timor); Becker et al. 2003 (E*); Trari et al. 2002: 331 (distr.; Morocco); Darsie & Ward 2005 (F*, L*; keys, distr.); Rattanarithikul et al. 2010 (F*, L*; keys, bion., distr.; Thailand); Becker et al. 2010: 194 (M*, F*, L*; keys, tax., distr., bion.); Ahmed et al. 2011 (distr.; Saudi Arabia); Namazov 2014 (distr.; Azerbaijan); Harrison et al. 2016 (F*, L*; keys, distr.); Robert et al. 2019 (distr.; Western Palearctic). **Ety.:** n.s. [to vex, annoy, injure, harass (L)]. **Informal name:** Vexing German Pointy Mosquito. Official ESA common name: vexans mosquito. (Vol. 1, p. 404)

syn. *parvus* Macquart
1834: 36 (*Culex*). **Type loc.:** Bordeaux [Gironde], France (NE).

syn. *articulatus* Rondani
1872: 30 (*Culex*). **Type loc.:** Italy (LU).

syn. *malariae* Grassi
1898: 168 (M; *Culex*). **Type loc.:** Italy (LU).

syn. *sylvestris* Theobald
1901a: 406 (M*, F*; *Culex*). **Type loc.:** Rondeau Provincial Park, Kent County, Lake Simcoe, Ontario; Stony Mountain, Manitoba, all Canada (NHMUK).

syn. *montcalmi* Blanchard
1905: 307 (new name for *sylvestris* Theobald, not *silvestris* Ross 1898; *Culex*).

syn. *minuta* Theobald
1907: 338 (F*; *Culicada*; not *minutus* (Theobald 1901a)). **Type loc.:** India (NHMUK).

syn. *eruthrosops* Theobald
1910c: 299 (F*; *Culicada*). **Type loc.:** Trincomalee, Ceylon [Sri Lanka] (NHMUK). **Refs.:** Knight & Hull 1953: 463 (syn.).

syn. *sudanensis* Theobald
1911c: 154 (M*, F*; *Culex*). **Type loc.:** Sudan (NHMUK). **Refs.:** Edwards 1941: 195 (syn.); White 1975: 309 (syn.; lectotype desig.).

syn. *euochrus* Howard, Dyar, & Knab
1917: 716 (F). **Type loc.:** Popcum, British Columbia, Canada (USNM).

ssp. ***nocturnus*** (Theobald)
1903a: 159 (F; *Culex*; as sp.). **Type loc.:** Ba, Fiji Islands (NHMUK). **Distr.:** Australia, Cook Isl. (Polynesia), Fiji, Indonesia, Kiribati, Malaysia, Mariana Isl., Marshall Isl., New Caledonia, Palau, Papua New Guinea, Samoa (Ind. State of Samoa, American Samoa, Western Samoa), Solomon Isl., Timor, Tonga, Tuvalu, Vanuatu. **Refs.:** Bohart & Ingram 1946b: 15, 41 (M*, F, P*, L*; to ssp. of *vexans*); Yamaguti & LaCasse 1950a: 73 (M*, F*, L*); Knight & Hull 1953: 460 (M*, F, L*); Belkin 1962: 427 (M*, F, P*, L*; to sp. status); Basio 1971b: 12 (M*; as ssp. of *vexans*; bion., vector, distr.); Reinert 1973b: 66 (syn.); Baisas 1974: 6 (M*, F, P*, L*; tax., bion., distr.; Philippines); Lee et al. 1982: 32 (tax., review; from syn. with *vexans* as sp.; distr.). **Ety.:** nocturnal (L); females active at night. **Informal name:** Nocturnal Fijian Pointy Mosquito.

ssp. syn. *niger* Theobald
1913b: 164 (A; *Culex nocturnus* var.). **Type loc.:** Canala, New Caledonia (B). **Refs.:** Reinert 1973b: 66 (syn.).

ssp. ***arabiensis*** (Patton)
1905: 633 (M*, F, L; *Culex*). **Type loc.:** Ulub Camp and Crater, West Aden Protectorate [Yemen] (NE). **Distr.:** Gambia, Mauritania, Mauritius, Nigeria, Saudi Arabia, Somalia, South Africa, Sudan and South Sudan, Tanzania, Yemen. **Refs.:** Edwards 1941: 195 (M*, F); Lewis 1945: 11 (P, L); Hopkins 1952: 198 (L*); White 1975: 309 (resur. from syn. to ssp. of *vexans*; distr.). **Ety.:** Arabia. **Informal name:** Arabia Yemeni Pointy Mosquito.

ssp. ***nipponii*** (Theobald)
1907: 337 (F*; *Culicada*; as sp.). **Type loc.:** Karnizana, Japan (NHMUK). **Distr.:** Georgia, Japan, Mongolia, People's Republic of China, Russia, South Korea. **Refs.:** Bohart & Ingram 1946b: 69 (M*, F, L*; bion., distr., tax.; ssp. of *vexans*); La Casse & Yamaguti 1950: 125 (M*, F*, L*; bion., distr., tax.); Yoshimeki 1955: 81 (L*); Hara 1957: 66 (F*); Matsuo, Yoshida, & Kunou 1972: 361 (E*); Reinert 1973b: 79 (M*, F*, P*, L*); Moriya et al. 1973 (E*); Sheremet 1975: 100 (distr.); Tanaka et al. 1975c: 224 (bion., distr.); Minar 1976: 335 (distr.); Tanaka et al. 1979: 410 M*, F*, L*); Mgeladze 1991: 331 (distr.; Georgia); Gornostaeva 2000: 478 (distr.; Russia); Ree 2003: 41 (distr.; Korea); Tanaka 2003a: 15 (P*; tax., key). **Ety.:** Japanese name for Japan. **Informal name:** Nippon Japanese Pointy Mosquito.

wigglesworthi Edwards
1941: 182, 394, 432 (M*, F, P*). **Type loc.:** Sapele, Nigeria (NHMUK). **Distr.:** Côte d'Ivoire, Nigeria, Uganda. **Refs.:** Hopkins 1952: 183 (L*); Mattingly 1956a: 39 (lectotype desig.). **Ety.:** Vincent Brian Wigglesworth. **Informal name:** Wigglesworth Nigerian Pointy Mosquito.

Subgenus *Alanstonea* Mattingly

1960: 170 (as subg. of genus *Aedes*). Type species: *Scutomyia treubi* de Meijere.

Name derivation: Alan Stone.

Important references: Mattingly 1971e: 133 (E*); Mogi & Chan 1996 (bion.); Reinert 2000e: 14 (F gen.*); Reinert, Harbach, & Kitching 2004: 289 (phyl., class.; to genus); Reinert, Harbach, & Kitching 2009 (phyl., class.); Rattanarithikul et al. 2010 (F*, L*; keys; Thailand); Wilkerson et al. 2015 (phyl., class.; to subg. of genus *Aedes*).

brevitibia (Edwards)
1914c: 125 (F; *Armigeres*). **Type loc.:** Tabuan Swamp, Kuching, Sarawak, Malaysia [Malaysian Borneo] (NHMUK). **Distr.:** Indonesia, Malaysia. **Refs.:** Brug 1934b (L*); Brug & Bonne-Wepster 1947: 185 (new comb.); Mattingly 1954a: 269 (tax.); Mattingly 1960: 170 (subg. transfer); Ramalingam & Ramakrishnan 1971: 231 (M*, F, P*, L*); Reinert 2000e: Fig. 11 (F*). **Ety.:** short tibia (L). **Informal name:** Short-legged Malaysian Pointy Mosquito.

treubi (de Meijere)
1910: 918 (M*, F*, P*, L*, E*; *Scutomyia*). **Type loc.:** Tjibodas (1,500 m) near Buitenzorg, Java [Indonesia] (AM, NHMUK). **Distr.:** Indonesia, Thailand. **Refs.:** Bonne-Wepster & Brug 1932: 109 (*Aedes* (*Stegomyia*)); Mattingly 1960: 170 (transfer); Mattingly 1971e (E*); Townsend 1990: 143 (syntype, NHMUK); Rattanarithikul et al. 2010 (F*, L*; keys, bion., distr.; Thailand). **Ety.:** n.s. [poss. ref. to M.W.F. Treub, politician (Harbach 2018c)]. **Informal name:** Treub Indonesian Pointy Mosquito.

Subgenus *Albuginosus* Reinert

1987: 308. Type species: *Stegomyia marshalli* Theobald (as subg. of genus *Aedes*).

Name derivation: white spotted (L).

Important references: Reinert 2000e: 15 (F gen.*); Reinert, Harbach, & Kitching 2004: 289 (phyl., morph.; to genus); Reinert, Harbach, & Kitching 2009 (phyl., class.); Huang & Rueda 2015 (F*; key; as subg. of genus *Aedes*); Wilkerson et al. 2015 (phyl., class.; to subg. of genus *Aedes*).

capensis Edwards
1924a: 162 (F). **Type loc.:** Cape Province, Caledon Division, South Africa (SA). **Distr.:** Central African Republic, Kenya, Malawi, Mozambique, Senegal, South Africa, Uganda. **Refs.:** Edwards 1941: 161 (M*, F*); Hopkins 1952: 163 (L*); Muspratt 1953b: 89 (A, L); Worth & De Meillon 1960: 241 (distr.); Pao & Knight 1970a: 108 (L*); White 1975: 306 (distr.); Reinert 1987 (from subg. *Aedimorphus*); Jupp 1996 (M*, F*; key). **Ety.:** poss. ref. to Cape Prov., South Africa. **Informal name:** Cape South African Pointy Mosquito.

gilliesi van Someren
1962: 24 (M, F). **Type loc.:** South Pare Hills, Tonto, Amani, Tanzania (NHMUK). **Distr.:** Tanzania. **Refs.:** Mattingly 1971a: Pl. 21 (P cephalothorax*); Reinert 1987 (from subg. *Aedimorphus*). **Ety.:** M.T. Gillies. **Informal name:** Gillies Tanzanian Pointy Mosquito.

haworthi Edwards
1923d: 499 (M*, F). **Type loc.:** Dar-es-Salaam, East Africa [Tanzania] (NHMUK). **Distr.:** Benin, Burkina Faso, Côte d'Ivoire, Kenya, Mali, South Africa, Tanzania, Zambia. **Refs.:** Edwards 1941: 165 (M*, F*); De Meillon et al. 1945: 95 (P, L); van Someren 1950a: 9 (A); Hopkins 1952: 167 (L); Pao & Knight 1970a: 109 (L*); Reinert 1987 (from subg. *Aedimorphus*); Jupp 1996 (M*, F*; key). **Ety.:** W.E. Haworth. **Informal name:** Haworth Tanzanian Pointy Mosquito.

kapretwae Edwards
1941: 162 (M, F). **Type loc.:** Mount Elgon, 8,000 ft above Kapretwa, Kenya (NHMUK). **Distr.:** Kenya. **Refs.:** van Someren 1946a: 118 (L*); Hopkins 1952: 163 (L*); Mattingly 1956a: 38 (lectotype desig.); Reinert 1987 (from subg. *Aedimorphus*). **Ety.:** Mt. Elgon, forest zone above Kapretwa [Kenya? Uganda?]. **Informal name:** Kapretwa Kenyan Pointy Mosquito.

kennethi Muspratt
1956b: 38 (M, F, L). **Type loc.:** Along Pongola River, Tongaland, Natal [KwaZulu-Natal, South Africa] (SAIM). **Distr.:** South Africa. **Refs.:** Pao & Knight 1970a: 109 (L*); Reinert 1987 (from subg. *Aedimorphus*); Jupp 1996 (M*, F*; key). **Ety.:** Kenneth C. Smithburn. **Informal name:** Smithburn South African Pointy Mosquito.

marshallii (Theobald)
1901a: 310 (M*, F*; *Stegomyia*). **Type loc.:** Salisbury, Mashonaland [Harare, Zimbabwe] (NHMUK). **Distr.:** Angola, Botswana, Democratic Republic of the Congo, Ethiopia, Kenya, Mozambique, Nigeria, South Africa, Uganda, Zambia, Zimbabwe. **Refs.:** Edwards 1941: 160 (M, F); De Meillon et al. 1945: 96 (P*); Hopkins 1952: 162 (L); Worth & De Meillon 1960: 240 (distr.); Pao & Knight 1970a: 110 (L*); Reinert 1987 (M*, F*; from subg. *Aedimorphus*); Jupp 1996 (M*, F*; key); Reinert 2000e: Fig. 12 (F*). **Ety.:** Guy Anstruther Knox Marshall. **Informal name:** Marshall Zimbabwean Pointy Mosquito.

ngong van Someren
1950a: 7 (M, F, P, L). **Type loc.:** Langata Forest, Ngong, Nairobi [Central], Kenya (NHMUK). **Distr.:** Kenya, Tanzania. **Refs.:** Reinert 1987 (from subg. *Aedimorphus*). **Ety.:** Ngong, Nairobi, Kenya. **Informal name:** Ngong Kenyan Pointy Mosquito.

stokesi Evans
1929c: 522 (new name for *occidentalis* Evans 1926, not Skuse 1889). **Type loc.:** Freetown, Sierra Leone (NHMUK). **Distr.:** Benin, Burkina Faso, Cameroon, Central African Republic, Côte d'Ivoire, Ghana, Mali, Nigeria, Senegal, Sierra Leone, Sudan and South Sudan, Togo, Uganda. **Refs.:** Edwards 1941: 163 (M, F, P); van Someren 1950a: 9 (M, F); Hopkins 1952: 165 (L*); Hamon

& Rickenbach 1954: 934 (M*); Reinert 1987 (from subg. *Aedimorphus*). **Ety.:** Adrian Stokes. **Informal name:** Stokes Sierra Leonean Pointy Mosquito.

syn. *occidentalis* Evans
1926: 99 (M*, L). **Type loc.:** Freetown, Sierra Leone (NHMUK). **Refs.:** Lee et al. 1982: 216 (F; key, tax., bion., distr., review).

teesdalei van Someren
1954a: 104 (M, F, P*, L). **Type loc.:** Kwale, Coast Province, Kenya (NHMUK). **Distr.:** Kenya. **Refs.:** Reinert 1987 (from subg. *Aedimorphus*). **Ety.:** C. Teesdale. **Informal name:** Teesdale Kenyan Pointy Mosquito.

Subgenus *Ayurakitia* Thurman

1954a: 198. Type species: *Ayurakitia griffithi* Thurman (as genus).

Name derivation: Luang Ayurakit Kosol.

Important references: Mattingly 1971a: 31, note 28 (to subg. of genus *Aedes*); Reinert 1972c (tax., keys, revision; SE Asia); Reinert 2000i (tax.; to genus); Reinert 2001a (F gen.*); Reinert, Harbach, & Kitching 2009 (phyl., class.); Rattanarithikul et al. 2010 (F*, L*; keys, bion.; Thailand); Wilkerson et al. 2015 (phyl., class.; subg. of genus *Aedes*).

griffithi (Thurman)
1954a: 198 (M*, F; *Ayurakitia*). **Type loc.:** Doi Chom Cheng, a peak of the Doi Suthep Range, Chiengmai Province, Thailand (USNM). **Distr.:** Thailand. **Refs.:** Stone & Knight 1957c: 196 (type info.); Reinert 1972c: 12 (M*, F*, P, L); Reinert 2000i (M*, F*, P*, L*); Reinert 2001a: Fig. 1 (F*); Rattanarithikul et al. 2010 (F*, L*; keys, bion., distr.; Thailand). **Ety.:** Melvin E. Griffith. **Informal name:** Griffith Thai Pointy Mosquito.

peytoni Reinert
1972c: 7 (M*, F*, P, L; *Aedes* (*Ayurakitia*)). **Type loc.:** Sip Si Hon, Phangnga, Thailand (USNM). **Distr.:** Thailand. **Refs.:** Rattanarithikul et al. 2010 (F*, L*; keys, bion., distr.; Thailand). **Ety.:** EL Peyton. **Informal name:** Peyton Thai Pointy Mosquito.

Subgenus *Aztecaedes* Zavortink

1972: 66. Type species: *Aedes* (*Gualteria*) *ramirezi* Vargas & Downs (as subg. of genus *Aedes*).

Name derivation: n.s. [Aztec culture, Aztec Empire, and *Aedes*].

Important references: Reinert 2000h (to subg. of genus *Ochlerotatus*); Reinert, Harbach, & Kitching 2004: 357 (to genus); Reinert, Harbach, & Kitching 2008 (phyl., class); Reinert, Harbach, & Kitching 2009 (phyl., class); Wilkerson et al. 2015 (phyl., class.; to subg. of genus *Aedes*).

ramirezi Vargas & Downs
1950: 164 (M*, F*, P*, L*). **Type loc.:** Morelos, Mexico (ISET). **Distr.:** Mexico. **Refs.:** Zavortink 1972: 67 (M*, F*, P*, L*); Reinert 2000e (F*); Reinert 2002e: Fig. 41 (F*). **Ety.:** Eliseo Ramírez. **Informal name:** Ramírez Mexican Pointy Mosquito.

Subgenus *Belkinius* Reinert

1982: 102 (as subg. of genus *Aedes*). Type-species: *Aedes aurotaeniatus* Edwards (new name for *Stegomyia aureostriata*).

Name derivation: John N. Belkin.

Important references: Reinert 2000e (F gen.*); Reinert, Harbach, & Kitching 2004: 289 (phl., class.; to genus); Reinert, Harbach, & Kitching 2009 (phyl., class.); Wilkerson et al. 2015 (phyl., class.; to subg. of genus *Aedes*).

aurotaeniatus Edwards
1922a: 256, 1922b: 464 (new name for *aurostriata* Banks, not Doleschall 1857). **Distr.:** Philippines. **Refs.:** Knight & Hull 1952: 169 (F); Huang 1970b: 281 (M*, F*, P*; subg. transfer); Basio 1971b: 25 (M*; bion.); Reinert 1982: 105 (from subg. *Paraedes*); Reinert 2000e: Fig. 13 (F*). **Ety.:** n.s. [gold ribbon (L, Gr); "mesonotum with a sublateral pair of rather broad golden stripes"]. **Informal name:** Gold-banded Philippine Pointy Mosquito.

syn. *aurostriata* Banks
1906b: 995 (F; *Stegomyia*). **Type loc.:** Canlaon volcano, Mount Siya-Siya, Negros Occidental, Philippines (NHMUK). **Refs.:** Edwards 1922b (syn.); Knight & Hull 1952: 169 (type info., lectotype desig.).

Subgenus *Bifidistylus* Reinert, Harbach, & Kitching

2009: 766 (as genus). Type species: *Aedes* (*Aedimorphus*) *lamborni* Edwards.

Name derivation: bifurcated gonostylus; bifida = split in two; stilus = Roman writing instrument.

Important references: Reinert 2010g (F gen.*); Wilkerson et al. 2015 (phyl., class.; to subg. of genus *Aedes*).

boneti Gil Collado
1936: 323 (M*). **Type loc.:** Rebola, Fernando Po [Equatorial Guinea] (MU). **Distr.:** Cameroon, Equatorial Guinea. **Refs.:** Edwards 1941: 187 (M). **Ety.:** F. Bonet.

Informal name: Bonet Equatoguinean Pointy Mosquito.

ssp. ***kumbae*** Chwatt

1948: 184 (M*, P, L*). **Type loc.:** Mambonjeze stream, Kumba, British Cameroons [Republic of Cameroon] (NHMUK). **Distr.:** Cameroon. **Refs.:** Hopkins 1952: 191 (L*); White 1975: 306 (distr.); Reinert 2010g: Fig. 2 (F*). **Ety.:** Kumba, Cameroon. **Informal name:** Kumba Cameroonian Pointy Mosquito.

lamborni Edwards

1923a: 398 (M). **Type loc.:** [Zomba], Mulungushi River, Malawi (NHMUK). **Distr.:** Cameroon, Congo, Democratic Republic of the Congo, Kenya, Malawi, South Africa, Tanzania, Zambia. **Refs.:** Edwards 1941: 186 (M*, F); Hopkins 1952: 189 (L*); Mattingly 1953c: 325 (P); Pao & Knight 1970a: 107 (L*); Jupp 1996 (F*; key); Reinert 2010g: Fig. 1 (F*). **Ety.:** W.A. Lamborn. **Informal name:** Lamborn Malawian Pointy Mosquito.

Subgenus *Borichinda* Harbach & Rattanarithikul

2007 (as genus). In Harbach, Rattanarithikul et al. 2007: 471. Type species: *Borichinda cavernicola* Rattanarithikul & Harbach.

Name derivation: Borichinda Cave, Chiangmai Prov., Thailand.

Important references: Reinert, Harbach, & Kitching 2009 (phyl., class.); Rattanarithikul et al. 2010 (F*, L*; keys, bion.; Thailand); Wilkerson et al. 2015 (phyl., class.; to subg. of genus *Aedes*).

cavernicola (Rattanarithikul & Harbach)

2007 (M*, F*, P*, L*; as *Borichinda*). In Harbach, Rattanarithikul et al. 2007: 469. **Type loc.:** Borichinda Cave, Chomthong District, Chiang Mai Province, Thailand (NHMUK). **Distr.:** Thailand. **Refs.:** Rattanarithikul et al. 2010 (F*, L*; keys, bion., distr.; Thailand). **Ety.:** cave inhabitant (L); ref. to larval habitat. **Informal name:** Cave-dwelling Thai Pointy Mosquito.

Subgenus *Bothaella* Reinert

1973e: 2 (as subg. of genus *Aedes*). Type species: *Aedes (Bothaella) helenae* Reinert.

Name derivation: Botha De Meillon.

Important references: Reinert, Harbach, & Kitching 2004 (phyl., class.; to genus); Reinert, Harbach, & Kitching 2009 (phyl., class.); Cook et al. 2010 (mol. phyl.); Rattanarithikul et al. 2010 (F*, L*; keys, bion.; Thailand); Wilkerson et al. 2015 (phyl., class.; to subg. of genus *Aedes*).

alongi Galliard & Ngu

1947: 77 (L; as *halongi*, emend. by Galliard & Ngu 1950: 67; L*). **Type loc.:** Baie d'Along, Tonkin, Indochina [Ha Long Bay, Hanoi province, Vietnam] (NE). **Distr.:** Vietnam. **Refs.:** Reinert, Harbach, & Huong 2006 (F, L; to genus *Bothaella*). **Ety.:** n.s. [prob. ref. to Bay of Along]. **Informal name:** Halong Vietnamese Pointy Mosquito.

brownscutumus Dong, Zhou, & Dong

1999: 137 (M*, P*, L*). **Type loc.:** Jilushan, Jinghong County, Yunnan Province, China [People's Republic of China] (YIM). **Distr.:** People's Republic of China. **Ety.:** n.s. [poss. ref. to brown scutum]. **Informal name:** Brown-scaled Chinese Pointy Mosquito.

eldridgei Reinert

1973e: 14 (M*, F*, P*, L*). **Type loc.:** Ban Tham Kraeb, Amphoe Chiang Dao, Chiang Mai, Thailand (USNM). **Distr.:** Thailand. **Refs.:** Rattanarithikul et al. 2010 (F*, L*; keys, bion., distr.; Thailand). **Ety.:** Bruce Frederick Eldridge. **Informal name:** Eldridge Thai Pointy Mosquito.

helenae Reinert

1973e: 9 (M*, F*, P*, L*). **Type loc.:** Ban Tham Kraeb, Amphoe Chiang Dao, Chiang Mai, Thailand (USNM). **Distr.:** People's Republic of China, Thailand. **Refs.:** Dong et al. 1999: 137 (distr.; People's Republic of China); Reinert 2000e: Fig. 14 (F*); Rattanarithikul et al. 2010 (F*, L*; keys, bion., distr.; Thailand). **Ety.:** Mary Helen (Mollie) Reinert. **Informal name:** Helen Thai Pointy Mosquito.

kleini Reinert

1973e: 18 (M*, F*). **Type loc.:** Oudong, Kompong Speu, Cambodia (USNM). **Distr.:** Cambodia. **Ety.:** Terry Allen Klein. **Informal name:** Klein Cambodian Pointy Mosquito.

manhi (Harbach & Cook)

2010 (M*, F, P*, L*; DNA, phyl.; *Bothaella*.). In Cook et al. 2010: 34. **Type loc.:** Cuc Phuong National Park, Ninh Binh Province, Vietnam (NHMUK). **Distr.:** Vietnam. **Ety.:** Nguyen Duc Manh. **Informal name:** Manh Vietnamese Pointy Mosquito.

Subgenus *Bruceharrisonius* Reinert

2003a: 309 (as subg. of genus *Ochlerotatus*). Type species: *Howardina greenii* Theobald.

Name derivation: Bruce A. Harrison.

Important references: Reinert, Harbach, & Kitching 2006: 89 (phyl., class.; to genus); Reinert, Harbach, & Kitching 2009 (phyl., class.); Rattanarithikul et al. 2010 (F*, L*; keys, bion., Thailand); Wilkerson et al. 2015 (phyl., class.; to subg. of genus *Aedes*); Soghigian et al. 2017 (mol. phyl.)

alektorovi Stackelberg
1943: 311 (M). **Type loc.:** Kamenush-ka, Shkotovo District, Ussuri Land, Southeastern Siberia [USSR] (ZIL). **Distr.:** Japan, Russia, South Korea. **Refs.:** Monchadskii 1951: 209 (L*; as *alectorovi*); Gutsevich 1955: 321 (M*); Gutsevich et al. 1974: 306 (M*, F, L*); Danilov 1977: 181 (distr.). **Ety.:** n.s. **Informal name:** Alektorov Russian Pointy Mosquito.

aureostriatus (Doleschall)
1857: 385 (F*; *Culex*). **Type loc.:** Amboina Island [Moluccas] (NE). **Distr.:** Cambodia, Indonesia (includes West Papua), Japan, Thailand, Timor. **Refs.:** Knight & Hull 1951b: 238 (M*, F, L); Basio 1971b: 16 (M*; bion.; as *leucopleurus*); Tanaka et al. 1975c: 221 (tax., distr.); Tanaka et al. 1979: 344 (M*, F*, L*); Lee et al. 1982: 118 (F; key, tax., bion., distr., review); Whelan & Hapgood 2000 (bion, distr.; East Timor). **Ety.:** n.s. [golden grooved (L); scutum with five golden stripes]. **Informal name:** Gold-striped Indonesian Pointy Mosquito.
syn. *leucopleurus* Rozeboom
1946: 588 (M*, L*). **Type loc.:** Irahuan River, Palawan, Philippines (USNM). **Refs.:** Reinert 1990 (syn.).

christophersi Edwards
1922a: 267 (F). **Type loc.:** Simla [East Punjab, India] (NHMUK). **Distr.:** India, Japan, Nepal, Pakistan, Thailand. **Refs.:** Barraud 1934: 195 (M*, F*, L*); Aslamkhan 1971b (distr.; Pakistan); Darsie et al. 1992 (distr.; Nepal); Rattanarithikul et al. 2010 (F*, L*; keys, bion., distr.; Thailand). **Ety.:** Samuel Rickard Christophers. **Informal name:** Christophers Indian Pointy Mosquito.

doonii Wattal, Bhatia, & Kalra
1958: 220 (M*, L*). **Type loc.:** Kalsi, Dehra Dun [India] (MSI). **Distr.:** India. **Ety.:** n.s. [poss. ref. to type locality, Dehra Dun]. **Informal name:** Dehra Dun Indian Pointy Mosquito.

greenii (Theobald)
1903a: 289 (F*; *Howardina*). **Type loc.:** Peradeniya [Central], Ceylon [Sri Lanka] (NHMUK). **Distr.:** India, Indonesia, Malaysia, Nepal, Philippines, Sri Lanka, Thailand. **Refs.:** Barraud 1923h (L*); Barraud 1934: 184 (M*, F*, L*); Joshi et al. 1965 (distr.; Nepal); Darsie & Pradhan 1990, 1991 (F, L; tax., keys, bion., distr.; Nepal); Harrison et al. 1991: 211 (to sp. from var. of *aureostriatus*); Reinert 2002e: Fig. 4 (F*); Reinert 2003a: 313 (M*, F*, P*, L*; tax., proposed type sp.); Rattanarithikul et al. 2010 (F*, L*; keys, bion., distr.; Thailand). **Ety.:** Christopher A. Green. **Informal name:** Green Sri Lankan Pointy Mosquito.
syn. *kanaranus* Barraud
1924a: 850 (F*; as *Finlaya greenii* var. *kanarana*). **Type loc.:** Karwar, North Kanara [Bombay], India (NHMUK). **Refs.:** Townsend 1990: 90 (type dep. from LU to NHMUK); Harrison et al. 1991: 211 (from var. *aureostriatus*; syn.).

hurlbuti Lien
1967: 179 (M*, F*, P*, L*). **Type loc.:** Between Hsinkaokou and Alishan, Wufeng, Chiai County, Taiwan (TMRI). **Distr.:** Taiwan. **Ety.:** Herbert S. Hurlbut. **Informal name:** Hurlbut Taiwanese Pointy Mosquito.

kobayashii Nakata
1956: 135 (M, F, P, L). **Type loc.:** Mount Kurama, Kyoto [Honshu], Japan (KLPH). **Distr.:** Japan, Korea. **Refs.:** Nakata 1959: 16 (M*, F*, P*, L*); Chow 1973 (distr.); Danilov 1977: 181 (syn.); Tanaka 2002b: 153 (P*; tax., key; from syn. with *alektorovi*). **Ety.:** n.k. [Harujir Kobayashi (Harbach 2018c)]. **Informal name:** Kobayashi Japanese Pointy Mosquito.

okinawanus Bohart
1946b: 39 (M*, F, L*). **Type loc.:** Okuma, Okinawa [Ryukyu-Retto, Japan] (USNM). **Distr.:** Japan. **Refs.:** Bohart & Ingram 1946b: 67 (M*, F*, P*, L*; bion., distr., tax.); Lien 1968d: 225 (as *aureostriatus* ssp.); Tanaka 2002b: 137 (as valid sp.), 156 (P*; tax., key). **Ety.:** Okinawa. **Informal name:** Okinawa Japanese Pointy Mosquito.

taiwanus Lien
1968d: 223 (M*, F*, P*, L*; *Aedes* (*Finlaya*) *aureostriatus* ssp.). **Type loc.:** Mashihshih, Taiwu, Pingtung Hsien, Taiwan (TMRI). **Distr.:** Japan, Taiwan. **Refs.:** Saugstad 1973: 284 (distr.); Tanaka 2002b: 137 (from ssp. of *aureostriatus*), 159 (P*; tax., key); Reinert 2003 (to sp.). **Ety.:** Taiwan. **Informal name:** Hoklo Taiwanese Pointy Mosquito [the Hoklo are a major ethnic group in Taiwan].

Subgenus *Cancraedes* Edwards

1929c: 342 (as subg. of genus *Aedes*). Type species: *Aedes* (*Skusea*) *cancricomes* Edwards.

Name derivation: Cancer = a crab (Gr) and *Aedes* [type species larvae found in crabholes].

Important references: Knight & Hull 1953: 477 (tax.; Philippines); Mattingly 1958a (keys); Mattingly 1958b (review); Reinert 2000e: 20 (F gen.*); Reinert, Harbach, & Kitching 2009 (phyl., class.; to genus); Rattanarithikul et al. 2010 (F*, L*; keys, bion.; Thailand); Wilkerson et al. 2015 (phyl., class.; to subg. of genus *Aedes*).

cancricomes Edwards
1922a: 272 (M*, F). **Type loc.:** Port Blair, Andaman Islands [India] (NHMUK). **Distr.:** India, Indonesia. **Refs.:** Barraud 1934: 298 (M, F); Mattingly 1958b: 80 (F*; tax.); Mattingly 1958c: 107 (type info.); Lien et al. 1977 (distr.); Reinert 2000e: Fig. 15 (F*). **Ety.:** crab companion

(L); ref. to larvae in crabholes. **Informal name:** Crabhole Indian Pointy Mosquito.

curtipes Edwards
1915c: 283 (F). **Type loc.:** Kuching, Sarawak, Borneo [Malaysia] (NHMUK). **Distr.:** Indonesia, Malaysia, Philippines, Singapore. **Refs.:** Edwards & Given 1928: 346 (L*); Knight & Hull 1953: 478 (M*, F, L); Mattingly 1958a: 49 (F*); Mattingly 1958c: 107 (type info.); Basio 1971b: 12 (M*; bion.); Basio 1971b: 13 (M*; bion.; as *miachaetessa*); Lien et al. 1977 (distr.). **Ety.:** short foot (L); legs shorter than other members of genus. **Informal name:** Short-legged Malaysian Pointy Mosquito.
syn. *miachaetessa* Dyar & Shannon
1925d: 78 (F). **Type loc.:** Camp Stotsenberg [Angeles], Pampanga, Luzon, Philippines (USNM). **Refs.:** Knight & Hull 1953: 477 (M*, F*; type info., lectotype desig.); Stone & Knight 1956a: 221 (type info.); Mattingly 1958a: 49 (syn.).

indonesiae Mattingly
1958a: 53 (M*, F*, P*, L*). **Type loc.:** Island in Bay of Batavia, Java [Indonesia] (NHMUK). **Distr.:** Indonesia, Malaysia, Thailand. **Refs.:** Mattingly 1974a: 41 (E*; distr.); Harrison et al. 1991: 223 (distr.; Thailand); Rattanarithikul et al. 2010 (F*, L*; keys, bion., distr.; Thailand). **Ety.:** Indonesia. **Informal name:** Srivijaya Indonesian Pointy Mosquito [Srivijaya was an Indonesian medieval state].

kohkutensis Mattingly
1958a: 55 (F*). **Type loc.:** Koh Kut Island, east Gulf of Siam [Thailand] (NHMUK). **Distr.:** Thailand. **Refs.:** Rattanarithikul et al. 2010 (F*, L*; keys, bion., distr.; Thailand). **Ety.:** Koh Kut Isl., Thailand. **Informal name:** Koh Kut Thai Pointy Mosquito.

mamoedjoensis Mattingly
1958a: 58 (F*). **Type loc.:** Mamoedjoe, Celebes [Sulawesi] (NHMUK). **Distr.:** Indonesia. **Ety.:** named for type loc. **Informal name:** Mamoedjoe Indonesian Pointy Mosquito.

masculinus Mattingly
1958a: 51 (M*, F*, P*, L*). **Type loc.:** Rantau Panjang, Selangor, Malaya [Malaysia] (NHMUK). **Distr.:** Malaysia, Philippines, Singapore. **Refs.:** Basio 1971b: 13 (M*; bion.). **Ety.:** n.s. **Informal name:** Masculine Malaysian Pointy Mosquito.

palawanicus Mattingly
1958a: 56 (F*). **Type loc.:** Puerto Princesa, Palawan, Philippines (USNM). **Distr.:** Philippines. **Refs.:** Basio 1971b: 13 (M*; bion.). **Ety.:** n.s. [for type loc. Palawan, Philippines]. **Informal name:** Palawan Philippine Pointy Mosquito.

penghuensis Lien
1968b: 96 (M*, F*, P*, L*). **Type loc.:** Kunpei, Huhsi Penghu Hsien, Taiwan (TMRI). **Distr.:** Taiwan. **Refs.:** Matsuo, Yoshida, & Lien 1974: 188 (E*). **Ety.:** named for type loc. **Informal name:** Penghu Taiwanese Pointy Mosquito.

simplex (Theobald)
1903a: 297 (M; *Ficalbia*). **Type loc.:** Kurunegala, [Northwestern] Ceylon [Sri Lanka] (NHMUK). **Distr.:** India, Sri Lanka. **Refs.:** Edwards 1928a: 273 (M*); Mattingly 1958b: 80 (M; tax.); Mattingly 1958c: 107 (type info.); Kaur 2003: 451 (distr.; India). **Ety.:** n.s. [single, simple (L)]. **Informal name:** Simple Sri Lankan Pointy Mosquito.

thurmanae Mattingly
1958a: 55 (F*). **Type loc.:** Kabaena Island [near Sulawesi, Indonesia] (NHMUK). **Distr.:** Indonesia. **Ety.:** n.s. [prob. ref. to Ernestine Hogan Basham Thurman (Harbach 2018c)]. **Informal name:** Thurman Indonesian Pointy Mosquito.

Subgenus *Catageiomyia* Theobald

1903d: i (Nov.) (as genus). Type species: *Stegomyia irritans* Theobald.

Name derivation: subterranean fly (Gr); larvae in crabholes.

syn. *Myxosquamus* Theobald 1909b: 7 (as genus). Type species: *Myxosquamosus confusus* Theobald.
syn. *Neopecomyia* Theobald 1909b: 12 (as genus). Type species: *Neopecomyia uniannulata* Theobald.
syn. *Stenoscutus* Theobald 1909a: 7 (as genus). Type species: *Stenoscutus africanus* Theobald.

Important references: Reinert, Harbach, & Kitching 2009 (phyl., class.; to genus); Reinert 2010d (F gen.*); Wilkerson et al. 2015 (to subg. of genus *Aedes*); Soghigian et al. 2017 (mol. phyl.).

adami Geoffroy
1971b: 273 (M*). **Type loc.:** Central African Republic (IERT). **Distr.:** Central African Republic. **Ety.:** Jean-Paul Adam. **Informal name:** Adam Central African Pointy Mosquito.

argenteopunctatus (Theobald)
1901a: 316 (M*, F; *Stegomyia*). **Type loc.:** Salisbury, Mashonaland [Harare, Zimbabwe] (NHMUK). **Distr.:** Benin, Burkina Faso, Cameroon, Central African Republic, Côte d'Ivoire, Democratic Republic of the Congo, Ethiopia, Gabon, Kenya, Madagascar (includes Glorioso and Juan De Nova Isl.), Mali, Mozambique, Senegal, South Africa, Sudan and South Sudan, Tanzania, Togo, Uganda, Zambia, Zimbabwe. **Refs.:** Edwards 1941: 167 (M*, F*); De Meillon et al. 1945: 97 (P*); Hopkins 1952: 170 (L*); Pao & Knight 1970a: 105 (L*); McIntosh 1975: 259 (M, F*); Jupp 1996 (M*, F*; key); Reinert 2000e: Fig. 10 (F*); Reinert 2010d: Fig. 2 (F*).

Ety.: n.s. [silvery ornamented (L); poss. ref. to silver spots on scutum]. **Informal name:** Silver-spotted Zimbabwean Pointy Mosquito.

syn. *quinquepunctata* Theobald
1913d: 598 (F; *Aedimorphus*). **Type loc.:** Lado, Sudan (NHMUK).

bedfordi Edwards
1936: 50 (F). **Type loc.:** Ntambanana, Zululand [uThungulu District, KwaZulu-Natal, South Africa] (NHMUK). **Distr.:** South Africa. **Refs.:** Edwards 1941: 171 (F); Muspratt 1959: 67 (M*; tax.); McIntosh 1975: 259 (M*, F); Jupp 1996 (M*, F*; key). **Ety.:** Bedford [poss. ref. to coll.]. **Informal name:** Bedford South African Pointy Mosquito.

chamboni Cornet
1968: 398 (M*, F*). **Type loc.:** Saboya, Nioro du Rip, Senegal (IERT). **Distr.:** Senegal. **Ety.:** L. Chambon. **Informal name:** Chambon Senegalese Pointy Mosquito.

dialloi Hamon & Brengues
1965: 103 (M*). **Type loc.:** Tchakarkou, Natitingou Prefecture, Dahomey (IERT). **Distr.:** Benin, Mali. **Ety.:** M. Bellan Diallo. **Informal name:** Diallo Beninese Pointy Mosquito.

falabreguesi Hamon
1957: 685 (M*). **Type loc.:** Abidjan lagoon, near Adiopodoume, Ivory Coast [Côte d'Ivoire] (IERT). **Distr.:** Cameroon, Côte d'Ivoire. **Ety.:** F. Falabrègues. **Informal name:** Falabrègues Ivorian Pointy Mosquito.

filicis Ingram & De Meillon
1927: 58 (M, F*, P*, L*). **Type loc.:** Eshowe, Zululand [KwaZulu-Natal, South Africa] (NHMUK). **Distr.:** Cameroon, Nigeria, South Africa. **Refs.:** Edwards 1941: 174 (M*, F), 393 (P); Hopkins 1952: 174 (L*); McIntosh 1975: 261 (M*, F*, L*; lectotype desig.); Jupp 1996 (M*, F*; key). **Ety.:** of the fern (L); larvae found in a small pool overshadowed by ferns. **Informal name:** Fern-haven South African Pointy Mosquito.

grenieri Hamon, Service, Adam & Taufflieb
1961a: 376 (M*). **Type loc.:** Daine, Man, Ivory Coast [Côte d'Ivoire] (IERT). **Distr.:** Cameroon, Côte d'Ivoire. **Ety.:** Paul Fernand Grenier. **Informal name:** Grenier Ivorian Pointy Mosquito.

hopkinsi Edwards
1936: 49 (M). **Type loc.:** Lira, Uganda (NHMUK). **Distr.:** Angola, Cameroon, Madagascar (includes Glorioso and Juan De Nova Isl.), Uganda. **Refs.:** Edwards 1941: 169 (M*, F), 393 (P); Haddow et al. 1951: 222 (M); Hopkins 1952: 171 (L*); Ribeiro & da Cunha Ramos 1973: 130 (distr.). **Ety.:** George Henry Evans Hopkins. **Informal name:** Hopkins Ugandan Pointy Mosquito.

insolens Edwards
1936: 50 (F). **Type loc.:** Arua, [Northern] Uganda (NHMUK). **Distr.:** Angola, Burkina Faso, Cameroon, Sierra Leone, Uganda. **Refs.:** Edwards 1941: 171 (F); Ribeiro & da Cunha Ramos 1973: 130 (distr.). **Ety.:** n.s. [arrogant? (L)]. **Informal name:** Proud Ugandan Pointy Mosquito.

irritans (Theobald)
1901d: iii (M*, F; *Stegomyia*). **Type loc.:** Bonny [Eastern Provinces], Nigeria (NHMUK). **Distr.:** Angola, Benin, Burkina Faso, Cameroon, Central African Republic, Congo, Democratic Republic of the Congo, Gambia, Ghana, Guinea Bissau, Kenya, Nigeria, Senegal, Uganda. **Refs.:** Macfie & Ingram 1922 (F*); Edwards 1941: 184 (M*, F*), 396 (P*); Hopkins 1952: 186 (L*); Reinert 2010d: Fig. 1 (F*). **Ety.:** n.s. [annoy, irritate (L)]. **Informal name:** Persistent Nigerian Pointy Mosquito.

syn. *senegalensis* Theobald
1903d: i (M*, F; *Catageiomyia*). **Type loc.:** Saint Louis, Senegal (NHMUK).

syn. *confusus* Theobald
1909b: 7 (F*; *Myxosquamus*). **Type loc.:** Accra [Ghana] (NHMUK).

syn. *albotaeniatus* Theobald
1909b: 5 (M*, F*; *Aedimorphus*). **Type loc.:** Accra, Ashanti, Ghana (NHMUK).

lokojensis Service
1959c: 244 (M*, P, L). **Type loc.:** Lokoja, Northern Nigeria (NHMUK). **Distr.:** Central African Republic, Nigeria. **Ety.:** Lokoja, Nigeria. **Informal name:** Lokoja Nigerian Pointy Mosquito.

lottei Hamon & Brengues
1965: 101 (M*). **Type loc.:** Tiassale, Ivory Coast [Côte d'Ivoire] (IERT). **Distr.:** Central African Republic, Côte d'Ivoire. **Ety.:** Michel Lotte. **Informal name:** Lotte Ivorian Pointy Mosquito.

microstictus Edwards
1936: 50 (M). **Type loc.:** Pretoria, Transvaal [South Africa] (NHMUK). **Distr.:** South Africa. **Refs.:** Edwards 1941: 170 (M*); McIntosh 1975: 258 (M*, F, L); Jupp 1996 (M*, F*; key). **Ety.:** n.s. [poss. ref. to small hit, blow (L)]. **Informal name:** Small-spotted South African Pointy Mosquito.

minutus (Theobald)
1901a: 319 (F*; *Stegomyia*). **Type loc.:** Salisbury, Mashonaland [Southern Rhodesia] (NHMUK). **Distr.:** Angola, Benin, Burkina Faso, Cameroon, Central African Republic, Congo, Côte d'Ivoire, Democratic Republic of the Congo, Ethiopia, Gambia, Ghana, Guinea, Liberia, Malawi, Mali, Mozambique, Namibia, Nigeria, Senegal, Sierra Leone, South Africa, Sudan and South Sudan, Togo, Zimbabwe. **Refs.:** Ingram & Macfie 1917b: 142 (L*); Edwards 1941: 177 (M*, F); Hopkins 1952: 178 (L*); McIntosh 1975: 260 (M*, F*, L; distr.); Jupp 1996 (M*, F*; key). **Ety.:** small, minute (L). **Informal name:** Small Zimbabwean Pointy Mosquito.

mixtus Edwards
1936: 49 (M). **Type loc.:** Kumasi [Ashanti], Ghana (NHMUK). **Distr.:** Benin, Burkina Faso, Côte d'Ivoire, Ghana, Mali, South Africa, Tanzania, Zimbabwe. **Refs.:** Edwards 1941: 170 (M*); Muspratt 1955: 164 (L); McIntosh 1975: 256 (M*, F*, L*); Jupp 1996 (M*, F*; key). **Ety.:** n.s. [mix, confound (L); poss. ref. to similarity with *punctothoracis*]. **Informal name:** Mixed Ghanaian Pointy Mosquito.

mutilus Edwards
1936: 50 (M). **Type loc.:** Mpanga Forest [Toro District], Uganda (NHMUK). **Distr.:** Cameroon, Central African Republic, Democratic Republic of the Congo, Uganda. **Refs.:** Edwards 1941: 170 (M*, F); van Someren 1950d: 65 (P*, L). **Ety.:** lost or stunted horns (L); ref. to gonostylus without a horn. **Informal name:** Hornless Ugandan Pointy Mosquito.

nyounae Hamon & Adam
1959b: 531 (M*). **Type loc.:** Adiopodoume, Abidjan, Ivory Coast [Côte d'Ivoire], French West Africa (IERT). **Distr.:** Cameroon, Central African Republic, Côte d'Ivoire. **Refs.:** Brengues & Hamon 1965: 92 (P*, L*); Harrison 1973c: 277 (type info.). **Ety.:** n.s. **Informal name:** Nyouna Ivorian Pointy Mosquito.

phyllolabis Edwards
1929c: 324 (M*, F). **Type loc.:** Stanleyville [Orientale], Zaire [Democratic Republic of the Congo] (NHMUK). **Distr.:** Cameroon, Congo, Côte d'Ivoire, Democratic Republic of the Congo, Kenya, Liberia, Uganda. **Refs.:** Edwards 1941: 176 (M*, F), 394 (P*), 430 (L*); Pao & Knight 1970a: 112 (L*). **Ety.:** n.s. [leaf lip (G); poss. ref. to a "leaf-like appendage" on the gonostylus]. **Informal name:** Leaf-lipped Congolese Pointy Mosquito.

pseudotarsalis van Someren
1946a: 119 (M*, F, P*, L*). **Type loc.:** Nairobi [Central], Kenya (NHMUK). **Distr.:** Cameroon, Congo, Democratic Republic of the Congo, Kenya. **Refs.:** Hopkins 1952: 179 (L*); Pao & Knight 1970a: 113 (L*). **Ety.:** false (Gr); similar to *tarsalis*. **Informal name:** False-legged Kenyan Pointy Mosquito.

punctothoracis (Theobald)
1909b: 4 (M, F*; *Aedimorphus*). **Type loc.:** Accra, Ashanti, Ghana (NHMUK). **Distr.:** Central African Republic, Côte d'Ivoire, Democratic Republic of the Congo, Gabon, Gambia, Ghana, Guinea Bissau, Liberia, Mali, Nigeria, Senegal, Sierra Leone, South Africa. **Refs.:** Macfie & Ingram 1922 (F); Edwards 1941: 168 (M*, F); Hamon, Abonnenc, & Noel 1955: 296 (M*); Service 1976a (distr.; Gabon). **Ety.:** spotted thorax (L); ref. to "six spots of small, flat, silvery white scales." **Informal name:** Spotty Ghanaian Pointy Mosquito.
syn. *punctithorax* Theobald
1910c: 205 (M*, F*; *Aedimorphus*). **Type loc.:** Accra [Ghana] (NHMUK).

reali Hamon & Adam
1959b: 531 (M*). **Type loc.:** Adiopodoume, Abidjan, Ivory Coast [Côte d'Ivoire], French West Africa (IERT). **Distr.:** Central African Republic, Côte d'Ivoire, Liberia. **Refs.:** Harrison 1973c: 277 (type info.). **Ety.:** P. Real. **Informal name:** Real Ivorian Pointy Mosquito.

smithburni van Someren
1950b: 39 (M*, F). **Type loc.:** Bwamba, Toro District, Uganda (NHMUK). **Distr.:** Uganda. **Ety.:** K. Smithburn. **Informal name:** Smithburn Ugandan Pointy Mosquito.

stenoscutus (Edwards)
1912c: 22 (*Ochlerotatus minutus* var.; new name for *africanus* Theobald, not Neveu-Lemaire 1906). **Distr.:** Cameroon, Ghana, Uganda, Zaire. **Refs.:** Edwards 1941: 178 (? = *congolensis*); White 1975: 305, 308 (from syn. with *neobiannulatus*; lectotype desig.); Reinert, Harbach, & Kitching 2009 (trans. from *Aedimorphus* to *Catageiomyia*). **Ety.:** n.s. [narrow scutum (Gr); poss. ref. to "scutellum bears a considerable number of narrow scales"]. **Informal name:** Narrow-scaled Ghanaian Pointy Mosquito.
syn. *africanus* Theobald
1909b: 8 (F*; *Stenoscutus*; not *Stegomyia africana* Theobald 1901d). **Type loc.:** Obuasi, Ashanti, Ghana (NHMUK). **Refs.:** White 1976 (syn.).
syn. *congolensis* Edwards
1927a: 351 (M*, F). **Type loc.:** Stanleyville, Belgian Congo [Kisangani, Tshopo Province, Democratic Republic of the Congo] (NHMUK & CMT). **Refs.:** Edwards 1941: 178 (M*, F); White 1975: 306 (syn.; lectotype desig.).

tarsalis (Newstead)
1907 (M*, F; *Duttonia*). In Newstead et al. 1907: 18. **Type loc.:** Kisui, Zaire [Democratic Republic of the Congo] (NHMUK). **Distr.:** Angola, Benin, Burkina Faso, Cameroon, Central African Republic, Congo, Côte d'Ivoire, Democratic Republic of the Congo, Ethiopia, Ghana, Guinea, Kenya, Liberia, Mali, Nigeria, Sierra Leone, South Africa, Sudan and South Sudan, Uganda, Zambia. **Refs.:** Edwards 1941: 174 (M*, F), 393 (P); Hopkins 1952: 176 (L); Pao & Knight 1970a: 114 (L*); McIntosh 1975: 265 (M*, F*, L*; distr.; holotype not found in NHMUK); Jupp 1996 (M*, F*; key). **Ety.:** n.s. [foot; flat surface between toes and heel (Gr); prob. ref. to tarsal claw characters]. **Informal name:** Flat-footed Congolese Pointy Mosquito.
syn. *biannulata* Theobald
1907: 263 (M*; *Reedomyia*). **Type loc.:** Nichol Brook, Mount Aureol, Sierra Leone (NHMUK) [This type male later incorrectly used as type male for *neobiannulata*]. **Refs.:** White 1975: 308 (M; holotype desig. as lectotype of *neobiannutatus*, tax.); Townsend 1990: 52 (lectotype desig.).
syn. *africana* Newstead
1907 (F*; *Duttonia*). In Newstead et al. 1907: 20. **Type loc.:** Kasongo, Zaire [Democratic Republic of the Congo] (NHMUK).

syn. *uniannulata* Theobald
1909b: 12 (F; *Neopecomyia*). **Type loc.:** Kumasi, Ashanti, Ghana (NHMUK).

syn. *bipunctata* Theobald
1909b: 5 (F*; *Reedomyia*). **Type loc.:** Obuasi, Ashanti, Ghana (NHMUK).

syn. *neobiannulata* Theobald
1910c: 255 (F; *Reedomyia*). **Type loc.:** Accra, Obuasi [Ghana] (NHMUK). **Refs.:** Edwards 1941: 178 (? = *congolensis*; type M is type of *biannulata* (= *tarsalis*)); White 1975: 305, 307 (M; holotype of *biannulata* desig. as lectotype, tax.).

syn. *sudanensis* Theobald
1913d: 595 (F; *Reedomyia*). **Type loc.:** Lado, Sudan (NHMUK).

veeniae McIntosh
1975: 263 (M*, F*, L*). **Type loc.:** Oslo Beach, South Africa (NIV). **Distr.:** South Africa. **Refs.:** Segerman 1990: 53 (type dep. from SAIM to NIV); Jupp 1996 (M*, F*; key). **Ety.:** Author's spouse, Veenia. **Informal name:** Veenia South African Pointy Mosquito.

wendyae Service
1959b: 73 (M*, F, P, L*). **Type loc.:** Ikarodu, near Lagos, S. Nigeria (NHMUK). **Distr.:** Nigeria. **Refs.:** Service 1960c: 90 (lectotype). **Ety.:** Wendy Service. **Informal name:** Wendy Nigerian Pointy Mosquito.

yangambiensis De Meillon & Lavoipierre
1944: 50 (M*). **Type loc.:** Yangambi [Orientale], Zaire [Democratic Republic of the Congo] (SAIM). **Distr.:** Angola, Benin, Cameroon, Central African Republic, Côte d'Ivoire, Democratic Republic of the Congo, Liberia, Nigeria. **Refs.:** De Meillon et al. 1945: 94 (P*, L*); Hopkins 1952: 179 (L); Ribeiro & da Cunha Ramos 1973: 134 (distr.). **Ety.:** Yangambi, Democratic Republic of the Congo. **Informal name:** Yangambi Congolese Pointy Mosquito.

yvonneae Edwards
1941: 176 (M*). **Type loc.:** Popakobaka, Kwango Province [Leopoldville], Zaire [Democratic Republic of the Congo] (NHMUK). **Distr.:** Cameroon, Democratic Republic of the Congo. **Ety.:** Yvonne Schwetz. **Informal name:** Yvonne Congolese Pointy Mosquito.

Subgenus *Catatassomyia* Dyar & Shannon

1925a: 71 (as genus). Type species: *Catatassomyia meronephada* Dyar & Shannon.

Name derivation: n.s. [poss. ref. to sublateral mesonotal sclerite completely covered with scales].

Important references: Edwards 1929d: 4 (syn. with subg. *Stegomyia* of genus *Aedes*); Knight & Hull 1952 (bion.); Baisas et al. 1962 (bion.); Huang 1978a (to syn. in subg. *Diceroymia*); Reinert, Harbach, & Kitching 2009 (phyl., class.; to genus); Wilkerson et al. 2015 (phyl., class.; to subg. of genus *Aedes*).

meronephada (Dyar & Shannon)
1925a: 71 (F; *Catatassomyia*). **Type loc.:** [Mount Makiling], Los Banos, Laguna, Luzon, Philippines (USNM). **Distr.:** Philippines. **Refs.:** Knight & Hull 1952: 174 (F, L*; type info., lectotype desig.); Stone & Knight 1956a: 221 (type info., lectotype desig.); Basio 1971b: 30 (M*; bion.), 12 (M*; bion.; as *wainwrighti*); Reinert Mattingly 1971a: Pl. 14 (P terminal segs.*); Huang 1978a: 335 (from subg. *Stegomyia*). **Ety.:** n.s. [poss. ref. to sublateral mesonotal sclerite covered with scales]. **Informal name:** Indecisive Philippine Pointy Mosquito [present authors undecided as to meaning].

syn. *wainwrighti* Baisas
1946c: 34 (F). **Type loc.:** Llavac, Infanta, Tayabas [= Quezon], Luzon, Philippines (NE). **Refs.:** Knight & Hull 1953: 464 (F); Reinert 1973a: 27 (syn.); Huang 1978b: 355 (from subg. *Stegomyia*).

Subgenus *Christophersiomyia* Barraud

1923b: 786 (as genus). Type species: *Stegomyia thomsoni* Theobald.

Name derivation: Samuel Rickart Christophers.

Important references: Barraud 1931a: 610 (as subg. of genus *Aedes*); Knight 1947b: 75 (key); Mattingly 1959: 51 (tax., keys; Indomalayan Area); Abercrombie 1977 (revision; as subg. of genus *Aedes*); Reinert 2000e (F gen.*); Reinert, Harbach, & Kitching 2004 (phyl., class.; to genus); Reinert, Harbach, & Kitching 2009 (phyl., class.); Rattanarithikul et al. 2010 (F*, L*; keys, bion.; Thailand); Wilkerson et al. 2015 (phyl., class.; to subg. of genus *Aedes*).

annulirostris (Theobald)
1905h: 239 (F; *Stegomyia*). **Type loc.:** Peradeniya [Central], Ceylon [Sri Lanka] (NHMUK). **Distr.:** India, Nepal, Sri Lanka, Thailand. **Refs.:** Barraud 1934: 215 (M*, F*); Peters & Dewar 1956: 39 (P*, L*); Abercrombie 1977: 7 (M*, F*, P*, L*; distr.); Darsie & Pradhan 1990, 1991 (F, L; tax., keys, bion., distr.; Nepal); Rattanarithikul et al. 2010 (F*, L*; keys, bion., distr.; Thailand). **Ety.:** ringed beak (L); proboscis. **Informal name:** Ring-snouted Sri Lankan Pointy Mosquito.

chionodes Belkin
1962: 434 (M*, F). **Type loc.:** Torokina, Bougainville, Solomon Islands (USNM). **Distr.:** Indonesia (West Papua), Solomon Isl. **Refs.:** Abercrombie 1977: 11 (M*, F*, P*, L*; distr.); Lee et al. 1982: 66 (tax., review, distr.). **Ety.:** snow white (Gr); ref. to thoracic scaling. **Informal name:** Snowy Solomon Islander Pointy Mosquito.

gombakensis Mattingly
1959: 56 (M*, F*, P*, L*). **Type loc.:** 16th mile, Ulu Gombak, Selangor, Malaya [Malaysia] (NHMUK). **Distr.:** Malaysia. **Refs.:** Mattingly 1971a: Pl. 32 (L head*); Abercrombie 1977: 14 (M*, F*, P*, L*). **Ety.:** Ulu Gombak, Selangor, Malaysia. **Informal name:** Gombak Malaysian Pointy Mosquito.

ibis Barraud
1931a: 610 (F). **Type loc.:** Sukna [Darjeeling District, West], Bengal, India (NHMUK). **Distr.:** Cambodia, India, Malaysia, People's Republic of China, Philippines, Thailand, Vietnam. **Refs.:** Basio 1971b: 13 (M*; bion.; as *brayi*); Abercrombie 1977: 17 (M*, F*, P*, L*; distr.); Rattanarithikul et al. 2010 (F*, L*; keys, bion., distr.; Thailand). **Ety.:** n.s. [wading bird? (L)]. **Informal name:** Ibis Indian Pointy Mosquito.
syn. *brayi* Knight
1947b: 73 (M*, F). **Type loc.:** San Jose, Nueva Ecija Province, Luzon, Philippines (USNM). **Refs.:** Abercrombie 1977: 17 (syn.).

thomsoni (Theobald)
1905e: 18 (A; *Stegomyia*). **Type loc.:** North West Provinces [NFWP, Pakistan] (NHMUK). **Distr.:** Bangladesh, India, Nepal, Pakistan, Philippines, Sri Lanka, Thailand. **Refs.:** Barraud 1923h (L*); Barraud 1934: 213 (M*, F*, L*); Peters & Dewar 1956: 40 (P*, L*); Aslamkhan 1971b (distr.; Pakistan); Abercrombie 1977: 21 (M*, F*, P*, L*; distr.); Darsie & Pradhan 1990, 1991 (F, L; tax., keys, bion., distr.; Nepal); Reinert 2000e: Fig. 16 (F*); Rattanarithikul et al. 2010 (F*, L*; keys, bion., distr.; Thailand). **Ety.:** n.s. [poss. ref. to Carl Gustaf Thomson]. **Informal name:** Thomson Pakistani Pointy Mosquito.

Subgenus *Coetzeemyia* Huang, Mathis, & Wilkerson

2010: 14 (as subg. of genus *Aedes*). Type species: *Culiseta fryeri* Theobald.

Name derivation: Maureen Coetzee.

Important references: Huang & Rueda 2014 (key to subgenera *Ochlerotatus* and *Coetzeemyia*; Afrotropical); Wilkerson et al. 2015 (phyl., class.).

fryeri (Theobald)
1912b: 84 (F; *Culicelsa*). **Type loc.:** Takamaka, Aldabra Island [Madagascar] (NHMUK). **Distr.:** Eritrea, French Southern Territories, Kenya, Madagascar (includes Glorioso and Juan De Nova Isl.), Mozambique, Seychelles, Tanzania. **Refs.:** Edwards 1941: 116 (M*, F); Hopkins 1952: 123 (L); Mattingly 1963: 165 (L*); Danilov 1981a: 86 (from subg. *Ochlerotatus*); Jupp 1996 (M*, F*; key); Huang et al. 2010: 1 (M*, F*; from subg. *Levua*; type info.). **Ety.:** J.C.F. Fryer. **Informal name:** Fryer Malagasy Pointy Mosquito.
syn. *mombasaensis* Mattingly
1963: 165 (M, F, L*). **Type loc.:** Mombasa, Kenya (NHMUK). **Refs.:** van Someren 1972: 90 (syn.).

Subgenus *Collessius* Reinert, Harbach, & Kitching

2006: 64 (as genus). Type species: *Ochlerotatus macfarlanei* Edwards.

Name derivation: Donald Henry Colless.

syn. *Alloeomyia* Reinert, Harbach, & Kitching 2008: 97, as subg. of genus *Collessius*. Type species: *Culex pseudotaeniatus* Giles.

Important references: Reinert 2008l (F gen.*); Reinert, Harbach, & Kitching 2009 (phyl., class.); Rattanarithikul et al. 2010 (F*, L*; keys, bion.; Thailand); Wilkerson et al. 2015 (phyl., class.; to subg. of genus *Aedes*); Soghigian et al. 2017 (mol. phyl.).

banksi Edwards
1922a: 270 (M, F). **Type loc.:** Montalban, Rizal, Luzon, Philippines (NHMUK). **Distr.:** Philippines. **Refs.:** Knight & Hull 1951b: 243 (M*, F, L*); Basio 1971b: 14 (M*; bion.). **Ety.:** Charles Summer Banks. **Informal name:** Banks Philippine Pointy Mosquito.

elsiae (Barraud)
1923a: 406 (M, F; *Finlaya*). **Type loc.:** Shiilong, Assam, India (NHMUK). **Distr.:** Cambodia, India, Malaysia, Nepal, People's Republic of China, Taiwan, Thailand, Vietnam. **Refs.:** Barraud 1923h (L*); Barraud 1934: 180 (M*, F*, L*); Ramalingam & Pillai 1973 (distr.); Darsie et al. 1993 (distr.; Nepal); Darsie 2010 (P*); Rattanarithikul et al. 2010 (F*, L*; keys, bion., distr.; Thailand). **Ety.:** n.s. **Informal name:** Elsie Indian Pointy Mosquito.
syn. *simulatus* Barraud
1931a: 611 (F). **Type loc.:** Haflong, Cachar Hills, Assam, India (MSI). **Refs.:** Barraud 1934: 180 (syn.).
ssp. ***vicarius*** Lien
1968d: 218 (M*, F*, P*, L*). **Type loc.:** Chuchi, Chiai Hsien, Taiwan (TMRI). **Distr.:** Taiwan. **Ety.:** n.s. [substitute, in place of (L); poss. ref. to similarity to *elsiae*]. **Informal name:** Substitute Taiwanese Pointy Mosquito.

hatorii Yamada
1921: 70 (M). **Type loc.:** Taihoku, Formosa [Taiwan, Republic of China] (IID). **Distr.:** China, Japan, South Korea, Taiwan, Thailand. **Refs.:** La Casse & Yamaguti 1950: 32 (M*, F*, P*, L*; bion., distr., tax.); Hara 1957: 60 (F*); Moriya et al. 1972 (E*); Moriya et al. 1973 (E*); Tanaka et al. 1979: 327 (M*, F*, L*); Tanaka 2002b: 141

(P*; tax., key). **Ety.:** Juro Hatori. **Informal name:** Hatori Taiwanese Pointy Mosquito.

macdougalli Edwards
1922a: 271 (M, F). **Type loc.:** Diyatalawa, Ceylon [Sri Lanka] (NHMUK). **Distr.:** India, Malaysia, Sri Lanka. **Refs.:** Barraud 1934: 177 (M, F*, L); Harrison et al. 1991: 212 (restr., distr.; Thailand); Darsie et al. 1993 (distr.; Nepal). **Ety.:** A.J. MacDougall. **Informal name:** MacDougall Sri Lankan Pointy Mosquito.

macfarlanei (Edwards)
1914b: 78 (M, F; *Ochlerotatus*). **Type loc.:** Hong Kong [People's Republic of China] (NHMUK). **Distr.:** Cambodia, Indonesia, Laos, Malaysia, People's Republic of China, Thailand, Vietnam. **Refs.:** Borel 1930: 251 (M*, F, L*); Barraud 1934: 180 (A), 181 (L); Abu-Hassan et al. 1992: 107 (distr.); Reinert 2008l: Fig. 1 (F*); Rattanarithikul et al. 2010 (F*, L*; keys, bion., distr.; Thailand). **Ety.:** H. Macfarlane. **Informal name:** Macfarlane Chinese Pointy Mosquito.
syn. *hegneri* Causey
1937a: 543 (M*, F). **Type loc.:** Chang, Thailand (LU). **Refs.:** Harrison et al. 1991: 212 (syn.).

ningheensis Lei
1989b: 100 (M*, F, L*) [type not designated in manuscript]. **Type loc.:** Wu Shan County (Hsien), Sichuan Province, China [People's Republic of China] (Chinese Academy of Medical Sciences, Institute of Parasitic Diseases and Control, Sichuan Province). **Distr.:** People's Republic of China. **Ety.:** n.s [prob. ref. to "Small Ninghe River" of Wu Shan Co. (Hsien), Si Chuan Prov. (Harbach 2018c)]. **Informal name:** Ninghe Chinese Pointy Mosquito.

pseudotaeniatus (Giles)
1901a: 607 (A; *Culex*). **Type loc.:** Bakloh [Punjab Hill States] and Naini Tal [United Provinces], India (NHMUK). **Distr.:** Bangladesh, Cambodia, India, Myanmar, Nepal, Pakistan, Sri Lanka, Thailand. **Refs.:** Barraud 1923h (L*); Barraud 1934: 178 (M*, F*, L*); Joshi et al. 1965 (distr.; Nepal); Aslamkhan 1971b (distr.; Pakistan); Ahmed 1987 (distr.; Bangladesh); Darsie & Pradhan 1990, 1991 (F, L; tax., keys, bion., distr.; Nepal); Suleman et al. 1993 (distr.; Pakistan); Darsie 1997: 286 (P*); Rattanarithikul et al. 2010 (F*, L*; keys, bion., distr.; Thailand). **Ety.:** false (Gr); similar to *taeniatus*. **Informal name:** False-ribboned Indian Pointy Mosquito.

shortti (Barraud)
1923a: 405 (M, F; *Finlaya*). **Type loc.:** Shillong, Khasi Hills, Assam, India (NHMUK). **Distr.:** India, Indonesia, Malaysia, Nepal, Pakistan, Thailand. **Refs.:** Barraud 1934: 183 (M*, F, L*); Aslamkhan 1971b (distr.; Pakistan); Darsie et al. 1993 (distr.; Nepal); Darsie 2010 (P*); Rattanarithikul et al. 2010 (F*, L*; keys, bion., distr.; Thailand). **Ety.:** n.s. **Informal name:** Shortt Indian Pointy Mosquito.

tonkinensis Galliard & Ngu
1947: 77 (M, L). **Type loc.:** Baie d'Along, Tonkin, Indochina [Vietnam] (NE). **Distr.:** Vietnam. **Refs.:** Galliard & Ngu 1950: 64 (M*, L*). **Ety.:** Tonkin, Vietnam. **Informal name:** Tonkin Vietnamese Pointy Mosquito.

Subgenus *Cornetius* Huang

2005 (as subg. of genus *Aedes*). Type species: *Aedes* (*Stegomyia*) *cozi* Cornet.

Name derivation: Michel Cornet.

Important references: Reinert, Harbach, & Kitching 2009 (phyl., class.; to genus); Wilkerson et al. 2015 (phyl., class.; to subg. of genus *Aedes*).

cozi Cornet
1973: 175 (M*, F*, P*, L*). **Type loc.:** Kédougou, Eastern Senegal, Senegal (IERT). **Distr.:** Senegal. **Refs.:** Huang 2005 (M*, F*, P*, L*; transfer from *Aedes* (*Stegomyia*) to new subg. *Cornetius*). **Ety.:** Jean Coz. **Informal name:** Coz Senegalese Pointy Mosquito.

Subgenus *Dahliana* Reinert, Harbach, & Kitching

2006: 66 (as genus). Type species: *Culex geniculatus* Olivier.

Name derivation: Christine Dahl.

Important references: Reinert 2008c (F gen.*); Reinert, Harbach, & Kitching 2009 (phyl., class.); Wilkerson et al. 2015 (phyl., class.; to subg. of genus *Aedes*).

echinus (Edwards)
1920a: 133 (F, L; *Ochlerotatus*). **Type loc.:** Stavros, near Salonica, Macedonia, Greece (NHMUK). **Distr.:** Albania, Algeria, Bulgaria, Cyprus, France, Greece, Iran, Israel, Italy (includes Sardinia and Sicily), Jordan, Macedonia, Montenegro, Morocco, Portugal, Spain, Turkey. **Refs.:** Edwards 1921d: 320 (L*); Martini 1930: 287 (M*, F, L*); Senevet & Andarelli 1958: 276 (P*); Mohrig 1967 (F*); Gutsevich et al. 1974: 299 (M, F, L*; distr.); Zaim 1987: 568 (distr.); Melero-Alcíbar et al. 2009: 2 (P; key); Becker et al. 2010: 203 (M*, F*, L*; keys, tax., distr., bion.); Robert et al. 2019 (distr.; Western Palearctic). **Ety.:** prickly, spiny (L); stellate setae on larva. **Informal name:** Urchin-like Greek Pointy Mosquito.

geniculatus (Olivier)
1791: 134 (A; *Culex*). **Type loc.:** Paris, France (NE). **Distr.:** Albania, Armenia, Austria, Azerbaijan, Belarus, Belgium, Bulgaria, Corsica, Crimean Peninsula, Croatia,

Czech Republic, Denmark, Finland, France (includes Corsica), FYRO Macedonia, Georgia, Germany, Greece, Hungary, Iran, Italy (includes Sardinia and Sicily), Kaliningrad Oblast, Kazakhstan, Lebanon, Lithuania, Luxembourg, Moldova, Montenegro, Morocco, Netherlands, Norway, Poland, Portugal, Romania, Russia (Central, northwestern, Southern, and Volga regions), Serbia, Slovakia, Slovenia, Spain, Sweden, Switzerland, Turkey, Turkmenistan, Ukraine, United Kingdom. **Refs.:** Brolemann 1919b (F*); Edwards 1921d: 319 (F*, L*); Séguy 1924: Pl. XXI (L*), 71 (M gen.*), 83 (F*); MacGregor 1932: 183 (E; bion.); Marshall 1938: 150 (M*, F*, L*, E*); Clavero 1946: 20 (M*, P*, L*); Natvig 1948: 389 (M*, F, L*; tax., bion.); Mohrig 1967 (F*); Sicart & Larrouy 1968 (F*); Mohrig 1969 (F*); Rjazantzeva 1970 (F*); Gutsevich et al. 1974: 295 (M*, F*, L*); Skierska 1977 (F*); Mgeladze 1991: 329 (distr.; Georgia); Alten et al. 2000 (distr.; Turkey); Eritja et al. 2000: 11 (distr.; Spain); Dahl & Blackmore 2001: 14 (distr.; Denmark, Norway, and Sweden); Reinert 2002e: Fig. 3 (F*); Trari et al. 2002: 331 (distr.; Morocco); Beck et al. 2003: 24 (distr.; Luxembourg); Reinert 2008c: Fig. 1 (F*); Becker et al. 2010: 206 (M*, F*, L*; keys, tax., distr., bion.); Namazov 2014 (distr.; Azerbaijan); Robert et al. 2019 (distr.; Western Palearctic). **Ety.:** knee, jointed (L); ref. to white tibiofemoral joint. **Informal name:** White-kneed French Pointy Mosquito. (Vol. 1, p. 406)

syn. *guttatus* Meigen
1818: 5 (A; *Culex*). **Type loc.:** Europe (NE).

syn. *lateralis* Meigen
1818: 5 (A; *Culex*). **Type loc.:** Austria (NE).

syn. *ornatus* Meigen
1818: 5 (A; *Culex*). **Type loc.:** Europe (NE).

syn. *guttatus* Curtis
1839: 537 (M*, F*; *Culex*). **Type loc.:** Cobham, Surrey, England (NMM).

syn. *albo-punctatus* Rondani
1872: 31 (M; *Culex*). **Type loc.:** Italy (LU).

gilcolladoi (Sánchez-Covisa, Rodríguez Rodríguez, & Guillén Llera)
1985: 441 (L; holotype not desig.). **Type loc.:** Casa de Campo, Madrid, Spain (NE). **Distr.:** Spain. **Refs.:** Merlero-Alcíbar & Molina 2007 (P*; tax.). **Ety.:** Juan Gil Collado. **Informal name:** Gil Collado Spanish Pointy Mosquito.

Subgenus *Danielsia* Theobald

1904c: 78 (as genus). Type species: *Danielsia albotaeniata* Leicester.

Name derivation: n.s. [poss. Charles Wilberforce Daniels (see also Harbach 2018c)]

Important references: Reinert 2008e (F gen.*); Reinert, Harbach, & Kitching 2008 (phyl., class.; from syn. of *Aedes* subg. *Finlaya* to genus); Reinert 2009e (redescr.); Reinert, Harbach, & Kitching 2009 (phyl., class.); Rattanarithikul et al. 2010 (F*, L*; keys, tax., bion.; Thailand); Wilkerson et al. 2015 (phyl., class.; to subg. of genus *Aedes*); Soghigian et al. 2017 (mol. phyl.).

albotaeniatus (Leicester)
1904c: 111 (M, F; *Danielsia*). In Theobald 1904c. **Type loc.:** Chang Road, Kuala Lumpur [Selangor], Malaya [Malaysia] (NHMUK). **Distr.:** India, Indonesia, Malaysia, Nepal, Sri Lanka, Thailand. **Refs.:** Barraud 1923h (F*; as *mikirana*) Barraud 1934: 174 (M*, F); Chow & Mattingly 1951: 87 (M, P*, L*; as var. *mikiranus*); Darsie et al. 1993 (distr.; Nepal); Reinert 2008e: Fig. 1 (F*); Reinert 2009e (F); Rattanarithikul et al. 2010 (F*, L*; keys, bion., distr.; Thailand). **Ety.:** n.s. [white ribbon (L); prob. ref. to white stripes on legs]. **Informal name:** White-banded Malaysian Pointy Mosquito.

syn. *mikiranus* Edwards
1922a: 269 (F; *Aedes* (*Finlaya*)). **Type loc.:** Mikir Hills, Sibsagar District, Assam, India (NHMUK). **Refs.:** Barraud 1934: 176 (F*; to var.); Harrison et al. 1991: 211 (syn.).

harperi Knight
1948a: 4 (M*, F, L*). **Type loc.:** Zig-Zag Pass, Subic Bay, Zambales, Luzon, Philippines (USNM). **Distr.:** Philippines. **Refs.:** Basio 1971b: 15 (M*; bion.); Baisas 1974: 10 (M, F*, P, L; tax., bion., distr.; Philippines). **Ety.:** Paul A. Harper. **Informal name:** Harper Philippine Pointy Mosquito.

lepchanus (Barraud)
1923d: 217 (M; as *Finlaya lepchana*). **Type loc.:** Tindharia, Darjeeling Hills [West Bengal], India (NHMUK). **Distr.:** India. **Refs.:** Barraud 1923h (M*); Knight 1948a: 2 (from syn. with *albotaeniatus*); Reinert 2008m (tax., type info.). **Ety.:** n.s. [Lepcha people; Lepchas of Darjeeling, India (Harbach 2018c)]. **Informal name:** Harvestman Indian Pointy Mosquito [*Lepchana* is a genus of Opiliones arachnids].

Subgenus *Dendroskusea* Edwards

1929c: 341 (as genus). Type species: *Culex micropterus* Giles.

Name derivation: n.s.; dendro- = a tree (Gr) and patronymic Skuse; meaning unclear.

Important references: Edwards 1932a: 172 (syn. with subg. *Diceromyia* of genus *Aedes*); Reinert, Harbach, & Kitching 2009 (phyl., class.; to genus); Wilkerson et al. 2015 (phyl., class.; to subg. of genus *Aedes*).

kanarensis Edwards
1934 (F). In Barraud 1934: 444. **Type loc.:** Yellapur, N Kanara [Bombay], India (NHMUK). **Distr.:** India. **Refs.:**

Knight & Hull 1953: 477 (tax.); Mattingly 1959: 35; Mattingly 1965b: 66 (provisional subgeneric placement); Reinert 1970a (tax.); Tewari et al. 1990 (M*, P*, L*). **Ety.:** Kanara, India. **Informal name:** Kanara Indian Pointy Mosquito.

micropterus (Giles)
1901a: 609 (A; *Culex*). **Type loc.:** Allahabad and Lucknow [United Provinces], India (NHMUK). **Distr.:** India, Nepal, Pakistan. **Refs.:** Barraud 1934: 275 (M*, F, L*); Aslamkhan 1971b (distr.; Pakistan); Darsie et al. 1996 (distr.; Nepal). **Ety.:** small wings (Gr); "wings proportionately small." **Informal name:** Small-winged Indian Pointy Mosquito.

periskelatus (Giles)
1902: 371 (M*; *Stegomyia*). **Type loc.:** Shahjahanpur [United Provinces], India (NE). **Distr.:** India, Pakistan. **Refs.:** Barraud 1934: 272 (F); Mattingly 1971a: Pl. 15 (P terminal segs.*); Aslamkhan 1971b (distr.; Pakistan). **Ety.:** n.s. [dried up, withered, mummified (Gr); prob. ref. to condition of specimen (Harbach 2018c)]. **Informal name:** Skeletal Indian Pointy Mosquito [author interpretation].
syn. *annulifemur* Edwards
1914b: 77 (M; *Ochlerotatus*). **Type loc.:** Jhansi [United Provinces], India (NHMUK). **Refs.:** Qutubuddin 1945: 34 (F); Khokhar & Tariq 1966: 117 (M*, F, P*, L*).

ramachandrai Reuben
1967: 234 (M*, F*). **Type loc.:** Musalimadugu Forest, Chittoor District, Andhra Pradesh, India (NHMUK). **Distr.:** India, Sri Lanka. **Refs.:** Reinert 2000e: Fig. 19 (F*). **Ety.:** T. Ramachandra. **Informal name:** Ramachandra Indian Pointy Mosquito.

reginae Edwards
1922a: 272 (M). **Type loc.:** Colombo, Ceylon [Sri Lanka] (NHMUK). **Distr.:** Bangladesh, India, Sri Lanka. **Refs.:** Barraud 1934: 277 (M, F, L); Reinert 1970a: 15 (M*, F*, P*, L*); Aslamkhan 1971b (distr.; Pakistan). **Ety.:** n.s. [queen (L)]. **Informal name:** Regina Sri Lankan Pointy Mosquito.

Subgenus *Diceromyia* Theobald

1911c: 151. Type species: *Diceromyia africana* Theobald (Edwards 1932a) (as genus).

Name derivation: Two-horned fly (Gr).

Important references: Barraud 1934: 271 (tax., keys; British India); Mattingly 1959: 34 (tax., keys; Indomalayan Area); Reinert 1970a (revision; SE Asia); Reinert 1970l: 129 (zoogeography); Huang 1986a (tax.; as subg. of genus *Aedes*; Furcifer Group); Jupp 1996 (tax.; southern Africa); Reinert 2000e (in part, F gen.*); Reinert, Harbach, &Kitching 2004: 289 (phyl., class.; to genus); Reinert, Harbach, & Kitching 2009 (phyl., class.); Brunhes et al. 2017 (in part, as genus *Paulianius*); Wilkerson et al. 2015 (phyl., class.; to subg. of genus *Aedes*).

adersi (Edwards)
1917: 214 (M, F; *Ochlerotatus* (*Diceromyia*)). **Type loc.:** Mnazi Moja, Zanzibar (NHMUK). **Distr.:** Kenya, Mozambique, South Africa, Tanzania. **Refs.:** Edwards 1941: 216 (M*, F); Hopkins 1952: 215 (L*); Worth & De Meillon 1960: 244 (distr.); Jupp 1996 (M*, F*; key); Huang & Rueda 2016 (F*; key). **Ety.:** W.M. Aders. **Informal name:** Aders Tanzanian Pointy Mosquito.

bananea Wolfs
1958: 298 (F). **Type loc.:** Banana, Bas-Congo, Zaire [Democratic Republic of the Congo] (CMT). **Distr.:** Democratic Republic of the Congo. **Refs.:** Huang & Rueda 2016 (F*; key). **Ety.:** Banana, Bas-Congo, Zaire. **Informal name:** Banana Congolese Pointy Mosquito.

cordellieri Huang
1986a: 636 (M*, F*, P*, L*). **Type loc.:** Dezidougou, M'Bahiakro, Centre Departement, Ivory Coast [Côte d'Ivoire] (USNM). **Distr.:** Côte d'Ivoire, Kenya, Republic of South Africa, Tanzania, Uganda. **Refs.:** Jupp 1996 (M*, F*; key); Jupp 1998 (tax., comp. morph.); Huang & Rueda 2016 (F*; key). **Ety.:** Roger Cordellier. **Informal name:** Cordellier Ivorian Pointy Mosquito.

fascipalpis (Edwards)
1912c: 19 (F*; *Ochlerotatus*). **Type loc.:** Little Ruaha River, Susangu District, German East Africa [Tanzania] (NHMUK). **Distr.:** Angola, Botswana, Malawi, Mozambique, Republic of South Africa, Tanzania, Zambia, Zimbabwe. **Refs.:** Edwards 1941: 217 (M*, F*), 402 (P); Hopkins 1952: 217 (L*); Muspratt 1959: 70 (M*); Worth & De Meillon 1960: 244 (distr.); Ribeiro & da Cunha Ramos 1973: 135 (distr.); Jupp 1996 (M*, F*; key); Huang & Rueda 2016 (F*; key). **Ety.:** n.s. [bundled feeler (L)]. **Informal name:** Bundle-lipped Tanzanian Pointy Mosquito.

flavicollis Edwards
1928a: 269 (M, F, P, L). **Type loc.:** Lagos, Nigeria (NHMUK). **Distr.:** Nigeria. **Refs.:** Edwards 1941: 217 (M*, F*), 401 (P); Hopkins 1952: 219 (L); Huang & Rueda 2016 (F*; key). **Ety.:** n.s. [yellow neck? (L); poss. ref. to yellow scales on head]. **Informal name:** Yellow-collared Nigerian Pointy Mosquito.

furcifer (Edwards)
1913b: 48 (new name for *nigra* Theobald, not Giles 1904b; *Ochlerotatus*). **Distr.:** Burkina Faso, Central African Republic, Côte d'Ivoire, Ethiopia, Gambia, Ghana, Guinea Bissau, Kenya, Mali, Nigeria, Senegal, Sudan and South Sudan, Tanzania, Togo, Uganda. **Refs.:** Edwards 1941: 214 (M*, F), 401 (P*); Lewis 1942a: 153 (P, L); Hopkins 1952: 213 (L); Reinert 1972b: 87 (E*); Port et al. 1981: 163 (distr., the Gambia); Ferrara et al. 1985: 179 (F*;

furcifer/taylori comparison); Hervy et al. 1985 (L*; *furcifer/taylori* comparison); Huang 1986a: 634 (tax.; Furcifer Group); Jupp et al. 1993 (*s.l.*; tax., distr.); Jupp 1996 (M*, F*; key); Jupp 1998 (tax., comp. morph.); Reinert 2000e: Fig. 17 (F*); Huang & Rueda 2016 (F*; key). **Ety.:** fork bearing (L); ref. to "forked male claspers." **Informal name:** Forked Sudanese Pointy Mosquito. (Vol. 1, p. 408)

syn. *nigra* Theobald
1906b: 80 (F*; *Mansonia*?). **Type loc.:** Blue Nile, Sudan (NHMUK).

syn. *africana* Theobald
1911c: 151 (M*, F*; *Diceromyia*; not *Stegomyia africana* Theobald 1901d). **Type loc.:** Nasser, Sobat River, Sudan (NHMUK).

mefouensis Ferrara
1974: 3 (M*, F). **Type loc.:** Zoatoupi, District of Mbankomo, Department of Mefou (03°49′ N, 011°23′ E), Cameroon (IERT). **Distr.:** Cameroon. **Refs.:** Huang & Rueda 2016 (F*; key). **Ety.:** Department of Méfou, Cameroon. **Informal name:** Méfou Cameroonian Pointy Mosquito.

taylori Edwards
1936: 55 (M). **Type loc.:** Gadau, Nigeria (NHMUK). **Distr.:** Burkina Faso, Côte d'Ivoire, Mali, Namibia, Nigeria, Senegal, Sudan and South Sudan, Zambia, Zimbabwe. **Refs.:** Edwards 1941: 216 (M*, F); Lewis 1942a: 153 (P, L); Hopkins 1952: 213 (L*); Muspratt 1955: 175 (M, L); van Someren, Teesdale, & Furlong 1955: 481 (tax.); Port et al. 1981: 163 (distr.; the Gambia); Ferrara et al. 1985: 179–183 (F*; *furcifer/taylori* comparison); Hervy et al. 1985: 17–24 (L*; *furcifer/taylori* comparison); Huang 1986a: 634 (M*; tax.; Furcifer Group); Jupp 1996 (M*, F*; key); Huang & Rueda 2016 (F*; key). **Ety.:** A.W. Taylor. **Informal name:** Taylor Nigerian Pointy Mosquito. (Vol. 1, p. 408)

Subgenus *Dobrotworskyius* Reinert, Harbach, & Kitching

2006: 68 (as genus). Type species: *Aedes tubbutiensis* Dobrotworsky.

Name derivation: Nikolas Vladimir Dobrotworsky.

Important references: Reinert 2008k (F gen.*); Reinert, Harbach, & Kitching 2009 (phyl., class.); Wilkerson et al. 2015 (phyl., class.; to subg. of genus *Aedes*); Soghigian et al. 2017 (mol. phyl.).

alboannulatus (Macquart)
1850: 314 (F; as *Culex*). **Type loc.:** Côte Orientale, Nouvelle-Hollande [Australia] (MNHP). **Distr.:** Australia. **Refs.:** Theobald 1901c: 389 (M*, F); Bancroft 1908: 35 (M, F); Lee 1944a: 60 (L*); Dobrotworsky 1959: 132 (M*, F, L*, E*); Pillai 1962 (E*); Dobrotworsky 1965: 137 (M*, F*, L*); Linley et al. 1991c (E*). **Ety.:** n.s. [white ringed (L); poss. ref. to white band on hind legs]. **Informal name:** White-ringed Australian Pointy Mosquito.

milsoni (Taylor)
1915: 179 (M, F; *Culicada*). **Type loc.:** Milson Island [Hawkesbury River], New South Wales, Australia (US). **Distr.:** Australia. **Refs.:** Dobrotworsky 1962: 301 (P*, L*; as sp.); Dobrotworsky 1965: 150 (M*, F*, L*); Lee et al. 1982: 189 (F; key, tax., bion., distr., review). **Ety.:** Milson Isl., Australia. **Informal name:** Milson Australian Pointy Mosquito.

syn. *milsoni* Taylor
1916: 566 (F; *Hulecoeteomyia*). **Type loc.:** Milson Island [Hawkesbury River], New South Wales, Australia (US).

occidentalis (Skuse)
1889: 1729 (F; *Culex*). **Type loc.:** King Georges Sound, Western Australia (MM). **Distr.:** Australia. **Refs.:** Bancroft 1908: 39 (M, F); Lee 1944a: 60 (L*). **Ety.:** n.s. [coming from the West (L); from Western Australia]. **Informal name:** West Australian Pointy Mosquito.

rubrithorax (Macquart)
1850: 313 (F; as *Culex*). **Type loc.:** Tasmania, Australia (MNHP). **Distr.:** Australia, Borneo, Brunei. **Refs.:** Bancroft 1908: 36 (F; *rubithorax* [*sic*]); Mackerras 1927: 297 (M*; tax., type loc. corrected to New South Wales); Dobrotworsky 1959: 134 (M*, F, L*, E*; as *queenslandis*); Klein & Marks 1960: 109 (F*; transfer from subg. *Ochlerotatus*; syn.); Dobrotworsky 1965: 140 (M*, F*, L*, E*); Lee et al. 1982: 247 (F; key, tax., bion., distr., review); Linley et al. 1991c (E*). **Ety.:** red thorax (L); color of thorax. **Informal name:** Ruby-chested Australian Pointy Mosquito.

syn. *similis* Strickland
1911a: 132 (F; *Culicelsa*). **Type loc.:** Mount Lofty, South Australia, Australia (NHMUK).

syn. *queenslandis* Strickland
1911a: 179 (F; *Culicelsa*). **Type loc.:** South Queensland, Australia (NHMUK). **Refs.:** Lee et al. 1957: 363 (to sp.; bion.); Lee et al. 1982 (syn.).

syn. *demansis* Strickland
1911a: 202 (F; *Culicada*). **Type loc.:** Tasmania, Australia (NHMUK).

syn. *cumpstoni* Taylor
1914c: 692 (F; *Culicada*). **Type loc.:** Melbourne, Victoria, Australia (US).

syn. *hybrida* Taylor
1916: 568 (F; *Culicada*). **Type loc.:** Milson Island [Hawkesbury River], New South Wales, Australia (US).

rupestris Dobrotworsky
1959: 136 (M*, F, P*, L*, E*). **Type loc.:** Lorne, Victoria, Australia (NMM). **Distr.:** Australia. **Refs.:** Dobrotworsky 1965: 143 (M*, F*, L*); Lee et al. 1982: 255 (F; key, tax.,

bion., distr., review). **Ety.:** n.s. [rock (L); prob. ref. to larvae in rock pools]. **Informal name:** Stone-reared Australian Pointy Mosquito.

subbasalis Dobrotworsky
1962: 298 (M*, F, P*, L*). **Type loc.:** Ginninderra Falls, New South Wales, Australia (CSIR). **Distr.:** Australia, Indonesia. **Refs.:** Dobrotworsky 1965: 148 (M*, F*, L*); Lee et al. 1982: 276 (F; key, tax., bion., distr., review). **Ety.:** nearly (L); near *basalis*. **Informal name:** Moth Australian Pointy Mosquito [many moths share the same name].

tubbutiensis Dobrotworsky
1959: 139 (M*, F, P*, L*). **Type loc.:** Tubbut, E. Gippsland, Victoria, Australia (NMM). **Distr.:** Australia. **Refs.:** Dobrotworsky 1965: 146 (M*, F, L*); Lee et al. 1982: 280 (F; key, tax., bion., distr., review); Reinert 2008k: Fig. 1 (F*). **Ety.:** Tubbut, Victoria, Australia. **Informal name:** Tubbut Australian Pointy Mosquito.

Subgenus *Downsiomyia* Vargas

1950c: 63 (as subg. of genus *Aedes*). Type species: *Stegomyia nivea* Ludlow.

Name derivation: Wilbur George Downs.

Important references: Knight 1946 (as Niveus Subgroup); Colless 1958 (as Niveus Subgroup); Colless 1959 (as Niveus Subgroup); Stone et al. 1959: 158 (as syn. of subg. *Finlaya* of genus *Aedes*); Reinert, Harbach, & Kitching 2004: 326, 357 (to genus from syn. with *Aedes* subg. *Finlaya*); Reinert & Harbach 2006 (revision, redescr.); Reinert 2008h (F gen.*); Reinert, Harbach, & Kitching 2009 (phyl., class.); Rattanarithikul et al. 2010 (F*, L*; keys, bion.; Thailand); Wilkerson et al. 2015 (phyl., class.; to subg. of genus *Aedes*); Soghigian et al. 2017 (mol. phyl.).

albolateralis (Theobald)
1908b: 289 (F; *Stegomyia*). **Type loc.:** Sylhet and Lungleh, Lushai Hills, Assam [Bangladesh] and India (IM). **Distr.:** Bangladesh, Cambodia, India, Indonesia (includes West Papua), Japan, Malaysia, Myanmar, Nepal, Pakistan, People's Republic of China, Philippines, South Korea, Sri Lanka, Taiwan, Thailand, Vietnam. **Refs.:** Barraud 1934: 205 (M*, F, L*; from syn. with *niveus*); Knight 1946: 279 (M*); Colless 1958: 479 (M*, L*); Colless 1959: 173 (M, F, L); Aslamkhan 1971b (distr.; Pakistan); Matsuo, Yoshida, & Lien 1974: 187 (E*); Lu & Li 1982a (F*); Ahmed 1987 (distr.; Bangladesh); Darsie & Pradhan 1990, 1991 (F, L; tax., keys, bion., distr.; Nepal); Rattanarithikul et al. 2010 (F*, L*; keys, bion., distr.; Thailand); Darsie 2012 (P*). **Ety.:** n.s. [white sides (L); prob. ref. to "broad silvery white patch on each side in front"]. **Informal name:** White-sided Bangladeshi Pointy Mosquito.

alboniveus Barraud
1934: 210 (M*, F, L*). **Type loc.:** Kurseong, Darjeeling District [West Bengal], Eastern Himalayas, India (NHMUK). **Distr.:** India, Indonesia, Malaysia, Philippines, Thailand. **Refs.:** Brug 1939: 105 (L*); Reinert 2002e: Fig. 7 (F*); Rattanarithikul et al. 2010 (F*, L*; keys, bion., distr.; Thailand). **Ety.:** n.s. [snow white (L); prob. ref. to complete white patch on scutum]. **Informal name:** Snow-white Indian Pointy Mosquito.

axitiosus Kulasekera, Knight, & Harbach
1990: 26 (M*, F, P*, L*). **Type loc.:** Rayoh, Tenom, Sabah [Borneo], Malaysia (USNM). **Distr.:** Malaysia. **Ety.:** acting together harmoniously (L); ref. to author teamwork. **Informal name:** Cooperative Malaysian Pointy Mosquito.

dorseyi Knight
1946: 277 (M*, F, L*). **Type loc.:** Garakayo, Palau Islands (USNM). **Distr.:** Palau. **Refs.:** Bohart 1956b: 30 (M*, F*, L*); Lee et al. 1982: 140 (F; key, tax., bion., distr., review). **Ety.:** C.K. Dorsey. **Informal name:** Dorsey Palauan Pointy Mosquito.

ganapathi Colless
1958: 474 (M*, F, L). **Type loc.:** Sian Kee Estate, Kuantan, Pahang, Malaya (NHMUK). **Distr.:** Malaysia, Thailand. **Refs.:** Bram 1967a: 90 (M*, L); Bram 1967b; Rattanarithikul et al. 2010 (F*, L*; keys, bion., distr.; Thailand). **Ety.:** A. Ganapathipillai. **Informal name:** Ganapathipillai Malaysian Pointy Mosquito.

harinasutai Knight
1978a: 106 (M*, F*, P*, L*). **Type loc.:** Ban Nithae, Kanchanaburi Province, Thailand (USNM). **Distr.:** Thailand. **Ety.:** Chamlong Harinasuta. **Informal name:** Harinasuta Thai Pointy Mosquito.

idjenensis Brug
1934a: 513 (A; as *Aedes niveus* var.). **Type loc.:** Ongop Ongop, Idjen, Java [Indonesia] (ITH). **Distr.:** Indonesia. **Refs.:** Knight 1946: 280 (to sp.). **Ety.:** n.s. [poss. ref. to Idjen, Java]. **Informal name:** Idjen Indonesian Pointy Mosquito.

inermis Colless
1958: 473 (M*, F, L*). **Type loc.:** MacRitchie Reservoir, Singapore (NHMUK). **Distr.:** Malaysia, Singapore, Thailand. **Refs.:** Rattanarithikul et al. 2010 (F*, L*; keys, bion., distr.; Thailand). **Ety.:** n.s. [unarmed (L)]. **Informal name:** Harmless Singaporean Pointy Mosquito [author interpretation; biting habits not stated].

lacteus Knight
1946: 275 (M*, F, L*). **Type loc.:** Cape Melville, Balabac Island, Philippines (USNM). **Distr.:** Philippines. **Refs.:**

Knight & Hull 1951b: 247 (tax.); Basio 1971b: 16 (M*; bion.). **Ety.:** n.s. [milky (L)]. **Informal name:** Milky Philippine Pointy Mosquito.

laoagensis Knight
1946: 276 (M*, F, L*). **Type loc.:** Laoag, Ilocos Norte, Luzon, Philippines (USNM). **Distr.:** Philippines. **Refs.:** Knight & Hull 1951b: 248 (tax.); Basio 1971b: 16 (M*; bion.). **Ety.:** Laoag, Ilocos Norte Prov., Luzon, Philippines. **Informal name:** Laoag Philippine Pointy Mosquito.

leonis Colless
1958: 476 (M*, F, L*). **Type loc.:** Pulau Blakang Mati, Singapore (NHMUK). **Distr.:** Malaysia, Singapore, Thailand. **Refs.:** Harrison et al. 1991: 199 (to sp. from ssp. of *niveus*); Rattanarithikul et al. 2010 (F*, L*; keys, bion., distr.; Thailand). **Ety.:** n.s. [lion (L)]. **Informal name:** Lion Singaporean Pointy Mosquito.

litoreus Colless
1958: 475 (M*, F, L*). **Type loc.:** Pulau Blakang Mati, Singapore (NHMUK). **Distr.:** Malaysia, Singapore, Thailand. **Refs.:** Rattanarithikul et al. 2010 (F*, L*; keys, bion., distr.; Thailand). **Ety.:** beach or seashore (L). **Informal name:** Seaside Singaporean Pointy Mosquito.

mikrokopion Knight & Harrison
1988: 215 (M*, F*, P*, L*). **Type loc.:** Sangkhla Buri, Kanchanaburi Province, Thailand (USNM). **Distr.:** Malaysia, Thailand. **Refs.:** Rattanarithikul et al. 2010 (F*, L*; keys, bion., distr.; Thailand). **Ety.:** very little oar (Gr); ref. to unusually small pupal paddles. **Informal name:** Small-oared Thai Pointy Mosquito.

mjobergi (Edwards)
1926b: 248 (F; *Armigeres*). **Type loc.:** Mount Murud, 7,000 ft, Borneo (NHMUK). **Distr.:** Borneo (island). **Refs.:** Stone 1957a: 338 (new comb.). **Ety.:** Eric Georg Mjöberg. **Informal name:** Mjöberg Bornean Pointy Mosquito.

mohani Knight
1969: 382 (M*, F, P, L*). **Type loc.:** Coonoor, Nilgiri Hills, Madras State, India (USNM). **Distr.:** India, Sri Lanka. **Ety.:** B.N. Mohan. **Informal name:** Mohan Indian Pointy Mosquito.

nipponicus La Casse & Yamaguti
1948: 79 (M*, F*, P*, L*; as *Aedes niveus* ssp.). **Type loc.:** Nagasaki, Fukuoka, Kagoshima, Kyushu, Maizuru, Wakayama, Yodo, Honshu, Japan (NE). **Distr.:** Japan, Russia, South Korea. **Refs.:** La Casse & Yamaguti 1950: 146 (M*, F*, P*, L*; bion., distr., tax.; from ssp.); Stone & Knight 1956a: 222 (type info.); Hara 1957 (F*?); Gutsevich et al. 1971: 167 (distr.); Tanaka et al. 1975c: 222 (tax., distr.); Tanaka et al. 1979: 351 (M*, F*, L*); Tanaka 2002b: 164 (P*; tax., key); Rattanarithikul et al. 2010 (F*, L*; bion., keys, distr.; Thailand). **Ety.:** Japan. **Informal name:** Nara Japanese Pointy Mosquito [Nara was the capital city of Japan from 710 to 794, during the Nara period].

nippononiveus Sasa & Nakahashi
1952: 258 (M*, F, L*). **Type loc.:** Oimachi, Tokyo [Honshu], Japan (IID). **Distr.:** Japan, Vietnam. **Refs.:** Hara 1957 (F*?). **Ety.:** Japan and white/snowy (L). **Informal name:** Yuki-onna Japanese Pointy Mosquito [the Yuki-onna is a snow spirit from Japanese folklore that lures humans to their deaths by exposure].

nishikawai Tanaka, Mizusawa, & Saugstad
1979: 356 (M*, F*, L*). **Type loc.:** Mount Yuwan, Mami Oshima, Ryukyus, Japan (NMST). **Distr.:** Japan. **Refs.:** Tanaka 2002b: 167 (P*; tax., key). **Ety.:** Masaru Nishikawa. **Informal name:** Nishikawa Japanese Pointy Mosquito.

niveoides Barraud
1934: 211 (M*). **Type loc.:** Nagargali [Belgaum District], Bombay, India (NHMUK). **Distr.:** Cambodia, India, Indonesia, Malaysia, Nepal, Thailand, Vietnam. **Refs.:** Bonne-Wepster & Brug 1939: 1244 (L*); Colless 1958: 479 (M*, L*); Colless 1959: 175 (M, F, L); Darsie et al. 1993 (distr.; Nepal); Rattanarithikul et al. 2010 (F*, L*; keys, bion., distr.; Thailand). **Ety.:** like *niveus*. **Informal name:** Snowy Indian Pointy Mosquito.

niveus (Ludlow)
1903 (F; as *Stegomyia nivea*). In Theobald 1903a: 139. **Type loc.:** Oras, Samar, Philippines (NHMUK). **Distr.:** Bangladesh, India (includes Andaman Isl.), Indonesia (Java, Flores, Timor), Japan, Laos, Malaysia, Pakistan, People's Republic of China, Philippines, Singapore, Sri Lanka, Thailand, Vietnam. **Refs.:** Barraud 1923g (M*, F*); Feng 1938: 512 (M*, F*, L*); Knight 1946: 278 (M*, F, L*); Knight & Chamberlain 1948: Fig. 33 (P*); Knight & Hull 1951b: 246 (M, F, L*; tax.); Basio 1971b: 17 (M*; bion.); Aslamkhan 1971b (distr.; Pakistan); Lu 1981 (tax.; *s.l.*); Ahmed 1987 (distr.; Bangladesh); Tewari & Hiriyan 1995: 167 (M*, F, P*, L*); Huang & Rueda 1998: 824 (F*; type info.); Reinert 2002e: Fig. 8 (F*); Reinert, Harbach, & Kitching 2006 (phyl., class., tax); Reinert & Harbach 2006 (F); Reinert 2008h: Fig. 1 (F*). **Ety.:** n.s. [snow, snowy (L)]. **Informal name:** Snowy Philippine Pointy Mosquito. (Vol. 1, p. 410)

novoniveus Barraud
1934: 211 (M*, F, L*). **Type loc.:** Mungpoo, Darjeeling District, Eastern Himalayas [West Bengal], India (NHMUK). **Distr.:** India, Malaysia, Nepal, Thailand. **Refs.:** Knight 1946: 279 (M); Colless 1958: 479 (M*), 480 (F*), 481 (L*); Colless 1959: 174 (M, F, L); Darsie et al. 1992 (distr.; Nepal); Rattanarithikul et al. 2010 (F*, L*; keys, bion., distr.; Thailand). **Ety.:** near/new *niveus*. **Informal name:** Snowfall Indian Pointy Mosquito [*novoniveus* = "new snow" = Snowfall].

omorii Lien
1968b: 105 (M*, F, P*, L*). **Type loc.:** Chungmin, Yuchih, Nantou Hsien, Taiwan (TMRI). **Distr.:** Taiwan. **Ety.:** Nanzaburo Omori. **Informal name:** Omori Taiwanese Pointy Mosquito.

pexus Colless
1958: 469 (M*, F, L*). **Type loc.:** MacRitchie Reservoir, Singapore (NHMUK). **Distr.:** Malaysia, Singapore, Thailand. **Refs.:** Rattanarithikul et al. 2010 (F*, L*; keys, bion., distr.; Thailand). **Ety.:** n.s. [comb, card (L); for wool]. **Informal name:** Combed Singaporean Pointy Mosquito.

pseudoniveus (Theobald)
1905f: 75 (F). **Type loc.:** Singapore (HNM). **Distr.:** Indonesia, Malaysia, People's Republic of China, Singapore, Thailand. **Refs.:** Knight 1946: 270 (F); Colless 1958: 478 (M*, L*); Colless 1959: 169 (M, F, L); Rattanarithikul et al. 2010 (F*, L*; keys, bion., distr.; Thailand). **Ety.:** false (Gr); similar to *niveus*. **Informal name:** Sleet Singaporean Pointy Mosquito [*pseudoniveus* = "false snow" = Sleet].

saperoi Knight
1946: 271 (M*, F, L*). **Type loc.:** Subic Bay [Zambales], Luzon, Philippines (USNM). **Distr.:** Philippines. **Refs.:** Knight & Hull 1951b: 247 (L*; tax.); Basio 1971b: 18 (M*; bion.); Rattanarithikul et al. 2010 (F*, L*; keys, bion., distr.; Thailand). **Ety.:** James Joseph Sapero. **Informal name:** Sapero Philippine Pointy Mosquito.

shehzadae Qutubuddin
1972: 119 (F). **Type loc.:** Mafulo, Papua (NHMUK). **Distr.:** Papua New Guinea. **Refs.:** Lee et al. 1982: 268 (F; key, tax., bion., distr., review). **Ety.:** Shehzad Qutubuddin. **Informal name:** Shehzad Papuan Pointy Mosquito.

sinensis Chow
1950: 283 (M*, F, P*, L*). **Type loc.:** Ta-Keng, Taichung Hsien, Formosa [Taiwan, Republic of China] (USNM). **Distr.:** People's Republic of China, Taiwan. **Ety.:** China. **Informal name:** China Taiwanese Pointy Mosquito.

subniveus Edwards
1922a: 269 (M). **Type loc.:** Kuching (?), Sarawak, Borneo [Malaysia] (NHMUK). **Distr.:** Australia, Malaysia, Singapore, Thailand. **Refs.:** Colless 1958: 478 (M*, L*; from syn. with *pseudoniveus* Theobald); Colless 1959: 171 (M, F, L); Ramalingam & Pillai 1973 (distr.); Rattanarithikul et al. 2010 (F*, L*; keys, bion., distr.; Thailand). **Ety.:** nearly (L); near *niveus*. **Informal name:** Frost Malaysian Pointy Mosquito [*subniveus* = "nearly snow" = Frost].

vanus Colless
1958: 471 (M*, F, L*). **Type loc.:** MacRitchie Reservoir, Singapore (NHMUK). **Distr.:** Indonesia, Malaysia, Singapore, Thailand. **Refs.:** Rattanarithikul et al. 2010 (F*, L*; keys, bion., distr.; Thailand). **Ety.:** n.s. [empty, useless (?) (L)]. **Informal name:** Pointless Singaporean Pointy Mosquito.

watteni Lien
1968b: 103 (M*, F, P*, L*). **Type loc.:** Chuanfanshih, Wolan, Hengchung, Pingtung Hsien, Taiwan (TMRI). **Distr.:** Taiwan. **Ety.:** R.H. Watten. **Informal name:** Watten Taiwanese Pointy Mosquito.

Subgenus *Edwardsaedes* Belkin

1962: 408 (as subg. of genus *Aedes*). Type species: *Culex imprimens* Walker.

Name derivation: Frederick Wallace Edwards.

Important references: Reinert 1976c: 9 (redescr., tax., bion.); Lee et al. 1982: 67 (tax., key, bion.; Australasia); Lee et al. 1984: 104 (key); Reinert, Harbach, & Kitching 2004: 289 (to genus); Reinert, Harbach, & Kitching 2009 (phyl., class.); Rattanarithikul et al. 2010 (F*, L*; keys, bion.; Thailand); Wilkerson et al. 2015 (phyl., class.; to subg. of genus *Aedes*); Soghigian et al. 2017 (mol. phyl.).

bekkui Mogi
1977: 129 (M*, F*, P*, L*). **Type loc.:** Sao-zaki, Kami-agata, Tsushima Island, Japan (NMST). **Distr.:** Japan, Russia, South Korea. **Refs.:** Tanaka et al. 1979: 423 (F*, L*; as *imprimens*). **Ety.:** n.s. [H. Bekku, collector of type specimen (Harbach 2018c)]. **Informal name:** Bekku Japanese Pointy Mosquito.

imprimens (Walker)
1860a: 144 (F; *Culex*). **Type loc.:** Amboina Island [Moluccas] (NHMUK). **Distr.:** Australia, Bangladesh, Cambodia, India, Indonesia, Malaysia, Papua New Guinea, Philippines, Solomon Isl., Thailand, Vietnam. **Refs.:** Hsiao & Bohart 1946: 22 (from *Aedimorphus*); La Casse & Yamaguti 1950: 102 (M*, F*; bion., distr., tax.); Waku 1950 (redescr.); Knight & Hull 1953: 465 (M*, F, L*); Hara 1957: 65 (F*); Mattingly 1961: 53 (M*, F*, P*, L*); Belkin 1962: 409 (M*, F, P*, L*); Basio 1971b: 24 (M*; bion.); Moriya et al. 1973 (E*); Arnell 1976: 44 (M*, F, L*; as *auratus*); Reinert 1976c: 12 (M*, F*, P*, L*, E*; distr.); Tanaka et al. 1979 (F*); Lee et al. 1982: 69 (tax., review); Ahmed 1987 (distr.; Bangladesh); Reinert 2000e: Fig. 20 (F*); Rattanarithikul et al. 2010 (F*, L*; bion., keys, distr.; Thailand). **Ety.:** n.s. [poss. ref. to printing]. **Informal name:** Gutenberg Indonesian Pointy Mosquito [*imprimens* = printing = Johannes Gutenberg (1398–1468), who introduced printing to Europe].
syn. *auratus* Leicester
1908: 153 (F; *Culex*). **Type loc.:** Malaya [Malaysia] (NE). **Refs.:** Belkin et al. 1970: 168 (M*, F, L*; from syn. with *tortilis* Theobald).

syn. *suknaensis* Theobald
1910a: 21 (F; *Culicada*). **Type loc.:** Sukna, Darjeeling District (Eastern Himalayas), [West Bengal], India (IM). **Refs.:** Barraud 1934: 257 (tax.); Reinert 1973b: 28 (subgeneric transfer).
syn. *brugi* Edwards
1924b: 371 (M). **Type loc.:** Merauke, New Guinea [West Papua] (NHMUK). **Refs.:** Knight & Hull 1953: 465 (? = *imprimens*); Mattingly 1961: 53 (syn.).

pingpaensis Chang
1965: 397 (M*, F*, L*). **Type loc.:** Pingpa, Kweichow Province, China [People's Republic of China] (IPD). **Distr.:** People's Republic of China. **Ety.:** Pingpa, Kweichow Prov., People's Republic of China. **Informal name:** Pingpa Chinese Pointy Mosquito.
syn. *antuensis* Su, Wang, & Li
1978: 439 (M*, F, L*). **Type loc.:** Antu Xian, Jilin Province, China [People's Republic of China] (LU). **Refs.:** Danilov 1985: 380 (syn.).

Subgenus *Elpeytonius* Reinert, Harbach, & Kitching

2009: 768 (as genus). Type species: *Ochlerotatus apicoannulatus* Edwards (new name for *Aedimorphus alboannulatus* Theobald).

Name derivation: EL Peyton.

Important references: Reinert 2010e (F gen.*); Wilkerson et al. 2015 (phyl., class.; to subg. of genus *Aedes*).

apicoannulatus (Edwards)
1912c: 18 (new name for *alboannulatus* Theobald). **Distr.:** Nigeria, Sierra Leone. **Refs.:** Ingram & Macfie 1917b: 144 (L*); Macfie & Ingram 1922 (F*); Reinert 2010e (F*). **Ety.:** ringed tip (L); apical white leg bands. **Informal name:** Ring-tipped Sierra Leonean Pointy Mosquito.
syn. *alboannulatus* Theobald
1905b: 154 (M, F; *Aedimorphus*). **Type loc.:** Sierra Leone (NHMUK). **Refs.:** Lee et al. 1982: 100 (F; key, tax., bion., distr., review).

simulans (Newstead & Carter)
1911b: 240 (M*; *Reedomyia*). **Type loc.:** Broomassie, Ashanti, West Africa [Ghana] (NHMUK). **Distr.:** Cameroon, Central African Republic, Congo, Côte d'Ivoire, Democratic Republic of the Congo, Ghana, Liberia, Nigeria, Sierra Leone, Sudan and South Sudan, Uganda. **Refs.:** Edwards 1941: 165 (M*, F*), 392 (P); Haddow et al. 1951: 221 (M, L); Hopkins 1952: 157 (L*); Pao & Knight 1970a: 111 (L*); Reinert 2010e (F*). **Ety.:** n.s. [imitate, copy (L); poss. ref. to similarity to *albopunctata*]. **Informal name:** Imitative Ghanaian Pointy Mosquito.

Subgenus *Finlaya* Theobald

1903a: 281 (as genus). Type species: *Culex kochi* Dönitz.

Name derivation: Carlos Finlay.

syn. *Popea* Ludlow 1905a: 95 (as genus). Type species: *Popea lutea* Ludlow.

Important references: Giles 1904a: 365 (as *Finlayia*; emend., tax., key); Blanchard 1905: 415, 630 (tax.); Knight & Laffoon 1946: 203 (keys; Kochi Group; Oriental); Marks 1947a (keys; Kochi Group; Australasia); Knight & Marks 1952: 513 (checklist); Belkin 1962 (tax., bion.; *s.l.*); Pao & Knight 1970a (Marshallii Group); Lee et al. 1982: 73 (tax., key, bion.; *s.l.*; Australasia); Reinert 2000h (tax.; as subg. of genus *Ochlerotatus*); Reinert, Harbach, & Kitching 2004 (phyl., class.; to genus; restricted to Kochi Group); Reinert & Harbach 2005b (tax., redescr.); Reinert, Harbach, & Kitching 2009 (phyl., class.); Rattanarithikul et al. 2010 (F*, L*; keys, bion.; Thailand); Wilkerson et al. 2015 (phyl., class.; to subg. of genus *Aedes*).

alocasicola Marks
1947a: 35 (M*, F, P*, L*). **Type loc.:** Mount Glorious, Queensland, Australia (UQIC). **Distr.:** Australia, Fiji. **Refs.:** Lee et al. 1982: 108 (F; key, tax., bion., distr., review). **Ety.:** from *Alocasia macrorrhiza* [plant bract]. **Informal name:** Elephant Ear-dwelling Australian Pointy Mosquito.

ananae Knight & Laffoon
1946: 218 (M*, F, P*, L*). **Type loc.:** Osmena, Basey Municipality, Samar, Philippines (USNM). **Distr.:** Philippines. **Refs.:** Knight & Hull 1951b: 230 (L*); Basio 1971b: 14 (M*, P*, L*; bion.). **Ety.:** pineapple [axil] (L). **Informal name:** Pineapple Philippine Pointy Mosquito.

avistylus Brug
1939: 107 (M*, L*; as *Aedes flavipennis* var.). **Type loc.:** Baoebaoe, Boeton [= Butung Island, Sulawesi] (NHMUK). **Distr.:** Indonesia, Malaysia. **Refs.:** Stone & Bohart 1944: 208 (M*; to sp.); Marks 1947a: 63 (M, F, L*); Lee et al. 1982: 125 (F; key, tax., bion., distr., review). **Ety.:** n.s. [bird beak (L); birdlike outline of the gonostylus]. **Informal name:** Bird-beaked Indonesian Pointy Mosquito.

bougainvillensis Marks
1947a: 19 (M*, F, P*, L*). **Type loc.:** Bougainville, Solomon Islands (UQIC). **Distr.:** Papua New Guinea, Solomon Isl. **Refs.:** Belkin 1962: 362 (M*, F, P*, L*). **Ety.:** Bougainville, Solomon Isl. **Informal name:** Bougainville Solomon Islander Pointy Mosquito.

burnetti Belkin
1962: 363 (M*, F, P*, L*). **Type loc.:** Tholo-i-Suva, Viti Levu, Fiji Islands (USNM). **Distr.:** Fiji. **Refs.:** Lee et al. 1982: 131 (F; key, tax., bion., distr., review). **Ety.:** G.F.

Burnett. **Informal name:** Burnett Fijian Pointy Mosquito.

croceus Knight & Laffoon
1946: 213 (M*, F, P*, L*); Basio 1971b: 15 (M*, P*, L*; bion.). **Type loc.:** Santa Rita, Olongapo, Subic Bay, Zambales, Luzon, Philippines (USNM). **Distr.:** Philippines. **Ety.:** yellow (L); ref. to predominantly yellow color. **Informal name:** Yellowed Philippine Pointy Mosquito.

dobrotworskyi Marks
1958a: 65 (M*, F, P*, L*). **Type loc.:** Barham Valley, Apollo Bay, Otway, Victoria, Australia (UQ). **Distr.:** Australia. **Refs.:** Dobrotworsky 1965: 128 (M*, F*, L*, E*); Lee et al. 1982: 139 (F; key, tax., bion., distr., review). **Ety.:** Nikolas Vladimir Dobrotworsky. **Informal name:** Dobrotworsky Australian Pointy Mosquito.

fijiensis Marks
1947a: 26 (M*, F, L*). **Type loc.:** Samabula, Viti Levu, Fiji Islands (IQ). **Distr.:** Fiji. **Refs.:** Paine 1943 (bion.); Amos 1947 (bion., distr.); Symes 1955 (bion., med.imp.); Laird 1957 (tax., bion., syst.; Kochi Group); Belkin 1962: 365 (M*, F, P*, L*); Huang 1977c (F*, L*; keys, distr.); Lee et al. 1982: 141 (F; key, tax., bion., distr., review). **Ety.:** Fiji. **Informal name:** Ratu Fijian Pointy Mosquito [*ratu* is an ancient Fijian term for a chief or ruler]. (Vol. 1, p. 412)

flavipennis Giles
1904a: 366 (M, F; *Finlaya*). **Type loc.:** Camp Stotsenberg, Angeles, Pampanga, Luzon, Philippines (NHMUK). **Distr.:** Indonesia, Malaysia, New Caledonia, Philippines, Thailand. **Refs.:** Stone & Bohart 1944: 207 (M*; as *aranetana*); Baisas 1946c: 30 (M*, F, P*, L*; as *aranetana*); Knight & Laffoon 1946: 214 (P, L*; as *aranetana*); Knight & Hull 1951b: 233 (tax.); Basio 1971b: 15 (M*, L*; bion.), 14 (M*, L*; bion.; as *aranetanus*); Ramalingam & Pillai 1973 (distr.); Baisas 1974: 8 (M*, F, P*, L*; tax., bion., distr.; Philippines); Lee et al. 1982: 146 (F; key, tax., bion., distr., review); Rattanarithikul et al. 2010 (F*, L*; keys, bion., distr.; Thailand). **Ety.:** yellow wing (L); "Wing generally yellow, especially on the costa." **Informal name:** Yellow-winged Philippine Pointy Mosquito.
syn. *aranetana* Banks
1906b: 1001 (M, F; *Finlaya*). **Type loc.:** Negros Occidental, Negros Island, Philippines (USNM). **Refs.:** Knight & Marks 1952: 543 (type info., lectotype desig.); Stone & Knight 1956a: 214 (type info.).

franclemonti Belkin
1962: 367 (M*, F, P*, L*). **Type loc.:** Munda, New Georgia, Solomon Islands (USNM). **Distr.:** Solomon Isl. **Refs.:** Lee et al. 1982: 148 (F; key, tax., bion., distr., review). **Ety.:** John G. Franclemont. **Informal name:** Franclemont Solomon Islander Pointy Mosquito.

freycinetiae Laird
1957: 343 (M*, F, L*). **Type loc.:** Glen Pool, Suva, Viti Levu, Fiji Islands (DMNZ). **Distr.:** Fiji. **Refs.:** Belkin 1962: 368 (M*, F, P*, L*); Lee et al. 1982: 150 (F; key, tax., bion., distr., review). **Ety.:** ref. to larvae found in *Freycinetia* leaf axils. **Informal name:** Freycinetia Fijian Pointy Mosquito.

fuscipalpis Belkin
1962: 370 (M*, F, P*, L*). **Type loc.:** Tenaru area, Guadalcanal, Solomon Islands (USNM). **Distr.:** Solomon Isl. **Refs.:** Lee et al. 1982: 152 (F; key, tax., bion., distr., review). **Ety.:** dark feeler (L); palpus. **Informal name:** Black-lipped Solomon Islander Pointy Mosquito.

fuscitarsis Belkin
1962: 371 (M*, F, P*, L*). **Type loc.:** Poha area, Guadalcanal, Solomon Islands (USNM). **Distr.:** Solomon Isl. **Ety.:** dark (L); tarsi. **Informal name:** Black-footed Solomon Islander Pointy Mosquito.

gahnicola Marks
1947a: 43 (M*, F, P*, L*). **Type loc.:** Caloundra, Queensland, Australia (UQIC). **Distr.:** Australia. **Refs.:** Lee et al. 1982: 154 (F; key, tax., bion., distr., review). **Ety.:** "Found breeding in axils of sword grass (*Gahnia* sp.)." **Informal name:** Grass-dwelling Australian Pointy Mosquito.

gani Bonne-Wepster
1940: 158 (M*, F*). **Type loc.:** Japero, Western New Guinea (Papua) (ITH). **Distr.:** Indonesia (West Papua), Papua New Guinea. **Refs.:** Lee et al. 1982: 156 (F; key, tax., bion., distr., review). **Ety.:** n.s. [Gan Houv Hwat; coll. of type specimen (Harbach 2018c)]. **Informal name:** Gan Papuan Pointy Mosquito.

gressitti Bohart
1957: 23 (M*, F*). **Type loc.:** Ngergoi Island, Palau Islands (CM). **Distr.:** Palau. **Ety.:** Judson Linsley Gressitt. **Informal name:** Gressitt Palauan Pointy Mosquito.

hollingsheadi Belkin
1962: 372 (M*, F, P*, L*). **Type loc.:** Poha area, Guadalcanal, Solomon Islands (USNM). **Distr.:** Papua New Guinea, Solomon Isl. **Refs.:** Lee et al. 1982: 163 (F; key, tax., bion., distr., review). **Ety.:** Charles Hollingshead. **Informal name:** Hollingshead Solomon Islander Pointy Mosquito.

horotoi Taylor
1972: 317 (F). **Type loc.:** Manitawaniuhi, San Cristobal, Solomon Islands (LU). **Distr.:** Solomon Isl. **Refs.:** Lee et al. 1982: 166 (F; key, tax., bion., distr., review; change type dep. from BBM to LU). **Ety.:** E. Sosimo Horoto. **Informal name:** Horoto Solomon Islander Pointy Mosquito.

hui Bohart
1957: 25 (M*, F, L). **Type loc.:** Koror Island, Palau Islands (USNM). **Distr.:** Palau. **Refs.:** Lee et al. 1982: 166 (tax., distr.). **Ety.:** S.M.K. Hu. **Informal name:** Hu Palauan Pointy Mosquito.

josephinae Marks
1958a: 58 (M*, F, P*, L*). **Type loc.:** Skull Creek, Cape York Peninsula, Queensland, Australia (UQIC). **Distr.:** Australia, Indonesia (West Papua). **Refs.:** Maffi et al. 1979: 212 (distr.); Lee et al. 1982: 168 (F; key, tax., bion., distr., review). **Ety.:** Mabel Josephine Mackerras. **Informal name:** Josephine Australian Pointy Mosquito.

knighti Stone & Bohart
1944: 210 (M*). **Type loc.:** Rendova Island, New Georgia Group, Solomon Islands (USNM). **Distr.:** Indonesia (West Papua), Solomon Isl. **Refs.:** Belkin 1962: 374 (M*, F, P*, L*). **Ety.:** Kenneth Lee Knight. **Informal name:** Knight Solomon Islander Pointy Mosquito.

kochi (Dönitz)
1901: 38 (F; *Culex*). **Type loc.:** Dreyer Hafen by Cap Cretin, New Guinea (ZM). **Distr.:** Australia, Indonesia (Kiriwina Isl., West Papua), Philippines, Samoa (Ind. State of Samoa, American Samoa, Western Samoa), Solomon Isl., Tonga. **Refs.:** Marks 1947a: 10 (M*, F, P*, L*; keys, tax., distr.); Marks 1958a: 73 (*s.l.*; tax.); Lee et al. 1982: 174 (F; key, tax., bion., distr., review); Lee et al. 1987b: 217 (F; key, tax., bion., distr., review); Townsend 1990: 91 (syntype, NHMUK); Reinert 2002e: Fig. 6 (F*); Reinert & Harbach 2005b: 160 (F; redescr.). **Ety.:** n.s. [Heinrich Hermann Robert Koch (Harbach 2018c; Kitzmiller 1982)]. **Informal name:** Koch Papuan Pointy Mosquito. (Vol. 1, p. 414)

lewelleni Starkey & Webb
1946: 179 (M*, F, L*). **Type loc.:** Angaur Island, Palau Islands (USNM). **Distr.:** Palau. **Refs.:** Bohart 1957: 23 (M*, F*, L*); Lee et al. 1982: 183 (tax., distr.). **Ety.:** William B. Lewellen. **Informal name:** Lewellen Palauan Pointy Mosquito.

luteus (Ludlow)
1905a: 96 (M; *Popea*). **Type loc.:** Camp Stotsenberg, Angeles, Pampanga, Luzon, Philippines (NE). **Distr.:** Philippines, Singapore. **Refs.:** Knight & Laffoon 1946: 216 (M, F, P, L*; as *flavipennis*); Knight & Hull 1951b: 234 (from syn. with *flavipennis*). **Ety.:** yellow (L); general color. **Informal name:** Yellow Philippine Pointy Mosquito.

maffii Taylor & Tenorio
1974: 577 (M*, F*, P*, L*). **Type loc.:** Tegano, Rennell Island, Solomon Islands (BBM). **Distr.:** Solomon Isl. **Refs.:** Lee et al. 1982: 186 (F; key, tax., bion., distr., review). **Ety.:** Mario Maffi. **Informal name:** Maffi Solomon Islander Pointy Mosquito.

medleri Knight & Laffoon
1946: 211 (M*, F, P*, L*). **Type loc.:** Leyte-Samar area, Jinamoc Island, Philippines (USNM). **Distr.:** Philippines. **Refs.:** Knight & Hull 1951b: 234 (tax.); Basio 1971b: 17 (M*, L*; bion.). **Ety.:** J.T. Medler. **Informal name:** Medler Philippine Pointy Mosquito.

neogeorgianus Belkin
1962: 375 (M*, F, P*, L*). **Type loc.:** Munda, New Georgia, Solomon Islands (USNM). **Distr.:** Papua New Guinea, Solomon Isl. **Refs.:** Lee et al. 1982: 194 (F; key, tax., bion., distr., review). **Ety.:** New Georgia, Solomon Isl. **Informal name:** New Georgia Solomon Islander Pointy Mosquito.

oceanicus Belkin
1962: 377 (M*, F, P*, L*). **Type loc.:** Amouli, Pago Pago, Tutuila, Samoa Islands (USNM). **Distr.:** Samoa (Ind. State of Samoa, American Samoa, Western Samoa), Tonga. **Refs.:** Ramalingam 1976: 303 (tax., bion., keys., distr.; Samoa & Tonga); Lee et al. 1982: 219 (F; key, tax., bion., distr., review). **Ety.:** from Oceania. **Informal name:** Oceania Samoan Pointy Mosquito.

poicilius (Theobald)
1903a: 283 (F*; as *Finlaya poicilia*). **Type loc.:** Pulau Jerejak, Penang, Malaya [Malaysia] (NHMUK). **Distr.:** Borneo, Brunei, India, Indonesia, Malaysia, Myanmar, Nepal, Philippines, Thailand. **Refs.:** Bancroft 1908: 53 (F; as genus *Finlaya*); Stone & Bohart 1944: 211 (M*); Knight & Laffoon 1946: 221 (M*, F, P*, L*; tax.); Knight & Hull 1951b: 231 (tax.); Stone et al. 1959: 169 (as *poecilus*); Mattingly 1971a: Pl. 32 (L thorax*); Basio 1971b: 18 (M*, L*; bion.); Baisas 1974: 20 (M, F, P, L; tax., bion., distr.; Philippines); Lien et al. 1977 (distr.); Lee et al. 1982: 231 (F; key, tax., bion., distr., review); Darsie et al. 1992 (distr.; Nepal); Rattanarithikul et al. 2010 (F*, L*; bion., keys, distr.; Thailand). **Ety.:** n.s. [spotted, variegated? (L, Gr)]. **Informal name:** Variegated Malaysian Pointy Mosquito. (Vol. 1, p. 416)

samoanus (Grünberg)
1913a: 130 (F*; *Finlaya*). **Type loc.:** Apia, Upolu, Samoa Islands (ZM). **Distr.:** Samoa (Ind. State of Samoa, American Samoa, Western Samoa). **Refs.:** Stone & Bohart 1944: 212 (M*, L); Bohart & Ingram 1946b: 14 (M*, F*, L*; bion., tax.); Marks 1947a: 32 (M, F, L); Laird 1957 (tax., bion., syst.; Kochi Group); Ramalingam & Belkin 1964: 105 (distr.); Ramalingam & Belkin 1976: 194 (P*, L*); Ramalingam 1976: 301 (tax., bion., keys., distr.; Samoa & Tonga); Lee et al. 1982: 258 (F; key, tax., bion., distr., review). **Ety.:** Samoa. **Informal name:** Apia Samoan Pointy Mosquito.

schlosseri Belkin
1962: 382 (M*, F, P*, L*). **Type loc.:** Poha area, Guadalcanal, Solomon Islands (USNM). **Distr.:** Solomon Isl. **Refs.:** Lee et al. 1982: 266 (F; key, tax., bion., distr.,

review). **Ety.:** Ralph J. Schlosser. **Informal name:** Schlosser Solomon Islander Pointy Mosquito.

solomonis Stone & Bohart
1944: 208 (M*, F, L). **Type loc.:** Guadalcanal, Solomon Islands (USNM). **Distr.:** Indonesia (West Papua), Solomon Isl. **Refs.:** Marks 1947a: 57 (M, F, P*, L*); Belkin 1962: 383 (M*, F, P*, L*); Lee et al. 1982: 269 (F; key, tax., bion., distr., review). **Ety.:** Solomon Isl. **Informal name:** Adaro Solomon Islander Pointy Mosquito [the *adaro* is a malevolent oceanic ghost in Solomon Islander mythology].

sorsogonensis Bañez & Jueco
1966: 194 (M*, F*, L*). **Type loc.:** Sorsogon Province [Luzon], Philippines (IHP). **Distr.:** Philippines. **Refs.**: Basio 1971b: 19 (M*; bion.); **Ety.:** Sorsogon Prov., Philippines. **Informal name:** Sorsogon Philippine Pointy Mosquito.

stonei Knight & Laffoon
1946: 208 (M*, F, P*, L*). **Type loc.:** Ducong, Basey Municipality, Samar, Philippines (USNM). **Distr.:** Philippines. **Refs**.: Basio 1971b: 20 (M*, P*, L*; bion.). **Ety.:** Alan Stone. **Informal name:** Stone Philippine Pointy Mosquito.

tutuilae Ramalingam & Belkin
1965: 3 (M*, F, P*, L*). **Type loc.:** Aoloau, Tutuila, American Samoa (USNM). **Distr.:** Samoa (Ind. State of Samoa, American Samoa, Western Samoa). **Refs.:** Ramalingam 1976: 302 (tax., bion., keys., distr.; Samoa & Tonga); Lee et al. 1982: 281 (F; key, tax., bion., distr., review). **Ety.:** Aoloau, Tutuila, American Samoa. **Informal name:** Tutuila Samoan Pointy Mosquito.

wallacei Edwards
1926a: 105 (F). **Type loc.:** Kavieng (Kaewung), New Ireland [Bismarck Archipelago] (NHMUK). **Distr.:** Australia, Indonesia (West Papua). **Refs.:** Marks 1947a: 51 (M*, F, L*); Penn 1949b: 46 (P*); Lee et al. 1982: 283 (F; key, tax., bion., distr., review); Peyton et al. 1999: 530 (distr.). **Ety.:** H.G. Wallace. **Informal name:** Wallace Papuan Pointy Mosquito.

Subgenus *Fredwardsius* Reinert

2000g (as subg. of genus *Aedes*). Type species: *Culex vittatus* Bigot.

Name derivation: Frederick Wallace Edwards.

Important references: Reinert, Harbach, & Kitching 2004 (phyl., class.; to genus); Reinert, Harbach, & Kitching 2009 (phyl., class.); Rattanarithikul et al. 2010 (F*, L*; keys, bion.; Thailand); Wilkerson et al. 2015 (phyl., class.; to subg. of genus *Aedes*); Soghigian et al. 2017 (mol. phyl.).

vittatus (Bigot)
1861: 227 (F; *Culex*). **Type loc.:** Corsica (NE). **Distr.:** Algeria, Angola, Bangladesh, Benin, Botswana, Burkina Faso, Cambodia, Cameroon, Central African Republic, Comoros, Côte d'Ivoire, Democratic Republic of the Congo, Cuba, Djibouti, Dominican Republic, Ethiopia, France (includes Corsica), Gabon, Gambia, Ghana, Guinea, India, Iran, Italy (includes Sardinia), Kenya, Laos, Liberia, Madagascar (includes Glorioso and Juan De Nova Isl.), Malawi, Malaysia, Mali, Mozambique, Namibia, Nepal, Niger, Nigeria, Pakistan, People's Republic of China, Portugal, Republic of South Africa, Saudi Arabia, Senegal, Sierra Leone, Somalia, Spain (includes Balearic Isl.), Sri Lanka, Sudan and South Sudan, Tanzania, Thailand, Tunisia, Uganda, Vietnam, Yemen, Zambia, Zimbabwe. **Refs.:** Patton 1905 (E*; as subg. *Stegomyia*); Theobald 1907: 199 (E; bion.; as *Scutomyia sugens*); Theobald 1910c (M*, F, L*, E*; after Patton 1905; as *Scutomyia sugens*); Macfie & Ingram 1922 (F*); Barraud 1923h (L*); Barraud 1934: 245 (M*, F*, L*); Edwards 1941: 155 (M*, F*), 391 (P*); Hopkins 1952: 160 (L*); Mattingly 1952: 291 (zoogeography); Senevet & Andarelli 1954b: 320 (L*); Mattingly & Knight 1956: 100 (L*; tax.); Senevet & Andarelli 1958: 273 (P*); Boorman 1961: 709 (bion.); Joshi et al. 1965 (distr.; Nepal); Hinton & Service 1969 (E*); Mattingly 1970g: 35 (L*); Service 1970c: 128 (M*, F*, P*, L*); Aslamkhan 1971b (distr.; Pakistan); Sharma & Chaudhry 1974: 163 (as variant); Huang 1977b: 114 (M*, F*, P*, L*; transfer; distr.); Brunhes 1977a (distr.; Comoros Isl.); Rodhain et al. 1977 (distr.); Huang & Ward 1981 (F*; key, Africa); Ahmed 1987 (distr.; Bangladesh); Zaim 1987: 568 (distr.); Darsie & Pradhan 1990, 1991 (F, L; tax., keys, bion., distr.; Nepal); Suleman et al. 1993 (distr.; Pakistan); Jupp 1996 (F*; key); Reinert 2000e (F*); Melero-Alcíbar 2006 (P; tax.); Rattanarithikul et al. 2010 (F*, L*; bion., keys; distr.; Thailand); Becker et al. 2010: 196 (M*, F*, L*; keys, tax., distr., bion.); Robert et al. 2019 (distr.; Western Palearctic). **Ety.:** bound with ribbon or band (L); white ring on tibiae. **Informal name:** Banded French Pointy Mosquito.

syn. *brumpti* Neveu-Lemaire
1905b: 9 (M, F*; *Stegomyia*). **Type loc.:** Harar [Hararghe], Ethiopia (PIP). **Refs.:** Rodhain & Boutonnier 1985: 275 (type dep. from NE to PIP).

syn. *albopunctata* Theobald
1907: 262 (F; *Reedomyia*). **Type loc.:** Sierra Leone (NHMUK).

Subgenus *Georgecraigius* Reinert, Harbach, & Kitching

2006: 70 (as genus). Type species: *Culex atropalpus* Coquillett.

Name derivation: George Brownlee Craig Jr.

syn. *Horsfallius* Reinert, Harbach, & Kitching 2006: 72 (phyl., class.; as subg. of genus *Georgecraigius*). Type species: *Culex fluviatilis* Lutz.

Important references: Reinert 2008a (F gen.*); Reinert, Harbach, & Kitching 2009 (phyl., class.); Wilkerson et al. 2015 (phyl., class.; to syn of subg. *Georgecraigius* of genus *Aedes*); Soghigian et al. 2017 (mol. phyl.).

atropalpus (Coquillett)
1902d: 292 (M, F; *Culex*). **Type loc.:** Plummers Island, Montgomery County, Maryland, United States (USNM). **Distr.:** Canada, Costa Rica, El Salvador, Italy, Mexico, Netherlands, Nicaragua, Paraguay, United States (continental), Venezuela. **Refs.:** Howard et al. 1913 (E*); Howard et al. 1917 (E); Trembley 1947: 244 (laboratory bion.); Ross 1947 (F*); Darsie 1951: 15 (P*); Yamaguti & LaCasse 1951d: 230 (M*, F*, L*); Hedeen 1953: 1 (bion.); Carpenter & LaCasse 1955: 253 (M*, F*, L*; keys); Craig 1956 (E*); Stone & Knight 1956a: 215 (type info., lectotype desig.); Craig & Horsfall 1958 (E*); Craig & Horsfall 1960 (E*); Dodge 1966: 349 (first instar L*; key); Mohrig 1967 (F*); Horsfall et al. 1970: 1710 (E*); O'Meara & Craig 1970a: 476 (tax., distr.); Zavortink 1972: 102 (M*, F*, P*, L*; subgeneric transfer; tax.); Brust 1974: 459 (E*; tax.); Lunt 1977a: 470 (mol. tax.); Bohart & Washino 1978: 42 (M, F, P, L; keys, tax., bion., distr.); Romi et al. 1997: 245 (distr.; Italy); Juliano & Lounibos 2005 (bion.; invasive); Darsie & Ward 2005 (F*, L*; keys, distr.); Reinert, Harbach, & Kitching 2006 (tax., phyl.; to *Georgecraigius* (*Georgecraigius*)); Reinert 2008a: Fig. 1 (F*); Becker et al. 2012b (review, invasive); Medlock et al. 2012 (bion., invasive review; Europe); Scholte et al. 2012 (bion., review); Wilkerson et al. 2015 (phyl.; to *Aedes* (*Georgecraigius*)); Harrison et al. 2016 (F*, L*; keys, distr.); Robert et al. 2019 (distr.; Western Palearctic). **Ety.:** dark feeler (L); palpus. **Informal name:** Black-lipped American Pointy Mosquito. (Vol. 1, p. 418)

epactius Dyar & Knab
1908: 53 (M, F). **Type loc.:** Córdoba [Veracruz], Mexico (USNM). **Distr.:** Costa Rica, El Salvador, Guatemala, Honduras, Mexico, Panama, United States (continental). **Refs.:** Dyar 1928: 221 (F*, P; as *perichares*); Aitken 1942: 166 (syn.); Lane 1953: 691 (M*, F, L*); Stone & Knight 1956a: 217 (type info., lectotype desig.); O'Meara & Craig 1970b: 1398 (from syn. to ssp.); Brust 1974: 459 (E*; tax.); Lunt 1977a: 470 (mol. evidence for a species complex); Darsie & Ward 2005 (F*, L*; keys, distr.). **Ety.:** n.s. [prob. *epactio* (Gr); on the shore]. **Informal name:** Shoreline Mexican Pointy Mosquito.

syn. *perichares* Dyar
1921b: 36 (M, F). **Type loc.:** Ciruelas, Costa Rica (USNM). **Refs.:** Kumm et al. 1940: 416 (syn.); Stone & Knight 1956a: 223 (type info., lectotype desig.); O'Meara & Craig 1970b (from syn. to ssp.); Zavortink 1972: 97 (syn.).

syn. *nielseni* O'Meara & Craig
1970a: 476 (M, F, L). **Type loc.:** Colorado River, near Moab, Grand County, Utah, United States (USNM). **Refs.:** O'Meara & Craig 1970b (tax.); Zavortink 1972: 98 (syn.).

fluviatilis (Lutz)
1904 (A; *Culex*). In Bourroul 1904: 72. **Type loc.:** Rio Grande, near Franca, São Paulo, Brazil (NHMUK). **Distr.:** Argentina, Belize, Bolivia, Brazil, Colombia, Costa Rica, Ecuador, El Salvador, French Guiana, Guatemala, Guyana, Honduras, Mexico, Nicaragua, Panama, Suriname, Uruguay, Venezuela. **Refs.:** Lane 1953: 692 (M*, F, L*), 697 (M*, F, L*; as *lithoecetor*); Forattini & Rabello 1960: 87 (P*, L*); Belkin et al. 1971: 20 (lectotype desig.); Zavortink 1972: 106 (M*, F*, P*, L*; subgeneric transfer; tax.); Pecor et al. 2002: 249 (tax., distr.; Belize); Forattini 2002 (E*); Rossi & Martínez 2003: 472 (distr.; Uruguay); Reinert 2008a (F). **Ety.:** n.s. [of rivers (L)]. **Informal name:** Riverine Brazilian Pointy Mosquito.

syn. *lithoecetor* Dyar & Knab
1907c: 201 (M, F). **Type loc.:** Chagres River, Panama (USNM). **Refs.:** Stone & Knight 1956a: 220 (type info., lectotype desig.); Zavortink 1972: 106 (syn.).

syn. *mediomaculata* Theobald
1907: 245 (M, F*; *Danielsia*). **Type loc.:** Pará, Brazil (NHMUK). **Refs.:** Belkin 1968b: 6 (lectotype desig.).

syn. *tripunctata* Theobald
1907: 247 (F*; *Danielsia*). **Type loc.:** Rio Grande, Brazil (NHMUK). **Refs.:** Belkin 1968b: 8 (type info.).

syn. *draconarius* Dyar
1922g: 194 (M, F). **Type loc.:** Saint Laurent, Maroni River, French Guiana [Guiana] (USNM). **Refs.:** Stone & Knight 1957c: 201 (type info., lectotype desig.).

Subgenus *Geoskusea* Edwards

1929c: 342 (as subg. of genus *Aedes*). Type species: *Aedes* (*Geoskusea*) *fimbripes* Edwards.

Name derivation: n.s. (geo—earth and Skuse or *Skusea*?]

Important references: Mattingly 1959: 47 (tax., keys, bion.; as subg. of genus *Aedes*); Belkin 1962 (tax., keys, bion.); Lee et al. 1984: 1 (tax., key, bion.; Australasia); Reinert 2000h: 177 (to subg. of genus *Ochlerotatus*); Reinert 2002e: 11 (F gen.*); Reinert, Harbach, &Kitching 2004: 289 (phyl., class.; to genus); Reinert, Harbach, & Kitching 2008 (phyl., class.; to subg. of genus *Ochlerotatus*); Reinert, Harbach, & Kitching 2009 (phyl., class.; to genus); Wilkerson et al. 2015 (phyl., class.; to subg. of genus *Aedes*); Soghigian et al. 2017 (mol. phyl.).

baisasi Knight & Hull
1951a: 197 (M*, F; as *Aedes*). **Type loc.:** Iwahig Penal Colony, Palawan, Philippines (USNM). **Distr.:** Japan,

Philippines. **Refs.:** Knight & Hull 1952: 163 (tax.); Mattingly 1959: 50 (M*, F*); Basio 1971b: 20 (M*; bion.); Tanaka et al. 1974: 308 (distr.); Tanaka et al. 1975c: 222 (bion., distr.); Tanaka et al. 1979: 415 M*, F*, L*); Tanaka 2003a: 18 (P*; tax.). **Ety.:** Francisco Edlagan Baisas. **Informal name:** Baisas Philippine Pointy Mosquito.

becki Belkin
1962: 334 (M*, F). **Type loc.:** Roviana, Solomon Islands (USNM). **Distr.:** Solomon Isl. **Refs.:** Lee et al. 1984: 6 (F; key, tax., bion., distr., review). **Ety.:** D. Elden Beck. **Informal name:** Beck Solomon Islander Pointy Mosquito.

daggyi Stone & Bohart
1944: 215 (M*, F, L). **Type loc.:** Espiritu Santo, New Hebrides Islands (USNM). **Distr.:** Solomon Isl., Vanuatu. **Refs.:** Belkin 1962: 335 (M*, F, P*, L*); Lee et al. 1984: 7 (F; key, tax., bion., distr., review). **Ety.:** R.H. Daggy. **Informal name:** Daggy Ni-Vanuatu Pointy Mosquito.

fimbripes Edwards
1924b: 390 (M, F). **Type loc.:** Rabaul, New Britain, Bismarck Archipelago (NHMUK). **Distr.:** Papua New Guinea, Philippines, Solomon Isl. **Refs.:** Lee et al. 1984: 10 (distr.); Reinert 2002e: Fig. 10 (F*). **Ety.:** fringed foot (L); fringe on hind tibia of male. **Informal name:** Fringe-footed Papuan Pointy Mosquito.

kabaenensis Brug
1939: 108 (M*, F*, L*). **Type loc.:** Kabaena Island [Sulawesi] (NHMUK). **Distr.:** Indonesia. **Refs.:** Knight & Hull 1951a: 198 (tax.); Mattingly 1959: 48 (M*, F*, P*, L*). **Ety.:** Kabaena, Indonesia. **Informal name:** Kabaena Indonesian Pointy Mosquito.

longiforceps Edwards
1929 (M). In Paine & Edwards 1929: 315. **Type loc.:** Cape Marsh, Pavuvu, Solomon Islands (NHMUK). **Distr.:** Indonesia (West Papua), Solomon Isl. **Refs.:** Belkin 1962: 337 (M*, F, P*, L*); Lee et al. 1984: 11 (F; key, tax., bion., distr., review). **Ety.:** n.s. [long pincers (L); poss. ref. to "long terminal spine of the male clasper"]. **Informal name:** Long-spined Solomon Islander Pointy Mosquito.

lunulatus King & Hoogstraal
1946c: 95 (M*, F, L*). **Type loc.:** Hollandia [Western New Guinea (Papua)] (USNM). **Distr.:** Indonesia (West Papua). **Refs.:** Lee et al. 1984: 13 (F; key, tax., bion., distr., review). **Ety.:** little moon (L); ref. to dark crescent on posterior pronotum. **Informal name:** Crescent Papuan Pointy Mosquito [*lunulatus* = "little moon" = Crescent].

perryi Belkin
1962: 336 (M*, F). **Type loc.:** Naro village, Guadalcanal, Solomon Islands (USNM). **Distr.:** Indonesia (West Papua), Philippines, Solomon Isl. **Refs.:** Lee et al. 1984: 14 (F; key, tax., bion., distr., review). **Ety.:** William J. Perry. **Informal name:** Perry Solomon Islander Pointy Mosquito.

timorensis (Miyagi, Toma, & Lien)
2004: 107 (M*, F*, P*, L*; *Ochlerotatus*). **Type loc.:** Sulamu Kupang, Timor, Indonesia (USNM). **Distr.:** Indonesia. **Ety.:** Timor Isl. (Kupang, Belu), Indonesia. **Informal name:** Timor Indonesian Pointy Mosquito.

tonsus Edwards
1924b: 390 (M). **Type loc.:** Saparoea, Amboina Island, Moluccas (NHMUK). **Distr.:** Indonesia. **Refs.:** Brug 1934a: 514 (tax.); Mattingly 1965a: 23 (M*). **Ety.:** n.s. [shorn, clipped, cut (L); poss. ref. to "hindtibia with only a few scattered bristles, no fringe"]. **Informal name:** Clipped Indonesian Pointy Mosquito.

Subgenus *Gilesius* Reinert, Harbach, & Kitching

2006: 73 (phyl., class.; as genus). Type species: *Culex pulchriventer* Giles.

Name derivation: George Michael James Giles.

Important references: Reinert 2009d (F gen.*); Reinert, Harbach, & Kitching 2009 (phyl., class.); Rattanarithikul et al. 2010 (F*, L*; keys; Thailand); Wilkerson et al. 2015 (phyl., class.; to subg. of genus *Aedes*).

alius Lien
1968d: 221 (M*, F*, P*, L*). **Type loc.:** Lishan, Hoping Taichung Hsien, Taiwan (TMRI). **Distr.:** Taiwan. **Refs.:** Meng 1981: 246 (from ssp. of *pulchriventer*). **Ety.:** n.s. [other, imposter (L); poss. ref. to difference from *pulchriventer*]. **Informal name:** Imposter Taiwanese Pointy Mosquito.

pulchriventer (Giles)
1901a: 608 (A; *Culex*). **Type loc.:** Naini Tal [United Provinces], India (NHMUK). **Distr.:** India, Pakistan, People's Republic of China, Thailand. **Refs.:** Christophers 1923b (F*); Barraud 1934: 199 (M*, F, L*); Aslamkhan 1971b (distr.; Pakistan); Darsie & Pradhan 1990, 1991 (F, L; tax., keys, bion., distr.; Nepal); Suleman et al. 1993 (distr.; Pakistan); Reinert 2009d: Fig. 1 (F*); Darsie 2009 (P*); Rattanarithikul et al. 2010 (F*, L*; bion, keys, distr.; Thailand). **Ety.:** beautiful venter (L); abdominal venter "elaborately adorned with golden snowy white and black markings." **Informal name:** Beautiful-bellied Indian Pointy Mosquito.
syn. *himalayana* Giles
1904b: 384 (M; *Howardina*). **Type loc.:** Naini Tal, Western Himalayas [United Provinces, India] (NHMUK).

(phyl., class.); Wilkerson et al. 2015 (phyl., class.; to subg. of genus *Aedes*); Soghigian et al. 2017 (mol. phyl.).

albonotatus (Coquillett)
1906a: 183 (M, F; *Gymnometopa*). **Type loc.:** San Francisco mines, Santo Domingo [Dominican Republic] (USNM). **Distr.:** Bahamas, Dominican Republic, Haiti, Venezuela. **Refs.:** Gerry 1932: 40 (F*); Lane 1953: 717 (M*, F, P*, L*); Stone & Knight 1956a: 213 (type info., lectotype desig.); Berlin 1969a: 59 (M*, F*, P*, L*); Belkin & Heinemann 1973: 218 (type loc. info.). **Ety.:** n.s. [observed in white (L); poss. ref. to white markings]. **Informal name:** White-robed Dominican Pointy Mosquito.

allotecnon Kumm, Komp & Ruiz
1940: 417 (A). **Type loc.:** Poás Volcano, Costa Rica (USNM). **Distr.:** Costa Rica, El Salvador, Guatemala, Mexico, Nicaragua, Panama. **Refs.:** Kumm & Komp 1941: 18 (M*, F*, L*); Lane 1953: 713 (M*, F, L*); Stone & Knight 1956a: 214 (type info., lectotype desig.); Berlin 1969a: 99 (M*, F*, P*, L*). **Ety.:** n.s. [of another kind (Gr) and that which is born, child, young, another born (Gr)]. **Informal name:** Newborn Costa Rican Pointy Mosquito.

arborealis Bonne-Wepster & Bonne
1920: 178 (M, F, L). **Type loc.:** Dam, Surinam (ITH). **Distr.:** Brazil, French Guiana, Guyana, Suriname. **Refs.:** Lane 1953: 708 (M*, F, P*, L*); Berlin 1969a: 22 (M*, F*, P*, L*); Reinert 2002e: Fig. 44 (F*). **Ety.:** of trees (L); larvae found in tree holes. **Informal name:** Treehole Surinamese Pointy Mosquito.

argyrites Dyar & Núñez Tovar
1927a: 190 (F). **Type loc.:** Venezuela (USNM). **Distr.:** Venezuela. **Refs.:** Lane 1953: 708 (F); Berlin 1969a: 37 (F*). **Ety.:** silver (Gr); with silvery lines and patches. **Informal name:** Silver Venezuelan Pointy Mosquito.

aureolineatus Berlin
1969a: 25 (M*, F*). **Type loc.:** Pirajá, Bahia, Brazil (USNM). **Distr.:** Brazil. **Ety.:** gold lined (L); golden lines on scutum. **Informal name:** Gold-lined Brazilian Pointy Mosquito.

aurites (Theobald)
1907: 216 (F*; *Howardina*). **Type loc.:** Newcastle [Surrey], Jamaica (NHMUK). **Distr.:** Jamaica. **Refs.:** Lane 1953: 720 (M*, F, P*, L*); Thompson 1956: 199 (syn.); Belkin 1968b: 4 (lectotype desig.); Berlin 1969a: 40 (M*, F*, P*, L*; from syn.); Belkin et al. 1970: 178 (M*, F*, P*, L*). **Ety.:** gild or overlay with gold (L, Gr); gold lines on scutum. **Informal name:** Gilded Jamaican Pointy Mosquito.

aurivittatus Cerqueira
1943a: 7 (F). **Type loc.:** Pampagrande, Florida, Santa Cruz, Bolivia (IOC). **Distr.:** Bolivia. **Refs.:** Lane 1953: 711 (F); Martínez & Prosen 1955: 25 (F*); Martínez et al. 1964 (M*); Berlin 1969a: 118 (M, F*). **Ety.:** gold ribboned (lines) (L); ref. to longitudinal stripes on the scutum. **Informal name:** Gold-banded Bolivian Pointy Mosquito.

bahamensis Berlin
1969a: 61 (M*, F*, P*, L*). **Type loc.:** Hunter, Grand Bahama, Bahama Islands (USNM). **Distr.:** Antigua and Barbuda, Bahamas, Dominican Republic, Guadeloupe, Haiti, United States (continental). **Refs.:** Mattingly 1974a: 43 (E*); Linley 1989b (M*); Darsie & Ward 2005 (F*, L*; keys, distr.). **Ety.:** Bahamas. **Informal name:** Lucayan Bahamian Pointy Mosquito [original inhabitants of the Bahamas].

brevis Berlin
1969a: 82 (M*, F*, P*, L*). **Type loc.:** La Lorra, Chiriquisito, Chiriquí Grande, Bocas del Toro, Panama (USNM). **Distr.:** Costa Rica, Panama. **Ety.:** n.s. [short (L)]. **Informal name:** Short Panamanian Pointy Mosquito.

brevivittatus Berlin
1969a: 116 (M*, F*). **Type loc.:** Zorrocucho, Azuay, Ecuador (USNM). **Distr.:** Ecuador. **Ety.:** n.s. [short ribboned or lined (L); poss. ref. to lines of white scales on scutum]. **Informal name:** Short-striped Ecuadorian Pointy Mosquito.

busckii (Coquillett)
1906e: 60 (M, F; *Stegomyia*). **Type loc.:** [Dominica Island, Lesser Antilles] (erroneously given as Santo Domingo) (USNM). **Distr.:** Lesser Antilles (includes Dominica, Grenada, Guadeloupe, Martinique, Montserrat, Saint Kitts and Nevis, Saint Lucia). **Refs.:** Lane 1953: 722 (M*, F, P*, L*); Stone & Knight 1956a: 215 (type info., lectotype desig.); Belkin, Schick, & Heinemann 1965: 33 (syn.); Berlin 1969a: 65 (M*, F*, P*, L); Mattingly 1974a: 42 (E*). **Ety.:** August Busck. **Informal name:** Busck Dominican Pointy Mosquito.
syn. *thaxteri* Dyar & Knab
1919: 8 (F, L). **Type loc.:** Grand Etang, Grenada Island, Lesser Antilles (USNM).
syn. *tracei* Senevet & Quievreux
1941: 251 (M*, F, P*, L*). **Type loc.:** Colson, Martinique Island [Lesser Antilles] (LPFM).

cozumelensis Díaz Nájera
1966: 334 (M*, F*, P*, L*). **Type loc.:** San Miguel de Cozumel, Quintana Roo, Mexico (ISET). **Distr.:** Belize, Mexico. **Refs.:** Berlin 1969a: 56 (M*, F*, P*, L*). **Ety.:** San Miguel de Cozumel, Quintana Roo, Mexico. **Informal name:** Cozumel Mexican Pointy Mosquito.

ecuadoriensis Berlin
1969a: 115 (M*, F*). **Type loc.:** Leon River, Saraguro, Azuay, Ecuador (USNM). **Distr.:** Ecuador. **Ety.:** Ecuador. **Informal name:** Quichua Ecuadorian Pointy Mosquito [one of the 13 indigenous languages recognized in modern Ecuador].

eleanorae Berlin
1969a: 105 (M*, F*, P*, L*). **Type loc.:** Buriticá, Antioquia, Colombia (USNM). **Distr.:** Colombia. **Ety.:** Eleanor Berlin. **Informal name:** Eleanor Colombian Pointy Mosquito.

fulvithorax (Lutz)
1904 (A; as *Haemagogus*). In Bourroul 1904: 47, 66. **Type loc.:** Ponte Ipe Arcado, Goiás, Brazil (NE). **Distr.:** Bolivia, Brazil, Colombia, French Guiana, Guyana, Peru, Suriname, Trinidad and Tobago, Venezuela. **Refs.:** Lane 1953: 706 (M*, F, L*); Martínez & Prosen 1955: 29 (F*); Berlin 1969a: 15 (M*, F*, P*, L*); Belkin et al. 1971: 20 (type info.); Reinert 2002e: Fig. 43 (F*). **Ety.:** tawny thorax (L); ref. to golden-brown scutum. **Informal name:** Tawny Brazilian Pointy Mosquito.
syn. *palliatus* Coquillett
1906e: 61 (F; *Taeniorhynchus*). **Type loc.:** Trinidad [Trinidad & Tobago] (USNM).
syn. *tachirensis* Anduze
1947: 355 (M). **Type loc.:** Las Mesas de Seboruco, Department Jaurequi Tachira, Venezuela (A). **Refs.:** Lane 1951 (syn.).

grabhami Berlin
1969a: 42 (M*, F*, P*, L*; syn.). **Type loc.:** Near Hope Farm, Porus, Manchester, Jamaica (USNM). **Distr.:** Jamaica. **Refs.:** Belkin et al. 1970: 180 (M*, F*, P*, L*). **Ety.:** Michael Grabham. **Informal name:** Michael Jamaican Pointy Mosquito.
syn. *aureostriata* Grabham
1906a: 171 (M, F*, P, L; *Howardina*; not Doleschall 1857). **Type loc.:** Newcastle [Surrey], Jamaica (NE). **Refs.:** Stone & Knight 1956a: 215 (type info.).

guatemala Berlin
1969a: 103 (F*). **Type loc.:** Yepocapa, Chimaltenango, Guatemala (USNM). **Distr.:** Guatemala. **Ety.:** Guatemala. **Informal name:** Maya Guatemalan Pointy Mosquito [Guatemala was the center of the Maya civilization].

guerrero Berlin
1969a: 102 (F*, L*). **Type loc.:** 21 mi south of Chilpancingo, Guerrero, Mexico (USNM). **Distr.:** Guatemala, Mexico. **Ety.:** Guerrero, Mexico. **Informal name:** Guerrero Mexican Pointy Mosquito.

inaequalis (Grabham)
1907: 25 (F, L; *Howardina*). **Type loc.:** Kingston [Surrey], Jamaica (USNM). **Distr.:** Jamaica. **Refs.:** Thompson 1956: 199 (M, F, L*; from syn. with *aurites*); Stone & Knight 1956a: 219 (type info., lectotype desig.); Berlin 1969a: 46 (M*, F*, P*, L*); Belkin et al. 1970: 182 (M*, F*, P*, L*); Reinert 2002e: Fig. 45 (F*). **Ety.:** n.s. [unequal (L); poss. ref. to near *aureostriata*]. **Informal name:** Unequal Jamaican Pointy Mosquito.

ioliota Dyar & Knab
1913: 77 (F). **Type loc.:** Trinidad [Trinidad & Tobago] (USNM). **Distr.:** Trinidad and Tobago, Venezuela. **Refs.:** Lane 1953: 719 (F); Stone & Knight 1956a: 219 (type info.); Suárez & Cova García 1961: 17 (M*, F*, P*, L*, E*); Berlin 1969a: 54 (M*, F*, P*, L*); Amarista 1996: 78 (distr.; Venezuela). **Ety.:** n.k. **Informal name:** Ioliote Trinbagonian Pointy Mosquito.

leei Berlin
1969a: 110 (M*, F*, P*, L*). **Type loc.:** Parace, Huila, Colombia (USNM). **Distr.:** Bolivia, Colombia, Peru. **Refs.:** Parker et al. 2012 (bion., distr.; Peru). **Ety.:** Vernon Harold Lee. **Informal name:** Lee Colombian Pointy Mosquito.

lorraineae Berlin
1969a: 80 (M*, L*). **Type loc.:** Vivero Rancho Teja, Ixtlán de Juárez, Oaxaca, Mexico (USNM). **Distr.:** Mexico. **Refs.:** Berlin 1975: 127 (M*, P*, L). **Ety.:** Lorraine Belkin. **Informal name:** Lorraine Mexican Pointy Mosquito.

marinkellei Berlin
1969a: 107 (F*, P*, L*). **Type loc.:** Suba, Bogotá, Cundinamarca, Colombia (USNM). **Distr.:** Colombia. **Ety.:** Cornelis Johannes Marinkelle. **Informal name:** Marinkelle Colombian Pointy Mosquito.

martinezi Berlin
1969a: 119 (F*). **Type loc.:** La Caldera, Salta, Argentina (USNM). **Distr.:** Anguilla, Argentina, Bolivia. **Ety.:** Antonia Martínez. **Informal name:** Martínez Argentinian Pointy Mosquito.

osornoi Berlin
1969a: 112 (M*, F*, P*, L*). **Type loc.:** "Chorro de Padilla," Rio San Francisco, Bogotá, Cundinamarca, Colombia (USNM). **Distr.:** Colombia, Venezuela. **Ety.:** Ernesto Osorno Mesa. **Informal name:** Osorno Colombian Pointy Mosquito.

pseudodominicii Komp
1936a: 75 (M*). **Type loc.:** Restrepo, Meta, Colombia (NE). **Distr.:** Argentina, Bolivia, Colombia. **Refs.:** Antunes 1937b: 16 (syn.); Berlin 1969a: 96 (M*, F*; from syn.). **Ety.:** false (Gr); similar to *dominicii*. **Informal name:** Pretender Colombian Pointy Mosquito [*pseudodominicii* = "false lord" = Pretender to the throne].
syn. *delpontei* Martínez & Prosen
1955: 35 (M*, F*). **Type loc.:** Chapare Province, Cochabamba, Bolivia (BA). **Refs.:** Berlin 1969a: 96 (syn.).

quadrivittatus (Coquillett)
1902d: 293 (F; *Culex*). **Type loc.:** Chacula, Guatemala (USNM). **Distr.:** Colombia, Costa Rica, Guatemala, Honduras, Mexico, Panama. **Refs.:** Kumm & Komp 1941: 22 (M, F*, L*); Lane 1953: 711 (M*, F, L*); Stone &

Knight 1956a: 224 (type info.); Berlin 1969a: 77 (M*, F*, P*, L*). **Ety.:** four bands (L); "scales of mesonotum black and with four vittae." **Informal name:** Four-banded Guatemalan Pointy Mosquito.

septemstriatus Dyar & Knab
1907a: 10 (F). **Type loc.:** Bluefields, Nicaragua (USNM). **Distr.:** Colombia, Costa Rica, Guyana, Nicaragua, Panama. **Refs.:** Lane 1953: 717 (M*, F); Stone & Knight 1956a: 225 (type info.); Berlin 1969a: 19 (M*, F*, P*, L*). **Ety.:** seven stria (L); thorax with seven narrow golden lines. **Informal name:** Seven-lined Nicaraguan Pointy Mosquito.

sexlineatus (Theobald)
1901a: 308 (F*; *Stegomyia*). **Type loc.:** Aqua Santa, Trinidad [Trinidad & Tobago] (NHMUK). **Distr.:** Colombia, Costa Rica, Ecuador, Honduras, Mexico, Nicaragua, Panama, Trinidad and Tobago, Venezuela. **Refs.:** Lane 1953: 715 (F); Berlin 1969a: 87 (M*, F*, P*, L*, E; syn.); Mattingly 1974a: 43 (E*). **Ety.:** six lines (L); ref. to six pale lines on the scutum. **Informal name:** Six-lined Trinbagonian Pointy Mosquito.
syn. *dominicii* Rangel & Romero Sierra
1907 (F*; *Stegomyia*). In Romero Sierra 1907: 48. **Type loc.:** "Rio Ananco y de la hacienda 'El Paraiso,' situada a sus orillas," Venezuela (LU). **Refs.:** Lane 1953: 700 (M*, F).

spinosus Berlin
1969a: 84 (M*, F, P*, L*). **Type loc.:** Near Cervantes, Cartago, Costa Rica (USNM). **Distr.:** Costa Rica, Honduras, Panama. **Ety.:** n.s. [thorny, prickly (L)]. **Informal name:** Spiny Costa Rican Pointy Mosquito.

stenei Thompson
1956: 196 (M, F, L*). **Type loc.:** Hope Farm, Clarendon Parish, Jamaica (USNM). **Distr.:** Jamaica. **Refs.:** Berlin 1969a: 49 (M*, F*, P*, L*); Belkin et al. 1970: 183 (M*, F, P*, L*). **Ety.:** Andrew E. Stene. **Informal name:** Stene Jamaican Pointy Mosquito.

vanemdeni Martini
1931a: 208 (M*). **Type loc.:** Yunga Coroico [La Paz], Bolivia (NHMUK). **Distr.:** Bolivia. **Refs.:** Lane 1953: 716 (M*; as *Aedes*); Belkin 1968b: 8 (lectotype desig.); Berlin 1969a: 120 (M*). **Ety.:** n.s. [Wolfgang van Emden (Harbach 2018c)]. **Informal name:** van Emden Bolivian Pointy Mosquito.

walkeri (Theobald)
1901a: 424 (A; *Culex*). **Type loc.:** Jamaica (NHMUK). **Distr.:** Jamaica. **Refs.:** Lane 1953: 725 (M*, F*, P*, L*); Berlin 1969a: 33 (M*, F*, P*, L*); Belkin et al. 1970: 175 (M*, F*, P*, L*); Reinert 2002e: Fig. 46 (F*). **Ety.:** Edmund Murton Walker. **Informal name:** Walker Jamaican Pointy Mosquito.

whitmorei Dunn
1918: 128 (F). **Type loc.:** Muzo, Colombia (USNM). **Distr.:** Colombia, Venezuela. **Refs.:** Komp 1936a: 73 (M*, F, L*; as *dominicii*); Antunes 1937b: 16 (tax.); Lane 1953: 701 (M*, F, L); Stone & Knight 1956a: 226 (type info., lectotype desig.); Berlin 1969a: 94 (M*, F*, P*, L*). **Ety.:** Eugene R. Whitmore. **Informal name:** Whitmore Colombian Pointy Mosquito.

Subgenus *Huaedes* Huang

1968b: 169 (as subg. of genus *Aedes*). Type species: *Huaedes wauensis* Huang.

Name derivation: Stephen M.K. Hu.

Important references: Lee et al. 1984: 30 (tax., key, bion.; Australasia); Reinert 2000h (as subg. of genus *Aedes*); Reinert, Harbach, & Kitching 2004 (phyl., class.; to genus); Reinert, Harbach, & Kitching 2009 (phyl., class.); Wilkerson et al. 2015 (phyl., class.; to subg. of genus *Aedes*).

medialis (Brug)
1932: 79 (M*, F; *Leptosomatomyia*). **Type loc.:** Upper Digoel River [Papua] (NHMUK). **Distr.:** Indonesia. **Refs.:** Huang 1968b: 169 (tax.); Lee et al. 1984: 33 (F; key, tax., bion., distr., review). **Ety.:** n.s. [medium, middle (L)]. **Informal name:** Midline Papuan Pointy Mosquito.

variepictus King & Hoogstraal
1946c: 98 (F; as *Aedes* subg. *Leptosomatomyia*). **Type loc.:** Mount Dafonsero, Cyclops Range, Hollandia Area [Western New Guinea (Papua)] (USNM). **Distr.:** Indonesia. **Refs.:** Huang 1968b: 169 (tax.); Lee et al. 1984: 35 (F; key, tax., bion., distr., review). **Ety.:** n.s. [different or diversely marked (L)]. **Informal name:** Patterned Papuan Pointy Mosquito.

wauensis Huang
1968b: 170 (M*, F*, P*, L*). **Type loc.:** Wau, NE New Guinea [Indonesia (West Papua)] (BBM). **Distr.:** Indonesia (West Papua), Papua New Guinea. **Refs.:** Lee et al. 1984: 37 (F; key, tax., bion., distr., review); Reinert 2000e: Fig. 22 (F*). **Ety.:** Wau, Papua New Guinea. **Informal name:** Wau Papuan Pointy Mosquito.

Subgenus *Hulecoeteomyia* Theobald

1904c: 163 (as genus). Type species: *Hulecoeteomyia trilineata* Leicester.

Name derivation: "True forest fly" (Gr) (Harbach 2018c).

Important references: Knight 1968 (tax., class.; Chysolineatus Subgroup of subg. *Finlaya* of genus *Aedes*); Reinert 2000e (subg. *Finlaya*, in part, of genus *Ochlerotatus*); Reinert 2002e (F gen.*; in part, Chrysolineatus Assemblage of subg. *Finlaya* of genus *Ochlerotatus*); Reinert, Harbach, & Kitching 2006: 92

(phyl., class.; to genus); Reinert 2008d (F gen.*); Reinert, Harbach, & Kitching 2009 (phyl., class.); Rattanarithikul et al. 2010 (F*, L*; keys, bion.; Thailand); Wilkerson et al. 2015 (phyl., class.; to subg. of genus *Aedes*); Soghigian et al. 2017 (phyl.).

bhutanensis Somboon & Harbach
2020 (F*, P*, L*; keys, mol. tax., phyl.). In Somboon et al. 2020. **Type loc.:** Sarpang District, Bhutan (26°58′ 57″ N; 90 °20′ 02″ E, 1200 m AMSL), about 4 km north of Dolumgang on the way to the village of Lhayul (QSBG). **Distr.:** Bhutan. **Ety.:** from Bhutan. **Informal name:** Sarpang Bhutanese Pointy Mosquito.

cherrapunjiensis (Natarajan, Rajavel, & Jambulingam)
2016: 330 (M*, F*, P, L*; keys; *Hulecoeteomyia*; distr.; India). **Type loc.:** Meghalaya (25°16624' N, 91°41558' E), Cherrapunji, India (VCRC). **Distr.:** India. **Ety.:** Meghalaya, Cherrapunji, India. **Informal name:** Cherrapunji Indian Pointy Mosquito.

chrysolineatus (Theobald)
1907: 218 (F*; *Howardina*). **Type loc.:** Pundabroya [Pundaluoya], Ceylon [Sri Lanka] (NHMUK). **Distr.:** Bangladesh, Cambodia, India, Indonesia, Laos, Malaysia, Nepal, Philippines, Sri Lanka, Thailand, Vietnam. **Refs.:** Barraud 1923g (L*; as *trilineata*); Barraud 1934: 185 (M*, F*, L*); Knight 1948c: 636 (M*, F, L); Joshi et al. 1965 (distr.; Nepal); Knight 1968: 8 (M*, F*, P, L*); Ahmed 1987 (distr.; Bangladesh); Darsie & Pradhan 1990 (F, L; tax., keys, bion., distr.; Nepal); Reinert 2002e: Fig. 2 (F*); Reinert 2008d: Fig. 1 (F*); Rattanarithikul et al. 2010 (F*, L*; keys, bion., distr.; Thailand). **Ety.:** gold lined (Gr, L); golden stripes on scutum. **Informal name:** Gold-banded Sri Lankan Pointy Mosquito.

syn. *trilineata* Leicester
1904 (M, F; *Hulecoeteomyia*; not Theobald 1901c). In Theobald 1904c: 163. **Type loc.:** Kuala Lumpur, Malaya [Malaysia] (NHMUK).

syn. *ceylonica* Theobald
1910c: 391 (F; *Culex japonicus* var.). **Type loc.:** Peradeniya [Central], Sri Lanka (NHMUK). **Refs.:** Knight 1948c: 636 (syn.).

formosensis Yamada
1921: 67 (M). **Type loc.:** Kakubanzan, Formosa [Taiwan, Republic of China] (IID). **Distr.:** Bangladesh, India, Indonesia, Malaysia, Nepal, People's Republic of China, Taiwan, Thailand, Vietnam. **Refs.:** Knight 1948c: 639 (M, F, L; tax.); Knight 1968: 10 (M*, F*, P, L*; syn.); Matsuo, Yoshida, & Lien 1974: 295 (E*); Ahmed 1987 (distr.; Bangladesh); Rattanarithikul & Harrison 1988: 82 (lectotype desig.); Darsie, Pradhan, & Vaidya 1991 (distr.; Nepal); Rattanarithikul et al. 2010 (F*, L*; keys, bion., distr.; Thailand). **Ety.:** Formosa. **Informal name:** Formosa Taiwanese Pointy Mosquito.

harveyi (Barraud)
1923a: 407 (M, F; *Finlaya*). **Type loc.:** Kurseong, Darjeeling [District, West Bengal], India (NHMUK). **Distr.:** Cambodia, India, Indonesia, Malaysia, Nepal, People's Republic of China, Sri Lanka, Taiwan, Thailand, Vietnam. **Refs.:** Barraud 1934: 188 (M*, F*); Carter & Wijesundara 1948: 141 (L*); Knight 1948c: 638 (tax.); Darsie et al. 1993 (distr.; Nepal); Rattanarithikul et al. 2010 (F*, L*; keys, bion., distr.; Thailand). **Ety.:** n.s. **Informal name:** Harvey Indian Pointy Mosquito.

japonicus (Theobald)
1901a: 385 (F; *Culex*). **Type loc.:** Tokyo, [Honshu], Japan (NHMUK). **Distr.:** Austria, Belgium, Canada, Croatia, France, Germany, Hungary, Israel, Italy, Japan, Liechtenstein, Luxembourg, Netherlands, Panama, People's Republic of China, Russia, Slovenia, South Korea, Switzerland, Taiwan, Turkey, United States (continental, Hawaiʻi). **Refs.:** Knight 1948c: 631 (M, F, L*); La Casse & Yamaguti 1950: 151 (M*, F*, L*; bion., distr., tax.); Hara 1957: 61 (F*); Mohrig 1967 (F*); Knight 1968: 15 (M*, F, L*); Miyagi 1971: 142 (A, L; distr.); Matsuo, Yoshida, & Kunou 1972: 360 (E*); Moriya et al. 1973 (E*); Tanaka et al. 1975c: 220 (tax., distr.); Tanaka et al. 1979: 308 (M*, F*, P*, L*; keys, tax., review, bion.); Peyton et al. 1999: 238 (F*, L*; distr.); Fonseca et al. 2001 (mol. tax.; *s.l.*); Tanaka 2002a: 64 (P*; tax., key); Schaffner et al. 2003 (distr.; France); Darsie & Ward 2005 (F*, L*; keys, distr.); Haddow et al. 2009 (E*); Harrison et al. 2016 (F*, L*; keys, distr., tax.); Robert et al. 2019 (distr.; Western Palearctic). **Ety.:** Japan. **Informal name:** Tokyo Japanese Pointy Mosquito. (Vol. 1, p. 420)

syn. *eucleptes* Dyar
1921i: 147 (M, F). **Type loc.:** Canton (Christian Coll), China [People's Republic of China] (USNM). **Refs.:** Edwards 1922b: 465 (syn. with *japonicus*); Tanaka et al. 1979: 308 (*japonicus* review; type info.).

ssp. ***shintienensis*** Tsai & Lien
1950: 177 (M*, F*, P*, L*). **Type loc.:** Chuchih, Shintien (Chen), Taipeh (Hsien), Formosa [Taiwan, Republic of China] (NTU?). **Distr.:** Taiwan. **Refs.:** Lien 1962: 623 (to ssp.); Miyagi 1971: 142 (syn.); Tanaka et al. 1979: 317 (M*, F*; from syn. with *japonicus*); Tanaka 2002a: 72 (P*; tax., key). **Ety.:** Shintien, Taiwan. **Informal name:** Shintien Taiwanese Pointy Mosquito.

syn. *tokushimaensis* Tanimura
1952: 176 (A; *Aedes*). **Type loc.:** Tokshima, Shikoku, Japan (ND). **Refs.:** Sasa & Kamimura 1971: 16 (syn.).

syn. *bisanensis* Suzuki, Tanimura, Miyagawa, & Murata
1953: 8 (A). **Type loc.:** Tokushima, Shikoku, Japan (ND). **Refs.:** Sasa & Kamimura 1971: 16 (syn.).

ssp. ***amamiensis*** Tanaka, Mizusawa, & Saugstad
1979: 312 (M, F*, L). **Type loc.:** Mount Yuwan, Amami Ōshima, Ryukyus, Japan (NMST). **Distr.:** Japan. **Refs.:**

Tanaka 2002a: 67 (P*; tax., key). **Ety.:** Amami, Tanzania. **Informal name:** Amami Japanese Pointy Mosquito.

ssp. ***yaeyamensis*** Tanaka, Mizusawa, & Saugstad 1979: 313 (M, F*, L). **Type loc.:** Near Yonehara, Ishigaki Island, Ryukyus, Japan (NMST). **Distr.:** Japan. **Refs.:** Tanaka 2002a: 70 (P*; tax., key). **Ety.:** Yaeyama Isl., Japan. **Informal name:** Yaeyama Japanese Pointy Mosquito.

jugraensis (Leicester)
1908: 109 (M, F; *Hulecoeteomyia*). **Type loc.:** Jugra, Malaya [Malaysia] (NE). **Distr.:** Indonesia, Malaysia, Philippines, Singapore, Thailand. **Refs.:** Knight 1948c: 647 (M, F, L); Knight & Hull 1951b: 242 (L*); Mattingly 1956b: 25 (lectotype desig.); Knight 1968: 19 (M*, F*, P, L*); Basio 1971b: 15 (M*; bion.); Rattanarithikul et al. 2010 (F*, L*; keys, bion., distr.; Thailand). **Ety.:** Jugra, Malaysia. **Informal name:** Jugra Malaysian Pointy Mosquito.

koreicus (Edwards)
1917: 212 (M, F; *Ochlerotatus*). **Type loc.:** Korea (NHMUK). **Distr.:** Belgium, Germany, Hungary, Italy, Japan, People's Republic of China, Russia (Southern Dists.), Slovenia, South Korea, Switzerland, Taiwan. **Refs.:** Knight 1948c: 633 (M*, F, L*; tax.); La Casse & Yamaguti 1950: 156 (M*, F*, L*; bion., distr., tax.); Knight 1968 (tax.; Chrysolineatus Group); Miyagi 1971: 147 (A, L; tax.); Gutsevich et al. 1974: 315 (M*, F, L*); Tanaka et al. 1979: 322 (M*, F*, P*, L*; keys, tax., review, bion.); Capelli et al. 2011 (distr.; Italy); Versteirt et al. 2012a, b (bion, distr.; Belgium); Robert et al. 2019 (distr.; Western Palearctic). **Ety.:** Korea. **Informal name:** Joseon Korean Pointy Mosquito [Joseon was the official name of Korea between 1392 and 1897]. (Vol. 1, p. 422)

nigrorhynchus Brug
1931c: 28 (M, F, L*). **Type loc.:** Djajasana, Preanger Regentschappen, Java [Indonesia] (NE). **Distr.:** Indonesia. **Refs.:** Knight 1968: 20 (M, F; new). **Ety.:** n.s. [black beak (Gr)]. **Informal name:** Black-beaked Indonesian Pointy Mosquito.

pallirostris Edwards
1922a: 270 (F). **Type loc.:** Golaghat, Sibsagar District, Assam [India] (NHMUK). **Distr.:** India, Thailand. **Refs.:** Barraud 1924a: 854 (M*, F*; as *khasiana*); Barraud 1934: 190 (F*; Pl. III, Fig. 7); Knight 1948c: 640 (F); Knight 1968: 10 (syn.); Tewari & Hiriyan 1996: 713 (M*, F*, P*, L*; from syn. with *formosensis*); Rattanarithikul et al. 2010 (F*, L*; keys, bion., distr.; Thailand). **Ety.:** covered beak (L); ref. to all-pale proboscis. **Informal name:** White-beaked Indian Pointy Mosquito.

syn. *khasiana* Barraud
1923a: 407 (M, F; *Finlaya*). **Type loc.:** Shillong, Khasi Hills, Assam, India (NHMUK). **Refs.:** Rattanarithikul & Harrison 1988: 83 (lectotype desig.); Tewari & Hiriyan 1996: 713 (transfer from syn. with *formosensis*).

reinerti Rattanarithikul & Harrison
1988: 78 (M*, F*, P*, L*). **Type loc.:** Doi Inthanon, Chiang Mai Province, Thailand (USNM). **Distr.:** Thailand. **Refs.:** Rattanarithikul et al. 2010 (F*, L*; keys, bion., distr.; Thailand). **Ety.:** John Francis Reinert. **Informal name:** Reinert Thai Pointy Mosquito.

rizali (Banks)
1906b: 999 (F; *Culex*). **Type loc.:** Canlaon volcano, Mount Siya-Siya, Negros Occidental, Negros Island, Philippines (NE). **Distr.:** Philippines. **Refs.:** Knight & Hull 1951b: 241 (tax.); Knight 1968: 22 (M*, F*, P, L*); Basio 1971b: 18 (M*; bion.). **Ety.:** José Rizal. **Informal name:** Rizal Philippine Pointy Mosquito.

syn. *abadsantosi* Baisas
1946c: 25 (M, F, P*, L*). **Type loc.:** Llavac, Infanta, Tayabas [= Quezon] Luzon, Philippines (PBH). **Refs.:** Knight 1948c: 641 (M, F*, L*); Knight 1968: 22 (syn.).

syn. *burgosi* Baisas
1946c: 27 (M, F, P, L*). **Type loc.:** Titunod Creek, Kolambugan, Lanao, Mindanao, Philippines (PBH). **Refs.:** Knight & Hull 1951b: 240 (M, L*; tax.); Knight 1968: 22 (syn.).

saxicola Edwards
1922a: 262, 1922b: 466 (new name for *fluviatilis* Leicester, not Lutz 1904). **Distr.:** Cambodia, India, Indonesia, Malaysia, People's Republic of China, Philippines, Thailand, Vietnam. **Refs.:** Barraud 1923g (L*; as *greigi*); Barraud 1934: 191 (M*, F*, L*); Knight 1948c: 628 (M*, F, L*); Knight & Hull 1951b: 239 (tax.); Knight 1968: 25 (M*, F, P*, L*); Basio 1971b: 19 (M*; bion.); Rattanarithikul et al. 2010 (F*, L*; bion., distr., keys). **Ety.:** n.s. [rock, large stone inhabitant (L); prob. ref. to larvae found in rock holes]. **Informal name:** Rock-dwelling Malaysian Pointy Mosquito.

syn. *fluviatilis* Leicester
1908: 111 (M, F; *Hulecoeteomyia*). **Type loc.:** Ulu Gombak Jungle [Selangor], Malaya (NE).

syn. *greigi* Barraud
1923a: 406 (M, F; *Finlaya*). **Type loc.:** Haflong, Cachar Hills, Assam, India (NHMUK).

sherki Knight
1948c: 645 (M*, F, L*). **Type loc.:** City of Baguio, Baguio Province, Luzon, Philippines (USNM). **Distr.:** Malaysia, Philippines. **Refs.:** Basio 1971b: 19 (M*; bion.). **Ety.:** Herman D. Sherk. **Informal name:** Sherk Philippine Pointy Mosquito.

yunnanensis (Gaschen)
1934: 331 (M*, F*, L*; *Finlaya*). **Type loc.:** Mountains of Si-Chang, Yunnan, China [People's Republic of China] (PIH). **Distr.:** People's Republic of China. **Ety.:** Yunnan Prov., People's Republic of China. **Informal name:** Yunnan Chinese Pointy Mosquito.

Subgenus *Indusius* Edwards

1934 (as subg. of genus *Aedes*). In Barraud 1934: 268, 445. Type species: *Aedes pulverulentus* Edwards.

Name derivation: Indus Valley, India.

Important references: Reinert 1976c (tax., revision); Reinert 2000e: 30 (F gen.*); Reinert, Harbach, & Kitching 2004 (phyl., class.; to genus); Reinert, Harbach, & Kitching 2009 (phyl., class.); Wilkerson et al. 2015 (phyl., class.; to subg. of genus *Aedes*).

pulverulentus Edwards
1922a: 273 (F). **Type loc.:** Nowshera, NW [Frontier Province, Pakistan] (NHMUK). **Distr.:** India, Pakistan. **Refs.:** Barraud 1934: 268 (M, F); Aslamkhan 1971b (distr.; Pakistan); Reinert 1976c: 4 (M*, F*; tax., distr.); Reinert 2000e: Fig. 23 (F*). **Ety.:** n.s. [dusty (L)]. **Informal name:** Dusty Pakistani Pointy Mosquito.

Subgenus *Isoaedes* Reinert

1979: 144 (as subg. of genus *Aedes*). Type species: *Aedes cavaticus* Reinert.

Name derivation: n.s. [prob. equal, similar to *Aedes* (Gr)].

Important references: Reinert 2000e (F gen.*); Reinert, Harbach, & Kitching 2004: 289 (to genus); Harbach, Rattanarithikul et al. 2007 (tax.; comp. to *Borichinda* and *Ayurakitia*); Reinert, Harbach, & Kitching 2009 (phyl., class.); Rattanarithikul et al. 2010 (F*, L*; keys, bion.; Thailand); Wilkerson et al. 2015 (phyl., class.; to subg. of genus *Aedes*).

cavaticus Reinert
1979: 148 (M*, F*, P*, L*). **Type loc.:** Gang Lawa Cave, Huai Bong Ti, Kanchanaburi, Thailand (USNM). **Distr.:** Thailand. **Refs.:** Reinert 2000e: Fig. 24 (F*); Rattanarithikul et al. 2010 (F*, L*; keys, bion., distr.; Thailand). **Ety.:** dwelling in caves (L); prob. ref. to larval habitat. **Informal name:** Cave-dwelling Thai Pointy Mosquito.

Subgenus *Jarnellius* Reinert, Harbach, & Kitching

2006: 76 (as genus). Type species: *Culex varipalpus* Coquillett.

Name derivation: J. Hal Arnell.

Important references: Arnell & Nielsen 1972 (as Varipalpus Group of *Aedes* subg. *Ochlerotatus*; tax., keys, revision); Reinert, Harbach, & Kitching 2009 (phyl., class.); Reinert 2010b (F gen.*); Wilkerson et al. 2015 (phyl., class.; to subg. of genus *Aedes*).

deserticola Zavortink
1970a: 1 (M*, F, P*, L*). **Type loc.:** 5 mi southwest of Littlerock, San Gabriel Mountains, Los Angeles County, California, United States (USNM). **Distr.:** United States (continental). **Refs.:** Arnell & Nielsen 1972: 21 (M*, F*, P*, L*); Bohart & Washino 1978: 46 (M, F*, L*; keys, tax., bion., distr.); Darsie & Ward 2005 (F*, L*; keys, distr.). **Ety.:** waste (L); desert inhabiting. **Informal name:** Desert-dwelling American Pointy Mosquito.

laguna Arnell & Nielsen
1972: 12 (M*, F*, P*, L*). **Type loc.:** Sierra Laguna, 20 mi ENE of Todos Santos, Baja California Sur, Mexico (USNM). **Distr.:** Mexico, United States (continental). **Ety.:** Sierra Laguna [La Laguna meadow?], Baja California Sur, Mexico. **Informal name:** Laguna Mexican Pointy Mosquito.

monticola Belkin & McDonald
1957: 179 (M*, F*, P*, L*). **Type loc.:** Madera Canyon, Santa Cruz County, Arizona, United States (USNM). **Distr.:** Mexico, United States (continental). **Refs.:** Arnell & Nielsen 1972: 14 (M*, F*, P*, L*); Lunt 1976: 200 (tax.); Heinemann & Belkin 1977c: 527 (distr.); Darsie & Ward 2005 (F*, L*; keys, distr.). **Ety.:** n.s. [mountain inhabitant (L)]. **Informal name:** Mountain-dwelling American Pointy Mosquito.

sierrensis (Ludlow)
1905b: 231 (M, F; *Taeniorhynchus*). **Type loc.:** Three Rivers (?), Sierra Nevada, California, United States (USNM). **Distr.:** Canada, Cuba, Puerto Rico, United States (continental). **Refs.:** Belkin & McDonald 1956a: 54 (from syn. with *varipalpus*); Belkin & McDonald 1956c (tax.; as western tree hole mosquito); Stone & Knight 1956a: 225 (type info., lectotype desig.); Belkin & McDonald 1957: 190 (tax.); Dodge 1966: 356 (first instar L; key); Myers 1967: 805 (E*); Cupp & Horsfall 1969: 51 (as *Finlaya*); Horsfall et al. 1970: 1710 (E*); Arnell & Nielsen 1972: 23 (M*, F*, P*, L*; as *Ochlerotatus*); Lunt 1976 (tax.); Bohart & Washino 1978: 115 (M, F*, L*; keys, tax., bion., distr.); Darsie & Ward 2005 (F*, L*; keys, distr.). **Ety.:** Sierra Nevada, California, United States. **Informal name:** Sierra Nevada American Pointy Mosquito. Official ESA common name: western treehole mosquito.

varipalpus (Coquillett)
1902d: 292 (F; *Culex*). **Type loc.:** Williams, Arizona, United States (USNM). **Distr.:** Canada, United States (continental). **Refs.:** Blasidell 1906; Gjullin 1937 (F*); Yamaguti & LaCasse 1951d: 238 (M*, F*, L*); Belkin & McDonald 1956a: 54 (tax., type info.); Craig 1956 (E*); Belkin & McDonald 1957: 190 (subgeneric transfer); Craig & Horsfall 1958 (E*); Myers 1967: 806 (E*); Laffoon

& Knight 1971 (F*); Arnell & Nielsen 1972: 17 (M*, F*, P*, L*); Lunt 1976: 200 (tax.); Harbach & Knight 1980 (F*); Darsie & Ward 2005 (F*, L*; keys, distr.); Reinert 2010b: Fig. 1 (F*). **Ety.:** n.s. [contrasting palpus (L); "same as *atropalpus*: Scales on apices of palpi . . . near the middle." **Informal name:** Spot-lipped American Pointy Mosquito.

Subgenus *Jihlienius* Reinert, Harbach, & Kitching

2006: 79 (as genus). Type species: *Aedes* (*Finlaya*) *chungi* Lien.

Name derivation: Jih Ching Lien.

Important references: Reinert 2009 (F gen.*); Reinert, Harbach, & Kitching 2009d (phyl., class.); Rattanarithikul et al. 2010 (F*, L*; keys, bion.; Thailand); Wilkerson et al. 2015 (phyl., class.; to subg. of genus *Aedes*).

chungi Lien
1968b: 100 (M*, F, P*, L*). **Type loc.:** Fenchihu, Chuchi, Chiai Hsien, Taiwan (TMRI). **Distr.:** Taiwan. **Refs.:** Reinert 2009c: Fig. 1 (F*). **Ety.:** Chao Lin Chung. **Informal name:** Chung Taiwanese Pointy Mosquito.

gonguoensis Gong & Lu
1986: 1 (M*, F, L*). **Type loc.:** Gonguo, Yunnan Province, China [People's Republic of China] (YIE). **Distr.:** People's Republic of China. **Ety.:** Gonguo, Yunnan Prov., China. **Informal name:** Gonguo Chinese Pointy Mosquito.

unicinctus Edwards
1922a: 266 (M). **Type loc.:** Simla [East Punjab], India (NHMUK). **Distr.:** India, Nepal, Thailand. **Refs.:** Barraud 1923h (L*); Barraud 1923g (M*, F*); Barraud 1934: 170 (M*, F*, L*); Darsie et al. 1992 (distr.; Nepal); Rattanarithikul et al. 2010 (F*, L*; keys, bion., distr.; Thailand); Darsie 2011b (P*). **Ety.:** one ring (L); ref. to single white ring at base of tarsomere 1 of all legs. **Informal name:** Single-banded Indian Pointy Mosquito.

Subgenus *Kenknightia* Reinert

1990: 4 (as subg. of genus *Aedes*). Type species: *Stegomyia dissimilis* Leicester.

Name derivation: Kenneth Lee Knight.

Important references: Reinert 2000h (to subg. of genus *Ochlerotatus*); Reinert 2002e: 13 (F gen.*); Reinert, Harbach, & Kitching 2004 (to genus); Reinert, Harbach, & Kitching 2006 (phyl., class.); Reinert, Harbach, & Kitching 2008 (phyl., class.); Reinert, Harbach, & Kitching 2009 (phyl., class.); Rattanarithikul et al. 2010 (F*, L*; keys, bion.; Thailand); Wilkerson et al. 2015 (phyl., class.; to subg. of genus *Aedes*); Soghigian et al. 2017 (mol. phyl.).

dissimilierodes Dong, Zhou, & Dong
2002b: 521 (M*, F*, P*, L*). **Type loc.:** Jinghong, Yunnan Province, China [People's Republic of China] (YIMC). **Distr.:** People's Republic of China. **Ety.:** similar (L); similar to *dissimilis*. **Informal name:** Dissimilar Chinese Pointy Mosquito.

dissimilis (Leicester)
1908: 91 (M, F; *Stegomyia*). **Type loc.:** Ampang & Ulu Klang, near Kuala Lumpur [Selangor], Malaya [Malaysia] (NHMUK). **Distr.:** Bangladesh, India, Malaysia, Nepal, People's Republic of China, Thailand, Vietnam. **Refs.:** Barraud 1934: 204 (M*, F*, L); Rozeboom 1946: 589 (tax.); Joshi et al. 1965 (distr.; Nepal); Aslamkhan 1971b (distr.; Pakistan); Ahmed 1987 (distr.; Bangladesh); Reinert 1990: 18 (M*, F*, P*, L*; keys, tax., bion.); Darsie & Pradhan 1990, 1991 (F, L; tax., keys, bion., distr.; Nepal); Reinert 2002e: Fig. 12 (F*); Rattanarithikul et al. 2010 (F*, L*; keys, bion., distr.; Thailand). **Ety.:** not near, different (L); dissimilar to all other *Stegomyia*. **Informal name:** Strange Malaysian Pointy Mosquito [*dissimilis* = "dissimilar" = Strange].

gaffigani Reinert
1990: 24 (M*, F*, P*, L*; keys, tax., bion.). **Type loc.:** Pensiangan, Sabah, Malaysia [Malaysian Borneo] (USNM). **Distr.:** Malaysia. **Ety.:** Thomas Vincent Gaffigan. **Informal name:** Gaffigan Malaysian Pointy Mosquito.

harbachi Reinert
1990: 27 (M*, F*, P*, L*; keys, tax., bion.). **Type loc.:** Doi Suthep, Ban Choeng, Chiang Mai Province, Thailand (USNM). **Distr.:** India, People's Republic of China, Thailand. **Refs.:** Dong, Zhou, & Dong 2002b: 516 (distr.; People's Republic of China); Rattanarithikul et al. 2010 (F*, L*; keys, bion., distr.; Thailand). **Ety.:** Ralph Edward Harbach. **Informal name:** Harbach Thai Pointy Mosquito.

karwari (Barraud)
1924a: 865 (F; *Finlaya*). **Type loc.:** Karwar, North Kanara [Bombay], India (NHMUK). **Distr.:** India. **Refs.:** Reinert 1990: 32 (M*, F*, P*, L*; keys, tax., bion.). **Ety.:** Karwar, India. **Informal name:** Karwar Indian Pointy Mosquito.

lerozeboomi Reinert
1990: 33 (M*, F*, P*, L*; keys, tax., bion.). **Type loc.:** Calso, Dalton Pass, Nueva Vizcaya, Philippines (USNM). **Distr.:** Philippines. **Ety.:** Lloyd Eugene Rozeboom. **Informal name:** Lerozeboom Philippine Pointy Mosquito.

leucomeres (Giles)
1904a: 367 (F; *Stegomyia*). **Type loc.:** Camp Stotsenberg, Angeles, Pampanga, Luzon, Philippines

(NHMUK). **Distr.:** Philippines. **Refs.:** Knight & Hull 1951b: 249 (tax.); Basio 1971b: 16 (M*; bion.); Reinert 1990: 37 (M*, F*, P*, L*; keys, tax., bion.). **Ety.:** n.s. [white part (Gr); poss. ref. to abdominal venter with brilliant broad white basal bands in contrast to *albocephalus*]. **Informal name:** White-banded Philippine Pointy Mosquito.

litwakae Reinert
1990: 39 (M*, F*, P*, L*; keys, tax., bion.). **Type loc.:** Caburan, Davao, Mindanao, Philippines (USNM). **Distr.:** Philippines. **Ety.:** Taina Rose Litwak. **Informal name:** Litwak Philippine Pointy Mosquito.

luzonensis Rozeboom
1946: 589 (M*, F, L*). **Type loc.:** Subic Bay [Zambales], Luzon, Philippines (USNM). **Distr.:** Philippines. **Refs.:** Knight & Hull 1951b: 249 (tax.); Basio 1971b: 16 (M*; bion.); Baisas 1974: 13 (M, F, P*, L*; tax., bion., distr.; Philippines); Reinert 1990: 43 (M*, F*, P*, L*; keys, tax., bion.); Reinert 2002e: Fig. 13 (F*). **Ety.:** Luzon, Philippines. **Informal name:** Luzon Philippine Pointy Mosquito.

paradissimilis Rozeboom
1946: 587 (M*, F, L*; as *Aedes*). **Type loc.:** Cape Melville, Balabac Island, Philippines (USNM). **Distr.:** Philippines. **Refs.:** Basio 1971b: 17 (M*; bion.); Reinert 1990: 47 (M*, F*, P*, L*; keys, tax., bion.); Dong, Zhou, & Dong 2002b: 516 (distr.; People's Republic of China). **Ety.:** like *dissimilis* (L). **Informal name:** Unusual Philippine Pointy Mosquito [*paradissimilis* = "almost dissimilar" = Unusual].

pecori Reinert
1990: 52 (M*, F*, P*, L*; keys, tax., bion.). **Type loc.:** Gunung Jasar, Cameron Highlands, Penang, Malaysia (USNM). **Distr.:** Malaysia, Philippines. **Ety.:** James Pecor. **Informal name:** Pecor Malaysian Pointy Mosquito.

wilkersoni Reinert
1990: 56 (M*, F*, P*, L*; keys, tax., bion.). **Type loc.:** Ulu Gombak, Selangor, Malaysia (USNM). **Distr.:** Malaysia. **Ety.:** Richard Charles Wilkerson. **Informal name:** Wilkerson Malaysian Pointy Mosquito.

Subgenus *Kompia* Aitken

1941: 81 (as subg. of genus *Aedes*). Type species: *Aedes* (*Kompia*) *purpureipes* Aitken.

Name derivation: W.H.W. Komp.

Important references: Belkin & McDonald 1957: 190 (tax., key); Mattingly 1961: 16 (to subg. [not clear where it was made a genus]); Zavortink 1972: 55 (tax., redescr.); Reinert 2000h (to subg. of genus *Ochlerotatus*); Reinert 2002e: 45 (F gen.*); Reinert, Harbach, & Kitching 2004 (to genus); Reinert, Harbach, & Kitching 2009 (phyl., class.); Wilkerson et al. 2015 (phyl., class.; to subg. of genus *Aedes*); Soghigian et al. 2017 (mol. phyl.).

purpureipes Aitken
1941: 82 (F). **Type loc.:** Triunfo, Distrito del Sur, Baja California, Mexico (CAS). **Distr.:** Mexico, United States (continental). **Refs.:** Gjullin 1946: 235 (F); Vargas 1949b: 261 (L*); Carpenter & LaCasse 1955: 220 (M, F, L; keys); McDonald 1957b: 529 (M*, F, P*, L*); Zavortink 1972: 55 (M*, F*, P*, L*); Reinert 2002e: Fig. 47 (F*); Darsie & Ward 2005 (F*, L*; keys, distr.). **Ety.:** purple, dark red (L); head, legs, wings, and abdomen with purple reflections. **Informal name:** Purple-footed Mexican Pointy Mosquito.

Subgenus *Leptosomatomyia* Theobald

Theobald 1905f: 110 (as genus). Type species: *Leptosomatomyia lateralis* Theobald.

Name derivation: n.s. [lepto = fine, small, thin, delicate; somato = body].

Important references: Lee et al. 1984: 38 (as subg. of genus *Aedes*; tax., descr.; Australasia); Reinert 2000h (tax.; as subg. of genus *Aedes*); Reinert 2000e: 32 (F gen.*); Reinert, Harbach, & Kitching 2004 (phyl., class.; to genus); Reinert, Harbach, & Kitching 2009 (phyl., class.); Wilkerson et al. 2015 (phyl., class.; to subg. of genus *Aedes*).

aurimargo Edwards
1922d: 94 (F). **Type loc.:** Moa Island, Northern Australia (US). **Distr.:** Australia, Indonesia, Papua New Guinea. **Refs.:** Penn 1947 (L*); Penn 1949b: 63 (P*); Lee et al. 1984: 40 (tax., review); Reinert 2000e: Fig. 25 (F*). **Ety.:** n.s. [gold margin (L); poss. ref. to golden scales on head]. **Informal name:** Gold-edged Australian Pointy Mosquito.
syn. *lateralis* Theobald
1905f: 110 (not Meigen 1818) (M*; *Leptosomatomyia*). **Type loc.:** Muina, [Northeast] New Guinea (HNM).

Subgenus *Levua* Stone & Bohart

1944: 214 (as subg. of genus *Aedes*). Type species: *Aedes* (*Levua*) *suvae* Stone & Bohart.

Name derivation: Viti Levu Isl., Fiji.

Important references: Lee et al. 1984: 44 (tax., descr., bion.; Australasia); Reinert 2000h (to subg. of genus *Ochlerotatus*); Reinert, Harbach, & Kitching 2004 (phyl., class.; to genus); Reinert, Harbach, & Kitching 2008 (phyl., class.; to subg. of genus *Ochlerotatus*); Reinert, Harbach, & Kitching 2009

syn. *sylvestris* Theobald
1910c: 196 (F*; *Chaetocruiomyia*). **Type loc.:** Kuranda, Queensland, Australia (NHMUK).

stoneorum Marks
1977: 33 (F*). **Type loc.:** Moa (Banks) Island, North Queensland, Australia (UQIC). **Distr.:** Australia. **Refs.:** Lee et al. 1984: 63 (F; key, tax., bion., distr., review). **Ety.:** Alan and Louise Stone. **Informal name:** Stone Australian Pointy Mosquito.

tremulus (Theobald)
1903e: 155 (M, F; *Macleaya*). **Type loc.:** South Queensland, Australia (NHMUK). **Distr.:** Australia, Indonesia, Papua New Guinea. **Refs.:** Bancroft 1908: 27 (M, F; as genus *Macleaya*); Edwards 1924b: 385 (A); Lee 1944a: 66 (L*); Penn 1949b: 54 (P*); Dobrotworsky 1965: 154 (M*, F*, L*); Lee et al. 1984: 64 (F; key, tax., bion., distr., review); Reinert 2002e: Fig. 15 (F*). **Ety.:** n.s. [trembling (L)]. **Informal name:** Trembling Australian Pointy Mosquito.
syn. *minuta* Taylor
1912a: 30 (M; *Danielsia*). **Type loc.:** Wandi, Northern Territory, Australia (US).
syn. *alboannulata* Taylor
1912a: 31 (F; *Danielsia*). **Type loc.:** Wandi, Northern Territory, Australia (NE).
syn. *australis* Taylor
1914b: 457 (F; *Aedimorphus*). **Type loc.:** Townsville, Queensland, Australia (US).
syn. *darwini* Taylor
1914b: 458 (F*; *Aedimorphus australis* var.). **Type loc.:** Darwin, Northern Territory, Australia (US).
syn. *doddi* Taylor
1919: 831 (M, F; *Mimeteomyia*). **Type loc.:** Itikinumu Plantation, Papua, New Guinea (US).

tulliae (Taylor)
1929: 275 (F; *Chaetocruiomyia*). **Type loc.:** Tully, Queensland, Australia (US). **Distr.:** Australia. **Refs.:** Marks 1949: 38 (M*; as *stricklandi* Edwards); Marks 1963a: 199 (M*, F*, P*, L*); Lee et al. 1982: 60 (F; key, tax., bion., distr., review). **Ety.:** Tully, Queensland, Australia. **Informal name:** Tully Australian Pointy Mosquito.

wattensis Taylor
1929: 275 (F). **Type loc.:** Watten, Queensland, Australia (US). **Distr.:** Australia. **Refs.:** Marks 1963a: 197 (F*); Dobrotworsky 1965: 157 (M*, F*); Lee et al. 1982: 63 (distr.), 62 (F; key, tax., bion., distr., review). **Ety.:** Watten, Queensland, Australia. **Informal name:** Watten Australian Pointy Mosquito.

Subgenus *Molpemyia* Theobald

1910c: 479 (as genus). Type species: *Molpemyia purpurea* Theobald.

Name derivation: n.s. [poss. ref. to molpe = song, myia = fly (Gr)].

syn. *Calomyia* Taylor 1914c: 684 (as genus) (not *Calomyia* Rossi 1848). Type species: *Calomyia priestly* Taylor.

Important references: Reinert 1993 (tax., redescr.; from syn. with subg. *Finlaya* to subg. of genus *Aedes*); Reinert 2000h (to subg. of genus *Ochlerotatus*); Reinert 2002e: 18 (F gen.*); Reinert, Harbach, & Kitching 2006 (phyl., class.; to genus); Reinert, Harbach, & Kitching 2009 (phyl., class.); Wilkerson et al. 2015 (phyl., class.; to subg. of genus *Aedes*).

auridorsum Edwards
1922d: 93 (M, F). **Type loc.:** Eidsvold, Queensland, Australia (LU). **Distr.:** Australia. **Refs.:** Edwards 1924b: 381 (tax.; change of type loc. from Sydney); Marks 1948: 21 (M*, F, P*, L*); Lee et al. 1982: 121 (F; key, tax., bion., distr., review); Townsend 1990: 48 (type dep. NHMUK to LU); Reinert 1993: 46 (from subg. *Finlaya*; M*, F*). **Ety.:** n.s. [gold back (L); prob. ref. to gold scales dorsally on thorax and abdomen]. **Informal name:** Gold-backed Australian Pointy Mosquito.

pecuniosus Edwards
1922d: 94 (F). **Type loc.:** Port Darwin, Northern Territory, Australia (NHMUK). **Distr.:** Australia. **Refs.:** Cooling 1924: 24 (M*, F*, L*; bion.); Lee 1944a: 65 (L*); Taylor 1944: 122 (syn.); Reinert 1988: 59 (F*; from syn. with *purpureus*); Reinert 1993: 51 (M*, F*, P*, L*; from subg. *Finlaya*). **Ety.:** n.s. [rich, wealthy (L)]. **Informal name:** Wealthy Australian Pointy Mosquito.

purpureus (Theobald)
1910c: 479 (F; *Molpemyia*). **Type loc.:** Stannary Hills, Queensland, Australia (NHMUK). **Distr.:** Australia. **Refs.:** Taylor 1944: 122 (F*; tax.); Lee et al. 1982: 236 (F; key, tax., bion., distr., review); Reinert 1988 (F*); Reinert 1993: 57 (F*; from subg. *Finlaya*); Reinert 2002e: Fig. 16 (F*). **Ety.:** purple, dark red (L); ref. to purple reflections on proboscis, abdomen, and legs. **Informal name:** Purple Australian Pointy Mosquito.
syn. *priestleyi* Taylor
1914c: 684 (F; *Calomyia*). **Type loc.:** Townsville, Queensland, Australia (US). **Refs.:** Taylor 1944: 122 (syn.).

Subgenus *Mucidus* Theobald

1901b: 235 (July 15), 1901d: 2 (Sept.), 1901a: 268 (Nov. 23) (as genus). Type species *Culex alternans* Westwood (Neveu-Lamaire 1902c: 219).

Name derivation: moldy, musty (L) [poss. ref. to overall look].

syn. *Pardomyia* Theobald 1907: 280 (as genus). Type species: *Pardomyia aurantia* Theobald. Reinert 2006b (phyl., class.; from syn. with subg. *Mucidus* to subg. of genus *Mucidus*); Wilkerson et al. 2015 (to syn. of subg. *Mucidus* of genus *Aedes*, as informal group).

syn. *Ekrinomyia* Leicester 1908: 65, 71 (as genus; syn. of *Pardomyia* Theobald). Type species: *Ekrinomyia aureostriata* Leicester.

Important references: Neveu-Lemaire 1902c (type species desig.); Edwards 1932a: 133 (subg. *Mucidus* of genus *Aedes*); Knight 1947a: 315 (keys); Mattingly 1961: 17 (keys; Indomalayan area); Tyson 1970b (revision; as Group B *Pardomyia* of subg. *Mucidus* of genus *Aedes*; SE Asia); Reinert 2000h (to subg. of genus *Ochlerotatus*); Reinert, Harbach, & Kitching 2004 (phyl., class.; to genus); Reinert, Harbach, & Kitching 2009 (phyl., class.); Wilkerson et al. 2015 (phyl., class.; to subg. of genus *Aedes*).

alternans (Westwood)
1835: 681 (A; *Culex*). **Type loc.:** Nova Hollandia [Australia] (NHMUK). **Distr.:** Australia, Indonesia, New Caledonia, Papua New Guinea, Timor. **Refs.:** Felt 1905b: 473 (M*); Bancroft 1908: 18 (M, F, E; "Scotch Grey" or "Hexham Grey" mosquito); Woodhill & Pasfield 1941: 208 (L*); Lee 1944a: 48 (L*); Knight 1947a: 321 (M, F, L); Penn 1949b: 40 (P*); Rageau & Hamon 1957 (M*; tax.); Belkin 1962: 402 (M*, F, P*, L*); Pillai 1962: 212 (E*); Dobrotworsky 1965: 74 (M*, F*, L*, E*); Mohrig 1967 (F*); Mattingly 1970d (E*); Lee et al. 1984: 73 (F; key, tax., bion., distr., review); Linley et al. 1991b (E*); Whelan & Hapgood 2000 (bion., distr.; East Timor); Reinert 2002e: Fig. 17 (F*); Becker et al. 2010: 357 (F*, L*; key, tax., distr., bion.). **Ety.:** n.s. [poss. ref. to alternating white and black on legs (L)]. **Informal name:** Alternating Australian Pointy Mosquito.

syn. *commovens* Walker
1856a: 432 (F; *Culex*). **Type loc.:** New Holland [Australia] (NHMUK).

syn. *hispidosus* Skuse
1889: 1726 (M, F; *Culex*). **Type loc.:** Hexham Swamp (near Newcastle, Richmond), Mount Kembla and Illawarra, New South Wales, Australia (MM).

syn. *kermorganti* Laveran
1901a: 568 (F; *Culex*). **Type loc.:** Nouméa, New Caledonia (PIP). **Refs.:** Knight 1947a: 322 (tax.); Reid 1947: 87 (tax.); Rageau & Hamon 1957 (syn.).

aurantius (Theobald)
1907: 280 (F; *Pardomyia*). **Type loc.:** Kuching, Sarawak [Borneo, Malaysia] (NHMUK). **Distr.:** Australia, Indonesia, Malaysia, Papua New Guinea, Singapore. **Refs.:** Edwards & Given 1928: 341 (P*, L*); Lee 1944a: 48 (L); Knight 1947a: 323 (M, F, L); Knight & Hull 1951b: 225 (tax.); Mattingly 1961: 43 (M*, F*, P*, L*); Mattingly 1970d: 88 (E*); Tyson 1970b: 46 (M*, F*, P*, L*); Mattingly 1971a: Pl. 21 (P cephalothorax*); Mattingly 1971b (E*); Lee et al. 1984: 83 (F; key, tax., bion., distr., review); Reinert 2002e: Fig. 18 (F*). **Ety.:** n.s. [orange (L); poss. ref. to golden and golden-yellow markings]. **Informal name:** Orange Malaysian Pointy Mosquito.

syn. *aureostriata* Leicester
1908: 71 (M, F; *Ekrinomyia*). **Type loc.:** Klang, near Kuala Lumpur [Selangor], Malaya [Malaysia] (NHMUK).

ssp. ***chrysogaster*** (Taylor)
1927: 67 (F; as *Mucidus* sp.). **Type loc.:** Berner Creek, near Innisfail, Queensland, Australia (US). **Distr.:** Australia, Papua New Guinea. **Refs.:** Knight et al. 1944: 50 (to ssp.); Knight 1948b: 68 (tax.); Penn 1949b: 41 (P*); Bick 1951: 406 (bion.); Lee et al. 1984: 87 (F; key, tax., bion., distr., review). **Ety.:** gold abdomen (L); abdomen with gold markings. **Informal name:** Gold-bellied Australian Pointy Mosquito.

ferinus Knight
1947a: 316 (M*, F, L*). **Type loc.:** San Ramon (Penal Farm), City of Zamboanga Province, Mindanao, Philippines (USNM). **Distr.:** Philippines. **Refs.:** Knight & Chamberlain 1948: Fig. 25 (P*); Knight & Hull 1951b: 224 (tax.); Mattingly 1961: 26 (M*, F*, P*, L*); Tyson 1970b: 37 (M*, F*, P*, L*); Basio 1971b: 20 (M*; bion.). **Ety.:** n.s. [of wild beasts (L)]. **Informal name:** Wild Philippine Pointy Mosquito.

grahamii (Theobald)
1909b: 2 (F; *Mucidus*). **Type loc.:** Obuasi, Ashanti, Ghana (NHMUK). **Distr.:** Burkina Faso, Cameroon, Central African Republic, Côte d'Ivoire, Democratic Republic of the Congo (Zaire), Ghana, Kenya, Liberia, Mali, Nigeria, Uganda. **Refs.:** Edwards 1941: 114 (M*, F*), 385 (P*); Wolfs 1945a: 64 (P*, L*); Mattingly 1947: 244 (L); Hopkins 1952: 123 (L); Mohrig 1967 (F*). **Ety.:** n.s. [poss. ref. to Walter Mayne Graham]. **Informal name:** Graham Ghanaian Pointy Mosquito.

syn. *africanus* Theobald
1901d: 2 (F; *Mucidus*). **Type loc.:** Asaba [Eastern Provinces], Nigeria (NHMUK). *Aedes (Stegomyia) africanus* (Theobald 1901d: 3) is a homonym, but because of its economic importance, it is here [Knight & Stone 1977] maintained, and *Mucidus africanus* is regarded as preoccupied.

laniger (Wiedemann)
1820: 9 (F; *Culex*). **Type loc.:** Java (ZMC). **Distr.:** Cambodia, India, Indonesia, Malaysia, Philippines, Singapore, Thailand, Vietnam. **Refs.:** Barraud 1934: 147 (M*, F); Knight 1947a: 320 (M, F; tax.); Knight & Hull 1951b: 225 (tax.); Mattingly 1961: 26 (M*, F*); Tyson 1970b: 39 (M*, F*, P*, L*); Basio 1971b: 20 (M*; bion.); Kaur 2003: 454 (distr.; India); Rattanarithikul et al. 2010 (F*, L*; keys, bion., distr.; Thailand). **Ety.:** n.s. [wool bearing (L)]. **Informal name:** Woolly Indonesian Pointy Mosquito.

lucianus Musspratt
1959: 64 (M*, F*; tax., distr.). **Type loc.:** M'canduzi, west side of Lake Saint Lucia, Zululand, Natal [KwaZulu-Natal, Republic of South Africa] (SAIM). **Distr.:** Republic of South Africa. **Refs.:** Jupp 1996 (M*, F*; key). **Ety.:** Lake Saint Lucia, Zululand, Natal. **Informal name:** Lucia South African Pointy Mosquito.

mucidus (Karsch)
1887: 25 (F; *Culex*). **Type loc.:** Delagoa Bay, Mozambique (ZM). **Distr.:** Burkina Faso, Cameroon, Gabon, Kenya, Madagascar (includes Glorioso and Juan De Nova Isl.), Mali, Mozambique, Nigeria, Republic of South Africa, Togo, Uganda, Zambia. **Refs.:** Edwards 1941: 112 (M*, F*), 385 (P*?); Hopkins 1952: 122 (L); Muspratt 1959: 64 (tax.); Mohrig 1967 (F*); Mattingly 1970d (E*); Service 1976a (distr.; Gabon); Jupp 1996 (M*, F*; key). **Ety.:** n.s. [moldy, musty (L)]. **Informal name:** Moldy Mozambican Pointy Mosquito.

nigerrimus (Theobald)
1913d: 591 (M*; as *Mucidus*). **Type loc.:** Wadelai, Lado District, Sudan (NHMUK). **Distr.:** Central African Republic, Democratic Republic of the Congo, Kenya, Mozambique, Sudan and South Sudan, Uganda. **Refs.:** Edwards 1941: 113 (M*, F), 385 (P*); Hopkins 1952: 122 (L); Darsie & Pradhan 1990, 1991 (F, L; tax., keys, bion., distr.; Nepal); Jupp 1996 (M*, F*; key). **Ety.:** blackest (L); a generally dark species. **Informal name:** Blackest Sudanese Pointy Mosquito.

nigrescens (Edwards)
1929 (F, P, L; *Pardomyia aurantia* var.). In Paine & Edwards 1929: 314. **Type loc.:** Lady Lever Plantation, Kulambangra [= Kolombangara], Solomon Islands (NHMUK). **Distr.:** Papua New Guinea, Solomon Isl. **Refs.:** Knight 1947a: 324 (M*, F, L); Knight & Hull 1951b: 226 (tax.; as *painei*); Belkin 1952: Pl. 1, Fig. 4 (L*); Belkin 1962: 405 (M*, F, P*, L*; to sp.; as *painei*); Lee et al. 1984: 89 (F; key, tax., bion., distr., review; as *painei*). **Ety.:** n.s. [become black, grow dark (L)]. **Informal name:** Dark Solomon Islander Pointy Mosquito.

syn. *painei* Knight
1948b: 68 (new name for *nigrescens* Edwards 1929, not Theobald 1907; *Aedes* (*Mucidus*) *aurantius* ssp.) [new name no longer needed; see Harbach 2018c: 267].

quadripunctis (Ludlow)
1910: 608 (F; as *Pardomyia* sp.). In Theobald 1910c. **Type loc.:** Parang [Cotabato], Mindanao, Philippines (USNM). **Distr.:** Philippines. **Refs.:** Bohart 1945: 55 (to sp.); Knight 1947a: 322 (M, F, L*); Mattingly 1961: 36 (M*, F, P*, L; to sp.); Tyson 1970b: 49 (M*, F*, P*, L*); Basio 1971b: 20 (M*; bion.; as *aurantius* ssp.). **Ety.:** four spots (L); "four dark spots on the wing." **Informal name:** Four-spotted Philippine Pointy Mosquito.

quasiferinus Mattingly
1961: 32 (M*, F*, P*, L*). **Type loc.:** Sungai Besi, Kuala Lumpur area, Malaya [Malaysia] (NHMUK). **Distr.:** Cambodia, India, Indonesia, Malaysia, Singapore, Sri Lanka, Thailand. **Refs.:** Tyson 1970b: 41 (M*, F*, P*, L*); Choochote et al. 1987 (E*); Rattanarithikul et al. 2010 (F*, L*; keys, bion., distr.; Thailand). **Ety.:** nearly (L); similar to *ferinus*. **Informal name:** Feral Malaysian Pointy Mosquito [*quasiferinus* = "half wild" = Feral].

scatophagoides (Theobald)
1901a: 277 (F*; as *Mucidus scataphagoides*, emendation by Edwards 1911b: 246). **Type loc.:** Moradad [Moradabad], Northwest Provinces (= Moradabad, United Provinces], India and Myingyan [Mandalay], Burma [Myanmar] (NHMUK). **Distr.:** Bangladesh, Gabon, Ghana, India, Madagascar (includes Glorioso and Juan De Nova Isl.), Mali, Mauritania, Mozambique, Myanmar, Nepal, Pakistan, People's Republic of China, Senegal, Sri Lanka, Sudan and South Sudan, Thailand, Vietnam. **Refs.:** Macfie & Ingram 1922 (F*); Barraud 1934: 145 (M*, F, L*); Edwards 1941: 110 (M*, F), 384 (P); Knight 1947a: 319 (M*, F; tax.); Hopkins 1952: 121 (L*); Worth & De Meillon 1960: 237 (distr.); Joshi et al. 1965 (distr.; Nepal); Tyson 1970b, c (F*); Aslamkhan 1971b (distr.; Pakistan); Service 1976a: 260 (distr.; Gabon); Ahmed 1987 (distr.; Bangladesh); Darsie & Pradhan 1990, 1991 (F, L; tax., keys, bion., distr.; Nepal). **Ety.:** n.s. [dung eater (?) (Gr)]. **Informal name:** Dung-eating Indian Pointy Mosquito.

sudanensis (Theobald)
1908c: 252 (F*; *Mucidus*). **Type loc.:** Upper White Nile; also at Sangikia, Kodok, and Khartoum, Sudan (NHMUK). **Distr.:** Burkina Faso, Congo, Ethiopia, Ghana, Kenya, Malawi, Mali, Mauritania, Mozambique, Namibia, Nigeria, Republic of South Africa, Senegal, Somalia, Sudan and South Sudan, Tanzania, Zambia, Zimbabwe. **Refs.:** Tyson 1970c: 81 (M, F*, P, L); Jupp 1996 (M*, F*; key). **Ety.:** Sudan. **Informal name:** Kush Sudanese Pointy Mosquito [ancient kingdom centered in modern Sudan].

tonkingi Gebert
1948: 96 (M, F; *Aedes*). **Type loc.:** Black River District, Mauritius (CLM). **Distr.:** Mauritius. **Refs.:** Mattingly 1970d: 87 (E*); Tyson 1970c: 83 (M*, F*, P*, L*, E*). **Ety.:** H.D. Tonking. **Informal name:** Tonking Mauritian Pointy Mosquito.

Subgenus *Neomelaniconion* Newstead

1907 (as genus). In Newstead, Dutton & Todd 1907: 31 (Feb. 1). Type species: *Neomelaniconion palpale* Newstead.

Name derivation: n.k. [new *melanoconion*?].

syn. *Banksinella* Theobald 1907 (Feb. 23): 468 (as genus). Type species: *Culex luteolateralis* Theobald.

Important references: Edwards 1932a: 171 (as *Banksinella*; tax., key, descr.); Edwards 1941: 201 (as *Banksinella*; tax., key, revision; Ethiopean Region); Le Berre & Hamon 1961: 1060 (keys; tropical Africa); Mattingly 1961 (key, descr.; Indomalayan Area); McIntosh 1971 (keys, revision; southern Africa); Lee et al. 1984: 92 (revision; Australasia); Jupp 1996 (tax.; southern Africa); Tanaka 2003a (P; Japan); Reinert, Harbach, & Kitching 2004 (phyl., class.; to genus); Le Goff et al. 2007 (revision; Madagascar); Reinert, Harbach, & Kitching 2009 (phyl., class.); Rattanarithikul et al. 2010 (F*, L*; keys, bion.; Thailand); Wilkerson et al. 2015 (phyl., class.; to subg. of genus *Aedes*).

albicosta (Edwards)
1913b: 47 (F; *Banksinella luteolateralis* var.). **Type loc.:** Wangi, British East Africa [Kenya] (NHMUK). **Distr.:** Ethiopia, Kenya, Somalia, Sudan and South Sudan, Tanzania. **Refs.:** Edwards 1941: 206 (M*, F); van Someren 1954b: 119 (P*, L*); van Someren, Teesdale, & Furlong 1955: 481 (A); Zavortink 1987: 2 (distr.). **Ety.:** white costa (L); "apical two-fifths of the costa with whitish-yellow scales." **Informal name:** White-margined Kenyan Pointy Mosquito.

albiradius (Le Goff, Boussès, & Brunhes)
2007: 197 (F; *Neomelaniconion*). **Type loc.:** Ampijoroa Forest Station, Province de Mahajanga, Madagascar (IRD). **Distr.:** Madagascar (includes Glorioso and Juan De Nova Isl.). **Ety.:** white (L); white scales at base of radial vein. **Informal name:** White-rayed Malagasy Pointy Mosquito.

albothorax (Theobald)
1907: 470 (F; *Banksinella luteolateralis* var.). **Type loc.:** Coast Province, 2 km north of Kakuyuni, Lake Mbaratuma, Kenya (CAS) (NHMUK). **Distr.:** Angola, Democratic Republic of the Congo, Ethiopia, Kenya, Mozambique, Nigeria, Republic of South Africa, Sudan and South Sudan, Tanzania. **Refs.:** Edwards 1941: 205 (M, F*,); van Someren 1954b: 121 (P, L); van Someren, Teesdale, & Furlong 1955: 480 (A); McIntosh 1971: 330 (M*, F); Zavortink 1987: 2 (distr.); Zavortink 2004: 198 (neotype desig., neotype loc.: Lake Mbaratumu, 2 km north of Kakuyuni, Coast Prov.). **Ety.:** n.s. [white thorax (L); prob. ref. to "thorax black, with white narrow-curved scales on each side"]. **Informal name:** White-chested Gambian Pointy Mosquito.

aurovenatus Worth
1960: 312 (F). **Type loc.:** Ndumu, Tongaland, Natal [KwaZulu-Natal, Republic of South Africa] (SAIM). **Distr.:** Republic of South Africa. **Refs.:** McIntosh 1971: 329 (F*); Cornel 1991 (M*, F*, P*, L*; key, bion.); Jupp 1996 (M*, F*; key). **Ety.:** n.s. [gold "venatus" (L); prob. ref. to yellow scaling of wings]. **Informal name:** Gold-winged South African Pointy Mosquito.

belleci (Le Goff, Boussès, & Brunhes)
2007: 191 (M*, F, P*, L*; *Neomelaniconion*). **Type loc.:** 10 km below Ranomafana thermal plant, Province de Fianarantsoa, Madagascar (IRD). **Distr.:** Madagascar (includes Glorioso and Juan De Nova Isl.). **Ety.:** Christian Bellec. **Informal name:** Bellec Malagasy Pointy Mosquito.

bequaerti Wolfs
1947a: 40 (M*, F). **Type loc.:** Kimilolo, Elisabethville [Katanga], Zaire [Democratic Republic of the Congo] (CMT). **Distr.:** Democratic Republic of the Congo, Zambia. **Refs.:** Zavortink 1987: 2 (distr.). **Ety.:** Joseph Charles Corneille Bequaert. **Informal name:** Bequaert Congolese Pointy Mosquito.

bergerardi Pajot & Geoffroy
1971: 269 (M*). **Type loc.:** Botambi forest, Ombella-M'Poko department, Central African Republic (IERT). **Distr.:** Central African Republic. **Ety.:** J. Bergerard. **Informal name:** Bergerard Central African Pointy Mosquito.

bolense Edwards
1936: 52 (M). **Type loc.:** Bole [Northern Territories], Ghana (NHMUK). **Distr.:** Bulgaria, Burkina Faso, Central African Republic, Côte d'Ivoire, Ghana, Senegal, Togo. **Refs.:** Edwards 1941: 213 (M*); Hamon & Rickenbach 1954: 934 (M*, F, P*, L*); Zavortink 1987: 2 (distr.). **Ety.:** Bole, Ghana. **Informal name:** Bole Ghanaian Pointy Mosquito.

carteri Edwards
1936: 52 (M*; *palpalis* var.). **Type loc.:** Lagos, Nigeria (NHMUK). **Distr.:** Central African Republic, Côte d'Ivoire, Democratic Republic of the Congo, Ghana, Liberia, Mozambique, Nigeria, Uganda. **Refs.:** Edwards 1941: 210 (M*, F), 400 (P); Haddow et al. 1951: 226 (to sp.); Worth & De Meillon 1960: 244 (distr.); Zavortink 1987: 2 (distr.). **Ety.:** "Recorded by Carter." **Informal name:** Carter Nigerian Pointy Mosquito.
syn. *maculicosta* Edwards
1936: 52 (F; *Aedes palpalis* var.). **Type loc.:** Lagos, Nigeria (NHMUK). **Refs.:** Zavortink 1987: 2 (distr.).

circumluteolus (Theobald)
1908a: 107 (F; *Banksiella* [*sic*] *luteolateralis* var. *circumluteola*). **Type loc.:** Transvaal [Republic of South Africa] (NHMUK). **Distr.:** Angola, Benin, Botswana, Burkina Faso, Cameroon, Central African Republic, Comoros, Congo, Côte d'Ivoire, Ethiopia, Gambia, Ghana, Kenya, Liberia, Madagascar (includes Glorioso and Juan De Nova Isl.), Mali, Mozambique, Niger, Nigeria, Republic of South Africa, Rwanda, São Tomé and Príncipe, Senegal, Sudan and South Sudan, Uganda, Zambia, Zimbabwe. **Refs.:** Edwards 1941: 204 (M*, F); Hopkins 1952: 208 (L*); Gillett 1955 (E); McIntosh 1971: 326 (M*, F*); Brunhes 1977a (distr.; Comoros Isl.); Zavortink 1987: 2 (distr.); Linley & Turell 1994: 1 (E*); Jupp 1996 (M*, F*;

key); Le Goff et al. 2007: 188 (F*; tax, distr.). **Ety.:** n.s. [around, yellow (L); poss. ref. to yellow scale pattern]. **Informal name:** Yellow-ringed South African Pointy Mosquito.

crassiforceps Edwards
1927a: 352 (M*, F). **Type loc.:** Stanleyville, Zaire [Kisangani, Tshopo Province, Democratic Republic of the Congo] (CMT). **Distr.:** Central African Republic, Democratic Republic of the Congo. **Refs.:** Edwards 1941: 212 (M*, F); Zavortink 1987: 2 (distr.). **Ety.:** n.s. [thick pincers (L)]. **Informal name:** Thick-clawed Congolese Pointy Mosquito.

ellinorae Edwards
1941: 212 (M, F). **Type loc.:** Gede, near Mombasa, Kenya (NHMUK). **Distr.:** Burkina Faso, Kenya. **Refs.:** Mattingly 1956a: 40 (lectotype desig.). **Ety.:** Ellinor C. MacDonald. **Informal name:** Ellinor Kenyan Pointy Mosquito.

flavimargo Edwards
1941: 209 (F). **Type loc.:** Gede, near Mombasa, Kenya (NHMUK). **Distr.:** Côte d'Ivoire, Kenya. **Refs.:** Mattingly 1956a: 40 (lectotype desig.). **Ety.:** n.s. [yellow border (L); ref. to yellow markings (esp. yellow costa?)]. **Informal name:** Yellow-edged Kenyan Pointy Mosquito.

fontenillei (Le Goff, Boussès, & Brunhes)
2007: 196 (M*, F; *Neomelaniconion*). **Type loc.:** Perinet/Andasibe forest station, Province de Toamasina, Madagascar (IRD). **Distr.:** Madagascar (includes Glorioso and Juan De Nova Isl.). **Ety.:** Didier Fontenille. **Informal name:** Fontenille Malagasy Pointy Mosquito.

fuscinervis (Edwards)
1914b: 73 (M; *Banksinella*). **Type loc.:** Accra, Ghana (NHMUK). **Distr.:** Benin, Burkina Faso, Central African Republic, Côte d'Ivoire, Ghana, Liberia, Nigeria, Senegal. **Refs.:** Edwards 1941: 213 (M*, F); Peters 1955c: 122 (L?); Zavortink 1987: 2 (distr.). **Ety.:** n.s. [dark wing (L); poss. ref. to color of veins]. **Informal name:** Dark-winged Ghanaian Pointy Mosquito.

hylaius Harbach
2018c: 3 (new name for *sylvaticum* Le Goff, Boussès, & Brunhes, not *sylvaticus* Meigen 1818). **Distr.:** Madagascar (includes Glorioso and Juan De Nova Isl.). **Ety.:** a wood, forest (L); ref. to found in forest habitats. **Informal name:** Jungle Malagasy Pointy Mosquito.
syn. *sylvaticum* Le Goff, Boussès, & Brunhes
2007: 195 (M*, F). **Type loc.:** Perinet/Andasibe forest station, Province de Toamasina, Madagascar (IRD). **Refs.:** Harbach 2018c: 3 (syn.).

jamoti Hamon & Rickenbach
1954: 932 (M*). **Type loc.:** Toussiana, Nasso et Peni, Subdivision et Cercle de Bobo Dioulasso, Haute Volta [Burkina Faso] (IERT). **Distr.:** Benin, Burkina Faso, Central African Republic, Côte d'Ivoire, Democratic Republic of the Congo, Ghana, Nigeria, Senegal. **Refs.:** Le Berre & Hamon 1961: 1054 (M, P*, L*); Zavortink 1987: 2 (distr.). **Ety.:** Dr. Jamot. **Informal name:** Jamot Burkinabe Pointy Mosquito.

lineatopennis (Ludlow)
1905a: 133 (F; *Taeniorhynchus*). **Type loc.:** Camp Gregg, Bayambang, Pangasinan, Luzon, Philippines (USNM). **Distr.:** Australia, Bangladesh, Borneo (island), Cambodia, India, Indonesia, Laos, Malaysia, Nepal, Pakistan, People's Republic of China, Philippines, Russia, South Korea, Sri Lanka, Thailand, Timor, Vietnam. **Refs.:** Macfie & Ingram 1922 (F*); Edwards 1941: 202 (M*, F*), 400 (P); Knight & Hull 1953: 468 (M*, F, L*; distr.); Stone & Knight 1956a: 220 (type info.); Mattingly 1961: 49 (M*, F*, P*, L*); Joshi et al. 1965 (distr.; Nepal); Aslamkhan 1971b (distr.; Pakistan); Basio 1971b: 24 (M*; bion.); McIntosh 1971: 322 (M*, F*, L); Mattingly 1974a: 42 (E*); Baisas 1974: 23 (M, F, P, L*; tax., bion., distr.; Philippines); Service 1976a: 260, 261 (distr.; Gabon); Tanaka et al. 1979: 419 M*, F*, L*); Lee et al. 1984: 94 (F; key, tax., bion., distr., review); Zavortink 1987: 2 (distr.); Darsie & Pradhan 1990, 1991 (F, L; tax., keys, bion., distr.; Nepal); Reinert 2000e: Fig. 28 (F*); Choochote et al. 2001 (E*); Tanaka 2003a: 22 (P*; tax.); Rattanarithikul et al. 2010 (F*, L*; keys, bion., distr.; Thailand). **Ety.:** n.s. [lined wings (L)]. **Informal name:** Thread-winged Philippine Pointy Mosquito.
syn. *linealis* Taylor
1913: 57 (F; *Pseudohowardina*). **Type loc.:** Ching Do and Townsville, Queensland, Australia (US).
ssp. ***aureus*** Gutsevich
1955: 320 (F). **Type loc.:** Kraskino, Maritime Territory, USSR [Russia] (LU). **Distr.:** Russia, Russian Federation. **Refs.:** Gutsevich et al. 1974: 339 (F); Danilov 1977: 182 (to ssp.). **Ety.:** n.s. [golden (L); poss. ref. to English transl.; "intense golden scales on lateral part of mesonotum"]. **Informal name:** Gold Russian Pointy Mosquito.

luridus McIntosh
1971: 327 (M*, F, P, L). **Type loc.:** Bethulie, South Africa [Republic of South Africa] (NIV). **Distr.:** Kenya, Republic of South Africa, Uganda. **Refs.:** Segerman 1990: 53 (type dep. change from SAIM to NIV); Jupp 1996 (M*, F*; key). **Ety.:** n.s. [sallow, wan (L); prob. ref. to pale yellow-and-white coloration]. **Informal name:** Sallow South African Pointy Mosquito.

luteolateralis (Theobald)
1901c: 71 (type F*; *Culex*). **Type loc.:** Durban, Natal [KwaZulu-Natal, Republic of South Africa] (NHMUK). **Distr.:** Kenya, Republic of South Africa, Sudan and South Sudan, Zimbabwe. **Refs.:** Edwards 1941: 204 (M, F); Muspratt 1953b: 90 (L); Mohrig 1967 (F*); McIntosh 1971: 329 (M*, F*); Jupp 1996 (M*, F*; key). **Ety.:** n.s. [yellow sided (L); prob. ref. to broad yellow to orange lines on sides of thorax]. **Informal name:** Yellow-sided South African Pointy Mosquito.

mcintoshi Huang
1985: 109 (M*, F*). **Type loc.:** Onderstepoort, Transvaal, Republic of South Africa (USNM). **Distr.:** Angola, Benin, Botswana, Burkina Faso, Congo, Democratic Republic of the Congo, Ethiopia, Gambia, Ghana, Kenya, Mali, Namibia, Nigeria, Republic of South Africa, Senegal, Sudan and South Sudan, Tanzania, Togo, Uganda, Zambia, Zimbabwe. **Refs.:** Knight & Hull 1953 (M*, F, L*; bion., distr.; as *lineatopennis*); Zavortink 1987: 2 (distr.); Linley & Turell 1994: 1 (E*); Jupp 1996 (M*, F*; key); Zavortink 2004 (tax.). **Ety.:** Bruce M. McIntosh. **Informal name:** McIntosh South African Pointy Mosquito. (Vol. 1, p. 424)

monotrichus Edwards
1936: 52 (M, F). **Type loc.:** Onitsha [Eastern Provinces], Nigeria (NHMUK). **Distr.:** Benin, Cameroon, Côte d'Ivoire, Liberia, Nigeria. **Refs.:** Edwards 1941: 209 (M*, F). **Ety.:** one seta (Gr); gonocoxite with single curved seta. **Informal name:** One-haired Nigerian Pointy Mosquito.

nigropterum (Le Goff, Boussès, & Brunhes)
2007: 194 (M*, F; *Neomelaniconion*). **Type loc.:** Perinet/Andasibe forest station, Province de Toamasina, Madagascar (IRD). **Distr.:** Madagascar. **Ety.:** dark wing (Gr); ref. to wing entirely dark. **Informal name:** Black-eyed Malagasy Pointy Mosquito.

palpalis (Newstead)
1907 (M*; as *Neomelaniconion palpale*). In Newstead et al. 1907: 31. **Type loc.:** Basoko, Haut Zaire, Zaire [Democratic Republic of the Congo] (NHMUK). **Distr.:** Cameroon, Central African Republic, Congo, Democratic Republic of the Congo, Madagascar (includes Glorioso and Juan De Nova Isl.), Nigeria, Republic of South Africa, Sudan and South Sudan, Uganda. **Refs.:** Edwards 1941: 209 (M*, F); Hopkins 1952: 211 (L*); McIntosh 1971: 331 (M*, F; key); Reinert 2000e: Fig. 27 (F*). **Ety.:** n.s. [pertaining to palpi (L)]. **Informal name:** Sensitive Congolese Pointy Mosquito.

pogonurus Edwards
1936: 52 (M). **Type loc.:** Stanleyville [Kisangani, Tshopo Province, Democratic Republic of the Congo] (NHMUK). **Distr.:** Central African Republic, Democratic Republic of the Congo. **Refs.:** Edwards 1941: 211 (M*, F); Zavortink 1987: 2 (distr.). **Ety.:** n.s. [bearded (Gr); poss. ref. to "coxite with dense tufts of long hair both above and below"]. **Informal name:** Hairy Congolese Pointy Mosquito.

punctocostalis (Theobald)
1909b: 14 (F; *Banksinella*). **Type loc.:** Obuasi, Ashanti, Ghana (NHMUK). **Distr.:** Côte d'Ivoire, Democratic Republic of the Congo, Ghana, Liberia, Nigeria. **Refs.:** Edwards 1941: 206 (F), 400 (P); Hopkins 1952: 210 (tax.); Zavortink 1987: 2 (distr.). **Ety.:** n.s. [spotted costa (L)]. **Informal name:** Spot-winged Ghanaian Pointy Mosquito.

taeniarostris (Theobald)
1909b: 13 (F*; *Banksinella*). **Type loc.:** Obuasi, Ashanti, Ghana (NHMUK). **Distr.:** Cameroon, Central African Republic, Côte d'Ivoire, Democratic Republic of the Congo, Ghana, Nigeria, Uganda. **Refs.:** Bailly-Choumara 1966b: 671 (P*, L*); Zavortink 1987: 2 (distr.). **Ety.:** banded beak (L); proboscis with broad median pale band. **Informal name:** Band-beaked Ghanaian Pointy Mosquito.

syn. *chrysothorax* Theobald
1909b: 13 (M*; *Banksinella*). **Type loc.:** Obuasi and Kumasi, Ashanti, Ghana (NHMUK). **Refs.:** Edwards 1941: 207 (syn.).

unidentatus McIntosh
1971: 325 (M*, F*, P, L). **Type loc.:** Olifantsvlei, Johannesburg, South Africa (NIV). **Distr.:** Lesotho, Republic of South Africa, Zimbabwe. **Refs.:** Zavortink 1987: 2 (unpublished annual report; distr.); Segerman 1990: 53 (type dep. change from SAIM to NIV); Jupp 1996 (M*, F*; key). **Ety.:** single tooth (L); ref. to unidentate hind tarsal claws. **Informal name:** Single-toothed South African Pointy Mosquito.

Subgenus *Nyctomyia* Harbach

2013a: 199 (new name for genus *Nyx* Harbach & Linton 2013, not Heppner 1982). Type species: *Aedes pholeocola* Linton & Harbach.

Name derivation: Greek goddess Nyx and fly (Gr) (Harbach 2018c).

syn. *Nyx* Harbach & Linton 2013 (as genus). In Harbach et al. 2013: 160. Type species: *Nyx pholeocola* Linton & Harbach.

Important references: Wilkerson et al. 2015 (phyl., class.; to subg. of genus *Aedes*).

biunguiculatus (Harbach)
2014 (M*; *Nyctomyia*). In Harbach & Taai 2014: 428. **Type loc.:** Wat Sumano cave, Phatthalung Province, Thailand (NHMUK). **Distr.:** Thailand. **Ety.:** double little claw (L); ref. to pair of gonostylar claws. **Informal name:** Two-clawed Thai Pointy Mosquito.

pholeocola (Linton & Harbach)
2013 (M*, F*, P*, L*; bion., mol. tax.; as *Nyx*). In Harbach et al. 2013. **Type loc.:** Wat Tham Phanturat cave, Khlong Sok (near Khao Sok National Park), Surat Thani Province, Thailand (08°53′ 596″ N, 98°31′ 312″ E) (NHMUK). **Distr.:** Thailand. **Ety.:** hole, cave, den inhabitant (Gr) [larvae]. **Informal name:** Cavern Thai Pointy Mosquito.

Subgenus *Ochlerotatus* Lynch Arribálzaga

1891a: 374, 1891b: 143 (as genus). Type species: *Ochlerotatus confirmatus* Lynch Arribálzaga.

Name derivation: n.s. [very troublesome; troublesome (Gr); -tatos is a superlative (Gr)].

syn. *Taeniorhynchus* Lynch Arribálzaga 1891a: 374, 1891b: 143 (as genus). Type species: *Culex taeniorhynchus* Wiedemann. Genus name suppressed, International Commission on Zoological Nomenclature 1959, Opinion 550.

syn. *Gilesia* Theobald 1903a: 233 (as genus). Type species: *Gilesia aculeata* Theobald.

Important references: Marks 1957a: 73 (tax.; Aculeatus Section of subg. *Ochlerotatus* of genus *Aedes*); Lee et al. 1984: 102 (as Aculeatus Section of *Aedes* subg. *Ochlerotatus*; tax., keys; Australasia); Reinert, Harbach, & Kitching 2008 (phyl., class.; to subg. of genus *Ochlerotatus*); Reinert, Harbach, & Kitching 2009 (phyl., class); Wilkerson et al. 2015 (phyl., class.; to syn with subg. *Ochlerotatus* of genus *Aedes*, as informal sp. group (Aculeatus Group)).

syn. *Culicada* Felt 1904: 391b (as genus). Type species: *Culex canadensis* Theobald.

Important references: Reinert, Harbach, & Kitching 2009 (phyl., class.; from syn. of subg. *Ochlerotatus* to subg. of genus *Ochlerotatus*); Wilkerson et al. 2015 (to syn. of subg. *Ochlerotatus* of genus *Aedes* as informal sp. group).

syn. *Culicelsa* Felt 1904: 391b (as genus). Type species: *Culex taeniorhynchus* Wiedemann.

Important references: Edwards 1932a: 136 (Group A, *Taeniorhynchus* group: *Culicelsa*); Reinert, Harbach, & Kitching 2008 (phyl., class.; from syn. of subg. *Ochlerotatus* to subg. of genus *Aedes*); Reinert, Harbach, & Kitching 2009 (phyl., class.); Wilkerson et al. 2015 (phyl., class.; to syn. with subg. *Ochlerotatus* of genus *Aedes*, as informal sp. group).

syn. *Protoculex* Felt 1904: 391d (as genus). Type species: *Culex serratus* Theobald.

Important references: Reinert, Harbach, & Kitching 2008 (phyl., class.; from syn. to subg. of genus *Ochlerotatus*); Reinert, Harbach, & Kitching 2009 (phyl., class.); Wilkerson et al. 2015 (phyl., class.; to syn. of subg. *Ochlerotatus* of genus *Aedes*).

syn. *Chrysoconops* Goeldi 1905: 114 (as genus). Type species: *Culex fulvus* Wiedemann.

Important references: Edwards 1932a: 136 (Group C, fulvus group: *Chrysoconops*); Reinert, Harbach, & Kitching 2008 (phyl., class.; as subg. of genus *Ochlerotatus*); Reinert, Harbach, & Kitching 2009 (phyl., class.); Wilkerson et al. 2015 (phyl., class.; to syn. of subg. *Ochlerotatus* of genus *Aedes* as informal sp. group).

syn. *Pseudoculex* Dyar 1905a: 47 (as genus). Type species: *Culex squamiger* Coquillett.

syn. *Lepidoplatys* Coquillett 1906b: 314 (as genus). Type species: *Culex squamiger* Coquillett.

syn. *Pseudohowardina* Theobald 1907: 223 (as genus). Type species: *Culex trivittatus* Coquillett.

syn. *Andersonia* Strickland 1911a: 250 (as genus). Type species: *Andersonia tasmaniensis* Strickland.

Important references: Edwards 1932a: 135 (type species desig.).

syn. *Hyparctius* Martini 1930: 264 (globus of [has general characteristics of subgenus] *Ochlerotatus*). Type species: *Culex pullatus* Coquillett.

Important references: Edwards 1932a: 135 (type species desig.).

syn. *Kummyia* Vargas 1950b: 155 (as subg. of genus *Aedes*). Type species: *Culex serratus* Theobald.

syn. *Empihals* Reinert, Harbach, & Kitching 2008: 105 (as subg. of genus *Ochlerotatus*). Type species: *Culex vigilax* Skuse.

Important references: Reinert, Harbach, & Kitching 2009 (phyl., class.); Rattanarithikul et al. 2010 (F*, L*; key, bion.; Thailand); Wilkerson et al. 2015 (phyl., class.; to syn. of subg. *Ochlerotatus* of genus *Aedes*, as informal sp. group).

syn. *Buvirilia* Reinert, Harbach, & Kitching 2008: 104 (as subg. of genus *Ochlerotatus*). Type species: *Aedes (Ochlerotatus) edgari* Stone & Rosen.

Important references: Reinert, Harbach, & Kitching 2009 (phyl., class.); Wilkerson et al. 2015 (phyl., class.; to syn. of *Aedes* subg. *Ochlerotatus*, as informal sp. group).

syn. *Pholeomyia* Reinert, Harbach, & Kitching 2008: 107 (as subg. of genus *Ochlerotatus*). Type species: *Aedes (Ochlerotatus) calcariae* Marks.

Important references: Dobrotworsky 1960b (tax.; Flavifrons Section); Lee et al. 1984 (tax.; Flavifrons Section); Reinert, Harbach, & Kitching 2009 (phyl., class.); Wilkerson et al. 2015 (phyl., class.; to syn. with subg. *Ochlerotatus* of genus *Aedes*, as informal sp. group Pholeomyia Group (= Flavifrons Section of Marks 1957a; Dobrotworsky 1960b; Dobrotworsky 1965; Lee et al.1984)).

syn. *Woodius* Reinert, Harbach, & Kitching 2009: 775 (as subg. of genus *Ochlerotatus*). Type species: *Aedes intruders* Dyar.

Important references: Reinert, Harbach, & Kitching 2009 (phyl., class.; to syn. with subg. *Ochlerotatus* of genus *Aedes*); Wilkerson et al. 2015 (phyl., class.; to syn. of subg. *Ochlerotatus* of genus *Aedes*).

syn. *Lepidokeneon* Reinert, Harbach, & Kitching 2009: 773 (as subg. of genus *Ochlerotatus*). Type species: *Aedes* (*Ochlerotatus*) *spilotus* Marks.

Important references: Marks 1963b (as the Stricklandi Section of *Aedes* subg. *Ochlerotatus*; revision); Dobrotworsky 1965: 122 (as the Stricklandi Section of *Aedes* subg. *Ochlerotatus*; keys, redescr.); Lee et al. 1984 (as Stricklandi Section of *Aedes* subg. *Ochlerotatus*; keys, tax.); Wilkerson et al. 2015 (phyl., class.; as syn. of subg. *Ochlerotatus* of genus *Aedes*).

syn. *Juppius* Reinert, Harbach, & Kitching 2009: 770 (as subg. of genus *Ochlerotatus*). Type species: *Grabhamia caballa* Theobald.

Important references: McIntosh 1973 (as subg. *Ochlerotatus*; tax.); Huang & Rueda 2014 (as *Aedes* subg. *Ochlerotatus*; key); Wilkerson et al. 2015 (phyl., class.; to syn. of subg. *Ochlerotatus* of genus *Aedes*, as informal sp. group).

Important references: Coquillett 1910: 577 (desig. of type species); da Costa Lima 1930e (tax.; as genus *Taeniorhynchus*); Peus 1933: 145 (F; key; Communis Group); Asanuma & Takahasi 1952 (M gen.); Marks 1957a: 71 (subg. review, tax.; Australasia); Belkin 1962 (tax., bion., distr.); Zavortink 1970a: 4 (keys; Varipalpus Complex); Arnell & Nielsen 1972: 1 (revision; Varipalpus Group); Horsfall & Voorhees 1972: 123 (E; Nearctic); Rjazantseva 1972: 35 (F gen.*; tax.); Arnell 1976 (revision; Scapularis Group); Reinert 2000h (tax.; to genus); Reinert 2002e (F gen.*); Reinert, Harbach, & Kitching 2008 (phyl., class.); Reinert, Harbach, & Kitching 2009 (phyl., class.); Rattanarithikul et al. 2010 (F*, L*; keys, bion.; Thailand); Huang & Rueda 2014 (F*; key, six sp. of *Och.* and one *Coe.*; Afrotropical); Wilkerson et al. 2015 (phyl., class.; to subg. of genus *Aedes*); Soghigian et al. 2017 (phyl., class.).

aboriginis Dyar
1917b: 99 (M*, F, L*). **Type loc.:** Longmire Springs, Mount Ranier National Park, Washington, United States (USNM). **Distr.:** Canada, United States (Alaska, continental). **Refs.:** Gjullin 1937 (F*); Knight 1951a: 96 (M*, F, L); Yamaguti & LaCasse 1951d: 26 (M*, F*, L*); Carpenter & LaCasse 1955: 147 (M*, F*, L*; keys); Stone & Knight 1956a: 213 (type info., lectotype desig.); Darsie 1957: 615 (P*); Bickley 1976 (tax.); Darsie & Ward 2005 (F*, L*; keys, distr.). **Ety.:** n.s. [by the Aboriginals? (L?)]. **Informal name:** Original American Pointy Mosquito. Official ESA common name: northwest coast mosquito.

abserratus (Felt & Young)
1904: 312 (A, L; *Culex*). **Type loc.:** Elizabethtown, New York, United States (NY). **Distr.:** Canada, United States (continental). **Refs.:** Darsie 1951: 23 (P*; as *implacabilis*); Knight 1951a: 98 (M*, F, L); Vockeroth 1954a: 114 (from syn. with *implacabilis*); Carpenter & LaCasse 1955: 149 (M*, F*, L*; keys); Kalpage & Brust 1968 (E*); Pickavance et al. 1970 (distr.); Darsie & Ward 2005 (F*, L*; keys, distr.); Harrison et al. 2016 (F*, L*; keys, distr., tax.). **Ety.:** n.s. [away from or not the species *serratus*? (L?)]. **Informal name:** Dull American Pointy Mosquito [*abserratus* = "not sharp" = Dull].

syn. *centrotus* Howard, Dyar, & Knab
1917: 747 (F). **Type loc.:** White River, north of Lake Superior, Ontario, Canada (USNM). **Refs.:** Knight 1951a: 98 (type info., lectotype desig.); Stone & Knight 1956a: 216 (type info.).

syn. *dysanor* Dyar
1921e: 70 (M*). **Type loc.:** Plattsburg, New York, United States (USNM).

aculeatus (Theobald)
1903a: 233 (F; *Gilesia aculeata*). **Type loc.:** Deception Bay, near Burpengary, Queensland, Australia (NHMUK). **Distr.:** Australia. **Refs.:** Bancroft 1908: 29 (F; as genus *Gilesia*); Marks 1949: 2 (M*, F, P*, L*, E*; bion.); Lee et al. 1984: 122 (F; key, tax., bion., distr., review). **Ety.:** n.s. [sharp, pointed (L)]. **Informal name:** Stinging Australian Pointy Mosquito.

aenigmaticus Cerqueira & Costa
1946: 133 (M*). **Type loc.:** Maracaju, Mato Grosso, Brazil (INER). **Distr.:** Brazil, Colombia. **Refs.:** Lane 1953: 661 (M*); Xavier 1973: 160 (type info.). **Ety.:** n.s. [enigmatic (L); poss. ref. to similarity with *serratus* and *nubilus*]. **Informal name:** Enigmatic Brazilian Pointy Mosquito.

akkeshiensis Tanaka
1998: 215–225 (M*, F, P, L*). **Type loc.:** Akkeshi-cho, Hokkaido, Japan (National Science Museum of Japan). **Distr.:** Japan. **Refs.:** Tanaka 1999: 120 (P*; tax.). **Ety.:** Akkeshi-cho, Japan. **Informal name:** Akkeshi-cho Japanese Pointy Mosquito.

albifasciatus (Macquart)
1838b: 35 (F; *Culex*). **Type loc.:** Rio de Janeiro, Guanabara, Brazil (MNHP). **Distr.:** Argentina, Bolivia, Brazil, Chile, Paraguay, Uruguay. **Refs.:** Lane 1953: 642 (M*, F, L*); Ronderos & García 1963a: 35 (P*); Mohrig 1967 (F*); Belkin 1968b: 4 (type info.); Belkin et al. 1971: 19 (type loc. info.); dos Santos-Mallet et al. 2009 (E*); Becker et al. 2010: 382 (F*, L*; key, tax., distr., bion.). **Ety.:** n.s. [white striped (L); prob. ref. to a dorsal abdominal band formed by contiguous triangular patches on each segment]. **Informal name:** White-bundled Brazilian Pointy Mosquito.

syn. *flavipes* Macquart
1838b: 35 (F; *Culex*). **Type loc.:** Concepción, Chile (MNHP). **Refs.:** Belkin 1968b: 5 (type info.).

syn. *annuliferus* Blanchard
1852 (A; *Culex*). In Gay 1852: 333. **Type loc.:** Near Coquimbo, Illapel, Chile (MNHP). **Refs.:** Belkin 1968b: 4 (syn.; lectotype desig.).

syn. *vittatus* Philippi
1865: 596 (F; *Culex*). **Type loc.:** Santiago, Chile (NE).
syn. *philippii* Dyar
1924g: 129 (new name for *vittatus* Philippi, not Bigot).
syn. *colonarius* Dyar
1924g: 130 (F). **Type loc.:** Azapa Valley, Province Tacana (*sic*, Tacna), Chile (USNM).

albineus Séguy
1923: 206 (M, F). **Type loc.:** Djamaa, Touggourt, Algeria (MNHP, NHMUK). **Distr.:** Algeria, Morocco, Tunisia. **Refs.:** Séguy 1924: 105 (M gen.*); Danilov 1979: 29 (from syn. with *caspius*); Townsend 1990: 39 (syntypes, NHMUK). **Ety.:** n.s. [white (L); ref. to white markings]. **Informal name:** Albino Algerian Pointy Mosquito.
syn. *flavidorsalis* Luh & Lee
1975: 428 (M*, F, L*). **Type loc.:** Ningxia Province, China [People's Republic of China] (ND). **Refs.:** Danilov 1979: 29 (syn.).

aloponotum Dyar
1917b: 98 (F). **Type loc.:** Lake Cushman, Washington, United States (USNM). **Distr.:** Canada, United States (continental). **Refs.:** Gjullin 1937 (F*); Wood 1977: 78 (tax.); Darsie & Ward 2005 (F*, L*; keys, distr.); Darsie 2011c (P*). **Ety.:** n.s. **Informal name:** Fox-marked American Pointy Mosquito.

amateuri Ortega & Zavortink
2019 (F*, M*, P*, L*). In Ortega et al. 2019. **Type loc.:** Sierra la Camotera, Santiago, NuevoLeón, Mexico (n.s., Bohart Museum?). **Distr.:** Mexico. **Ety.:** amateur, "because it was collected by personnel that were inexperienced in mosquito collections." **Informal name:** Amateur Mexican Pointy Mosquito.

ambreensis Rodhain & Boutonnier
1983c: 826 (F). **Type loc.:** Mount Ambre forest, 35 km south of Antsiranana, Madagascar (PIP). **Distr.:** Madagascar (includes Glorioso & Juan De Nova Isl.). **Ety.:** Montagne d'Ambre, Madagascar. **Informal name:** Ambre Malagasy Pointy Mosquito.

andersoni Edwards
1926a: 112 (new name for *tasmaniensis* Strickland, not Strickland 1911a (May); as *Aedes*). **Distr.:** Australia. **Refs.:** Dobrotworsky 1960a: 68 (M*, F, P*, L*); Dobrotworsky 1965: 117 (M*, F*, L*); Lee et al. 1984: 124 (F; key, tax., bion., distr., review). **Ety.:** n.s. [someone named Anderson]. **Informal name:** Anderson Australian Pointy Mosquito.
syn. *tasmaniensis* Strickland
1911a: 250 (July) (F*; *Andersonia*). **Type loc.:** Tasmania, Australia (NHMUK).
syn. *tasmaniensis* Taylor
1914b: 466 (F*; *Menolepsis*; not Strickland). **Type loc.:** Wedge Bay, Tasmania, Australia (US).

angustivittatus Dyar & Knab
1907a: 9 (M, F). **Type loc.:** Port Limon [Puerto Limón], Costa Rica (USNM). **Distr.:** Argentina, Belize, Bolivia, Brazil, Colombia, Costa Rica, Ecuador, El Salvador, Guatemala, Honduras, Mexico, Nicaragua, Panama, Peru, Venezuela. **Refs.:** Lane 1953: 670 (M*, F, L*); Stone & Knight 1956a: 214 (type info.); Ronderos & García 1963a: 37 (P*); Mohrig 1967 (F*); Arnell 1976: 29 (M*, F*, P*, L*; distr.); Pecor et al. 2000 (tax., distr.; Peru); Berti et al. 2015 (distr.; Venezuela). **Ety.:** narrow stripes (L); ref. to stripes on thorax. **Informal name:** Narrow-banded Costa Rican Pointy Mosquito. (Vol. 1, p. 426)
syn. *cuneatus* Dyar & Knab
1908: 54 (M, F). **Type loc.:** Córdoba, Veracruz, Mexico (USNM). **Refs.:** Stone & Knight 1956a: 216 (type info., lectotype desig.).
syn. *argentescens* Dyar & Knab
1908: 55 (M, F). **Type loc.:** Córdoba, Vera Cruz, Mexico (USNM). **Refs.:** Stone & Knight 1956a: 214 (type info., lectotype desig.).
syn. *traversus* Dyar
1925g: 215 (F). **Type loc.:** Zulia River, Venezuela (USNM). **Refs.:** Stone & Knight 1956a: 226 (type info., lectotype desig.).

annulipes (Meigen)
1830b: 241 (M, F; *Culex*). **Type loc.:** Europe (MNHP, NMW). **Distr.:** Armenia, Austria, Belarus, Belgium, Bulgaria, Crimean Peninsula, Croatia, Czech Republic, Denmark, Estonia, Finland, France, Germany, Greece, Hungary, Italy, Japan, Kaliningrad Oblast, Latvia, Lithuania, Luxembourg, Macedonia, Moldova, Montenegro, Netherlands, Norway, Poland, Romania, Russia (Northwestern), Serbia, Slovakia, Slovenia, Sweden, Switzerland, Turkey, Ukraine, United Kingdom. **Refs.:** Séguy 1924: 114 (L*), 111 (M gen.*); Marshall 1938: 194 (M*, F*, L*); Natvig 1948: 233 (M*, F, L*); Van Aken 1961: 259 (P*); Gutsevich et al. 1974: 225 (M*, F, L*); Dahl 1975: 82, 88 (tax, distr.); Alten et al. 2000 (distr.; Turkey); Spungis 2000: 9 (distr.; Latvia); Beck et al. 2003: 24 (distr.; Luxembourg); Becker et al. 2010: 209 (M*, F*, L*; keys, tax., distr., bion.); Robert et al. 2019 (distr.; Western Palearctic). **Ety.:** ringed feet (L); tarsi. **Informal name:** Ring-footed European Pointy Mosquito.
syn. *quartus* Martini
1920a: 128 (M*). **Type loc.:** Borsteler Jaeger, Germany (NHMUK).

antipodeus (Edwards)
1920a: 132 (F; *Ochlerotatus*). **Type loc.:** Karikari Bay, New Zealand (NHMUK). **Distr.:** New Zealand. **Refs.:** Belkin 1962: 389 (M*, F, P*, L*); Marks & Nye 1963: 50 (M, F*, P*, L*); Belkin 1968a: 98 (M, P, L; distr.); Lee et al. 1984: 126 (F; key, tax., bion., distr., review); Derraik 2004 (distr.; New Zealand). **Ety.:** n.s. [against feet (Gr); inferred, having to do with the feet of inhabitants that are against

each other on opposite sides of the world; in reference to Australia and New Zealand being on the other side of the earth from Britain (Harbach 2018c)]. **Informal name:** Antipodes New Zealander Pointy Mosquito.

atactavittatus Arnell
1976: 26 (M*, F*, P*, L*). **Type loc.:** Lerma, Campeche, Mexico (USNM). **Distr.:** Mexico. **Ety.:** n.s. [disorganized ribbon (Gr, L); poss. ref. to line of scutal scales]. **Informal name:** Rag Bag Mexican Pointy Mosquito [*atactavittatus* = "disordered ribbons" = Rag Bag].

atlanticus Dyar & Knab
1906d: 198 (L*). **Type loc.:** New Jersey and Sanford, Florida, United States (NE). **Distr.:** Guatemala, Mexico, United States (continental). **Refs.:** Yamaguti & LaCasse 1951d: 29 (M*, F*, L*); Carpenter & LaCasse 1955: 151 (M*, F*, L*; keys, bion.); Stone & Knight 1956a: 214 (type info.); Craig & Horsfall 1960 (E*); King et al. 1960: 66 (M, F*, L; keys); Ross & Horsfall 1965 (M*, F*, L*, E*; keys); Dodge 1966: 348 (first instar L*; key); Horsfall et al. 1970: 1711 (E*); Scanlon & Yates 1970 (tax.); Mattingly 1971a: Pl. 33 (L head*, terminal segs.*); Reinert 1971: 171 (P*); Roberts & Scanlon 1979 (F*; tax.); Roberts & Scanlon 1975 (bion.); Darsie & Ward 2005 (F*, L*; keys, distr.); Harrison et al. 2016 (F*, L*; keys, distr.). **Ety.:** Atlantic; region of North America. **Informal name:** Atlantic American Pointy Mosquito. (Vol. 1, p. 428)

auratus Grabham
1906b: 313 (M, F, L*). **Type loc.:** Kingston, Jamaica (USNM). **Distr.:** Jamaica. **Refs.:** Belkin, Heinemann, & Page 1970: 168 (M*, F, L*; from syn. with *tortilis* Theobald); Arnell 1976: 44 (M*, F, L*). **Ety.:** n.s. [golden (L); poss. ref. to "thorax rich golden yellow"]. **Informal name:** Theia Jamaican Pointy Mosquito [named for Theia, the ancient Greek goddess of gold].

aurifer (Coquillett)
1903a: 255 (M, F; *Culex*). **Type loc.:** Center Harbor, New Hampshire, United States (USNM). **Distr.:** Canada, United States (continental). **Refs.:** Ross 1947: 76 (M*, F*, L*); Rempel 1950: 23i (L*); Darsie 1951: 21 (P*); Yamaguti & LaCasse 1951d: 33 (M*, F*, L*); Carpenter & LaCasse 1955: 153 (M*, F*, L*; keys); Craig 1956 (E*); Stone & Knight 1956a: 215 (type info., lectotype desig.); Ross & Horsfall 1965 (M*, F*, L*, E*; keys); Lake 1968 (first instar L*); Darsie & Ward 2005 (F*, L*; keys, distr.); Harrison et al. 2016 (F*, L*; keys, distr., tax.). **Ety.:** n.s. [gold bearing (L); poss. ref. to "sides of mesonotum golden yellow" in contrast to white on *triseriatus*]. **Informal name:** Goldminer American Pointy Mosquito [*aurifer* = "gold bearing" = Goldminer].

behningi Martini
1926: 200 (F, L). **Type loc.:** Saton, Saratov Area, USSR (NHMUK). **Distr.:** Belarus, Croatia, Kaliningrad Oblast, Lithuania, Moldova, Poland, Romania, Russia (Northwestern, Central, Southern, and Volga regions), Slovakia, Ukraine. **Refs.:** Martini 1930: 266 (M*, F, L*); Mohrig 1967 (F*); Rjazantzeva 1972 (F*); Gutsevich et al. 1974: 216 (M*, F, L*); Becker et al. 2010: 211 (M*, F*, L*; keys, tax., distr., bion.); Robert et al. 2019 (distr.; Western Palearctic). **Ety.:** Arvid Liberovich Behning. **Informal name:** Behning Russian Pointy Mosquito.

bejaranoi Martínez, Carcavallo & Prosen
1960: 26 (M, F). **Type loc.:** Yungas del Palmar, Chapare Province, Cochabamba Department, Bolivia (A. Martínez) (LU). **Distr.:** Bolivia. **Ety.:** Don Juan Francisco Ricardo Bejarano. **Informal name:** Bejarano Bolivian Pointy Mosquito.

berlandi Séguy
1921a (M*, F*). **Type loc.:** Paris and Asnières, France (MNHP). **Distr.:** Algeria, France (includes Corsica), Greece, Italy (includes Sardinia and Sicily), Morocco, Portugal, Spain, Tunisia. **Refs.:** Séguy 1924: 92 (M gen.*); Séguy 1925 (E*); Clavero 1946: 9 (M*, F*, L*; as *longitubus*); Senevet et al. 1954: 266 (M*, F*, P*, L*; tax.; as *longitubus*); 1954: 267 (tax.); Rioux & Arnold 1955: 261 (to sp.; L*); Rioux 1958: 167 (M*, F*, L*); Senevet & Andarelli 1958: 280 (P*; as *longitubus*); Samanidou-Voyadjoglou & Darsie 1993: 465 (distr.); Eritja et al. 2000: 11 (distr.); Melero-Alcíbar 2005d (P; tax.); Becker et al. 2010: 212 (M*, F*, L*; keys, tax., distr., bion.); Robert et al. 2019 (distr.; Western Palearctic). **Ety.:** M.L. Berland. **Informal name:** Berland French Pointy Mosquito.
syn. *praeteritus* Séguy
1923: 207 (F). **Type loc.:** Le Pradet, Var, France (MNHP). **Refs.:** Rioux & Arnold 1955: 261 (syn.); [Coluzzi (*in litt.*) as synonym of *pulchritarsis* (Rondani)]; Stone 1967: 207.
syn. *longitubus* Cambournac
1938: 76 (M*, F*, L*). **Type loc.:** Aguas de Moura, 20 km east of Setubal, Portugal (NHMUK). **Refs.:** Rioux & Arnold 1955: 261 (syn.).
syn. *heracleensis* Callot
1944c: 56 (P, L*). **Type loc.:** Cavalaire, Var, France (IPS). **Refs.:** Callot 1944a (syn.); Rioux & Arnold 1955: 261 (tax.).

bimaculatus (Coquillett)
1902c: 84 (F; *Culex*). **Type loc.:** Brownsville, Texas, United States (USNM). **Distr.:** El Salvador, Mexico, United States (continental). **Refs.:** Rozeboom 1942b: 31 (in part; tax.); Ross 1943: 143 (M*, F, L*; tax.); Yamaguti & LaCasse 1951d: 41 (M*, F*, L*); Carpenter & LaCasse 1955: 157 (M*, F*, L*; keys); Darsie & Ward 2005 (F*, L*; keys, distr.). **Ety.:** n.s. [two-spotted (L)]. **Informal name:** Two-spotted American Pointy Mosquito.
syn. *rozeboomi* Vargas
1941b: 394 (M, F). **Type loc.:** Campeche, Mexico (ISET). **Refs.:** Ross 1943: 143 (syn.).

biskraensis Brunhes
1999: 25 (M*, F, P*, L*). **Type loc.:** Seriana/Biskra, Algeria (MNHP). **Distr.:** Algeria, Morocco, Tunisia. **Ety.:** Biskra, Algeria. **Informal name:** Biskra Algerian Pointy Mosquito.

bogotanus Arnell
1976: 22 (M*, F*, P*, L*). **Type loc.:** Ogamora, near Soacha, Cundinamarca, Colombia (USNM). **Distr.:** Colombia. **Ety.:** Bogotá, Colombia. **Informal name:** Bogotá Colombian Pointy Mosquito.

breedensis Muspratt
1953a: 55 (M, F, L*). **Type loc.:** Worcester, Breede River Valley, Cape Province, South Africa [Republic of South Africa] (SAIM). **Distr.:** Republic of South Africa. **Refs.:** Jupp 1996 (M*, F*; key). **Ety.:** Breede River Valley, Cape Prov., Republic of South Africa. **Informal name:** Breede South African Pointy Mosquito.

burjaticus (Kuchartshuk)
1973: 98 (M*, F, L). **Type loc.:** Khuzir [Khuzhir], Buryat Mongol ASSR [Russia] (LU). **Distr.:** Russia. **Ety.:** n.s. [for type locality; Khuzir, Buryat Mongol ASSR, USSR (Harbach 2018c)]. **Informal name:** Buryat Russian Pointy Mosquito.

burpengaryensis (Theobald)
1905g: 26 (M*, F*; *Culicada*). **Type loc.:** Burpengary, Queensland, Australia (NHMUK). **Distr.:** Australia. **Refs.:** Bancroft 1908: 45 (M, F); Mackerras 1927: 294 (tax.); Lee et al. 1984: 128 (F; key, tax., bion., distr., review). **Ety.:** Burpengary, Queensland, Australia. **Informal name:** Burpengary Australian Pointy Mosquito.

caballus (Theobald)
1912a: 93 (F; *Grabhamia*). **Type loc.:** Onderstepoort, Transvaal, Union of South Africa [Republic of South Africa] (NHMUK). **Distr.:** Angola, Ethiopia, Iran, Kenya, Namibia, Republic of South Africa, Saudi Arabia, Sudan and South Sudan, Yemen, Zimbabwe. **Refs.:** Edwards 1941: 117 (M*, F); Hopkins 1952: 124 (L*); McIntosh 1973: 265 (M*, F*, P, L); Jupp 1996 (M*, F*; key); Reinert 2002e: Fig. 23 (F*); Ahmed et al. 2011 (distr.; Saudi Arabia). **Ety.:** n.s. [horse (L)]. **Informal name:** Equine South African Pointy Mosquito.

cacozelus Marks
1963b: 45 (F). **Type loc.:** Darkan (110 mi south-southwest of Perth), Western Australia, Australia (CSIR). **Distr.:** Australia. **Refs.:** Lee et al. 1984: 130 (F; key, tax., bion., distr., review). **Ety.:** bad imitator (L); "refers to its partial resemblance to *A. stricklandi*." **Informal name:** Hammy Australian Pointy Mosquito [*cacozelus* = "bad imitator" = bad actor = Hammy].

calcariae Marks
1957a: 74 (M*, F*, P*, L*). **Type loc.:** "Calcaria," near Lake Eliza, circa 7 mi south of Robe, South Australia, Australia (CSIR). **Distr.:** Australia. **Refs.:** Dobrotworsky 1965: 97 (M*, F*, L*); Lee et al. 1984: 131 (F; key, tax., bion., distr., review). **Ety.:** Calcaria (informal place name belonging to Mr. Lines). **Informal name:** Calcaria Australian Pointy Mosquito.

calumnior Belkin, Heinemann, & Page
1970: 169 (M*, F, P*, L*). **Type loc.:** Mocho, near May Pen (Clarendon), Jamaica (USNM). **Distr.:** Jamaica. **Ety.:** n.s. [misrepresent (L); poss. ref. to close to *obturbator*]. **Informal name:** Liar Jamaican Pointy Mosquito.

campestris Dyar & Knab
1907d: 213 (M, F). **Type loc.:** Oxbow, Saskatchewan, Canada (USNM). **Distr.:** Canada, Mexico, Russia, United States (continental). **Refs.:** Dyar 1920m: 120 (L); Ross 1947: 77 (M*); Rempel 1950: 223 (L*); Rempel 1953: 467 (M*, F*); Yamaguti & LaCasse 1951d: 45 (M*, F*, L*); Vockeroth 1954b: 245 (F*); Carpenter & LaCasse 1955: 159 (M*, F*, L*; keys); Stone & Knight 1956a: 216 (type info.); Myers 1967: 799 (E*); Kalpage & Brust 1968 (E*); Bohart & Washino 1978: 44 (M, F*, L*; keys, tax., bion., distr.); Gornostaeva 2000: 479 (distr.; Russia); Darsie & Ward 2005 (F*, L*; keys, distr.); Darsie 2011c (P*). **Ety.:** n.s. [of the fields (L)]. **Informal name:** Field Canadian Pointy Mosquito.

syn. *callithotrys* Dyar
1920a: 16 (M, F). **Type loc.:** White Horse, Yukon Territory, Canada (USNM). **Refs.:** Stone & Knight 1956a: 216 (type info.).

camptorhynchus (Thomson)
1869: 443 (F; *Culex*). **Type loc.:** Sydney, New South Wales, Australia (SM). **Distr.:** Australia, New Zealand. **Refs.:** Mackerras 1927: 290 (M*); Lee 1944a: 52 (L*); Dobrotworsky 1960a: 63 (M*, F, L*); Dobrotworsky 1965: 115 (M*, F*, L*); Lee et al. 1984: 133 (F; key, tax., bion., distr., review); Linley et al. 1992b (E*); Reinert 2002e: Fig. 30 (F*); Barton et al. 2004 (bion.); Derraik 2004 (distr.; New Zealand); Becker et al. 2010: 360 (F*, L*; key, tax., distr., bion.). **Ety.:** bent nose (Gr); ref. to curved proboscis. **Informal name:** Bent-beaked Australian Pointy Mosquito.

syn. *labeculosus* Coquillett
1905: 116 (M, F; *Culex*). **Type loc.:** Goode Island, Victoria, Australia (USNM). **Refs.:** Stone & Knight 1956a: 220 (type info., lectotype desig.).

syn. *westralis* Strickland
1911a: 130 (F; *Culicelsa*). **Type loc.:** King's Park, Perth, Western Australia and Port River, Adelaide, South Australia, Australia (NHMUK).

syn. *inornata* Strickland
1911a: 201 (F; *Culicada*). **Type loc.:** Tasmania, Australia (NHMUK).

syn. *victoriensis* Taylor
1914b: 460 (F*; *Culicada*). **Type loc.:** Toorak, Victoria, Australia (US).

syn. *nigra* Taylor
1914c: 688 (M; *Culicada*). **Type loc.:** Launceston, Tasmania, Australia (US).

syn. *annulipes* Taylor
1914c: 693 (M*; *Culicada*). **Type loc.:** Bass Straits, Flinders Island [near Tasmania]; Launceston, Tasmania; Melbourne, Victoria, Australia (US).

canadensis (Theobald)

1901c: 3 (M*, F*; *Culex*). **Type loc.:** Front Creek, near Toronto, Ontario, Canada (NHMUK). **Distr.:** Canada, Dominican Republic, Mexico, United States (Alaska, continental). **Refs.:** Mitchell 1907b (E*); Howard et al. 1917 (E); Craig & Peinkowski 1931 (distr.; Alaska, United States); Ross 1947: 78 (M*, F*, L*); Darsie 1951: 20 (P*); Yamaguti & LaCasse 1951d: 49 (M*, F*, L*); Carpenter & LaCasse 1955: 161 (M*, F*, L*; keys); Craig 1956 (E*); Horsfall & Craig 1956: 372 (E*); Craig & Horsfall 1958 (E*); Price 1960: 550 (first instar L*); Ross & Horsfall 1965 (E*); Dodge 1966: 350 (first instar L*; key); Mohrig 1967 (F*); Belkin 1968b: 4 (lectotype desig.); Kalpage & Brust 1968 (E*); Horsfall et al. 1970: 1710 (E*); Darsie & Ward 1981 (F*, L*; keys, distr.); Darsie & Ward 2005 (F*, L*; keys, distr.); Harrison et al. 2016 (F*, L*; keys, distr.; note 14). **Ety.:** Canada. **Informal name:** Iroquois Canadian Pointy Mosquito [the word "Canada" is taken from the Saint Lawrence Iroquoian word for "village"]. (Vol. 1, p. 430)

syn. *nivitarsis* Coquillett

1904c: 168 (M, F; *Culex*). **Type loc.:** Paterson, New Jersey, United States (USNM). **Refs.:** Stone & Knight 1956a: 222 (type info., lectotype desig.).

ssp. ***mathesoni*** Middlekauff

1944: 42 (M*, F; as *Aedes* sp.). **Type loc.:** Kissimmee, Florida, United States (USNM). **Distr.:** Canada, United States (continental). **Refs.:** Rings & Hill 1946: 237 (L*); Rings & Hill 1948: 41 (F*; tax.; to ssp.); Yamaguti & LaCasse 1951d: 128 (M*, F*, L*); Carpenter & LaCasse 1955 (keys); Pickavance et al. 1970 (distr.); Bickley 1981 (tax., distr.; ssp. *mathesoni*); Darsie & Ward 1981 (F*, L*; keys, distr.); Darsie & Ward 2005 (F*, L*; keys, distr.); Harrison et al. 2016 (F*, L*; keys, distr.). **Ety.:** Robert Matheson. **Informal name:** Matheson Canadian Pointy Mosquito.

cantans (Meigen)

1818: 6 (F*; *Culex*). **Type loc.:** Europe (MNHP). **Distr.:** Austria, Belarus, Belgium, Bulgaria, Canada, Crimean Peninsula, Croatia, Czech Republic, Denmark, Estonia, Finland, France (includes Corsica), Germany, Greece, Hungary, Ireland, Italy (includes Sicily), Kaliningrad Oblast, Latvia, Lithuania, Luxembourg, Moldova, Mongolia, Netherlands, Norway, People's Republic of China, Poland, Romania, Russia (northwestern, Central, Southern, and Volga regions), Serbia, Slovakia, Slovenia, Spain, Sweden, Switzerland, Ukraine, United Kingdom. **Refs.:** Brolemann 1919b (F*); Séguy 1924: 114 (L*), 15, 111 (M gen.*); Marshall 1938: 199 (M*, F*, L*, E*; bion.; from syn. with *maculatus*); Clavero 1946: 10 (M*, F*, L*); Natvig 1948: 221 (M*, F*, L*; bion., tax.); Nielsen & Greve 1950: 233 (bion.); Senevet & Andarelli 1958: 282 (P*); Sicart & Larrouy 1968 (F*); Hinton & Service 1969 (E*); Rjazantzeva 1970 (F*); Minar 1971: 92 (distr.); Rjazantzeva 1972 (F*); Gutsevich et al. 1974: 210 (M*, F, L*); Dahl 1975: 83, 88 (distr.); Utrio 1976: 131,133 (L*; distr.); Bozicic 1980: 93 (distr.); Dahl 1997 (E*); Eritja et al. 2000: 11 (distr.; Spain); Spungis 2000: 9 (distr.; Latvia); Beck et al. 2003: 24 (distr.; Luxembourg); Melero-Alcíbar & Salom-Bonet 2003: 18 (P*; tax.); Becker et al. 2003 (E*); Becker et al. 2010: 214 (M*, F*, L*; keys, tax., distr., bion.); Robert et al. 2019 (distr.; Western Palearctic). **Ety.:** n.s. [singer, singing (?) (L)]. **Informal name:** Singing European Pointy Mosquito.

syn. *reptans* Meigen

1804: 3 (M, F; *Culex*; not Linnaeus). **Type loc.:** Europe (NE).

syn. *waterhousei* Theobald

1905 (M, F; *Culex*). In Waterhouse 1905: 674. **Type loc.:** New Forest, Brockenhurst [Hampshire], England (NHMUK).

syn. *subvexans* Martini

1922: 159 (A, L; *Aedes cantans* var.). **Type loc.:** Denmark (LU). **Refs.:** Natvig 1948 (syn.).

cantator (Coquillett)

1903a: 255 (M, F; *Culex*). **Type loc.:** New Jersey, United States (USNM). **Distr.:** Canada, United States (continental). **Refs.:** Headlee 1931: 40 (bion.); Darsie 1951: 17 (P*); Yamaguti & LaCasse 1951d: 55 (M*, F*, L*); Carpenter & LaCasse 1955: 164 (M*, F*, L*; keys); Craig 1956 (E*); Craig & Horsfall 1960 (E*); Lake 1963 (first instar L*); Dodge 1966: 350 (first instar L; key); Darsie & Ward 2005 (F*, L*; keys, distr.); Harrison et al. 2016 (F*, L*; keys, distr., tax.). **Ety.:** n.s. [singer (L)]. **Informal name:** Singer American Pointy Mosquito. Official ESA common name: brown saltmarsh mosquito.

caspius (Pallas)

1771: 475 (A; *Culex*). **Type loc.:** Caspian Sea (NE). **Distr.:** Afghanistan, Albania, Algeria, Armenia, Austria, Azerbaijan, Bahrain, Belarus, Belgium, Bosnia and Herzegovina, Bulgaria, Canada, Croatia, Cyprus, Czech Republic, Denmark, Djibouti, Egypt, Estonia, Finland, France, FYRO Macedonia, Georgia, Germany, Greece, Hungary, Iran, Iraq, Ireland, Israel (Gaza Strip and West Bank), Italy (includes Sardinia and Sicily), Kosovo, Japan, Jordan, Kazakhstan, Latvia, Lebanon, Libya, Lithuania, Malta, Mauritania, Moldova, Mongolia, Montenegro, Morocco, Netherlands, Norway, Pakistan, Palestine, People's Republic of China, Poland, Portugal, Qatar, Romania, Russia (Northwestern, Central, Southern and Volga Dists.), Saudi Arabia, Serbia, Slovakia, Slovenia, Spain (includes Balearic and Canary Isl.), Sri Lanka, Sudan and South Sudan, Sweden, Switzerland, Tajikistan, Tunisia, Turkey, Turkmenistan, Ukraine, United Arab Emirates, United Kingdom, United States, Uzbekistan, Yemen. **Refs.:** Kirkpatrick 1925b: 79 (M*, F*, P*, L*; bion.); Barraud 1934: 148 (M*, F, P, L); Marshall 1938 (E*); Clavero 1946: 8 (M*, F*, L*); Natvig 1948: 211 (M*, F*, L*; bion.), 202 (tax.); Senevet et al. 1949 (aberrant M*); Monchadskii 1951: 160, 164 (L*; tax.); Hopkins 1952: 125 (L*);

Mattingly & Knight 1956: 98 (tax.); Richards 1956: 264 (M*, F*, L*); Nielsen & Nielsen 1958: 282 (bion.); Senevet & Andarelli 1958: 279 (P*); Abul-Hab 1967: 279 (distr.); Aslamkhan 1971b (distr.; Pakistan); Minar 1971 (distr.); Rjazantzeva 1972 (F*); Gutsevich et al. 1974: 197 (M, F, L*); Minar 1976: 344 (M*, F*; tax.); Utrio 1976: 131, 134 (L*); Gutsevich 1976: 48 (polymorphism); Rodhain et al. 1977 (distr.); Moussiegt 1981 (annotated bib.); Zaim 1987 (distr.); Moussiegt 1988a (annotated bib.); Lambert et al. 1990 (E*); Mgeladze 1991: 329 (distr.; Georgia); Linley et al. 1993c (E*); Suleman et al. 1993 (distr.; Pakistan); Abdullah & Merdan 1995 (distr.; Saudi Arabia); Amr et al. 1997 (distr.; Jordan); Darsie & Samandou-Voyadjoglou 1997 (keys); Alten et al. 2000 (distr.; Turkey); Eritja et al. 2000: 11 (distr.); Hadjivassilis 2000 (distr.; Cyprus); Spungis 2000 (distr.; Latvia); Samanidou-Voyadjoglou & Harbach 2001 (key); Trari et al. 2002 (distr.; Morocco); Milankov et al. 2009 (tax., bion., review); Becker et al. 2010: 216 (M*, F*, L*; keys, tax., distr., bion.); Osório et al. 2010 (distr.; Portugal); Ahmed et al. 2011 (distr.; Saudi Arabia); Namazov 2014 (distr.; Azerbaijan); Robert et al. 2019 (distr.; Western Palearctic). **Ety.:** Caspian Sea. **Informal name:** Caspian Old World Pointy Mosquito. (Vol. 1, p. 432)

syn. *punctatus* Meigen
1804: 6 (M, F; *Culex*). **Type loc.:** Aden (MNHP).

syn. *siculus* Robineau-Desvoidy
1827: 406 (M, F; *Culex*). **Type loc.:** Sicily [Italy] (NE).

syn. *maculiventris* Macquart
1846a: 135 (F; *Culex*). **Type loc.:** Algeria (MNHP).

syn. *penicillaris* Rondani
1872: 31 (A; *Culex*). **Type loc.:** Italy (LU).

syn. *longisquamosa* Theobald
1905f: 103 (F*; *Grabhamia*). **Type loc.:** Sousse, Tunis [Tunisia] (HNM). **Refs.:** Edwards 1921d: 299 (syn.).

syn. *subtilis* Sergent & Sergent
1905: 673 (Apr. 14) (A, L; *Grabhamia*). **Type loc.:** Biskra [Constantine], Algeria (LU).

syn. *arabica* Giles
1906: 130 (M, F; *Mansonia*). **Type loc.:** Bahrain, Persian Gulf (NHMUK).

syn. *africanus* Neveu-Lemaire
1906: 271 (F*; *Taeniorhynchus*). **Type loc.:** Port Said, Suez and Cairo [Egypt] (NE).

syn. *willcocksii* Theobald
1907: 296 (M*, F*; *Grabhamia*). **Type loc.:** Kafr El Dawar, Egypt (NHMUK).

syn. *arabicus* Becker
1910: 140 (F; *Culex*). **Type loc.:** Ras Shoab, Socotra [Yemen] (ZM).

ssp. ***hargreavesi*** Edwards
1920a: 130 (F). **Type loc.:** Taranto [Puglia], Italy (NHMUK). **Distr.:** France, Italy. **Refs.:** Harbach & Howard 2007: 43 (to ssp.). **Ety.:** E. Hargreaves. **Informal name:** Hargreaves Italian Pointy Mosquito.

syn. *epsilonn* Séguy
1924: Pl. XXIII, 97 (L*). **Type loc.:** Region de Djamaa, Tiguedidine [Touggourt], Algeria (NE). **Refs.:** Monchadskii 1951: 160 (syn.).

ssp. ***meirai*** Ribeiro, Ramos, Capela, & Pires
1980: 69 (M, F, L, P). **Type loc.:** Curral Velho, Ilha da Boa Vista, Cape Verde (IHMT). **Distr.:** Cape Verde. **Ety.:** Manuel T.V. de Meira. **Informal name:** Meira Cabo Verdean Pointy Mosquito.

cataphylla Dyar
1916b: 86 (F). **Type loc.:** Fallen Leaf, Lake Tahoe, California, United States (USNM). **Distr.:** Austria, Azerbaijan, Belarus, Bulgaria, Canada, Crimean Peninsula, Croatia, Czech Republic, Denmark, Estonia, Finland, France, Germany, Greece, Hungary, Italy, Kaliningrad Oblast, Latvia, Lithuania, Moldova, Mongolia, Norway, Pakistan, People's Republic of China, Poland, Romania, Russia (Central, northwestern, Southern, and Volga regions), Serbia, Slovakia, Spain, Sweden, Switzerland, Turkey, Ukraine, United States (Alaska, continental). **Refs.:** Gjullin 1937: 262 (F*; as *Aedes*); Natvig 1948: 288 (M*, F, L*); Yamaguti & LaCasse 1951d: 59 (M*, F*, L*); Bohart 1954: 358 (L*); Carpenter & LaCasse 1955: 166 (M*, F*, L*; keys); Craig 1956 (E*); Dodge 1966: 350 (first instar L*; key); Myers 1967 (E*); Mohrig 1969 (F*); Polyakova 1970b: 135 (distr.); Minar 1971: 92 (distr.); Rjazantzeva 1972 (F*); Gutsevich et al. 1974 (M*, F, L*); Dahl 1975: 82, 87 (distr.); Bickley 1976 (distr.); Utrio 1976: 131, 134 (L*); Wolff & Nielsen 1976 (bion., distr.); Bohart & Washino 1978: 45 (M, F, L*; keys, tax., bion., distr.); Eritja et al. 2000: 11 (distr.; Spain); Spungis 2000: 9 (distr.; Latvia); Kaiser et al. 2001 (distr.; Greece); Nicolescu et al. 2003a: 2 (distr.; Romania); Darsie & Ward 2005 (F*, L*; keys, distr.); Becker et al. 2010: 218 (M*, F*, L*; keys, tax., distr., bion.); Namazov 2014 (distr.; Azerbaijan); Robert et al. 2019 (distr.; Western Palearctic). **Ety.:** down leaf (Gr); Fallen Leaf, Lake Tahoe, California, United States. **Informal name:** Leaf American Pointy Mosquito.

syn. *prodotes* Dyar
1917c: 118 (M*). **Type loc.:** Bozeman, Montana, United States (USNM).

syn. *rostochiensis* Martini
1920b: 204 (M*, F, L*). **Type loc.:** Mueritz area [Mecklenburg], Germany (NHMUK). **Refs.:** Dahl & White 1978 (syn.).

syn. *pacificensis* Hearle
1927: 101 (M, F). **Type loc.:** Discovery Island, British Columbia, Canada (CNC). **Refs.:** Matheson 1944: 260 (? to var.); Stone et al. 1959: 142 (to ssp.); Curtis 1967: 47 (syn.).

chelli (Edwards)
1915b: 275 (F; *Ochlerotatus*). **Type loc.:** Dido, Kenya (NHMUK). **Distr.:** Iran, Kenya, Namibia, Sudan and South Sudan, Yemen. **Refs.:** McIntosh 1973: 267 (M*, F;

from syn. with *caballus*; lectotype desig.). **Ety.:** G.R.H. Chell. **Informal name:** Chell Kenyan Pointy Mosquito.

churchillensis Ellis & Brust
1973: 922 (M*, F*, L*, E*). **Type loc.:** Churchill, Manitoba, Canada (CNC). **Distr.:** Canada. **Refs.:** Darsie & Ward 2005 (F*, L*; keys, distr.). **Ety.:** Churchill, Manitoba, Canada. **Informal name:** Churchill Canadian Pointy Mosquito.

clelandi (Taylor)
1914c: 690 (F*; *Culicada*). **Type loc.:** Bass Straits, Flinders Island [near Tasmania, Australia] (US). **Distr.:** Australia, Philippines. **Refs.:** Mackerras 1927: 294 (tax.); Dobrotworsky 1960b: 185 (M*, F*, P*, L*); Dobrotworsky 1965: 102 (M*, F*, L*); Lee et al. 1984: 138 (F; key, tax., bion., distr., review). **Ety.:** John Burton Cleland. **Informal name:** Cleland Australian Pointy Mosquito.

clivis Lanzaro & Eldridge
1992: 89 (M, F, P, L). **Type loc.:** Strawberry Lake, Tuolumne County, California, United States (USNM). **Distr.:** United States (continental). **Refs.:** Darsie & Ward 2005 (F*, L*; keys, distr.). **Ety.:** sloping, inclining (L); ref. to mountainous habitat. **Informal name:** Hillside American Pointy Mosquito.

coluzzii Rioux, Guilvard, & Pasteur
1998: 357 (mol. tax.). **Type loc.:** Salins de Giraud (43°24'N, 04°44'W), Basse-Camargue, Bouches-du-Rhone, France (MNHP). **Distr.:** France, Morocco, Portugal. **Refs.:** Trari et al. 2002: 331 (distr.; Morocco); Robert et al. 2019 (distr.). **Ety.:** Mario Coluzzi. **Informal name:** Coluzzi French Pointy Mosquito.

comitatus Arnell
1976: 66 (M*). **Type loc.:** Bosque Ocoa, east of Villavicencio, Meta, Colombia (USNM). **Distr.:** Colombia. **Ety.:** n.s. [accompanied (L)]. **Informal name:** Follower Colombian Pointy Mosquito.

communis (de Geer)
1776: 316 (M*, F*, P*, L*; *Culex*). **Type loc.:** Europe (SM). **Distr.:** Algeria, Austria, Belarus, Belgium, Bulgaria, Canada, Crimean Peninsula, Croatia, Czech Republic, Denmark, Estonia, Finland, France (includes Corsica), FYRO Macedonia, Germany, Greece, Hungary, Italy, Japan, Latvia, Lithuania, Moldova, Mongolia, Montenegro, Netherlands, Norway, Poland, Romania, Russia (Central, northwestern, Southern, and Volga regions), Serbia, Slovakia, Sweden, Switzerland, Syria, Togo, Turkey, Ukraine, United Kingdom, United States (Alaska, continental). **Refs.:** Brolemann 1919b (F*); Séguy 1924: 133 (L*), 127 (F), 71, 127 (M gen.*); Gjullin 1937: 261 (F*); Natvig 1948: 319 (M*, F, L*; bion., tax.); Darsie 1951: 22 (P*); Monchadskii 1951: 167 (L*); Yamaguti & LaCasse 1951d: 63 (M*, F*, L*); Asanuma et al. 1952 (M*); Bohart 1954: 360 (L*); Vockeroth 1954b: 246 (tax.); Carpenter & LaCasse 1955: 168 (M*, F*, L*; keys); Craig 1956 (E*); Craig & Horsfall 1958: 855 (E*); Price 1960: 552 (first instar L*); Brust & Horsfall 1965 (E*); Dodge 1966: 352 (first instar L; key); Myers 1967 (E*); Kalpage & Brust 1968 (E*); Minar 1971: 92 (distr.); Horsfall & Voorhees 1972 (E*); Rjazantzeva 1972 (F*); Ellis & Brust 1973: 917 (M*, F*, L*, E*); Gardner et al. 1973: 163 (L* mouthparts); Gutsevich et al. 1974: 241 (M*, F, L*); Dahl 1975: 82, 87 (tax., distr.); Utrio 1976: 131, 134 (L*); Wolff & Nielsen 1976 (bion., distr.); Bohart & Washino 1978: 45 (M, F*, L*; keys, tax., bion., distr.); Tanaka et al. 1979: 278 (M*, F*, L*); Belton 1982 (cuticular vestiture); Moussiegt 1993 (annotated bib.); Dahl 1997 (E*); Tanaka 1999: 112 (P*; tax.); Alten et al. 2000 (distr.; Turkey); Spungis 2000: 10 (distr.; Latvia); Kaiser et al. 2001 (distr.; Greece); Darsie & Ward 2005 (F*, L*; keys, distr.); Becker et al. 2010: 220 (M*, F*, L*; keys, tax., distr., bion.); Harrison et al. 2016 (F*, L*; keys, distr.); Robert et al. 2019 (distr.; Western Palearctic). **Ety.:** n.s. [common (L)]. **Informal name:** Communal European Pointy Mosquito.

syn. *fasciatus* Meigen
1804: 4 (M; *Culex*). **Type loc.:** Europe (NE).

syn. *nemorosus* Meigen
1818: 4 (M, F?; *Culex*). **Type loc.:** Europe (MNHP).

syn. *sylvaticus* Meigen
1818: 6 (M, F; *Culex*). **Type loc.:** Europe (NE).

syn. *obscurus* Meigen
1830a: Pl. 3, Fig. 2 (F*). **Type loc.:** Europe (MNHP).

syn. *lazarensis* Felt & Young
1904: 312 (M, F, L; *Culex*). **Type loc.:** Elizabethtown, New York, United States (NY).

syn. *borealis* Ludlow
1911a: 178 (F; *Culex*). **Type loc.:** Alaska, United States (USNM). **Refs.:** Stone & Knight 1956a: 215 (type info., lectotype desig.).

syn. *diplolineata* Schneider
1913: 37 (M, F; *Culicada nemorosa* form). **Type loc.:** Bonn, Germany (LU).

syn. *altiusculus* Dyar
1917b: 100 (M*, F, L*; *Culex*). **Type loc.:** "Indian Henry's," Mount Rainier National Park, Washington, United States (USNM). **Refs.:** Stone & Knight 1956a: 214 (type info., lectotype desig.).

syn. *palmeni* Edwards
1921b: 52 (M; *Ochlerotatus*). **Type loc.:** Seitjaur, Finland (NHMUK).

syn. *prolixus* Dyar
1922a: 2 (M, F). **Type loc.:** Alaska, United States (USNM). **Refs.:** Stone & Knight 1956a: 224 (type info., lectotype desig.).

condolescens Dyar & Knab
1907a: 11 (F). **Type loc.:** Nassau [New Providence Island], Bahama Islands (USNM). **Distr.:** Bahamas, Cayman Isl., Cuba, Mexico, United States (continental). **Refs.:** Lane 1953: 656 (M*, F; as *Aedes*); Perez Vigueras 1956: 278 (M*, P,

L*); Stone & Knight 1956a: 216 (type info.); Mohrig 1967 (F*); Arnell 1976: 46 (M*, F*; distr.); Darsie 2003: 1067 (distr.; United States). **Ety.:** n.s. [becoming painful (L)]. **Informal name:** Painful Bahamian Pointy Mosquito.

continentalis Dobrotworsky
1960a: 71 (M*, F, P*, L*). **Type loc.:** Carpendeit, Victoria, Australia (NMM). **Distr.:** Australia. **Refs.:** Dobrotworsky 1965: 120 (M*, F*, L*); Lee et al. 1984: 140 (F; key, tax., bion., distr., review). **Ety.:** n.s. [of the mainland (L)]. **Informal name:** Continental Australian Pointy Mosquito.

crinifer (Theobald)
1903a: 209 (F*; *Culex*). **Type loc.:** São Paulo, Brazil (NHMUK). **Distr.:** Argentina, Brazil, Paraguay, Uruguay, Venezuela. **Refs.:** Antunes & Lane 1934: 35 (M*, F*, L*); Lane 1953: 668 (M*, F, P*, L); Mohrig 1967 (F*); Belkin 1968b: 4 (lectotype desig.); Arnell 1976 (F*). **Ety.:** n.s. [hair bearer (L)]. **Informal name:** Furry Brazilian Pointy Mosquito.
syn. *lynchii* Brèthes
1910: 470 (F; *Culex*). **Type loc.:** Islas del Paraná, Argentina (BA). **Refs.:** Belkin et al. 1968: 13 (lectotype desig. according to Casal).
syn. *tapinops* Brèthes
1917: 227 (M*, F, L*; *Culex*). **Type loc.:** San Isidro, 40 km northeast of Buenos Aires, Argentina (BA). **Refs.:** Belkin et al. 1968: 13 (lectotype desig. according to Casal).
syn. *iguazu* Shannon & Del Ponte
1928 (M, F). In Dyar 1928: 168. **Type loc.:** Misiones, Argentina (LU).

cunabulanus Edwards
1924b: 378 (F). **Type loc.:** Cradle Valley, Tasmania, Australia (NHMUK). **Distr.:** Australia. **Refs.:** Dobrotworsky 1960a: 65 (M*, F, P*, L*); Lee et al. 1984: 141 (F; key, tax., bion., distr., review). **Ety.:** cradle (L); Cradle Valley, Tasmania, Australia. **Informal name:** Original Australian Pointy Mosquito.

cyprioides Danilov & Stupin
1982: 824 (M*, F*, L*). **Type loc.:** Urte-Kargane (12 km southwest of Kubukhay), Onon District, Chita Province, Siberia, USSR (IMP). **Distr.:** Russia. **Ety.:** n.s. [island of Cyprus?]. **Informal name:** Aphrodite Russian Pointy Mosquito [according to Greek myth, the goddess Aphrodite was born on the island of Cyprus].

cyprius Ludlow
1920a: 158 (F). **Type loc.:** Verkhne-Udinsk and Selenga, Siberia [USSR] (USNM). **Distr.:** Belarus, Estonia, Finland, Germany, Latvia, Lithuania, People's Republic of China, Poland, Russia (Central, northwestern, Southern, and Volga regions), Slovakia, Sweden, Turkey, Ukraine. **Refs.:** Peus 1937 (L*); Natvig 1948: 259 (M*, F, L*); Monchadskii 1951: 169 (L*); Stone & Knight 1956a: 216 (type info., lectotype desig.); Rjazantzeva 1972 (F*); Gutsevich et al. 1974: 229 (M*, F, L*); Dahl 1975: 83, 90 (tax., distr.); Utrio 1976: 130, 134 (L*); Spungis 2000: 10 (distr.; Latvia); Orszagh et al. 2001: 11 (distr.; Slovakia); Becker et al. 2010: 221 (M*, F*, L*; keys, tax., distr., bion.); Robert et al. 2019 (distr.; Western Palearctic). **Ety.:** coppery (L); adult appearance. **Informal name:** Coppery Russian Pointy Mosquito.
syn. *freyi* Edwards
1921d: 306 (M only). **Type loc.:** Eriksberg, Finland (HM).

dahlae (Nielsen)
2009 (F*, L*). **Type loc.:** 1 mi west of Mountain Green, Morgan County, Utah, United States (UCD). **Distr.:** United States (continental). **Ety.:** Christine Ida Dahl. **Informal name:** Dahl American Pointy Mosquito.

decticus Howard, Dyar, & Knab
1917: 737 (F). **Type loc.:** White River, north of Lake Superior, Ontario, Canada (USNM). **Distr.:** Canada, United States (Alaska, continental). **Refs.:** Dyar 1921I (*Aedes impiger* (*decticus* group)); Carpenter & LaCasse 1955: 170 (M*, F*, L*; keys; from syn. with *impiger*); Craig 1956 (E*); Kalpage & Brust 1968 (E*); Darsie & Ward 2005 (F*, L*; keys, distr.); Harrison et al. 2016 (F*, L*; keys, distr.). **Ety.:** n.s. [able to bite (Gr)]. **Informal name:** Biting American Pointy Mosquito.
syn. *pseudodiantaeus* Smith
1952: 21 (M*, F, P*, L*). **Type loc.:** Belchertown, Massachusetts, United States (USNM). **Refs.:** Frohne & Frohne 1954: 97 (bion.); Carpenter & LaCasse 1955: 170 (syn.).

deficiens Arnell
1976: 24 (F*, P*, L*). **Type loc.:** Suba, Cundinamarca, Colombia (USNM). **Distr.:** Colombia. **Ety.:** n.s. [deficient (L)]. **Informal name:** Deficient Colombian Pointy Mosquito.

detritus Haliday
1833: 151 (M). **Type loc.:** Holywood, Downshire, England (LU). **Distr.:** Albania, Algeria, Belarus, Belgium, Bulgaria, Croatia, Cyprus, Denmark, Egypt, France (includes Corsica), FYRO Macedonia, Germany, Greece, Hungary, Iran, Ireland, Israel (Gaza Strip and West Bank), Italy (includes Sardinia and Sicily), Malta, Mongolia, Montenegro, Morocco, Netherlands, Norway, Poland, Portugal, Romania, Russia (Northwestern Region), Serbia, Slovakia, Spain, Sweden, Tajikistan, Tunisia, Turkey, Ukraine, United Kingdom, United States. **Refs.:** Séguy 1924: Pl. XXVIII (L*), 136 (M gen.*); Marshall 1938: 170 (M*, F*, L*, E*; bion.); Natvig 1948: 279 (M*, F*, L*; bion.); Monchadskii 1951: 171 (L*); Horsfall 1955: 429 (review); Gutsevich et al. 1974: 276 (M*, F*, L*); Minar 1976: 337 (distr.); Gad et al. 1987: 207 (distr.); Moussiegt 1987 (annotated bib.); Zaim 1987: 568 (distr.); Dahl 1997 (E*); Agoulon et al. 1999: 35 (suppl.), 263 (distr.; France); Alten et al. 2000 (distr.; Turkey); Eritja et al. 2000: 11 (distr.; Spain); Hadjivassilis 2000: 38 (distr.; Cyprus); Orszagh et al. 2001: 11 (distr.; Slovakia); Trari et al. 2002: 331 (distr.; Morocco); Osório et al. 2010 (distr.; Portugal); Becker et al. 2010: 223 (M*, F*, L*; keys, tax.,

distr., bion.); Robert et al. 2019 (distr.; Western Palearctic). **Ety.:** rubbed away (L); scutum with little vestiture. **Informal name:** Rubbed British Pointy Mosquito.

syn. *salinus* Ficalbi
1896: 29 (M, F; *Culex*). **Type loc.:** Cagliari, Italy (LU). **Refs.:** Martini 1931c: 328 (M*, F, L*).

syn. *terriei* Theobald
1903a: 193 (F; *Culex*). **Type loc.:** Dartford, Kent, England (NHMUK).

syn. *maculosa* Theobald
1905f: 105 (F*; *Grabhamia*). **Type loc.:** Sfax, Tunis [Tunisia] (HNM). **Refs.:** Stone 1957b: 173 (syn.); Townsend 1990: 99 (syntypes, NHMUK).

diantaeus Howard, Dyar, & Knab
1913: Pl. 24, Fig. 167 (M*). **Type loc.:** Dublin, New Hampshire, United States (USNM). **Distr.:** Austria, Belarus, Canada, Czech Republic, Denmark, Estonia, Finland, France, Germany, Japan, Latvia, Lithuania, Moldova, Norway, People's Republic of China, Poland, Russia (Central, northwestern, Southern, and Volga regions), Slovakia, Sweden, Ukraine, United States (Alaska, continental). **Refs.:** Howard et al. 1917: 758; Natvig 1948: 358 (M*, F, L*; bion.); Monchadskii 1951: 174 (L*); Yamaguti & LaCasse 1951d: 67 (M*, F*, L*); Frohne & Frohne 1954: 95 (bion.); Carpenter & LaCasse 1955: 172 (M*, F*, L*; keys); Craig 1956 (E*); Stone & Knight 1956a: 217 (type info., lectotype desig.); Darsie 1957: 620 (P); Price 1960: 552 (first instar L*); Dodge 1966: 352 (first instar L; key); Kalpage & Brust 1968 (E*); Mohrig 1969 (F*); Polyakova 1970b: 135 (distr.); Mattingly 1971a: Pl. 33 (L head*, terminal segs.*); Horsfall & Voorhees 1972 (E*); Gutsevich et al. 1974: 258 (M*, F, L*); Utrio 1976: 129, 130, 134 (L*); Tanaka et al. 1979: 299 (M*, F*, L*); Dahl 1997 (E*); Tanaka 1999: 123 (distr.); Darsie & Ward 2005 (F*, L*; keys, distr.); Becker et al. 2010: 224 (M*, F*, L*; keys, tax., distr., bion.); Harrison et al. 2016 (F*, L*; keys, distr.); Robert et al. 2019 (distr.; Western Palearctic). **Ety.:** two columns, pillars (Gr, L); ref. to blunt-looking palpus. **Informal name:** Pillared American Pointy Mosquito.

syn. *serus* Martini
1920a: 96 (M*, F, L*). **Type loc.:** None designated (NHMUK).

dorsalis (Meigen)
1830b: 242 (F; *Culex*). **Type loc.:** Berlin, Germany (MNHP). **Distr.:** Albania, Austria, Azerbaijan, Belarus, Belgium, Bulgaria, Canada, Crimean Peninsula, Croatia, Czech Republic, Denmark, Estonia, Finland, France (includes Corsica), FYRO Macedonia, Germany, Greece, Hungary, Iraq, Ireland, Italy, Japan, Kazakhstan, Kosovo, Latvia, Lithuania, Mexico, Moldova, Mongolia, Netherlands, Norway, People's Republic of China, Poland, Romania, Russia (Central, northwestern, Southern, and Volga regions), Serbia, Slovakia, Slovenia, South Korea, Spain, Sweden, Switzerland, Taiwan, Tajikistan, Turkestan, Turkey, Ukraine, United Kingdom, United States (continental). **Refs.:** Howard et al. 1917 (E; as *currie*); Waterston 1918 (F*); Gjullin 1937: 261 (F*); Marshall 1938: 185 (M*, F*, L, E*); Natvig 1948: 201 (M*, F*, L; bion.); Bohart 1948 (P*, L*); La Casse & Yamaguti 1950 (F*); Monchadskii 1951: 161 (L*; as ssp. of *caspius*); Yamaguti & LaCasse 1951d: 71 (M*, F*, L*); Asanuma et al. 1952 (M*); Bohart 1954: 360 (L*); Barr 1955: 170 (M*; tax.); Carpenter & LaCasse 1955: 174 (M*, F*, L; keys); Craig 1956 (E*); Hara 1957: 59 (F*); Craig & Horsfall 1958 (E*); Price 1960: 554 (first instar L*); Chapman & Grodhaus 1963 (tax.); Ross & Horsfall 1965 (M*, F*, L*, E*; keys); Dodge 1966: 352 (first instar L; key); Myers 1967: 799 (E*); Kalpage & Brust 1968 (E*); Rjazantzeva 1972 (F*); Gutsevich et al. 1974: 201 (M*, F, L*; change in status); Dahl 1975: 81, 86 (distr.); Minar 1976: 344 (M*, F*; tax.); Utrio 1976: 131, 134 (L*); Bohart & Washino 1978: 47 (M*, F*, P*, L*; keys, tax., bion., distr.); Tanaka et al. 1979: 264 (M*, F*, L*); Lambert et al. 1990 (E*); Tanaka 1999: 107 (P*; tax., distr.; Japan); Gimnig & Eldridge 1999 (mol. tax., morph.); Spungis 2000: 10 (distr.; Latvia); Alten et al. 2000 (distr.; Turkey); Reinert 2002e: Fig. 22 (F*); Darsie & Ward 2005 (F*, L*; keys, distr.); Becker et al. 2010: 226 (M*, F*, L*; keys, tax., distr., bion.); Darsie 2011a (P*); Harrison et al. 2016 (F*, L*; keys, distr.); Robert et al. 2019 (distr.; Western Palearctic). **Ety.:** n.s. [back (L); prob. ref. to scutum patterned with two black stripes]. **Informal name:** Ridged German Pointy Mosquito [*dorsalis* has a high ridge along its back]. (Vol. 1, p. 434)

syn. *curriei* Coquillett
1901: 259 (F; *Culex*). **Type loc.:** Boise, Idaho, United States (USNM). **Refs.:** Stone & Knight 1956a: 216 (type info.).

syn. *onondagensis* Felt
1904: 304 (F*; *Culex*). **Type loc.:** Lake Onondaga, Syracuse, New York, United States (NY). **Refs.:** Howard et al. 1917: 629 (M, F, L*).

syn. *lativittatus* Coquillett
1906c: 109 (A; *Culex*). **Type loc.:** Arden, Alameda County, California, United States (USNM). **Refs.:** Stone & Knight 1956a: 220 (type info., lectotype desig.); Wood et al. 1979 (syn.).

syn. *quaylei* Dyar & Knab
1906d: 202 (A). **Type loc.:** California, United States (NE). **Refs.:** Stone & Knight 1956a: 224 (type info.).

syn. *broquetii* Theobald
1913c: 179 (F; *Grabhamia*). **Type loc.:** Tet-chili, northern China and Tamsui, Formosa [Taiwan, Republic of China] (NHMUK, GDZ).

syn. *grahami* Ludlow
1920a: 154 (M, F). **Type loc.:** Mostovoi, Siberia [USSR] (USNM). **Refs.:** Stone & Knight 1956a: 218 (type info., lectotype desig.).

dufouri Hamon
1953: 40 (M*, F*, P*, L*). **Type loc.:** Ile de la Reunion, Madagascar (MNHP). **Distr.:** French Southern Territories, Madagascar (includes Glorioso and Juan De Nova Isl.), Reunion. **Refs.:** Danilov 1981a: 86 (from subg.

Ochlerotatus); Huang et al. 2010: 1 (from subg. *Levua*). **Ety.:** M. le Dr. Dufour. **Informal name:** Dufour Malagasy Pointy Mosquito.

duplex Martini
1926: 203 (M). **Type loc.:** Elschanka, USSR [Russia] (NHMUK). **Distr.:** Moldova, Romania, Russia. **Refs.:** Martini 1930: 285 (M*); Mattingly 1955c: 28 (tax.). **Ety.:** double (L); basal lobe with two similar fine setae (L). **Informal name:** Double Russian Pointy Mosquito.

dupreei (Coquillett)
1904a: 10 (M, F; *Culex*). **Type loc.:** Baton Rouge, Louisiana, United States (USNM). **Distr.:** Brazil, Colombia, Costa Rica, French Guiana, Guyana, Mexico, Panama, United States (continental), Venezuela. **Refs.:** Mitchell 1907b (E*); Yamaguti & LaCasse 1951d: 75 (M*, F*, L*); Carpenter & LaCasse 1955: 177 (M*, F*, L*; keys); Craig 1956 (E*); Horsfall & Craig 1956: 372 (E*); Stone & Knight 1956a: 217 (type info., lectotype desig.); Craig & Horsfall 1958 (E*); Ross & Horsfall 1965 (M*, F*, L*, E*; keys); Dodge 1966: 352 (first instar L*; key); Reinert 1970b: 243 (P*); Horsfall et al. 1970: 1711 (E*); Darsie & Ward 2005 (F*, L*; keys, distr.); Harrison et al. 2016 (F*, L*; keys, distr.; note 6). **Ety.:** J.W. Dupree. **Informal name:** Dupree American Pointy Mosquito.

dzeta Séguy
1924: 98 (L*). **Type loc.:** Tanja El Balia, Morocco (NE). **Distr.:** Morocco. **Ety.:** n.s. [d + zeta; sixth letter of the Greek alphabet]. **Informal name:** Musical Moroccan Pointy Mosquito [DZETA is the name of a rock music band].

edgari Stone & Rosen
1952: 425 (M*, F, L*). **Type loc.:** Punaauia, Tahiti, Society Islands (USNM). **Distr.:** French Polynesia. **Refs.:** Belkin 1962: 391 (M*, F, P*, L*); Lee et al. 1984: 143 (F; key, tax., bion., distr., review). **Ety.:** S.A. Edgar. **Informal name:** Edgar French Polynesian Pointy Mosquito.

eidsvoldensis Mackerras
1927: 295 (F; as *Aedes theobaldi* form). **Type loc.:** Eidsvold, Queensland, Australia (MM). **Distr.:** Australia. **Refs.:** Marks 1949: 17 (F; to ssp.); Lee et al. 1984: 145 (F; key, tax., bion., distr., review; from ssp. of *theobaldi*). **Ety.:** Eidsvold, South Queensland, Australia. **Informal name:** Eidsvold Australian Pointy Mosquito.

eucephalaeus Dyar
1918d: 127 (M, F, L). **Type loc.:** Suriname (USNM). **Distr.:** French Guiana, Suriname. **Refs.:** Lane 1953: 663 (M*, F, P, L*); Stone & Knight 1956a: 217 (type info., lectotype desig.). **Ety.:** n.s. [original (true?) head (Gr)]. **Informal name:** Big-headed Surinamese Pointy Mosquito.

euedes Howard, Dyar, & Knab
1913: Pl. 28, Fig. 191 (M*) and 1917: 714 (M*, F; as *Aedes*). **Type loc.:** Trenton, Ontario, Canada (USNM). **Distr.:** Belarus, Canada, Finland, Latvia, Lithuania, Poland, Russia (Central, northwestern, Southern, and Volga regions), Sweden, United States (Alaska, continental). **Refs.:** Stone & Knight 1956a: 217 (type info., lectotype desig.); Price 1960: 550 (first instar L*; as *barri*); Dodge 1966: 350 (first instar L; key; as *barri*); Kalpage & Brust 1968 (E*; as *barri*); Skierska 1970: 614 (L*; distr.; as *beklemishevi*); Horsfall & Voorhees 1972 (E*; as *barri*); Nielsen & Horsfall 1973: 243 (tax., distr.; as *barri*); Gutsevich et al. 1974: 223 (M*, L*; as *beklemishevi*); Danilov 1975: 62 (distr. map; as *beklemishevi*); Utrio 1975: 63 (distr.; as *beklemishevi*); Bickley 1976 (tax.; as *beklemishevi*); Utrio 1976: 131, 133 (L*; distr.; as *beklemishevi*); Wood 1977: 72 (M*, F*, L; from syn. with *excrucians*; distr.); Dahl 1997 (E*); Spungis 2000: 10 (distr.; Latvia); Ramsdale & Snow 2000 (distr.; Europe); Darsie & Ward 2005 (F*, L*; keys, distr.); Becker et al. 2010: 228 (M*, F*, L*; keys, tax., distr., bion.); Robert et al. 2019 (distr.; Western Palearctic). **Ety.:** n.s. [*eueides* (Gr, L); well shaped or comely, good, true]. **Informal name:** Well-shaped Canadian Pointy Mosquito.

syn. *beklemishevi* Denisova
1955: 58 (L*). **Type loc.:** L'gov, Kursk, Dmitriev and Ivaninsk, USSR (LU). **Refs.:** Wood 1977: 72 (syn.).

syn. *grandilarva* Sazanova
1955: 99 (M*, L*). **Type loc.:** Oblasts of Vologod, Voronezh, Ryazan, Moscow, and Brest; Irtysh River (Khanty-Mansy District); Far East; USSR (ZIL). **Refs.:** Sazanova 1956: 145 (M*, F, L*; ? = *beklemishevi*); Wood 1977 (syn.).

syn. *barri* Rueger
1958: 34 (M*, F, P*, L*). **Type loc.:** Itasca State Park, Clearwater County, Minnesota, United States (USNM). **Refs.:** Danilov 1975: 61 (syn.); Wood 1977: 72 (syn.).

euiris Dyar
1922g: 193 (F). **Type loc.:** Bogotá, Cundinamarca, Colombia (USNM). **Distr.:** Bolivia, Colombia, Venezuela. **Refs.:** Anduze 1944 (M*, F); Lane 1953: 654 (M*, F); Stone & Knight 1957c: 201 (type info., lectotype desig.). **Ety.:** n.s. [original rainbow (Gr, L); poss. ref. to iridescent wings]. **Informal name:** Rainbow Colombian Pointy Mosquito.

euplocamus Dyar & Knab
1906d: 199 (L*). **Type loc.:** Zent, near Port Limon [Puerto Limón], Costa Rica (NE). **Distr.:** Colombia, Costa Rica, El Salvador, Guatemala, Guyana, Honduras, Mexico, Panama, Peru, Venezuela. **Refs.:** Dyar 1928: 167 (M*, F, L*); Stone & Knight 1956a: 217 (type info.); Arnell 1976: 48 (M*, F*, P*, L*; distr.). **Ety.:** n.s. **Informal name:** Hairy Costa Rican Pointy Mosquito.

excrucians (Walker)
1856a: 429 (A; *Culex*). **Type loc.:** Nova Scotia, Canada (NHMUK). **Distr.:** Austria, Belarus, Bulgaria, Canada,

Crimean Peninsula, Croatia, Czech Republic, Denmark, Estonia, Finland, France, Georgia, Germany, Hungary, Japan, Kaliningrad Oblast, Latvia, Lithuania, Moldova, Mongolia, Netherlands, Norway, Poland, Romania, Russia (Central, northwestern, Southern, and Volga regions), Serbia, Slovakia, Slovenia, Spain, Sweden, Switzerland, Turkey, Ukraine, United States (Alaska, continental). **Refs.:** Séguy 1924: 120 (L*); Clavero 1946: 12 (M*, F*, L*); Natvig 1948: 244 (M*, F, L*; bion.); La Casse & Yamaguti 1950: 138 (M*, F*; bion., distr., tax.); Monchadskii 1951: 176 (L*); Yamaguti & LaCasse 1951d: 79 (M*, F*, L*); Asanuma et al. 1952 (M*); Carpenter & LaCasse 1955: 178 (M*, F*, L*; keys); Craig 1956 (E*); Darsie 1957: 620 (P*); Hara 1957: 59 (F*); Craig & Horsfall 1958 (E*); Price 1960: 548 (first instar L*); Ross & Horsfall 1965 (M*, F*, L*, E*; keys); Dodge 1966: 353 (first instar L; key); Kalpage & Brust 1968 (E*); Sicart & Larrouy 1968 (F*); Horsfall & Voorhees 1972 (E*); Moriya et al. 1972 (E*); Rjazantzeva 1972 (F*); Moriya et al. 1973 (E*); Gutsevich et al. 1974: 219 (M*, F, L*); McDaniel & Webb 1974 (tax.); Dahl 1975: 83, 89 (tax., distr.); Matsuo 1975 (E*); Minar 1976: 335 (distr.); Utrio 1976: 130, 134 (L*); Wolff & Nielsen 1976 (bion., distr.); Wood 1977: 73 (M*, F*); Tanaka et al. 1979: 267 (M*, F*, L*); Mgeladze 1991: 330 (distr.; Georgia); Dahl 1997 (E*); Tanaka 1999: 109 (P*; tax., distr.; Japan); Alten et al. 2000 (distr.; Turkey); Spungis 2000: 10 (distr.; Latvia); Darsie & Ward 2005 (F*, L*; keys, distr.); Becker et al. 2010: 229 (M*, F*, L*; keys, tax., distr., bion.); Harrison et al. 2016 (F*, L*; keys, distr.); Robert et al. 2019 (distr.; Western Palearctic). **Ety.:** n.s. [torture, torment (L); prob. ref. to biting behavior]. **Informal name:** Excruciating Canadian Pointy Mosquito.

syn. *siphonalis* Grossbeck
1904a: 332 (M, F; *Culex*). **Type loc.:** Livingston Park, New Jersey, United States (USNM). **Refs.:** Stone & Knight 1956a: 225 (type info., lectotype desig.).

syn. *abfitchii* Felt
1904: 381 (L*; *Culex*). **Type loc.:** Karner, New York, United States (NY). **Refs.:** McCabe & Johnson 1980: 27 (L; lectotype desig. credited to D.M. Wood).

syn. *sansoni* Dyar & Knab
1909c: 102 (A). **Type loc.:** Banff, Alberta, Canada (USNM). **Refs.:** Howard et al. 1917: 686 (M, F, L*); Dyar 1917c: 114 (lectotype desig.); Stone & Knight 1956a: 225 (type info.).

syn. *dytes* Martini
1922: 164 (L; *Aedes excrucians* var.). **Type loc.:** Europe (LU).

explorator Marks
1964a: 60 (F*, L*). **Type loc.:** Mount Olga, 300 mi southwest of Alice Springs, Northern Territory, Australia (UO). **Distr.:** Australia. **Refs.:** Lee et al. 1984: 148 (F; key, tax., bion., distr., review). **Ety.:** a diligent searcher (L); ref. to difficult mosquito habitat. **Informal name:** Adventurous Australian Pointy Mosquito [*explorator* = "explorer" = Adventurous].

fitchii (Felt & Young)
1904: 312 (A, L; *Culex*). **Type loc.:** Karner, New York, United States (NY). **Distr.:** Canada, Russia, United States (Alaska, continental). **Refs.:** Felt 1904 (M*, F*, L*); Gjullin 1937: 259 (F*); Vockeroth 1950: 161 (F*); Darsie 1951: 18 (P*); Yamaguti & LaCasse 1951d: 83 (M*, F*, L*); Bohart 1954: 360 (L*); Carpenter & LaCasse 1955: 181 (M*, F, L*; keys); Craig 1956 (E*); Craig & Horsfall 1958: 855 (E*); Price 1960: 552 (first instar L*); Ross & Horsfall 1965 (M*, F*, L*, E*; keys); Dodge 1966: 353 (first instar L; key); Mohrig 1967 (F*); Myers 1967 (E*); Kalpage & Brust 1968 (E*); Gutsevich et al. 1971: 377 (distr.); Polyakova 1970a: 389 (distr.); Horsfall & Voorhees 1972 (E*); Gutsevich et al. 1974: 287 (M, F, L); McDaniel & Webb 1974 (tax.); Wolff & Nielsen 1976 (bion., distr.); Wood 1977: 73 (F*); Bohart & Washino 1978: 48 (M, F*, L*; keys, tax., bion., distr.); Darsie & Ward 2005 (F*, L*; keys, distr.); Harrison et al. 2016 (F*, L*; keys, distr.). **Ety.:** Asa Fitch. **Informal name:** Fitch American Pointy Mosquito.

syn. *palustris* Dyar
1916b: 89 (M, F, L). **Type loc.:** Fallen Leaf, Lake Tahoe, California, United States (USNM). **Refs.:** Stone & Knight 1956a: 223 (type info., lectotype desig.).

syn. *mimesis* Dyar
1917c: 116 (M, L). **Type loc.:** Drummond, Montana, United States (USNM).

syn. *pricei* Dyar
1917a: 16 (A; *palustris* var.). **Type loc.:** California, United States (USNM). **Refs.:** Stone & Knight 1956a: 223 (type info.).

flavescens (Müller)
1764: 87 (A; *Culex*). **Type loc.:** None designated (NHMUK). **Distr.:** Austria, Belarus, Belgium, Bulgaria, Canada, Crimean Peninsula, Croatia, Czech Republic, Denmark, Estonia, Finland, France, Germany, Hungary, Iran, Latvia, Lithuania, Moldova, Netherlands, Poland, Romania, Russia (Central, northwestern, Southern, and Volga regions), Serbia, Slovakia, Slovenia, Spain, Sweden, Switzerland, Turkey, Ukraine, United Kingdom, United States (Alaska, continental). **Refs.:** Hearle 1929 (first instar L*); Gjullin 1937: 261 (F*); Marshall 1938: 190 (M*, F*, L*, E*); Clavero 1946: 13 (M*, F*); Ross 1947 (F*); Natvig 1948: 265 (M*, F, L; bion., tax.); Yamaguti & LaCasse 1951d: 87 (M*, F*, L*); Vockeroth 1954b: 248 (F*); Craig 1956 (E*); Ross & Horsfall 1965 (M*, F*, L*, E*; keys); Dodge 1966: 353 (first instar L; key); Myers 1967 (E*); Kalpage & Brust 1968 (E*); Mattingly 1971b (E*); Rjazantzeva 1972 (F*); Gutsevich et al. 1974: 226 (M*, F, L*); Dahl 1975: 83, 89 (tax., distr.); Utrio 1976: 132, 134 (L*); Bohart & Washino 1978: 49 (M, F, L*; keys, tax., bion., distr.); Zaim 1987: 568 (distr.); Townsend 1990: 73 (syntype; type depository changed from Oxford Museum to NHMUK); Dahl 1997 (E*); Alten et al. 2000 (distr.; Turkey); Eritja et al.

2000: 11 (distr.; Spain); Spungis 2000: 10 (distr.; Latvia); Nicolescu et al. 2003a: 2 (distr.; Romania); Darsie & Ward 2005 (F*, L*; keys, distr.); Becker et al. 2010: 231 (M*, F*, L*; keys, tax., distr., bion.); Darsie 2011d (P*); Robert et al. 2019 (distr.; Western Palearctic). **Ety.:** becoming yellow (L); distinct yellow appearance. **Informal name:** Yellowing International Pointy Mosquito.

syn. *lutescens* Fabricius
1775: 800 (A; *Culex*). **Type loc.:** Havniae [= Copenhagen, Denmark] (NE).

syn. *variegatus* Schrank
1781: 482 (A; *Culex*). **Type loc.:** Augarten, Vienna, Austria (LU).

syn. *bipunctatus* Robineau-Desvoidy
1827: 405 (M; *Culex*). **Type loc.:** France (MNHP).

syn. *flavus* Motschulsky
1859: 503 (A; *Culex*). **Type loc.:** Somewhere along Amur River, Eastern Siberia [USSR] (LU). **Refs.:** Natvig 1948: 259 (syn.).

syn. *flavescens* Theobald
1901a: 410 (F; *Culex*). **Type loc.:** None designated (NHMUK). **Refs.:** Townsend 1990: 73 (syntype; type depository changed from Oxford Museum to NHMUK).

syn. *fletcheri* Coquillett
1902c: 84 (F; *Culex*). **Type loc.:** Carnduff, Assiniboia, Saskatchewan, Canada (USNM). **Refs.:** Stone & Knight 1956a: 218 (type info., lectotype desig.).

syn. *arcanus* Blanchard
1905: 303 (F; *Culex*; new name for *flavescens* Theobald, not Fabricius).

flavifrons (Skuse)
1889: 1735 (M, F; *Culex*). **Type loc.:** Blue Mountains, New South Wales, and Victoria, Brisbane, Queensland, Australia (MM). **Distr.:** Australia. **Refs.:** Bancroft 1908: 30 (F; as genus *Grabhamia*); Mackerras 1927: 289 (M*; tax.); Dobrotworsky 1960b: 180 (M*, F*, P*, L*); Dobrotworsky 1965: 95 (M*, F*, L*); Lee et al. 1984: 149 (F; key, tax., bion., distr., review). **Ety.:** yellow face (L); "Head densely covered with pale golden-yellow scales." **Informal name:** Yellow-headed Australian Pointy Mosquito.

syn. *vandema* Strickland
1911a: 202 (F*; *Culicada*). **Type loc.:** Tasmania, Australia (NHMUK).

syn. *variegatans* Strickland
1911a: 204 (F*; *Culicada vandema* var.). **Type loc.:** Tasmania, Australia (NHMUK).

fulvus (Wiedemann)
1828: 546 (F; *Culex*). **Type loc.:** Salvador, Bahia, Brazil (SNG). **Distr.:** Argentina, Belize, Bolivia, Brazil, Colombia, Costa Rica, Cuba, Ecuador, French Guiana, Guatemala, Guyana, Mexico, Panama, Peru, South Korea, Suriname, Trinidad and Tobago, United States (continental), Venezuela. **Refs.:** Goeldi 1905 (E*); Theobald 1907 (E*); Forattini 1965a (E*); Mohrig 1967 (F*); Belkin 1968b: 5 (tax.); Belkin et al. 1971: 19 (type loc. info.); Linton et al. 2013 (bion., tax., distr.; Ecuador). **Ety.:** n.s. [tawny (L)]. **Informal name:** Tawny Brazilian Pointy Mosquito.

syn. *flavicosta* Walker
1856a: 431 (A; *Culex*). **Type loc.:** Manaus, Amazonas, Brazil (NHMUK). **Refs.:** Belkin et al. 1971: 19 (type loc. info.).

ssp. ***pallens*** Ross
1943: 148 (M*, F, L*). **Type loc.:** New Orleans, Louisiana, United States (USNM). **Distr.:** Cuba, Mexico, United States (continental). **Refs.:** Yamaguti & LaCasse 1951d: 91 (M*, F*, L*); Carpenter & LaCasse 1955: 185 (M*, F*, L*; keys); Reinert 1970c: 47 (P*); Darsie & Ward 2005 (F*, L*; keys, distr.); Harrison et al. 2016 (F*, L*; keys, distr.). **Ety.:** n.s. [pale (L); poss. ref. to "can be separated at once from typical *fulvus* by the almost complete absence of pleural maculation of the thorax"]. **Informal name:** Pale American Pointy Mosquito.

grossbecki Dyar & Knab
1906d: 201 (L). **Type loc.:** New Jersey, United States (NE). **Distr.:** Canada, United States (continental). **Refs.:** Mitchell 1907b (E*; as syn. *sylvicola*); Howard et al. 1913 (E*); Howard et al. 1917: 708 (M*, F, L*, E*); Ross 1947: 74 (M*, F*, L*); Yamaguti & LaCasse 1951d: 96 (M*, F*, L*); Carpenter & LaCasse 1955: 187 (M*, F*, L*; keys); Stone & Knight 1956a: 218 (type info.); Ross & Horsfall 1965 (M*, F*, L*, E*; keys); Horsfall et al. 1970: 1710 (E*); Laffoon & Knight 1971 (F*); Harbach & Knight 1980 (F*); Bickley & Harison 1992: 205 (P*); Clements 1999 (F*); Reinert 2002e: Fig. 25 (F*); Darsie & Ward 2005 (F*, L*; keys, distr.); Harrison et al. 2016 (F*, L*; keys, distr., tax.). **Ety.:** John Arthur Grossbeck. **Informal name:** Grossbeck American Pointy Mosquito.

syn. *sylvicola* Grossbeck
1906: 129 (M*, F; *Culex*). **Type loc.:** Livingston Park, New Jersey, United States (USNM). **Refs.:** Stone & Knight 1956a: 225 (type info., lectotype desig.).

gutzevichi Dubitzky & Deshevykh
1978: 177 (M*, F, L*). **Type loc.:** Upper course of Ili River, near Borokhudzir, Kazakhstan, USSR (KZI). **Distr.:** Kazakhstan. **Ety.:** Aleksandr Vasilyevich Gutsevich. **Informal name:** Gutsevich Kazakhstani Pointy Mosquito.

hakusanensis Yamaguti & Tamaboko
1954: 414 (M*, F*, P, L*). **Type loc.:** Mount Hakusan, Ishikawa Prefecture [Honshu], Japan (LU). **Distr.:** Japan. **Refs.:** Nakagawa & Asanuma 1955: 103 (P*); Tanaka et al. 1979: 292 (M*, F*, L*); Tanaka 1999: 119 (P*; tax.). **Ety.:** n.s. [Mt. Hakusan, Japan]. **Informal name:** Hakusan Japanese Pointy Mosquito.

harrisoni Muspratt
1953a: 51 (M*, F, L*). **Type loc.:** Hex River Valley, Cape Province, Union of South Africa [Republic of South Africa] (SAIM). **Distr.:** Republic of South Africa.

Refs.: Jupp 1996 (M*, F*; key). **Ety.:** H.A. Harrison. **Informal name:** Harrison South African Pointy Mosquito.

hastatus Dyar
1922f: 160 (M). **Type loc.:** Paitilla, Canal Zone, Panama (USNM). **Distr.:** Argentina, Belize, Bolivia, Brazil, Colombia, Costa Rica, French Guiana, Guatemala, Guyana, Honduras, Mexico, Panama, Peru. **Refs.:** Lane 1953: 661 (M*, F, L*); Pecor et al. 2002: 249 (tax., distr.; Belize). **Ety.:** n.s. [speared (L); poss. ref. to claspette filament with sharp retrose point]. **Informal name:** Speared Panamanian Pointy Mosquito.

hesperonotius Marks
1959a: 131 (F). **Type loc.:** 2 mi northeast of Bullsbrook, Western Australia, Australia (CSIR). **Distr.:** Australia. **Refs.:** Lee et al. 1984: 152 (F; key, tax., bion., distr., review). **Ety.:** evening star, well known (L); ref. to distribution in Southwest Australia. **Informal name:** Evening Australian Pointy Mosquito.

hexodontus Dyar
1916b: 83 (M, F, L). **Type loc.:** Fallen Leaf, Lake Tahoe, California, United States (USNM). **Distr.:** Austria, Belarus, Canada, Finland, Norway, Poland, Russia (Northwestern Dist.), Sweden, United States (Alaska, continental). **Refs.:** Gjullin 1937 (F*); Knight 1951a: 93 (M*, F, L*; type info., lectotype desig.); Yamaguti & LaCasse 1951d: 100 (M*, F*, L*); Bohart 1954: 362 (L*); Vockeroth 1954b: 248 (tax.); Carpenter & LaCasse 1955: 188 (M*, F*, L*; keys); Stone & Knight 1956a: 218 (type info.); Darsie 1957: 613 (P*); Craig & Horsfall 1958: 854 (E*); Suzuki 1959: 291 (M*, F*, P*, L*); Dodge 1966: 353 (first instar L*; key); Myers 1967 (E*); Kalpage & Brust 1968 (E*); Gutsevich et al. 1971: 240 (distr.); Horsfall & Voorhees 1972 (E*); Dahl 1974: 59 (M*; distr.); Gutsevich et al. 1974: 250 (M*, F, L*); Matsuo 1975: 50 (E*); Utrio 1976: 130, 134 (L*; distr.); Wolff & Nielsen 1976 (bion., distr.); Dahl 1977: 63 (distr.); Wood 1977: 74 (tax.); Bohart & Washino 1978: 50 (M, F*, L*; keys, tax., bion., distr.); Tanaka 1999: 116 (distr.); Darsie & Ward 2005 (F*, L*; keys, distr.); Becker et al. 2010: 233 (M*, F*, L*; keys, tax., distr., bion.); Robert et al. 2019 (distr.; Western Palearctic). **Ety.:** six teeth (Gr); number of comb scales. **Informal name:** Six-toothed American Pointy Mosquito.
syn. *masamae* Dyar
1920f: 166 (F). **Type loc.:** Crater Lake, Oregon, United States (USNM). **Refs.:** Wood 1977: 74 (syn.).
syn. *cyclocerculus* Dyar
1920a: 23 (M, F, L). **Type loc.:** Prince Rupert, British Columbia, Canada (USNM). **Refs.:** Knight 1951a: 93 (type info., lectotype desig.); Stone & Knight 1956a: 216 (type info.).
syn. *leuconotips* Dyar
1920a: 24 (M, F, L). **Type loc.:** Prince Rupert, British Columbia, Canada (USNM). **Refs.:** Knight 1951a: 93 (type info., lectotype desig.); Stone & Knight 1956a: 220 (type info.).
syn. *labradorensis* Dyar & Shannon
1925d: 78 (F). **Type loc.:** Hawk's Harbor, Labrador, Canada (USNM). **Refs.:** Vockeroth 1954a: 115 (syn.).

hodgkini Marks
1959a: 112 (M*, F, P*, L*). **Type loc.:** Woodanilling, 94 mi north of Albany, Western Australia, Australia (CSIR). **Distr.:** Australia. **Refs.:** Lee et al. 1984: 153 (F; key, tax., bion., distr., review). **Ety.:** Ernest Pease Hodgkin. **Informal name:** Hodgkin Australian Pointy Mosquito.

hokkaidensis Tanaka, Mizusawa, & Saugstad
1979: 286 (M, F*, L). **Type loc.:** Onuma Koen, Hokkaido, Japan (NMST). **Distr.:** Japan. **Refs.:** King & Hoogstraal 1946d: 141 (M*, F*, L*); Tanaka 1999: 116 (P*; from ssp. of *hexodontus*). **Ety.:** Hokkaido, Japan. **Informal name:** Hokkaido Japanese Pointy Mosquito.

hungaricus Mihályi
1955a: 343 (M*, F). **Type loc.:** Budofok, Haros-sziget, Hungary (HNM). **Distr.:** Austria, Hungary, Moldova, Romania, Slovakia. **Refs.:** Mihályi 1961 (L*); Becker et al. 2010: 234 (M*, F*, L*; keys, tax., distr., bion.); Török et al. 2016 (distr.; Romania); Robert et al. 2019 (distr.; Western Palearctic). **Ety.:** Hungary. **Informal name:** Árpád Hungarian Pointy Mosquito [a ninth-century ruler considered to be the founder of modern Hungary].

imperfectus Dobrotworsky
1962: 296 (M*, F, P*, L*). **Type loc.:** Woori Yallock, Victoria, Australia (NMM). **Distr.:** Australia. **Refs.:** Dobrotworsky 1965: 93 (M*, F*, L*); Lee et al. 1984: 154 (F; key, tax., bion., distr., review; type dep. from USNM to NMM). **Ety.:** n.s. [imperfect (L)]. **Informal name:** Imperfect Australian Pointy Mosquito.

impiger (Walker)
1848: 6 (F; *Culex*). **Type loc.:** Saint Martin's Falls, Albany River, Hudson's Bay [Ontario, Canada] (NHMUK). **Distr.:** Canada, Finland, Japan, Norway, Russia (Northwestern Region), Sweden, United States (Alaska, continental). **Refs.:** Mitchell 1907b (E*); Gjullin 1937 (F*); Natvig 1948: 311 (M*, F, L*; tax.; as *nearcticus*); Rempel 1950: 225 (L*; as *nearcticus*); Yamaguti & LaCasse 1951d: 108 (M*, F*, L*), 136 (M*, F*, L*); Rempel 1953: 486 (M*, F*; as *nearcticus*); Carpenter & LaCasse 1955: 192 (M*, F*, L*; keys); Craig 1956 (E*); Sazonova 1958: 642 (distr.); Belkin 1968b: 5 (lectotype desig.); Kalpage & Brust 1968 (E*); Tanaka 1972: 229 (distr.); Danks & Corbet 1973: 367 (M*, F*, P*, L*, E*); Gutsevich et al. 1974: 269 (M*, F, L*); Utrio 1976: 130, 134 (L*; distr.); Darsie & Ward 2005 (F*, L*; keys, distr.); Becker et al. 2010: 236 (M*, F*, L*; keys, tax., distr., bion.). **Ety.:** n.s. [diligent, active, quick, energetic (L)]. **Informal name:** Diligent Canadian Pointy Mosquito.

syn. *nearcticus* Dyar
1919f: 32 (M, F, L). **Type loc.:** Bernard Harbour, Northwest Territories, Canada (CNC). **Refs.:** Vockeroth 1954a: 109 (syn.).

syn. *parvulus* Edwards
1921d: 314 (M*, F). **Type loc.:** Kittila, Finland (HM).

ssp. ***daisetsuzanus*** Tanaka, Mizusawa, & Saugstad
1979: 271 (M*, F*, L*). **Type loc.:** Yokomanbetsu, Mount Daisetsu, Hokkaido, Japan (NMST). **Distr.:** Japan. **Refs.:** Tanaka 1999: 110 (P*; tax.). **Ety.:** Mt. Daisetsu, Hokkaido, Japan. **Informal name:** Daisetsu Japanese Pointy Mosquito.

implicatus Vockeroth
1954a: 110 (M, F, L). **Type loc.:** Rupert House, Quebec, Canada (CNC). **Distr.:** Canada, Russia, United States (Alaska, continental). **Refs.:** Carpenter & LaCasse 1955: 195 (M*, F*, L*; keys); Craig 1956 (E*); Craig & Horsfall 1958 (E*); Dodge 1966: 353 (first instar L; key); Kalpage & Brust 1968 (E*); Danilov 1974a: 133 (M*, F, L*; distr., tax.); Wolff & Nielsen 1976 (bion., distr.); Mezenev 1980: 206 (syn.); Ward 1987: 95 (from syn. with *leucomelas*); Darsie & Ward 2000 (distr.; United States); Darsie & Ward 2005 (F*, L*; keys, distr.). **Ety.:** n.s. [involved (L)]. **Informal name:** Implicated Canadian Pointy Mosquito.

incomptus Arnell
1976: 20 (M*, F*, P*, L*). **Type loc.:** Cerro Mali, Serrania del Darien, Darien, Panama (USNM). **Distr.:** Costa Rica, Panama. **Ety.:** n.s. [unadorned (L); adult]. **Informal name:** Unadorned Panamanian Pointy Mosquito.

increpitus Dyar
1916b: 87 (M, F, L). **Type loc.:** Fallen Leaf, Lake Tahoe, California, United States (USNM). **Distr.:** Canada, Russia, United States (continental). **Refs.:** Gjullin 1937: 262 (F*); Bohart 1950a: 111 (L; bion.); Yamaguti & LaCasse 1951d: 116 (M*, F*, L*); Bohart 1954: 362 (L*); Carpenter & LaCasse 1955: 197 (M*, F*, L*; keys); Stone & Knight 1956a: 219 (type info., lectotype desig.); Dodge 1966: 354 (first instar L; key); Myers 1967: 799 (E*); Danilov 1976a: 71 (M*, L*; distr.); Wolff & Nielsen 1976 (bion., distr.); Bohart & Washino 1978: 51 (M, F*, L*; keys, tax., bion., distr.); Lanzaro & Eldridge 1992: 88 (M, F, P, L); Darsie & Ward 2005 (F*, L*; keys, distr.). **Ety.:** n.s. [rattle, snap, make noise (L)]. **Informal name:** Rattling American Pointy Mosquito.

syn. *vittata* Theobald
1903h: 313 (M, F, L*; *Grabhamia*; not Bigot 1861). **Type loc.:** Pecos Canon [*sic*; Cañón in Spanish, Canyon in English], New Mexico, United States (NHMUK). **Refs.:** Belkin 1968b: 8 (lectotype desig.).

syn. *mutatus* Dyar
1919b: 24 (M). **Type loc.:** Missoula, Montana, United States (USNM). **Refs.:** Stone & Knight 1956a: 222 (type info.).

syn. *hewitti* Hearle
1923: 5 (M, F). **Type loc.:** Yale, British Columbia, Canada (CNC).

inexpectatus Bonne-Wepster
1948: 318 (F*; as *Aedes* (*Aedimorphus*)). **Type loc.:** Baliem Valley, [Western] New Guinea (LM). **Distr.:** Indonesia (West Papua), Papua New Guinea. **Refs.:** Marks 1957a: 72 (subgeneric change); Lee et al. 1984: 156 (F; key, tax., bion., distr., review). **Ety.:** n.s. [unforeseen, unexpected (L)]. **Informal name:** Unexpected Indonesian Pointy Mosquito.

infirmatus Dyar & Knab
1906d: 197 (L*). **Type loc.:** Baton Rouge, Louisiana, United States (NE). **Distr.:** Honduras, Mexico, United States (continental). **Refs.:** Mitchell 1907b (E*); Howard et al. 1913: Fig. 680 (E*); Howard et al. 1917 (E*); Coher 1949: 102 (P*); Breland 1951: 362 (L*); Yamaguti & LaCasse 1951d: 120 (M*, F*, L*); Carpenter & LaCasse 1955: 199 (M*, F*, L*; keys); Stone & Knight 1956a: 219 (type info.); Craig & Horsfall 1960 (E*); Dodge 1966: 354 (first instar L*; key); Reinert 1970d: 147 (P*); Arnell 1976: 53 (M*, F*, P*, L*); Linley 1990 (E*); Darsie & Ward 2005 (F*, L*; keys, distr.); Harrison et al. 2016 (F*, L*; keys). **Ety.:** n.s. [weaken, diminish, be ill (L)]. **Informal name:** Infirm American Pointy Mosquito. (Vol. 1, p. 436)

intermedius Danilov & Gornostaeva
1987: 612 (M*, F*, L*). **Type loc.:** Abakan, Khakas [Khakass] Autonomous Province [Oblast], Krasnoyarsk Territory [Kray], USSR [Russian SFSR] (IMP). **Distr.:** Russia (includes Volga Region). **Ety.:** n.s. [in the middle (L); prob. ref. to "intermediate position between *Ae. leucomelas* & *Ae. implicatus*"]. **Informal name:** Intermediate Russian Pointy Mosquito.

intrudens Dyar
1919b: 23 (A, E). **Type loc.:** White River, Ontario, Canada (USNM). **Distr.:** Austria, Belarus, Canada, Czech Republic, Denmark, Estonia, Finland, Georgia, Germany, Japan, Kaliningrad Oblast, Latvia, Lithuania, Norway, Poland, Romania, Russia (Central, northwestern, Southern, and Volga regions), Slovakia, Sweden, Switzerland, Ukraine, United States (Alaska, continental). **Refs.:** Natvig 1948: 365 (M*, F*, L*; bion.); Darsie 1951: 23 (P*); Yamaguti & LaCasse 1951d: 124 (M*, F*, L*); Carpenter & LaCasse 1955: 200 (M*, F*, L*; keys); Stone & Knight 1956a: 219 (type info., lectotype desig.); Suzuki 1959: 293 (M*, F*, P*, L*); Price 1960: 554 (first instar L*); Dodge 1966: 354 (first instar L; key); Kalpage & Brust 1968 (E*); Mohrig 1969 (F*); Pickavance et al. 1970 (distr.); Polyakova 1970b: 134 (distr.); Horsfall & Voorhees 1972 (E*); Rjazantzeva 1972 (F*); Gutsevich et al. 1974: 261 (M*, F, L*); Dahl 1975: 81, 86 (distr.); Utrio 1976: 131, 134 (L*; distr.); Tanaka et al. 1979: 296 (M*, F*, L*); Mgeladze 1991: 331

(distr.; Georgia); Dahl 1997 (E*); Tanaka 1999: 122 (P*; tax.); Spungis 2000: 10 (distr.; Latvia); Reinert 2002e: Fig. 20 (F*); Nicolescu et al. 2003a: 2 (distr.; Romania); Darsie & Ward 2005 (F*, L*; keys, distr.); Becker et al. 2010: 237 (M*, F*, L*; keys, tax., distr., bion.); Harrison et al. 2016 (F*, L*; keys, distr.); Robert et al. 2019 (distr.; Western Palearctic). **Ety.:** thrust or force in (L); enters houses. **Informal name:** Trespasser Canadian Pointy Mosquito.

jacobinae Serafim & Davis
1933: 14 (M, F). **Type loc.:** Jacobina, Bahia, Brazil (USNM). **Distr.:** Brazil. **Refs.:** Lane 1953: 681 (M*, F); Stone & Knight 1956a: 220 (type info.). **Ety.:** Jacobina, Bahia, Brazil. **Informal name:** Jacobina Brazilian Pointy Mosquito.

jorgi Carpintero & Leguizamón
2000: 20 (F). **Type loc.:** Punta Lara, Berazategui District, Buenos Aires Province, Argentina (USNM). **Distr.:** Argentina. **Refs.:** Hribar & Hribar 2006 (type dep.). **Ety.:** Miguel E. Jorg. **Informal name:** Jorg Argentinian Pointy Mosquito.

juppi McIntosh
1973: 262 (M*, F*, P*, L*). **Type loc.:** Onderstepoort, South Africa [Republic of South Africa] (NIV). **Distr.:** Lesotho, Republic of South Africa. **Refs.:** Jupp 1996 (M*, F*; key). **Ety.:** Peter G. Jupp. **Informal name:** Jupp South African Pointy Mosquito.

kasachstanicus Gutsevich
1962: 886 (M*, F). **Type loc.:** Ili, Alma-Ata Oblast, Kazakhstan (ZIL). **Distr.:** Kazakhstan. **Ety.:** n.s. [pertaining to the type locality in Kazakhstan (Harbach 2018c)]. **Informal name:** Botai Kazakhstani Pointy Mosquito [culture of ancient Kazakhstan credited with domesticating the horse].

lasaensis Meng
1962: 159 (F*, L*). **Type loc.:** Lasa [Tibet, China] (LU). **Distr.:** People's Republic of China. **Ety.:** n.s. [pertaining to the type locality of Lasa (Lhasa), Tibet (Harbach 2018c)]. **Informal name:** Lhasa Chinese Pointy Mosquito.

ssp. ***gyirongensis*** Ma
1982: 162 (M*, F). **Type loc.:** Gyirong, Xizang Autonomous Region, China [People's Republic of China] (IZ). **Distr.:** People's Republic of China. **Ety.:** n.s. [prob. from type locality of Gyirong, Tibet Autonomous Region]. **Informal name:** Gyirong Chinese Pointy Mosquito.

lepidus Cerqueira & Paraense
1945: 12 (M*, F). **Type loc.:** Belo Horizonte Minas Gerais, Brazil (INER). **Distr.:** Brazil. **Refs.:** Lane 1953: 656 (M*, F); Cerqueira 1957 (P*, L*); Belkin et al. 1971: 19 (type info.); Xavier 1973: 160 (type info.). **Ety.:** n.s. [scaled (L); poss. ref. to differences in scales in comp. to *jacobinae*]. **Informal name:** Scaly Brazilian Pointy Mosquito.

leucomelas (Meigen)
1804: 3 (M; *Culex*). **Type loc.:** Europe (MNHP). **Distr.:** Albania, Austria, Belarus, Bosnia and Herzegovina, Crimean Peninsula, Croatia, Czech Republic, Denmark, Estonia, Finland, Germany, Hungary, Iran, Latvia, Lithuania, Moldova, Mongolia, Netherlands, Norway, Poland, Romania, Russia (Central, northwestern, Southern, and Volga regions), Slovakia, Spain, Sweden, Turkey, Ukraine, United Kingdom. **Refs.:** Martini 1920a: 112 (M*, L*; as *terriei* Theobald); Natvig 1948: 295 (M*, F, L*); Monchadskii 1951: 183 (L*); Rjazantzeva 1972 (F*); Gutsevich et al. 1974: 275 (A, L*); Dahl 1975: 82, 87 (tax., distr.); Minar 1976: 336 (distr.); Utrio 1976: 134 (distr.); Zaim 1987: 568 (distr.); Spungis 2000: 10 (distr.; Latvia); Becker et al. 2010: 238 (M*, F*, L*; keys, tax., distr., bion.); Robert et al. 2019 (distr.; Western Palearctic). **Ety.:** white, black (Gr); abdomen black with white rings. **Informal name:** Chanel European Pointy Mosquito [Gabrielle Bonheur "Coco" Chanel (1883–1971) was a French fashion designer who first popularized the black-and-white clothing combination].

syn. *salinellus* Edwards
1921b: 52 (A; *Ochlerotatus*). **Type loc.:** None designated (LU).

linesi Marks
1964a: 63 (F). **Type loc.:** Robe, South Australia, Australia (CSIR). **Distr.:** Australia. **Refs.:** Lee et al. 1984: 157 (F; key, tax., bion., distr., review). **Ety.:** E.W. Lines. **Informal name:** Lines Australian Pointy Mosquito.

longifilamentus Su & Zhang
1988: 13 (M*, F). **Type loc.:** Yumin, Bacju, and Kaiping of Xinjiang Uygur Autonomous Region, China [People's Republic of China] (IMEA). **Distr.:** People's Republic of China. **Ety.:** n.s. [long filament (L); prob. ref. to apical lobe of gonocoxite with three to six long setae]. **Informal name:** Long-threaded Chinese Pointy Mosquito.

luteifemur Edwards
1926a: 112 (F). **Type loc.:** Advent Bay, Strahan and Mount Farrell, Tasmania-Tarwin Lower, Victoria, Australia (CSIR). **Distr.:** Australia. **Refs.:** Dobrotworsky 1960a: 54 (M*, F, P*, L*); Dobrotworsky 1965: 106 (M*, F*, L*); Lee et al. 1984: 158 (F; key, tax., bion., distr., review). **Ety.:** yellow thigh (L); "hind femora almost entirely ochreous." **Informal name:** Yellow-legged Australian Pointy Mosquito.

macintoshi Marks
1959a: 117 (M*, F, P*, L*). **Type loc.:** Albany, Western Australia, Australia (CSIR). **Distr.:** Australia. **Refs.:** Lee et al. 1984: 160 (F; key, tax., bion., distr., review). **Ety.:** D.L. McIntosh. **Informal name:** McIntosh Australian Pointy Mosquito.

martineti Senevet
1937: 364 (M*, P, L). **Type loc.:** Cayenne [Guyane, French Guiana] (LPFM). **Distr.:** French Guiana. **Ety.:** Mme. Martinet. **Informal name:** Martinet French Guianese Pointy Mosquito.

mcdonaldi Belkin
1962: 396 (M*, F, P*, L*). **Type loc.:** West Paha swamp, Guadalcanal, Solomon Islands (USNM). **Distr.:** Papua New Guinea, Solomon Isl. **Refs.:** Lee et al. 1984: 161 (F; key, tax., bion., distr., review). **Ety.:** William A. McDonald. **Informal name:** McDonald Solomon Islander Pointy Mosquito.

melanimon Dyar
1924e: 126 (M, F). **Type loc.:** Bakersfield, California, United States (USNM). **Distr.:** Canada, Panama, United States (continental). **Refs.:** Dyar 1928: 201 (M*, F, L*); Yamaguti & LaCasse 1951d: 127 (M*, F*, L*; as *klotsi*); Barr 1955: 170 (from syn. with *dorsalis*); Carpenter & LaCasse 1955: 202 (M*, F; keys); Richards 1956: 261 (M*, F*, L*); Bohart 1956a (L*; tax.; *melanimon/dorsalis* comparison); Stone & Knight 1956a: 221 (type info., lectotype desig.); Chapman & Grodhaus 1963 (tax.); Dodge 1966: 354 (first instar L; key); Myers 1967: 803 (E*); Bohart & Washino 1978 (M*, F*, P*, L*; keys, tax., bion.); Bohart & Washino 1978: 51 (M, F*, L*; keys, tax., bion., distr.); Darsie & Ward 2005 (F*, L*; keys, distr.); Darsie 2011d (P*). **Ety.:** n.s. [dark (Gr); prob. ref. to all-dark wing scales]. **Informal name:** Encephalitis American Pointy Mosquito [*melanimon* is a major vector of western equine and California group encephalitis]. (Vol. 1, p. 438)

syn. *mediolineata* Ludlow
1907: 129 (F; *Grabhamia*). **Type loc.:** Fort Lincoln, North Dakota, United States (USNM). **Refs.:** Stone & Knight 1956a: 220 (type info., lectotype desig.); Wood 1977: 76 (syn.).

syn. *klotsi* Matheson
1933: 69 (M*, F). **Type loc.:** Mountain Home Lake, Fort Garland, Colorado, United States (USNM). **Refs.:** Carpenter & LaCasse 1955: 202 (syn.).

meprai Martínez & Prosen
1953b: 1 (F). **Type loc.:** Reserva Nac "Finca El Rey," Departamento de Anta, Salta, Argentina (BA). **Distr.:** Argentina, Bolivia, Ecuador, Peru. **Refs.:** Ronderos & García 1963b: 38 (syn.); Arnell 1976: 28 (F*; from syn. with *angustivittatus*; distr.). **Ety.:** acronym: Misión de Estudios de Patología Regional Argentina. **Informal name:** Mepra Argentinian Pointy Mosquito.

mercurator Dyar
1920a: 13 (M, F, L). **Type loc.:** Dawson, Yukon Territory, Canada (USNM). **Distr.:** Canada, Russia (Northwestern Dist.), United States (Alaska). **Refs.:** Stone & Knight 1956a: 221 (type info., lectotype desig.); Danilov 1974b: 322 (M*, F, L*; from syn. with *stimulans*; distr.); Wood 1977: 76 (F*; tax.); Darsie & Ward 2005 (F*, L*; keys, distr.); Becker et al. 2010: 242 (M*, F*, L*; keys, tax., distr., bion.); Robert et al. 2019 (distr.; Western Palearctic). **Ety.:** n.s. [Mercury, Grecian messenger of the gods]. **Informal name:** Mercury Canadian Pointy Mosquito.

syn. *albertae* Dyar
1920d: 115 (M, F; *stimulans* ssp.). **Type loc.:** Edmonton, Alberta, Canada (USNM). **Refs.:** Stone & Knight 1956a: 213 (type info., lectotype desig.); Wood 1977: 76 (syn.).

syn. *ater* Gutsevich
1955: 321 (M*; *riparius* ssp.). **Type loc.:** Baranovka, Khabarovsk, USSR (LU). **Refs.:** Danilov 1974b: 322 (syn.); Gutsevich et al. 1974: 208 (M*, F; tax.).

milleri Dyar
1922g: 194 (F). **Type loc.:** Bogota, Cundinamarca, Colombia (USNM). **Distr.:** Argentina, Bolivia, Colombia, Ecuador, Honduras, Peru. **Refs.:** Knight & Marks 1952: 517 (tax.; exclusion from *Finlaya*); Lane 1953: 676 (M*, F, L*). **Ety.:** F.A. Miller. **Informal name:** Miller Colombian Pointy Mosquito.

syn. *araozi* Shannon & Del Ponte
1928: 74 (A). **Type loc.:** Córdoba; Zapla, Jujuy; Quebrada de San Lorenzo, Salta; all in Argentina (INM). **Refs.:** Belkin et al. 1968: 14 (lectotype desig. according to Casal).

syn. *oroecetor* Martini
1931a: 204 (M*, F). **Type loc.:** Sorata, 2,300 m, Bolivia (NHMUK). **Refs.:** Lane 1953: 676 (syn.); Mattingly 1955c: 29 (lectotype desig.).

mitchellae (Dyar)
1905b: 74 (M, F; *Culex*). **Type loc.:** Jacksonville, Florida, United States (USNM). **Distr.:** Mexico, United States (continental). **Refs.:** Yamaguti & LaCasse 1951d: 132 (M*, F*, L*); Carpenter & LaCasse 1955: 204 (M*, F*, L*; keys); Stone & Knight 1956a: 222 (type info.); Craig & Horsfall 1960 (E*); Ross & Horsfall 1965 (M*, F*, L*, E*; keys); Dodge 1966: 354 (first instar L*; key); Reinert 1970e: 73 (P*); Darsie & Ward 2005 (F*, L*; keys, distr.); Harrison et al. 2016 (F*, L*; keys, distr., tax.). **Ety.:** Evelyn G. Mitchell. **Informal name:** Mitchell American Pointy Mosquito.

montchadskyi Dubitzky
1968: 218 (M*, F*, L*). **Type loc.:** Ili River Basin, Alma Ata, Kazakh, USSR [Kazakhstan] (KZI). **Distr.:** Kazakhstan. **Ety.:** n.s. [Alexander S. Montchadsky (Harbach 2018c)]. **Informal name:** Montchadsky Kazakhstani Pointy Mosquito.

nevadensis Chapman & Barr
1964: 444 (M*, F, P*, L; as *communis* ssp.). **Type loc.:** Lamoille Canyon, Elko County, Utah, United States (USNM). **Distr.:** United States (continental). **Refs.:** Ellis & Brust 1973: 922 (M*, F*, L*; to sp.); Belton 1982 (cuticular vestiture); Darsie & Ward 2005 (F*, L*; keys,

distr.). **Ety.:** from Nevada, United States. **Informal name:** Nevada American Pointy Mosquito.

nigrinus (Eckstein)
1918: 67 (M, F, L; *Culicada*). **Type loc.:** Ackenheim, Wolfisheim, and Oberschäffolsheim, Strasburg, Germany (W). **Distr.:** Austria, Croatia, Czech Republic, Denmark, Estonia, Finland, France (includes Corsica), Germany, Hungary, Lithuania, Netherlands, Norway, Poland, Russia (Northwestern Region), Slovakia, Sweden. **Refs.:** Natvig 1948: 354 (M*, F, L*); Gutsevich et al. 1974: 255 (M, F, L); Utrio 1976: 134 (distr.); Becker et al. 2010: 243 (M*, F*, L*; keys, tax., distr., bion.); Robert et al. 2019 (distr.; Western Palearctic). **Ety.:** black, dark (L); overall color of adult. **Informal name:** Black German Pointy Mosquito.

nigripes (Zetterstedt)
1838: 807 (M, F; *Culex*). **Type loc.:** Greenland (LS). **Distr.:** Canada, Greenland, Iceland, Norway, Poland, Russia, Sweden, United States (Alaska, continental). **Refs.:** Natvig 1948: 302 (M*, F*, L*; tax.); Yamaguti & LaCasse 1951d: 140 (M*, F*, L*); Natvig 1955: 86 (M*, F*); Vockeroth 1954b: 251 (F*); Carpenter & LaCasse 1955: 206 (M*, F*, L*; keys); Mohrig 1967 (F*); Kalpage & Brust 1968 (E*); Danks & Corbet 1973: 367 (M*, F*, P*, L*, E*); Gutsevich et al. 1974: 266 (M*, F, L*); Corbet & Danks 1975 (E*); Utrio 1976: 130 (L*); Dahl 1997 (E*); Darsie & Ward 2005 (F*, L*; keys, distr.); Becker et al. 2010: 245 (M*, F*, L*; keys, tax., distr., bion.); Robert et al. 2019 (distr.; Western Palearctic). **Ety.:** black foot (L); all dark feet. **Informal name:** Black-footed Greenlandic Pointy Mosquito.
syn. *innuitus* Dyar & Knab
1918b: 166 (M, F). **Type loc.:** Kangerdlooksoah, Inglefield Gulf, Greenland (USNM). **Refs.:** Stone & Knight 1956a: 219 (type info., lectotype desig.).

nigrithorax (Macquart)
1847: 25 (M; *Culex*). **Type loc.:** Tasmania, Australia (BC). **Distr.:** Australia. **Refs.:** Edwards 1924b: 377 (M, F); Marks 1960a: 117 (M*, F*; syn.); Dobrotworsky 1965: 86 (M*, F*, L*, E*); Dobrotworsky 1966: 135 (M*, L); Lee et al. 1984: 162 (F; key, tax., bion., distr., review). **Ety.:** black thorax (L). **Informal name:** Black-bodied Australian Pointy Mosquito.
syn. *macleayanus* Mackerras
1927: 291 (F). **Type loc.:** Launceston, Tasmania, Australia (SAM).

nigrocanus Martini
1927: 390 (M*). **Type loc.:** Anatolischen Bahn, Turkey (NHMUK). **Distr.:** Turkey. **Refs.:** Lane 1992a: 12 (M*; type info., tax.); Lane 1994: 35 (type loc. info.). **Ety.:** black, whitish gray (L); adult colors. **Informal name:** Black-gray Turkish Pointy Mosquito.

nigromaculis (Ludlow)
1906a: 83 (A; *Grabhamia*). **Type loc.:** Fort Keogh, Montana, United States (USNM). **Distr.:** Canada, Mexico, United States (continental). **Refs.:** Gjullin 1937: 257 (F*); Rempel 1950: 220 (L*); Yamaguti & LaCasse 1951d: 144 (M*, F*, L*); Arnold 1953: 37 (embryology); Rempel 1953: 470 (M*, F*); Bohart 1954: 364 (L*); Carpenter & LaCasse 1955: 208 (M*, F*, L*; keys); Craig 1956 (E*); Stone & Knight 1956a: 222 (type info., lectotype desig.); Price 1960: 556 (first instar L*); Ross & Horsfall 1965 (M*, F*, L*, E*; keys); Dodge 1966: 355 (first instar L*; key); Myers 1967: 798 (E*); Mohrig 1967 (F*); Kalpage & Brust 1968 (E*); Bohart & Washino 1978: 53 (M, F*, L*; keys, tax., bion., distr.); Darsie & Ward 2005 (F*, L*; keys, distr.). **Ety.:** n.s. [black spot (L)]. **Informal name:** Black-spotted American Pointy Mosquito.
syn. *grisea* Ludlow
1907: 130 (F; *Grabhamia*). **Type loc.:** Boise Barracks, Idaho, United States (USNM). **Refs.:** Stone & Knight 1956a: 218 (type info.).

niphadopsis Dyar & Knab
1918b: 166 (F). **Type loc.:** Salt Lake County, Utah, United States (USNM). **Distr.:** United States (continental). **Refs.:** Yamaguti & LaCasse 1951d: 148 (M*, F*, L*); Rees & Collett 1954: 207 (M*, F, L*; bion.); Carpenter & LaCasse 1955: 210 (M*, F*, L*; keys); Stone & Knight 1956a: 222 (type info., lectotype desig.); Myers 1967: 804 (E*); Bohart & Washino 1978: 114 (M, F, L*; keys, tax., bion., distr.); Darsie & Ward 2005 (F*, L*; keys, distr.). **Ety.:** n.s. [white appearing (Gr); poss. ref. to white morphological features]. **Informal name:** Snowflake American Pointy Mosquito.

nivalis Edwards
1926a: 112 (F). **Type loc.:** Marysville, Victoria, Australia (NHMUK). **Distr.:** Australia. **Refs.:** Mackerras 1927: 293 (F); Dobrotworsky 1960a: 60 (M*, F, P*, L*); Dobrotworsky 1965: 112 (M*, F*, L*); Lee et al. 1984: 165 (F; key, tax., bion., distr., review). **Ety.:** n.s. [snowy, snow covered, snow-like (L)]. **Informal name:** Snowy Australian Pointy Mosquito.

normanensis (Taylor)
1915: 182 (M, F; *Culex*). **Type loc.:** Normanton, Queensland, Australia (US). **Distr.:** Australia, Papua New Guinea. **Refs.:** Marks 1949: 18 (M*, F, P*, L*); Lee et al. 1984: 169 (F; key, tax., bion., distr., review); Townsend 1990: 112 (syntypes, NHMUK); Becker et al. 2010: 360 (F*, L*; key, tax., distr., bion.). **Ety.:** Normanton, Queensland, Australia. **Informal name:** Normanton Australian Pointy Mosquito.

nubilus (Theobald)
1903a: 208 (F*; *Culex*). **Type loc.:** Pomeroon Mission and Christianburg River, British Guiana [Guyana] (NHMUK). **Distr.:** Argentina, Brazil, Colombia, Ecuador, French Guiana, Guyana, Panama, Paraguay, Peru. **Refs.:** Antunes & Lane 1934: 38 (tax.); Senevet & Abonnenc 1939c: 474 (M*); Komp 1949: 105 (syn.); Mohrig 1967 (F*); Belkin 1968b: 6 (lectotype desig.); Belkin et al.

1970: 159 (to sp.); Lane 1992b: 16 (distr.). **Ety.:** n.s. [cloudy (L); poss. ref. to overall deep dusky brown color]. **Informal name:** Overcast Guyanese Pointy Mosquito.

syn. *polyagrus* Dyar

1918a: 77 (M, L). **Type loc.:** Taboga Island, Panama (USNM). **Refs.:** Komp 1949: 107 (M*; syn.); Belkin et al. 1970: 159 (? syn.).

obturbator Dyar & Knab

1907a: 9 (F). **Type loc.:** Tarpon Bay [Eleuthera], Bahama Islands (USNM). **Distr.:** Bahamas, Puerto Rico. **Refs.:** Lane 1953: 653 (F); Stone & Knight 1956a: 222 (type info.); Maldonado-Capriles et al. 1958: 65 (M*, L*); Mohrig 1967 (F*); Arnell 1976: 72 (M*, F*, L*). **Ety.:** n.s. [disturb, agitate (L)]. **Informal name:** Disturbing Bahamian Pointy Mosquito.

oligopistus Dyar

1918a: 76 (F). **Type loc.:** Trinidad [Trinidad & Tobago] (USNM). **Distr.:** Argentina, Bolivia, Brazil, Colombia, French Guiana, Guyana, Honduras, Panama, Suriname, Trinidad and Tobago, Venezuela. **Refs.:** Lane 1953: 664 (M*, F); Xavier 1967: 227 (P*, L*); Panday 1975a, b (distr.; Suriname). **Ety.:** n.s. [small, faithful, genuine (Gr)]. **Informal name:** Faithful Trinbagonian Pointy Mosquito.

patersoni Shannon & Del Ponte

1928 (M, F). In Dyar 1928: 73. **Type loc.:** San Pedro, Jujuy, Argentina (INM). **Distr.:** Argentina, Bolivia. **Refs.:** Ronderos & García 1963a: 35 (P; from syn.); Belkin et al. 1968: 14 (lectotype desig. by Casal (INM)); Arnell 1976: 50 (M*, F*, P*, L*; distr.). **Ety.:** Guillermo Paterson. **Informal name:** Paterson Argentinian Pointy Mosquito.

pectinatus Arnell

1976: 74 (M*, F*, P*, L*). **Type loc.:** Rio Cuja, near Fusagasugá, Cundinamarca, Colombia (USNM). **Distr.:** Colombia. **Ety.:** pecten; ref. to large number of pecten spines. **Informal name:** Combed Colombian Pointy Mosquito.

pennai Antunes & Lane

1938: 607 (M*, F, L*). **Type loc.:** Cabreúva and Rocinha, São Paulo, Brazil (FH). **Distr.:** Argentina, Bolivia, Brazil. **Refs.:** Lane 1953: 644 (M*, F, P*, L*). **Ety.:** Henrique de Azevedo Penna. **Informal name:** Penna Brazilian Pointy Mosquito.

perkinsi Marks

1949: 32 (M*, P*, L*). **Type loc.:** Caloundra, Queensland, Australia (UQIC). **Distr.:** Australia. **Refs.:** Dobrotworsky 1965: 104 (M*, F*, L*); Lee et al. 1984: 174 (F; key, tax., bion., distr., review). **Ety.:** F.A. Perkins. **Informal name:** Perkins Australian Pointy Mosquito.

pertinax Grabham

1906b: 316 (M, F, L*). **Type loc.:** Jamaica (USNM). **Distr.:** Bahamas, Dominican Republic, Haiti, Jamaica, United States (continental). **Refs.:** Stone & Knight 1956a: 223 (type info., lectotype desig.); Belkin et al. 1970: 157 (M*, F, P*, L*; from syn. with *serratus* Theobald); Shroyer et al. 2015: 97 (distr.; United States). **Ety.:** n.s. [persevering, obstinate (L)]. **Informal name:** Persistent Jamaican Pointy Mosquito.

syn. *quasiserratus* Theobald

1907: 465 (F; *Protoculex*). **Type loc.:** Kingston, Jamaica (NHMUK). **Refs.:** Howard et al. 1917 (syn.); Belkin 1968b: 7 (lectotype desig.); Belkin et al. 1970: 157 (change in syn.).

phaecasiatus Marks

1964a: 64 (F). **Type loc.:** 5 mi from Maningrida, Arnhem Land, Northern Territory, Australia (UQIC). **Distr.:** Australia. **Refs.:** Lee et al. 1984: 175 (F; key, tax., bion., distr., review). **Ety.:** a wearer of white shoes (Gr); ref. to white hind tarsal segment 5. **Informal name:** Sneaker Australian Pointy Mosquito [*phaecasiatus* = "wearer of white shoes" = Sneaker].

phaeonotus Arnell

1976: 56 (M*, F*, P*, L*). **Type loc.:** Chantimelle Village, near Sauteurs, Saint Patrick Parish, Grenada, Lesser Antilles (USNM). **Distr.:** Grenada. **Ety.:** dusky back (Gr); scutum. **Informal name:** Dusky Grenadian Pointy Mosquito.

pionips Dyar

1919b: 19 (M, F, L). **Type loc.:** White River, Ontario, Canada (USNM). **Distr.:** Belarus, Canada, Finland, Montenegro, Norway, Poland, Russia (Northwestern Region), Sweden, United States (Alaska, continental). **Refs.:** Rempel 1950: 225 (L*); Yamaguti & LaCasse 1951d: 152 (M*, F*, L*); Vockeroth 1952: 243 (M*, F*, L*; tax.); Rempel 1953: 474 (M*, F*); Beckel 1954: 329 (tax.); Carpenter & LaCasse 1955: 212 (M*, F*, L*; keys); Craig 1956; Stone & Knight 1956a: 223 (type info., lectotype desig.); Darsie 1957: 620 (P); Mohrig 1967 (F*); Kalpage & Brust 1968 (E*); Gutsevich et al. 1971: 235 (distr.); Horsfall & Voorhees 1972 (E*); Dahl 1974: 62 (L*; distr.); Gutsevich et al. 1974: 245 (M*, F, L*); Utrio 1975: 63 (distr.); Utrio 1976: 129, 131, 134 (L*; distr.); Dahl 1997 (E*); Darsie & Ward 2005 (F*, L*; keys, distr.); Becker et al. 2010: 246 (M*, F*, L*; keys, tax., distr., bion.); Robert et al. 2019 (distr.; Western Palearctic). **Ety.:** fat worm (Gr); ref. to large larvae. **Informal name:** Fat Canadian Pointy Mosquito.

procax (Skuse)

1889: 1742 (F; *Culex*). **Type loc.:** Gosford and South Clifton, New South Wales, Australia (MM). **Distr.:** Australia. **Refs.:** Bancroft 1908: 46 (F); Klein & Marks 1960: 112 (*rubrithorax* of authors, not Macquart; from syn.); Dobrotworsky 1965: 81 (M*, F*, L*); Lee et al. 1984: 176 (F; key, tax., bion., distr., review); Linley et al. 1991a (E*). **Ety.:** n.s. [impudent, undisciplined (L)]. **Informal name:** Bold Australian Pointy Mosquito.

pseudonormanensis Marks
1949: 26 (M*, F, L*). **Type loc.:** Eidsvold, Queensland, Australia (MM). **Distr.:** Australia. **Refs.:** Lee et al. 1984: 181 (F; key, tax., bion., distr., review). **Ety.:** false (Gr); similar to *normanensis*. **Informal name:** Saxon Australian Pointy Mosquito [*pseudonormanensis* = "false Norman" = Saxon]

pulcritarsis (Rondani)
1872: 31 (A; *Culex*). **Type loc.:** Italy (FM). **Distr.:** Albania, Afghanistan, Azerbaijan, Bulgaria, Crimean Peninsula, Croatia, Cyprus, Czech Republic, France (includes Corsica), Georgia, Greece, Hungary, India, Iran, Israel (Gaza Strip and West Bank), Italy (includes Sicily), Kazakhstan, Kosovo, Lebanon, Macedonia, Moldova, Morocco, Pakistan, Portugal, Romania, Russia (Southern Dists.), Serbia, Slovakia, Spain (includes Balearic Isl.), Tunisia, Turkey, Ukraine, Uzbekistan. **Refs.:** Martini 1930: 316 (M*, F, L*); Barraud 1934: 150 (M, F; tax.); Edwards 1934: 442 (appendix); Barraud 1934; Sicart 1952a: 95 (M*, F, P, L*; bion.); Sicart & Larrouy 1968 (F*); Aslamkhan 1971b (distr.; Pakistan); Rjazantzeva 1972 (F*); Gutsevich et al. 1974: 206 (M*, F, L); Margalit & Tahori 1974: 84 (distr.); Snow 1986: 361 (corr. specific epithet from *pulchitarsis*, Article 32, International Commission on Zoological Nomenclature 1999; Mgeladze 1991: 329 (distr.; Georgia); Harrison et al. 1991: 213 (removed Thailand from distr.); Eritja et al. 2000: 11 (distr.; Spain); Ramsdale et al. 2001: 19 (distr.; Turkey); Rueda, Pecor et al. 2008: 397 (distr.); Becker et al. 2010: 248 (M*, F*, L*; keys, tax., distr., bion.); Namazov 2014 (distr.; Azerbaijan); Robert et al. 2019 (distr.; Western Palearctic). **Ety.:** beautiful ankle (L); ref. to white tarsal bands. **Informal name:** Pretty-footed Italian Pointy Mosquito.

syn. *leucacanthus* Loew
1873 (M; *Culex*). **Type loc.:** Kasan, Southern Russia [USSR] (LU).

syn. *simici* Baranoff
1916: 15 (M*, F, L*). **Type loc.:** Schischevo bei Skopje, Serbia [Republic of Macedonia] (CNC). **Refs.:** Peterson 1971: 1199 (lectotype desig.).

ssp. ***asiaticus*** Edwards
1926c: 73 (A). **Type loc.:** Drosh, Chitral [NW Frontier Provinces, Pakistan] (NHMUK). **Distr.:** Bangladesh, India, Nepal, Pakistan, People's Republic of China, Sri Lanka. **Refs.:** Barraud 1934: 151 (F, L); Monchadskii 1951: 193 (L*; ssp.); Gutsevich et al. 1974: 208 (A, L*). **Ety.:** Asian. **Informal name:** Chitral Pakistani Pointy Mosquito.

ssp. syn. *stegomyina* Stackelberg & Monchadskii
1926 (M, L*; *pulcritarsis* var.). In Montchadsky 1926: 151. **Type loc.:** Katta-Kourgan, Turkestan [Turkistan] (LU).

pullatus (Coquillett)
1904c: 168 (M, F; *Culex*). **Type loc.:** Kaslo, British Columbia, Canada (USNM). **Distr.:** Austria, Bosnia and Herzegovina, Bulgaria, Canada, Czech Republic, Finland, France (includes Corsica), Germany, Greece, Hungary, India, Italy, Lithuania, Mongolia, Montenegro, Netherlands, Norway, Pakistan, Poland, Portugal, Romania, Russia (Central, northwestern, Southern, and Volga regions), Serbia, Slovakia, Spain, Sweden, Switzerland, Turkey, United States (Alaska, continental). **Refs.:** Séguy 1921b (M gen.*; as *jugorum*); Barraud 1934: 152 (M), 442 (F); Gjullin 1937: 258 (F*); Natvig 1948: 372 (M*, F, L*); Yamaguti & LaCasse 1951d: 156 (M*, F*, L*); Vockeroth 1954b: 252 (distr.); Carpenter & LaCasse 1955: 214 (M*, F*, L*; keys); Craig 1956 (E*); Stone & Knight 1956a: 224 (type info., lectotype desig.); Myers 1967: 798 (E*); Minar 1971: 92 (distr.); Horsfall & Voorhees 1972 (E*); Dahl 1974: 63 (L*; distr.); Gutsevich et al. 1974: 263 (M*, F, L*); Utrio 1976: 129, 131, 134 (L*); Wolff & Nielsen 1976 (bion., distr.); Bohart & Washino 1978: 114 (M, F, L*; keys, tax., bion., distr.); Darsie & Ward 2000 (distr.; United States); Eritja et al. 2000: 11 (distr.; Spain); Kaiser et al. 2001 (distr.; Greece); Kaur 2003: 454 (distr.; India); Darsie & Ward 2005 (F*, L*; keys, distr.); Becker et al. 2010: 249 (M*, F*, L*; keys, tax., distr., bion.); Robert et al. 2019 (distr.; Western Palearctic). **Ety.:** clad in black (L); overall coloration of adult. **Informal name:** Black Canadian Pointy Mosquito.

syn. *acrophilus* Dyar
1917d: 127 (M, L). **Type loc.:** Lake Louise, Laggan, Alberta, Canada (USNM).

syn. *jugorum* Villeneuve
1919: 58 (M, F; *Culex*). **Type loc.:** Lautaret, Hautes Alpes, France (ISN).

syn. *gallii* Martini
1920a: 110 (M*, L*). **Type loc.:** Lake Rogneda (2,390 m), Valtellina, Italy (NHMUK) [Revised type-loc. data from Peus, *in litt.* 1972]. **Refs.:** Mattingly 1955c: 29 (tax.).

syn. *metalepticus* Dyar
1920b: 51 (M, F). **Type loc.:** Scais, Prov Sondrio, [Lombardy], Italy (USNM). **Refs.:** Stone & Knight 1956a: 221 (type info., lectotype desig.).

syn. *pearyi* Dyar & Shannon
1925d: 78 (F). **Type loc.:** Hawk's Harbor, Labrador, Canada (USNM). **Refs.:** Vockeroth 1954a: 115 (syn.).

syn. *seguyi* Apfelbeck
1930: 288 (M*, F, L*). **Type loc.:** Sarajevo, Yugoslavia [Bosnia & Herzegovina] (LU). **Refs.:** Townsend 1990: 132 (syntypes, NHMUK).

punctodes Dyar
1922a: 1 (M, F). **Type loc.:** Alaska, United States (USNM). **Distr.:** Belgium, Finland, Norway, Russia, Sweden, United States (Alaska, continental). **Refs.:** Knight 1951a: 97 (from syn. with *punctor*; type info., lectotype desig.); Carpenter & LaCasse 1955: 216 (M*, L*; keys); Darsie 1957: 616 (P*); Dahl 1974: 64 (L*; distr.); Utrio 1976: 130, 134 (L*; distr.); Gornostaeva 2000: 479 (distr.; Russia); Darsie & Ward 2005 (F*, L*; keys, distr.); Becker et al. 2010: 250 (M*, F*, L*; keys, tax.,

distr., bion.); Robert et al. 2019 (distr.; Western Palearctic). **Ety.:** resembling *punctor* (Gr); Punctor Group. **Informal name:** Spotlike American Pointy Mosquito.

punctor (Kirby)
1837: 309 (A; *Culex*). **Type loc.:** Mackenzie River Valley, near Fort Norman, Northwest Territories, Canada (LU). **Distr.:** Austria, Belarus, Belgium, Bulgaria, Canada, Crimean Peninsula, Croatia, Czech Republic, Denmark, Estonia, Finland, France, Georgia, Germany, Greece, Hungary, Ireland, Italy, Japan, Latvia, Lithuania, Luxembourg, Moldova, Montenegro, Netherlands, Norway, Poland, Portugal, Romania, Russia, Slovakia, Slovenia, Spain, Sweden, Switzerland, Turkey, Ukraine, United Kingdom, United States (Alaska, continental). **Refs.:** Séguy 1924: 127, Pl. XXVII (L*), 127 (M gen.*, F*); Natvig 1948: 333 (M*, F, L*, E*); Monchadskii 1951: 196 (L*); Darsie 1951: 24 (P*); Knight 1951a: 90 (M*, F, L; tax.); Yamaguti & LaCasse 1951d: 112 (M*, F*, L*; as *implacabilis*); Yamaguti & LaCasse 1951d: 160 (M*, F*, L*); Asanuma et al. 1952 (M*); Vockeroth 1954b: 248 (tax.); Carpenter & LaCasse 1955: 217 (M*, F*, L*; keys); Hara 1957: 59 (F*); Price 1960: 556 (first instar L*); Ross & Horsfall 1965 (M*, F*, L*, E*; keys); Dodge 1966: 356 (first instar L; key); Kalpage & Brust 1968 (E*); Sicart & Larrouy 1968 (F*); Horsfall & Voorhees 1972 (E*); Rjazantzeva 1972 (F*); Dahl 1974: 64 (M*); Gutsevich et al. 1974: 247 (M*, F, L*); Dahl 1975: 82, 87 (distr.); Utrio 1976: 130, 134 (L*); Tanaka et al. 1979: 281 (M*, F*, L*; tax.); Mgeladze 1991: 331 (distr.; Georgia); Dahl 1997 (E*); Tanaka 1999: 113 (P*; tax., distr.; Japan); Eritja et al. 2000: 11 (distr.; Spain); Spungis 2000: 10 (distr.; Latvia); Kaiser et al. 2001 (distr.; Greece); Beck et al. 2003: 24 (distr.; Luxembourg); Melero-Alcíbar 2005c (P; tax.); Darsie & Ward 2005 (F*, L*; keys, distr.); Becker et al. 2010: 251 (M*, F*, L*; keys, tax., distr., bion.); Harrison et al. 2016 (F*, L*; keys, distr.; note 12); Robert et al. 2019 (distr.; Western Palearctic). **Ety.:** stings or punctures (L); ref. to its persistent biting. **Informal name:** Stinging Canadian Pointy Mosquito.
syn. *implacabilis* Walker
1848: 7 (F; *Culex*). **Type loc.:** Saint Martin's Falls, Albany River, Hudson's Bay [Ontario, Canada] (NHMUK). **Refs.:** Vockeroth 1954a: 114 (syn.); Belkin 1968b: 5 (type info.).
syn. *fusculus* Zetterstedt
1850: 3459 (M, F; *Culex*). **Type loc.:** Areskutan, Mulfiellen Jämtland, and Sparresater, Westrogothia; Sweden; and Toien, Christiania [Oslo], Norway (LS). **Refs.:** Natvig 1948: 392 (syn.).
syn. *auroides* Felt
1905b: 449 (F, L; *Culicelsa*). **Type loc.:** Elizabethtown, New York, United States (NY).
syn. *wahlgreni* Theobald
1910c: 396 (M, F; *Culex*; for *fusculus* Zetterstedt Wahlgren 1904). **Type loc.:** Areskutan, Mulfiellen Jämtland, and Sparresater, Westrogothia, Sweden; and Toien, Christiania [Oslo], Norway (LS).
syn. *haplolineata* Schneider
1913: 37 (A; *Culicada nemorosa* form). **Type loc.:** Bonn, Germany (LU). **Refs.:** Belkin et al. 1971: 8 (info. on type); Harbach & Howard 2007: 44 (to ssp.).
syn. *alineata* Schneider
1913: 37 (F; *Culicada nemorosa* form). **Type loc.:** Bonn, Germany (LU).
syn. *meigenanus* Dyar
1921e: 72 (A). **Type loc.:** Europe (NE).

purpuraceus Brug
1932: 77 (M*, F). **Type loc.:** Upper Digoel River, Western New Guinea (Papua) (NHMUK). **Distr.:** Indonesia, Papua New Guinea. **Refs.:** Lee et al. 1984: 183 (F; key, tax., bion., distr., review). **Ety.:** n.s. [purple (L); ref. to abdominal terga II–VII as "metallic violet with creamy-yellow, basal, lateral spots"]. **Informal name:** Purple Indonesian Pointy Mosquito.

purpureifemur Marks
1959a: 132 (F). **Type loc.:** Forrestdale, 15 mi southeast of Perth, Western Australia, Australia (CSIR). **Distr.:** Australia. **Refs.:** Lee et al. 1984: 185 (F; key, tax., bion., distr., review). **Ety.:** purple, dark red femur (L); ref. to "entirely purplish-black anterior surface of the hindfemur." **Informal name:** Purple-legged Australian Pointy Mosquito.

purpuriventris Edwards
1926a: 113 (F). **Type loc.:** Eaglehawk, N Tasmania, Australia (LU). **Distr.:** Australia. **Refs.:** Dobrotworsky 1960b: 183 (M*, F*, P*, L*); Dobrotworsky 1965: 100 (M*, F*, L*); Lee et al. 1984: 186 (F; key, tax., bion., distr., review). **Ety.:** purple venter (L); abdominal venter color. **Informal name:** Purple-bellied Australian Pointy Mosquito.

ratcliffei Marks
1959a: 123 (M*, F, P*, L*). **Type loc.:** Gnangara, 12 mi north of Perth, Western Australia, Australia (CSIR). **Distr.:** Australia. **Refs.:** Lee et al. 1984: 188 (F; key, tax., bion., distr., review). **Ety.:** F.N. Ratcliffe. **Informal name:** Ratcliffe Australian Pointy Mosquito.

raymondi Del Ponte, Castro, & García
1951: 235 (F). **Type loc.:** San Pedro, Jujuy, Argentina (INM). **Distr.:** Argentina. **Refs.:** Arnell 1976: 52 (F*). **Ety.:** Raymond Corbett Shannon. **Informal name:** Raymond Argentinian Pointy Mosquito.

rempeli Vockeroth
1954a: 112 (M*, F*, L*). **Type loc.:** Great Whale River, Quebec, Canada (CNC). **Distr.:** Canada, Russia. **Refs.:** Dubitzky & Tupitzin 1970: 171 (L*); Gutsevich et al. 1974: 288 (M, F, L); Polyakova & Kukharchuk 1970: 130 (M*); Dubitzky 1977: 72 (M*, F); Gornostaeva 2000: 479 (distr.; Russia); Darsie & Ward 2005 (F*, L*; keys, distr.). **Ety.:** n.s. [prob. ref. to J.G. Rempel]. **Informal name:** Rempel Canadian Pointy Mosquito.

rhyacophilus da Costa Lima
1933: 404 (M*, F, L*). **Type loc.:** São João de Petrópolis, Vale do Canaã, Brazil (LU). **Distr.:** Argentina, Brazil. **Refs.:** Lane 1953 (M*, F, P*, L*); Belkin et al. 1971: 19 (type loc. info.); Arnell 1976: 58 (syn.); Sallum et al. 1988: (M*, F*, P*, L*; from syn. with *Aedes scapularis*. **Ety.:** n.s. [rushing stream lover (Gr); prob. ref. to larvae in rock holes next to streams]. **Informal name:** Torrent Brazilian Pointy Mosquito.

riparioides Su & Zhang
1987: 450 (M*, F, L*). **Type loc.:** Jingpeng, Keshiketeng Banner, Inner Mongolia Autonomous Region, China [People's Republic of China] (BMC). **Distr.:** People's Republic of China. **Ety.:** like (Gr); similar to *riparius*. **Informal name:** Jingpeng Chinese Pointy Mosquito.

riparius Dyar & Knab
1907d: 213 (M, F). **Type loc.:** Winnipeg, Manitoba, Canada (USNM). **Distr.:** Belarus, Bosnia and Herzegovina, Bulgaria, Canada, Crimean Peninsula, Croatia, Czech Republic, Denmark, Estonia, Finland, Germany, Hungary, Latvia, Lithuania, Moldova, Mongolia, Montenegro, Netherlands, Norway, Poland, Romania, Russia, Serbia, Slovenia, Sweden, Ukraine, United States (Alaska, continental). **Refs.:** Monchadskii 1951: 199 (L*); Yamaguti & LaCasse 1951d: 164 (M*, F*, L*); Vockeroth 1954b: 253 (F*); Carpenter & LaCasse 1955: 222 (M*, F*, L*; keys); Darsie 1955: 23 (P*); Craig 1956 (E*); Stone & Knight 1956a: 225 (type info., lectotype desig.); Kalpage & Brust 1968 (E*); Gutsevich et al. 1974: 213 (M*, F, L*); Dahl 1975: 83, 89 (tax., distr.); Utrio 1976: 130, 134 (L*); Minar 1976: 337 (distr.); Wood 1977: 73 (F*); Merdic 1992: 13 (distr.; Croatia); Dahl 1997 (E*); Spungis 2000: 10 (distr.; Latvia); Reinert 2002e: Fig. 21 (F*); Darsie & Ward 2005 (F*, L*; keys, distr.); Becker et al. 2010: 253 (M*, F*, L*; keys, tax., distr., bion.); Robert et al. 2019 (distr.; Western Palearctic). **Ety.:** of river banks (L); along banks of Assiniboine River. **Informal name:** Streamside Canadian Pointy Mosquito.
syn. *semicantans* Martini
1920b: 205 (M*, L*). **Type loc.:** Gr-Mueritz, Mecklenburg, Germany (NHMUK). **Refs.:** Martini 1931c: 330 (M*, F, L*).

sagax (Skuse)
1889: 1744 (F; *Culex*). **Type loc.:** Murrumbidgee, New South Wales (MM). **Distr.:** Australia. **Refs.:** Mackerras 1927: 292 (M*, F); Dobrotworsky 1965: 88 (M*, F*, L*); Lee et al. 1984: 189 (F; key, tax., bion., distr., review); Linley et al. 1991a (E*); Becker et al. 2010: 362 (F*, L*; key, tax., distr., bion.). **Ety.:** n.s. [keen senses, acute, clever (L)]. **Informal name:** Wise Australian Pointy Mosquito.
syn. *wilsoni* Taylor
1919: 833 (M*, F; *Culicada*). **Type loc.:** Kyabram, Echuca, Bamawm, Swan Hill & Mildura, Victoria, Australia (US). **Refs.:** Townsend 1990: 151 (syntypes, NHMUK).

sallumae González & Reyes
2017 (M*, F*, P*, L*; mol. tax., bion., distr.). In González et al. 2017: 35. **Type loc.:** Huasco wetland, Región de Copiapó, Huasco Province, Chile (MHNC). **Distr.:** Chile. **Refs.:** Danilov & Stupin 1986 (morph., tax., bion., distr.); **Ety.:** Maria Anice Mureb Sallum. **Informal name:** Sallum Chilean Pointy Mosquito.

sapiens Marks
1964a: 66 (M*, F*). **Type loc.:** Wilcannia and Bourke, western New South Wales, and Richmond, Australia (CSIR). **Distr.:** Australia. **Refs.:** Lee et al. 1984: 193 (F; key, tax., bion., distr., review). **Ety.:** wise (L); named to suggest affinities with *sagax*. **Informal name:** Prudent Australian Pointy Mosquito.

scapularis (Rondani)
1848: 109 (F; *Culex*). **Type loc.:** Vicinity of Belém, Pará, Brazil (LU). **Distr.:** Argentina, Bahamas, Belize, Bolivia, Brazil, Colombia, Costa Rica, Cuba, Dominican Republic, Ecuador, El Salvador, French Guiana, Guatemala, Guyana, Haiti, Honduras, Jamaica, Lesser Antilles (includes Grenada, Trinidad and Tobago), Mexico, Nicaragua, Panama, Paraguay, Peru, Puerto Rico, Suriname, United States (continental), Uruguay, Venezuela. **Refs.:** Séguy 1924: 71 (M gen.*); Coher 1949: 101 (F*); Breland 1951: 369 (L*); Yamaguti & LaCasse 1951d: 170 (M*, F*, L*); Leví-Castillo 1952a: 262 (M*; as *camposanus*); Lane 1953: 665 (M*, F, P*, L*); Forattini 1965a (E*); Belkin et al. 1971: 19 (type loc. info.; type loc. restricted to vicinity of Rio de Janeiro (Guanabara)); Arnell 1976: 57 (M*, F*, P*, L*; distr.; type loc. corrected to vicinity of Belém, Pará, Brazil); Forattini 1996 (E*); Belkin et al. 1970: 161 (M*, F, P*, L*; from syn. with *scapularis* Rondani; as *hemisurus*); Reinert 2000e (F*); Forattini 2002 (E*); Reinert 2002e: Fig. 26 (F*); Rossi & Martínez 2003: 472 (distr.; Uruguay); Darsie & Ward 2005 (F*, L*; keys, distr.); dos Santos-Mallet et al. 2010 (E*); Becker et al. 2010: 383 (F*, L*; key, tax., distr., bion.); Berti et al. 2015 (distr.; Venezuela). **Ety.:** n.s. [shoulder, back (L); poss. ref. to appearance of wearing a white scapular]. **Informal name:** Scapular Brazilian Pointy Mosquito. (Vol. 1, p. 440)
syn. *confirmatus* Lynch Arribálzaga
1891b: 146 (F; *Ochlerotatus*). **Type loc.:** Navarro, Buenos Aires, Argentina (BA). **Refs.:** Arnell 1976 (syn.).
syn. *hemisurus* Dyar & Knab
1906d: 199 (L). **Type loc.:** Jamaica (NE). **Refs.:** Stone & Knight 1956a: 218 (type info.); Arnell 1976 (syn.).
syn. *indolescens* Dyar & Knab
1907a: 11 (F). **Type loc.:** Cayamas, Cuba (USNM). **Refs.:** Stone & Knight 1956a: 219 (type info.); Belkin et al. 1970: 161 (syn.); Arnell 1976: 58 (syn.).
syn. *camposanus* Dyar
1918d: 128 (M, F). **Type loc.:** Guayaquil, Ecuador (USNM). **Refs.:** Leví-Castillo 1951b: 383 (M*; from syn.

with *euplocamus*); Stone & Knight 1956a: 216 (type info., lectotype desig.); Arnell 1976: 58 (syn.).

schizopinax Dyar
1929a: 1 (M, F, L). **Type loc.:** Story Creek railway crossing, Montana, United States (USNM). **Distr.:** Canada, United States (continental). **Refs.:** Yamaguti & LaCasse 1951d: 174 (M*, F*, L*); Carpenter & LaCasse 1955: 226 (M*, F*, L*; keys); Myers 1967: 804 (E*); Wolff & Nielsen 1976 (bion., distr.); Enfield 1977: 82 (distr.); Bohart & Washino 1978: 115 (M, F, L*; keys, tax., bion., distr.); Darsie & Ward 2005 (F*, L*; keys, distr.). **Ety.:** n.s. [split board, plank (Gr); prob. ref. to "anal segment with the plate . . . divided to base, with irregular edge"]. **Informal name:** Splintered American Pointy Mosquito.

schtakelbergi Shingarev
1928: 52 (M*, F). **Type loc.:** Vladimir, USSR [Russia] (VMS). **Distr.:** Russia. **Ety.:** n.s. [A.A. Shtakel'berg (Harbach 2018c)]. **Informal name:** Schtakelberg Russian Pointy Mosquito.

scutellalbum Boshell-Manrique
1939: 309 (M*, F, L*). **Type loc.:** Rio Upín Municipality of Restrepo, Meta, Colombia (USNM). **Distr.:** Colombia, Venezuela. **Refs.:** Lane 1953: 699 (M*, F, L*); Zavortink 1972: 3 (new combination). **Ety.:** n.s. [scutellum white (L); prob. ref. to all-white scutellum]. **Informal name:** White-shielded Colombian Pointy Mosquito.

sedaensis Lei
1989a: 6 (M, L). **Type loc.:** Seda and Songpan Counties, Sichuan Province, China [on plateau > 3,000 m] (SIP). **Distr.:** People's Republic of China. **Refs.:** Lei 1989b: 111 (M*, F, L*; as *Aedes*). **Ety.:** Seda Co., Sichuan Prov., China. **Informal name:** Seda Chinese Pointy Mosquito.

sergievi Danilov, Markovich, & Proskuryakova
1978: 44 (M*, F*, L*). **Type loc.:** Shagopar, Tuva, ASSR (South Siberia), USSR [Russia] (ZIL). **Distr.:** Russia. **Ety.:** n.s. [serrated (L); poss. ref. to a zigzag line where contrasting colors meet on the abdomen]. **Informal name:** Sergiev Russian Pointy Mosquito.

serratus (Theobald)
1901c: 45 (M*, F*; *Culex*). **Type loc.:** Rio de Janeiro, Brazil (NHMUK). **Distr.:** Antigua and Barbuda, Argentina, Bahamas, Belize, Bolivia, Brazil, Colombia, Costa Rica, Cuba, Dominican Republic, Ecuador, El Salvador, French Guiana, Guadeloupe, Guatemala, Guyana, Honduras, Mexico, Nicaragua, Panama, Paraguay, Peru, Puerto Rico, Suriname, Trinidad and Tobago, Venezuela. **Refs.:** Mitchell 1907b (E*); Theobald 1910c: 401 (E*); Antunes & Lane 1934: 38 (F; tax.); Komp 1949: 105 (tax.); Lane 1953: 658 (M*, F, L*); Xavier & Mattos 1967: 617 (P*); Belkin 1968b: 7 (lectotype desig.); Panday 1975a, b (distr.; Suriname); Belkin & Heinemann 1976c: 187 (distr.); Pecor et al. 2000 (tax.); Reinert 2002e: Fig. 27 (F*); Linton et al. 2013 (bion., mol. tax., distr.; Ecuador); Berti et al. 2015 (distr.; Venezuela). **Ety.:** n.s. [serrated (L); poss. ref. to abdominal markings]. **Informal name:** Serrated Brazilian Pointy Mosquito.

syn. *mathisi* Neveu-Lemaire
1902b: 13 (F*; *Culex*). **Type loc.:** Cunani, Amapá, Brazil (NE). **Refs.:** Belkin 1968b: 6 (tax.); Belkin et al. 1971: 20 (type loc. info., tax.).

syn. *meridionalis* Dyar & Knab
1906d: 195 (L*). **Type loc.:** Las Loras, near Puntarenas, Costa Rica (USNM). **Refs.:** Stone & Knight 1956a: 221 (type info., lectotype desig.); Belkin et al. 1970: 159 (? = serratus).

shannoni Vargas & Downs
1950: 161 (M*, F*, P*, L). **Type loc.:** Gabriel Mariaca, Morelos, Mexico (ISET). **Distr.:** Mexico. **Ety.:** Raymond Corbett Shannon. **Informal name:** Shannon Mexican Pointy Mosquito.

silvestris (Dobrotworsky)
1961b: 248 (new name for *Aedes waterhousei* Dobrotworsky 1960a, not Theobald; *Culex*). **Distr.:** Australia. **Refs.:** Dobrotworsky 1965: 109 (M*, F*, L*); Lee et al. 1984: 194 (F; key, tax., bion., distr., review). **Ety.:** n.s. [wild, wooded, untamed (L)]. **Informal name:** Wild Australian Pointy Mosquito.

syn. *waterhousei* Dobrotworsky
1960a: 57 (M*, F, P*, L*). **Type loc.:** Wattle Glen, Victoria (NMM).

simanini Gutsevich
1966: 457 (M*, F). **Type loc.:** Kokand, Fergana Valley, Uzbekistan (ZIL). **Distr.:** Uzbekistan. **Refs.:** Danilov 1981b: 36 (syn.); Ward 1987: 95 (from syn. with *niphadopsis*). **Ety.:** P.I. Simanin. **Informal name:** Simanin Uzbek Pointy Mosquito.

sinkiangensis Hsiao
1977: 451 (M*, F*, L*). **Type loc.:** Ake Su Region, Ha La Ku Leh (IZ). **Distr.:** China. **Ety.:** Xinjiang Uyghur Autonomous Region, China. **Informal name:** Xinjiang Chinese Pointy Mosquito.

sollicitans (Walker)
1856a: 427 (F; *Culex*). **Type loc.:** Charleston, South Carolina, United States (NHMUK). **Distr.:** Anguilla, Bahamas, Bermuda, Canada, Cayman Isl., Cuba, Dominican Republic, Haiti, Jamaica, Mexico, People's Republic of China, Puerto Rico, United States (continental), Virgin Isl. **Refs.:** Mitchell 1907b (E*); Howard et al. 1913 (E*); Howard et al. 1917: 658 (M*, F, L*, E*; bion.); Gerry 1932 (F*); Ross 1947: 69 (M*, F*, L*); Darsie 1951: 15 (P*); Yamaguti & LaCasse 1951d: 176 (M*, F*, L*); Decoursey & Webster 1952 (E*); Carpenter & LaCasse 1955: 227 (M*, F*, L*; keys); Craig 1956 (E*); Craig & Horsfall 1958 (E*); Craig & Horsfall 1960 (E*); Ross & Horsfall 1965 (M*, F*, L*, E*; keys); Belkin et al. 1966: 3 (restriction of type loc.); Dodge 1966: 356 (first instar L;

key); Mohrig 1967 (F*); Belkin 1968b: 7 (lectotype desig.); Belkin et al. 1970: 155 (M*, F, P*, L*); Horsfall et al. 1970: 1710 (E*); Olson & Meola 1976 (E*); Scotton & Axtell 1979 (E*); Reinert 2002e: Fig. 28 (F*); Darsie & Ward 2005 (F*, L*; keys, distr.); Becker et al. 2010: 391 (F*, L*; key, tax., distr., bion.); Harrison et al. 2016 (F*, L*; keys, distr.). **Ety.:** n.s. [disturbing, harassing (L)]. **Informal name:** Disturbing American Pointy Mosquito. Official ESA common name: saltmarsh mosquito. (Vol. 1, p. 442)

spencerii (Theobald)
1901c: 99 (F*; *Culex*). **Type loc.:** Stony Mountain, Manitoba, Canada (NHMUK). **Distr.:** Canada, Mexico, United States (continental). **Refs.:** Ross 1947: 79 (M*, F*, L*); Yamaguti & LaCasse 1951d: 180 (M*, F*, L*); Carpenter & LaCasse 1955: 229 (M*, F, L*; keys); Nielsen & Rees 1959: 46; Belkin 1968b: 7 (lectotype desig.); Kalpage & Brust 1968 (E*); Reinert 2002e: Fig. 29 (F*); Darsie & Ward 2005 (F*, L*; keys, distr.). **Ety.:** W.I. Spencer. **Informal name:** Spencer Canadian Pointy Mosquito.

ssp. ***idahoensis*** (Theobald)
1903a: 250 (A; as *Grabhamia spencerii* var.). **Type loc.:** Market Lake, Idaho, United States (NHMUK). **Distr.:** Canada, United States (continental). **Refs.:** Dyar 1917e (L); Gjullin 1937: 260 (M*); Yamaguti & LaCasse 1951d: 104 (M*, F*, L*); Carpenter & LaCasse 1955: 190 (M*, F*, L*; keys); Myers 1967 (E*); Darsie & Ward 2005 (F*, L*; keys, distr.). **Ety.:** Idaho, United States. **Informal name:** Idaho American Pointy Mosquito.

spilotus Marks
1963b: 31 (M*, F*, P*, L*). **Type loc.:** Cavendish, Victoria, Australia (NMM). **Distr.:** Australia. **Refs.:** Dobrotworsky 1965: 154 (M*, F*, L*); Lee et al. 1984: 196 (F; key, tax., bion., distr., review). **Ety.:** spotted, speckled (Gr); ref. to "extensive mottling of wings, femora, tibiae and tarsi." **Informal name:** Spotted Australian Pointy Mosquito.

squamiger (Coquillett)
1902c: 85 (F; *Culex*). **Type loc.:** Palo Alto, California, United States (USNM). **Distr.:** Mexico, United States (continental). **Refs.:** Howard et al. 1917 (E*); Gjullin 1937: 258 (F; key); Bohart 1948: 216 (P*, L*); Yamaguti & LaCasse 1951d: 184 (M*, F*, L*); Freeborn & Bohart 1951 (M*, F, L; keys, bion.); Yamaguiti 1952 (M*, F; keys); Bohart 1954: 364 (first instar L*); Carpenter & LaCasse 1955: 231 (M*, F*, L*; keys); Craig 1956 (E*); Stone & Knight 1956a: 225 (type info., lectotype desig.); Dodge 1966: 357 (first instar L; key); Myers 1967: 798 (E*; key; California, Nevada); Mohrig 1967 (F*); Bohart & Washino 1978 (M, F*, L*; keys, bion., tax.), 116 (M, F*, L*; keys, tax., bion., distr.); Eldridge et al. 1991 (mol. evol.); Darsie & Ward 2005 (F*, L*; keys, distr.); Darsie 2011a (P*). **Ety.:** n.s. [scale bearer (L)]. **Informal name:** Scaly American Pointy Mosquito. Official ESA common name: California saltmarsh mosquito. (Vol. 1, p. 444)

syn. *deniedmannii* Ludlow
1904b: 234 (F; *Grabhamia*). **Type loc.:** Benicia, California, United States (USNM). **Refs.:** Stone & Knight 1956a: 217 (type info., lectotype desig.).

sticticus (Meigen)
1838: 1 (F; *Culex*). **Type loc.:** Bavaria, Germany (MNHP). **Distr.:** Albania, Austria, Belarus, Belgium, Bosnia and Herzegovina, Bulgaria, Canada, Croatia, Czech Republic, Denmark, France, FYRO Macedonia, Georgia, Germany, Greece, Hungary, Italy, Japan, Kosovo, Lithuania, Luxembourg, Moldova, Mongolia, Montenegro, Netherlands, Norway, Poland, Romania, Russia, Serbia, Slovakia, Slovenia, Spain, Sweden, Switzerland, Ukraine, United Kingdom, United States (continental). **Refs.:** Mitchell 1907b (E*; as *pretans*); Séguy 1924: 135 (L*); Edwards 1932a: 144 (catalog; as *lateralis* Meigen); Peus 1933: 153 (tax.); Gjullin 1937 (F*); Ross 1947: 18 (M*, F*, L*); Natvig 1948: 347 (M*, F, L*); Gjullin, Yates, & Stage 1950: 262 (bion.); Rempel 1950: 225 (L*); Darsie 1951: 25 (P*); Yamaguti & LaCasse 1951d: 188 (M*, F*, L*); Asanuma et al. 1952 (M*); Rempel 1953: 484 (M*, F*); Bohart 1954: 364 (L*); Beckel 1954: 330 (tax.); Carpenter & LaCasse 1955: 233 (M*, F*, L*; keys); Horsfall & Craig 1956: 374 (E*); Craig 1956 (E*); Hara 1957: 59 (F*); Craig & Horsfall 1958 (E*); Price 1960: 557 (first instar L*); Ross & Horsfall 1965 (M*, F*, L*, E*; keys); Dodge 1966: 358 (first instar L; key); Myers 1967 (E*); Kalpage & Brust 1968 (E*); Sicart & Larrouy 1968 (F*); Horsfall et al. 1970: 1712 (E*); Rjazantzeva 1972 (F*); Gutsevich et al. 1974: 253 (M*, F, L*); Minar 1976: 338 (distr.); Enfield 1977: 84 (distr.); Bohart & Washino 1978: 117 (M, F*, L*; keys, tax., bion., distr.); Tanaka et al. 1979: 275 (M*, F*, L*); Hnton 1981 (E*); Mgeladze 1991: 330 (distr.; Georgia); Dahl 1997 (E*); Eritja et al. 2000: 11 (distr.; Spain); Kaiser et al. 2001 (distr.; Greece); Beck et al. 2003: 24 (distr.; Luxembourg); Darsie & Ward 2005 (F*, L*; keys, distr.); Becker et al. 2010: 255 (M*, F*, L*; keys, tax., distr., bion.); Harrison et al. 2016 (F*, L*; keys, distr.); Robert et al. 2019 (distr.; Western Palearctic). **Ety.:** being dappled, spotted (Gr, L); abdomen with white lateral spots. **Informal name:** Spotted German Pointy Mosquito. Official ESA common name: floodwater mosquito.

syn. *concinnus* Stephens
1846: 19 (A; *Culex*). **Type loc.:** Scotland (NHMUK).

syn. *sylvae* Theobald
1901c: 96 (M*, F; *Culex nigripes* var.). **Type loc.:** New Forest, Hampshire, England (NHMUK).

syn. *hirsuteron* Theobald
1901c: 98 (F*; *Culex*). **Type loc.:** Woodstock, Virginia, United States (NHMUK). **Refs.:** Belkin 1968b: 5 (lectotype desig.).

syn. *aestivalis* Dyar
1904b: 245 (L; *Culex*). **Type loc.:** British Columbia, Canada (USNM). **Refs.:** Stone & Knight 1956a: 213 (type info., lectotype desig.).

syn. *pretans* Grossbeck
1904b: 332 (M, F; *Culex*). **Type loc.:** Great Piece Meadow, New Jersey, United States (USNM). **Refs.:** Stone & Knight 1956a: 224 (type info., lectotype desig.).
syn. *aldrichi* Dyar & Knab
1908: 57 (F). **Type loc.:** Market Lake, Idaho, United States (USNM). **Refs.:** Stone & Knight 1956a: 213 (type info., lectotype desig.).
syn. *gonimus* Dyar & Knab
1918b: 165 (F). **Type loc.:** Kerrville, Texas, United States (USNM). **Refs.:** Natvig 1948: 347 (syn.); Stone & Knight 1956a: 218 (type info., lectotype desig.).
syn. *vinnipegensis* Dyar
1919b: 34 (F). **Type loc.:** Winnipeg Beach, Manitoba, Canada (USNM). **Refs.:** Stone & Knight 1956a: 226 (type info.).
syn. *dorso-vittatus* Villeneuve
1919: 57 (M, F; *Culex nemorosus* var.). **Type loc.:** Lyon, Rhone, and Creuse, France (ISN).
syn. *lesnei* Séguy
1920: 328 (M*; *Ochlerotatus*). **Type loc.:** Samois-sur-Seine, Seine-et-Marne, France (MNHP). **Refs.:** Stone 1963: 125 (Peus [*in litt.*] says this is related to *annulipes* (Meigen)).
syn. *paradiantaeus* Apfelbeck
1930: 290 (M*, L*). **Type loc.:** Sarajevo, Bosnia, Yugoslavia [Bosnia & Herzegovina] (NHMUK). **Refs.:** Townsend 1990: 118 (syntypes; change type dep. from LU to NHMUK).

stigmaticus Edwards
1922d: 78 (F). **Type loc.:** Asuncion, Paraguay (NHMUK). **Distr.:** Argentina, Bolivia, Brazil, Mexico, Paraguay. **Refs.:** Lane 1953: 647 (F); Martínez et al. 1959: 112 (F*); Belkin 1968b: 7 (lectotype desig.). **Ety.:** n.s. [branded, marked (L)]. **Informal name:** Spotted Paraguayan Pointy Mosquito.

stimulans (Walker)
1848: 4 (F; *Culex*). **Type loc.:** Nova Scotia, Canada (NHMUK). **Distr.:** Canada, Nigeria, Uganda, United States (continental). **Refs.:** Mitchell 1907b (E*; as *subcantans*); Ross 1947: 72 (M*, F, L*); Rempel 1950: 219 (L*); Darsie 1951: 19 (P*); Yamaguti & LaCasse 1951d: 192 (M*, F*, L*); Rempel 1953: 462 (M*, F*); Carpenter & LaCasse 1955: 236 (M*, F*, L*; keys); Horsfall & Craig 1956: 370 (E*); Craig 1956 (E*); Craig & Horsfall 1958 (E*); Price 1960: 557 (first instar L*); Ross & Horsfall 1965 (M*, F*, L*, E*; keys); Dodge 1966: 358 (first instar L; key); Kalpage & Brust 1968 (E*); Horsfall et al. 1970: 1710 (E*); Pickavance et al. 1970 (distr.); McDaniel & Webb 1974 (tax.); Enfield 1977: 84 (distr.); Wood 1977: 73 (F*); Darsie & Ward 2005 (F*, L*; keys, distr.); Harrison et al. 2016 (F*, L*; keys, distr., tax.). **Ety.:** n.s. [goad, torment (L)]. **Informal name:** Goading Canadian Pointy Mosquito.
syn. *subcantans* Felt
1905b: 448, 474 (M*, F*, L; *Culicada*). **Type loc.:** None designated (NY).
syn. *classicus* Dyar
1920d: 113 (M, F; *stimulans* ssp.). **Type loc.:** Hudson County, New Jersey, United States (USNM). **Refs.:** Stone & Knight 1956a: 216 (type info., lectotype desig.).
syn. *mississippii* Dyar
1920d: 113 (M, F; *stimulans* ssp.). **Type loc.:** Electric Mills, Mississippi, United States (USNM). **Refs.:** Stone & Knight 1956a: 221 (type info., lectotype desig.).

stramineus Dubitzky
1970a: 408 (M*, F*, L*; as *Aedes*). **Type loc.:** Upper Ili River, near Borokhudzir, Kazakh SSR, USSR (IZAA). **Distr.:** Kazakhstan, Mongolia, Russia. **Refs.:** Gutsevich et al. 1974: 286 (M, F, L); Minar 1976: 355 (distr.); Danilov 1979: 29 (syn.); Ward 1987: 95 (from syn. with *albineus*); Gornostaeva 2000: 479 (distr.; Russia); Robert et al. 2019 (distr.; Western Palearctic). **Ety.:** n.s. [straw colored (L)]. **Informal name:** Straw-colored Kazakhstani Pointy Mosquito.

stricklandi (Edwards)
1912b: 523 (new name, for *australis* Strickland, not Erichson 1842; *Ochlerotatus*). **Distr.:** Australia. **Refs.:** Marks 1949: 38 (F only; M = *turneri*; see Marks 1963b: 42); Marks 1963b: 38 (M*, F*, P*, L*); Dobrotworsky 1965: 126 (M*, F*, L*); Lee et al. 1984: 197 (F; key, tax., bion., distr., review). **Ety.:** Cyril Strickland. **Informal name:** Strickland Australian Pointy Mosquito.
syn. *australis* Strickland
1911a: 133 (F*; *Grabhamia*). **Type loc.:** Serpentine, Darling Range, near Perth, Western Australia, Australia (NHMUK).
syn. *flindersi* Taylor
1914c: 686 (F; *Grabhamia*). **Type loc.:** Bass Straits, Flinders Island [near Tasmania], Australia (US).

subalbirostris Klein & Marks
1960: 115 (F). **Type loc.:** Invercargill, New Zealand (NHMUK). **Distr.:** New Zealand. **Refs.:** Marks & Nye 1963: 56 (M*, F*, P*, L*); Belkin 1968a: 100 (M*, F, P*, L*); Lee et al. 1984: 199 (F; key, tax., bion., distr., review); Derraik 2004 (distr.; New Zealand). **Ety.:** nearly (L); near *albirostris*. **Informal name:** Whitish-beaked New Zealander Pointy Mosquito [*subalbirostris* = "sub-white nosed" = Whitish-beaked].

surcoufi (Theobald)
1912 (M, F; *Culicada*). In Surcouf 1912: 59. **Type loc.:** Chaville, Seine-et-Oise and Nemours, Seine-et-Marne, France (MNHP, GDZ). **Distr.:** France, Hungary, Italy, Spain. **Refs.:** Arnaud et al. 1976 (E*; bion.); Eritja et al. 2000: 12 (distr.; Spain); Becker et al. 2010: 231 (M*, F*, L*; keys, tax., distr., bion.); Harbach 2018c: 46, 254 (*lapsus*, as syn. of *excrucians* (Walker)); Robert et al. 2019 (distr.;

Western Palearctic). **Ety.:** Jacques M.R. Surcouf. **Informal name:** Surcouf French Pointy Mosquito.

synchytus Arnell
1976: 71 (M*, F*). **Type loc.:** Cataratas del Iquazú, Misiones, Argentina (USNM). **Distr.:** Argentina. **Ety.:** n.s. **Informal name:** Shod Argentinian Pointy Mosquito [*synchytus* = "shoe-like" = Shod].

taeniorhynchus (Wiedemann)
1821: 43 (A; *Culex*). **Type loc.:** Mexico (NMW). **Distr.:** Argentina, Bahamas, Belize, Bermuda, Brazil, Cayman Isl., Colombia, Costa Rica, Cuba, Dominican Republic, Ecuador, El Salvador, French Guiana, Guatemala, Guyana, Haiti, Honduras, Jamaica, Lesser Antilles (Anguilla, Antigua and Barbuda, Barbados, Curaçao, Dominica, Grenada, Guadeloupe, Martinique, Montserrat, Saba, Saint Kitts and Nevis, Saint Lucia, Saint Vincent and the Grenadines, Trinidad and Tobago, Virgin Isl.), Mexico, Nicaragua, Panama, Peru, Puerto Rico, Suriname, United States (continental), Venezuela. **Refs.:** Mitchell 1907b (E*); Howard et al. 1913 (E*); Howard et al. 1917: 667 (M*, F, L, E*; bion.); Gerry 1932: 39 (F*); Gjullin 1937: 256 (F*); Darsie 1951: 16 (P*); Yamaguti & LaCasse 1951d: 198 (M*, F*, L*); Lane 1953: 678 (M*, F, L*); Nielsen & Nielsen 1953: 141 (bion.); Bohart 1954: 365 (L*); Carpenter & LaCasse 1955: 238 (M*, F*, L*); Craig 1956 (E*); Forattini 1958: 176 (P*); Craig & Horsfall 1960 (E*); Dodge 1966: 358 (first instar L*; key); Myers 1967 (E*); Belkin et al. 1970: 152 (M*, F, P*, L*); Bohart & Washino 1978: 118 (M, F*, L*; keys, tax., bion., distr.); Scotton & Axtell 1979 (E*); Darsie & Ward 2005 (F*, L*; keys, distr.); Becker et al. 2010: 393 (F*, L*; key, tax., distr., bion.); Marcondes et al. 2012 (E*); Harrison et al. 2016 (F*, L*; keys, distr.; note 4). **Ety.:** banded beak (Gr); white bands on a black proboscis. **Informal name:** Band-beaked Mexican Pointy Mosquito. (Vol. 1, p. 446)

syn. *damnosus* Say
1823: 11 (A; *Culex*). **Type loc.:** Pennsylvania, United States (NE).

syn. *niger* Giles
1904b: 384 (F; *Taeniorhynchus*). **Type loc.:** Antigua Island [Lesser Antilles] (NHMUK). **Refs.:** Belkin 1968b: 6 (lectotype desig.).

syn. *portoricensis* Ludlow
1905c: 386 (F; *Culex*). **Type loc.:** San Juan, Puerto Rico (USNM). **Refs.:** Stone & Knight 1956a: 224 (type info., lectotype desig.).

syn. *epinolus* Dyar & Knab
1914: 61 (F). **Type loc.:** Ventanillas, Peru (USNM). **Refs.:** Stone & Knight 1956a: 217 (type info., lectotype desig.).

syn. *pix* Martini
1935: 55 (F). **Type loc.:** Belize, British Honduras (USNM). **Refs.:** Lane 1953: 652 (F); Mattingly 1955c: 29 (type info.); Stone & Knight 1956a: 224 (type info.); Belkin 1977a (syn.).

tahoensis Dyar
1916b: 82 (M, F, L). **Type loc.:** Fallen Leaf, Lake Tahoe, California, United States (USNM). **Distr.:** United States (continental). **Refs.:** Stone & Knight 1956a: 225 (type info., lectotype desig.); Brust & Munstermann 1992: 2 (from syn. with *communis*); Darsie & Ward 2005 (F*, L*; keys, distr.). **Ety.:** Lake Tahoe, California, United States. **Informal name:** Tahoe American Pointy Mosquito.

thelcter Dyar
1918d: 129 (F). **Type loc.:** Brownsville, Texas, United States (USNM). **Distr.:** United States (continental). **Refs.:** Yamaguti & LaCasse 1951d: 202 (M*, F*, L*); Carpenter & LaCasse 1955: 240 (M*, F*, L*); Arnell 1976: 76 (M*, F*, P*, L; distr.); Darsie & Ward 2005 (F*, L*; keys, distr.). **Ety.:** n.s. **Informal name:** Butterfly American Pointy Mosquito [butterfly = closest meaning found].

syn. *keyensis* Buren
1947: 228 (F). **Type loc.:** Key West, Florida, United States (A). **Refs.:** Thurman et al. 1949: 171 (syn.).

theobaldi (Taylor)
1914a: 751 (A; *Grabhamia*). **Type loc.:** South Queensland, Australia (NHMUK). **Distr.:** Australia. **Refs.:** Marks 1949: 11 (M*, F, P*, L*); Dobrotworsky 1965: 83 (M*, F*, L*); Lee et al. 1984: 201 (F; key, tax., bion., distr., review); Linley et al. 1991a (E*); Becker et al. 2010: 363 (F*, L*; key, tax., distr., bion.). **Ety.:** Frederic Vincent Theobald. **Informal name:** Theobald Australian Pointy Mosquito.

thibaulti Dyar & Knab
1910: 174 (M, F). **Type loc.:** Scott, Arkansas, United States (USNM). **Distr.:** Canada, Russia, United States (continental). **Refs.:** Jenkins & Carpenter 1946: 36 (bion.); Ross 1947: 73 (M*, F*, L*); Yamaguti & LaCasse 1951d: 206 (M*, F*, L*); Carpenter & LaCasse 1955: 242 (M*, F*, L*; keys); Craig 1956 (E*); Stone & Knight 1956a: 225 (type info., lectotype desig.); Craig & Horsfall 1958 (E*); Craig & Horsfall 1960 (E*); Ross & Horsfall 1965 (M*, F*, L*, E*; keys); Dodge 1966: 358 (first instar L*; key); Gutsevich & Goritskaya 1970: 72 (M*; distr.); Lacey & Lake 1972: 89 (P*); Darsie & Ward 2005 (F*, L*; keys, distr.); Harrison et al. 2016 (F*, L*; keys, distr.). **Ety.:** J.K. Thibault Jr. **Informal name:** Thibault American Pointy Mosquito.

tormentor Dyar & Knab
1906d: 191 (L*). **Type loc.:** Baton Rouge, Louisiana, United States (NE). **Distr.:** Guatemala, Honduras, Mexico, United States (continental). **Refs.:** Howard et al. 1917 (E); Yamaguti & LaCasse 1951d: 210 (M*, F*, L*); Carpenter & LaCasse 1955: 243 (M*, F*, L*; keys, bion.); Stone & Knight 1956a: 225 (type info.); King et al. 1960: 81 (M, F*, L; keys); Dodge 1966: 358 (first instar L*; key); Mohrig 1967 (F*); Scanlon & Yates 1970

(tax.); Reinert 1971: 176 (P*); Roberts & Scanlon 1979 (F*; tax.); Darsie & Ward 2005 (F*, L*; keys, distr.); Harrison et al. 2016 (F*, L*; keys, distr., tax.). **Ety.:** n.s. [tormenter (L)]. **Informal name:** Tormenting American Pointy Mosquito.

tortilis (Theobald)
1903g: 281 (F; *Culex*). **Type loc.:** Kingston [Surrey], Jamaica (NHMUK). **Distr.:** Argentina, Bahamas, Brazil, Cuba, Dominican Republic, Guatemala, Haiti, Jamaica, Lesser Antilles (Antigua and Barbuda, Cayman Isl., Dominica, Guadeloupe, Montserrat, Saint Kitts and Nevis, Saint Lucia, Saint Vincent and the Grenadines, Virgin Isl.), Mexico, Puerto Rico, Turks and Caicos Isl., United States (continental). **Refs.:** Gerry 1932: 38 (F*); Yamaguti & LaCasse 1951d: 214 (M*, F*, L*); Lane 1953: 653 (M*, F, L*); Carpenter & LaCasse 1955: 245 (M*, F*, L*; keys); Mohrig 1967 (F*); Belkin 1968b: 8 (lectotype desig.); Belkin et al. 1970: 165 (M*, F, P*, L*); Arnell 1976: 39 (M*, F*, P*, L*; distr.); Darsie & Ward 2005 (F*, L*; keys, distr.). **Ety.:** n.s. [twisted, coiled (L); poss. ref. to habit of "carrying their hind legs twisted right forward over their head when settled"]. **Informal name:** Twisted Jamaican Pointy Mosquito.
syn. *bracteatus* Coquillett
1906a: 184 (F; *Culex*). **Type loc.:** Havana, Cuba (USNM). **Refs.:** Stone & Knight 1956a: 215 (type info.).
syn. *habanicus* Dyar & Knab
1906d: 198 (L*). **Type loc.:** Havana, Cuba (USNM). **Refs.:** Stone & Knight 1956a: 218 (type info., lectotype desig.).
syn. *plutocraticus* Dyar & Knab
1907a: 11 (F). **Type loc.:** Nassau [New Providence Island], Bahama Islands (USNM). **Refs.:** Stone & Knight 1956a: 224 (type info.).
syn. *balteatus* Dyar & Knab
1907a: 9 (F). **Type loc.:** Santo Domingo [Dominican Republic] (USNM). **Refs.:** Stone & Knight 1956a: 215 (type info.).
syn. *virginensis* Dyar
1922c: 56 (F; *tortilis* ssp.). **Type loc.:** Saint Thomas Island [Lesser Antilles] (USNM).

trivittatus (Coquillett)
1902b: 193 (F; *Culex*). **Type loc.:** Chester, New Jersey, United States (USNM). **Distr.:** Canada, Honduras, Mexico, Panama, United States (continental), Venezuela. **Refs.:** Ross 1947: 75 (M*, F*, L*); Abdel-Malek 1948b: 51, 951 (bion.); Abdel-Malek 1949: 19 (P*, L*, E*); Breland 1951: 369 (L*); Darsie 1951: 21 (P*); Yamaguti & LaCasse 1951d: 222 (M*, F*, L*); Carpenter & LaCasse 1955: 249 (M*, F*, L*; keys); Horsfall & Craig 1956: 372 (E*); Craig 1956 (E*); Stone & Knight 1956a: 226 (type info.); Price 1960: 558 (first instar L*); Ross & Horsfall 1965 (M*, F*, L*, E*; keys); Dodge 1966: 358 (first instar L*; key); Horsfall et al. 1970: 1711 (E*); Arnell 1976: 35 (M*, F*, P*, L*); Darsie & Ward 2005 (F*, L*; keys, distr.); Harrison et al. 2016 (F*, L*; keys, distr.). **Ety.:** carrying three bands, ribbons, vittae (L); ref. to "near *triseriatus*, but with three vittae of blackish scales on the mesonotum." **Informal name:** Three-banded American Pointy Mosquito. (Vol. 1, p. 448)
syn. *inconspicuus* Grossbeck
1904b: 332 (M, F; *Culex*). **Type loc.:** Garret Mountains, Paterson, New Jersey, United States (USNM). **Refs.:** Stone & Knight 1956a: 219 (type info., lectotype desig.).

turneri Marks
1963b: 42 (M*, F*). **Type loc.:** 4 mi west of Piawaning (80 mi north-northwest of Perth), Western Australia (CSIR). **Distr.:** Australia. **Refs.:** Lee et al. 1984: 206 (F; key, tax., bion., distr., review). **Ety.:** A.J. Turner. **Informal name:** Turner Australian Pointy Mosquito.

upatensis Anduze & Hecht
1943: 185 (M*). **Type loc.:** Region de Upata, Distrito Piar, Bolivar, Venezuela (FH). **Distr.:** Colombia, Venezuela. **Refs.:** Lane 1953: 679 (M); Zavortink 1972: 3 (new comb.). **Ety.:** Upata Region, Piar Dist., Bolivar State, Venezuela. **Informal name:** Upata Venezuelan Pointy Mosquito.

ventrovittis Dyar
1916b: 84 (M). **Type loc.:** Fallen Leaf, Lake Tahoe, California, United States (USNM). **Distr.:** United States (Alaska, continental). **Refs.:** Gjullin 1937: 260 (F*); Yamaguti & LaCasse 1951d: 226 (M*, F*, L*); Bohart 1954: 365 (first instar L*); Carpenter & LaCasse 1955: 251 (M*, F*, L*; keys); Dodge 1966: 359 (first instar L; key); Myers 1967: 806 (E*); Nielsen 1970: 530 (syn.); Bickley 1976 (distr.); Bohart & Washino 1978: 119 (M, F*, L*; keys, tax., bion., distr.); Darsie & Ward 2005 (F*, L*; keys, distr.). **Ety.:** ventral band (L); ref. to abdominal "venter white-scaled, with a median black band." **Informal name:** Band-bellied American Pointy Mosquito.
syn. *fisheri* Dyar
1917a: 19 (F). **Type loc.:** Tahoe Tavern, Lake Tahoe, California, United States (USNM). **Refs.:** Stone & Knight 1956a: 217 (type info., lectotype desig.).
syn. *cacothius* Dyar
1923a: 44 (F). **Type loc.:** Shoshone Point, Yellowstone National Park, Wyoming, United States (USNM). **Refs.:** Carpenter & LaCasse 1955: 159 (F); Stone & Knight 1956a: 215 (type info., lectotype desig.).

vigilax (Skuse)
1889: 1731 (F; *Culex*). **Type loc.:** Gosford, Kiama, and National Park, New South Wales, and Brisbane, Queensland, Australia (MM). **Distr.:** Australia, Fiji, Indonesia (Java, Flores, Kalimantan, Sulawesi, Sumatra [includes Ketulauan Riau Archipelago, Timor]), Malaysia, New Caledonia, New Hebrides, Papua New Guinea, Philippines, Seychelles, Solomon Isl., Taiwan, Thailand, Timor, Tonga, Vanuatu, Vietnam. **Refs.:** Bancroft 1908: 41 (M, F; "Black Bush-Mosquito"); Edwards 1924b: 375

(F; as *albirostris*); Mackerras 1927: 296 (M*); Borel 1930: 280 (M*, F, L*); Hamlyn-Harris 1933: 229 (F*; bion.); Taylor 1943b: 74 (M, F, L*); Lee 1944a: 53 (L*); Penn 1949b: 42 (P*); Klein & Marks 1960: 112 (F*; as *albirostris*; syn.), 115 (syn.); Mattingly 1961: 39 (M*, F*, P*, L*); Belkin 1962 (M*, F, P*, L*; tax., bion., distr.); Dobrotworsky 1965: 79 (M*, F*, L*); Mohrig 1967 (F*); Marks 1969b: 81 (tax.; as *albirostris*); Basio 1971b: 25 (M*; bion.); Tanaka et al. 1975c: 220 (distr.); Tanaka et al. 1979: 262 (L*); Lee et al. 1984: 207 (F; key, tax., bion., distr., review); Kay & Jorgensen 1986 (E*); Linley et al. 1992a (E*); Whelan & Hapgood 2000 (bion, distr.; East Timor); Rattanarithikul et al. 2010 (F*, L*; keys, bion., distr.; Thailand); Becker et al. 2010: 364 (F*, L*; key, tax., distr., bion.); Puslednik et al. 2012 (distr., bion., mol. tax.). **Ety.:** n.s. [vigilant, watchful (L)]. **Informal name:** Watchful Australian Pointy Mosquito. (Vol. 1, p. 450)

syn. *marinus* Theobald
1901a: 396 (M*, F*; *Culex*). **Type loc.:** Queensland, Australia (NHMUK).

ssp. ***ludlowae*** (Blanchard)
1905: 630 (as *Culex ludlowii*; new name for *annulifera* Ludlow, not Blanchard 1852). **Distr.:** Philippines. **Refs.:** Knight & Hull 1951b: 226 (M*, F, L*; to ssp.; type info., lectotype desig.); Laffoon & Knight 1971 (F*); Baisas 1974: 25 (M*, F, P*, L*; tax., bion., distr.; Philippines); Harbach & Knight 1980 (F*). **Ety.:** Clara Southmayd Ludlow. **Informal name:** Ludlow Philippine Pointy Mosquito.

ssp. syn. *annulifera* Ludlow
1903: 141 (F; *Culex annulifera*). **Type loc.:** Mangarin, Mindoro, Philippines (USNM). **Refs.:** Stone & Knight 1956a: 214 (type info.).

syn. *pseudovigilax* Theobald
1907: 382 (M*, F*; *Culicelsa*). **Type loc.:** Burpengary, Queensland, Australia (NHMUK).

syn. *uniformis* Strickland
1911a: 131 (F; *Culicelsa*). **Type loc.:** Kings Park, Perth, Western Australia, Australia (NHMUK). **Refs.:** Belkin 1962: 393 (lectotype).

ssp. ***vansomerenae*** Mattingly
1955: 78 (M*, F*, P*, L*). In Mattingly & Brown 1955. **Type loc.:** Anse Lascar, Silhouette Island, Seychelles (NHMUK). **Distr.:** Seychelles. **Refs.:** Le Goff et al. 2012 (F; key, bion., distr.; Seychelles). **Ety.:** Ellinor Catharine Cunningham-van Someren. **Informal name:** Someren Seychellois Pointy Mosquito.

vittiger (Skuse)
1889: 1728 (F; *Culex*). **Type loc.:** Port Denison and Wide Bay, Queensland and Gosford, New South Wales, Australia (MM). **Distr.:** Australia, Papua New Guinea. **Refs.:** Bancroft 1908: 33 (F); Mackerras 1927: 289 (M*); Lee 1944a: 55 (L*); Dobrotworsky 1965: 91 (M*, F*, L*); Lee et al. 1984: 234 (F; key, tax., bion., distr., review); Linley et al. 1992a (E*). **Ety.:** n.s. [ribbon, band bearer (L); prob. ref. to "thorax black, with five vittae of whitish scales"]. **Informal name:** Ribboned Australian Pointy Mosquito.

washinoi Lanzaro & Eldridge
1992: 93 (M, F, P, L). **Type loc.:** Union City, Alameda County, California, United States (USNM). **Distr.:** United States (continental). **Refs.:** Darsie & Ward 2005 (F*, L*; keys, distr.). **Ety.:** Robert K. Washino. **Informal name:** Washino American Pointy Mosquito.

Subgenus *Paraedes* Edwards

Edwards 1934 (subg. of *Aedes*). In Barraud 1934: 466. Type species: *Paraedes barraudi* Edwards.

Name derivation: n.s. [near *Aedes* (Gr)].

Important references: Mattingly 1958a: 19 (keys; to subg. of genus *Aedes*); Reinert 1981 (revision); Reinert 2000e (F gen.*); Reinert, Harbach, &Kitching 2004: 289 (phyl., class.; to genus); Reinert, Harbach, & Kitching 2009 (phyl., class.); Rattanarithikul et al. 2010 (F*, L*; keys, bion.; Thailand); Wilkerson et al. 2015 (phyl., class.; to subg. of genus *Aedes*); Soghigian et al. 2017 (mol. phyl.).

barraudi (Edwards)
1934 (M*; *Paraedes*). In Barraud 1934: 447. **Type loc.:** Virajpet, Coorg, India (NHMUK). **Distr.:** India. **Refs.:** Reinert 1981: 10 (M*; keys, distr., bion. tax.); Tewari & Hiriyan 1994b (F*); Reinert 2000e: Fig. 29 (F*). **Ety.:** Philip James Barraud. **Informal name:** Barraud Indian Pointy Mosquito.

bonneae Mattingly
1958a: 32 (M*, F*, P*, L*). **Type loc.:** Rantau Panjang, Selangor, Malaya (NHMUK). **Distr.:** Malaysia. **Refs.:** Reinert 1981: 13 (M*, F*, P*, L*; keys, distr., bion., tax.). **Ety.:** Johanna Bonne-Wepster. **Informal name:** Bonne Malaysian Pointy Mosquito.

chrysoscuta (Theobald)
1910c: 228 (F*; *Pseudohowardina*). **Type loc.:** Peradeniya [Central], Ceylon [Sri Lanka] (NHMUK). **Distr.:** Sri Lanka. **Refs.:** Reinert 1981: 17 (M*, F*, P*, L*; keys, distr., bion. tax.; from syn. with *ostentatio*). **Ety.:** gold shield (Gr, L); ref. to gold markings on scutum. **Informal name:** Gold-plated Sri Lankan Pointy Mosquito.

collessi Mattingly
1958a: 28 (M*, F*, P*, L*). **Type loc.:** Rantau Panjang, Selangor, Malaya [Malaysia] (NHMUK). **Distr.:** Malaysia, Singapore, Taiwan. **Refs.:** Reinert 1981: 22 (M*, F*, P*, L*; keys, distr., bion. tax.); Jeffery et al. 2010 (distr.; Singapore). **Ety.:** Donald Henry Colless. **Informal name:** Colless Malaysian Pointy Mosquito.

jambulingami Natarajan
2019 (M*, F). **Type loc.:** Dibrugarh District, Soraipung, India. (Vector Control Museum, Puducherry). **Distr.:** India. **Ety.:** Purushothaman Jambulingam. **Informal name:** Jambulingam Indian Pointy Mosquito.

menoni Mattingly
1958b: 77 (M*, F*). **Type loc.:** Travancore, India (NHMUK). **Distr.:** India. **Refs.:** Reinert 1981: 28 (M*, F*; keys, distr., bion., tax.). **Ety.:** M.A.U. Menon. **Informal name:** Menon Indian Pointy Mosquito.

ostentatio (Leicester)
1908: 193 (F; *Aioretomyia*). **Type loc.:** [Pahang Road, Kuala Lumpur, Selangor], Malaya [Malaysia] (NHMUK). **Distr.:** Cambodia, India, Indonesia, Laos, Malaysia, Philippines, Sri Lanka, Thailand, Vietnam. **Refs.:** Knight & Hull 1953: 463 (F; tax.); Mattingly 1958a: 25 (F); Mattingly 1958c: 106 (lectotype desig.); Basio 1971b: 26 (M*; bion.); Harrison et al. 1974: 148 (tax.); Reinert 1981: 31 (M*, F*, P*, L*; keys, distr., bion., tax.); Lee et al. 1984: 240 (tax., review); Reinert 2000e (F*; Fig. 30); Rattanarithikul et al. 2010 (F*, L*; keys, bion., distr.; Thailand). **Ety.:** n.s. [display, showing off (L); prob. ref. to conspicuous bright markings]. **Informal name:** Ostentatious Malaysian Pointy Mosquito.

pagei (Ludlow)
1911b: 128 (F; *Danielsia*). **Type loc.:** Fort Pikit [Malabang, Lanao], Mindanao, Philippines (USNM). **Distr.:** Philippines. **Refs.:** Knight & Hull 1953: 463 (type info., lectotype desig.); Stone & Knight 1956a: 223 (type info.); Stone & Knight 1957c: 198 (type info., lectotype desig.); Mattingly 1958c: 106; Reinert 1981: 38 (F*; key, distr., bion., tax.). **Ety.:** Henry Page. **Informal name:** Page Philippine Pointy Mosquito.

thailandensis Reinert
1976b: 319 (M*, F*, P*, L*). **Type loc.:** Ban Bu Phram, Prachin Buri, Thailand (USNM). **Distr.:** Thailand, Vietnam. **Refs.:** Reinert 1981: 40 (M*, F*, P*, L*; keys, distr., bion., tax.); Rattanarithikul et al. 2010 (F*, L*; keys, bion., distr.; Thailand). **Ety.:** Thailand. **Informal name:** Sukhothai Thai Pointy Mosquito [the Sukhothai Kingdom (1238–1583) was based around the city of Sukhothai in north central Thailand].

Subgenus *Patmarksia* Reinert, Harbach, & Kitching

2006: 81 (as genus). Type species: *Leucomyia australiensis* var. *papuensis* Taylor.

Name derivation: Elisabeth Nesta (Pat) Marks.

Important references: King & Hoogstraal 1946b (Papuensis Group/Subgroup of *Finlaya*); Lee et al. 1982: 96 (tax., bion., key; as Papuensis Subgroup; Australasia); Reinert, Harbach, & Kitching 2008 (phyl., class.); Reinert 2008g (F gen.*); Reinert, Harbach, & Kitching 2009 (phyl., class.); Wilkerson et al. 2015 (phyl., class.; to subg. of genus *Aedes*).

anggiensis Bonne-Wepster
1937: 97 (M*, F). **Type loc.:** Anggi Lakes, 2,000 m, Western New Guinea (Papua) [Indonesia] (ITH). **Distr.:** Indonesia (Papua), Papua New Guinea. **Refs.:** Lee et al. 1982: 112 (F; key, tax., bion., distr., review). **Ety.:** Anggi Lakes, Indonesia (Western New Guinea). **Informal name:** Anggi Indonesian Pointy Mosquito.

argenteitarsis Brug
1932: 76 (F). **Type loc.:** Upper Digoel River, Western New Guinea (Papua) (NHMUK). **Distr.:** Indonesia (West Papua), Papua New Guinea. **Refs.:** King & Hoogstraal 1946d: 154 (M*, F, L); Lee et al. 1982: 114 (F; key, tax., bion., distr., review); Peyton et al. 1999: 529 (distr.). **Ety.:** silvery foot (Gr); tarsi. **Informal name:** Silver-footed Indonesian Pointy Mosquito.

argyronotum Belkin
1962: 354 (M*, F, P*, L*). **Type loc.:** Matanikau Valley, Guadalcanal, Solomon Islands (USNM). **Distr.:** Solomon Isl. **Refs.:** Lee et al. 1982: 116 (F; key, tax., bion., distr., review). **Ety.:** silvery notum (Gr); scutum. **Informal name:** Silver-marked Solomon Islander Pointy Mosquito.

buxtoni Belkin
1962: 356 (F). **Type loc.:** Bougainville?, Solomon Islands (US). **Distr.:** Solomon Isl., Vanuatu. **Refs.:** Lee et al. 1982: 132 (F; key, tax., bion., distr., review). **Ety.:** Patrick Alfred Buxton. **Informal name:** Buxton Solomon Islander Pointy Mosquito.

clintoni Taylor
1946: 211 (M*). **Type loc.:** Lae, Northeast New Guinea (US). **Distr.:** Papua New Guinea. **Refs.:** Knight 1948a: 2 (? = *hollandius*); Lee et al. 1982: 135 (tax., review). **Ety.:** K.J. Clinton. **Informal name:** Clinton Papuan Pointy Mosquito.

derooki Brug
1932: 75 (F). **Type loc.:** Ternate, Moluccas (NHMUK). **Distr.:** Indonesia. **Refs.:** Lee et al. 1982: 136 (F; key, tax., bion., distr., review). **Ety.:** Hendrik DeRook. **Informal name:** DeRook Indonesian Pointy Mosquito.

dobodurus King & Hoogstraal
1946b: 37 (M, F). **Type loc.:** Dobodura, Papua, New Guinea (USNM). **Distr.:** Indonesia, Papua New Guinea. **Refs.:** King & Hoogstraal 1946d: 138 (M*, F*, L*); Lee et al. 1982: 137 (F; key, tax., bion., distr., review). **Ety.:** Dobodura, Papua New Guinea. **Informal name:** Dobodura Papuan Pointy Mosquito.

hollandius King & Hoogstraal
1946b: 38 (M, F). **Type loc.:** Mount Dafonsero, Cyclops Range, Hollandia Area, Western New Guinea (Papua)

(USNM). **Distr.:** Indonesia, Papua New Guinea. **Refs.:** Penn 1949b: 51 (P*); Lee et al. 1982: 161 (F; key, tax., bion., distr., review); Peyton et al. 1999: 529 (distr.). **Ety.:** Mt. Dafonsero, Cyclops Range, Hollandia area, Papua, Indonesia. **Informal name:** Hollandia Indonesian Pointy Mosquito.

mackerrasi Taylor
1927: 68 (M, F). **Type loc.:** Berner Creek, Innisfail District, Queensland, Australia (US). **Distr.:** Australia. **Refs.:** Lee et al. 1982: 184 (F; key, tax., bion., distr., review). **Ety.:** Ian Murray Creagh Mackerras. **Informal name:** Mackerras Australian Pointy Mosquito.

novalbitarsis King & Hoogstraal
1946d: 146 (new name for *albitarsis* Taylor, not Ludlow 1905a). **Distr.:** Australia, Indonesia, Papua New Guinea. **Refs.:** Penn 1949b: 52 (P*); Bick 1951: 408 (bion.); Lee et al. 1982: 213 (F; key, tax., bion., distr., review); Peyton et al. 1999: 529 (distr.). **Ety.:** new (L) [near *albitarsis*]. **Informal name:** Snow-legged Papuan Pointy Mosquito [*novalbitaris* = "new white legged" = Snow-legged].
syn. *albitarsis* Taylor
1914d: 194 (F; *Leucomyia*?). **Type loc.:** Lakekamu Gold Field, Papua, New Guinea (US).

palmarum Edwards
1924b: 382 (F). **Type loc.:** Palm Island, Queensland, Australia (NHMUK). **Distr.:** Australia. **Refs.:** Marks 1948: 34 (M*, F, P*, L*; bion.); Lee et al. 1982: 224 (F; key, tax., bion., distr., review). **Ety.:** n.s. [palm (of hand) wide? (L)]. **Informal name:** Palm Australian Pointy Mosquito.

papuensis (Taylor)
1914d: 193 (F*; as *Leucomyia australiensis* var.). **Type loc.:** Milne Bay, Papua, New Guinea (US). **Distr.:** Indonesia, Papua New Guinea. **Refs.:** King & Hoogstraal 1946d: 152 (M*, F*, L*); Penn 1949b: 50 (P*); Lee et al. 1982: 226 (F; key, tax., bion., distr., review); Peyton et al. 1999: 530 (distr.); Reinert 2002e: Fig. 9 (F*); Reinert 2008g: Fig. 1 (F*). **Ety.:** Papua New Guinea. **Informal name:** New Guinea Papuan Pointy Mosquito.

subalbitarsis King & Hoogstraal
1946d: 144 (M*, F*, L*). **Type loc.:** Mount Dafonsero, Cyclops Mountains, Hollandia Area, Western New Guinea [West Papu] (USNM). **Distr.:** Indonesia (West Papua). **Refs.:** Lee et al. 1982: 272 (F; key, tax., bion., distr., review); Peyton et al. 1999: 530 (distr.). **Ety.:** nearly (L); near *albitarsis*. **Informal name:** Frost-legged Indonesian Pointy Mosquito [*subalbitaris* = "sub-white legged" = Frost-legged].

Subgenus *Paulianius* Brunhes & Boussès

2017 (tax., bion., keys; as genus). In Brunhes et al. 2017: 345. Type species: *Aedes tiptoni* Grejbine.

Name derivation: Renaud Paulian.

ambremontis (Brunhes & Boussès)
2017: 356 (M*, F*; distr.; as *Paulianius ambremontis*). In Brunhes et al. 2017. **Type loc.:** Ambre Mountain, Province of Antsiranana, Madagascar (PIP). **Distr.:** Madagascar (includes Glorioso and Juan De Nova Isl.). **Ety.:** Mt. Ambre. **Informal name:** Ambre Malagasy Pointy Mosquito.

brucei Wilkerson
2020. In Wilkerson, Linton, & Strickman 2020. New name for *Aedes* (*Paulianius*) *hirsutus* (Brunhes & Boussès 2017), not *Aedes* (*Aedimorphus*) *hirsutus* (Theobald 1901a) [original combination *Culex hirsutum*]. **Ety.:** Bruce Arthur Harrison. **Informal name:** Harrison Malagasy Pointy Mosquito.
syn. *hirsutus* Brunhes & Boussès
2017 (P*, L*; distr.; original comb. *Paulianius hirsutus*). In Brunhes et al. 2017: 352. **Type loc.:** Bealoy, Province de Mahajanga, Madagascar (PIP). **Distr.:** Madagascar (includes Glorioso and Juan De Nova Isl.).

coulangesi Rodhain & Boutonnier
1983b (F). **Type loc.:** Ampijora, Madagascar (PIP). **Distr.:** Madagascar (includes Glorioso and Juan De Nova Isl.). **Refs.:** Rodhain & Boutonnier 1983d (distr.; Madagascar); Fontenille 1987 (M*); Brunhes et al. 2017: 17 (M*, F*; bion., distr.; as genus *Paulianius*). **Ety.:** Pierre Coulanges. **Informal name:** Coulanges Malagasy Pointy Mosquito.

grassei Doucet
1951a: 69 (M*, F). **Type loc.:** Perinet, Tananarive, Madagascar (IRSM). **Distr.:** Madagascar (includes Glorioso and Juan De Nova Isl.). **Refs.:** Rodhain & Boutonnier 1983d (distr.; Madagascar); Brunhes et al. 2017: 18 (M*, F; tax., distr.; in genus *Paulianius*). **Ety.:** Pierre-Paule Grassé. **Informal name:** Grassé Malagasy Pointy Mosquito.

madagascarensis van Someren
1949a: 7 (F). **Type loc.:** Sakaramy, near Diego Suarez, Madagascar (NHMUK). **Distr.:** Madagascar (includes Glorioso and Juan De Nova Isl.). **Refs.:** Rodhain & Boutonnier 1983a: 197 (from subg. *Finlaya*); Rodhain & Boutonnier 1983d (distr.; Madagascar); Brunhes et al. 2017: 24 (M*, F*; tax., bion., distr.; in gen. *Paulianius*). **Ety.:** Madagascar. **Informal name:** Vazimba Malagasy Pointy Mosquito [the Vazimba are popularly believed to have been the original inhabitants of Madagascar].
syn. *sylvaticus* Brunhes
1983: 316 (M*). **Type loc.:** Edge of Onive River, about 10 km upstream from Ambohitralanana, Diego Suarez, Madagascar (SSC). **Refs.:** Brunhes et al. 2017 (syn.).

rodhaini (Brunhes & Boussès)
2017 (M*, F*; *Paulianius rodhaini*). In Brunhes et al. 2017: 364. **Type loc.:** Ambodimanga, S.P. de Moramanga, Province of Toamasina, Madagascar (IRD). **Distr.:**

Madagascar (includes Glorioso and Juan De Nova Isl.). **Ety.:** Jerome Rodhain. **Informal name:** Rodhain Malagasy Pointy Mosquito.

tiptoni Grjebine
1953a: 469 (L*). **Type loc.:** Ampijora, Ambato-Boeni District, Majunga, Madagascar (IRSM). **Distr.:** Madagascar (includes Glorioso and Juan De Nova Isl.). **Refs.:** Ravaonjanahary 1978 (E*), 103 (transfer from subg. *Aedimorphus*), (E*); Rodhain & Boutonnier 1983d (distr.; Madagascar); Brunhes et al. 2017 (M*, F*, P*, L*; tax., bion., distr.; in gen. *Paulianius*). **Ety.:** n.s. [Vernon John Tipton]. **Informal name:** Tipton Malagasy Pointy Mosquito.

zethus De Meillon & Lavoipierre
1944: 57 (new name for *niveus* De Meillon, not Ludlow 1903). **Distr.:** Zambia, Zimbabwe. **Ety.:** n.s. [Zethus, one of the twin sons of Zeus]. **Informal name:** Zethus Zambian Pointy Mosquito.
syn. *niveus* De Meillon
1943: 94 (M*, F, L*). **Type loc.:** Livingstone, Northern Rhodesia [Zambia] (SAIM). **Refs.:** Baisas 1974: 17 (M, F, P*, L*; tax., bion., distr.; Philippines).

Subgenus *Petermattinglyius* Reinert, Harbach, & Kitching

2009: 778 (phyl., class; as genus). Type species: *Aedes* (*Skusea*) *iyengari* Edwards.

Name derivation: Peter Frederick Mattingly.

syn. *Aglaonotus* Reinert, Harbach, & Kitching 2009: 779. Type species: *Aedes* (*Diceromyia*) *whartoni* Mattingly (as subg. of genus *Petermattinglyius*).

Important references: Rattanarithikul et al. 2010 (F*, L*; keys, bion.; Thailand); Reinert 2010h (F gen.*); Wilkerson et al. 2015 (phyl., class.; syn. with *Aedes* subg. *Petermattinglyius*).

franciscoi Mattingly
1959: 42 (M, F*, P, L). **Type loc.:** P Blakang Mati, Singapore (NHMUK). **Distr.:** Malaysia, Singapore, Thailand. **Refs.:** Mattingly 1961: 59 (M*, P, L); Reinert 1970a: 8 (M*, F*, P*, L*); Mattingly 1971a: Pl. 15 (P terminal segs.*); Harrison et al. 1991: 223 (distr.); Rattanarithikul et al. 2010 (F*, L*; keys, bion., distr.; Thailand). **Ety.:** Francisco Baisas. **Informal name:** Francisco Singaporean Pointy Mosquito.

iyengari Edwards
1923b: 4 (F). **Type loc.:** Meenglas and Jalpaiguri [West Bengal], India (NHMUK). **Distr.:** Bangladesh, India, Indonesia, Laos, Myanmar, Nepal, Thailand. **Refs.:** Brug 1932: 78 (L*); Barraud 1934: 273 (M*, F, L); Mattingly 1959: 38 (M*, F*, P, L*); Mattingly 1961: 59 (M*); Reinert 1970a: 10 (M*, F*, P*, L*); Aslamkhan 1971b (distr.; Pakistan); Darsie et al. 1992 (distr.; Nepal); Rattanarithikul et al. 2010 (F*, L*; keys, bion., distr.; Thailand); Reinert 2010h: Fig. 1 (F*). **Ety.:** M.O.T. Iyengar. **Informal name:** Iyengar Indian Pointy Mosquito.
syn. *punctissimus* Barraud
1928b: 360 (M*, F). **Type loc.:** Karwar [Bombay], India (NHMUK).

punctipes Edwards
1921c: 77 (F). **Type loc.:** Maymyo [Mandalay], Upper Burma [Myanmar] (IM). **Distr.:** Myanmar. **Refs.:** Barraud 1934: 273 (F); Reinert 1970a: 14 (F). **Ety.:** n.s. [foot spot (L); distinctive leg markings]. **Informal name:** Spot-footed Burmese Pointy Mosquito.

scanloni Reinert
1970f: 17 (M*, F*). **Type loc.:** Nakhon Ratchasima, Thailand (USNM). **Distr.:** Cambodia, Thailand. **Refs.:** Rattanarithikul et al. 2010 (F*, L*; keys, bion., distr.; Thailand). **Ety.:** John E. Scanlon. **Informal name:** Nakhon Thai Pointy Mosquito.

whartoni Mattingly
1965b: 65 (M*, P*, L*). **Type loc.:** Ulu Gombak, Malaya [Malaysia] (NHMUK). **Distr.:** Malaysia, Thailand. **Refs.:** Reinert 1970f: 19 (M*, F*, P*, L*); Rattanarithikul et al. 2010 (F*, L*; keys, bion., distr.; Thailand); Reinert 2010h: Fig. 1 (F*). **Ety.:** Ronald Harry Wharton. **Informal name:** Wharton Malaysian Pointy Mosquito.

Subgenus *Phagomyia* Theobald

1905e: 21 (as genus). Type species: *Culex gubernatoris* Giles.

Name derivation: n.s. [eat, devour, consume (Gr)].

syn. *Lepidotomyia* Theobald 1905e: 22 (as genus). Type species: *Lepidotomyia magna* Theobald.
syn. *Pseudocarrollia* Theobald 1910a: 12 (as genus). Type species: *Pseudocarrollia lophoventralis* Theobald.

Important references: Edwards 1932a: 149 (type species desig.); Knight & Marks 1952: 517, 522, 523 (*Aedes* subg. *Finlaya*, *terrens-gubernatoris* group: Gualteria, Subgroup III, *gubernatoris*; key, tax); Reinert, Harbach, & Kitching 2006 (phyl., class.; to genus from syn. with subg. *Finlaya* of genus *Aedes*); Reinert 2008b (F gen.*); Reinert, Harbach, & Kitching 2009 (phyl., class.); Rattanarithikul et al. 2010 (F*, L*; keys, bion; Thailand); Wilkerson et al. 2015 (phyl., class.; to subg. of genus *Aedes*); Soghigian et al. 2017 (mol. phyl.).

assamensis (Theobald)
1908b: 290 (F; *Stegomyia*). **Type loc.:** Sylhet, Assam, India [Bangladesh] (IM). **Distr.:** Bangladesh, Cambodia, India, Indonesia, Japan, Nepal, People's Republic of China, Thailand, Vietnam. **Refs.:** Barraud 1923e (L*); Barraud

1923g (M*, F*); Barraud 1934: 166 (M*, F*, L*); Joshi et al. 1965 (distr.; Nepal); Aslamkhan 1971b (distr.; Pakistan); Ahmed 1987 (distr.; Bangladesh); Darsie & Pradhan 1990, 1991 (F, L; tax., keys, bion., distr.; Nepal); Rattanarithikul et al. 2010 (F*, L*; keys, bion., distr.; Thailand). **Ety.:** Assam, India. **Informal name:** Assam Bangladeshi Pointy Mosquito.

cacharanus (Barraud)
1923a: 406 (M, F; *Finlaya*). **Type loc.:** Haflong, Cachar Hills, Assam, India (NHMUK). **Distr.:** India, Nepal. **Refs.:** Barraud 1923g (M*, F*); Barraud 1934: 166 (M*, F*); Darsie et al. 1993 (distr.; Nepal). **Ety.:** Cachar Dist., Assam, India. **Informal name:** Cachar Indian Pointy Mosquito.

cogilli Edwards
1922d: 92 (F). **Type loc.:** Karwar, North Kanara [Bombay], India (NHMUK). **Distr.:** India, Nepal. **Refs.:** Barraud 1923g (M*, F*); Barraud 1934: 165 (M, F*); Darsie et al. 1993 (distr.; Nepal). **Ety.:** n.s. [Harry Cogill?]. **Informal name:** Cogill Indian Pointy Mosquito.

deccanus (Barraud)
1923d: 217 (M, F; as *Finlaya deccana*). **Type loc.:** Belgaum, Bombay, Deccan, India (NHMUK). **Distr.:** India, Nepal. **Refs.:** Barraud 1923e (L*); Barraud 1923h (F*); Barraud 1934: 163 (M*, L*); Darsie et al. 1993 (distr.; Nepal); Darsie 2012 (P*). **Ety.:** n.s. [dean, leader? (L)]. **Informal name:** Deccan Indian Pointy Mosquito.

feegradei Barraud
1934: 164 (M, F, L). **Type loc.:** Rangoon [Pegu], Burma [Myanmar] (NHMUK). **Distr.:** Japan, Myanmar, Nepal, Thailand. **Refs.:** Darsie et al. 1993 (distr.; Nepal); Rattanarithikul et al. 2010 (F*, L*; keys, bion., distr.; Thailand). **Ety.:** n.s. [Feegrade (collector?)]. **Informal name:** Feegrade Burmese Pointy Mosquito.

gubernatoris (Giles)
1901a: 607 (May 18) (A; *Culex*). **Type loc.:** Allahabad [United Provinces], India (NHMUK). **Distr.:** Bangladesh, India, Nepal, People's Republic of China, Sri Lanka, Thailand. **Refs.:** Barraud 1923e (L*; as *magna*); Barraud 1934: 159 (M*, F*, L*); Joshi et al. 1965 (distr.; Nepal); Aslamkhan 1971b (distr.; Pakistan); Ahmed 1987 (distr.; Bangladesh); Darsie & Pradhan 1990, 1991 (F, L; tax., keys, bion., distr.; Nepal); Reinert 2002e: Fig. 5 (F*); Reinert 2008b: Fig. 1 (F*); Darsie 2011b (P*). **Ety.:** n.s. [poss. ref. to Allahabad Government House Garden]. **Informal name:** Gubernatorial Indian Pointy Mosquito.

syn. *magna* Theobald
1905e: 22 (M, F; *Lepidotomyia*). **Type loc.:** Bombay, India (NHMUK).

ssp. ***kotiensis*** Barraud
1934: 161 (M, F, L*; *Aedes (Finlaya) gubernatoris* var.). **Type loc.:** Koti, near Kalka [Punjab Hill States], Western Himalayas [India] (NHMUK). **Distr.:** India. **Ety.:** Koti, India. **Informal name:** Koti Indian Pointy Mosquito.

inquinatus Edwards
1922a: 267 (M, F). **Type loc.:** Simla [East Punjab], India (NHMUK). **Distr.:** India. **Refs.:** Barraud 1923h (M*); Barraud 1934: 169 (M*, F). **Ety.:** n.s. [stained? (L)]. **Informal name:** Messy Indian Pointy Mosquito.

iwi Marks
1955: 11 (F). **Type loc.:** Badu (Mulgrave) Island, Torres Strait, Northern Queensland, Australia (UQIC). **Distr.:** Australia. **Refs.:** Lee et al. 1982: 167 (F; key, tax., bion., distr., review). **Ety.:** word for "mosquito" in language spoken in Badu and other western islands of the Torres Strait. **Informal name:** Iwi Australian Pointy Mosquito.

khazani Edwards
1922a: 265 (M, F). **Type loc.:** Pudupadi, Malabar Coast [Madras], India (NHMUK). **Distr.:** Bangladesh, Cambodia, India, Malaysia, Nepal, Thailand, Vietnam. **Refs.:** Barraud 1934: 168 (M*, F*, L); Aslamkhan 1971b (distr.; Pakistan); Ahmed 1987 (distr.; Bangladesh); Darsie & Pradhan 1990, 1991 (F, L; tax., keys, bion., distr.; Nepal); Darsie et al. 1996 (distr.; Nepal); Rattanarithikul et al. 2010 (F*, L*; keys, bion., distr.; Thailand). **Ety.:** Khazan Chand. **Informal name:** Khazan Indian Pointy Mosquito.

kiangsiensis Tung
1955: 313 (M*, F). **Type loc.:** Kanhsien, Kiangsi, China [People's Republic of China] (CUMC). **Distr.:** People's Republic of China. **Ety.:** Kanhsien, Kiangsi, People's Republic of China. **Informal name:** Kiangsi Chinese Pointy Mosquito.

lophoventralis (Theobald)
1910a: 13 (F; *Pseudocarrollia*). **Type loc.:** Purniah, Behar (Northern Bengal), India (IM). **Distr.:** Bangladesh, India, Nepal, Thailand, Vietnam. **Refs.:** Barraud 1923g (M*, F*); Barraud 1934: 167 (M, F*); Sen 1938: 359 (P*, L*); Qutubuddin 1954: 135 (M*, P*, L*); Aslamkhan 1971b (distr.; Pakistan); Ahmed 1987 (distr.; Bangladesh); Darsie et al. 1993 (distr.; Nepal); Rattanarithikul et al. 2010 (F*, L*; keys, bion., distr.; Thailand). **Ety.:** tufted belly (Gr); ref. to marked ventral abdominal tufts. **Informal name:** Tuft-bellied Indian Pointy Mosquito.

melanopterus (Giles)
1904a: 367 (F*; *Finlaya*). **Type loc.:** Camp Stotsenberg, Angeles, Pampanga, Luzon, Philippines (NHMUK). **Distr.:** Philippines, Taiwan. **Refs.:** Knight & Hull 1951b: 235 (M*, F, L*); Basio 1971b: 17 (M*; bion.); Matsuo, Yoshida, & Lien 1974: 187 (E*; distr.); Baisas 1974: 16 (M*, F*, P*, L*; tax., bion., distr.; Philippines). **Ety.:** black wing (Gr); "Wings quite unspotted; its veins clothed with uniformly sooty scales." **Informal name:** Black-winged Philippine Pointy Mosquito.

syn. *palawanensis* Ludlow
1914a: 31 (F; *Popea*). **Type loc.:** Puerto Princesa, Palawan, Philippines (USNM).

plumiferus King & Hoogstraal
1946f: 306 (M*, F*, L). **Type loc.:** Hollandia, Western New Guinea (Papua) (USNM). **Distr.:** Indonesia, Papua New Guinea. **Refs.:** Lee et al. 1982: 230 (F; key, tax., bion., distr., review). **Ety.:** feathered (L); ref. to "large tufts of scales on the abdominal sternites." **Informal name:** Plumed Indonesian Pointy Mosquito.

prominens (Barraud)
1923f: 228 (M, F; *Finlaya*). **Type loc.:** Sukna [Darjeeling District, West] Bengal, India (NHMUK). **Distr.:** Cambodia, India, Indonesia, Malaysia, Nepal, People's Republic of China, Thailand, Vietnam. **Refs.:** Barraud 1923h; Borel 1930: 262 (M*, F, L*); Barraud 1934: 169 (M, F, L); Darsie et al. 1992 (distr.; Nepal); Rattanarithikul et al. 2010 (F*, L*; keys, bion., distr.; Thailand). **Ety.:** n.s. [prominent, projecting (L)]. **Informal name:** Prominent Indian Pointy Mosquito.

stevensoni (Barraud)
1923d: 219 (M, F; *Finlaya*). **Type loc.:** Nagargali [Belgaum District], Bombay, India (NHMUK). **Distr.:** India, Indonesia. **Refs.:** Barraud 1934: 174 (M*, F*); Lien et al. 1977 (distr.). **Ety.:** n.s. **Informal name:** Stevenson Indian Pointy Mosquito.

watasei Yamada
1921: 64 (F). **Type loc.:** Ōmura, Kyushu, Japan (IID). **Distr.:** Japan. **Refs.:** Omori 1962: 10 (M*, P*, L*); Tanaka et al. 1975c: 222 (tax., distr.); Tanaka et al. 1979: 364 (M*, F*, L*); Tanaka 2002b: 173 (P*; tax., key). **Ety.:** S. Watasé. **Informal name:** Watasé Japanese Pointy Mosquito.

Subgenus *Polyleptiomyia* Theobald

Theobald 1905e: 21 (as genus). Type species: *Stegomyia albocephala* Theobald.

Name derivation: n.s. [lept- = slender, thin, fine (Gr); poss. ref. to narrow curved scales on the nape].

Important references: Reinert, Harbach, & Kitching 2009 (phyl., class.; to genus from syn. with genus *Aedimorphus*); Reinert 2010f (F gen.*; tax.); Wilkerson et al. 2015 (phyl., class.; to subg. of genus *Aedes*).

albocephalus (Theobald)
1903c: 4 (M; *Stegomyia*). **Type loc.:** The Gambia (NHMUK). **Distr.:** Angola, Benin, Cameroon, Comoros, Gambia, Kenya, Liberia, Madagascar (includes Glorioso and Juan De Nova Isl.), Mozambique, Nigeria, Republic of South Africa, Senegal, Seychelles, Sierra Leone, Tanzania, Togo, Uganda, Zimbabwe. **Refs.:** Theobald 1910c: 205 (M only of type series; as *punctithorax*; syn. by Edwards1941: 177); Ingram & Macfie 1917a: 82 (P*); Ingram & Macfie 1917b: 141 (L*); Macfie & Ingram 1922: 169 (F*); Edwards 1941: 177 (M*, F; syn.), 396 (P); Haddow et al. 1951: 223 (tax.); Hopkins 1952: 181 (L*); Hamon, Abonnenc, & Noel 1955: 296 (M*); van Someren, Teesdale, & Furlong 1955: 478 (M, F); Muspratt 1961: 100 (bion.); Pao & Knight 1970a: 112 (L*); Ribeiro & da Cunha Ramos 1973: 126 (distr.); McIntosh 1975: 266 (M*, F*, L*; distr.); Jupp 1996 (M*, F*; key); Reinert 2000e (F); Reinert 2010f: Fig. 1 (F*). **Ety.:** white head (L); "head covered with flat dull white scales." **Informal name:** White-headed Gambian Pointy Mosquito.

syn. *seychellensis* Theobald
1912b: 83 (F*; *Reedomyia*). **Type loc.:** Dennis Island, Silhouette, Mahe, near Morne Blanc, Takamaka and Aldabra, Seychelles (NHMUK). **Refs.:** Edwards 1941: 177 (? = *albocephalus*).

gandarai da Cunha Ramos, Capela, & Ribeiro
1995: 98 (M*, F*, P*, L*). **Type loc.:** Praia das Conchas, São Tomé (IHMT). **Distr.:** São Tomé and Príncipe. **Ety.:** Álvaro F. Gândara. **Informal name:** Gândara São Toméan Pointy Mosquito.

Subgenus *Protomacleaya* Theobald

1907: 149, 253 (as genus). Type species: *Culex triseriatus* Say.

Name derivation: n.s. [false (Gr) and *Macleaya*].

syn. *Gualteria* Lutz (not Agassiz 1872). In Bourroul 1904: 47, 54. Type species: *Haemagogus oswaldi* Lutz (Brunetti 1914: 61). Zavortink 1972: 10 (syn.).

syn. *Promacleaya* Surcouf & González-Rincones 1912: 277 (unjust. emend. of *Protomacleaya* Theobald).

Important references: Surcouf & González-Rincones 1912: 277 (as *Promacleaya*: emend.); Schick 1970a (keys; Terrens Group); Schick 1970b: 13 (keys; Terrens Group); Zavortink 1972: 10 (tax.; resurrected from syn. with *Finlaya*); Reinert 2000h (to subg. of genus *Ochlerotatus*); Reinert, Harbach, & Kitching 2009 (as "*Ochlerotatus* (*Protomacleaya*)" *sensu auctorum*); Wilkerson et al. 2015 (to syn. of subg. *Ochlerotatus* of genus *Aedes*); Wilkerson & Linton 2015 (to subg. of genus *Aedes*).

aitkeni Schick
1970a: 61 (F*, P*). **Type loc.:** La Sierra, 10 km southeast on Hwy 2, San Jose, Costa Rica (USNM). **Distr.:** Costa Rica, Panama. **Ety.:** Thomas Henry Gardiner Aitken. **Informal name:** Aitken Costa Rican Pointy Mosquito.

alboapicus Schick
1970a: 50 (M*, F*, P*, L*). **Type loc.:** Barro Colorado Island, Canal Zone, Panama (USNM). **Distr.:** Costa Rica, Panama. **Ety.:** white tip (L); ref. to "silvered fifth tarsal segments." **Informal name:** White-tipped Panamanian Pointy Mosquito.

amabilis Schick
1970a: 68 (F*). **Type loc.:** Cueva del Nacimiento del Agua, near La Penuela, Mexico (USNM). **Distr.:** Mexico. **Ety.:** n.s. [lovely (L)]. **Informal name:** Lovely Mexican Pointy Mosquito.

argyrothorax Bonne-Wepster & Bonne
1920: 179 (M). **Type loc.:** Paramaribo, Suriname (USNM, ITH). **Distr.:** Brazil, Colombia, Ecuador, French Guiana, Guyana, Peru, Suriname, Venezuela. **Refs.:** Cerqueira 1950: 173 (M*, F*, P, L*); Lane 1953: 688 (M*, F, P*, L*); Stone & Knight 1956a: 214 (type info.; type depos. change USNM to ITH); Belkin 1968b: 4 (tax.); Schick 1970a: 36 (M*, F*, P*, L*); Schick 1970b: 14 (distr.); Linton et al. 2013 (bion., mol. tax., distr.; Ecuador). **Ety.:** silver thorax (Gr). **Informal name:** Silver-chested Surinamese Pointy Mosquito.

berlini Schick
1970a: 48 (F*, P*, L*). **Type loc.:** Parrott Hall, 155 mi post, near Parlatuvier, Tobago Island (USNM). **Distr.:** Colombia, Trinidad and Tobago, Venezuela. **Refs.:** Schick 1970b: 22 (M*, F*, P*, L*; distr.). **Ety.:** Olavil George William Berlin. **Informal name:** Berlin Trinbagonian Pointy Mosquito.

bertrami Schick
1970a: 38 (P). **Type loc.:** New capital site, near Cayo, British Honduras [Belize] (USNM). **Distr.:** Belize. **Refs.:** Pecor et al. 2002: 249 (tax., distr.; Belize). **Ety.:** Douglas Somerville Bertram. **Informal name:** Bertram Belizean Pointy Mosquito.

braziliensis Gordon & Evans
1922: 329 (M; as *Aedes oswaldi* var.). **Type loc.:** Macapá, Amazonas, Brazil (NHMUK). **Distr.:** Brazil, Colombia, French Guiana, Venezuela. **Refs.:** Belkin 1968b: 4 (from syn. with *terrens*); Schick 1970a: 43 (M*); Schick 1970b: 17 (M*, F*, P*, L*); Berti et al. 2015 (distr.; Venezuela). **Ety.:** Brazil. **Informal name:** Pindorama Brazilian Pointy Mosquito [Pindorama is the original Guaraní name for the region of Brazil, meaning "land of the palm trees"].

brelandi Zavortink
1972: 25 (M*, F*, P*, L*). **Type loc.:** Chisos Mountains, Big Bend National Park, Brewster County, Texas, United States (USNM). **Distr.:** Mexico, United States (continental). **Refs.:** Darsie & Ward 2005 (F*, L*; keys, distr.); Sánchez-Trinidad et al. 2014 (bion., distr.; Mexico). **Ety.:** Osmond Philip Breland. **Informal name:** Breland American Pointy Mosquito.

buenaventura Schick
1970a: 52 (M*, F*, P*, L*). **Type loc.:** Circa 4 km west of Rio Raposo Virus Field Station, near Buenaventura, Valle, Colombia (USNM). **Distr.:** Colombia. **Ety.:** Buenaventura, Colombia. **Informal name:** Buenaventura Colombian Pointy Mosquito.

burgeri Zavortink
1972: 42 (M*, F*, P*, L*). **Type loc.:** Bodie Canyon, east of Lochiel, Santa Cruz County, Arizona, United States (USNM). **Distr.:** Mexico, United States (continental). **Refs.:** Darsie & Ward 2005 (F*, L*; keys, distr.). **Ety.:** John Frederick Burger. **Informal name:** Burger American Pointy Mosquito.

campana Schick
1970a: 79 (M*, F, P*, L*). **Type loc.:** Cerro Campana, Panama (USNM). **Distr.:** Panama. **Ety.:** Cerro Campana, Panama. **Informal name:** Campana Panamanian Pointy Mosquito.

casali Schick
1970b: 21 (M*, F*, P*, L*). **Type loc.:** Las Tablillas, near General Enrique Mosconi (Vespucio), Argentina (USNM). **Distr.:** Argentina. **Refs.:** Zavortink 1972: 3 (transfer). **Ety.:** Osvaldo Hugo Casal. **Informal name:** Casal Argentinian Pointy Mosquito.

chionotum Zavortink
1972: 47 (M*, F*, P*, L*). **Type loc.:** Near Cuernavaca, Morelos, Mexico (USNM). **Distr.:** Mexico. **Ety.:** n.s. [snowy notum [scutum] (Gr); poss. ref. to scattered light scales on scutum]. **Informal name:** Snow-backed Mexican Pointy Mosquito.

daryi Schick
1970: 80 (M*, F*, P*, L*). **Type loc.:** Finca Trece Aguas, near Senahu, Alta Vera Paz, Guatemala (USNM). **Distr.:** Costa Rica, Guatemala, Panama. **Ety.:** Mario Dary R. **Informal name:** Dary Guatemalan Pointy Mosquito.

diazi Schick
1970: 89 (M*). **Type loc.:** Cueva del Nacimiento del Agua, near Penuela, Veracruz, Mexico (USNM). **Distr.:** Mexico. **Ety.:** Alfonso Díaz Nájera. **Informal name:** Díaz Mexican Pointy Mosquito.

gabriel Schick
1970: 69 (M*, F*, P*, L*). **Type loc.:** Gabriel Mariaca, 0.5 mi east of Morelos, Mexico (USNM). **Distr.:** Mexico, Panama. **Ety.:** Gabriel Mariaca, Morelos, Mexico. **Informal name:** Gabriel Mexican Pointy Mosquito.

galindoi Schick
1970: 77 (M*, F*, P*, L*). **Type loc.:** Barro Colorado Island, Canal Zone, Panama (USNM). **Distr.:** Costa Rica, Panama. **Ety.:** Pedro Galindo. **Informal name:** Galindo Panamanian Pointy Mosquito.

hendersoni Cockerell
1918: 199 (F; as *Aedes triseriatus* var.). **Type loc.:** Box Elder Creek, 19 mi west of Douglas, Wyoming, United States (USNM). **Distr.:** Canada, Greece, United States (continental). **Refs.:** Jenkins & Carpenter 1946: 39 (syn.); Stone & Knight 1956a: 218 (type info., lectotype desig.); Breland 1960: 601 (M, F*, L*; to sp., from syn.); Harmston 1969: 490 (F*); Zavortink 1972: 20 (M*, F*, P*, L*); Grimstad et al. 1974: 795 (tax.); Lunt 1977b: 654 (ident.); Saul et al. 1977: 705 (mol. tax.); Zaim et al. 1977: 489 (E*); Linley & Craig 1993: 65 (E*); Darsie & Ward 2005 (F*, L*; keys, distr.); Harrison et al. 2016 (F*, L*; keys, distr.). **Ety.:** J. Henderson. **Informal name:** Henderson American Pointy Mosquito.

heteropus Dyar
1921j: 152 (M, F). **Type loc.:** Alajuela, Costa Rica (USNM). **Distr.:** Costa Rica, El Salvador. **Refs.:** Stone & Knight 1956a: 218 (type info., lectotype desig.); Schick 1970a: 74 (M*, F*, P*, L*; from syn. with *terrens*). **Ety.:** n.s. [different foot (L); poss. ref. to broad black ring on midtarsomere 2 in contrast to *podographicus*]. **Informal name:** Odd-footed Costa Rican Pointy Mosquito.

homoeopus Dyar
1922d: 92 (M, F; *terrens* ssp.). **Type loc.:** Alajuela, Costa Rica (USNM). **Distr.:** Costa Rica, El Salvador, Guatemala, Mexico, Panama. **Refs.:** Lane 1951 (syn.); Stone & Knight 1956a: 219 (type info., lectotype desig.); Schick 1970a: 63 (M*, F*, P*, L*; from syn. with *argyrothorax*). **Ety.:** n.s. [same work (Gr, L); poss. ref. to similarity with *terrens*]. **Informal name:** Uniform Costa Rican Pointy Mosquito.

idanus Schick
1970a: 70 (M*, F, P*, L*). **Type loc.:** Viejo de Agua de Obispo, near Chilpancingo, Guerrero, Mexico (USNM). **Distr.:** Mexico. **Ety.:** n.s. [fair, beautiful (Gr)]. **Informal name:** Beautiful Mexican Pointy Mosquito.

impostor Schick
1970a: 66 (F*, P*, L*). **Type loc.:** Finca Trece Aguas, Alta Vera Paz, Guatemala (USNM). **Distr.:** Guatemala. **Ety.:** n.s. [imposter (L); poss. ref. to superficial resemblance to members of subg. *Eumelanomyia*]. **Informal name:** Disguised Guatemalan Pointy Mosquito [*impostor* = Disguised].

insolitus (Coquillett)
1906e: 62 (F; *Verrallina*). **Type loc.:** Trinidad [Trinidad & Tobago] (USNM). **Distr.:** Colombia, Costa Rica, Guatemala, Panama, Trinidad and Tobago, Venezuela. **Refs.:** Schick 1970a: 57 (M*, F*, P*, L*; from syn. with *terrens*); Schick 1970b: 20 (distr.). **Ety.:** n.s. [unusual (L)]. **Informal name:** Unusual Trinbagonian Pointy Mosquito.
syn. *laternaria* Coquillett
1906a: 184 (M; *Verrallina*). **Type loc.:** Trinidad [Trinidad & Tobago] (USNM). **Refs.:** Stone & Knight 1956a: 220 (type info.); Schick 1970a: 57 (syn.).

knabi (Coquillett)
1906a: 183 (F; *Culex*). **Type loc.:** Tehuantepec, Oaxaca, Mexico (USNM). **Distr.:** Mexico. **Refs.:** Dyar 1928: 227 (L*); Stone & Knight 1956a: 220 (type info., lectotype desig.); Zavortink 1972: 53 (M*, F, P*, L*); Reinert 2002e: 31 (F*). **Ety.:** Frederick Knab. **Informal name:** Knab Mexican Pointy Mosquito.

kompi Vargas & Downs
1950: 167 (M*, F*, P*, L). **Type loc.:** Tepoztlan, Morelos, Mexico (ISET). **Distr.:** Mexico, Panama, United States (continental). **Refs.:** Zavortink 1972: 44 (M*, F*, P*, L*); Reinert 2002e: Fig. 31 (F*). **Ety.:** William H. Wood Komp. **Informal name:** Komp Mexican Pointy Mosquito.

lewnielseni Ortega & Zavortink
2019 (M*, F*, P*, L*). In Ortega et al. 2019. **Type loc.:** Sierra la Camotera, Santiago, Nuevo León, Mexico (n.s., Bohart Museum?). **Ety.:** Lewis T. Nielsen. **Informal name:** Nielsen Mexican Pointy Mosquito.

metoecopus Dyar
1925d: 30 (M, F). **Type loc.:** Ecuador (USNM). **Distr.:** Ecuador. **Refs.:** Lane 1953: 686 (syn.); Stone & Knight 1956a: 221 (type info., lectotype desig.); Schick 1970a: 54 (M*, F*, P*, L*; from syn. with *terrens*). **Ety.:** n.k. **Informal name:** Active Ecuadorian Pointy Mosquito.

niveoscutum Zavortink
1972: 49 (M*, F*, P*, L*). **Type loc.:** El Mirador, Jalisco, Mexico (USNM). **Distr.:** Mexico. **Ety.:** n.s. [snowy scutum (L); however, only midlobe of scutellum pale]. **Informal name:** Snow-shield Mexican Pointy Mosquito.

podographicus Dyar & Knab
1906b: 165 (M, F, L). **Type loc.:** Sonsonate, El Salvador (USNM). **Distr.:** Belize, Costa Rica, Ecuador, El Salvador, Guatemala, Honduras, Mexico, Nicaragua, Panama, Venezuela. **Refs.:** da Costa Lima 1930a: 257 (M*); Lane 1936a: 11 (M*, L*); Kumm et al. 1940: 416 (syn.); Stone & Knight 1956a: 224 (type info., lectotype desig.); Schick 1970a: 82 (M*, F*, P*, L*; from syn. with *terrens*); Schick 1970b: 23 (distr.); Heinemann & Belkin 1977b: 414 (distr.). **Ety.:** n.k. **Informal name:** Footnote Salvadoranian Pointy Mosquito [*podographicus* = "foot writing" = Footnote].

sandrae Zavortink
1972: 51 (M*, F*, P*, L*). **Type loc.:** Hipodromo del Norte, Guatemala City, Guatemala (USNM). **Distr.:** Guatemala, Mexico. **Ety.:** Sandra J. Heinemann. **Informal name:** Sandra Guatemalan Pointy Mosquito.

schicki Zavortink
1972: 46 (M*, F*, P*, L*). **Type loc.:** La Emerta, Durango, Mexico (USNM). **Distr.:** Mexico. **Ety.:** Robert Xavier Schick. **Informal name:** Schick Mexican Pointy Mosquito.

schroederi Schick
1970a: 88 (M*, P*, L*). **Type loc.:** Tehuantepec, Oaxaca, Mexico (USNM). **Distr.:** Mexico. **Ety.:** D. Schroeder. **Informal name:** Schroeder Mexican Pointy Mosquito.

sumidero Schick
1970a: 72 (M*, F*). **Type loc.:** Sumidero, Chiapas, Mexico (USNM). **Distr.:** Mexico. **Ety.:** Sumidero, Chiapas, Mexico. **Informal name:** Sumidero Mexican Pointy Mosquito.

tehuantepec Schick
1970a: 87 (M*, F*, L*). **Type loc.:** Tehuantepec, Oaxaca, Mexico (USNM). **Distr.:** Mexico. **Ety.:** Tehuantepec, Oaxaca, Mexico. **Informal name:** Tehuantepec Mexican Pointy Mosquito.

terrens (Walker)
1856a: 429 (M; *Culex*). **Type loc.:** Rio de Janeiro, Guanabara, Brazil (NHMUK). **Distr.:** Argentina, Belize, Bolivia, Brazil, Colombia, Costa Rica, Ecuador, French Guiana, Guatemala, Guyana, Mexico, Panama, Paraguay, Suriname, Venezuela. **Refs.:** Lane 1953: 686 (M*, F, P*, L*); Schick 1970a: 41 (M*, F*, P*, L*); Schick 1970b: 15 (M*, F*, P*, L*; redefined); Belkin et al. 1971: 20 (type loc. info); Panday 1975a, b (distr.; Suriname); Reinert 2002e: Fig. 33 (F*); Alencar et al. 2005a (E*); Berti et al. 2015 (distr.; Venezuela). **Ety.:** n.s. [alarming, scaring (L) (Harbach 2018c)]. **Informal name:** Earthy Brazilian Pointy Mosquito.

syn. *oswaldi* Lutz
1904: 47, 66 (A; *Gualteria*). In Bourroul 1904. **Type loc.:** Fazenda do Bonito, Serra da Bocaina, São Paulo, Brazil (NHMUK). **Refs.:** da Costa Lima 1930a: 256 (P*); Belkin 1968b: 6 (tax.); Belkin et al. 1971: 20 (lectotype desig.).

syn. *apollo* Schick
1970a: 46 (M*, F*, P*, L*). **Type loc.:** Finca Vanguardia (on Villavicencio-Restrepo road, closer to Villavicencio), Colombia (USNM). **Refs.:** Schick 1970b: 15 (syn.).

thorntoni Dyar & Knab
1907a: 10 (F). **Type loc.:** Bluefields, Nicaragua (USNM). **Distr.:** Belize, Costa Rica, Nicaragua, Panama. **Refs.:** Kumm et al. 1940: 416 (syn.); Stone & Knight 1956a: 225 (type info.); Schick 1970a: 33 (M*, F*, P*, L*; from syn. with *terrens*); Pecor et al. 2002: 249 (tax., distr.; Belize). **Ety.:** W.F. Thornton. **Informal name:** Thornton Nicaraguan Pointy Mosquito.

triseriatus (Say)
1823: 12 (F; *Culex*). **Type loc.:** Pennsylvania, United States (NE). **Distr.:** Canada, Cuba, Greenland, Mexico, United States (continental). **Refs.:** Mitchell 1907b (E*); Howard et al. 1913 (E*); Howard et al. 1917 (E); Ross 1947: 65 (M*, F*, L*); Darsie 1951: 14 (P*); Yamaguti & LaCasse 1951d: 234 (M*, F*, L*); Rempel 1953: 483 (M*, F*); Carpenter & LaCasse 1955: 255 (M*, F*, L*; keys); Craig 1956 (E*); Horsfall & Craig 1956: 370 (E*); Price 1960: 558 (first instar L*); Ross & Horsfall 1965 (M*, F*, L*, E*; keys); Dodge 1966: 358 (first instar L*; key); Mohrig 1967 (F*); Kalpage & Brust 1968 (E*); Harmston 1969: 490 (F*); Pratt & Kidwell 1969 (E*); Horsfall et al. 1970: 1710 (E*); Zavortink 1972: 27 (M*, F*, P*, L*; distr.); Grimstad et al. 1974: 795 (tax.); Lunt 1977b: 654 (tax.); Saul et al. 1977: 705 (mol. tax.); Zaim et al. 1977: 489 (E*); Linley 1989a (E*); Linley & Craig 1993: 65 (E*); Reinert 2002e: Fig. 34 (F*); Juliano & Lounibos 2005 (bion., review); Darsie & Ward 2005 (F*, L*; keys, distr.); Sánchez-Trinidad et al. 2014 (bion., distr.; Mexico); Harrison et al. 2016 (F*, L*; keys, distr.; note 13). **Ety.:** n.s. [three serrations (L); prob. ref. to three white spots on abdominal segments]. **Informal name:** Three-spotted American Pointy Mosquito. Official ESA common name: eastern treehole mosquito. (Vol. 1, p. 452)

syn. *nigra* Ludlow
1905c: 387 (F; *Finlaya*?). **Type loc.:** Rock Island Arsenal, Illinois, United States (USNM).

vargasi Schick
1970a: 73 (M*, F*, P*, L*). **Type loc.:** Cordoba, 5–15 mi south, Veracruz, Mexico (USNM). **Distr.:** Mexico. **Ety.:** Luís Vargas. **Informal name:** Vargas Mexican Pointy Mosquito.

zavortinki Schick
1970a: 45 (M*, F*, P*, L*). **Type loc.:** Barro Colorado Island, Canal Zone, Panama (USNM). **Distr.:** Colombia, Panama. **Ety.:** Thomas James Zavortink. **Informal name:** Zavortink Panamanian Pointy Mosquito.

zoosophus Dyar & Knab
1918b: 165 (F). **Type loc.:** Kerrville, Texas, United States (USNM). **Distr.:** Mexico, United States (continental). **Refs.:** Breland 1949b (morph., tax.; as *alleni*); Yamaguti & LaCasse 1951d: 243 (M*, F*, L*); Carpenter & LaCasse 1955: 259 (M*, F*, L*; keys); Dodge 1966: 359 (first instar L; key); Zavortink 1972: 34 (M*, F, P*, L*); Reinert 2002e: Fig. 35 (F*); Darsie & Ward 2005 (F*, L*; keys, distr.); Sánchez-Trinidad et al. 2014 (bion., distr.; Mexico). **Ety.:** n.s. [animal wise? (L)]. **Informal name:** Sophisticated American Pointy Mosquito.

syn. *alleni* Turner
1924: 84 (M). **Refs.:** Gjullin 1946 (syn.). **Type loc.:** Mission, Texas, United States (USNM).

Subgenus *Pseudalbuginosus* Huang & Rueda

2015: 382 (as subg. of genus *Aedes*). Type species: *Aedes* (*Aedimorphus*) *grjebinei* Hamon, Taufflieb, & Maillot.

Name derivation: false *Albuginosus* (Gr).

Important references: Huang & Rueda 2015 (key; as Grjebinei Group of subg. *Aedimorphus*).

grjebinei Hamon, Taufflieb, & Maillot
1957: 695 (M). **Type loc.:** Bangou, Gabon (IERT). **Distr.:** Cameroon, Congo, Gabon. **Ety.:** Alexid Grjebine. **Informal name:** Grjebine Gabonese Pointy Mosquito.

Subgenus *Pseudarmigeres* Stone & Knight

1956c: 34 (as subg. of genus *Aedes*; new name for *Dunnius* Edwards). Type species: *Aedes albomarginatus* var. *dunni* Evans.

Name derivation: false *Armigeres* (Gr).

syn. *Dunnius* Edwards 1930a: 297 (not Distant 1902). Type species: *Aedes albomarginatus* var. *dunni* Evans.

Important references: Edwards 1941 (as subg. *Dunnius*; tax., key); Hopkins 1952 (as subg. *Dunnius*; tax.; Ethiopian Region); Jupp 1996 (tax.); Reinert 2000e (F gen.*); Reinert, Harbach, & Kitching 2004: 289 (phyl., class.; to genus); Reinert, Harbach, & Kitching 2009 (phyl., class.); Wilkerson et al. 2015 (phyl., class.; to subg. of genus *Aedes*).

albomarginatus (Newstead)
1907 (F*; *Stegomyia*). In Newstead et al. 1907: 16. **Type loc.:** Kasongo, Belgian Congo [Democratic Republic of the Congo] (NHMUK). **Distr.:** Democratic Republic of the Congo, Kenya, Uganda. **Refs.:** Edwards 1932a: 174 (as *albomaginatus*, *lapsus*); Edwards 1941: 221 (M); van Someren 1946d (tax.); van Someren 1947d: 132 (M, F). **Ety.:** n.s. [white bordered (L)]. **Informal name:** White-edged Congolese Pointy Mosquito.

argenteoventralis (Theobald)
1909b: 24 (F; *Dendromyia*). **Type loc.:** Obuasi, Ashanti, Gold Coast [Ghana] (NHMUK). **Distr.:** Cameroon, Central African Republic, Côte d'Ivoire, Ghana, Liberia, Nigeria, Sierra Leone, Togo. **Refs.:** Edwards 1941: 219 (M*, F*). **Ety.:** silvery venter (L); abdominal venter silver. **Informal name:** Silver-bellied Ghanaian Pointy Mosquito.
syn. *affinis* Theobald
1909b: 25 (F; *Dendromyia*). **Type loc.:** Obuasi, Ashanti, Gold Coast [Ghana] (NHMUK).
ssp. ***dunni*** (Evans)
1928: 39 (M*, F, P*, L*; as *Armigeres albomarginata* var.). **Type loc.:** Lagos, Nigeria (NHMUK). **Distr.:** Cameroon, Democratic Republic of the Congo, Nigeria. **Refs.:** Edwards 1941: 220 (M, F, P*); Hopkins 1952: 219 (L*); White 1975: 310 (from var. to ssp.; lectotype desig.); Reinert 2000e (F*; Fig. 30). **Ety.:** L.H. Dunn. **Informal name:** Dunn Nigerian Pointy Mosquito.

kummi Edwards
1930a: 298 (M*, F). **Type loc.:** Ibadan [Western Provinces], Nigeria (NHMUK). **Distr.:** Cameroon, Central African Republic, Congo, Côte d'Ivoire, Nigeria. **Refs.:** Edwards 1941: 220 (M*, F), 403 (P); Hopkins 1952: 220 (L). **Ety.:** Henry William Kumm. **Informal name:** Kumm Nigerian Pointy Mosquito.

michaelikati van Someren
1946c: 3 (M*, F, P). **Type loc.:** Gede [near Mombasa, Coast], Kenya (NHMUK). **Distr.:** Kenya. **Refs.:** Hopkins 1952: 221 (L*). **Ety.:** Michael Ikata. **Informal name:** Ikata Kenyan Pointy Mosquito.
ssp. ***gurneri*** van Someren
1946c: 5 (M*, F, P). **Type loc.:** Nairobi [Central], Kenya (NHMUK). **Distr.:** Democratic Republic of the Congo, Kenya. **Refs.:** Hopkins 1952: 222 (L*). **Ety.:** n.s. [prob. ref. to: "Nairobi Gurner R. Cunningham van Somerens"]. **Informal name:** Gurner Kenyan Pointy Mosquito.

natalensis Edwards
1930a: 298 (M*). **Type loc.:** Durban, Natal [KwaZulu-Natal, Republic of South Africa] (NHMUK). **Distr.:** Eswatini, Kenya, Mozambique, Republic of South Africa, Tanzania. **Refs.:** Edwards 1941: 221 (M*, F); Muspratt 1953b: 90 (L*); Worth & De Meillon 1960: 244 (distr.); Jupp 1996 (M*, F*; key). **Ety.:** Natal, Republic of South Africa. **Informal name:** Natal South African Pointy Mosquito.

Subgenus *Pseudoskusea* Theobald

1907: 192 (as genus). Type species: *Skusea multiplex* Theobald.

Name derivation: false *Skusea* (Gr).

syn. *Caenocephalus* Taylor 1914c: 700 (not *Caenocephalus* van der Wulp 1898).

Important references: Dobrotworsky 1961a (tax., bion.; Australia); Dobrotworsky 1965 (tax., bion.; Australia); Lee et al. 1984: 241 (tax., key, distr., bion.; Australia); Reinert 2000h (to subg. of genus *Ochlerotatus*); Reinert 2002e: 32 (F gen.*); Reinert, Harbach, & Kitching 2006 (phyl., class.; to genus); Reinert, Harbach, & Kitching 2008 (to subg. of genus *Ochlerotatus*); Wilkerson et al. 2015 (phyl., class.; syn. of subgenus *Ochlerotatus* of genus *Aedes*); Wilkerson & Linton 2015 (to subg. of genus *Aedes*).

bancroftianus Edwards
1921c: 74 (M, F). **Type loc.:** Eidsvold, Queensland, Australia (NHMUK). **Distr.:** Australia. **Refs.:** Dobrotworsky 1961a: 257 (M*, F*, P*, L*); Dobrotworsky 1965: 161 (M*, F*, L*); Lee et al. 1984: 244 (F; key, tax., bion., distr., review). **Ety.:** Thomas Lane Bancroft. **Informal name:** Bancroft Australian Pointy Mosquito.

culiciformis (Theobald)
1905f: 77 (F*; *Skusea*). **Type loc.:** Paumomu River [Papua], New Guinea (HNM). **Distr.:** Australia, Papua New Guinea. **Refs.:** Edwards 1924b: 386 (A); Lee et al. 1984: 247 (F; key, tax., bion., distr., review). **Ety.:** n.s. [gnat appearing (L)]. **Informal name:** Gnat-like Papuan Pointy Mosquito.

multiplex (Theobald)
1903a: 293 (F*; *Skusea*). **Type loc.:** Queensland, Australia (NHMUK). **Distr.:** Australia. **Refs.:** Bancroft 1908: 23 (M, F; as genus *Pseudoskusea*); Edwards 1924b: 386 (A); Dobrotworsky 1961a: 264 (M*, F*, P*, L*); Dobrotworsky 1965: 166 (M*, F*, L*); Lee et al. 1984: 249 (F; key, tax., bion., distr., review); Reinert 2002e: Fig. 36 (F*). **Ety.:** n.s. [multitudinous (L); poss. ref. to observed intraspecific variability]. **Informal name:** Manifold Australian Pointy Mosquito.

postspiraculosus Dobrotworsky
1961a: 261 (M*, F*, P*, L*). **Type loc.:** Wattle Glen, Victoria [Australia] (NMM). **Distr.:** Australia. **Refs.:** Dobrotworsky 1965: 163 (M*, F*, L*); Mattingly 1971a: Pl. 7 (M head*); Lee et al. 1984: 252(F; key, tax., bion., distr., review). **Ety.:** after spiracle (L); ref. to two patches of scales on the postspiracular area. **Informal name:** Spiracle Australian Pointy Mosquito.

Subgenus *Rampamyia* Reinert, Harbach, & Kitching

2006: 83 (as genus). Type species: *Culex notoscriptus* Skuse.

Name derivation: Rampa Rattanarithikul.

Important references: Lee et al. 1982 (keys, tax., bion.; as group E (Notoscriptus Group), in part); Reinert 2008f (F gen.*); Reinert, Harbach, & Kitching 2009 (phyl., class.); Wilkerson et al. 2015 (phyl., class.; to subg. of genus *Aedes*); Soghigian et al. 2017 (mol. phyl.).

albilabris Edwards
1925: 258 (F). **Type loc.:** Tulagi [Florida], Solomon Islands (NHMUK). **Distr.:** Indonesia, Papua New Guinea, Solomon Isl. **Refs.:** Edwards 1926a: 110 (M); Paine & Edwards 1929: 315 (L); Belkin 1962: 350 (M*, F, P*, L*); Lee et al. 1982: 98 (F; key, tax., bion., distr., review). **Ety.:** n.s. [white lip (L); poss. ref. to white markings on head or proboscis]. **Informal name:** White-lipped Solomon Islander Pointy Mosquito.

notoscriptus (Skuse)
1889: 1738 (M*, F; *Culex*). **Type loc.:** Sydney, New South Wales, Australia (MM). **Distr.:** Australia, Indonesia, New Caledonia, New Zealand, Papua New Guinea, Philippines, Solomon Isl., United States (continental). **Refs.:** Bancroft 1908: 24 (M, F; as genus *Scutomyia*); Brug 1934a: 513 (M*); Lee 1944a: 57 (L*); Penn 1949b: 43 (P*); Bick 1951: 408 (bion.); Belkin 1962: 348 (M*, F, P*, L*); Dobrotworsky 1965: 131 (M*, F*, L*); Mohrig 1967 (F*); Belkin 1968a: 103 (distr.); Maffi 1973b: 46 (distr.); Lee et al. 1982: 196 (F; key, tax., bion., distr., review); Linley et al. 1991b (E*); Derraik 2004 (distr.; New Zealand); Foley et al. 2004 (bion., mol. gen.); Reinert 2008f: Fig. 1 (F*); Becker et al. 2010: 358 (F*, L*; key, tax., distr., bion.); Endersby et al. 2013 (*s.l.*; mol. tax.); Peterson & Campbell 2015 (US invasive, niche model). **Ety.:** back (Gr), to write (L); ref. to thorax "with several fine longitudinal lines." **Informal name:** Scriptwriting Australian Pointy Mosquito. (Vol. 1, p. 454)

ssp. ***montanus*** Brug
1939: 104 (M*; as *Aedes* (*Finlaya*) *notoscriptus* var. *montana*). **Type loc.:** Lembang, Java (NHMUK). **Distr.:** Indonesia. **Refs.:** Stone et al. 1959: 167 (to ssp.). **Ety.:** from a mountain (L). **Informal name:** Montane Indonesian Pointy Mosquito.

quinquelineatus Edwards
1922d: 93 (F). **Type loc.:** Queensland, Australia (NHMUK). **Distr.:** Australia. **Refs.:** Marks 1955: 21 (M, P*, L*); Marks 1959b: 23 (M*); Lee et al. 1982: 243 (F; key, tax., bion., distr., review). **Ety.:** five lines (L); five lines of narrow golden scales on scutum. **Informal name:** Five-lined Australian Pointy Mosquito.

Subgenus *Rhinoskusea* Edwards

1929c: 342 (as subg. of genus *Aedes*). Type species: *Ficalbia longirostris* Leicester.

Name derivation: n.k. [rhino = nose, but Edwards names it that because "first hindtarsal segment as long as the tibia"(?); one could assume named for type species instead].

Important references: Mattingly 1953a: 35 (keys); Mattingly 1958a (tax., in part); Reinert 1976d (subg. revision); Reinert 2000h: 177 (to subg. of genus *Ochlerotatus*); Reinert 2002e: 34 (F gen.*); Reinert, Harbach, & Kitching 2004: 289 (phyl., class.; to genus); Reinert, Harbach, & Kitching 2008 (phyl., class.; to subg. of genus *Ochlerotatus*); Wilkerson et al. 2015 (phyl., class.; to subg. of genus *Aedes*); Soghigian et al. 2017 (mol. phyl.).

longirostris (Leicester)
1908: 228 (M, F; *Ficalbia*). **Type loc.:** Kuala Klang, Malaya [Malaysia] (NHMUK). **Distr.:** Australia, India, Indonesia, Malaysia, Papua New Guinea, Philippines, Singapore, Thailand, Vietnam. **Refs.:** Edwards 1926a: 120 (L*); Barraud 1934: 216 (M*, F, L); Mattingly 1958a: 38 (M*, F*, P*, L*); Basio 1971b: 26 (M*; bion.); Mattingly 1971a: Pl. 21

(P cephalothorax*); Mattingly 1974a: 41 (E*); Baisas 1974: 27 (M*, F*, P*, L*; tax., bion., distr.; Philippines); Lien et al. 1975 (distr.); Reinert 1976d: 8 (M*, F*, P*, L*; distr.); Lee et al. 1984: 255 (tax., review); Reinert 2002e: Fig. 37 (F*); Rattanarithikul et al. 2010 (F*, L*; keys, bion., distr.; Thailand). **Ety.:** long snout (L); proboscis. **Informal name:** Long-beaked Malaysian Pointy Mosquito.

syn. *hilli* Taylor
1919: 841 (M*; *Uranotaenia*). **Type loc.:** Darwin, Northern Territory, Australia (US).

pillaii Mattingly
1958a: 40 (M*, F*, P, L*). **Type loc.:** Kampong Sireh, Selangor, Malaya [Malaysia] (NHMUK). **Distr.:** Malaysia, Singapore. **Refs.:** Reinert 1976d: 14 (M*, F*, P*, L*; distr.). **Ety.:** n.s. [for Mr. Ganapathipillai (Harbach 2018c)]. **Informal name:** Pilla Malaysian Pointy Mosquito.

portonovoensis Tewari & Hiriyan
1992a: 123 (M*, F*, P*, L*). **Type loc.:** Pichavaram mangrove swamp, South Arcot District, India (USNM). **Distr.:** India. **Ety.:** Porto Novo, India. **Informal name:** Pichavaram Indian Pointy Mosquito [scientific name names "From Porto Novo," but the type locality is the Pichavaram mangrove swamp].

wardi Reinert
1976d: 18 (M*, F*, P*, L*). **Type loc.:** Subic US Naval Base, Zambales, Luzon, Philippines (USNM). **Distr.:** Indonesia, Malaysia, Philippines, Sri Lanka. **Ety.:** Ronald Anthony Ward. **Informal name:** Ward Philippine Pointy Mosquito.

Subgenus *Rusticoidus* Shevchenko & Prudkina

1973: 43, 46 (as subg. of genus *Aedes*; M*; key). Type species: *Aedes refiki* Medschid.

Name derivation: like *rusticus* (Gr); *rusticus* = a species of *Aedes*, -oides = resembling.

syn. *Feltianus* Martini 1930 (as *Aedes* subg. *Ochlerotatus* globus Felitanus). In Lindner 1930: 264. Type species: *Culex diversus* Theobald.

Important references: Reinert 1999d (tax., review); Reinert 2000h: (to subg. of genus *Ochlerotatus*); Reinert 2002e: 35 (F gen.*); Reinert, Harbach, & Kitching 2008 (phyl., class.); Reinert, Harbach, & Kitching 2009 (phyl., class.); Becker et al. 1910 (tax., bion.; Europe); Wilkerson et al. 2015 (to syn. subg. *Ochlerotatus* of genus *Aedes*); Wilkerson & Linton 2015 (class.; to subg. of genus *Aedes*); Soghigian et al. 2017 (mol. phyl.).

albescens Edwards
1921d: 308 (F). **Type loc.:** Omsk, Western Siberia [Russia] (HM). **Distr.:** Russia (Southern Dists.). **Refs.:** Gutsevich & Pendikova 1972a: 401 (revalidation); Gutsevich & Pendikova 1972b: 241–242; Robert et al. 2019 (distr.; Western Palearctic). **Ety.:** becoming white (L); ref. to "nearly all scales of the body whitish." **Informal name:** Whitening Russian Pointy Mosquito.

bicristatus Thurman & Winkler
1950: 239 (M*, F*, P*, L*). **Type loc.:** Kelseyville (Township 13 North, Range 9 West, Sections 11 and 12, elevation 1,390 ft), Lake County, California, United States (USNM). **Distr.:** United States (continental). **Refs.:** Yamaguti & LaCasse 1951d: 37 (M*, F*, L*); Bohart 1954: 358 (first instar L*); Carpenter & LaCasse 1955: 155 (M*, F*, L*; keys); Stone & Knight 1956a: 215 (type info., lectotype desig.); Dodge 1966: 350 (first instar L; key); Myers 1967: 798 (E*); Bohart & Washino 1978: 43 (M, F*, L*; keys, tax., bion., distr.); Reinert 2000f (from subg. *Ochlerotatus*); Darsie & Ward 2005 (F*, L*; keys, distr.). **Ety.:** n.s. [double plumed (L); poss. ref. to two groups of setae on the gonocoxite]. **Informal name:** Two-crested American Pointy Mosquito.

krymmontanus Alekseev
1989: 173 (M*, F*, L*). **Type loc.:** Southern slope of Crimean Mountains, Crimea, USSR [Crimean Peninsula]. **Distr.:** Crimean Peninsula. **Ety.:** Crimean Mountains. **Informal name:** Crimea Ukrainian Pointy Mosquito.

lepidonotus Edwards
1920a: 132 (M, F). **Type loc.:** Galiko River, Salonica, Macedonia, Greece (NHMUK). **Distr.:** FYRO Macedonia, Greece, Turkey. **Refs.:** Edwards 1921d: 308 (M*, F*); Medschid 1928: 306 (M*, F, L*; as *Aedes*); Martini 1930: 297 (M*, F, L*); Mohrig 1967 (F*); Reinert 1999d: 2 (from subg. *Ochlerotatus*); Becker et al. 2010: 257 (M*, F*, L*; keys, tax., distr., bion.); Robert et al. 2019 (distr.; Western Palearctic). **Ety.:** scaly back (Gr); scales on postnotum. **Informal name:** Scaly Greek Pointy Mosquito.

provocans (Walker)
1848: 7 (M; *Culex*). **Type loc.:** Nova Scotia, Canada (NHMUK). **Distr.:** Canada, United States (continental). **Refs.:** Howard et al. 1917 (E*; as *trichurus*); Rempel 1950: 221 (L*; as *trichurus*); Darsie 1951: 25 (P*; as *trichurus*); Yamaguti & LaCasse 1951d: 218 (M*, F*, L*; as *trichurus*); Rempel 1953: 479 (M*, F*; as *trichurus*); Carpenter & LaCasse 1955: 247 (M*, F*, L*; keys; as *trichurus*); Craig 1956 (E*; as *trichurus*); Price 1960: 557 (first instar L*; as *trichurus*); Dodge 1966: 358 (first instar L; key; as *trichurus*); Mohrig 1967 (F*); Belkin 1968b: 6 (lectotype desig.); Kalpage & Brust 1968 (E*; as *trichurus*); Wood 1977: 77 (M; distr.; from syn. with *punctor*); Reinert 2000f (from subg. *Ochlerotatus*); Darsie & Ward 2005 (F*, L*; keys, distr.); Harrison et al. 2016 (F*, L*; keys, distr.). **Ety.:** n.s. [challenge, provoke

(L)]. **Informal name:** Provocative Canadian Pointy Mosquito.

syn. *trichurus* Dyar
1904a: 170 (L*, E; *Culex*). **Type loc.:** British Columbia, Canada (NE). **Refs.:** Stone & Knight 1956a: 226 (type info.); Wood 1977: 77 (syn.).

syn. *cinereoborealis* Felt & Young
1904: 312 (A, L; *Culex*). **Type loc.:** [New York, United States] (NY). **Refs.:** Wood 1977 (tax., syn.).

syn. *pagetonotum* Dyar & Knab
1909d: 253 (M, F). **Type loc.:** Ottawa, Ontario, Canada (USNM). **Refs.:** Stone & Knight 1956a: 223 (type info., lectotype desig.). **Refs.:** Wood 1977 (tax., syn.).

syn. *poliochros* Dyar
1919b: 35 (F). **Type loc.:** Aweme, Manitoba, Canada (USNM). **Refs.:** Wood 1977 (tax., syn.).

quasirusticus Torres Cañamares
1951: 79 (M*, F*, P*, L*). **Type loc.:** Majadas, Cuenca, Spain (NHMUK). **Distr.:** Algeria, Morocco, Spain. **Refs.:** Townsend 1990: 126 (syntypes, NHMUK); Reinert 1999b: 35 (lectotype desig.); Reinert 1999d: 2 (from subg. *Ochlerotatus*; tax.); Melero-Alcíbar & Salom 2002 (distr.; Algeria); Melero-Alcíbar 2004 (P*; tax.); Becker et al. 2010: 258 (M*, F*, L*; keys, tax., distr., bion.). **Ety.:** nearly (L); similar to *rusticus*. **Informal name:** Suburban Spanish Pointy Mosquito [*quasirusticus* = "half rustic" = Suburban].

refiki Medschid
1928: 310 (M*, F, L*). **Type loc.:** Angora, Ankara, Turkey (NHMUK). **Distr.:** Austria, Bosnia and Herzegovina, Crimean Peninsula, Czech Republic, France, Germany, Greece, Hungary, Italy, Romania, Russia, Slovakia, Slovenia, Spain, Sweden, Switzerland, Turkey, Ukraine. **Refs.:** Martini 1931c: 326 (M*, F, L*); Mohrig 1969 (F*); Gutsevich et al. 1971: 225 (distr.); Townsend 1990: 128 (syntypes; change type dep. from LU to NHMUK); Reinert 2000j (M, F, L; lectotype desig.); Reinert 2002e: Fig. 38 (F*); Becker et al. 2010: 259 (M*, F*, L*; keys, tax., distr., bion.); Robert et al. 2019 (distr.; Western Palearctic). **Ety.:** Refik Saydam. **Informal name:** Refik Turkish Pointy Mosquito.

syn. *stampari* Apfelbeck
1930: 286 (M*, F*, L*). **Type loc.:** Sarajevo, Yugoslavia [Bosnia & Herzegovina] (LU). **Refs.:** Stone 1963: 125 (Peus (*in litt.*) = syn. of *refiki* Medschid).

syn. *pseudorusticus* Callot
1945: 31 (M*, L*). **Type loc.:** Clermont-Perraud (Puy de Dome), France (IPS). **Refs.:** Stone et al. 1959: 52 (syn. by Callot, *in litt.*).

rusticus (Rossi)
1790: 333 (A; *Culex*). **Type loc.:** Pisa [Tuscany], Italy (LU). **Distr.:** Albania, Algeria, Austria, Belgium, Bulgaria, Crimean Peninsula, Croatia, Czech Republic, Denmark, Estonia, France (includes Corsica), FYRO Macedonia, Germany, Greece, Hungary, Ireland, Italy (includes Sardinia and Sicily), Latvia, Lithuania, Luxembourg, Montenegro, Morocco, Netherlands, Norway, Poland, Portugal, Serbia, Slovakia, Spain, Sweden, Switzerland, Turkey, Ukraine, United Kingdom. **Refs.:** Brolemann 1919b (F*); Séguy 1924: Pl. XXIX (L*), 139, 140 (F*), 15, 139 (M gen.*); Marshall 1938 (E*); Clavero 1946: 14 (M*, F*, L*); Natvig 1948: 381 (M*, F, L*); Monchadskii 1951: 201 (L*); Gillett 1955 (E); Senevet & Andarelli 1958: 285 (P*); Dodge 1966: 356 (first instar L; key); Sicart & Larrouy 1968 (F*); Mohrig 1969 (F*); Gutsevich et al. 1974: 232 (M*, F, L*); Bozicic 1980: 93 (distr.); Reinert 1999d (from subg. *Ochlerotatus*); Dahl 1997 (E*); Reinert 1999d (to subg. *Rusticoidus*); Alten et al. 2000 (distr.; Turkey); Eritja et al. 2000 (distr.; Spain); Spungis 2000 (distr.; Latvia); Beck et al. 2003 (distr.; Luxembourg); Becker et al. 2003 (E*); Melero-Alcíbar 2005b (P; tax.); Becker et al. 2010: 261 (M*, F*, L*; keys, tax., distr., bion.); Robert et al. 2019 (distr.; Western Palearctic). **Ety.:** n.s. [rural, of the country (L)]. **Informal name:** Rustic Italian Pointy Mosquito.

syn. *maculatus* Meigen
1804: 4 (M; *Culex*). **Type loc.:** Europe (NE). **Refs.:** Marshall 1938: 107, 199 (tax.).

syn. *musicus* Leach
1825: 293 (A; *Culex*). **Type loc.:** Nice [Alpes-Maritimes], France (LU).

syn. *pungens* Robineau-Desvoidy
1827: 407 (M, F; *Culex*). **Type loc.:** Gallia [France] (NE).

syn. *quadratimaculatus* Macquart
1834: 34 (M, F; *Culex*). **Type loc.:** France (NE).

syn. *diversus* Theobald
1901c: 73 (F*; *Culex*). **Type loc.:** Tunbridge Wells, Kent, England (NHMUK). **Refs.:** Martini 1930: 277 (M*, F, L*).

syn. *luteovittata* Theobald
1901c: 85 (A; *Culex nemorosus* var.). **Type loc.:** None designated (NHMUK).

ssp. ***subtrichurus*** Martini
1927: 386 (M*, F). **Type loc.:** Eastern end of Gulf of Ismid, Turkey (NHMUK). **Distr.:** Turkey. **Refs.:** Mattingly 1955c (ssp. desig.); Harbach & Howard 2007: 43 (correction to ssp.). **Ety.:** nearly (L); similar to *trichurus*. **Informal name:** Ismid Turkish Pointy Mosquito.

subdiversus Martini
1926: 204 (M). **Type loc.:** Nowodewidschje (Saratov), USSR [Russia] (NHMUK). **Distr.:** FYRO Macedonia, Russia. **Refs.:** Martini 1931c: 332 (M*); Mattingly 1955c: 30 (tax.); Gutsevich et al. 1974: 239 (M*, F, L); Reinert 1999d: 2 (from subg. *Ochlerotatus*); Becker et al. 2010: 263 (M*, F*, L*; keys, tax., distr., bion.); Robert et al. 2019 (distr.; Western Palearctic). **Ety.:** nearly (L); smaller than diversus [ref. to basal lobe of the male hypopygium]. **Informal name:** Variable Russian Pointy Mosquito [*subdiversus* = "subdifferent" = Variable].

Subgenus *Sallumia* Reinert, Harbach, & Kitching

2008: 110 (phyl., class.; as subg. of genus *Ochlerotatus*). Type species: *Aedes hortator* Dyar & Knab.

Name derivation: Maria Anice Mureb Sallum.

Important references: Reinert, Harbach, & Kitching 2009 (phyl., class.; to genus); Reinert 2010a (F gen.*); Wilkerson et al. 2015 (phyl., class.; to subg. of genus *Aedes*).

hortator Dyar & Knab
1907a: 12 (F). **Type loc.:** Trinidad [Trinidad & Tobago] (USNM). **Distr.:** Bolivia, Brazil, Colombia, Ecuador, French Guiana, Guyana, Panama, Peru, Suriname, Trinidad and Tobago. **Refs.:** Lane 1953: 649 (F*, P, L*); Stone & Knight 1956a: 219 (type info.); Pecor et al. 2000 (distr.; Peru); Reinert 2010a: Fig. 1 (F*). **Ety.:** n.s. [encourager (L)]. **Informal name:** Encouraging Trinbagonian Pointy Mosquito.

perventor Cerqueira & Costa
1946: 135 (M*, F). **Type loc.:** Mangaratiba, Rio de Janeiro, Brazil (INER). **Distr.:** Brazil, French Guiana. **Refs.:** Lane 1953: 650 (M*, F); Forattini 1958: 177 (P*); Belkin et al. 1971: 19 (type info.); Xavier 1973: 160 (type info.); Fauran & Pajot 1974a: 101 (distr.). **Ety.:** n.s. [come to, reach, arrive (L)]. **Informal name:** Arrival Brazilian Pointy Mosquito.

Subgenus *Scutomyia* Theobald

1904c: 77 (as genus). Type species: *Scutomyia albolineata* Theobald.

Name derivation: n.s. [poss. ref. to all flat scutellar scales].

Important references: Knight & Rozeboom 1946 (as Albolineatus Group of subg. *Stegomyia*); Huang 1979a (tax., keys; as Albolineatus Group of subg. *Stegomyia*); Reinert 1985 (revision; from syn. with subg. *Stegomyia* to subg. of genus *Aedes*); Reinert 2000e (F gen.*); Reinert, Harbach, & Kitching 2004: 289 (phyl., class.; to genus); Reinert, Harbach, & Kitching 2009 (phyl., class.); Rattanarithikul et al. 2010 (F*, L*; keys, bion.; Thailand); Wilkerson et al. 2015 (phyl., class.; to subg. of genus *Aedes*); Soghigian et al. 2017 (mol. phyl.).

albolineatus (Theobald)
1904c: 77 (F; *Scutomyia*). **Type loc.:** [Ampang Jungle, near] Kuala Lumpur [Selangor], Malaya [Malaysia] (NHMUK). **Distr.:** Cambodia, India, Indonesia, Malaysia, Mariana Isl., Papua New Guinea, People's Republic of China, Philippines, Singapore, Solomon Isl., Taiwan, Thailand, Vietnam. **Refs.:** Knight & Rozeboom 1946: 84 (M*, F, P*, L*; tax.); Penn 1949b: 59 (P*); Belkin 1962: 444 (M*, F, P*, L*); Basio 1971b: 27 (M*, P*; bion.); Lien et al. 1977 (distr.); Taylor & Maffi 1978: 216 (distr.); Huang 1979a: 7 (M*, F*, P*, L*; keys, distr.; as subg. *Stegomyia*); Reinert 1985 (from subg. *Stegomyia*); Lee et al. 1987a: 70 (F; key, tax., bion., distr., review); Reinert 2000e: Fig. 32 (F*); Jeffery et al. 2010 (distr.; Singapore); Rattanarithikul et al. 2010 (F*, L*; keys, bion., distr.; Thailand). **Ety.:** n.s. [white lined (L); prob. ref. to "thorax black, with a broad median silvery white line in front"]. **Informal name:** White-lined Malaysian Pointy Mosquito.

arboricolus Knight & Rozeboom
1946: 88 (M*, F, P, L*). **Type loc.:** Sohoton Springs, Basey River, Samar, Philippines (USNM). **Distr.:** Philippines. **Refs.:** Knight & Hull 1952: 185 (L*; tax.); Basio 1971b: 29 (M*; bion.); Huang 1979a (M*, P*, L*; keys, distr.; as subg. *Stegomyia*); Reinert 1985 (from subg. *Stegomyia*). **Ety.:** tree inhabitant (L). **Informal name:** Tree-dwelling Philippine Pointy Mosquito.

bambusicolus Knight & Rozeboom
1946: 94 (M*). **Type loc.:** Culion Island [Calamian Group], Philippines (USNM). **Distr.:** Philippines. **Refs.:** Ross 1950: 79 (M*, F*, P*, L*); Basio 1971b: 29 (M*; bion.); Huang 1979a (M*, P*, L*; keys, distr.; as subg. *Stegomyia*); Reinert 1985 (from subg. *Stegomyia*). **Ety.:** bamboo inhabitant (L). **Informal name:** Bamboo-dwelling Philippine Pointy Mosquito.

boharti Knight & Rozeboom
1946: 90 (M*, F, P, L). **Type loc.:** Osmena, Samar, Philippines (USNM). **Distr.:** Malaysia, Philippines. **Refs.:** Knight & Hull 1952: 185 (L*); Basio 1971b: 29 (M*; bion.); Baisas 1974: 35 (M, F*, P*, L; tax., bion., distr.; Philippines); Huang 1979a (M*, P*, L*; keys, distr.; as subg. *Stegomyia*); Reinert 1985 (from subg. *Stegomyia*). **Ety.:** George Edward Bohart. **Informal name:** Bohart Philippine Pointy Mosquito.

hoogstraali Knight & Rozeboom
1946: 92 (M*, F, P*, L*). **Type loc.:** Subic Bay [Zambales], Luzon, Philippines (USNM). **Distr.:** Philippines. **Refs.:** Knight & Hull 1952: 187 (? = *impatibilis*); Basio 1971b: 30 (M*, P*; bion.); Huang 1978c (F*); Huang 1979a (M*, P*, L*; keys, distr.; as subg. *Stegomyia*); Reinert 1985 (from subg. *Stegomyia*). **Ety.:** Harry Hoogstraal. **Informal name:** Hoogstraal Philippine Pointy Mosquito.

impatibilis (Walker)
1859: 91 (M; *Culex*). **Type loc.:** Makassar, Celebes [Sulawesi] (NHMUK). **Distr.:** Indonesia, Philippines. **Refs.:** Knight & Hull 1952: 187 (from syn. with *aegypti*); Huang 1978c (F*); Huang 1979a (M*, P*; keys, distr.; as subg. *Stegomyia*); Reinert 1985 (from subg. *Stegomyia*). **Ety.:** n.k. **Informal name:** Unendurable Indonesian Pointy Mosquito.

laffooni Knight & Rozeboom
1946: 94 (M, F). **Type loc.:** San Ramon [Zamboanga], Mindanao, Philippines (USNM). **Distr.:** Philippines, Vietnam. **Refs.:** Knight & Hull 1951b: 221 (L); Knight & Hull 1952: 186 (? = *pseudalbolineatus* ssp.); Basio 1971b: 30 (M*; bion.); Huang 1978b (F*); Huang 1979a (M*, F*, P*, L*; keys, distr.; as subg. *Stegomyia*); Reinert 1985 (from subg. *Stegomyia*). **Ety.:** Jean Luther Laffoon. **Informal name:** Laffoon Philippine Pointy Mosquito.

platylepidus Knight & Hull
1951a: 201 (F). **Type loc.:** Puerto Princesa, Palawan, Philippines (USNM). **Distr.:** Philippines. **Refs.:** Mattingly 1959: 41 (F; to subg. *Diceromyia*); Reinert 1970a: 12 (M*, F*); Basio 1971b: 13 (M*; bion.); Huang 1979a (M*; from subg. *Diceromyia* to subg. *Stegomyia*); Reinert & Ramalingam 1983 (P*, L*); Reinert 1985 (from subg. *Stegomyia*). **Ety.:** n.s. [broad scale (Gr); prob. ref. to "broad plate-like dark (with metallic reflections) scutal and scutellar scales"]. **Informal name:** Broad-scaled Philippine Pointy Mosquito.

pseudalbolineatus Brug
1939: 103 (M*, F, L). **Type loc.:** Malino, Lamboean, Beroko, and Kabaena Island, Celebes [Sulawesi]; Taroena Islands; and Sangir Islands (NHMUK). **Distr.:** Indonesia. **Refs.:** Knight & Rozeboom 1946: 88 (M, F, L); Huang 1978b (F*); Huang 1979a (M*, F*; keys, distr.; as subg. *Stegomyia*); Reinert 1985 (from subg. *Stegomyia*). **Ety.:** false (Gr); similar to *albolineatus*. **Informal name:** False-white Indonesian Pointy Mosquito.

Subgenus *Skusea* Theobald

1903a: 291 (as genus). Type species: *Aedes pembaensis* Theobald (Blanchard 1905: 416).

Name derivation: n.s., prob. Frederick Arthur Askew Skuse.

Important references: Brunhes 1977b: 165 (M gen*; tax.; western Indian Ocean); Reinert 2000e: 42 (F gen.*); Reinert, Harbach, & Kitching 2004: 289 (phyl., class.; to genus); Reinert, Harbach, & Kitching 2009 (phyl., class.); Wilkerson et al. 2015 (phyl., class.; to subg. of genus *Aedes*).

cartroni (Ventrillon)
1906b: 143 (F; *Stegomyia*). **Type loc.:** Morondava, Madagascar (NHMUK, MNHP). **Distr.:** Comoros, Madagascar (includes Glorioso and Juan De Nova Isl.). **Refs.:** Edwards 1941: 223 (? = *pembaensis*); Brunhes 1977a (distr.; Comoros Isl.); Brunhes 1977b: 166 (from syn. with *pembaensis*). **Ety.:** n.s. [Dr. Cartron (Harbach 2018c)]. **Informal name:** Cartron Malagasy Pointy Mosquito.

lambrechti van Someren
1971: 21 (M*, F, P*, L). **Type loc.:** Seychelles (NE). **Distr.:** Madagascar (includes Glorioso and Juan De Nova Isl.), Seychelles. **Refs.:** Townsend 1990: 92 (change type dep. from NHMUK to NE). **Ety.:** Frank Laurent Lambrecht. **Informal name:** Lambrecht Seychellois Pointy Mosquito.

moucheti Ravaonjanahary & Brunhes
1977: 213 (M*). **Type loc.:** Maradoka, Nosy Be Island, Madagascar (IRSM). **Distr.:** Madagascar (includes Glorioso and Juan De Nova Isl.). **Ety.:** René Libert Joseph Mouchet. **Informal name:** Mouchet Malagasy Pointy Mosquito.

pembaensis Theobald
1901c: 235 (F*). **Type loc.:** Pemba Island [Tanzania] (NHMUK). **Distr.:** Comoros, Kenya, Madagascar (includes Glorioso and Juan De Nova Isl.), Mozambique, Seychelles, Tanzania. **Refs.:** Edwards 1941: 223 (M*, F), 403 (P); Hopkins 1952: 222 (L*); Goiny et al. 1957: 1 (E*); Mattingly 1958c: 108 (type info.); Mattingly 1959 (E*); Worth et al. 1961: 257 (bion.); Mohrig 1967 (F*); da Cunha Ramos & Ribeiro 1995 (P*); Jupp 1996 (M*, F*; key); Reinert 2000e: Fig. 33 (F*). **Ety.:** Pemba Isl., Tanzania. **Informal name:** Pemba Tanzanian Pointy Mosquito.
syn. *mutica* Enderlein
1923: 28 (F; *Phagomyia*). **Type loc.:** Patta Island [Kenya] (ZM).

Subgenus *Stegomyia* Theobald

1901 (as genus). In Howard 1901: 235. Type species: *Culex aegypti* Linnaeus 1762 (Neveu-Lemaire 1902c: 211).

Name derivation: Egypt.

syn. *Quasistegomyia* Theobald 1906b: 69 (as genus). Type species: *Quasistegomyia unilineata* Theobald.
syn. *Pseudostegomyia* Ludlow 1908b: 10 (as genus). Type species: [*Stegomyia*] *gardnerii* Ludlow.
Important references: Theobald 1910c: 135.

syn. *Kingia* Theobald 1910c: 135 (as genus) (not Schloenbach 1866). Type species: *Stegomyia luteocephala* Newstead.
Important references: Brunetti 1914: 63 (type species desig.).

syn. *Aniella* Enderlein 1923: 26 (as genus). Type species: *Stegomyia africana* Theobald.
syn. *Actinothrix* Reinert, Harbach, & Kitching 2009: 781 (as subg. of genus *Stegomyia*). Type species: *Stegomyia edwardsi* Barraud.
Important references: Wilkerson et al. 2015 (phyl., class.; to syn. of subg. *Stegomyia* of genus *Aedes*).

syn. *Bohartius* Reinert, Harbach, & Kitching 2009: 781. Type species: *Aedes* (*Stegomyia*) *pandani* Stone (as subg. of genus *Stegomyia*).
Important references: Wilkerson et al. 2015 (phyl., class.; to syn. with subg. *Stegomyia* of genus *Aedes*).

syn. *Heteraspidion* Reinert, Harbach, & Kitching 2009: 781 (as subg. of genus *Stegomyia*). Type species: *Stegomyia annandalei* Theobald.

Important references: Huang 1977a: 15 (as Annandalei Subgroup; tax., keys; SE Asia); Huang 1979a (as Annandalei Subgroup; tax., keys; Oriental Region); Rattanarithikul et al. 2010 (F*, L*; keys; Thailand); Wilkerson et al. 2015 (phyl., tax.; to syn. of subg. *Stegomyia* of genus *Aedes*).

syn. *Huangmyia* Reinert, Harbach, & Kitching 2009: 782 (phyl., class.; as subg. of genus *Stegomyia*).

Important references: Huang 1977a: 31 (tax.; as Mediopunctatus Subgroup); Huang 1979a (tax., keys; as Mediopunctatus Subgroup; Oriental Region); Rattanarithikul et al. 2010 (F*, L*; keys; Thailand); Wilkerson et al. 2015 (phyl., class.; to syn. with subg. *Stegomyia* of genus *Aedes*).

syn. *Mukwaya* Reinert, Harbach, & Kitching 2009: 782 (phyl., class.; as subg. of genus *Stegomyia*). Type species: *Stegomyia simpsoni* Theobald.

Important references: Huang 2004 (tax., keys; Simpsoni Group of subg. *Stegomyia* of genus *Aedes*); Wilkerson et al. 2015 (phyl., class.; to syn. of subg. *Stegomyia* of genus *Aedes*).

syn. *Xyele* Reinert, Harbach, & Kitching 2009: 783 (as subg. of genus *Stegomyia*). Type species: *Stegomyia desmotes* Giles.

Important references: Huang 1977a: 25 (as Desmotes Subgroup of subg. *Stegomyia*); Reinert 2000e: 46 (F gen.*; as Desmotes Assemblage); Rattanarithikul et al. 2010 (F*, L*; keys, bion.; Thailand); Wilkerson et al. 2015 (phyl., class.; to syn. of subg. *Stegomyia* of genus *Aedes*).

syn. *Zoromorphus* Reinert, Harbach, & Kitching 2009: 783 (as subg. of genus *Stegomyia*). Type species: *Aedes (Stegomyia) futunae* Belkin.

Important references: Wilkerson et al. 2015 (phyl., class.; to syn. of subg. *Stegomyia* of genus *Aedes*).

Important references: Theobald 1901a: 283 (Nov. 23); Theobald 1901b: 235 (July 15); Theobald 1901d: ii (Sept.); Neveu-Lemaire 1902c: 211 (desig. of type species); Mattingly 1952: 235 (keys; Ethiopian Region); Mattingly 1953a (Ethiopian Region); Marks 1954: 349 (keys; Scutellaris Subgroup); Belkin 1962 (keys, tax., bion; South Pacific); Mattingly et al. 1962 (proposal to the International Commission on Zoological Nomenclature to validate name *Culex aegypti* Linnaeus 1762 for the yellow fever mosquito); International Commission on Zoological Nomenclature 1964 (Opinion 711; validation of *Culex aegypti* Linnaeus 1762 as name for the yellow fever mosquito); Gerberg & van Someren 1970 (keys; East Africa); Huang 1970c (overview; SE Asia); Huang 1972c (Scutellaris Group; SE Asia); Reinert 1973a: 28 (tax.); Huang 1977a (SE Asia); Huang 1979a (tax., keys; Oriental Region); Huang & Hitchcock 1980 (keys; Scutellaris Group in Fiji, Tonga, Samoa); Lee et al. 1987a (tax., key, bion.; Australasia); Huang 1988b (Pseudonigeria Group; Afrotropical Region); Huang 1990 (tax.; Africanus Group; Afrotropical Region); Huang 1997 (tax.; Dendrophilus Group; Afrotropical Region); Reinert 2000e (F gen.*); Huang 2004 (tax., keys; Afrotropical Region); Reinert, Harbach, & Kitching 2004 (phyl., tax.; to genus); Reinert, Harbach, & Kitching 2009 (phyl., tax.); Rattanarithikul et al. 2010 (F*, L*; keys, bion.; Thailand); Wilkerson et al. 2015 (phyl., tax.; to subg. of genus *Aedes*); Soghigian et al. 2017 (mol. phyl.).

aegypti (Linnaeus)
1762: 470 (A; *Culex*). **Type loc.:** Kuala Lumpur, Selangor, Malaya [Malaysia] (NHMUK). **Distr.:** Afghanistan, Algeria, Angola, Anguilla, Argentina, Aruba, Australia, Azerbaijan, Bahamas, Bangladesh, Belize, Benin, Bermuda, Bhutan, Bolivia, Bonaire, Bosnia and Herzegovina, Botswana, Brazil, Brunei, Burkina Faso, Cambodia, Cameroon, Canada, Cape Verde Isl., Cayman Isl., Central African Republic, Chad, Chile, Colombia, Comoros, Congo, Cook Isl. (Polynesia), Costa Rica, Côte d'Ivoire, Croatia, Cuba, Democratic Republic of the Congo, Djibouti, Dominican Republic, Ecuador, Egypt, El Salvador, Equatorial Guinea, Eritrea, Ethiopia, Fiji, France, French Guiana, French Polynesia, French Southern Territories, Gabon, Gambia, Georgia, Ghana, Guam, Guatemala, Guinea, Guinea Bissau, Guyana, Haiti, Honduras, Hong Kong, India (includes Andaman Isl.), Indonesia (includes Flores, Java, Kalimantan, Moluccas, Sumatra [includes Ketulauan Riau Archipelago, Sulawesi, Timor]), Iran, Iraq, Israel (Gaza Strip and West Bank), Italy, Jamaica, Japan, Kenya, Kiribati, Korea, Republic of Laos, Lebanon, Lesser Antilles (Antigua and Barbuda, Barbados, Curaçao, Dominica, Grenada, Guadeloupe, Martinique, Montserrat, Saba, Saint Kitts and Nevis, Saint Lucia, Saint Martin, Saint Vincent and the Grenadines, Trinidad and Tobago, Virgin Isl.), Liberia, Libya, Madagascar (includes Glorioso and Juan De Nova Isl.), Malawi, Malaysia, Maldives, Mali, Mariana Isl., Marshall Isl., Mauritania, Mauritius, Mayotte, Mexico, Micronesia (Caroline Isl.), Micronesia (Wake Isl.), Morocco, Mozambique, Myanmar, Namibia, Nauru, Nepal, Netherlands, New Caledonia, Nicaragua, Niger, Nigeria, Niue, Oman, Pakistan, Panama, Papua New Guinea, Paraguay, Peru, People's Republic of China (includes Hainan), Philippines, Portugal, Puerto Rico, Republic of South Africa, Reunion, Russia (Southern Dists.), Rwanda, Samoa (Ind. State of Samoa, American Samoa, Western Samoa), São Tomé and Príncipe, Saudi Arabia, Senegal, Seychelles, Sierra Leone, Singapore, Solomon Isl., Somalia, Spain, Sri Lanka, Sudan and South Sudan, Suriname, Syria, Tahiti, Taiwan, Tanzania, Thailand, Timor, Togo,

Tokelau, Tonga, Tunisia, Turkey, Tuvalu, Uganda, United Kingdom, United States (continental, Hawaiʻi), Uruguay, Vanuatu, Venezuela, Vietnam, Wallis and Futuna, Yemen, Zambia, Zimbabwe. **Refs.:** Goeldi 1905 (*E; as syn. *fasciata* Fabricius 1805); Banks 1908 (M*, F*; as *Stegomyia persistans*); Bancroft 1908: 20 (M, F; "Tiger Mosquito"; *Stegomyia fasciata*); Howard et al. 1913 (E*; as syn. *calopus* Meigen 1818); Howard et al. 1917: 827 (M*, F, P*, L*, E*; as syn. *calopus* Meigen 1818); Dyar 1920i: 204 (tax.); Macfie & Ingram 1922 (F*); Barraud 1923g (L*; as *Stegomyia argentea*); Martini 1929, 1930, 1931a, 1931b, 1931c (*E; as syn. *fasciata* Fabricius 1805); Gerry 1932 (F*); Patton 1933: 182 (tax.); Barraud 1934: 221 (M*, F, L*); Edwards 1941: 128 (M*, F, P*; bion.); Bohart & Ingram 1946b: 6, 37, 66 (M*, F*, P*, L*; distr., bion.); Ross 1947 (F*); La Casse & Yamaguti 1950 (F*); Darsie 1951: 10 (P*); Yamaguti & LaCasse 1951d: 247 (M*, F*, L*); Hopkins 1952: 134 (L*; bion.); Knight & Hull 1952: 167 (M, F, L*); Carpenter & LaCasse 1955: 261 (M*, F*, L*; keys); Teesdale 1955: 711 (bion.); Horsfall 1955: 476 (review); Iyengar & Menon 1955 (distr.); Bohart 1957: 6 (distr.; Micronesia); Craig 1956 (E*); Mattingly 1957b: 392 (tax., bion.); Hara 1957 (F*); Craig & Horsfall 1958 (E*); Christophers 1960: 1 (E*; bion., morph.); Craig & Horsfall 1960 (E*); Belkin 1962: 8, 443 (distr.); Mattingly et al. 1962: 208 (tax., neotype desig., type loc. desig.); Clements 1963 (F*); International Commission on Zoological Nomenclature 1964 (validation of name); Spielman 1964 (F*); Mattingly 1965b (F*); Ross & Horsfall 1965 (M*, F*, L*, E*; keys); Dodge 1966: 348 (first instar L*; key); Mohrig 1967 (F*); Hinton & Service 1969 (E*); Pratt & Kidwell 1969 (E*); Belkin et al. 1970: 184 (M*, F*, P*, L*); Horsfall et al. 1970: 1710 (E*); Rjazantzeva 1970 (F*); Aslamkhan 1971b (distr.; Pakistan); Basio 1971b: 27 (M*; bion.); Lambrecht & van Someren 1971: 483 (distr.); Mattingly 1971a: Pl. 18 (P*), Pl. 30 (L head*), Pl. 31 (L head*, terminal segs.*); Laffoon & Knight 1971 (F*); Matsuo, Yoshida, & Kunou 1972: 360 (E*); Harrison & Rattanarithikul 1973: 280 (L*; first and third instars); Hartberg & Craig 1973: 206 (genetic tax.); Moriya et al. 1973 (E*); Matsuo, Yoshida, & Lien 1974: 180 (E*); Baisas 1974: 30 (M, F, P, L; tax., bion., distr.; Philippines); Tanaka et al. 1975c: 222 (bion., distr.); Paterson et al. 1976: 252 (Africa; *s.l.*); Ramalingam 1976: 304 (tax., bion., keys., distr.; Samoa & Tonga); Brunhes 1977a (distr.; Comoros Isl.); Huang 1977c: 292 (distr.); Rodhain et al. 1977 (distr.); Rossignol & McIver 1977 (F*); Taylor & Maffi 1978: 216 (distr.); Harbach & Knight 1978 (E*); Huang 1979a (M*, F*, P*, L*; keys, distr.); Tanaka et al. 1979: 396 (M*, F*, L*); White 1980: 140 (distr.); Harbach & Knight 1980 (F*); Hinton 1981 (E); Ahmed 1987 (distr.; Bangladesh); Lee et al. 1987a: 14 (F; key, tax., bion., distr., review); Linley 1989b (E*); Jupp 1996 (M*, F*; key); Clements 1999 (F*); Darsie & Ward 2000 (distr.; United States); Reinert 2000e: Fig. 34 (F*); Jarial 2001 (E*); Chadee 2003: 199 (distr.; Trinidad & Tobago); Pena et al. 2003 (distr.; Dominican Republic); Rossi & Martínez 2003: 471 (distr.; Uruguay); Toto et al. 2003: 344 (distr.; Equatorial Guinea); Trari et al. 2002: 331 (distr.; Morocco); Carpintero & Leguizamón 2004: 501 (distr.; Argentina); Huang 2004 (M*, F*; distr., keys); Darsie & Ward 2005 (F*, L*; keys, distr.); McClelland 2009 (tax.); Rattanarithikul et al. 2010 (F*, L*; keys, bion., distr.; Thailand); Becker et al. 2010: 198 (M*, F*, L*; keys, tax., distr., bion.); Berti et al. 2015 (distr.; Venezuela); Kraemer et al. 2015 (distr.); Harrison et al. 2016 (F*, L*; keys, distr.); Robert et al. 2019 (distr.; Western Palearctic). **Ety.:** Egypt. **Informal name:** Yellow Fever Malaysian Pointy Mosquito. Official ESA common name: yellow fever mosquito. (Vol. 1, p. 456)

syn. *argenteus* Poiret
1787: 245 (F; *Culex*). **Type loc.:** Barbary [Barbary Coast?] (NE). **Refs.:** Knab 1916a: 59 (tax.); Kirkpatrick 1925b: 87 (M*, F, P*, L*).

syn. *fasciatus* Fabricius
1805: 36 (A; *Culex*). **Type loc.:** Americae Insulis [Antilles] (ZMC). **Refs.:** Theobald 1901a: 289 (M*, F*; tax.); Barraud 1928c: 377 (comprehensive review); Martini 1930: 245 (M*, F*, P*, L*, E*; bion.); Bonne-Wepster & Brug 1932: 47 (M*, F, L*; bion.); Belkin 1968b: 5 (lectotype desig.).

syn. *calopus* Meigen
1818: 3 (M, F; *Culex*). **Type loc.:** Portugal (NE).

syn. *mosquito* Robineau-Desvoidy
1827: 407 (M; *Culex*). **Type loc.:** Cuba (NE). **Refs.:** Theobald 1901a: 295 (tax.).

syn. *frater* Robineau-Desvoidy
1827: 408 (M*; *Culex*). **Type loc.:** West Indies [Antilles] (NE).

syn. *taeniatus* Wiedemann
1828: 10 (M, F; *Culex*). **Type loc.:** Savannah [Georgia, United States] (NMW?). **Refs.:** Belkin 1968b: 7 (tax.).

syn. *sugens* Wiedemann
1828: 545 (F; *Culex*). **Type loc.:** Nubia [Northeast Africa] (SNG).

syn. *kounoupi* Brullé
1833: 289 (A; *Culex*). **Type loc.:** Morea, Greece (NE).

syn. *toxorhynchus* Macquart
1838a: 39 (F; *Culex*). **Type loc.:** Rio de Janeiro, Guanabara, Brazil (MNHP). **Refs.:** Lane 1944: 183 (? = *Toxorhynchites theobaldi*); Belkin 1968b: 8 (change in syn.); Belkin et al. 1971: 20 (type loc.info.).

syn. *annulitarsis* Macquart
1846a: 136 (F; *Culex*). **Type loc.:** De l'ile de France [Mauritius] (NE). **Refs.:** Mattingly & Bruce-Chwatt 1954: 191 (tax., syn.).

syn. *viridifrons* Walker
1848: 3 (F; *Culex*). **Type loc.:** Type loc. not stated: "Presented by Captain Lord Byron" (NHMUK).

syn. *excitans* Walker
1848: 4 (F; *Culex*). **Type loc.:** Georgia [United States] (NHMUK).
ssp. ***formosus*** (Walker)
1848: 4 (F; *Culex*). **Type loc.:** Sierra Leone (NHMUK). **Distr.:** Angola, Côte d'Ivoire, Gabon, Kenya, Nigeria, Republic of South Africa, Sierra Leone, Senegal, Tanzania, Uganda. **Refs.:** Mattingly 1957b: 395 (to ssp.); Service 1976a (distr.; Gabon); Huang 2004 (F*; tax., keys, distr.). **Ety.:** beautiful (L); [not named for island of Formosa]. **Informal name:** Beautiful Sierra Leonean Pointy Mosquito.
syn. *inexorabilis* Walker
1848: 4 (F; *Culex*). **Type loc.:** West Africa (NHMUK).
syn. *exagitans* Walker
1856a: 430 (F; *Culex*). **Type loc.:** Para [Brazil] (NHMUK). **Refs.:** Belkin et al. 1971: 20 (type info.).
syn. *insatiabilis* Bigot
1859: 118 (M*, F*; *Culex*). **Type loc.:** Madagascar (NE).
syn. *bancrofti* Skuse
1889: 1740 (M*; *Culex*). **Type loc.:** Brisbane, Queensland, Australia (MM).
syn. *elegans* Ficalbi
1890b: 95 (M, F; *Culex*). **Type loc.:** Italy (LU).
syn. *rossii* Giles
1899: 64 (M, F; *Culex*). **Type loc.:** Calcutta [West Bengal], India (LU).
syn. *queenslandensis* Theobald
1901a: 297 (F; *Stegomyia fasciata* var.). **Type loc.:** Burpengary, Queensland, Australia (NHMUK). **Refs.:** Lewis 1945: 9 (M, F); Mattingly 1953a: 46, 60 (tax.); van Someren, Teesdale, & Furlong 1955: 473 (M, F); Mattingly & Knight 1956: 100, 119, 135 (tax.); Mattingly 1957b: 395 (to var.); Huang 1979a: 2, 41 (syn.); Lee et al. 1987b: 14 (F; key, tax., bion., distr., review; to syn. with *aegypti*).
syn. *luciensis* Theobald
1901a: 297 (M, F; *Stegomyia fasciata* var.). **Type loc.:** Saint Lucia [Lesser Antilles] (NHMUK). **Refs.:** Barraud 1934: 222 (A); Belkin 1968b: 6 (lectotype desig.).
syn. *nigeria* Theobald
1901a: 303 (F*; *Stegomyia*). **Type loc.:** Bonny [Eastern Provinces, Nigeria] (NHMUK).
syn. *lamberti* Ventrillon
1904: 552 (M, F; *Stegomyia*). **Type loc.:** Ankazobe, Diego Suarez, Majunga, Madagascar (NHMUK). **Refs.:** Rodhain & Boutonnier 1985: 275 (from syn. with *albopictus*; syntypes PIP; syn. with *aegypti*); Townsend 1990: 92 (syntypes; change type dep. from LU to NHMUK); Ward 1992 (as syn. of *aegypti*).
syn. *canariensis* Pittaluga
1905: 432 (M*, F*; *Stegomyia calopus* var.). **Type loc.:** Canary Islands (LU).
syn. *persistans* Banks
1906b: 996 (M, F; *Stegomyia fasciata* ssp.). **Type loc.:** Manila, Negros Occidental, and Iloilo, Philippines (NE).
syn. *anguste-alatus* Becker
1908: 79 (M; *Culex*). **Type loc.:** Tenerife, Canary Islands (ZM).
syn. *albopalposus* Becker
1908: 80 (F; *Culex*). **Type loc.:** Santa Cruz, Tenerife, Canary Islands (ZM).
syn. *alboannulis* Ludlow
1911b: 132 (F; *Duttonia*). **Type loc.:** Mindanao, Philippines (USNM).
syn. *pulcherrima* Taylor
1919: 830 (M, *Mimeteomyia*). **Type loc.:** Cairns, Queensland, Australia (US). **Refs.:** Knight & Marks 1952: 517 (syn.).
syn. *atritarsis* Edwards
1920a: 129 (M; *Stegomyia fasciata* var.). **Type loc.:** Accra, Ghana (NHMUK). **Refs.:** Edwards 1941: 132 (M, F).

africanus (Theobald)
1901d: 3 (F; *Stegomyia*). **Type loc.:** Duke Town and Bonny [Eastern Provinces], Nigeria (NHMUK). **Distr.:** Angola, Benin, Burkina Faso, Cameroon, Central African Republic, Congo, Côte d'Ivoire, Democratic Republic of the Congo, Equatorial Guinea, Ethiopia, Gabon, Ghana, Guinea, Kenya, Liberia, Mali, Mozambique, Nigeria, São Tomé and Príncipe, Senegal, Sierra Leone, Sudan and South Sudan, Thailand, Gambia, Uganda, Zambia. **Refs.:** Edwards 1941: 150 (M*, F*), 391 (P*); Hopkins 1952: 156 (L*); Mattingly 1952: 241 (tax.); Mattingly & Bruce-Chwatt 1954: 187 (M*, F*); Mohrig 1967 (F*); Service 1976a (distr.; Gabon); Huang 1990: 14 (M*, F*, P*, L*; lectotype desig., distr.); Linley et al. 1993c (E*); Reinert 2000e: Fig. 35 (F*); Toto et al. 2003: 344 (distr.; Equatorial Guinea); Huang 2004: (M*, F*; tax., adult key, distr.); Rueda 2004 (F*, L*; keys for African dengue vectors). **Ety.:** Africa. **Informal name:** Local Nigerian Pointy Mosquito [Nigeria is "Local" to Africa]. (Vol. 1, p. 458)
syn. *dubia* Theobald
1910c: 170 (F*; *Stegomyia*). **Type loc.:** Katemar, Bihe, Angola (NHMUK).

agrihanensis Bohart
1957: 39 (M*, F*, P*, L*). **Type loc.:** Southwest village, Agrihan, Mariana Islands (USNM). **Distr.:** Mariana Isl. **Refs.:** Lee et al. 1987a: 69 (tax., distr.). **Ety.:** Agrihan, Mariana Isl. **Informal name:** Agrihan Mariana Islander Pointy Mosquito.

albopictus (Skuse)
1895: 20 (F*; *Culex*). **Type loc.:** Calcutta [West Bengal], India (USNM). **Distr.:** Albania, Algeria, Argentina, Armenia, Australia, Austria, Bangladesh, Belgium, Belize, Bermuda, Bhutan, Bolivia, Borneo, Bosnia and Herzegovina, Brazil, British Indian Ocean Territory, Brunei, Bulgaria, Cambodia, Cameroon, Canada, Cayman Isl., Central African Republic, Cocos (Keeling) Isl., Colombia, Comoros, Congo, Cook Isl. (Polynesia),

Costa Rica, Côte d'Ivoire, Croatia, Cuba, Czech Republic, Democratic Republic of the Congo, Dominican Republic, Ecuador, El Salvador, Equatorial Guinea, Fiji, France (includes Corsica), French Polynesia, French Southern Territories, FYRO Macedonia, Gabon, Georgia, Germany, Greece (includes Crete), Guam, Guatemala, Haiti, Honduras, Hong Kong, Hungary, India (includes Andaman Isl.), Indonesia (Flores, Java, Ketulauan Riau Archipelago, Moluccas, Sumatra, Timor), Israel (Gaza Strip and West Bank), Iran, Italy (includes Giglo, Sardinia, Sicily, Ventotene, Ustica Isl.), Japan, Jordan, Kosovo, Laos, Lebanon, Lesser Antilles (Barbados, Trinidad and Tobago), Madagascar (includes Glorioso and Juan De Nova Isl.), Madeira, Malaysia, Maldives, Mali, Malta, Mariana Isl., Marshall Isl., Mauritius, Mexico, Micronesia (Wake Isl.), Montenegro, Morocco, Mozambique, Myanmar, Nepal, Netherlands, New Caledonia, New Zealand, Nicaragua, Nigeria, Pakistan, Palau, Palestine, Panama, Papua New Guinea, Paraguay, People's Republic of China (includes Hong Kong), Philippines, Polynesian Isl. (Howland, Jarvis, Johnston Atoll, Pitcairn, Wallis and Futuna), Puerto Rico, Republic of the Congo, Republic of South Africa, Reunion, Romania, Russia, Samoa (Ind. State of Samoa, American Samoa, Western Samoa), San Marino, São Tomé and Principe, Serbia, Seychelles, Singapore, Slovakia, Slovenia, Solomon Isl., South Korea, Spain (includes Balearic Isl., Ibiza, Minorca), Sri Lanka, Switzerland, Syria, Taiwan, Thailand, Timor, Tonga, Turkey, Tuvalu, United States (continental, Hawai'i), Uruguay, Vanuatu, Venezuela, Vietnam. **Refs.:** Banks 1908 (E*, L*); Barraud 1923h (L*); Barraud 1934: 233 (M*, F*, L*); Edwards 1941: 153 (A), 391 (P*); Bohart & Ingram 1946b (M*, F*, L*); La Casse & Yamaguti 1950: 111 (M*, F*, P*, L*; keys, bion.); Knight & Hull 1952: 176 (M*, F, L); Mattingly 1953a: 17 (tax.), 49 (distr.); Horsfall 1955: 506 (review); Iyengar & Menon 1955 (distr.); Hara 1957: 65 (F*); Bullock 1960 (E*); Joshi et al. 1965 (distr.; Nepal); Mohrig 1967 (F*); Huang 1968c: 297 (M*, F*, P*; neotype desig.); Pratt & Kidwell 1969 (E*); Aslamkhan 1971b (distr.; Pakistan); Mattingly 1971a: Pl. 15 (P abdomen*); Lambrecht & van Someren 1971: 483 (distr.); Basio 1971b: 27 (M*; bion.); Huang 1971 (F*); Huang 1972c: 13 (M*, F*, P*, L*; distr.); Matsuo, Yoshida, & Kunou 1972 (E*); Moriya et al. 1972 (E*); Colless 1973: 226 (distr.); Moriya et al. 1973 (E*); Matsuo, Yoshida, & Lien 1974: 180 (E*); Baisas 1974: 33 (M, F*, P*, L*; tax., bion., distr.; Philippines); Tanaka et al. 1975c: 223 (bion., distr.); Huang 1979a (F*; keys, distr.); Tanaka et al. 1979: 380 (M*, F*, L*); Lu 1985 (tax.; *s.l.*); Ahmed 1987 (distr.; Bangladesh); Lee et al. 1987a: 75 (F; key, tax., bion., distr., review); Linley 1989b (E*); Darsie & Pradhan 1990, 1991 (F, L; tax., keys, bion., distr.; Nepal); Pozza & Mojori 1992: 318 (distr.; Italy); Suleman et al. 1993 (distr.; Pakistan); Ibáñez-Bernal & Martínez-Campos 1994: 231 (distr.; Mexico); Knudsen 1995: 93 (distr.); Ogata & Samayoa 1996: 503 (distr.; Guatemala); Jupp 1996 (M*, F*; key); Adhami & Reiter 1998: 340 (distr.; Albania); Darsie 1999: 614 (distr.); Albuquerque et al. 2000 (distr.; Brazil); Delaunay et al. 2000: 17 (distr.; France); Tanaka 2000b: 232 (P*; tax., key); Whelan & Hapgood 2000: (bion., distr.; East Timor); Chadee et al. 2003: 438 (distr.; Trinidad); Pena et al. 2003 (distr.; Dominican Republic); Rossi & Martínez 2003: 471 (distr.; Uruguay); Toto et al. 2003: 343 (distr.; Equatorial Guinea); Girod 2004: 74 (distr.; Comoros, Reunion); Huang 2004: 14 (M*, F*; tax., keys, distr.); Darsie & Ward 2005 (F*, L*; keys, distr.); Scholte et al. 2007 (distr.; Netherlands); Rattanarithikul et al. 2010 (F*, L*; keys, bion., distr.; Thailand); Becker et al. 2010: 201 (M*, F*, L*; keys, tax., distr., bion.); Oter et al. 2013 (bion., distr.; Turkey); Ngoagouni et al. 2015 (distr.; Africa); Kutateladze et al. 2016 (distr.; Republic of Georgia); Harrison et al. 2016 (F*, L*; keys, distr., tax.); Giordano 2019 (distr.; Canada); Robert et al. 2019 (distr.; Western Palearctic). **Ety.:** n.s. [painted white (L); ref. to distinctive white and silvery-white scales]. **Informal name:** Asian Tiger Mosquito. Official ESA common name: Asian tiger mosquito. (Vol. 1, p. 460)

syn. *samarensis* Ludlow
1903: 138 (A; *Stegomyia scutellaris* ssp.). **Type loc.:** Samar, Philippines (USNM). **Refs.:** Stone & Knight 1956a: 225 (type info., lectotype desig.).

syn. *nigritia* Ludlow
1910: 194 (F; *Stegomyia*). **Type loc.:** Cotabato, Mindanao, Philippines (USNM). **Refs.:** Stone & Knight 1956a: 222 (type info., lectotype desig.).

syn. *quasinigritia* Ludlow
1911b: 129 (M; *Stegomyia*). **Type loc.:** Turucan, Mindanao, Philippines (USNM).

alcasidi Huang
1972c: 37 (M*, F*, P*, L*). **Type loc.:** Dalton Pass, Nueva Ecija, Luzon, Philippines (USNM). **Distr.:** Malaysia, Philippines, Taiwan. **Refs.:** Huang 1973a: 125 (distr.); Matsuo, Yoshida, & Lien 1974: 180 (E*; distr.); Huang 1979a (L*; keys, distr.). **Ety.:** G.L Alcasid. **Informal name:** Alcasid Philippine Pointy Mosquito.

alorensis Bonne-Wepster & Brug
1932: 92 (M*; *variegatus* var.). **Type loc.:** Alor Island [near Timor] (NHMUK). **Distr.:** Indonesia. **Refs.:** Stone & Farner 1945: 161 (to sp.); Marks 1954: 383 (M*, F*); Huang 1972c: 41 (M*, F*); Huang 1979a: (tax., key, distr.); Lee et al. 1987a: 90 (tax., distr.). **Ety.:** Alor Isl. **Informal name:** Alor Indonesian Pointy Mosquito.

amaltheus De Meillon & Lavoipierre
1944: 52 (M*, F, L*). **Type loc.:** Livingstone, Northern Rhodesia [Zambia] (SAIM). **Distr.:** Botswana, Zambia, Zimbabwe. **Refs.:** De Meillon et al. 1945: 98 (P*); Hopkins 1952: 154 (L*); Mattingly 1952: 246; Huang 1974b: 27 (M*, F*); Jupp 1996 (M*, F*; key); Huang 1997: 17 (M*, F*, P, L; keys, tax., bion., distr.); Huang

2004 (tax., key to adults, distr.); Mattingly 1953a: 6 (tax.). **Ety.:** n.s. [soft (Gr)]. **Informal name:** Gentle Zambian Pointy Mosquito.

andrewsi Edwards
1926a: 103 (M*; *variegatus* var.). **Type loc.:** Christmas Island, south of Java (NHMUK). **Distr.:** Australia, Indonesia. **Refs.:** Stone & Farner 1945: 159 (to sp.); Marks 1954: 383 (M*, F*); Huang 1972c: 42 (M*, F*); Huang 1979a (tax., key to adults, distr.); Lee et al. 1987a: 91 (F; key, tax., bion., distr., review). **Ety.:** C.W. Andrews. **Informal name:** Andrews Australian Pointy Mosquito.

angustus Edwards
1935a: 135 (M, F). **Type loc.:** Between Mount Mgahinga and Mount Sabinio, Congo–Uganda border, and at Muko, southwest of Kabale, Uganda (NHMUK). **Distr.:** Angola, Democratic Republic of the Congo, Uganda. **Refs.:** Edwards 1941: 150 (M*, F*); Hopkins 1952: 156 (L*); Mattingly 1952 (tax.); Mattingly 1953a (tax.); Ribeiro & da Cunha Ramos 1973: 120 (distr.); Huang 2004 (M*, F*; tax., key to adults, distr.). **Ety.:** narrow (L) (laterally compressed scutum). **Informal name:** Narrow Ugandan Pointy Mosquito.

annandalei (Theobald)
1910a: 10 (F; *Stegomyia*). **Type loc.:** Sukna, Darjeeling District (base of Eastern Himalayas, 500 ft) [West Bengal], India (IM). **Distr.:** Bangladesh, Cambodia, India, Indonesia, Myanmar, Nepal, Papua New Guinea, People's Republic of China, Taiwan, Thailand, Vietnam. **Refs.:** Barraud 1934: 227 (M*, F*, L*); Mathis 1938: 493 (E*, bion.); Mattingly 1965b: 39 (M*, F*, P*, L*); Mattingly 1970g: 34 (L*); Aslamkhan 1971b (distr.; Pakistan); Mattingly 1971a: Pl. 14 (P terminal segs.*), Pl. 30 (L head*, terminal segs.*); Matsuo, Yoshida, & Lien 1974: 180 (E*); Huang 1977a: 17 (M*, F*, P*, L*; distr.); Huang 1979a (F, P, L; distr.); Ahmed 1987 (distr.; Bangladesh); Lee et al. 1987a: 92 (F; key, tax., bion., distr., review); Darsie et al. 1992 (distr.; Nepal); Rattanarithikul et al. 2010 (F*, L*; keys, bion., distr.; Thailand). **Ety.:** Thomas Nelson Annandale. **Informal name:** Sukna Indian Pointy Mosquito.
syn. *horishensis* Yamada
1921: 58 (F). **Type loc.:** Horisha, Formosa [Taiwan, Republic of China] (IID). **Refs.:** Lien 1962: 626 (to ssp.); Mattingly 1965b: 39 (syn.); Huang 1977a: 16 (syn.).
ssp. ***quadricinctus*** Barraud
1923f: 227 (F; *Stegomyia annandalei* var. *quadricincta*). **Type loc.:** Nongpoh, Assam, India (NHMUK). **Distr.:** India. **Refs.:** Barraud 1934: 228 (A); Huang 1977a: 17 (syn.); Harbach & Howard 2007: 43 (to ssp.). **Ety.:** four rings (L); four basal white rings on all tarsi. **Informal name:** Four-ringed Taiwanese Pointy Mosquito.

aobae Belkin
1962: 452 (M*, F*, P*, L*). **Type loc.:** Near Crater Lake, Aoba Island, New Hebrides (USNM). **Distr.:** Vanuatu. **Refs.:** Lee et al. 1987a: 94 (F; key, tax., bion., distr., review). **Ety.:** Aoba Isl., New Hebrides [Vanuatu]. **Informal name:** Aoba Ni-Vanuatu Pointy Mosquito.

apicoargenteus (Theobald)
1909b: 3 (F; *Stegomyia*). **Type loc.:** Obuasi and Kumasi, Ashanti, Gold Coast [Ghana] (NHMUK). **Distr.:** Angola, Benin, Burkina Faso, Cameroon, Central African Republic, Congo, Côte d'Ivoire, Democratic Republic of the Congo, Gabon, Ghana, Kenya, Liberia, Nigeria, Senegal, Sierra Leone, Sudan and South Sudan, Togo, Uganda. **Refs.:** Macfie & Ingram 1922 (F*); Edwards 1941: 136 (M*, F*), 390 (P); Hopkins 1952: 143 (L*); Mattingly 1952 (tax.); Ribeiro & da Cunha Ramos 1973: 121 (distr.); Townsend 1990: 44 (lectotype desig.); Huang 2004 (M*, F*; tax., key, distr.). **Ety.:** silver tip (L); palpus. **Informal name:** Silver-tipped Ghanaian Pointy Mosquito.
syn. *albertii* Theobald
1912 (F; *Kingia*). In Schouteden 1912: 78. **Type loc.:** [Ponthierville], Belgian Congo [Democratic Republic of the Congo] (CMT).
syn. *maculoabdominalis* Theobald
1913d: 597 (F; *Kingia*). **Type loc.:** Khor Kobwa, Lado District, Sudan (NHMUK). **Refs.:** Edwards 1941: 136 (syn.).
syn. *ziemanni* Enderlein
1923: 27 (F; *Aniella*). **Type loc.:** South Cameroons [Cameroon] (ZM). **Refs.:** Edwards 1941: 136 (syn.?).
syn. *togoensis* Enderlein
1923: 28 (F; *Aniella*). **Type loc.:** Klein Popo, Togo, West Africa (ZM).

bambusae Edwards
1935a: 134 (M, F). **Type loc.:** Mount Mgahinga and Mount Sabinio, Congo–Uganda border, and at Muko, southwest of Kabale, Uganda (NHMUK). **Distr.:** Central African Republic, Congo, Democratic Republic of the Congo, Uganda. **Refs.:** Edwards 1941: 148 (M, F*), 390 (P); Hopkins 1952: 155 (L); Mattingly 1953a: 14 (tax.); Huang 1997: 19 (M*, F*, P, L; keys, tax., bion., distr.); Huang 2004 (tax., key to adults, distr.). **Ety.:** bamboo (L); found in bamboo stems. **Informal name:** Bamboo Ugandan Pointy Mosquito.

blacklocki Evans
1925a: 121 (M*, F*). **Type loc.:** Daru, Sierra Leone (NHMUK). **Distr.:** Cameroon, Nigeria, Sierra Leone. **Refs.:** Edwards 1941: 139 (syn.); Huang 2004 (M*, F*; tax., key to adults, distr.; to sp.). **Ety.:** B. Blacklock. **Informal name:** Blacklock Sierra Leonean Pointy Mosquito.

bromeliae (Theobald)
1911a: 10 (M, F*; *Stegomyia*). **Type loc.:** Kampala Swamp, Uganda (NHMUK). **Distr.:** Angola, Benin, Burkina Faso, Cameroon, Central African Republic, Comoros, Congo, Côte d'Ivoire, Democratic Republic of the Congo, Ethiopia, Ghana, Guinea, Kenya, Liberia, Madagascar

(includes Glorioso and Juan De Nova Isl.), Malawi, Nigeria, Republic of South Africa, Senegal, Sierra Leone, Sudan and South Sudan, Tanzania, Uganda, Zimbabwe. **Refs.:** Huang 1979b: 226 (from syn. with *simpsoni*); Huang 1986c (distr., bion.); Jupp 1996 (M*, F*; key); Huang 2004 (M*, F*; tax., key to adults, distr.); Bennett et al. 2015 (mol. tax.; *bromeliae*, *lilii* comparison); **Ety.:** n.s. [pineapple (L); poss. ref. to larvae found in pineapple bracts]. **Informal name:** Bromeliad Ugandan Pointy Mosquito. (Vol. 1, p. 462)

burnsi Basio & Reisen
1971: 58 (M*, L*). **Type loc.:** Andersen Air Force Base, Guam (PNM). **Distr.:** Guam. **Ety.:** J.P. Burns. **Informal name:** Burns Guamanian Pointy Mosquito.

calceatus Edwards
1924c (M*, F*). In Haworth 1924: 197. **Type loc.:** Lindi, Tanganyika [Tanzania] (NHMUK). **Distr.:** Angola, Cameroon, Kenya, Mozambique, Tanzania, Zimbabwe, Zambia. **Refs.:** Edwards 1941: 142 (F*); De Meillon et al. 1945: 99 (P*); Hopkins 1952: 148 (L*); Mattingly 1952 (tax.); Mattingly 1953d: 296 (distr.; Cameroon); Worth & De Meillon 1960: 240 (distr.); Ribeiro & da Cunha Ramos 1973: 121 (distr.); Huang 1979a (F*); Huang 1981: 93 (M*, F*; distr.); Townsend 1990: 56 (corrects attribution of Knight & Stone 1977: 158); Jupp 1996 (M*, F*; key); Huang 2004 (M*, F*; tax., keys, distr.). **Ety.:** shoe-like (L); "black (or shoe-like) tip of the hindtarsi." **Informal name:** Shoe-like Tanzanian Pointy Mosquito.

chaussieri Edwards
1923a: 397 (F). **Type loc.:** Sandoa, Belgian Congo [Democratic Republic of the Congo] (NHMUK). **Distr.:** Congo, Democratic Republic of the Congo, Kenya, Zambia. **Refs.:** Edwards 1941: 136 (F*); Rodhain & Boutonnier 1985: 275 (syntypes, PIP); Huang 2004 (F*; tax., keys, distr.). **Ety.:** Dr. Chaussier. **Informal name:** Chaussier Congolese Pointy Mosquito.

chemulpoensis Yamada
1921: 54 (M, F). **Type loc.:** Jinsen (Chemulpo), Korea (IID). **Distr.:** People's Republic of China, South Korea. **Refs.:** Feng 1938: 506 (M*, F*, L*, E*); La Casse & Yamaguti 1950: 111 (F*; key); Gutsevich 1955: 324 (M*); Bullock 1960 (E*); Huang 1974a: 208 (M*; lectotype desig.); Tanaka et al. 1979: 403 (M*, F*, L*); Huang 1988b (M*; tax., keys); Reinert 2000e: Fig. 37 (F*); Tanaka 2001: 189 (P*; tax.); Huang 2004 (tax., keys). **Ety.:** Chemulpo; near Incheon, South Korea. **Informal name:** Chemulpo Korean Pointy Mosquito.

contiguus Edwards
1936: 55 (M). **Type loc.:** Salisbury [Mashonaland], Southern Rhodesia [Zimbabwe] (NHMUK). **Distr.:** Ethiopia, Republic of South Africa, Tanzania, Zimbabwe. **Refs.:** Edwards 1941: 145 (M*, F*), 390 (P); Hopkins 1952: 149 (L*); Mattingly 1952: 247 (tax.?); Mattingly 1953a: 11 (tax.); Muspratt 1956a: 70 (M*, F*, L*); White 1975: 310 (distr.); Huang 1981: 111 (M*, F*); Jupp 1996 (M*, F*; key); Huang 2004 (M*, F*; tax., keys, distr.). **Ety.:** bordering (L); lobes of male ninth tergite contiguous. **Informal name:** Bordering Zimbabwean Pointy Mosquito.

cooki Belkin
1962: 454 (M*, F, P*, L*). **Type loc.:** Niue Island (NHMUK). **Distr.:** Cook Isl. (Polynesia), India, Niue, Thailand, Tonga. **Refs.:** Huang 1977c: 291 (distr.); Huang & Hitchcock 1980: 10 (M*, F*, P*, L*; distr.); Lee et al. 1987a: 96 (F; key, tax., bion., distr., review). **Ety.:** James Cook. **Informal name:** Cook Niuean Pointy Mosquito.

corneti Huang
1986b: 765 (M*, F*, P*, L*). **Type loc.:** Tiwai Island (Moa River), Potoro, Southern Province, Sierra Leone (USNM). **Distr.:** Cameroon, Côte d'Ivoire, Sierra Leone. **Refs.:** Huang 1990: 22 (M*, F*, P*, L*; distr.); Huang 2004 (tax., keys, distr.). **Ety.:** Michel Cornet. **Informal name:** Cornet Sierra Leonean Pointy Mosquito.

craggi (Barraud)
1923f: 227 (M*; *Stegomyia*). **Type loc.:** Haflong, Assam, India (NHMUK). **Distr.:** India, Nepal, Thailand. **Refs.:** Barraud 1934: 229 (M*); Huang 1977a: 22 (M*, F*, P*, L*; distr.); Darsie et al. 1992 (distr.; Nepal); Rattanarithikul et al. 2010 (F*, L*; keys, bion., distr.; Thailand). **Ety.:** F.W. Cragg. **Informal name:** Cragg Indian Pointy Mosquito.
syn. *purii* Barraud
1931b: 226 (M*). **Type loc.:** Marianbarrie Tea Estate, near Sukna, North Bengal, India (NHMUK, MSI). **Refs.:** Barraud 1934: 229 (syn.).

cretinus Edwards
1921d: 325 (F). **Type loc.:** Crete, Greece (DEI). **Distr.:** Cyprus, FYRO Macedonia, Georgia, Greece (includes Crete), Lebanon, Turkey, Ukraine, Russia (USSR). **Refs.:** Séguy 1924: Pl. XX (L*), as *Ae. delta*; Mattingly 1954a: 265 (M, F, L); Mohrig 1967 (F*); Gutsevich et al. 1974: 321 (M*, F, L); Lane 1982 (F*); Darsie 1999: 615 (P*; tax., key, distr.); Alten et al. 2000 (distr.; Turkey); Becker et al. 2010: 203 (M*, F*, L*; keys, tax., distr., bion.). **Ety.:** Crete. **Informal name:** Minotaur Greek Pointy Mosquito [the Minotaur is a legendary monster from the island of Crete].
syn. *delta* Séguy
1924: 80 (L*). **Type loc.:** Jagladzlik, Macedonia (MNHP). **Refs.:** Monchadskii 1951: 223 (syn.).
syn. *lindtropi* Shingarev
1927: 545 (F). **Type loc.:** Black Sea Coast of the Caucasus, near Gudauty, USSR (LU).

daitensis Miyagi & Toma
1981: 428 (M*, F*, P*, L*, E*). **Type loc.:** Zaisho, Minami Daito Islands, Ryukyus, Japan (NMST). **Distr.:** Japan. **Ety.:** Minami Daito Isl., Japan. **Informal name:** Daito Japanese Pointy Mosquito.

deboeri Edwards
1926a: 128 (M). **Type loc.:** Nairobi [Central], Kenya (NHMUK). **Distr.:** Central African Republic, Kenya, Tanzania. **Refs.:** Edwards 1941: 147 (M*, F), 390 (P); van Someren 1946a: 112 (P, L); Mattingly 1952: 247; Mattingly 1953a: 14 (tax.); Huang 1997: 22 (M*, F*, P*, L*; keys, tax., bion., distr.); Huang 2004 (tax., keys, distr.). **Ety.:** n.k. **Informal name:** Deboer Kenyan Pointy Mosquito.

demeilloni Edwards
1936: 55 (A; *deboeri* var.). **Type loc.:** Eshowe, Zululand [KwaZulu-Natal, Republic of South Africa] (NHMUK). **Distr.:** Republic of South Africa. **Refs.:** van Someren 1946a: 113 (P*, L*, E*); Hopkins 1952: 147 (L); Mattingly 1952: 247 (tax; to sp.); Mattingly 1953a: 8 (tax.; to sp.); Muspratt 1956a: 56 (M*, F*, L*); Jupp 1996 (M*, F*; key); Huang 1997: 25 (M*, F*, P*, L*; keys, tax., bion., distr.); Huang 2004 (tax., keys, distr.). **Ety.:** Botha De Meillon. **Informal name:** De Meillon South African Pointy Mosquito.

denderensis Wolfs
1949: 190 (M*, F; as *apicoargenteus* var.). **Type loc.:** Dendere, Costermansville [Kivu], Belgian Congo [Democratic Republic of the Congo] (CMT). **Distr.:** Burkina Faso, Cameroon, Central African Republic, Congo, Côte d'Ivoire, Democratic Republic of the Congo, Ghana, Kenya, Liberia, Nigeria, Senegal, Sierra Leone, Uganda. **Refs.:** Mattingly 1952: 249 and 1953: 13 (L; to ssp.); Mattingly 1953a: 13 (L; to ssp.); Townsend 1990: 66 (syntypes, NHMUK); Huang 2004: (M*, F*; tax., keys, distr.; to sp.). **Ety.:** Dendere, Bukavu, Republic of the Congo. **Informal name:** Dendere Congolese Pointy Mosquito.

dendrophilus Edwards
1921c: 74 (M). **Type loc.:** Oblogo, Ghana (NHMUK). **Distr.:** Angola, Burkina Faso, Cameroon, Central African Republic, Côte d'Ivoire, Democratic Republic of the Congo, Equatorial Guinea, Ghana, Nigeria, Republic of South Africa, Senegal, Togo, Zambia. **Refs.:** van Someren 1946a: 113 (M*, F*, P*, L*, E*; as *demeilloni)*; Hopkins 1952: 145 (L*); Mattingly 1952: 240, 283 (tax.); Ribeiro & da Cunha Ramos 1973: 121 (distr.); Jupp 1996 (M*, F*; key); Huang 1997: 29 (M*, F*, P*, L*; keys, tax., bion., distr.); Huang 2004 (tax., keys, distr.). **Ety.:** n.s. [tree loving (Gr); found in tree holes]. **Informal name:** Tree-loving Ghanaian Pointy Mosquito.
syn. *trinidad* Gil Collado
1936: 315, 326 (M*, P*). **Type loc.:** Santa Isabel, Fernando Po Island (MU). **Refs.:** Mattingly 1952: 240 (syn.).

desmotes (Giles)
1904a: 367 (F; *Stegomyia*). **Type loc.:** [Camp Stotsenberg, Angeles, Pampanga, Luzon], Philippines (NHMUK). **Distr.:** Bangladesh, Cambodia, India, Indonesia, Malaysia, Philippines, Singapore, Taiwan, Thailand, Vietnam. **Refs.:** Edwards & Given 1928: 339 (L*; bion.; as *gracilis*); Barraud 1934: 225 (M*, F, L); Knight & Hull 1952: 170 (M*, F, L*); Mattingly 1965b: 43 (M*, F*, P*, L*; lectotype desig.); Basio 1971b: 29 (M*; bion.); Mattingly 1971a: Pl. 15 (P abdomen*), Pl. 2 (A thorax*; as *gracilis*); Matsuo, Yoshida, & Lien 1974: 187 (E*); Baisas 1974: 38 (M*, F*, P*, L*; tax., bion., distr.; Philippines); Huang 1977a: 26 (M*, F*, P*, L*; distr.); Huang 1979a (tax., keys, distr.); Benjaphong & Rattanarithikul 1991: 53 (distr.; as *gracilis*); Reinert 2000e: Fig. 41 (F*), Fig. 1 (F*; as *gracilis*); Jeffery et al. 2010 (distr.; Singapore); Rattanarithikul et al. 2010 (F*, L*; keys, bion., distr.; Thailand; as *gracilis*). **Ety.:** n.s. [one bound, a prisoner (Gr)]. **Informal name:** Captive Philippine Pointy Mosquito.
syn. *albitarsis* Ludlow
1905a: 131 (F; *Anisocheleomyia*). **Type loc.:** Camp Stotsenberg, Angeles, Pampanga, Luzon, Philippines (USNM).
syn. *gracilis* Leicester
1908: 81 (M, F; *Stegomyia*). **Type loc.:** Ulu Klang jungle, 8 mi from Kuala Lumpur, Malaya [Malaysia] (NHMUK).
syn. *albipes* Theobald
1910a: 11 (F; *Stegomyia*). **Type loc.:** Maddathorai, west base of Western Ghats, Travancore, India (IM).
syn. *montana* Koidzumi
1918: 141 (M; *Stegomyia*). **Type loc.:** Chikutoki, Kagi District, Formosa [Taiwan, Republic of China] (LU). **Refs.:** Lien 1962: 626 (syn.).

dybasi Bohart
1957: 53 (M*, F*, P*, L*). **Type loc.:** Koror Island, Palau Islands (USNM). **Distr.:** Palau, Philippines. **Refs.:** Lee et al. 1987a: 103 (tax., distr.). **Ety.:** H.S. Dybas. **Informal name:** Dybas Palauan Pointy Mosquito.

ealaensis Huang
2004: 36 (M*, F*). **Type loc.:** Eala, Cuvette-Centrale, Zaire [Democratic Republic of the Congo] (CMT). **Distr.:** Congo, Democratic Republic of the Congo. **Ety.:** Eala, Zaire. **Informal name:** Eala Congolese Pointy Mosquito.

edwardsi (Barraud)
1923b: 784 (M*, F*; *Stegomyia*). **Type loc.:** Andaman Islands (MSI). **Distr.:** Andaman Isl., India. **Refs.:** Barraud 1934: 232 (M*, F*); Huang 1977a: 6 (F*); Tewari et al. 2008 (P*, L*). **Ety.:** Frederick Wallace Edwards. **Informal name:** Edwards Andamanese Pointy Mosquito.

ethiopiensis Huang
2004: 39 (M*). **Type loc.:** Ethiopia (PIP). **Distr.:** Ethiopia. **Ety.:** Ethiopia. **Informal name:** Abyssinia Ethiopian Pointy Mosquito [Abyssinia is a historical exonym for Ethiopia].

flavopictus Yamada
1921: 52 (M*, F). **Type loc.:** Shiba, Tokyo, Honshu, Japan (IID). **Distr.:** Guam, India, Japan, Russia, South Korea. **Refs.:** La Casse & Yamaguti 1950: 1116 (M*, F*, P*, L*; bion., distr., tax.); Sasa & Kano 1951: 112 (L*); Asanuma &

Nakagawa 1953: 87 (P*); Hara 1957: 65 (F*); Gutsevich et al. 1971: 305 (distr.); Huang 1972c: 21 (M*, F, P*, L*; lectotype desig.); Moriya et al. 1973 (E*); Matsuo 1975: 49 (E*); Tanaka et al. 1975c: 223 (tax., distr.); Danilov 1976c: 353 (tax.); Tanaka et al. 1979: 386 (M*, F*, L*); Tanaka 2000b: 236 (P*; tax., key); Kaur 2003: 452 (distr.; India). **Ety.:** painted yellow (L); yellow scutal markings instead of the silver markings of *albopictus*. **Informal name:** Yellow-painted Japanese Pointy Mosquito.

ssp. ***downsi*** Bohart & Ingram
1946a: 51 (M*, F, L*; as sp.). **Type loc.:** Chizuka, Okinawa [Ryukyu-Retto, Japan] (USNM). **Distr.:** Japan. **Refs.:** Bohart & Ingram 1946b: 64 (M, F, L*; bion., distr., tax.); Bohart 1953: 184 (to ssp.); Huang 1972c: 17 (M*, F, P*, L*; to sp.); Tanaka et al. 1979: 389 (M*, F*, L; to ssp. of *flavopictus*); Huang 1979a (L*; keys, distr.; as sp.); Tanaka 2000b: 239 (P*; tax., key). **Ety.:** W.G. Downs. **Informal name:** Downs Japanese Pointy Mosquito.

ssp. ***miyarai*** Tanaka, Mizusawa, & Saugstad
1979: 390 (M*, F*, L*). **Type loc.:** Mount Bana, Ishigaki Island, Ryukyus, Japan (NMST). **Distr.:** Japan. **Refs.:** Tanaka 2000b: 242 (P*; tax., key). **Ety.:** n.s. [Kobo Miyara]. **Informal name:** Miyara Japanese Pointy Mosquito.

fraseri (Edwards)
1912c: 11 (F*; *Stegomyia*). **Type loc.:** Mpumu Forest, Uganda (NHMUK). **Distr.:** Angola, Cameroon, Côte d'Ivoire, Ghana, Kenya, Nigeria, Sierra Leone, Uganda. **Refs.:** van Someren 1946a: 109 (M, F, L*); Mattingly 1952: 246, 250, 282 (tax.); Huang 2004 (M*, F*; tax., keys, distr.). **Ety.:** A.D. Fraser. **Informal name:** Fraser Ugandan Pointy Mosquito.

futunae Belkin
1962: 455 (M*, F, P*, L*). **Type loc.:** Mu'a, Alofi Island, Horne Islands (USNM). **Distr.:** Polynesian Isl. (Howland, Jarvis, Johnston Atoll, Pitcairn, Wallis and Futuna). **Refs.:** Huang & Hitchcock 1980 (M*, F*, P*, L*; keys); Lee et al. 1987a: 104 (F; key, tax., bion., distr., review). **Ety.:** Mu'a, Alofi Isl., Futuna (Horne, Hoorn) group, Wallis & Futuna. **Informal name:** Futuna Polynesian Pointy Mosquito.

galloisi Yamada
1921: 47 (M*, F). **Type loc.:** Sapporo, Honshu, Japan (IID). **Distr.:** Japan, People's Republic of China, Russia, South Korea. **Refs.:** Sasa & Kano 1951: 110 (L*); Hara 1957: 64 (M*); Huang 1972a: 253 (F*; lectotype desig.); Tupitsyn & Dubitzky 1972: 106 (distr.); Chow 1973: 44 (distr.); Gutsevich et al. 1974: 323 (M*, F, L*; distr.); Tanaka et al. 1979: 376 (M*, F*, L*); Tanaka 2000b: 230 (P*; tax., key); Huang 2004: 8 (tax.). **Ety.:** Edme Gallois. **Informal name:** Gallois Japanese Pointy Mosquito.

galloisiodes Liu & Lu
1984: 76 (M*, F, L*). **Type loc.:** Kunming, Yunnan Province, China [People's Republic of China] (KIM). **Distr.:** People's Republic of China. **Refs.**: Lu & Yu 2000 (P). **Ety.:** similar to *galloisi*. **Informal name:** Imitator Chinese Pointy Mosquito [*galloisiodes* = "like Gallois" = imitating Gallois = Imitator].

gandaensis Huang
2004: 41 (F*). **Type loc.:** Ganda, Coast Region, Kenya (NHMUK). **Distr.:** Kenya. **Ety.:** Ganda, Kenya. **Informal name:** Ganda Kenyan Pointy Mosquito.

gardnerii (Ludlow)
1905a: 99 (M, F; *Stegomyia*). **Type loc.:** Bulacao, Mindoro Island, Philippines (USNM). **Distr.:** Bangladesh, Cambodia, Indonesia, Malaysia, Philippines, Singapore, Thailand, Vietnam. **Refs.:** Knight & Hull 1952: 172 (M*, F, L; type info., lectotype desig.); Stone & Knight 1956a: 218 (type info.); Mattingly 1965b: 32 (M, F*, P, L*); Basio 1971b: 29 (M*; bion.); Baisas 1974: 40 (M, F*, P*, L*; tax., bion., distr.; Philippines); Huang 1977a: 46 (M*, F*, P*, L*; distr.); Huang 1979a (M, F, P, L; tax., keys, distr.); Lee et al. 1987a: 106 (tax., distr.); Jeffery et al. 2010 (distr.; Singapore). **Ety.:** Fletcher Gardner. **Informal name:** Gardner Philippine Pointy Mosquito.

ssp. ***imitator*** (Leicester)
1908: 89 (F; *Stegomyia*). **Type loc.:** Gombak Road, Kuala Lumpur [Selangor], Malaya [Malaysia] (NHMUK). **Distr.:** Bangladesh, Cambodia, India, Malaysia, Nepal, People's Republic of China, Taiwan, Thailand, Vietnam. **Refs.:** Mattingly 1965b: 36 (M*, F*, P*, L; lectotype desig.; from syn.); Matsuo, Yoshida, & Lien 1974: 187 (E*; distr.); Huang 1977a: 50 (M*, F*, P, L; distr.); Huang 1979a (F, P, L; tax., keys, distr.); Darsie & Pradhan 1990, 1991 (F, L; tax., keys, bion., distr.; Nepal); Rattanarithikul et al. 2010 (F*, L*; keys, bion., distr.; Thailand). **Ety.:** n.s. [imitator (L); poss. ref. to similarity with other species]. **Informal name:** Imitative Malaysian Pointy Mosquito.

ssp. syn. *argenteomaculata* Theobald
1907: 184 (F; *Stegomyia*). **Type loc.:** Narcondam Island (80 mi from Andaman & Nicobar Islands), Bay of Bengal (NHMUK). **Refs.:** Huang 1977a: 50 (syn. of *gardnerii* imitator?).

ssp. syn. *minutissima* Theobald
1910a: 9 (F; *Stegomyia*). **Type loc.:** Sukna, Darjeeling District [West Bengal], India (IM). **Refs.:** Huang 1977a: 50 (from syn. of *w-albus*).

ssp. syn. *christianus* Dyar
1921i: 148 (M, F). **Type loc.:** Canton and Honan, China [People's Republic of China] (USNM). **Refs.:** Stone & Knight 1956a: 216 (type info., lectotype desig.); Mattingly 1965b: 36 (syn.).

ssp. syn. *indosinensis* Borel
1928: 93 (M, F; *Stegomyia*). **Type loc.:** Indochina (LU). **Refs.:** Huang 1977a: 50 (syn.).

grantii (Theobald)
1901a: 306 (F*; *Stegomyia*). **Type loc.:** Socotra [Yemen] (NHMUK). **Distr.:** Yemen. **Refs.:** Edwards 1941 (F); Leeson & Theodor 1948: 226 (M, L*); Mattingly 1952: 244; Mattingly 1953a: 17 (tax.); Marks 1954: 353 (tax.);

Mattingly 1954a: 268 (M*); Mattingly & Knight 1956: 100 (tax.); Huang 2004 (M*, F*; tax., keys, distr.). **Ety.:** Grant [poss. ref. to a collector]. **Informal name:** Grant Yemeni Pointy Mosquito.

guamensis Farner & Bohart
1944: 117 (M*, F). **Type loc.:** Mata, Guam [Mariana Islands] (USNM). **Distr.:** Guam, Mariana Isl., Papua New Guinea. **Refs.:** Bohart & Ingram 1946b: 35 (M, F*, L*; bion., distr., tax.); Yamaguti & LaCasse 1950a: 54 (M*, F*, P*, L*); Marks 1954: 382 (M*, F*); Lee et al. 1987a: 107 (tax., distr.). **Ety.:** Guam. **Informal name:** Guam Mariana Islander Pointy Mosquito.

gurneyi Stone & Bohart
1944: 217 (M*, F, L). **Type loc.:** Bougainville, Solomon Islands (USNM). **Distr.:** Papua New Guinea, Solomon Isl. **Refs.:** Marks 1954: 382, Pl. 18 (M*, F*); Belkin 1962: 456 (M*, F, P*, L*); Lee et al. 1987a: 108 (F; key, tax., bion., distr., review). **Ety.:** Ashley B. Gurney. **Informal name:** Gurney Solomon Islander Pointy Mosquito.

hakanssoni Knight & Hurlbut
1949: 23 (M*, F, P*, L*). **Type loc.:** Sokas Island, Ponape Island, Caroline Islands (USNM). **Distr.:** Micronesia (Wake Isl.). **Refs.:** Marks 1954: 382, Pl. 18 (M*, F*); Lee et al. 1987a: 110 (tax., distr.). **Ety.:** E.G. Hakansson. **Informal name:** Hakansson Micronesian Pointy Mosquito.

hansfordi Huang
1997: 33 (M*, F*, P*, L*; keys, tax., bion., distr.). **Type loc.:** Bishop's Seat, Eshowe district, Natal [KwaZulu-Natal], South Africa [Republic of South Africa] (USNM). **Distr.:** Burkina Faso, Cameroon, Central African Republic, Côte d'Ivoire, Kenya, Republic of South Africa, Senegal, Tanzania, Uganda. **Refs.:** Huang 2004 (M, F; keys, tax., distr.). **Ety.:** Clifford F. Hansford. **Informal name:** Hansford South African Pointy Mosquito.

hebrideus Edwards
1926a: 102 (M*; *variegatus* var.). **Type loc.:** Hog Harbor, Espiritu Santo, New Hebrides Islands (NHMUK). **Distr.:** Indonesia, Papua New Guinea, Solomon Isl., Vanuatu. **Refs.:** Farner & Bohart 1945: 46 (M*, F); Stone 1947: 85 (syn.); Belkin 1962: 457 (M*, F, P*, L*; from syn.); Maffi 1973b: 46 (distr.); Rodhain & Gaxotte 1977 (distr.); Lee et al. 1987a: 111 (F; key, tax., bion., distr., review). **Ety.:** New Hebrides. **Informal name:** Hebrides Ni-Vanuatu Pointy Mosquito.

heischi van Someren
1951: 1 (M, F, P, L). **Type loc.:** Taveta, Kenya (NHMUK). **Distr.:** Kenya, Mozambique, Republic of South Africa, Tanzania. **Refs.:** Mattingly 1952: 246; Mattingly 1953a: 8, 44 (tax.); Mattingly & Bruce-Chwatt 1954: 192 (tax.); Worth & De Meillon 1960: 240 (distr.); Jupp 1996: 70 (M, F*; key); Huang 1997: 38 (M*, F*, P*, L*; keys, tax., bion., distr.); Huang 2004 (M, F; tax., keys, distr.). **Ety.:** R.B. Heisch. **Informal name:** Heisch Kenyan Pointy Mosquito.

hensilli Farner
1945: 59 (M, F, L). **Type loc.:** Ulithi Islands, Western Caroline Islands (USNM). **Distr.:** Micronesia (Wake Isl.), Palau, Singapore. **Refs.:** Bohart & Ingram 1946b: 25 (M*, F*, L; bion., distr., tax.); Marks 1954: 383, Pl. 18 (M*, F*); Bohart 1957: 48 (M*, F*, L*); Colless 1957b (M*; bion.); Colless 1962: 314 (to ssp.); Huang 1972c: 43 (M*, F; revalidation); Lee et al. 1987a: 118 (F; key, tax., bion., distr., review). **Ety.:** George S. Hensill. **Informal name:** Hensill Micronesian Pointy Mosquito.

hogsbackensis Huang
2004: 43 (M*, F*, P*, L*). **Type loc.:** Hogsback, Cape Province, Republic of South Africa (USNM). **Distr.:** Republic of South Africa. **Ety.:** Hogsback, Cape Prov., Republic of South Africa. **Informal name:** Hogsback South African Pointy Mosquito.

hoguei Belkin
1962: 460 (M*, F, P*, L*). **Type loc.:** Hutuna, Rennell Island, Solomon Islands (NHMUK). **Distr.:** Solomon Isl. **Refs.:** Taylor & Maffi 1971: 119 (F); Lee et al. 1987a: 120 (F; key, tax., bion., distr., review). **Ety.:** Charles L. Hogue. **Informal name:** Hogue Solomon Islander Pointy Mosquito.

horrescens Edwards
1935a: 128 (M, F, L). **Type loc.:** Ura, Taveuni, Fiji Islands (NHMUK). **Distr.:** Fiji. **Refs.:** Stone & Farner 1945: 160 (M*); Marks 1954: 383, Pl. 18 (M*, F*); Belkin 1962: 461 (M*, F, P*, L*); Huang & Hitchcock 1980 (M*, F*, P*, L*; keys, tax., distr.); Lee et al. 1987a: 121 (F; key, tax., bion., distr., review). **Ety.:** n.s. [dread, become terrified (L)]. **Informal name:** Horrendous Fijian Pointy Mosquito.

josiahae Huang
1988a: 155 (M*, F*). **Type loc.:** 60–70 km south of Ifakara, Tanzania (VBDK). **Distr.:** Tanzania. **Refs.:** Huang 2004 (M*, F*; tax., keys, distr.). **Ety.:** Phoebe A.O. Josiah. **Informal name:** Josiah Tanzanian Pointy Mosquito.

katherinensis Woodhill
1949: 141 (M*, F). **Type loc.:** Katherine, Northern Territory, Australia (MM). **Distr.:** Australia. **Refs.:** Woodhill 1950: 251 (hybridization); Marks 1954: 383, Pl. 18 (M*, F*); Lee et al. 1987a: 124 (F; key, tax., bion., distr., review; from ssp. of *scutellaris*); Linley & Service 1995: 148 (E*). **Ety.:** Katherine, Northern Territory, Australia. **Informal name:** Katherine Australian Pointy Mosquito.

keniensis van Someren
1946a: 111 (M, F, L*). **Type loc.:** City Park, Nairobi, Kenya (NHMUK). **Distr.:** Kenya, Tanzania. **Refs.:** Hopkins 1952: 151 (L*); Huang 1997: 42 (M*, F*, L; keys, tax., bion., distr.); Huang 2004 (M, F; tax., keys, distr.).

Ety.: Kenya. **Informal name:** Bantu Kenyan Pointy Mosquito [historically, the Bantu had a large effect on the area that is now Kenya].

kenyae van Someren
1946c: 2 (M, F). **Type loc.:** Kaimosi, North Kavirondo, Kenya (NHMUK). **Distr.:** Kenya. **Refs.:** Mattingly 1953a: 14 (tax.); Huang 1997: 44 (M*, F*, L; keys, tax., bion., distr.; from ssp. of *bambusae*); Huang 2004 (M, F; tax., keys, distr.). **Ety.:** Kenya. **Informal name:** Kilwa Kenyan Pointy Mosquito [the medieval Kilwa Sultanate ruled over much of the Swahili Coast].

kesseli Huang & Hitchcock
1980: 23 (M*, F*, P*, L*; distr.). **Type loc.:** Tafahi Island, Tonga (USNM). **Distr.:** Tonga. **Refs.:** Lee et al. 1987a: 131 (F; key, tax., bion., distr., review). **Ety.:** John F. Kessel. **Informal name:** Kessel Tongan Pointy Mosquito.

kivuensis Edwards
1941: 132 (F; *subargenteus* var.). **Type loc.:** Kibati lava plain, Kivu, Belgian Congo [Democratic Republic of the Congo] (CMT). **Distr.:** Democratic Republic of the Congo. **Refs.:** Mattingly 1953a: 9 (F; to sp.); Huang 2004 (F*; tax., keys, distr.). **Ety.:** North Kivu, Democratic Republic of the Congo. **Informal name:** Kivu Congolese Pointy Mosquito.

krombeini Huang
1975b: 345 (M*, F*, P*, L*). **Type loc.:** Udawattekele, Kandy District, Central Province, Sri Lanka (USNM). **Distr.:** India, Sri Lanka. **Refs.:** Huang 1979a (M*, F*, P*, L*; keys, distr.); Kaur 2003: 452 (distr.; India). **Ety.:** Karl V. Krombein. **Informal name:** Krombein Sri Lankan Pointy Mosquito.

langata van Someren
1946c: 1 (M*, F, P). **Type loc.:** Langata Forest, Nairobi, Kenya (NHMUK). **Distr.:** Kenya, Malawi, Republic of South Africa, Zambia, Zimbabwe. **Refs.:** Hopkins 1952: 150 (L*); Mattingly 1952: 247; Mattingly 1953a: 11, 37 (tax.); Huang 2004 (M*, F*; tax., keys, distr.). **Ety.:** Langata Forest, Nairobi, Kenya. **Informal name:** Langata Kenyan Pointy Mosquito.

ledgeri Huang
1981: 96 (M*, F*, P*, L*; distr.). **Type loc.:** Makonde, Sibasa, Republic of South Africa (USNM). **Distr.:** Botswana, Kenya, Republic of South Africa, Tanzania, Zambia. **Refs.:** Jupp 1996 (M*, F*; key); Huang 2004 (M*, F*; tax., keys, distr.). **Ety.:** John A. Ledger. **Informal name:** Ledger South African Pointy Mosquito.

lilii (Theobald)
1910c: 160 (M, F*; *Stegomyia*). **Type loc.:** Bor, Sudan (NHMUK). **Distr.:** Burkina Faso, Central African Republic, Côte d'Ivoire, Ethiopia, Ghana, Liberia, Nigeria, Senegal, Sierra Leone, Sudan and South Sudan, Uganda. **Refs.:** Huang 1979b: 226 (from syn. with *simpsoni*); Tanaka et al. 1979: 251 (M*, F*, P, L); Huang 1986c (distr., bion.); Huang 2004 (M*, F*; tax., keys, distr.); Bennett et al. 2015 (mol. tax.; comparison with *bromeliae*; distr.). **Ety.:** larvae at base of lily leaves. **Informal name:** Lily Sudanese Pointy Mosquito.

luteocephalus (Newstead)
1907 (F*; *Stegomyia*). In Newstead et al. 1907: 15. **Type loc.:** Kumba, Belgian Congo [Democratic Republic of the Congo] (NHMUK). **Distr.:** Angola, Benin, Burkina Faso, Cameroon, Central African Republic, Congo, Côte d'Ivoire, Democratic Republic of the Congo, Eritrea, Ethiopia, Ghana, Guinea, Liberia, Mali, Nigeria, Senegal, Sierra Leone, Sudan and South Sudan, Tanzania, Uganda, Zambia, Zimbabwe. **Refs.:** Macfie & Ingram 1922 (F); Edwards 1941: 151 (M*, F*), 391 (P); Hopkins 1952: 157 (L*); Mattingly 1952: 245, 286 (tax.); Huang 1990: 25 (M*, F*, P*, L*; lectotype desig., distr.); Jupp 1996 (M*, F*; key); Huang 2004 (M., F; tax., keys, distr.). **Ety.:** n.s. [yellow headed (L)]. **Informal name:** Yellow-faced Congolese Pointy Mosquito. (Vol. 1, p. 462)

maehleri Bohart
1957: 40 (M*, P, L*). **Type loc.:** Guror, Yap Island, Caroline Islands (USNM). **Distr.:** Micronesia (Wake Isl.). **Ety.:** K.L. Maehler. **Informal name:** Maehler Micronesian Pointy Mosquito.

malayensis Colless
1962: 314 (M*, F). **Type loc.:** Pulau Hantu, Keppel Harbor, Singapore (CSIR). **Distr.:** Cambodia, India, Malaysia, Singapore, Taiwan, Thailand, Vietnam. **Refs.:** Huang 1971: 2 (M*, F*, P*, L*); Huang 1972c: 45 (M*, F*, P*, L*; to sp.); Colless 1973: 225 (maintains ssp.); Huang 1979a (M, F, P, L; tax., keys, distr.); Kaur 2003: 452 (distr.; India); Rattanarithikul et al. 2010 (F*, L*; keys, bion., distr.; Thailand). **Ety.:** Malaya [Malaysia]. **Informal name:** Malay Singaporean Pointy Mosquito.

malikuli Huang
1973b: 225 (M*, P*, L*). **Type loc.:** Huai Phrao, Chiang Mai, Thailand (USNM). **Distr.:** Taiwan, Thailand. **Refs.:** Huang 1977a: 33 (F*, P*, L*); Huang 1979a (M, F, P, L; tax., keys, distr.); Rattanarithikul et al. 2010 (F*, L*; keys, bion., distr.; Thailand). **Ety.:** Vichai Malikul. **Informal name:** Malikul Thai Pointy Mosquito.

marshallensis Stone & Bohart
1944: 218 (M*, F, L). **Type loc.:** Airok Island, Ailinglaplap Atoll, Marshall Islands (USNM). **Distr.:** Kiribati, Marshall Isl., Micronesia (Wake Isl.). **Refs.:** Farner & Bohart 1945: 49 (M*); Bohart & Ingram 1946b: 21 (M*, F*, L; bion., distr., tax.); Marks 1954: 383, Pl. 18 (M*, F*); Belkin 1962: 463 (M*, F, P*, L*); Belkin 1962: 8 (distr.); Lee et al. 1987a: 137 (F; key, tax., bion., distr., review). **Ety.:** Marshall Isl. **Informal name:** Nitijela Marshallese Pointy Mosquito [the Nitijela is the legislature of the Marshall Isl.].

mascarensis MacGregor
1924: 409 (M, F, L*). **Type loc.:** Mauritius (NHMUK). **Distr.:** Mauritius. **Refs.:** Edwards 1941: 153 (M*, F); Hopkins 1952: 159 (L); Mattingly 1952: 244; Mattingly 1953a: 15 (tax.); Hartberg & Craig 1973: 206 (genetic tax.); McClelland et al. 1973: 570 (tax.); Huang 2004 (M*, F*; tax., keys, distr.). **Ety.:** Mascarene Isl. **Informal name:** Abel Mauritian Pointy Mosquito [Abel Tasman (1603–1659) set out from Dutch Mauritius to Western Australia].

masseyi Edwards
1923a: 397 (F). **Type loc.:** Ruwe, Belgian Congo [Democratic Republic of the Congo] (NHMUK). **Distr.:** Congo, Democratic Republic of the Congo, Tanzania, Zambia. **Refs.:** Edwards 1941: 146 (F*); Mattingly 1953c: 322 (L*?); van Someren 1972: 91 (M, F); Huang 1997: 47 (M*, F*; keys, tax., bion., distr.); Huang 2004 (M, F; tax., keys, distr.). **Ety.:** A. Yale Massey. **Informal name:** Massey Congolese Pointy Mosquito.

mattinglyorum Huang
1994: 11 (M*, F*, P*, L*). **Type loc.:** Tiwai Island, Potoru, Southern Province, Sierra Leone (USNM). **Distr.:** Côte d'Ivoire, Ghana, Nigeria, Sierra Leone. **Refs.:** Huang 1997: 49 (M*, F*, P*, L*; keys, tax., bion., distr.); Huang 2004 (M, F; tax., keys, distr.). **Ety.:** Peter Frederick and Christine Mattingly. **Informal name:** Mattingly Sierra Leonean Pointy Mosquito.

maxgermaini Huang
1990: 31 (F*). **Type loc.:** Mamfe, West Cameroon, Cameroon (USNM). **Distr.:** Cameroon. **Refs.:** Huang 2004 (M, F; tax., keys, distr.). **Ety.:** Max Germain. **Informal name:** Germain Cameroonian Pointy Mosquito.

mediopunctatus (Theobald)
1905h: 240 (F; *Stegomyia*). **Type loc.:** Peradeniya [Central], Ceylon [Sri Lanka] (NHMUK). **Distr.:** India, Philippines, Sri Lanka, Thailand. **Refs.:** Borel 1928: 87 (M*, F, L*); Borel 1930: 222 (M*, F, L*); Barraud 1934: 230 (M*, F*, L); Huang 1973b: 230 (tax.); Huang 1977a: 36 (M*, F*, P*, L*; distr.); Huang 1979a (M, F, P, L; tax., keys, distr.). **Ety.:** n.s. [spotted in the middle (L)]. **Informal name:** Spot-waisted Sri Lankan Pointy Mosquito.
syn. *submediopunctata* Barraud
1923b: 781 (F; *Stegomyia mediopunctata* var.). **Refs.:** Huang 1977a: 36 (syn.). **Type loc.:** Nagargali [Belgaum District], Bombay, India (NHMUK).
ssp. ***sureilensis*** Barraud
1934: 231 (F). **Type loc.:** Sureil, Darjeeling District [West Bengal], India (NHMUK). **Distr.:** India. **Refs.:** Harbach & Howard 2007: 44 (to ssp.). **Ety.:** Sureil, Darjeeling Dist., India. **Informal name:** Sureil Indian Pointy Mosquito.

metallicus (Edwards)
1912c: 12 (new name for *quasistegomyia dubia* Theobald, not *Stegomyia dubia* Theobald 1910c; *Stegomyia*). **Distr.:** Angola, Botswana, Burkina Faso, Ghana, Kenya, Namibia, Nigeria, Republic of South Africa, Senegal, Sudan and South Sudan, Tanzania, Togo, Uganda, Zimbabwe. **Refs.:** Macfie & Ingram 1922 (F); Edwards 1941: 134 (M*, F*), 389 (P); Hopkins 1952: 141 (L*); Muspratt 1956a: 64 (M*, F*, L*); Jupp 1996 (M*, F*; key); Huang 2004 (M*, F*; tax., keys, distr.). **Ety.:** n.s. [metallic (L)]. **Informal name:** Metallic Sudanese Pointy Mosquito.
syn. *dubia* Theobald
1910c: 133 (M*; *Quasistegomyia*). **Type loc.:** Bor, Sudan (NHMUK).

mickevichae Huang
1988b: 6 (F*). **Type loc.:** Kaunro, Kenya (NHMUK). **Distr.:** Kenya. **Refs.:** Huang 2004 (M, F; tax., keys, distr.). **Ety.:** Mary F. Mickevich. **Informal name:** Mickevich Kenyan Pointy Mosquito.

mpusiensis Huang
2004: 47 (M*). **Type loc.:** Mount Mpuse, Zaire [Democratic Republic of the Congo] (CMT). **Distr.:** Congo, Democratic Republic of the Congo. **Ety.:** Mt. Mpuse, Democratic Republic of the Congo. **Informal name:** Mpuse Congolese Pointy Mosquito.

muroafcete Huang
1997: 54 (F*; keys, tax., bion., distr.). **Type loc.:** Mount Karisimwa, Kivu-Central, Zaire [Democratic Republic of the Congo] (CMT). **Distr.:** Congo, Democratic Republic of the Congo. **Refs.:** Huang 2004 (M, F; tax., keys, distr.). **Ety.:** acronym: Musee Royale de l'Afrique Centrale at Tervuren, Belgium. **Informal name:** Karisimwa Congolese Pointy Mosquito.

neoafricanus Cornet, Valade & Dieng
1978: 228 (M*, F*). **Type loc.:** 10 km northwest of Departement de Kédougou, Senegal (SSC). **Distr.:** Senegal. **Refs.:** Huang 1990: 33 (M*, F*, L*; distr.); Huang 2004 (M, F; tax., keys, distr.). **Ety.:** n.s. [near *africanus*?]. **Informal name:** Carthaginian Senegalese Pointy Mosquito ["Afri" was a Latin term for the Carthaginians].

neogalloisi Chen & Chen
2000: 343 (M*, F, L*). **Type loc.:** Taihangshan, Jiyuan City, Henan Province, 1,200 m [China] (HGMC). **Distr.:** People's Republic of China. **Ety.:** n.s. [near *galloisi*]. **Informal name:** New-Gallois Chinese Pointy Mosquito.

neopandani Bohart
1957: 36 (M*, F*, P*, L*). **Type loc.:** Mount Tapotchau, Saipan, Mariana Islands (USNM). **Distr.:** Guam, Mariana Isl. **Refs.:** Lee et al. 1987a: 139 (tax., distr.). **Ety.:** near *pandani* [larvae found in *Pandanus* and taro axils]. **Informal name:** Tapotchau Mariana Islander Pointy Mosquito.

njombiensis Huang
1997: 56 (F*; keys, tax., bion., distr.). **Type loc.:** Njombe, Iringa Region, Tanzania (NHMUK). **Distr.:** Kenya,

Tanzania. **Refs.:** Huang 2004 (M, F; tax., keys, distr.). **Ety.:** Njombe, Tanzania. **Informal name:** Iringa Tanzanian Pointy Mosquito [named for wider region that encompasses Njombe Dist.].

novalbopictus Barraud
1931b: 224 (M*). **Type loc.:** Pusa, Bihar, India (NHMUK). **Distr.:** India, Thailand. **Refs.:** Barraud 1934: 237 (M*, L*); Huang 1972c: 24 (M*, P*, L*); Bhat 1976: 266 (chrom.*); Huang 1979a (M, F, P, L; tax., keys, distr.); Lu 1982 (F*); Amerasinghe 1983 (E*); Rattanarithikul et al. 2010 (F*, L*; keys, bion., distr.; Thailand). **Ety.:** near *albopictus*. **Informal name:** Whitewashed Indian Pointy Mosquito [*novalbopictus* = "new white painted" = Whitewashed].

opok Corbet & van Someren
1962: 73 (M*, F). **Type loc.:** Awere, Acholi, Uganda (NHMUK). **Distr.:** Burkina Faso, Central African Republic, Côte d'Ivoire, Mali, Senegal, Uganda. **Refs.:** Germain et al. 1975: 41.; Huang 1990: 36 (M*, F*; distr.); Huang 2004 (M, F; tax., keys, distr.). **Ety.:** opok: a local common name for the tree *Terminalia velutina*; adults captured in grasses under this tree species. **Informal name:** Opok Ugandan Pointy Mosquito.

palauensis Bohart
1957: 51 (M*, F*, L*). **Type loc.:** Ngergoi Island, Palau Islands (USNM). **Distr.:** Micronesia (Wake Isl.), Palau. **Refs.:** Lee et al. 1987a: 140 (tax., distr.). **Ety.:** Palau Isl., Palau. **Informal name:** Ngergoi Palauan Pointy Mosquito.

pandani Stone
1939: 162 (M*, F). **Type loc.:** Barrigada Area, Guam [Mariana Islands] (USNM). **Distr.:** Guam, Mariana Isl. **Refs.:** Bohart & Ingram 1946b: 38 (M*, F*, L*; bion., distr., tax.); Yamaguti & LaCasse 1950a: 60 (M*, F*, P*, L*); Bohart 1957: 34 (M*, F*, P*, L*); Lee et al. 1987a: 141 (tax., distr.); Reinert 2000e: Fig. 40 (F*). **Ety.:** "in axils of *Pandanus* leaves." **Informal name:** Pandanus Mariana Islander Pointy Mosquito.

patriciae Mattingly
1954a: 262 (M, F, P*, L). **Type loc.:** Krol Mountains, near Solan (Pun jab [Punjab] States), Western Himalayas [India] (NHMUK). **Distr.:** India, Malaysia, Pakistan, People's Republic of China, Thailand, Vietnam. **Refs.:** Aslamkhan 1971b (distr.; Pakistan); Huang 1972c: 26 (M*, F, P*, L*); Danilov 1976c: 353 (L; distr.); Huang 1977d: 625 (distr.); Huang 1979a (L*; keys, distr.); Rattanarithikul et al. 2010 (F*, L*; keys, bion., distr.; Thailand). **Ety.:** Elisabeth Nesta [Patricia, "Pat"] Marks. **Informal name:** Patricia Indian Pointy Mosquito.

paullusi Stone & Farner
1945: 155 (M*, F). **Type loc.:** San Antonio, Samar, Philippines (USNM). **Distr.:** Indonesia, Malaysia, Philippines. **Refs.:** Knight & Hull 1952: 178 (M*, F, L); Marks 1954: 376, 383, Pl. 18 (M*, F*); Basio 1971b: 31 (M*; bion.); Huang 1972c: 49 (M*, F*, P*, L*); Ramalingam & Pillai 1973 (distr.); Huang 1979a (M, F, P, L; tax., keys, distr.); Lee et al. 1987a: 142 (F; key, tax., bion., distr., review). **Ety.:** J.H. Paullus. **Informal name:** Paullus Philippine Pointy Mosquito.

pernotatus Farner & Bohart
1944: 118 (M*, F). **Type loc.:** Segond Channel Area, Espiritu Santo, New Hebrides Islands (USNM). **Distr.:** Vanuatu. **Refs.:** Perry 1950: 126 (M*, F*, P*, L*); Marks 1954: 383, Pl. 18 (M*, F*); Belkin 1962: 464 (M*, F, P*, L*); Lee et al. 1987a: 144 (F; key, tax., bion., distr., review). **Ety.:** n.s. [thoroughly inscribed? (L)]. **Informal name:** Spotty Ni-Vanuatu Pointy Mosquito.

perplexus (Leicester)
1908: 83 (M, F; *Stegomyia*). **Type loc.:** Kuala Lumpur and the Gap [Selangor], Malaya [Malaysia] (NHMUK). **Distr.:** India, Malaysia, Taiwan, Thailand. **Refs.:** Knight & Hull 1952: 182 (M*, F, L*); Mattingly 1965b: 46 (M, F*, P*, L*; to ssp.; lectotype desig.); Basio 1971b: 30 (M*; bion.; as var. of *mediopunctatus*); Matsuo, Yoshida, & Lien 1974: 187 (E*; distr.); Huang 1975a: 93 (M*, F*, P*, L*); Huang 1977a: 40 (resurrection from var. of *mediopunctatus*; M*, F*, P*, L*; distr.); Huang 1979a (M, F, P, L; tax., keys, distr.); Rattanarithikul et al. 2010 (F*, L*; keys, bion., distr.; Thailand). **Ety.:** perplexing (L); "mosquito somewhat difficult to place." **Informal name:** Weird Malaysian Pointy Mosquito.

pia Le Goff & Robert
2013 (M*, F*, P, L*). In Le Goff et al. 2013. **Type loc.:** Réserve Forestière de Majimbini, 3.5 km west of Mamoudzou, Mayotte, Comoros Archipelago, northern Mozambique Channel, western Indian Ocean (IRD). **Distr.:** Madagascar (includes Glorioso and Juan De Nova Isl.). **Ety.:** *pia* means fresh or new in the Shimaore language of Mayotte and is first name of the stepdaughter of author V. Robert. **Informal name:** Pia Comoran Pointy Mosquito.

polynesiensis Marks
1951a: 138 (M, L). **Type loc.:** Taveuni, Fiji Islands (NHMUK). **Distr.:** Australia, Cook Isl. (Polynesia), Fiji, French Polynesia, Niue, Polynesian Isl. (Howland, Jarvis, Johnston Atoll, Pitcairn, Wallis and Futuna), Samoa (Ind. State of Samoa, American Samoa, Western Samoa), South China Sea Isl. (Paracel, Spratly, Tokelau), Tahiti, Tuvalu. **Refs.:** Marks 1954: 383, Pl. 18 (M*, F*); Rosen & Rozeboom 1954: 531 (L*); Belkin 1962: 468 (M*, F, P*, L*); Mattingly 1970g: 35 (L*); Huang 1975a (M*, F*, P*, L*; tax.); Ramalingam 1976: 304 (tax., bion., keys., distr.; Samoa & Tonga); Huang & Hitchcock 1980 (M*, F*, P*, L*; tax., keys, distr.); Lee et al. 1987a: 147 (F; key, tax., bion., distr., review); Linley & Service 1995 (E*). **Ety.:** Polynesia. **Informal name:** Polynesia Fijian Pointy Mosquito. (Vol. 1, p. 466)

poweri (Theobald)
1905g: 19 (F*; *Stegomyia*). **Type loc.:** Natal [KwaZulu-Natal, Republic of South Africa] (NHMUK). **Distr.:** Republic of South Africa. **Refs.:** Mattingly 1952: 246; Mattingly 1953a: 10 (tax.); Muspratt 1953b: 83 (M*, F*, L); Jupp 1996 (M*, F*; key); Huang 2004 (M, F; tax., keys, distr.). **Ety.:** Dr. Power. **Informal name:** Power South African Pointy Mosquito.

pseudalbopictus (Borel)
1928: 85 (M*, F, L*; *Stegomyia*). **Type loc.:** Terres Rouges, Cochin China, Indochina (LU). **Distr.:** India, Indonesia, Laos, Malaysia, Myanmar, Nepal, Taiwan, Thailand, Vietnam. **Refs.:** Bonne-Wepster & Brug 1932: 81 (M*, L*); Barraud 1934: 235 (M*, F, L); Huang 1972c: 28 (M*, F*, P*, L*); Matsuo, Yoshida, & Lien 1974: 180 (E*; distr.); Huang 1979a (F*; keys, distr.); Lee et al. 1987a: 185 (tax., distr.); Darsie & Pradhan 1990, 1991 (F, L; tax., keys, bion., distr.; Nepal); Darsie, Pradhan, & Vaidya 1991 (distr.; Nepal); Rattanarithikul et al. 2010 (F*, L*; keys, bion., distr.; Thailand). **Ety.:** false (Gr); similar to *albopictus*. **Informal name:** Trang Vietnamese Pointy Mosquito [*trang* is the Vietnamese word for "white"].

pseudoafricanus Chwatt
1949: 808 (M, F, P, L). **Type loc.:** [Gbohun-Gbohun], Lagos, Nigeria (NHMUK). **Distr.:** Cameroon, Congo, Côte d'Ivoire, Democratic Republic of the Congo, Nigeria, Sierra Leone. **Refs.:** Mattingly 1952: 242, 245 (tax.); Mattingly & Bruce-Chwatt 1954: 183 (M*, F*, P*, L*; bion.); Huang 1990: 39 (M*, F*; distr.); Huang 2004 (M, F; tax., keys, distr.). **Ety.:** false (Gr); similar to *africanus*. **Informal name:** Punic Nigerian Pointy Mosquito [Roman rule of North Africa was established after the Punic Wars (264 B.C.–146 B.C.)].

pseudonigeria (Theobald)
1910c: 166 (F*; *Stegomyia*). **Type loc.:** Wambu, Angola (NHMUK). **Distr.:** Angola, Botswana, Namibia, Republic of South Africa. **Refs.:** Edwards 1941: 147 (F); Mattingly 1953a (tax.); Huang 1988b: 8 (F*; distr.); Jupp 1996 (M*, F*; key); Huang 2004 (M, F*; tax., keys, distr.). **Ety.:** n.s. [false (Gr); poss. ref. to similarity to *Ae. nigeria*, a syn. of *Ae. aegypti*]. **Informal name:** Nigeria Angolan Pointy Mosquito.

syn. *wellmanii* Theobald
1905b: 103 (F; as *Danielsia*). **Type loc.:** Bailunder, Angola (NHMUK). **Refs.:** Edwards 1941: 122 (F), 387 (tax.); Robinson 1948: 66 (M, F, L); Hopkins 1952: 129 (tax.); White 1980: 128 (emend. from *wellmani* to orig. spell.).

pseudoscutellaris (Theobald)
1910b: 156 (F*; *Stegomyia*). **Type loc.:** Suva [Viti Levu], Fiji Islands (NHMUK). **Distr.:** Fiji, Solomon Isl. **Refs.:** Paine 1943 (E); Bohart & Ingram 1946b: 12 (M*, F*, L*; distr., bion., tax); Marks 1954: 383 (M*, F*; var.); Rosen & Rozeboom 1954: 529 (L); Woodhill 1954: 19 (tax.); Belkin 1962: 469 (M*, F, P*, L*); Huang 1975a: 88 (M*, F*, P*, L*); Huang & Hitchcock 1980 (M*, F*, P*, L*; tax., keys, distr.); Lee et al. 1987a: 186 (F; key, tax., bion., distr., review). **Ety.:** false (Gr); similar to *scutellaris*. **Informal name:** False-plated Fijian Pointy Mosquito.

quasiscutellaris Farner & Bohart
1944: 120 (M*, F). **Type loc.:** Guadalcanal, Solomon Islands (USNM). **Distr.:** Papua New Guinea, Solomon Isl. **Refs.:** Perry 1949: 754 (L*); Marks 1954: 383, Pl. 18 (M*, F*); Belkin 1962: 471 (M*, F, P*, L*); Lee et al. 1987a: 202 (F; key, tax., bion., distr., review). **Ety.:** nearly (L); similar to *scutellaris*. **Informal name:** Half-plated Solomon Islander Pointy Mosquito.

rhungkiangensis Chang & Chang
1974: 347 (M*, F, L*). **Type loc.:** Rhungkiang, Kweichow Province, China [People's Republic of China] (Provincial Station of Sanitary & Diseases Control, Kweichow, China). **Distr.:** People's Republic of China. **Ety.:** from type locality of Rhungkiang. **Informal name:** Rhungkiang Chinese Pointy Mosquito.

riversi Bohart & Ingram
1946a: 50 (M*, F, L*). **Type loc.:** Chizuka, Okinawa [Ryukyu-Retto, Japan] (USNM). **Distr.:** Japan. **Refs.:** Bohart & Ingram 1946b: 65 (M*, F, P*, L*; bion., distr., tax.); Marks 1954: 383, Pl. 18 (M*, F*); Huang 1972c: 52 (M*, F*, P*, L*); Tanaka et al. 1975c: 223 (bion., distr.); Mogi 1976: 347 (distr.); Huang 1979a (M, F, P, L; tax., keys, distr.); Tanaka et al. 1979: 372 (M*, F*, L*); Tanaka 2000b: 227 (P*; tax., key). **Ety.:** T.M. Rivers. **Informal name:** Rivers Japanese Pointy Mosquito.

robinsoni Belkin
1962: 447 (M*, F, P*, L*). **Type loc.:** Near Government House, Peuo, Vanikoro, Santa Cruz Islands (USNM). **Distr.:** Solomon Isl. **Refs.:** Lee et al. 1987a: 205 (F; key, tax., bion., distr., review). **Ety.:** W.A. Robinson. **Informal name:** Robinson Solomon Islander Pointy Mosquito.

rotanus Bohart & Ingram
1946b: 39 (M*, F, L*; bion., distr., tax.). **Type loc.:** Rota Island, Mariana Islands (USNM). **Distr.:** Guam, Mariana Isl. **Refs.:** Bohart 1957: 39 (M*, F*, P*, L*); Lee et al. 1987a: 206 (tax., distr.). **Ety.:** Rota Isl., Marianas. **Informal name:** Rota Mariana Islander Pointy Mosquito.

rotumae Belkin
1962: 472 (M*, F, P*, L*). **Type loc.:** Oinafa village, Rotuma Island (USNM). **Distr.:** Fiji. **Refs.:** Huang & Hitchcock 1980 (M*, F*, P*, L*; tax., keys, distr.); Lee et al. 1987a: 207 (F; key, tax., bion., distr., review). **Ety.:** Rotuma Isl., Fiji. **Informal name:** Rotuma Fijian Pointy Mosquito.

ruwenzori Haddow & van Someren
1950: 281 (F*). **Type loc.:** [Toro], Uganda (NHMUK). **Distr.:** Uganda. **Refs.:** Gillett 1951: 195 (M, P, L*); Mattingly 1952: 245; Mattingly 1953a: 15 (tax.); Mat-

tingly 1971a: Pl. 13 (P cephalothorax*, terminal segs.*); Huang 1990: 42 (M*, F*; distr.); Huang 2004 (M, F; tax., keys, distr.). **Ety.:** Rwenzori Mountains [formerly called Ruwenzori Range), Uganda and Democratic Republic of the Congo]. **Informal name:** Ruwenzor Ugandan Pointy Mosquito.

saimedres Huang
1988: 10 (F*). **Type loc.:** Okokarara, Namibia (NHMUK). **Distr.:** Botswana, Namibia. **Refs.:** Jupp 1996 (M*, F*; key); Huang 2004 (M, F; tax., keys, distr.). **Ety.:** acronym: South African Institute for Medical Research, Johannesburg, Republic of South Africa. **Informal name:** Okokara Namibian Pointy Mosquito.

saipanensis Stone
1945b: 66 (M*, F, L*). **Type loc.:** Saipan, Mariana Islands (USNM). **Distr.:** Mariana Isl. **Refs.:** Bohart & Ingram 1946b: 40 (M*, F, L*; bion., distr., tax.); Bohart 1957: 38 (F*, M*, L*); Lee et al. 1987a: 210 (tax., distr.). **Ety.:** Saipan, Mariana Isl. **Informal name:** Saipan Mariana Islander Pointy Mosquito.

sampi Huang
2004: 49 (F*). **Type loc.:** Mataara (Matahara), Central Region, Kenya (VBDK). **Distr.:** Kenya. **Ety.:** acronym: Systematics of *Aedes* Mosquitoes Project. **Informal name:** SAMP Kenyan Pointy Mosquito.

schwetzi Edwards
1926a: 127 (M, F). **Type loc.:** Elizabethville (Katanga), Belgian Congo [Democratic Republic of the Congo] (NHMUK). **Distr.:** Angola, Congo, Democratic Republic of the Congo, Zambia. **Refs.:** Edwards 1941: 139 (M*, F*); Hopkins 1952: 145 (L); Mattingly 1952: 246 (tax.); Mattingly 1953a: 14 (tax.); Mohrig 1967 (F*); Danilov 1988 (tax., distr.). **Ety.:** Jacques Schwetz. **Informal name:** Katanga Congolese Pointy Mosquito.

scutellaris (Walker)
1859: 77 (F; *Culex*). **Type loc.:** Aroe Islands [New Guinea] (NHMUK). **Distr.:** Admiralty Isl., Australia (Christmas Isl. or Kiritimati), Cambodia, Fiji, Guam, Hong Kong, India (includes Andaman Isl.), Indonesia (Moluccas, Sumatra [includes Ketulauan Riau Archipelago, West Papua]), Malaysia, Mariana Isl., Micronesia (Wake Isl.), Myanmar, New Britain, New Hebrides, People's Republic of China, Philippines, Sri Lanka, Taiwan, Tuvalu, Vanuatu. **Refs.:** Buxton & Hopkins 1925 (E*; as syn. *variegatus* Doleschall 1858); Stone & Farner 1946 (*Ae. scutellaris* group); Penn 1949b: 56 (P*); Knight & Hull 1952: 180 (M*, F, L*); Bonne-Wepster 1954: 79 (M, F*, L*); Marks 1954: 383, Pl. 18 (M*, F*); Smith-White & Woodhill 1954: 163 (genetics); Belkin 1962: 474 (M*, F); Colless 1962: 313 (F*); Mohrig 1967 (F*); Huang 1970a: 471 (M*; topotype); Basio 1971b: 31 (M*; bion.); Huang 1972c: 55 (M*, F, P*, L*); Lien et al. 1977 (distr.); Lee et al. 1987a: 211 (F; key, tax., bion., distr., review); Reinert 2000e: Fig. 38 (F*); Kaur 2003: 452 (distr.; India). **Ety.:** n.s. **Informal name:** Plated Papuan Pointy Mosquito. (Vol. 1, p. 468)
syn. *variegatus* Doleschall
1858: 77 (not Schrank 1781) (A; *Culex*). **Type loc.:** Amboina Island [Moluccas] (NE). **Refs.:** Bonne-Wepster & Brug 1932: 83 (M*, F, L*; bion.).
syn. *zonatipes* Walker
1861a: 229 (F; *Culex*). **Type loc.:** Dorey, New Guinea (NHMUK). **Refs.:** Bohart & Ingram 1946b: 26 (M*, F, L*; bion., distr., tax.).

scutoscriptus Bohart & Ingram
1946b: 27 (M*, F*, L*; bion., distr., tax.). **Type loc.:** Moen Island, Truk Islands, Caroline Islands (USNM). **Distr.:** Micronesia (Wake Isl.). **Refs.:** Marks 1954: 382, Pl. 18 (M*, F*); Bohart 1957: 46 (M*, F*, P, L*); Lee et al. 1987a: 228 (tax., distr.). **Ety.:** scutum, script (L); ref. to median and lateral lines on scutum. **Informal name:** Lined Micronesian Pointy Mosquito.

seampi Huang
1974c: 137 (F*). **Type loc.:** Andaman Islands, India (NHMUK). **Distr.:** India. **Refs.:** Huang 1977a: 7 (F*); Huang 1979a (F; tax., keys, distr.). **Ety.:** acronym: [Southeast Asia Mosquito Project]. **Informal name:** SEAMP Indian Pointy Mosquito.

seatoi Huang
1969: 234 (M*, F, P, L*). **Type loc.:** Bangphra, Chon Buri, Thailand (USNM). **Distr.:** Thailand. **Refs.:** Huang 1972c: 32 (M*, F, P*, L*); Harrison & Rattanarithikul 1973: 280 (L*, first and third instars); Huang 1979a (L*; keys, distr.); Rattanarithikul et al. 2010 (F*, L*; keys, bion., distr.; Thailand). **Ety.:** acronym: [Southeast Asia Treaty Organization (Medical Research Laboratory)]. **Informal name:** SEATO Thai Pointy Mosquito.

segermanae Huang
1997: 57 (M*, F*; keys, tax., bion., distr.). **Type loc.:** Port Saint Johns, Pondoland, South Africa (SAIMR). **Distr.:** Republic of South Africa. **Refs.:** Huang 2004 (M, F; tax., keys, distr.). **Ety.:** Joyce Segerman. **Informal name:** Segerman South African Pointy Mosquito.

sibiricus Danilov & Filippova
1978: 170 (M*, F, L*). **Type loc.:** Kedrovaya Pad Reserve, Maritime Territory, USSR (ZIL). **Distr.:** Russia. **Ety.:** Siberia. **Informal name:** Siberia Russian Pointy Mosquito.

simpsoni (Theobald)
1905c: 224 (M, F; *Stegomyia*). **Type loc.:** Nelspruit, Transvaal [Republic of South Africa] (NHMUK). **Distr.:** Democratic Republic of the Congo, Eswatini, Ethiopia, Mali, Republic of South Africa, Sudan and South Sudan. **Refs.:** Macfie & Ingram 1922 (F); Edwards 1941: 133 (M*, F*), 389 (P*); Hopkins 1952: 139 (L*); Mattingly 1952: 245, 285 (tax.); Horsfall 1955: 514 (review); Muspratt

1956a: 47 (M*, F*, L); Gerberg 1972: 9 (change in type loc.); Gerberg & Hartberg 1975: 137 (bibliography); White 1975: 310 (distr.); Pajot 1976: 201 (A*); Brunhes 1977a (distr.; Comoros Isl.); Huang 1979b: 222 (lectotype desig.); Huang 1986c (distr., bion.); Lutwama & Mukwaya 1994 (tax., morph. comp.; *s.l.*); Jupp 1996 (M*, F*; key); Reinert 2000e: Fig. 36 (F*); Huang 2004 (M*, F*; tax., keys, distr.). **Ety.:** n.s. [C.B. Simpson]. **Informal name:** Simpson South African Pointy Mosquito.

soleatus Edwards
1924c (M*, F*). In Haworth 1924: 198. **Type loc.:** Lindi, Tanganyika [Tanzania] (NHMUK). **Distr.:** Angola, Kenya, Republic of South Africa, Tanzania, Zimbabwe. **Refs.:** van Someren 1946d (tax.); van Someren 1947d: 131 (P); Hopkins 1952: 149 (L); Mattingly 1952: 246; Mattingly 1953a: 12 (tax.); Muspratt 1956a: 68 (M*, F*, L*); Ribeiro & da Cunha Ramos 1973: 124 (distr.); Townsend 1990: 135 (corrects attribution of Knight & Stone 1977: 165); Jupp 1996 (M*, F*; key); Huang 2004 (M*, F*; tax., keys, distr.). **Ety.:** having sandals (L); "name alludes to the black 'sandals' of the hind feet which form naked-eye distinction from *A.* (*S.*) *calceatus*." **Informal name:** Sandal Tanzanian Pointy Mosquito.

strelitziae Muspratt
1950: 73 (M*, F, P, L*). **Type loc.:** Margate, near Port Shepstone, Natal [KwaZulu-Natal, Republic of South Africa] (SAIM). **Distr.:** Eswatini, Republic of South Africa. **Refs.:** Mattingly 1952: 245 (tax.); Muspratt 1956a: 51 (M, F*, L*); Segerman 1990: 54 (type dep.); Jupp 1996 (M*, F*; key); Huang 2004 (M*, F*; tax., keys, distr.). **Ety.:** n.s. [larvae in axils of *Strelitzia nicolai* (Harbach 2018c)]. **Informal name:** Strelitzia South African Pointy Mosquito.

subalbopictus Barraud
1931b: 225 (M*). **Type loc.:** Belgaum, Bombay, Deccan, India (NHMUK). **Distr.:** India, Malaysia, Nepal, People's Republic of China, Thailand, Vietnam. **Refs.:** Barraud 1934: 238 (M*); Huang 1972c: 35 (M*, F*, P*, L*); Bhat 1976: 266 (chrom.*); Huang 1979a (F*; keys, distr.); Darsie & Pradhan 1990, 1991 (F, L; tax., keys, bion., distr.; Nepal). **Ety.:** nearly (L); near *albopictus*. **Informal name:** Sita Indian Pointy Mosquito [*sita* is a Hindi word for "white"].

subargenteus Edwards
1925: 262 (M, F). **Type loc.:** Fort Johnston, Nyasaland [Malawi] (NHMUK). **Distr.:** Kenya, Malawi, Mozambique, Republic of South Africa. **Refs.:** Edwards 1941: 132 (M*, F*); Hopkins 1952: 138 (L*); Mattingly 1952: 248; Mattingly 1953a: 8 (tax.); Worth & De Meillon 1960: 238 (distr.); Jupp 1996 (M*, F*; key); Huang 2004 (M*, F*; tax., keys, distr.). **Ety.:** nearly (L); near *argenteus*. **Informal name:** Stainless Malawian Pointy Mosquito [*subargenteus* = "subsilver" = stainless steel = Stainless].

tabu Ramalingam & Belkin
1965: 1 (M*, F, P*, L). **Type loc.:** Eua Island, Tonga (USNM). **Distr.:** Tonga. **Refs.:** Ramalingam 1976: 306 (tax., bion., keys., distr.; Samoa & Tonga); Huang & Hitchcock 1980: 45 (M*, F*, P*, L*; to ssp. of *tongae*; distr.); Lee et al. 1987a: 229 (F; key, tax., bion., distr., review; from ssp. of *tongae*). **Ety.:** Tongatabu Isl. (Tonga). **Informal name:** Eua Tongan Pointy Mosquito.

tongae Edwards
1926a: 103 (M*, F). **Type loc.:** Haapai, Tonga Islands (NHMUK). **Distr.:** Tonga. **Refs.:** Marks 1954: 383, Pl. 18 (M*, F*); Huang 1972b: 338 (M*, F; tax.); Ramalingam 1976: 305 (tax., bion., keys., distr.; Samoa & Tonga); Huang & Hitchcock 1980: 35 (M*, F*, P*, L*; distr.); Lee et al. 1987a: 236 (F; key, tax., bion., distr., review). **Ety.:** Tonga Isl. **Informal name:** Friendly Tongan Pointy Mosquito [the Tonga Isl. were widely known as the Friendly Isl. after the Tongans' welcoming reception of Captain Cook in 1773].

tulagiensis Edwards
1926a: 109 (F). **Type loc.:** Tulagi Hospital (NHMUK). **Distr.:** Solomon Isl. **Refs.:** Belkin 1962: 446 (M*, F, P*, L*; to sp.); Lee et al. 1987a: 242 (F; key, tax., bion., distr., review). **Ety.:** Tulagi Hospital, Santa Cruz Isl. [Solomons?]. **Informal name:** Tulagi Solomon Islander Pointy Mosquito.

unilineatus (Theobald)
1906b: 70 (F*; *Quasistegomyia*). **Type loc.:** Bahr-el-Ghazal, Sudan (NHMUK). **Distr.:** Angola, Botswana, Burkina Faso, Central African Republic, Côte d'Ivoire, Gabon, Ghana, India, Kenya, Malawi, Nepal, Nigeria, Pakistan, Republic of South Africa, Saudi Arabia, Senegal, Sudan and South Sudan, Uganda, Zambia, Zimbabwe. **Refs.:** Macfie & Ingram 1922 (F*); Barraud 1934: 242 (M*, F*, L*); Edwards 1941: 152 (M*, F*), 391 (P*); Hopkins 1952: 158 (L); Mattingly 1952: 243 (tax.); Aslamkhan 1971b (distr.; Pakistan); Ribeiro & da Cunha Ramos 1973: 125 (distr.); Service 1976a (distr.; Gabon); Huang 1979a (M*, F*; keys, distr.); Darsie & Pradhan 1990 (F, L; tax., keys, bion., distr.; Nepal); Suleman et al. 1993 (distr.; Pakistan); Jupp 1996 (M*, F*; key); Godsey et al. 2003: 84 (distr.; Saudi Arabia); Huang 2004 (M*, F*; tax., keys, distr.). **Ety.:** n.s. [one line (L); poss. ref. to median white line on head or line of white scales on scutum]. **Informal name:** One-lined Sudanese Pointy Mosquito.

syn. *gebeleinensis* Theobald
1910c: 157 (F*; *Stegomyia*). **Type loc.:** Gebelein, Sudan (NHMUK).

upolensis Marks
1957b: 51 (M*, F, P, L*). **Type loc.:** Afiamalu, Upolu Island, Samoa Islands (NHMUK). **Distr.:** Samoa (Ind. State of Samoa, American Samoa, Western Samoa). **Refs.:** Belkin 1962: 477 (M*, F, P*, L*); Ramalingam 1976: 305

(tax., bion., keys., distr.; Samoa & Tonga); Huang & Hitchcock 1980 (M*, F*, P*, L*; tax., keys, distr.); Lee et al. 1987a: 244 (F; key, tax., bion., distr., review). **Ety.:** Upolu Isl., Western Samoa. **Informal name:** Upolu Samoan Pointy Mosquito.

usambara Mattingly
1953d: 294 (F*). **Type loc.:** Amani Bustani, Tanganyika [Tanzania] (NHMUK). **Distr.:** Côte d'Ivoire, Tanzania. **Refs.:** van Someren 1962: 25 (M, F); van Someren & Hamon 1964: 80 (P*); Huang 2004 (M*, F*; tax., keys, distr.). **Ety.:** n.s. [Usambara Mountains, Tanzania]. **Informal name:** Usambara Tanzanian Pointy Mosquito.

varuae Belkin
1962: 478 (M*, F, P*, L*). **Type loc.:** Lawaii, Graciosa Bay, Santa Cruz Island (USNM). **Distr.:** Indonesia (West Papua), Papua New Guinea, Solomon Isl. **Refs.:** Lee et al. 1987a: 247 (F; key, tax., bion., distr., review). **Ety.:** ref. to the "brigantine 'Varua' of the Robinson-Peabody Museum Expedition to the Southwest Pacific, 1956." **Informal name:** Varua Solomon Islander Pointy Mosquito.

vinsoni Mattingly
1953a: 16 (F). **Type loc.:** Moka, Mauritius (NHMUK). **Distr.:** Mauritius. **Refs.:** Halcrow 1954: 235 (? = fertile hybrid between *aegypti* and *mascarensis*); Mattingly 1963: 168 (F). **Ety.:** n.s. [Monsieur Vinson (Harbach 2018c)]. **Informal name:** Vinson Mauritian Pointy Mosquito.

wadai Tanaka, Mizusawa & Saugstad
1979: 400 (M*, F*, L*). **Type loc.:** Kitafukurozawa, Chichijima, Ogasawara Islands, Japan (NMST). **Distr.:** Japan. **Refs.:** Tanaka 2000b: 244 (P*; tax., key). **Ety.:** Yoshitake Wada. **Informal name:** Wada Japanese Pointy Mosquito.

w-albus (Theobald)
1905f: 74 (F*; *Stegomyia*). **Type loc.:** Matheran, India Orientalis [Maharashtra State, India] (HNM). **Distr.:** Bangladesh, India, Indonesia, Malaysia, Nepal, Pakistan, People's Republic of China, Sri Lanka, Thailand, Vietnam. **Refs.:** Barraud 1934: 232 (M*, F*); Rajagopalan 1956: 481 (P*, L*); Aslamkhan 1971b (distr.; Pakistan); Huang 1977a: 55 (M*, F*, P*, L*; distr.); Huang 1979a (M, F, P, L; tax., keys, distr.); Ahmed 1987 (distr.; Bangladesh); Darsie & Pradhan 1990 (F, L; tax., keys, bion., distr.; Nepal); Reinert 2000e: Fig. 39 (F*); Rattanarithikul et al. 2010 (F*, L*; keys, bion., distr.; Thailand). **Ety.:** white (L); ref. to white W-shaped area in front of scutum. **Informal name:** Dumbledore Indian Pointy Mosquito [Albus Dumbledore is a character in the Harry Potter book series by J.K. Rowling (1965–)].

woodi Edwards
1922d: 82 (F). **Type loc.:** Cholo, Nyasaland [Malawi] (NHMUK). **Distr.:** Kenya, Malawi, Mozambique, Republic of South Africa, Tanzania. **Refs.:** Mattingly 1952: 245; Mattingly 1953a: 9 (M, F); van Someren & Harper 1955: 77 (P*, L*); Mattingly 1970g: 34 (L*); Linley & Service 1994: 64 (E*); Jupp 1996 (M*, F*; key); Huang 2004 (M*, F*; tax., keys, distr.). **Ety.:** R.C. Wood. **Informal name:** Wood Malawian Pointy Mosquito.

Subgenus *Tanakaius* Reinert, Harbach, & Kitching

Reinert, Harbach, & Kitching 2004: 349 (phyl., class.; as genus). Type species: *Culicelsa togoi* Theobald.

Name derivation: Kazuo Tanaka.

Important references: Reinert 2008i (F gen.*); Reinert, Harbach, & Kitching 2009 (phyl., class.); Rattanarithikul et al. 2010 (F*, L*; keys, bion.; Thailand); Wilkerson et al. 2015 (phyl., class.; to subg. of genus *Aedes*); Soghigian et al. 2017 (mol. phyl.).

savoryi Bohart
1957: 31 (F*). **Type loc.:** Chichi Jima, Bonin Islands, Ogasawara Gunto [Japan] (USNM). **Distr.:** Japan. **Refs.:** Wada et al. 1973: 129 (M*, F*, P*, L*); Tanaka et al. 1979: 334 (M*, F*, L*); Lee et al. 1982: 265 (tax., distr.); Tanaka 2002b: 147 (P*; tax., key). **Ety.:** Nathaniel Savory. **Informal name:** Savory Japanese Pointy Mosquito.

togoi (Theobald)
1907: 379 (F; *Culicelsa*). **Type loc.:** Osaka [Honshu], Japan (NHMUK). **Distr.:** Cambodia, Canada, Japan, Malaysia, People's Republic of China (includes Hong Kong), Russia, South Korea, Taiwan, Thailand, United States (continental), Vietnam. **Refs.:** Ho 1931: 123 (M*, F); Bohart & Ingram 1946b: 66 (M*, F, P*, L*; bion., distr., tax.); La Casse & Yamaguti 1950 (M*, F*, P*, L*; keys, bion.); Asanuma & Nakagawa 1953: 89 (P*); Omori & Fujii 1953: 23 (bion.); Hara 1957: 61 (F*); Bullock 1960 (E*); Mohrig 1967 (F*); Hinton & Service 1969 (E*); Ramalingam 1969 (distr.; Malaysia); Gutsevich et al. 1971: 288 (distr.); Matsuo, Yoshida, & Kunou 1972: 360 (E*); Moriya et al. 1972 (E*); Moriya et al. 1973 (E*); Gutsevich et al. 1974: 310 (M*, F, L*); Matsuo, Yoshida, & Lien 1974: 187 (E*); Tanaka et al. 1975c: 220 (distr.); Tanaka et al. 1979 (M*, F*, L*; keys, tax, bion.); Belton 1980: 625 (distr.; United States); Linley & Chan 1991 (E*); Gornostaeva 2000: 479 (distr.; Russia); Tanaka 2002b: 144 (P*; tax., key); Tanaka 2003b (P; key, tax.); Sames et al. 2004 (distr.; United States); Reinert, Harbach, & Kitching 2004 (phyl., tax.; to genus *Tanakaius*); Darsie & Ward 2005 (F*, L*; keys, distr.); Reinert 2008i: Fig. 1 (F*); Sames et al. 2008 (review); Rattanarithikul et al. 2010 (F*, L*; keys, bion., distr.; Thailand); Wilkerson et al. 2015 (phyl., tax.; to *Aedes* subg. *Tanakaius*). **Ety.:** Togo [Togolese Republic]. **Informal name:** Togo Japanese Pointy Mosquito. (Vol. 1, p. 470)

Subgenus *Tewarius* Reinert

2006a (as genus). Type species: *Aedes agastyai* Tewari & Hiriyan.

Name derivation: Satish Chandra Tewari.

Important references: Reinert 1973d (tax.); Reinert, Harbach, & Kitching 2008 (phyl., class.); Reinert, Harbach, & Kitching 2009 (phyl., class.); Rattanarithikul et al. 2010 (F*, L*; bion., keys; Thailand); Wilkerson et al. 2015 (phyl., class.; to subg. of genus *Aedes*).

agastyai Tewari & Hiriyan
1992b: 154 (M*, F*, P*, L*). **Type loc.:** Kannidatti, Ambasamudram, Tirunelveli District, Tamil Nadu, India (USNM). **Distr.:** India. **Ety.:** Agastya Hills, India. **Informal name:** Agastya Indian Pointy Mosquito.

nummatus Edwards
1923b: 4 (M, F). **Type loc.:** Meenglas, Bengal, India (NHMUK). **Distr.:** India. **Refs.:** Barraud 1934: 265 (M*, F, L*); Reinert 1973d: 23 (M*, F*, L*; transfer). **Ety.:** n.s. [poss. having to do with coins or money (L)]. **Informal name:** Wealthy Indian Pointy Mosquito [*nummatus* = "having money" = Wealthy].

pseudonummatus Reinert
1973d: 29 (F*). **Type loc.:** Doi Jom Jang, near Wat Phra Bat Mon Kaeo, Amphoe Sanpaton, Chiang Mai, Thailand (USNM). **Distr.:** Thailand. **Refs.:** Reinert 2000e: Fig. 18 (F*); Rattanarithikul et al. 2010 (F*, L*; keys, bion., distr.; Thailand). **Ety.:** false (Gr); similar to *nummatus*. **Informal name:** Counterfeit Thai Pointy Mosquito [*pseudonummatus* = "false money" = Counterfeit].

reubenae Tewari & Hiriyan
1992b: 164 (M*, F*, P, L). **Type loc.:** Panthanthodu, Silent Valley, Palghat District, Kerala, India (USNM). **Distr.:** India. **Ety.:** Rachel Reuben. **Informal name:** Reuben Indian Pointy Mosquito.

Subgenus *Vansomerenis* Reinert, Harbach, & Kitching

2006: 85 (as genus). Type species: *Aedes* (*Finlaya*) *pulchrithorax* Edwards.

Name derivation: Ellinor Catherine Cunningham van Someren.

Important references: Reinert 2008j (F gen.*); Reinert, Harbach, & Kitching 2009 (phyl., class.); Wilkerson et al. 2015 (phyl., class.; to subg. of genus *Aedes*).

hancocki van Someren
1962: 21 (M*, F, L*). **Type loc.:** Amani, Tanzania (NHMUK). **Distr.:** Tanzania. **Ety.:** George Leonard Rhys Hancock. **Informal name:** Hancock Tanzanian Pointy Mosquito.

luteostriatus Robinson
1950: 80 (M*, F*, P, L*). **Type loc.:** Ndola, Northern Rhodesia [Zambia] (SAIM). **Distr.:** Angola, Democratic Republic of the Congo, Zambia. **Ety.:** yellow grooves (L); ref. to lateral yellow lines on the scutum. **Informal name:** Yellow-striped Zambian Pointy Mosquito.

pulchrithorax Edwards
1939a: 17 (A, L*). **Type loc.:** Nairobi, Kenya (NHMUK). **Distr.:** Kenya, Uganda. **Refs.:** Edwards 1941: 124 (M, F), 387 (P), 429 (L*); Clements 1963 (F*); Reinert 2008j: Fig. 1 (F*); Huang & Rueda 2017a (M*; lectotype desig.). **Ety.:** n.s. [beautiful thorax (L)]. **Informal name:** Beautiful Kenyan Pointy Mosquito.

Subgenus *Zavortinkius* Reinert

Reinert 1999a: 7 (as subg. of genus *Aedes*). Type species: *Stegomyia longipalpis* Grünberg.

Name derivation: Thomas James Zavortink.

Important references: Reinert 2000h: 177 (to subg. of genus *Ochlerotatus*); Reinert 2002e (F gen.*); Reinert, Harbach, & Kitching 2004 (phyl., class.; to genus); Reinert, Harbach, & Kitching 2009 (phyl., class.); Wilkerson et al. 2015 (phyl., class.; to subg. of genus *Aedes*); Huang & Rueda 2015 (Longipalpis Group; Afrotropics).

brunhesi Reinert
1999a: 55 (M*; key, tax.). **Type loc.:** Correfour Demberi, Sada Mayotte, Comoros (IRD). **Distr.:** Comoros. **Ety.:** Jacques Brunhes. **Informal name:** Brunhes Comoran Pointy Mosquito.

brygooi Brunhes
1971: 335 (M, F, P*, L*). **Type loc.:** Ambovombe, Province de Tulear, Madagascar (IERT). **Distr.:** Madagascar (includes Glorioso and Juan De Nova Isl.). **Refs.:** Reinert 1999a: 43 (M*, F*, P*, L*; keys, tax.). **Ety.:** Édouard-Raoul Brygoo. **Informal name:** Brygoo Malagasy Pointy Mosquito.

fulgens (Edwards)
1917: 213 (M, F; *Ochlerotatus*). **Type loc.:** Zanzibar [Tanzania] (NHMUK). **Distr.:** Angola, Botswana, Kenya, Malawi, Republic of South Africa, Senegal, Tanzania (includes Zanzibar), Zambia, Zimbabwe. **Refs.:** De Meillon et al. 1945: 99 (P; tax.); Hopkins 1952: 129 (L);

Ribeiro & da Cunha Ramos 1973: 117 (distr.); Jupp 1996 (M*, F*; key); Reinert 1999a: 26 (M*, F*, P*, L*; keys, tax.). **Ety.:** n.s. [flashing (L)]. **Informal name:** Flashing Tanzanian Pointy Mosquito.

geoffroyi Reinert
1999a: 31 (M*, F*, P*, L*; keys, tax.). **Type loc.:** Adiopodoume, Inst. Past. Forest, Abidjan Department, Ivory Coast [Côte d'Ivoire] (USNM). **Distr.:** Cameroon, Côte d'Ivoire, Ghana, Liberia, Nigeria, Sierra Leone. **Ety.:** Bernard Geoffroy. **Informal name:** Geoffroy Ivorian Pointy Mosquito.

huangae Reinert
1999a: 34 (M*, F*, P*, L*; keys, tax.). **Type loc.:** Center Department, Dezidougou, Ivory Coast [Côte d'Ivoire] (USNM). **Distr.:** Cameroon, Côte d'Ivoire, Liberia, Sierra Leone, Uganda. **Ety.:** Yiau-Min Huang. **Informal name:** Huang Ivorian Pointy Mosquito.

interruptus Reinert
1999a: 47 (M*, F*; keys, tax.). **Type loc.:** Province Tamatave, Madagascar (IRD). **Distr.:** Madagascar (includes Glorioso and Juan De Nova Isl.). **Ety.:** n.s. [interrupted (L); poss. ref. to scutum with "dorsocentral white scaled stripe interrupted by a dark-scaled area"]. **Informal name:** Interrupted Malagasy Pointy Mosquito.

longipalpis (Grünberg)
1905: 383 (F*; *Stegomyia*). **Type loc.:** Duala (Hafen), Cameroon (ZM). **Distr.:** Burkina Faso, Cameroon, Central African Republic, Congo, Côte d'Ivoire, Democratic Republic of the Congo, Kenya, Liberia, Mali, Nigeria, Senegal, Uganda, Zambia. **Refs.:** Edwards 1941: 120 (M, F*), 386 (P*), 429 (L); Haddow et al. 1951: 218 (A); Hopkins 1952: 127 (L*); Gillett 1955 (E); van Someren 1962: 19 (M*); Reinert 1999a: 19 (M*, F*, P*, L*; keys, tax.); Reinert 2002e: Fig. 39 (F*). **Ety.:** long feeler (L); ref. to long palpus. **Informal name:** Long-lipped Cameroonian Pointy Mosquito.

monetus Edwards
1935a: 132 (F). **Type loc.:** Maevatanana, Madagascar (NHMUK). **Distr.:** Comoros, Madagascar (includes Glorioso and Juan De Nova Isl.). **Refs.:** Brunhes 1971: 344 (M*, F*, L*); Brunhes 1977a (distr.; Comoros Isl.); Reinert 1999a: 51 (M*, F*, P*, L*; keys, tax.). **Ety.:** n.s. [money? (L)]. **Informal name:** Juno Malagasy Pointy Mosquito [the Roman goddess Juno was known as Juno Moneta, the origin of the English word "money"].

mzooi van Someren
1962: 19 (M*, F). **Type loc.:** Sigi River bridge, near Amani, Usambara, Tanganyika [Tanzania] (NHMUK). **Distr.:** Tanzania. **Refs.:** Reinert 1999a: 38 (M*, F*; keys, tax.). **Ety.:** Fabian Mzoo. **Informal name:** Mzoo Tanzanian Pointy Mosquito.

phillipi van Someren
1949a: 7 (M). **Type loc.:** Sakaramy, near Diego Suarez, Madagascar (NHMUK). **Distr.:** Madagascar (includes Glorioso and Juan De Nova Isl.). **Refs.:** Brunhes 1971: 340 (M*, F, P*, L*); Reinert 1999a: 49 (M*, F*, P*, L*; keys, tax.). **Ety.:** n.s. [Captain Phillip Glasgow (Harbach 2018c)]. **Informal name:** Phillip Malagasy Pointy Mosquito.

pollinctor (Graham)
1910: 271 (M, F; *Stegomyia*). **Type loc.:** Lagos, Nigeria (NHMUK). **Distr.:** Côte d'Ivoire, Ghana, Nigeria, Sierra Leone, Uganda. **Refs.:** Reinert 1999a: 40 (M*, F*, P*, L*; keys, tax.). **Ety.:** n.s. [undertaker (L)]. **Informal name:** Undertaker Nigerian Pointy Mosquito.

Subgenus Uncertain *Aedes*

alticola Bonne-Wepster
1948: 313 (F). **Type loc.:** Scree Valley, at foot of Mount Wilhelmina, 3,800 m, Western New Guinea (Papua) (ITH). **Distr.:** Indonesia (West Papua). **Refs.:** Bonne-Wepster 1954: 71 (F*); Lee et al. 1982: 110 (F; key, tax., bion., distr., review). **Ety.:** n.s. [high dweller (L): prob. ref. to found at a high altitude (3,800 m)]. **Informal name:** High Papuan Pointy Mosquito.

auronitens Edwards
1922a: 268 (M, F). **Type loc.:** Simla [East Punjab], India (NHMUK). **Distr.:** India. **Refs.:** Barraud 1934: 202 (M, F, L). **Ety.:** shining gold (L); ref. to golden scales on head and thorax. **Informal name:** Shining Indian Pointy Mosquito.

australiensis (Theobald)
1910c: 313 (F; *Leucomyia*). **Type loc.:** Stannary Hills, Queensland, Australia (NHMUK). **Distr.:** Australia. **Refs.:** Marks 1948: 3 (F, P*, L*); Marks 1955: 24 (M, F); Lee et al. 1982: 123 (F; key, tax., bion., distr., review). **Ety.:** Australia. **Informal name:** Stannary Australian Pointy Mosquito.

biocellatus (Taylor)
1914b: 463 (F; *Culex*). **Type loc.:** Milson Island, Hawkesbury River, New South Wales, Australia (US). **Distr.:** Australia. **Refs.:** Marks 1948: 6 (M*, F, P*, L*); Lee et al. 1982: 126 (F; key, tax., bion., distr., review). **Ety.:** n.s. [two little eyes (L); poss. ref. to wings with two spots]. **Informal name:** Two-eyed Australian Pointy Mosquito.

britteni Marks & Hodgkin
1958: 33 (M*, F, P*, L*). **Type loc.:** Kalumburu (Drysdale River Mission), Western Australia, Australia (UQIC). **Distr.:** Australia. **Refs.:** Lee et al. 1982: 130 (F; key, tax., bion., distr., review). **Ety.:** E.J. Britten. **Informal name:** Britten Australian Pointy Mosquito.

candidoscutellum Marks
1947b: 1 (M*, F, P*, L*). **Type loc.:** Binna Burra, Lamington National Park, Queensland, Australia (UQIC). **Distr.:** Australia, Indonesia (includes West Papua), Papua New Guinea. **Refs.:** Penn 1949b: 48 (P*); Lee et al. 1982: 133 (F; key, tax., bion., distr., review). **Ety.:** bright saucer (L); scutellum shining white. **Informal name:** Bright-plated Australian Pointy Mosquito.

crossi Lien
1967: 177 (M*, F*, P*, L*). **Type loc.:** Between Tashan and Alishan, Wufeng, Chiai County, Taiwan (TMRI). **Distr.:** Taiwan. **Ety.:** John H. Cross. **Informal name:** Cross Taiwanese Pointy Mosquito.

daliensis (Taylor)
1916: 565 (F; *Stegomyia*). **Type loc.:** Daly River, Northern Territory, Australia (US). **Distr.:** Australia. **Refs.:** Edwards 1924b: 391 (tax.); Lee et al. 1984: 46 (tax., review). **Ety.:** Daly River, Australia. **Informal name:** Daly Australian Pointy Mosquito.

eatoni (Edwards)
1916: 358 (M; *Ochlerotatus*). **Type loc.:** Monte Funchal (2,000 ft), Madeira Islands (NHMUK). **Distr.:** Portugal (Madeira), Spain (Canary Isl.). **Refs.:** Clavero 1946: 20 (P*, L*); Senevet & Andarelli 1958: 277 (P*); Reinert 2005a: 1-2 (M; holotype desc.). **Ety.:** A.E. Eaton. **Informal name:** Eaton Portuguese Pointy Mosquito.

gracilelineatus Bonne-Wepster
1937: 93 (M*, F*). **Type loc.:** Anggi Lakes, Western New Guinea (ITH). **Distr.:** Papua New Guinea. **Refs.:** Lee et al. 1982: 158 (F; key, tax., bion., distr., review). **Ety.:** slender lined (L); ref. to slender lines of scales on the scutum. **Informal name:** Thin-lined Papuan Pointy Mosquito.

keefei King & Hoogstraal
1946f: 311 (M*, F, L*). **Type loc.:** K.B. Mission, Milne Bay, Papua, New Guinea (USNM). **Distr.:** Australia, Papua New Guinea. **Refs.:** Penn 1949b: 44 (P*); Lee et al. 1982: 170 (F; key, tax., bion., distr., review). **Ety.:** A.M. Keefe. **Informal name:** Keefe Papuan Pointy Mosquito.

koreicoides Sasa, Kano, & Hayashi
1950: 627 (M*, F). **Type loc.:** Jozankei, Hokkaido, Japan (IID). **Distr.:** Japan. **Refs.:** Hara 1957: 61 (M*); Sakakibara & Omori 1962: 15 (M*, F, P*, L*); Miyagi 1971: 148 (A, L); Tanaka et al. 1979: 348 (M*, F*, L*); Tanaka 2002b: 161 (P*; tax., key). **Ety.:** similar (Gr); like *koreicus*. **Informal name:** Kankoku Japanese Pointy Mosquito [Kankoku is the Japanese name for South Korea].

lauriei (Carter)
1920: 623 (M*, F, P*, L*; as *Ochlerotatus* laurei, emend. by Edwards 1924b: 383). **Type loc.:** Lord Howe Island, South Pacific (NHMUK). **Distr.:** Australia. **Refs.:** Knight & Marks 1952: 573 (A); Belkin 1962: 352 (M*, F, P*, L*); Lee et al. 1982: 182 (F; key, tax., bion., distr., review). **Ety.:** R.D. Laurie. **Informal name:** Laurie Australian Pointy Mosquito.

mallochi Taylor
1944: 121 (F). **Type loc.:** Eidsvold, Queensland, Australia (US). **Distr.:** Australia. **Refs.:** Knight & Marks 1952: 588 (neotype desig.); Marks 1955: 13 (M*, F*, P*, L*; bion.); Dobrotworsky 1965: 133 (M*, F*, L*); Lee et al. 1982: 187 (F; key, tax., bion., distr., review); Townsend 1990: 100 (syntype, NHMUK). **Ety.:** J.R. Malloch. **Informal name:** Malloch Australian Pointy Mosquito.

monocellatus Marks
1948: 14 (M*, F, P*, L*). **Type loc.:** Upper Cedar Creek, Queensland, Australia (UQIC). **Distr.:** Australia. **Refs.:** Lee et al. 1982: 192 (F; key, tax., bion., distr., review). **Ety.:** n.s. ["having one spot on the wing" (L); derivation not clear: single eye?]. **Informal name:** One-eyed Australian Pointy Mosquito.

oreophilus (Edwards)
1916: 357 (F; *Ochlerotatus*). **Type loc.:** Gharial, Murree Hills, Western Himalayas [Punjab, Pakistan] (NHMUK). **Distr.:** India, Japan, Nepal, Pakistan, South Korea. **Refs.:** Barraud 1934: 192 (M*, F*, L*; as *Aedes*); Hara 1957 (F*; as *bunanoki*); Aslamkhan 1971b (distr.; Pakistan); Chow 1973: 43 (distr.); Tanaka et al. 1979: 361 (M*, F*, L*); Darsie et al. 1996: 130 (distr.; Nepal)); Tanaka 2002b: 169 (P*; tax., key). **Ety.:** mountain loving (Gr); first found in a mountainous area. **Informal name:** Montane Pakistani Pointy Mosquito.
syn. *bunanoki* Sasa & Ishimura
1951: 103 (M*, F*, P*, L*). **Type loc.:** Bonjusa Aomori, Honshu, Japan (IID). **Refs.:** Kamimura 1968: 28 (syn.).

peipingensis Feng
1938: 520 (M*, F*). **Type loc.:** Peiping, Hopeh, China [People's Republic of China] (CUMC). **Distr.:** People's Republic of China. **Ety.:** Peiping [Beijing, People's Republic of China]. **Informal name:** Peiping Chinese Pointy Mosquito.

plagosus Marks
1959b: 21 (M*). **Type loc.:** Ben Lomond, 35 mi north of Armidale, New South Wales, Australia (CSIR). **Distr.:** Australia. **Refs.:** Dobrotworsky 1965: 135 (F*); Lee et al. 1982: 229 (F; key, tax., bion., distr., review). **Ety.:** "Full of stripes" (L); ref. to lines on scutum and legs. **Informal name:** Stripey Australian Pointy Mosquito.

quasirubithorax (Theobald)
1910c: 327, 348 (F*; *Culex*). **Type loc.:** Kuranda, Queensland, Australia (NHMUK). **Distr.:** Australia, Indonesia, Papua New Guinea. **Refs.:** Lee 1944a: 62 (L*); King & Hoogstraal 1946f: 311 (M, F, L); Penn 1949b: 46 (P); Lee et al. 1982: 240 (F; key, tax., bion., distr., review). **Ety.:** nearly (L); similar to *rubithorax*. **Informal name:**

Reddish Australian Pointy Mosquito [quasirubi = "half red" = Reddish].

roai Belkin
1962: 353 (M*, F). **Type loc.:** Wrights Creek, Matanikau Valley, Guadalcanal, Solomon Islands (USNM). **Distr.:** Solomon Isl. **Refs.:** Lee et al. 1982: 245 (F; key, tax., bion., distr., review). **Ety.:** Vernon R. Roa. **Informal name:** Roa Solomon Islander Pointy Mosquito.

rubiginosus Belkin
1962: 346 (M*, F). **Type loc.:** "Apparently from the Solomons?," Guadalcanal (USNM). **Distr.:** Papua New Guinea, Solomon Isl. **Refs.:** Lee et al. 1982: 246 (F; key, tax., bion., distr., review). **Ety.:** rusty color (L); overall color. **Informal name:** Rusty Solomon Islander Pointy Mosquito.

sintoni (Barraud)
1924b: 967 (M*, F; *Finlaya*). **Type loc.:** Ferozepore Nullah, near Tangmarg [Pakistan] (NHMUK). **Distr.:** India, Pakistan. **Refs.:** Aslamkhan 1971b (distr.; Pakistan); Kaur 2003: 454 (distr.; India). **Ety.:** John Alexander Sinton. **Informal name:** Sinton Pakistani Pointy Mosquito.

stanleyi Peters
1963a: 1 (M, F, P*, L*). **Type loc.:** Minj, Western Highlands District, New Guinea (CSIR). **Distr.:** Indonesia, Papua New Guinea. **Refs.:** Lee et al. 1982: 271 (F; key, tax., bion., distr., review). **Ety.:** Stanley Hinton Christian. **Informal name:** Stanley Papuan Pointy Mosquito.

subauridorsum Marks
1948: 28 (M*, F, P*, L*; as *Aedes*). **Type loc.:** Mount Mowbullan, Bunya Mountains, Queensland, Australia (UQIC). **Distr.:** Australia. **Refs.:** Dobrotworsky 1965: 152 (M*, F*, L*); Lee et al. 1982: 274 (F; key, tax., bion., distr., review). **Ety.:** nearly (L); near *auridorsum*. **Informal name:** Brass-backed Australian Pointy Mosquito [subauri = "subgold" = Brass].

suffusus Edwards
1922a: 270 (F). **Type loc.:** Simla, Western Himalayas [East Punjab], India (NHMUK). **Distr.:** India, Nepal. **Refs.:** Darsie et al. 1996 (distr.; Nepal). **Ety.:** suffused (L); ref. to indefinite markings on scutum. **Informal name:** Suffused Indian Pointy Mosquito.

toxopeusi Bonne-Wepster
1948: 315 (M*, F). **Type loc.:** Scree Valley, at foot of Mount Wilhelmina, Western New Guinea (Papua) [West Papua] (ITH). **Distr.:** Indonesia (West Papua). **Refs.:** Bonne-Wepster 1954: 71 (M, F*); Lee et al. 1982: 278 (F; key, tax., bion., distr., review). **Ety.:** L.J. Toxopeus. **Informal name:** Toxopeus Indonesian Pointy Mosquito.

tsiliensis King & Hoogstraal
1946f: 305 (M*). **Type loc.:** Tsili Tsili on Watut River, branch of Markham River, [Northeast] New Guinea (USNM). **Distr.:** Papua New Guinea. **Refs.:** Lee et al. 1982: 279 (tax., review). **Ety.:** Tsili Tsili, Papua New Guinea. **Informal name:** Tsili Tsili Papuan Pointy Mosquito.

versicolor (Barraud)
1924k: 73 (F; *Finlaya*). **Type loc.:** Yusmarg, Kashmir [Pakistan] (NHMUK). **Distr.:** Azerbaijan, India, Iran, Pakistan, Turkey. **Refs.:** Barraud 1934: 151 (F); Aslamkhan 1971b (distr.; Pakistan); Danilov 1978: 82 (from subg. *Ochlerotatus*, from var. *pulchritarsis*). **Ety.:** n.s. [having colors that change (L)]. **Informal name:** Colorful Pakistani Pointy Mosquito.

wasselli Marks
1947b: 9 (F). **Type loc.:** Mountain Creek, Buderim Mountain, Queensland, Australia (UQIC). **Distr.:** Australia. **Refs.:** Lee et al. 1982: 286 (F; key, tax., bion., distr., review). **Ety.:** J.L. Wassell. **Informal name:** Wassell Australian Pointy Mosquito.

Tribe Aedini

Genus *Armigeres* Theobald

(Vol. 1, p. 222)

Subgenus *Armigeres* Theobald

1901b: 235 (July 15), 1901a: 98, 322 (Nov. 23). Type species: *Culex obturbans* Walker.

Name derivation: armor bearer (L).

syn. *Blanchardiomyia* Brunetti 1912: 440 (as genus). Type species: *Culex obturbans* Walker.
syn. *Desvoidya* Blanchard 1901b: 1046 (Dec. 6). Type species: *Culex obturbans* Walker. Theobald 1903a: 134 (as *Desvoidea*, *lapsus*).
syn. *Desvoidea* Theobald 1903a: 134 (variant spelling of *Desvoidya* Blanchard).
syn. *Neosquamomyia* Taylor 1914d: 186 (as genus). Type species: *Neosquamomyia breinli* Taylor.

Important references: Brunetti 1914: 55 (desig. type species); Barraud 1934: 310, 313 (tax., keys, morph.; India, Sri Lanka, Myanmar); Thurman 1959b (keys; Thailand); Delfinado 1966a: (keys; Philippines); Steffan 1968b: 135 (keys; Papuan Subregion); Tanaka et al. 1979 (tax., keys, bion.; Japan); Lee et al. 1988a (tax., key; Australasia); Darsie & Pradhan 1990 (tax., keys; Nepal); Lu et al. 1997 (People's Republic of China); Darsie 2000a (P; tax., key; Nepal); Reinert, Harbach, & Kitching 2004 (phyl., class.); Rattanarithikul & Harrison et al. 2005 (F, L; tax., keys, bion.; Thailand); Reinert, Harbach, & Kitching 2009 (phyl., class.); Rattanarithikul et al. 2010 (F*, L*; keys, bion.; Thailand); Wilkerson et al. 2015 (phyl., class.); Soghigian et al. 2017 (mol. phyl.).

alkatirii Toma, Miyagi, & Syafruddin
1995: 110 (M*, F*, P*, L*). **Type loc.:** Matayangan, North Sulawesi, Indonesia (USNM). **Distr.:** Indonesia. **Ety.:** Junas Alkatiri. **Informal name:** Alkatiri Indonesian Cord Mosquito.

apoensis Bohart & Farner
1944: 69 (M*, F). **Type loc.:** Sibulan River, Mount Apo [Davao], Mindanao, Philippines (MCZ). **Distr.:** Indonesia, Philippines. **Refs.:** Delfinado 1966a: 79 (M*); Basio 1971b: 47 (M*; bion.). **Ety.:** Mt. Apo, Mindanao, Philippine Isl. **Informal name:** Apo Philippine Cord Mosquito.

aureolineatus (Leicester)
1908: 79 (F; *Desvoidya*). **Type loc.:** Ampang, Malaya [Selangor, Malaysia] (NHMUK). **Distr.:** Cambodia, India, Laos, Malaysia, Nepal, People's Republic of China, Philippines, Sri Lanka, Thailand, Vietnam. **Refs.:** Senior-White 1927: 67, Pl. VII, Fig. 6 (L*); Barraud 1934: 319 (M*, F, L); Thurman 1959b: 83 (M*, L); Delfinado 1966a: 80 (M*, distr.); Basio 1971b: 48 (M*; bion.); Darsie & Pradhan 1990 (F, L; tax., keys, bion., distr.; Nepal); Darsie 2000b: 109 (P*; tax., key); Rattanarithikul et al. 2010 (F*, L*; keys, bion., distr.; Thailand). **Ety.:** n.s. [gold lined (L); ref. to golden-yellow lines on thorax]. **Informal name:** Gold-lined Malaysian Cord Mosquito.

azurini Basio
1971a: 54 (M*). **Type loc.:** Mount Makiling, Laguna, Luzon, Philippines (PNM). **Distr.:** Philippines. **Ety.:** Jesus C. Azurin. **Informal name:** Azurin Philippine Cord Mosquito.

baisasi Stone & Thurman
1958: 240 (M*, F). **Type loc.:** Abuyog, Leyte [Eastern Visayas], Philippines (USNM). **Distr.:** Philippines. **Refs.:** Baisas 1935a: 488 (M*, F; as *kuchingensis*); Delfinado 1966a: 80 (M*, F, P*, L*); Basio 1971b: 48 (M*; bion.); Baisas 1974: 75 (M, F, P*, L*; tax., bion., distr.; Philippines); Darsie 2000b (P*; tax., key). **Ety.:** Francisco Edlagan Baisas. **Informal name:** Baisas Philippine Cord Mosquito.

bhayungi Thurman & Thurman
1958: 186 (F). **Type loc.:** Tad Muey Falls, Doi Suthep mountain, Chiengmai Province, Thailand (USNM). **Distr.:** Thailand. **Refs.:** Baisas 1965 (M*, F*, P*, L*); Rattanarithikul et al. 2010 (F*, L*; keys, bion., distr.; Thailand). **Ety.:** Luang Bhayung Vejjasastr. **Informal name:** Bhayung Thai Cord Mosquito.

breinli (Taylor)
1914d: 186 (F*; *Neosquamomyia*). **Type loc.:** Milne Bay, Papua New Guinea (ANIC). **Distr.:** Australia, Indonesia, Papua New Guinea, Solomon Island. **Refs.:** Lee 1944c: 217 (M*, F, L*); Penn 1949a: 65 (P*); Belkin 1962: 482 (M*, F, P*, L*); Steffan 1968b: 143 (M*, F, P*, L*); Rodhain & Gaxotte 1977 (distr.); Maffi et al. 1979: 214 (distr.); Lee et al. 1988a: 6 (F; lectotype F; type info., key, tax., bion., distr., review). **Ety.:** Anton Breinl. **Informal name:** Breinl Papuan Cord Mosquito.

candelabrifer Brug
1939: 99 (M*, F, L*). **Type loc.:** Near Lindoe Lake, Celebes [Sulawesi] (NHMUK). **Distr.:** Indonesia (Sulawesi). **Ety.:** candlestick bearing (L); three candle-shaped setae on gonocoxite. **Informal name:** Candlestick Indonesian Cord Mosquito.

confusus Edwards
1915c: 283 (M, F). **Type loc.:** Kuching, Sarawak, [Malaysian] Borneo (NHMUK). **Distr.:** Indonesia, Malaysia, Singapore, Thailand. **Refs.:** Bonne-Wepster & Brug 1937: 529 (M, F*); Bonne-Wepster & Brug 1939: 1234 (L*); Mattingly 1971d: 122 (E*); Jeffery et al. 2010 (distr.; Singapore); Rattanarithikul et al. 2010 (F*, L*; keys, bion.,

distr.; Thailand). **Ety.:** n.s. [confused (L)]. **Informal name:** Puzzling Malaysian Cord Mosquito.

conjungens Edwards
1914c: 126 (M, F). **Type loc.:** Kuching, Sarawak, [Malaysian] Borneo (NHMUK). **Distr.:** Malaysia, Singapore, Thailand. **Refs.:** Mattingly & Qutubuddin 1952: 92 (M*); Toma & Miyagi 2005b (M*, P*, L*; redescr.); Hirao et al. 2010 (M*, P*, L*). **Ety.:** connect, unite (L); ref. to being intermediate between *Armigeres* and *Leicesteria*. **Informal name:** United Malaysian Cord Mosquito.

denbesteni Brug
1925: 345 (M*, F, L*). **Type loc.:** Piroe, Ceram, Moluccas [Seram, Maluku, Indonesia] (NHMUK). **Distr.:** Indonesia, Malaysia, Papua New Guinea. **Refs.:** Steffan 1968b: 151 (M*, F, P*, L*); Lee et al. 1988a: 11 (F; key, tax., bion., distr., review). **Ety.:** Dr. den Besten. **Informal name:** den Besten Indonesian Cord Mosquito.
syn. *spathulatus* Brug
1939: 101 (M*). **Type loc.:** Ralio Sawalit, Ceram, Moluccas [Seram, Maluku, Indonesia] (NHMUK). **Refs.:** Lee 1944c: 215 (syn.); Townsend 1990: 136 (type info.).

durhami (Edwards)
1917: 206 (M, F). **Type loc.:** Bukit Kutu, Malaya [Malaysia] (NHMUK). **Distr.:** Cambodia, India, Indonesia, Laos, Malaysia, Nepal, Thailand, Vietnam. **Refs.:** Thurman 1959a: 391 (M*; to sp.); Thurman 1959b: 85 (M*, L); Darsie & Pradhan 1990 (F, L; tax., keys, bion., distr.; Nepal); Darsie 2000b (P*; tax., key). **Ety.:** Dr. Durham. **Informal name:** Durham Malaysian Cord Mosquito.
syn. *dibrugarhensis* Barraud
1927b: 539 (A; *Armigeres* (*Armigeres*) *kuchingensis* form). **Type loc.:** Upper Assam Valley [India] (LU). **Refs.:** Mattingly 1956b: 26 (type info.); Thurman 1959b: 85 (syn.).
syn. *shillongensis* Barraud
1927b: 539 (A; *Armigeres* (*Armigeres*) *kuchingensis* form). **Type loc.:** Assam Hills [India] (LU). **Refs.:** Mattingly 1956b: 26 (type info.); Thurman 1959b: 85 (syn.).
syn. *javanensis* Brug
1939: 100 (M, F, L*; *Armigeres* (*Armigeres*) *kuchingensis* var.). **Type loc.:** Tangkoeban Prahoe, Java [Indonesia] (NHMUK). **Refs.:** Thurman 1959b: 87 (syn.?); Ramalingam 1987: 64 (syn.).

ejercitoi Baisas
1935a: 486 (M*, F). **Type loc.:** Kolambugan, Lanao, Mindanao, Philippines (NE). **Distr.:** Philippines. **Refs.:** Delfinado 1966a: 82 (M*, F); Basio 1971b: 48 (M*; bion.). **Ety.:** Antonio Ejercito. **Informal name:** Ejercito Philippine Cord Mosquito.

fimbriatus Edwards
1930a: 305 (M). **Type loc.:** Bettatan, near Sandakan, North Borneo [Sabah, Malaysian Borneo] (NHMUK). **Distr.:** Malaysia. **Ety.:** fringed (L); "The somewhat fringed appearance of the tip of the abdomen is distinctive." **Informal name:** Fringed Malaysian Cord Mosquito.

foliatus Brug
1931c: 32 (M*). **Type loc.:** Ranau-See [Danau Ranau], Sumatra [Indonesia] (NHMUK). **Distr.:** Indonesia, Thailand. **Refs.:** Rattanarithikul et al. 2010 (F*, L*; keys, bion., distr.; Thailand). **Ety.:** n.s. [leafy (L)]. **Informal name:** Leaf-like Indonesian Cord Mosquito.
syn. *shrivastavii* Wattal, Kalra, & Krishnan
1966: 159 (M*). **Type loc.:** Ferrargunj, S Andaman Islands (NCID). **Refs.:** Sirivanakarn 1972: 27 (syn.).
syn. *chungkiangensis* Chang & Chang
1974: 347 (M*). **Type loc.:** Chungkiang, Kweichow Province, China [People's Republic of China] (Provincial Station of Sanitary & Diseases Control, Kweichow, China). **Refs.:** Meng & Chen 1980: 161 (syn.).

giveni Edwards
1926a: 118 (M, L). **Type loc.:** Singapore (NHMUK). **Distr.:** Singapore. **Refs.:** Toma & Miyagi 2002: 264 (M*). **Ety.:** D.H.C. Given. **Informal name:** Given Singaporean Cord Mosquito.

hybridus Edwards
1914c: 126 (M, F). **Type loc.:** Kuching, Sarawak, [Malaysian] Borneo (NHMUK). **Distr.:** Malaysia, People's Republic of China, Singapore. **Refs.:** Toma et al. 2011 (P*, L*). **Ety.:** n.s. [hybrid (L); poss. ref. to similarity with *obturbans*]. **Informal name:** Hybrid Malaysian Cord Mosquito.

joloensis (Ludlow)
1904b: 236 (M, F; *Desvoidea fusca* var.). **Type loc.:** Jolo, Jolo Island [Sulu Island], Philippines (USNM). **Distr.:** India, Philippines. **Refs.:** Banks 1908 (P*, L*, E*); Stone & Knight 1955: 282 (lectotype desig.); Stone & Thurman 1958: 243 (M*; from syn. with *obturbans*); Baisas 1965 (M*, F*, P*, L*); Basio 1971b: 49 (M*; bion.); Bhattacharyya et al. 2000 (distr.; India). **Ety.:** Jolo Jolo, Philippine Isl. **Informal name:** Jolo Philippine Cord Mosquito.

jugraensis (Leicester)
1908: 77 (M, F; *Desvoidya*). **Type loc.:** Ampang Jungle, Kuala Lumpur [Selangor], Malaya [Malaysia] (NHMUK). **Distr.:** Indonesia, Malaysia, Thailand. **Refs.:** Thurman 1959b: 86 (M*, L); Hirao et al. 2010 (M*, P*, L*); Rattanarithikul et al. 2010 (F*, L*; keys, bion., distr.; Thailand). **Ety.:** Jugra, Malaysia. **Informal name:** Jugra Malaysian Cord Mosquito.

kesseli Ramalingam
1987: 56 (M*, F*, P*, L*). **Type loc.:** Ulu Gombak, Selangor, Malaysia (USNM). **Distr.:** Bangladesh, India, Malaysia, Nepal, Thailand. **Refs.:** Darsie & Pradhan 1990 (F, L; tax., keys, bion., distr.; Nepal); Darsie et al. 1996 (distr.; Nepal); Rattanarithikul et al. 2010 (F*, L*; keys, bion., distr.; Thailand). **Ety.:** John F. Kessel. **Informal name:** Kessel Malaysian Cord Mosquito.

kinabaluensis Ramalingam
1972: 460 (M*, F, P*, L*). **Type loc.:** Mount Kinabalu, Sabah, East Malaysia [Malaysian Borneo] (USNM). **Distr.:** Malaysia. **Ety.:** Mt. Kinabalu, Sabah, Malaysia. **Informal name:** Kinabalu Malaysian Cord Mosquito.

kuchingensis Edwards
1915c: 283 (M, F). **Type loc.:** Kuching Reservoir, Sarawak, [Malaysian] Borneo (NHMUK). **Distr.:** Bangladesh, Cambodia, India, Indonesia, Laos, Malaysia (includes Borneo-Sarawak), Nepal, Pakistan, Philippines, Thailand, Vietnam. **Refs.:** Borel 1930: 186 (M*, F, L*); Barraud 1934: 314 (tax.), 317 (M*, F, L); Stone & Thurman 1958: 240 (M*; from syn. with *obturbans*); Thurman 1959b: 86 (M*, L); Joshi et al. 1965 (distr.; Nepal); Ahmed 1987 (distr.; Bangladesh); Darsie & Pradhan 1990 (F, L; tax., keys, bion., distr.; Nepal); Darsie 2000b (P*; tax., key); Rattanarithikul et al. 2010 (F*, L*; keys, bion., distr.; Thailand). **Ety.:** Kuching Reservoir, Sarawak. **Informal name:** Kuching Malaysian Cord Mosquito.
syn. *nongpohensis* Barraud
1927b: 539 (A; *Armigeres* (*Armigeres*) *kuchingensis* form). **Type loc.:** Assam Hills and E Himalayas [India] (LU). **Refs.:** Mattingly 1956b: 26 (type info.); Thurman 1959b: 87 (syn.).

lacuum Edwards
1922d: 97 (M, F). **Type loc.:** Ile des Lacs, Bismarck Archipelago [Garove, Vitu Islands] (HNM). **Distr.:** Indonesia, Papua New Guinea. **Refs.:** Hill 1925: 70 (P*, L*); Lee 1944c: 219 (M*, F, L); Steffan 1968b: 147 (M*, F, P*, L*; distr.); Lee et al. 1988a: 13 (F; key, tax., bion., distr., review). **Ety.:** n.s. [of lakes (L)]. **Informal name:** Lakeside Papuan Cord Mosquito.

laoensis Toma & Miyagi
2003: 169 (M*). **Type loc.:** Thapachon, Khammouane Province, Lao People's Democratic Republic [Laos] (USNHM). **Distr.:** Laos. **Ety.:** Lao People's Democratic Republic [Laos]. **Informal name:** Muang Laotian Cord Mosquito [Muang Lao, meaning "Lao Country," is an indigenous exonym for Laos].

mahantai Bhattacharyya, Prakash, Mohapatra, & Sarma
2009 (M*, F*). **Type loc.:** Baghmara, South Garo Hills, Meghalaya, NE India (NHMUK). **Distr.:** India. **Ety.:** J. Mahanta. **Informal name:** Mahanta Indian Cord Mosquito.

maiae Edwards
1917: 208 (M*). **Type loc.:** Jungle, Pahang Road, 6 mi from Kuala Lumpur [Selangor], Malaya [Malaysia] (NHMUK). **Distr.:** Malaysia. **Refs.:** Mattingly 1971d: 123 (E). **Ety.:** Maia Edwards (?). **Informal name:** Maia Malaysian Cord Mosquito.

malayi (Theobald)
1901c: 258 (F*; *Uranotaenia*). **Type loc.:** Klang Jungle, Selangor, Malaya [Malaysia] (NHMUK). **Distr.:** Bangladesh, Indonesia, Malaysia, Papua New Guinea, People's Republic of China, Philippines, Thailand, Timor, Vietnam. **Refs.:** Edwards 1917: 207 (M*); Brug 1924: 439 (L*; as *jugraensis*); Bonne-Wepster & Brug 1937b: 532 (M, F*); Bonne-Wepster & Brug 1939: 1234 (L*); Knight & Chamberlain 1948: 10 (P*); Thurman 1959b: 88 (M*, F, P*, L*); Steffan 1968b: 141 (M*, F, P*, L*); Basio 1971b: 49 (M*, L*; bion.); Mattingly 1971a: Pl. 13 (P terminal segs.*); Baisas 1974: 77 (M, F, P*, L*; tax., bion., distr.; Philippines); Lee et al. 1988a: 16 (F; key, tax., bion., distr., review); Hirao et al. 2010 (M*, P*, L*); Rattanarithikul et al. 2010 (F*, L*; keys, bion., distr.; Thailand); Anderson & Davis 2014 (distr.; Timor-Leste). **Ety.:** Malaya [Malaysia]. **Informal name:** Tanah Malaysian Cord Mosquito [Tanah Malayu, meaning "Malay Land," is an indigenous exonym for the Malaysian Peninsula].
syn. *fusca* Theobald
1903a: 135 (M*, F, P*, L*; *Desvoidea*). **Type loc.:** Jugra, Malaya [Malaysia] (NHMUK).
syn. *russelli* Baisas
1935a: 490 (M*, F). **Type loc.:** Masiit, Calauan, Laguna, Luzon, Philippines (NE). **Refs.:** Bohart 1945: 47 (syn.).

manalangi Baisas
1935a: 492 (M*, F). **Type loc.:** Tungkong Mangga, San Jose, Bulacan Luzon, Philippines (PBH). **Distr.:** Philippines. **Refs.:** Delfinado 1966a: 86 (M*, F, P*, L*); Basio 1971b: 49 (M*; bion.); Baisas 1974: 79 (M*, F*, P*, L*; tax., bion., distr.; Philippines). **Ety.:** Cristobal Manalang. **Informal name:** Manalang Philippine Cord Mosquito.

maximus Edwards
1922a: 292 (F). **Type loc.:** Suban Ajam [Soebanajam, Benkoelen], Sumatra [Indonesia] (AM). **Distr.:** Indonesia, Malaysia, Thailand. **Refs.:** Harrison et al. 1991: 224 (distr.; Thailand); Rattanarithikul et al. 2010 (F*, L*; keys, bion., distr.; Thailand). **Ety.:** largest (L); ref. to general appearance of a large *Armigeres*. **Informal name:** Great Indonesian Cord Mosquito.

milnensis Lee
1944c: 215 (M*, F, L*). **Type loc.:** Milne Bay, Papua New Guinea (CSIR). **Distr.:** Indonesia, Papua New Guinea, Solomon Isl., Timor. **Refs.:** Taylor 1914d: 188 (M only, as *breinli*; syn. by Lee 1944c: 218); Penn 1949b: 66 (P*); Steffan 1968b: 153 (M*, F, P*, L*); Lee et al. 1988a: 19 (F; key, tax., bion., distr., review); Anderson & Davis 2014 (distr.). **Ety.:** Milne Bay, Australia. **Informal name:** Milne Papuan Cord Mosquito.
syn. *ventralis* Walker
1865: 103 (F; *Culex*; not Walker 1860b). **Type loc.:** New Guinea (NMM). **Refs.:** Lee et al. 1988a: 123 (syn. by Marks; from *nomen dubium* of genus *Culex*; syn.; type info.).

moultoni Edwards
1914c: 125 (M, F). **Type loc.:** Kuching, Sarawak, [Malaysian] Borneo (NHMUK). **Distr.:** Cambodia, Indonesia, Laos, Malaysia, Thailand, Vietnam. **Refs.:** Hirao et al.

2010 (M*, P*, L*); Rattanarithikul et al. 2010 (F*, L*; keys, bion., distr.; Thailand). **Ety.:** J.C. Moulton. **Informal name:** Moulton Malaysian Cord Mosquito.

obturbans (Walker)
1859: 91 (F; *Culex*). **Type loc.:** Makassar, Celebes [Sulawesi] (NMM). **Distr.:** China (includes Hong Kong), India, Indonesia, Malaysia, Maldives, Philippines, Singapore, Sri Lanka, Thailand. **Refs.:** Edwards 1932a: 185 (*Armigeres*); Ty 1945: 304 (bion.); Coher 1949 (F*); Iyengar & Menon 1955 (distr.); Lee et al. 1988a (to sp.; type info.), 22, 27 (F; key, tax., bion., distr., review). **Ety.:** n.s. [disturb, agitate (L)]. **Informal name:** Agitated Indonesian Cord Mosquito.

pallithorax Dong, Zhou, & Dong
2004: 125 (M*, F). **Type loc.:** Xishuangbanna (elev. 780 m), Yunnan Province, China [People's Republic of China] (IPD). **Distr.:** People's Republic of China. **Ety.:** pale thorax (L) (in Chinese, English summary) [prob. ref. to same as *durhami* "but mesonotum has brown scales, abdominal sternite [*sic*] almost is covered with pale scales"]. **Informal name:** Ashen Chinese Cord Mosquito.

papuensis Peters
1963a: 4 (M*, F*, P*, L*). **Type loc.:** Maprik, [East] Sepik District, New Guinea [Papua New Guinea] (ANIC). **Distr.:** Indonesia, Papua New Guinea. **Refs.:** Steffan 1968b (F); Lee et al. 1988a (F; key, tax., bion., distr., review). **Ety.:** Papua New Guinea. **Informal name:** Hiri Papuan Cord Mosquito [Hiri is an official language in Papua New Guinea].

sembeli Toma & Miyagi
2002: 257 (M*, F*, P*, L*). **Type loc.:** Langowan, North Sulawesi, Indonesia (USNM). **Distr.:** Indonesia. **Ety.:** Dante T. Sembel. **Informal name:** Sembel Indonesian Cord Mosquito.

seticoxitus Luh & Li
1981: 188 (M*, F). **Type loc.:** Guangxi Autonomous Region, China [People's Republic of China] (ANC). **Distr.:** People's Republic of China. **Refs.**: Dong, Zhou, Dong, & Wang 1994 (P*, L*). **Ety.:** coxite setae (Gr); ref. to large patch of setae on the gonocoxite. **Informal name:** Hairy-hipped Chinese Cord Mosquito.

setifer Delfinado
1966a: 87 (M*, F). **Type loc.:** Pinigisan, Mantalingajan, Palawan, Philippines (USNM). **Distr.:** Laos, Philippines. **Refs.**: Basio 1971b: 50 (M*; bion.); Toma et al. 2007 (M*, F*, P*, L*); Hirao et al. 2010 (M*, P*, L*). **Ety.:** n.s. **Informal name:** Hairy Philippine Cord Mosquito.

subalbatus (Coquillett)
1898: 302 (F; *Culex*). **Type loc.:** Japan (USNM). **Distr.:** Bangladesh, Cambodia, India, Indonesia, Japan, Laos, Malaysia, Myanmar, Nepal, Pakistan, People's Republic of China, Philippines, South Korea, Sri Lanka, Taiwan, Thailand, Vietnam. **Refs.:** Edwards 1934 (from syn. with *obturbans*), in Barraud 1934 (p. 314); Bohart & Ingram 1946b: 62 (M*, F, P*, L*; bion., distr., tax.); La Casse & Yamaguti 1950: 53 (M*, F*, P*, L*; bion., distr., tax.); Stone & Knight 1955: 287 (type info.); Hara 1957: 50 (F*); Thurman 1959a: 390 (M*; tax.); Joshi et al. 1965 (distr.; Nepal); Aslamkhan 1971b (distr.; Pakistan); Mattingly 1971a: Pl. 4 (M, F heads*), Pl. 13 (P terminal segs.*), Pl. 29 (L head*, terminal segs.*); Mattingly 1971d: 122 (E*); Matsuo et al. 1972: 361 (E*); Moriya et al. 1972 (E*); Moriya et al. 1973 (E*); Baisas 1974: 80 (M, F*, P*, L*; tax., bion., distr.; Philippines); Tanaka et al. 1975c: 224 (distr.); Tanaka et al. 1979: 447 (M*, F*, L*); Darsie & Pradhan 1990 (F, L; tax., keys, bion., distr.; Nepal); Miyagi et al. 1994: 21 (distr.); Darsie 2000b: 111 (P*; tax., key); Reinert 2002d: Fig. 1 (F*); Rattanarithikul et al. 2010 (F*, L*; keys, bion., distr.; Thailand). **Ety.:** n.s. **Informal name:** White-bellied Japanese Cord Mosquito. (Vol. 1, p. 472)
syn. *panalectoris* Giles
1901a: 608 (A; *Culex*). **Type loc.:** Calcutta, West Bengal, India (NHMUK). **Refs.:** Thurman 1959b: 78 (syn.).
ssp. ***chrysocorporis*** Hsieh & Liao
1956: 126 (M*, F*, L*; *obturbans* var.). **Type loc.:** Amoy, China [People's Republic of China] (LU). **Distr.:** People's Republic of China. **Refs.:** Harbach & Howard 2007: 41 (to ssp.). **Ety.:** gold body (L); "mesonotum golden brown." **Informal name:** Gold-bodied Chinese Cord Mosquito.

theobaldi Barraud
1934: 319 (new name for *apicalis* Theobald 1910a, not Theobald 1908b). **Distr.:** Bangladesh, India, Laos, Myanmar, Thailand. **Refs.:** Mattingly & Qutubuddin 1952: 92 (M*); Thurman 1959b: 91 (M*, F); Toma et al. 1994: 11 (M*, F*, P*, L*); Dong, Zhou, Dong, & Wang 1994 (P*, L*); Rattanarithikul et al. 2010 (F*, L*; keys, bion., distr.; Thailand). **Ety.:** Frederic Vincent Theobald. **Informal name:** Theobald Indian Cord Mosquito.
syn. *apicalis* Theobald
1910a: 5 (F; *Desvoidea*). **Type loc.:** Balighai, near Puri, Orissa, India (IM).

yunnanensis Dong, Zhou, & Dong
1995b: 97 (M*, F, P*, L*). **Type loc.:** Jinhong, Yunnan Province, China [People's Republic of China] (Institute of Malaria Prevention & Treatment of Yunnan Province). **Distr.:** People's Republic of China. **Ety.:** Yunnan Prov., China. **Informal name:** Yunnan Chinese Cord Mosquito.

Subgenus *Leicesteria* Theobald

1904c: 211 (as genus). Type species: *Leicesteria longipalpis* Leicester.

Name derivation: George Frederick Leicester.

syn. *Brevirhynchus* Theobald 1908b: 293. Type species: *Brevirhynchus magnus* Theobald.

syn. *Chaetomyia* Leicester 1908: 100 (not Brauer & Bergenstamm 1891). Type species: *Chaetomyia flava* Leicester.

syn. *Leicesteriomyia* Brunetti 1912: 452 (as genus; new name for *Chaetomyia*). Type species: *Chaetomyia flava* Leicester. Macdonald 1960: 112 (syn.).

Important references: Barraud 1934: 310, 313 (tax., keys, morph.; India, Sri Lanka, Myanmar); Thurman 1959b: 76 (keys; Thailand); Macdonald 1960: 110 (rev., bion.); Delfinado 1966a: 75 (keys; Philippines); Lee et al. 1988a (tax., keys; Australasia); Darsie & Pradhan 1990 (tax., keys; Nepal); Lu et al. 1997 (People's Republic of China); Darsie 2000a (P; tax., key; Nepal); Reinert, Harbach, & Kitching 2004 (phyl., class.); Reinert, Harbach, & Kitching 2009 (phyl., class.); Rattanarithikul et al. 2010 (F*, L*; keys, bion.; Thailand); Wilkerson et al. 2015 (phyl., class.); Soghigian et al. 2017 (mol. phyl.).

annulipalpis (Theobald)
1910a: 6 (F; *Brevirhynchus*). **Type loc.:** Maddathorai, west base of Western Ghats, Travancore [India] (IM). **Distr.:** Cambodia, India, Indonesia, Myanmar, People's Republic of China, Thailand. **Refs.:** Bonne-Wepster 1934: 272 (M*); Brug 1939: 102 (L); Macdonald 1960: 134 (M*, F, L); Lee et al. 1988a: 33 (F; key, tax., bion., distr., review); Rattanarithikul et al. 2010 (F*, L*; keys, bion., distr.; Thailand); Toma et al. 2012 (M*, F*, P*, L*; redescr.). **Ety.:** ringed feeler (L); ref. to long black palpi with a snow-white band. **Informal name:** Ring-lipped Indian Cord Mosquito.

annulitarsis (Leicester)
1908: 99 (M, F; *Leicesteria*). **Type loc.:** Pahang Road, 5 3/4 mi from Kuala Lumpur [Selangor], Malaya [Malaysia] (NHMUK). **Distr.:** Bangladesh, Cambodia, India, Indonesia, Laos, Malaysia, Nepal, Taiwan, Thailand, Vietnam. **Refs.:** Borel 1930: 198 (M*, F, L*); Barraud 1934: 325 (M*, F, L); Macdonald 1960: 119 (M*, F, L); Ahmed 1987 (distr.; Bangladesh); Darsie & Pradhan 1990 (F, L; tax., keys, bion., distr.; Nepal); Darsie et al. 1996 (distr.; Nepal); Darsie 1998 (P*); Rattanarithikul et al. 2010 (F*, L*; keys, bion., distr.; Thailand). **Ety.:** ringed foot (L, Gr); tarsi. **Informal name:** Ring-legged Malaysian Cord Mosquito.

balteatus Macdonald
1960: 128 (M*, F, L*). **Type loc.:** Ulu Gombak, Selangor, Malaya [Malaysia] (NHMUK). **Distr.:** Indonesia, Malaysia, Philippines, Thailand. **Refs.:** Rattanarithikul et al. 2010 (F*, L*; keys, bion., distr.; Thailand). **Ety.:** belted (L); ref. to apical abdominal banding. **Informal name:** Banded Malaysian Cord Mosquito.

cingulatus (Leicester)
1908: 97 (F; *Leicesteria*). **Type loc.:** The Gap [Selangor], Malaya [Malaysia] (NHMUK). **Distr.:** Cambodia, India, Malaysia, Myanmar, Thailand, Vietnam. **Refs.:** Thurman 1959b: 100 (syn.); Rattanarithikul et al. 2010 (F*, L*; keys, bion., distr.; Thailand). **Ety.:** n.s. [girdled, ringed (L)]. **Informal name:** Girdled Malaysian Cord Mosquito.

dentatus Barraud
1927b: 547 (M*). **Type loc.:** Nongpoh, Assam, India (NHMUK). **Distr.:** Bangladesh, Cambodia, India, Malaysia, Nepal, Thailand, Vietnam. **Refs.:** Macdonald 1960: 124 (M*, F, L*); Mattingly 1971a: Pl. 7 (M head*); Mattingly 1971d: 125 (E*); Ahmed 1987 (distr.; Bangladesh); Darsie & Pradhan 1990 (F, L; tax., keys, bion., distr.; Nepal); Darsie et al. 1996 (distr.; Nepal); Darsie 1998 (P*); Rattanarithikul et al. 2010 (F*, L*; keys, bion., distr.; Thailand). **Ety.:** n.s. [toothed (L)]. **Informal name:** Toothy Indian Cord Mosquito.

digitatus (Edwards)
1914a: 262 (M*, F; *Leicesteria*). **Type loc.:** Ulu Gombak, Selangor, Malaya [Malaysia] (NHMUK). **Distr.:** Bangladesh, Cambodia, India, Indonesia, Malaysia, Nepal, Philippines, Taiwan, Thailand, Vietnam. **Refs.:** Barraud 1934: 330 (M*, F); Baisas 1935a: 493 (M*; as *degitatus*, *lapsus*); Coher 1949: 105 (F*); Macdonald 1960: 130 (M*, F, L*); Delfinado 1966a: 88 (M*, F, P*, L*); Basio 1971b: 50 (M*; bion.); Mattingly 1971e: 130 (E*); Matsuo, Yoshida, & Lien 1974 (E*); Baisas 1974: 81 (M, F, P*, L*; tax., bion., distr.; Philippines); Ahmed 1987 (distr.; Bangladesh); Okazawa et al. 1991 (E*); Darsie et al. 1992 (distr.; Nepal); Darsie 1998 (P*); Rattanarithikul et al. 2010 (F*, L*; keys, bion., distr.; Thailand). **Ety.:** finger (L); ref. to five-finger near-terminal spines on gonostylus. **Informal name:** Fingered Malaysian Cord Mosquito.

dolichocephalus (Leicester)
1908: 95 (M, F; as *Leicesteria dolicocephala*). **Type loc.:** Kuala Lumpur [Selangor], Malaya [Malaysia] (NHMUK). **Distr.:** Cambodia, India, Indonesia, Laos, Malaysia, Nepal, Thailand, Vietnam. **Refs.:** Thurman 1959b: 99 (M*, F); Macdonald 1960: 121 (M*, F, L*); Mattingly 1971e: 131 (E*); Darsie & Pradhan 1990 (F, L; tax., keys, bion., distr.; Nepal); Darsie et al. 1996 (distr.; Nepal); Darsie 1998 (P*); Kaur 2003: 453 (distr.; India); Rattanarithikul et al. 2010 (F*, L*; keys, bion., distr.; Thailand). **Ety.:** long head (Gr); head compressed laterally. **Informal name:** Long-headed Malaysian Cord Mosquito.

flavus (Leicester)
1908: 101 (M, F; *Chaetomyia*). **Type loc.:** Ulu Klang Jungle, Kuala Lumpur, Selangor, Malaya (NHMUK). **Distr.:** Bangladesh, Cambodia, India, Indonesia, Laos, Malaysia, Myanmar, Nepal, People's Republic of China, Philippines, Taiwan, Thailand, Vietnam. **Refs.:** Barraud 1934: 321 (M*, F, L*); Mattingly 1956b: 25 (lectotype desig.); Macdonald 1960: 116 (M*, F, L); Delfinado 1966a: 89 (M*, F, P*, L*); Mohrig 1967 (F*); Aslamkhan 1971b (distr.; Pakistan); Basio 1971b: 50 (M*; bion.); Mattingly 1971d: 124 (E*); Baisas 1974: 84 (M, F, P*, L*; tax., bion., distr.; Philippines); Rattanarithikul et al. 2010 (F*, L*; keys,

bion., distr.; Thailand); Toma et al. 2012: 13 (M*, F*, P*, L*; redescr.). **Ety.:** n.s. [yellow (L); poss. ref. to large creamy-yellow lateral abdominal bands]. **Informal name:** Yellow Malaysian Cord Mosquito.

syn. *apicalis* Theobald
1908b: 291 (M, F; *Leicesteria*). **Type loc.:** Lushai Hills, Assam, India (IM).

syn. *apicalis* Theobald
1910a: 7 (F; *Brevirhynchus*). **Type loc.:** Sylhet, Assam [India] (IM).

inchoatus Barraud
1927b: 544 (M*, F). **Type loc.:** Tindharia, Eastern Himalayas, West Bengal [India] (NHMUK). **Distr.:** Bangladesh, India, Malaysia, Nepal, Thailand. **Refs.:** Macdonald 1960: 122 (M*, F, L); Mattingly 1971e: 132 (E*); Ahmed 1987 (distr.; Bangladesh); Darsie et al. 1992 (distr.; Nepal); Darsie 1998 (P*); Rattanarithikul et al. 2010 (F*, L*; keys, bion., distr.; Thailand). **Ety.:** n.s. [established (L)]. **Informal name:** Established Indian Cord Mosquito.

lepidocoxitus Dong, Zhou, & Dong
1995b: 286 (M*, F, P*, L*). **Type loc.:** Jinghong Xian (altitude 780 m), Yunnan Province, China [People's Republic of China] (YIM). **Distr.:** People's Republic of China. **Ety.:** scaly coxa (L); ref. to a row of broad scales on inner border of the basimere (gonocoxite). **Informal name:** Scaly-hipped Chinese Cord Mosquito.

longipalpis (Leicester)
1904 (M, F; *Leicesteria*). In Theobald 1904c: 211. **Type loc.:** Kuala Lumpur [Selangor], Malaya [Malaysia] (NHMUK). **Distr.:** Cambodia, India, Indonesia, Laos, Malaysia, Thailand, Vietnam. **Refs.:** Barraud 1927b: 545 (M*); Borel 1930: 202 (M*, L*); Barraud 1934: 329 (M*, F; as *cingulatus*); Macdonald 1960: 126 (M*, F, L); Dong et al. 1993b (P*, L*); Reinert 2002d: Fig. 2 (F*); Rattanarithikul et al. 2010 (F*, L*; keys, bion., distr.; Thailand). **Ety.:** long palpus (L); ref. to palpus a half length of proboscis [third segment very long]. **Informal name:** Long-lipped Malaysian Cord Mosquito.

magnus (Theobald)
1908b: 293 (M, F; *Brevirhynchus*). **Type loc.:** Sylhet, Assam [India] (NHMUK). **Distr.:** Bangladesh, Cambodia, India, Indonesia, Laos, Macau, Malaysia, Myanmar, Nepal, People's Republic of China, Philippines, Sri Lanka, Taiwan, Thailand, Vietnam. **Refs.:** Barraud 1923g (M*, F*); Borel 1930: 194 (M*, F, L*); Barraud 1934: 324 (M*, F); Macdonald 1960: 117 (M*, F, L); Delfinado 1966a: 91 (M*, F, P*, L*); Aslamkhan 1971b (distr.; Pakistan); Basio 1971b: 50 (M*; bion.); Baisas 1974: 83 (M, F, P*, L*; tax., bion., distr.; Philippines); Ahmed 1987 (distr.; Bangladesh); Townsend 1990: 100 (type info.); Darsie & Pradhan 1990 (F, L; tax., keys, bion., distr.; Nepal); Darsie et al. 1996 (distr.; Nepal); Darsie 1998 (P*); Rattanarithikul et al. 2010 (F*, L*; keys, bion., distr.; Thailand). **Ety.:** n.s. [large (L); poss. it is a large mosquito]. **Informal name:** Magnificent Indian Cord Mosquito.

menglaensis Dong, Zhou, & Dong
2002a: 159 (M*, F, L*). **Type loc.:** Mengla County, Yunnan, China [People's Republic of China] (YIM). **Distr.:** People's Republic of China. **Ety.:** n.s. [prob. for type locality]. **Informal name:** Mengla Chinese Cord Mosquito.

omissus (Edwards)
1914b: 76 (M, F, L; *Leicesteria*). **Type loc.:** Colombo, Ceylon [Sri Lanka] (NHMUK). **Distr.:** Bangladesh, Cambodia, India, Indonesia, Malaysia, Nepal, People's Republic of China, Philippines, Sri Lanka, Taiwan, Thailand. **Refs.:** Barraud 1934: 330 (M*, F); Macdonald 1960: 126 (M*, F, L*); Delfinado 1966a: 92 (M*, F, P*, L*; distr.); Aslamkhan 1971b (distr.; Pakistan); Basio 1971b: 51 (M*; bion.); Matsuo, Yoshida, & Lien 1974 (E*); Ahmed 1987 (distr.; Bangladesh); Darsie 2000a: 964 (P*; tax., key, distr.); Rattanarithikul et al. 2010 (F*, L*; keys, bion., distr.; Thailand). **Ety.:** omit (L) [prob. ref. to similarity with *dolicocephala* but lacking yellow lateral spots on abdominal segments II–V]. **Informal name:** Spotless Sri Lankan Cord Mosquito.

pectinatus (Edwards)
1914a: 263 (M*, F; *Leicesteria*). **Type loc.:** Jugra, Malaya [Malaysia] (NHMUK). **Distr.:** Cambodia, Indonesia, Laos, Malaysia, Philippines, Thailand, Vietnam. **Refs.:** Macdonald 1960: 123 (M*, F, L*); Rattanarithikul et al. 2010 (F*, L*; keys, bion., distr.; Thailand). **Ety.:** pectinate (L); ref. to comb-like row of nine teeth on gonostylus. **Informal name:** Comb-like Malaysian Cord Mosquito.

pendulus (Edwards)
1914a: 261 (M*; *Leicesteria*). **Type loc.:** The Gap [Selangor], Malaya [Malaysia] (NHMUK). **Distr.:** Malaysia, Thailand. **Refs.:** Macdonald 1960: 131 (M*, F, L); Mohrig 1967 (F*); Rattanarithikul et al. 2010 (F*, L*; keys, bion., distr.; Thailand). **Ety.:** hanging down (L); ref. to "genitalia very small, completely exserted and pendulous (whence the specific name)." **Informal name:** Dangling Malaysian Cord Mosquito.

traubi Macdonald
1960: 132 (M*, F, L*). **Type loc.:** Ulu Gombak, Selangor, Malaya [Malaysia] (NHMUK). **Distr.:** Malaysia, Thailand. **Refs.:** Mattingly 1971e: 132 (E*); Rattanarithikul et al. 2010 (F*, L*; keys, bion., distr.; Thailand). **Ety.:** Robert Traub. **Informal name:** Traub Malaysian Cord Mosquito.

vimoli Thurman & Thurman
1958: 187 (F). **Type loc.:** Doi Suthep mountain, Chiengmai Province, Thailand (USNM). **Distr.:** Thailand. **Refs.:** Rattanarithikul et al. 2010 (F*, L*; keys, bion., distr.; Thailand). **Ety.:** Vimol Notananda. **Informal name:** Vimol Thai Cord Mosquito.

Subfamily Culicinae
Tribe Aedini

Important references: Reinert 2005b (Aedini egg lit. review); Reinert 2010i (F gen.; generic level summary); Reinert 2010j (F gen.; generic level summary).

Genus *Eretmapodites* Theobald

(Vol. 1, p. 224)

1901d: ii (as genus). Type species: *Eretmapodites quinquevittatus* Theobald.

Name derivation: eretmon = oar and pod, foot (Gr): oar footed.

Important references: Edwards 1912c: 46 (as *Eretmopodites*; unjust. emend.); Brunetti 1914: 60 (desig. type species); Edwards 1941: 224, 403 (tax., adults, pupae, keys; Africa); Haddow 1946: 57 (keys); van Someren 1949b: 119 (key to males); Hopkins 1952: 224 (tax., bion., keys, larvae; Africa); Hamon 1961: 892 (rev. *Eretmapodites oedipodeios* ssp.); Hamon & van Someren 1961b: 907 (key; Oedipodeios Group); Service 1990 (Africa); Reinert 2001d (F gen.*); Reinert et al. 2004 (phyl., class.); Reinert et al. 2009 (phyl., class.); Wilkerson et al. 2015 (phyl., class.).

adami Ferrara & Eouzan
1974: 261 (M*). **Type loc.:** On road from Yaoundé to Kribi, Cameroon (IERT). **Distr.:** Cameroon. **Ety.:** Jean-Paul Adam. **Informal name:** Adam Cameroonian African Mosquito.

angolensis da Cunha Ramos & Ribeiro
1992: 32 (M*). **Type loc.:** Dundo, Carrisso Park, northeastern Angola (IHMT). **Distr.:** Angola. **Ety.:** Angola. **Informal name:** Bantu Angolan African Mosquito [the Bantu are an indigenous people of Angola].

argyrurus Edwards
1936: 53 (A). **Type loc.:** Lagos, Nigeria (NHMUK). **Distr.:** Cameroon, Côte d'Ivoire, Nigeria. **Refs.:** Edwards 1941: 234 (M*, F), 406 (P); Hopkins 1952: 237 (L); Mattingly 1958c: 106 (lectotype desig.); Mattingly 1971a: Pl. 14 (P terminal segs.*); Pl. 31 (L head*, terminal segs.*); Rickenbach et al. 1976: 94 (distr.). **Ety.:** silver (Gr); tergum VII. **Informal name:** Silver Nigerian African Mosquito.

brenguesi Rickenbach & Lombrici
1975: 39 (M*). **Type loc.:** Nkadip (Mefou), Cameroon (IERT). **Distr.:** Cameroon. **Ety.:** Jacques Brengues. **Informal name:** Brengues Cameroonian African Mosquito.

brottesi Rickenbach
1967: 374 (M*). **Type loc.:** Ayene, Yaoundé, Cameroun [Cameroon] (IERT). **Distr.:** Cameroon. **Ety.:** Henri Brottes. **Informal name:** Brottes Cameroonian African Mosquito.

caillardi Rickenbach, Ferrara, & Eouzan
1968: 297 (M*). **Type loc.:** Ototomo, Cameroun [Cameroon] (IERT). **Distr.:** Cameroon. **Ety.:** M. le Médecin-Général Caillard. **Informal name:** Caillard Cameroonian African Mosquito.

chrysogaster Graham
1909b: 157 (M, F). **Type loc.:** Obuasi, Ashanti, Gold Coast [Ghana] (NHMUK). **Distr.:** Benin, Burkina Faso, Cameroon, Central African Republic, Côte d'Ivoire, Democratic Republic of the Congo, Ethiopia, Gabon, Ghana, Guyana, Kenya, Liberia, Mali, Nigeria, Republic of the Congo, Republic of South Africa, São Tomé and Príncipe, Senegal, Sierra Leone, Uganda, Zambia. **Refs.:** Edwards 1941: 226 (M*, F; distr.), 405 (P*); Hopkins 1952: 226 (L*); Mattingly 1956b: 28 (type info.); Mattingly 1971a: Pl. 31 (L*); da Cunha Ramos & Ribeiro 1975: 136 (tax., bion., distr.; Angola); Reinert 2001d: Fig. 2 (F*). **Ety.:** n.s. [gold stomach (Gr); ref. to golden scales on abdomen]. **Informal name:** Gold-bellied Ghanaian African Mosquito.

corbeti Hamon
1962: 256 (M*). **Type loc.:** Nyanga Forest, near Kampala, Uganda (IERT). **Distr.:** Uganda. **Ety.:** P. Corbet [Philip Steven?]. **Informal name:** Corbet Ugandan African Mosquito.

dracaenae Edwards
1916: 362 (M, F). **Type loc.:** Freetown, Sierra Leone (NHMUK). **Distr.:** Benin, Burkina Faso, Cameroon, Central African Republic, Côte d'Ivoire, Ethiopia, Ghana, Mali, Nigeria, Senegal, Sierra Leone, Uganda. **Refs.:** Edwards 1941: 237 (M*, F*), 408 (P*); Haddow et al. 1951: 229 (M); Hopkins 1952: 239 (L); Mattingly 1971a: Pl. 13 (P*); Rickenbach et al. 1976: 94 (distr.). **Ety.:** immature habitat in leaf axils of the "Cocked Hat" plant, *Dracaena* sp. **Informal name:** Dracaena Sierra Leonean African Mosquito.

syn. *ferox* Haddow
1946: 69 (M*, F, P*, L*). **Type loc.:** Bwamba, Toro District, Uganda (NHMUK). **Refs.:** Ovazza et al. 1956: 166 (syn.).

dundo da Cunha Ramos & Ribeiro
1992 (M*). **Type loc.:** Luachimo River, Dundo (Carrisso Park), Angola (IHMT). **Distr.:** Angola. **Ety.:** Dundo, Angola. **Informal name:** Dundo Angolan African Mosquito.

eouzani Rickenbach & Lombrici
1974: 218 (M*). **Type loc.:** Ebogo (Nyong and So), Yaoundé, Cameroon (IERT). **Distr.:** Cameroon. **Refs.:**

Rickenbach et al. 1976: 96 (distr.). **Ety.:** Jean-Pierre Eouzan. **Informal name:** Eouzan Cameroonian African Mosquito.

ferrarai Rickenbach & Eouzan
1970: 131 (M*). **Type loc.:** Ototomo, Yaoundé, Cameroon (IERT). **Distr.:** Cameroon. **Ety.:** Léo Ferrara. **Informal name:** Ferrara Cameroonian African Mosquito.

forcipulatus Edwards
1936: 53 (M). **Type loc.:** Sunyani, Ashanti, Gold Coast [Ghana] (NHMUK). **Distr.:** Burkina Faso, Cameroon, Côte d'Ivoire, Ghana, Liberia. **Refs.:** Edwards 1941: 234 (M*, F); Adam & Hamon 1961: 198 (P*, L*); Rickenbach et al. 1976: 94 (distr.). **Ety.:** n.s. [small forceps? (Gr)]. **Informal name:** Small-clawed Ghanaian African Mosquito.

germaini Rickenbach & Eouzan
1970: 133 (M*). **Type loc.:** Ofoumselek, Yaoundé, Cameroon (IERT). **Distr.:** Cameroon. **Ety.:** Max Germain. **Informal name:** Germain Cameroonian African Mosquito.

gilletti van Someren
1949b: 125 (M*). **Type loc.:** Bwamba, Toro District, Uganda (NHMUK). **Distr.:** Benin, Cameroon, Côte d'Ivoire, Democratic Republic of the Congo, Liberia, Republic of the Congo, Uganda. **Refs.:** Peters 1955c: 123 (F, P*, L*); Rickenbach et al. 1976: 93 (distr.). **Ety.:** Mr. Gillett. **Informal name:** Gillett Ugandan African Mosquito.

grahami Edwards
1911c: 71 (M). **Type loc.:** Obuasi, Ashanti, Gold Coast [Ghana] (NHMUK). **Distr.:** Cameroon, Côte d'Ivoire, Democratic Republic of the Congo, Ghana, Nigeria, Republic of the Congo, Sierra Leone, Tanzania, Togo, Uganda. **Refs.:** Edwards 1941: 230 (M*, F), 406 (P); van Someren 1949b: 126 (M); Hopkins 1952: 232 (L); Rickenbach et al. 1976: 96 (distr.). **Ety.:** n.s. [poss. Walter Mayne Graham]. **Informal name:** Graham Ghanaian African Mosquito.

grenieri Hamon & van Someren
1961b: 907 (M*). **Type loc.:** Amani, Tanganyika [Tanzania] (IERT). **Distr.:** Tanzania. **Ety.:** Paul Fernand Grenier. **Informal name:** Grenier Tanzanian African Mosquito.

haddowi van Someren
1949b: 122 (M*). **Type loc.:** Mongiro, Bwamba County, Toro District, Uganda (NHMUK). **Distr.:** Cameroon, Uganda. **Refs.:** Rickenbach et al. 1976: 96 (distr.). **Ety.:** Alexander John Haddow. **Informal name:** Haddow Ugandan African Mosquito.

hamoni Grjebine
1972: 308 (M*, F). **Type loc.:** Djili River, Brazzaville, Congo [Republic of the Congo] (MNHP). **Distr.:** Republic of Congo. **Ety.:** Jacques Pierre Jean Hamon. **Informal name:** Hamon Congolese African Mosquito.

harperi van Someren
1949b: 123 (M*). **Type loc.:** Mongiro, Bwamba County, Toro District, Uganda (NHMUK). **Distr.:** Cameroon, Uganda. **Refs.:** Rickenbach et al. 1976: 93 (distr.). **Ety.:** John Osmond Harper. **Informal name:** Harper Ugandan African Mosquito.

hightoni van Someren
1947c: 13 (M*, F). **Type loc.:** Kaimosi Forest, North Kavirondo, Kenya (NHMUK). **Distr.:** Kenya. **Refs.:** Hopkins 1952: 241 (L). **Ety.:** R.B. Highton. **Informal name:** Highton Kenyan African Mosquito.

inornatus Newstead
1907 (M, F*). In Newstead et al. 1907: 12. **Type loc.:** Coquilhatville and Lusambo, Équateur, Belgian Congo [Democratic Republic of the Congo] (NHMUK). **Distr.:** Cameroon, Central African Republic, Democratic Republic of the Congo, Kenya, Liberia, Nigeria, Republic of the Congo, Uganda. **Refs.:** Edwards 1941: 232 (M*, F*); Haddow 1946: 63 (M*, F*, P*, L*); Mattingly 1971a: Pl. 13 (P cephalothorax*, terminal segs.*); Rickenbach et al. 1976: 94 (distr.). **Ety.:** n.s. [unadorned (L)]. **Informal name:** Inornate Congolese African Mosquito.

intermedius Edwards
1936: 53 (M). **Type loc.:** Stanleyville, Orientale, Belgian Congo [Kisangani, Tshopo Province, Democratic Republic of the Congo] (NHMUK). **Distr.:** Cameroon, Democratic Republic of the Congo, Ethiopia, Kenya, Mali, Republic of the Congo, Sudan and South Sudan, Tanzania, Uganda. **Refs.:** Edwards 1941: 228 (M*, F), 405 (P); van Someren 1949b: 121 (M*); Rickenbach et al. 1976: 93 (distr.). **Ety.:** n.s. [in the middle (L), intermediate; prob. ref. to similarity to two other species]. **Informal name:** Intermediate Congolese African Mosquito.

jani Rickenbach & Lombrici
1975: 42 (M*). **Type loc.:** Nkolbisson, Cameroon (IERT) [originally in Collections des Services Scientifiques Centraux de l'O.R.S.T.O.M]. **Distr.:** Cameroon. **Ety.:** Charles Jan. **Informal name:** Jan Cameroonian African Mosquito.

lacani Rickenbach & Eouzan
1970: 135 (M*). **Type loc.:** Nkolbisson, Yaoundé, Cameroon (IERT). **Distr.:** Cameroon. **Ety.:** Andre Marie Maurice Lacan. **Informal name:** Lacan Cameroonian African Mosquito.

leucopous Graham
1909a: 88 (M, F). **Type loc.:** Obuasi, Ashanti, Gold Coast [Ghana] (NHMUK). **Distr.:** Cameroon, Central African Republic, Côte d'Ivoire, Democratic Republic of the Congo, Gabon, Ghana, Nigeria, Republic of the Congo, Sierra Leone. **Refs.:** Edwards 1941: 241 (M*, F); Mattingly

1956b: 28 (lectotype desig.); Rickenbach & Lombrici 1975 (M*; as *leucopus*); Rickenbach et al. 1976: 94 (distr.); Service 1976a (distr.; Gabon). **Ety.:** white foot (Gr); "the two last segments of the tarsi are pure white." **Informal name:** White-footed Ghanaian African Mosquito.

mahaffyi van Someren
1949b: 122 (M*). **Type loc.:** Mongiro, Bwamba County, Toro District, Uganda (NHMUK). **Distr.:** Cameroon, Uganda. **Refs.:** Rickenbach et al. 1976: 93 (distr.). **Ety.:** Dr. Mahaffy. **Informal name:** Mahaffy Ugandan African Mosquito.

marcelleae Adam & Hamon
1959: 524 (M*; as *marcellei*). **Type loc.:** Banco Forest, near Abidjan, Ivory Coast [Côte d'Ivoire] (IERT). **Distr.:** Central African Republic, Côte d'Ivoire, Liberia. **Refs.:** Doucet 1960: 812 (*oedipodeios* ssp.); Hamon 1961: 906 (to sp.); White 1975: 324 (emend.). **Ety.:** Marcelle Hamon. **Informal name:** Marcelle Ivorian African Mosquito.

mattinglyi Hamon & van Someren
1961a: 728 (M*). **Type loc.:** Amani, Tanganyika [Tanzania] (IERT). **Distr.:** Tanzania. **Ety.:** Peter Frederick Mattingly. **Informal name:** Mattingly Tanzanian African Mosquito.

melanopous Graham
1909b: 158 (M, F). **Type loc.:** Obuasi, Ashanti, Gold Coast [Ghana] (NHMUK). **Distr.:** Ghana. **Refs.:** Mattingly 1956b: 28 (lectotype desig.). **Ety.:** n.s. [black foot (Gr)]. **Informal name:** Black-footed Ghanaian African Mosquito.

mortiauxi da Cunha Ramos & Ribeiro
1990: 139 (M*). **Type loc.:** Dundo (Carisso Park), Angola (IHMT). **Distr.:** Angola. **Ety.:** Mons(eigneur) Mortiaux. **Informal name:** Mortiaux Angolan African Mosquito.

oedipodeios Graham
1909a: 86 (M, F). **Type loc.:** Obuasi, Ashanti, Gold Coast [Ghana] (NHMUK). **Distr.:** Cameroon, Central African Republic, Côte d'Ivoire, Democratic Republic of the Congo, Ghana, Kenya, Liberia, Madagascar (includes Glorioso and Juan De Nova Isl.), Nigeria, Republic of the Congo, Sierra Leone, Togo, Uganda. **Refs.:** Edwards 1912c: 48 (emend.); Edwards 1941: 238 (M*, F); White 1980: 142 (unjust. emend. to *oedipodius* noted). **Ety.:** n.s. [swollen feet (Gr); poss. ref. to "hind tarsi of abnormal form and densely plumed"]. **Informal name:** Swell-footed Ghanaian African Mosquito.

syn. *stanleyi* Edwards
1941: 239 (M*, F; *oedipodeios* ssp.), 408 (P). **Type loc.:** Stanleyville, Orientale, Belgian Congo [Kisangani, Tshopo Province, Democratic Republic of the Congo] (NHMUK). **Refs.:** Mattingly 1956b: 26 (lectotype desig.); Hamon 1961: 900 (M*; syn.).

parvipluma Edwards
1941: 240 (M*, F), 408 (P*, in part). **Type loc.:** Kasuku, Uganda (NHMUK). **Distr.:** Democratic Republic of the Congo, Kenya, Republic of the Congo, Uganda. **Refs.:** Hopkins 1952: 242 (L*); Mattingly 1956b: 27 (type info.); Hamon 1961: 906 (M*; to sp.). **Ety.:** n.s. [small feather, plume (L); poss. ref. to difference with *oedipodius*]. **Informal name:** Small-plumed Ugandan African Mosquito.

pauliani Grjebine
1950: 45 (M*, P*, L*). **Type loc.:** Banco Forest, 4 km from Abidjan, Ivory Coast [Côte d'Ivoire] (IERT). **Distr.:** Cameroon, Central African Republic, Côte d'Ivoire. **Refs.:** Hopkins 1952: 230 (L); Hamon & Adam 1959a: 527 (M*); Harrison 1973c: 279 (type info.); Rickenbach et al. 1976: 93 (distr.). **Ety.:** Renaud Paulian. **Informal name:** Paulian Ivorian African Mosquito.

penicillatus Edwards
1941: 234 (M*). **Type loc.:** Ilesha, Western Provinces, Nigeria (NHMUK). **Distr.:** Cameroon, Côte d'Ivoire, Nigeria, Republic of the Congo, Sierra Leone, Uganda. **Refs.:** Haddow 1946: 66 (M*, F*, P*, L*); Mattingly 1956b: 26 (type info.). **Ety.:** having a painter's brush (L); ref. to differing by "having the proximal claspettes well developed, ending in a brush of hairs." **Informal name:** Tufty Nigerian African Mosquito.

plioleucus Edwards
1941: 241 (M*, F). **Type loc.:** Kumasi, Ashanti, Gold Coast [Ghana] (NHMUK). **Distr.:** Côte d'Ivoire, Democratic Republic of the Congo, Ghana, Madagascar (includes Glorioso and Juan De Nova Isl.). **Refs.:** Mattingly 1956b: 27 (lectotype desig.). **Ety.:** n.s. [more white (Gr); poss. ref to white leg markings]. **Informal name:** White-legged Ghanaian African Mosquito.

ssp. ***brevis*** Edwards
1941: 242 (M*). **Type loc.:** Stanleyville, Orientale, Belgian Congo [Kisangani, Tshopo Province, Democratic Republic of the Congo] (NHMUK). **Distr.:** Democratic Republic of the Congo. **Refs.:** Mattingly 1956b: 27 (type info.). **Ety.:** n.s. [short (L); ref. to several characters on male genitalia shorter than the nominotypical species]. **Informal name:** Short Congolese African Mosquito.

productus Edwards
1941: 241 (M*, F), 408 (P). **Type loc.:** Stanleyville, Orientale, Belgian Congo [Kisangani, Tshopo Province, Democratic Republic of the Congo] (NHMUK). **Distr.:** Cameroon, Central African Republic, Democratic Republic of the Congo, Kenya, Uganda. **Refs.:** Haddow 1946: 73 (A, P*, L*); Knight & Chamberlain 1948: Fig. 27 (P*); Mattingly 1956b: 27 (type info.); Rickenbach & Lombrici 1975 (M*; to valid sp.); White 1975: 324 tax.; Rickenbach et al. 1976: 94 (distr.). **Ety.:** n.s. [lead forward, bring out (L)]. **Informal name:** Productive Congolese African Mosquito.

quinquevittatus Theobald
1901d: ii (F*). **Type loc.:** Freetown, Sierra Leone (NHMUK). **Distr.:** Central African Republic, Comoros, Democratic Republic of the Congo, Equatorial Guinea, Ghana, Kenya, Madagascar (includes Glorioso and Juan De Nova Isl.), Nigeria, Republic of South Africa, Senegal, Sierra Leone, Sudan and South Sudan, Tanzania, Togo, Uganda, Zimbabwe. **Refs.:** Edwards 1941: 236 (M*, F*), 407 (P*); Hopkins 1952: 237 (L*); Mattingly 1956b: 30 (lectotype desig.); Mohrig 1967 (F*); Mattingly 1970b: 17 (E*); Brunhes 1977a (distr.; Comoros Isl.); Linley & Service 1994: 67 (E*); Jupp 1996 (M*, F*; key); Reinert 2001d: Fig. 1 (F*); Toto et al. 2003: 344 (distr.; Equatorial Guinea). **Ety.:** five lines (L); five dark ferruginous lines on scutum. **Informal name:** Five-banded Sierra Leonean African Mosquito.

syn. *condei* Ventrillon
1905a: 444 (M, F). **Type loc.:** Mayotte Island, Madagascar (MNHP).

syn. *austenii* Theobald
1910c: 572 (M, F). **Type loc.:** Freetown, Sierra Leone (NHMUK).

ravissei Rickenbach & Eouzan
1970: 134 (M*). **Type loc.:** Yaoundé, Cameroon (IERT). **Distr.:** Cameroon. **Ety.:** Pierre Ravisse. **Informal name:** Ravisse Cameroonian African Mosquito.

rickenbachi Ferrara & Eouzan
1974: 259 (M*). **Type loc.:** Ofoumselek, Cameroon (IERT). **Distr.:** Cameroon. **Ety.:** André Rickenbach. **Informal name:** Rickenbach Cameroonian African Mosquito.

salauni Rickenbach, Ferrara, & Eouzan
1968: 299 (M*). **Type loc.:** Ototomo, Cameroun [Cameroon] (IERT). **Distr.:** Cameroon. **Refs.:** Ferrara 1970: 137 (M*). **Ety.:** Jean-Jacques Salaün. **Informal name:** Salaün Cameroonian African Mosquito.

semisimplicipes Edwards
1914b: 76 (M; as *chrysogaster* var.). **Type loc.:** Akrokerri, Ashanti, Gold Coast [Ghana] (NHMUK). **Distr.:** Cameroon, Central African Republic, Côte d'Ivoire, Democratic Republic of the Congo, Ghana, Kenya, Nigeria, Republic of the Congo, Sierra Leone, Togo, Uganda. **Refs.:** Edwards 1941: 229 (M*), 406 (P); Haddow 1946: 62 (A, P, L). **Ety.:** n.s. [nearly simple foot (L); prob. ref. to lesser paddle-like fringe on male Ta-$III_{4,5}$ than on *chrysogaster*]. **Informal name:** Semi-simple Ghanaian African Mosquito.

silvestris Ingram & De Meillon
1927: 63 (F*, P*, L*). **Type loc.:** Eshowe, Zululand [KwaZulu Natal, Republic of South Africa] (NHMUK). **Distr.:** Ethiopia, Kenya, Nigeria, Republic of South Africa, Sudan and South Sudan, Tanzania. **Refs.:** Edwards 1941: 231 (M*, F*), 406 (P); Hopkins 1952: 232 (L*); Townsend 1990: 133 (lectotype desig.); Jupp 1996 (M*, F*; key). **Ety.:** n.s. [wild, wooded, untamed (L)]. **Informal name:** Wild South African African Mosquito.

ssp. ***conchobius*** Edwards
1941: 231 (M, F). **Type loc.:** Tiwi, Kenya (NHMUK). **Distr.:** Ethiopia, Kenya, Nigeria, Sudan and South Sudan. **Refs.:** Hoogstraal & Knight 1951: 659 (M, F, P, L*); Mattingly 1956b: 26 (type info.). **Ety.:** shell (L); ref. to immatures found in snail shells. **Informal name:** Snail-shell Kenyan African Mosquito.

subsimplicipes Edwards
1914b: 76 (M, L; as *chrysogaster* var.). **Type loc.:** Sebeleni Swamp, Zanzibar [Tanzania] (NHMUK). **Distr.:** Comoros, Kenya, Madagascar (includes Glorioso and Juan De Nova Isl.), Malawi, Mayotte, Mozambique, Republic of South Africa, Tanzania (includes Zanzibar). **Refs.:** Edwards 1941: 229 (M*, F; to sp.), 405 (P); Hopkins 1952: 231 (L); Worth & De Meillon 1960: 245 (distr.); Brunhes 1977a (distr.; Comoros Isl.); Jupp 1996 (M*, F*; key). **Ety.:** n.s. [nearly simple foot (L); prob. ref. to comp. to much less paddle-like fringe on male Ta-$III_{4,5}$ of *chrysogaster*]. **Informal name:** Simple-footed Tanzanian African Mosquito.

tendeiroi da Cunha Ramos, Ribeiro, & de Barros Machado
1992a (M, F; key). **Type loc.:** Dundo, Carrisso Park, Angola (IHMT). **Distr.:** Angola. **Ety.:** J. Leal da Silva Tendeiro. **Informal name:** Tendeiro Angolan African Mosquito.

tonsus Edwards
1941: 235 (M*). **Type loc.:** Kakamega, Kenya (NHMUK). **Distr.:** Cameroon, Ethiopia, Kenya, Tanzania, Uganda. **Refs.:** Haddow et al. 1951: 229 (M); van Someren 1954b: 121 (M, F); van Someren & Hamon 1964: 82 (P*, L*); Rickenbach et al. 1976: 94 (distr.). **Ety.:** n.s. [shorn, clipped, cut (L); poss. postnotum completely bare]. **Informal name:** Tonsured Kenyan African Mosquito.

vansomereni Hamon
1962: 253 (M*). **Type loc.:** Zika Forest, near Entebbe, Uganda (IERT). **Distr.:** Uganda. **Ety.:** Ellinor Catharine Cunningham van Someren. **Informal name:** van Someren Ugandan African Mosquito.

wansoni Edwards
1941: 240 (M*). **Type loc.:** Matadi, Belgian Congo [Democratic Republic of the Congo] (NHMUK). **Distr.:** Burkina Faso, Cameroon, Central African Republic, Côte d'Ivoire, Democratic Republic of the Congo, Republic of the Congo, Togo. **Refs.:** Mattingly 1956b: 27 (lectotype desig.); Hamon 1961: 902 (M*; to sp.); Rickenbach et al. 1976: 94 (distr.). **Ety.:** n.s. [poss. Marcel Jules Emil Marie Wanson]. **Informal name:** Wanson Congolese African Mosquito.

ssp. ***douceti*** Adam & Hamon
1959: 525 (M*, P, L). **Type loc.:** Nasso Forest, Bobo Dioulasso, Upper Volta, French West Africa [Burkina Faso] (IERT). **Distr.:** Burkina Faso, Côte d'Ivoire. **Refs.:** Hamon 1961: 905 (tax.). **Ety.:** Jean-Paul Doucet. **Informal name:** Doucet Burkinabe African Mosquito.

Genus *Haemagogus* Williston

(Vol. 1, p. 226)

Important references: Martínez et al. 1961a (tax.; Argentina); Arnell 1973 (revision); Reinert et al. 2009 (phyl.); Maestre-Serrano et al. 2013 (distr.); Wilkerson et al. 2015 (phyl.); Soghigian et al. 2017 (mol. phyl.).

Subgenus *Conopostegus* Dyar

1925f: 141 (as subg. of *Aedes*). Type species: *Aedes leucocelaenus* Dyar & Shannon.

Name derivation: conopo = gnat, mosquito; stego = roof, covered place (Gr) [?].

Important references: Edwards 1932a: 154 (syn. with *Aedes* (*Finlaya*)); Zavortink 1972: 111 (to subg. of genus *Haemagogus* from syn. with *Aedes* (*Finlaya*).

clarki (Galindo, Carpenter, & Trapido)
1953: 531 (M*, F, P*, L*; *leucocelaenus* ssp.). **Type loc.:** Coclé Province, Panama (USNM). **Distr.:** Costa Rica, Guyana, Honduras, Nicaragua, Panama, Venezuela. **Refs.:** Zavortink 1972: 127 (M*, F, P*, L*; to sp. and new comb.). **Ety.:** Herbert Charles Clark. **Informal name:** Clark Panamanian Shoulder Mosquito.

leucocelaenus (Dyar & Shannon)
1924d: 484 (*Aedes leucocelaenus*; new name for *leucomelas* Lutz, not Meigen 1804). **Type loc.:** Franca, São Paulo, Brazil (NHMUK) (as *leucomelas*). **Distr.:** Argentina, Bolivia, Brazil, Colombia, French Guiana, Guyana, Lesser Antilles (includes Trinidad and Tobago), Panama, Paraguay, Peru, Suriname, Uruguay, Venezuela. **Refs.:** Galindo et al. 1953: 529 (tax.); Lane 1953: 682 (M*, F, P*, L*); Zavortink 1972: 124 (M*, F*, P*, L*; tax.); Panday 1975a (distr.); Reinert 2002c: Fig. 1 (F*); Rossi & Martínez 2003: 472 (distr.; Uruguay); Alencar et al. 2003 (E*). **Ety.:** n.s. [murky white (Gr)]. **Informal name:** Murky-white Brazilian Shoulder Mosquito.
syn. *leucomelas* Lutz,
1904 (F; *Haemagogus*). In Bourroul 1904: 47, 66. **Type loc.:** Franca, São Paulo, Brazil (NHMUK). **Refs.:** Dyar & Shannon 1924d (syn.); Belkin et al. 1971: 20 (lectotype desig.).

leucophoebus (Galindo, Carpenter, & Trapido)
1953: 535 (M*, F, P*, L*; *Aedes*). **Type loc.:** Feijo, Acre, Brazil (FH). **Distr.:** Brazil. **Refs.:** Zavortink 1972: 122 (M*, F, P*, L*; tax.). **Ety.:** n.s. [*leucophoebus* = bright white (Gr) = Alabaster]. **Informal name:** Alabaster Brazilian Shoulder Mosquito

leucotaeniatus (Komp)
1938: 261 (M*, F; *Aedes*). **Type loc.:** Comacho, Canal Zone, Panama (USNM). **Distr.:** Costa Rica, Panama. **Refs.:** Galindo et al. 1953: 536 (M*, P*, L*); Barreto-Reyes 1957 (F); Zavortink 1972: 119 (M*, F, P*, L*; tax.). **Ety.:** n.s. [white ribboned (Gr); poss. ref. to median line of silvery-white scales on scutum. or femora white lined beneath]. **Informal name:** White-ribboned Panamanian Shoulder Mosquito.

Subgenus *Haemagogus* Williston

1896: 271 (as genus). Type species: *Haemagogus splendens* Williston.

Name derivation: n.k. [haemo = blood and feeder? (Gr)].

syn. *Stegoconops* Lutz 1905: 83 (as genus). Type species: *Stegoconops capricornii* Lutz.
Important references: Howard et al. 1917: 863 (syn. with *Haemagogus*); Dyar 1921g: 102 (from syn. to subg. *Haemagogus*); Arnell 1973: 20 (syn. with *Haemagogus*).

syn. *Cacomyia* Coquillett 1906d: 16, 25. Type species: *Haemagogus albomaculatus* Theobald.
Important references: Coquillett 1910: 516 (desig. of type species); Howard et al. 1917: 863 (syn. with *Haemagogus*).

syn. *Longipalpifer* Leví-Castillo 1951a: 10 (as subg. of genus *Haemagogus*) [pp. 7, 11 as *Masculongipalpi* but changed to *Longipalpifer* in accompanying errata p. 10]. Type species: *Haemagogus panarchys* Dyar. Important reference: Arnell 1973: 20 (syn.).

Important references: Lutz 1904 (*in* Bourroul 1904) (subfamily Haemagoginae); Lane 1953: 775 (tax., bion., keys); Forattini 1965b (review, tax., bion.); Cova García et al. 1966: 366 (key., tax.; Venezuela); Belkin et al. 1970 (tax., bion.; Jamaica); Arnell 1973 (revision); Clark-Gil & Darsie 1983 (keys, bion.; Guatemala); Darsie 1985 (F, L; keys; Argentina); Mitchell & Darsie 1985 (review, bibliography., distr.; Argentina); Reinert 2002c (F gen.*); Reinert et al. 2004 (phyl., class.); Liria & Navarro 2009 (key, bion., distr.; Venezuela); Reinert et al. 2009 (phyl., class.); Marcondes & Alencar 2010 (tax., key, distr.); Wilkerson et al. 2015 (phyl., class.); Soghigian et al. 2017 (mol. phyl.).

acutisentis Arnell
1973: 58 (M*, F, P*, L*). **Type loc.:** Valencia, Los Rios, Ecuador (USNM). **Distr.:** Ecuador. **Ety.:** n.s. [sharp thorn (L); poss. ref. to sharply pointed comb scales]. **Informal name:** Thorn-like Ecuadorian Shoulder Mosquito.

aeritinctus Galindo & Trapido
1967: 103 (M*, F*, P*, L*). **Type loc.:** Stann Creek, British Honduras [Belize] (USNM). **Distr.:** Belize, Guatemala, Honduras. **Refs.:** Arnell 1973: 80 (M*, F, P*, L*). **Ety.:** n.s. [copper, bronze, brass (L); prob. ref. to "peculiar coppery

sheen of the mesonotal scales"]. **Informal name:** Coppery Belizean Shoulder Mosquito.

albomaculatus Theobald
1903a: 308 (F*). **Type loc.:** Cara Cara, Demerara River, British Guiana [Guyana] (NHMUK). **Distr.:** Brazil, Ecuador, French Guiana, Guyana, Lesser Antilles (includes Trinidad and Tobago), Nicaragua, Panama, Suriname, Venezuela. **Refs.:** Lane 1953: 800 (M*, F); Komp 1954b: 148 (M*); Mohrig 1967 (F*); Belkin 1968b: 22 (type info.); Arnell 1973: 50 (M*, F, P*, L*). **Ety.:** n.s [white spotted (L); prob. ref. to abdomen with various white spots]. **Informal name:** White-spotted Guyanese Shoulder Mosquito.

anastasionis Dyar
1921k: 155 (M, F). **Type loc.:** Puntarenas, Costa Rica (USNM). **Distr.:** Aruba, Colombia, Costa Rica, Curaçao, Ecuador, El Salvador, Guatemala, Honduras, Mexico, Nicaragua, Venezuela. **Refs.:** Dyar 1928: 137 (M*, F; as *anastationis*, emend.); Leví-Castillo 1951a: 23 (M*, F, L*); Lane 1953: 798 (M*, F, L*); Stone & Knight 1955: 287 (lectotype desig.); Arnell 1973: 24 (M*, F, P*, L*); Navarro et al. 2013a (distr.; Ecuador); Berti et al. 2015 (distr.; Venezuela). **Ety.:** n.s. [prob. derived from Anastasio, first name of collector A. Alfaro (Harbach 2018c)]. **Informal name:** Puntarenas Costa Rican Shoulder Mosquito.
syn. *dominguezi* Duret
1971b: 83 (M*). **Type loc.:** Departamento de Zelaya, Nicaragua (USNM). **Refs.:** Arnell 1973: 24 (syn.).

andinus Osorno-Mesa
1944b: 170 (M*, F, L*). **Type loc.:** Fusagasugá, Cundinamarca, Colombia (USNM). **Distr.:** Colombia. **Refs.:** Lane 1953: 794 (M*, F, L*); Stone & Knight 1955: 287 (type info.); Arnell 1973: 33 (M*, F, P*, L*). **Ety.:** Andes mountains. **Informal name:** Andes Colombian Shoulder Mosquito.

argyromeris Dyar & Ludlow
1921b: 679 (M). **Type loc.:** Corozal, Canal Zone, Panama (USNM). **Distr.:** Panama. **Refs.:** Galindo et al. 1951a: 118 (bion.); Lane 1953: 786 (M*, F, L*); Stone & Knight 1955: 287 (lectotype desig.); Arnell 1973: 90 (M*, F, P*, L*). **Ety.:** n.s. [part silver? (Gr)]. **Informal name:** Silvery Panamanian Shoulder Mosquito.
syn. *gladiator* Dyar
1921g: 108 (M, F, L). **Type loc.:** Corozal, Canal Zone, Panama (USNM). **Refs.:** Dyar 1925f: 139 (syn.); Stone & Knight 1955: 288 (type info.).

baresi Cerqueira
1960a: 1 (M*, F, P*, L*). **Type loc.:** Igarapé do Tarumã, Manaus, Amazonas, Brazil (FH). **Distr.:** Brazil, Peru. **Refs.:** Belkin et al. 1971: 21 (type info.); Arnell 1973: 31 (M*, F, P*, L*). **Ety.:** Barés indigenous tribe from near Manaus. **Informal name:** Barés Brazilian Shoulder Mosquito.

boshelli Osorno-Mesa
1944b: 165 (M*, F, L*). **Type loc.:** Bahía de Solano, Chocó, Colombia (USNM). **Distr.:** Colombia, Ecuador, Panama. **Refs.:** Lane 1953: 784 (M*, F, L*); Stone & Knight 1955: 287 (type info.); Arnell 1973: 97 (M*, F, P*, L*). **Ety.:** Jorge Boshell-Manrique. **Informal name:** Boshell Colombian Shoulder Mosquito.
syn. *garciai* Leví-Castillo
1955a: 361 (M*, F). **Type loc.:** Isla de Changuaral, Esmeraldas, Ecuador (USNM). **Refs.:** Arnell 1973: 97 (syn.).

capricornii Lutz
1904 (F). In Bourroul 1904: 66. **Type loc.:** Horto Florestal, Serrada Cantareira, São Paulo, Brazil (CPRR). **Distr.:** Argentina, Bolivia, Brazil, Colombia, Guyana, Nicaragua, Paraguay. **Refs.:** Antunes 1939: 106 (from syn. with *equinus*); Cerqueira 1943a (F*); Cerqueira & Lane 1945: 279 (M*, F, P*, L*; tax.); Knight & Chamberlain 1948: 10 (P*); Coher 1949 (F); Kumm & Cerqueira 1951: 173 (M*); Lane 1953: 790 (M*, F*, P*, L*); Martínez et al. 1961a: 76 (tax.); Forattini 1965b (E*); Belkin et al. 1971: 21 (type info.); Mattingly 1971a: Pl. 18 (P*); Pl. 33 (L head*, terminal segs.*); Arnell 1973: 44 (M*, F, P*, L*); Xavier & Mattos 1989: 50 (neotype desig.); Alencar et al. 2005b (E*); Alencar et al. 2009 (tax.). **Ety.:** n.s. [Tropic of Capricorn?]. **Informal name:** Capricorn Brazilian Shoulder Mosquito.

celeste Dyar & Núñez Tovar
1927a: 152 (M). **Type loc.:** Maracay [Aragua], Venezuela (USNM). **Distr.:** Brazil, Colombia, Lesser Antilles (includes Trinidad and Tobago), Venezuela. **Refs.:** Edwards 1932a: 179 (syn. with splendens); Stone & Knight 1955: 287 (lectotype desig.); Arnell 1973: 73 (M*, F, P*, L*; to sp.); Chadee & Bennett 1990 (E*); Berti et al. 2015 (distr.; Venezuela). **Ety.:** Celeste Núñez Tovar. **Informal name:** Celeste Venezuelan Shoulder Mosquito.

chalcospilans Dyar
1921g: 110 (M, F, L). **Type loc.:** Caldera Island, Porto Bello Bay, Panama (USNM). **Distr.:** Colombia, Costa Rica, Ecuador, Honduras, Nicaragua, Panama. **Refs.:** Galindo et al. 1951a: 119 (bion.); Lane 1953: 782 (M*, F, L*); Stone & Knight 1955: 287 (type info.); Arnell 1973: 93 (M*, F, P*, L*). **Ety.:** n.s. [copper? (Gr)]. **Informal name:** Coppery Panamanian Shoulder Mosquito.

chrysochlorus Arnell
1973: 27 (M*, F). **Type loc.:** Curaçao, Netherlands Antilles (USNM). **Distr.:** Curaçao. **Ety.:** n.s. [gold green (Gr)]. **Informal name:** Gold-green Curaçaoan Shoulder Mosquito.

equinus Theobald
1903g: 282 (F). **Type loc.:** Kingston, Jamaica (NHMUK). **Distr.:** Belize, Colombia, Costa Rica, El Salvador, Guatemala, Guyana, Honduras, Jamaica, Lesser Antilles

(includes Trinidad and Tobago), Mexico, Nicaragua, Panama, United States (continental), Venezuela. **Refs.:** Osorno-Mesa 1944a: 39 (bion.); Galindo et al. 1951a: 116 (bion.); Lane 1953: 802 (M*, F, P*, L*); Komp 1954a: 51 (tax.); Komp 1955a: 163 (bion.); Breland 1958: 218 (L*); Mohrig 1967 (F*); Belkin 1968b: 22 (type info.); Belkin et al. 1970: 188 (M*, F, P*, L*); Díaz Nájera 1971: 82 (M*); Díaz Nájera 1972: 113 (M*); Arnell 1973: 60 (M*, F, P*, L*); Chadee & Bennett 1990 (E*); Darsie & Ward 2005 (F*, L*; keys, distr.). **Ety.:** n.s. [of horses, equine (L); ref. to captured while feeding on a horse]. **Informal name:** Equine Jamaican Shoulder Mosquito.

syn. *affirmatus* Dyar & Knab
1906b: 164 (F; *Aedes*). **Type loc.:** Salina Cruz, Oaxaca, Mexico (USNM). **Refs.:** Howard et al. 1917: 817 (syn.); Stone & Knight 1955: 287 (type info.)

syn. *philosophicus* Dyar & Knab
1906d: 190, 195 (L; *Aedes*). **Type loc.:** Tehuantepec, Oaxaca, Mexico (USNM). **Refs.:** Howard et al. 1917: 874 (syn.); Stone & Knight 1955: 288 (type info.)

iridicolor Dyar
1921g: 106 (M, F). **Type loc.:** Higuito, San Mateo, Costa Rica (USNM). **Distr.:** Costa Rica, Honduras, Nicaragua, Panama. **Refs.:** Lane 1953: 789 (M*, F, L*); Stone & Knight 1955: 288 (type info.); Komp 1955b: 29 (L; tax., lectotype desig.); Mohrig 1967 (F*); Arnell 1973: 76 (M*, F, P*, L*). **Ety.:** rainbow colored (L). **Informal name:** Rainbow Costa Rican Shoulder Mosquito.

janthinomys Dyar
1921g: 112 (M, F). **Type loc.:** Trinidad [Trinidad & Tobago] (USNM). **Distr.:** Argentina, Bolivia, Brazil, Chile, Colombia, Costa Rica, Ecuador, French Guiana, Guyana, Honduras, Lesser Antilles (includes Trinidad and Tobago), Nicaragua, Panama, Peru, Suriname, Venezuela. **Refs.:** Cerqueira 1943a: 10 (syn.); Bates 1947: 1 (bion.; as *falco*); Lane 1953: 794 (M*; as *falco*); Komp 1953 (bion., tax.); Komp 1955d (tax.); Stone & Knight 1955: 288 (lectotype desig.); Galindo 1957b: 122 (M*; tax.; as *falco*); Martínez et al. 1961a: 76 (to ssp.); Arnell 1973: 36 (M*, F*, P*, L*; keys, tax., distr.); Mattingly 1973a (E*; as *Haemagogus spegazzinii*); Alencar et al. 2009 (tax.); Becker et al. 2010: 381 (F*, L*; key, tax., distr., bion.); Liria & Navarro 2010 (niche model); Linton et al. 2013 (bion., mol. tax., distr.; Ecuador); Berti et al. 2015 (distr.; Venezuela). **Ety.:** n.s. [color of violets (Gr); poss. ref. to overall purple aspect]. **Informal name:** Violet Trinbagonian Shoulder Mosquito. (Vol. 1, p. 474)

syn. *obscurescens* Martini
1931a: 212 (F; *Haemagogus uriartei* var.). **Type loc.:** Ucayali River, Peru (NHMUK). **Refs.:** Stone 1957a: 339 (syn. with *anastasionis*); Arnell 1973: 37 (syn.).

syn. *falco* Kumm, Osorno-Mesa, & Boshell-Manrique
1946: 14 (M*, F*, L*; *spegazzinii* ssp.). **Type loc.:** Volcanes Forest, Caparrapi, Cundinamarca, Colombia (USNM). **Refs.:** Stone & Knight 1955: 283 (type info.); Martínez et al. 1961a: 76 (tax.); Arnell 1973: 37 (syn.).

syn. *petrocchiae* Martínez, Carcavallo, & Prosen
1961: 79 (M, F; *Haemagogus* (*Stegoconops*) *capricornii* ssp.). **Type loc.:** Salvador Mazza (Pocitos), Departamento de General José de San Martín, Salta Province, Argentina (USNM). **Refs.:** Belkin et al. 1968: 14 (type info.); Arnell 1973: 37 (syn.).

lucifer (Howard, Dyar, & Knab)
1913: Pl. 23, Fig. 164 (M*; *Stegoconops*). **Type loc.:** Tabernilla, Canal Zone, Panama (USNM). **Distr.:** Colombia, Costa Rica, Ecuador, Panama, Venezuela. **Refs.:** Dyar 1921g: 107 (lectotype desig.); Galindo et al. 1951a: 117 (bion.); Lane 1953: 781 (M*, F*, L*); Komp 1954c: 193 (M*; tax. = *regalis*); Mohrig 1967 (F*); Arnell 1973: 84 (M*, F, P*, L*); Mattingly 1973a: 24 (E*); Townsend 1990: 96 (type info.). **Ety.:** n.s [poss. ref. to Satan]. **Informal name:** Lucifer Panamanian Shoulder Mosquito.

mesodentatus Komp & Kumm
1938: 253 (M*, F*, L*). **Type loc.:** San Jose, Costa Rica (LU). **Distr.:** Belize, Costa Rica, El Salvador, Guatemala, Mexico, Panama. **Refs.:** Lane 1953: 796 (M*, F, L*); Stone & Knight 1955: 288 (type info.); Arnell 1973: 46 (M*, F, P*, L*); Liria & Navarro 2010 (niche model). **Ety.:** mesosome [aedeagus] with teeth (Gr, L). **Informal name:** Toothy Costa Rican Shoulder Mosquito. (Vol. 1, p. 476)

syn. *gorgasi* Galindo & Trapido
1956: 228 (M, F, P, L; *mesodentatus* ssp.). **Type loc.:** Tapachula, Chiapas, Mexico (USNM). **Refs.:** Arnell 1973: 46 (syn.).

syn. *alticola* Galindo, Trapido, & Boshell-Manrique
1956 (M, F, P, L; *mesodentatus* ssp.). In Galindo & Trapido 1956: 231. **Type loc.:** Sumidero Canyon, north of Tuxtla Gutierrez, Chiapas, Mexico (USNM). **Refs.:** Arnell 1973: 47 (syn.).

nebulosus Arnell
1973: 35 (F, P*, L*). **Type loc.:** Rancho Grande, Aragua, Venezuela (USNM). **Distr.:** Venezuela. **Ety.:** n.s. [misty, foggy (L); poss. ref. to habitat]. **Informal name:** Nebulous Venezuelan Shoulder Mosquito.

panarchys Dyar
1921g: 104 (M, F). **Type loc.:** El Salado, Guayaquil, Ecuador (USNM). **Distr.:** Ecuador. **Refs.:** Lane 1953: 806 (M*, F, P*, L*); Komp 1955c: 237 (L); Stone & Knight 1955: 288 (type info.); Arnell 1973: 53 (M*, F*, P*, L*). **Ety.:** n.s. **Informal name:** Overlord Ecuadorian Shoulder Mosquito [*panarchys* = "all governing" = Overlord].

regalis Dyar & Knab
1906b: 167 (A). **Type loc.:** Sonsonate, El Salvador (USNM). **Distr.:** Colombia, El Salvador, Guatemala,

Honduras, Mexico, Panama. **Refs.:** Lane 1953: 787 (M*, F, L*); Stone & Knight 1955: 289 (type info.); Mohrig 1967 (F*); Arnell 1973: 82 (M*, F, P*, L*). **Ety.:** n.s. [royal (L); prob. ref. to metallic colors]. **Informal name:** Regal Salvadoranian Shoulder Mosquito.

soperi Leví-Castillo
1955c: 480 (M*, F, P*, L*). **Type loc.:** Juan Montalvo, Los Rios, Ecuador (USNM). **Distr.:** Ecuador. **Refs.:** Arnell 1973: 56 (M*, F, P*, L*; keys, tax., distr.). **Ety.:** Frederick Lowe Soper. **Informal name:** Soper Ecuadorian Shoulder Mosquito.

spegazzinii Brèthes
1912: 39 (F). **Type loc.:** Jujuy, Argentina (BA). **Distr.:** Argentina, Bolivia, Brazil, Colombia, Ecuador, French Guiana, Honduras, Lesser Antilles (includes Trinidad and Tobago), Paraguay, Peru, Venezuela. **Refs.:** Cerqueira 1943a: 10 (F*; from syn. with *equinus*); Cerqueira & Boshell-Manrique 1946 (M*, F, L*; tax.); Coher 1949 (F); Kumm & Cerqueira 1951: 174 (M*, F, L*; as *uriartei*); Lane 1953: 800 (M*, F, P*, L*; as *uriartei*); Leví-Castillo 1956 (tax.); Belkin et al. 1968: 14 (type info.); Mattingly 1971a: Pl. 17 (P*); Arnell 1973: 28 (M*, F, P*, L*); Mattingly 1973a (E*); Liria & Navarro 2010 (niche model). **Ety.:** Carlos Luigi Spegazzini. **Informal name:** Spegazzini Argentinian Shoulder Mosquito. (Vol. 1, p. 478)

syn. *uriartei* Shannon & Del Ponte
1928: 68 (A). **Type loc.:** Vipos, Tucumán, Argentina (USNM). **Refs.:** Stone & Knight 1955: 289 (type info.). Martínez et al. 1961a: 71 (syn.).

syn. *lindneri* Martini
1931b: 118 (F?). **Type loc.:** Chiquitos, Bolivia (SMNS). **Refs.:** Lane 1939 (syn. with *uriartei*).

splendens Williston
1896: 272 (F). **Type loc.:** Saint Vincent Island [Lesser Antilles] (NHMUK). **Distr.:** Colombia, Lesser Antilles (includes Grenada, Martinique, Saint Vincent and the Grenadines), Panama, Venezuela. **Refs.:** Lane 1953: 779 (M*, F, P*, L*, in part); Komp 1954a: 49 (tax.); Belkin 1968b: 22 (lectotype desig.); Mattingly 1971a: Pl. 1 (M, F head*, wing*), Pl. 8 (P paddle*), Pl. 22 (L head*, terminal segs.*); Arnell 1973: 71 (M*, F*, P*, L*); Reinert 2002c: Fig. 2 (F*). **Ety.:** n.s. [shine, gleam, glitter (L); ref. to brilliant metallic coloration]. **Informal name:** Resplendent Vincentian Shoulder Mosquito.

tropicalis Cerqueira & Antunes
1938 (M*, F, P*, L*). **Type loc.:** Curralinho, Pará, Brazil (LU). **Distr.:** Brazil. **Refs.:** Lane 1953: 807 (M*, F, P*, L*); Belkin et al. 1971: 20 (type info.); Arnell 1973: 68 (M*, F, L*). **Ety.:** n.s. [tropical (L]. **Informal name:** Tropical Brazilian Shoulder Mosquito.

Genus *Heizmannia* Ludlow

(Vol. 1, p. 228)

Important references: Barraud 1934 (tax.; SE Asia); Mattingly 1957d (keys; Indomalayan region); Thurman 1959b (tax.; Thailand); Mattingly 1970a (revision; SE Asia); Reinert 1973f (tax., class.); Lee et al. 1988b (tax., class., bion.; Australasia); Reinert 2002b (F gen.*); Rattanarithikul & Harrison et al. 2005 (F, L; tax., keys, bion.; Thailand); Reinert et al. 2009 (class., phyl.); Rattanarithikul et al. 2010 (tax., keys, bion.; Thailand); Wilkerson et al. 2015 (class., phyl.); Soghigian et al. 2017 (phyl.).

Subgenus *Heizmannia* Ludlow

1905a: 130 (as genus). Type species: *Heizmannia scintillans* Ludlow. Brunetti 1914: 61 (*Heizmannia*; emend.).

Name derivation: C.L. Heizmann.

syn. *Bolbodeomyia* Theobald 1910a: 31 (as genus). Type species: *Bolbodeomyia complex* Theobald.

Important references: Brunetti 1914: 61 (*Heinzmannia*; emend.); Barraud 1934: 310, 313 (tax., keys, morph.; India, Sri Lanka, Myanmar); Mattingly 1957 (keys; Indomalayan area); Thurman 1959b (tax.; Thailand); Delfinado 1966a (tax., keys; Philippines); Mattingly 1970a (revision; SE Asia); Reinert 1973f (tax.); Tanaka et al. 1979: 247 (keys; Japan); Lee et al. 1988b: 80 (keys, tax.; Australasia); Reinert 2002b (F gen.*); Reinert et al. 2004 (phyl., class.); Rattanarithikul et al. 2010 (F*, L*; keys, bion.; Thailand); Wilkerson et al. 2015 (phyl., class.).

aurea Brug
1932: 75 (F). **Type loc.:** Halmahera, Moluccas (NHMUK). **Distr.:** Indonesia. **Ety.:** n.s. [golden (L); poss. ref. to pale yellow marking on hind femur]. **Informal name:** Golden Indonesian Metallic Mosquito.

aureochaeta (Leicester)
1908: 255 (F; *Dendromyia*). **Type loc.:** Near Kuala Lumpur [Selangor], Malaya [Malaysia] (NHMUK). **Distr.:** Cambodia, India, Malaysia, Thailand. **Refs.:** Mattingly 1957d: 50 (F*); Mattingly 1970a: 11 (M*, F*, P*, L*; distr.); Mattingly 1971a: Pl. 14 (P abdomen*); Kaur 2003: 453 (distr.; India); Rattanarithikul et al. 2010 (F*, L*; keys, bion., distr.; Thailand). **Ety.:** n.s. [golden setae (L)]. **Informal name:** Gold-maned Malaysian Metallic Mosquito.

carteri Amerasinghe
1993: 110 (F*). **Type loc.:** Udawattakele Forest, Central Province, Sri Lanka (USNM). **Distr.:** Sri Lanka. **Ety.:** Henry Francis Carter. **Informal name:** Carter Sri Lankan Metallic Mosquito.

chandi Edwards
1922a: 291 (M*, F) and 1922b: Figs. 31, 32 (M*). **Type loc.:** Pudupadi, Malabar Coast [Madras], India (NHMUK). **Distr.:** India. **Refs.:** Barraud 1934: 308 (M*, F); Tewari et al. 2012 (M, F). **Ety.:** Khazan Chand. **Informal name:** Chand Indian Metallic Mosquito.

chengi Lien
1968c: 137 (M*, F*, P*, L*). **Type loc.:** Peiyuan, Tungho Township, Taitung Hsien, Taiwan (Republic of China) (TMRI). **Distr.:** India, Malaysia, Taiwan, Thailand. **Refs.:** Mattingly 1970a: 13 (M*, F*, P*, L*; distr.); Rattanarithikul et al. 2010 (F*, L*; keys, bion., distr.; Thailand). **Ety.:** Che I. Cheng. **Informal name:** Cheng Taiwanese Metallic Mosquito.

communis (Leicester)
1908: 254 (M, F; *Dendromyia*). **Type loc.:** The Gap [Selangor], Malaya [Malaysia] (NHMUK). **Distr.:** Indonesia, Malaysia, Thailand, Vietnam. **Refs.:** Mattingly 1957d: 43 (M*, F*, L*?; tax.); Mattingly 1970a: 15 (M*, F*, P*, L*; distr.); Mattingly 1971f: 197 (E*); Rattanarithikul et al. 2010 (F*, L*; keys, bion., distr.; Thailand). **Ety.:** n.s. [common (L)]. **Informal name:** Common Malaysian Metallic Mosquito.

complex (Theobald)
1910a: 31 (M, F; *Bolbodeomyia*). **Type loc.:** Dawna Hills, Lower Burma [Myanmar] (NHMUK). **Distr.:** Cambodia, India, Laos, Malaysia, Myanmar, Thailand, Vietnam. **Refs.:** Edwards 1922b: 447 (M*); Borel 1930: 158 (M*, F, L*); Barraud 1934: 302 (M, F); Mattingly 1970a: 17 (M*, F*, P*, L*; distr.); Mattingly 1971a: Pl. 14 (P abdomen*); Rattanarithikul et al. 2010 (F*, L*; keys, bion., distr.; Thailand). **Ety.:** n.s. [complex, entwined (L); poss. ref. to complex male genitalia]. **Informal name:** Entwined Burmese Metallic Mosquito.
syn. *stonei* Mattingly
1957d: 51 (F*). **Type loc.:** Ulu Langat, Selangor, Malaya [Malaysia] (NHMUK). **Refs.:** Mattingly 1970a: 17 (syn.).

covelli Barraud
1929b: 265 (M*, F). **Type loc.:** Andaman Islands. (NHMUK). **Distr.:** Bangladesh, India, Myanmar, Thailand, Vietnam. **Refs.:** Barraud 1934: 304 (M*, F, L?); Mattingly 1970a: 19 (M*, F*, L*); Ahmed 1987 (distr.; Bangladesh); Rattanarithikul et al. 2010 (F*, L*; keys, bion., distr.; Thailand). **Ety.:** G. Covell. **Informal name:** Covell Indian Metallic Mosquito.

demeilloni Mattingly
1970a: 21 (F*, P*, L). **Type loc.:** Kalaw, Burma [Myanmar] (USNM). **Distr.:** Myanmar, Nepal, Thailand. **Refs.:** Darsie et al. 1992: 132 (distr.); Rattanarithikul et al. 2010 (F*, L*; keys, bion., distr.; Thailand). **Ety.:** Botha De Meillon. **Informal name:** De Meillon Burmese Metallic Mosquito.

funerea (Leicester)
1908: 252 (F; *Wyeomyia*). **Type loc.:** 6 mi from Kuala Lumpur [Selangor], Malaya [Malaysia] (NHMUK). **Distr.:** Australia, India, Malaysia, Singapore. **Refs.:** Mattingly 1957d: 28 (syn.); Mattingly 1970a: 23 (M*, F*, L*; from syn. with *scintillans*); Kaur 2003: 453 (distr.; India). **Ety.:** n.s. [funereal (L); poss. ref. to dark color]. **Informal name:** Funereal Malaysian Metallic Mosquito.

greenii (Theobald)
1905h: 247 (M*, F; *Wyeomyia*). **Type loc.:** Peradeniya [Central], Ceylon [Sri Lanka] (NHMUK). **Distr.:** India, Sri Lanka. **Refs.:** Edwards 1922b: 449 (M*); Carter & Wijesundara 1948: 142 (L*); Mattingly 1957d: 6 (tax.); Amerasinghe 1989: 83 (M*, F*, P*, L*; lectotype desig., distr.). **Ety.:** Christopher A. Green. **Informal name:** Green Sri Lankan Metallic Mosquito.

heterospina Gong & Lu
1986: 183 (M*) [in Lu & Gong 1986]. **Type loc.:** Xianguan, Yunnan Province, China [People's Republic of China] (YIE). **Distr.:** People's Republic of China. **Ety.:** different thorn (L); "Distimere [gonostylus] and (?) claw-like subapical spine are unique among the genus." **Informal name:** Thorny Chinese Metallic Mosquito.

himalayensis Edwards
1922a: Figs. 28, 29 (M*), 1922b: 290 (M*, F). **Type loc.:** Darjeeling, Eastern Himalayas [West Bengal], India (IM). **Distr.:** India, Nepal. **Refs.:** Barraud 1934: 306 (M*, F, L); Mattingly 1949c: 221 (P*); Joshi et al. 1965 (distr.; Nepal); Darsie & Pradhan 1990, 1991 (F, L; tax., keys, bion., distr.; Nepal). **Ety.:** Himalayas. **Informal name:** Himalaya Indian Metallic Mosquito.

indica (Theobald)
1905f: 115 (M; *Phoniomyia*). **Type loc.:** Singapore (HNM). **Distr.:** India, Malaysia, Nepal, Singapore, Thailand. **Refs.:** Mattingly1957d: 34 (M*, F*, L*; tax.), 41 (M*?, F*; tax.; as *metallica*); Joshi et al. 1965: 139 (distr.); Mattingly 1970a: 25 (M*, F*); Mattingly 1971f: 198 (E*); Kaur 2003: 453 (distr.; India). **Ety.:** India. **Informal name:** India Singaporean Metallic Mosquito.
syn. *metallica* Leicester
1908: 251; (F; *Wyeomyia*). **Type loc.:** Bukit Kutu, Malaya [Malaysia] (NHMUK). **Refs.:** Mattingly 1970a: 25 (syn.).

kana Tanaka, Mizusawa, & Saugstad
1979: 249 (F*). **Type loc.:** Mount Yuwan, Amami Ōshima, Ryukyus, Japan (NMST). **Distr.:** Japan. **Refs.:** Miyagi, Toma, Tamashiro et al. 2012 (M*, F, P*, L*, E*). **Ety.:** Taisuke Kanao. **Informal name:** Kanao Japanese Metallic Mosquito.

kanhsienensis Tung
1955: 309 (M*). **Type loc.:** Kanhsien, Kiangsi [Jiangxi] Province, China [People's Republic of China] (CUMC). **Distr.:** People's Republic of China. **Ety.:** Kanhsien, China. **Informal name:** Kanhsien Chinese Metallic Mosquito.

lii Wu
1936: 46 (M, F*). **Type loc.:** Yun-Shi, Hangchow, Chekiang, China [Zhejiang Province, People's Republic of China] (BEH, NHMUK). **Distr.:** People's Republic of China, South Korea, Taiwan. **Refs.:** Lu & Li 1982b (L*); Townsend 1990: 44 (type info.). **Ety.:** Mr. F.S. Li (in Harbach 2018c). **Informal name:** Li Chinese Metallic Mosquito.

macdonaldi Mattingly
1957d: 47 (M*, F*, P*, L*). **Type loc.:** Ulu Langat Forest Reserve, Selangor, Malaya [Malaysia] (NHMUK). **Distr.:** Malaysia, Taiwan, Thailand. **Refs.:** Mattingly 1970a: 26 (M*, F*, P*, L*); Knight 1978b: 40 (distr.); Rattanarithikul et al. 2010 (F*, L*; keys, bion., distr.; Thailand). **Ety.:** W.W. Macdonald. **Informal name:** Macdonald Malaysian Metallic Mosquito.
syn. *nivirostris* Lien
1968c: 139 (M*, F*, P*, L*). **Type loc.:** Laitou, Fanlu, Chiai Hsien, Taiwan (TMRI). **Refs.:** Mattingly 1970a: 26 (syn.).

mattinglyi Thurman
1959b: 70 (F). **Type loc.:** Rong Kwang, Prae Province, Thailand (USNM). **Distr.:** Cambodia, Thailand. **Refs.:** Mattingly 1970a: 28 (F*; distr.); Rattanarithikul et al. 2010 (F*, L*; keys, bion., distr.; Thailand). **Ety.:** Peter Frederick Mattingly. **Informal name:** Mattingly Thai Metallic Mosquito.

maximalepido Dong, Zhou, & Dong
2005. **Type loc.:** Mengla County, Yunnan Province, China. **Refs.:** Dong, Zhou, & Dong 2004. **Ety.:** n.s. [maximally agreeable/charming (L)]. **Informal name:** Charming Chinese Metallic Mosquito.

menglianensis Lu & Gong
1986: 184 (M*, F*, L; [*sic*] *memglianensis*). **Type loc.:** Menglian, Yunnan Province, China [People's Republic of China] (YIE). **Distr.:** People's Republic of China. **Ety.:** Menglian, Yunnan Province, China. **Informal name:** Menglian Chinese Metallic Mosquito.

menglianeroides Dong, Dong, & Zhou
2003 (M*, P*, L*). In Dong, Dong, Zhou, Wang, & Lu 2003: 523. **Type loc.:** Jiluoshong (680 m), Xishuangbanna, Yunnan Province, China [People's Republic of China] (YIM). **Distr.:** China. **Ety.:** similar to *menglianensis*. **Informal name:** Jiluoshong Chinese Metallic Mosquito.

persimilis Mattingly
1970a: 30 (M*, F*, P*, L*). **Type loc.:** 15th mile, Ulu Gombak, Selangor, Malaysia (NHMUK). **Distr.:** Malaysia, Thailand, Vietnam. **Refs.:** Rattanarithikul et al. 2010 (F*, L*; keys, bion., distr.; Thailand). **Ety.:** n.s. [very similar (L); prob. ref. to similarity with *propinqua*

and *proxima*]. **Informal name:** Familiar Malaysian Metallic Mosquito [*persimilis* = "very similar" = Familiar].

propinqua Mattingly
1970a: 32 (M*, F*, P*, L*). **Type loc.:** Ulu Gombak, Selangor, Malaysia (NHMUK). **Distr.:** Malaysia, Thailand. **Refs.:** Rattanarithikul et al. 2010 (F*, L*; keys, bion., distr.; Thailand). **Ety.:** n.s. [near, neighboring (L); prob. ref. to similarity with *persimilis* and *proxima*]. **Informal name:** Nearby Malaysian Metallic Mosquito.

proxima Mattingly
1970a: 33 (M*, P*, L*). **Type loc.:** Doi Inthanon, Chiang Mai, Thailand (USNM). **Distr.:** Thailand. **Refs.:** Lu & Gong 1986 (F*); Rattanarithikul et al. 2010 (F*, L*; keys, bion., distr.; Thailand). **Ety.:** n.s. [near (L); prob. ref. to near *propinqua*, *persimilis*, and other species]. **Informal name:** Approximate Thai Metallic Mosquito.

rajagopalani Natarajan, Eapen, & Jambulingam
2020 (M*, F*; keys). **Type loc.:** Kerala, Idukki District, Pathinaramkandam, India (Zoological Survey of India, Chennai, India [IM?]). **Distr.:** India. **Refs.:** Edwards 1922b: 448–449, 462 (as *metallica*); Barraud 1929b: 267 (as *metallica*); Barraud 1934: 299, 301, 307–308. **Ety.:** P.K. Rajagopalan. **Informal name:** Rajagopalani Indian Metallic Mosquito.

reidi Mattingly
1957d: 38 (M*, F*). **Type loc.:** Sukna [Darjeeling District, West] Bengal, India (NHMUK). **Distr.:** Cambodia, India, Malaysia, Myanmar, Nepal, Taiwan, Thailand, Vietnam. **Refs.:** Lien 1968c (F); Mattingly 1970a: 35 (M*, F*, P*, L*; distr.); Darsie & Pradhan 1990, 1991 (F, L; tax., keys, bion., distr.; Nepal); Rattanarithikul et al. 2010 (F*, L*; keys, bion., distr.; Thailand). **Ety.:** John Alexander Reid. **Informal name:** Reid Indian Metallic Mosquito.
syn. *cheni* Lien
1968c: 135 (M*, F*, P*, L*). **Type loc.:** Fenchihu, Chuchi Township, Chiai Hsien, Taiwan (TMRI). **Refs.:** Mattingly 1970a: 35 (syn.).

ruiliensis Dong, Zhou, & Wang
1997 (M*, F*, L*). In Dong, Zhou, Wang, & Wang 1997: 27. **Type loc.:** Ruili County (elevation 780 m), Yunnan Province, China [People's Republic of China] (YIM). **Distr.:** People's Republic of China. **Ety.:** Ruili Co., Yunnan Province, China. **Informal name:** Ruili Chinese Metallic Mosquito.

scanloni Mattingly
1970a: 38 (M*). **Type loc.:** Ban Chang Kian, Chiang Mai, Thailand (USNM). **Distr.:** Thailand. **Refs.:** Rattanarithikul et al. 2010 (F*, L*; keys, bion., distr.; Thailand). **Ety.:** John E. Scanlon. **Informal name:** Scanlon Thai Metallic Mosquito.

scintillans Ludlow
1905a: 130 (F). **Type loc.:** Camp Stotsenberg, Angeles, Pampanga, Luzon, Philippines (USNM). **Distr.:** Cambodia, Indonesia, Malaysia, Philippines, Singapore, Thailand, Vietnam. **Refs.:** Mattingly 1957d: 28 (M*, F*, P*, L*; tax.); Mattingly 1970a: 39 (M*, F*, P*, L*); Basio 1971b: 74 (M*; bion.); Mattingly 1971a: Pl. 6 (M head*), Pl. 14 (P terminal segs.*), Pl. 32 (L head*, terminal segs.*); Baisas 1974: 130 (M, F, P, L; tax., bion., distr.; Philippines); Reinert 2002b: Fig. 1 (F*); Rattanarithikul et al. 2010 (F*, L*; keys, bion., distr.; Thailand). **Ety.:** n.s. [send out sparks (L); prob. ref. to body "with iridescent and silvery reflections"]. **Informal name:** Glittering Philippine Metallic Mosquito.
syn. *pilosa* Brug
1931a: 245 (M*, F, P*, L*). **Type loc.:** Dermajoe, near Benkoelen, Sumatra [Indonesia] (NHMUK). **Refs.:** Mattingly 1957d: 28 (syn.).

taiwanensis Lien
1968c: 141 (M*, F*, P*, L*). **Type loc.:** Kuantzulin, Paiho Township, Tainan Hsien, Taiwan [Republic of China] (TMRI). **Distr.:** Malaysia, Taiwan, Thailand. **Refs.:** Mattingly 1970a: 41 (M*, F*, P*, L*); Rattanarithikul et al. 2010 (F*, L*; keys, bion., distr.; Thailand). **Ety.:** Taiwan. **Informal name:** Paiho Taiwanese Metallic Mosquito.

tengchongensis Dong, Wang, & Zhou
2002: 133 (M*). **Type loc.:** Gudong (1,798 m), Tengchong County, Yunnan, China [People's Republic of China] (YIM). **Distr.:** People's Republic of China. **Ety.:** Tengchong Co., Yunnan Prov., China. **Informal name:** Tengchong Chinese Metallic Mosquito.

thelmae Mattingly
1970a: 48 (M*, P*, L*). **Type loc.:** Ban Chatri, Ranong, Thailand (USNM). **Distr.:** Thailand. **Refs.:** Reinert 2003b: 433 (tax.; from subg. *Mattinglyia*); Rattanarithikul et al. 2010 (F*, L*; keys, bion., distr.; Thailand). **Ety.:** n.s. **Informal name:** Thelma Thai Metallic Mosquito.

viridis Barraud
1929b: 266 (F). **Type loc.:** Yellapur, North Kanara [Bombay], India (NHMUK). **Distr.:** India. **Refs.:** Barraud 1934: 306 (F). **Ety.:** n.s. [fresh, green (L); poss. ref. to metallic green scales on scutum and scutellum]. **Informal name:** Viridian Indian Metallic Mosquito.

Subgenus *Mattinglyia* Lien

1968c: 128 (as genus). Type species *Mattinglyia catesi* Lien. Mattingly 1970a: 2 (to subg. of genus *Heizmannia*).

Name derivation: Peter Frederick Mattingly.

syn. *Sinoaedes* Gong & Lu 1991: 55 (as subg. of genus *Aedes*). Type species: *Aedes* (*Sinoaedes*) *occidentayunnanus*

Genus *Psorophora* Robineau-Desvoidy

(Vol. 1, p. 232)

Important references: Horsfall et al. 1952: 618 (E; key; North America); Lane 1953 (tax., keys; Neotropics); Forattini 1965a (tax., keys, bion.; Neotropics); Cova García et al. 1966 (tax., keys, bion.; Venezuela); Barr & Barr 1969a (P; key; North America, Puerto Rico); Belkin et al. 1970: 115 (tax, keys, bion.; Jamaica); Wood et al. 1979 (tax., bion.; Canada); Clark-Gil & Darsie 1983 (keys; Guatemala); Darsie 1985 (keys; Argentina); Reinert 2000b (F* gen.); Reinert, Harbach, & Kitching 2004 (phyl., class.); Darsie & Ward 2005 (keys, distr.; North America); Reinert, Harbach, & Kitching 2006 (phyl., class.); Reinert, Harbach, & Kitching 2008 (phyl., class.); Reinert, Harbach, & Kitching 2009 (phyl., class.); Liria & Navarro 2014 (phyl.); Wilkerson et al. 2015 (phyl., class.); Soghigian et al. 2017 (mol. phyl.).

Subgenus *Grabhamia* Theobald

1903a: 243 (as genus). Type species: *Culex jamaicensis* Theobald. Felt 1904: 391b (type species desig.).

Name derivation: Michael Grabham.

syn. *Feltidia* Dyar 1905a: 47. Type species: *Culex jamaicensis* Theobald.
syn. *Ceratocystia* Dyar & Knab 1906d: 183 (as genus). Type species: *Culex discolor* Coquillett.
syn. *Lanesia* Leví-Castillo 1953f: 102 (as subg. of genus *Aedes*). Type species: *Aedes garciai* Leví-Castillo.

Important references: see under genus *Psorophora*. Reinert 2000b (F* gen.).

cingulata (Fabricius)
1805: 36 (A; *Culex*). **Type loc.:** Cayenne, French Guiana [Guyana] (ZMC). **Distr.:** Argentina, Belize, Bolivia, Brazil, Colombia, Costa Rica, Ecuador, French Guiana, Guatemala, Guyana, Haiti, Lesser Antilles (includes Antigua and Barbuda, Grenada, Montserrat, Saint Lucia, Trinidad and Tobago), Mexico, Nicaragua, Panama, Paraguay, Peru, Suriname, Venezuela. **Refs.:** Dyar & Knab 1906d: 181 (L; as *scholasticus* Theobald); Lane 1953: 762 (M*, F, P*, L*); da Silva Maciel 1964 (P*); Guedes et al. 1965: 21 (M*, F*); Belkin 1968b: 26 (lectotype desig.); Linley & Chadee 1990 (E*); Liria & Navarro 2007 (morph.; cibarial armature*); Linton et al. 2013 (bion., mol. tax., distr.; Ecuador); Berti et al. 2015 (distr.; Venezuela). **Ety.:** n.s. [girdled (L)]. **Informal name:** Girdled French Guianese Gang Mosquito.
syn. *apicalis* Theobald
1903a: 171 (M, F; *Culex*). **Type loc.:** Rio de Janeiro, Brazil (NHMUK). **Refs.:** Belkin 1968b: 25 (lectotype desig.).
syn. *indoctum* Dyar & Knab
1906b: 161 (A, L; *Janthinosoma*). **Type loc.:** Trinidad [Trinidad & Tobago, Lesser Antilles] (USNM). **Refs.:** Stone & Knight 1955: 285 (type info.).
syn. *neoapicalis* Theobald
1910c: 336 (*Culex*; new name for *apicalis* Theobald 1903a, not Adams 1903).
syn. *garciai* Leví-Castillo
1953g: 102 (M; *Aedes*). **Type loc.:** Región de Dos Ríos, jurisdicción de Tena, provincia de Napo-Pastaza, Ecuador (NE). **Refs.:** Stone 1957a: 336 (syn.).

columbiae (Dyar & Knab)
1906a: 135 (F; *Janthinosoma*). **Type loc.:** Grassymead, Virginia, United States (USNM). **Distr.:** Bahamas, Belize, Canada, Cayman Isl., Cuba, Honduras, Mexico, Panama, United States (continental). **Refs.:** Howard et al. 1913 (E*); Howard et al. 1917: 590 (M*, F, P*, L*, E*); Gerry 1932: 41 (F*); Aitken 1940: 677 (tax.); Horsfall 1942: 10 (bion.); Ross 1947: 91 (M*, F*, L*); Yamaguti & LaCasse 1951b (F*); Horsfall et al. 1952: 618 (E*); Carpenter & LaCasse 1955: 129 (M*, F*, L*; keys; as *confinnis*); Stone & Knight 1955: 283 (type info.); Craig 1956 (E*; as *confinnis*); Ross & Horsfall 1965 (M*, F*, L*, E*; keys; as *confinnis*); Belkin et al. 1970: 137 (provisional resurrection); Horsfall et al. 1970: 1714 (E*); Bohart & Washino 1978: 37 (M*, F*, P*, L*; keys, tax., bion., distr.); Bosworth et al. 1983 (E*); Linley & Chadee 1990 (E*); Darsie & Ward 2005 (F*, L*; keys, distr.); Becker et al. 2010: 394 (F*, L*; key, tax., distr., bion.); Lanzaro et al. 2015 (tax.); Harrison et al. 2016 (F*, L*; keys, distr.). **Ety.:** n.s. [of Columbia; poss. the New World or District of Columbia]. **Informal name:** Columbia American Gang Mosquito. (Vol. 1, p. 480)
syn. *floridense* Dyar & Knab
1906a: 135 (A; *Janthinosoma*). **Type loc.:** Alligator Creek, Florida, United States (USNM). **Refs.:** Stone & Knight 1955: 283 (type info.); Lanzaro et al. 2015 (type loc. restricted).
syn. *texanum* Dyar & Knab
1906a: 135 (A; *Janthinosoma*). **Type loc.:** Brownsville, Texas, United States (USNM). Refs.: Stone & Knight 1955: 286 (type info.).

confinnis (Lynch Arribálzaga)
1891b: 149 (A; *Taeniorhynchus*). **Type loc.:** Formosa, Chaco, Argentina (BA). **Distr.:** Argentina, Bahamas, Belize, Bolivia, Bonaire, Brazil, Cayman Isl., Colombia, Costa Rica, Cuba, Dominican Republic, El Salvador, French Guiana, Guatemala, Guyana, Haiti, Honduras, Jamaica, Lesser Antilles (includes Aruba, Curaçao, Virgin Isl.), Mexico, Nicaragua, Panama, Paraguay, Peru, Puerto Rico, United States (continental), Uruguay, Venezuela. **Refs.:** Lynch Arribálzaga 1891b: index (as *confinis*, *lapsus*); Yamaguti & LaCasse 1951b: 71 (M*, F*, L*); Lane 1953: 768 (M*, F, P, L*); Guedes et al. 1965: 23 (M*, F*); Dodge 1966: 380 (first instar L*; key); Belkin

et al. 1968: 13 (lectotype desig. according to Casal); Belkin et al. 1970: 137 (tax.); Bickley 1984 (as *confinnis* complex); Rossi & Martínez 2003: 472 (distr.; Uruguay); Liria & Navarro 2007 (morph.; cibarial armature*); Lanzaro et al. 2015 (tax., distr.; Argentina, Amazonas?). **Ety.:** n.s. [neighboring? (L); poss. close to another species]. **Informal name:** Neighboring Argentinian Gang Mosquito.

syn. *scutipunctatus* Lutz & Neiva
1911: 298 (F; *Culex*). **Type loc.:** [Rio] Tietê, immediately above Itapura, São Paulo, Brazil (IOC). **Refs.:** Belkin et al. 1971: 19 (type info.).

syn. *chiquitana* Pinto
1932: 291 (F). **Type loc.:** Los Naranjos, Oriente da Bolivia, Bolivia (LU). **Refs.:** Lane 1953: 768 (syn.); Belkin et al. 1968: 20 (type loc. info.).

dimidiata Cerqueira
1943a: 5 (M*, F). **Type loc.:** Lagunillas, Valle Grande, Santa Cruz, Bolivia (IOC). **Distr.:** Argentina, Bolivia, Paraguay. **Refs.:** Lane 1953: 766 (M*, F); Guedes et al. 1965: 23 (M*, F*). **Ety.:** n.s. [halved? (L)]. **Informal name:** Halved Bolivian Gang Mosquito.

discolor (Coquillett)
1903a: 256 (F; *Culex*). **Type loc.:** Delair, New Jersey, United States (USNM). **Distr.:** Mexico, United States (continental). **Refs.:** Mitchell 1907b (E*); Howard et al. 1913 (E*); Howard et al. 1917: 578 (M*, F, L*, E); Yamaguti & LaCasse 1951b: 77 (M*, F*, L*); Horsfall et al. 1952: 618 (E*); Carpenter & LaCasse 1955: 131 (M*, F*, L*; keys); Stone & Knight 1955: 283 (type info.); Ross & Horsfall 1965 (M*, F*, L*, E*; keys); Dodge 1966: 380 (first instar L; key); Barr & Barr 1969a: 9 (P*); Darsie & Ward 2005 (F*, L*; keys, distr.); Harrison et al. 2016 (F*, L*; keys, distr.). **Ety.:** n.s. [of different colors (L)]. **Informal name:** Varicolored American Gang Mosquito.

funiculus Dyar
1920e: 141 (M, F). **Type loc.:** Rio Frio, Magdalena, Colombia (USNM). **Distr.:** Colombia. **Refs.:** Dyar 1928: 127 (syn.); Stone & Knight 1955: 284 (lectotype desig.); Lanzaro et al. 2015 (from syn. with *confinnis*). **Ety.:** n.s. [thin rope, cord, string (L)]. **Informal name:** Threadlike Colombian Gang Mosquito.

syn. *trigonophorus* Lutz
1928: 32 (M, F*; *Taeniorhynchus*). **Type loc.:** San Jacinto, Aragua, Venezuela (LU). **Refs.:** Edwards 1932a: 126 (syn. with *funiculus*); Lanzaro et al. 2015 (syn.).

infinis (Dyar & Knab)
1906d: 182 (L*; *Janthinosoma*; as *infine*). **Type loc.:** Santo Domingo [Dominican Republic] (USNM). **Distr.:** Cuba, Dominican Republic, Ecuador, Haiti, Jamaica, Mexico, United States (continental). **Refs.:** Dyar 1928: 131 (M*, F, L*); Lane 1953: 762 (syn. with *cingulata*); Stone & Knight 1955: 285 (lectotype desig.; as *infine*); Belkin et al. 1970: 131 (M*, F, P*, L*; from syn.); Mattingly 1971a: Pl. 17 (P segment II*). **Ety.:** n.s. [unlimited (L)]. **Informal name:** Unlimited Dominican Gang Mosquito.

insularius (Dyar & Knab)
1906a: 135 (A; *Janthinosoma*). **Type loc.:** Santo Domingo [Dominican Republic] (USNM). **Distr.:** Bahamas, Cuba, Dominican Republic, Haiti, Jamaica, Mexico, Puerto Rico. **Refs.:** Lane 1953: 770 (M*, F, L*); Stone & Knight 1955: 285 (type info.); Barr & Barr 1969a: 10 (P*); Belkin et al. 1970: 138 (M*, F, P*, L*); Harbach 2018c: 64 (spelling corrected from *insularia*). **Ety.:** n.s. [of islands (L)]. **Informal name:** Islander Dominican Gang Mosquito.

syn. *haruspicus* Dyar & Knab
1908: 56 (M, F; *Aedes*). **Type loc.:** Port Antonio, Jamaica (USNM). **Refs.:** Stone & Knight 1955: 284 (lectotype desig.).

jamaicensis (Theobald)
1901a: 345 (F; *Culex*). **Type loc.:** Spanish Road, Kingston, Jamaica (NHMUK). **Distr.:** Argentina, Cayman Isl., Colombia, Cuba, Dominican Republic, Haiti, Jamaica, Lesser Antilles (Virgin Isl.), Puerto Rico. **Refs.:** Aitken 1940: 677 (syn. with *confinnis*); Belkin 1968b: 27 (lectotype desig.); Belkin et al. 1970: 134 (M*, F, P*, L*; from syn.); Reinert 2000e: Fig. 1 (F*). **Ety.:** Jamaica. **Informal name:** Jamrock Jamaican Gang Mosquito [Jamrock is a local slang term for the island of Jamaica].

syn. *walsinghamii* Theobald
1907: 484 (F*; *Taeniorhynchus*). **Type loc.:** Runaway Bay, Jamaica (NHMUK). **Refs.:** Howard et al. 1917: 581 (syn. with *jamaicensis*); Belkin 1968b: 29 (type info.).

leucocnemis Martini
1931a: 214 (F). **Type loc.:** Montevideo, Uruguay (NHMUK). **Distr.:** Uruguay. **Refs.:** Lane 1953: 764 (? = *varinervis*); Mattingly 1955c: 28 (type info.). **Ety.:** n.s. [white leg (L)]. **Informal name:** White-legged Uruguayan Gang Mosquito.

paulli Paterson & Shannon
1927: 4, 9 (F). **Type loc.:** Tres Pozos, Embarcación, Salta, Argentina (INM). **Distr.:** Argentina, Bolivia, Paraguay. **Refs.:** Lane 1953: 768 (F); Guedes et al. 1965: 18 (F*); Belkin et al. 1968: 13 (lectotype desig. according to Casal); Stein et al. 2013 (M, P, L; redescr.). **Ety.:** Archibaldo Paul. **Informal name:** Paul Argentinian Gang Mosquito.

pygmaea (Theobald)
1903a: 245 (F; *Grabhamia*). **Type loc.:** Antigua Island [Lesser Antilles] (NHMUK). **Distr.:** Argentina, Bahamas, Cayman Isl., Cuba, Dominican Republic, Haiti, Jamaica, Lesser Antilles (includes Anguilla, Antigua and Barbuda, Barbados, Saint Kitts and Nevis, Saint Martin, Turks and Caicos Isl., Virgin Isl.), Mexico, Panama, Puerto Rico, United States (continental). **Refs.:** Theobald 1907 (E*);

Yamaguti & LaCasse 1951b: 81 (M*, F*, L*); Lane 1953: 771 (M*, F, L*); Carpenter & LaCasse 1955: 133 (M*, F*, L*; keys); Belkin 1968b: 28 (lectotype desig.); Barr & Barr 1969a: 10 (P*); Belkin et al. 1970: 140 (M*, F, P*, L*); Darsie & Ward 2005 (F*, L*; keys, distr.). **Ety.:** n.s. [small (Gr); poss. ref. to small size]. **Informal name:** Pygmy Antiguan Gang Mosquito.

syn. *nanus* Coquillett
1903a: 256 (F; *Culex*). **Type loc.:** Key West, Florida, United States (USNM). **Refs.:** Stone & Knight 1955: 285 (lectotype desig.)

syn. *antiguae* Giles
1904b: 384 (F; *Taeniorhynchus*). **Type loc.:** Antigua Island [Lesser Antilles] (NHMUK). **Refs.:** Belkin 1968b: 25 (lectotype desig.).

santamarinai Broche
1971: Pl. 33 (L head*, terminal segs.*). **Type loc.:** San Antonio de Cabezas, Municipio Unión de Reyes, Provincia de Matanzas, Cuba (IMTC). **Distr.:** Cuba. **Ety.:** Alberto Santamarina. **Informal name:** Santamarina Cuban Gang Mosquito.

signipennis (Coquillett)
1904c: 167 (M, F; *Taeniorhynchus*). **Type loc.:** Monterrey [Nuevo Leon], Mexico (USNM). **Distr.:** Canada, Mexico, United States (continental). **Refs.:** Carpenter et al. 1946: 162 (M*, F, L*); Yamaguti & LaCasse 1951b: 85 (M*, F*, P*, L*); Hedeen 1954: 143 (L); Stone & Knight 1955: 286 (lectotype desig.); Dodge 1966: 382 (first instar L; key); Barr & Barr 1969a: 9 (P); Bohart & Washino 1978: 38 (M, F, L*; keys, tax., bion., distr.); Darsie & Ward 2005 (F*, L*; keys, distr.). **Ety.:** sign, wing (L); ref. to apical half of wing with light-yellow scales except for two black patches. **Informal name:** Mark-winged Mexican Gang Mosquito.

syn. *pruinosa* Martini
1935: 47 (F). **Type loc.:** Torreón, Coahuila, Mexico (NHMUK). **Distr.:** Mexico. **Refs.:** Diaz Najera 1965a, 1965b (syn.)

tolteca (Dyar & Knab)
1906a: 135 (A; *Janthinosoma*). **Type loc.:** Tehuantepec, Oaxaca, Mexico (USNM). **Distr.:** Mexico, United States (continental). **Refs.:** Aitken 1940: 677 (syn.); Stone & Knight 1955: 286 (type info.); Lanzaro et al. 2015 (from syn. with *columbiae*). **Ety.:** n.s. **Informal name:** Toltec Mexican Gang Mosquito.

varinervis Edwards
1922d: 78 (F). **Type loc.:** Paraguay (HNM). **Distr.:** Argentina, Bolivia, Brazil, Colombia, Paraguay, Uruguay. **Refs.:** Lane 1953: 764 (F); Martínez et al. 1959: 115 (M*); da Silva Maciel 1964 (M*); Guedes et al. 1965: 24 (M*, F*). **Ety.:** n.s. [contrasting veins? (L)]. **Informal name:** Guarani Paraguayan Gang Mosquito [named for the dominant indigenous people of Paraguay].

Subgenus *Janthinosoma* Lynch Arribálzaga

1891a: 374, 1891b: 152 (as genus). Type species: *Culex discrucians* Walker. Dyar 1905a: 47 (type species desig.).

Name derivation: violet covered body (L).

syn. *Conchyliastes* Theobald 1901. In Howard 1901: 155, 235. Type species: *Culex musicus* Say. Coquillett 1910: 526 (type species desig.).

syn. *Ianthinosoma* Lynch Arribálzaga 1902 (unjust. emend.). In Goeldi 1902: 9.

syn. *Lepidosia* Coquillett 1906b: 314 (as genus). Type species: *Culex cyanescens* Coquillett.

Important references: see under genus *Psorophora*; Guedes & Souza 1964: 471 (key); Reinert 2000b (F* gen.); Harrison et al. 2008 (subg. *Janthinosoma*; key, tax.; United States).

albigenu (Lutz)
1908 (F; *Janthinosoma*). In Peryassú 1908: 155. **Type loc.:** Chanaan [Canaan], São Paulo, Brazil (IOC). **Distr.:** Argentina, Bolivia, Brazil, Colombia, Costa Rica, Ecuador, El Salvador, Haiti, Nicaragua, Panama, Paraguay, Peru, Suriname, Uruguay, Venezuela. **Refs.:** Guedes & Souza 1964: 474 (M*, F*, P*, L*; from syn.); Guedes et al. 1965 (F*); Belkin et al. 1971: 18 (type info.); Forattini 1996 (E*); Forattini 2002 (E*); Rossi & Martínez 2003: 472 (distr.; Uruguay); Pacheco et al. 2012 (E*); Linton et al. 2013 (bion., mol. tax., distr.; Ecuador). **Ety.:** n.s. [white knees (L); hind legs with white "knees"]. **Informal name:** White-kneed Brazilian Gang Mosquito.

syn. *paraguayensis* Strickland
1911b: 268 (F; *Janthinosoma*). **Type loc.:** Puerto Max, Paraguay (LU). **Refs.:** Belkin 1968b: 28 (tax.).

syn. *bruchi* Petrocchi
1927 (M, F). In Shannon & Del Ponte 1927: 728. **Type loc.:** Provincia de Buenos Aires, Argentina (INM). **Refs.:** Stone 1967: 206 (syn.); Casal & García 1970: 12 (lectotype desig.).

albipes (Theobald)
1907: 157 (F; *Janthinosoma*). **Type loc.:** Aqua [Acqua] Santa, Trinidad [Trinidad & Tobago, Lesser Antilles] (NHMUK). **Distr.:** Argentina, Belize, Bolivia, Brazil, Colombia, Costa Rica, Ecuador, French Guiana, Guatemala, Guyana, Honduras, Lesser Antilles (includes Trinidad and Tobago), Mexico, Nicaragua, Panama, Paraguay, Peru, Suriname, Venezuela. **Refs.:** da Costa Lima 1938: 257 (from syn. with *lutzii*); Lane 1953: 757 (M*, F); Guedes & Souza 1964: 476 (M*, P*, L*); Guedes et al. 1965 (F*); Heinemann & Belkin 1977b: 443 (distr.); Heinemann & Belkin 1977c: 530 (distr.); Liria & Navarro 2007 (morph.; cibarial armature*); Linton et al. 2013 (bion., mol. tax., distr.; Ecuador); Berti et al. 2015 (distr.; Venezuela). **Ety.:** n.s. [white foot (L); prob. ref. to "apex of

the second hind tarsal also white"]. **Informal name:** White-footed Trinbagonian Gang Mosquito.

amazonica Cerqueira
1960b: 1 (M*, F, P*, L*). **Type loc.:** Igarapé do Tarumã, Manaus, Amazonas, Brazil (FH). **Distr.:** Brazil. **Refs.:** Belkin et al. 1971: 18 (type info.). **Ety.:** Amazon. **Informal name:** Amazon Brazilian Gang Mosquito.

champerico (Dyar & Knab)
1906a: 134 (F; *Janthinosoma*). **Type loc.:** Champerico, Guatemala (USNM). **Distr.:** Argentina, Brazil, Costa Rica, El Salvador, Guatemala, Mexico, Panama, Venezuela. **Refs.:** Lane 1953: 758 (M*, F); Stone & Knight 1955: 283 (type info.); Guedes et al. 1965 (F*); Muller et al. 2012 (bion., distr.; Brazil). **Ety.:** Champerico, Guatemala. **Informal name:** Champerico Guatemalan Gang Mosquito.
syn. *pisces* Lassmann
1944: 4 (M*, F). **Type loc.:** Tempoal, Estado de Veracruz, Mexico (USNM). **Refs.:** Lane 1951 (syn.); Stone & Knight 1955: 285 (lectotype desig.).

circumflava Cerqueira
1943a: 3 (F). **Type loc.:** Axiamas [Ixiamos?] Caupolican, La Paz, Bolivia (LU). **Distr.:** Bolivia, Brazil, Nicaragua, Peru. **Refs.:** Lane 1953: 760 (F); Cerqueira 1960b: 5 (M*); Guedes et al. 1965: 18 (F*). **Ety.:** n.s. [around, yellow (L); poss. ref. to yellow scale pattern]. **Informal name:** Yellow Bolivian Gang Mosquito.

cyanescens (Coquillett)
1902a: 137 (F; *Culex*). **Type loc.:** Brownsville, Texas, United States (USNM). **Distr.:** Argentina, Bolivia, Colombia, Guatemala, Guyana, Honduras, Lesser Antilles (includes Aruba, Trinidad and Tobago), Mexico, Nicaragua, Paraguay, Peru, United States (continental), Uruguay, Venezuela. **Refs.:** Schwardt 1939 (E*); Carpenter et al. 1946: 139 (M*, F, L*); Yamaguti & LaCasse 1951b: 49 (M*, F*, L*); Lane 1953: 747 (M*, F, L*); Carpenter & LaCasse 1955: 118 (M*, F*, L*; keys); Stone & Knight 1955: 283 (lectotype desig.); Guedes et al. 1965: 18 (M*, F*); Ross & Horsfall 1965 (M*, F*, L*, E*; keys); Dodge 1966: 380 (first instar L*; key); Barr & Barr 1969a: 6 (P*); Horsfall et al. 1970: 1714 (E*); Heinemann & Belkin 1977b: 453 (distr.); Heinemann & Belkin 1977c: 530 (distr.); Darsie 1994: 51–514 (distr.; Guatemala); Darsie & Ward 2005 (F*, L*; keys, distr.); Liria & Navarro 2007 (morph.; cibarial armature*); Harrison et al. 2008 (tax., key, distr.; United States); Berti et al. 2015 (distr.; Venezuela); Harrison et al. 2016 (F*, L*; keys, distr.). **Ety.:** n.s. [becoming dark blue (Gr); various features dark blue]. **Informal name:** Dark-blue American Gang Mosquito.
syn. *tovari* Evans
1922: 218 (F). **Type loc.:** Maracay [Aragua], Venezuela (NHMUK). **Refs.:** Edwards 1932a: 125 (syn.); Belkin 1968b: 29 (tax., lectotype desig.).

syn. *purpurascens* Edwards
1922d: 77 (F). **Type loc.:** Paraguay (HNM). **Refs.:** Dyar 1928: 122 (syn.).

discrucians (Walker)
1856a: 430 (M, F; *Culex*). **Type loc.:** Rio de Janeiro, Guanabara, Brazil (NHMUK). **Distr.:** Argentina, Bolivia, Brazil, Panama, Paraguay, Uruguay, Venezuela. **Refs.:** Howard et al. 1917 (E); Lane 1953: 749 (M*, F, L*); Xavier 1964: 237 (P*); Guedes et al. 1965: 19 (M*, F*); Belkin et al. 1971: 18 (type loc. info.); Forattini 1996 (E*); Reinert 2000e: Fig. 2 (F*); Forattini 2002 (E*). **Ety.:** n.s. [torturing (L)]. **Informal name:** Torturous Brazilian Gang Mosquito.
syn. *arribalzagae* Giles
1902: 341 (M, F*; *Janthinosoma*). **Type loc.:** São Paulo, Brazil (NHMUK). **Refs.:** Belkin 1968b: 25 (type info.).

ferox (von Humboldt)
1819: 340 (A; *Culex*). **Type loc.:** Borodon [Borondón?], Ecuador (NE). **Distr.:** Argentina, Bahamas, Belize, Bolivia, Brazil, Canada, Colombia, Costa Rica, Cuba, Dominican Republic, Ecuador, El Salvador, French Guiana, Guatemala, Guyana, Haiti, Honduras, Jamaica, Lesser Antilles (includes Dominica, Guadeloupe, Saint Lucia, Trinidad and Tobago), Mexico, Nicaragua, Panama, Paraguay, Peru, Suriname, United States (continental), Uruguay, Venezuela. **Refs.:** Goeldi 1905 (E*; as *musica*); Mitchell 1907b (E*; as *posticatus*); Theobald 1907 (E*; as *sayi*); Howard et al. 1913 (E*; as *sayi*); Howard et al. 1917 (E; as *sayi*); Pawan 1922 (E*; as *posticata*); Gerry 1932 (F*); Senevet 1946: 327 (L*); Yamaguti & LaCasse 1951b: 53 (M*, F*, L*); Horsfall et al. 1952: 618 (E*); Lane 1953: 745 (M*, F, P*, L*); Carpenter & LaCasse 1955: 120 (M*, F*, L*; keys); Guedes et al. 1965: 19 (M*, F*); Forattini 1965a (E*); Ross & Horsfall 1965 (M*, F*, L*, E*; keys); Dodge 1966: 382 (first instar L*; key); Belkin 1968b: 26 (tax.); Belkin et al. 1970: 126 (M*, F, P*, L*); Horsfall et al. 1970: 1714 (E*); Wallis & Whitman 1971 (bion.); Mattingly 1971a: Pl. 17 (P segment IV*); Magnarelli 1980 (bion.); Linley & Chadee 1990 (E*); Rossi & Martínez 2003: 472 (distr.; Uruguay); Darsie & Ward 2005 (F*, L*; keys, distr.); Liria & Navarro 2007 (morph.; cibarial armature*); Harrison et al. 2008 (tax., key; United States); Becker et al. 2010: 395 (F*, L*; key, tax., distr., bion.); Berti et al. 2015 (distr.; Venezuela); Harrison et al. 2016 (F*, L*; keys, distr.). **Ety.:** n.s. [wild (L)]. **Informal name:** Wild Ecuadorian Gang Mosquito. (Vol. 1, p. 482)
syn. *posticatus* Wiedemann
1821: 43 (A; *Culex*). **Type loc.:** Mexico (NMW). **Refs.:** Belkin 1968b: 28 (type info.).
syn. *musicus* Say
1829: 149 (A; *Culex*). **Type loc.:** Indiana, United States (NE). **Refs.:** Goeldi 1905: 117 (F*, E*).
syn. *vanhalli* Dyar & Knab
1906a: 134 (A; *Janthinosoma*). **Type loc.:** Paramaribo, Suriname (USNM). **Refs.:** Stone & Knight 1955: 286 (type info.).

syn. *sayi* Dyar & Knab
1906d: 181 (new name for *musicus* Say, not Leach; *Janthinosoma*). **Refs.:** Howard et al. 1917: 554 (M*, F, P*, L*, E*; bion.).

syn. *echinata* Grabham
1906b: 311 (M, F, L*; *Janthinosoma*). **Type loc.:** Molynes Road, near Kingston [Surrey], Jamaica (USNM). **Refs.:** Stone & Knight 1955: 283 (type info.); Belkin et al. 1970: 126 (lectotype desig.).

syn. *terminalis* Coquillett
1906d: 8, 17 (A; *Janthinosoma*). **Type loc.:** Saint Lucia Island [Lesser Antilles] (NHMUK). **Refs.:** Belkin 1968b: 29 (lectotype desig.).

syn. *coquillettii* Theobald
1907: 153 (A; *Janthinosoma*). **Type loc.:** None designated (NE).

syn. *sayi* Theobald
1907: 155 (new name for *musicus* Say, not Leach; *Janthinosoma*). **Refs.:** Stone et al. 1959: 127 (syn.).

syn. *jamaicensis* Theobald
1907: 157 (F; *Janthinosoma sayi* var.). **Type loc.:** Runaway Bay and Kingston [Surrey], Jamaica (NHMUK). **Refs.:** Belkin 1968b: 27 (lectotype desig.).

syn. *pazosi* Pazos
1908: 432 (F*; figure only; *Aedes*). **Type loc.:** Vuelta Abajo (Las Villas), Cuba (NE).

syn. *centrale* Brèthes
1910: 469 (A; *Janthinosoma*). **Type loc.:** Islas del Paraná, Buenos Aires, Argentina (BA). **Refs.:** Belkin et al. 1968: 13 (lectotype desig. according to Casal).

fiebrigi Edwards
1922d: 77 (M, F). **Type loc.:** Paraguay (NHMUK). **Distr.:** Bolivia, Paraguay. **Refs.:** Lane 1951 (syn.); Vargas 1951: 56 (M*); de Souza & Guedes 1962: 249 (M*, F*; from syn.); Guedes et al. 1965 (F*); Belkin 1968b: 26 (lectotype desig.). **Ety.:** n.s. [prob. ref. to Karl (Carlos) August Gustav Fiebrig (Gertz)]. **Informal name:** Fiebrig Paraguayan Gang Mosquito.

forceps Cerqueira
1939: 83 (M*, F). **Type loc.:** Jacarepaguá, Rio de Janeiro, Brazil (IOC). **Distr.:** Brazil, Colombia, Ecuador, Panama. **Refs.:** Lane 1953: 758 (M*, F); Xavier 1964: 235 (P*, L*); Guedes et al. 1965: 11 (M*, F). **Ety.:** n.s. [pincers (L); poss. ref. to the unusual gonostylus]. **Informal name:** Pincer Brazilian Gang Mosquito.

horrida (Dyar & Knab)
1908: 56 (F, in part; *Aedes*). **Type loc.:** Corinth, Mississippi, United States (USNM). **Distr.:** United States (continental). **Refs.:** Howard et al. 1913 (E*); Howard et al. 1917 (E); Matheson 1934a: 42 (from syn. with *champerico*); Rozeboom 1939 (L*); Roth 1945: 1 (M*, F*, L*; bion., tax.); Yamaguti & LaCasse 1951b: 57 (M*, F*, L*); Horsfall et al. 1952: 618 (E*); Carpenter & LaCasse 1955: 122 (M*, F*, L*; keys); Stone & Knight 1955: 284 (type info.); Ross & Horsfall 1965 (M*, F*, L*, E*; keys); Dodge 1966: 382 (first instar L*; key); Barr & Barr 1969a: 7 (P); Horsfall et al. 1970: 1714 (E*); Darsie & Ward 2005 (F*, L*; keys, distr.); Harrison et al. 2008 (tax., key); Harrison et al. 2016 (F*, L*; keys, distr., tax.). **Ety.:** n.s. [dreadful (L)]. **Informal name:** Horrid American Gang Mosquito.

johnstonii (Grabham)
1905: 410 (F; *Janthinosoma*). **Type loc.:** 5 1/2 mi along Molynes Road, Kingston [Surrey], Jamaica (LU). **Distr.:** Bahamas, Cayman Isl., Cuba, Dominican Republic, Haiti, Jamaica, Leeser Antilles (includes Virgin Isl.), Puerto Rico, United States (continental). **Refs.:** Thurman et al. 1951: 144 (M*, F, P*, L, E*; bion., tax.); Yamaguti & LaCasse 1951b: 61 (M*, F*, L*); Lane 1953: 751 (M*, F, P*, L*); Carpenter & LaCasse 1955: 124 (M*, F*, L*; keys); Belkin et al. 1970: 124 (M*, F, P*, L*); Belkin & Heinemann 1972: 67 (distr.); Darsie & Ward 2005 (F*, L*; keys, distr.); Harrison et al. 2008 (tax., key, US). **Ety.:** n.s. [someone named Johnston]. **Informal name:** Johnston Jamaican Gang Mosquito.

syn. *coffini* Dyar & Knab
1906a: 134 (F; *Janthinosoma*). **Type loc.:** Nassau [New Providence Island], Bahama Islands (USNM). **Refs.:** Dyar 1921m: 31 (M); Thurman et al. 1951: 144 (syn.); Stone & Knight 1955: 283 (type info.).

syn. *schwarzi* Dyar & Knab
1906a: 135 (F; *Janthinosoma*). **Type loc.:** Cayamas, Cuba (USNM). **Refs.:** Dyar 1928: 1212 (syn.); Stone & Knight 1955: 286 (type info.).

lanei Shannon & Cerqueira
1943: 135 (F). **Type loc.:** Maracaju, Mato Grosso, Brazil (LU). **Distr.:** Bolivia, Brazil, Ecuador, Peru, Venezuela. **Refs.:** Lane 1953: 760 (F); Guedes et al. 1965: 18 (F*); Belkin et al. 1971: 18 (type info.); Linton et al. 2013 (bion., mol. tax., distr.; Ecuador). **Ety.:** John Lane. **Informal name:** Lane Brazilian Gang Mosquito.

longipalpus Randolph & O'Neill
1944: 88 (F). **Type loc.:** Texas, United States (USNM). **Distr.:** Mexico, United States (continental). **Refs.:** Yamaguti & LaCasse 1951b: 65 (M*, F*, L*; as *longipalpis* Roth); Carpenter & LaCasse 1955: 125 (M*, F*, L*; keys; as *longipalpis* Roth); Stone 1961b: 246 (nomenclature); Dodge 1966: 382 (first instar L; key; as *longipalpis* Roth); Barr & Barr 1969a: 7 (P*); Darsie & Ward 2005 (F*, L*; keys, distr.); Harrison et al. 2008 (tax., key, distr.; United States); Dia et al. 2013 (tax., bion., distr.; Africa). **Ety.:** long palpus (L); ref. to palpus about one-third length of proboscis. **Informal name:** Long-lipped American Gang Mosquito.

syn. *longipalpis* Roth
1945: 13 (M*, F*, L*). **Type loc.:** Fayetteville, Arkansas, United States (USNM). **Refs.:** Stone & Knight 1955: 285 (type info.); Stone 1963: 124 (syn.).

Navarro 2007 (morph.; cibarial armature*); Berti et al. 2015 (distr.; Venezuela). **Ety.:** n.s. [lined (L)]. **Informal name:** Lined Colombian Gang Mosquito.

syn. *blanchardi* Surcouf & González-Rincones
1911: 120 (M, F*). **Type loc.:** Maturin [Monagas], Venezuela (MNHP). **Refs.:** Belkin 1968b: 25 (lectotype desig.).

ochripes (Macquart)
1850: 315 (F; *Culex*). **Type loc.:** Panará River, Buenos Aires, Argentina (MNHP). **Distr.:** Argentina. **Refs.:** Belkin 1968b: 27 (from syn. with *Aedes fulvus*). **Ety.:** yellow feet (Gr); ref. to reddish-yellow "feet." **Informal name:** Yellow-footed Argentinian Gang Mosquito.

pallescens Edwards
1922d: 76 (M, F). **Type loc.:** Paraguay (NHMUK). **Distr.:** Argentina, Bolivia, Paraguay. **Refs.:** Del Ponte & Castro 1952: 221 (M*, F*; tax.); Lane 1953: 737 (M*, F); Stone 1957a: 337 (M*); Guedes et al. 1965: 18 (F*); Belkin 1968b: 28 (lectotype desig.); Stein et al. 2009 (M*, F, P*, L*). **Ety.:** n.s. [grow pale, fade (L)]. **Informal name:** Pallid Paraguayan Gang Mosquito.

syn. *stigmatephora* Dyar
1922e: 116 (M, F*). **Type loc.:** Rio Tapenagá, Colonie Florencia, Gran Chaco, Argentina (USNM). **Refs.:** Stone & Knight 1955: 286 (lectotype desig.).

pilipes (Macquart)
1834: 36 (M, F; *Culex*). **Type loc.:** Brazil (NE; originally in Macquart collection). **Distr.:** Brazil. **Refs.:** Belkin 1968b: 28 (tax.). **Ety.:** n.s. [single hair, foot (L); poss. ref. to character on hind legs]. **Informal name:** Hair-footed Brazilian Gang Mosquito.

saeva Dyar & Knab
1906a: 133 (F). **Type loc.:** Trinidad [Trinidad & Tobago, Lesser Antilles] (USNM). **Distr.:** Argentina, Bolivia, Brazil, Lesser Antilles (includes Trinidad and Tobago), Peru, Suriname, Venezuela. **Refs.:** Pinto 1930: 179 (M*, F*, L*; tax.; as *genumaculata*); Stone & Knight 1955: 286 (type info.); Forattini et al. 1956: 86 (P*; as *lineata*); Stone 1967: 206 (from syn.); Peyton et al. 1983: 74 (distr., tax.); Liria & Navarro 2003a (tax.). **Ety.:** n.s. [acute, sharp (L)]. **Informal name:** Sharp Trinbagonian Gang Mosquito.

syn. *genumaculata* Cruz
1907b: 329 (M, F). **Type loc.:** Santos, São Paulo, Brazil (IOC). **Refs.:** Lane 1951 (syn.); Belkin et al. 1971: 18 (type info.).

stonei Vargas
1956a: 15 (M, F, L). **Type loc.:** Mazatlán, Sinaloa, Mexico (ISET). **Distr.:** Mexico. **Ety.:** Alan Stone. **Informal name:** Stone Mexican Gang Mosquito.

Genus *Udaya* Thurman

(Vol. 1, p. 234)

1954b: 84 (as subg. of genus *Paraedes* Edwards). Type species: *Paraedes argyrurus* Edwards.

Name derivation: Udaya Sandhinand.

Important references: Mattingly 1958a: 4 (to genus); Thurman 1959b (Thailand); Macdonald & Mattingly 1960 (tax.); Reinert 1990 (tax.); Reinert 2000c (F gen.*); Reinert, Harbach, & Kitching 2004 (class., phyl.); Rattanarithikul & Harrison et al. 2005 (F, L; tax., keys, bion.; Thailand); Reinert & Harbach 2005a: 227 (gender of *Udaya* corrected to masculine); Reinert, Harbach, & Kitching 2009 (class., phyl.); Rattanarithikul et al. 2010 (F*, L*; keys, tax.; Thailand); Wilkerson et al. 2015 (phyl., class.); Soghigian et al. 2017 (mol. phyl.).

argyrurus (Edwards)
1934 (F; *Paraedes*). In Barraud 1934: 488. **Type loc.:** Nongpoh, Assam, India (NHMUK). **Distr.:** India, Malaysia, Singapore, Thailand. **Refs.:** Thurman 1954b: 83 (M*, F); Macdonald & Mattingly 1960: 26 (P*, L*); Reinert 2000c: Fig. 1 (F*); Rattanarithikul et al. 2010 (F*, L*; keys, bion., distr.; Thailand). **Ety.:** silver (Gr); thoracic pleura and sides of abdomen. **Informal name:** Silver Indian Compressed Mosquito.

lucaris Macdonald & Mattingly
1960: 22 (M*, F, P*, L*). **Type loc.:** Ulu Gombak Forest Reserve, Selangor, Malaya [Malaysia] (NHMUK). **Distr.:** Malaysia. **Refs.:** Mattingly 1971a: Pl. 5 (A wing*), Pl. 31 (L head*, terminal segs.*); Reinert 1989 (type info., descr.). **Ety.:** n.s. [sacred grove (L)]. **Informal name:** Sacred Malaysian Compressed Mosquito.

subsimilis (Barraud)
1927c: 552 (M*; *Finlaya*). **Type loc.:** Sukna, Darjeeling District (base of Eastern Himalayas), [West] Bengal, India (NHMUK). **Distr.:** India. **Ety.:** nearly (L); near *dissimilis*. **Informal name:** Similar Indian Compressed Mosquito.

Genus *Verrallina* Theobald

(Vol. 1, p. 238)

Important references: Blanchard 1905: 417; Leicester 1908: 196 (as genus); Laffoon 1946: 228 (keys; Philippines); King & Hoogstraal 1947a: 113 (keys; New Guinea); Belkin 1962: 412 (from syn.; keys; Papuan Subregion); Delfinado 1967 (keys); Delfinado 1968 (keys); Huang 1968a (Papuan Subregion); Huang 1968b: 169 (subg. keys; Papuan Subregion); Reinert 1974a (review; as subg. of genus *Aedes*); Reinert 1984 (tax.; Sri Lanka); Lee et al. 1987b: 250 (tax., keys, bion., distr.; Australasian Region); Reinert 1999c (to genus; tax., class.); Reinert 2001c (F gen.); Reinert, Harbach, & Kitching 2004 (class., phyl.); Rattanarithikul & Harrison et al. 2005 (F, L; tax., keys, bion.; Thailand); Reinert, Harbach, & Kitching 2009 (class., phyl.); Rattanarithikul et al. 2010 (tax., keys, bion.; Thailand); Wilkerson et al. 2015 (phyl., class.); Soghigian et al. 2017 (mol. phyl.).

Subgenus *Harbachius* Reinert

1999c: 18. Type species: *Aedes yusafi* Barraud.

Name derivation: Ralph E. Harbach.

Important references: Reinert 1974a (review; subg. of genus *Aedes*); Reinert 2001c (F gen.*); Reinert, Harbach, & Kitching 2004 (class., phyl.); Reinert, Harbach, & Kitching 2009 (class., phyl.); Rattanarithikul et al. 2010 (tax., keys, bion.; Thailand); Soghigian et al. 2017 (mol. phyl.).

abdita (Barraud)
1931a: 614 (F*; *Aedes* (*Aedes*) *abditus*). **Type loc.:** Sukna [Darjeeling District, West] Bengal, India (NHMUK). **Distr.:** Cambodia, India, Thailand. **Refs.:** Barraud 1934 (F*); Reinert 1974a: 18 (F*; transfer); Reinert 1999c: 22 (to genus *Verrallina*). **Ety.:** n.s. [hidden, concealed (L)]. **Informal name:** Concealed Indian Broadtail Mosquito.

consonensis (Reinert)
1973c: 252 (M*, P*, L*; *Aedes*). **Type loc.:** Con Son, Con Son Province, South Vietnam (USNM). **Distr.:** Vietnam. **Refs.:** Reinert 1974a: 38 (M*, P*, L*; from *Neomacleaya*); Reinert 1999c: 22 (to genus *Verrallina*). **Ety.:** Con Son Prov., Vietnam. **Informal name:** Con Son Vietnamese Broadtail Mosquito.

fragilis Leicester
1908: 199 (M). **Type loc.:** Kuala Lumpur [Selangor], Malaya [Malaysia] (NHMUK). **Distr.:** Cambodia, Malaysia. **Refs.:** Edwards 1917: 222 (M*); Delfinado 1968: 15 (M*, F*; syn.); Basio 1971b: 74 (M*; bion.; as *malayi*); Klein 1973: 5 (F*); Reinert 1974a: 43 (M*, F*; from *Neomacleaya*); Reinert 1999c: 22 (to genus *Verrallina*). **Ety.:** n.s. [fragile (L)]. **Informal name:** Fragile Malaysian Broadtail Mosquito.
syn. *malayi* Leicester
1908: 198 (F). **Type loc.:** Gombak Road, Kuala Lumpur, Malaya [Malaysia] (NHMUK). **Refs.:** Laffoon 1946: 238 (tax.).
syn. imitator Leicester
1908: 201 (F). **Type loc.:** Kuala Lumpur [Selangor], Malaya [Malaysia] (NHMUK). **Refs.:** Stone et al. 1959 (syn.).

hamistylus (Laffoon)
1946: 232 (M*, F*, L*; *Aedes*). **Type loc.:** Cape Melville, Balabac Island, Philippines (USNM). **Distr.:** Indonesia, Malaysia, Philippines. **Refs.:** Delfinado 1968: 16 (M*, F*, L*); Basio 1971b: 22 (M*; bion.); Reinert 1974a: 45 (M*, F*; from *Neomacleaya*; distr.); Reinert 1999c: 22 (to genus *Verrallina*). **Ety.:** n.s. [hooked stylus (L)]. **Informal name:** Hooked Philippine Broadtail Mosquito.

indecorabilis Leicester
1908: 200 (M, F). **Type loc.:** Kuala Lumpur, Malaya [Malaysia] (NHMUK). **Distr.:** Cambodia, Malaysia, Thailand. **Refs.:** Delfinado 1967: 25 (M*, F*); Reinert 1974a: 52 (F*; from *Neomacleaya*; distr.); Reinert 1999c: 22 (to genus *Verrallina*); Rattanarithikul et al. 2010 (F*, L*; keys, bion., distr.; Thailand). **Ety.:** n.s. [unbecoming (L); prob. ref. to overall dark, dingy appearance]. **Informal name:** Dingy Malaysian Broadtail Mosquito.

nobukonis (Yamada)
1932: 228 (F; *Aedes*). **Type loc.:** Ōmura, Kyushu, Japan (IID). **Distr.:** Japan, Russia. **Refs.:** Sasa 1949 (M, F*); La Casse & Yamaguti 1950: 100 (F*); Hara 1957: 63 (F*); Delfinado 1968 (F*); Delfinado 1968: 18 (M*, F*, P*, L*; as *ishigakiensis*); Gutsevich et al. 1971: 313 (distr.); Moriya et al. 1973 (E*); Reinert 1974a: 69 (M, F*, E; from *Aedes*); Tanaka et al. 1975c: 224 (distr.); Tanaka et al. 1979: 439 (M*, F*, L*); Reinert 1999c: 22 (to genus *Verrallina*). **Ety.:** [Nobuko (author's daughter) (Harbach 2018c)]. **Informal name:** Omura Japanese Broadtail Mosquito.
syn. *ishigakiensis* Bohart
1956b: 32 (M*, F*, P, L*; *Aedes* (*Aedes*)). **Type loc.:** Mount Banna, Ishigaki Island [Ryukyu-Retto, Japan] (USNM). **Refs.:** Reinert 1974a: 69 (syn.).

pahangi (Delfinado)
1968: 31 (F*; *Aedes*). **Type loc.:** Kuala Terengan, Pahang, Malaya [Malaysia] (BBM). **Distr.:** Malaysia. **Refs.:** Reinert 1974a: 71 (F*; from *Neomacleaya*); Reinert 1999c: 22 (to genus *Verrallina*). **Ety.:** Pahang, Malaysia. **Informal name:** Pahang Malaysian Broadtail Mosquito.

ramalingami (Reinert)
1974: 78 (M*; *Aedes*). **Type loc.:** Kampung Maraba, Beaufort, Sabah [Malaysian Borneo] (USNM). **Distr.:** Malaysia. **Refs.:** Reinert 1999c: 22 (to genus *Verrallina*). **Ety.:** Shivaji Ramalingam. **Informal name:** Ramalingam Malaysian Broadtail Mosquito.

robertsi (Laffoon)
1946: 230 (M*, F*, L; *Aedes*). **Type loc.:** Tacloban, Leyte, Philippines (ANS). **Distr.:** Philippines. **Refs.:** Delfinado 1968: 35 (M*, F*, L*); Basio 1971b: 23 (M*; bion.); Reinert 1974a: 17 (from *Neomacleaya*); Reinert 1999c: 22 (to genus *Verrallina*). **Ety.:** H.R. Roberts. **Informal name:** Roberts Philippine Broadtail Mosquito.

srilankensis (Reinert)
1977: 366 (M*; *Aedes*). **Type loc.:** 18 mi northeast of Colombo, Yakkala, Western Province, Sri Lanka (USNM). **Distr.:** Sri Lanka. **Refs.:** Reinert 1984 (F*); Reinert 1999c: 22 (to genus *Verrallina*). **Ety.:** Sri Lanka. **Informal name:** Vedda Sri Lankan Broadtail Mosquito [the Vedda are an indigenous people of Sri Lanka].

stunga (Klein)
1973: 10 (F*; as *Aedes stungus*). **Type loc.:** Stung Chral, Kompong Sela, Cambodia (USNM). **Distr.:** Cambodia. **Refs.:** Reinert 1974a: 17 (from *Neomacleaya*); Reinert 1999c: 22 (to genus *Verrallina*). **Ety.:** Stung Chral, Kompong Sela, Cambodia. **Informal name:** Stung Chral Cambodian Broadtail Mosquito.

uniformis (Theobald)
1910a: 33 (F; *Skusea*). **Type loc.:** Pallode, 20 mi northeast of Trivandrum, Travancore, India (IM). **Distr.:** India, Nepal, Sri Lanka. **Refs.:** Barraud 1928a (F*); Barraud 1934: 281 (M*, F*); Harrison et al. 1974: 153 (distr.); Reinert 1984 (F*); Lee et al. 1987a: 321 (F; key, tax., bion., distr., review); Darsie et al. 1992 (distr.; Nepal); Reinert 1999c: 22 (to genus *Verrallina*). **Ety.:** n.s. [uniform appearance (L); poss. ref. to "legs uniformly deep brown"]. **Informal name:** Uniform Indian Broadtail Mosquito.

yusafi (Barraud)
1931a: 612 (F*; *Aedes*). **Type loc.:** Roshanara Gardens, Delhi, India (NHMUK). **Distr.:** India, Sri Lanka, Thailand. **Refs.:** Barraud 1934: 282 (F*); Reinert 1974a: 96 (M*, F*, P*, L*; from *Neomacleaya*; distr.); Reinert 1999c: 19 (M*, F*, P*, L*; to *Verrallina* (*Neomacleaya*)); Reinert 2001c: Fig. 1 (F*); Rattanarithikul et al. 2010 (F*, L*; keys, bion., distr.; Thailand). **Ety.:** Mohamed Yusaf. **Informal name:** Yusaf Indian Broadtail Mosquito.

syn. *siamensis* Delfinado
1968: 36 (M*, F*, P, L*; *Aedes* (*Neomacleaya*)). **Type loc.:** Lampang, Ban Pong Tao, Thailand (USNM). **Refs.:** Reinert 1974a: 96 (syn.); Reinert 1999c: 22 (to genus *Verrallina*).

Subgenus *Neomacleaya* Theobald

1907: 238 (as genus). Type species: *Neomacleaya indica* Theobald.

Name derivation: New (Gr) and *Macleaya*.

syn. *Aioretomyia* Leicester 1908: 185. Type species: *Aioretomyia varietas* Leicester (Brunetti 1914: 55).

Important references: Laffoon 1946: 228 (keys; Philippines); Delfinado 1967 (from syn.); Delfinado 1968 (keys; SE Asia); Reinert 1974a (review; subg. of genus *Aedes*); Reinert 1984 (tax.; Sri Lanka); Reinert 1999c: 23 (tax., rev.; from subg. of genus *Aedes* to subg. of *Verrallina*); Reinert 2001c (F gen.*); Reinert, Harbach, & Kitching 2004 (class., phyl.); Reinert, Harbach, & Kitching 2009 (class., phyl.); Rattanarithikul et al. 2010 (tax., keys, bion.; Thailand); Soghigian et al. 2017 (mol. phyl.).

adusta (Laffoon)
1946: 236 (M*, F; as *Aedes adustus*). **Type loc.:** Cape Melville, Balabac Island, Philippines (USNM). **Distr.:** Cambodia, Malaysia, Philippines, Thailand. **Refs.:** Knight & Hull 1953: 472 (tax.); Reinert 1974a: 18 (M*, P*, L*; from syn. with *andamanensis* and transfer; distr.); Reinert 1999c: 22 (to genus *Verrallina*); Rattanarithikul et al. 2010 (F*, L*; keys, bion., distr.; Thailand). **Ety.:** n.s. [brown (L)]. **Informal name:** Brown Philippine Broadtail Mosquito.

agrestis (Barraud)
1931a: 613 (F*; *Aedes*). **Type loc.:** Nilgiri Hills [Madras], India (NHMUK). **Distr.:** India. **Refs.:** Barraud 1934 (F*); Reinert 1974a: 20 (M*; from *Aedes*); Tewari & Hiriyan 1988 (M*); Tewari & Hiriyan 1994a: 36 (P*, L*); Reinert 1999c: 22 (to genus *Verrallina*). **Ety.:** n.s. [of the land, rural (L)]. **Informal name:** Rural Indian Broadtail Mosquito.

andamanensis (Edwards)
1922a: 272 (M*; *Aedes*). **Type loc.:** Andaman Islands (NHMUK). **Distr.:** Andaman Isl., Bangladesh, Cambodia, India, Indonesia, Malaysia, Philippines, Singapore, Thailand, Vietnam. **Refs.:** Barraud 1928b: 368 (M*, F* L; distr.); Barraud 1934: 290 (M*, F*; distr.); Knight & Hull 1953: 473 (tax.); Delfinado 1967: 9 (M*, F*, P*, L*; syn.?); Basio 1971b: 21 (M*; bion.); Reinert 1974a: 20 (M*, F*; from *Neomacleaya*; distr.); Ahmed 1987 (distr.; Bangladesh); Reinert 1999c: 22 (to genus *Verrallina*); Rattanarithikul et al. 2010 (F*, L*; keys, bion., distr.; Thailand). **Ety.:** Andaman Isl. **Informal name:** Rajendra Andamanese Broadtail Mosquito [Rajendra Chola I (1014–1042 C.E.) was an Indian Tamil ruler who added the Andaman Isl. to his empire].

syn. *simplex* Theobald
1908b: 291 (A; *Neomocleaya indica* var.; not Theobald 1903a). **Type loc.:** Sylhet, Assam, India (IM).

assamensis Bhattacharyya, Tewari, Prakash, Mohapatra, & Mahanta
2004 (M*, F*, P*, L*). **Type loc.:** Dihing Patkai Reserve Forest, Dibrugarh District, Assam, India (NHMUK). **Distr.:** Bangladesh, Cambodia, India, Indonesia, Japan, Nepal, People's Republic of China, Thailand, Vietnam.

Ety.: Assam, India. **Informal name:** Assam Indian Broadtail Mosquito.

atria (Barraud)
1928b: 371 (F*; *Aedes*). **Type loc.:** Nongpoh, Assam, India (NHMUK). **Distr.:** India, Thailand. **Refs.:** Barraud 1934 (F*); Delfinado 1967: 11 (M*, F*, P*, L*); Reinert 1999c: 22 (to genus *Verrallina*); Reinert & Harbach 2005a (*atrius* corr. to *atria*); Rattanarithikul et al. 2010 (F*, L*; keys, bion., distr.; Thailand). **Ety.:** n.s. [atrium (L); atrial plates distinct (= upper vaginal sclerites)]. **Informal name:** Atrium Indian Broadtail Mosquito.

atriisimilis (Tanaka & Mizusawa)
1973: 625 (M*, F*, L*; *Aedes*). **Type loc.:** Itokawarindo, Iriomote Island, Yaeyama Gunto, Ryukyu Islands [Japan] (NMST). **Distr.:** Japan. **Refs.:** Reinert 1974a: 17 (from *Neomacleaya*); Tanaka et al. 1975c: 224 (bion., distr.); Tanaka et al. 1979: 443 (M*, F*, L*); Reinert 1999c: 22 (to genus *Verrallina*); Tanaka 2003a: 25 (P*; tax.). **Ety.:** n.s. [resembling darkness (L); poss. ref. to overall dark color]. **Informal name:** Shaded Japanese Broadtail Mosquito [*atriisimilis* = "resembling darkness" = Shaded].

campylostylus (Laffoon)
1946: 240 (M*, F*, L*; *Aedes*). **Type loc.:** Balsahan River (near Iwahig), Palawan Island, Philippines (USNM). **Distr.:** Indonesia, Philippines. **Refs.:** Delfinado 1968: 13 (M*, F*, P*, L*); Basio 1971b: 21 (M*; bion.); Reinert 1974a: 28 (M*, F*; from *Neomacleaya*; distr.); Lee et al. 1987a: 264 (F; key, tax., bion., distr., review); Reinert 1999c: 28 (to genus *Verrallina*). **Ety.:** n.s. [crooked staff (Gr, L)]. **Informal name:** Crooked-staff Philippine Broadtail Mosquito.

cauta (Barraud)
1928b: 368 (M*, F*; as *Aedes cautus*). **Type loc.:** Tavargatti, Bombay, Deccan, India (NHMUK). **Distr.:** India. **Refs.:** Barraud 1934 (F*); Delfinado 1967: 12 (M*, F*); Reinert 1974a: 34 (M, F*; transfer; distr.). **Ety.:** n.s. [cautious (L)]. **Informal name:** Cautious Indian Broadtail Mosquito.

ceylonica (Edwards)
1917: 221 (M*; *Aedes*). **Type loc.:** Colombo, Ceylon [Sri Lanka] (NHMUK). **Distr.:** Sri Lanka. **Refs.:** Wijesundara 1951: 174 (F*); Reinert 1974a: 61(F*; transfer from genus *Aedes*); Reinert 1984 (F*); Reinert 1999c: 28 (to genus *Verrallina*). **Ety.:** Ceylon [Sri Lanka]. **Informal name:** Ceylon Sri Lankan Broadtail Mosquito.
syn. *lankaensis* Stone & Knight
1958: 69 (when both were included in genus *Aedes*, new name for *ceylonicus* Edwards, not *ceylonica* Theobald 1910; *Aedes* (*Aedes*) *lankaensis*). **Type loc.:** Colombo, Ceylon [Sri Lanka] (NHMUK).

clavata (Barraud)
1931a: 614 (M*; as *Aedes clavatus*). **Type loc.:** Sukna [Darjeeling District, West] Bengal, India (NHMUK). **Distr.:** Cambodia, India, Thailand, Vietnam. **Refs.:** Delfinado 1967: 13 (M*, F*); Klein 1973: 2 (distr.); Reinert 1974a: 34 (M*, F*, P*; from *Neomacleaya*; distr.); Reinert 1999c: 38 (M*, F*, P*, L*; to genus *Verrallina*); Rattanarithikul et al. 2010 (F*, L*; keys, bion., distr.; Thailand). **Ety.:** clubbed (L); proctiger with clubbed armlike processes]. **Informal name:** Clubbed Indian Broadtail Mosquito.

comata (Barraud)
1931a: 615 (F*; as *Aedes comatus*). **Type loc.:** Sukna [Darjeeling District, West Bengal, India (NHMUK). **Distr.:** India. **Refs.:** Barraud 1934 (F*); Reinert 1974a: 35 (M*; transfer from (*Aedes*)); Reinert 1999c: 38 (M*, F*, P*, L*; to genus *Verrallina*). **Ety.:** n.s. [shaggy (L); prob. ref. to setae on lower mesepimeron]. **Informal name:** Shaggy Indian Broadtail Mosquito.

comosa (Reinert)
1974: 36 (F*; as *Aedes comosus*). **Type loc.:** Kota Belud, Sabah [Malaysian Borneo] (USNM). **Distr.:** Malaysia. **Refs.:** Reinert 1999c: 38 (M*, F*, P*, L*; to genus *Verrallina*). **Ety.:** n.s. [having long hair? (L)]. **Informal name:** Long-haired Malaysian Broadtail Mosquito.

cretata (Delfinado)
1967: 14 (M*, F*, P*, L*; as *Aedes cretatus*). **Type loc.:** Tham Pho Ti Yan, Khon Kaen, Thailand (USNM). **Distr.:** Cambodia, Malaysia, Thailand. **Refs.:** Klein 1973: 3 (distr.); Reinert 1974a: 38 (M*, F*; from *Neomacleaya*; distr.); Reinert 1999c: 28 (to genus *Verrallina*); Rattanarithikul et al. 2010 (F*, L*; keys, bion., distr.; Thailand). **Ety.:** n.s. [chalked (L)]. **Informal name:** Chalky Thai Broadtail Mosquito.

cyrtolabis (Edwards)
1928a: 273 (M*; *Aedes*). **Type loc.:** Singapore (NHMUK). **Distr.:** Cambodia, Malaysia, Singapore, Thailand. **Refs.:** Delfinado 1967: 16 (M*, P*, L); Klein 1973: 3 (F*); Reinert 1974a: 39 (M*, F, P*, L*; from *Neomacleaya*); Reinert 1999c: 28 (to genus *Verrallina*); Rattanarithikul et al. 2010 (F*, L*; keys, bion., distr.; Thailand). **Ety.:** n.s. [curved lips? (Gr)]. **Informal name:** Curve-lipped Singaporean Broadtail Mosquito.

gibbosa (Delfinado)
1967: 20 (M*, F*, P*; as *Aedes gibbosus*). **Type loc.:** Hat Yai, Songkhla, Thailand (USNM). **Distr.:** Indonesia, Malaysia, Thailand. **Refs.:** Reinert 1974a: 44 (M*, F*, P, L*; type data amended; from *Neomacleaya*); Reinert 1999c: 28 (to genus *Verrallina*); Rattanarithikul et al. 2010 (F*, L*; keys, bion., distr.; Thailand). **Ety.:** n.s. [protuberance (L); poss. ref. to male genitalia character]. **Informal name:** Gibbous Thai Broadtail Mosquito.

harrisonica (Reinert)
1974: 46 (F*; *Aedes*). **Type loc.:** Megamendg Mountains, Java, Indonesia (USNM). **Distr.:** Indonesia. **Refs.:** Reinert

1999c: 28 (to genus *Verrallina*); Reinert & Harbach 2005a (*harrisonicus* corr. to *harrisonica*). **Ety.:** Bruce Arthur Harrison. **Informal name:** Harrison Indonesian Broadtail Mosquito.

hispida (Delfinado)
1967: 21 (F*; as *Aedes hispidus*). **Type loc.:** Nong Bua Sampor, Udorn, Thailand (USNM). **Distr.:** Thailand. **Refs.:** Reinert 1974a: 17 (from *Neomacleaya*); Reinert 1999c: 28 (to genus *Verrallina*); Rattanarithikul et al. 2010 (F*, L*; keys, bion., distr.; Thailand). **Ety.:** n.s. [hairy (L)]. **Informal name:** Hairy Thai Broadtail Mosquito.

indica (Theobald)
1907: 238 (F; *Neomacleaya indicus*). **Type loc.:** India (NHMUK). **Distr.:** India, Nepal, Sri Lanka. **Refs.:** Barraud 1928a (F*); Barraud 1934: 283 (M*, F*, L*); Reinert 1974a: 53 (M*, F*, P*, L*; lectotype desig.; from *Neomacleaya*; distr.); Reinert 1984 (F*); Darsie & Pradhan 1990, 1991 (F, L; tax., keys, bion., distr.; Nepal); Reinert 1999c: 28 (M*, F*, P*, L*; to genus *Verrallina*); Reinert 2001c: Fig. 2 (F*). **Ety.:** India. **Informal name:** Harappa Indian Broadtail Mosquito [Harappa was a major center for the ancient Harappan civilization of modern-day India, emerging circa 2600 B.C.E.].
syn. *mediofasciata* Theobald
1907: 544 (M*, F*; *Skusea*). **Type loc.:** India (NHMUK).

johnsoni (Laffoon)
1946: 235 (M*, F*, L; *Aedes*). **Type loc.:** Pasananca, near Zamboanga, Mindanao, Philippines (USNM). **Distr.:** Philippines. **Refs.:** Delfinado 1968: 19 (M*, F*, P*, L*); Basio 1971b: 22 (M*; bion.); Reinert 1974a: 58 (M*, F*; from *Neomacleaya*); Reinert 1999c: 28 (to genus *Verrallina*). **Ety.:** D.R. Johnson. **Informal name:** Johnson Philippine Broadtail Mosquito.

johorensis (Reinert)
1974: 58 (F*; *Aedes*). **Type loc.:** Pontian Besar, Johore, Malaysia (USNM). **Distr.:** Indonesia, Malaysia. **Refs.:** Reinert 1999c: 28 (to genus *Verrallina*). **Ety.:** Johore, Malaysia. **Informal name:** Johore Malaysian Broadtail Mosquito.

komponga (Klein)
1973: 8 (M*; as *Aedes kompongus*). **Type loc.:** Kirirom Hills, Kompong Speu, Cambodia (USNM). **Distr.:** Cambodia. **Refs.:** Reinert 1974a: 17 (from *Neomacleaya*); Reinert 1999c: 28 (to genus *Verrallina*). **Ety.:** Kirirom Hills, Kompong Speu, Cambodia. **Informal name:** Kompong Cambodian Broadtail Mosquito.

latipennis (Delfinado)
1967: 26 (M*, F*, P*, L*; *Aedes*). **Type loc.:** Nga Chang, Songkhla, Thailand (USNM). **Distr.:** Thailand. **Refs.:** Reinert 1974a: 62 (F*, P, L; type data amended; from *Neomacleaya*); Reinert 1999c: 28 (to genus *Verrallina*); Rattanarithikul et al. 2010 (F*, L*; keys, bion., distr.; Thailand). **Ety.:** n.s. [wide wing (L); poss. ref. to broad scales on alula]. **Informal name:** Wide-winged Thai Broadtail Mosquito.

leicesteri (Edwards)
1917: 222 (M*; *Aedes*). **Type loc.:** Kuala Lumpur [Selangor], Malaya [Malaysia] (NHMUK). **Distr.:** Indonesia, Malaysia. **Refs.:** Delfinado 1968: 21 (M*); Reinert 1974a: 63 (M*, F*, P*, L*; from *Neomacleaya*); Reinert 1999c: 28 (to genus *Verrallina*). **Ety.:** George Frederick Leicester. **Informal name:** Leicester Malaysian Broadtail Mosquito.

macrodixoa (Dyar & Shannon)
1925a: 79 (M; *Aedes*). **Type loc.:** Infanta, Tayabas [= Quezon], Luzon, Philippines (USNM). **Distr.:** Philippines. **Refs.:** Laffoon 1946: 233 (M*, F*); Delfinado 1968: 23 (M*, F*, L*); Basio 1971b: 22 (M*; bion.); Reinert 1974a: 67 (M*, F*; from *Neomacleaya*); Lee et al. 1987a: 293 (tax., review); Reinert 1999c: 28 (to genus *Verrallina*). **Ety.:** n.s. **Informal name:** Tayabas Philippine Broadtail Mosquito.

margarsen (Dyar & Shannon)
1925a: 80 (M, F; *Aedes*). **Type loc.:** Camp Eldridge, Laguna, Luzon, Philippines (USNM). **Distr.:** Philippines. **Refs.:** Bohart 1945 (F*); Laffoon 1946: 237 (M*, F; distr., keys; lectotype desig.); Stone & Knight 1956a: 220 (type info.); Delfinado 1968: 25 (M*, F*, P*, L*); Basio 1971b: 22 (M*; bion.); Reinert 1974a: 67 (M*, F*; from *Neomacleaya*); Reinert 1999c: 28 (to genus *Verrallina*). **Ety.:** n.s. **Informal name:** Margarsen Philippine Broadtail Mosquito.

neomacrodixoa (King & Hoogstraal)
1947a: 124 (M*, F; *Aedes*). **Type loc.:** Hollandia, Western New Guinea (Papua) (USNM). **Distr.:** Indonesia (West Papua), Papua New Guinea, Solomon Isl. **Refs.:** Huang 1968a: 41 (M*, F, P*, L*; keys); Delfinado 1968 (F*); Reinert 1974a: 68 (F*; distr.); Lee et al. 1987a: 298 (F; key, tax., bion., distr., review); Reinert 1999c: 28 (to genus *Verrallina*). **Ety.:** similar to *macrodixoa*. **Informal name:** Hollandia Indonesian Broadtail Mosquito.

nigrotarsis (Ludlow)
1908a: 52 (F; *Pseudoskusea*). **Type loc.:** Infanta, Tayabas [= Quezon], Philippines (USNM). **Distr.:** Philippines, Vietnam. **Refs.:** Bohart 1945 (F*); Laffoon 1946: 239 (M*, F*, L*); Delfinado 1968: 28 (M*, F*, P*, L*); Basio 1971b: 23 (M*; bion.); Reinert 1974a: 68 (M*, F*; from *Neomacleaya*; distr.); Reinert 1999c: 28 (to genus *Verrallina*). **Ety.:** black feet (L); ref. to tibiae and tarsi entirely dark brown. **Informal name:** Black-footed Philippine Broadtail Mosquito.

notabilis (Delfinado)
1967: 28 (F*; *Aedes*). **Type loc.:** Bang Lamung, Chon Buri, Thailand (USNM). **Distr.:** Cambodia, Thailand. **Refs.:** Klein 1973: 9 (distr.); Reinert 1974a: 70 (F*; holotype data amended; from *Neomacleaya*); Rattanarithikul et al. 2010 (F*, L*; keys, bion., distr.; Thailand). **Ety.:** n.s. [notable,

remarkable (L); poss. ref. to distinctive male terminalia]. **Informal name:** Noteworthy Thai Broadtail Mosquito.

nubicola (Laffoon)
1946: 237 (M*; *Aedes*). **Type loc.:** Sibulan River on Mount Apo, near Davao, Mindanao, Philippines (MCZ). **Distr.:** Philippines. **Refs.:** Basio 1971b: 23 (M*; bion.); Reinert 1974a: 17 (from *Neomacleaya*); Reinert 1999c: 28 (to genus *Verrallina*). **Ety.:** n.s. [cloud dweller (L); prob. ref. to collection at high altitude (7,000–8,000 ft)]. **Informal name:** Cloud-dwelling Philippine Broadtail Mosquito.

panayensis (Ludlow)
1914c: 159 (M, F; *Aedes*). **Type loc.:** Iloilo, Panay, Philippines (USNM). **Distr.:** Indonesia, Malaysia, Papua New Guinea, Philippines. **Refs.:** Bohart 1945 (F*); Laffoon 1946: 242 (M*, F*; lectotype desig); Delfinado 1968: 32 (M*, F*); Huang 1968a: 47 (M, F); Basio 1971b: 32 (M*; bion., keys); Reinert 1974a: 71 (M*, F*; distr.); Lien et al. 1977 (distr.); Lee et al. 1987a: 301 (F; key, tax., bion., distr., review); Reinert 1999c: 28 (to genus *Verrallina*). **Ety.:** Iloilo, Panay, Philippines. **Informal name:** Panay Philippine Broadtail Mosquito.

petroelephantus (Wijesundara)
1951: 176 (F*; *Aedes*). **Type loc.:** Kurunegala, North Western, Ceylon [Sri Lanka] (NHMUK). **Distr.:** Sri Lanka. **Refs.:** Reinert 1974a: 72 (transfer from (*Aedes*); lectotype desig.); Reinert 1984 (F*); Reinert 1999c: 28 (to genus *Verrallina*). **Ety.:** n.s. [elephant (L)]. **Informal name:** Elephant Sri Lankan Broadtail Mosquito.

philippinensis (Delfinado)
1968: 33 (M*, F*; *Aedes*). **Type loc.:** Mahaplag, Leyte, Philippines (BBM). **Distr.:** Japan, Philippines. **Refs.:** Basio 1971b: 23 (M*; bion.); Reinert 1974a: 73 (type not found in USNM; from *Neomacleaya*); Reinert 1999c: 28 (to genus *Verrallina*). **Ety.:** Philippine Isl. **Informal name:** Villalobos Philippine Broadtail Mosquito [Spanish explorer Ruy Lopez de Villalobos (1500–1544) named the archipelago Las Islas Filipinas in honor of King Philip II].

phnoma (Klein)
1973: 9 (M*; as *Aedes phnomus*). **Type loc.:** Chrui Changvar, Kandal, Cambodia (USNM). **Distr.:** Cambodia, Thailand. **Refs.:** Reinert 1974a: 73 (transfer; distr.); Reinert 1999c: 28 (to genus *Verrallina*); Rattanarithikul et al. 2010 (F*, L*; keys, bion., distr.; Thailand). **Ety.:** n.s. [prob. ref. to Kompong Sela, Phnom Praung rain forest, Cambodia]. **Informal name:** Phnom Praung Cambodian Broadtail Mosquito.

prioekanensis (Brug)
1931a: 249 (M*; *Aedes*). **Type loc.:** Air Prioekan, Sumatra [Indonesia] (NHMUK). **Distr.:** Indonesia. **Refs.:** Delfinado 1968 (F*); Reinert 1974a: 74 (M*; from *Neomacleaya*); Reinert 1999c: 28 (to genus *Verrallina*). **Ety.:** Air Prioekan, Sumatra. **Informal name:** Prioekan Indonesian Broadtail Mosquito.

protuberans (Delfinado)
1967: 29 (M*; *Aedes*). **Type loc.:** Luang, Nakhon Si Thammarat, Thailand (USNM). **Distr.:** Thailand. **Refs.:** Reinert 1974a: 17 (from *Neomacleaya*); Reinert 1999c: 28 (to genus *Verrallina*); Rattanarithikul et al. 2010 (F*, L*; keys, bion., distr.; Thailand). **Ety.:** n.s. [swell, bulge out (L); poss. ref. to "distimere swollen near middle"]. **Informal name:** Protuberant Thai Broadtail Mosquito.

pseudodiurna (Theobald)
1910a: 32 (M; *Skusea*). **Type loc.:** Sukna, Darjeeling District, base of Eastern Himalayas, 500 ft, [West Bengal], India (IM). **Distr.:** India, Thailand. **Refs.:** Barraud 1934: 295 (M*); Reinert 1974a: 16 (from *Neomacleaya*); Reinert 1999c: 28 (to genus *Verrallina*); Rattanarithikul et al. 2010 (F*, L*; keys, bion., distr.; Thailand). **Ety.:** false (Gr); similar to *diurna* (*Skusea*). **Informal name:** Twilight Indian Broadtail Mosquito [*pseudodiurna* = "false day" = Twilight].

pseudomediofasciata (Theobald)
1910c: 489 (M*; *Skusea*). **Type loc.:** Peradeniya [Central] and Hakgala, Ceylon [Sri Lanka] (NHMUK). **Distr.:** India, Pakistan, Sri Lanka, Thailand. **Refs.:** Senior-White 1927: 76 (L*); Barraud 1928a (F*); Barraud 1934: 286 (M*, F*); Wijesundara 1951 (F); Reinert 1974a: 75 (M*, F*; transfer from *Aedes*); Reinert 1978 (P*, L*; bion.); Reinert 1984: 45 (F*; lectotype desig.); Lee et al. 1987a: 305 (tax., review); Reinert 1999c: 28 (to genus *Verrallina*). **Ety.:** false (Gr); similar to *mediofasciata*. **Informal name:** False-banded Sri Lankan Broadtail Mosquito.

pseudovarietas (Reinert)
1974: 76 (M*; *Aedes*). **Type loc.:** Matunggong, Kudat, Sabah [Malaysian Borneo] (USNM). **Distr.:** Malaysia. **Refs.:** Reinert 1999c: 28 (to genus *Verrallina*). **Ety.:** false (Gr); similar to *varietas*. **Informal name:** Hypocritical Malaysian Broadtail Mosquito [false varied = "false variety" = Hypocritical].

rami (Barraud)
1928b: 370 (M*; *Aedes*). **Type loc.:** Golaghat, Upper Assam, India (NHMUK). **Distr.:** India. **Refs.:** Barraud 1934: 295 (M*); Reinert 1974a: 80 (F*; from *Aedes*); Reinert 1999c: 28 (to genus *Verrallina*). **Ety.:** n.s. [a branch (L)]. **Informal name:** Ram Indian Broadtail Mosquito.

rara (Delfinado)
1968: 35 (M*; as *Aedes rarus*). **Type loc.:** Tawau, North Borneo [Sabah, Malaysian Borneo] (NHMUK). **Distr.:** Cambodia, Indonesia, Malaysia. **Refs.:** Klein 1973 (F*); Reinert 1974a: 81 (M, F*; from *Neomacleaya*; distr.). **Ety.:** n.s. [rare, few, infrequent (L); prob. ref. to an apparently rare species]. **Informal name:** Rare Malaysian Broadtail Mosquito.

syn. *khmerus* Klein
1973: 7 (M*, F*; *Aedes*). **Type loc.:** Sre Klong, Kompong Speu, Cambodia (USNM). **Refs.:** Reinert 1974a: 81 (syn.); Reinert 1999c: 28 (to genus *Verrallina*).

sabahensis (Reinert)
1974: 82 (F*; *Aedes*). **Type loc.:** Kampung Bingul, Beaufort, Sabah [Malaysian Borneo] (USNM). **Distr.:** Malaysia. **Refs.:** Reinert 1999c: 28 (to genus *Verrallina*). **Ety.:** Sabah, Malaysia. **Informal name:** Sabah Malaysian Broadtail Mosquito.

seculata (Menon)
1950: 139 (M*; *Aedes*). **Type loc.:** Pathanamthitta, Travancore, India (NHMUK). **Distr.:** India, Sri Lanka. **Refs.:** Wijesundara 1951 (F*); Reinert 1974a: 84 (transfer from (*Aedes*); Reinert 1999c: 28 (to genus *Verrallina*). **Ety.:** sickle (L); sickle shaped gonostylus. **Informal name:** Sickle Indian Broadtail Mosquito.

singularis (Leicester)
1908: 188 (M; *Aioretomyia*). **Type loc.:** Kuala Lumpur [Selangor], Malaya [Malaysia] (NHMUK). **Distr.:** Borneo (island), Malaysia. **Refs.:** Edwards 1922a: 265 (M*); Delfinado 1968: 38 (F*, M*); Reinert 1974a: 17 (from *Neomacleaya*); Reinert 1999c: 28 (to genus *Verrallina*). **Ety.:** n.s. [alone, unique, single (L); poss. ref. to unique male genitalia]. **Informal name:** Singular Malaysian Broadtail Mosquito.

sohni (Reinert)
1974: 85 (M*; *Aedes*). **Type loc.:** Pensiangan, Sapulut, Sabah, Malaysia (USNM). **Distr.:** Malaysia, Thailand. **Refs.:** Reinert 1999c: 28 (to genus *Verrallina*); Rattanarithikul et al. 2010 (F*, L*; keys, bion., distr.; Thailand). **Ety.:** Young T. Sohn. **Informal name:** Sohn Malaysian Broadtail Mosquito.

spermatheca (Wijesundara)
1951: 175 (F*; *Aedes*). **Type loc.:** Weligama, Southern, Ceylon [Sri Lanka] (NHMUK). **Distr.:** Sri Lanka. **Refs.:** Reinert 1974a: 17 (from subg. *Aedes*); Reinert 1984 (F*); Townsend 1990: 136 (syntype; change MRI to NHMUK). **Ety.:** n.s. [spermatheca (L); morph. ref.?]. **Informal name:** Seed-like Sri Lankan Broadtail Mosquito.

taeniata (Leicester)
1908: 190 (F; *Aioretomyia*). **Type loc.:** Kuala Lumpur [Selangor], Malaya [Malaysia] (NHMUK). **Distr.:** Cambodia, Indonesia, Malaysia, Thailand. **Refs.:** King & Hoogstraal 1947a: 115 (tax.; as *incertus*); Delfinado 1967: 22 (F*, P*, L*; as *incertus*), 17 (M*, P*; as *dermajoensis*); Huang 1968a: 68 (distr.; as *incertus*); Klein 1973: 4 (M*, F*; as *dermajoensis*); Reinert 1974a: 48 (M, F*, P, L; from *Neomacleaya*; distr.; as *incertus*); Lee et al. 1987a: 283 (F; key, tax., bion., distr., review; as *incertus*); Rattanarithikul et al. 2010 (F*, L*; keys, bion., distr.; Thailand; as *incerta* (Edwards)). **Ety.:** ribbon, band (L); banded abdomen. **Informal name:** Ribboned Malaysian Broadtail Mosquito.

syn. *incertus* Edwards
1922a: 264, 1922b: 468 (new name for *taeniata* Leicester 1908, not Wiedemann 1828; *Aedes (Aedes) incertus* [replacement name needed when both were in genus *Aedes*]). **Refs.:** Reinert 1999c: 28 (to genus *Verrallina*).

syn. *dermajoensis* Brug
1931a: 250 (M*; *Aedes*). **Type loc.:** Dermajoe [Benkoelen], Sumatra [Indonesia] (NHMUK). **Refs.:** Reinert 1974a: 48 (syn.); Reinert 1999c: 28 (to genus *Verrallina*).

torosa (Delfinado)
1967: 30 (M*; as *Aedes torosus*). **Type loc.:** Tham Pho Ti Yan, Khon Kaen, Thailand (USNM). **Distr.:** Thailand. **Refs.:** Reinert 1974a: 87 (M*, P*, L*; from *Neomacleaya*); Reinert 1999c: 28 (to genus *Verrallina*); Rattanarithikul et al. 2010 (F*, L*; keys, bion., distr.; Thailand). **Ety.:** n.s. [muscular, brawny (L)]. **Informal name:** Muscular Thai Broadtail Mosquito.

unca (Theobald)
1901c: 53 (F*; as *Culex uncus*). **Type loc.:** Klang jungle, Selangor, Malaya [Malaysia] (NHMUK). **Distr.:** Cambodia, India, Indonesia, Malaysia, Philippines, Thailand, Vietnam. **Refs.:** Leicester 1908: 185 (F; only of type series; as *varietas*); Barraud 1928a (F*); Barraud 1934: 291 (M*, F*; as *hirsutipleura*); Bohart 1945 (F*); Laffoon 1946: 237 (M*, F*); Knight & Hull 1951b: 222 (L); Knight & Hull 1953: 474 (syn.); Delfinado 1967: 30 (M*, F*, P*); Basio 1971b: 23 (M*; bion.; as *uncus*); Klein 1973: 11 (distr.); Reinert 1974a: 88 (M*, F*, L*; from *Neomacleaya*); Reinert 1999c: 28 (L*; to genus *Verrallina*); Reinert & Harbach 2005a (name corrected to *uncus*); Rattanarithikul et al. 2010 (F*, L*; keys, bion., distr.; Thailand; as *uncus*). **Ety.:** (n.s.) [hooked, curved (L); poss. ref. to curved scales on head]. **Informal name:** Hooked Malaysian Broadtail Mosquito.

syn. *hirsutipleura* Barraud
1928b: 370 (M*, F*; *Aedes (Aedes)*). **Type loc.:** Golaghat, Upper Assam, India (NHMUK). **Refs.:** Reinert 1974a: 88 (syn.; lectotype desig.).

vallistris (Barraud)
1928b: 369 (M*, F*; *Aedes*). **Type loc.:** Golaghat, Upper Assam, India (NHMUK). **Distr.:** Cambodia, India, Myanmar, Thailand. **Refs.:** Barraud 1934: 290 (M*, F*); Iyengar & Menon 1956: 791 (M*); Delfinado 1967: 32 (M*, F*; key); Klein 1973: 11 (distr.); Reinert 1974a: 91 (M*, F*, P*, L*; from *Neomacleaya*); Reinert 1999c: 28 (to genus *Verrallina*); Rattanarithikul et al. 2010 (F*, L*; keys, bion., distr.; Thailand). **Ety.:** n.s. [valley?]. **Informal name:** Valley Indian Broadtail Mosquito.

varietas (Leicester)
1908: 185 (M only; *Aioretomyia*). **Type loc.:** Ulu Gombak, Pangkor Laut and Klang, all near Kuala Lumpur, Malaysia. **Distr.:** Indonesia, Malaysia, Singapore, Thailand. **Refs.:** Edwards 1917: 222 (M*); Knight & Hull 1953: 474 (type info.; tax.); Delfinado 1968: 39 (M*, F*; keys); Reinert 1974a:

92 (F*; from *Neomacleaya*; distr.); Townsend 1990: 147 (unjustified lectotype desig.); Reinert 1999c: 28 (to genus *Verrallina*). **Ety.:** n.s. [variety, difference, mottled (L)]. **Informal name:** Different Malaysian Broadtail Mosquito.

virilis Leicester
1908: 197 (M). **Type loc.:** Kuala Lumpur [Selangor], Malaya [Malaysia] (NHMUK). **Distr.:** Indonesia, Malaysia. **Refs.:** Edwards 1917: 222 (M*); Reinert 1974a: 92 (M*, F*; from *Neomacleaya*; distr.); Reinert 1999c: 28 (to genus *Verrallina*). **Ety.:** n.s. [manly, virile, mature (L)]. **Informal name:** Manly Malaysian Broadtail Mosquito.

yerburyi (Edwards)
1917: 222 (M*; *Aedes*). **Type loc.:** Kitli Station, Ceylon [Sri Lanka] (NHMUK). **Distr.:** India, Sri Lanka. **Refs.:** Barraud 1934: 293 (M*); Wijesundara 1951: 174 (F*); Reinert 1974a: 95 (M*, F*; transfer from subg. *Aedes* to subg. *Verrallina*); Reinert 1984 (F*); Reinert 1999c: 28 (to genus *Verrallina*); Kaur 2003: 455 (distr.; India). **Ety.:** Lt. Col. Yerbury. **Informal name:** Yerbury Sri Lankan Broadtail Mosquito.

Subgenus *Verrallina* Theobald

Theobald 1903a: 295 (as genus, subg.). Type species *Aedes butleri* Theobald.

Name derivation: n.s. [prob. for George Henry Verrall].

syn. *Verralina* Lutz 1904 (synoptic table; incorrect spelling of *Verrallina*). In Bourroul 1904: 45.

Important references: Blanchard 1905: 417 (type species desig.); Leicester 1908: 196 (as genus); Laffoon 1946: 228 (keys; Philippines); King & Hoogstraal 1947a: 113 (keys; New Guinea); Belkin 1962: 412 (tax., bion., keys; South Pacific); Huang 1968a (*Aedes* (*Verrallina*) spp., keys, tax.; Papuan Subregion); Huang 1968b (key to *Aedes* subg.; Papuan Subregion); Reinert 1974a (tax., keys); Reinert 1984 (review, tax.; Sri Lanka); Lee et al. 1987b: 250 (tax., keys, bion., distr.; Australasia); Reinert 1999c (tax., to genus); Reinert 2001c (F gen.*); Reinert et al. 2004 (phyl., class.); Reinert et al. 2009 (phyl., class.); Rattanarithikul et al. 2010 (L*, F*; tax. keys, bion.; Thailand); Wilkerson et al. 2015 (phyl., class.); Soghigian et al. 2017 (phyl., class.).

azureosquamata (Bonne-Wepster)
1948: 320 (F*; *Aedes*). **Type loc.:** Bernhard Camp, Western New Guinea (Papua) (ITH). **Distr.:** Indonesia (includes West Papua). **Refs.:** Huang 1968a: 15 (M, F; keys); Lee et al. 1987a: 259 (F; key, tax., bion., distr., review; type dep. change from LM to ITH); Reinert 1999c: 38 (to genus *Verrallina*). **Ety.:** n.s. [prob. ref. to blue scaled (L); wings]. **Informal name:** Blue-scaled Indonesian Broadtail Mosquito.

bancrofti (Taylor)
1914b: 465 (M, F*; *Skusea*). **Type loc.:** Eidsvold, Queensland, Australia (US). **Distr.:** Australia. **Refs.:** Reinert 1974a: 17 (from subg. *Aedes*); Lee et al. 1987a: 271 (F; key, tax., bion., distr., review; as *cunninghami*); Reinert 1999c: 34 (to genus *Verrallina*). **Ety.:** Thomas Lane Bancroft. **Informal name:** Bancroft Australian Broadtail Mosquito.
syn. *cunninghami* Taylor
1944: 123 (new name for *bancrofti* Taylor, not Skuse 1889; *Aedes* (*Aedes*) *cunninghami*). **Type loc.:** Eidsvold, Queensland, Australia (US).

bifoliata (King & Hoogstraal)
1947: 120 (M*; as *Aedes bifoliatus*). **Type loc.:** Hollandia, Western New Guinea (Papua) (USNM). **Distr.:** Indonesia (West Papua). **Refs.:** Huang 1968a: 16 (M, F; keys); Lee et al. 1987a: 260 (tax., review); Reinert 1999c: 38 (to genus *Verrallina*). **Ety.:** two-leafed (L); ref. to gonocoxite with two broadened leaflets. **Informal name:** Two-leafed Indonesian Broadtail Mosquito.

butleri (Theobald)
1901c: 230 (F; *Aedes*). **Type loc.:** Klang jungle, Selangor, Malaya [Malaysia] (NHMUK). **Distr.:** Australia, Borneo, Brunei, Cambodia, India, Indonesia, Malaysia, Myanmar, Philippines, Singapore, Sri Lanka, Thailand, Vietnam. **Refs.:** Barraud 1928a (F*); Barraud 1934: 296 (F*; only); Laffoon 1946: 244 (tax.), 243 (M*, F*, L*; as *umbrosus*); Wijesundara 1951 (F); Knight & Hull 1953: 476 (tax.); Iyengar & Menon 1956: 790 (M*, F*); Huang 1968a: 16 (M, F, P, L; keys); Basio 1971b: 32 (M*; bion.); Reinert 1974a: 22 (M*, F*, P*, L*; distr.); Matsuo, Yoshida, & Kunou 1974 (E*); Lien et al. 1975 (distr.); Reinert 1984 (F*); Lee et al. 1987a: 261 (F; key, tax., bion., distr., review); Reinert 1999c: 34 (M*, F*, P*, L*; to genus *Verrallina*); Reinert 2001c: Fig. 3 (F*); Rattanarithikul et al. 2010 (F*, L*; keys, bion., distr.; Thailand). **Ety.:** A.L. Butler. **Informal name:** Butler Malaysian Broadtail Mosquito.
syn. *diurna* Theobald
1903f: 259 (F; *Skusea*). **Type loc.:** Jugra, Kuala Lumpur [Selangor], Malaya (NHMUK). **Refs.:** Edwards 1913c: 229 (syn.).
syn. *hatiensis* Carter
1910b: 275 (F*; *Stegomyia*). **Type loc.:** Saigon and Ha-tien [Cochin China], Indochina [Vietnam] (NHMUK). **Refs.:** Edwards 1913c: 229 (syn.).
syn. *umbrosus* Brug
1924: 437 (M*, F; *Aedes* (*Skusea*)). **Type loc.:** Tanah Grogot, South East Division, Borneo (NHMUK). **Refs.:** Knight & Hull 1953: 476 (syn.).
syn. *carteri* Wijesundara
1951: 176 (M*, F*; not Edwards 1936; *Aedes* (*Aedes*) *carteri*). **Type loc.:** Matugama, Kalutara District, Sri Lanka (NHMUK). **Refs.:** Stone 1957a: 339 (syn. of *seculatus*); Reinert 1974a: 22 (to syn. with *butleri*).

carmenti (Edwards)

1924b: 388 (F; *Aedes*). **Type loc.:** Maravovo, Guadalcanal, Solomon Islands (NHMUK). **Distr.:** Australia, Indonesia, Papua New Guinea, Solomon Isl. **Refs.:** King & Hoogstraal 1947a: 115 (M*, F*, L*; lectotype desig.); Belkin 1962: 416 (M*, F, P*, L*); Huang 1968a: 18 (M*, F, P*, L*, E*); Reinert 1974a: 28 (M*, F*, P*, L*; distr.); Lee et al. 1987a: 265 (F; key, tax., bion., distr., review); Reinert 1999c: 38 (M*, F*, P*, L*; to genus *Verrallina*); Reinert 2001c: Fig. 4 (F*). **Ety.:** A.G. Carment. **Informal name:** Carment Solomon Islander Broadtail Mosquito.

syn. *ceramensis* Brug

1934a: 512 (M*; *Aedes* (*Aedes*)). **Type loc.:** Oewin and Warasiwa, Ceram, Moluccas (NHMUK). **Refs.:** King & Hoogstraal 1947a: 115 (syn.); Belkin 1962: 417 (from syn.); Reinert 1974a: 28 (syn.; lectotype desig.).

cuccioi (Belkin)

1962: 414 (M*, F, P*, L*; *Aedes*). **Type loc.:** Matanikau Valley, Guadalcanal, Solomon Islands (USNM). **Distr.:** Papua New Guinea, Solomon Isl. **Refs.:** Huang 1968a: 23 (M, F, P, L; distr.; *Aedes*); Lee et al. 1987a: 270 (F; key, tax., bion., distr., review); Reinert 1999c: 38 (to genus *Verrallina*). **Ety.:** J. Cuccio. **Informal name:** Cuccio Solomon Islander Broadtail Mosquito.

dux (Dyar & Shannon)

1925a: 81 (M, F; *Aedes*). **Type loc.:** Camp Stotsenberg [Angeles], Pampanga, Luzon, Philippines (USNM). **Distr.:** Cambodia, India, Laos, Malaysia, People's Republic of China, Philippines, Thailand, Vietnam. **Refs.:** Barraud 1928a (F*); Barraud 1934 (F*); Bohart 1945 (F*); Laffoon 1946: 233 (M*, F*, L*); Iyengar & Menon 1956 (F*); Chu 1957 (F*); Delfinado 1967: 18 (M*, F*, P*, L*); Delfinado 1968 (F*); Basio 1971b: 21 (M*; bion.); Reinert 1974a: 41 (M*, F*, E*; from *Neomacleaya*; distr.); Lien et al. 1977 (distr.); Reinert 1999c: 34 (to genus *Verrallina*); Rattanarithikul et al. 2010 (F*, L*; keys, bion., distr.; Thailand). **Ety.:** n.s. [leader? (L)]. **Informal name:** Leader Philippine Broadtail Mosquito.

syn. *sigmoides* Barraud

1928b: 373 (F*; *Aedes* (*Aedes*)). **Type loc.:** Port Blair, Andaman Islands (NHMUK). **Refs.:** Causey 1937b: 414 (syn.); Iyengar & Menon 1956: 789 (M*).

embiensis (Huang)

1968a: 24 (M*; *Aedes*). **Type loc.:** Embi Lakes, Popondetta, New Guinea (BBM). **Distr.:** Papua New Guinea. **Refs.:** Reinert 1974a: 42 (M*; tax.); Lee et al. 1987a: 273 (tax., review); Reinert 1999c: 38 (to genus *Verrallina*). **Ety.:** Embi Lakes, Popondetta, SE New Guinea. **Informal name:** Embi Papuan Broadtail Mosquito.

foliformis (King & Hoogstraal)

1947: 121 (M*, F; *Aedes*). **Type loc.:** Gusika, Finschhafen area, NE New Guinea (USNM). **Distr.:** Indonesia, Papua New Guinea. **Refs.:** Huang 1968a: 26 (M*, F, P*, L*); Lee et al. 1987a: 274 (F; key, tax., bion., distr.); Reinert 1999c: 38 (to genus *Verrallina*). **Ety.:** n.s. [leaf form (L)]. **Informal name:** Leaflike Papuan Broadtail Mosquito.

funerea (Theobald)

1903a: 292 (F; as *Skusea funereus*). **Type loc.:** Queensland, Australia (NHMUK). **Distr.:** Australia, Indonesia, Papua New Guinea. **Refs.:** Bancroft 1908: 55 (F; as genus *Skusea*); Lee 1944a: 76 (L*); King & Hoogstraal 1947a: 117 (M*, F); Penn 1949b: 62 (P); Huang 1968a: 29 (M, F, P, L, E*; keys); Lee et al. 1987a: 276 (F; key, tax., bion., distr., review); Reinert 1999c: 22 (to genus *Verrallina*). **Ety.:** n.s. [funereal (L); poss. ref. to dark color]. **Informal name:** Funereal Australian Broadtail Mosquito.

syn. *basalis* Taylor

1912b: 27 (F; *Pseudoskusea*). **Type loc.:** Cooktown, Queensland, Australia (US). **Refs.:** Edwards 1924b: 388 (syn.).

iriomotensis (Tanaka & Mizusawa)

1973: 633 (F*; *Aedes*). **Type loc.:** Funaura, Iriomote Island, Yaeyama Gunto, Ryukyu Islands [Japan] (NSMT). **Distr.:** Japan. **Refs.:** Reinert 1974a: 17 (from *Neomacleaya*); Tanaka et al. 1975c: 224 (bion., distr.); Miyagi & Toma 1979 (M*, P*, L*); Tanaka et al. 1979: 442 (F*); Reinert 1999c: 34 (to genus *Verrallina*). **Ety.:** Iriomote Isl., Japan. **Informal name:** Iriomote Japanese Broadtail Mosquito.

killertonis (Huang)

1968a: 31 (M*, F*; keys; *Aedes*). **Type loc.:** Cape Killerton, SE New Guinea (BBM). **Distr.:** Papua New Guinea. **Refs.:** Lee et al. 1987a: 284 (F; key, tax., bion., distr., review); Reinert 1999c: 38 (to genus *Verrallina*). **Ety.:** Cape Killerton, SE New Guinea. **Informal name:** Killerton Papuan Broadtail Mosquito.

leilae (King & Hoogstraal)

1947: 121 (M*; keys; *Aedes*). **Type loc.:** Hollandia, Western New Guinea (Papua) (USNM). **Distr.:** Australia, Indonesia (West Papua), Papua New Guinea. **Refs.:** Lee et al. 1987a: 285 (F; key, tax., bion., distr., review); Reinert 1999c: 38 (to genus *Verrallina*). **Ety.:** n.s. **Informal name:** Leila Indonesian Broadtail Mosquito.

lineata (Taylor)

1914d: 191 (F; as *Lepidotomyia lineatus*). **Type loc.:** Lakekamu Gold Field, Mekeo District, Papua, New Guinea (US). **Distr.:** Australia, Indonesia, Papua New Guinea, Solomon Isl., Vanuatu. **Refs.:** King & Hoogstraal 1947a: 118 (M*, F, L*); Penn 1949b: 62 (P*); Bonne-Wepster 1954: 94 (M, F*, L*); Belkin 1962: 418 (M*, F, P*,

L*); Huang 1968a: 33 (M*, F, P*, L*, E*; keys); Rodhain & Gaxotte 1977 (distr.); Lee et al. 1987a: 290 (F; key, tax., bion., distr., review); Reinert 1999c: 38 (to genus *Verrallina*). **Ety.:** lines (L); ref. to a median and two lateral lines on scutum. **Informal name:** Lined Papuan Broadtail Mosquito.

syn. *ornata* Theobald
1905f: 79 (F; not Meigen 1818; *Skusea funerea* var.). **Type loc.:** Sattelberg, Huon Gulf and Friedrich-Wilhelmshafen, [NE] New Guinea (HNM). **Refs.:** King & Hoogstraal 1947a: 118 (syn.).

lugubris (Barraud)
1928b: 372 (F*; *Aedes*). **Type loc.:** Rangoon [Pegu], Burma [Myanmar] (NHMUK). **Distr.:** India, Malaysia, Myanmar, Thailand. **Refs.:** Barraud 1934 (F*); Laffoon 1946: 244 (tax.); Delfinado 1968: 22 (M*, F*; Thailand); Reinert 1974a: 66 (F*, P*; transfer; distr.); Reinert 1984 (lectotype desig.); Reinert & Harbach 1992 (L*); Reinert 1999c: 34 (to genus *Verrallina*); Rattanarithikul et al. 2010 (F*, L*; keys, bion., distr.; Thailand). **Ety.:** n.s. [poss. ref. to mournful]. **Informal name:** Mournful Burmese Broadtail Mosquito.

mccormicki (Belkin)
1962: 420 (M*, F, P*, L*; *Aedes*). **Type loc.:** Lunga area, Guadalcanal, Solomon Islands (USNM). **Distr.:** Papua New Guinea, Solomon Isl. **Refs.:** Huang 1968a: 37 (M, F, P, L; keys); Lee et al. 1987a: 294 (F; key, tax., bion., distr., review); Reinert 1999c: 38 (to genus *Verrallina*). **Ety.:** E.J. McCormick Jr. **Informal name:** McCormick Solomon Islander Broadtail Mosquito.

milnensis (King & Hoogstraal)
1947: 122 (M*; *Aedes*). **Type loc.:** Milne Bay [Papua], New Guinea (USNM). **Distr.:** Papua New Guinea. **Refs.:** Huang 1968a: 38 (M*, F; keys); Lee et al. 1987a: 295 (F; key, tax., bion., distr., review); Reinert 1999c: 38 (to genus *Verrallina*). **Ety.:** Milne Bay, Australia. **Informal name:** Milne Papuan Broadtail Mosquito.

multifolium (King & Hoogstraal)
1947: 122 (M*; *Aedes*). **Type loc.:** Hollandia, Western New Guinea (Papua) (USNM). **Distr.:** Australia, Indonesia (West Papua), Papua New Guinea. **Refs.:** Huang 1968a: 38 (M*, F, P*, L*; keys); Lee et al. 1987a: 296 (F; key, tax., bion., distr., review); Reinert 1999c: 38 (to genus *Verrallina*). **Ety.:** many leaves (L); ref. to four subapical leaflets on the gonocoxite. **Informal name:** Multi-leafed Indonesian Broadtail Mosquito.

obsoleta (Huang)
1968a: 45 (M*, F*; keys). **Type loc.:** 3 km south of the Vanapa River, Brown River Road, SE New Guinea (BBM). **Distr.:** Papua New Guinea. **Refs.:** Lee et al. 1987a: 300 (F; key, tax., bion., distr., review); Reinert 1999c: 38 (to genus *Verrallina*). **Ety.:** n.s. [obsolete (L)]. **Informal name:** Obsolete Papuan Broadtail Mosquito.

parasimilis (King & Hoogstraal)
1947: 125 (M*, F, L*; *Aedes*). **Type loc.:** Hollandia, Western New Guinea (Papua) (USNM). **Distr.:** Indonesia (West Papua), Papua New Guinea. **Refs.:** Huang 1968a: 48 (F, P, L, E*; keys); Lee et al. 1987a: 303 (F; key, tax., bion., distr., review); Reinert 1999c: 38 (to genus *Verrallina*). **Ety.:** close (Gr); close to *similis*. **Informal name:** Fraternal Indonesian Broadtail Mosquito [*parasimilis* = "almost-similar" = Fraternal twins].

pipkini (Bohart)
1957: 65 (F*; *Aedes*). **Type loc.:** Wena (Moen), Truk Islands, Caroline Islands (USNM). **Distr.:** Micronesia (Wake Isl.), Palau. **Refs.:** Reinert 1974a: 73 (F*; from *Aedes*); Lee et al. 1987a: 305 (tax., review); Reinert 1999c: 38 (to genus *Verrallina*). **Ety.:** A.C. Pipkin. **Informal name:** Pipkin Micronesian Broadtail Mosquito.

quadrifolium (Brug)
1934a: 512 (M*; *Aedes*). **Type loc.:** Torpedo-boot River, [Western] New Guinea (NHMUK). **Distr.:** Australia, Indonesia (West Papua), Papua New Guinea. **Refs.:** King & Hoogstraal 1947a: 119 (A); Huang 1968a: 50 (M*, F, P*; keys); Lee et al. 1987a: 307 (F; key, tax., bion., distr., review); Reinert 1999c: 38 (to genus *Verrallina*). **Ety.:** four leaves (L); four leaflike appendages near apex of gonocoxite. **Informal name:** Four-leafed Indonesian Broadtail Mosquito.

quadrispinata (King & Hoogstraal)
1947: 126 (M*, F; *Aedes*). **Type loc.:** Toem, Western New Guinea (Papua) (USNM). **Distr.:** Australia, Indonesia (West Papua), Papua New Guinea. **Refs.:** Huang 1968a: 52 (M*, F, P, L; keys); Lee et al. 1987a: 309 (F; key, tax., bion., distr., review); Reinert 1999c: 38 (to genus *Verrallina*). **Ety.:** four spines (L); "Coxite with four heavy, bluntly pointed subapical spines below base of style." **Informal name:** Four-spined Indonesian Broadtail Mosquito.

reesi (King & Hoogstraal)
1947: 127 (M*, F; *Aedes*). **Type loc.:** Vivigani, Goodenough Islands, Papua, New Guinea (USNM). **Distr.:** Australia, Indonesia, Papua New Guinea. **Refs.:** Huang 1968a: 54 (M, F; keys); Lee et al. 1987a: 311 (F; key, tax., bion., distr., review); Reinert 1999c: 38 (to genus *Verrallina*). **Ety.:** B.E. Rees. **Informal name:** Rees Papuan Broadtail Mosquito.

sentania (King & Hoogstraal)
1947: 128 (M*, F, L*; *Aedes*). **Type loc.:** Poee Village, south side of Lake Sentani, Hollandia area, Western New Guinea (Papua) (USNM). **Distr.:** Indonesia (West Papua), Papua New Guinea. **Refs.:** Huang 1968a: 54 (M*, F, P, L; keys; as *sentanius*); Reinert 1974a (F*; as *sentanius*); Lee et al. 1987a: 314 (F; key, tax., bion., distr., review); Reinert 1999c: 38 (to genus *Verrallina*). **Ety.:** Lake Sentani, Netherlands New Guinea (Papua, Indonesia). **Informal name:** Sentani Indonesian Broadtail Mosquito.

similis (Theobald)
1910c: 189 (F*; *Pseudoskusea*). **Type loc.:** Kuranda, Queensland, Australia (NHMUK). **Distr.:** Australia, Indonesia, Papua New Guinea. **Refs.:** King & Hoogstraal 1947a: 119 (F); Huang 1968a: 56 (M*, F, P*, L*; keys); Lee et al. 1987a: 316 (F; key, tax., bion., distr., review); Reinert 1999c: 38 (to genus *Verrallina*). **Ety.:** n.s. [like, resembling (L); poss. ref. to "It comes very near *Pseudoskusea multiplex*"]. **Informal name:** Similar Australian Broadtail Mosquito.

simpla (King & Hoogstraal)
1947: 130 (M*; as *Aedes simplus*). **Type loc.:** Hollandia area, Western New Guinea (Papua) (USNM). **Distr.:** Indonesia. **Refs.:** Huang 1968: 59 (M; keys); Lee et al. 1987a: 319 (F; key, tax., bion., distr., review); Reinert 1999c: 38 (to genus *Verrallina*); Reinert & Harbach 2005a (*simplus* corr. to *simpla*). **Ety.:** simple (L); "Coxite unusually simple for this group." **Informal name:** Simple Indonesian Broadtail Mosquito.

trispinata (King & Hoogstraal)
1947: 130 (M*, F; *Aedes*). **Type loc.:** Milne Bay, Papua New Guinea (USNM). **Distr.:** Papua New Guinea. **Refs.:** Huang 1968a: 60 (M*, F, P*, L*; keys); Lee et al. 1987a: 320 (F; key, tax., bion., distr., review); Reinert 1999c: 34 (to genus *Verrallina*). **Ety.:** three spines (L); "Coxite with three strong, rather long, tapered subapical spines." **Informal name:** Three-spined Papuan Broadtail Mosquito.

vanapa (Huang)
1968a: 63 (M*, F, P*; keys; *Aedes*). **Type loc.:** 3 km south of the Vanapa River, Brown River Road, SE New Guinea (BBM). **Distr.:** Papua New Guinea. **Refs.:** Lee et al. 1987a: 322 (F; key, tax., bion., distr., review); Reinert 1999c: 38 (to genus *Verrallina*). **Ety.:** Vanapa River, SE New Guinea. **Informal name:** Vanapa Papuan Broadtail Mosquito.

variabilis (Huang)
1968a: 65 (M*, F, P*, L*; keys; *Aedes*). **Type loc.:** 3 km south of the Vanapa River, Brown River Road, SE New Guinea (BBM). **Distr.:** Papua New Guinea. **Refs.:** Lee et al. 1987a: 323 (F; key, tax., bion., distr., review); Reinert 1999c: 38 (to genus *Verrallina*). **Ety.:** variable (L); ref. to adult variability. **Informal name:** Variable Papuan Broadtail Mosquito.

Genus *Zeugnomyia* Leicester

(Vol. 1, p. 238)

1908: 231 (as genus). Type species: *Zeugnomyia gracilis* Leicester.

Name derivation: "It links on the *Uranotaenias* [*sic*] to *Colonemyia* [*sic*] and *Skeiromyia* [*sic*] and through them to the *Wyeomyia* [*sic*]; hence its name." zygos (Gr) = yoke; zeugnymi = to connect, join; myia (Gr) = fly (Harbach 2018c).

Important references: Baisas & Feliciano 1953: 161 (tax.; Philippines); Delfinado 1966a: 60 (tax., keys; Philippines); Reinert 2000d (F gen.*); Reinert et al. 2004 (phyl., class.); Rattanarithikul & Harrison et al. 2005 (F, L; tax., keys, bion.; Thailand); Reinert et al. 2009 (phyl., class.); Rattanarithikul et al. 2010 (F*, L*; keys, bion.; Thailand); Wilkerson et al. 2015 (phyl., class.).

aguilari Baisas & Feliciano
1953: 164 (M*, F, P*, L). **Type loc.:** Mount Apo, Davao, Mindanao, Philippines (USNM). **Distr.:** Philippines. **Refs.:** Basio 1971b: 93 (M*; bion.); Mattingly 1971a: Pl. 32 (L head*, terminal segs.*). **Ety.:** Eusebio Aguilar. **Informal name:** Aguilar Philippine Silver Mosquito.

fajardoi Baisas & Feliciano
1953: 172 (M*, P*, L). **Type loc.:** Mount Apo, Davao, Mindanao, Philippines (USNM). **Distr.:** Philippines. **Refs.**: Basio 1971b: 93 (M*; bion.). **Ety.:** Francisco Fajardo. **Informal name:** Fajardo Philippine Silver Mosquito.

gracilis Leicester
1908: 232 (M, F). **Type loc.:** Kuala Lumpur, Jugra, Malaya [Malaysia] (NHMUK). **Distr.:** Malaysia, Philippines, Singapore, Thailand. **Refs.:** Edwards & Given 1928: 339 (L*; bion.); Basio 1971b: 94 (M*; bion.); Mattingly 1971a: Pl. 2 (A thorax*); Benjaphong & Rattanarithikul 1991: 53 (distr.); Reinert 2000c: Fig. 1 (F*); Rattanarithikul et al. 2010 (F*, L*; keys, bion., distr.; Thailand). **Ety.:** n.s. [slender (L)]. **Informal name:** Slender Malaysian Silver Mosquito.

lawtoni Baisas
1946b: 27 (M*, F, P*, L*). **Type loc.:** Wenes Calinan, Davao, Mindanao, Philippines (USNM). **Distr.:** Philippines. **Refs.:** Knight & Chamberiain 1948: 14 (tax.); Baisas & Feliciano 1953: 162 (M*; bion.); Stone & Knight 1957a: Stone & Knight 1957c: 201 (type info., lectotype desig.); Basio 1971b: 94 (M*; bion.); Mattingly 1971a: Pl. 12 (P abdomen*). **Ety.:** Henry Ware Lawton. **Informal name:** Lawton Philippine Silver Mosquito.

Tribe Culicini

Genus *Culex* Linnaeus

(Vol. 1, p. 240)

Important references: Coquillett 1906b (tax.); Edwards 1911b (*Culex*; tax; Africa); Barraud 1934 (M*, F*, L*; keys, tax.; S Asia [British India]); Edwards 1941 (M*, F*, P*; keys; Afrotropics); Michener 1944: 263 (cibarial armature); Senevet 1947b: 212 (L; key; N Africa); Hopkins 1952 (L*; key; Afrotropics); Lane 1953 (revision; Neotropics); Sicart 1954: 27 (P; key; Tunisia); Senevet & Andarelli 1954a: 36 (key; N Africa); Mattingly & Knight 1956 (tax.; Arabia); Belkin 1962 (South Pacific); DuBose & Curtin 1965 (F, L; keys; Mediterranean); Forattini 1965a (Neotropics); Cova García et al. 1966 (Venezuela); Delfinado 1966a (Philippines); Bram 1967a (tax., subg. *Culex*; New World); Bram 1967b (Thailand); Belkin 1968a (revision; New Zealand); Sirivanakarn 1968 (subg. *Lophoceraomyia*); Berlin 1969a (revision; subg. *Howardina*); Berlin 1969c (revision; subg. *Micraedes*); Belkin et al. 1970 (Jamaica); Sirivanakarn 1971b (revision; subg. *Neoculex*; SE Asia); Sirivanakarn 1972 (revision; subg. *Eumelanomyia*; SE Asia); Adames & Galindo 1973 (tax.; subg. *Belkinomyia*); Valencia 1973 (revision; subg. *Carrollia*); Gutsevich et al. 1974 (revision; former Union of Soviet Socialist Republics [USSR]); Sirivanakarn 1976 (revision; subg. *Culex*; Oriental Region); Sirivanakarn 1977a (revision; various subg.; Oriental Region); Tanaka et al. 1979 (revision; Japan, Korea); Wood et al. 1979 (tax., keys; Canada); Berlin & Belkin 1980 (revision; subg. *Aedinus*, *Tinolestes*, *Anoedioporpa*; Neotropics); Ribeiro & da Cunha Ramos 1980 (tax.; Angola); Darsie & Ward 1981 (F, L; keys; North America); Lu & Li 1982a (tax.; People's Republic of China); Clark-Gil & Darsie 1983 (F, L; keys; Guatemala); Sirivanakarn 1983 (revision; subg. *Melanoconion*); Darsie 1985 (F, L; keys; Argentina); Lu & Su 1987 (tax.; People's Republic of China,); Lee et al. 1988b (F; review; Australasia); Harbach 1988 (revision; subg. *Culex*; SW Asia and Egypt); Lee et al. 1989a (F; review; Australasia); Darsie & Pradhan 1990, 1991 (F, L; keys; Nepal); Harbach & Peyton 1992b (tax.; subg. *Phenacomyia*); Pecor et al. 1992 (tax., catalog; subg. *Melanoconion*); Rattanarithikul & Panthusiri 1994 (keys, med. imp. spp.; Thailand); Reuben et al. 1994 (keys, JE vectors; Southeast Asia); Sallum & Forattini 1996 (revision; Spissipes Section, subg. *Melanoconion*); Miller et al. 1996 (mol. phyl.); Lu et al. 1997 (tax.; People's Republic of China); Navarro & Liria 2000 (L mouthparts; tax., phyl.); González Broche & Rodríguez R. 2001 (subg. *Nicaromyia*); Tanaka 2004b (subg. *Oculeomyia*, *Sirivanakarnius*); Darsie & Ward 2005 (F, L; keys, distr.; North America); Rattanarithikul & Harrison et al. 2005 (F, L; keys, distr.; Thailand); Rattanarithikul & Harrison et al. 2005 (tax. background; Thailand); St John 2007 (morph., phyl.); Rossi & Harbach 2008 (subg. *Phytotelmatomyia*); Harbach 2011 (class., tax., *Culex*); Vesgueiro et al. 2011 (mol. phyl.); Demari-Silva et al. 2011 (mol. phyl.); Harbach et al. 2012 (morph., phyl.); Laurito et al. 2013 (morph., phyl.); Harbach, Culverwell, & Kitching 2016 (phyl.).

Subgenus *Acalleomyia* Leicester

1908: 194. Type species: *Acalleomyia obscurus* Leicester.

Name derivation: n.s. [acallo = ugly (Gr); ugly fly?].

Important references: Bram 1968 (M*, L*; redescr.); Harbach et al. 2012 (phyl.).

obscurus (Leicester)
1908: 194 (M, F; *Acalleomyia*). **Type loc.:** Malaya [Malaysia] (NHMUK). **Distr.:** Indonesia, Malaysia, Singapore. **Refs.:** Edwards 1922a: 285 (M*); Bram 1967c (type info., lectotype desig.); Bram 1968: 53 (M*, F, P*, L*); Lien et al. 1975 (distr.). **Ety.:** n.s. [dark, obscure (L); poss. ref. to "pale colourless pleura with black scaled legs and abdomen"]. **Informal name:** Dark Malaysian Typical Mosquito.

Subgenus *Acallyntrum* Stone & Penn

1948:109 (as subg.). Type species: *Culex* (*Acallyntrum*) *perkinsi* Stone & Penn.

Name derivation: callyntro = broom or brush (Gr); a- = without [ref. to "The most striking character of the subgenus *Acallyntrum* is the great reduction of the ventral hairs of the anal segment of the larva"].

Important references: Harbach et al. 2012 (phyl.).

axillicolus Steffan
1979: 133 (M*, F, P*, L*; as *axillicolus*). **Type loc.:** Pukamil Hamlet, Minj, Western Highlands District, Papua New Guinea (BBM). **Distr.:** Papua New Guinea. **Refs.:** Ward 1984b: 240 (emend.); Lee et al. 1989a: 7 (F; key, tax., bion., distr., review). **Ety.:** from plant axils (taro) (L). **Informal name:** Axil-dwelling Papuan Typical Mosquito.

belkini Stone & Penn
1948: 114 (M*, F, P*, L*). **Type loc.:** West Cape, Guadalcanal, Solomon Islands (USNM). **Distr.:** Papua New Guinea, Solomon Isl. **Refs.:** Belkin 1962: 225 (M*, F, P*, L*); Mattingly 1971a: Pl. 25 (L head*, terminal segs.*); Lee et al. 1989a: 8 (F; key, tax., bion., distr., review). **Ety.:** John Nicholas Belkin. **Informal name:** Belkin Solomon Islander Typical Mosquito.

bicki Stone & Penn
1947: 89 (M*, F, P*, L*). **Type loc.:** Toem, Western New Guinea (Papua) (USNM). **Distr.:** Indonesia (West

Papua), Papua New Guinea. **Refs.:** Lee et al. 1989a: 9 (F; key, tax., bion., distr., review). **Ety.:** George H. Bick. **Informal name:** Bick Indonesian Typical Mosquito.

binigrolineatus Knight & Rozeboom
1945: 289 (M*, F*, P*, L*). **Type loc.:** Lake Sentani, Hollandia, Western New Guinea (Papua) (USNM). **Distr.:** Indonesia (West Papua), Papua New Guinea. **Refs.:** Knight & Chamberlain 1948: 18 (P*); Lee et al. 1989a: 10 (F; key, tax., bion., distr., review). **Ety.:** twice black lined (L); ref. to "black stripe on each side of middorsal line" of scutum. **Informal name:** Ebon-lined Indonesian Typical Mosquito [*binigro* = "twice black" = Ebon].

bougainvillensis Steffan
1979: 138 (M*). **Type loc.:** Bougainville Island, Solomon Islands (USNM). **Distr.:** New Guinea, Solomon Isl. **Refs.:** Lee et al. 1982: 128 (F; key, tax., bion., distr., review); Lee et al. 1989a: 12 (tax., bion., distr., review). **Ety.:** Bougainville, Solomon Isl. **Informal name:** Bougainville Solomon Islander Typical Mosquito.

miyagii Mogi & Toma
1999: 477 (M*, F, P*, L*). **Type loc.:** Tanah Merah, Pasahari, North Seram, Indonesia (USNM). **Distr.:** Indonesia. **Ety.:** Ichiro Miyagi. **Informal name:** Miyagi Indonesian Typical Mosquito.

pallidiceps (Theobald)
1905f: 101 (M; *Melanoconion*). **Type loc.:** Friedrich-Wilhelmshafen, [NE] New Guinea (HNM). **Distr.:** Papua New Guinea, Solomon Isl. **Refs.:** Lee et al. 1989a: 13 (tax., bion., distr., review). **Ety.:** n.s. [pale headed (L)]. **Informal name:** Pale-headed Papuan Typical Mosquito.

perkinsi Stone & Penn
1948: 110 (M*, F, P*, L*). **Type loc.:** Umasini River, Guadalcanal, Solomon Islands (USNM). **Distr.:** Papua New Guinea, Solomon Isl. **Refs.:** Laird 1955: 285 (L*); Stone 1957b: 172 (syn.); Belkin 1962: 226 (M*, F, P*, L*; from syn.); Lee et al. 1989a: 15 (F; key, tax., bion., distr., review). **Ety.:** F.A. Perkins. **Informal name:** Perkins Solomon Islander Typical Mosquito.

Subgenus *Aedinus* Lutz

1904 (as genus). In Bourroul 1904: 43. Type species: *Aedinus amazonensis* Lutz 1905: 103.

Name derivation: n.s. [-inus = likeness, belonging to, derived, coming from (L); prob. ref. to being like *Aedes* or coming from a bad, distasteful thing (Gr)].

syn. *Eubonnea* Dyar 1919c: 150. Type species: *Culex* (*Eubonnea*) *tapena* Dyar. (Belkin 1968b: 11 (syn.)).

Important references: Edwards 1932a: 220 (as subg.); Belkin 1968b: 51 (attrib. author to Lutz); Berlin & Belkin 1980: 6 (revision).

accelerans Root
1927c: 581 (M*). **Type loc.:** Porto das Caixas, Rio de Janeiro, Brazil (USNM). **Distr.:** Brazil, Colombia, French Guiana, Guyana, Lesser Antilles (includes Trinidad and Tobago), Panama. **Refs.:** Lane 1953: 386 (M*); Belkin et al. 1971: 27 (type loc. info.); Berlin & Belkin 1980: 17 (M*, F, P*, L*; keys, tax., bion., distr.). **Ety.:** n.s. [accelerating (L)]. **Informal name:** Speedy Brazilian Typical Mosquito.

amazonensis (Lutz)
1905: 103 (M, F; *Aedinus*). **Type loc.:** Manaus, Amazonas, Brazil (LU). **Distr.:** Argentina, Brazil, Colombia, Ecuador, French Guiana, Guatemala, Guyana, Lesser Antilles (includes Trinidad and Tobago), Nicaragua, Panama, Paraguay, Peru, Suriname, Venezuela. **Ety.:** Amazon. **Informal name:** Amazon Brazilian Typical Mosquito.
syn. *tapena* Dyar
1919e: 150 (M). **Type loc.:** Paramaribo, Suriname (USNM). **Refs.:** Dyar 1923e: 189 (syn.).
syn. *paraplesia* Dyar
1922g: 192 (M*). **Type loc.:** Puerto Nino, Colombia (USNM). **Refs.:** Dyar 1923e: 189 (syn.).
syn. *hildebrandi* Evans
1923c: 377 (M*). **Type loc.:** Itacoatiara, Amazonas, Brazil (NHMUK). **Refs.:** Dyar 1923e: 190 (syn.); Belkin 1968b: 15 (type info.); Belkin et al. 1971: 27 (type loc. info.).

clastrieri Casal & García
1968b: 119 (M*). **Type loc.:** Belém, Pará, Brazil (INM). **Distr.:** Brazil, French Guiana. **Refs.:** Belkin et al. 1971: 27 (type loc. info.); Berlin & Belkin 1980: 15 (M*; key, tax., bion., distr.). **Ety.:** Jean Clastrier. **Informal name:** Clastrier Brazilian Typical Mosquito.

guyanensis Clastrier
1970a: 116 (M*). **Type loc.:** Forêt de Cabassou, Cayenne, French Guiana [Guiana] (MNHP). **Distr.:** French Guiana. **Refs.:** Berlin & Belkin 1980: 16 (M*; key, tax., bion., distr.). **Ety.:** French Guiana. **Informal name:** Arawak French Guianese Typical Mosquito [the Arawak were among the first inhabitants of modern French Guiana].

Subgenus *Afroculex* Danilov

1989: 790. Type species *Pseudohowardina lineata* Theobald.

Name derivation: African.

Important references: Edwards 1941 (as *Cx. pulchrithorax*); Danilov 1989 (tax., distr.); Harbach et al. 2012 (morph., phyl.).

lineatus (Theobald)
1912a: 92 (M, F; *Pseudohowardina*). **Type loc.:** Onderstepoort, Transvaal [Republic of South Africa] (VRL, NHMUK). **Distr.:** Mozambique, Republic of South Africa. **Refs.:** Edwards 1941: 253 (M*, F*; tax.; as *pulchrithorax*); Kok 1979 (L*; bion.; as *pulchrithorax*); Danilov 1989 (from subg. *Maillotia*); Townsend 1990: 94 (type info.); Jupp 1996 (M*, F*; key). **Ety.:** lined (L); three thin pale lines on the thorax. **Informal name:** Striped South African Typical Mosquito.
syn. *pulchrithorax* Edwards
1914b: 73 (new name for *lineata* Theobald, not von Humboldt 1819). **Refs.:** Danilov 1989 (from subg. *Maillotia*; syn.).

Subgenus *Allimanta* Casal & García

1968a: 331 (as subg.). Type species: *Culex* (*Culex*) *tramazayguesi* Duret.

Name derivation: n.s. [alli = big toe; manta = cloak (L, Sp)].

Important references: Harbach 2011 (phyl.).

tramazayguesi Duret
1954b: 23 (M*, F). **Type loc.:** Monte Comán, Mendoza, Argentina (USNM). **Distr.:** Argentina. **Refs.:** Belkin et al. 1968: 15 (type loc. info.); Casal & García 1968a: 333 (M*, P*, L*; to subg. *Allimanta*); Harbach et al. 1991: 194 (type info.). **Ety.:** M.M. Tramazaygues. **Informal name:** Tramazaygues Argentinian Typical Mosquito.

Subgenus *Anoedioporpa* Dyar

1923e: 190. Type species: *Culex conservator* Dyar & Knab.

Name derivation: n.s. [poss. ref. to genitalia character: an = without (Gr); oedi = swollen (Gr); porpa = pin, buckle, clasp (Gr)].

Important references: Belkin 1968b: 11 (resurrected from syn.); Berlin & Belkin 1980: 26 (revision); Harbach et al. 2012 (phyl.).

bamborum Rozeboom & Komp
1948: 399 (M*, F, L*). **Type loc.:** Acacias [Meta], Colombia (USNM). **Distr.:** Colombia, Ecuador. **Refs.:** Belkin, Schick, & Heinemann 1965: 10 (type loc. info.); Mattingly 1971a: Pl. 20 (P cephalothorax*); Berlin & Belkin 1980: 45 (M*, F, P*, L*; keys, tax., bion., distr.); Morales & Cura 1983 (P*); Linton et al. 2013 (bion., mol. tax., distr.; Ecuador). **Ety.:** n.s. [from bamboo?]. **Informal name:** Bamboo Colombian Typical Mosquito.

belemensis Duret & Damasceno
1955: 404 (M*). **Type loc.:** Belém, Pará, Brazil (USNM). **Distr.:** Brazil, French Guiana. **Refs.:** Belkin et al. 1971: 27 (type loc. info.); Fauran & Pajot 1974a: 102 (distr.); Berlin & Belkin 1980: 47 (M*; key, tax., bion., distr.); Harbach et al. 1991: 192 (type info.). **Ety.:** Belém, Brazil. **Informal name:** Belém Brazilian Typical Mosquito.

browni Komp
1936c: 326 (M*). **Type loc.:** Gatun, Canal Zone, Panama (USNM). **Distr.:** Colombia, Ecuador, Panama. **Refs.:** Lane 1953: 395 (M*); Belkin, Schick, & Heinemann 1965: 55 (type loc. info.); Berlin & Belkin 1980: 43 (M*, F, P*, L*; keys, tax., bion., distr.); Linton et al. 2013 (bion., distr.; Ecuador). **Ety.:** C.G. Brown. **Informal name:** Brown Panamanian Typical Mosquito.

canaanensis Lane & Whitman
1943: 398 (M*, F, P*, L*). **Type loc.:** Vale do Canaã, S Joao de Petropolis, Espirito Santo, Brazil (INER). **Distr.:** Argentina, Brazil. **Refs.:** Belkin et al. 1971: 27 (type loc. info.); Xavier 1973: 160 (type info.); Berlin & Belkin 1980: 40 (M*, P, L; keys, tax., bion., distr.). **Ety.:** Valley of Canaã, Espirito Santo, Brazil. **Informal name:** Canaã Brazilian Typical Mosquito.

chaguanco Casal, García, & Fernández
1968: 217 (M*). **Type loc.:** Tablillas, near Campamento Vespucio, Departamento San Martin, Salta, Argentina (INM). **Distr.:** Argentina. **Refs.:** Berlin & Belkin 1980: 48 (M*, F, P*, L*; keys, tax., bion., distr.). **Ety.:** Chaguanaco (Amerindian tribe of SE Bolivia?). **Informal name:** Chaguanaco Argentinian Typical Mosquito.

conservator Dyar & Knab
1906d: 221 (L*). **Type loc.:** Saint Joseph, Trinidad [Trinidad & Tobago, Lesser Antilles] (USNM). **Distr.:** Argentina, Belize, Bolivia, Brazil, Colombia, Costa Rica, Ecuador, French Guiana, Guatemala, Guyana, Honduras, Lesser Antilles (Saint Kitts and Nevis, Saint Lucia, Saint Vincent and the Grenadines, Trinidad and Tobago), Mexico, Nicaragua, Panama, Peru, Suriname, United States (continental), Venezuela. **Refs.:** Dyar 1928: 345 (M*, F, L*); Lane 1953: 391 (M*, F, L*); Stone & Knight 1957a: 46 (lectotype desig.); Belkin, Schick, & Heinemann 1965 (type loc. info.); Heinemann & Belkin 1977b: 415, 427, 442 (distr.; Belize, Guatemala, Honduras); Heinemann & Belkin 1977c: 528 (distr.; Costa Rica); Berlin & Belkin 1980: 35 (M*, F*, P*, L*; keys, tax., bion., distr.); Pecor et al. 2000 (tax., bion., distr.; Peru). **Ety.:** n.s. [keeper, defender, savior (L)]. **Informal name:** Defensive Trinbagonian Typical Mosquito.
syn. *divisior* Dyar & Knab
1906d: 222 (L*). **Type loc.:** Trinidad [Trinidad & Tobago, Lesser Antilles] (USNM). **Refs.:** Lane 1951: 334 (syn.); Stone & Knight 1957a: 47 (lectotype desig.); Belkin, Schick, & Heinemann 1965: 70 (type loc. info.).

syn. *bifoliata* Dyar
1922d: 94 (M). **Type loc.:** Miraflores, Canal Zone, Panama (USNM). **Refs.:** Dyar 1928: 345 (syn.); Stone & Knight 1957a: 44 (lectotype desig.); Belkin, Schick, & Heinemann 1965: 56 (type loc. info.).

syn. *paganus* Evans
1923b: 104 (M*, F). **Type loc.:** Aragua, Venezuela (NHMUK). **Refs.:** Lane & Whitman 1943: 397 (from syn. with *conservator*); Lane 1953: 394 (M*, F); Belkin 1968b: 18 (lectotype desig.); Berlin & Belkin 1980 (syn.).

syn. *surukumensis* Anduze
1941b: 833 (M*). **Type loc.:** Rio Surukum, Bolivar, Venezuela (DPFA). **Refs.:** Rozeboom & Komp 1950a: 96 (syn.); Belkin, Schick, & Heinemann 1965: 75 (type loc. info.).

corrigani Dyar & Knab
1907c: 203 (F, L). **Type loc.:** Tabernilla, Canal Zone, Panama (USNM). **Distr.:** Colombia, Costa Rica, Nicaragua, Panama. **Refs.:** Dyar 1928: 347 (M*, F, L*); Lane 1953: 392 (M*, F, L*); Belkin, Schick, & Heinemann 1965: 56 (type loc. info.); Belkin 1968b: 14 (n. comb.); Heinemann & Belkin 1977b: 452 (distr.); Berlin & Belkin 1980: 56 (M*, F, P*, L*; keys, tax., bion., distr.). **Ety.:** John Corrigan. **Informal name:** Corrigan Panamanian Typical Mosquito.

syn. *chalcocorystes* Martini
1914: 70 (M, F). **Type loc.:** Porto Bello, Panama (NHMUK). **Refs.:** Mattingly 1955c: 31 (lectotype desig.).

damascenoi Duret
1969c: 143 (M*). **Type loc.:** Rio Preto, Municipio de João Goulart, Amazonas, Brazil (USNM). **Distr.:** Brazil, French Guiana. **Refs.:** Belkin et al. 1971: 27 (type loc. info.); Berlin & Belkin 1980: 42 (M*, P; keys, tax., bion., distr.); Harbach et al. 1991: 193 (type info.). **Ety.:** Reinaldo G. Damasceno. **Informal name:** Damasceno Brazilian Typical Mosquito.

syn. *menui* Clastrier
1971: 649 (M*, P*) (MNHP). **Type loc.:** Maripasoula, French Guiana [Guiana] (A). **Refs.:** Berlin & Belkin 1980: 42 (syn.).

luteopleurus (Theobald)
1903a: 239 (F; *Melanoconion*). **Type loc.:** Pará, Brazil (NHMUK). **Distr.:** Brazil. **Refs.:** Duret & Damasceno 1955: 397 (M*; tax.); Belkin 1968b: 17 (to subg. *Micraedes*); Belkin et al. 1971: 27 (n. comb.); Berlin & Belkin 1980: 54 (M*; key, tax., bion., distr.). **Ety.:** yellow sided (L); pleura bright yellow. **Informal name:** Yellow-sided Brazilian Typical Mosquito.

originator Gordon & Evans
1922: 323 (M*, F, L*). **Type loc.:** Macapá, near Manaus, Amazonas, Brazil (NHMUK). **Distr.:** Argentina, Brazil, French Guiana, Guyana, Lesser Antilles (includes Grenada, Martinique, Trinidad and Tobago). **Refs.:** Senevet & Abonnenc 1939a: 112 (M*, L*); Lane 1953: 394 (M*, F, L); Fauran 1961a: 12 (P*); Belkin 1968b: 18 (lectotype desig.); Belkin et al. 1971: 27 (type loc. info.); Belkin & Heinemann 1976a: 272 (distr.; Grenada); Berlin & Belkin 1980: 50 (M*, F, P*, L*; keys, tax., bion., distr.). **Ety.:** n.s. [source, beginning (L)]. **Informal name:** Original Brazilian Typical Mosquito.

quasioriginator Duret
1972: 3 (M*). **Type loc.:** Pirelli, Belém, Pará, Brazil (USNM). **Distr.:** Brazil. **Refs.:** Berlin & Belkin 1980: 53 (tax., bion., distr.); Harbach et al. 1991: 194 (type info.). **Ety.:** nearly (L); similar to *originator*. **Informal name:** Beginner Brazilian Typical Mosquito [*quasiorginator* = "almost original" = Beginner].

restrictor Dyar & Knab
1906d: 222 (L*). **Type loc.:** Almoloya, Oaxaca, Mexico (USNM). **Distr.:** Belize, Brazil, Colombia, Costa Rica, El Salvador, Guatemala, Honduras, Mexico, Nicaragua, Panama, Venezuela. **Refs.:** Dyar 1928: 348 (M*, F, L*); Lane 1953: 527 (M*, F, P*, L); Stone & Knight 1957a: 55 (type info.); Berlin & Belkin 1980: 26 (M*, F*, P*, L*; keys, tax., bion., distr.; from subg. *Microculex*); Pecor et al. 2002: 242 (tax., distr.; Belize); Mendenhall et. al. 2011: 148 (distr.; Colombia). **Ety.:** n.s. [restrict, tie up (L)]. **Informal name:** Restrictive Mexican Typical Mosquito.

syn. *consternator* Dyar & Knab
1908: 59 (M, F). **Type loc.:** Córdoba [Veracruz], Mexico (USNM). **Refs.:** Howard et al. 1915: 333 (syn.); Stone & Knight 1957a: 46 (lectotype desig.); Belkin, Schick, & Heinemann 1965: 38 (type loc. info.).

Subgenus *Barraudius* Edwards

1921d: 332 (as subg. of genus *Culex*). Type species: *Culex pusillus* Macquart.

Name derivation: Philip James Barraud.

Important references: Gutsevich et al. 1974 (tax., bion., distr.); Tanaka et al. 1979 (M, F, L; subg. descry.); Becker et al. 2010 (tax., bion.; Europe); Harbach 2011 (phyl.).

inatomii Kamimura & Wada
1974: 13 (M*, F*, P*, L*). **Type loc.:** Utoma, Kojima City, Okayama Prefecture, Honshu, Japan (NMST). **Distr.:** Japan, South Korea. **Refs.:** Tanaka et al. 1979: 234 (M*, F*, L*; from ssp. *modestus*). **Ety.:** Seiiti Inatomi. **Informal name:** Inatomi Japanese Typical Mosquito.

modestus Ficalbi
1890a: 93 (F). **Type loc.:** Ravenna [Emilia], Italy (LU). **Distr.:** Afghanistan, Albania, Algeria, Austria, Azerbaijan, Belarus, Bulgaria, Crimean Peninsula, Croatia, Czech Republic, Denmark, France (includes Corsica), FYRO Macedonia, Georgia, Germany, Greece, Hungary, Iran, Iraq, Israel (and Gaza Strip and West Bank), Italy

(includes Sicily), Moldova, Mongolia, Montenegro, Morocco, Netherlands, Pakistan, People's Republic of China, Poland, Portugal, Romania, Russia, Serbia, Slovakia, Slovenia, Spain, Sweden, Switzerland, Tajikistan, Turkey, Ukraine, United Kingdom. **Refs.:** Séguy 1924: Pl. XII (L*); Barraud 1934: 346 (M*, F, L); Stackelberg 1937: 219 (M*); Senevet 1947a: 109 (L*); Senevet 1949: 52 (M*); Monchadskii 1951: 245 (L*); Senevet & Andarelli 1959: 80 (M*, F, P*, L*); Brunhes & Venard 1966 (tax.); Polyakova 1970b: 136 (distr.); Aslamkhan 1971b (distr.; Pakistan); Lotfi 1973: 206 (distr.); Gutsevich et al. 1974: 351 (M*, F, L*); Margalit & Tahori 1974: 85 (distr.); Mattingly 1975: 22 (E*); Minar 1976: 335 (distr.); Moussiegt 1990 (complete annotated bib.); Alten et al. 2000 (distr.; Turkey); Eritja et al. 2000: 12 (distr.; Spain); Orszagh et al. 2001: 19 (distr.; Slovakia); Rueda, Pecor et al. 2008: 397 (distr.); Becker et al. 2010: 265 (M*, F*, L*; keys, tax., distr., bion.); Osório et al. 2010 (distr.; Portugal); Namazov 2014 (distr.; Azerbaijan); Gunay et al. 2015 (mol. tax., distr.; Turkey); Robert et al. 2019 (distr.; Western Palearctic). **Ety.:** unassuming (L); nondescript overall color. **Informal name:** Modest Italian Typical Mosquito.

syn. *eadithae* Barraud
1924f: 1004 (M*). **Type loc.:** Sopor village, Jhelum River near Wular Lake, Kashmir [Pakistan] (NHMUK).

syn. *tanajcus* Stschelkanovzev
1926: 133 (M*, L*). **Type loc.:** Asov [Rostov], Union of Soviet Socialist Republics (LU). **Refs.:** Monchadsky 1936: 332 (syn.).

syn. *nudipalpis* Shingarev
1927: 550 (M*). **Type loc.:** Moscow District, Union of Soviet Socialist Republics (MTI). **Refs.:** Stackelberg 1937: 219 (syn.).

pusillus Macquart
1850: 313 (F). **Type loc.:** Egypt (BC). **Distr.:** Afghanistan, Algeria, Dubai, Egypt, Greece, Iran, Iraq, Israel (and Gaza Strip and West Bank), Jordan, Kuwait, Libya, Morocco, Qatar, Saudi Arabia, Sudan and South Sudan, Syria, Tajikistan, Tunisia, Turkey, Turkmenistan, United Arab Emirates, Uzbekistan. **Refs.:** Séguy 1924: Pl. XIV (L*), Pl. XIII (M gen.*); Stackelberg 1937: 221 (M*); Senevet 1947a: 107 (L*); Senevet 1949: 50 (M*); Monchadskii 1951: 246 (L*); Sicart 1954: 28 (P); Mattingly & Knight 1956: 109 (F), 113 (L) (distr.; Saudi Arabia); Senevet & Andarelli 1959: 85 (M*, F, P*, L*); Gutsevich et al. 1974: 353 (M*, F, L*); Margalit & Tahori 1974: 86 (distr.); Harbach 1998: 14 (distr.; United Arab Emirates); Al-Houty 1997 (distr.; Kuwait, as *pisullus* [*sic*]); Rueda, Pecor et al. 2008 (distr.); Becker et al. 2010: 267 (M*, F*, L*; keys, tax., distr., bion.); Ahmed et al. 2011 (distr.; Saudi Arabia); Gunay et al. 2015 (mol. tax., distr.; Turkey); Robert et al. 2019 (distr.; Western Palearctic). **Ety.:** n.s. [tiny (L)]. **Informal name:** Tiny Egyptian Typical Mosquito.

richeti Brunhes & Venard
1966: 75 (M*, F, P*, L*). **Type loc.:** Lake Chad, Nigeria (IERT). **Distr.:** Niger, Nigeria. **Ety.:** P. Richet. **Informal name:** Richet Nigerian Typical Mosquito.

Subgenus *Belkinomyia* Adames & Galindo

1973: 55. Type species: *Culex* (*Belkinomyia*) *eldridgei* Adames & Galindo.

Name derivation: John Nicholas Belkin.

Important references: Harbach et al. 2012 (phyl.).

eldridgei Adames & Galindo
1973: 56 (M*, F, P*, L*). **Type loc.:** Curiche, El Chocó, Colombia (USNM). **Distr.:** Colombia. **Ety.:** Bruce Frederick Eldridge. **Informal name:** Eldridge Colombian Typical Mosquito.

Subgenus *Carrollia* Lutz

1905: 81 (as genus). Type species: *Carrollia iridescens* Lutz.

Name derivation: n.s. [prob. ref. to James Carroll, who researched yellow fever in Cuba].

syn. *Carrollella* Lutz 1921 (as genus): 163. Type species: *Carrollia iridescens* Lutz.
Edwards 1932a: 220 (syn.).

Important references: Dyar 1918b: 108 (to subg. of *Culex*); Casal & García 1968f: 97–98; Valencia 1973 (revision).

anduzei Cerqueira & Lane
1944: 219 (M*). **Type loc.:** Rio Maués, Amazonas, Brazil (FH). **Distr.:** Brazil, Venezuela. **Refs.:** Barata & Cotrim 1972 (M*, F, P*, L*); Valencia 1973: 52 (M*, F*, P*, L*); Berti et al. 2015 (distr.; Venezuela). **Ety.:** Pablo Anduze. **Informal name:** Anduze Brazilian Typical Mosquito.

antunesi Lane & Whitman
1943: 389 (M*, F, P*, L*). **Type loc.:** Vale do Canaã, S João de Petropolis, Espirito Santo, Brazil (INER). **Distr.:** Brazil, Colombia, Costa Rica, French Guiana, Panama. **Refs.:** Valencia 1973: 73 (M*, F*, P*, L*); Xavier 1973: 160 (type info.). **Ety.:** Paulo César de Azevedo Antunes. **Informal name:** Antunes Brazilian Typical Mosquito.

syn. *manaensis* Floch & Fauran
1955: 3 (M*, P). **Type loc.:** Haute-Mana, French Guiana [Guiana] (PIG). **Refs.:** Valencia 1973: 73 (syn.).

babahoyensis Leví-Castillo
1953e: 92 (M*, F, P*). **Type loc.:** Hacienda Mora, Juan Montalvo, Los Rios, Ecuador (USNM). **Distr.:** Bahamas, Cayman Isl., Costa Rica, Dominican Republic, Ecuador,

Haiti, Jamaica, Lesser Antilles (Antigua and Barbuda, Guadeloupe), Nicaragua, United States (continental), Virgin Isl. **Refs.:** Valencia 1973: 70 (M*, F*, P*, L*). **Ety.:** Rio Babahoyo, Ecuador. **Informal name:** Babahoyo Ecuadorian Typical Mosquito.

bihaicolus Dyar & Núñez Tovar
1928: 91 (M, F). **Type loc.:** Ocumare de la Costa, Aragua, Venezuela (USNM). **Distr.:** Colombia, Costa Rica, Ecuador, Guatemala, Mexico, Panama, Venezuela. **Refs.:** Dyar 1928: 91 (M, F; republ. of original description? [not stated]); Stone & Knight 1957a: 44 (type info.); Lane 1953: 509 (M*, F, L*); Valencia 1973: 29 (M*, F*, P*, L*; lectotype desig.). **Ety.:** *Heliconia bihai* bract; larval habitat. **Informal name:** Bihai Venezuelan Typical Mosquito.

bonnei Dyar
1921k: 155 (A). **Type loc.:** Suriname (USNM). **Distr.:** Brazil, Colombia, Ecuador, French Guiana, Guyana, Peru, Suriname. **Refs.:** Antunes & Ramos 1939: 381 (M*, F*, L*); Lane 1953: 500 (M*, F*, P*, L); Stone & Knight 1957a: 44 (lectotype desig.); Valencia 1973: 57 (M*, F*, P*, L*). **Ety.:** Cornelis Bonne. **Informal name:** Bonne Surinamese Typical Mosquito.

cerqueirai Valencia
1973: 76 (M*, F*, P*, L*). **Type loc.:** Quebrada Grande, Bocas del Toro, Panama (USNM). **Distr.:** Panama. **Ety.:** n.s. [Nelson Leander Cerqueira]. **Informal name:** Cerqueira Panamanian Typical Mosquito.

guerreroi Cova García, Sutil, & Pulido
1971: 203 (M*, P*, L*). **Type loc.:** Colonia Tovar, Municipio Tovar, Estado Aragua, Venezuela (DERM). **Distr.:** Venezuela. **Refs.:** Valencia 1973: 33 (M*, F*, P*, L*). **Ety.:** Lacenio Guerrero. **Informal name:** Guerrero Venezuelan Typical Mosquito.

infoliatus Bonne-Wepster & Bonne
1920: 170 (A, L). **Type loc.:** Dam, Suriname (ITH). **Distr.:** Brazil, Colombia, Ecuador, French Guiana, Guyana, Peru, Suriname, Venezuela. **Refs.:** Senevet & Abonnenc 1939a: 126 (M*, L*); Senevet & Abonnenc 1958a: 288 (P*); Belkin 1968b: 16 (lectotype desig.); Valencia 1973: 41 (M*, F*, P*, L*). **Ety.:** n.s. [not leafy (L)]. **Informal name:** Cactus Surinamese Typical Mosquito [*infoliatus* = "not leafy" = Cactus].

insigniforceps Clastrier & Claustre
1978a: 359 (M*). **Type loc.:** Forêt du Gallion, Guyane Française (MNHP). **Distr.:** French Guiana. **Ety.:** remarkable pincers (L); ref. to notable male genitalia characters. **Informal name:** Pincered French Guianese Typical Mosquito.

iridescens (Lutz)
1905: 81 (M, F; *Carrollia*). **Type loc.:** Serra da Cantareira, São Paulo, Brazil (FH). **Distr.:** Brazil, Colombia, Ecuador. **Refs.:** Antunes & Ramos 1939: 380 (M*, F*, L*; neotype desig.); Lane 1953: 498 (M*, F*, P*, L*); Belkin et al. 1971: 27 (type loc. info., tax.); Valencia 1973: 64 (M*, F, P*, L*); Linton et al. 2013 (bion., mol. tax., distr.; Ecuador). **Ety.:** n.s. [becoming iridescent (L); prob. ref. to it being a very beautiful species ("*muito bonita e caracteristica*")]. **Informal name:** Iridescent Brazilian Typical Mosquito.

kompi Valencia
1973: 81 (M*, F, P*, L*). **Type loc.:** Salinas, Restrepo, Meta, Colombia (USNM). **Distr.:** Colombia. **Ety.:** William H. Wood Komp. **Informal name:** Komp Colombian Typical Mosquito.

metempsytus Dyar
1921k: 154 (M, F). **Type loc.:** Alajuela, Costa Rica (USNM). **Distr.:** Colombia, Costa Rica, Ecuador, Guatemala, Panama. **Refs.:** Lane 1953: 510 (M*, F); Stone & Knight 1957a: 53 (lectotype desig.); Valencia 1973: 39 (M*, F*, P*, L*). **Ety.:** n.s. **Informal name:** Mid-spit Costa Rican Typical Mosquito.

rausseoi Cova García, Sutil O., & Pulido F.
1972: 207 (M*, F, P*, L*). **Type loc.:** Campamento Rangel, Tejerias, Aragua, Venezuela (DERM). **Distr.:** Venezuela. **Refs.:** Valencia 1973: 36 (M*, F*, P*, L*). **Ety.:** José Antonio Rausseo Aguilera. **Informal name:** Rausseo Venezuelan Typical Mosquito.

secundus Bonne-Wepster & Bonne
1920: 170 (M, L; as *secunda*). **Type loc.:** Canal Zone, Panama (USNM). **Distr.:** Brazil, Colombia, Costa Rica, Ecuador, Panama. **Refs.:** Dyar 1928: 281 (M*, F, L*; corrected spelling); Lane 1953: 506 (M*, F, P*, L*); Belkin 1968b: 20 (type info.); Barata & Cotrim 1972 (M*, F, P*, L*); Valencia 1973: 60 (M*, F*, P*, L*); Linton et al. 2013 (bion., distr.; Ecuador). **Ety.:** n.s. [second, next (L)]. **Informal name:** Second Panamanian Typical Mosquito.

soperi Antunes & Lane
1937: 21 (M*, F). **Type loc.:** Perus, São Paulo, Brazil (FH). **Distr.:** Argentina, Brazil. **Refs.:** Antunes & Ramos 1939: 383 (M*, F*, L*); Lane 1953: 503 (M*, F*, P*, L*); Belkin et al. 1971: 28 (type info.); Valencia 1973: 67 (M*, F, P*, L*). **Ety.:** Frederick Lowe Soper. **Informal name:** Soper Brazilian Typical Mosquito.

urichii (Coquillett)
1906e: 61 (F; *Melanoconion*). **Type loc.:** Trinidad [Trinidad & Tobago, Lesser Antilles] (USNM). **Distr.:** Bolivia, Brazil, Colombia, Costa Rica, Ecuador, French Guiana, Guyana, Lesser Antilles (includes Trinidad and Tobago), Panama, Peru, Suriname, Venezuela. **Refs.:** Dyar 1928: 283 (M*, F, L*); Senevet & Abonnenc 1939a: 128 (M*, L*); Lane 1953: 511 (M*, F*, P*, L*), 513 (M*, F; as *mathesoni*); Senevet & Abonnenc 1958a: 290 (P*); Valencia 1973: 48 (M*, F*, P*, L*); Chadee & Tikasingh 1986 (E*); Linton et al. 2013 (bion., mol. tax., distr.; Ecuador); Berti et al.

2015 (distr.; Venezuela). **Ety.:** F.W. Urich. **Informal name:** Urich Trinbagonian Typical Mosquito.

syn. *mathesoni* Anduze
1942a: 47 (M*). **Type loc.:** Peñón de Paraitepui, Venezuela (FH). **Refs.:** Forattini et al. 1970: 43 (type info.); Valencia 1973: 48 (syn.).

wannonii Cova García & Sutil O
1976b: 354 (M*, P*, L*). **Type loc.:** La Florida, Táchira, Venezuela (ND). **Distr.:** Venezuela. **Ety.:** Luis Wannoni Lander. **Informal name:** Wannoni Venezuelan Typical Mosquito.

wilsoni Lane & Whitman
1943: 393 (M*). **Type loc.:** Vale do Canaã, São João de Petrópolis, Espirito Santo, Brazil (INER). **Distr.:** Brazil, Colombia, Ecuador. **Refs.:** Valencia 1973: 79 (M*, P*, L*); Xavier 1973: 283 (type info.); Linton et al. 2013 (bion., mol. tax., distr.; Ecuador). **Ety.:** D. Bruce Wilson. **Informal name:** Wilson Brazilian Typical Mosquito.

Subgenus *Culex* Linnaeus

1758: 602. Type species: *Culex pipiens* Linneaus (Latreille 1810: 442).

Name derivation: gnat, midge (L).

syn. *Heteronycha* Lynch Arribálzaga 1891a: 373, 1891b: 155 (as genus). Type species: *Heteronycha dolosa* Lynch Arribálzaga.

syn. *Lasioconops* Theobald 1903c: viii (as genus). Type species: *Lasioconops poicilipes* Theobald. Edwards 1911b: 256 (syn.).

syn. *Heptaphlebomyia* Theobald 1903a: 336 (as genus). Type species: *Heptaphlebomyia simplex* Theobald. Edwards 1911b: 262 (syn. of *simplex* with *univittatus* Theobald).

syn. *Pseudoheptaphlebomyia* Ventrillon 1905b: 427 (as genus). Type species: *Pseudoheptaphlebomyia madagascariensis* Ventrillon. Edwards 1932a: 200 (syn.).

syn. *Trichopronomyia* Theobald 1905f: 98 (as genus). Type species: *Trichopronomyia annulata* Theobald. Edwards 1932a: 200 (syn.).

syn. *Aporoculex* Theobald 1907: 316. Type species: *Aporoculex punctipes* Theobald.

syn. *Leucomyia* Theobald 1907: 372 (as genus) (not Brauer & Bergenstam 1891). Type species: *Culex gelidus* Theobald. Edwards 1932a: 200 (syn.).

syn. *Theobaldiomyia* Brunetti 1912: 462 (as genus). Type species: *Culex gelidus* Theobald. Theobald 1901c: 20 (fixed type of replacement name). Edwards 1932a: 200 (syn.)

syn. *Phalangomyia* Dyar & Knab 1914: 58 (as genus). Type species: *Phalangomyia debilis* Dyar & Knab. Edwards 1932a: 200 (syn.).

syn. *Laiomyia* Izquierdo 1916: 65 (as genus). Type species: *Culex stigmatosoma* Dyar. Edwards 1932a: 200 (syn.).

syn. *Cacoculex* Dyar 1918b (as subg. of genus *Culex*): 100. Type species: *Culex habilitator* Dyar & Knab. Edwards 1932a: 200 (syn.).

syn. *Transculicia* Dyar 1918e (as subg. of genus *Culex*): 184. Type species: *Culex eleuthera* Dyar. Edwards 1932a: 205 (syn.).

Important references: Belkin 1962 (revision; South Pacific); Bram 1967a (revision; Neotropics); Belkin, Schick, & Heinemann 1968 (type loc. info.; South America); Belkin, Schick, & Heinemann 1971 (type loc. info.; Brazil); Chen 1972: 282 (cibarium key to some *Culex* spp.); Cupp & Ibrahim 1973: 277 (Pipiens Complex); Matsuo et al. 1974: 51 (Mimeticus Group); Sirivanakarn 1975a: 69 (Vishnui Complex); Sirivanakarn 1976 (revision; Oriental Region); Harbach 1988 (revision; SW Asia); Lee et al. 1989a (revision; Australasia); Deus 2009 (mol. phyl.); Vesgueiro et al. 2011 (mol. phyl.); Demari-Silva et al. 2011 (mol. phyl.); Harbach et al. 2012 (phyl.); Laurito et al. 2013 (phyl.); Demari-Silva et al. 2015 (mol. phyl.); Harbach et al. 2016 (phyl.).

abnormalis Lane
1936b: 189 (M*, F, L*). **Type loc.:** Coronel Ponce, Mato Grosso, Brazil (FH). **Distr.:** Brazil, Colombia. **Refs.:** Lane 1953: 368 (M*, F, L*); Bram 1967a: 20 (M*, F, L); Cotrim et al. 1974: 9 (M*). **Ety.:** n.s. [abnormal (L)]. **Informal name:** Abnormal Brazilian Typical Mosquito.

syn. *scutatus* Rozeboom & Komp
1948: 396 (M*, F*, L*). **Type loc.:** Villavicencio [Meta], Colombia (USNM). **Refs.:** Bram 1967a: 20 (syn.).

acharistus Root
1927c: 578 (M*, F). **Type loc.:** Agua Limpa, Minas Gerais, Brazil (USNM). **Distr.:** Argentina, Brazil, Chile, Colombia. **Refs.:** Lane 1953: 347 (M*, F, P*); Stone & Knight 1957a: 42 (lectotype desig.); Bachmann & Casal 1963 (F*); Bram 1967a: 22 (M*); Laurito et al. 2009 (M*, F, P*, L*). **Ety.:** n.s. [unpleasant, disagreeable (L)]. **Informal name:** Unpleasant Brazilian Typical Mosquito.

alani Forattini
1965a: 143 (M*). **Type loc.:** Catera a Gachancipa, Colombia (USNM). **Distr.:** Colombia. **Refs.:** Bram 1967a: 23 (M*). **Ety.:** Alan Stone. **Informal name:** Alan Colombian Typical Mosquito.

alienus Colless
1957a: 97 (M, F, L*). **Type loc.:** Long Tebangan, Fourth Division, Sarawak, Borneo [Malaysia] (NHMUK). **Distr.:** Cambodia, Malaysia, Singapore, Thailand, Vietnam. **Refs.:** Bram 1967b: 202 (M*, F, P, L*; bion., distr., tax.); Sirivanakarn 1976: 122 (M*, F, P*, L*; distr.); Rattanarithikul et al. 2005b (F*, L*; bion., distr., keys). **Ety.:** n.s. [from somewhere else, foreign (L); poss. ref. to non-type-locality specimens in a museum collection]. **Informal name:** Alien Malaysian Typical Mosquito.

alis Theobald
1903a: 167 (M*, F). **Type loc.:** Christmas Island (NHMUK). **Distr.:** Cambodia, Guam, Indonesia, Malaysia, Papua New Guinea, Philippines, Singapore, Taiwan, Thailand, Vietnam. **Refs.:** Brug 1934a: 517 (M, F*); Colless 1957b: 465 (L*); Bram 1967c (type info., lectotype desig.); Sirivanakarn 1976: 100 (M*, F, P*, L*; distr.); Lee et al. 1989a: 33 (F; key, tax., bion., distr., review); Rattanarithikul et al. 2005b (F*, L*; bion., distr., keys). **Ety.:** wing (attribute) (L); "Wing with the fork-cells short, their stems longer than their cells." **Informal name:** Wing Christmas Islander Typical Mosquito.
syn. *neolitoralis* Bram
1967b: 236 (M*, F, L*; bion., distr., tax.). **Type loc.:** Ban Laem Sing, Chanthaburi, Thailand (USNM). **Refs.:** Sirivanakarn 1976: 100 (syn.).

ameliae Casal
1967: 29 (M*). **Type loc.:** San Pedro and El Dorado Road, 10 km from San Pedro, Misiones, Argentina (INM). **Distr.:** Argentina. **Refs.:** Laurito et al. 2011 (P*, L*). **Ety.:** Amelia Casal. **Informal name:** Amelia Argentinian Typical Mosquito.

andersoni Edwards
1914b: 65 (M*, L). **Type loc.:** Kabete [Kenya] (NHMUK). **Distr.:** Comoros, Democratic Republic of the Congo, Eritrea, Ethiopia, Kenya, Malawi, Republic of South Africa, Rwanda, Tanzania, Uganda. **Refs.:** Edwards 1941: 324 (M*, F*), 422 (P); Hopkins 1952: 310 (L). **Ety.:** Anderson (collector). **Informal name:** Anderson Kenyan Typical Mosquito.
ssp. ***abyssinicus*** Edwards
1941: 325 (M*). **Type loc.:** Addis Ababa [Shawa], Ethiopia (NHMUK). **Distr.:** Ethiopia. **Refs.:** Hopkins 1952: 310 (L); Mattingly 1956a: 42 (lectotype desig.). **Ety.:** Abyssinia, Ethiopia. **Informal name:** Abyssinia Ethiopian Typical Mosquito.
ssp. ***bwambanus*** Edwards
1941: 325 (M*). **Type loc.:** Bwamba Pass, Ruwenzori [Toro District], Uganda (NHMUK). **Distr.:** Lesotho, Republic of South Africa, Uganda. **Refs.:** De Meillon 1943: 99 (M, F, L*); De Meillon et al. 1945: 93 (P; as *andersoni* var.); Hopkins 1952: 310 (L*); Mattingly 1956a: 42 (type info.); Jupp 1996 (M*, F*; key). **Ety.:** Bwamba Pass, Uganda. **Informal name:** Bwamba Ugandan Typical Mosquito.

annulirostris Skuse
1889: 1737 (F). **Type loc.:** Blue Mountains and Berowra, New South Wales, Australia (MM). **Distr.:** Australia, Bismarck Archipelago, Cook Isl. (Polynesia), Easter Isl., Fiji, French Polynesia, Guam, Indonesia, Kiribati, Mariana Isl., Micronesia (Wake Isl.), Nauru, New Caledonia, Palau, Papua New Guinea, Philippines, Samoa (Ind. State of Samoa, American Samoa, Western Samoa), Solomon Isl., Tahiti, Timor, Tonga, Tuvalu, Vanuatu. **Refs.:** Bancroft 1908: 43 (M, F); Bohart & Ingram 1946b: 18 (M*, F, L*; bion., distr., tax.); Knight & Hurlbut 1949: 30 (M*, F, P, L*; tax.); Penn 1949b: 81 (P*); Miller & Phillips 1952: 23 (F*, L*); Bohart 1957: 80 (M*, F, L*; distr.; Micronesia); Belkin 1962: 207 (M*, F, P*, L*); Dobrotworsky 1965: 207 (M*, F*, L*); Delfinado 1966a: 139 (M*, F, L*; syn.); Basio 1971b: 53 (M*; bion.); Baisas 1974: 87 (M*, F*, P, L*; tax., bion., distr.; Philippines); Ramalingam 1976: 308 (tax., bion., keys., distr.; Samoa and Tonga); Sirivanakarn 1976: 107 (M*, F*, P*, L*; distr.); Kay 1978 (tax.); Marks 1982 (tax.); Russell 1986a (bion.); Russell 1987 (bion.); Lee et al. 1989a: 36 (F; key, tax., bion., distr., review); Whelan & Hapgood 2000 (bion., distr.; East Timor); Becker et al. 2010: 365 (F*, L*; key, tax., distr., bion.); Jansen et al. 2013 (morph., mol. tax.). **Ety.:** ringed beak (L); proboscis. **Informal name:** Ring-beaked Australian Typical Mosquito. (Vol. 1, p. 486)
syn. *jepsoni* Bahr
1912: 18 (F*). **Type loc.:** Fiji Islands (NHMUK).
syn. *somerseti* Taylor
1912b: 28 (F). **Type loc.:** Somerset, Queensland, Australia (ANIC). **Refs.:** Lee et al. 1989a: 36 (type info.).
syn. *consimilis* Taylor
1913: 55 [alternate pub., p. 8] (M*, F; *Culicelsa*). **Type loc.:** Ayr, Queensland, Australia (ANIC). **Refs.:** Lee et al. 1989a: 36 (type info.).
syn. *simplex* Taylor
1914c: 698 (F*; *Culicelsa*). **Type loc.:** Townsville, Queensland, Australia (ANIC).
syn. *palmi* Baisas
1938: 207 (M*, F*, P*, L). **Type loc.:** Paranaque, Rizal, Luzon, Philippines (NE). **Refs.:** Lee et al. 1989a: 36 (type info.).
ssp. ***marianae*** Bohart & Ingram
1946b: 42 (M*, L*; bion., distr., tax.). **Type loc.:** Charan Kanoa, Saipan, Mariana Islands (USNM). **Distr.:** Mariana Isl. **Refs.:** Bohart & Ingram 1946b: 32 (M*, L*; bion., distr.); Yamaguti & LaCasse 1950a: 18 (M*, F*, P*, L*). **Ety.:** Mariana Isl. **Informal name:** Saipan Mariana Islander Typical Mosquito.

annuliventris (Blanchard)
1852: 334 (A; *Anopheles*). **Type loc.:** Valdivia, Chile (MNHP). **Distr.:** Chile. **Refs.:** Belkin 1968b: 13 (lectotype desig; from syn. with *albifasciatus* (Macquart)). **Ety.:** ringed belly (L); abdominal venter. **Informal name:** Ring-bellied Chilean Typical Mosquito.

annulus Theobald
1901a: 358 (F). **Type loc.:** Tai Po, Pokfulam, Hong Kong [People's Republic of China] (NHMUK). **Distr.:** Bangladesh, Cambodia, Indonesia, Malaysia, Myanmar, People's Republic of China (includes Hong Kong), Philippines, Vietnam. **Refs.:** Colless 1957a: 93 (M, F, L*; from syn. with *tritaeniorhynchus*); Bram 1967b: 205

(M*, F*, P*, L*; bion., distr., tax.); Reuben 1969: 643 (syn. with *vishnui*); Basio 1971b: 54 (M*; bion.); Baisas 1974: 86 (M*, F*, P, L*; tax., bion., distr.; Philippines); Lien et al. 1975 (distr.; Sumatra; as valid sp.); Chen 1980b: 436 (from syn. with *vishnui*). **Ety.:** ringed (L); femur. **Informal name:** Ringed Chinese Typical Mosquito.

syn. *pseudoinfula* Theobald
1911d: 237 (F). **Type loc.:** Pasuruan and Semarang, Java [Indonesia] (AM). **Refs.:** Colless 1957a: 93 (syn.).

syn. *adelae* Baisas
1938: 200 (M*, F*, P*, L). **Type loc.:** Tungkong Mangga [San Jose, Bulacan, Luzon] Philippines (NE). **Refs.:** Colless 1957a: 95 (syn.).

syn. *permixtus* Hsieh & Liao
1956: 123 (M*, L*). **Type loc.:** Amoy, China [People's Republic of China] (LU). **Refs.:** Chen 1980b: 436 (syn.).

syn. *cheni* Ho
1963: 368 (M*, F, L*). **Type loc.:** Canton, China [People's Republic of China] (CMC). **Refs.:** Chen 1980b: 436 (syn.).

antennatus (Becker)
1903: 68 (M; *Anopheles*). **Type loc.:** Cairo, Egypt (ZM). **Distr.:** Algeria, Angola, Benin, Botswana, Burkina Faso, Cameroon, Côte d'Ivoire, Democratic Republic of the Congo, Egypt, Ethiopia, Iran, Iraq, Israel (and Gaza Strip and West Bank), Jordan, Kenya, Liberia, Madagascar (includes Glorioso and Juan De Nova Isl.), Mali, Mauritania, Morocco, Nigeria, São Tomé and Príncipe, Senegal, Seychelles, Sudan and South Sudan, Tanzania, Tunisia, Turkey, Uganda. **Refs.:** Séguy 1924: Pl. V (M gen.*; as *laurenti*); Kirkpatrick 1925b: 129 (M*, F*, P*, L*; as *laurenti*); Hopkins 1936: 224 (L; new comb.); Edwards 1941: 333 (M*, F*), 425 (P); Larivière & Abonnenc 1956: 1195 (M*, L*, E*; bion., morph.); Lotfi 1970: 400 (distr.); Margalit & Tahori 1974: 86 (distr.); Gad et al. 1987 (bion., distr.; Egypt); Harbach 1988: 48 (M*, F, P*, L*; tax., keys, bion., distr., lit. review); da Cunha Ramos et al. 1992b: 163 (distr.; Príncipe Isl.); Jupp 1996 (M*, F*; key); Amr et al. 1997 (distr.; Jordan); Alten et al. 2000: 28 (distr.); Becker et al. 2010: 328 (F*, L*; key, tax., distr., bion.); Robert et al. 2019 (distr.; Western Palaearctic). **Ety.:** antenna (L); various antennal chars. **Informal name:** Antenna Egyptian Typical Mosquito. (Vol. 1, p. 488)

syn. *laurenti* Newstead
1907 (M, F). In Newstead et al. 1907: 24. **Type loc.:** Leopoldville, Belgian Congo [Kinshasa, Democratic Republic of the Congo] (NHMUK). **Refs.:** Hopkins 1936: 224 (syn.); Harbach 1988: 48 (tax., review; as syn. of *antennatus*).

apicinus Philippi
1865: 596 (A). **Type loc.:** Santiago, Chile (NE). **Distr.:** Argentina, Bolivia, Chile, Peru. **Refs.:** Lane 1953: 341 (M*, F, L*); Bram 1967a: 24 (M*, F, L); Rossi et al. 2008 (M*, F*, P*, L*). **Ety.:** n.s. [tip, top (apical?) (L)]. **Informal name:** Apex Chilean Typical Mosquito.

syn. *debilis* Dyar & Knab
1914: 58 (M, F; *Phalangomyia*). **Type loc.:** Matucana [Lima], Peru (USNM). **Refs.:** Stone & Knight 1957a: 47 (lectotype desig.).

syn. *escomeli* Brèthes
1920 (M*, F). **Type loc.:** Arequipa, Peru (BA).

syn. *alticola* Martini
1931a: 216 (A). **Type loc.:** Serata (2,300 m), Bolivia (NHMUK). **Refs.:** Mattingly 1955c: 31 (lectotype desig.); Stone 1957a: 340 (M*); Belkin 1968b: 12 (tax.).

aquarius Strickman
1990: 150 (M*, F*, P*, L*). **Type loc.:** San Isidro de Coronado, San Jose Province, Costa Rica (USNM). **Distr.:** Costa Rica, Panama. **Ety.:** of water (L); contrast to *salinarius*—of salt water. **Informal name:** Watery Costa Rican Typical Mosquito.

archegus Dyar
1929b: 511 (M). **Type loc.:** Tarma [Junin], Peru (USNM). **Distr.:** Bolivia, Chile, Colombia, Ecuador, Peru. **Refs.:** Lane 1951: 334 (syn.); Leví-Castillo 1952b: 901 (M*, F*, P*, L*); Leví-Castillo 1953a: 187 (M*, F, P*, L*); Bram 1967a: 28 (M*, F, L; from syn. with *articularis*). **Ety.:** n.s. [originator, author, founder? (Gr)]. **Informal name:** Original Peruvian Typical Mosquito.

argenteopunctatus (Ventrillon)
1905a: 446 (M, F; *Heptaphlebomyia*). **Type loc.:** Tananarive, Madagascar (MNHP, NHMUK). **Distr.:** Benin, Burkina Faso, Cameroon, Central African Republic, Côte d'Ivoire, Kenya, Madagascar (includes Glorioso and Juan De Nova Isl.), Mali, Mozambique, Nigeria, Senegal, Togo. **Refs.:** Edwards 1941: 303 (M*, F*); Worth & De Meillon 1960: 248 (distr.); Brunhes 1967: 213 (M*, P*, L*); Townsend 1990: 45 (type info.). **Ety.:** n.s. [silvery ornamented (L)]. **Informal name:** Silver-spotted Malagasy Typical Mosquito.

ssp. ***kingii*** (Theobald)
1913d: 601 (F; *Heptaphlebomyia*; as sp.). **Type loc.:** Nyumbe and Alenga, Lado District, Sudan [South Sudan] (NHMUK). **Distr.:** Angola, Burkina Faso, Cameroon, Central African Republic, Democratic Republic of the Congo, Ghana, Kenya, Namibia, Nigeria, Republic of the Congo, Republic of South Africa, Senegal, Sierra Leone, Sudan and South Sudan. **Refs.:** Edwards 1941: 304 (F; to ssp.), 419 (P*), 432 (L*); Lewis 1949: 63 (L*); Rickenbach et al. 1976: 96 (distr.); Jupp 1996 (M*, F*; key). **Ety.:** n.s. [poss. Mr. Harold King, collector (Harbach 2018c)]. **Informal name:** King Sudanese Typical Mosquito.

articularis Philippi
1865: 596 (A). **Type loc.:** Casa Pangue, Llanquihue, Chile (USNM). **Distr.:** Argentina, Chile, Ecuador, Peru. **Refs.:** Dyar 1928: 358 (M*, F); Lane 1953: 362 (M*, F); Bram 1967a: 30 (M*, F; neotype desig.); Laurito et al. 2011

(P*, L*). **Ety.:** n.s. [jointed? (L)]. **Informal name:** Jointed Chilean Typical Mosquito.

asteliae Belkin
1968a: 66 (M*, F, P*, L*). **Type loc.:** Rangitoto Island, North Auckland, New Zealand (NELS). **Distr.:** New Zealand. **Refs.:** Lee et al. 1989a: 87 (F; key, tax., bion., distr., review); Derraik 2004 (distr.; New Zealand). **Ety.:** in axils of broad-leaf *Astelia* spp. **Informal name:** Astelia New Zealander Typical Mosquito.

astridianus De Meillon
1942: 89 (M*, F). **Type loc.:** Astrida, Ruanda Urundi, Belgian Congo [Democratic Republic of the Congo] (SAIM). **Distr.:** Democratic Republic of the Congo. **Refs.:** White 1975: 316 (distr.). **Ety.:** Astrida (Zaire). **Informal name:** Astrida Congolese Typical Mosquito.

atriceps Edwards
1926a: 105 (F*). **Type loc.:** Papeari [Tahiti-nui District], Tahiti [Society Islands] (NHMUK). **Distr.:** French Polynesia. **Refs.:** Edwards 1928a: 279 (M*); Marks 1951b: 129 (A*); Stone & Rosen 1953: 356 (L*); Belkin 1962: 198 (M*, F, P*, L*); Lee et al. 1989a: 88 (F; key, tax., bion., distr., review). **Ety.:** dark head (L); head with blackish integument and scales. **Informal name:** Dark-headed Tahitian Typical Mosquito.
syn. *nigriceps* Buxton
1927: 307 (*lapsus* for *atriceps*).

australicus Dobrotworsky & Drummond
1953: 143 (M*, F, P, L*; as *pipiens* ssp.). **Type loc.:** Melbourne, Victoria, Australia (NMM). **Distr.:** Australia, New Caledonia, Vanuatu. **Refs.:** Dobrotworsky 1955b: 42 (M*); Belkin 1962: 194 (M*, F, L*; to sp.); Dobrotworsky 1965: 213 (M*, F*, L*); Lee et al. 1989a: 90 (F; key, tax., bion., distr., review). **Ety.:** Australia. **Informal name:** Gunai Australian Typical Mosquito [Gunai is an indigenous Australian nation of southeastern Australia].

bahamensis Dyar & Knab
1906d: 210 (L*). **Type loc.:** Bahama Islands (USNM). **Distr.:** Bahamas, Cayman Isl., Cuba, Dominican Republic, Haiti, Jamaica, Lesser Antilles (includes Anguilla, Antigua and Barbuda, Guadeloupe, Montserrat, Saint Kitts and Nevis, Saint Barts, Saint Croix, Saint Johns, Saint Thomas, Sint Maarten, Trinidad and Tobago, Virgin Isl.), Mexico, Puerto Rico, United States (continental). **Refs.:** Gerry 1932 (F*); Yamaguti & LaCasse 1951c: 17 (M*, F*, L*); van der Kuyp 1954: 77 (P); Carpenter & LaCasse 1955: 274 (M*, F*, L*; keys); Stone & Knight 1957a: 43 (lectotype desig.); Bram 1967a: 32 (M*, F, L); Belkin et al. 1970: 59 (M*, F, P*, L*); Darsie & Ward 2005 (F*, L*; keys, distr.). **Ety.:** Bahamas. **Informal name:** Lucayan Bahamian Typical Mosquito [the Lucayans were the indigenous people of the Bahamas prior to the arrival of Europeans].
syn. *eleuthera* Dyar
1918e: 184 (M, F). **Type loc.:** Governor's Harbor, Eleuthera Island, Bahama Islands (USNM). **Refs.:** Stone & Knight 1957a: 48 (lectotype desig.).
syn. *petersoni* Dyar
1920k: 27 (M, F, L). **Type loc.:** Leinster Bay, Saint John Island [Lesser Antilles] (USNM).

banksensis Maffi & Tenorio
1977: 503 (M*, F, P*, L*). **Type loc.:** Qeteqavite, Gava Island, Banks Island, New Hebrides. **Distr.:** Vanuatu. **Refs.:** Lee et al. 1989a: 99 (F; key, tax., bion., distr., review). **Ety.:** Banks Isl., New Hebrides [Vanatu]. **Informal name:** Banks Ni-Vanuatu Typical Mosquito.

barraudi Edwards
1922a: 284 (M*). **Type loc.:** Madhopur [Punjab Hill States], India (NHMUK). **Distr.:** Cambodia, India, Nepal, Pakistan, People's Republic of China, Sri Lanka, Thailand. **Refs.:** Barraud 1923c (L*); Barraud 1934: 403 (M*, F*); Peters & Dewar 1956: 46 (M*, L*); Mattingly 1958c: 106 (lectotype desig.); Bram 1967b: 209 (M*, F, L*; bion., distr., tax.); Aslamkhan 1971b (distr.; Pakistan); Marks 1971b (tax.); Sirivanakarn 1976: 138 (M*, F, P, L*; distr.); Lee et al. 1989a: 100 (tax.); Darsie & Pradhan 1990, 1991 (F, L; tax., keys, bion., distr.; Nepal); Rattanarithikul et al. 2005b (F*, L*; bion., distr., keys). **Ety.:** Philip James Barraud. **Informal name:** Barraud Indian Typical Mosquito.

beta Séguy
1924: Pl. VIII, 41 (L*). **Type loc.:** Algeria (MNHP). **Distr.:** Algeria. **Refs.:** Senevet 1947a: 128 (tax.). **Ety.:** n.s. [second letter of the Greek alphabet]. **Informal name:** Beta Algerian Typical Mosquito.

bickleyi Forattini
1965a: 146 (M*). **Type loc.:** Bosque Calderón, Bogotá, Colombia (USNM). **Distr.:** Colombia, Ecuador. **Refs.:** Bram 1967a: 33 (M*). **Ety.:** William E. Bickley. **Informal name:** Bickley Colombian Typical Mosquito.

bidens Dyar
1922g: 190 (M*, F). **Type loc.:** Rosario [Beni], Bolivia (USNM). **Distr.:** Argentina, Bolivia, Brazil, Costa Rica, Mexico, Nicaragua, Paraguay, Uruguay, Venezuela. **Refs.:** Lane 1953: 351 (syn. with *virgultus* = *Cx. declarator*); Stone & Knight 1957a: 44 (lectotype desig.); Bram 1967a: 35 (M*, F, L; to valid sp.); Harbach et al. 1986a: 139 (M*; tax.); Rossi & Martínez 2003: 471 (distr.; Uruguay). **Ety.:** n.s. [two toothed (L); poss. ref. to "mesosome stout, with two long teeth from the apex"]. **Informal name:** Two-toothed Bolivian Typical Mosquito.

bihamatus Edwards
1926a: 107 (M). **Type loc.:** Atamboea, Timor [Indonesia] (NHMUK). **Distr.:** Indonesia. **Refs.:** Sirivanakarn 1974: 259 (M*); Lee et al. 1989a: 101 (tax., review). **Ety.:** two-hooked (L); two pairs of equal curved hooks on

mesosome. **Informal name:** Two-hooked Indonesian Typical Mosquito.

bilineatus Theobald
1903a: 196 (M, F). **Type loc.:** São Paulo, São Paulo State, Brazil (NHMUK). **Distr.:** Brazil. **Refs.:** Lane 1951: 334 (syn.); Belkin 1968b: 13 (lectotype desig.); Belkin et al. 1971: 21 (type loc. info.); Laurito et al. 2013 (from syn. with *dolosus*). **Ety.:** two-lined (L); "thorax with two black parallel median lines." **Informal name:** Two-lined Brazilian Typical Mosquito.

bonneae Dyar & Knab
1919: 3 (M, F, L). **Type loc.:** Compagnie des Mines d'Or, Lawa River, Suriname (USNM). **Distr.:** Brazil, Colombia, Costa Rica, Ecuador, French Guiana, Guyana, Panama, Suriname. **Refs.:** Dyar 1928: 385 (M*, F, L*); Senevet & Abonnenc 1939a: 78 (M*, L*); Lane 1953: 330 (M*, F, L*); Stone & Knight 1957a: 44 (lectotype desig.); Bram 1967a: 36 (M*, F, L); Linton et al. 2013 (bion., mol. tax., distr.; Ecuador). **Ety.:** Johanna Bonne-Webster. **Informal name:** Lawa Surinamese Typical Mosquito.

brami Forattini, Rabello, & Souza Lopes
1967: 13 (M*, F*). **Type loc.:** Boraceia, Casa Grande, Municipio de Salesopolis, São Paulo, Brazil (FH). **Distr.:** Brazil. **Ety.:** Ralph Albert Bram. **Informal name:** Bram Brazilian Typical Mosquito.

brethesi Dyar
1919c: 86 (M). **Type loc.:** [San Isidro, Buenos Aires] Argentina (BA). **Distr.:** Argentina, Bolivia, Uruguay, Venezuela. **Refs.:** Dyar 1928: 378 (M*, F); Bachmann & Casal 1962a: 78 (M*, P*, L*); Bram 1967a; 38 (M*, F, L); Darsie, Becnel et al. 1991 (distr.; Argentina, Uruguay); Rossi & Martínez 2003: 471 (distr.; Uruguay); Rossi 2006 (M*, P*, L*). **Ety.:** Juan Brèthes. **Informal name:** Brèthes Argentinian Typical Mosquito.

brevispinosus Bonne-Wepster & Bonne
1920: 171 (M, F, L). **Type loc.:** Kwakoegron, Saramacca, Suriname (ITH). **Distr.:** Brazil, Colombia, French Guiana, Lesser Antilles (includes Trinidad and Tobago), Suriname, Venezuela. **Refs.:** Dyar 1928: 389 (M*, F, L*); Lane 1953: 334 (M*, F, P, L*); Bram 1967a: 40 (M*, F, L); Belkin 1968b: 14 (lectotype desig.); Berti et al. 2015 (distr.; Venezuela). **Ety.:** short spine (L); short spines at tip of siphon, comp. to *coronator*. **Informal name:** Short-spined Surinamese Typical Mosquito.

brumpti Galliard
1931a: 134 (M*, F, L*). **Type loc.:** Several localities on Corsica (NHMUK). **Distr.:** Algeria, France, Italy, Morocco, Spain (includes Corsica). **Refs.:** Aitken 1954: 483 (M, L); Townsend 1990: 55 (type info.); Becker et al. 2010: 269 (M*, F*, L*; keys, tax., distr., bion.); Robert et al. 2019 (distr.; Western Palaearctic). **Ety.:** Alexandre Joseph Émile Brumpt. **Informal name:** Brumpt Corsican Typical Mosquito.

bukavuensis Wolfs
1947b: 159 (M*, F, P*, L*). **Type loc.:** Kawa, Kivu, Belgian Congo [Democratic Republic of the Congo] (CMT). **Distr.:** Democratic Republic of the Congo. **Ety.:** Bukavu, Democratic Republic of the Congo. **Informal name:** Bukavu Congolese Typical Mosquito.

calurus Edwards
1935a: 136 (M). **Type loc.:** Mount Kinangop, Aberdare Mountains, Kenya (NHMUK). **Distr.:** Kenya. **Refs.:** Edwards 1941: 320 (M*). **Ety.:** n.s. [country, farm (L)]. **Informal name:** Rural Kenyan Typical Mosquito.

camposi Dyar
1925d: 28 (M; as race of *coronator*). **Type loc.:** Ecuador (USNM). **Distr.:** Colombia, Ecuador, Peru, Venezuela. **Refs.:** Lane 1953: 333 (M; from syn. with *coronator*); Bram 1967a: 49 (M*, F; to sp.); Linton et al. 2013 (bion., mol. tax., distr.; Ecuador). **Ety.:** Francisco O. Campos Rivadeneira. **Informal name:** Campos Ecuadorian Typical Mosquito.

carcinoxenus de Oliveira Castro
1932 (M*, F, P*, L*). **Type loc.:** Bertioga, São Paulo, Brazil (IOC). **Distr.:** Brazil. **Refs.:** Forattini et al. 1956: 85 (P*); Bram 1967a: 40 (M*, F, L); Belkin et al. 1971: 21 (lectotype desig.). **Ety.:** crab guest (Gr); larvae in crabholes. **Informal name:** Crab Brazilian Typical Mosquito.

carleti Brunhes & Ravaonjanahary
1971 (M*, F, P*, L*). **Type loc.:** Ifaho, Sous-Prefecture de Manakara, Province de Fianarantsoa, Madagascar (IERT). **Distr.:** Comoros, Madagascar (includes Glorioso and Juan De Nova Isl.). **Refs.:** Brunhes 1977a (distr.; Comoros Isl.). **Ety.:** n.s. **Informal name:** Carlet Malagasy Typical Mosquito.

castelli Hamon
1957: 686 (M*). **Type loc.:** Bouake, Ivory Coast [Côte d'Ivoire] (IERT). **Distr.:** Côte d'Ivoire. **Ety.:** n.s. [fortress (L)]. **Informal name:** Castell Ivorian Typical Mosquito.

chidesteri Dyar
1921h: 117 (M*). **Type loc.:** Colon, Canal Zone, Panama (USNM). **Distr.:** Bahamas, Belize, Bolivia, Brazil, Chile, Colombia, Costa Rica, Cuba, Ecuador, El Salvador, Guatemala, Jamaica, Lesser Antilles (Guadeloupe, Saint Kitts and Nevis, Saint Lucia, Saint Vincent and the Grenadines), Mexico, Panama, Paraguay, Puerto Rico, United States (continental), Uruguay, Venezuela. **Refs.:** Yamaguti & LaCasse 1951c: 21 (M*, F*, L*); Lane 1953: 375 (M*, F, P*, L*); Carpenter & LaCasse 1955: 275 (M*, F*, L*; keys); Stone & Knight 1957a: 45 (lectotype desig.); Bram 1967a: 42 (M*, L; syn.); Belkin et al. 1970: 64 (M*, F, P*, L*); Pecor et al. 2002: 242 (tax., distr.; Belize); Rossi & Martínez 2003: 472 (distr.; Uruguay); Darsie & Ward 2005 (F*, L*; keys, distr.). **Ety.:** W.S. Chidester. **Informal name:** Chidester Panamanian Typical Mosquito.

syn. *finlayi* Pérez Vigueras
1956: 382 (M*, F*, L*). **Type loc.:** Finca "El Pino," Casiguas, Barrio de Jaruco (Habana), Cuba (LU). **Refs.:** Montchadsky & García Avila 1966: 43 (syn.).
syn. *deanei* Corrêa & Ramalho
1959b: 141 (M*, F, P, L*). **Type loc.:** Campo de Marte, Bairro de Santana, São Paulo, Brazil (FH).

chitae Duret
1967b: 113 (M*). **Type loc.:** Monserrat, near Bogotá, Colombia (USNM). **Distr.:** Colombia. **Refs.:** Harbach et al. 1991: 193 (type info.). **Ety.:** named for spouse, full name n.s. **Informal name:** Chita Colombian Typical Mosquito.

chorleyi Edwards
1941: 330, 423, 433 (M*, P, L). **Type loc.:** Masaka, Uganda (NHMUK). **Distr.:** Burundi, Cameroon, Democratic Republic of the Congo, Ethiopia, Republic of the Congo, Republic of South Africa, Uganda. **Refs.:** Hopkins 1952: 316 (L*); Mattingly 1956a: 43 (lectotype desig.); Jupp 1996 (M*, F*; key). **Ety.:** n.s. [T.W. Chorley]. **Informal name:** Chorley Ugandan Typical Mosquito.

comorensis Brunhes
1977b: 136 (M*, F, P*, L*). **Type loc.:** Near M'jamaoue, Anjouan (SSC). **Distr.:** Comoros, Madagascar (includes Glorioso and Juan De Nova Isl.). **Ety.:** Comoros Isl. **Informal name:** Lunar Comoran Typical Mosquito [the name Comoros is taken from the Arabic word for moon].
ssp. ***kartalae*** Brunhes
1977b: 140 (M*, F, L). **Type loc.:** Karthala Volcano, Grande Comore Island, Comoros (SSC). **Distr.:** Comoros. **Ety.:** Karthala Volcano, Comoros Island. **Informal name:** Karthala Comoran Typical Mosquito.

coronator Dyar & Knab
1906d: 215 (L*). **Type loc.:** Saint Joseph, Trinidad [Trinidad & Tobago, Lesser Antilles] (USNM). **Distr.:** Antigua and Barbuda, Argentina, Belize, Bolivia, Brazil, Colombia, Costa Rica, Dominican Republic, Ecuador, El Salvador, French Guiana, Guatemala, Guyana, Honduras, Lesser Antilles (includes Trinidad and Tobago), Mexico, Nicaragua, Panama, Paraguay, Peru, Saint Lucia, Suriname, United States (continental), Uruguay, Venezuela. **Refs.:** Yamaguti & LaCasse 1951c: 25 (M*, F*, L*); Lane 1953: 331 (M*, F, P*, L*); Carpenter & LaCasse 1955: 277 (M*, F*, L*); Stone & Knight 1957a: 46 (lectotype desig.); Bram 1967a: 47 (M*, F, L); Belkin & Heinemann 1976a: 288 (distr.; Saint Lucia); Rossi & Martínez 2003: 471 (distr.; Uruguay); Darsie & Ward 2005 (F*, L*; keys, distr.); Berti et al. 2015 (distr.; Venezuela); Harrison et al. 2016 (F*, L*; keys, distr., tax.); Laurito et al. 2017 (phyl.). **Ety.:** crown (L); ref. to crown of short "spikes" on the larval siphon. **Informal name:** Crowned Trinbagonian Typical Mosquito.
syn. *mooseri* Vargas & Martínez Palacios
1954: 33 (M*, F; *coronator* ssp.). **Type loc.:** Campeche, Mexico (ISET). **Refs.:** Bram 1967a: 46 (syn. with *coronator*).

covagarciai Forattini
1965a: 128 (M). **Type loc.:** San Antonio de Caparo, Municipal Capital, Venezuela (USNM). **Distr.:** Venezuela. **Refs.:** Bram 1967a: 53 (M*, L*). **Ety.:** Pablo Cova-García. **Informal name:** Cova-García Venezuelan Typical Mosquito.

crinicauda Edwards
1921c: 77 (new name for *parvus* Taylor, not Macquart 1834). **Distr.:** Australia. **Refs.:** Knight et al. 1944: 43 (M); Lee et al. 1989a: 114 (F; key, tax., bion., distr., review). **Ety.:** n.s. [hairy tail (L); poss. ref. to male genitalia character]. **Informal name:** Hairy-tailed Australian Typical Mosquito.
syn. *parvus* Taylor
1912a: 27 (M, F). **Type loc.:** Umbrawarra Creek, Northern Territory, Australia (US, NHMUK). **Refs.:** Townsend 1990: 118 (type info.).

curvibrachius Angulo
1993 (M*, F). In Angulo & Olivares 1993: 33. **Type loc.:** Concepción, Chile (UCCC). **Distr.:** Chile. **Ety.:** curved arm (L, Gr); ref. to recurved basal arms of the mesosome. **Informal name:** Curve-armed Chilean Typical Mosquito.

cuyanus Duret
1968e: 323 (F*). **Type loc.:** Jáchal, San Juan, Argentina (USNM). **Distr.:** Argentina. **Refs.:** Rossi et al. 2006 (F, P*, L*). **Ety.:** Cuyo Region, Argentina. **Informal name:** Cuyo Argentinian Typical Mosquito.

decens Theobald
1901d: vii (M*, F). **Type loc.:** Bonny [Eastern Provinces], Nigeria (NHMUK). **Distr.:** Angola, Benin, Burkina Faso, Cameroon, Central African Republic, Comoros, Côte d'Ivoire, Gabon, Ghana, Ethiopia, Liberia, Madagascar (includes Glorioso and Juan De Nova Isl.), Mali, Mauritania, Nigeria, Republic of South Africa, São Tomé and Príncipe, Saudi Arabia, Senegal, Sierra Leone, Sudan and South Sudan, Tanzania, The Gambia, Yemen. **Refs.:** Macfie & Ingram 1922 (F*); Edwards 1941: 336 (M*, F*; tax.), 425 (P*); Lewis 1949: 63 (L*); Hopkins 1952: 320 (L*); Brunhes 1977a (distr.; Comoros Isl.); Harbach 1988: 44 (M*, F, P*, L*; tax., keys, bion., distr.); Jupp 1996 (M*, F*; key); Abdel-Mohsin 2004: (distr.; Saudi Arabia). **Ety.:** n.s. [fitting, decent? (L)]. **Informal name:** Decent Nigerian Typical Mosquito.
syn. *masculus* Theobald
1901c: 125 (M*, F). **Type loc.:** Wilberforce, Freetown, Sierra Leone (NHMUK). **Refs.:** Harbach 1988: 44 (lectotype desig.).
syn. *minutus* Theobald
1905g: 30 (syntype M* only). **Type loc.:** Transvaal [Republic of South Africa] (LU). **Refs.:** Edwards 1941: 300 (syn.); Harbach 1988: 44 (lectotype desig.; syn.).
syn. *nigrocostalis* Theobald
1909b: 16 (M*, F*). **Type loc.:** Accra, Ashanti, Ghana (NHMUK). **Refs.:** Edwards 1941: 336 (syn.); Harbach 1988: 44 (lectotype desig.).

syn. *lividocostalis* Graham
1910: 269 (M, F). **Type loc.:** [Lagos, Nigeria], West Africa (NHMUK). **Refs.:** Harbach 1988: 45 (lectotype desig.).

declarator Dyar & Knab
1906d: 211 (L*). **Type loc.:** Trinidad [Trinidad & Tobago, Lesser Antilles] (USNM). **Distr.:** Argentina, Belize, Bolivia, Brazil, Colombia, Costa Rica, Ecuador, El Salvador, French Guiana, Guatemala, Guyana, Haiti, Honduras, Lesser Antilles (Antigua and Barbuda, Dominica, Montserrat, Saint Kitts and Nevis, Saint Lucia, Saint Vincent and the Grenadines, Trinidad and Tobago, Virgin Isl.), Mexico, Nicaragua, Panama, Paraguay, Peru, Suriname, United States (continental), Uruguay, Venezuela. **Refs.:** Yamaguti & LaCasse 1951c: 29 (M*, F*, L*); Lane 1953: 351 (M*, F, P*, L*); Breland 1954b (M*, L*; tax.); Carpenter & LaCasse 1955: 298 (M*, F*, L*; keys; as *virgultus*); Stone 1957a: 341 (tax.; from syn. with *virgultus*); Stone & Knight 1957a: 47 (lectotype desig.); Dodge 1966: 370 (first instar L; key); Bram 1967b: 58 (M*, L; key, distr., syn.); Pecor et al. 2002: 243 (tax., distr.; Belize); Darsie & Day 2005 (P*); Darsie & Ward 2005 (F*, L*; keys, distr.). **Ety.:** n.s. [announcer (L)]. **Informal name:** Announcer Trinbagonian Typical Mosquito.
syn. *inquisitor* Dyar & Knab
1906d: 211 (L*). **Type loc.:** [Montserrat], Trinidad [Trinidad & Tobago, Lesser Antilles] (USNM). **Refs.:** Stone & Knight 1957a: 51 (lectotype desig.).
syn. *proclamator* Dyar & Knab
1906d: 211 (L*). **Type loc.:** Santa Lucrecia [Veracruz], Mexico (USNM). **Refs.:** Stone & Knight 1957a: 55 (lectotype desig.).
syn. *jubilator* Dyar & Knab
1907c: 201 (M, F, L). **Type loc.:** Taboga Island, Panama (USNM). **Refs.:** Stone & Knight 1957a: 52 (lectotype desig.).
syn. *revelator* Dyar & Knab
1907c: 202 (M, F, L). **Type loc.:** Taboga Island, Panama (USNM). **Refs.:** Stone & Knight 1957a: 55 (lectotype desig.).
syn. *vindicator* Dyar & Knab
1909d: 255 (A, L). **Type loc.:** Dominica Island [Lesser Antilles] (USNM). **Refs.:** Stone & Knight 1957a: 58 (type info., lectotype desig.).
syn. *dictator* Dyar & Knab
1909d: 255 (A). **Type loc.:** Dominica Island [Lesser Antilles] (USNM) **Refs.:** Stone & Knight 1957a: 47 (lectotype desig.).
syn. *forattinii* Corrêa & Ramalho
1959b: 55 (M*). **Type loc.:** Santa Cruz de Rio Pardo, São Paulo, Brazil (FH). **Refs.:** Belkin et al. 1971: 21 (type loc. info.).

delys Howard, Dyar & Knab
1915: 317 (F). **Type loc.:** Tabernilla, Canal Zone, Panama (USNM). **Distr.:** Panama. **Refs.:** Lane 1951: 334 (from syn. with *mollis*); Lane 1953: 360 (F); Bram 1967a: 57 (F). **Ety.:** n.s. [fleur de lys?]. **Informal name:** Tabernilla Panamanian Typical Mosquito.

demeilloni Doucet
1950: 53 (L*). **Type loc.:** Ambositra [Fianarantsoa], Madagascar (IRSM). **Distr.:** Madagascar (includes Glorioso and Juan De Nova Isl.). **Ety.:** Botha De Meillon. **Informal name:** Botha Malagasy Typical Mosquito.

diengensis Brug
1931c: 35 (M*, F). **Type loc.:** Merdodo Lake and Telaga Warna (Dienghochebene), Java [Indonesia] (NHMUK). **Distr.:** Indonesia. **Ety.:** Dieng Plateau, Indonesia (Java). **Informal name:** Dieng Indonesian Typical Mosquito.

diplophyllum Dyar
1929b: 509 (M, F, L). **Type loc.:** Verrugas Canyon, Lima, Peru (USNM). **Distr.:** Peru. **Refs.:** Lane 1953: 343 (M*, F, L*); Stone & Knight 1957a: 46 (lectotype desig.); Bram 1967a: 57 (M*, F, L). **Ety.:** n.s. [double leaf (Gr); prob. ref. to two leaflike setae on the gonostylus]. **Informal name:** Two-leafed Peruvian Typical Mosquito.

dohenyi Hogue
1975: 357 (M*, F, P*, L*). **Type loc.:** Wafer Bay, Cocos Island, Costa Rica (USNM). **Distr.:** Costa Rica. **Ety.:** Timothy M. Doheny. **Informal name:** Doheny Costa Rican Typical Mosquito.

dolosus (Lynch Arribálzaga)
1891b: 156 (M*, F*; *Heteronycha*). **Type loc.:** Buenos Aires, Argentina (BA). **Distr.:** Antigua and Barbuda, Argentina, Bolivia, Brazil, Chile, Ecuador, Panama, Paraguay, Uruguay. **Refs.:** Lane 1953: 362 (M*, F, P*, L*); Duret & Barreto 1956: 84 (tax.); Bram 1967a: 59 (M*, F, L); Belkin et al. 1968: 15 (lectotype desig. according to Casal); Senise & Sallum 2008 (M*, F, P*, L*; redescr.); Demari-Silva et al. 2015 (mol. tax.; as *s.l*). **Ety.:** n.s. [crafty (L)]. **Informal name:** Crafty Argentinian Typical Mosquito.
syn. *bonariensis* Brèthes
1916: 213 (M*, F). **Type loc.:** San Isidro [Buenos Aires], Argentina (BA). **Refs.:** Belkin et al. 1968: 15 (lectotype desig. according to Casal).

duplicator Dyar & Knab
1909d: 258 (A). **Type loc.:** San Francisco Mountains, Santo Domingo [Dominican Republic] (USNM). **Distr.:** Dominican Republic, Haiti. **Refs.:** Dyar 1928: 384 (M*, F); Lane 1953: 330 (M*, F); Stone & Knight 1957a: 48 (lectotype desig.); Bram 1967b: 61 (M*, L); Belkin & Heinemann 1973: 218 (type loc. info.). **Ety.:** n.s. [duplicator (L)]. **Informal name:** Duplicator Dominican Typical Mosquito.

duttoni Theobald
1901d: 5 (M*, F*). **Type loc.:** Duke Town [Eastern Provinces], Nigeria (NHMUK). **Distr.:** Angola, Benin, Burkina Faso, Cameroon, Côte d'Ivoire, Equatorial

Guinea, Ethiopia, Gabon, Ghana, Kenya, Madagascar (includes Glorioso and Juan De Nova Isl.), Malawi, Mali, Morocco, Nigeria, Republic of the Congo, Republic of South Africa, Saudi Arabia, Senegal, Sierra Leone, Somalia, Sudan and South Sudan, Tanzania, Uganda, Yemen, Zimbabwe. **Refs.:** Macfie & Ingram 1922 (F*); Edwards 1941: 300 (M*, F*), 418 (P*); Hopkins 1952: 287 (L*); Harbach 1988: 82 (M*, F, P*, L*; tax., keys, bion., distr.); Jupp 1996 (M*, F*; key); Toto et al. 2003: 344 (distr.; Equatorial Guinea); Abdel-Mohsin 2004 (distr.; Saudi Arabia); Robert et al. 2019 (distr.; Western Palaearctic). **Ety.:** n.s. [poss. ref. to Joseph Everett Dutton]. **Informal name:** Dutton Nigerian Typical Mosquito.

syn. *dissimilis* Theobald
1901a: 376 (M*, F*). **Type loc.:** Mount Clureol [= Mount Aureol?], Freetown, Sierra Leone (NHMUK). **Refs.:** Harbach 1988: 82 (lectotype desig.).

syn. *hirsutipalpis* Theobald
1901a: 378 (M*, F*). **Type loc.:** Salisbury, Mashonaland [Zimbabwe] (NHMUK). **Refs.:** Harbach 1988: 82 (lectotype desig.).

syn. *anarmostus* Theobald
1903a: 170 (F). **Type loc.:** Freetown, Sierra Leone (NHMUK).

syn. *bifoliata* Theobald
1905g: 31 (M*). **Type loc.:** Transvaal [Republic of South Africa] (NHMUK). **Refs.:** Edwards 1941: 300 (syn.); Harbach 1988: 82 (lectotype desig.).

syn. *condylodesmus* Grünberg
1905: 385 (M*, F). **Type loc.:** Victoria, Kamerun [Nigeria] (ZM).

syn. *albovirgatus* Graham
1910: 264 (M, F). **Type loc.:** [Lagos, Nigeria], West Africa (NHMUK). **Refs.:** Harbach 1988: 82 (lectotype desig.).

eduardoi Casal & García
1968e: 105 (M*, F, P*, L*). **Type loc.:** Canal 6 and Paraná de las Palmas, Estación Agropecuaria Delta, Buenos Aires, Argentina (INM). **Distr.:** Argentina, Costa Rica, Lesser Antilles (includes Anguilla). **Ety.:** Eduardo Casal. **Informal name:** Eduardo Argentinian Typical Mosquito.

edwardsi Barraud
1923i: 507 (F). **Type loc.:** Shillong, Assam, India (LU). **Distr.:** Australia, India, Nepal, Papua New Guinea, Philippines, Sri Lanka, Thailand. **Refs.:** Barraud 1934: 397 (M*, F,); Carter & Wijesundara 1948: 148 (L*); Bram 1967b: 209 (*barraudi* to syn. with *edwardsi*); Marks 1971b: 195 (tax.; from syn. with *barraudi*); Sirivanakarn 1976: 141 (M*, F, P, L; distr.); Lee et al. 1989a: 117 (F; key, tax., bion., distr., review); Darsie & Pradhan 1990, 1991 (F, L; tax., keys, bion., distr.; Nepal); Rattanarithikul et al. 2005b (F*, L*; bion., distr., keys). **Ety.:** Frederick Wallace Edwards. **Informal name:** Edwards Indian Typical Mosquito.

erythrothorax Dyar
1907: 124 (A). **Type loc.:** Jack Slough, California, United States (USNM). **Distr.:** Colombia, Honduras, Mexico, Panama, United States (continental). **Refs.:** Yamaguti & LaCasse 1951c: 33 (M*, F*, L*); Carpenter & LaCasse 1955: 279 (M*, F*, L*; keys); Bohart & Washino 1957: 463 (L*); Stone & Knight 1957a: 49 (type info.); Bram 1967a: 62 (M*, L; keys); Bohart & Washino 1978: 124 (M, F*, L; keys, tax., bion., distr.); Darsie & Ward 2005 (F*, L*; keys, distr.). **Ety.:** red thorax (Gr); ref. to "thorax light red." **Informal name:** Red-chested American Typical Mosquito.

syn. *federalis* Dyar
1923d: 186 (M). **Type loc.:** [Regino Balzanario], Xochimilco, Distrito Pederal, Mexico (USNM). **Refs.:** Stone & Knight 1957a: 50 (lectotype desig.).

syn. *badgeri* Dyar
1924e: 125 (M, F). **Type loc.:** Bakersfield, California, United States (USNM). **Refs.**: Stone & Knight 1957a: 43 (lectotype desig.).

fasyi Baisas
1938: 215 (M*, F). **Type loc.:** Baguio [Mountain], Luzon, Philippines (NE). **Distr.:** Philippines. **Refs.:** Basio 1971b: 55 (M*; bion.); Sirivanakarn 1976: 149 (M*, F, P, L). **Ety.:** n.s. **Informal name:** Fasy Philippine Typical Mosquito.

fernandezi Casal, García, & Cavalieri
1966: 185 (M*, F, P*, L*). **Type loc.:** Quebrada de Piquirenda, Salta, Argentina (INM). **Distr.:** Argentina. **Refs.:** Bram 1967a: 120 (M, L; tax.). **Ety.:** Héctor I. Fernández. **Informal name:** Fernández Argentinian Typical Mosquito.

foliaceus Lane
1945b: 206 (M*, F). **Type loc.:** Ares, Rio Grande do Norte, Brazil (FH). **Distr.:** Brazil. **Refs.:** Stone 1950: 237 (L*); Lane 1953: 339 (M*, F, L*); Bram 1967a: 64 (M*, L); Belkin et al. 1971: 21 (type info.); Rozeboom & Komp 1950a: 93 (M*; as sp. A). **Ety.:** n.s. [leafy (L); poss. ref. to two large leaves on the male genitalia]. **Informal name:** Leafy Brazilian Typical Mosquito.

fuscocephala Theobald
1907: 420 (F). **Type loc.:** Peradeniya [Central], Ceylon [Sri Lanka] (NHMUK). **Distr.:** Bangladesh, Cambodia, Guam, Hong Kong, India, Indonesia, Japan, Laos, Malaysia, Myanmar, Nepal, Pakistan, People's Republic of China, Philippines, Singapore, Sri Lanka, Taiwan, Thailand, Timor, Vietnam. **Refs.:** Barraud 1924m (L*); Barraud 1934: 424 (M*, F, L*); Bonne-Wepster & Brug 1937a: 80 (M, F*); Bonne-Wepster & Brug 1939: 1277 (L*); Bohart 1945: 78 (M*); Bram 1967b: 184 (M*, F*, P*, L*; bion., distr., tax.); Aslamkhan 1971b (distr.; Pakistan); Basio 1971b: 55 (M*; bion.); Pinhão 1974 (distr.; Timor); Baisas 1974: 90 (M, F, P, L; tax., bion., distr.; Philippines); Tanaka et al. 1975c: 219 (bion., distr.);

Sirivanakarn 1976: 51 (M*, F*, P*, L*; distr.); Tanaka et al. 1979: 135 (M*, F*, L*); Ahmed 1987 (distr.; Bangladesh); Lee et al. 1989a: 119 (F; key, tax., bion., distr., review), 163 (tax., bion., distr., review; as *uniformis*); Darsie & Pradhan 1990, 1991 (F, L; tax., keys, bion., distr.; Nepal); Suleman et al. 1993 (distr.; Pakistan); Tanaka 2004a: 12 (P*; tax., key); Rattanarithikul et al. 2005b (F*, L*; bion., distr., keys). **Ety.:** dark head (Gr); "head deep blackish-brown." **Informal name:** Dark-headed Sri Lankan Typical Mosquito. (Vol. 1, p. 490)

syn. *uniformis* Leicester
1908: 159 (F). **Type loc.:** Batu Gajah [Perak], Malaya [Malaysia] (NE).

syn. *minimus* Leicester
1908: 160 (M, F). **Type loc.:** Kuala Lumpur [Selangor], Malaya [Malaysia] (NE).

syn. *taytayensis* Banks
1909: 545 (M, F). **Type loc.:** Taytay, Rizal [Luzon], Philippines (NE).

syn. *luteola* Theobald
1910c: 378 (F*). **Type loc.:** Peradeniya [Central], Ceylon [Sri Lanka] (NHMUK).

syn. *inelegans* Dyar
1920h: 179 (M). **Type loc.:** Los Banos [Laguna, Luzon], Philippines (USNM).

syn. *fuscitarsis* Barraud
1924g: 1272 (M*, F). **Type loc.:** Pachmarhi, Central Provinces, India (NHMUK).

gameti Bailly-Choumara
1966c: 148 (M*). **Type loc.:** Mbol, Yokadouma, Cameroon (IERT). **Distr.:** Cameroon. **Ety.:** Médecin-Colonel Gamet. **Informal name:** Gamet Cameroonian Typical Mosquito.

garciai Broche
2000: 30 (M*, F, P*, L*). **Type loc.:** Alto de la Majagua, Turquino, Sierra Maestra, Provincia de Santiago de Cuba, Cuba (IMTC). **Distr.:** Cuba. **Ety.:** Israel García Avila. **Informal name:** García Cuban Typical Mosquito.

gelidus Theobald
1901c: 20 (F*). **Type loc.:** Selangor, Malaya [Malaysia] (NHMUK). **Distr.:** Bangladesh, Cambodia, India, Indonesia, Japan, Laos, Malaysia, Myanmar, Nepal, Pakistan, Papua New Guinea, People's Republic of China (includes Hong Kong), Philippines, Singapore, Sri Lanka, Taiwan, Thailand, Vietnam. **Refs.:** Barraud 1924m (L*); Barraud 1934: 407 (M*, F, L*); Baisas 1938: 213 (M*, F, P*, L*); Bonne-Wepster 1954: 119 (M, F*, L*); Bram 1967b: 247 (M*, F*, P*, L*; bion., distr., tax.); Aslamkhan 1971b (distr.; Pakistan); Basio 1971b: 56 (M*; bion.); Berlin & Subramanian 1974: 263 (E*); Baisas 1974: 92 (M, F, P, L; tax., bion., distr.; Philippines); Sirivanakarn 1976: 58 (M*, F*, P*, L*); Ahmed 1987 (distr.; Bangladesh); Lee et al. 1989a: 122 (F; key, tax., bion., distr., review); Darsie & Pradhan 1990, 1991 (F, L; tax., keys, bion., distr.; Nepal); Rattanarithikul et al. 2005b (F*, L*; bion., distr., keys). **Ety.:** ice cold (L); ref. to frosty-white scales on the head and scutum. **Informal name:** Icy Malaysian Typical Mosquito. (Vol. 1, p. 492)

syn. *cuneatus* Theobald
1901c: 22 (F*; as var.). **Type loc.:** Taiping, Perak, Malaya [Malaysia], and Quilon, Travancore, India (NHMUK).

syn. *bipunctata* Theobald
1907: 374 (M; *Leucomyia*; as var.). **Type loc.:** India and Sarawak [Malaysian Borneo] (NHMUK).

globocoxitus Dobrotworsky
1953: 357 (M*, F*, P, L*). **Type loc.:** Williamstown, Victoria, Australia (NMM). **Distr.:** Australia. **Refs.:** Dobrotworsky 1955b: 41 (M*; hybridization studies); Dobrotworsky 1965: 211 (M*, F*, L*); Lee et al. 1989a: 126 (F; key, tax., bion., distr., review); Duret 1953a, b (M*; key, distr.; Argentina; as *intrincatus*). **Ety.:** ball, sphere (L); ref. to swollen gonocoxites. **Informal name:** Ball-hipped Australian Typical Mosquito.

grahamii Theobald
1910c: 628 (new name for *pullatus* Graham, not Coquillett 1904). **Distr.:** Angola, Benin, Burkina Faso, Cameroon, Central African Republic, Côte d'Ivoire, Democratic Republic of the Congo, Ghana, Guinea, Kenya, Liberia, Madagascar (includes Glorioso and Juan De Nova Isl.), Mali, Nigeria, Republic of the Congo, Senegal, Sudan and South Sudan, The Gambia, Togo, Uganda. **Refs.:** Edwards 1941: 347 (M*, F), 426 (P*); Hopkins 1952: 334 (L*); Rickenbach et al. 1976: 97 (distr.). **Ety.:** n.s. [Walter Mayne Graham]. **Informal name:** Graham Nigerian Typical Mosquito.

syn. *pullatus* Graham
1910: 265 (M, F). **Type loc.:** [Lagos, Nigeria], West Africa (NHMUK).

ssp. ***farakoensis*** Hamon
1955: 674 (M, F, L*). **Type loc.:** Farako, Sikasso, French Sudan [Mali] (IERT, NHMUK). **Distr.:** Mali, Nigeria. **Refs.:** Townsend 1990: 72 (type info.); Harbach & Howard 2007: 41 (to ssp.). **Ety.:** Cascade of Farako, French Sudan (Mali). **Informal name:** Farako Malian Typical Mosquito.

guayasi Leví-Castillo
1953c: 15 (M*, F). **Type loc.:** Babahoyo, Los Rios, Ecuador (NE). **Distr.:** Ecuador. **Refs.:** Bram 1967a: 111 (unrecognized sp.). **Ety.:** Province of Guayas. **Informal name:** Guayas Ecuadorian Typical Mosquito.

guiarti Blanchard
1905: 629 (new name for *viridis* Theobald, not Robineau-Desvoidy 1827). **Distr.:** Benin, Burkina Faso, Cameroon, Central African Republic, Côte d'Ivoire, Democratic Republic of the Congo, Gabon, Ghana, Kenya, Liberia,

Madagascar (includes Glorioso and Juan De Nova Isl.), Mali, Mozambique, Nigeria, Republic of the Congo, Republic of South Africa, Senegal, Sudan and South Sudan, Tanzania, The Gambia, Uganda. **Refs.:** Ingram & Macfie 1917a: 87 (P*); Edwards 1941: 342 (M*, F*); Hopkins 1952: 329 (L*); Muspratt 1959: 71 (M; tax.); Service 1976a: 262 (distr.); Rickenbach et al. 1976: 97 (distr.); Jupp 1996 (M*, F*; key). **Ety.:** n.s. [Jules Guiart]. **Informal name:** Guiart Ugandan Typical Mosquito.

syn. *viridis* Theobald
1903a: 212 (M*, F*). **Type loc.:** Buse, Uganda (NHMUK).

ssp. ***sudanicus*** Edwards
1941: 345 (M; as var. of *Culex guiarti*). **Type loc.:** Bole [Northern Territories], Gold Coast [Ghana] (NHMUK). **Distr.:** Ghana, Nigeria. **Refs.:** Mattingly 1956a: 44 (lectotype desig.); Harbach & Howard 2007 (to ssp.). **Ety.:** n.s. [Sudan?]. **Informal name:** Sudan Ghanaian Typical Mosquito.

guizhouensis Chen & Zhao
1985: 265 (M*, F*, P*, L*; holotype not desig.). **Type loc.:** Guiyang, Guizhou Province, China [People's Republic of China] (type series at GMC). **Distr.:** People's Republic of China. **Ety.:** Guiyang, Guizhou Prov., China. **Informal name:** Guizhou Chinese Typical Mosquito.

habilitator Dyar & Knab
1906d: 212 (L*). **Type loc.:** San Domingo [Ciudad Trujillo, Dominican Republic] (USNM). **Distr.:** Argentina, Dominican Republic, Haiti, Lesser Antilles (includes Antigua and Barbuda, Montserrat, Saint Kitts and Nevis, Saint Lucia, Saint Vincent and the Grenadines, Sint Maarten, Trinidad and Tobago, Virgin Isl.), Puerto Rico, Peru. **Refs.:** Howard et al. 1915: 261 (M*, F, L*; as *eremita*); Dyar 1928: 361 (M*, F, L*); Lane 1953: 325 (M*, F, P*, L*); Stone & Knight 1957a: 50 (lectotype desig.); Bram 1967a: 66 (M*, F, L; distr.; Peru). **Ety.:** n.s. [habilitator (L)]. **Informal name:** Habilitated Dominican Typical Mosquito.

syn. *eremita* Howard, Dyar, & Knab
1913: 95, 331; 1915: 261 (M*, F, L*). **Type loc.:** Santo Domingo [National District, Dominican Republic] (USNM). **Refs.:** Stone & Knight 1957a: 49 (lectotype desig.).

hancocki Edwards
1930a: 294 (M, F*). **Type loc.:** Bulambuli, Uganda (NHMUK). **Distr.:** Kenya, Uganda. **Refs.:** Edwards 1941: 330 (M*, F*), 424 (P*); Hopkins 1952: 317 (L*). **Ety.:** George Leonard Rhys Hancock. **Informal name:** Hancock Ugandan Typical Mosquito.

hopkinsi Edwards
1932b: 562 (M, F). **Type loc.:** Kameranjoka, Rwenzori foothills [Toro District], Uganda (NHMUK). **Distr.:** Democratic Republic of the Congo, Liberia, Uganda. **Refs.:** Edwards 1941: 326 (M*, F), 422 (P); Hopkins 1952: 312 (L*). **Ety.:** George Henry Evans Hopkins. **Informal name:** Hopkins Ugandan Typical Mosquito.

huangae Meng
1958: 351 (M*, F, L). **Type loc.:** Kunming [Yunnan], China [People's Republic of China] (KMC). **Distr.:** People's Republic of China. **Refs.**: Chen & Zhao 1987 (P*). **Ety.:** W.H. Huang. **Informal name:** Huang Chinese Typical Mosquito.

hutchinsoni Barraud
1924g: 1261 (M*, F). **Type loc.:** Nongpoh [Khasi Hills District], Assam, India (NHMUK). **Distr.:** Bangladesh, Cambodia, India, Laos, Malaysia, Myanmar, Nepal, Pakistan, Philippines, Singapore, Thailand, Vietnam. **Refs.:** Colless 1955b: 316 (M, F, L*); Bram 1967b: 188 (M*, F*, P*, L*; bion., distr., tax.); Basio 1971b: 56 (M*; bion.); Basio 1972: 205 (M*, F*); Sirivanakarn 1976: 40 (M*, F*, P*, L*; distr.); Ahmed 1987 (distr.; Bangladesh); Darsie & Pradhan 1990, 1991 (F, L; tax., keys, bion., distr.; Nepal); Rattanarithikul et al. 2005b (F*, L*; bion., distr., keys). **Ety.:** n.s. **Informal name:** Hutchinson Indian Typical Mosquito.

incognitus Baisas
1938: 203 (M*, F*, P*, L). **Type loc.:** "Tanque" at Calauan, Laguna, Luzon, Philippines (NE). **Distr.:** Indonesia, Philippines, Solomon Isl. **Refs.:** Delfinado 1966a: 145 (M*, F, P, L*); Basio 1971b: 57 (M*; bion.); Baisas 1974: 94 (M*, F*, P, L*; tax., bion., distr.; Philippines); Sirivanakarn 1976: 126 (M*, F*, P*, L*; distr.); Lee et al. 1989a: 130 (F; key, tax., bion., distr., review). **Ety.:** n.s. [unknown (L)]. **Informal name:** Incognito Philippine Typical Mosquito.

inflictus Theobald
1901c: 115 (F). **Type loc.:** Belize, Colombia, Costa Rica, Cuba, El Salvador, Guatemala, Honduras, Lesser Antilles (includes Barbados, Dominica, Grenada, Martinique, Saint Kitts and Nevis, Saint Lucia, Saint Vincent and the Grenadines, Trinidad and Tobago), Mexico, Nicaragua, Panama, Venezuela. **Refs.:** Dyar 1928: 390 (M*, F, L*); Senevet & Quievreux 1941: 259 (M*); Martínez Palacios 1952a: 89 (M*, F, L); Lane 1953: 367 (M*, F, L); Bram 1967a: 68 (M*, L); Belkin 1968b: 16 (lectotype desig.); Pecor et al. 2002: 243 (tax., distr.; Belize). **Ety.:** n.s. [dashed against, inflicted (L); poss. ref. to pallid posterior portion of thorax with strongly contrasted black setae in two rows]. **Informal name:** Dashed Grenadian Typical Mosquito.

syn. *scholasticus* Theobald
1901c: 120 (M*, P*). **Type loc.:** Grenada Island [Lesser Antilles] (NHMUK). **Refs.:** Belkin 1968b: 20 (lectotype desig.).

syn. *extricator* Dyar & Knab
1906d: 211 (L*). **Type loc.:** Cedros, Trinidad [Trinidad & Tobago, Lesser Antilles] (USNM). **Refs.:** Stone & Knight 1957a: 49 (type info.).

ingrami Edwards
1916: 360 (M*). **Type loc.:** Sunyani [Ashanti], Gold Coast [Ghana] (NHMUK). **Distr.:** Cameroon, Central African Republic, Comoros, Côte d'Ivoire, Democratic Republic

of the Congo, Gabon, Ghana, Liberia, Nigeria, Republic of the Congo, Sierra Leone, Uganda, Sudan and South Sudan. **Refs.:** Edwards 1941: 346 (M*, F*), 426 (P*); van Someren 1946d (tax.); Hopkins 1952: 332 (L*); Peters 1956: 546 (tax.); Service 1959a: 1 (L*); Service 1976a: 262 (distr.). **Ety.:** Alexander Ingram. **Informal name:** Ingram Ghanaian Typical Mosquito.

interfor Dyar
1928: 372 (M, F). **Type loc.:** between Tucuman and Jujuy, Argentina (USNM). **Distr.:** Argentina, Paraguay. **Refs.:** Lane 1953: 346 (M*, F); Stone & Knight 1957a: 50 (lectotype desig.); Bram 1967a: 35 (syn. with *bidens*); Harbach et al. 1986a: 139 (M*; from syn.); Sallum et al. 1996 (F*, P*, L*). **Ety.:** n.s. [interrupt (L)]. **Informal name:** Interrupted Argentinian Typical Mosquito.

interrogator Dyar & Knab
1906d: 209 (L*). **Type loc.:** Rincon Antonio, Mexico (USNM). **Distr.:** Belize, Costa Rica, El Salvador, Guatemala, Honduras, Mexico, Nicaragua, Panama, Peru, United States (continental). **Refs.:** Yamaguti & LaCasse 1951c: 37 (M*, F*, L*); Carpenter & LaCasse 1955: 280 (M*, F*, L*; keys); Stone & Knight 1957a: 51 (lectotype desig.); Bram 1967a: 69 (M*, F, L); Pecor et al. 2000 (distr.; Belize); Darsie & Ward 2005 (F*, L*; keys, distr.). **Ety.:** n.s. [interrogator (L)]. **Informal name:** Interrogator Mexican Typical Mosquito.
syn. *reflector* Dyar & Knab
1909d: 256 (A). **Type loc.:** Ancon, Canal Zone, Panama (USNM). **Refs.:** Stone & Knight 1957a: 55 (lectotype desig.).

invidiosus Theobald
1901d: xi (F*). **Type loc.:** Bonny [Eastern Provinces], Nigeria (NHMUK). **Distr.:** Benin, Burkina Faso, Cameroon, Côte d'Ivoire, Democratic Republic of the Congo, Gabon, Ghana, Kenya, Liberia, Mali, Nigeria, Republic of the Congo, São Tomé and Príncipe, Senegal, Sierra Leone, Sudan and South Sudan, Tanzania, The Gambia, Uganda. **Refs.:** Ingram & Macfie 1917a: 85 (P*); Edwards 1941: 338 (M*, F; to sp.), 425 (P); Hopkins 1952: 323 (L*); Rickenbach et al. 1976: 97 (distr.); Service 1976a: 262 (distr.). **Ety.:** n.s. [arousing hatred (L)]. **Informal name:** Hateful Nigerian Typical Mosquito.
syn. *euclastus* Theobald
1903a: 8 (F*). **Type loc.:** Gambia (NHMUK).
syn. *chloroventer* Theobald
1909b: 16 (M*, F*). **Type loc.:** Accra, Ashanti, Gold Coast [Ghana] (NHMUK). **Refs.:** Townsend 1990: 59 (type info.).
syn. *aquilus* Graham
1910: 266 (M). **Type loc.:** West Africa [Lagos, Nigeria] (NHMUK).
ssp. ***vexillatus*** Edwards
1941: 339 (M*). **Type loc.:** Kampala, Uganda (NHMUK). **Distr.:** Cameroon, Democratic Republic of the Congo, Nigeria, Republic of the Congo, Uganda. **Refs.:** Mattingly 1956a: 43 (lectotype desig.); Harbach & Howard 2007: 42 (to ssp.). **Ety.:** having a flag, banner (L); ref. to "shape of appendage of the coxite lobe . . . expanded at the tip like a small flag." **Informal name:** Banner Ugandan Typical Mosquito.
ssp. ***vicinalis*** De Meillon & Lavoipierre
1944: 49 (M*). **Type loc.:** Yangambi [Orientale], Belgian Congo [Democratic Republic of the Congo] (SAIM). **Distr.:** Democratic Republic of the Congo. **Ety.:** n.s. [neighbor (L); poss. ref. to near *invidiosus*]. **Informal name:** Neighboring Congolese Typical Mosquito.

iyengari Mattingly & Rageau
1958: 244 (M*, F*, L*). **Type loc.:** We, Lifu Island, Loyalty Islands (NHMUK). **Distr.:** Australia, New Caledonia. **Refs.:** Belkin 1962: 193 (M*, F, L*); Lee et al. 1989a: 131 (F; key, tax., bion., distr., review). **Ety.:** M.O.T. Iyengar. **Informal name:** Iyengar Australian Typical Mosquito.

jacksoni Edwards
1934 (M, F, L). In Barraud 1934: 452. **Type loc.:** Shonson Hill, Hong Kong [People's Republic of China] (NHMUK). **Distr.:** People's Republic of China (includes Hong Kong), India, Nepal, Russia, South Korea, Sri Lanka, Taiwan, Thailand. **Refs.:** Edwards 1935a: 131 (A); Carter & Wijesundara 1948: 149 (L*; as *fuscifurcatus*); Joshi et al. 1965 (distr.; Nepal; as *fuscifurcatus*); Gutsevich et al. 1974: 378 (A, L); Sirivanakarn 1976: 150 (M*, F*, P*, L*; lectotype desig., distr.); Tanaka et al. 1979: 161 (M*, F*, L*); Darsie & Pradhan 1990, 1991 (F, L; tax., keys, bion., distr.; Nepal); Tanaka 2004a: 32 (P*; tax., key); Rattanarithikul et al. 2005b (F*, L*; bion., distr., keys). **Ety.:** R.B. Jackson. **Informal name:** Jackson Chinese Typical Mosquito.
syn. *fuscifurcatus* Edwards
1934 (M, F). In Barraud 1934: 452. **Type loc.:** Ootacamund, Nilgiri Hills [Madras], India (NHMUK). **Refs.:** Sirivanakarn 1976 (syn.); Townsend 1990: 77 (type info., invalidate lectotype desig.).
syn. *kangi* Lien
1968d: 235 (M*, F*, P*, L*). **Type loc.:** Chienchi, Hsinchu, Hsinchu Hsien, Taiwan [Republic of China] (TMRI).

janitor Theobald
1903a: 183 (M*, F*). **Type loc.:** Kingston [Surrey], Jamaica (NHMUK). **Distr.:** Cayman Isl., Colombia, Dominican Republic, Haiti, Jamaica, Puerto Rico. **Refs.:** Lane 1953: 360 (M*, F, L*); Bram 1967a: 71 (M*, F, L); Belkin 1968b: 16 (lectotype desig.); Belkin et al. 1970: 66 (M*, F, P*, L*). **Ety.:** doorkeeper (L); "found congregated with *De. cancer* at the entrance to crabholes." **Informal name:** Doorkeeper Jamaican Typical Mosquito.

kesseli Belkin
1962: 200 (L*). **Type loc.:** Teahupoo, Tahiti (USNM). **Distr.:** French Polynesia, Tonga. **Refs.:** Belkin 1964: 236 (M*, F, P*); Lee et al. 1989a: 133 (F; key, tax., bion., distr.,

review). **Ety.:** John F. Kessel. **Informal name:** Kessel Tahitian Typical Mosquito.

lahillei Bachmann & Casal
1962b: 267 (M*, F*, P*). **Type loc.:** Achiras, Cordoba, Argentina (INM). **Distr.:** Argentina. **Refs.:** Bram 1967a: 72 (M*, F); Laurito et al. 2015 (M, F, P*, L*). **Ety.:** Fernando Lahille. **Informal name:** Lahille Argentinian Typical Mosquito.

laticinctus Edwards
1913a: 49 (M, F). **Type loc.:** Tiberias, Palestine [Israel] (NHMUK). **Distr.:** Albania, Algeria, Azerbaijan, Croatia, Djibouti, Egypt, Ethiopia, Greece, Iran, Iraq, Israel (and Gaza Strip and West Bank), Italy (includes Sardinia and Sicily), Jordan, Lebanon, Liberia, Libya, Malta, Morocco, Oman, Palestine, Portugal, Romania, Saudi Arabia, Somalia, Spain (includes Balearic Isl., Canary Isl.), Sudan and South Sudan, Syria, Tunisia, Turkey, Yemen. **Refs.:** Séguy 1924: Pl. II (L*), Pl. I (M gen.*); Kirkpatrick 1925b: 119 (M*, F, P*, L*); Edwards 1941: 313 (M*, F*), 420 (P); Senevet 1946: 317 (L*); Hopkins 1952: 298 (L*); Lotfi 1973: 206 (distr.); Rodhain et al. 1977 (distr.); Zaim 1987: 568 (distr.); Harbach 1988: 67 (M*, F, P*, L*; tax., keys, bion., distr.); Ribeiro et al. 2002 (distr.; Portugal); Trari et al. 2002: 331 (distr.; Morocco); Becker et al. 2010: 270 (M*, F*, L*; keys, tax., distr., bion.); Osório et al. 2010 (distr.; Portugal); Gunay et al. 2015 (mol. tax., distr.; Turkey); Robert et al. 2019 (distr.; Western Palearctic). **Ety.:** wide girdle (L) [comp. to *pipiens*; broad, white abdominal bands]. **Informal name:** Wide-belted Israeli Typical Mosquito.

laticlasper Galindo & Blanton
1954: 246 (M*, F). **Type loc.:** Cerro Punta, Panama (USNM). **Distr.:** Panama. **Refs.:** Bram 1967a: 72 (M*, F). **Ety.:** broad clasper (L); ref. to gonocoxopodite. **Informal name:** Wide-clasped Panamanian Typical Mosquito.

levicastilloi Lane
1945b: 209 (M*). **Type loc.:** Ecuador (FH). **Distr.:** Colombia, Ecuador, Venezuela. **Refs.:** Lane 1953: 340 (M*); Bram 1967a: 74 (M*, L; syn.). **Ety.:** Jose Roberto Leví-Castillo. **Informal name:** Leví-Castillo Ecuadorian Typical Mosquito.
syn. *tejerai* Cova García
1962: 312 (M*, F, P*, L*). **Type loc.:** near La Cumbre de Choroni, Aragua, Venezuela (DERM).

litoralis Bohart
1946b: 43 (M*, F, L*). **Type loc.:** north shore, Rota Island, Mariana Islands (USNM). **Distr.:** French Polynesia, Guam, Indonesia, Mariana Isl., Polynesian Isl. (Howland, Jarvis, Johnston atoll, Pitcairn, Wallis and Futuna), Thailand. **Refs.:** Bohart & Ingram 1946b: 43 (M*, F, L*; bion., distr., tax.); Yamaguti & LaCasse 1950a: 24 (M*, F*, L*); Colless 1957b: 465 (L*); Lee et al. 1989a: 134 (tax., review). **Ety.:** beach or seashore (L). **Informal name:** Seaside Mariana Islander Typical Mosquito.

litwakae Harbach
1985b: 255 (M*, F, P*, L*). **Type loc.:** Mazeras, Mambasa District, Coast Region, Kenya (USNM). **Distr.:** Kenya, Tanzania. **Ety.:** Taina Rose Litwak. **Informal name:** Litwak Kenyan Typical Mosquito.

lygrus Root
1927c: 579 (M*, F). **Type loc.:** [Mage], Rio de Janeiro, Brazil (USNM). **Distr.:** Argentina, Brazil. **Refs.:** Lane 1953: 354 (M*, F); Stone & Knight 1957a: 53 (lectotype desig.); Bram 1967a: 76 (M*, F). **Ety.:** baneful (Gr). **Informal name:** Baneful Brazilian Typical Mosquito.

maracayensis Evans
1923b: 102 (M*). **Type loc.:** Maracay [Aragua], Venezuela (NHMUK). **Distr.:** Colombia, Curaçao, Lesser Antilles (includes Trinidad and Tobago), Venezuela. **Refs.:** Dyar 1928: 362 (M*, F, L*); Lane 1953: 327 (M*, F, L*); van der Kuyp 1954: 75 (L*), 77 (P); Bram 1967a: 76 (M*, F, L; distr.; Trinidad); Belkin 1968b: 17 (type info.). **Ety.:** Maracay, Venezuela. **Informal name:** Maracay Venezuelan Typical Mosquito.
syn. *aglischrus* Dyar
1924d: 124 (M*, L). **Type loc.:** Barranquilla [Magdalena], Colombia (USNM). **Refs.:** Edwards 1932a: 206 (syn.); Stone & Knight 1957a: 42 (lectotype desig).

marquesensis Stone & Rosen
1953: 354 (M*, F, L*). **Type loc.:** Atuona Bay, Hivaoa, Marquesas Islands (USNM). **Distr.:** French Polynesia. **Refs.:** Belkin 1962: 200 (M*, F, P*, L*); Lee et al. 1989a: 136 (F; key, tax., bion., distr., review). **Ety.:** Marquesas Isl. **Informal name:** Marquesas French Polynesian Typical Mosquito.

mattinglyi Knight
1953b: 320 (M*, F, L*). **Type loc.:** Birket Shiekh Kunnaf, San'a, Yemen (USNM). **Distr.:** Saudi Arabia, Syria, Yemen. **Refs.:** Harbach 1988: 72 (M*, F, P*, L*; tax., keys, bion., distr.). **Ety.:** Peter Frederick Mattingly. **Informal name:** Yetagaini Yemeni Typical Mosquito [again named after Mattingly].

mauesensis Lane
1945b: 208 (M*, F). **Type loc.:** Maués, Amazonas, Brazil (FH). **Distr.:** Brazil. **Refs.:** Lane 1953: 366 (M*, F); Bram 1967a: 78 (M*). **Ety.:** Maués, Amazonas, Brazil. **Informal name:** Maués Brazilian Typical Mosquito.

maxi Dyar
1928: 386 (M). **Type loc.:** San Pedro [Jujuy], Argentina (USNM). **Distr.:** Argentina, Brazil, Paraguay, Uruguay. **Refs.:** Lane 1953: 337 (M*, F); Stone & Knight 1957a: 53 (lectotype desig.); Bachmann & Casal 1963 (F*); Bram 1967b: 79 (M*, F); Harbach, Peyton, & Jakob 1984: 310 (M*); Almirón & Harbach 1996 (M*, F, P, L*; tax., bion.); Rossi & Martínez 2003: 472 (distr.; Uruguay). **Ety.:** n.s. **Informal name:** Max Argentinian Typical Mosquito.

syn. *oswaldoi* Forattini
1965a: 167 (M*). **Type loc.:** Macaphyba [Macafba?], Natal, Brazil (USNM). **Refs.:** Bram 1967a: 86 (M*); Harbach, Peyton, & Jakob 1984: 317 (syn.).

mimeticus Noè
1899: 240 (F). **Type loc.:** Grassano in Basilicata [Italy] (LU). **Distr.:** Afghanistan, Albania, Algeria, Azerbaijan, Bulgaria, Crimean Peninsula, Cyprus, Egypt, France (includes Corsica), FYRO Macedonia, Georgia, Greece, Hong Kong, Hungary, India, Indonesia, Iran, Iraq, Israel (and Gaza Strip and West Bank), Italy (includes Sardinia and Sicily), Japan, Jordan, Kosovo, Lebanon, Malaysia, Montenegro, Morocco, Myanmar, Nepal, Pakistan, People's Republic of China, Philippines, Portugal, Russia, Saudi Arabia, Singapore, South Korea, Spain, Sri Lanka, Syria, Taiwan, Thailand, Tunisia, Turkey, Ukraine, Vietnam. **Refs.:** Barraud 1923c (L*); Séguy 1924: 34 (L*); Bohart & Ingram 1946b: 80 (M*, F, L*; bion., distr., tax.); La Casse & Yamaguti 1950: 241 (M*, F*, P*, L*; bion., distr., tax.); Abul-Hab 1967: 279 (distr.); Aslamkhan 1971b (distr.; Pakistan); Gutsevich et al. 1974: 375 (M*, F*, L*); Tanaka et al. 1975c: 220 (distr.); Sirivanakarn 1976: 145 (M*, F*, P*, L*); Tanaka et al. 1979: 164 (M*, F*, L*); Zaim 1987: 568 (distr.); Harbach 1988: 97 (M*, F, P*, L*; tax., keys, bion., distr.); Darsie & Pradhan 1990, 1991 (F, L; tax., keys, bion., distr.; Nepal); Suleman et al. 1993 (distr.; Pakistan); Amr et al. 1997 (distr.; Jordan); Eritja et al. 2000: 12 (distr.; Spain); Ribeiro et al. 2002 (distr.; Portugal); Trari et al. 2002: 331 (distr.; Morocco); Tanaka 2004a: 35 (P*; tax., key); Rattanarithikul et al. 2005b (F*, L*; bion., distr., keys); Rueda, Pecor et al. 2008: 397 (distr.); Becker et al. 2010: 271 (M*, F*, L*; keys, tax., distr., bion.); Santiago & Claveria 2012 (distr.; Philippines); Namazov 2014 (distr.; Azerbaijan); Gunay et al. 2015 (mol. tax., distr.; Turkey); Robert et al. 2019 (distr.; Western Palearctic). **Ety.:** imitation, mimic (L); similar to *superpictus*. **Informal name:** Mimic Italian Typical Mosquito.
syn. *pseudomimeticus* Sergent
1909: 445 (A). **Type loc.:** Oued Chiffa, Algeria (NE). **Refs.:** syn. by F.W. Edwards in Barraud 1934: 451.

mimuloides Barraud
1924e: 992 (M*, F; *mimeticus* var.). **Type loc.:** Nilgiri Hills [Madras], India (NHMUK). **Distr.:** India, People's Republic of China. **Refs.:** Barraud 1934: 451 (M*, F; to sp.); Mattingly 1956a: 37 (lectotype desig.); Sirivanakarn 1976: 153 (M*, F*, P*, L*; distr.). **Ety.:** resemblance (Gr); close to *mimulus*. **Informal name:** Imitative Indian Typical Mosquito.

mimulus Edwards
1915c: 284 (M, F). **Type loc.:** Sarawak, [Malaysian] Borneo (NHMUK). **Distr.:** Australia, Bangladesh, Cambodia, India, Indonesia, Malaysia, Myanmar, Nepal, Pakistan, Papua New Guinea, People's Republic of China (includes Hong Kong), Philippines, Singapore, Sri Lanka, Taiwan, Thailand, Vietnam. **Refs.:** Barraud 1923c (L*); Barraud 1934: 412 (M*, F, L*); Bonne-Wepster & Brug 1937a: 66 (M, F*); Baisas 1938: 214 (M*); Bonne-Wepster & Brug 1939: 1275 (L*); Delfinado 1966a: 150 (M*, F; syn.); Bram 1967b: 213 (M*, F*, P*, L*; bion., distr., tax.); Aslamkhan 1971b (distr.; Pakistan); Basio 1971b: 57 (M*; bion.); Sirivanakarn 1976: 157 (M*, F*, P*, L*; lectotype desig., distr.); Lee et al. 1989a: 138 (F; key, tax., bion., distr., review); Townsend 1990: 105 (invalidation of lectotype desig.); Darsie & Pradhan 1990, 1991 (F, L; tax., keys, bion., distr.; Nepal); Rattanarithikul et al. 2005b (F*, L*; bion., distr., keys). **Ety.:** little, small (L) [ref. to smaller than *mimeticus*]. **Informal name:** Little Malaysian Typical Mosquito.
syn. *mossmani* Taylor
1915: 181 (M*, F). **Type loc.:** Mossman, Queensland, Australia (ANIC). **Refs.:** Lee et al. 1989a: 138 (type info.).
syn. *confusus* Baisas
1938: 216 (M*, F). **Type loc.:** Balabac, Balabac Island, Palawan, Philippines (NE).
syn. *neomimulus* Lien
1968d: 240 (M*, F*, P*, L*). **Type loc.:** Pinglin, Shuangchi, Taipei Hsien, Taiwan [Republic of China] (TMRI). **Refs.:** Sirivanakarn 1976 (syn.).

miraculosus Bonne-Wepster
1937: 92 (F*). **Type loc.:** Anggi Lakes (2,000 m), Western New Guinea (Papua) (LU). **Distr.:** Indonesia (West Papua), Papua New Guinea. **Refs.:** Bonne-Wepster 1938: 396 (M*); Bonne-Wepster 1954: 117 (M, F*, L); Lee et al. 1989a: 141 (F; key, tax., bion., distr., review). **Ety.:** n.s. [wonder, marvel, miracle (L)]. **Informal name:** Miraculous Indonesian Typical Mosquito.

mirificus Edwards
1913b: 57 (M*). **Type loc.:** Njoro, Kenya (NHMUK). **Distr.:** Kenya. **Refs.:** Edwards 1941: 319 (M*, F*); Hopkins 1952: 306 (L*). **Ety.:** n.s. [wonderful, amazing (L)]. **Informal name:** Wonderful Kenyan Typical Mosquito.

mollis Dyar & Knab
1906b: 171 (A; *carmodyae* ssp.). **Type loc.:** Sangre Grande, Trinidad [Trinidad & Tobago, Lesser Antilles] (USNM). **Distr.:** Argentina, Belize, Bolivia, Brazil, Colombia, Costa Rica, Ecuador, French Guiana, Guatemala, Guyana, Honduras, Lesser Antilles (includes Trinidad and Tobago), Mexico, Nicaragua, Panama, Paraguay, Peru, Suriname, United States (continental), Venezuela. **Refs.:** Senevet & Abonnenc 1939a: 77 (M*, L*); Lane 1953: 358 (M*, F, L*); Stone & Knight 1957a: 53 (lectotype desig.); Bram 1967a: 80 (M*, F, L; tax., distr.); Heinemann & Belkin 1977b: 452 (distr.; Nicaragua); Linton et al. 2013 (bion., mol. tax., distr.; Ecuador); Demari-Silva et al. 2015 (mol. tax; as *s.l*). **Ety.:** n.s. [soft, flexible (L)]. **Informal name:** Soft Trinbagonian Typical Mosquito.

syn. *equivocator* Dyar & Knab
1907c: 203 (M, F). **Type loc.:** Tabernilla, Canal Zone, Panama (USNM). **Refs.:** Stone & Knight 1957a: 48 (lectotype desig.).
syn. *lateropunctata* Theobald
1907: 458 (F*). **Type loc.:** Supenaam, British Guiana [Guyana] (NHMUK).
syn. *elocutilis* Dyar & Knab
1909d: 255 (M, F). **Type loc.:** Cascajal River, Porto Bello Bay, Panama (USNM). **Refs.:** Stone & Knight 1957a: 48 (lectotype desig.).
syn. *lepostenis* Dyar
1923b: 70 (M). **Type loc.:** Cascajal River [Porto Bello Bay], Panama (USNM). **Refs.:** Komp 1935: 10 (syn.).
syn. *tisseuli* Senevet
1937: 375 (M*; as *tisseuilli*). **Type loc.:** Cayenne [Guyane], French Guiana [Guiana] (FMP; originally in LPFM). **Refs.:** Belkin, Schick, & Heinemann 1965: 24 (emend.); Belkin 1968b: 21 (type info.).

murrelli Lien
1968d: 243 (M*, F*, P*, L*). **Type loc.:** Hsinhsien, Wulai, Taipei Hsien, Taiwan [Republic of China] (TMRI). **Distr.:** Malaysia, People's Republic of China, Taiwan, Thailand, Vietnam. **Refs.:** Sirivanakarn 1976: 162 (M*, F*, P*, L*; distr.); Rattanarithikul et al. 2005b (F*, L*; bion., distr., keys). **Ety.:** K.D. Murrell. **Informal name:** Murrell Taiwanese Typical Mosquito.

musarum Edwards
1932b: 562 (M, F). **Type loc.:** [Fort Portal, Western] Uganda (NHMUK). **Distr.:** Democratic Republic of the Congo, Ethiopia, Kenya, Nigeria, Uganda. **Refs.:** Edwards 1941: 331 (M*, F), 425 (P*); Hopkins 1952: 319 (L*); White 1975: 316 (distr.). **Ety.:** n.s. [goddesses of science/poetry (L)]. **Informal name:** Muse Ugandan Typical Mosquito.

nakuruensis Mattingly
1951c: 44 (M*). **Type loc.:** Nakuru, Kenya (NHMUK). **Distr.:** Kenya. **Refs.:** van Someren 1967: 11 (M, F, P*, L*). **Ety.:** Nakuru, Kenya. **Informal name:** Nakuru Kenyan Typical Mosquito.

neavei Theobald
1906b: 76 (F*). **Type loc.:** Luala, Upper White Nile, Sudan (NHMUK). **Distr.:** Angola, Benin, Burkina Faso, Chad, Democratic Republic of the Congo, Ethiopia, Gabon, Kenya, Liberia, Madagascar (includes Glorioso and Juan De Nova Isl.), Mali, Nigeria, Republic of the Congo, Republic of South Africa, Senegal, Sudan and South Sudan, Tanzania, Uganda. **Refs.:** Theobald 1901b: 58 (M*, only of type series; as *cinereus*; syn. by Edwards 1941: 272). Edwards 1941: 308 (M*, F; tax.; as *univittatus* var.); Muspratt 1959: 70 (tax.); Ribeiro 1966: 164 (M*); Jupp 1971: 339 (to sp.; distr.); Jupp 1972: 103 (M*; tax.); White 1975: 317 (lectotype desig.); Service 1976a (distr.; Gabon); Fontenille & Jupp 1989: 287 (distr.); Jupp 1996 (M*, F*; key). **Ety.:** Sheffield Neave. **Informal name:** Neave Sudanese Typical Mosquito.

nigripalpus Theobald
1901c: 322 (M*, F*). **Type loc.:** Saint Lucia Island [Lesser Antilles] (NE?). **Distr.:** Bahamas, Belize, Bolivia, Brazil, Cayman Isl., Chile, Colombia, Costa Rica, Cuba, Dominican Republic, Ecuador, El Salvador, French Guiana, Guatemala, Guyana, Haiti, Honduras, Jamaica, Lesser Antilles (includes Anguilla, Antigua and Barbuda, Barbados, Curaçao, Dominica, Grenada, Guadeloupe, Martinique, Montserrat, Saint Kitts and Nevis, Saint Lucia, Saint Vincent and the Grenadines, Trinidad and Tobago, Virgin Isl.), Mexico, Nicaragua, Panama, Paraguay, Peru, Puerto Rico, Suriname, United States (continental), Venezuela. **Refs.:** Gerry 1932 (F*); Coutinho 1941 (L anomaly); Dodge 1945: 166 (L*); Yamaguti & LaCasse 1951c: 41 (M*, F*, L*); Lane 1953: 348 (M*, F, P*, L*); Carpenter & LaCasse 1955: 282 (M*, F*, L*; keys); Dodge 1966: 370 (first instar L*; key); Bram 1967a: 84 (M*, F, L); Belkin 1968b: 18 (type info.); Belkin et al. 1970: 70 (M*, F, P*, L*); Nayar 1982 (bion., phys.); Chadee & Haeger 1986 (E*); Darsie & Day 2003: 101 (P*; tax., key); Darsie & Day 2004: 110 (L*; tax.); Darsie & Ward 2005 (F*, L*; keys, distr.); Becker et al. 2010: 384 (F*, L*; key, tax., distr., bion.); Berti et al. 2015 (distr.; Venezuela); Harrison et al. 2016 (F*, L*; keys, distr.). **Ety.:** black palpus (L). **Informal name:** Black-lipped Saint Lucian Typical Mosquito. (Vol. 1, p. 494)
syn. *palus* Theobald
1903a: 194 (M*, F). **Type loc.:** Barbados [Lesser Antilles] (NHMUK). **Refs.:** Belkin 1968b: 19 (lectotype desig.).
syn. *similis* Theobald
1903a: 207 (F). **Type loc.:** Red Hills, Kingston [Surrey], Jamaica (NE?). **Refs.:** Belkin 1968b: 20 (type info.).
syn. *biocellatus* Theobald
1903a: 224 (F; *flavipes* var.). **Type loc.:** Trinidad [Trinidad & Tobago, Lesser Antilles] (NHMUK). **Refs.:** Stone 1957a: 340 (tax.); Stone et al. 1959: 252 (syn.); Belkin 1968b: 13 (tax.).
syn. *microsquamosus* Theobald
1905 (M*, F, L*). In Grabham 1905: 407. **Type loc.:** Rio Cobre Canal Dam, near Spanish Town [Middlesex], Jamaica (NHMUK). **Refs.:** Belkin 1968b: 17 (lectotype desig.).
syn. *carmodyae* Dyar & Knab
1906d: 210 (L*). **Type loc.:** [San Francisco Mines], Santo Domingo [Dominican Republic] (USNM). **Refs.:** Stone & Knight 1957a: 44 (lectotype desig.).
syn. *mortificator* Dyar & Knab
1906d: 210 (L*). **Type loc.:** Zent, Costa Rica (NE). **Refs.:** Stone & Knight 1957a: 53 (type info.).
syn. *factor* Dyar & Knab
1906d: 212 (L*). **Type loc.:** Tehuantepec [Oaxaca], Mexico (USNM). **Refs.:** Stone & Knight 1957a: 49 (lectotype desig.).

syn. *regulator* Dyar & Knab
1906d: 213 (L*). **Type loc.:** Santo Domingo [Dominican Republic] (USNM). **Refs.:** Stone & Knight 1957a: 55 (lectotype desig.).

syn. *microannulata* Theobald
1907: 481 (M*; *Trichopronomyia*). **Type loc.:** Stanley Town, New Amsterdam [Berbice], British Guiana [Guyana] (NHMUK). **Refs.:** Belkin 1968b: 17 (type info.).

syn. *proximus* Dyar & Knab
1909a: 38 (M*). **Type loc.:** [Taboga Island], Panama (USNM). **Refs.:** Stone & Knight 1957a: 55 (lectotype desig.).

syn. *caraibeus* Howard, Dyar & Knab
1913: Fig. 333; 1915: 257 (F, L*). **Type loc.:** Barbados [Lesser Antilles] (USNM). **Refs.:** Stone & Knight 1957a: 45 (lectotype desig.).

syn. *prasinopleurus* Martini
1914: 68 (M*, F). **Type loc.:** Santiago de Cuba [Oriente], Cuba (NHMUK). **Refs.:** Mattingly 1955c: 31 (lectotype desig.).

syn. *azuayus* Leví-Castillo
1954c: 264 (M*, F, P*, L*). **Type loc.:** Zhurucuchu, Azuay, Ecuador (USNM?). **Refs.:** Bram 1967a: 84 (syn.).

nilgiricus Edwards
1916: 358 (M*, F). **Type loc.:** Utakamand (Ootacamund), Nilgiri Hills, Madras, India (NHMUK). **Distr.:** India. **Refs.:** 1934: 413 (M*, F, L*). **Ety.:** Nilgiri Hills, Madras, India. **Informal name:** Nilgiri Indian Typical Mosquito.

ninagongoensis Edwards
1928b: 324 (M, F). **Type loc.:** Mount Ninagongo, Kivu Volcanoes, Kivu, Belgian Congo [Democratic Republic of the Congo] (NHMUK). **Distr.:** Democratic Republic of the Congo, Uganda. **Refs.:** Edwards 1941: 319 (M*, F), 421 (P); Hopkins 1952: 308 (L). **Ety.:** Mt. Ninagongo, Kivu Volcanoes. **Informal name:** Ninagongo Congolese Typical Mosquito.

omani Belkin
1962: 210 (M*, F, P*, L*). **Type loc.:** Chacon swamp, Guadalcanal, Solomon Islands (USNM). **Distr.:** Indonesia, Solomon Isl. **Refs.:** Maffi et al. 1979: 215 (distr.); Lee et al. 1989a: 148 (F; key, tax., bion., distr., review). **Ety.:** Paul W. Oman. **Informal name:** Oman Solomon Islander Typical Mosquito.

orientalis Edwards
1921d: 338 (M, F). **Type loc.:** Tokyo [Honshu], Japan (NHMUK). **Distr.:** India, Japan, People's Republic of China (includes Hong Kong), Philippines, Russia, South Korea, Taiwan. **Refs.:** La Casse & Yamaguti 1950: 246 (M*, F*, P*, L*; bion., distr., tax.); Hara 1957: 56 (F*); Gutsevich et al. 1974: 380 (M*, F, L*); Sirivanakarn 1976: 165 (M*, F, L*); Tanaka et al. 1979: 167 (M*, F*, L*); Tanaka 2004a: 38 (P*; tax., key). **Ety.:** n.s. [eastern (L)]. **Informal name:** Jimmu Japanese Typical Mosquito [according to legend, Jimmu (accession date 660 B.C.E.) was the first emperor of Japan].

ornatothoracis Theobald
1909b: 15 (M*, F*). **Type loc.:** Accra, Ashanti, Gold Coast [Ghana] (NHMUK). **Distr.:** Benin, Côte d'Ivoire, Ghana, Kenya, Senegal, Uganda. **Refs.:** Edwards 1941: 340 (M*, F). **Ety.:** ornate thorax (L); thorax distinctly marked. **Informal name:** Fancy-chested Ghanaian Typical Mosquito.

ousqua Dyar
1918b: 99 (M, L). **Type loc.:** Panama (USNM). **Distr.:** Belize, Colombia, Costa Rica, El Salvador, Guatemala, Honduras, Mexico, Nicaragua, Panama. **Refs.:** Bram 1967a: 50 (M*; from syn. with *coronator*; syn.). **Ety.:** n.k. [*ousqua* = *ousia* = Lively?]. **Informal name:** Lively Panamanian Typical Mosquito.

syn. *albertoi* Anduze
1943b: 193 (M*). **Type loc.:** Mesa de Escalante, near Río Guaruries, District Tovar, Merida, Venezuela (FH). **Refs.:** Lane 1951: 334 (syn.); Bram 1967a (syn.); Forattini et al. 1970: 34 (type info.).

pacificus Edwards
1916: 360 (M*, F). **Type loc.:** Port Vila, Zagabe [Efate Island], New Hebrides Islands (NHMUK). **Distr.:** New Caledonia, Vanuatu. **Refs.:** Lee 1944a: 108 (L*); Mattingly & Rageau 1958: 241 (M*, F, L*; tax.); Belkin 1962: 191 (M*, F, P*, L*); Lee et al. 1989a: 149 (F; key, tax., bion., distr., review). **Ety.:** n.s. [poss. ref. to peace (L) or Pacific Ocean]. **Informal name:** Pacific Ni-Vanuatu Typical Mosquito.

pajoti da Cunha Ramos & Ribiero
1982 (M*, F*, P*, L). **Type loc.:** Mumbondo, Angola (IHMT). **Distr.:** Angola. **Ety.:** F.-X. Pajot. **Informal name:** Pajot Angolan Typical Mosquito.

palpalis Taylor
1912a: 29 (F). **Type loc.:** Umbrawarra Creek, Northern Territory, Australia (ANIC). **Distr.:** Australia, Papua New Guinea. **Refs.:** Lee et al. 1989a: 151 (F; key, tax., bion., distr., review; from syn. with *annulirostris*); Jansen et al. 2013 (morphological and DNA ident.). **Ety.:** n.s. [pertaining to palpi (L)]. **Informal name:** Palpi Australian Typical Mosquito.

paramaxi Duret
1968b: 57 (M*). **Type loc.:** Ingeniero Dolabella, Minas Gerais, Brazil (USNM). **Distr.:** Brazil. **Refs.:** Belkin et al. 1971: 22 (type loc. info.); Harbach et al. 1991: 194 (type info.). **Ety.:** near *maxi*. **Informal name:** Paramax Brazilian Typical Mosquito.

perexiguus Theobald
1903a: 199 (M, F). **Type loc.:** Sidon, Palestine [Lebanon] (NHMUK). **Distr.:** Afghanistan, Albania, Algeria, Bulgaria, Burkina Faso, Cabo Verde, Cyprus, Egypt,

France, FYRO Macedonia, Ghana, Greece, Iran, Iraq, Israel (includes Gaza Strip and West Bank), Italy (includes Sicily), Jordan, Kenya, Kuwait, Lebanon, Mali, Malta, Mauritania, Morocco, Oman, Pakistan, Palestine, Poland, Portugal, Romania, Saudi Arabia, Spain, Sudan and South Sudan, Syria, Tanzania, Tunisia, Turkey. **Refs.:** Barraud 1924m (L*); Séguy 1924: Pl. III (L*); Kirkpatrick 1925b: 125 (M*, F, P*, L*); Mattingly & Knight 1956: 104 (to var.); White 1975: 318 (tax., distr.; from syn. with *univittatus*; lectotype desig.); Zaim 1987: 568 (distr.); Harbach 1988: 57 (M*, F, P*, L*; review, tax., keys, bion., distr.); Jupp & Harbach 1990 (morph., crossmating); Amr et al. 1997 (distr.; Jordan); Harbach 1999 (distr.; East Africa, Med. Subregion); Eritja et al. 2000: 12 (distr.; Spain, as *univittatus*); Ribiero et al. 2002 (distr.; Portugal, as *univittatus*); Trari et al. 2002: 331 (distr.; Morocco); Rueda, Pecor et al. 2008: 397 (distr.); Alves et al. 2010 (distr.; Cabo Verde); Becker et al. 2010: 273 (M*, F*, L*; bion., keys, tax.; distr.); Osório et al. 2010 (distr.; Portugal); Ahmed et al. 2011 (distr.; Saudi Arabia); Gunay et al. 2015 (mol. tax., distr.; Turkey); Robert et al. 2019 (distr.; Western Palearctic). **Ety.:** n.s. [lesser, very small (L); prob. ref. to "small delicate mosquito"]. **Informal name:** Tiny Lebanese Typical Mosquito. (Vol. 1, p. 496)

perfidiosus Edwards
1914b: 72 (M*). **Type loc.:** Ilesha [Western Provinces], Nigeria (NHMUK). **Distr.:** Benin, Cameroon, Central African Republic, Democratic Republic of the Congo, Gabon, Ghana, Liberia, Madagascar (includes Glorioso and Juan De Nova Isl.), Nigeria, Republic of the Congo, Zambia. **Refs.:** Edwards 1941: 342 (M*, F); Hopkins 1952: 327 (L*). **Ety.:** deceitful, untrue (L); ref. to "deceptive resemblance to *C. invidiosus* and *C. perfuscus*." **Informal name:** Perfidious Nigerian Typical Mosquito.

perfuscus Edwards
1914b: 71 (M*). **Type loc.:** Port Herald, Nyasaland [Malawi] (NHMUK). **Distr.:** Benin, Burkina Faso, Cameroon, Central African Republic, Côte d'Ivoire, Democratic Republic of the Congo, Ethiopia, Gabon, Ghana, Kenya, Liberia, Madagascar (includes Glorioso and Juan De Nova Isl.), Malawi, Mali, Mauritania, Mozambique, Nigeria, Republic of South Africa, Senegal, Sierra Leone, Sudan and South Sudan, Tanzania, The Gambia, Uganda. **Refs.:** Edwards 1941: 340 (M*, F*); Hopkins 1952: 324 (L*); Rickenbach et al. 1976: 97 (distr.); Service 1976a: 260 (distr.; Gabon); Jupp 1996 (M*, F*; key). **Ety.:** completely dark (L). **Informal name:** Pitch-black Malawian Typical Mosquito [*perfuscus* = "completely dark" = Pitch black].

perplexus Leicester
1908: 150 (M, F). **Type loc.:** Kuala Lumpur [Selangor], Malaya [Malaysia] (NHMUK). **Distr.:** Cambodia, India, Indonesia, Malaysia, Papua New Guinea, Philippines, Singapore, Thailand. **Refs.:** Edwards 1917: 225 (to syn. of *vishnui*); Colless 1957a: 95 (M, F, L*; tax; to sp.); Bram 1967b: 221 (M*, F, L*; bion., distr., tax.); Sirivanakarn 1976: 120 (M*, F, P*, L*; distr.); Lee et al. 1989a: 155 (F; key, tax., bion., distr., review); Rattanarithikul et al. 2005b (F*, L*; bion., distr., keys). **Ety.:** n.s. [perplexing (L)]. **Informal name:** Perplexing Malaysian Typical Mosquito.

pervigilans von Bergroth
1889: 295 (F). **Type loc.:** Greymouth [Westland], New Zealand (LU). **Distr.:** Australia, Papua New Guinea, New Zealand. **Refs.:** Lee 1944a: 109 (L*); Miller & Phillips 1952: 19 (F*, L*); Mattingly & Rageau 1958: 241 (M*, F, L; tax.); Belkin 1962: 190 (M*, F, P*, L*); Belkin 1968a: 56 (M*, F, P, L; tax., distr.); Lee et al. 1989a: 157 (F; key, tax., bion., distr., review); Derraik 2004 (distr.; New Zealand). **Ety.:** n.s. [remain awake all night (L)]. **Informal name:** Nocturnal New Zealander Typical Mosquito.

philipi Edwards
1929c: 327 (M*, F). **Type loc.:** Lagos, Nigeria (NHMUK). **Distr.:** Algeria, Benin, Cameroon, Côte d'Ivoire, Ghana, Nigeria, Senegal, Sierra Leone, The Gambia. **Refs.:** Edwards 1941: 350 (M*, F), 427 (P); Hopkins 1952: 337 (L); Rickenbach et al. 1976: 97 (distr.). **Ety.:** C.B. Philip. **Informal name:** Philip Nigerian Typical Mosquito.

philippinensis Sirivanakarn
1976: 124 (M*, F, P*, L*). **Type loc.:** Tay-Tay, Rizal, Luzon, Philippines (USNM). **Distr.:** Philippines. **Ety.:** Philippines. **Informal name:** King Philippine Typical Mosquito [the Philippines were named after King Philip II of Spain (1527–1598)].

pinarocampa Dyar & Knab
1908: 59 (M, F). **Type loc.:** Cordoba [Veracruz], Mexico (USNM). **Distr.:** Costa Rica, Guatemala, Honduras, Mexico, Panama, Venezuela. **Refs.:** Dyar 1928: 379 (M*, F, L*); Martínez Palacios 1950: 187 (M*); Lane 1953: 352 (M*, F, L*); Stone & Knight 1957a: 55 (lectotype desig.); Bram 1967a: 88 (M*, F, L); Heinemann & Belkin 1977a: 283 (distr.; Costa Rica); Berti et al. 2015 (distr.; Venezuela). **Ety.:** n.s. [dirty caterpillar? (Gr)]. **Informal name:** Earthy Mexican Typical Mosquito [*pinaros* + *campa* = "dirty field" = Earthy].

pipiens Linnaeus
1758: 602 (A). **Type loc.:** near Lake Krankesjo, Silvakra farm, Veberod, Scania, Sweden (USNM). **Distr.:** Afghanistan, Albania, Algeria, Angola, Argentina, Armenia, Australia, Austria, Azerbaijan, Belarus, Belgium, Bosnia and Herzegovina, Bulgaria, Burundi, Burkina Faso, Cameroon, Canada, Canary Isl., Central African Republic, Chad, Chile, Comoros, Côte d'Ivoire, Crimean Peninsula, Croatia, Cyprus, Czech Republic, Denmark, Democratic Republic of the Congo, Djibouti, Egypt, Estonia, Ethiopia, Finland, France (includes Corsica),

FYRO Macedonia, Georgia, Germany, Ghana, Greece (includes Crete), Guam, Guinea, Guyana, Honduras, Hungary, Iran, Iraq, Ireland, Israel (and Gaza Strip and West Bank), Italy (includes Sardina and Sicily), Japan, Jordan, Kazakhstan, Kenya, Kosovo, Kuwait, Latvia, Lebanon, Lesser Antilles (includes Trinidad and Tobago), Liberia, Libya, Liechtenstein, Lithuania, Luxembourg, Madagascar (includes Glorioso and Juan De Nova Isl.), Malaysia, Mali, Malta, Mauritania, Mexico, Moldova, Montenegro, Morocco, Netherlands, New Caledonia, Nigeria, Norway, Pakistan, Palestine, People's Republic of China, Philippines, Poland, Portugal (includes Azores and Madeira), Puerto Rico, Qatar, Republic of South Africa, Romania, Russia, Rwanda, Saudi Arabia, Serbia, Slovakia, Slovenia, South Korea, Spain (includes Balearic Isl. and Canary Isl.), Sudan and South Sudan, Sweden, Switzerland, Syria, Tajikistan, Tanzania, Thailand, Tunisia, Turkey, Turkmenistan, Uganda, Ukraine, United Kingdom, United States (continental), Uruguay, Uzbekistan, Vietnam, Yemen, Zimbabwe. **Refs.:** Blanchard 1905 (F*); Brolemann 1920 (F*); Séguy 1924: 36 (M gen.*, L*); La Casse & Yamaguti 1950 (F*); Darsie 1951: 34 (P*); Christophers 1951 (F*); Mattingly 1951b: 331 (tax., review); Yamaguti & LaCasse 1951c: 45 (M*, F*, L*); Carpenter & LaCasse 1955: 284 (M*, F*, L*; keys); Horsfall 1955: 563 (review); Lewis 1956b: 710 (tax.); Mattingly & Knight 1956: 102,120 (tax.); Barr 1957b: 153 (tax.); Bohart & Washino 1957: 463 (L*); Hara 1957 (F*); Mattingly 1957c: 686 (tax.); Rioux 1958: 244 (M*, L*; tax.); Dodge 1966: 370 (first instar L; key); Ribeiro & Mexia: 1966: 176 (tax.); Laffoon & Knight 1971 (F*); Mattingly 1971a: Pl. 19 (P*); Tewfik & Barr 1974: 216 (chrom.; *molestus*?); Rickenbach et al. 1976: 96 (distr.); Utrio 1976: 129, 135 (L*); Iglisch 1977: 267 (M*); Bohart & Washino 1978: 126 (M*, F, L*; keys, tax., bion., distr.); Tanaka et al. 1979: 146 (M*, F*; as *pipiens molestus* form); Harbach & Knight 1980 (F*); Brogdon 1981 (*s.l.*; siphonal index for ident.); Brogdon 1984a (*s.l.*; siphonal index for id.); Brogdon 1984b (*Cx. pipiens s.l.*; Memphis, TN, United States); Harbach, Harrison, & Gad 1984 (F*) (as *molestus* Forskål); Mitchell et al. 1984: 308 (distr.; Argentina); Harbach et al. 1985: 9 (neotype desig.); Rjazantzeva 1985 (F*); Harbach 1988: 23 (M*, F, P*, L*; complete syn.; tax., keys, bion., distr.); Abdullah & Merdan 1995 (distr.; Saudi Arabia); Jupp 1996 (M*, F*; key); Amr et al. 1997 (distr.; Jordan); Clements 1999 (F*); Alten et al. 2000 (distr.; Turkey); Fonseca et al. 2000 (mol. tax.; *s.l.*); Hadjivassilis 2000: 38 (distr.; Cyprus); Spungis 2000: 10 (distr.; Latvia); Orszagh et al. 2001: 20 (distr.; Slovakia); Trari et al. 2002: 331 (distr.; Morocco); Beck et al. 2003: 24 (distr.; Luxembourg); Nicolescu et al. 2003b: 20 (distr.; Romania); Rossi & Martínez 2003: 472 (distr.; Uruguay); Fonseca et al. 2004 (mol. tax.; *s.l.*); Darsie & Ward 2005 (F*, L*; keys, distr.); Fonseca et al. 2006 (mol. tax.; *s.l.*); Becker et al. 2010: 275 (M*, F*, L*; keys, tax., distr., bion.); Fonseca et al. 2010 (mol. tax.; *s.l.*); Osório et al. 2010 (distr.; Portugal); Becker et al. 2012a (Pipiens Complex; rev., Europe); Gunay et al. 2015 (mol. tax., distr.; Turkey); Harrison et al. 2016 (F*, L*; keys, distr.; notes 10, 24); Robert et al. 2019 (distr.; Western Palearctic). **Ety.:** n.s. [piping (L); shrill sounding (wing beat?)]. **Informal name:** Piping Swedish Typical Mosquito. Official ESA common name: northern house mosquito. (Vol. 1, p. 498)

syn. *bifurcatus* Linnaeus
1758: 603 (A; type = *An.* (*Ano.*) *claviger*). **Type loc.:** Europe (LSL). **Refs.:** Natvig 1948: 438 (tax.); Harbach, Dahl, & White 1985: 7 (lectotype desig.).

syn. *fasciatus* Mueller
1764: 87 (A). **Type loc.:** none designated (LU).

syn. *molestus* Forskål
1775: 85 (A). **Type loc.:** Rosetta, Káhira, and Alexandria, Egypt (LU). **Refs.:** Natvig 1948: 452 (M, F, L; tax.); Knight 1951b: 354 (M, F, L, E; bion., tax.); Knight & Abdel-Malek 1951: 183 (M, F, L, E; bion.); Mattingly 1951b: 340 (to ssp.; tax.); Kitzmiller & Laven 1954: 767 (hybridization); Dobrotworsky 1955a: 193 (bion., tax.); Dobrotworsky 1955b: 41 (M*; tax.); Hara 1957: 54 (F*); Dobrotworsky 1965: 217 (M*, F*, L*); Harbach, Harrison, & Gad 1984 (neotype desig; syn.); Lee et al. 1989a: 143 (improper sp. desig.).

syn. *trifurcatus* Fabricius
1794: 401 (A). **Type loc.:** "Europae paludosis" [malarious Europe] (NE)

syn. *luteus* Meigen
1804: 6 (F). **Type loc.:** Europe (NE).

syn. *domesticus* Germar
1817: 290 (A). **Type loc.:** Dalmatia [Croatia] (LU). **Refs.:** Marshall & Staley 1937: 23 (syn.); Harbach 1988: 229 (from syn. of *molestus*).

syn. *rufus* Meigen
1818: 7 (M). **Type loc.:** none designated (MNHP).

syn. *bicolor* Meigen
1818: 9 (M, F). **Type loc.:** Africa? (MNHP).

syn. *meridionalis* Leach
1825: 292 (A). **Type loc.:** Nice [Alpes Maritimes], France (LU).

syn. *marginalis* Stephens
1825: 455 (F). **Type loc.:** London, England (NHMUK).

syn. *consobrinus* Robineau-Desvoidy
1827: 408 (A). **Type loc.:** Pennsylvania, United States (NE).

syn. *calcitrans* Robineau-Desvoidy
1827: 409 (F). **Type loc.:** Germany and Paris, France (NE).

syn. *thoracicus* Robineau-Desvoidy
1827: 409 (F). **Type loc.:** Paris, France (NE).

syn. *pallipes* Waltl
1835: 110 (F). **Type loc.:** Spain (MNHP). **Refs.:** Meigen 1838: 1 (F).

syn. *unistriatus* Curtis
1837: 231 (A). **Type loc.:** England (NMM). **Refs.:** Stone et al. 1959: 254 (syn.).

syn. *pallipes* Macquart
1838a: 37 (A). **Type loc.:** Egypt (NE).
syn. *rufinus* Bigot
1888: 7 (F). **Type loc.:** Chott El Djerid, Tunisia (NE).
syn. *agilis* Bigot
1889: 122 (F). **Type loc.:** Algeria (BC).
syn. *phytophagus* Ficalbi
1890c: 126 (M, F). **Type loc.:** Toscana, Romagna, Sardinia, Italy (LU).
syn. *haematophagus* Ficalbi
1893: 143 (A). **Type loc.:** Europa e d'Italia [Europe and Italy] (LU). **Refs.:** Marshall & Stanley 1937: 23 (syn.); Harbach 1988: 230 (from syn. with *molestus*).
ssp. ***pallens*** Coquillett
1899: 303 (M, F). **Type loc.:** Japan (USNM). **Distr.:** Japan, People's Republic of China, Mexico, South Korea, United States (continental). **Refs.:** La Casse & Yamaguti 1950: 225 (M*, F*, P*, L*; bion., distr., tax.; as *pipiens*); Mattingly 1951b: 334 (tax.); Bekku 1956: 963 (tax.); Stone & Knight 1957a: 54 (type info.); Ishii 1969: 204 (F); Chow 1973: 44 (distr.); Ishii 1977: 31 (tax., review); Tanaka et al. 1979: 144 (M*, F*); Tanaka 2004a: 19 (P*; tax., key). **Ety.:** n.s. [pale (L); prob. ref. to pleura without maculation in comp. to *fulvus*]. **Informal name:** Pale Japanese Typical Mosquito.
ssp. syn. *osakaensis* Theobald
1907: 439 (M*, F). **Type loc.:** Osaka [Honshu], Japan (NHMUK).
ssp. syn. *comitatus* Dyar & Knab
1909a: 35 (M*). **Type loc.:** [Nat City], California, United States (USNM). **Refs.:** Mattingly 1951b: 334 (tax.); Rozeboom 1951b: 350 (tax.); Stone & Knight 1957a: 46 (lectotype desig.).
ssp. syn. *dipseticus* Dyar & Knab
1909a: 34 (M*; *quinquefasciatus* race). **Type loc.:** Indio and Coachella, California, United States, and La Paz (Baja California), Salina Cruz (Oaxaca), and Acapulco (Guerrero), Mexico (USNM). **Refs.:** Mattingly 1951b: 335 (syn.?); Belkin et al. 1966: 20 (type info.); Roubaud & Ghelelovitch 1957: 7 (M*).
syn. *melanorhinus* Giles
1900a: 342 (new name for *pallipes* Macquart 1838a, not Waltl).
syn. *varioannulatus* Theobald
1903a: 198 (F*). **Type loc.:** Saint Michaels, Azores (NHMUK).
syn. *azoriensis* Theobald
1903a: 210 (M, F*). **Type loc.:** Saint Michaels, Azores (NHMUK).
syn. *longefurcatus* Becker
1903: 68 (F). **Type loc.:** Cairo, Egypt (ZM).
syn. *quasimodestus* Theobald
1905f: 88 (F; *bicolor* var.?). **Type loc.:** Sfax, Tunis [Tunisia] (HNM). **Refs.:** Edwards 1921: 246 (syn. of *pipiens*); Stone 1957b: 172 (syn. of *molestus*); Harbach 1988: 230 (from syn. of *molestus*).
syn. *doliorum* Edwards
1912a: 263 (M; *pipiens* var.). **Type loc.:** Kingswear, Devon, England (NHMUK).
syn. *autogenicus* Roubaud
1935b: 444 (E*; *pipiens* ssp.). **Type loc.:** France (PIP). **Refs.:** Marshall & Staley 1937: 18 (syn.); Gaschen 1955: 125 (tax.); Harbach 1988: 230 (from syn. of *molestus*).
syn. *berbericus* Roubaud
1935b: 445 (E*; *pipiens* ssp.). **Type loc.:** North Africa [Algeria] (PIP). **Refs.:** Knight & Abdel-Malek 1951: 184 (syn. with *molestus*); Roubaud & Ghelelovitch 1956: 2900 (tax.); Harbach 1988: 230 (from syn. of *molestus*).
syn. *sternopallidus* Roubaud
1945: 54 (A*; *Culex autogenicus* ssp.). **Type loc.:** Aigues-Mortes [Bouches du Rhone] and Bages, Pyrenees-Orientales, France (PIP). **Refs.:** Harbach, Harrison, & Gad 1984 (syn.); Harbach 1988: 230 (tax.).
syn. *sternopunctatus* Roubaud
1945: 54 (A*; *autogenicus* ssp.). **Type loc.:** Arles [Bouches du Rhone] and Paris, France (PIP). **Refs.:** Harbach, Harrison, & Gad 1984 (syn.); Harbach 1988: 230 (tax.).
syn. *disjunctus* Roubaud
1957: 3116 (*Culex pipiens* ssp.). **Type loc.:** Conde-sur-Ifs, Normandy, France (NE). **Refs.:** Stone et al. 1959: 254 (syn.).
syn. *calloti* Rioux & Pech
1959: 117 (A; *pipiens* ssp.). **Type loc.:** France and Tunisia (LU). **Refs.:** Harbach 1988: 231 (syn.).
syn. *erectus* Iglisch
1977: 269 (M*; *Culex pipiens* ssp.). **Type loc.:** Krummer Fenn, Berlin-Zehlendorf, West Germany (ND). **Refs.:** Harbach 1988: 231 (syn.).
syn. *torridus* Iglisch
1977: 269 (M*). **Type loc.:** Krummer Fenn, Berlin-Zehlendorf, West Germany (ND). **Refs.:** Harbach 1988: 231 (syn.).

plicatus Olivares
1993 (M*, F). In Angulo & Olivares 1993: 1993. **Type loc.:** Concepción, Chile (UCCC). **Distr.:** Chile. **Ety.:** folded, corrugated (L); ref. to apex of dististyle. **Informal name:** Folded Chilean Typical Mosquito.

propinquus Colless
1955b: 315 (F*). **Type loc.:** Pulau Blakang Mati, Singapore (NHMUK). **Distr.:** Singapore. **Ety.:** near, neighboring (L); near *mimulus*. **Informal name:** Nearby Singaporean Typical Mosquito.

prosecutor Séguy
1927: 27 (new name for *pseudomimeticus* Séguy, not Sergent 1909). **Distr.:** France. **Ety.:** n.s. [pursuer (L)]. **Informal name:** Prosecutor French Typical Mosquito.
syn. *pseudomimeticus* Séguy
1925: 15 (L*). **Type loc.:** Camargue, Bouches-du-Rhône, France (LU).

pruina Theobald
1901d: 8 (M*, F*). **Type loc.:** [Agumi, Nigeria], West Africa (NHMUK). **Distr.:** Cameroon, Central African Republic, Côte d'Ivoire, Democratic Republic of the Congo, Gabon, Ghana, Liberia, Nigeria, Republic of the Congo, Sierra Leone, Sudan and South Sudan, Uganda. **Refs.:** Macfie & Ingram 1922 (F*); Edwards 1941: 349 (M*, F*), 427 (P); Hopkins 1952: 335 (L*). **Ety.:** n.s. [hoarfrost, gray (L); prob. ref. to "thorax covered with frosty-grey scales"]. **Informal name:** Frosty Nigerian Typical Mosquito.

syn. *pallidothoracis* Theobald
1909b: 17 (M*, F*). **Type loc.:** Obuasi, Ashanti, Gold Coast [Ghana] (NHMUK).

ssp. ***eschirasi*** Galliard
1931b: 227 (M, L). **Type loc.:** Sainte-Croix des Eschiras, Gabon (NHMUK). **Distr.:** Burkina Faso, Cameroon, Central African Republic, Gabon, Côte d'Ivoire, Nigeria, Republic of the Congo, Uganda. **Refs.:** Edwards 1941: 350 (tax.); Hopkins 1952: 337 (L*); Townsend 1990: 71 (type info.). **Ety.:** after type locality. **Informal name:** Eschiras Gabonese Typical Mosquito.

pseudojanthinosoma Senevet & Abonnenc
1946: 139 (F, L*). **Type loc.:** French Guiana [Guiana] (FMP; originally in PIA). **Distr.:** Colombia, French Guiana. **Refs.:** Senevet & Abonnenc 1958b: 345 (P*, L*); Belkin 1968b: 19 (type info.). **Ety.:** false (Gr); similar to *janthinosoma*. **Informal name:** Janus French Guianese Typical Mosquito [*pseudojanthinosoma* = "false *janthinosoma*" = imposter = treacherous = Janus].

pseudopruina van Someren
1951: 3 (M*). **Type loc.:** Bwamba County, Toro District, Uganda (NHMUK). **Distr.:** Uganda. **Ety.:** false (Gr); similar to *pruina*. **Informal name:** Hoarfrost Ugandan Typical Mosquito [*pseudopruina* = "false frost" = Hoarfrost].

pseudostigmatosoma Strickman
1990: 144 (M*, F*, P*, L*). **Type loc.:** Llano Guagololo, La Paz Department, Honduras (USNM). **Distr.:** Costa Rica, El Salvador, Guatemala, Honduras, Mexico, Panama. **Ety.:** false (Gr); similar to *stigmatosoma*. **Informal name:** Pretender Honduran Typical Mosquito.

pseudovishnui Colless
1957a: 88 (M*, F, L*). **Type loc.:** Singapore (NHMUK). **Distr.:** Bangladesh, Cambodia, India, Indonesia, Iran, Iraq, Japan, Laos, Macau, Malaysia, Nepal, Pakistan, Papua New Guinea, People's Republic of China (includes Hong Kong), Philippines, Singapore, South Korea, Sri Lanka, Taiwan, Thailand, Vietnam. **Refs.:** Delfinado 1966a: 147 (P, L*); Bram 1967b: 222 (M*, F, L*; bion., distr., tax.); Reuben 1969: 648 (M*, F*, L*); Miyagi & Iha 1970: 71 (tax., distr.; as *neovishnui*); Aslamkhan 1971b (distr.; Pakistan); Basio 1971b: 58 (M*; bion.); Matsuo & Iwaki 1972: 349 (tax.); Matsuo & Ramalingam 1972: 55 (tax.); Baisas 1974: 100 (M, F, P, L; tax., bion., distr.; Philippines); Lien et al. 1975 (distr.; as *neovishnui*); Sirivanakarn 1975a: 72 (M*, F, P*, L*); Tanaka et al. 1975c: 218 (distr.); Sirivanakarn 1976: 116 (M*, F, P*, L*; distr.); Tanaka et al. 1979: 152 (M*, F*, L*); Zaim 1987: 568 (distr.); Harbach 1988: 111 (M*, F, P*, L*; tax., keys, bion., distr.); Lee et al. 1989a: 162 (F; key, tax., bion., distr., review); Darsie & Pradhan 1990, 1991 (F, L; tax., keys, bion., distr.; Nepal); Tanaka 2004a: 26 (P*; tax., key); Rattanarithikul et al. 2005b (F*, L*; bion., distr., keys); Robert et al. 2019 (distr.; Western Palearctic). **Ety.:** false (Gr); similar to *vishnui*. **Informal name:** Narayana Singaporean Typical Mosquito [Narayana is another name for the Hindu god Vishnu].

syn. *neovishnui* Lien
1968d: 230 (M*, F, P*, L*). **Type loc.:** Peiyuan, Tungho, Taitung, Taiwan [Republic of China] (TMRI). **Refs.:** Sirivanakarn 1976: 116 (syn.).

quasiguiarti Theobald
1910c: 374 (M, F*). **Type loc.:** Mpumu, Uganda (NHMUK). **Distr.:** Benin, Burkina Faso, Cameroon, Côte d'Ivoire, Democratic Republic of the Congo, Ethiopia, Kenya, Madagascar (includes Glorioso and Juan De Nova Isl.), Mali, Rwanda, Uganda. **Refs.:** Edwards 1941: 334 (M*, F; tax.; from syn. with *pallidocephalus* and *pipiens*); van Someren 1956: 10 (P, L; tax.); Rickenbach et al. 1976: 97 (distr.). **Ety.:** nearly (L); similar to *guiarti*. **Informal name:** In-law Ugandan Typical Mosquito [*quasiguiarti* = "half Guiart" = In-law].

quinquefasciatus Say
1823: 10 (A). **Type loc.:** Mississippi River, United States (NE). **Distr.:** Afghanistan, Algeria, Angola, Argentina, Australia, Bahamas, Bahrain, Bangladesh, Belize, Bolivia, Bonaire, Brazil, British Indian Ocean Territory, Borneo, Burkina Faso, Cambodia, Cameroon, Cayman Isl., Central African Republic, Chile, Colombia, Comoros, Cook Isl. (Polynesia), Costa Rica, Côte d'Ivoire, Croatia, Cuba, Djibouti, Dominican Republic, Ecuador, Egypt, El Salvador, Equatorial Guinea, Ethiopia, Fiji, France, French Guiana, French Polynesia, Gabon, Ghana, Greece, Guam, Guatemala, Guyana, Haiti, Honduras, Hong Kong, India, Indonesia, Iran, Iraq, Israel (and Gaza Strip and West Bank), Jamaica, Japan, Jordan, Kenya, Kiribati, Kuwait, Laos, Lesser Antilles (includes Anguilla, Antigua and Barbuda, Aruba, Barbados, Curaçao, Dominica, Grenada, Guadeloupe, Martinique, Montserrat, Saint Kitts and Nevis, Saint Lucia, Saint Vincent and the Grenadines, Sint Maarten, Trinidad and Tobago, Virgin Isl.), Liberia, Madagascar (includes Glorioso and Juan De Nova Isl.), Malaysia, Maldives, Mali, Mariana Isl., Marshall Isl., Mauritania, Mauritius, Mexico, Micronesia (Wake Isl.), Morocco, Mozambique, Myanmar, Nauru, Nepal, New Caledonia, New Hebrides, New Zealand, Nicaragua, Niger, Nigeria, Oman, Pakistan, Palau, Panama, Papua New Guinea, Paraguay, People's Republic

of China, Peru, Philippines, Puerto Rico, Republic of the Congo, Republic of South Africa, Réunion, Russia, Samoa (Ind. State of Samoa, American Samoa, Western Samoa), São Tomé and Príncipe, Saudi Arabia, Senegal, Singapore, Solomon Isl., Somalia, South Korea, Sri Lanka, Sudan and South Sudan, Suriname, Taiwan, Tanzania, Thailand, The Gambia, Timor, Togo, Tonga, Turkey, Tuvalu, Uganda, United States (continental, Hawai'i), Uruguay, Vanuatu, Venezuela, Vietnam, Yemen, Zambia. **Refs.:** Giles 1902 (F*) (as *fatigans* Wiedemann); Bancroft 1908: 31 (M, F; "House Mosquito"; as *fatigans*); Bancroft 1908: 50 (M, F; as genus *Taeniorhynchus acer*); Banks 1908 (F*, P*, L*, E*; as *fatigans*); Macfie & Ingram 1922 (F*) (as *fatigans*); Barraud 1924m (L*; as *fatigans*); Séguy 1924: 36 (M gen.*; as *fatigans*); Gerry 1932 (F*); Barraud 1934: 420 (M*, F, L*; as *fatigans*); Edwards 1941: 316 (M*, F*; as *fatigans*), 421 (P); Bohart & Ingram 1946b (M*, L*; bion.); La Casse & Yamaguti 1950: 220 (M*, F*, L*; bion., distr., tax.); Darsie 1951: 35 (P*); Christophers 1951 (F*); Yamaguti & LaCasse 1951c: 49 (M*, F*, L*); Lane 1953: 344 (M*, F*, P*, L*; as *fatigans*); Carpenter & LaCasse 1955: 286 (M*, F*, L*; keys); Dobrotworsky 1955b: 41 (hybridization studies; as *fatigans*); Iyengar & Menon 1955 (distr.); Mattingly & Brown 1955: 86 (tax.); Barr 1957: 153 (tax.); Bohart 1957: 77 (M*, F, L*); Bohart & Washino 1957: 463 (L*); Hara 1957: 55 (F*); Stone 1957a: 342 (tax.); Belkin 1962: 195 (M*, F, P*, L*; tax.), 8 (distr.); Dobrotworsky 1965: 215 (M*, F, L*; as *fatigans*); Belkin et al. 1966: 4 (restriction of type loc.); Dodge 1966: 370 (first instar L*; key); Bram 1967b: 192 (M*, F*, P*, L*; bion., distr., tax.; as *pipiens* ssp.); Belkin 1968a: 68 (M, F, P, L; tax., distr.); Belkin et al. 1970: 73 (M*, F, P*, L*; as sp.); Basio 1971b: 57 (M*; bion.; as *pipiens* ssp.); Lambrecht & van Someren 1971: 483 (distr.); Baisas 1974: 95 (M, F*, P, L*; tax., bion., distr.; Philippines); Cheng & Barr 1976: 347 (genetics [*pipiens* or *quinquefasciatus*?]); Tanaka et al. 1975c: 219 (distr.); White 1975: 319 (tax.); Ramalingam 1976: 307 (tax., bion., keys., distr.; Samoa and Tonga); Sirivanakarn 1976: 30 (M*, F*, P*, L*; distr.; as valid sp.); Belkin 1977b: 45 (nomenclatural review); Brunhes 1977a (distr.; Comoros Isl.); Iglisch 1977: 271 (M*); Rodhain et al. 1977 (distr.); Bohart & Washino 1978: 126 (M*, F, L*; keys, distr.); Sirivanakarn & White 1978: 362 (M*, F*, P*, L*; neotype desig.); Taylor & Maffi 1978: 182 (distr.); Tanaka et al. 1979: 141 (M*, F*, L*; as *pipiens* ssp.); Harbach, Dahl, & White 1985 (F*); Ahmed 1987 (distr.; Bangladesh); Zaim 1987: 568 (distr.); Harbach 1988: 31 (M*, F, P*, L*; complete syn.; tax., keys, bion., distr.); Lee et al. 1989a: 166 (F; key, tax., bion., distr., review); Darsie & Pradhan 1990, 1991 (F, L; tax., keys, bion., distr.; Nepal); Suleman et al. 1993 (distr.; Pakistan); Abdullah & Merdan 1995 (distr.; Saudi Arabia); Jupp 1996 (M*, F*; key); Whelan & Hapgood 2000 (bion., distr.; East Timor); Pena et al. 2003 (distr.; Dominican Republic); Ree 2003: 40 (distr.; remove Korea); Rossi & Martínez 2003: 472 (distr.; Uruguay); Toto et al. 2003: 344 (distr.; Equatorial Guinea); Derraik 2004 (distr.; New Zealand); Tanaka 2004a: 17 (P*; tax., key); Rattanarithikul et al. 2005b (F*, L*; bion., distr., keys); Darsie & Ward 2005 (F*, L*; keys, distr.); Suman et al. 2008 (E*); Becker et al. 2010: 278 (M*, F*, L*; keys, tax., distr., bion.); Berti et al. 2015 (distr.; Venezuela); Harrison et al. 2016 (F*, L*; keys, distr.; notes 10, 24). **Ety.:** five banded (L); "abdomen cinereous; tergum with five black, broad, fasciae." **Informal name:** Five-banded American Typical Mosquito. Official ESA common name: southern house mosquito. (Vol. 1, p. 500)

syn. *fatigans* Wiedemann
1828: 10 (M, F). **Type loc.:** East Indies [Indonesia] (NMW). **Refs.:** Stone 1957a: 342 (syn.); Belkin 1968a: 68 (lectotype desig.); Sirivanakarn & White 1978 (syn.).

syn. *aestuans* Wiedemann
1828: 11 (M, F). **Type loc.:** Salvador, Bahia, Brazil (NMW). **Refs.:** Stone 1958a: 186 (lectotype desig.; syn.); Belkin 1968b: 12 (tax.); Belkin et al. 1971: 22 (type loc. info.).

syn. *pungens* Wiedemann
1828: 9 (F). **Type loc.:** New Orleans [Louisiana, United States] (NMW). **Refs.:** Belkin 1968b: 19 (lectotype desig.).

syn. *acer* Walker
1848: 8 (F). **Type loc.:** New Zealand (NHMUK).

syn. *cingulatus* Doleschall
1856: 405 (A). **Type loc.:** Ambarawa, Java [Indonesia] (LU).

syn. *cubensis* Bigot
1857: 329 (A). **Type loc.:** Cuba (MNHP). **Refs.:** Edwards 1932a: 208 (syn. with *fatigans*); Belkin 1968b: 15 (tax., lectotype desig.).

syn. *anxifer* Bigot
1859: 117 (M*, F*). **Type loc.:** Madagascar (LU).

syn. *serotinus* Philippi
1865: 595 (M, F). **Type loc.:** Santiago and Valdivia, Chile (NE).

syn. *autumnalis* Weyenbergh
1882: 23 (M*, F*, P*, L*). **Type loc.:** Primero River, Cordoba, Argentina (Universidad de Cordoba?).

syn. *penafieli* Sánchez
1885: 213 (M, F; descr.). **Type loc.:** Mexico City, Mexico (LU). **Refs.:** Sánchez 1886: 326 (name given).

syn. *macleayi* Skuse
1889: 1746 (F). **Type loc.:** King George's Sound, Western Australia, Australia (ANIC). **Refs.:** Lee et al. 1989a: 166 (type info).

syn. *skusii* Giles
1900a: 292 (M, F). **Type loc.:** Australia (LU).

syn. *doleschallii* Giles
1900a: 338 (new name for *cingulatus* Doleschall, not Fabricius 1805).

syn. *quasipipiens* Theobald
1901c: 136 (F*). **Type loc.:** Sambalpur, Central Provinces, India (NHMUK). **Refs.:** Bram 1967c (type info., lectotype desig.).

syn. *fouchowensis* Theobald
1901c: 137 (M*, F*). **Type loc.:** Foochow [Fukien], China [People's Republic of China] (NHMUK).

syn. *reesii* Theobald
1901c: 145 (M*, F*). **Type loc.:** Hong Kong [People's Republic of China] (NHMUK).

syn. *sericeus* Theobald
1901c: 147 (F*). **Type loc.:** Hong Kong [People's Republic of China] (LU).

syn. *luteoannulatus* Theobald
1901c: 159 (A; *fatigans* ssp.). **Type loc.:** none designated (NHMUK).

syn. *trilineatus* Theobald
1901c: 159 (A; *fatigans* ssp.). **Type loc.:** none designated (LU).

syn. *albolineatus* Giles
1901a: 609 (A). **Type loc.:** Shahjahanpur [United Provinces], India (NHMUK). **Refs.:** Sirivanakarn 1976: 33 (lectotype desig.).

syn. *pallidocephala* Theobald
1904b: 73 (F). **Type loc.:** Sennar, Blue Nile, Sudan (NHMUK). **Refs.:** Edwards 1941: 316 (syn.?).

syn. *cartroni* Ventrillon
1905b: 429 (M, F). **Type loc.:** Majunga, Moroudara, Madagascar (MNHP). **Refs.:** Harbach 1988: 31 (lectotype desig.).

syn. *barbarus* Dyar & Knab
1906d: 210 (L*). **Type loc.:** Trinidad [Trinidad & Tobago. Lesser Antilles] (USNM).

syn. *zeltneri* Neveu-Lemaire
1906: 251 (M*, F*). **Type loc.:** Harar [Hararge], Ethiopia (NE). **Refs.:** Edwards 1941: 353 (tax.); White 1975: 322 (syn.).

syn. *pygmaeus* Neveu-Lemaire
1906: 256 (M*, F*). **Type loc.:** Imi [Ogaden], Ethiopia (NE).

syn. *didieri* Neveu-Lemaire
1906: 257 (F*). **Type loc.:** Leopoldville, Belgian Congo [Kinshasa, Democratic Republic of the Congo] (NE).

syn. *raymondii* Tamayo
1907 (M*, F*, P*, L*). In Tamayo & García 1907: 36. **Type loc.:** Huacachina [desert oasis] [Ica], Peru (LU).

syn. *quasilinealis* Theobald
1907: 415 (F). **Type loc.:** Adelaide [South Australia], Australia (NHMUK).

syn. *stoehri* Theobald
1907: 419 (F*). **Type loc.:** British Central Africa [Malawi] (NHMUK).

syn. *christophersii* Theobald
1907: 453 (M, F). **Type loc.:** India (NHMUK). **Refs.:** Bram 1967c (type info., lectotype desig.).

syn. *minor* Theobald
1908b: 298 (M, F). **Type loc.:** Sylhet and Lungleh, Lushai Hills, Assam and Calcutta, Bengal, India (NHMUK).

syn. *aikenii* Dyar & Knab
1908: 61 (M). **Type loc.:** New Amsterdam, Suriname (USNM). **Refs.:** Edwards 1932a: 208 (syn. with *fatigans*); Stone & Knight 1957a: 43 (lectotype desig.).

syn. *revocator* Dyar & Knab
1909d: 256 (A). **Type loc.:** Hope Gardens, Newcastle [Surrey], Jamaica (USNM). **Refs.:** Stone & Knight 1957a: 55 (lectotype desig.).

syn. *lachrimans* Dyar & Knab
1909d: 259 (new name for *aikenii* Dyar & Knab, not Aiken & Rowland 1906).

syn. *goughii* Theobald
1911b: 268 (syntype M only). **Type loc.:** Onderstepoort, Transvaal [Republic of South Africa] (NHMUK). **Refs.:** White 1975: 321 (F; lectotype desig.; paralectotype = *univittatus*; syn.).

syn. *fuscus* Taylor
1914c: 699 (M; *Culicelsa*). **Type loc.:** Townsville, Queensland, Australia (NE?).

syn. *aseyehae* Dyar & Knab
1915: 112 (M, F). **Type loc.:** New Providence Island, Bahama Islands (USNM). **Refs.**: Stone & Knight 1957a: 43 (lectotype desig.).

syn. *townsvillensis* Taylor
1919: 836 (new name for *fuscus* Taylor 1914c, not Theobald 1909b).

syn. *hensemaeon* Dyar
1920h: 178 (F). **Type loc.:** Los Banos [Laguna, Luzon], Philippines (USNM).

syn. *nigrirostris* Enderlein
1920: 51 (F; *fatigans* var.). **Type loc.:** Tananarive, Madagascar (ZM).

quitensis Leví-Castillo
1953a: 190 (M*, F, P*, L*). **Type loc.:** Quito [Pichincha], Ecuador (NE). **Distr.:** Ecuador. **Refs.:** Bram 1967a: 111 (unrecognized sp.). **Ety.:** Quito, Ecuador. **Informal name:** Quito Ecuadorian Typical Mosquito.

restuans Theobald
1901c: 142 (F*). **Type loc.:** Toronto, Ontario, Canada (NHMUK). **Distr.:** Bahamas, Canada, Guatemala, Honduras, Mexico, United States (continental). **Refs.:** Ross 1947: 47 (M*, F*, L*); Darsie 1951: 33 (P*); Yamaguti & LaCasse 1951c: 54 (M*, F*, L*); Carpenter & LaCasse 1955: 288 (M*, F, L*; keys); Bohart & Washino 1957: 463 (L*); Dodge 1966: 370 (first instar L*; key); Bram 1967a: 94 (M*, F, L); Belkin 1968b: 19 (type info.); Bohart & Washino 1978: 129 (M, F*, L*; keys, tax., bion., distr.); Strickman & Darsie 1988: 21 (F*, L*; distr.; Guatemala); Darsie & Ward 2005 (F*, L*; keys, distr.); Darsie & Day 2007 (P*); Becker et al. 2010: 396 (F*, L*; key, tax., distr., bion.); Harrison et al. 2016 (F*, L*; keys, distr.). **Ety.:** n.s. [see, look at, protect, watch? (L)]. **Informal name:** Watchful Canadian Typical Mosquito. (Vol. 1, p. 502)

syn. *brehmei* Knab
1916b: 161 (M, F, L). **Type loc.:** [Newark], New Jersey, United States (USNM). **Refs.:** Edwards 1932a: 207 (syn.); Stone & Knight 1957a: 45 (lectotype desig., type loc. info).

riojanus Duret
1968e: 326 (M*). **Type loc.:** Chamical, La Rioja, Argentina (USNM). **Distr.:** Argentina. **Refs.:** Harbach et al. 1991: 194 (type info.). **Ety.:** La Rioja Prov., Argentina. **Informal name:** Rioja Argentinian Typical Mosquito.

roseni Belkin
1962: 203 (M*, F, P*, L*). **Type loc.:** Hitiaa, Tahiti (USNM). **Distr.:** French Polynesia, Tonga. **Refs.:** Lee et al. 1989a: 217 (F; key, tax., bion., distr., review). **Ety.:** L. Rosen. **Informal name:** Rosen Tahitian Typical Mosquito.

rotoruae Belkin
1968a: 64 (M*, F, P*, L). **Type loc.:** Kuirau Hot Springs Reserve, Rotorua, South Auckland, New Zealand (NELS). **Distr.:** New Zealand. **Refs.:** Lee et al. 1989a: 220 (F; key, tax., bion., distr., review); Derraik 2004 (distr.; New Zealand). **Ety.:** Kuirau Hot Springs Reserve, Rotorua, South Auckland, New Zealand. **Informal name:** Rotorua New Zealander Typical Mosquito.

salinarius Coquillett
1904b: 73 (M, F, L). **Type loc.:** [Elizabeth], New Jersey, United States (USNM). **Distr.:** Bahamas, Bermuda, Canada, Mexico, United States (continental). **Refs.:** Ross 1947: 49 (M*, F, L*); Darsie 1951: 34 (P*); Yamaguti & LaCasse 1951c: 58 (M*, F*, L*); Carpenter & LaCasse 1955: 290 (M*, F*, L*; keys); Stone & Knight 1957a: 56 (lectotype desig.); Dodge 1966: 370 (first instar L*; key); Bram 1967a: 95 (M*, F, L); Darsie & Ward 2005 (F*, L*; keys, distr.); Darsie et al. 2006 (L*; tax.); Darsie & Day 2006 (P*); Becker et al. 2010: 397 (F*, L*; key, tax., distr., bion.); Harrison et al. 2016 (F*, L*; keys, distr.). **Ety.:** pertaining to salt (L); in contrast to *nigritulus*. **Informal name:** Saline American Typical Mosquito. (Vol. 1, p. 504)

saltanensis Dyar
1928: 386 (M). **Type loc.:** Campo Santa de Salta [Salta], Argentina (USNM). **Distr.:** Argentina, Brazil, Colombia, Panama, Paraguay, Venezuela. **Refs.:** Lane 1953: 329 (M*); Bram 1967a: 97 (M*, F, L); Cotrim et al. 1974: 14 (M*, P*, L*); Laurito et al. 2008 (M*, F, P*, L*). **Ety.:** n.s. [Campo Santa de Salta, Argentina]. **Informal name:** Salta Argentinian Typical Mosquito.
syn. *beauperthuyi* Anduze
1943a: 459 (M*, F, P, L). **Type loc.:** Borburata, Carabobo, Venezuela (IHC).

scheuberi Carpintero & Leguizamón
2004: 502 (M*). **Type loc.:** "Cuesta del Sur" ranch, Bajo Giulani, Toay District, La Pampa Province, Argentina (USNM?). **Distr.:** Argentina. **Refs.**: Hribar & Hribar 2006 (type dep.). **Ety.:** Luis Scheuber. **Informal name:** Scheuber Argentinian Typical Mosquito.

schwetzi Edwards
1929c: 326 (M*, F). **Type loc.:** Stanleyville, Belgian Congo [Kisangani, Tshopo Province, Democratic Republic of the Congo] (NHMUK). **Distr.:** Benin, Cameroon, Côte d'Ivoire, Democratic Republic of the Congo, Liberia. **Refs.:** Edwards 1941: 347 (M*, F*); Peters 1955c: 126 (P*, L*). **Ety.:** Jacques Schwetz. **Informal name:** Schwetz Congolese Typical Mosquito.

scimitar Branch & Seabrook
1959: 217 (M*, F). **Type loc.:** Hog Island, Bahama Islands (USNM). **Distr.:** Bahamas, Cuba. **Refs.:** Bram 1967a: 98 (M*, F, L). **Ety.:** scimitar (English); ref. to shape of gonostylus. **Informal name:** Scimitar Bahamian Typical Mosquito.

scottii Theobald
1912b: 86 (M*, F*). **Type loc.:** Mahe Island and Silhouette Island, Seychelles (NHMUK). **Distr.:** São Tomé and Príncipe, Seychelles. **Refs.:** Edwards 1941: 332 (M*, F; from syn. with *fatigans*); Mattingly & Brown 1955: 86, 89, 93 (tax.). **Ety.:** H. Scott. **Informal name:** Scott Seychellois Typical Mosquito.

sechani Brunhes & Boussès
2009: 21 (M*, F, P*, L*; key). **Type loc.:** Iles de la Société, on way to Mount Marau, Tahiti (IRD). **Distr.:** French Polynesia. **Ety.:** Yves Séchan. **Informal name:** Séchan Tahitian Typical Mosquito.

secutor Theobald
1901c: 321 (M, F*). **Type loc.:** Cinchona, Jamaica (NHMUK). **Distr.:** Dominican Republic, Haiti, Jamaica, Lesser Antilles (includes Dominica, Saint Kitts and Nevis, Saint Lucia, Saint Vincent and the Grenadines), Puerto Rico, Venezuela. **Refs.:** Lane 1953: 355 (M*, F, L*); Bram 1967a: 100 (M*, F, L); Belkin 1968b: 20 (lectotype desig.); Belkin et al. 1970: 68 (M*, F, P*, L*). **Ety.:** n.s. [pursuer, attendant (L); ref. to "The insects follow one about in a regular cloud"]. **Informal name:** Follower Jamaican Typical Mosquito.
syn. *lamentator* Dyar & Knab
1906d: 219 (L*). **Type loc.:** San Francisco Mines, Santo Domingo [Dominican Republic] (USNM). **Refs.:** Stone & Knight 1957a: 52 (lectotype desig.); Belkin & Heinemann 1973: 218 (type loc. info.).
syn. *toweri* Dyar & Knab
1907a: 13 (A). **Type loc.:** Mayaguez, Puerto Rico (USNM). **Refs.:** Dyar 1918b: 97 (syn.); Stone & Knight 1957a: 56 (type info.).
syn. *quasisecutor* Theobald
1907: 398 (M*, F*). **Type loc.:** Newcastle [Surrey], Jamaica (NHMUK). **Refs.:** Belkin 1968b: 19 (lectotype desig.).

seldeslachtsi Wolfs
1947c: 224 (F, P*, L*). **Type loc.:** Costermansville [Bukavu], Kivu, Belgian Congo [Democratic Republic of the Congo] (CMT). **Distr.:** Democratic Republic of the Congo. **Ety.:** Dr. Seldeslachts. **Informal name:** Seldeslachts Congolese Typical Mosquito.

shoae Hamon & Ovazza
1954: 416 (M*, F*, P*, L*). **Type loc.:** Shoa [Shewa], Ethiopia (IERT). **Distr.:** Ethiopia. **Refs.:** Harrison 1973c: 278 (type info.). **Ety.:** Shewa, Ethiopia (romanized formerly as Shoa). **Informal name:** Shoa Ethiopian Typical Mosquito.
ssp. ***ugandae*** van Someren
1967: 14 (M*, F, P, L). **Type loc.:** Ngogwe (33°′ E, 0°15′ N), Uganda (NHMUK). **Distr.:** Uganda. **Ety.:** Uganda. **Informal name:** Ngogwe Ugandan Typical Mosquito.

simpsoni Theobald
1905g: 28 (M*, F*). **Type loc.:** Transvaal [Republic of South Africa] (NHMUK). **Distr.:** Burkina Faso, Comoros, Democratic Republic of the Congo, Djibouti, Ethiopia, Kenya, Madagascar (includes Glorioso and Juan De Nova Isl.), Mali, Mauritania, Mauritius, Morocco, Nigeria, Republic of South Africa, Saudi Arabia, Senegal, Seychelles, Somalia, Sudan and South Sudan, Tanzania, Uganda, Yemen, Zimbabwe. **Refs.:** Edwards 1941: 309 (M*, F), 420 (P); Hopkins 1952: 293 (L*); Mattingly & Brown 1955: 85 (L; tax.); Gerberg & Arnett 1976: 343 (distr.); Brunhes 1977a (distr.; Comoros Isl.); Rodhain et al. 1977 (distr.); Harbach 1988: 75 (M*, F, P*, L*; tax., keys, bion., distr.); Jupp 1996 (M*, F*; key); Ahmed et al. 2011 (distr.; Saudi Arabia); Robert et al. 2019 (distr.; Western Palearctic). **Ety.:** C.B. Simpson. **Informal name:** Simpson South African Typical Mosquito.
syn. *richteri* Ingram & De Meillon
1927: 72 (M*, F, P*, L*). **Type loc.:** Begamuzi River, Zululand [KwaZulu-Natal, Republic of South Africa] (NHMUK). **Refs.:** Harbach 1988: 75 (lectotype desig.).
syn. *mauritanicus* Callot
1940: 362 (L*). **Type loc.:** Taghjicht [Grand Atlas Region], Morocco (IPS). **Refs.:** Senevet et al. 1949: 42 (M*, P, L; tax.); Harbach 1988: 75 (syn.).

sinaiticus Kirkpatrick
1925a: 383 (M*, F, L*). **Type loc.:** Several localities in Sinai, Egypt (DAC, NHMUK). **Distr.:** Egypt, Eritrea, Iran, Israel (and Gaza Strip and West Bank), Jordan, Oman, Saudi Arabia, Somalia, Sudan and South Sudan, Yemen. **Refs.:** Kirkpatrick 1925b: 122 (M*, F, P*, L*); Edwards 1941: 311 (M*, F), 420 (P); Lewis 1949: 63 (L*); Hopkins 1952: 295 (L); Mattingly & Knight 1956: 103 (L; tax.); Lotfi 1973: 206 (distr.); Harbach 1988: 78 (M*, F, P*, L*; tax., keys, bion., distr.); Townsend 1990: 135 (type info.); Robert et al. 2019 (distr.; Western Palearctic). **Ety.:** Sinai Peninsula. **Informal name:** Sinai Egyptian Typical Mosquito.

sitiens Wiedemann
1828: 542 (F). **Type loc.:** Sumatra [Indonesia] (ZMC). **Distr.:** Australia, Bangladesh, Borneo, Cambodia, Cameroon, Comoros, Djibouti, Egypt, Ethiopia, Fiji, French Polynesia, Guam, Hong Kong, India, Indonesia, Iran, Japan, Kenya, Kuwait, Madagascar (includes Glorioso and Juan De Nova Isl.), Malaysia, Maldives, Mariana Isl., Morocco, Mozambique, Myanmar, Nauru, New Caledonia, New Hebrides, Nigeria, Oman, Pakistan, Palau, Papua New Guinea, People's Republic of China, Philippines, Polynesian Isl. (Howland, Jarvis, Johnston Atol, Pitcairn, Wallis and Futuna), Samoa (Ind. State of Samoa, American Samoa, Western Samoa), Saudi Arabia, Senegal, Seychelles, Sierra Leone, Singapore, Solomon Isl., South Korea, Sri Lanka, Sudan and South Sudan, Taiwan, Tanzania, Thailand, Timor, Tonga, Tuvalu, United Arab Emirates, Vanuatu, Vietnam, Yemen, Zambia. **Refs.:** Barraud 1924m (L*); Edwards 1941: 296 (M*, F), 417 (P); Williams 1943: 215 (F*, L*); Bohart & Ingram 1946b: 18, 32, 81 (M*, F, L*; bion., distr., tax.); Penn 1949b: 80 (P*); Yamaguti & LaCasse 1950a: 30, 85–86 (M*, F*, L*; tax., keys, bion., distr.; Guam); Iyengar & Menon 1955: 7 (M*, L*); Mattingly & Knight 1956: 104 (F, L; tax.); Worth & De Meillon 1960: 247 (distr.); Delfinado 1966a: 151 (M*, F, L*; syn.); Bram 1967b: 239 (M*, F*, P*, L*; bion., distr., tax.); Lotfi 1970: 401 (distr.); Aslamkhan 1971b: (distr.); Basio 1971b: 59 (M*; bion.); Baisas 1974: 101 (M, F, P, L; tax., bion., distr.; Philippines); Maffi & Taylor 1974: 207 (distr.); Tanaka et al. 1975c: 218 (tax., bion., distr.); Ramalingam 1976: 308 (tax., bion., keys., distr.; Samoa and Tonga); Rickenbach et al. 1976: 96 (distr.); Sirivanakarn 1976: 95 (M*, F*, P*, L*; distr.); Brunhes 1977a (distr.; Comoros Isl.); Tanaka et al. 1979: 155 (M*, F*, L*); Chadee & Bennett 1988 (E*); Harbach 1988: 87 (M*, F, P*, L*; complete syn.; tax., keys, bion., distr.); Lee et al. 1989a: 225 (F; key, tax., bion., distr., review); Jupp 1996 (M*, F*; key); Whelan & Hapgood 2000; Trari et al. 2002: 331 (distr.; Morocco; as *mauritanicus*); Tanaka 2004a: 28 (P*; tax., key); Rattanarithikul et al. 2005b (F*, L*; bion., distr., keys); Becker et al. 2010: 366 (F*, L*; key, tax., distr., bion.); Robert et al. 2019 (distr.; Western Palearctic). **Ety.:** n.s. [thirsty, arid, dry (L)]. **Informal name:** Thirsty Indonesian Typical Mosquito. (Vol. 1, p. 506)
syn. *impellens* Walker
1859: 91 (F). **Type loc.:** Makassar, Celebes [Sulawesi] (NHMUK).
syn. *microannulatus* Theobald
1901a: 353 (M*, F*). **Type loc.:** Quilon, Travancore, Madras and Shahjahanpur [United] Provinces, [India] (NHMUK).
syn. *bancroftii* Theobald
1901a: 367 (F*; *annulirostris* ssp.). **Type loc.:** Burpengary, Queensland, Australia (NHMUK). **Refs.:** Stone et al. 1959: 241 (syn.); Lee et al. 1989a: 225 (to syn. with *sitiens* by E.N. Marks).
syn. *gnophodes* Theobald
1903a: 163 (F). **Type loc.:** Bruas, Dindings, Malaya [Malaysia] (NHMUK).
syn. *somaliensis* Neveu-Lemaire
1906: 254 (M*, F*). **Type loc.:** Djibouti, French Somaliland (NE?).

syn. *nigricephala* Leicester
1908: 149 (M, F). **Type loc.:** Batu Gajah [Perak], Malaya [Malaysia] (NE).
syn. *salus* Theobald
1908c: 256 (M*, F*). **Type loc.:** Port Sudan [Kassala], Sudan (NHMUK).
syn. *jepsoni* Theobald
1910b: 158 (F*). **Type loc.:** Suva [Viti Levu], Fiji (NHMUK).
syn. *saibaii* Taylor
1912b: 28 (F). **Type loc.:** Saibai Island, Torres Strait (ANIC). **Refs.:** Lee et al. 1989a: 225 (type info.).
syn. *paludis* Taylor
1913: 56 (F; *Culicelsa*). **Type loc.:** Townsville, Queensland, Australia (ANIC). **Refs.:** Lee et al. 1989a: 225 (type info.).
syn. *milni* Taylor
1914d: 196 (M*, F; *Culicelsa annulirostris* var.). **Type loc.:** Milne Bay, Papua, New Guinea (ANIC). **Refs.:** Lee et al. 1989a: 225 (type info.).
syn. *annulata* Taylor
1914c: 689 (M, F*; *Culicada*). **Type loc.:** Townsville, Queensland, Australia (ANIC). **Refs.:** Lee et al. 1989a: 225 (type info.).
syn. *salinus* Baisas
1938: 204 (M*, F*, P*, L). **Type loc.:** Caloocan, Rizal, Luzon, Philippines (NE).
syn. *mauritanicus* Callot
1940: 362 (L*). **Type loc.:** Taghjicht [Grand Atlas Region], Morocco (IPS). **Refs.:** Harbach 1988: 87 (lectotype desig.).

solitarius Bonne-Wepster
1938: 394 (M*, F*). **Type loc.:** Anggi Lakes, Western New Guinea (Papua) (LU). **Distr.:** Indonesia (West Papua), Papua New Guinea. **Refs.:** Bonne-Wepster 1954: 129 (M, F*, L); Lee et al. 1989a: 243 (F; key, tax., bion., distr., review). **Ety.:** n.s. [solitary (L); poss. ref. to relationship to *mimeticus* and *mimulus*]. **Informal name:** Solitary Indonesian Typical Mosquito.

sphinx Howard, Dyar, & Knab
1913: Fig. 134, 1915: 301 (M*, F). **Type loc.:** Nassau [New Providence Island], Bahama Islands (USNM). **Distr.:** Bahamas, Cuba. **Refs.:** Lane 1953: 335 (M*, F, L*); Stone & Knight 1957a: 56 (lectotype desig.); Bram 1967a: 101 (M*, F, L); Gutsevich & García-Avila 1969: 6 (distr.; Cuba). **Ety.:** n.s. [sphinx]. **Informal name:** Sphinx Bahamian Typical Mosquito.

spinosus Lutz
1905: 26 (M, F). **Type loc.:** São Paulo, Brazil (FSP). **Distr.:** Argentina, Brazil, Colombia, Venezuela. **Refs.:** Lane 1953: 369 (M*, F, P*, L*); Bram 1967a: 103 (M*, F, L); Belkin 1968b: 20 (type info.); Belkin et al. 1971: 22 (lectotype desig.); Lourenço-de-Oliveira & Townsend 1989: 431 (M*, L*; invalid lectotype desig., neotype desig.). **Ety.:** thorny (L); "spinose legs." **Informal name:** Spiny Brazilian Typical Mosquito.

stenolepis Dyar & Knab
1908: 60 (M, F). **Type loc.:** Córdoba [Veracruz], Mexico (USNM). **Distr.:** Costa Rica, Cuba, Dominican Republic, Guatemala, Haiti, Honduras, Mexico, Panama. **Refs.:** Martínez Palacios 1950: 187 (M*); Lane 1953: 357 (M*, F, L); Stone & Knight 1957a: 56 (lectotype desig.); Bram 1967a: 104 (M*, F, L). **Ety.:** n.s. [narrow scale (Gr); poss. very narrow scales on wings]. **Informal name:** Narrow-scaled Mexican Typical Mosquito.

stigmatosoma Dyar
1907: 123 (A, L). **Type loc.:** Pasadena, California, United States (USNM). **Distr.:** Colombia, Costa Rica, El Salvador, Guatemala, Honduras, Mexico, United States (continental), Venezuela. **Refs.:** Yamaguti & LaCasse 1951c: 62 (M*, F*, L*); Carpenter & LaCasse 1955: 292 (M*, F*, L*; keys); Bohart & Washino 1957: 463 (L*); Breland 1957: 175 (L*); Stone & Knight 1957a: 56 (type info.); Stone 1958b: 236 (syn. with *peus*); Bohart & Washino 1978: 125 (M, F, L*; keys, tax., bion., distr.; as *peus*, a *nom. nudum*); Strickman 1988a: 490 (from syn.); Eldridge & Harbach 1989: 247 (conservation of *stigmatosoma*); International Commission on Zoological Nomenclature 1991 (Opinion 1644; conservation of name); Eldridge & Harbach 1992 (conservation of *stigmatosoma* and *thriambus*); Darsie & Ward 2005 (F*, L*; keys, distr.); Ortega-Morales et al. 2011 (tax., distr.). **Ety.:** n.s. [spot body (Gr); prob. ref. to "coloration of the underside of the abdomen, which has only a row of round black spots"]. **Informal name:** Dotted American Typical Mosquito.
syn. *eumimetes* Dyar & Knab
1908: 61 (M, F). **Type loc.:** Orizaba [Veracruz], Mexico (USNM). **Refs.:** Stone & Knight 1957a: 49 (lectotype desig.); Stone et al. 1959: 253 (syn.); Strickman 1988a: 490 (syn.).

striatipes Edwards
1941: 311 (M*, F). **Type loc.:** Nairobi [Central], Kenya (NHMUK). **Distr.:** Burkina Faso, Democratic Republic of the Congo, Ethiopia, Ivory Coast, Kenya, Madagascar (includes Glorioso and Juan De Nova Isl.), Zambia, Zimbabwe. **Refs.:** De Meillon et al. 1945: 90 (P*, L*); Hopkins 1952: 296 (L*); Mattingly 1956a: 42 (type info.). **Ety.:** n.s. [groove, wrinkle foot (L); poss. ref. to striped middle legs]. **Informal name:** Line-footed Kenyan Typical Mosquito.
ssp. ***joanae*** Muspratt
1955: 185 (M, F, L). **Type loc.:** Whiskey Spruit Stream, near Lydenburg Sabie Road, Transvaal [Republic of South Africa] (SAIM). **Distr.:** Republic of South Africa. **Refs.:** Jupp 1996 (M*, F*; key). **Ety.:** Joan Wright. **Informal name:** Joan South African Typical Mosquito.

summorosus Dyar
1920h: 180 (M). **Type loc.:** Los Banos, [Laguna, Luzon], Philippines (USNM). **Refs.:** Colless 1957a: 98 (M*, F, L*;

tax., to ssp.); Bram 1967b: 225 (syn. with *tritaeniorhynchus*); Sirivanakarn 1976: 129 (tax.); Airi & Kaur 2015 (to sp.).

syn. *siamensis* Barraud & Christophers
1931: 283 (M; *Culex tritaeniorhynchus* var.). **Type loc.:** Chieng mai [Chiang Mai], Thailand (NHMUK). **Refs.:** Mattingly 1956a: 37 (lectotype desig.); Colless 1957a: 98 (syn.).

surinamensis Dyar
1918d: 121 (M, F, L). **Type loc.:** Suriname (USNM). **Distr.:** Bolivia, Brazil, Colombia, French Guiana, Guyana, Suriname, Venezuela. **Refs.:** Senevet & Abonnenc 1939a: 74 (M*, L*); Lane 1953: 336 (M*, F, L*); Stone & Knight 1957a: 56 (lectotype desig.); Bram 1967a: 106 (M*, F, L). **Ety.:** Suriname. **Informal name:** Sranang Surinamese Typical Mosquito [Sranang, an English-based Creole language, is widely used in Suriname].

tamsi Edwards
1934: 327 (M, F). **Type loc.:** São Tomé (NHMUK). **Distr.:** Angola, São Tomé and Príncipe. **Refs.:** Edwards 1941: 323 (M*, F), 422 (P); Hopkins 1952: 309 (L). **Ety.:** n.s. **Informal name:** Tams São Tomean Typical Mosquito.

tarsalis Coquillett
1896: 43 (M, F). **Type loc.:** Argus Mountains, California, United States (USNM). **Distr.:** Canada, Mexico, United States (Alaska, continental). **Refs.:** Ross 1947: 49 (M*, F*, L*); Yamaguti & LaCasse 1951c: 66 (M*, F*, L*); Lungstrom 1954: 86 (bion.); Carpenter & LaCasse 1955: 294 (M*, F*, L*; keys); Horsfall 1955: 592 (bion., disease relations); Lungstrom 1955: 68 (P*, L*); Bohart & Washino 1957: 463 (L*); Dow et al. 1957: 294 (bion., disease relations); Stone & Knight 1957a: 57 (lectotype desig.); Dodge 1966: 371 (first instar L*; key); Bram 1967a: 106 (M*, L); Bohart & Washino 1978: 130 (M*, F*, L*; keys, tax., bion., distr.); Darsie & Day 2003: 101 (P*; tax., key); Darsie & Ward 2005 (F*, L*; keys, distr.); Becker et al. 2010: 398 (F*, L*; key, tax., distr., bion.); Larson et al. 2010 (bion.); Harrison et al. 2016 (F*, L*; keys, distr., tax.). **Ety.:** foot (Gr) [prob. ref. to tarsal claw characters]. **Informal name:** Leggy American Typical Mosquito. (Vol. 1, p. 508)

syn. *willistoni* Giles
1900a: 281 (F). **Type loc.:** Argus Mountains, California, United States (USNM). **Refs.:** Belkin 1968b: 21 (type info.).

syn. *kelloggii* Theobald
1903i: 211 (M, F). **Type loc.:** Stanford University, California, United States (NHMUK). **Refs.:** Stone 1957a: 341 (lectotype desig.); Belkin 1968b: 17 (type info.).

tatoi Casal & García
1971: 631 (M*, F, P*, L*). **Type loc.:** Salta, Laguna de Knutzen, 3 km S. de Campamento Vespucio, Departamento San Martín, Argentina (INM). **Distr.:** Argentina. **Ety.:** Tato Pavlovsky. **Informal name:** Tato Argentinian Typical Mosquito.

telesilla De Meillon & Lavoipierre
1945 (P*, L*). In De Meillon et al. 1945: 91. **Type loc.:** Yangambi [Orientale], Belgian Congo [Democratic Republic of the Congo] (SAIM, NHMUK). **Distr.:** Angola, Cameroon, Central African Republic, Democratic Republic of the Congo, Ghana, Liberia, Mozambique, Nigeria, Sudan. **Refs.:** De Meillon 1947a: 110 (M*, F); Hopkins 1952: 326 (L*); Rickenbach et al. 1976: 97 (distr.); Segerman 1990: 54 (invalid holotype); Townsend 1990: 141 (type info.); Jupp 1996 (M*, F*; key). **Ety.:** n.s. [poss. ref. to Telesilla, Greek poetess, native of Argos]. **Informal name:** Telesilla Congolese Typical Mosquito.

tenagius van Someren
1954b: 122 (M*, F, P*, L*). **Type loc.:** Bwamba County, Toro District, Uganda (NHMUK). **Distr.:** Djibouti, Kenya, Uganda. **Refs.:** Rodhain et al. 1977 (distr.). **Ety.:** n.s. [shoals, shallows (*tenagos*) (Gr)]. **Informal name:** Shallow Ugandan Typical Mosquito.

terzii Edwards
1941: 312 (M*, F). **Type loc.:** Salisbury [Mashonaland], Southern Rhodesia [Zimbabwe] (NHMUK). **Distr.:** Algeria, Democratic Republic of the Congo, Ethiopia, Kenya, Liberia, Republic of South Africa, Uganda, Zimbabwe. **Refs.:** Mattingly 1956a: 42 (type info.); van Someren 1958: 84 (P*, L*); Muspratt 1961: 98 (L); Jupp 1996 (M*, F*; key). **Ety.:** J.E. Terzi. **Informal name:** Terzi Zimbabwean Typical Mosquito.

thalassius Theobald
1903a: 168 (F*). **Type loc.:** The Gambia (NHMUK). **Distr.:** Angola, Benin, Cameroon, Comoros, Côte d'Ivoire, Democratic Republic of the Congo, Djibouti, Gabon, Ghana, Guinea-Bissau, Kenya, Liberia, Madagascar (includes Glorioso and Juan De Nova Isl.), Mauritius, Mozambique, Nigeria, Republic of South Africa, São Tomé and Príncipe, Senegal, Syria, Tanzania, The Gambia, Togo. **Refs.:** Ingram & Macfie 1917a: 86 (P*); Macfie & Ingram 1922 (F*); Edwards 1941: 298 (M*, F), 417 (P); Hopkins 1952: 284 (L*); Rickenbach et al. 1976: 96 (distr.); Service 1976a (distr.; Gabon); Jupp 1996 (M*, F*; key). **Ety.:** sea water, salt water (Gr); ref. to larvae in mangrove swamp and in tidal pool. **Informal name:** Oceanic Gambian Typical Mosquito.

syn. *ronaldi* de Charmoy
1908: 259 (M, F, L*). **Type loc.:** Fanfara Bastian, Mauritius (LSTM). **Refs.:** Edwards 1941: 298 (syn.).

syn. *accraensis* Theobald
1909b: 19 (M, F*; *Culicelsa*). **Type loc.:** Accra, Ashanti, Gold Coast [Ghana] (NHMUK).

syn. *fuscus* Theobald
1909b: 20 (F*; *Culicelsa accraensis* var.). **Type loc.:** Accra, Ashanti, Gold Coast [Ghana] (NHMUK).

syn. *neotaeniorhynchus* Theobald
1910c: 320 (M*, F*; *Culicelsa*). **Type loc.:** Delagoa Bay [Mozambique] (NHMUK).

theileri Theobald
1903a: 187 (M*, F*). **Type loc.:** Pretoria, Transvaal [Republic of South Africa] (NHMUK). **Distr.:** Afghanistan, Albania, Algeria, Armenia, Azerbaijan, Bangladesh, Belarus, Bulgaria, Crimean Peninsula, Cyprus, Egypt, Eritrea, Ethiopia, France (includes Corsica), FYRO Macedonia, Georgia, Greece, Hungary, India, Iran, Iraq, Israel (and Gaza Strip and West Bank), Italy (includes Sardinia and Sicily), Jordan, Kenya, Kuwait, Lebanon, Liberia, Libya, Madagascar (includes Glorioso and Juan De Nova Isl.), Madiera, Moldova, Mongolia, Morocco, Mozambique, Myanmar, Namibia, Nepal, Nigeria, Pakistan, People's Republic of China, Portugal (includes Madeira), Republic of South Africa, Romania, Russia, Saudi Arabia, Slovakia, Spain (includes Canary Isl.), Switzerland, Syria, Tajikistan, Tanzania, Tunisia, Turkey, Ukraine, Yemen, Zimbabwe. **Refs.:** Barraud 1924m (L*); Kirkpatrick 1925b: 115 (M*, F, P*, L*; as *tipuliformis*); Edwards 1941: 305 (M, F*), 419 (P); Senevet 1949: 57 (M*); Hopkins 1952: 289 (L*); Noèldner 1953: 10 (M*, L*); Rioux 1958: 251 (M*, F*, L*); Aslamkhan 1971b (distr.; Pakistan); Minar 1976: 336 (distr.); Sirivanakarn 1976: 46 (M*, F*, P, L*; distr.); McIntosh et al. 1980 (Rift Valley virus incrimination); Rjazantzeva 1985 (F); Ahmed 1987 (distr.; Bangladesh); Harbach 1988: 61 (M*, F, P*, L*; distr.); Darsie & Pradhan 1990, 1991 (F, L; tax., keys, bion., distr.; Nepal); Jupp 1996 (M*, F*; key); Eritja et al. 2000: 12 (distr.; Spain); Ribeiro et al. 2002 (distr.; Portugal); Trari et al. 2002: 331 (distr.; Morocco); Becker et al. 2010: 280 (M*, F*, L*; keys, tax., distr., bion.); Namazov 2014 (distr.; Azerbaijan); Gunay et al. 2015 (mol. tax., distr.; Turkey); Robert et al. 2019 (distr.; Western Palearctic). **Ety.:** Arnold Theiler. **Informal name:** Theiler South African Typical Mosquito. (Vol. 1, p. 510)
syn. *creticus* Theobald
1903a: 189 (F). **Type loc.:** Crete, Greece (NHMUK).
syn. *pettigrewii* Theobald
1910a: 15 (F). **Type loc.:** Ukhrul, Manipur, 6,400 ft [Assam], India (NHMUK). **Refs.:** Townsend 1990: 120 (type info.).
syn. *onderstepoortensis* Theobald
1911b: 265 (F*). **Type loc.:** Onderstepoort, Transvaal [Republic of South Africa] (NHMUK). **Refs.:** Townsend 1990: 115 (type info.).
syn. *annulata* Theobald
1913a: 321 (F; *Culex theileri* var.). **Type loc.:** Onderstepoort, Transvaal [Republic of South Africa] (NHMUK).
syn. *alpha* Séguy
1924: 41 (L*). **Type loc.:** Algeria (MNHP). **Refs.:** Senevet 1947a: 128 (tax.); Harbach 1988: 61 (syn.; lectotype L illustration designated; specimen NE?).

thriambus Dyar
1921a: 33 (M, F, L). **Type loc.:** Kerrville, Texas, United States (USNM). **Distr.:** Colombia, Costa Rica, Dominican Republic, Guatemala, Honduras, Mexico, Panama, United States (continental), Venezuela. **Refs.:** Galindo & Kelley 1943: 87 (M, F, L; to sp.); Yamaguti & LaCasse 1951c: 70 (M*, F*, L*); Carpenter & LaCasse 1955: 296 (M*, F*, L*; keys); Bohart & Washino 1957: 463 (L*); Stone & Knight 1957a: 57 (lectotype desig.); Dodge 1966: 372 (first instar L; key); Bram 1967a: 107 (M*, F, L); Bohart & Washino 1978: 134 (M, F*, L*; keys, tax., bion., distr.); Strickman 1988a: 490 (syn. with *peus*); Eldridge & Harbach 1989: 247 (conservation of *thriambus*); International Commission on Zoological Nomenclature 1991 (Opinion 1644; conservation of name); Eldridge & Harbach 1992 (conservation of *stigmatosoma* and *thriambus*); Darsie & Ward 2005 (F*, L*; keys, distr.); Ortega-Morales et al. 2011 (tax., distr.). **Ety.:** n.s. [poss. ref. to Thria, nurse of Apollo, inventer of divination]. **Informal name:** Thria American Typical Mosquito.
syn. *affinis* Adams
1903: 25 (F). **Type loc.:** [Oak Creek Canyon], Arizona, United States (USNM). **Refs.:** Strickman 1988a: 490 (syn.).

tianpingensis Chen
1981: 426 (M*, F*, L*). **Type loc.:** Tianpingsham, Guangxi Zhuang Autonomous Region, South China, China [People's Republic of China] (GMC). **Distr.:** People's Republic of China. **Refs.**: Chen & Zhao 1987 (P*). **Ety.:** Tianpingsham, Guangxi Zhuang Autonomous Region, South China. **Informal name:** Tianpingsham Chinese Typical Mosquito.

toroensis Edwards & Gibbins
1939: 32 (M; *vansomereni* ssp.). **Type loc.:** Kisomoro, Ruwenzori [Toro District], Uganda (NHMUK). **Distr.:** Burundi, Cameroon, Democratic Republic of the Congo, Ethiopia, Kenya, Malawi, Republic of South Africa, Rwanda, Sudan and South Sudan, Uganda. **Refs.:** Edwards 1941: 329 (M*, P, L; to sp.); Hopkins 1952: 314 (L*); Mattingly 1956a: 43 (lectotype desig.); Service 1959a (L); Jupp 1996 (M*, F*; key). **Ety.:** n.s. [poss. Toro Kingdom of Uganda]. **Informal name:** Toro Ugandan Typical Mosquito.
ssp. ***macrophyllus*** Edwards & Gibbins
1939: 31 (M; *vansomereni* ssp.). **Type loc.:** Lugezi Camp, between Mount Mgahinga and Mount Sabinio, Uganda (NHMUK). **Distr.:** Cameroon, Uganda. **Refs.:** Edwards 1941: 329 (M*); Mattingly 1956a: 43 (lectotype desig.). **Ety.:** large leaf (Gr); ref. to leaf on gonocoxite larger than on *toroensis*. **Informal name:** Large-leafed Ugandan Typical Mosquito.

torrentium Martini
1925: 336 (M, F, L). **Type loc.:** Schwartza River [Thuringia], Germany (NHMUK). **Distr.:** Algeria, Armenia, Austria, Belarus, Belgium, Crimean Peninsular, Czech Republic, Denmark, Estonia, Finland, France (includes Corsica), Germany, Greece, Hungary, Iran, Iraq, Ireland, Italy, Latvia, Libya, Liechtenstein, Lithuania, Luxembourg, Moldova, Netherlands, Norway, Poland,

Portugal, Romania, Russia, Serbia, Shetland Isl., Slovakia, Spain, Sweden, Switzerland, Turkey, Ukraine, United Kingdom. **Refs.:** Natvig 1948: 455 (M*, F, L); Mattingly 1951a: 172 (M*; tax.); Noèldner 1953: 7 (M*, F, L*; as *exilis*); Senevet & Andarelli 1959: 203 (M*, F, P, L); Sicart & Ruffie 1960: 647 (P*); Service 1968: 313 (M*; tax.); Lotfi 1970: 401 (distr.); Iversen 1971: 235 (distr.); Gutsevich et al. 1974: 392 (M, F, L); Iglisch 1977: 270 (M*); Harbach 1988: 40 (M*, F, P*, L*; tax., keys, bion., distr.); Alten et al. 2000 (distr.; Turkey); Aranda et al. 2000: 7 (distr.; Spain); Ribeiro et al. 2002 (distr.; Portugal); Beck et al. 2003: 24 (distr.; Luxembourg); Becker et al. 2010: 279 (M*, F*, L*; keys, tax., distr., bion.); Ahmed et al. 2011 (distr.; Saudi Arabia); Danabalan et al. 2012 (mol. tax., United Kingdom); Gunay et al. 2015 (mol. tax., distr.; Turkey); Robert et al. 2019 (distr.; Western Palearctic). **Ety.:** rapid, a torrent (L); ref. to being found in river water left after flooding. **Informal name:** Torrential German Typical Mosquito.

syn. *pavlovsky* Shingarev
1928: 52 (M*). **Type loc.:** Naryn District, Tomsk, Siberia, former Union of Soviet Socialist Republics (USSR) [Russia] (TMS).

toviiensis Klein, Rivière, & Séchan
1984: 72 (P*, L*). **Type loc.:** Tovii Plateau, Nuku-Hivi Island, Marquesas Islands (SSC). **Distr.:** French Polynesia. **Refs.:** Lee et al. 1989a: 258 (tax., bion., distr., review; Australasia). **Ety.:** Plateau de Tovii, Marquesas Isl. **Informal name:** Tovii French Polynesian Typical Mosquito.

trifilatus Edwards
1914b: 65 (M*, F). **Type loc.:** Kabete [Kenya] (NHMUK). **Distr.:** Angola, Cameroon, Democratic Republic of the Congo, Ethiopia, Gabon, Guinea, Kenya, Madagascar (includes Glorioso and Juan De Nova Isl.), Malawi, Mozambique, Republic of the Congo, Republic of South Africa, Sudan and South Sudan, Tanzania, Uganda, Zimbabwe, Zambia. **Refs.:** Edwards 1941: 321 (M*, F); Worth & De Meillon 1960: 250 (distr.); Service 1976a (distr.; Gabon); Ribeiro et al. 1982 (L*) Jupp 1996 (M*, F*; key). **Ety.:** three filaments (L); ref. to "three filament-like divisions of the harpagones" [claspette]. **Informal name:** Filamented Kenyan Typical Mosquito.

ssp. ***aenescens*** Edwards
1941: 323, 421 (M*, P). **Type loc.:** Kameranjoka, Ruwenzori foothills [Toro District], Uganda (NHMUK). **Distr.:** Uganda. **Refs.:** Hopkins 1952: 308 (L*); Mattingly 1956a: 42 (lectotype desig.). **Ety.:** n.s. [copper, bronze colored) (L); ref. to scutum with a "brassy-yellow tint"]. **Informal name:** Copper Ugandan Typical Mosquito.

trifoliatus Edwards
1914b: 70 (M*). **Type loc.:** Kasala, Uganda (NHMUK). **Distr.:** Benin, Burkina Faso, Cameroon, Central African Republic, Côte d'Ivoire, Democratic Republic of the Congo, Ethiopia, Ghana, Kenya, Mali, Namibia, Nigeria, Republic of South Africa, Sudan and South Sudan, Tanzania, Uganda. **Refs.:** Edwards 1941: 339 (M*, F); Ovazza et al. 1956: 176 (P*, L*); van Someren 1956: 11 (L); Worth & De Meillon 1960: 250 (distr.); Rickenbach et al. 1976: 97 (distr.); Jupp 1996 (M*, F*; key). **Ety.:** three leaves (L); ref. to three leaflike plates on the gonocoxite. **Informal name:** Three-leafed Ugandan Typical Mosquito.

tritaeniorhynchus Giles
1901a: 606 (A). **Type loc.:** Travancore, India (NHMUK). **Distr.:** Afghanistan, Albania, Angola, Azerbaijan, Bangladesh, Benin, Borneo, Brunei, Cambodia, Cameroon, Central African Republic, Djibouti, Egypt, Ethiopia, Fiji, Gabon, Georgia, Ghana, Greece, Guam, Hong Kong, India, Indonesia, Iran, Iraq, Israel (and Gaza Strip and West Bank), Japan, Jordan, Kenya, Kuwait, La Réunion, Laos, Lebanon, Liberia, Madagascar (includes Glorioso and Juan De Nova Isl.), Malaysia, Maldives, Mariana Isl., Mauritius, Micronesia (Wake Isl.), Mozambique, Myanmar, Nepal, Nigeria, Oman, Pakistan, Papua New Guinea, People's Republic of China, Philippines, Russia, Saudi Arabia, Senegal, Singapore, Somalia, South Korea, Sri Lanka, Syria, Taiwan, Tanzania, Thailand, The Gambia, Timor, Togo, Turkey, Turkmenistan, Uzbekistan, Vietnam, Yemen. **Refs.:** Macfie & Ingram 1922 (F*); Barraud 1923c (L*); Kirkpatrick 1925b: 111 (M*, F, P*, L*); Barraud 1934: 404 (M*, F, L*); Bohart & Ingram 1946b: 81 (M*, F, L*; bion., distr., tax.); La Casse & Yamaguti 1950: 230 (M*, F*, P*, L*; bion., distr., tax.; as *summorosus*); Ovazza et al. 1956: 169 (L*); Colless 1957a: 100 (tax.); Hara 1957 (F*), 56 (F*); Joshi et al. 1965 (distr.; Nepal); Bram 1967b: 225 (M*, F*, P*, L*; bion., distr., tax.); Reuben 1969: 650 (M*, F*, L*); Aslamkhan 1971b (distr.; Pakistan); Basio 1971b: 60 (M*; bion.); Matsuo & Iwaki 1972: 349 (tax.); Baisas 1974: 104 (M, F, P, L*; tax., bion., distr.; Philippines; as ssp. *summorosus*); Gutsevich et al. 1974: 382 (M*, F*, L*); Sirivanakarn 1975a: 73 (M*, F*, P*, L*); Tanaka et al. 1975c: 219 (bion., distr.); Sirivanakarn 1976: 129 (M*, F*, P*, L*; distr.); Rodhain et al. 1977 (distr.); Tanaka et al. 1979: 100 (M*, F*, L*); Ahmed 1987 (distr.; Bangladesh); Harbach 1988: 106 (M*, F, P*, L*; tax., keys, bion., distr.); Lee et al. 1989a: 259 (F; key, tax., bion., distr., review); Darsie & Pradhan 1990, 1991 (F, L; tax., keys, bion., distr.; Nepal); Suleman et al. 1993 (distr.; Pakistan); Jupp 1996 (M*, F*; key); Al-Houty 1997 (distr.; Kuwait); Amr et al. 1997 (distr.; Jordan); Whelan & Hapgood 2000 (bion., distr.; East Timor); Samanidou & Harbach 2003: 15 (distr.; Greece); Tanaka 2004a: 23 (P*; tax., key); Rattanarithikul et al. 2005b (F*, L*; bion., distr., keys); Rueda et al. 2008: 398 (distr.); Suman et al. 2008 (E*); Becker et al. 2010: 349 (F*, L*; key, tax., distr., bion.); Namazov 2014 (distr.; Azerbaijan); Gunay et al. 2015 (mol. tax., distr.; Turkey); Robert et al. 2019 (distr.; Western Palearctic). **Ety.:**

three-banded, snout (L); proboscis with three yellow bands. **Informal name:** Three-banded Indian Typical Mosquito. (Vol. 1, p. 512)

syn. *biroi* Theobald
1905f: 82 (M*, F). **Type loc.:** Bombay, India (HNM).

tsengi Lien
1968d: 237 (M*, F*, P*, L*). **Type loc.:** Chihtuan, Loshui, Tatung, Ilan Hsien, Taiwan [Republic of China] (TMRI). **Distr.:** People's Republic of China (includes Taiwan), Thailand. **Refs.:** Sirivanakarn 1976: 152 (M*, F*, P, L*). **Ety.:** Po Tsun Tseng. **Informal name:** Tseng Taiwanese Typical Mosquito.

umbripes Edwards
1941: 321 (F). **Type loc.:** Pashie stream, Kundelungu Mountains, Katanga, Belgian Congo [Democratic Republic of the Congo] (NHMUK). **Distr.:** Democratic Republic of the Congo. **Refs.:** Mattingly 1953c: 339 (M*); Mattingly 1956a: 42 (lectotype desig.). **Ety.:** n.s. [dark foot (L); ref. to very dark legs]. **Informal name:** Dark-footed Congolese Typical Mosquito.

univittatus Theobald
1901c: 29 (M*, F*). **Type loc.:** Salisbury, Rhodesia [Zimbabwe] (NHMUK). **Distr.:** Algeria, Angola, Benin, Burkina Faso, Cameroon, Central African Republic, Democratic Republic of the Congo, Djibouti, Egypt, Ethiopia, Gabon, Ghana, India, Iran, Iraq, Israel (and Gaza Strip and West Bank), Kenya, Kuwait, Lebanon, Liberia, Madagascar (includes Glorioso and Juan De Nova Isl.), Mali, Mauritania, Morocco, Mozambique, Namibia, Niger, Nigeria, Oman, Pakistan, Qatar, Republic of the Congo, Republic of South Africa, Saudi Arabia, Senegal, Somalia, Sudan and South Sudan, Syria, Tanzania, The Gambia, Tunisia, Turkmenistan, Uganda, United States (continental), Yemen, Zambia, Zimbabwe. **Refs.:** Séguy 1924: Pl. IV (M gen.*); Edwards 1941: 306 (M*, F*), 419 (P); Hopkins 1952: 291 (L*); Mattingly & Knight 1956: 104 (tax.); Senevet et al. 1957: 52 (M*; tax.); Jupp 1970: 9 (M*; tax.); Aslamkhan 1971b (distr.; Pakistan); Jupp 1971: 339 (tax.); Jupp 1972: 103 (M*; tax.); Gutsevich et al. 1974: 385 (M*, F, L*); White 1975: 316, 320 (tax., lectotype desig., distr.); Sirivanakarn 1976: 48 (M*, F*, P, L*; distr.); Rodhain et al. 1977 (distr.); Beier et al. 1986: 164 (distr.); Harbach 1988: 52 (M*, F, P*, L*; tax., keys, bion., distr.); Suleman et al. 1993 (distr.; Pakistan); Jupp 1996 (M*, F*; key); Al-Houty 1997 (checklist; Kuwait); Ahmed et al. 2011 (distr.; Saudi Arabia); Robert et al. 2019 (distr.; Western Palearctic). **Ety.:** n.s. [one banded (L); prob. ref. to single bands on tibiae]. **Informal name:** One-banded Zimbabwean Typical Mosquito. (Vol. 1, p. 514)

syn. *simplex* Theobald
1903a: 337 (F*; *Heptaphlebomyia*). **Type loc.:** Salisbury, Mashonaland [Zimbabwe] (NHMUK).

syn. *montforti* Ventrillon
1905a: 448 (M*, F; *Heptaphlebomyia*). **Type loc.:** Ankajobe, Madagascar (NHMUK). **Refs.:** White 1975: 321 (lectotype desig.); Townsend 1990: 106 (type info.).

syn. *ataeniatus* Theobald
1911b: 261 (F; as *ataeniata*). **Type loc.:** Onderstepoort, Transvaal [Republic of South Africa] (NHMUK). **Refs.:** Edwards 1941: 353 (tax.); White 1980: 137 (syn.).

syn. *goughii* Theobald
1911b: 268 (syntype F* only). **Type loc.:** Onderstepoort, Transvaal [Republic of South Africa] (NHMUK). **Refs.:** White 1975: 320 (tax.).

usquatissimus Dyar
1922b: 19 (M*, F). **Type loc.:** Toro Point, Canal Zone, Panama (USNM). **Distr.:** Argentina, Brazil, Colombia, Costa Rica, Ecuador, Guyana, Panama, Venezuela. **Refs.:** Stone & Knight 1957a: 57 (lectotype desig.); Bram 1967a: 51 (M*; tax., from syn. with *coronator*). **Ety.:** n.s. [large (L); larger than *usquatus*]. **Informal name:** Large Panamanian Typical Mosquito.

usquatus Dyar
1922d: 122 (M, L). **Type loc.:** Paramaribo, Suriname (USNM). **Distr.:** Argentina, Belize, Brazil, Colombia, Costa Rica, Ecuador, Guatemala, Lesser Antilles (includes Trinidad and Tobago), Mexico, Panama, Paraguay, Peru, Suriname. **Refs.:** Stone & Knight 1957a: 57 (lectotype desig.); Bram 1967a: 49 (M*; tax.; from syn. with *coronator*); Linton et al. 2013 (bion., mol. tax., distr.; Ecuador). **Ety.:** n.s. **Informal name:** Whisky Surinamese Typical Mosquito [*usquatus* = "usquebaugh-like" = Whisky].

vagans Wiedemann
1828: 545 (M, F). **Type loc.:** China [People's Republic of China] (NE). **Distr.:** Afghanistan, Bangladesh, Egypt, Guam, Hong Kong, India, Iran, Iraq, Israel (and Gaza Strip and West Bank), Japan, Macau, Mongolia, Nepal, Pakistan, People's Republic of China, Russia, South Korea, Turkey, Vietnam. **Refs.:** Séguy 1924: Pl. VI (L*; as *Cx. tipuliformis*); La Casse & Yamaguti 1950: 216 (M*, F*, P*, L*; bion., distr., tax.); Hara 1957: 55 (F*); Joshi et al. 1965 (distr.; Nepal); Ishii 1969: 204 (F); Lotfi 1970: 400 (distr.); Aslamkhan 1971b (distr.; Pakistan); Gutsevich et al. 1974: 390 (M*, F, L*); Tanaka et al. 1975c: 219 (distr.); Minar 1976: 335 (distr.); Sirivanakarn 1976: 37 (M*, F*, P*, L*; distr.); Tanaka et al. 1979: 138 (M*, F*, L*); Harbach 1985a: 86 (distr.; Israel); Harbach 1988: 36 (M*, F, P*, L*; tax., keys, bion., distr.); Darsie & Pradhan 1990, 1991 (F, L; tax., keys, bion., distr.; Nepal); Tanaka 2004a: 14 (P*; tax., key); Rueda, Pecor et al. 2008: 397 (distr.); Robert et al. 2019 (distr.; Western Palearctic). **Ety.:** n.s. [ranging, wandering (L)]. **Informal name:** Wandering Chinese Typical Mosquito.

syn. *tipuliformis* Theobald
1901c: 325 (F*). **Type loc.:** Bakloh, Northwest Provinces [India] (NHMUK).

syn. *virgatipes* Edwards
1914c: 126 (M, F). **Type loc.:** Hong Kong [People's Republic of China] (NHMUK).
syn. *exilis* Dyar
1924f: 127 (M). **Type loc.:** Vladivostok [Maritime Territory], former Union of Soviet Socialist Republics (USSR) (USNM). **Refs.:** Stone & Knight 1957a: 49 (lectotype desig.).

vansomereni Edwards
1926a: 129 (M*, F). **Type loc.:** Nairobi [Central], Kenya (NHMUK). **Distr.:** Democratic Republic of the Congo, Ethiopia, Kenya, Liberia, Mozambique, Republic of South Africa, Sudan and South Sudan, Tanzania, Uganda, Zimbabwe. **Refs.:** Edwards 1941: 327 (M*, F*), 422 (P), 434 (L); Hopkins 1952: 312 (L*); Lewis 1956b: 715 (L); Worth & De Meillon 1960: 250 (distr.). **Ety.:** Ellinor Catharine Cunningham van Someren. **Informal name:** van Someren Kenyan Typical Mosquito.
ssp. ***draconis*** Ingram & De Meillon
1927: 67 (M*, F*; as sp.). **Type loc.:** Pietermaritzburg, Natal [KwaZulu-Natal, Republic of South Africa] (NHMUK). **Distr.:** Republic of South Africa. **Refs.:** Edwards 1941: 328 (M; to ssp.), 423 (P); Hopkins 1952: 313 (L); Jupp 1996 (M*, F*; key). **Ety.:** n.s. [dragon (Gr); poss. ref. to characteristics of the male genitalia]. **Informal name:** Draconic South African Typical Mosquito.
ssp. ***elgonicus*** Edwards
1941: 328 (M*). **Type loc.:** Buginyanya, Mount Elgon, Uganda (NHMUK). **Distr.:** Ethiopia, Uganda. **Refs.:** Mattingly 1956a: 43 (lectotype desig.). **Ety.:** Mt. Elgon, Uganda/Kenya. **Informal name:** Elgon Ugandan Typical Mosquito.

ventrilloni Edwards
1920b: 135 (F). **Type loc.:** Tananarive, Madagascar (MNHP). **Distr.:** Madagascar (includes Glorioso and Juan De Nova Isl.). **Refs.:** Edwards 1941: 300 (M, F). **Ety.:** Édouard Ventrillon. **Informal name:** Ventrillon Malagasy Typical Mosquito.
syn. *albigenu* Enderlein
1920: 50 (M, F). **Type loc.:** Tananarive, Madagascar (ZM).

verutus Harbach
1987: 216 (M*, F, P*, L*). **Type loc.:** Kasawe Forest Reserve, Moyamba District, Southern Province, Sierra Leone (USNM). **Distr.:** Côte d'Ivoire, Liberia, Sierra Leone. **Ety.:** "Armed with a dart or javelin" (L); ref. to posterior median process of spiracular apparatus with a strong erect spine. **Informal name:** Javelin Sierra Leonean Typical Mosquito.

vicinus (Taylor)
1916: 569 (M, F; *Leucomyia*). **Type loc.:** Stapleton, Northern Territory, Australia (ANIC). **Distr.:** Australia, Papua New Guinea. **Refs.:** Lee et al. 1989a: 265 (F; key, tax., bion., distr., review; type info). **Ety.:** neighboring (L); ref. to resemblance to two other species. **Informal name:** Neighboring Australian Typical Mosquito.
syn. *annulata* Taylor
1914c: 695 (M, F; *Leucomyia*). **Type loc.:** Townsville, Queensland, Australia (ANIC). **Refs.:** Lee et al. 1989a: 265 (type info.).
syn. *basicinctus* Edwards
1921c: 78 (new name for *annulata* Taylor, not Schrank 1776). **Refs.:** Stone & Knight 1958: 69 (syn.).

vishnui Theobald
1901a: 355 (type F* only). **Type loc.:** Madras, India (NHMUK). **Distr.:** Bangladesh, Borneo, Cambodia, India, Indonesia, Japan, Laos, Malaysia, Maldives, Myanmar, Nepal, Pakistan, Papua New Guinea, People's Republic of China (includes Hong Kong), Philippines, Singapore, South Korea, Sri Lanka, Taiwan, Thailand, Timor, Vietnam. **Refs.:** Barraud 1923c (L*); Barraud 1934: 400 (M*, F, L*); Bohart & Ingram 1946b: 83 (M*, F, L*; bion., distr., tax.); La Casse & Yamaguti 1950: 236 (M*, F*, P*, L*; bion., distr., tax.); Iyengar & Menon 1955 (distr.); Colless 1957a: 87 (M; tax.); Joshi et al. 1965 (distr.; Nepal); Reuben 1969: 643 (M*, F*, L*; syn.); Aslamkhan 1971b (distr.; Pakistan); Lotfi 1973: 204 (distr.); Pinhão 1974 (distr.; Timor); Sirivanakarn 1975a: 71 (M*, F*, P*, L*); Sirivanakarn 1976: 111 (M*, F*, P*, L*; distr.); Lee et al. 1989a: 269 (F; key, tax., bion., distr., review); Darsie & Pradhan 1990, 1991 (F, L; tax., keys, bion., distr.; Nepal); Suleman et al. 1993 (distr.; Pakistan); Whelan & Hapgood 2000 (bion, distr.; East Timor); Tanaka 2004a: 41 (P; tax., key); Rattanarithikul et al. 2005b (F*, L*; bion., distr., keys). **Ety.:** n.s. [Vishnu (a Hindu deity)]. **Informal name:** Vishnu Indian Typical Mosquito. (Vol. 1, p. 516)

watti Edwards
1920a: 135 (M, F). **Type loc.:** Kumasi [Ashanti], Gold Coast [Ghana] (NHMUK). **Distr.:** Angola, Cameroon, Democratic Republic of the Congo, Ghana, Kenya, Madagascar (includes Glorioso and Juan De Nova Isl.), Mali, Tanzania, Uganda. **Refs.:** Edwards 1941: 302 (M*, F); Ravaonjanahary 1979 (P*, L*). **Ety.:** W.G. Watt. **Informal name:** Watt Ghanaian Typical Mosquito.

weschei Edwards
1935a: 136 (M, F). **Type loc.:** Sunyani [Ashanti], Gold Coast [Ghana] (NHMUK). **Distr.:** Burkina Faso, Cameroon, Democratic Republic of the Congo, Ghana, Guinea, Madagascar (includes Glorioso and Juan De Nova Isl.), Mali, Mozambique, Nigeria, Senegal, Seychelles, Sudan and South Sudan, Togo. **Refs.:** Edwards 1941: 345 (M*, F); Hopkins 1952: 330 (L*); Rickenbach et al. 1976: 97 (distr.); Jupp 1996 (M*, F*; key). **Ety.:** n.s. **Informal name:** Wesche Ghanaian Typical Mosquito.

ssp. ***gediensis*** Edwards
1941: 345 (M, F). **Type loc.:** Gede, near Mombasa [Coast], Kenya (NHMUK). **Distr.:** Kenya. **Refs.:** van Someren, Teesdale, & Furlong 1955: 490 (L?); Mattingly 1956a: 44 (lectotype desig.). **Ety.:** Gede, near Mombasa, Kenya. **Informal name:** Gede Kenyan Typical Mosquito.

whitei Barraud
1923i: 508 (M). **Type loc.:** Haflong, Cachar Hills, Assam, India (NHMUK). **Distr.:** Bangladesh, Cambodia, India, Indonesia, Malaysia, Nepal, Philippines, Thailand, Vietnam. **Refs.:** Peters & Dewar 1956: 46 (P*, L*); Bram 1967b: 233 (M*, F, L*; bion., distr., tax.); Lien et al. 1975 (distr.); Sirivanakarn 1976: 135 (M*, F, P*, L*; distr.); Lee et al. 1989a: 274 (F; key, tax., bion., distr., review); Darsie & Pradhan 1990, 1991 (F, L; tax., keys, bion., distr.; Nepal); Rattanarithikul et al. 2005b (F*, L*; bion., distr., keys). **Ety.:** R. Senior-White. **Informal name:** Senior-White Indian Typical Mosquito.

whitmorei (Giles)
1904a: 367 (F; *Taeniorhynchus*). **Type loc.:** Camp Stotsenburg [Angeles], Pampanga, [Luzon], Philippines (NHMUK). **Distr.:** Australia, Bangladesh, Cambodia, India, Indonesia, Japan, Laos, Malaysia, Nepal, Pakistan, Papua New Guinea, People's Republic of China, Philippines, Russia, South Korea, Sri Lanka, Taiwan, Thailand, Vietnam. **Refs.:** Barraud 1923c (L*); La Casse & Yamaguti 1950: 211 (M*, F*, P*, L*; bion., distr., tax.); Hara 1957: 56 (F*); Bram 1967b: 253 (M*, F*, L*; bion., distr., tax.); Aslamkhan 1971b (distr.; Pakistan); Basio 1971b: 61 (M*; bion.); Gutsevich et al. 1974: 372 (M*, F, L*); Baisas 1974: 107 (M, F, P, L; tax., bion., distr.; Philippines); Tanaka et al. 1975c: 219 (distr.); Sirivanakarn 1976: 103 (M*, F*, P*, L*; distr.); Tanaka et al. 1979: 158 (M*, F*, L*); Lee et al. 1989a: 275 (F; key, tax., bion., distr., review); Townsend 1990: 151 (lectotype desig.); Darsie & Pradhan 1990, 1991 (F, L; tax., keys, bion., distr.; Nepal); Tanaka 2004a: 40 (P; tax., key); Rattanarithikul et al. 2005b (F*, L*; bion., distr., keys). **Ety.:** n.s. [someone named Whitmore]. **Informal name:** Whitmore Philippine Typical Mosquito.
syn. *argenteus* Ludlow
1905a: 98 (F; *Taeniorhynchus*). **Type loc.:** Angeles, Pampanga, Luzon, Philippines (USNM).
syn. *plegepennis* Theobald
1907: 375 (F; *Leucomyia*). **Type loc.:** Kobe [Honshu], Japan (NHMUK).
syn. *albus* Leicester
1908: 148 (F). **Type loc.:** Kuala Lumpur [Selangor], Malaya [Malaysia] (NHMUK).
syn. *loricatus* Leicester
1908: 151 (F). **Type loc.:** Kuala Lumpur [Selangor], Malaya [Malaysia] (NHMUK).

whittingtoni Belkin
1962: 204 (L*). **Type loc.:** Munda, New Georgia [Solomon Islands] (USNM). **Distr.:** Solomon Isl. **Refs.:** Lee et al. 1989a: 279 (tax., bion., distr., review). **Ety.:** Forrest B. Whittington. **Informal name:** Whittington Solomon Islander Typical Mosquito.

yojoae Strickman
1990: 148 (M*, F*, P*, L*). **Type loc.:** La Joya, Cortes Department, Honduras (USNM). **Distr.:** Belize, Honduras. **Refs.:** Pecor et al. 2002: 243 (tax., distr.; Belize). **Ety.:** Lake Yojoa, Honduras. **Informal name:** Yojoa Honduran Typical Mosquito.

zombaensis Theobald
1901c: 143 (F*). **Type loc.:** Zomba [Malawi] (NHMUK). **Distr.:** Angola, Democratic Republic of the Congo, Ethiopia, Kenya, Malawi, Mozambique, Nigeria, Republic of South Africa, Sudan and South Sudan, Tanzania, Uganda, Zambia. **Refs.:** Edwards 1941: 317 (M*, F; from syn. with *pipiens*); Wolfs 1948c: 87 (P*, L*); Ovazza et al. 1956: 172 (L*); Jupp 1996 (M*, F*; key). **Ety.:** Zomba, Malawi. **Informal name:** Zomba Malawian Typical Mosquito.

Subgenus *Culiciomyia* Theobald

Name derivation: n.s. [-culis = gnat, mosquito (L); myia = small fly (Gr)].

syn. *Trichorynchus* Theobald 1905h: 241 (not Balbiani 1887) (as genus). Type species: *Trichorynchus fuscus* Theobald. Edwards 1932a: 198 (syn.).
Culiciomyia Theobald 1907: 227 (as genus). Type species: *Culiciomyia inornata* Theobald (Edwards 1912c: 33). Stone 1961c: 45 (tax.); Borel 1926 (to subg.).
syn. *Neomelanoconion* Theobald 1907: 514 (as genus). Type species: *Culex rima* Theobald [This is based upon a misidentification of *Culex nebulosus* Theobald, which is a true *Culiciomyia*. The true *Culex rima* has been assigned to the subgenus *Eumelanomyia*]. Stone 1961c: 45 (tax.). Edwards 1932a: 198 (syn.).
syn. *Pectinopalpus* Theobald 1909b: 11 (as genus). Type species: *Pectinopalpus fuscus* Theobald. Edwards 1932a: 198 (syn.).
syn. *Trichorhynchomyia* Brunetti 1912: 477 (as genus). Type species: *Trichorhynchus fuscus* Theobald. Edwards 1932a:198 (syn.).
syn. *Thaiomyia* Bram 1966: 73 (as subg. of genus *Culex*). Type species: *Culex dispectus* Bram. Harrison 1987b: 111 (syn.)

Important references: King & Hoogstraal 1946a: 143 (keys; New Guinea); Vattier & Hamon 1962: 251 (L; key; Africa); Belkin 1962 (revision; South Pacific); Bram 1967b (revision;

Thailand); Harrison 1987a (book review); Lee et al. 1989b (revision; Australasia).

azurini Toma, Miyagi, & Cabrera
1984: 172 (M*, F*, P*, L*). **Type loc.:** White Beach, Puerto Princesa, Palawan Island, Philippines (NMST). **Distr.:** Philippines. **Ety.:** Jesus C. Azurin. **Informal name:** Azurin Philippine Typical Mosquito.

bahri (Edwards)
1914b: 79 (M, F; *Culiciomyia*). **Type loc.:** Badulla, Ceylon [Sri Lanka] (NHMUK). **Distr.:** Indonesia, Sri Lanka. **Refs.:** Sirivanakarn 1977c: 94 (M*, F). **Ety.:** P.H. Bahr. **Informal name:** Bahr Sri Lankan Typical Mosquito.

bailyi Barraud
1934: 382 (M*, F). **Type loc.:** Virajpet, Coorg, India (NHMUK). **Distr.:** Cambodia, India, Indonesia, Malaysia, Myanmar, Papua New Guinea, Philippines, Sri Lanka, Thailand. **Refs.:** Carter & Wijesundara 1948: 147 (L*); Bram 1967b: 123 (M*, F*, P*, L*; bion., distr., tax.); Maffi et al. 1979: 216 (distr.); Lee et al. 1989b: 5 (F; key, tax., bion., distr., review); Rattanarithikul et al. 2005b (F*, L*; bion., distr., keys). **Ety.:** J.D. Baily. **Informal name:** Baily Indian Typical Mosquito.

barrinus Bram
1967b: 125 (M*, F, P, L*; bion., distr., tax.). **Type loc.:** Ban Pa Chi, Mae Hong Son, Thailand (USNM). **Distr.:** Thailand. **Refs.:** Kalpage & Brust 1968 (E*); Rattanarithikul et al. 2005b (F*, L*; bion., distr., keys). **Ety.:** elephant (L); larvae in elephant tracks. **Informal name:** Elephant Thai Typical Mosquito.

cambournaci Hamon & Gandara
1955: 866 (M*, L*). **Type loc.:** São Tomé (IERT). **Distr.:** Angola, São Tomé and Príncipe. **Ety.:** Francisco José Carrasqueiro Cambournac. **Informal name:** Cambournac São Tomean Typical Mosquito.

ceramensis Sirivanakarn & Kurihara
1973: 220 (M*, F, P, L*). **Type loc.:** Piru, Ceram, Maluku Province, Indonesia (USNM). **Distr.:** Indonesia. **Refs.:** Lee et al. 1989b: 7 (F; key, tax., bion., distr., review). **Ety.:** Ceram, Indonesia. **Informal name:** Ceram Indonesian Typical Mosquito.

cheni Dong, Wang, & Lu
2003 (M*, F, L*). In Dong, Dong, Zhou, Wang, & Lu 2003: 528. **Type loc.:** Tenchong County (1,795 m), Yunnan, China [People's Republic of China] (YIM). **Distr.:** People's Republic of China. **Ety.:** n.s. [H.B. Chen (Harbach 2018c)]. **Informal name:** Chen Chinese Typical Mosquito.

cinerellus Edwards
1922d: 90 (M). **Type loc.:** Kasala, Uganda (NHMUK). **Distr.:** Angola, Benin, Burkina Faso, Cameroon, Central African Republic, Comoros, Côte d'Ivoire, Democratic Republic of the Congo, Kenya, Liberia, Madagascar (includes Glorioso and Juan De Nova Isl.), Mali, Nigeria, Republic of the Congo, Republic of South Africa, São Tomé and Príncipe, Sierra Leone, Sudan and South Sudan, Tanzania, Uganda. **Refs.:** Evans 1929a: 411 (P*, L*); Edwards 1941: 275 (M*, F*), 413 (P); Hopkins 1952: 273 (L*); Brunhes 1977a (distr.; Comoros Isl.); Jupp 1996 (M*, F*; key). **Ety.:** n.s. [ash colored (L); poss. ref. to color]. **Informal name:** Ashy Ugandan Typical Mosquito.

cinereus Theobald
1901c: 58 (type F*, only). **Type loc.:** Freetown, Sierra Leone (NHMUK). **Distr.:** Angola, Benin, Burkina Faso, Burundi, Cameroon, Central African Republic, Côte d'Ivoire, Democratic Republic of the Congo, Ghana, Kenya, Liberia, Madagascar (includes Glorioso and Juan De Nova Isl.), Mali, Nigeria, Republic of South Africa, Sierra Leone, Sudan and South Sudan, Tanzania, Uganda, Zambia, Zimbabwe. **Refs.:** Edwards 1941: 274 (M*, F*), 413 (P); Hopkins 1952: 271 (L*); Jupp 1996 (M*, F*; key). **Ety.:** ash colored (L) (ref. to dense gray scales on thorax). **Informal name:** Ashy Sierra Leonean Typical Mosquito.

syn. *freetownensis* Theobald
1901c: 69 (M* syntype; only of type series) [F syntype syn. of *nebulosus*]. **Type loc.:** [Freetown], Sierra Leone (NHMUK).

syn. *uniformis* Theobald
1909b: 8 (M*, F*; *Culiciomyia*). **Type loc.:** Obuasi, Ashanti, Gold Coast [Ghana] (NHMUK). **Refs.:** Edwards 1941: 274 (A; to ssp.); Knight & Stone 1977 (syn.).

delfinadoae Sirivanakarn
1973b: 118 (M*, F). **Type loc.:** Tawi Tawi, Sangir Island, Philippines (USNM). **Distr.:** Malaysia, Philippines. **Ety.:** Mercedes Delfino Delfinado. **Informal name:** Delfinado Philippine Typical Mosquito.

dispectus Bram
1966: 75 (M*, F*, P*, L*). **Type loc.:** Doi Sam Sao, Tak Province, Thailand (USNM). **Distr.:** Malaysia, People's Republic of China, Thailand. **Refs.:** Bram 1967b: 113 (M*, F*, P*, L*; bion., distr., tax.); Harrison 1987b: 113 (from *Thaiomyia*); Rattanarithikul et al. 2005b (F*, L*; bion., distr., keys). **Ety.:** n.s. [discovered (L)]. **Informal name:** Discovered Thai Typical Mosquito.

eouzani Geoffroy
1971a: 279 (M*). **Type loc.:** La Mingui, near the village of Batalimo, Central African Republic (IERT). **Distr.:** Cameroon, Central African Republic. **Refs.:** Rickenbach et al. 1976: 96 (distr.). **Ety.:** Jean-Pierre Eouzan. **Informal name:** Eouzan Central African Typical Mosquito.

fragilis Ludlow
1903: 142 (M, F). **Type loc.:** Oras, Samar, Philippines (USNM). **Distr.:** Bangladesh, Cambodia, India, Indonesia, Malaysia (includes Sabah and Sarawak), Papua New Guinea, Philippines, Solomon Isl., Sri Lanka, Thailand,

Vietnam. **Refs.:** King & Hoogstraal 1946a: 148 (M*, L*); Penn 1949b: 72 (P*); Stone & Knight 1957a: 50 (lectotype desig.); Bram 1967b: 129 (M*, F*, P*, L*; bion., distr., tax.); Basio 1971b: 62 (M*; bion.); Lee et al. 1989b: 9 (F; key, tax., bion., distr., review); Rattanarithikul et al. 2005b (F*, L*; bion., distr., keys). **Ety.:** fragile (L); "A very delicate looking insect." **Informal name:** Fragile Philippine Typical Mosquito.

syn. *fuscus* Theobald
1905h: 242 (F*; *Trichorhynchus*). **Type loc.:** Peradeniya [Central], Ceylon [Sri Lanka] (NHMUK).

syn. *inornata* Theobald
1907: 227 (M*, F*; *Culiciomyia*). **Type loc.:** Kuching, Sarawak [Malaysian Borneo] (NHMUK).

syn. *ceylonica* Theobald
1907: 236 (M, F*; *Culiciomyia*). **Type loc.:** Peradeniya [Central] and Maskeliya, Ceylon [Sri Lanka] (NHMUK).

syn. *graminis* Leicester
1908: 158 (M, F). **Type loc.:** [Kuala Lumpur, Selangor], Malaya [Malaysia] (NE).

furlongi van Someren
1954a: 107 (M*, F, P, L*). **Type loc.:** Kwale [Coast], Kenya (NHMUK). **Distr.:** Kenya. **Ety.:** n.s. [long, tall, tedious? (L)]. **Informal name:** Furlong Kenyan Typical Mosquito.

fuscicinctus King & Hoogstraal
1946a: 145 (M*). **Type loc.:** Hollandia, Western New Guinea (Papua) (USNM). **Distr.:** Indonesia. **Refs.:** Lee et al. 1989b: 13 (F; key, tax., bion., distr., review). **Ety.:** n.s. [dark ring, belt (L); poss. ref. to two brownish stripes on pleuron]. **Informal name:** Dark-belted Indonesian Typical Mosquito.

gilliesi Hamon & van Someren
1961a: 724 (M*, F). **Type loc.:** Amani, Tanganyika [Tanzania] (IERT). **Distr.:** Tanzania. **Refs.:** Vattier & Hamon 1962: 246 (P*, L*). **Ety.:** M.T. Gillies. **Informal name:** Gillies Tanzanian Typical Mosquito.

grenieri Eouzan
1969: 163 (M*). **Type loc.:** Nkolbisson, Yaoundé, Cameroon (SSC). **Distr.:** Cameroon. **Ety.:** Paul Fernand Grenier. **Informal name:** Grenier Cameroonian Typical Mosquito.

hainanensis Chen
1977: 95 (M*, F, L*). **Type loc.:** Diaoluoshan, Hainan Island [People's Republic of China] (KMC). **Distr.:** People's Republic of China. **Refs.:** Harrison 1987b: 113 (from *Thaiomyia*). **Ety.:** Hainan Isl., China. **Informal name:** Hainan Chinese Typical Mosquito.

harleyi Peters
1955a: 81 (M*, F, P*, L*). **Type loc.:** Kpain, Liberia (NHMUK). **Distr.:** Cameroon, Liberia. **Refs.:** Rickenbach et al. 1976: 96 (distr.). **Ety.:** Dr. and Mrs. George Way Harley. **Informal name:** Harley Liberian Typical Mosquito.

harrisoni Sirivanakarn
1977c: 102 (M*, F, P, L*). **Type loc.:** Chiang Dao, Chiang Mai, Thailand (USNM). **Distr.:** People's Republic of China, Thailand. **Refs.:** Zhang & Zhang 1982 (distr.; People's Republic of China); Rattanarithikul et al. 2005b (F*, L*; bion., distr., keys). **Ety.:** Bruce Arthur Harrison. **Informal name:** Harrison Thai Typical Mosquito.

javanensis Bonne-Wepster
1934: 274 (M*, F*). **Type loc.:** Tjisarua (near Buitenzorg), Java [Indonesia] (NHMUK). **Distr.:** Indonesia, Malaysia (includes Sabah). **Refs.:** Sirivanakarn 1977c: 100 (M*, F, P, L*); Tsukamoto et al. 1989 (L; keys; *Culiciomyia*). **Ety.:** Java. **Informal name:** Java Indonesian Typical Mosquito.

kyotoensis Yamaguti & LaCasse
1952 (M*, F*, L*). **Type loc.:** Japan (LU). **Distr.:** Japan, South Korea, Taiwan. **Refs.:** Asanuma & Nakagawa 1953: 94 (P*); Hara 1957: 53 (F*); Tanaka et al. 1979: 218 (M*, F*, L*). **Ety.:** Kyoto, Japan. **Informal name:** Kyoto Japanese Typical Mosquito.

lampangensis Sirivanakarn
1973b: 113 (M*, F, P, L*). **Type loc.:** Ban Pha Daeng, Lampang, Thailand (USNM). **Distr.:** Thailand. **Refs.:** Rattanarithikul et al. 2005b (F*, L*; bion., distr., keys). **Ety.:** Lampang Prov., Thailand. **Informal name:** Lampang Thai Typical Mosquito.

liberiensis Peters
1955a: 85 (M*, F, P*, L*). **Type loc.:** Kpain, Liberia (NHMUK). **Distr.:** Côte d'Ivoire, Democratic Republic of the Congo, Liberia. **Refs.:** Dyar 1924i: 183 (syn.). **Ety.:** Liberia. **Informal name:** Free Liberian Typical Mosquito [Liberia means "Land of the Free"].

macfiei Edwards
1923a: 399 (M, F). **Type loc.:** Ofako, Gold Coast [Ghana] (NHMUK). **Distr.:** Benin, Burkina Faso, Cameroon, Central African Republic, Côte d'Ivoire, Democratic Republic of the Congo, Gabon, Ghana, Kenya, Liberia, Mali, Nigeria, São Tomé and Príncipe, Senegal, Sierra Leone, Sudan and South Sudan, Tanzania, Uganda. **Refs.:** Edwards 1941: 276 (M*, F*), 413 (P*); Hopkins 1952: 275 (L*); Service 1976a (distr.; Gabon). **Ety.:** Dr. Macfie. **Informal name:** Macfie Ghanaian Typical Mosquito.

maplei Knight & Hurlbut
1949: 31 (M*, F, P*, L*; as *pullus* ssp.). **Type loc.:** Sokas Island, Ponape Island, Caroline Islands (USNM). **Distr.:** Micronesia (Wake Isl.). **Refs.:** Bohart 1957: 73 (M, P, L*; to sp.); Lee et al. 1989b: 14 (tax., distr.). **Ety.:** John D. Maple II. **Informal name:** Maple Micronesian Typical Mosquito.

megaonychus Yang, Li, & Chen
1993: 225 (M*, F, L*). **Type loc.:** Lushui County, Yunnan Province, China [People's Republic of China] (Biology Department of Guiyang Medical College). **Distr.:**

People's Republic of China. **Ety.:** large claw (Gr); ref. to large subapical claw on the gonostylus. **Informal name:** Large-clawed Chinese Typical Mosquito.

milloti Doucet
1949b: 331 (L*). **Type loc.:** Tsimbazaza, Tananarive, Madagascar (IRSM). **Distr.:** Madagascar (includes Glorioso and Juan De Nova Isl.). **Refs.:** Hopkins 1952: 270 (L). **Ety.:** Jacques Millot. **Informal name:** Millot Malagasy Typical Mosquito.

mongiro van Someren
1951: 2 (M*, F). **Type loc.:** Mongiro, Bwamba County, Toro District, Uganda (NHMUK). **Distr.:** Uganda. **Ety.:** Mongiro, Bwamba, Toro District, Uganda. **Informal name:** Mongiro Ugandan Typical Mosquito.

muspratti Hamon & Lambrecht
1959: 582 (M*). **Type loc.:** Irangi Forest, Kivu, Belgian Congo [Democratic Republic of the Congo] (IERT). **Distr.:** Central African Republic, Democratic Republic of the Congo. **Ety.:** J. Muspratt. **Informal name:** Muspratt Congolese Typical Mosquito.

nailoni King & Hoogstraal
1946a: 143 (M*, F, L*). **Type loc.:** Hollandia, Western New Guinea (Papua) (USNM). **Distr.:** Indonesia (West Papua); Papua New Guinea. **Refs.:** Lee et al. 1989b: 15 (F; key, tax., bion., distr., review). **Ety.:** T. Nailon. **Informal name:** Nailon Indonesian Typical Mosquito.

nebulosus Theobald
1901d: x (M*). **Type loc.:** Old Calabar [Eastern Provinces], Nigeria (LU). **Distr.:** Benin, Burkina Faso, Cameroon, Central African Republic, Comoros, Côte d'Ivoire, Democratic Republic of the Congo, Ethiopia, Gabon, Ghana, Kenya, Liberia, Madagascar (includes Glorioso and Juan De Nova Isl.), Malaysia, Mali, Niger, Nigeria, Republic of the Congo, Republic of South Africa, Saudi Arabia, Senegal, Sierra Leone, Tanzania, Togo, Uganda, Yemen, Zambia. **Refs.:** Macfie & Ingram 1922 (F); Edwards 1941: 272 (M*, F*), 413 (P); Hopkins 1952: 268 (L*); Peters 1955c: 125 (L*); Lewis 1956b: 704 (L); Brunhes 1977a (distr.; Comoros Isl.); Townsend 1990: 108 (type info.); Ahmed et al. 2011 (distr.; Saudi Arabia). **Ety.:** n.s. [misty, foggy (L)]. **Informal name:** Nebulous Nigerian Typical Mosquito.

syn. *nigrochaetae* Theobald
1901c: 60 (F of type series [M of type series syn. of *Coquillettidia* (*Coquillettidia*) *metalica*]). **Type loc.:** Lagos, Nigeria (NHMUK). **Refs.:** Townsend 1990: 111 (type info.).

ssp. ***pseudocinereus*** Theobald
1901c: 62 (M*). **Type loc.:** Salisbury, Mashonaland [Zimbabwe] (NHMUK). **Distr.:** Botswana, Democratic Republic of the Congo, Lesotho, Malawi, Namibia, Nigeria, Republic of South Africa, Tanzania, Zimbabwe. **Refs.:** Edwards 1941: 273 (M, F; to ssp.); Peters 1955c: 125 (L*; tax.); Jupp 1996 (M*, F*; key). **Ety.:** false (Gr); similar to *cinereus*. **Informal name:** Cinderella Zimbabwean Typical Mosquito [*pseudocinereus* = "false ashy" = Cinderella, who wasn't truly a cinder-girl after all].

syn. *freetownensis* Theobald
1901c: 69 (F syntype [M syntype syn. of *cinereus*]). **Type loc.:** [Freetown], Sierra Leone (NHMUK). **Refs.:** Edwards 1941: 272 (syn.).

syn. *invenustus* Theobald
1901d: ix (F*). **Type loc.:** Degama [Eastern Provinces], Nigeria (NHMUK).

syn. *fuscus* Theobald
1909b: 11 (M; *Pectinopalpus*). **Type loc.:** Obuasi, Ashanti, Gold Coast [Ghana] (NHMUK).

nigropunctatus Edwards
1926a: 121 (new name for *annulata* Theobald, not Schrank 1776). **Distr.:** Bangladesh, Cambodia, India, Indonesia, Japan, Laos, Malaysia (includes Sabah and Sarawak), Micronesia (Wake Isl.), Myanmar, Nepal, Palau, People's Republic of China (includes Hong Kong), Philippines, Singapore, Sri Lanka, Taiwan, Thailand, Vietnam. **Refs.:** Barraud 1934: 383 (M*, F, L; bion., distr.); Bohart 1957: 70 (M*, F, L*); Bram 1967b: 133 (M*, F*, P*, L*; bion., distr., tax.); Aslamkhan 1971b (distr.; Pakistan); Basio 1971b: 62 (M*; bion.); Baisas 1974: 109 (M, F, P, L; tax., bion., distr.; Philippines); Tanaka et al. 1975c: 218 (distr.); Tanaka et al. 1979: 220 (M*, F*, L*); Lee et al. 1989b: 16 (tax., distr.); Darsie & Pradhan 1990, 1991 (F, L; tax., keys, bion., distr.; Nepal); Rueda et al. 2004: 205 (distr.; Hong Kong); Rattanarithikul et al. 2005b (F*, L*; bion., distr., keys). **Ety.:** n.s. [black point (L)]. **Informal name:** Black-spotted Malaysian Typical Mosquito.

syn. *annulata* Theobald
1907: 230 (M*, F*; *Culiciomyia*). **Type loc.:** Kuching, Sarawak [Malaysian Borneo] (NHMUK).

pallidothorax Theobald
1905g: 32 (M, F). **Type loc.:** India (NHMUK). **Distr.:** Bangladesh, Cambodia, India, Indonesia, Japan, Macau, Malaysia, Myanmar, Nepal, People's Republic of China (includes Hong Kong), Philippines, Sri Lanka, Taiwan, Thailand, Timor, Vietnam. **Refs.:** Bohart & Ingram 1946b: 75 (M*, F*, L*, E*; bion., distr., tax.); La Casse & Yamaguti 1950: 182 (M*, F*, P*, L*; bion., distr., tax.); Hara 1957: 52 (F*); Bram 1967b: 137 (M*, F*, P*, L*; bion., distr., tax.); Basio 1971b: 63 (M*; bion.); Pinhão 1974 (distr.; Timor); Tanaka et al. 1975c: 218 (distr.); Tanaka et al. 1979: 123 (M*, F*, L*); Ahmed 1987 (distr.; Bangladesh); Lee et al. 1989b: 17 (F; key, tax., bion., distr., review); Darsie & Pradhan 1990, 1991 (F, L; tax., keys, bion., distr.; Nepal); Rattanarithikul et al. 2005b (F*, L*; bion., distr., keys). **Ety.:** pale thorax (L); "Head and thorax uniformly pale fawn coloured." **Informal name:** Pale-chested Indian Typical Mosquito.

syn. *albopleura* Theobald
1907: 456 (F). **Type loc.:** India (NHMUK).

syn. *annuloabdominalis* Theobald
1910c: 236 (M*, F*; *Culiciomyia*). **Type loc.:** Peradeniya [Central] and Hakgala, Ceylon [Sri Lanka] (NE).

pandani Brunhes
1969: 175 (M*, F, P*, L*). **Type loc.:** Manakara, Fianarantsoa, Madagascar (IERT). **Distr.:** Comoros, Madagascar (includes Glorioso and Juan De Nova Isl.). **Ety.:** immatures in *Pandanus* leaf axils. **Informal name:** Pandanus Malagasy Typical Mosquito.

papuensis (Taylor)
1914d: 201 (F*; *Melanoconion*). **Type loc.:** Lakekamu Gold Field, Papua New Guinea (ANIC). **Distr.:** Cambodia, Indonesia, Malaysia (includes Sabah and Sarawak), Papua New Guinea, Philippines, Solomon Isl., Thailand. **Refs.:** King & Hoogstraal 1946a: 146 (M*, L*); Penn 1949b: 73 (P*); Laird 1955: 286 (L*; tax.); Belkin 1962: 230 (M*, F, P*, L*); Bram 1967b: 123 (M*, F*, P*, L*; bion., distr., tax.); Basio 1971b: 63 (M*; bion.); Ramalingam & Pillai 1973 (distr.); Lee et al. 1989b: 21 (F; key, tax., bion., distr., review, type info.); Peyton et al. 1999: 541 (distr.); Rattanarithikul et al. 2005b (F*, L*; bion., distr., keys). **Ety.:** Papua New Guinea. **Informal name:** Lakekamu Papuan Typical Mosquito.

pullus Theobald
1905f: 87 (F*). **Type loc.:** Muina, [NE] New Guinea (HNM). **Distr.:** Australia, Bangladesh, Indonesia, Micronesia (Wake Isl.), Papua New Guinea, Solomon Isl., Timor. **Refs.:** Lee 1944a: 94 (L*); King & Hoogstraal 1946a: 149 (M*); Bohart & Ingram 1946b: 31 (bion., distr., tax.); Penn 1949b: 75 (P*); Belkin 1962: 232 (M*, F, P*, L*); Pinhão 1974 (distr.; Timor); Ahmed 1987 (distr.; Bangladesh); Lee et al. 1989b: 26 (F; key, tax., bion., distr., review). **Ety.:** dark, dusky (L); overall color. **Informal name:** Dusky Papuan Typical Mosquito.
syn. *muticus* Edwards
1923b: 6 (M, F). **Type loc.:** Rabaul, New Britain [Bismarck Archipelago] (NHMUK).

rajah Tsukamoto
1989: 216 (M, F, P*, L*). **Type loc.:** Mount Kinabalu, Sabah, Malaysia [Malaysian Borneo] (NMST). **Distr.:** Malaysia. **Ety.:** larvae in pitcher plant *Nepenthes rajah*. **Informal name:** Rajah Malaysian Typical Mosquito.

ramakrishnii Wattal & Kalra
1965: 309 (M*, F). **Type loc.:** Lachhiwala Reserve Forest, Dehra Dun, India (NICD). **Distr.:** India. **Refs.:** Stone 1967: 215 (not emendable under Article 31a, International Commission on Zoological Nomenclature 1964). **Ety.:** S.P. Ramakrishnan. **Informal name:** Ramakrishnan Indian Typical Mosquito.

ramalingami Sirivanakarn
1973b: 119 (M*, F, P*, L*). **Type loc.:** Tanjong, Tualang, Perak, West Malaysia (USNM). **Distr.:** Malaysia. **Ety.:** Shivaji Ramalingam. **Informal name:** Ramalingam Malaysian Typical Mosquito.

ruthae Peters
1958: 91 (M*, F; as *ruthi*). **Type loc.:** Mapamoiwa, Fergusson Island, Papua New Guinea (CSIR). **Distr.:** Papua New Guinea. **Refs.:** Peters 1959: 153 (emend.); Lee et al. 1989b: 35 (F; key, tax., bion., distr., review). **Ety.:** Ruth Peters. **Informal name:** Ruth Papuan Typical Mosquito.

ryukyensis Bohart
1946b: 41 (M*, F, L*). **Type loc.:** Chizuka, Okinawa, Ryukyu-Retto [Japan] (USNM). **Distr.:** Japan. **Refs.:** Bohart & Ingram 1946b: 77 (M*, F*, L*; bion., distr., tax.); La Casse & Yamaguti 1950 (F*); Tanaka et al. 1975c: 213 (bion., distr.); Tanaka et al. 1979: 217 (M*, F*, L*). **Ety.:** n.s. [prob. Ryukyu Isl.]. **Informal name:** Ryukyu Japanese Typical Mosquito.

sasai Kano, Nitahara, & Awaya
1954: 14 (M*, F*, L*; bion.). **Type loc.:** foot of Mount Fuji, Honshu, Japan (TMDU). **Distr.:** Japan, South Korea, Taiwan, Thailand. **Refs.:** Rattanarithikul et al. 2005b (F*, L*; bion., distr., keys). **Ety.:** Manabu Sasa. **Informal name:** Sasa Japanese Typical Mosquito.
syn. *spiculothorax* Bram
1967b: 155 (L*; bion., distr., tax.). **Type loc.:** Ban Khum Klang, Chiang Mai, Thailand (USNM). **Refs.:** Ramalingam & Pillai 1973 (distr.); Phanitchakun et al. 2017 (syn.).

scanloni Bram
1967b: 147 (M*, F*, P*, L*). **Type loc.:** Quam Ga Lang, Satun, Thailand (USNM). **Distr.:** Cambodia, Indonesia, Malaysia, Philippines, Sri Lanka, Thailand, Vietnam. **Refs.:** Ramalingam & Pillai 1973 (distr.); Lien et al. 1977 (distr.); Rattanarithikul et al. 2005b (F*, L*; bion., distr., keys). **Ety.:** John E. Scanlon. **Informal name:** Scanlon Thai Typical Mosquito.

semibrunneus Edwards
1927a: 353 (M*, F). **Type loc.:** Stanleyville, Belgian Congo [Kisangani, Tshopo Province, Democratic Republic of the Congo] (CMT, NHMUK). **Distr.:** Burkina Faso, Cameroon, Central African Republic, Côte d'Ivoire, Democratic Republic of the Congo, Japan, Kenya, Madagascar (includes Glorioso and Juan De Nova Isl.), Republic of the Congo, Uganda. **Refs.:** Edwards 1941: 275 (M*, F); van Someren 1956: 7 (M, P*, L*); Rickenbach et al. 1976: 95 (distr.). **Ety.:** n.s. [semibrown (L)]. **Informal name:** Brownish Congolese Typical Mosquito.

shebbearei Barraud
1924i: 19 (M*, F). **Type loc.:** Kurseong [Darjeeling District, West Bengal], India (NHMUK). **Distr.:** India, Malaysia, People's Republic of China, Sri Lanka, Thailand. **Refs.:** Barraud 1934: 380 (M*, F, L); Bram 1967c (type info., lectotype desig.); Sirivanakarn 1977c: 98 (M*, F, P, L*). **Ety.:** E.O. Shebbeare. **Informal name:** Shebbeare Indian Typical Mosquito.

spathifurca (Edwards)
1915c: 284 (M; *Culiciomyia*). **Type loc.:** Kuching, Sarawak, Borneo [Malaysian Borneo] (NHMUK). **Distr.:** Cambodia, India, Indonesia, Malaysia, Maldives,

Philippines, Singapore, Sri Lanka, Thailand. **Refs.:** Edwards 1926a: 120 (L); Delfinado 1966a: 123 (M*, F, P, L*); Bram 1967b: 123 (M*, F*, P*, L*; bion., distr., tax.); Basio 1971b: 64 (M*; bion.); Lien et al. 1975 (distr.); Lee et al. 1989b: 36 (F; key, tax., bion., distr., review); Rajavel et al. 1998: 346 (distr.; India); Rattanarithikul et al. 2005b (F*, L*; bion., distr., keys). **Ety.:** n.s. [broad blade, two-prong fork (Gr); poss. ref. to morphology of the male genitalia]. **Informal name:** Broad-forked Malaysian Typical Mosquito.

syn. *stylifurcatus* Carter & Wijesundara
1948: 145 (M*, F, L*). **Type loc.:** Ratmalana and Pelwatte Ganga River at Denipitiya, Ceylon [Sri Lanka] (NHMUK). **Refs.:** Mattingly 1955b: 10 (syn.); Townsend 1990: 137 (lectotype desig.).

spiculostylus Chen
1989: 86 (M*, F*, L*). **Type loc.:** Mengla Xian, Xishuangbanna, Yunnan Province, China [People's Republic of China] (GMC). **Distr.:** People's Republic of China. **Ety.:** n.s. [spiculose style (L); poss. ref. to "with several spiculate [*sic*] on basal half of distimere"]. **Informal name:** Spiculate Chinese Typical Mosquito.

subaequalis Edwards
1941: 276 (M*). **Type loc.:** Uluku, Belgian Congo [Democratic Republic of the Congo] (NHMUK). **Distr.:** Cameroon, Côte d'Ivoire, Democratic Republic of the Congo, Gabon, Kenya, Madagascar (includes Glorioso and Juan De Nova Isl.), Uganda. **Refs.:** Hopkins 1952: 274 (L*); Mattingly 1956a: 41 (type info.); Rickenbach et al. 1976: 96 (distr.); Service 1976a (distr.; Gabon). **Ety.:** n.s. [subequal, equivalent (L)]. **Informal name:** Subequal Congolese Typical Mosquito.

termi Thurman
1955: 18 (M*, F, P, L*). **Type loc.:** New Kong Lom, Lampang, Thailand (USNM). **Distr.:** Cambodia, Thailand. **Refs.:** Bram 1967b: 157 (M*, F*, P*, L*; bion., distr., tax.); Rattanarithikul et al. 2005b (F*, L*; bion., distr., keys). **Ety.:** Term Vejarasthira. **Informal name:** Term Thai Typical Mosquito.

thurmanorum Bram
1967b: 159 (M*, F, P*, L*; bion., distr., tax.). **Type loc.:** Doi Chang, Mae Hong Son, Thailand (USNM). **Distr.:** Cambodia, Thailand. **Refs.:** Rattanarithikul et al. 2005b (F*, L*; bion., distr., keys). **Ety.:** D.C. and E.B. Thurman. **Informal name:** Thurman Thai Typical Mosquito.

tricuspis Edwards
1930a: 294 (new name for *trifidus* Edwards, not Dyar 1921). **Distr.:** Indonesia. **Refs.:** Mattingly & Marks 1955: 172 (M*; tax.); Sirivanakarn 1971a: 67 (n. comb.); Sirivanakarn 1973b: 116 (M*); Lee et al. 1989b: 38 (tax., distr.). **Ety.:** n.s. [having three prongs (L)]. **Informal name:** Three-pointed Indonesian Typical Mosquito.

syn. *trifidus* Edwards
1926a: 108 (M). **Type loc.:** Alor Island [near Timor] (NHMUK).

viridiventer Giles
1901a: 609 (A). **Type loc.:** [Naini Tal, United Provinces], India (NHMUK). **Distr.:** Bangladesh, India, Nepal, Pakistan, People's Republic of China, Thailand, Vietnam. **Refs.:** Barraud 1934: 378 (M*, F, L); Bram 1967c (type info., lectotype desig.); Aslamkhan 1971b (distr.; Pakistan); Sirivanakarn 1977c: 95 (M*, F, P*, L*; distr.); Darsie & Pradhan 1990, 1991 (F, L; tax., keys, bion., distr.; Nepal); Rattanarithikul et al. 2005b (F*, L*; bion., distr., keys). **Ety.:** green venter (L); abdominal venter "green in fresh specimens." **Informal name:** Green-bellied Indian Typical Mosquito.

syn. *angulatus* Theobald
1901c: 324 (F*). **Type loc.:** Naini Tal [United Provinces], India (NHMUK). **Refs.:** Bram 1967c (type info., lectotype desig.).

syn. *longifurcatus* Theobald
1910a: 19 (M, F). **Type loc.:** Dahawangahary Hill, near Bengal frontier, Nepal (NHMUK).

syn. *pseudolongifurcatus* Theobald
1910c: 366 (new name for *longifurcatus* Theobald, not Becker 1903).

yaoi Tung
1955: 311 (M*, F). **Type loc.:** Lushan, Kiangsi, China [People's Republic of China] (CUMC). **Distr.:** People's Republic of China. **Ety.:** K.F. Yao. **Informal name:** Yao Chinese Typical Mosquito.

Subgenus *Eumelanomyia* Theobald

1909b: 10 (as genus). Type species: *Eumelanomyia inconspicuosa* Theobald.

Name derivation: true (Gr) *Melanomyia*.

syn. *Protomelanoconion* Theobald 1909b: 20 (as genus). Type species: *Protomelanoconion fusca* Theobald. Sirivanakarn 1971a: 72 (tax.; syn.).

syn. *Mochthogenes* Edwards 1930a (as subg. of genus *Culex*): 305. Type species: *Culex malayi* Leicester. Sirivanakarn 1971a: 66 (syn.).

Important references: Sirivanakarn 1971a (revision); Sirivanakarn 1972 (revision; SE Asia); Tanaka et al. 1979 (revision; Japan, Korea); Lee et al. 1989b (revision; Australasia); Harbach 2011 (phyl.).

acrostichalis Edwards
1941: 268 (F*). **Type loc.:** Kasala, Uganda (NHMUK). **Distr.:** Democratic Republic of the Congo, Liberia, Uganda. **Refs.:** Mattingly 1956a: 41 (lectotype desig.).

Ety.: n.s. [poss. ref. to presence of acrostichal setae]. **Informal name:** Acrostic Ugandan Typical Mosquito.

adami (Hamon & Mouchet)
1955: 860 (M*; *Neoculex*). **Type loc.:** Younde, Cameroon (IERT). **Distr.:** Cameroon. **Ety.:** Jean-Paul Adam. **Informal name:** Adam Cameroonian Typical Mosquito.

adersianus Edwards
1941: 268 (M*, F). **Type loc.:** Mombasa [Coast], Kenya (NHMUK). **Distr.:** Côte d'Ivoire, Kenya, Tanzania. **Refs.:** Hopkins 1952: 264 (L); Mattingly 1956a: 41 (lectotype desig.). **Ety.:** W.M. Aders. **Informal name:** Aders Kenyan Typical Mosquito.

albertianus Edwards
1941: 265 (M*, F*). **Type loc.:** Blukwa, Lake Albert [Orientale], Belgian Congo [Democratic Republic of the Congo] (NHMUK). **Distr.:** Democratic Republic of the Congo, Kenya, Liberia. **Refs.:** Mattingly 1956a: 41 (type info.). **Ety.:** Lake Albert, Kenya. **Informal name:** Albert Congolese Typical Mosquito.

albiventris Edwards
1922d: 88 (new name for *inconspicuosa* Theobald, not Theobald 1908a). **Distr.:** Benin, Cameroon, Central African Republic, Côte d'Ivoire, Ghana, Kenya, Liberia, Mali, Nigeria, Republic of the Congo, Sierra Leone, The Gambia, Uganda. **Refs.:** Macfie & Ingram 1922 (F) (as *inconspicuosa* Theobald); Edwards 1941: 267 (M*, F*), 411 (P); Hopkins 1952: 263 (L*). **Ety.:** n.s. [white venter (L); ref. to pure white abdominal venter]. **Informal name:** White-bellied Ghanaian Typical Mosquito.
syn. *inconspicuosa* Theobald
1909b: 10 (M*, F*; *Eumelanomyia*). **Type loc.:** Obuasi, Ashanti, Gold Coast [Ghana] (NHMUK).

amaniensis van Someren & Hamon
1964: 78 (M*, F*). **Type loc.:** Amani, Tanganyika [Tanzania] (NHMUK). **Distr.:** Tanzania. **Ety.:** Amani, Tanzania. **Informal name:** Amani Tanzanian Typical Mosquito.

andreanus Edwards
1927a: 354 (M*, F). **Type loc.:** Irebu [Équateur], Belgian Congo [Zaire] (CMT, NHMUK). **Distr.:** Cameroon, Democratic Republic of the Congo, Ghana, Liberia, Nigeria, Republic of the Congo, Uganda. **Refs.:** Edwards 1941: 258 (M*, F*); Hamon, Taufflieb, & Maillot 1957: 694 (M*); Rickenbach et al. 1976: 95 (distr.). **Ety.:** Major André. **Informal name:** André Congolese Typical Mosquito.

baisasi Sirivanakarn
1972: 35 (M*, P). **Type loc.:** Kolambugan, Lanao, Mindanao, Philippines (USNM). **Distr.:** Philippines. **Ety.:** Francisco Edlagan Baisas. **Informal name:** Baisas Philippine Typical Mosquito.

bokorensis Klein & Sirivanakarn
1970: 587 (M*). **Type loc.:** Bokor Hill, Kampot, Cambodia (USNM). **Distr.:** Cambodia. **Refs.:** Sirivanakarn 1972: 57 (M*). **Ety.:** Bokor Hill, Cambodia. **Informal name:** Bokor Cambodian Typical Mosquito.

brenguesi Brunhes & Ravaonjanahary
1973 (M*, F, P*, L*). **Type loc.:** Moramanga, Tamatave Province, Madagascar (IERT). **Distr.:** Madagascar (includes Glorioso and Juan De Nova Isl.). **Ety.:** Jacques Brengues. **Informal name:** Brengues Malagasy Typical Mosquito.

brevipalpis (Giles)
1902: 384 (M*, F*; *Stegomyia*). **Type loc.:** Shahjahanpur [United Provinces], India (NHMUK). **Distr.:** Bangladesh, Cambodia, India, Indonesia, Japan, Malaysia, Myanmar, Nepal, Pakistan, Papua New Guinea, People's Republic of China, Philippines, Singapore, Sri Lanka, Taiwan, Thailand, Vietnam. **Refs.:** Barraud 1924m: 432 (L*); Barraud 1934: 348 (M*, F, L*); Baisas 1935d: 177 (M*, F); Lee 1944a: 111 (L*); Bohart & Ingram 1946b: 72 (M*, F, L*; bion., distr., tax.); Joshi et al. 1965: 139 (distr.); Delfinado 1966a: 125 (M*, F, P*, L*); Bram 1967b: 25 (M*, F*, P*, L*; bion., distr., tax.); Aslamkhan 1971b (distr.; Pakistan); Basio 1971b: 71 (M*; bion.); Sirivanakarn 1972: 13 (M*, F*, P*, L*; distr.); Baisas 1974: 127 (M, F, P, L; tax., bion., distr.; Philippines); Mattingly 1975: 22 (E*); Tanaka et al. 1975c: 216 (distr.); Tanaka et al. 1979: 193 (M*, F*, L*); Lee et al. 1989b: 42 (F; key, tax., bion., distr., review); Darsie & Pradhan 1990, 1991 (F, L; tax., keys, bion., distr.; Nepal); Rattanarithikul et al. 2005b (F*, L*; bion., distr., keys). **Ety.:** short feeler (L); short palpus. **Informal name:** Short-lipped Indian Typical Mosquito.
syn. *longipes* Theobald
1901c: 68 (F). **Type loc.:** Singapore (NHMUK).
syn. *macropus* Blanchard
1905: 327 (new name for *longipes* Theobald, not Fabricius 1850).
syn. *uniformis* Leicester
1908: 136 (M, F; *Melanoconion*). **Type loc.:** Malaya [Malaysia] (NHMUK). **Refs.:** Baisas 1974 (M*, L*).
syn. *fidelis* Dyar
1920h: 180 (M). **Type loc.:** Los Banos [Laguna, Luzon], Philippines (USNM). **Refs.:** Stone & Knight 1957a: 50 (lectotype desig.).

calabarensis Edwards
1941: 265 (M*, F*). **Type loc.:** Calabar, [Eastern Provinces], Nigeria (NHMUK). **Distr.:** Benin, Nigeria. **Refs.:** Hamon & Rickenbach 1955b: 853 (M*); Mattingly 1956a: 41 (type info.). **Ety.:** Calabar, Nigeria. **Informal name:** Calabar Nigerian Typical Mosquito.

campilunati Carter & Wijesundara
1948: 143 (M*, F). **Type loc.:** Moon Plains, Nuwara Eliya, Ceylon [Sri Lanka] (NHMUK). **Distr.:** Sri Lanka. **Refs.:** Sirivanakarn 1972: 43 (M*, F; lectotype desig.). **Ety.:** field

and crescent shaped (L) [prob. ref. to type loc. Moon Plains]. **Informal name:** Crescent Sri Lankan Typical Mosquito.

castor De Meillon & Lavoipierre
1944: 49 (M*). **Type loc.:** Bili [Orientale], Belgian Congo [Democratic Republic of the Congo] (SAIM). **Distr.:** Democratic Republic of the Congo, Republic of the Congo. **Refs.:** De Meillon et al. 1945: 86 (P*); Harrison 1973c: 279 (type info.); Segerman 1990: 54 (type info.). **Ety.:** n.s. [beaver (Gr)]. **Informal name:** Beaver Congolese Typical Mosquito.

castrensis Edwards
1922a: 285 (M*, F). **Type loc.:** Castle Rock [North Kanara, Bombay], India (NHMUK). **Distr.:** India, Sri Lanka. **Refs.:** Barraud 1934: 355 (M*, F, L*); Sirivanakarn 1972: 33 (M*, F; lectotype desig.); Lee et al. 1989b: 45 (tax., distr.); Townsend 1990: 110 (incorrect desig. of lectotype of *castrensis* to *nigrescens*); Darsie et al. 1992 (distr.; Nepal). **Ety.:** Castle Rock, India. **Informal name:** Castle Indian Typical Mosquito.

cataractarum Edwards
1923b: 7 (M, F). **Type loc.:** Rabaul, New Britain [Bismarck Archipelago] (NHMUK). **Distr.:** Papua New Guinea, Philippines. **Refs.:** Basio 1971b: 70 (M*; bion.; as *tricontus*); Sirivanakarn 1972: 34 (M*, F; lectotype desig.); Lee et al. 1989b: 46 (F; key, tax., bion., distr., review). **Ety.:** waterfalls (L); on rocks under waterfalls. **Informal name:** Cataract Papuan Typical Mosquito.

syn. *tricontus* Delfinado
1966a: 133 (M*). **Type loc.:** Molawin Creek, College, Laguna, Philippines (USNM). **Refs.:** Sirivanakarn 1972: 34 (syn.).

chauveti Brunhes & Rambelo
1968 (M*, F, P*, L*). **Type loc.:** Ambalakindresy, S/P Ambohimahasoa, Fianarantsoa, Madagascar (IERT). **Distr.:** Comoros, Madagascar (includes Glorioso and Juan De Nova Isl.). **Refs.:** Brunhes 1977a (distr.; Comoros Isl.). **Ety.:** Guy Chauvet. **Informal name:** Chauvet Malagasy Typical Mosquito.

femineus Edwards
1926a: 107 (M*, F). **Type loc.:** Hog Harbour, Espiritu Santo, New Hebrides (NHMUK). **Distr.:** French Polynesia, Vanuatu. **Refs.:** Lee 1944a: 89 (L*); Belkin 1962: 235 (M*, F, P*, L*); Lee et al. 1989b: 48 (F; key, tax., bion., distr., review). **Ety.:** n.s. [feminine? (L)]. **Informal name:** Feminine Ni-Vanuatu Typical Mosquito.

fimbriforceps Edwards
1935a: 135 (M). **Type loc.:** Jinja Road, Uganda (NHMUK). **Distr.:** Cameroon, Democratic Republic of the Congo, Uganda. **Refs.:** Edwards 1941: 279 (M*, F*); De Meillon et al. 1945: 86 (P*); Hopkins 1952: 278 (L*); Rickenbach et al. 1976: 96 (distr.). **Ety.:** fringed pincers (L); "distal portion of style [gonostylus] slender and provided with a fine fringe." **Informal name:** Fringe-clawed Ugandan Typical Mosquito.

foliatus Brug
1932: 82 (M*, P*, L*). **Type loc.:** Bandoeng, Java [Indonesia] (NHMUK). **Distr.:** Cambodia, India, Indonesia, Malaysia, Nepal, People's Republic of China (includes Hong Kong), Philippines, Sri Lanka, Taiwan, Thailand, Timor, Vietnam. **Refs.:** Bohart 1945: 73 (M*; as *chiyutoi*); Peters & Dewar 1956: 46 (M*); Delfinado 1966a: 129 (M*, P, L; as *chiyutoi*); Bram 1967b: 34 (M*, F, L*; bion., distr., tax., syn.); Basio 1971b: 70 (M*; bion.); Sirivanakarn 1972: 27 (M*, F*, P*, L*); Lee et al. 1989b: 50 (F; key, tax., bion., distr., review); Darsie & Pradhan 1990, 1991 (F, L; tax., keys, bion., distr.; Nepal); Townsend 1990: 74 (lectotype desig.); Whelan & Hapgood 2000 (distr.; East Timor); Rattanarithikul et al. 2005b (F*, L*; bion., distr., keys). **Ety.:** leafy (L); in contrast to *casterensis* gonocoxite with six broad leaves. **Informal name:** Foliaceus Indonesian Typical Mosquito.

syn. *chiyutoi* Baisas
1935d: 176 (M*, F). **Type loc.:** Kolambugan, Lanao, Mindanao, Philippines (NE). **Refs.:** Bram 1967b: 34 (syn. with *foliatus*; type depos. change NE to USNM).

galliardi Edwards
1941: 266 (M*, F*). **Type loc.:** Forcados [Western Provinces], Nigeria (NHMUK). **Distr.:** Democratic Republic of the Congo, Gabon, Liberia, Nigeria, Sierra Leone, The Gambia. **Refs.:** Mattingly 1956a: 41 (type info.); Service 1976a (distr.; Gabon). **Ety.:** Henri Galliard. **Informal name:** Galliard Nigerian Typical Mosquito.

garioui Bailly-Choumara & Rickenbach
1966: 144 (M*). **Type loc.:** Mbol, Yokadouma, E Cameroon (IERT). **Distr.:** Cameroon, Central African Republic, Sierra Leone. **Ety.:** Médecin-Colonel Gariou. **Informal name:** Gariou Cameroonian Typical Mosquito.

germaini Geoffroy
1974: 93 (M*). **Type loc.:** Mbaere River, on road from Boda to Bambio (3°56′ N, 17°00′ E), Central African Republic (IERT). **Distr.:** Central African Republic, Republic of South Africa. **Ety.:** Max Germain. **Informal name:** Germain Central African Typical Mosquito.

hackeri Edwards
1923b: 4 (M, F). **Type loc.:** Cameron's Highlands [near Tapah, Perak], Malaya [Malaysia] (NHMUK). **Distr.:** Malaysia. **Refs.:** Sirivanakarn 1972: 52 (M*, F*; lectotype desig.); Sirivanakarn & Ramalingam 1976: 214 (tax., distr.); Townsend 1990: 81 (type info.; invalidates lectotype desig.). **Ety.:** Henry Pollard Hacker. **Informal name:** Hacker Malaysian Typical Mosquito.

hamoni Brunhes, Adam, & Bailly-Choumara
1967: 48 (M*). **Type loc.:** Bangou Forest, Congo [Republic of the Congo] (IERT). **Distr.:** Republic of the Congo.

Ety.: Jacques Pierre Jean Hamon. **Informal name:** Hamon Congolese Typical Mosquito.

hayashii Yamada
1917: 67 (M, F, P, L, E). **Type loc.:** Tokyo [Honshu], Japan (NHMUK). **Distr.:** Japan, People's Republic of China, Russia, South Korea, Taiwan. **Refs.:** Bohart & Ingram 1946b: 71 (M*, F, L*, E*; bion., distr., tax.); La Casse & Yamaguti 1950: 257 (M*, F*, P*, L*; bion., distr., tax.); Monchadskii 1951: 250 (L*); Hara 1957: 57 (F*); Sirivanakarn 1972: 50 (M*, F*, P*, L*); Gutsevich et al. 1974: 363 (M*, F, L*); Tanaka et al. 1975c: 216 (distr.); Tanaka et al. 1979: 187 (M*, F*, L*); Townsend 1990: 82 (type info.). **Ety.:** n.k. **Informal name:** Hayashi Japanese Typical Mosquito.
ssp. ***ryukyuanus*** Tanaka, Mizusawa, & Saugstad
1979: 190 (M, F). **Type loc.:** near Shiraham, Iriomote Island, Ryukyu Archipelago, Japan (NMST). **Distr.:** Japan. **Ety.:** Ryukyu Isl., Japan. **Informal name:** Ryukyu Japanese Typical Mosquito.

helenae Brunhes, Adam, & Bailly-Choumara
1967: 50 (M*). **Type loc.:** Limie, Cameroon (IERT). **Distr.:** Cameroon. **Ety.:** n.s. [Helena da Cunha Ramos]. **Informal name:** Helena Cameroonian Typical Mosquito.

hinglungensis Chu
1957: 159, 163 (M). **Type loc.:** Hing-lung, Hainan Island, China [People's Republic of China] (SMMU). **Distr.:** Cambodia, India, People's Republic of China, Philippines, Thailand. **Refs.:** Chu 1958: 111 (M*); Bram 1967b: 37 (M*; bion., distr., tax.); Basio 1971b: 69 (M*; bion.; as *culionicus*); Sirivanakarn 1972: 31 (M*); Rattanarithikul et al. 2005b (F*, L*; bion., distr., keys). **Ety.:** Hing-lung, Hainan Isl., China. **Informal name:** Hing-lung Chinese Typical Mosquito.
syn. *culionicus* Delfinado
1966a: 130 (M*, F). **Type loc.:** San Pedro, Culion Island, Palawan, Philippines (USNM). **Refs.:** Sirivanakarn 1972: 31 (syn.).

horridus Edwards
1922d: 88 (new name for *fusca* Theobald 1909b, not Theobald 1908b). **Distr.:** Angola, Benin, Burkina Faso, Cameroon, Central African Republic, Comoros, Côte d'Ivoire, Democratic Republic of the Congo, Ghana, Kenya, Liberia, Madagascar (includes Glorioso and Juan De Nova Isl.), Mali, Mozambique, Nigeria, Republic of the Congo, Republic of South Africa, Sudan and South Sudan, Tanzania, Togo, Uganda, Zambia, Zimbabwe. **Refs.:** Edwards 1941: 269 (M*, F*), 411 (P); Hopkins 1952: 265 (L*); Hamon & Rickenbach 1955b: 847 (M*); Brunhes 1977a (distr.; Comoros Isl.); Jupp 1996 (M*, F*; key). **Ety.:** n.s. [dreadful (L)]. **Informal name:** Horrid Ghanaian Typical Mosquito.
syn. *fusca* Theobald
1909b: 20 (M*, F*; *Protomelanoconion*). **Type loc.:** Accra, Gold Coast [Ghana] (NHMUK).

ssp. ***rageaui*** (Hamon & Rickenbach)
1955b: 859 (M*; *Neoculex*). **Type loc.:** Nkolbisson, Yaoundé Region, Cameroon (IERT). **Distr.:** Benin, Burkina Faso, Cameroon. **Refs.:** Harbach & Howard 2007: 42 (to ssp.). **Ety.:** J. Rageau. **Informal name:** Rageau Cameroonian Typical Mosquito.

inconspicuosus (Theobald)
1908a: 109 (M, F; *Aedes*). **Type loc.:** Transvaal [Republic of South Africa] (NHMUK). **Distr.:** Angola, Benin, Burkina Faso, Cameroon, Central African Republic, Côte d'Ivoire, Democratic Republic of the Congo, Gabon, Ghana, Kenya, Liberia, Mali, Mozambique, Nigeria, Republic of the Congo, Republic of South Africa, São Tomé and Príncipe, Senegal, Sudan and South Sudan, The Gambia, Uganda, Zambia, Zimbabwe. **Refs.:** Edwards 1941: 278 (M*, F*), 412 (P*); De Meillon et al. 1945: 88 (P*); Hopkins 1952: 276 (L*); Service 1976a: 260 (distr.; Gabon); Jupp 1996 (M*, F*; key). **Ety.:** n.s. [not visible (L); ref. to "a small, brown, inconspicuous mosquito"]. **Informal name:** Inconspicuous South African Typical Mosquito.
syn. *nyangae* Galliard
1931b: 230 (L*). **Type loc.:** Tchibanga (Nyanga), Gabon (LU). **Refs.:** Edwards 1941: 278 (? = *inconspicuosus* Theobald); White 1980: 139 (syn. with *inconspicuosus*).

insignis (Carter)
1911: 37 (F*; *Culiciomyia*). **Type loc.:** Entebbe, Uganda (NHMUK). **Distr.:** Benin, Burkina Faso, Cameroon, Côte d'Ivoire, Democratic Republic of the Congo, Ghana, Kenya, Liberia, Madagascar (includes Glorioso and Juan De Nova Isl.), Malawi, Mali, Mauritius, Mozambique, Nigeria, Republic of the Congo, Republic of South Africa, Senegal, Sierra Leone, Sudan and South Sudan, Tanzania, The Gambia, Uganda. **Refs.:** Macfie & Ingram 1922 (F*); Edwards 1941: 263 (M*, F*; from syn. with *rima*); Hamon & Rickenbach 1955b: 851 (M*); van Someren 1956: 6 (P*, L); Rickenbach et al. 1976: 95 (distr.); Jupp 1996 (M*, F*; key). **Ety.:** n.s. [remarkable (L); prob. ref. to notable abdominal markings]. **Informal name:** Remarkable Ugandan Typical Mosquito.

iphis Barraud
1924h: 1279 (M*, F). **Type loc.:** Nilgiri Hills [Madras], India (NHMUK). **Distr.:** India. **Refs.:** Sirivanakarn 1972: 56 (M*, F*; lectotype desig.); Sirivanakarn 1977d: 541 (P*, L*). **Ety.:** n.s. [strongly (Gr)]. **Informal name:** Strong Indian Typical Mosquito.

jefferyi Sirivanakarn
1977b: 77 (M*, F, P*, L). **Type loc.:** Carey Island, Selangor, Peninsular Malaysia, Malaysia (USNM). **Distr.:** Malaysia. **Ety.:** John Jeffery. **Informal name:** Jeffery Malaysian Typical Mosquito.

kanyamwerima van Someren
1951: 4 (M*, F). **Type loc.:** Kanyamwerima, Bwamba County, Toro District, Uganda (NHMUK). **Distr.:**

Democratic Republic of the Congo, Uganda. **Ety.:** Kanyamwerima, Bwamba, Toro District, Uganda. **Informal name:** Kanyamwerima Ugandan Typical Mosquito.

khazani Edwards
1922a: 286 (M*). **Type loc.:** Pudupadi [Madras], India (NHMUK). **Distr.:** India, Thailand. **Refs.:** Barraud 1923g (M*, F*); Barraud 1934: 357 (M*, F, L?); Galliard & Ngu 1949: 500 (L*); Sirivanakarn 1972: 45 (M*, F, L). **Ety.:** Khazan Chand. **Informal name:** Khazan Indian Typical Mosquito.

kilara van Someren
1951: 5 (M*, F). **Type loc.:** Kilara, Bwamba County, Toro District, Uganda (NHMUK). **Distr.:** Uganda. **Ety.:** Kilara, Bwamba, Toro District, Uganda. **Informal name:** Kilara Ugandan Typical Mosquito.

kingianus Edwards
1922d: 86 (M*, F). **Type loc.:** River Menzi, Sudan (NHMUK). **Distr.:** Burkina Faso, Cameroon, Central African Republic, Côte d'Ivoire, Democratic Republic of the Congo, Liberia, Madagascar (includes Glorioso and Juan De Nova Isl.), Nigeria, Sudan and South Sudan, Uganda. **Refs.:** Edwards 1941: 259 (M*, F*); Mattingly 1953c: 331 (M*, P*, L*). **Ety.:** Harold Holmes King. **Informal name:** King Sudanese Typical Mosquito.

kiriensis Klein & Sirivanakarn
1970: 583 (M*). **Type loc.:** Bokor Hill, Kampot, Cambodia (USNM). **Distr.:** Cambodia, Thailand. **Refs.:** Sirivanakarn 1972: 53 (M*, F*, P); Rattanarithikul et al. 2005b (F*, L*; bion., distr., keys). **Ety.:** n.s. [poss. ref. to Kirirom Hills, Kompong Speu, Cambodia]. **Informal name:** Kirirom Cambodian Typical Mosquito.

laplantei (Hamon, Adam, & Mouchet)
1955: 862 (M*; *Neoculex*). **Type loc.:** Ebogo, Mbalmayo, Cameroon (IERT). **Distr.:** Cameroon, Côte d'Ivoire. **Ety.:** Laplante; a colleague. **Informal name:** Laplante Cameroonian Typical Mosquito.

latifoliatus Delfinado
1966a: 131 (M*, F). **Type loc.:** Molawin Creek, College, Laguna, Philippines (USNM). **Distr.:** Philippines. **Refs.:** Basio 1971b: 70 (M*; bion.); Sirivanakarn 1972: 30 (M*, F, P, L). **Ety.:** n.s. [leafy on the side (L)]. **Informal name:** Leaf-sided Philippine Typical Mosquito.

laureli Baisas
1935d: 176 (M*). **Type loc.:** Malaybalay, Bukidnon, Mindanao, Philippines (NE). **Distr.:** Philippines, Timor. **Refs.:** Bohart 1945: 73 (syn.); Delfinado 1966a: 132 (M*, F, P*, L*; from syn. with *malayia*); Basio 1971b: 70 (M*; bion.); Baisas 1974: 126 (M*, F, P*, L*; tax., bion., distr.; Philippines); Lee et al. 1989b: 52 (F; key, tax., bion., distr., review); Whelan & Hapgood 2000 (distr.; East Timor). **Ety.:** n.s. **Informal name:** Laurel Philippine Typical Mosquito.

macrostylus Sirivanakarn & Ramalingam
1976: 210 (M*, F, P, L*). **Type loc.:** Genting Heights, Pahang, Malaysia (USNM). **Distr.:** Malaysia. **Ety.:** large stylus (Gr); ref. to unique modifications of the male genitalia. **Informal name:** Large-styled Malaysian Typical Mosquito.

malayensis Sirivanakarn
1972: 10 (M*, F*). **Type loc.:** Rantau, Panjang, Kelantan, West Malaysia (USNM). **Distr.:** Malaysia. **Refs.:** Sirivanakarn 1977b: 74 (M*, F, P*, L*; distr.). **Ety.:** Malaya [Malaysia]. **Informal name:** Langkasuka Malaysian Typical Mosquito [Langkasuka = one of the earliest ancient kingdoms on the Malay Peninsula].

malayi (Leicester)
1908: 184 (M, F; *Aedes*). **Type loc.:** Kuala Lumpur [Selangor], Malaya [Malaysia] (NHMUK). **Distr.:** Bangladesh, Cambodia, India, Indonesia, Malaysia, Maldives, Myanmar, Nepal, Pakistan, People's Republic of China, Sri Lanka, Taiwan, Thailand, Timor, Vietnam. **Refs.:** Barraud 1934: 358 (M*, F, L); Bohart 1945: 73 (M*); Galliard & Ngu 1949: 502 (M*, L*); Bram 1967a: 39 (M*, F, L*; bion., distr., tax.); Aslamkhan 1971b (distr.; Pakistan); Sirivanakarn 1972: 37 (M*, F*, P*, L*); Ahmed 1987 (distr.; Bangladesh); Lee et al. 1989b: 53 (F; key, tax., bion., distr., review); Darsie & Pradhan 1990, 1991 (F, L; tax., keys, bion., distr.; Nepal); Whelan & Hapgood 2000 (distr.; East Timor); Rattanarithikul et al. 2005b (F*, L*; bion., distr., keys). **Ety.:** Malaya [Malaysia]. **Informal name:** Malaya Malaysian Typical Mosquito.
syn. *nigrescens* Theobald
1907: 540 (1 M* only of type series; *Aedes*; not Theobald 1907). **Type loc.:** Castle Rock [North Kanara, Bombay], India (NHMUK). **Refs.:** Sirivanakarn 1972: 38 (M in type series recognized as the holotype; F in type series is *Culex* (*Eumelanomyia*) *castrensis* Edwards 1922a).
syn. *aedes* Leicester
1908: 189 (M; *Aioretomyia*). **Type loc.:** Kuala Lumpur [Selangor], Malaya [Malaysia] (NHMUK).

manusensis Sirivanakarn
1975b: 394 (M*, F*, P*, L*). **Type loc.:** Lorengau Village, Manus District, Manus Island, Papua New Guinea (BBM). **Distr.:** Papua New Guinea. **Refs.:** Lee et al. 1989b: 55 (F; key, tax., bion., distr., review). **Ety.:** Manus Isl., Papua New Guinea. **Informal name:** Manus Papuan Typical Mosquito.

megafolius Chen & Dong
1992: 78 (M*, P*, L*). **Type loc.:** Gengma, Ruili, and Mengla (800–1,000 m), Yunnan Province, China [People's Republic of China] (YIM). **Distr.:** People's Republic of China. **Ety.:** large leaf (Gr); ref. to prominent broad leaflet on gonocoxite. **Informal name:** Large-leafed Chinese Typical Mosquito.

miaolingensis Chen
1982: 312 (M*, F*, P*, L*). **Type loc.:** Guiyang, Guizhou Province, China [People's Republic of China] (GMC).

Distr.: People's Republic of China. **Ety.:** n.s. [type loc. is in the Miaoling mountains (Harbach 2018c)]. **Informal name:** Miaoling Chinese Typical Mosquito.

micolo Ribeiro, da Cunha Ramos, Capela, & Pires 1998: 12 (M*). **Type loc.:** Micóló, Ilha de São Tomé (IHMT). **Distr.:** São Tomé and Príncipe. **Ety.:** Micóló, Ilha de São Tomé. **Informal name:** Micóló São Tomean Typical Mosquito.

mijanae Brunhes, Adam, & Bailly-Choumara 1967: 51 (M*). **Type loc.:** Ntui-Mbam, Cameroun [Cameroon] (IERT). **Distr.:** Cameroon. **Ety.:** n.s. **Informal name:** Mijana Cameroonian Typical Mosquito.

mohani Sirivanakarn 1977d: 537 (M*, F, P*, L*). **Type loc.:** Coonoor, Nilgiri Hills, Madras, India (USNM). **Distr.:** India. **Ety.:** B.N. Mohan. **Informal name:** Mohan Indian Typical Mosquito.

mundulus Grünberg 1905: 388 (F). **Type loc.:** Kamerun [Cameroons, Nigeria] (ZM). **Distr.:** Nigeria. **Refs.:** Edwards 1941: 353 (tax.); White 1975: 322 (tax.). **Ety.:** n.s. [somewhat clean, somewhat neat (L)]. **Informal name:** Semi-clean Nigerian Typical Mosquito.

okinawae Bohart 1953: 187 (M*, F, P, L*). **Type loc.:** East Taira, Okinawa, Ryukyu-Retto [Japan] (USNM). **Distr.:** Japan, Philippines, Taiwan. **Refs.:** Sirivanakarn 1972: 54 (M*, F*, P*, L*); Tanaka et al. 1975c: 216 (bion., distr.); Tanaka et al. 1979: 191 (M*, F*, L*). **Ety.:** Okinawa. **Informal name:** Okinawa Japanese Typical Mosquito.

syn. *lini* Lien 1968d: 225 (M*, F, P*, L*). **Type loc.:** Chulu, Peinan, Taitung Hsien, Taiwan (TMRI). **Refs.:** Sirivanakarn 1972: 54 (syn.).

oresbius Harbach & Rattanarithikul 1988: 69 (M*, F, P*, L*). **Type loc.:** Doi Inthanon, Chom Thong, Chiang Mai Province, Thailand (USNM). **Distr.:** Cambodia, People's Republic of China, Thailand. **Refs.:** Dong et al. 1993a: 161 (distr.; People's Republic of China); Rattanarithikul et al. 2005b (F*, L*; bion., distr., keys). **Ety.:** from *oresbios* (Gr); ref. to living on mountains. **Informal name:** Montane Thai Typical Mosquito.

orstom Brunhes, Adam, & Bailly-Choumara 1967: 49 (M*). **Type loc.:** Irangi Forest, Kivu, Congo [Democratic Republic of the Congo] (IERT). **Distr.:** Côte d'Ivoire, Democratic Republic of the Congo, Republic of the Congo. **Ety.:** acronym: Office de la Recherche Scientifique et Technique Outre-Mer. **Informal name:** ORSTOM Congolese Typical Mosquito.

otachati Klein & Sirivanakarn 1970: 585 (M*). **Type loc.:** Kirirom Hills, O Tachat, Kompong Speu, Cambodia (USNM). **Distr.:** Cambodia, Thailand. **Refs.:** Sirivanakarn 1972: 59 (M*); Rattanarithikul et al. 2005b (F*, L*; bion., distr., keys). **Ety.:** Kirirom Hills, O Tachat, Kompong Speu, Cambodia. **Informal name:** O Tachat Cambodian Typical Mosquito.

phangngae Sirivanakarn 1972: 17 (M*, F, P, L). **Type loc.:** Nam Tai, Phangnga, Thailand (USNM). **Distr.:** Thailand. **Refs.:** Rattanarithikul et al. 2005b (F*, L*; bion., distr., keys). **Ety.:** Nam Tai, Phangnga, Thailand. **Informal name:** Phangnga Thai Typical Mosquito.

pluvialis Barraud 1924h: 1281 (M*, F). **Type loc.:** Kadra, North Kanara [Bombay], India (NHMUK). **Distr.:** India, Malaysia, Sri Lanka. **Refs.:** Sirivanakarn 1972: 42 (M*; lectotype desig.); Sirivanakarn & Ramalingam 1976: 214 (tax.); Sirivanakarn 1977d: 540 (P*, L*; distr.). **Ety.:** n.s. [rain bringing, rainy (L)]. **Informal name:** Rainy Indian Typical Mosquito.

pseudoandreanus Bailly-Choumara 1966a: 660 (M*). **Type loc.:** Ebogo, Arrondissement de Mbalmayo, 100 km south of Yaoundé, Cameroon (IERT). **Distr.:** Cameroon. **Refs.:** Rickenbach et al. 1976: 95 (distr.). **Ety.:** false (Gr); similar to *andreanus*. **Informal name:** False-André Cameroonian Typical Mosquito.

quintetti Brunhes, Adam, & Bailly-Choumara 1967: 49 (M*). **Type loc.:** Man, Ivory Coast (IERT). **Distr.:** Angola, Côte d'Ivoire. **Ety.:** n.s. **Informal name:** Quintet Ivorian Typical Mosquito.

richardgarciai Jeffery, Oothuman, & Rudnick 1987: 140 (M*). **Type loc.:** Pasoh Forest Reserve, Negeri Sembilan, Malaysia (USNM). **Distr.:** Malaysia. **Ety.:** Richard García. **Informal name:** Richard Malaysian Typical Mosquito.

richei Klein 1970: 504 (M*). **Type loc.:** Popork Vil, Bokor Hills, Kampot Province, Cambodia (USNM). **Distr.:** Cambodia, Thailand. **Refs.:** Chen 1980a (F*, L*); Chen & Zhao 1987 (P*). **Ety.:** Armand Riche. **Informal name:** Riche Cambodian Typical Mosquito.

rima Theobald 1901d: xi (F). **Type loc.:** Old Calabar [Eastern Provinces], Nigeria (NHMUK). **Distr.:** Benin, Burkina Faso, Cameroon, Central African Republic, Côte d'Ivoire, Democratic Republic of the Congo, Gabon, Ghana, Liberia, Mali, Mozambique, Niger, Nigeria, Republic of the Congo, Republic of South Africa, São Tomé and Príncipe, Sierra Leone, Sudan and South Sudan, The Gambia. **Refs.:** Macfie & Ingram 1922 (F*); Edwards 1941: 260 (M*, F*); De Meillon et al. 1945: 89 (P*); Klein 1956: 921 (M*); Rickenbach et al. 1976: 95 (distr.); Jupp 1996 (M*, F*; key). **Ety.:** cleft (L); ref. to "clypeus black, with a transverse sulcus". **Informal name:** Fissure Nigerian Typical Mosquito.

syn. *koumbai* Galliard

1931c: 514 (M*; *rima* var.). **Type loc.:** Port-Gentil, Gabon (NHMUK). **Refs.:** Edwards 1941: 260 (syn.); Townsend 1990: 91 (type info.).

rubinotus Theobald

1906b: 78 (F*). **Type loc.:** Luala, Upper White Nile, Sudan (NHMUK). **Distr.:** Angola, Burkina Faso, Cameroon, Central African Republic, Chad, Côte d'Ivoire, Democratic Republic of the Congo, Ethiopia, Gabon, Kenya, Liberia, Madagascar (includes Glorioso and Juan De Nova Isl.), Mozambique, Nigeria, Republic of the Congo, Republic of South Africa, Senegal, Sudan and South Sudan, Tanzania, Uganda, Zambia. **Refs.:** Edwards 1941: 260 (M, F*), 410 (P*); Hopkins 1952: 257 (L*); Service 1976a (distr.; Gabon); Jupp 1996 (M*, F*; key). **Ety.:** red back (L); ref. to "thorax bright reddish-brown." **Informal name:** Red-backed Sudanese Typical Mosquito.

selai Klein & Sirivanakarn

1970: 589 (M*). **Type loc.:** Kompong Sela, Koh-Kong, Cambodia (USNM). **Distr.:** Cambodia, Malaysia. **Refs.:** Sirivanakarn 1972: 44 (M*); Sirivanakarn 1978b: 78 (P*, L*; distr.). **Ety.:** Kompong Sela, Koh-Kong, Cambodia. **Informal name:** Sela Cambodian Typical Mosquito.

simplicicornis Edwards

1930a: 306 (M). **Type loc.:** Samawang jungle, near Sandakan, North Borneo, [Malaysian] Borneo (NHMUK). **Distr.:** Malaysia. **Refs.:** Sirivanakarn 1972: 9 (M*, F*; lectotype desig.); Sirivanakarn 1977b: 78 (P*, L*; distr.). **Ety.:** n.s. [single horn (L); poss. ref. to it being the only other oriental *Culex* with a "simple male antenna" other than *nilgiricus*]. **Informal name:** Single-horned Malaysian Typical Mosquito.

simpliciforceps Edwards

1941: 279 (M*, F). **Type loc.:** Bundle to Hierallah, Upper White Nile, Sudan (NHMUK). **Distr.:** Cameroon, Central African Republic, Côte d'Ivoire, Democratic Republic of the Congo, Guinea, Kenya, Liberia, Republic of the Congo, Republic of South Africa, Sudan and South Sudan, Tanzania, Uganda. **Refs.:** De Meillon et al. 1945: 86 (P*); Hopkins 1952: 280 (L); Muspratt 1955: 181 (M*); Mattingly 1956a: 41 (lectotype desig.); Rickenbach et al. 1976: 96 (distr.); Jupp 1996 (M*, F*; key). **Ety.:** n.s. [simple, single forceps (L); poss. ref. to lack of tooth and fringe on gonostylus]. **Informal name:** Simple-clawed Sudanese Typical Mosquito.

stellatus van Someren

1947a: 29 (M*, F, P*, L*). **Type loc.:** Mahe Island, Seychelles (NHMUK). **Distr.:** Seychelles. **Refs.:** Hopkins 1952: 266 (L*); Mattingly & Brown 1955: 85 (tax.). **Ety.:** provided with stars (L); ref. to many stellate larval setae. **Informal name:** Starry Seychellois Typical Mosquito.

subrima Edwards

1941: 262 (M*, F*). **Type loc.:** Onitsha [Eastern Provinces], Nigeria (NHMUK). **Distr.:** Cameroon, Côte d'Ivoire, Democratic Republic of the Congo, Liberia, Nigeria, Republic of the Congo, Uganda. **Refs.:** Hopkins 1952: 258 (L); Hamon & Rickenbach 1955b: 849 (M*); Mattingly 1956a: 40 (type info.). **Ety.:** nearly (L); near *rima*. **Informal name:** Cavity Nigerian Typical Mosquito [*subrima* = "subfissure" = Cavity].

sunyaniensis Edwards

1941: 263, 411 (M*, P). **Type loc.:** Sunyani [Ashanti], Gold Coast (NHMUK). **Distr.:** Benin, Cameroon, Central African Republic, Côte d'Ivoire, Democratic Republic of the Congo, Gabon, Ghana, Liberia, Madagascar (includes Glorioso and Juan De Nova Isl.), Mozambique, Nigeria, Senegal, Sierra Leone, Sudan and South Sudan, The Gambia. **Refs.:** Hopkins 1952: 259 (L*); Hamon & Rickenbach 1955b: 851 (M*); Mattingly 1956a: 40 (type info.); Worth & De Meillon 1960: 246 (distr.); Service 1976a (distr.; Gabon). **Ety.:** Sunyani, Ghana. **Informal name:** Sunyani Ghanaian Typical Mosquito.

tauffliebi Geoffroy & Hervé

1976: 151 (M*). **Type loc.:** Bozo, Central African Republic (SSC). **Distr.:** Central African Republic. **Ety.:** R. Taufflieb. **Informal name:** Taufflieb Central African Typical Mosquito.

tenuipalpis Barraud

1924h: 1278 (M*, F). **Type loc.:** Mungpoo, Darjeeling Hills [West Bengal], India (NHMUK). **Distr.:** Cambodia, India, Indonesia, Malaysia, Thailand. **Refs.:** Barraud 1924m (L*); Barraud 1934: 351 (M*, F, L*; tax.); Bram 1967b: 29 (M*, F, L*; bion., distr., tax.); Sirivanakarn 1972: 48 (M*, F*, P*, L*); Ramalingam & Pillai 1973 (distr.); Sirivanakarn & Ramalingam 1976: 213 (tax.); Rattanarithikul et al. 2005b (F*, L*; bion., distr., keys). **Ety.:** n.s. **Informal name:** Thin-lipped Indian Typical Mosquito.

uncinatus Delfinado

1966a: 134 (M*, F). **Type loc.:** Osmena, Samar [Philippines] (USNM). **Distr.:** Philippines. **Refs.:** Basio 1971b: 71 (M*; bion.); Sirivanakarn 1972: 25 (M*, F). **Ety.:** n.s. [hooked (L)]. **Informal name:** Barbed Philippine Typical Mosquito.

vattieri Geoffroy

1971c: 265 (M*). **Type loc.:** Central African Republic; type location of holotype not specifically designated (IERT). **Distr.:** Central African Republic. **Ety.:** Mme. Vattier-Bernard. **Informal name:** Vattier Central African Typical Mosquito.

vinckei Hamon, Holstein, & Rivola

1957: 681 (M*). **Type loc.:** Nyakalunge (Kalimbi), Terr Kaleha, Kivu, Belgian Congo [Zaire] (IERT). **Distr.:** Côte d'Ivoire, Democratic Republic of the Congo, Republic of the Congo. **Refs.:** Doucet 1962: 1160 (M*, P*, L*). **Ety.:** Dr. Vincke. **Informal name:** Vincke Congolese Typical Mosquito.

wansoni Wolfs
1945c: 221 (M*, F, P*, L*). **Type loc.:** Coquilhatville, Équateur, Belgian Congo [Zaire] (CMT, NHMUK). **Distr.:** Cameroon, Democratic Republic of the Congo, Sierra Leone. **Refs.:** Townsend 1990: 149 (type info.). **Ety.:** n.s. [prob. Marcel Jules Emil Marie Wanson]. **Informal name:** Wanson Congolese Typical Mosquito.

wigglesworthi Edwards
1941: 264 (M*, F*). **Type loc.:** Freetown, Sierra Leone (NHMUK). **Distr.:** Burkina Faso, Cameroon, Central African Republic, Comoros, Côte d'Ivoire, Democratic Republic of the Congo, Ghana, Kenya, Liberia, Madagascar (includes Glorioso and Juan De Nova Isl.), Mali, Nigeria, Republic of South Africa, Sierra Leone, Sudan and South Sudan, Togo, Uganda. **Refs.:** De Meillon et al. 1945: 8 (P*, L*); Hopkins 1952: 261 (L*); Hamon & Rickenbach 1955b: 851 (M*); Mattingly 1956a: 40 (type info.); Brunhes 1977a (distr.; Comoros Isl.); Jupp 1996 (M*, F*; key). **Ety.:** Vincent Brian Wigglesworth. **Informal name:** Wigglesworth Sierra Leonean Typical Mosquito.

yeageri Baisas
1935d: 175 (M*). **Type loc.:** Iwahig, Palawan, Philippines (NE). **Distr.:** Philippines. **Refs.:** Basio 1971b: 71 (M*; bion.); Sirivanakarn 1972: 41 (M*, F*). **Ety.:** n.s. **Informal name:** Yeager Philippine Typical Mosquito.

Subgenus *Kitzmilleria* Danilov

Danilov 1989: 792 [114 in translation] (as subg.). Type species: *Culex moucheti* Evans.

Name derivation: James B. Kitzmiller.

Important references: Edwards 1941: 351 (revision (adults); as *Culex* (*Culex*)); Hopkins 1952: 338 (L; revision; as *Culex* (*Culex*)); Danilov 1989 (tax., distr.); Harbach 2011 (phyl.).

moucheti Evans
1923a: 89 (M*, F). **Type loc.:** Stanleyville, Belgian Congo [Kisangani, Tshopo Province, Democratic Republic of the Congo] (CMT). **Distr.:** Benin, Burkina Faso, Cameroon, Central African Republic, Comoros, Democratic Republic of the Congo, Kenya, Liberia, Madagascar (includes Glorioso and Juan De Nova Isl.), Nigeria, Republic of the Congo, Sierra Leone, Sudan and South Sudan, Uganda. **Refs.:** Edwards 1941: 351 (M*, F*), 428 (P*); Hopkins 1952: 338 (L*); Danilov 1989 (from subg. *Culex*). **Ety.:** René Libert Joseph Mouchet. **Informal name:** Mouchet Congolese Typical Mosquito.

Subgenus *Lasiosiphon* Kirkpatrick

1925a: 372 (as subg.). Type species: *Culex* (*Lasiosiphon*) *adairi Kirkpatrick* [replacement name for *Culex pluvialis* Kirkpatrick].

Name derivation: lasio- = shaggy with hair or wool (Gr); -siphon; siphon with numerous tufts of long and abundant hairs in a single ventral row.

Important references: Harbach et al. 2012 (phyl.).

adairi Kirkpatrick
1926: 112 (new name for *pluvialis* Kirkpatrick, not Barraud 1924). **Distr.:** Egypt, Israel (and Gaza Strip and West Bank), Oman. **Refs.:** Rioux & Quezel 1957: 461 (P*, L*); Margalit & Tahori 1974: 85 (distr.); Harbach et al. 1988: 320 (syntype info.); Townsend 1990: 122 (syntype info.); Robert et al. 2019 (distr.; Western Palaearctic). **Ety.:** E.W. Adair. **Informal name:** Adair Egyptian Typical Mosquito.
syn. *pluvialis* Kirkpatrick
1925a: 373 (M*, F, L*). **Type loc.:** Wadi No'oth, North Galala Mountains and Wadi Hof, 12 km southeast of Helouan [Helewan]; Egypt (DAC, ESE, NHMUK). **Refs.:** Kirkpatrick 1926 (syn.); Harbach et al. 1988: 320 (type info.); Townsend 1990: 122 (type info.).
syn. *kirkpatriki* Edwards
1926c: 123 (new name for *pluvialis* Kirkpatrick, not Barraud 1924g).
syn. *kirkpatricki* Stackelberg
1927: 148 (new name for *pluvialis* Kirkpatrick, not Barraud 1924g).

Subgenus *Lophoceraomyia* Theobald

1905h: 245 (April 15). Type species: *Lophoceraomyia uniformis* Theobald.

Name derivation: back of neck, crest of hill, crest/ridge; cerao = animal's horn [unusual setae on male antennae mentioned (?); Fig. 3 shows scape with a dorsal protuberance not mentioned in the text].
Lophoceratomyia Theobald 1905f: 93 (June, lapsus of *Lophoceraomyia*).

syn. *Philodendromyia* Theobald 1907: 623 (as genus). Type species: *Philodendromyia barkerii* Theobald. Edwards 1932a: 196 (syn.).
syn. *Cyathomyia* de Meijere 1910: 921. Type species: *Cyathomyia jenseni* de Meijere. Edwards 1932a: 196 (syn.).

Important references: Belkin 1962 (revision; South Pacific); Bram 1967b (revision; Thailand); Colless 1965: 216 (keys; Malaya); Sirivanakarn 1968: 75 (keys; New Guinea and

Bismarck Archipelago); Sirivanakarn 1977a (revision; Oriental Region); Lee et al. 1989b (revision; Australasia); Harbach 2011 (phyl.).

aculeatus Colless
1965: 286 (M*, F, L*). **Type loc.:** Bukit Tunggal, Kampong Gajan Forest Reserve, Perak, Malaya [Malaysia] (ANIC). **Distr.:** Cambodia, Malaysia, Thailand. **Refs.:** Bram 1967b: 50 (M*, F, L*; bion., distr., tax.); Sirivanakarn 1977a: 48 (M*, F, P, L; keys, tax., bion., distr.); Rattanarithikul et al. 2005b (F*, L*; bion., distr., keys). **Ety.:** n.s. [sharp, pointed (L)]. **Informal name:** Sharp-pointed Malaysian Typical Mosquito.

acutipalus Colless
1965: 299 (M*, F, L). **Type loc.:** Kampong Chantek Bahru, Singapore (ANIC). **Distr.:** Malaysia, Singapore, Thailand. **Refs.:** Sirivanakarn 1977a: 140 (M*, F, P, L; distr.). **Ety.:** n.s. [sharp stick (L); poss. ref. to shape of palpus]. **Informal name:** Sharp Thai Typical Mosquito.

aestivus Sirivanakarn
1977a: 50 (M*). **Type loc.:** Long Tebangan, Sarawak, Malaysia (NHMUK). **Distr.:** Malaysia. **Ety.:** n.s. [summer (L); poss. ref. to time of year of capture]. **Informal name:** Summer Singaporean Typical Mosquito.

alorensis Sirivanakarn
1977a: 25 (M*, F, P, L; keys, tax., bion., distr.). **Type loc.:** Alor, Indonesia (USNM). **Distr.:** Indonesia, Timor. **Refs.:** Lee et al. 1989b: 77 (F; key, tax., bion., distr., review). **Ety.:** Alor Isl. **Informal name:** Alor Indonesian Typical Mosquito.

alphus Colless
1965: 283 (M*, F, L*). **Type loc.:** Kg-Sijankang, Selangor, Malaya (ANIC). **Distr.:** Malaysia, Singapore, Thailand. **Refs.:** Bram 1967b: 53 (M*, F, L*; bion., distr., tax.); Sirivanakarn 1977a: 65 (M*, F*, P*, L*; keys, tax., bion., distr.); Rattanarithikul et al. 2005b (F*, L*; bion., distr., keys). **Ety.:** n.s. [white spot on skin (Gr); poss. ref. to white-scaled sternum VIII]. **Informal name:** White Malaysian Typical Mosquito.

atracus Colless
1960: 385 (M*). **Type loc.:** Kavieng, New Ireland (CSIR). **Distr.:** Papua New Guinea, Solomon Isl. **Refs.:** Sirivanakarn 1968: 161 (M*, F, P, L; keys, tax., bion., distr.; syn.); Lee et al. 1989b: 78 (F; key, tax., bion., distr., review). **Ety.:** n.s. [dark (L)]. **Informal name:** Dark Papuan Typical Mosquito.
syn. *franclemonti* Belkin
1962: 256 (M*, F, P*, L*). **Type loc.:** Munda, New Georgia, Solomon Islands (USNM). **Refs.:** Sirivanakarn 1968 (syn.).

bandoengensis Brug
1939: 112 (M*). **Type loc.:** Bandoeng, Java (NHMUK). **Distr.:** Indonesia, Malaysia. **Refs.:** Colless 1965: 292 (M*, F, L; to sp.); Sirivanakarn 1977a: 104 (M*, F, P, L; keys, tax., bion., distr.); Lee et al. 1989b: 80 (F; key, tax., bion., distr., review). **Ety.:** Bandoeng, Java. **Informal name:** Bandoeng Indonesian Typical Mosquito.

becki Belkin
1962: 265 (M*, F, P*, L*). **Type loc.:** Sprague swamp, Guadalcanal, Solomon Islands (USNM). **Distr.:** Solomon Isl. **Refs.:** Lee et al. 1989b: 81 (F; key, tax., bion., distr., review). **Ety.:** D. Elden Beck. **Informal name:** Beck Solomon Islander Typical Mosquito.

bengalensis Barraud
1934: 371 (M). **Type loc.:** Nongpoh, Assam, India (NHMUK). **Distr.:** Bhutan, India, Indonesia, Malaysia, People's Republic of China, Thailand. **Refs.:** Colless 1965: 293 (M*, F, L*; to sp.); Bram 1967b: 72 (M*, F, L*; bion., distr., tax.); Sirivanakarn 1977a: 112 (M*, F, P*, L; keys, tax., bion., distr.); Townsend 1990: 51 (lectotype desig.; LU to NHMUK); Rattanarithikul et al. 2005b (F*, L*; bion., distr., keys); Namgay et al. 2018 (bion., distr.; Bhutan). **Ety.:** Bengal region, India. **Informal name:** Bengal Indian Typical Mosquito.

bergi Belkin
1962: 258 (M*, F, P*, L*). **Type loc.:** Poha River, about 3 mi from coast, Guadalcanal, Solomon Islands (USNM). **Distr.:** Papua New Guinea, Solomon Isl. **Refs.:** Lee et al. 1989b: 82 (F; key, tax., bion., distr., review). **Ety.:** Clifford Osburn Berg. **Informal name:** Berg Solomon Islander Typical Mosquito.

bicornutus (Theobald)
1910a: 25 (M; *Lophoceratomyia*). **Type loc.:** Dawna Hills (base), near Kawkareik, Lower Burma [Myanmar] (NHMUK). **Distr.:** India, Japan, Malaysia, Myanmar, People's Republic of China, Singapore, Sri Lanka, Thailand, Vietnam. **Refs.:** Colless 1965: 291 (from syn. with *minor*); Sirivanakarn 1977a: 102 (M, F, P, L; keys, tax., bion., distr.; from syn. with *minor*); Tanaka et al. 1979: 207 (M*, F*, L*); Rattanarithikul et al. 2005b (F*, L*; bion., distr., keys). **Ety.:** two horned (L); ref. to "two prominent lateral horn-like processes on the basal segments of the antenna." **Informal name:** Two-horned Burmese Typical Mosquito.

bolii Sirivanakarn
1968: 100 (M*, F*). **Type loc.:** Marani Creek, Kundiawa, in Chimbu, Eastern Highlands, NE New Guinea (BBM). **Distr.:** Papua New Guinea. **Refs.:** Lee et al. 1989b: 83 (F; key, tax., bion., distr., review). **Ety.:** Mr. Boli. **Informal name:** Boli Papuan Typical Mosquito.

brevipalpus (Theobald)
1905f: 96 (not Giles 1902) (M*; *Lophoceratomyia*). **Type loc.:** Singapore (HNM). **Distr.:** Indonesia, Malaysia, Singapore. **Refs.:** Stone 1957b: 172 (from syn.); Colless 1965: 301 (M*, F, L); Sirivanakarn 1977a: 135 (M*, F*, P*, L*; keys, tax., bion., distr.). **Ety.:** short feeler (L); short

palpus. **Informal name:** Short-lipped Singaporean Typical Mosquito.

buxtoni Edwards
1926a: 106 (M*; as *hilli* var.). **Type loc.:** Vila, Efate Island, New Hebrides (NHMUK). **Distr.:** Vanuatu. **Refs.:** Lee 1944a: 93 (L*); Perry 1946: 15 (syn.); Mattingly 1949c: 227 (L*); Belkin 1962: 253 (M*, F, P*, L*; from syn.); Lee et al. 1989b: 84 (F; key, tax., bion., distr., review). **Ety.:** Patrick Alfred Buxton. **Informal name:** Buxton Ni-Vanuatu Typical Mosquito.

carolinensis Bohart & Ingram
1946b: 30 (M*, F, P*, L*; bion., distr., tax.). **Type loc.:** Fala Island, Truk Islands, [Caroline Islands] (USNM). **Distr.:** Micronesia (Wake Isl.). **Refs.:** Bohart 1957: 66 (M*, F, P, L*); Lee et al. 1989b: 87 (tax., distr.). **Ety.:** Caroline Isl. **Informal name:** Caroline Micronesian Typical Mosquito.

castaneus Sirivanakarn
1973a: 212 (M*). **Type loc.:** Gusika, Finschhafen, Madang, NE New Guinea (USNM). **Distr.:** Papua New Guinea. **Refs.:** Lee et al. 1989b: 88 (tax., distr.). **Ety.:** n.s. [chestnut colored (L); prob. ref. to "brownish species"]. **Informal name:** Chestnut Papuan Typical Mosquito.

christiani Colless
1960: 389 (M*, F, L*). **Type loc.:** Minj, Western Highlands, New Guinea (CSIR). **Distr.:** Indonesia, Papua New Guinea. **Refs.:** Sirivanakarn 1968: 147 (M*, F, P*, L*; keys, tax., bion., distr.); Lee et al. 1989b: 89 (F; key, tax., bion., distr., review). **Ety.:** S.H. Christian. **Informal name:** Christian Indonesian Typical Mosquito.

cinctellus Edwards
1922a: 287 (M*; new name for *taeniata* Leicester, not Wiedemann 1828). **Distr.:** Cambodia, India, Indonesia, Japan, Malaysia, People's Republic of China, Philippines, Singapore, Thailand, Vietnam. **Refs.:** Edwards & Given 1928: 353 (L*?); Barraud 1934: 366 (M*, F, L); Mattingly 1949c: 226 (L*?); Bram 1967b: 55 (M*, F, L*; bion., distr., tax.); Basio 1971b: 64 (M*; bion.); Tanaka et al. 1975c: 217 (distr.); Sirivanakarn 1977a: 33 (M*, F*, P*, L*; keys, tax., bion., distr.); Tanaka et al. 1979: 202 (M*, F*, L); Rattanarithikul et al. 2005b (F*, L*; bion., distr., keys). **Ety.:** n.s. [girded (L)]. **Informal name:** Girded Malaysian Typical Mosquito.
syn. *taeniata* Leicester
1908: 127 (M, F; *Lophoceratomyia*). **Type loc.:** Kuala Lumpur and Klang [Selangor], Malaya [Malaysia] (NHMUK).

coerulescens Edwards
1928a: 277 (M*, F). **Type loc.:** Kuching, Sarawak, Borneo (NHMUK). **Distr.:** Malaysia, Singapore. **Refs.:** Mattingly 1949c: 226 (L); Sirivanakarn 1977a: 129 (M*, F, P, L*; keys, tax., bion., distr.). **Ety.:** n.s. [becoming sky blue (L)]. **Informal name:** Sky-blue Malaysian Typical Mosquito.

collessi Sirivanakarn
1968: 170 (M*, F, P*, L*). **Type loc.:** Keravat, New Britain, Bismarck Archipelago (BBM). **Distr.:** Papua New Guinea. **Refs.:** Lee et al. 1989b: 91 (F; key, tax., bion., distr., review). **Ety.:** Donald Henry Colless. **Informal name:** Colless Papuan Typical Mosquito.

cottlei Sirivanakarn
1968: 180 (M*, F, P*, L*). **Type loc.:** Lorengau, Manus Island, Bismarck Archipelago (BBM). **Distr.:** Papua New Guinea. **Refs.:** Lee et al. 1989b: 92 (F; key, tax., bion., distr., review). **Ety.:** Robert Cottle. **Informal name:** Cottle Papuan Typical Mosquito.

crassicomus Colless
1965: 296 (M*). **Type loc.:** Ulu Gombak, Selangor, Malaya [Malaysia] (ANIC). **Distr.:** Malaysia. **Refs.:** Sirivanakarn 1977a: 110 (M*; key, tax., bion., distr.). **Ety.:** n.s. [thick? (L)]. **Informal name:** Thick Malaysian Typical Mosquito.

crowei Sirivanakarn
1968: 115 (M*, F, P, L). **Type loc.:** Daunda Village 9, Vanimo, Sepik District, NE New Guinea (BBM). **Distr.:** Indonesia, Papua New Guinea. **Refs.:** Lee et al. 1989b: 93 (F; key, tax., bion., distr., review). **Ety.:** Lockland Crowe. **Informal name:** Crowe Papuan Typical Mosquito.

cubiculi Marks
1989 (new name for *annulata* Taylor 1916, not Schrank 1776). In Lee et al. 1989b: 94. **Distr.:** Australia. **Refs.:** Lee et al. 1989b: 94 (F; key, tax., bion., distr., review). **Ety.:** bedroom (L); type and cotype captured in a bedroom. **Informal name:** Bedroom Australian Typical Mosquito.
syn. *annulata* Taylor
1916: 571 (M, F; *Lophoceratomyia*; not Schrank 1776). **Type loc.:** Stapleton and Daly River, Northern Territory, Australia (NHMUK). **Refs.:** Knight et al. 1944: 47 (syn.); Lee et al. 1989b (syn.).

cubitatus Colless
1965: 273 (M*, F, L). **Type loc.:** Mandai Road, Singapore (ANIC). **Distr.:** Indonesia, Malaysia, Papua New Guinea, Philippines, Singapore. **Refs.:** Sirivanakarn 1973b: 215 (distr.); Sirivanakarn 1977a: 57 (M*, F, P, L; keys, tax., bion., distr.); Lee et al. 1989b: 97 (tax., distr.). **Ety.:** n.s. [reclined (L)]. **Informal name:** Reclining Singaporean Typical Mosquito.

curtipalpis (Edwards)
1914c: 127 (M; *Lophoceratomyia*). **Type loc.:** Kuching Reservoir, Sarawak, Borneo (NHMUK). **Distr.:** Cambodia, Indonesia, Malaysia, Singapore, Thailand, Vietnam. **Refs.:** Edwards 1928a: 279 (M*); Edwards & Given 1928: 357 (P, L*); Colless 1965: 301 (M*, F, L); Bram 1967b: 76 (M*, F, L*; bion., distr., tax.); Sirivanakarn 1977a: 142 (M*, F*, P*, L*; keys, tax., bion., distr.); Rattanarithikul et al. 2005b (F*, L*; bion., distr., keys). **Ety.:** short feeler (L); short palpus. **Informal name:** Small-lipped Malaysian Typical Mosquito.

ssp. ***sumatranus*** Brug
1931a: 248 (M*, L*). **Type loc.:** Dermajoe, Benkoelen, Sumatra (NHMUK). **Distr.:** Cambodia, Indonesia, Macau, People's Republic of China (includes Hong Kong), Vietnam. **Refs.:** Sirivanakarn 1971a: 67 (n. comb.); Sirivanakarn 1977a: 145 (M*, F, P, L*); Meng & Chen 1980 (syn.). **Ety.:** Sumatra, Indonesia. **Informal name:** Sumatra Indonesian Typical Mosquito.

ssp. syn. *gudouensis* Chang, Zhao, Hang, & Chen
1975 (M*, F, L*). **Type loc.:** Xin-hui Shien, Kwangtung Province, China [People's Republic of China] (Kwangtung Health Station, Kwangchow). **Refs.:** Meng & Chen 1980 (syn.).

cylindricus Theobald
1903a: 202 (M*, F*). **Type loc.:** South Queensland, Australia (NHMUK). **Distr.:** Australia, Malaysia, Papua New Guinea. **Refs.:** Bancroft 1908: 47 (M, F); Dobrotworsky 1958a: 319 (M*, F, P*, L*); Dobrotworsky 1965: 202 (M*, F, L*); Lee et al. 1989b: 98 (F; key, tax., bion., distr., review). **Ety.:** cylindrical (L); ref. to narrow cylindrical abdomen of adult female. **Informal name:** Cylindrical Australian Typical Mosquito.

demissus Colless
1965: 296 (M*). **Type loc.:** Ulu Gombak, Selangor, Malaya [Malaysia] (ANIC). **Distr.:** Malaysia, Thailand. **Refs.:** Bram 1967b: 87 (M*, F, L*; bion., distr., tax.; as *fuscosiphonus*); Sirivanakarn 1977a: 91 (M*, F, P*, L*; keys, tax., bion., distr.); Rattanarithikul et al. 2005b (F*, L*; bion., distr., keys). **Ety.:** n.s. [low lying or low altitude (L)]. **Informal name:** Lowland Malaysian Typical Mosquito.

syn. *fuscosiphonis* Bram & Rattanarithikul
1967: 14 (M*, F, L*). **Type loc.:** Muang, Phatthalung Province, Thailand (USNM). **Refs.:** Sirivanakarn 1977a: 91 (syn.).

digoelensis Brug
1932: 81 (M, F). **Type loc.:** Upper Digoel River, [Western] New Guinea (NHMUK). **Distr.:** Indonesia (West Papua), Papua New Guinea. **Refs.:** King & Hoogstraal 1955: 11 (M*); Sirivanakarn 1968: 101 (M*, F, P*, L*; keys, tax., bion., distr.); Lee et al. 1989b: 102 (F; key, tax., bion., distr., review). **Ety.:** Digoel (Digul) River, New Guinea. **Informal name:** Digoel Indonesian Typical Mosquito.

syn. *caeruleus* King & Hoogstraal
1947b: 67 (M*, F, L*). **Type loc.:** Samboga River, Dobodura [Papua], New Guinea (USNM). **Refs.:** Sirivanakarn 1973a: 215 (syn.).

durhami Sirivanakarn
1968: 177 (M*). **Type loc.:** Upau Stream, Maprik, Sepik, NE New Guinea (BBM). **Distr.:** Papua New Guinea. **Refs.:** Lee et al. 1989b: 104 (tax., bion., distr.). **Ety.:** S. Durham. **Informal name:** Durham Papuan Typical Mosquito.

eminentia (Leicester)
1908: 131 (M; *Lophoceratomyia*). **Type loc.:** Kuala Lumpur [Selangor], Malaya [Malaysia] (NHMUK). **Distr.:** Malaysia, Singapore. **Refs.:** Edwards & Given 1928: 355 (L*); Mattingly 1949c: 226 (L*); Colless 1965: 298 (M*, L); Sirivanakarn 1977a: 139 (M*, F, P, L; keys, tax., bion., distr.). **Ety.:** standing out (L); ref. to a very distinct and easily recognized species. **Informal name:** Eminent Malaysian Typical Mosquito.

eukrines Bram & Rattanarithikul
1967: 11 (M*, F, L*). **Type loc.:** Huai Bong Ti, Kanchanaburi Province, Thailand (USNM). **Distr.:** Cambodia, Thailand. **Refs.:** Bram 1967b: 85 (M*, F, L*; bion., distr., tax.); Sirivanakarn 1977a: 116 (M*, F, P, L*; keys, tax., bion., distr.); Rattanarithikul et al. 2005b (F*, L*; bion., distr., keys). **Ety.:** "clear, distinct" (Gr). **Informal name:** Distinct Thai Typical Mosquito.

flavicornis Barraud
1924j: 45 (M*, F). **Type loc.:** Nilgiri Hills [Madras], India (NHMUK). **Distr.:** India. **Refs.:** Barraud 1934: 375 (M*, F); Bram 1967c (type info., lectotype desig.); Sirivanakarn 1977a: 120 (M*, F, P*, L*; keys, tax., bion., distr.). **Ety.:** n.s. [yellow horned (L); poss. ref. to long bright yellow setae on sixth segment of male antenna]. **Informal name:** Yellow-horned Indian Typical Mosquito.

fraudatrix (Theobald)
1905f: 94 (M*, F*; *Lophoceratomyia*). **Type loc.:** Friedrich-Wilhelmshafen, [NE] New Guinea (HNM). **Distr.:** Australia, Indonesia, Papua New Guinea, Philippines, Solomon Islands, Thailand. **Refs.:** Barraud 1934: 368 (M*, F, L); Lee 1944a: 91 (L*); Bohart 1945: 74 (M*, L); Mattingly 1949c: 227 (L*; key); Penn 1949b: 69 (P*); Laird 1954: 289 (L*; tax.); Colless 1960: 382 (M*); Sirivanakarn 1968: 156 (M*, F, P*, L*; keys, tax., bion., distr., lectotype desig.); Lee et al. 1989b: 105 (F; key, tax., bion., distr., review). **Ety.:** n.s [fraud (L); prob. ref. to "pseudo-vein" between cubital and anal veins]. **Informal name:** Fraudulent Papuan Typical Mosquito.

syn. *molestus* Wiedemann
1828: 542 (F; not Wiedemann 1821). **Type loc.:** Sumatra (LU).

fulleri (Ludlow)
1909b: 97 (F; *Oculiomyia*). **Type loc.:** Parang [Cotabato], Mindanao, Philippines (USNM). **Distr.:** Philippines. **Refs.:** Bohart 1945: 74 (M*); Stone & Knight 1957a: 50 (lectotype desig.); Basio 1971b: 65 (M*; bion.); Sirivanakarn 1977a: 36 (M*, F; keys, tax., bion., distr.). **Ety.:** Major Fuller. **Informal name:** Fuller Philippine Typical Mosquito.

gagnei Evenhuis
1989 (new name for *ornatus* Theobald 1905f, not Meigen 1818). In Evenhuis & Gon 1989: 209. **Distr.:** Indonesia, Papua New Guinea. **Refs.:** King & Hoogstraal 1955: 7

(M*, F, L; as *ornatus*); Colless 1960: 386 (M; as *ornatus*); Sirivanakarn 1968: 141 (M*, F, P*, L*; as *ornatus*); Peyton et al. 1999: 543 (distr.). **Ety.:** Raymond Joseph Gagné. **Informal name:** Gagné Papuan Typical Mosquito.

syn. *ornatus* Theobald
1905f: 100 (F*; *Melanoconion*). **Type loc.:** Friedrich-Wilhelmshafen, [NE] New Guinea (HNM). **Refs.:** N.L. Evenhuis 1989 (syn.). In Evenhuis & Gon 1989.

ganapathi Colless
1965: 294 (M*, F, L*). **Type loc.:** Ulu Langat, Selangor, Malaya [Malaysia] (ANIC). **Distr.:** Cambodia, Malaysia, Thailand. **Refs.:** Bram 1967b: 90 (M*, F, L*; bion., distr., tax.); Sirivanakarn 1977a: 93 (M*, F, P*, L*; keys, tax., bion., distr.); Rattanarithikul et al. 2005b (F*, L*; bion., distr., keys). **Ety.:** A. Ganapathipillai. **Informal name:** Ganapathipillai Malaysian Typical Mosquito.

gibbulus Delfinado
1966a: 105 (M*). **Type loc.:** San Jose, Mindoro, Philippines (USNM). **Distr.:** Philippines. **Refs.:** Basio 1971b: 65 (M*; bion.); Sirivanakarn 1977a: 41 (M*; key, tax., bion., distr.). **Ety.:** n.s. [protuberance, lump on the body (L)]. **Informal name:** Gibbous Philippine Typical Mosquito.

gossi Bohart
1957: 68 (M*, F, L). **Type loc.:** Kolonia, Yap Island, Caroline Islands (USNM). **Distr.:** Micronesia (Wake Isl.). **Refs.:** Lee et al. 1989b: 113 (tax., distr.). **Ety.:** R.J. Goss. **Informal name:** Goss Micronesian Typical Mosquito.

gracicornis Sirivanakarn
1977a: 58 (M*). **Type loc.:** Kampung Ayer Mati, Perak, Malaysia (ANIC). **Distr.:** Malaysia, Thailand. **Refs.:** Rattanarithikul et al. 2005b (F*, L*; bion., distr., keys). **Ety.:** n.s. [slender horn (L); poss. ref. to characteristics of the phallosome]. **Informal name:** Slender-horned Malaysian Typical Mosquito.

gressitti Sirivanakarn
1968: 153 (M*, F, P*, L*). **Type loc.:** Edie Creek, Wau, Morobe, Papua New Guinea (BBM). **Distr.:** Papua New Guinea. **Refs.:** Lee et al. 1982: 160 (tax., distr.); Lee et al. 1989b: 114 (F; key, tax., bion., distr., review). **Ety.:** Judson Linsley Gressitt. **Informal name:** Gressitt Papuan Typical Mosquito.

hewitti (Edwards)
1914b: 80 (M; *Lophoceratomyia*). **Type loc.:** Matang, Sarawak, Borneo (NHMUK). **Distr.:** Indonesia, Malaysia (includes Borneo (Sarawak)), Singapore. **Refs.:** Edwards & Given 1928: 355 (L*); Colless 1965: 303 (M*, F, L); Sirivanakarn 1977a: 131 (M*, F*, P*, L*; keys, tax., bion., distr.; lectotype desig.); Townsend 1990: 82 (holotype M; NHMUK, invalidates lectotype desig.). **Ety.:** J. Hewitt [poss. ref. to collector]. **Informal name:** Hewitt Malaysian Typical Mosquito.

hilli Edwards
1922d: 95 (M, F). **Type loc.:** 70 mi south of Darwin, Northern Territory, Australia (NHMUK). **Distr.:** Australia. **Refs.:** Knight et al. 1944: 46 (M); Lee et al. 1989b: 115 (F; key, tax., bion., distr., review). **Ety.:** G.F. Hill. **Informal name:** Hill Australian Typical Mosquito.

syn. *australis* Taylor
1915: 178 (F*; *Neomacleaya*; not Erichson 1842). **Type loc.:** Halifax, Queensland, Australia (US).

hirtipalpis Sirivanakarn
1977a: 152 (M*; key, tax., bion., distr.). **Type loc.:** Doi Suthep, near temple, Chiang Mai, Thailand (USNM). **Distr.:** Thailand. **Refs.:** Dong, Zhou, & Dong 1995c (F*, P*, L*); Rattanarithikul et al. 2005b (F*, L*; bion., distr., keys). **Ety.:** n.s. [shaggy feeler (L); prob. ref. to "prominent rows of long setae on palpal segment 3"]. **Informal name:** Shaggy-palped Thai Typical Mosquito.

hurlbuti Belkin
1962: 266 (M*, F, P*, L*). **Type loc.:** Tenaru area, Guadalcanal, Solomon Islands (USNM). **Distr.:** Papua New Guinea, Solomon Isl. **Refs.:** Sirivanakarn 1968: 177 (distr.); Lee et al. 1989b: 119 (tax., bion., distr., review). **Ety.:** Herbert S. Hurlbut. **Informal name:** Hurlbut Solomon Islander Typical Mosquito.

impostor Sirivanakarn
1977b: 80 (M*, F, P*, L). **Type loc.:** Sapulut, Pensiangan, Sabah, [Malaysian] Malaysia (USNM). **Distr.:** Malaysia (Borneo (Sabah)). **Ety.:** imposter [similar to *homeopus*) (L)]. **Informal name:** Impostor Malaysian Typical Mosquito.

incomptus Bram & Rattanarithikul
1967: 16 (M*, F). **Type loc.:** Doi Suthep, Chiang Mai Province, Thailand (USNM). **Distr.:** Thailand. **Refs.:** Bram 1967b: 81 (M*; bion., distr., tax.); Sirivanakarn 1977a: 111 (M*; key, tax., bion., distr.); Rattanarithikul et al. 2005b (F*, L*; bion., distr., keys). **Ety.:** n.s. [disheveled, untidy, unpolished (L)]. **Informal name:** Disheveled Thai Typical Mosquito.

inculus Colless
1965: 280 (M*, F, L*). **Type loc.:** Kampong Sireh, Selangor, Malaya [Malaysia] (ANIC). **Distr.:** Cambodia, Indonesia, Malaysia. **Refs.:** Lien et al. 1975 (distr.); Sirivanakarn 1977a: 43 (M*, F, P*, L*; keys, tax., bion., distr.). **Ety.:** n.s. **Informal name:** Rough Malaysian Typical Mosquito [*inculus* is similar to "incultus" = Rough].

infantulus Edwards
1922a: 287 (M*). **Type loc.:** Hong Kong [People's Republic of China] (NHMUK). **Distr.:** Cambodia, India, Indonesia, Japan, Malaysia, Maldives, Myanmar, Nepal, People's Republic of China (includes Hong Kong), Philippines, South Korea, Sri Lanka, Thailand, Vietnam. **Refs.:** Bohart & Ingram 1946b: 73 (M*, F, P*, L*, E*; bion., distr., tax.); La Casse & Yamaguti 1950: 197 (M*, F*, P*, L*; bion., distr.,

tax.); Iyengar & Menon 1955 (L*); Bram 1967b: 58 (M*, F, L*; bion., distr., tax.); Basio 1971b: 65 (M*; bion.); Ramalingam & Pillai 1973 (distr.); Baisas 1974: 111 (M, F, P, L; tax., bion., distr.; Philippines); Mattingly 1975: 19 (E*); Tanaka et al. 1975c: 217 (bion., distr.); Sirivanakarn 1977a: 26 (M*, F*, P*, L*; keys, tax., bion., distr.); Tanaka et al. 1979: 199 (M*, F*, L*); Darsie & Pradhan 1990, 1991 (F, L; tax., keys, bion., distr.; Nepal); Rattanarithikul et al. 2005b (F*, L*; bion., distr., keys). **Ety.:** n.s. [little infant (L)]. **Informal name:** Infant Chinese Typical Mosquito.

syn. *parainfantulus* Menon
1944: 389 (M*, F*; *Lophoceratomyia*). **Type loc.:** Trivandrum, Travancore, India (LU). **Refs.:** Mattingly 1949c: 224 (syn.).

insequens Marks
1989 (new name for *cairnsensis* Taylor 1919). In Lee et al. 1989b: 120. **Distr.:** Australia. **Refs.:** Lee et al. 1989b: 120 (F; key, tax., bion., distr., review). **Ety.:** following closely (L); ref. to close publication of two sp. named *cairnsensis* that led to homonymy and need for this new name. **Informal name:** Pursuer Australian Typical Mosquito.

syn. *cairnsensis* Taylor
1919: 837 (M*, F; *Lophoceratomyia*; not Taylor 1919: 829). **Type loc.:** Cairns, Queensland, Australia (US). **Refs.:** Lee et al. 1989b: 120.

insularis Sirivanakarn
1968: 166 (M*, F, P, L). **Type loc.:** Kerevat, New Britain, Bismarck Archipelago (BBM). **Distr.:** Papua New Guinea. **Refs.:** Lee et al. 1989b: 122 (F; key, tax., bion., distr., review). **Ety.:** n.s. [of islands (L)]. **Informal name:** Islander Papuan Typical Mosquito.

jenseni (de Meijere)
1910: 922 (M*, F, P*, L*; *Cyathomyia*). **Type loc.:** Tjibodas (1,500 m), near Buitenzorg, Java (AM). **Distr.:** Indonesia, Malaysia, Singapore. **Refs.:** Edwards & Given 1928: 356 (L); Mattingly 1949c: 226 (L); Bram 1967c (type info., lectotype desig.); Mattingly 1975: 19 (E*); Sirivanakarn 1977a: 133 (M*, F*, P*, L*; keys, tax., bion., distr.). **Ety.:** Dr. Jensen. **Informal name:** Jensen Indonesian Typical Mosquito.

josephineae Baisas
1935d: 172 (M*, F). **Type loc.:** Del Carmen, Pampanga Province, Luzon, Philippines (NE). **Distr.:** Philippines. **Refs.:** Bohart 1945: 74 (syn.); Delfinado 1966a: 106 (M*; from syn.); Basio 1971b: 66 (M*; bion.); Sirivanakarn 1977a: 56 (M*, F*, P, L; keys, tax., bion., distr.). **Ety.:** n.s. [someone named Josephine]. **Informal name:** Josephine Philippine Typical Mosquito.

kaviengensis Sirivanakarn
1968: 167 (M*, F, P*, L*). **Type loc.:** Kavieng, New Ireland, Bismarck Archipelago (BBM). **Distr.:** Papua New Guinea. **Refs.:** Lee et al. 1989b: 123 (tax., distr.). **Ety.:** Kavieng, New Ireland Prov., Papua New Guinea. **Informal name:** Kavieng Papuan Typical Mosquito.

kowiroensis Sirivanakarn
1968: 133 (M*, F, P*, L*). **Type loc.:** Kowiro, inland of Wewak, Sepik, NE New Guinea (BBM). **Distr.:** Papua New Guinea. **Refs.:** Lee et al. 1989b: 124 (F; key, tax., bion., distr., review). **Ety.:** Kowiro, Sepik Prov., Papua New Guinea. **Informal name:** Kowiro Papuan Typical Mosquito.

kuhnsi King & Hoogstraal
1955: 1 (M*, F, L*). **Type loc.:** Doromena, Hollandia area, Western New Guinea (Papua) (USNM). **Distr.:** Indonesia (includes West Papua), Malaysia, Papua New Guinea, Philippines. **Refs.:** Sirivanakarn 1968: 96 (M*, F, P*, L*); Sirivanakarn 1977a: 107 (M*, F, P, L*; keys, tax., bion., distr.); Lee et al. 1989b: 125 (F; key, tax., bion., distr., review). **Ety.:** Dwight M. Kuhns. **Informal name:** Kuhns Papuan Typical Mosquito.

syn. *mercedesae* Baisas
1974: 121 (M, F, P, L; tax., bion., distr.; Philippines; *uniformis* ssp.). **Type loc.:** Mount Makiling, Luzon, Philippines (LU). **Refs.:** Sirivanakarn 1977a: 107 (syn.).

kusaiensis Bohart
1957: 69 (M*, F, P, L*). **Type loc.:** Weye Cave, Kusaie Island, Caroline Islands (USNM). **Distr.:** Micronesia (Kusaie Isl., Wake Isl.). **Refs.:** Lee et al. 1989b: 127 (tax., distr.). **Ety.:** Kusaie Isl. [now Kosrae], Micronesia. **Informal name:** Kusaie Micronesian Typical Mosquito.

laffooni Belkin
1962: 261 (M*, F, L*). **Type loc.:** Halavo, Florida, Solomon Islands (USNM). **Distr.:** Solomon Isl. **Refs.:** Lee et al. 1989b: 128 (F; key, tax., bion., distr., review). **Ety.:** Jean Luther Laffoon. **Informal name:** Laffoon Solomon Islander Typical Mosquito.

lairdi Belkin
1962: 257 (M*, F, P*, L*). **Type loc.:** Poha swamp, Guadalcanal, Solomon Islands (USNM). **Distr.:** Papua New Guinea, Solomon Isl. **Refs.:** Lee et al. 1989b: 129 (F; key, tax., bion., distr., review). **Ety.:** Marshall Laird. **Informal name:** Laird Solomon Islander Typical Mosquito.

lakei Sirivanakarn
1968: 120 (M*, F, P*, L*). **Type loc.:** Rigo Road, Port Moresby, SE New Guinea (BBM). **Distr.:** Papua New Guinea. **Refs.:** Lee et al. 1989b: 131 (F; key, tax., bion., distr., review). **Ety.:** Harry Lake. **Informal name:** Lake Papuan Typical Mosquito.

lasiopalpis Sirivanakarn
1977a: 123 (M*, F, P*, L). **Type loc.:** Matale, Matale District, Central Province, Sri Lanka (USNM). **Distr.:** India, Sri Lanka. **Refs.:** Howard et al. 1913, Figs. 277, 466, 549 (L*; as *trinidadensis*). **Ety.:** n.s. [hairy feeler (L); poss. ref. to male with "numerous densely packed hairlike setae on lateral, ventral and mesal surfaces of palpal segments 2 and 3"]. **Informal name:** Hairy-palped Sri Lankan Typical Mosquito.

lavatae Stone & Bohart
1944: 220 (M*). **Type loc.:** Los Banos [Laguna], Luzon, Philippines (USNM). **Distr.:** Malaysia, Philippines. **Refs.:** Bohart 1945: 75 (M*); Basio 1971b: 66 (M*; bion.); Baisas 1974: 112 (M*, F, P*, L*; tax., bion., distr.; Philippines); Sirivanakarn 1977a: 86 (M*, F, P, L; keys, tax., bion., distr.). **Ety.:** n.s. [wash, bathe (L); poss. ref. to type loc., Los Banos, Philippines]. **Informal name:** Bath Philippine Typical Mosquito.

leei King & Hoogstraal
1955: 4 (M*, F, L). **Type loc.:** "Roseboompond," Hollandia, Western New Guinea (Papua) (USNM). **Distr.:** Indonesia. **Refs.:** Sirivanakarn 1968: 138 (M*, F, P*, L*; keys, tax., bion., distr.); Lee et al. 1989b: 132 (F; key, tax., bion., distr., review). **Ety.:** Vernon Harold Lee. **Informal name:** Lee Indonesian Typical Mosquito.

lucaris Colless
1965: 299 (M*). **Type loc.:** Ulu Pandan, Singapore (ANIC). **Distr.:** Malaysia, Singapore, Thailand. **Refs.:** Bram 1967b: 79 (M*, F, L*; bion., distr., tax.); Sirivanakarn 1977a: 138 (M*, F, P, L; keys, tax., bion., distr.); Rattanarithikul et al. 2005b (F*, L*; bion., distr., keys). **Ety.:** n.s. [sacred grove (L)]. **Informal name:** Sacred Singaporean Typical Mosquito.

macdonaldi Colless
1965: 276 (M*, L*). **Type loc.:** Singapore (ANIC). **Distr.:** Cambodia, India, Indonesia, Malaysia, Philippines, Singapore, Thailand, Vietnam. **Refs.:** Delfinado 1966a: 108 (M*, P*, L*); Bram 1967b: 61 (M*, L*; bion., distr., tax.); Basio 1971b: 66 (M*; bion.); Sirivanakarn 1977a: 61 (M*, F*, P*, L*; keys, tax., bion., distr.); Rattanarithikul et al. 2005b (F*, L*; bion., distr., keys). **Ety.:** W.W. Macdonald. **Informal name:** Macdonald Singaporean Typical Mosquito.

mammilifer (Leicester)
1908: 128 (M, F; *Lophoceratomyia*). **Type loc.:** Raub [Pahang] and Kuala Lumpur [Selangor], Malaya [Malaysia] (NHMUK). **Distr.:** Cambodia, India, Indonesia, Malaysia, People's Republic of China, Philippines, Sri Lanka, Thailand. **Refs.:** Barraud 1934: 374 (M*, F, L*); Baisas 1935d: 174 (M*); Colless 1965: 287 (M*, F, L*; syn.); Delfinado 1966a: 109 (M*, P*, L*); Bram 1967b: 93 (M*, F, L*; bion., distr., tax.); Basio 1971b: 66 (M*; bion.); Baisas 1974: 114 (M, F, P, L; tax., bion., distr.; Philippines); Sirivanakarn 1977a: 87 (M*, F*, P*, L*; keys, tax., bion., distr.); Bhattacharyya et al. 2003 (distr.; India); Rattanarithikul et al. 2005b (F*, L*; bion., distr., keys). **Ety.:** mammal (L); ref. to teat-like projection on upper angle on the inner side of the antenna. **Informal name:** Mammal Malaysian Typical Mosquito.
syn. *chiungchungensis* Hsu
1963: 229 (M*, F, L*). **Type loc.:** Chiungchungsang, Hainan Island, Kwangtung Province, China [People's Republic of China] (LU).

marksae King & Hoogstraal
1955: 6 (M*). **Type loc.:** Hollandia, Western New Guinea (Papua) (USNM). **Distr.:** Indonesia (includes West Papua), Singapore. **Refs.:** van den Assem 1958c: 218 (F, L*); Sirivanakarn 1968: 123 (M*, L); Lee et al. 1989b: 133 (tax., distr.). **Ety.:** Elisabeth Nesta Marks. **Informal name:** Marks Indonesian Typical Mosquito.

minjensis Sirivanakarn
1968: 151 (M*, F). **Type loc.:** Tengen stream, Minj, Western Highlands, NE New Guinea (BBM). **Distr.:** Papua New Guinea. **Refs.:** Lee et al. 1989b: 135 (F; key, tax., bion., distr., review). **Ety.:** Minj, Western Highlands, Papua New Guinea. **Informal name:** Minj Papuan Typical Mosquito.

minor (Leicester)
1908: 126 (M, F; *Lophoceratomyia*). **Type loc.:** [Kuala Lumpur, Selangor], Malaya [Malaysia] (NHMUK). **Distr.:** Cambodia, India, Indonesia, Japan, Malaysia, Myanmar, Nepal, People's Republic of China, Philippines, Thailand, Vietnam. **Refs.:** Barraud 1934: 370 (M*, L); Mattingly 1949c: 227 (L*; tax.); Colless 1965: 289 (M*, F, L*; syn.); Bram 1967b: 96 (M*, F, L*; bion., distr., tax.); Basio 1971b: 67 (M*; bion.); Baisas 1974: 115 (M*, F, P*, L*; tax., bion., distr.; Philippines); Sirivanakarn 1977a: 98 (M*, F, P*, L*; keys, tax., bion., distr.); Lee et al. 1989b: 136 (F; key, tax., bion., distr., review); Rattanarithikul et al. 2005b (F*, L*; bion., distr., keys). **Ety.:** smaller (L); small species. **Informal name:** Minor Malaysian Typical Mosquito.
syn. *plantaginis* Barraud
1924j: 46 (M*). **Type loc.:** Koti, near Kasauli [Punjab Hill States], India (NHMUK). **Refs.:** Peters & Dewar 1956: 44 (L*?); Sirivanakarn 1977a: 98 (syn.; lectotype desig.).
syn. *nolledoi* Baisas
1935d: 170 (M*, F). **Type loc.:** Kolambugan, Lanao Province, Mindanao, Philippines (NE; paratypes USNM). **Refs.:** Colless 1965: 289 (syn.).
syn. *harpagophallus* Wang & Feng
1964: 38 (M*). **Type loc.:** Szemao, Yunnan, China [People's Republic of China] (MMC). **Distr.:** People's Republic of China,. **Refs.** Sirivanakarn 1977a: 9 (tax.); Meng & Chen 1980 (syn.).

minutissimus (Theobald)
1907: 235 (F; *Culiciomyia*). **Type loc.:** Peradeniya [Central], Ceylon [Sri Lanka] (NHMUK). **Distr.:** Bangladesh, India, Indonesia, Malaysia, Maldives, Pakistan, People's Republic of China, Sri Lanka, Thailand. **Refs.:** Barraud 1934: 370 (M*, L); Mattingly 1949c: 227 (L*); Iyengar & Menon 1955 (L*); Joshi et al. 1965 (distr.; Nepal); Aslamkhan 1971b (distr.; Pakistan); Lien et al. 1977 (distr.); Sirivanakarn 1977a: 22 (M*, F, P, L; keys, tax., bion., distr.); Ahmed 1987 (distr.; Bangladesh); Darsie & Pradhan 1990: 72 (distr.); Rattanarithikul et al. 2005b (F*, L*; bion., distr., keys). **Ety.:** very small (L). **Informal name:** Minute Sri Lankan Typical Mosquito.

syn. *nigerrima* Theobald
1910c: 233 (F*; *Culiciomyia*). **Type loc.:** Trincomalee, Ceylon [Sri Lanka] (NHMUK).
syn. *juxtapallidiceps* Theobald
1910c: 456 (F; *Melanoconion*). **Type loc.:** Trincomalee, Ceylon [Sri Lanka] (NHMUK).

muruae Sirivanakarn
1968: 130 (M*, F, P*, L*). **Type loc.:** Murua, Papua New Guinea (BBM). **Distr.:** Papua New Guinea. **Refs.:** Lee et al. 1989b: 138 (F; key, tax., bion., distr., review). **Ety.:** Murua, Papua New Guinea. **Informal name:** Murua Papuan Typical Mosquito.

navalis Edwards
1926a: 121 (M, L). **Type loc.:** Singapore (NHMUK). **Distr.:** Indonesia, Malaysia, Singapore. **Refs.:** Edwards & Given 1928: 353 (L*); Mattingly 1949c: 226 (L*); Colless 1965: 304 (M*, F, L); Sirivanakarn 1977a: 127 (M*, F*, P*, L*; keys, tax., bion., distr.; lectotype desig.). **Ety.:** n.s. [naval? (L)]. **Informal name:** Naval Singaporean Typical Mosquito.

niger (Leicester)
1908: 123 (M, F; *Lophoceratomyia*). **Type loc.:** Kuala Lumpur [Selangor], Malaya [Malaysia] (NHMUK). **Distr.:** Malaysia. **Refs.:** Colless 1965: 282 (M*, F, L); Sirivanakarn 1977a: 40 (M*, F, P*, L*; keys, tax., bion., distr.). **Ety.:** black, dark (L); "A very sombre-coloured mosquito." **Informal name:** Somber Malaysian Typical Mosquito.
syn. *atratulus* Edwards
1922a: 287 (new name for *niger* Leicester 1908, in error).

orbostiensis Dobrotworsky
1958a: 317 (M*, F, P*, L*, E*). **Type loc.:** Orbost, Victoria, Australia (NMM). **Distr.:** Australia. **Refs.:** Dobrotworsky 1965: 204 (M*, F*, L*); Lee et al. 1989b: 139 (F; key, tax., bion., distr., review). **Ety.:** Orbost, Shire of East Gippsland, Victoria, Australia. **Informal name:** Orbost Australian Typical Mosquito.

oweni Belkin
1962: 259 (M*, F, P*, L*). **Type loc.:** Tributary of Matanikau River, Guadalcanal, Solomon Islands (USNM). **Distr.:** Solomon Isl. **Refs.:** Lee et al. 1989b: 145 (F; key, tax., bion., distr., review). **Ety.:** William B. Owen. **Informal name:** Owen Solomon Islander Typical Mosquito.

pairoji Sirivanakarn
1977a: 63 (M, F, P, L*). **Type loc.:** Khao Yai National Park, Nakhon Ratchasima, Thailand (USNM). **Distr.:** Indonesia, Malaysia, Singapore, Thailand. **Refs.:** Rattanarithikul et al. 2005b (F*, L*; bion., distr., keys). **Ety.:** Pairoj Suwanakorn. **Informal name:** Pairoj Thai Typical Mosquito.

paraculeatus Sirivanakarn
1977a: 49 (M*). **Type loc.:** Tawau, Sabah, Malaysia [Malaysian Borneo] (NHMUK). **Distr.:** Malaysia (Borneo (Sabah)), Philippines. **Ety.:** like *aculeatus*. **Informal name:** Bag-like Malaysian Typical Mosquito [*paraculeatus* = "near bag" = Bag-like].

perryi Belkin
1962: 268 (M*, F, P*, L*). **Type loc.:** Balasuma area, Guadalcanal, Solomon Islands (USNM). **Distr.:** Solomon Isl. **Refs.:** Lee et al. 1989b: 146 (F; key, tax., bion., distr., review). **Ety.:** William J. Perry. **Informal name:** Perry Solomon Islander Typical Mosquito.

petersi Colless
1960: 388 (M*, F, L*). **Type loc.:** Minj, Western Highlands, New Guinea (CSIR). **Distr.:** Indonesia, Papua New Guinea. **Refs.:** Sirivanakarn 1968: 109 (M*, F, P*, L*); Lee et al. 1989b: 147 (F; key, tax., bion., distr., review). **Ety.:** W. Peters. **Informal name:** Peters Indonesian Typical Mosquito.

peytoni Bram & Rattanarithikul
1967: 7 (M*, F, L*). **Type loc.:** Phatthalung-Trang Road, Phatthalung Province, Thailand (USNM). **Distr.:** Cambodia, India, Indonesia, Malaysia, Thailand, Vietnam. **Refs.:** Bram 1967b: 100b (M*, F, L*; bion., distr., tax.); Sirivanakarn 1977a: 114 (M*, F, P*, L*; keys, tax., bion., distr.); Rattanarithikul et al. 2005b (F*, L*; bion., distr., keys). **Ety.:** EL Peyton. **Informal name:** Peyton Thai Typical Mosquito.

pholeter Bram & Rattanarithikul
1967: 13 (M*, F, L*). **Type loc.:** Khao Mai Ha Wa, Chon Buri Province, Thailand (USNM). **Distr.:** India, Thailand. **Refs.:** Bram 1967b: 103 (M*, F, L*; bion., distr., tax.); Sirivanakarn 1977a: 118 (M, F, P*, L*; keys, tax., bion., distr.). **Ety.:** one who lurks in a hole (Gr); ref. to larval habitat (crabholes). **Informal name:** Hole-dwelling Thai Typical Mosquito.

pilifemoralis Wang & Feng
1964: 37 (M*). **Type loc.:** Szemao, Yunnan, China [People's Republic of China] (MMC). **Distr.:** India, People's Republic of China, Thailand. **Refs.:** Sirivanakarn 1977a: 150 (M*, F, P, L; keys, tax., bion., distr.); Rattanarithikul et al. 2005b (F*, L*; bion., distr., keys). **Ety.:** hairy femur (L); distal dorsal portion of femur of foreleg with 10–12 setae arranged closely in a row. **Informal name:** Hairy-legged Chinese Typical Mosquito.

pseudornatus Colless
1960: 386 (M*, F). **Type loc.:** Edie Creek, New Guinea (CSIR). **Distr.:** Papua New Guinea. **Refs.:** Sirivanakarn 1968: 149 (M*, F, L*); Lee et al. 1989b: 148 (F; key, tax., bion., distr., review). **Ety.:** false (Gr); similar to *ornatus*. **Informal name:** Gilded Papuan Typical Mosquito [*pseudornatus* = "false ornate" = Gilded].

pseudorubithoracis Sirivanakarn
1968: 178 (M*). **Type loc.:** Korogo, Sepik River, NE New Guinea (LU). **Distr.:** Papua New Guinea. **Refs.:** Lee et al. 1989b: 150 (tax., bion., distr., review). **Ety.:** false (Gr);

similar to *rubithoracis*. **Informal name:** Garnet Papuan Typical Mosquito [*pseudorubi* = "false red" = false ruby = Garnet].

quadripalpis (Edwards)
1914b: 80 (M; *Lophoceraomyia*). **Type loc.:** Sarawak, Borneo [Malaysia] (NHMUK). **Distr.:** Cambodia, India, Indonesia, Malaysia (includes Borneo (Sarawak)), Philippines, Singapore, Sri Lanka, Thailand, Vietnam. **Refs.:** Mattingly 1949c: 227 (L*; tax.); Bram 1967b: 65 (M*, F, L*; bion., distr., tax.); Basio 1971b: 67 (M*; bion.; as *pachecoi*); Sirivanakarn 1977a: 44 (M*, F*, P*, L*; keys, tax., bion., distr.); Bhattacharyya et al. 2003 (distr.; India); Rattanarithikul et al. 2005b (F*, L*; bion., distr., keys). **Ety.:** n.s. [four palpi (L); prob. ref. to "distinct finger-like process at the base of each palp" poss. giving the impression of four palpi]. **Informal name:** Four-lipped Malaysian Typical Mosquito.
syn. *sylvestris* Leicester
1908: 125 (M, F; *Lophoceratomyia*; not Theobald 1901a). **Type loc.:** Malaya [Malaysia] (NHMUK). **Refs.:** Townsend 1990: 139 (type info.).
syn. *roubaudi* Borel
1927: 112 (M*, L*; *Lophoceratomyia*). **Type loc.:** Terres Rouges, Cochin-China, Indochina [southern Vietnam] (PIP). **Refs.:** Harrison 1973c: 278 (type info.).
syn. *pachecoi* Baisas
1935d: 171 (M*, F). **Type loc.:** College of Agriculture, Los Banos, Laguna Province, Luzon, Philippines (NE; paratypes USNM). **Refs.:** Sirivanakarn 1977a: 44 (syn.).

raghavanii Rahman, Chowdhury, & Kalra
1969: 28 (M*, F, P*, L*). **Type loc.:** Coonoor, Singara Estate, Nilgiri Hills, South India (NICD). **Distr.:** India. **Ety.:** N.G.S. Raghavan. **Informal name:** Raghavan Indian Typical Mosquito.

rajaneeae Sirivanakarn
1968: 174 (M*). **Type loc.:** Lorengau, Manus Island, Bismarck Archipelago (BBM). **Distr.:** Papua New Guinea. **Refs.:** Lee et al. 1989b: 151 (tax., bion., distr., review). **Ety.:** Rajanee Sirivanakarn. **Informal name:** Rajanee Papuan Typical Mosquito.

reidi Colless
1965: 279 (M*, F, L*). **Type loc.:** West Coast Road, Singapore (ANIC). **Distr.:** Cambodia, Indonesia, Malaysia, Philippines, Singapore, Thailand. **Refs.:** Bram 1967b: 66 (M*, F, L*; bion., distr., tax.); Basio 1971b: 68 (M*; bion.); Baisas 1974: 117 (M, F, P, L; tax., bion., distr.; Philippines); Lien et al. 1975 (distr.); Sirivanakarn 1977a: 51 (M*, F, P, L*; keys, tax., bion., distr.); Rattanarithikul et al. 2005b (F*, L*; bion., distr., keys). **Ety.:** John Alexander Reid. **Informal name:** Reid Singaporean Typical Mosquito.

rubithoracis (Leicester)
1908: 119 (M, F; *Lophoceratomyia*). **Type loc.:** Kuala Lumpur [Selangor], Malaya [Malaysia] (NE). **Distr.:** Cambodia, Guam, India, Indonesia, Japan, Malaysia, Myanmar, People's Republic of China (includes Hong Kong), Philippines, Singapore, Sri Lanka, Taiwan, Thailand, Vietnam. **Refs.:** Baisas 1947: 255 (P, L); Dantis 1947; Mattingly 1949c: 224 (L*); La Casse & Yamaguti 1950: 192 (M*, F*, P*, L*; bion., distr., tax.); Hara 1957: 57 (F*); Bram 1967a: 69 (M*, F, L*; bion., distr., tax.); Basio 1971b: 68 (M*; bion.); Baisas 1974: 119 (M, F, P, L; tax., bion., distr.; Philippines); Tanaka et al. 1975c: 217 (distr.); Sirivanakarn 1977a: 37 (M*, F*, P*, L*; keys, tax., bion., distr.); Tanaka et al. 1979: 204 (M*, F*, L*); Lee et al. 1989b: 152 (tax., bion., distr., review); Rattanarithikul et al. 2005b (F*, L*; bion., distr., keys). **Ety.:** red thorax (L); ref. to "a conspicuously reddish-brown thorax." **Informal name:** Ruby-chested Malaysian Typical Mosquito.

sangengluoensis Wang
1984: 217 (M*, F*, L*). **Type loc.:** Wanning, Hainan, Sangenluo Commune, China [People's Republic of China] (TMMC). **Distr.:** People's Republic of China. **Refs.:** Wang 1984 (M*, F, L*). **Ety.:** Wanning, Hainan, Sangengluo Commune, China. **Informal name:** Sangengluo Chinese Typical Mosquito.

schilfgaardei Sirivanakarn
1968: 163 (M*, F, P*, L*). **Type loc.:** Singuawa River area, Morobe, NE New Guinea (BBM). **Distr.:** Papua New Guinea. **Refs.:** Lee et al. 1989b: 154 (F; key, tax., bion., distr., review). **Ety.:** R. Schilfgaarde. **Informal name:** Schilfgaarde Papuan Typical Mosquito.

sedlacekae Sirivanakarn
1968: 180 (M*). **Type loc.:** Lae, Morobe, New Guinea (BBM). **Distr.:** Papua New Guinea. **Refs.:** Lee et al. 1989b: 155 (tax., distr.). **Ety.:** Marie Sedlacek. **Informal name:** Sedlacek Papuan Typical Mosquito.

seniori Barraud
1934: 365 (M*). **Type loc.:** Garden Reach, Calcutta [Bengal], India (NHMUK). **Distr.:** India. **Refs.:** Sirivanakarn 1977a: 31 (M*; key, tax., bion., distr.). **Ety.:** Ronald Senior-White. **Informal name:** Senior-White Indian Typical Mosquito.

shanahani Sirivanakarn
1968: 117 (M*, F*). **Type loc.:** Cape Killerton, Popondetta, SE New Guinea (BBM). **Distr.:** Indonesia, Papua New Guinea. **Refs.:** Lee et al. 1989b: 156 (F; key, tax., bion., distr., review). **Ety.:** P. Shanahan. **Informal name:** Shanahan Papuan Typical Mosquito.

singhbhumensis Natarajan & Rajavel
2009: 403 (M*, F, P*, L*). **Type loc.:** Orissa, Keonjhar District, Mamalapusi, India (USNM). **Distr.:** India. **Ety.:** Singhbhum Hills, India. **Informal name:** Singhbhum Indian Typical Mosquito.

singuawaensis Sirivanakarn
1969: 119 (new name for *confusus* Sirivanakarn, not Baisas 1938). **Distr.:** Papua New Guinea. **Refs.:** Lee et al.

1989b: 157 (F; key, tax., bion., distr., review). **Ety.:** Singuawa River area, New Guinea. **Informal name:** Singuawa Papuan Typical Mosquito.

syn. *confusus* Sirivanakarn
1968: 105 (M*). **Type loc.:** Singuawa River area, Morobe, Lae, NE New Guinea (BBM).

solomonis Edwards
1929 (M, F, L; *fraudatrix* var.). In Paine & Edwards 1929: 316. **Type loc.:** Manaba, Malaita, Solomon Islands. (NHMUK). **Distr.:** Indonesia, Solomon Isl. **Refs.:** Mattingly 1949c: 227 (L*); Belkin 1962: 262 (M*, F, P*, L*; lectotype desig.; to sp.); Sirivanakarn 1973a: 215 (distr.); Lee et al. 1989b: 159 (tax., bion., distr., review). **Ety.:** Solomon Isl. **Informal name:** Malaita Solomon Islander Typical Mosquito.

spiculosus Bram & Rattanarithikul
1967: 3 (M*, F, L*). **Type loc.:** Doi Sam Sao, Tak Province, Thailand (USNM). **Distr.:** Cambodia, Indonesia, Malaysia, Myanmar, Taiwan, Thailand. **Refs.:** Bram 1967b: 105 (M*, F, L*; bion., distr., tax.); Ramalingam & Pillai 1973 (distr.); Sirivanakarn 1977a: 96 (M*, F, P, L*; keys, tax., bion., distr.); Miyagi et al. 1994: 21 (distr.); Rattanarithikul et al. 2005b (F*, L*; bion., distr., keys). **Ety.:** small spikes (L); ref. to spiculate larval thorax and abdomen. **Informal name:** Spiky Thai Typical Mosquito.

syn. *hui* Lien
1968d: 227 (M*, F, P*, L*). **Type loc.:** Chukou, Fanlu, Chiai Hsien, Taiwan (TMRI). **Refs.:** Sirivanakarn 1977a: 96 (syn.).

steffani Sirivanakarn
1968: 112 (M*, F, P*, L*). **Type loc.:** Rigo Road, Port Moresby, Central District, SE New Guinea (BBM). **Distr.:** Papua New Guinea. **Refs.:** Lee et al. 1989b: 161 (F; key, tax., bion., distr., review). **Ety.:** Wallace Allan Steffan. **Informal name:** Steffan Papuan Typical Mosquito.

submarginalis Sirivanakarn
1973a: 212 (M*). **Type loc.:** Gusika, Finschhafen, Madang, NE New Guinea (USNM). **Distr.:** Papua New Guinea. **Refs.:** Lee et al. 1989b: 162 (tax., distr.). **Ety.:** near margin (L); ref. to "an irregular row of 5–6 strong submarginal setae on the sidepiece [gonocoxite]." **Informal name:** Submarginal Papuan Typical Mosquito.

szemaonensis Wang & Feng
1964: 37 (M*). **Type loc.:** Szemao, Yunnan, China [People's Republic of China] (MMC). **Distr.:** People's Republic of China. **Refs.:** Belkin 1962: 475 (M*, F, P*, L*; as *tongae*). **Ety.:** Szemao, People's Republic of China. **Informal name:** Szemao Chinese Typical Mosquito.

traubi Colless
1965: 295 (M*, F, L*). **Type loc.:** Ulu Gombak, 16th mile, Selangor, Malaya [Malaysia] (ANIC). **Distr.:** Indonesia, Malaysia, Thailand. **Refs.:** Bram 1967b: 108 (M*, F, L*; bion., distr., tax.); Sirivanakarn 1977a: 82 (M*, F, P*, L*; keys, tax., bion., distr.); Rattanarithikul et al. 2005b (F*, L*; bion., distr., keys). **Ety.:** Robert Traub. **Informal name:** Traub Malaysian Typical Mosquito.

tuberis Bohart
1946b: 42 (M*, L). **Type loc.:** Chizuka, Okinawa, Ryukyu-Retto, Japan (USNM). **Distr.:** Japan, Thailand. **Refs.:** Bohart & Ingram 1946b: 74 (M*, L*; bion., distr., tax.); Bohart 1953: 185 (F, P, L); Tanaka et al. 1975c: 217 (bion., distr.); Sirivanakarn 1977a: 105 (M*, F, P, L; keys, tax., bion., distr.); Tanaka et al. 1979: 21 (M*, F*, L*); Rattanarithikul et al. 2005b (F*, L*; bion., distr., keys). **Ety.:** (n.s.) [protuberance (L); prob. ref. to torus (pedicel) "with a pronounced knob at upper inner angle"]. **Informal name:** Knobby Japanese Typical Mosquito.

uniformis (Theobald)
1905h: 245 (M*, F; *Lophoceraomyia*). **Type loc.:** Peradeniya [Central], Ceylon [Sri Lanka] (NHMUK). **Distr.:** India, Malaysia, Philippines, Sri Lanka. **Refs.:** Barraud 1934: 373 (M*, F, L); Mattingly 1949c: 226 (L*); Penn 1949b: 71 (P*); Basio 1971b: 68 (M*; bion.); Baisas 1974 (M*, L*); Sirivanakarn 1977a: 84 (M*, F, P, L*; keys, tax., bion., distr., lectotype desig); Lee et al. 1989b: 163 (tax., bion., distr., review). **Ety.:** n.s. [uniform (L)]. **Informal name:** Uniform Sri Lankan Typical Mosquito.

variatus (Leicester)
1908: 121 (M, F; *Lophoceratomyia*). **Type loc.:** [Kuala Lumpur, Selangor], Malaya [Malaysia] (NHMUK). **Distr.:** Cambodia, India, Indonesia, Malaysia, People's Republic of China, Singapore, Thailand, Vietnam. **Refs.:** Colless 1965: 270 (M*, F, L*; from syn.; lectotype desig.); Bram 1967b: 72 (M*, F, L*; bion., distr., tax.); Lien et al. 1975 (distr.); Sirivanakarn 1977a: 53 (M*, F*, P*, L*; keys, tax., bion., distr.); Rattanarithikul et al. 2005b (F*, L*; bion., distr., keys). **Ety.:** vary, variegate (L); "A rather variable mosquito." **Informal name:** Variant Malaysian Typical Mosquito.

versabilis Sirivanakarn
1968: 126 (M*, F, P*, L*). **Type loc.:** Madang, Gum River area, NE New Guinea (BBM). **Distr.:** Papua New Guinea. **Refs.:** Lee et al. 1989b: 165 (F; key, tax., bion., distr., review). **Ety.:** n.s. [changeable, adaptable (L); prob. ref. to "considerable variation in all stages"]. **Informal name:** Versatile Papuan Typical Mosquito.

walukasi Belkin
1962: 264 (M*, F, P*, L*). **Type loc.:** Sprague swamp, Guadalcanal, Solomon Islands (USNM). **Distr.:** Papua New Guinea, Solomon Isl. **Refs.:** Lee et al. 1989b: 166 (F; key, tax., bion., distr., review). **Ety.:** Leo Walukas. **Informal name:** Walukas Solomon Islander Typical Mosquito.

wamanguae Sirivanakarn
1968: 135 (M*, F, P*, L*). **Type loc.:** Wamangu, inland of Wewak, Sepik, NE New Guinea (BBM). **Distr.:** Papua New

Guinea. **Refs.:** Lee et al. 1989b: 168 (F; key, tax., bion., distr., review). **Ety.:** Wamangu, Sepik Prov., Papua New Guinea. **Informal name:** Wamangu Papuan Typical Mosquito.

wardi Sirivanakarn
1977a: 90 (M*, F, P, L). **Type loc.:** Kanneliya, Galle District, Sabaragamuwa Province, Sri Lanka (USNM). **Distr.:** Sri Lanka. **Ety.:** Ronald Anthony Ward. **Informal name:** Ward Sri Lankan Typical Mosquito.

whartoni Colless
1965: 275 (M*, F, L*). **Type loc.:** Singapore (ANIC). **Distr.:** Indonesia, Malaysia, Singapore, Thailand. **Refs.:** Sirivanakarn 1977a: 59 (M*, F*, P, L; keys, tax., bion., distr.); Rattanarithikul et al. 2005b (F*, L*; bion., distr., keys). **Ety.:** Ronald Harry Wharton. **Informal name:** Wharton Singaporean Typical Mosquito.

wilfredi Colless
1965: 297 (M*). **Type loc.:** Malaya [Malaysia] (ANIC). **Distr.:** Cambodia, India, Malaysia, People's Republic of China, Thailand, Vietnam. **Refs.:** Bram 1967b: 110 (M*, F, L*; bion., distr., tax.); Sirivanakarn 1977a: 147 (M*, F*, P*, L*; keys, tax., bion., distr.); Rattanarithikul et al. 2005b (F*, L*; bion., distr., keys). **Ety.:** Wilfred Chellapah. **Informal name:** Wilfred Malaysian Typical Mosquito.

winkleri Belkin
1962: 260 (M*, F, P*, L*). **Type loc.:** tributary of Matanikau River, Guadalcanal, Solomon Islands (USNM). **Distr.:** Australia, Indonesia (West Papua), Solomon Isl. **Refs.:** Lee et al. 1989b: 169 (tax., bion., distr., review). **Ety.:** Eric C. Winkler. **Informal name:** Winkler Solomon Islander Typical Mosquito.

Subgenus *Maillotia* Theobald

1907: 274. Type species: *Maillotia pilifera* Theobald.

Name derivation: n.s. [poss. ref. to Lionel Maillot; ref. in Evenhuis (2013) for Diptera author names.

Important references: Sirivanakarn 1971a: 71 (to subg.); Mallampalli 1995 (phyl.); Harbach et al. 2012 (phyl.); Becker et al. 2010 (tax., bion., distr.; Europe).

arbieeni Salem
1938: 20 (M*, F, P*, L*). **Type loc.:** Sinai, Egypt (LU). **Distr.:** Algeria, Egypt, Iran, Israel, Jordan, Spain (includes Canary Isl.), Sudan and South Sudan, Yemen. **Refs.:** Hopkins 1952: 251 (L*); Mattingly 1955a: 381 (M*, P, L); Lotfi 1973: 206 (distr.). **Ety.:** Valley of El-Arbieen, Southern Sinai. **Informal name:** El-Arbieen Egyptian Typical Mosquito.

syn. *jenkinsi* Knight
1953b: 323 (M*, F). **Type loc.:** El-Hauban, Wadi el Malah, Yemen (USNM). **Refs.:** Mattingly 1955a: 381 (syn.).

avianus De Meillon
1943: 96 (M*, F*, L*). **Type loc.:** Cathedral Peak, Drakensberg, Natal [KwaZulu-Natal, Republic of South Africa] (SAIM). **Distr.:** Madagascar (includes Glorioso and Juan De Nova Isl.), Republic of South Africa. **Refs.:** Jupp 1996 (M*, F*; key). **Ety.:** n.s. [pertaining to a bird (L)]. **Informal name:** Avian South African Typical Mosquito.

deserticola Kirkpatrick
1925a: 378 (M*, F, L*). **Type loc.:** Wadi No'oth, North Galala Mountains, Eastern Desert, Egypt (DAC, ESE, NHMUK). **Distr.:** Afghanistan, Algeria, Egypt, Iran, Iraq, Israel (and Gaza Strip and West Bank), Jordan, Libya, Morocco, Spain, Sudan and South Sudan, Syria, Tunisia, Turkey. **Refs.:** Senevet 1949: 53 (M*); Sicart 1952b: 149 (L*); Mattingly 1955a: 385 (M*; tax.); Senevet et al. 1955: 51 (P*; tax.); Senevet & Andarelli 1959: 107 (M*, F*, P*, L*); Lotfi 1970: 401 (distr.); Margalit & Tahori 1974: 86 (distr.); Harbach 1985a: 84 (from subg. *Neoculex*); Harbach et al. 1988: 320 (type info.); Townsend 1990: 67 (type info.); Amr et al. 1997 (distr.; Jordan); Alten et al. 2000 (distr.; Turkey); Eritja 2000: 12 (distr.; Spain); da Cunha Ramos, Lucientes et al. 2002 (M*; tax., distr.); Rueda, Pecor et al. 2008: 397 (distr.); Gunay et al. 2015 (mol. tax., distr. Turkey); Robert et al. 2019 (distr.; Western Palearctic). **Ety.:** waste (L); desert inhabiting. **Informal name:** Desert-dwelling Egyptian Typical Mosquito.

hortensis Ficalbi
1889a: 27 (M, F). **Type loc.:** Pisa, Siena, and Firenze, Tuscany, Italy (NHMUK). **Distr.:** Afghanistan, Albania, Algeria, Andorra, Armenia, Austria, Azerbaijan, Belgium, Bosnia and Herzegovina, Bulgaria, Crimean Peninsula, Cyprus, Czech Republic, France (includes Corsica), FYRO Macedonia, Georgia, Germany, Greece, Hungary, India, Iran, Iraq, Israel (and Gaza Strip and West Bank), Italy (includes Sardinia and Sicily), Kosovo, Lebanon, Liechtenstein, Malta, Morocco, Poland, Portugal (includes Madeira), Romania, Russia, Serbia, Slovakia, Slovenia, Spain (includes Balearic Isl. and Canary Isl.), Switzerland, Tajikistan, Tunisia, Turkey, Ukraine. **Refs.:** Séguy 1924: 47, Pl. X (L*), 46 (M gen.*); Martini 1931c: 366 (M*, F, L*); Senevet 1936b: 437 (P*); Callot 1943: 129 (F*, L*); Senevet 1949: 52 (M*); Sicart 1952b: 149 (L*); Mattingly 1955a: 385 (M*; tax.); Abul-Hab 1969: 246 (distr.); Bhat & Kulkarni 1971: 583 (distr.); Gutsevich et al. 1974: 359 (M*, F, L*); Townsend 1990: 84 (type info.); Alten et al. 2000 (distr.; Turkey); Trari et al. 2002: 331 (distr.; Morocco); Rueda, Pecor et al. 2008: 397 (distr.); Becker et al. 2010: 282 (M*, F*, L*; keys, tax., distr., bion.); Osório et al. 2010 (distr.; Portugal); Namazov 2014 (distr.; Azerbaijan); Gunay et al. 2015 (mol. tax., distr.; Turkey); Robert et al. 2019 (distr.; Western Palearctic). **Ety.:** garden (L); larva found in botanical gardens. **Informal name:** Garden Italian Typical Mosquito.

syn. *pilifera* Theobald
1907: 274 (F; *Maillotia*). **Type loc.:** Algeria (NHMUK).

syn. *lavieri* Larrousse
1925: 68 (L*). **Type loc.:** Bornand, Haute Savoie, France (IPS). **Refs.:** Callot 1944b: 196 (syn.).

ssp. ***maderensis*** Mattingly
1955a: 377 (M*, F, P, L). **Type loc.:** Monte, Madeira, Madeira Islands (NHMUK). **Distr.:** Portugal. **Ety.:** Madère (Madeira, Portugal). **Informal name:** Madeira Portuguese Typical Mosquito.

peringueyi Edwards
1924a: 162 (M). **Type loc.:** Cape Town, Cape Province [South Africa] (SA). **Distr.:** South Africa. **Refs.:** Edwards 1941: 255 (M*, F*); Muspratt 1955: 178 (M*, L*); Jupp 1996 (M*, F*; key). **Ety.:** Louis Péringuey. **Informal name:** Péringuey South African Typical Mosquito.

quettensis Mattingly
1955a: 374 (M*, F). **Type loc.:** Quetta, Baluchistan, Pakistan (NHMUK). **Distr.:** Afghanistan, Pakistan. **Refs.:** Rueda, Pecor et al. 2008: 397 (distr.). **Ety.:** Quetta, Pakistan. **Informal name:** Quetta Pakistani Typical Mosquito.

salisburiensis Theobald
1901c: 112 (F*). **Type loc.:** Salisbury, Mashonaland [Harare, Zimbabwe] (NHMUK). **Distr.:** Democratic Republic of the Congo, Kenya, Lesotho, Madagascar (includes Glorioso and Juan De Nova Isl.), Nigeria, Saudi Arabia, South Africa, Sudan and South Sudan, Uganda, Yemen, Zambia, Zimbabwe. **Refs.:** Edwards 1941: 257 (M*, F*); Hopkins 1952: 254 (L*); Knight 1953a: 226 (M*, F); Jupp 1996 (M*, F*; key); Ahmed et al. 2011 (distr.; Saudi Arabia). **Ety.:** Campo Santa de Salta, Argentina. **Informal name:** Salisbury Zimbabwean Typical Mosquito.

syn. *bostocki* Theobald
1905g: 29 (A). **Type loc.:** Transvaal [South Africa] (NHMUK). **Refs.:** Townsend 1990: 53 (type info.).

syn. *amboannulatus* Theobald
1913a: 320 (M; *salisburiensis* var.). **Type loc.:** Onderstepoort, Transvaal [South Africa] (GDZ).

ssp. ***capensis*** De Meillon
1935: 354 (F; as var.). **Type loc.:** Hout Bay, Pinelands and Palmiet River, Cape Province [South Africa] (LU). **Distr.:** South Africa. **Refs.:** Edwards 1941: 257 (syn.); White 1975: 323 (to ssp.). **Ety.:** n.s. [poss. after Cape Province, South Africa]. **Informal name:** Cape South African Typical Mosquito.

ssp. ***coursi*** Doucet
1949a: 144 (L*). **Type loc.:** Route between Bejofo–Amparafaravola [Tamatave], Madagascar (IRSM). **Distr.:** Madagascar (includes Glorioso and Juan De Nova Isl.). **Refs.:** Hopkins 1952: 253 (L; tax.). **Ety.:** n.s. **Informal name:** Cours Malagasy Typical Mosquito.

syn. *naudeanus* Muspratt
1961: 97 (M, F, L*; *Culex* (*Neoculex*) *salisburiensis* ssp.). **Type loc.:** Naudesberg Pass, 40 mi south of Middleberg, Cape Province [South Africa] (SIAM). **Refs.:** White 1975: 323 (syn.).

seyrigi Edwards
1941: 256 (M*). **Type loc.:** Ambositra [Fiarantsoa], Madagascar (MNHP). **Distr.:** Madagascar (includes Glorioso and Juan De Nova Isl.). **Refs.:** Brunhes 1968: 15 (M*, P*, L*; syn.). **Ety.:** A. Seyrig. **Informal name:** Seyrig Malagasy Typical Mosquito.

syn. *robici* Doucet
1950: 54 (L*). **Type loc.:** Ambositra [Fianarantsoa], Madagascar (IRSM).

subsalisburiensis Hervé & Geoffroy
1974: 191 (M*). **Type loc.:** Moundi (6°52′ N, 15°33′ E), Central African Republic (IERT). **Distr.:** Central African Republic. **Ety.:** nearly (L); near *salisburiensis*. **Informal name:** Sarum Central African Typical Mosquito [Sarum is a name for the Salisbury region in England].

Subgenus *Melanoconion* Theobald

1903a: 238 (as genus). Type species: *Culex atratus* Theobald (Dyar 1905a: 49). Dyar & Knab 1906d: 205 (syn. with genus *Culex*); Dyar 1918b: 103 (to subg. of genus *Culex*).

Name derivation: n.s. [melano- = dark; -conion (?); poss. ref. to overall color].

syn. *Melanoconium* Blanchard 1905: 395 (unjustified emend.). Dyar & Knab 1906d: 205 (syn.).
syn. *Gnophodeomyia* Theobald 1905g: 21 (as genus). Type species: *Gnophodeomyia inornata* Theobald.
syn. *Mochlostyrax* Dyar & Knab 1906d: 223 (as genus). Type species: *Mochlostyrax caudelli* Dyar & Knab. Howard, Dyar & Knab 1915: 216 (syn. with genus *Culex*); Dyar 1918b: 107 (to subg.); Rozeboom & Komp 1950b: 77 (syn. with subg. *Melanoconion*); Foote 1953 (morph., tax.); Foote 1954 (to subg.); Belkin et al. 1970: 92 (syn.); Stone & Haeger 1970 (L, E; tax.).
syn. *Choeroporpa* Dyar 1918b: 103 (as subg.). Type species: *Culex* (*Choeroporpa*) *anips* Dyar). Dyar 1928: 269 (syn. with *Mochlostyrax*); Edwards 1932a: 212 (syn. with subg. *Melanoconion*).
syn. *Helcoporpa* Dyar 1918d: 125 (as subg.). Type species: *Culex* (*Helcoporpa*) *menytes* Dyar. Dyar 1928: 269 (syn. with *Mochlostyrax*); Edwards 1932a: 213 (syn. with subg. *Melanoconion*).
syn. *Culex* (*Mochlostyrax*), section *Dinoporpa* Dyar 1928: 286. Type species: *Culex* (*Helcoporpa*) *trifidus* Dyar. Edwards 1932a: 213 (syn.).
syn. *Upsiloporpa* Komp & Curry 1932: 83 (section of *Culex*). Haplotype: *Culex* (*Upsiloporpa*) *haynei* Komp & Curry. Komp 1935: 2 (syn. of *haynei* with *menytes*).

Important references: Dyar 1918b: 103 (to subg.); Rozeboom & Komp 1950a: 75 (M gen.*); Foote 1952 (morph., tax.); Foote 1953 (morph., tax.); Belkin 1954a (L; chaetotaxy); Foote 1954 (P*, L*); Belkin, Schick, & Heinemann 1965 (type info., bion.); Belkin et al. 1966 (type info., bion.); Belkin et al. 1968 (type info., bion.); Berlin 1969c (phyl., tax.); Mattingly 1970b (E*); Belkin et al. 1971 (type info., bion.); Adames & Galindo 1973 (phyl.); Mattingly 1976 (E*); Sirivanakarn 1983: 153 (internal class.; M*, F*, P*, L*); Pecor et al. 1992 (M*; catalog, distr.); Sallum & Forattini 1996 (revision, Spissipes Section); Harbach 2011 (phyl.); Torres-Gutierrez & Sallum 2015 (tax., catalog; South America).

abominator Dyar & Knab
1909d: 257 (A). **Type loc.:** Plano, Texas, United States (USNM). **Distr.:** United States (continental). **Refs.:** Dyar & Knab 1918c: 170 (syn. with *erraticus*); King & Bradley 1937: 352 (tax.; from syn.); Rozeboom & Komp 1950a: 86 (M*); Yamaguti & LaCasse 1951c: 75 (M*, F*, L*); Foote 1954: 12 (L*); Carpenter & LaCasse 1955: 300 (M*, F*, L*; keys); Stone & Knight 1957a: 42 (type info.); Mattingly 1970b: 18 (E*); Pecor et al. 1992: 9 (M*; tax., catalog citation); Darsie & Ward 2005 (F*, L*; keys, distr.). **Ety.:** n.s. [you will loathe (L)]. **Informal name:** Abominable American Typical Mosquito.

abonnenci Clastrier
1970c: 468 (M*). **Type loc.:** Forêt du Gallion, French Guiana [Guiana] (MNHP). **Distr.:** French Guiana. **Refs.:** Pecor et al. 1992: 9 (M*; tax., catalog citation); Torres-Gutierrez & Sallum 2015: 10 (tax., bion., distr., catalog citation). **Ety.:** Émile Abonnenc. **Informal name:** Abonnenc French Guianese Typical Mosquito.

adamesi Sirivanakarn & Galindo
1980: 26 (M*, F*, P*, L*). **Type loc.:** Across canal from Summit Empire Firing Range, Canal Zone, Panama (USNM). **Distr.:** Brazil, Colombia, Ecuador, French Guiana, Panama, Peru. **Refs.:** Pecor et al. 1992: 9 (M*; tax., catalog citation); Sallum & Forattini 1996: 552 (M, F); Navarro & Weaver 2004 (phyl.); Torres-Gutierrez & Sallum 2015: 10 (tax., bion., distr., catalog citation). **Ety.:** Abdiel José Adames. **Informal name:** Adames Panamanian Typical Mosquito.

akritos Forattini & Sallum
1995: 125 (M*, F*). **Type loc.:** Experimental Station, Pariquera-Açu, São Paulo State, Brazil. **Distr.:** Brazil. **Refs.:** Sallum & Forattini 1996: 539 (M*, F); Torres-Gutierrez & Sallum 2015: 10 (tax., bion., distr., catalog citation). **Ety.:** confused (Gr); ref. to confused with *taeniopus*. **Informal name:** Confused Brazilian Typical Mosquito.

albinensis Bonne-Wepster & Bonne
1920: 173 (M, F, L). **Type loc.:** Paramaribo, Suriname (ITH). **Distr.:** Argentina, Brazil, Colombia, Ecuador, French Guiana, Guyana, Panama, Paraguay, Suriname, Venezuela. **Refs.:** Bonne & Bonne-Wepster 1925: 298 (M*); Dyar 1928: 311 (M*, F, L*); Rozeboom & Komp 1950a: 86 (M*; tax.); Rozeboom & Komp 1950a: 92 (M*; as *maroniensis*); Lane 1953: 425 (M*, F, L*); Foote 1954: 16 (P*, L*; tax; Colombia); Duret 1954a: 104 (M*); Galindo & Blanton 1955: 70 (distr.; Panama); Cova García et al. 1966: 217 (M*; Venezuela); Belkin 1968b: 12 (lectotype desig.); Duret 1969a (distr.; Paraguay); Sirivanakarn 1983: 280 (M*, F*); Mitchell & Darsie 1985: 284, 310 (distr.; Argentina); Forattini & Sallum 1987a: 153 (M*); Pecor et al. 1992: 9 (M*; tax., catalog citation); Torres-Gutierrez & Sallum 2015: 10 (tax., bion., distr., catalog citation). **Ety.:** n.s. [white (L); prob. ref. to lateral silvery-white abdominal patches]. **Informal name:** White Surinamese Typical Mosquito.
syn. *maroniensis* Bonne-Wepster & Bonne
1920: 175 (M). **Type loc.:** Suriname (ITH). **Refs.:** Lane 1951: 334 (syn.); Belkin 1968b: 17 (type info.).
syn. *gordoni* Evans
1924: 369 (M*). **Type loc.:** Bosque, Manáos [Amazonas], Brazil (NHMUK). **Refs.:** Belkin 1968b: 15 (type info.); Townsend 1990: 79 (type info.).

alcocki Bonne-Wepster & Bonne
1920: 171 (M, F, L; as *alcocci*). **Type loc.:** Zanderij, Suriname (ITH). **Distr.:** Brazil, French Guiana, Suriname. **Refs.:** Bonne & Bonne-Wepster 1925: 290 (M*, F, L*; as *alcocci*); Dyar 1928: 302 (M*, F, L*; as *alcocci*); Rozeboom & Komp 1950a: 86 (M*; as *alcocci*); Lane 1953: 456 (M, F, L*; as *alcocci*); Foote 1954: 18 (L*; as *alcocci*); Belkin 1968b: 12 (emend.; lectotype desig.); Fauran & Pajot 1974a: 103 (distr.); Pecor et al. 1992: 10 (M*; tax., catalog citation); Torres-Gutierrez & Sallum 2015: 11 (tax., bion., distr., catalog citation). **Ety.:** Alfred William Alcock. **Informal name:** Alcock Surinamese Typical Mosquito.

aliciae Duret
1953a: 75 (M*). **Type loc.:** Cerro Azul (Arroyo San Juan), Misiones, Argentina (USNM). **Distr.:** Argentina, Bolivia, Brazil, Panama, Paraguay. **Refs.:** Duret 1954a: 106 (M*; type info.); Duret & Barreto 1956: 87 (distr.; Brazil); Belkin et al. 1968: 16 (type info.); Duret 1969a (distr.; Paraguay); Peyton et al. 1983: 72 (distr.; Bolivia); Harbach et al. 1991: 192 (type info.); Pecor et al. 1992: 10 (M*; tax., catalog citation); Torres-Gutierrez & Sallum 2015: 11 (tax., bion., distr., catalog citation). **Ety.:** Alicia Diana Duret. **Informal name:** Alicia Argentinian Typical Mosquito.

alinkios Sallum & Hutchings
2003: 618 (M*). **Type loc.:** State of São Paulo, Cananéia County, Itapitangui, Itapuan Farm, Atlantic Tropical System, Brazil (FSP). **Distr.:** Brazil. **Refs.:** Torres-Gutierrez & Sallum 2015: 11 (tax., bion., distr., catalog citation). **Ety.:** resembling, like (Gr); ref. to similarity to *Cx. coppenamensis*. **Informal name:** Itapuan Brazilian Typical Mosquito.

alogistus Dyar

1918d: 126 (M, F). **Type loc.:** Suriname (USNM). **Distr.:** Brazil, Colombia, Costa Rica, French Guiana, Guyana, Panama, Suriname, Venezuela. **Refs.:** Dyar 1924i: 185 (distr.; Colombia); Bonne & Bonne-Wepster 1925: 308 (M*, F, L*); Dyar 1928: 288 (M*; as *megapus*); Dyar 1928: 302 (M*, F; distr.; Panama); Rozeboom & Komp 1950a: 87 (M*; tax., lectotype desig.); Lane 1953: 456 (M*, F); Foote 1954: 107 (P*, L*; tax.); Galindo & Blanton 1955: 70 (distr.; Costa Rica); Stone & Knight 1957a: 42 (type info.); Stone 1961c: 47 (distr.; French Guiana); Pecor et al. 1992: 10 (M*; tax., catalog citation); Torres-Gutierrez & Sallum 2015: 11 (tax., bion., distr., catalog citation). **Ety.:** n.s. ["silly," "thoughtlessly" (Gr)]. **Informal name:** Silly Surinamese Typical Mosquito.

syn. *megapus* Root

1927c: 595 (M*). **Type loc.:** Angra dos Reis, Rio de Janeiro, Brazil (USNM). **Refs.:** Belkin et al. 1971: 25 (type info.).

amitis Komp

1936c: 333 (M*). **Type loc.:** Quiriquire [Monagas], Venezuela (USNM). **Distr.:** Venezuela. **Refs.:** Rozeboom & Komp 1950a: 87 (M*); Lane 1953: 487 (M*); Stone & Knight 1957a: 43 (lectotype desig.); Cova García et al. 1966: 187 (M*); Pecor et al. 1992: 11 (M*; tax., catalog citation); Torres-Gutierrez & Sallum 2015: 11 (tax., bion., distr., catalog citation). **Ety.:** n.s. [supporting pole or fork, fence rail, crossbar (L)]. **Informal name:** Pole Venezuelan Typical Mosquito.

andricus Root

1927c: 592 (M*). **Type loc.:** Lassance, Minas Gerais, Brazil (USNM). **Distr.:** Brazil. **Refs.:** Dyar 1928: 330 (M*); Rozeboom & Komp 1950a: 87 (M*); Lane 1953: 469 (M*); Belkin et al. 1971: 22 (type info.); Sirivanakarn 1983: 280 (M*); Pecor et al. 1992: 11 (M*; tax., catalog citation); Torres-Gutierrez & Sallum 2015: 11 (tax., bion., distr., catalog citation). **Ety.:** n.s. [to or of a man, male; masculine (Gr)]. **Informal name:** Manlike Brazilian Typical Mosquito.

anips Dyar

1916a: 48 (M, F). **Type loc.:** San Diego, California, United States (USNM). **Distr.:** Mexico, United States (continental). **Refs.:** Dyar 1928: 299 (M*, F); Rozeboom & Komp 1950a: 87 (M*); Brookman & Reeves 1953: 231 (L*; Mexico); Foote 1954: 20 (L*); Carpenter & LaCasse 1955: 301 (M*, F, L*; keys); Bohart & Washino 1957: 462 (L*); Stone & Knight 1957a: 43 (lectotype desig.); Bohart & Washino 1978: 122 (M, F, L*; keys, tax., bion., distr.); Sirivanakarn 1983: 280 (M*); Pecor et al. 1992: 11 (M*; tax., catalog citation); Darsie & Ward 2005 (F*, L*; keys, distr.). **Ety.:** n.s. [unwashed (Gr)]. **Informal name:** Unwashed American Typical Mosquito.

anoplicitus Forattini & Sallum

1989a: 1 (M*). **Type loc.:** Vilarinho, Itapitangui, Cananéia, São Paulo, Brazil (FSP). **Distr.:** Brazil. **Refs.:** Pecor et al. 1992: 12 (M*; tax., catalog citation); Torres-Gutierrez & Sallum 2015: 11 (tax., bion., distr., catalog citation). **Ety.:** unarmed (Gr); gonostylus apex unusual. **Informal name:** Unarmed Brazilian Typical Mosquito.

arboricolus Galindo & Mendez

1961: 3 (M*, P*, L*). **Type loc.:** Cerro La Victoria, Panama (USNM). **Distr.:** Costa Rica, Panama. **Refs.:** Pecor et al. 1992: 60 (emend.), 12 (M*; tax., catalog citation). **Ety.:** n.s. [tree neck or hill? (L); larva from a tree hole]. **Informal name:** Tree-dwelling Panamanian Typical Mosquito.

atratus Theobald

1901c: 55 (M*, F*). **Type loc.:** Ferry Swamp, Jamaica (NHMUK). **Distr.:** Bahamas, Brazil, Cayman Isl., Colombia, Cuba, Dominican Republic, Guyana, Haiti, Jamaica, Lesser Antilles (includes Antigua and Barbuda, Barbados, Dominica, Guadeloupe, Martinique, Montserrat, Saint Kitts and Nevis, Saint Lucia, Saint Vincent and the Grenadines, Trinidad and Tobago, Virgin Isl.), Panama, Puerto Rico, Suriname, United States (continental), Venezuela. **Refs.:** Theobald 1903a: 239 (P*, L; distr.; Guyana, Brazil, Barbados, Saint Lucia); Dyar 1905a: 49 (type sp. desig. for *Melanoconion*); Dyar 1905c: 55 (M*); Theobald 1910c: 456 (distr.; United States); Howard et al. 1913: Fig. 92 (M*, L*); Howard et al. 1915: 388 (M, F, L); Root 1922b: 400 (distr.; Puerto Rico); Dyar 1923d: 187 (M*); Bonne & Bonne-Wepster 1925: 268 (M, F, L; distr.; Suriname); Root 1927a: 464 (distr.; Haiti); Dyar 1928: 338 (M*, F, L*; distr.; Cuba, Virgin Isl.); Matheson 1944: 216 (M*, F, L); Floch & Abonnenc 1945e: 29 (M*, L*; as *advieri*); Carpenter et al. 1946: 260 (M, F, L*); Rozeboom & Komp 1950a: 87 (M*); Yamaguti & LaCasse 1951c: 79 (M*, F*, L*); Lane 1953: 473 (M*, F, P, L*; distr.; Panama); Foote 1954: 21 (P*, L*; tax.); Carpenter & LaCasse 1955: 303 (M*, F*, L*; keys); King et al. 1960: 109 (A, L); Dodge 1966: 368 (first instar L*; key); Belkin 1968b: 13 (lectotype desig.); Stone 1969: 6 (distr.; Dominica); Belkin et al. 1970: 78 (M*, F*, P*, L*); Knight & Haeger 1971: 553 (M*; tax.); Belkin & Heinemann 1973: 219 (distr.; Dominican Republic); Belkin & Heinemann 1975a: 286, 296 (distr.; Puerto Rico, Virgin Isl.); Belkin & Heinemann 1975b: 372, 377, 393 (distr.; Bahamas, Cayman Isl., Nevis); Belkin & Heinemann 1976b (distr.; Antigua, Montserrat); Mattingly 1976: 225 (E*); Sirivanakarn 1983: 269 (M*, F*, P*, L*); Townsend 1990: 47 (type info.); Pecor et al. 1992: 12 (M*; tax., catalog citation); Darsie & Ward 2005 (F*, L*; keys, distr.); Torres-Gutierrez & Sallum 2015: 12 (tax., bion., distr., catalog citation). **Ety.:** n.s. [dark (L); prob. ref. to "a small dark mosquito"]. **Informal name:** Dark Jamaican Typical Mosquito.

syn. *falsificator* Dyar & Knab

1909d: 257 (A). **Type loc.:** Havana, Cuba (USNM). **Refs.:** Bonne & Bonne-Wepster 1925: 268 (syn.); Stone & Knight 1957a: 49 (lectotype desig.).

syn. *advieri* Senevet
1938: 185 (M*). **Type loc.:** Pointe-à-Pitre, Guadeloupe Island, Lesser Antilles (FMP). **Refs.:** Rozeboom & Komp 1950a: 87 (syn.); Belkin 1968b: 12 (tax.).

aureonotatus Duret & Barreto
1956: 91 (M*). **Type loc.:** Monte Alegre, Ribeirão Preto, São Paulo, Brazil (FMRP). **Distr.:** Brazil. **Refs.:** Belkin et al. 1971: 22 (type info.); Pecor et al. 1992: 13 (M*; tax., catalog citation); Torres-Gutierrez & Sallum 2015: 12 (tax., bion., distr., catalog citation). **Ety.:** branded gold (L); golden-yellow scales on scutum. **Informal name:** Gold-marked Brazilian Typical Mosquito.

bahiensis Duret
1969b: 40 (M*). **Type loc.:** Uruçuca, Bahia, Brazil (USNM). **Distr.:** Brazil. **Refs.:** Stone 1970: 162 (emend.); Harbach et al. 1991: 192 (type info.); Pecor et al. 1992: 13 (M*; tax., catalog citation); Lestani & Rossi 2012 (M, F, P, L); Torres-Gutierrez & Sallum 2015: 12 (tax., bion., distr., catalog citation). **Ety.:** Bahia, Brazil. **Informal name:** Bahia Brazilian Typical Mosquito.

bastagarius Dyar & Knab
1906b: 170 (M, L). **Type loc.:** Laventille, Trinidad [Trinidad & Tobago. Lesser Antilles] (USNM). **Distr.:** Argentina, Belize, Bolivia, Brazil, Colombia, Costa Rica, Ecuador, El Salvador, French Guiana, Guatemala, Guyana, Lesser Antilles (includes Guadeloupe, Trinidad and Tobago), Mexico, Nicaragua, Panama, Paraguay, Peru, Suriname, Venezuela. **Refs.:** Howard et al. 1913: Figs. 88, 364 (M*, L*); Howard et al. 1915: 424 (M, L); Bonne & Bonne-Wepster 1925: 302 (M, F, L); Dyar 1928: 331 (M*; as *xivilis*, *lapsus* for *xivylis*); Senevet & Abonnenc 1939a: 89 (M*, L*; as *innominatus*); Senevet & Abonnenc 1939a: 98 (M*, L; as *thomasi*); Kumm et al. 1940: 402 (distr.; Costa Rica); Anduze 1941a: 14 (distr.; Venezuela); Floch & Abonnenc 1945e: 34 (M*, L*; as *thomasi*); Rozeboom & Komp 1950a: 87 (M*; tax.); Martínez Palacios 1952b: 83 (distr.; Mexico); Duret 1953a, b (M*; key, distr.; Argentina); Lane 1953: 434 (M*, F, L*); Duret 1954a: 108 (M*); Foote 1954: 23 (P*, L*; tax.), 94 (L*; as *thomasi*); Stone & Knight 1957a: 44 (lectotype desig.); Stone et al. 1959: 270 (distr.; Peru); Stone 1961c: 46 (tax.; as *thomasi*); Cova García et al. 1966: 201 (M*); Duret 1969a: 7 (distr.; Paraguay); Clark-Gil & Darsie 1983: 254 (distr.; Guatemala); Sirivanakarn 1983: 280 (M*, F*); Pecor et al. 1992: 13 (M*; tax., catalog citation), 87 (distr.; Nicaragua); Torres-Gutierrez & Sallum 2015: 12 (tax., bion., distr., catalog citation). **Ety.:** n.s. [baggage master? (L)]. **Informal name:** Burdened Trinbagonian Typical Mosquito.
syn. *vapulans* Dyar
1920c: 69 (M). **Type loc.:** Paramaribo, Suriname (USNM). **Refs.:** Komp 1935: 7 (syn.); Rozeboom & Komp 1950a: 87 (lectotype desig.); Stone & Knight 1957a: 57 (type info.).
syn. *alfaroi* Dyar
1921b: 34 (M). **Type loc.:** Atirro, Costa Rica (USNM). **Refs.:** Dyar 1925f: 174 (as syn. of *mutator* without explanation); Komp 1935: 7 (syn.).
syn. *innominatus* Evans
1924: 363 (M*). **Type loc.:** Itacoatiara, Amazonas, Brazil (NHMUK). **Refs.:** Komp 1935: 6 (syn.); Belkin 1968b: 16 (lectotype desig.); Belkin et al. 1971: 22 (type loc. info.). Ma
syn. *thomasi* Evans
1924: 372 (M*). **Type loc.:** Manaus, Amazonas, Brazil (NHMUK). **Refs.:** Belkin et al. 1971: 25 (type loc. info.); Sirivanakarn 1983: 274 (syn.); Townsend 1990: 51 (type info.).
syn. *cuclyx* Dyar & Shannon
1924a: 48 (M). **Type loc.:** Cardenas River, Fort Clayton, Canal Zone, Panama (USNM). **Refs.:** Dyar 1928: 332 (syn.); Rozeboom & Komp 1950a: 87 (syn.).

batesi Rozeboom & Komp
1948: 403 (M*, F, L*). **Type loc.:** Villavicencio [Meta], Colombia (USNM). **Distr.:** Colombia, Costa Rica, Ecuador, French Guiana, Guyana, Nicaragua, Panama. **Refs.:** Rozeboom & Komp 1950a: 88 (M*); Lane 1953, 491 (M*, F, L*); Foote 1954: 26 (P*, L*); Fauran & Pajot 1974a: 103 (distr.; French Guiana); Sirivanakarn 1983: 280 (M, F*); Pecor et al. 1992: 14 (M*; tax., catalog citation, distr.; Costa Rica, Ecuador); Torres-Gutierrez & Sallum 2015: 13 (tax., bion., distr., catalog citation). **Ety.:** Marston Bates. **Informal name:** Bates Colombian Typical Mosquito.

bejaranoi Duret
1953a: 75 (M*); Duret 1954a: 109 (M*). **Type loc.:** Bernardo de Irigoyen, Misiones, Argentina (USNM). **Distr.:** Argentina. **Refs.:** Harbach et al. 1991: 192 (type info.); Pecor et al. 1992: 15 (M*; tax., catalog citation); Torres-Gutierrez & Sallum 2015: 13 (tax., bion., distr., catalog citation). **Ety.:** Don Juan Francisco Ricardo Bejarano. **Informal name:** Bejarano Argentinian Typical Mosquito.

bequaerti Dyar & Shannon
1925b: 39 (M, F). **Type loc.:** Sororoca, Rio Branca, Amazonas, Brazil (USNM). **Distr.:** Brazil. **Refs.:** Dyar 1928: 298 (M*, F); Rozeboom & Komp 1950a: 88 (M*); Lane 1953: 474 (M, F); Pecor et al. 1992: 15 (M*; tax., catalog citation); Torres-Gutierrez & Sallum 2015: 13 (tax., bion., distr., catalog citation). **Ety.:** Joseph Charles Corneille Bequaert. **Informal name:** Bequaert Brazilian Typical Mosquito.

bifoliolatus Duret & Barreto
1956: 96 (M*, F). **Type loc.:** Monte Alegre, Ribeirão Preto, São Paulo, Brazil (FMRP). **Distr.:** Brazil. **Refs.:** Belkin et al. 1971: 23 (type info.); Pecor et al. 1992: 15 (M*; tax., catalog citation); Torres-Gutierrez & Sallum 2015: 13 (tax., bion., distr., catalog citation). **Ety.:** two

broad leaves (L); ref. to "two equal leaves on the distal lobe of the subapical lobe of the basistyle." **Informal name:** Two-leafed Brazilian Typical Mosquito.

brachiatus Hutchings & Sallum
2008: 45 (M*). **Type loc.:** Amazonas State, Novo Airão municipality, Parque Nacional do Jaú, Jaú River, Miratucu Stream, Brazil (INPA). **Distr.:** Brazil. **Refs.:** Torres-Gutierrez & Sallum 2015: 13 (tax., bion., distr., catalog citation). **Ety.:** with arms, branches (L); distal division of subapical lobe subdivided into two arms. **Informal name:** Brachiated Brazilian Typical Mosquito.

carcinophilus Dyar & Knab
1906d: 220 (L*). **Type loc.:** San Domingo City [Cuidad Trujillo, Santo Domingo, Dominican Republic]. **Distr.:** Cuba, Dominican Republic, Guatemala, Haiti, Puerto Rico. **Refs.:** Howard et al. 1913: Figs. 85, 368, 592 (M*, L*); Howard et al. 1915: 412 (M, F, L); Root 1927a: 464 (distr.; Haiti); Dyar 1928: 319 (M*, F, L*); Martini 1935: 61 (distr.; Guatemala); Rozeboom & Komp 1950a: 88 (M*; tax.); Lane 1953: 475 (M*, F, P*, L*); Foote 1954: 29 (P*, L*); Stone & Knight 1957a: 45 (lectotype desig.); García Avila 1976: 9 (distr.; Cuba); Pecor et al. 1992: 15 (M*; tax., catalog citation). **Ety.:** crab loving (Gr); larvae in crab-holes. **Informal name:** Crab-loving Dominican Typical Mosquito.

caribeanus Galindo & Blanton
1954: 244 (M*). **Type loc.:** Mojinga Swamp, Canal Zone, Panama. **Distr.:** Panama. **Refs.:** Pecor et al. 1992: 16 (M*; tax., catalog citation). **Ety.:** Caribbean. **Informal name:** Caribbean Panamanian Typical Mosquito.

caudatus Clastrier
1970c: 470 (M*). **Type loc.:** Forêt de Cabassou, French Guiana. **Distr.:** Brazil, French Guiana, Lesser Antilles (includes Trinidad and Tobago). **Refs.:** Heinemann & Belkin 1979: 94 (distr.; Brazil); Pecor et al. 1992: 16 (M*; tax., catalog citation); Torres-Gutierrez & Sallum 2015: 13 (tax., bion., distr., catalog citation). **Ety.:** n.s. [tailed (L); poss. ref. to something caudal]. **Informal name:** Tailed French Guianese Typical Mosquito.

caudelli (Dyar & Knab)
1906d: 224 (L*; *Mochlostyrax*). **Type loc.:** Arima, Trinidad [Trinidad & Tobago, Lesser Antilles] (USNM). **Distr.:** Brazil, Colombia, French Guiana, Guyana, Lesser Antilles (includes Trinidad and Tobago), Panama, Peru, Suriname, Venezuela. **Refs.:** Howard et al. 1913: Figs. 74, 378, 590 (M*, L*); Howard et al. 1915: 395 (M, F, L); Bonne & Bonne-Wepster 1925: 312 (M*, L*; as *multispinosus*), 313 (M, L*); Dyar 1928: 287 (M*, F, L*); Kumm & Novis 1938: 503 (distr.; Brazil); Senevet & Abonnenc 1939a: 118 (M*; distr.; French Guiana); Anduze 1941a: 16 (distr.; Venezuela); Rozeboom & Komp 1950a: 88 (M*); Lane 1953: 422 (M*, F, L*); Foote 1954: 109 (P*, L*; tax., distr.; Colombia, Panama); Stone & Knight 1957a: 45 (lectotype desig.); Stone et al. 1959: 277 (distr.; Guyana); Forattini 1965a: 181 (M*, F, L); Cova García et al. 1966: 165 (M*); Sirivanakarn 1983: 280 (M*, F*); Chadee 1986: 196 (E*); Pecor et al. 1992: 16 (M*; tax., catalog citation); Pecor et al. 2000 (distr.; Peru); Torres-Gutierrez & Sallum 2015: 13 (tax., bion., distr., catalog citation). **Ety.:** A.N. Caudell. **Informal name:** Caudell Trinbagonian Typical Mosquito.

syn. *multispinosus* Bonne-Wepster & Bonne
1920: 177 (M, F, L). **Type loc.:** Kabelstation, Suriname (ITH). **Refs.:** Dyar 1928: 287 (syn.); Belkin 1968b: 18 (lectotype desig.).

cedecei Stone & Hair
1968: 39 (M*, F*). **Type loc.:** Mahogany Hammock, Dade, Florida, United States. **Distr.:** United States (continental). **Refs.:** Belkin 1969a: 68 (syn. with *opisthopus*); Belkin 1969b: 27 (syn. with *annulipes*); Belkin et al. 1970: 82 (syn.); Mattingly 1976: 228 (E*); Weaver et al. 1986: 619 (from syn.); Pecor et al. 1992: 16 (M*; tax., catalog citation); Sallum & Forattini 1996: 530 (M*); Darsie & Ward 2005 (F*, L*; keys, distr.). **Ety.:** acronym: Centers for Disease Control and Prevention, United States. **Informal name:** CDC American Typical Mosquito.

changuinolae Galindo & Blanton
1954: 238 (M*). **Type loc.:** Almirante, Bocas del Toro, Panama (USNM). **Distr.:** Panama. **Refs.:** Pecor et al. 1992: 17 (M*; tax., catalog citation). **Ety.:** Changuinola, Bocas del Toro, Panama. **Informal name:** Changuinola Panamanian Typical Mosquito.

clarki Evans
1924: 365 (M*). **Type loc.:** Manaus, Amazonas, Brazil (NHMUK). **Distr.:** Argentina, Brazil, Colombia, Paraguay, Uruguay, Venezuela. **Refs.:** Dyar 1928: 297 (M*); Edwards 1932a: 216 (syn. with *nigrescens*); Komp 1935: 4 (from syn.); Rozeboom & Komp 1950a: 93 (M*; syn. with *nigrescens*); Lane 1951: 334 (from syn.); Lane 1953: 480 (syn. with *nigrescens*); Casal 1963: 317 (M*; from syn.); Belkin 1968b: 14 (lectotype desig.); Duret 1969a (distr.; Paraguay); Belkin et al. 1971: 24 (type loc. info.); Sirivanakarn 1983: 279 (M*); Townsend 1990: 61 (type info.); Pecor et al. 1992: 17 (M*; tax., catalog citation), 87 (distr.; Venezuela); Rossi & Martínez 2003: 472 (distr.; Uruguay); Torres-Gutierrez & Sallum 2015: 13 (tax., bion., distr., catalog citation). **Ety.:** Herbert Charles Clark. **Informal name:** Clark Brazilian Typical Mosquito.

comatus Senevet & Abonnenc
1939a: 103 (M*, L*). **Type loc.:** Crique Mangue, Saut Tigre [Inini], French Guiana [Guiana] (NE). **Distr.:** Brazil, Colombia, French Guiana. **Refs.:** Rozeboom & Komp 1950a: 88 (M*); Foote 1954: 33 (L*; tax.); Belkin 1968b: 53 (type info.); Pecor et al. 1992: 17 (M*; tax., catalog citation); Torres-Gutierrez & Sallum 2015: 14 (tax., bion., distr., catalog citation). **Ety.:** n.s. [shaggy (L)].

Informal name: Shaggy French Guianese Typical Mosquito.

commevynensis Bonne-Wepster & Bonne
1920: 176 (M). **Type loc.:** Alkmaar [Commewijne] Suriname (ITH). **Distr.:** Belize, Brazil, Colombia, French Guiana, Panama, Suriname. **Refs.:** Dyar 1923e: 188 (M*); Bonne & Bonne-Wepster 1925: 271 (M*); Dyar 1928: 340 (M*); Rozeboom & Komp 1950a: 89 (M*; tax., distr.; Colombia, Panama); Lane 1953: 440 (M*, P*, L*); Foote 1954: 34 (P*, L*; tax.); Fauran 1961b: 35 (distr.; French Guiana); Belkin 1968b: 14 (tax.); Sirivanakarn 1983: 279 (M*); Pecor et al. 1992: 17 (M*; tax., catalog citation); Pecor et al. 2002: 247 (P, L; tax., distr.; Belize); Torres-Gutierrez & Sallum 2015: 14 (tax., bion., distr., catalog citation). **Ety.:** Commewijne Dist., Suriname. **Informal name:** Commewijne Surinamese Typical Mosquito.

comminutor Dyar
1920c: 70 (M). **Type loc.:** Suriname (USNM). **Distr.:** Brazil, Colombia, Ecuador, French Guiana, Lesser Antilles (includes Trinidad and Tobago), Panama, Suriname. **Refs.:** Dyar 1928: 322 (M*); Senevet & Abonnenc 1939a: 88 (M*, L); Rozeboom & Komp 1950a: 89 (M*; tax.); Lane 1953: 458 (M*, L); Foote 1954: 36 (L; tax.); Galindo & Blanton 1955: 70 (distr.; Panama); Duret & Barreto 1956: 86 (distr.; Brazil); Stone et al. 1959: 268 (distr.; Ecuador); Stone 1961c: 46 (distr.; Trinidad); Sirivanakarn 1983: 279 (M*); Pecor et al. 1992: 18 (M*; tax., catalog citation); Torres-Gutierrez & Sallum 2015: 14 (tax., bion., distr., catalog citation). **Ety.:** n.s. [crusher? (L)]. **Informal name:** Crusher Surinamese Typical Mosquito.

confundior Komp & Rozeboom
1951: 132 (M*). **Type loc.:** Paramaribo, Suriname (USNM). **Distr.:** Suriname. **Refs.:** Pecor et al. 1992: 18 (M*; tax., catalog citation); Torres-Gutierrez & Sallum 2015: 14 (tax., bion., distr., catalog citation). **Ety.:** n.s. [confounder? (L)]. **Informal name:** Confounding Surinamese Typical Mosquito.

conspirator Dyar & Knab
1906d: 217 (L*). **Type loc.:** Almoloya, Oaxaca, Mexico (USNM). **Distr.:** Belize, Colombia, Costa Rica, Ecuador, El Salvador, Guatemala, Honduras, Lesser Antilles (includes Trinidad and Tobago), Mexico, Nicaragua, Panama, Venezuela. **Refs.:** Howard et al. 1913: Figs. 70, 346, 591 (M*, L*); Howard et al. 1915: 412 (M, F, L); Dyar 1925d: 28 (distr.; Ecuador); Dyar 1928: 306 (M*, F, L*); Komp 1935: 5 (M); Rozeboom & Komp 1950a: 89 (M*); Lane 1953: 460 (M*, F, P*, L*); Foote 1954: 37 (L*; distr.; Honduras); Stone & Knight 1957a: 46 (lectotype desig.); Cova García et al. 1966: 149 (L*); Heinemann & Belkin 1977b: 428, 453 (distr.; Guatemala, Nicaragua); Sirivanakarn 1983: 280 (M*, F*); Pecor et al. 1992: 18 (M*; tax., catalog citation), 87 (distr.; Trinidad); Pecor et al. 2002: 247 (tax., distr.; Belize); Torres-Gutierrez & Sallum 2015: 14 (tax., bion., distr., catalog citation). **Ety.:** n.s. [*conspirator* (L)]. **Informal name:** Conspirator Mexican Typical Mosquito.

syn. *merodaemon* Dyar
1921f: 100 (M). **Type loc.:** Orotina, Costa Rica (USNM). **Refs.:** Dyar & Shannon 1924a: 47 (syn.); Dyar 1925f: 172 (syn. with *holoneus*); Stone & Knight 1957a: 51 (type info.).

syn. *pasadaemon* Dyar
1921f: 100 (M). **Type loc.:** San José, Costa Rica (USNM). **Refs.:** Dyar 1923b: 69 (syn.).

syn. *holoneus* Dyar
1921b: 35 (M). **Type loc.:** Orotina, Costa Rica (USNM). **Refs.:** Komp 1935: 4 (syn.); Belkin, Schick, & Heinemann 1965: 13 (type loc. info.).

syn. *dysmathes* Dyar & Ludlow
1921a: 47 (M, F). **Type loc.:** Cativa, Canal Zone, Panama (USNM). **Refs.:** Dyar 1923b: 69 (syn.); Stone & Knight 1957a: 48 (lectotype desig.).

syn. *fatuator* Dyar & Shannon
1924a: 47 (M). **Type loc.:** Cardenas River, Canal Zone, Panama (USNM). **Refs.:** Dyar 1928: 306 (M*, F); Komp 1935: 5 (syn.).

syn. *meroneus* Dyar
1925b: 22 (M). **Type loc.:** Barranquilla [Magdalena], Colombia (USNM).

syn. *macaronensis* Dyar & Núñez Tovar
1927a: 153 (M). **Type loc.:** Macaro, Venezuela (USNM). **Refs.:** Dyar 1928: 309 (M*); Komp 1935: 5 (syn.).

syn. *inducens* Root
1928 (M, F). In Dyar 1928: 307. **Type loc.:** [Maracay, Aragua], Venezuela (USNM). **Refs.:** Komp 1935: 5 (syn.); Stone & Knight 1957a: 51 (lectotype desig.).

contei Duret
1968c: 121 (M*). **Type loc.:** São Miguel do Guamá Pará, Brazil (USNM). **Distr.:** Brazil, French Guiana, Guyana, Lesser Antilles (includes Trinidad and Tobago), Panama. **Refs.:** Harbach et al. 1991: 193 (type info.); Pecor et al. 1992: 19 (M*; tax., catalog citation); Torres-Gutierrez & Sallum 2015: 14 (tax., bion., distr., catalog citation). **Ety.:** Juvenal O. Conte. **Informal name:** Conte Brazilian Typical Mosquito.

syn. *seneveti* Clastrier
1970c: 464 (M*). **Type loc.:** Forêt de Cabassou, Cayenne, French Guiana [Guiana] (MNHP). **Refs.:** Sirivanakarn 1983: 274 (syn.).

coppenamensis Bonne-Wepster & Bonne
1920: 173 (M, L). **Type loc.:** Kabelstation, Suriname (ITH). **Distr.:** Brazil, Colombia, French Guiana, Guyana, Suriname, Venezuela. **Refs.:** Bonne & Bonne-Wepster 1925: 300 (M*, L*); Dyar 1928: 314 (M*, L*); Anduze 1941a: 15 (distr.; Venezuela); Rozeboom & Komp 1950a: 89 (M*); Lane 1953: 413 (M*, L*); Foote 1954: 38 (L*); Cova García et al. 1966: 163 (M*); Belkin 1968b: 14 (lectotype desig.); Sirivanakarn 1983: 280 (M*); Pecor et al. 1992: 19 (M*; tax., catalog citation); Sallum &

Hutchings 2003: 615 (M*; tax.); Torres-Gutierrez & Sallum 2015: 15 (tax., bion., distr., catalog citation). **Ety.:** Coppename River, Suriname. **Informal name:** Coppename Surinamese Typical Mosquito.

corentynensis Dyar
1920c: 65 (M). **Type loc.:** Suriname (USNM). **Distr.:** Colombia, French Guiana, Suriname. **Refs.:** Dyar 1928: 315 (M*); Rozeboom & Komp 1950a: 89 (M*; lectotype desig.); Lane 1953: 477 (M*); Stone & Knight 1957a: 46 (type info.); Fauran 1961b: 35 (distr.; French Guiana); Pecor et al. 1992: 20 (M*; tax., catalog citation); Torres-Gutierrez & Sallum 2015: 15 (tax., bion., distr., catalog citation). **Ety.:** Corentyne, Suriname. **Informal name:** Corentyne Surinamese Typical Mosquito.

creole Anduze
1949a: 149 (M*). **Type loc.:** Caripito, Monagas, Venezuela (FH). **Distr.:** Brazil, Colombia, French Guiana, Venezuela. **Refs.:** Lane 1953: 413 (? = *coppenamensis*); Cova García et al. 1966: 219 (M*); Clastrier 1968: 225 (M*; French Guiana); Forattini et al. 1970: 39 (type info.); Sirivanakarn 1983: 280 (M*); Pecor et al. 1992: 20 (M*; tax., catalog citation); Torres-Gutierrez & Sallum 2015: 15 (tax., bion., distr., catalog citation). **Ety.:** Creole Petroleum Corp. **Informal name:** Creole Venezuelan Typical Mosquito.

cristovaoi Duret
1968a: 14 (M*). **Type loc.:** Caracaraí, Roraima, Brazil (USNM). **Distr.:** Brazil, French Guiana. **Refs.:** Fauran & Pajot 1974a: 104 (distr.; French Guiana); Harbach et al. 1991: 193 (type info.); Pecor et al. 1992: 20 (M*; tax., catalog citation); Torres-Gutierrez & Sallum 2015: 15 (tax., bion., distr., catalog citation). **Ety.:** Cristóvão da Silva Maciel. **Informal name:** Cristóvão Brazilian Typical Mosquito.

crybda Dyar
1924i: 184 (M). **Type loc.:** Atrato River, Murindo (Antioquia), Colombia (USNM). **Distr.:** Brazil, Colombia, Costa Rica, Ecuador, French Guiana, Lesser Antilles (includes Trinidad and Tobago), Mexico, Nicaragua, Panama, Suriname, Venezuela. **Refs.:** Dyar 1928: 293 (syn. with *taeniopus*); Rozeboom & Komp 1950a: 89 (A, P; tax.); Lane 1953: 403 (M); Stone 1961c: 46 (distr.; Trinidad); Belkin, Schick, & Heinemann 1965: 9 (type loc. info.); Galindo 1969: 87 (syn. with *epanastasis*); Sirivanakarn & Belkin 1980: 11 (M*; from syn.); Sirivanakarn 1983: 279 (F*); Pecor et al. 1992: 20 (M*; tax., catalog citation), 136 (M*); Sallum & Forattini 1996: 560 (M*, F; dist.); Torres-Gutierrez & Sallum 2015: 15 (tax., bion., distr., catalog citation). **Ety.:** n.s. **Informal name:** Hidden Colombian Typical Mosquito [*crybda* = "crybelo" = Hidden].

delpontei Duret
1969a: 8 (M*). **Type loc.:** Las Palmas, Chaco, Argentina (USNM). **Distr.:** Argentina, Brazil, Paraguay. **Refs.:** Forattini & Sallum 1989c: 473 (M*, F*; Brazil); Harbach et al. 1991: 193 (type info.); Pecor et al. 1992: 21 (M*; tax., catalog citation); Sallum & Forattini 1996: 574 (M*, F); Sallum et al. 2001: 932 (P*, L*; tax.); Torres-Gutierrez & Sallum 2015: 15 (tax., bion., distr., catalog citation). **Ety.:** Eduardo Francisco Del Ponte. **Informal name:** Del Ponte Argentinian Typical Mosquito.

diamphidius Peyton & Harbach
1991: 199 (M*). **Type loc.:** Talosita, Oaxaca, Mexico (17°00′ N, 96°30′ W) (USNM). **Distr.:** Mexico. **Ety.:** utterly different (Gr); ref. to gonostylus character. **Informal name:** Different Mexican Typical Mosquito.

distinguendus Dyar
1928: 305 (M, F). **Type loc.:** Mojinga Swamp, Canal Zone, Panama (USNM). **Distr.:** Brazil, Colombia, Costa Rica, Ecuador, French Guiana, Lesser Antilles (includes Trinidad and Tobago), Panama, Peru. **Refs.:** Rozeboom & Komp 1950a: 89 (M*; lectotype desig.); Lane 1951: 334 (syn. with *comminutor*); Foote 1954: 41 (L*; from syn.; distr.; Brazil); Stone & Knight 1957a: 47 (type info.); Stone et al. 1959: 269 (distr.; Costa Rica); Fauran & Pajot 1974a: 104 (distr.; French Guiana); Pecor et al. 1992: 21 (M*; tax., catalog citation); Pecor et al. 2000 (bion., distr.; Peru); Torres-Gutierrez & Sallum 2015: 16 (tax., bion., distr., catalog citation). **Ety.:** n.s. [distinguishing, separating (L)]. **Informal name:** Distinguished Panamanian Typical Mosquito.

dolichophyllus Clastrier
1970b: 858 (M*). **Type loc.:** Forêt de Cabassou, Cayenne, French Guiana [Guiana] (MNHP). **Distr.:** French Guiana. **Refs.:** Pecor et al. 1992: 21 (M*; tax., catalog citation); Torres-Gutierrez & Sallum 2015: 16 (tax., bion., distr., catalog citation). **Ety.:** n.s. [long leaf (L)]. **Informal name:** Long-leafed French Guianese Typical Mosquito.

dunni Dyar
1918d: 123 (M, F). **Type loc.:** Mandingo River, Panama [Canal Zone] (USNM). **Distr.:** Argentina, Belize, Brazil, Colombia, Costa Rica, Ecuador, El Salvador, French Guiana, Guatemala, Guyana, Lesser Antilles (includes-Trinidad and Tobago), Mexico, Nicaragua, Panama, Paraguay, Suriname, Venezuela. **Refs.:** Dyar 1923e: 188 (M*); Dyar 1924i: 183 (distr.; Colombia); Bonne & Bonne-Wepster 1925: 267 (M, F; distr.; Suriname); Dyar 1928: 340 (M*, F*, L*); Komp 1935: 8 (M; tax.); Senevet & Abonnenc 1939a: 80 (distr.; French Guiana); Anduze 1941a: 16 (distr.; Venezuela); Rozeboom & Komp 1950a: 89 (M*; type loc. info.); Martínez Palacios 1952b: 83 (distr.; Mexico); Foote 1954: 42 (P*, L*; tax.); Stone & Knight 1957a: 48 (lectotype desig.); Stone 1961c: 46 (distr.; Trinidad); Forattini 1965a: 182 (M*, L*); Cova García et al. 1966: 177 (M*); Duret 1969a: 7 (distr.); Bertram 1971: 745 (distr.; Belize); Sirivanakarn 1983: 279 (A*); Pecor et al. 1992: 87 (distr.; Nicaragua); Pecor et al. 1992: 21 (M*; tax., catalog citation); Berti et al. 2015 (distr.;

Venezuela); Torres-Gutierrez & Sallum 2015: 16 (tax., bion., distr., catalog citation). **Ety.:** L.H. Dunn. **Informal name:** Dunn Panamanian Typical Mosquito.

syn. *ruffinis* Dyar & Shannon
1924b: 144 (M). **Type loc.:** Barro Colorado Island, Gatun Lake, Canal Zone, Panama (USNM). **Refs.:** Komp 1935: 8 (syn.).

syn. *exedrus* Root
1927c: 580 (M*, F). **Type loc.:** Porto das Caixas, Rio de Janeiro, Brazil (USNM). **Refs.:** Dyar 1928: 341 (syn. with *ruffinis*); Stone & Knight 1957a: 49 (lectotype desig.).

dureti Casal & García
1968d: 123 (M*). **Type loc.:** San Ignacio, Misiones, Argentina (INM). **Distr.:** Argentina, Brazil, Paraguay, Venezuela. **Refs.:** Heinemann & Belkin 1979: 80 (distr.; Brazil); Pecor et al. 1992: 22 (M*; tax., catalog citation), 87 (distr.; Venezuela); Torres-Gutierrez & Sallum 2015: 16 (tax., bion., distr., catalog citation). **Ety.:** José Pedro Duret. **Informal name:** Duret Argentinian Typical Mosquito.

dyius Root
1927c: 587 (M*). **Type loc.:** Brazil ("Probably from the State of Rio de Janeiro") (USNM). **Distr.:** Brazil, Ecuador, French Guiana. **Refs.:** Dyar 1928 308 (M*; syn. of *elevator*); Rozeboom & Komp 1950a: 90 (M*; type info.; from syn.); Fauran & Pajot 1974a: 105 (distr.; French Guiana); Sirivanakarn 1983: 280 (M*); Pecor et al. 1992: 22 (M*; tax., catalog citation); Torres-Gutierrez & Sallum 2015: 16 (tax., bion., distr., catalog citation). **Ety.:** n.s. **Informal name:** Double Brazilian Typical Mosquito [*dyius* = "dyo" = Double].

eastor Dyar
1920c: 71 (M). **Type loc.:** Suriname (USNM). **Distr.:** Brazil, Colombia, Ecuador, French Guiana, Guatemala, Lesser Antilles (includes Trinidad and Tobago), Mexico, Nicaragua, Panama, Peru, Suriname, Venezuela. **Refs.:** Dyar 1924i: 184 (distr.; Colombia); Dyar 1928: 323 (M*, F); Komp 1935: 7 (distr.; Panama); Martini 1935: 61 (distr.; Guatemala); Rozeboom & Komp 1950a: 90 (M*); Lane 1953: 463 (M*, F); Foote 1954: 44 (P*, L*); Vargas 1956c: 28 (distr.; Mexico); Stone 1967: 217 (distr.; Trinidad); Sirivanakarn 1983: 279 (M*); Pecor et al. 1992: 22 (M*; tax., catalog citation); Pecor et al. 2000 (bion., distr.; Peru); Linton et al. 2013 (bion., mol. tax., distr.; Ecuador); Torres-Gutierrez & Sallum 2015: 17 (tax., bion., distr., catalog citation). **Ety.:** n.s. **Informal name:** Eastor Surinamese Typical Mosquito.

syn. *manaosensis* Evans
1924: 370 (M*). **Type loc.:** Manáos [Amazonas], Brazil (NHMUK). **Refs.:** Rozeboom & Komp 1950a: 90 (syn.); Belkin 1968b: 17 (type info.); Townsend 1990: 101 (type info.).

educator Dyar & Knab
1906d: 217 (L*). **Type loc.:** Rio Aranjuez, near Puntarenas, Costa Rica (USNM). **Distr.:** Argentina, Belize, Bolivia, Brazil, Colombia, Costa Rica, Ecuador, El Salvador, French Guiana, Guatemala, Honduras, Mexico, Nicaragua, Panama, Paraguay, Peru, Puerto Rico, Suriname, Venezuela. **Refs.:** Howard et al. 1915: 321 (M, F; as *apateticus*), 415 (L); Dyar 1918b: 105 (M*), 106 (syn.; type info; as *apateticus*); Dyar 1925b: 22 (distr.; Colombia); Root 1927c: 586 (distr.; Brazil); Dyar 1928: 325 (M*, F, L*; Venezuela); Rozeboom & Komp 1950a: 90 (M*); Martínez Palacios 1952b: 84 (distr.; Mexico); Duret 1953a, b (M*; key, distr.; Argentina); Lane 1953: 467 (M*, F, P*, L*); Duret 1954a: 112 (M*); Foote 1954: 46 (P*, L*); Galindo & Blanton 1955: 70 (distr.; Panama); Stone & Knight 1957a: 48 (type info.); Stone et al. 1959: 268 (distr.; Ecuador, Honduras); Cova García et al. 1966: 185 (M*); Bertram 1971: 745 (distr.; Belize); Heinemann & Belkin 1977b: 453 (distr.; Nicaragua); Sirivanakarn 1983: 279 (M*, F*); Pecor et al. 1992: 23 (M*; tax., catalog citation), 34 (M*; tax., catalog citation; as *keenani*), 87 (distr.; Peru); Forattini & Sallum 1993: 89 (M*, F*, P*, L*); Berti et al. 2015 (distr.; Venezuela); Torres-Gutierrez & Sallum 2015: 17 (tax., bion., distr., catalog citation). **Ety.:** n.s. [educator, tutor (L)]. **Informal name:** Educator Costa Rican Typical Mosquito.

syn. *apeteticus* Howard, Dyar, & Knab
1913: Fig. 91; 1915: 321 (M*, F; as *apateticus*). **Type loc.:** Upper Pequini River, Panama (USNM). **Refs.:** Stone & Knight 1957a: 43 (lectotype desig.; as *apateticus*); Stone et al. 1959: 269 (emend.).

syn. *aneles* Dyar & Ludlow
1922: 63 (M*, F). **Type loc.:** Cardenas River, Canal Zone, Panama (USNM). **Refs.:** Dyar 1928: 325 (syn.); Stone & Knight 1957a: 43 (lectotype desig.).

syn. *keenani* Galindo & Mendez
1961: 2 (M*). **Type loc.:** Pacora, Panama (USNM). **Refs.:** Forattini & Sallum 1993b: 89 (syn.).

egcymon Dyar
1923b: 67 (M*). **Type loc.:** Tabernilla, Canal Zone, Panama (USNM). **Distr.:** Colombia, Costa Rica, Guatemala, Panama. **Refs.:** Dyar 1928: 292 (M*, F, L*); Rozeboom & Komp 1950a: 90 (M*); Lane 1953: 417 (M*, F, L*); Foote 1954: 48 (L*); Stone & Knight 1957a: 48 (lectotype desig.); Sirivanakarn 1983: 280 (M*); Pecor et al. 1992: 24 (M*; tax., catalog citation), 87 (distr.; Colombia); Torres-Gutierrez & Sallum 2015: 17 (tax., bion., distr., catalog citation). **Ety.:** n.s. **Informal name:** Tabernilla Panamanian Typical Mosquito.

eknomios Forattini & Sallum
1992b: 273 (M*, F*, P*, L*). **Type loc.:** Guaporé River, Costa Marques Town, Rondônia, Brazil (USNM). **Distr.:** Brazil, Ecuador. **Refs.:** Torres-Gutierrez & Sallum 2015: 17 (tax., bion., distr., catalog citation). **Ety.:** n.s. [unusual, marvelous (Gr)]. **Informal name:** Marvelous Brazilian Typical Mosquito.

elephas Komp
1936c: 328 (M*). **Type loc.:** Juán Diaz, Panama (USNM). **Distr.:** Colombia, Panama, Venezuela. **Refs.:** Rozeboom

& Komp 1950a: 90 (M*); Lane 1953: 417 (M*); Sirivanakarn 1983: 280 (A*); Pecor et al. 1992: 24 (M*; tax., catalog citation); Torres-Gutierrez & Sallum 2015: 17 (tax., bion., distr., catalog citation). **Ety.:** elephant (L); "The very long, tapering, curved clasper bears a ludicrous resemblance to the proboscis of an elephant." **Informal name:** Elephant Panamanian Typical Mosquito.

elevator Dyar & Knab
1906d: 217 (L*). **Type loc.:** Port Limón, Costa Rica (USNM). **Distr.:** Argentina, Belize, Brazil, Colombia, Costa Rica, Ecuador, El Salvador, French Guiana, Guatemala, Honduras, Lesser Antilles (includes Dominica, Guadeloupe), Mexico, Nicaragua, Panama, Peru, Puerto Rico, Venezuela. **Refs.:** Howard et al. 1913: Figs. 76, 354, 593 (M*, L*); Howard et al. 1915: 414 (M, F, L); Dyar 1928: 308 (M*, F, L*); Komp 1935: 5 (tax.); Floch & Abonnenc 1945e: 31 (M*, F, L*; as *bonneti*); Rozeboom & Komp 1950a: 91 (M*); Yamaguti & LaCasse 1951c: 83 (M*, F*, L*); Martínez Palacios 1952b: 84 (distr.; Mexico); Duret 1953a (M; key, distr.; Argentina); Lane 1953: 447 (M*, F, L*); Duret 1954a: 113 (M); Foote 1954: 49 (P*, L*); Stone & Knight 1957a: 48 (lectotype desig.); Fauran 1961b: 35 (distr.; French Guiana); Cova García et al. 1966: 193 (M*); Stone 1969: 6 (distr.; Dominica); Heinemann & Belkin 1977b: 442, 453 (distr.; Honduras, Nicaragua); Sirivanakarn 1983: 280 (F*); Pecor et al. 1992: 24 (M*; tax., catalog citation), 87 (distr.; Guatemala); Pecor et al. 2002: 247 (tax., distr.; Belize); Linton et al. 2013 (bion., mol. tax., distr.; Ecuador); Torres-Gutierrez & Sallum 2015: 17 (tax., bion., distr., catalog citation). **Ety.:** n.s. **Informal name:** Elevator Costa Rican Typical Mosquito.

syn. *dornarum* Dyar & Shannon
1924a: 46 (M). **Type loc.:** Sweet Water Reservoir, Fort Sherman, Canal Zone, Panama (USNM). **Refs.:** Komp 1935: 5 (syn.).

syn. *curryi* Dyar
1926: 112 (M, F, L). **Type loc.:** Mojinga River Swamp, Canal Zone, Panama (USNM). **Refs.:** Komp 1935: 7 (syn.); Rozeboom & Komp 1950a: 91 (syn.); Stone & Knight 1957a: 46 (lectotype desig.).

syn. *bonneti* Senevet
1938: 187 (M*). **Type loc.:** Pointe-á-Pitre, Guadeloupe Island, Lesser Antilles (FMP; originally in LPFM). **Refs.:** Rozeboom & Komp 1950a: 91 (syn.); Belkin 1968b: 13 (type info.).

syn. *vogelsangi* Anduze
1949b: 61 (M*). **Type loc.:** Caripito, Monagas, Venezuela (FH). **Refs.:** Lane 1953: 448 (syn.); Forattini et al. 1970: 50 (type info.).

ensiformis Bonne-Wepster & Bonne
1920: 176 (M, F, L). **Type loc.:** Dam, Suriname (ITH). **Distr.:** Belize, Bolivia, Brazil, Colombia, Ecuador, French Guiana, Suriname. **Refs.:** Bonne & Bonne-Wepster 1925: 272 (M*, F, L); Senevet & Abonnenc 1939a: 81 (M*, L*; French Guiana); Rozeboom & Komp 1950a: 98 (syn. with *zeteki*); Foote 1954: 97 (P*, L*; as *zeteki*); Belkin 1968b: 15 (from syn.; type info., lectotype desig.); Peyton et al. 1983: 72 (distr.; Bolivia); Pecor et al. 1992: 25 (M*; tax., catalog citation); Pecor et al. 2002: 247 (P, L; tax., distr.; Belize); Torres-Gutierrez & Sallum 2015: 18 (tax., bion., distr., catalog citation). **Ety.:** n.s. [sword shape (L); poss. ref. to plate at base of gonocoxite narrow and bladelike]. **Informal name:** Swordlike Surinamese Typical Mosquito.

epanastasis Dyar
1922g: 191 (M*). **Type loc.:** Arenal River, Toro Point, Canal Zone, Panama (USNM). **Distr.:** Brazil, Colombia, French Guiana, Nicaragua, Panama, Venezuela. **Refs.:** Dyar 1928: 296 (M*); Komp 1935: 4 (syn. with *taeniopus*); Galindo 1969: 86 (from syn.); Heinemann & Belkin 1977b: 453 (distr.; Nicaragua); Sirivanakarn & Belkin 1980: 10 (tax.); Pecor et al. 1992: 25 (M*; tax., catalog citation); Sallum & Forattini 1996: 562 (M*); Torres-Gutierrez & Sallum 2015: 18 (tax., bion., distr., catalog citation). **Ety.:** n.s. **Informal name:** Arenal Panamanian Typical Mosquito.

syn. *pseudotaeniopus* Galindo & Blanton
1954: 240 (M*, P*, L*). **Type loc.:** Mojinga Swamp, Canal Zone, Panama (USNM). **Refs.:** Galindo 1969: 88 (tax.); Sirivanakarn & Belkin 1980: 11 (syn.).

equinoxialis Floch & Abonnenc
1945c: 3 (M*). **Type loc.:** Camp Rochambeau, French Guiana [Guiana] (PIP). **Distr.:** Brazil, French Guiana. **Refs.:** Rozeboom & Komp 1950a: 91 (M*); Belkin 1968b: 15 (tax.); Harrison 1973c: 277 (type info.); Pecor et al. 1992: 26 (M*; tax., catalog citation), 87 (distr.; Brazil); Torres-Gutierrez & Sallum 2015: 18 (tax., bion., distr., catalog citation). **Ety.:** n.s. ["Pertaining to the equinox" (L)]. **Informal name:** Equinox French Guianese Typical Mosquito.

ernanii Duret
1968f: 74 (M*). **Type loc.:** Boa Vista, Roraima, Brazil (USNM). **Distr.:** Brazil. **Refs.:** Harbach et al. 1991: 193 (type info.); Pecor et al. 1992: 26 (M*; tax., catalog citation); Torres-Gutierrez & Sallum 2015: 18 (tax., bion., distr., catalog citation). **Ety.:** Ernani Ferreira. **Informal name:** Ernani Brazilian Typical Mosquito.

ernsti Anduze
1949b: 63 (M*). **Type loc.:** Caripito, Monagas, Venezuela (FH). **Distr.:** Brazil, Venezuela. **Refs.:** Forattini et al. 1970: 40 (type info.); Sirivanakarn 1983: 280 (M*); Pecor et al. 1992: 26 (M*; tax., catalog citation); Torres-Gutierrez & Sallum 2015: 18 (tax., bion., distr., catalog citation). **Ety.:** A. Ernst. **Informal name:** Ernst Venezuelan Typical Mosquito.

erraticus (Dyar & Knab)
1906d: 224 (L*; *Mochlostyrax*). **Type loc.:** Baton Rouge, Louisiana, United States (USNM). **Distr.:** Bahamas,

Belize, Brazil, Canada, Colombia, Costa Rica, Cuba, Curaçao, Dominican Republic, Ecuador, El Salvador, French Guiana, Guatemala, Guyana, Haiti, Honduras, Jamaica, Lesser Antilles (includes Trinidad and Tobago, Virgin Isl.), Mexico, Nicaragua, Panama, Paraguay, Peru, Puerto Rico, Suriname, United States (continental), Venezuela. **Refs.:** Howard et al. 1913: Figs. 349, 588, 689 (L*, E*), Figs. 72, 363, 588, (M*, L*; as *trachycampa*), Figs. 84, 353 (M*, L*; as *leprincei*); Howard et al. 1915: 329 (M, F, L; as *trachycampa*), 397 (M, F, L; as *leprincei*); Dyar 1923e (tax.; as *leprincei*); Dyar 1928: 299 (M*, F, L*; M*, in part; as *abominator*); King & Bradley 1937: 345 (M*, L*; tax.); Rozeboom 1942b: 40 (M*, F, L); Michener 1944: 263 (F*); Carpenter et al. 1946: 262 (M*); Ross 1947 (F*); Rozeboom & Komp 1950a: 91 (M*); Yamaguti & LaCasse 1951c: 87 (M*, F*, L*); Lane 1953: 492 (M*, F, L*); Foote 1954: 51 (P*, L*; distr.; Brazil, Nicaragua, Honduras, Cuba, Mexico); van der Kuyp 1954: 39 (M, F, L*; bion.); Carpenter & LaCasse 1955: 305 (M*, F*, L*; keys); Stone & Knight 1957a: 49 (lectotype desig.); Stone et al. 1959: 270 (distr.; Dominican Republic); Cova García et al. 1966: 197 (M*); Dodge 1966: 369 (first instar L*; key); Duret 1969a (distr.; Paraguay); Belkin et al. 1970: 84 (M*, F, P*, L*); Bertram 1971: 745 (distr.; Belize); Knight & Haeger 1971: 553 (F*; tax.); Panday 1975a, b (distr.; Suriname); Mattingly 1976: 226 (E*); Heinemann & Belkin 1977b: 428, 433, (distr.; Guatemala, El Salvador); Sirivanakarn 1978a: 244 (M*); Sirivanakarn 1983: 279 (M*, F*); Townsend 1990: 143 (type info.); Pecor et al. 1992: 26 (M*; tax., catalog citation); Darsie & Ward 2005 (F*, L*; keys, distr.); Mendenhall et al. 2012 (mol. tax.); Torres-Gutierrez & Sallum 2015: 18 (tax., bion., distr., catalog citation); Harrison et al. 2016 (F*, L*; keys, distr.); Giordano 2019 (distr.; Canada). **Ety.:** erratic (L); ref. to comb scales in irregular row. **Informal name:** Erratic American Typical Mosquito. (Vol. 1, p. 518)

syn. *leprincei* Dyar & Knab
1907c: 202 (M, F). **Type loc.:** Tabernilla, Canal Zone, Panama (USNM). **Refs.:** King & Bradley 1937: 346 (syn.); Stone & Knight 1957a: 52 (lectotype desig.).

syn. *egberti* Dyar & Knab
1907d: 214 (F). **Type loc.:** Warner's Camp, Lake Okeechobee, Florida, United States (USNM). **Refs.:** King & Bradley 1937: 345 (syn.); Stone & Knight 1957a: 48 (lectotype desig.).

syn. *trachycampa* Dyar & Knab
1909c: 101 (M, P). **Type loc.:** Las Cascadas, Canal Zone, Panama (USNM). **Refs.:** Dyar 1918b: 105 (syn. with *leprincei*); King & Bradley 1937: 346 (tax.); Stone & Knight 1957a: 57 (lectotype desig.).

syn. *peribleptus* Dyar & Knab
1918c: 181 (M, F). **Type loc.:** Parr Shoals, South Carolina, United States (USNM). **Refs.:** Dyar 1920l: 161 (L); King & Bradley 1937: 346 (syn.); Stone & Knight 1957a: 54 (lectotype desig.).

syn. *pose* Dyar & Knab
1918c: 182 (F). **Type loc.:** Dallas, Texas, United States (USNM). **Refs.:** King & Bradley 1937: 346 (syn.); Stone & Knight 1957a: 53 (lectotype desig.).

syn. *moorei* Dyar
1918b: 108 (F). **Type loc.:** Plantation Plaisance, Georgetown [Demerara], British Guiana [Guyana] (USNM). **Refs.:** King & Bradley 1937: 346 (syn.); Stone & Knight 1957a: 53 (lectotype desig.).

syn. *degustator* Dyar
1921c: 39 (M). **Type loc.:** Scott, Arkansas, United States (USNM). **Refs.:** King & Bradley 1937: 346 (syn.); Stone & Knight 1957a: 47 (lectotype desig.).

syn. *homoeopas* Dyar & Ludlow
1921a: 46 (M). **Type loc.:** Jackson Barracks, New Orleans, Louisiana, United States (USNM). **Refs.:** King & Bradley 1937: 346 (syn.).

syn. *borinqueni* Root
1922b: 400 (M*, F, L*). **Type loc.:** Rio Piedras [San Juan], Puerto Rico (USNM). **Refs.:** King & Bradley 1937: 346 (syn.); Stone & Knight 1957a: 44 (lectotype desig.).

syn. *tovari* Evans
1924: 367 (M*). **Type loc.:** Palo Negro, Venezuela (NHMUK). **Refs.:** King & Bradley 1937: 346 (syn.); Belkin 1968b: 21 (type info.); Townsend 1990: 143 (type info.).

evansae Root
1927c: 593 (M*, L*). **Type loc.:** Mage, Rio de Janeiro, Brazil (USNM). **Distr.:** Brazil, Costa Rica, Ecuador, French Guiana, Honduras, Lesser Antilles (includes Trinidad and Tobago), Panama, Peru. **Refs.:** Dyar 1928: 320 (M*, F, L*); Senevet & Abonnenc 1939a: 86 (M*, L*; distr.; French Guiana); Rozeboom & Komp 1950a: 91 (M*; lectotype desig., type loc. info.); Lane 1953: 442 (M*, F, P*, L*); Foote 1954: 54 (L*; tax.); Galindo & Blanton 1955: 71 (distr.; Panama); Stone & Knight 1957a: 49 (type info.); Stone 1961c: 46 (distr.; Trinidad); Heinemann & Belkin 1979: 107 (distr.; Ecuador); Pecor et al. 1992: 28 (M*; tax., catalog citation), 61 (type info.); Linton et al. 2013 (bion., mol. tax., distr.; Ecuador); Torres-Gutierrez & Sallum 2015: 19 (tax., bion., distr., catalog citation). **Ety.:** Alwen Myfanwy Evans. **Informal name:** Evans Brazilian Typical Mosquito.

fairchildi Galindo & Blanton
1954: 234 (M*). **Type loc.:** Patino Point, Darien Province, Panama (USNM). **Distr.:** Brazil, Guatemala, Panama, Venezuela. **Refs.:** Pecor et al. 1992: 28 (M*; tax., catalog citation), 87 (distr.; Brazil); Torres-Gutierrez & Sallum 2015: 19 (tax., bion., distr., catalog citation). **Ety.:** Alexander Graham Bell Fairchild. **Informal name:** Fairchild Panamanian Typical Mosquito.

faurani Duret
1968f: 77 (M*). **Type loc.:** Manaus, Amazonas, Brazil (USNM). **Distr.:** Brazil, French Guiana. **Refs.:** Galindo

1969: 88 (tax.); Harbach et al. 1991: 193 (type info.); Pecor et al. 1992: 29 (M*; tax., catalog citation); Sallum & Forattini 1996: 592 (M*, F*); Torres-Gutierrez & Sallum 2015: 19 (tax., bion., distr., catalog citation). **Ety.:** P. Fauran. **Informal name:** Fauran Brazilian Typical Mosquito.

ferreri Duret
1968f: 79 (M*). **Type loc.:** Agualasal, Cucutá, Norte de Santander Colombia (USNM). **Distr.:** Colombia, Venezuela. **Refs.:** Harbach et al. 1991: 193 (type info.); Pecor et al. 1992: 29 (M*; tax., catalog citation); Torres-Gutierrez & Sallum 2015: 19 (tax., bion., distr., catalog citation). **Ety.:** H. Ferrer. **Informal name:** Ferrer Colombian Typical Mosquito.

flabellifer Komp
1936: 323 (M*). **Type loc.:** Santa Rosa, Colón Province, Panama (USNM). **Distr.:** Belize, French Guiana, Guatemala, Guyana, Honduras, Mexico, Nicaragua, Panama, Venezuela. **Refs.:** Rozeboom & Komp 1950a: 91 (M*; distr.; Mexico); Galindo & Blanton 1955: 71 (distr.; Honduras); Belkin et al. 1970: 89 (corrected distr.); Fauran & Pajot 1974a: 105 (distr.; French Guiana); Pecor et al. 1992: 29 (M*; tax., catalog citation), 87 (distr.; Venezuela, Guatemala); Pecor et al. 2002: 247 (tax., distr.; Belize); Torres-Gutierrez & Sallum 2015: 19 (tax., bion., distr., catalog citation). **Ety.:** small fan bearing (L); ref. to a large fanlike leaf on the male genitalia. **Informal name:** Fan-bearing Panamanian Typical Mosquito.

foliafer Komp & Rozeboom
1951: 121 (M*). **Type loc.:** Suriname (USNM). **Distr.:** Bolivia, French Guiana, Guyana, Panama, Suriname. **Refs.:** Foote 1954: 111 (P*, L*; tax.); Galindo & Blanton 1955: 71 (distr.; Panama); Clastrier 1968: 227 (M*; distr.; French Guiana); Pecor et al. 1992: 29 (M*; tax., catalog citation); Torres-Gutierrez & Sallum 2015: 20 (tax., bion., distr., catalog citation). **Ety.:** leaf bearing (L); ref. to a striate leaf on the base of the outer division of the ninth tergal lobe. **Informal name:** Leafy Surinamese Typical Mosquito.

galindoi Komp & Rozeboom
1951: 131 (M*). **Type loc.:** Quebrada Escondida, Rio Pequini, Colón Province, Panama (USNM). **Distr.:** Ecuador, Panama. **Refs.:** Pecor et al. 1992: 29 (M*; tax., catalog citation); Torres-Gutierrez & Sallum 2015: 20 (tax., bion., distr., catalog citation). **Ety.:** Pedro Galindo. **Informal name:** Galindo Panamanian Typical Mosquito.

galvaoi Duret
1968b: 60 (M*). **Type loc.:** Cheré, Rio de Janeiro, Brazil (USNM). **Distr.:** Brazil. **Refs.:** Harbach et al. 1991: 193 (type info.); Pecor et al. 1992: 30 (M*; tax., catalog citation); Torres-Gutierrez & Sallum 2015: 20 (tax., bion., distr., catalog citation). **Ety.:** Augusto Leopoldo Ayroza Galvão. **Informal name:** Galvão Brazilian Typical Mosquito.

garcesi Duret
1968b: 63 (M*). **Type loc.:** Quibdo, Chocó, Colombia (USNM). **Distr.:** Colombia, Costa Rica, Panama. **Refs.:** Strickman 1989: 210 (distr.; Panama); Harbach et al. 1991: 193 (type info.); Pecor et al. 1992: 30 (M*; tax., catalog citation); Torres-Gutierrez & Sallum 2015: 20 (tax., bion., distr., catalog citation). **Ety.:** Gilbert Garcés. **Informal name:** Garcés Colombian Typical Mosquito.

glyptosalpinx Harbach, Peyton, & Harrison
1984: 185 (M*, F*, P*, L*). **Type loc.:** Santa Cruz, Sandoval, Rincón del Tigre, Bolivia (USNM). **Distr.:** Argentina, Bolivia, Brazil, Paraguay. **Refs.:** Duret 1954a: 114 (M*; as *intrincatus*); Duret & Barreto 1956: 86 (M*; as *intrincatus*); Pecor et al. 1992: 30 (M*; tax., catalog citation); Torres-Gutierrez & Sallum 2015: 20 (tax., bion., distr., catalog citation). **Ety.:** notched trumpet (Gr); ref. to "lateral margin [of trumpet] with conspicuous funnel-shaped indentation or notch." **Informal name:** Notched Bolivian Typical Mosquito.

gnomatos Sallum, Hutchings, Leila, & Ferreira
1997: 215 (M*, F*). **Type loc.:** State of Amazonas, Novo Airão County, Parque Nacional do Jaú, Miratucu River, Brazil (INPA-CPEN). **Distr.:** Brazil, Peru. **Refs.:** Pecor et al. 2000 (bion., distr.; Peru). **Ety.:** small being (Gr, L) [prob. ref. to "characteristic aspect of the foliform setal on the distal division of the subapical lobe of gonocoxite"]. **Informal name:** Gnome Brazilian Typical Mosquito.

guedesi Mattos & Xavier
1991: 193 (M*). **Type loc.:** Lagoa dos Mares, Município de Lagoa Santa, Estado de Minas Gerais, Brazil (FSP). **Distr.:** Brazil. **Refs.:** Pecor et al. 1992: 30 (M*; tax., catalog citation); Torres-Gutierrez & Sallum 2015: 21 (tax., bion., distr., catalog citation). **Ety.:** Arino da Silveira Guedes. **Informal name:** Guedes Brazilian Typical Mosquito.

herrerai Sutil Oramas, Pulido Florenzano, & Amarista Meneses
1987: 81 (M*). **Type loc.:** Jabilla, Queseras del Medio, Apure, Venezuela (DERM). **Distr.:** Venezuela. **Refs.:** Pecor et al. 1992: 30 (M*; tax., catalog citation); Torres-Gutierrez & Sallum 2015: 21 (tax., bion., distr., catalog citation). **Ety.:** Jesús Herrera. **Informal name:** Herrera Venezuelan Typical Mosquito.

idottus Dyar
1920c: 77 (M). **Type loc.:** Suriname (USNM). **Distr.:** Argentina, Bolivia, Brazil, Colombia, French Guiana, Lesser Antilles (includes Dominica, Grenada, Guadeloupe, Martinique, Saint Lucia, Trinidad and Tobago), Guyana, Paraguay, Suriname, Venezuela. **Refs.:** Dyar 1928: 332 (M*); Townsend 1934: 491 (distr.; Brazil); Rozeboom & Komp 1950a: 91 (M*); Duret 1953a, b (M*; key, distr.;

Argentina); Lane 1953: 470 (M*); Duret 1954a: 113 (M*); Floch & Fauran 1954: 3 (M*; distr.; French Guiana); Foote 1954: 94 (L*; as *thomasi*); Fauran 1961a: 6 (tax.); Stone 1961c: 46 (tax., distr.; Trinidad); Cova García et al. 1966: 41 (M*); Duret 1969a (distr.; Paraguay); Stone 1969: 6 (distr.; Dominica); Belkin & Heinemann 1976a: 272, 288 (distr.; Grenada, Saint Lucia); Belkin & Heinemann 1976c: 188 (distr.; Guadeloupe); Sirivanakarn 1983: 279 (M*); Peyton et al. 1983: 72 (distr.; Bolivia); Pecor et al. 1992: 31 (M*; tax., catalog citation); Torres-Gutierrez & Sallum 2015: 21 (tax., bion., distr., catalog citation). **Ety.:** n.s. **Informal name:** Akurio Surinamese Typical Mosquito [Akurio = ethnic group of Suriname].

syn. *terepaima* Anduze
1949b: 65 (M*). **Type loc.:** El Valle, Distrito Federal, Venezuela (FH). **Refs.:** Lane 1953: 470 (syn.); Forattini et al. 1970: 49 (type info.).

ikelos Forattini & Sallum
1995: 132 (M*, F*, P*, L*). **Type loc.:** Brazil, São Paulo State, Sorocaba County, Campolim (FSP, USP). **Distr.:** Brazil. **Refs.:** Sallum & Forattini 1996: 544 (M*, F); Torres-Gutierrez & Sallum 2015: 21 (tax., bion., distr., catalog citation). **Ety.:** like (Gr); ref. to similarity to *taeniopus*. **Informal name:** Sorocaba Brazilian Typical Mosquito.

inadmirabilis Dyar
1928: 297 (M*). **Type loc.:** São Paulo, Brazil (USNM). **Distr.:** Argentina, Brazil, Colombia, French Guiana, Lesser Antilles (includes Trinidad and Tobago), Panama, Venezuela. **Refs.:** Rozeboom & Komp 1950a: 91 (M*; tax.); Lane 1953: 477 (M*); Fauran & Pajot 1974a: 106 (distr.; French Guiana); Pecor et al. 1992: 31 (M*; tax., catalog citation), 37 (M*; tax., catalog citation; as *mistura*); Forattini & Sallum 1993a: 88 (M*, F*, P*, L*); Berti et al. 2015 (distr.; Venezuela); Torres-Gutierrez & Sallum 2015: 21 (tax., bion., distr., catalog citation). **Ety.:** n.s. [not admirable (L)]. **Informal name:** Unadmirable Brazilian Typical Mosquito.

syn. *mistura* Komp & Rozeboom
1951: 124 (M*, F, L*). **Type loc.:** Laguna de la Palmita, Villavicencio, Meta, Colombia (USNM). **Refs.:** Foote 1954: 69 (P*, L*); Galindo & Blanton 1955: 71 (distr.; Panama); Cova García et al. 1966: 199 (M*); Sirivanakarn 1983: 279 (M*); Forattini & Sallum 1993a: 77 (syn.).

inhibitator Dyar & Knab
1906d: 216 (L*). **Type loc.:** San Francisco Mountains, Santo Domingo [Dominican Republic] (USNM). **Distr.:** Belize, Brazil, Colombia, Costa Rica, Dominican Republic, El Salvador, French Guiana, Guatemala, Guyana, Haiti, Jamaica, Mexico, Panama, Puerto Rico, Suriname, Venezuela. **Refs.:** Howard et al. 1913: Fig. 83 (M*); Howard et al. 1915: 425 (M, F); Dyar 1928: 317 (in part M*, F; misident. of L*; see *erraticus*); Gerry 1932 (F*); Martini 1935: 61 (distr.; Guatemala); King & Bradley 1937: 349 (M*, L; tax.); Kumm et al. 1940: 404 (distr.; Costa Rica); Anduze 1941a: 16 (distr.; Venezuela); Rozeboom & Komp 1950a: 91 (M*); Fox 1953: 179 (distr.; Puerto Rico); Lane 1953: 466 (M*, F, P*, L); Foote 1954: 59 (L*); Stone & Knight 1957a: 51 (type info.); Cova García et al. 1966: 191 (M*); Belkin et al. 1970: 88 (M*, F, P*, L*); Belkin & Heinemann 1973: 218 (corrected type loc.); Fauran & Pajot 1974a: 106 (distr.; French Guiana); Panday 1975a, b (distr.; Suriname); Sirivanakarn 1983: 280 (M*, F*); Pecor et al. 1992: 31 (M*; tax., catalog citation). **Ety.:** n.s. [inhibitor (L)]. **Informal name:** Inhibited Dominican Typical Mosquito.

syn. *investigator* Dyar & Knab
1906d: 216 (L*). **Type loc.:** Santa Lucrecia [Veracruz], Mexico (USNM?). **Refs.:** Dyar 1928: 317 (syn.); Stone & Knight 1957a: 51 (type info.).

innovator Evans
1924: 373 (M*). **Type loc.:** Itacoatiara, Amazonas, Brazil (NHMUK). **Distr.:** Brazil, French Guiana. **Refs.:** Dyar 1928: 292 (M*); Senevet & Abonnenc 1939a: 118 (M*; French Guiana); Rozeboom & Komp 1950a: 92 (M*); Foote 1954: 111 (L*; tax.); Belkin 1968b: 16 (tax., lectotype desig.); Belkin et al. 1971: 25 (type loc. info.); Pecor et al. 1992: 32 (M*; tax., catalog citation); Torres-Gutierrez & Sallum 2015: 22 (tax., bion., distr., catalog citation). **Ety.:** n.s. [innovator (L)]. **Informal name:** Innovative Brazilian Typical Mosquito.

intonsus Galindo & Blanton
1954: 231 (M*). **Type loc.:** Lancetilla Valley, Tela, Honduras (USNM). **Distr.:** Honduras. **Refs.:** Pecor et al. 1992: 32 (M*; tax., catalog citation). **Ety.:** n.s. [uncut, unshaven (L)]. **Informal name:** Shaggy Honduran Typical Mosquito.

intrincatus Brèthes
1916: 214 (M*). **Type loc.:** San Isidro [8 mi north of Buenos Aires], Argentina (BA). **Distr.:** Argentina, Bolivia, Brazil, Ecuador, Paraguay, Peru, Suriname. **Refs.:** Dyar 1928: 315 (M*); Rozeboom & Komp 1950a: 92 (M*); Lane 1953: 488 (M*); Duret 1954a: 114 (M*); Foote 1954: 60 (L*; tax.); Cova García et al. 1966: 207 (M*; Venezuela); Duret 1969a (distr.; Paraguay); Fauran & Pajot 1974a: 106 (distr.; French Guiana); Sirivanakarn 1983: 279 (M*); Harbach, Peyton, & Harrison 1984: 198 (M*; tax.); Pecor et al. 1992: 32 (M*; tax., catalog citation); Pecor et al. 2000 (distr.; Peru); Torres-Gutierrez & Sallum 2015: 22 (tax., bion., distr., catalog citation). **Ety.:** n.s. [entangled (L); poss. ref. to very complicated male genitalia]. **Informal name:** Intricate Argentinian Typical Mosquito.

syn. *xivylis* Dyar
1920c: 78 (M, L). **Type loc.:** Suriname (USNM). **Refs.:** Komp 1935: 7 (as *xivilis*, *lapsus* for *xivlys*; syn. with *bastagarius*); Rozeboom & Komp 1950a: 92 (syn.; corrected spelling; tax.); Stone & Knight 1957a: 58 (type info.).

syn. *cenus* Root
1927c: 590 (M*). **Type loc.:** Mage, Rio de Janeiro, Brazil (USNM). **Refs.:** Stone & Knight 1957a: 45 (lectotype desig.).

invocator Pazos
1908: 426 (A). **Type loc.:** San Antonio de los Banos [Havana], Cuba (USNM). **Distr.:** Brazil, Cuba. **Refs.:** Dyar & Knab 1909d: 258 (*invocator*; as new sp.); Howard et al. 1913: Fig. 81 (M*); Howard et al. 1915: 323 (M, F; corrected authorship); Dyar 1928: 317 (syn. with *inhibitator*); King & Bradley 1937: 345 (syn. with *erraticus*); Stone & Knight 1957a: 52 (lectotype desig.); Sirivanakarn 1978a: 239 (M*, F; from syn.); Pecor et al. 1992: 33 (M*; tax., catalog citation). **Ety.:** n.s. [one who invokes (L)]. **Informal name:** Priestly Cuban Typical Mosquito [*invocator* = "one who invokes" = a priest = Priestly].

iolambdis Dyar
1918b: 106 (M). **Type loc.:** Panama (USNM). **Distr.:** Bahamas, Belize, Brazil, Cayman Isl., Colombia, Costa Rica, Ecuador, El Salvador, Guatemala, Jamaica, Mexico, Nicaragua, Panama, Puerto Rico, United States (continental). **Refs.:** Dyar 1928: 328 (M*, F); Rozeboom & Komp 1950a: 92 (M*; as *iolanbdis*, *lapsus* for *iolambdis*); Martínez Palacios 1952b: 84 (distr.; Mexico); Pratt & Seabrook 1952: 27 (M*, F, P*, L*; distr.; United States, Puerto Rico); Lane 1953: 471 (M*, F, P*, L*; distr., Colombia); Foote 1954: 61 (P*, L*; tax.); Carpenter & LaCasse 1955: 307 (M*, F*, L*; keys); Belkin et al. 1970: 86 (M*, F, P*, L*); Knight & Haeger 1971: 553 (F*; tax.); Pecor et al. 1992: 33 (M*; tax., catalog citation), 88 (distr.; El Salvador); Darsie 2002d: 277 (P*; tax.); Pecor et al. 2002: 247 (tax., distr.; Belize); Darsie & Ward 2005 (F*, L*; keys, distr.); Linton et al. 2013 (bion., mol. tax., distr.; Ecuador); Torres-Gutierrez & Sallum 2015: 22 (tax., bion., distr., catalog citation). **Ety.:** n.s. **Informal name:** Lambda Panamanian Typical Mosquito [*iolambdis* = Lambda, the eleventh letter of the Greek alphabet].

isabelae Duret
1968f: 72 (M*). **Type loc.:** Caracaraí, Roraima, Brazil (USNM). **Distr.:** Brazil, Peru. **Refs.:** Harbach et al. 1991: 194 (type info.); Pecor et al. 1992: 34 (M*; tax., catalog citation); Pecor et al. 2000 (bion., distr.; Peru); Torres-Gutierrez & Sallum 2015: 22 (tax., bion., distr., catalog citation). **Ety.:** Isabel Duret. **Informal name:** Isabel Brazilian Typical Mosquito.

jocasta Komp & Rozeboom
1951: 128 (M*). **Type loc.:** Grenville, Grenada Island, Lesser Antilles (USNM). **Distr.:** Lesser Antilles (includes Grenada, Saint Vincent and the Grenadines). **Refs.:** Belkin & Heinemann 1976a: 296 (distr.; Saint Vincent); Pecor et al. 1992: 34 (M*; tax., catalog citation). **Ety.:** wife and mother of Oedipus (Gr) [close resemblance to *C. oedipus*]. **Informal name:** Jocasta Grenadian Typical Mosquito.

johnnyi Duret
1968d: 17 (M*). **Type loc.:** Rio Preto, Município de Joíão Goulard, Amazonas, Brazil (USNM). **Distr.:** Brazil. **Refs.:** Harbach et al. 1991: 194 (type info.); Pecor et al. 1992: 34 (M*; tax., catalog citation); Torres-Gutierrez & Sallum 2015: 22 (tax., bion., distr., catalog citation). **Ety.:** A.J. Walker. **Informal name:** Johnny Brazilian Typical Mosquito.

johnsoni Galindo & Mendez
1961a: 1 (M*). **Type loc.:** Pacora, Panama (USNM). **Distr.:** Colombia, Panama. **Refs.:** Pecor et al. 1992: 34 (M*; tax., catalog citation), 88 (distr.; Colombia); Torres-Gutierrez & Sallum 2015: 23 (tax., bion., distr., catalog citation). **Ety.:** Carl M. Johnson. **Informal name:** Johnson Panamanian Typical Mosquito.

jubifer Komp & Brown
1935: 254 (M*). **Type loc.:** Mojinga Swamp, Lower Chagres River, Canal Zone, Panama (USNM). **Distr.:** Guyana, Lesser Antilles (includes Trinidad and Tobago), Panama, Venezuela. **Refs.:** Rozeboom & Komp 1950a: 92 (M*); Lane 1953: 435 (M*); Forattini et al. 1970: 43 (distr.; Brazil); Sirivanakarn 1983: 279 (M*, F*); Pecor et al. 1992: 34 (M*; tax., catalog citation); Sallum & Forattini 1996: 589 (M*; distr.). **Ety.:** n.s. [name bearing (L)]. **Informal name:** Mojinga Panamanian Typical Mosquito.

kummi Komp & Rozeboom
1951: 130 (M*). **Type loc.:** Almirante, Bocas del Toro Province, Panama (USNM). **Distr.:** Colombia, Nicaragua, Panama. **Refs.:** Foote 1954: 63 (L*); Barreto & Lee 1969: 430 (distr.; Colombia); Pecor et al. 1992: 35 (M*; tax., catalog citation), 88 (distr.; Nicaragua); Torres-Gutierrez & Sallum 2015: 23 (tax., bion., distr., catalog citation). **Ety.:** Henry William Kumm. **Informal name:** Kumm Panamanian Typical Mosquito.

lacertosus Komp & Rozeboom
1951: 123 (M*). **Type loc.:** Almirante, Bocas Province, Panama (USNM). **Distr.:** French Guiana, Panama. **Refs.:** Clastrier 1972: 169 (M*, F, P*, L*; distr.; French Guiana); Pecor et al. 1992: 35 (M*; tax., catalog citation); Torres-Gutierrez & Sallum 2015: 23 (tax., bion., distr., catalog citation). **Ety.:** strong limbed (L); ref. to thickened rods on male genitalia. **Informal name:** Muscular Panamanian Typical Mosquito.

limacifer Komp
1936c: 325 (M*). **Type loc.:** Chase, Bocas del Toro Province, Costa Rica (USNM). **Distr.:** Belize, Costa Rica, El Salvador, Guatemala, Honduras, Mexico, Panama. **Refs.:** Rozeboom & Komp 1950a: 92 (M*); Martínez Palacios 1952b: 85 (distr.; Mexico); Lane 1953: 440 (M*); Pecor et al. 1992: 35 (M*; tax., catalog citation), 88 (distr.; El Salvador); Pecor et al. 2002: 247 (tax., distr.; Belize). **Ety.:** slug bearing (L); ref. to "a fanciful resemblance of the

lobes of the ninth tergite to a pair of garden slugs (Limax), with their antennae erect." **Informal name:** Slug Costa Rican Typical Mosquito.

lopesi Sirivanakarn & Jakob
1979: 139 (M*). **Type loc.:** Porto do Ribeira, Iguape, São Paulo, Brazil (USNM). **Distr.:** Brazil. **Refs.:** Sirivanakarn 1983: 279 (M*); Forattini & Sallum 1990 (F*); Pecor et al. 1992: 35 (M*; tax., catalog citation); Sallum & Forattini 1996: 591 (M*, F*); Torres-Gutierrez & Sallum 2015: 23 (tax., bion., distr., catalog citation). **Ety.:** Oscar de Souza Lopes. **Informal name:** Lopes Brazilian Typical Mosquito.

lucifugus Komp
1936c: 331 (M*). **Type loc.:** Quiriquire [Monagas], Venezuela (USNM). **Distr.:** Argentina, Colombia, Ecuador, Lesser Antilles (includes Trinidad & Tobago), Venezuela. **Refs.:** Rozeboom & Komp 1950a: 92 (M*; distr.; Trinidad); Lane 1953: 467 (M*); Foote 1954: 66 (P*, L*; Colombia); Cova García et al. 1966: 195 (M*); García & Casal 1968b: 108 (distr.; Argentina); Pecor et al. 1992: 26 (M*; tax., catalog citation); Torres-Gutierrez & Sallum 2015: 23 (tax., bion., distr., catalog citation). **Ety.:** n.s. [light fleeing (L)]. **Informal name:** Vampire Venezuelan Typical Mosquito [*lucifugus* = "light fleeing" = Vampire].

madininensis Senevet
1936a: 129 (M*). **Type loc.:** Trinité, Martinique Island [Lesser Antilles] (FMP; originally in LPFM). **Distr.:** Lesser Antilles (includes Dominica, Guadeloupe, Martinique, Montserrat, Saint Kitts and Nevis, Saint Lucia). **Refs.:** Rozeboom & Komp 1950a: 92 (M*); Lane 1953: 441 (M*); Belkin 1968b: 17 (type info.); Belkin & Heinemann 1975b: 389 (distr.; Dominica, Leeward Isl.); Belkin & Heinemann 1976a: 289 (distr.; Saint Lucia); Belkin & Heinemann 1976c: 188 (distr.; Guadeloupe); Pecor et al. 1992: 26 (M*; tax., catalog citation). **Ety.:** n.s. [Madinina = "island of flowers" = Martinique]. **Informal name:** Madinina Martinican Typical Mosquito.

martinezi Casal & García
1968c: 455 (M*, P*, L*). **Type loc.:** Kilometer 34 Vespucio–San Pedrito Road, Salta, Argentina (INM). **Distr.:** Argentina, Bolivia. **Refs.:** Pecor et al. 1992: 26 (M*; tax., catalog citation); Torres-Gutierrez & Sallum 2015: 23 (tax., bion., distr., catalog citation). **Ety.:** Antonia Martínez. **Informal name:** Martínez Argentinian Typical Mosquito.

maxinocca Dyar
1920c: 71 (M, L). **Type loc.:** Suriname (USNM). **Distr.:** Brazil, French Guiana, Guatemala, Suriname. **Refs.:** Dyar 1928: 323 (M*, L*); Rozeboom & Komp 1950a: 92 (M*; lectotype desig.); Lane 1953: 478 (M*, P*, L*); Foote 1954: 68 (L*; tax.); Stone & Knight 1957a: 53 (type info.); Fauran 1961b: 38 (distr.; French Guiana); Pecor et al. 1992: 26 (M*; tax., catalog citation); Torres-Gutierrez & Sallum 2015: 23 (tax., bion., distr., catalog citation). **Ety.:** n.s. **Informal name:** Maxinocca Surinamese Typical Mosquito.

syn. *tosimus* Dyar
1920c: 72 (M). **Type loc.:** Suriname (USNM). **Refs.:** Dyar 1928: 323 (syn.); Stone & Knight 1957a: 53 (type info.).

mesodenticulatus Galindo & Mendez
1961: 2 (M*). **Type loc.:** Almirante, Bocas del Toro Province, Panama (USNM). **Distr.:** Colombia, Panama. **Refs.:** Pecor et al. 1992: 37 (M*; tax., catalog citation); Torres-Gutierrez & Sallum 2015: 24 (tax., bion., distr., catalog citation). **Ety.:** n.s. [middle teeth or denticles (L)]. **Informal name:** Middle-toothed Panamanian Typical Mosquito.

milwardi Xavier & Mattos
1972: 569 (M*). **Type loc.:** Bairro Ressaca, Belo Horizonte, Minas Gerais, Brazil (FSP). **Distr.:** Brazil. **Refs.:** Forattini et al. 1988: 535 (type info.); Pecor et al. 1992: 37 (M*; tax., catalog citation); Torres-Gutierrez & Sallum 2015: 24 (tax., bion., distr., catalog citation). **Ety.:** Roberto Milward. **Informal name:** Milward Brazilian Typical Mosquito.

misionensis Duret
1953a: 75 (M*) & Duret 1954a: 115 (M*; type info.). **Type loc.:** Aristóbulo del Valle, Misiones, Argentina (USNM). **Distr.:** Argentina, Brazil. **Refs.:** Duret & Barreto 1956: 86 (distr.; Brazil); Harbach et al. 1991: 194 (type info.); Pecor et al. 1992: 37 (M*; tax., catalog citation); Torres-Gutierrez & Sallum 2015: 24 (tax., bion., distr., catalog citation). **Ety.:** Misiones, Argentina. **Informal name:** Misiones Argentinian Typical Mosquito.

mulrennani Basham
1948: 1 (M*, F, P*, L*). **Type loc.:** Big Pine Key, Monroe County, Florida, United States (USNM). **Distr.:** Bahamas, Cayman Isl., Cuba, United States (continental). **Refs.:** Rozeboom & Komp 1950a: 93 (M*); Yamaguti & LaCasse 1951c: 91 (M*, L*); Foote 1954: 71 (P*, L*); Carpenter & LaCasse 1955: 308 (M*, F, L*; keys); Stone et al. 1959: 273 (distr.; Cuba); King et al. 1960: 117 (A, L); Knight & Haeger 1971: 554 (F*; tax.); Belkin & Heinemann 1975b: 372, 377 (distr.; Bahamas, Cayman Isl.); Sirivanakarn 1983: 280 (M*); Pecor et al. 1992: 37 (M*; tax., catalog citation); Darsie & Ward 2005 (F*, L*; keys, distr.). **Ety.:** John A. Mulrennan. **Informal name:** Mulrennan American Typical Mosquito.

mutator Dyar & Knab
1906d: 216 (L*). **Type loc.:** Cordoba [Veracruz], Mexico (USNM). **Distr.:** Belize, Colombia, Costa Rica, El Salvador, Mexico, Panama. **Refs.:** Howard et al. 1913: Figs. 68, 361, 595 (M*, L*); Howard et al. 1915: 422 (M, F, L); Dyar 1923e (tax., type info.); Dyar 1928: 316 (M*, F, L*; distr.; Costa Rica); Komp 1935: 6 (tax.); Rozeboom & Komp 1950a: 93 (M*; lectotype desig.); Lane 1953: 461 (M*, L*);

Foote 1954: 73 (P*, L*); Galindo & Blanton 1955: 71 (distr.; Panama); Stone & Knight 1957a: 53 (type info.; clarification of lectotype); Pecor et al. 1992: 38 (M*; tax., catalog citation); Pecor et al. 2002: 248 (tax., distr.; Belize). **Ety.:** n.s. [move, change, shift (L)]. **Informal name:** Changeling Mexican Typical Mosquito.

nicceriensis Bonne-Wepster & Bonne
1920: 174 (M, F, L). **Type loc.:** Kabelstation, Suriname (ITH). **Distr.:** Lesser Antilles (includes Trinidad and Tobago), Suriname, Venezuela. **Refs.:** Bonne & Bonne-Wepster 1925: 288 (M*, F, L*); Dyar 1928: 303 (M*, F, L*); Rozeboom & Komp 1950a: 93 (M*); Foote 1954: 74 (L*); Cova García et al. 1966: 209 (M*); Belkin 1968b: 18 (lectotype desig.); Pecor et al. 1992: 38 (M*; tax., catalog citation); Torres-Gutierrez & Sallum 2015: 24 (tax., bion., distr., catalog citation). **Ety.:** n.s. **Informal name:** Niccer Surinamese Typical Mosquito.

ocossa Dyar & Knab
1919: 6 (M). **Type loc.:** Georgetown, British Guiana [Guyana] (USNM). **Distr.:** Argentina, Brazil, Colombia, Ecuador, Guyana, Honduras, Panama, Peru, Suriname, Venezuela. **Refs.:** Dyar 1923c: 120 (syn. with *aikenii*); Dyar 1928: 337 (in part; see *panocossa*; distr.; Colombia, Panama, Suriname, Venezuela; as *aikenii*); Stone & Knight 1957a: 54 (lectotype desig.); Belkin 1970b: 60 (from syn.); Belkin et al. 1970: 93 (M*, P*, L*); Kreutzer & Galindo 1980 (mol. tax.); Sirivanakarn & Jakob 1982: 195 (distr.; Argentina); Lee & Craig 1983 (F*; cibarium); Sirivanakarn 1983: 279 (M*, F*); Forattini & Sallum 1992b (F*; cibarium); Pecor et al. 1992: 38 (M*; tax., catalog citation); Sallum & Forattini 1996: 585 (M*, F*); Pecor et al. 2000 (bion., distr.; Peru); Sallum et al. 2001: 928 (P*, L*; tax.); Torres-Gutierrez & Sallum 2015: 24 (tax., bion., distr., catalog citation). **Ety.:** n.s. [for bone (L)]. **Informal name:** Bone Guyanese Typical Mosquito. (Vol. 1, p. 520)

oedipus Root
1927c: 588 (M*). **Type loc.:** Mage, Rio de Janeiro, Brazil (USNM). **Distr.:** Argentina, Brazil, Ecuador, Panama. **Refs.:** Dyar 1928: 309 (M*; syn. with *phlogistus*); Rozeboom & Komp 1950a: 94 (M*; lectotype desig.; from syn.); Duret 1953a, b (M*; key, distr.; Argentina); Lane 1953: 446 (M*); Duret 1954a: 117 (M*); Galindo & Blanton 1955: 71 (distr.; Panama); Stone & Knight 1957a: 54 (type info.); García & Casal 1965a: 9 (P*); Forattini & Sallum 1987a: 123 (M*, F*, P*, L*); Pecor et al. 1992: 39 (M*; tax., catalog citation), Pecor et al. 1992: 88 (distr.; Ecuador); Torres-Gutierrez & Sallum 2015: 25 (tax., bion., distr., catalog citation). **Ety.:** swollen foot (Gr); "swollen tip of the outer arm of the inner division of the lobe of the side-piece." **Informal name:** Oedipal Brazilian Typical Mosquito.

olimpioi Xavier, da Silva, & Mattos
1970: 183 (M*). **Type loc.:** Rio Branco, State of Acre, Brazil (INER). **Distr.:** Brazil, Peru. **Refs.:** Pecor et al. 1992: 39 (M*; tax., catalog citation); Pecor et al. 2000 (bion., distr.; Peru); Torres-Gutierrez & Sallum 2015: 25 (tax., bion., distr., catalog citation). **Ety.:** Olympio da Silva Pinto. **Informal name:** Olimpio Brazilian Typical Mosquito.

orfilai Duret
1953a: 74 (M*). **Type loc.:** Iguazu (Villa Tacuara), Misiones, Argentina (USNM). **Distr.:** Argentina. **Refs.:** Duret 1954a: 117 (M*; type info.); Harbach et al. 1991: 194 (type info.); Pecor et al. 1992: 39 (M*; tax., catalog citation); Torres-Gutierrez & Sallum 2015: 25 (tax., bion., distr., catalog citation). **Ety.:** Ricardo N. Orfila. **Informal name:** Orfila Argentinian Typical Mosquito.

palaciosi Duret
1968d: 15 (M*). **Type loc.:** Boa Vista, Roraima, Brazil (USNM). **Distr.:** Brazil, French Guiana. **Refs.:** Fauran & Pajot 1974a: 108 (distr.; French Guiana); Harbach et al. 1991: 194 (type info.); Pecor et al. 1992: 39 (M*; tax., catalog citation); Torres-Gutierrez & Sallum 2015: 25 (tax., bion., distr., catalog citation). **Ety.:** Silvio. **Informal name:** Brazilian Typical Mosquito.

panocossa Dyar
1923c: 120 (M). **Type loc.:** Bas Obispo, Canal Zone, Panama (USNM). **Distr.:** Belize, Brazil, Colombia, Costa Rica, El Salvador, Guatemala, Jamaica, Mexico, Panama, Venezuela. **Refs.:** Anduze 1941a: 15 (distr.; Venezuela; as *aikenii*); Stone & Knight 1957a: 54 (lectotype desig.); Cova García et al. 1966: 175 (M*; as *aikenii*); Belkin 1970b: 60 (from syn.; distr.; Costa Rica, Mexico); Belkin et al. 1970b: 93 (M*); Bertram 1971: 745 (distr.; Belize); Pecor et al. 1992: 40 (M*; tax., catalog citation), 88 (distr.; El Salvador); Sallum & Forattini 1996: 587 (M*); Torres-Gutierrez & Sallum 2015: 25 (tax., bion., distr., catalog citation). **Ety.:** n.s. [all or the whole (Gr) with sp. *ocossa*]. **Informal name:** Armored Panamanian Typical Mosquito [*panocossa* = "all over bone" = Armored].

paracrybda Komp
1936c: 330 (M*). **Type loc.:** Juán Díaz, Panama (USNM). **Distr.:** Argentina, Ecuador, Guatemala, Panama, Suriname. **Refs.:** Rozeboom & Komp 1950a: 94 (M*); Duret 1954a: 119 (M*); Duret 1969a: 10 (M*); Galindo 1969: 87 (tax.); Panday 1975a, b (distr.; Suriname); Sirivanakarn 1983: 279 (M*); Clark-Gil & Darsie 1983: 255 (distr.; Guatemala); Pecor et al. 1992: 40 (M*; tax., catalog citation); Sallum & Forattini 1996: 572 (M*, F); Torres-Gutierrez & Sallum 2015: 25 (tax., bion., distr., catalog citation). **Ety.:** near *crybda*. **Informal name:** Close Panamanian Typical Mosquito.

patientiae Floch & Fauran
1955: 1 (M*). **Type loc.:** Patience (Haute Mana), French Guiana [Guiana] (PIG). **Distr.:** French Guiana. **Refs.:** Pecor et al. 1992: 40 (M*; tax., catalog citation); Torres-Gutierrez & Sallum 2015: 26 (tax., bion., distr., catalog

citation). **Ety.:** Patience (Haute Mana), French Guiana. **Informal name:** Patience French Guianese Typical Mosquito.

pavlovskyi Casal & García
1967a: 19 (M*). **Type loc.:** Aeropuerto de Camba Punta, Corrientes, Argentina (INM). **Distr.:** Argentina. **Refs.:** Belkin et al. 1968: 16 (type info.); Pecor et al. 1992: 40 (M*; tax., catalog citation); Torres-Gutierrez & Sallum 2015: 26 (tax., bion., distr., catalog citation). **Ety.:** Eduardo Alejo Pavlovsky. **Informal name:** Pavlovsky Argentinian Typical Mosquito.

peccator Dyar & Knab
1909d: 256 (A). **Type loc.:** Scott, Lonoke County, Arkansas, United States (USNM). **Distr.:** Cuba, Mexico, Puerto Rico, United States (continental). **Refs.:** Howard et al. 1913: Fig. 67 (M*); Howard et al. 1915: 318 (M, F); Dyar 1928: 300 (in part, M*, F; misident. of L; see *erraticus*); King & Bradley 1937: 350 (L*; tax.); Rozeboom 1942b: 41 (M*, F, L); Matheson 1944: 217 (M*, F, L); Carpenter et al. 1946: 265 (M*, F, L*); Rozeboom & Komp 1950a: 94 (M*); Yamaguti & LaCasse 1951c: 98 (M*, F*, L*); Foote 1954: 79 (L*); Carpenter & LaCasse 1955: 312 (M*, F*, L*; keys); Vargas 1956b: 29 (distr.; Mexico); Stone & Knight 1957a: 54 (type info.); Belkin et al. 1966: 22 (type loc. info.); Dodge 1966: 370 (first instar L*; key); Knight & Haeger 1971: 554 (F*; tax.); Mattingly 1976: 227 (E*); Sirivanakarn 1983: 280 (F*); Pecor et al. 1992: 41 (M*; tax., catalog citation); Darsie & Ward 2005 (F*, L*; keys, distr.); Harrison et al. 2016 (F*, L*; keys, distr.). **Ety.:** n.s. **Informal name:** Criminal American Typical Mosquito [*peccator* = "pecco" = Criminal].
syn. *incriminator* Dyar & Knab
1909d: 257 (A). **Type loc.:** Agricultural College, Mississippi, United States (USNM). **Refs.:** Dyar & Knab 1918c: 179 (syn.); Stone & Knight 1957a: 51 (lectotype desig.).

pedroi Sirivanakarn & Belkin
1980: 12 (M*, F, *, P*, L*). **Type loc.:** Juán Mina (0.5 km south of Rio Chagres), Canal Zone, Panama. **Distr.:** Argentina, Bolivia, Brazil, Colombia, Costa Rica, Ecuador, French Guiana, Guatemala, Guyana, Lesser Antilles (includes Trinidad and Tobago), Mexico, Panama, Peru, Suriname, Venezuela. **Refs.:** Sirivanakarn 1983: 279 (M*); Sallum & Forattini 1996: 548 (M*, F*; Bolivia); Pecor et al. 2000 (tax., distr.; Peru). **Ety.:** Pedro Galindo. **Informal name:** Pedro Panamanian Typical Mosquito.

penai Sirivanakarn
1979: 135 (M*). **Type loc.:** Chapare, San Antonio, Bolivia (USNM). **Distr.:** Bolivia, Ecuador. **Refs.:** Sirivanakarn 1983: 280 (M*); Pecor et al. 1992: 42 (M*; tax., catalog citation); Torres-Gutierrez & Sallum 2015: 26 (tax., bion., distr., catalog citation). **Ety.:** Luis E. Peña G. **Informal name:** Peña Bolivian Typical Mosquito.

pereyrai Duret
1967a: 81 (M*). **Type loc.:** Cecilio Baez, Caaguazú, Paraguay (USNM). **Distr.:** Brazil, Paraguay. **Refs.:** Galindo 1969: 88 (tax.); Forattini & Sallum 1989c: 478 (M*, F*; Brazil); Harbach et al. 1991: 194 (type info.); Pecor et al. 1992: 42 (M*; tax., catalog citation); Sallum & Forattini 1996: 577 (M*, F*); Sallum et al. 2001: 937 (P*, L*; tax.); Torres-Gutierrez & Sallum 2015: 26 (tax., bion., distr., catalog citation). **Ety.:** Atilio Pereyra. **Informal name:** Pereyra Paraguayan Typical Mosquito.

phlabistus Dyar
1920c: 63 (M). **Type loc.:** Suriname (USNM). **Distr.:** Brazil, French Guiana, Suriname. **Refs.:** Dyar 1928: 313 (M*); Rozeboom & Komp 1950a: 94 (M*; tax.); Fauran & Pajot 1974a: 106 (distr.; French Guiana); Sirivanakarn 1983: 279 (M*); Pecor et al. 1992: 42 (M*; tax., catalog citation); Torres-Gutierrez & Sallum 2015: 27 (tax., bion., distr., catalog citation). **Ety.:** n.s. **Informal name:** Alabi Surinamese Typical Mosquito [Alabi (1783–1820) was a leader of the Maroon people of Suriname].
syn. *kerri* Duret
1968a: 11 (M*). **Type loc.:** Rio Preto, João Goulard, Amazonas, Brazil (USNM). **Refs.:** Sirivanakarn 1983: 274 (syn.); Harbach et al. 1991: 194 (type info.).

phlogistus Dyar
1920c: 61 (M, L). **Type loc.:** Suriname (USNM). **Distr.:** Brazil, Colombia, Ecuador, French Guiana, Guyana, Lesser Antilles (includes Trinidad & Tobago), Nicaragua, Panama, Suriname, Venezuela. **Refs.:** Dyar 1928: 309 (in part, L*; misident. of M*; see *oedipus*); Komp 1935: 5 (distr.; Panama); Rozeboom & Komp 1950a: 94 (M*; tax.); Lane 1953: 444 (M*, F, P*, L); Floch & Fauran 1954: 6 (M*; French Guiana); Foote 1954: 81 (P*, L*); Fauran 1961a: 4 (tax.); Cova García et al. 1966: 189 (M*); Forattini & Sallum 1987a: 149 (M*); Pecor et al. 1992: 42 (M*; tax., catalog citation); Torres-Gutierrez & Sallum 2015: 27 (tax., bion., distr., catalog citation). **Ety.:** n.s. [related to inflammation and fevers (Gr)]. **Informal name:** Inflamed Surinamese Typical Mosquito.

phyllados Hutchings & Sallum
2008: 42 (M*). **Type loc.:** Amazonas State, Novo Airão municipality, Parque Nacional do Jaú, Jaú River, Miratucu Stream, Brazil (INPA). **Distr.:** Brazil. **Refs.:** Torres-Gutierrez & Sallum 2015: 27 (tax., bion., distr., catalog citation). **Ety.:** leafy (Gr); ref. to distinguishing foliform setae on the male genitalia. **Informal name:** Leafy Brazilian Typical Mosquito.

pifanoi Anduze
1949b: 60 (M*). **Type loc.:** Caripito, Monagas, Venezuela (A). **Distr.:** Venezuela. **Refs.:** Cova García et al. 1966: 221 (M*); Forattini et al. 1970: 45 (type info.); Sirivanakarn 1983: 279 (M*); Pecor et al. 1992: 43 (M*; tax., catalog citation); Torres-Gutierrez & Sallum 2015: 27 (tax., bion.,

distr., catalog citation). **Ety.:** Félix Pifano C. **Informal name:** Pifano Venezuelan Typical Mosquito.

pilosus (Dyar & Knab)
1906d: 224 (L*; *Mochlostyrax*). **Type loc.:** Santa Lucrecia, [Veracruz], Mexico (USNM). **Distr.:** Argentina, Australia, Bahamas, Belize, Bolivia, Brazil, Colombia, Costa Rica, Cuba, Dominican Republic, Ecuador, El Salvador, French Guiana, Guatemala, Haiti, Honduras, Jamaica, Lesser Antilles (includes Trinidad and Tobago, Virgin Isl.), Mexico, Nicaragua, Panama, Paraguay, Peru, Puerto Rico, Suriname, United States (continental), Venezuela. **Refs.:** Howard et al. 1913: Figs. 80, 379 (M*, L*); Howard et al. 1915: 393 (M, F, L); Howard et al. 1915: 426 (M, F; as *mastigia*); Dyar & Barrett 1918 (L; as *floridanus*); Dyar 1924i: 186 (distr.; Colombia); Dyar 1925d: 29 (distr.; Ecuador); Dyar 1925f: 170 (distr.; Panama); Dyar 1928: 290 (M*, F, L*; distr.; Venezuela); Martini 1935: 61 (distr.; Belize); King & Bradley 1937: 353 (tax.); Kumm et al. 1940: 404 (distr.; Costa Rica); Matheson 1944: 218 (M*, F, L); Michener 1944: 263 (M*); Carpenter et al. 1946: 268 (M*, F, L*); Rozeboom & Komp 1950a: 94 (M*); Yamaguti & LaCasse 1951c: 102 (M*, F*, L*); Duret 1953a, b (M*; key, distr.; Argentina); Lane 1953: 452 (M*, F, P*, L*; distr.; Brazil); Duret 1954a: 99 (M); Foote 1954: 114 (P*, L*; distr.; Honduras, Bahamas, Nicaragua); Carpenter & LaCasse 1955: 313 (M*, F*, L*; keys); Stone & Knight 1957a: 54 (lectotype desig.); King et al. 1960: 115 (A, L); Stone 1961c: 47 (distr.; Trinidad); Cova García et al. 1966: 169 (M*); Dodge 1966: 370 (first instar L*; key); Duret 1969a (distr.; Paraguay); Belkin et al. 1970: 90 (M*, F, P*, L*); Knight & Haeger 1971: 554 (F*; tax.); Belkin & Heinemann 1972: 66 (distr.; Dominican Republic); Mattingly 1976: 223 (E*); Heinemann & Belkin 1977b: 428 (distr.; Guatemala); Peyton et al. 1983: 73 (distr.; Bolivia); Sirivanakarn 1983: 280 (M*, F*); Pecor et al. 1992: 43 (M*; tax., catalog citation); Pecor et al. 2000 (bion., distr.; Peru); Darsie & Ward 2005 (F*, L*; keys, distr.); Linton et al. 2013 (bion., mol. tax., distr.; Ecuador); Torres-Gutierrez & Sallum 2015: 27 (tax., bion., distr., catalog citation); Harrison et al. 2016 (F*, L*; keys, distr.). **Ety.:** n.s. [hairy, shaggy (L)]. **Informal name:** Hairy Mexican Typical Mosquito.
syn. *floridanus* Dyar & Knab
1906b: 171 (L; *Mochlostyrax*). **Type loc.:** Estero, Florida, United States (NE). **Refs.:** Dyar 1924i: 186 (syn.); Stone & Knight 1957a: 50 (type info.).
syn. *cubensis* Dyar & Knab
1906d: 225 (L*; *Mochlostyrax*). **Type loc.:** Havana, Cuba (NE). **Refs.:** Dyar & Knab 1907b: 100 (preoccupied name; see *agitator* Dyar & Knab); Stone & Knight 1957a: 46 (type info.).
syn. *jamaicensis* Grabham
1906b: 318 (M, F, L*; *Mochlostyrax*). **Type loc.:** Kingston, [Surrey], Jamaica (USNM). **Refs.:** Dyar & Knab 1909d: 257 (preoccupied name; see *reductor* Dyar & Knab); Dyar 1924i: 186 (syn.); Belkin et al. 1970: 90 (lectotype desig.).
syn. *agitator* Dyar & Knab
1907b: 100 (new name for *cubensis* Dyar & Knab, not Bigot 1857; *Mochlostyrax*). **Type loc.:** Havana Cuba (NE). **Refs.:** Dyar & Knab 1918c: 180 (syn. with *floridanus*).
syn. *hesitator* Dyar & Knab
1907c: 205 (M, F). **Type loc.:** Las Cascadas, Canal Zone, Panama (USNM). **Refs.:** Lane 1951: 334 (syn.); Stone & Knight 1957a: 50 (lectotype desig.).
syn. *deceptor* Dyar & Knab
1909d: 257 (A). **Type loc.:** Fort White, Florida, United States (USNM). **Refs.:** Dyar & Knab 1918c: 180 (syn. with *foridanus*); Stone & Knight 1957a: 47 (lectotype desig.).
syn. *reductor* Dyar & Knab
1909d: 257 (new name for *jamaicensis* Grabham, not Theobald 1901a). **Refs.:** Dyar 1924i: 186 (syn.).
syn. *ignobilis* Dyar & Knab
1909b: 39 (A). **Type loc.:** San Antonio de los Baños, [Havana], Cuba (USNM). **Refs.:** Dyar 1928: 94 (syn.); Stone & Knight 1957a: 51 (lectotype desig.).
syn. *mastigia* Howard, Dyar & Knab
1913: Fig. 90 (M*), 1915: 426 (M*, F). **Type loc.:** San Antonio de los Baños, [Havana], Cuba (USNM). **Refs.:** Dyar & Knab 1918c: 180 (syn. with *floridanus*); Stone & Knight 1957a: 53 (lectotype desig.).
syn. *curopinensis* Bonne-Wepster & Bonne
1920: 177 (M). **Type loc.:** Suriname (ITH). **Refs.:** Dyar 1924i: 186 (syn.); Belkin 1968b: 15 (lectotype desig.).
syn. *colombiensis* Dyar
1924i: 184 (M). **Type loc.:** Colombia (USNM). **Refs.:** Dyar 1928: 289 (syn. with *hesitator*).
syn. *radiatus* Senevet & Abonnenc
1939a: 120 (M*, L*). **Type loc.:** Cayenne [Guyane] and Saut Tigre [Inini], French Guiana [Guiana] (FMP; originally in LPFM). **Refs.:** Rozeboom & Komp 1950a: 94 (syn.); Belkin 1968b: 19 (type info.).

plectoporpe Root
1927c: 589 (M*). **Type loc.:** Bangu, Rio de Janeiro, Brazil (USNM). **Distr.:** Argentina, Brazil, Colombia, Ecuador, French Guiana, Panama. **Refs.:** Dyar 1928: 312 (M*); Komp 1935: 6 (distr.; Panama); Senevet & Abonnenc 1939a: 83 (M*, L*); Rozeboom & Komp 1950a: 95 (M*); Lane 1953: 447 (M*, L); Foote 1954: 82 (L*); Stone & Knight 1957a: 55 (lectotype desig.); Stone 1961c: 47 (syn. with *phlogistus*); Clastrier 1970c: 467 (from syn.); Sirivanakarn & Jakob 1982: 197 (distr.; Argentina); Forattini & Sallum 1987a: 127 (M*, F*, P*, L*); Pecor et al. 1992: 45 (M*; tax., catalog citation); Torres-Gutierrez & Sallum 2015: 28 (tax., bion., distr., catalog citation). **Ety.:** n.s. [twisted (pin or ring?) (Gr); poss. ref. to "peculiar lateral flap near the tip of the clasper"]. **Informal name:** Twisted Brazilian Typical Mosquito.

portesi Senevet & Abonnenc
1941: 41 (M*). **Type loc.:** French Guiana [Guiana] (LPFM). **Distr.:** Brazil, French Guiana, Guyana, Lesser

Antilles (includes Trinidad and Tobago), Peru, Suriname, Venezuela. **Refs.:** Rozeboom & Komp 1950a: 95 (M*; tax.); Lane 1951: 334 (syn.); Aitken & Galindo 1966: 202 (M*, F; from syn.; distr.; Brazil, Trinidad); Belkin 1968b: 53 (type info.); Galindo 1969: 87 (tax.); Panday 1975a, b (distr.; Suriname); Mattingly 1976: 224 (E*); Sirivanakarn & Dégallier 1982: 153 (M*, F, P*, L*; distr.; Guyana); Sirivanakarn 1983: 278 (M*); Sallum & Forattini 1996: 581 (M*, F*); Pecor et al. 2000 (distr.; Peru). **Ety.:** M. le Professeur Portes. **Informal name:** Portes French Guianese Typical Mosquito.

syn. *cayennensis* Floch & Abonnenc 1945b: 4 (M*). **Type loc.:** Cayenne [Guyane], French Guiana [Guiana] (NE). **Refs.:** Flock & Abonnenc 1947 (syn.); Lane 1953: 430 (syn.); Floch & Kramer 1965: 3 (M*; from syn.); Aitken & Galindo 1966 (syn. with *portesi*); Belkin 1968b: 14 (type info.); Harrison 1973c: 277 (type info.); Sirivanakarn & Dégallier 1982: 154 (syn.).

productus Senevet & Abonnenc
1939a: 107 (M*, L*). **Type loc.:** Saint-Élie [Inini], French Guiana [Guiana] (LPFM). **Distr.:** Brazil, French Guiana. **Refs.:** Rozeboom & Komp 1950a: 89 (syn. with *comminutor*); Foote 1954: 83 (L*; from syn.); Belkin 1968b: 53 (type info.); Clastrier 1971: 638 (M*, F, P*, L*); Pecor et al. 1992: 45 (M*; tax., catalog citation); Torres-Gutierrez & Sallum 2015: 28 (tax., bion., distr., catalog citation). **Ety.:** n.s. [lead forward, bring out (L); prob. ref. to "basal lobe more produced and pointed" (comp. to *leucopus*)]. **Informal name:** Pointed French Guianese Typical Mosquito.

psatharus Dyar
1920g: 173 (M). **Type loc.:** Colón, Canal Zone, Panama (USNM). **Distr.:** Costa Rica, Ecuador, Panama. **Refs.:** Dyar 1928: 296 (M*, F); Rozeboom & Komp 1950a: 95 (M*; lectotype desig.); Lane 1953: 481 (M*, F, P*, L*); Foote 1954: 84 (P*, L*); Stone & Knight 1957a: 55 (type info.); Heinemann & Belkin 1979: 108 (distr.; Ecuador); Sirivanakarn 1983: 279 (M*); Pecor et al. 1992: 46 (M*; tax., catalog citation); Torres-Gutierrez & Sallum 2015: 28 (tax., bion., distr., catalog citation). **Ety.:** n.s. **Informal name:** Brittle Panamanian Typical Mosquito [*psatharus* = "psathyrus" = Brittle].

putumayensis Matheson
1934b: 121 (M*). **Type loc.:** Santo Antonio do Ica, Amazonas, Brazil (USNM). **Distr.:** Brazil, Colombia, Ecuador, French Guiana, Lesser Antilles (includes Trinidad and Tobago), Peru, Suriname. **Refs.:** Rozeboom & Komp 1950a: 95 (M*); Lane 1953: 431 (M*); Stone 1961c: 47 (distr.; Trinidad); Belkin et al. 1971: 24 (type loc. info.); Clastrier 1971: 630 (M*, F, P*, L*); Panday 1975a, b (distr.; Suriname); Sirivanakarn 1983: 279 (M*); Pecor et al. 1992: 46 (M*; tax., catalog citation); Torres-Gutierrez & Sallum 2015: 28 (tax., bion., distr., catalog citation). **Ety.:** Putumayo, Colombia. **Informal name:** Putumayo Brazilian Typical Mosquito.

syn. *cavernicolus* Floch & Abonnenc 1945c: 1 (M*; as *cavernicolus*). **Type loc.:** Cayenne [Guyane], French Guiana [Guiana] (PIP). **Refs.:** Rozeboom & Komp 1950a: 95 (syn., as *cavernicolus*); Belkin 1968b: 14 (tax., as *cavernicolus*); Harrison 1973c: 277 (type info.); Knight & Stone 1977: 273 (emend.).

quadrifoliatus Komp
1936c: 322 (M*). **Type loc.:** Mojinga Swamp, lower Chagres River, Canal Zone, Panama (USNM). **Distr.:** Panama. **Refs.:** Rozeboom & Komp 1950a: 96 (M*); Pecor et al. 1992: 46 (M*; tax., catalog citation). **Ety.:** n.s. [four leaves (L)]. **Informal name:** Shamrock Panamanian Typical Mosquito [*quadrifoliatus* = "four-leafed" = Shamrock, a four-leafed clover].

quasihibridus Galindo & Blanton
1954: 232 (M*). **Type loc.:** Puerto Pilón, Panama (USNM). **Distr.:** Colombia, El Salvador, Panama. **Refs.:** Barreto & Lee 1969: 430 (distr.; Colombia); Pecor et al. 1992: 46 (M*; tax., catalog citation); Torres-Gutierrez & Sallum 2015: 29 (tax., bion., distr., catalog citation). **Ety.:** nearly (L); similar to a hybrid (*educator* and *iolambdis*). **Informal name:** Near-hybrid Panamanian Typical Mosquito.

rabanicolus Floch & Abonnenc
1946a: 1 (M*). **Type loc.:** Raban, French Guiana [Guiana] (PIP). **Distr.:** French Guiana, Guyana, Lesser Antilles (includes Trinidad and Tobago). **Refs.:** Rozeboom & Komp 1950a: 96 (M*); Lane 1953: 451 (M*); Stone 1967: 217 (emend.); Harrison 1973c: 278 (type info.); Heinemann & Belkin 1978: 445 (distr.; Guyana); Heinemann et al. 1980: 214 (distr.; Trinidad); Sirivanakarn 1983: 279 (M*); Rodhain & Boutonnier 1984a: 277 (type info.); Pecor et al. 1992: 47 (M*; tax., catalog citation); Torres-Gutierrez & Sallum 2015: 29 (tax., bion., distr., catalog citation). **Ety.:** Raban, French Guiana. **Informal name:** Raban French Guianese Typical Mosquito.

rabelloi Forattini & Sallum
1987a: 130 (M*, F*, P*, L*). **Type loc.:** Pariquera-Mirím, Pariquera-Açu, São Paulo, Brazil (FSP). **Distr.:** Argentina, Brazil. **Refs.:** Forattini et al. 1988: 537 (type info.); Pecor et al. 1992: 47 (M*; tax., catalog citation); Torres-Gutierrez & Sallum 2015: 29 (tax., bion., distr., catalog citation). **Ety.:** Ernesto X. Rabello. **Informal name:** Rabello Brazilian Typical Mosquito.

rachoui Duret
1968b: 58 (M*). **Type loc.:** Paragominas, Município de Capim, Pará, Brazil (USNM). **Distr.:** Brazil. **Refs.:** Harbach et al. 1991: 194 (type info.); Pecor et al. 1992: 47 (M*; tax., catalog citation); Torres-Gutierrez & Sallum 2015: 29 (tax., bion., distr., catalog citation). **Ety.:** René Guimarães Rachou. **Informal name:** Rachou Brazilian Typical Mosquito.

ribeirensis Forattini & Sallum
1985: 171 (M*, F*, P*, L*). **Type loc.:** Ribeira Valley, Pariquera-Açu, São Paulo, Brazil (FSP). **Distr.:** Argentina, Brazil. **Refs.:** Forattini et al. 1988: 537 (type info.); Pecor et al. 1992: 47 (M*; tax., catalog citation); Forattini 1996 (F*); Sallum & Forattini 1996: 564 (M, F*); Navarro & Weaver 2004 (phyl.); Torres-Gutierrez & Sallum 2015: 29 (tax., bion., distr., catalog citation). **Ety.:** Ribeira Valley, São Paulo State, Brazil. **Informal name:** Ribeira Brazilian Typical Mosquito.

ronderosi de Linero
1967: 289 (M*, F*). **Type loc.:** Oricopiche, Ciudad Bolivar, Bolivar, Venezuela (EUO). **Distr.:** Venezuela. **Refs.:** Pecor et al. 1992: 48 (M*; tax., catalog citation); Torres-Gutierrez & Sallum 2015: 29 (tax., bion., distr., catalog citation). **Ety.:** Ricardo Ronderos. **Informal name:** Ronderos Venezuelan Typical Mosquito.

rooti Rozeboom
1935: 251 (M*). **Type loc.:** Panama City, Panama (USNM). **Distr.:** Argentina, Belize, Brazil, Colombia, French Guiana, Guyana, Mexico, Panama, Venezuela. **Refs.:** Rozeboom 1936 (L*); Rozeboom & Komp 1950a: 96 (M*); Martínez 1952b: 85 (distr.; Mexico); Duret 1953a, b (M*; key, distr.; Argentina); Duret 1954a: 102 (M*); Foote 1954: 116 (P*, L*; tax.); Stone & Knight 1957a: 55 (lectotype desig.); Sutil Oramas 1980: 15 (distr.; Venezuela); Pecor et al. 1992: 48 (M*; tax., catalog citation); Pecor et al. 2002: 248 (tax., distr.; Belize); Torres-Gutierrez & Sallum 2015: 30 (tax., bion., distr., catalog citation). **Ety.:** Francis Metcalf Root. **Informal name:** Root Panamanian Typical Mosquito.

rorotaensis Floch & Abonnenc
1946a: 3 (M*). **Type loc.:** Rorota [Guyane], French Guiana [Guiana] (PIG). **Distr.:** Brazil, Colombia, French Guiana, Suriname. **Refs.:** Rozeboom & Komp 1950a: 96 (M*); Lane 1953: 485 (M*); Belkin 1968b: 47 (type info.); Forattini et al. 1988: 537 (distr.; Brazil); Pecor et al. 1992: 48 (M*; tax., catalog citation), 88 (distr.; Suriname); Torres-Gutierrez & Sallum 2015: 30 (tax., bion., distr., catalog citation). **Ety.:** Rorota, French Guiana. **Informal name:** Rorota French Guianese Typical Mosquito.

sacchettae Sirivanakarn & Jakob
1982: 192 (M*, F). **Type loc.:** Cananéia (Brucouha), São Paulo, Brazil (FH). **Distr.:** Brazil. **Refs.:** Forattini et al. 1988: 537 (type info.); Forattini & Sallum 1989d: 117 (M*, F*, P*, L*); Forattini 1996 (F*); Sallum & Forattini 1996: 582 (M*, F*). **Ety.:** Lia de Abreu Sacchetta. **Informal name:** Sacchetta Brazilian Typical Mosquito.

saramaccensis Bonne-Wepster & Bonne
1920: 172 (M, F, L). **Type loc.:** Kabelstation, Suriname River, Suriname (ITH). **Distr.:** Ecuador, French Guiana, Suriname. **Refs.:** Bonne & Bonne-Wepster 1925: 305 (M*, F, L); Dyar 1928: 304 (M*, F, L*); Rozeboom & Komp 1950a: 96 (M*); Lane 1953: 454 (M, F, L); Foote 1954: 86 (L*); Foote 1954: 58 (L*; as *implicatus*); Belkin 1968b: 20 (lectotype desig.); Sirivanakarn 1983: 279 (M*); Pecor et al. 1992: 48 (M*; tax., catalog citation); Torres-Gutierrez & Sallum 2015: 30 (tax., bion., distr., catalog citation). **Ety.:** Saramacca River, Suriname. **Informal name:** Saramacca Surinamese Typical Mosquito.
syn. *implicatus* Senevet & Abonnenc
1939a: 99 (M*, L*). **Type loc.:** Sinnamary [Guyane], French Guiana [Guiana] (NE). **Refs.:** Lane 1951: 334 (syn. with *nigrescens*); Belkin 1968b: 53 (type info.); Sirivanakarn 1983: 274 (from syn. with *nigrescens*).

sardinerae Fox
1953: 179 (M*, F, P*, L*). **Type loc.:** Dorado (Finca Sardinera) [Arecibo], Puerto Rico (STMPR). **Distr.:** Cuba, Guatemala, Mexico, Panama, Puerto Rico. **Refs.:** Sirivanakarn 1983: 279 (M); Clark-Gil & Darsie 1983: 257 (distr.; Guatemala); Pecor et al. 1992: 49 (M*; tax., catalog citation). **Ety.:** Finca Sardinera, Puerto Rico. **Informal name:** Sardinera Puerto Rican Typical Mosquito.
syn. *bilobatus* Galindo & Blanton
1954: 236 (M*). **Type loc.:** Rio Banana, near Almirante, Panama (USNM). **Refs.:** Galindo & Blanton 1955: 72 (syn.).

serratimarge Root
1927c: 589 (M*). **Type loc.:** Sant'Anna [de Japin], Rio de Janeiro, Brazil (USNM). **Distr.:** Argentina, Bolivia, Brazil, Colombia, Ecuador, French Guiana, Guatemala, Guyana, Lesser Antilles (includes Trinidad and Tobago), Nicaragua, Panama, Paraguay, Peru, Venezuela. **Refs.:** Dyar 1928: 312 (M*; unjustified emend. to *serratamargo*); Komp 1935: 6 (distr.; Panama; as *serratamargo*); Senevet & Abonnenc 1939a: 85 (M*, L*; distr.; French Guiana, as *serratimargo*); Rozeboom & Komp 1950a: 96 (M*; original spelling); Duret 1953a, b (M*; key, distr.; Argentina); Lane 1953: 419 (M*, L); Duret 1954a: 120 (M*); Foote 1954: 87 (P*, L*; distr.; Colombia); Duret 1969a: 8 (distr.; Paraguay); Pecor et al. 1992: 49 (M*; tax., catalog citation), 88 (distr.; Trinidad); Pecor et al. 2000 (distr.; Peru); Linton et al. 2013 (bion., mol. tax., distr.; Ecuador); Torres-Gutierrez & Sallum 2015: 30 (tax., bion., distr., catalog citation). **Ety.:** n.s. [saw, saw up (L)]. **Informal name:** Sharp-edged Brazilian Typical Mosquito.

silvai Duret
1968f: 69 (M*). **Type loc.:** Caracaraí, Roraima, Brazil (A). **Distr.:** Brazil. **Refs.:** Harbach et al. 1991: 194 (type info.); Pecor et al. 1992: 49 (M*; tax., catalog citation); Torres-Gutierrez & Sallum 2015: 30 (tax., bion., distr., catalog citation). **Ety.:** Tito López da Silva. **Informal name:** Silva Brazilian Typical Mosquito.

simulator Dyar & Knab
1906d: 218 (L*). **Type loc.:** Arima, Trinidad [Trinidad & Tobago] (USNM). **Distr.:** Brazil, Lesser Antilles (includes

Trinidad and Tobago), Panama, Venezuela. **Refs.:** Howard et al. 1913: Figs. 352, 575 (L*); Howard et al. 1915: 302 (L); Dyar 1928: 333 (L*); Rozeboom & Komp 1950a: 98 (tax.); Foote 1954: 89 (L*; tax.); Stone & Knight 1957a: 56 (lectotype desig.); Sirivanakarn & Heinemann 1980: 41 (M*, F*, P*, L*; distr.; Panama); Pecor et al. 1992: 50 (M*; tax., catalog citation); Sallum & Forattini 1996: 588 (M*; Brazil); Torres-Gutierrez & Sallum 2015: 31 (tax., bion., distr., catalog citation). **Ety.:** n.s. [simulator (L)]. **Informal name:** Simulator Trinbagonian Typical Mosquito.

syn. *venezuelensis* Anduze
1949b: 64 (M*). **Type loc.:** Caripito, Monagas, Venezuela (FH). **Refs.:** Forattini et al. 1970: 50 (type info.); Sirivanakarn 1983: 274 (syn.).

spathulatus Forattini & Sallum
1987b: 167 (M*). **Type loc.:** Santa Helena Farm, São João da Boa Vista, São Paulo, Brazil (FSP). **Distr.:** Brazil. **Refs.:** Forattini et al. 1988: 538 (type info.); Pecor et al. 1992: 50 (M*; tax., catalog citation); Torres-Gutierrez & Sallum 2015: 31 (tax., bion., distr., catalog citation). **Ety.:** flat piece (L); ref. to distal division of gonocoxite subapical lobe with a "remarkable large spatula-shaped seta." **Informal name:** Spatula Brazilian Typical Mosquito.

spissipes (Theobald)
1903a: 242 (F*; *Melanoconion*). **Type loc.:** Trinidad [Trinidad & Tobago, Lesser Antilles] (NHMUK). **Distr.:** Belize, Bolivia, Brazil, Colombia, Ecuador, French Guiana, Guatemala, Guyana, Honduras, Lesser Antilles (includes Trinidad and Tobago), Mexico, Nicaragua, Panama, Paraguay, Peru, Suriname, Venezuela. **Refs.:** Bourroul 1904: 70 (distr.; Brazil); Howard et al. 1915: 312 (M; tax.); Dyar 1925f: 169 (distr.; Panama); Dyar 1925g: 214 (distr.; Venezuela); Bonne & Bonne-Wepster 1925: 268 (in part, Suriname; misident. of M; see Takahashi 1968); Martini 1935: 60 (distr.; Belize, Mexico); Foote 1954: 94 (tax.); Barreto 1955: 60 (distr.; Colombia); Prosen et al. 1963: 110 (Lane cited for Bolivian record); Takahashi 1968: 329 (M*, F*; tax.); Galindo 1969: 87 (tax.; Spissipes Group); Stone 1970: 164 (distr.; Honduras); Heinemann & Belkin 1979: 108 (distr.; Ecuador); Darsie & Hobbs 1982: 73 (distr.; Guatemala); Sirivanakarn 1983: 278 (M*, F*); Pecor et al. 1992: 9 (M*; tax., catalog citation), 11 (M*; tax., catalog citation; as *alvarezi*), 41 (M*; tax., catalog citation), 45 (M*; tax., catalog citation), 48 (M*; tax., catalog citation), 50 (M*; tax., catalog citation), 51 (M*; tax., catalog citation), 55 (M*; tax., catalog citation); Sallum & Forattini 1996: 522 (M*, F*); Pecor et al. 2000 (bion., distr.; Peru); Navarro & Weaver 2004 (phyl.); Berti et al. 2015 (distr.; Venezuela); Torres-Gutierrez & Sallum 2015 (tax., bion., distr., catalog citation). **Ety.:** thick footed (L); ref. to thick, flattened midlegs. **Informal name:** Thick-footed Trinbagonian Typical Mosquito. (Vol. 1, p. 522)

syn. *fur* Dyar & Knab
1907a: 13 (F). **Type loc.:** Colón, Canal Zone, Panama (USNM). **Refs.:** Bonne-Wepster & Bonne 1921a: 20 (syn. with *spissipes*); Lane 1951: 334 (from syn.); Foote 1954: 94 (syn.); Takahashi 1968: 329 (tax.).

syn. *menytes* Dyar
1918d: 125 (M). **Type loc.:** Trinidad River, Panama (USNM). **Refs.:** Rozeboom & Komp 1950a: 93 (M*; type loc. info.); Takahashi 1968: 331 (syn.).

syn. *haynei* Komp & Curry
1932: 82 (M*). **Type loc.:** Mojinga Swamp, lower Chagres River, Canal Zone, Panama (USNM). **Refs.:** Komp 1935: 3 (syn.); Stone & Knight 1957a: 59 (type info.).

syn. *alvarezi* Sutil Oramas, Pulido Florenzano, & Amarista Meneses
1987: 83 (M*). **Type loc.:** Chiricoa, San Camillo, Apure, Venezuela (DERM). **Refs.:** Sallum & Forattini 1996: 523 (syn.).

sursumptor Dyar
1924d: 123 (M). **Type loc.:** Barranquilla [Magdalena], Colombia (USNM). **Distr.:** Colombia, Ecuador, Panama, Venezuela. **Refs.:** Dyar 1924i: 183 (tax.); Dyar 1928: 329 (M*, F, L*); Anduze 1941a: 16 (distr.; Venezuela); Rozeboom & Komp 1950a: 96 (M*; tax.); Foote 1954: 90 (P*, L*; tax.); Galindo & Blanton 1955: 72 (distr.; Panama); Stone & Knight 1957a: 56 (lectotype desig.); Cova García et al. 1966: 209 (M*); Pecor et al. 1992: 51,88 (M*; tax., catalog citation, distr.; Ecuador); Torres-Gutierrez & Sallum 2015: 31 (tax., bion., distr., catalog citation). **Ety.:** n.s. [upward, on high (L); poss. ref. to "occiput dark brown, the scales oval, not very broad, but this type reaching the vertex"]. **Informal name:** Upward Colombian Typical Mosquito.

syn. *ligator* Dyar
1924d: 123 (M). **Type loc.:** Barranquilla [Magdalena], Colombia (USNM). **Refs.:** Stone & Knight 1957a: 52 (lectotype desig.).

symbletos Sallum & Hutchings
2003: 620 (M*). **Type loc.:** State of Amazonas, Novo Airão County, Parque Nacional do Jaú, Carabinani River (01°59′ S, 61°32′ W), Brazil (INPA). **Distr.:** Brazil, Ecuador, Peru. **Refs.:** Linton et al. 2013 (bion., mol. tax., distr.; Ecuador); Torres-Gutierrez & Sallum 2015: 32 (tax., bion., distr., catalog citation). **Ety.:** comparable (Gr); ref. to similarity with *mesodenticulatus*. **Informal name:** Carabinani Brazilian Typical Mosquito.

taeniopus Dyar & Knab
1907b: 100 (F). **Type loc.:** Bluefields, Nicaragua (USNM). **Distr.:** Argentina, Bahamas, Belize, Bolivia, Brazil, Cayman Isl., Colombia, Costa Rica, Dominican Republic, Ecuador, French Guiana, Guatemala, Guyana, Honduras, Jamaica, Lesser Antilles (includes Trinidad and Tobago), Mexico, Nicaragua, Panama, Paraguay,

Peru, Puerto Rico, Suriname, United States (continental), Venezuela. **Refs.:** Komp 1935: 4 (M; tax.); Rozeboom & Komp 1950a: 96 (in part, tax.; misident. of M*; see *pedroi*); Yamaguti & LaCasse 1951c (F*) (as *opisthopus* Komp); Lane 1953: 403 (M*, F*; as *opisthopus*); Duret 1954a: 102 (M*); Carpenter & LaCasse 1955: 310 (in part, M*; misident. of L*); Galindo 1969: 83 (tax.); Belkin et al. 1970: 82 (M*, F*, P*L*; as *opisthopus*); Morales-Ayala 1971: 143 (distr.; Peru); Sirivanakarn & Belkin 1980: 8 (M, F; tax.); Sirivanakarn 1983: 265 (M*, F*); Mitchell & Darsie 1985: 285, 314 (distr.; Argentina); Cupp et al. 1989 (bion.); Forattini & Sallum 1992a: 72 (distr.; Brazil); Pecor et al. 1992: 51 (tax., review); Sallum & Forattini 1996: 526 (M*, F); Navarro & Weaver 2004 (phyl.); Torres-Gutierrez & Sallum 2015: 32 (tax., bion., distr., catalog citation). **Ety.:** banded legs or feet (Gr); legs "black, with broad white rings on the tarsi." **Informal name:** Ribbon-footed Nicaraguan Typical Mosquito. (Vol. 1, p. 524)

syn. *annulipes* Theobald
1907: 512 (F*; *Melanoconion*). **Type loc.:** Red Hills [Kingston, Surrey], Jamaica (NHMUK). **Refs.:** Belkin 1968b: 13 (tax.); Belkin 1969a: 68 (tax., syn.); Galindo 1969: 87 (tax.).

syn. *opisthopus* Komp
1926: 44 (M, F). **Type loc.:** Puerto Castillo, Honduras (USNM). **Refs.:** Yamaguti & LaCasse 1951c: 91 (M*, F*, L*); Stone & Knight 1957a: 54 (lectotype desig.); Sirivanakarn & Belkin 1980: 7 (syn.); Pecor et al. 1992: 52 (tax. review).

syn. *mychonde* Komp
1928 (M). In Dyar 1928: 295. **Type loc.:** Almirante, Panama (USNM). **Refs.:** Komp 1935: 3 (syn.); Pecor et al. 1992: 52 (tax., review); Sallum & Forattini 1996 (syn.).

tecmarsis Dyar
1918d: 124 (M, F). **Type loc.:** Trinidad River, Panama (USNM). **Distr.:** Brazil, Colombia, Costa Rica, Ecuador, Panama, Venezuela. **Refs.:** Dyar 1925g: 214 (distr.; Venezuela); Dyar 1928: 313 (M*, F); Rozeboom & Komp 1950a: 97 (M*; lectotype desig.); Lane 1953: 420 (M*, F); Stone & Knight 1957a: 57 (type info.); Cova García et al. 1966: 167 (M*); Sirivanakarn 1983: 280 (M*); Pecor et al. 1992: 52 (M*; tax., catalog citation), Pecor et al. 1992: 88 (distr.; Colombia); Torres-Gutierrez & Sallum 2015: 32 (tax., bion., distr., catalog citation). **Ety.:** n.s. **Informal name:** Driven Panamanian Typical Mosquito [*tecmarsis* = tecmarto = "goal" = Driven].

terebor Dyar
1920c: 56 (M). **Type loc.:** Suriname (USNM). **Distr.:** Suriname. **Refs.:** Dyar 1928: 304 (M*); Rozeboom & Komp 1950a: 97 (M*); Lane 1953: 449 (M*); Pecor et al. 1992: 52 (M*; tax., catalog citation); Torres-Gutierrez & Sallum 2015: 33 (tax., bion., distr., catalog citation). **Ety.:** n.s. **Informal name:** Terebor Surinamese Typical Mosquito.

theobaldi (Lutz)
1904 (F; *Melanoconion*). In Bourroul 1904: 70. **Type loc.:** Lagoa, São Paulo, Brazil (NHMUK, IOC?). **Distr.:** Argentina, Belize, Bolivia, Brazil, Colombia, Costa Rica, Ecuador, French Guiana, Guatemala, Guyana, Honduras, Lesser Antilles (includes Trinidad and Tobago), Mexico, Nicaragua, Panama, Paraguay, Peru, Suriname, Uruguay, Venezuela. **Refs.:** Bonne & Bonne-Wepster 1925: 293 (M*, F, L); Cerqueira 1943b: 34 (distr.; Bolivia); Rozeboom & Komp 1950a: 97 (M*; tax.); Duret 1954a: 103 (M*); Foote 1954: 92 (P*, L*; tax.); Belkin 1968b: 20 (type info.); Duret 1969a: 8 (distr.); Belkin et al. 1971: 25 (type info., lectotype desig.); Morales-Ayala 1971: 143 (distr.; Peru); Mitchell & Darsie 1985: 285, 314 (distr.; Argentina); Forattini & Sallum 1989b: 201 (M*, F*, P*, L*); Pecor et al. 1992: 52 (M*; tax., catalog citation); Rossi & Martínez 2003: 472 (distr.; Uruguay); Linton et al. 2013 (bion., mol. tax., distr.; Ecuador); Torres-Gutierrez & Sallum 2015: 33 (tax., bion., distr., catalog citation). **Ety.:** Frederic Vincent Theobald. **Informal name:** Theobald Brazilian Typical Mosquito.

syn. *chrysonotum* Dyar & Knab
1908: 57 (M, F). **Type loc.:** Ancon, Canal Zone, Panama (USNM). **Refs.:** Rozeboom & Komp 1950a: 88 (in part, lectotype desig.; misident. of M; see Forattini & Sallum 1989b); Lane 1951: 410 (syn. with *spissipes*); Foote 1954: 31 (P*, L*; from syn.); Stone & Knight 1957a: 45 (type info.); Forattini & Sallum 1989b: 201 (syn.).

syn. *aurilatus* Senevet & Abonnenc
1939a: 94 (M*, L*). **Type loc.:** French Guiana [Guiana] (NE). **Refs.:** Rozeboom & Komp 1950a: 88 (syn.); Belkin 1968b: 53 (type info.); Forattini & Sallum 1989b (syn.).

tournieri Senevet & Abonnenc
1939a: 105 (M*, L*). **Type loc.:** Crique Mangue, Saut Tigre [Inini], French Guiana [Guiana] (NE). **Distr.:** French Guiana, Guyana. **Refs.:** Rozeboom & Komp 1950a: 97 (M*); Foote 1954: 95 (L*); Belkin 1968b: 53 (type info.); Pecor et al. 1992: 53 (M*; tax., catalog citation); Torres-Gutierrez & Sallum 2015: 33 (tax., bion., distr., catalog citation). **Ety.:** Monsieur le Médecin Colonel Tournier. **Informal name:** Tournier French Guianese Typical Mosquito.

trifidus Dyar
1921h: 115 (M*, F). **Type loc.:** Tiribí, Costa Rica (USNM). **Distr.:** Costa Rica, El Salvador, Guatemala, Honduras, Mexico, Panama. **Refs.:** Bonne & Bonne-Wepster 1925: 282 (M, F); Dyar 1928: 286 (M*, F); Dampf 1941: 253 (M*; distr.; Mexico); Rozeboom & Komp 1950a: 97 (M*; lectotype desig., distr.; Panama); Lane 1953: 412 (M*, F); Stone & Knight 1957a: 57 (type info.); Stone et al. 1959: 276 (distr.; Honduras); Clark-Gil & Darsie 1983: 256 (distr.; Guatemala); Sirivanakarn 1983: 279 (M*, F*); Pecor et al. 1992: 53 (M*; tax., catalog citation). **Ety.:** three branched (L); "clasper thick, divided into three

erect parallel lobes." **Informal name:** Three-branched Costa Rican Typical Mosquito.

trigeminatus Clastrier
1970c: 473 (M*). **Type loc.:** Forêt du Gallion, French Guiana [Guiana] (MNHP). **Distr.:** Brazil, French Guiana. **Refs.:** Heinemann & Belkin 1979: 94 (distr.; Brazil); Pecor et al. 1992: 54 (M*; tax., catalog citation); Torres-Gutierrez & Sallum 2015: 33 (tax., bion., distr., catalog citation). **Ety.:** three pairs (L); ref. to the subapical lobe of the gonocoxite with three groups of two elements each. **Informal name:** Triplet French Guianese Typical Mosquito [*trigeminatus* = "three twin" = Triplet].

trilobulatus Duret & Barreto
1956: 93 (M*). **Type loc.:** Rio Tamandua, Ribeirão Preto, São Paulo, Brazil (FMRP). **Distr.:** Brazil. **Refs.:** Pecor et al. 1992: 54 (M*; tax., catalog citation); Torres-Gutierrez & Sallum 2015: 33 (tax., bion., distr., catalog citation). **Ety.:** n.s. [three lobed (L); prob. ref. to apical lobe of basistyle (gonocoxite) with three lobes]. **Informal name:** Three-lobed Brazilian Typical Mosquito.

trisetosus Fauran
1961a: 1 (M*). **Type loc.:** French Guiana; confluence of Oyac and Conte Rivers, Saint Antoine (MNHP). **Distr.:** French Guiana, Guyana. **Refs.:** Pecor et al. 1992: 54 (M*; tax., catalog citation); Torres-Gutierrez & Sallum 2015: 33 (tax., bion., distr., catalog citation). **Ety.:** three setae (L); ref. to three long filamentous striated setae inserted between the two lobes of the gonocoxite. **Informal name:** Three-threaded French Guianese Typical Mosquito.

unicornis Root
1928 (M, F, L). In Dyar 1928: 291. **Type loc.:** [Maracay, Aragua], Venezuela (USNM). **Distr.:** Brazil, French Guiana, Guyana, Venezuela. **Refs.:** Rozeboom & Komp 1950a: 97 (M*; lectotype desig.); Lane 1953: 483 (M*, F, L); Foote 1954: 118 (P*, L*); Stone & Knight 1957a: 57 (M gen.; type info., desig. as lectotype); Stone 1961c: 47 (distr.; French Guiana); Cova García et al. 1966: 171 (M*); Pecor et al. 1992: 54 (M*; tax., catalog citation); Torres-Gutierrez & Sallum 2015: 33 (tax., bion., distr., catalog citation). **Ety.:** n.s. [single horn (L); poss. ref. to male genitalia character]. **Informal name:** Unicorn Venezuelan Typical Mosquito.

vaxus Dyar
1920c: 73 (M, L). **Type loc.:** Suriname (USNM). **Distr.:** Argentina, Brazil, French Guiana, Suriname. **Refs.:** Dyar 1928: 328 (M*, L*; as *bibulus*); Stone & Knight 1957a: 58 (lectotype desig.); Forattini & Sallum 1993: 98 (from syn.; M*, F*, P*, L*); Torres-Gutierrez & Sallum 2015: 34 (tax., bion., distr., catalog citation). **Ety.:** n.s. **Informal name:** Vexing Surinamese Typical Mosquito.

syn. *bibulus* Dyar
1920c: 74 (M, L). **Type loc.:** Suriname (USNM). **Refs.:** Komp 1935: 7 (syn. with *educator*); Forattini & Sallum 1993b: 98 (transferred syn.).

vexillifer Komp
1936c: 320 (M*, F, L). **Type loc.:** Barro Colorado Island, Gatun Lake, Canal Zone, Panama (USNM). **Distr.:** Belize, Honduras, Panama. **Refs.:** Rozeboom & Komp 1950a: 97 (M*; lectotype desig.); Lane 1953: 421 (M*, F, L*); Foote 1954: 120 (L*); Stone & Knight 1957a: 58 (type info.); Heinemann & Belkin 1977b: 443 (distr.; Honduras); Pecor et al. 1992: 54 (M*; tax., catalog citation); Pecor et al. 2002: 248 (tax., distr.; Belize). **Ety.:** flag, banner (L); ref. to "the conspicuous long pennon-like projection from the tips of the rods of the inner division of the sidepiece." **Informal name:** Flagged Panamanian Typical Mosquito.

vidali Floch & Fauran
1954: 4 (M*). **Type loc.:** Moulin-de-Vidal, Ile de Cayenne, French Guiana [Guiana] (PIG). **Distr.:** French Guiana. **Refs.:** Dégallier & Claustre 1980: 144 (type info.); Pecor et al. 1992: 55 (M*; tax., catalog citation); Torres-Gutierrez & Sallum 2015: 34 (tax., bion., distr., catalog citation). **Ety.:** Moulin-de-Vidal, Cayenne, French Guiana. **Informal name:** Vidal French Guianese Typical Mosquito.

vomerifer Komp
1932: 79 (M*). **Type loc.:** Almirante, Panama (USNM). **Distr.:** Brazil, Colombia, Ecuador, French Guiana, Lesser Antilles (includes Trinidad and Tobago), Panama, Peru, Venezuela. **Refs.:** Rozeboom & Komp 1950a: 97 (M*); Stone 1961c: 47 (distr.; Trinidad); Forattini 1965a: 185 (M*); Floch & Kramer 1965: 1 (tax.); Aitken & Galindo 1966: 202 (tax.; French Guiana); Barreto & Lee 1969: 431 (distr.; Colombia); Galindo 1969: 87 (tax.); Sirivanakarn 1979: 139 (distr.; Brazil); Sirivanakarn & Jakob 1982: 194 (M*); Sirivanakarn 1983: 278 (M*); Pecor et al. 1992: 55 (M*; tax., catalog citation); Sallum & Forattini 1996: 579 (M*, F*); Pecor et al. 2000 (distr.; Peru); Navarro & Weaver 2004 (phyl.); Linton et al. 2013 (bion., mol. tax., distr.; Ecuador); Torres-Gutierrez & Sallum 2015: 34 (tax., bion., distr., catalog citation). **Ety.:** plough bearing (L); ref. to structure on male genitalia that resembles a plough, with a heavily sclerotized tip that forms the share. **Informal name:** Plowman Panamanian Typical Mosquito. (Vol. 1, p. 526)

wepsterae Komp & Rozeboom
1951: 127 (M*; as *wepsteri*). **Type loc.:** Paramaribo, Suriname (USNM). **Distr.:** Suriname. **Refs.:** Stone & Knight 1957a: 59 (emend.); Pecor et al. 1992: 55 (M*; tax., catalog citation); Torres-Gutierrez & Sallum 2015: 35 (tax., bion., distr., catalog citation). **Ety.:** Mrs. J. Bonne-Wepster. **Informal name:** Wepster Surinamese Typical Mosquito.

ybarmis Dyar
1920c: 57 (M, L). **Type loc.:** Paramaribo, Suriname (USNM). **Distr.:** Brazil, French Guiana, Guyana, Lesser Antilles (includes Trinidad and Tobago), Suriname, Venezuela. **Refs.:** Dyar 1928: 302 (M*, L*); Senevet & Abonnenc 1939a: 131 (distr.; French Guiana); Rozeboom & Komp 1950a: 97 (M*); Lane 1953: 433 (M*, L; distr.;

Brazil); Foote 1954: 96 (L*); Stone 1967: 217 (distr.; Trinidad); Forattini et al. 1970: 50 (distr.; Venezuela); Pecor et al. 1992: 55 (M*; tax., catalog citation); Torres-Gutierrez & Sallum 2015: 35 (tax., bion., distr., catalog citation). **Ety.:** n.s. **Informal name:** Paramaribo Surinamese Typical Mosquito.

syn. *jonistes* Dyar
1920c: 76 (M). **Type loc.:** Suriname (USNM). **Refs.:** Komp 1935: 7 (syn.).

zeteki Dyar
1918d: 122 (M). **Type loc.:** Gatun, Canal Zone, Panama (USNM). **Distr.:** Belize, Brazil, Colombia, French Guiana, Guyana, Honduras, Lesser Antilles (includes Trinidad and Tobago), Nicaragua, Panama, Paraguay, Suriname, Venezuela. **Refs.:** Dyar 1923e: 188 (M*; as *zeteci*); Dyar 1928: 339 (M*; as *zeteci*); Komp 1935: 8 (M; as *zeteci*); Rozeboom & Komp 1950a: 98 (M; tax., emend.); Lane 1953: 438 (M*; distr.; French Guiana); Foote 1954: 97 (P*, L*; tax.); Cova García et al. 1966 (M*); Stone 1967: 217 (distr.; Trinidad); Barreto & Lee 1969: 431 (distr.; Colombia); Bertram 1971: 745 (distr.; Belize); Panday 1975a, b (distr.; Suriname); Sirivanakarn 1983: 279 (F*); Pecor et al. 1992: 56 (M*; tax., catalog citation), 88 (distr.; Nicaragua, Paraguay); Torres-Gutierrez & Sallum 2015: 35 (tax., bion., distr., catalog citation). **Ety.:** J. Zetek. **Informal name:** Zetek Panamanian Typical Mosquito.

syn. *loturus* Dyar
1925g: 214 (M). **Type loc.:** Catatumbo River [Zulia], Venezuela (USNM). **Refs.:** Komp 1935: 9 (syn.).

Subgenus *Micraedes* Coquillett

1906d (as genus): 16, 24. Type species: *Micraedes bisulcatus* Coquillett.

Name derivation: small and *Aedes* (Gr).

Important references: Howard et al. 1915: 216 (syn. with genus *Culex*); Dyar 1918b: 102 (syn. with subg. *Culex*); Dyar 1928 (syn. with subg. *Melanoconion*); Edwards 1932a: 218 (to subg. of *Culex*); Lane 1953: 387 (syn. with subg. *Tinolestes*); Stone et al. 1959: 281 (as syn. subg. *Aedinus*); Belkin 1968b: 11 (resurrected from syn.); Berlin 1969c: 21 (revision); Harbach et al. 2012 (phyl.).

antillummagnorum Dyar
1928: 344 (M*, F). **Type loc.:** San Antonio de los Baños [Havana] Cuba (USNM). **Distr.:** Cuba, Dominican Republic, Haiti, Lesser Antilles (includes Virgin Isl.), Puerto Rico. **Refs.:** Gerry 1932 (F*); Rozeboom & Komp 1950a: 77 (syn.); Stone & Knight 1957a: 43 (lectotype desig.); Belkin 1968b: 11 (from syn.); Berlin 1969c: 34 (M*, F, P*, L*); Mattingly 1971a: Pl. 19 (P*); Mattingly 1975: 21 (E*). **Ety.:** Greater Antilles. **Informal name:** Antilles Cuban Typical Mosquito.

arawak Berlin
1969c: 37 (M*, L*). **Type loc.:** Moneague (Saint Ann), Jamaica (USNM). **Distr.:** Jamaica. **Refs.:** Belkin et al. 1970: 95 (M*, L*). **Ety.:** Caribbean indigenous peoples. **Informal name:** Arawak Jamaican Typical Mosquito.

biscaynensis Zavortink & O'Meara
1999: 263 (M*, F, P*, L*). **Type loc.:** Dade County, Florida, United States (CAS). **Distr.:** United States (continental). **Refs.:** Darsie & Ward 2005 (F*, L*; keys, distr.). **Ety.:** Biscayne Bay, Florida, United States. **Informal name:** Biscayne American Typical Mosquito.

bisulcatus (Coquillett)
1906a: 185 (M, F; *Micraedes*). **Type loc.:** Guadeloupe Island [Lesser Antilles] (USNM). **Distr.:** Argentina, Bahamas, Cuba, Dominican Republic, Haiti, Lesser Antilles (includes Antigua and Barbuda, Dominica, Guadeloupe, Martinique, Montserrat, Saint Kitts and Nevis, Saint Lucia, Sint Maarten, Saba, Trinidad and Tobago, Windward Isl.). **Refs.:** Floch & Abonnenc 1945e: 37 (M*, F, L*); Stone & Knight 1957a: 44 (lectotype desig.); Berlin 1969c: 30 (M*, F*, P*, L*). **Ety.:** n.s. [double farrowed (L)]. **Informal name:** Double-farrowed Guadeloupean Typical Mosquito.

erethyzonfer Galindo & Blanton
1954: 244 (M*, L*). **Type loc.:** Palo Santo, Panama (USNM). **Distr.:** Costa Rica, Guatemala, Honduras, Mexico, Panama. **Refs.:** Berlin 1969c: 43 (M*, F, P*, L*). **Ety.:** n.s. [porcupine bearing? (L)]. **Informal name:** Porcupine Panamanian Typical Mosquito.

jalisco Berlin
1974: 273 (M*, F, P*, L*). **Type loc.:** near El Tuito, 16 km from Boca de Tomatlàn (on highway from Puerto Vallarta to Manzanillo), Jalisco, Mexico (USNM). **Distr.:** Mexico. **Ety.:** Jalisco, Mexico. **Informal name:** Jalisco Mexican Typical Mosquito.

sandrae Berlin
1969c: 41 (F, P*, L*). **Type loc.:** Omilteme, Guerrero, Mexico (USNM). **Distr.:** Mexico. **Ety.:** Sandra June Heinemann. **Informal name:** Sandra Mexican Typical Mosquito.

schicki Berlin
1969c: 39 (M*, F, P*, L*). **Type loc.:** Rancho Viejo de Agua de Obispo, Chilpancingo, Guerrero, Mexico (USNM). **Distr.:** Mexico. **Ety.:** Robert Xavier Schick. **Informal name:** Schick Mexican Typical Mosquito.

Subgenus *Microculex* Theobald

1907: 461 (as genus). Type species: *Microculex arenteoumbrosus* Theobald.

Name derivation: very small and *Culex* (Gr).

Important references: Dyar 1918b: 92 (to subg. of *Culex*); Edwards 1932a: 219 (tax., bion., distr.); Lane & Whitman 1943 (tax., class., bion.; Brazil); Rozeboom & Komp 1950b: 147 (M; key); Lane 1953 (F, L; keys, tax., class., distr.); Coutinho & Forattini 1962: 209 (tax.); Cotrim & Bianchi Galati 1977 (revision; Pleuristriatus Series); Harbach et al. 2012 (phyl.).

albipes Lutz
1904 (F). In Bourroul 1904: 6. **Type loc.:** Itaparica, Bahia, Brazil (NE). **Distr.:** Brazil, Panama. **Refs.:** Kumm 1933: 568 (M*, F*, L*); Rozeboom & Komp 1950b: 156 (M*; tax.); Belkin et al. 1971: 25 (type info.); Cotrim et al. 1977: 177 (L*). **Ety.:** n.s [white foot (L); ref. to tarsal banding]. **Informal name:** White-footed Brazilian Typical Mosquito.

aphylactus Root
1927c: 584 (M*). **Type loc.:** Rio Soberbo, Rio de Janeiro, Brazil (USNM). **Distr.:** Brazil. **Refs.:** Rozeboom & Komp 1950b: 156 (M*); Lane & Whitman 1951: 363 (M, F, P*, L*); Stone & Knight 1957a: 43 (lectotype desig.); Belkin et al. 1971: 25 (type loc. info.). **Ety.:** n.s. [without guard (Gr)]. **Informal name:** Unguarded Brazilian Typical Mosquito.

aureus Lane & Whitman
1951: 349 (M*, F, P*, L). **Type loc.:** Rio de Janeiro, Guanabara, Brazil (INER). **Distr.:** Brazil. **Refs.:** Forattini & Coutinho 1963: 481 (P*, L*); Belkin et al. 1971: 25 (type loc. info.); Xavier 1973: 159 (type info.); Cotrim & Bianchi Galati 1977: 177 (M*, F, P*, L*). **Ety.:** golden (L); ref. to "broad and continuous golden band over occiput, mesonotum and scutellum." **Informal name:** Golden Brazilian Typical Mosquito.

azymus Dyar & Knab
1906b: 169 (F, L). **Type loc.:** Arima, Trinidad [Trinidad & Tobago] (USNM). **Distr.:** Lesser Antilles (Trinidad and Tobago). **Refs.:** Dyar 1928: 350 (F, L*); Belkin, Schick, & Heinemann 1965: 70 (type loc. info.); Cotrim et al. 1977: 185 (F, P*, L*); Leví-Castillo 1953b: 161 (M*, F, P). **Ety.:** n.s. [uncleaved (L)]. **Informal name:** Uncloven Trinbagonian Typical Mosquito.

carioca Lane & Whitman
1951: 358 (M*, F, P*, L*). **Type loc.:** Rio de Janeiro, Guanabara, Brazil (INER). **Distr.:** Brazil, Colombia. **Refs.:** Barreto & Lee 1969: 278 (distr.); Belkin et al. 1971: 25 (type loc. info.); Xavier 1973: 159 (type info.). **Ety.:** Carioca [a person from Rio de Janeiro]. **Informal name:** Carioca Brazilian Typical Mosquito.

chryselatus Dyar & Knab
1919: 5 (M, F, P, L). **Type loc.:** Compagnie des Mines d'Or, Lawa River, Suriname (USNM). **Distr.:** Brazil, Colombia, Ecuador, French Guiana, Guyana, Panama, Suriname, Venezuela. **Refs.:** Dyar 1928: 352 (M*, F, L*); Lane 1953: 541 (M*, F, P*, L*); Stone & Knight 1957a: 45 (lectotype desig.); Fauran & Pajot 1974a: 107 (distr.; French Guiana); Linton et al. 2013 (bion., mol. tax., distr.; Ecuador). **Ety.:** n.s. [gold side (Gr, L)]. **Informal name:** Gold-sided Surinamese Typical Mosquito.

consolator Dyar & Knab
1906b: 169 (M, L). **Type loc.:** Arima, Trinidad [Trinidad & Tobago] (USNM). **Distr.:** Argentina, Brazil, Lesser Antilles (Trinidad and Tobago). **Refs.:** Rozeboom & Komp 1950b: 155 (M*); Lane & Whitman 1951: 352 (M, F, P*, L*). **Ety.:** n.s. [comforter (L)]. **Informal name:** Consoling Trinbagonian Typical Mosquito.
syn. *trychnus* Root
1927c: 585 (M*). **Type loc.:** Rio Soberbo, Rio de Janeiro, Brazil (USNM). **Refs.:** Stone & Knight 1957a: 57 (lectotype desig.); Belkin et al. 1971: 26 (type loc. info.).

daumastocampa Dyar & Knab
1908: 58 (M, F). **Type loc.:** Porto Bello, Fort San Felipe, Panama (USNM). **Distr.:** Colombia, Costa Rica, Guatemala, Nicaragua, Panama, Venezuela. **Refs.:** Dyar 1928: 355 (M*, F, L*); Rozeboom & Komp 1950b: 155 (M*); Stone & Knight 1957a: 46 (lectotype desig.); Barreto & Lee 1969: 432 (distr.; Colombia). **Ety.:** n.s. **Informal name:** San Felipe Panamanian Typical Mosquito.

davisi Kumm
1933: 565 (M*, F*, L*). **Type loc.:** São Salvador, Bahia, Brazil (NHMUK). **Distr.:** Argentina, Brazil, Paraguay. **Refs.:** Lane 1953: 528 (M*, F, P, L); Mattingly 1971a: Pl. 20 (P cephalothorax*); Belkin et al. 1971: 26 (type loc. info.); Cotrim et al. 1977: 185 (M*, L*); Townsend 1990: 65 (type info.). **Ety.:** Nelson Caryl Davis. **Informal name:** Davis Brazilian Typical Mosquito.

dubitans Lane & Whitman
1951: 357 (M*, F, P*, L*). **Type loc.:** Rio de Janeiro, Guanabara, Brazil (INER). **Distr.:** Brazil. **Refs.:** Belkin et al. 1971: 26 (type loc. info.); Xavier 1973: 159 (type info.). **Ety.:** n.s. [doubting (L); ref. to author doubt that it might already be named]. **Informal name:** Dubious Brazilian Typical Mosquito.

elongatus Rozeboom & Komp
1950b: 147 (M*, F, P*, L*). **Type loc.:** Acacias, Meta, Colombia (USNM). **Distr.:** Colombia, Peru. **Refs.:** Pecor et al. 2000 (tax., bion., distr.; Peru). **Ety.:** n.s. [prolonged (L); poss. ref. to very long larval siphon]. **Informal name:** Elongated Colombian Typical Mosquito.

gairus Root
1927c: 583 (M*, L*). **Type loc.:** Botanical Garden, Rio de Janeiro, Guanabara, Brazil (USNM). **Distr.:** Brazil. **Refs.:** Lane 1953: 543 (M*, F, P*, L*); Stone & Knight 1957a: 50 (lectotype desig.); Belkin et al. 1971: 26 (type loc. info.); Cotrim & Bianchi Galati 1977: 188 (M*, L). **Ety.:** n.s. [country or farm? (L)]. **Informal name:** Rural Brazilian Typical Mosquito.

gaudeator Dyar & Knab
1907c: 204 (M, F; *jenningsi* var.). **Type loc.:** Tabernilla, Canal Zone, Panama (USNM). **Distr.:** Colombia, Costa Rica, Panama, Venezuela. **Refs.:** Lane 1953: 533 (M, F; to sp.); Stone & Knight 1957a: 50 (type info.). **Ety.:** n.s. [rejoicer (L)]. **Informal name:** Rejoicing Panamanian Typical Mosquito.

hedys Root
1927c: 584 (M*). **Type loc.:** Angra dos Reis, Rio de Janeiro, Brazil (USNM). **Distr.:** Brazil, Colombia. **Refs.:** Pessoa & Galvão 1936: 84 (M, L*); Rozeboom & Komp 1950b: 151 (M*; tax.); Belkin et al. 1971: 26 (type loc. info.). **Ety.:** n.s. [sweet (Gr)]. **Informal name:** Sweet Brazilian Typical Mosquito.

imitator Theobald
1903a: 175 (M*). **Type loc.:** São Paulo, São Paulo State, Brazil (NHMUK). **Distr.:** Argentina, Brazil, Colombia, Ecuador, French Guiana, Guyana, Honduras, Lesser Antilles (includes Trinidad and Tobago), Mexico, Panama, Suriname, Uruguay, Venezuela. **Refs.:** Kumm 1933: 564 (M*, F*, L*); Senevet & Abonnenc 1939a: 115 (L*); Rozeboom & Komp 1950b: 155 (M*); Forattini & Coutinho 1963: 488 (P*, L*); Belkin et al. 1971: 26 (type loc. info., tax.); Rossi & Martínez 2003: 472 (distr.; Uruguay); Demari-Silva et al. 2015 (mol.tax; as *s.l*). **Ety.:** n.s. [imitator (L)]. **Informal name:** Imitative Brazilian Typical Mosquito.

syn. *daumasturus* Dyar & Knab
1906d: 220 (L*). **Type loc.:** La Brea, Trinidad [Trinidad & Tobago, Lesser Antilles] (USNM). **Refs.:** Stone & Knight 1957a: 46 (lectotype desig.); Belkin et al. 1965: 70 (type loc. info.).

syn. *vector* Dyar & Knab
1906d: 220 (L*). **Type loc.:** Trinidad [Trinidad & Tobago, Lesser Antilles] (USNM). **Refs.:** Stone & Knight 1957a: 58 (lectotype desig.).

syn. *argenteoumbrosus* Theobald
1907: 461 (M, F; *Microculex*). **Type loc.:** Rio de Janeiro, Guanabara, Brazil (NHMUK). **Refs.:** Belkin 1968b: 13 (lectotype desig.); Belkin et al. 1971: 26 (type loc. info.).

ssp. ***retrosus*** Lane & Whitman
1951: 356 (M*, F, P*, L). **Type loc.:** Rio de Janeiro, Guanabara, Brazil (INER). **Distr.:** Brazil. **Refs.:** Belkin et al. 1971: 26 (type loc. info.); Xavier 1973: 159 (type info.). **Ety.:** turned back (L); ref. to "retrose hooks of mesosome." **Informal name:** Retrose Brazilian Typical Mosquito.

inimitabilis Dyar & Knab
1906d: 221 (L*). **Type loc.:** Trinidad [Trinidad & Tobago, Lesser Antilles] (NE). **Distr.:** Brazil, Colombia, Ecuador, French Guiana, Guyana, Lesser Antilles (includes Grenada, Trinidad and Tobago), Panama, Suriname, Venezuela. **Refs.:** Pessoa & Galvão 1936: 86 (L*); Lane 1953: 525 (M*, F, P, L); Stone & Knight 1957a: 51 (type info.); Belkin & Heinemann 1976a: 272 (distr.; Grenada). **Ety.:** n.s. [not able to be imitated (L)]. **Informal name:** Inimitable Trinbagonian Typical Mosquito.

ssp. ***fuscatus*** Lane & Whitman
1951: 361 (M, F, P*, L*). **Type loc.:** Rio de Janeiro, Guanabara, Brazil (INER). **Distr.:** Brazil. **Refs.:** Belkin et al. 1971: 26 (type loc. info.); Xavier 1973: 159 (type info.). **Ety.:** darkened (L); similar to *inimitabilis* except no white scales on abdominal dorsum. **Informal name:** Darkened Brazilian Typical Mosquito.

intermedius Lane & Whitman
1951: 348 (M*, F, P, L). **Type loc.:** Rio de Janeiro, Guanabara, Brazil (INER). **Distr.:** Brazil. **Refs.:** Coutinho & Forattini 1962: 217 (M*, F, P*, L*); Belkin et al. 1971: 26 (type loc. info.); Xavier 1973: 159 (type info.); Cotrim & Bianchi Galati 1977: 188 (M*, F, P, L*). **Ety.:** n.s. [in the middle (L); poss. ref. to large median wing spot]. **Informal name:** Middle Brazilian Typical Mosquito.

jenningsi Dyar & Knab
1907c: 204 (M, F). **Type loc.:** Tabernilla, Canal Zone, Panama (USNM). **Distr.:** Colombia, Costa Rica, Mexico, Nicaragua, Panama. **Refs.:** Dyar 1928: 356 (M*, F, L*); Rozeboom & Komp 1950b: 155 (M*); Stone & Knight 1957a: 52 (lectotype desig.). **Ety.:** A.H. Jennings. **Informal name:** Jennings Panamanian Typical Mosquito.

kukenan Anduze
1942c: 558 (M*, F). **Type loc.:** La Gran Sabana, Surukum [Bolivar], Venezuela (FH). **Distr.:** Colombia, Venezuela. **Refs.:** Barreto & Lee 1969: 432 (distr.). **Ety.:** n.s. [poss. ref. to Kukenán Tepui, Venezuela]. **Informal name:** Kukenán Venezuelan Typical Mosquito.

lanei Coutinho & Forattini
1962: 210 (M*, F, P*, L*). **Type loc.:** Bertioga (Itaguare), Santos, São Paulo, Brazil (FH). **Distr.:** Brazil. **Refs.:** Belkin et al. 1971: 26 (type loc. info.); Cotrim et al. 1977: 192 (M*, F, P, L*). **Ety.:** John Lane. **Informal name:** Lane Brazilian Typical Mosquito.

microphyllus Root
1927c: 586 (M*). **Type loc.:** Mage, Rio de Janeiro, Brazil (USNM). **Distr.:** Brazil. **Refs.:** Lane & Whitman 1951: 362 (M, F, P*, L*; from syn. with *inimitabilis*); Stone & Knight 1957a: 53 (lectotype desig.); Belkin et al. 1971: 26 (type loc. info.). **Ety.:** small leaf (Gr); ref. to a small leaf on male hypopygium. **Informal name:** Small-leafed Brazilian Typical Mosquito.

neglectus Lutz
1904 (A, L). In Bourroul 1904: 27. **Type loc.:** Serra da Cantareira, São Paulo, Brazil (USNM). **Distr.:** Argentina, Brazil. **Refs.:** Rozeboom & Komp 1950b: 156 (M*); Lane & Whitman 1951: 364 (M*, F, P*, L*; lectotype desig); Stone & Knight 1957a: 54 (type info.; credit Bourroul as author); Belkin et al. 1971: 7 (tax.). **Ety.:** n.s.

[neglected, disregarded (L)]. **Informal name:** Neglected Brazilian Typical Mosquito.

pleuristriatus Theobald
1903a: 177 (F). **Type loc.:** São Paulo, Brazil (NHMUK). **Distr.:** Argentina, Bolivia, Brazil, Colombia, Costa Rica, Ecuador, French Guiana, Guyana, Lesser Antilles (includes Grenada, Trinidad and Tobago), Suriname, Venezuela. **Refs.:** Kumm 1933: 563 (M*, F*, L*); Senevet & Abonnenc 1939a: 114 (M*, L*); Lane & Whitman 1951: 347 (P*, L); Coutinho & Forattini 1962: 223 (M*, P*, L*); Belkin 1968b: 19 (lectotype desig.); Cotrim et al. 1977: 171 (M*, F, P*, L*); Linton et al. 2013 (bion., mol. tax., distr.; Ecuador). **Ety.:** n.s. [side striae (Gr); prob. ref. to "pleurae very pale, with two broken parallel brown lines"]. **Informal name:** Striped Brazilian Typical Mosquito.

pulidoi Cova García & Sutil Oramas
1974: 29 (M*). **Type loc.:** La Soledad, Municipio Altamira, Estado Barinas, Venezuela (DERM). **Distr.:** Venezuela. **Ety.:** Juan Pulido F. **Informal name:** Pulido Venezuelan Typical Mosquito.

reducens Lane & Whitman
1951: 354 (M*, P*, L*). **Type loc.:** Rio de Janeiro, Guanabara, Brazil (INER). **Distr.:** Brazil. **Refs.:** Belkin et al. 1971: 26 (type loc. info.); Xavier 1973: 159 (type info.). **Ety.:** n.s. [bring back, restore (L)]. **Informal name:** Restored Brazilian Typical Mosquito.

reginae Floch & Fauran
1955: 5 (M*). **Type loc.:** Régina (Approuague) [Guyane], French Guiana [Guiana] (PIG). **Distr.:** French Guiana. **Refs.:** Belkin 1968b: 47 (type info.). **Ety.:** Régina (Approuague), French Guiana. **Informal name:** Régina French Guianese Typical Mosquito.

rejector Dyar & Knab
1906d: 221 (L*). **Type loc.:** Córdoba [Veracruz], Mexico (NE). **Distr.:** Belize, Costa Rica, Guatemala, Honduras, Mexico, Nicaragua. **Refs.:** Dyar 1928: 356 (M*, F, L*); Rozeboom & Komp 1950b: 155 (M*); Stone & Knight 1957a: 55 (type info.); Heinemann & Belkin 1977b: 427, 453 (distr.; Guatemala, Nicaragua); Pecor et al. 2002: 248 (tax., distr.; Belize). **Ety.:** n.s. [rejector (L)]. **Informal name:** Reject Mexican Typical Mosquito.

shopei Forattini & Toda
1966: 507 (M*, F*, P*, L*). **Type loc.:** Utinga, Belém, Pará, Brazil (FH). **Distr.:** Brazil. **Refs.:** Belkin et al. 1971: 27 (type info., type loc. info.). **Ety.:** Robert E. Shope. **Informal name:** Shope Brazilian Typical Mosquito.

siphanulatus Lourenço-de-Oliveira & Silva-do-Nascimento
1987: 101 (M*, F*, P*, L*). **Type loc.:** Granjas Calabria, Jacarepaguá, Rio de Janeiro, Brazil (IOC). **Distr.:** Brazil. **Ety.:** siphon, anulus (L); ref. to siphon with two dark rings. **Informal name:** Siphon Brazilian Typical Mosquito.

stonei Lane & Whitman
1943: 405 (M*, F, P, L). **Type loc.:** Trinidad [Trinidad & Tobago] (USNM). **Distr.:** Brazil, Colombia, Ecuador, French Guiana, Guyana, Lesser Antilles (includes Trinidad and Tobago), Peru, Suriname, Venezuela. **Refs.:** Rozeboom & Komp 1950b: 155 (M*); Lane 1953: 519 (M*, F, P*, L*); Stone & Knight 1957a: 56 (lectotype desig.); Pecor et al. 2000 (distr.; Peru); Linton et al. 2013 (bion., mol. tax., distr.; Ecuador). **Ety.:** Alan Stone. **Informal name:** Stone Trinbagonian Typical Mosquito.

sutili Cova García & Pulido F.
1974: 27 (M*, L*). **Type loc.:** Hacienda de La Maravilla, Municipio Tácata, Estado Miranda, Venezuela (DERM). **Distr.:** Venezuela. **Refs.:** Cova García & Pulido F. 1974: 227–228 (M*, L*; illustration inadvertently published early in Vol. XIV (1–2): 27–28). **Ety.:** Ezequiel Sutil Oramas. **Informal name:** Sutil Venezuelan Typical Mosquito.

worontzowi Pessoa & Galvão
1936: 86 (M*, F, L*). **Type loc.:** Tabatinguera, São Paulo, Brazil (FMSP). **Distr.:** Brazil. **Refs.:** Lane & Whitman 1951: 353 (M, F, P*, L*). **Ety.:** Cesar Worontzowi. **Informal name:** Worontzowi Brazilian Typical Mosquito.

xenophobus Ronderos
1965: 249 (M*, F*, P*, L*). **Type loc.:** San Bonifacio, Sucre, Venezuela (INM). **Distr.:** Venezuela. **Refs.:** Cotrim & Galati Bianchi 1977: 177 (M*, F, P, L*). **Ety.:** n.s. [stranger, fear? (Gr)]. **Informal name:** Xenophobic Venezuelan Typical Mosquito.

Subgenus *Neoculex* Dyar

Dyar 1905a: 48 (as genus). Type species: *Culex territans* Walker.

Name derivation: new *Culex*.

Important references: Belkin 1962 (tax., keys, bion., distr.; South Pacific); Bram 1967b (tax., keys, bion.; Thailand); Linam & Nielsen 1970: 149 (tax.); Mattingly 1970h (oviposition); Sirivanakarn 1971a: 69 (tax., keys); Lee et al. 1988b (tax., key, distr.; Australasia); Becker 2010 (tax., bion.; Europe); Harbach et al. 2012 (phyl.).

apicalis Adams
1903: 26 (F). **Type loc.:** Arizona, United States (USNM). **Distr.:** Mexico, United States (continental). **Refs.:** Ross 1947 (F*); Bohart 1949: 336 (M*, F, P, L*; tax.); Yamaguti & LaCasse 1951c: 107 (M*, F*, L*); Carpenter & LaCasse 1955: 315 (M*, F*, L*; keys); Bohart & Washino 1957: 462 (L*); Stone 1958b (lectotype desig.); Bohart & Washino 1978: 123 (M*, F, P*, L*; keys, tax., bion., distr.); Darsie & Ward 2005 (F*, L*; keys, distr.). **Ety.:** n.s. [apex, apical (L)]. **Informal name:** Apex American Typical Mosquito.

arizonensis Bohart
1949: 341 (M*, F, P, L*). **Type loc.:** Prescott, Arizona, United States (USNM). **Distr.:** Mexico, United States (continental). **Refs.:** Yamaguti & LaCasse 1951c: 109 (M*, F*, L*); Carpenter & LaCasse 1955: 317 (M*, F*, L*; keys); Darsie & Ward 2005 (F*, L*; keys, distr.). **Ety.:** Arizona, United States. **Informal name:** Arizona American Typical Mosquito.

boharti Brookman & Reeves
1950: 159 (new name for *reevesi* Bohart, not Wirth 1948). **Distr.:** United States (continental). **Refs.:** Yamaguti & LaCasse 1951c: 113 (M*, F*, L*); Carpenter & LaCasse 1955: 319 (M*, F*, L*; keys); Bohart & Washino 1957: 462 (L*); Baisas 1974: 35 (M, F*, P*, L; tax., bion., distr.; Philippines); Bohart & Washino 1978: 123 (M, F, L*; keys, tax., bion., distr.); Darsie & Ward 2005 (F*, L*; keys, distr.). **Ety.:** George Edward Bohart. **Informal name:** Bohart American Typical Mosquito.
syn. *reevesi* Bohart
1949: 342 (M*, F, P, L*). **Type loc.:** Monticello, Napa County, California, United States (USNM). **Refs.:** Brookman & Reeves 1950 (syn.).

chaetoventralis (Theobald)
1910c: 461 (F*; *Neomelanoconion*). **Type loc.:** Kumanda [= Kuranda], Queensland, Australia (NHMUK). **Distr.:** Australia. **Refs.:** Mattingly & Marks 1955: 172 (M*, F, P*, L*); Lee et al. 1988b: 19 (F; key, tax., bion., distr., review). **Ety.:** hairy belly (L); apex of adult abdomen with ventral setae. **Informal name:** Hairy-bellied Australian Typical Mosquito.

cheesmanae Mattingly & Marks
1955: 166 (M*, F*, P*, L*). **Type loc.:** Nassirah, New Caledonia (NHMUK). **Distr.:** New Caledonia. **Refs.:** Belkin 1962: 240 (M*, F, P, L*); Lee et al. 1988b: 21 (F; key, tax., bion., distr., review). **Ety.:** Evelyn Cheesman. **Informal name:** Cheesman New Caledonian Typical Mosquito.

crassistylus Brug
1934a: 516 (M*). **Type loc.:** Inawattan, New Guinea (NHMUK). **Distr.:** Indonesia. **Refs.:** Lee et al. 1988b: 23 (type loc., tax., bion., distr., review). **Ety.:** thick pillar (L) [poss. ref. to structures on gonocoxite]. **Informal name:** Trunklike Indonesian Typical Mosquito [*crassistylus* = "thick pillar" = Trunklike].

derivator Dyar & Knab
1906d: 216 (L*). **Type loc.:** Cordoba [Veracruz], Mexico (NE). **Distr.:** Costa Rica, Guatemala, Honduras, Mexico, Panama. **Refs.:** Dyar 1928: 393 (M*, F, L*); Bohart 1949: 345 (M, F, L); Stone & Knight 1957a: 46 (type info.). **Ety.:** n.s. [derived? (L)]. **Informal name:** Derivative Mexican Typical Mosquito.

douglasi Dobrotworsky
1956: 110 (M*, F*, P*, L*, E*). **Type loc.:** Meredith, Victoria, Australia (NMM). **Distr.:** Australia. **Refs.:** Dobrotworsky 1965: 196 (M*, F, L*); Lee et al. 1988b: 24 (F; key, tax., bion., distr., review; type info). **Ety.:** G.W. Douglas. **Informal name:** Douglas Australian Typical Mosquito.

dumbletoni Belkin
1962: 244 (M, L*). **Type loc.:** Saint Louis, New Caledonia (USNM). **Distr.:** New Caledonia. **Refs.:** Lee et al. 1988b: 26 (tax., bion., distr., review). **Ety.:** L.J. Dumbleton. **Informal name:** Dumbleton New Caledonian Typical Mosquito.

europaeus da Cunha Ramos, Ribeiro, & Harrison
2003: 7 (M*, F*, P, L; key). **Type loc.:** Avo, Portugal. **Distr.:** Albania, Algeria, Austria, Azerbaijan, Belarus, Belgium, Bulgaria, Crimean Peninsula, Croatia, Czech Republic, Denmark, Estonia, Finland, France (includes Corsica), FYRO Macedonia, Georgia, Germany, Greece (includes Crete), Hungary, Iran, Iraq, Israel (and Gaza Strip and West Bank), Italy, Jordan, Kosovo, Liechtenstein, Lithuania, Luxembourg, Moldova, Netherlands, Norway, Poland, Portugal, Romania, Russia, Serbia, Slovakia, Slovenia, Spain, Sweden, Switzerland, Tunisia, Turkey, Ukraine, United Kingdom. **Refs.:** Brolemann 1920 (F*; as *pyrenaicus* Brolemann); Séguy 1924: 45, Pl. IX (L*), 44 (M gen.*; as *pyrenaicus*); Mattingly 1955a: 385 (M*; tax.; as *territans*); Derwesh 1965: 44 (distr.; Iraq; as *apicalis*); Lotfi 1973: 205 (distr.; Iran; as *territans*); Gutsevich et al. 1974: 357 (M*, F, L*; as *territans*); Bozicic 1980: 93 (Serbia; as *territans*); Rjazantzeva 1985 (F; as *territans*); Amr et al. 1997 (distr.; Jordan; as *territans*); Clements 1999 (F*; as *territans*); Eritja et al. 2000: 12 (distr.; Spain; as *territans*); Beck et al. 2003: 24 (Luxembourg; as *territans*); Becker et al. 2010: 286 (M*, F*, L*; keys, tax., distr., bion.; as *territans*); Namazov 2014 (distr.; Azerbaijan; as *territans*); Robert et al. 2019 (distr.; Western Palearctic). **Ety.:** Europe [the continent of Europe was named after Europa, a mythological figure abducted by Zeus]. **Informal name:** Europa Portuguese Typical Mosquito.

fergusoni (Taylor)
1914b: 459 (M, F; *Culicada*). **Type loc.:** Milson Island, Hawkesbury River, New South Wales, Australia (ANIC). **Distr.:** Australia. **Refs.:** Lee 1944a: 110 (L*); Dobrotworsky 1956: 105 (M*, F*, P*, L, E*); Dobrotworsky 1965: 194 (M*, F*, L*); Lee et al. 1988b: 26 (F; key, tax., bion., distr., review; type info.). **Ety.:** Dr. Ferguson. **Informal name:** Ferguson Australian Typical Mosquito.

gamma Séguy
1924: Pl. XI, 47 (L*). **Type loc.:** Biskra [Constantine], Algeria (MNHP). **Distr.:** Algeria. **Refs.:** Senevet 1947a: 128 (tax.). **Ety.:** n.s. [third letter of the Greek alphabet]. **Informal name:** Gamma Algerian Typical Mosquito.

gaufini Belkin
1962: 243 (M*, F, L*). **Type loc.:** APO [Army Post Office] 25, New Caledonia (USNM). **Distr.:** New Caledonia.

Refs.: Lee et al. 1988b: 30 (F; key, tax., bion., distr., review). **Ety.:** A.R. Gaufin. **Informal name:** Gaufin New Caledonian Typical Mosquito.

impudicus Ficalbi
1890d: 81 (M, F). **Type loc.:** Sardinia (LU). **Distr.:** Albania, Algeria, Cyprus, France (includes Corsica), Greece, Iran, Italy (includes Sardinia and Sicily), Morocco, Portugal, Spain, Tunisia, Turkey. **Refs.:** Séguy 1924: 28 (M*), 29 (M gen.*); Sicart 1940: 471 (M*, F, P*, L*); Senevet 1947a: 116 (L*); Senevet 1949: 53 (M*); Senevet & Andarelli 1954a: 43 (M, F*); Mattingly 1955a: 385 (M*; tax.); Lotfi 1970: 401 (distr.); Lotfi 1973: 204 (distr.); Eritja et al. 2000: 12 (distr.; Spain); Trari et al. 2002: 331 (distr.; Morocco); Becker et al. 2010: 284 (M*, F*, L*; keys, tax., distr., bion.); Gunay et al. 2015 (mol. tax.; distr.; Turkey); Robert et al. 2019 (distr.; Western Palearctic). **Ety.:** lewd, impudent, shameless (L); ref. to prominent gonocoxites. **Informal name:** Impudent Italian Typical Mosquito.
syn. *sergentii* Theobald
1903a: 218 (F*). **Type loc.:** Algeria (NHMUK). **Refs.:** Mattingly 1955c: 28 (syn.).

johni Cova García, Pulido F., & Escalante de Ugueto
1979: 21 (M*, P*, L*). **Type loc.:** La Mucuy, Municipio Tabay, Merida, Venezuela (DERM). **Distr.:** Venezuela. **Ety.:** n.s. [poss. ref. to John Lane]. **Informal name:** John Venezuelan Typical Mosquito.

judaicus Edwards
1926c: 127 (M, F; as *apicalis* var.). **Type loc.:** Palestine (NHMUK). **Distr.:** Israel (and Gaza Strip and West Bank), Jordan, Lebanon, Palestine. **Refs.:** Mattingly 1953b: 108 (M, F; to sp.; tax.); Senevet & Andarelli 1959: 117 (M*); Margalit & Tahori 1974: 86 (distr.); Robert et al. 2019 (distr.; Western Palaearctic). **Ety.:** n.s. [relating to the Jews (L)]. **Informal name:** Judea Israeli Typical Mosquito [Judea was a Roman province in what is now Israel].

latus Dobrotworsky
1956: 108 (M*, F*, L*). **Type loc.:** Kewdale, Western Australia, Australia (CSIR). **Distr.:** Australia. **Refs.:** Lee et al. 1988b: 31 (F; key, tax., bion., distr., review). **Ety.:** wide, broad (L); "broad longitudinal stripe of pale scales on the scutum." **Informal name:** Broad Australian Typical Mosquito.

leonardi Belkin
1962: 245 (M*, F, P*, L*). **Type loc.:** West Poha swamp, Guadalcanal, Solomon Islands (USNM). **Distr.:** Australia, Solomon Isl. **Refs.:** Lee et al. 1988b: 32 (F; key, tax., bion., distr., review). **Ety.:** Justin W. Leonard. **Informal name:** Leonard Solomon Islander Typical Mosquito.

martinii Medschid
1930: 364 (M*, F, L*). **Type loc.:** Adana [Seyhan], Turkey (LU). **Distr.:** Armenia, Austria, Crimean Peninsula, Croatia, Cyprus, Czech Republic, Egypt, France (includes Corsica), Germany, Greece, Hungary, Israel (and Gaza Strip and West Bank), Italy, Lebanon, Morocco, Romania, Slovakia, Switzerland, Tajikistan, Turkey, Ukraine, United Kingdom, Uzbekistan. **Refs.:** Monchadskii 1951: 256 (L*); Cervone 1957: 236 (M*, F*, L); Bailly-Choumara 1968: 140 (distr.); Gutsevich et al. 1974: 361 (M*, F, L*); Becker et al. 2010: 285 (M*, F*, L*; keys, tax., distr., bion.); Gunay et al. 2015 (mol. tax.; distr.; Turkey); Robert et al. 2019 (distr.; Western Palearctic). **Ety.:** Erich Martini. **Informal name:** Martini Turkish Typical Mosquito.

millironi Belkin
1962: 242 (M*, F, L*). **Type loc.:** Poindimié, New Caledonia (USNM). **Distr.:** New Caledonia. **Refs.:** Lee et al. 1988b: 33 (F; key, tax., bion., distr., review). **Ety.:** H.E. Milliron. **Informal name:** Milliron New Caledonian Typical Mosquito.

pedicellus King & Hoogstraal
1947b: 65 (M*). **Type loc.:** Hollandia, Western New Guinea (Papua) (USNM). **Distr.:** Indonesia. **Refs.:** Lee et al. 1988b: 35 (tax., bion., distr., review). **Ety.:** n.s. **Informal name:** Sneaker Indonesian Typical Mosquito [*pedicellus* = "of the foot" = Sneaker].

postspiraculosus Lee
1944c: 221 (M*, F, L*). **Type loc.:** Narrabeen, New South Wales, Australia (CSIR). **Distr.:** Australia. **Refs.:** Dobrotworsky 1965: 198 (M*, F*, L*); Lee et al. 1988b: 36 (F; key, tax., bion., distr., review). **Ety.:** postspiracular (setae) [ref. to a *Culex* with the unique character of having postspiracular setae]. **Informal name:** Postspiracle Australian Typical Mosquito.

pseudomelanoconia Theobald
1907: 416 (F). **Type loc.:** South Queensland, Australia (NHMUK). **Distr.:** Australia. **Refs.:** Bancroft 1908: 44 (F); Lee 1944a: 112 (L*); Mattingly & Marks 1955: 171 (M*; tax.); Dobrotworsky 1956: 113 (F*, P*, L*); Dobrotworsky 1965: 200 (M*, F, L*); Lee et al. 1988b: 38 (F; key, tax., bion., distr., review). **Ety.:** false (Gr); similar to *Melanoconion*. **Informal name:** Fake Australian Typical Mosquito.

reevesi Wirth
1948: 230 (F, L). **Type loc.:** Monterey, California, United States (LU). **Distr.:** Mexico, United States (continental). **Refs.:** Brookman & Reeves 1953: 226 (M*, F, P, L*; tax.); Carpenter & LaCasse 1955: 320 (M*, F, L*; keys); Bohart & Washino 1957: 426 (L*); Bohart & Washino 1978: 129 (M, F*, L*; keys, tax., bion., distr.); Darsie & Ward 2005 (F*, L*; keys, distr.). **Ety.:** n.s. [poss. ref. to W.C. Reeves]. **Informal name:** Reeves American Typical Mosquito.

rubensis Sasa & Takahasi
1948: 51 (M*, F, L). **Type loc.:** Rubeshibe, Hokkaido, Japan (IID). **Distr.:** Japan, People's Republic of China, Russia, South Korea. **Refs.:** La Casse & Yamaguti 1950: 261 (M*, F*, P*, L*; bion., distr., tax.); Hara 1957: 57 (F*); Danilov & Kupriyanova 1976: 928 (L*; distr); Tanaka

et al. 1979: 180 (M*, F*, L*); Shu-Po & Chen 1986 (distr., People's Republic of China). **Ety.:** n.s. [prob. ref. to type loc.]. **Informal name:** Rubeshibe Japanese Typical Mosquito.

territans Walker
1856a: 428 (F). **Type loc.:** United States (NE). **Distr.:** Albania, Algeria, Armenia, Austria, Azerbaijan, Belarus, Belgium, Bulgaria, Canada, Croatia, Czech Republic, Denmark, Estonia, Finland, France (includes Corsica), FYRO Macedonia, Georgia, Germany, Greece (includes Crete), Hungary, Iran, Iraq, Israel (and Gaza Strip and West Bank), Italy, Jordan, Liechtenstein, Lithuania, Luxembourg, Moldova, Morocco, Netherlands, Norway, Poland, Portugal, Romania, Russia, Serbia, Slovakia, Slovenia, Spain, Sweden, Switzerland, Tunisia, Turkey, Ukraine, United Kingdom (includes Guernsey; Jersey; Isle of Man), United States (includes Alaska, continental). **Refs.:** Knab 1904 (E); Bohart 1949: 332 (M*, F, P, L*; tax.); Darsie 1951: 32 (P*); Yamaguti & LaCasse 1951c: 117 (M*, F*, L*); Mattingly 1953b: 106 (M, F, P, L; tax.); Carpenter & LaCasse 1955: 322 (M*, F*, L*; keys); Bohart & Washino 1957: 462 (L*); Belkin et al. 1966: 3 (restriction of type loc.); Dodge 1966: 368 (first instar L*; key); Belkin 1968b: 20 (tax.); Basio 1971b: 71 (M*; bion.; as *nematoides*); Laffoon & Knight 1971 (F*); Bohart & Washino 1978: 134 (M, F, L*; keys, tax., bion., distr.); Tanaka et al. 1979: 183 (tax.); Harbach & Knight 1980 (F*); Bickley & Harrison 1989; Darsie & Ward 2005 (F*, L*; keys, distr.); Harrison et al. 2016 (F*, L*; keys, distr.); Robert et al. 2019 (distr.; Western Palearctic). **Ety.:** n.s. [alarming, terrifying (L)]. **Informal name:** Terrifying American Typical Mosquito.

syn. *saxatilis* Grossbeck
1905: 360 (F). **Type loc.:** Garret Mountain, New Jersey, United States (USNM). **Refs.:** Stone & Knight 1957a: 56 (lectotype desig.); Belkin et al. 1966: 20 (type loc. info).

syn. *frickii* Ludlow
1906b: 132 (F). **Type loc.:** Fort Snelling, Minnesota, United States (USNM).

syn. *pyrenaicus* Brolemann
1919a: 427 (M*, F). **Type loc.:** several localities in Pyrenees, France (NE).

syn. *nematoides* Dyar & Shannon
1925a: 84 (M, F). **Type loc.:** Hagthorpe, East Riding, Yorkshire, England (USNM). **Refs.:** Stone & Knight 1957a: 54 (lectotype desig.); Sirivanakarn 1971b: 1 (syn.).

Subgenus *Nicaromyia* González & Rodriquez

2001: 65 (as subg. of genus *Culex*). Type species: *Culex nicaroensis* Duret.

Name derivation: Nicaro, Cuba (type loc.).

nicaroensis Duret
1967a: 77 (M*, F). **Type loc.**: Nicaro, Municipio de Mayari, Oriente, Cuba (USNM). **Distr.**: Cuba. **Refs.**: Sirivanakarn 1983: 279 (M*); Harbach et al. 1991: 194 (type info.); Pecor et al. 1992: 38 (M*; tax., catalog citation; in subg. *Melanoconion*); Sallum & Forattini 1996: 518 (subg. uncertain); Broche & Rodríguez 2001: 65 (F*, P*, L*; to subg. *Nicaromyia*). **Ety.**: Nicaro, Municipio de Mayarí, Provincia de Oriente, Cuba. **Informal name:** Nicaro Cuban Typical Mosquito.

Subgenus *Oculeomyia* Theobald

1907: 515 (as genus). Type species: *Oculeomyia sarawaki* Theobald.

Name derivation: n.s. [eye (L) and myia (fly); prob. ref. to "Eyes very large and completely united in the middle line"].

Important references: Sirivanakarn 1976 (as Bitaeniorhynchus Subgroup of Sitiens Group of subg. *Culex*; tax., bion., distr.); Tanaka 2004b (from syn. to subg. of genus *Culex*; tax, distr.; Japan); Harbach 2011 (phyl.).

albinervis Edwards
1929c: 338 (M, F, P, L). **Type loc.:** Suva [Viti Levu], Fiji Islands (NHMUK). **Distr.:** Fiji, Tonga. **Refs.:** Amos 1947: 14 (L*), 22 (A*); Belkin 1962: 212 (M*, F, P*, L*); Lee et al. 1989a: 31 (F; key, tax., bion., distr., review); Tanaka 2004b (from subg. *Culex*). **Ety.:** n.s. [white veins (L); ref. to white on some crossveins]. **Informal name:** White-veined Fijian Typical Mosquito.

annulioris Theobald
1901a: 371 (F*). **Type loc.:** Salisbury, Mashonaland [Southern Rhodesia] (NHMUK). **Distr.:** Angola, Benin, Burkina Faso, Cameroon, Central African Republic, Côte d'Ivoire, Democratic Republic of the Congo, Ethiopia, Ghana, Guinea, Kenya, Liberia, Madagascar (includes Glorioso and Juan De Nova Isl.), Malawi, Mali, Nigeria, Republic of the Congo, São Tomé and Príncipe, Senegal, Seychelles, South Africa, Sudan and South Sudan, Tanzania, Uganda, Zimbabwe. **Refs.:** Macfie & Ingram 1922 (F*); Macfie & Ingram 1923: 434 (P*, L); Edwards 1941: 293 (M*, F*), 417 (P*); Hopkins 1952: 283 (L*); Hamon & Ovazza 1956: 89 (M*; tax.); Jupp 1996 (M*, F*; key). **Ety.:** ringed mouth (L); proboscis. **Informal name:** Ring-mouthed Zimbabwean Typical Mosquito.

syn. *gambiensis* Theobald
1903a: 5 (F; as var.). **Type loc.:** [Bathurst], Gambia (NHMUK). **Refs.:** Edwards 1941: 293 (syn.).

ssp. ***consimilis*** Newstead
1907 (F; as *tigripes* var.). In Newstead et al. 1907: 23. **Type loc.:** several localities in Belgian Congo (NHMUK). **Distr.:** Cameroon, Democratic Republic of the Congo, Gabon, Ghana, Kenya, Liberia, Mozambique, Nigeria,

Sierra Leone, Sudan and South Sudan, Tanzania, Uganda. **Refs.:** Edwards 1941: 295 (M*; to ssp.); Hamon & Ovazza 1956: 89 (M*; tax.). **Ety.:** n.s. **Informal name:** Similar Congolese Typical Mosquito.

ssp. syn. *congolensis* Evans
1923a: 91 (M; *annulioris* var.). **Type loc.:** Leopoldville, Belgian Congo (NHMUK). **Refs.:** Edwards 1941: 295 (syn.).

ssp. syn. *mayumbae* Galliard
1931b: 228 (M*; *bitaeniorhynchus* var.). **Type loc.:** Tchibanga and Mouila, Gabon (NHMUK). **Refs.:** Edwards 1941: 295 (syn.).

ssp. syn. *major* Edwards
1935a: 135 (M, F; *annulioris* var.). **Type loc.:** Nairobi [Central], Kenya (NHMUK). **Refs.:** Hamon & Ovazza 1956: 89 (syn.).

syn. *pseudoannulioris* Theobald
1909b: 15 (F*). **Type loc.:** Obuasi, Ashanti, Gold Coast (NHMUK). **Refs.:** Townsend 1990: 123 (type info, lectotype desig).

aurantapex Edwards
1914b: 74 (F). **Type loc.:** Nairobi [Central], Kenya (NHMUK). **Distr.:** Democratic Republic of the Congo, Kenya, Madagascar (includes Glorioso and Juan De Nova Isl.), Mozambique, South Africa, Zambia. **Refs.:** Edwards 1941: 292 (M*, F*), 416 (P); Hopkins 1952: 282 (L); Worth & De Meillon 1960: 247 (distr.); Jupp 1996 (M*, F*; key). **Ety.:** n.s. [orange coloration? (L, Gr)]. **Informal name:** Orange Kenyan Typical Mosquito.

ssp. ***jinjaensis*** Edwards
1941: 293 (M*, F). **Type loc.:** Jinja, Uganda (NHMUK). **Distr.:** Uganda, Zambia. **Refs.:** Mattingly 1956a: 41 (lectotype desig.); Muspratt 1959: 70 (tax.). **Ety.:** Jinja, Uganda. **Informal name:** Jinja Ugandan Typical Mosquito.

ssp. ***ellinorae*** Ovazza, Hamon, & Neri
1956: 167 (new name for *abyssinicus* van Someren, not Edwards 1941). **Distr.:** Ethiopia. **Ety.:** n.s. [Ellinor C. MacDonald, collector (Harbach 2018c)]. **Informal name:** Ellinora Kenyan Typical Mosquito.

ssp. syn. *abyssinicus* van Someren
1945: 79 (F; as *aurantapex* ssp.). **Type loc.:** Sciasciamanna [Arusi], Ethiopia (NHMUK).

bitaeniorhynchus Giles
1901a: 607 (A). **Type loc.:** Travancore, India (NHMUK). **Distr.:** Australia, Bangladesh, Benin, Burkina Faso, Cambodia, Cameroon, Democratic Republic of the Congo, Djibouti, Egypt, Ethiopia, Gabon, Ghana, Hong Kong, India, Indonesia, Iran, Japan, Kenya, Laos, Lesotho, Liberia, Madagascar (includes Glorioso and Juan De Nova Isl.), Malaysia, Mali, Micronesia (Wake Isl.), Mozambique, Myanmar, Namibia, Nepal, New Caledonia, Nigeria, Pakistan, Palau, Papua New Guinea, People's Republic of China, Philippines, Russia, Saudi Arabia, Senegal, Singapore, Solomon Isl., Somalia, South Africa, South Korea, Sri Lanka, Sudan and South Sudan, Taiwan, Tanzania, Thailand, The Gambia, Uganda, Vietnam, Yemen, Zambia, Zimbabwe. **Refs.:** Barraud 1923c (L*); Barraud 1934: 391 (M*, F, L*); Bohart & Ingram 1946b: 77 (M*, F, L*; bion., distr., tax.); La Casse & Yamaguti 1950 (M*,F*, P*, L*; tax., keys, bion., distr.); Asanuma & Nakagawa 1953: 93 (P*); Hara 1957: 55 (F*); Belkin 1962: 215 (M*, F, P*, L*); Bram 1967b: 258 (M*, F*, P*, L*; bion., distr., tax.); Aslamkhan 1971b (distr.; Pakistan); Basio 1971b: 54 (M*; bion.); Lotfi 1973: 206 (distr.); Sirivanakarn 1973c: 235 (M*, F*, P, L; tax.); Baisas 1974: 88 (M*, F*, P*, L*; tax., bion., distr.; Philippines); Gutsevich et al. 1974: 367 (M*, F, L*); Tanaka et al. 1975c: 220 (distr.); Rickenbach et al. 1976: 96 (distr.); Service 1976a: 262 (distr); Sirivanakarn 1976: 65 (M*, F*, P*, L*; distr.); Rodhain et al. 1977 (distr.); Tanaka et al. 1979: 173 (M*, F*, L*; as *Culex* (*Culex*)); Harbach 1988: 101 (M*, F, P*, L*; tax., keys, bion., distr.); Lee et al. 1989a: 102 (F; key, tax., bion., distr., review); Darsie & Pradhan 1990, 1991 (F, L; tax., keys, bion., distr.; Nepal); Jupp 1996 (M*, F*; key); Tanaka 2004b: 221 (P*, L mandible*, maxilla*; from subg. *Culex*); Abdel-Mohsin 2004 (distr.; Saudi Arabia); Rattanarithikul et al. 2005b (F*, L*; bion., distr., keys). **Ety.:** n.s. [double banded (L, Gr); prob. ref. to proboscis with two bands]. **Informal name:** Double-banded Indian Typical Mosquito. (Vol. 1, p. 528)

syn. *ager* Giles
1901b: 196 (F; *Taeniorhynchus*). **Type loc.:** Madras Presidency, India (NHMUK). **Refs.:** Townsend 1990: 38 (type info.).

syn. *sarawaki* Theobald
1907: 515 (F*; *Oculeomyia*). **Type loc.:** Sarawak [Malaysian Borneo] (NHMUK). **Refs.:** Sirivanakarn 1976: 71 (syn.).

syn. *maculipesarabiensis* Patton
1905: 635 (*Taeniorhynchus tenax* var.). **Refs.:** White 1980 (syn.).

syn. *ethiopicus* Edwards
1912c: 30 (A*; *ager* var.). **Type loc.:** [Bole, Northern Territories, Ghana] (NHMUK). **Refs.:** Edwards 1941: 291 (M*, F*; tax.; from syn. with *bitaeniorhynchus*), 415 (P*); Hopkins 1952: 282 (L*); Harbach 1988: 101 (syn.); Jupp 1996 (M*, F*; key).

syn. *karatsuensis* Mochizuki
1913: 28 (M*, F*, E). **Type loc.:** Karatsu, Kyushu, Japan (LU).

syn. *abdominalis* Taylor
1913: 53 (F*; *Culicelsa*). **Type loc.:** Ayr and Townsville, Queensland, Australia (NE).

syn. *arabiensis* Theobald
1913a: 318 (A; *Pseudotaeniorhynchus quasigelidus* var.). **Type loc.:** Transvaal [Republic of South Africa] (NE). **Refs.:** Edwards 1941: 291 (syn.).

cornutus Edwards
1922a: 283 (M*, F). **Type loc.:** Tavargatti jungle, Bombay, India (NHMUK). **Distr.:** India, Thailand. **Refs.:** Barraud

1923c (L*); Barraud 1934: 396 (M*, F, L*); Sirivanakarn 1976: 86 (M*, F, P, L; lectotype desig., distr.); Tanaka 2004b (from subg. *Culex*). **Ety.:** n.s. [horned (L)]. **Informal name:** Horned Indian Typical Mosquito.

epidesmus (Theobald)
1910a: 22 (F; *Taeniorhynchus*). **Type loc.:** Bhogaon, Purneah District [Bihar], India (LU). **Distr.:** Bangladesh, India, Nepal, Thailand, Sri Lanka. **Refs.:** Barraud 1934: 389 (M*, F); Darsie & Pradhan 1990, 1991 (F, L; tax., keys, bion., distr.; Nepal). **Ety.:** n.s. [common (Gr) or on, upon, band (Gr); "marked abdominal banding"]. **Informal name:** Common Indian Typical Mosquito.
syn. *ochracea* Theobald
1905g: 25 (F; *Grabhamia*; not Theobald 1901c). **Type loc.:** India (NHMUK). **Refs.:** Sirivanakarn 1976: 89 (lectotype desig.).
syn. *luteoabdominalis* Theobald
1910a: 23 (F; *Taeniorhynchus*). **Type loc.:** Katihar, Purneah District [Bihar] India (IM).

geminus Colless
1955: 314b (M*, F, L*). **Type loc.:** Singapore (NHMUK). **Distr.:** Malaysia, Singapore. **Refs.:** Sirivanakarn 1976: 90 (M*, F, P, L*); Tanaka 2004b (from subg. *Culex*). **Ety.:** twin of [close to] (L) *pseudosinensis*. **Informal name:** Twin Singaporean Typical Mosquito.

giganteus Ventrillon
1906a: 100 (M, F). **Type loc.:** Tananarive, Madagascar (MNHP). **Distr.:** Madagascar (includes Glorioso and Juan De Nova Isl.). **Refs.:** Edwards 1941: 296 (M*, F); Brunhes 1967: 208 (M*, F, P*, L*). **Ety.:** n.s. [giant (L)]. **Informal name:** Gigantic Malagasy Typical Mosquito.
syn. *neireti* Ventrillon
1906a: 103 (M, F). **Type loc.:** Tananarive, Madagascar (MNHP). **Refs.:** White 1975: 317 (tax.).

infula Theobald
1901a: 370 (F*). **Type loc.:** Taiping, Perak, Malaya [Malaysia] (NHMUK). **Distr.:** Bangladesh, India, Indonesia, Malaysia, Myanmar, Nepal, Philippines, Singapore, Sri Lanka, Thailand, Vietnam. **Refs.:** Sirivanakarn 1976: 71 (M*, F*, P, L*; distr.; from syn. with *bitaeniorhynchus*); Darsie & Pradhan 1990, 1991 (F, L; tax., keys, bion., distr.; Nepal); Tanaka 2004b (from subg. *Culex*); Rattanarithikul et al. 2005b (F*, L*; bion., distr., keys). **Ety.:** n.s. [headband (L); prob. ref. to banded proboscis]. **Informal name:** Headband Malaysian Typical Mosquito.
syn. *tenax* Theobald
1901c: 198 (F*; *Taeniorhynchus*). **Type loc.:** Perak, Malaya [Malaysia] (NHMUK). **Refs.:** Bohart 1957: 84 (syn.); Sirivanakarn 1976: 71 (syn.).
syn. *ambiguus* Theobald
1903a: 248 (M; *Grabhamia*). **Type loc.:** Quilon [Travancore], India (NHMUK). **Refs.:** Bohart 1957: 84 (syn.); Sirivanakarn 1976: 71 (syn.).
syn. *taeniarostris* Theobald
1907: 299 (F; *Grabhamia*). **Type loc.:** Peradeniya [Central], Ceylon [Sri Lanka] (NHMUK). **Refs.:** Sirivanakarn 1976: 71 (syn.).
syn. *ocellata* Theobald
1907: 488 (F*; *Taeniorhynchus tenax* var.). **Type loc.:** Kuching, Sarawak [Malaysian Borneo] (NHMUK). **Refs.:** Sirivanakarn 1976: 71 (syn.).
syn. *domesticus* Leicester
1908: 169 (M, F; *Taeniorhynchus*; not Germar 1817). **Type loc.:** Kuala Lumpur [Selangor], Malaya [Malaysia] (NHMUK). **Refs.:** Sirivanakarn 1976: 71 (syn.).
syn. *afridii* Qutubuddin
1956: 140 (M*). **Type loc.:** Dacca, Pakistan [Bangladesh] (NHMUK). **Refs.:** Cotrim et al. 1974: 12 (M*); Sirivanakarn 1976: 71 (syn.).

kinabaluensis Sirivanakarn
1976: 91 (M*, F, P*, L*). **Type loc.:** Mount Kinabalu, Sabah, Malaysia [Malaysian Borneo] (USNM). **Distr.:** Malaysia. **Refs.:** Tanaka 2004b (from subg. *Culex*). **Ety.:** Mt. Kinabalu, Sabah, Borneo, Malaysia. **Informal name:** Kinabalu Malaysian Typical Mosquito.

longicornis Sirivanakarn
1976: 81 (M*, F). **Type loc.:** Doi Suthep, Chiang Mai, Thailand (USNM). **Distr.:** Thailand. **Refs.:** Rattanarithikul et al. 2005b (F*, L*; bion., distr., keys). **Ety.:** long horned (L); ref. to phallosome with "slender and long hornlike apical sternal process." **Informal name:** Long-horned Thai Typical Mosquito.

luzonensis Sirivanakarn
1976: 75 (M*, F*, P, L). **Type loc.:** Imugan, Dalton, N Vizcaya, Luzon, Philippines (USNM). **Distr.:** Philippines, Timor. **Refs.:** Whelan & Hapgood 2000 (distr.; East Timor); Tanaka 2004b (from subg. *Culex*). **Ety.:** Luzon, Philippine Isl. **Informal name:** Luzon Philippine Typical Mosquito.

poicilipes (Theobald)
1903a: 236 (F; *Lasioconops*). **Type loc.:** Bonny [Eastern Provinces], Nigeria (NHMUK). **Distr.:** Angola, Benin, Burkina Faso, Cameroon, Central African Republic, Egypt, Ethiopia, Gabon, Ghana, Guinea-Bissau, Israel (and Gaza Strip and West Bank), Kenya, Madagascar (includes Glorioso and Juan De Nova Isl.), Malawi, Mali, Mauritania, Nigeria, Senegal, Somalia, South Africa, Sudan and South Sudan, Tanzania, The Gambia, Uganda, Zimbabwe. **Refs.:** Macfie & Ingram 1922 (F*) (as *quasigelidus* Theobald); Edwards 1941: 289 (M*, F*); Hopkins 1952: 280 (L*); Ribeiro & da Cunha Ramos 1980 (tax., key; distr.; Angola); Harbach 1988: 92 (M*, F, P*, L*; tax., keys, bion., distr.); Jupp 1996 (M*, F*; key); Capela et al. 1997 (distr.; Guinea-Bissau); Robert et al. 2019 (distr.; Western Palearctic). **Ety.:** n.s. [spotted hair footed (Gr); poss. ref. to leg markings]. **Informal name:** Sock Nigerian Typical

Mosquito [*poicilipes* = "spot hair footed" = woolly sock]. (Vol. 1, p. 530)

syn. *quasigelidus* Theobald
1903a: 181 (M, F*). **Type loc.:** Entebbe, Uganda (NHMUK). **Refs.:** Kirkpatrick 1925b: 107 (M*, F, P*, L*); Harbach 1988: 92 (lectotype desig.).

syn. *taeniorhynchoides* Giles
1904a: 369 (F). **Type loc.:** Benguela, Angola (NHMUK).

syn. *maculipes* Theobald
1904b: 79 (A*; *Taeniorhynchus tenax* var.). **Type loc.:** Kenissa and Middle Sobat, Upper Nile, Sudan (NHMUK).

syn. *madagascariensis* Ventrillon
1905b: 427 (F; *Pseudoheptaphlebomyia*). **Type loc.:** Tananarive, Madagascar (MMHP). **Refs.:** Harbach 1988: 92 (lectotype desig.).

syn. *punctipes* Theobald
1907: 316 (F*; *Aporoculex*). **Type loc.:** Chinde [Zambezia, Mozambique] (NHMUK).

syn. *par* Newstead
1907 (F*). In Newstead et al. 1907: 25. **Type loc.:** Tshumbiri [Leopoldville], Belgian Congo [Democratic Republic of the Congo] (NHMUK).

syn. *auritaenia* Enderlein
1920: 49 (M, F). **Type loc.:** Tananarive, Madagascar (NHMUK). **Refs.:** Harbach 1988: 92 (lectotype desig.).

pseudosinensis Colless
1955b: 311 (M*, F, L*). **Type loc.:** Singapore (NHMUK). **Distr.:** Cambodia, Indonesia, Laos, Malaysia, Singapore, Thailand, Vietnam. **Refs.:** Bram 1967b: 265 (M*, F*, L*; bion., distr., tax.); Sirivanakarn 1976: 79 (M*, F*, P, L*; distr.); Kohn 1990: 165 (distr.); Tanaka 2004b (from subg. *Culex*); Rattanarithikul et al. 2005b (F*, L*; bion., distr., keys). **Ety.:** false (Gr); similar to *sinensis*. **Informal name:** False-China Singaporean Typical Mosquito.

samoaensis (Theobald)
1914: 36 (F; *Pseudotaeniorhynchus*). **Type loc.:** Apia [Upolu], Samoa Islands (NHMUK). **Distr.:** Samoa (Ind. State of Samoa, American Samoa, Western Samoa). **Refs.:** Bohart & Ingram 1946b: 17 (F); Belkin 1962: 210 (F); Ramalingam 1976: 308 (distr.), 309 (tax., bion., keys., distr.; Samoa and Tonga); Lee et al. 1989a: 221 (F; key, tax., bion., distr., review). **Ety.:** Samoa. **Informal name:** Apia Samoan Typical Mosquito.

selangorensis Sirivanakarn
1976: 77 (M*, F*, P, L). **Type loc.:** Bukit Kutu, Selangor, Malaysia (USNM). **Distr.:** Malaysia. **Refs.:** Tanaka 2004b (from subg. *Culex*). **Ety.:** Selangor, Malaysia. **Informal name:** Selangor Malaysian Typical Mosquito.

sinensis Theobald
1903a: 180 (F; *gelidus* var.). **Type loc.:** Shaohyling, China [People's Republic of China] (NHMUK). **Distr.:** Australia, Bangladesh, Cambodia, India, Indonesia, Japan, Laos, Malaysia, Myanmar, Nepal, Papua New Guinea, People's Republic of China (includes Hong Kong), Philippines, Russia, Sri Lanka, South Korea, Sudan and South Sudan, Taiwan, Thailand, Vietnam. **Refs.:** Baisas 1938: 209 (F, E); Bohart & Ingram 1946b: 79 (M*, F, L*; bion., distr., tax.); La Casse & Yamaguti 1950: 206 (M*, F*, P*, L*; bion., distr., tax.); Hara 1957: 55 (F*); Joshi et al. 1965 (distr.; Nepal); Bram 1967b: 268 (M*, F*, L*; bion., distr., tax.); Aslamkhan 1971b (distr.; Pakistan); Basio 1971b: 59 (M*; bion.); Gutsevich et al. 1974: 371 (M*, F, L*); Tanaka et al. 1975c: 220 (distr.); Sirivanakarn 1976 (F*); Tanaka et al. 1979: 177 (M*, F*, L*; as *Culex* (*Culex*)); Lee et al. 1989a: 223 (F; key, tax., bion., distr., review); Darsie & Pradhan 1990, 1991 (F, L; tax., keys, bion., distr.; Nepal); Gornostaeva 2000: 480 (distr.; Russia); Tanaka 2004b: 224 (from subg. *Culex*); Rattanarithikul et al. 2005b (F*, L*; bion., distr., keys). **Ety.:** China. **Informal name:** Local Chinese Typical Mosquito [*sinensis* = "from China" = Local].

syn. *sepositus* Leicester
1908: 152 (F). **Type loc.:** Kuala Lumpur [Selangor], Malaya [Malaysia] (NE).

syn. *tenax* Leicester
1908: 167 (F; *Taeniorhynchus*). **Type loc.:** Kuala Lumpur [Selangor], Malaya [Malaysia] (NE).

syn. *tripunctatus* Mochizuki
1913: 24 (M*, F*, E). **Type loc.:** Fukuoka, Kyushu, Japan (LU).

squamosus (Taylor)
1914c: 691 (F*; *Culicada*). **Type loc.:** Townsville, Queensland, Australia (ANIC). **Distr.:** Australia, Indonesia, Papua New Guinea, Solomon Islands. **Refs.:** Brug 1934a: 515 (M*, F, P*, L*); Lee 1944a: 98 (L*); Penn 1949b: 79 (P*); Belkin 1962: 214 (M*, F, P*, L*); Lee et al. 1989a: 245 (F; key, tax., bion., distr., review, type info); Tanaka 2004b (from subg. *Culex*). **Ety.:** n.s. [squamous, scaly (L); ref. to distinguishing character of "pleurae brown with white flat scales"]. **Informal name:** Scaly Australian Typical Mosquito.

syn. *annulata* Theobald
1905f: 98 (M*; *Trichopronomyia*; not Schrank 1776). **Type loc.:** Friedrich-Wilhelmshafen, [NE] New Guinea (HNM).

syn. *annulirostris* Taylor
1914c: 696 (M, F; *Leucomyia*). **Type loc.:** Townsville and Ching Do, Queensland, Australia (ANIC). **Refs.:** Lee et al. 1989a: 245 (type info.).

syn. *taylori* Edwards
1921c: 78 (new name for *annulirostris* Taylor, not Skuse 1889).

starckeae Stone & Knight
1958: 69 (new name for *basicinctus* Edwards 1922d, not Edwards 1921c). **Distr.:** Australia, Indonesia, New Caledonia, Vanuatu. **Refs.:** Belkin 1962: 217 (M*, F, P*, L*); Pinhão 1974 (distr.; Timor); Lee et al. 1989a: 252 (F; key, tax., bion., distr., review); Tanaka 2004b (from

subg. *Culex*). **Ety.:** Gordon Sylvester Starkey. **Informal name:** Starkey Australian Typical Mosquito.

syn. *basicinctus* Edwards
1922d: 96 (M, F). **Type loc.:** Townsville, Queensland, Australia (NHMUK). **Refs.:** Lee 1944a: 99 (L*).

Subgenus *Phenacomyia* Harbach & Peyton

1992b: 242 (as subg. of genus *Culex*). Type species: *Culex corniger* Theobald.

Name derivation: phenax = deceiver, cheat, imposter (Gr), in reference to the included species having been mistaken for a member of subgenus *Culex*.

Important references: Strickman & Pratt 1989 (tax.; *Cx. corniger* and *Cx. lactator*); Deus 2009 (mol. phyl.); Demari-Silva et al. 2011 (mol. phyl.); Vesgueiro et al. 2011 (mol. phyl.); Harbach et al. 2012 (phyl.).

airozai Lane
1945b: 204 (M*, F). **Type loc.:** Rio Paruari, Amazonas, Brazil (FH). **Distr.:** Argentina, Brazil, Chile, Colombia. **Refs.:** Lane 1953: 338 (M*, F); Bram 1967a: 23 (M*; distr.); Cotrim et al. 1974: 12 (M*); Harbach & Peyton 1992b: 242 (from subg. *Culex*). **Ety.:** Augusto Leopoldo Ayrosa Galvão. **Informal name:** Ayrosa Brazilian Typical Mosquito.

corniger Theobald
1903a: 173 (M, F*). **Type loc.:** Pará, Brazil (NHMUK). **Distr.:** Argentina, Belize, Bolivia, Brazil, Cayman Isl., Colombia, Costa Rica, Cuba, Dominican Republic, Ecuador, El Salvador, French Guiana, Guatemala, Guyana, Haiti, Honduras, Jamaica, Lesser Antilles (includes Dominica, Guadeloupe, Trinidad and Tobago), Mexico, Nicaragua, Panama, Paraguay, Peru, Suriname, Uruguay, Venezuela. **Refs.:** Martínez 1950: 187 (M*); Lane 1953: 374 (M*, F, L*); Senevet & Abonnenc 1958b: 344 (P*, L*); Bram 1967a: 43 (M*, F, L); Belkin 1968b: 15 (lectotype desig.); Belkin et al. 1970: 61 (M*, F, P*, L*); Heinemann & Belkin 1977b: 427, 442 (distr.; Guyana, Honduras); Strickman & Pratt 1989: 552 (M*, F*, P*, L*); Harbach & Peyton 1992b: 242 (from subg. *Culex*); Pecor et al. 2002: 248 (tax., distr.; Belize); Pena et al. 2003 (distr.; Dominican Republic); Berti et al. 2015 (distr.; Venezuela). **Ety.:** n.s. [horn bearing (L)]. **Informal name:** Horned Brazilian Typical Mosquito.

syn. *hassardii* Grabham
1906a: 167 (M*, F*, P, L*). **Type loc.:** Newcastle [Surrey], Jamaica (USNM). **Refs.:** Belkin et al. 1970: 61 (lectotype desig.).

syn. *basilicus* Dyar & Knab
1906b: 169 (F, L). **Type loc.:** Arima, Trinidad [Trinidad & Tobago] (USNM). **Refs.:** Stone & Knight 1957a: 43 (lectotype desig.).

syn. *subfuscus* Theobald
1907: 403 (M*). **Type loc.:** Moneague [Middlesex], Jamaica (NHMUK). **Refs.:** Belkin 1968b: 20 (type info.).

syn. *loquaculus* Dyar & Knab
1909d: 254 (A; *lactator* var.). **Type loc.:** [Corozal], Canal Zone, Panama (USNM). **Refs.:** Stone & Knight 1957a: 52 (lectotype desig.).

syn. *rigidus* Senevet & Abonnenc
1939a: 68 (M*, F, L*). **Type loc.:** Saut Tigre [Inini], French Guiana [Guyana] (NE). **Refs.:** Lane 1951: 334 (syn.).

lactator Dyar & Knab
1906d: 209 (L*). **Type loc.:** Rincon Antonio, Mexico (USNM). **Distr.:** Belize, Colombia, Guatemala, Honduras, Mexico, Panama, Venezuela. **Refs.:** Stone & Knight 1957a: 52 (lectotype desig.); Strickman & Pratt 1989: 562 (M*, F*, P*, L*; from syn. with *corniger*); Harbach & Peyton 1992b: 242 (from subg. *Culex*). **Ety.:** n.s. [entice (L)]. **Informal name:** Enticing Mexican Typical Mosquito.

Subgenus *Phytotelmatomyia* Rossi & Harbach

2008: 2 (as subg. of genus *Culex*). Type species: *Culex renatoi* Lane & Ramalho.

Name derivation: "Derived from three Greek words: Phyton (n. plant, combining form phyto-); telma, -tos (n. standing water, combining form telmato-), and myia (f. fly)."

Important references: Harbach et al. 2012 (phyl.).

castroi Casal & García
1967b: 451 (M*, F, P*, L*). **Type loc.:** Canal 6 and Parana de las Palmas, Estación Agropecuaria Delta, Buenos Aires, Argentina (INM). **Distr.:** Argentina, Lesser Antilles (includes Anguilla, Antigua and Barbuda), Uruguay. **Refs.:** Rossi & Martínez 2003: 471 (distr.; Uruguay); Rossi & Harbach 2008: 1 (L*; to subg. *Phytotelmatomyia*). **Ety.:** Manuel P. Castro. **Informal name:** Castro Argentinian Typical Mosquito.

hepperi Casal & García
1967c: 88 (M*, F, P*, L*). **Type loc.:** Canal 6 and Parana de las Palmas, Estación Agropecuaria Delta, Buenos Aires, Argentina (INM). **Distr.:** Argentina, Lesser Antilles (includes Anguilla, Antigua and Barbuda), Uruguay. **Refs.:** Rossi & Martínez 2003: 472 (distr.; Uruguay); Rossi & Harbach 2008 (M*, P*, L*; to subg. *Phytotelmatomyia*). **Ety.:** Héctor C. Hepper. **Informal name:** Hepper Argentinian Typical Mosquito.

machadoi Mattos, Guedes & Xavier
1978: 31 (M*, P*, L*). **Type loc.:** "Sitio do Sr Alexandre Ribiero da Silva," Florestal County, Minas Gerais, Brazil (CPRR). **Distr.:** Brazil. **Refs.:** Xavier et al. 1980 (F*); Rossi & Harbach 2008 (to subg. *Phytotelmatomyia*). **Ety.:**

A.B.M. Machado. **Informal name:** Machado Brazilian Typical Mosquito.

renatoi Lane & Ramalho
1960: 173 (M*, F, P*, L*). **Type loc.:** Bairro de São Miguel Paulista, São Paulo, Brazil (FH). **Distr.:** Argentina, Brazil. **Refs.:** Bram 1967a: 92 (M*, F, L); Belkin et al. 1971: 22 (type info.); Rossi & Harbach 2008 (L*; to subg. *Phytotelmatomyia*). **Ety.:** Renato R. Corrêa. **Informal name:** Renato Brazilian Typical Mosquito.

Subgenus *Sirivanakarnius* Tanaka

2004b: 224 (as subg. of genus *Culex*). Type species: *Culex* (*Culex*) *boninensis* Bohart.

Name derivation: Sunthorn Sirivanakarn.

Important references: Bohart 1957 (bion., distr.); Tanaka et al. 1979 (bion., distr.); Harbach et al. 2012 (phyl.).

boninensis Bohart
1957: 75 (M*, F, P, L*). **Type loc.:** Chichi Jima, Bonin Islands (USNM). **Distr.:** Japan. **Refs.:** Tanaka et al. 1979: 171 (M*, F*, L*); Toma & Miyagi 2005a (bion.); Lee et al. 1989a: 113 (tax.); Tanaka 2004b (from subg. *Culex*). **Ety.:** Bonin Isl., Japan. **Informal name:** Bonin Japanese Typical Mosquito.

Subgenus *Tinolestes* Coquillett

1906d: 16, 24; Coquillett 1906a: 185 (as genus). Type species: *Tinolestes latisquama* Coquillett.

Name derivation: n.s [possibly: tino- = repay, take vengeance (Gr), plus robber, plunderer, pirate (Gr)].

Important references: Belkin 1968b: 11 (resurrected from syn.); Berlin & Belkin 1980 (revision); Harbach 2011 (phyl.).

breviculus Senevet & Abonnenc
1939a: 110 (M*). **Type loc.:** Saut Tigre [Inini], French Guiana [Guyana] (FMP; originally LPFM). **Distr.:** Brazil, French Guiana. **Refs.:** Rozeboom & Komp 1950a: 88 (M*); Belkin 1968b: 14 (tax.); Sirivanakarn 1983: 272 (transfer from subg. *Melanoconion*). **Ety.:** n.s. [short posterior (L)]. **Informal name:** Small French Guianese Typical Mosquito.
syn. *mojuensis* Duret & Damasceno
1955: 401 (M*, F). **Type loc.:** Oriboca, Rio Guajará, Pará, Brazil (USNM). **Refs.:** Belkin 1968b: 12, 14 (syn.); Belkin et al. 1971: 23 (type loc. info.); Harbach et al. 1991: 194 (type info.).

cauchensis Floch & Abonnenc
1945b: 1 (M*). **Type loc.:** Caux, French Guiana [Guyana] (PIP). **Distr.:** Brazil, French Guiana, Guyana. **Refs.:** Belkin 1968b: 12, 14 (transfer; type info.); Harrison 1973c: 277 (type info.); Sirivanakarn 1983: 272 (transfer from subg. *Melanoconion*). **Ety.:** Caux, French Guiana. **Informal name:** Caux French Guianese Typical Mosquito.

latisquama (Coquillett)
1906a: 185 (M, F; *Tinolestes*). **Type loc.:** Port Limón, Costa Rica (USNM). **Distr.:** Colombia, Costa Rica, Guatemala, Honduras, Nicaragua, Panama, Suriname, United States (continental). **Refs.:** Rozeboom & Komp 1950a: 92 (M*; lectotype desig.); Lane 1953: 387 (M*, F, L*); Foote 1954: 65 (L; tax.); Stone & Knight 1957a: 52 (type info.); Berlin & Belkin 1980: 22 (M*, F*, P*, L*; keys, tax., bion., distr.). **Ety.:** n.s. [scaled on the side (L)]. **Informal name:** Scale-sided Costa Rican Typical Mosquito.

Subgenus Uncertain *Culex*

cairnsensis (Taylor)
1919: 829 (F; *Pseudoskusea*). **Type loc.:** Cairns, Queensland, Australia (ANIC). **Distr.:** Australia. **Refs.:** Mattingly & Marks 1955: 166 (tax.); Lee et al. 1988b: 35 (tax., distr., review). **Ety.:** Cairns, Australia. **Informal name:** Cairns Australian Typical Mosquito.

flochi Duret
1969d: 119 (M*, F). **Type loc.:** Rio Moju, Belém, Pará, Brazil (USNM). **Distr.:** Brazil, Colombia, French Guiana, Guyana. **Refs.:** Harbach et al. 1991: 193 (type info); Pecor et al. 1992: 3 (to subg. uncertain). **Ety.:** Hervé Floch. **Informal name:** Floch Brazilian Typical Mosquito.

inornatus (Theobald)
1905g: 21 (F*; *Gnophodeomyia*; not Williston 1893). **Type loc.:** New Amsterdam [Berbice], British Guiana [Guyana] (NHMUK). **Distr.:** Guyana, Suriname. **Refs.:** Dyar & Knab 1919: 7 (rejected name); Belkin 1968b: 16 (lectotype desig.); Belkin 1970b: 59 (tax.). **Ety.:** n.s. [unadorned (L); poss. ref. to unmarked thorax, abdomen, and legs]. **Informal name:** Unadorned Guyanese Typical Mosquito.

nigrimacula Lane & Whitman
1943: 403 (M*, F, P*, L*). **Type loc.:** Vicinity of Rio de Janeiro, Guanabara, Brazil (INER). **Distr.:** Brazil, French Guiana. **Refs.:** Belkin et al. 1971: 24 (type loc. info.); Xavier 1973: 159 (type info.); Pecor et al. 1992: 3 (to subg. uncertain). **Ety.:** n.s. [black spot (L)]. **Informal name:** Black-dotted Brazilian Typical Mosquito.

ocellatus Theobald
1903a: 222 (F*). **Type loc.:** São Paulo, Brazil (LU). **Distr.:** Bolivia, Brazil, Colombia, French Guiana, Guyana. **Refs.:** Rozeboom & Komp 1950a: 94 (M*; tax.); Lane 1953: 405 (M*, F, P*, L*); Foote 1954: 76 (P*, L*; tax.); Belkin 1968b: 18 (tax.); Townsend 1990: 114 (type info.); Pecor et al. 1992: 3 (to subg. uncertain). **Ety.:** n.s. [having eyes? (L);

poss. ref. to "eye-like spots at the roots of the wings"]. **Informal name:** Eye-spotted Brazilian Typical Mosquito.

syn. *automartus* Root
1927c: 591 (M*). **Type loc.:** Botanical Garden, Rio de Janeiro, Guanabara, Brazil (USNM). **Refs.:** Lane & Whitman 1943: 402 (syn.); Belkin et al. 1971: 24 (type loc. info.).

punctiscapularis Floch & Abonnenc
1946b: 1 (M*). **Type loc.:** Crique Anguille, French Guiana [Guyana] (PIP). **Distr.:** French Guiana, Guyana. **Refs.:** Rozeboom & Komp 1950a: 95 (M*; tax.); Lane 1951: 334 (syn.); Duret 1969d: 122 (M*; from syn.); Harrison 1973c: 278 (type info.); Pecor et al. 1992: 3 (to subg. uncertain). **Ety.:** n.s. [marked scapula (L); poss. ref. to dark circular spots at the wing roots]. **Informal name:** Spot-shouldered French Guianese Typical Mosquito.

romeroi Surcouf & González-Rincones
1912: 295 (new name for *nigritulus* Zetterstedt of Romero Sierra 1907). **Type loc.:** Caracas, Venezuela (LU). **Distr.:** Venezuela. **Ety.:** n.s. [Bachiller Jesús Maria Romero Sierra (Harbach 2018c)]. **Informal name:** Romero Venezuelan Typical Mosquito.

Genus *Deinocerites* Theobald

(Vol. 1, p. 242)

1901a: 98, 1901b: 235 (July 15), 1901c: 215 (Nov. 23) (as genus). Type species: *Deinocerites cancer* Theobald.

Name derivation: n.s. [terrible, fearful horn (Gr); poss. ref. to stout setae on gonocoxite].

syn. *Brachiosoma* Theobald 1901b: 235 (as genus). Type species: *Deinocerites cancer* Theobald (Coquillett 1910: 515). Blanchard 1905: 414 (syn.).
syn. *Brachiomyia* Theobald 1901c: 343. Type species: *Brachiomyia magna* Theobald. Theobald 1903a: 275 (syn.).
syn. *Deinokerides* Giles 1902: 472 (unjustified emend.).
syn. *Dinocerites* Blanchard 1905: 413 (unjust. emend. of *Deinocerites*).
syn. *Dinomimetes* Knab 1907b: 120. Type species: *Dinomimetes epitedeus* Knab. Dyar 1918b: 101 (syn.).
syn. *Dinanamesus* Dyar & Knab 1909d: 259 (as genus). Type species: *Dinanamesus spanius* Dyar & Knab.

Important references: Belkin & Hogue 1959 (revision); Belkin et al. 1970 (tax., bion.; Jamaica); Adames 1971 (revision, tax., keys, bion.); Navarro & Liria 2000 (L mouthparts*; phyl.; as subg. of genus *Culex*); Harbach et al. 2012 (phyl.).

atlanticus Adames
1971: 28 (M*, F*, P*, L*). **Type loc.:** La Boca, Ocumare de la Costa, Aragua, Venezuela (USNM). **Distr.:** Colombia, Panama, Venezuela. **Ety.:** Atlantic Region; part of Central and South America. **Informal name:** Atlantic Venezuelan Crabhole Mosquito.

barretoi Adames
1971: 40 (M*, F*). **Type loc.:** mouth of Rio Raposo, Valle, Colombia (USNM). **Distr.:** Colombia. **Refs.:** Barreto & Lee 1969: 435 (as *dyari*). **Ety.:** Pablo Barreto. **Informal name:** Barreto Colombian Crabhole Mosquito.

belkini Adames
Adames 1971: 34 (M*, F*, P*, L*). **Type loc.:** Barra de Navidad, Jalisco, Mexico (USNM). **Distr.:** El Salvador, Mexico. **Ety.:** John Nicholas Belkin. **Informal name:** Belkin Mexican Crabhole Mosquito.

cancer Theobald
1901c: 215 (F*). **Type loc.:** Kingston, Jamaica (NHMUK). **Distr.:** Bahamas, Belize, Cayman Isl., Colombia, Costa Rica, Cuba, Dominican Republic, Guatemala, Haiti, Honduras, Jamaica, Lesser Antilles (includes Aruba, Bonaire, Curaçao, Sint Maarten, Trinidad and Tobago), Mexico, Nicaragua, Panama, Puerto Rico, United States (continental), Venezuela. **Refs.:** Gerry 1932 (F*); Cerqueira 1938 (F*); Yamaguti & LaCasse 1951c: 124 (M*, F*, L*); Carpenter & LaCasse 1955: 325 (M*, F*, L*; keys); Belkin & Hogue 1959: 432 (M*, F*, P, L; lectotype desig.); Dodge 1966: 375 (first instar L*; key); Belkin et al. 1970: 46 (M*, F, P*, L*); Adames 1971: 44 (M*, F*, P*, L*); Mattingly 1971a: Pl. 3 (M, F head*), Pl. 29 (L head*, terminal segs.*); Mattingly 1973c: 222 (E*); Liria & Navarro 1999 (L mouthparts); Darsie & Ward 2005 (F*, L*; keys, distr.; United States). **Ety.:** crab (L); larvae in crabholes. **Informal name:** Cancer Jamaican Crabhole Mosquito. Official ESA common name: crabhole mosquito.
syn. *tetraspathus* Dyar & Knab
1909d: 260 (F). **Type loc.:** Bluefields, Nicaragua (USNM). **Refs.:** Dyar 1928: 262 (syn.); Stone & Knight 1957c: 197 (type info., lectotype desig.).

colombianus Adames
1971: 64 (M*, F*, P*, L*). **Type loc.:** La Punta, Turbo, Antioquia, Colombia (USNM). **Distr.:** Colombia, Panama. **Ety.:** Colombia. **Informal name:** Quimbaya Colombian Crabhole Mosquito [the Quimbaya were one of the original civilizations of Colombia].

costaricensis Adames & Hogue
1970: 12 (M*, F*, P*, L*). **Type loc.:** 1 km north of Boca del Rio Barranca, Hacienda Bonilla, Puntarenas Province, Costa Rica (USNM). **Distr.:** Costa Rica. **Refs.:** Adames 1971 (F*). **Ety.:** Costa Rica. **Informal name:** Puntarenas Costa Rican Crabhole Mosquito.

curiche Adames
1971: 68 (M*, F*, P*, L*). **Type loc.:** El Naranjo, Curiche, El Chocó, Colombia (USNM). **Distr.:** Colombia, Costa Rica. **Ety.:** El Naranjo, Curiche, Colombia. **Informal name:** Curiche Colombian Crabhole Mosquito.

dyari Belkin & Hogue
1959: 428 (M*, F*). **Type loc.:** Corozal, Canal Zone, Panama (USNM). **Distr.:** Colombia, Costa Rica, Panama. **Refs.:** Adames 1971: 37 (M*, F*, P*, L*). **Ety.:** Harrison G. Dyar. **Informal name:** Dyar Panamanian Crabhole Mosquito.

epitedeus (Knab)
1907b: 120 (M, F; *Dinomimetes*). **Type loc.:** Port Limón, Costa Rica (USNM). **Distr.:** Costa Rica, Guatemala, Honduras, Nicaragua, Panama. **Refs.:** Lane 1953: 556 (M*, F, P*, L*); Stone & Knight 1957c: 197 (type info., lectotype desig.); Belkin & Hogue 1959: 438 (M*, F*, P, L); Adames 1971: 59 (M*, F*, P*, L*); Taylor & Turner 1998: 215 (distr.; Honduras). **Ety.:** n.s. **Informal name:** Limón Costa Rican Crabhole Mosquito.

howardi Belkin & Hogue
1959: 441 (M*, F*). **Type loc.:** Las Peñas (Puerto Vallarta), Jalisco, Mexico (USNM). **Distr.:** El Salvador, Mexico, Nicaragua. **Refs.:** Adames 1971: 77 (M*, F*, P*, L*). **Ety.:** Leland Ossian Howard. **Informal name:** Howard Mexican Crabhole Mosquito.

magnus (Theobald)
1901c: 344 (F; *Brachiomyia*). **Type loc.:** Saint Lucia Island, Lesser Antilles (NE). **Distr.:** Argentina, Brazil, French Guiana, Guyana, Lesser Antilles (includes Antigua and Barbuda, Barbados, Dominica, Grenada, Guadeloupe, Martinique, Montserrat, Saint Kitts and Nevis, Saint Lucia, Saint Vincent and the Grenadines, Trinidad and Tobago, Virgin Isl.), Puerto Rico, Suriname, Venezuela. **Refs.:** Floch & Abonnenc 1945e: 38 (M*, L*; tax.; as *troglodytus*); Belkin & Hogue 1959: 431 (M*, F*, P, L; from syn. with *cancer* Theobald); Belkin 1968b: 22 (type info.); Adames 1971: 52 (M*, F*, P*, L*). **Ety.:** n.s. [large? (L)]. **Informal name:** Large Saint Lucian Crabhole Mosquito.

syn. *troglodytus* Dyar & Knab
1909d: 260 (M*, F*). **Type loc.:** Trinidad [Trinidad and Tobago, Lesser Antilles] (USNM). **Refs.:** Stone & Knight 1957c: 197 (type info.); Belkin & Hogue 1959: 430 (syn.).

mathesoni Belkin & Hogue
1959: 426 (M*, F*, P, L). **Type loc.:** Brownsville, Texas (USNM). **Distr.:** Mexico, United States (continental). **Refs.:** Knight & Chamberlain 1948: 18 (P*; as *spanius*); Yamaguti & LaCasse 1951c (F*; as *spanius* (Dyar & Knab)); Carpenter & LaCasse 1955: 327 (M*, F*, L*; keys; as *spanius*, in part); Dodge 1966: 376 (first instar L; key); Adames 1971: 31 (M*, F*, P*, L*); Heinemann & Belkin 1977c: 534 (distr.; Mexico); Darsie & Ward 2005 (F*, L*; keys, distr.; United States). **Ety.:** Robert Matheson. **Informal name:** Matheson American Crabhole Mosquito.

mcdonaldi Belkin & Hogue
1959: 437 (M*, F*, P*, L*). **Type loc.:** San Blas, Nayarit, Mexico (USNM). **Distr.:** Mexico. **Refs.:** Adames 1971: 79 (M*, F*, P*, L*); Mattingly 1971a: Pl. 11 (P abdomen*); Cheng & Hogue 1974: 211 (bion., distr.; Mexico). **Ety.:** William A. McDonald. **Informal name:** McDonald Mexican Crabhole Mosquito.

melanophylum Dyar & Knab
1907c: 200 (M, F; as *cancer* var.). **Type loc.:** Colón, Canal Zone, Panama (USNM). **Distr.:** Colombia, Panama, Venezuela. **Refs.:** Stone & Knight 1957c: 197 (type info., lectotype desig.); Belkin & Hogue 1959: 434 (M*, F*; to sp.); Adames 1971: 49 (M*, F*, P*, L*); Liria & Navarro 1999 (L mouthparts). **Ety.:** black type (Gr); ref. to dark color in contrast to cancer. **Informal name:** Darkened Panamanian Crabhole Mosquito.

syn. *monospathus* Dyar & Knab
1925 (F). In Dyar 1925f: 155. **Type loc.:** Fort Sherman, Canal Zone, Panama (USNM). **Refs.:** Belkin & Hogue 1959: 434 (syn.).

nicoyae Adames & Hogue
1970: 9 (M*, F*, P*, L*). **Type loc.:** Estero El Mero, Boca del Rio Barranca, Puntarenas Province, Costa Rica (USNM). **Distr.:** Costa Rica, Nicaragua. **Refs.:** Adames 1971: 41 (M*, F*, P*, L*); Heinemann & Belkin 1977b: 453 (distr.; Nicaragua). **Ety.:** Nicoya Peninsula, Costa Rica. **Informal name:** Nicoya Costa Rican Crabhole Mosquito.

panamensis Adames
1971: 61 (M*, F*, P*, L*). **Type loc.:** Portobelo, Colón, Panama (USNM). **Distr.:** Panama. **Ety.:** Panama. **Informal name:** Cueva Panamanian Crabhole Mosquito [the Cueva were some of the original inhabitants of Panama].

pseudes Dyar & Knab
1909d: 260 (M, F). **Type loc.:** Ancón, Canal Zone, Panama (USNM). **Distr.:** Colombia, Costa Rica, Ecuador, El Salvador, Guatemala, Honduras, Mexico, Nicaragua, Panama, Peru, United States (continental). **Refs.:** Matheson 1944: 276 (A*); Lane 1953: 556 (M*, F, L*); Stone & Knight 1957c: 197 (type info., lectotype desig.); Belkin & Hogue 1959: 440 (M*, F*, L); Reinert 1970f: 17 (P*); Adames 1971: 72 (M*, F*, P*, L*); Darsie & Ward 2005 (F*, L*; keys, distr.; United States). **Ety.:** n.s. **Informal name:** False Panamanian Crabhole Mosquito.

spanius (Dyar & Knab)
1909d: 259 (M, F; *Dinanamesus*). **Type loc.:** Corozal, Canal Zone, Panama (USNM). **Distr.:** El Salvador, Panama, Venezuela. **Refs.:** Yamaguti & LaCasse 1951c: 128 (M*, F*, L*); Carpenter & LaCasse 1955: 327 (distr.; in part); Stone & Knight 1957c: 197 (type info., lectotype desig.); Belkin & Hogue 1959: 427 (M*, F*); Adames 1971: 27 (M*, F*). **Ety.:** n.s. **Informal name:** Spaniard Panamanian Crabhole Mosquito.

Genus *Galindomyia* Stone & Barreto

(Vol. 1, p. 244)

1969: 143. Type species: *Galindomyia leei* Stone & Barreto.

Name derivation: Pedro Galindo.

Important references: Adames 1971 (phyl.); Harbach & Kitching 1998 (phyl.); Harbach et al. 2012 (phyl.).

leei Stone & Barreto
1969: 145 (M*, F). **Type loc.:** near mouth of Rio Raposo, Departamento de Valle, Colombia (USNM). **Distr.:** Colombia, Ecuador. **Refs.:** Mattingly 1971a: Pl. 4 (M, F heads*), Pl. 6 (M tarsal claw*); Adames & Arzube 1975: 113 (distr.; Ecuador); Adames & Galindo 1975: 132 (P*, L*). **Ety.:** Vernon Harold Lee. **Informal name:** Lee Colombian Galindo Mosquito.

Genus *Lutzia* Theobald

(Vol. 1, p. 246)

Important references: Edwards 1932a: 190 (to subg. of *Culex*); Edwards 1941 (M, F, P; keys, tax.; as subg. of *Culex*; Afrotropics); Hopkins 1952 (L; tax.; as subg. of *Culex*; Afrotropical Region); Belkin 1962 (as subg. of *Culex*; South Pacific); Bram 1967b (as subg. of *Culex*; Thailand); Tanaka et al. 1979 (as subg. of *Culex*; Japan); Lee et al. 1988b (tax., bion.; as subg. of *Culex*; Australasian Region); Miller et al. 1996 (mol. phyl.); Navarro & Liria 2000 (morph. phyl.); Liria & Navarro 2003b (tax., class.; L mouthparts); Tanaka 2003c: 159 (tax.; as genus; Japan); Rattanarithikul, Harrison et al. 2005 (F, L; tax., keys, bion.; Thailand); Deus 2009 (mol. phyl.); Demari-Silva et al. 2011 (mol. phyl.); Vesgueiro et al. 2011 (mol. phyl.); Harbach et al. 2012 (morph. phyl.); Kitching et al. 2015 (phyl.; review of status).

Subgenus *Insulalutzia* Tanaka

2003: 160 (*Lutzia* subg.). Type species: *Culex* (*Lutzia*) *shinonagai* Tanaka, Mizusawa, & Saugstad.

Name derivation: n.s. [poss. ref. to occurrence on an island].

Important references: Tanaka et al. 1979 (as *Culex* subg. *Lutzia*; Japan, Korea).

shinonagai (Tanaka, Mizusawa, & Saugstad)
1979: 245 (M*, F, L*; as *Culex*). **Type loc.:** Kitafukuro-zawa, Chichijima, Ogasawara Islands [Japan] (NMST). **Distr.:** Japan. **Refs.:** Tanaka 2003c: 166 (P*; transfer from genus *Culex*). **Ety.:** Satoshi Shinonaga. **Informal name:** Shinonaga Japanese Predatory Mosquito.

Subgenus *Lutzia* Theobald

1903a: 155. Type species: *Culex bigoti* Bellardi.

Name derivation: Adolfo Lutz.

Important references: Lane 1953 (tax., keys, bion.; as *Culex* subg. *Lutzia*; Neotropics); Harbach et al. 2012 (phyl., class.); Kitching et al. (phyl.).

allostigma Howard, Dyar, & Knab
1915: 471 (M, P, L, E). **Type loc.:** [Las Cascadas, Canal Zone, Panama] (USNM). **Distr.:** Belize, Brazil, Colombia, Costa Rica, Ecuador, French Guiana, Guyana, Mexico, Nicaragua, Panama, Paraguay, Peru, Suriname, Venezuela. **Refs.:** Senevet & Abonnenc 1939a: 65 (M*, L*); Stone & Knight 1957a: 43 (lectotype desig.); Pecor et al. 2002: 246 (tax., distr.; Belize); Tanaka 2003c: 159 (transfer from genus *Culex*); Darsie 2006 (P*); Linton et al. 2013 (bion., mol. tax., distr.; Ecuador). **Ety.:** n.s. [another mark (Gr); poss. ref. to diverse male genitalia]. **Informal name:** Marked Panamanian Predatory Mosquito.

bigoti (Bellardi)
1862: 3. **Type loc.:** Mexico (BC). **Distr.:** Argentina, Belize, Bolivia, Brazil, Colombia, El Salvador, Guatemala, Mexico, Panama, Paraguay, Peru, Venezuela. **Refs.:** Lane 1953: 311 (M*, F*, P*, L*); Tanaka 2003c: 159 (transfer from genus *Culex*); Darsie 2006 (P*); Berti et al. 2015 (distr.; Venezuela). **Ety.:** Jacques Marie Frangile Bigot. **Informal name:** Bigot Mexican Predatory Mosquito.
syn. *brasiliae* Dyar
1923b: 67 (M*). **Type loc.:** São Paulo, Brazil (USNM). **Refs.:** Lane 1936c: 429 (syn.); Stone & Knight 1957a: 44 (lectotype desig.).
syn. *patersoni* Shannon & Del Ponte
1928 (M, F). In Dyar 1928: 268. **Type loc.:** San Pedro de Jujuy, Argentina (USNM). **Refs.:** Lane 1951: 334 (syn.); Stone & Knight 1957a: 54 (lectotype desig.).

Subgenus *Metalutzia* Tanaka

2003: 160 (*Lutzia*: subgen.). Type species: *Lutzia vorax* Edwards.

Name derivation: n.s. [near (Gr) *Lutzia*].

syn. *Jamesia* Christophers 1906: 12 (not Jeckel 1861). Type species: *Culex concolor* Robineau-Desvoidy (Edwards 1932a: 190; as *Culex fuscanus* Wiedemann).

Important references: Edwards 1941 (M, F, P; keys, tax.; Afrotropics); Hopkins 1952 (L; tax.; Afrotropics); Belkin 1962 (South Pacific); Bram 1967b (Thailand); Tanaka et al. 1979 (tax.; Japan); Lee et al. 1988b (tax.; Australasian Region); Tanaka 2003c: 159 (tax.; Japan); Rattanarithikul et al. 2005b (tax.; Thailand); Harbach et al. 2012 (morph. phyl.); Kitching et al. 2015 (phyl., tax.).

agranensis Singh & Prakash
2008 (M*, L*). **Type loc.:** Dayalbagh at Agra, India (USNM?). **Distr.:** India. **Ety.:** n.s. [named for Dayalbagh at Agra, Indian name plus -nensis, meaning from the type locality of Agra] (Harbach 2018c). **Informal name:** Agra Indian Predatory Mosquito.

chiangmaiensis Somboon & Harbach
2019. (M*, F*, L*, P*). **Type loc.:** Chiang Mai Province, Mae Taeng District, Ban Pang Mai Daeng, Thailand. **Distr.:** Thailand. **Refs.:** Bram 1967b: 11–23 (in part, as *halifaxii*); Apiwathanasorn 1986: 26, 47 (in part, as *halifaxii*); Harrison et al. 1991: 226 (in part, as *halifaxii*); Rattanarithikul et al. 2005: 16, 25, 31 (in part, as *vorax*);

Phanitchakun et al. 2019 (as *Lutzia* CM form). **Ety.:** from Chiang Mai Province. **Informal name:** Chaing Mai Thai Predatory Mosquito.

fuscana (Wiedemann)
1820: 9 (M, F; *Culex*). **Type loc.:** India (NMW). **Distr.:** Bangladesh, Cambodia, India, Indonesia, Japan, Macau, Malaysia, Mariana Isl., Micronesia (Wake Island), Myanmar, Nepal, Pakistan, Palau, People's Republic of China, Philippines, Russia, Singapore, South Korea, Sri Lanka, Taiwan, Thailand, Timor, Vietnam. **Refs.:** Barraud 1924c (M*); Barraud 1924l (L*); Barraud 1934: 341 (M*, F, L); T'an 1945: 46 (L*); Bohart 1957: 75 (M*, F); Joshi et al. 1965: 139 (distr.; Nepal); Delfinado 1966a: 97 (M*, F, P*, L*); Bram 1967b: 13 (M*, F*; tax., distr.; Thailand); Aslamkhan 1971b (distr.; Pakistan); Basio 1971b: 69 (M*; bion.); Pratt & Siren 1971 (distr.; Mariana Isl.); Tanaka 1971: 4 (distr.; Ryuku Isl.); Berlin & Pandian 1973b: 227 (E*); Pinhão 1974 (distr.; Timor); Gutsevich et al. 1974: 347 (M*, F, L*); Baisas 1974: 123 (M, F, P, L; tax., bion., distr.; Philippines); Tanaka et al. 1975c: 215 (distr.); Tanaka et al. 1979: 240 (M*, F*, L*; as *Cx.* (*Lut.*) *fuscanus*); Ahmed 1987 (distr.; Bangladesh); Lee et al. 1988b: 3 (F; key, tax., bion., review, distr.; Australasia); Darsie & Pradhan 1990 (F, L; tax., keys, bion., distr.; Nepal); Tanaka 2003c: 165 (P; transfer from genus *Culex*); Kirti & Kaur 2004: 70 (F*); Rattanarithikul et al. 2005b (F*, L*; bion., keys; distr.; Thailand). **Ety.:** n.s. [dark (L)]. **Informal name:** Dark Indian Predatory Mosquito.
syn. *concolor* Robineau-Desvoidy
1827: 405 (M). **Type loc.:** None designated (ND).
syn. *luridus* Doleschall
1857: 384 (F*; *Culex*). **Type loc.:** Gombong, Java [Indonesia] (LU).
syn. *setulosus* Doleschall
1857: 384 (M, F*; *Culex*). **Type loc.:** Gombong, Java [Indonesia] (NMW).

halifaxii (Theobald)
1903a: 231 (F). **Type loc.:** Dindings, Malaya [Malaysia] (NHMUK). **Distr.:** Australia, Bangladesh, Cambodia, India, Indonesia, Japan, Malaysia, Nepal, Pakistan, Papua New Guinea, People's Republic of China, Philippines, Russia, Singapore, Solomon Isl., South Korea, Sri Lanka, Taiwan, Thailand, Timor, Vietnam. **Refs.:** Barraud 1924d: 977 (L*; as *raptor*); Barraud 1924c (M*; as *raptor*); Barraud 1924l (L*; as *raptor*); Barraud 1934: 343 (M*, F, L; as *raptor*); Cook 1944: 47 (L*); Lee 1944a: 88 (L*); Knight & Chamberlain 1948: 18 (P*; Fig. 29); Belkin 1962: 220 (M*, F, P*, L*); Bram 1967b: 17 (M*, F*, L*; bion., tax., distr.; Thailand); Aslamkhan 1971b (distr.; Pakistan); Basio 1971b: 69 (M*; bion.); Chow 1973: 44 (distr.); Baisas 1974: 125 (M, F, P, L; tax., bion., distr.; Philippines); Tanaka et al. 1975c: 216 (distr.); Tanaka et al. 1979: 242 (M*, F*, L; as *Cx.* (*Lut.*)); Lee et al. 1988b: 5 (F; key, tax., bion., review, distr.; Australasia); Darsie & Pradhan 1990 (F, L; tax., keys, bion., distr.; Nepal); Whelan & Hapgood 2000 (bion.); Tanaka 2003c: 159 (transfer from genus *Culex*); Kirti & Kaur 2004: 70 (F*); Rattanarithikul et al. 2005b (F*, L*; keys, bion., distr.; Thailand). **Ety.:** n.s. [poss. ref. to F.J. Halifax]. **Informal name:** Halifax Malaysian Predatory Mosquito.
syn. *multimaculosus* Leicester
1908: 155 (M, F; *Culex*). **Type loc.:** Malaya [Malaysia] (NHMUK).
syn. *aureopunctis* Ludlow
1910: 195 (F; *Culex*). **Type loc.:** Cottabato [Cotabato], Mindanao, Philippines (USNM).
syn. *raptor* Edwards
1922a: 275 (M*). **Type loc.:** Amritsar [Punjab Hill States], India (NHMUK). **Refs.:** Bram 1967b: 17 (syn.).

tigripes (de Grandpre & de Charmoy)
1901: 6 (M, F). **Type loc.:** Mauritius (MNHP). **Distr.:** Angola, Benin, Burkina Faso, Cameroon, Central African Republic, Comoros, Côte d'Ivoire, Democratic Republic of the Congo, Djibouti, Equatorial Guinea, Ethiopia, Gabon, Ghana, Kenya, Liberia, Madagascar (includes Glorioso and Juan De Nova Isl.), Malawi, Mali, Mauritius, Mozambique, Nigeria, Réunion, São Tomé and Príncipe, Saudi Arabia, Senegal, Sierra Leone, South Africa, Sudan and South Sudan, Tanzania, The Gambia, Uganda. **Refs.:** Bancroft 1908: 37 (M, F, L); Macfie & Ingram 1922 (F*); Edwards 1941: 247 (M*, F*; tax.), 409 (P*); Hopkins 1952: 249 (L*); Jackson 1953: 153 (bion.); Rodhain et al. 1977 (distr.); Lee et al. 1988b: 14 (distr., review); Townsend 1990: 142 (2 syntypes: NHMUK); Jupp 1996 (M*, F*; key); Tanaka 2003c: 159 (transfer from gen. *Culex*); Toto et al. 2003: 344 (distr.; Equatorial Guinea); Darsie 2006 (P*); Ahmed et al. 2011 (distr.; Saudi Arabia). **Ety.:** n.s. **Informal name:** Tiger-footed Mauritian Predatory Mosquito.
syn. *sierraleonis* Theobald
1901c: 36 (F; *Culex maculicrura* var.). **Type loc.:** Freetown, Sierra Leone (NHMUK). **Refs.:** Edwards 1941: 247 (syn.).
syn. *mombasaensis* Theobald
1901c: 36 (F*; *Culex maculicrura* var.). **Type loc.:** Mombasa [Coast, Kenya] (NHMUK). **Refs.:** Edwards 1941: 247 (syn.).
syn. *maculicrures* Theobald
1901d: 8 (F). **Type loc.:** Bonny [Eastern Provinces, Nigeria] (NHMUK).
syn. *fusca* Theobald
1909b: 17 (A; as var.). **Type loc.:** Obuasi, Ashanti, Gold Coast [Ghana] (NHMUK). **Refs.:** Edwards 1941: 247 (syn.).
syn. *bimaculata* Theobald
1910c: 393 (F; *Culex trigripes* var.). **Type loc.:** Meshra el Firaffe [Sudan] (NHMUK). **Refs.:** Edwards 1941: 247 (syn.).

vorax Edwards
1921d: 327 (M*). **Type loc.:** Tokyo [Honshu], Japan (NHMUK). **Distr.:** India, Japan, Malaysia, Mariana Isl., Myanmar, Nepal, Sri Lanka, Taiwan, Thailand. **Refs.:** Barraud 1924c (M*); Barraud 1924l (L*); Bohart & Ingram 1946b: 70 (M*, F, L*; bion., distr., tax.); La Casse & Yamaguti 1950: 251 (M*, F*, P*, L*; bion., distr., tax.); Asanuma & Nakagawa 1953: 91 (P*); Bohart 1957: 73 (M*, F, L*); Hara 1957: 52 (F*); Bram 1967b: 17 (syn.); Tanaka 2003c: 161 (from syn. with *halifaxii*; P*); Rattanarithikul et al. 2005b (F*, L*; bion., distr., keys). **Ety.:** n.s. [ravenous, insatiable (L)]. **Informal name:** Voracious Japanese Predatory Mosquito.

Tribe Culisetini

Genus *Culiseta* Felt

(Vol. 1, p. 248)

Important references: Belkin 1962 (tax.; South Pacific); Barr 1963 (P; key, tax.; Nearctic); Maslov 1964: 193 (tax.); DuBose & Curtin 1965 (tax., keys; Mediterranean); Forattini 1965b (tax.; Neotropics); Maslov 1967 (revision; world fauna) [see Maslov 1989, for English transl.]; Belkin 1968a (review; New Zealand); Dobrotworsky 1971: 43 (tax., keys to subg.; SE Asia); Mattingly 1972c (E*); Gutsevich et al. 1974 (former USSR); Tanaka et al. 1979 (tax., keys, revision; Japan, Korea); Wood et al. 1979 (tax., keys, revision; Canada); Lu & Li 1982a (People's Republic of China); Lee et al. 1988b (Australasian Region); Service 1990 (Afrotropics); Lu et al. 1997 (People's Republic of China); Harbach & Kitching 1998 (phyl.); Darsie & Ward 2005 (keys; North America); Reidenbach et al. 2009 (phyl.); Becker et al. 2010 (tax., keys, bion.; Europe).

Subgenus *Allotheobaldia* Brolemann

1919b: 91 (as subg. *Allotheobaldia* of genus *Theobaldia*). Type species: *Culex spathipalpis* Rondani.

Name derivation: false (Gr), or the other (Gr), *Theobaldia*.

Important references: Edwards 1941 (M, F, P; tax., keys, distr.; as subg. of genus *Theobaldia*); Hopkins 1952 (L; tax., keys, bion., distr.; as subg. of genus *Theobaldia*); Maslov 1964: 214 (to genus); DuBose & Curtin 1965 (Mediterranean area; keys); Maslov 1967 (revision; world fauna) [see Maslov 1989, for English transl.]; Dobrotworsky 1971: 39 (to subg.); Gutsevich et al. 1974 (F, L; keys, tax., bion., distr.; former USSR); Service 1990 (Afrotropics); Becker et al. 2010 (tax., keys, bion.; Europe).

longiareolata (Macquart)
1838a: 38 (F; *Culex*). **Type loc.:** Canary Islands (NE). **Distr.:** Afghanistan, Albania, Algeria, Austria, Azerbaijan, Botswana, Bulgaria, Crimean Peninsula, Croatia, Cyprus, Djibouti, Egypt, Eritrea, Ethiopia, France (includes Corsica), FYRO Macedonia, Georgia, Germany, Greece (includes Crete), Hungary, India, Iran, Iraq, Israel (includes Gaza Strip and West Bank), Italy (includes Sardinia and Sicily), Jordan, Kosovo, Lebanon, Lesotho, Malta, Mauritania, Moldova, Morocco, Namibia, Pakistan, Portugal (includes Madeira), Romania, Russia, Saudi Arabia, Serbia, Slovakia, Slovenia, Somalia, South Africa, Spain (includes Balearic Isl. and Canary Isl.), Sudan and South Sudan, Switzerland, Syria, Tajikistan, Tunisia, Turkey, Ukraine, United Kingdom, Yemen. **Refs.:** Séguy 1924: Pl. XV (L*), 54 (M*); Edwards 1941: 68 (M, F*), 373 (P*); Senevet 1946: 318 (L*); Mattingly 1950: 113 (M*, F*, L); Hopkins 1952: 77 (L*); Maslov 1967: 176 (M*, F, L*) [see Maslov 1989, for English transl.]; Mattingly 1971a: Pl. 20 (P*), Pl. 29 (L head*, terminal segs.*); Gutsevich et al. 1974: 142 (M*, F*, L*); da Cunha Ramos & Ribeiro 1975: 136 (tax., bion., distr.; Angola); Rodhain et al. 1977 (distr.); Jupp 1996 (M*, F*; key); Al-Houty 1997 (distr.; Kuwait); Alten et al. 2000 (distr.; Turkey); Hadjivassilis 2000: 38 (distr.; Cyprus); Becker et al. 2010: 289 (M*, F*, L*; keys, tax., distr., bion.); Osório et al. 2010 (distr.; Portugal); Becker & Hoffmann 2011 (distr.; Germany); Namazov 2014 (distr.; Azerbaijan). **Ety.:** long, open spaces (L); ref. to elongated wing cells. **Informal name:** Long Spanish Cool Mosquito.
syn. *spathipalpis* Rondani
1872: 31 (A, L; *Culex*). **Type loc.:** Italy (FM).
syn. *leucogrammus* Loew
1874: 413 (F; *Culex*). **Type loc.:** Schahrud [Shahrud County, Semnan Province, Iran] (LU).
syn. *serratipes* Becker
1908: 78 (M; *Culex*). **Type loc.:** Tenerife, Canary Islands (ZM).
syn. *marocanus* D'Anfreville
1916: 141 (M*, F*; *Culex annulatus* var.). **Type loc.:** Sale, French Morocco [Morocco] (NE).

Subgenus *Austrotheobaldia* Dobrotworsky

1954: 68 (as subg. *Austrotheobaldia* of genus *Theobaldia*). Type species: *Chrysoconops littleri* Taylor.

Name derivation: n.s. [prob. ref. to a *Theobaldia* from Australia].

Important references: Maslov 1964: 213 (to genus); Dobrotworsky 1965 (tax., bion., distr.; Australia); Maslov 1967 (revision; world fauna) [see Maslov 1989, for English transl.]; Dobrotworsky 1971: 39 (to subg.); Lee et al. 1988b (F; key, tax., bion., distr.; Australia).

littleri (Taylor)
1914c: 702 (F; *Chrysoconops*). **Type loc.:** Mount Arthur, near Launceston, Tasmania, Australia (ANIC). **Distr.:** Australia. **Refs.:** Dobrotworsky 1954: 75 (M*, F, P*, L*); Dobrotworsky 1965: 172 (M*, F, L*); Mattingly 1971a: Pl. 30 (L head*, terminal segs.*); Lee et al. 1988b: 47 (F; key, tax., bion., distr., review). **Ety.:** F.M. Littler. **Informal name:** Littler Australian Cool Mosquito.

Subgenus *Climacura* Howard, Dyar, & Knab

1915: 452 (as *Culex* subg. *Climacura*). Type species: *Culex melanurus* Coquillett.

Name derivation: n.s. [clima- = ladder, staircase (Gr); cura = care, attention (Gr)?].

Important references: Dobrotworsky 1965 (tax., bion., distr; Australia); Maslov 1967 (revision; world fauna) [see Maslov 1989, for English transl.]; Dobrotworsky 1971 (tax., descr.); Gutsevich et al. 1974 (former USSR; descr.); Wood et al. 1979 (tax.); Lee et al. 1988b (F; key, tax., bion., distr.; Australia).

antipodea Dobrotworsky
1962: 291 (M*, F, P*, L*). **Type loc.:** Cann River, Victoria [Australia] (NMM). **Distr.:** Australia. **Refs.:** Dobrotworsky 1965: 185 (M*, F, L*); Lee et al. 1988b: 50 (F; key, tax., bion., distr., review). **Ety.:** n.s. [against feet (Gr); inferred, having to do with the feet of inhabitants that are against each other on opposite sides of the world; in reference to Australia and New Zealand being on the other side of the earth from Britain (Harbach 2018c)]. **Informal name:** Antipodean Australian Cool Mosquito.

marchettei Garcia, Jeffery, & Rudnick
1969: 252 (M*, F, P*, L*). **Type loc.:** Tanjong Rabok, Kuala Langat Forest Reserve, Selangor, Malaysia (USNM). **Distr.:** Cambodia, Malaysia. **Refs.:** Dobrotworsky 1971: 47 (M*, F*, P*, L*). **Ety.:** Nyven J. Marchette. **Informal name:** Marchette Malaysian Cool Mosquito.

melanura (Coquillett)
1902b: 193 (M, F; *Culex*). **Type loc.:** Center Harbor, New Hampshire, United States (USNM). **Distr.:** Canada, Mexico, United States (continental). **Refs.:** Yamaguti & LaCasse 1951a: 79 (M*, F*, L*); Carpenter & LaCasse 1955: 95 (M*, F*, L*; keys); Chamberlain et al. 1955: 18 (E*; bion.); Stone & Knight 1957c: 196 (type info., lectotype desig.); Darsie et al. 1962: 167 (P*); Barr 1963: 327 (P*); Dodge 1966: 373 (first instar L*; key); Andreadis & Munstermann 1997 (F; key; char. morph.); Maslov 1989: 163 (M*, F, L*; key, distr., bion.); Darsie & Ward 2005 (F*, L*; keys, distr.); Ortega-Morales et al. 2009: 100 (distr.); Harrison et al. 2016 (F*, L*; keys, distr., tax.). **Ety.:** black tail (Gr); ref. "given in allusion to the distinctly blackish anal tube of the larva." **Informal name:** Black-tailed American Cool Mosquito. (Vol. 1, p. 532)

novaezealandiae Pillai
1966: 127 (M*, F, P*, L*). **Type loc.:** Tahakopa, South Otago, New Zealand (NSIR). **Distr.:** New Zealand. **Refs.:** Lee et al. 1988b: 52 (F; key, tax., bion., distr., review); Derraik 2004 (distr.; New Zealand). **Ety.:** New Zealand. **Informal name:** Maori New Zealander Cool Mosquito [the Maori are the indigenous Polynesian people of New Zealand].

taiwanica Lien, Lin, & Weng
1999. **Type loc.:** n.k. **Distr.:** n.k. **Ety.:** from Taiwan. **Informal name:** Taiwan Chinese Cool Mosquito.

tonnoiri (Edwards)
1925: 258 (F; *Theobaldia*). **Type loc.:** Waiho, New Zealand (NSIR). **Distr.:** New Zealand. **Refs.:** Miller & Phillipps 1952: 26 (F*, L*); Belkin 1962: 284 (F, L*); Dumbleton 1965 (M*, P*, L*); Belkin 1968a: 71 (M, F, P, L); Lee et al. 1988b: 53 (F; key, tax., bion., distr., review); Pillai 1968 (M*; bion.); Derraik 2004 (distr.; New Zealand). **Ety.:** A. Tonnoir. **Informal name:** Tonnoir New Zealander Cool Mosquito.

Subgenus *Culicella* Felt

1904: 391c. Type species: *Culex dyari* Coquillett.

Name derivation: n.s. [culi- = midge (L); -cella = small (L)].

Important references: Dobrotworsky 1965 (tax., subg. characters; Australia); Gutsevich et al. 1974 (former USSR; descr.); Maslov 1967 (revision; world fauna) [see Maslov 1989, for English transl.]; Wood et al. 1979 (tax.; Canada); Lee et al. 1988b (F; key, tax., bion., distr.; Australia).

amurensis Maslov
1964: 210 (M*, F, L). **Type loc.:** Central Amur region, Ussuri Basin, Southern Maritime Province, Northeast China [People's Republic of China] (LU). **Distr.:** People's Republic of China, Russia. **Refs.:** Maslov 1967: 157 (M*, F, L*) [see Maslov 1989, for English transl.]. **Ety.:** Amur Region, China. **Informal name:** Amur Chinese Cool Mosquito.

atra (Lee)
1944c: 220 (M*, F; *Theobaldia*). **Type loc.:** Perth, Western Australia, Australia (WAM). **Distr.:** Australia. **Refs.:** Dobrotworsky 1968: 11 (M*, F, P*, L*); Lee et al. 1988b: 58 (F; key, tax., bion., distr., review). **Ety.:** n.s. [dark (L); prob. ref. to overall darker integument]. **Informal name:** Dark Australian Cool Mosquito.

drummondi (Dobrotworsky)
1960c: 241 (M*, F, P*, L*; *Theobaldia*). **Type loc.:** Sylvan, Victoria [Australia] (NMM). **Distr.:** Australia. **Refs.:** Maslov 1963: 801 (placed in *Neotheobaldia*); Dobrotworsky 1965: 177 (M*, F, L*); Lee et al. 1988b: 60 (F; key, tax., bion., distr., review). **Ety.:** n.s. [prob. ref. to Frank Harvey N. Drummond]. **Informal name:** Drummond Australian Cool Mosquito.

fumipennis (Stephens)
1825: 453 (A; *Culex*). **Type loc.:** Ripley, Surrey, England (NHMUK). **Distr.:** Albania, Algeria, Armenia, Azerbaijan, Belarus, Belgium, Bulgaria, Croatia, Czech Republic, Denmark, Estonia, France (includes Corsica), Georgia, Germany, Greece, Hungary, Iran, Iraq, Ireland, Italy (includes Sardinia), Macedonia, Morocco, Netherlands, Norway, Poland, Portugal, Romania, Russia, Serbia, Spain, Sweden, Switzerland, Syria, Tunisia, Turkey,

Ukraine, United Kingdom. **Refs.:** Séguy 1924: 56 (L*); Stackelberg 1937: 110 (M*, F; as *setivalva*); Marshall 1938: 220 (M*, F*, L*, E*; tax.); Natvig 1948: 180 (M*, F, L; tax.); Monchadskii 1951: 133 (L*); Sicart & Ruffie 1960: 641 (P); Maslov 1967: 149 (M*, F, L*; as *setivalva*) [see Maslov 1989, for English transl.]; Gutsevich et al. 1974: 164 (M*, F, L*; distr.), 166 (M*, F, L; as *setivalva*); Townsend 1990: 76 (lectotype desig.); Becker et al. 2010: 291 (M*, F*, L*; keys, tax., distr., bion.); Robert et al. 2019 (distr., Western Palearctic). **Ety.:** smoky wing (L); ref. to dusky wing. **Informal name:** Smoke-winged British Cool Mosquito.

syn. *ficalbii* Noè
1899: 238 (M*, F; *Culex*). **Type loc.:** Maccarese [Latium] and Porto, Italy (LU).

syn. *theobaldi* de Meijere
1911: 142 (M*, F, P*, L*; *Culicada*). **Type loc.:** several localities in [Utrecht], Holland (LU).

syn. *setivalva* Monchadskii
1936: 211 (L; *Theobaldia*). **Type loc.:** Anatolia [Turkey] and south shores of Crimea [Former Union of Soviet Socialist Republics (USSR)] [Ukraine] (LU). **Refs.:** Danilov 1984: 313 (syn.).

inconspicua (Lee)
1937: 294 (M*, F, L*; *Theobaldia* (*Culicella*)). **Type loc.:** Tinderry, New South Wales, Australia (CSIR). **Distr.:** Australia. **Refs.:** Dobrotworsky 1954: 75 (M, F, P*, L, E*); Dobrotworsky 1965: 183 (M*, F, L*); Lee et al. 1988b: 61 (F; key, tax., bion., distr., review). **Ety.:** n.s. [not visible, not clearly seen (L)]. **Informal name:** Inconspicuous Australian Cool Mosquito.

litorea (Shute)
1928: 186 (M, P, L; *Culicella morsitans* var.). **Type loc.:** Walton-on the-Naze, Essex, England (NHMUK). **Distr.:** Albania, Algeria, France (includes Corsica), Ireland, Italy (includes Sardinia and Sicily), Morocco, Portugal, Spain, United Kingdom. **Refs.:** Marshall & Staley 1933: 119 (M*, F, L; tax.; to sp.); Rioux & Arnold 1955: 268 (M*, F*, L*); Mattingly 1956a: 38 (lectotype desig.); Senevet & Andarelli 1956b: 400 (M*, P*, L); Rioux 1958: 208 (M*, F*, P*, L*); Service 1970d: 481 (M, F, P, L); Trari et al. 2002: 331 (distr.; Morocco); Becker et al. 2010: 292 (M*, F*, L*; keys, tax., distr., bion.); Robert et al. 2019 (distr.; Western Palearctic). **Ety.:** seashore (L); ref. to larvae in a coastal marsh. **Informal name:** Coastal British Cool Mosquito.

minnesotae Barr
1957: 163 (M*, F). **Type loc.:** Olcott Park, City of Virginia, Saint Louis County, Minnesota, United States (USNM). **Distr.:** Canada, United States (Alaska, continental). **Refs.:** Price 1958 (P*, L*; tax.); Maslov 1964: 209 (103) (to ssp.); Darsie & Ward 2005 (F*, L*; keys, distr.); Harrison et al. 2016 (F*, L*; keys, distr., tax.). **Ety.:** Minnesota, United States. **Informal name:** Minnesota American Cool Mosquito.

morsitans (Theobald)
1901c: 8 (M*, F*; *Culex*). **Type loc.:** England (NHMUK). **Distr.:** Armenia, Austria, Azerbaijan, Belarus, Belgium, Bulgaria, Canada, Crimean Peninsula, Croatia, Czech Republic, Denmark, Estonia, Finland, France (includes Corsica), Georgia, Germany, Greece, Hungary, Iran, Ireland, Israel (and Gaza Strip and West Bank), Italy, Kazakhstan, Latvia, Lithuania, Macedonia, Morocco, Netherlands, Norway, Poland, Romania, Russia, Serbia, Slovakia, Slovenia, Spain, Sweden, Switzerland, Turkey, Ukraine, United Kingdom, United States (Alaska, continental). **Refs.:** Marshall 1938: 227 (M*, F*, L*, E*; first instar L*); Natvig 1948: 172 (M*, F*, L*; bion.); Darsie 1951: 38 (P*); Yamaguti & LaCasse 1951a: 75 (M*, F*, L*); Carpenter & LaCasse 1955: 93 (M*, F*, L*; keys); Dodge 1966: 374 (first instar L; key); Maslov 1967: 153 (M*, F, L*) [see Maslov 1989, for English transl.]; Pickavance et al. 1970 (distr.); Service 1970d: 481 (M, F, P*, L); Gutsevich et al. 1974: 158 (M*, F, L*); Margalit & Tahori 1974: 85 (distr.); Utrio 1976 (L*; key); Danilov 1983b (tax., bion.); Alten et al. 2000 (distr.; Turkey); Spungis 2000: 8 (distr.; Latvia); Darsie & Ward 2005 (F*, L*; keys, distr.); Becker et al. 2010: 294 (M*, F*, L*; keys, tax., distr., bion.); Harrison et al. 2016 (F*, L*; keys, distr.); Robert et al. 2019 (distr., Western Palearctic). **Ety.:** n.s. [biting (L)]. **Informal name:** Biting British Cool Mosquito.

syn. *dyari* Coquillett
1902b: 192 (M, F; *Culex*). **Type loc.:** Center Harbor, New Hampshire, United States (USNM). **Refs.:** Stone & Knight 1957c: 196 (type info., lectotype desig.); Maslov 1964: 209, 103 (to ssp.); Wood et al. 1979 (syn.).

syn. *brittoni* Felt
1905a: 79 (F; *Culex*). **Type loc.:** Branford, Connecticut, United States (NY).

syn. *parodites* Dyar
1928: 244 (M*, F; *Culicella* (*Culicella*)). **Type loc.:** Saxeville, Wisconsin, United States (USNM). **Refs.:** Stone & Knight 1957c: 196 (type info., lectotype desig.); Maslov 1964 (syn.); Maslov 1967: 154 (syn. with *morsitans dyari*) [see Maslov 1989, for English transl.].

nipponica La Casse & Yamaguti
1950: 84 (L*). **Type loc.:** Lake Akan, Hokkaido, Japan (USNM). **Distr.:** Japan, South Korea. **Refs.:** Tanaka et al. 1979: 100 (M*, F*, L*). **Ety.:** Japan. **Informal name:** Nippon Japanese Cool Mosquito.

ochroptera (Peus)
1935: 113 (M*, F, P*, L*; *Theobaldia*). **Type loc.:** Luchsee bei Krausnick, Mark Brandenburg, Germany (NE). **Distr.:** Belarus, Belgium, Czech Republic, Estonia, Finland, Germany, Hungary, Lithuania, Netherlands, People's Republic of China, Poland, Romania, Russia, Slovakia, Sweden, Ukraine. **Refs.:** Monchadskii 1951: 139 (L*); Maslov 1964: 209 (103) (to ssp.); Gutsevich et al. 1974: 160 (M*, F, L*); Utrio 1976: 135 (distr.); Spungis 2000: 8

(distr.; Latvia); Becker et al. 2010: 296 (M*, F*, L*; keys, tax., distr., bion.). **Ety.:** yellow wing (Gr). **Informal name:** Ochre-winged German Cool Mosquito.

otwayensis (Dobrotworsky)
1960c: 246 (M*, F, P*, L*; *Theobaldia*). **Type loc.:** Cape Horn, Otway, Victoria (NMM). **Distr.:** Australia. **Refs.:** Dobrotworsky 1965: 181 (M*, F, L*); Maslov 1967: 170 (in *Neotheobaldia*) [see Maslov 1989, for English transl.]; Lee et al. 1988b: 63 (F; key, tax., bion., distr., review). **Ety.:** Cape Horn, Otway, Victoria, Australia. **Informal name:** Otway Australian Cool Mosquito.

sylvanensis (Dobrotworsky)
1960c: 245 (M*, F, P*, L*; *Theobaldia*). **Type loc.:** Sylvan, Victoria (NMM). **Distr.:** Australia. **Refs.:** Dobrotworsky 1965: 179 (M*, F, L*); Maslov 1967: 171 (*Neotheobaldia*) [see Maslov 1989, for English transl.]; Lee et al. 1988b: 64 (F; key, tax., bion., distr., review). **Ety.:** n.s. [from the woods (L)]. **Informal name:** Sylvan Australian Cool Mosquito.

victoriensis (Dobrotworsky)
1954: 73 (M*, F, P*, L*, E*; *Theobaldia*). **Type loc.:** Ringwood, Victoria, Australia (NMM). **Distr.:** Australia. **Refs.:** Dobrotworsky 1965: 175 (M*, F*, L*); Maslov 1967: 172 (*Neotheobaldia*) [see Maslov 1989, for English transl.]; Lee et al. 1988b: 65 (F; key, tax., bion., distr., review). **Ety.:** Ringwood, Victoria, Australia. **Informal name:** Victoria Australian Cool Mosquito.

weindorferi (Edwards)
1926a: 111 (M, F; *Theobaldia*). **Type loc.:** Cradle Valley, Harz Mount, Tasmania, Australia (ANIC). **Distr.:** Australia. **Refs.:** Lee 1937: 296 (M*, F); Dobrotworsky 1954: 72 (M*; of Lee 1937 transposed); Dobrotworsky 1966: 143 (M*, F, P*, L*); Maslov 1967: 173 (*Neotheobaldia*) [see Maslov 1989, for English transl.]; Lee et al. 1988b: 66 (F; key, tax., bion., distr., review). **Ety.:** n.s. [Gustav Weindorfer]. **Informal name:** Weindorf Australian Cool Mosquito.

Subgenus *Culiseta* Felt

syn. *Theobaldia* Neveu-Lemaire 1902a: 1331 (preoccupied senior synonym of *Culiseta* Felt) (not Fischer 1885). Type species: *Culex annulatus* Schrank.

1904: 391c. Type species: *Culex absobrinus* Felt 1904.

Name derivation: n.s. [culi- = posterior; -seta = hair (L, Gr); poss. ref. to "Larvae with pecten prolonged into setae."]

syn. *Theobaldinella* Blanchard 1905: 390 (as genus). Type species: *Culex annulatus* Schrank.

syn. *Pseudotheobaldia* Theobald 1907: 271 (as genus). Type species: *Pseudotheobaldia niveitaeniata* Theobald.

Important references: Maslov 1967 (revision; world fauna) [see Maslov 1989, for English transl.]; Gutsevich et al. 1974 (descr.; former USSR); Wood et al. 1979 (tax.; Canada); Reidenbach et al. 2009 (mol. phyl.); Becker et al. 2010 (revision, keys, bion; Europe); Harbach & Greenwalt 2012 (phyl.).

alaskaensis (Ludlow)
1906c: 326 (M, F; *Theobaldia*). **Type loc.:** Fort Egbert [Eagle], Alaska, United States (USNM). **Distr.:** Armenia, Austria, Belarus, Canada, Crimean Peninsula, Czech Republic, Denmark, Estonia, Finland, France, Georgia, Germany, Hungary, Iran, Ireland, Latvia, Lithuania, Moldova, Mongolia, Netherlands, Norway, Poland, Romania, Russia, Slovakia, Slovenia, Sweden, Switzerland, Turkey, Ukraine, United Kingdom, United States (Alaska, continental). **Refs.:** Natvig 1948: 139 (M*, F*, L; bion.); Monchadskii 1951: 126 (L*); Yamaguti & LaCasse 1951a: 55 (M*, F*, L*); Frohne 1954: 9 (bion.); Carpenter & LaCasse 1955: 83 (M*, F*, L*; keys); Stone & Knight 1957c: 196 (type info., lectotype desig.); Barr 1963: 328 (P*); Maslov 1967: 131 (M*, L*) [see Maslov 1989, for English transl.]; Minar 1971: 92 (distr.); Utrio 1976: 129, 131, 135 (L*); Zaim 1987: 568 (distr.); Darsie & Ward 2005 (F*, L*; keys, distr.); Becker et al. 2010: 298 (M*, F*, L*; keys, tax., distr., bion.); Robert et al. 2019 (distr., Western Palearctic). **Ety.:** Alaska, United States. **Informal name:** Alaska American Cool Mosquito.

syn. *arctica* Edwards
1920a: 136 (M; *Theobaldia*). **Type loc.:** Arkhangel [Arkhangelsk Oblast], USSR (NHMUK).

syn. *siberiensis* Ludlow
1920a: 151 (F; *Culex*). **Type loc.:** Mostovoi, Selenga, and Verkhne Udinsk [= Ulan-Ude], USSR (USNM). **Refs.:** Stone & Knight 1957c: 197 (type info., lectotype desig.).

ssp. ***indica*** (Edwards)
1920a: 137 (M; *Theobaldia*). **Type loc.:** Bakloh [Gurdaspur District, Punjab Hill States], India (NHMUK). **Distr.:** Armenia, Azerbaijan, Georgia, India, Kazakhstan, Kyrgyzstan, Pakistan, Russia, Tajikistan, Turkmenistan, Ukraine, Uzbekistan. **Refs.:** Barraud 1934: 93 (M*, F); Qutubuddin 1952: 37 (L*); Maslov 1964: 203 (Ent. Rev. p. 100) (to ssp.); Maslov 1967: 132 (tax.) [see Maslov 1989, for English transl.]; Gutsevich et al. 1971: 147 (distr.). **Ety.:** Indian (L). **Informal name:** Bakloh Indian Cool Mosquito.

ssp. syn. *wassilievi* Shingarev
1927c: 548 (M*, F; *Theobaldia*). **Type loc.:** Turkestan [USSR] (MTI). **Refs.:** Stackelberg 1937: 100 (M; to ssp., emend.); Maslov 1967: 132 (syn.) [see Maslov 1989, for English transl.]; Knight & Stone 1977: 280 (emend.; returned to original transliteration according to the International Commission on Zoological Nomenclature).

annulata (Schrank)
1776: 97 (A; *Culex*). **Type loc.:** Linz [Ober-Oesterreich], Austria (LU). **Distr.:** Albania, Algeria, Armenia, Austria,

Azerbaijan, Belarus, Belgium, Bulgaria, Crimean Peninsula, Croatia, Cyprus, Czech Republic, Denmark, Estonia, Finland, France (includes Corsica), FYRO Macedonia, Georgia, Germany, Greece, Hungary, Iran, Iraq, Ireland, Israel (and Gaza Strip and West Bank), Italy (includes Sardinia and Sicily), Jordan, Kosovo, Latvia, Lebanon, Lithuania, Luxembourg, Macedonia, Moldova, Montenegro, Morocco, Netherlands, Norway, Poland, Portugal, Romania, Russia, Saudi Arabia, Serbia, Slovakia, Slovenia, Spain (includes Balearic Isl.), Sweden, Switzerland, Syria, Tunisia, Turkey, Ukraine, United Kingdom. **Refs.:** Séguy 1924: 56 (L*); Marshall 1938: 209 (M*, F*, P*, L*, E*; bion.); Natvig 1948: 146 (M*, F*, L*); Rioux 1958: 199 (M*, P*, L*); Maslov 1967: 134 (A, L*; distr.); Maslov 1989 [English transl. of Maslov 1967]; Amr et al. 1997 (distr.; Jordan); Hadjivassilis 2000: 38 (distr.; Cyprus); Beck et al. 2003: 24 (distr.; Luxembourg); Becker et al. 2010: 299 (M*, F*, L*; keys, tax., distr., bion.); Osório et al. 2010 (distr.; Portugal); Harrison et al. 2016 (F*, L*; keys, distr., tax.); Robert et al. 2019 (distr.; Western Palearctic). **Ety.:** ringed (L); abdomen, legs. **Informal name:** Ringed Austrian Cool Mosquito.

syn. *annulatus* Fourcroy
1785: 516 [Vol I] (A; *Culex*). **Type loc**. Paris, France (NE).

syn. *annulatus* Fabricius
1787: 363 (A; *Culex*). **Type loc.:** Copenhagen, Denmark (K).

syn. *nicaensis* Leach
1825: 292 (A; *Culex*). **Type loc.:** Nice [Alpes-Maritimes], France (LU).

syn. *affinis* Stephens
1825: 452 (A; *Culex*). **Type loc.:** Hertford [Hertfordshire], England (NHMUK).

atlantica (Edwards)
1932b: 560 (F; *Theobaldia*). **Type loc.:** Serra Gorda, Ilha da Pico, Azores (NHMUK). **Distr.:** Portugal (Azores). **Refs.:** Maslov 1967: 136 (F) [see Maslov 1989, for English transl.]; Townsend 1990: 47 (lectotype desig.); Robert et al. 2019 (distr.; Western Palearctic). **Ety.:** Atlantic. **Informal name:** Atlantic Portuguese Cool Mosquito.

bergrothi (Edwards)
1921b: 50 (A; *Theobaldia*). **Type loc.:** Sweden (NHMUK, SM, HM). **Distr.:** Denmark, Finland, Japan, Mongolia, Norway, People's Republic of China, Russia, South Korea, Sweden. **Refs.:** Natvig 1942: 2 (from syn. with *glaphyroptera*); Natvig 1948: 160 (M*, F*, P*, L*; tax.); LaCasse & Yamaguti 1950: 80 (M*, F*, P*, L*; bion., distr., tax.; as *kanayamensis*); Monchadskii 1951: 123 (L*); Maslov 1967: 136 (M*, F, L) [see Maslov 1989, for English transl.]; Gutsevich et al. 1974: 155 (M*, F, L*); Minar 1976: 335 (distr.); Utrio 1976: 129, 135 (L*; distr.); Tanaka et al. 1979: 104 (M*, F*, L*; as *kanayamensis*); da Cunha Ramos & Ribeiro 1980 (M*, P*, L*); Townsend 1990: 51 (type info.); Becker et al. 2010: 301 (M*, F*, L*; keys, tax., distr., bion.); Robert et al. 2019 (distr.; Western Palearctic). **Ety.:** Ernst Evald Bergroth. **Informal name:** Bergroth Swedish Cool Mosquito.

syn. *borealis* Shingarev
1927c: 548 (M*, F; *Theobaldia*). **Type loc.:** Kola Peninsula, Lake Imandra, and environs of Murmansk [Karelo-Finnish SSR] (MTI). **Refs.:** Natvig 1942: 2 (syn.).

syn. *kogievnikovi* Shingarev
1927c: 547 (L; *Theobaldia*). **Type loc.:** Kola Peninsula, Polyarny Krug [Karelo-Finnish SSR] (MTI). **Refs.:** Stone et al. 1959: 219 (syn.).

syn. *kanayamensis* Yamada
1932: 218 (F; *Theobaldia*). **Type loc.:** Kanayama, Hokkaido, Japan (IID). **Refs.:** Danilov 1976b: 185 (syn.).

glaphyroptera (Schiner)
1864: 628 (A; *Culex*). **Type loc.:** Gmunden [Ober-Oesterreich], Austria (NMW). **Distr.:** Austria, Bosnia and Herzegovina, Bulgaria, Croatia, Czech Republic, Finland, France, FYRO Macedonia, Germany, Greece, Hungary, Italy, Moldova, Poland, Romania, Russia, Slovakia, Slovenia, Ukraine. **Refs.:** Natvig 1948: 160 (M, F*, P*, L; tax.); Monchadskii 1951: 122 (L*); Noèldner 1953: 21 (M*, F, L*); Maslov 1967: 138 (M*, F, L*; distr.) [see Maslov 1989, for English transl.]; Gutsevich et al. 1971: 142 (distr.); Gutsevich et al. 1974: 146 (M*, F, L*); Becker et al. 2010: 303 (M*, F*, L*; keys, tax., distr., bion.); Robert et al. 2019 (distr.; Western Palearctic). **Ety.:** n.s. [elegant wings (Gr)]. **Informal name:** Elegant Austrian Cool Mosquito.

syn. *zottae* Ungureanu
1956: 4 (M*, F*, P*, L*; *Theobaldia glaphyroptera* var.). **Type loc.:** Neamt, Varatec, Romania (LU).

impatiens (Walker)
1848: 5 (M, F; *Culex*). **Type loc.:** St Martin's Falls, Albany River [Ontario], Canada (NHMUK). **Distr.:** Canada, United States (Alaska, continental). **Refs.:** Yamaguti & LaCasse 1951a: 59 (M*, F*, L*); Frohne 1953: 103 (bion.); Carpenter & LaCasse 1955: 85 (M*, F*, L*; keys); Barr 1963: 329 (P); Maslov 1967: 139 (M*, F, L) [see Maslov 1989, for English transl.]; Bohart & Washino 1978: 31 (M, F, L*; keys, tax., bion., distr.); Darsie & Ward 2005 (F*, L*; keys, distr.); Harrison et al. 2016 (F*, L*; keys, distr., tax.). **Ety.:** n.s. [impatient, desiring immediate action (L)]. **Informal name:** Impatient Canadian Cool Mosquito.

syn. *pinguis* Walker
1866: 337 (F; *Culex*). **Type loc.:** British Columbia, Canada (LU).

syn. *absobrinus* Felt
1904: 318 (M*, F*, L*; *Culex*). **Type loc.:** Elizabethtown, New York, United States (NY).

incidens (Thomson)

1869: 443 (F; *Culex*). **Type loc.:** California, United States (LU, SM?). **Distr.:** Canada, Mexico, United States (Alaska, continental). **Refs.:** Cook 1944: 40 (L*); Knight & Chamberlain 1948: 15 (P*); Yamaguti & LaCasse 1951a: 63 (M*, F*, L*); Carpenter & LaCasse 1955: 87 (M*, F*, L*; keys, bion.); Barr 1963: 328 (P*); Maslov 1967: 141 (M*, F, L*) [see Maslov 1989, for English transl.]; Belkin 1968b: 21 (tax.); Heinemann & Belkin 1977c: 529 (distr.); Bohart & Washino 1978:31 (M, F*, P*, L*; keys, tax., bion., distr.); Darsie & Ward 2005 (F*, L*; keys, distr.). **Ety.:** n.s. [incidental (L)]. **Informal name:** Incidental American Cool Mosquito.

inornata (Williston)

1893: 253 (F; *Culex*). **Type loc.:** Argus mountains, California, United States (USNM). **Distr.:** Canada, Cuba, Mexico, United States (continental). **Refs.:** Leicester 1908: 99 (M; as *annulitarsis* (*Leicester*)); Howard et al. 1913: Figs. 691–692 (E*); Darsie 1951: 39 (P*); Rees & Onishi 1951: 233 (M*, F*); Yamaguti & LaCasse 1951a: 67 (M*, F*, L*); Carpenter & LaCasse 1955: 89 (M*, F*, L*; keys); Dodge 1966: 372 (first instar L*; key); Maslov 1967: 195: (M*, F, L*; keys, distr., bion.) [see Maslov 1989, for English transl.]; Owen 1977 (morph.); Bohart & Washino 1978: 32 (M*, F*, L*; keys, tax., bion., distr.); Owen 1980 (morph.); Darsie & Ward 2005 (F*, L*; keys, distr.); Harrison et al. 2016 (F*, L*; keys, distr.). **Ety.:** n.s. [unadorned (L); poss. ref. to lack of banding and other diagnostic characters]. **Informal name:** Unadorned American Cool Mosquito. (Vol. 1, p. 534)

syn. *magnipennis* Felt

1904: 322 (M*, F*, L*; *Culex*). **Type loc.:** Sodus Point, New York, United States (NY).

megaloba Luh, Chao, & Xu

1974 (M*, F*, L*). **Type loc.:** Yinchwan, Ninghsia Hui Autonomous Region, China [People's Republic of China] (ND). **Distr.:** People's Republic of China. **Ety.:** large lobe (G); ref. to "a pair of projected prominent lobes on the 9th abdominal tergite." **Informal name:** Large-lobed Chinese Cool Mosquito.

niveitaeniata (Theobald)

1907: 272 (M*; *Pseudotheobaldia*). **Type loc.:** Dehra Dun [United Provinces], India (NHMUK). **Distr.:** India, Nepal, Pakistan, People's Republic of China, Taiwan. **Refs.:** Barraud 1934: 91 (M*, F, L*); Qutubuddin 1952: 39 (L); Joshi et al. 1965: 139 (distr.); Maslov 1967: 145 (M*, F, L*) [see Maslov 1989, for English transl.]; Dobrotworsky 1971: 45 (M*, F*, P*, L*); Darsie & Pradhan 1990 (F, L; tax., keys, bion., distr.; Nepal); Darsie et al. 1996 (distr.; Nepal). **Ety.:** n.s. [snowy white band (L); poss. ref. to basal silvery white scales on abdomen]. **Informal name:** Snow-ribbon Indian Cool Mosquito.

syn. *sinensis* Meng & Wu

1962: 382 (M*, F, P, L*; *Theobaldia* (*Theobaldia*)). **Type loc.:** China [People's Republic of China] (LU). **Refs.:** Maslov 1967: 145 (syn.) [see Maslov 1989, for English transl.].

syn. *lishanensis* Lien

1968a: 6 (M*, F, P*, L*). **Type loc.:** Lishan, Hoping Township, Taichung Hsien, Taiwan [Republic of China] (TMRI). **Refs.:** Dobrotworsky 1971: 45 (syn.).

particeps (Adams)

1903: 26 (M, F; *Culex*). **Type loc.:** Arizona, United States (USNM). **Distr.:** Costa Rica, El Salvador, Guatemala, Mexico, Panama, United States (Alaska, continental). **Refs.:** Yamaguti & LaCasse 1951a: 70 (M*, F*, L*; as *maccrackenae*); Carpenter & La Casse 1955: 91 (M*, F*, L*; as *maccrackenae*); Stone 1958b: 236 (tax.; from syn. with *incidens*; lectotype desig.); Barr 1963: 328 (P*); Dodge 1966: 372 (first instar L*; key); Maslov 1967: 147 (M*, F, L*) [see Maslov 1989, for English transl.]; Bohart & Washino 1978: 33 (M, F*, L*; keys, tax., bion., distr.); Strickman 1989: 210 (bion.; larval movement); Darsie & Ward 2005 (F*, L*; keys, distr.). **Ety.:** n.s. [sharer, partaker (L)]. **Informal name:** Comradely American Cool Mosquito.

syn. *maccrackenae* Dyar & Knab

1906a: 133 (M, F). **Type loc.:** Stanford University, California, United States (USNM). **Refs.:** Stone & Knight 1957c: 196 (type info., lectotype desig.); Stone 1958b: 236 (syn.).

syn. *dugesi* Dyar & Knab

1906a: 134 (F). **Type loc.:** Guanajuato, Mexico (USNM). **Refs.:** Stone & Knight 1957c: 196 (type info., lectotype desig.); Stone 1958b: 236 (syn.).

subochrea (Edwards)

1921 (A; *Theobaldia annulata* var.). In Wesenberg-Lund 1921: 198. **Type loc.:** Amager, Copenhagen, Denmark (NHMUK). **Distr.:** Afghanistan, Algeria, Belgium, Croatia, Czech Republic, Denmark, Egypt, Finland, France (includes Corsica), Germany, Greece, Hungary, Iran, Iraq, Ireland, Israel (and Gaza Strip and West Bank), Italy (includes Sardinia and Sicily), Kazakhstan, Kyrgyzstan, Lebanon, Morocco, Netherlands, Norway, Poland, Portugal, Romania, Russia, Saudi Arabia, Serbia, Slovakia, Spain, Sweden, Syria, Tajikistan, Tunisia, Turkmenistan, United Kingdom, Uzbekistan. **Refs.:** Marshall 1938: 213 (M*, F*, P*, L*, E*); Natvig 1948: 158 (M*, F, L*; tax.); Senevet & Andareili 1956a: 223 (M*, P*, L); Rioux 1958: 201 (M*, P*, L*); Maslov 1964: 204 (100) (to ssp.); Maslov 1967: 135 (tax.) [see Maslov 1989, for English transl.]; Abul-Hab 1967: 279 (distr.); Utrio 1976: 135 (distr.); Ribeiro et al. 1977: 55 (from ssp. of *annulata*); Harbach et al. 1988: 320 (distr.; Egypt); Abdullah & Merdan 1995 (distr.; Saudi Arabia); Eritja et al. 2000: 12 (distr.; Spain); Rueda, Pecor et al. 2008: 397 (distr.); Becker et al. 2010: 305 (M*, F*, L*; keys, tax., distr., bion.); Robert et al. 2019 (distr.; Western Palearctic). **Ety.:** nearly yellow (L, Gr);

abdomen color. **Informal name:** Pale-yellow Danish Cool Mosquito [*subochrea* = "subyellow" = Pale-yellow].

syn. *ferruginata* Martini

1924: 438 (A; *Theobaldia annulata* var.). **Type loc.:** Wohldorf [near Hamburg, Germany] (NHMUK). **Refs.:** Peus 1930: 52 (tax., syn.).

Subgenus *Neotheobaldia* Dobrotworsky

1958b: 186 (as subg. *Neotheobaldia* in genus *Theobaldia*). Type species: *Theobaldia hilli* Edwards 1926. (Original description by Dobrotworsky 1954: 68 invalid, type species subsequently designated by Dobrotworsky 1958b.)

Name derivation: n.s. [a new *Theobaldia*].

Important references: Dobrotworsky 1965 (tax.; subg. characters; Australia); Lee et al. 1988b (F; key, tax., bion., distr.; Australia); Maslov 1967 (revision; world fauna) [see Maslov 1989, for English transl.].

frenchii (Theobald)

1901c: 66 (F*; *Culex*). **Type loc.:** Victoria, Australia (NHMUK). **Distr.:** Australia. **Refs.:** Dobrotworsky 1954: 72 (M*, F, P, L, E*); Dobrotworsky 1965: 189 (M*, F*, L*); Maslov 1967: 168 (M*, F, L) [see Maslov 1989, for English transl.]; Lee et al. 1988b: 70 (F; key, tax., bion., distr., review). **Ety.:** Mr. French. **Informal name:** French Australian Cool Mosquito.

ssp. ***atritarsalis*** (Dobrotworsky)

1954: 73 (M, F; *Theobaldia*). **Type loc.:** Stony Creek, Victoria, Australia (NMM). **Distr.:** Australia. **Refs.:** Dobrotworsky 1965: 191 (A, L); Lee et al. 1988b: 69 (F; key, tax., bion., distr., review). **Ety.:** dark tarsi (L, Gr); "legs also are dark apically." **Informal name:** Dark-legged Australian Cool Mosquito.

hilli (Edwards)

1926a: 111 (M; *Theobaldia*). **Type loc.:** Beaconsfield, Victoria, Australia (NHMUK). **Distr.:** Australia. **Refs.:** Dobrotworsky 1954: 69 (M*, F, P*, L*, E*); Dobrotworsky 1965: 187 (M*, F, L*); Maslov 1967: 168 (M*, F, L*) [see Maslov 1989, for English transl.]; Lee et al. 1988b: 72 (F; key, tax., bion., distr., review). **Ety.:** G.F. Hill. **Informal name:** Hill Australian Cool Mosquito.

Subgenus *Theomyia* Edwards

1930a: 303 (as subg. *Theomyia* of genus *Theobaldia*). Type species: *Leptosomatomyia fraseri* Edwards.

Name derivation: n.s. [theo- = god, spirit (Gr); -myia = a small fly].

Important references: Edwards 1941 (F, P; keys, bion.; as subg. of *Theobaldia*; distr.); Hopkins 1952 (P; key, bion.; as subg. of *Theobaldia*; distr.); Maslov 1964: 214 (to genus); Maslov 1967 (revision; as genus; world fauna) [see Maslov 1989, for English transl.]; Dobrotworsky 1971: 39 (to subg.); Service 1990 (tax., bion., distr.; Afrotropics).

fraseri (Edwards)

1914b: 74 (M; *Leptosomatomyia*). **Type loc.:** Kasala Stream, Uganda (NHMUK). **Distr.:** Cameroon, Central African Republic, Congo, Côte d'Ivoire, Democratic Republic of the Congo, Gabon, Mozambique, Nigeria, São Tomé and Príncipe, Sierra Leone, Sudan and South Sudan, Tanzania, Togo, Uganda, Zimbabwe. **Refs.:** Edwards 1941: 70 (M*, F); Hopkins 1952: 79 (L*); van Someren 1956: 3 (P*); Maslov 1967: 175 (F, L*) [see Maslov 1989, for English transl.]; Service 1976a: 262 (distr.); Jupp 1996 (M*, F*; key). **Ety.:** A.D. Fraser. **Informal name:** Fraser Ugandan Cool Mosquito.

Subgenus Uncertain *Culiseta*

arenivaga Marks

1968: 51 (F*). **Type loc.:** Forest Station, Fraser Island, Queensland, Australia (UQIC). **Distr.:** Australia. **Refs.:** Lee et al. 1988b: 74 (F; key, tax., bion., distr., review). **Ety.:** a wanderer of sandy places (L). **Informal name:** Sand-wandering Australian Cool Mosquito.

wui Lin, Tseng, & Lien

2008 (M*, F*, P*, L*). **Type loc.:** Boa Esperança, São Tomé Island (EMIP). **Distr.:** São Tomé. **Ety.:** Cheng-Hua Wu. **Informal name:** Wu Taiwanese Cool Mosquito.

Tribe Ficalbiini

Genus *Ficalbia* Theobald

(Vol. 1, p. 250)

1903a: 296. Type species: *Uranotaenia minima* Theobald (Brunetti 1914: 60).

Name derivation: Eugenio Ficalbi.

Important references: Edwards 1941 (M, F, P; tax., keys (includes *Mimomyia*); Afrotropical Region); Hopkins 1952 (L; tax., key (includes *Mimomyia*); Afrotropical Region); Mattingly 1957a: 6 (keys; Indomalayan area); Mattingly & Grjebine 1958: 259 (keys); Belkin 1962 (tax. (includes *Mimomyia*), bion.; South Pacific); Mattingly 1971a: 1 (generic restriction); da Cunha Ramos & Ribeiro 1975: 116 (keys; Angola); Mattingly 1981 (phyl.); Grjebine 1986 (review, tax., bion.; Madagascar); Lee et al. 1988b: 75 (review (includes *Mimomyia*), tax., bion., distr.); Service 1990 (review, tax., bion.; Afrotropical Region); Harbach & Kitching 1998 (phyl.); Rattanarithikul, Harrison et al. 2005 (F, L; tax., keys, bion.; Thailand); Rattanarithikul et al. 2006b (tax., keys, bion.; Thailand).

circumtestacea (Theobald)
1908c: 264 (M*, F*; *Mimomyia*). **Type loc.:** Upper White Nile, Sudan (NHMUK). **Distr.:** Benin, Burkina Faso, Chad, Madagascar (includes Glorioso and Juan De Nova Isl.), Senegal, Sierra Leone, South Africa, Sudan and South Sudan. **Refs.:** Edwards 1941: 88 (A); Hamon 1954a: 584 (L*); Hamon 1954b: 591 (P*); Mattingly 1971a: Pl. 15 (P trumpet*, terminal segs*); Grjebine 1986: 43 (M*, F*, L*, P*; keys); Jupp 1996 (M*, F*; key). **Ety.:** n.s. [around brick (L); poss. ref. to brownish coloration]. **Informal name:** Brickwork Sudanese Small Mosquito [*circumtestacea* = "around brick" = Brickwork].

ichiromiyagii Toma & Higa
2004: 195 (M*). **Type loc.:** Funaura, Iriomote Island, Ryukyu Archipelago, Japan (NMST). **Distr.:** Japan. **Refs.:** Higa et al. 2012: 279 (F, P*, L*). **Ety.:** Ichiro Miyagi. **Informal name:** Miyagi Japanese Small Mosquito.

jacksoni Mattingly
1949b: 9 (M, F, P, L*). **Type loc.:** Hong Kong [People's Republic of China] (NHMUK). **Distr.:** People's Republic of China. **Refs.:** Mattingly 1957a: 30 (tax.). **Ety.:** R.B. Jackson. **Informal name:** Jackson Chinese Small Mosquito.

ludlowae Brunetti
1920: 173 (as *ludlowi*, new name for *minima* Ludlow, not Theobald 1901c). **Distr.:** Indonesia, Philippines. **Refs.:** Bohart 1945: 39 (tax.; from syn. with *hybrida*); Mattingly 1957a: 29 (F*; tax.); Basio 1971b: 72 (M*; bion.). **Ety.:** Clara Ludlow. **Informal name:** Ludlow Philippine Small Mosquito.
syn. *minima* Ludlow
1907: 413 (M, F; *Ludlowia*). **Type loc.:** Carandangan [Cudarangan], Mindanao, Philippines (USNM). **Refs.:** Stone & Knight 1957c: 197 (type info., lectotype desig.).

malfeyti (Newstead)
1907 (M*, F*; *Mimomyia*). In Newstead et al. 1907: 29. **Type loc.:** Boma [Leopoldville], Belgian Congo [Democratic Republic of the Congo] (NHMUK). **Distr.:** Angola, Benin, Cameroon, Côte d'Ivoire, Democratic Republic of the Congo, Gabon, Ghana, Liberia, Malawi, Mozambique, Nigeria, Senegal, South Africa, Sudan and South Sudan, Togo, Uganda. **Refs.:** Edwards 1941: 381 (P*; as *uniformis*); Hopkins 1952: 98 (L*); Hamon 1954a: 584 (L*; as *uniformis*); Hamon 1954c: 836 (M*, F*; as *uniformis*); Mattingly & Hamon 1955: 490 (M*, F*, L*; tax.; to sp.); Mattingly 1971a: Pl. 15 (P trumpet*, terminal segs*); da Cunha Ramos & Ribeiro 1975: 117 (L*; tax., distr.; Angola); Service 1976a (distr.; Gabon). **Ety.:** n.s. [Major Malfeyt (Harbach 2018c)]. **Informal name:** Malfeyt Congolese Small Mosquito.

minima (Theobald)
1901c: 262 (M*; *Uranotaenia*). **Type loc.:** Quilon, Travancore, India (NHMUK). **Distr.:** Australia, Bangladesh, Cambodia, India, Indonesia, Laos, Malaysia, Myanmar, Papua New Guinea, People's Republic of China, Singapore, Sri Lanka, Thailand, Vietnam. **Refs.:** Iyengar 1935: 423 (E*); Pratt 1951: 300 (P*, L*); Mattingly 1957a: 26 (M*, F*, P*, L*, E*); Mattingly 1971a: Pl. 4 (F head*), Pl. 7 (M head*), Pl. 28 (L head*, terminal segs.*); Yin Yin May 1972: 15 (distr.); Ahmed 1987 (distr.; Bangladesh); Lee et al. 1988b: 79 (distr.); Townsend 1990: 105 (lectotype desig.); Rattanarithikul et al. 2006b (F*, L*; bion., distr., keys). **Ety.:** smallest (L); ref. to being a very small species. **Informal name:** Mini Indian Small Mosquito.
syn. *minuta* Theobald
1908b: 301 (M; *Mimomyia*). **Type loc.:** Sylhet, Assam, India (IM).

nigra (Theobald)
1901d: 14 (A; *Aedes*). **Type loc.:** Old Calabar [Eastern Provinces], Nigeria (NHMUK). **Distr.:** Benin, Democratic Republic of the Congo, Liberia, Nigeria, Uganda. **Refs.:** Edwards 1941: 88 (M, F). **Ety.:** black (L). **Informal name:** Black Nigerian Small Mosquito.
syn. *africana* Newstead
1907 (F*; *Mimomyia*). In Newstead et al. 1907: 28. **Type loc.:** Nouvelle Anvers [Equateur], Belgian Congo [Democratic Republic of the Congo] (NHMUK).
syn. *fusca* Theobald
1911a: 32 (F; *Mimomyia*). **Type loc.:** Kampala swamp, Uganda (NHMUK).

uniformis (Theobald)
1904b: 80 (M*, F*; *Mimomyia*). **Type loc.:** Lado, Sudan [South Sudan] (NHMUK). **Distr.:** Angola, Benin, Burkina Faso, Cameroon, Central African Republic, Côte d'Ivoire, Democratic Republic of the Congo, Guinea-Bissau, Kenya, Liberia, Madagascar (includes Glorioso and Juan De Nova Isl.), Mali, Mozambique, Niger, Nigeria, Republic of the Congo, Senegal, Sudan and South Sudan, Tanzania, Uganda, Zambia, Zimbabwe. **Refs.:** Mattingly & Hamon 1955: 488 (M*, F*, L*; tax.); Correia 1968: 164 (distr.); da Cunha Ramos & Ribeiro 1975: 117 (L*; tax., bion., distr.; Angola); Grjebine 1986: 37 (M*, F*, P*, L*; keys); Jupp 1996 (M*, F*; key). **Ety.:** n.s. **Informal name:** Uniform Sudanese Small Mosquito.

syn. *grenieri* Hamon
1954c: 836 (M*, F*, P*, L). **Type loc.:** Lemouroudougou, Cercle de Banfora, Haute Volta [Burkina Faso] (IERT). **Refs.:** Mattingly & Hamon 1955: 488 (syn.).

Genus *Mimomyia* Theobald

(Vol. 1, p. 252)

Important references: Edwards 1932a: 111 (to subg. of *Ficalbia* Theobald); Barraud 1934 (tax., bion.; as subg. of *Ficalbia*; South Asia); Edwards 1941 (M, F, P; tax., keys, bion.; as subg. of *Ficalbia*; Afrotropics); Hopkins 1952 (L; tax., key, bion.; as subg. of *Ficalbia*; Afrotropics); Mattingly & Grjebine 1958 (tax., keys, bion.; as subg. *Mimomyia* of genus *Ficalbia*); Belkin 1962 (tax., bion., keys; South Pacific); Mattingly 1971a (to genus); Tanaka et al. 1979 (tax., keys, bion.; Japan); Mattingly 1981 (class., phyl.); Grjebine 1986 (tax., keys, bion.; Madagascar); Lee et al. 1988c (F; tax., key, bion.; Australasia); Service 1990 (tax., keys, bion.; Afrotropics);); Harbach & Kitching 1998 (phyl.); Rattanarithikul et al. 2006a (F, L; keys, tax., bion.; Thailand).

Subgenus *Etorleptiomyia* Theobald

1904b: 71. Type species: *Etorleptiomyia mediolineata* Theobald.

Name derivation: n.s. [etor- = heart, seat of life; leptos? = fine, small, delicate].

syn. *O'Reillia* Ludlow 1905a:101 (as genus). Type species: *O'Reillia luzonensis* Ludlow.
syn. *Dixomyia* Taylor 1914c: 702. Type species: *Dixomyia elegans* Taylor.
syn. *Luzonus* Stone & Bohart 1944: 212 as *Aedes* subg. *Luzonus*. Type species: *Aedes* (*Luzonensis*) *clavirostris* Stone & Bohart.

Important references: Mattingly 1957a (tax., bion.; Indomalaya; as subg. of genus *Ficalbia*); Mattingly & Grjebine 1958 (tax., keys, bion.; as subg. *Mimomyia* of genus *Ficalbia*); Belkin 1962 (tax., bion., keys; South Pacific; as subg. of *Ficalbia*); Delfinado 1966a (tax., bion.; Philippines; as subg. of *Ficalbia*); Tanaka et al. 1979 (tax., keys, bion.; Japan); Grjebine 1986 (tax., keys, bion.; Madagascar); Lee et al. 1988c (F; tax., key, bion.; Australasia); Service 1990 (tax., keys, bion.; Afrotropics).

bougainvillensis (Belkin)
1962: 294 (M*, F, P*, L*; *Ficalbia*). **Type loc.:** Bougainville, Solomon Islands (USNM). **Distr.:** Solomon Isl. **Refs.:** Lee et al. 1988c: 10 (F; key, tax., bion., distr., review). **Ety.:** Bougainville, Solomon Isl. **Informal name:** Bougainville Solomon Islander Indistinct Mosquito.

elegans (Taylor)
1914c: 703 (F*; *Dixomyia*). **Type loc.:** Townsville, Queensland, Australia (US). **Distr.:** Australia, Cambodia, Indonesia, Japan, Malaysia, Papua New Guinea, Philippines, Solomon Isl., Thailand, United States (Hawai'i), Vietnam. **Refs.:** Penn 1949b: 35 (P*); Mattingly 1957a: 46 (M, F*, P, L*); Basio 1971b: 72 (M*, L*; bion.); Saugstad 1973: 284 (distr.); Tanaka et al. 1975c: 214 (tax., distr.); Tanaka et al. 1979: 95 (M*, F*); Lee et al. 1988c: 12 (F; key, tax., bion., distr., review); Rattanarithikul et al. 2006a (F*, L*; bion., distr., keys). **Ety.:** n.s. [fine, elegant (L)]. **Informal name:** Elegant Australian Indistinct Mosquito.

luzonensis (Ludlow)
1905a: 101 (F; *O'Reillia*). **Type loc.:** Bayambang [Pangasinan], Luzon, Philippines (USNM). **Distr.:** Bangladesh, Cambodia, India, Indonesia, Japan, Malaysia, Myanmar, Nepal, Pakistan, People's Republic of China (includes Hong Kong), Philippines, Singapore, Sri Lanka, Taiwan, Thailand, Vietnam. **Refs.:** Bohart & Ingram 1946b: 61 (M, F, L*; bion., distr., tax.); Chen & Lien 1956: 199 (M*, F*, P*, L*); Mattingly 1957a: 42 (M*, F*, P*, L*; distr.); Basio 1971b: 72 (M*; bion.); Mattingly 1971a: Pl. 5 (A wing*), Pl. 12 (P paddle*, trumpet*); Baisas 1974: 129 (M, F, P, L; tax., bion., distr.; Philippines; as *Ficalbia*); Tanaka et al. 1975c: 213 (distr.); Tanaka et al. 1979: 92 (M*, F, L*); Lee et al. 1988c: 17 (F; key, tax., bion., distr., review); Darsie & Pradhan 1990 (F, L; tax., keys, bion., distr.; Nepal); Darsie 2000c: 626 (P*; tax., key); Rattanarithikul et al. 2006a (F*, L*; bion., distr., keys). **Ety.:** Luzon, Philippine Isl. **Informal name:** Luzon Philippine Indistinct Mosquito.
syn. *completiva* Leicester
1908: 178 (M; *Etorleptiomyia*). **Type loc.:** Singapore (NHMUK).
syn. *clavirostris* Stone & Bohart
1944: 212 (M*, F; *Aedes* (*Luzonus*)). **Type loc.:** Camp Nichols, Rizal, Luzon, Philippines (USNM). **Refs.:** Stone 1945a: 39 (syn.).

martinei (Doucet)
1951a: 68 (F; *Ficalbia*). **Type loc.:** Perinet, Tananarive, Madagascar (IRSM). **Distr.:** Madagascar (includes Glorioso and Juan De Nova Isl.). **Refs.:** Mattingly & Grjebine 1958: 271 (F*); White 1974: 240 (tax.); Service 1990: 106 (from subg. *Ingramia*). **Ety.:** n.s. **Informal name:** Martine Malagasy Indistinct Mosquito.

mediolineata (Theobald)
1904b: 71 (F*; *Etorleptiomyia*). **Type loc.:** Pibor River [Upper Nile], Sudan (NHMUK). **Distr.:** Benin, Botswana, Burkina Faso, Cameroon, Chad, Democratic Republic of the Congo, Djibouti, Gabon, Ghana, Kenya, Liberia, Mali, Mozambique, Namibia, Nigeria, South Africa, Sudan and South Sudan, Tanzania, The Gambia, Uganda, Zimbabwe. **Refs.:** Theobald 1905e: 44 (as *mediopunctata*, *lapsus*); Edwards 1941: 84 (M, F*), 380 (P*); Hopkins 1952: 96 (L*); Rodhain et al. 1977 (distr.); Grjebine 1986: 88 (M*, F*, P*, L*; keys); Jupp 1996 (M*, F*; key). **Ety.:** n.k. [yellow (Gr) plus middle lined or belted (L)]. **Informal name:** Midline Sudanese Indistinct Mosquito.

syn. *quadrimaculata* Newstead
1907 (F*; *Anisocheleomyia*). In Newstead et al. 1907: 32. **Type loc.:** Boma, Belgian Congo [Kinshasa, Democratic Republic of the Congo] (NHMUK).

solomonis (Belkin)
1962: 292 (M*, F, P*, L*; *Ficalbia*). **Type loc.:** Chacon swamp, Lunga area, Guadalcanal, Solomon Islands (USNM). **Distr.:** Solomon Isl. **Refs.:** Lee et al. 1988b: 98 (F; key, tax., bion., distr., review). **Ety.:** Solomon Isl. **Informal name:** Alvaro Solomon Islander Indistinct Mosquito [Álvaro de Mendaña de Neira (1542–1595) was the first European to visit the Solomon Isl. in 1568].

xanthozona (van Someren)
1948: 132 (F; *Ficalbia*). **Type loc.:** Mongiro, Bwamba County, Toro District, Uganda (NHMUK). **Distr.:** Cameroon, Uganda. **Refs.:** Eouzan et al. 1967: 223 (M*, F, P*, L*). **Ety.:** n.s. [yellow belt, zone (Gr)]. **Informal name:** Yellow-belted Ugandan Indistinct Mosquito.

Subgenus *Ingramia* Edwards

1912c: 43 (as genus). Type species: *Dasymyia fusca* Leicester.

Name derivation: Alexander Ingram.

syn. *Dasymyia* Leicester 1908: 102 (not Egger 1858). Type species: *Dasymyia fusca* Leicester.
syn. *Ravenalites* Doucet 1957 (*Ficalbia* subg.). In Mattingly 1957a: 2, 18. Type species: *Ravenalites roubaudi* Doucet.

Important references: Mattingly 1957a (tax., bion.; Indomalaya; as subg. *Ravenalites* of genus *Ficalbia*); Delfinado 1966a (tax., bion.; Philippines; as subg. *Ravenalites* of genus *Ficalbia*); White 1974: 240 (syn. of *Ravenalites*; revalidation of *Ingramia*); Grjebine 1986 (tax., keys, bion.; Madagascar); Service 1990 (tax., keys, bion.; Afrotropics).

aurata (Doucet)
1951a: 70 (M, F*, P*, L*; *Ravenalites*). **Type loc.:** Perinet, Tananarive, Madagascar (IRSM). **Distr.:** Madagascar (includes Glorioso and Juan De Nova Isl.). **Refs.:** White 1974: 240 (tax.); Grjebine 1986: 98 (M*, F*, P*, L*; keys). **Ety.:** n.s. [golden (L); prob. ref. to a small golden mosquito with black legs]. **Informal name:** Golden Malagasy Indistinct Mosquito.

bernardi (Doucet)
1950: 49 (L*; *Ravenalites*). **Type loc.:** Tampolo, Fort Dauphin, and Perinet, Madagascar (IRSM). **Distr.:** Madagascar (includes Glorioso and Juan De Nova Isl.). **Refs.:** White 1974: 240 (tax.); Grjebine 1986: 106 (M*, F*, P*, L*; keys). **Ety.:** n.s. **Informal name:** Bernard Malagasy Indistinct Mosquito.

beytouti (Doucet)
1951c: 112 (L*; *Ravenalites*). **Type loc.:** Amparihy [Tulear], Madagascar (IRSM). **Distr.:** Madagascar (includes Glorioso and Juan De Nova Isl.). **Refs.:** White 1974: 240 (tax.); Grjebine 1986: 113 (M*, F*, P*, L*; keys). **Ety.:** Le médecin-lieutenant Beytout. **Informal name:** Beytout Malagasy Indistinct Mosquito.

brygooi Grjebine
1986: 119 (M*, P*, L*). **Type loc.:** Manakara, Fianarantsoa, Madagascar (SSC). **Distr.:** Madagascar (includes Glorioso and Juan De Nova Isl.). **Ety.:** Édouard-Raoul Brygoo. **Informal name:** Brygoo Malagasy Indistinct Mosquito.

collessi Grjebine
1986: 124 (M*, F*, P*, L*). **Type loc.:** Manakara, East Coast, Madagascar (SSC). **Distr.:** Madagascar (includes Glorioso and Juan De Nova Isl.). **Ety.:** Donald Henry Colless. **Informal name:** Colless Malagasy Indistinct Mosquito.

deguzmanae (Mattingly)
1957a: 53 (M, F, P*, L*; *Ficalbia*). **Type loc.:** Llavac, Sierra Madre Range, Infanta of Quezon Province, Luzon, Philippines (NHMUK). **Distr.:** Philippines, Thailand. **Refs.:** Delfinado 1966a: 20 (M*, F, P*, L*); Basio 1971b: 73 (M*, L*; bion.); White 1974: 240 (tax.). **Ety.:** Milagros de Guzman. **Informal name:** de Guzman Philippine Indistinct Mosquito.

fusca (Leicester)
1908: 102 (M, F; *Dasymyia*). **Type loc.:** Kuala Lumpur [Selangor], Malaya [Malaysia] (NHMUK). **Distr.:** Cambodia, Indonesia, Malaysia, Philippines, Singapore, Taiwan, Thailand. **Refs.:** Chen & Lien 1956: 203 (M*, F*, P*); Mattingly 1957a: 49 (M*, F*, P*, L*); Mattingly 1971a: Pl. 5 (A wing*), Pl. 12 (P paddle*, trumpet*, L head*, terminal segs.*), Pl. 28 (L head*); White 1974: 240 (tax.); Grjebine 1986: 203 (M*, F*, P*, L*; keys); Rattanarithikul et al. 2006a (F*, L*; bion., distr., keys). **Ety.:** n.s. [dark (L); overall color of adult]. **Informal name:** Dark Malaysian Indistinct Mosquito.

grjebinei (Brunhes)
1977b: 153 (M*, F*, P*, L*; *Ficalbia*). **Type loc.:** near Gouloue River, Mayotte Island, Comoros (SSG). **Distr.:** Comoros, Madagascar (includes Glorioso and Juan De Nova Isl.). **Refs.:** Grjebine 1986: 130 (M*, F*, P*, L*; keys). **Ety.:** Alexis Grjebine. **Informal name:** Grjebine Comoran Indistinct Mosquito.

jeansottei (Doucet)
1950: 49 (L*; *Ravenalites*). **Type loc.:** Tampolo, East Madagascar (IRSM). **Distr.:** Madagascar (includes Glorioso and Juan De Nova Isl.). **Refs.:** White 1974: 240 (tax.); Grjebine 1986: 136 (M*, F*, P*, L*; keys). **Ety.:** n.s. **Informal name:** Sotte Malagasy Indistinct Mosquito.

kiriromi (Klein)
1969: 215 (M*, F*; *Ficalbia*). **Type loc.:** O Tachat, Kirirom Hills, Kompong-Speu Province, Cambodia (USNM). **Distr.:** Cambodia. **Refs.:** White 1974: 240 (tax.). **Ety.:** Kirirom Hills, Kompong Speu, Cambodia. **Informal name:** Kirirom Cambodian Indistinct Mosquito.

levicastilloi Grjebine
1986: 145 (P*, L*; keys). **Type loc.:** Manakara, Fianarantsoa, Madagascar (SSC). **Distr.:** Madagascar (includes Glorioso and Juan De Nova Isl.). **Ety.:** Jose Roberto Leví-Castillo. **Informal name:** Leví-Castillo Malagasy Indistinct Mosquito.

longicornis Grjebine
1986: 149 (M*, F*, P*, L*; keys). **Type loc.:** Ambodiriana (forest), Tamatave Province, Madagascar (SSC). **Distr.:** Madagascar (includes Glorioso and Juan De Nova Isl.). **Ety.:** long horned (L); ref. to very long pupal trumpet. **Informal name:** Long-horned Malagasy Indistinct Mosquito.

marksae Grjebine
1986: 154 (M*, F*, P*, L*; keys). **Type loc.:** Manakara, Fianarantsoa Province, Madagascar (SSC). **Distr.:** Madagascar (includes Glorioso and Juan De Nova Isl.). **Ety.:** Elisabeth Nesta Marks. **Informal name:** Marks Malagasy Indistinct Mosquito.

mattinglyi Grjebine
1986: 158 (M, P*, L*; keys). **Type loc.:** Analamazaotra (forest), Perinet, Tamatave Province, Madagascar (SSC). **Distr.:** Madagascar (includes Glorioso and Juan De Nova Isl.). **Ety.:** Peter Frederick Mattingly. **Informal name:** Mattingly Malagasy Indistinct Mosquito.

milloti Grjebine
1986: 163 (M*, F*, P*, L*; keys). **Type loc.:** Manombo, km 745 route Tananarive-Tulear, Madagascar (SSC). **Distr.:** Madagascar (includes Glorioso and Juan De Nova Isl.). **Ety.:** Jacques Millot. **Informal name:** Millot Malagasy Indistinct Mosquito.

mogii Miyagi, Toma, & Higa
2004: 11 (M*, F*, P*, L*). **Type loc.:** Kelimutu, Flores Island, Indonesia (USNM). **Distr.:** Indonesia. **Ety.:** Motoyoshi Mogi. **Informal name:** Mogi Indonesian Indistinct Mosquito.

ramalai Grjebine
1986: 169 (F*, P*, L*; keys). **Type loc.:** La Mandraka, Tananarive Province, Madagascar (SSC). **Distr.:** Madagascar (includes Glorioso and Juan De Nova Isl.). **Ety.:** Ramala [assistant's name]. **Informal name:** Ramala Malagasy Indistinct Mosquito.

roubaudi (Doucet)
1950: 51 (L*; *Ravenalites*). **Type loc.:** [Nosy Be Island, Lokobe Reserve] Madagascar (IRSM). **Distr.:** Madagascar (includes Glorioso and Juan De Nova Isl.). **Refs.:** Mattingly & Grjebine 1958: 285 (M, F, P); White 1974: 240 (tax.); Grjebine 1986: 174 (M*, F*, P*, L*; keys). **Ety.:** Emile Roubaud. **Informal name:** Roubaud Malagasy Indistinct Mosquito.

spinosa (Doucet)
1951a: 77 (L*; *Ravenalites*). **Type loc.:** Perinet, Tananarive, Madagascar (IRSM). **Distr.:** Madagascar (includes Glorioso and Juan De Nova Isl.). **Refs.:** White 1974: 240 (tax.); Grjebine 1986: 183 (M*, F*, P*, L*; keys). **Ety.:** spine (L); ref. to very long spinelike comb scales. **Informal name:** Spiny Malagasy Indistinct Mosquito.

stellata Grjebine
1986: 192 (M*, F, P*, L*; keys). **Type loc.:** Perinet (forest), Moramanga region, Madagascar (SSC). **Distr.:** Madagascar (includes Glorioso and Juan De Nova Isl.). **Ety.:** starlike (L); ref. to the numerous larval stellate setae. **Informal name:** Starry Malagasy Indistinct Mosquito.

vansomerenae Grjebine
1986: 199 (L*; key). **Type loc.:** Lokobe Reserve, Nosy Be Island, Madagascar (SSC). **Distr.:** Madagascar (includes Glorioso and Juan De Nova Isl.). **Ety.:** Ellinor Catharine Cunningham van Someren. **Informal name:** van Someren Malagasy Indistinct Mosquito.

Subgenus *Mimomyia* Theobald

1903a: 304 (as genus, subgenus). Type species: *Mimomyia splendens* Theobald.

Name derivation: n.s. [mimo- = imitator, actor, plus small fly (Gr)].

syn. *Boycia* Newstead 1907 (as genus). In Newstead et al. 1907: 33. Type species: *Boycia mimomyiaformis* Newstead.
syn. *Ludlowia* Theobald 1907: 193 (as genus). Type species: *Mimomyia chamberlaini* Theobald.
syn. *Megaculex* Theobald 1907: 282 (as genus). Type species: *Culex albitarsis* Theobald.
syn. *Radioculex* Theobald 1908b: 295 (as genus). Type species: *Radioculex clavipalpus* Theobald.
syn. *Conopomyia* Leicester 1908: 113 (as genus). Type species: *Conopomyia metallica* Leicester (Brunetti 1914: 68).
syn. *Hispidimyia* Theobald 1910c: 245 (as genus). Type species: *Hispidimyia hispida* Theobald.

Important references: Mattingly 1957a (tax., bion.; Indomalaya; as subg. of genus *Ficalbia*); Mattingly & Grjebine 1958 (tax., keys, bion.; as subg. of *Ficalbia*); Belkin 1962 (tax., keys, bion.; South Pacific; as subg. of *Ficalbia*); Delfinado 1966a (tax., bion.; Philippines; as subg. of *Ficalbia*); Grjebine 1986 (tax., bion., keys; Madagascar); Lee et al. 1988c (F; key, tax., bion.; Australasia); Service 1990 (tax., keys, bion.; Afrotropics).

aurea (Leicester)
1908: 116 (M, F; *Conopomyia*). **Type loc.:** Raub [Pahang], Malaya [Malaysia] (NHMUK). **Distr.:** Cambodia, India, Malaysia, Singapore, Thailand, Vietnam. **Refs.:** Mattingly 1957a: 37 (M*, F*, P*, L*); Mattingly 1970f: 161 (E*); Mattingly 1971c: 66 (E*); Rattanarithikul et al. 2006a (F*, L*; bion., distr., keys). **Ety.:** n.s. [golden (L); prob. ref. to golden markings on the head and abdomen]. **Informal name:** Gold Malaysian Indistinct Mosquito.

chamberlaini Ludlow
1904c: 297 (M). **Type loc.:** Bayambang, Pangasinan, Luzon, Philippines (NHMUK). **Distr.:** Australia, Bangladesh, Cambodia, India, Indonesia, Malaysia, Myanmar, Nepal, Papua New Guinea, People's Republic of China, Philippines, Sri Lanka, Thailand, Vietnam. **Refs.:** Barraud 1923h (L*); Galliard & Ngu 1949: 497 (L*?); Mattingly 1957a: 31 (M*, F*, P*, L*; tax.); Menon & Tampi 1959: 13 (E*); Basio 1971b: 73 (M*, L*; bion.); Mattingly 1971a: Pl. 11 (P paddle*, trumpet*), Pl. 28 (L head*); Ameen & Talukdar 1974a: 206 (P*); Lien et al. 1975 (distr.); Ahmed 1987 (distr.; Bangladesh); Darsie & Pradhan 1990 (F, L; tax., keys, bion., distr.; Nepal); Darsie 2000c: 626 (P*; tax., key); Rattanarithikul et al. 2006a (F*, L*; bion., distr., keys). **Ety.:** W.P. Chamberlain. **Informal name:** Chamberlain Philippine Indistinct Mosquito.

ssp. ***metallica*** (Leicester)
1908: 113 (M, F; *Conopomyia*). **Type loc.:** Barrack Pool, Kuala Lumpur, Malaya [Malaysia] (NHMUK). **Distr.:** Australia, Cambodia, Indonesia, Laos, Malaysia, Papua New Guinea, Philippines, Singapore, Taiwan, Thailand, Vietnam. **Refs.:** Galliard & Ngu 1949: 498 (M*, L*); Chen & Lien 1956: 204 (M*, F*, P*, L*); Mattingly 1957a: 34 (A, L*; tax.; as variety); Lee et al. 1988c: 23 (F; key, tax., bion., distr., review); Rattanarithikul et al. 2006a (F*, L*; bion., distr., keys). **Ety.:** metallic (L); ref. to abundant metallic scale features. **Informal name:** Metallic Malaysian Indistinct Mosquito.

ssp. ***clavipalpus*** (Theobald)
1908b: 295 (M, F; *Radioculex*). **Type loc.:** Calcutta, West Bengal, India (IM, NHMUK). **Distr.:** India, Pakistan, Sri Lanka. **Refs.:** Mattingly 1957a: 34 (L*; to ssp.); Townsend 1990: 61 (type info.). **Ety.:** club palpus (L); ref. to palpal "apical segment large and swollen but short." **Informal name:** Club-lipped Indian Indistinct Mosquito.

femorata (Edwards)
1936: 54 (F; *Ficalbia*). **Type loc.:** Kibanga, Uganda (NHMUK). **Distr.:** Uganda. **Refs.:** Edwards 1941: 80 (F). **Ety.:** n.s. [femur, thigh (L); ref. to "all femora . . . with anterior surfaces almost entirely black-scaled," differing from all other African species of the subgenus]. **Informal name:** Leggy Ugandan Indistinct Mosquito.

flavens (King & Hoogstraal)
1946e: 188 (M; *Ficalbia*). **Type loc.:** Hollandia, Western New Guinea (Papua) (USNM). **Distr.:** Indonesia. **Refs.:** Mattingly 1957a: 36 (? = *chamberlaini*); Lee et al. 1988c: 29 (F; key, tax., bion., distr., review). **Ety.:** n.s. [being yellow (L)]. **Informal name:** Yellow Indonesian Indistinct Mosquito.

flavopicta (Edwards)
1936: 54 (A; *Ficalbia*). **Type loc.:** Kisumu, Kenya (NHMUK). **Distr.:** Central African Republic, Côte d'Ivoire, Kenya, Uganda. **Refs.:** Edwards 1941: 80 (M, F). **Ety.:** painted yellow (L); yellow markings on scutum and abdomen. **Informal name:** Yellow-painted Kenyan Indistinct Mosquito.

gurneyi (Belkin)
1962: 296 (M*, F, P*, L*; *Ficalbia*). **Type loc.:** Bougainville, Solomon Islands (USNM). **Distr.:** Papua New Guinea, Solomon Isl. **Refs.:** Lee et al. 1988c: 31 (F; key, tax., bion., distr., review). **Ety.:** Ashley B. Gurney. **Informal name:** Gurney Solomon Islander Indistinct Mosquito.

hispida (Theobald)
1910c: 245 (M*, F*, P; *Hispidimyia*). **Type loc.:** Gebel Ahmed Aga [Jebel Ahmed Agha], Sudan (NHMUK). **Distr.:** Benin, Burkina Faso, Democratic Republic of the Congo, Ethiopia, Ghana, Guinea, Kenya, Liberia, Madagascar (includes Glorioso and Juan De Nova Isl.), Mali, Mozambique, Niger, Nigeria, Republic of the Congo, Senegal, South Africa, Sudan and South Sudan, Tanzania, The Gambia, Uganda, Zimbabwe. **Refs.:** Edwards 1941: 77 (M, F), 376 (P*); Hopkins 1952: 87 (L); Mattingly & Grjebine 1958: 275 (L*); Correia 1968: 164 (distr.); Grjebine 1986: 62 (M*, F*, P*, L*; keys); Jupp 1996 (M*, F*; key). **Ety.:** n.s. [hairy (L)]. **Informal name:** Hairy Sudanese Indistinct Mosquito.

syn. *palustris* Theobald
1911a: 13 (M*, F*; *Megaculex*). **Type loc.:** Kampala swamp, Uganda (NHMUK). **Refs.:** Edwards 1941: 78 (A), 377 (P); Stone et al. 1959: 98 (syn.).

syn. *sunyaniensis* Edwards
1935a: 134 (M, F; *Ficalbia* (*Mimomyia*) *hispida* var.). **Type loc.:** Sunyani [Ashanti], Gold Coast [Ghana] (NHMUK). **Refs.:** Edwards 1941: 79 (A); Hopkins 1952: 87 (L*); Stone et al. 1959: 98 (syn.).

hybrida (Leicester)
1908: 115 (M, F; *Conopomyia*). **Type loc.:** Kuala Lumpur [Selangor, Malaysia] (NHMUK). **Distr.:** Bangladesh, Cambodia, India, Indonesia, Laos, Malaysia, Myanmar, Nepal, Philippines, Singapore, Sri Lanka, Thailand, Vietnam. **Refs.:** Menon 1938: 185 (E*); Galliard & Ngu 1949: 495 (L*); Mattingly 1957a: 38 (M*, F, P*, L*; tax.); Basio 1971b: 73 (M*, P*; bion.); Mattingly 1971a: Pl. 11 (P paddle*, trumpet*), Pl. 26 (L terminal segs.*); Yin Yin May 1972: 12 (distr.); Ahmed 1987 (distr.; Bangladesh); Lee et al. 1988c: 32 (tax., review); Darsie & Pradhan 1990 (F, L; tax., keys, bion., distr.; Nepal); Darsie 2000c: 628 (P*; tax., key); Rattanarithikul et al. 2006a (F*, L*; bion., distr., keys). **Ety.:** n.s. [hybrid (L)]. **Informal name:** Hybrid Malaysian Indistinct Mosquito.

intermedia Barraud
1929a: 1055 (M; *chamberlaini* var.). **Type loc.:** Golaghat, Assam, India (NHMUK). **Distr.:** India, Nepal, Sri Lanka, Vietnam. **Refs.:** Mattingly 1957a: 34 (F; to sp.); Darsie et al. 1992 (distr.; Nepal). **Ety.:** in the middle (L); male characters intermediate between two closely related species. **Informal name:** Middle Indian Indistinct Mosquito.

lacustris (Edwards)
1935a: 134 (M, F; *Ficalbia*). **Type loc.:** Entebbe, Uganda (NHMUK). **Distr.:** Benin, Burkina Faso, Central African Republic, Chad, Democratic Republic of the Congo, Kenya, Mozambique, Nigeria, Republic of the Congo, Senegal, South Africa, Sudan and South Sudan, Tanzania, Togo, Uganda. **Refs.:** Edwards 1941: 79 (M, F), 377 (P*); Hopkins 1952: 88 (L*); Mattingly & Grjebine 1958: 271 (F*); Grjebine 1986: 69 (M*, F*, P*, L*; keys); Jupp 1996 (M*, F*; key). **Ety.:** n.s. [lake (L)]. **Informal name:** Lake Ugandan Indistinct Mosquito.

mimomyiaformis (Newstead)
1907 (M*, F*; *Boycia*). In Newstead et al. 1907: 34. **Type loc.:** Boma, Belgian Congo [Kinshasa, Democratic Republic of the Congo] (NHMUK). **Distr.:** Benin, Burkina Faso, Cameroon, Chad, Côte d'Ivoire, Democratic Republic of the Congo, Djibouti, Ethiopia, Gabon, Ghana, Kenya, Liberia, Madagascar (includes Glorioso and Juan De Nova Isl.), Mali, Niger, Nigeria, Republic of the Congo, Senegal, South Africa, Sudan and South Sudan, Tanzania, Uganda, Zimbabwe. **Refs.:** Edwards 1941: 81 (M, F; tax.), 378 (P)); Hopkins 1952: 93 (L*); Ribeiro 1966: 165 (L*); da Cunha Ramos & Ribeiro 1975: 119 (tax., bion., distr.; Angola); Rodhain et al. 1977 (distr.); Grjebine 1986: 74 (M*, F*, P*, L*; keys); Jupp 1996 (M*, F*; key). **Ety.:** n.s. [poss. ref. to similarity with *Mimomyia*]. **Informal name:** Boma Congolese Indistinct Mosquito.

syn. *sudanensis* Theobald
1907: 195 (M; *Ludlowia*). **Type loc.:** Bahr-el-Jebel, Sudan [Upper Nile, South Sudan] (NE).

syn. *pincerna* Graham
1910: 267 (M, F; *Megaculex*). **Type loc.:** [Yaba, Lagos, Nigeria], West Africa (NHMUK). **Refs.:** Peters 1956: 538 (L).

modesta (King & Hoogstraal)
1946e: 186 (M*, F; *Ficalbia*). **Type loc.:** Hollandia, Western New Guinea (Papua) (USNM). **Distr.:** Indonesia (West Papua). **Refs.:** Mattingly 1957a: 38 (? = *hybrida*); van den Assem 1958a: 131 (L*); Lee et al. 1988c: 33 (F; key, tax., bion., distr., review). **Ety.:** n.s. [modest, reserved (L)]. **Informal name:** Modest Indonesian Indistinct Mosquito.

pallida Edwards
1925: 263 (M, F). **Type loc.:** Fort Johnston, Nyasaland [Mangochi, Southern Region, Malawi] (NHMUK). **Distr.:** Benin, Cameroon, Democratic Republic of the Congo, Ghana, Kenya, Nigeria, Republic of the Congo, South Africa, The Gambia, Uganda. **Refs.:** Edwards 1941: 80 (M, F), 378 (P*); Hopkins 1952: 91 (L*); Jupp 1996 (M*, F*; key). **Ety.:** n.s. [pale, ashen (L)]. **Informal name:** Pallid Malawian Indistinct Mosquito.

parenti (De Meillon & Lavoipierre)
1944: 49 (M, F; *Ficalbia*). **Type loc.:** Yangambi, Belgian Congo [Isangi Territory, Tshopo District, Orientale Province, Democratic Republic of the Congo] (SAIM). **Distr.:** Democratic Republic of the Congo. **Refs.:** De Meillon et al. 1945: 99 (P*). **Ety.:** Dr. Parent, coll. [poss. ref. to Octave Parent]. **Informal name:** Parent Congolese Indistinct Mosquito.

perplexens (Edwards)
1932b: 561 (F; *Ficalbia*). **Type loc.:** Kampala, Uganda (NHMUK). **Distr.:** Cameroon, Côte d'Ivoire, Uganda. **Refs.:** Edwards 1941: 79 (F), 375 (P*); Hopkins 1952: 89 (L*). **Ety.:** n.s. [perplexing (L)]. **Informal name:** Perplexing Ugandan Indistinct Mosquito.

plumosa (Theobald)
1901a: 373 (M*; *Culex*). **Type loc.:** Salisbury, Mashonaland [Zimbabwe] (NHMUK). **Distr.:** Angola, Benin, Burkina Faso, Cameroon, Côte d'Ivoire, Democratic Republic of the Congo, Ethiopia, Gabon, Ghana, Kenya, Madagascar (includes Glorioso and Juan De Nova Isl.), Mali, Mozambique, Namibia, Nigeria, Republic of the Congo, Senegal, Seychelles, South Africa, Sudan and South Sudan, Tanzania, Uganda, Zambia, Zimbabwe. **Refs.:** Edwards 1941: 83 (M*, F), 379 (P*); Hopkins 1952: 94 (L*); Mattingly & Grjebine 1958: 273 (M*); Mattingly 1971a: Pl. 28 (L head*); da Cunha Ramos & Ribeiro 1975: 120 (tax., bion., distr.; Angola); Service 1976a (distr.; Gabon); Grjebine 1986: 80 (M*, F*, P*, L*; keys); Jupp 1996 (M*, F*; key). **Ety.:** n.s. (plumose (L)). **Informal name:** Feathery Zimbabwean Indistinct Mosquito.

syn. *albitarsis* Theobald
1901c: 25 (M*; *Culex*). **Type loc.:** Bonny [Eastern Provinces], Nigeria (NHMUK).

splendens Theobald
1903a: 304 (F*). **Type loc.:** Entebbe, Uganda (NHMUK). **Distr.:** Benin, Burkina Faso, Cameroon, Central African Republic, Democratic Republic of the Congo, Ethiopia, Ghana, Kenya, Madagascar (includes Glorioso and Juan De Nova Isl.), Mali, Mauritania, Mozambique, Nigeria, Senegal, South Africa, Sudan and South Sudan, Tanzania, Uganda. **Refs.:** Edwards 1941: 77 (M, F), 375 (P*); Hopkins 1952: 85 (L*); Mattingly & Grjebine 1958: 273 (M*); da Cunha Ramos & Ribeiro 1975: 121 (L*; tax., bion., distr.; Angola); Grjebine 1986: 57 (M*, F*, P*, L*; keys); Jupp 1996 (M*, F*; key). **Ety.:** shine, be bright (L); "a very marked and beautiful little species." **Informal name:** Splendid Ugandan Indistinct Mosquito.

Tribe Hodgesiini

Genus *Hodgesia* Theobald

(Vol. 1, p. 254)

1904a: 17 (as genus). Type species: *Hodgesia sanguinae* Theobald.

Name derivation: Aubrey Hodges.

Important references: Barraud 1934 (tax.; South Asia); Edwards 1941 (M, F, P; keys, tax., bion.; Afrotropics); Hopkins 1952 (L; key, tax., bion.; Afrotropics); Lee et al. 1988b (F; key, tax., bion., distr.; Australasia); Service 1990 (review; Afrotropics); Lu et al. 1997 (People's Republic of China); Rattanarithikul et al. 2006a (F, L; tax., keys, bion.; Thailand); Harbach & Kitching 1998 (phyl.); Rattanarithikul et al. 2010 (F*, L*; keys, bion., distr.; Thailand).

bailyi Barraud
1929a: 1062 (F). **Type loc.:** Virajpet, Coorg, India (NHMUK). **Distr.:** Cambodia, India, Sri Lanka, Thailand, Vietnam. **Refs.:** Barraud 1934: 55 (F); Carter & Wijesundara 1948: 137 (M*, P*, L*); Bram 1967b: 123 (M*, F*, P* L*; bion., distr., tax.); Rattanarithikul et al. 2006a (F*, L*; bion., distr., keys). **Ety.:** J.D. Baily. **Informal name:** Baily Indian Tiny Mosquito.

cairnsensis Taylor
1919: 842 (F). **Type loc.:** Cairns, Queensland, Australia (ANIC). **Distr.:** Australia, Papua New Guinea, Solomon Isl. **Refs.:** Lee et al. 1988b: 90 (F; key, tax., bion., distr., review). **Ety.:** Cairns, Australia. **Informal name:** Cairns Australian Tiny Mosquito.

cyptopus Theobald
1909b: 25 (F*; as *cuptopous*, emend. by Edwards 1912c: 35). **Type loc.:** Obuasi, Ashanti, Gold Coast [Ghana] (NHMUK). **Distr.:** Benin, Cameroon, Democratic Republic of the Congo, Gabon, Ghana, Liberia, Nigeria, Uganda. **Refs.:** Edwards 1912c: 35 (emend., to *cyptopus*); Edwards 1913b: 59 (M orig. described, not F); Edwards 1941: 40 (M*, F); De Meillon et al. 1945: 101 (P, L); Hopkins 1952: 48 (L*); Service 1976a: 260 (distr.; Gabon); White 1980: 140 (unjustified emend. of Edwards 1912c, restored to original spelling *cyptopous*). **Ety.:** n.s. [stooped foot (Gr)]. **Informal name:** Stoop-footed Ghanaian Tiny Mosquito.

lampangensis Thurman
1959b: 52 (M*, P*, L*). **Type loc.:** Ngao, Lampang Province, Thailand (USNM). **Distr.:** Thailand. **Refs.:** Rattanarithikul et al. 2006a (F*, L*; bion., distr., keys). **Ety.:** Lampang Prov., Thailand. **Informal name:** Lampang Thai Tiny Mosquito.

malayi Leicester
1908: 231 (M, F). **Type loc.:** near Kuala Lumpur [Selangor], Malaya [Malaysia] (NHMUK). **Distr.:** Cambodia, Indonesia, Laos, Malaysia, Philippines, Sri Lanka, Thailand, Vietnam. **Refs.:** Borel 1930: 136 (M*, F*, L*; as *ampyx*); Barraud 1934: 54 (M, F*, L*); Delfinado 1966a: 58 (M*, P*, L*); Mattingly 1971a: Pl. 2 (A wing*, wing scales*), Pl. 8 (P trumpet*), Pl. 28 (L head*, terminal segs.*); Baisas 1974: 131 (M, F, P*, L*; tax., bion., distr.; Philippines); Lien et al. 1975 (distr.); Lee et al. 1988b: 92 (tax., bion., distr., review); Rattanarithikul et al. 2006a (F*, L*; bion., distr., keys). **Ety.:** Malaya [Malaysia]. **Informal name:** Parameswara Malaysian Tiny Mosquito [Prince Parameswara (1344–1414) established the Malacca Sultanate in Malaysia in the early 15th century].
syn. *ampyx* Dyar
1920h: 176 (F). **Type loc.:** College of Agriculture, Los Banos [Laguna, Luzon], Philippines (USNM). **Refs.:** Stone & Knight 1957c: 197 (type info., lectotype desig.).

nigeriae Edwards
1930a: 300 (M, F). **Type loc.:** Ebute Metta [Western Provinces, Nigeria] (NHMUK). **Distr.:** Benin, Cameroon, Ghana, Liberia, Nigeria, Republic of the Congo. **Refs.:** Edwards 1941: 40 (M, F), 364 (P*); Hopkins 1952: 47 (L*). **Ety.:** Nigeria. **Informal name:** Metta Nigerian Tiny Mosquito.

psectropus Edwards
1930a: 300 (M*, F). **Type loc.:** Stanleyville, Belgian Congo [Kisangani, Tshopo Province, Democratic Republic of the Congo] (NHMUK). **Distr.:** Cameroon, Democratic Republic of the Congo, Republic of the Congo. **Refs.:** Edwards 1941: 40 (M, F), 364 (P); Hopkins 1952: 47 (L). **Ety.:** n.s. [curry comb, a scraper (Gr)]. **Informal name:** Comb-footed Congolese Tiny Mosquito.

quasisanguinae Leicester
1908: 230 (F). **Type loc.:** near Kuala Lumpur [Selangor], Malaya [Malaysia] (NHMUK). **Distr.:** Australia, Cambodia, Indonesia, Malaysia, Papua New Guinea, Philippines. **Refs.:** Bohart 1945: 31 (A); Basio 1971b: 74 (M*; bion.); Lee et al. 1988b: 94 (F; key, tax., bion., distr., review). **Ety.:** nearly (L); similar to *sanguinae*. **Informal name:** Ketchup Malaysian Tiny Mosquito [*quasisanguinae* = "similar to blood" = Ketchup].
syn. *niveocaputis* Ludlow
1911b: 130 (F). **Type loc.:** Fort Pikit, Malabang [Lanao, Mindanao], Philippines (USNM). **Refs.:** Stone & Knight 1957c: 198 (type info., lectotype desig.).
syn. *triangulatus* Taylor
1914d: 204 (F). **Type loc.:** Lakekamu Gold Field, Papua New Guinea (ANIC). **Refs.:** Lee et al. 1988b: 94 (type info.).

sanguinae Theobald
1904a: 17 (F*). **Type loc.:** Entebbe, Uganda (NHMUK). **Distr.:** Uganda. **Refs.:** Edwards 1912c: 35 (as *sanguinis*, emend.); Edwards 1941: 39 (M*, F), 364 (P); Hopkins

1952: 46 (L); Townsend 1990: 130 (lectotype desig.). **Ety.:** n.s. [blood (L)]. **Informal name:** Bloody Ugandan Tiny Mosquito.

solomonis Belkin
1962: 280 (M*, F, P*, L*). **Type loc.:** West Poha swamp, Guadalcanal, Solomon Islands (USNM). **Distr.:** Australia, Indonesia, Papua New Guinea, Solomon Isl. **Refs.:** Knight & Chamberlain 1948: 15 (P*; as *elegans*); Penn 1949b: 35 (P*; as *elegans*); Lee et al. 1988b: 98 (F; key, tax., bion., distr., review). **Ety.:** Solomon Isl. **Informal name:** Poha Solomon Islander Tiny Mosquito.

spoliata Edwards
1923b: 8 (F). **Type loc.:** Mekeo District, [Papua] New Guinea (NHMUK, ANIC). **Distr.:** Indonesia, Papua New Guinea. **Refs.:** Knight & Chamberlain 1948: 17 (P*); Lien et al. 1977 (distr.); Lee et al. 1988b: 100 (F; key, tax., bion., distr., review, type info.). **Ety.:** n.s. [rob, plunder, deprive (L)]. **Informal name:** Plundering Papuan Tiny Mosquito.

Tribe Mansoniini

Ronderos & Bachmann 1964 (generic group revision).

Genus *Coquillettidia* Dyar

(Vol. 1, p. 256)

Important references: Lane 1953 (subg. *Rhynchotaenia* of genus *Mansonia*; keys, tax., distr.); Belkin 1962 (as *Mansonia* subg.; tax.; South Pacific); Wharton 1962: 1 (Malaya); Ronderos & Bachman 1963a: 48 (to genus) [but see Mattingly 1971a]; Cova García et al. 1966 (tax., keys; Venezuela); Delfinado 1966a (subg. *Coquillettidia*, tax., keys; Philippines); Belkin 1968a (subg. *Austromansonia* and *Coquillettidia*; tax., distr., bion.; New Zealand); Belkin et al. 1970: 100 (tax.); Tanaka et al. 1979 (as *Mansonia* subg.; tax., keys, bion., distr.; Japan); Lee et al. 1988a (tax., key, bion., distr.; Australasia); Service 1990 (tax., keys, bion., distr.; Afrotropics); Rattanarithikul, Harrison et al. 2005 (F, L; tax., keys, bion.; Thailand); Rattanarithikul et al. 2006a (tax., bion., keys; Thailand).

Subgenus *Austromansonia* Belkin

1968a: 76. Type species: *Taeniorhynchus* (*Coquillettidia*) *tenuipalpis* Edwards.

Name derivation: n.s. [ref. to Australasia].

Important references: Belkin 1968a (bion., distr., tax.; New Zealand); Lee et al. 1988a (descr., tax.; New Zealand).

tenuipalpis (Edwards)
1924b: 366 (M, F; *Taeniorhynchus*). **Type loc.:** Ohakune [Wellington], New Zealand (NHMUK). **Distr.:** New Zealand. **Refs.:** Belkin 1962: 305 (M*, F); Belkin 1968a: 76 (M*, F*, P*); Lee et al. 1988a: 74 (F; key, tax., bion., distr., review); Derraik 2004 (distr.; New Zealand). **Ety.:** n.s. [thin, slender, slight palpus (L); poss. ref. to comparison with *iracundus*]. **Informal name:** Thin-lipped New Zealander Underwater Mosquito.

Subgenus *Coquillettidia* Dyar

1905a: 47. Type species: *Culex perturbans* Walker.

Name derivation: Daniel William Coquillett.

Important references: Gillett 1946: 425 (keys; Ethiopian Region); Lane 1953 (as subg. *Rhynchotaenia* of genus *Mansonia*; keys, tax., distr.); Belkin 1962: 301 (as *Mansonia* subg.; tax.; South Pacific); Wharton 1962: 1 (Malayasia); Ronderos & Bachman 1963a: 48 (to genus) [but see Mattingly 1971a]; Cova García et al. 1966 (tax., keys; Venezuela); Delfinado 1966a (subg. *Coquillettidia*; tax., keys; Philippines); Belkin 1968a (subg. *Austromansonia* and *Coquillettidia*; tax., distr., bion.; New Zealand); Belkin et al. 1970: 100 (tax.); Tanaka et al. 1979 (as *Mansonia* subg.; tax., keys, bion., distr.; Japan, Korea); Lee et al. 1988a (tax., key, bion., distr.; Australasia); Service 1990 (tax., keys, bion., distr.; Afrotropics); Rattanarithikul et al. 2006a (tax., bion., keys; Thailand).

annettii (Theobald)
1901d: 13 (M, F; *Taeniorhynchus*). **Type loc.:** Old Calabar and Bonny [Eastern Provinces], Nigeria (NHMUK). **Distr.:** Benin, Cameroon, Central African Republic, Democratic Republic of the Congo, Liberia, Mozambique, Nigeria, Republic of the Congo. **Refs.:** Edwards 1941: 94 (M*); Worth & De Meillon 1960: 236 (distr.). **Ety.:** Dr. Annett. **Informal name:** Annett Nigerian Underwater Mosquito.

aurata (Dobrotworsky)
1962: 295 (F*; *Mansonia*). **Type loc.:** Cabbage Tree Creek, Victoria, Australia (NMM). **Distr.:** Australia. **Refs.:** Dobrotworsky 1965: 63 (M*, F*, E*); Mattingly 1971g (E*); Lee et al. 1988a: 79 (F; key, tax., bion., distr., review). **Ety.:** n.s. [golden (L); prob. ref. to light golden scales on scutum]. **Informal name:** Golden Australian Underwater Mosquito.

aurea (Edwards)
1915b: 281 (M; *Taeniorhynchus*). **Type loc.:** Umbilo, Durban, Natal [KwaZulu-Natal, South Africa] (NHMUK). **Distr.:** Angola, Democratic Republic of the Congo, Mozambique, South Africa, Sudan and South Sudan, Tanzania, Uganda, Zimbabwe. **Refs.:** Edwards 1941: 100 (M*); Worth & De Meillon 1960: 236 (distr.); da Cunha Ramos & Ribeiro 1975: 122 (tax., bion., keys, distr.; Angola); Jupp 1996 (M*, F*; key). **Ety.:** n.s. [golden (L)]. **Informal name:** Gold South African Underwater Mosquito.

aureosquammata (Ludlow)
1909c: 234 (F; *Taeniorhynchus*?). **Type loc.:** Parang [Cotabato, Mindanao], Philippines (USNM). **Distr.:** Indonesia, Malaysia, Philippines. **Refs.:** Bohart 1945: 43 (M*); Stone & Knight 1957c: 198 (type info., lectotype desig.); Wharton 1962 (E); Basio 1971b: 51 (M*; bion.); Mattingly 1971g (E). **Ety.:** n.s. [golden scaled (L); poss. ref. to head with "light yellowish-white flat scales laterally"]. **Informal name:** Gold-scaled Philippine Underwater Mosquito.
syn. *pagei* Ludlow
1910 (M, F; *Taeniorhynchus*). In Theobald 1910c: 618. **Type loc.:** Parang [Cotabato, Mindanao], Philippines (USNM).

aurites (Theobald)
1901d: 13 (M; *Taeniorhynchus*). **Type loc.:** Bonny and Ogugumanga [Eastern Provinces], Nigeria (NHMUK).

Distr.: Angola, Benin, Cameroon, Central African Republic, Côte d'Ivoire, Gabon, Kenya, Liberia, Mali, Mozambique, Nigeria, Republic of the Congo, Tanzania, Uganda, Zambia, Zimbabwe. **Refs.:** Edwards 1941: 99 (M*, F); Gillett 1946: 434 (P*, L*); Gillett 1949: 99 (M, F; tax.); Hopkins 1952: 109 (L*); Worth & De Meillon 1960: 236 (distr.); Mattingly 1961 (E*); Mattingly 1971g (E*); da Cunha Ramos & Ribeiro 1975: 122 (tax., bion., keys, distr.; Angola). **Ety.:** gold agent (L, Gr); gold scales on scutum and abdomen. **Informal name:** Jeweler Nigerian Underwater Mosquito [*aurites* = "gold agent" = Jeweler].

buxtoni (Edwards)
1923b: 9 (F; *Taeniorhynchus*). **Type loc.:** Huleh, Israel (NHMUK). **Distr.:** Boznia & Herzegovina, France (includes Corsica), Germany, Greece (includes Samos), Israel (and Gaza Strip and West Bank), Italy (includes Sardinia and Sicily), Kosovo, Moldova, Montenegro, Morocco, Portugal (Madeira), Romania, Russia, Spain, Switzerland, Syria, Turkey, Ukraine. **Refs.:** Aitken 1954: 466 (M*, F); Coluzzi & Contini 1962: 215 (P*, L*); Mandoul et al. 1968: 285 (L*); Gutsevich et al. 1974: 173 (M, F, L*); Ramsdale & Snow 2001 (distr.; France, Italy, Romania, Spain, Ukraine); Linton et al. 2007 (distr.; Samos Isl.); Becker et al. 2010: 307 (M*, F*, L*; keys, tax., distr., bion.); Robert et al. 2019 (distr.; Western Palearctic). **Ety.:** Patrick Alfred Buxton. **Informal name:** Buxton Israeli Underwater Mosquito.

chrysosoma (Edwards)
1915b: 281 (M*; *Taeniorhynchus*). **Type loc.:** Karonga, Nyasaland [Malawi] (NHMUK). **Distr.:** Ethiopia, Kenya, Liberia, Malawi, Mozambique, South Africa, Sudan and South Sudan, Tanzania. **Refs.:** Edwards 1941: 99 (M*); Jupp 1996 (M*, F*; key). **Ety.:** n.s. [gold body (Gr); comp. to *aurites*]. **Informal name:** Gold-bodied Malawian Underwater Mosquito.

crassipes (van der Wulp)
1881: 9 (F*; *Culex*). **Type loc.:** Soeroelangoen [Benkoelen], Sumatra [Indonesia] (LM). **Distr.:** Australia, Bangladesh, Cambodia, Fiji, Guam, India, Indonesia, Japan, Laos, Malaysia, Mariana Isl., Myanmar, Nepal, Pakistan, Papua New Guinea, People's Republic of China (includes Hong Kong), Philippines, Sri Lanka, Taiwan, Thailand, Vietnam. **Refs.:** Barraud 1934: 120 (M*, F); Bohart 1945: 43 (M*); Bonne-Wepster 1954: 40 (M, F, L*); Basio 1971b: 51 (M*; bion.); Mattingly 1971a: Pl. 5 (A* wing*), Pl. 7 (M head*); Belkin 1962: 308 (distr.); Wharton 1962: 6 (M*, F, P*, L*, E); Baisas 1974: 136 (M, F, P, L; tax., bion., distr.; Philippines); Tanaka et al. 1975c: 215 (distr.); Tanaka et al. 1979: 122 (M*, F*); Ahmed 1987 (distr.; Bangladesh); Lee et al. 1988a: 80 (F; key, tax., bion., distr., review); Darsie et al. 1996 (distr.; Nepal); Rattanarithikul et al. 2006a (F*, L*; bion., distr., keys). **Ety.:** n.s. [thick foot (L)]. **Informal name:** Thick-footed Indonesian Underwater Mosquito.

syn. *brevicellulus* Theobald
1901c: 212 (M*, F*; *Taeniorhynchus*). **Type loc.:** Selangor, Malaya, and Thayetmyo [Magwe], Burma [Myanmar] (NHMUK).

syn. *pygmaeus* Theobald
1908b: 300 (F; *Chrysoconops*). **Type loc.:** Sylhet, Assam, India (IM). **Refs.:** Bohart 1945 (syn.).

syn. *diaeretus* Dyar
1920h: 181 (F; *Mansonia* (*Coquillettidia*)). **Type loc.:** Los Banos [Laguna, Luzon], Philippines (USNM). **Refs.:** Bohart 1945: 43 (syn.); Stone & Knight 1957c: 198 (type info., lectotype desig.).

cristata (Theobald)
1904b: 78 (F; *Taeniorhynchus*). **Type loc.:** Pibor, Upper Nile, Sudan [Pibor Post, South Sudan] (NHMUK). **Distr.:** Angola, Burkina Faso, Central African Republic, Côte d'Ivoire, Democratic Republic of the Congo, Ethiopia, Gabon, Liberia, Malawi, Mozambique, Nigeria, Republic of the Congo, Sierra Leone, South Africa, Sudan and South Sudan, The Gambia, Uganda, Zambia, Zimbabwe. **Refs.:** Bonne-Wepster 1939 (L; distr.; Borneo); Edwards 1941: 96 (M*, F); Worth & De Meillon 1960: 236 (distr.); da Cunha Ramos & Ribeiro 1975: 122 (M*; tax., bion., keys, distr.; Angola); Jupp 1996 (M*, F*; key). **Ety.:** n.s. [crested (L)]. **Informal name:** Crested Sudanese Underwater Mosquito.

syn. *nocturnus* Theobald
1913d: 593 (M*, F*; *Chrysoconops*). **Type loc.:** Several localities in Bahr-el-Ghazal, Sudan [South Sudan] (NHMUK).

fijiensis (Belkin)
1962: 308 (M*, F, P*, L*; *Mansonia*). **Type loc.:** Koronivia, Viti Levu, Fiji (USNM). **Distr.:** Fiji. **Refs.:** Lee et al. 1988a: 88 (F; key, tax., bion., distr., review). **Ety.:** Fiji. **Informal name:** Viti Fijian Underwater Mosquito.

flavocincta (Edwards)
1936: 54 (A; *Taeniorhynchus*). **Type loc.:** Lubumbashi, Belgian Congo [Democratic Republic of the Congo] (NHMUK). **Distr.:** Angola, Democratic Republic of the Congo, Tanzania, Zambia, Zimbabwe. **Refs.:** Edwards 1941: 96 (M); Lips & Hamon 1956: 86 (P*, L*); da Cunha Ramos & Ribeiro 1975: 123 (tax., bion., keys, distr.; Angola); Jupp 1996 (M*, F*; key). **Ety.:** yellow girded (L); hind tibia with yellow ring. **Informal name:** Yellow-girded Congolese Underwater Mosquito.

fraseri (Theobald)
1911a: 22 (M*, F*; *Chrysoconops*). **Type loc.:** Kampala swamp, Uganda (NHMUK). **Distr.:** Angola, Cameroon, Central African Republic, Democratic Republic of the Congo, Kenya, Nigeria, Sudan and South Sudan, Uganda. **Refs.:** Gillett 1949: 100 (M, F; from syn. with *aurites*); Mattingly 1971g (E*); da Cunha Ramos & Ribeiro 1975: 124 (tax., bion., keys, distr.; Angola). **Ety.:** A.D. Fraser. **Informal name:** Fraser Ugandan Underwater Mosquito.

syn. *atroapicalis* Gillett
1946: 436 (M*, F, P*, L*; *Taeniorhynchus* (*Coquillettidia*)). **Type loc.:** Uganda (NHMUK). **Refs.:** Gillett 1949: 100 (syn.).

fuscopennata (Theobald)
1903a: 265 (M*, F*; *Taeniorhynchus*). **Type loc.:** Entebbe, Uganda (NHMUK). **Distr.:** Angola, Democratic Republic of the Congo, Ethiopia, Ghana, Kenya, Liberia, Rwanda, South Africa, Tanzania, Uganda. **Refs.:** Edwards 1941: 97 (M*, F); Gillett 1946: 433 (P*, L*); Mattingly 1971g (E*); Jupp 1996 (M*, F*; key). **Ety.:** n.s. [dark winged (L); wing with dusky-brown scales]. **Informal name:** Dark-winged Ugandan Underwater Mosquito.
syn. *drymoecius* Speiser
1909: 42 (F; *Culex*). **Type loc.:** Kilimanjaro, near Kibonoto and Momo, Usambara, Kenya (SM).
syn. *bakeri* Theobald
1911a: 19 (M*, F*; *Chrysoconops*). **Type loc.:** Kampala swamp, Uganda (NHMUK).

fuscopteron (Theobald)
1911d: 239 (F; *Chrysoconops*). **Type loc.:** Digoel River, [Western] New Guinea (AM). **Distr.:** Indonesia (West Papua), Papua New Guinea. **Refs.:** Lee et al. 1988a: 91 (F; key, tax., bion., distr., review). **Ety.:** n.s. [dark winged (L, Gr)]. **Informal name:** Black-winged Indonesian Underwater Mosquito.

giblini (Taylor)
1914d: 198 (M, F; *Pseudotaeniorhynchus conopas* var.). **Type loc.:** Lakekamu Gold Field, Papua New Guinea (ANIC). **Distr.:** Australia, Indonesia, Malaysia, Papua New Guinea. **Refs.:** Taylor 1944: 123 (M*); Bohart 1945: 43 (M*); Bonne-Wepster 1954: 42 (M, F); Lee et al. 1988a: 92 (F; key, tax., bion., distr., review). **Ety.:** Dr. Giblin. **Informal name:** Giblin Papuan Underwater Mosquito.

grandidieri (Blanchard)
1905: 627 (new name for *flavus* Ventrillon, not Motschulsky 1859; *Culex*). **Distr.:** Liberia, Madagascar (includes Glorioso and Juan De Nova Isl.). **Refs.:** Edwards 1941: 98 (M, F); Le Goff et al. 2002 (L; tax., review). **Ety.:** n.s. [poss. Alfred Grandidier]. **Informal name:** Grandidier Malagasy Underwater Mosquito.
syn. *flavus* Ventrillon
1904: 550 (M, F; *Culex*). **Type loc.:** Ankazobe, Madagascar (NE).

hodgkini (Wharton)
1962: 12 (M*, F*, E; *Mansonia*). **Type loc.:** Lamir (Pekan), Pahang, Malaya [Malaysia] (NHMUK). **Distr.:** Indonesia, Malaysia. **Refs.:** Mattingly 1971g (E). **Ety.:** Ernest Pease Hodgkin. **Informal name:** Hodgkin Malaysian Underwater Mosquito.

iracunda (Walker)
1848: 6 (F; *Culex*). **Type loc.:** New Zealand (NHMUK). **Distr.:** New Zealand. **Refs.:** Miller & Phillipps 1952: 23 (M, F*); Belkin 1962: 306 (M*, F); Belkin 1968a: 78 (M, F, P*); Lee et al. 1988a: 95 (F; key, tax., bion., distr., review); Derraik 2004 (distr.; New Zealand). **Ety.:** n.s. [angry, hot tempered (L)]. **Informal name:** Angry New Zealander Underwater Mosquito.

karandalaensis (Wolfs)
1951: 12 (M*, F, P*, L*; *Taeniorhynchus*). **Type loc.:** Karandala [Ruanda-Urundi], Belgian Congo [Democratic Republic of the Congo] (CMT). **Distr.:** Democratic Republic of the Congo. **Ety.:** Karandala, India. **Informal name:** Karandala Congolese Underwater Mosquito.

linealis (Skuse)
1889: 1747 (F; *Culex*). **Type loc.:** several localities in New South Wales, Australia (ANIC). **Distr.:** Australia, Papua New Guinea. **Refs.:** Bancroft 1908: 48 (F); Edwards 1924b: 366 (M); Dobrotworsky 1965: 61 (M*, F*); Lee et al. 1988a: 97 (F; key, tax., bion., distr., review). **Ety.:** n.s. [string, line (L); poss. ref. to scutum with "four distinct lines of golden-yellow scales"]. **Informal name:** Linear Australian Underwater Mosquito.

lutea (Belkin)
1962: 309 (M*, F, P*, L*; *Mansonia*). **Type loc.:** Burns Creek, Lunga area, Guadalcanal, Solomon Islands (USNM). **Distr.:** Solomon Isl. **Refs.:** Lee et al. 1988a: 103 (F; key, tax., bion., distr., review). **Ety.:** yellow (L); overall colors. **Informal name:** Yellow Solomon Islander Underwater Mosquito.

maculipennis (Theobald)
1911a: 27 (M*, F*; *Chrysoconops*). **Type loc.:** Kampala swamp, Uganda (NHMUK). **Distr.:** Benin, Burkina Faso, Cameroon, Central African Republic, Côte d'Ivoire, Democratic Republic of the Congo, Kenya, Liberia, Malawi, Mali, Mozambique, Nigeria, Republic of the Congo, Rwanda, South Africa, Sudan and South Sudan, Tanzania, Uganda, Zambia, Zimbabwe. **Refs.:** Edwards 1941: 95 (M*); Gillett 1945: 395 (P*, L*); Gillett 1946: 433 (P*, L*); Corbet 1958: 127 (F*); Worth & De Meillon 1960: 236 (distr.). **Ety.:** spotted wings (L). **Informal name:** Spot-winged Ugandan Underwater Mosquito.

memorans (Bonne-Wepster)
1930: 393 (F; *Taeniorhynchus*). **Type loc.:** Upper Digoel, [Western] New Guinea (NHMUK). **Distr.:** Australia, Indonesia (includes West Papua), Papua New Guinea. **Refs.:** Lee et al. 1988a: 105 (F; key, tax., bion., distr., review); Townsend 1990: 103 (type info.). **Ety.:** n.s. [be mindful of, relate, speak of (L)]. **Informal name:** Mindful Indonesian Underwater Mosquito.

metallica (Theobald)
1901d: 8 (M*, F*; *Culex*). **Type loc.:** Bonny [Eastern Provinces], Nigeria (NHMUK). **Distr.:** Angola, Benin, Botswana, Burkina Faso, Cameroon, Central African Republic, Democratic Republic of the Congo, Ethiopia, Ghana, Kenya, Liberia, Madagascar (includes Glorioso

and Juan De Nova Isl.), Mali, Mozambique, Nigeria, Senegal, Sierra Leone, South Africa, Sudan and South Sudan, Tanzania, Uganda. **Refs.:** Edwards 1941: 91 (M*, F); Gillett 1946: 430 (P*, L*); da Cunha Ramos & Ribeiro 1975: 124 (tax., bion., keys, distr.; Angola); Jupp 1996 (M*, F*; key). **Ety.:** metallic (L) [prob. ref. to metallic purple scales on abdomen]. **Informal name:** Metallic Nigerian Underwater Mosquito.

syn. *nigrochaetae* Theobald
1901c: 60 (M* of type series; *Culex* [F of type series syn. of Culex (Culiciomyia) *nebulosus*]). **Type loc.:** Lagos, Nigeria (NHMUK?).

syn. *violaceus* Theobald
1908c: 262 (M*, F*; *Taeniorhynchus*). **Type loc.:** Taufikia and Kodok, Upper Nile, Sudan (NHMUK).

syn. *nigra* Theobald
1910c: 434 (F; *Chrysoconops*). **Type loc.:** Angola (NHMUK).

microannulata (Theobald)
1911a: 26 (F*; *Chrysoconops*). **Type loc.:** Kampala swamp, Uganda (NHMUK). **Distr.:** Côte d'Ivoire, Democratic Republic of the Congo, Ethiopia, Kenya, Mozambique, South Africa, Sudan and South Sudan, Tanzania, Uganda, Zambia, Zimbabwe. **Refs.:** Edwards 1941: 101 (M*, F); Gillett 1946: 435 (P*, L*); Mattingly 1971a: Pl. 8 (P trumpet*); Jupp 1996 (M*, F*; key). **Ety.:** small rings (L); ref. to "Legs golden yellow with minute dark apical tarsal bands." **Informal name:** Small-ringed Ugandan Underwater Mosquito.

syn. *auripennis* Edwards
1915b: 279 (M, F; *Taeniorhynchus*). **Type loc.:** Entebbe, Uganda (NHMUK). **Refs.:** Edwards 1941: 101 (tax.); Stone et al. 1959: 103 (syn.).

syn. *chubbi* Edwards
1915b: 280 (M*, F; *Taeniorhynchus*). **Type loc.:** Umbilo, Durban, Natal [KwaZulu-Natal, South Africa] (NHMUK).

nigritarsis (Wolfs)
1958: 297 (M; *Mansonia*). **Type loc.:** Kikwit, Kwango, Belgian Congo [Democratic Republic of the Congo] (CMT). **Distr.:** Democratic Republic of the Congo. **Ety.:** black tarsi (L). **Informal name:** Black-footed Congolese Underwater Mosquito.

nigrithorax (Theobald)
1910c: 439 (M*, F*; *Chrysoconops*). **Type loc.:** Bailundu, Angola (NHMUK). **Distr.:** Angola, Central African Republic, Democratic Republic of the Congo. **Refs.:** Edwards 1941: 95 (M*, F); da Cunha Ramos & Ribeiro 1975: 124 (tax., bion., keys, distr.; Angola). **Ety.:** black thorax (L). **Informal name:** Black-chested Angolan Underwater Mosquito.

nigrochracea (Bonne-Wepster)
1930: 392 (F; *Taeniorhynchus*). **Type loc.:** Upper Digoel, [Western] New Guinea, and Dramajoe [= Dermajoe, Benkoelen], Sumatra (LM, ITH). **Distr.:** Indonesia. **Refs.:** Lee et al. 1988a: 107 (F; key, tax., bion., distr., review). **Ety.:** n.s. [black, yellow (Gr); poss. color of proboscis]. **Informal name:** Black-sheathed Indonesian Underwater Mosquito.

nigrosignata (Edwards)
1917: 224 (F*; *Taeniorhynchus*). **Type loc.:** several localities in Malaya [Malaysia] and Formosa [Taiwan, Republic of China] (NHMUK). **Distr.:** Cambodia, Indonesia, Malaysia, Philippines, Singapore, Taiwan, Thailand. **Refs.:** Wharton 1962: 7 (M*, F*, L*, E; from syn.); Basio 1971b: 52 (M*; bion.); Lee et al. 1988a: 109 (F; key, tax., bion., distr., review); Townsend 1990: 111 (type info.); Rattanarithikul et al. 2006a (F*, L*; bion., distr., keys); Jeffery et al. 2010 (distr.; Singapore). **Ety.:** n.s. [black mark (L)]. **Informal name:** Black-marked Malaysian Underwater Mosquito.

novochracea (Barraud)
1927d: 558 (M*; *Taeniorhynchus*). **Type loc.:** Nongpoh, Assam, India (NHMUK). **Distr.:** India, Indonesia, Thailand. **Refs.:** Barraud 1934: 121 (M*, F); Rattanarithikul et al. 2006a (F*, L*; bion., distr., keys). **Ety.:** near *ochracea*. **Informal name:** Yellowish Underwater Mosquito.

ochracea (Theobald)
1903a: 263 (F*; *Taeniorhynchus*). **Type loc.:** Kuala Lumpur [Selangor], Malaya [Malaysia] (NHMUK). **Distr.:** Bangladesh, Cambodia, India, Indonesia, Japan, Laos, Malaysia, Papua New Guinea, Philippines, South Korea, Thailand, Vietnam. **Refs.:** Barraud 1934: 122 (M*, F); Kano & Hayashi 1949: 23 (M*, F*, E*); LaCasse & Yamaguti 1950: 72 (M*, F*, P*, L*; bion., distr., tax.); Hara 1957: 50 (F*); Basio 1971b: 52 (M*; bion.); Baisas 1974: 137 (M, F, P, L; tax., bion., distr.; Philippines); Tanaka et al. 1975c: 215 (distr.); Tanaka et al. 1979: 119 (M*, F*, L*); Ahmed 1987 (distr.; Bangladesh); Lee et al. 1988a: 110 (F; key, tax., bion., distr., review); Rattanarithikul et al. 2006a (F*, L*; bion., distr., keys). **Ety.:** ocher-colored (L); ref. to abdomen, legs, and wings with "ochraceous yellow scales." **Informal name:** Ocher Malaysian Underwater Mosquito.

syn. *chrysogona* Knab
1909: 386 (M, F; *Mansonia*). **Type loc.:** Parang [Cotabato, Mindanao], Philippines (USNM). **Refs.:** Stone & Knight 1957c: 198 (type info., lectotype desig.).

syn. *shakujiiensis* Ogasawara
1939a: 238 (F*; *Culex*). **Type loc.:** Shakujii, Tokyo, Honshu, Japan (LU). **Refs.:** Kano & Hayashi 1949: 23 (syn.).

perturbans (Walker)
1856a: 428 (F; *Culex*). **Type loc.:** United States (NHMUK). **Distr.:** Canada, Mexico, United States (continental). **Refs.:** Dyar & Currie 1903 (L, E); Dyar & Barrett 1918 (L); Darsie 1951: 43 (P*); Yamaguti & LaCasse 1951b: 31 (M*, F*, L*); Asanuma & Nakagawa 1954a: 373 (P*); Carpenter &

LaCasse 1955: 109 (M*, F*, L*; keys); Belkin et al. 1966: 3 (type loc. info); Dodge 1966: 377 (first instar L*; key); Mattingly 1971g: 203 (E*); Lee & Craig 1983 (F*; cibarium); Darsie & Ward 2005 (F*, L*; keys, distr.); Becker et al. 2010: 400 (F*, L*; distr., keys); Harrison et al. 2016 (F*, L*; keys, distr.). **Ety.:** n.s. [disturb, confuse (L)]. **Informal name:** Agitated American Underwater Mosquito. (Vol. 1, p. 536)

syn. *testaceus* van der Wulp
1867: 128 (M*; *Culex*). **Type loc.:** Kumlien, Wisconsin, United States (LM). **Refs.:** Belkin 1968b: 11 (type info.).

syn. *ochropus* Dyar & Knab
1907b: 100 (F; *Culex*). **Type loc.:** Center Harbor, New Hampshire, United States (USNM). **Refs.:** Belkin et al. 1966: 7 (type loc. info.).

pseudoconopas (Theobald)
1910c: 443 (F*; *Chrysoconops*). **Type loc.:** Mpumu, Uganda (NHMUK). **Distr.:** Cameroon, Central African Republic, Côte d'Ivoire, Democratic Republic of the Congo, Liberia, Republic of the Congo, Sudan and South Sudan, Uganda. **Refs.:** Edwards 1941: 93 (M*, F; from syn. with *annettii*); Gillett 1946: 432 (P*, L*); Corbet 1958: 127 (F*). **Ety.:** false (Gr); similar to *conopas*. **Informal name:** False-gnat Ugandan Underwater Mosquito.

richiardii (Ficalbi)
1889b: 50 (M, F; *Culex*). **Type loc.:** Ravenna [Emilia], Italy (LU). **Distr.:** Afghanistan, Albania, Austria, Azerbaijan, Belarus, Belgium, Bulgaria, Crimean Peninsula, Croatia, Czech Republic, Denmark, Estonia, Finland, France (includes Corsica), FYRO Macedonia, Georgia, Germany, Greece, Hungary, Iran, Ireland, Italy (includes Sardinia and Sicily), Kazakhstan, Latvia, Lithuania, Luxembourg, Moldova, Montenegro, Morocco, Netherlands, Norway, Poland, Portugal, Romania, Russia, Serbia, Slovakia, Slovenia, Spain, Sweden, Switzerland, Syria, Tajikistan, Turkey, Ukraine, United Kingdom, Uzbekistan. **Refs.:** Edwards 1919 (P, L*); Séguy 1924: 51 (M gen.*, L*); Marshall 1938: 264 (M*, F*, P*, L*, E); Colas-Belcour 1943: 101 (E*); Natvig 1948: 184 (M*, F*, P*, L*; bion.); Mattingly 1971g: 202 (E*); Gutsevich et al. 1974: 169 (M*, F, L*); Utrio 1976: 130 (L*); Ramsdale & Snow 2001 (distr.; Europe); Trari et al. 2002: 331 (distr.; Morocco); Beck et al. 2003: 24 (distr.; Luxembourg); Rueda, Pecor et al. 2008: 397 (distr.); Osório et al. 2010 (distr.; Portugal); Becker et al. 2010: 308 (M*, F*, L*; keys, tax., distr., bion.); Namazov 2014 (distr.; Azerbaijan); Robert et al. 2019 (distr.; Western Palearctic). **Ety.:** Sebastiano Richiardi. **Informal name:** Richiardi Italian Underwater Mosquito.

syn. *nikolskyi* Shingarev
1927: 546 (F; *Taeniorhynchus*). **Type loc.:** Environs of Lake Nor-Zaisan, Siberia and Turkestan [USSR] (MTI). **Refs.:** Natvig 1948: 184 (syn.).

rochei (Doucet)
1951c: 109 (M*, F; *Taeniorhynchus nikolsky* [*sic*]). **Type loc.:** Vangaindrano, Madagascar (IRSM). **Distr.:** Madagascar (includes Glorioso and Juan De Nova Isl.). **Ety.:** n.k. **Informal name:** Roche Malagasy Underwater Mosquito.

samoaensis Stone
1966: 331 (M*, F). **Type loc.:** Tabalogi Ridge, Tau Island, Manua Group, Eastern Samoa (USNM). **Distr.:** Samoa (Ind. State of Samoa, American Samoa, Western Samoa). **Refs.:** Ramalingam 1976: 309 (distr.); Lee et al. 1988a: 112 (F; key, tax., bion., distr., review). **Ety.:** Samoa. **Informal name:** Tabalogi Samoan Underwater Mosquito.

saotomensis Lien, Lin, Lin, & Tseng
2008 (M*). **Type loc.:** Taiwan Medical Mission Compound in Agua Grande, São Tomé Island (EMIP). **Distr.:** São Tomé. **Ety.:** São Tomé. **Informal name:** São Tomé Island Underwater Mosquito.

schoutedeni (Wolfs)
1948a: 244 (F, P, L*; *Taeniorhynchus*). **Type loc.:** Ruzizi, Ruanda, Urundi, Belgian Congo [Democratic Republic of the Congo] (CMT). **Distr.:** Rwanda. **Ety.:** Dr. Schouteden. **Informal name:** Schouteden Congolese Underwater Mosquito.

vanoyei (Wolfs)
1948b: 83 (M*, F, P, L; *Taeniorhynchus*). **Type loc.:** Ruzizi, Ruanda, Urundi, Belgian Congo [Democratic Republic of the Congo] (CMT). **Distr.:** Rwanda. **Ety.:** Paul van Oye. **Informal name:** Van Oye Congolese Underwater Mosquito.

variegata (Dobrotworsky)
1962: 293 (M*, F*; *Mansonia*). **Type loc.:** Cann River, Victoria, Australia (NMM). **Distr.:** Australia. **Refs.:** Dobrotworsky 1965: 63 (M*, F*; assigned here?); Lee et al. 1988a: 113 (F; key, tax., bion., distr., review). **Ety.:** n.s. [variegated (L); poss. ref. to "proboscis more or less mottled"]. **Informal name:** Variegated Australian Underwater Mosquito.

versicolor (Edwards)
1913b: 50 (M, F; *Taeniorhynchus*). **Type loc.:** Nairobi [Central], Kenya (NHMUK). **Distr.:** Cameroon, Democratic Republic of the Congo, Ethiopia, Kenya, Liberia, Rwanda, Tanzania, Uganda. **Refs.:** Edwards 1941: 92 (M*, F); Gillett 1946: 431 (P*, L*). **Ety.:** n.s. [having colors that change (L)]. **Informal name:** Colorful Kenyan Underwater Mosquito.

voltaensis (Danilov)
1982: 26 (F*; *Mansonia*). **Type loc.:** Boromo, Upper Volta [Burkina Faso] (LU). **Distr.:** Burkina Faso. **Ety.:** Republic of Upper Volta. **Informal name:** Volta Burkinabe Underwater Mosquito.

wahlbergi (Edwards)
1936: 55 (M; *Taeniorhynchus*). **Type loc.:** "Caffraria," South Africa [Cape Province, South Africa] (SM). **Distr.:** Mozambique, South Africa. **Refs.:** Edwards 1941:

101 (M*); Worth & De Meillon 1960: 236 (distr.); Jupp 1996 (M*, F*; key). **Ety.:** Wahlberg (coll.?). **Informal name:** Wahlberg South African Underwater Mosquito.

xanthogaster (Edwards)
1924b: 366 (M, F; *Taeniorhynchus*). **Type loc.:** Burpengary, Queensland, Australia (NHMUK). **Distr.:** Australia, Indonesia, New Caledonia, Papua New Guinea, Vanuatu. **Refs.:** Taylor 1944: 124 (M*, P*, L*); Knight & Chamberlain 1948: 16 (P*); Perry 1949: 379 (P*, L*); Belkin 1962: 311 (M*, F, P*, L*); Lee et al. 1988a: 114 (F; key, tax., bion., distr., review); Becker et al. 2010: 367 (F*, L*; key, tax., distr., bion.). **Ety.:** yellow abdomen (Gr); ref. to "entirely yellow abdomen." **Informal name:** Yellow-bellied Australian Underwater Mosquito.

Subgenus *Rhynchotaenia* Brèthes

1910: 470 (as genus). Type species: *Taeniorhynchus fasciolatus* Lynch Arribálzaga (Edwards 1932a: 116).

Name derivation: rhyncho- = snout, muzzle, beak (Gr); -taenia = ribbon (Gr) [proboscis with a broad median band].

syn. *Pseudotaeniorhynchus* Theobald 1911a: 19 (as genus). Type species: *Taeniorhynchus fasciolatus* Lynch Arribálzaga (Brunetti 1914: 69).

Important references: Dyar 1928 (to subg. of genus *Mansonia* Blanchard); Edwards 1932a: 116 (to subg. of *Mansonia*); de Oliviera Castro & Bressanello 1952: 229 (keys, tax.); Lane 1953 (as subg. of *Mansonia*; distr., keys, tax.; Neotropical Region); Belkin 1962 (as subg. of *Mansonia*; tax.; South Pacific); Ronderos & Bachman 1963: 43 (to subg. of *Coquillettidia*); Cova García et al 1966 (keys, tax.; Venezuela); Delfinado 1966a (bion., distr., keys, tax; Philippines); Belkin 1968a (bion., distr., tax.; New Zealand); Belkin et al. 1970 (M*, F*, P*, L*; tax., bion.; Jamaica); Mattingly 1972b: 45 (E*); Tanaka et al 1979 (as subg. of *Mansonia*; bion., distr., keys, tax.; Japan, Korea); Lee et al. 1988a (bion., distr., keys, tax.; Australasia); Service 1990 (bion., distr., keys, tax.; Afrotropics); Rattanarithikul et al. 2006a (bion., keys; Thailand).

albicosta (Chagas)
1908 (F*; *Taeniorhynchus*). In Peryassú 1908: 220. **Type loc.:** Xerem, Rio de Janeiro, Brazil (IOC). **Distr.:** Argentina, Bolivia, Brazil, Colombia, Ecuador, French Guiana, Guyana, Paraguay, Peru, Suriname, Venezuela. **Refs.:** Shannon 1934: 108 (F*); Lane & Coutinho 1940: 590 (M*, F); Belkin et al. 1971: 16 (lectotype desig.); Panday 1975a, b (distr.; Suriname). **Ety.:** n.s. [white costa (L); prob. ref. to dense yellowish-white scales on the costal and subcostal veins]. **Informal name:** White-ribbed Brazilian Underwater Mosquito.

albifera (Prado)
1931a: 193 (M*, F; *Mansonia*). **Type loc.:** Rio Pinheiros, Butantan, São Paulo, Brazil (IB). **Distr.:** Brazil, Ecuador. **Refs.:** Lane & Coutinho 1940: 590 (M*, F; tax.). **Ety.:** n.k. [carrying white (L)]. **Informal name:** White-cloaked Brazilian Underwater Mosquito [*albifera* = "white bearing" = White-cloaked].

arribalzagae (Theobald)
1903a: 261 (F*; *Taeniorhynchus*). **Type loc.:** Pará, Brazil (NHMUK). **Distr.:** Bolivia, Brazil, Colombia, Costa Rica, Ecuador, French Guiana, Guyana, Nicaragua, Panama, Peru, Suriname, Venezuela. **Refs.:** Shannon 1931: 152 (M*); Lane 1953: 616 (M*, F); Stone 1957a: 334 (unjustified emend.); Belkin 1968b: 11 (lectotype desig.); Heinemann & Belkin 1977b: 452 (distr.). **Ety.:** Félix Lynch-Arribálzaga. **Informal name:** Arribálzaga Brazilian Underwater Mosquito.
syn. *coticula* Dyar & Knab
1907b: 101 (F; *Taeniorhynchus*). **Type loc.:** Bocas del Toro, Panama (USNM). **Distr.:** Panama. **Refs.:** Stone & Knight 1957c: 198 (type info.); Belkin et al. 1966: 49 (type loc. info.).

chrysonotum (Peryassú)
1922a: 117 (M, F; *Taeniorhynchus*). **Type loc.:** Valle do Rio Doce, Espírito Santo, Brazil (MNRT, IOC). **Distr.:** Argentina, Bolivia, Brazil, Nicaragua. **Refs.:** Davis 1933a: 280 (M*, F); Davis 1933b: 623 (L*); Lane & Coutinho 1940: 590 (M*, F; tax.); Belkin et al. 1971: 16 (type info.). **Ety.:** marked gold (Gr, L); ref. to three gold stripes on scutum. **Informal name:** Gold-spotted Brazilian Underwater Mosquito.

fasciolata (Lynch Arribálzaga)
1891b: 150 (F; *Taeniorhynchus*). **Type loc.:** Parque Lezama, Capital Federal, Buenos Aires, Argentina (INM). **Distr.:** Argentina, Belize, Brazil, Colombia, Costa Rica, Ecuador, El Salvador, French Guiana, Guyana, Honduras, Lesser Antilles (includes Trinidad and Tobago), Nicaragua, Panama, Peru, Suriname, Uruguay, Venezuela. **Refs.:** Castro & Bressanello 1952: 230 (neotype F); Belkin et al. 1968: 12 (type info.). **Ety.:** n.s. [little bandage (L)]. **Informal name:** Bandaged Argentinian Underwater Mosquito.

hermanoi (Lane & Coutinho)
1940: 593 (F; *Mansonia*). **Type loc.:** Acampamento dos Morros Azues, Mato Grosso, Brazil (FH). **Distr.:** Argentina, Bolivia, Brazil, Colombia, Peru. **Refs.:** Barreto-Reyes 1957 (F); Morales-Ayala 1971: 140 (distr.). **Ety.:** Hermano Ribeira da Silva. **Informal name:** Hermano Brazilian Underwater Mosquito.

juxtamansonia (Chagas)
1907: 23 (F*; *Taeniorhynchus*, as *justamansonia*). **Type loc.:** Juiz de Fora, Minas Gerais, Brazil (IOC). **Distr.:** Argentina, Bolivia, Brazil, Colombia, Ecuador, Paraguay, Peru, Venezuela. **Refs.:** Davis 1933a: 281 (M*); Davis

1933b: 620 (M*, P*, L*, E*); Lane 1953: 613 (M*, F, P*, L*); Belkin et al. 1971: 16 (lectotype desig., type loc. info.); de Mello et al. 2014 (E*); Berti et al. 2015 (distr.; Venezuela). **Ety.:** n.s. [near Mansonia (L)]. **Informal name:** Gerais Brazilian Underwater Mosquito.

syn. *hypocrindyna* Dyar
1918c: 115 (F; *Mansonia*). **Type loc.:** São Paulo, Brazil (USNM).

lynchi (Shannon)
1931: 152 (M*; *Mansonia*). **Type loc.:** Pará, Brazil (USNM). **Distr.:** Bolivia, Brazil, Colombia, Ecuador, French Guiana, Guyana, Nicaragua, Panama, Peru. **Refs.:** Lane 1953: 617 (M*, F). **Ety.:** Félix Lynch-Arribálzaga. **Informal name:** Lynch Brazilian Underwater Mosquito.

neivai (Lane & Coutinho)
1940: 592 (M*; *Mansonia*). **Type loc.:** Juquia, São Paulo, Brazil (FH). **Distr.:** Brazil. **Refs.:** Lane 1951 (to syn. with *nigricans*); Belkin et al. 1970: 103 (from syn.). **Ety.:** Arthur Neiva. **Informal name:** Neiva Brazilian Underwater Mosquito.

nigricans (Coquillett)
1904c: 166 (F; *Taeniorhynchus*). **Type loc.:** Panama (USNM). **Distr.:** Argentina, Belize, Bolivia, Brazil, Colombia, Costa Rica, Cuba, Ecuador, El Salvador, French Guiana, Guatemala, Honduras, Jamaica, Lesser Antilles (includes Trinidad and Tobago), Mexico, Nicaragua, Panama, Paraguay, Peru, Venezuela. **Refs.:** Lane 1953: 615 (M*, F); Stone & Knight 1957c: 198 (type info., lectotype desig.); Belkin et al. 1970: 100 (M*, F, P*, L*; distr.); Heinemann & Belkin 1977b: 452 (distr.); Berti et al. 2015 (distr.; Venezuela). **Ety.:** black, dark colored (L); ref. to predominant color. **Informal name:** Black Panamanian Underwater Mosquito.

syn. *persephassa* Dyar & Knab
1909d: 254 (F; *Bancroftia*). **Type loc.:** San Antonio de Los Baños [Havana], Cuba (USNM).

nitens (Cerqueira)
1943a: 2 (F; *Taeniorhynchus*). **Type loc.:** Puerto Suárez, Chiquitos, Santa Cruz, Bolivia (IOC). **Distr.:** Bolivia. **Ety.:** n.s. [shine, look bright (L); ref. to metallic coloration, esp. proboscis and legs]. **Informal name:** Shiny Bolivian Underwater Mosquito.

shannoni (Lane & Antunes)
1937: 227 (F; *Mansonia*). **Type loc.:** Cuiabá, Mato Grosso, Brazil (FH). **Distr.:** Argentina, Bolivia, Brazil, El Salvador, Paraguay, Venezuela. **Refs.:** Coutinho 1942: 210 (M*). **Ety.:** Raymond Corbett Shannon. **Informal name:** Shannon Brazilian Underwater Mosquito.

venezuelensis (Theobald)
1912 (F; *Pseudotaeniorhynchus*). In Surcouf 1912: 61. **Type loc.:** Cano de la Viuda, Venezuela (MNHP). **Distr.:** Argentina, Belize, Bolivia, Brazil, Colombia, Costa Rica, Ecuador, El Salvador, French Guiana, Guatemala, Guyana, Honduras, Lesser Antilles (includes Trinidad and Tobago), Mexico, Nicaragua, Panama, Paraguay, Peru, Suriname, Uruguay, Venezuela. **Refs.:** Davis 1933b: 623 (M*, L*; as *fascioliata*); Shannon 1934: 108 (F*; as *fascioliata*); Castro & Bressanello 1952: 235 (tax.); Belkin 1968b: 11 (type info.); Berti et al. 2015 (distr.; Venezuela). **Ety.:** Venezuela. **Informal name:** Mariche Venezuelan Underwater Mosquito [the Mariche (now a lost people) were an indigenous tribe of what is now Venezuela].

syn. *araozi* Shannon & Del Ponte
1928 (F; *Mansonia* (*Rynchotaenia*)). In Dyar 1928: 257. **Type loc.:** Ledesma, Jujuy, Argentina (USNM). **Refs.:** Belkin et al. 1968: 17 (type info.).

Genus *Mansonia* Blanchard

(Vol. 1, p. 258)

Important references: Gillett 1949 (key; Ethiopian Region); Lane 1953 (as subg. *Rhynchotaenia* of *Coquillettidia*; Neotropical Region); Belkin 1962 (tax.; South Pacific); Wharton 1962 (Malaysia); Ronderos & Bachmann 1963a: 46 (generic gp. revision; Neotropical Region); Forattini 1965b (Neotropical Region); Cova García et al. 1966 (Venezuela); Delfinado 1966a (Philippines); Belkin et al. 1970 (Jamaica); Tanaka et al. 1979 (Japan, Korea); Danilov 1983a (tax., key); Darsie 1985 (keys; Argentina); Lee et al. 1988b (Australasian Region); Service 1990 (Afrotropical Region); Rattanarithikul & Panthusiri 1994 (keys; Thailand); Rattanarithikul, Harrison et al. 2005 (F, L; tax., keys, bion.; Thailand).

Subgenus *Mansonia* Blanchard

Panoplites Theobald 1900a: 261 (as genus). (Theobald 1900a: 5 (not Gould 1853)). Type species: *Culex titillans* Walker 1848 (Blanchard 1901: 1046).
Mansonia Blanchard 1901: 1046 (as genus and subg.; new name for *Panoplites* Theobald). Type species: *Culex titillans* Walker 1848. On Official List of Generic Names in Zoology, International Commission on Zoological Nomenclature 1959 (Opinion 550 protects generic name *Mansonia* Blanchard).

Name derivation: Patrick Manson.

Important references: Lane 1953 (tax., keys; Neotropics); Ronderos & Bachmann 1963b (descr., keys; Neotropics); Forattini 1965b (tax., keys, bion., distr.; Neotropics); Cova García et al. 1966 (distr., keys; Venezuela); Belkin et al. 1970 (bion. distr., keys, tax.; Jamaica); Mattingly 1972b: 46 (E*); Clark-Gil & Darsie 1983 (tax., keys, bion.; Guatemala); Darsie 1985 (keys; Argentina).

amazonensis (Theobald)
1901c: 182 (F; *Panoplites*). **Type loc.:** Gurupá, Pará, Brazil (NHMUK). **Distr.:** Bolivia, Brazil, Colombia, Peru, Venezuela. **Refs.:** da Costa Lima 1929b (M*; tax.); Barretto & Coutinho 1944: 61 (M*, F); Pratt 1953: 15 (F*); Belkin et al. 1971: 16 (type loc. info.). **Ety.:** Amazon. **Informal name:** Amazon Brazilian Submarine Mosquito.

cerqueirai (Barretto & Coutinho)
1944: 73 (M*; *Taeniorhynchus*). **Type loc.:** Maracajú, Mato Grosso, Brazil (INER). **Distr.:** Brazil. **Refs.:** Xavier 1973: 160 (type info.). **Ety.:** Nelson Leander Cerqueira. **Informal name:** Cerqueira Brazilian Submarine Mosquito.

chagasi (da Costa Lima)
1935a: 464 (M*; *Taeniorhynchus*). **Type loc.:** Bicudos, Minas Gerais, Brazil (IOC). **Distr.:** Brazil. **Refs.:** Belkin et al. 1971: 17 (type info.). **Ety.:** Evandro Chagas. **Informal name:** Chagas Brazilian Submarine Mosquito.

dyari Belkin, Heinemann & Page
1970: 109 (M*, F*, P*, L*). **Type loc.:** [Fort Simonds], Clarendon, Jamaica (USNM). **Distr.:** Belize, Colombia, Costa Rica, Dominican Republic, Ecuador, El Salvador, Guatemala, Honduras, Jamaica, Mexico, Nicaragua, Panama, Puerto Rico, United States (continental). **Refs.:** Heinemann & Belkin 1977b: 415, 433, 443 (distr.); Linley 1989c E*); Pena & Chadee 2004 (distr.; Dominican Republic); Darsie & Ward 2005 (F*, L*; keys, distr.); Harrison et al. 2016 (F*, L*; keys, distr., tax.). **Ety.:** Harrison G. Dyar. **Informal name:** Dyar Jamaican Submarine Mosquito.

flaveola (Coquillett)
1906a: 182 (M; *Taeniorhynchus*). **Type loc.:** Saint Thomas Island, Lesser Antilles (USNM). **Distr.:** Argentina, Bolivia, Brazil, Dominican Republic, Ecuador, Jamaica, Lesser Antilles (includes Trinidad and Tobago, Virgin Isl.), Nicaragua, Panama, Peru, Puerto Rico, Suriname, Venezuela. **Refs.:** Shannon 1934: 107 (from syn. with *titillans*); Barretto & Coutinho 1944: 62 (M*, F); Pratt 1945: 122 (M*, F*, P*, L*); Belkin et al. 1970: 107 (M*, F*, P*, L*; tax.); Mitchell & Darsie 1985: 317 (distr., tax.); Pena & Chadee 2004 (distr.; Dominican Republic). **Ety.:** n.s. [yellow (L)]. **Informal name:** Yellow Lesser Antillean Submarine Mosquito.

fonsecai (Pinto)
1932: 289 (P*; *Taeniorhynchus*). **Type loc.:** Los Naranjos, Bolivia (IOC). **Distr.:** Bolivia, Brazil. **Refs.:** da Costa Lima 1935a: 467 (syn.); Barbosa et al. 2005 (M*, F*, P*, L*; Brazil). **Ety.:** O. da Fonseca. **Informal name:** Fonseca Bolivian Submarine Mosquito.

humeralis Dyar & Knab
1916: 65 (F, E*). **Type loc.:** Georgetown [Demerara], British Guiana [Guyana] (USNM). **Distr.:** Argentina, Bolivia, Brazil, Colombia, French Guiana, Guyana, Lesser Antilles (includes Trinidad and Tobago, Virgin Isl.), Panama, Paraguay, Peru, Suriname, Venezuela. **Refs.:** Barretto & Coutinho 1944: 64 (M*, F); Pratt 1953: 14 (F*); Stone & Knight 1957c: 198 (type info.). **Ety.:** n.s. [robe, cape (L); poss. ref. to "three large patches of golden scales upon the mesonotum"]. **Informal name:** Robed Guyanese Submarine Mosquito.

iguassuensis Barbosa, Navarro da Silva, & Sallum
2007: 46 (M*, F*, P*, L*). **Type loc.:** Curitiba (25°25′04″ S, 49°14′30″ W), Paraná, Brazil. **Distr.:** Brazil. **Ety.:** Parque Regional do Iguaçu. **Informal name:** Iguaçu Brazilian Submarine Mosquito.

indubitans Dyar & Shannon
1925b: 41 (F). **Type loc.:** Belém, Pará, Brazil (USNM). **Distr.:** Argentina, Belize, Bolivia, Brazil, Colombia, Cuba, Ecuador, French Guiana, Guatemala, Haiti, Jamaica, Lesser Antilles (includes Trinidad and Tobago), Mexico, Panama, Peru, Puerto Rico, Suriname, Uruguay. **Refs.:** da Costa Lima 1935a: 467 (M*, F*); Yamaguti & LaCasse 1951b: 22 (M*, F*, L*); Kutz & Darsie 1963 (P*); Dodge 1966: 376 (first instar L*; key); Belkin et al. 1971: 17 (type loc. info.); Mitchell & Darsie 1985: 317 (distr.). **Ety.:** n.s. [having misgivings (L)]. **Informal name:** Doubtful Brazilian Submarine Mosquito.

leberi Boreham
1970: 384 (M*, F*, P*, L*, E*). **Type loc.:** Chilibre River, approximately 350 m from its confluence with Chagres River, Canal Zone, Panama (USNM). **Distr.:** Costa Rica, Ecuador, Guatemala, Nicaragua, Panama. **Ety.:** Walter P. Leber. **Informal name:** Leber Panamanian Submarine Mosquito.

pessoai (Barreto & Coutinho)
1944: 69 (M*; *Taeniorhynchus*). **Type loc.:** Curitiba, Paraná, Brazil (FH). **Distr.:** Brazil. **Refs.:** Belkin et al. 1971: 17 (type info.). **Ety.:** Samuel B. Pessoa. **Informal name:** Pessoa Brazilian Submarine Mosquito.

pseudotitillans (Theobald)
1901c: 178 (F; *Panoplites*). **Type loc.:** Lower Amazon, Brazil (NHMUK). **Distr.:** Argentina, Bolivia, Brazil, Cayman Isl., Colombia, Ecuador, French Guiana, Guyana, Haiti, Lesser Antilles (includes Trinidad and Tobago), Panama, Peru, Suriname, Venezuela. **Refs.:** da Costa Lima 1929b (M*; tax.); Barreto & Coutinho 1944: 59 (M*, F); Pratt 1953: 15 (F*); Ronderos & Bachmann 1963b: 61 (M*, F*, P*, L*); Belkin 1968b: 23 (lectotype desig.); Mitchell & Darsie 1985: 318 (distr.); Berti et al. 2015 (distr.; Venezuela). **Ety.:** false (Gr); similar to *titillans*. **Informal name:** Insensitive Brazilian Submarine Mosquito [*pseudotitillans* = "false tickler" = Insensitive].

suarezi Cova García & Sutil Oramas
1976c: 360 (M*). **Type loc.:** La Juanita, Municipality of Miranda, State of Carabobo, Venezuela (ND). **Distr.:** Venezuela. **Ety.:** Miguel A. Suarez F. **Informal name:** Suarez Venezuelan Submarine Mosquito.

titillans (Walker)
1848: 5 (F; *Culex*). **Type loc.:** Belém, Pará, Brazil (NHMUK). **Distr.:** Antigua and Barbuda, Argentina, Belize, Bolivia, Brazil, Colombia, Costa Rica, Cuba, Dominican Republic, Ecuador, El Salvador, French Guiana, Guatemala, Guyana, Haiti, Honduras, Jamaica, Lesser Antilles (includes Guadeloupe, Saint Kitts and Nevis; Saint Lucia; St Vincent and the Grenadines, Trinidad and Tobago), Mexico, Nicaragua, Panama, Paraguay, Peru, Puerto Rico, Romania, Suriname, United States (continental), Uruguay, Venezuela. **Refs.:** da Costa Lima 1929b (M*; tax.); Lane 1939: 89 (distr.); Peréz Vigueras 1948: 298 (M*, F*, P*, L*); Yamaguti & LaCasse 1951b: 27 (M*, F*, L*); Pratt 1953: 15 (F*, P*); Carpenter & LaCasse 1955: 107 (M*, F*, L*; keys); Kutz & Darsie 1963 (P*); Ronderos & Bachmann 1963b: 61 (M*, P*, L*, E*); Belkin et al. 1970: 103 (M*, F*, P*, L*); Belkin et al. 1971: 17 (type loc. info.); Heinemann & Belkin 1977c: 529 (distr.); Linley 1989c (E*); Rossi & Martínez 2003: 472 (distr.; Uruguay); Darsie & Ward 2005 (F*, L*; keys, distr.); Becker et al. 2010: 385 (F*, L*; key, tax., distr., bion.); Berti et al. 2015 (distr.; Venezuela); Harrison et al. 2016 (F*, L*; keys, distr., tax.). **Ety.:** n.s. [tipulid-like; "Several species of tipulidae [*sic*] closely resemble it on the wing"]. **Informal name:** Ticklish Brazilian Submarine Mosquito. (Vol. 1, p. 538)

wilsoni (Barretto & Coutinho)
1944: 71 (M*, F; *Taeniorhynchus*). **Type loc.:** São Paulo, Brazil (LU). **Distr.:** Brazil, Colombia, Paraguay. **Refs.:** Barretto & Lee 1969: 106 (distr.); Belkin et al. 1971: 17 (type info.). **Ety.:** n.s. **Informal name:** Wilson Brazilian Submarine Mosquito.

Subgenus *Mansonioides* Theobald

1907: 498 (as genus). Type species: *Mansonioides septemguttata* Theobald 1907.

Name derivation: similar to *Mansonia* (Gr).

Important references: Gillett 1949 (key; Ethiopian Region); Carter 1950 (key; Sri Lanka); Wharton 1962 (Malaysia); Belkin 1962 (tax., bion; South Pacific); Ronderos & Bachmann 1963a, b (tax., to subg.); Delfinado 1966a (tax., keys; Philippines); Mattingly 1972b (E*); Tanaka et al. 1979 (tax., keys, bion.; Japan, Korea); Lee et al. 1988b (tax., key, bion.; Australasian Region); Service 1990 (tax., keys, bion.; Afrotropical Region); Rattanarithikul & Panthusiri 1994 (keys; Thailand); Rattanarithikul et al. 2006a (tax., keys, bion.; Thailand).

africana (Theobald)
1901d: 12 (F*; *Panoplites*). **Type loc.:** Asaba [Eastern Provinces], Nigeria (NHMUK). **Distr.:** Angola, Benin, Botswana, Burkina Faso, Cameroon, Central African Republic, Côte d'Ivoire, Democratic Republic of the Congo, Equatorial Guinea, Ethiopia, Gabon, Ghana, Kenya, Liberia, Mali, Mozambique, Nigeria, Senegal, Sierra Leone, South Africa, Sudan and South Sudan, Tanzania, The Gambia, Uganda, Zambia. **Refs.:** Edwards 1941: 103 (M, F*), 384 (P*); Hopkins 1952: 110 (L*); Laurence 1960: 491 (bion.); Mattingly 1972a (E*; bion.); da Cunha Ramos & Ribeiro 1975: 125 (tax., bion., keys, distr.; Angola); Lee et al. 1988b: 117 (tax., distr., review); Jupp 1996 (M*, F*; key); Toto et al. 2003: 344

(distr.; Equatorial Guinea); Becker et al. 2010: 330 (F*, L*; key, tax., distr., bion.). **Ety.:** Africa. **Informal name:** Continental Nigerian Submarine Mosquito [Africa is a continent = Continental].

syn. *major* Theobald
1903a: 270 (F*). **Type loc.:** Bahr-el-Ghazal [Sudan] (NHMUK). **Refs.:** Edwards 1932a: 120 (syn.).

ssp. ***nigerrima*** Theobald
1910c: 450 (F; as sp.). **Type loc.:** Mpumu, Uganda (NHMUK). **Distr.:** Senegal, Tanzania, Uganda. **Refs.:** Edwards: 1941: 105 (F); Harbach & Howard 2007: 42 (to ssp.). **Ety.:** dark, black (L); "Very much like *M. nigra*, Theobald, but with black forked scales." **Informal name:** Dark Ugandan Submarine Mosquito.

annulata Leicester
1908: 174 (F). **Type loc.:** Kuala Lumpur [Selangor], Malaya [Malaysia] (NHMUK). **Distr.:** Bangladesh, Indonesia, Malaysia, Philippines, Thailand, Vietnam. **Refs.:** Bonne-Wepster 1954: 46 (M, F*); Wharton 1962: 15 (M*, F*, P*, L*, E); Basio 1971b: 75 (M*; bion.); Iwaki & Choochote 1991 (E*); Rattanarithikul et al. 2006a (F*, L*; bion., distr., keys); Rueanghiran et al. 2011 (mol. tax.). **Ety.:** ringed (L); legs, proboscis. **Informal name:** Ringed Malaysian Submarine Mosquito.

annulifera (Theobald)
1901c: 183 (F*; *Panoplites*). **Type loc.:** Madras and Quilon, Travancore, India (NHMUK). **Distr.:** Australia, Bangladesh, Cambodia, India, Indonesia, Laos, Malaysia, Myanmar, Nepal, Papua New Guinea, Philippines, Sri Lanka, Thailand, Vietnam. **Refs.:** Barraud 1934: 127 (M*, F*, P*, L*); Menon 1940: 181 (L*); Knight et al. 1944: 64 (distr.); Carter 1950: 23 (M*, F*, L*; bion.); Delfinado 1966a: 23 (M*, F*); Basio 1971b: 76 (M*; bion.); Mattingly 1972a (E*; bion.); Baisas 1974: 138 (M, F, P, L; tax., bion., distr.; Philippines); Lien et al. 1977 (distr.); Ahmed 1987 (distr.; Bangladesh); Lee et al. 1988b: 119 (F; key, tax., bion., distr., review); Darsie & Pradhan 1990 (F, L; tax., keys, bion., distr.; Nepal); Iwaki & Choochote 1991 (E*); Rattanarithikul et al. 2006a (F*, L*; bion., distr., keys). **Ety.:** ring bearing (L); legs with many white bands. **Informal name:** Ring-bearing Indian Submarine Mosquito.

syn. *seguini* Laveran
1901b: 992 (F; *Panoplites*). **Type loc.:** Hanoi, Tonkin, Indochina [Vietnam] (PIP).

syn. *septemguttata* Theobald
1907: 499 (F*; *Mansonioides*). **Type loc.:** Sarawak, Borneo [Malaysian Borneo] (NHMUK).

bonneae Edwards
1930b: 542 (M). **Type loc.:** Kuala Lumpur [Selangor], Malaya [Malaysia] (NHMUK). **Distr.:** Cambodia, Indonesia, Malaysia, Philippines, Singapore, Thailand. **Refs.:** Stone 1957a: 334 (M*; tax.); Wharton 1962: 17 (M*, F*, P*, L*, E); Basio 1971b: 76 (M*; bion.); Lien et al. 1977 (distr.); Iwaki & Choochote 1991 (E*); Rattanarithikul et al. 2006a (F*, L*; bion., distr., keys); Jeffery et al. 2010 (distr.; Singapore). **Ety.:** Johanna Bonne-Wepster. **Informal name:** Bonne Malaysian Submarine Mosquito.

dives (Schiner)
1868: 31 (*Culex*; new name for *annulipes* Walker, not Meigen 1830b). **Distr.:** Bangladesh, Cambodia, India, Indonesia, Malaysia, People's Republic of China, Philippines, Singapore, Thailand, Vietnam. **Refs.:** Bonne-Wepster 1939 (L; distr.; Borneo; as *longipalpis*); Bonne-Wepster 1954: 50 (M, F*, L*; as *longipalpis*); Stone 1957a: 336 (tax.); Wharton 1962: 20 (M*, F*, P*, L*, E); Basio 1971b: 77 (M*; bion.); Ahmed 1987 (distr.; Bangladesh); Lee et al. 1988b: 123 (F; key, tax., bion., distr., review); Iwaki & Choochote 1991 (E*); Rattanarithikul et al. 2006a (F*, L*; bion., distr., keys). **Ety.:** n.s. [rich (L)]. **Informal name:** Rich Singaporean Submarine Mosquito.

syn. *annulipes* Walker
1856b: 6 (F; *Culex*). **Type loc.:** Singapore (NHMUK).

syn. *longipalpis* van der Wulp
1881: 9 (F*; *Culex*). **Type loc.:** Alahan Panjang, near Soeroelangoen [Benkoelen], Sumatra [Indonesia] (LM). **Refs.:** Stone 1957a: 336 (syn.).

indiana Edwards
1930b: 541 (M, F). **Type loc.:** Bandjaran, South Preanger, Java [Indonesia] (NHMUK). **Distr.:** Bangladesh, Cambodia, India, Indonesia, Laos, Malaysia, Myanmar, Nepal, Papua New Guinea, Philippines, Singapore, Sri Lanka, Thailand, Vietnam. **Refs.:** Barraud 1934: 130 (M*, F*, P*, L*); Carter 1950: 21 (M*, F*, L); Joshi et al. 1965: 139 (distr.); Mattingly 1972a (E*; bion.); Lee et al. 1988b: 128 (tax., distr., review); Eng et al. 1990 (E*); Darsie & Pradhan 1990 (F, L; tax., keys, bion., distr.; Nepal); Iwaki & Choochote 1991 (E*); Rattanarithikul et al. 2006a (F*, L*; bion., distr., keys). **Ety.:** India. **Informal name:** Yawadvipa Indonesian Submarine Mosquito [Yawadvipa (circa 200 B.C.E.) was a Hindu kingdom in modern Indonesia]. (Vol. 1, p. 540)

melanesiensis Belkin
1962: 314 (M*, F*, P*, L*). **Type loc.:** Doma Cove areas, Guadalcanal, Solomon Islands (USNM). **Distr.:** Australia, Papua New Guinea, Solomon Isl. **Refs.:** Lee et al. 1988b: 130 (F; key, tax., bion., distr., review). **Ety.:** Melanesia. **Informal name:** Melanesia Solomon Islander Submarine Mosquito.

papuensis (Taylor)
1914d: 200 (F; *Taeniorhynchus*). **Type loc.:** Lakekamu Gold Field, Papua New Guinea (ANIC, NHMUK). **Distr.:** Australia, Indonesia, Papua New Guinea. **Refs.:** Bonne-Wepster 1954: 52 (F); Peters 1958: 89 (M*); Lee et al. 1988b: 133 (F; key, type info., tax., bion., distr., review); Townsend 1990: 118 (type info.); Peyton et al. 1999: 546 (distr.). **Ety.:** Papua New Guinea. **Informal name:** Lakekamu Papuan Submarine Mosquito.

septempunctata Theobald
1905f: 107 (F). **Type loc.:** Friedrich-Wilhelmshafen [NE New Guinea] (HNM). **Distr.:** Australia, Indonesia, Papua New Guinea. **Refs.:** Knight et al. 1944: 23 (F; tax.); Lee et al. 1988b: 137 (F; key, tax., bion., distr., review). **Ety.:** seven punctuated (L); ref. to thorax "with seven pale puncta." **Informal name:** Seven-spotted Papuan Submarine Mosquito.

syn. *bonnewepsterae* van den Assem
1958b: 205 (M*, F, P, L*). **Type loc.:** Lake Sentani, North New Guinea (LM). **Refs.:** Lee et al. 1988b: 137 (syn.).

uniformis (Theobald)
1901c: 180 (F; *Panoplites*). **Type loc.:** Quilon, Travancore, India (NHMUK). **Distr.:** Angola, Australia, Bangladesh, Benin, Botswana, Burkina Faso, Cambodia, Central African Republic, Comoros, Côte d'Ivoire, Ethiopia, Gabon, Ghana, Guam, Hong Kong, India, Indonesia, Japan, Kenya, Liberia, Madagascar (includes Glorioso and Juan De Nova Isl.), Malawi, Malaysia, Mali, Mozambique, Myanmar, Nepal, Niger, Nigeria, Pakistan, Papua New Guinea, People's Republic of China, Philippines, Senegal, Sierra Leone, Solomon Isl., Somalia, South Africa, South Korea, Sri Lanka, Sudan and South Sudan, Taiwan, Tanzania, Thailand, The Gambia, Timor, Uganda, Vietnam, Zambia. **Refs.:** Bancroft 1908: 52 (F); Bohart & Ingram 1946b: 61 (M*, F, L*; bion., distr., tax.); Knight & Chamberlain 1948: 16 (P*); Carter 1950: 14 (M*, F*, P*, L*, E*; bion.); LaCasse & Yamaguti 1950: 66 (M*, F*, P*, L*; bion., distr., tax.); Hara 1957: 50 (F*); Laurence 1960: 491 (bion.); Wharton 1962: 24 (M*, F*, P*, L*); Dobrotworsky 1965: 66 (M*, F*, L*); Basio 1971b: 77 (M*; bion.); Mattingly 1971a: Pl. 4 (A wing*), Pl. 7 (M head*), Pl. 8 (P trumpet*), Pl. 26 (L terminal segs.*); Mattingly 1972a (E*; bion.); Baisas 1974: 139 (M, F, P, L; tax., bion., distr.; Philippines); da Cunha Ramos & Ribeiro 1975: 125 (tax., bion., keys, distr.; Angola); Tanaka et al. 1975c: 214 (distr.); Brunhes 1977a (distr.; Comoros Isl.); Tanaka et al. 1979: 116 (M*, F*, L*); Ahmed 1987 (distr.; Bangladesh); Lee et al. 1988b: 145 (F; key, tax., bion., distr., review); Darsie & Pradhan 1990 (F, L; tax., keys, bion., distr.; Nepal); Kohn 1990: 165 (distr.); Iwaki & Choochote 1991 (E*); Suleman et al. 1993 (distr.; Pakistan); Jupp 1996 (M*, F*; key); Whelan & Hapgood 2000 (bion., distr.; East Timor); Rattanarithikul et al. 2006a (F*, L*; bion., distr., keys); Becker et al. 2010: 330 (F*, L*; key, tax., distr., bion.). **Ety.:** uniform appearance (L); ref. to more uniform appearance compared to related species—i.e., no white thoracic spots. **Informal name:** Uniform Indian Submarine Mosquito. (Vol. 1, p. 542)

syn. *reversus* Theobald
1901c: 189 (F; *Panoplites africanus* var.). **Type loc.:** Zomba, Nyasaland [Malawi] (NHMUK).

syn. *australiensis* Giles
1902: 355 (F; *Panoplites*). **Type loc.:** Burpengary, Queensland, Australia (NHMUK).

syn. *marquesensis* Dyar
1925e: 43 (M*). **Type loc.:** Lourenço Marques, Mozambique (USNM).

Tribe Orthopodomyiini

Genus *Orthopodomyia* Theobald

(Vol. 1, p. 260)

1904c: 236 (as genus). Type species: *Orthopodomyia albipes* Leicester.

Name derivation: straight-footed fly (Gr).

syn. *Bancroftia* Lutz 1904. In Bourroul 1904: 40, 59. Type species: *Bancroftia albipes* Lutz. Edwards 1913c: 239 (syn.).
syn. *Pneumaculex* Dyar 1905a: 46 (as genus). Type species: *Culex signifer* Coquillett. Edwards 1913c: 239 (syn.).
syn. *Newsteadina* Theobald 1909a: 297 (as genus). Type species: *Culex arboricollis* Charmoy. Edwards 1920b: 135 (syn.).
syn. *Thomasina* Newstead & Carter 1911a: 553. Type species: *Mansonia longipalpis* Newstead & Carter. Howard et al. 1917: 878 (syn.).

Important references: Barraud 1934 (tax., bion.; southern Asia); Knight & Mattingly 1950: 1 (tax., keys; *Anopheloides* subg); Lane 1953 (tax., keys, bion.; Neotropics); Thurman 1959b (tax., bion., review; Thailand); Delfinado 1966a (tax., key; bion.; Philippines); Zavortink 1968 (world revision); Belkin et al. 1970 (tax., bion.; Jamaica); Mattingly 1971c: 66 (E); Zavortink 1971 (review, tax., bion., keys; SE Asia); Tanaka et al. 1979 (tax., bion.; Japan); Lee et al. 1988c (tax., bion., key; Australasia); Service 1990 (tax., keys; bion.; Afrotropics); Brunhes & Hervy 1995 (tax., keys, bion.; Afrotropics); Harbach & Kitching 1998 (phyl.); Darsie & Ward 2005 (F, L; keys; North America); Rattanarithikul, Harrison et al. 2005 (F, L; tax., keys, bion.; Thailand); Rattanarithikul et al. 2007 (F, L; keys, bion.; Thailand); Byrd et al. 2012 (mol. phyl.; Signifera Group; North America).

alba Baker
1936: 1 (M, F, P*, L*). **Type loc.:** Ithaca, New York, United States (USNM). **Distr.:** Canada, United States (continental). **Refs.:** Darsie 1951: 41 (P*); Wilkins & Breland 1951: 225 (L*; bion.); Yamaguti & LaCasse 1951b: 8 (M*, F*, L*); Carpenter & LaCasse 1955: 98 (M, F, L*; keys); Breland 1959: 137 (first instar L*); Dodge 1966: 378 (first instar L*; key); Smith & Trimble 1973 (distr.; Canada); Darsie & Ward 2005 (F*, L*; keys, distr.); Harrison et al. 2016 (F*, L*; keys, distr., tax.). **Ety.:** n.s. [white (L); prob. ref. to extensive white markings]. **Informal name:** White American Red Mosquito.

albicosta (Lutz)
1904 (M, F; *Bancroftia*). In Bourroul 1904: 40, 59. **Type loc.:** Serra da Cantareira, São Paulo, Brazil (IOC). **Distr.:** Brazil, Colombia, Venezuela. **Refs.:** Dyar 1928: 398 (M*, F, L); Edwards 1939b: 122 (M, F*); Lane 1953: 624 (M*, F, P*); Stone 1957a: 334 (tax.); Belkin 1968b: 23 (tax.); Zavortink 1968: 66 (M*, F*, P*, L*); Belkin et al. 1971: 18 (lectotype desig.). **Ety.:** n.s [white costa (L)]. **Informal name:** White-edged Brazilian Red Mosquito.

albipes Leicester
1904 (M, F). In Theobald 1904c: 237 (Sept.). **Type loc.:** Kuala Lumpur, Malaya [Malaysia] (NHMUK). **Distr.:** India, Malaysia, Philippines, Singapore, Thailand, Vietnam. **Refs.:** Knight & Mattingly 1950: 16 (M*, F, P*, L*); Mattingly 1971c: 66 (E); Zavortink 1971: 12 (M*, F*, P*, L*; tax., keys); Rattanarithikul et al. 2007 (F*, P*, L*; bion., distr., keys). **Ety.:** n.s. [white foot (L); prob. ref. to "last three tarsal joints creamy white"]. **Informal name:** White-shod Malaysian Red Mosquito.

ambremontis Brunhes & Hervy
1995: 21 (M* F, P*, L*). **Type loc.:** Montagne d'Ambre, Province de Diego Suarez (Antsiranana), Madagascar (IRD). **Distr.:** Madagascar (includes Glorioso and Juan De Nova Isl.). **Informal name:** Mt. Ambre Malagasy Red Mosquito.

andamanensis Barraud
1934: 102 (M*, F*; *anopheloides* var.). **Type loc.:** Andaman Islands (NHMUK). **Distr.:** Andaman Isl., Cambodia, Indonesia, Malaysia, Philippines, Thailand, Vietnam. **Refs.:** Knight & Mattingly 1950: 15 (M*, P*, L*; description (elevation?) of *Or. anopheleloides* var. *andamanensis* as sp. *andamanensis*); Bonne-Wepster 1954: 28 (M, F*, L*); Basio 1971b: 78 (M*; bion.); Zavortink 1971: 14 (M*, F*, P*, L*; tax., keys); Lee et al. 1988c: 45 (tax., distr., review); Rattanarithikul et al. 2007 (F*, P*, L*; bion., distr., keys). **Ety.:** Andaman Isl. **Informal name:** Hanuman Andamanese Red Mosquito [the name Andaman is thought to have been taken from the Malay name for the Hindu god Hanuman].

ankaratrensis Brunhes & Hervy
1995: 27 (L*). **Type loc.:** Manjakatompo (1,800 m), Sous-Prefecture d'Ambatolampy, Province de Tananarive (Antananarivo), Madagascar (LTV). **Distr.:** Madagascar (includes Glorioso and Juan De Nova Isl,). **Ety.:** Ankaratra Volcano. **Informal name:** Angaratra Malagasy Red Mosquito.

anopheloides (Giles)
1903 (M, F*, L; *Mansonia*). In Thomson 1903: 315. **Type loc.:** Dehra Dun [United Provinces], India (NHMUK). **Distr.:** Bangladesh, Cambodia, China, India, Indonesia, Japan, Malaysia, Nepal, Pakistan, Philippines, Singapore, Sri Lanka, Taiwan, Thailand, Vietnam. **Refs.:** Barraud 1932b (P*, L*); Knight & Mattingly 1950: 7 (M*, F, P*, L; tax.); Asanuma & Nakagawa 1954a: 373 (P*); Hara 1957: 50 (F*); Zavortink 1968: 146 (M*, F*, P*, L*; syn.); Basio 1971b: 78 (M*; bion.; as *maculipes*); Mattingly 1971a: (as

lemmonae), Pl. 5 (F leg*), Pl. 7 (M head*); Zavortink 1971: 16 (M*, F*, P*, L*); Baisas 1974: 142 (M*, F*, P*, L*; tax., bion., distr.; Philippines); Tanaka et al. 1975c: 215 (distr.); Tanaka et al. 1979: 109 (M*, F*, L*); Darsie & Pradhan 1990 (F, L; tax., keys, bion., distr.; Nepal); Rattanarithikul et al. 2007 (F*, P*, L*; bion., distr., keys). **Ety.:** *Anopheles*-like (Gr). **Informal name:** *Anopheles* Indian Red Mosquito.

syn. *nigritarsis* Leicester
1908: 177 (M, F; *albipes* var.). **Type loc.:** Pangkor Laut [Perak], Malaya [Malaysia] (NHMUK). **Refs.:** Barraud 1934: 98 (syn.).

syn. *maculata* Theobald
1910a: 29 (F). **Type loc.:** Maddathorai, Travancore, India (IM). **Refs.:** Barraud 1934: 101 (M, F, L; to var.); Knight & Mattingly 1950: 9 (M*, P, L; to ssp.); Zavortink 1968: 146 (syn.).

syn. *maculipes* Theobald
1910c: 470 (F*). **Type loc.:** Andaman Islands (NHMUK). **Refs.:** Knight & Mattingly 1950: 10 (M*, F, P*, L; to ssp.); Zavortink 1968: 149 (syn.).

syn. *manganus* Baisas
1946a: 34 (M*, F, P*, L*). **Type loc.:** Llavac, Laguna, Luzon, Philippine Islands (PBH). **Refs.:** Knight & Mattingly 1950: 7 (syn.).

syn. *nipponica* La Casse & Yamaguti
1948: 264 (M*, F*, P*, L*). **Type loc.:** Kyoto, Honshu, Japan (USNM). **Refs.:** Knight & Mattingly 1950: 7 (syn.); Stone & Knight 1957c: 198 (type info., lectotype desig.).

syn. *lemmonae* Thurman
1959b: 58 (L*). **Type loc.:** Doi Chom Cheng of Doi Suthep range, Chiengmai Province, Thailand (USNM). **Refs.:** Zavortink 1968: 147 (syn.).

arboricollis (de Charmoy)
1908: 257 (M*, F*; *Culex*). **Type loc.:** Vacoas, Mauritius (NHMUK). **Distr.:** Mauritius. **Refs.:** Edwards 1941: 72 (M, F*); Hopkins 1952: 82 (L); Zavortink 1968: 102 (M*, F, L*; lectotype desig.). **Ety.:** tree inhabitant (L). **Informal name:** Tree-dwelling Mauritian Red Mosquito.

aureoantennata Ferrara
1973: 199 (M*). **Type loc.:** Nkolbisson, Yaoundé, Cameroon (IERT). **Distr.:** Cameroon. **Ety.:** golden antenna (L); golden antennal hairs. **Informal name:** Gold-browed Cameroonian Red Mosquito.

comorensis Brunhes
1977b: 145 (M*, F*, P*, L*). **Type loc.:** near Dembeni, Mayotte Island, Comoros (SSC). **Distr.:** Comoros, Madagascar (includes Glorioso and Juan De Nova Isl.), Mayotte. **Ety.:** Comoros Isl. **Informal name:** Phoenicia Comoran Red Mosquito [according to legend, the Comoros were first discovered by Phoenician sailors].

fascipes (Coquillett)
1906a: 182 (F; *Mansonia*). **Type loc.:** Puntarenas, Costa Rica (USNM). **Distr.:** Argentina, Bolivia, Brazil, Colombia, Costa Rica, Ecuador, French Guiana, Guyana, Lesser Antilles (includes Trinidad and Tobago), Nicaragua, Panama, Peru, Suriname, Venezuela. **Refs.:** Senevet 1946: 326 (L*); Lane 1953: 620 (M*, F, P*, L*); Stone & Knight 1957c: 198 (type info., lectotype desig.); Zavortink 1968: 75 (M*, F, P*, L*; syn.); Linton et al. 2013 (bion., mol. tax., distr.; Ecuador). **Ety.:** banded foot (L); "Distinguished by the color of the hindtarsi." **Informal name:** Black-footed Costa Rican Red Mosquito.

syn. *longipalpis* Newstead & Thomas
1910: 145 (F*; *Mansonia*). **Type loc.:** Manaus, Amazonas, Brazil (NHMUK). **Refs.:** Howard et al. 1917: 882 (syn.); Zavortink 1968: 75 (lectotype desig.).

syn. *townsendi* da Costa Lima
1935b: 178 (F*). **Type loc.:** Boa Vista, Pará, Brazil (IOC). **Refs.:** Lane 1951 (syn.).

syn. *bacigalupoi* Martínez & Prosen
1958: 17 (M*, F). **Type loc.:** Buenavista (Tacu), Province de Ichilo, Santa Cruz, Bolivia (BA). **Refs.:** Zavortink 1968: 75 (syn.).

flavicosta Barraud
1927a: 531 (M*, F, L). **Type loc.:** Karwar [Bombay], India (NHMUK). **Distr.:** India. **Refs.:** Barraud 1932b (P*, L*); Barraud 1934: 104 (M, F, P*, L*); Zavortink 1968: 116 (M*, F, P*, L*; lectotype desig.); Mattingly 1971a: Pl. 16 (P cephalothorax*, terminal segs.*). **Ety.:** yellow side (L); ref. to yellow scales on basal half of costa. **Informal name:** Yellow-sided Indian Red Mosquito.

flavithorax Barraud
1927a: 529 (M*, F, L). **Type loc.:** Karwar [Bombay], India (NHMUK). **Distr.:** India, Sri Lanka. **Refs.:** Barraud 1932b (P*, L*); Barraud 1934: 103 (M, F, P*, L*); Zavortink 1968: 119 (M*, F, P*, L; lectotype desig.). **Ety.:** yellow thorax (L); ref. to bright golden scutal scales. **Informal name:** Yellow-chested Indian Red Mosquito.

fontenillei Brunhes & Hervy
1995: 46 (M*, P*, L*). **Type loc.:** Moramanga, Province de Tamatave (Toamasina), Madagascar (LTV). **Distr.:** Madagascar (includes Glorioso and Juan De Nova Isl.). **Ety.:** Didier Fontenille. **Informal name:** Fontenille Malagasy Red Mosquito.

joyoni Brunhes
1977b: 140 (M*, F*, P*, L*). **Type loc.:** Grille, Grande Comore Island, Comoros (SSC). **Distr.:** Comoros. **Ety.:** n.s. **Informal name:** Joyon Comoran Red Mosquito.

kummi Edwards
1939b: 121 (M, F*). **Type loc.:** Orosi, Costa Rica (NHMUK). **Distr.:** Costa Rica, El Salvador, Guatemala, Mexico, Panama, United States (continental). **Refs.:** de Buen 1953a: 243 (P*, L*); Lane 1953: 626 (M*, F); Zavortink 1968: 68 (M*, F*, P*, L*); Darsie & Ward 2005 (F*, L*; keys, distr.). **Ety.:** Henry William Kumm. **Informal name:** Kumm Costa Rican Red Mosquito.

lanyuensis Lien
1968a: 4 (M*, F, P*, L*). **Type loc.:** south of Hongtou, Lanyu Island, Taitung Hsien, Taiwan [Republic of China] (TMRI). **Distr.:** Taiwan. **Refs.:** Zavortink 1971: 8 (M, F; tax., keys). **Ety.:** Lanyu Isl., Taiwan. **Informal name:** Lanyu Taiwanese Red Mosquito.

madrensis Baisas
1946a: 41 (M*, F, P*, L*). **Type loc.:** Llavac, Laguna, Luzon, Philippines (NE). **Distr.:** Philippines. **Refs.:** Knight & Mattingly 1950: 17 (M, F, P, L); Basio 1971b: 79 (M*; bion.); Zavortink 1971: 6 (M*, F*, P*, L*; tax., keys). **Ety.:** n.s. [poss. ref. to the Sierra Madre, Philippines]. **Informal name:** Madre Philippine Red Mosquito.

milloti Doucet
1951a: 66 (M*). **Type loc.:** Perinet, Tananarive, Madagascar (IRSM). **Distr.:** Madagascar (includes Glorioso and Juan De Nova Isl.). **Refs.:** Zavortink 1968: 97 (M*, F, L*). **Ety.:** Jacques Millot. **Informal name:** Millot Malagasy Red Mosquito.

nkolbissonensis Rickenbach & Hamon
1966: 1112 (M*, F*). **Type loc.:** Nkolbisson, 8 km from Yaoundé, Cameroon (IERT). **Distr.:** Cameroon. **Refs.:** Ferrara 1968: 161 (L*); Zavortink 1968: 100 (M, F). **Ety.:** Recherches Agronomiques de Nkolbisson, Cameroon. **Informal name:** Nkolbisson Cameroonian Red Mosquito.

ototomoensis Huang & Rueda
2013 (M*). **Type loc.:** Ototomo (03°39′ N, 11°19′ E), Cameroon (IRD). **Distr.:** Cameroon. **Ety.:** Ototomo, Cameroon. **Informal name:** Ototomo Cameroonian Red Mosquito.

papuensis Zavortink
1968: 139 (M*, F, P*, L*). **Type loc.:** Finschhafen, [Papua] New Guinea (USNM). **Distr.:** Australia, Indonesia, Papua New Guinea. **Refs.:** Lee et al. 1988c: 47 (tax., bion., distr., review). **Ety.:** Papua New Guinea. **Informal name:** Retez Papuan Red Mosquito [the 16th-century Spanish explorer Yñigo Ortiz de Retez is credited with bestowing the island's name, Nueva Guinea].

peytoni Leguizamón & Carpintero
2004: 204 (M*). **Type loc.:** Finca 1, Colonia Frias, Departamento de General Consesa, Provincia de Río Negro, Argentina (APCP). **Distr.:** Argentina. **Refs.:** Hribar & Hribar 2006 (type dep.). **Ety.:** EL Peyton. **Informal name:** Peyton Argentinian Red Mosquito.

phyllozoa (Dyar & Knab)
1907c: 199 (M; *Mansonia*). **Type loc.:** Tabernilla, Canal Zone, Panama (USNM). **Distr.:** Colombia, Costa Rica, Ecuador, Nicaragua, Panama, Venezuela. **Refs.:** Lane 1953: 626 (M*, F, L*); Zavortink 1968a: 84 (M*, F*, P*, L*). **Ety.:** n.s. [leaf bearing (Gr)]. **Informal name:** Leafy Panamanian Red Mosquito.

pulcripalpis (Rondani)
1872: 31 (A; as *Culex pulcripalpus*). **Type loc.:** Italy (FM). **Distr.:** Albania, Algeria, Austria, Azerbaijan, Belgium, Bulgaria, Crimean Peninsula, Croatia, France, FYRO Macedonia, Greece, Hungary, Italy, Morocco, Portugal (includes Madeira), Romania, Russia, Serbia, Spain, Switzerland, Tunisia, Turkey, Ukraine, United Kingdom. **Refs.:** Séguy 1924: Pl. XXI (L*), Pl. XVII (M gen.*; as *albionensis*); Tate 1932 (L*); Marshall 1938: 258 (M*, F*, P*, L*, E*; bion.); Dodge 1966: 379 (first instar L; key); Zavortink 1968: 52 (M*, F*, P*, L*; lectotype desig.; as *pulchripalpis*); Gutsevich et al. 1974: 134 (M*, F, L*); Snow 1986: 361 (emend.; corrected to original spelling, *pulcripalpis*); Ramsdale & Snow 2001 (distr.; Europe); Becker et al. 2010: 310 (M*, F*, L*; keys, tax., distr., bion.); Namazov 2014 (distr.; Azerbaijan); Azari-Hamidian et al. 2017 (bion., distr.; Iran); Robert et al. 2019 (distr.; Western Palearctic). **Ety.:** beautiful palpus (L). **Informal name:** Lipstick Italian Red Mosquito [*pulcripalpis* = "lovely lips" = Lipstick].
syn. *albionensis* MacGregor
1919: 451 (M, F*, P, L*). **Type loc.:** Epping Forest, Essex, England (NHMUK). **Refs.:** Zavortink 1968: 52 (lectotype desig.); Townsend 1990: 39 (type info.).

rajaonariveloi Brunhes & Hervy
1995: 75 (F*); 1995: 78 (M*, F). **Type loc.:** Sakondroely, Canton de Fenerive, Province de Tamatave (Toamasina), Madagascar (LTV). **Distr.:** Madagascar (includes Glorioso and Juan De Nova Isl.). **Ety.:** E. Rajaonarivelo. **Informal name:** Rajaonarivelo Malagasy Red Mosquito.

ravaonjanaharyi Brunhes & Hervy
1995: 78 (M*, F). **Type loc.:** Province de Diego Suarez (Antsiranana), Madagascar (LTV). **Distr.:** Madagascar (includes Glorioso and Juan De Nova Isl.). **Ety.:** Ch. Ravaonjanahary. **Informal name:** Ravaonjanahary Malagasy Red Mosquito.

reunionensis Brunhes & Hervy
1995: 81 (M*, F, P*, L*). **Type loc.:** Réunion Island (LTV). **Distr.:** Réunion. **Ety.:** Réunion Isl. **Informal name:** Morgabin Réunionnais Red Mosquito [medieval Arab traders referred to the island by the name of Dina Morgabin].

rodhaini Brunhes & Hervy
1995: 89 (L*). **Type loc.:** D'Ambinanitelo, Sous-Prefecture de Maroantsetra, Province de Tamatave (Toamasina), Madagascar (LTV). **Distr.:** Madagascar (includes Glorioso and Juan De Nova Isl.). **Ety.:** F. Rodhain. **Informal name:** Rodhain Malagasy Red Mosquito.

sampaioi da Costa Lima
1935b: 176 (M*, F*, L*). **Type loc.:** Tijuca, Rio de Janeiro, Brazil (IOC). **Distr.:** Argentina, Brazil. **Refs.:** Lane 1953: 623 (M*, F, L); Zavortink 1968: 82 (M*, F, L). **Ety.:** Paulo C.

Sampaio. **Informal name:** Sampaio Brazilian Red Mosquito.

siamensis Zavortink
1968: 126 (M*, F, P*, L*). **Type loc.:** Thailand (USNM). **Distr.:** Thailand. **Refs.:** Zavortink 1971: 9 (M*, F, P*, L*; tax., keys); Rattanarithikul et al. 2007 (F*, P*, L*; bion., distr., keys). **Ety.:** Siam ["an exonym that was formerly used as the name of Thailand" (*Wikipedia*)]. **Informal name:** Siam Thai Red Mosquito.

signifera (Coquillett)
1896: 43 (F; *Culex signifer*). **Type loc.:** Washington, District of Columbia, United States (USNM). **Distr.:** Canada, Cuba, Dominican Republic, Haiti, Lesser Antilles (Virgin Isl.), Mexico, Puerto Rico, United States (continental). **Refs.:** Dyar 1928: 397 (M*, F, L*; corrected spelling); Martini 1930: 249 (E*); Darsie 1951: 40 (P*); Yamaguti & LaCasse 1951b: 12 (M*, F*, L*; as *californica*), 17 (M*, F*, L*); Carpenter & LaCasse 1955: 100 (M, F, L*; keys; as *californica*), 101 (M*, F*, L*; keys); Breland 1959: 137 (first instar L*); Dodge 1966: 378 (first instar L*; key); Zavortink 1968: 35 (M*, F*, P*, L*; syn.); Belkin & Heinemann 1972: 67 (distr.); Smith & Trimble 1973 (distr.; Canada); Bohart & Washino 1978: 35 (M, F*, P*, L*; keys, tax., bion., distr.); Darsie & Ward 2005 (F*, L*; keys, distr.); Harrison et al. 2016 (F*, L*; keys, distr.). **Ety.:** n.s. [sign, signal (L); poss. ref. to silvery-white lines on scutum]. **Informal name:** Marked American Red Mosquito.
syn. *californica* Bohart
1950b: 399 (M, F, P, L*). **Type loc.:** Elkhorn Ferry, near Sacramento, Yolo County, California, United States (USNM). **Refs.:** Zavortink 1968: 35 (syn.).

vernoni van Someren
1949a: 5 (M, L). **Type loc.:** Sakaramy, near Diego Suarez, Madagascar (NHMUK). **Distr.:** Madagascar (includes Glorioso and Juan De Nova Isl.). **Refs.:** Doucet 1950: 57 (L*); Hopkins 1952: 83 (L); Zavortink 1968: 96 (M*, L). **Ety.:** n.s. **Informal name:** Vernon Malagasy Red Mosquito.
syn. *geberti* Grjebine
1953a: 465 (L*). **Type loc.:** Ampijoroa, Ambato-Boeni District [Majunga], Madagascar (IRSM). **Refs.:** Zavortink 1968: 100 (to *nomen dubium*); White 1980: 146 (from *nomen dubium* to valid sp.); Brunhes & Hervy 1995 (syn.).
syn. *antanosyorum* Rodhain & Boutonnier
1984b: 220 (M*, F*, P, L). **Type loc.:** Andaka plain, near Fort Dauphin, Madagascar (PIP). **Refs.:** Brunhes & Hervy 1995 (syn.).

wanxianensis Lei & Li
1990: 10 (M*, L*). **Type loc.:** Wanxian County, Sichuan Province, China [People's Republic of China] (Sichuan Academy of Medical Sciences, Wanxian District, Sichuan Province, China). **Distr.:** People's Republic of China. **Ety.:** Wanxian Dist., Sichuan Prov., China. **Informal name:** Wanxian Chinese Red Mosquito.

waverleyi (Grabham)
1907: 25 (A*, L; *Mansonia*). **Type loc.:** Waverley [Maverly?] Estate, Constant Spring [Surrey], Jamaica (USNM). **Distr.:** Jamaica. **Refs.:** Zavortink 1968: 31 (M*, F*, P*, L*; from syn. with *signifera*; lectotype desig.); Belkin et al. 1970: 113 (M*, F, P*, L*; type loc. info.). **Ety.:** Waverley Estate, Constant Spring, Jamaica. **Informal name:** Waverley Jamaican Red Mosquito.

wilsoni Macdonald
1958: 121 (M*, F, P*, L*). **Type loc.:** Ulu Gombak Forest Reserve, Selangor, Malaya [Malaysia] (NHMUK). **Distr.:** Malaysia, Thailand. **Refs.:** Mattingly 1970f: 160 (E*); Mattingly 1971a: Pl. 16 (P cephalothorax*, terminal segs.*), Pl. 29 (L head*, terminal segs.*); Zavortink 1971: 11 (M*, F*, P*, L*; tax., keys); Rattanarithikul et al. 2007 (F*, P*, L*; bion., distr., keys). **Ety.:** T. Wilson. **Informal name:** Wilson Malaysian Red Mosquito.

Tribe Sabethini

Important references: Lane & Cerqueira 1942 (rev.: New World); Belkin et al. 1965 (type info.; New World); Belkin 1968b (type info.); Belkin et al. 1968 (type info.; New World); Barr & Barr 1969b (E); Mattingly 1969d (E*; syst.); Belkin et al. 1971 (type info.; New World).

Genus *Isostomyia* Coquillett

(Vol. 1, p. 262)

1906d: 16, 24. Type species: *Aedes perturbans* Williston.

Name derivation: n.s. [poss. ref. to iso- = the same: "Proboscis, feet and abdomen wholly black-scaled"].

syn. *Goeldia*, subg. *Isogoeldia* Edwards 1930a: 301. Type species: *Aedes perturbans* Williston.

Important references: Coquillett 1910 (tax., type species); Zavortink 1979b: 12 (as subg. of genus *Runchomyia*); Harbach & Peyton 1993 (L mouthparts; class., phyl.; to genus); Judd 1996 (phyl.); Harbach & Kitching 1998 (phyl.); Harbach & Peyton 2000 (phyl.).

espini (Martini)
1914: 65 (F; *Lesticocampa*). **Type loc.:** Miraflores Lake, Canal Zone, Panama (NHMUK). **Distr.:** Argentina, Bolivia, Brazil, Costa Rica, Guatemala, Honduras, Nicaragua, Panama. **Refs.:** Lane & Cerqueira 1942: 508 (M, F, L); Lane 1953: 833 (M*, F, L*); Belkin 1968b: 35 (tax.); Zavortink 1979b: 12 (from genus *Trichoprosopon*). **Ety.:** n.s. [Dr. Espin]. **Informal name:** Espin Panamanian Grasping Mosquito.
syn. *shropshirei* Ludlow
1919: 168 (F; *Trichoprosopon* (*Joblotia*)). **Type loc.:** Camp Gaillard, Canal Zone, Panama (USNM). **Refs.:** Stone & Knight 1957b: 119 (type info., lectotype desig.); Zavortink 1979b: 12 (to sp. from genus *Trichoprosopon*).

lunata (Theobald)
1901c: 279 (F*; *Wyeomyia*). **Type loc.:** Rio de Janeiro, Brazil (NHMUK). **Distr.:** Argentina, Brazil, Ecuador, Lesser Antilles (includes Trinidad and Tobago), Venezuela. **Refs.:** Lane & Cerqueira 1942: 515 (M*, F, P*, L*); Belkin 1968b: 36 (lectotype desig.); Zavortink 1979b: 12 (from genus *Trichoprosopon*). **Ety.:** bent into a crescent (L); "with a semicircular band of black upright fork scales across the nape." **Informal name:** Crescent Brazilian Grasping Mosquito.

paranensis (Brèthes)
1910: 471 (F; *Lynchiaria*). **Type loc.:** Islas del Paraná, Argentina (BA). **Distr.:** Argentina, Uruguay. **Refs.:** Lane & Cerqueira 1942: 515 (syn.); García & Casal 1965b: 14 (M*; from syn.); Belkin et al. 1968: 11 (lectotype desig. according to Casal); Zavortink 1979b: 12 (from genus *Trichoprosopon*); Rossi & Martínez 2003: 472 (distr.; Uruguay); Campos & Zavortink 2010 (M*, F*, P*, L*; descr., tax.). **Ety.:** Islas del Paraná. **Informal name:** Paraná Argentinian Grasping Mosquito.

perturbans (Williston)
1896: 271 (M, F*; *Aedes*). **Type loc.:** Saint Vincent Island [Lesser Antilles] (NHMUK). **Distr.:** Belize, Guatemala, Honduras, Lesser Antilles (includes Dominica, Grenada, Guadeloupe, Martinique, Saint Vincent and Saint Lucia), Venezuela. **Refs.:** Martini 1935 (distr.; Belize); Lane 1953: 833 (M, F); Belkin 1968b: 37 (tax.); Mattingly 1971a: Pl. 6 (M head*); Belkin & Heinemann 1976a: 262, 272, 289 (distr.); Bohart & Washino 1978: 29 (M, F*, P*, L*; keys, tax., bion., distr.); Zavortink 1979b: 12 (from genus *Trichoprosopon*). **Ety.:** n.s. [disturb, confuse (L)]. **Informal name:** Disturbing Vincentian Grasping Mosquito.

Genus *Johnbelkinia* Zavortink

(Vol. 1, p. 264)

1979b: 18 (as genus). Type species: *Culex longipes* Fabricius.

Name derivation: John Nicholas Belkin.

Important references: Harbach & Peyton 1993 (L mouthparts; phyl.).

leucopus (Dyar & Knab)
1906a: 137 (F; *Lesticocampa*). **Type loc.:** Bluefields, Nicaragua (USNM). **Distr.:** Brazil, Costa Rica, Ecuador, Nicaragua, Panama. **Refs.:** Lane 1953: 857 (M*, F); Stone & Knight 1957b: 118 (type info., lectotype desig.); Barreto & Lee 1969: 415 (distr.); Zavortink 1979b: 314 (M*, F; tax., distr.; from genus *Trichoprosopon*). **Ety.:** n.s. [white foot (Gr); hind tarsi white at tip]. **Informal name:** White-footed Nicaraguan Brushy Mosquito.

longipes (Fabricius)
1805: 34 (A; *Culex*). **Type loc.:** French Guiana [Guiana] (ZMC). **Distr.:** Bolivia, Brazil, Colombia, Costa Rica, Ecuador, French Guiana, Guatemala, Guyana, Nicaragua, Panama, Paraguay, Peru, Suriname, Venezuela. **Refs.:** Townsend 1934: 492 (M; as *trichopus*); Evans & Walker 1935: 465 (M*; as *trichopus*); Kumm & Novis 1938: 512 (L*; as *trichopus*); Senevet & Abonnenc 1939b: 248 (M*, L*); Lane 1939: 162 (syn.; as *trichopus*); Lane 1953: 855 (M*, F, L*); Cerqueira 1961a: 460 (M*, F, P, L*; as *edwardsianum*); Belkin et al. 1965: 41 (type loc. info.); Zavortink 1979b: 35 (M*, F, P*, L*; from genus *Trichoprosopon*). **Ety.:** n.s. [long foot (L); poss. ref. to long tibia]. **Informal name:** Long-footed French Guianese Brushy Mosquito.

syn. *trichopus* Dyar
1919a: 10 (F; *Lesticocampa*). **Type loc.:** Tefe, Amazonas, Brazil (USNM). **Refs.:** da Costa Lima 1931c: 66 (syn.); Lane 1939 (syn.); Lane & Cerqueira 1942: 525 (tax.).

syn. *edwardsianus* Lane & Cerqueira
1942: 528 (F; *Trichoprosopon* (*Hyloconops*)). **Type loc.:** Belterra, Santarem, Pará, Brazil (IOC). **Refs.:** Lane 1953: 860 (M, F); Zavortink 1979b: 35 (syn.).

ulopus (Dyar & Knab)
1906a: 137 (F; *Lesticocampa*). **Type loc.:** Bluefields, Nicaragua (USNM). **Distr.:** Brazil, Colombia, Costa Rica, Ecuador, French Guiana, Guatemala, Guyana, Honduras, Lesser Antilles (includes Trinidad and Tobago), Mexico, Nicaragua, Panama, Suriname, Venezuela. **Refs.:** Howard et al. 1915: 182 (syn. with *longipes*); Zavortink 1979b: 24 (M*, F*, P*, L*; to sp.; from genus *Trichoprosopon*); Berti et al. 2015 (distr.; Venezuela). **Ety.:** n.s. **Informal name:** Bluefields Nicaraguan Brushy Mosquito.

syn. *culicivora* Dyar & Knab
1907c: 207 (F; *Lesticocampa*). **Type loc.:** Tabernilla, Canal Zone, Panama (USNM). **Refs.:** Zavortink 1979b: 25 (from genus *Trichoprosopon*; syn.).

Genus *Kimia* Vu & Harbach

(Vol. 1, p. 266)

2007. In Harbach, Vu, & Kitching 2007: 103. Type species: *Topomyia decorabilis* Leicester.

Name derivation: Tran Thi Bach Kim (wife of first author).

Important references: Harbach, Vu, & Kitching 2007 (tax., phyl., class., bion.).

decorabilis (Leicester)
1908: 239 (M, F; *Topomyia*). **Type loc.:** The Gap [Selangor], Malaya [Malaysia] (NHMUK). **Distr.:** Malaysia, Thailand. **Refs.:** Edwards 1922b: 440 (M*, F); Thurman 1959b: 45 (n. comb.); Mattingly 1971a: Pl. 9 (P terminal segs.*), Pl. 26 (L head*); Harbach, Vu, & Kitching 2007 (transfer to *Kimia*); Miyagi et al 2007 (M*, F*, P*, L*; as *Topomyia*). **Ety.:** n.s. [ornamented (L)]. **Informal name:** Ornamented Malaysian Ridged Mosquito.

imitata (Baisas)
1946d: 32 (M*, P*, L*; as *Topomyia imitatus*). **Type loc.:** Kidapawan, Cotabato, Mindanao, Philippines (USNM). **Distr.:** Philippines. **Refs.:** Thurman 1959b: 45 (n. comb.); Basio 1971b: 80 (M*; bion.); Harbach, Vu, & Kitching 2007 (transfer to *Kimia*). **Ety.:** n.s. [imitated (L)]. **Informal name:** Imitated Philippine Ridged Mosquito.

miyagii (Toma & Mogi)
2003: 25 (*Topomyiia* (*Suaymyia*)); M*, F, P*, L*). **Type loc.:** Tomberab, Ende, Flores, Indonesia (USNM). **Distr.:** Indonesia. **Refs.:** Harbach, Vu, & Kitching 2007 (transfer to *Kimia*). **Ety.:** Ichiro Miyagi. **Informal name:** Miyagi Indonesian Ridged Mosquito.

nemorosa (Gong)
1996: 365 (M*, F, L). **Type loc.:** Ruili City, Yunnan Province, China [People's Republic of China] (Institute of Epidemiology of Yunnan Province). **Distr.:** China. **Refs.:** Harbach, Vu, & Kitching 2007 (transfer to *Kimia*). **Ety.:** n.s. [well wooded (L); poss. ref. to larval habitat]. **Informal name:** Sylvan Chinese Ridged Mosquito.

suchariti (Miyagi & Toma)
1989: 16 (*Topomyia* (*Suaymyia*)); M*, P*, L*). **Type loc.:** Trak Nong, Chanthaburi, Thailand (USNM). **Distr.:** Thailand. **Refs.:** Harbach, Vu, & Kitching 2007 (transfer to *Kimia*); Rattanarithikul et al. 2007 (F*, P*, L*; bion., distr., keys). **Ety.:** Supan Sucharit. **Informal name:** Sucharit Thai Ridged Mosquito.

Genus *Limatus* Theobald

(Vol. 1, p. 268)

1901c: 349 (as genus). Type species: *Limatus durhamii* Theobald. Theobald 1907: 22 (in subfam. Limatinae).

Name derivation: n.s. [filed, polished (L)].

syn. *Simondella* Laveran 1902c: 1160 (as genus). Type species: *Simondella curvirostris* Laveran.

syn. *Lemmamyia* Dyar 1919d: 140 (as genus). Type species: *Limatus methysticus* Dyar & Knab.

Important references: Lane 1953 (revision, keys, tax., bion., distr.); Forattini 1965b (review, tax., bion., distr.); Cova García et al. 1966 (tax.; Venezuela); Belkin et al. 1970 (tax.; Jamaica); Harbach & Peyton 1993 (L mouthparts; tax., phyl.); Harbach & Kitching 1998 (phyl.); Harbach & Peyton 2000 (phyl.); Harbach, Vu, & Kitching 2007 (phyl.); Motta et al. 2007 (phyl.).

andinus Leví-Castillo
1954b: 252 (M*, F). **Type loc.:** Valencia, Los Rios, Ecuador (CEIE). **Distr.:** Ecuador. **Ety.:** Andes Mountains. **Informal name:** Andes Ecuadorian Royal Mosquito.

asulleptus (Theobald)
1903a: 315 (F; *Dendromyia*). **Type loc.:** Demerara River, British Guiana [Guyana] (NHMUK). **Distr.:** Belize, Brazil, Colombia, Costa Rica, Ecuador, French Guiana, Guyana, Honduras, Lesser Antilles (includes Trinidad and Tobago), Nicaragua, Panama, Peru, Suriname, Venezuela. **Refs.:** Dyar 1928: 26 (M*, F, L*); De Oliveira Castro 1935: 148 (tax.); Senevet & Abonnenc 1939b: 272 (M*, L*); Belkin 1968b: 22 (lectotype desig.); Mattingly 1971a: Pl. 6 (M head*); Heinemann & Belkin 1977b: 415 (distr.); Berti et al. 2015 (distr.; Venezuela). **Ety.:** n.s. [poss. ref. to "prothoracic lobes yellow scaled"]. **Informal name:** Yellow-scaled Guyanese Royal Mosquito.

syn. *methysticus* Dyar & Knab
1909d: 266 (M, F). **Type loc.:** Port Limon [Puerto Limón], Costa Rica (USNM). **Refs.:** Stone & Knight 1957b: 117 (type info., lectotype desig.).

durhamii Theobald
1901c: 350 (F). **Type loc.:** Pará, Brazil (NHMUK). **Distr.:** Argentina, Belize, Bolivia, Brazil, Colombia, Costa Rica, Ecuador, El Salvador, French Guiana, Guatemala, Guyana, Honduras, Lesser Antilles (includes Dominica, Grenada, Guadeloupe, Trinidad and Tobago), Mexico, Nicaragua, Panama, Peru, Suriname, Venezuela. **Refs.:** Senevet & Abonnenc 1939b: 273 (M*, L*); Lane & Cerqueira 1942: 652 (M*, F, P*, L*); Knight & Chamberlain 1948: 13 (P*); Lane 1953: 1046 (M*, F, P*, L*); Stone 1969: 4 (distr.); Mattingly 1971a: Pl. 6 (M head*), Pl. 9 (P terminal segs.*), Pl. 10 (P trumpet*), Pl. 23 (L head*, terminal segs.*); Belkin & Heinemann 1976a: 272 (distr.); Belkin & Heinemann 1976c: 188 (distr.); Heinemann & Belkin 1977b: 415 (distr.); Linton et al. 2013 (bion., mol. tax., distr.; Ecuador); Berti et al. 2015 (distr.; Venezuela). **Ety.:** Dr. Durham. **Informal name:** Durham Brazilian Royal Mosquito.

syn. *curvirostris* Laveran
1902c: 1160 (M, F; *Simondella*). **Type loc.:** Rio de Janeiro, Brazil (NHMUK). **Refs.:** Belkin 1968b: 22 (type info.); Townsend 1990: 65 (type info.).

syn. *exhibitor* Shannon & Del Ponte
1928: 92 (M, F, L). **Type loc.:** Cataratas del Iguazu, Misiones, Argentina (INM). **Refs.:** Edwards 1932a (syn.); Belkin et al. 1968: 11 (lectotype desig. according to Casal).

flavisetosus de Oliveira Castro
1935: 143 (M*, F, P*, L*). **Type loc.:** Cubatão, São Paulo, Brazil (LU). **Distr.:** Brazil, Colombia, Ecuador, French Guiana, Guyana, Haiti, Nicaragua, Panama, Peru, Suriname. **Refs.:** Lane & Cerqueira 1942: 655 (M*, F); Belkin et al. 1971: 13 (type info.); Panday 1975a, b (distr.; Suriname); Linton et al. 2013 (bion., mol. tax., distr.; Ecuador). **Ety.:** n.s. [yellow haired (L); poss. ref. to golden setae toward the apex of the male abdomen]. **Informal name:** Yellow-haired Brazilian Royal Mosquito.

guayasi Leví-Castillo
1954b: 251 (M*, F). **Type loc.:** El Empalme, Guayas, Ecuador (NE). **Distr.:** Ecuador. **Refs.:** Giles 1901a: 194 (July) (F). **Ety.:** Prov. of Guayas. **Informal name:** Guayas Ecuadorian Royal Mosquito.

hoffmani Root
1927a: 465 (M*). **Type loc.:** Rivière Froide, Haiti (USNM). **Distr.:** Brazil, Dominican Republic, Ecuador, Haiti, Jamaica. **Refs.:** Dyar 1928: 25 (M*, L*); Lane 1953: 1050 (M, L*); Stone & Knight 1957b: 117 (type info.); Belkin et al. 1970: 230 (F). **Ety.:** William A. Hoffman. **Informal name:** Hoffman Haitian Royal Mosquito.

martiali Senevet & Abonnenc
1939b: 276 (M*). **Type loc.:** Saut Tigre [Inini], French Guiana [Guyana] (LPFM, now in FMP). **Distr.:** French Guiana. **Refs.:** Lane 1953: 1052 (M*); Belkin 1968b: 23 (tax.). **Ety.:** Médecin-Colonel Martial. **Informal name:** Martial French Guianese Royal Mosquito.

paraensis (Theobald)
1903a: 316 (F; *Dendromyia*). **Type loc.:** Pará, Brazil (NHMUK). **Distr.:** Brazil. **Refs.:** Lane & Cerqueira 1942: 654 (M, F; from syn. with *durhamii*). **Ety.:** State of Pará, Brazil. Informal name: Pará Brazilian Royal Mosquito.

syn. *cacophrades* Dyar & Knab 1909d: 266 (M, F). **Type loc.:** Tabernilla, Canal Zone, Panama (USNM). **Refs.:** Lane & Cerqueira 1942 (syn.); Stone & Knight 1957b: 117 (type info., lectotype desig.).

pseudomethysticus (Bonne-Wepster & Bonne) 1920: 166 (F, L; *Lemmamyia*). **Type loc.:** Suriname (ITH). **Distr.:** Brazil, French Guiana, Suriname. **Refs.:** Lane & Cerqueira 1942: 656 (from syn. with *asuleptus*); Lane 1953: 1052 (F, L); Belkin 1968b: 23 (lectotype desig.); Fauran & Pajot 1974a: 100 (distr.). **Ety.:** false (Gr); similar to *methysticus*. **Informal name:** Plonk French Surinamese Royal Mosquito [*pseudomethysticus* = "false wine" = Plonk].

Genus *Malaya* Leicester

(Vol. 1, p. 270)

1908: 258 (as genus). Type species: *Malaya genurostris* Leicester.

Name derivation: Malaya.

syn. *Grahamia* Theobald 1909b: 23 (May) (as genus). Type species: *Grahamia trichorostris* Theobald.
syn. *Harpagomyia* de Meijere 1909: 165 (Dec.) (as genus). Type species: *Harpagomyia splendens* de Meijere.

Important references: Barraud 1934 (tax., bion.; as genus *Harpagomyia*; distr.; South Asia); Wharton 1947 (keys; Australasian and Oriental regions); Thurman 1959b (tax.; Thailand); Belkin 1962 (tax., bion.; South Pacific); Tanaka et al. 1979 (tax., bion.; Japan); Lee et al. 1988b (tax., keys, bion.; Australasian Region); Service 1990 (tax., keys, bion.; Afrotropical Region); Harbach & Peyton 1993 (comparative morphology of larval maxillae); Judd 1996 (phyl.); Harbach & Kitching 1998 (phyl.); Harbach & Peyton 2000 (phyl.); Rattanarithikul, Harrison et al. 2005 (F, L; tax., keys, bion.; Thailand); Rattanarithikul et al. 2007 (tax., bion.; Thailand); Harbach, Vu, & Kitching 2007 (phyl.).

farquharsoni (Edwards)
1922c: 500 (M*, F; *Harpagomyia*). **Type loc.:** Ibadan [Western Provinces], Nigeria (NHMUK). **Distr.:** Benin, Kenya, Nigeria, Togo. **Refs.:** Edwards 1941: 36 (M), 362 (P); Hopkins 1952: 45 (L). **Ety.:** C.O. Farquharson. **Informal name:** Farquharson Nigerian Ant Mosquito.

fraseri (Edwards)
1922c: 499 (M*; *Harpagomyia*). **Type loc.:** Mpumu Forest, Uganda (NHMUK). **Distr.:** Uganda. **Refs.:** Edwards 1941: 37 (M). **Ety.:** A.D. Fraser. **Informal name:** Fraser Ugandan Ant Mosquito.

genurostris Leicester
1908: 258 (M). **Type loc.:** Malaya [Malaysia] (NHMUK). **Distr.:** Australia, Bangladesh, Cambodia, India, Indonesia, Japan, Malaysia, Maldives, Myanmar, Nepal, Papua New Guinea, People's Republic of China, Philippines, Singapore, Sri Lanka, Taiwan, Thailand, Vietnam. **Refs.:** Barraud 1934: 48 (M*, F, P, L*); Bohart & Ingram 1946b: 54 (M, F*, P*, L*; bion., distr., tax.; as *Harpagomyia*); Wharton 1947: 64 (M, F, L*); Iyengar & Menon 1956 (A*, P*, L*); Knight & Chamberlain 1948: 13 (P*); Thurman 1959b: 49 (M*, L); Basio 1971b: 75 (M*, L*; bion.); Mattingly 1971a: Pl. 2 (F head*), Pl. 9 (P seg. VII*), Pl. 24 (L head*, thorax*, terminal segs.*); Lane 1972: 411 (P*); Baisas 1974: 134 (M*, F, P, L*; tax., bion., distr.; Philippines); Tanaka et al. 1975c: 225 (distr.); Tanaka et al. 1979: 487 (M*, F*, L*); Ahmed 1987 (distr.; Bangladesh); Lee et al. 1988b: 105 (F; key, tax., bion., distr., review); Darsie & Pradhan 1990 (F, L; tax., keys, bion., distr.; Nepal); Rattanarithikul et al. 2007 (F*, P*, L*; bion., distr., keys). **Ety.:** knee beak (L); ref. to articulated proboscis. **Informal name:** Knee-beaked Malaysian Ant Mosquito.
syn. *caeruleovittata* Ludlow
1911b: 131 (F; *Harpagomyia*). **Type loc.:** Ludlow Barrack [Parang, Cotabato], Mindanao, Philippines (USNM). **Refs.:** Bohart 1945: 25 (syn.); Stone & Knight 1957b: 117 (type info.).

incomptas Ramalingam & Pillai
1972: 451 (M*, F, P*, L*). **Type loc.:** Limau Kasturi Kelantan, West Malaysia [Malaysia] (USNM). **Distr.:** Malaysia. **Ety.:** n.s. [unadorned (L); adult]. **Informal name:** Disheveled Malaysian Ant Mosquito.

jacobsoni (Edwards)
1930b: 543 (F; *Harpagomyia*). **Type loc.:** Fort de Kock [West Coast], Sumatra [Indonesia] (NHMUK). **Distr.:** Bangladesh, Cambodia, India, Indonesia, Malaysia, Nepal, Taiwan, Thailand, Vietnam. **Refs.:** Barraud 1934: 51 (A, L*); Thurman 1958: 15 (A); Thurman 1959b: 50 (M*, L); Lane 1972: 411 (P*); Ramalingam & Pillai 1972: 458 (distr.); Ahmed 1987 (distr.; Bangladesh); Darsie, Pradhan & Vaidya 1991 (distr.; Nepal); Rattanarithikul et al. 2007 (F*, P*, L*; bion., distr., keys). **Ety.:** E. Jacobson. **Informal name:** Jacobson Indonesian Ant Mosquito.

leei (Wharton)
1947: 60 (M*, F*, L*; *Harpagomyia*). **Type loc.:** Madang, [Papua] New Guinea (CSIR). **Distr.:** Australia, Indonesia, Papua New Guinea, Philippines. **Refs.:** Lane 1972: 413 (distr.); Lee et al. 1988b: 108 (F; key, tax., bion., distr., review); Peyton et al. 1999: 545 (distr.). **Ety.:** David Joseph Lee. **Informal name:** Lee Papuan Ant Mosquito.

marceli (Mattingly)
1953c: 315 (M*, F; as *Harpagomyia marcelli*). **Type loc.:** source of River Lofoi, Katanga, Belgian Congo [Democratic Republic of the Congo] (NHMUK). **Distr.:** Democratic Republic of the Congo. **Ety.:** Marcel Lips. **Informal name:** Marcel Congolese Ant Mosquito.

moucheti (Hamon & Adam)
1955: 882 (M*; *Harpagomyia*). **Type loc.:** Yaoundé, Cameroun [Cameroon] (IERT). **Distr.:** Cameroon. **Ety.:** René Libert Joseph Mouchet. **Informal name:** Mouchet Cameroonian Ant Mosquito.

solomonis (Wharton)
1947: 63 (M, F, L; *Harpagomyia*). **Type loc.:** Bougainville, Solomon Islands (UQIC). **Distr.:** Papua New Guinea, Solomon Isl. **Refs.:** Belkin 1962: 490 (M*, F, P*, L*); Lee et al. 1988b: 111 (F; key, tax., bion., distr., review). **Ety.:** Solomon Isl. **Informal name:** Agunua Solomon Islander Ant Mosquito [Agunua is the chief god of the people of San Cristoval Island in the Solomon Isl.].

splendens (de Meijere)
1909: 167 (M*, F*, *Harpagomyia*). **Type loc.:** Batavia and Semarang, Java [Indonesia] (AM). **Distr.:** Indonesia, Malaysia. **Refs.:** Wharton 1947: 66 (M, F, L). **Ety.:** shine, be bright (L); "a very marked and beautiful little species." **Informal name:** Shining Indonesian Ant Mosquito.

taeniarostris (Theobald)
1911a: 34 (M; *Harpagomyia*). **Type loc.:** Kampala swamp, Uganda (NHMUK). **Distr.:** Cameroon, Central African Republic, Côte d'Ivoire, Democratic Republic of the Congo, Ethiopia, Gabon, Liberia, Nigeria, South Africa, Sudan and South Sudan, Tanzania, Uganda. **Refs.:** Edwards 1941: 36 (M, F*), 362 (P*); Hopkins 1952: 43 (L*); Jupp 1996 (M*, F*; key). **Ety.:** banded beak (L); proboscis "banded dark brown at the swollen apex." **Informal name:** Ribbon-beaked Ugandan Ant Mosquito.

trichorostris (Theobald)
1909b: 23 (M*, F*; *Grahamia*). **Type loc.:** Obuasi, Ashanti, Gold Coast [Ghana] (NHMUK). **Distr.:** Benin, Cameroon, Central African Republic, Ghana, Liberia, Sierra Leone. **Refs.:** Edwards 1941: 36 (M); Hamon 1955: 676 (L*); da Cunha Ramos & Ribeiro 1975: 114 (tax., bion., distr.; Angola). **Ety.:** n.s. [hairy beak (Gr)]. **Informal name:** Hairy-beaked Ghanaian Ant Mosquito.

Genus *Maorigoeldia* Edwards

(Vol. 1, p. 272)

Rachionotomyia, subg. *Maorigoeldia* Edwards 1930a: 302. Type species: *Culex argyropus* Walker 1848.

Name derivation: Maori people of New Zealand and Emilio Augusto Goeldi.

Important references: Belkin 1962: 492 (to genus); Harbach & Petyton 1993 (L mouthparts*; phyl.); Judd 1996 (phyl.); Harbach & Kitching 1998 (phyl.).

argyropus (Walker)
1848: 2 (F; *Culex*). **Type loc.:** New Zealand (NHMUK). **Distr.:** New Zealand. **Refs.:** Graham 1929: 227 (M*, F*, P*, L*, E*; bion.); Lee 1946: 229 (M*, F*, L*); Belkin 1962: 493 (M*, F, P*, L*); Belkin 1968a: 106 (M*, F*, P, L; bion.); Lee et al. 1988c: 3 (tax., bion., distr., review); Derraik 2004 (distr.; New Zealand). **Ety.:** silver foot (Gr); tarsi. **Informal name:** Silver-footed New Zealander Maori Mosquito.

Genus *Onirion* Peyton & Harbach

(Vol. 1, p. 274)

2000. In Harbach & Peyton 2000: 2000. Type species: *Dendromyia personata* Lutz (in Bourroul 1904).

Name derivation: "*Onirion* is derived from the Greek Oneiros (m.), dream or vision. Addition of the diminutive suffix -ion results in a neuter name meaning 'littledream'" (Harbach & Peyton 2000).

Important references: Harbach & Peyton 2000 (phyl.); Harbach, Vu, & Kitching 2007 (phyl.); Motta et al. 2007 (mol. phyl.).

aenigma Harbach
2000 (M*, F, P*, L*). In Harbach & Peyton 2000: 132. **Type loc.:** Pakitza, Rio Manu, Madre de Dios, Peru (USNM). **Distr.:** Peru. **Ety.:** a riddle, mystery (L); ref. to "the mystifying similarity of the male genitalia to those of *On. brucei*." **Informal name:** Enigmatic Peruvian Dream Mosquito.

brucei (Del Ponte & Cerqueira)
1938: 231 (F; as *Wyeomyia*). **Type loc.:** Cuiabá, Mato Grosso, Brazil (LU). **Distr.:** Argentina, Brazil, Peru. **Refs.:** da Costa Lima 1943: 306 (syn.); Belkin et al. 1971: 11, 33 (type info.); Harbach & Peyton 2000: 127 (M*, F*, P*, L*). **Ety.:** D. Bruce Wilson. **Informal name:** Bruce Brazilian Dream Mosquito.
syn. *belkini* Casal & García
1966: 155 (M*, F*, P*, L*; *Wyeomyia* (*Dendromyia*)). **Type loc.:** Igu, Misiones, Argentina (INM). **Refs.:** Harbach & Peyton 2000: 127 (to syn. with *brucei*).

celatum Peyton & Harbach
2000: 124 (M*, F*, P*, L*). **Type loc.:** Pakitza, Rio Manu, Madre de Dios, Peru. **Distr.:** Peru. **Ety.:** concealed, hidden (L); similar to *brucei*. **Informal name:** Concealed Peruvian Dream Mosquito.

imparis Peyton & Harbach
2000 (M*, F*, P*, L*). In Harbach & Peyton 2000: 130. **Type loc.:** Carabobo, Puerto Cabello, Borburata, Venezuela (USNM). **Distr.:** Ecuador, Venezuela. **Refs.:** Linton et al. 2013 (bion., mol. tax., distr.; Ecuador). **Ety.:** n.s. [unequal, unlike, discordant (L)]. **Informal name:** Discordant Venezuelan Dream Mosquito.

personatum (Lutz)
1904 (M, F; *Dendromyia*). In Bourroul 1904: 22, 68. **Type loc.:** São Paulo, Brazil (NHMUK). **Distr.:** Argentina, Bolivia, Brazil, Colombia, Venezuela. **Refs.:** da Costa Lima 1943: 305 (tax.); Lane 1953: 962 (M*, F, P*, L*); Forattini et al. 1960: 99 (P*, L*); Belkin 1968b: 41 (tax.); Barreto & Lee 1969: 418 (distr.); Belkin et al. 1971: 11 (lectotype desig.); Motta & Lourenço-de-Oliveira 1995: 375 (to subg. uncertain); Harbach & Peyton 2000: 124 (M*, F*, P*, L*; tax.). **Ety.:** n.s. [masked (L); poss. ref. to scales on the clypeus]. **Informal name:** Masked Brazilian Dream Mosquito.
syn. *luederwaldti* Lane
1936a: 6 (F; *Goeldia* (*Isogoeldia*)). **Type loc.:** Avaré, São Paulo, Brazil (FH). **Refs.:** Lane 1953: 385 (F); Zavortink 1979b: 12 (from *Trichoprosopon* (*Isostomyia*) to *Wyeomyia* (*Dendromyia*)); Harbach & Peyton 2000 (syn.).

regale Peyton & Harbach
2000 (M*, F*, L*). In Harbach & Peyton 2000: 132. **Type loc.:** Finca La Selva, Puerto Viejo, Heredia, Costa Rica (USNM). **Distr.:** Costa Rica. **Ety.:** of a king, royal, regal (L); name "chosen for no reason other than it is an appealing name." **Informal name:** Regal Costa Rican Dream Mosquito.

sirivanakarni (Duret)
1982: 167 (M*). **Type loc.:** Darién, Panama (USNM). **Distr.:** Panama. **Refs.:** Motta et al. 1995 (to subg. uncertain); Harbach & Peyton 2000: 129 (M*, F*; bion., distr., tax.; generic transfer). **Ety.:** Sunthorn Sirivanakarn. **Informal name:** Sirivanakarn Panamanian Dream Mosquito.

Genus *Runchomyia* Theobald

(Vol. 1, p. 276)

Important references: Zavortink 1979b: 12 (keys; from subg. of genus *Trichoprosopon*); Harbach & Peyton 1993 (L mouthparts; class.).

Subgenus *Ctenogoeldia* Edwards

Goeldia, subg. *Ctenogoeldia* 1930a: 301. Type species: *Phoniomyia magna* Theobald.

Name derivation: cteno- = comb (Gr) and Emilio Augusto Goeldi (*Goeldia*).

Important references: Lane 1953 (tax., bion.; as subg. of genus *Trichoprosopon*); Zavortink 1979b:12 (keys; from subg. of genus *Trichoprosopon*); Harbach & Peyton 1993 (L mouthparts; class.); Judd 1996 (phyl.).

magna (Theobald)
1905f: 117 (F; *Phoniomyia*). **Type loc.:** San Antonio, Bolivia (HNM). **Distr.:** Belize, Bolivia, Brazil, Colombia, Costa Rica, Ecuador, French Guiana, Guatemala, Honduras, Nicaragua, Panama, Paraguay, Suriname, Venezuela. **Refs.:** Lane 1953: 836 (M*, F, P*, L); Mattingly 1971a: Pl. 9 (P terminal segs.*), Pl. 10 (P trumpet*); Heinemann & Belkin 1977b: 415 (distr.); Zavortink 1979b: 12 (from genus *Trichoprosopon*); Linton et al. 2013 (bion., mol. tax., distr.; Ecuador). **Ety.:** large (L); ref. to "one of the largest of the genus"]. **Informal name:** Large Bolivian Filament Mosquito.

syn. *homotina* Dyar & Knab
1906a: 141 (A; *Phoniomyia*). **Type loc.:** Port Limón, Costa Rica (USNM). **Refs.:** Stone & Knight 1957b: 118 (type info.); Zavortink 1979b: 12 (from genus *Trichoprosopon*).

syn. *dicellaphora* Howard, Dyar, & Knab
1913: Figs. 43, 305; 1915: 166 (M*, F, L*; *Lesticocampa*). **Type loc.:** Miraflores, Canal Zone, Panama (USNM). **Refs.:** Stone & Knight 1957b: 118 (type info., lectotype desig.).

walcotti (Lane & Cerqueira)
1942: 510 (F). **Type loc.:** Caravelas, Bahia, Brazil (IOC). **Distr.:** Brazil. **Refs.:** Zavortink 1979b: 12 (from genus *Trichoprosopon*). **Ety.:** A.M. Walcott. **Informal name:** Walcott Brazilian Filament Mosquito.

Subgenus *Runchomyia* Theobald

1903a: 319 (as genus, subgenus). Type species: *Runchomyia frontosa* Theobald.

Name derivation: n.s. [runcho = ?; "The character upon which the genus is founded is the peculiar projecting blunt frontal process in the (female)"].

syn. *Rhynchomyia* Lutz 1904 (unjust. emend. for *Runchomyia* Theobald). In Bourroul 1904: 55.
syn. *Binotia* Blanchard 1904: 478. Type species: *Runchomyia frontosa* Theobald. Proposed as replacement name for *Runchomyia*.
syn. *Lesticocampa* Dyar & Knab 1906d: 226 (as genus). Type species: *Wyeomyia lunata* Theobald (Bruneti 1914: 63, *Lestiocampa*, unjust. emend.). Type species: *Wyeomyia lunata* Theobald.
syn. *Lynchiaria* Brèthes 1910: 471 (as genus). Type species: *Lynchiaria paraensis* Brèthes.

Important references: Lane 1953 (tax., bion.; as subg. of genus *Trichoprosopon*); Zavortink 1979b: 12 (keys; from subg. of genus *Trichoprosopon*); Darsie 1985 (keys; Argentina); Harbach & Peyton 1993 (L mouthparts; class.); Judd 1996 (phyl.).

cerqueirai (Stone)
1944: 339 (M, F). **Type loc.:** Rio de Janeiro, Distrito Federal, Brazil (USNM). **Distr.:** Brazil, Panama. **Refs.:** Lane 1953: 844 (M*, F, P*, L*); Zavortink 1979b: 12 (from genus *Trichoprosopon*). **Ety.:** Nelson Leander Cerqueira. **Informal name:** Cerqueira Brazilian Filament Mosquito.

frontosa Theobald
1903a: 319 (F*; *Runchomyia*). **Type loc.:** Barima River, Pomeroon Mission [Essequibo], British Guiana [Guyana] (NHMUK). **Distr.:** Argentina, Brazil, Guyana, Lesser Antilles (includes Trinidad and Tobago), Panama, Venezuela. **Refs.:** Lane 1953: 846 (M*, F, P*, L*); Belkin 1968b: 35 (lectotype desig.); Mattingly 1971a: Pl. 23 (L head*, terminal segs.*); Zavortink 1979b: 12 (from genus *Trichoprosopon*); Zavortink 1986 (bion.); Berti et al. 2015 (distr.; Venezuela). **Ety.:** n.s. [bulging forehead (L); prob. ref. to a yellow frontal tubercle]. **Informal name:** Bulge-headed Guyanese Filament Mosquito.

humboldti (Lane & Cerqueira)
1942: 520 (M*, F, P*, L*). **Type loc.:** Paineiras, Distrito Federal, Brazil (IOC). **Distr.:** Brazil. **Refs.:** Zavortink 1979b: 12 (from genus *Trichoprosopon*). **Ety.:** n.s. [poss. ref. to Friedrich Wilhelm Heinrich Alexander (Freiherr) von Humboldt]. **Informal name:** Humboldt Brazilian Filament Mosquito.

rapax Dyar & Knab
1906a: 137 (M, F; *Lesticocampa*). **Type loc.:** Trinidad [Trinidad and Tobago, Lesser Antilles] (USNM). **Refs.:** Stone 1944: 339 (syn. with *frontosum*); Stone & Knight 1957b: 118 (type info.); Zavortink 1979b: 12 (from syn., in genus *Trichoprosopon*). **Ety.:** n.s. [grasping, greedy (L)]. **Informal name:** Greedy Trinidadian Filament Mosquito.

reversa (Lane & Cerqueira)
1942: 522 (M*, F, P*, L*). **Type loc.:** Mage, Rio de Janeiro, Brazil (IOC). **Distr.:** Argentina, Brazil, Paraguay. **Refs.:** Zavortink 1979b: 12 (from genus *Trichoprosopon*). **Ety.:** n.s. **Informal name:** Reverse Brazilian Filament Mosquito.

theobaldi (Lane & Cerqueira)
1942: 523 (M*, F, P*, L*). **Type loc.:** Mage, Xerém, Rio de Janeiro, Brazil (IOC). **Distr.:** Brazil, Lesser Antilles (includes Trinidad and Tobago). **Refs.:** Zavortink 1979b: 12 (from genus *Trichoprosopon*); Marcondes et al. 2003: 1 (distr.; Brazil). **Ety.:** Frederic Vincent Theobald. **Informal name:** Theobald Brazilian Filament Mosquito.

Genus *Sabethes* Robineau-Desvoidy

(Vol. 1, p. 278)

Important references: Belkin 1968b: 29 (tax.).

Subgenus *Davismyia* Lane & Cerqueira

1942: 582 (as subg. of genus *Wyeomyia*). Type species: *Miamyia petrocchiae* Shannon & Del Ponte.

Name derivation: Nelson C. Davis.

Important references: Lane & Causey 1955 (P*, L*; tax.); Harbach & Peyton 1991b (from genus *Wyeomyia*); Harbach 1991a (morph. comp. to other genus *Sabethes*); Huang 2002 (proposed move to subg. of *Wyeomyia*).

petrocchiae (Shannon & Del Ponte)
1928: 94 (F; *Miamyia*). **Type loc.:** Raco, Tucuman, Argentina (NE). **Distr.:** Argentina, Bolivia, Brazil, Paraguay. **Refs.:** Del Ponte & Cerqueira 1938: 227 (from syn. with *arthrostigma*); Lane 1953: 936 (F); Lane & Causey 1955: 12 (M*, P*, L*); Harbach & Peyton 1991: 157 (M*, F*, P*, L*; from genus *Wyeomyia*). **Ety.:** Juana Miguela Petrocchi? **Informal name:** Petrocchi Argentinian Canopy Mosquito.
syn. *monoleua* Martini
1931b: 116 (F; *Miamyia*). **Type loc.:** San Jose, Argentina (SMNS). **Refs.:** Lane & Cerqueira 1942: 583 (syn.); Belkin 1968b: 41 (type info.).

Subgenus *Peytonulus* Harbach

1991a (as subg. of genus *Sabethes*). Type species: *Sabethinus aurescens* Lutz.

Name derivation: EL Peyton.

Important references: Judd 1996 (phyl.).

aurescens (Lutz)
1905: 350 (M, F; *Sabethinus*). **Type loc.:** Cantareira, São Paulo, Brazil (NHMUK). **Distr.:** Argentina, Brazil, Panama, Venezuela. **Refs.:** Dyar 1922h (M*); Lane & Cerqueira 1942: 684 (M*, F, P*, L*); Knight & Chamberlain 1948: 12 (P*); Lane 1953: 1089 (M*, F, P*, L*); Forattini et al. 1965: 455 (P*, L*); Belkin 1968b: 30 (tax.); Belkin et al. 1971: 15 (lectotype desig.), 42 (elimination of combination *aurescens* Theobald 1907: 622). **Ety.:** n.s. [becoming golden (scales) (L); prob. ref. to general golden color]. **Informal name:** Alchemy Brazilian Canopy Mosquito [*aurescens* = "becoming golden"= Alchemy, the medieval theory of turning base metals into gold].

fabricii Lane & Cerqueira
1942: 687 (M*, F, P*, L*). **Type loc.:** Tingua, Rio de Janeiro, Brazil (IOC). **Distr.:** Brazil, Panama. **Refs.:** Lane 1953 (M*, F, P*, L*). **Ety.:** Johann Christian Fabricius. **Informal name:** Fabricius Brazilian Canopy Mosquito.

gorgasi Duret
1971a: 15 (M*). **Type loc.:** Juan Mina, Rio Chagres, Provincia de Panama, Panama (A). **Distr.:** Panama. **Ety.:** Instituto Gorgas. **Informal name:** Gorgas Panamanian Canopy Mosquito.

hadrognathus Harbach
1995a: 87 (M*, F*, P*, L*). **Type loc.:** Finca La Selva, Puerto Viejo Province, Costa Rica (USNM). **Distr.:** Costa Rica, French Guiana, Panama, Peru. **Ety.:** bulky jaw (Gr); ref. to "the enormous, protruding maxillae of the larva." **Informal name:** Bulky-jawed Costa Rican Canopy Mosquito.

identicus Dyar & Knab
1907c: 207 (F, L). **Type loc.:** Tabernilla, Canal Zone, Panama (USNM). **Distr.:** Argentina, Bolivia, Brazil, Colombia, Costa Rica, Panama. **Refs.:** Lane 1951: 335 (from syn. with *aurescens* (Lutz); Lane 1953: 1083 (M*, F, P*, L*); Stone & Knight 1957b: 118 (type info.); Forattini et al. 1965: 460 (P*, L*). **Ety.:** identical (L); identical to *Sabethes undosus*. **Informal name:** Identical Panamanian Canopy Mosquito.
syn. *lutzianus* Lane & Cerqueira
1942: 679 (M*, F, P*, L*). **Type loc.:** Tingua, Brazil (IOC). **Refs.:** Lane 1951 (syn.); Belkin et al. 1971: 15 (type loc. info.).

ignotus Harbach
1995b: 583 (M*, F). **Type loc.:** Villavicencio, Meta, Colombia (USNM). **Distr.:** Colombia. **Ety.:** unknown, obscure, strange (L). **Informal name:** Unknown Colombian Canopy Mosquito.

luxodens Hall, Howard & Harbach
1999: 329 (M*, P*, L*). **Type loc.:** 12 km west on Route 3 (17MPT 1358) 5 m, Guayas, Guayaquil, Ecuador (USNM). **Distr.:** Colombia, Ecuador. **Ety.:** dislocated tooth (L); ref. to "pronounced dorsal displacement of the first (distal) lateral tooth of the larval maxilla." **Informal name:** Braces Ecuadorian Canopy Mosquito [*luxodens* = "dislocated tooth" = orthodontic Braces].

paradoxus Harbach
2002 (M*, P*, L*). In Harbach & Howard 2002: 363. **Type loc.:** Darien, Panama (USNM). **Distr.:** Panama, French Guiana. **Ety.:** contrary to expectation (L); ref. to the unusual coloration of the male and "enigmatic presence of a delineated membranous area in one pupa of the type series." **Informal name:** Odd Panamanian Canopy Mosquito.

soperi Lane & Cerqueira

1942: 682 (M*, F, P*, L*). **Type loc.:** Pirajá, Bahia, Brazil (IOC). **Distr.:** Argentina, Bolivia, Brazil, French Guiana. **Refs.:** Lane 1953: 1088 (M*, F, P*, L*); Mattingly 1971a: Pl. 10 (P trumpet*). **Ety.:** Frederick Lowe Soper. **Informal name:** Soper Brazilian Canopy Mosquito.

undosus (Coquillett)

1906a: 186 (A; *Sabethoides*). **Type loc.:** Trinidad [Trinidad and Tobago, Lesser Antilles] (USNM). **Distr.:** Argentina, Belize, Bolivia, Brazil, Colombia, Costa Rica, Ecuador, French Guiana, Guyana, Honduras, Lesser Antilles (includes Trinidad and Tobago), Nicaragua, Panama, Paraguay, Peru, Venezuela. **Refs.:** Howard et al. 1913: Figs. 4, 463, 468, 520, 547, Pl. 42 (M*, L*); Dyar 1922h (M*); Lane 1953: 1084 (M*, F, P*, L*); Stone & Knight 1957a, b: 118 (type info.); Heinemann & Belkin 1977b: 443 (distr.); Pecor et al. 2002: 249 (tax., distr.; Belize). **Ety.:** abounding in waves (L); ref. to wavy upper border of abdominal venter markings. **Informal name:** Wavy Trinbagonian Canopy Mosquito.

whitmani Lane & Cerqueira

1942: 685 (M*, F, P*, L*). **Type loc.:** Santa Tereza, Espírito Santo, Brazil (IOC). **Distr.:** Brazil. **Refs.:** Lane 1953: 1092 (M*, F, P*, L*). **Ety.:** Loring Whitman. **Informal name:** Whitman Brazilian Canopy Mosquito.

xenismus Harbach

1995b: 583 (M*, F). **Type loc.:** Chichimene, Colombia (NHMUK). **Distr.:** Colombia. **Ety.:** strangeness, novelty (Gr). **Informal name:** Strange Colombian Canopy Mosquito.

Subgenus *Sabethes* Robineau-Desvoidy

1827: 411 (as genus, subgenus). Type species: *Sabethes locuples* Robineau-Desvoidy.

Name derivation: n.s.

syn. *Sabettus* Agassiz 1846: 329 [Harbach 2018c]; Scudder 1882: 297 (unjust. emend. of *Sabethes*).

Important references: Harbach 1991c (tax., comp. morph.); Judd 1996 (phyl.).

albiprivus Theobald

1903a: 323 (M, F). **Type loc.:** São Paulo and Rio de Janeiro, Brazil (NHMUK). **Distr.:** Argentina, Bolivia, Brazil, Colombia, French Guiana, Guyana, Paraguay, Suriname. **Refs.:** Coher 1949: 95 (F*); Lane 1953: 1065 (M*, F); Forattini et al. 1965: 449 (P*, L*; syn.); Belkin 1968b: 29 (lectotype desig.). **Ety.:** n.s. ["white" and "single"? (L); poss. ref. to a lack of white on paddle of midleg]. **Informal name:** Unicorn Brazilian Canopy Mosquito [*albiprivus* = "white one" = Unicorn, the mythical white horse with a single horn].

syn. *albiprivatus* Lutz

1904 (A). In Bourroul 1904: 67. **Type loc.:** São Paulo, Brazil (NHMUK). **Refs.:** Belkin 1968b: 30 (type info.); Belkin et al. 1971: 13 (lectotype desig.).

syn. *albiprivatus* Theobald

1907: 595 (F). **Type loc.:** São Paulo, Brazil (NHMUK). **Refs.:** Belkin 1968b: 30 (lectotype desig.; same specimen as lectotype of *albiprivus* Theobald 1903 and *albiprivatus* Lutz 1904).

syn. *neivai* Petrocchi

1927 (F). In Shannon & del Ponte 1927: 725. **Type loc.:** Santa Clara, Jujuy, Argentina (INM). **Refs.:** Casal & García 1970: 11 (lectotype desig.).

amazonicus Gordon & Evans

1922: 316 (F). **Type loc.:** Macapá, Amazonas, Brazil (NHMUK). **Distr.:** Bolivia, Brazil, Ecuador, Peru, Suriname. **Refs.:** Lane 1953: 1061 (F); Forattini 1959: 286 (M); Belkin 1968b: 30 (tax.); Pecor et al. 2000 (distr.; Peru); Navarro et al. 2013a (distr.; Ecuador). **Ety.:** Amazon. **Informal name:** Amazon Brazilian Canopy Mosquito.

syn. *kappleri* Bonne

1923a: 122 (F). **Type loc.:** Moengo, Suriname (ITH). **Refs.:** Belkin 1968b: 31 (tax.).

syn. *longfieldae* Edwards

1928a: 283 (F). **Type loc.:** Melguierra, Rio Amolar, Mato Grosso, Brazil (NHMUK). **Refs.:** Lane & Cerqueira 1942: 664 (syn.).

batesi Lane & Cerqueira

1942: 669 (M*). **Type loc.:** Tingua, Rio de Janeiro, Brazil (IOC). **Distr.:** Brazil. **Ety.:** Henry Walter Bates. **Informal name:** Bates Brazilian Canopy Mosquito.

belisarioi Neiva

1908a: 351 (F*). **Type loc.:** Bicudos, Minas Gerais, Brazil (LU). **Distr.:** Argentina, Bolivia, Brazil, Colombia, French Guiana, Guyana, Lesser Antilles (includes Trinidad and Tobago), Panama, Peru, Suriname, Venezuela. **Refs.:** Coher 1949: 93 (F*); Lane 1953: 1060 (F); Lane & Causey 1955: 16 (M*); Vargas & Diaz Nájera 1959 (M*); Mattingly 1971a: Pl. 2 (F leg*); Belkin et al. 1971: 14 (type info.). **Ety.:** Belisario Penna (coll.). **Informal name:** Belisario Brazilian Canopy Mosquito.

syn. *schausi* Dyar & Knab

1908: 63 (F). **Type loc.:** Omai [Essequibo], British Guiana [Guyana] (USNM).

syn. *goeldii* Howard, Dyar & Knab

1915: 24 (F). **Type loc.:** Trinidad [Trinidad and Tobago, Lesser Antilles] (USNM). **Refs.:** Howard et al. 1917: 1040 (tax.).

syn. *argyronotum* Edwards

1928a: 283 (F). **Type loc.:** Melguierra, Rio Amolar, Mato Grosso, Brazil (NHMUK). **Refs.:** Belkin 1968b: 30 (lectotype desig.).

bipartipes Dyar & Knab
1906a: 136 (A). **Type loc.:** Santo Domingo [Dominican Republic] (USNM). **Distr.:** Dominican Republic, Ecuador, French Guiana, Panama, Suriname. **Refs.:** Bonne-Wepster & Bonne 1921b (first recorded *Sabethes* larva); Dyar 1922h (M*); Dyar 1928: 15 (M*, F, L*); Lane 1953: 1070 (M*, F, L); Stone & Knight 1957b: 117 (type info., lectotype desig.). **Ety.:** n.s. [double-parted foot (L)]. **Informal name:** Split-footed Dominican Canopy Mosquito.
syn. *chroiopus* Dyar & Knab
1913: 76 (M). **Type loc.:** Potaro Highlands [Essequibo], British Guiana [Guyana] (USNM).

cyaneus (Fabricius)
1805: 35 (A; *Culex*). **Type loc.:** Cayenne, French Guiana [Guiana] (ZMC). **Distr.:** Argentina, Australia, Belize, Brazil, Colombia, Costa Rica, French Guiana, Guyana, Honduras, Lesser Antilles (Saint Kitts and Nevis, Saint Lucia, Saint Vincent and the Grenadines, Trinidad and Tobago), Mexico, Nicaragua, Panama, Peru, Suriname, Venezuela. **Refs.:** Dyar 1922h (M*); Shannon 1931: 155 (L*); Lane 1953: 1063 (M*, F, L); Belkin 1968b: 30 (tax.); Pecor et al. 2002: 249 (tax., distr.; Belize); dos Santos-Mallet et al. 2013 (E*). **Ety.:** n.s. [dark blue (Gr)]. **Informal name:** Blue French Guianese Canopy Mosquito.
syn. *locuples* Robineau-Desvoidy
1827: 412 (A). **Type loc.:** Rio de Janeiro, Guanabara, Brazil (NE). **Refs.:** Belkin et al. 1971: 14 (type loc. info.).
syn. *remipes* Wiedemann
1828: 573 (M; *Culex*). **Type loc.:** Salvador, Bahia, Brazil (ZMC). **Refs.:** Belkin 1968b: 32 (type info.); Belkin et al. 1971: 14 (type loc. info.).

forattinii Cerqueira
1961b: 37 (M*, F). **Type loc.:** road from Manaus–Caracaraí (km 23), Manaus, Amazonas, Brazil (FH). **Distr.:** Brazil. **Refs.:** Forattini & Toda 1967: 103 (P*, L*). **Ety.:** Oswaldo Paulo Forattini. **Informal name:** Forattini Brazilian Canopy Mosquito.

gymnothorax Harbach & Petersen
1992: 102 (M*, F*, P*, L*). **Type loc.:** Bayano reservoir, Maje Island, Panama (USNM). **Distr.:** Belize, Costa Rica, Honduras, Mexico, Panama. **Ety.:** bare thorax (Gr); ref. to "the character of the proepisterna which readily distinguishes this species from *Sa. tarsopus*." **Informal name:** Shirtless Panamanian Canopy Mosquito [*gymnothorax* = "bare chested" = Shirtless].

lanei Cerqueira
1961b: 46 (M*, F). **Type loc.:** road from Manaus–Caracaraí (km 23), Manaus, Amazonas, Brazil (FH). **Distr.:** Brazil. **Ety.:** John Lane. **Informal name:** Lane Brazilian Canopy Mosquito.

nitidus Theobald
1901c: 347 (M, only of type series). **Type loc.:** Pará, Brazil (NHMUK). **Distr.:** Brazil. **Refs.:** Mattingly 1958c: 105 (type info.); Belkin 1968b: 31 (type info., lectotype desig.); Belkin et al. 1971: 14 (from syn. with *bipartipes*; type loc. restr.). **Ety.:** n.s. [shining, bright (L); prob. ref. to its overall bright coloration]. **Informal name:** Shining Brazilian Canopy Mosquito.

ortizi Vargas & Diaz Nájera
1961: 74 (M*, F). **Type loc.:** Aldea El Naranjo, Chiapas, Mexico (ISET). **Distr.:** Mexico. **Ety.:** Carlos Ortiz Mariotte. **Informal name:** Ortiz Mexican Canopy Mosquito.

paraitepuyensis Anduze
1941b: 824 (F). **Type loc.:** Parai-tepuí, Bolivar, Venezuela (DPFA). **Distr.:** Venezuela. **Ety.:** Peñón de Paraitepuy, Venezuela. **Informal name:** Paraitepuy Venezuelan Canopy Mosquito.

purpureus (Theobald)
1907: 617 (F; *Sabethoides*). **Type loc.:** Rio de Janeiro, Brazil (NHMUK). **Distr.:** Argentina, Bolivia, Brazil, French Guiana, Venezuela,. **Refs.:** Lane 1953: 1071 (M*, F, P*, L*); Belkin 1968b: 31 (lectotype desig.); Mattingly 1971a: Pl. 10 (P terminal segs.*), Pl. 23 (L terminal segs.*); Berti et al. 2015 (distr.; Venezuela). **Ety.:** purple, dark red (L); abdomen metallic purple. **Informal name:** Purple Brazilian Canopy Mosquito.
syn. *purpureus* Peryassú
1908: 287 (A). **Type loc.:** Juiz de Fora, Minas Gerais, Brazil (IOC). **Refs.:** Belkin et al. 1971: 14 (lectotype desig.).
syn. *remipusculus* Dyar
1924b: 100 (new name for *purpureus* Peryassú).

quasicyaneus Peryassú
1922b: 179 (F*). **Type loc.:** Matta do Utinga, Belém, Pará, Brazil (LU). **Distr.:** Brazil, Colombia, French Guiana, Peru. **Refs.:** Lane & Cerqueira 1942: 668 (M*, F; from syn. with *albiprivus*); Barreto-Reyes 1957 (F). **Ety.:** nearly (L); similar to *cyaneus*. **Informal name:** Bluish Brazilian Canopy Mosquito.

schnusei (Martini)
1931a: 202 (A; *Dendromyia*). **Type loc.:** San Carlos, Mapiri (La Paz), Bolivia (NE). **Distr.:** Bolivia, Peru. **Refs.:** Lane & Cerqueira 1942: 584 (F); Belkin 1968b: 42 (tax.); Belkin 1971b: 26 (type loc. info.); Harbach & Peyton 1991: 150 (to *Wyeomyia* subg. uncertain). **Ety.:** n.s. **Informal name:** Schnuse Bolivian Canopy Mosquito.

shannoni Cerqueira
1961b: 40 (M*, F, P*, L*). **Type loc.:** Igarapé do Leão, Manaus, Amazonas, Brazil (FH). **Distr.:** Brazil. **Ety.:** Raymond Corbett Shannon. **Informal name:** Shannon Brazilian Canopy Mosquito.

spixi Cerqueira
1961b: 44 (M*, F). **Type loc.:** Estrada Manaus–Caracaraí (km 23), Manaus, Amazonas, Brazil (FH). **Distr.:** Brazil. **Ety.:** Johann Batist von Spix. **Informal name:** Spix Brazilian Canopy Mosquito.

tarsopus Dyar & Knab
1908: 62 (F). **Type loc.:** Bocas del Toro, Panama (USNM). **Distr.:** Belize, Costa Rica, Guatemala, Mexico, Nicaragua, Panama. **Refs.:** Galindo et al. 1951b: 91 (M*); Lane 1953: 1061 (M*, F); Stone & Knight 1957b: 118 (type info., lectotype desig.); Harbach & Peterson 1992: 115 (M*, F*, P*, L*; tax.). **Ety.:** n.s. [having to do with the ankle joint (Gr); poss. ref. to outstanding scales ("ciliation") on the legs ("paddles")]. **Informal name:** Ankle Panamanian Canopy Mosquito.

Subgenus *Sabethinus* Lutz

1904 (as genus). In Bourroul 1904: 48, 57. Type species: *Sabethinus intermedius* Lutz.

syn. Sabettinus Blanchard 1905: 634 (unjust. emend. of *Sabethinus*).

Name derivation: n.s.

Important references: Lane & Cerqueira 1942 (to subg. of genus *Sabethes*); Harbach 1991a (tax., comp. morph.); Harbach 1994b (revision).

andreae Harbach
2018b (M*, F, P*, L*; keys). **Type loc.:** Costa Rica, Province Puntarenas, Golfito, Jiménez, ACOSA, del Cerro Mueller, 0.2 km north Hacia Cerro Rincón, 680 m (MNCR). **Distr.:** Costa Rica. **Ety.:** Andrea Lynn (daughter of author). **Informal name:** Andrea Costa Rican Canopy Mosquito.

idiogenes Harbach
1994b: 227 (M*, F*, P*, L*). **Type loc.:** Pakitza, Rio Manu, Madre de Dios, Peru (USNM). **Distr.:** French Guiana, Peru. **Ety.:** distinctive, peculiar (Gr). **Informal name:** Peculiar Peruvian Canopy Mosquito.

intermedius (Lutz)
1904 (A; *Sabethinus*). In Bourroul 1904: 48, 57. **Type loc.:** near São Paulo, Brazil (NHMUK). **Distr.:** Argentina, Bolivia, Brazil, Colombia, French Guiana, Panama. **Refs.:** Lane & Cerqueira 1942: 676 (M*, F, P*, L*); Lane 1953: 1077 (M*, F, P*, L*); Belkin 1968b: 31 (tax.); Belkin et al. 1971: 15 (lectotype desig.); Harbach 1994b: 211 (M*, F*,P*, L*). **Ety.:** n.s. [in the middle, intermediate (L)]. **Informal name:** Middling Brazilian Canopy Mosquito.

melanonymphe Dyar
1924b: 100 (new name for *albiprivatus* Theobald, not Lutz 1904). **Distr.:** Argentina, Brazil. **Refs.:** Lane & Cerqueira 1942: 678 (M*, F, P*, L*; from syn. with *intermedius*); Lane 1953: 1081 (M*, P*, L*); García & Casal 1968a: 63 (distr.); Harbach 1994b: 217 (M*, F*, P*, L*). **Ety.:** n.s. [black nymph (Gr)]. **Informal name:** Echo Brazilian Canopy Mosquito [*melanonymphe* = "black nymph" = sad nymph = Echo, cursed for her love for Zeus in Greek mythology].
syn. *albiprivatus* Theobald
1907: 620 (M, F*; *Sabethinus*). **Type loc.:** Cantareira [São Paulo], Brazil (NHMUK). **Refs.:** Belkin 1968b: 29 (lectotype desig.); Lane & Cerqueira 1942 (syn.).

theresae Harbach
2018b (M*, F, P*, L*; keys). **Type loc.:** Costa Rica, Province Puntarenas, Golfito, Jiménez, ACOSA, del Cerro Mueller, 0.2 km north hacia [toward] Cerro Rincón, 680 m (MNCR). **Distr.:** Costa Rica. **Ety.:** Theresa M. Howard. **Informal name:** Theresa Costa Rican Canopy Mosquito.

xhyphydes Harbach
1994b: 222 (M*, F*, P*, L*). **Type loc.:** Boracéia Field Station, Salesópolis, São Paulo, Brazil (USNM). **Distr.:** Brazil, Colombia. **Ety.:** "An arbitrary combination of letters." **Informal name:** Arbitrary Brazilian Canopy Mosquito [name was composed Arbitrarily].

Subgenus *Sabethoides* Theobald

1903a: 328 (as genus). Type species: *Sabethoides confusus* Theobald.

syn. *Sabettoides* Blanchard 1905: 423 (unjust. emend.).

Name derivation: n.s.

Important references: Lane & Cerqueira 1942 (to subg.); Harbach 1991a (tax., comp. morph.); Harbach 1991c (larval ontogeny); Judd 1996 (phyl.).

chloropterus (von Humboldt)
1819: 340 (A; *Culex*). **Type loc.:** Guayaquil River, near Borodan, Ecuador (NE). **Distr.:** Argentina, Belize, Bolivia, Brazil, Colombia, Costa Rica, Ecuador, El Salvador, French Guiana, Guatemala, Guyana, Honduras, Lesser Antilles (includes Trinidad and Tobago), Mexico, Nicaragua, Panama, Peru, Suriname, Venezuela. **Refs.:** Anduze 1941b: 825 (F; as *imperfectus*); Lane & Cerqueira 1942: 675 (F; tax., keys, bion., distr.); Lane 1953: 1076 (M*, F, L); Leví-Castillo 1953f: 96 (M*, F, P*, L*); Lane & Causey 1955: 17 (M, P*, L); Galindo 1957a (behav.); Galindo 1958: 429 (bion.); Belkin 1968b: 30 (type info.); Harbach 1991c (L*, all instars; morph., ontogeny); Navarro & Machado-Allison 1995 (bion.); Judd 1996 (M, F, P, L; phyl.); Zsemlye et al. 2005 (behav.). **Ety.:** n.s. [green wing (Gr)]. **Informal name:** Green-winged Ecuadorian Canopy Mosquito. (Vol. 1, p. 544)
syn. *confusus* Theobald
1903a: 328 (F*; *Sabethoides*). **Type loc.:** Brazil (NHMUK). **Refs.:** Mattingly 1958c: 105 (type info.).
syn. *rangeli* Surcouf & Gonzales-Rincones
1911: 251 (F*; *Sabethoides*). **Type loc.:** Maturin, Monagas, Venezuela (MNHP).

syn. *imperfectus* Bonne-Wepster & Bonne 1920: 165 (F; *Sabethoides*). **Type loc.:** Suriname (ITH). **Refs.:** Lane 1951 (syn.); Belkin 1968b: 30 (type info.).

conditus Moses, Howard & Harbach 2000: 992 (M*, P*, L*). **Type loc.:** Aragua, Ocumare de la Costa, Puerto Ocumare, East of (19PFM3459), Venezuela (USNM). **Distr.:** Brazil, Venezuela. **Ety.:** hidden, unseen (L); close to *chloropterus*. **Informal name:** Concealed Venezuelan Canopy Mosquito.

glaucodaemon (Dyar & Shannon) 1925b: 39 (F; *Sabethoides*). **Type loc.:** near San Alberto, Rio Branco, Amazonas, Brazil (USNM). **Distr.:** Bolivia, Brazil, Guyana, Suriname. **Refs.:** da Costa Lima 1931b: 57 (M*); Lane & Cerqueira 1942: 673 (M*, F); Lane & Causey 1955: 16 (P*, L*). **Ety.:** n.s. [bluish-gray spirit (L, Gr)]. **Informal name:** Egoi Brazilian Canopy Mosquito [Egoi was a minor deity of the south wind in Basque mythology].

tridentatus Cerqueira 1961b: 48 (M*, F, P*, L*). **Type loc.:** Igarapé da Bolivia, Manaus, Amazonas, Brazil (FH). **Distr.:** Brazil. **Ety.:** n.s. [three teeth (L)]. **Informal name:** Three-toothed Brazilian Canopy Mosquito.

Genus *Shannoniana* Lane & Cerqueira

(Vol. 1, p. 280)

1942: 502 (as subg. of genus *Trichoprosopon*). Type species: *Goeldia fluviatilis* Theobald.

Name derivation: Raymond Corbett Shannon.

syn. *Goeldia* Theobald 1903a: 330 (not *Goeldia* Keyserling 1891). Type species: *Goeldia fluviatilis* Theobald.

Important references: Lane 1953 (tax., review, keys; as subg. of *Trichoprosopon*); Zavortink 1979b: 11 (from subg. of *Trichoprosopon*); Clark-Gil & Darsie 1983 (keys; Guatemala); Harbach & Peyton 1993 (L mouthparts, phyl.); Judd 1996 (phyl.); Harbach & Kitching 1998 (phyl.); Harbach & Peyton 2000 (phyl.); Harbach, Vu, & Kitching 2007 (phyl.).

fluviatilis (Theobald)
1903a: 331 (F; *Goeldia*). **Type loc.:** São Paulo, Brazil (NHMUK). **Distr.:** Argentina, Bolivia, Brazil, Colombia, Ecuador, French Guiana, Guatemala, Guyana, Mexico, Nicaragua, Panama. **Refs.:** Lane & Cerqueira 1942: 503 (M*, F, P*, L*); Lane 1953: 829 (M*, F, P*, L*); Díaz Nájera 1961: 201 (L*), 203 (L*; as *longipalpis*); Belkin 1968b: 35 (type info., tax.); Belkin et al. 1971: 9 (type info.); Zavortink 1979b: 11 (from genus *Trichoprosopon*). **Ety.:** n.s. [of rivers (L); from the Demerara River]. **Informal name:** Riverine Brazilian Toothy Mosquito.
syn. *longipalpis* Lutz
1905: 127 (M, F; *Hyloconops*). **Type loc.:** Brazil (IOC). **Refs.:** Lane & Cerqueira 1942: 503 (syn.); Belkin 1968b: 36 (tax.); Belkin et al. 1971: 9 (type info.).

moralesi (Dyar & Knab)
1919: 3 (F; *Lesticocampa*). **Type loc.:** San Felipe, Departamento Retalhuela, Guatemala (USNM). **Distr.:** Belize, Guatemala, Mexico, Panama. **Refs.:** Vargas & Martínez 1953: 302 (M*, F, L*; from syn. with *fluviatilis*); Stone & Knight 1957b: 118 (type info., lectotype desig.); Díaz Nájera 1961: 203 (L*); Zavortink 1979b: 12 (from genus *Trichoprosopon*). **Ety.:** Dr. Morales. **Informal name:** Morales Guatemalan Toothy Mosquito.

schedocyclia (Dyar & Knab)
1908: 64 (F; *Lesticocampa*). **Type loc.:** [Bluefields], Interior, Nicaragua (USNM). **Distr.:** Bolivia, Brazil, French Guiana, Guatemala, Mexico, Nicaragua, Panama, Venezuela. **Refs.:** Lane & Cerqueira 1942: 505 (F, P, L); Vargas & Martínez 1953: 300 (M*); Stone & Knight 1957b: 119 (type info., lectotype desig.); Zavortink 1979b: 12 (from genus *Trichoprosopon*). **Ety.:** n.s. **Informal name:** Motorcycle Nicaraguan Toothy Mosquito [*schedocyclia* = "swift cycle" = Motorcycle].

Genus *Topomyia* Leicester

(Vol. 1, p. 282)

Important references: Thurman 1959b (tax., keys, bion.; Thailand); Marks 1971a (metapleural generic character); Tanaka et al. 1979 (tax., bion.; Japan); Lee et al. 1988c (tax., key, bion., distr.; Australasia); Harbach & Peyton 1993 (L mouthparts, phyl.); Judd 1996 (phyl.); Harbach & Kitching 1998 (phyl.); Harbach & Peyton 2000 (phyl.); Rattanarithikul, Harrison et al. 2005 (F, L; tax., keys, bion.; Thailand); Harbach, Vu, & Kitching 2007 (phyl., comp. morph.); Miyagi & Toma 2010a, 2010b (morph.); Harbach & Culverwell 2014 (tax., bion.; subg. *Miyagiella*; subgeneric comp.).

Subgenus *Miyagiella* Harbach

2014 (*Topomyia* subg.). In Harbach & Culverwell 2014: 576. Type species: *Topomyia* (*Miyagiella*) *discors* Harbach.

Name derivation: Dr. Ichiro Miyagi.

Important references: Thurman 1959b (morph. comp.); Harbach & Culverwell 2014 (tax., bion.; subg. *Miyagiella*; subgeneric comp.).

discors Harbach
2014 (M*; comp. of *Topomyia* subg.). In Harbach & Culverwell 2014. **Type loc.:** Kampung Paradason (06°46135' N, 116°47183' W), Sabah, Malaysian Borneo. **Distr.:** Malaysia. **Ety.:** different, not near (L); ref. to unusual male genitalia. **Informal name:** Different Malaysian Omnivore Mosquito.

Subgenus *Suaymyia* Thurman

1959b: 44 (as subg. of genus *Topomyia*). Type species: *Topomyia* (*Suayamia*) *cristata* Thurman.

Name derivation: pretty fly (Thai word for pretty and myia [fly (L)]).

Important references: Thurman 1959b (morph. comp.); Tanaka et al. 1979 (tax., bion.; Japan); Lee et al. 1988 (tax., key, bion.; Australasia); Harbach & Peyton 1993 (L mouthparts; phyl.); Harbach 2007 (tax., phyl.; five species moved to genus *Kimia*); Harbach & Culverwell 2014 (tax., bion.; subg. *Miyagiella*; subgeneric comp.).

apsarsae Klein
1977a: 126 (M*, F*). **Type loc.:** Stung Charal forest, Kompong Speu Province, Cambodia (SSC). **Distr.:** Cambodia, Philippines, Thailand. **Refs.:** Rattanarithikul et al. 2007 (F*, P*, L*; bion., distr., keys) as *apsarae* [*sic*]. **Ety.:** n.s. **Informal name:** Apsarsa Cambodian Omnivore Mosquito.
syn. *rausai* Miyagi
1980: 178 (M*, F, P*). **Type loc.:** Montigble Subcolony, Iwahig, Palawan Island, Philippines (USNM). **Refs.:** Miyagi, Toma, & Cabrera 1983: 112 (syn.).

argenteoventralis Leicester
1908: 240 (M, F; as *argentoventralis* [*lapsus*?]). **Type loc.:** The Gap [Selangor], Malaya [Malaysia] (NHMUK). **Distr.:** Indonesia, Malaysia. **Refs.:** Edwards 1922b: 442 (M*); Bonne-Wepster 1954: 17 (M, F, L*); Thurman 1959b: 45 (n. comb.); Miyagi et al. 2006 (redescr.). **Ety.:** silvery venter (L). **Informal name:** Silver-bellied Malaysian Omnivore Mosquito.

auriceps Brug
1939: 96 (M*, F). **Type loc.:** Kalawara, Celebes [Sulawesi] (NHMUK). **Distr.:** Indonesia, Malaysia (includes Sabah and Sarawak). **Refs.:** Thurman 1959b: 45 (n. comb.); Miyagi & Toma 2010a (M*, F, P, L). **Ety.:** gold head (L); broad golden ventral border on head, gold stripe on proboscis. **Informal name:** Gold-headed Indonesian Omnivore Mosquito.

bambusaihole Dong, Zhou, & Dong
1997a: 251 (M*, P*, L*). **Type loc.:** Jinghong, Yunnan Province, China [People's Republic of China] (YIM). **Distr.:** People's Republic of China. **Ety.:** n.s. [poss. ref. to holes in bamboo]. **Informal name:** Bamboo Chinese Omnivore Mosquito.

bannaensis Gong & Lu
1995: 478 (F). **Type loc.:** Mengla County (21°30′ N, 101°35′ E, 600 m), Yunnan Province, China [People's Republic of China] (IEYP). **Distr.:** People's Republic of China. **Ety.:** Xishuangbanna, Yunnan Prov., China. **Informal name:** Yunnan Chinese Omnivore Mosquito.

cristata Thurman
1959b: 46 (M*). **Type loc.:** Doi Suthep, Chiengmai Province, Thailand (USNM). **Distr.:** India, People's Republic of China, Thailand. **Refs.:** Rattanarithikul et al. 2007 (F*, P*, L*; bion., distr., keys). **Ety.:** n.s. [crested (L)]. **Informal name:** Crested Thai Omnivore Mosquito.

dulongensis Gong & Lu
1995: 485 (F, L*). **Type loc.:** Dulong River, Gongshan County (27°69′ N, 98°33′ E; 1,600 m), China [Yunnan Province, People's Republic of China] (IEYP). **Distr.:** People's Republic of China. **Refs.:** Gong et al. 1999: 227 (M*; tax.). **Ety.:** Dulong River, China. **Informal name:** Dulong Chinese Omnivore Mosquito.

houghtoni Feng
1941: 244 (M*, F, P*, L*). **Type loc.:** Peng Lung Chai, near Chefang, Yunnan Province, China [People's Republic of China] (LU). **Distr.:** People's Republic of China, Thailand. **Refs.:** Thurman 1959b: 45 (n. comb.); Rattanarithikul et al.

2007 (F*, P*, L*; bion., distr., keys). **Ety.:** Henry. S. Houghton. **Informal name:** Houghton Chinese Omnivore Mosquito.

kelabitensis Miyagi & Toma
2010b: 354 (M*, F*, P*, L*). **Type loc.:** Arur Dalan (elevation 1,000–1,200 m above sea level), Bario, Sarawak, Malaysian Borneo. **Distr.:** Malaysia. **Refs.:** Harbach 2010 (justified emendation from *kelabitense*). **Ety.:** after the Kelabat, one of the smallest tribes in Sarawak, Malaysia. **Informal name:** Kelabat Malaysian Omnivore Mosquito.

lehcharlesi Miyagi & Toma
2008: 163 (M*, F, P*, L*). **Type loc.:** Penrissen Highlands, Padawan Sarawak, Malaysia (USNM). **Distr.:** Malaysia. **Ety.:** Moi Ung Charles Leh. **Informal name:** Leh Malaysian Omnivore Mosquito.

leucotarsis Thurman
1959b: 47 (M*). **Type loc.:** Doi Suthep, Chiengmai Province, Thailand (USNM). **Distr.:** Thailand. **Refs.:** Rattanarithikul et al. 2007 (F*, P*, L*; bion., distr., keys). **Ety.:** white tarsus (Gr). **Informal name:** White-ankled Thai Omnivore Mosquito.

mengi Dong, Wang, & Lu
1990: 64 (M*). **Type loc.:** Mengla County, Yunnan Province, China [People's Republic of China] (YIM). **Distr.:** People's Republic of China. **Ety.:** Menglang, China. **Informal name:** Meng Chinese Omnivore Mosquito.

nepenthicola Miyagi & Toma
2007: 167 (M*, F*, P*, L*). **Type loc.:** Bario Pa Umor (approximately 1,200 m), Sarawak, Malaysia [Malaysian Borneo]. **Distr.:** Malaysia. **Ety.:** larvae associated with *Nepenthes* pitcher plants. **Informal name:** Nepenthes Malaysian Omnivore Mosquito.

papuensis Marks
1960b: 91 (M*, F, P*, L*). **Type loc.:** Koipa, 4–5 mi from Saiho (08°50′ S, 148°05′ E) on the Popondetta road, Northern District, Papua (UQIC). **Distr.:** Indonesia, Papua New Guinea. **Refs.:** Lee et al. 1988c: 53 (tax., bion., distr., review); Miyagi & Toma 1997: 134 (distr.). **Ety.:** Papua New Guinea. **Informal name:** Koipa Papuan Omnivore Mosquito.

pseudoauriceps Miyagi & Toma
2010a: 35 (M*, F, P, L). **Type loc.:** Sarawak, Malaysia [Malaysian Borneo]. **Distr.:** Malaysia (Borneo, Sarawak). **Ety.:** false (Gr); similar to *auriceps*. **Informal name:** Gilt-headed Malaysian Omnivore Mosquito [*pseudoauriceps* = "false gold headed" = Gilt-headed].

pseudoleucotarsis Thurman
1959b: 48 (M*). **Type loc.:** Doi Suthep, Chiengmai Province, Thailand (USNM). **Distr.:** Thailand. **Ety.:** false (Gr); similar to *leucotarsis*. **Informal name:** Dirty-ankled Thai Omnivore Mosquito [*pseudoleucotarsis* = "false white ankled" = Dirty-ankled].

roslihashimi Miyagi & Toma
2005: 275 (M*, P*, L*). **Type loc.:** Ulu Gombak, Malaysia (USNM). **Distr.:** Malaysia. **Ety.:** Rosli Hashim. **Informal name:** Hashim Malaysian Omnivore Mosquito.

spathulirostris Edwards
1923b: 2 (M, F). **Type loc.:** Cameron's Highlands, near Tapah [Perak], Malaya [Malaysia] (NHMUK). **Distr.:** Malaysia, People's Republic of China, Thailand. **Refs.:** Mattingly 1971a: Pl. 8 (P trumpet*), Pl. 9 (P abdominal segment VII*), Pl. 26 (L terminal segs.*); Miyagi, Toma, & Ramalingam 1989: 41 (from subg. *Topomyia*); Dong, Zhou, & Dong 1997a: 251 (distr.; People's Republic of China); Rattanarithikul et al. 2007 (F*, P*, L*; bion., distr., keys). **Ety.:** flat beak (L); "In both sexes the apical fifth, or rather more of the proboscis . . . is very broad and flat." **Informal name:** Flat-beaked Malaysian Omnivore Mosquito.

spinophallus Zhou, Zhu, & Lu
1999: 107 (M*, P*). **Type loc.:** Xishuangbanna, Yunnan Province, China [People's Republic of China] (Yunnan Institute of Malaria Control). **Distr.:** People's Republic of China. **Ety.:** spiny phallus (L, Gr); "Phallosome very well developed and divided 2 lateral plates; and with teeth on it." **Informal name:** Spiny Chinese Omnivore Mosquito.

tumetarsalis Chen & Zhang
1988: 213 (M*, F*, P*, L*). **Type loc.:** Luo-tian County, Guizhou Province, China [People's Republic of China] (GMC). **Distr.:** People's Republic of China. **Ety.:** n.s. [begin to swell, tarsus (L)]. **Informal name:** Swell-footed Chinese Omnivore Mosquito.

yanbarensis Miyagi
1976: 202 (M*, F, P*, L*). **Type loc.:** Yona, Kunigami, Northern Okinawa-Honto, Ryukyu Islands, Japan (EIHU). **Distr.:** Japan, Thailand. **Refs.:** Tanaka et al. 1979: 478 (M*, F*, L*); Okazawa et al. 1986b (E*); Rattanarithikul et al. 2007 (F*, P*, L*; bion., distr., keys). **Ety.:** *yanbaru* is a local name for "the northern sylvan region of Okinawa-honto." **Informal name:** Yanbaru Japanese Omnivore Mosquito.

yanbareroides Dong & Miyagi
1995: 343 (M*, P*, L*). **Type loc.:** China [People's Republic of China] (YIM). **Distr.:** People's Republic of China. **Ety.:** similar, resembling (Gr); like *yanbarensis*. **Informal name:** Huli Jing Chinese Omnivore Mosquito [Huli jing are fox spirits that inhabit woods and fields in Chinese folklore].

Subgenus *Topomyia* Leicester

1908: 238 (as genus, subgenus). Type species: *Topomyia minor* Leicester (Brunetti 1914: 72).

Name derivation: n.s. [topo- = place or spot (Gr) plus small fly (Gr)].

syn. *Pseudograhamia* Theobald 1910a: 26 (as genus). Type species: *Pseudograhamia aureoventer* Theobald.

Important references: Thurman 1959b (morph. comp.); Lee et al. 1988c (tax., key, bion.; Australasia); Harbach & Peyton 1993 (L mouthparts; phyl.); Harbach & Culverwell 2014 (tax., bion.; subg. *Miyagiella*; subgeneric comp.).

aenea Thurman
1959b: 39 (M*, F). **Type loc.:** Ngao, Lampang Province, Thailand (USNM). **Distr.:** Thailand. **Refs.:** Rattanarithikul et al. 2007 (F*, P*, L*; bion., distr., keys). **Ety.:** n.s. [bronze colored (L); poss. ref. to "posterior pronotal lobe with coppery scales" and other coppery scutal features]. **Informal name:** Bronze Thai Omnivore Mosquito.

aliyusopi Miyagi & Toma
2014 (M*, F, P*, L*; bion.). In Miyagi et al. 2014: 28. **Type loc.:** Mount Murud, Sarawak, Malaysia [Malaysian Borneo]. **Distr.:** Malaysia. **Ety.:** Haji Ali Bin Yusop. **Informal name:** Yusop Malaysian Omnivore Mosquito.

angkoris Klein
1977a: 123 (M*, F). **Type loc.:** Mount Bokor, km 15 on Kampot–Bokor road, Kampot Province, Cambodia (SSC). **Distr.:** Cambodia, Thailand. **Refs.:** Ward 1984b: 251 (as *ankoris*, *lapsus*); Rattanarithikul et al. 2007 (F*, P*, L*; bion., distr., keys). **Ety.:** Angkor [Archaeological Park]. **Informal name:** Angkor Cambodian Omnivore Mosquito.

argyropalpis Leicester
1908: 242 (M, F). **Type loc.:** Kuala Lumpur and the Gap [Selangor], Malaya (NHMUK). **Distr.:** Cambodia, Indonesia, Malaysia, Philippines, Thailand. **Refs.:** Edwards 1922b: 441 (M*, F); Baisas & Feliciano 1953: 174 (M*); Basio 1971b: 79 (M*; bion.); Rattanarithikul et al. 2007 (F*, P*, L*; bion., distr., keys). **Ety.:** silver feeler (Gr); palpus. **Informal name:** Silver-lipped Malaysian Omnivore Mosquito.

syn. *gregoryi* Ludlow
1911b: 128 (F; *Kingia*). **Type loc.:** Ludlow Barracks, Parang [Cotabato], Mindanao, Philippines (USNM). **Refs.:** Bohart 1945: 24 (syn.).

aureoventer (Theobald)
1910a: 27 (F; *Pseudograhamia*). **Type loc.:** Palode, 20 mi northeast of Trivandrum, Travancore, India (IM). **Distr.:** India, Malaysia, Nepal. **Refs.:** Barraud 1934: 46 (F); Joshi et al. 1965: 139 (distr.); Darsie & Pradhan 1990 (F, L; tax., keys, bion., distr.; Nepal); Natarajan et al. 2019 (M*, P*, L*). **Ety.:** golden belly (L); abdominal venter pale dull golden. **Informal name:** Gold-bellied Indian Omnivore Mosquito.

baolini Gong
1989b: 220 (M*, F*). **Type loc.:** Yingjiang County, Yunnan Province, China [People's Republic of China] (YIE). **Distr.:** People's Republic of China. **Ety.:** Lu Baolin. **Informal name:** Baolin Chinese Omnivore Mosquito.

barbus Baisas
1946d: 41 (M*, F, P*, L*). **Type loc.:** Mindanao, Philippines (USNM). **Distr.:** Philippines. **Refs.:** Knight & Chamberlain 1948: 13 (P*); Mattingly 1971a: Pl. 9 (P terminal segs.*); Basio 1971b: 80 (M*; bion.); Baisas 1974: 144 (M*, F, P*, L*; tax., bion., distr.; Philippines). **Ety.:** n.s. [beard or bearded (L)]. **Informal name:** Bearded Philippine Omnivore Mosquito.

bifurcata Dong, Wang, & Lu
1995: 484 (M*). **Type loc.:** Mengla, Yunnan Province, China [People's Republic of China] (YIM). **Distr.:** India, People's Republic of China, Thailand. **Refs.:** Rattanarithikul et al. 2007 (F*, P*, L*; bion., distr., keys). **Ety.:** "Distimere [gonostylus] bifurcated." **Informal name:** Bifurcated Chinese Omnivore Mosquito.

cabrerai Miyagi, Toma, & Rivera
1983: 1 (M*). **Type loc.:** Mindanao Island, Philippines (PNM). **Distr.:** Philippines. **Ety.:** Benjamin Cabrera. **Informal name:** Cabrera Philippine Omnivore Mosquito.

chaii Miyagi & Toma
2012 (M*, F, P*, L*). In Miyagi, Toma, Okazawa et al. 2012: 104. **Type loc.:** Malaysia Headquarters of Lanjak-Entimau Wildlife Sanctuary (01°38777' N, 112°167709' E), Matang National Park, Katibas, Lanjak-Entimau, Sarawak, Borneo High, Bario [Sarawak, Malaysian Borneo] (USNM). **Distr.:** Malaysia. **Ety.:** Paul Chai. **Informal name:** Chai Malaysian Omnivore Mosquito.

danaraji Ramalingam
1988: 245 (M*, P*). **Type loc.:** The Gap, Selangor, Peninsular Malaysia (USNM). **Distr.:** Malaysia. **Ety.:** Tan Sri T.J. Danaraj. **Informal name:** Danaraj Malaysian Omnivore Mosquito.

dejesusi Baisas & Feliciano
1953: 173 (M*, F). **Type loc.:** Mount Apo, Davao, Mindanao, Philippines (USNM). **Distr.:** Philippines. **Refs.:** Basio 1971b: 80 (M*; bion.). **Ety.:** Vicente de Jesus. **Informal name:** De Jesus Philippine Omnivore Mosquito.

dubitans Leicester
1908: 246 (M). **Type loc.:** The Gap [Selangor], Malaya [Malaysia] (NE). **Distr.:** Indonesia, Malaysia. **Refs.:** Brug 1939: 97 (M*, F, L*?); Lien et al. 1977 (distr.). **Ety.:** n.s. [doubting (L)]. **Informal name:** Dubious Malaysian Omnivore Mosquito.

gracilis Leicester
1908: 244 (M, F). **Type loc.:** The Gap [Selangor], Malaya [Malaysia] (NHMUK). **Distr.:** Cambodia, Indonesia, Laos, Malaysia, Thailand, Vietnam. **Refs.:** Edwards 1922b: 443 (M*); Borel 1930: 145 (M*, P, L*); Brug 1931c: 9 (P*, L*); Mattingly 1971a: Pl. 24 (L head*, thorax*, terminal segs.*); Rattanarithikul et al. 2007 (F*, P*, L*; bion., distr., keys). **Ety.:** slender (L); proboscis, legs. **Informal name:** Gracile Malaysian Omnivore Mosquito.

hardini Miyagi, Toma, & Ramalingam
1989: 91 (M*, F, P*, L*). **Type loc.:** Mount Serapi, Kaman, Kuching, Sarawak, Malaysia [Malaysian Borneo] (USNM). **Distr.:** Malaysia. **Ety.:** Stalin Hardin. **Informal name:** Hardin Malaysian Omnivore Mosquito.

hernandoi Baisas & Feliciano
1953: 177 (M*). **Type loc.:** Mount Apo, Davao, Mindanao, Philippines (USNM). **Distr.:** Philippines. **Refs.:** Basio 1971b: 80 (M*; bion.). **Ety.:** Eugenio Hernando. **Informal name:** Hernando Philippine Omnivore Mosquito.

hirtusa Gong
1989a: 90 (M*). **Type loc.:** Yunnan, China [People's Republic of China] (YIE). **Distr.:** India, People's Republic of China. **Ety.:** corrupted form of "hirtus" meaning shaggy (L) [poss. ref. to specialized tufts of setae on the gonocoxite and gonostylus)]. **Informal name:** Shaggy Chinese Omnivore Mosquito.

inclinata Thurman
1959b: 41 (M*). **Type loc.:** Buker Cabin Area, Doi Suthep, Chiengmai Province, Thailand (USNM). **Distr.:** Thailand. **Refs.:** Rattanarithikul et al. 2007 (F*, P*, L*; bion., distr., keys). **Ety.:** n.s. [inclined (L)]. **Informal name:** Inclined Thai Omnivore Mosquito.

irianensis Miyagi & Toma
1997: 134 (M*, F, P*, L*). **Type loc.:** Arso, Irian Jaya [Indonesia: West Papua] (USNM). **Distr.:** Indonesia. **Ety.:** Irian Jaya, Indonesia. **Informal name:** Irian Jaya Indonesian Omnivore Mosquito.

javaensis Miyagi & Toma
1995: 179 (M*, F, P*, L*). **Type loc.:** Mount Cibodas (1,000 m), Bogor Botanical Garden, West Java, Indonesia (USNM). **Distr.:** Indonesia. **Ety.:** Java. **Informal name:** Java Indonesian Omnivore Mosquito.

katibasensis Miyagi & Toma
2012 (M*, P*, L*). In Miyagi, Toma, Okazawa et al. 2012: 98. **Type loc.:** Katibas, Lanjak-Entimau, Sarawak, Malaysia [Headquarters (01°38777' N, 112°167709' E) of Lanjak-Entemau Wildlife Sanctuary, Matang National Park, Borneo High, Bario] (USNM). **Distr.:** Malaysia (Borneo, Sarawak). **Ety.:** ref. to river Sg. Katibas. **Informal name:** Katibas Malaysian Omnivore Mosquito.

lindsayi Thurman
1959b: 42 (M*). **Type loc.:** Doi Suthep, Chiengmai Province, Thailand (USNM). **Distr.:** Thailand. **Refs.:** Dong & Wang 1988 (F, P*, L*); Rattanarithikul et al. 2007 (F*, P*, L*; bion., distr., keys). **Ety.:** Dale R. Lindsay. **Informal name:** Lindsay Thai Omnivore Mosquito.

longisetosa Gong
1994: 210 (M*, L*). **Type loc.:** Yingjiang County, Yunnan Province, China [People's Republic of China] (Research Institute of Epidemic Diseases of Yunnan Province). **Distr.:** People's Republic of China. **Ety.:** long setae (L); ref. to "ventro-medial of basimere band-like and with two rows of very long flat setae." **Informal name:** Long-bristled Chinese Omnivore Mosquito.

malaysiensis Ramalingam & Banu
1987: 119 (M*, F*, P*, L*). **Type loc.:** Tambunan, Sabah, Malaysia [Malaysian Borneo] (USNM). **Distr.:** Malaysia (Borneo, Sabah). **Ety.:** Malaysia. **Informal name:** Tambunan Malaysian Omnivore Mosquito.

margina Gong & Lu
1995: 476, 484 (F*, L*). **Type loc.:** Jinghong County (29°98′ N, 100°77′ E, elevation 660 m), Yunnan Province, China [People's Republic of China] (IEYP). **Distr.:** People's Republic of China. **Ety.:** n.s. [margin, border (L)]. **Informal name:** Marginal Chinese Omnivore Mosquito.

minor Leicester
1908: 238 (M, F). **Type loc.:** near Kuala Lumpur, Malaya [Malaysia] (NHMUK). **Distr.:** Malaysia. **Refs.:** Edwards 1922b: 440 (M*). **Ety.:** smaller (L); smallest *Topomyia*. **Informal name:** Small Malaysian Omnivore Mosquito.

nicksoni Miyagi & Toma
2012 (M*, F*, P*, L*). In Miyagi, Toma, Okazawa et al. 2012: 109. **Type loc.:** Katibas, Lanjak-Entimau, Sarawak [Malaysia Headquarters (01°38777' N, 112°167709' E) of Lanjak-Entemau Wildlife Sanctuary, Matang National Park, Borneo High, Bario] (USNM). **Distr.:** Malaysia (Borneo, Sarawak). **Ety.:** Nickson Joseph Robi. **Informal name:** Nickson Malaysian Omnivore Mosquito.

nigra Leicester
1908: 245 (M, F). **Type loc.:** Ampang, near Kuala Lumpur [Selangor], Malaya (NHMUK). **Distr.:** Indonesia, Malaysia. **Refs.:** Edwards 1922b: 443 (M*, P); Brug 1931c: 9 (L*). **Ety.:** n.s. [black (L); overall color: "Male very black without white spot on the scutellum"]. **Informal name:** Black Malaysian Omnivore Mosquito.

pilosa Brug
1931c: 12 (M*, F, L*). **Type loc.:** Ranau-See [Ranau Danau], Sumatra [Indonesia] (NHMUK). **Distr.:** Indonesia. **Ety.:** n.s. **Informal name:** Hairy Indonesian Omnivore Mosquito.

pseudobarbus Baisas
1946d: 46 (M*, F). **Type loc.:** San Miguel, Bulacan, Philippines (NE; paratype USNM). **Distr.:** Philippines. **Refs.:** Basio 1971b: 80 (M*; bion.); Baisas 1974: 146 (M*, F, P, L; tax., bion., distr.; Philippines). **Ety.:** false (Gr); similar to *barbus*. **Informal name:** False-bearded Philippine Omnivore Mosquito.

rubithoracis Leicester
1908: 243 (M, F). **Type loc.:** The Gap, Malaya [Malaysia] (NHMUK). **Distr.:** Malaysia. **Refs.:** Edwards 1922b: 442 (M*); Miyagi et al. 2011 (P*, L*). **Ety.:** red thorax (L); ref. to "the light reddish-brown thorax and pleurae." **Informal name:** Red-chested Malaysian Omnivore Mosquito.

sabahensis Ramalingam & Ramakrishna
1988: 34 (M*, F*, P*, L*). **Type loc.:** Semporna, Sabah, Malaysia [Malaysian Borneo] (USNM). **Distr.:** Malaysia (Borneo, Sarawak). **Refs.:** Delfinado 1967 (F*). **Ety.:** Sabah, Malaysia. **Informal name:** Sabah Malaysian Omnivore Mosquito.

svastii Thurman
1959b: 43 (M*). **Type loc.:** Doi Suthep, Chiengmai Province, Thailand (USNM). **Distr.:** Thailand. **Refs.:** Rattanarithikul et al. 2007 (F*, P*, L*; bion., distr., keys); Dong & Guo 2008 (F*, P*, L*). **Ety.:** Svasti Daengsvang. **Informal name:** Svasti Thai Omnivore Mosquito.

sylvatica Lu, Dong, & Wang
1986: 406 (M*). **Type loc.:** Mengla, Yunnan Province, China [People's Republic of China] (YIM). **Distr.:** People's Republic of China. **Ety.:** n.s. [sylvan (L)]. **Informal name:** Sylvan Chinese Omnivore Mosquito.

tenuis Edwards
1922a: 290 (M*). **Type loc.:** The Gap [Selangor], Malaya [Malaysia] (NHMUK). **Distr.:** Malaysia, Thailand. **Refs.:** Mattingly 1971a: Pl. 26 (L abdomen segment V*); Rattanarithikul et al. 2007 (F*, P*, L*; bion., distr., keys). **Ety.:** n.s. [fine, delicate (L); poss. ref. to fine filament on aedeagus]. **Informal name:** Delicate Malaysian Omnivore Mosquito.

tipuliformis Leicester
1908: 247 (M). **Type loc.:** Raub [Pahang] and the Gap [Selangor], Malaya [Malaysia] (NHMUK). **Distr.:** Indonesia, Malaysia, Thailand. **Refs.:** Edwards 1922b: 441 (M*, F); Barraud 1924m (L*); Brug 1931c: 9 (L*); Mattingly 1971a: Pl. 9 (P trumpet*); Rattanarithikul et al. 2007 (F*, P*, L*; bion., distr., keys). **Ety.:** n.s. [tipulid-like; "Several species of tipulidae [*sic*] closely resemble it on the wing"]. **Informal name:** Tipulid Malaysian Omnivore Mosquito.

trifida Edwards
1922a: 289 (M*). **Type loc.:** Kuching?, Sarawak, Borneo [Malaysian Borneo] (NHMUK). **Distr.:** Malaysia (Borneo, Sarawak). **Ety.:** n.s. [three prongs (L); poss. ref. to male genitalia character]. **Informal name:** Three-pronged Malaysian Omnivore Mosquito.

unispinosa Thurman
1959b: 44 (M*). **Type loc.:** Doi Suthep, Chiengmai Province, Thailand (USNM). **Distr.:** Malaysia, Thailand. **Refs.:** Rattanarithikul et al. 2007 (F*, P*, L*; bion., distr., keys). **Ety.:** n.s. [one spine (L)]. **Informal name:** One-spined Thai Omnivore Mosquito.

vijayae Ramalingam
1975: 185 (M*, F, P*, L*). **Type loc.:** Ulu Langat Forest Reserve, Selangor, Malaysia (USNM). **Distr.:** Malaysia, Thailand. **Refs.:** Rattanarithikul et al. 2007 (F*, P*, L*; bion., distr., keys). **Ety.:** Vijaya Ramalingam. **Informal name:** Vijaya Malaysian Omnivore Mosquito.

winter Dong, Wu, & Mao
2006: 217 (M*, L*). **Type loc.:** Nandao River, Simao County, Yunnan Province [People's Republic of China]. **Distr.:** People's Republic of China. **Ety.:** n.s. [transl. from Chinese; type specimens were collected in the winter season (Harbach 2018c)]. **Informal name:** Winter Chinese Omnivore Mosquito.

yongi Miyagi, Toma, & Ramalingam
1991: 185 (M*, F, P*, L*). **Type loc.:** Ulu Gombak Forest Reserve, Selangor, Dani Ehsan, Malaysia (USNM). **Distr.:** Malaysia. **Ety.:** Yong Hoi Sen. **Informal name:** Yong Malaysian Omnivore Mosquito.

zhangi Gong
1991: 228 (M*, F, L*). **Type loc.:** Dulong River, Gongshan County, Yunnan Province, China [People's Republic of China] (YIE). **Distr.:** People's Republic of China. **Ety.:** n.s. [Benhua Zhang, entomologist (Harbach 2018c)]. **Informal name:** Zhang Chinese Omnivore Mosquito.

Genus *Trichoprosopon* Theobald

(Vol. 1, p. 284)

1901a: 98; 1901c: 283 (as genus). Type species: *Trichoprosopon nivipes* Theobald.

Name derivation: hair and face or front (Gr.) [ref. to setulae on clypeus].

syn. *Joblotia* Blanchard 1901: 1046. Type species: *Trichoprosopon nivipes* Theobald.
syn. *Hyloconops* Lutz 1904 (as genus). In Bourroul 1904: 49, 55. Type species: *Hyloconops pallidiventer* Lutz. Belkin 1968b: 35 (syn.).
syn. *Trichoprosopon*, subg. *Vonplessenia* Lane & Cerqueira 1942: 501. Type species: *Lesticocampa vonplesseni* Dyar & Knab. Zavortink 1979b: 11 (syn.).
syn. *Trichoprosopon*, subg. *Limamyia* Lane & Cerqueira 1942: 500. Type species: *Isostomyia brevipes* da Costa Lima. Zavortink 1979b: 11 (syn.).

Important references: Stone 1944: 335 (tax.); Forattini 1965b (tax., bion.); Belkin 1968b: 34 (tax.); Zavortink 1979a (tax., review); Zavortink 1981 (tax., review); Darsie 1985 (keys; Argentina); Mitchell & Darsie 1985 (review, bibliography, distr.; Argentina); Harbach & Peyton 1993 (L mouthparts; phyl.); Judd 1996 (phyl.); Harbach & Kitching 1998 (phyl.); Harbach & Peyton 2000 (phyl.); Harbach, Vu, & Kitching 2007 (phyl.).

andinum Leví-Castillo
1953d: 63 (M*, F, P*, L*). **Type loc.:** Macuchi, Cotopaxi, Ecuador (LU). **Distr.:** Colombia, Ecuador. **Ety.:** Andes Mountains. **Informal name:** Apu Ecuadorian Hairy-lipped Mosquito [Apu were Incan gods of the mountains].

brevipes (da Costa Lima)
1931c: 68 (M*, F, P*, L*; *Isostomyia*). **Type loc.:** Laranjeiras, Rio de Janeiro, Brazil (IOC). **Distr.:** Brazil. **Refs.:** Belkin et al. 1971: 9 (lectotype desig.); Zavortink 1979b: 11 (from subg. *Limamyia*). **Ety.:** short foot (L); legs shorter, in comp. to pallidiventer. **Informal name:** Short-footed Brazilian Hairy-lipped Mosquito.

castroi Lane & Cerqueira
1942: 514 (M*, F). **Type loc.:** Terezópolis, Rio de Janeiro, Brazil (IOC). **Distr.:** Argentina, Brazil, Panama. **Refs.:** Zavortink 1979b: 11 (from subg. *Runchomyia*). **Ety.:** Gustavo Mendes de Oliveira Castro. **Informal name:** Castro Brazilian Hairy-lipped Mosquito.

compressum Lutz
1905: 171 (M, F). **Type loc.:** São Paulo, Brazil (IOC, FMP?, NHMUK). **Distr.:** Argentina, Bolivia, Brazil, Colombia, Ecuador, French Guiana, Guyana, Panama, Paraguay, Venezuela. **Refs.:** Knight & Chamberlain 1948: 12 (P*); Coher 1949: 89 (F*); Lane 1953: 823 (M*, F, P*, L*); Belkin 1968b: 35 (tax.); Townsend 1990: 62 (type info.); Linley 1991 (E*). **Ety.:** compressed (L); laterally compressed abdomen. **Informal name:** Compressed Brazilian Hairy-lipped Mosquito.

ssp. ***mogilasium*** (Dyar & Knab)
1907c: 206 (F; *Joblotia*; as sp.). **Type loc.:** Tabernilla, Canal Zone, Panama (USNM). **Distr.:** Brazil, Ecuador, French Guiana, Panama, Venezuela. **Refs.:** Stone 1944: 338 (to ssp.); Stone & Knight 1957b: 118 (type info., lectotype desig.); Stone et al. 1959: 73 (to var.); Harbach & Howard 2007: 44 (to ssp.). **Ety.:** n.s. [hardly or scarcely? (Gr)]. **Informal name:** Tabernilla Panamanian Hairy-lipped Mosquito.

syn. *trichorryes* Dyar & Knab
1907c: 206 (M, F; *Joblotia*). **Type loc.:** Tabernilla, Canal Zone, Panama (USNM). **Refs.:** Stone 1944: 338 (lectotype desig.); Stone & Knight 1957b: 119 (type info.).

digitatum (Rondani)
1848: 109 (F; *Culex*). **Type loc.:** Rio de Janeiro, Guanabara, Brazil (LU). **Distr.:** Argentina, Belize, Bolivia, Brazil, Colombia, Costa Rica, Ecuador, El Salvador, French Guiana, Guatemala, Guyana, Honduras, Lesser Antilles (includes Trinidad and Tobago), Mexico, Nicaragua, Panama, Paraguay, Peru, Suriname, Venezuela. **Refs.:** Howard et al. 1913 (M*, L*; tax., bion.); Senevet & Abonnenc 1939b: 250 (M*, L*); Lane & Cerqueira 1942: 494 (M*, F, P*, L*; keys, bion.); Belkin 1952: 122 (P*, L*); Lane 1953: 820 (M*, F, P*, L*); Belkin et al. 1971: 8 (type loc. info.); Mattingly 1971a: Pl. 1 (A scutum*), Pl. 24 (L head*, terminal segs.*); Zavortink 1981 (tax.); Zavortink et al. 1983 (M*, F*, P*, L*; tax., bion.); Linley et al. 1990 (E*); Linton et al. 2013 (bion., mol. tax., distr.; Ecuador); Berti et al. 2015 (distr.; Venezuela). **Ety.:** n.s. [finger (L)]. **Informal name:** Digital Brazilian Hairy-lipped Mosquito. (Vol. 1, p. 546)

syn. *nivipes* Theobald
1901c: 285 (M*, F*). **Type loc.:** Aqua Santa, Trinidad [Trinidad and Tobago, Lesser Antilles] (NHMUK). **Refs.:** Belkin 1968b: 36 (tax., lectotype desig.).

syn. *splendens* Lutz
1904 (A). In Bourroul 1904: 68. **Type loc.:** Manaus [Amazonas], Brazil (NE). **Refs.:** Lane & Cerqueira 1942: 494 (syn.); Belkin et al. 1971: 8 (type info.).

syn. *wilsoni* Ludlow
1918: 66 (M, F). **Type loc.:** Chagres Camp, Las Cascades, Canal Zone, Panama (USNM?). **Refs.:** Stone 1944: 335 (tax.); Stone & Knight 1957b: 119 (type info.).

syn. *subsplendens* Martini
1931a: 200 (A; *Joblotia splendens* var.). **Type loc.:** Pto. [Puerto] Bermudas, Pichis, Peru (SMT). **Refs.:** Lane & Cerqueira 1942: 494 (syn.); Belkin 1971a: 31 (type info.).

ssp. ***townsendi*** Stone
1944: 337 (M; as var.). **Type loc.:** Boa Vista [previously in Fordlandia, currently Belterra], Rio Tapajos, Pará, Brazil (USNM). **Distr.:** Brazil, Lesser Antilles (includes

Trinidad and Tobago), Panama. **Refs.:** Belkin et al. 1971: 8 (type info.); Harbach & Howard 2007: 44 (to ssp.). **Ety.:** C.H.T. Townsend. **Informal name:** Townsend Brazilian Hairy-lipped Mosquito.

evansae Antunes
1942 (M*, F). In Lane & Cerqueira 1942: 529. **Type loc.:** Restrepo, Meta, El Caibe, Colombia (IOC). **Distr.:** Colombia, Venezuela. **Refs.:** Zavortink 1979b: 11 (from subg. *Runchomyia*). **Ety.:** Alwen Myfanwy Evans. **Informal name:** Evans Colombian Hairy-lipped Mosquito.

lampropus (Howard, Dyar, & Knab)
1913: Fig. 44; 1915: 167 (M*, F; *Lesticocampa*). **Type loc.:** Upper Pequini River, Panama (USNM). **Distr.:** Brazil, Costa Rica, Ecuador, Panama. **Refs.:** Stone 1944: 340 (tax.); Lane 1953: 858 (M*, F); Stone & Knight 1957b: 118 (type info.); Zavortink 1979b: 11 (from subg. *Runchomyia*). **Ety.:** n.s. [bright foot (Gr); poss. ref. to leg coloration)]. **Informal name:** Bright-footed Panamanian Hairy-lipped Mosquito.

lanei (Antunes)
1937a: 71 (M*, F; *Goeldia*). **Type loc.:** Caney, Restrepo, Meta, Colombia (IOC). **Distr.:** Colombia, Ecuador. **Refs.:** Lane 1953: 859 (M*, F); Zavortink 1979b: 11 (from subg. *Runchomyia*). **Ety.:** John Lane. **Informal name:** Lane Colombian Hairy-lipped Mosquito.

obscurum Lane & Cerqueira
1942: 498 (M*, F). **Type loc.:** Mangaratiba, Rio de Janeiro, Brazil (IOC). **Distr.:** Argentina, Brazil. **Ety.:** dim, dark (L); thorax in lateral view generally darker than other species in the genus. **Informal name:** Obscure Brazilian Hairy-lipped Mosquito.

pallidiventer (Lutz)
1905: 125 (M, F; *Hyloeonops*). **Type loc.:** Brazil (NHMUK). **Distr.:** Argentina, Bolivia, Brazil, Colombia, French Guiana, Nicaragua, Panama, Paraguay, Peru, Venezuela. **Refs.:** Lane & Cerqueira 1942: 511 (M*, F, P*, L*); Coher 1949: 92 (F*); Lane 1953: 838 (M*, F, P*, L*); Belkin 1968b: 36 (tax.); Mattingly 1971a: Pl. 2 (A wing*); Belkin et al. 1971: 8 (lectotype desig.); Townsend 1990: 116 (type info.); Marcondes et al. 2003: 1 (distr.; Brazil). **Ety.:** pale venter (L); ref. to golden to pearly-white color of the abdominal venter. **Informal name:** Pale-bellied Brazilian Hairy-lipped Mosquito.

simile Lane & Cerqueira
1942: 514 (M*, F). **Type loc.:** Campso de Jordão, São Paulo, Brazil (IOC). **Distr.:** Argentina, Brazil. **Refs.:** Xavier & Mattos 1967: 618 (P*, L*); Zavortink 1979b: 11 (from subg. *Runchomyia*). **Ety.:** similar (L); to pallidiventer. **Informal name:** Simile Brazilian Hairy-lipped Mosquito.

soaresi Lane & Cerqueira
1942: 499 (M*, F, P*). **Type loc.:** São João de Petropolis, Espírito Santo, Brazil (IOC). **Distr.:** Brazil, French Guiana, Mexico. **Refs.:** Mattingly 1971a: Pl. 10 (P trumpet*, terminal segs.*). **Ety.:** Gabriel Soares de Souza. **Informal name:** Soares Brazilian Hairy-lipped Mosquito.

vonplesseni (Dyar & Knab)
1906a: 137 (F; *Lesticocampa*). **Type loc.:** Pastaza River, Ecuador (USNM). **Distr.:** Ecuador. **Refs.:** Lane & Cerqueira 1942: 502 (F; from syn. with *longipalpis*); Stone & Knight 1957b: 119 (type info., lectotype desig.); Zavortink 1979b: 11 (from subg. *Vonplessenia*). **Ety.:** Baron von Plessen. **Informal name:** von Plessen Ecuadorian Hairy-lipped Mosquito.

Genus *Tripteroides* Giles

(Vol. 1, p. 286)

Important references: Barraud 1934 (tax., keys, bion.; South Asia); Lee 1946: 219 (tax., keys; Australasia); Belkin 1950a: 201 (keys; Solomon Isl.); Baisas & Ubaldo-Pagayon 1953 (keys; Philippines); Peters 1959: 135 (keys; New Guinea); Thurman 1959b (tax., keys, bion.; Thailand); Belkin 1962 (tax., keys, bion.; South Pacific); Peters 1963c: 89 (P; key; New Guinea); Tanaka et al. 1979 (tax., bion.; Japan); Mattingly 1980 (class.); Mattingly 1981 (tax., keys, bion.; Oriental Region; subg. *Rachionotomyia* and *Tricholeptomyia*); Lee et al. 1989c (tax., key, bion.; Australasia); Harbach & Peyton 1993 (L mouthparts; phyl.); Judd 1996 (phyl.); Rattanarithikul, Harrison et al. 2005 (F, L; tax., keys, bion.; Thailand); Rattanarithikul et al. 2007 (tax., keys, bion.; Thailand).

Subgenus *Polylepidomyia* Theobald

1905f: 118 (as subg. of genus *Tripteroides*). Type species: *Polylepidomyia argenteiventris* Theobald. Mattingly 1980: 165 (from syn. with subg. *Rachionotomyia*).

Name derivation: n.s. [many-scaled little fly; poss. ref. to "Head . . . clothed with flat scales over all the front sides and centre"].

syn. *Mimeteomyia* Theobald 1910c: 210 (as genus). Type species: *Mimeteomyia apicotriangulata* Theobald.

Important references: Lee 1946: 219 (tax., keys; Australasia); Belkin 1962 (tax., keys, bion.; South Pacific); Peters 1963c: 89 (P; key; New Guinea); Mattingly 1980 (class.); Mattingly 1981 (tax., keys, bion.); Lee et al. 1989c (tax., key, bion.; Australasia).

altivallis Bonne-Wepster
1948: 311 (F). **Type loc.:** Moss Forest camp, near Lake Habbema, [Western] New Guinea (ITH). **Distr.:** Indonesia, Papua New Guinea. **Refs.:** Mattingly 1980: 165 (from subg. *Rachionotomyia*); Lee et al. 1989c: 25 (F; key, tax., bion., distr., review, type info.). **Ety.:** n.s. [high valley (L); poss. ref. to collection locality at 2,800 m]. **Informal name:** High Indonesian Pestle Mosquito.

apicotriangulatus (Theobald)
1910c: 211 (F*; *Mimeteomyia*). **Type loc.:** Kuranda, Queensland, Australia (NHMUK). **Distr.:** Australia. **Refs.:** Lee et al. 1989c: 26 (F; key, tax., bion., distr., review; from syn. with *atripes*). **Ety.:** triangle tipped (abdomen with apicolateral triangular white spots) (L). **Informal name:** Triangular Australian Pestle Mosquito.

argenteiventris (Theobald)
1905f: 118 (F; *Polylepidomyia*). **Type loc.:** Paumomu River, [Papua] New Guinea (HNM). **Distr.:** Australia, Indonesia, Papua New Guinea. **Refs.**: Lee et al. 1989c: 28 (F; key, tax., bion., distr., review). **Ety.:** silvery venter (L). **Informal name:** Silver-bellied Papuan Pestle Mosquito.
syn. *atra* Taylor
1914d: 190 (F; *Stegomyia*). **Type loc.:** Lakekamu Gold Field, Papua, New Guinea (IJS). **Refs.:** Lee 1946: 270 (F); Stone et al. 1959: 72 (as *ater*).
syn. *brugi* Edwards
1927b: 356 (F; *Rachionotomyia*). **Type loc.:** Nassau Mountains, New Guinea [West Papua] (NHMUK). **Refs.:** Lee 1946: 270 (syn.).

atripes (Skuse)
1889: 1750 (M, F; *Culex*). **Type loc.:** Knapsack Gully, New South Wales, Australia (ANIC). **Distr.:** Australia. **Refs.:** Lee 1946: 261 (M*, F*, L*); Dobrotworsky 1965: 52 (M*, F*, L*); Mattingly 1980: 165 (from subg. *Rachionotomyia*); Lee et al. 1989c: 33 (F; key, tax., bion., distr., review, type info.). **Ety.:** dark foot (L). **Informal name:** Dark-footed Australian Pestle Mosquito.

caledonicus (Edwards)
1922d: 100 (M, F; *Rachionotomyia*). **Type loc.:** Houailou, New Caledonia (NHMUK). **Distr.:** New Caledonia. **Refs.:** Williams 1943: 207 (F*, L*); Lee 1946: 265 (M*, F, L*); Belkin 1955: 227 (M, F, P, L*); Belkin 1962: 514 (M*, F, P*, L*); Iyengar 1969: 214 (E*); Mattingly 1980: 165 (from subg. *Rachionotomyia*); Lee et al. 1989c: 38 (F; key, tax., bion., distr., review). **Ety.:** New Caledonia. **Informal name:** Nautilus New Caledonian Pestle Mosquito [the nautilus is featured on the New Caledonia emblem].

coheni Belkin
1950a: 250 (M*, F, P*, L*). **Type loc.:** White River Valley, Guadalcanal, Solomon Islands (USNM). **Distr.:** Solomon Isl. **Refs.:** Taylor 1973: 65 (Rennell Isl.); Mattingly 1980: 165 (from subg. *Rachionotomyia*); Lee et al. 1989c: 41 (F; key, tax., bion., distr., review). **Ety.:** Murray Cohen. **Informal name:** Cohen Solomon Islander Pestle Mosquito.

collessi Lee
1946: 271 (M*, F, L*). **Type loc.:** Upper Barron, Queensland, Australia (CSIR). **Distr.:** Australia. **Refs.:** Mattingly 1980: 165 (from subg. *Rachionotomyia*); Lee et al. 1989c: 43 (F; key, tax., bion., distr., review). **Ety.:** Donald Henry Colless. **Informal name:** Colless Australian Pestle Mosquito.

digoelensis Brug
1934a: 506 (M*). **Type loc.:** Tanah Merah, Upper Digoel River, [Western] New Guinea (NHMUK). **Distr.:** Indonesia. **Refs.:** Lee 1946: 264 (M); Mattingly 1980: 165 (from subg. *Rachionotomyia*); Lee et al. 1989c: 45

(tax., bion., distr., review). **Ety.:** Digoel (Digul) River, New Guinea. **Informal name:** Digoel Indonesian Pestle Mosquito.

floridensis Belkin
1950a: 269 (L*). **Type loc.:** Halavo, Florida, Solomon Islands (USNM). **Distr.:** Solomon Isl. **Refs.:** Belkin 1962: 511 (L*; to subg. *Rachionotomyia*); Mattingly 1980: 165 (from subg. *Rachionotomyia*); Lee et al. 1989c: 46 (tax., bion., distr., review). **Ety.:** Florida [ref. to type loc. Halavo, Florida Isl., Solomon Isl.]. **Informal name:** Florida Solomon Islander Pestle Mosquito.

folicola Belkin
1955: 230 (A, P, L*). **Type loc.:** Bomber 3, Renee River area, Espiritu Santo, New Hebrides Islands (USNM). **Distr.:** Vanuatu. **Refs.:** Belkin 1962: 517 (M*, F, P*, L*); Mattingly 1980: 165 (from subg. *Rachionotomyia*); Lee et al. 1989c: 47 (F; key, tax., bion., distr., review). **Ety.:** leaf inhabitant (L); larvae found in leaf axils. **Informal name:** Leaf-dwelling Ni Vanuatu Pestle Mosquito.

marksae Dobrotworsky
1965: 58 (M*, F, L*). **Type loc.:** Cabbage Tree Creek, Victoria, Australia (NMM). **Distr.:** Australia. **Refs.:** Mattingly 1980: 165 (from subg. *Rachionotomyia*); Lee et al. 1989c: 49 (F; key, tax., bion., distr., review). **Ety.:** Elisabeth Nesta Marks. **Informal name:** Marks Australian Pestle Mosquito.

melanesiensis Belkin
1955: 233 (M*, F, P, L*). **Type loc.:** Segond Channel area, Espiritu Santo, New Hebrides (USNM). **Distr.:** Australia, New Caledonia, Solomon Isl., Vanuatu. **Refs.:** Knight & Chamberlain 1948: 12 (P*; as *caledonica*); Coher 1949: 99 (F*); Belkin 1962: 515 (M*, F, P*, L*); Iyengar 1969: 214 (E*); Mattingly 1980: 165 (from subg. *Rachionotomyia*). **Ety.:** Melanesia. **Informal name:** Melanesia Ni Vanuatu Pestle Mosquito.

microlepis (Edwards)
1927b: 353 (F; *Rachionotomyia*). **Type loc.:** Nassau Mountains, [Western] New Guinea (NHMUK). **Distr.:** Indonesia, Papua New Guinea. **Refs.:** Lee 1946: 270 (F*); van den Assem 1960: 13 (M*, L*); Mattingly 1980: 165 (from subg. *Rachionotomyia*); Lee et al. 1989c: 55 (F; key, tax., bion., distr., review). **Ety.:** small scale (Gr); ref. to "the very small mesonotal scales." **Informal name:** Small-scaled Indonesian Pestle Mosquito.

perplexus Peters
1963c: 91 (F, P*). **Type loc.:** Maprik, [East] Sepik District, New Guinea [Papua New Guinea] (CSIR). **Distr.:** Papua New Guinea. **Refs.:** Mattingly 1980: 165 (from subg. *Rachionotomyia*); Lee et al. 1989c: 58 (F; key, tax., bion., distr., review). **Ety.:** n.s. [perplexing (L)]. **Informal name:** Perplexing Papuan Pestle Mosquito.

punctolateralis (Theobald)
1903e: 156 (M, F; *Stegomyia*). **Type loc.:** South Queensland, Australia (NHMUK). **Distr.:** Australia, Indonesia, Timor. **Refs.:** Bancroft 1908: 22 (M, F; as genus *Stegomyia*); Lee 1946: 263 (M, L; from var.); Mattingly 1980: 165 (from subg. *Rachionotomyia*); Lee et al. 1989c: 59 (F; key, tax., bion., distr., review); Whelan & Hapgood 2000 (bion., distr.; East Timor). **Ety.:** n.s. [lateral spots (L); "Abdomen black, with apical white lateral spots"]. **Informal name:** Spot-sided Australian Pestle Mosquito.
syn. *occidentalis* Brug
1934a: 506 (M*; *Tripteroides atripes* var.). **Type loc.:** Atapoepoe,
Oebe, Timor [Indonesia] (NHMUK). **Refs.:** Lee 1946: 263 (syn.).

rotumanus (Edwards)
1929c: 337 (M, F, P, L; *Rachionotomyia*). **Type loc.:** Rotuma Island (NHMUK). **Distr.:** Fiji. **Refs.:** Belkin 1955: 243 (A, P, L*); Belkin 1962: 513 (M*, F, P*, L*); Mattingly 1980: 165 (from subg. *Rachionotomyia*); Lee et al. 1989c: 64 (F; key, tax., bion., distr., review). **Ety.:** Rotuma Isl., Fiji. **Informal name:** Rotuma Fijian Pestle Mosquito.

solomonis (Edwards)
1924b: 363 (F; *Rachionotomyia*). **Type loc.:** Tulagi [Florida], Solomon Islands (NHMUK). **Distr.:** New Caledonia, Papua New Guinea, Solomon Isl. **Refs.:** Lee 1944a: 20 (L*); Belkin 1950a: 240 (M*, F, P*, L*); Mattingly 1980: 165 (from subg. *Rachionotomyia*); Lee et al. 1989c: 66 (F; key, tax., bion., distr., review). **Ety.:** Solomon Isl. **Informal name:** Nggela Solomon Islander Pestle Mosquito [The species' type locality, Tulagi, is in the Nggela Island group].

standfasti Peters
1959: 148 (M*, F, P*, L*). **Type loc.:** Maprik, [East] Sepik District, New Guinea [Papua New Guinea] (CSIR). **Distr.:** Papua New Guinea. **Refs.:** Mattingly 1980: 165 (from subg. *Rachionotomyia*); Lee et al. 1989c: 69 (F; key, tax., bion., distr., review). **Ety.:** H.A. Standfast. **Informal name:** Standfast Papuan Pestle Mosquito.

tasmaniensis (Strickland)
1911a: 249 (F; *Stegomyia*). **Type loc.:** Tasmania, Australia (NHMUK). **Distr.:** Australia. **Refs.:** Lee 1946: 267 (M*, L*); Dobrotworsky 1965: 56 (M*, F*, L*); Mattingly 1980: 165 (from subg. *Rachionotomyia*); Lee et al. 1989c: 70 (F; key, tax., bion., distr., review). **Ety.:** Tasmania. **Informal name:** Tasman Australian Pestle Mosquito.
syn. *cephasi* Edwards
1923b: 8 (F; *Rachionotomyia*). **Type loc.:** Mole Creek, Tasmania, Australia (NHMUK).

Subgenus *Rachionotomyia* Theobald

1905h: 248 (as genus). Type species: *Rachionotomyia ceylonensis* Theobald.

Name derivation: n.s. [rachio- = spine, backbone, anything ridged; poss. ref. to "scutellum drawn out into a large thick backward projection spine"].

syn. *Skeiromyia* Leicester 1908: 248 (as genus). Type species: *Skeiromyia fusea* Leicester.
syn. *Squamomyia* Theobald 1910a: 28 (as genus). Type species: *Squamomyia inornata* Theobald.
syn. *Mimeteomyia* Theobald 1910c: 210. Type species: *Mimeteomyia apicotriangulata* Theobald.

Important references: Barraud 1934 (tax., keys, bion.; South Asia); Belkin 1955: 221 (keys; *caledonica* complex; Melanesia); Thurman 1959b (tax., keys, bion.; Thailand); Belkin 1962: 506 (resurrected from syn.); Mattingly 1980: 164 (tax.); Mattingly 1981 (tax., keys, bion.); Rattanarithikul et al. 2007 (tax., keys, bion.; Thailand).

affinis (Edwards)
1913c: 241 (new name for *caeruleocephala* Theobald, not Leicester 1908; *Rachionotomyia*). **Distr.:** India, Sri Lanka, Thailand. **Refs.:** Barraud 1934: 44 (M, P, L); Carter & Wijesundara 1948: 136 (M*; tax.); Mattingly 1980: 168 (from subg. *Tripteroides*); Mattingly 1981: 43 (M*, F*, P*, L*; keys, tax., bion., distr.); Rattanarithikul et al. 2007 (F*, P*, L*; bion., distr., keys). **Ety.:** n.s. [near, neighboring (L); poss. ref. to affinity with *aranoides* or *argentiventris*]. **Informal name:** Neighboring Sri Lankan Pestle Mosquito.
syn. *caeruleocephala* Theobald
1910c: 577 (M, F*; *Phoniomyia*). **Type loc.:** Hakgala, Ceylon [Sri Lanka] (NHMUK). **Refs.:** Mattingly 1980: 168 (from subg. *Tripteroides*).

aranoides (Theobald)
1901c: 274 (F*; *Wyeomyia*). **Type loc.:** Taiping, Perak, Malaya [Malaysia] (NHMUK). **Distr.:** Bangladesh, Cambodia, India, Indonesia, Laos, Malaysia, Myanmar, Nepal, People's Republic of China, Singapore, Sri Lanka, Taiwan, Thailand, Vietnam. **Refs.:** Barraud & Covell 1928: Fig. 44 (F*); Borel 1930: 153 (M*, F, L*); Brug 1931c: 15 (L*); Barraud 1934: 42 (M*, F, P, L*); Thurman 1959b: 30 (A, P, L*); Mattingly 1981: 20 (M*, F*, P*, L*; keys, tax., bion., distr.); Okazawa et al. 1986b (E*); Ahmed 1987 (distr.; Bangladesh); Darsie, Pradham, & Vaidya 1990 (F, L; tax., keys, bion., distr.; Nepal); Darsie, Becnel et al. 1991 (distr.; Nepal); Rattanarithikul et al. 2007 (F*, P*, L*; bion., distr., keys). **Ety.:** resembling a spider (L); ref. to long legs. **Informal name:** Spidery Malaysian Pestle Mosquito.
syn. *fusca* Leicester
1908: 248 (M, F; *Skeiromyia*). **Type loc.:** Malaya [Malaysia] (NHMUK). **Refs.:** Mattingly 1981: 22 (lectotype desig.).
syn. *inornata* Theobald
1910a: 28 (M; *Squamomyia*). **Type loc.:** West slopes of Dawna Hills, Lower Burma [Myanmar] (NHMUK). **Refs.:** Townsend 1990: 87 (type info.).
syn. *szechwanensis* Hsu
1964: 278 (M*, L*). **Type loc.:** Chengtu, Szechwan [Sichuan] Province, China [People's Republic of China] (SMC). **Refs.:** Mattingly 1980: 168 (from subg. *Rachisoura*; syn.).

ceylonensis (Theobald)
1905h: 248 (F; *Rachionotomyia*). **Type loc.:** Peradeniya [Central], Ceylon [Sri Lanka] (NHMUK). **Distr.:** Bangladesh, Sri Lanka. **Refs.:** Mattingly 1981: 26 (M*, F, P*, L*; keys, bion., distr., tax.; from syn. with *aranoides*). **Ety.:** Ceylon [Sri Lanka]. **Informal name:** Ceylon Sri Lankan Pestle Mosquito.

coonorensis Mattingly
1981: 30 (M*, F*, P*, L*; keys, tax., bion., distr.). **Type loc.:** Sim's Park, Coonor [*sic*, Coonoor], Madras State, India (USNM). **Distr.:** India. **Ety.:** Coonoor, Madras State, India. **Informal name:** Coonoor Indian Pestle Mosquito.

dofleini (Guenther)
1913: 204, 259 (P*, L*; *Ficalbia*). **Type loc.:** near Udugama [Southern], Ceylon [Sri Lanka] (LU). **Distr.:** Sri Lanka. **Refs.:** Barraud 1934: 45 (L; from syn. with *affinis*); Carter & Wijesundara 1948: 135 (M*, F, L); Mattingly 1980: 168 (from subg. *Tripteroides*); Mattingly 1981: 46 (M*, F*, P*, L*; keys, tax., bion. ,distr.). **Ety.:** n.s. [Franz Doflein (Harbach 2018c)]. **Informal name:** Doflein Sri Lankan Pestle Mosquito.

edwardsi (Barraud)
1929b: 264 (F; *Heizmannia*). **Type loc.:** Yellapur [northern Kanara District, Bombay], India (NHMUK, MSI). **Distr.:** India. **Refs.:** Mattingly 1980: 168 (from *Tripteroides*); Mattingly 1981: 49 (F*; key, distr., tax., lectotype desig.); Kaur 2003: 453 (distr.; India). **Ety.:** Frederick Wallace Edwards. **Informal name:** Edwards Indian Pestle Mosquito.

longipalpis Dong, Zhou, & Dong
1997b: 199 (M*, P*, L*). **Type loc.:** Jinghong County (21°51′ N, 101°64′ E, 700 m), Yunnan Province, China [People's Republic of China] (YIM). **Distr.:** People's Republic of China. **Refs.**: Dong 2000 (F). **Ety.:** n.s. [long palpus (L); prob. ref. to male palpus five-sixths length of proboscis]. **Informal name:** Long-lipped Chinese Pestle Mosquito.

nepenthis (Edwards)
1915c: 285 (M, F; *Rachionotomyia*). **Type loc.:** Kuching, Sarawak, [Malaysian] Borneo (NHMUK). **Distr.:** Malaysia, Singapore. **Refs.:** Edwards 1926a: 117 (L*); Mattingly 1981: 38 (M*, F*, P*, L*; keys, tax., bion., distr.); Townsend 1990: 109 (lectotype desig.). **Ety.:** n.s. [prob. ref. to larvae in pitcher plants, genus *Nepenthes*; painless (Gr)]. **Informal name:** Nepenthes Malaysian Pestle Mosquito.

nepenthisimilis Mattingly
1981: 41 (M*, F*, P*, L*; keys, tax., bion., distr.). **Type loc.:** Selangor; Batang, Berjuntai (K. [Kualala] Selangor),

Malaysia (USNM). **Distr.:** Malaysia, Singapore. **Ety.:** near *Nepenthes*. **Informal name:** Comfortable Malaysian Pestle Mosquito [*nepenthisimilis* = "similar to painless" = Comfortable].

pallidothorax Dong, Dong, & Wu
2008: 187 (M*, F, P, L*). **Type loc.:** Jiangchen County (24.4° N, 102.8° E), Yunnan Province [People's Republic of China] (Yunnan Institute of Parasitic Diseases). **Distr.:** People's Republic of China. **Ety.:** (in Chinese) [pale thorax (L); pale coloration of postpronotum and scutum]. **Informal name:** White-chested Chinese Pestle Mosquito.

ponmeki Miyagi & Toma
2001: 269 (M*, F*, P*, L*). **Type loc.:** Hinboon, Khammouane Province (105°46′ E, 17°17′ N), Lao PDR (USNM). **Distr.:** Laos. **Ety.:** Ponmek Dalaloy. **Informal name:** Ponmek Laotian Pestle Mosquito.

rozeboomi Baisas & Ubaldo-Pagayon
1953: 109 (M*). **Type loc.:** Pasonanca, Zamboanga City, Zamboanga, Mindanao, Philippines (USNM). **Distr.:** Philippines. **Refs.:** Basio 1971b: 88 (M*; bion.); Mattingly 1980: 168 (from subg. *Tripteroides*); Mattingly 1981: 50 (M*; key, distr., tax.). **Ety.:** Lloyd Eugene Rozeboom. **Informal name:** Rozeboom Philippine Pestle Mosquito.

serratus (Barraud)
1929a: 1059 (A; *Rachionotomyia*). **Type loc.:** Nagargali, Belgaum District [Bombay], India (MSI). **Distr.:** India, Thailand. **Refs.:** Barraud 1934: 43 (F); Mattingly 1981: 32 (M*, F*; keys, bion., distr., tax.; from var. of *aranoides*; lectotype desig.); Rattanarithikul et al. 2007 (F*, P*, L*; bion., distr., keys). **Ety.:** n.s. [prob. ref. to scales of lateral border of abdomen, which are not in a straight line and are irregular triangular patches instead]. **Informal name:** Serrated Indian Pestle Mosquito.

tenax (de Meijere)
1910: 928 (M*, F, P*, L*; *Ficalbia*). **Type loc.:** Tjibodas (1,500 m), near Buitenzorg, Java [Indonesia] (AM). **Distr.:** Indonesia, Malaysia, Singapore, Thailand, Vietnam. **Refs.:** Mattingly 1981: 33 (M*, F, P*, L*; keys, bion., distr., tax.; from syn. with *aranoides*; lectotype desig.); Rattanarithikul et al. 2007 (F*, P*, L*; bion., distr., keys). **Ety.:** n.s. [steadfast, persistent (L)]. **Informal name:** Tenacious Indonesian Pestle Mosquito.

Subgenus *Rachisoura* Theobald

1910c: 207 (as genus). Type species: *Rachisoura sylvestris* Theobald.

Name derivation: n.s. [etymology not determined; poss. ref. to "Abdomen with the apex very bristly"].

Important references: van den Assem 1959: 35 (bion.; New Guinea); Belkin 1962 (tax., keys, bion.; South Pacific); Mattingly 1981 (tax., keys, bion.); Lee et al. 1989c (tax., key, bion.; Australasia); Harbach & Peyton 1993 (L mouthparts; phyl.).

adentata van den Assem
1959: 48 (M*, F, L*). **Type loc.:** Cyclops Mountains, near Ifar, [Netherlands] New Guinea (ITH). **Distr.:** Indonesia. **Refs.:** Lee et al. 1989c: 78 (F; key, tax., bion., distr., review, type info.). **Ety.:** n.s. [no teeth (L); "absence of a lateral comb on the 8th segment"]. **Informal name:** Toothless Indonesian Pestle Mosquito.

bisquamatus Lee
1946: 252 (M*, F, L*). **Type loc.:** Hollandia, Western New Guinea (Papua) (CSIR). **Distr.:** Australia, Indonesia, Papua New Guinea. **Refs.:** Peters 1959: 137 (P*); van den Assem 1959: 41, 50 (F, L*, E; bion.); Lee et al. 1989c: 79 (F; key, tax., bion., distr., review). **Ety.:** n.s. [two scaled (L)]. **Informal name:** Two-scaled Indonesian Pestle Mosquito.

brevirhynchus Brug
1934a: 503 (M*, F, P, L*). **Type loc.:** Tanah Merah, Upper Digoel River, [Western] New Guinea (NHMUK). **Distr.:** Australia, Indonesia, Papua New Guinea. **Refs.:** Lee 1946: 259 (M*, F, L*); Peters 1959: 140 (P*); Lee et al. 1989c: 82 (F; key, tax., bion., distr., review). **Ety.:** n.s. [short nose (Gr); poss. ref. to short, stout proboscis]. **Informal name:** Short-nosed Indonesian Pestle Mosquito.

concinnus Lee
1946: 260 (F*). **Type loc.:** Mount Dafonsero, Western New Guinea (Papua) (CSIR). **Distr.:** Indonesia, Papua New Guinea. **Refs.:** Lee et al. 1989c: 84 (F; key, tax., bion., distr., review). **Ety.:** n.s. [well arranged, skillfully joined, beautiful, striking (L)]. **Informal name:** Beautiful Indonesian Pestle Mosquito.

confusus Lee
1946: 251 (M*, F, L*). **Type loc.:** Milne Bay, Papua, New Guinea (CSIR). **Distr.:** Indonesia, Papua New Guinea. **Refs.:** Lee et al. 1989c: 84 (F; key, tax., bion., distr., review). **Ety.:** n.s. [confused (L)]. **Informal name:** Confused Papuan Pestle Mosquito.

cuttsi van den Assem
1959: 46 (M*, F, L*). **Type loc.:** Homejo, Kemabu Valley, New Guinea (ITH). **Distr.:** Indonesia. **Refs.:** Lee et al. 1989c: 87 (F; key, tax., bion., distr., review, type info.). **Ety.:** W. Cutts. **Informal name:** Cutts Indonesian Pestle Mosquito.

exnebulis Bonne-Wepster
1948: 308 (F). **Type loc.:** Mist Camp, New Guinea [Central Highlands, Western New Guinea (Papua)] (ITH). **Distr.:** Indonesia. **Refs.:** Lee et al. 1989c: 88 (F; key, tax., bion., distr., review, type info.). **Ety.:** n.s. [from

the clouds (L); poss. ref. to capture in a cloud forest ("mist camp")]. **Informal name:** Cloudy Indonesian Pestle Mosquito [*exnebulis* = "from the clouds" = Cloudy].

felicitatis Bonne-Wepster
1948: 307 (F). **Type loc.:** Mist Camp, New Guinea [Central Highlands, Western New Guinea (Papua)] (ITH). **Distr.:** Indonesia. **Refs.:** Lee et al. 1989c: 89 (F; key, tax., bion., distr., review, type info.). **Ety.:** n.s. [happiness (L)]. **Informal name:** Happy Indonesian Pestle Mosquito.

filipes (Walker)
1861a: 229 (F; *Culex*). **Type loc.:** Dorey, New Guinea [Western New Guinea (Papua)] (NHMUK). **Distr.:** Australia, Indonesia, Papua New Guinea. **Refs.:** Lee 1946: 242 (M*, F*, L*); Peters 1959: 139 (P*); van den Assem 1959: 37 (M*, F); Lee et al. 1989c: 90 (F; key, tax., bion., distr., review). **Ety.:** n.s. [fern foot (L)]. **Informal name:** Fern-footed Indonesian Pestle Mosquito.
syn. *papua* Brug
1934a: 505 (M*, F). **Type loc.:** Tanah Merah, Upper Digoel River, New Guinea [Indonesia: West Papua] (NHMUK). **Refs.:** Lee 1946: 246 (tax.); Lee et al. 1989c: 90 (syn.).

flabelliger Bonne-Wepster
1948: 308 (F*). **Type loc.:** Mist Camp, Central Highlands, Irian Jaya [Indonesia: West Papua] (ITH). **Distr.:** Indonesia (West Papua). **Refs.:** van den Assem 1959: 38 (M*, F, L*); Lee et al. 1989c: 96 (F; key, tax., bion., distr., review). **Ety.:** n.s. [small fan bearing (L); poss. ref. to setae and scales on postnotum]. **Informal name:** Fan-bearing Indonesian Pestle Mosquito.

fuliginosus Lee
1946: 248 (M*). **Type loc.:** Hollandia, Western New Guinea [Indonesia: West Papua] (CSIR). **Distr.:** Indonesia. **Refs.:** Lee et al. 1989c: 98 (tax., bion., distr., review). **Ety.:** sooty, blackish (L); general color. **Informal name:** Sooty Indonesian Pestle Mosquito.

fuscipleura Lee
1946: 246 (M*, F, L*). **Type loc.:** Doromena, near Hollandia, Western New Guinea [Indonesia: West Papua] (CSIR). **Distr.:** Indonesia (West Papua), Papua New Guinea. **Refs.:** Peters 1959: 145 (P*); Mattingly 1971a: Pl. 9 (P trumpet*); Lee et al. 1989c: 99 (F; key, tax., bion., distr., review). **Ety.:** dark side (Gr); dark thoracic pleura. **Informal name:** Sith Indonesian Pestle Mosquito [*fuscipleura* = "dark side" = Sith (wielders of the Dark Side of the Force in George Lucas' [1944–] *Star Wars* movie franchise)].

kingi Lee
1946: 256 (M*, F, L*). **Type loc.:** Mount Dafonsero, Western New Guinea (Papua) (CSIR). **Distr.:** Indonesia (West Papua). **Refs.:** van den Assem 1959: 45 (M*, F, L*); Lee et al. 1989c: 101 (F; key, tax., bion., distr., review). **Ety.:** W.V. King. **Informal name:** King Indonesian Pestle Mosquito.

latisquama (Edwards)
1927b: 354 (F; *Rachionotomyia*). **Type loc.:** Nassau Mountains, Western New Guinea [West Papua] (NHMUK). **Distr.:** Indonesia (West Papua). **Refs.:** Lee 1946: 249 (F); Lee et al. 1989c: 103 (F; key, tax., bion., distr., review). **Ety.:** n.s. [scaled on the side (L)]. **Informal name:** Side-scaled Indonesian Pestle Mosquito.

leei Peters
1959: 141 (M*, F, P*, L*). **Type loc.:** Maprik, [East] Sepik District, New Guinea [Papua New Guinea] (CSIR). **Distr.:** Indonesia, Papua New Guinea. **Refs.:** Lee et al. 1989c: 105 (F; key, tax., bion., distr., review); Peyton et al. 1999: 549 (distr.). **Ety.:** David Joseph Lee. **Informal name:** Lee Papuan Pestle Mosquito.

longipalpatus Lee
1946: 249 (M*, L*). **Type loc.:** Hollandia, Western New Guinea (Papua) (CSIR). **Distr.:** Indonesia (West Papua), Papua New Guinea. **Refs.:** Peters 1959: 144 (P*, L*); van den Assem 1959: 40 (F, L*); Lee et al. 1989c: 106 (F; key, tax., bion., distr., review). **Ety.:** long palpus (L); ref. to long male palpus, as in *confusus*. **Informal name:** Long-lipped Indonesian Pestle Mosquito.

mathesoni Belkin
1950a: 219 (M*, F*, P*, L*). **Type loc.:** Matanikau River Valley, Guadalcanal, Solomon Islands (USNM). **Distr.:** Papua New Guinea, Solomon Isl. **Refs.:** Paine & Edwards 1929: 310 (P, L*; as *filipes*); Belkin 1952: 124 (P*); Belkin 1962: 523 (M*, F, P*, L*); Lee et al. 1989c: 108 (F; key, tax., bion., distr., review). **Ety.:** Robert Matheson. **Informal name:** Matheson Solomon Islander Pestle Mosquito.

obscurus Brug
1934a: 504 (M*). **Type loc.:** Tanah Merah, Upper Digoel River, [Western] New Guinea (NHMUK). **Distr.:** Australia, Indonesia. **Refs.:** Lee 1946: 273 (tax.); Mattingly 1980: 168 (from subg. *Rachionotomyia*); Lee et al. 1989c: 111 (F; key, tax., bion., distr., review). **Ety.:** n.s. [obscure (L)]. **Informal name:** Dark Indonesian Pestle Mosquito.
syn. *subobscura* Lee
1946: 273 (M*, F; *Tripteroides* (*Mimeteomyia*)). **Type loc.:** Jacky Jacky, Cape York, Queensland, Australia (CSIR). **Refs.:** Mattingly 1980: 168 (from subg. *Rachionotomyia*); E.N. Marks 1989 (syn.). In Lee et al. 1989b: 111 (syn.).

pallidus Lee
1946: 255 (M*, F, L*). **Type loc.:** Hollandia, Western New Guinea [West Papua] (CSIR). **Distr.:** Indonesia (West Papua). **Refs.:** Lee et al. 1989c: 114 (F; key, tax., bion., distr., review). **Ety.:** n.s. [pale, ashen (L)]. **Informal name:** Pale Indonesian Pestle Mosquito.

pilosus Lee
1946: 255 (F). **Type loc.:** Mount Dafonsero, near Hollandia, Western New Guinea [West Papua] (CSIR). **Distr.:** Indonesia (West Papua). **Refs.:** Lee et al. 1989c: 115 (F; key, tax., bion., distr., review). **Ety.:** n.s. [hairy, shaggy (L)]. **Informal name:** Furry Indonesian Pestle Mosquito.

plumiger Bonne-Wepster
1948: 310 (F). **Type loc.:** Rattan Camp [Central Highlands], New Guinea [Indonesia: West Papua] (ITH). **Distr.:** Indonesia (West Papua). **Refs.:** Peters 1959: 146 (M*, P*, L*); Lee et al. 1989c: 116 (F; key, tax., bion., distr., review, type info.). **Ety.:** n.s. [lead (L); poss. ref. to lead because of overall dark color]. **Informal name:** Leaden Indonesian Pestle Mosquito.

simplex Brug
1934a: 507 (M*). **Type loc.:** Tanah Merah, Upper Digoel River, [Western] New Guinea (NHMUK). **Distr.:** Indonesia. **Refs.:** Lee 1946: 259 (tax.); Lee et al. 1989c: 118 (tax., bion., distr., review). **Ety.:** n.s. [simple (L)]. **Informal name:** Simple Indonesian Pestle Mosquito.

stonei Belkin
1950a: 228 (M*, F, P*, L*). **Type loc.:** Poha River Valley, Guadalcanal, Solomon Islands (USNM). **Distr.:** Papua New Guinea, Solomon Isl. **Refs.:** Mattingly 1971a: Pl. 25 (L head*, thorax*, terminal segs.*), Pl. 9 (P terminal segs.*); Lee et al. 1989c: 119 (F; key, tax., bion., distr., review). **Ety.:** Alan Stone. **Informal name:** Stone Solomon Islander Pestle Mosquito.

subnudipennis (Edwards)
1927b: 355 (F; *Rachionotomyia*). **Type loc.:** Nassau Mountains, New Guinea [West Papua] (NHMUK). **Distr.:** Indonesia (West Papua). **Refs.:** Lee 1946: 261 (tax.); Lee et al. 1989c: 121 (F; key, tax., bion., distr., review; from subg. uncert.). **Ety.:** nearly bare wing (L); ref. to "wing much more scantily scaled, scales smaller and narrower" [comp. to *vanleeuweni*]. **Informal name:** Bare-winged Indonesian Pestle Mosquito.

sylvestris (Theobald)
1910c: 208 (M*, F*; *Rachisoura*). **Type loc.:** Kuranda, Queensland, Australia (NHMUK). **Distr.:** Australia. **Refs.:** Lee 1946: 242 (syn.; F only); Belkin 1962: 520 (to sp.); Lee et al. 1989c: 122 (F; key, tax., bion., distr., review). **Ety.:** n.s. [a wood or forest (L); poss. habitat)]. **Informal name:** Forest Australian Pestle Mosquito.
syn. *hilli* Taylor
1914b: 456 (F*; *Stegomyia*). **Type loc.:** Melville Island, Australia (LU). **Refs.:** Edwards 1924b: 362 (syn. with *filipes*); Lee et al. 1989c: 122 (from syn. of filipes).

tityae Slooff
1961: 22 (F*). **Type loc.:** Wasirawa River, Arfak Mountains, Vogelkop Peninsula, Netherlands New Guinea (NHMUK). **Distr.:** Indonesia (West Papua). **Refs.:** Lee et al. 1989c: 126 (F; key, tax., bion., distr., review). **Ety.:** n.s. [tickle, provoke (L)]. **Informal name:** Provoking Indonesian Pestle Mosquito.

torokinae Belkin
1950a: 236 (M*, F, L). **Type loc.:** Torokina, Empress Augusta Bay, Bougainville, Solomon Islands (USNM). **Distr.:** Papua New Guinea, Solomon Isl. **Refs.:** Belkin 1962: 525 (M*, F, P*, L*); Lee et al. 1989c: 127 (F; key, tax., bion., distr., review). **Ety.:** Torokina, Empress Augusta Bay, Bougainville. **Informal name:** Torokina Solomon Islander Pestle Mosquito.

vanleeuweni (Edwards)
1927b: 355 (F; *Rachionotomyia*). **Type loc.:** Rouffaer River, tributary of Mamberamo River, [Western] New Guinea (NHMUK). **Distr.:** Australia, Indonesia, Papua New Guinea. **Refs.:** Lee 1946: 255 (F); van den Assem 1959: 43 (M*, F, L*); Lee et al. 1989c: 128 (F; key, tax., bion., distr., review). **Ety.:** Dr. van Leeuwen. **Informal name:** Van Leeuwen Indonesian Pestle Mosquito.

Subgenus *Tricholeptomyia* Dyar & Shannon

1925a: 72 (as genus). Type species: *Wyeomyia nepenthicola* Banks.

Name derivation: n.s. [trich- = hair, lepto- = thin; poss. ref. to having "hair-like wing scales"].

Important references: Mattingly 1980:164 (from syn. with *Rachionotomyia*); Mattingly 1981 (tax., keys, bion.).

apoensis Baisas & Ubaldo-Pagayon
1953: 68 (M*, F, P*, L*). **Type loc.:** Taglawig, Tagum, Davao, Mindanao, Philippines (USNM). **Distr.:** Philippines. **Refs.:** Basio 1971b: 82 (M*; bion.); Mattingly 1980: 168 (from subg. *Tripteroides*); Mattingly 1981: 57 (M*, F*, P*, L*; keys, tax., bion., distr.). **Ety.:** Mt. Apo, Mindanao, Philippine Isl. **Informal name:** Apo Philippine Pestle Mosquito.

barraudi Baisas & Ubaldo-Pagayon
1953: 75 (M*, F, P*, L*). **Type loc.:** road to Banaue Mountain, Luzon, Philippines (USNM). **Distr.:** Philippines. **Refs.:** Basio 1971b: 83 (M*; bion.); Mattingly 1980: 168 (from subg. *Tripteroides*); Mattingly 1981: 59 (M*, F*, P*, L*; keys, tax., bion., distr.). **Ety.:** Philip James Barraud. **Informal name:** Barraud Philippine Pestle Mosquito.

belkini Baisas & Ubaldo-Pagayon
1953 (1952: 80 (M*, F, P*, L*). **Type loc.:** Bacungan, Puerto Princesa, Palawan, Philippines (USNM). **Distr.:** Philippines. **Refs.:** Basio 1971b: 83 (M*; bion.); Mattingly 1980: 168 (from subg. *Tripteroides*); Mattingly 1981: 74 (M*, F*, P*, L*; keys, tax., bion., distr.). **Ety.:** John

Nicholas Belkin. **Informal name:** Belkin Philippine Pestle Mosquito.

christophersi Baisas & Ubaldo-Pagayon
1953: 86 (M*, F, P*, L*). **Type loc.:** road to Banaue Mountain, Luzon, Philippines (USNM). **Distr.:** Philippines. **Refs.:** Basio 1971b: 83 (M*; bion.); Mattingly 1980: 168 (from subg. *Tripteroides*); Mattingly 1981: 61 (M*, F*, P*, L*; keys, tax., bion., distr.). **Ety.:** Samuel Rickard Christophers. **Informal name:** Christophers Philippine Pestle Mosquito.

delpilari Baisas & Ubaldo-Pagayon
1953: 91 (M*, F, P*, L*). **Type loc.:** Sierra Madre, near Famy, Laguna, Luzon, Philippines (USNM). **Distr.:** Philippines. **Refs.:** Basio 1971b: 83 (M*; bion.); Mattingly 1980: 168 (from subg. *Tripteroides*); Mattingly 1981: 76 (M*, F*, P*, L*; keys, tax., bion., distr.). **Ety.:** Gregoria del Pilar. **Informal name:** Del Pilar Philippine Pestle Mosquito.

microcala (Dyar)
1929c: 61 (M, F; *Rachionotomyia*). **Type loc.:** Bamban River, Pampanga, Luzon, Philippines (USNM). **Distr.:** Philippines. **Refs.:** Baisas & Ubaldo-Pagayon 1949: 71 (P*, L*); Baisas & Ubaldo-Pagayon 1952: 97 (M*, F, P*, L*, E*?); Basio 1971b: 85 (M*; bion.); Mattingly 1980: 168 (from subg. *Tripteroides*); Mattingly 1981: 63 (M*, F*, P*, L*; keys, tax., bion., distr.). **Ety.:** n.s. [poss. ref. to small proboscis]. **Informal name:** Bamban Philippine Pestle Mosquito.

nepenthicola (Banks)
1909: 550 (M, F; *Wyeomyia*). **Type loc.:** Trinidad, Benguet [Mountain, Luzon], Philippines (NE). **Distr.:** Philippines. **Refs.:** Baisas & Ubaldo-Pagayon 1953: 103 (M*, F, P*, L*); Basio 1971b: 86 (M*; bion.); Mattingly 1980: 168 (from subg. *Rachionotomyia*); Mattingly 1981: 67 (M*, F, P*, L*; keys, bion., distr., tax.; lectotype desig.). **Ety.:** larvae associated with *Nepenthes* pitcher plants. **Informal name:** Lotus-eater Philippine Pestle Mosquito [*nepenthicola* = "living in painlessness" = dweller of the paradisical land of the lotus-eaters in Homer's *Odyssey*].
syn. *mus* Dyar
1920h: 175 (M, F; *Wyeomyia* (*Dodecamyia*)). **Type loc.:** Los Baños [Laguna, Luzon], Philippines (USNM). **Refs.:** Stone & Knight 1957b: 119 (type info., lectotype desig.); Mattingly 1980: 168 (from subg. *Rachionotomyia*).

roxasi Baisas & Ubaldo-Pagayon
1953: 103 (M*, F, P*, L*). **Type loc.:** Sumilao, Bukidnon, Mindanao, Philippines (USNM). **Distr.:** Philippines. **Refs.:** Basio 1971b: 87 (M*; bion.); Mattingly 1980: 168 (from subg. *Tripteroides*); Mattingly 1981: 70 (M*, F*, P*, L*; keys, tax., bion., distr.). **Ety.:** Hilario Roxas. **Informal name:** Roxas Philippine Pestle Mosquito.

werneri Baisas & Ubaldo-Pagayon
1953: 111 (M*, F, P, L). **Type loc.:** Mount McKinley, Davao, Mindanao, Philippines (USNM). **Distr.:** Philippines. **Refs.:** Basio 1971b: 89 (M*; bion.); Mattingly 1980: 168 (from subg. *Tripteroides*); Mattingly 1981: 72 (M*, F*, P*, L*; keys, tax., bion., distr.). **Ety.:** F.G. Werner. **Informal name:** Werner Philippine Pestle Mosquito.

Subgenus *Tripteroides* Giles

1904a: 369 (as genus, subgenus). Type species: *Runchomyia philippinensis* Giles.

Name derivation: "tri-, L. (Gr. treis, trion) three; ptero-, pteron, Gr. feather, wing; -eides, Gr. (L. -aides) like, resembling, having the form of" (Harbach 2018c).

syn. Colonemyia Leicester 1908: 233. Type species: *Colonemyia caeruleocephala* Leicester (in Brunetti 1914: 58).

Important references: Thurman 1959b (tax., keys, bion.; Thailand); Belkin 1962: 497 (tax.; South Pacific); Tanaka et al. 1979 (tax., bion.; Japan); Mattingly 1980:164 (class.); Mattingly 1981 (tax., keys, bion.); Lee et al. 1989c (tax., key, bion.; Australasia); Rattanarithikul et al. 2007 (tax., keys, bion.; Thailand).

aeneus (Edwards)
1921c: 79 (F; *Rachionotomyia*). **Type loc.:** Ampang Jungle, Kuala Lumpur, Malaya [Malaysia] (NHMUK). **Distr.:** Cambodia, Malaysia, Thailand. **Refs.:** Thurman 1959b: 34 (M*); Rattanarithikul et al. 2007 (F*, P*, L*; bion., distr., keys). **Ety.:** n.s. [bronze colored (L); poss. ref. to bronzy-green scutum]. **Informal name:** Bronze Malaysian Pestle Mosquito.

alboscutellatus Lee
1946: 235 (M*, F, L*). **Type loc.:** Hollandia, Western New Guinea (Papua) (CSIR). **Distr.:** Australia, Indonesia, Papua New Guinea. **Refs.:** Peters 1959: 135 (P*); Lee et al. 1989c: 137 (F; key, tax., bion., distr., review). **Ety.:** white scutellum (L); "silvery scaled scutellum." **Informal name:** White-shield Indonesian Pestle Mosquito.

antennalis Bohart & Farner
1944: 70 (M*). **Type loc.:** Sibulan River, Mount Apo [Davao], Mindanao, Philippines (MCZ). **Distr.:** Philippines. **Refs.:** Baisas & Ubaldo-Pagayon 1953: 113 (M*); Basio 1971b: 82 (M*; bion.). **Ety.:** antenna (L); unusual male antenna. **Informal name:** Antenna Philippine Pestle Mosquito.

bambusa (Yamada)
1917: 61 (M*, F, P*, L*, E; *Rachionotomyia*). **Type loc.:** Tokyo [Honshu], Japan (IID, NHMUK). **Distr.:** Japan, People's Republic of China, South Korea, Taiwan. **Refs.:** LaCasse & Yamaguti 1950: 47 (M*, F*, P*, L*; bion., distr.,

tax.); Nakagawa & Asanuma 1954: 89 (P*, L*); Hara 1957: 49 (F*); Delfinado & Hodges 1968: 371 (M*); Chow 1973: 43 (distr.); Mattingly 1974b: 231 (E*); Tanaka et al. 1975c: 224 (distr.); Tanaka et al. 1979: 478 (M*, F*, L*); Townsend 1990: 50 (type info.); Linley & Chan 1991 (E*). **Ety.:** from bamboo (L). **Informal name:** Bamboo Japanese Pestle Mosquito.

ssp. ***yaeyamensis*** Tanaka, Mizusawa, & Saugstad 1979: 481 (M, F*). **Type loc.:** Mount Banna, Ishigaki Island, Ryukus, Japan (NMST). **Distr.:** Japan. **Ety.:** Yaeyama Isl., Japan. **Informal name:** Yaeyama Japanese Pestle Mosquito.

bimaculipes (Theobald)
1905f: 114 (F; as *Phoniomyia*). **Type loc.:** Moroka [Papua] and Friedrich Wilhelmshafen, [NE] New Guinea (NHMUK). **Distr.:** Australia, Indonesia (West Papua), Papua New Guinea. **Refs.:** Brug 1934a: 509 (M*, L*); Lee 1946: 231 (M*, F*, L*); Penn 1949b: 25 (P*); Bonne-Wepster 1954: 13 (M, F, L*); Lee et al. 1989c: 139 (F; key, tax., bion., distr., review); Townsend 1990: 52 (lectotype desig.). **Ety.:** n.s. [two spotted (L); ref. to two silvery spots on the femora]. **Informal name:** Two-spot-footed Papuan Pestle Mosquito.

syn. *ornata* Taylor
1914d: 189 (M; *Stegomyia*). **Type loc.:** Milne Bay, Papua, New Guinea (ANIC). **Refs.:** Lee 1946: 231 (syn.); Lee et al. 1989c: 130 (type info.).

binotatus Belkin
1950a: 264 (M, F, L). **Type loc.:** Empress Augusta Bay, Bougainville, Solomon Islands [Papua New Guinea] (USNM). **Distr.:** Papua New Guinea, Solomon Isl. **Refs.:** Belkin 1962: 504 (M*, F, P*, L*); Lee et al. 1989c: 143 (F; key, tax., bion., distr., review). **Ety.:** two spots (L); ref. to patches of dark scales in front of wing roots. **Informal name:** Two-spotted Papuan Pestle Mosquito.

bonneti Belkin
1962: 501 (M*, P*). **Type loc.:** Temotu (Malo), Santa Cruz Islands (USNM). **Distr.:** Solomon Isl. **Refs.:** Lee et al. 1989c: 145 (F; key, tax., bion., distr., review). **Ety.:** David D. Bonnet. **Informal name:** Bonnet Solomon Islander Pestle Mosquito.

brevipalpis Brug
1934a: 508 (M*, F, P, L*). **Type loc.:** Warasiwa, Oewin, and Hoelong, Ceram, Moluccas (NHMUK). **Distr.:** Indonesia, Papua New Guinea. **Refs.:** Lee 1946: 240 (tax.); Penn 1949b: 26 (P*); Lee et al. 1989c: 146 (F; key, tax., bion., distr., review). **Ety.:** short feeler (L); short palpus. **Informal name:** Short-lipped Indonesian Pestle Mosquito.

caeruleocephalus (Leicester)
1908: 233 (M, F; *Colonemyia*). **Type loc.:** Bukit Kutu and Ulu Klang [Selangor], Malaya [Malaysia] (NHMUK). **Distr.:** Indonesia, Malaysia, Thailand. **Refs.:** Thurman 1959b: 34 (A, P*, L*); Rattanarithikul et al. 2007 (F*, P*, L*; bion., distr., keys). **Ety.:** sky blue (L, Gr); ref. to color of head. **Informal name:** Blue-headed Malaysian Pestle Mosquito.

cheni Lien
1968a: 2 (M*, F, P*, L*). **Type loc.:** Yayu, Hongtou Village, Lanyu Island, Taiwan (TMRI). **Distr.:** Taiwan. **Ety.:** H.H. Chen. **Informal name:** Chen Taiwanese Pestle Mosquito.

claggi Bohart & Farner
1944: 71 (M*). **Type loc.:** Sibulan River, Mount Apo [Davao], Mindanao, Philippines (MCZ). **Distr.:** Philippines. **Refs.:** Baisas & Ubaldo-Pagayon 1953: 117 (tax.); Basio 1971b: 83 (M*; bion.). **Ety.:** C.S. Clagg. **Informal name:** Clagg Philippine Pestle Mosquito.

denticulatus Delfinado & Hodges
1968: 362 (M*, P*, L*). **Type loc.:** Pulau Tiga, Lower Perak, Malaya [Malaysia] (NHMUK). **Distr.:** Malaysia, Singapore, Thailand. **Refs.:** Rattanarithikul et al. 2007 (F*, P*, L*; bion., distr., keys). **Ety.:** n.s. [small toothed (L)]. **Informal name:** Toothy Malaysian Pestle Mosquito.

distigma (Edwards)
1925: 257 (F; *Rachionotomyia*). **Type loc.:** Tulagi [Florida], Solomon Islands (NHMUK). **Distr.:** Solomon Isl. **Refs.:** Belkin 1950a: 268 (tax.); Belkin 1962: 502 (M*, F, P*, L*); Lee et al. 1989c: 148 (F; key, tax., bion., distr., review). **Ety.:** double spotted (L); ref. to dark-brown spots in front of wing roots. **Informal name:** Double-spotted Solomon Islander Pestle Mosquito.

dyari Bohart & Farner
1944: 72 (M*; syn.). **Type loc.:** Los Baños, Laguna, Luzon, Philippines (USNM). **Distr.:** Philippines. **Refs.:** Baisas & Ubaldo Pagayon 1949: 71 (P*); Baisas & Ubaldo Pagayon 1952: 118 (M*, F, P, L*); Basio 1971b: 84 (M*; bion.); Baisas 1974: 150 (M*, F, P, L*; tax., bion., distr.; Philippines). **Ety.:** Harrison G. Dyar. **Informal name:** Dyar Philippine Pestle Mosquito.

dyi Baisas & Ubaldo-Pagayon
1953: 128 (M*, F). **Type loc.:** Mount McKinley, Davao, Mindanao, Philippines (USNM). **Distr.:** Philippines. **Refs.:** Basio 1971b: 84 (M*; bion.). **Ety.:** Francisco Dy. **Informal name:** Dy Philippine Pestle Mosquito.

elegans Brug
1934a: 507 (F). **Type loc.:** Torpedoboot River, [Western] New Guinea (NHMUK). **Distr.:** Australia, Indonesia, Papua New Guinea. **Refs.:** Lee 1946: 235 (M*, F); Peters 1963c: 89 (P*, L*); Lee et al. 1989c: 150 (F; key, tax., bion., distr., review). **Ety.:** n.s. [fine, elegant (L)]. **Informal name:** Elegant Indonesian Pestle Mosquito.

erlindae Baisas & Ubaldo-Pagayon
1953: 130 (M*, F). **Type loc.:** Mount McKinley, Davao, Mindanao, Philippines (USNM). **Distr.:** Philippines.

Refs.: Basio 1971b: 84 (M*; bion.). **Ety.:** Erlind Ubaldo-Pagayon. **Informal name:** Erlind Philippine Pestle Mosquito.

hoogstraali Baisas
1947a: 121 (M*, F). **Type loc.:** Mount Data, northern Luzon, Philippines (USNM). **Distr.:** Philippines. **Refs.:** Baisas & Ubaldo-Pagayon 1953: 133 (M*, F); Basio 1971b: 84 (M*; bion.). **Ety.:** Harry Hoogstraal. **Informal name:** Hoogstraal Philippine Pestle Mosquito.

hybridus (Leicester)
1908: 237 (F; *Colonemyia*). **Type loc.:** Bukit Kutu, Malaya [Malaysia] (NHMUK). **Distr.:** Malaysia, Thailand. **Refs.:** Thurman 1959b: 35 (M*). **Ety.:** n.s. [hybrid (L)]. **Informal name:** Hybrid Malaysian Pestle Mosquito.

indeterminatus Baisas & Ubaldo-Pagayon
1953 (1952): 134 (M). **Type loc.:** Mount McKinley, Davao, Mindanao, Philippines (USNM). **Distr.:** Philippines. Refs.: Basio 1971b: 85 (M*; bion.). **Ety.:** n.s. [not confined within limits (L)]. **Informal name:** Indeterminate Philippine Pestle Mosquito.

indicus (Barraud)
1929a: 1061 (M*; *Rachionotomyia*; as var.). **Type loc.:** Sukna [Darjeeling District, West] Bengal, India (MSI, NHMUK). **Distr.:** India, Indonesia, Myanmar, Nepal, Thailand, Vietnam. **Refs.:** Barraud 1934: 39 (M*, F); Brug 1939: 94 (L; to ssp.); Thurman 1959b: 33 (M*, P, L*; to sp.); Townsend 1990: 85 (type info.); Darsie et al. 1993 (distr.; Nepal); Rattanarithikul et al. 2007 (F*, P*, L*; bion., distr., keys). **Ety.:** India. **Informal name:** Maurya Indian Pestle Mosquito [the Maurya Empire (322–185 B.C.E.) was an extensive Iron Age empire located in modern India].

intermediatus Baisas & Ubaldo-Pagayon
1953: 135 (M*, P, L). **Type loc.:** Taglawig, Tagum, Davao, Mindanao, Philippines (USNM). **Distr.:** Philippines. **Refs.:** Basio 1971b: 85 (M*; bion.). **Ety.:** n.s. [intermediate, half (L)]. **Informal name:** Intermediate Philippine Pestle Mosquito.

knighti Baisas & Ubaldo-Pagayon
1953: 139 (M*). **Type loc.:** Pasonanca, Zamboanga City, Zamboanga, Mindanao, Philippines (USNM). **Distr.:** Philippines. **Refs.:** Basio 1971b: 85 (M*; bion.); Lee et al. 1982: 172 (F; key, tax., bion., distr., review). **Ety.:** Kenneth Lee Knight. **Informal name:** Knight Philippine Pestle Mosquito.

latispinus Ghong & Ji
1989: 482 (M*, F). **Type loc.:** Dehong Zhou, Yunnan Province, China [People's Republic of China] (YIE). **Distr.:** People's Republic of China. **Ety.:** wide, broad spine (L) [prob. ref. to "apex of basal lobe of IXth tergum of male broad and with a tuft of short flat spines."]. **Informal name:** Broad-spined Chinese Pestle Mosquito.

lipovskyi Belkin
1950a: 256 (M*, F, P*, L*). **Type loc.:** Sprague swamp, Guadalcanal, Solomon Islands (USNM). **Distr.:** Solomon Isl. **Refs.:** Lee et al. 1989c: 152 (F; key, tax., bion., distr., review). **Ety.:** Louis J. Lipovsky. **Informal name:** Lipovsky Solomon Islander Pestle Mosquito.

littlechildi (Edwards)
1930b: 544 (F; *Rachionotomyia*). **Type loc.:** Tauri River, Tapala, Papua, New Guinea (NHMUK). **Distr.:** Indonesia, Papua New Guinea. **Refs.:** Lee 1946: 234 (tax.); Maffi et al. 1979: 217 (distr.); Lee et al. 1989c: 154 (F; key, tax., bion., distr., review). **Ety.:** H.O.C. Littlechild. **Informal name:** Littlechild Papuan Pestle Mosquito.

longisiphonus Dong, Zhou, & Dong
2001 (M*, F*, P*, L*). **Type loc.:** Dong, Jinghong County, Yunnan Province, China [People's Republic of China] (YIM). **Distr.:** People's Republic of China. **Ety.:** long siphon (L); ref. to very long siphon. **Informal name:** Long-siphoned Chinese Pestle Mosquito.

lorengaui Peters
1963b: 67 (M*, F, P*, L*). **Type loc.:** Lorengau River, Manus Island, New Guinea (CSIR). **Distr.:** Papua New Guinea. **Refs.:** Mattingly 1971a: Pl. 9 (P trumpet*); Lee et al. 1989c: 156 (F; key, tax., bion., distr., review). **Ety.:** Lorengau River, Manus Isl., New Guinea. **Informal name:** Lorengau Papuan Pestle Mosquito.

mabinii Baisas & Ubaldo-Pagayon
1953: 60 (M*, F, P*, L*). **Type loc.:** Parang-Malabang Highway [Cotabato Lanao], Mindanao, Philippines (USNM). **Distr.:** Philippines. **Refs.:** Basio 1971b: 82 (M*; bion.); Mattingly 1980: 168 (from subg. *Rachisoura*); Mattingly 1981: 81 (M*, F, P*, L*; keys, tax., bion., distr.). **Ety.:** Apolinario Mabini. **Informal name:** Mabini Philippine Pestle Mosquito.

magnesianus (Edwards)
1924b: 361 (F; *Rachionotomyia*). **Type loc.:** Magnetic Island, Queensland, Australia (NHMUK). **Distr.:** Australia, Papua New Guinea, Timor. **Refs.:** Lee 1946: 236 (F); Lee et al. 1989c: 158 (F; key, tax., bion., distr., review); Whelan & Hapgood 2000 (distr.; East Timor). **Ety.:** Magnetic Isl., Queensland, Australia. **Informal name:** Magnetic Australian Pestle Mosquito.

malayi Delfinado & Hodges
1968: 368 (M*, P*, L*). **Type loc.:** 20 mi Pahang Road, Kuanta, Gunong Tabu, Terengganu, Malaya [Malaysia] (NHMUK). **Distr.:** Malaysia. **Ety.:** Malaya [Malaysia]. **Informal name:** Malacca Malaysian Pestle Mosquito [Malacca was the most prominent of the 15th-century Muslim sultanates of Malaysia].

malvari Baisas & Ubaldo-Pagayon
1953: 141 (M*, F, P, L*). **Type loc.:** Sierra Madre, near Famy, Laguna, Luzon, Philippines (USNM). **Distr.:** Philippines.

Refs.: Basio 1971b: 85 (M*; bion.). **Ety.:** Miguel Malvar. **Informal name:** Malvar Philippine Pestle Mosquito.

mendacis (Daniels)
1908: 266 (M, F; *Colonemyia*). **Type loc.:** East Coast, Malaya [Malaysia] (NE). **Distr.:** Malaysia, Singapore. **Ety.:** n.s. [false, counterfeit (L)]. **Informal name:** False Malaysian Pestle Mosquito.

monetifer (Dyar)
1920h: 176 (type F only; *Rachionotomyia*). **Type loc.:** Los Baños [Laguna, Luzon], Philippines (USNM). **Distr.:** Philippines. **Refs.:** Bohart & Farner 1944: 72 (tax.); Baisas & Ubaldo-Pagayon 1949: 71 (P*, L*); Baisas & Ubaldo-Pagayon 1953: 146 (M*, F, P, L*); Stone & Knight 1957b: 119 (type info.); Basio 1971b: 86 (M*; bion.); Baisas 1974: 152 (M, F, P, L; tax., bion., distr.; Philippines). **Ety.:** n.s. [money bearing? (L)]. **Informal name:** Wealthy Philippine Pestle Mosquito [*monetifer* = "bearer of money" = Wealthy].

nissanensis Lee
1946: 239 (M*, F, L*). **Type loc.:** Nissan Island, Green Islands, Bismarck Archipelago (CSIR). **Distr.:** Papua New Guinea. **Refs.:** Peters 1963b: 70 (P*); Lee et al. 1989c: 162 (F; key, tax., bion., distr., review). **Ety.:** Nissan, Green Isl., Papua New Guinea. **Informal name:** Nissan Papuan Pestle Mosquito.

nitidoventer (Giles)
1904a: 368 (F; *Uranotaenia*). **Type loc.:** Philippines (NHMUK). **Distr.:** Indonesia, Philippines. **Refs.:** Brug 1939: 95 (M*, F, L*); Baisas & Ubaldo-Pagayon 1949: 71 (P*, L*); Baisas & Ubaldo-Pagayon 1953: 153 (M*, F, P*, L*); Basio 1971b: 86 (M*; bion.). **Ety.:** shining, bright venter (L); abdominal "venter marked with white apical bands, which in certain lights are brilliantly luminous." **Informal name:** Shiny-bellied Philippine Pestle Mosquito.
syn. *philippinensis* Giles
1904a: 368 (F*; *Runchomyia*). **Type loc.:** Philippines (NHMUK).

novohanoverae Peters
1963b: 63 (M*, F, P*, L*). **Type loc.:** Noipuas, New Hanover, New Guinea (CSIR). **Distr.:** Papua New Guinea. **Refs.:** Lee et al. 1989c: 164 (F; key, tax., bion., distr., review). **Ety.:** New Hanover, Papua New Guinea. **Informal name:** New Hanover Papuan Pestle Mosquito.

plumosus (Brug)
1931c: 18 (M*, F, P*, L*; *Rachionotomyia*). **Type loc.:** Tjoeroep [Benkoelen], Sumatra; Petjet, Java, and Batoeriti, Bali [all Indonesia] (NHMUK). **Distr.:** Indonesia. **Refs.:** Townsend 1990: 122 (type info.). **Ety.:** n.s. [plumose (L)]. **Informal name:** Plumed Indonesian Pestle Mosquito.

powelli (Ludlow)
1909c: 235 (F; *Uranotaenia*). **Type loc.:** Camp Wilheim, Tayabas [= Quezon], Luzon, Philippines (USNM). **Distr.:** Cambodia, Indonesia, Malaysia, People's Republic of China, Philippines, Thailand, Vietnam. **Refs.:** Brug 1939: 93 (L*); Baisas & Ubaldo-Pagayon 1953: 158 (M*, F, P, L); Thurman 1959b: 32 (M*, P, L*); Basio 1971b: 86 (M*; bion.); Baisas 1974: 153 (M, F, P, L; tax., bion., distr.; Philippines); Rattanarithikul et al. 2007 (F*, P*, L*; bion., distr., keys). **Ety.:** W.A. Powell. **Informal name:** Powell Philippine Pestle Mosquito.
ssp. ***escodae*** Baisas & Ubaldo-Pagayon
1953: 167 (M*, L). **Type loc.:** Puerto Princesa, Palawan, Philippines (USNM). **Distr.:** Philippines. Refs.: Basio 1971b: 86 (M*; bion.). **Ety.:** Josepha Escoda. **Informal name:** Escoda Philippine Pestle Mosquito.
ssp. ***laffooni*** Baisas & Ubaldo-Pagayon
1953: 168 (M*). **Type loc.:** Banta, Leyte, Philippines (USNM). **Distr.:** Philippines. Refs.: Basio 1971b: 87 (M*; bion.). **Ety.:** Jean Luther Laffoon. **Informal name:** Laffoon Philippine Pestle Mosquito.
ssp. ***mattinglyi*** Baisas & Ubaldo-Pagayon
1953: 169 (M*). **Type loc.:** Culion Island [Calamian Group], Philippines (USNM). **Distr.:** Philippines. Refs.: Basio 1971b: 87 (M*; bion.). **Ety.:** Peter Frederick Mattingly. **Informal name:** Mattingly Philippine Pestle Mosquito.

proximus (Edwards)
1915c: 285 (M, F; *Rachionotomyia*). **Type loc.:** Kuching, Sarawak, [Malaysian] Borneo (NHMUK). **Distr.:** Indonesia, Malaysia, Thailand, Vietnam. **Refs.:** Borel 1930: 150 (M*, F, L*); Brug 1932: 74 (P*, L*); Rattanarithikul et al. 2007 (F*, P*, L*; bion., distr., keys). **Ety.:** close (L); close to *coeruleocephala*. **Informal name:** Approximate Malaysian Pestle Mosquito.

purpuratus (Edwards)
1921c: 79 (F; *Rachionotomyia*). **Type loc.:** Suva [Viti Levu], Fiji (NHMUK). **Distr.:** Fiji. **Refs.:** Amos 1944: 24 (A*, L*); Lee 1946: 241 (F, L); Belkin 1962: 498 (M*, F, P*, L*); Lee et al. 1989c: 165 (F; key, tax., bion., distr., review). **Ety.:** n.s. [dressed in purple (L); ref. to purple abdominal scales]. **Informal name:** Purple Fijian Pestle Mosquito.

quasiornatus (Taylor)
1915: 177 (M, F*; *Stegomyia*). **Type loc.:** Innisfail, Queensland, Australia (ANIC). **Distr.:** Australia, Papua New Guinea. **Refs.:** Lee 1946: 237 (M*, F, L*); Peters 1959: 136 (P*); Lee et al. 1989c: 168 (F; key, tax., bion., distr., review, type info.). **Ety.:** nearly (L); similar to *ornatus*. **Informal name:** Glittery Australian Pestle Mosquito [*quasiornatus* = "half ornamented" = Glittery].

reiseni Basio & Basio
1971b: 56 (M*). **Type loc.:** Mount Makiling, Laguna, Luzon, Philippines (PNM). **Distr.:** Philippines. **Ety.:** William Kenneth Reisen. **Informal name:** Reisen Philippine Pestle Mosquito.

riverai Miyagi, Toma, & Tsukamoto
1983: 5 (M*, F*, P*, L*). **Type loc.:** Iwahig, Palawan Island, Philippines (NMST). **Distr.:** Philippines. **Ety.:** Delfin G. Rivera. **Informal name:** Rivera Philippine Pestle Mosquito.

similis (Leicester)
1908: 235 (M, F; *Colonemyia*). **Type loc.:** Bukit Kutu, Malaya [Malaysia] (NHMUK). **Distr.:** Cambodia, India, Indonesia, Malaysia, People's Republic of China, Thailand. **Refs.:** Barraud 1934: 38 (M*, F); Delfinado & Hodges 1968: 371 (M*); Rattanarithikul et al. 2007 (F*, P*, L*; bion., distr., keys). **Ety.:** similar (L); to *caeruleocephala*. **Informal name:** Similar Malaysian Pestle Mosquito.

simulatus Baisas & Ubaldo-Pagayon
1953: 170 (M*, F; as *simulata*; notes F unknown but is described). **Type loc.:** Banaue Mountain, Luzon, Philippines (USNM). **Distr.:** Philippines. Refs.: Basio 1971b: 88 (M*; bion.). **Ety.:** n.s. [simulate (L); poss. ref. to similarity with *dyari*, *malvari*, and *intermediata*]. **Informal name:** Simulating Philippine Pestle Mosquito.

splendens Lee
1946: 233 (M*, F*). **Type loc.:** Dobodura, [Papua] New Guinea (CSIR). **Distr.:** Australia, Papua New Guinea. **Refs.:** Lee et al. 1989c: 174 (F; key, tax., bion., distr., review). **Ety.:** n.s. [shine, be bright (L); poss. ref. to overall coloration]. **Informal name:** Splendid Papuan Pestle Mosquito.

sullivanae Baisas & Ubaldo-Pagayon
1953: 175 (M*, F, P*, L*). **Type loc.:** Matanao, Davao, Mindanao, Philippines (USNM). **Distr.:** Philippines. **Refs.**: Basio 1971b: 88 (M*; bion.). **Ety.:** Dorothy B. Sullivan. **Informal name:** Sullivan Philippine Pestle Mosquito.

tarsalis Delfinado & Hodges
1968: 372 (M*, F, P*, L*). **Type loc.:** Frasers Hill, West Pahang, Malaya [Malaysia] (NHMUK). **Distr.:** Malaysia, Thailand. **Refs.:** Rattanarithikul et al. 2007 (F*, P*, L*; bion., distr., keys). **Ety.:** tarsus (L); ref. to male tarsal and claw modifications. **Informal name:** Tarsus Malaysian Pestle Mosquito.

toffaletii Baisas & Ubaldo-Pagayon
1953: 182 (M*, F, P, L). **Type loc.:** Iwahig, Palawan, Philippines (USNM). **Distr.:** Australia, Philippines. **Refs**.: Basio 1971b: 88 (M*; bion.). **Ety.:** Mrs. F.M.J. Metselaar-van Sluys [a nickname?]. **Informal name:** Toffaleti Philippine Pestle Mosquito.

uichancoi Baisas & Ubaldo-Pagayon
1953: 189 (M*). **Type loc.:** Mount McKinley, Davao, Mindanao, Philippines (USNM). **Distr.:** Philippines. **Refs**.: Basio 1971b: 88 (M*; bion.). **Ety.:** Leopoldo B. Uichanco. **Informal name:** Uichanco Philippine Pestle Mosquito.

vicinus (Edwards)
1914c: 128 (M, F; *Rachionotomyia*). **Type loc.:** Kuching Reservoir, Sarawak, [Malaysian] Borneo (NHMUK). **Distr.:** Indonesia, Malaysia, People's Republic of China. **Refs.:** Barraud 1934: 40 (M); Delfinado & Hodges 1968: 371 (M*). **Ety.:** neighboring (L); ref. to resemblance to several other species. **Informal name:** Neighboring Malaysian Pestle Mosquito.

Genus *Wyeomyia* Theobald

(Vol. 1, p. 288)

Subgenus *Antunesmyia* Lane & Cerqueira

1942: 587 (as subg. of genus *Wyeomyia*). Type species: *Wyeomyia* (*Antunesmyia*) *rooti* Lane & Cerqueira.

Name derivation: P.C.A. Antunes.

Important references: Lane 1953 (tax., keys, bion., distr.); Motta et al. 2007 (phyl.).

alani Lane & Cerqueira
1957: 244 (new name for *rooti*, Lane & Cerqueira, not Del Ponte 1939). **Distr.:** Brazil. **Ety.:** Alan Stone. **Informal name:** Alan Brazilian Bromeliad Mosquito.
syn. *rooti* Lane & Cerqueira
1942: 587 (M*, F, P*, L*). **Type loc.:** Mangaratiba, Rio de Janeiro, Brazil (IOC). **Refs.:** Lane 1953: 941 (M*, F, P*, L*).

colombiana Lane
1945a: 142 (F). **Type loc.:** Restrepo, Meta, Colombia (FH). **Distr.:** Colombia. **Ety.:** Colombia. **Informal name:** Tairona Colombian Bromeliad Mosquito [the Tairona were an ancient civilization of what is now Colombia].

flavifacies Edwards
1922d: 82 (F). **Type loc.:** Aruka River, British Guiana [Guyana] (NHMUK). **Distr.:** Brazil, Colombia, Guyana. **Refs.:** Lane & Cerqueira 1942: 581 (F); Belkin 1968b: 40 (tax.). **Ety.:** yellow face (L); ref. to pale-yellow clypeus. **Informal name:** Yellow-faced Guyanese Bromeliad Mosquito.

Subgenus *Caenomyiella* Harbach & Peyton

1990: 16 (as subg. of *Wyeomyia*). Type species: *Sabethes fernandezyepezi* Cova García, Sutil Oramas & Pulido.

Name derivation: "Latinized diminutive of the Greek word Caenomyia, from kaines, new, and myia, fly."

Important references: Harbach & Peyton 1990 (from genus *Sabethes*; morph. comp. to subg. *Decamyia*, *Wyeomyia* & *Zinzala*).

fernandezyepezi (Cova García, Sutil Oramas, & Pulido F.)
1974: 23 (M*, P*, L*; *Sabethes*). **Type loc.:** La Raya, Município Altamira, Estado Barinas, Venezuela (DERM). **Distr.:** Panama, Venezuela. **Refs.:** Harbach & Peyton 1990: 16 (M*, P*, L*; from genus *Sabethes*). **Ety.:** Francisco Fernández Yépez. **Informal name:** Yépez Venezuelan Bromeliad Mosquito.

Subgenus *Cruzmyia* Lane & Cerqueira

1942: 579 (as subg. of *Wyeomyia*). Type species: *Wyeomyia* (*Cruzmyia*) *dyari* Lane & Cerqueira.

Name derivation: Oswaldo Cruz.

Important references: Lane 1953 (tax., keys, bion., distr. Neotropics); Judd 1996 (phyl.); Motta et al. 2007 (class., phyl.).

dyari Lane & Cerqueira
1942: 580 (M*, F, P*, L*). **Type loc.:** Petrópolis, Rio de Janeiro, Brazil (FH). **Distr.:** Brazil. **Refs.:** Lane 1953: 933 (M*, F, P*, L*); Lane & Cerqueira 1958 (type info.). **Ety.:** Harrison G. Dyar. **Informal name:** Dyar Brazilian Bromeliad Mosquito.

forattinii Clastrier
1974: 155 (M*, F, P*, L*). **Type loc.:** Matoury, French Guiana [Guiana] (A). **Distr.:** French Guiana. **Ety.:** Oswaldo Paulo Forattini. **Informal name:** Forattini French Guianese Bromeliad Mosquito.

kummi Lane & Cerqueira
1942: 582 (F). **Type loc.:** Curralinho, Pará, Brazil (IOC). **Distr.:** Brazil, Colombia. **Refs.:** Barreto & Lee 1969: 416 (distr.). **Ety.:** Henry William Kumm. **Informal name:** Kumm Brazilian Bromeliad Mosquito.

mattinglyi Lane
1953: 935 (F). **Type loc.:** Salvador, Brazil (LU). **Distr.:** Brazil, Colombia. **Refs.:** Barreto & Lee 1969: 417 (distr.); Belkin et al. 1971: 10 (type loc. info.). **Ety.:** Peter Frederick Mattingly. **Informal name:** Mattingly Brazilian Bromeliad Mosquito.

Subgenus *Decamyia* Dyar

1919d: 135 (as genus). Type species: *Wyeomyia onidus* Dyar & Knab.

Name derivation: n.s. [etymology not known; 10 (Gr.), little fly].

syn. *Heliconiamyia* Dyar 1919d: 123 (as genus). Type species: *Wyeomyia galoa* Dyar & Knab.

Important references: Edwards 1932a: 86 (syn. with *Dendromyia*); Harbach & Peyton 1990: 16 (from syn with *Dendromyia*; morph. comp.); Motta et al. 2007 (phyl.).

felicia (Dyar & Núñez Tovar)
1927b: 3 (M, F; *Dendromyia*). **Type loc.:** Tío Julian, Aragua, Venezuela (USNM). **Distr.:** Brazil, Lesser Antilles (includes Trinidad and Tobago), Nicaragua, Panama, Venezuela. **Refs.:** Lane 1945a: 148 (M*, F, P*, L*); Lane 1953: 979 (M*, F, P*, L*); Stone & Knight 1957b: 123 (type

info., lectotype desig.); Mattingly 1971a: Pl. 10 (P terminal segs.*); Harbach & Peyton 1990: 16 (from subg. *Dendromyia*). **Ety.:** n.s. [happy, blessed, lucky (L)]. **Informal name:** Lucky Venezuelan Bromeliad Mosquito.

pseudopecten Dyar & Knab
1906a: 139 (A). **Type loc.:** Trinidad [Trinidad and Tobago, Lesser Antilles] (USNM). **Distr.:** Brazil, Colombia, Costa Rica, Ecuador, French Guiana, Guatemala, Guyana, Honduras, Lesser Antilles (includes Trinidad and Tobago), Mexico, Panama, Peru, Suriname, Venezuela. **Refs.:** Lane 1945a: 150 (M*, P*, L*); Vargas & Downs 1948: 96 (P*); Lane 1953: 972 (M*, F, P*, L*); Stone & Knight 1957b: 125 (type info.); Harbach & Peyton 1990: 16 (from subg. *Dendromyia*); Pecor et al. 2000 (distr.; Peru); Linton et al. 2013 (bion., mol. tax., distr.; Ecuador). **Ety.:** false (Gr) pecten; having false pecten on siphon. **Informal name:** False-pecten Trinbagonian Bromeliad Mosquito.
syn. *galoa* Dyar & Knab
1906a: 140 (A). **Type loc.:** Cacao, Trece Aguas, Alta Vera Paz, Guatemala (USNM). **Refs.:** Stone & Knight 1957b: 123 (type info.).
syn. *cara* Dyar & Knab
1909d: 264 (A). **Type loc.:** Trinidad [Trinidad and Tobago, Lesser Antilles] (USNM).
syn. *eloisa* Howard, Dyar, & Knab
1913: Figs. 36, 287; 1915: 121 (M*, F, L*). **Type loc.:** Miraflores, Canal Zone, Panama (USNM). **Refs.:** Lane 1953: 972 (syn.); Stone & Knight 1957b: 122 (type info., lectotype desig.).
syn. *bicompressa* Lutz
1928: 33 (F; *Dendromyia*). **Type loc.:** Tumero, Aragua, Venezuela (LU). **Refs.:** Lane 1953: 972 (syn.).

ulocoma (Theobald)
1903a: 313 (F*; *Dendromyia*). **Type loc.:** Demerara River, British Guiana [Guyana] (NHMUK). **Distr.:** Colombia, Costa Rica, Dominican Republic, Ecuador, Guyana, Lesser Antilles (includes Trinidad and Tobago), Panama, Peru, Suriname, Venezuela. **Refs.:** Dyar 1928: 79 (M*, F, L*); Lane 1953: 975 (M*, F, L); Belkin 1968b: 43 (lectotype desig.); Panday 1975a, b (distr.; Suriname); Harbach & Peyton 1990: 16 (from subg. *Dendromyia*); Pecor et al. 2000 (bion., distr.; Peru); Linton et al. 2013 (bion., mol. tax., distr.; Ecuador). **Ety.:** n.s. [destructive hair? (Gr)]. **Informal name:** Bedhead Guyanese Bromeliad Mosquito [*ulocoma* = "destructive hair" = Bedhead].
syn. *onidus* Dyar & Knab
1909d: 261 (M, F). **Type loc.:** Tabernilla, Canal Zone, Panama (USNM). **Refs.:** Stone & Knight 1957b: 125 (type info., lectotype desig.).
syn. *pantoia* Dyar & Knab
1909d: 262 (M, F). **Type loc.:** Tabernilla, Canal Zone, Panama (USNM). **Refs.:** Stone & Knight 1957b: 125 (type info., lectotype desig.).
syn. *cacodela* Dyar & Knab
1909d: 265 (M, F). **Type loc.:** Gorgona, Canal Zone, Panama (USNM). **Refs.:** Stone & Knight 1957a, b: 121 (type info., lectotype desig.).

Subgenus *Dendromyia* Theobald

1903a: 313 (genus). Type species: *Dendromyia luteoventralis* Theobald (Blanchard 1905: 426).

Name derivation: n.s. [dendr- = tree; little tree fly].

syn. *Wyeomyia*, subg. *Shropshirea* Dyar 1922d: 97. Type species: *Wyeomyia* (*Shropshirea*) *ypsipola* Dyar.
syn. *Dendromyia*, subg. *Melanolepis* Dyar 1928: 69, 72. Type species: *Prosopolepis jocosa* Dyar & Knab (Edwards 1932a: 87).

Important references: Motta & Lourenço-de-Oliveira 1995: 375 (tax.); Motta et al. 1998 (phyl., class.); Motta & Lourenço-de-Oliveira 2000 (revision, class., keys); Motta & Lourenço-de-Oliveira 2005 (comp. morph.); Motta et al. 2007 (phyl.).

complosa (Dyar)
1928: 74 (M*, F, L*; *Dendromyia*). **Type loc.:** San Juan de Pequini, Panama (USNM). **Distr.:** Brazil, Colombia, Costa Rica, Ecuador, French Guiana, Guyana, Honduras, Nicaragua, Panama, Venezuela. **Refs.:** Lane 1953: 995 (M*, F, L); Stone & Knight 1957b: 122 (type info., lectotype desig.); Motta & Lourenço-de-Oliveira 2000: 672 (M*, F*, P*, L*; tax.); Navarro et al. 2013b (distr.; Ecuador). **Ety.:** n.s. [struck, applaud, clap (L)]. **Informal name:** Struck Panamanian Bromeliad Mosquito.

jocosa (Dyar & Knab)
1908: 64 (F; *Prosopolepis*). **Type loc.:** Caldera, Panama (USNM). **Distr.:** Colombia, Costa Rica, Ecuador, Guatemala, Mexico, Nicaragua, Panama, Venezuela. **Refs.:** Dyar 1928: 73 (M*, F, L*; as *prolepidis*); Lane & Cerqueira 1942: 616 (tax.); Lane 1953: 996 (M*); Barreto & Lee 1969: 417 (distr.); Motta & Lourenço-de-Oliveira 2000: 657 (M*, F*, P*, L*; tax.). **Ety.:** n.s. [facetious (L)]. **Informal name:** Facetious Panamanian Bromeliad Mosquito.
syn. *prolepidis* Dyar & Knab
1919: 1 (F). **Type loc.:** Culebra, Canal Zone, Panama (USNM). **Refs.:** Lane 1953: 996 (syn.); Stone & Knight 1957b: 125 (type info., lectotype desig.).
syn. *favor* Dyar & Núñez Tovar
1928 (F; *Dendromyia* (*Melanolepis*)). In Dyar 1928: 73. **Type loc.:** Rancho Grande, Venezuela (USNM). **Refs.:** Lane 1953: 996 (syn.).

luteoventralis Theobald
1901c: 348 (F). **Type loc.:** Pará, Brazil (NHMUK). **Distr.:** Argentina, Brazil, Colombia, Guyana, Nicaragua, Venezuela. **Refs.:** da Costa Lima 1930d (M*, F*, P*, L*);

da Costa Lima 1931a: 367 (M*); Lane 1953: 946 (M*, F); Belkin 1968b: 40 (lectotype desig.); Motta & Lourenço-de-Oliveira 1995: 375 (M*, F*, P*, L*). **Ety.:** yellow belly (L); ref. to abdomen "ochraceous to nearly white ventrally." **Informal name:** Yellow-bellied Brazilian Bromeliad Mosquito.

syn. *quasiluteoventralis* Theobald
1903a: 317 (F; *Dendromyia*). **Type loc.:** Moranhauna and Barima Rivers and Demarara River, British Guiana [Guyana] (NE). **Refs.:** Belkin 1968b: 42 (tax., syn.).

testei Senevet & Abonnenc
1939b: 269 (M*, L*). **Type loc.:** Sautigre [Inini], French Guiana [Guiana] (FMP, originally LPFM). **Distr.:** French Guiana, Guyana. **Refs.:** Belkin 1968b: 43 (tax.); Motta & Lourenço-de-Oliveira 2000: 664 (M*, F*, P*, L*; tax.). **Ety.:** M. le Médecin-Colonel Teste. **Informal name:** Teste French Guianese Bromeliad Mosquito.

trifurcata Clastrier
1973: 39 (M*, F, P*, L*). **Type loc.:** Tonnegrande, French Guiana [Guiana] (A). **Distr.:** French Guiana. **Ety.:** n.s. [three pronged (L); prob. ref. to gonostylus with three branches, as in *testei*]. **Informal name:** Three-forked French Guianese Bromeliad Mosquito.

ypsipola Dyar
1922d: 97 (M*). **Type loc.:** Comacho [Canal Zone, Panama] (USNM). **Distr.:** Bolivia, Brazil, Colombia, Costa Rica, French Guiana, Guatemala, Guyana, Lesser Antilles (includes Trinidad and Tobago), Nicaragua, Panama, Suriname, Venezuela. **Refs.:** Lane 1953: 959 (M*, F, P*, L*); Bruijning 1959: 139 (distr.); Motta & Lourenço-de-Oliveira 2000: 651 (M*, F*, P, L; tax.). **Ety.:** n.s. [Y-shaped (Gr)]. **Informal name:** Y-shaped Panamanian Bromeliad Mosquito.

Subgenus *Dodecamyia* Dyar

1919d: 138. Type species: *Wyeomyia aphobema* Dyar.

Name derivation: n.s. [etymology not known; dodeca- = 12 (Gr.), little fly].

Important references: Belkin & Heinemann 1978 (change from syn. of *Wyeomyia* subg. *Wyeomyia* to *Wyeomyia* subg. *Dodecamyia*: a revalidation without comment); Peyton et al. 1983: 75 (repeats Belkin & Heinemann 1978 revalidation without comment); Judd 1996 (phyl.); Motta & Lourenço-de-Oliveira 2005 (comp. morph.); Motta et al. 2007 (phyl.).

aphobema Dyar
1918d: 120 (M). **Type loc.:** Lawa River, Suriname (USNM). **Distr.:** Argentina, Bolivia, Brazil, Colombia, Ecuador, French Guiana, Guyana, Peru, Suriname. **Refs.:** Bonne-Wepster & Bonne 1919; Senevet & Abonnenc 1939b: 252 (M*, L*); Lane 1953: 887 (M*, F, P*, L*); Peyton et al. 1983: 75 (from subg. *Wyeomyia*); Linton et al. 2013 (bion., mol. tax., distr.; Ecuador). **Ety.:** n.s. [fearless (Gr)]. **Informal name:** Fearless Surinamese Bromeliad Mosquito.

syn. *bodkini* Edwards
1922d: 81 (M*, F). **Type loc.:** Issororo, NW district, British Guiana [Guyana] (NHMUK). **Refs.:** Lane & Cerqueira 1942: 560 (syn.); Belkin 1968b: 39 (lectotype desig.).

syn. *aequatorialis* Leví-Castillo
1952c: 131 (M*, F; *aphobema* var.). **Type loc.:** Tena, Napo-Pastaza, Ecuador (NE). **Refs.:** Lane 1953: 887 (syn.).

Subgenus *Exallomyia* Harbach & Peyton

1992a: 92 (as subg.). Type species: *Wyeomyia tarsata* Lane & Cerqueira.

Name derivation: "Exuallos (quite different) prefixed to the stem Myia (fly) in reference to the distinctness of the group."

Important references: Motta & Lourenço-de-Oliveira 2005 (comp. morph.); Motta et al. 2007 (phyl.).

arborea Galindo, Carpenter, & Trapido
1951b: 86 (M*, F, P*, L*). **Type loc.:** Bijao, Chiriquí Volcano, Chiriquí Province, Panama (USNM). **Distr.:** Nicaragua, Panama. **Refs.:** Lane 1953: 936 (M*, F, P*, L*); Harbach & Peyton 1992a: 103 (M*, F, P*, L*; from subg. *Davismyia*). **Ety.:** of trees (L); found in canopy. **Informal name:** Arboreal Panamanian Bromeliad Mosquito.

carrilloi (Sutil Oramas & Pulido F.)
1978: 199 (M*, P*, L*; *Sabethes*). **Type loc.:** Guaca, Municipio San Antonio de Caparo, Táchira, Venezuela (DERM). **Distr.:** Venezuela. **Refs.:** Harbach & Peyton 1992a: 96 (M*, F*, P*, L*; from *Sabethes*). **Ety.:** Salvador José Carrillo. **Informal name:** Carrillo Venezuelan Bromeliad Mosquito.

tarsata Lane & Cerqueira
1942: 612 (M*, F). **Type loc.:** Anápolis, Goiás, Brazil (IOC). **Distr.:** Bolivia, Brazil, Colombia. **Refs.:** Barreto & Lee 1969: 418 (distr.); Harbach & Peyton 1992a: 92 (M*, F*; from subg. *Dendromyia*). **Ety.:** n.s. [pertaining to tarsi (L)]. **Informal name:** Foot-like Brazilian Bromeliad Mosquito.

Subgenus *Hystatomyia* Dyar

1919d: 140 (as genus). Type species: *Wyeomyia circumcincta* Dyar & Knab.

Name derivation: n.s. [etymology not determined].

Important references: Dyar & Shannon 1924d: 481 (to subg. of *Wyeomyia*); Dyar 1928: 67 (syn. with *Dendromyia* Theobald); Judd 1998: 572 (to subg. of *Wyeomyia*).

autocratica Dyar & Knab
1906d: 230 (L*). **Type loc.:** Trinidad [Trinidad and Tobago, Lesser Antilles] (USNM). **Distr.:** Bahamas, Brazil, Ecuador, Lesser Antilles (includes Trinidad and Tobago), Nicaragua, Venezuela. **Refs.:** Dyar 1928: 86 (M*, F, L*); Lane 1945a: 143 (M*, F, L*); Lane 1953: 946 (M*, F, P, L*); Stone & Knight 1957b: 121 (type info.); Judd 1998: 579 (M*, P*, L*; from subg. uncertain); Linton et al. 2013 (bion., mol. tax., distr.; Ecuador). **Ety.:** n.s. [self (Gr)]. **Informal name:** Autocratic Trinbagonian Bromeliad Mosquito.

baltae Porter
2009 (M*, F*, P*, L*). **Type loc.:** Cusco, Provincia Paucartambo, Distrito Kosñipata Quitacalzón (about 1.5 km west of Chontachaca) (13°156' S, 71°2998' W, elevation 1,019 m), Peru. **Distr.:** Peru. **Refs.:** Edwards 1914a: 260 (M*). **Ety.:** Mercedes Rosario Balta Leon. **Informal name:** Balta Peruvian Bromeliad Mosquito.

chocoensis Porter & Wolff
2004: 3 (M*, F*, P*, L*). **Type loc.:** Ensenada de Utria (6°013' N, 77°210' W, elevation 0–10 m), Chocó , Colombia (USNM). **Distr.:** Colombia. **Ety.:** Chocó Department, Colombia. **Informal name:** Chocó Colombian Bromeliad Mosquito.

circumcincta Dyar & Knab
1907c: 210 (M; *Wyeomyia*). **Type loc.:** Tabernilla, Canal Zone, Panama (USNM). **Distr.:** Belize, Bolivia, Brazil, Costa Rica, Panama. **Refs.:** Dyar 1928: 85 (M*, F, L*); Lane 1953: 975 (M*, F, P*, L*); Stone & Knight 1957b: 122 (type info., lectotype desig.); Mattingly 1971a: Pl. 10 (P trumpet*); Judd 1998: 572 (M*, F*, P*, L*; from subg. uncertain); Pecor et al. 2002: 250 (tax., distr.; Belize). **Ety.:** n.s. [girded around (L)]. **Informal name:** Girdled Panamanian Bromeliad Mosquito.

syn. *macrotus* Dyar & Knab
1907c: 212 (A, P). **Type loc.:** Boquerón River, Panama (USNM). **Refs.:** Dyar 1928: 85 (syn.); Stone & Knight 1957b: 124 (type info., lectotype desig.).

syn. *andropus* Dyar & Knab
1908: 68 (M, F). **Type loc.:** Porto Bello, Fort San Felipe, Panama (USNM). **Refs.:** Dyar 1928: 85 (syn.); Stone & Knight 1957b: 120 (type info., lectotype desig.).

coenonus Howard, Dyar & Knab
1913: Fig. 38; 1915: 153 (M*, F; key). **Type loc.:** Tabernilla, Canal Zone, Panama (USNM). **Distr.:** Panama. **Refs.:** Dyar 1928: 86 (M*, F, L*); Lane 1953: 990 (M*, F, L); Stone & Knight 1957b: 122 (type info., lectotype desig.); Judd 1998 (from subg. uncertain). **Ety.:** n.s. [common (Gr)]. **Informal name:** Common Panamanian Bromeliad Mosquito.

esmeraldasi (Leví-Castillo)
1955b: 390 (M*; *Phoniomyia*). **Type loc.:** Changuaral Island, Ancón de Sardinas Bay, Esmeraldas, Ecuador (NE). **Distr.:** Ecuador. **Refs.:** Judd 1998: 997 (type info.; from genus *Phoniomyia*). **Ety.:** Esmeraldas Prov., Ecuador. **Informal name:** Esmeraldas Ecuadorian Bromeliad Mosquito.

intonca Dyar & Knab
1910: 173 (M). **Type loc.:** Empire, Canal Zone, Panama (USNM). **Distr.:** Panama. **Refs.:** Lane 1945a: 146 (syn. with *circumcincta*); Judd 1998: 577 (to valid sp.); Porter & Wolff 2004: 17 (M*, F*, P*, L*; tax.). **Ety.:** n.s. **Informal name:** Empire Panamanian Bromeliad Mosquito.

lamellata (Bonne-Wepster & Bonne)
1920: 168 (M; *Hystatomyia*). **Type loc.:** Lawa River, Suriname (ITH). **Distr.:** French Guiana, Suriname. **Refs.:** Dyar 1928: 87 (M*, F); Lane 1953: 949 (M*); Belkin 1968b: 40 (tax.); Judd 1998 (from subg. uncertain). **Ety.:** n.s. [small plates, leaves, lamellae (L)]. **Informal name:** Small-leafed Surinamese Bromeliad Mosquito.

lopezii Cova García, Sutil Oramas, & Pulido F.
1974: 21 (M*, P*, L*). **Type loc.:** La Soledad, Municipio Altamira, Estado Barinas, Venezuela. **Distr.:** Venezuela. **Refs.:** Judd 1998 (from subg. uncertain). **Ety.:** José Antonio López. **Informal name:** López Venezuelan Bromeliad Mosquito.

Subgenus *Menolepis* Lutz

1905: 269 (as genus). Type species: *Menolepis leucostigma* Lutz.

Name derivation: n.s. [courage, spirit, strength (Gr)].

Important references: Motta et al. 2007 (phyl.).

leucostigma Lutz
1904 (A). In Bourroul 1904: 48, 67. **Type loc.:** São Paulo, Brazil (NE). **Distr.:** Argentina, Bolivia, Brazil. **Refs.:** Lane & Cerqueira 1942: 584 (M*, F, P*, L*); Belkin 1968b: 40 (tax.); Belkin et al. 1971: 10 (type info.). **Ety.:** n.s. [white mark (Gr); poss. ref. to scutum covered with white scales]. **Informal name:** White-spotted Brazilian Bromeliad Mosquito.

syn. *leontiniae* Brèthes
1910: 471 (A; *Limatus*). **Type loc.:** Islas del Paraná, Argentina (BA). **Refs.:** Belkin et al. 1968: 11 (lectotype desig. according to Casal).

Subgenus *Miamyia* Dyar

1919d: 116 (as genus). Type species: *Wyeomyia symmachus* Dyar & Knab.

Name derivation: n.s. [mia = my (L); myia = fly (Gr); my fly].

Important references: Judd 1996 (phyl.); Motta et al. 2007 (from syn. with subg. *Wyeomyia*; phyl.).

codiocampa Dyar & Knab
1907c: 209 (M, F). **Type loc.:** Tabernilla, Canal Zone, Panama (USNM). **Distr.:** Argentina, Brazil, Colombia, Ecuador, Lesser Antilles (includes Trinidad and Tobago), Panama, Paraguay. **Refs.:** Dyar 1928: 57 (M*, F, L*); da Costa Lima 1930b: 73 (tax.); Lane 1953: 923 (M*, F, P, L); Stone & Knight 1957b: 122 (type info., lectotype desig.); Mattingly 1971a: Pl. 10 (P terminal segs.*); Motta et al. 2007 (transfer to subg. *Miamyia*); Linton et al. 2013 (bion., mol. tax., distr.; Ecuador). **Ety.:** n.s. [sheepskin field? (Gr)]. **Informal name:** Sheepfold Panamanian Bromeliad Mosquito.

hosautos Dyar & Knab
1907c: 211 (A; as *hosautus*). **Type loc.:** Tabernilla, Canal Zone, Panama (USNM). **Distr.:** Brazil, Colombia, Panama. **Refs.:** Dyar 1928: 59 (M*, F, L*); Lane 1953: 894 (M*, F, L); Motta et al. 2007 (transfer to subg. *Miamyia*). **Ety.:** n.s. [in like manner, likewise (Gr) (Harbach 2018c)]. **Informal name:** Conformity Panamanian Bromeliad Mosquito [*hosautos* = "in the same way" = Conformity].

syn. *euethes* Dyar & Knab
1909d: 263 (F). **Type loc.:** Tabernilla, Canal Zone, Panama (USNM).

syn. *symmachus* Dyar & Knab
1909d: 262 (F). **Type loc.:** Tabernilla, Canal Zone, Panama (USNM). **Refs.:** Stone & Knight 1957b: 125 (type info., lectotype desig.).

ssp. ***leucotarsis*** Lane
1936b: 180 (F). **Type loc.:** Boa Esperança and Pocinho, Mato Grosso, Brazil (LU). **Distr.:** Brazil. **Refs.:** Belkin et al. 1971: 10 (type info.); Harbach & Howard 2007: 45 (to ssp.). **Ety.:** white tarsi (Gr); ref. to more extensive white on Ta-III$_{2-5}$) in comp. to *hosautos* (Ta-III$_{4-5}$ only). **Informal name:** White-legged Brazilian Bromeliad Mosquito.

limai Lane & Cerqueira
1942: 571 (M*, F). **Type loc.:** Londrina, Paraná, Brazil (IOC). **Distr.:** Argentina, Bolivia, Brazil, Colombia. **Refs.:** Forattini et al. 1960: 95 (P*, L*); Motta et al. 2007 (to subg. *Miamyia*). **Ety.:** Angelo Moreiro da Costa Lima. **Informal name:** Lima Brazilian Bromeliad Mosquito.

lutzi (da Costa Lima)
1930b: 75 (M*, F, L*; *Miamyia*). **Type loc.:** Pindamonhangaba, São Paulo, Brazil (IOC). **Distr.:** Brazil. **Refs.:** Lane & Cerqueira 1942: 572 (M*, F, P*, L*); Knight & Chamberlain 1948: 12 (P*); Motta et al. 2007 (transfer to subg. *Miamyia*). **Ety.:** Adolfo Lutz. **Informal name:** Lutz Brazilian Canopy Mosquito.

oblita (Lutz)
1904 (M, F; *Dendromyia*). In Bourroul 1904: 68. **Type loc.:** São Paulo and Ponte Ipê Arcado, Goiás, Brazil (IOC). **Distr.:** Argentina, Bolivia, Brazil, Colombia, Ecuador, French Guiana, Suriname. **Refs.:** Lane & Cerqueira 1942: 562 (M*, F, P*, L*); Lane 1953: 906 (M*, F, P*, L*); Belkin 1968b: 41 (tax.); Belkin et al. 1971: 10 (type info.); Motta et al. 2007 (transfer to subg. *Miamyia*); Linton et al. 2013 (bion., mol. tax., distr.; Ecuador). **Ety.:** n.s. [forget (L)]. **Informal name:** Forgetful Brazilian Bromeliad Mosquito.

syn. *fallax* Bonne-Wepster & Bonne
1919: 110 (F, L). **Type loc.:** Paramaribo, Suriname (ITH). **Refs.:** Belkin 1968b: 39 (lectotype desig.).

syn. *pintoi* da Costa Lima
1930c: 191 (M*, F, L*; *Miamyia* (*Miamyia*)). **Type loc.:** Mendes, "E do Rio," Brazil (IOC). **Refs.:** Lane & Cerqueira 1942: 562 (syn.).

sabethea Lane & Cerqueira
1942: 573 (M*, F, P*, L*). **Type loc.:** Teresópolis, Rio de Janeiro, Brazil (IOC). **Distr.:** Argentina, Brazil. **Refs.:** Lane 1953: 913 (M*, F, P*, L*); García & Casal 1968a: 61 (distr.); Motta et al. 2007 (transfer to subg. *Miamyia*). **Ety.:** n.s. **Informal name:** Teresópolis Brazilian Bromeliad Mosquito.

serrata (Lutz)
1905: 287 (M, F; *Dendromyia*). **Type loc.:** Rio Frio, Pindamonhangaba, São Paulo, Brazil (IOC). **Distr.:** Brazil, Lesser Antilles (includes Trinidad and Tobago). **Refs.:** da Costa Lima 1930b: 73 (M*, F*; tax.); Lane & Cerqueira 1942: 566 (M*, F, P*, L*); Belkin 1968b: 42 (tax.); Belkin et al. 1971: 10 (lectotype desig.); Motta et al. 2007 (transfer to subg. *Miamyia*). **Ety.:** n.s. [toothed margin (Gr); poss. ref. to "tenth sternites comb-like, with about 13 teeth" or "peculiar form of the lobes of the ninth tergite"]. **Informal name:** Serrated Brazilian Bromeliad Mosquito.

Subgenus *Nunezia* Dyar

1928: 28, 50 (as subg. of genus *Wyeomyia*). Type species: *Dendromyia* (*Eunicemyia*) *bicornis* Root.

Name derivation: M. Núñez Tovar.

Important references: Judd 1996 (phyl.); Judd 1998 (phyl.); Porter 2014: 332 (bion., morph., tax.).

bicornis (Root)
1928 (M, L; *Dendromyia*). In Dyar 1928: 50. **Type loc.:** [Ocumare, Araqua], Venezuela (USNM). **Distr.:** Bolivia, Brazil, Colombia, Peru, Venezuela. **Refs.:** Lane 1953: 931 (M*, F, L*); Stone & Knight 1957b: 121 (type info., lectotype desig.); Navarro & Liria 2007 (M*, P*, L*; syn. of *trujilloi*); Parker et al. 2012 (bion., distr.; Peru). **Ety.:** n.s. [two horns (L)]. **Informal name:** Two-horned Venezuelan Bromeliad Mosquito.

syn. *trujilloi* Pulido & Sutil
1981 (M*, P*, L*). **Type loc.:** Altamira, Municipio Altamira, Barinas, Venezuela (DERM). **Refs.:** Navarro & Liria 2007: 89 (syn.).

lateralis Petrocchi
1927 (F). In Shannon & Del Ponte 1927: 726. **Type loc.:** Zapla, Jujuy, Argentina (NE). **Distr.:** Argentina, Bolivia. **Refs.:** Lane & Cerqueira 1942: 578 (F); Belkin et al. 1968: 11 (type info.). **Ety.:** n.s. **Informal name:** Lateral Argentinian Bromeliad Mosquito.

paucartamboensis Porter
2014: 311 (M*, F*, P*, L*). **Type loc.:** San Pedro (13°0385' S, 71°3330' W, elevation 1,415 m), Kosnipata District, Paucartambo Province, Cusco, Peru (MNL). **Distr.:** Peru. **Ety.:** Paucartambo Prov., Peru. **Informal name:** Paucartambo Peruvian Bromeliad Mosquito.

Subgenus *Phoniomyia* Theobald

1903a: 311 (as genus). Type species: *Wyeomyia longirostris* Theobald (Dyar 1905a: 49).

Name derivation: n.s. [phono- = sound, tone, voice (Gr) plus little fly; poss. unknown ref. to "Proboscis very long, longer than the whole body"].

syn. *Dyarina* Bonne-Wepster & Bonne 1921: 6. Type species: *Dyarina tripartita* Bonne-Wepster & Bonne.

Important references: Corrêa & Ramalho 1956 (review); Forattini, Rabello, & Cotrim 1970 (type info.); Judd 1996 (phyl.); Judd 1998: 572 (to subg. of *Wyeomyia*); Motta et al. 2007 (phyl.); Porter 2014 (comp. morph.).

antunesi Lane & Guimarães
1937: 213 (M*, F, L*). **Type loc.:** Campos do Jordão [São Paulo], Brazil (FH). **Distr.:** Brazil. **Refs.:** Lane 1953: 1020 (M*, F, P*, L*); Corrêa & Ramalho 1956: 13 (M*, F*, P*, L*); Judd 1998 (from genus *Phoniomyia*). **Ety.:** Paulo César de Azevedo Antunes. **Informal name:** Antunes Brazilian Bromeliad Mosquito.

bonnei (Lane & Cerqueira)
1942: 630 (M*, F, P*, L*; *Phoniomyia*). **Type loc.:** Rio de Janeiro, Guanabara, Brazil (IOC). **Distr.:** Brazil. **Refs.:** Corrêa & Ramalho 1956: 15 (M*, F*, P*, L*); Belkin et al. 1971: 12, 37 (type info.); Judd 1998 (from genus *Phoniomyia*). **Ety.:** Johanna Bonne-Wepster. **Informal name:** Bonne Brazilian Bromeliad Mosquito.

davisi (Lane & Cerqueira)
1942: 647 (M*, F). **Type loc.:** Mangaratiba, Rio de Janeiro, Brazil (IOC). **Distr.:** Brazil. **Refs.:** Corrêa & Ramalho 1956: 17 (M*, F*, P*, L*); Belkin et al. 1971: 12, 37 (type info.); Mattingly 1971a: Pl. 6 (M head*); Judd 1998 (from genus *Phoniomyia*). **Ety.:** Nelson Caryl Davis. **Informal name:** Davis Brazilian Bromeliad Mosquito.

deanei (Lourenço-de-Oliveira)
1983 (M*, F*). **Type loc.:** Granjas Calabria Jacarepaguá, Rio de Janeiro, Brazil (IOC). **Distr.:** Brazil. **Refs.:** Lourenço-de-Olivera et al. 1991 (P, L); Judd 1998 (from genus *Phoniomyia*). **Ety.:** Leonidas M. Deane. **Informal name:** Deane Brazilian Bromeliad Mosquito.

diabolica (Lane & Forattini)
1952: 254 (F*, P*, L*; *Phoniomyia*). **Type loc.:** Presidente Wenceslau, Cuiabá River, Serra do Diabo, São Paulo, Brazil (FH). **Distr.:** Argentina, Brazil. **Refs.:** Corrêa & Ramalho 1956: 20 (M*, P*, L*); Judd 1998 (from genus *Phoniomyia*). **Ety.:** Região da Serra do Diablo, São Paulo State, Brazil. **Informal name:** Diablo Brazilian Bromeliad Mosquito.

edwardsi (Lane & Cerqueira)
1942: 637 (M*, F, P*, L*). **Type loc.:** Porto das Caixas, Rio de Janeiro, Brazil (USNM). **Distr.:** Brazil. **Refs.:** Knight & Chamberlain 1948: 13 (P*); Corrêa & Ramalho 1956: 22 (M*, F*, P*, L*); Belkin et al. 1971: 12 (tax., lectotype desig.); Judd 1998 (from genus *Phoniomyia*). **Ety.:** Frederick Wallace Edwards. **Informal name:** Edwards Brazilian Bromeliad Mosquito.

flabellata (Lane & Cerqueira)
1942: 641 (M*, F). **Type loc.:** Muriqueira, Bahia, Brazil (IOC). **Distr.:** Brazil. **Refs.:** Corrêa & Ramalho 1956: 24 (M*, F); Judd 1998 (from genus *Phoniomyia*). **Ety.:** small fanned (L); ref. to male sternum VIII very characteristic, in the form of a fan. **Informal name:** Small-fanned Brazilian Bromeliad Mosquito.

fuscipes Edwards
1922d: 76 (F). **Type loc.:** Paraguay (NHMUK). **Distr.:** Brazil, Paraguay. **Refs.:** Lane & Cerqueira 1942: 639 (M*, F, P*, L*; from syn. with *tripartita*); Coher 1949: 98 (F*); Lane 1953: 1026 (M*, F, P*, L*); Corrêa & Ramalho 1956: 25 (tax.); Belkin 1968b: 24 (lectotype desig.); Mattingly 1971a: Pl. 23 (L terminal segs.*); Judd 1998 (from genus *Phoniomyia*). **Ety.:** n.s. [dark foot (L)]. **Informal name:** Dark-footed Paraguayan Bromeliad Mosquito.

galvaoi (Corrêa & Ramalho)
1956: 26 (M*, F*, P*, L*; *Phoniomyia*). **Type loc.:** Ilha de Santo Amaro, Guarujá, São Paulo, Brazil (FH). **Distr.:** Brazil. **Refs.:** Judd 1998 (from genus *Phoniomyia*). **Ety.:** Augusto Leopoldo Ayroza Galvão. **Informal name:** Galvão Brazilian Bromeliad Mosquito.

incaudata Root
1928 (M*). In Dyar 1928: 54. **Type loc.:** Rio de Janeiro, Brazil (USNM). **Distr.:** Brazil. **Refs.:** Lane & Cerqueira 1942: 634 (M*, F, P*, L*); Lane 1953: 1018 (M*, F, P*, L*); Corrêa & Ramalho 1956: 30 (M*, F*, P*, L*); Stone & Knight 1957b: 117 (type info., lectotype desig.); Judd 1998 (from genus *Phoniomyia*). **Ety.:** n.s. [not tailed (L)]. **Informal name:** Tailless Brazilian Bromeliad Mosquito.

lassalli (Bonne-Wepster & Bonne)
1921a: 8 (M; *Dyarina*). **Type loc.:** Trinidad [Trinidad and Tobago, Lesser Antilles] (USNM). **Distr.:** Brazil, Lesser Antilles (includes Trinidad and Tobago). **Refs.:** Lane & Cerqueira 1942: 645 (M*, F; from syn. with *trinidadensis*); Coher 1949: 96 (F*); Lane 1953: 1037 (M*, F, L*); Stone 1967: 202 (lectotype desig.); Judd 1998 (from genus *Phoniomyia*). **Ety.:** n.s. **Informal name:** Lassall Trinbagonian Bromeliad Mosquito.

longirostris Theobald
1901c: 275 (F*). **Type loc.:** Rio de Janeiro, Brazil (NHMUK). **Distr.:** Brazil. **Refs.:** Lane & Cerqueira 1942: 626 (M*, F, P*, L); Lane 1953: 1008 (M*, F, P*, L); Corrêa & Ramalho 1956: 33 (M*, F, P*, L*; tax.); Belkin 1968b: 24 (lectotype desig.); Judd 1998: 587 (M*; from genus *Phoniomyia*). **Ety.:** long snout (L); proboscis. **Informal name:** Long-nosed Brazilian Bromeliad Mosquito.

lopesi (Corrêa & Ramalho)
1956: 37 (M*, F, P*, L*; *Phoniomyia*). **Type loc.:** Ilha de Santo Amaro, Guarujá, São Paulo, Brazil (FH). **Distr.:** Brazil. **Ety.:** Tito Lopes da Silva. **Informal name:** Lopes Brazilian Bromeliad Mosquito.

muehlensi Petrocchi
1927 (M, F). In Shannon & Del Ponte 1927: 727. **Type loc.:** Sáenz Peña, Resistencia and Charadai, Chaco, Argentina (INM). **Distr.:** Argentina, Bolivia, Brazil. **Refs.:** Lane & Cerqueira 1942: 646 (M*, F; from syn. with *tripartita*); Lane 1953: 1039 (M*, F); Corrêa & Ramalho 1956: 40 (M*, F*, P*, L*); de Souza & Xavier 1964: 459 (P*, L*); Casal & García 1970: 12 (lectotype desig.); Judd 1998 (from genus *Phoniomyia*). **Ety.:** n.s. [Pedro Mühlens (Harbach 2018c)] **Informal name:** Mühlens Argentinian Bromeliad Mosquito.

pallidoventer (Theobald)
1907: 598 (M; *Phoniomyia*). **Type loc.:** Rio de Janeiro, Brazil (NHMUK). **Distr.:** Brazil. **Refs.:** Lane & Cerqueira 1942: 643 (M*, F, P*); Lane 1953: 1033 (M*, F, P*, L*); Corrêa & Ramalho 1956: 43 (M*, F*, P*, L*); Mattingly 1971a: Pl. 9 (P terminal segs.*); Judd 1998 (from genus *Phoniomyia*). **Ety.:** pale venter (L); abdomen with "venter deep ochraceous." **Informal name:** Pale-bellied Brazilian Bromeliad Mosquito.

palmata (Lane & Cerqueira)
1942: 648 (M*, F, P*, L*). **Type loc.:** Rio de Janeiro, Federal District, Brazil (IOC). **Distr.:** Brazil. **Refs.:** Corrêa & Ramalho 1956: 45 (M*, F*, P*, L*); Judd 1998 (from genus *Phoniomyia*). **Ety.:** n.s. **Informal name:** Palm Brazilian Bromeliad Mosquito.

pilicauda Root
1928 (M*, F, L*). In Dyar 1928: 55. **Type loc.:** [Porto das Caixas], Rio de Janeiro, Brazil (USNM). **Distr.:** Brazil. **Refs.:** Lane & Cerqueira 1942: 632 (M*, F, P*, L*); Lane 1953: 1016 (M*, F, P*, L*); Corrêa & Ramalho 1956: 58 (M*, F*, P*, L*); Stone & Knight 1957b: 117 (type info., lectotype desig.); Judd 1998 (from genus *Phoniomyia*). **Ety.:** n.s. [single hair, caudal (L)]. **Informal name:** Ball-tailed Brazilian Bromeliad Mosquito.

quasilongirostris (Theobald)
1907: 598 (F; *Phoniomyia*). **Type loc.:** Mana, Rio de Janeiro, Brazil (NHMUK). **Distr.:** Argentina, Brazil. **Refs.:** Lane & Cerqueira 1942: 628 (M*, F, P*, L*; from syn. with *longirostris*); Lane 1953: 1011 (M*, F, P*, L*); Corrêa & Ramalho 1956: 50 (M*, F*, P*, L*); Belkin 1968b: 24 (lectotype desig.); Judd 1998 (from genus *Phoniomyia*). **Ety.:** nearly (L); similar to *longirostris*. **Informal name:** Medium-nosed Brazilian Bromeliad Mosquito [*quasilongirostris* = "half-long nose" = Medium-nosed].
syn. *neivai* Lane & Cerqueira
1942: 649 (F; *Phoniomyia*). **Type loc.:** Londrina, Paraná, Brazil (IOC). **Refs.:** Corrêa & Ramalho 1956: 50 (syn.).

splendida Bonne-Wepster & Bonne
1919: 111 (M, F, P, L). **Type loc.:** Lawa River, Sarah Creek, Suriname (ITH). **Distr.:** Brazil, French Guiana, Guyana, Lesser Antilles (includes Trinidad and Tobago), Suriname, United States (continental). **Refs.:** Lane 1953: 1014 (M*, F, P*, L*); Corrêa & Ramalho 1956: 52 (M*, F, P*, L*); Belkin 1968b: 24 (lectotype desig.); Judd 1998: 587 (L*; from genus *Phoniomyia*). **Ety.:** n.s. [shine, gleam, glitter (L); ref. to metallic coloration]. **Informal name:** Splendid Surinamese Bromeliad Mosquito.

theobaldi (Lane & Cerqueira)
1942: 638 (F*). **Type loc.:** Distrito Federal, Brazil (IOC). **Distr.:** Brazil. **Refs.:** Lane 1953: 1024 (M*, F, P*); Corrêa & Ramalho 1956: 54 (M*, F*, P*, L*); Belkin et al. 1971: 13 (type info.); Judd 1998 (from genus *Phoniomyia*). **Ety.:** Frederic Vincent Theobald. **Informal name:** Theobald Brazilian Bromeliad Mosquito.

trinidadensis Theobald
1901c: 277 (F*). **Type loc.:** Trinidad [Trinidad and Tobago, Lesser Antilles] (NHMUK). **Distr.:** Lesser Antilles (includes Trinidad and Tobago), Venezuela. **Refs.:** Corrêa & Ramalho 1956: 56 (M*, F; tax.); Belkin

1968b: 24 (lectotype desig.); Judd 1998: 587 (M*; from genus *Phoniomyia*). **Ety.:** Trinidad. **Informal name:** Arawak Trinbagonian Bromeliad Mosquito [the Arawak were an indigenous, pre-Columbian people of Trinidad and Tobago].

tripartita (Bonne-Wepster & Bonne)
1921a: 7 (M; *Dyarina*). **Type loc.:** Brazil (NE). **Distr.:** Argentina, Brazil. **Refs.:** Lane 1953: 1035 (M*, F); Belkin 1968b: 24 (type info.); Judd 1998 (from genus *Phoniomyia*). **Ety.:** n.s. [three parts (L)]. **Informal name:** Three-part Brazilian Bromeliad Mosquito.

Subgenus *Prosopolepis* Lutz

1905: 312. Type species: *Prosopolepis confusus* Lutz.

Name derivation: n.k.

Important references: Edwards 1932a: 86 (syn. with subg. *Dendromyia*); Motta & Lourenço-de-Oliveira 1995: 375 (to subg. uncertain in genus *Wyeomyia*); Lourenço-de-Oliveira et al. 1999: 200 (to subg. of genus *Wyeomyia*); Motta & Lourenço-de-Oliveira 2005 (comp. morph.); Motta et al. 2007 (phyl.).

confusa (Lutz)
1905: 312 (F; *Prosopolepis*). **Type loc.:** Brazil (IOC). **Distr.:** Bolivia, Brazil, Suriname, Venezuela. **Refs.:** Lane & Cerqueira 1942: 616 (M*, F, P*, L*; tax.); Lane 1953: 997 (M*, F, P*, L*); Belkin et al. 1971: 11 (type info.); Mattingly 1971a: Pl. 24 (L head*, terminal segs.*); Motta & Lourenço-de-Oliveira 1995: 375 (to subg. uncertain in *Wyeomyia*); Lourenço-de-Oliveira et al. 1999: 200 (M*, F*, P*, L*; from subg. uncertain). **Ety.:** n.s. [confused (L)]. **Informal name:** Confusing Brazilian Bromeliad Mosquito.
syn. *pusillum* Lutz
1928: 33 (F*, P; *Trichoprosopon*). **Type loc.:** Guayabita near Turmero, Aragua, Venezuela (LU). **Refs.:** Lane & Cerqueira 1942: 616 (syn., tax.); Lourenço-de-Oliveira et al. 1999: 200 (as *nomen dubium*) [not followed here].

Subgenus *Spilonympha* Motta & Lourenço-de-Oliveira

2005 (as subg.). Type species: *Wyeomyia aningae* Motta & Lourenço-de-Oliveira.

Name derivation: "Derived from the Greek spilus (spotted) and nympha (immature stage) in reference to the dark pigmented pattern on the integument of the pupa of the majority of species."

Important references: Motta et al. 2007 (phyl.).

airosai Lane & Cerqueira
1942: 593 (M*, F, P*, L*). **Type loc.:** Santa Tereza, Espírito Santo, Brazil (IOC). **Distr.:** Brazil. **Refs.:** Lane 1953: 957 (M*, F, P*, L*); Motta & Lourenço-de-Oliveira 1995: 375 (to subg. uncertain); Motta & Lourenço-de-Oliveira 2005: 838 (to subg. *Spilonympha*). **Ety.:** Augusto Leopoldo Ayrosa Galvão. **Informal name:** Ayrosa Brazilian Bromeliad Mosquito.

aningae Motta & Lourenço-de-Oliveira
2005: 847 (M*, F, P*, L*). **Type loc.:** near Paricatuba River, Estrada do Mosqueiro (1 1 5S48 13 W S48 13 W), Belém, Pará, Brazil (IOC). **Distr.:** Brazil. **Ety.:** aninga plant (*Montrichardia* sp. [larval habitat]). **Informal name:** Aninga Brazilian Bromeliad Mosquito.

bourrouli (Lutz)
1905: 347 (M, F; *Dendromyia*). **Type loc.:** Itaci [São Paulo], Brazil (NE). **Distr.:** Brazil, French Guiana, Nicaragua, Suriname. **Refs.:** Lane & Cerqueira 1942: 591 (M*, F, P*, L*); Lane 1953: 950 (M*, F, P*, L*); Belkin et al. 1971: 11 (type info.); Clastrier 1974: 153 (M*; distr.); Motta et al. 1995 (to subg. uncertain). **Ety.:** n.s. [Celestino Bourroul?]. **Informal name:** Bourroul Brazilian Bromeliad Mosquito.

finlayi Lane & Cerqueira
1942: 594 (M*, F, P*, L*). **Type loc.:** Xerém or Tigua, Rio de Janeiro, Brazil (IOC). **Distr.:** Brazil. **Refs.:** Lane 1953: 955 (M*, F, P*, L*); Motta et al. 1995 (to subg. uncertain); Motta & Lourenço-de-Oliviera 2005 (tax.; to subg. *Spilonympha*). **Ety.:** Carlos Finlay. **Informal name:** Finlay Brazilian Bromeliad Mosquito.

forcipenis Lourenço-de-Oliveira & Silva-do-Nascimento
1985: 321 (M*, F, P*, L*). **Type loc.:** Granjas Calabria, Jacarepaguá, Rio de Janeiro, Brazil (IOC). **Distr.:** Brazil. **Refs.:** Motta et al. 1995 (to subg. uncertain). **Ety.:** forceps, pincers (L); "hind plate of phallosome, forcep-like." **Informal name:** Forceps Brazilian Bromeliad Mosquito.

howardi Lane & Cerqueira
1942: 595 (M*, F). **Type loc.:** Muriqueira, Bahia, Brazil (IOC). **Distr.:** Brazil, Lesser Antilles (includes Trinidad and Tobago). **Refs.:** Motta et al. 1995 (to subg. uncertain). **Ety.:** Leland Ossian Howard. **Informal name:** Howard Brazilian Bromeliad Mosquito.

mystes Dyar
1924a: 92 (M). **Type loc.:** [Rio de Janeiro], Brazil (USNM). **Distr.:** Bolivia, Brazil, French Guiana. **Refs.:** Lane & Cerqueira 1942: 592 (M*, F, P*, L*); Lane 1953: 953 (M*, F, P*, L*); Stone & Knight 1957b: 124 (type info., lectotype desig.); Motta et al. 1995 (to subg. uncertain). **Ety.:** n.s. [initiated in secret rites (L)]. **Informal name:** Mysterious Brazilian Bromeliad Mosquito.

Subgenus *Triamyia* Dyar

1919d: 120 (as genus). Type species: *Wyeomyia aporonoma* Dyar & Knab.

Name derivation: n.s [tri- = three plus little fly (Gr)].

Important references: Motta et al. 2007 (from syn. with *Wyeomyia* subg. *Dendomyia*).

aporonoma Dyar & Knab
1906d: 230 (L*). **Type loc.:** Sonsonate, El Salvador (USNM). **Distr.:** Argentina, Belize, Bolivia, Brazil, Colombia, Costa Rica, Ecuador, El Salvador, French Guiana, Guatemala, Guyana, Honduras, Lesser Antilles (includes Grenada, Saint Vincent and the Grenadines, Trinidad and Tobago), Mexico, Nicaragua, Panama, Peru, Suriname, Venezuela. **Refs.:** da Costa Lima 1943: 307 (tax.); Lane 1953: 964 (M*, F, P*, L*); Stone & Knight 1957b: 121 (type info., lectotype desig.); García & Casal 1968a: 62 (distr.); Mattingly 1971a: Pl. 1 (A scutum*); Belkin & Heinemann 1976a: 272, 296 (distr.); Heinemann & Belkin 1977b: 429, 443 (distr.); Motta et al. 1995 (to subg. uncertain); Motta et al. 2007 (in subg. *Triamyia*). **Ety.:** doubt, question (Gr); ref. to author doubt. **Informal name:** Doubtful Salvadoranian Bromeliad Mosquito.

staminifera Lourenço-de-Oliveira, Motta & de Castro
1992: 115 (M*, F*, P*, L*). **Type loc.:** Branco River (near Fazenda Rio Branco) (9° 54′ S, 62° 58′ W), Ariquemes, Rondônia state, Brazil (IOC). **Distr.:** Brazil. **Refs.:** Motta & Lourenço-de-Oliveira 1995: 384 (exclusion from subg. *Dendromyia*); Motta et al. 2007 (placement in subg. *Triamyia*). **Ety.:** stamen bearing (L); ref. to gonocoxite with "subapical lobe covered by minute setae, bearing three stamen-like strong setae." **Informal name:** Stamen Brazilian Bromeliad Mosquito.

Subgenus *Wyeomyia* Theobald

1901b: 235; 1901a: 98; 1901c: 267 (as genus, subgenus). Type species: *Wyeomyia grayii* Theobald (Neveu-Lemaire 1902c: 223).

Name derivation: n.s. [etymology not determined].

syn. *Diphalangarpe* Dyar 1919d: 126. Type species: *Wyeomyia abascanta* Dyar & Knab.
syn. *Caltadimyia* Dyar 1919d: 137. Type species: *Wyeomyia pandora* Dyar & Knab.
syn. *Pentemyia* Dyar 1919d: 122 (as genus). Type species: *Wyeomyia drapetes* Dyar & Knab.
syn. *Wyeomyia*, subg. *Phyllozomyia* Dyar 1924c: 112. Type species: [*Aedes*] *smithii* Coquillett.
syn. *Wyeomyia*, subg. *Technicomyia* Dyar 1925a: 20. Type species: *Wyeomyia florestan* Dyar.

Important references: Bruijning 1959: 99 (tax.); Stone 1969: 2 (tax.); Belkin, Heinemann, & Page 1970: 205 (tax.).

abebela Dyar & Knab
1908: 67 (M, F). **Type loc.:** Córdoba [Veracruz], Mexico (USNM). **Distr.:** Belize, Costa Rica, Guatemala, Honduras, Mexico, Panama. **Refs.:** Dyar 1928: 35 (M*, F, L*); de Buen 1953b: 253 (P*); Lane 1953: 911 (M*, F, L); Stone & Knight 1957b: 120 (type info., lectotype desig.); Heinemann & Belkin 1977b: 429 (distr.); Pecor et al. 2002: 250 (tax., distr.; Belize). **Ety.:** n.s. [a thing that must be stepped on? (Gr)]. **Informal name:** Córboda Mexican Bromeliad Mosquito.

ablechra Dyar & Knab
1908: 66 (F). **Type loc.:** Sonsonate, El Salvador (USNM). **Distr.:** El Salvador. **Refs.:** Lane & Cerqueira 1942: 556 (syn.); Stone & Knight 1957b: 120 (type info., lectotype desig.); Stone 1969: 3 (to sp.). **Ety.:** n.s. [weak? (L)]. **Informal name:** Weak Salvadoranian Bromeliad Mosquito.

adelpha Dyar & Knab
1906a: 140 (A). **Type loc.:** Esparta, Costa Rica (USNM). **Distr.:** Belize, Costa Rica, Mexico, Panama. **Refs.:** Lane & Cerqueira 1942: 556 (syn.); Stone & Knight 1957b: 120 (type info., lectotype desig.); Stone 1969: 3 (to sp.); Bertram 1971: 642 (distr.; Belize). **Ety.:** n.s. [sibling (Gr)]. **Informal name:** Sibling Costa Rican Bromeliad Mosquito.
syn. *ablabes* Dyar & Knab
1908: 66 (M, F). **Type loc.:** Córdoba [Veracruz], Mexico (USNM). **Refs.:** Stone & Knight 1957a, b: 120 (type info., lectotype desig.); Stone 1969: 3 (syn.).
syn. *rolonca* Dyar & Knab
1910: 173 (M). **Type loc.:** Upper Pequini River, Panama (USNM). **Refs.:** Stone 1969: 3 (syn.).

arthrostigma (Lutz)
1905: 311 (M, F; *Dendromyia*). **Type loc.:** Brazil (IOC). **Distr.:** Belize, Brazil, Colombia, Costa Rica, Ecuador, El Salvador, French Guiana, Guatemala, Honduras, Lesser Antilles (includes Trinidad and Tobago), Mexico, Nicaragua, Panama, Venezuela. **Refs.:** Lane & Cerqueira 1942: 561 (M*, F, P*, L*); da Costa Lima 1943: 297 (tax.); Lane 1953: 909 (M*, F, P*, L*); Belkin et al. 1971: 10 (lectotype desig.); Bertram 1971: 642 (distr.; Belize); Heinemann & Belkin 1977b: 429, 433, 443 (distr.). **Ety.:** joint mark (Gr, L); pale spots on leg articulations. **Informal name:** Joint-marked Brazilian Bromeliad Mosquito.
syn. *bromeliarum* Dyar & Knab
1906a: 138 (L). **Type loc.:** Trinidad [Trinidad and Tobago, Lesser Antilles] (USNM). **Refs.:** Lane &

Cerqueira 1942: 561 (syn.); Stone & Knight 1957b: 121 (type info., lectotype desig.).

syn. *espartana* Dyar & Knab

1906a: 140 (A). **Type loc.:** Esparta, Costa Rica (USNM). **Refs.:** Lane & Cerqueira 1942: 561 (syn.).

syn. *panamena* Dyar & Knab

1907c: 210 (A). **Type loc.:** Tavernilla, Canal Zone, Panama (USNM). **Refs.:** Lane & Cerqueira 1942: 561 (syn.).

syn. *drapetes* Dyar & Knab

1909d: 264 (M, F). **Type loc.:** San Juan, Trinidad [Trinidad and Tobago, Lesser Antilles] (USNM). **Refs.:** Lane & Cerqueira 1942: 561 (syn.); Stone & Knight 1957b: 122 (type info., lectotype desig.).

atrata Belkin, Heinemann, & Page

1970: 227 (M*, F, P*, L*). **Type loc.:** Dunns River (Saint Ann), Jamaica (USNM). **Distr.:** Jamaica. **Ety.:** dark (L); spiculation of the larva. **Informal name:** Dark Jamaican Bromeliad Mosquito.

bahama Dyar & Knab

1906a: 138 (A). **Type loc.:** Tarpon Bay [Eleuthera Island], Bahama Islands (USNM). **Distr.:** Bahamas, Cuba. **Refs.:** Lane 1953: 881 (F); Stone & Knight 1957b: 121 (type info.). **Ety.:** Bahamas. **Informal name:** Taino Bahamian Bromeliad Mosquito [the name Bahamas is taken from the Taino *ba ha ma*, meaning "big upper middle land"].

syn. *minor* Dyar & Knab

1906a: 138 (A). **Type loc.:** Baracca [Oriente], Cuba (USNM). **Refs.:** Stone & Knight 1957b: 124 (type info.).

caracula Dyar & Núñez Tovar

1928: 90 (M). **Type loc.:** Mariara, Carabobo, Venezuela (USNM). **Distr.:** Brazil, Panama, Venezuela. **Refs.:** Hill & Hill 1945: 296 (M*, F, P*, L*); Belkin et al. 1970: 191 (distr.). **Ety.:** small amount of water (L); ref. to larvae found in a bromeliad. **Informal name:** Droplet Venezuelan Bromeliad Mosquito.

celaenocephala Dyar & Knab

1906a: 140 (A). **Type loc.:** Cacao, Trece Aguas, Alta Vera Paz, Guatemala (USNM). **Distr.:** Belize, Colombia, Costa Rica, Ecuador, El Salvador, Guatemala, Honduras, Lesser Antilles (includes Curaçao, Trinidad and Tobago), Mexico, Nicaragua, Panama, Venezuela. **Refs.:** Lane 1953: 884 (M*, F, P*, L*); Berti et al. 2015 (distr.; Venezuela). **Ety.:** n.s. [dark head (Gr); prob. ref. to head color]. **Informal name:** Dark-headed Guatemalan Bromeliad Mosquito.

syn. *chrysomus* Dyar & Knab

1907c: 208 (M; *Phoniomyia chrysomus*). **Type loc.:** Tabernilla, Canal Zone, Panama (USNM).

syn. *philophone* Dyar & Knab

1907c: 209 (F; *Phoniomyia*). **Type loc.:** Tabernilla, Canal Zone, Panama (USNM). **Refs.:** Stone & Knight 1957b: 125 (type info.).

syn. *megalodora* Dyar & Knab

1908: 69 (F). **Type loc.:** Sonsonate, El Salvador (USNM). **Refs.:** Stone & Knight 1957b: 124 (type info., lectotype desig.).

syn. *mataea* Dyar & Knab

1908: 70 (F). **Type loc.:** Sonsonate, El Salvador (USNM). **Refs.:** Stone & Knight 1957b: 124 (type info., lectotype desig.).

charmion Dyar

1928: Errata & Addenda (M). **Type loc.:** Agua Clara Reservoir, near Gatún, Canal Zone, Panama (USNM). **Distr.:** Costa Rica, Panama. **Refs.:** Lane 1953: 928 (M*, P*, L*). **Ety.:** n.s. [source of delight (Gr); servant and advisor to Cleopatra VII]. **Informal name:** Charmion Panamanian Bromeliad Mosquito.

corona Belkin, Heinemann, & Page

1970: 218 (M*, P*, L*). **Type loc.:** Hermitage Dam (Saint Andrew), Jamaica (USNM). **Distr.:** Jamaica. **Ety.:** crown (L); ref. to crown of short setae on the larval siphon. **Informal name:** Crowned Jamaican Bromeliad Mosquito.

downsi Lane

1945a: 140 (F). **Type loc.:** Mirassol, São Paulo, Brazil (FH). **Distr.:** Brazil. **Ety.:** Wilbur George Downs. **Informal name:** Downs Brazilian Bromeliad Mosquito.

florestan Dyar

1925a: 20 (M). **Type loc.:** Fort Clayton, Canal Zone, Panama (USNM). **Distr.:** Brazil, Panama. **Refs.:** Lane 1953: 929 (M*). **Ety.:** n.s. **Informal name:** Floral Panamanian Bromeliad Mosquito.

gaudians Dyar & Núñez Tovar

1927b: 5 (M, F). **Type loc.:** Mariara, Carabobo, Venezuela (USNM). **Distr.:** Venezuela. **Refs.:** Dyar 1928: 38 (M*, F); Lane 1953: 903 (M*, F); Stone & Knight 1957b: 123 (type info., lectotype desig.). **Ety.:** n.s. [rejoicing? (L)]. **Informal name:** Joyful Venezuelan Bromeliad Mosquito.

gausapata Dyar & Núñez Tovar

1927b: 7 (M). **Type loc.:** Ocumare de la Costa, Aragua, Venezuela (USNM). **Distr.:** Brazil, Guyana, Lesser Antilles (includes Trinidad and Tobago), Venezuela. **Refs.:** Dyar 1928: 39 (M*, F); Lane 1953: 905 (M*, F). **Ety.:** n.s. [shaggy (L)]. **Informal name:** Shaggy Venezuelan Bromeliad Mosquito.

grayii Theobald

1901c: 269 (F*). **Type loc.:** Castries, Saint Lucia Island [Lesser Antilles] (NHMUK). **Distr.:** Dominican Republic, Lesser Antilles (includes Barbados, Dominica, Grenada, Guadeloupe, Martinique, Montserrat, Saint Lucia, Saint Vincent and the Grenadines, Trinidad and Tobago). **Refs.:** Belkin 1968b: 40 (to sp.; lectotype desig.); Belkin et al. 1970: 208 (tax.). **Ety.:** n.s. [poss. George

Gray]. **Informal name:** Gray Saint Lucian Bromeliad Mosquito.

syn. *colsoni* Senevet & Quiévreux
1941: 249 (M*, L; as var.). **Type loc.:** Colson, Martinique Island [Lesser Antilles] (FMP, originally LPFM). **Refs.:** Belkin 1968b: 39 (syn.).

syn. *antillarum* Floch & Abonnenc
1945e: 4 (M*, F, L*). **Type loc.:** Guadeloupe Island [Lesser Antilles] (PIP). **Refs.:** Belkin 1968b: 39 (syn.; lectotype desig.).

guatemala Dyar & Knab
1906a: 139 (A). **Type loc.:** Cacao, Trece Aguas, Alta Vera Paz, Guatemala (USNM). **Distr.:** Costa Rica, Guatemala, Mexico. **Refs.:** Lane 1953: 902 (syn.); Stone & Knight 1957b: 123 (type info.); Stone 1969: 3 (to sp.); Belkin et al. 1970: 195 (tax.). **Ety.:** Guatemala. **Informal name:** Cacao Guatemalan Bromeliad Mosquito.

hemisagnosta Dyar & Knab
1906d: 230 (L*). **Type loc.:** Sonsonate, El Salvador (USNM). **Distr.:** Argentina, Bolivia, Brazil, Costa Rica, El Salvador, Panama. **Refs.:** Dyar 1928: 64 (F, L*); Lane 1953: 889 (F, L); Stone & Knight 1957b: 123 (type info., lectotype desig.). **Ety.:** n.s. [half unknown (Gr)]. **Informal name:** Obscure Salvadoranian Bromeliad Mosquito [*hemisagnosta* = "half unknown" = Obscure].

syn. *gynaecopus* Dyar & Knab
1908: 66 (F). **Type loc.:** Esparta, Costa Rica (USNM).

syn. *baria* Dyar & Knab
1908: 69 (F). **Type loc.:** Sonsonate, El Salvador (USNM).

hirsuta (Hill & Hill)
1946: 39 (M*, F, P*, L*; *Phoniomyia*). **Type loc.:** Hermitage Dam [Saint Andrew, Surrey], Jamaica (USNM). **Distr.:** Jamaica. **Refs.:** Belkin et al. 1970: 222 (M*, F, P*, L*). **Ety.:** hairy (L); ref. to hairy appearance of the larva). **Informal name:** Hirsute Jamaican Bromeliad Mosquito.

juxtahirsuta Belkin, Heinemann, & Page
1970: 220 (M*, F, P*, L*). **Type loc.:** "St. James, Montego Bay, Reading, roadside near Five mile (SR894413), 30 m," Jamaica (USNM). **Distr.:** Jamaica. **Ety.:** nearly hairy (L); similar to morphology and bromeliad habitat of *hirsuta*. **Informal name:** Downy Jamaican Bromeliad Mosquito [*juxtahirsuta* = "almost hairy" = Downy].

labesba Howard, Dyar, & Knab
1913: Pl. 89, Fig. 28; 1915: 106 (F, L*). **Type loc.:** Tabernilla, Canal Zone, Panama (USNM). **Distr.:** Panama. **Refs.:** Lane 1953: 902 (syn.); Stone & Knight 1957b: 124 (type info., lectotype desig.); Stone 1969: 3 (from syn.). **Ety.:** n.s. [poss. anagram for *ablabes* (Harbach 2018c)]. **Informal name:** Stained Panamanian Bromeliad Mosquito.

luna Belkin, Heinemann, & Page
1970: 213 (M*, F, P*, L*). **Type loc.:** track leading south into Unity Valley (15 mi south of Moneague) [Saint Ann], Jamaica (USNM). **Distr.:** Jamaica. **Ety.:** moon (L); ref. to pearly iridescence of pupa. **Informal name:** Lunar Jamaican Bromeliad Mosquito.

medioalbipes Lutz
1904 (A). In Bourroul 1904: 68. **Type loc.:** Bahia, Brazil (NHMUK). **Distr.:** Argentina, Belize, Brazil, Colombia, Cuba, Dominican Republic, Guyana, Haiti, Jamaica, Lesser Antilles (includes Dominica,Trinidad and Tobago), Nicaragua, Paraguay, Suriname, United States (continental), Venezuela. **Refs.:** da Costa Lima 1943: 302 (tax.); Lane 1953: 893 (M*, F, P*, L*); Bruijning 1959: 103 (M*, F, P*, L*; syn.); Belkin 1968b: 40 (tax.); Belkin et al. 1970: 208 (tax.); Belkin et al. 1971: 10 (lectotype desig.). **Ety.:** middle-white foot (L); ref. to middle of legs (femora and tibiae) white below. **Informal name:** Middle-footed Brazilian Bromeliad Mosquito.

syn. *telestica* Dyar & Knab
1906d: 230 (L*). **Type loc.:** Trinidad [Trinidad and Tobago, Lesser Antilles] (USNM). **Refs.:** da Costa Lima 1943: 302 (from syn. with *quasiluteoventralis*); Lane 1953: 927 (M*, P, L); Stone & Knight 1957b: 126 (type info.); Bruijning 1959: 103 (syn.).

syn. *abascanta* Dyar & Knab
1908: 65 (M, F). **Type loc.:** Trinidad [Trinidad and Tobago, Lesser Antilles] (USNM). **Refs.:** Lane 1953: 893 (syn.); Stone & Knight 1957b: 120 (type info., lectotype desig.).

melanopus Dyar
1919d: 130 (M*). **Type loc.:** Porto Bello, Panama (USNM). **Distr.:** Belize, Brazil, Colombia, Costa Rica, Guatemala, Honduras, Lesser Antilles (includes Trinidad and Tobago), Mexico, Nicaragua, Panama, Venezuela. **Refs.:** Lane 1947: 97 (M*, F, P*, L*); Heinemann & Belkin 1977b: 429, 443 (distr.); Pecor et al. 2002: 250 (tax., distr.; Belize). **Ety.:** n.s. [black wing (Gr)]. **Informal name:** Black-winged Panamanian Bromeliad Mosquito.

syn. *culebrae* Dyar
1923b: 65 (F). **Type loc.:** Culebra, Canal Zone, Panama (USNM).

mitchellii (Theobald)
1905d: 37 (F; *Dendromyia*). **Type loc.:** Jamaica (NHMUK). **Distr.:** Brazil, Colombia, Cuba, Dominican Republic, Guam, Haiti, Honduras, Jamaica, Mexico, Nicaragua, Panama, Puerto Rico, United States (continental, Hawaiʻi), Venezuela. **Refs.:** Carpenter et al. 1946: 90 (M*, L*); Roth 1946a: 296 (F*); Lane 1953: 902 (M*, F, P*, L*); Carpenter & LaCasse 1955: 66 (M*, F*, L*; keys); Bruijning 1959: 103 (syn.); Belkin 1968b: 41 (tax.); Stone 1969: 3 (tax.); Belkin et al. 1970: 205 (M*, F*, P*, L*; tax.); Reinert 1970h: 166 (P*); Lee et al. 1989c: 301 (tax., bion.,

distr., review); Darsie & Ward 2005 (F*, L*; keys, distr.). **Ety.:** Mrs. Alfred Mitchell. **Informal name:** Mitchell Jamaican Bromeliad Mosquito.

syn. *jamaicensis* Theobald
1905d: 11 (*Dendromyia*; different name in key for specimen described under the name *mitchelli* in text).

syn. *violescens* Dyar & Knab
1906a: 138 (A). **Type loc.:** Cayamas, Cuba (USNM). **Refs.:** Stone & Knight 1957b: 126 (type info.).

syn. *glaucocephala* Dyar & Knab
1906a: 140 (A). **Type loc.:** near Santo Domingo City, Santo Domingo [Dominican Republic] (USNM). **Refs.:** Stone & Knight 1957b: 123 (type info.).

syn. *ochrura* Dyar & Knab
1906d: 229 (L*). **Type loc.:** near Santo Domingo City, Santo Domingo [Dominican Republic] (USNM). **Refs.:** Stone & Knight 1957b: 124 (type info., lectotype desig.).

syn. *abia* Dyar & Knab
1908: 67 (M, F). **Type loc.:** Santo Domingo, Dominican Republic (USNM). **Refs.:** Lane 1953: 893 (syn.); Stone & Knight 1957b: 120 (type info., lectotype desig.); Belkin 1970a: 58 (syn., type info., type loc. info.).

syn. *antoinetta* Dyar & Knab
1909d: 263 (A). **Type loc.:** Estero, Florida, United States (USNM).

nigritubus Galindo, Carpenter, & Trapido
1951b: 89 (M*, F, P*, L*). **Type loc.:** La Victoria, Panama Province, Panama (USNM). **Distr.:** Dominican Republic, Haiti, Jamaica, Lesser Antilles (includes Trinidad and Tobago), Mexico, Panama. **Refs.:** Lane 1953: 890 (M*, F, P*, L*); Belkin et al. 1970: 200 (M*, F*, P*, L*); Belkin & Heinemann 1972: 69 (distr.). **Ety.:** black tube (L); ref. to heavily sclerotized, dark larval siphon. **Informal name:** Black-tubed Panamanian Bromeliad Mosquito.

pertinans (Williston)
1896: 271 (M*, F; *Aedes*). **Type loc.:** Saint Vincent Island [Lesser Antilles] (NHMUK). **Distr.:** Belize, Brazil, Colombia, Costa Rica, El Salvador, French Guiana, Guatemala, Guyana, Honduras, Lesser Antilles (includes Barbados, Dominica, Grenada, Saint Lucia, Saint Vincent and the Grenadines, Trinidad and Tobago), Mexico, Nicaragua, Panama, Suriname, Venezuela. **Refs.:** Lane & Cerqueira 1942: 549 (syn.); Belkin 1968b: 41 (lectotype desig.); Belkin et al. 1970: 209 (distr.). **Ety.:** n.s. [pertain to, relate to (L)]. **Informal name:** Pertinent Vincentian Bromeliad Mosquito.

pseudorobusta Pajot & Fauran
1975: 125 (M*, P*). **Type loc.:** Pasteur Institute Experiment Station of l'Inini, Cayenne, French Guiana [Guiana] (PIG). **Distr.:** French Guiana. **Ety.:** false (Gr); similar to *robusta*. **Informal name:** Steroid French Guianese Bromeliad Mosquito [*pseudorobusta* = "false robust" = Steroid, a drug used to achieve muscular development falsely].

robusta Senevet & Abonnenc
1939b: 253 (M*, L*). **Type loc.:** Saut Tigre [Inini], French Guiana [Guiana] (FMP, originally LPFM). **Distr.:** French Guiana. **Refs.:** Belkin 1968b: 42 (tax.). **Ety.:** n.s. [hardy, robust (L)]. **Informal name:** Robust French Guianese Bromeliad Mosquito.

scotinomus (Dyar & Knab)
1907c: 209 (M, F; *Phoniomyia*). **Type loc.:** Boquerón River, Panama (USNM). **Distr.:** Argentina, Brazil, Colombia, Costa Rica, Ecuador, Guyana, Saint Lucia, Saint Vincent and the Grenadines, Nicaragua, Panama, Paraguay. **Refs.:** Lane 1953: 900 (M*, F, P*, L*); Stone & Knight 1957b: 125 (type info., lectotype desig.). **Ety.:** n.s. [darkness, gloom (Gr)]. **Informal name:** Shade-wandering Panamanian Bromeliad Mosquito.

syn. *homothe* Dyar & Knab
1907c: 211 (A). **Type loc.:** Tabernilla, Canal Zone, Panama (USNM). **Refs.:** Lane 1953: 900 (syn.).

syn. *leucopisthepus* Dyar & Knab
1907c: 212 (A). **Type loc.:** Tabernilla, Canal Zone, Panama (USNM).

syn. *abrachys* Dyar & Knab
1909d: 262 (F). **Type loc.:** Caldera Island, Porto Bello Bay, Panama (USNM). **Refs.:** Stone & Knight 1957b: 120 (type info., lectotype desig.).

syn. *chresta* Dyar & Knab
1909d: 263 (F). **Type loc.:** Tabernilla, Canal Zone, Panama (USNM). **Refs.:** Stone & Knight 1957b: 122 (type info., lectotype desig.).

syn. *hapla* Dyar & Knab
1909d: 265 (F). **Type loc.:** Caldera Island, Porto Bello Bay, Panama (USNM).

syn. *incana* Dyar
1922g: 189 (M*). **Type loc.:** Margarita, Canal Zone, Panama (USNM). **Refs.:** Stone & Knight 1957b: 123 (type info., lectotype desig.).

syn. *camptocomma* Dyar
1924d: 120 (M, L). **Type loc.:** Barranquilla [Magdalena], Colombia (USNM). **Refs.:** Lane & Cerqueira 1942: 552 (syn.); Stone & Knight 1957b: 121 (type info., lectotype desig.).

simmsi (Dyar & Knab)
1908: 65 (F; *Phoniomyia*). **Type loc.:** Porto Bello, San Felipe, Panama (USNM). **Distr.:** Colombia, Costa Rica, Nicaragua, Panama, Venezuela. **Refs.:** Dyar 1928: 45 (M*, F, L*); Lane 1953: 888 (M*, F, L). **Ety.:** H. Simms. **Informal name:** Simms Panamanian Bromeliad Mosquito.

syn. *dymodora* Dyar & Knab
1908: 68 (F). **Type loc.:** Fort San Felipe, Porto Bello, Panama (USNM). **Refs.:** Lane 1953: 888 (syn.); Stone & Knight 1957b: 122 (type info.).

syn. *roloncetta* Dyar
1919d: 131 (M*). **Type loc.:** Fort San Felipe, Porto Bello, Panama (USNM). **Refs.:** Romero Sierra 1907: 39 (F*; as *nigritulus* Zetterstedt).

smithii (Coquillett)
1901: 260 (A; *Aedes*). **Type loc.:** Lahaway, New Jersey, United States (USNM). **Distr.:** Canada, Jamaica, United States (continental). **Refs.:** Roth 1946a: 294 (F*); Ross 1947: 34 (M*, F, L*); Darsie 1951: 45 (P*); Yamaguti & LaCasse 1951a: 18 (M*, F*, L*; as *haynei*); Yamaguti & LaCasse 1951a: 26 (M*, F*, L*); Lane 1953: 882 (M*, F, P*, L*); Carpenter & LaCasse 1955: 64 (M*, F*, L*; keys; as *haynei*); Carpenter & LaCasse 1955: 68 (M*, F*, L*; keys), 64 (M*, F*, L*; as *haynei*); Stone & Knight 1957b: 125 (type info.); Dodge 1966: 388 (first instar L*; key; as *haynei*); Barr & Barr 1969b: 193 (E*); Bradshaw & Lounibos 1977 (bion.; syn. of *haynei*); Bradshaw 1986 (bion.); Darsie & Ward 2005 (F*, L*; keys, distr.); Harrison et al. 2016 (F*, L*; keys, distr.; note 2). **Ety.:** J.B. Smith. **Informal name:** Smith American Bromeliad Mosquito. Official ESA common name: pitcherplant mosquito.
syn. *haynei* Dodge
1947: 118 (M*, F, P*, L*). **Type loc.:** Columbia, South Carolina, United States (USNM). **Refs.:** Bradshaw & Lounibos 1977: 547 (syn.).

sororcula Dyar & Knab
1906a: 139 (A). **Type loc.:** San Francisco Mines, Santo Domingo [Dominican Republic] (USNM). **Distr.:** Bahamas, Dominican Republic, Martinique. **Refs.:** Stone & Knight 1957b: 125 (type info.); Belkin 1970a: 58 (from syn. with *vanduzeei*); Belkin & Heinemann 1973: 218 (type loc. info). **Ety.:** n.s. [little sister (L)]. **Informal name:** Sisterly Dominican Bromeliad Mosquito.
syn. *fratercula* Dyar & Knab
1906a: 139 (A). **Type loc.:** Dominican Republic (USNM). **Refs.:** Belkin 1970a: 58 (type loc., syn.).

stellata Belkin, Heinemann, & Page
1970: 215 (M*, P*, L*). **Type loc.:** Hermitage Dam Road [Saint Andrew], Jamaica (USNM). **Distr.:** Jamaica. **Ety.:** provided with stars (L); ref. to stellate setae of larva. **Informal name:** Stellar Jamaican Bromeliad Mosquito.

stonei Vargas & Martínez Palacios
1953: 297 (M*, F). **Type loc.:** Poana or Teapa, Tabasco, Mexico (ISET). **Distr.:** Mexico. **Ety.:** Alan Stone. **Informal name:** Stone Mexican Bromeliad Mosquito.

vanduzeei Dyar & Knab
1906a: 138 (A). **Type loc.:** Estero, Florida, United States (USNM). **Distr.:** Cayman Isl., Cuba, Dominican Republic, Haiti, Jamaica, Puerto Rico, United States (continental). **Refs.:** Gerry 1932: 37 (F*); Roth 1946a: 296 (P*); Yamaguti & LaCasse 1951a: 31 (M*, F*, L*); Lane 1953: 880 (M*, F, L*); Carpenter & LaCasse 1955: 70 (M*, F*, L*; keys); Stone & Knight 1957b: 126 (type info., lectotype desig.); Dodge 1966: 388 (first instar L; key); Belkin et al. 1970: 210 (M*, F, P*, L*); Reinert 1970h: 163 (P*); Darsie & Ward 2005 (F*, L*; keys, distr.). **Ety.:** J.B. Van Duzee. **Informal name:** Van Duzee American Bromeliad Mosquito.
syn. *argyrura* Dyar & Knab
1908: 70 (F). **Type loc.:** San Antonio de Los Baños [Habana], Cuba (USNM).
syn. *conchita* Dyar & Knab
1909d: 264 (A). **Type loc.:** San Antonio de Los Baños [Habana], Cuba (USNM). **Refs.:** Stone & Knight 1957b: 122 (type info., lectotype desig.).

Subgenus *Zinzala* Zavortink

1986:46 (as subg. of genus *Wyeomyia*). Type species: *Wyeomyia* (*Zinzala*) *zinzala* Zavortink.

Name derivation: a kind of gnat (L).

Important references: Judd 1996 (phyl.).

fishi Zavortink
1986: 55 (M*, F, P*, L*). **Type loc.:** Santa Elena Road (km 146), Estado Bolivar, Venezuela (CAS). **Distr.:** Venezuela. **Ety.:** Durland Fish. **Informal name:** Fish Venezuelan Bromeliad Mosquito.

zinzala Zavortink
1986: 50 (M*, F, P*, L*). **Type loc.:** summit of Mount Roraima, Estado Bolívar, Venezuela (CAS). **Distr.:** Venezuela. **Ety.:** a kind of a gnat (L). **Informal name:** Gnat Venezuelan Bromeliad Mosquito.

Subgenus Uncertain *Wyeomyia*

syn. *Dinomyia* Dyar 1919d: 117. Type species: *Dinomyia proviolans* Dyar.
syn. *Calladimyia* Dyar 1919d: 137. Type species: *Wyeomyia pandora* Dyar & Knab.
syn. *Cleobonnea* Dyar 1919d: 134 (formerly syn. of *Dendromyia* Theobald 1903a). Type species: *Wyeomyia occulta* Bonne-Wepster & Bonne.
syn. *Prosopolepis*, subg. *Eunicemyia* Dyar & Shannon 1924d: 481. Type species: *Wyeomyia albosquamata* Bonne-Wepster & Bonne.
syn. *Prosopolepis*, subg. *Janicemyia* Dyar & Shannon 1924d: 481. *Type species: Wyeomyia* clasoleuca Dyar & Knab.

Important references: Motta et al. 1995 (tax.).

aequatorianna Leví-Castillo
1954b: 249 (M*, P*, L*). **Type loc.:** Hacienda Pichilingue, Los Rios, Ecuador (NE). **Distr.:** Ecuador. **Refs.:** Motta et al. 1995 (to subg. uncertain). **Ety.:** n.s. [equator? (L)]. **Informal name:** Equatorial Ecuadorian Bromeliad Mosquito.

albosquamata Bonne-Wepster & Bonne
1919: 107 (M, F, L). **Type loc.:** Lawa River, Suriname (ITH). **Distr.:** French Guiana, Panama, Suriname. **Refs.:**

Dyar 1928: 83 (M*, F, L*); Senevet & Abonnenc 1939b: 257 (M*, L*); Belkin 1968b: 38 (lectotype desig.); Motta et al. 1995 (to subg. uncertain). **Ety.:** n.s. [white scaled (L); presence of broad white scales on the postnotum (postpronotum)]. **Informal name:** White-scaled Surinamese Bromeliad Mosquito.

amazonica Leví-Castillo
1954b: 247 (M*). **Type loc.:** Tena, Napo-Pastaza, Ecuador (CEIE). **Distr.:** Ecuador. **Refs.:** Motta et al. 1995 (to subg. uncertain). **Ety.:** Amazon. **Informal name:** Amazon Ecuadorian Bromeliad Mosquito.

argenteorostris (Bonne-Wepster & Bonne)
1920: 167 (M, F, L; *Cleobonnea*). **Type loc.:** Lawa River, Suriname (ITH). **Distr.:** Brazil, French Guiana, Suriname. **Refs.:** Dyar 1928: 61 (M*, F, L*); Senevet & Abonnenc 1939b: 262 (M*, L*); Lane 1953: 990 (M*, F, L); Fauran 1961a: 13 (P*); Belkin 1968b: 39 (lectotype desig.); Motta et al. 1995 (to subg. uncertain). **Ety.:** silvery beak (L); ref. to white scale markings on proboscis. **Informal name:** Silver-beaked Surinamese Bromeliad Mosquito.
syn. *albocaerulea* Senevet & Abonnenc
1939b: 259 (F, L*). **Type loc.:** Saut Tigre [Sinnamary River] [Inini], French Guiana [Guiana] (FMP). **Refs.:** Lane 1951 (syn.); Belkin 1968b: 38 (tax., lectotype desig.).
syn. *compta* Senevet & Abonnenc
1939b: 264 (A, L*). **Type loc.:** Saut-Tigre, [Inini], French Guiana [Guiana] (FMP, originally LPFM). **Refs.:** Lane 1953: 981 (F, L); Belkin 1968b: 39 (tax.); Motta et al. 1995 (to subg. uncertain); Nascimento-Pereira et al. 2019 (syn. with *compta*).

cesari Del Ponte & Cerqueira
1938: 226 (F). **Type loc.:** Cuiabá, Mato Grosso, Brazil (IOC). **Distr.:** Brazil. **Refs.:** Motta et al. 1995 (to subg. uncertain). **Ety.:** Gastão Cesar. **Informal name:** Cesar Brazilian Bromeliad Mosquito.

chalcocephala Dyar & Knab
1906a: 140 (A). **Type loc.:** Cacao, Trece Aguas, Alta Vera Paz, Guatemala (USNM). **Distr.:** Belize, Brazil, Colombia, Costa Rica, Ecuador, French Guiana, Guatemala, Honduras, Nicaragua, Panama. **Refs.:** Dyar 1928: 77 (M*, F, L*); Lane 1953: 977 (M*, F, L); Stone & Knight 1957b: 121 (type info.); Barreto & Lee 1969: 417 (distr.); Heinemann & Belkin 1977b: 443 (distr.); Motta et al. 1995 (to subg. uncertain); Pecor et al. 2002: 250 (tax., distr.; Belize). **Ety.:** n.s. [copper head (Gr)]. **Informal name:** Copper-headed Guatemalan Bromeliad Mosquito.
syn. *luciae* Senevet, Chabelard, & Abonnenc
1942: 343 (M*, P*, L*). **Type loc.:** Port-Inini, French Guiana [Guiana] (FMP, originally LPFM). **Refs.:** Lane 1951 (syn.); Belkin 1968b: 40 (tax., status?).

clasoleuca Dyar & Knab
1908: 68 (F). **Type loc.:** Caldera Island, Panama (USNM). **Distr.:** Colombia, Costa Rica, French Guiana, Lesser Antilles (includes Grenada, Saint Lucia, Saint Vincent and the Grenadines, Trinidad and Tobago), Nicaragua, Panama. **Refs.:** Dyar 1928: 88 (M*, F); Lane 1953: 984 (M*, F); Stone & Knight 1957b: 122 (type info., lectotype desig.); Motta et al. 1995 (to subg. uncertain). **Ety.:** n.s. **Informal name:** Caldera Panamanian Bromeliad Mosquito.
syn. *agyrtes* Dyar & Knab
1909d: 265 (F). **Type loc.:** Tabernilla, Canal Zone, Panama (USNM). **Refs.:** Lane 1951 (syn.).

covagarciai Sutil Oramas & Pulido F.
1974: 24 (M*, P*). **Type loc.:** La Chimenea, Municipio Altamira, Estado Barinas, Venezuela (DERM). **Distr.:** Venezuela. **Refs.:** Motta et al. 1995 (to subg. uncertain). **Ety.:** Pablo Cova-García. **Informal name:** Cova-García Venezuelan Bromeliad Mosquito.

exallos Rocha, Lourenço-de-Oliveira & Motta
2012 (M*, F, P*, L*). **Type loc.:** PARNA, Itatiaia (22°25'4304 S, 44°37'2546 W) [Brazil] (Laboratório de Transmissores de Hematozoários, CCULI, IOC). **Distr.:** Brazil. **Ety.:** quite distinct (Gr); different from all other *Wyeomyia* spp. in all stages. **Informal name:** Distinct Brazilian Bromeliad Mosquito.

flui (Bonne-Wepster & Bonne)
1920: 169 (F; *Prosopolepis*). **Type loc.:** Albina and Dam, Suriname (ITH). **Distr.:** Bolivia, Brazil, Ecuador, Guyana, Peru, Suriname. **Refs.:** Lane 1951: 335 (syn. with *confusa*); Lane 1953: 1000 (M*, F; as *kerri*); Belkin 1968b: 40 (lectotype desig.); Lourenço-de-Olivera et al. 1999: 209 (to valid sp. in subg. uncertain; distr.); Pecor et al. 2000 (distr.; Peru). **Ety.:** n.s. [Paul Christian Flu?]. **Informal name:** Flu Surinamese Bromeliad Mosquito.
syn. *kerri* Del Ponte & Cerqueira
1938: 228 (M*, F). **Type loc.:** Cuiabá, Mato Grosso, Brazil (IOC). **Refs.:** Lourenço-de-Olivera et al. 1999: 209 (syn.).

gutierrezi Duret
1971a: 17 (M*). **Type loc.:** El Paso, Distrito Savegre, Cantón de Aguirre, Costa Rica (USNM). **Distr.:** Costa Rica. **Refs.:** Motta et al. 1995 (to subg. uncertain). **Ety.:** José Gutiérrez. **Informal name:** Gutiérrez Costa Rican Bromeliad Mosquito.

ininicola Fauran & Pajot
1974b: 145 (M*). **Type loc.:** Inini River at l'Itany (Maroni) about 6 km southeast of Maripasoula (3°38′ N, 54°02′ W), French Guiana [Guiana] (PIG). **Distr.:** French Guiana, Guyana, Lesser Antilles (includes Trinidad and Tobago). **Refs.:** Harbach & Peyton 1991: 150 (from subg. *Davismyia*). **Ety.:** inhabitant of River Inini, French Guiana. **Informal name:** Inini French Guianese Bromeliad Mosquito.

knabi Lane & Cerqueira
1942: 600 (M*, F). **Type loc.:** Cachoeira, Rio de Janeiro, Brazil (IOC). **Distr.:** Brazil. **Refs.:** Lane & Causey 1955:

14 (P*, L*); Motta et al. 1995 (to subg. uncertain). **Ety.:** Frederick Knab. **Informal name:** Knab Brazilian Bromeliad Mosquito.

melanocephala Dyar & Knab
1906a: 140 (A). **Type loc.:** Trinidad [Trinidad and Tobago, Lesser Antilles] (USNM). **Distr.:** Argentina, Bolivia, Brazil, Colombia, Costa Rica, Ecuador, French Guiana, Guyana, Lesser Antilles (includes Saint Kitts and Nevis, Saint Lucia, Saint Vincent and the Grenadines, Trinidad and Tobago), Panama, Paraguay, Peru, Suriname, Venezuela. **Refs.:** Senevet & Abonnenc 1939b: 267 (M*, L*); Lane 1953: 970 (M*, F, P*, L*); Motta et al. 1995 (to subg. uncertain). **Ety.:** n.s. [black head (Gr)]. **Informal name:** Black-headed Trinbagonian Bromeliad Mosquito.
syn. *canfieldi* Dyar & Knab
1907c: 207 (F; *Sabethes*). **Type loc.:** Lion Hill, Canal Zone, Panama (USNM). **Refs.:** Stone & Knight 1957b: 121 (type info.).
syn. *agnostips* Dyar & Knab
1907c: 211 (A). **Type loc.:** Tabernilla, Canal Zone, Panama (USNM). **Refs.:** Bonne & Bonne-Wepster 1925: 100 (as *melanocephala* form).
syn. *pandora* Dyar & Knab
1909d: 261 (A). **Type loc.:** Corozal, Canal Zone, Panama (USNM). **Refs.:** Stone & Knight 1957b: 125 (type info., lectotype desig.).
syn. *grenadensis* Edwards
1916: 363 (F). **Type loc.:** Grenada Island [Lesser Antilles] (NHMUK). **Refs.:** Lane 1951 (syn.); Belkin 1968b: 40 (valid sp.?).
syn. *fauna* Dyar & Knab
1919: 2 (A). **Type loc.:** Bas Obispo, Canal Zone, Panama (USNM). **Refs.:** Stone & Knight 1957b: 123 (type info.).
syn. *modalma* Dyar
1922d: 97 (F). **Type loc.:** Gatún, Canal Zone, Panama (USNM). **Refs.:** Stone & Knight 1957b: 124 (type info., lectotype desig.).
syn. *hemisiris* Dyar & Shannon
1925c: 234 (F; *Prosopolepis*). **Type loc.:** France Field, Canal Zone, Panama (USNM).
syn. *typharum* Shannon & Del Ponte
1928: 96 (F; *Dendromyia* (*Calladimyia?*)). **Type loc.:** Ledesma, Jujuy, Argentina (BA). **Refs.:** Lane & Cerqueira 1942: 603 (syn.); Belkin et al. 1968: 11 (lectotype desig. according to Casal).
syn. *melanoides* Root
1928 (M*, F, L*; *Dendromyia* (*Calladimyia*)). In Dyar 1928: 79. **Type loc.:** Mage [Rio de Janeiro], Brazil (USNM). **Refs.:** Lane & Cerqueira 1942: 603 (syn.); Stone & Knight 1957b: 124 (type info., lectotype desig.).

moerbista (Dyar & Knab)
1919: 2 (F; *Sabethinus*). **Type loc.:** Rupununi [Essequibo], British Guiana [Guyana] (USNM). **Distr.:** Brazil, Colombia, Guyana, Suriname, Venezuela. **Refs.:** Lane & Cerqueira 1942: 615 (F); Mattingly 1971a: Pl. 2 (wing*); Motta et al. 1995 (to subg. uncertain). **Ety.:** n.s. **Informal name:** Rupununi Guyanese Bromeliad Mosquito.
syn. *ocumarensis* Lutz
1928: 34 (F; *Dendromyia*). **Type loc.:** Ocumare de la Costa, Aragua, Venezuela (LU).

negrensis Gordon & Evans
1922: 319 (M*, F, P*, L*). **Type loc.:** Macapá, Amazonas, Brazil (NHMUK). **Distr.:** Brazil. **Refs.:** Dyar 1928: 61 (M*, F, L*); Lane 1953: 988 (M*, F, L); Belkin 1968b: 41 (tax., lectotype desig.); Motta et al. 1995 (to subg. uncertain). **Ety.:** n.s. [black?]. **Informal name:** Black Brazilian Bromeliad Mosquito.

nigricephala Clastrier & Claustre
1978b: 543 (M*). **Type loc.:** Forêt du Galion, French Guiana [Guiana] (MNHP). **Distr.:** French Guiana. **Ety.:** black head (L). **Informal name:** Black-headed French Guianese Bromeliad Mosquito.

occulta Bonne-Wepster & Bonne
1919: 105 (M, F, P, L). **Type loc.:** Suriname (ITH). **Distr.:** Argentina, Bolivia, Brazil, Ecuador, French Guiana, Guyana, Suriname. **Refs.:** Dyar 1928: 60 (M*, F, L*); Senevet & Abonnenc 1939b: 267 (L*); Lane 1953: 987 (M*, P, L); Belkin 1968b: 41 (lectotype desig.); Motta et al. 1995 (to subg. uncertain). **Ety.:** n.s. [hidden, concealed (L); poss. ref. to "larvae found in a jelly-like mass at the base of *Heliconia* leaves"]. **Informal name:** Occult Surinamese Bromeliad Mosquito.

pampithes (Dyar & Núñez Tovar)
1928 (M; *Miamyia*). In Dyar 1928: 64. **Type loc.:** Rancho Grande, Venezuela (USNM). **Distr.:** Jamaica, Venezuela. **Refs.:** Lane 1953: 988 (M*); Motta et al. 1995 (to subg. uncertain). **Ety.:** n.s. [all straight? (Gr)]. **Informal name:** Sprite Venezuelan Bromeliad Mosquito [Pampi is slang for a magical fairy or sprite].

phroso Howard, Dyar, & Knab
1915: 149 (F). **Type loc.:** Gatún, Canal Zone, Panama (USNM). **Distr.:** Colombia, Panama. **Refs.:** Dyar 1928: 70 (M*, F, L*); Lane 1953: 979 (M*, F, L); Motta et al. 1995 (to subg. uncertain). **Ety.:** n.s. **Informal name:** Gatún Panamanian Bromeliad Mosquito.
syn. *proviolans* Dyar
1919d: 117 (M*; *Dinomyia*). **Type loc.:** Porto Bello, Panama (USNM). **Refs.:** Stone & Knight 1957b: 125 (type info., lectotype desig.).

rooti (Del Ponte)
1939: 541 (F; *Dendromyia*). **Type loc.:** Goiás, Brazil (IOC). **Distr.:** Brazil. **Refs.:** Belkin et al. 1971: 11 (type loc. info.; lectotype desig.); Motta et al. 1995 (to subg.

uncertain). **Ety.:** Francis Metcalf Root. **Informal name:** Root Brazilian Bromeliad Mosquito.

syn. *delpontei* Lane & Cerqueira
1942: 613 (F). **Type loc.:** Cuiabá, Mato Grosso, Brazil (IOC). **Refs.:** Lane & Cerqueira 1957: 244 (syn.).

rorotai Senevet, Chabelard & Abonnenc
1942: 336 (M*, F, P*, L*). **Type loc.:** Rorota, near Cayenne, French Guiana [Guiana] (FMP, originally LPFM). **Distr.:** Colombia, French Guiana, Guatemala, Lesser Antilles (includes Trinidad and Tobago), Panama, Suriname. **Refs.:** Lane 1951 (syn.); Belkin 1968b: 42 (tax.; to sp.; lectotype desig.); Motta et al. 1995 (to subg. uncertain). **Ety.:** Rorota, French Guiana. **Informal name:** Rorota French Guianese Bromeliad Mosquito.

roucouyana (Bonne-Wepster & Bonne)
1920: 166 (M, L; *Dendromyia*). **Type loc.:** Lawa district, Suriname (ITH). **Distr.:** Brazil, Suriname. **Refs.:** Dyar 1928: 62 (F, L*); Lane 1953: 961 (F, L); Belkin 1968b: 42 (type info.); Motta et al. 1995 (to subg. uncertain). **Ety.:** n.s. [poss. ref. to Village Roucouyenne, French Guiana]. **Informal name:** Roucouyenne Surinamese Bromeliad Mosquito.

serratoria (Dyar & Núñez Tovar)
1927b: 7 (F; *Sabethoides*). **Type loc.:** Villegas, Aragua, Venezuela (USNM). **Distr.:** Argentina, Brazil, Colombia, Venezuela. **Refs.:** Lane 1953: 992 (F); Motta et al. 1995 (to subg. uncertain). **Ety.:** serrated (L); ref. to toothed tarsal claws. **Informal name:** Serrated Venezuelan Bromeliad Mosquito.

shannoni Lane & Cerqueira
1942: 599 (M*, F). **Type loc.:** Mangaratiba, Rio de Janeiro, Brazil (IOC). **Distr.:** Brazil, Ecuador, Nicaragua. **Ety.:** Raymond Corbett Shannon. **Informal name:** Shannon Brazilian Bromeliad Mosquito.

subcomplosa (Del Ponte)
1939: 541 (F; *Dendromyia*). **Type loc.:** Goiás, Brazil (LU). **Distr.:** Brazil. **Refs.:** Belkin et al. 1971: 12 (tax.); Motta & Lourenço-de-Oliveira 1995 (to subg. uncertain). **Ety.:** nearly (L); near *complosa*. **Informal name:** Sub-struck Brazilian Bromeliad Mosquito.

surinamensis Bruijning
1959: 135 (M*). **Type loc.:** Ornamibo, Suriname (LM). **Distr.:** French Guiana, Suriname. **Refs.:** Motta et al. 1995 (to subg. uncertain). **Ety.:** Suriname. **Informal name:** Ornamibo Surinamese Bromeliad Mosquito.

taurepana Anduze
1941b: 825 (M*). **Type loc.:** Cabeceras del Rio Surukum, Bolivar, Venezuela (DPFA). **Distr.:** Venezuela. **Refs.:** Motta et al. 1995 (to subg. uncertain). **Ety.:** Taurepan is the name of the indigenous peoples at the headwaters of the Rio Surukum. **Informal name:** Taurepan Venezuelan Bromeliad Mosquito.

undulata Del Ponte & Cerqueira
1938: 233 (F). **Type loc.:** Cuiabá, Mato Grosso, Brazil (IOC). **Distr.:** Bolivia, Brazil, Colombia, Costa Rica, Ecuador, Nicaragua, Panama, Paraguay. **Refs.:** Lane & Cerqueira 1942: 609 (M*, F, P*, L); Lane 1953: 984 (M*, F, P*, L); Motta et al. 1995 (to subg. uncertain). **Ety.:** n.s. **Informal name:** Wavy Brazilian Bromeliad Mosquito.

Tribe Toxorhynchitini

Genus *Toxorhynchites* Theobald

(Vol. 1, p. 290)

Important references: Barraud 1934 (as *Megarhinus*; Central Asia); Edwards 1941 (F, P; keys; as *Megarhinus*; Afrotropics); van Someren 1946b: 177 (key; Ethiopian Region); Stone 1948 (tax.; *Megarhinus* replaced by *Toxorhynchites*); Hopkins 1952 (L; key; Afrotropics); Vargas 1953b: 27 (key; North and Central America); Lane 1953 (tax., keys; Neotropical Region); Stone & Knight 1957d (International Commission on Zoological Nomenclature; *Toxorhynchites*); Thurman 1959b (Thailand); da Costa Lima et al. 1962: 225 (revision; key; *Ankylorhynchus* and *Lynchiella*); Belkin 1962 (review; South Pacific); Forattini 1965b (Neotropical Region); Cova García et al. 1966 (Venezuela); Belkin 1968b: 32 (type info.); Mattingly 1969b: 13 (E*); Belkin et al. 1970 (New World); Berlin & Pandian 1973a: 274 (cat.; India); Harbach 1977 (phyl.); Harbach 1978 (phyl.); Tanaka et al. 1979 (F, L; keys, tax., bion.; Japan); Steffan et al. 1980 (annotated bibliography); Steffan 1981 (type info.; NMNH); Steffan & Evenhuis 1981 (bion.); Manning et al. 1982 (cat. supplement); Clark-Gil & Darsie 1983 (keys; Guatemala); Darsie 1985 (keys; Argentina); Mitchell & Darsie 1985 (review, bibliography., distr.; Argentina); Steffan & Evenhuis 1985 (phyl.; subg. *Toxorhynchites*); Evenhuis & Steffan 1986 (Acaudatus Group; Oriental Region); Lee et al. 1988c (subg. *Toxorhynchites*; Australasia); Service 1990 (Afrotropics); Ribeiro 1991 (tax.; Brevipalpis Group; Afrotropics); Ribeiro 1992b (subg. *Afrorhynchus*); Harbach & Kitching 1998 (phyl.); Collins & Blackwell 2000 (bion.); Mitchell et al. 2002 (mol. phyl.; as tribe in subfamily Culicinae); Ribeiro 2004 (tax.; subg. *Afrorhynchus*; Madagascar); Ribeiro 2005b (tax., class.); Rattanarithikul, Harrison et al. 2005 (F, L; tax., keys, bion.; Thailand); Reidenbach et al. 2009 (mol. phyl.; as member of subfamily Culicinae); Tyagi et al. 2015 (tax., keys, bion.; SE Asia).

Subgenus *Afrorhynchus* Ribeiro

1992b: 195. Type species: *Megarhinus lutescens* Theobald.

Name derivation: n.k.

Important references: see under genus *Toxorhynchites*.

aeneus (Evans)
1926: 102 (F, P; *Megarhinus*). **Type loc.:** Hill Station, Freetown, Sierra Leone (NHMUK). **Distr.:** Côte d'Ivoire, Democratic Republic of the Congo, Nigeria, Sierra Leone. **Refs.:** Wolfs 1947d: 69 (P*, L*); Doucet 1960: 814 (distr.); Ribeiro 1992b (transfer from subg. *Toxorhynchites*). **Ety.:** n.s. [bronze colored (L); poss. ref. to bronzy reflections on scutum]. **Informal name:** Bronze Sierra Leonean Carnivorous Mosquito.

angolensis Ribeiro
1992c: 188 (M*). **Type loc.:** Dundo area, northeastern Angola (IHMT). **Distr.:** Angola, Côte d'Ivoire. **Ety.:** Angola. **Informal name:** Khoi Angolan Carnivorous Mosquito [the Khoi are Angolan indigenous pastoralists].

brunhesi Ribeiro
2004: 250 (M*, F). **Type loc.:** Cant. Perinet, District Moramanga, Province Tamatave, Madagascar (IRD). **Distr.:** Madagascar (includes Glorioso and Juan De Nova Isl.). **Ety.:** Jacques Brunhes. **Informal name:** Brunhes Malagasy Carnivorous Mosquito.

capelai Ribeiro
1992a: 135 (M*, F). **Type loc.:** São Tomé (FCLM). **Distr.:** São Tomé and Príncipe. **Ety.:** Rúben Antunes Capela. **Informal name:** Capela São Tomean Carnivorous Mosquito.

erythrurus (Edwards)
1941: 30, 361 (M, F, P; *Megarhinus*). **Type loc.:** Ibadan [Western Provinces], Nigeria (NHMUK). **Distr.:** Côte d'Ivoire, Nigeria. **Refs.:** Mattingly 1956a: 38 (lectotype desig.); Surtees 1958: 63 (L); Ribeiro 1992b (transfer from subg. *Toxorhynchites*). **Ety.:** red (Gr); ref. to large red caudal tufts. **Informal name:** Red Nigerian Carnivorous Mosquito.

fontenillei Ribeiro
2004: 252 (M*). **Type loc.:** Madagascar (IRD). **Distr.:** Madagascar (includes Glorioso and Juan De Nova Isl.). **Ety.:** Didier Fontenille. **Informal name:** Fontenille Malagasy Carnivorous Mosquito.

grjebinei Ribeiro
2004: 251 (M*, F). **Type loc.:** Sainte-Luce, Madagascar (IRD). **Distr.:** Madagascar (includes Glorioso and Juan De Nova Isl.). **Ety.:** Alexis Grjebine. **Informal name:** Grjebine Malagasy Carnivorous Mosquito.

helenae Ribeiro
1992c: 191 (M*). **Type loc.:** Dundo area, northeastern Angola (IHMT). **Distr.:** Angola. **Ety.:** Helena da Cunha Ramos. **Informal name:** Helena Angolan Carnivorous Mosquito.

kaimosi (van Someren)
1946b: 184 (M, F, P, L; *Megarhinus*). **Type loc.:** Kaimosi Forest, North Kavirondo, Kenya (NHMUK). **Distr.:** Cameroon, Central African Republic, Kenya. **Refs.:** Corbet 1963: 11 (P*); Steffan & White 1981: 118 (lectotype desig.); Ribeiro 1992b (transfer from subg. *Toxorhynchites*). **Ety.:** Kaimosi Forest, Kenya. **Informal name:** Kaimosi Kenyan Carnivorous Mosquito.

lemuriae Ribeiro
2004: 254 (F*). **Type loc.:** Province Fianarantsoa, Loharano, SP Manakara, Madagascar (IRD). **Distr.:** Madagascar (includes Glorioso and Juan De Nova Isl.). **Ety.:** Lemuriens, inhabitants of the hypothetical "lost land" located in the Indian and Pacific Oceans. **Informal name:** Lemuria Malagasy Carnivorous Mosquito.

lutescens (Theobald)
1901a: 233 (M*; *Megarhinus*). **Type loc.:** Salisbury, Mashonaland [Harare, Zimbabwe] (NHMUK). **Distr.:** Kenya, South Africa, Tanzania, Zimbabwe. **Refs.:** Edwards 1941: 29 (M); van Someren 1946b: 181 (F); Ribeiro 1992b (transfer from subg. *Toxorhynchites*); Jupp 1996 (M*, F*; key). **Ety.:** n.s. [becoming yellow (L); poss. ref. to yellow caudal tuft and yellow body). **Informal name:** Yellowing Zimbabwean Carnivorous Mosquito.

madagascarensis Ribeiro
2004: 248 (M*, F). **Type loc.:** Fort Dauphin, Madagascar (IRD). **Distr.:** Madagascar (includes Glorioso and Juan De Nova Isl.). **Ety.:** Madagascar. **Informal name:** Dauphin Malagasy Carnivorous Mosquito.

nairobiensis (van Someren)
1946b: 181 (M, F, L*; *Megarhinus*). **Type loc.:** City Park, Nairobi, Kenya (NHMUK). **Distr.:** Kenya. **Refs.:** Steffan & White 1981: 119 (lectotype desig.); Ribeiro 1992b (transfer from subg. *Toxorhynchites*). **Ety.:** Nairobi, Kenya. **Informal name:** Nairobi Kenyan Carnivorous Mosquito.

nigeriensis Ribeiro
2005a: 227 (M*). **Type loc.:** Shaki, Nigeria (NHMUK). **Distr.:** Nigeria. **Ety.:** Nigeria. **Informal name:** Nok Nigerian Carnivorous Mosquito [the Nok civilization of Northern Nigeria (500 B.C.–A.D. 200) produced some of the earliest known sub-Saharan sculptures].

pauliani (Doucet)
1951c: 107 (M*; *Megarhinus*). **Type loc.:** Vangaindrano, Madagascar (IRSM). **Distr.:** Madagascar (includes Glorioso and Juan De Nova Isl.). **Refs.:** Ribeiro 1992b (transfer from subg. *Toxorhynchites*); Ribeiro 2004: 247 (M*; key). **Ety.:** Renaud Paulian. **Informal name:** Paulian Malagasy Carnivorous Mosquito.

ruwenzori (van Someren)
1948: 128 (M, F, L; *Megarhinus*). **Type loc.:** Bwamba Pass, Ruwenzori [Toro District], Uganda (NHMUK). **Distr.:** Uganda. **Refs.:** Ribeiro 1992b (transfer from subg. *Toxorhynchites*). **Ety.:** Rwenzori Mountains [formerly Ruwenzori Range], Uganda, and Democratic Republic of the Congo. **Informal name:** Ruwenzori Ugandan Carnivorous Mosquito.

viridibasis (Edwards)
1935a: 133 (F; *Megarhinus aeneus* var.). **Type loc.:** Kampala, Uganda (NHMUK). **Distr.:** Burkina Faso, Central African Republic, Côte d'Ivoire, Mali, Nigeria, Sudan and South Sudan, Uganda. **Refs.:** Edwards 1941: 31 (M, F; to sp.); Wolfs 1947d: 66 (P*, L*); Hamon 1954a: 583 (L*); Ribeiro 1992b (transfer from subg. *Toxorhynchites*). **Ety.:** green base (L); ref. to metallic scales of first three abdominal segments. **Informal name:** Green-based Ugandan Carnivorous Mosquito.

ssp. ***voltaicus*** Ribeiro
2005a: 229 (M*). **Type loc.:** Banankeledage 18754, Bob-Dioulasso [District], Haut Volta [Burkina Faso] (Insect Collection of the Unite d'Ecologie Virale, Institut Pasteur, Paris, France). **Distr.:** Burkina Faso. **Ety.:** Haute-Volta [Upper Volta, Burkina Faso]. **Informal name:** Volta Burkinabe Carnivorous Mosquito.

wolfsi Ribeiro
2005a: 224 (M*). **Type loc.:** Coquilhatville, Congo [Mbandaka, Democratic Republic of the Congo] (CMT). **Distr.:** Congo. **Ety.:** J. Wolfs. **Informal name:** Wolfs Congolese Carnivorous Mosquito.

zairensis Ribeiro
2005a: 227 (M*). **Type loc.:** Mapanda Fuma, Bakali-Inzia, Kwanga Province, Zaire [Democratic Republic of the Congo] (CMT). **Distr.:** Republic of the Congo. **Ety.:** Zaire. **Informal name:** Zaire Congolese Carnivorous Mosquito.

Subgenus *Ankylorhynchus* Lutz

1904. In Bourroul 1904: 53, 65. Type species: *Megarhinus purpureus* Theobald (Stone 1961c: 31).

Name derivation: n.k.

Important references: see under genus *Toxorhynchites*; da Costa Lima et al. 1962: 225 (revision; key; *Ankylorhynchus* and *Lynchiella*).

catharinensis (da Costa Lima, Guitton, & Ferreira)
1962: 230 (M*, F, P*, L*; *Ankylorhynchus*). **Type loc.:** Brusque, Santa Catarina, Brazil (IOC). **Distr.:** Brazil. **Refs.:** Belkin et al. 1971: 7 (type loc. info.). **Ety.:** n.s. [poss. ref. to Catherine]. **Informal name:** Santa Catarina Brazilian Carnivorous Mosquito.

hexacis (Martini)
1931a: 217 (F; *Megarrhinus*). **Type loc.:** Yungas von Coroico [La Paz], Bolivia (SMT). **Distr.:** Bolivia, Brazil. **Refs.:** Belkin 1968b: 33 (type info.); Belkin 1971a: 31 (type info.). **Ety.:** n.s. **Informal name:** Hex Bolivian Carnivorous Mosquito.

purpureus (Theobald)
1901a: 230 (F*; *Megarhinus*). **Type loc.:** Manaus, Amazonas, Brazil (NHMUK). **Distr.:** Brazil, Uruguay. **Refs.:** Dyar 1928: 400 (M*, F, L*); Lane 1953: 118 (M*, F, P*, L); Belkin 1968b: 34 (lectotype desig.); Belkin et al. 1971: 7

(type loc. info.); Rossi & Martinez 2003: 473 (distr.; Uruguay). **Ety.:** purple, dark red (L); abdomen metallic purple. **Informal name:** Purple Brazilian Carnivorous Mosquito.

trichopygus (Wiedemann)
1828: 4 (A; *Culex*). **Type loc.:** Salvador, Bahia, Brazil (SNG). **Distr.:** Brazil, French Guiana. **Refs.:** Dyar 1928: 401 (F, L*); Lane 1953: 115 (M*, F, P*, L); Belkin 1968b: 34 (tax.); Belkin et al. 1971: 7 (type loc. info.). **Ety.:** n.s. [hair tail (Gr)]. **Informal name:** Hairy Brazilian Carnivorous Mosquito.
syn. *neglectus* Lutz
1904 (F; *Ankylorhynchus*). In Bourroul 1904: 14. **Type loc.:** São Paulo, Brazil (NE). **Refs.:** Belkin et al. 1971: 26 (type info.).

Subgenus *Lynchiella* Lahille

1904: 82 [14] (as genus; new name for *Megarhinus* Robineau-Desvoidy). Type species: *Culex haemorrhoidalis* Fabricius.

Name derivation: Félix Lynch Arribálzaga

syn. *Megarhinus* Robineau-Desvoidy 1827: 412 (not Rafinesque 1820). Type species: *Culex haemorrhoidalis* Fabricius.
syn. *Megarhina* Macquart 1838b: 32 (as genus). Harbach 2018c: 150: "Unjustified emendation of *Megarhinus* Robineau-Desvoidy 1827; preoccupied by *Megarhina* Le Peletier & Audinet-Serville 1828; senior [but invalid] synonym of *Lynchiella* Lahille 1904; synonymy by Evenhuis et al. 2016."
syn. *Megarrhina* Skuse 1889: 1720 (variant spelling of *Megarhinus* Robineau-Desvoidy).
syn. *Megarrhinus* Stackelberg 1926: 50 (incorrect spelling of *Megarhinus* Robineau-Desvoidy).

Important references: see under genus *Toxorhynchites*; Lane 1953 (to subg.); da Costa Lima et al. 1962: 225 (*Ankylorhynchus* and *Lynchiella*; revision, key).

bambusicola (Lutz & Neiva)
1913: 136 (M, F, P, L; *Megarhinus*). **Type loc.:** Petropolis [Rio de Janeiro], Brazil (IOC). **Distr.:** Argentina, Bolivia, Brazil, Colombia, Ecuador, Panama, Paraguay, Peru, Suriname, Venezuela. **Refs.:** Lane 1953: 131 (M*, F, P*, L); Linton et al. 2013 (bion., mol. tax.; distr.; Ecuador). **Ety.:** bamboo inhabitant (L). **Informal name:** Bamboo-dwelling Brazilian Carnivorous Mosquito.
syn. *aldrichanus* Bonne-Wepster & Bonne
1920: 179 (M; *Megarhinus*). **Type loc.:** Dam, Suriname (ITH). **Refs.:** Belkin 1968b: 32 (lectotype desig.).

cavalierii García & Casal
1967: 436 (F, P*, L*). **Type loc.:** Hosteria Ruffino, Puerto Iguazu Misiones, Argentina (INM). **Distr.:** Argentina. **Ety.:** Fiorello Cavalieri. **Informal name:** Cavalieri Argentinian Carnivorous Mosquito.

gerbergi Belkin
1977c: 329 (M*, F, P*, L*). **Type loc.:** Industry Village, Union, Saint Mark, Grenada, Southern Lesser Antilles (USNM). **Distr.:** Lesser Antilles (Grenada, Saint Vincent and the Grenadines). **Ety.:** Eugene Jordan Gerberg and Jo Betty Vick Gerberg. **Informal name:** Gerberg Grenadian Carnivorous Mosquito.

grandiosus (Williston)
1900: 224 (F; *Megarrhina* [*sic*]). **Type loc.:** Omilteme, Guerrero, Mexico (NHMUK). **Distr.:** Guatemala, Guyana, Mexico. **Refs.:** Lane 1953: 125 (F). **Ety.:** n.s. [large (L)]. **Informal name:** Grandiose Mexican Carnivorous Mosquito.
syn. *longipes* Theobald
1901c: 241 (F*; *Megarhinus*). **Type loc.:** Mexico (NHMUK).

guadeloupensis (Dyar & Knab)
1906c: 254 (M, F; *Megarhinus*). **Type loc.:** Guadeloupe Island, Lesser Antilles (USNM). **Distr.:** Argentina, Brazil, Colombia, French Guiana, Grenada, Haiti, Lesser Antilles (includes Dominica, Guadeloupe, Montserrat, Saba, Saint Kitts and Nevis, Saint Lucia, Trinidad and Tobago), Suriname, Venezuela. **Refs.:** Knight & Chamberlain 1948: 18 (P*; as *horei*); Lane 1953: 125 (M*, F, P, L); van der Kuyp 1954: 65 (M, F, P, L*, E); Stone & Knight 1957c: 199 (type info.); Belkin & Heinemann 1976a: 148 (distr.); Belkin & Heinemann 1976b: 272; Augier et al. 2003: 99 (P*, L*; tax.). **Ety.:** Guadeloupe Isl. **Informal name:** Karukera Guadeloupean Carnivorous Mosquito [Karukera, "Island of the Beautiful Waters," was the original name of Guadeloupe Island].
syn. *guianensis* Bonne-Wepster & Bonne
1920: 180 (M, F; *Megarhinus guadeloupensis*; ssp.). **Type loc.:** Kwakoegron (Saramacca), Suriname (ITH). **Refs.:** Lane 1953: 125 (syn.); Belkin 1968b: 33 (lectotype desig.).
syn. *horei* Gordon & Evans
1922: 330 (M*, F, P*, L*; *Megarhinus*). **Type loc.:** Macapá, Amazonas, Brazil (NHMUK). **Refs.:** Lane 1951: 333 (syn.); Vargas 1953a: 312 (M*); Belkin 1968b: 33 (tax, lectotype desig.).
syn. *tucumanus* Brèthes
1926a: 318 (M*, F; *Megarhinus*). **Type loc.:** Concepción, Tucumán, Argentina (BA). **Refs.:** Lane 1951: 333 (syn.); Belkin et al. 1968: 10 (lectotype desig. according to Casal).
syn. *arborealis* Shannon & Del Ponte
1928: 89 (P; *Megarhinus tucumanus* var.). **Type loc.:** Lules, Tucumán, Argentina (NE). **Refs.:** Lane 1953: 125 (syn.).

haemorrhoidalis (Fabricius)
1787: 364 (M; *Culex*). **Type loc.:** Cayenne [Guyane], French Guiana [Guyana] (NE). **Distr.:** Argentina, Belize,

Bolivia, Brazil, Colombia, Costa Rica, Ecuador, French Guiana, Guatemala, Guyana, Honduras, Lesser Antilles (includes Trinidad and Tobago), Mexico, Nicaragua, Panama, Paraguay, Peru, Suriname, Venezuela. **Refs.:** Dyar 1928: 403 (M*, F); Séguy 1950: 144 (M*); Lane 1953: 119 (M, F, P, L*); Belkin 1968b: 33 (type info.); Berti et al. 2015 (distr.; Venezuela). **Ety.:** n.s. (poss. ref. to red scaling at apex of abdomen). **Informal name:** Red-tailed French Guianese Carnivorous Mosquito.

ssp. ***separatus*** (Lynch Arribálzaga)
1891b: 133 (M; *Megarhina*; as sp.). **Type loc.:** Formosa, Argentina (BA). **Distr.:** Argentina, Bolivia, Brazil, Nicaragua, Paraguay. **Refs.:** Lane 1951 (to ssp.); Forattini & Lane 1952: 57 (P*, L*); Lane 1953: 123 (M, F, P, L); Belkin et al. 1968: 11 (lectotype desig. according to Casal). **Ety.:** n.s. **Informal name:** Separate Argentinian Carnivorous Mosquito.

ssp. syn. ***lynchi*** Dyar & Knab
1906c: 244 (M, F; *Megarhinus*). **Type loc.:** Argentina (BA). **Refs.:** Lane 1939: 42 (syn.); Belkin et al. 1968: 11 (lectotype desig. according to Casal).

ssp. ***superbus*** (Dyar & Knab)
1906c: 255 (M, F; *Megarhinus*; as sp.). **Type loc.:** Trinidad [Trinidad & Tobago, Lesser Antilles] (USNM). **Distr.:** Belize, Colombia, Costa Rica, Cuba, Ecuador, French Guiana, Guatemala, Honduras, Lesser Antilles (includes Trinidad and Tobago), Mexico, Nicaragua, Panama, Suriname, Venezuela. **Refs.:** Lane 1951 (to ssp.); Lane 1953: 121 (M*, F, L); Vargas 1953a: 311 (M*, L); Stone & Knight 1957c: 200 (type info.); Pecor et al. 2002: 249 (tax., distr.; Belize). **Ety.:** n.s. [overbearing, proud (L)]. **Informal name:** Proud Trinbagonian Carnivorous Mosquito.

hypoptes (Knab)
1907a: 50 (M; *Megarhinus*). **Type loc.:** Bluefields, Nicaragua (USNM). **Distr.:** Colombia, Costa Rica, Honduras, Nicaragua, Panama. **Refs.:** Lane 1951: 333 (syn.); Zavortink & Chaverri 2009 (from syn. with *theobaldi*; distr.). **Ety.:** n.s. [stranger shy? (Gr)]. **Informal name:** Shy Nicaraguan Carnivorous Mosquito.

mariae (Bourroul)
1904: 3 (F, L; *Megarhinus*). **Type loc.:** Ilha de Itaparica, Bahia, Brazil (NE). **Distr.:** Brazil, Lesser Antilles (includes Trinidad and Tobago). **Refs.:** Dyar 1928: 404 (L*; as *violaceus*); Lane 1944: 189 (M, F, P, L; from syn. with *violaceus*); Belkin et al. 1971: 7 (tax.). **Ety.:** n.s. **Informal name:** Maria Brazilian Carnivorous Mosquito.

syn. *iris* Knab
1913a: 35 (F; *Megarhinus*). **Type loc.:** Trinidad [Trinidad and Tobago, Lesser Antilles] (USNM). **Refs.:** Lane 1944: 189 (syn.).

moctezuma (Dyar & Knab)
1906c: 251 (M, F; *Megarhinus*). **Type loc.:** Rio Aranjuez, near Puntarenas, Costa Rica (USNM). **Distr.:** Belize, Colombia, Costa Rica, El Salvador, Guatemala, Honduras, Lesser Antilles (includes Trinidad and Tobago), Mexico, Nicaragua, Panama, United States (continental), Venezuela. **Refs.:** Lane 1953: 129 (syn.); Stone & Knight 1957c: 199 (type info.); Chadee et al. 1987 (E*); Darsie & Ward 2005 (F*, L*; keys, distr.); Zavortink & Chaverri 2009 (from syn. with *theobaldi*; distr.). **Ety.:** Moctezuma. **Informal name:** Moctezuma Costa Rican Carnivorous Mosquito.

syn. *trinidadensis* Dyar & Knab
1906c: 252 (M, F; *Megarhinus*). **Type loc.:** Trinidad [Trinidad and Tobago, Lesser Antilles] (USNM). **Refs.:** da Costa Lima 1931d: 312 (syn.); Stone & Knight 1957c: 200 (type info.); Mattingly 1970g: 34 (L*); Zavortink & Chaverri 2009 (syn.).

portoricensis (von Röder)
1885: 337 (M; *Megarrhina*). **Type loc.:** Puerto Rico (NE). **Distr.:** Cuba, Dominican Republic, Haiti, Jamaica, Lesser Antilles (includes Montserrat), Puerto Rico. **Refs.:** Lane 1953a: 128 (M*, F, L*); Vargas 1953: 313 (M*, L); Dodge 1966: 384 (first instar L; key); Belkin 1968b: 34 (tax.); Belkin et al. 1970: 233 (M*, F, P*, L*). **Ety.:** Puerto Rico. **Informal name:** Igneri Puerto Rican Carnivorous Mosquito [the Igneri were an Arawak people of the Caribbean, conquered by the Caribs in the 10th century].

syn. *haitiensis* Dyar & Knab
1906c: 253 (M, F; *Megarhinus*). **Type loc.:** San Francisco Mines, Santo Domingo [Dominican Republic] (USNM). **Refs.:** Stone & Knight 1957c: 199 (type info.); Belkin & Heinemann 1973: 218 (type loc. info.).

pusillus (da Costa Lima)
1931d: 313 (M, F, L*; *Megarhinus*). **Type loc.:** Alto da Boa Vista, Tijuca, Rio de Janeiro, Brazil (IOC). **Distr.:** Brazil. **Refs.:** Lane 1953: 130 (M, F, P*). **Ety.:** tiny (L); ref. to one of the smallest species in this subgenus. **Informal name:** Tiny Brazilian Carnivorous Mosquito.

rizzoi (de Deus Palma & Galvão)
1969: 3 (F, P*, L*). **Type loc.:** Goianápolis, Goiás, Brazil (IOC). **Distr.:** Brazil. **Ety.:** José Ângelo Rizzo. **Informal name:** Rizzo Brazilian Carnivorous Mosquito.

rutilus (Coquillett)
1896: 44 (M, F; *Megarhinus*). **Type loc.:** Georgiana, Florida, United States (USNM). **Distr.:** Canada, United States (continental). **Refs.:** Jenkins & Carpenter 1946: 44 (P, L; bion.); Yamaguti & LaCasse 1951a: 5 (M*, F*, L*; as *Megarhinus*); Vargas 1953a: 313 (M*, L*); Vargas 1956b: 33 (L*; tax.); Stone & Knight 1957c: 199 (type info., lectotype desig.); Reinert 1970g: 209 (P*); Trimble & Smith 1975: 115 (bibliography); Harbach & Knight 1978 (E*); Chadee et al. 1987 (E); Linley 1989d (E*); Darsie & Ward 2005 (F*, L*; keys, distr.); Obenauer et al. 2009 (E*); Harrison et al. 2016 (F*, L*; keys, distr.; note 1). **Ety.:** n.s. [red, golden red, reddish yellow (L); poss. ref. to golden accents on palpus, thorax, pleura, abdomen, and legs].

Informal name: Red-gold American Carnivorous Mosquito.

ssp. ***septentrionalis*** (Dyar & Knab)

1906c: 249 (M, F; *Megarhinus*; as sp.). **Type loc.:** Woodstock, Virginia, United States (USNM). **Distr.:** Canada, United States (continental). **Refs.:** Jenkins & Carpenter 1946: 43 (P, L; bion.; to ssp.); Breland 1949a: 38 (L*, E*); Yamaguti & LaCasse 1951a: 10 (M*, F*, L*; as *Megarhinus*); Vargas 1953a: 313 (M*, L*); Carpenter & LaCasse 1955: 60 (M*, F*, L*; keys); Vargas 1956b: 33 (L*; tax.); Stone & Knight 1957c: 199 (type info.); Dodge 1964: 46 (L*); Dodge 1966: 384 (first instar L*; key); Parker 1977: 93 (distr.); Lamb & Smith 1980 (E*); Steffan & Evenhuis 1980 (P*); Darsie & Ward 2005 (F*, L*; keys, distr.); Harrison et al. 2016 (F*, L*; keys, distr.). **Ety.:** n.s. [northern part of a country (L)]. **Informal name:** Northern American Carnivorous Mosquito.

ssp. syn. *herrickii* Theobald

1906a: 241 (M, P, L; *Megarhinus*). **Type loc.:** Mississippi, United States (USNM). **Refs.:** Belkin 1968b: 33 (tax.).

solstitialis (Lutz)

1904 (M, F; *Megarhinus*). In Bourroul 1904: 10. **Type loc.:** São Paulo, Brazil (NHMUK). **Distr.:** Argentina, Brazil. **Refs.:** Lane 1953a: 123 (M*, F, P*, L); Vargas 1953: 312 (M*); Belkin 1968b: 34 (tax.); Belkin et al. 1971: 8 (lectotype desig.). **Ety.:** n.s. [of or belonging to summer (L)]. **Informal name:** Summer Brazilian Carnivorous Mosquito.

syn. *chrysocephalus* Theobald

1907: 136 (M*; *Mergarhinus* [*sic*]). **Type loc.:** São Paulo, Brazil (NHMUK).

syn. *mara* Anduze

1942a: 43 (M*; *Megarhinus*). **Type loc.:** La Rivera (Rio Escalante), Zulia, Venezuela (IHC). **Refs.:** Lane 1944: 180 (syn.); da Costa Lima et al. 1962: 243 (syn. suggested); Belkin et al. 1965: 72 (syn.).

theobaldi (Dyar & Knab)

1906c: 246 (M, F; *Megarhinus*). **Type loc.:** Bogotá, Colombia (O). **Distr.:** Argentina, Belize, Bolivia, Brazil, Colombia, Costa Rica, Ecuador, El Salvador, Guatemala, Honduras, Lesser Antilles (includes Trinidad and Tobago), Mexico, Nicaragua, Panama, Paraguay, Venezuela. **Refs.:** Lane 1953a: 129 (M*, F, P*, L); Vargas 1953: 313 (M*, L); Belkin 1968b: 34 (tax., lectotype desig.); Mattinglyi 1969b (E*; as *theobaldi*?); Lee et al. 1988c: 59 (tax., review); Pecor et al. 2002: 249 (tax., distr.; Belize); Rossi & Martínez 2003: 473 (distr.; Uruguay); Zavortink & Chaverri 2009 (distr. restricted to type loc.; Colombia); Berti et al. 2015 (distr.; Venezuela). **Ety.:** Frederic Vincent Theobald. **Informal name:** Theobald Colombian Carnivorous Mosquito.

syn. *ferox* Wiedemann

1828: 1 (M; *Culex*). **Type loc.:** Salvador, Bahia, Brazil (SNG, NMW). **Refs.:** Belkin 1968b: 32 (tax.); Belkin et al. 1971: 8 (type loc. info.; type info.).

syn. *ambiguus* Dyar & Knab

1906c: 246 (M; *Megarhinus*; new name for *ferox* Wiedemann 1828 (in part), not Humboldt 1819). **Type loc.:** Salvador, Bahia, Brazil (LU, NMW?). **Refs.:** Lane 1953: 129 (syn.); Belkin 1968: 32 (type info.); Belkin et al. 1971: 8 (type info., type loc. info.).

syn. *wiedemanni* Dyar & Knab

1906c: 246 (*Megarhinus*; new name for *ferox* Wiedemann 1828 (in part), not Humboldt 1819). **Refs.:** Lane 1953: 129 (syn.).

syn. *fluminensis* Neiva

1908 (M, F, L*; *Megarhinus*). In Peryassú 1908: 128. **Type loc.:** Rio de Janeiro, Brazil (IOC?). **Refs.:** da Costa Lima 1931d: 312 (M*; syn.).

syn. *posticatus* Lutz & Neiva

1913: 139 (M, F, P; *Megarrhinus*). **Type loc.:** Petrópolis, Rio de Janeiro, Brazil (IOC). **Refs.:** Belkin et al. 1971: 8 (tax.).

syn. *moengoensis* Bonne-Wepster & Bonne

1923: 7 (M, F, L; *Megarhinus*). **Type loc.:** Moengo, Suriname (ITH). **Refs.:** Lane 1939: 43 (syn.); Belkin 1968b: 33 (lectotype desig.).

violaceus (Wiedemann)

1821: 7 (M; *Culex*). **Type loc.:** Bahia, Brazil (NMW). **Distr.:** Brazil. **Refs.:** da Costa Lima 1931d: 310 (M*); Lane 1944: 187 (F; tax.); Belkin 1968b: 34 (type info., lectotype desig.). **Ety.:** n.s. [purple, violet (L)]. **Informal name:** Violet Brazilian Carnivorous Mosquito.

Subgenus *Toxorhynchites* Theobald

1901b: 234 (July 15), 1901a: 244 (Nov. 23) (as genus, subgenus). Type species: *Toxorhynchites brevipalpis* Theobald (International Commission on Zoological Nomenclature 1959, Opinion 548; validates name *Toxorhynchites*).

Name derivation: n.s. [toxo- = curved (Gr), snout.; ref. to downward-curved proboscis].

syn. *Worcesteria* Banks 1906a: 779 (as genus). Type species: *Worcesteria grata* Banks.

syn. *Teromyia* Leicester 1908: 49 (as genus). Type species: *Teromyia acaudata* Leicester (Brunetti 1914: 72).

Important references: see under genus *Toxorhynchites*.

acaudatus (Leicester)

1908: 49 (M, F; *Teromyia*). **Type loc.:** Singapore (NHMUK). **Distr.:** Indonesia, Malaysia, Singapore. **Refs.:** Edwards 1926a: 114 (P*, L); Steffan & White 1981: 116 (lectotype desig.); Tyagi et al. 2015 (tax., keys, bion.; SE Asia). **Ety.:** n.s. [prob. ref. to similarity with *ferox* except lacks a caudal tuft]. **Informal name:** Tailless Singaporean Carnivorous Mosquito.

albipes (Edwards)
1922a: 287 (M, F; *Megarhinus*). **Type loc.:** Simla [East Punjab], India (NHMUK). **Distr.:** Cambodia, India, Laos, Thailand, Vietnam. **Refs.:** Borel 1930: 116 (F*, L*); Barraud 1931c (P*, L*); Barraud 1934: 16 (M, F, P*, L*); Rattanarithikul et al. 2007 (F*, P*, L*; bion., distr., keys); Tyagi et al. 2015 (tax., keys, bion.; SE Asia). **Ety.:** n.s. [white foot (L); ref. to white markings on tarsomeres]. **Informal name:** White-footed Indian Carnivorous Mosquito.

amboinensis (Doleschall)
1857: 381 (A; *Culex*). **Type loc.:** Amboina Island, Moluccas (NMW). **Distr.:** Fiji, French Polynesia, Indonesia, Malaysia, Papua New Guinea, Philippines, Samoa (Ind. State of Samoa, American Samoa, Western Samoa), Sri Lanka, Thailand. **Refs.:** Banks 1908 (M*, F*, P*, L*,E*; as *Worcesteria grata* Banks); Knight & Chamberlain 1948: 11 (P*); Steffan 1968a: 150 (M*, F, P*, L*; syn.); Steffan 1968c (M*, F; tax., redescr.); Basio 1971b: 81 (M*, P*; bion.); Mattingly 1973b: 197 (E*); Ramalingam & Belkin 1976: 194 (distr.); Ramalingam 1976: 309 (tax., bion., keys., distr.; Samoa & Tonga); Harbach & Knight 1978 (E*); Harbach & Knight 1980 (E*); Steffan et al. 1982 (M*, F*); Chadee et al. 1987 (E*); Lee et al. 1988c: 64 (F; key, tax., bion., distr., review); Linley 1989d (E*); Linley & Seabury 1990 (E*); Rattanarithikul et al. 2007 (F*, P*, L*; bion., distr., keys); Tyagi et al. 2015 (tax., keys, bion.; SE Asia). **Ety.:** Amboina (Ambon) Isl., Indonesia. **Informal name:** Amboina Indonesian Carnivorous Mosquito.

syn. *subulifer* Doleschall
1857: 382 (F; *Culex*). **Type loc.:** Amboina Island [Moluccas] (NMW). **Refs.:** Knight et al. 1944: 12 (M; to ssp.); Bonne-Wepster 1954: 11 (A); Steffan 1977 (syn.).

syn. *immisericors* Walker
1859: 90 (M; *Megarhina*). **Type loc.:** Makessar, Celebes [Sulawesi] (NHMUK). **Refs.:** Steffan & White 1981: 118 (lectotype desig.).

syn. *lewaldii* Ludlow
1904b: 233 (M; *Megarhinus*). **Type loc.:** Salog, Gui Maras Island [Iloilo], Philippines (USNM). **Refs.:** Blanchard 1905: 625 (as *waldi*; emend.).

syn. *argenteotarsis* Ludlow
1906d: 367 (F). **Type loc.:** Margosatubig [Zamboanga], Mindanao, Philippines (USNM). **Refs.:** Bohart 1945: 23 (syn.); Stone & Knight 1957c: 199 (type info., lectotype desig.).

syn. *grata* Banks
1906a: 780 (M*, F*; *Worcesteria*). **Type loc.:** Hacienda "Louisiana," Bago; Mailum, Negros Occidental; and Cebu, Philippines (USNM). **Refs.:** Bohart 1945: 23 (syn.); Stone & Knight 1957c: 199 (type info., lectotype desig.).

angustiplatus Evenhuis & Steffan
1986: 548 (M*, F, P*, L*). **Type loc.:** Pacific Tin, Berjuntai, Batang, Selangor, Malaysia (USNM). **Distr.:** Malaysia. **Ety.:** n.s. [narrow and flat (L, Gr); poss. ref. to narrow pupal paddle]. **Informal name:** Narrow Malaysian Carnivorous Mosquito.

ater (Daniels)
1908: 265 (M, F, P; *Teromyia*). **Type loc.:** Pahang, Malaya [Malaysia] (NE). **Distr.:** Malaysia. **Ety.:** n.s. [dark (L); habitus]. **Informal name:** Dark Malaysian Carnivorous Mosquito.

auranticauda Lane
1992c: 301 (F*, P*, L*). **Type loc.:** Dumoga-Bone National Park, North Sulawesi Province, Indonesia (NHMUK). **Refs.:** Tyagi et al. 2015 (tax., keys, bion.; SE Asia). **Distr.:** Indonesia. **Ety.:** orange caudal scale tufts (L). **Informal name:** Orange Indonesian Carnivorous Mosquito.

auripes (Edwards)
1935b: 579 (F; *Megarhinus*). **Type loc.:** Mount Dulit, Sarawak, [Malaysian] Borneo (NHMUK). **Distr.:** Malaysia. **Ety.:** n.s. [gold foot (L); ref. to mostly golden tarsi]. **Informal name:** Gold-footed Malaysian Carnivorous Mosquito.

barbipes Edwards
1913b: 47 (M). **Type loc.:** Mpanga Forest, Toro District, Uganda (NHMUK). **Distr.:** Kenya, Liberia, Uganda. **Refs.:** Edwards 1941: 28 (M*); van Someren 1946b: 179 (M, F, P, L). **Ety.:** bearded foot (L); "first joint of the hind tarsus is thickly clothed with long black hair." **Informal name:** Beard-footed Ugandan Carnivorous Mosquito.

bengalensis Rosenberg & Evenhuis
1985: 36 (M*, P*). **Type loc.:** Chunarughat, Sylhet District, Bangladesh (USNM). **Distr.:** Bangladesh. **Refs.:** Tyagi et al. 2015 (tax., keys, bion.; SE Asia). **Ety.:** Bengal region, India. **Informal name:** Bengal Bangladeshi Carnivorous Mosquito.

bickleyi Thurman
1959b: 14 (M). **Type loc.:** Doi Suthep, Chiengmai Province, Thailand (USNM). **Distr.:** Thailand. **Refs.:** Rattanarithikul et al. 2007 (F*, P*, L*; bion., distr., keys); Tyagi et al. 2015 (tax., keys, bion.; SE Asia). **Ety.:** William E. Bickley. **Informal name:** Bickley Thai Carnivorous Mosquito.

brevipalpis Theobald
1901a: 245 (F*). **Type loc.:** [Durban], Natal [KwaZulu Natal, South Africa] NHMUK). **Distr.:** Angola, Benin, Burkina Faso, Cameroon, Central African Republic, Côte d'Ivoire, Democratic Republic of the Congo, Ghana, Kenya, Liberia, Mozambique, Nigeria, Senegal, South Africa, Tanzania, Uganda, Zambia, Zimbabwe. **Refs.:** Edwards 1941: 26 (M, F), 360 (P); Doucet 1951a: 82 (L*); Muspratt 1951: 355 (L*, E*; bion.); Hopkins 1952: 35 (L*); Vargas 1953a: 312 (M*); Worth & De Meillon 1960: 234 (distr.); Belkin 1962: 530 (M*, F, P*, L*); Mattingly 1969b (E*); Mattingly 1971b (E*); da Cunha Ramos & Ribeiro 1975: 114 (distr.); Harbach & Knight 1978 (E*); Harbach &

Knight 1980 (E*); Lamb & Smith 1980 (E*); Steffan & White 1981: 117 (lectotype desig.); Chadee et al. 1987 (E*); Lee et al. 1988c: 77 (F; key, tax., bion., distr., review); Jupp 1996 (M*, F*; key). **Ety.:** short feeler (L); short palpus. **Informal name:** Short-lipped South African Carnivorous Mosquito.

syn. *marshallii* Theobald
1903a: 121 (M*). **Type loc.:** Salisbury, Mashonaland [Zimbabwe] (NHMUK).

ssp. ***conradti*** Grünberg
1907a: 405 (M, F; as sp.). **Type loc.:** Johann-Albrechts Hohe [station], Cameroons [Republic of Cameroon] (ZM). **Distr.:** Burkina Faso, Cameroon, Côte d'Ivoire, Democratic Republic of the Congo, Equatorial Guinea, Gabon, Ghana, Liberia, Mali, Nigeria, Republic of the Congo, Sierra Leone, Sudan and South Sudan, The Gambia, Uganda. **Refs.:** Hopkins 1936: 32 (L*; to ssp.); Edwards 1941: 27 (M, F), 360 (P*); Hopkins 1952: 35 (L*); Vargas 1953a: 312 (M*); Corbet 1963: 9 (P*, L*, E). **Ety.:** n.s. **Informal name:** Conradt Cameroonian Carnivorous Mosquito.

ssp. syn. *tessmanni* Enderlein
1931: 124 (F). **Type loc.:** Alen Benito, Uam Region [Rio Muni, Equatorial Guinea] (ZM). **Refs.:** Edwards 1941: 27 (syn.).

syn. *schultzei* Enderlein
1931: 123 (F). **Type loc.:** Mongumba "am Uban Fluss." **Refs.:** Edwards 1941: 27 (syn.).

ssp. ***abyssinicus*** Ribeiro
1991: 43 (M*). **Type loc.:** Keffa Province, Ethiopia (NHMUK). **Distr.:** Ethiopia. **Ety.:** Abyssinia, Ethiopia. **Informal name:** Abyssinia Ethiopian Carnivorous Mosquito.

camaronis Ribeiro
1991: 50 (M*, F). **Type loc.:** Ototomo, Cameroon [Republic of Cameroon] (SSC). **Distr.:** Cameroon. **Ety.:** Cameroon. **Informal name:** Sao Cameroonian Carnivorous Mosquito [the Sao civilization (6th century B.C.–16th century A.D.) was the earliest civilization to leave traces in Cameroon].

christophi (Portschinsky)
1884: 122 (F; *Megarhina*). **Type loc.:** Amur, USSR (LU). **Distr.:** Japan, Russia, South Korea. **Refs.:** Stackelberg 1927 (tax., transl.); Monchadskii 1951: 103 (L*); Chow 1973: 43 (distr.); Shamrai & Gutsevich 1974: 427 (M*); Gutsevich et al. 1974: 120 (F*, L*); Filippova 1978 (P*); Tanaka et al. 1979: 502 (M*, F*, L*); Tyagi et al. 2015 (tax., keys, bion.; SE Asia). **Ety.:** n.s. **Informal name:** Christoph Russian Carnivorous Mosquito.

ssp. ***aurifluus*** (Edwards)
1921a: 631 (M, F; *Megarhinus*). **Type loc.:** Several localities on Formosa [Taiwan, Republic of China] (DEI). **Distr.:** Indonesia, Taiwan. **Refs.:** Brug 1932: 73 (P*); Brug 1939: 92 (F?); Lien 1965: 9 (M*, F*, P*, L*). **Ety.:** n.s. [gold, flowing (L); ref. to golden scale patches on thorax, abdomen, and legs]. **Informal name:** Gold-flowing Taiwanese Carnivorous Mosquito.

syn. *formosaensis* Ogasawara
1939b: 242 (M; as var.) [syn. of *aurifluus*]. **Type loc.:** Ikenohata (Shinchiku), Formosa [Chiehtuan, Loswei Village, Tatung Township, Ilan Hsien] (LU). **Refs.:** Lien 1962: 617 (syn.).

syn. *changbaiensis* Su & Wang
1981: 327 (M*, F, P*, L*). **Type loc.:** Huang-Song-Pu forest area, Antu Xian, Jilin Province, China [People's Republic of China] (BMC). **Refs.:** Danilov 1987a: 154 (syn.).

coeruleus (Brug)
1934a: 501 (M, F; *Megarhinus*). **Type loc.:** Batang Palupuh [West Coast], Sumatra [Indonesia] (NHMUK). **Distr.:** Indonesia. **Refs.:** Steffan & White 1981: 117 (lectotype desig.); Tyagi et al. 2015 (tax., keys, bion.; SE Asia). **Ety.:** n.s. [bluish, poss. ref. to bluish abdominal dorsum]. **Informal name:** Bluish Indonesian Carnivorous Mosquito.

darjeelingensis Tyagi, Munirathinam, Krishnamoorthy, Baskaran, Govindarajan, Krishnamoorthi, Mariappan, Dhananjeyan, & Venkatesh
2015: 16 (M*, F*, P*, L*; mol. tax., key). **Type loc**: West Bengal, Jalpaiguri District, Darjeeling Hills, India (CRME). **Distr**.: India. **Ety**.: ref. to type loc., Darjeeling Hills. **Informal name**: Darjeeling Indian Carnivorous Mosquito.

dundo Ribeiro
1991: 51 (M*, F). **Type loc.:** Dundo, Angola (IHMT). **Distr.:** Angola. **Ety.:** Dundo, Angola. **Informal name:** Dundo Angolan Carnivorous Mosquito.

edwardsi (Barraud)
1924d: 999 (M, F; *Megarhinus*). **Type loc.:** Krol Mountain, near Solon, Western Himalayas, India (NHMUK). **Distr.:** India. **Refs.:** Barraud 1931c (P*, L*); Barraud 1934: 23 (M, F, P*, L*); Mattingly 1971a: Pl. 10 (P trumpet*); Steffan & White 1981: 117 (lectotype desig.). **Ety.:** Frederick Wallace Edwards. **Informal name:** Edwards Indian Carnivorous Mosquito.

evansae (Edwards)
1936: 53 (M, F; *Megarhinus*). **Type loc.:** Freetown, Sierra Leone (NHMUK). **Distr.:** Cameroon, Sierra Leone. **Ety.:** Alwen Myfanwy Evans. **Informal name:** Evans Sierra Leonean Carnivorous Mosquito.

funestus (Leicester)
1908: 58 (M; *Teromyia*). **Type loc.:** Kuala Lumpur [Selangor], Malaya [Malaysia] (NHMUK). **Distr.:** Brunei, Malaysia. **Ety.:** funereal (L); "To the naked eye it is a very funereal looking mosquito." **Informal name:** Funereal Malaysian Carnivorous Mosquito.

gigantulus (Dyar & Shannon)
1925a: 67 (M; *Megarhinus*). **Type loc.:** Limay, Bataan, Luzon, Philippines (USNM). **Distr.:** Philippines. **Refs.:** Bohart 1945: 23 (M); Berlin 1969b: 21 (M*, F, P*, L*);

Basio 1971b: 81 (M*; bion.). **Ety.:** n.s. [diminutive of giant (Gr); however, it is a "small slender species"]. **Informal name:** Troll Philippine Carnivorous Mosquito [*gigantulus* = "little giant" = Troll, a mythical monster from Norse mythology that is smaller than the giant].

gravelyi (Edwards)
1921c: 73 (M; *Megarhinus*). **Type loc.:** Pashok [West Bengal], India (IM). **Distr.:** India, Malaysia, Thailand. **Refs.:** Barraud 1931c (P*, L*); Barraud 1934: 20 (M*, F, P*, L*); Rattanarithikul et al. 2007 (F*, P*, L*; bion., distr., keys); Tyagi et al. 2015 (tax., keys, bion.; SE Asia). **Ety.:** F.H. Gravely. **Informal name:** Gravely Indian Carnivorous Mosquito.

indicus Evenhuis & Steffan
1986: 558 (M*, F). **Type loc.:** Kuching, Sarawak, Malaysia (NHMUK). **Distr.:** Malaysia. **Ety.:** India. **Informal name:** India Malaysian Carnivorous Mosquito.

inornatus (Walker)
1865: 102 (M; *Megarhina*). **Type loc.:** New Guinea (NHMUK). **Distr.:** Australia, Fiji, Indonesia, Papua New Guinea. **Refs.:** Hill 1925: 67 (P*, L*); Laird 1947: 455 (L*; bion.); Belkin 1962: 533 (M*, F, P*, L*); Steffan & White 1981: 118 (lectotype desig.); Lee et al. 1988c: 82 (F; key, tax., bion., distr., review); Tyagi et al. 2015 (tax., keys, bion.; SE Asia). **Ety.:** n.s. [unadorned (L)]. **Informal name:** Inornate Papuan Carnivorous Mosquito.
ssp. ***albitarsis*** (Brug)
1939: 91 (M; *Megarhinus*; as var.). **Type loc.:** Kalawara, Celebes [Sulawesi] (RM) [to ssp. status]. **Distr.:** Indonesia. **Ety.:** white foot (L, Gr); tarsi marked with white. **Informal name:** White-ankled Indonesian Carnivorous Mosquito.

kempi (Edwards)
1921c: 72 (M, F; *Megarhinus*). **Type loc.:** Talewadi, North Kanara [Bombay], India (IM). **Distr.:** Cambodia, India, Indonesia, Laos, Philippines, Vietnam. **Refs.:** Borel 1930: 119 (M*, F*, L*); Barraud 1931c (P*, L*); Barraud 1934: 18 (M*, F*, P*, L*); Tyagi et al. 2015 (tax., keys, bion.; SE Asia). **Ety.:** S. Kemp. **Informal name:** Kemp Indian Carnivorous Mosquito.

klossi (Edwards)
1921c: 72 (M; *Megarhinus*). **Type loc.:** Kedah Peak (Gunong Jerai) [Kedah], Malaya [Malaysia] (NHMUK). **Distr.:** India, Malaysia. **Refs.:** Tyagi et al. 2015 (tax., keys, bion.; SE Asia). **Ety.:** C. Boden Kloss. **Informal name:** Kloss Malaysian Carnivorous Mosquito.

leicesteri Theobald
1904c: 36 (M, F). **Type loc.:** Kuala Lumpur, Malaya [Malaysia] (NHMUK). **Distr.:** Malaysia, Philippines, Singapore, Thailand. **Refs.:** Edwards 1926a: 114 (P*, L); Steffan & White 1981: 118 (lectotype desig.); Lee et al. 1988c: 88 (tax., distr., review); Rattanarithikul et al. 2007 (F*, P*, L*; bion., distr., keys); Tyagi et al. 2015 (tax., keys, bion.; SE Asia). **Ety.:** George Frederick Leicester. **Informal name:** Leicester Malaysian Carnivorous Mosquito.

lewisi Ribeiro
1991: 53 (M*, F). **Type loc.:** Farangbaia, Sierra Leone (NHMUK). **Distr.:** Sierra Leone. **Ety.:** David James Lewis. **Informal name:** Lewis Sierra Leonean Carnivorous Mosquito.

macaensis Ribeiro
1997: 213 (M*, F). **Type loc.:** Paque Seac Pai Van, Ilha de Coloane, Macau, China. **Distr.:** People's Republic of China. **Ety.:** Macau. **Informal name:** Macau Chinese Carnivorous Mosquito.

magnificus (Leicester)
1908: 54 (M, F; *Teromyia*). **Type loc.:** Malaya [Malaysia] (NHMUK). **Distr.:** Malaysia, Singapore, Thailand. **Refs.:** Edwards 1926a: 114 (P*); Steffan & White 1981: 119 (lectotype desig.); Rattanarithikul et al. 2007 (F*, P*, L*; bion., distr., keys); Tyagi et al. 2015 (tax., keys, bion.; SE Asia). **Ety.:** n.s. [noble, magnificent (L); prob. ref. to overall coloration]. **Informal name:** Magnificent Malaysian Carnivorous Mosquito.

manicatus (Edwards)
1921a: 630 (F; *Megarhinus*). **Type loc.:** Toa Tsui Kutsu, Formosa [Taiwan, Republic of China] (DEI). **Distr.:** Japan, Taiwan. **Refs.:** Lien 1965: 2 (M*, F*, P*, L*); Tanaka et al. 1979 (M*, F*, L*). **Ety.:** n.s. [having long sleeves (L); poss. ref. to tarsal banding (L); or rise in the morning (L); poss. ref. to active in the morning]. **Informal name:** Long-sleeved Taiwanese Carnivorous Mosquito.
ssp. ***yamadai*** (Ôuchi)
1939: 223 (M*; *Megarhinus*). **Type loc.:** Mount Yuwan [Kyushu], Japan (SCL). **Distr.:** Japan. **Refs.:** Tanaka et al. 1979: 494 (to ssp.). **Ety.:** Shinichiro Yamada. **Informal name:** Yamada Japanese Carnivorous Mosquito.
ssp. ***yaeyamae*** Bohart
1956b: 29 (M, F, P, L). **Type loc.:** Nakara River, Iriomote-shima, Ryukyu-Retto [Japan] (USNM). **Distr.:** Japan. **Refs.:** Tanaka et al. 1975c: 225 (tax., distr.); Tanaka et al. 1979: 496 (to ssp.). **Ety.:** Yaeyama Isl., Japan. **Informal name:** Yaeyama Japanese Carnivorous Mosquito.

manopi Thurman
1959b: 16 (M). **Type loc.:** Doi Suthep, Chiengmai Province, Thailand (USNM). **Distr.:** Thailand. **Refs.:** Rattanarithikul et al. 2007 (F*, P*, L*; bion., distr., keys; as *apsarae* [*sic*]). **Ety.:** Manop Rattanopradith. **Informal name:** Manop Thai Carnivorous Mosquito.

metallicus Leicester
1904 (M, F). In Theobald 1904c: 37. **Type loc.:** Kuala Lumpur, Malaya [Malaysia] (NHMUK). **Distr.:** India, Indonesia, Malaysia. **Refs.:** Steffan & White 1981: 119 (lectotype desig.); Tyagi et al. 2015 (tax., keys, bion.; SE Asia). **Ety.:** metallic (L); ref. to "Thorax brilliant metallic

green." **Informal name:** Metallic Malaysian Carnivorous Mosquito.

minimus (Theobald)
1905h: 237 (M*; *Megarhinus*). **Type loc.:** Yatiyantota, Ceylon [Sri Lanka] (NHMUK). **Distr.:** India, Indonesia, Malaysia, Philippines, Sri Lanka. **Refs.:** Barraud 1931c (P*, L*); Barraud 1934: 22 (M, F, P*, L*); Baisas 1974: 147 (M, F, P, L; tax., bion., distr.; Philippines); Tyagi et al. 2015 (tax., keys, bion.; SE Asia). **Ety.:** smallest (L); ref. to being a very small species. **Informal name:** Mini Sri Lankan Carnivorous Mosquito.

nepenthicola Steffan & Evenhuis
1982: 2 (M*, F*, P*, L*). **Type loc.:** near Olosobip, Upper Fly River, Western District, Papua New Guinea (BBM). **Distr.:** Papua New Guinea. **Refs.:** Lee et al. 1988c: 89 (F; key, tax., bion., distr., review). **Ety.:** larvae associated with *Nepenthes* pitcher plants. **Informal name:** Nepenthes Papuan Carnivorous Mosquito.

nepenthis (Dyar & Shannon)
1925a: 66 (F; *Megarhinus*). **Type loc.:** Los Baños, Laguna, Luzon, Philippines (USNM). **Distr.:** Philippines. Refs.: Basio 1971b: 81 (M*; bion.). **Ety.:** n.s. [poss. ref. to association with *Nepenthes* pitcher plants]. **Informal name:** Pitcher Plant Philippine Carnivorous Mosquito.

nigripes (Edwards)
1935b: 580 (M, F; *Megarhinus*). **Type loc.:** Mount Dulit, Sarawak, Borneo [Malaysia] (NHMUK). **Distr.:** Indonesia, Malaysia. **Refs.:** Steffan & White 1981: 119 (lectotype desig.). **Ety.:** black foot (L); "Legs (except for undersides of femora), wholly blue-black." **Informal name:** Black-footed Malaysian Carnivorous Mosquito.

okinawensis Toma, Miyagi, & Tanaka
1990: 345 (M*, F, P*, L*). **Type loc.:** Ada, Okinawa Jima, Japan (HUJ). **Distr.:** Japan. **Ety.:** Okinawa. **Informal name:** Okinawa Japanese Carnivorous Mosquito.

pendleburyi (Edwards)
1930a: 305 (F; *Megarhinus*). **Type loc.:** Kenokok, Mount Kinabalu, North Borneo, [Malaysian] Borneo (NHMUK). **Distr.:** Malaysia. **Ety.:** H.M. Pendlebury. **Informal name:** Pendlebury Malaysian Carnivorous Mosquito.

phytophagus Theobald
1909b: 1 (F; as *phytophygus*). **Type loc.:** Obuasi, Ashanti, Gold Coast [Ghana] (NHMUK). **Distr.:** Cameroon, Côte d'Ivoire, Democratic Republic of the Congo, Ghana, Liberia, Sierra Leone. **Refs.:** Edwards 1912c: 3 (emend.); Edwards 1941: 28 (M, F); Wolfs 1947d: 63 (P*, L*). **Ety.:** n.s. [plant eater? (Gr)]. **Informal name:** Herbivorous Ghanaian Carnivorous Mosquito.

quasiferox (Leicester)
1908: 51 (M, F; *Teromyia*). **Type loc.:** Jugra, Malaya [Malaysia] and Singapore (NHMUK). **Distr.:** Indonesia, Malaysia, Philippines, Singapore. **Refs.:** Edwards 1926a: 114 (P*, L); Steffan & White 1981: 120 (lectotype desig.); Tyagi et al. 2015 (tax., keys, bion.; SE Asia). **Ety.:** nearly (L); similar to *ferox*. **Informal name:** Feral Malaysian Carnivorous Mosquito [*quasiferox* = "half savage" = Feral].
syn. *javaensis* Theobald
1911d: 233 (F). **Type loc.:** Java [Indonesia] (AM).

rajah Tsukamoto
1989: 221 (M, F, P*, L*). **Type loc.:** Mount Kinabalu, Sabah, Malaysia [Malaysian Borneo] (NMST). **Distr.:** Malaysia, Philippines. **Ety.:** larvae in pitcher plant *Nepenthes rajah*. **Informal name:** Rajah Malaysian Carnivorous Mosquito.

ramalingami Evenhuis & Steffan
1986: 566 (M*, F, P*, L*). **Type loc.:** Bukit Puan Forest Reserve, Belait District, Brunei [Borneo] (USNM). **Distr.:** Borneo; Brunei. **Ety.:** Shivaji Ramalingam. **Informal name:** Ramalingam Bornean Carnivorous Mosquito.

raris (Leicester)
1908: 56 (M, L; *Teromyia*). **Type loc.:** Ulu Klang, Malaya [Malaysian Borneo] (NE). **Distr.:** Malaysia. **Ety.:** n.s. [(rare (L); poss. ref. to "I presume it is a rare insect"]. **Informal name:** Rare Malaysian Carnivorous Mosquito.

rickenbachi Ribeiro
1991: 54 (M*, F); 1991: 57 (F*). **Type loc.:** Elouden, Cameroon (SSC, P). **Distr.:** Cameroon. **Ety.:** André Rickenbach. **Informal name:** Rickenbach Cameroonian Carnivorous Mosquito.

rodhaini Ribeiro
1991: 57 (F*). **Type loc.:** Tai Forest, Ivory Coast [Côte d'Ivoire] (PIP). **Distr.:** Côte d'Ivoire. **Ety.:** François Rodhain. **Informal name:** Rodhain Ivorian Carnivorous Mosquito.

speciosus (Skuse)
1889: 1722 (M*; *Megarrhina*). **Type loc.:** Port Denison, Queensland, Australia (ANIC). **Distr.:** Australia, Fiji, Indonesia, Papua New Guinea, Singapore. **Refs.:** Bancroft 1908: 16 (M); Colledge 1911: 121 (M, F*, P, L*, E*); Lee 1944a: 14 (L*); Lee et al. 1988c: 90 (F; key, tax., bion., distr., review, type info.); Tyagi et al. 2015 (tax., keys, bion.; SE Asia). **Ety.:** beautiful, handsome (L); are "extremely magnificent insects." **Informal name:** Beautiful Australian Carnivorous Mosquito.

splendens (Wiedemann)
1819: 2 (M; *Culex*). **Type loc.:** Java [Indonesia] (ZMC, NMW). **Distr.:** Australia, Bangladesh, Cambodia, Fiji, India, Indonesia, Malaysia, Myanmar, Nepal, Papua New Guinea, People's Republic of China, Philippines, Sri Lanka, Thailand, Vietnam. **Refs.:** Green 1905 (E*; as *immisericors* Walker); Barraud 1931c (P*, L*); Barraud

1934: 24 (M*, F*, P*, L*, E); Paine 1943 (E*); Newkirk 1947: 522 (E*; bion.); Bonne-Wepster 1954: 10 (M*, F*, L*); Belkin 1962: 532 (M*, F, P*, L*); Mattingly 1969b (E*); Basio 1971b: 81 (M*; bion.); Mattingly 1971b (E*); Baisas 1974: 148 (M, F, P, L; tax., bion., distr.; Philippines); Ameen & Talukdar 1974b: 201 (P*); Bhat 1976: 266 (chrom.*); Furumizo & Rudnick 1978 (E*); Ahmed 1987 (distr.; Bangladesh); Chadee et al. 1987 (E*); Lee et al. 1988c: 95 (F; key, tax., bion., distr., review); Darsie & Pradhan 1990 (F, L; tax., keys, bion., distr.; Nepal); Linley & Seabury 1990 (E*); Rattanarithikul et al. 2007 (F*, P*, L*; bion., distr., keys); Tyagi et al. 2015 (tax., keys, bion.; SE Asia). **Ety.:** n.s. [shine, be bright (L)]. **Informal name:** Bright Indonesian Carnivorous Mosquito.

syn. *regius* Tennent
1859: 268 (A; *Culex*). **Type loc.:** Kandy [Central], Ceylon [Sri Lanka] (NE).

syn. *gilesii* Theobald
1901a: 227 (M, F*; *Megarhina*). **Type loc.:** Upper Burma [Myanmar] and Sikkim, India (NHMUK). **Refs.:** Steffan & White 1981: 118 (lectotype desig.).

syn. *sikkimensis* Giles
1901a: 604 (F; *Megarhina*). **Type loc.:** Sikkim, India (NHMUK). **Refs.:** Barraud 1934: 24 (syn.).

sumatranus (Brug)
1939: 93 (F; *Megarhinus*). **Type loc.:** Benkoelen, Sumatra [Indonesia] (NHMUK). **Distr.:** Indonesia. **Refs.:** Tyagi et al. 2015 (tax., keys, bion.; SE Asia). **Ety.:** Sumatra. **Informal name:** Sumatra Indonesian Carnivorous Mosquito.

sunthorni Thurman
1959b: 19 (F). **Type loc.:** Doi Suthep, Chiengmai Province, Thailand (USNM). **Distr.:** People's Republic of China, Thailand. **Refs.:** Rattanarithikul et al. 2007 (F*, P*, L*; bion., distr., keys); Tyagi et al. 2015 (tax., keys, bion.; SE Asia). **Ety.:** Sunthorn Srihongse. **Informal name:** Sunthorn Thai Carnivorous Mosquito.

towadensis (Matsumura)
1916: 445 (M*; *Megarrhina*). **Type loc.:** Towada, Honshu, Japan (LU). **Distr.:** Japan, South Korea. **Refs.:** LaCasse & Yamaguti 1950: 38 (M, F*; bion., distr., tax.); Ishimura 1952: 12 (P*, L*); Asanuma & Nakagawa 1954b: 82 (P*); Hara 1957: 48 (F*); Tanaka et al. 1979: 498 (M*, F*, L*). **Ety.:** n.s. [prob. ref. to Towada, Honshu, Japan]. **Informal name:** Towada Japanese Carnivorous Mosquito.

tyagii Krishnamoorthy, Munirathinam, Dhananjeyan, Hiriyan, Mariappan, Samuel & Venkatesh
2013: 448 (M*, F*, P*, L*). **Type loc.:** Tamil Nadu, Nilgiri District, Gudalur, India (CRME). **Refs.:** Tyagi et al. 2015 (tax., keys, bion.; SE Asia). **Distr.:** India. **Ety.:** B.K. Tyagi. **Informal name:** Tyagi Indian Carnivorous Mosquito.

Tribe Uranotaeniini

Genus *Uranotaenia* Lynch Arribálzaga

(Vol. 1, p. 292)

Important references: Barraud 1934 (tax., bion., distr.; southern Asia); Edwards 1941 (F, P; tax., keys; Afrotropics); Hopkins 1952 (L; tax., key; Afrotropics); Belkin 1953a: 312 (tax., keys; Solomon Isl.); Lane 1953 (tax., bion., keys; Neotropics); Galindo et al. 1954 (tax., keys, bion.; Panama), 107 (keys; Americas); Belkin 1962 (review, tax., bion., keys; South Pacific); Delfinado 1966a (tax., keys, bion.; Philippines); Belkin et al. 1970 (tax., keys, bion.; Jamaica); Peyton 1972: 16 (subg. class.); Peyton 1977 (*Pseudoficalbia* revision; SE Asia); Tanaka et al. 1979 (tax., bion., keys; Japan); Clark-Gil & Darsie 1983 (keys; Guatemala); Darsie 1985 (keys; Argentina); Mitchell & Darsie 1985 (review, bibliography., distr.; Argentina); Lee et al. 1989c (tax., bion., key; Australasia); Darsie & Pradhan 1990 (tax., bion. keys; Nepal); Service 1990 (tax., bion., keys; Afrotropics); da Cunha Ramos 1993 (revision; Afrotropics); da Cunha Ramos & Brunhes 2004 (tax., bion., keys; Madagascar); Rattanarithikul, Harrison et al. 2005 (F, L; tax., keys, bion.; Thailand); Rattanarithikul et al. 2006a (tax., bion., keys; Thailand).

Subgenus *Pseudoficalbia* Theobald

1912b: 89 (as genus). Type species: *Pseudoficalbia pandani* Theobald (Howard, Dyar, & Knab 1917: 899).

Name derivation: false *Ficalbia*.

Important references: Peyton 1972: 21 (revalidation); Peyton 1977 (*Pseudoficalbia* revision; SE Asia); Tanaka et al. 1979 (tax., bion., keys; Japan); Lee et al. 1989c (tax., bion., key; Australasia); da Cunha Ramos 1993 (keys; Afrotropical spp.); da Cunha Ramos & Brunhes 2004 (tax., bion., keys; Madagascar); Rattanarithikul et al. 2006a (tax., bion., keys; Thailand).

abdita Peyton
1977: 125 (M*, F, P*, L*). **Type loc.:** Ban Rong Na, Lampang, Thailand (USNM). **Distr.:** Cambodia, Thailand. **Refs.:** Rattanarithikul et al. 2006a (F*, L*; bion., distr., keys). **Ety.:** n.s. [hidden, concealed (L); poss. ref. to similarity to several other species or ref. to different larval habits]. **Informal name:** Concealed Thai Heaven Mosquito.

abstrusa Peyton
1977: 35 (M*, F, P*, L*). **Type loc.:** San Jose, Mindoro, Philippines (USNM). **Distr.:** Philippines. **Refs.:** Rattanarithikul et al. 2006a (F*, L*; bion., distr., keys). **Ety.:** n.s. [secret (L); prob. ref. to similarity with several other species]. **Informal name:** Secret Philippine Heaven Mosquito.

albimanus da Cunha Ramos & Brunhes
2004: 91 (M*, F, P*). **Type loc.:** Manakara, Madagascar (LTV). **Distr.:** Madagascar (includes Glorioso and Juan De Nova Isl.). **Ety.:** white feet (L); ref. to hind tarsomeres 3–5 white scaled [white extremities of the legs of adults]. **Informal name:** White-footed Malagasy Heaven Mosquito.

albinotata da Cunha Ramos & Brunhes
2004: 220 (M*, F, P*, L*). **Type loc.:** Bords de la rivière Manana, au nord du marais d'Ambila (S-P de Manakara), Madagascar (LTV). **Distr.:** Madagascar (includes Glorioso and Juan De Nova Isl.). **Ety.:** white notum (L); three patches of white scales on anterior part of scutum. **Informal name:** White-spotted Malagasy Heaven Mosquito.

albipes Peyton
1977: 64 (F). **Type loc.:** Khao Yai National Park, Nakhon Nayok, Thailand (USNM). **Distr.:** Thailand. **Refs.:** Rattanarithikul et al. 2006a (F*, L*; bion., distr., keys). **Ety.:** n.s. [white foot (L); poss. ref. to tarsomere 5 of all legs white]. **Informal name:** White-footed Thai Heaven Mosquito.

ambodimanga da Cunha Ramos & Brunhes
2004: 94 (M*, F). **Type loc.:** Ambodimanga, Moramanga, Madagascar (LTV). **Distr.:** Madagascar (includes Glorioso and Juan De Nova Isl.). **Ety.:** Ambodimanga, Madagascar. **Informal name:** Ambodimanga Malagasy Heaven Mosquito.

andreae Doucet
1962: 1157 (M*, F, P*, L). **Type loc.:** Banco Forest, Abidjan, Ivory Coast [Côte d'Ivoire] (IERT, Adiopodoume). **Distr.:** Cameroon, Côte d'Ivoire. **Refs.:** da Cunha Ramos 1993: 226 (M*, F, P*, L*; keys, tax., bion., distr.); **Ety.:** Andrée Doucet. **Informal name:** Andrée Ivorian Heaven Mosquito.

anhydor Dyar
1907: 128 (L). **Type loc.:** Sweetwater Junction, near San Diego, California, United States (NE). **Distr.:** Mexico, United States (continental). **Refs.:** Yamaguti & LaCasse 1951a: 36 (M*, F*, L*); Brookman & Reeves 1953: 234 (M*); Galindo et al. 1954: 118 (M, P*, L*); Carpenter & LaCasse 1955: 74 (M*, F*, L*; keys); Belkin & McDonald 1956b: 105 (M, F*, P*, L*, E*; tax.); Mattingly 1970c (E*); Bohart & Washino 1978: 40 (M, F*, P*, L*; keys, tax., bion., distr.); Darsie & Ward 2005 (F*, L*; keys, distr.). **Ety.:** dry (Gr); larval habitat dry when revisited. **Informal name:** Dry American Heaven Mosquito.

ssp. ***syntheta*** Dyar & Shannon
1924c: 189 (F; as sp.). **Type loc.:** Mission, Texas, United States (USNM). **Distr.:** Mexico, United States (continental). **Refs.:** Dampf 1943: 147 (M*, F); Porter 1946 (L*);

Yamaguti & LaCasse 1951a: 48 (M*, F*, L*); Breland 1954a (L*); Carpenter & LaCasse 1955: 79 (M*, F*, L*; keys); Belkin & McDonald 1956b: 105 (M, F*, P*, L; tax.; to ssp.); Dodge 1966: 385 (first instar L; key); Darsie & Ward 2005 (F*, L*; keys, distr.). **Ety.:** n.s. [together? (Gr)]. **Informal name:** Synthetic American Heaven Mosquito.

annulata Theobald
1901d: 14 (M, F). **Type loc.:** Bonny (Eastern Provinces), Nigeria. **Distr.:** Benin, Cameroon, Equatorial Guinea, Ghana, Guinea, Liberia, Nigeria, Senegal, Sierra Leone, The Gambia, Uganda. **Refs.:** da Cunha Ramos 1993: 89 (M*, F, L*; keys, tax., bion., distr., lectotype desig.). **Ety.:** n.s. [ringed (L); poss. ref. to apical white bands on hind legs]. **Informal name:** Ringed Nigerian Heaven Mosquito.

antalahaensis da Cunha Ramos & Brunhes
2004: 291 (F*, P*). **Type loc.:** Onive River, Parc National de Masoala, Madagascar (LTV). **Distr.:** Madagascar (includes Glorioso and Juan De Nova Isl.). **Ety.:** named for village closest to the type loc. **Informal name:** Antalaha Malagasy Heaven Mosquito.

apicosquamata da Cunha Ramos & Brunhes
2004: 97 (M*, F, P*, L*). **Type loc.:** Manakara, Madagascar (LTV). **Distr.:** Madagascar (includes Glorioso and Juan De Nova Isl.). **Ety.:** scaled tip (L); pecten reach apex of siphon. **Informal name:** Scale-tipped Malagasy Heaven Mosquito.

apicotaeniata Theobald
1909b: 21 (M, F; as sp.). **Type loc.:** Obuasi, Ashanti, Gold Coast [Ghana] (NHMUK). **Distr.:** Benin, Cameroon, Côte d'Ivoire, Democratic Republic of the Congo, Equatorial Guinea, Ghana, Liberia, Nigeria, Republic of the Congo, Sudan and South Sudan. **Refs.:** Edwards 1941: 56 (M, F; to var.); da Cunha Ramos 1993: 98 (M*, F; keys, tax., bion., distr.; to sp.; lectotype desig.). **Ety.:** ribboned tip (L); hind tibia with white tip. **Informal name:** Ribbon-tipped Ghanaian Heaven Mosquito.

approximata Peyton
1977: 38 (M*, F, P*, L*). **Type loc.:** Ban Sop O Kok, Chiang Dao, Chiang Mai, Thailand (USNM). **Distr.:** Thailand. **Refs.:** Rattanarithikul et al. 2006a (F*, L*; bion., distr., keys). **Ety.:** approaching, approximating (L); very similar to *demeilloni*. **Informal name:** Approximate Thai Heaven Mosquito.

ascidiicola de Meijere
1910: 925 (M*, F, P*, L*, E*). **Type loc.:** Tjibodas (1,500 m), near Buitenzorg [Bogor], Java [Indonesia] (AM). **Distr.:** Indonesia. **Refs.:** Mattingly 1970c (E*); Mattingly 1971a: Pl. 12 (P terminal segs.*); Peyton 1977: 66 (M*, F, P*, L*, E; type info.); Townsend 1990: 46 (type info.). **Ety.:** n.s. [bag inhabitant (Gr); poss. ref. to larval habitat]. **Informal name:** Bag-dwelling Indonesian Heaven Mosquito.

atra Theobald
1905f: 114 (F). **Type loc.:** Muina, [NE] New Guinea (HNM). **Distr.:** Papua New Guinea, Solomon Isl. **Refs.:** Lee 1944a: 35 (L*; as *nigerrima*); Knight & Chamberlain 1948: 14 (P*; as *nigerrima*); Penn 1949b: 30 (P*; as *nigerrima*); Stone 1957b: 173 (tax.); Lee et al. 1989c: 192 (tax., bion., distr., review). **Ety.:** n.s. [dark (L); prob. ref. to "general blackish brown hue"]. **Informal name:** Dark Papuan Heaven Mosquito.

syn. *nigerrima* Taylor
1914d: 203 (M*, F). **Type loc.:** Milne Bay and Lakekamu Gold Field, Papua, New Guinea (ANIC, NHMUK). **Refs.:** Stone 1957b: 173 (syn.); Lee et al. 1989c: 192 (type info.); Townsend 1990: 109 (type info.).

bambusicola da Cunha Ramos & Brunhes
2004: 225 (F, P*, L*). **Type loc.:** Mantadia, Moramango, Madagascar (LTV). **Distr.:** Madagascar (includes Glorioso and Juan De Nova Isl.). **Ety.:** bamboo inhabitant (L). **Informal name:** Bamboo-dwelling Malagasy Heaven Mosquito.

belkini Grjebine
1979: 65 (L*). **Type loc.:** Manakara, Fianarantsoa Province, Madagascar (SSC). **Distr.:** Madagascar (includes Glorioso and Juan De Nova Isl.). **Ety.:** John Nicholas Belkin. **Informal name:** Belkin Malagasy Heaven Mosquito.

bicincta da Cunha Ramos & Brunhes
2004: 237 (M*, P*, L*). **Type loc.:** Lokobe, Nosy Be Island, Madagascar (LTV). **Distr.:** Madagascar (includes Glorioso and Juan De Nova Isl.). **Ety.:** two rings (L); two distinctive submedian scutal bands. **Informal name:** Double-belted Malagasy Heaven Mosquito.

bicolor Leicester
1908: 225 (M, F). **Type loc.:** Kuala Lumpur, Malaya [Malaysia] (NHMUK). **Distr.:** Cambodia, India, Indonesia, Malaysia, Nepal, People's Republic of China, Philippines, Sri Lanka, Thailand, Vietnam. **Refs.:** Basio 1971b: 90 (M*; bion.; as *lagunensis*); Baisas 1974: 157 (M*, F, P, L*; tax., bion., distr.; Philippines; as *lagunensis*); Peyton 1977: 41 (M*, F*, P*, L*; distr.); Darsie et al. 1992 (distr.; Nepal); Qu & Qian 1993 (cibarial armature*); Rattanarithikul et al. 2006a (F*, L*; bion., distr., keys). **Ety.:** n.s. [two colors (L); poss. ref. to regular banding of the abdomen]. **Informal name:** Bicolored Malaysian Heaven Mosquito.

syn. *fusca* Leicester
1908: 227 (M, F). **Type loc.:** Sungei Lembing, Malaya [Malaysia] (NHMUK). **Refs.:** Edwards 1922b: 436 (syn.).

syn. *leicesteri* Edwards
1913c: 239 (new name for *fusca* Leicester, not Theobald 1907). **Refs.:** Edwards 1922b: 460 (syn.).

syn. *kalabahensis* Haga
1925: 43 (F). **Type loc.:** Kalabahi, Alor Island [East Nusa Tenggara Province, Indonesia] (NHMUK). **Refs.:** Peyton 1977: 41 (syn.).

syn. *lagunensis* Baisas
1935b: 70 (M*, F, L*). **Type loc.:** College of Agriculture, Los Baños, Laguna, Luzon, Philippines (NE). **Refs.:** Peyton 1977: 41 (syn.).

bifasciata da Cunha Ramos & Brunhes
2004: 102 (F, P*, L*). **Type loc.:** Nosy Mangabe, Madagascar (LTV). **Distr.:** Madagascar (includes Glorioso and Juan De Nova Isl.). **Ety.:** two banded (L); ref. to two bands of white scales on border of scutum. **Informal name:** Double-banded Malagasy Heaven Mosquito.

bimaculata Leicester
1908: 226 (M, F). **Type loc.:** The Gap, Malaya [Malaysia] (NHMUK). **Distr.:** Japan, Malaysia, Thailand. **Refs.:** Roth 1946b: 67 (P*, L*; tax.); Bohart & Ingram 1946b: 57 (M, F, P*, L*, E*; bion., distr., tax.); LaCasse & Yamaguti 1950: 42 (M*, F*, P*, L*; bion., distr., tax.); Asanuma & Nakagawa 1954a: 372 (P*); Hara 1957: 49 (F*); Mattingly, in Delfinado 1966a: 41 (lectotype desig.); Mattingly 1970c (E*); Moriya et al. 1973 (E*); Peyton 1977: 69 (M*, F*, P*, L*; distr.); Rattanarithikul et al. 2006a (F*, L*; bion., distr., keys). **Ety.:** two spotted (L); ref. to two costal spots. **Informal name:** Double-spotted Malaysian Heaven Mosquito.

bosseri Grjebine
1979: 57 (L*). **Type loc.:** Manakara, Fianarantsoa Province, Madagascar (SSC). **Distr.:** Madagascar (includes Glorioso and Juan De Nova Isl.). **Ety.:** Jean Bosser. **Informal name:** Bosser Malagasy Heaven Mosquito.

boussesi da Cunha Ramos & Brunhes
2004: 151 (L*). **Type loc.:** "Brickaville/Ampasimanolotra & Tananarive/Antananarivo (15km a partir de Brickaville)," Madagascar (LTV). **Distr.:** Madagascar (includes Glorioso and Juan De Nova Isl.). **Ety.:** Philippe Boussès. **Informal name:** Boussès Malagasy Heaven Mosquito.

breviseta da Cunha Ramos & Brunhes
2004: 107 (M*, P*). **Type loc.:** Manakara, Madagascar (LTV). **Distr.:** Madagascar (includes Glorioso and Juan De Nova Isl.). **Ety.:** short hair (L); ref. to weak setae 5-IV–V of pupa. **Informal name:** Short-haired Malagasy Heaven Mosquito.

browni Mattingly
1955: 75 (M, F, P*, L*). In Mattingly & Brown 1955. **Type loc.:** Vallee de Mai, Praslin Island, Seychelles (NHMUK). **Distr.:** Seychelles. **Ety.:** E.S. Brown. **Informal name:** Brown Seychellois Heaven Mosquito.

brumpti Doucet
1951c: 111 (L*). **Type loc.:** Ampisopiso, Amparihy, and Vohimalazo, Madagascar (IRSM). **Distr.:** Madagascar (includes Glorioso and Juan De Nova Isl.), France (Corsica). **Ety.:** Alexandre Joseph Emile Brumpt. **Informal name:** Brumpt Malagasy Heaven Mosquito.

brunhesi Grjebine
1979: 61 (L*). **Type loc.:** Sainte-Luce, District Fort Dauphin, Madagascar (SSC). **Distr.:** Madagascar (includes Glorioso and Juan De Nova Isl.). **Refs.:** Robert et al. 2019 (distr.; Western Palearctic). **Ety.:** Jacques Brunhes. **Informal name:** Brunhes Malagasy Heaven Mosquito.

cachani (Doucet)
1950: 45 (L*; *Aedes*). **Type loc.:** Fenerive-Anosibe, Tampolo, Maroantsetra and Mangoky, Madagascar (IRSM). **Distr.:** Madagascar (includes Glorioso and Juan De Nova Isl.). **Refs.:** Mattingly & Brown 1955: 84 (n. comb.). **Ety.:** P. Cachan. **Informal name:** Cachan Malagasy Heaven Mosquito.

capelai da Cunha Ramos
1993: 235 (M*, F, P*, L*; keys, tax., bion., distr.). **Type loc.:** Rio Ió Grande, Ilha de São Tomé, São Tomé (IHMT). **Distr.:** São Tomé and Príncipe. **Ety.:** Rúben Antunes Capela. **Informal name:** Capela São Tomean Heaven Mosquito.

carcinicola da Cunha Ramos & Brunhes
2004: 111 (M*, F, P*, L*). **Type loc.:** Nosy Mangabe Island, Madagascar (LTV). **Distr.:** Madagascar (includes Glorioso and Juan De Nova Isl.). **Ety.:** crab inhabitant (Gr, L); larvae found in crabholes. **Informal name:** Crabhole Malagasy Heaven Mosquito.

cavernicola Mattingly
1954b: 167 (M*, F, L*). **Type loc.:** Yolohafiri Cave on Mount Hoye [Orientale], Belgian Congo [Democratic Republic of the Congo (NHMUK). **Distr.:** Angola, Burkina Faso, Cameroon, Côte d'Ivoire, Democratic Republic of the Congo, Ghana, Guinea, Liberia, Mali, Republic of the Congo, South Africa, Tanzania, Togo, Uganda, Zambia. **Refs.:** Vattier & Adam 1963: 911 (M*, P*, L*); da Cunha Ramos 1993: 102 (M*, F, P*, L*; keys, tax., bion., distr.). **Ety.:** cave inhabitant (L); ref. to larval habitat. **Informal name:** Cave-dwelling Congolese Heaven Mosquito.

colocasiae Edwards
1928a: 281 (M, F, P, L*). **Type loc.:** Ovalau Islands, Fiji (NHMUK). **Distr.:** Fiji. **Refs.:** Paine 1943: 17 (A, P*, L*); Belkin 1962: 157 (M*, F, P*, L*); Lee et al. 1989c: 198 (tax., bion., distr., review). **Ety.:** *Colocasia* [taro]. **Informal name:** Colocasia Fijian Heaven Mosquito.

combesi Doucet
1950: 39 (L*). **Type loc.:** Miandrivazo, Tsinjoarivo-Onive, Ambositra, Maroantsetra, and Tampolo, Madagascar (IRSM, NHMUK?). **Distr.:** Madagascar (includes Glorioso and Juan De Nova Isl.). **Refs.:** Townsend 1990: 62 (type info.). **Ety.:** n.s. [strongly (L); poss. ref. to thickness of head hair B (6-C?)]. **Informal name:** Combes Malagasy Heaven Mosquito.

comorensis da Cunha Ramos & Brunhes
2004: 170 (M*, F, P*, L*). **Type loc.:** Massif de la Grille (north of Grande Comore), Madagascar (LTV). **Distr.:** Comoros, Mayotte. **Ety.:** Comoros Isl. **Informal name:** Comore Malagasy Heaven Mosquito.

confusa Peyton
1977: 129 (M*, F, L*). **Type loc.:** San Jose, Mindoro, Mindoro Island, Philippines (USNM). **Distr.:** Indonesia, Philippines. **Ety.:** n.s. [confused (L); poss. ref. to similarity to *philippinensis*]. **Informal name:** Confusing Philippine Heaven Mosquito.

contrastata da Cunha Ramos & Brunhes
2004: 247 (P*, L). **Type loc.:** Ambodimanga, Moramanga, Madagascar (LTV). **Distr.:** Madagascar (includes Glorioso and Juan De Nova Isl.). **Ety.:** stood against (L); ref. to five patches of white scales on the dark-brown lateral margin of the scutum. **Informal name:** Contrasting Malagasy Heaven Mosquito.

cornuta da Cunha Ramos & Brunhes
2004: 295 (F*, P*, L*). **Type loc.:** 5 km from Sainte-Luce/Manafiafy, Madagascar (LTV). **Distr.:** Madagascar (includes Glorioso and Juan De Nova Isl.). **Ety.:** horned (L) (ref. to hornlike trumpet of the pupa). **Informal name:** Horned Malagasy Heaven Mosquito.

damasei Grjebine
1979: 70 (L*). **Type loc.:** Manakara, km 5, route Vohipeno, Fianarantsoa Province, Madagascar (SSC). **Distr.:** Madagascar (includes Glorioso and Juan De Nova Isl.). **Ety.:** Damase Randriamanantena. **Informal name:** Damase Malagasy Heaven Mosquito.

dandakaranyensis Natarajan, Rajavel, & Jambulingam
2017 (M, F, P*, L*). **Type loc.:** Andhra Pradesh, Araku District, Araku-Maliguda-Bora section (Mosquito Museum, Vector Control Research Centre, Puducherry, India). **Distr.:** India. **Ety.:** Dandakaranya Forest, Odisha, India. **Informal name:** Dandakaranya Indian Heaven Mosquito.

demeilloni Peyton & Rattanarithikul
1970: 404 (M, F, P, L). **Type loc.:** Khlong Bang Man, Ranong, Thailand (USNM). **Distr.:** Malaysia, Philippines, Thailand, Vietnam. **Refs.:** Peyton 1977: 47 (M*, F, P*, L*; distr.); Rattanarithikul et al. 2006a (F*, L*; bion., distr., keys). **Ety.:** Botha De Meillon. **Informal name:** De Meillon Thai Heaven Mosquito.

devemyi Hamon
1955: 672 (M*). **Type loc.:** Kolda, Casamance [Senegal] (IERT). **Distr.:** Senegal. **Refs.:** da Cunha Ramos 1993: 252 (M*; key, tax., bion., distr.). **Ety.:** P. Devemy. **Informal name:** Devemy Senegalese Heaven Mosquito.

diagonalis Brug
1934a: 510 (M*, F, L*). **Type loc.:** Tanah Merah, Upper Digoel River, [Western] New Guinea (NHMUK). **Distr.:** Australia, Indonesia, Papua New Guinea. **Refs.:** Marks 1970: 165 (distr.); Lee et al. 1989c: 201 (F; key, tax., bion., distr., review). **Ety.:** diagonal (L) [oblique dark-brown stripe on pleura]. **Informal name:** Diagonal Indonesian Heaven Mosquito.

dibrugarhensis Bhattacharyya, Prakash, Mohapatra, & Mahanta
2004 (M*, F, P*, L*). **Type loc.:** Dihing Reserve Forest (elevation 152 m), Soraipung, Dibrugarh District, Assam, Northeast India, India (NHMUK). **Distr.:** India. **Ety.:** Dibrugarh Dist., Assam, India. **Informal name:** Dibrugarh Indian Heaven Mosquito.

donai da Cunha Ramos & Brunhes
2004: 254 (L*). **Type loc.:** Sainte-Luce/Manafiafy, Fort Dauphin/Tôlanaro, Madagascar (LTV). **Distr.:** Madagascar (includes Glorioso and Juan De Nova Isl.). **Ety.:** Donadieu Randrianambinina. **Informal name:** Donadieu Malagasy Heaven Mosquito.

douceti Grjebine
1953a: 459 (L*). **Type loc.:** Androhibe, Majunga, Madagascar (IRSM). **Distr.:** Comoros, Madagascar (includes Glorioso and Juan De Nova Isl.). **Refs.:** Brunhes 1977a (distr.; Comoros Isl.). **Ety.:** Jean-Paul Doucet. **Informal name:** Doucet Malagasy Heaven Mosquito.

enigmatica Peyton
1977: 133 (M*, F, P*, L*). **Type loc.:** Ban Wang Mo, Nan, Thailand (USNM). **Distr.:** Thailand. **Refs.:** Rattanarithikul et al. 2006a (F*, L*; bion., distr., keys). **Ety.:** n.s. [enigmatic, puzzle (L); prob. ref. to similarity to *abdita*]. **Informal name:** Enigmatic Thai Heaven Mosquito.

fulgens da Cunha Ramos & Brunhes
2004: 257 (F, P*, L*). **Type loc.:** On route from Moramanga to Anosibe-An'ala [Anosibe An'ala region, Madagascar] (LTV). **Distr.:** Madagascar (includes Glorioso and Juan De Nova Isl.). **Ety.:** flashing (L); particularly brilliant silvery scale patches on the pleuron. **Informal name:** Flashing Malagasy Heaven Mosquito.

fusca Theobald
1907: 564 (M*, F*). **Type loc.:** Mount Aureol, Sierra Leone (NHMUK). **Distr.:** Angola, Burkina Faso, Cameroon, Central African Republic, Democratic Republic of the Congo, Gabon, Ghana, Guinea, Kenya, Liberia, Mali, Mozambique, Nigeria, Republic of the Congo, Senegal, Sierra Leone, South Africa, Sudan and South Sudan, Tanzania, Uganda, Zambia, Zimbabwe. **Refs.:** Edwards 1941: 61 (M, F), 370 (P*); Hopkins 1952: 67 (L*); Mattingly 1954b: 171 (L*); Worth & De Meillon 1960: 235 (distr.); da Cunha Ramos & Ribeiro 1975: 130 (L*; tax., bion., keys, distr.; Angola); da Cunha Ramos 1993: 146 (M, F, P*, L*; keys, tax., bion., distr.); Jupp 1996 (M*, F*; key). **Ety.:** n.s. [dark (L); unmarked species]. **Informal name:** Dark Sierra Leonean Heaven Mosquito.

syn. *inornata* Theobald
1908a: 108 (M, F; *Ficalbia*). **Type loc.:** Transvaal [South Africa] (NHMUK). **Refs.:** Edwards 1941: 61 (syn.).

garnhami van Someren
1948: 130 (M*, P*, L*). **Type loc.:** Kigezi, Uganda (NHMUK). **Distr.:** Uganda. **Refs.**: da Cunha Ramos 1993: 254 (M*, P*, L*; keys, tax., bion., distr.). **Ety.:** Percy Cyril Claude Garnham. **Informal name:** Garnham Ugandan Heaven Mosquito.

gigantea Brug
1931c: 6 (A, L*). **Type loc.:** Balige, Siborong-borong and Samosir [Tapanuli], Hoeta Gindjang, Mount Toba Lake, N Sumatra [Indonesia] (NHMUK?). **Distr.:** Indonesia. **Refs.:** Peyton 1977: 72 (M*, F, L*; type loc. info). **Ety.:** n.s. [giant (L)]. **Informal name:** Gigantic Indonesian Heaven Mosquito.

gouldi Peyton & Klein
1970: 248 (M, F, P, L). **Type loc.:** Khao Sung, Phangnga, Thailand (USNM). **Distr.:** Cambodia, Thailand. **Refs.:** Peyton 1977: 50 (M*, F, P*, L*); Rattanarithikul et al. 2006a (F*, L*; bion., distr., keys). **Ety.:** Douglas J. Gould. **Informal name:** Gould Thai Heaven Mosquito.

grenieri Doucet
1951c: 110 (L*). **Type loc.:** Ampataka, Madagascar (IRSM). **Distr.:** Madagascar (includes Glorioso and Juan De Nova Isl.). **Ety.:** Paul Fernand Grenier. **Informal name:** Grenier Malagasy Heaven Mosquito.

grjebinei da Cunha Ramos & Brunhes
2004: 300 (M*, F*, P*, L*). **Type loc.:** Manakara, Madagascar (LTV). **Distr.:** Madagascar (includes Glorioso and Juan De Nova Isl.). **Ety.:** Alexis Grjebine. **Informal name:** Grjebine Malagasy Heaven Mosquito.

haddowi da Cunha Ramos & Brunhes
2004: 116 (M*, P*, L*). **Type loc.:** Nosy Be (Reserve Naturelle), Lokobe, Madagascar (LTV). **Distr.:** Madagascar (includes Glorioso and Juan De Nova Isl.). **Ety.:** Alexander John Haddow. **Informal name:** Haddow Malagasy Heaven Mosquito.

harrisoni Peyton
1977: 29 (M*, F, P*, L*). **Type loc.:** San Antonio, Victoria District, Oriental, Mindoro Island, Philippines (USNM). **Distr.:** Malaysia, Philippines. **Ety.:** Bruce Arthur Harrison. **Informal name:** Harrison Philippine Heaven Mosquito.

henrardi Edwards
1935c: 96 (M). **Type loc.:** Leopoldville, Belgian Congo [Kinshasa, Democratic Republic of the Congo] (CMT). **Distr.:** Cameroon, Côte d'Ivoire, Democratic Republic of the Congo, Republic of the Congo, Tanzania, Uganda. **Refs.:** Edwards 1941: 63 (M*); Doucet 1960: 810 (P, L*); van Someren 1962: 25 (F); White 1975: 325 (distr.); da Cunha Ramos 1987 (M*); da Cunha Ramos 1993: 257 (M*; key, tax., bion., distr.). **Ety.:** Constant Louis Joseph Ghislain Henrard. **Informal name:** Henrard Congolese Heaven Mosquito.

henriquei da Cunha Ramos
1993: 158 (M*, F, P*, L*; keys, tax., bion., distr.). **Type loc.:** Dundo, Angola (IHMT). **Distr.:** Angola, Cameroon, Sierra Leone. **Ety.:** Henrique Ribeiro. **Informal name:** Henrique Angolan Heaven Mosquito.

hervyi da Cunha Ramos & Brunhes
2004: 262 (M*, P*, L*). **Type loc.:** Beforona, Madagascar (LTV). **Distr.:** Madagascar (includes Glorioso and Juan De Nova Isl.). **Ety.:** John-Paul Hervy. **Informal name:** Hervy Malagasy Heaven Mosquito.

hirsuta Boussès & Brunhes
2013: 403 (L*). **Type loc.:** Forest d'Ambovonamby, Réserve de Namoroka, Mahajanga Province, Madagascar (16°28.2'S - 45°20.9'E) (IRD). **Distr.:** Madagascar (includes Glorioso and Juan De Nova Isl.). **Ety.:** hairy (L); ref. to larvae. **Informal name:** Hairy Malagasy Heaven Mosquito.

hirsutifemora Peters
1964: 21 (M*, F*). **Type loc.:** Port Moresby, Central District, Papua New Guinea (CSIR). **Distr.:** Australia, Cambodia, Indonesia, Malaysia, Papua New Guinea, Singapore, Solomon Isl., Thailand. **Refs.:** Marks 1970: 166 (distr.); Peyton 1977: 136 (M*, F, P*; distr.); Harbach et al. 1986b (L; distr.; Thailand); Lee et al. 1989c: 204 (F; key, tax., bion., distr., review); Rattanarithikul et al. 2006a (F*, L*; bion., distr., keys). **Ety.:** hairy leg (L); ref. to rows of long hairs on femora. **Informal name:** Hairy-legged Papuan Heaven Mosquito.

hongayi Galliard & Ngu
1947: 77 (A, L). **Type loc.:** Baie d'Along, Tonkin, Indochina [Halong Bay, Hanoi province, Vietnam] (LU). **Distr.:** Vietnam. **Refs.:** Galliard & Ngu 1950: 67 (M*, P*, L*); Peyton 1977: 139 (F; distr.). **Ety.:** n.s. [Hongay (Hon Gai), Vietnam]. **Informal name:** Hongay Vietnamese Heaven Mosquito.

husaini Qutubuddin
1947: 117 (M*, F). **Type loc.:** Hyderabad, India (IM). **Distr.:** India. **Ety.:** Mian Afzal Husain. **Informal name:** Husain Indian Heaven Mosquito.

jacksoni Edwards
1935a: 130 (M). **Type loc.:** Hong Kong [People's Republic of China] (NHMUK). **Distr.:** Japan, People's Republic of China (includes Hong Kong). **Refs.:** Bohart & Ingram 1946b: 59 (M, F, P*, L*; bion., distr., tax.; as *stonei*); Tanaka et al. 1975c: 213 (bion., distr.); Peyton 1977: 141 (M*, F, P*, L*); Tanaka et al. 1979: 455 (M*, F*, L*). **Ety.:** R.B. Jackson. **Informal name:** Jackson Chinese Heaven Mosquito.

syn. *stonei* Bohart & Ingram
1946a: 47 (M*, F, L*). **Type loc.:** Chizuka, Okinawa, Ryukyu-Retto [Japan] (USNM). **Refs.:** Peyton 1972: 36 (syn.).

jinhongensis Dong, Dong, & Zhou
2003 (M*, L*). **Type loc.:** Jinhong County (elevation 760 m), Yunnan, China [People's Republic of China] (IPD). **Distr.:** People's Republic of China. **Ety.:** Jinhong County, China. **Informal name:** Jinhong Chinese Heaven Mosquito.

koli Peyton & Klein
1970: 246 (M, F, P, L). **Type loc.:** Khao Sai Dao Mountain, Chanthaburi, Thailand (USNM). **Distr.:** Cambodia, Thailand, Vietnam. **Refs.:** Peyton 1977: 145 (M*, F*, P*, L*; distr.); Rattanarithikul et al. 2006a (F*, L*; bion., distr., keys). **Ety.:** Kol Mongkolpanya. **Informal name:** Kol Thai Heaven Mosquito.

kraussi Grjebine
1953a: 459 (M*, F*, P*, L*). **Type loc.:** La Mandraka, Manjakandriana District [Tananarive], Madagascar (IRSM). **Distr.:** Madagascar (includes Glorioso and Juan De Nova Isl.). **Ety.:** M. Krauss. **Informal name:** Krauss Malagasy Heaven Mosquito.

laffosseae da Cunha Ramos & Brunhes
2004: 193 (M*, F, P*, L*). **Type loc.:** Beforona, Moramanga, Madagascar (LTV). **Distr.:** Madagascar (includes Glorioso and Juan De Nova Isl.). **Ety.:** Josée Laffosse. **Informal name:** Laffosse Malagasy Heaven Mosquito.

lavieri Doucet
1950: 40 (L*). **Type loc.:** Montagne d'Ambre, Diego Suarez, Madagascar (IRSM). **Distr.:** Madagascar (includes Glorioso and Juan De Nova Isl.). **Refs.:** Doucet 1951a: 82 (L*). **Ety.:** n.s. **Informal name:** Lavier Malagasy Heaven Mosquito.

legoffi da Cunha Ramos & Brunhes
2004: 305 (M*, P*, L*). **Type loc.:** Beforona, Madagascar (LTV). **Distr.:** Madagascar (includes Glorioso and Juan De Nova Isl.). **Ety.:** Gilbert Le Goff. **Informal name:** Le Goff Malagasy Heaven Mosquito.

leiboensis Chu
1981: 273 (L*). **Type loc.:** Leibo County, Sichuan Province, China [People's Republic of China] (AMC). **Distr.:** People's Republic of China. **Ety.:** Leibo County, Sichuan Prov., China. **Informal name:** Leibo Chinese Heaven Mosquito.

longitubus da Cunha Ramos & Brunhes
2004: 126 (L*). **Type loc.:** Onive River (altitude 150 m), Masoala, Antalaha, Madagascar (LTV). **Distr.:** Madagascar (includes Glorioso and Juan De Nova Isl.). **Ety.:** long tube, pipe (L); ref to long siphon. **Informal name:** Long-piped Malagasy Heaven Mosquito.

lousthei Boussès & Brunhes
2013: 406 (L*). **Type loc.:** near Ambila Lemaitso village, Toamasina Province, Madagascar (18°51'S, 49°09'E) (IRD). **Distr.:** Madagascar (includes Glorioso and Juan De Nova Isl.). **Ety.:** collected by Vincent Robert and his son Mathieu Lousthe-Robert [states named after the collector; poss. named after both]. **Informal name:** Lousthe Malagasy Heaven Mosquito.

lucyae van Someren
1954a: 103 (M*, F). **Type loc.:** Kwale, Kenya (NHMUK). **Distr.:** Kenya, Tanzania. **Refs.**: da Cunha Ramos 1993: 109 (M*, F; keys, tax., bion., distr.). **Ety.:** n.s. [Lucy Ellinor Tunstall, author's mother (Harbach 2018c)]. **Informal name:** Lucy Kenyan Heaven Mosquito.

lui Lien
1968a: 9 (M*, F, P*, L*). **Type loc.:** Chihtuan of Loswei Village, Tatung, Ilan Hsien, Taiwan [Republic of China] (TMRI). **Distr.:** Taiwan. **Refs.:** Peyton 1977: 76 (M*, F, P*, L*). **Ety.:** Liang Cheng Lu. **Informal name:** Lu Taiwanese Heaven Mosquito.

lunda da Cunha Ramos
1993: 180 (M*, F, P*, L*; keys, tax., bion., distr.). **Type loc.:** Dundo, Angola (IHMT). **Distr.:** Angola. **Ety.:** Lunda [Norte?] Prov., Angola. **Informal name:** Lunda Angolan Heaven Mosquito.

luteola Edwards
1934 (M*, F). In Barraud 1934: 80. **Type loc.:** Malabar Coast [Madras], India (NHMUK). **Distr.:** India, Nepal, People's Republic of China, Taiwan, Thailand. **Refs.:** Joshi et al. 1965: 139 (distr.); Darsie & Pradhan 1990 (F, L; tax., keys, bion., distr.; Nepal). **Ety.:** n.s. [yellow (L); poss. ref. to yellow markings on abdomen]. **Informal name:** Yellow Indian Heaven Mosquito.

lutescens Leicester
1908: 222 (M, F). **Type loc.:** Malaya [Malaysia] (NHMUK). **Distr.:** Cambodia, Malaysia, Thailand, Vietnam. **Refs.:** Barraud 1926: 344 (M*, F); Borel 1930: 129 (M*, F, L*); Barraud 1934: 80 (distr.); Peyton 1977: 53 (M*, F, P*, L*; distr.); Rattanarithikul et al. 2006a (F*, L*; bion., distr., keys). **Ety.:** becoming yellow (L); ref. to "pale ochraceous mesonotum and pleura." **Informal name:** Yellowing Malaysian Heaven Mosquito.

maculipleura Leicester
1908: 223 (F). **Type loc.:** Kuala Lumpur, Malaya [Malaysia] (NHMUK). **Distr.:** Malaysia, Thailand. **Refs.:** Joshi et al. 1965: 139 (distr.); Wattal & Kalra 1965: 312 (M*); Peyton 1977: 79 (M*, F*, P*, L*; distr.); Darsie & Pradhan 1990 (F, L; tax., keys, bion., distr.; Nepal). **Ety.:** spotted side (L); ref. to "pleura is creamy with three dark brown spots." **Informal name:** Spot-sided Malaysian Heaven Mosquito.

madagascarensis da Cunha Ramos & Brunhes
2004: 267 (M*, P*, L*). **Type loc.:** Ambodimanga, Madagascar (LTV). **Distr.:** Madagascar (includes Glorioso and Juan De Nova Isl.). **Ety.:** Madagascar. **Informal name:** Imerina Malagasy Heaven Mosquito [Imerina (1540–1897) was a precolonial kingdom of Madagascar].

maikalensis Natarajan, Rajavel, & Jambulingam
2017: 251 (M, F, P*, L*). **Type loc.:** Chhattisgarh, Korba District, Lapha, India (Mosquito Museum, Vector Control Research Centre, Puducherry, India). **Distr.:** India. **Ety.:** Maikal Hills, India. **Informal name:** Maikal Indian Heaven Mosquito.

manakaraensis da Cunha Ramos & Brunhes
2004 (M*, F, P*). **Type loc.:** Mankara, Madagascar (LTV). **Distr.:** Madagascar (includes Glorioso and Juan De Nova Isl.). **Ety.:** Mankara, Madagascar. **Informal name:** Mankara Malagasy Heaven Mosquito.

mashonaensis Theobald
1901c: 259 (M*, F*). **Type loc.:** Salisbury, Mashonaland [Harare, Zimbabwe] (NHMUK). **Distr.:** Angola, Benin, Burkina Faso, Cameroon, Central African Republic, Côte d'Ivoire, Democratic Republic of the Congo, Equatorial Guinea, Gabon, Ghana, Guinea, Israel, Kenya, Liberia, Malawi, Mali, Mozambique, Niger, Nigeria, Republic of the Congo, Sierra Leone, Somalia, South Africa, Sudan and South Sudan, Tanzania, Togo, Uganda, Zimbabwe. **Refs.:** Ingram & De Meillon 1927: 70 (P*, L*); Edwards 1941: 59 (M*, F), 369 (P); Hopkins 1952: 65 (L*); Worth & De Meillon 1960: 235 (distr.); da Cunha Ramos & Ribeiro 1975: 131 (L*; tax., bion., keys, distr.; Angola); Service 1976a (distr.; Gabon); da Cunha Ramos 1993: 124 (M*, F, P*, L*; keys, tax., bion., distr.); Jupp 1996 (M*, F*; key). **Ety.:** Mashonaland, Zimbabwe. **Informal name:** Mashonaland Zimbabwean Heaven Mosquito.

mattinglyi Qutubuddin
1951: 107 (M*, F). **Type loc.:** Hyderabad, India (NHMUK). **Distr.:** India. **Ety.:** Peter Frederick Mattingly. **Informal name:** Mattingly Indian Heaven Mosquito.

maxima Leicester
1908: 221 (F). **Type loc.:** The Gap, Malaya [Malaysia] (NHMUK). **Distr.:** India, Malaysia, People's Republic of China, Thailand. **Refs.:** Barraud 1934: 77 (M*, F, L*); Peyton 1977: 178 (M*, F, P*, L*; lectotype desig.); Qu & Qian 1993 (cibarial armature*); Rattanarithikul et al. 2006a (F*, L*; bion., distr., keys). **Ety.:** largest (L); the largest *Uranotaenia* the author had ever found. **Informal name:** Largest Malaysian Heaven Mosquito.

mengi Chen, Wang, & Zhao
1989: 224 (M*, F*, P*, L*). **Type loc.:** Chishui County, Guizhou Province, China [People's Republic of China] (GMC). **Distr.:** People's Republic of China. **Ety.:** n.s. **Informal name:** Meng Chinese Heaven Mosquito.

micromelas Edwards
1934: 326 (M, F; as *micromeles*). **Type loc.:** São Tomé (NHMUK). **Distr.:** São Tomé and Príncipe. **Refs.:** Edwards 1941: 61 (M*; emend.); da Cunha Ramos & Antunes Capela 1988 (P*, L*); da Cunha Ramos 1993: 134 (M*, F, P, L; keys, tax., bion., distr.). **Ety.:** n.s. **Informal name:** Honey Sao Tomean Heaven Mosquito [melas = "mellis" = Honey].

modesta Leicester
1908: 218 (M, F). **Type loc.:** Ampang Jungle [Selangor], Malaya [Malaysia] (NHMUK). **Distr.:** Malaysia, Philippines, Thailand,. **Refs.:** Basio 1971b: 93 (M*; bion.; as *tubanguii*); Mattingly 1971a: Pl. 12 (P terminal segs.*); Peyton 1977: 83 (M*, F, P*, L*); Rattanarithikul et al. 2006a (F*, L*; bion., distr., keys). **Ety.:** n.s. [modest, reserved (L); prob. ref. to "a very small dull-coloured mosquito"]. **Informal name:** Modest Malaysian Heaven Mosquito.

syn. *tubanguii* Baisas
1935b: 69 (M*, F, L*). **Type loc.:** Kolambugan, Lanao, Mindanao, Philippines (NE; paratypes USNM). **Refs.:** Peyton 1972: 37 (syn.).

montana Ingram & De Meillon
1927: 80 (M, F). **Type loc.:** Eshowe, Matubatuba, and Gingindhlovu, Natal Province [KwaZulu Natal, South Africa] (NHMUK). **Distr.:** South Africa. **Refs.:** Edwards 1941: 54 (M, F); De Meillon et al. 1945: 100 (P*); Hopkins 1952: 58 (L); da Cunha Ramos 1993: 112 (M*, F, P*, L; keys, tax., bion., distr.); Jupp 1996 (M*, F*; key). **Ety.:** n.s. [mountain, mountainous (L)]. **Informal name:** Montane South African Heaven Mosquito.

moufiedi Peyton
1977: 86 (M*, F*). **Type loc.:** Mount Kinabalu, Sabah, Malaysia [Malaysian Borneo] (USNM). **Distr.:** Malaysia. **Ety.:** Moufied A. Moussa. **Informal name:** Moufied Malaysian Heaven Mosquito.

moultoni Edwards
1914c: 128 (M, F). **Type loc.:** Kuching Reservoir, Sarawak [Malaysian Borneo] (NHMUK). **Distr.:** Indonesia, Malaysia, Singapore. **Refs.:** Edwards 1926a: 115 (L*; as *brevitrostris*); Peyton & Hochman 1968: 379 (M*; as *brevirostris*); Peyton 1977: 88 (M*, F, P*, L*; distr.). **Ety.:** J.C. Moulton. **Informal name:** Moulton Malaysian Heaven Mosquito.

syn. *brevirostris* Edwards
1915c: 284 (M, F). **Type loc.:** Kuching, Sarawak [Malaysian Borneo] (NHMUK). **Refs.:** Peyton 1972: 37 (syn.).

musarum Edwards
1936: 54 (A; as var.). **Type loc.:** Fort Portal, Uganda (NHMUK). **Distr.:** Cameroon, Uganda. **Refs.:** Edwards 1941: 58 (M); van Someren 1951: 9 (M, F, P, L); White 1975: 325 (to ssp.); da Cunha Ramos 1993: 195 (M*, F, P*, L*; keys, tax., bion., distr.; to sp.). **Ety.:** goddesses of

science/poetry (L). **Informal name:** Muse Ugandan Heaven Mosquito.

nepenthes (Theobald)
1912b: 92 (F*, P*?, L*?; *Pseudoficalbia*). **Type loc.:** Silhouette Island, Seychelles (NHMUK). **Distr.:** Seychelles. **Refs.:** Paulian 1950: 120 (L*); Mattingly & Brown 1955: 73 (M, P*, L*; tax.). **Ety.:** larvae in a "Nepenthes pitcher." **Informal name:** Nepenthes Seychellois Heaven Mosquito.

nigricephala da Cunha Ramos & Brunhes
2004: 199 (M*, F, P*, L*). **Type loc.:** near Antalaha, Madagascar. **Distr.:** Madagascar (includes Glorioso and Juan De Nova Isl.). **Ety.:** black head (L); dark headed. **Informal name:** Black-headed Malagasy Heaven Mosquito.

nigripes (Theobald)
1905i: 199 (M; *Ficalbia*). **Type loc.:** Kortright, Freetown, Sierra Leone (NHMUK). **Distr.:** Cameroon, Democratic Republic of the Congo, Equatorial Guinea, Ghana, Liberia, Nigeria, Republic of the Congo, Sierra Leone, Togo, Uganda. **Refs.:** Edwards 1941: 58 (M, F), 368 (P*); Hopkins 1952: 62 (L); da Cunha Ramos 1993: 170 (M*, F, P*, L*; keys, tax., bion., distr.). **Ety.:** black foot (L); legs deep blackish brown, coxae pale ochraceous. **Informal name:** Black-footed Sierra Leonean Heaven Mosquito.

nigripleura da Cunha Ramos & Brunhes
2004: 310 (M*, F*, P*). **Type loc.:** Ifaho, Madagascar [Manakara Sud, Fianarantsoa, 22°1′0″ S, 47°53′0″ E; original name is Ifaho] (LTV). **Distr.:** Madagascar (includes Glorioso and Juan De Nova Isl.). **Ety.:** n.s. [dark sides (Gr); poss. ref. to dark markings on pleuron]. **Informal name:** Black-sided Malagasy Heaven Mosquito.

nigromaculata Edwards
1941: 60 (new name for *bimaculata* Theobald, not Leicester 1908 (M*, F*)). **Distr.:** Angola, Benin, Burkina Faso, Cameroon, Central African Republic, Côte d'Ivoire, Democratic Republic of the Congo, Ethiopia, Gabon, Ghana, Liberia, Mozambique, Nigeria, Republic of the Congo, South Africa, Tanzania, Uganda. **Refs.:** Peters 1955c: 120 (P*, L*); Worth & De Meillon 1960: 235 (distr.); Service 1976a (distr.; Gabon); da Cunha Ramos 1993: 138 (M*, F, P*, L*; keys, tax., bion., distr.); Jupp 1996 (M*, F*; key). **Ety.:** n.s. [black spot (L); prob. ref. to black spots above wing bases]. **Informal name:** Black-spotted Ghanaian Heaven Mosquito.

syn. *bimaculata* Theobald
1909b: 22 (M*). **Type loc.:** Obuasi, Ashanti, Gold Coast [Ghana] (NHMUK).

nivipleura Leicester
1908: 219 (M, F). **Type loc.:** The Gap [Selangor], Malaya [Malaysia] (NHMUK). **Distr.:** Cambodia, India, Indonesia, Japan, Laos, Malaysia, Nepal, People's Republic of China (includes Hong Kong), Singapore, Sri Lanka, Taiwan, Thailand, Vietnam. **Refs.:** Bohart & Ingram 1946b: 57 (M; bion., distr., tax.); Peyton & Hochman 1968: 380 (M*); Tanaka et al. 1975c: 213 (distr.); Peyton 1977: 92 (M*, F, P*, L*; type info., distr.); Tanaka et al. 1979: 467 (M*, F*, L*); Darsie et al. 1992 (distr.; Nepal); Qu & Qian 1993 (cibarial armature*); Rattanarithikul et al. 2006a (F*, L*; bion., distr., keys). **Ety.:** snow-white side (L, Gr); "Pleura and coxae silvery-white." **Informal name:** Snow-sided Malaysian Heaven Mosquito.

syn. *loshanensis* Wang & Feng
1965: 206 (M*, F, L*). **Type loc.:** Loshan, Szechuan Province, China [People's Republic of China] (MMC). **Refs.:** Chu 1980: 238 (syn.; as *loashanensis*, *lapsus*).

nivipous Theobald
1912a: 93 (F). **Type loc.:** Onderstepoort, Transvaal [South Africa] (NHMUK). **Distr.:** Cameroon, Democratic Republic of the Congo, Guinea, Kenya, Liberia, Mozambique, Nigeria, Republic of the Congo, Senegal, South Africa, Sudan and South Sudan, Uganda, Zimbabwe. **Refs.:** Edwards 1941: 56 (F), 367 (P*; as *candidipes*); Hopkins 1952: 59 (L*; as *candidipes*); Mattingly 1954b: 169 (M*, L; tax.; as *candidipes*); Worth & De Meillon 1960: 235 (distr.); da Cunha Ramos 1993: 116 (M*, F, P*, L*; keys, tax., bion., distr.); Jupp 1996 (M*, F*; key). **Ety.:** n.s. [snowy feet (L); hind tarsomeres 3–5 creamy white]. **Informal name:** Snow-footed South African Heaven Mosquito.

syn. *candidipes* Edwards
1912c: 42 (new name for *nivipous* Theobald).

nocticola Peyton
1977: 149 (M*, F, P*, L*). **Type loc.:** Chiang Dao, Chiang Mai, Thailand (USNM). **Distr.:** Thailand. **Refs.:** Rattanarithikul et al. 2006a (F*, L*; bion., distr., keys). **Ety.:** n.s. [night lover (L)]. **Informal name:** Nocturnal Thai Heaven Mosquito.

novobscura Barraud
1934: 84 (M, F, L). **Type loc.:** Sukna, Darjeeling District [West Bengal], India (NHMUK). **Distr.:** Bangladesh, Cambodia, India, Japan, Laos, Malaysia, People's Republic of China (includes Hong Kong), Taiwan, Thailand. **Refs.:** Bohart & Ingram 1946b (E*; as *bimaculata* Leicester); Qutubuddin 1951 (M*); Mattingly 1970c (E*; as *bimaculata* Leicester, in part); Peyton 1977: 96 (M*, F*, P*, L*, E*; distr.); Tanaka et al. 1979: 460 (M*, F*, L*); Ahmed 1987 (distr.; Bangladesh); Qu & Qian 1993 (cibarial armature*); Bhattacharyya et al. 2003 (distr.; India); Rattanarithikul et al. 2006a (F*, L*; bion., distr., keys). **Ety.:** near *obscura*. **Informal name:** Newly obscured Indian Heaven Mosquito.

ssp. ***ryukyuana*** Tanaka, Mizusawa, & Saugstad
1979: 465 (M*, F*, L). **Type loc.:** Okinawa Island, Ryukyus, Japan (NMST). **Distr.:** Japan. **Ety.:** Ryukyu Isl., Japan. **Informal name:** Ryukyu Japanese Heaven Mosquito.

obscura Edwards
1915c: 285 (A). **Type loc.:** Kuching, Sarawak [Malaysian] Borneo (NHMUK). **Distr.:** Australia, Cambodia, India, Indonesia, Malaysia, Papua New Guinea, Philippines, Singapore, Sri Lanka, Thailand. **Refs.:** Edwards & Given 1928: 338 (L); Penn 1949b: 29 (P*; as *papua*); Bonne-Wepster 1954: 26 (F, L*; as *papua*); Peyton & Hochman 1968: 380 (M*); Basio 1971b: 92 (M*; bion.; as *philippinensis*); Lee et al. 1989c: 207 (tax., bion., distr., review); Rattanarithikul et al. 2006a (F*, L*; bion., distr., keys). **Ety.:** n.s. [black, indistinct (L); prob. ref. to "integument and scales almost uniformly dark brownish"]. **Informal name:** Shadowed Malaysian Heaven Mosquito.
syn. *papua* Brug
1924: 441 (F). **Type loc.:** Pioneer Bivak, Mamberamo River, Western New Guinea (Papua) (NHMUK). **Refs.:** Peyton 1972: 37 (syn.).
syn. *philippinensis* Delfinado
1966b: 36 (M*, F, P*, L*). **Type loc.:** San Jose, Mindoro, Philippines (USNM). **Refs.:** Peyton 1972: 37 (syn.).

ohamai Tanaka, Mizusawa, & Saugstad
1975a: 27 (M*, F, L*). **Type loc.:** Yashigawa, Iriomote Island, Ryukyu Islands [Japan] (USNM). **Distr.:** Japan. **Refs.:** Peyton 1977: 152 (M*, F, P*, L*); Tanaka et al. 1979: 457 (M*, F*, L). **Ety.:** Shinken Ohama. **Informal name:** Ohama Japanese Heaven Mosquito.

ornata Theobald
1909b: 22 (M). **Type loc.:** Obuasi, Ashanti, Gold Coast [Ghana] (NHMUK). **Distr.:** Angola, Benin, Burkina Faso, Cameroon, Central African Republic, Côte d'Ivoire, Democratic Republic of the Congo, Ethiopia, Gabon, Ghana, Guinea, Kenya, Liberia, Mali, Nigeria, Republic of the Congo, Sierra Leone, Uganda, Zimbabwe. **Refs.:** Summers Connal 1931: 459 (P*, L*); Edwards 1941: 56 (M, F*), 368 (P*); Hopkins 1952: 60 (L*); da Cunha Ramos & Ribeiro 1975: 132 (tax., bion., keys, distr.; Angola); Service 1976a (distr.; Gabon); da Cunha Ramos 1993: 205 (M*, F, P*, L*; keys, tax., bion., distr.); Jupp 1996 (M*, F*; key). **Ety.:** n.s. [ornate (L)]. **Informal name:** Ornate Ghanaian Heaven Mosquito.

ototomo da Cunha Ramos
1993: 261 (M*; key, tax., bion., distr.). **Type loc.:** Ototomo, Cameroon (SSC). **Distr.:** Cameroon. **Ety.:** Ototomo, Cameroon. **Informal name:** Ototomo Cameroonian Heaven Mosquito.

painei Edwards
1935a: 127 (M, F, L). **Type loc.:** Salialevu, Taveuni Island, Fiji (NHMUK). **Distr.:** Fiji, Solomon Isl. **Refs.:** Belkin 1962: 162 (M*, F, P*, L*); Lee et al. 1989c: 211 (tax., bion., distr., review). **Ety.:** R.W. Paine. **Informal name:** Paine Fijian Heaven Mosquito.

pallidipleura da Cunha Ramos & Brunhes
2004: 273 (M*, F, P*). **Type loc.:** Loharano, Manakara, Madagascar (LTV). **Distr.:** Madagascar (includes Glorioso and Juan De Nova Isl.). **Ety.:** pale side (Gr); ref. to pale color of the lateral parts of the thorax. **Informal name:** Pale-sided Malagasy Heaven Mosquito.

pandani (Theobald)
1912b: 90 (M*, F*; *Pseudoficalbia*). **Type loc.:** Cascade, Mahe Island, Seychelles (NHMUK). **Distr.:** Comoros, Madagascar (includes Glorioso and Juan De Nova Isl.), Seychelles. **Refs.:** Hopkins 1952: 63 (L*); Mattingly & Brown 1955: 70 (M*, P*, L*); Brunhes 1977a (distr.; Comoros Isl.). **Ety.:** n.s. ["Amongst dead leaves of *Pandanus hornei*"]. **Informal name:** Pandanus Seychellois Heaven Mosquito.

patriciae Peyton
1977: 103 (M*, F*, P*, L*). **Type loc.:** The Gap, Selangor, Malaya [Malaysia] (USNM). **Distr.:** Malaysia, Singapore, Thailand. **Refs.:** Rattanarithikul et al. 2006a (F*, L*; bion., distr., keys). **Ety.:** Patricia Ann Peyton. **Informal name:** Patricia Malaysian Heaven Mosquito.

pauliani Doucet
1949b: 325 (L*). **Type loc.:** Ambohiby, Tsiroanomandidy District, Madagascar (IRSM). **Distr.:** Madagascar (includes Glorioso and Juan De Nova Isl.). **Refs.:** Hopkins 1952: 64 (L; tax.); Mattingly & Brown 1955: 70 (syn.). **Ety.:** Renaud Paulian. **Informal name:** Paulian Malagasy Heaven Mosquito.

pilosa da Cunha Ramos & Brunhes
2004: 313 (M*, P*, L*). **Type loc.:** Sainte-Luce/Manafiafy, Madagascar (LTV). **Distr.:** Madagascar (includes Glorioso and Juan De Nova Isl.). **Ety.:** n.s. [hairy, shaggy (L)]. **Informal name:** Furry Malagasy Heaven Mosquito.

principensis da Cunha Ramos
1993: 264 (M*, F, P*, L*; keys, tax., bion., distr.). **Type loc.:** Rio Papagaio, Bela Vista, Príncipe, Republica de S Tome e Príncipe [São Tomé & Príncipe] (IHMT). **Distr.:** Cameroon, São Tomé and Príncipe. **Ety.:** Ilha do Príncipe. **Informal name:** Príncipe São Tomean Heaven Mosquito.

propinqua Peyton
1977: 106 (M*, F*, P*, L*). **Type loc.:** Ulu Bakau, Selangor, Malaya [Malaysia] (USNM). **Distr.:** Malaysia. **Ety.:** n.s. [near, neighboring (L); poss. ref. to similarity to *modesta*]. **Informal name:** Nearly Malaysian Heaven Mosquito.

pseudoalbimanus da Cunha Ramos & Brunhes
2004: 132 (M*, F, P*, L*). **Type loc.:** Manakara, Madagascar (LTV). **Distr.:** Madagascar (includes Glorioso and Juan De Nova Isl.). **Ety.:** false (Gr); similar to *albimanus*. **Informal name:** Cardsharp Malagasy Heaven Mosquito [*albimanus* = "false white handed" = false handed = Cardsharp].

pseudohenrardi Peters
1955c: 117 (M*, F). **Type loc.:** Kpayan, Liberia (NHMUK). **Distr.:** Côte d'Ivoire, Democratic Republic of the Congo, Liberia. **Refs.**: da Cunha Ramos 1993: 274 (M*, F; keys, tax., bion., distr.). **Ety.:** false (Gr); similar to *henrardi*. **Informal name:** Alias Liberian Heaven Mosquito [*pseudohenrardi* = "false Henrard" = Alias].

pseudomaculipleura Peyton & Rattanarithikul
1970: 408 (M, F, P, L). **Type loc.:** Nam Tai, Phangnga, Thailand (USNM). **Distr.:** Malaysia, Thailand. **Refs.:** Peyton 1977: 109 (M*, F*, P*, L*; distr.); Rattanarithikul et al. 2006a (F*, L*; bion., distr., keys). **Ety.:** false (Gr); similar to *maculipleura*. **Informal name:** False-spotted Thai Heaven Mosquito.

pseudoshillitonis da Cunha Ramos & Brunhes
2004: 276 (M*, P*, L*). **Type loc.:** Masoala, Ambohitralanana, Madagascar (LTV). **Distr.:** Madagascar (includes Glorioso and Juan De Nova Isl.). **Ety.:** false (Gr); similar to *shillitonis*. **Informal name:** Masoala Malagasy Heaven Mosquito.

pseudostricklandi Natarajan, Rajavel, & Jambulingam
2018 (M, F*, P*, L*). **Type loc.:** India, Kerala, Kani forest, Molamedu, 08° 36.361' N, 77° 12.187' E (Mosquito Museum of the Vector Control Research Centre, Puducherry, India). **Distr.:** India. **Ety.:** near *stricklandi*. **Informal name:** Kerala Indian Heaven Mosquito.

pylei Baisas
1946a: 45 (M*, F, P*, L*). **Type loc.:** Llavac, Laguna, Luzon, Philippines (NE). **Distr.:** Philippines. **Refs.:** Basio 1971b: 92 (M*; bion.); Peyton 1977: 155 (M*, F*, P*, L*). **Ety.:** n.s. **Informal name:** Pyle Philippine Heaven Mosquito.

quadrimaculata Edwards
1929 (M, F, P, L). In Paine & Edwards 1929: 313. **Type loc.:** Rere, Guadalcanal, Solomon Islands (NHMUK). **Distr.:** Papua New Guinea, Solomon Isl. **Refs.:** Belkin 1953a: 373 (M*, F*, P*, L*); Laird 1955: 285 (L*; tax.); Belkin 1962: 159 (M*, F*, P*, L*; lectotype desig.); Lee et al. 1989c: 213 (tax., bion., distr., review). **Ety.:** four spots (L); ref. to four conspicuous black spots on scutum and postpronotum. **Informal name:** Four-spotted Solomon Islander Heaven Mosquito.

quasimodesta Peyton
1977: 112 (M*, F, P*, L*). **Type loc.:** Mount Kinabalu, Sabah, [Malaysian] Borneo (USNM). **Distr.:** Malaysia. **Ety.:** nearly (L); similar to *modesta*. **Informal name:** Half-modest Malaysian Heaven Mosquito.

qui Dong, Dong, & Zhou
2003 (M*, P*, L*). In Dong, Dong, Zhou, Wang, & Lu 2003: 519. **Type loc.:** Jinhong County (780 m), Yunnan Province, China [People's Republic of China] (YIM). **Distr.:** People's Republic of China. **Ety.:** n.s. [prob. for Qu Feng-yi (Harbach 2018c)]. **Informal name:** Qui Chinese Heaven Mosquito.

quinquemaculata Bonne-Wepster
1934: 276 (F*). **Type loc.:** Tjisarua (near Buitenzorg), Java [Indonesia] (NHMUK). **Distr.:** Indonesia, Malaysia. **Refs.:** Peyton 1977: 116 (F*; distr.). **Ety.:** five spots (L); "five conspicuous dark brown (almost black) spots on the integument of the mesonotum." **Informal name:** Five-spotted Indonesian Heaven Mosquito.

ramosa da Cunha Ramos
1993: 214 (L*; key, tax., bion., distr.). **Type loc.:** Dundo, Angola (IHMT). **Distr.:** Angola. **Ety.:** n.s. [poss. ref. to Helena da Cunha Ramos]. **Informal name:** Ramos Angolan Heaven Mosquito.

ravenalaphila Boussès & Brunhes
2013: 400 (L*). **Type loc.:** near Ifaho village, Manakara District, Fianarantsoa Province, Madagascar (22°10′ S, 45°04′ E) (IRD). **Distr.:** Madagascar (includes Glorioso and Juan De Nova Isl.). **Ety.:** larvae found in "le stipe sectionne et creux de *Ravenala madagascariensis*." **Informal name:** Ravenala Malagasy Heaven Mosquito.

ravenalicola da Cunha Ramos & Brunhes
2004: 207 (M*, F, P*, L*). **Type loc.:** La base des Feuilles de Ravenala Province de Diégo-Suarez/Antsiranana Parc National de Masoala, Antalaha, Madagascar (?) (LTV). **Distr.:** Madagascar (includes Glorioso and Juan De Nova Isl.). **Ety.:** palm leaves; water in palm leaves of the genus *Ravenala*. **Informal name:** Palm-dwelling Malagasy Heaven Mosquito [*Ravenala* is a genus of palm tree].

recondita Edwards
1922d: 91 (M, F). **Type loc.:** Karwar, N Kanara District [Bombay], India (NHMUK). **Distr.:** India, Nepal, Sri Lanka, Thailand. **Refs.:** Barraud & Covell 1928: 676 (F*); Barraud 1934: 83 (M*, F, L*); Qutubuddin 1951 (M*); Darsie & Pradhan 1990 (F, L; tax., keys, bion., distr.; Nepal). **Ety.:** hidden, concealed (L); not easy to recognize. **Informal name:** Hidden Indian Heaven Mosquito.

reinerti Peyton
1977: 116 (M*, F*). **Type loc.:** Mount Kinabalu, Sabah [Malaysian Borneo] (USNM). **Distr.:** Malaysia (includes Sabah and Sarawak). **Ety.:** John Francis Reinert. **Informal name:** Reinert Malaysian Heaven Mosquito.

rickenbachi da Cunha Ramos
1993: 275 (M*; key, tax., bion., distr.). **Type loc.:** Ototomo, Cameroon (SSC). **Distr.:** Cameroon. **Ety.:** André Rickenbach. **Informal name:** Rickenbach Cameroonian Heaven Mosquito.

rossi Delfinado
1966b: 36 (M*, F). **Type loc.:** San Jose, Mindoro, Philippines (USNM). **Distr.:** Malaysia, Philippines. **Refs.:** Basio 1971b: 92 (M*; bion.); Peyton 1977: 159 (M*, F*,

P*, L*; distr.). **Ety.:** Rossiya [Russia], from Russia. **Informal name:** Rossiya Philippine Heaven Mosquito.

satpuraensis Natarajan, Rajavel, & Jambulingam 2017: 259 (M, F, P*, L*). **Type loc.:** Chhattisgarh, Korba District, Lapha, India (Mosquito Museum, Vector Control Research Centre, Puducherry, India). **Distr.:** India. **Ety.:** Satpura Hill Range, Chhattisgarh State, India. **Informal name:** Satpura Indian Heaven Mosquito.

scutostriata da Cunha Ramos & Brunhes 2004: 137 (M*, F, P*, L*). **Type loc.:** Ambodimanga, Moramanga, Madagascar (LTV). **Distr.:** Madagascar (includes Glorioso and Juan De Nova Isl.). **Ety.:** scutum striated (L); ref. to longitudinal stripes on scutum. **Informal name:** Striated Malagasy Heaven Mosquito.

shillitonis Edwards 1932b: 561 (M, F). **Type loc.:** Nyakasura, Uganda (NHMUK). **Distr.:** Angola, Cameroon, Côte d'Ivoire, Democratic Republic of the Congo, Guinea, Liberia, Madagascar (includes Glorioso and Juan De Nova Isl.), Mozambique, Republic of the Congo, Sudan and South Sudan, Tanzania, Togo, Uganda. **Refs.:** Edwards 1941: 62 (M*, F), 371 (P*); Doucet 1951a: 82 (L*); Hopkins 1952: 68 (L*); Worth & De Meillon 1960: 235 (distr.); da Cunha Ramos 1993: 278 (M*, F, P*, L*; keys, tax., bion., distr.). **Ety.:** J.F. Shillito. **Informal name:** Shillito Ugandan Heaven Mosquito.

signata da Cunha Ramos 1993: 219 (P*; key, tax., bion., distr.). **Type loc.:** Dundo, Angola (IHMT). **Distr.:** Angola. **Ety.:** mark (L); distinguishing mark on pupal paddle. **Informal name:** Marked Angolan Heaven Mosquito.

spiculosa Peyton & Rattanarithikul 1970: 409 (M, F, P, L). **Type loc.:** Khao Mai Ha Wa, Chon Buri, Thailand (USNM). **Distr.:** Cambodia, Thailand, Vietnam. **Refs.:** Peyton 1977: 162 (M*, F*, P*, L*; distr.); Rattanarithikul et al. 2006a (F*, L*; bion., distr., keys). **Ety.:** n.s. [spiculose (L)]. **Informal name:** Spiny Thai Heaven Mosquito.

spinitubus da Cunha Ramos & Brunhes 2004: 142 (L*). **Type loc.:** Onive (150 m), Madagascar (LTV). **Distr.:** Madagascar (includes Glorioso and Juan De Nova Isl.). **Ety.:** spine tube (L); spiny siphon and posterior segments. **Informal name:** Spike-piped Malagasy Heaven Mosquito.

spinosa da Cunha Ramos & Brunhes 2004: 317 (M*, P*, L*). **Type loc.:** Antalaha? [Madagascar] (LTV). **Distr.:** Madagascar (includes Glorioso and Juan De Nova Isl.). **Ety.:** spinose (L); ref. to very spinose larvae. **Informal name:** Spiky Malagasy Heaven Mosquito.

spiraculata da Cunha Ramos & Brunhes 2004: 281 (M*, F, P*, L*). **Type loc.:** Beforona (29 km à l'est de Périnet/Andasibe (LTV). **Distr.:** Madagascar (includes Glorioso and Juan De Nova Isl.). **Ety.:** spiracle (L); ref. to large patch of white scales on postspiracular area. **Informal name:** Spiracle Malagasy Heaven Mosquito.

srilankensis Peyton 1974: 222 (M*, F). **Type loc.:** Wilpattu, Hunuwilagama, Anuradhapura District, North Central Province, Sri Lanka (USNM). **Distr.:** Sri Lanka. **Refs.:** Peyton et al. 1979 (M*, F*, P*, L*; bion.). **Ety.:** Sri Lanka. **Informal name:** Ravana Sri Lankan Heaven Mosquito [in the Indian epic *The Ramayana*, Ravana is the demon king of Lanka (Sri Lanka)].

stricklandi Barraud 1926: 345 (M*, F). **Type loc.:** Balasum River, near Kurseong, Eastern Himalayas [West Bengal], India (NHMUK). **Distr.:** India, Nepal, People's Republic of China, Thailand. **Refs.:** Peyton 1977: 166 (M*, F, P*, L*; distr.); Darsie & Pradhan 1990 (F, L; tax., keys, bion., distr.; Nepal); Rattanarithikul et al. 2006a (F*, L*; bion., distr., keys). **Ety.:** Cyril Strickland. **Informal name:** Strickland Indian Heaven Mosquito.

sumethi Peyton & Rattanarithikul 1970: 411 (M, F, P, L). **Type loc.:** Gang Lawa Cave, Huey Bong Ti, Kanchanaburi, Thailand (USNM). **Distr.:** Thailand. **Refs.:** Peyton 1977: 170 (M*, F*, P*, L*); Rattanarithikul et al. 2006a (F*, L*; bion., distr., keys). **Ety.:** Sumeth Chunchulcherm. **Informal name:** Sumeth Thai Heaven Mosquito.

tanakai Miyagi & Toma 2013 (M*, F, P*, L*; key to sp. of *Ur.* (*Pseudoficalbia*) of the Ryukyu Archipelago). **Type loc.:** Otomi, Iriomote Island, Ryukyu Archipelago, Japan (elevation 50–100 m above sea level). **Distr.:** Japan. **Ety.:** Azusa Tanaka. **Informal name:** Tanaka Japanese Heaven Mosquito.

tanzaniae da Cunha Ramos 1993: 287 (M*, F; keys, tax., bion., distr.). **Type loc.:** Amani, Tanzania (OEATOM-Bondy, France). **Distr.:** Tanzania. **Ety.:** Tanzania. **Informal name:** Hadza Tanzanian Heaven Mosquito [the Hadza are thought to be the earliest inhabitants of what is modern Tanzania].

tricolor da Cunha Ramos & Brunhes 2004: 322 (L*). **Type loc.:** Sainte-Luce and Manafiafy, Madagascar (LTV). **Distr.:** Madagascar (includes Glorioso and Juan De Nova Isl.). **Ety.:** three colored (L); ref. to red, blue, and white dorsal larval coloration. **Informal name:** Tricolored Malagasy Heaven Mosquito.

tridentata da Cunha Ramos & Brunhes 2004: 287 (L*). **Type loc.:** Tananarive, Antananarivo, Tamatave, and Toamasina, Lakato, Madagascar (LTV). **Distr.:** Madagascar (includes Glorioso and Juan De Nova Isl.). **Ety.:** three teeth (L); three well-developed teeth on mentum of larva. **Informal name:** Three-toothed Malagasy Heaven Mosquito.

tsaratananae Doucet
1950: 42 (L*). **Type loc.:** Mont Tsaratanana, Sambirano, Madagascar (IRSM). **Distr.:** Madagascar (includes Glorioso and Juan De Nova Isl.). **Ety.:** Mt. Tsaratanana, Madagascar. **Informal name:** Tsaratanana Malagasy Heaven Mosquito.

ugandae da Cunha Ramos
1993: 290 (M*; key, tax., bion., distr.). **Type loc.:** Floresta Kalinzu, Uganda (NHMUK). **Distr.:** Uganda. **Ety.:** Uganda. **Informal name:** Kalinzu Ugandan Heaven Mosquito.

unguiculata Edwards
1913a: 51 (M). **Type loc.:** Tiberias, Israel (IM). **Distr.:** Afghanistan, Albania, Algeria, Armenia, Austria, Azerbaijan, Bulgaria, Crimea, Croatia, Czech Republic, Egypt, France (includes Corsica), FYRO Macedonia, Georgia, Germany, Greece, Hungary, India, Iran, Iraq, Israel (and Gaza Strip and West Bank), Italy (includes Sardinia and Sicily), Jordan, Kazakhstan, Kosovo, Kyrgyzstan, Lebanon, Malta, Moldova, Montenegro, Morocco, Pakistan, Portugal, Romania, Russia, Saudi Arabia, Serbia, Slovakia, Spain, Syria, Tajikistan, Tunisia, Turkey, Turkmenistan, Ukraine, Uzbekistan. **Refs.:** Séguy 1924: 64 (L*); Barraud 1934: 67 (M*, F, P, L*); Stackelberg 1937: 85 (M*, F*); Monchadskii 1951: 112 (L*); Harant et al. 1952: 407 (A*, L*; bion.); Dow 1953: 689 (distr.; Iran); Senevet & Andarelli 1959: 222 (M, F, P*, L*); Gutsevich et al. 1974: 130 (M*, F, L*); Ryba et al. 1974: 142 (distr.; Czechoslovakia); Amr et al. 1997 (distr.; Jordan); Ramsdale & Snow 2001: 29 (distr.; Europe); Rueda, Pecor et al. 2008: 397 (distr.); Becker et al. 2010: 312 (M*, F*, L*; keys, tax., distr., bion.); Melero-Alcíbar et al. 2010 (P; key); Osório et al. 2010 (distr.; Portugal); Robert et al. 2019 (distr.; Western Palearctic). **Ety.:** fingernail or toenail (L); ref to unusual front tarsal claws. **Informal name:** Fingernail Israeli Heaven Mosquito.

ssp. ***pefflyi*** Stone
1961a: 249 (M, F). **Type loc.:** Qatif Oasis, Al Hasa Province, Saudi Arabia (USNM). **Distr.:** Saudi Arabia. **Ety.:** R.L. Peffly. **Informal name:** Peffly Saudi Heaven Mosquito.

xanthomelaena Edwards
1925: 259 (M, L). **Type loc.:** Kuala Lumpur [Selangor], Malaya [Malaysia] (NHMUK). **Distr.:** Malaysia, Singapore. **Refs.:** Peyton 1977: 118 (M*, F, P*, L*). **Ety.:** n.s. [yellowish and dark (Gr); poss. ref. to overall yellow coloration]. **Informal name:** Yellow-black Malaysian Heaven Mosquito.

yaeyamana Tanaka, Mizusawa & Saugstad
1975a: 31 (M*, F, L*). **Type loc.:** Funaura, Iriomote Island, Ryukyu Islands [Japan] (NMST). **Distr.:** Japan. **Refs.:** Peyton 1977: 174 (M*, F*, P*, L*); Tanaka et al. 1979: 460 (M*, F, L*); Qu & Qian 1993 (cibarial armature*). **Ety.:** Yaeyama Guntô, Ryukyu Isl., Japan. **Informal name:** Yaeyama Japanese Heaven Mosquito.

yovani van Someren
1951: 6 (M*, F, P*, L*). **Type loc.:** Bwamba, Toro District, Uganda (NHMUK). **Distr.:** Cameroon, Democratic Republic of the Congo, Liberia, Uganda. **Refs.:** da Cunha Ramos 1993: 222 (M*, F, P*, L*; keys, tax., bion., distr.). **Ety.:** Yovani Senkubuge. **Informal name:** Yovani Ugandan Heaven Mosquito.

Subgenus *Uranotaenia* Lynch Arribálzaga

1891a: 375, 1891b: 163 (as genus, subgenus). Type species: *Uranotaenia pulcherrima* Lynch Arribálzaga (Neveu-Lemaire 1902c: 227).

Name derivation: [heaven or sky ribbon (Gr); prob. ref. to bright ornamentation].

syn. *Anisocheleomyia* Theobald 1905a: 52. Type species: *Anisocheleomyia nivipes* Theobald (Brunetti 1914: 55).
syn. *Pseudouranotaenia* Theobald 1905g: 33 (as genus). Type species: *Pseudouranotaenia rowlandi* Theobald.

Important references: Belkin, Schick, & Heinemann 1965 (type info.; New World); Belkin, Schick, & Heinemann 1966 (type info.; New World); Belkin, Schick, & Heinemann 1968 (type info.; New World); Belkin, Schick, & Heinemann 1971 (type info.; New World); Lee et al. 1989c (tax., bion., key; Australasia); Service 1990 (tax., bion., keys; Afrotropics); da Cunha Ramos 1993 (keys; Afrotropics); da Cunha Ramos & Brunhes 2004 (tax., bion., keys; Madagascar); Rattanarithikul et al. 2006a (tax., bion., keys; Thailand).

aequatorianna Leví-Castillo
1953c: 14 (M, F). **Type loc.:** Babahoyo, Los Rios, Ecuador (NE). **Distr.:** Ecuador. **Refs.:** Galindo et al. 1954: 118 (tax.); Leví-Castillo 1954a: 84 (M*). **Ety.:** n.s. [equator? (L)]. **Informal name:** Equatorial Ecuadorian Heaven Mosquito.

alba Theobald
1901c: 262 (M; *mashonaenis* var.). **Type loc.:** Salisbury, Mashonaland [Harare, Zimbabwe] (NHMUK). **Distr.:** Angola, Burkina Faso, Cameroon, Democratic Republic of the Congo, Equatorial Guinea, Ethiopia, Kenya, Madagascar (includes Glorioso and Juan De Nova Isl.), Nigeria, Republic of the Congo, South Africa, Sudan and South Sudan, Uganda, Zambia, Zimbabwe. **Refs.:** Edwards 1941: 49 (M, F), 365 (P*); Hopkins 1952: 54 (L); da Cunha Ramos & Ribeiro 1975: 128 (L*; tax., bion., keys, distr.; Angola); da Cunha Ramos 1993: 294 (M*, F, P*, L*; keys, tax., bion., distr.); Jupp 1996 (M*, F*; key). **Ety.:** n.s. [white (L)]. **Informal name:** White Zimbabwean Heaven Mosquito.

albescens Taylor
1914c: 705 (M*, F*). **Type loc.:** Townsville, Queensland, Australia (ANIC). **Distr.:** Australia, Papua New Guinea.

Refs.: Lee 1944a: 32 (L*); Peters 1964: 27 (A, P*, L); Lee et al. 1989c: 223 (F; key, tax., bion., distr., review, type info.); . **Ety.:** n.s. [becoming white (L)]. **Informal name:** White Australian Heaven Mosquito.

alboabdominalis Theobald
1910c: 508 (M*, F*). **Type loc.:** "Bor to Mongalla," Sudan (NHMUK). **Distr.:** Angola, Benin, Burkina Faso, Cameroon, Comoros, Democratic Republic of the Congo, Ghana, Kenya, Liberia, Madagascar (includes Glorioso and Juan De Nova Isl.), Mozambique, Nigeria, Republic of the Congo, Sudan and South Sudan, Tanzania, Uganda. **Refs.:** Edwards 1941: 48 (M, F), 365 (P); Hopkins 1952: 53 (L*); Worth & De Meillon 1960: 234 (distr.); Brunhes 1977a (distr.; Comoros Isl.); da Cunha Ramos 1993: 307 (M*, F, P*, L*; keys, tax., bion., distr.); Jupp 1996 (M*, F*; key). **Ety.:** white abdomen (L); "known by the white basal area of the abdomen." **Informal name:** White-bellied Sudanese Heaven Mosquito.

alboannulata (Theobald)
1905a: 54 (M*; *Anisocheleomyia*). **Type loc.:** India (NHMUK). **Distr.:** India. **Refs.:** Barraud 1934: 62 (M*, F). **Ety.:** n.s. [white ringed (L); prob. ref. to either a narrow white line around the eyes or to the hind legs with broad apical white bands]. **Informal name:** White-ringed Indian Heaven Mosquito.

albocephala da Cunha Ramos & Brunhes
2004: 334 (F, P*, L*). **Type loc.:** Onive River (altitude 150 m), Antalaha, Madagascar (LTV). **Distr.:** Madagascar (includes Glorioso and Juan De Nova Isl.). **Ety.:** white head (L); ref. to white scales on the vertex of the head. **Informal name:** White-headed Malagasy Heaven Mosquito.

albosternopleura Peters
1963e: 203 (M, F*). **Type loc.:** Maprik, [East] Sepik District, New Guinea [Papua New Guinea] (CSIR). **Distr.:** Australia, Papua New Guinea. **Refs.:** Marks 1970: 165 (distr.); Lee et al. 1989c: 229 (F; key, tax., bion., distr., review). **Ety.:** n.s. [white sterna and pleura (L); ref. to "distinct patch of silvery-white scales on upper sternopleuron"]. **Informal name:** White-sided Papuan Heaven Mosquito.

amiensis Peters
1963e: 207 (M*, F*). **Type loc.:** Ami, 10 mi northwest of Maprik, [East] Sepik District, New Guinea [Papua New Guinea] (CSIR). **Distr.:** Australia, Indonesia, Papua New Guinea. **Refs.:** Marks 1970: 165 (distr.); Lee et al. 1989c: 231 (F; key, tax., bion., distr., review). **Ety.:** Ami, Papua New Guinea. **Informal name:** Ami Papuan Heaven Mosquito.

andavakae Doucet
1950: 44 (L*). **Type loc.:** Mont Tsaratanana, Sambirano, Madagascar (IRSM). **Distr.:** Comoros, Madagascar (includes Glorioso and Juan De Nova Isl.). **Refs.:** Brunhes 1977a (distr.; Comoros Isl.). **Ety.:** abbrev. for village of Andavakinimenarana, Madagascar. **Informal name:** Andavaka Malagasy Heaven Mosquito.

angolensis da Cunha Ramos
1985: 74 (M*, F*). **Type loc.:** Dundo, Angola, (IHMT). **Distr.:** Angola. **Refs.**: da Cunha Ramos 1993: 359 (M*, F; keys, tax., bion., distr.); **Ety.:** Angola. **Informal name:** Ndongo Angolan Heaven Mosquito [Angola derives its name from the ancient Bantu kingdom of Ndongo].

annandalei Barraud
1926: 343 (M*, F). **Type loc.:** Golaghat, Assam, India (NHMUK). **Distr.:** Cambodia, India, Japan, Myanmar, Nepal, People's Republic of China, Philippines, Taiwan, Thailand, Vietnam. **Refs.:** Edwards 1932b: 559 (L*); Baisas 1935b: 64 (M*, P, L*); Peters & Dewar 1956: 43 (P*?); Delfinado 1966a: 39 (M*, F, P*, L*); Basio 1971b: 89 (M*; bion.); Baisas 1974: 155 (M*, F, P*, L*; tax., bion., distr.; Philippines); Tanaka et al. 1975c: 213 (distr.); Tanaka et al. 1979: 469 (M*, F*, L*); Darsie & Pradhan 1990 (F, L; tax., keys, bion., distr.; Nepal); Qu & Qian 1993 (cibarial armature*); Rattanarithikul et al. 2006a (F*, L*; bion., distr., keys). **Ety.:** Thomas Nelson Annandale. **Informal name:** Annandale Indian Heaven Mosquito.

syn. *nanseica* Bohart & Ingram
1946b: 56 (M; tax.). **Type loc.:** Chizuka, Okinawa, Ryukyu Retto [Japan] (USNM). **Refs.:** Bohart 1953: 184 (syn.).

anopheloides Brunhes & Razafindrasolo
1975 (M*, F, P*, L*). **Type loc.:** Maevetanana, Majunga Province, Madagascar (IERT). **Distr.:** Madagascar (includes Glorioso and Juan De Nova Isl.). **Ety.:** *Anopheles*-like (Gr). **Informal name:** *Anopheles*-like Malagasy Heaven Mosquito.

antennalis Taylor
1919: 840 (F). **Type loc.:** Cairns, Queensland, Australia (US). **Distr.:** Australia, Papua New Guinea. **Refs.:** King & Hoogstraal 1947c: 595 (? = *tibialis*); Lee et al. 1989c: 233 (tax., bion., distr., review). **Ety.:** n.s. [antenna (L); attribute?]. **Informal name:** Antenna Australian Heaven Mosquito.

apicalis Theobald
1903a: 298 (M, F*). **Type loc.:** Antigua [Lesser Antilles] (NHMUK). **Distr.:** Argentina, Brazil, Colombia, Costa Rica, Ecuador, French Guiana, Guatemala, Guyana, Honduras, Lesser Antilles (includes Antigua and Barbuda, Trinidad and Tobago), Mexico, Nicaragua, Panama, Paraguay, Peru, Suriname, Venezuela. **Refs.:** Lane 1943: 142 (M*, F); Galindo et al. 1954: 119 (M*, F, P*, L*); Belkin 1968b: 37 (lectotype desig.). **Ety.:** n.s. [apex (L); poss. ref. to abdominal apical white bands or apical white spots on hind legs]. **Informal name:** Apex Antiguan Heaven Mosquito.

argentipleura da Cunha Ramos & Brunhes
2004: 349 (M*, F). **Type loc.:** Nosy Be, Madagascar (LTV). **Distr.:** Madagascar (includes Glorioso and Juan De Nova Isl.). **Ety.:** silvery pleura (Gr); thorax. **Informal name:** Silver-sided Malagasy Heaven Mosquito.

arguellesi Baisas
1935b: 68 (M*, F, L*). **Type loc.:** Calauan, Laguna, Luzon, Philippines (NE). **Distr.:** Philippines. **Refs.:** Delfinado 1966a: 40 (M*, F, P*, L*); Basio 1971b: 89 (M*; bion.); Baisas 1974: 156 (M*, F, P*, L*; tax., bion., distr.; Philippines). **Ety.:** Angel S. Argüelles. **Informal name:** Argüelles Philippine Heaven Mosquito.

argyrotarsis Leicester
1908: 214 (M, F). **Type loc.:** Gombak Road, Kuala Lumpur [Selangor], Malaya [Malaysia] (NHMUK). **Distr.:** Australia, Indonesia, Malaysia, Papua New Guinea, Philippines. **Refs.:** Baisas 1935b: 65 (M*, L*); Knight & Chamberlain 1948: 14 (P*); Penn 1949b: 33 (P*); Peters 1964: 25 (M*, F, P, L); Basio 1971b: 90 (M*, L*; bion.); Lee et al. 1989c: 235 (F; key, tax., bion., distr., review). **Ety.:** silver foot (Gr); tarsi. **Informal name:** Silver-footed Malaysian Heaven Mosquito.
syn. *parangensis* Ludlow
1909a: 24 (F; *Pseudouranotaenia*). **Type loc.:** Parang [Cotabato], Mindanao, Philippines (USNM).

balfouri Theobald
1904b: 82 (F*). **Type loc.:** Pibor River [Upper Nile], Sudan (NHMUK). **Distr.:** Angola, Benin, Burkina Faso, Cameroon, Central African Republic, Côte d'Ivoire, Democratic Republic of the Congo, Djibouti, Ethiopia, Gabon, Ghana, Guinea, Kenya, Liberia, Madagascar (includes Glorioso and Juan De Nova Isl.), Malawi, Mali, Mauritania, Morocco, Mozambique, Namibia, Niger, Nigeria, Republic of the Congo, São Tomé and Príncipe, Senegal, Sierra Leone, South Africa, Sudan and South Sudan, Tanzania, The Gambia, Togo, Uganda, Zambia. **Refs.:** Edwards 1941: 52 (M*, F), 366 (P*); Hopkins 1952: 54 (L*); Bailly-Choumara 1968: 142 (distr.); Correia 1968: 164 (distr.); da Cunha Ramos & Ribeiro 1975: 128 (tax., bion., keys, distr.; Angola); Service 1976a (distr.; Gabon); Rodhain et al. 1977 (distr.); da Cunha Ramos 1993: 327 (M, F, P*, L*; keys, tax., bion., distr.); Jupp 1996 (M*, F*; key). **Ety.:** n.s. [for Dr. Andrew Balfour (Harbach 2018c)]. **Informal name:** Balfour Sudanese Heaven Mosquito.

barnesi Belkin
1953a: 332 (M*, F*, P*, L*). **Type loc.:** Togiri swamp, Guadalcanal, Solomon Islands (USNM). **Distr.:** Papua New Guinea, Solomon Isl., Vanuatu. **Refs.:** Perry 1946: 11 (F, L; as *tibialis*); Belkin 1962: 172 (M*, F*, P*, L*); Maffi & Taylor 1974: 207 (distr.); Lee et al. 1989c: 240 (tax., bion., distr., review). **Ety.:** Arthur W. Barnes Jr. **Informal name:** Barnes Solomon Islander Heaven Mosquito.

benoiti Wolfs
1964 (L). In Benoit 1964: 309. **Type loc.:** Mwendua River, Kasongo, Congo [Democratic Republic of the Congo] (CMT). **Distr.:** Democratic Republic of the Congo, Republic of the Congo. **Refs.**: da Cunha Ramos 1993: 363 (F, L*; keys, tax., bion., distr.). **Ety.:** P.L.G. Benoit. **Informal name:** Benoit Congolese Heaven Mosquito.

bertii Cova García & Rausseo
1964: 221 (M*, F, P*, L*). **Type loc.:** Cumbre de Choroni, Aragua, Venezuela (DERM). **Distr.:** Lesser Antilles (includes Trinidad and Tobago), Venezuela. **Ety.:** Arturo Luis Berti. **Informal name:** Berti Venezuelan Heaven Mosquito.

bidentata da Cunha Ramos & Brunhes
2004: 355 (L*). **Type loc.:** Manambolosy, Tamatave/Toamasina Province, Madagascar (LTV). **Distr.:** Madagascar (includes Glorioso and Juan De Nova Isl.). **Ety.:** n.s. [two toothed]. **Informal name:** Two-toothed Malagasy Heaven Mosquito.

bilineata Theobald
1909b: 21 (M, F). **Type loc.:** Obuasi, Ashanti, Gold Coast [Ghana] (NHMUK). **Distr.:** Angola, Cameroon, Côte d'Ivoire, Democratic Republic of the Congo, Ghana, Guinea, Liberia, Mozambique, Nigeria, Republic of the Congo, South Africa, Uganda, Zambia. **Refs.:** Edwards 1941: 50 (M*, F), 366 (P); Hopkins 1952: 54 (L); Worth & De Meillon 1960: 234 (distr.); da Cunha Ramos & Ribeiro 1975: 128 (tax., bion., keys, distr.; Angola); da Cunha Ramos 1993: 336 (M*, F, P*, L*; keys, tax., bion., distr.); Jupp 1996 (M*, F*; key). **Ety.:** two lined (L); "two pale lateral parallel lines on the sides of the thorax." **Informal name:** Bilinear Ghanaian Heaven Mosquito.

bimaculiala Leicester
1908: 208 (M, F). **Type loc.:** Raub [Pahang], Malaya [Malaysia] (NHMUK). **Distr.:** Cambodia, Indonesia, Malaysia, Thailand. **Refs.:** Iyengar & Menon 1956: 785 (A*, L*); Mattingly 1970c: 62 (E*); Rattanarithikul et al. 2006a (F*, L*; bion., distr., keys). **Ety.:** two spotted (L); wing. **Informal name:** Two-spotted Malaysian Heaven Mosquito.

bricenoi Cova García, Pulido, de Ugueto, Amarista, & Mora
1987 (M*, P*, L*). **Type loc.:** Cata Public Beach, Ocurare de las Costa Municipality, Venezuela (DERM). **Distr.:** Venezuela. **Ety.:** Leopoldo Briceño Iragorry. **Informal name:** Briceño Venezuelan Heaven Mosquito.

briseis Dyar
1925g: 216 (F). **Type loc.:** Catatumbo River [Zulia], Venezuela (USNM). **Distr.:** Brazil, Lesser Antilles (includes Trinidad and Tobago), Panama, Venezuela. **Ety.:** n.s. **Informal name:** Briseis Venezuelan Heaven Mosquito [Briseis appears as a character in Homer's *Iliad*].

caeruleocephala Theobald
1901d: 14 (F). **Type loc.:** Old Calabar [Eastern Provinces], Nigeria (NHMUK). **Distr.:** Angola, Cameroon, Democratic Republic of the Congo, Liberia, Nigeria, Republic of the Congo, Sudan and South Sudan, Uganda. **Refs.:** Edwards 1941: 47 (M, F*; as *coeruleocephala*); da Cunha Ramos 1993: 367 (M*, F; keys, tax., bion., distr.). **Ety.:** sky blue (L, Gr); ref. to color of head. **Informal name:** Blue-headed Nigerian Heaven Mosquito.

caliginosa Philip
1931: 190 (M, F). **Type loc.:** Ebute Metta [Western Provinces], Nigeria (NHMUK). **Distr.:** Angola, Benin, Cameroon, Côte d'Ivoire, Democratic Republic of the Congo, Liberia, Nigeria, Republic of the Congo. **Refs.**: da Cunha Ramos 1993: 372 (M*, F; keys, tax., bion., distr.). **Ety.:** n.s. [dark (L)]. **Informal name:** Murky Nigerian Heaven Mosquito.

calosomata Dyar & Knab
1907c: 200 (A). **Type loc.:** Tabernilla, Canal Zone, Panama (USNM). **Distr.:** Bolivia, Brazil, Colombia, Costa Rica, Ecuador, French Guiana, Guyana, Lesser Antilles (includes Trinidad and Tobago), Panama, Peru, Venezuela. **Refs.:** Lane 1953: 572 (M*, F, L*); Galindo et al. 1954: 132 (M*, F; tax.); Stone & Knight 1957c: 200 (type info., lectotype desig.); Linton et al. 2013 (bion., mol. tax., distr.; Ecuador); Berti et al. 2015 (distr.; Venezuela). **Ety.:** n.s. [beautiful body (Gr)]. **Informal name:** Beautiful Panamanian Heaven Mosquito.
syn. *albitarsis* Gordon & Evans
1922: 335 (M, F; *calosomata* var.). **Type loc.:** Macapá, Amazonas, Brazil (NHMUK). **Refs.:** Lane 1943: 155 (syn.); Belkin 1968b: 37 (lectotype desig.).

campestris Leicester
1908: 213 (M, F). **Type loc.:** Malaya [Malaysia] (NHMUK). **Distr.:** Bangladesh, Cambodia, India, Indonesia, Malaysia, Nepal, Sri Lanka, Thailand, Timor, Vietnam. **Refs.:** Senior-White 1927: 66 (L*); Barraud 1934: 68 (M*, F, L); Peters & Dewar 1956: 42 (P*, L); Ahmed 1987 (distr.; Bangladesh); Lee et al. 1989c: 243 (tax., bion., distr., review); Darsie & Pradhan 1990 (F, L; tax., keys, bion., distr.; Nepal); Rattanarithikul et al. 2006a (F*, L*; bion., distr., keys). **Ety.:** n.s. [swampy open ground (L); poss. ref. to larval habitat]. **Informal name:** Field Malaysian Heaven Mosquito.

chorleyi Edwards
1936: 54 (A). **Type loc.:** Kasala Forest, Uganda (NHMUK). **Distr.:** Angola, Burkina Faso, Cameroon, Côte d'Ivoire, Democratic Republic of the Congo, Ghana, Guinea, Liberia, Mali, Nigeria, Republic of the Congo, South Africa, Sudan and South Sudan, Tanzania, Togo, Uganda, Zambia. **Refs.:** Edwards 1941: 53 (M*), 367 (P); Hopkins 1952: 57 (L*); Peters 1955c: 120 (P*, L*); da Cunha Ramos & Ribeiro 1975: 130 (tax., bion., keys, distr.; Angola); da Cunha Ramos 1993: 407 (M*, F, P*, L*; keys, tax., bion., distr.); Jupp 1996 (M*, F*; key). **Ety.:** T.W. Chorley. **Informal name:** Chorley Ugandan Heaven Mosquito.

christophersi Barraud
1926: 336 (M*, F). **Type loc.:** Andaman Islands (NHMUK). **Distr.:** India. **Refs.:** Barraud 1934: 64 (M*, F). **Ety.:** Samuel Rickard Christophers. **Informal name:** Christophers Indian Heaven Mosquito.

civinskii Belkin
1953a: 341 (M*, F*, P*, L*). **Type loc.:** Bonegi River Valley, Guadalcanal, Solomon Islands (USNM). **Distr.:** Papua New Guinea, Solomon Isl. **Refs.:** Belkin 1962: 168 (M*, F*, P*, L*); Lee et al. 1989c: 245 (tax., bion., distr., review). **Ety.:** Stanley Civinksi. **Informal name:** Civinski Solomon Islander Heaven Mosquito.

clara Dyar & Shannon
1925a: 68 (M). **Type loc.:** Ludlow Barracks [Parang, Cotabato], Mindanao, Philippines (USNM). **Distr.:** Indonesia, Philippines. **Refs.:** Belkin 1953a: 340 (from syn. with *ludlowae*); Stone & Knight 1957c: 200 (type info., lectotype desig.); Delfinado 1966a: 42 (M*, P*, L*); Basio 1971b: 90 (M*; bion.); Lee et al. 1989c: 248 (distr.). **Ety.:** n.s. [clear (L), or poss. named for Clara Ludlow]. **Informal name:** Clara Philippine Heaven Mosquito.
syn. *delae* Baisas
1935b: 73 (F*, L*). **Type loc.:** Salimbao, Cotabato, Mindanao, Philippines (NE). **Refs.:** Belkin 1953a: 340 (M; syn.).

coatzacoalcos Dyar & Knab
1906d: 186 (L). **Type loc.:** Santa Lucrecia [Veracruz], Mexico (USNM). **Distr.:** Colombia, Costa Rica, El Salvador, Guatemala, Honduras, Mexico, Panama, Venezuela. **Refs.:** Lane 1953: 572 (M*, F, L); Galindo et al. 1954: 136 (M, F, L*); Stone & Knight 1957c: 200 (type info.). **Ety.:** Coatzacoalcos, Mexico. **Informal name:** Coatzacoalcos Mexican Heaven Mosquito.
syn. *basalis* Howard, Dyar, & Knab
1913: Pl. 37, Fig. 251; 1917: 917 (M*, L). **Type loc.:** Cordoba [Veracruz], Mexico (USNM).

connali Edwards
1912c: 39 (M*, F). **Type loc.:** Accra, Gold Coast (NHMUK). **Distr.:** Angola, Benin, Cameroon, Democratic Republic of the Congo, Ghana, Liberia, Nigeria, Republic of the Congo, Sierra Leone, Uganda, Zambia, Zimbabwe. **Refs.:** Edwards 1941: 51 (M*; to var.); da Cunha Ramos 1993: 345 (from var. of *bilineata*); Jupp 1996 (M*, F*; key). **Ety.:** A.C. Connal. **Informal name:** Connal Ghanaian Heaven Mosquito.

cooki Root
1937: 98 (M*, F). **Type loc.:** Port au Prince [Ouest], Haiti (USNM). **Distr.:** Cuba, Haiti, Jamaica, Puerto Rico, Virgin Isl. **Refs.:** Pratt 1947: 578 (M*, P*, L*); Lane 1953: 581 (M*, F); Stone & Knight 1957c: 200 (type info., lectotype desig.);

Gutsevitch & García-Avila 1969: 4 (distr.); Belkin et al. 1970: 41 (M*, F*, P*, L*); da Cunha Ramos 1993: 345 (M*, F; keys, tax., bion., distr.). **Ety.:** James Cook. **Informal name:** Cook Haitian Heaven Mosquito.

davisi Lane
1943: 158 (M*, F). **Type loc.:** Salvador, Bahia, Brazil (FH). **Distr.:** Argentina, Bolivia, Brazil, Colombia. **Ety.:** Nelson Caryl Davis. **Informal name:** Davis Brazilian Heaven Mosquito.

diraphati Peyton & Klein
1970: 245 (M, F). **Type loc.:** Turien Nok, Narathiwat, Thailand (USNM). **Distr.:** Cambodia, Malaysia, Thailand. **Refs.:** Rattanarithikul et al. 2006a (F*, L*; bion., distr., keys). **Ety.:** Chaliou Diraphat. **Informal name:** Diraphat Thai Heaven Mosquito.

ditaenionota Prado
1931b: 209 (M*, F). **Type loc.:** Instituto Butantan Garden, São Paulo, Brazil (IB). **Distr.:** Argentina, Bolivia, Brazil, Colombia, Panama, Paraguay, Venezuela. **Refs.:** Lane 1943: 158 (M*, F, L*); Galindo et al. 1954: 138 (M*, F). **Ety.:** n.s. **Informal name:** Butantan Brazilian Heaven Mosquito.
syn. *burkii* Lane
1936b: 194 (M*, F, L*). **Type loc.:** near Ponce, Mato Grosso, Brazil (FH). **Refs.:** Lane 1943: 158 (syn.).

dumonti Doucet
1949b: 327 (L*). **Type loc.:** Ambohiby, Tsiroanomandidy District, Madagascar (IRSM). **Distr.:** Madagascar (includes Glorioso and Juan De Nova Isl.). **Ety.:** M. l'Administrateur Dumont. **Informal name:** Dumont Malagasy Heaven Mosquito.

edwardsi Barraud
1926: 337 (F). **Type loc.:** Golaghat, Assam, India (NHMUK). **Distr.:** India, Malaysia, Nepal, People's Republic of China, Thailand, Vietnam. **Refs.:** Barraud 1934: 65 (M*, F); Darsie & Pradhan 1990 (F, L; tax., keys, bion., distr.; Nepal); da Cunha Ramos, Ribeiro et al. 2002: 124 (distr.; People's Republic of China); Rattanarithikul et al. 2006a (F*, L*; bion., distr., keys). **Ety.:** Frederick Wallace Edwards. **Informal name:** Edwards Indian Heaven Mosquito.

falcipes Banks
1906: 1004 (M, F). **Type loc.:** Camp 320, Manila Waterworks, Rizal, Luzon, Philippines (NHMUK). **Distr.:** Philippines. **Refs.:** Peyton 1972: 34 (to sp.); Townsend 1990: 72 (type info.). **Ety.:** n.s. [sickle foot (L); poss. ref. to "peculiar falcate ungues in both male and female"]. **Informal name:** Sickle-footed Philippine Heaven Mosquito.

fimbriata King & Hoogstraal
1947c: 592 (M*). **Type loc.:** Hollandia, Western New Guinea (Papua) (USNM). **Distr.:** Indonesia, Papua New Guinea. **Refs.:** Lee et al. 1989c: 249 (tax., bion., distr., review). **Ety.:** fringed (L); ref. to fringes of setae on foretibia and some tarsal segments. **Informal name:** Fringed Indonesian Heaven Mosquito.

fraseri Edwards
1912c: 39 (F). **Type loc.:** Mpumu Forest, Uganda (NHMUK). **Distr.:** Angola, Cameroon, Democratic Republic of the Congo, Ghana, Kenya, Nigeria, Republic of the Congo, Sudan and South Sudan, Tanzania, Uganda. **Refs.:** Ingram & De Meillon 1929: 78 (P*, L*); da Cunha Ramos 1993: 350 (M*, F, P*, L*, E*; keys, tax., bion., distr.; to sp.); Jupp 1996 (M*, F*; key). **Ety.:** A.D. Fraser. **Informal name:** Fraser Ugandan Heaven Mosquito.
syn. *obsoleta* Edwards
1936: 54 (A; *bilineata* var.). **Type loc.:** Kasakiro, Uganda (NHMUK). **Refs.:** Edwards 1941: 51(M); da Cunha Ramos et al. 1993 (syn.).

gabaldoni Cova García, Pulido F., Escalante de Ugueto, & Mora R.
1987: 94 (M*, P*, L*). **Type loc.:** Capital (Caicara River), Apure State, Venezuela (DERM). **Distr.:** Venezuela. **Ety.:** Arnaldo Gabaldón. **Informal name:** Gabaldón Venezuelan Heaven Mosquito.

geniculata da Cunha Ramos & Brunhes
2004: 361 (M*, P*, L*). **Type loc.:** Onive River (altitude 150 m), Antalaha, Madagascar (LTV). **Distr.:** Madagascar (includes Glorioso and Juan De Nova Isl.). **Ety.:** knee, jointed (L); ref. to white tibio-femoral joint. **Informal name:** Jointed Malagasy Heaven Mosquito.

geometrica Theobald
1901c: 247 (F*). **Type loc.:** Cubatão, near Santos [São Paulo], Brazil (NHMUK). **Distr.:** Argentina, Belize, Bolivia, Brazil, Colombia, Costa Rica, Ecuador, El Salvador, French Guiana, Guatemala, Guyana, Honduras, Lesser Antilles (includes Trinidad and Tobago), Mexico, Nicaragua, Panama, Paraguay, Peru, Suriname, Uruguay, Venezuela. **Refs.:** Lane 1943: 144 (M*, F, P*, L*); Knight & Chamberlain 1948: 18 (P*); Galindo et al. 1954: 142 (M*, F, P*, L*); Belkin 1968b: 37 (lectotype desig.); Heinemann & Belkin 1977b: 453 (distr.); Rossi & Martinez 2003: 473 (distr.; Uruguay); Berti et al. 2015 (distr.; Venezuela). **Ety.:** n.s. [geometrical (L)]. **Informal name:** Geometric Brazilian Heaven Mosquito.

gerdae Slooff
1963b: 84 (M*, F, P*, L*). **Type loc.:** Sorong, Klademak, West New Guinea (USNM). **Distr.:** Indonesia (West Papua). **Refs.:** Lee et al. 1989c: 251 (tax., bion., distr., review). **Ety.:** Gerda Sloof. **Informal name:** Gerda Indonesian Heaven Mosquito.

grassei da Cunha Ramos & Brunhes
2004: 366 (F, P*, L*). **Type loc.:** Perinet-Andasibe Reserve Naturelle (1,000 m), Madagascar (LTV). **Distr.:** Madagascar

(includes Glorioso and Juan De Nova Isl.). **Ety.:** Pierre-Paule Grassé. **Informal name:** Grassé Malagasy Heaven Mosquito.

hamoni Grjebine
1953a: 465 (L; as var. of *chorleyi*). **Type loc.:** La Mandraka (District Manjakandriana) [Tananarive], Madagascar (IRSM). **Distr.:** Madagascar (includes Glorioso and Juan De Nova Isl.). **Refs.:** White 1975: 326 (to sp.). **Ety.:** Jacques Pierre Jean Hamon. **Informal name:** Hamon Malagasy Heaven Mosquito.

hebes Barraud
1931a: 609 (F). **Type loc.:** Nongpoh, Assam, India (NHMUK). **Distr.:** India, Taiwan, Thailand. **Refs.:** Qu & Qian 1993 (cibarial armature*); Rattanarithikul et al. 2006a (F*, L*; bion., distr., keys). **Ety.:** n.s. [blunt (L)]. **Informal name:** Blunt Indian Heaven Mosquito.
syn. *nii* Lien
1968a: 10 (M*, F, P*, L*). **Type loc.:** Aoshenshui, Yanchao, Kaohsiung. **Refs.:** Peyton 1972: 34 (syn.).

hebrardi da Cunha Ramos & Brunhes
2004: 376 (L*). **Type loc.:** Manakara, Madagascar (PIP). **Distr.:** Côte d'Ivoire, Democratic Republic of the Congo, Madagascar (includes Glorioso and Juan De Nova Isl.). **Ety.:** Georges Hebrard. **Informal name:** Hebrard Malagasy Heaven Mosquito.

heiseri Baisas
1935b: 72 (F, L*). **Type loc.:** Parang, Cotabato, Mindanao, Philippines (NE). **Distr.:** Philippines. **Refs.:** Basio 1971b: 90 (M*; bion.). **Ety.:** Victor G. Heiser. **Informal name:** Heiser Philippine Heaven Mosquito.

hopkinsi Edwards
1932b: 561 (A). **Type loc.:** Kampala, Uganda (NHMUK). **Distr.:** Democratic Republic of the Congo, Guinea, Mozambique, Nigeria, Republic of the Congo, Togo, Uganda. **Refs.:** Edwards 1941: 54 (F), 368 (P); Hopkins 1952: 57 (L); da Cunha Ramos 1993: 417 (F, P*, L; keys, tax., bion., distr.); Jupp 1996 (M*, F*; key). **Ety.:** George Henry Evans Hopkins. **Informal name:** Hopkins Ugandan Heaven Mosquito.

hystera Dyar & Knab
1913: 78 (M). **Type loc.:** Manoa, Orinoco River, Venezuela (USNM). **Distr.:** Belize, Bolivia, Brazil, Colombia, Costa Rica, Ecuador, French Guiana, Guatemala, Guyana, Nicaragua, Panama, Peru, Venezuela. **Refs.:** Cerqueira 1943b: 24 (M*); Galindo et al. 1954: 146 (M*, F; tax.). **Ety.:** n.s. [womb (Gr)]. **Informal name:** Manoa Venezuelan Heaven Mosquito.
syn. *bicolor* Martini
1935: 29 (F). **Type loc.:** Rio Cacao, British Honduras [Belize] (NHMUK). **Refs.:** Galindo et al. 1954: 146 (syn.).
syn. *martinii* Lane
1943: 152 (new name for *bicolor* Martini, not Leicester 1908).

incognita Galindo, Blanton, & Peyton
1954: 150 (M*, F). **Type loc.:** David, Chiriquí Province, Panama (USNM). **Distr.:** Argentina, Colombia, Ecuador, French Guiana, Guatemala, Lesser Antilles (includes Trinidad and Tobago), Panama. **Ety.:** n.s. [unknown (L)]. **Informal name:** Unknown Panamanian Heaven Mosquito.

iriartei Cova-García, Pulido F., Escalante de Ugueto, & Mora R.
1987: 45 (M*, P*, L*). **Type loc.:** El Yaguel and Mantecal, Apure State, Venezuela (DERM). **Distr.:** Venezuela. **Ety.:** David lriarte. **Informal name:** Iriarte Venezuelan Heaven Mosquito.

joucouri da Cunha Ramos & Brunhes
2004: 379 (L*). **Type loc.:** Foulpointe, Tamatave-Toamasina Province, Madagascar (LTV). **Distr.:** Madagascar (includes Glorioso and Juan De Nova Isl.). **Ety.:** n.s. [Joucour?]. **Informal name:** Joucour Malagasy Heaven Mosquito.

lanei Martínez & Prosen
1953a: 4 (F). **Type loc.:** Formosa, Argentina (BA). **Distr.:** Argentina. **Ety.:** John Lane. **Informal name:** Lane Argentinian Heaven Mosquito.

lateralis Ludlow
1905c: 385 (F; *caeruleocephala* var.). **Type loc.:** Cotabato [Mindanao], Philippines (USNM). **Distr.:** Australia, India, Indonesia, Malaysia, Papua New Guinea, People's Republic of China, Philippines, Solomon Isl., Sri Lanka, Thailand, Timor, Vietnam. **Refs.:** Belkin 1953a: 324 (M*, F*, P*, L*; tax.; as *atra*); Stone 1957b: 173 (tax.; from syn. with *atra*); Stone & Knight 1957c: 200 (type info., lectotype desig.); Belkin 1962: 170 (M*, F*, P*, L*); Mattingly 1970c: 62 (E*); Basio 1971b: 91 (M*; bion.); Baisas 1974: 159 (M*, F, P*, L*; tax., bion., distr.; Philippines); Lee et al. 1989c: 252 (F; key, tax., bion., distr., review); da Cunha Ramos 1993 (E*); Whelan & Hapgood 2000 (distr.; East Timor); Rattanarithikul et al. 2006a (F*, L*; bion., distr., keys). **Ety.:** n.s. [of the side (L); poss. ref. to "well marked lateral spots"]. **Informal name:** Lateral Philippine Heaven Mosquito.
syn. *cancer* Leicester
1908: 215 (M, F). **Type loc.:** Port Swettenham and Klang [Selangor], Malaya [Malaysia] (NHMUK).
syn. *ceylonica* Theobald
1910c: 503 (F*). **Type loc.:** Galle, Ceylon [Sri Lanka] (NHMUK).
syn. *propria* Taylor
1914c: 704 (M*). **Type loc.:** Townsville, Queensland, Australia (ANIC). **Refs.:** Lee et al. 1989c: 252 (type info.).
syn. *cairnsensis* Taylor
1919: 839 (F*). **Type loc.:** Cairns, Queensland, Australia (ANIC). **Refs.:** Lee et al. 1989c: 252 (type info.).

lebiedi da Cunha Ramos & Brunhes
2004: 381 (M*). **Type loc.:** Parc National de Masoala, Antalaha, Madagascar (LTV). **Distr.:** Madagascar (includes Glorioso and Juan De Nova Isl.). **Ety.:** Ber Lebied. **Informal name:** Lebied Malagasy Heaven Mosquito.

leucoptera (Theobald)
1907: 575 (F; *Anisocheleomyia*). **Type loc.:** Stanley Town, New Amsterdam [Berbice], British Guiana [Guyana] (NHMUK). **Distr.:** Argentina, Bolivia, Brazil, Colombia, Ecuador, French Guiana, Guatemala, Guyana, Honduras, Lesser Antilles (includes Trinidad and Tobago), Mexico, Panama, Suriname, Venezuela. **Refs.:** Galindo et al. 1954: 152 (M*, F, P*, L*); Belkin 1968b: 37 (tax.); Berti et al. 2015 (distr.; Venezuela). **Ety.:** white winged (Gr); wings white scaled. **Informal name:** White-winged Guyanese Heaven Mosquito.

longirostris Leicester
1908: 217 (M). **Type loc.:** Malaya [Malaysia] (NHMUK). **Distr.:** Cambodia, India, Indonesia, Malaysia, Thailand, Vietnam. **Refs.:** Barraud 1934: 71 (M, F, L); Iyengar & Menon 1956: 787 (L*); Rattanarithikul et al. 2006a (F*, L*; bion., distr., keys). **Ety.:** long snout (L); proboscis. **Informal name:** Long-nosed Malaysian Heaven Mosquito.

lowii Theobald
1901c: 339 (M). **Type loc.:** Saint Lucia Island [Lesser Antilles] (NHMUK). **Distr.:** Argentina, Bahamas, Belize, Bolivia, Brazil, Cayman Isl., Colombia, Costa Rica, Cuba, Dominican Republic, Ecuador, El Salvador, French Guiana, Guatemala, Guyana, Haiti, Honduras, Jamaica, Lesser Antilles (includes Bonaire, Grenada, Saint Lucia, Trinidad and Tobago, Virgin Isl.), Mexico, Nicaragua, Panama, Paraguay, Peru, Puerto Rico, Suriname, United States (continental), Uruguay, Venezuela. **Refs.:** Lane 1943: 149 (M*, F); Pratt 1947: 581 (M*, P*, L*); Yamaguti & LaCasse 1951a: 40 (M*, F*, L*); Lane 1953: 568 (M*, F, L*); Galindo et al. 1954: 155 (M*, F, P*, L*); van der Kuyp 1954 (E); Carpenter & LaCasse 1955: 76 (M*, F*, L*; keys); Dodge 1966: 386 (first instar L*; key); Belkin 1968b: 37 (lectotype desig.); Belkin et al. 1970: 43 (M*, F*, P*, L*); Rossi & Martínez 2003: 473 (distr.; Uruguay); Darsie & Ward 2005 (F*, L*; keys, distr.); Berti et al. 2015 (distr.; Venezuela); Harrison et al. 2016 (F*, L*; keys, distr.). **Ety.:** Dr. Low. **Informal name:** Low Saint Lucian Heaven Mosquito.

syn. *continentalis* Dyar & Knab
1906d: 187 (L). **Type loc.:** Baton Rouge, Louisiana, United States (LU).

syn. *minuta* Theobald
1907: 559 (F). **Type loc.:** New Amsterdam [Berbice], British Guiana [Guyana] (NHMUK).

syn. *monilis* Shannon & Del Ponte
1928: 84 (A). **Type loc.:** Ledesma, San Pedro, Jujuy, Corrientes; Corrientes, Resistencia; Chaco, Cataratas del Iguaçu; Misiones Argentina (INM). **Refs.:** Lane 1943: 149 (syn.); Belkin et al. 1968: 12 (lectotype desig. according to Casal).

ludlowae Dyar & Shannon
1925a: 68 (F). **Type loc.:** Ludlow Barracks [Parang, Cotabato], Mindanao, Philippines (USNM). **Distr.:** Philippines. **Refs.:** Baisas 1935b: 66 (F, L); Stone & Knight 1957c: 200 (type info., lectotype desig.); Delfinado 1966a: 47 (M, F, L; syn.); Basio 1971b: 91 (M*, F*; bion.). **Ety.:** Clara Ludlow. **Informal name:** Ludlow Philippine Heaven Mosquito.

macfarlanei Edwards
1914c: 127 (F). **Type loc.:** Hong Kong (NHMUK). **Distr.:** Cambodia, India, Indonesia, Japan, Malaysia, Nepal, People's Republic of China, (includes Hong Kong), Taiwan, Thailand, Vietnam. **Refs.:** Bohart & Ingram 1946b: 55 (M, F*, P*, L*; bion., distr., tax.); Tanaka et al. 1975c: 214 (tax., bion., distr.); Tanaka et al. 1979: 472 (M*, F*, L*); Darsie & Pradhan 1990 (F, L; tax., keys, bion., distr.; Nepal); Qu & Qian 1993 (cibarial armature*); Rattanarithikul et al. 2006a (F*, L*; bion., distr., keys). **Ety.:** H. Macfarlane. **Informal name:** Macfarlane Chinese Heaven Mosquito.

syn. *zelena* Barraud
1934: 69 (M, F; *Uranotaenia campestris* var.). **Type loc.:** Pir Pao, Bombay, India (NHMUK). **Refs.:** Peyton 1972: 34 (syn.).

machadoi da Cunha Ramos
1986: 110 (M*, F, P*, L*). **Type loc.:** Dundo, Angola (NML). **Distr.:** Angola. **Refs.**: da Cunha Ramos 1993: 375 (M*, F, P*, L*; keys, tax., bion., distr.). **Ety.:** António de Barros Machado. **Informal name:** Machado Angolan Heaven Mosquito.

madagascarica da Cunha Ramos & Brunhes
2004: 383 (M*, P*, L*). **Type loc.:** Mandena, Fort Dauphin/Tôlanaro, Madagascar (LTV). **Distr.:** Madagascar (includes Glorioso and Juan De Nova Isl.). **Ety.:** Madagascar. **Informal name:** Mandena Malagasy Heaven Mosquito.

mathesoni Lane
1943: 153 (M*). **Type loc.:** Juquia, São Paulo, Brazil (FH). **Distr.:** Brazil, French Guiana. **Refs.:** Fauran & Pajot 1974a: 100 (distr.). **Ety.:** Robert Matheson. **Informal name:** Matheson Brazilian Heaven Mosquito.

mayeri Edwards
1912c: 40 (F). **Type loc.:** Oshogbo [Western Provinces], Nigeria (NHMUK). **Distr.:** Angola, Cameroon, Democratic Republic of the Congo, Ghana, Kenya, Mozambique, Nigeria, Republic of the Congo, Sierra Leone. **Refs.:** Edwards 1941: 50 (F); Worth & De Meillon 1960: 234 (distr.); da Cunha Ramos 1993: 319 (M, F, P*; keys, tax., bion., distr.). **Ety.:** T.F.G. Mayer. **Informal name:** Mayer Nigerian Heaven Mosquito.

mayottensis Brunhes
1977b: 158 (M*, F, P*, L*). **Type loc.:** Vahibé, route between Passamainti-Combani, Mayotte Island, Comores (SSC). **Distr.:** Comoros, Madagascar (includes Glorioso and Juan De Nova Isl.). **Ety.:** Mayotte [an insular department of France near Madagascar]. **Informal name:** Mayotte Comoran Heaven Mosquito.

mendiolai Baisas
1935b: 71 (M*, F, L*). **Type loc.:** College of Agriculture, Los Baños, Laguna, Luzon, Philippines (NE; paratypes USNM). **Distr.:** Malaysia, Philippines. **Refs.:** Delfinado 1966a: 48 (M*, F, P*, L*); Basio 1971b: 91 (M*; bion.); Mattingly 1971a: Pl. 12 (P terminal segs.*); Baisas 1974: 160 (M*, F, P, L*; tax., bion., distr.; Philippines). **Ety.:** J. Mendiola. **Informal name:** Mendiola Philippine Heaven Mosquito.

metatarsata Edwards
1914b: 81 (M). **Type loc.:** Circular Road, Kuala Lumpur [Selangor], Malaya [Malaysia] (NHMUK). **Distr.:** Cambodia, Indonesia, Malaysia, Philippines, Thailand, Vietnam. **Refs.:** Rattanarithikul et al. 2006a (F*, L*; bion., distr., keys). **Ety.:** n.s. [among, between (Gr), with tarsus (L); prob. ref. to being another species of this genus with modified male front tarsi]. **Informal name:** Middle-legged Malaysian Heaven Mosquito.
syn. *innotata* Dyar & Shannon
1925a: 69 (F). **Type loc.:** Camp Nichols, Rizal, Luzon, Philippines (USNM). **Refs.:** Bohart 1945: 34 (syn.); Peyton 1972: 34 (syn.).

micans Leicester
1908: 206 (M, F). **Type loc.:** [Circular Road, Kuala Lumpur, Selangor], Malaya [Malaysia] (NHMUK). **Distr.:** Cambodia, Indonesia, Malaysia, Thailand, Vietnam. **Refs.:** Bonne-Wepster 1954: 25 (M, F); Rattanarithikul et al. 2006a (F*, L*; bion., distr., keys). **Ety.:** n.s. [flashing, gleaming, sparkling (L)]. **Informal name:** Shining Malaysian Heaven Mosquito.

moramangae da Cunha Ramos & Brunhes
2004: 393 (M*, P*, L*). **Type loc.:** Ambodimanga, Beparasy, Moramanga, Madagascar (LTV). **Distr.:** Madagascar (includes Glorioso and Juan De Nova Isl.). **Ety.:** Moramanga, Madagascar. **Informal name:** Moramanga Malagasy Heaven Mosquito.

moresbyensis Peters
1963e: 204 (M*). **Type loc.:** Port Moresby, Central District, Papua (CSIR). **Distr.:** Australia, Papua New Guinea. **Refs.:** Marks 1970: 166 (distr.); Lee et al. 1989c: 260 (F; key, tax., bion., distr., review). **Ety.:** Port Moresby, Papua New Guinea. **Informal name:** Moresby Papuan Heaven Mosquito.

nataliae Lynch Arribálzaga
1891b: 164 (A). **Type loc.:** Baradero, Buenos Aires, Argentina (BA). **Distr.:** Argentina, Bolivia, Brazil, Colombia, Ecuador, French Guiana, Guatemala, Guyana, Honduras, Lesser Antilles (includes Trinidad and Tobago), Mexico, Nicaragua, Panama, Paraguay, Peru, Suriname, Uruguay, Venezuela. **Refs.:** Lane 1943: 147 (M*, F); Lane 1953: 577 (M*, F); Galindo et al. 1954: 157 (M*, F, P*, L*); Belkin et al. 1968: 12 (lectotype desig. according to Casal); Heinemann & Belkin 1977b: 453 (distr.); Rossi & Martínez 2003: 473 (distr.; Uruguay); Berti et al. 2015 (distr.; Venezuela). **Ety.:** n.s. [a woman's or girl's name (Harbach 2018c)]. **Informal name:** Natalia Argentinian Heaven Mosquito.
syn. *rowlandii* Theobald
1905g: 34 (M*, F*; *Pseudouranotaenia*). **Type loc.:** Stanley Town, New Amsterdam [Berbice], British Guiana [Guyana] (NHMUK). **Refs.:** Belkin 1968b: 38 (lectotype desig.).
syn. *noctivaga* Neiva & Pinto
1922c: 374 (F). **Type loc.:** Rio de Janeiro, Brazil (LU). **Refs.:** Belkin et al. 1971: 17 (tax.).
syn. *argenteopennis* Peryassú
1923a: 2 (M, F). **Type loc.:** "Distrito Federal e Baixada Fluminense," Brazil (LU).
syn. *capitis* Shannon & Del Ponte
1928: 85 (F). **Type loc.:** Ledesma, Jujuy, Argentina (USNM). **Refs.:** Lane 1953: 577 (syn.).

neireti Edwards
1920b: 138 (M, F). **Type loc.:** Tananarive?, Madagascar (MNHP). **Distr.:** Madagascar (includes Glorioso and Juan De Nova Isl.). **Refs.:** Edwards 1941: 53 (M, F); Mattingly 1954b: 170 (M*). **Ety.:** Dr. Neiret. **Informal name:** Neiret Malagasy Heaven Mosquito.

neotibialis King & Hoogstraal
1947c: 590 (M*, P*, L*). **Type loc.:** Hollandia, Western New Guinea (Papua) (USNM). **Distr.:** Papua New Guinea. **Refs.:** Lee et al. 1989c: 262 (tax., bion., distr., review). **Ety.:** near (L) *tibialis*. **Informal name:** Near-tibialis Indonesian Heaven Mosquito.

nivea Leicester
1908: 211 (M). **Type loc.:** Kuala Lumpur [Selangor], Malaya [Malaysia] (NHMUK). **Distr.:** Malaysia, Philippines. **Refs.:** Bohart 1945: 36 (to sp.); Delfinado 1966a: 50 (M*, F, L*); Basio 1971b: 92 (M*; bion.). **Ety.:** snowy white (L); band of snowy-white scales in front of wings. **Informal name:** Snowy Malaysian Heaven Mosquito.
syn. *triangulata* Ludlow
1908a: 331 (M; *Pseudo-uranotaenia*). **Type loc.:** Reina Regente [Cotabato] Mindanao, Philippines (USNM).

nivipes (Theobald)
1905a: 53 (M*; *Anisocheleomyia*). **Type loc.:** Queensland, Australia (NHMUK). **Distr.:** Australia, Indonesia, Papua New Guinea. **Refs.:** Bancroft 1908: 58 (M; as genus *Anisocheleomyia*); Lee et al. 1989c: 263 (F; key, tax., bion., distr., review). **Ety.:** white foot (L); ref. to "last two and apex of the antepenultimate hind tarsi white." **Informal name:** Snow-footed Australian Heaven Mosquito.

syn. *albofasciata* Taylor
1920: 164 (A). **Type loc.:** near Darwin, Northern Territory, Australia (SAM).

novaguinensis Peters
1963d: 141 (M*, F, P*, L). **Type loc.:** Ami, 10 mi northwest of Maprik, [East] Sepik District, New Guinea [Papua New Guinea] (CSIR). **Distr.:** Australia, Papua New Guinea. **Refs.:** Marks 1970: 166 (distr.); Lee et al. 1989c: 271 (F; key, tax., bion., distr., review). **Ety.:** New Guinea. **Informal name:** Sepik Papuan Heaven Mosquito.

ssp. ***alticola*** Peters
1963d: 142 (M, F). **Type loc.:** Goroka, Eastern Highlands District, New Guinea (CSIR). **Distr.:** Papua New Guinea. **Refs.:** Lee et al. 1989c: 269 (F; key, tax., bion., distr., review). **Ety.:** n.s. [high-altitude dweller (L)]. **Informal name:** Altitude Papuan Nail Mosquito.

orientalis Barraud
1926: 339 (M*). **Type loc.:** Golaghat, Assam, India (NHMUK). **Distr.:** India, Thailand. **Refs.:** Barraud 1934: 66 (M*); Rattanarithikul et al. 2006a (F*, L*; bion., distr., keys). **Ety.:** n.s. [eastern (L)]. **Informal name:** Eastern Indian Heaven Mosquito.

orthodoxa Dyar
1921h: 118 (F). **Type loc.:** Tiribi, Costa Rica (USNM). **Distr.:** Colombia, Costa Rica, Ecuador, Guatemala, Honduras, Panama. **Refs.:** Galindo et al. 1954: 159 (M*, F). **Ety.:** n.s. [straight opinion, judgment (Gr)]. **Informal name:** Orthodox Costa Rican Heaven Mosquito.

oteizai Perez Vigueras
1956: 443 (M, F, L*). **Type loc.:** Guanimar, Cuba (LU). **Distr.:** Cuba. **Ety.:** n.s. [Dr. Alberto Oteiza y Zetién (Harbach 2018c)]. **Informal name:** Oteiza Cuban Heaven Mosquito.

pallidocephala Theobald
1908c: 266 (F*). **Type loc.:** Upper White Nile, Sudan (NHMUK). **Distr.:** Angola, Benin, Burkina Faso, Cameroon, Central African Republic, Democratic Republic of the Congo, Gabon, Kenya, Liberia, Mozambique, Nigeria, Republic of the Congo, Sudan and South Sudan, The Gambia, Uganda, Zambia. **Refs.:** Edwards 1941: 45 (M*, F); Hopkins 1952: 52 (L*); Peters 1955c: 118 (P*); Service 1976a (distr.; Gabon); da Cunha Ramos 1993: 384 (M*, F, P*, L*; keys, tax., bion., distr.); Jupp 1996 (M*, F*; key). **Ety.:** n.s. [pale head (Gr)]. **Informal name:** Pale-headed Sudanese Heaven Mosquito.

syn. *similis* Theobald
1908c: 257 (F*). **Type loc.:** Sudan (NHMUK).

syn. *caerulea* Theobald
1908c: 267 (F*; *pallidocephala* ssp.; original type M in NHMUK). **Type loc.:** Upper White Nile, Sudan (NHMUK?).

syn. *abnormalis* Theobald
1910c: 512 (M*, F*). **Type loc.:** "Bor to Mongalla," Sudan (NHMUK).

pallidoventer Theobald
1903a: 300 (F*). **Type loc.:** Pará, Brazil (NHMUK). **Distr.:** Argentina, Brazil, Colombia, French Guiana, Guatemala, Guyana, Lesser Antilles (includes Trinidad and Tobago), Panama, Peru, Suriname, Venezuela. **Refs.:** Dyar 1928: 419 (M*, F); Galindo et al. 1954: 160 (M*, F, P*, L*). **Ety.:** pale venter (L); "venter deep ochraceus." **Informal name:** Pale-bellied Brazilian Heaven Mosquito.

palmeirimi De Meillon & Rebêlo
1941: 76 (M, F). **Type loc.:** Pebane, Quelimane District, Mozambique (SAIM). **Distr.:** Angola, Mozambique. **Refs.:** da Cunha Ramos 1993: 427 (F; key, tax., bion., distr.); Jupp 1996 (M*, F*; key). **Ety.:** Vasco Palmeirim. **Informal name:** Palmeirim Mozambican Heaven Mosquito.

ssp. ***dundo*** da Cunha Ramos
1993 (F*; key, tax., bion., distr.). **Type loc.:** Dundo, Angola (IHMT). **Distr.:** Angola. **Ety.:** Dundo, Angola. **Informal name:** Dundo Angolan Heaven Mosquito.

paludosa Galindo, Blanton, & Peyton
1954: 162 (M*, F). **Type loc.:** Almirante, Bocas del Toro, Panama (USNM). **Distr.:** Panama. **Ety.:** n.s. [swampy, marshy (L)]. **Informal name:** Boggy Panamanian Heaven Mosquito.

paralateralis Peters
1964: 19 (M*, F). **Type loc.:** Kavieng, New Ireland District, New Guinea (CSIR). **Distr.:** Australia, Papua New Guinea. **Refs.:** Marks 1970: 166 (distr.); Lee et al. 1989c: 275 (F; key, tax., bion., distr., review). **Ety.:** paired with (L); close to *lateralis*. **Informal name:** Kavieng Papuan Heaven Mosquito.

paranovaguinensis Peters
1963d: 135 (M*, P*, L*). **Type loc.:** Maprik, [East] Sepik District, New Guinea [Papua New Guinea] (CSIR). **Distr.:** Papua New Guinea. **Refs.:** Lee et al. 1989c: 277 (tax., bion., distr., review). **Ety.:** close to *novaguinensis*. **Informal name:** Maprik Papuan Heaven Mosquito.

philonuxia Philip
1931: 188 (M*, F). **Type loc.:** Ebute Metta [Western Provinces], Nigeria (NHMUK, CAS). **Distr.:** Benin, Cameroon, Côte d'Ivoire, Liberia, Mozambique, Nigeria, Uganda. **Refs.:** Hamon 1954b: 591 (P*); Peters 1955c: 118 (P*, L*); Worth & De Meillon 1960: 234 (distr.); Arnaud 1979: 58 (type info.); da Cunha Ramos 1993: 396 (M*, F, P*, L*; keys, tax., bion., distr.); Jupp 1996 (M*, F*; key). **Ety.:** n.s. **Informal name:** Night-loving Nigerian Heaven Mosquito.

pifanoi Cova García, Pulido F., & Escalante de Ugueto
1981: 210 (M*, P*, L*). **Type loc.:** La Toma, Municipio Salom, San Esteban River, Carabobo, Venezuela

(DERM). **Distr.:** Venezuela. **Ety.:** Félix Pifano C. **Informal name:** Pifano Venezuelan Heaven Mosquito.

prajimi Peyton & Rattanarithikul
1970: 406 (M, F, P, L). **Type loc.:** Ban Pa Daeng, Lampang, Thailand (USNM). **Distr.:** Thailand. **Refs.:** Rattanarithikul et al. 2006a (F*, L*; bion., distr., keys). **Ety.:** Prajim Boonyakanist. **Informal name:** Prajim Thai Heaven Mosquito.

pulcherrima Lynch Arribálzaga
1891b: 165 (M*). **Type loc.:** Las Conchas, Buenos Aires, Argentina (NE). **Distr.:** Argentina, Belize, Bolivia, Brazil, Colombia, Costa Rica, Ecuador, El Salvador, French Guiana, Guatemala, Guyana, Honduras, Lesser Antilles (includes Trinidad and Tobago), Mexico, Nicaragua, Panama, Paraguay, Peru, Suriname, Uruguay, Venezuela. **Refs.:** Lane 1953: 563 (M*, F, P*, L*); Galindo et al. 1954: 163 (M*, F, P*, L*); Duret & Barreto 1956: 87 (M*; tax.); Rossi & Martínez 2003: 473 (distr.; Uruguay); Berti et al. 2015 (distr.; Venezuela). **Ety.:** n.s. [very, very pretty (prettiest?) (L)]. **Informal name:** Beautiful Argentinian Heaven Mosquito.

ssp. ***elnora*** Paterson & Shannon
1927: 10 (F). **Type loc.:** Tres Pozos, Embarcación, Salta, Argentina (USNM). **Distr.:** Argentina. **Ety.:** Elnora S. Shannon. **Informal name:** Elnora Argentinian Heaven Mosquito.

syn. *urania* Shannon & Del Ponte
1928: 83 (M, F). **Type loc.:** Resistencia, Chaco, Argentina (USNM). **Refs.:** Lane 1951: 334 (syn.); Stone & Knight 1957c: 200 (type info., lectotype desig.).

syn. *modesta* Martini
1935: 29 (F; *pulcherrima* var.). **Type loc.:** near Rio Cacao, British Honduras [Belize] (LU). **Refs.:** Lane 1953: 564 (syn.).

pygmaea Theobald
1901c: 254 (F*). **Type loc.:** Burpengary, Queensland, Australia (NHMUK). **Distr.:** Australia. **Refs.:** Bancroft 1908: 56 (F); Dyar & Shannon 1925a: 69 (= *arguellesi*); Lee 1944a: 31 (L*); Lee et al. 1989c: 279 (F; key, tax., bion., distr., review). **Ety.:** n.s. [small (Gr); author notes its small size]. **Informal name:** Pygmy Australian Heaven Mosquito.

rachoui Xavier, Mattos, & da Silva
1970: 411 (M*). **Type loc.:** Rio Bananal, Linhares District, Espírito Santo, Brazil (INER). **Distr.:** Brazil. **Ety.:** René Guimarães Rachou. **Informal name:** Rachou Brazilian Heaven Mosquito.

rampae Peyton & Klein
1970: 243 (M, F). **Type loc.:** Svay Chrum, Kampong Chhnang, Cambodia (USNM). **Distr.:** Bangladesh, Cambodia, Malaysia, Thailand, Vietnam. **Refs.:** Rattanarithikul et al. 2006a (F*, L*; bion., distr., keys). **Ety.:** Rampa Rattanarithikul. **Informal name:** Rampa Cambodian Heaven Mosquito.

reyi Baisas
1935b: 74 (F*, L*). **Type loc.:** Simoay, Cotabato, Mindanao, Philippines (NE). **Distr.:** Philippines. **Refs.:** King & Hoogstraal 1947c: 592 (M); Delfinado 1966a: 47 (M, F, L; syn.); Peyton 1972: 35 (to sp.). **Ety.:** n.s. **Informal name:** Rey Philippine Heaven Mosquito.

riverai Duret
1970: 43 (M*, F*). **Type loc.:** Bogotà City, Colombia (USNM). **Distr.:** Colombia. **Refs.:** Harbach et al. (1991) 1990: 194 (type info.). **Ety.:** Hernado Rivera. **Informal name:** Rivera Colombian Heaven Mosquito.

roberti da Cunha Ramos & Brunhes
2004: 404 (L*). **Type loc.:** Lakato, Moramanga, Madagascar (LTV). **Distr.:** Madagascar (includes Glorioso and Juan De Nova Isl.). **Ety.:** Vincent Robert. **Informal name:** Robert Malagasy Heaven Mosquito.

rutherfordi Edwards
1922a: 288 (F). **Type loc.:** Peradeniya [Central], Ceylon [Sri Lanka] (NHMUK). **Distr.:** Sri Lanka. **Refs.:** Carter & Wijesundara 1948: 139 (M*). **Ety.:** A. Rutherford. **Informal name:** Rutherford Sri Lankan Heaven Mosquito.

sapphirina (Osten Sacken)
1868: 47 (M, F; *Aedes*). **Type loc.:** Washington, DC, and Brooklyn, New York, United States (MCZ). **Distr.:** Canada, Dominican Republic, El Salvador, Haiti, Mexico, Puerto Rico, Suriname, United States (continental). **Refs.:** Blanchard 1905 (E*); Howard et al. 1917 (E*); Gerry 1932: 38 (F*); Darsie 1951: 44 (P*); Yamaguti & LaCasse 1951a: 44 (M*, F*, L*); Carpenter & LaCasse 1955: 77 (M*, F*, L*; keys); Ross & Horsfall 1965 (M*, F*, L*, E*; keys); Dodge 1966: 386 (first instar L*; key); Belkin et al. 1970: 37 (M*, F*, P*, L*; keys); Mattingly 1970c (E*); Mattingly 1971a: Pl. 30 (L head*, terminal segs.*); Darsie & Ward 2005 (F*, L*; keys, distr.); Harrison et al. 2016 (F*, L*; keys, distr.); Reeves et al. 2018 (bion.). **Ety.:** n.s. [blue gem, sapphire (L); prob. ref. to having bluish reflections]. **Informal name:** Sapphire American Heaven Mosquito.

syn. *coquilletti* Dyar & Knab
1906d: 187 (L). **Type loc.:** United States (LU).

setosa King & Hoogstraal
1947c: 585 (M*, F, P*, L*). **Type loc.:** Hollandia, Western New Guinea (Papua) (USNM). **Distr.:** Indonesia, Papua New Guinea. **Refs.:** Lee et al. 1989c: 282 (tax., bion., distr., review). **Ety.:** n.s. [setose (L)]. **Informal name:** Setose Indonesian Heaven Mosquito.

sexaueri Belkin
1953a: 359 (M*, F*, P*, L*). **Type loc.:** West Poha swamp, Guadalcanal, Solomon Islands (USNM). **Distr.:** Papua New Guinea, Solomon Isl. **Refs.:** Belkin 1962: 167 (M*, F*, P*, L*); Lee et al. 1989c: 284 (tax., bion., distr., review). **Ety.:** Henry F. Sexauer. **Informal name:** Sexauer Solomon Islander Heaven Mosquito.

socialis Theobald
1901c: 340 (M*, F*). **Type loc.:** Kingston [Surrey], Jamaica (NHMUK). **Distr.:** Belize, Brazil, Colombia, Costa Rica, Cuba, Dominican Republic, Ecuador, El Salvador, French Guiana, Guatemala, Guyana, Haiti, Honduras, Jamaica, Lesser Antilles (includes Trinidad and Tobago, Virgin Isl.), Mexico, Nicaragua, Panama, Puerto Rico. **Refs.:** Galindo et al. 1954: 164 (M*, F, P*, L*; tax.; from syn. with *sapphirina*); Belkin 1968b: 38 (lectotype desig.); Belkin et al. 1970: 39 (M*, F*, P*, L*); Heinemann & Belkin 1977b: 415 (distr.). **Ety.:** n.s. [allied, in partnership (L); poss. ref. to "Found in association with *Anopheles* larvae"]. **Informal name:** Allied Jamaican Heaven Mosquito.

solomonis Belkin
1953a: 350 (M*, F*, P*, L*). **Type loc.:** Lunga, Guadalcanal, Solomon Islands (USNM). **Distr.:** Indonesia, Papua New Guinea, Solomon Isl. **Refs.:** Belkin 1962: 174 (M*, F*, P*, L*); Maffi et al. 1979: 218 (distr.); Lee et al. 1989c: 286 (tax., bion., distr., review). **Ety.:** Solomon Isl. **Informal name:** Lapita Solomon Islander Heaven Mosquito [the Lapita (1600–500 B.C.E.) were a prehistoric Polynesian people believed to have settled on the Solomon Isl.].

sombooni Peyton & Klein
1970: 249 (M, F, P, L). **Type loc.:** Khao Sai Dao Mountain, Chanthaburi, Thailand (USNM). **Distr.:** Cambodia, Malaysia, Thailand. **Refs.:** Qu & Qian 1993 (cibarial armature*); Rattanarithikul et al. 2006a (F*, L*; bion., distr., keys). **Ety.:** Somboon Maneechai. **Informal name:** Somboon Thai Heaven Mosquito.

subnormalis Martini
1920a, Pl. II, Fig. 6b (M*). **Type loc.:** Singapore (LU). **Distr.:** Cambodia, Indonesia, Malaysia, Singapore, Thailand, Vietnam. **Refs.:** Edwards 1922b: 436 (M, F); Edwards 1926a: 116 (L*); Peyton & Hochman 1968: 379 (M*); Rattanarithikul et al. 2006a (F*, L*; bion., distr., keys). **Ety.:** n.s. **Informal name:** Subnormal Singaporean Heaven Mosquito.
syn. *roperi* Edwards
1922a: 289 (F). **Type loc.:** Membakut [Sabah, Malaysian] Borneo (NHMUK).

subtibioclada King & Hoogstraal
1947c: 595 (M*). **Type loc.:** Hollandia, Western New Guinea (Papua) (USNM). **Distr.:** Indonesia. **Refs.:** Lee et al. 1989c: 289 (tax., bion., distr., review). **Ety.:** nearly (L); near *tibioclada* [foretibia appears branched in both]. **Informal name:** Hollandia Indonesian Heaven Mosquito.

telmatophila Galindo, Blanton, & Peyton
1954: 168 (M*, F, P*, L*). **Type loc.:** Tocumen, Panama Province, Panama (USNM). **Distr.:** Ecuador, Panama. **Ety.:** marsh, pool lover (Gr); ref. to larval habitat. **Informal name:** Marsh-loving Panamanian Heaven Mosquito.

testacea Theobald
1905f: 113 (F*). **Type loc.:** Singapore (HNM). **Distr.:** Cambodia, India, Malaysia, Myanmar, Nepal, People's Republic of China, Singapore, Thailand. **Refs.:** Edwards 1932b: 559 (L*; as *testacea*?); Barraud 1934: 74 (M*, F); Baisas 1935d: 169 (M*); Delfinado 1966a: 53 (=*falcipes*); Basio 1971b: 93 (M*; bion.); Baisas 1974: 161 (M*, F, P, L*; tax., bion., distr.; Philippines); Darsie et al. 1992 (distr.; Nepal); Qu & Qian 1993 (cibarial armature*); Rattanarithikul et al. 2006a (F*, L*; bion., distr., keys). **Ety.:** resembling brick (L); ref. to "Thorax bright testaceous." **Informal name:** Brick-like Singaporean Heaven Mosquito.
syn. *unilineata* Leicester
1908: 220 (M, F). **Type loc.:** Malaya [Malaysia] (NHMUK).

tibialis Taylor
1919: 839 (M, F*). **Type loc.:** Cairns, Queensland, Australia (ANIC). **Distr.:** Australia, Papua New Guinea, Vanuatu. **Refs.:** Perry 1946: 11 (F, L); King & Hoogstraal 1947c: 595 (tax.); Lee et al. 1989c: 290 (F; key, tax., bion., distr., review, type info.). **Ety.:** "Apex of foretibia with a tuft of long, brown, hair-like scales." **Informal name:** Leggy Australian Heaven Mosquito.

tibioclada King & Hoogstraal
1947c: 593 (M*, F, P*, L*). **Type loc.:** Samboga River, Dobodura, Papua, New Guinea (USNM). **Distr.:** Papua New Guinea. **Refs.:** Lee et al. 1989c: 294 (tax., bion., distr., review). **Ety.:** spotted tibia (L); hind tibia with apical white spot. **Informal name:** Spot-legged Papuan Heaven Mosquito.

trapidoi Galindo, Blanton, & Peyton
1954: 170 (M*, F, P*, L*). **Type loc.:** El Hato, Chiriquí Province, Panama (USNM). **Distr.:** Costa Rica, Panama. **Ety.:** Harold Trapido. **Informal name:** Trapido Panamanian Heaven Mosquito.

trilineata Leicester
1908: 204 (M, F). **Type loc.:** Malaya [Malaysia] (NHMUK). **Distr.:** Cambodia, Malaysia, Thailand. **Refs.:** Rattanarithikul et al. 2006a (F*, L*; bion., distr., keys). **Ety.:** n.s. [three lines (L)]. **Informal name:** Three-lined Malaysian Heaven Mosquito.

typhlosomata Dyar & Knab
1907c: 200 (M). **Type loc.:** Taboga Island, Panama (USNM). **Distr.:** Brazil, Colombia, Costa Rica, Ecuador, Guatemala, Lesser Antilles (includes Trinidad and Tobago), Nicaragua, Panama, Paraguay, Venezuela. **Refs.:** Galindo et al. 1954: 175 (M*, F, P*, L*; tax.; from syn. with *coatzacoalcos*); Berti et al. 2015 (distr.; Venezuela). **Ety.:** n.s. [blind or blind ended, body (Gr)]. **Informal name:** Blind Panamanian Heaven Mosquito.

unimaculiala Leicester
1908: 210 (F). **Type loc.:** Malaya [Malaysia] (NHMUK). **Distr.:** Malaysia. **Ety.:** n.s. [one spot (L); poss. ref. to single costal spot]. **Informal name:** One-spotted Malaysian Heaven Mosquito.

wysockii Belkin
1953a: 366 (M*, F*, P*, L*). **Type loc.:** Tenaru, Guadalcanal, Solomon Islands (USNM). **Distr.:** Papua New Guinea, Solomon Isl. **Refs.:** Lee et al. 1989c: 296 (tax., bion., distr., review). **Ety.:** F.B. Wysocki. **Informal name:** Wysocki Solomon Islander Heaven Mosquito.

yunnanensis Dong, Dong, & Wu
2004: 568 (M*, F, P*, L*). **Type loc.:** Jindong County, Yunnan Province (1,160 m), China [People's Republic of China] (IPD). **Distr.:** People's Republic of China. **Ety.:** Yunnan Prov., China. **Informal name:** Nanzhao Chinese Heaven Mosquito [Nanzhao (eighth through ninth centuries C.E.) was a kingdom located in present-day Yunnan].

Fossil Culicidae

Important references: Brodie 1845 (fossil insects; England); Scudder 1877 (fossil insects); Scudder 1882 (fossil insect names); Meunier 1902 (fossil, amber); Statz 1943 (fossils); Nei Menggu Zizhi Qu Dizhi & Dongbei Dizhi Kexue 1976 (fossils); Lutz 1985 (fossil); Evenhuis 1994 (fossil fly catalog); Wichard et al. 2009 (fossils, Baltic amber); Kentzer 2011 (Baltic amber mosquito inclusion).

Subfamily Anophelinae

Genus *Anopheles* Meigen

1818: 10. Type species: *Anopheles maculipennis* Meigen (International Commission on Zoological Nomenclature [ICZN] 1959, Opinion 547).

dominicanus Zavortink & Poinar
2000 (F). **Type loc.:** El Mamey Formation, northern mountain ranges, Dominican Republic (Dominican amber) (PAC). **Distr.:** Eocene–Miocene. **Refs.:** Zavortink & Poinar 2000.

rottensis Statz [*Anopheles*?]
1944: 109 (M*). **Type loc.:** Rott [near Meissner, Germany] (LACM). **Distr.:** Upper Oligocene. **Refs.:** Poinar et al. 2000; Zavortink & Poinar 2000.

Subfamily Burmaculicinae

Genus *Burmaculex* Borkent & Grimaldi

2004: 883 (fossil genus).

antiquus Borkent & Grimaldi
2004 (holotype F). Ref.: Borkent & Grimaldi 2016 (Burmese amber). **Type loc.:** near Tanai, Kachin Province, Myanmar (Burmese amber) (AMNH). **Distr.:** Cretaceous.

Subfamily Culicinae

Genus *Aedes* Meigen

serafini (Szadziewski) [*Ochlerotatus*]
1998 (as *Aedes* (*Finlaya*)). **Type loc.:** not stated (MAI). **Distr.:** Eocene. **Refs.:** Szadziewski & Gilka 2011 (to *Ochlerotatus*).

Genus *Aetheapnomyia* Harbach & Greenwalt

2012: 33 (F; fossil genus).

hoffeinsorum (Szadziewski)
1998 (F; as *Aedes*). **Type loc.:** Poland (Baltic amber). **Distr.:** Eocene. **Refs.:** Podenas 1999; Harbach & Greenwalt 2012 (from genus *Aedes*).

Genus *Culex* Linnaeus

1758: 602.

ciliaris Linnaeus [extant *Culex*]
1767. **Type loc.:** Sweden (copal) (LU). **Distr.:** Holocene. **Refs.:** Bloch 1776; Edwards 1932a (as *cinereus*); Natvig 1948 (as *cinereus*); Poinar et al. 2000 (as *cinereus*).

damnatorum Scudder [*Culex*]
1890: 582 (F*). **Type loc.:** Green River, Wyoming, United States (MCZ). **Distr.:** Eocene. **Refs.:** Edwards 1923c (*Culex*?).

erikae Szadziewski & Szadziewska [*Culex*]
1985: 513 (F). **Type loc.:** Poland (Baltic amber) (MNB). **Distr.:** Eocene.
syn. *perkunas* Podėnas [*Culex*]
1999 (F; *Aedes*). **Type loc.:** Poland (Baltic amber) (ZMVU). **Distr.:** Eocene. **Refs.:** Szadziewski & Gilka 2011 (syn.).

flavus Gistel [*Culex*]
1831: 247 (A). **Type loc.:** Brazil? **Distr.:** Quaternary (copal).

malariager Poinar [*Culex*]
2005a (F). **Type loc.:** La Toca amber mine, Dominican Republic (Dominican amber) (PAC). **Distr.:** Eocene–Miocene. **Refs.:** Poinar 2005a, b.

pipiens Linnaeus [extant *Culex*]
1758. **Type loc.:** no type (LU). **Distr.:** Eocene. **Refs.:** Anecdotal.

protorhinus Cockerell [*Culex*]
1916: 488 (M*). **Type loc.:** Gurnet Bay, Isle of Wight, England (USNM). **Distr.:** Middle Oligocene. **Refs.:** Edwards 1923c: 148 (generic placement queried).

vectensis Edwards [*Culex*]
1923c: 151 (F*). **Type loc.:** Gurnet Bay, Isle of Wight, England (NHMUK). **Distr.:** Middle Oligocene.

winchesteri Cockerell [*Culex*]
1919: 44 (F). **Type loc.:** Smith's Ranch, near Cathedral Bluffs, Colorado, United States (USNM). **Distr.:** Eocene. **Refs.:** Edwards 1923c: 140 (as "true" *Culex*).

Genus *Culiseta* Felt

gedanica Szadziewski & Gilka [*Culiseta*]
2011. **Type loc.:** not stated; Germany (SDEI). **Distr.:** Eocene.

kishenehn Harbach & Greenwalt [*Culiseta*]
2012 (F). **Type loc.:** Constenius Park (W 113° 42.085', N 48° 23.517'), Kishenehn Formation, Montana, United States (USNM). **Distr.:** Eocene. **Refs.:** Harbach & Greenwalt 2012.

lemniscata Harbach & Greenwalt [*Culiseta*]
2012a (F). **Type loc.:** Dakin (W 113° 42.173', N 48° 23.476'), Kishenehn Formation, Montana, United States (USNM). **Distr.:** Eocene. **Refs.:** Harbach & Greenwalt 2012b (erratum).

Genus *Eoaedes* Harbach & Greenwalt

2012: 33 (fossil genus).

damzeni (Szadziewski) [*Eoaedes*]
1998 (M). **Type loc.:** not stated (Baltic amber) (GPIUH). **Distr.:** Eocene. **Refs.:** Podenas 1999; Harbach & Greenwalt 2012 (from genus *Aedes*).

Genus *Mansonia* Blanchard

cockerelli (Edwards) [*Mansonia*]
1923c: 151 (F*; as *Taeniorhynchus*). **Type loc.:** Gurnet Bay, Isle of Wight, England (NHMUK). **Distr.:** Middle Oligocene. **Refs.:** Edwards 1932a: 7 (states probably a *Mansonia*); International Commission on Zoological Nomenclature 1959, Opinion 550 (name *Taeniorhynchus* suppressed); Evenhuis 1994 (as *Ochlerotatus*); Poinar et al. 2000 (as *Mansonia*).

martinii Statz [*Mansonia*]
1944: 110 (F*). **Type loc.:** Rott [near Meissner, Germany] (LACM). **Distr.:** Upper Oligocene. **Refs.:** Statz 1941 (as *Mansonia martinii*, a *nomen nudum*).

varivestita Statz [*Mansonia*]
1944: 111 (F*). **Type loc.:** Rott [near Meissner, Germany] (KCB). **Distr.:** Upper Oligocene. **Refs.:** Statz 1941: 14 (as *Mansonia varivesta*, a *nomen nudum*).

Genus *Neoculicites* Evenhuis

1994: 234 (genus of fossil Culicidae, subf. Culicinae; new name for *Culicites* Meunier 1915).

syn. *Culicites* Meunier 1915 (as genus). Evenhuis 1994 (syn.).

arvernensis Piton [*Neoculicites*]
1936 (M*, P*). **Type loc.:** Lac Chambon, Puy-de-Dôme, France (A). **Distr.:** Mio-Pliocene. **Refs.:** Meunier 1915 (as *Culicites*, not von Heyden 1862); Evenhuis 1994 (replacement name *Neoculicites*).

ceyx von Heyden [*Neoculicites*]
1870: 252 (F*; as *Culex*). **Type loc.:** Rott, near Meissner, Germany (LU). **Distr.:** Upper Oligocene. **Refs.:** Meunier 1915 (as *Culicites*, not von Heyden 1862); Evenhuis 1994 (replacement name *Neoculicites*).

depereti Meunier [*Neoculicites*]
1915: 17 (A). **Type loc.:** Aix-en-Provence [Bouches-du-Rhone], France (LGUL). **Distr.:** Oligocene. **Refs.:** Meunier 1915 (to *Culicites*, not von Heyden 1862); Evenhuis 1994 (replacement name *Neoculicites*).

Genus *Paleoculicis* Poinar, Zavortink, Pike, & Johnston

2000: 121 (fossil genus).

minutus Poinar, Zavortink, Pike, & Johnston [*Paleoculicis*]
2000: 121 (M*). **Type loc.:** Grassy Lake, Alberta, Canada. (RTMP). **Distr.:** Cretaceous (amber).

Genus *Toxorhynchites* Theobald

mexicanus Zavortink & Poinar [*Toxorhynchites*]
2008 (F). **Type loc.:** near Simojovel, Chiapas, Mexico (in Mexican amber) (PAC). **Distr.:** Eocene–Miocene. **Refs.:** Zavortink & Poinar 2008.

Genus Not Determined

protolepis (Cockerell) [genus not det.]
1916: 488 (A*; *Culex*). **Type loc.:** Gurnet Bay, Isle of Wight, England (USNM). **Distr.:** Middle Oligocene. **Refs.:** Edwards 1923c: 142 (M*, F*).
syn. *petrifactellus* Cockerell [genus not det.]
1916: 489 (A*; *Culex*). **Type loc.:** Gurnet Bay, Isle of Wight, England (USNM). **Distr.:** Oligocene. **Refs.:** Edwards 1923c (syn.).

Nomina Dubia

The following names comply with the ICZN definition for designation of *nomen dubium* (pl. *nomina dubia*): "A Latin term meaning 'a name of unknown or doubtful application.'"

aequatorianus Leví-Castillo [*Toxorhynchites*]
1953: 72 (M*, F, P*, L*; *Toxorhynchites*). **Type loc.:** Hacienda Pichilingue, Los Rios, Ecuador (CEIE). **Refs.:** Stone et al. 1959: 65 (to *nomen dubium*).

africanus Roque [*Anopheles*]
1903: 110 (M, F*; *superpictus* ssp.). **Type loc.:** Mossamedes, Angola (LU). **Refs.:** Evans 1938: 394 (M, F; tax.); Edwards 1941: 178 (? = *congolensis*); De Meillon 1947b: 9 (unrecognized sp.).

aikenii Aiken & Rowland [*Culex*]
1906: 34 (L*; *Gnophodeomyia*). **Type loc.:** New Amsterdam [Berbice], British Guiana [Guyana] (NE). **Refs.:** Stone et al. 1959: 266 (catalog); Belkin 1968b: 12 (tax.); Belkin 1970b: 59 (to *nomen dubium*).

albipes (Giles) [*Orthopodomyia*]
1904c (Dec. 1): 366 (A; *Finlaya*). **Type loc.:** not stated (LU). **Refs.:** Stone 1967: 205.

allophus (Lutz & Peryassú) [*Anopheles*]
1921 (M*, P*; as *Cellia allopha*). In Peryassú 1921a: 70. **Type loc.:** several localities in Rio de Janeiro and Distrito Federal, Brazil (MNRS). **Refs.:** Lane 1939: 22 (syn.); Belkin et al. 1971: 4 (type info.); Lourenço-de-Oliveira & Deane 1984 (to *nomen dubium*?); Linthicum 1988 (*nomen dubium*).

americanus (Neveu-Lemaire) [*Culex*]
1902b: 23 (F*; *Aedeomyia*). **Type loc.:** Counani, French Guiana [Guiana] (NE). **Refs.:** Lane 1953: 389 (M*, F, P, L*); van der Kuyp 1954: 58 (M, F, P, L*); Fauran 1961b: 45 (tax.); Belkin 1968b: 12 (tax.); Belkin et al. 1971: 27 (type info.).

arnoulti Grjebine [*Anopheles*]
1966: 273 (L*). **Type loc.:** Andamazaotra Forest, near Perinet, eastern Madagascar (MNHP). **Refs.:** Brunhes, Le Goff, & Geoffroy 1998: 140 (to *nomen dubium*).

asebeomyia Aiken [*Culex* subgenus]
1911: 193 (A; new genus). Type species: *Culex epirus* Aiken. **Refs.**: Edwards 1932a: 212 (syn. with *Melanoconion*); Sirivanakarn 1983: 273 (to *nomen dubium*).

barkerii (Theobald) [*Culex*]
1907: 623 (M*; *Philodendromyia*). **Type loc.:** Sarawak [Borneo] (NHMUK). **Refs.:** Stone et al. 1959: 232 (from ? syn. with *quadripalpis*); Sirivanakarn 1977a: 9 (to *nomen dubium*).

bernardi (Borel) [*Culex*]
1927. **Type loc.:** Terres Rouges, Cochin China, Indochina [South Vietnam] (LU). **Refs.:** Borkent & Grimaldi 2004: 111 (M*, L*; *Lophoceratomyia*).

brachypus Dönitz [*Anopheles*]
1902: 52 (A). ***Type loc.:*** Indonesia (LU). **Refs.:** Edwards 1932a: 40 (? = *Anopheles albotaeniatus*). ***Ety.:*** n.s. [short foot (Gr)].

chrysothorax Newstead & Thomas [*Culex*]
1910: 147 (M, F; *Neomelaniconion*; not Peryassú 1908). **Type loc.:** Inner Flores swamp, Pensador, near Manaos [Manaus] (Amazonas), Brazil (NHMUK). **Refs.:** Dyar 1928: 327 (syn. with *theobaldi*); Senevet 1937: 374 (syn. with *chrysonotum*); Rozeboom & Komp 1950a: 97 (syn. with *theobaldi*); Belkin 1968b: 14 (lectotype desig.); Townsend 1990: 60 (type info.).

chrysothorax (Peryassú) [*Culex*]
1908: 244 (M, F; *Melanoconion*). **Type loc.:** Copacabana, Rio de Janeiro, Guanabara, Brazil (IOC). **Refs.:** Dyar 1928: 326 (M, F); Belkin et al. 1971: 23 (lectotype desig., type info.); Sirivanakarn 1983: 273 (to *nomen dubium*).

ciliaris (Linnaeus) [*Aedes*]
1767: 1002 (A; *Culex*). **Type loc.:** Sweden (NE). **Refs.:** Natvig 1948: 401 (? = *Aedes cinereus*).

conopas (von Frauenfeld) [*Mansonia*]
1867: 451 (F; *Culex*). **Type loc.:** China [People's Republic of China] (LU). **Refs.:** Edwards 1932a: 120 (? = *Mansonia uniformis*).

costalis Loew [*Anopheles*]
1866: 55 (F). **Type loc.:** Caffrerei [Cape Province, South Africa] (NE). **Refs.:** Evans 1938: 340 (? = *listeri* De Meillon), 343 (tax.); Gillies & De Meillon 1968: 263 (*nomen dubium*).

cotopaxensis Leví-Castillo [*Trichoprosopon*]
1953: 66 (M*, F, P*). **Type loc.:** Macuchi, Cotopaxi, Ecuador (NE?). **Refs.:** Zavortink 1979b: 11 (type prob. nonexistent; composite description of several sp.; from *Trichoprosopon* (*Runchomyia*)).

courdurieri Grjebine [*Anopheles*]
1966: 378 (L*). **Type loc.:** Bevilamy, Fort Dauphin, eastern Madagascar (MNHP). **Refs.:** Brunhes et al. 1998: 140 (to *nomen dubium*).

decorator Dyar & Knab [*Culex*]
1906d: 218 (L*). **Type loc.:** Tobago (NE). **Refs.:** Stone & Knight 1957a: 47 (type info.); Sirivanakarn 1983: 273 (type info.; *taxon dubium* [= *nomen dubium*]).

epirus Aiken [*Culex*]
1909: 8 (F). **Type loc.:** Corentyne [Courantyne] River, [Berbice], British Guiana [Guyana] (NE). **Refs.:** Belkin 1968b: 15 (tax.); Sirivanakarn 1983: 273 (to *nomen dubium*).

fasciolatus (Lutz) [*Culex*]
1904 (A; *Melanoconion*). In Bourroul 1904: 70. **Type loc.:** São Paulo, Brazil (IOC?). **Refs.:** Dyar 1928: 335 (F); Belkin et al. 1971: 23 (type info.); Sirivanakarn 1983: 273 (type info.; *taxon dubium* [= *nomen dubium*]).

flavirostris (Meigen) [*Culiseta*]
1830b: 242 (M; *Culex*). **Type loc.:** Europe (MNHP, NMW). **Refs.:** Edwards 1932a: 105 (? = *Culiseta morsitans*).

goeldii (Giles) [*Psorophora*]
1904b: 383 (A; *Taeniorhynchus*). **Type loc.:** none designated (LU). **Refs.:** Belkin 1968b: 26.

gravitator Dyar & Knab [*Culex*]
1906d: 218 (L*). **Type loc.:** Cordoba [Veracruz], Mexico (NE). **Refs.:** Foote 1954: 55 (L*; tax.); Stone & Knight 1957a: 50 (type info.); Sirivanakarn 1983: 273 (type info.; *taxon dubium* [= *nomen dubium*]).

humilis Theobald [*Culex*]
1901: 336 (M, F). **Type loc.:** São Paulo, Brazil (NHMUK). **Refs.:** Foote 1954: 56 (L*; tax.); Belkin 1968b: 16 (lectotype desig.); Sirivanakarn 1983: 273 (type info.; *taxon dubium* [= *nomen dubium*]).

hyperleucum (Martini) [*Trichoprosopon*]
1931a: 201 (A; *Goeldia leucopus* var.). **Type loc.:** Urubamba River [Cusco], Peru (NE). **Refs.:** Lane 1953: 858 (F); Belkin 1968b: 36 (tax.); Zavortink 1979b: 12 (type lost; inadequately described; from *Trichoprosopon* (*Runchomyia*)).

indecorabilis (Theobald) [*Culex*]
1903: 241 (?M, F*; *Melanoconion*). **Type loc.:** Pará, Brazil (NHMUK). **Refs.:** Dyar 1928: 332 (M, F); Belkin 1968b: 16 (lectotype desig.); Sirivanakarn 1983: 273 (to *nomen dubium*).

jacobi (Hill & Haydon) [*Anopheles*]
1907: 144 (F*, L*; *Cellia*). **Type loc.:** Natal [KwaZulu-Natal, South Africa] (LU). **Refs.:** Evans 1938: 388 (F); De Meillon 1947b: 9 (tax.); Gillies & De Meillon 1968: 263 (*nomen dubium*).

javanensis Swellengrebel & Swellengrebel de Graaf [*Anopheles*]
1920a: 91.

lugens Lutz [*Culex*]
1905: 28 (M, F). **Type loc.:** Lagoa, São Paulo, Brazil (IOC). **Refs.:** Dyar 1928: 336 (M, F); Belkin et al. 1971: 24 (type info.); Sirivanakarn 1983: 273 (type info.; *taxon dubium* [= *nomen dubium*]).

lutzii Theobald [*Sabethes*]
1903: 323 (A). **Type loc.:** Manaos [Amazonas], Brazil (IOC?). **Refs.:** Lane & Cerqueira 1942: 670 (tax.); Belkin 1968b: 31 (type info.; to *nomen dubium*); Belkin et al. 1971: 14 (type info.).

maculatus von Humboldt [*Culex*]
1819: 340 (A; *Culex*; not Meigen 1804). **Type loc.:** Near Babaoyo, [Los Rios], Ecuador (NE). **Refs.:** Belkin 1968b: 17 (tax.).

madecassorum Rodhain & Boutonnier [*Orthopodomyia*]
1984b: 228 (F). **Type loc.:** Perinet forest, Madagascar (PIP). **Refs.:** Brunhes & Hervy 1995: 15 (to *nomen dubium*).

marmorata (Philippi) [*Psorophora*]
1865: 597 (M, F; *Culex*). **Type loc.:** Chile (NE). **Refs.:** Edwards 1932a: 124 (subg. *Psorophora*?).

mcgregori (Banks) [*Orthopodomyia*]
1909: 548 (M; *Kerteszia*). **Type loc.:** Basilan Island [Zamboanga, Mindanao], Philippines (NE). **Refs.:** Knight & Chamberlain 1948: 15 (P*; from syn. with *albipes*); Knight & Mattingly 1950: 12 (M, P, L; tax.); Zavortink 1968: 154 (to *nomen dubium*); Basio 1971b: 79 (M*; bion.); Zavortink 1971: 20 (as *nomen dubium*).

microtaeniata Theobald [*Culex*]
1911: 236 (F; *Culex*). **Type loc.:** Samarang, Java (AM). **Refs.:** Colless 1957a: 100 (tax.).

mindanaoensis Baisas [*Culex*]
1935d: 168 (M*). **Type loc.:** Cotabato, Mindanao, Philippines (NE). **Refs.:** Basio 1971b: 67 (M*; bion.); Sirivanakarn 1977a: 9 (to *nomen dubium*).

minutus Macquart [*Anopheles*]
1834: 33 (F). **Type loc.:** Senegal (NE). **Refs.:** Edwards 1932a: 51 (? = *funestus* Giles); Gillies & De Meillon 1968: 263 (*nomen dubium*).

molestus Kollar [*Culex*]
1832: 18 (M, F*; not Forskål 1775). In Pohl & Kollar 1832. **Type loc.:** Rio de Janeiro, Guanabara, Brazil (NE). **Refs.:** Belkin 1968b: 17 (tax.; to sp.); Belkin et al. 1971: 28 (tax.); Lee et al. 1989a: 143 (F; key, tax., bion., distr., review).

nero (Doleschall) [*Anopheles*]
1857: 383 (M; *Culex*). **Type loc.:** Gombong, Java [Indonesia] (LU). **Refs.:** Reid 1953: 27 (? = *nigerrimus*).

nigrescens Theobald [*Culex*]
1907: 540 (M, F*, only of type series; *Aedes*) (not Theobald 1907: 248). **Type loc.:** Santo Amaro [southwest of São Paulo], Brazil (NHMUK). **Refs.:** Bonne & Bonne-Wepster 1925: 288 (M*, F, L*); Dyar 1928: 303 (M*, F, L*); Rozeboom & Komp 1950a: 93 (M*); Lane 1953: 480 (M*; tax.); Casal 1963: 317 (M*); Belkin et al. 1971: 24 (type loc. info.); Sirivanakarn 1983: 273 (*taxon dubium* [= *nomen dubium*]).

nigricorpus Theobald [*Culex*]
1901: 231 (F; *Aedes*). **Type loc.:** Itacoatiara, Amazonas, Brazil (NHMUK). **Refs.:** Dyar 1928: 334 (F); Belkin 1968b: 18 (lectotype desig.); Sirivanakarn 1983: 273 (*taxon dubium* [= *nomen dubium*]); Townsend 1990: 110 (type info.).

niveus (von Eichwald) [*Aedes*]
1837. **Type loc.:** Baku [Azerbaijan] (LU).

novaeguineae Evenhuis [*Culex*]
1989 (F; *Culex ventralis*, not Walker 1860b). In Evenhuis & Gon 1989 (F; *Culex ventralis*, not Walker 1860). **Refs.:** Evenhuis & Gon 1989: 210 (new name for *ventralis* Walker 1865; preoccupied by *ventralis* Walker 1860b).

oblitus (Lynch Arribálzaga) [*Culex*]
1891: 154 (M; *Janthinosoma*?). **Type loc.:** Navarro, Province Bonaerense [Buenos Aires], Argentina (BA). **Refs.:** Belkin et al. 1968: 16 (n. comb.).

pallipes Robineau-Desvoidy [*Culex*]
1827: 410 (F). **Type loc.:** Rio de Janeiro, Guanabara, Brazil (NE, originally in MNHP). **Refs.:** Belkin 1968b: 19 (*nomen dubium*; type info.); Belkin et al. 1971: 28 (type loc. info.).

penetrans (Robineau-Desvoidy) [*Culiseta*]
1827: 407 (M, F; *Culex*). **Type loc.:** "Gallia" [France] (NE). **Refs.:** Edwards 1932a: 104 (? = *Culiseta subochrea*).

perdita Leicester [*Verrallina*]
1908: 192 (M; *Aioretomyia*). **Type loc.:** Malaya, Gombak Road, Kuala Lumpur [Selangor] (NE). **Refs.:** Reinert et al. 2004: 354 (to *nomen dubium*).

perterrens (Walker) [*Psorophora*]
1856a: 431 (F; *Culex*). **Type loc.:** Rio de Janeiro, Guanabara, Brazil (NE). **Refs.:** Stone 1957a: 337 (from syn. with *ciliata*; tax.); Belkin et al. 1971: 19 (type loc. info.).

rimandoi (Basio) [*Aedes*]
1971a: 52 (M*, F). **Type loc.:** University of Philippines College of Forestry Campus, Los Baños, Laguna, Luzon, Philippines (PNM). **Refs.:** Huang 1979a: 27 (to *nomen dubium* from *Stegomyia*).

silvestris (Shingarev) [*Culiseta*]
1928: 51 (M*, F; *Theobaldia*). **Type loc.:** Moskva, Union of Soviet Socialist Republics (USSR) (LU). **Refs.:** Maslov 1967: 161 (M*, F, L); Maslov 1989 [English translation of Maslov 1967]; Dahl & White 1978: 393 (to *nomina dubia* in footnote without explanation).

soalalaensis Grjebine [*Anopheles*]
1953a: 501 (F; as *fuscicolor* var.). **Type loc.:** Namoroka, Soalala District, Madagascar (IRSM). **Refs.:** White 1980: 116 (from *fuscicolor* var.); Brunhes et al. 1998: 139 (to *nomen dubium*).

striocrurus (Giles) [*Armigeres*]
1904a: 367 (F?). **Type loc.:** Philippines (NE).

suborientalis Baisas [*Culex*]
1938: 217 (F). **Type loc.:** Baguio, Luzon, Philippines (NE). **Refs.:** Delfinado 1966a: 156 (to *nomen dubium* from *Culex* (*Culex*)).

upemba Lips [*Anopheles*]
1960b: 303 (F; validated by reference to Mattingly 1955: 50). **Type loc.:** Lusinga, Upemba National Park, Katanga, Congo [Democratic Republic of the Congo] (CMT). **Refs.:** White 1975: 316 (type info.); Brunhes et al. 1997: 180 (to *nomen dubium*).

ventralis (Walker) [*Armigeres*]
1860b: 144 (F; *Culex*). **Type loc.:** Amboina Island, Moluccas [Indonesia] (NHMUK). **Refs.:** Edwards 1932a: 185 (? = *Armigeres obturbans*); Lee et al. 1988a: 30 (tax., review).

ventralis Walker [*Culex*]
1865: 103 (F; *Culex*; not Walker 1860b). **Type loc.:** New Guinea (NE).

virgultus Theobald [*Culex*]
1901c: 123 (M*). **Type loc.:** Parque do Museu, Rio de Janeiro, Brazil (NHMUK). **Refs.:** Stone 1957a: 341 (tax.); Belkin 1968b: 21 (type info.; *nomen dubium*).

vulgaris Hatori [*Anopheles*]
1901: 274 (A*). **Type loc.:** Taihoku, Kielun, Tansui, Formosa [Taiwan, Republic of China] (LU).

vulgaris Linnaeus [*Culex*]
1792: 380 (A; *Culex*). **Type loc.:** Lapland [Sweden] (NE) Howard, Dyar, & Knab 1915:218 (mixture of a simuliid and perhaps an *Aedes* sp.).

Nomina Nuda and Rejected Names

The following names comply with the ICZN definition for designation of *nomen nudum* (pl. *nomina nuda*): "A Latin term referring to a name that, if published before 1931, fails to conform to Article 12; or, if published after 1930, fails to conform to Article 13. A *nomen nudum* is not an available name, and therefore the same name may be made available later for the same or a different concept; in such a case it would take authorship and date [Articles 21, 50] from that act of establishment, not from any earlier publication as a *nomen nudum*."

africanus Theobald [*Anopheles*]
1901d: 1 (*Anopheles barbirostris* var.). **Refs.:** International Commission on Zoological Nomenclature 1968, Opinion 850 (name suppressed).

ainshamsi Gad, El Said, Hassan, & Shoukry [*Anopheles*]
1987: 214. **Refs.:** Harbach 2018c (without comment).

albirostris Macquart [*Aedes*]
1850: 10 (F; *Culex*). **Type loc.:** New South Wales, Australia (MNHP). **Refs.:** Belkin 1968b: 102 (type loc. info.); Marks & Brown 1970: 227 (request for suspension of rules); International Commission on Zoological Nomenclature 1972, Opinion 979 (name suppressed).

alboannulatus Giles [*Anopheles*]
1904c: 38 (*Myzorhynchus*). **Refs.**: Stone et al. 1959: 287 (as *alboannulus* Giles).

alboaurata Petrocchi [*Psorophora*]
1923: 92 (in a list as "n. sp." but without further information).

albus La Casse & Yamaguti [*Aedes*]
1950: 136 (*Aedes* (*Ochlerotatus*) *dorsalis* ssp.). **Refs.:** Harbach 2018c (without comment).

aranoideoides Dong, Zhou, Gong et al. [*sic*] [*Tripteroides*]
2004: 187. **Refs.:** Harbach 2018c (without comment).

argentina Petrocchi [*Uranotaenia*]
1923: 92 (in a list as "n. sp." but without further information).

argyrotarsis Lutz [*Wyeomyia*]
1904. In Bourroul 1904: 77. **Refs.:** Knight & Stone 1977 (without comment); Harbach 2018c (without comment).

bambusae Dyar [*Culex*]
Dyar 1922j: 419 (in a list of spp. identified by Dyar from Costa Rica).

bifoliata Osorno-Mesa & Muñoz-Sarmiento [*Anopheles*]
1948: 105 (M*, F, L*, E*; *An. pseudopunctipennis* var. *bifoliata* [*sic*]). **Type loc.**: Florida, Valle del Cauca, Colombia (DMB). **Refs.**: Harbach & Howard 2007 (to *nomen nudum*).

bironelli Christophers
1924a: 16. New name for *gracilis* Theobald 1905f; as *Anopheles* (*Bironella*) *bironelli*. Rejected name (ICZN 1965, Opinion 737).

caucasicus Stackelberg [*Anopheles*]
1937: 52, table (*An. maculipennis maculipennis caucasicus*). **Refs.:** Knight & Stone 1977 (table); Harbach 2018c (in a table).

centaurus Theobald [*Psorophora*]
1905e: 16 (a Walker manuscript name).

chaquense Petrocchi [*Culex*]
1923: 91 (in a list as "n. sp." but without further information).

circumvolens Walker [*Culex*]
1861b. In Tennent 1861: 434. **Refs.:** Knight & Stone 1977 (without comment); Harbach 2018c (without comment).

cleopatrae Willcocks [*Anopheles*]
1910: 586 (*Pyretophorus*). **Refs.:** Knight & Stone 1977 (without comment); Harbach 2018c (without comment).

colon Harris [*Culex*]
1835. In Hitchcock 1835: 595. **Refs.:** Knight & Stone 1977 (without comment); Harbach 2018c (without comment).

contrahens Walker [*Culex*]
1861b. In Tennent 1861: 462.

cucphuongensis Anonymous [*Anopheles*]
1987. In IMPE (pp. 8, 36, 49). **Refs.:** Harbach 2018c (unavailable name).

cyanoconops Antunes
1939: 106 (*Haemagogus* subg.). **Refs.:** Knight & Stone 1977 (without comment); Harbach 2018c (without comment).

dahomeensis Holstein [*Anopheles*]
1951: 464, 465 (*Anopheles funestus* var.; in list only).

dangi Anonymous [*Anopheles*]
1987. In IMPE (pp. 8, 36, 49). Harbach 2018c (unavailable name; therefore not a synonym of *crawfordi*, as described by Ngo et al. (2013).

dubitans Christophers [*Anopheles*]
1924a: 30 (name only listed as from a Leicester label on specimens of *An. separatus*).

dubius Blanchard [*Anopheles*]
1905: 205 (name listed without comment in a synonymy of *Nyssorhynchus cubensis* Agramonte (= *An. albimanus*).

dyari Petrocchi [*Psorophora*]
1923: 92 (in a list as "n. sp." but without further information).

errabunda Swellengrebel [*Anopheles*]
1925: 1913 (*Cellia*). **Refs.:** International Commission on Zoological Nomenclature: 1969, Opinion 874 (name suppressed).

flavinervis Edwards [*Aedes*]
1915a: 161 (*Banksinella luteolateralis* var. refers to co-occurring nongenetic variant of species *luteolateralis*).

flavus Yamada [*Aedes*]
1927: 575 (*Aedes esoensis* var.). **Refs.:** Knight & Stone 1977 (without comment); Harbach 2018c (without comment).

florense Petrocchi [*Culex*]
1923: 91 (in a list as "n. sp." but without further information).

funerearoides Dong, Zhou, Gong et al. [*sic*] [*Heizmannia*]
2004: 187. **Refs.:** Knight & Stone 1977 (Harbach 2018c (without comment).

furtivus Dyar [*Culex*]
1922j: 419 (in a list of spp. identified by Dyar from Costa Rica).

fusco Petrocchi [*Culex*]
1923: 91 (in a list as "n. sp." but without further information).

guttulatus Harris [*Culex*]
1835: 595 (*Anopheles quadrimaculatus*? var.). **Refs.:** Knight & Stone 1977 (without comment); Harbach 2018c (without comment). [not seen]

haddowi Gillett [*Anopheles*]
1975: 366 (a suggestion for a name for *Anopheles* (*Cellia*) *gambiae* species D, = *Anopheles* (*Cellia*) *bwambae* White 1985).

hamadryadis Cooling [*Aedes*]
1924: 24 (*Aedes* (*Finlaya*) *priestleyi* ssp.). **Refs.:** Lee et al. 1982: 236 (a manuscript name; therefore, transferred from synonymy of *purpureus* to *nomen nudum*); Ward 1984b

juppi Dumas, Atyame, Malcolm, Le Goff, Unal, Makoundou, Pasteur, Weill, & Duron [*Culex*]
2016: 807. **Refs.:** Harbach 2018c (to *nomen nudum*: not compliant with ICZN).

keybergi Lips [*Anopheles*]
1960a: 49. **Refs.:** White 1975: 313 (tax.; to *nomen dubium*); White 1980: 118 (as *nomen dubium* in catalog entry for *An.* (*Cel.*) *dureni millecampsi* Lips).

kolhapuriensis Sathe & Girhe [*Aedes*]
2002: 197. **Refs.:** Harbach 2018c: 69 (to *nomen nudum*: not compliant with ICZN; no name-bearing type or type depository designated).

krishnai Sathe & Japtag
2002. **Refs.:** Harbach 2018c: 70 (fails to conform to Articles 13.1 and 16.4 of the Code of the ICZN (1999) [status by Harbach, 27 June 2012, http://mosquito-taxonomic-inventory.info/whats-new].

leucotelus McCormack [*Culex*]
1919: 29. **Refs.:** Knight & Stone 1977 (without comment); Harbach 2018c (without comment).

loewi (Giebel) [*Toxorhynchites*, extant and fossil]
1862: 317 (A; *Culex* [a synonym of *Toxorhynchites brevipalpis* Theobald 1901]). **Type loc.:** East Africa? (NMC). **Distr.:** Upper Oligocene. **Refs.:** Hennig 1966a: 5; Stone 1967: 218 (tax.); International Commission on Zoological Nomenclature 1982, Opinion 1213 (name suppressed).

marajoensis Peryassú [*Culex*]
1923c: 76. **Refs.:** Knight & Stone 1977 (without comment); Harbach 2018c (without comment).

martinii Statz [*Mansonia* fossil]
1941: 8 (see *martinii* Statz 1944).

maximalepido Dong, Zhou, Gong et al. [*sic*] [*Heizmannia*]
2004: 187. **Refs.:** Harbach 2018c (without comment).

melanotarsis Anonymous. Director General Medical Services [*Anopheles*]
1943: 8. **Refs.:** Lee et al. 1988a: 3, 14 (not validly published).

minutus Eversmann [*Culex*]
1834: 420. **Refs.:** Knight & Stone 1977 (without comment); Harbach 2018c (without comment).

moenensis Iyengar [*Aedes*]
1955: 30. **Refs.:** International Commission on Zoological Nomenclature 1985 (unavailable name; Article 12a(i)).

monstrosus Anonymous. [*Anopheles*]
1987. In IMPE (pp. 11, 33, 57). **Refs.:** Harbach 2018c (unavailable name).

multimaculatus Gastiaburú [*Anopheles*]
1945. In Aitken 1945: 327. **Refs.:** Knight & Stone 1977 (without comment); Harbach 2018c (without comment).

neivai Petrocchi [*Megarhynus*]
1923: 92 (in a list as "n. sp." but without further information; name later used by Pettrochi 1926: 725).

niansimbii Gillett [*Anopheles*]
1975: 366 (a suggestion for a name for *Anopheles* (*Cellia*) *gambiae* species D, = *Anopheles* (*Cellia*) *bwambae* White 1985).

nielseni O'Meara & Craig [*Aedes*]
1970: 53 (*Ae.* (*Finlaya*) *atropalpus* ssp.; *nielseni*; name inadvertently (?) published (Carpenter 1970) in advance of formal description of *Aedes nielseni*).

niger Herrick [*Anopheles*]
1905: 283. Ward 1979 (as *lapsus calami*); Harbach 2018c (as *nomen nudum*).

nigerrimus Dong, Zhou, Gong et al. [*sic*] [*Toxorhynchites*]
2004: 187. Harbach 2018c (*nomen nudum*).

Osornomyia Leví-Castillo
1955a: 361 (in *Haemagogus* subg.). **Refs.:** Knight & Stone 1977 (without comment); Harbach 2018c (without comment).

pallida Theobald
1907: 470 (*Banksinella luteolateralis* var.). Rejected name, suppression by the International Commission on Zoological Nomenclature, 1994, Opinion 1775.

pallidabdomen Dong, Zhou, Gong et al. [*sic*] [*Armigeres*]
2004: 187. Harbach 2018c (*nomen nudum*).

panchgangee Sathe & Girhe [*Aedes*]
2002: 197. **Refs.:** Harbach 2018c (to *nomen nudum*: not compliant with ICZN; no name-bearing type or type depository designated).

papuensis De Rook [*Anopheles*]
1929: 29, 39, 45 [pp. 34, 48, 55 in 1930 English translation] (as *Anopheles*). **Refs.:** Lee et al. 1988a: 67 (probably a *Bironella* species; to *nomen nudum*).

perlisiensis Yong, Cheong, Chiang, Dhaliwal, Loong, & Sarjan [*Anopheles*]
1983: 613 (*An. balabacensis* ssp.). **Refs.:** Ward 1984b (to *nomen nudum*); Sallum et al. 2005a (as syn. of *An.* (*Cel.*) *cracens* Sallum & Peyton); Harbach 2018c (corrected to *nomen nudum*).

peus Speiser [*Culex*]
1904: 148 (name suppression by International Commission on Zoological Nomenclature 1991, Opinion 1644).

portoricensis Weinbren & Weinbren [*Wyeomyia*]
1970: E-161 (name only in a footnote as "*Wyeomyia portoricensis* Pratt").

pseudoconfirmatus Giles [*Coquillettidia*?]
1902: Pl. XIII. **Refs.:** Harbach 2018c (to *nomen nudum*; "name informally used to indicate resemblance of a '*Panoplites* sp.' to '*C.* (*Ochlerotatus*) *confirmatus*'").

putaneus Theobald [*Psorophora*]
1901a: 257 (*Janthinosoma*). **Refs.:** Harbach 2018c.

pygmaeus Curtis [*Anopheles*]
1831, column 202. **Refs.:** Knight & Stone 1977 (without comment); Harbach 2018c (without comment).

rabaiensis Lounibos [*Anopheles*]
1979: 28. **Refs.:** Ward 1984b (to *nomen nudum*; given as "White, in press" but not subsequently published).

rectirostris Theobald [*Toxorhynchites*]
1910c: 214. **Refs.:** Knight & Stone 1977 (without comment); Harbach 2018c (without comment).

saheliensis Senevet & Andarelli [*Anopheles*]
1955b: 128 (*Anopheles claviger petragnanii* [*sic*] race). **Refs.:** Knight & Stone 1977 (without comment); Harbach 2018c (without comment).

salteno Petrocchi [*Culex*]
1923: 91 (in a list as "n. sp." but without further information).

sangitee Sathe & Girhe [*Aedes*]
2002: 197. **Refs.:** Harbach 2018c (to *nomen nudum*: not compliant with ICZN; no name-bearing type or type depository designated).

sangiti Girhe & Sathe [*Aedes*]
2001: 46. **Refs.:** Harbach 2018c (to *nomen nudum*: not compliant with ICZN; no name-bearing type or type depository designated).

silvestris Ross [*Culex*]
1898: 134. **Refs.:** Knight & Stone 1977 (without comment); Harbach 2018c (without comment).

simplex Baisas & Catipon [*Uranotaenia*]
1958: 54. **Refs.:** Knight & Stone 1977 (without comment); Harbach 2018c (without comment).

singalesi Marlatt [*Culex*]
1903: 122. **Refs.:** Knight & Stone 1977 (without comment); Harbach 2018c (without comment).

singalesi Theobald [*Culex*]
1901. In Green 1901: 368. **Refs.:** Knight & Stone 1977 (without comment); Harbach 2018c (without comment).

smudges Dong, Zhou, Gong et al. [*sic*] [*Uranotaenia*]
2004: 187. **Refs.:** Harbach 2018c (without comment).

solomonensis Cumpston [*Anopheles*]
1924: 1402. **Refs.:** Harbach 2018c (without comment).

spinophallus Dong, Zhou, Gong et al. [*sic*] [*Uranotaenia*]
2004: 187. **Refs.:** Harbach 2018c (without comment).

splendenroides Dong, Zhou, Gong et al. [*sic*] [*Toxorhynchites*]
2004: 187. **Refs.:** Harbach 2018c (without comment).

tanzaniae Capasso [*Culex* fossil]
1991: 173. **Refs.:** Harbach 2018c (without comment).

timendus Walker [*Culex*]
1874: 25. **Refs.:** Knight & Stone 1977 (without comment); Harbach 2018c (without comment).

varivestita Statz [*Mansonia*]
1941: 11. **Refs.:** Harbach 2018c (without comment).

vadakadiensis Doraisamy [*Anopheles subpictus* var.]
1963: 42 (unavailable name: published as a variety after 1960). **Refs.:** Harbach 2018c: 135.

varovum Breeland [*Anopheles*]
1953: 2, 7 (*An. crucians* ssp.).

vietnamensis Anonymous. [*Anopheles*]
1987. In IMPE (pp. 13, 33, 57). Refs.: Harbach 2018c (unavailable name).

yunquensis Weinbren & Weinbren [*Wyeomyia*]
1970: E-161 (name only in a footnote as "*Wyeomyia yunquensis* Pratt").

Family-Group and Other Suprageneric Names

The following is a list of names compiled by Harbach (2018c, pp. 157–160). The names and primary citations are given to point the reader to names that for the most part are not covered in the present catalog. See Harbach (2018c) for further explanations and taxonomic histories. Names in bold are currently in use as valid.

Aedeomyiinae. Corrected spelling of Aedeomyinae Neveu-Lemaire 1902c.

Aedeomyiini Theobald 1901b: 235.

Aedeomyina Theobald 1901b: 235.

Aedeomyinae Neveu-Lemaire 1902a: 201.

Aedina Lahille 1904: 79, 84.

Aedinae Lutz 1904: 45, 53, 65. In Bourroul 1904.

Aedinae Neveu-Lemaire 1902c: 201, 220.

Aedini Neveu-Lemaire 1902c: 220.

Aedomina Giles 1902: 255, 256, 475.

Ankylorhynchae Lutz 1904: 53. In Bourroul 1904.

Anopheleinae Eysell 1905: 24, 1905b: 55

Anophelidae Eysell 1905: 24, 1905b: 55.

Anophelina Theobald 1901a: 97, 115.

Anophelinae Grassi 1900: 359.

Anophelini Edwards 1912c: 53.

Anophilini Pearse 1936: 53. In Pearse et al. 1936, 45–59.

Asiphonatae Lutz 1904: 51. In Bourroul 1904.

Branchiopodiformes Bouché 1834: 5.

Burmaculicinae Borkent & Grimaldi 2016: 465.

Campestrae Lynch Arribálzaga 1891a: 367.

Culicales Alcock 1911: 241, 247.

Culices Griffith & Pidgeon 1832: 726.

Culicida Perty 1841: 888.

Culicidae Meigen 1818: xxxiii.

Culicidea Bellardi 1859: 5.

Culicides Billberg 1820: 122.

Culicidi Lioy 1863: 202.

Culiciformes Meigen 1818: xxxiii.

Culicinae Meigen 1818: xxxiii.

Culicini Meigen 1818: xxxiii.

Culisetina Maslov 1964: 198.

Culisetini Belkin 1962: 282.

Culisetini Stone 1957a: 333 (*nomen nudum*).

Deinoceritinae Mitchell 1906: 19.

Dendromyiinae. Corrected spelling of Dendromyinae Lutz: 1904.

Dendromyinae Lutz 1904: 47, 56, 67. In Bourroul 1904.

Dinoceratinae Alcock 1911: 241.

Domesticae Lynch Arribálzaga 1891a: 367.

Epialurgi Alcock 1911: 241, 247.

Euculicidae Lutz 1904: 51, 63. In Bourroul 1904.

Ficalbiini Belkin 1962: 287.

Haemagoginae Lutz 1904: 47, 66. In Bourroul 1904.

Harpagomyiinae. Corrected spelling of Harpagomyinae Ludlow.

Harpagomyinae Ludlow 1911b: 127, 132.

Heptaphlebomyiinae. Corrected spelling of Heptaphlebomyinae Lutz.

Heptaphlebomyiini Handlirsch 1924: 973.

Heptaphlebomyina Theobald 1903a: 336.

Heptaphlebomyinae Lutz 1904. In Bourroul 1904.

Heteropalpae Lutz 1904: 55. In Bourroul 1904.

Hétéropsélaphes Lahille 1904: 79.

Hodgesiini Belkin 1962: 278.

Hyloconopinae Lutz 1904: 49, 55, 48. In Bourroul 1904.

Joblotiinae Grünberg 1907b: 63, 107.

Joblotina Theobald 1903a: 334.

Joblotinae Theobald 1905e: 5, 33.

Limatinae Theobald 1907: 18, 627.

Lynchiellina Lahille 1904: 79, 81.

Lynchinellinae Peryassú 1908: 29.

Macropsélaphes Lahille 1904: 79.

Mansoniini Belkin 1962: 299.

Mansonini Ronderos & Bachmann 1963a: 45.

Megalorhini Alcock 1911: 241.

Megalorhinina Marshall 1923: 99. In Neveu-Lemaire 1923.

Megalorrhini Brunetti 1920: 90.

Megarhinae Giles 1902: 265.

Megarhini Brunetti 1914: 32, 34.

Megarhinina Theobald 1901d: 97, 215.

Megarhininae Neveu-Lemaire 1902c: 201, 206.

Megarhinini Edwards 1932a: 8, 29, 57.

Megarinina Theobald 1904c: 36.

Megarrhini Bertrand 1954: 200.

Megarrhininae Bezzi 1906: 209.

Megharinae González-Rincones & Guyon 1953: 169, 182, 199.

Metanotopsilae Lutz 1904: 53. In Bourroul 1904.

Metanototricha Alcock 1911: 241, 249.

Metanototrichae Lutz 1904.

Metanototrichina Marshall 1923: 99. Footnote in Neveu-Lemaire 1923.

Metanotrichae Lutz 1904: 55. In Bourroul 1904.

Micropalpae Lutz 1904: 53, 56. In Bourroul 1904.

Micropsélaphes Lahille 1904: 79.

Opificinae Miller 1922: 115.

Orthopodomyiini Belkin, Heinemann, & Page 1970: 112.

Orthorhynchae Lutz 1904. In Bourroul 1904.

Palustrae Lynch Arribálzaga 1891a: 367.

Protoptères Lahille 1904: 78, 79.

Psorophorinae Mitchell 1906: 21.

Sabethinae Dyar 1906: 191.

Sabethini Blanchard 1905: 420.

Sabethinini Shannon 1927: 12.

Sabettinae Blanchard 1905: 420.

Silvicolae Lynch Arribálzaga 1891a: 367.

Siphonatae Lutz 1904: 53. In Bourroul 1904.

Stegomyiinae. Corrected spelling of Stegomyinae Peryassú.

Stegomyina Christophers 1906: 16, 17.

Stegomyinae Peryassú 1908: 364.

Téléoptères Lahille 1904: 78, 79.

Toxorhynchina Lahille 1904: 79, 84.

Toxorhynchitina da Costa Lima, Guitton, & Ferreira 1962: 226.

Toxorhynchitinae Theobald 1905e: 5, 13.

Toxorhynchitini Lahille 1904: 79.

Trichoprosopinae. Corrected spelling of Trichoprosoponinae Coquillett.

Trichoprosopini. Corrected spelling of Trichoprosoponina.

Trichoprosoponina Theobald 1901a: 98.

Trichoprosoponinae Coquillett 1906d: 11, 26 (but see Sabrosky 1999: 313).

Uranotaeniani Pearse 1936: 53.

Uranotaeniinae Coquillett 1906d: 11, 26.

Uranotaeniini Lahiille 1904: 79.

Uranotaenina Lahille 1904: 79, 88.

Uranotaeninae Mitchell 1907: 297.

Wyeomiinae Lahille 1904: 77.

Wyeomyiinae (Corrected spelling of Wyeomyinae Leicester 1908).

Wyeomyinae Leicester 1908: 250.

Informal Species Groups

Genera:

Aedes
Anopheles
Culex
Mimomyia
Orthopodomyia
Toxorhynchites
Tripteroides
Uranotaenia Old World/New World
Verrallina

Genus *Aedes*

The classification used here follows Wilkerson et al. (2015) and Wilkerson & Linton (2015), which is a modified version of the schema proposed by Reinert, Harbach, & Kitching (2004, 2006, 2008, 2009). The Niveus Section of Brunhes et al. (2017) is not used here since *niveus* De Meillon is now in genus *Aedes*, not genus *Paulianus*. Species *niveus* De Meillon is therefore preoccupied, and *zethus* is the replacement species name, and for purposes of this list is also used for the section name.

Subgenus *Acartomyia* Theobald (Mariae Complex of Coluzzi et al. 1976)
mariae (Sergent & Sergent 1903)
phoeniciae Coluzzi & Sabatini 1968
zammitii (Theobald 1903a)
Subgenus *Collessius*
Alloeomyia Group (Wilkerson et al. 2015)
banksi Edwards 1922a
pseudotaeniatus (Giles 1901a)
tonkinensis Galliard & Ngu 1947
Collessius Group (Wilkerson et al. 2015)
elsiae (Barraud 1923a)
elsiae vicarious Lien 1968d
hatorii Yamada 1921
macdougalli Edwards 1922a
macfarlanei (Edwards 1914b)
ningheensis Lei 1989b
shortti (Barraud 1923a)
Subgenus *Downsiomyia*
Alboniveus Group (Reinert & Harbach 2006)
alboniveus Barraud 1934
saperoi Knight 1946
Niveus Group (Reinert & Harbach 2006)
albolateralis (Theobald 1908b)
axitiosus Kulasekera, Knight, & Harbach 1990
dorseyi Knight 1946
ganapathi Colless 1958
harinasutai Knight 1978a
idjenensis Brug 1934a
inermis Colless 1958
lacteus Knight 1946
laoagensis Knight 1946
leonis Colless 1958
litoreus Colless 1958
mikrokopion Knight, & Harrison 1988
mjobergi (Edwards 1926b)
mohani Knight 1969
nipponicus La Casse, & Yamaguti 1948
nippononiveus Sasa, & Nakahashi 1952
nishikawai Tanaka, Mizusawa, & Saugstad 1979
niveoides Barraud 1934
niveus (Ludlow 1903)
novoniveus Barraud 1934
omorii Lien 1968b
pexus Colless 1958
pseudoniveus (Theobald 1905f)
shehzadae Qutubuddin 1972
sinensis Chow 1950
subniveus Edwards 1922a
vanus Colless 1958
watteni Lien 1968b
Subgenus *Georgecraigius*
Georgecraigius Group (Wilkerson et al. 2015)
atropalpus (Coquillett 1902d)
epactius Dyar & Knab 1908
Horsfallius Group (Wilkerson et al. 2015)
fluviatilis (Lutz 1904)
Subgenus *Hopkinsius*
Hopkinsius Group (Wilkerson et al. 2015)
barnardi Edwards 1924a
embuensis Edwards 1930a
ingrami Edwards 1930a
nyasae Edwards 1930a
wellmanii (Theobald 1905b)
Yamada Group (Wilkerson et al. 2015)
albocinctus (Barraud 1924f)
seoulensis Yamada 1921
Subgenus *Macleaya*
Chaetocruiomyia Group (Wilkerson et al. 2015)
calabyi Marks 1963a
elchoensis Taylor 1929
humeralis Edwards 1922d
macmillani Marks 1964b
moloiensis Taylor 1929
spinosipes Edwards 1922d
tulliae (Taylor 1929)
wattensis Taylor 1929
Macleaya Group (Wilkerson et al. 2015)
littlechildi Taylor 1933
stoneorum Marks 1977
tremulus (Theobald 1903e)

Subgenus *Mucidus*
Mucidus Group (Wilkerson et al. 2015)
alternans (Westwood 1835)
ferinus Knight 1947a
grahamii (Theobald 1909b)
laniger (Wiedemann 1820)
lucianus Muspratt 1959
mucidus (Karsch 1887)
nigerrimus (Theobald 1913d)
quasiferinus Mattingly 1961
scatophagoides (Theobald 1901a)
sudanensis (Theobald 1908c)
tonkingi Gebert 1948
Pardomyia Group (Wilkerson et al. 2015)
aurantius (Theobald 1907)
aurantius chrysogaster (Taylor 1927)
nigrescens (Edwards 1929b)
quadripunctis (Ludlow 1910)
Subgenus *Ochlerotatus*
Buvirilia Group (Wilkerson et al. 2015)
edgari Stone & Rosen 1952
Chrysoconops Group (Wilkerson et al. 2015)
bimaculatus (Coquillett 1902c)
fulvus (Wiedemann 1828)
fulvus pallens (Ross 1943)
jorgi Carpintero & Lequizamón 2000
pennai Antunes & Lane 1938
stigmaticus Edwards 1922d
Culicada Group (Wilkerson et al. 2015)
canadensis (Theobald 1901c)
canadensis mathesoni Middlekauff 1944
Culicelsa Group (Wilkerson et al. 2015)
mitchellae (Dyar 1905b)
nigromaculis (Ludlow 1906a)
sollicitans (Walker 1856a)
taeniorhynchus (Wiedemann 1821)
Empihals Group (Wilkerson et al. 2015)
vigilax (Skuse 1889)
vigilax ludlowae (Blanchard 1905)
vigilax vansomerenae Mattingly 1955
Gilesia Group (Wilkerson et al. 2015; Aculeatus Group of Marks 1957a)
aculeatus (Theobald 1903a)
mcdonaldi Belkin 1962
purpuraceus Brug 1932
Juppius Group (Wilkerson et al. 2015)
caballus (Theobald 1912a)
chelli (Edwards 1915b)
juppi McIntosh 1973
Lepidokeneon Group (Wilkerson et al. 2015; Stricklandi Section of Marks 1957a)
spilotus Marks 1963b
stricklandi (Edwards 1912b)
turneri Marks 1963b
Ochlerotatus Group (Wilkerson et al. 2015; Scapularis Group of Arnell 1976)
angustivittatus Dyar & Knab 1907a
atactavittatus Arnell 1976
auratus Grabham 1906b
bogotanus Arnell 1976
comitatus Arnell 1976
condolescens Dyar & Knab 1907a
crinifer (Theobald 1903a)
deficiens Arnell 1976
euplocamus Dyar & Knab 1906d
incomptus Arnell 1976
infirmatus Dyar & Knab 1906d
meprai Martínez & Prosen 1953b
obturbator Dyar & Knab 1907a
patersoni Shannon & Del Ponte 1928
pectinatus Arnell 1976
phaeonotus Arnell 1976
raymondi Del Ponte, Castro, & García 1951
rhyacophilus da Costa Lima 1933
scapularis (Rondani 1848)
synchytus Arnell 1976
thelcter Dyar 1918d
tortilis (Theobald 1903g)
trivittatus (Coquillett 1902b)
Pholeomyia Group (Wilkerson et al. 2015; Flavifrons Section of Marks 1957)
calcariae Marks 1957a
clelandi (Taylor 1914c)
flavifrons (Skuse 1889)
purpuriventris Edwards 1926a
Protoculex Group (Wilkerson et al. 2015; Serratus Group of Belkin et al. 1970)
aenigmaticus Cerqueira & Costa 1946
atlanticus Dyar & Knab 1906d
dupreei (Coquillett 1904a)
eucephalaeus Dyar 1918d
hastatus Dyar 1922f
nubilus Theobald 1903a
oligopistus Dyar 1918a
pertinax Grabham 1906b
serratus (Theobald 1901c)
tormentor Dyar & Knab 1906d
Woodius Group (Wilkerson et al. 2015)
diantaeus Howard, Dyar, & Knab 1913
intrudens Dyar 1919b
Subgenus *Paulianius*
Coulangesi Section (Brunhes et al. 2017)
ambremontis (Brunhes & Boussès 2017).
coulangesi Rodhain & Boutonnier 1983b
grassei Doucet 1951a
madagascarensis van Someren 1949a
rodhaini (Brunhes & Boussès 2017). In Brunhes et al. 2017
Tiptoni Section (Brunhes et al. 2017)

brucei Wilkerson 2020. In Wilkerson, Linton, & Strickman 2020 [replacement name for *hirsutus* Brunhes & Boussès 2017]
tiptoni Grjebine 1953a
Zethus Section
zethus De Meillon & Lavoipierre 1944
Subgenus *Petermattinglyius*
Aglaonotus Group (Wilkerson et al. 2015)
whartoni Mattingly 1965
Petermattinglyius Group (Wilkerson et al. 2015)
franciscoi Mattingly 1965b
iyengari Edwards 1923b
punctipes Edwards 1921c
scanloni Reinert 1970f
Subgenus *Stegomyia*
Actinothrix Group (Wilkerson et al. 2015)
edwardsi (Barraud 1923b)
robinsoni Belkin 1962
seampi Huang 1974
tulagiensis Edwards 1926a
Aegypti Group (Huang 2004)
aegypti (Linnaeus 1762)
aegypti formosus (Walker 1848)
mascarensis (MacGregor 1924)
pia Le Goff et al. 2013
Africanus Group (Huang 2004)
africanus (Theobald 1901d)
corneti Huang 1986b
luteocephalus (Newstead 1907). In Newstead et al. 1907
maxgermaini Huang 1990
neoafricanus Cornet, Valade, & Dieng 1978
opok Corbet & van Someren 1962
pseudoafricanus Chwatt 1949
ruwenzori Haddow & van Someren 1950
Annandalei Group (Huang 1977a; as subgroup)
annandalei (Theobald 1910a)
craggi (Barraud 1923f)
Apicoargenteus Group (Huang 2004)
apicoargenteus (Theobald 1909b)
blacklocki Evans 1925a
denderensis Wolfs 1949
ealaensis Huang 2004
fraseri (Edwards 1912c)
schwetzi Edwards 1926a
soleatus Edwards 1924c
Dendrophilus Group (Huang 2004)
amaltheus De Meillon & Lavoipierre 1944
bambusae Edwards 1935a
deboeri Edwards 1926a
demeilloni Edwards 1936
dendrophilus Edwards 1921c
hansfordi Huang 1997
heischi van Someren 1951
keniensis van Someren 1946a
kenyae van Someren 1946c
masseyi Edwards 1923a
mattinglyorum Huang 1994
muroafcete Huang 1997
njombiensis Huang 1997
segermanae Huang 1997
Desmotes Group (Huang 1977a; as subgroup)
desmotes (Giles 1904a)
Edwardsi Group (Huang 1970)
edwardsi Barraud 1923b
robinsoni Belkin 1962
seampi Huang 1974
tulagiensis Edwards 1926a
Granti Group (Huang 2004)
granti (Theobald 1901a)
Maehleri Group (Huang1970)
burnsi Basio & Reisen 1971
maehleri Bohart 1957
Mediopunctatus Group (Huang 1977a; as subgroup)
malikuli Huang 1973
mediopunctatus (Theobald 1905h)
mediopunctatus sureilensis Barraud 1934
perplexus (Leicester 1908)
Metallicus Group (Huang 2004)
metallicus (Edwards 1912c)
Pandani Group (Bohart 1957; as subgroup)
agrihanensis Bohart 1957
neopandani Bohart 1957
pandani Stone 1939
rotanus Bohart & Ingram 1946b
saipanensis Stone 1945b
Poweri Group (Huang 2004)
angustus Edwards 1935a
calceatus Edwards 1924c
chaussieri Edwards 1923a
contiguus Edwards 1936
ethiopiensis Huang 2004
hogsbackensis Huang 2004
langata van Someren 1946c
ledgeri Huang 1981
mpusiensis Huang 2004
poweri (Theobald 1905g)
usambara Mattingly 1953d
Pseudonigeria Group (Huang 2004)
chemulpoensis Yamada 1921
mickevichae Huang 1988
pseudonigeria (Theobald 1910c)
saimedres Huang 1988
Scutellaris Group (Huang 2004)
Albopictus Subgroup (Huang 1972)
albopictus (Skuse 1895)
flavopictus Yamada 1921
flavopictus downsi Bohart & Ingram 1946a
flavopictus miyarai Tanaka, Mizusawa, & Saugstad 1979

galloisi Yamada 1921
galloisiodes Liu & Lu 1984
neogalloisi Chen & Chen 2000
novalbopictus Barraud 1931b
patriciae Mattingly 1954a
pseudalbopictus (Borel 1928)
seatoi Huang 1969
sibiricus Danilov & Filippova 1978
subalbopictus Barraud 1931b
Scutellaris Subgroup (Huang 1972)
alcasidi Huang 1972
alorensis Bonne-Wepster & Brug 1932
andrewsi Edwards 1926a
aobae Belkin 1962
cooki Belkin 1962
daitensis Miyagi & Toma 1981
dybasi Bohart 1957
guamensis Farner & Bohart 1944
gurneyi Stone & Bohart 1944
hakanssoni Knight & Hurlbut 1949
hebrideus Edwards 1926a
hensilli Farner 1945
hoguei Belkin 1962
horrescens Edwards 1935a
katherinensis Woodhill 1949
kesseli Huang & Hitchcock 1980
krombeini Huang 1975
malayensis Colless 1962
marshallensis Stone & Bohart 1944
palauensis Bohart 1957
paullusi Stone & Farner 1945
pernotatus Farner & Bohart 1944
polynesiensis Marks 1951a
pseudoscutellaris (Theobald 1910b)
quasiscutellaris Farner & Bohart 1944
riversi Bohart & Ingram 1946a
rotumae Belkin 1962
scutellaris (Walker 1859)
scutoscriptus Bohart & Ingram 1946b
tabu Ramalingam & Belkin 1965
tongae Edwards 1926a
upolensis Marks 1957b
varuae Belkin 1962
Simpsoni Group (Huang 2004)
bromeliae (Theobald 1911a)
gandaensis Huang 2004
josiahae Huang 1988
kivuensis Edwards 1941
lilii (Theobald 1910c)
sampi Huang 2004
simpsoni (Theobald 1905c)
strelitziae Muspratt 1950
subargenteus Edwards 1925
woodi Edwards 1922d
Unilineatus Group (Huang 2004)
unilineatus (Theobald 1903b)
W-albus Group (Huang 1970c)
gardnerii (Ludlow 1905a)
gardnerii imitator (Leicester 1908)
w-albus (Theobald 1905f)
Zoromorphus Group (Wilkerson et al. 2015)

Genus *Anopheles*

Informal names proposed for members of genus *Anopheles* (six subgenera: *Anopheles*, *Baimia*, *Cellia*, *Cristya*, *Kerteszia*, and *Nyssorhynchus*) are based on Christophers (1924a), Edwards (1932), Reid & Knight (1961), Grjebine (1966), Gillies & De Meillon (1968), Reid (1968), Faran (1980), Linthicum (1988), and others. Harbach (1994a, 2004, 2013b) summarized use of these names and their component species. The list below is modified from Harbach et al. (2016) [*Anopheles* classification http://mosquito-taxonomic-inventory.info/node/11358, accessed July 24, 2016]. Refer to the above and to the citations given with each group for details.

Genus *Anopheles*
Subgenus *Anopheles*
Angusticorn Section (Reid & Knight 1961)
annulipalpis Lynch Arribálzaga 1878 (see Harbach & Kitching 2015)
Anopheles Series (Edwards 1932)
algeriensis Theobald 1903a
concolor Edwards 1938
marteri Senevet & Prunnelle 1927
Claviger Complex (Coluzzi et al. 1965)
claviger (Meigen 1804)
petragnani Del Vecchio 1939
Aitkenii Group (Reid & Knight 1961)
aberrans Harrison & Scanlon 1975
acaci Baisas 1946a
aitkenii James 1903. In Theobald 1903a
bengalensis Puri 1930
borneensis McArthur 1949
fragilis (Theobald 1903f)
insulaeflorum (Swellengrebel & Swellengrebel de Graaf 1920)
palmatus (Rodenwaldt 1926)
peytoni Kulasekera, Harrison, & Amerasinghe 1989
pilinotum Harrison & Scanlon 1974
pinjaurensis Barraud 1932
stricklandi Reid 1965
tigertti Scanlon & Peyton 1967
Alongensis Group (Phan et al. 1991)
alongensis Venhuis 1940
cucphuongensis Vu, Nguyen, Tran, & Nguyen 1991
Atratipes Group (Lee et al. 1987b)
atratipes Skuse 1889
tasmaniensis Dobrotworsky 1966

Culiciformis Group (Reid & Knight 1961)
- *culiciformis* Cogill 1903
- *sintoni* Puri 1929b
- *sintonoides* Ho 1938a

Lindesayi Group (Reid & Knight 1961)

Gigas Complex (Harrison et al. 1991)
- *baileyi* Edwards 1929c
- *gigas* Giles 1901b
  - *gigas crockeri* Colless 1955a
  - *gigas danaubento* Mochtar & Walandouw 1934
  - *gigas formosus* Ludlow 1909a
  - *gigas oedjalikalah* Nainggolan 1939
  - *gigas pantjarbatu* Waktoedi Koesoemawinangoen 1954
  - *gigas refutans* Alcock 1913
  - *gigas simlensis* (James 1911). In James & Liston 1911
- *prachongae* Rattanarithikul & Harrison 2017. In Harbach et al. 2017
- *sumatranus* Swellengrebel & Rodenwaldt 1932

Lindesayi Complex (Harrison et al. 1991)
- *lindesayi* Giles 1900a
  - *lindesayi benguetensis* King 1931
  - *lindesayi cameronensis* Edwards 1929c
  - *lindesayi japonicus* Yamada 1918
  - *lindesayi pleccau* Koidzumi 1924
- *menglangensis* Ma 1981a
- *nilgiricus* Christophers 1924b
- *wellingtonianus* Alcock 1912

Maculipennis Group (Reid & Knight 1961)
- *atropos* Dyar & Knab 1906b
- *aztecus* Hoffmann 1935
- *lewisi* Ludlow 1920b
- *walkeri* Theobald 1901a

Freeborni Subgroup (Linton 2004)
- *earlei* Vargas 1943
- *freeborni* Aitken 1939
- *hermsi* Barr & Guptavanij 1989
- *occidentalis* Dyar & Knab 1906b

Maculipennis Subgroup (Linton 2004)
- *artemievi* Gordeyev, Zvantsov, Goryacheva, Shaĭkevich, & Yezhov 2005
- *atroparvus* van Thiel 1927
- *daciae* Linton, Nicolescu, & Harbach 2004. In Nicolescu et al. 2004
- *labranchiae* Falleroni 1926
- *maculipennis* Meigen 1918
- *martinius* Shingarev 1926
- *melanoon* Hackett 1934
- *messeae* Falleroni 1926
- *persiensis* Linton, Sedaghat, & Harbach 2003. In Sedaghat et al. 2003
- *sacharovi* Favre 1903

Quadrimaculatus Subgroup (Linton 2004)
- *beklemishevi* Stegnii & Kabanova 1976
- *diluvialis* Reinert 1997
- *inundatus* Reinert 1997
- *maverlius* Reinert 1997
- *quadrimaculatus* Say 1824
- *smaragdinus* Reinert 1997

Plumbeus Group (Reid & Knight 1961)
- *arboricola* Zavortink 1970b
- *barberi* Coquillett 1903b
- *barianensis* James 1911
- *fausti* Vargas 1943a
- *judithae* Zavortink 1969
- *omorii* Sakakibara 1959
- *plumbeus* Stephens 1828
- *powderi* Zavortink 1970b
- *xelajuensis* de León 1938

Pseudopunctipennis Group (Reid & Knight 1961)
- *chiriquiensis* Komp 1936b
- *eiseni* Coquillett 1902b
  - *eiseni geometricus* Corrêa 1944
- *franciscanus* McCracken 1904
- *hectoris* Giaquinto-Mira 1931
- *parapunctipennis* Martini 1932b
  - *parapunctipennis guatemalensis* de León 1938
- *pseudopunctipennis* Theobald 1901c
  - *pseudopunctipennis levicastilloi* Leví-Castillo 1944
  - *pseudopunctipennis neghmei* Mann 1950
  - *pseudopunctipennis noei* Mann 1950
  - *pseudopunctipennis patersoni* Alvarado & Heredia 1947
  - *pseudopunctipennis rivadeneirai* Leví-Castillo 1945
- *tibiamaculatus* (Neiva 1906)

Punctipennis Group (Reid & Knight 1961)
- *perplexens* Ludlow 1907
- *punctipennis* (Say 1823)

Crucians Complex (Wilkerson et al. 2004)
- *bradleyi* King 1939
- *crucians* Wiedemann 1828 (species A, B, C, D, & E)
- *georgianus* King 1939

Stigmaticus Group (Reid & Knight 1961)
- *colledgei* Marks 1956
- *corethroides* Theobald 1907
- *papuensis* Dobrotworsky 1957
- *powelli* Lee 1944b
- *pseudostigmaticus* Dobrotworsky 1957
- *stigmaticus* Skuse 1889

Cycloleppteron Series (Edwards 1932a)
- *grabhamii* Theobald 1901a

Lophoscelomyia Series (Edwards 1932a)
- *bulkleyi* Causey 1937a

Asiaticus Group (Reid 1968)
annandalei Prashad 1918
noniae Reid 1963a
Asiaticus Subgroup (Rattanarithikul et al. 2006b)
asiaticus Leicester 1903
Interruptus Subgroup (Rattanarithikul et al. 2006b)
interruptus Puri 1929a
Laticorn Section (Reid & Knight 1961)
Arribalzagia Series (Root 1922a)
anchietai Corrêa & Ramalho 1968
apicimacula Dyar & Knab 1906a
bustamentei Galvão 1955
calderoni Wilkerson 1991
costai da Fonseca & da Silva Ramos 1940a
costai s.l. (Bourke et al. 2018)
evandroi da Costa Lima 1937
fluminensis Root 1927d
fluminensis s.l. (Bourke et al. 2018)
forattinii Wilkerson & Sallum 1999
gabaldoni Vargas 1941c
guarao Anduze & Capdevielle 1949
maculipes (Theobald 1903a)
malefactor Dyar & Knab 1907c
malefactor s.l. (Bourke et al. 2018)
mattogrossensis Lutz & Neiva 1911
medialis Harbach 2018c
mediopunctatus (Lutz 1903)
minor da Costa Lima 1929c
neomaculipalpus Curry 1931
peryassui Dyar & Knab 1908
peryassui s.l. (Bourke et al. 2018)
pseudomaculipes (Chagas 1908). In Peryassú 1908
punctimacula Dyar & Knab 1906a
punctimacula s.l. (Bourke et al. 2018)
rachoui Galvão 1952
shannoni Davis 1931
veruslanei Vargas 1979a
vestitipennis Dyar & Knab 1906a
Myzorhynchus Series (Edwards 1932a)
obscurus (Grünberg 1905)
Albotaeniatus Group (Reid & Knight 1961)
albotaeniatus (Theobald 1903a)
balerensis Mendoza 1947
ejercitoi Mendoza 1947
montanus Stanton & Hacker 1917
saperoi Bohart & Ingram 1946a
saperoi ohamai Ohama 1947
Bancroftii Group (Reid & Knight 1961)
bancroftii Giles 1902
bancroftii barbiventris Brug 1938
pseudobarbirostris Ludlow 1902a
Barbirostris Group (Reid & Knight 1961)
freyi Meng 1957
koreicus Yamada & Watanabe 1918
Barbirostris Subgroup (Reid 1968)
donaldi Reid 1962
franciscoi Reid 1962
hodgkini Reid 1962
pollicaris Reid 1962
Barbirostris Complex (Satoto 2001)
barbirostris van der Wulp 1884
campestris Reid 1962
dissidens Taai & Harbach 2015 (species III of Paredes-Esquivel et al. 2009)
saeungae Taai & Harbach 2015 (species IV of Paredes-Esquivel et al. 2009)
vanderwulpi Townson & Harbach 2013. In Townson et al. 2013
wejchoochotei Taai & Harbach 2015
Vanus Subgroup (Reid 1968)
ahomi Chowdhury 1929
barbumbrosus Strickland & Chowdhury 1927a
manalangi Mendoza 1940
reidi Harrison 1973a
vanus Walker 1859
Coustani Group (Reid & Knight 1961)
caliginosus De Meillon 1943
coustani Laveran 1900
crypticus Coetzee 1994a
fuscicolor van Someren 1947b
namibiensis Coetzee 1984
paludis Theobald 1900b
symesi Edwards 1928a
tenebrosus Dönitz 1902
ziemanni Grünberg 1902
Hyrcanus Group (Reid 1953)
argyropus (Swellengrebel 1914)
belenrae Rueda 2005
chodukini Martini 1929
engarensis Kanda & Oguma 1978
hailarensis Xu & Luo 1998
heiheensis Ma 1981b
hyrcanus (Pallas 1771)
hyrcanus spIR (Djadid et al. 2009)
kleini Rueda 2005
kweiyangensis Yao & Wu 1944
liangshanensis Kang, Tan, Cao, Cheng, Yang, & Huang 1984
nimpe Nguyen, Tran, & Harbach 2000. In Nguyen et al. 2000
pseudopictus Grassi 1899
pullus Yamada 1937
sinensis Wiedemann 1828
sineroides Yamada 1924
xui Dong, Zhou, Dong, & Mao 2007
Lesteri Subgroup (Harrison 1972)

crawfordi Reid 1953
lesteri Baisas & Hu 1936
peditaeniatus (Leicester 1908)
vietnamensis Nguyen, Tran, & Nguyen 1993
Nigerrimus Subgroup (Harrison 1972)
nigerrimus Giles 1900a
nitidus Harrison, Scanlon, & Reid 1973
pseudosinensis Baisas 1935c
pursati Laveran 1902a
Umbrosus Group (Reid 1950)
brevipalpis Roper 1914
brevirostris Reid 1950
hunteri (Strickland 1916a)
samarensis Rozeboom 1951a
similissimus Strickland & Chowdhury 1927a
Baezai Subgroup (Rattanarithikul et al. 2006b)
baezai Gater 1933
Letifer Subgroup (Reid 1968)
collessi Reid 1963a
letifer Sandosham 1944
roperi Reid 1950
whartoni Reid 1963a
Separatus Subgroup (Rattanarithikul et al. 2006b)
separatus (Leicester 1908)
Umbrosus Subgroup (Rattanarithikul et al. 2006b)
umbrosus (Theobald 1903a)
Subgenus *Baimaia*
kyondawensis Abraham 1947
Subgenus *Cellia*
Cellia Series (Christophers 1924a)
argenteolobatus (Gough 1910)
brumpti Hamon & Rickenbach 1955a
cristipalpis Service 1977
murphyi Gillies & De Meillon 1968
pharoensis Theobald 1901a
swahilicus Gillies 1964b
Squamosus Group (Grjebine 1966)
cydippis De Meillon 1931a
squamosus Theobald 1901a
Myzomyia Series (Christophers 1924a)
apoci Marsh 1933
azaniae Bailly-Choumara 1960
barberellus Evans 1932
bervoetsi D'Haenens 1961
brunnipes (Theobald 1910c)
domicolus Edwards 1916
dthali Patton 1905
erythraeus Corradetti 1939c
ethiopicus Gillies & Coetzee 1987
flavicosta Edwards 1911a
fontinalis Gillies & De Meillon 1968
gabonensis Rahola, Makanga, & Paupy 2014. In Rahola et al. 2014
majidi Young & Majid 1928
moucheti Evans 1925b
schwetzi Evans 1934
tchekedii De Meillon & Leeson 1940
walravensi Edwards 1930a
Demeilloni Group (Gillies & De Meillon 1968)
carteri Evans & De Meillon 1933
demeilloni Evans 1933
freetownensis Evans 1925c
garnhami Edwards 1930a
keniensis Evans 1931a
lloreti Gil Collado 1936
sergentii (Theobald 1907)
sergentii macmahoni Evans 1936
Funestus Group (Garros et al. 2005a)
jeyporiensis James 1902
Aconitus Subgroup (Chen et al. 2003)
aconitus Dönitz 1902
filipinae Manalang 1930
mangyanus (Banks 1906b)
pampanai Büttiker & Beales 1959
varuna Iyengar 1924
Culicifacies Subgroup (Garros et al. 2005b)
culicifacies Giles 1901b (species A, B, C, D, & E; Kar et al. 1999)
Funestus Subgroup (Garros et al. 2005b)
aruni Sobti 1968
confusus Evans & Leeson 1935
funestus Giles 1900b
funestus-like species (Spillings et al. 2009)
longipalpis (Theobald 1903a) (Type C) (Koekemoer et al. 2009)
parensis Gillies 1962
vaneedeni Gillies & Coetzee 1987
Minimus Subgroup (Chen et al. 2003)
flavirostris (Ludlow 1914a)
leesoni Evans 1931b
longipalpis (Theobald 1903a) (Type A) (Koekemoer et al. 2009)
Fluviatilis Complex (Sarala et al. 1994)
fluviatilis James 1902 (species S, T, U, X, Y; Chen et al. 2006)
Minimus Complex (Green et al. 1990)
harrisoni Harbach & Manguin 2007 (species C)
minimus Theobald 1901a (species A)
yaeyamaensis Somboon & Harbach 2010. In Somboon et al. 2010
Rivulorum Subgroup (Garros et al. 2005b)
brucei Service 1960c
fuscivenosus Leeson 1930
rivulorum Leeson 1935

rivulorum-like species (Cohuet et al. 2003)
Marshallii Group (Gillies & De Meillon 1968)
austenii (Theobald 1905b)
berghei Vincke & Leleup 1949c
brohieri Edwards 1929c
gibbinsi Evans 1935
hancocki Edwards 1929c
hargreavesi Evans 1927
harperi Evans 1936
mortiauxi Edwards 1938
mousinhoi De Meillon & Pereira 1940
njombiensis Peters 1955b
seydeli Edwards 1929c
Marshallii Complex (Gillies & Coetzee 1987)
hughi Lambert & Coetzee 1982
kosiensis Coetzee, Segerman, & Hunt 1987
letabensis Lambert & Coetzee 1982
marshallii (Theobald 1903a)
Wellcomei Group (Gillies & De Meillon 1968)
distinctus (Newstead & Carter 1911b)
erepens Gillies 1958
theileri Edwards 1912d
wellcomei Theobald 1904b
wellcomei ugandae Evans 1934
wellcomei ungujae White 1975
Neocellia Series (Christophers 1924a)
dancalicus Corradetti 1939b
hervyi Brunhes, Le Goff, & Geoffroy 1999
karwari (James 1903)
maculipalpis Giles 1902
moghulensis Christophers 1924c
paltrinierii Shidrawi & Gillies 1988
pattoni Christophers 1926
pretoriensis (Theobald 1903a)
pulcherrimus Theobald 1902a
rufipes (Gough 1910)
rufipes broussesi Edwards 1929a
salbaii Maffi & Coluzzi 1958
stephensi Liston 1901
superpictus Grassi 1899 (species A & B; Oshaghi et al. 2007, 2008)
theobaldi Giles 1901b
Annularis Group (Reid 1968)
Annularis Complex (Atrie et al. 1999)
annularis van der Wulp 1884 (species A & B)
pallidus Theobald 1901a
philippinensis Ludlow 1902b
schueffneri Stanton 1915
Nivipes Complex (Green et al. 1985b; Harrison et al. 1991)
nivipes (Theobald 1903f) (two cytogenetic species in Thailand)
Jamesii Group (Rattanarithikul et al. 2006b)
jamesii Theobald 1901a
pseudojamesi Strickland & Chowdhury 1927b
splendidus Koidzumi 1920
Maculatus Group (Rattanarithikul & Green 1987)
dispar Rattanarithikul & Harbach 1991
greeni Rattanarithikul & Harbach 1991
pseudowillmori (Theobald 1910c)
willmori (James 1903)
Maculatus Subgroup (Rattanarithikul et al. 2006b)
dravidicus Christophers 1924c
maculatus Theobald 1901a
Sawadwongporni Subgroup (Rattanarithikul et al. 2006b)
notanandai Rattanarithikul & Green 1987
rampae Harbach & Somboon 2011. In Somboon et al. 2011
sawadwongporni Rattanarithikul & Green 1987
Neomyzomyia Series (Christophers 1924a)
amictus Edwards 1921c
annulatus Haga 1930
aurirostris (Watson 1910)
dualaensis Brunhes, Le Goff & Geoffroy 1999
hilli Woodhill & Lee 1944
incognitus Brug 1931b
kokhani Vythilingam, Jeffery & Harbach 2007. In Vythilingam et al. 2007
kolambuganensis Baisas 1932
longirostris Brug 1928a
meraukensis Venhuis 1932
novaguinensis Venhuis 1933
saungi Colless 1955a
stookesi Colless 1955a
watsonii (Leicester 1908)
Annulipes Complex (Green 1972)
annulipes Walker 1956a (species A–Q; Foley et al. 2007a,c)
Lungae Complex (Belkin 1962)
lungae Belkin & Schlosser 1944
nataliae Belkin 1945
solomonis Belkin, Knight, & Rozeboom 1945
Ardensis Group (Gillies & De Meillon 1968)
ardensis (Theobald 1905g)
buxtoni Service 1958
cinctus (Newstead & Carter 1910)
deemingi Service 1970a
dureni Edwards 1938
eouzani Brunhes, Le Goff & Boussès 2003
kingi Christophers 1923a
machardyi Edwards 1930b
maliensis Bailly-Choumara & Adam 1959
millecampsi Lips 1960a

multicinctus Edwards 1930a
natalensis (Hill & Haydon 1907)
vernus Gillies & De Meillon 1968
vinckei De Meillon 1942
Nili Complex (Gillies & De Meillon 1968)
carnevalei Brunhes, Le Goff, & Geoffroy 1999
nili (Theobald 1904b)
ovengensis Awono-Ambene, Kengne, Simard, Antonio-Nkondjio, & Fontenille 2004
somalicus Rivola & Holstein 1957
Kochi Group (Rattanarithikul et al. 2006b)
kochi Dönitz 1901
Leucosphyrus Group (Reid 1949)
Hackeri Subgroup (Sallum et al. 2005a)
hackeri Edwards 1921c
mirans Sallum & Peyton 1905
pujutensis Colless 1948
recens Sallum & Peyton 2005
sulawesi Waktoedi Koesoemawinangoen 1954
Leucosphyrus Subgroup (Peyton 1990; Sallum et al. 2005b)
baisasi Colless 1957c
Dirus Complex (Sallum et al. 2005b)
aff. *takasagoensis* (Takano et al. 2010)
baimaii Sallum & Peyton 2005
cracens Sallum & Peyton 2005
dirus Peyton & Harrison 1979
elegans (James 1903)
nemophilous Peyton & Ramalingam 1988
scanloni Sallum & Peyton 2005
takasagoensis Morishita 1946
Leucosphyrus Complex (Sallum et al. 2005a)
balabacensis Baisas 1936b
introlatus Colless 1957c
latens Sallum & Peyton 2005
leucosphyrus Dönitz 1901
Riparis Subgroup (Peyton 1990)
cristatus King & Baisas 1936
macarthuri Colless 1956
riparis King & Baisas 1936
Mascarensis Group (Harbach 1994a)
mascarensis De Meillon 1947a
Pauliani Group (Grjebine 1966)
grassei Grjebine 1953b
grenieri Grjebine 1964
milloti Grjebine & Lacan 1953
pauliani Grjebine 1953a
radama De Meillon 1943
Punctulatus Group (Schmidt et al. 2001)
clowi Rozeboom & Knight 1946
farauti 4 (Foley et al. 1993)
koliensis Owen 1945
punctulatus Dönitz 1901
rennellensis Taylor & Maffi 1991
sp. near *punctulatus* (Foley et al. 1995)
Farauti Complex (Schmidt et al. 2003)
farauti Laveran 1902
farauti 5 (Foley et al. 1993)
farauti 8 (Bower et al. 2008)
hinesorum Schmidt 2001. In Schmidt et al. 2001
irenicus Schmidt 2003. In Schmidt et al. 2003
oreios Bangs & Harbach 2014. In Bangs et al. 2014
torresiensis Schmidt 2001
Ranci Group (Grjebine 1966)
griveaudi Grjebine 1961
Ranci Subgroup (Grjebine 1966)
ranci Grjebine 1953a
Roubaudi Subgroup (Grjebine 1966)
lacani Grjebine 1953a
notleyi van Someren 1949a
roubaudi Grjebine 1953a
Rhodesiensis Group (Gillies & De Meillon 1968)
cameroni De Meillon & Evans 1935
lounibosi Gillies & Coetzee 1987
rhodesiensis Theobald 1901a
rhodesiensis rupicolus Lewis 1937
rodhaini Leleup & Lips 1950
ruarinus Edwards 1940
Smithii Group (Gillies & De Meillon 1968)
caroni Adam 1961
faini Leleup 1952
hamoni Adam 1962
jebudensis Froud 1944
lovettae Evans 1934
rageaui Mattingly & Adam 1954
smithii Theobald 1905b
vanhoofi Wanson & Lebied 1945
wilsoni Evans
Tessellatus Group (Rattanarithikul et al. 2006b)
tessellatus Theobald 1901a
tessellatus kalawara Stoker & Waktoedi Koesoemawinangoen 1949
tessellatus orientalis (Swellengrebel & Swellengrebel de Graaf 1920)
Paramyzomyia Series (Christophers & Barraud 1931)
Cinereus Group (Gillies & De Meillon 1968)
azevedoi Ribeiro 1969
cinereus Theobald 1901a
cinereus hispaniola (Theobald 1903a)
turkhudi Liston 1901
turkhudi telamali Saliternik & Theodor 1942
Listeri Group (Gillies & De Meillon 1968)
listeri De Meillon 1931b
multicolor Cambouliu 1902

seretsei Abdulla-Khan, Coetzee, & Hunt 1998
Pyretophorus Series (Edwards 1932a)
christyi (Newstead & Carter 1911b)
comorensis Brunhes, Le Goff, & Geoffroy 1997
daudi Coluzzi 1958
indefinitus (Ludlow 1904c)
limosus King 1932a
litoralis King 1932
ludlowae (Theobald 1903a)
ludlowae torakala Stoker & Waktoedi Koesoemawinangoen 1949
parangensis (Ludlow 1914b)
pseudosundaicus Tyagi, Hiriyan, Tewari, Ayanar, Samuel, Arunachalam, Paramasivan, Krishnamoorthy, Dhananjeyan, Leo, & Rajendran 2009
vagus Dönitz 1902
Gambiae Complex (White 1985)
amharicus Hunt, Wilkerson, & Coetzee 2013. In Coetzee et al. 2013
arabiensis Patton 1905
bwambae White 1985
coluzzii Coetzee & Wilkerson 2013. In Coetzee et al. 2013
fontenillei Barrón, Paupy, Rahola, Akone-Ella, Ngangue, Wilson-Bahun, Pombi, Kengne, Costantini, Simard, González, & Ayala 2019
gambiae Giles 1902
melas (Theobald 1903a)
merus Dönitz 1902
quadriannulatus (Theobald 1911b)
Subpictus Complex (Suguna et al. 1994)
subpictus Grassi 1899 (species A, B, C, & D)
Sundaicus Complex (Sukowati et al. 1999)
epiroticus Linton & Harbach 2005. In Linton et al. 2005
sundaicus (Rodenwaldt 1925)
sundaicus s.l. (species B, C, D, & E; Dusfour, Londeau, et al. 2007)
Subgenus *Christya*
implexus (Theobald 1903b)
okuensis Brunhes, Le Goff, & Geoffroy 1997
Subgenus *Kerteszia*
auyantepuiensis Harbach & Navarro 1996
bambusicolus Komp 1937
bellator Dyar & Knab 1906b
boliviensis (Theobald 1905f)
gonzalezrinconesi Cova García, Pulido F., & Escalante de Ugueto 1977
homunculus Komp 1937
laneanus Corrêa & Cerqueira 1944
lepidotus Zavortink 1973
neivai Howard, Dyar, & Knab 1913
pholidotus Zavortink 1973
rollai Cova García, Pulido F., & Escalante de Ugueto 1977
Cruzii Complex (Ramirez & Dessen 2000a)
cruzii Dyar & Knab 1908 (species A, B, & C; Ramirez & Dessen 2000b)*
Subgenus *Lophopodomyia*
gilesi (Neiva 1908b)
gomezdelatorrei Leví-Castillo 1955
oiketorakras Osorno-Mesa 1947
pseudotibiamaculatus Galvão & Barretto 1941
squamifemur Antunes 1937a
vargasi Gabaldón, Cova-García, & Lopez 1941b
Subgenus *Nyssorhynchus*
Albimanus Section (Leví Castillo 1949)
Albimanus Series (Faran 1980)
albimanus Wiedemann 1821
Oswaldoi Series (Faran 1980)
Oswaldoi Group (Faran 1980)
Oswaldoi Subgroup (Faran 1980)
anomalophyllus Komp
aquasalis Curry 1932
evansae (Brèthes 1926b)
galvaoi Causey 1943b
ininii Senevet & Abonnenc 1938
Konderi Complex (Ruiz-Lopez et al. 2013)
konderi Galvão & Damasceno 1942a
sp. nr. *konderi* (Ruiz-Lopez et al. 2013)
konderi s.l. (Bourke et al. 2018)
Nuneztovari Complex (Mirabello & Conn 2008; Foster et al. 2013; Scarpassa et al. 2016)
dunhami Causey 1945
goeldii Rozeboom & Gabaldón 1941
nuneztovari Gabaldón 1940
Oswaldoi Complex (Ruiz-Lopez et al. 2013)
oswaldoi (Peryassú 1922b)
oswaldoi A (Ruiz-Lopez et al. 2013)
oswaldoi B (Ruiz et al. 2005; Ruiz et al. 2010)
oswaldoi s.l. (Bourke et al. 2018)
rangeli Gabaldón, Cova García, & Lopez 1940a
sanctielii Senevet & Abonnenc 1938
trinkae Faran 1979
Strodei Subgroup (Faran 1980)
albertoi Unti 1941
arthuri Unti 1941
arthuri s.l. (Bourke et al. 2018)

* de Carvalho-Pinto & Lourenço-de-Oliveira (2004) and Rona et al. (2009, 2010a, b, 2013) suggest the complex consists of four or five species, but how these relate to the three chromosomal species of Ramirez & Dessen (2000b) is not clear.

rondoni (Neiva & Pinto 1922a)
striatus Sant'Ana & Sallum 2017
strodei Root 1926
Benarrochi Complex (Ruiz-Lopez et al. 2005)
benarrochi Gabaldón, Cova García, & Lopez 1941a
benarrochi B (Ruiz et al. 2005)
benarrochi s.l. (Bourke et al. 2018)
Triannulatus Group (Faran 1980)
halophylus Silva-do-Nascimento & Lourenço-de-Oliveira 2002
triannulatus (Neiva & Pinto 1922b)
triannulatus s.l. (species C; Silva-do-Nascimento & Lourenço-de-Oliveira 2007)
Argyritarsis Section (Leví Castillo 1949)
Albitarsis Series (Linthicum 1988)
Albitarsis Group (Linthicum 1988)
Albitarsis Complex (Wilkerson, Parsons et al. 1995)
albitarsis Lynch Arribálzaga 1878
albitarsis s.l. (species F, G, H, & I; Brochero et al. 2007; Ruiz et al. 2012)
albitarsis s.l. (Bourke et al. 2018)
deaneorum Rosa-Freitas 1989a
janconnae Wilkerson & Sallum 2009
lineage nr. *janconnae* (Gutiérrez et al. 2010)
marajoara Galvão & Damasceno 1942b
oryzalimnetes Wilkerson & Motoki 2009. In Motoki et al. 2009a
Braziliensis Group (Linthicum 1988)
braziliensis (Chagas 1907)
Argyritarsis Series (Linthicum 1988)
Argyritarsis Group (Linthicum 1988)
argyritarsis Robineau-Desvoidy 1827
sawyeri Causey, Deane, Deane, & Sampaio 1943
Darlingi Group (Linthicum 1988)
darlingi Root 1926
Lanei Group (Linthicum 1988)
lanei Galvão & Franco do Amaral 1938
Pictipennis Group (Linthicum 1988)
atacamensis González & Sallum 2010
pictipennis (Philippi 1865)
Myzorhynchella Section (Peyton et al. 1992)
antunesi Galvão & Franco do Amaral 1940
guarani Shannon 1928. In Dyar 1928
lutzii Cruz 1901
nigritarsis (Chagas 1907)
parvus (Chagas 1907)
pristinus Nagaki & Sallum 2010. In Nagaki et al. 2010

Subgenus *Stethomyia*
acanthotorynus Komp 1937
canorii Flock & Abonnenc 1945d
kompi Edwards 1930a
nimbus (Theobald 1902b)
thomasi Shannon 1933
Nomina dubia
africanus Roque 1903
allopha (Lutz & Peryassú 1921). In Peryassú 1921a
arnoulti Grjebine 1966
brachypus Dönitz 1902
costalis Loew 1866
courdurieri Grjebine 1866
futunae Belkin 1962
jacobi (Hill & Haydon 1903)
minutus Macquart 1834
nero (Doleschall 1857)
pallida (Ludlow 1905a)
soalalaensis Grjebine 1955a
upemba Lips 1906b
vulgaris Hatori 1792

Genus *Culex*

Organization of content follows Harbach, R.E. [http://mosquito-taxonomic-inventory.info/ltemgtculexltemgt-classification] and Harbach et al. (2016).

In summary, informal group names within subgenera of genus *Culex* were proposed by Edwards 1932a (*Culex*, *Culiciomyia*, *Lophoceraomyia*); Edwards 1934 (*Lophoceraomyia*); Edwards 1941 (*Culex*); Lane & Whitman 1951 (*Microculex*); Mattingly & Rageau 1958 (*Culex*); Belkin 1962 (*Acallyntrum*, *Culex*, *Lophoceraomyia*); Colless 1965 (*Lophoceraomyia*); Bram 1967a, b (*Culex*), Bram 1967b (*Culiciomyia*); Sirivanakarn 1968 (*Lophoceraomyia*); Berlin 1969c (*Micraedes*); Sirivanakarn 1971a (*Eumelanomyia*, *Maillotia*, *Neoculex*); Sirivanakarn 1972 (*Eumelanomyia*); Valencia 1973 (*Carrollia*); Chen 1974 (cibarium morphology); Sirivanakarn 1976 (*Culex*, *Oculeomyia*); Sirivanakarn 1977a (*Lophoceraomyia*); Berlin & Belkin 1980 (*Anoedioporpa*); Sirivanakarn 1983 (*Melanoconion*); Harbach 1988 (*Culex*); Sallum & Forattini 1996 (*Melanoconion*); Forattini 2002 (*Culex*); Natarajan & Rajavel 2009 (*Lophoceraomyia*).

Subgenus *Acalleomyia*
obscurus (Leicester 1908)
Subgenus *Acallyntrum*
Bicki Group (Belkin 1962)
bicki Stone & Penn 1947
binigrolineatus Knight & Rozeboom 1945
miyagii Mogi & Toma 1999
Perkinsi Group (Belkin 1962)
axillicola Steffan 1979
belkini Stone & Penn 1948

bougainvillensis Steffan 1979
pallidiceps (Theobald 1905f)
perkinsi Stone & Penn 1948
Subgenus *Aedinus*
accelerans Root 1927c
amazonensis (Lutz 1905)
clastrieri Casal & García 1968b
guyanensis Clastrier 1970a
Subgenus *Afroculex*
lineatus (Theobald 1912a)
Subgenus *Allimanta*
tramazayguesi Duret 1954b
Subgenus *Anoedioporpa*
Conservator Group (Berlin & Belkin 1980)
bamborum Rozeboom & Komp 1948
belemensis Duret & Damasceno 1955
browni Komp 1936
canaanensis Lane & Whitman 1943
chaguanco Casal, García, & Fernández 1968
conservator Dyar & Knab 1906d
corrigani Dyar & Knab 1907c
damascenoi Duret 1969c
luteopleurus (Theobald 1903a)
originator Gordon & Evans 1922
quasioriginator Duret 1972
Restrictor Group (Berlin & Belkin 1980)
restrictor Dyar & Knab 1906d
Subgenus *Barraudius*
inatomii Kamimura & Wada 1974
modestus Ficalbi 1890a
pusillus Macquart 1850
richeti Brunhes & Venard 1966
Subgenus *Belkinomyia*
eldridgei Adames & Galindo 1973
Subgenus *Carrollia*
Bihaicola Group (Valencia 1973)
bihaicolus Dyar & Núñez Tovar 1928
guerreroi Cova García, Sutil, & Pulido 1971
infoliatus Bonne-Wepster & Bonne 1920
metempsytus Dyar 1921k
rausseoi Cova García, Sutil O, & Pulido F. 1972
Iridescens Group (Valencia 1973)
Iridescens Subgroup (Valencia 1973)
antunesi Lane &Whitman 1943
babahoyensis Leví-Castillo 1953e
bonnei Dyar 1921k
cerqueirai Valencia 1973
insigniforceps Clastrier & Claustre 1978a
iridescens (Lutz 1905)
kompi Valencia 1973
secundus Bonne-Wepster & Bonne 1920
soperi Antunes & Lane 1937
wannonii Cova García & Sutil O. 1976b
wilsoni Lane & Whitman 1943
Urichii Subgroup (Valencia 1973)
anduzei Cerqueira & Lane 1944
urichii (Coquillett 1906e)
Subgenus *Culex*
Atriceps Group (Belkin 1962)
atriceps Edwards 1926a
kesseli Belkin 1962
marquesensis Stone & Rosen 1953
Duttoni Group (Harbach 1988)
duttoni Theobald 1901d
Guiarti Group (Edwards 1941)
grahamii Theobald 1910c
grahamii farakoensis Hamon 1955
guiarti Blanchard 1905
guiarti sudanicus Edwards 1941
ingrami Edwards 1916
pajoti da Cunha Ramos & Ribeiro 1982
schwetzi Edwards 1929c
verutus Harbach 1987
weschei Edwards 1935a
weschei gediensis Edwards 1941
Pipiens Group (Edwards 1932a, in part)
abnormalis Lane 1936b
andersoni Edwards 1914b
andersoni bwambanus Edwards 1941
andersoni abyssinicus Edwards 1941
argenteopunctatus (Ventrillon 1905a)
argenteopunctatus kingii (Theobald 1913d)
astridianus De Meillon 1942
bickleyi Forattini 1965
bukavuensis Wolfs 1947b
calurus Edwards 1935a
carleti Brunhes & Ravaonjanahary 1971
chorleyi Edwards 1941
comorensis Brunhes 1977b
comorensis kartalae Brunhes 1977b
covagarciai Forattini 1965
demeilloni Doucet 1950
guayasi Leví-Castillo 1953c
hancocki Edwards 1930a
hopkinsi Edwards 1932b
huangae Meng 1958
levicastilloi Lane 1945
mirificus Edwards 1913b
musarum Edwards 1932b
nakuruensis Mattingly 1951c
nilgiricus Edwards 1916
ninagongoensis Edwards 1928b
ornatothoracis Theobald 1909b
perfidiosus Edwards 1914b
perfuscus Edwards 1914b
philipi Edwards 1929c
prosecutor Séguy 1927
pruina Theobald 1901d

pruina eschirasi Galliard 1931b
pseudopruina van Someren 1951
quasiguiarti Theobald 1910c
riojanus Duret 1968e
scottii Theobald 1912b
seldeslachtsi Wolfs 1947c
shoae Hamon & Ovazza 1954
shoae ugandae van Someren 1967
striatipes Edwards 1941
striatipes joanae Muspratt 1955
telesilla De Meillon & Lavoipierre 1945. In De Meillon et al. 1945
terzii Edwards 1941
toroensis Edwards & Gibbins 1939
toroensis macrophyllus Edwards & Gibbins 1939
trifoliatus Edwards 1914b
umbripes Edwards 1941
vansomereni Edwards 1926a
vansomereni draconis Ingram & De Meillon 1927
vansomereni elgonicus Edwards 1941
ventrilloni Edwards 1920b
watti Edwards 1920a
yojoae Strickman 1990
zombaensis Theobald 1901c

Coronator Complex (Forattini 2002)
camposi Dyar 1925d
coronator Dyar & Knab 1906d
ousqua Dyar 1918b
usquatissimus Dyar 1922b
usquatus Dyar 1922b

Restuans Complex (Bram 1967a; as *restuans-laticlasper-acharistus* complex)
acharistus Root 1927c
brethesi Dyar 1919c
laticlasper Galindo & Blanton 1954
restuans Theobald 1901c

Salinarius Complex (Bram 1967a)
alani Forattini 1965
archegus Dyar 1929b
bilineatus Theobald 1903a
dolosus (Lynch Arribálzaga 1891)
salinarius Coquillett 1904b
spinosus Lutz 1905

Apicinus Subgroup (Edwards 1932a; as *salinarius-apicinus* series)
ameliae Casal 1967
apicinus Philippi 1865
aquarius Strickman 1990
articularis Philippi 1865
bonneae Dyar & Knab 1919
brami Forattini, Rabello, & Lopes 1967
carcinoxenus de Oliveira Castro 1932
chidesteri Dyar 1921h
curvibrachius Angulo 1993. In Angulo & Olivares 1993
delys Howard, Dyar, & Knab 1915
diplophyllum Dyar 1929b
dohenyi Hogue 1975
eduardoi Casal & García 1968e
erythrothorax Dyar 1907
foliaceus Lane 1945
inflictus Theobald 1901c
lahillei Bachmann & Casal 1962b
mollis Dyar & Knab 1906b
nigripalpus Theobald 1901c
plicatus Olivares 1993
scimitar Branch & Seabrook 1959
sphinx Howard, Dyar, & Knab 1913
tatoi Casal & García 1971

Decens Subgroup (Harbach 1988)
antennatus (Becker 1903)
decens Theobald 1901d
invidiosus Theobald 1901d
invidiosus vicinalis De Meillon & Lavoipierre 1944
invidiosus vexillatus Edwards 1941
litwakae Harbach 1985b

Pipiens Subgroup (Sirivanakarn 1976)
australicus Dobrotworsky & Drummond 1953
globocoxitus Dobrotworsky 1953
pipiens Linnaeus 1758
pipiens pallens (Coquillett 1898)
quinquefasciatus Say 1823

Simpsoni Subgroup (Harbach 1988)
simpsoni Theobald 1905g
sinaiticus Kirkpatrick 1925a

Tarsalis Subgroup (Edwards 1932a; as *tarsalis* series)
bahamensis Dyar & Knab 1906d
bidens Dyar 1922g
brevispinosus Bonne-Wepster & Bonne 1920
chitae Duret 1967b
cuyanus Duret 1968e
declarator Dyar & Knab 1906d
duplicator Dyar & Knab 1909d
garciai Broche 2000
habilitator Dyar & Knab 1906d
interfor Dyar 1928
interrogator Dyar & Knab 1906d
janitor Theobald 1903a
lygrus Root 1927c
maracayensis Evans 1923b
maxi Dyar 1928
paramaxi Duret 1968b
pinarocampa Dyar & Knab 1908
pseudostigmatosoma Strickman 1990
saltanensis Dyar 1928

secutor Theobald 1901c
stenolepis Dyar & Knab 1908
stigmatosoma Dyar 1907
surinamensis Dyar 1918d
tarsalis Coquillett 1896
thriambus Dyar 1921a
Theileri Subgroup (Sirivanakarn 1976)
laticinctus Edwards 1913a
mattinglyi Knight 1953
tenagius van Someren 1954b
theileri Theobald 1903a
Trifilatus Subgroup (Mattingly & Rageau 1958)
asteliae Belkin 1968a
banksensis Maffi & Tenorio 1977
guizhouensis Chen & Zhao 1985
hutchinsoni Barraud 1924g
iyengari Mattingly & Rageau 1958
miraculosus Bonne-Wepster 1937
pacificus Edwards 1916
pervigilans von Bergroth 1889
rotoruae Belkin 1968a
tamsi Edwards 1934
torrentium Martini 1925
trifilatus Edwards 1914b
trifilatus aenescens Edwards 1941
vagans Wiedemann 1828
Univittatus Subgroup (Sirivanakarn 1976)
fuscocephala Theobald 1907
neavei Theobald 1906b
perexiguus Theobald 1903a
univittatus Theobald 1901c
Sitiens Group (Edwards 1932a, in part)
bihamatus Edwards 1926a
castelli Hamon 1957
crinicauda Edwards 1921c
omani Belkin 1962
roseni Belkin 1962
sechani Brunhes & Boussès 2009
thalassius Theobald 1903a
toviiensis Klein, Rivière, & Séchan 1984
vicinus (Taylor 1916)
whittingtoni Belkin 1962
Barraudi Subgroup (Sirivanakarn 1976)
barraudi Edwards 1922a
edwardsi Barraud 1923i
Gelidus Subgroup (Sirivanakarn 1976)
gelidus Theobald 1901c
Mimeticus Subgroup (Mimeticus Series of Edwards 1932a)
diengensis Brug 1931c
fasyi Baisas 1938
jacksoni Edwards 1934
mimeticus Noè 1899
mimuloides Barraud 1924e
mimulus Edwards 1915c
murrelli Lien 1968d
orientalis Edwards 1921d
propinquus Colless 1955b
solitarius Bonne-Wepster 1938
tianpingensis Chen 1981
tsengi Lien 1968d
Sitiens Subgroup (Bram 1967a; Sitiens Series of Edwards 1932a, in part)
alis Theobald 1903a
annulirostris Skuse 1889
annulirostris marianae Bohart & Ingram 1946b
litoralis Bohart 1946b
palpalis Taylor 1912a
sitiens Wiedemann 1828
whitmorei (Giles 1904a)
Vishnui Subgroup (Bram 1967a)
alienus Colless 1957a
annulus Theobald 1901a
incognitus Baisas 1938
perplexus Leicester 1908
philippinensis Sirivanakarn 1976
pseudovishnui Colless 1957a
tritaeniorhynchus Giles 1901a
vishnui Theobald 1901a
whitei Barraud 1923i
Uncertain Placement
annuliventris (Blanchard 1852)
beta Séguy 1924
brumpti Gailiard 1931a
fernandezi Casal, García, & Cavalieri 1966
gameti Bailly-Choumara 1966c
mauesensis Lane 1945
pseudojanthinosoma Senevet & Abonnenc 1946
quitensis Leví-Castillo 1953a
scheuberi Carpintero & Leguizamón 2004
Subgenus *Culiciomyia*
Dispectus Group (Bram 1969)
cheni Dong, Wang, & Lu 2003
dispectus Bram 1966
hainanensis Chen 1977
Fragilis Group (Edwards 1932a) (Oriental and Australasian species)
bahri (Edwards 1914b)
barrinus Bram 1967b
ceramensis Sirivanakarn & Kurihara 1973
fragilis Ludlow 1903
fuscicinctus King & Hoogstraal 1946a
lampangensis Sirivanakarn 1973b
maplei Knight & Hurlbut 1949
nailoni King & Hoogstraal 1946a
nigropunctatus Edwards 1926a
pallidothorax Theobald 1905g
papuensis (Taylor 1914d)
pullus Theobald 1905f

ramakrishnii Wattal & Kalra 1965
ramalingami Sirivanakarn 1973b
ryukyensis Bohart 1946b
scanloni Bram 1967b
spathifurca (Edwards 1915c)
spiculothorax Bram 1967b
termi Thurman 1955
thurmanorum Bram 1967b
viridiventer Giles 1901a
yaoi Tung 1955

Nebulosus Group (Edwards 1932a)
cambournaci Hamon & Candara 1955
cinerellus Edwards 1922d
cinereus Theobald 1901c
eouzani Geoffroy 1971a
furlongi van Someren 1954a
gilliesi Hamon & van Someren 1961a
grenieri Eouzan 1969
harleyi Peters 1955a
liberiensis Peters 1955a: 85
macfiei Edwards 1923a
milloti Doucet 1949b
mongiro van Someren 1951
muspratti Hamon & Lambrecht 1959
nebulosus Theobald 1901d
nebulosu pseudocinereus Theobald 1901c
pandani Brunhes 1969
ruthae Peters 1958
semibrunneus Edwards 1927a
subaequalis Edwards 1941

Shebbearei Group (unnamed "group or complex" of Sirivanakarn 1977c)
bailyi Barraud 1934
harrisoni Sirivanakarn 1977c
javanensis Bonne-Wepster 1934
kyotoensis Yamaguti & LaCasse 1952
megaonychus Yang, Li, & Chen 1993
rajah Tsukamoto 1989
sasai Kano, Nitahara & Awaya 1954
shebbearei Barraud 1924i
spiculostylus Chen 1989

Tricuspis Group (Harrison 1987)
azurini Toma, Miyagi, & Cabrera 1984
delfinadoae Sirivanakarn 1973b
tricuspis Edwards 1930a

Subgenus *Eumelanomyia*

Eumelanomyia Group (Sirivanakarn 1971a)
acrostichalis Edwards 1941
adersianus Edwards 1941
albiventris Edwards 1922d
garioui Bailly-Choumara & Rickenbach 1966
kanyamwerima van Someren 1951
kilara van Someren 1951
taufliebi Geoffroy & Hervé 1976
vinckei Hamon, Holstein, & Rivola 1957

Mochthogenes Group (Sirivanakarn 1971a)
mohani Sirivanakarn 1977d

Bokorensis Subgroup (Sirivanakarn 1972)
bokorensis Klein & Sirivanakarn 1970

Femineus Subgroup (Sirivanakarn 1971a)
femineus Edwards 1926a

Foliatus Subgroup (Sirivanakarn 1972)
foliatus Brug 1932
latifoliatus Delfinado 1966a

Hinglungensis Subgroup (Sirivanakarn 1971a)
baisasi Sirivanakarn 1972
castrensis Edwards 1922a
cataractarum Edwards 1923b
hinglungensis Chu 1957
manusensis Sirivanakarn 1975b

Inconspicuosus Subgroup (Sirivanakarn 1971a)
castor De Meillon & Lavoipierre 1944
fimbriforceps Edwards 1935a
germaini Geoffroy 1974
hamoni Brunhes, Adam, & Bailly-Choumara 1967
helenae Brunhes, Adam, & Bailly-Choumara 1967
inconspicuosus (Theobald 1908a)
micolo Ribeiro, da Cunha Ramos, Capela, & Pires 1998
mijanae Brunhes, Adam, & Bailly-Choumara 1967
orstom Brunhes, Adam, & Bailly-Choumara 1967
quintetti Brunhes, Adam, & Bailly-Choumara 1967
simpliciforceps Edwards 1930a

Iphis Subgroup (Sirivanakarn1972)
iphis Barraud 1924h

Khazani Subgroup (Sirivanakarn 1972)
khazani Edwards 1922a

Malayi Subgroup (Sirivanakarn 1971a)
laureli Baisas 1935d
malayi (Leicester 1908)
yeageri Baisas 1935d

Okinawae Subgroup (Sirivanakarn 1972)
miaolingensis Chen 1982
okinawae Bohart 1953

Otachati Subgroup (Sirivanakarn 1971a)
otachati Klein & Sirivanakarn 1970
richardgarciai Jeffery, Oothuman, & Rudnick 1987

Pluvialis Subgroup (Sirivanakarn 1972)
campilunati Carter & Wijesundara 1948
pluvialis Barraud 1924h
selai Klein & Sirivanakarn 1970

Tenuipalpis Subgroup (Sirivanakarn 1972)
hackeri Edwards 1923b

hayashii Yamada 1917
hayashii ryukyuanus Tanaka, Mizusawa, & Saugstad 1979
kiriensis Klein & Sirivanakarn 1970
macrostylus Sirivanakarn & Ramalingam 1976
megafolius Chen & Dong 1992
oresbius Harbach & Rattanarithikul 1988
richei Klein 1970
tenuipalpis Barraud 1924h
Uncinatus Subgroup (Sirivanakarn 1971a)
uncinatus Delfinado 1966a
Protomelanoconion Group (Sirivanakarn 1971a)
brevipalpis (Giles 1902)
horridus Edwards 1922d
horridus rageaui (Hamon & Rickenbach 1955b)
phangngae Sirivanakarn 1972
stellatus van Someren 1947a
Rubinotus-rima Group (Sirivanakarn 1971a)
brenguesi Brunhes & Ravaonjanahary 1973
jefferyi Sirivanakarn 1977b
Rima Subgroup (Sirivanakarn 1971a)
adami (Hamon & Mouchet 1955)
albertianus Edwards 1941
amaniensis van Someren & Hamon 1964
calabarensis Edwards 1941
chauveti Brunhes & Rambelo 1968
galliardi Edwards 1941
insignis (Carter 1911)
laplantei Hamon, Adam, & Mouchet 1955
mundulus Grünberg 1905
rima Theobald 1901d
subrima Edwards 1941
sunyaniensis Edwards 1941
vattieri Geoffroy 1971c
wansoni Wolfs 1945c
wigglesworthi Edwards 1941
Rubinotus Subgroup (Sirivanakarn 1971a)
andreanus Edwards 1927a
kingianus Edwards 1922d
malayensis Sirivanakarn 1972
pseudoandreanus Bailly-Choumara 1966a
rubinotus Theobald 1906b
simplicicornis Edwards 1930a
Subgenus *Kitzmilleria*
moucheti Evans 1923a
Subgenus *Lasiosiphon*
adairi Kirkpatrick 1926
Subgenus *Lophoceraomyia*
Fraudatrix Group (Edwards 1934. In Barraud 1934)
cubiculi Marks 1989
gossi Bohart 1957
kusaiensis Bohart 1957
orbostiensis Dobrotworsky 1958a
Fraudatrix Subgroup (Sirivanakarn 1977a)
Alphus Complex (Sirivanakarn 1977a)
alphus Colless 1965
Bergi Complex (Belkin 1962)
bergi Belkin 1962
laffooni Belkin 1962
oweni Belkin 1962
winkleri Belkin 1962
Buxtoni Complex (Belkin 1962)
buxtoni Edwards 1926a
lairdi Belkin 1962
Christiani Complex (Sirivanakarn 1968)
christiani Colless 1960
gressitti Sirivanakarn 1968
minjensis Sirivanakarn 1968
pseudornatus Colless 1960
Cinctellus Complex (Sirivanakarn 1977a)
cinctellus Edwards 1922a
fulleri (Ludlow 1909b)
Cottlei Complex (Sirivanakarn 1968)
cottlei Sirivanakarn 1968
Fraudatrix Complex (Sirivanakarn 1968)
atracus Colless 1960
collessi Sirivanakarn 1968
fraudatrix (Theobald 1905f)
insequens Marks 1989
insularis Sirivanakarn 1968
kaviengensis Sirivanakarn 1968
rajaneeae Sirivanakarn 1968
schilfgaardei Sirivanakarn 1968
submarginalis Sirivanakarn 1973a
Hilli Complex (Sirivanakarn 1968)
carolinensis Bohart & Ingram 1946b
hilli Edwards 1922d
lakei Sirivanakarn 1968
Hurlbuti Complex (Belkin 1962)
hurlbuti Belkin 1962
perryi Belkin 1962
Inculus Complex (Sirivanakarn 1977a)
inculus Colless 1965
Marksae Complex (Sirivanakarn 1968)
kowiroensis Sirivanakarn 1968
leei King & Hoogstraal 1955
marksae King & Hoogstraal 1955
muruae Sirivanakarn 1968
versabilis Sirivanakarn 1968
wamanguae Sirivanakarn 1968
Ornatus Complex (Sirivanakarn 1968)
gagnei Evenhuis 1989. In Evenhuis & Gon 1989
Petersi Complex (Sirivanakarn 1968)
crowei Sirivanakarn 1968
petersi Colless 1960
shanahani Sirivanakarn 1968
steffani Sirivanakarn 1968

Pseudorubithoracis Complex (Sirivanakarn 1968)
castaneus Sirivanakarn 1973a
pseudorubithoracis Sirivanakarn 1968
sedlacekae Sirivanakarn 1968
Quadripalpis Complex (Sirivanakarn 1977a)
aculeatus Colless 1965
aestivus Sirivanakarn 1977a
paraculeatus Sirivanakarn 1977a
quadripalpis (Edwards 1914b)
reidi Colless 1965
Rubithoracis Complex (Sirivanakarn 1977a)
gibbulus Delfinado 1966a
niger (Leicester 1908)
rubithoracis (Leicester 1908)
sangenluoensis Wang 1984
Seniori Complex (Sirivanakarn 1977a)
seniori Barraud 1934
Solomonis Complex (Belkin 1962)
becki Belkin 1962
durhami Sirivanakarn 1968
solomonis Edwards 1929
walukasi Belkin 1962
Variatus Complex (Sirivanakarn 1977a)
cubitatus Colless 1965
gracicornis Sirivanakarn 1977a
josephineae Baisas 1935d
macdonaldi Colless 1965
pairoji Sirivanakarn 1977a
variatus (Leicester 1908)
whartoni Colless 1965
Minutissimus Subgroup (Sirivanakarn 1977a)
alorensis Sirivanakarn 1977a
cylindricus Theobald 1903a
infantulus Edwards 1922a
minutissimus (Theobald 1907)
Mammilifer Group (Edwards 1932a)
szemaonensis Wang & Feng 1964
Bolii Subgroup (Sirivanakarn 1968)
bolii Sirivanakarn 1968
Brevipalpus Subgroup (Colless 1965)
Brevipalpus Complex (Sirivanakarn 1977a)
acutipalus Colless 1965
brevipalpus (Theobald 1905f)
eminentia (Leicester 1908)
lucaris Colless 1965
Curtipalpis Complex (Sirivanakarn 1977a)
curtipalpis (Edwards 1914c)
curtipalpis sumatranus Brug
Hewitti Complex (Sirivanakarn 1977a)
hewitti (Edwards 1914b)
Jenseni Complex (Sirivanakarn 1977a)
jenseni (de Meijere 1910)
Navalis Complex (Sirivanakarn 1977a)
coerulescens Edwards 1928a
navalis Edwards 1926a
Digoelensis Subgroup
digoelensis Brug 1932
singuawaensis Sirivanakarn 1969
Mammilifer Subgroup (Colless 1965)
kuhnsi King & Hoogstraal 1955
Flavicornis Complex (Sirivanakarn 1977a)
flavicornis Barraud 1924j
lasiopalpis Sirivanakarn 1977a
raghavanii Rahman, Chowdhury, & Kalra 1969
Ganapathi Complex (Sirivanakarn 1977a)
ganapathi Colless 1965
spiculosus Bram & Rattanarithikul 1967
Impostor Complex (Sirivanakarn 1977a)
impostor Sirivanakarn 1977b
Mammilifer Line (Sirivanakarn 1977a)
Mammilifer Complex (Sirivanakarn 1977a)
demissus Colless 1965
mammilifer (Leicester 1908)
wardi Sirivanakarn 1977
Minor Line (Sirivanakarn 1977a)
Minor Complex (Sirivanakarn 1977a)
bandoengensis Brug 1939
bengalensis Barraud 1934
bicornutus (Theobald 1910a)
crassicomus Colless 1965
incomptus Bram & Rattanarithikul 1967
kuhnsi King & Hoogstraal 1955
minor (Leicester 1908)
tuberis Bohart 1946b
Peytoni Complex (Sirivanakarn 1977a)
eukrines Bram & Rattanarithikul 1967
peytoni Bram & Rattanarithikul 1967
Pholeter Complex (Sirivanakarn 1977a)
pholeter Bram & Rattanarithikul 1967
Singhbhumensis Complex (Natarajan & Rajavel 2009)
singhbhumensis Natarajan & Rajavel 2009
Traubi Complex (Sirivanakarn 1977a)
lavatae Stone & Bohart 1944
traubi Colless 1965
uniformis (Theobald 1905h)
Wilfredi Group (Sirivanakarn 1977a)
hirtipalpis Sirivanakarn 1977a
pilifemoralis Wang & Feng 1964
wilfredi Colless 1965
Subgenus *Maillotia*
Hortensis Group (Sirivanakarn 1971a)
arbieeni Salem 1938
hortensis Ficalbi 1889a
hortensis maderensis Mattingly 1955a
quettensis Mattingly 1955a
Seyrigi Group (Sirivanakarn 1971a)
avianus De Meillon 1943
peringueyi Edwards 1924a
salisburiensis Theobald 1901c

salisburiensis capensis De Meillon 1935
salisburiensis coursi Doucet 1949a
seyrigi Edwards 1941
subsalisburiensis Hervé & Geoffroy 1974
Unplaced Species
deserticola Kirkpatrick 1925
Subgenus *Melanoconion*
Melanoconion Section (Sirivanakarn 1983)
guedesi da Silva Mattos & Xavier 1991
herrerai Sutil Oramas, Pulido, F., & Amarista, M. 1987
Atratus Group (Sirivanakarn 1983)
atratus Theobald 1901c
caribeanus Galindo & Blanton 1954
commevynensis Bonne-Wepster & Bonne 1920
dunni Dyar 1918d
ensiformis Bonne-Wepster & Bonne 1920
trigeminatus Clastrier 1970c
zeteki Dyar 1918d
Bastagarius Group (Sirivanakarn 1983)
diamphidius Peyton & Harbach 1991
Bastagarius Subgroup (Sirivanakarn 1983)
alinkios Sallum & Hutchings 2003
bastagarius Dyar & Knab 1906b
brachiatus Hutchings & Sallum 2008
comatus Senevet & Abonnenc 1939a
coppenamensis Bonne-Wepster & Bonne 1920
creole Anduze 1949a
intonsus Galindo & Blanton 1954
phyllados Hutchings & Sallum 2008
tournieri Senevet & Abonnenc 1939a
Iolambdis Subgroup (Sirivanakarn 1983)
bifoliolatus Duret & Barreto 1956
confundior Komp & Rozeboom 1951
corentynensis Dyar 1920c
dolichophyllus Clastrier 1970b
dureti Casal & García 1968d
iolambdis Dyar 1918b
limacifer Komp 1936
quasihibridus Galindo & Blanton 1954
Conspirator Group (Sirivanakarn 1983)
aliciae Duret 1953
conspirator Dyar & Knab 1906d
dyius Root 1927c
elevator Dyar & Knab 1906d
jocasta Komp & Rozeboom 1951
lucifugus Komp 1936
madininensis Senevet 1936a
martinezi Casal & García 1968c
olimpioi Xavier, da Silva & da Silva Mattos 1970
terebor Dyar 1920c
Distinguendus Group (Sirivanakarn 1983)
Distinguendus Subgroup (Sirivanakarn 1983)
alcocki Bonne-Wepster & Bonne 1920
comminutor Dyar 1920c
distinguendus Dyar 1928
maxinocca Dyar 1920c
nicceriensis Bonne-Wepster & Bonne 1920
patientiae Floch & Fauran 1955
productus Senevet & Abonnenc 1939a
Galindoi Subgroup (Sirivanakarn 1983)
galindoi Komp & Rozeboom 1951
Putumayensis Subgroup (Sirivanakarn 1983)
bahiensis Duret 1969b
phlabistus Dyar 1920c
putumayensis Matheson 1934b
Rorotaensis Subgroup (Sirivanakarn 1983)
rorotaensis Floch & Abonnenc 1946a
Educator Group (Sirivanakarn 1983)
cristovaoi Duret 1968a
educator Dyar & Knab 1906d
eknomios Forattini & Sallum 1992b
inadmirabilis Dyar 1928
rachoui Duret 1968b
theobaldi (Lutz 1904)
vaxus Dyar 1920c
Erraticus Group (Sirivanakarn 1983)
Clarki Subgroup (Sirivanakarn 1983)
clarki Evans 1924
Erraticus Subgroup (Sirivanakarn 1983)
aureonotatus Duret & Barreto 1956
erraticus (Dyar & Knab 1906d)
invocator Pazos 1908
Psatharus Subgroup (Sirivanakarn 1983)
psatharus Dyar 1920g
Evansae Group (Sirivanakarn 1983)
batesi Rozeboom & Komp 1948
changuinolae Galindo & Blanton 1954
evansae Root 1927c
johnnyi Duret 1968d
Inhibitator Group (Sirivanakarn 1983)
spathulatus Forattini & Sallum 1987b
Egcymon Subgroup (Sirivanakarn 1983)
caudatus Clastrier 1970c
egcymon Dyar 1923b
elephas Komp 1936
isabelae Duret 1968f
serratimarge Root 1927c
Inhibitator Subgroup (Sirivanakarn 1983)
abonnenci Clastrier 1970c
albinensis Bonne-Wepster & Bonne 1920
amitis Komp 1936
bejaranoi Duret 1953
carcinophilus Dyar & Knab 1906d
contei Duret 1968c
ernanii Duret 1968f
ernsti Anduze 1949b
flabellifer Komp 1936
inhibitator Dyar & Knab 1906d

kummi Komp & Rozeboom 1951
mesodenticulatus Galindo & Mendez 1961
oedipus Root 1927c
orfilai Duret 1953
pavlovskyi Casal & García 1967a
phlogistus Dyar 1920c
plectoporpe Root 1927c
rabelloi Forattini & Sallum 1987a
symbletos Sallum & Hutchings 2003
vidali Floch & Fauran 1954
wepsterae Komp & Rozeboom 1951
Mulrennani Subgroup (Sirivanakarn 1983)
mulrennani Basham 1948
Intrincatus Group (Sirivanakarn 1983)
Andricus Subgroup (Sirivanakarn 1983)
andricus Root 1927c
Eastor Subgroup (Sirivanakarn 1983)
eastor Dyar 1920c
Idottus Subgroup (Sirivanakarn 1983)
fairchildi Galindo & Blanton 1954
ferreri Duret 1968f
idottus Dyar 1920c
ronderosi de Linero 1967
sardinerae Fox 1953
Intrincatus Subgroup (Sirivanakarn 1983)
bequaerti Dyar & Shannon 1925b
equinoxialis Floch & Abonnenc 1945c
glyptosalpinx Harbach, Peyton, & Harrison 1984
intrincatus Brèthes 1916
johnsoni Galindo & Mendez 1961a
milwardi Xavier & da Silva Mattos 1972
misionensis Duret 1953
mutator Dyar & Knab 1906d
pifanoi Anduze 1949b
quadrifoliatus Komp 1936
rabanicolus Floch & Abonnenc 1946a
silvai Duret 1968f
sursumptor Dyar 1924d
trilobulatus Duret & Barreto 1956
trisetosus Fauran 1961a
ybarmis Dyar 1920c
Penai Subgroup (Sirivanakarn 1983)
penai Sirivanakarn 1979
Tecmarsis Subgroup (Sirivanakarn 1983)
tecmarsis Dyar 1918d
Peccator Group (Sirivanakarn 1983)
abominator Dyar & Knab 1909d
anips Dyar 1916a
peccator Dyar & Knab 1909d
Pilosus Group (Sirivanakarn 1983)
Caudelli Subgroup (Sirivanakarn 1983)
alogistus Dyar 1918d
arboricolus Galindo & Mendez 1961
caudelli (Dyar & Knab 1906d)
foliafer Komp & Rozeboom 1951
galvaoi Duret 1968b
garcesi Duret 1968b
lacertosus Komp & Rozeboom 1951
palaciosi Duret 1968d
vexillifer Komp 1936
Pilosus Subgroup (Sirivanakarn 1983)
innovator Evans 1924
pilosus (Dyar & Knab 1906d)
rooti Rozeboom 1935
unicornis Root 1928. In Dyar 1928
Saramaccensis Group (Sirivanakarn 1983)
saramaccensis Bonne-Wepster & Bonne 1920
Trifidus Group (Sirivanakarn 1983)
trifidus Dyar 1921h
Spissipes Section (Sirivanakarn 1983)
Crybda Group (Sallum & Forattini 1996)
Paracrybda Subgroup (Sirivanakarn 1983)
delpontei Duret 1969a
paracrybda Komp 1936
Pedroi Subgroup (Sirivanakarn 1983)
adamesi Sirivanakarn & Galindo 1980
crybda Dyar 1924i
epanastasis Dyar 1922g
pedroi Sirivanakarn & Belkin 1980
ribeirensis Forattini & Sallum 1985
Pereyrai Subgroup (Sirivanakarn 1983)
pereyrai Duret 1967a
Faurani Group (Sirivanakarn 1983)
faurani Duret 1968f
Jubifer Group (Sirivanakarn 1983)
jubifer Komp & Brown 1935
simulator Dyar & Knab 1906d
Lopesi Group (Sirivanakarn 1983)
lopesi Sirivanakarn & Jakob 1979
Ocossa Group (Sirivanakarn 1983)
ocossa Dyar & Knab 1919
panocossa Dyar 1923c
Spissipes Group (Sirivanakarn 1983)
spissipes (Theobald 1903a)
Taeniopus Group (Sallum & Forattini 1996)
akritos Forattini & Sallum 1995
cedecei Stone & Hair 1968
ikelos Forattini & Sallum 1995
taeniopus Dyar & Knab 1907b
Vomerifer Group (Sallum & Forattini 1996)
gnomatos Sallum, Hutchings, Leila, & Ferreira 1997
portesi Senevet & Abonnenc 1941
sacchettae Sirivanakarn & Jakob 1982
vomerifer Komp 1932
Subgenus *Micraedes*
Bisulcatus Group (Berlin 1969c)
antillummagnorum Dyar 1928
arawak Berlin 1969c

biscaynensis Zavortink & O'Meara 1999
bisulcatus (Coquillett 1906a)
Erethyzonfer Group (Berlin 1969c)
erethyzonfer Galindo & Blanton 1954
Schicki Group (Berlin 1969c)
jalisco Berlin 1974
sandrae Berlin 1969c
schicki Berlin 1969c
Subgenus *Microculex*
Consolator Series (Lane & Whitman 1951)
consolator Dyar & Knab 1906b
hedys Root 1927c
reducens Lane & Whitman 1951
worontzowi Pessoa & Galvão 1936
Imitator Series (Lane & Whitman 1951)
carioca Lane & Whitman 1951
dubitans Lane & Whitman 1951
imitator Theobald 1903a
imitator retrosus Lane & Whitman 1951
Inimitabilis Series (Lane & Whitman 1951)
aphylactus Root 1927c
inimitabilis Dyar & Knab 1906d
inimitabilis fuscatus Lane & Whitman 1951
microphyllus Root 1927c
neglectus Lutz 1904
Pleuristriatus Series (Lane & Whitman 1951)
albipes Lutz 1904
aureus Lane & Whitman 1951
azymus Dyar & Knab 1906b
davisi Kumm 1933
gairus Root 1927c
intermedius Lane & Whitman 1951
lanei Coutinho & Forattini 1962
pleuristriatus Theobald 1903a
xenophobus Ronderos 1965
Unplaced Species
chryselatus Dyar & Knab 1919
daumastocampa Dyar & Knab 1908
elongatus Rozeboom & Komp 1950b
gaudeator Dyar & Knab 1907c
jenningsi Dyar & Knab 1907c
kukenan Anduze 1942c
pulidoi Cova García & Sutil Oramas 1974
reginae Floch & Fauran 1955
rejector Dyar & Knab 1906d
shopei Forattini & Toda 1966
siphanulatus Lourenço-de-Oliveira & Silva-do-Nascimento 1987
stonei Lane & Whitman 1943
sutili Cova García & Pulido F. 1974
Subgenus *Neoculex*
Crassistylus Group (Sirivanakarn 1971a)
crassistylus Brug 1934a
leonardi Belkin 1962
pedicellus King & Hoogstraal 1947
Pseudomelanoconia Group (Sirivanakarn 1971a)
chaetoventralis (Theobald 1910c)
cheesmanae Mattingly & Marks 1955
douglasi Dobrotworsky 1956
dumbletoni Belkin 1962
fergusoni (Taylor 1914b)
gaufini Belkin 1962
latus Dobrotworsky 1956
millironi Belkin 1962
postspiraculosus Lee 1944c
pseudomelanoconia Theobald
Territans Group (Sirivanakarn 1971a)
apicalis Adams 1903
arizonensis Bohart 1949
boharti Brookman & Reeves 1950
derivator Dyar & Knab 1906d
europaeus da Cunha Ramos, Ribeiro & Harrison 2003
impudicus Ficalbi 1890d
judaicus Edwards 1926c
martinii Medschid 1930
reevesi Wirth 1948
rubensis Sasa & Takahasi 1948
territans Walker 1856a
Unplaced Species
gamma Séguy 1924
johni Cova García, Pulido F., & Escalante de Ugueto 1979
Subgenus *Oculeomyia*
albinervis Edwards 1929c
annulioris Theobald 1901a
annulioris consimilis Newstead 1907
aurantapex Edwards 1914b
aurantapex jinjaensis Edwards 1941
aurantapex ellinorae Ovazza, Hamon & Neri 1956
giganteus Ventrillon 1906a
poicilipes (Theobald 1903a)
samoaensis (Theobald 1914)
squamosus (Taylor 1914c)
starckeae Stone & Knight 1958
Bitaeniorhynchus Complex (Sirivanakarn 1976)
bitaeniorhynchus Giles 1901a
infula Theobald 1901a
longicornis Sirivanakarn 1976
luzonensis Sirivanakarn 1976
pseudosinensis Colless 1955b
selangorensis Sirivanakarn 1976
Geminus Complex (Sirivanakarn 1976)
geminus Colless 1955
kinabaluensis Sirivanakarn 1976
Sinensis Complex (Sirivanakarn 1976)
cornutus Edwards 1922a
epidesmus (Theobald 1910a)
sinensis Theobald 1903a

Mimomyia Subgenus *Mimomyia*

Informal group names are available for *Mimomyia* (*Mimomyia*) [17 sp./ssp.], but not for *Mi.* (*Etorleptiomyia*) [seven spp.] or *Mi.* (*Ingramia*) [21 spp.].

According to Mattingly & Grjebine (1958) and Grjebine (1986) (as *Ficalbia* subgenus *Mimomyia*).

Group A
- *splendens* Theobald 1903a

Group B
- *aurea* (Leicester 1908)
- *chamberlaini* Ludlow 1904c
- *chamberlaini clavipalpus* (Theobald 1908b)
- *chamberlaini metallica* (Leicester 1908)
- *femorata* (Edwards 1936)
- *flavens* (King & Hoogstraal 1946e)
- *hispida* (Theobald 1910c)
- *hybrida* (Leicester 1908)
- *intermedia* Barraud 1929a
- *lacustris* (Edwards 1935a)
- *modesta* (King & Hoogstraal 1946e)
- *pallida* Edwards 1925
- *perplexans* (Edwards 1932b)

Group C
- *mimomyiaformis* (Newstead 1907)

Group D
- *plumosa* (Theobald 1901a)

Group E
- *flavopicta* (Edwards 1936)

Genus *Orthopodomyia*

From Zavortink (1968).

Bancroftia Section
Albicosta Group
albicosta (Lutz 1904)
Signifera Group
Kummi Subgroup
kummi Edwards 1939b
species 5
Pulchripalpis Subgroup
pulcripalpis (Rondani 1872)
Signifera Subgroup
alba Baker 1936
signifera (Coquillett 1896)
waverleyi (Grabham 1907)
Folicolae Section
phyllozoa (Dyar & Knab 1907c)
Orthopodomyia Section
Albipes Group
Albipes Subgroup
albipes Leicester 1904
Anopheloides Subgroup
andamanensis Barraud 1934
anopheloides (Giles 1903). In Thomson 1903
papuensis Zavortink 1968
Flavicosta Subgroup
flavicosta Barraud 1927a
flavithorax Barraud 1927a
madrensis Baisas 1946a
siamensis Zavortink 1968
Wilsoni Subgroup
wilsoni Macdonald 1958
Arboricolus Group
arboricollis (de Charmoy 1908)
Nkolbissonensis Group
nkolbissonensis Rickenbach & Hamon 1966 (1965)
species 14
Vernoni Group
milloti Doucet 1951a
vernoni van Someran 1949a
Thomasina Section
fascipes (Coquillett 1906a)
sampaioi da Costa Lima 1935b

Described since Zavortink (1968), not yet placed.
- *ambremontis* Brunhes & Hervy 1995
- *ankaratrensis* Brunhes & Hervy 1995
- *aureoantennata* Ferrara 1973
- *comorensis* Brunhes 1977b
- *fontenillei* Brunhes & Hervy 1995
- *joyoni* Brunhes 1977b
- *lanyuensis* Lien 1968a
- *ototomoensis* Huang & Rueda 2013
- *peytoni* Leguizamón & Carpintero 2004
- *rajaonariveloi* Brunhes & Hervy 1995
- *ravaonjanaharyi* Brunhes & Hervy 1995
- *reunionensis* Brunhes & Hervy 1995
- *rodhaini* Brunhes & Hervy 1995
- *wanxianensis* Lei & Li 1990

Genus *Toxorhynchites* Theobald

Edwards (1932) judged that there were three groups in genus *Toxorhynchites* based on the length of the female palpus. Subgenus *Lynchiella* was Edward's Group A (Megarhinus

Group), subgenus *Ankylorhynchus* corresponded to Edward's Group B (Ankylorhynchus Group), and subgenus *Toxorhynchites* corresponded to his Group C (Toxorhynchites Group). Seven species included by Edwards in subgenus *Toxorhynchites* were later placed in the Lutescens Group by Edwards (1941). These were separated (Ribeiro 1991) from subgenus *Toxorhynchites* and included in subgenus *Afrorhynchus*, which now contains 20 species.

Steffan & Evenhuis (1985), following Moss et al. (1979) and Simon (1983), proposed informal groupings for subgenus *Toxorhynchites* as follows. Also included below are species in Edward's Brevipalpis Group proposed by Ribeiro (1991).

Evenhuis & Steffan (1986) further treated a group of nepenthic subgenus *Toxorhynchites* species they had previously designated in 1985, the Acaudatus Group.

Acaudatus Group (Steffan & Evenhuis 1985)
- Acaudatus Subgroup (Steffan & Evenhuis 1985)
  - *acaudatus* (Leicester 1908)
  - *angustiplatus* Evenhuis & Steffan 1986
  - *coeruleus* (Brug 1934a)
  - *indicus* Evenhuis & Steffan 1986
  - *ramalingami* Evenhuis & Steffan 1986
- Nepenthis Subgroup (Steffan & Evenhuis 1985)
  - *nepenthis* (Dyar & Shannon 1925a)

Ater Group (Steffan & Evenhuis 1985)
- *ater* (Daniels 1908)
- *nigripes* (Edwards 1935b)
- *sumatranus* (Brug 1939)

Brevipalpis Group (Edwards 1941)
- Brevipalpis Subgroup (Ribeiro 1991)
  - *brevipalpis* Theobald 1901a
    - *brevipalpis conradti* Grünberg 1907a
    - *brevipalpis abyssinicus* Ribeiro 1991

Christophi Group (Steffan & Evenhuis 1985)
- *bickleyi* Thurman 1959b
- *christophi* (Portschinsky 1884)
  - *christophi aurifluus* (Edwards 1921a)
- *edwardsi* (Barraud 1924d)
- *manopi* Thurman 1959b
- *okinawensis* Toma, Miyagi, & Tanaka 1990
- *quasiferox* (Leicester 1908)
- *sunthorni* Thurman 1959b
- *towadensis* (Matsumura 1916)

Leicesteri Group (Steffan & Evenhuis 1985)
- Leicesteri Subgroup (Steffan & Evenhuis 1985)
  - *gravelyi* (Edwards 1921c)
  - *kempi* (Edwards 1921c)
  - *klossi* (Edwards 1921c)
  - *leicesteri* Theobald 1904c
  - *metallicus* Leicester 1904
- Manicatus Subgroup (Steffan & Evenhuis 1985)
  - *albipes* (Edwards 1922a)
  - *auripes* (Edwards 1935b)
  - *manicatus* (Edwards 1921a)
    - *manicatus yamadai* (Ouchi 1939)
  - *yaeyamae* Bohart 1956b

Magnificus Group (Steffan & Evenhuis 1985)
- *bengalensis* Rosenberg & Evenhuis 1985
- *funestus* (Leicester 1908)
- *magnificus* (Leicester 1908)
- *raris* (Leicester 1908)
- Phytophygus Subgroup (Ribeiro 1991)
  - *barbipes* (Edwards 1913b)
  - *camaronis* Ribeiro 1991
  - *dundo* Ribeiro 1991
  - *evansae* (Edwards 1936)
  - *lewisi* Ribeiro 1991
  - *phytophagus* Theobald 1909b
  - *rickenbachi* Ribeiro 1991

Minimus Group (Steffan & Evenhuis 1985)
- *gigantulus* (Dyar & Shannon 1925a)

Splendens Group (Steffan & Evenhuis 1985)
- *amboinensis* (Doleschall 1857)
- *inornatus* (Walker 1865)
  - *inornatus albitarsis* (Brug 1939)
- *nepenthicola* Steffan & Evenhuis 1982
- *pendleburyi* (Edwards 1930a)
- *speciosus* (Skuse 1889)
- *splendens* (Wiedemann 1819)
- *tyagii* Krishnamoorthy et al. 2013

Genus *Tripteroides*

From Mattingly (1981).

Subgenus *Polylepidomyia*
(not subdivided)

Subgenus *Rachionotomyia*
- Aranoides Group (Mattingly 1981)
  - Affinis Subgroup (Mattingly 1981)
    - *affinis* (Edwards 1913c)
    - *dofleini* (Guenther 1913)
  - Aranoides Subgroup (Mattingly 1981)
    - *aranoides* (Theobald 1901c)
    - *ceylonensis* (Theobald 1905h)
    - *coonorensis* Mattingly 1981
    - *pallidothorax* Dong, Dong, & Wu 2008
    - *serratus* (Barraud 1929a)
    - *tenax* (de Meijere 1910)
  - Nepenthis Subgroup (Mattingly 1981)
    - *nepenthis* (Edwards 1915c)
    - *nepenthisimilis* Mattingly 1981
- Edwardsi Group (Mattingly 1981)
  - *edwardsi* (Barraud 1929b)
  - *longipalpis* Dong, Zhou, & Dong 1997b

ponmeki Miyagi & Toma 2001
rozeboomi Baisas & Ubaldo-Pagayon 1952
Subgenus *Rachisoura* (not subdivided)
Subgenus *Tricholeptomyia*
Delpilari Group (Mattingly 1981)
delpilari Baisas & Ubaldo-Pagayon 1952
Nepenthicola Group (Mattingly 1981)
Belkini Subgroup (Mattingly 1981)
belkini Baisas & Ubaldo-Pagayon 1952
Nepenthicola Subgroup (Mattingly 1981)
apoensis Baisas & Ubaldo-Pagayon 1952
barraudi Baisas & Ubaldo-Pagayon 1952
christophersi Baisas & Ubaldo-Pagayon 1952
microcala (Dyar 1929c)
nepenthicola (Banks 1909)
roxasi Baisas & Ubaldo-Pagayon 1952
werneri Baisas & Ubaldo-Pagayon 1952
Subgenus *Tripteroides*
Mabinii Group (Belkin 1962)
mabinii Baisas & Ubaldo-Pagayon 1952
Nitidoventer Group (Mattingly 1981?)
aeneus (Edwards 1921c)
alboscutellatus Lee 1946
antennalis Bohart & Farner 1944
bambusa (Yamada 1917)
bambusa yaemamensis Tanaka, Mizusawa & Saugstad 1979
bimaculipes (Theobald 1905f)
binotatus Belkin 1950a
bonneti Belkin 1962
brevipalpis Brug 1934a
caeruleocephalus (Leicester 1908)
cheni Lien 1968a
claggi Bohart & Farner 1944
denticulatus Delfinado & Hodges 1968
distigma (Edwards 1925)
dyari Bohart & Farner 1944
dyi Baisas & Ubaldo-Pagayon 1952
elegans Brug 1934a
erlindae Baisas & Ubaldo-Pagayon 1952
hoogstraali Baisas 1947a
hybridus (Leicester 1908)
indeterminatus Baisas & Ubaldo-Pagayon 1952
indicus (Barraud 1929a)
intermediatus Baisas & Ubaldo-Pagayon 1952
knighti Baisas & Ubaldo-Pagayon 1952
latispinus Gong & Ji 1989
lipovskyi Belkin 1950a
littlechildi (Edwards 1930b)
longisiphonus Dong, Zhou, & Dong 2001
lorengaui Peters 1963b
mabinii Baisas & Ubaldo-Pagayon 1952
magnesianus (Edwards 1924b)
malayi Delfinado & Hodges 1968
malvari Baisas & Ubaldo-Pagayon 1952
mendacis (Daniels 1908)
monetifer (Dyar 1920h)
nissanensis Lee 1946
nitidoventer (Giles 1904a)
novohanoverae Peters 1963b
plumosus (Brug 1931c)
powelli (Ludlow 1909c)
powelli escodae Baisas & Ubaldo-Pagayon 1952
powelli laffooni Baisas & Ubaldo-Pagayon 1952
powelli mattinglyi Baisas & Ubaldo-Pagayon 1952
proximus (Edwards 1915c)
purpuratus (Edwards 1921c)
quasiornatus (Taylor 1915)
reiseni Basio 1971b
riverai Miyagi, Toma, & Tsukamoto 1983
similis (Leicester 1908)
simulatus Baisas & Ubaldo-Pagayon 1952
splendens Lee 1946
sullivanae Baisas & Ubaldo-Pagayon 1952
tarsalis Delfinado & Hodges 1968
toffaletii Baisas & Ubaldo-Pagayon 1952
uichancoi Baisas & Ubaldo-Pagayon 1952
vicinus (Edwards 1914c)

Genus *Uranotaenia*

Subgenus *Pseudoficalbia* Old World
Annulata Section (da Cunha Ramos 1993/da Cunha Ramos & Brunhes 2004)
Annulata Group (da Cunha Ramos 1993/da Cunha Ramos & Brunhes 2004)
albimanus da Cunha Ramos & Brunhes 2004
annulata Theobald 1901d (in Annulata Series of Peyton 1977)
apicotaeniata Theobald 1909b
carcinicola da Cunha Ramos & Brunhes 2004
Subgroup A (da Cunha Ramos 1993)
cavernicola Mattingly 1954b
lucyae van Someren 1954a
Subgroup B (da Cunha Ramos 1993)
montana Ingram & De Meillon 1927
nivipous Theobald 1912a
Lavieri Group (da Cunha Ramos 1993/da Cunha Ramos & Brunhes 2004)
ambodimanga da Cunha Ramos & Brunhes 2004

apicosquamata da Cunha Ramos & Brunhes 2004
bifasciata da Cunha Ramos & Brunhes 2004
breviseta da Cunha Ramos & Brunhes 2004
haddowi da Cunha Ramos & Brunhes 2004
lavieri Doucet 1950
longitubus da Cunha Ramos & Brunhes 2004
manakaraensis da Cunha Ramos & Brunhes 2004
pseudoalbimanus da Cunha Ramos & Brunhes 2004
scutostriata da Cunha Ramos & Brunhes 2004
spinitubus da Cunha Ramos & Brunhes 2004

Mashoensis Section (da Cunha Ramos 1993/da Cunha Ramos & Brunhes 2004)
Mashoensis Group (da Cunha Ramos 1993/da Cunha Ramos & Brunhes 2004)
mashonaensis Theobald 1901c (in Bicolor Series of Peyton 1977)
micromelas Edwards 1934
nigromaculata Edwards 1941 (in Bicolor Series of Peyton 1977)

Nigripes Section (da Cunha Ramos 1993/da Cunha Ramos & Brunhes 2004)
Fusca Group (da Cunha Ramos 1993/da Cunha Ramos & Brunhes 2004)
fusca Theobald 1907 (in Bicolor Series of Peyton 1977)
Ornata Group (da Cunha Ramos 1993/da Cunha Ramos & Brunhes 2004)
lunda da Cunha Ramos 1993
musarum Edwards 1936
ornata Theobald 1909b (in Bicolor Series of Peyton 1977)
ramosa da Cunha Ramos 1993
signata da Cunha Ramos 1993
yovani van Someren 1951
Nigripes Group (da Cunha Ramos 1993/da Cunha Ramos & Brunhes 2004)
bosseri Grjebine 1979
boussesi da Cunha Ramos & Brunhes 2004
brumpti Doucet 1951c
cachani (Doucet 1950)
combesi Doucet 1950
comorensis da Cunha Ramos & Brunhes 2004
confusa Peyton 1977 (in Recondita Series of Peyton 1977)
douceti Grjebine 1953a
grenieri Doucet 1951c
henriquei da Cunha Ramos 1993
kraussi Grjebine 1953a
laffosseae da Cunha Ramos & Brunhes 2004
nigricephala da Cunha Ramos & Brunhes 2004
nigripes (Theobald 1905i) (in Bicolor Series of Peyton 1977)
pauliani Doucet 1949b
ravenalicola da Cunha Ramos & Brunhes 2004
tsaratananae Doucet 1950

Shillitonis Section (da Cunha Ramos 1993/da Cunha Ramos & Brunhes 2004)
Shillitonis Group (da Cunha Ramos 1993/da Cunha Ramos & Brunhes 2004)
albinotata da Cunha Ramos & Brunhes 2004
bambusicola da Cunha Ramos & Brunhes 2004
belkini Grjebine 1979
brunhesi Grjebine 1979
contrastata da Cunha Ramos & Brunhes 2004
damasei Grjebine 1979
devemyi Hamon 1955
donai da Cunha Ramos & Brunhes 2004
fulgens da Cunha Ramos & Brunhes 2004
henrardi Edwards 1935c
hervyi da Cunha Ramos & Brunhes 2004
madagascarensis da Cunha Ramos & Brunhes 2004
ototomo da Cunha Ramos 1993
pallidipleura da Cunha Ramos & Brunhes 2004
pseudohenrardi Peters 1955c
pseudoshillitonis da Cunha Ramos & Brunhes 2004
rickenbachi da Cunha Ramos 1993
spiraculata da Cunha Ramos & Brunhes 2004
tanzaniae da Cunha Ramos 1993
tridentata da Cunha Ramos & Brunhes 2004
ugandae da Cunha Ramos 1993
Subgroup A (da Cunha Ramos 1993)
garnhami van Someren 1948
shillitonis Edwards 1932b
Subgroup B (da Cunha Ramos 1993)
andreae Doucet 1962
capelai da Cunha Ramos 1993
principensis da Cunha Ramos 1993
Bicincta Group (da Cunha Ramos 1993/da Cunha Ramos & Brunhes 2004) [not verified]

bicincta da Cunha Ramos & Brunhes 2004
Spinosa Section (da Cunha Ramos 1993/da Cunha Ramos & Brunhes 2004)
Spinosa Group (da Cunha Ramos 1993/da Cunha Ramos & Brunhes 2004)
antalahaensis da Cunha Ramos & Brunhes 2004
cornuta da Cunha Ramos & Brunhes 2004
grjebinei da Cunha Ramos & Brunhes 2004
legoffi da Cunha Ramos & Brunhes 2004
nigripleura da Cunha Ramos & Brunhes 2004
pilosa da Cunha Ramos & Brunhes 2004
spinosa da Cunha Ramos & Brunhes 2004
tricolor da Cunha Ramos & Brunhes 2004
Bicolor Series (Peyton 1977; see five additional species above)
abstrusa Peyton 1977
approximata Peyton 1977
atra Theobald 1905f (in Atra Section of Belkin 1962)
bicolor Leicester 1908
browni Mattingly & Brown 1955
colocasiae Edwards 1928a (in Colocasiae Section of Belkin 1962)
demeilloni Peyton & Rattanarithikul 1970
diagonalis Brug 1934a
gouldi Peyton & Klein 1970
lutescens Leicester 1908
nepenthes (Theobald 1912b)
obscura Edwards 1915c
painei Edwards 1935a (in Painei Section of Belkin 1962)
pandani (Theobald 1912b)
quadrimaculata Edwards 1929 (in Atra Section of Belkin 1962)
Bimaculata Series (Peyton 1977)
albipes Peyton 1977
ascidiicola de Meijere 1910
bimaculata Leicester 1908
gigantea Brug 1931c
leiboensis Chu 1981
lui Lien 1968a
maculipleura Leicester 1908
modesta Leicester 1908
moufiedi Peyton 1977
moultoni Edwards 1914c
nivipleura Leicester 1908
novobscura Barraud 1934
novobscura ryukyuana Tanaka, Mizusawa & Saugstad 1979
patriciae Peyton 1977
propinqua Peyton 1977
pseudomaculipleura Peyton & Rattanarithikul 1970
quasimodesta Peyton 1977
qui Dong, Dong, & Zhou 2003 [uncertain.; see orig. descr.]
quinquemaculata Bonne-Wepster 1934
reinerti Peyton 1977
xanthomelaena Edwards 1925
Harrisoni Series (Peyton 1977)
harrisoni Peyton 1977
Maxima Series (Peyton 1977)
maxima Leicester 1908
unguiculata Edwards 1913a
unguiculata pefflyi Stone 1961a
Recondita Series (Peyton 1977; see one additional species above)
abdita Peyton 1977
dandakaranyensis Natarajan, Rajavel, & Jambulingam 2017
dibrugarhensis Bhattacharyya, Prakash, Mohapatra, & Mahanta 2004
enigmatica Peyton 1977
hirsutifemora Peters 1964
hongayi Galliard & Ngu 1947
husaini Qutubuddin 1947
jacksoni Edwards 1935a
jinhongensis Dong, Dong, & Zhou 2003
koli Peyton & Klein 1970
luteola Edwards 1934
maikalensis Natarajan, Rajavel, & Jambulingam 2017
mattinglyi Qutubuddin 1951
nocticola Peyton 1977
ohamai Tanaka, Mizusawa, & Saugstad 1975a
pylei Baisas 1946a
recondita Edwards 1922d
rossi Delfinado 1966b
satpuraensis Natarajan, Rajavel, & Jambulingam 2017
spiculosa Peyton & Rattanarithikul 1970
srilankensis Peyton 1974
stricklandi Barraud 1926
sumethi Peyton & Rattanarithikul 1970
yaeyamana Tanaka, Mizusawa, & Saugstad 1975a
Subgenus *Pseudoficalbia* New World
Anhydor Series (Galindo, Blanton, & Peyton 1954)
anhydor Dyar 1907
anhydor syntheta Dyar & Shannon 1924c
Maxima Series (Peyton 1977)
Subgenus *Uranotaenia* Old World
Alba Section (da Cunha Ramos & Brunhes 2004 and/or da Cunha Ramos 1993)

Alba Group (da Cunha Ramos & Brunhes 2004 and/or da Cunha Ramos 1993)
alba Theobald 1901c
alboabdominalis Theobald 1910c
joucouri da Cunha Ramos & Brunhes 2004
mayeri Edwards 1912c
Balfouri Group (da Cunha Ramos & Brunhes 2004 and/or da Cunha Ramos 1993)
balfouri Theobald 1904b
hebrardi da Cunha Ramos & Brunhes 2004
Bilineata Group (da Cunha Ramos & Brunhes 2004 and/or da Cunha Ramos 1993)
bilineata Theobald 1909b
connali Edwards 1912c
fraseri Edwards 1912c
Anopheloides Section (da Cunha Ramos & Brunhes 2004 and/or da Cunha Ramos 1993)
Anopheleoides Group (da Cunha Ramos & Brunhes 2004 and/or da Cunha Ramos 1993)
albocephala da Cunha Ramos & Brunhes 2004
anopheloides Brunhes & Razafindrasolo 1975
Neireti Group (da Cunha Ramos & Brunhes 2004 and/or da Cunha Ramos 1993)
andavakae Doucet 1950
neireti Edwards 1920b
Caeruleocephala Section (da Cunha Ramos & Brunhes 2004 and/or Ramos 1993)
Caeruleocephala Group (da Cunha Ramos & Brunhes 2004 and/or da Cunha Ramos 1993)
angolensis da Cunha Ramos 1985
benoiti Wolfs 1964
caeruleocephala Theobald 1901d
caliginosa Philip 1931
machadoi da Cunha Ramos 1986
pallidocephala Theobald 1908c
philonuxia Philip 1931
Madagascarica Group (da Cunha Ramos & Brunhes 2004 and/or da Cunha Ramos 1993)
argentipleura da Cunha Ramos & Brunhes 2004
geniculata da Cunha Ramos & Brunhes 2004
madagascarica da Cunha Ramos & Brunhes 2004
Chorleyi Section (da Cunha Ramos & Brunhes 2004 and/or da Cunha Ramos 1993)
Chorleyi Group (da Cunha Ramos & Brunhes 2004 and/or da Cunha Ramos 1993)
chorleyi Edwards 1936
Hamoni Group (da Cunha Ramos & Brunhes 2004 and/or da Cunha Ramos 1993)
bidentata da Cunha Ramos & Brunhes 2004
hamoni Grjebine 1953a
moramangae da Cunha Ramos & Brunhes 2004
roberti da Cunha Ramos & Brunhes 2004
Dumonti Section (da Cunha Ramos & Brunhes 2004 and/or da Cunha Ramos 1993)
Dumonti Group (da Cunha Ramos & Brunhes 2004 and/or da Cunha Ramos 1993)
dumonti Doucet 1949b
grassei da Cunha Ramos & Brunhes 2004
lebiedi da Cunha Ramos & Brunhes 2004
mayottensis Brunhes 1977b
Hopkinsi Section (da Cunha Ramos & Brunhes 2004 and/or da Cunha Ramos 1993)
Hopkinsi Group (da Cunha Ramos & Brunhes 2004 and/or da Cunha Ramos 1993)
hopkinsi Edwards 1932b
palmeirimi De Meillon & Rebelo 1941
palmeirimi dundo da Cunha Ramos 1993
Argyrotarsis Series (EL Peyton, *in. lit.*)
argyrotarsis Leicester 1908
Campestris Series (EL Peyton, *in. lit.*)
arguellesi Baisas 1935b
campestris Leicester 1908
christophersi Barraud 1926
heiseri Baisas 1935b
macfarlanei Edwards 1914c
mendiolai Baisas 1935b
metatarsata Edwards 1914b
rampae Peyton & Klein 1970
subnormalis Martini 1920a
Ludlowae Series (EL Peyton, *in. lit.*)
clara Dyar & Shannon 1925a
lateralis Ludlow 1905c
ludlowae Dyar & Shannon 1925a
reyi Baisas 1935b
Micans Series (EL Peyton, *in. lit.*)
bimaculiala Leicester 1908
longirostris Leicester 1908
micans Leicester 1908
Nivea Series (EL Peyton, *in. lit.*)
nivea Leicester 1908
orientalis Barraud 1926
Testacea Series (EL Peyton, *in. lit.*)
annandalei Barraud 1926
falcipes Banks 1906
hebes Barraud 1931a
prajimi Peyton & Rattanarithikul 1970
somblooni Peyton & Klein 1970
testacea Theobald 1905f
yunnanensis Dong, Dong, & Wu 2004
Trilineata Series (EL Peyton, *in. lit.*)
rutherfordi Edwards 1922a
trilineata Leicester 1908
Unimaculiala Series (EL Peyton, *in. lit.*)

diraphati Peyton & Klein 1970
unimaculiala Leicester 1908
Anisocheleomyia Section (Belkin 1962)
albescens Taylor 1914c
barnesi Belkin 1953a
civinskii Belkin 1953a
nivipes (Theobald 1905a)
sexaueri Belkin 1953a
solomonis Belkin 1953a
Wysockii Section (Belkin 1962)
wysockii Belkin 1953a
Tibialis Group (King & Hoogstraal 1946)
antennalis Taylor 1919
fimbriata King & Hoogstraal 1947
gerdae Slooff 1963b
neotibialis King & Hoogstraal 1947
setosa King & Hoogstraal 1947
subtibioclada King & Hoogstraal 1947
tibialis Taylor 1919
tibioclada King & Hoogstraal 1947
Unplaced
alboannulata (Theobald 1905a)
albosternopleura Peters 1963e
amiensis Peters 1963e
edwardsi Barraud 1926
moresbyensis Peters 1963e
novaguinensis Peters 1963d
novaguinensis alticola Peters 1963d
paralateralis Peters 1964
paranovaguinensis Peters 1963d
pygmaea Theobald 1901c
Subgenus *Uranotaenia* New World
Leucoptera Series (Galindo, Blanton, & Peyton 1953)
hystera Dyar & Knab 1913
lanei Martínez & Prosen 1953a [presumptive, not placed by authors]
leucoptera (Theobald 1907)
nataliae Lynch Arribálzaga 1891
Lowii Series (Galindo, Blanton, & Peyton 1953)
bertii Cova García & Rausseo 1964 [presumptive, not placed by authors]
briseis Dyar 1925g
lowii Theobald 1901c
orthodoxa Dyar 1921h
paludosa Galindo 1954
riverai Duret 1970
telmatophila Galindo, Blanton, & Peyton 1954
Pallidoventer Series (Galindo, Blanton, & Peyton 1954)
calosomata Dyar & Knab 1907c
coatzacoalcos Dyar & Knab 1906d
cooki Root 1937
davisi Lane 1943
ditaenionota Prado 1931b
incognita Galindo, Blanton, & Peyton 1954
mathesoni Lane 1943
pallidoventer Theobald 1903a
trapidoi Galindo, Blanton, & Peyton 1954
typhlosomata Dyar & Knab 1907c
Pulcherrima Series (Galindo, Blanton, & Peyton 1954)
aequatorianna Leví-Castillo 1953c
apicalis Theobald 1903a
geometrica Theobald 1901c
pulcherrima Lynch-Arribalzaga 1891
pulcherrima elnora Paterson & Shannon 1927
rachoui Xavier, da Silva Mattos, & da Silva 1970 [presumptive, not placed by authors]
sapphirina (Osten Sacken 1868)
socialis Theobald 1901c
Unplaced (Galindo, Blanton, & Peyton 1954)
bricenoi Cova García, Pulido F., Escalante de Ugueto, & Mora 1987
gabaldoni Cova García, Pulido F., Escalante de Ugueto, & Mora 1987
iriartei Cova García, Pulido F., Escalante de Ugueto, & Mora 1987
oteizai Perez Vigueras 1956
pifanoi Cova García, Pulido F., Escalante de Ugueto, & Mora 1987

Genus *Verrallina*

After Reinert (1999c).

Subgenus *Harbachius*
13 species, no further subdivisions.

Subgenus *Neomacleaya*
50 species, no further subdivisions.

Subgenus *Verrallina*
Butleri Series (Reinert 1999)
bancrofti (Taylor 1914b)
butleri (Theobald 1901c)
dux (Dyar and Shannon 1925a)
iriomotensis (Tanaka & Mizusawa 1973)
lugubris (Barraud 1928b)
Carmenti Series (Reinert 1999)
azureosquamata (Bonne-Wepster 1948)
bifoliata (King & Hoogstraal 1947)
carmenti (Edwards 1924b)
cuccioi (Belkin 1962)
embiensis (Huang 1968)
foliformis (King & Hoogstraal 1947)
funerea (Theobald 1903a)

killertonis (Huang 1968)
leilae (King & Hoogstraal 1947)
lineata (Taylor 1914d)
mccormicki (Belkin 1962)
milnensis (King & Hoogstraal 1947)
multifolium (King & Hoogstraal 1947)
obsoleta (Huang)
parasimilis (King & Hoogstraal 1947)
pipkini (Bohart 1968)
quadrifolium (Brug 1934a)
quadrispinata (King & Hoogstraal 1947)
reesi (King & Hoogstraal 1947)
sentania (King Hoogstraal 1947)
similis (Theobald 1910c)
simpla (King & Hoogstraal 1947)
trispinata (King & Hoogstraal 1947)
vanapa (Huang 1968)
variabilis (Huang 1968)

Literature Cited

Abdel-Malek, A. (1948a). Plant hormones (auxins) as a factor in the hatching of *Aedes trivittatus* (Coquillett) eggs. *Annals of the Entomological Society of America, 41*(1), 51–57.

Abdel-Malek, A. (1948b). The biology of *Aedes trivittatus. Journal of Economic Entomology, 41*(6), 951–954.

Abdel-Malek, A. (1949). A study of the morphology of the immature stages of *Aedes trivittatus* (Coquillett) (Diptera: Culicidae). *Annals of the Entomological Society of America, 42*(1), 19–37.

Abdel-Malek, A. (1960). The culicine mosquitoes of the northern region of the United Arab Republic. *Bulletin of the Egyptian Entomological Society, 44*, 111–128.

Abdel-Malek, A. A., & Baldwin, W.F. (1961). Specificity of plant feeding in mosquitoes as determined by radioactive phosphorus. *Nature, 192*(4798), 178–179.

Abdel-Mohsin, M.O.A. (2004). First record of three Afrotropical *Culex* species (Diptera: Culicidae) in Saudi Arabia. *Annals of Medical Entomology, 13*(1–2), 1–9.

Abdul-Ghani, R., Al-Mekhlafi, A.M., & Alabsi, M.S. (2012). Microbial control of malaria: Biological warfare against the parasite and its vector. *Acta Tropica, 121*(2), 71–84.

Abdullah, M.A.R., & Merdan, A.I. (1995). Distribution and ecology of the mosquito fauna in the southwestern Saudi Arabia. *Journal of the Egyptian Society of Parasitology, 25*(3), 815–837.

Abdulla-Khan, R., Coetzee, M., & Hunt, R.H. (1998). Description of *Anopheles* (*Cellia*) *seretsei* sp. nov. from Kasane, Botswana. *Journal of the American Mosquito Control Association, 14*(3), 248–252.

Abercrombie, J. (1977). Medical entomology studies. IX. The subgenus *Christophersiomyia* of the genus *Aedes* (Diptera: Culicidae). *Contributions of the American Entomological Institute, 14*(2), 1–72.

Abonnenc, E. (1954). Sur un anophèle cavernicole de la Guinée: *Anopheles cavernicolus* n. sp. (Diptera-Culicidae). *Bulletin et mémoires de l'École Préparatoire de Médecine et de Pharmacie de Dakar. Parasitologie, 2*, 288–290.

Abonnenc, E. (1956). Speologica africana. Sur un anophèle cavernicole de la Guinée; *Anopheles cavernicolus* n. sp. (Diptera-Culicidae). *Bulletin de l'Institut Français d'Afrique Noire, 18*(3), 802–805.

Abraham, E.G., & Jacobs-Lorena, M. (2004). Mosquito midgut barriers to malaria parasite development. *Insect Biochemistry and Molecular Biology, 34*(7), 667–671.

Abraham, E.G., Donnelly-Doman, M., Fujioka, H., Ghosh, A., Moreira, L., & Jacobs-Lorena, M. (2005). Driving midgut-specific expression and secretion of a foreign protein in transgenic mosquitoes with *AgAper1* regulatory elements. *Insect Molecular Biology, 14*(3), 271–279.

Abraham, O.K. (1947). *Anopheles* (*Anopheles*) *kyondawensis* n. sp., from lower Burma. *Medical Journal of Malaya, 1*(3), 173–176.

Abu-Hassan, A., Rahman, A., Adanan, C.R., & Rashid, M.Z.A. (1992). *Aedes* (*Finlaya*) *macfarlanei* (Edwards), a new record from peninsular Malaysia and notes on mosquitoes of rock pools in stream beds from Endau Forest Reserve, Johore, Malaysia. *Malaysian Applied Biology, 21*(2), 107–111.

Abul-Hab, J. (1967). Larvae of culicine mosquitos in north Iraq (Diptera, Culicidae). *Bulletin of Entomological Research, 57*(2), 279–284.

Abul-Hab, J. (1969). Larvae of culicine mosquitoes of Iraq with a key for their identification. *Bulletin of Endemic Diseases, 10*(for 1968) (1–4), 243–255.

Adam, J.P. (1961). *Anopheles caroni* n. sp. (Diptera, Culicidae) anophèle cavernicole nouveau de la République du Congo. *Bulletin de la Société de pathologie exotique, 54*(4), 714–717.

Adam, J.P. (1962). Un anophèle cavernicole nouveau de la République du Congo (Brazzaville): *Anopheles* (*Neomyzomyia*) *hamoni* n. sp. (Diptera-Culicidae). *Bulletin de la Société de pathologie exotique, 55*(1), 153–165.

Adam, J.P., & Hamon, J. (1959). Description de deux nouvelles sous-espèces d'*Eretmapodites oedipodius* Graham, capturées en Afrique Occidentale. *Bulletin de la Société de pathologie exotique, 51*(for 1958)(4), 524–527.

Adam, J.P., & Hamon, J. (1961). Les stades pre-imaginaux *d'Eretmapodites forcipulatus* Edwards, 1936. *Bulletin de la Société de pathologie exotique, 54*(2), 198–202.

Adam, J.-P., & Mattingly, P.F. (1956). Note sur la morphologie et la biologie d'*Anopheles* (*Neomyzomyia*) *smithi* var. *rageaui* Mattingly

(P.F.) et Adam (J.-P.) 1954. *Bulletin de la Société de pathologie exotique, 49*(4), 734–747.
Adam, J.-P., Chwatt, L.J.B., & Hamon, J. (1956). Contribution à l'étude des Culicidae (Diptères) de la région éthiopienne: Description du mâle et de quelques variations de la femelle et des premiers stades d'*Anopheles domicolus* Edwards, 1916. *Bulletin de la Société de pathologie exotique, 49*(1), 100–108.
Adam, J.-P., Hamon, J., Rickenbach, A., & Lips, M. (1956). I. Description du mâle et du pharynx de la femelle *d'Anopheles brohieri* Edwards 1929 et *A. hancocki* var. *masseguini* Hamon 1954. II. Étude des affinités existant entre *Anopheles hancocki, A. hancocki* var. *messeguini, A. brohieri, A. theileri, A. seydeli. Bulletin de la Société de pathologie exotique, 49*(4), 747–758.
Adames, A.J. (1971). Mosquito studies (Diptera, Culicidae). XXIV. A revision of the crabhole mosquitoes of the genus *Deinocerites. Contributions of the American Entomological Institute, 7*(2), 1–154.
Adames, A.J., & Arzube, M.E. (1975). Geographical extension of *Galindomyia leei* Stone and Barreto to Ecuador (Diptera, Culicidae). *Mosquito Systematics, 7*(2), 113–114.
Adames, A.J., & Galindo, P. (1973). Mosquito studies (Diptera, Culicidae). XXX. A new subgenus and species of *Culex* from Colombia. *Contributions of the American Entomological Institute, 9*(3), 55–61.
Adames, A.J., & Galindo, P. (1975). Description of the immature stages of *Galindomyia leei* Stone and Barreto, 1970. *Mosquito Systematics, 7*(2), 132–136.
Adames, A.J., & Hogue, C.L. (1970). Mosquito studies (Diptera, Culicidae). XVII. Two new species of *Deinocerites* from Costa Rica. *Contributions of the American Entomological Institute, 5*(for 1969) (1), 9–20.
Adams, C.F. (1903). *Dipterological contributions* (Vol. 2). Lawrence: University of Kansas.
Adams, P.C.G. (1940). Some observations on the flight of stained anophelines at N'Kana, Northern Rhodesia. *Annals of Tropical Medicine and Parasitology, 34*(1), 35–43.
Adelmann, H. B. (1966). *Marcello Malpighi and the evolution of embryology.* Ithaca, NY: Cornell University Press.
Adhami, J., & Murati, N. (1987). Prani e mushkonjës *Aedes albopictus* në shqipëri. *Revista mjekësore, 1,* 13–16.
Adhami, J., & Reiter, P. (1998). Introduction and establishment of *Aedes* (*Stegomyia*) *albopictus* Skuse (Diptera: Culicidae) in Albania. *Journal of the American Mosquito Control Association, 14*(3), 340–343.
Adis, J., & Junk, W.J. (2002). Terrestrial invertebrates inhabiting lowland river floodplains of Central Amazonia and Central Europe: A review. *Freshwater Biology, 47*(4), 711–731.
Afify, A., & Galizia, C.G. (2015). Chemosensory cues for mosquito oviposition site selection. *Journal of Medical Entomology, 52*(2), 120–130.
Agassiz, L. (1846). *Nomenclatoris zoologici index universalis, continens nomina systematica classium, ordinum, familiarum et generum animalium omnium, tam viventium quam fossilium, secundum ordinem alphabeticum unicum disposita, adjectis homonymiis plantarum, nec non variis adnotionibus et emendationibus.* Jent et Gassmann, Soloduri [Solothrun, Switzerland].
Agoulon, A., Guilloteau, J., & Marjolet, M. (1999). The taxon *Aedes detritus* (Diptera: Culicidae) on the French Atlantic Coast. *Annales de la Société Entomologique de France, 35*(suppl.), 263–267.
Agramonte, A. (1900). Los mosquitos malaricos. *El progreso médico, 10*(12), 455–465.
Agre, P., King, L.S., Yasui, M., Guggino, W.B., Ottersen, O.P., Fujiyashi, Y., . . . Nielsen, S. (2002). Aquaporin water channels—from atomic structure to clinical medicine. *The Journal of Physiology, 542*(1), 3–16.
Aguilar, R., Jedlicka, A.E., Mintz, M., Mahairaki, V., Scott, A.L., & Dimopoulos, G. (2005). Global gene expression analysis of *Anopheles gambiae* responses to microbial challenge. *Insect Biochemistry and Molecular Biology, 35*(7), 709–719.
Ahmadi, A., & McClelland, G.A.H. (1985). Mosquito-mediated attraction of female mosquitoes to a host. *Physiological Entomology, 10*(3), 251–255.
Ahmed, A.M., Shaalan, E.A., Aboul-Soud, M.A.M., Tripet, F., & Al-Khedhairy, A.A. (2011). Mosquito vectors survey in the Al-Ahsaa district of eastern Saudi Arabia. *Journal of Insect Science, 11,* 176.
Ahmed, T.U. (1987). Checklist of the mosquitoes of Bangladesh. *Mosquito Systematics, 19*(3), 187–200.
Ahumada, J.A., Lapointe, D., & Samuel, M.D. (2004). Modeling the population dynamics of *Culex quinquefasciatus* (Diptera: Culicidae) along an elevational gradient in Hawaii. *Journal of Medical Entomology, 41*(6), 1157–1170.
Aiken, J. (1909). Notes of the mosquitoes of British Guiana. *The British Guiana Medical Annual, 1908,* 1–25.
Aiken, J. (1911). A synoptical view of the mosquitoes of British Guiana. *Timehri—the Journal of the Royal Agricultural and Commercial Society of British Guiana, 1*(11), 187–204.
Aiken, J., & Rowland, E.D. (1906). Preliminary notes on the mosquitoes of British Guiana. *The British Guiana Medical Annual, 1905,* 13–38.
Airi, M., & Kaur, S. (2015). Confirmation of *Culex* (*Culex*) *tritaeniorhynchus summorosus* (Diptera: Culicidae) as a separate species. *Journal of Vector Borne Diseases, 52*(3), 219–223.
Aitken, T.H.G. (1939). The *Anopheles maculipennis* complex of western America (Diptera, Culicidae). *The Pan-Pacific Entomologist, 15*(4), 191–192.
Aitken, T.H.G. (1940). The genus *Psorophora* in California (Diptera, Culicidae). *Revista de entomología, 11*(3), 672–682.
Aitken, T.H.G. (1941). A new American subgenus and species of *Aedes* (Diptera, Culicidae). *The Pan-Pacific Entomologist, 17*(2), 81–84.
Aitken, T.H.G. (1942). Contributions toward a knowledge of the insect fauna of lower California. *Proceedings of the California Academy of Sciences, 24*(6), 161–170.
Aitken, T.H.G. (1945). Studies on the anopheline complex of western America. *University of California Publications in Entomology, 7*(11), 273–354.
Aitken, T.H.G. (1953). The anopheline fauna of Sardinia. *American Journal of Hygiene Monograph Series, 20*, 303–352.
Aitken, T.H.G. (1954). The Culicidae of Sardinia and Corsica (Diptera). *Bulletin of Entomological Research, 45*(3), 436–494.
Aitken, T.H.G., & Galindo, P. (1966). On the identity of *Culex* (*Melanoconion*) *portesi* Senevet and Abonnenc 1941. *Proceedings of the Entomological Society of Washington, 68*(3), 198–208.
Akoh, J.I., Aigbodion, F.I., & Kumbak, D. (1992). Studies on the effect of larval diet, adult body weight, size of blood-meal, and age on fecundity of *Culex quinquefasciatus* (Diptera: Culicidae). *Insect Science and Its Application, 13*(2), 177–181.
Alam, M.T., Das, M.K., Dev, V., Ansari, M.A., & Sharma, Y.D. (2007). Identification of two cryptic species in the *Anopheles* (*Cellia*) *annularis* complex using ribosomal DNA PCR-RFLP. *Parasitology Research, 100*(5), 943–948.
Albeny-Simões, D., Murrell, E.G., Elliot, S.L., Andrade, M.R., Lima, E., Juliano, S.A., & Vilela, E.F. (2014). Attracted to the enemy: *Aedes aegypti* prefers oviposition sites with predator-killed conspecifics. *Oecologia, 175*(2), 481–492.
Albuquerque, C., Melo-Santos, M.A.V., Bezerra, M.A.S., Barbosa, R., Silva, D.F., & Silva, E. (2000). First report of *Aedes albopictus* in areas of rain forest in Brazil. *Revista de Saúde Pública, 34*(3), 314–315.

Alcock, A. (1911). Remarks on the classification of the Culicidae, with particular reference to the constitution of the genus *Anopheles*. *Annals and Magazine of Natural History, 8*(44), 240–250.

Alcock, A. (1912). Descriptions of a new species of *Anopheles* from the Malay Peninsula. *Journal of the London School of Tropical Medicine, 2*(1), 1–2.

Alcock, A. (1913). Synopsis of the anopheline mosquitos of Africa and of the Oriental region. *Journal of the London School of Tropical Medicine, 2*(3), 153–166.

Aldersley, A.J. (2015). *The swarm and the mosquito: Investigations into acoustic interactions in the frequency domain*. Publication no. 0000 0004 5920 0226 [Doctoral dissertation, University of Bristol]. EthOS uk.bl.ethos.689693.

Alekseev, A.N., Rasnitsyn, S.P., & Vitlin, L.M. (1976). Group behavior of female blood-sucking mosquitoes (Diptera, Culicidae, *Aedes*). I. Demonstration of the "invitation effect." *Meditsinskaia parazitologiia i parazitarnye bolezni, 46*(1), 23–24.

Alekseev, E.V. (1989). Bloodsucking mosquito *Aedes* (*Rusticoidus*) *krymmontanus* Alekseev, sp. n., a relic of the entomofauna of the Crimea. *Parazitologyia Leningrad, 23*(2), 173–178.

Alencar, J., Guimarães, A.É., Mello, R.P., Lopes, C. M., Dégallier, N., & Santos-Mallett, J.R. (2003). Scanning electron microscopy of eggs of *Haemagogus leucocelaenus* (Diptera: Culicidae). *Revista de saúde pública, 37*(5), 1–7.

Alencar, J., Guimarães, A.É., Gil-Santana, H.R., & Santos-Mallet, J.R. (2005a). Scanning electron microscopy of eggs of *Ochlerotatus* (*Protomacleaya*) *terrens* Walker. *Journal of the American Mosquito Control Association, 21*(4), 355–359.

Alencar, J., Guimarães, A.É., Mello, R.P., Lopes, C.M., Dégallier, N., & Santos-Mallet, J.R. (2005b). Scanning electron microscopy study of the egg of *Haemagogus* (*Haemagogus*) *capricornii* Lutz, 1904 (Diptera: Culicidae). *Journal of Medical Entomology, 42*(1), 1–6.

Alencar, J., Rodriguez-Fernández, J., Dégallier, N., Marcondes, C.B., Costa, J.M., & Guimarães, A.É. (2009). Multivariate discrimination between two cryptic *Haemagogus* species associated with the transmission of yellow fever virus in the Americas. *Journal of the American Mosquito Control Association, 25*(1), 18–24.

Alfonso-Parra, C., Avila, F.W., Deewatthanawong, P., Sirot, L.K., Wolfner, M.F., & Harrington, L.C. (2014). Synthesis, depletion and cell-type expression of a protein from the male accessory glands of the dengue vector mosquito *Aedes aegypti*. *Journal of Insect Physiology, 70*, 117–124.

Alfonso-Parra, C., Ahmed-Braimah, Y.H., Degner, E.C., Avila, F.W., Villarreal, S.M., Pleiss, J.A., . . . Harrington, L.C. (2016). Mating-induced transcriptome changes in the reproductive tract of female *Aedes aegypti*. *PLOS Neglected Tropical Diseases, 10*(2), e0004451.

Ali, S.R., & Rozeboom, L.E. (1973). Comparative laboratory observations on selective mating of *Aedes* (*Stegomyia*) *albopictus* Skuse and *A.* (*S.*) *polynesiensis* Marks. *Mosquito News, 33*(1), 23–28.

Aliabadi, B.W., & Juliano, S.A. (2002). Escape from gregarine parasites affects the competitive interactions of an invasive mosquito. *Biological Invasions, 4*(3), 283–297.

Allen, M.L, O'Brochta, D.A., Atkinson, P.W., & Leresque, C.S. (2001). Stable, germ-line transformation of *Culex quinquefasciatus* (Diptera: Culicidae). *Journal of Medical Entomology, 38*(5), 701–710.

Alley, R.B. (2000). *The two-mile time machine: Ice cores, abrupt climate change, and our future*. Princeton, NJ: Princeton University Press.

Almeida, A.P.G., Conçalves, Y.M., Novo, M.T., Sousa, C.A., Melim, M., & Grácio, A.J.S. (2007). Vector monitoring of *Aedes aegypti* in the Autonomous Region of Madeira, Portugal. *Euro Surveillance, 12*(46), 3311.

Almirón, W.R., & Harbach, R.E. (1996). Taxonomy and biology of *Culex* (*Culex*) *maxi* Dyar (Diptera: Culicidae) in South America. *Memórias do Instituto Oswaldo Cruz* (Rio de Janeiro), *91*(5), 579–588.

Alten, B., Caglar, S.S., & Ozer, N. (2000). Malaria and its vectors in Turkey. *European Mosquito Bulletin, 7*, 27–33.

Al-Houty, W. (1997). Checklist of the insect fauna of Kuwait. *Kuwait Journal of Science and Engineering, 24*(1), 145–162.

Al-Tikrity, A.B. (1963). Recent study of the ecology of *Anopheles apoci* in Iraq. Part I. The geographical distribution of the breeding places and the morphology of *Anopheles apoci* pupa in Iraq. *Bulletin of Endemic Diseases, 5*(1–2), 73–81.

Al-Tikrity, A.B. (1969). Addition to the *Anopheles* species record in Iraq—*A. turkhudi*. *Bulletin of Endemic Diseases* (Baghdad), *11*, 74–81.

Alvarado, C.A., & Heredia, R.L. (1947). Observaciones sobre una nueva variedad del *Anopheles* (*A.*) *pseudopunctipennis* Theobald 1901, encontrada en la provincia de Tucumán (Nota previa). *Anales del Instituta de Medicina Regional Universidad Tucumán, 2*(1), 73–78.

Alvarado, C.A., & Ponte, E.D. (1935). Sobre la existencia de *Anopheles annulipalpis* en la provincia de Mendoza (R. A.).—descripción de su larva. *Folia biologica* (90–93), 388–392.

Alves, J., Gomes, B., Rodrigues, R., Silva, J., Arez, A.P., Pinto, J., & Sousa, C.A. (2010). Mosquito fauna on the Cape Verde Islands (West Africa): An update on species distribution and a new finding. *Journal of Vector Ecology, 35*(2), 307–312.

Amarista, J. (1996). Nuevo registro sobre la distribución geográfica de *Aedes* (*Howardina*) *ioliota* Dyar & Knab 1913. (Diptera: Culicidae) en Venezuela. *Boletín de la Dirección de Malariología y Saneamiento Ambiental, 78*(1, 2), 361–362.

Ameen, M.-U., & Talukdar, M.Z.I. (1974a). Pupa of *Ficalbia* (*Mimomyia*) *chamberlaini* (Ludlow) (Diptera: Culicidae). *Mosquito Systematics, 6*(3), 206–210.

Ameen, M.-U., & Talukdar, M.Z.I. (1974b). Pupa of *Toxorhynchites splendens* (Wiedemann) (Diptera: Culicidae). *Mosquito Systematics, 6*(3), 201–205.

Ameen, M.-U., & Talukdar, M.Z.I. (1974c). Pupal chaetotaxy of the common mosquitoes of Dacca. *Beiträge zur Entomologie, 24*(1, 4), 87–95.

Amerasinghe, F.P. (1983). A note on *Aedes* (*Stegomyia*) *novalbopictus* Barraud from Sri Lanka, including first descriptions of the female and egg (Diptera: Culicidae). *Journal of Natural History, 17*(2), 203–212.

Amerasinghe, F.P. (1989). Redescription of *Heizmannia* (*Heizmannia*) *greenii* Theobald from Sri Lanka (Diptera: Culicidae). *Mosquito Systematics, 21*(2), 83–99.

Amerasinghe, F.P. (1990). A guide to the identification of the anopheline mosquitoes (Diptera: Culicidae) of Sri Lanka. Vol. 1. Adult females. *Ceylon Journal of Science (Biological Sciences), 21*(1), 1–16.

Amerasinghe, F.P. (1991). A guide to the identification of the anopheline mosquitoes (Diptera: Culicidae) of Sri Lanka. 2. Larvae. *Ceylon Journal of Science (Biological Sciences), 22*(1), 1–13.

Amerasinghe, F.P. (1993). A new species of *Heizmannia* from Sri Lanka (Diptera: Culicidae). *Mosquito Systematics, 25*(2), 110–115.

Amerasinghe, F.P. (1995). A guide to the identification of the anopheline mosquitoes of Sri Lanka. 3. Pupae. *Journal of the National Science Council of Sri Lanka, 23*(3), 115–129.

Amos, D.W. (1944). Mosquito control training manual. *Fiji Times and Herald*, Suva, Fiji.

Amr, Z.S., Al-Khalili, Y.H., & Arbaji, A. (1997). Larval mosquitoes collected from northern Jordan and the Jordan Valley. *Journal of the American Mosquito Control Association, 13*(4), 375–378.

An, S., Dong, S., Wang, Q., Li, S., Gilbert, L.I., Stanley, D., & Song, Q. (2012). Insect neuropeptide bursicon homodimers induce innate immune and stress genes during molting by activating the NF-κB transcription factor *relish*. *PLOS ONE, 7*(3), e34510.

Anderson, E.M., & Davis, J.A. (2014). First records of *Armigeres malayi* and *Armigeres milnensis* in Timor-Leste. *Journal of the American Mosquito Control Association, 30*(1), 51–53.

Anderson, J.F., Andreadis, T.G., Main, A.J., Ferrandino, F.J., & Vossbrinck, C.R. (2006). West Nile virus from female and male mosquitoes (Diptera: Culicidae) in subterranean, ground and canopy habitats in Connecticut. *Journal of Medical Entomology, 43*(5), 1010–1019.

Anderson, R.A., Knols, B.G.J., & Koella, J.C. (2000). *Plasmodium falciparum* sporozoites increase feeding-associated mortality of their mosquito hosts *Anopheles gambiae* s.l. *Parasitology, 120*(4), 329–333.

Andreadis, T.G., & Munstermann, L.E. (1997). Intraspecific variation in key morphological characters of *Culiseta melanura* (Diptera: Culicidae). *Journal of the American Mosquito Control Association, 13*(2), 127–133.

Andreadis, T.G., Anderson, J.F., Munstermann, L.E., Wolfe, R.J., & Florin, D.A. (2001). Discovery, distribution, and abundance of the newly introduced mosquito *Ochlerotatus japonicus* (Diptera: Culicidae) in Connecticut, USA. *Journal of Medical Entomology, 38*(6), 774–779.

Anduze, P.J. (1941a). Lista provisional de los zancudos hematófagos de Venezuela (Diptera: Culicidae). *Boletín de entomología Venezolana, 1*(1), 6–18.

Anduze, P.J. (1941b). Primer informe ô entomologica médica del estado Bolivar (Venezuela). Vol. 1. La fauna culicidiana. Descripción de tres especies nuevas. *Revista Venezolana de sanidad y asistencia social, 6*, 812–839.

Anduze, P.J. (1942a). Fauna culicidiana de Venezuela. Descripción de dos especies nuevas. *Boletín de entomología Venezolana, 1*(2–3), 43–49.

Anduze, P.J. (1942b). Sobre la morfología de la armadura bucofaríngea de algunos representantes Venezolanos del subgénero *Kerteszia* (Diptera: Culicidae). *Revista Venezolana de sanidad y asistencia social, 6*, 435–436.

Anduze, P.J. (1942c). La fauna culicidiana de Venezuela. Part II. Descripción de una especie nueva (Diptera: Culicidae). *Revista Venezolana de sanidad y asistencia social, 7*(4), 557–560.

Anduze, P.J. (1942d). Sobre la fauna culicidiana de Venezuela. Descripción del huevo del *Anopheles* (*Kerteszia*) *homunculus* Komp. (Diptera: Culicidae). *Revista de sanidad y asistencia social, 7*(3), 433–434.

Anduze, P.J. (1943a). *Culex* (*Culex*) *beauperthuyi* sp. nov. (Diptera: Culicidae). *Revista Venezolana de sanidad y asistencia social, 8*(3), 459–461.

Anduze, P.J. (1943b). Estudios de entomología médica en el estado Merida-Venezuela. La fauna culicidiana-descripción del *Culex* (*Culex*) *albertoi* sp. n. *Boletín de entomología Venezolana, 2*(4), 189–196.

Anduze, P.J. (1944). *Aedes* (*Ochlerotatus*) *euiris* Dyar. *Boletín de entomología Venezolana, 3*(3), 161–163.

Anduze, P.J. (1947). Contribución al estudio de los vectores de la fiebre amarilla en Venezuela. *Boletín de la Academia de Ciencias Físicas, Matemátias y Naturales, 10*(30), 331–373.

Anduze, P.J. (1948). Notas entomológicas. I. Dos variedades nuevas de *Anopheles aquasalis*. *Boletín Médico, 1*(1), 17–19.

Anduze, P.J. (1949a). Fauna culicina de Venezuela. Descripción de una especie nueva. *Boletín de entomología Venezolana, 7*(for 1948) (3–4), 149–150.

Anduze, P.J. (1949b). Fauna culicina de Venezuela. Descripción de cinco nuevas especies. *Boletín de entomología Venezolana, 7*(for 1948)(1–2), 60–66.

Anduze, P.J., & Capdevielle, A. (1949). *Anopheles guarao* sp. nov. *Boletín de entomología Venezolana, 8*(3–4), 119–124.

Anduze, P.J., & Hecht, O. (1943). *Aedes* (*Finlaya*) *upatensis* sp. nov. *Boletín de entomología Venezolana, 2*(4), 185–188.

Angarita-Jaimes, N.C., Parker, J.E.A., Abe, M., Mashauri, F., Martine, J., Towers, C.E., . . . Towers, D.P. (2016). A novel video-tracking system to quantify the behaviour of nocturnal mosquitoes attacking human hosts in the field. *Journal of the Royal Society Interface, 13*(117), 20150974.

Angelon, K.A., & Petranka, J.W. (2002). Chemicals of predatory mosquitofish (*Gambusia affinis*) influence selection of oviposition site by *Culex* mosquitoes. *Journal of Chemical Ecology, 28*(4), 797–806.

Angulo, A.O., & Olivares, T.S. (1993). Catalogo de los culicidos de Chile (Diptera, Culicidae) y dos especies nuevas de *Culex* (*Culex*) Linnaeus. *Boletín de la Sociedad Biologico de Concepción, Chile, 64*, 21–26.

Anthony, T.G., Harbach, R.E., & Kitching, I.J. (2002). Phylogeny of the Pyretophorus Series of *Anopheles* subgenus *Cellia* (Diptera: Culicidae). *Systematic Entomology, 24*(2), 193–205.

Antonio-Nkondjio, C., & Simard, F. (2013). Highlights on *Anopheles nili* and *Anopheles moucheti*, malaria vectors in Africa. In S. Manguin (Ed.), Anopheles *mosquitoes: New insights into malaria vectors* (pp. 221–238). Rijeka, Croatia: InTech.

Antonio-Nkondjio, C., Simard, F., Cohuet, A., & Fontenille, D. (2002). Morphological variability in the malaria vector, *Anopheles moucheti*, is not indicative of speciation: Evidences from sympatric south Cameroon populations. *Infection, Genetics and Evolution, 2*(1), 69–72.

Antonio-Nkondjio, C., Ndo, C., Kengne, P., Mukwaya, L., Awono-Ambene, P., Fontenille, D., & Simard, F. (2008). Population structure of the malaria vector *Anopheles moucheti* in the equatorial forest region of Africa. *Malaria Journal, 7*, 120.

Antonio-Nkondjio, C., Ndo, C., Costantini, C., Awono-Ambene, P., Fontenille, D., & Simard, F. (2009). Distribution and larval habitat characterization of *Anopheles moucheti*, *Anopheles nili*, and other malaria vectors in river networks of southern Cameroon. *Acta Tropica, 112*(3), 270–276.

Antunes, P.C.A. (1937a). A new *Anopheles* and a new *Goeldia* from Colombia (Dip. Culic.). *Bulletin of Entomological Research, 28*(1), 69–73.

Antunes, P.C.A. (1937b). Informe sobre una investigación entomológica realizada en Colombia. *Revista de la Facultad de Medicina* (Universidad Nacional de Colombia, Bogotá), *6*(2), 1–29.

Antunes, P.C.A. (1939). Nota sobre o genero *Haemagogus* Williston (Diptera, Culicidae). *Revista Paulista de medicina* (São Paulo), *14*, 106.

Antunes, P.C.A., & Lane, J. (1934). Notes on some species of *Aedes* (*Ochlerotatus*) (Diptera, Culicidae). *Revista de biologia e higiene* (São Paulo), *5*, 35–40.

Antunes, P.C.A., & Lane, J. (1937). Um novo *Culex, Culex* (*Carrollia*) *soperi*, encontrado em São Paulo (Diptera, Culicidae). *Revista de biologia e higiene* (São Paulo), *8*(1), 21–23.

Antunes, P.C.A., & Lane, J. (1938). Um novo *Aedes, Aedes* (*Ochlerotatus*) *pennai*, encontrado em São Paulo. *Revista do Museu Paulista, 23*, 605–614.

Antunes, P.C.A., & Ramos, A.S. (1939). *Culex* (*Carrollia*) *iridescens, bonnei e soperi* (Diptera, Culicidae). *Boletim biólogico, 4*(3), 374–385.

Apfelbeck, V.V. (1930). Neue culiciden aus Bosnien und Dalmatien. *Konowia: Zeitschrift für systematische Insektenkunde, 8*(for 1929)(3), 286–292.

Apiwathnasorn, C. (1986). *A list of mosquito species in southeast Asia.* Bangkok, Thailand: Museum and Reference Centre, SEAMEO-TROPMED National Centre of Thailand, Mahidol University.

Appawu, M.A., Dadzie, S.K. & Quartey, S.Q. (2000). Studies on the feeding behaviour of larvae of the predaceous mosquito *Culex* (*Lutzia*) *tigripes* Grandpré and Charmoy (Diptera: Culicidae). *Insect Science and Its Application, 20*(4), 245–250.

Apperson, C.S., Engber, B., & Lerine, J.F. (1989). Relative suitability of *Aedes albopictus* and *Aedes aegypti* in North Carolina to support development of *Dirofilaria immitis*. *Journal of the American Mosquito Control Association, 5*(3), 377–382.

Apperson, C.S., Hassan, H.K., Harrison, B.A., Savage, H.M., Aspen, S.E., Farajollahi, A., . . . Anderson, M. (2004). Host feeding patterns of established and potential mosquito vectors of West Nile virus in the eastern United States. *Vector-Borne and Zoonotic Diseases, 4*(1), 71–82.

Apte-Deshpande, A., Paingankar, M., Gokhale, M.D., & Deobagkar, D.N. (2012). *Serratia odorifera,* a midgut inhabitant of *Aedes aegypti* mosquito enhances its susceptibility to dengue-2 virus. *PLOS ONE, 7*(7), e40401.

Aranda, C., Eritja, R., Schaffner, F., & Escosa, R. (2000). *Culex* (*Culex*) *torrentium* Martini (Diptera: Culicidae), a new species from Spain. *European Mosquito Bulletin, 8*, 7–9.

Arbogast, B.S., Edwards, S.V., Wakely, J., Beerli, P., & Slowinski, J.B. (2002). Estimating divergence times from molecular data on phylogenetic and population genetic timescales. *Annual Review of Ecology and Systematics, 33*(1), 707–740.

Arcà, B., Lombardo, F., Valenzuela, J.G., Francischetti, I.M.B., Marinotti, O., Coluzzi, M., & Ribeiro, J.M.C. (2005). An updated catalogue of salivary gland transcripts in the adult female mosquito, *Anopheles gambiae*. *Journal of Experimental Biology, 208*, 3971–3986.

Arensburger, P., Megy, K., Waterhouse, R.M., Abrudan, J., Amedeo, P., Antelo, B., . . . Atkinson, P.W. (2010). Sequencing of *Culex quinquefasciatus* establishes a platform for mosquito comparative genomics. *Science, 330*(6000), 86–88.

Arik, A.J., Rasgon, J.L., Quicke, K.M., & Riehle, M.A. (2009). Manipulating insulin signaling to enhance mosquito reproduction. *BMC Physiology, 9*(1), 15.

Armbruster, P., & Conn, J.E. (2006). Geographic variation of larval growth in North American *Aedes albopictus* (Diptera: Culicidae). *Annals of the Entomological Society of America, 99*(6), 1234–1243.

Arnaud, J.-D., Rioux, J.-A., Croset, H., & Guilvard, E. (1976). *Aedes* (*Ochlerotatus*) *surcoufi* (Theobald, 1912). Rétablissement du binome; analyse morphologique position au sein du complexe holarctique *«excrucians»*. *Annales de Parasitologie Humaine et Comparée, 51*(4), 477–494.

Arnaud, P.H., Jr. (1979). *A catalog of the types of Diptera in the collection of the California Academy of Sciences* (Vol. 1). San Francisco, CA: Insect Associates.

Arnell, J.H. (1973). Mosquito studies (Diptera, Culicidae). XXXII. A revision of the genus *Haemagogus*. *Contributions of the American Entomological Institute, 10*(2), 1–174.

Arnell, J.H. (1976). Mosquito studies (Diptera, Culicidae). XXXII. A revision of the Scapularis Group of *Aedes* (*Ochlerotatus*). *Contributions of the American Entomological Institute, 13*(3).

Arnell, J.H., & Nielsen, L.T. (1972). Mosquito studies (Diptera, Culicidae). XXVII. The Varipalpus Group of *Aedes* (*Ochlerotatus*). *Contributions of the American Entomological Institute, 8*(2), 1–48.

Arnold, J.R. (1953). *A few observations from the project on the embryology of Aedes nigromaculis (Ludlow).* Paper presented at the 21st Annual Conference of the California Mosquito Control Association.

Arnqvist, G. (1998). Comparative evidence for the evolution of genitalia by sexual selection. *Nature, 393*(6687), 784–786.

Arredondo-Bernal, H.C., & Reyes-Villanueva, F. (1989). Diurnal pattern and behavior of oviposition of *Toxorhynchites theobaldi* in the field. *Journal of the American Mosquito Control Association, 5*(1), 25–28.

Arsic, D., & Guerin, P.M. (2008). Nutrient content of diet affects the signaling activity of the insulin/target of rapamycin/p70 S6 kinase pathway in the African malaria mosquito *Anopheles gambiae*. *Journal of Insect Physiology, 64*(8), 1226–1235.

Asanuma, K., Kano, R., & Takahasi, H. (1952). Notes on Culicidae of Hokkaido. I. Descriptions of the male terminalia of subgenus *Ochlerotatus* Arrib. (*Aedes*, Culicidae) *Report of the Hokkaido Institute of Public Health, 3*, 34–40.

Asanuma, K., & Nakagawa, H. (1953). Morphological and taxonomic studies of some mosquito pupae belonging to the genera *Aedes* and *Culex* from Japan, with special regards to the chaetotaxy. *Miscellaneous Reports of the Research Institute for Natural Resources, 31*, 86–98.

Asanuma, K., & Nakagawa, H. (1954a). A key to the genera of Japanese mosquitoes, based on pupal chaetotaxy. *Japan Society of Medical Entomology and Zoology, 4*, 362–375.

Asanuma, K., & Nakagawa, H. (1954b). Pupal chaetotaxy of mosquitoes belonging to the tribe Megarhinini. Studies on the pupal stage of Japanese mosquitoes. Part X. *Miscellaneous Reports of the Research Institute for Natural Resources, 35*, 82–85.

Ascunce, M.S., Yang, C.C., Oakey, J., Calcaterra, L., Wu, W.J., Shih, C.J., . . . Shoemaker, D. (2011). Global invasion history of the fire ant *Solenopsis invicta*. *Science, 331*(6020), 1066–1068.

Ashburner, M. (2005). The genomes of Diptera. In D.K. Yeates & B.M. Wiegman (Eds.), *The evolutionary biology of flies* (pp. 65–125). New York, NY: Columbia University Press.

Aslamkhan, M. (1971a). A redescription of *Aedes* (*Aedimorphus*) *punctifemoris* (Ludlow) from Pakistan. *Biologia* (Pakistan), *17*(1), 61–65.

Aslamkhan, M. (1971b). The mosquitoes of Pakistan. I. A checklist. *Mosquito Systematics, 3*(4), 147–159.

Aslamkhan, M. (1972). The mosquitoes of Pakistan. II. Mosquitoes originally described from Pakistan. *Mosquito Systematics, 4*(4), 98–102.

Aslamkhan, M., Aaqil, M., & Hafeez, M. (1972). Genetical and morphological variations in a natural population of the malaria mosquito, *Anopheles stephensi* from Karachi, Pakistan. *Biologia, 18*(1), 29–41.

Astaiza, V.R., Murillo, B.C., & Fajardo, O.P. (1988). Biology of *Anopheles neivai* H.D. and K., 1913 (Diptera: Culicidae) on the Pacific coast of Colombia. II. Adult population fluctuation. *Revista de saúde pública, 22*(3), 101–108.

Atrie, B., Subbarao, S.K., Pillai, M.K.K., Rao, S.R.V., & Sharma, V.P. (1999). Population cytogenetic evidence for sibling species in *Anopheles annularis* (Diptera: Culicidae). *Annals of the Entomological Society of America, 92*(2), 243–249.

Attardo, G.M., Hansen, I.A., & Raikhel, A.S. (2005). Nutritional regulation of vitellogenesis in mosquitoes: Implications for anautogeny. *Insect Biochemistry and Molecular Biology, 35*(7), 661–675.

Attardo, G.M., Hansen, I.A., Shiao, S.H., & Raikhel, A.S. (2006). Identification of two cationic amino acid transporters required for nutritional signaling during mosquito reproduction. *Journal of Experimental Biology, 209*(16), 3071–3078.

Augier, L.M., Dantur Juri, M.J., & Molina, G.A. (2003). Redescripción de la larva y la pupa de *Toxorhynchites* (*Lynchiella*) *guadeloupensis* (Diptera: Culicidae). *Revista de la Sociedad Entomológica Argentina, 62*(1–2), 99–106.

Avila, F.W., Sirot, L.K., LaFlamme, B.A., Rubinstein, C.D., & Wolfner, M.F. (2011). Insect seminal fluid proteins: Identification and function. *Annual Review of Entomology, 56*, 21–40.

Awasthi, A.K., Wu, C.H., & Hwang, J.S. (2012). Diving as an antipredator behavior in mosquito pupae. *Zoological Studies, 51*(8), 1225–1234.

Awono-Ambene, H.P., Kengne, P., Simard, F., Antonio-Nkondjio, C., & Fontenille, D. (2004). Description and bionomics of *Anopheles* (*Cellia*) *ovengensis* (Diptera: Culicidae), a new malaria vector species of the *Anopheles nili* group from South Cameroon. *Journal of Medical Entomology, 41*(4), 561–568.

Awono-Ambene, H.P., Simard, F., Antonio-Nkondjio, C., Cohuet, A., Kengne, P., & Fontenille, D. (2006). Multilocus enzyme electrophoresis supports speciation within the *Anopheles nili* group of malaria vectors in Cameroon. *The American Journal of Tropical Medicine and Hygiene, 75*(4), 656–658.

Ayala, D., Costatini, C., Ose, K., Kamden, G.C., Antonio-Nkondjio, C., Agbor, J.P., . . . Simard, F. (2009). Habitat suitability and ecological niche profile of major malaria vectors in Cameroon. *Malaria Journal, 8*(1), 307.

Azambuja, P., García, E.S., & Ratcliffe, N.A. (2005). Gut microbiota and parasite transmission by insect vectors. *Trends in Parasitology, 21*(12), 568–572.

Azari-Hamidian, S., & Harbach, R.E. (2009). Keys to the adult females and fourth-instar larvae of the mosquitoes of Iran (Diptera: Culicidae). *Zootaxa, 2078*, 1–33.

Azari-Hamidian, S., Abai, M.R., Ladonni, H., Vatandoost, H., & Akbarzadeh, K. (2006). *Anopheles peditaeniatus* (Leicester) new to the Iranian mosquito fauna with notes on *Anopheles hyrcanus* group in Iran. *Journal of the American Mosquito Control Association, 22*(1), 144–146.

Azari-Hamidian, S., Norouzi, B., & Noorallahi, A. (2017). *Orthopodomyia pulcripalpis* (Diptera: Culicidae), a genus and species new to the Iranian mosquito fauna, with a review of bionomical information. *Zootaxa, 4299*(1), 141–145.

Azrag, R.S., Ibrahim, K., Malcolm, C., El Rayah, E., & El-Sayed, B. (2016). Laboratory rearing of *Anopheles arabiensis*: Impact on genetic variability and implications for Sterile Insect Technique (SIT) based mosquito control in northern Sudan. *Malaria Journal, 15*(1), 432.

Baba, K. (1950). A study of anopheline mosquitoes in relation to the epidemiology of malaria in Canton delta, South China. I. Species of anopheline mosquitoes recorded from Kwangtong Province, South China and their breeding habits. *Transactions of the Kansai Entomological Society* (China), *15*(2), 1–11.

Baber, I., Keita, M., Sogoba, N., Konate, M., Diallo, M., Doumbia, S., . . . Manoukis, N.C. (2010). Population size and migration of *Anopheles gambiae* in the Bancoumana Region of Mali and their significance for efficient vector control. *PLOS ONE, 5*(4), e10270.

Bachmann, A.O., & Casal, O.H. (1962a). Notas sobre *Culex* (*Culex*) Argentinos (Diptera: Culicidae). *Anales del Instituto Nacional de Microbiología, 1*, 77–81.

Bachmann, A.O., & Casal, O.H. (1962b). *Culex* (*Culex*) *lahillei* una nueva especie de la Argentina (Diptera: Culicidae). *Physis, 23*(65), 267–270.

Bachman, A.O., & Casal, O.H. (1963). Notas sobre *Culex* (*Culex*) Argentinos. II. *Revista de la Sociedad Entomologica Argentina, 25*(for 1962), 39–42.

Bader, C.A., & Williams, C.R. (2012). Mating, ovariole number and sperm production of the dengue vector mosquito *Aedes aegypti* (L.) in Australia: Broad thermal optima provide the capacity for survival in a changing climate. *Physiological Entomology, 37*(2), 136–144.

Baerg, D.C. (1968). Overwintering habits of *Orthopodomyia californica* Bohart (Diptera: Culicidae). *Mosquito News, 28*(2), 234–235.

Baerg, D.C., & Boreham, M.M. (1974a). *Anopheles neivai* Howard, Dyar and Knab: Laboratory observations on the life cycle and description of the egg stage (Diptera: Culicidae). *Journal of Medical Entomology, 11*(5), 629–630.

Baerg, D.C., & Boreham, M.M. (1974b). Experimental rearing of *Chagasia bathana* (Dyar) using induced mating, and description of the egg stage (Diptera: Culicidae). *Journal of Medical Entomology, 11*(5), 631–632.

Baeza Cuéllar, M. (1933). Estudio médico de los Culicidos hematófagos. Madrid: Talleres Gráficos Herrera.

Bahr, P.H. (1912). Filariasis and elephantiasis in Fiji; being a report to the London School of Tropical Medicine. Dysentery [sic] in Fiji during the year 1910. [Series] *Journal of the London School of Tropical Medicine* (Suppl. 1), 1–192.

Bai, H., Ramaseshadri, P., & Palli, S.R. (2007). Identification and characterization of juvenile hormone esterase gene from the yellow fever mosquito, *Aedes aegypti*. *Insect Biochemistry and Molecular Biology, 37*(8), 829–837.

Bailey, C.L., Vongpradist S., & Mongkolpanya, K. (1975). Observations of an unusual oviposition habit in *Armigeres* (*Leicesteria*) *annulitarsis* (Leicester). *Mosquito News, 35*(4), 564–565.

Bailly-Choumara, H. (1960). Une espèce nouvelle d'*Anopheles* du golfe d'Aden, *Anopheles* (*Myzomyia*) *azaniae*. Notes morphologiques, systématiques et écologiques. *Bulletin de la Société de pathologie exotique, 53*(3), 531–542.

Bailly-Choumara, H. (1966a). Contribution a l'étude des *Culex* (*Neoculex*) (Diptera, Culicidae) de la région Éthiopienne. Description de *Culex* (*Neoculex*) *pseudoandreanus* sp. n. *Bulletin de la Société de pathologie exotique, 58*(for 1965)(4), 660–664.

Bailly-Choumara, H. (1966b). Description de la larve et de la nymphe d'*Aedes* (*Neomelaniconion*) *taeniarostris* Theobald 1910. Observation sur une variation de coloration chez l'adulte. *Bulletin de la Société de pathologie exotique, 58*(for 1965)(4), 671–676.

Bailly-Choumara, H. (1966c). *Culex* (*Culex*) *gameti* (Diptera, Culicidae), moustique nouveau du Cameroun description du mâle. *Bulletin de la Société de pathologie exotique, 59*(1), 148–150.

Bailly-Choumara, H. (1968). Contribution à l'étude des moustiques du Maroc (Diptera, Culicidae). Six espèces nouvelles pour le pays. *Cahiers ORSTOM. Série entomologie médicale et parasitologie, 6*(2), 139–144.

Bailly-Choumara, H., & Adam, J.P. (1959). *Anopheles maliensis* n. sp. (Diptera Culicidae), une espèce nouvelle d'anophélinés du Fouta-Djalon, République de Guinée. *Academie des Sciences* (Paris), *248*(26), 3741–3743.

Bailly-Choumara, H., & Adam, J.P. (1960). Notes complémentaires sur la morphologie, la position systematique et la biologie d'*Anopheles* (*Neomyzomyia*) *maliensis* Bailly-Choumara et Adam, 1959. *Bulletin de la Société de pathologie exotique, 52*, 110–119.

Bailly-Choumara, H., & Pajot, F.X. (1966). Contribution a l'étude des *Culex* (*Neoculex*) de la région Éthiopienne (Diptera, Culicidae). Description de la larve et de la nymphe de *Culex* (*Neoculex*) *pseudoandreanus* Bailly-Choumara. *Bulletin de la Société de pathologie exotique, 58*(for 1965)(4), 665–671.

Bailly-Choumara, H., & Rickenbach, A. (1966). Contribution a l'étude des *Culex* (Diptera, Culicidae) de la région Éthiopienne. Description de *Culex* (*Neoculex*) *garioui* sp. n., moustique nouveau du Cameroun. *Bulletin de la Société de pathologie exotique, 59*(1), 144–148.

Baimai, V., & Harrison, B.A. (1980). *Evidence of sibling speciation in the* Balabacensis *complex of Southeast Asia (Diptera: Culicidae).* Paper presented at the 10th International Congress of Tropical Medicine and Malaria, Manila, Philippines.

Baimai, V., Andre, R.G., Harrison, B.A., Kijchalao, U., & Panthusiri, L. (1987). Crossing and chromosomal evidence for two additional sibling species within the taxon *Anopheles dirus* Peyton and Harrsion (Diptera: Culicidae) in Thailand. *Proceedings of the Entomological Society of Washington, 89*(1), 157–166.

Baimai, V., Harbach, R.E., & Kijchalao, U. (1988a). Cytogenetic evidence for a fifth species within the taxon *Anopheles dirus* in Thailand. *Journal of the American Mosquito Control Association, 4*(3), 333–338.

Baimai, V., Harbach, R.E., & Sukowati, S. (1988b). Cytogenetic evidence for two species within the current concept of the malaria vector *Anopheles leucosphyrus* in southeast Asia. *Journal of the American Mosquito Control Association, 4*(1), 44–50.

Baimai, V., Kijchalao, U., & Rattanarithikul, R. (1996). Metaphase karyotypes of *Anopheles* of Thailand and southeast Asia. V. The Myzomyia series, subgenus *Cellia* (Diptera: Culicidae). *Journal of the American Mosquito Control Association, 12*(1), 97–105

Baisas, F.E. (1932). *The morphology of some Philippine* Anopheles. Paper presented at the Transactions of the Eighth Congress of the Far Eastern Association of Tropical Medicine, 1930. Bangkok, Siam.

Baisas, F.E. (1935a). Notes on Philippine mosquitoes. I. The *Armigeres* group. *Philippine Journal of Science, 56*(4), 485–497.

Baisas, F.E. (1935b). Notes on Philippine mosquitoes. II. *Uranotaenia* group. *Philippine Journal of Science, 57*(1), 63–80.

Baisas, F.E. (1935c). Notes on Philippine mosquitoes. V. The pupal characters of anophelines under the *Myzorhynchus* series and group *Neocellia*, with further comments on the larvae and adults of *sinensis*. *Monthly Bulletin of the Bureau of Health* (Manila), *15*(9), 291–339.

Baisas, F.E. (1935d). Notes on Philippine mosquitoes. III. Genus *Culex*: groups *Lophoceratomyia*, *Mochthogenes*, and *Neoculex*. *Philippine Journal of Science, 57*(2), 167–179.

Baisas, F.E. (1936a). Notes of Philippine mosquitoes. VI. The pupal characters on anophelines of the subgenus *Myzomyia*. *Philippine Journal of Science, 61*(2), 205–245.

Baisas, F.E. (1936b). Notes on Philippine mosquitoes. IV. The pupal and certain adult characters of some rare species of *Anopheles*. *Philippine Journal of Science, 59*(1), 65–98.

Baisas, F.E. (1938). Notes on Philippine mosquitoes. VII. A. *Culex* (*Culex*) with banded proboscis and tarsi. B. *Anopheles*: The pupae of three rare species; the Leucosphyrus-subgroup. *Monthly Bulletin of the Bureau of Health* (Manila), *18*(5), 175–232.

Baisas, F.E. (1946a). Notes on Philippine mosquitoes. VIII. Species found in the jungles of Llavac. *Monthly Bulletin of the Bureau of Health* (Manila), *22*(1), 27–49.

Baisas, F.E. (1946b). Notes on Philippine mosquitoes. IX. A new species of *Zeugnomyia*. *Monthly Bulletin of the Bureau of Health* (Manila), *22*(2), 27–35.

Baisas, F.E. (1946c). Notes on Philippine mosquitoes. X. Some species of *Aedes* (*Finlaya*) and (*Aedimorphus*). *Monthly Bulletin of the Bureau of Health* (Manila), *22*(1), 21–37.

Baisas, F.E. (1946d). Notes of Philippine mosquitoes, XII. *Topomyia*. *Monthly Bulletin of the Bureau of Health* (Manila), *22*(4), 31–47.

Baisas, F.E. (1947a). Notes of Philippine mosquitoes. XI. A new species of *Tripteroides*. *Fieldiana: Zoology, 31*(15), 121–124.

Baisas, F.E. (1947b). Notes on Philippine mosquitoes. XIV. The larval instars of *Anopheles*. *Monthly Bulletin of the Bureau of Health* (Manila), *23*(3), 197–207.

Baisas, F.E. (1957). Notes on Philippine mosquitoes. XVIII. The malaria vector, *flavirostris*: Summary of recent findings. *Bulletin of the Department of Health, 29*(2), 68–90.

Baisas, F.E. (1965). Additional notes on Philippine mosquitoes. Part I: *Armigeres baisasi* Stone and Thurman, 1958 and *Armigeres joloensis* (Ludlow), 1904. *PACAF Epidemiological Laboratory Technical Report 17-65, 5th Epidemiological Flight.*

Baisas, F.E. (1974). *The mosquito fauna of Subic Bay Naval Reservation, Republic of the Philippines.* San Francisco, CA: First Medical Service Wing (PACAF), Technical Report 72-2.

Baisas, F.E., & Catipon, R. (1958). Notes on Philippine mosquitoes. XX. Daytime observations in houses of two barrios in Laguna Province. *Philippine Journal of Science, 87*(1), 47–63.

Baisas, F.E., & Dowell, F.H. (1965). Keys to the adult female and larval anopheline mosquitoes of the Philippines. *PACAF Epidemiological Laboratory Technical Report*, 13–65.

Baisas, F.E., & Dowell, F.H. (1967). Keys to the adult female and larval anopheline mosquitoes of the Philippines. *Journal of Medical Entomology, 4*(1), 11–24.

Baisas, F.E., & Feliciano, P. (1953). Notes on Philippine mosquitoes. XIII. Four new species of *Zeugnomyia* and *Topomyia*. *Fieldiana: Zoology, 33*(3), 161–179.

Baisas, F.E., & Hu, S. M.K. (1936). *Anopheles hyrcanus* var. *sinensis* of the Philippines and certain parts of China, with some comments on *Anopheles hyrcanus* var. *nigerrimus* of the Philippines. *Monthly Bulletin of the Bureau of Health* (Manila), *16*(6), 205–242.

Baisas, F.E., & Ubaldo-Pagayon, A. (1949). Notes on Philippine mosquitoes. XV. The chaetotaxy of the pupae and larvae of *Tripteroides*. *Philippine Journal of Science, 78*(1), 43–72.

Baisas, F.E., & Ubaldo-Pagayon, A. (1953). *Notes on Philippine mosquitoes. XVI. Genus Tripteroides.* Manila, Philippines (for 1952): Bureau of Printing.

Baisas, F.E., & Ubaldo-Pagayon, A. (1956). Notes on Philippine mosquitoes. XVII. The eggs and first-instar larvae of some *Neomyzomyias*. *Philippine Journal of Science, 85*(2), 215–230.

Baisas, F.E., Bañez, L.F.L., & Leuenberger, N. (1962). Notes on Philippine mosquitoes. XXII. The axil-breeding species. *Philippine Journal of Science, 8*(for 1960)(2), 183–199.

Baker, F.C. (1936). A new species of *Orthopodomyia*, *O. alba* sp. n. (Diptera, Culicidae). *Proceedings of the Entomological Society of Washington, 38*(1), 1–7.

Baker, M. (1961). The altitudinal distribution of mosquito larvae in the Colorado Front Range. *Transactions of the American Entomological Society, 87*(4), 231–246.

Baker, R.H., & Kitzmiller, J.B. (1964). The salivary gland chromosomes of *Anopheles aztecus*. *Revista del Instituto de Salubridad y Enfermedades Tropicales* (México), *24*, 43–54.

Baker, R.H., Drench, W., & Kitzmiller, J.B. (1962). Induced copulation in *Anopheles* mosquitoes. *Mosquito News, 22*(1), 16–17.

Baker, R.H., Kitzmiller, J.B., & Chowdaiah, B.N. (1966). The salivary gland chromosomes of *Anopheles hectoris*. *Chromosoma, 19*(1), 126–136.

Baker, R.H., Nasir, A.S., & Aslamkhan, M. (1968). The salivary gland chromosomes of *Anopheles pulcherrimus* Theobald. *Parassitologia, 10*, 167–177.

Baldini, F., Gabrieli, P., Rogers, D.W., & Catteruccia, F. (2012). Function and composition of male accessory gland secretions in *Anopheles gambiae*: A comparison with other insect vectors of infectious diseases. *Pathogens and Global Health, 106*(2), 82–93.

Baldini, F., Gabrieli, P., South, A., Valim, C., Mancini, F., & Catteruccia, F. (2013). The interaction between a sexually transferred steroid hormone and a female protein regulates oogenesis in the malaria mosquito *Anopheles gambiae*. *PLOS Biology, 11*(10), e1001695.

Ballard, J.W.O., Melvin, R.G., Katewa, S.D., & Mass, K. (2007). Mitochondrial DNA variation is associated with measurable differences in life-history traits and mitochondrial metabolism in *Drosophilia simulans*. *Evolution, 61*(7), 1735–1747.

Bancroft, T.L. (1908). List of the mosquitoes of Queensland, with the original descriptions and notes on the life-history of a number. *Annals of the Queensland Museum, 8*, 1–64.

Bañez, L.F., & Jueco, N.L. (1966). Adult and larva of *Aedes* (*Finlaya*) *sorsogonensis* Bañez, 1963. Comparison with *Aedes* (*Finlaya*) *ananae* Knight and Laffoon, 1946 and *Aedes* (*Finlaya*) *poecilus* (Theobald), 1903. *Acta Médica Philippina, 2*(4), 194–200.

Bangs, M.J., Taai, K., Howard, T.M., Cook, S., & Harbach, R.E. (2014). The mosquito *Anopheles* (*Cellia*) *oreios* sp. n., formerly species 6 of the Australasian *Anopheles farauti* complex, and a critical review of its biology and relation to disease. *Medical and Veterinary Entomology, 29*(1), 68–81.

Banks, C.S. (1906a). A new genus and species of Culicidae. *Philippine Journal of Science, 1*(7), 779–784.

Banks, C.S. (1906b). A list of Philippine Culicidae with descriptions of some new species. *Philippine Journal of Science, 1*(9), 977–1005.

Banks, C.S. (1908). Biology of Philippine Culicidae. *Philippine Journal of Science, 3*(4), 235–258.

Banks, C.S. (1909). Four new Culicidae from the Philippines. *Philippine Journal of Science, 4*(6), 275–285.

Banks, C.S. (1914). A new Philippine malaria mosquito. *Philippine Journal of Science, 9*(4), 405–407.

Baranoff, N. (1927). Eine neue *Aedes*-Art aus Sud-Serbien (Dipt.: Culic). *Neue Beiträge zur systematischen Insektenkunde, 4*(1–2), 15–16.

Barata, J.M.S., & Cotrim, M.D. (1971). Uma nova espécie de *Psorophora* A (Diptera, Culicidae). *Revista Brasileira de entomología, 15*(6), 35–40.

Barata, J.M.S., & Cotrim, M.D. (1972). Redescrição de *Culex* (*Carrollia*) *secundus* Bonne-Wepster & Bonne, 1920 e *Culex* (*C.*) *anduzei* Cerqueira & Lane, 1944, com dados sobre formas imaturas (Diptera, Culicidae). *Revista Brasileira de entomología, 16*(4), 17–30.

Barbosa, A.A., Navarro-Silva, M.A., & Sallum, M.A.M. (2005). Description and revalidation of *Mansonia* (*Mansonia*) *fonsecai* (Pinto) (Diptera: Culicidae). *Zootaxa, 905*, 1–11.

Barbosa, A.A., Navarro-Silva, M.A., & Sallum, M.A.M. (2007). *Mansonia* (*Mansonia*) *iguassuensis* sp. nov. (Diptera: Culicidae) from Brasil. *Zootaxa, 1527*, 45–52.

Barbosa, R.M.R., Furtado, A., Regis, L., & Leal, W.S. (2010). Evaluation of an oviposition-stimulating kairomone for the yellow fever mosquito, *Aedes aegypti*, in Recife, Brazil. *Journal of Vector Ecology, 35*(1), 204–207.

Bargielowski, I., Alphey, L., & Koella, J.C. (2011). Cost of mating and insemination capacity of a genetically modified mosquito *Aedes aegypti* OX513A compared to its wild type counterpart. *PLOS ONE, 6*(10), e26086.

Bargielowski, I., Blosser, E., & Lounibos, L.P. (2015). The effects of interspecific courtship on the mating success of *Aedes aegypti* and *Aedes albopictus* (Diptera: Culicidae) males. *Annals of the Entomological Society of America, 108*(4), 513–518.

Bargielowski, I.E., & Lounibos, L.P. (2016). Satyrization and satyrization-resistance in competitive displacements of invasive mosquito species. *Insect Science, 23*(2), 162–174.

Barmak, D.H., Dorso, C.O., Otero, M., & Solari, H.G. (2011). Dengue epidemics and human mobility. *Physical Review E, 84*(1), 011901.

Barr, A.R. (1954). Hybridization of experiments with some American dark-winged anophelines. *Experimental Parasitology, 3*(5), 445–457.

Barr, A.R. (1955). The resurrection of *Aedes melanimon* Dyar. *Mosquito News, 15*(3), 170–172.

Barr, A.R. (1957a). A new species of *Culiseta* (Diptera: Culicidae) from North America. *Proceedings of the Entomological Society of Washington, 59*(4), 163–167.

Barr, A.R. (1957b). The distribution of *Culex p. pipiens* and *C. p. quinquefasciatus* in North America. *The American Journal of Tropical Medicine and Hygiene, 6*(1), 153–165.

Barr, A.R. (1958). The mosquitoes of Minnesota (Diptera: Culicidae: Culicinae). *Technical Bulletin 228, University of Minnesota Agricultural Experiment Station*, 1–154.

Barr, A.R. (1963). Pupae of the genus *Culiseta* Felt. II. Descriptions and a key to the north American species (Diptera, Culicidae). *Annals of the Entomological Society of America, 56*(3), 324–330.

Barr, A.R. (1974). New concepts of mosquito taxonomy. *Mosquito Systematics, 6*(2), 134–136.

Barr, A.R., & Barr, S. (1969a). Mosquito studies (Diptera, Culicidae). XIII. Pupae of the genus *Psorophora* in North America and Puerto Rico. *Contributions of the American Entomological Institute, 4*(4), 1–20.

Barr, A.R., & Barr, S. (1969b). The egg of *Wyeomyia smithii* (Coquillett) and a review of knowledge of the eggs of the Sabethini. *Mosquito News, 29*(2), 193–198.

Barr, A.R., & Guptavanij, P. (1989). *Anopheles hermsi* n. sp., an unrecognized American species of the *Anopheles maculipennis* group (Diptera: Culicidae). *Mosquito Systematics, 20*(for 1988)(3), 352–356.

Barr, J.F. (1996). Aspects of common loon (*Garia immer*) feeding biology on its breeding ground. *Hydrobiologia, 321*(2), 119–144.

Barraud, P.J. (1923a). Some new culicine mosquitoes found in India, and a note on *Finlaya assamensis* (Theo.). *Bulletin of Entomological Research, 13*(4), 405–408.

Barraud, P.J. (1923b). A revision of the culicine mosquitoes of India. Part I. The genera *Stegomyia*, Theo. and *Christophersiomyia*, gen. n. adult stage. *Indian Journal of Medical Research* (Calcutta), *10*(3), 772–788.

Barraud, P.J. (1923c). A revision of the culicine mosquitoes of India. Part III. Notes on certain Indian species of the genus *Finlaya*, Theo. and descriptions of new species. *Indian Journal of Medical Research* (Calcutta), *11*(1), 214–219.

Barraud, P.J. (1923d). A revision of the culicine mosquitoes of India. Part V. Further notes on the genera *Stegomyia*, Theo. and *Finlaya*, Theo. with descriptions of new species. *Indian Journal of Medical Research* (Calcutta), *11*(1), 224–228.

Barraud, P.J. (1923e). Two new species of *Culex* (Diptera, Culicidae) from Assam. *Indian Journal of Medical Research* (Calcutta), *11*(2), 507–509.

Barraud, P.J. (1923f). A revision of the culicine mosquitoes of India. Part IV. The larvae of some Indian species of *Finlaya*, Theo. *Indian Journal of Medical Research* (Calcutta), *11*(1), 220–223.

Barraud, P.J. (1923g). A revision of the culicine mosquitoes of India. Part VII. The larvae of some species of *Stegomyia*; the larvae of some species of *Finlaya*; the larva of *Christophersiomyia thomsoni*; the larva of *Mimomyia chamberlaini*; the larva of *Aedomyia catasticta*. *Indian Journal of Medical Research* (Calcutta), *11*(2), 495–505.

Barraud, P.J. (1923h). A revision of the culicine mosquitoes of India. Part VI. Some Indian species of the genus *Finlaya* Theo. adult stage. *Indian Journal of Medical Research (Calcutta)*, *11*(2), 475–493.

Barraud, P.J. (1923i). A revision of the culicine mosquitoes of India. Part II. The larvae of some Indian species of *Culex*. *Indian Journal of Medical Research* (Calcutta), *10*(4), 934–493.

Barraud, P.J. (1924a). A revision of the culicine mosquitoes of India. Part VIII. Further descriptions of Indian species of *Finlaya* Theo. *Indian Journal of Medical Research* (Calcutta), *11*(3), 845–865.

Barraud, P.J. (1924b). A revision of the culicine mosquitoes of India. Part X. The larvae of Indian species of *Lutzia* Theo. *Indian Journal of Medical Research* (Calcutta), *11*(4), 977–978.

Barraud, P.J. (1924c). A new mosquito from Kashmir. *Indian Journal of Medical Research* (Calcutta), *11*, 967–968.

Barraud, P.J. (1924d). Four new mosquitoes from the western Himalayas. *Indian Journal of Medical Research* (Calcutta), *11*(4), 999–1006.

Barraud, P.J. (1924e). A revision of the culicine mosquitoes of India. Part XI. Some Indian species of *Culex* L. *Indian Journal of Medical Research* (Calcutta), *11*(4), 979–998.

Barraud, P.J. (1924f). A revision of the culicine mosquitoes of India. Part XII. Further descriptions of Indian species of *Culex* L. including two new species. *Indian Journal of Medical Research* (Calcutta), *11*(4), 1259–1274.

Barraud, P.J. (1924g). A revision of the culicine mosquitoes of India. Part XIII. Further descriptions of Indian species of *Culex* L. including three new species. *Indian Journal of Medical Research* (Calcutta), *11*, 1275–1282.

Barraud, P.J. (1924h). A revision of the culicine mosquitoes of India. Part XIV. The Indian species of the subgenus *Culiciomyia* (Theo.) Edw., including one new species. *Indian Journal of Medical Research (Calcutta)*, *12*(1), 15–22.

Barraud, P.J. (1924i). A revision of the culicine mosquitoes of India. Part XV. The Indian species of the subgenus *Lophoceratomyia* (Theo.) Edw., including two new species. *Indian Journal of Medical Research* (Calcutta), *12*(1), 39–46.

Barraud, P.J. (1924j). A new mosquito from Kashmir and the northwest frontier province. *Indian Journal of Medical Research* (Calcutta), *12*(1), 73–74.

Barraud, P.J. (1924k). A revision of the culicine mosquitoes of India. Part IX. The Indian species of *Lutzia* Theo. *Indian Journal of Medical Research* (Calcutta), *11*(4), 971–976.

Barraud, P.J. (1924l). A revision of the culicine mosquitoes of India. Part X. The larvae of Indian species of *Lutzia* Theo. *Indian Journal of Medical Research* (Calcutta), *11*(4), 977–978.

Barraud, P.J. (1924m). A revision of the culicine mosquitoes of India. Part XVII. *Indian Journal of Medical Research* (Calcutta), *12*(2), 427–434.

Barraud, P.J. (1926). A revision of the culicine mosquitoes of India. Part XVIII. The Indian species of *Uranotaenia* and *Harpagomyia*, with descriptions of five new species. *Indian Journal of Medical Research* (Calcutta), *14*(2), 331–350.

Barraud, P.J. (1927a). A revision of the culicine mosquitoes of India. Part XIX. The Indian species of *Aedeomyia* and *Orthopodomyia* with descriptions of two new species. *Indian Journal of Medical Research* (Calcutta), *14*(3), 523–532.

Barraud, P.J. (1927b). A revision of the culicine mosquitoes of India. Part XX. The Indian species of *Armigeres* (including *Leicesteria*), with descriptions of two new species. *Indian Journal of Medical Research* (Calcutta), *14*(3), 533–548.

Barraud, P.J. (1927c). A revision of the culicine mosquitoes of India. Part XXI. Descriptions of new species of *Aedimorphus* and *Finlaya*, and notes on *Stegomyia albolineata* (Theo.). *Indian Journal of Medical Research* (Calcutta), *14*(3), 549–554.

Barraud, P.J. (1927d). A revision of the culicine mosquitoes of India. Part XXII. The Indian species of the genus *Taeniorhynchus* (including *Mansonioides*), with a description of one new species. *Indian Journal of Medical Research* (Calcutta), *14*(3), 555–563.

Barraud, P.J. (1928a). A revision of the culicine mosquitoes of India. Part XXIV. The Indian species of the subgenera *Skusea* and *Aedes*, with descriptions of eight new species, and remarks on a new method for identifying the females of the subgenus *Aedes*. *Indian Journal of Medical Research* (Calcutta), *16*(2), 357–375.

Barraud, P.J. (1928b). The distribution of "*Stegomyia fasciata*" in India, with remarks on dengue and yellow fever. *Indian Journal of Medical Research* (Calcutta), *16*(2), 377–388.

Barraud, P.J. (1928c). A revision of the culicine mosquitoes of India. Part XXIII. The genus *Aedes* and the classification of the subgenera. Descriptions of the Indian species of *Aedes* (*Aedimorphus*), *Aedes* (*Ochlerotatus*) and *Aedes* (*Banksinella*) with notes on *Aedes* (*Stegomyia*) *variegatus*. *Indian Journal of Medical Research* (Calcutta), *15*, 653–669.

Barraud, P.J. (1929a). A revision of the culicine mosquitoes of India. Part XXV. The genera *Mucidus*, *Mimomyia*, *Ficalbia*, *Rachionotomyia*, and *Hodgesia*. *Indian Journal of Medical Research* (Calcutta), *16*, 1052–1063.

Barraud, P.J. (1929b). A revision of the culicine mosquitoes of India. Part XXVI. The genera *Heizmannia*, *Haemagogus*, *Topomyia*, and *Megarhinus*. *Indian Journal of Medical Research* (Calcutta), *17*(1), 259–280.

Barraud, P.J. (1931a). Descriptions of eight new species of Indian culicine mosquitoes. *Indian Journal of Medical Research* (Calcutta), *19*(2), 609–616.

Barraud, P.J. (1931b). Notes on some Indian mosquitoes of the subgenus *Stegomyia*, with descriptions of new species. *Indian Journal of Medical Research* (Calcutta), *19*(1), 221–229.

Barraud, P.J. (1931c). The early stages of some Indian mosquitoes: *Megarhinus*. *Indian Journal of Medical Research* (Calcutta), *18*(4), 1127–1137.

Barraud, P.J. (1932a). *Anopheles aitkeni* James var. *n. pinjaurensis*. *Records of the Malaria Survey of India*, *3*(2), 353–355.

Barraud, P.J. (1932b). The early stages of some Indian mosquitoes: *Orthopodomyia*. *Indian Journal of Medical Research* (Calcutta), *19*(4), 1013–1017.

Barraud, P.J. (1934). *The fauna of British India, including Ceylon and Burma. Diptera: Vol. 5. Family Culicidae, tribes Megarhinini and Culicini*. London, England: Taylor and Francis.

Barraud, P.J., & Christophers, S.R. (1931). On a collection of anopheline and culicine mosquitoes from Siam. *Records of the Malaria Survey of India*, *2*(2), 269–285.

Barraud, P.J., & Covell, G. (1928). The morphology of the buccal cavity in anopheline and culicine mosquitoes. *Indian Journal of Medical Research* (Calcutta), *15*, 671–679.

Barrera, R. (2008). Competition and resistance to starvation in larvae of container-inhabiting *Aedes*. *Ecological Entomology*, *21*(2), 117–127.

Barreto, M.P., & Coutinho, J.O. (1944). Sôbre o gênero *Taeniorhynchus* Arribálzaga, 1891, com a descrição de três novas espécies do subgênero *Taeniorhynchus* (Diptera, Culicidae). *Arquivos de higiene e saúde pública* (São Paulo), *9*(21), 53–85.

Barreto, M.P., & Coutinho, J.O. (1951). Sôbre o *Anopheles* (*Kerteszia*) *cruzii* Dyar e Knab e *Anopheles* (*Kerteszia*) *Ianeanus* Corrêa e Cerqueira (Diptera, Culicidae). *Arquivos de higiene e saúde pública* (São Paulo), *15*(for 1950), 177–180.

Barreto, P., & Lee, V.H. (1969). Arthropodos hematófagos del Rio Raposo, Valle, Colombia. II. Culicidae. *Caldasia, 10*(49), 407–440.

Barreto-Reyes, P. (1955). Lista de mosquitos de Colombia, S.A. *Revista anales de la Sociedad de Biologia de Bogota, 7*(2), 45–94.

Barreto-Reyes, P. (1957). Mosquitos de Colombia. Nuevos registros. *Sociedad de Biologia de Bogata, 7*, 190–196.

Barrón, M.G., Paupy, C., Rahola, N., Akone-Ella, O., Ngangue, M.F., Wilson-Bahun, T.A., . . . Ayala, D. (2019). A new species in the major malaria vector complex sheds light on reticulated species evolution. *Scientific Reports, 9*, 14753.

Barros de Carvalho, C.J., & de Mello-Patiu, C.A. (2008). Key to the adults of the most common forensic species of Diptera in South America. *Revista Brasileira de entomología, 52*(3), 390–406.

Bartholomew, J.G. (1913). *A Literary and Historical Atlas of Africa and Australia*. London, England: J.M. Dent & Sons.

Bartlett-Healy, K.B., Unlu, I., Obenauer, P., Hughes, T., Healy, S., Crepeau, T., . . . Strickman, D. (2012). Larval mosquito habitat utilization and community dynamics of *Aedes albopictus* and *Aedes japonicus* (Diptera: Culicidae). *Journal of Medical Entomology, 49*(4), 813–824.

Barton, P.S., Aberton, J.G., & Kay, B.H. (2004). Spatial and temporal definition of *Ochlerotatus camptorhynchus* (Thomson) (Diptera: Culicidae) in the Gippsland Lakes system of eastern Victoria. *Australian Journal of Entomology, 43*(1), 16–22.

Basham, E.H. (1948). *Culex* (*Melanoconion*) *mulrennani*, a new species from Florida (Diptera: Culicidae). *Annals of the Entomological Society of America, 41*(1), 1–7.

Basio, R.G. (1971a). On Philippine mosquitoes. III. Some new species from Mt. Makiling and its vicinity in Luzon (Diptera: Culicidae). *The Philippine Entomologist, 2*(1), 51–57.

Basio, R.G. (1971b). *The Mosquito Fauna of the Philippines* (*Diptera Culicidae*). Monograph No. 4. Manila, Philippines: National Museum of the Philippines.

Basio, R.G. (1972). On Philippine mosquitoes. IX. Notes and descriptions of *Culex* (*Culex*) *hutchinsoni* Barraud (Diptera: Culicidae). *The Philippine Entomologist, 2*(3), 205–208.

Basio, R.G., & Basio, L.S. (1971a). On Philippine mosquitoes. IV. A new species of *Armigeres*, *Armigeres azurini* subgenus *Armigeres* (Diptera: Culicidae). *Philippine Journal of Science, 100*(1), 53–56.

Basio, R.G., & Basio, L.S. (1971b). On Philippine mosquitoes. VI. *Tripteroides* (*Tripteroides*) *reiseni*, a new species (Diptera: Culicidae). *Philippine Journal of Science, 100*(2), 103–105.

Basio, R.G., & Reisen, W.K. (1971). On some mosquitoes of Guam, Marianas Islands (Diptera: Culicidae). *The Philippine Entomologist, 2*(1), 57–61.

Bataille, A., Cunningham, A.A., Cedeño, V., Cruz, M., Eastwood, G., Fonseca, D.M., . . . Goodman, S.J. (2009). Evidence for regular ongoing introductions of mosquito disease vectors into the Galápagos Islands. *Proceedings of the Royal Society B: Biological Sciences, 276*(1674), 3769–3775.

Bates, M. (1940). The nomenclature and taxonomic status of the mosquitoes of the *Anopheles maculipennis* complex. *Annals of the Entomological Society of America, 33*(2), 343–356.

Bates, M. (1947). The development and longevity of *Haemagogus* mosquitoes under laboratory conditions. *Annals of the Entomological Society of America, 40*(1), 1–12.

Bates, M. (1949). *The natural history of mosquitoes*. New York, NY: Macmillan.

Bates, M., Beklemishev, W.N., & La Face, L. (1949). Anophelines of the Palearctic region. In M.F. Boyd (Ed.), *Malariology: A comprehensive survey of all aspects of this group of diseases from a global standpoint, by 65 contributors* (pp. 419–442). Philadelphia, PA: Saunders.

Baton, L.A., & Ranford-Cartwright, L.C. (2005). How do malaria ookinetes cross the mosquito midgut wall? *Trends in Parasitology, 21*(1), 22–28.

Bauer, J.H. (1928). The transmission of yellow fever by mosquitoes other than *Aedes aegypti*. *American Journal of Tropical Medicine, 8*(4), 261–282.

Beach, R.F. (1978). The required day number and timely induction of diapause in geographic strains of the mosquito *Aedes atropalpus*. *Journal of Insect Physiology, 24*(5), 449–455.

Beach, R.F., & Craig, G.B., Jr. (1977). Night length measurements by the circadian clock controlling diapause induction in the mosquito *Aedes atropalpus*. *Journal of Insect Physiology, 23*(7), 865–870.

Beales, P.F. (1984). A review of the taxonomic status of *Anopheles sinensis* and its bionomics in relation to malaria transmission. (WHO/MAL/84.1007 [WHO/VBC/84.898]). Geneva, Switzerland: World Health Organization.

Beck, M., Galm, M., Weitzel, T., Fohlmeister, V., Kasiser, A., Arnold, A., & Becker, N. (2003). Preliminary studies on the mosquito fauna of Luxembourg. *European Mosquito Bulletin, 14*, 21–24.

Beckel, W.E. (1954). The identification of adult female *Aedes* mosquitoes (Diptera, Culicidae) of the black-legged group taken in the field at Churchill, Manitoba. *Canadian Journal of Zoology, 32*, 324–330.

Becker, N. (2006). Biological control of mosquitoes: Management of the upper Rhine mosquito population as a model programme. In J. Eilenberg & H.M.T. Hokkanen (Eds.), *An ecological and societal approach to biological control* (pp. 227–245). Dordrecht, The Netherlands: Springer.

Becker, N. (2008). Influence of climate change on mosquito development and mosquito-borne diseases in Europe. *Parasitology Research, 103*(Suppl. 1), 519–528.

Becker, N., & Hoffmann, D. (2011). First record of *Culiseta longiareolata* (Macquart) for Germany. *European Mosquito Bulletin, 29*, 143–150.

Becker, N., Petrić, D., Zgomba, M., Boase, C., Dahl, C., Lane, J., & Kaiser, A. (2003). *Mosquitoes and their control*. New York, NY: Kluwer Academic/Plenum.

Becker, N., Petrić, D., Zgomba, M., Boase, C., Madon, M., Dahl, C., & Kaiser, A. (2010). *Mosquitoes and their control* (2nd ed.). Berlin, Germany: Springer-Verlag.

Becker, N., Jöst, A., & Weitzel, T. (2012a). The *Culex pipiens* complex in Europe. *Journal of the American Mosquito Control Association, 28*(Suppl. 4), 53–67.

Becker, N., Pluskota, B., Kaiser, A., & Schaffner, F. (2012b). Exotic mosquitoes conquer the world. In H. Mehlhorn (Ed.), *Arthropods as vectors of emerging diseases* (Vol. 3, pp. 31–60). New York, NY: Springer.

Becker, T. (1903). Ägyptische Dipteren. *Mitteilungen ans dem Zoologischen Museum in Berlin, 2*, 67–195.

Becker, T. (1908). Dipteren der Kanarischen inseln. *Mitteilungen aus dem Zoologischen Museum in Berlin, 6*, 78–81.

Becker, T. (1910). Dipteren aus Südarabien und von der Insel Sokótra. *Denkschriften der Akademie der Wissenschaften 71*, 131–160.

Beckmann, J.F., & Fallon, A.M. (2013). Detection of the *Wolbachia* protein WPIP0282 in mosquito spermathecae: Implications for cytoplasmic incompatibility. *Insect Biochemistry and Molecular Biology, 43*(9), 867–878.

Beckmann, J.F., Markowski, T.W., Witthuhn, B.A., & Fallon, A.M. (2013). Detection of the *Wolbachia*-encoded DNA binding protein, HU beta, in mosquito gonads. *Insect Biochemistry and Molecular Biology, 43*(3), 272–279.

Beckmann, J.F., Ronau, J.A., & Hochstrasser, M. (2017). A *Wolbachia* deubiquitylating enzyme induces cytoplasmic incompatibility. *Nature Microbiology, 2*(5), 17007.

Beebe, N.W., & Cooper, R.D. (2002). Distribution and evolution of the *Anopheles punctulatus* group (Diptera: Culicidae) in Australia and Papua New Guinea. *International Journal for Parasitology, 32*(5), 563–574.

Beebe, N.W., Maung, J., van den Hurk, A.F., Ellis, J.T., & Cooper, R.D. (2001). Ribosomal DNA spacer genotypes of the *Anopheles bancroftii* group (Diptera: Culicidae) from Australia and Papua New Guinea. *Insect Molecular Biology, 10*(5), 407–413.

Beebe, N.W., Russell, T.L., Burkot, T.R., Lobo, N.F., & Cooper, R.D. (2013). The systematics and bionomics of malaria vectors in the Southwest Pacific. In S. Manguin (Ed.), Anopheles *mosquitoes: New insights into malaria vectors* (pp. 357–394). Rijeka, Croatia: InTech.

Beebe, N.W., Russell, T., Burkot, T.R., & Cooper, R.D. (2015). *Anopheles punctulatus* group: Evolution, distribution, and control. *Annual Review of Entomology, 60*, 335–350.

Beerntsen, B.T., James, A.A., & Christensen, B.M. (2000). Genetics of mosquito vector competence. *Microbiology and Molecular Biology Reviews, 64*(1), 115–137.

Bei-Bienko, G.I.A., & Steyskal, G.C. (1988). *Keys to the insects of the European part of the USSR. Vol. V, Diptera and Siphonaptera. Part I. Zoologicheskii (Akademii a nauk SSSR), Opredeliteli po fauna SSSR, no. 100.* [in English]. Washington, DC: Smithsonian Institution/National Science Foundation.

Beidas, M.F., & Gillies, M.T. (1980). The egg of *Anopheles (Cellia) culicifacies adenensis* Christophers. *Mosquito Systematics, 12*(2), 172–174.

Beier, J.C. (1996). Frequent blood-feeding and restrictive sugar-feeding behavior enhance the malaria vector potential of *Anopheles gambiae* sl and *An. funestus* (Diptera: Culicidae) in western Kenya. *Journal of Medical Entomology, 33*(4), 613–618.

Beier, J.C., Berry, W.J., & Craig, G.B., Jr. (1982). Horizontal distribution of adult *Aedes triseriatus* (Diptera: Culicidae) in relation to habitat structure, oviposition, and other mosquito species. *Journal of Medical Entomology, 19*(3), 239–247.

Beier, J.C., Kenawy, M.A., Said, S.E., & Merdan, A.I. (1986). Vector potential of culicine mosquitoes in Faiyum Governorate, Egypt. *Journal of the American Mosquito Control Association, 2*(2), 164–167.

Beier, J.C., Copeland, R.S., Oyaro, C., Masinya, A., Odago, W.O., Oduor, S., . . . Roberts, C.R. (1990). *Anopheles gambiae* complex egg stage survival in dry soil from larval development sites in western Kenya. *Journal of the American Mosquito Control Association, 6*(1), 105–109.

Bejarano, E.E. (2003). Occurrence of the malaria vector *Anopheles albimanus* Wiedemann (Diptera: Culicidae) in Isla Fuerte, Colombia. *Neotropical Entomology, 32*(3), 517–518.

Beketov, M.A., & Liess, M. (2007). Predation risk perception and food scarcity induce alterations of life-cycle traits of the mosquito *Culex pipiens. Ecological Entomology, 32*(4), 405–410.

Bekku, H. (1956). Studies on the *Culex pipiens* group of Japan. I. Comparative studies on the morphology of those obtained from various localities in the Far East. *Nagasaki Medical Journal, 31*(11), 956–966.

Belkin, J.N. (1945). *Anopheles nataliae*, a new species from Guadalcanal. *Journal of Parasitology, 31*(5), 315–318.

Belkin, J.N. (1950a). Mosquitoes of the genus *Tripteroides* in the Solomon Islands. *Proceedings of the United States National Museum, 100*(3262), 201–274.

Belkin, J.N. (1950b). A revised nomenclature for the chaetotaxy of the mosquito larva (Diptera: Culicidae). *American Midland Naturalist, 44*(3), 678–698.

Belkin, J.N. (1952). The homology of the chaetotaxy of immature mosquitoes and a revised nomenclature for the chaetotaxy of the pupa (Diptera, Culicidae). *Proceedings of the Entomological Society of Washington, 54*(3), 115–130.

Belkin, J.N. (1953a). Mosquitoes of the genus *Uranotaenia* in the Solomon Islands (Diptera: Culicidae). *Pacific Science* (Honolulu), *7*(3), 312–391.

Belkin, J.N. (1953b). Corrected interpretations of some elements of the abdominal chaetotaxy of the mosquito larva and pupa (Diptera, Culicidae). *Proceedings of the Entomological Society of Washington, 55*(6), 318–324.

Belkin, J.N. (1954a). Remarks on the larval chaetotaxy of the subgenus *Melanoconion* (Diptera: Culicidae). *Entomological News, 65*(2), 29–32.

Belkin, J.N. (1954b). The dorsal hairless setal ring of mosquito pupae (Diptera: Culicidae). *The Pan-Pacific Entomologist, 30*(3), 227–230.

Belkin, J.N. (1955). The *Tripteroides caledonica* complex of mosquitoes in Melanesia (Diptera: Culicidae). *Pacific Science* (Honolulu), *9*(2), 221–246.

Belkin, J.N. (1956). Mosquito studies (Diptera: Culicidae). IV. The mosquitoes of the Robinson-Peabody Museum of Salem expedition to the southwest Pacific, 1956. *Contributions of the American Entomological Institute, 1*(4), 11–34.

Belkin, J.N. (1960). Innervation as a criterion of homology of the elements of the larval and pupal chaetotaxy of mosquitoes (Diptera, Culicidae). *Proceedings of the Entomological Society of Washington, 62*(3), 197.

Belkin, J.N. (1962). *The mosquitoes of the South Pacific (Diptera, Culicidae)* (Vols. 1–2). Berkeley: University of California Press.

Belkin, J.N. (1964). The adults and pupa of *Culex (C.) kesseli* from Tahiti and remarks on the *atriceps* group (Diptera: Culicidae). *Annals of the Entomological Society of America, 57*(2), 236–239.

Belkin, J.N. (1968a). Mosquito studies (Diptera, Culicidae). VII. The Culicidae of New Zealand. *Contributions of the American Entomological Institute, 3*(1), 1–182.

Belkin, J.N. (1968b). Mosquito studies (Diptera, Culicidae). IX. The type specimens of New World mosquitoes in European museums. *Contributions of the American Entomological Institute, 3*(4), 1–69.

Belkin, J.N. (1969a). *Culex (Melanoconion) annulipes* invalid. *Mosquito Systematics, 1*(3), 68–70.

Belkin, J.N. (1969b). The problem of the identity of the species of *Culex (Melanoconion)* related to *Opisthopus. Mosquito Systematics, 1*(2), 26–27.

Belkin, J.N. (1970a). Corrected type localities for *Wyeomyia abia* D. and K., 1908 and *W. fratercula* D. and K., 1906. *Mosquito Systematics, 2*(2), 57–58.

Belkin, J.N. (1970b). *Culex (Melanoconion) aikenii* (A. & R., 1906) a *nomen dubium*; *ocossa* D. & K., 1919 and *panocossa* Dyar, 1923 both valid. *Mosquito Systematics Newsletter, 2*(2), 59–60.

Belkin, J.N. (1971a). Mosquito types in East Germany. *Mosquito Systematics Newsletter, 3*(2), 31.

Belkin, J.N. (1971b). Type locality restriction for *Wyeomyia schnusei. Mosquito Systematics Newsletter, 3*(2), 26.

Belkin, J.N. (1977a). *Aedes (Ochlerotatus) pix* Martini 1935, a synonym of *Aedes (O.) taeniorhynchus* (Wiedemann 1821). *Mosquito Systematics, 9*(4), 535.

Belkin, J.N. (1977b). *Quinquefasciatus* or *fatigans* for the tropical (southern) house mosquito (Diptera: Culicidae). *Proceedings of the Entomological Society of Washington, 79*(1), 45–52.

Belkin, J.N. (1977c). *Toxorhynchites (Lyn.) gerbergi*, a new species from the southern Lesser Antilles. *Mosquito Systematics, 9*(3), 329–332.

Belkin, J.N., & Heinemann, S.J. (1971). *Aedes vexans* in Guatemala. *Mosquito Systematics Newsletter, 3*(2), 27.

Belkin, J.N., & Heinemann, S.J. (1972). A tentative annotated list of the Culicidae of the island of Hispaniola. *Mosquito Systematics, 4*(2), 63–72.

Belkin, J.N., & Heinemann, S.J. (1973). Collection records of the project "Mosquitoes of Middle America" 1. Introduction; Dominican Republic (RDO). *Mosquito Systematics, 5*(3), 201–220.

Belkin, J.N., & Heinemann, S.J. (1975a). Collection records of the project "Mosquitoes of Middle America" 2. Puerto Rico (PR, PRA, PRX) and Virgin Is. (VI, VIA). *Mosquito Systematics, 7*(3), 269–296.

Belkin, J.N., & Heinemann, S.J. (1975b). Collection records of the project "Mosquitoes of Middle America" 3. Bahama Is. (BAH), Cayman Is. (CAY), Cuba (CUB), Haiti (HAC, HAR, HAT) and Lesser Antilles (LAR). *Mosquito Systematics, 7*(4), 367–393.

Belkin, J.N., & Heinemann, S.J. (1975c). *Psorophora (Janthinosoma) mathesoni*. sp. nov. for "*varipes*" of the Southeastern U.S.A. *Mosquito Systematics, 7*(4), 363–366.

Belkin, J.N., & Heinemann, S.J. (1976a). Collection records of the project "Mosquitoes of Middle America" 6. Southern Lesser Antilles: Barbados (BAR), Dominica (DOM), Grenada (GR, GRR), St. Lucia (LU), St. Vincent (VT). *Mosquito Systematics, 8*(3), 237–297.

Belkin, J.N., & Heinemann, S.J. (1976b). Collection records of the project "Mosquitoes of Middle America" 4. Leeward Islands: Anguilla (ANG), Antigua (ANT), Barbuda (BAB), Montserrat (MNT), Nevis (NVS), St.Kitts (KIT). *Mosquito Systematics, 8*(2), 123–162.

Belkin, J.N., & Heinemann, S.J. (1976c). Collection records of the project "Mosquitoes of Middle America" 5. French West Indies: Guadeloupe (FWI) and Martinique (FWIM, MAR). *Mosquito Systematics, 8*(2), 163–193.

Belkin, J.N., & Hogue, C.L. (1959). A review of the crabhole mosquitoes of the genus *Deinocerites* (Diptera, Culicidae). *University of California Publications in Entomology, 14*(6), 411–458.

Belkin, J.N., & McDonald, W.A. (1956a). A change to the scientific name of the western tree-hole mosquito. *California Vector Views, 3*(10), 54.

Belkin, J.N., & McDonald, W.A. (1956b). A population of *Uranotaenia anhydor* from Death Valley, with descriptions of all stages and discussion of the complex (Diptera, Culicidae). *Annals of the Entomological Society of America, 49*(2), 105–132.

Belkin, J.N., & McDonald, W.A. (1956c). *Aedes sierrensis* (Ludlow, 1905), a change in name for the western tree-hole mosquito of the pacific slope (Diptera: Culicidae). *Proceedings of the Entomological Society of Washington, 58*(6), 344.

Belkin, J.N., & McDonald, W.A. (1957). A new species of *Aedes (Ochlerotatus)* from tree-holes in southern Arizona and a discussion of the *varipalpus* complex (Diptera: Culicidae). *Annals of the Entomological Society of America, 50*(2), 179–191.

Belkin, J.N., & Schlosser, R.J. (1944). A new species of *Anopheles* from the Solomon Islands. *Journal of the Washington Academy of Sciences, 34*(8), 268–273.

Belkin, J.N., Knight, K.L., & Rozeboom, L.E. (1945). Anopheline mosquitoes of the Solomon Islands and New Hebrides. *The Journal of Parasitology, 31*(4), 241–265.

Belkin, J.N., Ehmann, N., & Heid, G. (1951). Preliminary field observations on the behaviour of the adults of *Anopheles franciscanus* McCracken in Southern California. *Mosquito News, 11*, 23–31.

Belkin, J.N., Schick, R.X., & Heinemann, S.J. (1965a). Mosquito studies (Diptera: Culicidae). V. Mosquitoes originally described from Middle America. *Contributions of the American Entomological Institute, 1*(5), 1–95.

Belkin, J.N., Schick, R.X., Galindo, P., & Aitken, T.H.G. (1965b). Mosquito studies (Diptera, Culicidae) I. A project for a systematic study of the mosquitoes of Middle America. II. Methods for the collection, rearing and preservation of mosquitoes. *Contributions of the American Entomological Institute, 1*(2), 1–78.

Belkin, J.N., Schick, R.X., & Heinemann, S.J. (1966). Mosquito studies (Diptera: Culicidae). VI. Mosquitoes originally described from North America. *Contributions of the American Entomological Institute, 1*(6), 1–39.

Belkin, J.N., Schick, R.X., Galindo, P., & Aitken, T.H.G. (1967). Estudios sobre mosquitos. Ia. Un proyecto para un estudio sistemático de los mosquitos de Meso-América. *Contributions of the American Entomological Institute, 1*(2a), 1–88.

Belkin, J.N., Schick, R.X., & Heinemann, S.J. (1968). Mosquito studies (Diptera, Culicidae). XI. Mosquitoes originally described from Argentina, Bolivia, Chile, Paraguay, Peru, and Uruguay. *Contributions of the American Entomological Institute, 4*(1), 9–29.

Belkin, J.N., Heinemann, S.J., & Page, W.A. (1970). The Culicidae of Jamaica. Mosquito studies. XXI. *Contributions of the American Entomological Institute, 6*(1), 458.

Belkin, J.N., Schick, R.X., & Heinemann, S.J. (1971). Mosquito studies (Diptera, Culicidae). XXV. Mosquitoes originally described from Brazil. *Contributions of the American Entomological Institute, 7*(5), 1–64.

Bellamy, R.E. (1939). An anopheline from inland Georgia resembling the brackish-water race of *Anopheles crucians*. *Journal of Parasitology, 25*(2), 186.

Bellamy, R.E. (1956). An investigation of the taxonomic status of *Anopheles perplexens* Ludlow, 1907. *Annals of the Entomological Society of America, 49*(6), 515–529.

Bellamy, R.E., & Repass, R.P. (1950). Notes of the ova of *Anopheles georgianus* King. *Journal of the National Malaria Society, 9*, 84–88.

Bellardi, L. (1859). *Saggio di ditterologia messicana. Part I.* Torino.

Bellardi, L. (1862). *Saggio di ditterologia messicana*. Appendice. Torino.

Belmont, M., Cazzamali, G., Williamson, M., Hauser, H., & Grimmelikhuijzen, C.J.P. (2006). Identification of four evolutionarily related G protein-coupled receptors from the malaria mosquito *Anopheles gambiae*. *Biochemical and Biophysical Research Communications, 344*(1), 160–165.

Belton, P. (1980). The first record of *Aedes togoi* (Theo.) in the United States—aboriginal or ferry passenger? *Mosquito News, 40*(4), 624–626.

Belton, P. (1982). The cuticular vestiture of larvae of *Aedes communis* and *A. nevadensis* (Diptera: Culicidae). *Canadian Journal of Research, 60*(7), 1642–1646.

Beltrán-Aguilar, A., Ibáñez-Bernal, S., Mendoza-Palmero, F., Sandoval-Ruiz, C.A., & Hernández-Xoliot, R.A. (2011). Taxonomy and distribution of the anopheline mosquitoes in the state of Veracruz, Mexico (Diptera: Culicidae, Anophelinae). *Acta zoologica Mexicana* (n.s.), *27*(3), 601–755.

Benabadji, M., & Larrouy, G. (1968). Observations concernant la faune culicidienne du massif du hoggar. *Bulletin de la Société d'Histoire Naturelle de l'Afrique du Nord, 58*(for 1967)(1–2), 15–23.

Ben Cheikh, R., Berticat, C., Berthomieu, A., Pasteur, N., Ben Cheikh, H., & Weill, M. (2008). Characterization of a novel high-activity esterase in Tunisian populations of the mosquito *Culex pipiens*. *Journal of Economic Entomology, 101*(2), 484–491.

Ben Cheikh, R., Berticat, C., Berthomieu, A., Pasteur, N., Ben Cheikh, H., & Weill, M. (2009). Genes conferring resistance to organophosphorus insecticides in *Culex pipiens* (Diptera: Culicidae) in Tunisia. *Journal of Medical Entomology, 46*(3), 523–530.

Benedict, M.Q., & Robinson, A.S. (2003). The first releases of transgenic mosquitoes: An argument for the sterile insect technique. *Trends in Parasitology, 19*(8), 349–355.

Benedict, M.Q., & Seawright, J.A. (1987). Changes in pigmentation in mosquitoes (Diptera: Culicidae) in response to color of environment. *Annals of the Entomological Society of America, 80*(1), 55–61.

Benedict, M.Q., Levine, R.S., Hawley, W.A., & Lounibos, L.P. (2007). Spread of the tiger: Global risk of invasion by the mosquito *Aedes albopictus. Vector-Borne and Zoonotic Diseases, 7*(1), 76–85.

Benedict, M.Q., Hood-Nowotny, R.C., Howell, P.I., & Wilkins, E.E. (2009). Methylparaben in *Anopheles gambiae* sl sugar meals increases longevity and malaria oocyst abundance but is not a preferred diet. *Journal of Insect Physiology, 55*(3), 197–204.

Benedict, M.Q., Sandve, S.R., Wilkins, E.E., & Roberts, J.M. (2010). Relationship of larval desiccation to *Anopheles gambiae* Giles and *An. arabiensis* Patton survival. *Journal of Vector Ecology, 35*(1), 116–123.

Benelli, G. (2015). The best time to have sex: Mating behaviour and effect of daylight time on male sexual competitiveness in the Asian tiger mosquito, *Aedes albopictus* (Diptera: Culicidae). *Parasitology Research, 114*(3), 887–894.

Benet, A., Mai, A., Bockarie, F., Lagog, M., Zimmerman, P., Alpers, M.P., Reeder, J.C., & M.J. Bockarie. (2004). Polymerase chain reaction diagnosis and the changing pattern of vector ecology and malaria transmission dynamics in Papua New Guinea. *The American Journal of Tropical Medicine and Hygiene, 71*(3), 277–284.

Benjaphong, N., & Rattanarithikul, R. (1991). *Zeugnomyia gracilis* Leicester (Diptera: Culicidae), a new genus and species occurrence recorded from Thailand. *Mosquito Systematics, 23*(1), 53.

Bennett, K.L., Linton, Y.-M., Shija, F., Kaddumukasa, M., Djouaka, R., Misinzo, G., . . . Walton, C. (2015). Molecular differentiation of the African Yellow Fever vector *Aedes bromeliae* (Diptera: Culicidae) from its sympatric non-vector sister species, *Aedes lilii. PLOS Neglected Tropical Diseases, 9*(12), e0004250.

Bennett, K.L., Kaddumukasa, M., Shija, F., Djouaka, R., Misinzo, G., Lutwama, J., . . . Walton, C. (2018). Comparative phylogeography of *Aedes* mosquitoes and the role of past climatic change for evolution in Africa. *Ecology and Evolution, 8*(5), 3019–3036.

Benoit, J.B., & Denlinger, D.L. (2007). Suppression of water loss during adult diapause in the northern house mosquito, *Culex pipiens. Journal of Experimental Biology, 210*(2), 217–226.

Benoit, J.B., & Denlinger, D.L. (2010). Meeting the challenges of on-host and off-host water balance in blood-feeding arthropods. *Journal of Insect Physiology, 56*(10), 1366–1376.

Benoit, J.B., Lopez-Martinez, G., Phillips, Z.P., Patrick, K.R., & Denlinger, D.L. (2010a). Heat shock proteins contribute to mosquito dehydration tolerance. *Journal of Insect Physiology, 56*(2), 151–156.

Benoit, J.B., Patrick, K.R., Desai, K., Hardesty, J.J., Krause, T.B., & Denlinger, D.L. (2010b). Repeated bouts of dehydration deplete nutrient reserves and reduce egg production in the mosquito *Culex pipiens. Journal of Experimental Biology, 213*(16), 2763–2769.

Benoit, J.B., Hansen, I.A., Szuter, E.M., Drake, L.L., Burnett, D.L., & Attardo, G.M. (2014). Emerging roles of aquaporins in relation to the physiology of blood-feeding arthropods. *Journal of Comparative Physiology B, 184*(7), 811–825.

Benoit, P.L.G. (1964). *Diptera—Culicidae. Observations biologiques et écologiques*, pp. 297–320. Mission de Zoologie médicale au Maniema (Congo, Leopoldville) (P.L.G. Benoit, 1959): Résultats scientifiques. Musée Royal de l'Afrique centrale -Tervuren, Belgique. Annales - Serie in-8° - Sciences Zoologiques - n° 132.

Bentley, C.A. (1902). *Anopheles* mosquitoes in Tezpur, Assam. *Indian Medical Gazette, 37*(1), 15–16.

Benzon, G.L., & Apperson, C.S. (1988). Reexamination of chemically mediated oviposition behaviour in *Aedes aegypti* (L.) (Diptera: Culicidae). *Journal of Medical Entomology, 25*(3), 158–164.

Bergo, E.S., Souto, R.N.P., Galardo, A.K.R., Nagaki, S.S., Calado, D.C., & Sallum, M.A.M. (2007). Systematic notes on *Anopheles* Meigen (Diptera: Culicidae) species in the state of Amapá, Brazil. *Memórias do Instituto Oswaldo Cruz, 102*(3), 373–376.

Berlin, O.G.W. (1969a). Mosquito studies (Diptera, Culicidae). XII. A revision of the Neotropical subgenus *Howardina* of *Aedes. Contributions of the American Entomological Institute, 4*(2), 1–190.

Berlin, O.G.W. (1969b). Mosquito studies (Diptera, Culicidae). XIV. Redescription of *Toxorhynchites* (*T.*) *gigantulus* from the Philippines. *Contributions of the American Entomological Institute, 4*(4), 21–26.

Berlin, O.G.W. (1969c). Mosquito studies (Diptera, Culicidae). XVIII. The subgenus *Micraedes* of *Culex. Contributions of the American Entomological Institute, 5*(1), 21–63.

Berlin, O.G.W. (1974). A new species of bromeliad-breeding *Culex* (*Micraedes*) from Mexico (Diptera, Culicidae). *Mosquito Systematics, 6*(4), 273–278.

Berlin, O.G.W. (1975). Description of the male and the pupa of *Aedes* (*Howardina*) *lorraineae* Berlin, with a note on the Quadrivittatus Group (Diptera, Culicidae). *Mosquito Systematics, 7*(2), 127–131.

Berlin, O.G.W., & Belkin, J.N. (1980). Mosquito studies (Diptera, Culicidae). 36. Subgenera *Aedinus*, *Tinolestes* and *Anoedioporpa* of *Culex. Contributions of the American Entomological Institute, 17*(2), 1–104.

Berlin, O.G.W., & Pandian, R.S. (1973a). Annotated list of nominal species of culicines of India. I. Introduction. Genus *Toxorhynchites. Mosquito Systematics, 5*(4), 274–276.

Berlin, O.G.W., & Pandian, R.S. (1973b). Description of the egg of *Culex* (*Lutzia*) *fuscanus* Wiedemann (Diptera, Culicidae). *Mosquito Systematics, 5*(3), 227–229.

Berlin, O.G.W., & Subramanian, M. (1974). Description of the egg of *Culex* (*Culex*) *gelidus* Theobald with a note on development (Diptera, Culicidae). *Mosquito Systematics, 6*(4), 263–265.

Bernáth, B., Horváth, G., & Meyer-Rochow, V.B. (2012). Polarotaxis in egg-laying yellow fever mosquitoes *Aedes* (*Stegomyia*) *aegypti* is masked due to infochemicals. *Journal of Insect Physiology, 58*(7), 1000–1006.

Bernier, U.R., Kline, D.L., Posey, K.H., Booth, M.M., Yost, R.A., & Barnard, D.R. (2003). Synergistic attraction of *Aedes aegypti* (L.) to binary blends of L-lactic acid and acetone, dichloromethane, or dimethyl disulfide. *Journal of Medical Entomology, 40*(5), 653–656.

Bernier, U.R., Kline, D.L., Allan, S.A., & Barnard, D.R. (2007a). Laboratory comparison of *Aedes aegypti* attraction to human odors and to synthetic human odor compounds and blends. *Journal of the American Mosquito Control Association, 23*(3), 288–293.

Bernier, U.R., Kline, D.L., & Posey, K.H. (2007b). Human emanations and related natural compounds that inhibit mosquito host-finding abilities. In M. Debboun, S.P. Frances, & D. Strickman (Eds.), *Insect repellents: Principles, methods and uses* (pp. 77–102). Boca Raton, FL: CRC Press.

Berti, J., Guzmán, H., Estrada, Y., & Ramírez, R. (2015). New records of mosquitoes (Diptera: Culicidae) from Bolívar State in South Eastern Venezuela, with 27 new species for the state and 5 of them new in the country. *Frontiers in Public Health, 2*, 10.

Berticat, C., Boquien, G., Raymond, M., & Chevillon, C. (2002). Insecticide resistance genes induce a mating competition cost in *Culex pipiens* mosquitoes. *Genetics Research, 79*(1), 41–47.

Bertone, M.A., Courtney, G.W., & Wiegmann, B.M. (2008). Phylogenetics and temporal diversification of the earliest true flies (Insecta: Diptera) based on multiple nuclear genes. *Systematic Entomology, 33*(4), 668–687.

Bertram, D.S. (1971). Mosquitoes of British Honduras, with some comments on malaria, and on arbovirus antibodies in man and equines. *Transactions of the Royal Society of Tropical Medicine and Hygiene, 65*(6), 742–762.

Bertrand, H. (1954). Les insectes aquatiques d'Europe. genres larvaes. nymphs, imagos. Vol. 2. *Trichoptères, Lépidoptères, Diptères, Hyménoptères. Encyclopedie Entomologique.* Série A. *Travaux généraux*, 31. Paul Lechevalier, Paris.

Besansky, N.J., & Fahey, G.T. (1997). Utility of the *white* gene in estimating phylogenetic relationships among mosquitoes (Diptera: Culicidae). *Molecular Biology and Evolution, 14*(4), 442–454.

Beyenbach, K.W. (2003). Transport mechanisms of diuresis in Malpighian tubules of insects. *Journal of Experimental Biology, 206*(21), 3845–3856.

Beyenbach, K.W. (2012). A dynamic paracellular pathway serves diuresis in mosquito Malpighian tubules. *Annals of the New York Academy of Sciences, 1258*(1), 166–176.

Beyenbach, K.W., & Masia, R. (2002). Membrane conductances of principal cells in Malpighian tubules of *Aedes aegypti. Journal of Insect Physiology, 48*(3), 375–386.

Bezzi, M. (1906). Ditteri Eritrei raccolti dal Dott. Andreini e dal Prof. Tellini. *Bollettino della Societa Entomologica Italiana, 37*(for 1905), 195–304.

Bhat, H.R., & Kulkarni, S.M. (1971). A report on the occurrence of *Culex* (*Neoculex*) *hortensis* Ficalbi, 1889, in Ladakh, India. *Oriental Insects, 5*(4), 583–584.

Bhat, U.K. (1976). Karyotypes of three species of mosquitoes from India. *Current Science, 45*(7), 266–267.

Bhatnagar, V.N., Bhatia, B.L., & Krishnan, K.S. (1958). On certain morphological abnormalities noted in *A. pallidus* Theobald, 1901 and *A. stephensi stephensi* Liston, 1901. *Indian Journal of Malariology, 12*(1), 39–42.

Bhatt, R.M., Sharma, R.C., Yadav, R.S., & Sharma, V.P. (1989). Resting of mosquitoes in outdoor pit shelters in Kheda district, Gujarat. *Indian Journal of Malariology, 26*(2), 75–81.

Bhattacharyya, D.R., Tewari, S.C., Dutta, P., & Mahanta, J. (1995). Description of the larva and pupa of *Aedes* (*Aedimorphus*) *nigrostriatus* (Diptera: Culicidae). *Mosquito Systematics, 27*(3), 191–196.

Bhattacharyya, D.R., Prakash, A., Tewari, S.C., Mohapatra, P.K., & Mahanta, J. (2000). *Armigeres joloensis* (Diptera: Culcidae), a rare mosquito in upper Assam: First report from India. *Entomon, 25*(1), 63–65.

Bhattacharyya, D.R., Mohapatra, A., & Mahanta, J. (2003). Occurrence of *Culex* (*Lophoceraomyia*) *quadripalpis, Culex* (*Lophoceraomyia*) *mammilifer*, and *Uranotaenia* (*Pseudoficalbia*) *novobscura* in Assam, India. *Journal of the American Mosquito Control Association, 19*(1), 13–14.

Bhattacharyya, D.R., Prakash, A., Mohapatra, P.K., & Mahanta, J. (2004). *Uranotaenia dibrugarhensis*, a new species in subgenus *Pseudoficalbia* from Assam, India. *Journal of the American Mosquito Control Association, 20*(1), 1–5.

Bhattacharyya, D.R., Tewari, S.C., Prakash, A., Mohapatra, P.K., & Mahanta, J. (2004). *Verrallina* (*Neomacleaya*) *assamensis*, a new species from Assam, India. *Journal of the American Mosquito Control Association, 20*(2), 115–120.

Bhattacharyya, D.R., Prakash, A., Mohapatra, P.K., & Sarma, D.K. (2009). *Armigeres* (*Armigeres*) *mahantai*, a new mosquito species from India. *Journal of the American Mosquito Control Association, 25*(1), 1–5.

Bhombore, S.R., Sitaraman, N.L., & Achuthan, C. (1954). A first record of *Anopheles theobaldi* Giles, 1901, from Mysore State. *Indian Journal of Malariology, 8*(1), 45–46.

Bhombore, S.R., Sitaraman, N.L., & Achuthan, C. (1956). Studies on the bionomics of *Anopheles fluviatilis* in Mysore State, India. II. Bionomics in Western Hill-tracts, Mysore State. *Indian Journal of Malariology, 10*(1), 23–33.

Bian, G., Raikhel, A.S., & Zhu, J. (2008). Characterization of a juvenile hormone-regulated chymotrypsin-like serine protease gene in *Aedes aegypti* mosquito. *Insect Biochemistry and Molecular Biology, 38*(2), 190–200.

Bianchi, U. (1968). Sulla tassonomia di *Anopheles labranchiae* et *Anopheles atroparvus. Rivista di malariologia, 29*(3), 221–226.

Bick, G.H. (1951). The ecology of the mosquito larvae of New Guinea. *Pacific Science, 5*(4), 392–431.

Bickley, W.E. (1945). The anal gills of mosquito larvae. *Mosquito News, 5*(1), 18.

Bickley, W.E. (1957). Notes on the distribution of mosquitoes in Maryland and Virginia. *Mosquito News, 17*(1), 22–25.

Bickley, W.E. (1976). Notes on the distribution of Alaskan mosquitoes. *Mosquito Systematics, 8*(3), 232–236.

Bickley, W.E. (1980). Notes on the status of *Aedes cinereus hemiteleus* Dyar. *Mosquito Systematics, 12*(3), 367–370.

Bickley, W.E. (1981). Notes on the geographical distribution of *Aedes canadensis mathesoni. Mosquito Systematics, 13*(2), 150–152.

Bickley, W.E. (1984). Notes on the *Psorophora confinnis* complex. *Mosquito Systematics, 16*(2), 162–167.

Bickley, W.E., & Harrison, B.A. (1989). Separation of variable *Culex territans* specimens from other *Culex* (*Neoculex*) in North America. *Mosquito Systematics, 21*(3), 188–196.

Bickley, W.E., & Harrison, B.A. (1992). Description of the pupa of *Aedes* (*Ochlerotatus*) *grossbecki* Dyar and Knab (Diptera: Culicidae). *Mosquito Systematics, 23*(for 1991)(3), 205–208.

Bidlingmayer, W.L. (1994). How mosquitoes see traps: Role of visual responses. *Journal of the American Mosquito Control Association, 10*(2), 272–279.

Bigot, J. (1857). Dipteros. In R. de la Sagra (Ed.), *Historia fisica, politica y natural de la Isla de Cuba. Secunda parte. Tomo VII. Crustaceos, Aragnides é Insectos.* Paris, France: Bertrand.

Bigot, J. (1859). Diptères de Madagascar. Première partie. *Annales de la Société Entomologique de France, 7*, 115–135.

Bigot, J. (1861). Trois diptères nouveaux de la Corse. *Annales de la Société Entomologique de France, 1*, 227–229.

Bigot, J.-M.-F. (1888). *Énumération des diptères recueillis en Tunisie dans la Mission de 1884 par M. Valéry Mayet, membre de la Mission de l'Exploration Scientifique de la Tunisie, et Description des Espèces Nouvelles, par J.-M.-F. Bigot.* Exploration Scientifique de la Tunisie series. Paris, France: Imprimerie Nationale.

Bigot, J.-M.-F. (1889). Bulletin entomologique. Bigot envoie, de Quincy, une note sur une nouvelle espècie du genre *Culex. Annales de la Société Entomologique de France, 9*, cxxii.

Billberg, G.J. (1820). *Enumeratio insectorum in Museo Gust. Joh. Billberg.* [Holmiae]: Typis Gadelianis.

Black, W.C., IV. (2004). Learning to use *Ochlerotatus* is just the beginning. *Journal of the American Mosquito Control Association, 20*(2), 215–216.

Black, W.C., IV, Hawley, W.A., Rai, K.S., & Craig, G.B. (1988). Breeding structure of a colonizing species: *Aedes albopictus* (Skuse) in peninsular Malaysia and Borneo. *Heredity, 61*(3), 439–446.

Black, W.C., IV, Karamjit, S.R., Turco, B.J. & Arroyo, D.C. (1989). Laboratory study of competition between United States strains of *Aedes albopictus* and *Aedes aegypti*. (Diptera: Culicidae). *Journal of Medical Entomology, 26*(4), 260–271.

Blanchard, C. (1918). Sur un nouveau type larvaire du groupe des Anophélines. *Bulletin de la Société de pathologie exotique, 11*(7), 669–677.

Blanchard, E. (1852). Orden. IX. Dipteros. In Gay, C. (Ed.), *Historia fisica y politica de Chile. Zoologia, 7*, 327–468.

Blanchard, R. (1901). Observations sur quelques moustiques. *Comptes rendus des séances de la Société de Biologie, 53*, 1045–1046.

Blanchard, R. (1902). Nouvelle note sur les moustiques. *Société de Biologie, 54*, 793–795.
Blanchard, R. (1904). Notes et informations. *Archives de parasitologie, 8*, 478.
Blanchard, R. (1905). *Les moustiques, histoire naturelle et médicale.* Paris, France: F. R. de Rudeval, Imprimeur-Editeur.
Blasidell, F.E. (1906). Notes and description of the larva of *Culex varipalpus* Coq. *Entomological News and Proceedings of the Entomological Section, Academy of Natural Sciences of Philadelphia, 17*(4), 107–110.
Blaustein, L., Blaustein, J., & Chase, J. (2005). Chemical detection of the predator *Notonecta irrorate* by ovipositing *Culex* mosquitoes. *Journal of Vector Ecology, 30*(2), 299–301.
Bloch, M.E. (1776). Beitrag zur Naturgeschichte des Kopals. *Beschaftigungen der Berlinischen Gesellschaft Naturforschender Freunde, 2*, 91–96.
Blow, J.A., Turell, M.J., Walker, E.D., & Silverman, A.L. (2002). Post-bloodmeal diuretic shedding of hepatitus B virus by mosquitoes (Diptera: Culicidae). *Journal of Medical Entomology, 39*(4), 605–612.
Boes, K.E., Ribeiro, J.M., Wong, A., Harrington, L.C., Wolfner, M.F., & Sirot, L.K. (2014). Identification and characterization of seminal fluid proteins in the Asian tiger mosquito, *Aedes albopictus. PLOS Neglected Tropical Diseases, 8*(6), e2946.
Boëte, C., & Beisel, U. (2013). Transgenic mosquitoes for malaria control: From the bench to the public opinion survey. In S. Manguin (Ed.), Anopheles *mosquitoes: New insights into malaria vectors* (pp. 797–813). Rijeka, Croatia: InTech.
Bohart, R.M. (Ed.) (1945). *A synopsis of the Philippine mosquitoes. US NAVMED* (Bureau of Medicine and Surgery, Navy Department), 580.
Bohart, R.M. (1946a). A key to the Chinese culicine mosquitoes. *US NAVMED* (Bureau of Medicine and Surgery, Navy Department), *961*, 1–23.
Bohart, R.M. (1946b). New species of mosquitoes from the Marianas and Okinawa (Diptera, Culicidae). *Proceedings of the Biological Society of Washington, 59*, 39–46.
Bohart, R.M. (1948). Differentiation of larvae and pupae of *Aedes dorsalis* and *Aedes squamiger* (Diptera, Culicidae). *Proceedings of the Entomological Society of Washington, 50*(8), 216–218.
Bohart, R.M. (1949). The subgenus *Neoculex* in America North of Mexico (Diptera, Culicidae). *Annals of the Entomological Society of America, 41*(for 1948)(3), 330–345.
Bohart, R.M. (1950a). Observations on snow mosquitoes in California (Diptera: Culicidae). *The Pan-Pacific Entomologist, 26*(3), 111–118.
Bohart, R.M. (1950b). A new species of *Orthopodomyia* from California (Diptera, Culicidae). *Annals of the Entomological Society of America, 43*(3), 399–404.
Bohart, R.M. (1953). A new species of *Culex* and notes on other species of mosquitoes from Okinawa (Diptera, Culicidae). *Proceedings of the Entomological Society of Washington, 55*(4), 183–188.
Bohart, R.M. (1954). Identification of first stage larvae of California *Aedes* (Diptera, Culicidae). *Annals of the Entomological Society of America, 47*(2), 355–366.
Bohart, R.M. (1956a). Identification and distribution of *Aedes melanimon* and *Aedes dorsalis* in California. *Proceedings and Papers of the Annual Conference of the California Mosquito Control Association, 24*, 81–83.
Bohart, R.M. (1956b). New species of mosquitoes from the southern Ryukyu Islands. *Bulletin of the Brooklyn Entomological Society, 51*(2), 29–34.
Bohart, R.M. (1957). *Insects of Micronesia Diptera: Culicidae.* 12(for 1956)(1). Honolulu, HI: Bernice P. Bishop Museum.
Bohart, R.M., & Farner, D.S. (1944). New culicine mosquitoes from the Philippine Islands (Diptera, Culicidae). *Proceedings of the Biological Society of Washington, 57*, 69–74.
Bohart, R.M., & Ingram, R.L. (1946a). Four new species of mosquitoes from Okinawa (Diptera: Culicidae). *Journal of the Washington Academy of Sciences, 36*(2), 46–52.
Bohart, R.M., & Ingram, R.L. (1946b). Mosquitoes of Okinawa and islands in the Central Pacific. In *US NAVMED* (Bureau of Medicine and Surgery, Navy Department).
Bohart, R.M., & Washino, R.K. (1957). Differentiation of second and third stage larvae of California *Culex* (Diptera: Culicidae). *Annals of the Entomological Society of America, 50*, 459–463.
Bohart, R.M., & Washino, R.K. (1978). *Mosquitoes of California* (3rd ed.). Berkeley: University of California Agriculture and Natural Resources.
Bohbot, J.D., & Dickens, J.C. (2012). Selectivity of odorant receptors in insects. *Frontiers in Cellular Neuroscience, 6*, Article 29.
Bohbot, J.D., Sparks, J.T., & Dickens, J.C. (2014). The maxillary palp of *Aedes aegypti*, a model of multisensory integration. *Insect Biochemistry and Molecular Biology, 48*, 29–39.
Boissière, A., Tchioffo, M.T., Bachar, D., Abate, L., Marie, A., Nsango, S.E., . . . Morlais, I. (2012). Midgut microbiota of the malaria mosquito vector *Anopheles gambiae* and interactions with *Plasmodium falciparum* infection. *PLOS Pathogens, 8*(5), e1002742.
Bond, H.A., Craig, G.B., Jr., & Foy, R.W. (1970). Field mating and movement of *Aedes aegypti. Mosquito News, 30*(3), 394–402.
Bond, J.G., Arredondo-Jiménez, J.I., Rodríguez, M.H., Quiroz-Martínez, H., & Williams, T. (2005). Oviposition habitat selection for a predator refuge and food source in a mosquito. *Ecological Entomology, 30*(3), 255–263.
Bonne, C. (1923a). A new *Sabethes* from Surinam (Diptera, Culicidae). *Insecutor Inscitiae Menstruus, 11*(7–9), 122–123.
Bonne, C. (1923b). The eggs of *Anopheles mediopunctatus* Theobald. *Tijdschrift voor Entomologie, 66*, 118.
Bonne, C. (1923c). The male hypopygium of *Anopheles mediopunctatus* Theobald. *Tijdschrift voor Entomologie, 66*, 115–117.
Bonne, C., & Bonne-Wepster, J. (Eds.). (1925). *Mosquitoes of Surinam. A study on Neotropical mosquitoes* (Vol. 21). Amsterdam, The Netherlands: Instituut Druk De Bussy.
Bonne-Wepster, J. (1930). The genus *Taeniorhynchus* Arribalzaga in the Dutch East Indies. Part II. Subgenus *Coquillettidia* Dyar. *Mededeelingen van den Dienst der Volksgezondheid in Nederlandsch-Indië, 19*(3), 387–399.
Bonne-Wepster, J. (1934). New mosquitos (Dipt.) from the Netherlands Indies. *Stylops, 3*(12), 272–276.
Bonne-Wepster, J. (1937). Notes on mosquitoes from the Netherlands Indies. A new *Culex* and two new *Finlaya*'s from New Guinea. *Mededeelingen van den Dienst der Volksgezondheid in Nederlandsch-Indië, 26*(1–2), 92–98.
Bonne-Wepster, J. (1938). Notes on mosquitoes from the Netherlands Indies. A new *Culex* from New Guinea: *Culex* (*Culex*) *solitarius* nov. sp. Description of male terminalia of *Culex* (*C.*) *miraculosus* B. -W. *Mededeelingen van den Dienst der Volksgezondheid in Nederlandsch-Indië, 27*(3), 394–396.
Bonne-Wepster, J. (1939). Notes on mosquitoes from the Netherlands Indies. Two new *Mansonia*-larvae from Borneo. *Mededeelingen van den Dienst der Volksgezondheid in Nederlandsch-Indië, 28*(1), 11–13.
Bonne-Wepster, J. (1940). Notes on mosquitoes from the Netherlands Indies: A new *Finlaya* from New-Guinea. *Mededeelingen van den Dienst der Volksgezondheid in Nederlandsch-Indië, 29*(3–4), 158–159.

Bonne-Wepster, J. (1948). Results of the third Archbold Expedition 1938–1939 (Diptera, Culicidae). Notes on the mosquitoes collected by the Neth. Indian-American expedition to Central and North New Guinea. *Treubia, 19*(2), 305–322.

Bonne-Wepster, J. (1951a). *Anopheles venhuisi* n. sp. *Documenta Neerlandica et Indonesica de Morbis Tropicis, 3,* 284.

Bonne-Wepster, J. (1951b). Notes on Oriental and Australasian mosquitoes of the Malay Archipelago. *Documenta Neerlandica et Indonesica de Morbis Tropicis, 3*(1), 67–74.

Bonne-Wepster, J. (1954). Synopsis of a hundred common non-anopheline mosquitoes of the Greater and Lesser Sundas, the Moluccas and New Guinea. *Documenta de medicina geographica et tropica, 6,* 1–29.

Bonne-Wepster, J., & Bonne, C. (1919). Description of the larvae of *Wyeomyia aphobema* Dyar (Diptera, Culicidae). *Insecutor Inscitiae Menstruus, 7*(7–9), 114.

Bonne-Wepster, J., & Bonne, C. (1919). Four new South American mosquitoes (Diptera, Culicidae). *Insecutor Inscitiae Menstruus, 7*(7–9), 105–113.

Bonne-Wepster, J., & Bonne, C. (1920). Diagnoses of new mosquitoes from Surinam, with a note on synonymy (Diptera, Culicidae). *Insecutor Inscitiae Menstruus, 7*(for 1919)(10–12), 165–180.

Bonne-Wepster, J., & Bonne, C. (1921a). Notes on South American mosquitoes in the British Museum (Diptera, Culicidae). *Insecutor Inscitiae Menstruus, 9*(1–3), 1–26.

Bonne-Wepster, J., & Bonne, C. (1921b). The first *Sabethes* larvae found (Diptera, Culicidae). *Insecutor Inscitiae Menstruus, 9*(4–6), 98–99.

Bonne-Wepster, J., & Bonne, C. (1923). A new *Megarhinus* from Surinam (Diptera, Culicidae). *Insecutor Inscitiae Menstruus, 11*(1–3), 7–9.

Bonne-Wepster, J., & Brug, S.L. (1932). The subgenus *Stegomyia* in Netherland India. *Geneeskundig tijdschrift voor Nederlandsch-Indië, 72,* 39–119.

Bonne-Wepster, J., & Brug, S.L. (1937a). Nederlandsch-Indische culicinen. *Geneeskundig tijdschrift voor Nederlandsch-Indië* [a reprint of 1937b with separate pagination], *7,* 1–105.

Bonne-Wepster, J., & Brug, S.L. (1937b). Nederlandsch-Indische culicinen. *Geneeskundig tijdschrift voor Nederlandsch-Indië, 77,* 515–617.

Bonne-Wepster, J., & Brug, S.L. (1939). Larven van Nederlandsch-Indische culicinen. *Geneeskundig tijdschrift voor Nederlandsch-Indië, 79,* 1218–1279.

Bonne-Wepster, J., & Swellengrebel, N.H. (1953). *The anopheline mosquitoes of the Indo-Australian Region.* Amsterdam, The Netherlands: de Bussy.

Boorman, J.P.T. (1961). Observations on the habits of mosquitos of Plateau Province, Northern Nigeria, with particular reference to *Aedes* (*Stegomyia*) *vittatus* (Bigot). *Bulletin of Entomological Research, 52*(4), 709–725.

Booth, D.R., & Bryan, J.H. (1986). Cytogenetic and crossbreeding evidence for additional cryptic species in the *Anopheles annulipes* Walker Complex (Diptera: Culicidae). *Journal of the Australian Entomological Society, 25*(4), 315–325.

Boreham, M.M. (1970). *Mansonia leberi*, a new mosquito from the Panama Canal Zone, with notes on its biology (Diptera: Culicidae). *Journal of Medical Entomology, 7*(3), 383–390.

Boreham, M.M., & Baerg, D.C. (1974). Description of the larva, pupa and egg of *Anopheles* (*Lophopodomyia*) *squamifemur* Antunes, with notes on development. (Diptera: Culicidae). *Journal of Medical Entomology, 11*(5), 564–569.

Borel, E. (1926). Les moustiques de la Cochinchine et du Sud-Annam. I. *Archives des Instituts Pasteur d'Indochine, 3–4,* 75–121.

Borel, E. (1928). Les moustiques de la Cochinchine et du Sud-Annam. II. *Archives des Instituts Pasteur d'Indochine, 7,* 75–106.

Borel, E. (1930). Les moustiques de la Cochinchine et du Sud-Annam. III. Paris, France: Société de Pathologie Exotique.

Borel, M. (1927). Description de deux espèces nouvelles de Culicides de Cochinchine *Lophoceratomya roubaudi* et *L. bernardi* n. sp. *Bulletin de la Société de pathologie exotique, 20*(1), 25–30.

Borkent, A., & Grimaldi, D.A. (2004). The earliest fossil mosquito (Diptera: Culicidae), in mid-Cretaceous amber. *Annals of the Entomological Society of America, 97*(5), 882–888.

Borkent, A., & Grimaldi, D.A. (2016). The cretaceous fossil *Burmaculex antiquus* confirmed as the earliest known lineage of mosquitoes (Diptera: Culicidae). *Zootaxa, 4079*(4), 457–456.

Borovsky, D. (2003). Trypsin-modulating oostatic factor: A potential new larvicide for mosquito control. *Journal of Experimental Biology, 206*(211), 3869–3875.

Bosch, O.J., Geier, M., & Boeckh, J. (2000). Contribution of fatty acids to olfactory host finding of female *Aedes aegypti*. *Chemical Senses, 25*(3), 323–330.

Boshell-Manrique, J. (1939). A new species of *Aedes* from Colombia (Dipt. Culicidae). *Revista de entomologia, 10*(2), 309–312.

Bosworth, A.B., Meola, S.M., & Olson, J.K. (1983). The chorionic morphology of eggs of the *Psorophora confinnis* complex in the United States. 1. Taxonomic considerations. *Mosquito Systematics, 15*(4), 285–309.

Bouché, P.F. (1834). *Naturgeschichte der lnsekten, besonders in Hinsicht ihrer ersten Zustände als Larven und Puppen.* Berlin: Erste Lieferung.

Boudko, D.Y., Moroz, L.L., Harvey, W.R., & Linser, P.J. (2001). Alkalinization by chloride/bicarbonate pathway in larval mosquito midgut. *Proceedings of the National Academy of Sciences of the United States of America, 98*(26), 15354–15359.

Bourke, B.P., Foster, P.G., Bergo, E.S., Calado, D.C., & Sallum, M.A.M. (2010). Phylogenetic relationships among species of *Anopheles* (*Nyssorhynchus*) (Diptera, Culicidae) based on nuclear and mitochondrial gene sequences. *Acta Tropica, 114,* 88–96.

Bourke, B.P., Nagaki, S.S., Bergo, E.S., da Cruz Cardoso, J., & Sallum, M.A.M. (2011). Molecular phylogeny of the Myzorhynchella Section of *Anopheles* (*Nyssorhynchus*) (Diptera: Culicidae): Genetic support for recently described and resurrected species. *Memórias do Instituto Oswaldo Cruz, 106*(6), 705–715.

Bourke, B.P., Conn, J.E., de Oliveira, T.M.P., Chaves, L.S.M., Bergo, E.S., Laporta, G.Z., & Sallum, M.A.M. (2018). Exploring malaria vector diversity on the Amazon Frontier. *Malaria Journal, 17,* 342.

Bourroul, C. (1904). *Mosquitos do Brasil.* Faculdade de Medicina da Bahia.

Boussès, P., & Brunhes, J. (2013). Trois nouveaux *Uranotaenia* Lynch Arribálzaga de Madagascar (Diptera, Culicidae). *Bulletin de la Société Entomologique de France, 118*(3), 399–410.

Bower, J.E., Dowton, M., Cooper, R.D., & Beebe, N.W. (2008). Intraspecific concerted evolution of the rDNA ITS1 in *Anopheles farauti* sensu stricto (Diptera: Culicidae) reveals recent patterns of population structure. *Journal of Molecular Evolution, 67*(4), 397–411.

Boyd, A.-M., & Foley, D.H. (2007). Distribution of sibling species of the *Anopheles annulipes* complex (Diptera: Culicidae) in the Townsville region of Australia. *Australian Journal of Entomology, 46*(2), 133–139.

Boyd, M.F. (Ed.) (1949). *Malariology: A comprehensive survey of all aspects of this group of diseases from a global standpoint. By 65 contributors* (Vol. I.). Philadelphia, PA: Saunders.

Boyer, S., Gilles, J., Merancienne, D., Lempérière, G., & Fontenille, D. (2011). Sexual performance of male mosquito *Aedes albopictus*. *Medical and Veterinary Entomology, 25*(4), 454–459.

Bozicic, B. (1980). Culicidae in Vojvodina Yugoslavia. *Acta entomologica Jugoslavica, 16*(1–2), 93–100.

Brackney, D.E., Foy, B.D., & Olson, K.E. (2008). The effects of midgut serine proteases on dengue virus type 2 infectivity of *Aedes aegypti*. *The American Journal of Tropical Medicine and Hygiene, 79*(2), 267–274.

Brackney, D.E., Isoe, J., Black, W.C., Zamora, J., Foy, B.D., Miesfeld, R.L., & Olson, K.E. (2010). Expression profiling and comparative analyses of seven midgut serine proteases from the yellow fever mosquito, *Aedes aegypti*. *Journal of Insect Physiology, 56*(7), 736–744.

Bradshaw, W.E. (1980). Thermoperiodism and the thermal environment of the pitcher-plant mosquito *Wyeomyia smithii*. *Oecologia, 46*(1), 13–17.

Bradshaw, W.E. (1986). Variable iteroparity as a life-history tactic in the pitcher-plant mosquito *Wyeomyia smithii*. *Evolution, 40*(3), 471–478.

Bradshaw, W.E., & Holzapfel, C.M. (1983). Predator-mediated, non-equilibrium coexistence of tree-hole mosquitoes in southeastern North America. *Oecologia, 57*(1–2), 239–256.

Bradshaw, W.E., & Holzapfel, C.M. (1984). Seasonal development of tree-hole mosquitoes (Diptera: Culicidae) and Chaoborids in relation to weather and predation. *Journal of Medical Entomology, 21*(4), 366–378.

Bradshaw, W.E., & Holzapfel, C.M. (1989). Life historical consequences of density-dependent selection in the pitcher-plant mosquito, *Wyeomyia smithii*. *The American Naturalist, 133*(6), 869–887.

Bradshaw, W.E., & Lounibos, L.P. (1977). Evolution of dormancy and its photoperiodic control in pitcher-plant mosquitoes. *Society for the Study of Evolution, 31*(3), 546–567.

Brahmachari Bahadur, U.N. (1912). On some new anophelines of Calcutta and on the seasonal prevalence and variations of anopheline *fuliginosus* of Calcutta. *Indian Medical Gazette, 47*, 186–187.

Braks, M.A.H., Meijerink, J., & Takken, W. (2001). The response of the malaria mosquito, *Anopheles gambiae*, to two components of human sweat, ammonia and L-lactic acid, in an olfactometer. *Physiological Entomology, 26*(2), 142–148.

Braks, M.A.H., Juliano, S.A., & Lounibos, L.P. (2006). Superior reproductive success on human blood without sugar is not limited to highly anthropophilic mosquito species. *Medical and Veterinary Entomology, 20*(1), 53–59.

Bram, R.A. (1966). *Culex* (*Thaiomyia*) *dispectus*, a new subgenus and species from Thailand (Diptera: Culicidae). *Proceedings of the Entomological Society of Washington, 68*(2), 73–77.

Bram, R.A. (1967a). Classification of *Culex* subgenus *Culex* in the New World (Diptera: Culicidae). *Proceedings of the United States National Museum, 120*(3557), 1–122.

Bram, R.A. (1967b). Contributions to the mosquito fauna of Southeast Asia. II. The genus *Culex* in Thailand (Diptera: Culicidae). *Contributions of the American Entomological Institute, 2*(1), 1–296.

Bram, R.A. (1967c). Lectotype assignments for several species of the genus *Culex* in Southeast Asia. *Proceedings of the Entomological Society of Washington, 69*(4), 327–328.

Bram, R.A. (1968). A redescription of *Culex* (*Acalleomyia*) *obscurus* (Leicester) (Diptera: Culicidae). *Proceedings of the Entomological Society of Washington, 70*(1), 52–57.

Bram, R.A. (1969). Relationships of adult and larval anatomy in supraspecific classification of the genus *Culex* in southeast Asia (Diptera: Culicidae). *Mosquito Systematics Newsletter, 1*(2), 9–12.

Bram, R.A., & Rattanarithikul, M. (1967). Six new species of the *Culex* (*Lophoceraomyia*) *mammilifer* group from Thailand (Diptera: Culicidae). *Proceedings of the Entomological Society of Washington, 69*(1), 1–17.

Branch, N., & Seabrook, E.L. (1959). *Culex* (*Culex*) *scimitar*, a new species of mosquito from the Bahama Islands (Diptera: Culicidae). *Proceedings of the Entomological Society of Washington, 61*(5), 216–218.

Breeland, S.G. (1951). The identification of the early larval instars of three common anophelines of southern Georgia, U.S.A. *Journal of the U.S.A. National Malaria Society, 10*, 224–232.

Breeland, S.G. (1953). *A morphological investigation of an anopheline of the Crucians complex reared from atypical ova (Diptera, Culicidae)*. Faculty publications and research, together with abstracts of doctoral dissertations and master's theses, North Carolina State University, Raleigh.

Breeland, S.G. (1980). A bibliography to the literature of *Anopheles albimanus* (Diptera: Culicidae). *Mosquito Systematics, 12*(1), 50–150.

Breeland, S.G. (1982). Bibliography and notes on Florida mosquitoes with limited distribution in the United States. *Mosquito Systematics, 14*(1), 53–72.

Breeland, S.G., Snow, W.E., & Pickard, E. (1961). Mosquitoes of the Tennessee Valley. *Journal of the Tennessee Academy of Science, 36*(4), 249–319.

Breland, O.P. (1949a). The biology and the immature stages of the mosquito, *Megarhinus septentrionalis* Dyar and Knab. *Annals of the Entomological Society of America, 42*, 38–47.

Breland, O.P. (1949b). Distinctive features of the larvae of *Aedes alleni* Turner (Diptera: Culicidae). *Journal of the New York Entomological Society, 57*, 93–100.

Breland, O.P. (1951). The immature stages of *Aedes infirmatus* Dyar and Knab with notes on related species (Diptera: Culicidae). *Annals of the Entomological Society of America, 44*(3), 362–371.

Breland, O.P. (1954a). Notes on the larvae of *Uranotaenia syntheta* (Diptera: Culicidae). *Journal of the Kansas Entomological Society, 27*, 156–158.

Breland, O.P. (1954b). Notes on the *Culex virgultus* complex (Diptera: Culicidae). *Mosquito News, 14*(2), 68–71.

Breland, O.P. (1957). Variations in the larvae of *Culex stigmatosoma* Dyar with notes on similar species (Diptera: Culicidae). *Annals of the Entomological Society of America, 50*(2), 175–178.

Breland, O.P. (1958). A report on *Haemagogus* mosquitoes in the United States with notes on identification (Diptera: Culicidae). *Annals of the Entomological Society of America, 51*(3), 217–221.

Breland, O.P. (1959). The first instar larvae of *Orthopodomyia alba* Baker and *Orthopodomyia signifera* (Coquillett) with comparative notes (Diptera: Culicidae). *Annals of the Entomological Society of America, 52*(2), 137–141.

Breland, O.P. (1960). Restoration of the name, *Aedes hendersoni* Cockerell, and its elevation to full specific rank (Diptera, Culicidae). *Annals of the Entomological Society of America, 53*(5), 600–606.

Brengues, J., & Hamon, J. (1965). Description de la larve et de la nymphe d'*Aedes* (*Aedimorphus*) *nyounae* Hamon et Adam, 1958. *Bulletin de la Société de pathologie exotique, 58*(1), 92–101.

Brèthes, J. (1910). Dípteros nuevos ó poco conocidos de Sud-América. *Anales del Museo Nacional de Buenos Aires, 13*, 469–472.

Brèthes, J. (1912). Los mosquitos de la República Argentina. *Boletín de Instituto Entomológia Patagonia Vegetal, 1*, 1–48.

Brèthes, J. (1916). Algunas notas sobre mosquitos Argentines, su relation con las Énfermedades palúdicas, etc., y description de très espècies nuevas. *Anales del Museo Nacional de Historia Natural de Buenos Aires, 28*, 193–218.

Brèthes, J. (1917). Description d'une nouvelle espèce de moustique de Buenos Aires. *Physis, 3*, 226–229.

Brèthes, J. (1920). Description d'un nouveau moustique du Peru. *Revista Chilena de historia natural, 24*, 41–43.

Brèthes, J. (1926a). Contribución para el conocimiento de los mosquitoes Argentinos. Descripción de un nuevo *Megarhinus*: *Megarhinus tucumanus*. *Boletín del Instituto de Clinica Quirurgica, 2*, 318–321.

Brèthes, J. (1926b). Description provisoire de deux espèces nouvelles d'Anophelinae argentins. *Prensa médica Argentina, 13*(3), 106–107.

Briegel, H. (1986). Protein catabolism and nitrogen partitioning during oogenesis in the mosquito *Aedes aegypti*. *Journal of Insect Physiology, 32*(5), 455–462.

Briegel, H. (2003). Physiological bases of mosquito ecology. *Journal of Vector Ecology, 28*(1), 1–11.

Briegel, H., Knüsel, I., & Timmermann, S.E. (2001). *Aedes aegypti*: Size, reserves, survival, and flight potential. *Journal of Vector Ecology, 26*(1), 21–31.

Bright, D.B., & Hogue, C.L. (1972). *A synopsis of the burrowing land crabs of the world and lists of their arthropod symbionts and burrow associates*. Los Angeles, CA: Natural History Museum of Los Angeles.

Broadie, K.S., & Bradshaw, W.E. (2008). Mechanisms of interference competition in the western tree-hole mosquito, *Aedes sierrensis*. *Ecological Entomology, 16*(2), 145–154.

Brochero, H.H.L., Li, C., & Wilkerson, R.C. (2007). A newly recognized species in the *Anopheles* (*Nyssorhynchus*) *albitarsis* complex (Diptera: Culicidae) from Puerto Carreño, Colombia. *The American Journal of Tropical Medicine and Hygiene, 76*(6), 1113–1117.

Brodie, P.B. (1845). *A history of the fossil insects of the secondary rocks of England; accompanied by a particular account of the strata in which they occur, and of the circumstances connected with their preservation*. London, England: van Voorst.

Brogdon, W.G. (1981). Use of the siphonal index to separate *Culex pipiens* subspecies and hybrids. *Mosquito Systematics, 13*(2), 129–137.

Brogdon, W.G. (1984a). The siphonal index. I. A method for evaluating *Culex pipiens* subspecies and intermediates. *Mosquito Systematics, 16*(2), 144–152.

Brogdon, W.G. (1984b). The siphonal index. II. Relative abundance of *Culex* species, subspecies, and intermediates in Memphis, Tennessee. *Mosquito Systematics, 16*(2), 153–161.

Brolemann, H.W. (1919a). Sur quelques *Culex* des Pyrenées et description d'une espèce nouvelle. I. Campagne 1917. *Annales de la Société Entomologique de France, 87*(3–4), 425–440.

Brolemann, H.W. (1919b). Sur quelques *Culex* des Pyrenées. II. Campagne 1918. *Annales de la Société Entomologique de France, 88*(1–2), 65–103.

Brolemann, H.W. (1920). Sur quelques *Culex* des Pyrenées III. *Annales de la Société Entomologique de France, 89*(1), 51–73.

Brookman, B., & Reeves, W.C. (1950). A new name for a California mosquito (Diptera, Culicidae). *The Pan-Pacific Entomologist, 26*(4), 159–160.

Brookman, B., & Reeves, W.C. (1953). New records of mosquitoes from Lower California, Mexico, with notes and descriptions (Diptera: Culicidae). *Annals of the Entomological Society of America, 46*(2), 225–236.

Brown, B.V., Borkent, A., Cumming, J.M., Wood, D.M., Woodley, N.E., & Zumbado, M.A. (Eds.). (2009). *Manual of Central American Diptera*, vol. 1. Ottawa, Ontario, Canada: NRC Research Press.

Brown, B.V., Borkent, A., Cumming, J.M., Wood, D.M., Woodley, N.E., & Zumbado, M.A. (Eds.). (2010). *Manual of Central American Diptera*, vol. 2. Ottawa, Ontario, Canada: NRC Research Press.

Brown, J.E., McBride, C.S., Johnson, P., Ritchie, S., Paupy, C., Bossin, H., . . . Powell, J.R. (2011). Worldwide patterns of genetic differentiation imply multiple "domestications" of *Aedes aegypti*, a major vector of human diseases. *Proceedings of the Royal Society B: Biological Sciences, 278*(1717), 2446–2454.

Brown, M.R., & Cao, C. (2001). Distribution of ovary ecdysteroidogenic hormone I in the nervous system and gut of mosquitoes. *Journal of Insect Science, 1*(3), 1–11.

Brown, R.W. (1956). *Composition of scientific words*. Washington, DC: Smithsonian Institution Press.

Brug, S.L. (1924). Notes on Dutch East-Indian mosquitos. *Bulletin of Entomological Research, 14*(4), 433–442.

Brug, S L. (1925). A new *Armigeres* (Diptera, Culicidae) from Ceram (Moluccas). *Bulletin of Entomological Research* (London), *15*(4), 345–347.

Brug, S.L. (1926). *Lophoscelomya annandalei* var. *djajasanensis* nov. var. (Diptera, Culicidae). *Bulletin de la Société de pathologie exotique, 19*(9), 805–806.

Brug, S.L. (1928a). *Anopheles* (*Neomyzomyia*) *longirostris* n. sp. (Diptera, Culicidae). *Geneeskundig tijdschrift voor Nederlandsch-Indië, 68*, 278–279.

Brug, S.L. (1928b). Aanteekeningen omtrent muskieten. IV. *Geneeskundig tijdschrift voor Nederlandsch-Indië, 68*, 921–927.

Brug, S.L. (1931a). New Culicidae from Sumatra. *Tijdschrift voor Entomologie, 74*, 245–250.

Brug, S.L. (1931b). *Anopheles incognitus* n. sp. *Geneeskundig tijdschrift voor Nederlandsch-Indië, 71*(2), 136–137.

Brug, S.L. (1931c). Culiciden der Deutschen Limnologischen Sunda-Expedition. *Archiv für Hydrobiologie (Tropische Binnengewässer), 2*(Suppl. Bd-9), 1–42.

Brug, S.L. (1932). Notes on Dutch East Indian mosquitos. *Bulletin of Entomological Research, 23*(1), 73–83.

Brug, S.L. (1934a). Notes on Dutch East Indian mosquitos. *Bulletin of Entomological Research, 25*(4), 501–519.

Brug, S.L. (1934b). Contribution a l'étude de la faune nepenthicole. Art. VI. Culicidae, collected from *Nepenthes* in Borneo. *Overdruk uit het Natuurhistorisch Maandblad, 23*(10), 149–150.

Brug, S.L. (1938). Waarnemingen bij anophelinen. *Geneeskundig tijdschrift voor Nederlandsch-Indië, 78*, 520–523.

Brug, S.L. (1939). Notes on Dutch East-Indian mosquitos. *Tijdschrift voor Entomologie, 82*, 91–113.

Brug, S.L., & Bonne-Wepster, J. (1947). The geographical distribution of the mosquitoes of the Malay Archipelago. *Overdruk Uit Chronica Naturae, 103*, 179–197.

Brug, S.L., & de Rook, H. (1922). *Bironella gracilis* Theobald 1905 (Sous famille *Anophelinae*). *Bulletin de la Société de pathologie exotique, 15*, 305–310.

Bruijning, C.F.A. (1959). Notes on *Wyeomyia* mosquitoes of Suriname, with a description of *Wyeomyia surinamensis* sp. n. *Studies on the Fauna of Suriname and Other Guyanas, 3*(8), 99–146.

Brullé, A. (1833). *Expédition scientifique de Morée*. Section des Sciences Physiques (Vol. 3, part 1). Zoologie. Deuxième Section: Des animaux articulés. Paris, France: Levrault.

Brunetti, E. (1912). Annotated catalogue of Oriental Culicidae. *Records of the Indian Museum, 4*(Suppl.), 403–517.

Brunetti, E. (1914). Critical review of "genera" in Culicidae. *Records of the Indian Museum, 10*(1), 15–73.

Brunetti, E. (1920). Catalogue of Oriental and South Asiatic Nemocerca. *Records of the Indian Museum, 17*, 1–300.

Brunhes, J. (1967). Contribution a l'étude des culicides de Madagascar. Description des larves et nymphes de *Culex* (*C.*) *giganteus* Ventrillon et *Culex* (*C.*) *argenteopunctatus* Ventrillon. Notes com-

plémentaires sur les adultes de ces deux espèces. *Cahiers ORSTOM. Série entomologie médicale et parasitologie, 5*(4), 207–216.
Brunhes, J. (1968). Contribution a l'étude des culicides de Madagascar. Synonymie entre *Culex (N.) seyrigi* Edwards 1941 et *Culex (N.) robici* Doucet 1960; description de la nymphe et de la femelle de *Culex (N.) seyrigi* Edwards. *Cahiers ORSTOM. Série entomologie médicale et parasitologie, 6*(1), 15–18.
Brunhes, J. (1969). Culicides de Madagascar. IV. Description des imagos, de la nymphe et de la larve de *Culex (Culiciomyia) pandani* sp. n. *Cahiers ORSTOM. Série entomologie médicale et parasitologie, 7*(3), 175–180.
Brunhes, J. (1971). Culicides de Madagascar. V. Quelques *Aedes* (sous-genre *Finlaya*) de Madagascar. *Cahiers ORSTOM. Série entomologie médicale et parasitologie, 9*(4), 335–349.
Brunhes, J. (1977a). Les moustiques de l'archipel des Comores. I. Inventaire, répartition et description de quatre espèces ou sous-espècies nouvelles. *Cahiers ORSTOM. Série entomologie médicale et parasitologie, 15*(2), 131–152.
Brunhes, J. (1977b). Les moustiques de l'archipel des Comores. II. Description de quatre espècies nouvelles ou peu connues. Répartition des membres du sous-genre *Skusea* dans l'océan Indien occidental. Affinités de la faune culicidienne des Comores. *Cahiers ORSTOM. Série entomologie médicale et parasitologie, 15*(2), 153–170.
Brunhes, J. (1983). Culicides de Madagascar. IX. Compléments à la description d'*Aedes (Diceromyia) grassei* Doucet, 1951 et description d'*Aedes* (*Diceromyia*) *sylvaticus* n.sp. (Diptères, Nematoceres). *Cahiers ORSTOM. Série entomologie médicale et parasitologie, 20*(for 1982)(4), 313–318.
Brunhes, J. (1999). Culicidae du Maghreb. Description d'*Aedes* (*Ochlerotatus*) *biskraensis* n. sp. d'Algérie (Diptera, Nematocera). *Bulletin de la Société Entomologique de France, 104*(1), 25–30.
Brunhes, J., & Boussès, P. (2009). Description de *Culex* (*Culex*) *sechani* n. sp. de Tahiti (iles de la Société) (Diptera, Culicidae). *Bulletin de la Société Entomologique de France, 114*(1), 21–28.
Brunhes, J., & Hervy, J.P. (1995). Insectes diptères: Culicidae Culicinae genre *Orthopodomyia* de la sous-région malgache et de la région afrotropicale. *Insectes diptères : Culicidae Culicinae genre* Orthopodomyia *de la sous-région malgache et de la région afrotropicale.* Paris: Muséum National d'Histoire Naturelle (Faune de Madagascar; 85).
Brunhes, J., & Rambelo, J. (1968). Contribution a l'étude des culicidés de Madagascar. Description des adultes, nymphe et larve de *Culex* (*Neoculex*) *chauveti*. *Cahiers ORSTOM. Série entomologie médicale et parasitologie, 6*(1), 113–118.
Brunhes, J., & Ravaonjanahary, C. (1971). Culicidés de Madagascar: VI. Description de *Culex* (*Culex*) *carleti* sp. n. *Cahiers ORSTOM. Série entomologie médicale et parasitologie, 9*(2), 177–182.
Brunhes, J., & Ravaonjanahary, C. (1973). Culicidés de Madagascar: VII. Description des imagos, de la nymphe et de la larve de *Culex* (*Eumelanomyia*) *brenguesi* sp. n. *Cahiers ORSTOM. Série entomologie médicale et parasitologie, 11*(3), 169–173.
Brunhes, J., & Razafindrasolo, E. (1975). Culicidés de Madagascar: VIII. Description des imagos, de la nymphe et de la larve de *Uranotaenia anopheloides* sp. n. *Cahiers ORSTOM. Série entomologie médicale et parasitologie, 13*(4), 223–227.
Brunhes, J., & Venard, P. (1966). Contribution a l'étude des culicidés de la région éthiopienne. Description de *Culex* (*Barraudius*) *richeti* n. sp. Notes complémentaires sur *Culex* (*B.*) *pusillus* Macquart et *Culex* (*B.*) *modestus* Ficalbi. *Cahiers ORSTOM. Série entomologie médicale et parasitologie, 4*(7), 75–84.
Brunhes, J., Adam, H.P., & Bailly-Choumara, H. (1967). Contribution a l'étude des *Culex* de la région éthiopienne appartenant au sous-genre *Mochthogenes* (Diptères, Culicides), avec description des males de cinq nouvelles espèces. *Cahiers ORSTOM. Série entomologie médicale et parasitologie, 5*(1), 43–52.
Brunhes, J., Le Goff, G., & Geoffroy, B. (1997). *Anophèles* afrotropicaux. I. Descriptions d'espèces nouvelles et changements de statuts taxonomiques (Diptera: Culicidae). *Annales de la Société Entomologique de France, 33*(2), 172–183.
Brunhes, J., Le Goff, G., Manga, L., & Geoffroy, B. (1998a). *Anophèles* Afrotropicaux. IV. Mise au point sur le complexe *Anopheles moucheti*, réhabilitation d'*An. multicinctus* et d'*An. garnhami basilewskyi* (Diptera: Culicidae). *Annales de la Société Entomologique de France, 34*(4), 397–405.
Brunhes, J., Le Goff, G., & Geoffroy, B. (1998b). *Anophèles* afrotropicaux. II. Mises au point sur les espèces de la sous-région malgache (Diptera: Culicidae). *Bulletin de la Société Entomologique de France, 103*(2), 139–152.
Brunhes, J., Le Goff, G., & Geoffroy, B. (1999). Afro-tropical anopheline mosquitoes. III. Description of three new species: *Anopheles carnevalei* sp. nov., *An. hervyi* sp. nov., and *An. dualaensis* sp. nov., and resurrection of *An. rageaui* Mattingly and Adam. *Journal of the American Mosquito Control Association, 15*(4), 552–558.
Brunhes, J., Hassaïne, K., Rhaim, A., & Hervy, J.P. (2000). Les Culicides de l'Afrique méditerranéenne: Espèces présentes et répartition (Diptera, Nematocera). *Bulletin de la Société Entomologique de France, 105*(2), 195–204.
Brunhes, J., Le Goff, G., & Boussès, P. (2003). *Anophèles* afrotropicaux. V. Description du mâle et des stades pré-imaginaux d'*An. deemingi* Service, 1970 et description d'*An. eouzani* n. sp. (Diptera: Culicidae). *Annales de la Société Entomologique de France, 39*(2), 179–185.
Brunhes, J., Boussès, P., & da Cunha Ramos, H. (2011). Les *Aedeomyia* Theobald, 1901, des régions afro-tropicale et malgache (Diptera, Culicidae). *Bulletin de la Société Entomologique de France, 116*(1), 99–128.
Brunhes, J., Boussès, P., Tantely, M.L., & Kengue, P. (2017). Un nouveau genre de Culicidae (Diptera), *Paulianius* n. gen., avec la description de trois nouvelles espèces malgaches. *Annales de la Société Entomologique de France* (n.s.), *53*(5), 344–373.
Brust, R.A. (1974). Reproductive isolation within the *Aedes atropalpus* group, and description of eggs. *Journal of Medical Entomology, 11*(4), 459–466.
Brust, R.A., & Horsfall, W.R. (1965). Thermal stress and anomalous development of mosquitoes (Diptera: Culicidae). IV. *Aedes communis* [*sic*]. *Canadian Journal of Zoology, 43*(1), 17–53.
Brust, R.A., & Mahon, R.J. (1997). Resurrection of *Aedes* (*Halaedes*) *ashworthi* Edwards 1921: Morphological characterisation and separation from its sibling *Aedes* (*Halaedes*) *australis* (Erichson) 1842 (Diptera: Culicidae). *Australian Journal of Entomology, 36*(2), 129–136.
Brust, R.A., & Munstermann, L.E. (1992). Morphological and genetic characterization of the *Aedes* (*Ochlerotatus*) *communis* complex (Diptera: Culicidae) in North America. *Annals of the Entomological Society of America, 85*(1), 1–10.
Brust, R.A., Ballard, J.W.O., Driver, F., Hartley, D.M., Galway, N.J., & Curran, J. (1998). Molecular systematics, morphological analysis, and hybrid crossing identify a third taxon, *Aedes* (*Halaedes*) *wardangensis* sp. nov., of the *Aedes* (*Halaedes*) *australis* species-group (Diptera: Culicidae). *Canadian Journal of Research, 76*, 1236–1246.
Bryan, J.H. (1973). Studies on the *Anopheles punctulatus* complex. I. Identfication by proboscis morphological criteria and by crossmating experiments. II. Hybridization of the member species. III. Mating behaviour of the F1 hybrid adults from crosses between *Anopheles farauti* no. 1 and *Anopheles farauti* no. 2. *Transactions of the Royal Society of Tropical Medicine and Hygiene, 67*(1), 64–91.

Bryan, J.H. (1974). Morphological studies on the *Anopheles punctulatus* Dönitz complex. *Transactions of the Royal Entomological Society of London, 125*(4), 413–435.

Bryan, J.H., & Coluzzi, M. (1971). Cytogenetic observations on *Anopheles farauti* Laveran. *Bulletin of the World Health Organization, 45*, 266–267.

Bryan, J.H., Foley, D.H., Geary, M., & Carven, C.T.J. (1991). *Anopheles annulipes* Walker (Diptera: Culicidae) at Griffith, New South Wales. 3. Dispersal of two sibling species. *Journal of the Australian Entomological Society, 30*(2), 119–121.

Bryant, B., Macdonald, W., & Raikhel, A.S. (2010). microRNA *miR-275* is indispensable for blood digestion and egg development in the mosquito *Aedes aegypti*. *Proceedings of the National Academy of Sciences of the United States of America, 107*(52), 22391–22398.

Bueno Marí, R., Bernués Bañeres, A., Chordá Olmos, F.A., & Jiménez Peydró, R. (2011). Contributions to the knowledge of the distribution and biology of *Anopheles algeriensis* Theobald, 1903 in Spain. Aportes al conocimiento de la distribución y biología de *Anopheles algeriensis* Theobald, 1903 en España. *Boletín de Malariología y Salud Ambiental, 51*(1), 93–96.

Bulichev, V.P. (1977). *Mosquito human attack activity at the virus foci of Tyagin in Tajikistan*. Paper presented at the Theoretical Basis of Prophylaxis and Reduction of Infectious Diseases conference, Leninobod [Khujand], Tajikistan.

Bullini, L., Bianchi-Bullini, A.P., Cianchi, R., Sabatini, A., & Coluzzi, M. (1980). Tassonomia biochimica del complesso *Anopheles maculipennis*. *Parassitology, 22*, 290–293.

Bullock, H.R. (1960). Chorionic pattern of *Aedes* eggs by SUMP method. *Transactions of the American Microscopical Society, 79*(2), 167–170.

Buonomini, G., & Mariani, M. (1946). Qual'è il valore sistematico del nome *maculipennis* Meig.? *Rivista di malariologia, 25*(3–4), 101–107.

Buonomini, G., & Mariani, M. (1953). World anophelines belonging to the subgenus *Maculipennia* Buonomini e Mariani. *Rivista di malariologia, 32*(4–6), 173–188.

Buren, W.F. (1947). A new *Aedes* from the Florida Keys (Diptera, Culicidae). *Proceedings of the Entomological Society of Washington, 49*(8), 228–229.

Burger, J.F. (1965). *Aedes kompi* Vargas and Downs 1950, new to the United States. *Mosquito News, 25*(4), 396–398.

Burger, J.F. (1981). New records of mosquitoes (Diptera: Culicidae) from New Hampshire. *Entomological News, 92*(1), 49–50.

Burgess, N.R.H., & Cowan, G.O. (1993). Mosquitoes. In *A colour atlas of medical entomology* (pp. 9–30). The Netherlands: Springer.

Burkett, D.A., Kline, D.L., & Carlson, D.A. (1999). Sugar meal composition of five North Central Florida mosquito species (Diptera: Culicidae) as determined by gas chromatography. *Journal of Medical Entomology, 36*(4), 462–467.

Burkett-Cadena, N.D., Eubanks, M.D., & Unnasch, T.R. (2008). Preference of female mosquitoes for natural and artificial resting sites. *Journal of the American Mosquito Control Association, 24*(2), 228–235.

Burkett-Cadena, N.D., Blosser, E.M., & Day, J. (2017). *Aedeomyia squamipennis* (Diptera: Culicidae) in Florida, USA, a new state and country record. *Journal of Medical Entomology, 54*(3), 788–792.

Burt, A., Coulibaly, M., Crisanti, A., Diabaté, A., & Kayonde, J.K. (2018). Gene drive to reduce malaria transmission in sub-Saharan Africa. *Journal of Responsible Innovation, 5*(Suppl. 1), 566–580.

Bush, J.W.M., Hu, D.L., & Prakesh, M. (2008). The integument of water-walking arthropods: Form and function. *Advances in Insect Physiology, 34*(147), 118–192.

Butail, S., Manoukis, N., Diallo, M., Yaro, A.S., Dao, A., Traoré, S.F., . . . Paley, D.A. (2011, August). 3D tracking of mating events in wild swarms of the malaria mosquito *Anopheles gambiae*. In *Engineering in Medicine and Biology Society, EMBC, 2011 Annual International Conference of the IEEE* (pp. 720–723). Boston, MA: Institute of Electrical and Electronics Engineers.

Butail, S., Manoukis, N., Diallo, M., Ribeiro, J.M., Lehmann, T., & Paley, D.A. (2012). Reconstructing the flight kinematics of swarming and mating in wild mosquitoes. *Journal of the Royal Society Interface, 9*(75), 2624–2638.

Büttiker, W., & Beales, P. (1959). *Anopheles pampanae*, a new species of mosquito from Cambodia. *Acta Tropica, 16*, 63–69.

Buxton, P.A. (1927). Sur les moustiques de Tahiti et du groupe des Îles de la Société. *Bulletin de la Société d'études océaniennes, Papeete, Tahiti, 2*(21), 306–308.

Buxton, P.A., & Hopkins, G.H.E. (1925). The early stages of Somoan [*sic*] mosquitos. *Bulletin of Entomological Research, 15*(3), 295–301.

Byrd, B.D., Harrison, B.A., Zavortink, T.J., & Wesson, D.M. (2012). Sequence, secondary structure, and phylogenetic analyses of the ribosomal Internal Transcribed Spacer 2 (ITS2) in members of the North American Signifera Group of *Orthopodomyia* (Diptera: Culicidae). *Journal of Medical Entomology, 49*(6), 1189–1197.

Cabrera, M., & Jaffe, K. (2007). An aggregation pheromone modulates lekking behavior in the vector mosquito *Aedes aegypti* (Diptera: Culicidae). *Journal of the American Mosquito Control Association, 23*(1), 1–10.

Cady, C., & Hagedorn, H.H. (1999). The effect of putative diuretic factors on in vivo urine production in the mosquito *Aedes aegypti*. *Journal of Insect Physiology, 45*(4), 317–325.

Cagampang-Ramos, A., & Darsie, R.F., Jr. (1970). Illustrated keys to the *Anopheles* mosquitoes of the Philippine Islands. *US Air Force 5th Epidemiology Flight, PACAF, Technical Report 70-1*, 1–49.

Cain, A.J. (1956). The genus in evolutionary taxonomy. *Systematic Biology, 5*(3), 97–109.

Calado, D.C., & Navarro-Silva, M.A. (2005). Identification of *Anopheles* (*Kerteszia*) *cruzii* Dyar & Knab and *Anopheles* (*Kerteszia*) *homunculus* Komp (Diptera, Culicidae, Anophelinae) by molecular markers (RAPD and RFLP). *Revista Brasileira de zoologia, 22*(4), 1127–1133.

Calado, D.C., Foster, P.G., Bergo, E.S., dos Santos, C.L.S., Galardo, A.K.R., & Sallum, M.A.M. (2008). Resurrection of *Anopheles goeldii* from synonymy with *Anopheles nuneztovari* (Diptera, Culicidae) and a new record for *Anopheles dunhami* in the Brazilian Amazon. *Memórias do Instituto Oswaldo Cruz, 103*(8), 791–799.

Calderón Falero, G. (1995). Clave para identificar especies de *Anopheles* (Diptera: Culicidae, Anophelinae) del Peru (adultos hembras). *Revista Peruana de entomología, 11*(for 1994), 31–40.

Calderón Falero, G., Fernández, R., & Valle, J. (1995). Especiés de la fauna anofelina, su distribución y algunas consideraciones sobre su abundancia e infectividad en el Perú. *Revista Peruana de epidmiología, 8*(1), 5–23.

Calle L., D.A., Quiñones, M.L., Erazo, H.F., & Jaramillo, N. (2002). Morphometric discrimination of females of five species of *Anopheles* of the subgenus *Nyssorhynchus* from Southern and Northwest Colombia. *Memórias do Instituto Oswaldo Cruz* (Rio de Janeiro), *97*(8), 1191–1195.

Callot, J. (1940). Sur quelques moustiques du Maroc. *Archives de l'Institut Pasteur du Maroc, 2*(2), 361–365.

Callot, J. (1943). Sur *Culex hortensis* et *Culex apicalis* a Richelieu (Indre-et Loire). *Annales de parasitologie humaine et comparée, 19*(for 1942–1943)(4–6), 129–141.

Callot, J. (1944a). A propos de *Aedes (Finlaya) heracleensis* Callot 1944. *Annales de parasitologie humaine et comparée, 20*(1944–1945), 93–94.

Callot, J. (1944b). Culicides des eaux acides d'Auvergne. II. A propos de *Culex lavieri* Larrousse. *Comptes rendus des séances de la Société de Physique Biologique de France, 18*(80), 196–197.

Callot, J. (1944c). Sur un nouveau moustique arboricole: *Aedes* (*Finlaya*) *heracleensis* sp. nov. *Bulletin de la Société de pathologie exotique, 37*(1–2), 56–59.

Callot, J. (1945). Diptères nématocères piqueurs d'Auvergne. *Revue des sciences naturelles d'Auvergne, 11*, 28–36.

Calvin, W.H. (1996). *How brains think*. New York, NY: HarperCollins.

Calvo, E., Pham, V.M., & Ribeiro, J.M.C. (2008). An insight into the sialotranscriptome of the non-blood feeding *Toxorhynchites amboinensis* mosquito. *Insect Biochemistry and Molecular Biology, 38*(5), 499–507.

Cambouliu, M. (1902). Contribution a l'étude des *Anopheles* de l'isthme de Suez. *Comptes rendus de l'Académie des Sciences, 135*(17), 704–706.

Cambournac, F.J. (1938). *Aedes* (*Ochlerotatus*) *longitubus*, a new species from Portugal (Diptera, Culicidae). *Proceedings of the Royal Entomological Society of London, Series B: Taxonomy, 7*(3), 74–80.

Cambournac, F.J.C., Petrarca, V., & Coluzzi, M. (1982). *Anopheles arabiensis* in the Cape Verde Archipelago. *Parassitologia, 24*(2–3), 265–267.

Campos, R.E., & Zavortink, T.J. (2010). Description of the larva and pupa and redescription of the adults of *Isostomyia paranensis* (Brèthes) (Diptera: Culicidae). *Zootaxa, 2689*, 27–36.

Campos, R.E., Fernandez, L.A., & Sy, V.E. (2004). Study of insects associated with the floodwater mosquito *Ochlerotatus abifasciatus* (Diptera: Culicidae) and their possible predators in Buenos Aires Province, Argentina. *Hydrobiologia, 524*(1), 91–102.

Campos, S.S., Fernandes, R.S., Cunha dos Santos, A.A., Moraes de Miranda, R., Telleria, E.L., Ferreira-de-Brito, A., . . . Lourenço-de-Oliveira, R. (2017). Zika virus can be venereally transmitted between *Aedes aegypti* mosquitoes. *Parasites and Vectors, 10*(1), 605.

Cano, J., Nzambo, S., Buatiche, J.N., Ondo-Esono, M., Micha, F., & Benito, A. (2003). *Anopheles* (*Cellia*) *carnevalei* in Equatorial Guinea (West-Central Africa). *Journal of the American Mosquito Control Association, 19*(4), 435–437.

Cantino, P.D., & de Queiroz, K. (2010). *International code of phylogenetic nomenclature*. (Version 4c). Retrieved from https://www.ohio.edu/phylocode/PhyloCode4c.pdf

Capasso, L. (1991). Fossil mosquitoes and the spread of infectious diseases in man's ancestors. *Journal of Paleopathology, 3*(3), 171–201.

Capelli, G., Drago, A., Martini, S., Montarsi, F., Soppelsa, M., Delai, N., . . . Russo, F. (2011). First report in Italy of the exotic mosquito species *Aedes* (*Finlaya*) *koreicus*, a potential vector of arboviruses and filariae. *Parasites and Vectors, 4*, 188.

Caplan, A.L., Parent, B., Shen, M., & Plunkett, C. (2015). No time to waste—the ethical challenges created by CRISPR. *European Molecular Biology Organisation Reports, 16*(11), 1421–1426.

Caragata, E.P., Rancès, E., Hedges, L.M., Gofton, A.W., Johnson, K.N., O'Neill, S.L., & McGraw, E.A. (2013). Dietary cholesterol modulates pathogen blocking by *Wolbachia*. *PLOS Pathology, 9*(6), e1003459.

Caragata, E.P., Rancès, E., O'Neill, S.L., & McGraw, E.A. (2014). Competition for amino acids between *Wolbachia* and the mosquito host, *Aedes aegypti*. *Microbial Ecology, 67*(1), 205–218.

Caragata, E.P., Dutra, H.L., & Moreira, L.A. (2016). Exploiting intimate relationships: Controlling mosquito-transmitted disease with *Wolbachia*. *Trends in Parasitology, 32*(3), 207–218.

Cardamatis, J.P. (1931). Les espèces de moustiques en Grèce et tout particulierement d'Athenes. *Bulletin de la Société de pathologie exotique, 24*, 122–131.

Cardoso, J.D.C., Bergo, E.S., Oliveira, T.M.P., Sant'Ana, D.C., Motoki, M.T., & Sallum, M.A.M. (2012). New records of *Anopheles homunculus* in Central and Serra Do Mar biodiversity corridors of the Atlantic Forest, Brazil. *Journal of the American Mosquito Control Association, 28*(1), 1–5.

Carey, J.R., & Liedo, P. (1999). Mortality dynamics of insects: General principles derived from aging research on the Mediterranean fruit fly. *American Entomologist, 45*(1), 49–55.

Carlson, D.A., & Service, M.W. (1980). Identification of mosquitoes of *Anopheles gambiae* species complex A and B by analysis of cuticular components. *Science, 207*(4435), 1089–1091.

Caroci, A.S., & Noriega, F.G. (2003). Free amino acids are important for the retention of protein and non-protein meals by the midgut of *Aedes aegypti* females. *Journal of Insect Physiology, 49*(9), 839–844.

Carpenter, S.J. (1970). Review of recent literature on mosquitoes of North America. Supplement I. *California Vector Views, 17*(6), 39–66.

Carpenter, S.J., & LaCasse, W.J. (1955). *Mosquitoes of North America (North of Mexico)*. Berkeley: University of California Press.

Carpenter, S J., Middlekauff, W.W., & Chamberlain, R.W. (1946). *The mosquitoes of the southern United States; east of Oklahoma and Texas*. Notre Dame, IN: University of Notre Dame Press.

Carpintero, D.J., & Leguizamón, M.N. (2000). Una nueva especie de mosquito en la la provincia de Buenos Aires, *Aedes* (*Ochlerotatus*) *jorgi*. *Mariposas del mundo, 5*, 20–23.

Carpintero, D.J., & Leguizamón, M.N. (2004). Description of a new *Culex* (*Culex*) species (Diptera: Culicidae) from La Pampa Province, Argentina, and a preliminary list of the mosquitoes of the province. *Studia dipterologica, 11*(2), 501–503.

Carrasquilla, M.C., & Lounibos, L.P. (2015). Satyrization without evidence of successful insemination from interspecific mating between invasive mosquitoes. *Biology Letters, 11*(9), 20150527.

Carter, H.F. (1910a). A new anopheline from South Africa; *Pyretophorus transvaalensi*, sp. nov. *The Entomologist, 43*(568), 237–238.

Carter, H.F. (1910b). Report on a collection of Culicidae from Cochin, China. *The Entomologist, 43*(569), 274–276.

Carter, H.F. (1911). A new mosquito from Uganda. *Bulletin of Entomological Research, 2*(1), 37–38.

Carter, H.F. (1920). Descriptions of the adult, larval and pupal stages of a new mosquito from Lord Howe Island, S. Pacific. *Proceedings of the general meeting for scientific business of the Zoological Society of London, 1920*(4), 623–628.

Carter, H.F. (1950). The genus *Taeniorhynchus* Lynch Arribálzaga (Diptera, Culicidae) with special reference to the bionomics and relation to disease of the species occurring in Ceylon. *Ceylon Journal of Science B, 24*(1), 1–26.

Carter, H.F., & Wijesundara, D.P. (1948). Notes on some Ceylon culicine mosquitoes. *Ceylon Journal of Science B, 23*(3), 135–151.

Carter, T.E., Yared, S., Gebresilassie, A., Bonnell, V., Damodaran, L., Lopez, K., . . . & Janies, D. (2018). First detection of *Anopheles stephensi* Liston, 1901 (Diptera: Culicidae) in Ethiopia using molecular and morphological approaches. *Acta Tropica, 188*, 180–186.

Carvajal, H., De Herrera, M.A., Quintero, J., Alzate, A., & Herrera, S. (1989). *Anopheles neivai*: A vector of malaria in the Pacific lowlands of Colombia. *Transactions of the Royal Society of Tropical Medicine and Hygiene, 83*, 609.

Carvalho, D.O., McKemey, A.R., Garziera, L., Lacroix, R., Donnelly, C.A., Alphey, L., . . . Capurro, M.L. (2015). Suppression of a field population of *Aedes aegypti* in Brazil by sustained release of transgenic male mosquitoes. *PLOS Neglected Tropical Diseases, 9*(7), e0003864.

Casal, O.H. (1963). Comentarios a propósito de la revalidación de *Culex (Melanoconion) clarki* Evans (Diptera: Culicidae). *Revista Brasileira de biologia, 23*(3), 317–320.

Casal, O.H. (1967). *Culex (Culex) ameliae*, una nueva especie de la República Argentina (Diptera, Culicidae). *Neotropica-La Plata, 13*(40), 29–30.

Casal, O.H., & García, M. (1966). Una nueva especie de *Wyeomyia (Dendromyia)* de la República Argentina (Diptera, Culicidae). *Physis, 26*(71), 155–162.

Casal, O.H., & García, M. (1967a). *Culex (Melanoconion) pavlovskyi*, nueva especie de la República Argentina (Diptera, Culicidae). *Neotropica-La Plata, 13*(40), 19–20.

Casal, O.H., & García, M. (1967b). *Culex (Culex) castroi,* nueva especie de la República Argentina (Diptera, Culicidae). *Physis, 26*(73), 451–457.

Casal, O.H., & García, M. (1967c). *Culex (Culex) hepperi*, nueva especie del delta Bonaerense del Rio Paraná (Diptera, Culicidae). *Physis, 27*(74), 87–94.

Casal, O.H., & García, M. (1968a). *Culex (Allimanta)* nuevo subgénero para *Culex (Culex) tramazayguesi* Duret, 1954 (Diptera, Culicidae). *Physis, 27*(75), 329–335.

Casal, O.H., & García, M. (1968b). *Culex (Eubonnea) clastrieri*, una nueva especie del Brasil (Diptera, Culicidae). *Neotropica-La Plata, 14*, 119–120.

Casal, O.H., & García, M. (1968c). *Culex (Melanoconion) martinezi*, una nueva especie de la República Argentina (Diptera, Culicidae). *Physis, 27*(75), 455–459.

Casal, O.H., & García, M. (1968d). *Culex (Melanoconion) dureti*, una nueva especie Argentino-Paraguaya (Diptera, Culicidae). *Physis, 28*(76), 123–125.

Casal, O.H., & García, M. (1968e). *Culex (Culex) eduardoi,* una nueva especie de la República Argentina (Diptera, Culicidae). *Revista de la Sociedad Entomologica Argentina, 30*(1–4), 105–111.

Casal, O.H., & García, M. (1968f). El subgenero *Culex (Carrollia)* Lutz 1905, nuevo para la entomofauna Argentina (Diptera, Culicida). *Revista de la Sociedad Entomológica Argentina, 30*(for 1967), 97–98.

Casal, O.H., & García, M. (1970). Los tipos de tres especies descriptas por Juana Petrocchi (Diptera: Culicidae). *Physis, 30*(80), 11–12.

Casal, O.H., & García, M. (1971). *Culex (Culex) tatoi*, una nueva especie de la República Argentina (Diptera, Culicidae). *Physis, 30*(81), 631–635.

Casal, O.H., García, M., & Cavalieri, F. (1966). *Culex (Culex) fernandezi*, una nueva especie bromelicola de la República Argentina (Diptera, Culicidae). *Physis, 26*(71), 185–192.

Casal, O.H., García, M., & Fernández, H.I. (1968). El subgénero *Culex (Aedinus)* Bourroul, 1904, nuevo para la entomofauna Argentina, con la descripción de una nueva especie (Diptera, Culicidae). *Physis, 28*(76), 217–218.

Castro, M., & Bressanello, M. (1952). Revisión de las espécies de "*Taeniohynchus (Rhynchotaenia)*" (Diptera, Culicidae). *Revista Brasileira de biologia, 12*(3), 229–246.

Cator, L. (2011). *The role of bioaccoustics in the mating behavior of medically important mosquitoes* [Doctoral dissertation, Cornell University]. eCommons Cornell Digital Repository. Retrieved from http://hdl.handle.net/1813/29394

Cator, L.J., & Harrington, L.C. (2011). The harmonic convergence of fathers predicts the mating success of sons in *Aedes aegypti*. *Animal Behaviour, 82*(4), 627–633.

Cator, L.J., Arthur, B.J., Harrington, L.C., & Hoy, R.R. (2009). Harmonic convergence in the love songs of the dengue vector mosquito. *Science, 323*(5917), 1077–1079.

Cator, L.J., Arthur, B.J., Ponlawat, A., & Harrington, L.C. (2011). Behavioral observations and sound recordings of free-flight mating swarms of *Ae. aegypti* (Diptera: Culicidae) in Thailand. *Journal of Medical Entomology, 48*(4), 941–946.

Causey, O.R. (1937a). New anopheline and culicine mosquitoes from Siam. *The Journal of Parasitology, 23*(5), 543–545.

Causey, O.R. (1937b). Some anopheline and culicine mosquitoes of Siam with remarks on malaria control. *The American Journal of Hygiene, 25*(2), 400–420.

Causey, O.R. (1945). Description of *Anopheles (Nyssorhynchus) dunhami*, a new species from the Upper Amazon Basin. *Journal of the National Malaria Society, 4*(3), 231–234.

Causey, O.R., Deane, L.M., & Deane, M.P. (1943a). Ecology of *Anopheles gambiae* in Brazil. *American Journal of Tropical Medicine, 23*(1), 73–94.

Causey, O.R., Deane, L.M., & Deane, M.P. (1943b). Descrição de um novo anofelino da parte alta do Vale do Amazonas *Anopheles (Nyssorhynchus) galvaoi* n. sp. *Revista Paulista médica, 23*, 293–296.

Causey, O.R., Deane, L.M., Deane, M.P., & Sampaio, M.M. (1943). *Anopheles (Nyssorhynchus) sawyeri,* a new anopheline mosquito from Ceará, Brazil. *Annals of the Entomological Society of America, 36*(1), 11–20.

Causey, O.R., Deane, L.M., & Deane, M.P. (1944). An illustrated key to the eggs of thirty species of Brazilian anophelines, with several new descriptions. *American Journal of Hygiene, 39*(1), 1–7.

Causey, O.R., Deane, L.M., & Deane, M.P. (1945a). *Anopheles aquasalis* vs. *Anopheles tarsimaculatus* as the name for the brackish water anopheline of Central and South America and the Caribbean Islands. *Journal of the National Malaria Society, 4*(3), 243–250.

Causey, O.R., Deane, L.M., & Deane, M.P. (1945b). Description of *Chagasia rozeboomi* an anopheline from Ceará, Brasil. *Journal of the National Malaria Society, 4*(4), 341–350.

Causey, O.R., Deane, L.M., & Deane, M.P. (1946). II. An illustrated key by male genitalic characteristics for the identification of thirty-four species of Anophelini from the northeast and Amazon regions of Brazil, with a note on dissection technique. *American Journal of Hygiene Monograph Series, 18*, 21–31.

Cavanagh, S., & Townson, H. (1986). Evidence for an aggregation pheromone in feeding mosquitoes. *Transactions of the Royal Society of Tropical Medicine and Hygiene, 80*(2), 334.

Cequina, E.Y., & Schoenig, E. (1976). An ecological study of *Anopheles* mosquitoes in Asuncion, Davao del Norte. *Philippine Scientist, 13*, 132–134.

Cerqueira, N.L. (1938). Nota sobre o genero *Deinocerites*: A sua presença no Estado do Maranhão, Brasil (Dipt. Culicidae). *Revista de entomología, 8*(3–4), 289–291.

Cerqueira, N.L. (1939). Sobre *Psorophora lutzi* Theobald, *Psorophora albipes* Theobald e *Psorophora forceps* n. sp. (Dipt. Culicidae). *Revista de entomología, 10*(1), 78–85.

Cerqueira, N.L. (1943a). Algumas espécies novas da Bolívia, e referência a três espécies de *Haemagogus* (Diptera, Culicidae). *Memórias do Instituto Oswaldo Cruz, 39*(1), 1–14.

Cerqueira, N.L. (1943b). Lista dos mosquitos da Bolívia (Diptera, Culicidae). *Memórias do Instituto Oswaldo Cruz, 39*(1), 15–36.

Cerqueira, N.L. (1950). Description of the larva, pupa and female of *Aedes (Finlaya) argyrothorax* Bonne-Wepster and Bonne (Diptera, Culicidae). *Proceedings of the Entomological Society of Washington, 52*(4), 173–178.

Cerqueira, N.L. (1957). Descrição da larva e pupa de *Aedes* (*Ochlerotatus*) *lepidus* Cerqueira & Paraense (Diptera: Culicidae, Culicini). *Boletim do Museu Paraense Emílio Goeldi, zoologia, 3*, 1–4.

Cerqueira, N.L. (1960a). Uma nova espécie de *Haemagogus* do Amazonas (Diptera, Culicidae). *Boletim do Museu Paraense Emilio Goeldi, Zoologia, 25*, 1–5.

Cerqueira, N.L. (1960b). Sôbre uma n. sp. de *Janthinosoma* da Série Lutzi e a descrição do macho de *Psorophora* (J.) *circumflava* Cerqueira, 1943 (Diptera Culicidae). *Boletim do Museu Paraense Emílio Goeldi, zoologia, 26*, 1–7.

Cerqueira, N.L. (1961a). Sôbre *Trichoprosopon* (*R.*) *edwardsianum* (Lane & Cerqueira, 1942) (Diptera, Culicidae). *Studia entomologica, 4*(1–4), 459–464.

Cerqueira, N.L. (1961b). Cinco novos sabetinos da Amazonia (Diptera, Culicidae). *Revista Brasileira de entomología*, 10, 37–52.

Cerqueira, N., & Antunes, P.C.A. (1938). *Haemagogus tropicalis*, a new species from Pará, Brazil (Diptera, Culicidae). *Proceedings of the Entomological Society of Washington, 40*(1), 1–9.

Cerqueira, N.L., & Boshell-Manrique, J. (1946). Note on *Haemagogus spegazzinii* Brèthes, 1912 (Diptera, Culicidae). *Proceedings of the Entomological Society of Washington, 48*(8), 191–200.

Cerqueira, N.L., & Costa, A.F. (1946). Duas novas espécies de *Aedes* (*Ochlerotatus*) (Diptera, Culicidae). *Livro de homenagem a R.F. d'Almeida, 11*, 133–140.

Cerqueira, N.L., & Lane, J. (1944). Sôbre algumas espécies de *Culex* (*Carrollia*) (Diptera, Culicidae). *Revista de entomología, 15*(1–2), 216–220.

Cerqueira, N.L., & Lane, J. (1945). Note on *Haemagogus capricornii* Lutz, 1904 (Diptera, Culicidae). *Proceedings of the Entomological Society of Washington, 47*(9), 279–288.

Cerqueira, N.L., & Lobato Paraense, W. (1945). Uma nova espécie de *Aedes*, transmissora de *Plasmodium gallinaceum* (Diptera, Culicidae). *Memórias do Instituto Oswaldo Cruz, 42*(1), 11–15.

Cervone, L. (1957). Sulla presenza di *Culex* (*Neoculex*) *martinii* Medschid in Provincia di Latina e contributo alla conoscenza della specie. *Rivista di parassitologia, 18*(4), 235–248.

Chadee, D.D. (1986). Description of the egg of *Culex* (*Melanoconion*) *caudelli* Dyar and (Diptera, Culicidae). *Mosquito Systematics, 18*(2), 196–198.

Chadee, D.D. (2003). Surveillance for the dengue vector *Aedes aegypti* in Tobago, West Indies. *Journal of the American Mosquito Control Association, 19*(3), 199–205.

Chadee, D.D., & Bennett, H. (1988). Description of *Culex sitiens* eggs from Oman (Diptera: Culicidae). *Mosquito Systematics, 20*(3), 370–373.

Chadee, D.D., & Bennett, H. (1990). Scanning electron microscopy of the eggs of *Haemagogus celeste* and *Haemagogus equinus* (Diptera: Culicidae) from Trinidad, West Indies. *Mosquito Systematics, 22*(1), 11–14.

Chadee, D.D., & Gilles, J.R. (2014). The diel copulation periodicity of the mosquito, *Aedes aegypti* (L.) (Diptera: Culicidae) at indoor and outdoor sites in Trinidad, West Indies. *Acta Tropica, 132*(Suppl.), S91–S95.

Chadee, D.D., & Haeger, J.S. (1986). A description of the egg of *Culex* (*Culex*) *nigripalpus* Theobald from Florida, with notes on five egg rafts (Diptera: Cullicae). *Mosquito Systematics, 18*(3, 4), 288–292.

Chadee, D.D., & Tikasingh, E.S. (1986). The eggs of *Culex* (*Carrollia*) *urichii* (Coquillett) (Diptera: Culicidae) in Trinidad, W.I. *Mosquito Systematics, 18*(3, 4), 284–287.

Chadee, D.D., & Wilkerson, R.C. (2005). *Anopheles triannulatus* (Neiva and Pinto): A new *Anopheles* record from Trinidad, West Indies. *Journal of the American Mosquito Control Association, 21*(3), 316–317.

Chadee, D.D., Small, G.J., & O'Malley, S.L.C. (1987). The description of the eggs of *Toxorhynchites moctezuma* (Diptera: Culicidae) from Trinidad, West Indies, with notes on other eggs. *Mosquito Systematics, 19*(3), 237–242.

Chadee, D.D., Corbet, P.S., & Greenwood, J.J.D. (1990). Egg-laying yellow fever mosquitoes avoid sites containing eggs laid by themselves or by conspecifics. *Entomología Experimentalis et Applicata, 57*(3), 295–298.

Chadee, D.D., Fat, F.H., & Persad, R.C. (2003). First record of *Aedes albopictus* from Trinidad, West Indies. *Journal of the American Mosquito Control Association, 19*(4), 438–439.

Chagas, C. (1907). *Novas especies de culicidos brasileiros* . Rio de Janeiro, Brazil: Besnard Frères.

Chalmers, A.J. (1900). Uncomplicated aestivo-autumnal fever in Europeans in the Gold Coast Colony, West Africa. *Lancet, 2*, 1262–1264.

Chamberlain, R.W., Sudia, W.D., & Nelson, D.B. (1955). Laboratory observations on a mosquito, *Culiseta melanura* (Coquillett). *Mosquito News, 15*(1), 18–21.

Chambers, G.M., & Klowden, M.J. (1990). Correlation of nutritional reserves with a critical weight for pupation in larval *Aedes aegypti* mosquitoes. *Journal of the American Mosquito Control Association, 6*(3), 394–399.

Chandra, G., Bhattacharjee, I., & Chatterjee, S. (2010). A review on *Anopheles subpictus* Grassi—a biological vector. *Acta Tropica, 115*(1), 142–154.

Chang, H. (1965). Description of a mosquito *Aedes* (*Edwardsaedes*) *pingpaensis*, sp. nov. [in Chinese]. *Acta parasitologica Sinica, 2*(4), 397–402.

Chang, H.-T., Zhao, S.-X., Hang, B.-X., & Chen, S.-F. (1975). A new species of *Culex*. *Acta entomologica Sinica, 18*(3), 338–340.

Chang, M.C., Teng, H.C., Chen, C.F., Chen, Y.C., & Jeng, C.R. (2008). The resting sites and blood-meal sources of *Anopheles minimus* in Taiwan. *Malaria Journal, 7*(1), 105.

Chang, P.-X., & Chang, S.-M. (1974). Two new species of Culicini from Kweichow. *Acta entomologica Sinica, 17*(3), 347–352.

Chapman, H.C., & Barr, A.R. (1964). *Aedes communis nevadensis*, a new subspecies of mosquito from western North America (Diptera: Culicidae). *Mosquito News, 24*(4), 439–447.

Chapman, H.C., & Grodhaus, G. (1963). The separation of adult females of *Aedes dorsalis* (Meigen) and *A. melanimon* Dyar in California. *California Vector Views, 10*(8), 53–56.

Chapman, T. (2009). Finding the right plugin: Mosquitoes have the answer. *PLOS Biology, 7*(12), e1000273.

Charlwood, J.D., Graves, P.M., & Marshall, T.D.C. (1988). Evidence for a "memorized" home range in *Anopheles farauti* females from Papua New Guinea. *Medical and Veterinary Entomology, 2*(2), 101–108.

Charlwood, J.D., Billingsley, P.F., & Hoc, T.Q. (1995a). Mosquito-mediated attraction of female European but not African mosquitoes. *Annals of Tropical Medicine and Parasitology, 89*(3), 327–329.

Charlwood, J.D., Kihonda, J., Sama, S., Billingsley, P.F., Hadji, H., Verhave, J.P., . . . Smith, T. (1995b). The rise and fall of *Anopheles arabiensis* (Diptera: Culicidae) in a Tanzanian village. *Bulletin of Entomological Research, 85*(1), 37–44.

Charlwood, J.D., Smith, T., Kihonda, J., Heiz, B., Billingsley, P.F., & Takken, W. (1995c). Density independent feeding success of malaria vectors (Diptera: Culicidae) in Tanzania. *Bulletin of Entomological Research, 85*(1), 29–35.

Charlwood, J.D., Vij, R., & Billingsley, P.F. (2000). Dry season refugia of malaria-transmitting mosquitoes in a dry savannah zone of east Africa. *The American Journal of Tropical Medicine and Hygiene, 62*(6), 726–732.

Charlwood, J.D., Tomás, E.V.E., Kelly-Hope, L., & Briët, O.J.T. (2014). Evidence of an "invitation" effect in feeding sylvatic *Stegomyia albopicta* from Cambodia. *Parasites and Vectors, 7*(1), 324.

Charrel, R.N., de Lamballerie, X., & Raoult, D. (2007). Chikungunya outbreaks—the globalization of vectorborne diseases. *New England Journal of Medicine, 356*(8), 769–771.

Chaudhry, S., Gupta, N.S., & Chhilar, J.S. (2005). Salivary polytene chromosome mapping of *Anopheles* (*Cellia*) *subpictus* Grassi (Culicidae: Diptera). *Genome, 48*(2), 241–246.

Chauvet, G. (1962). Variabilité geographique chez les femelles d'*Anopheles mascarensis* De Meillon, 1947. Absence d'*A. marshalli* Theobald, 1929 à Madagascar. *Bulletin de la Société de pathologie exotique, 55*(6), 1145–1156.

Chauvet, G., & Dejardin, J. (1968). Caractères chetotaxiques de distinction entre larves (stade IV) de l'espèce A et de l'espèce B du complexe anophèle *gambiae* à Madagascar. *Cahiers ORSTOM. Série entomologie médicale, 6*(1), 69–101.

Chauvet, G., Davidson, G., & Dejardin, J. (1969). Validité d'une méthode chétotaxique de distinction des larves des espèces *A* et *B* du complexe *Anopheles gambiae* Giles à Madagascar. *Cahiers ORSTOM. Série entomologie médicale, 7*(1), 51–60.

Chaverri, L.G. (2009). Culicidae (mosquitos [mosquitoes], zancudos). In B.V. Brown, A. Borkent, J.M. Cumming, D.M. Wood, N.E. Woodley, & M.A. Zumbado (Eds.), *Manual of Central American Diptera* (Vol. 1, pp. 369–388). Ottawa, Canada: NRC Research Press.

Chaves, L.F., Harrington, L.C., Keogh, C.L., Nguyen, A.M., & Kitron, U.D. (2010). Blood feeding patterns of mosquitoes: Random or structured? *Frontiers in Zoology, 7*(1), 3.

Chavshin, A.R., Oshaghi, M.A., Vatandoost, H., Pourmand, M.R., Raeisi, A., Enayati, A.A., . . . Ghoorchian, S. (2012). Identification of bacterial microflora in the midgut of the larvae and adult of wild caught *Anopheles stephensi*: A step toward finding suitable paratransgenesis candidates. *Acta Tropica, 121*(2), 129–134.

Chen, B., Harbach, R.E., & Butlin, R.K. (2002). Molecular and morphological studies on the *Anopheles minimus* group of mosquitoes in southern China: Taxonomic review, distribution and malaria vector status. *Medical and Veterinary Entomology, 16*(3), 253–265.

Chen, B., Butlin, R.K., & Harbach, R.E. (2003). Molecular phylogenetics of the Oriental members of the Myzomyia Series of *Anopheles* subgenus *Cellia* (Diptera: Culicidae) inferred from nuclear and mitochondrial DNA sequences. *Systematic Entomology, 28*(1), 57–69.

Chen, B., Butlin, R.K., Pedro, P.M., Wang, X.Z., & Harbach, R.E (2006). Molecular variation, systematics and distribution of the *Anopheles fluviatilis* complex in southern Asia. *Medical and Veterinary Entomology, 20*(1), 33–43.

Chen, B., Pedro, P.M., Harbach, R.E., Somboon, P., Walton, C., & Butlin, R.K. (2011). Mitochondrial DNA variation in the malaria vector *Anopheles minimus* across China, Thailand and Vietnam: Evolutionary hypothesis, population structure and population history. *Heredity, 106*(2), 241–252.

Chen, B., Harbach, R.E., Walton, C., He, Z., Zhong, D., Yan, G., & Butlin, R.K. (2012). Population genetics of the malaria vector *Anopheles aconitus* in China and Southeast Asia. *Infection, Genetics and Evolution, 12*(8), 1958–1967.

Chen, C.C., Jenkins, E., Epp, T., Waldner, C., Curry, P.S., & Soos, C. (2013). Climate change and West Nile virus in a highly endemic region of North America. *International Journal of Environmental Research and Public Health, 10*(7), 3052–3071.

Chen, C.Y. (1972). Studies on morphology of the cibarium in culicine mosquitos. I. Eight species of culicines common in the Taipei area, Taiwan. *Journal of the Formosan Medical Association, 71*, 282–291.

Chen, C.Y. (1974). Studies on morphology of the cibarium in culicine mosquitos. II. Some *Culex* mosquitos of Taiwan belonging to Subgenera *Lophoceraomyia, Eumelanomyia, Culiciomyia*. *Journal of the Formosan Medical Association, 73*, 511–525.

Chen, H.-B. (1980a). Descriptions of the female and larva of *Culex* (*Eumelanomyia*) *richei* Klein. *Acta zootaxonomica Sinica, 5*(3), 320–321.

Chen, H.-B. (1980b). Study on the *Culex vishnui* subgroup in China with discussions on the taxonomic status of *C. neovishnui* Lien, *C. permixtus* Hsieh & Liao and *C. cheni* Ho. *Acta entomologica Sinica, 23*(4), 434–440.

Chen, H.-B. (1981). A new species of *Culex* (Diptera: Culicidae). *Acta entomologica Sinica, 24*(4), 426–428.

Chen, H.-B. (1982). A new species of *Culex* from Guizhou, *Culex* (*Eumelanomyia*) *miaolingensis*, sp.nov. *Acta zootaxonomica Sinica, 7*(3), 312–316.

Chen, H.-B. (1989). A new species of *Culex* (*Culiciomyia*) from Yunnan (Diptera: Culicidae). *Acta zootaxonomica Sinica, 14*(1), 86–90.

Chen, H.-B., & Dong, X.-S. (1992). A new species of *Culex* (*Eumelanomyia*) from China (Diptera: Culicidae). *Acta zootaxonomica Sinica, 17*(1), 78–83.

Chen, H.-B., & Zhang, P.-X. (1988). *Topomyia* (*Suaymyia*) *tumetarsalis*, a new species from China (Diptera: Culicidae). *Acta entomologica Sinica, 31*(2), 213–218.

Chen, H.-B., & Zhao, H. (1984). Descriptions of the larva and pupa of *Aedes* (*Aedes*) *mubiensis* Luh & Shih. *Acta zootaxonomica Sinica, 9*(2), 222–224.

Chen, H.-B, & Zhao, H. (1985). A new species of the genus *Culex* from China (Diptera: Culicidae). *Entomotaxonomia, 7*(4), 265–269.

Chen, H.-B., & Zhao, H. (1987). Descriptions of three species of *Culex* pupae from China (Diptera: Culicidae). *Acta zootaxonomica Sinica, 12*(2), 222–224.

Chen, H.-B., Wang, S.H., & Zhao, H. (1989). A new species of *Uranotaenia* breeding in crab holes from Guizhou Province of China (Diptera: Culicidae). *Acta zootaxonomica Sinica, 14*(2), 224–228.

Chen, H.-H., & Lien, J.-C. (1956). Mosquitoes of genus *Ficalbia* in Taiwan (Formosa), China. *Journal of the Formosan Medical Association, 55*(5), 199–215.

Chen, H.-L., & Chen, H.-B. (2000). Description of a new species of the *Aedes* from Henan, China (Diptera: Culicidae). *Acta zootaxonomica Sinica, 25*(3), 341–344.

Chen, H.-P. (1977). A new species of *Culex* (Diptera: Culicidae). *Acta entomologica Sinica, 20*(1), 95–98.

Chen, L., Zhu, J., Sun, G., & Raikhel, A.S. (2004). The early gene *Broad* is involved in the ecysteroid hierarchy governing vitellogenesis of the mosquito *Aedes aegypti*. *Journal of Molecular Endocrinology, 33*(3), 743–761.

Cheng, L., & Hogue, C.L. (1974). New distribution and habitat records of biting midges and mangrove flies from the coasts of southern Baja California, Mexico (Diptera: Ceratopogonidae, Culicidae, Chironomidae, and Phoridae). *Entomological News, 85*(7–8), 211–218.

Cheng, M.L., & Barr, A.R. (1976). The inheritance of pigmented-paddle (du2) in *Culex pipiens* (Diptera: Culicidae). *Journal of Medical Entomology, 13*(3), 347–350.

Chevalier, V., de la Rocque, S., Baldet, T., Vial, L., & Roger, F. (2004). Epidemiological processes involved in the emergence of vectorborne diseases: West Nile fever, Rift Valley fever, Japanese encephalitis and Crimean-Congo haemorrhagic fever. *Revue scienti-*

fique et technique de l'Office International des Epizooties, 23(2), 535–555.

Chhilar, J.S., & Chaudhry, S. (2012). Phylogenetic analysis of *Anopheles (Cellia) subpictus* Grassi using rDNA-ITS2 sequence. *Proceedings of the Zoological Society* (Calcutta), *65*(1), 1–10.

Chilaka, N., Perkins, E., & Tripet, F. (2012). Visual and olfactory associative learning in the malaria vector *Anopheles gambiae* sensu stricto. *Malaria Journal, 11*(1), 27.

Chinayev, P.P. (1965). Range and epidemiological importance of the malarial mosquitoes *Anopheles maculipennis sacharovi* Fv., *A. superpictus* Fr. and *A. pulcherrimus* Theob. (Diptera, Culicidae) in Soviet central Asia. *Entomological Review, 44*(2), 164–169.

Choochote, W., & Saeung, A. (2013). Genetic and phenetic approaches to *Anopheles* systematics. In S. Manguin (Ed.), Anopheles Anopheles *mosquitoes: New insights into malaria vectors* (pp. 57–79). Rijeka, Croatia: InTech.

Choochote, W., Iwaki, M., & Suwanpanit, P. (1987). Scanning electron microscopy of the egg of *Aedes* (*Mucidus*) *quasiferinus* Mattingly, 1961. *Southeast Asian Journal of Tropical Medicine and Public Health, 18*, 569–571.

Choochote, W., Jitpakdi, A., Sukontason, K., Suntaravitun, T., Wongkamchai, S., & Pitasawat, B. (2001). Scanning electron microscopy of *Aedes lineatopennis* (Diptera: Culicidae) eggs. *Journal of Medical Entomology, 38*(5), 753–755.

Chow, C.Y. (1949a). The anopheline mosquitoes of Taiwan (Formosa), China. *Quarterly Journal of the Taiwan Museum, 2*(1), 1–8.

Chow, C.Y. (1949b). The identification and distribution of Chinese anopheline mosquitoes. *Journal of the National Malaria Society, 8*(2), 121–131.

Chow, C.Y. (1950). Collection of culicine mosquitoes (Diptera, Culicidae) in Taiwan (Formosa), China, with description of a new species. *Quarterly Journal of the Taiwan Museum, 3*(4), 281–287.

Chow, C.Y. (1970). Bionomics of malaria vectors in the western Pacific region. *Southeast Asian Journal of Tropical Medicine and Public Health, 1*(1), 40–57.

Chow, C.Y. (1973). Arthropods of public health importance in Korea. *Korean Journal of Entomology, 1*(1), 31–54.

Chow, C.Y., & Mattingly, P.F. (1951). The male genitalia and early stages of *Aedes* (*Finlaya*) *albocinctus* Barraud and *Aedes* (*Finlaya*) *albotaeniatus* var. *mikaranus* Edwards with some notes on related species. *Proceedings of the Royal Society of London, Series B: Taxonomy 20*(7–8), 80–90.

Chowdaiah, B.N., & Seetharam, P.L. (1975). Chromosome studies of oriental anophelines. IV. Polytene chromosomes of *Anopheles fluviatilis. Mosquito News, 35*(4), 495–500.

Chowdaiah, B.N., Avirachan, T.T., & Seetharam, P.L. (1970). Chromosome studies of oriental anophelines. I. The salivary gland chromosomes of *Anopheles barbirostris. Experientia, 26*, 315–317.

Chowdhury, K.L. (1929). A new variety of protanopheline, *A. barbirostris* van der Wulp, var. *ahomi,* found in Upper Assam. *Indian Journal of Medical Research* (Calcutta), *16*(4), 986–988.

Christophers, S.R. (1906). On the importance of larval characters in the classification of mosquitoes. *Scientific Memoirs by Officers of the Medical and Sanitary Departments of the Government of India* (Calcutta) (n.s.), *25*, 1–18.

Christophers, S.R. (1911a). [Notes on mosquitoes]. II. A new anopheline. *Neocellia fowleri* n. sp. *Paludism* (Simla), *2*, 64–68 [plate for *fowleri* in Chistophers 1911b, p. 130].

Christophers, S.R. (1911b). [Notes on mosquitoes]. III. A new culicine, *Leslieomyia taeniorhynchoides*, nov. gen. et sp. *Paludism* (Simla), *3*, 68–72, 130 [plate accompanying Christophers 1911a].

Christophers, S.R. (1915). The male genitalia of *Anopheles. Indian Journal of Medical Research* (Calcutta), *3*, 371–394.

Christophers, S.R. (1916a). A new anopheline with unspotted wings from Mesopotamia (*Anopheles lukisii*). *Indian Journal of Medical Research* (Calcutta), *4*(1), 120–122.

Christophers, S.R. (1916b). A revision of the nomenclature of Indian Anophelini. *Indian Journal of Medical Research* (Calcutta), *3*(3), 454–488.

Christophers, S.R. (1923a). A new east African anopheline (*A. kingi*) related to *A. natalensis*, Hill and Hayden, and *A. watsonii*, Leicester. *Indian Journal of Medical Research* (Calcutta), *10*(4), 1008–1019.

Christophers, S.R. (1923b). An *Anopheles* of the *Myzorhynchus* group (*Anopheles amazonicus* sp. n.) from South America. *Annals of Tropical Medicine and Parasitology, 17*(1), 71–78.

Christophers, S.R. (1924a). Provisional list and reference catalogue of the Anophelini. *Indian Medical Research Memoirs, 3*, 1–105.

Christophers, S.R. (1924b). Some Himalayan and peninsular varieties of Indian species of *Anopheles. Indian Journal of Medical Research* (Calcutta), *12*(1), 11–13.

Christophers, S.R. (1924c). Some further varieties of Indian species of *Anopheles* with notes on the species *A. pallidus* Theobald and *A. philippinensis* Ludlow. *Indian Journal of Medical Research* (Calcutta), *12*(2), 295–301.

Christophers, S.R. (1926). *A.* (*Mysomyia*) *pattoni*, a new *Anopheles* from Shan-tung, North China; with notes on some other species of *Anopheles* from the same locality. *Indian Journal of Medical Research* (Calcutta), *13*, 871–877.

Christophers, S.R. (1933). *The fauna of British India, including Ceylon and Burma. Diptera.* Vol. IV. *Family Culicidae. Tribe Anophelini.* London, England: Taylor and Francis.

Christophers, S.R. (1951). Note on morphological characters differentiating *Culex pipiens* L. from *Culex molestus* Forskål and the status of these forms. *Transactions of the Royal Entomological Society of London, 102*, 372–379.

Christophers, S.R. (1960). Aedes aegypti *(L.), the yellow fever mosquito: Its life history, bionomics and structure.* Cambridge, United Kingdom: Cambridge University Press.

Christophers, S.R., & Barraud, P.J. (1931). The eggs of Indian *Anopheles*, with descriptions of the hitherto undescribed eggs of a number of species. *Records of the Malaria Survey of India, 2*(1), 161–192.

Christophers, S.R., & Chand, K. (1915). Notes on some anophelines from Arabia and Mesopotamia. *Indian Journal of Medical Research* (Calcutta), *3*(1), 180–200.

Christophers, S.R., & Puri, I.M. (1931a). Species and varieties of the *funestus* series of *Anopheles. Records of the Malaria Survey of India, 2*(3), 481–493.

Christophers, S.R., & Puri, I.M. (1931b). Notes on some anopheline mosquitoes collected in Sierra Leone including differentiation of *Anopheles dthali* Patton (Mediterranean) as a distinct species from *Anopheles rhodesiensis* Theo. (Ethiopian). *Indian Journal of Medical Research* (Calcutta), *18*(4), 1133–1166.

Christophides, G.K. (2005). Transgenic mosquitoes and malaria transmission. *Cellular Microbiology, 7*(3), 325–333.

Chu, F.-I. (1957). Collection of megarhine and culicine mosquitoes from Hainan Island, South China, with description of a new species. *Acta zoologica Sinica, 9*(2), 145–164.

Chu, F.-I. (1958). Advances in the study of culicine mosquitoes of Hainan, South China. *Indian Journal of Malariology, 12*(2), 109–113.

Chu, F.-I. (1980). The synonomy, distribution and classification of Chinese *Uranotaenia* (Diptera: Culicidae). *Entomotaxonomia, 2*(3), 237–245.

Chu, F.-I. (1981). A new species of Chinese *Uranotaenia* (Diptera: Culicidae). *Entomotaxonomia, 3*(4), 273–275.

Chu, F.-I., & Chen, H. (1983). Two new records of *Uranotaenia* mosquitoes breeding in crab holes from Yimnan Province (Diptera: Culicidae). *Entomotaxonomia, 5*(4), 343–344.

Chua, T.H., Manin, B.O., Daim, S., Vythilingam, I., & Drakeley, C. (2017). Phylogenetic analysis of simian *Plasmodium* spp. infecting *Anopheles balabacensis* Baisas in Sabah, Malaysia. *PLOS Neglected Tropical Diseases, 11*(10), e0005991.

Chwatt, L.J. (1948). A new *Aedes* from the Cameroons, *A.* (*Aedimorphus*) *boneti* n. sp. *kumbae* s.sp. nov. *Annals of Tropical Medicine and Parasitology, 42*(2), 184–189.

Chwatt, L.J. (1949). *Aedes* (*Stegomyia*) *pseudoafricanus* sp. nov.: A new species of *Aedes* from the coast of Nigeria (British West Africa). *Nature, 163*(4151), 808.

Cianchi, R., Sabatini, A., Boccolini, D., Bullini, L., & Coluzzi, M. (1987). *Electrophoretic evidence of reproductive isolation between sympatric populations of* Anopheles melanoon *and* An. subalpinus. Paper presented at the Third International Congress on Malaria and Babiosis, International Laveran Foundation, Annecy, France, p. 1560.

Ciota, A.T., Styer, L.M., Meola, M.A., & Kramer, L.D. (2011). The costs of infection and resistance as determinants of West Nile virus susceptibility in *Culex* mosquitoes. *BMC Ecology, 11*(1), 23.

Cirimotich, C.M., Ramirez, J.L., & Dimopoulos, G. (2011). Native microbiota shape insect vector competence for human pathogens. *Cell Host and Microbe, 10*(4), 307–310.

Clark, M.A., Moran, N.A., Baumann, P., & Wernegreen, J.J. (2000). Cospeciation between bacterial endosymbionts (*Buchnera*) and a recent radiation of aphids (*Uroleucon*) and pitfalls of testing for phylogenetic congruence. *Evolution, 54*(2), 517–525.

Clark-Gil, S., & Darsie, R.F., Jr. (1983). The mosquitoes of Guatemala: Their identification, distribution and bionomics, with keys to adult females and larvae, in English and Spanish. *Mosquito Systematics, 15*(3), 151–284.

Clastrier, J. (1968). Présence, en Guyane Française, de *Culex* (*Melanoconion*) *creole* Anduze, 1948 et *Culex* (*Mochlostyrax*) *foliafer* Komp et Rozeboom, 1951 (Diptera, Culicidae). *Archives de l'Institut Pasteur de la Guyane Française et du Territoire de l'Inini, 21*(523), 225–230.

Clastrier, J. (1970a). *Culex* (*Eubonnea*) *guyanensis* n. sp. (Diptera, Culicidae) nouveau moustique de la Guyane Française. *Annales de parasitologie humaine et comparée, 45*(1), 115–118.

Clastrier, J. (1970b). *Culex* (*Melanoconion*) *dolichophyllus* n. sp. et *Culex* (*Microculex*) *stonei* Lane et Whitman, 1943 (Diptera, Culicidae) en Guyane Française. *Annales de parasitologie humaine et comparée, 45*(5), 857–861.

Clastrier, J. (1970c). Quatre nouveaux *Melanoconion* (Diptera, Culicidae) de la Guyane Française. *Annales de parasitologie humaine et comparée, 45*(4), 463–476.

Clastrier, J. (1971). Sur deux *Melanoconion* et un nouvel *Aedinus* de la Guyane Française (Dipt. Culicidae). *Annales de la Société Entomologique de France, 7*(3), 629–654.

Clastrier, J. (1972). Description des estades inconnus de *Culex* (*Mochlostyrax*) *lacertosus* Komp & Rozeboom, 1951 (Dipt. Culicidae). *Annales de la Société Entomologique de France, 8*(1), 169–178.

Clastrier, J. (1973). *Wyeomyia* (*Dendromyia*) *trifurcata* n. sp. nouveau moustique de la Guyane Française (Diptera, Culicidae). *Nouvelle revue d'entomologie, 3*(1), 39–47.

Clastrier, J. (1974). *Wyeomyia* (*Dendromyia*) *bourrouli* (Lutz, 1905) et *Wyeomyia* (*Cruzmyia*) *forattinii* n. sp. en Guyane Française (Diptera, Culicidae). *Cahiers ORSTOM. Serie entomologie médicale, 12*(3), 153–163.

Clastrier, J., & Claustre, J. (1978a). *Culex* (*Carrollia*) *insigniforceps* n.sp. de la Guyane Française (Diptera, Culicidae). *Nouvelle revue d'entomologie, 8*(3), 359–362.

Clastrier, J., & Claustre, J. (1978b). *Limatus martiali* Senevet et Abonnenc, 1939 et *Wyeomyia nigricephala* n.sp. (Diptera, Culicidae) de la Guyane Française. *Annales de parasitologie humaine et comparée, 53*(5), 539–546.

Clavero, G. (1946). Aedinos de España. *Revista de sanidad e higiene pública, 20*, 1–28.

Clements, A.N. (1963). *The physiology of mosquitoes*. New York, NY: Pergamon Press.

Clements, A.N. (1992). *The biology of mosquitoes: Vol. 1. Development, nutrition and reproduction*. Chapman & Hall, United Kingdom.

Clements, A.N. (1999). *The biology of mosquitoes: Vol. 2. Sensory reception and behaviour*. Wallingford, United Kingdom: CABI.

Clements, A.N. (2012). *The biology of mosquitoes: Vol. 3. Viral and bacterial pathogens and bacterial symbionts*. Cambridge University Press, Cambridge, United Kingdom.

Clifton, M.E., Correa, S., Rivera-Pérez, C., Nouzova, M., & Noriega, F.G. (2014). Male *Aedes aegypti* mosquitoes use JH III transferred during copulation to influence previtellogenic ovary physiology and affect the reproductive output of female mosquitoes. *Journal of Insect Physiology, 64*, 40–47.

Coast, G.M., Garside, C.S., Webster, S.G., Schegg, K.M., & Schooley, D.A. (2005). Mosquito natriuretic peptide identified as a calcitonin-like diuretic hormone in *Anopheles gambiae* (Giles). *Journal of Experimental Biology, 208*(17), 3281–3291.

Cockburn, A.F., Tarrant, C.A., & Mitchell, S. (1988). Use of DNA probes to distinguish sibling species of the *Anopheles quadrimaculatus* complex. *The Florida Entomologist, 71*(3), 299–302.

Cockerell, T.D.A. (1916). British fossil insects. *Proceedings of the United States National Museum, 49*(2119), 469–499.

Cockerell, T.D.A. (1918). The mosquitoes of Colorado. *Journal of Economic Entomology, 11*(2), 195–200.

Cockerell, T.D.A. (1919). The oldest mosquitoes. *Nature, 103*(2577), 43–46.

Coetzee, M. (1980). The *Anopheles coustani* group of species in South Africa. *Proceedings of the Third Entomological Congress of the Entomological Society of Southern Africa, Pretoria, 1980*, 17–18.

Coetzee, M. (1984). A new species of *Anopheles* (*Anopheles*) from Namibia (Diptera: Culicidae). *Systematic Entomology, 9*(1), 1–8.

Coetzee, M. (1987). A taxonomic description of *Anopheles quadriannulatus* Theobald (Diptera: Culicidae) from southern Africa. *Systematic Entomology, 12*(3), 305–312.

Coetzee, M. (1994a). *Anopheles crypticus*, new species from South Africa is distinguished from *Anopheles coustani* (Diptera: Culicidae). *Mosquito Systematics, 26*(3), 125–131.

Coetzee, M. (1994b). The eggs of two anopheline mosquitoes (Diptera: Culicidae) endemic to Madagascar. *African Entomology, 2*(1), 72–74.

Coetzee, M. (2017). Culicidae. In A.H. Kirk-Spriggs & B.J. Sinclair (Eds.), *Manual of Afrotropical Diptera: Nematocerous Diptera and lower Brachycera. Suricata 5* (Vol. 2, pp. 677–692). Pretoria: South African National Biodiversity Institute.

Coetzee, M., & du Toit, C.L.N. (1979). A full description of the adult, pupa and larva of *Anopheles* (*Cellia*) *cameroni* De Meillon and Evans, from the Cape Province, South Africa. *Mosquito Systematics, 11*(4), 291–299.

Coetzee, M., Segerman, J., & Hunt, R.H. (1987). Description of a new species *Anopheles* (*Cellia*) *kosiensis* (Diptera: Culicidae) from Zululand, South Africa. *Systematic Entomology, 12*(1), 23–28.

Coetzee, M., Estrada-Franco, J.G., Wunderlich, C.A., & Hunt, R.H. (1999). Cytogenetic evidence for a species complex within *Anopheles pseudopunctipennis* Theobald (Diptera: Culicidae). *American Society of Tropical Medicine and Hygiene, 60*(4), 649–653.

Coetzee, M., Hunt, R.H., Wilkerson, R., Della Torre, A., Coulibaly, M.B., & Besansky, N.J. (2013). *Anopheles coluzzii* and *Anopheles amharicus*, new members of the *Anopheles gambiae* complex. *Zootaxa, 3619*(3), 246–274.
Cogill, H. (1903). The *Anopheles* of Karwar (North Kanara). *Journal of the Bombay Natural History Society, 15*(2), 327–336.
Cohen, E. (2012). Roles of aquaporins in osmoregulation, desiccation and cold hardiness in insects. *Entomology, Ornithology and Herpetology, 1*(2012), 001.
Coher, E.I. (1949). A study of the female genitalia of Culicidae; with particular reference to characters of generic value. *Entomologica Americana, 28*(for 1948)(3), 75–112.
Cohuet, A., Simard, F., Toto, J.-C., Kenge, P., Coetzee, M., & Fontenille, D. (2003). Species identification within the *Anopheles funestus* group of malaria vectors in Cameroon and evidence for a new species. *The American Journal of Tropical Medicine and Hygiene, 69*(2), 200–205.
Colas-Belcour, J. (1943). L'oeuf et la ponte de *Taeniorhynchus* (*Coquillettidia*) *richiardii* Ficalbi. *Bulletin de la Société de pathologie exotique, 36*, 101–105.
Colledge, W.R. (1911). Notes on a brush-tongued mosquito. *Proceedings of the Royal Society of Queensland, 23*, 121–130.
Colless, D.H. (1948). The anopheline mosquitoes of north-west Borneo. *Proceedings of the Linnean Society of New South Wales, 73*(3–4), 71–119.
Colless, D.H. (1955a). New anopheline mosquitoes from North Borneo. *Sarawak Museum Journal, 6*(5), 331–342.
Colless, D.H. (1955b). Notes on the culicine mosquitoes of Singapore. I. Three new species of *Culex* (Diptera, Culicidae) and a redescription of *Culex hutchinsoni* Barraud. *Annals of Tropical Medicine and Parasitology, 49*(3), 311–319.
Colless, D.H. (1956). The *Anopheles leucosphyrus* group. *Transactions of the Royal Entomological Society of London, 108*(3), 37–116.
Colless, D.H. (1957a). Notes on the culicine mosquitoes of Singapore. II. The *Culex vishnui* group (Diptera, Culicidae), with descriptions of two new species. *Annals of Tropical Medicine and Parasitology, 51*(1), 87–101.
Colless, D.H. (1957b). Records of two Pacific island species of mosquito from Singapore harbour. *Medical Journal of Malaya, 12*(2), 464–467.
Colless, D.H. (1957c). Further notes on the systematics of the *Anopheles leucosphyrus* group (Diptera: Culicidae). *Proceedings of the Royal Entomological Society of London B: Taxonomy. 26*(7–8), 131–139.
Colless, D.H. (1958). Notes on the culicine mosquitoes of Singapore. IV. The *Aedes niveus* subgroup (Diptera, Culicidae): Introduction and description of five new species and of one new subspecies. *Annals of Tropical Medicine and Parasitology, 52*(4), 468–483.
Colless, D.H. (1959). Notes on the culicine mosquitoes of Singapore. V. The *Aedes niveus* subgroup (Diptera, Culicidae): Previously described species and keys to adults and larvae. *Annals of Tropical Medicine and Parasitology, 53*(2), 166–179.
Colless, D.H. (1960). Some species of *Culex* (*Lophoceraomyia*) from New Guinea and adjacent islands, with descriptions of four new species and notes on the male of *Culex fraudatrix* Theobald (Diptera: Culicidae). *Proceedings of the Linnean Society of New South Wales, 84*(for 1959)(3), 382–390.
Colless, D.H. (1962). Notes on the taxonomy of the *Aedes scutellaris* group, and new records of *A. paullusi* and *A. albopictus* (Diptera: Culicidae). *Proceedings of the Linnean Society of New South Wales, 87*(3), 312–315.
Colless, D.H. (1965). The genus *Culex* subgenus *Lophoceraomyia* in Malaya (Diptera: Culicidae). *Journal of Medical Entomology, 2*(3), 216–307.
Colless, D.H. (1973). A note on the status of *Aedes malayensis* and the distribution of *Aedes albopictus*. *Mosquito Systematics, 5*(3), 225–226.
Colless, D.H. (1976). *Anopheles amictus* Edwards and the subspecies concept. *Mosquito Systematics, 8*(3), 221–222.
Collins, F.H., & Paskewitz, S.M. (1996). A review of the use of ribosomal DNA (rDNA) to differentiate among cryptic *Anopheles* species. *Insect Molecular Biology, 5*(1), 1–9.
Collins, L.E., & Blackwell, A. (2000). The biology of *Toxorhynchites* mosquitoes and their potential for as biocontrol agents. *Biocontrol News and Information, 21*(4), 105N–116N.
Collucci, E., & Sallum, M.A.M. (2003). Phylogenetic analysis of the subgenus *Kerteszia* of *Anopheles* (Diptera: Culicidae: Anophelinae) based on morphological characters. *Insect Systematics and Evolution, 34*, 361–372.
Collucci, E., & Sallum, M.A.M. (2007). Cladistic analysis of the subgenus *Anopheles* (*Anopheles*) Meigen (Diptera: Culicidae) based on morphological characters. *Memórias do Instituto Oswaldo Cruz, 102*(3), 277–291.
Colton, L., Biggerstaff, B.J., Johnson, A., & Nasci, R.S. (2005). Quantification of West Nile virus in vector mosquito saliva. *Journal of the American Mosquito Control Association, 21*(1), 49–53.
Colton, Y.M., Chadee, D.D., & Severson, D.W. (2003). Natural skip oviposition of the mosquito *Aedes aegypti* indicated by codominant genetic markers. *Medical and Veterinary Entomology, 17*(2), 195–204.
Coluzzi, M. (1958). Descrizione di una nouva specie rinvenuta in Somalia. *Rivista di malariologia, 37*(1–3), 69–71.
Coluzzi, M. (1962). Le forme di *Anopheles claviger* Meigen indicate con i nomi *missirolii* e *petragnanii*, sono due specie riproduttivamente isolate. *Rendiconti Accademia dei Lincei, 32*(6), 1025–1030.
Coluzzi, M. (1964). Morphological divergences in the *Anopheles gambiae* complex. *Rivista di malariologia, 43*(4–6), 197–232.
Coluzzi, M., & Bullini, L. (1971). Enzyme variants as markers in the study of pre-copulatory isolating mechanisms. *Nature, 231*(5303), 455–456.
Coluzzi, M., & Cancrini, G. (1971). I cromosomi salivari di *Anopheles plumbeus* Stephens. *Parassitologia, 13*(1), 145–149.
Coluzzi, M., & Contini, C. (1962). The larva and pupa of *Mansonia buxtoni* (Edwards), 1923 (Diptera, Culicidae). *Bulletin of Entomological Research, 53*(2), 215–218.
Coluzzi, M., & Kitzmiller, J.B. (1975). Anopheline mosquitoes. In R.C. King (Ed.), *Handbook of Genetics*, Vol. 3. (pp. 285–307). New York, NY: Plenum.
Coluzzi, M., & Sabatini, A. (1968). Divergenze morfologiche e barriere di sterilitá nel complesso *Aedes mariae* (Diptera, Culicidae). *Rivista de parassitologia, 29*(1), 49–70.
Coluzzi, M., Saccà, G., & Feliciangefi, E.D. (1965). Il complesso *A. claviger* nella sotto-regione mediterranea. *Cahiers ORSTOM. Série entomologie médicale, 3*(3–4) 97–102.
Coluzzi, M., Gironi, A.M., & Muir, D.A. (1970). Ulteriori esperimenti d'incrocio tra le forme del complesso *mariae* del genere *Aedes*. *Parassitologia, 12*(2–3), 119–123.
Coluzzi, M., Sabatini, A., Bullini, L., & Ramsdale, C. (1974). Nuovi dati sulla distribuzione delle specie del complesso *mariae* del genere *Aedes*. *Revista di parassitologia, 35*(4), 321–330.
Coluzzi, M., Bianchi-Bullini, A.P., & Bullini, L. (1976). Speciation in the *mariae* complex of the genus *Aedes* (Diptera, Culicidae). *Atti Associazione Genetica Italiana* (Pavia), *21*, 221–223.
Coluzzi, M.A, Sabatini, A., Della Torre, A., Di Deco, M.A., & Petrarca, V. (2002). A polytene chromosome analysis of the *Anopheles gambiae* species complex. *Science, 298*(5597), 1415–1418.
Committee on Gene Drive Research in Non-Human Organisms. (2016). *Gene drives on the horizon: Advancing science, navigating*

uncertainty and aligning research with public values. Washington, DC: National Academic Press.

Condon, M.A., Scheffer, S.J., Lewis, M.L., & Swensen, S.M. (2008). Hidden neotropical diversity: Greater than the sum of its parts. *Science, 320*(5878), 928–931.

Conn, J. (1990). A genetic study of the malaria vector *Anopheles nuneztovari* from western Venezuela. *Journal of the American Mosquito Control Association, 6*(3), 400–406.

Conn, J. (1991). A cytogenetic analysis of the polytene chromosomes of *Anopheles triannulatus* (Diptera, Culicidae) from western Venezuela. *Genome, 34*(2), 267–272.

Conn, J., Rangel Puertas, Y., & Seawright, J.A. (1993). A new cytotype of *Anopheles nuneztovari* from western Venezuela and Colombia. *Journal of the American Mosquito Control Association, 9*(3), 294–301.

Conn, J.E., Wilkerson, R.C., Segura, M.N.O., Souza, R.T.L., Schlichting, C.D., Wirtz, R.A., & Povoa, M.M. (2002). Emergence of a new malaria vector facilitated by human migration and changes in land use. *The American Journal of Tropical Medicine and Hygiene, 66*(1), 18–22.

Conn, J., Quiñones, M., & Póvoa, M.M. (2013). Phylogeography, vectors, and transmission in Latin America. In S. Manguin (Ed.), Anopheles *mosquitoes: New insights into malaria vectors* (pp. 145–172). Rijeka, Croatia: InTech.

Conner, W.E., & Itagaki, H. (1984). Pupal attendance in the crabhole mosquito *Deinocerites cancer*: The effects of pupal sex and age. *Physiological Entomology, 9*(3), 263–267.

Consoli, R.A.G.B., & Lourenço-de-Oliveira, R. (1994). *Principais mosquitos de importância sanitária no Brasil.* Rio de Janeiro, Brazil: Editora Fiocruz.

Conti, B., Benelli, G., Flamini, G., Cioni, P.L., Profeti, R., Ceccarini, L., . . . Canale, A. (2012). Larvicidal and repellent activity of *Hyptis suaveolens* (Lamiaceae) essential oil against the mosquito *Aedes albopictus* Skuse (Diptera: Culicidae). *Parasitology Research, 110*(5), 2013–2021.

Conway, M.J., Watson, A.M., Colpitts, T.M., Dragovic, S.M., Li, Z., Wang, P., . . . Fikrig, E. (2014). Mosquito saliva serine protease enhances dissemination of dengue virus into the mammalian host. *Journal of Virology, 88*(1), 164–175.

Cook, E.F. (1944). The morphology of the larval heads of certain Culicidae (Diptera). *Microentomology, 9*, 38–68.

Cook, S., Lien, N.G., McAlister, E., & Harbach, R.E. (2010). *Bothaella manhi*, a new species of tribe Aedini (Diptera: Culicidae) from the Cuc Phuong National Park of Vietnam based on morphology and DNA sequence. *Zootaxa, 2661*, 33–46.

Cooling, L.E. (1924). A synonymic list of the more important species of Culicidae of the Australian region. Commonwealth of Australia, Department of Health Service Públication. *School of Public Health and Tropical Medicine, 2*, 1–61.

Coon, K.L., Vogel, K.J., Brown, M.R., & Strand, M.R. (2014). Mosquitoes rely on their gut microbiota for development. *Molecular Ecology, 23*(11), 2727–2739.

Cooper, R.D., Waterson, D.G.E., Bangs, M.J., & Beebe, N.W. (2000). Rediscovery of *Anopheles* (*Cellia*) *clowi* (Diptera: Culicidae), a rarely recorded member of the *Anopleles punctulatus* group. *Journal of Medical Entomology, 37*(6), 840–845.

Cooper, R.D., Waterson, D.G.E., Frances, S.P., Beebe, N.W., & Sweeney, A.W. (2002). Speciation and distribution of the members of the *Anopheles punctulatus* (Diptera: Culicidae) group in Papua New Guinea. *Journal of Medical Entomology, 39*(1), 16–27.

Cooper, R.D., Waterson, D.G.E, Frances, S.P., Beebe, N.W., Pluess, B., & Sweeeney, A.W. (2009). Malaria vectors of Papua New Guinea. *International Journal for Parasitology, 39*(13), 1495–1501.

Cooperband, M., Golden, F.V., Clark, G.G., Jany, W., & Allan, S.A. (2010). Prallethrin-induced excitation increases contact between sprayed ultralow volume droplets and flying mosquitoes (Diptera: Culicidae) in a wind tunnel. *Journal of Medical Entomology, 47*(6), 1099–1106.

Copeland, R.S., & Craig, G.B., Jr. (1990). Habitat segregation among treehole mosquitoes (Diptera: Culicidae) in the Great Lakes region of the United States. *Annals of the Entomological Society of America, 83*(6), 1063–1073.

Coquillett, D.W. (1896). New Culicidae from North America. *The Canadian Entomologist, 28*, 43–44.

Coquillett, D.W. (1898). Report on a collection of Japanese Diptera, presented to the U.S. National Museum by the Imperial University of Tokyo. *Proceedings United States National Museum, 21*(1146), 301–340.

Coquillett, D.W. (1901). Three new species of Culicidae. *The Canadian Entomologist, 33*(9), 258–260.

Coquillett, D.W. (1902a). New Orthorrhaphous Diptera from Mexico and Texas. *Journal of the New York Entomological Society, 10*(3), 136–141.

Coquillett, D.W. (1902b). New forms of Culicidae from North America. *Journal of the New York Entomological Society, 10*(3), 191–194.

Coquillett, D.W. (1902c). New Diptera from North America. *Proceedings of the United States National Museum, 25*(1280), 83–126.

Coquillett, D.W. (1902d). Three new species of *Culex. The Canadian Entomologist, 34*(11), 292–293.

Coquillett, D.W. (1903a). Four new species of *Culex. The Canadian Entomologist, 35*(9), 255–257.

Coquillett, D.W. (1903b). A new *Anopheles* with unspotted wings. *The Canadian Entomologist, 35*(11), 310.

Coquillett, D.W. (1904a). Several new Diptera from North America. *The Canadian Entomologist, 36*(1), 10–12.

Coquillett, D.W. (1904b). Notes on *Culex nigritulus. Entomological News, 15*(2), 73–74.

Coquillett, D.W. (1904c). New North American Diptera. *Proceedings of the Entomological Society of Washington, 6*(3), 166–192.

Coquillett, D.W. (1905). A new *Culex* from Australia. *Entomological News, 16*(4), 116.

Coquillett, D.W. (1906a). New Culicidae from the West Indies and Central America. *Proceedings of the Entomological Society of Washington, 7*(for 1905)(4), 182–186.

Coquillett, D.W. (1906b). On the breaking-up of the old genus *Culex. Science* (n.s.), *23*(582), 312–314.

Coquillett, D.W. (1906c). A new *Culex* near *curriei. Entomological News, 17*(4), 109.

Coquillett, D.W. (1906d). *A classification of the mosquitoes of North and Middle America* (Vol. 11). Washington, DC: Kessinger.

Coquillett, D.W. (1906e). Five new Culicidae from the West Indies. *The Canadian Entomologist, 38*(2), 58–62.

Coquillett, D.W. (1910). The type-species of the North American genera of Diptera. *Proceedings of the United States National Museum, 37*(1719), 499–647.

Corbel, V., Raymond, M., Chandre, F., Darriet, F., & Hougard, J.M. (2003). Efficiency of insecticide mixtures against larvae of *Culex quinquefasciatus* (Say) (Diptera: Culicidae) resistant to pyrethroids and carbamates. *Pest Management Science, 60*(4), 375–380.

Corbel, V., N'Guessan, R., Brengues, C., Chandre, F., Djogbenou, L., Martin, T., . . . Rowland, M. (2007). Multiple insecticide resistance mechanisms in *Anopheles gambiae* and *Culex quinquefasciatus* from Benin, West Africa. *Acta Tropica, 101*(3), 207–216.

Corbet, P.S. (1958). Identification of adult *Mansonia* (*Coquillettidia*) *pseudoconopas* Theobald and *M.* (*C.*) *maculipennis* Theobald

(Diptera: Culicidae). *Proceedings of the Royal Entomological Society of London, Series B: Taxonomy, 27*(7–8), 127–128.
Corbet, P.S. (1963). Observations on *Toxorhynchites brevipalpis conradti* Grünb. (Diptera, Culicidae) in Uganda. *Bulletin of Entomological Research, 54*(1), 9–17.
Corbet, P.S., & Chadee, D.D. (1993). An improved method for detecting substrate preferences shown by mosquitoes that exhibit "skip oviposition." *Physiological Entomology, 18*(2), 114–118.
Corbet, P.S., & Danks, H.V. (1975). Egg-laying habits of mosquitoes in the high arctic. *Mosquito News, 35*(1), 8–14.
Corbet, P.S., & Downe, A.E.R. (1966). Natural hosts of mosquitoes in northern Ellesmere Island. *Arctic, 19*(2), 153–161.
Corbet, P.S., & van Someren, E.C.C. (1962). *Aedes* (*Stegomyia*) *opok* sp. nov., a new species of mosquito from Uganda. *Annals of Tropical Medicine and Parasitology, 56*(1), 73–77.
Cornel, A.J. (1991). Description of *Aedes* (*Neomelaniconion*) *aurovenatus* Worth (Diptera: Culicidae). *Mosquito Systematics, 23*(1), 54–65.
Cornel, A.J., Mcabee, R.D., Rasgon, J., Stanich, M.A., Scott, T.W., & Coetzee, M. (2003). Differences in extent of genetic introgression between sympatric *Culex pipiens* and *Culex quinquefasciatus* (Diptera: Culicidae) in California and South Africa. *Journal of Medical Entomology, 40*(1), 36–51.
Cornel, A., Lee, Y., Fryxell, R.T., Siefert, S., Nieman, C., & Lanzaro, G. (2012). *Culex pipiens* sensu lato in California: A complex within a complex? *Journal of the American Mosquito Control Association, 28*(Suppl. 4), 113–121.
Cornet, M. (1968). *Aedes* (*Aedimorphus*) *chamboni* n. sp., une nouvelle espèce de Culicidae du Sénégal. *Bulletin de la Société de pathologie exotique, 60*(for 1967)(4), 398–402.
Cornet, M. (1973). *Aedes* (*Stegomyia*) *cozi* n. sp., une nouvelle espèce de Culicidae au Sénégal. *Cahiers ORSTOM. Série entomologie médicale, 11*(3), 175–180.
Cornet, M., Valade, M., & Dieng, P.Y. (1978). *Aedes* (*Stegomyia*) *neoafricanus* un nouvelle espèce de moustique capturée au Sénégal Oriental (Diptera: Culicidae). *Cahiers ORSTOM. Série entomologie médicale, 16*(3), 227–230.
Corradetti, A. (1939a). Studi morfologici sulla specie anofelica precedentemente identificata come *A. dthali* nel semien e sua classificazione come nuova varietà di *A. rhodesiensis*. *Rivista di parassitologia, 3*(1), 57–63.
Corradetti, A. (1939b). Una nuova specie di *Anopheles* rinvenuta in Dancalia: *Anopheles* (*Neocellia*) *dancalicus* n. sp. Nota preliminare. *Rivista di parassitologia, 3*(3), 277–278.
Corradetti, A. (1939c). Su alcune larve di *Anopheles* rinvenutea Ghinda (Eritrea) e descrizione di una nuova larva: *Anopheles* (*Myzomyia*) *erythraeus* n. sp. *Rivista di parassitologia, 3*(4), 287–292.
Corradetti, A. (1940a). Descrizione dell'*Anopheles* (*Neocellia*) *dancalicus*. *Rivista di parassitologia, 4*(1), 31–44.
Corradetti, A. (1940b). Note sull'*Anopheles garnhami* d'Abissinia. *Rivista di parassitologia, 4*(3), 141–146.
Corradetti, A., & Archetti, I. (1947). Gli anofelini raccolti dalla Missione Biologica Sagan-Omo. *Rendiconti Instituto Superiore de Sanità, 10*, 676–679.
Corrêa, R.R. (1942). Das formas evolutivas aquaticas do *Anopheles* (*Anopheles*) *eiseni* Coquillett, 1902. *Arquivos de higiene e saúde pública, 7*(15), 25–33.
Corrêa, R.R. (1944). Descrição de uma nova subespécie do subgênero *Anopheles* (Diptera, Culicidae). *Annais Paulistas de medicina e cirurgia, 47*(5), 405–408.
Corrêa, R.R. (1946). Do ciclo evolutivo em condições experiméntais do *Anopheles* (*Kerteszia*) *laneanus* (Diptera, Culicidae). *Arquivos de higiene e saúde pública, 11*(30), 605–611.
Corrêa, R.R. (1950a). Descrição de *Anopheles* (*Kerteszia*) *montemor* nova espécie de anofelino do Brasil (Diptera, Culicidae). *Arquivos de higiene e saúde pública, 14*(39–42), 53–55.
Corrêa, R.R. (1950b). Alguns informes sôbre *Anopheles* (*Arribalzagia*) *intermedius* (Chagas, 1908), (Diptera, Culicidae). *Arquivos de higiene e saúde pública* (São Paulo), *14*, 79–88.
Corrêa, R.R., & Cerqueira, F.M.C. (1944). Descrição de *Anopheles* (*Kerteszia*) *laneanus*, nova espécie de anofelino de Campos do Jordão (Diptera, Culicidae). *Arquivos de higiene e saúde pública* (São Paulo), *9*, 111–117.
Corrêa, R.R., & Rabelo, E.X. (1951). Observações sôbre o *Anopheles maculipes* Theobald no estado de São Paulo (Diptera, Culicidae). *Arquivos do Museu Nacional* (Rio de Janeiro), *42*, 167–173.
Corrêa, R.R., & Ramalho, G.R. (1956). Revisão de *Phoniomyia* Theobald, 1903 (Diptera, Culicidae, *Sabethini*). *Folia clinica et biologica* (São Paulo), *25*(1–6), 1–176.
Corrêa, R.R., & Ramalho, G.R. (1959a). Descrição de uma nova espécie de *Culex* (Diptera, Culicidae). *Revista Brasileira de malariologia e doenças tropicais* (Rio de Janeiro), *11*(1), 55–57.
Corrêa, R.R., & Ramalho, G.R. (1959b). *Culex* (*Culex*) *deanei*, nova espécie do gênero *Culex* Linnaeus 1758 (Diptera, Culicidae). *Revista do Instituto de Medicina Tropical de São Paulo, 1*(2), 141–143.
Corrêa, R.R., & Ramalho, G.R. (1968). *Anopheles* (*Arribalzagia*) *anchietai* nova espécie de anofelino de São Paulo, Brasil (Diptera, Culicidae). *Revista Brasileira de malariologia e doenças tropicais* (Rio de Janeiro), *20*(8–9), 115–132.
Corrêa, R.R., & Ramos, A.S. (1942). Os anofelinos da região meridional do Estado de São Paulo. *Arquivos de higiene e saúde pública, 7*, 37–57.
Corrêa, R.R., & Ramos, A.S. (1943). Descrição de uma nova subespécie de anofelino do sub-gênero *Nyssorhynchus* Blanchard, 1902 (Diptera, Culicidae). *Revista Paulista de medicina, 22*, 246–248.
Corrêa, R.R., & Ramos, A.S. (1944). Nota sistemática sôbre *Anopheles* (*N.*) *rondoni*—descrição do ovo (Diptera, Culicidae). *Anais Paulistas de medicina e cirugia, 47*(5), 407.
Corrêa, R.R., Lima, F.O., & Coda, D. (1950). Observations on the flight and longevity in nature of *Anopheles albitarsis domesticus*. *Journal of the National Malaria Society, 9*(3), 280–284.
Correia, M.G. (1968). Contribuição para o conhecimento dos Culicidae da Guiné Portuguesa (Ilha de Bissau). Estudo morfológico das larvas dos géneros *Ficalbia* e *Uranotaenia*. *Boletim cultural da Guiné Portuguesa, 23*(89–90), 163–180.
Costantini, C., Ayala, D., Guelbeogo, N.M., Pombi, M., Some, C.Y., Bassole, I.H.N., . . . Simard, F. (2009). Living at the edge: Biogeographic patterns of habitat segregation conform to speciation by niche expansion in *Anopheles gambiae*. *BMC Ecology, 9*(1), 16.
Costanzo, K.S., Kesavaraju, B., & Juliano, S.A. (2005). Condition-specific competition in container mosquitoes: The role of non-competing life-history stages. *Ecology, 86*(12), 3289–3295.
Cot & Hovasse (1917). Quelques remarques sur les Anophélines de Macédoine. *Bulletin de la Société de pathologie exotique, 10*(10), 890–895.
Cotrim, M.D., & Bianchi Galati, E.A. (1977). Revisão da série pleuristriatus do subgênera *Microculex* Theobald, 1907 (Diptera, Culicidae). *Revista Brasileira de entomología, 20*(3, 4), 169–205.
Cotrim, M.D., Bianchi, E.A., & von Atzingen, N.C.B. (1974). Redescrição de algumas espécies do subgênero *Culex* (Diptera, Culicidae). *Revista Brasileira de entomología, 18*(1), 9–19.
Coutinho, J.O. (1941). Sobre uma curiosa anomalia em uma larva de *Culex* (*C.*) *nigripalpus* Theobald, 1901 (Diptera, Culicidae). *Revista Brasileira de biologia* (Rio de Janiero), *1*(3), 343–344.

Coutinho, J.O. (1942). Sobre os mosquitos do estado de Mato Grosso, Brasil, com a descrição do macho de *Taeniorhynchus* (*Rhynchotaenia*) *shannoni* (Lane and Antunes, 1937) (Diptera, Culicidae). *Revista Brasileira de biologia* (Rio de Janeiro), *2*(2), 209–212.

Coutinho, J.O. (1944). *Anopheles* (*Shannoniezia*) *costalimai* Fonseca e Ramos, 1940, novo nome e a redescrição da espécie (Diptera-Culicidae). *Memórias do Instituto Oswaldo Cruz, 39*(for 1943)(3), 425–434.

Coutinho, J.O. (1946a). Contribuição para o estudo do sub-gênero *Kerteszia* com a descrição do macho de *Anopheles* (*Kerteszia*) *bambusicolus* Komp, 1937. In *Livro de Homenagem a R.F. d'Almeida* (Vol. 13, pp. 149–154).

Coutinho, J.O. (1946b). *Contribuição para o conhecimento das espécies do subgénero* Kerteszia (*Diptera, Culicidae*), *sua importância na transmissão da malaria* [Doctoral thesis, Universidade de São Paulo, Brazil].

Coutinho, J.O., & Farias, G.S. (1942). *Anopheles* (*Ayrozamyia*) *tibiamaculatus* (Neiva, 1906)—descrição do macho e criação de novo subgênero (Diptera, Culicidae). *Anais da Academia Brasileira de Ciencias, 14*(4), 343–348.

Coutinho, J.O., & Forattini, O.P. (1962). Notas sôbre Culicidae (Diptera). 4a-Contribuição ao estudo do subgênero *Microculex* com a descrição duma nova espécie. *Studia entomologica, 5*(1–4), 209–234.

Coutinho, J.O., & Verano, O.T. (1942). Estudo sôbre as cerdas clipeais das larvas de *Anopheles* (*Nyssorhynchus*) *oswaldoi* (Peryassú, 1922) e *Anopheles* (*Nyssorhynchus*) *argyritarsis* R. Desvoidy, 1827, da Baixada Fluminense (Diptera, Culicidae). *Revista Brasileira de biologia, 2*(3), 353–360.

Cova García, P. (1946). Notas sôbre los anofelinos de Venezuela y su identificacion. (CSP12/C1). Editorial Grafolit.

Cova García, P. (1961). *Notas sôbre los anofelinos de Venezuela y su identificación* (2nd ed.). Caracas, Venezuela: Editora Grafos, C.A.

Cova García, P. (1962). Hallazgo del insecto *Culex* (*Culex*) *tejerai* sp. n. en la cumbre del Choroni, Estado Aragua. *Revista Venezolana de sanidad y asistencia social, 27*(2), 312–316.

Cova García, P., & Pulido F., J. (1974). Nueva especie de *Culex*. (*Microculex*) (Diptera, Culicidae): *Culex* (*Microculex*) *sutili*. *Boletín de la Dirección de Malariología y Saneamiento Ambiental, 14*(5–6), 227–228 [illustration inadvertently pre-published in *14*(1–2): 27].

Cova García, P., & Rausseo, J.A. (1964). *Uranotaenia bertii*, n. sp. *Revista Venezolana de sanidad y asistencia social, 29*(2), 221–229.

Cova García, P., & Sutil Oramas, E. (1974). *Culex* (*Microculex*) *pulidoi* (Diptera, Culicidae). *Boletín de la Dirección de Malariología y Saneamiento Ambiental, 14*(1–2), 29–30.

Cova García, P., & Sutil O., E. (1975a). Clave para adultos hembras de anofelinos de Venezuela. *Boletín de la Dirección de Malariología y Saneamiento Ambiental, 15*(5), 201–216.

Cova García, P., & Sutil O., E. (1975b). Clave para larvas de anofelinos de Venezuela. *Boletín de la Dirección de Malariología y Saneamiento Ambiental, 15*(1–2), 6–24.

Cova García, P., & Sutil O., E. (1976a). Clave para la identificación de los anofelinos de Venezuela por las terminalia del macho. *Boletín de la Dirección de Malariología y Saneamiento Ambiental, 16*(1), 13–32.

Cova García, P., & Sutil Oramas, E. (1976b). *Culex* (*Carrollia*) *wannonii* (Diptera, Culicidae), una nueva especie de Venezuela. *Boletín de la Dirección de Malariología y Saneamiento Ambiental, 16*(4), 354–359.

Cova García, P., & Sutil Oramas, E. (1976c). Nueva especie para Venezuela *Mansonia* (*Mansonia*) *suarezi* (Diptera, Culicidae). *Boletín de la Dirección de Malariología y Saneamiento Ambiental, 16*(4), 360–362.

Cova García, P., & Sutil O., E. (1977). *Claves gráficas para la clasificación de anofelinos de Venezuela*. Maracay, Aragua, Venezuela: Division de Endemias Rurales.

Cova García, P., Sutil, E., & Rausseo, J.A. (1966). *Mosquitos (Culicinos) de Venezuela* (Vol. 1. [adults], Vol. 2. [larvae]). Caracas, Venezuela. Ministerio de Sanidad y Asistencia Social.

Cova García, P., Sutil, E., & Pulido, J. (1971). Descripción de nueva especie de *Culex*, del subgénero *Carrollia* (Diptera, Culicidae). *Boletín de la Sociedad Venezolana de Ciencias Naturales* (Caracas), *29*, 203–210.

Cova García, P., Sutil Oramas, E., & Pulido F., J. (1972). Neuva especie de *Culex* (*carrollia* [*sic*]) *ransseoi*. *Boletín de la Dirección de Malariología y Saneamiento Ambiental, 12*, 207–209.

Cova García, P., Sutil, E., & Pulido F., J. (1974a). *Wyeomyia lopezii* n. sp. (Diptera, Culicidae). *Boletín de la Dirección de Malariología y Saneamiento Ambiental, 14*(1–2), 21–22.

Cova García, P., Sutil Oramas, E., & Pulido F., J. (1974b). *Sabethes fernandez-yepezi* n. sp. *Boletín de la Dirección de Malariología y Saneamiento Ambiental, 14*(1–2), 23–24.

Cova García, P., Pulido F., J., & Escalante de Ugueto, C. (1976). *Anopheles* (*Kerteszia*) *hilli* (Diptera, Culicidae), una nueva especie de Venezuela. *Boletín de la Dirección de Malariología y Saneamiento Ambiental, 16*(4), 344–353.

Cova García, P., Pulido F., J., & Amarista M., J.R. (1977). *Anopheles* (*Nyssorhynchus*) *deltaorinoquensis* n.sp. (Diptera, Culicidae) de Venezuela. *Boletín de la Dirección de Malariología y Saneamiento Ambiental, 17*(2), 150–161.

Cova García, P., Pulido F., J., & Escalante de Ugueto, C. (1977). *Anopheles* (*Kerteszia*) *gonzálezrinconesi* n. sp. (Diptera, Culicidae) de Venezuela. *Boletín de la Dirección de Malariología y Saneamiento Ambiental, 17*(2), 140–149.

Cova García, P., Pulido F., J., & Escalante de Ugueto, C. (1977). *Anopheles* (*Kerteszia*) *rollai,* nomen novum *Anopheles* (*Kerteszia*) *hilli* Cova García, 1976 (non *Anopheles hilli* Woodhill & Lee 1944). *Boletín de la Dirección de Malariología y Saneamiento Ambiental, 17*(3), 241.

Cova García, P., Pulido F., J., & Escalante de Ugueto, C.E. (1979). *Culex* (*Neoculex*) *johni* (Diptera, Culicidae), nueva especie de Venezuela. *Boletín de malariología y salud ambiental, 19*(1), 21–27.

Cova García, P., Pulido F., J., & Escalante de Ugueto, C. (1981). *Uranotaenia pifanoi* (Diptera, Culicidae), nueva especie de Venezuela. *Boletín de la Dirección de Malariología y Saneamiento Ambiental, 21*(3–4), 210–218.

Cova García, P., Pulido F., J., Escalante de Ugueto, C., & Mora F., J.D. (1987a). *Uranotaenia iriartei* (Diptera, Culicidae), nueva especie de Venezuela. *Boletín de la Dirección de Malariología y Saneamiento Ambiental, 27*(1–4), 45–52.

Cova García, P., Pulido F., J., Escalante de Ugueto, C., & Mora R., J.D. (1987b). *Uranotaenia gabaldoni* (Diptera, Culicidae), nueva especie de Venezuela. *Boletín de la Dirección de Malariología y Saneamiento Ambiental, 27*(1–4), 94–104.

Cova García, P., Pulido F., J., Escalante de Ugueto, C., Amarista, J., & Mora, J.D. (1987). *Uranotaenia bricenoi* (Diptera, Culicidae), nueva especie de Venezuela. *Boletín de la Dirección de Malariología y Saneamiento Ambiental, 27*(1–4), 65–73.

Covell, G. (1927). A new species of *Anopheles* from Eastern India, *A.* (*Myzomyia*) *ramsayi*; with a new description of *A.* (*Myzomyia*) *jamesii* Theobald. *Indian Journal of Medical Research* (Calcutta), *14*(4), 1019–1025.

Coz, J., Grjebine, A., & Hamon, J. (1960). *Anopheles flavicosta* Edwards, 1911 de Madagascar et d'Afrique. *Bulletin de la Société de pathologie exotique, 53*(4), 742–751.

Craig, G.B., Jr. (1955). Preparation of the chorion of eggs of aedine mosquitoes for microscopy. *Mosquito News, 15*, 228–231.

Craig, G.B., Jr. (1956). Classification of eggs of nearctic Aedinae mosquitoes (Diptera: Culicidae) (Doctoral dissertation abstract, University of Illinois Urbana) (5).

Craig, G.B., Jr., & Horsfall, W.R. (1958). Taxonomic and ecological significance of eggs of aedine mosquitoes. *Proceedings of the Tenth International Congress of Entomology, 3*, 853–857.

Craig, G.B., Jr., & Horsfall, W.R. (1960). Eggs of floodwater mosquitoes. VII. Species of *Aedes* common in the Southeastern United States (Diptera: Culicidae). *Annals of the Entomological Society of America, 53*(1), 11–18.

Craig, G.B., Jr., & Peinkowski, R.L. (1931). The occurrence of *Aedes canadensis* (Theobald) in Alaska (Diptera, Culicidae). *Proceedings of the Entomological Society of Washington, 57*(6), 268.

Cranston, P. (2005). Biogeographic patterns in the evolution of Diptera. In D.K. Yeates & B.M. Wiegman (Eds.), *The evolutionary biology of flies* (pp. 274–311). New York, NY: Columbia University Press.

Cranston, P.S., Ramsdale, C.D., Snow, K.R., & White, G.B. (1987). *Keys to the adults, male hypopygia, fourth-instar larvae, and pupae of the British mosquitoes (Culicidae): With notes on their ecology and medical importance* (Vol. 48). Ambleside, England: Freshwater Biological Association.

Crawford, J.E., Riehle, M.M., Markianos, K., Bischoff, E., Guelbeogo, W.M., Gneme, A., . . . Lazzaro, B.P. (2016). Evolution of GOUNDRY, a cryptic subgroup of *Anopheles gambiae* sl, and its impact on susceptibility to *Plasmodium* infection. *Molecular Ecology, 25*(7), 1494–1510.

Crawford, R. (1938). *Some anopheline pupae of Malaya with a note on pupal structure*. Singapore: Government Printing Office.

Cross, H.A. (1983). A new distribution record of *Anopheles merus* Donitz. *Journal of the Entomological Society of Southern Africa, 46*(1), 155.

Crovello, T.J. (1969). Numerical taxonomy: Its value to mosquito systematics. *Mosquito Systematics, 1*(3), 63–67.

Cruse, H., Dürr, V., & Schmitz, J. (2007). Insect walking is based on a decentralized architecture revealing a simple and robust controller. *Philosophical Transactions of the Royal Society of London A: Mathematical, Physical and Engineering Sciences, 365*(1850), 221–250.

Cruz, J., Man-Padros, D., Zou, Z., & Raikhel, A.S. (2012). Distinct roles of isoforms of the hemeliganded nuclear receptor E75, an insect ortholog of the vertebrate Rev-erb, in mosquito reproduction. *Molecular and Cellular Endocrinology, 349*(2), 262–271.

Cruz, O.G. (1901). Contribuição para o estudo dos culicidios do Rio de Janeiro. *Brazil-Medico, 14*(43), 423–426.

Cruz, O.G. (1906). Um novo genero da sub-familia Anophelina. *Brazil-Medico, 20*(20), 199–200.

Cruz, O.G. (1907a). Um novo genero brasiliero da sub-familia "Anophelinae." *Brazil-Medico, 21*(28), 271–273.

Cruz, O.G. (1907b). Uma nova espécie do genero *Psorophora*. *Brazil-Medico, 21*, 329–330.

Culley, T.M., & Hardiman, N.A. (2009). The role of intraspecific hybridization in the evolution of invasiveness: A case study of the ornamental pear tree *Pyrus calleryana*. *Biological Invasions, 11*(5), 1107–1119.

Cumming, J.M., & Wood, D.M. (2009). Adult morphology and terminology. In B.V. Brown et al. (Eds.), *Manual of Central American Diptera* (Vol. 1, pp. 9–50).

Cumpston, J. H. H. (1923, 1924). *Disease distribution in the Pacific Basin*. Paper presented at the Pan-Pacific Science Congress, Australia.

Cupp, E.W., & Horsfall, W.R. (1969). Biological bases for placement of *Aedes sierrensis* (Ludlow) in the subgenus *Finlaya* (Theobald). *Mosquito Systematics, 1*(3), 51–52.

Cupp, E.W., & Ibrahim, A.N. (1973). Identification of members of the *Culex pipiens* complex by immunodiffusion and immunoelectrophoresis. *Journal of Medical Entomology, 10*(3), 277–281.

Cupp, E.W., Kreutzer, R.D., & Weaver, S.C. (1989). The biosystematics of *Culex (Melanoconion) taeniopus sensu lato* in relation to Venezuelan equine encephalomyelitis. *Mosquito Systematics, 21*(3), 216–221.

Curry, D.P. (1928). A new anopheline mosquito, *Anopheles (Chagasia) bathanus* Dyar, discovered in the Canal Zone. *The American Journal of Tropical Medicine and Hygiene, 8*(3), 243–248.

Curry, D.P. (1931). *Anopheles (Anopheles) neomaculipalpus*. A new species of the *Arribalzagia* group of *Anopheles* from Panama. *American Journal of Hygiene, 13*(2), 643–647.

Curry, D.P. (1932). Some observations on the *Nyssorhynchus* group of the *Anopheles* (Culicidae) of Panama. *American Journal of Hygiene, 15*(2), 566–572.

Curtis, D.P. (1831). *A guide to an arrangement of British insects; being a catalogue of all the named species hitherto discovered in Great Britain and Ireland* (for 1829). London, England: Richard Taylor (for the author).

Curtis, J. (1837). *A guide to an arrangement of British insects being a catalog of all named species hitherto discovered in Great Britain and Ireland* (2nd ed.). London, England: Pigot.

Curtis, J. (1839). 537. *Culex guttatus*. The white-spotted gnat or mosquito. In *British entomology: Being illustrations and descriptions of the genera of insects found in Great Britain and Ireland: Containing coloured figures from nature of the most rare and beautiful species, and in many instances of the plants upon which they are found* (Vol. XVI). London, England. Printed for the author, 1823–1840

Curtis, L.C. (1967). The mosquitoes of British Columbia. *Occasional Papers of the British Columbia Provincial Museum* (No. 15). Victoria, B.C.: Dept. of Recreation and Conservation.

Cywinska, A., Hunter, F.F., & Hebert, P.D.N. (2006). Identifying Canadian mosquito species through DNA barcodes. *Medical and Veterinary Entomology, 20*(4), 413–424.

da Costa Lima, A. (1929a). Sobre alguns anophelineos encontrados no Brasil. *Memórias do Instituto Oswaldo Cruz, 12*(Suppl.), 275–293.

da Costa Lima, A. (1929b). Sobre algumas espécies de *Mansonia* encontradas no Brasil. *Memórias do Instituto Oswaldo Cruz, 12*(Suppl. 2), 297–300.

da Costa Lima, A. (1929c). Sobre um novo *Anopheles* do Brasil. *Brazil-Medico, 43*(2), 1100–1101.

da Costa Lima, A. (1930a). Sobre os mosquitos que se criam em buracos de arvores. *Memórias do Instituto Oswaldo Cruz, 23*(5), 255–260.

da Costa Lima, A. (1930b). Sobre espécies do genero *Miamyia*, subgenero *Miamyia* (Diptera: Culicidae). *Memórias do Instituto Oswaldo Cruz, 24*(2), 73–78.

da Costa Lima, A. (1930c). Sobre espécies do genero *Miamyia*, subgenero *Miamyia* (2a. nota) (Diptera: Culicidae). *Memórias do Instituto Oswaldo Cruz, 24*(3), 187–194.

da Costa Lima, A. (1930d). Nota sobre a *Wyeomyia (Dendromyia) luteoventralis* Theobald, 1901 (Diptera: Culicidae). *Memórias do Instituto Oswaldo Cruz, 24*(1), 35–39.

da Costa Lima, A. (1930e). Sobre a revalidação do genero *Taeniorhynchus* F. Lch. A. (Diptera: Culicidae). *Memórias do Instituto Oswaldo Cruz, 23*(2), 105–108.

da Costa Lima, A. (1931a). Notas sobre Culicidae. *Anales del Museo Nacional de Historia Natural Bernardino Rivadavia* (Buenos Aires), *36*(29), 359–368.

da Costa Lima, A. (1931b). Sobre as espécies dos generos *Sabethes* e *Sabethoides* (Diptera: Culicidae). *Memórias do Instituto Oswaldo Cruz, 25*(1), 51–64.
da Costa Lima, A. (1931c). Nota sobre sabethineos do grupo *Joblotia* (Diptera: Culicidae). *Memórias do Instituto Oswaldo Cruz, 25*(1), 65–71.
da Costa Lima, A. (1931d). Sobre ás espécies de *Megarhinus* do Brasil (Diptera: Culicidae). *Memórias do Instituto Oswaldo Cruz, 25*(4), 307–315.
da Costa Lima, A. (1933). Sobre um novo *Aëdes* (*Ochlerotatus*) do Brasil. *Memórias do Instituto Oswaldo Cruz, 27*(4), 403–406.
da Costa Lima, A. (1935a). Espécies de *Taeniorhynchus* (*Taeniorhynchus*) (Diptera: Culicidae). *Memórias do Instituto Oswaldo Cruz, 30*(3), 453–469.
da Costa Lima, A. (1935b). Sobre ás espécies de *Orthopodomyia* encontradas no Brasil (Diptera: Culicidae). *Revista médico-cirurgica do Brasil, 43*(6), 175–179.
da Costa Lima, A. (1937). Um novo *Anopheles* da Baixada *fluminense* (Diptera: Culicidae). *Reimpresso da médico-cirurgica do Brasil, 45*(1), 5–7.
da Costa Lima, A. (1938). Nota sobre algumas especies de *Psorophora* (*Janthinosoma*) (Diptera: Culicidae). *Entomología médica, reimpresso de acta médica, 3*, 255–258.
da Costa Lima, A. (1943). Nota sobre alguns sabetineos. *Anais da Academia Brasileira de Ciencias, 15*(4), 295–308.
da Costa Lima, A.M., Guitton, N., & Ferreira, O. (1962). Comentários relativos às espécies da tribo Toxorhynchitini (Megarhinini) com a descrição de uma espécie nova de *Lynchiella* (Diptera, Culicidae). *Memórias do Instituto Oswaldo Cruz, 60*(2), 225–252.
da Cunha Ramos, H. (1985). Research on the mosquitoes of Angola (Diptera: Culicidae). XVII. Description of *Uranotaenia* (*Uranotaenia*) *angolensis* sp. nov. *Boletim da Sociedade Portuguesa de Entomologia, 1*(Suppl. 1), 73–82.
da Cunha Ramos, H. (1986). Research on the mosquitoes of Angola (Diptera: Culicidae). XVIII. Description of *Uranotaenia* (*Uranotaenia*) *machadoi* sp. nov. *Mosquito Systematics, 18*(1), 110–115.
da Cunha Ramos, H. (1987). Description of the male genitalia of *Uranotaenia henrardi* Edwards, 1935 (Diptera, Culicidae). *Mosquito Systematics, 19*(2), 135–138.
da Cunha Ramos, H. (1993). *Revisão do género Uranotaenia (Diptera, Culicidae) na região Afrotropical*. Lisbon, Portugal: Ministério do Planeamento e da Administração do Território, Secretaria de Estado da Ciência e Technologia, Instituto de Investigação Científica Tropical.
da Cunha Ramos, H., & Antunes Capela, R. (1988). On the terrestrial and riverine fauna of the Democratic Republic of São Tóme and Príncipe. II. Description of the fourth instar larva and pupa of *Uranotaenia* (*Pseudoficalbia*) *micromelas* Edwards, 1934 (Diptera, Culicidae). *Arquivos do Museu Bocage* [Universidade de Lisboa; Museu e Laboratório Zoológico e Antropológico], Nova Serie, *Vol. Especial*, 17–24.
da Cunha Ramos, H., & Brunhes, J. (2004). Insecta Diptera Culicidae *Uranotaenia*. Faune de Madagascar 91. Institut de recherche pour le developpement (Montpellier); Centre de cooperation internationale en recherché agronomique pour le développement. Paris, France: Museum national d'histoire naturelle.
da Cunha Ramos, H., & Ribeiro, H. (1975). Research on the mosquitoes of Angola (Diptera: Culicidae). IX. Genera *Toxorhynchites* Theo., 1901, *Malaya* Leicester, 1908, *Ficalbia* Theo., 1903, *Coquillettidia* Dyar, 1905, *Mansonia* Blanchard, 1901, *Uranotaenia* L. Arribálzaga, 1891, *Aedeomyia* Theo., 1901, *Eretmapodites* Theo., 1901 and *Culiseta* Felt, 1904. *Anais do Instituto de Higiene e Medicina Tropical, 2*(1–4), 111–141.
da Cunha Ramos, H., & Ribeiro, H. (1980). Description of the male, pupa and fourth instar larva of *Culiseta* (*Culiseta*) *atlantica* (Edwards, 1932) (Diptera: Culicidae), an endemic species of the Azores Archipelago. *Mosquito Systematics, 12*(3), 320–332.
da Cunha Ramos, H., & Ribeiro, H. (1982). Research on the mosquitoes of Angola (Diptera: Culicidae). XII. Description of *Culex* (*Culex*) *pajoti* sp. nov. *Mosquito Systematics, 13*(for 1981)(2), 176–190.
da Cunha Ramos, H., & Ribeiro, H. (1990). Research on the mosquitoes of Angola (Diptera, Culicidae). XX - *Eretmapodites mortiauxi*, a new species of the "*leucopus* group." *Boletim da Sociedade Portuguesa de Entomología, 4*(115), 138–144.
da Cunha Ramos, H., & Ribeiro, H. (1992). Research on the mosquitoes of Angola (Diptera, Culicidae). XXI - Description of *Eretmapodites angolensis* sp. nov. and *Eretmapodites dundo* sp. nov. of the "*oedipodeios* group." *Garcia De Orta Série de Zoologia, 17*(for 1990)(1–2), 31–35.
da Cunha Ramos, H., & Ribeiro, H. (1995). Redescription of the pupa of *Aedes* (*Skusea*) *pembaensis* (Diptera: Culicidae). *Mosquito Systematics, 27*(3), 187–190.
da Cunha Ramos, H., Ribeiro, H., & de Barros Machado, A. (1992a). Research on the mosquitoes of Angola (Diptera, Culicidae). XIX - *Eretmapodites tendeiroi*, a new species of the "*plioleucus* group." *Garcia De Orta Série de Zoologia, 16*(for 1989)(1–2), 133–136.
da Cunha Ramos, H., Ribeiro, H., Pires, C.A., & Capela, R.A. (1992b). Os mosquitos (Diptera, Culicidae) da Ilha do Príncipe. *Garcia De Orta Série de Zoologia, 16*(for 1989)(1–2), 163–170.
da Cunha Ramos, H., Ribeiro, H., & Capela, R.A. (1993). Description of the female, pupa and phallosome of *Culex* (*Culiciomyia*) *cambournaci* (Diptera: Culicidae) and redesdcription of the fourth-instar larva and male. *Mosquito Systematics, 25*(1), 35–40.
da Cunha Ramos, H., Capela, R.A., & Ribeiro, H. (1995). Description of *Aedes* (*Aedimorphus*) *gandarai*, a new species from the Island of São Tóme, with keys to all known species of the Tarsalis group (Diptera: Culicidae). *Mosquito Systematics, 26*(for 1994)(3), 97–104.
da Cunha Ramos, H., Ribeiro, H., Mitchell, C.J., Baptista, S., Anselmo, M.L., Novo, M.T., . . . Easton, E.R. (2002). Two new *Uranotaenia* records for Macau (Diptera: Culicidae). *Garcia De Orta Série de Zoologia, 24*(1–2), 123–124.
da Cunha Ramos, H., Lucientes, J., Blasco-Zumeta, J., Osacar, J.J., & Ribeiro, H. (2002). Some morphological characteristics of *Culex deserticola*, a recent addition to the Spanish fauna (Diptera: Culicidae). *European Mosquito Bulletin, 12*, 27–29.
da Cunha Ramos, H., Ribeiro, H., & Harrison, B.A. (2003). A new European mosquito species: *Culex* (*Neoculex*) *europaeus* (Diptera: Culicidae). *European Mosquito Bulletin [Journal of the European Mosquito Control Association], 15*, 6–11.
da Fonseca, F., & da Silva Ramos, A. (1940a). Novo subgenero e novas especies de anofelinas neotropicas (Diptera. Culicidae). *Memórias do Instituto Butantan, 13*(for 1939), 383–387.
da Fonseca, F., & Ramos, A.S. (1940b). *Shannonesia* nov. nom. (Dipt. Culicidae). *Revista de entomología* (Rio de Janeiro), *11*, 966.
da Silva Maciel, C. (1964). Descriação do macho de *Psorophora* (*Grabhamia*) *varinervis* Edwards, 1922, e da pupa de *Psorophora* (*Grabhamia*) *cingulata* (Fabricius, 1805) (Diptera, Culicidae). *Revista Brasileira de malariología e doenças tropicais* (Rio de Janeiro), *16*, 465–469.
Dabiré, K.R., Sawadodgo, S., Diabaté, A., Toe, K.H., Kengne, P., Ouari, A., . . . Lehmann, T. (2013). Assortative mating in mixed swarms of the mosquito *Anopheles gambiae* s.s. M and S molecular forms, in Burkina Faso, West Africa. *Medical and Veterinary Entomology, 27*(3), 298–312.

D'Abrera, V.S.E. (1944). The eggs of the Ceylon anopheline mosquitoes. *Journal of the Malaria Institute of India, 5*(3), 337–359.

Dadd, R.H., & Kleinjan, J.E. (1979). Essential fatty acid for the mosquito *Culex pipiens*: Arachidonic acid. *Journal of Insect Physiology, 25*(6), 495–502. doi:10.1016/S0022-1910(79)80008-6

Dahan, Y.L., & Koekemoer, L.L. (2014). Analysis of the genitalia rotation in the male *Anopheles funestus* (Diptera: Culicidae). *Acta Tropica, 132*(Suppl.), S20–S25.

Dahl, C. (1974). Circumpolar *Aedes* (*Ochlerotatus*) species in North Fennoscandia. *Mosquito Systematics, 6*(1), 57–73.

Dahl, C. (1975). Culicidae (Dipt. Nematocera) of the Baltic Island of Öland. *Entomologisk tidsskrift, 96*(3–4), 77–96.

Dahl, C. (1977). Taxonomy and geographic distribution of Swedish Culicidae (Diptera, Nematocera). *Insect Systematics and Evolution, 8*(1), 59–69.

Dahl, C. (1997). Diptera Culicidae, mosquitoes *Nilsson, Anders. Aquatic insects of North Europe. A taxonomic handbook: Vol. 2. Odonata - Diptera*. Stenstrup, Denmark: Apollo Books, 163–186.

Dahl, C., & Blackmore, M.S. (2001). The distribution and status of *Ochlerotatus geniculatus* (Olivier) in Fennoscandia. *European Mosquito Bulletin, 9*, 12–16.

Dahl, C., & White, G.B. (1978). Culicidae. In J. Illies (Ed.), *Limnofauna Europaea: A checklist of the animals inhabiting european inland waters, with accounts of their distribution and ecology (except Protozoa)* (pp. 390–395). Stuttgart, German: Fischer Verlag.

Dahl, C., Widahl, L.E., & Nilsson, C. (1988). Functional analysis of the suspension feeding system in mosquitoes (Diptera: Culicidea). *Annals of the Entomological Society of America, 81*(1), 105–127.

Dai, L., & Adams, M.E. (2009). Ecdysis triggering hormone signaling in the yellow fever mosquito, *Aedes aegypti*. *General and Comparative Endocrinology, 162*(1), 43–51.

Dale, P.E.R., Knight, J., Kay, B.H., Chapman, H., Ritchie, S.A., & Brown, M.D. (2008). Habitat characteristics and eggshell distribution of the salt marsh mosquito, *Aedes vigilax*, in marshes in subtropical eastern Australia. *Journal of Insect Science, 8*, Article 25.

Damiani, C., Ricci, I., Crotti, E., Rossi, P., Rizzi, A., Scuppa, P., . . . Favia, G. (2008). Paternal transmission of symbiotic bacteria in malaria vectors. *Current Biology, 18*(23), 1087–1088.

Damiani, C., Ricci, I., Crotti, E., Rossi, P., Rizzi, A., Scuppa, P., . . . Favia, G. (2010). Mosquito-bacteria symbiosis: The case of *Anopheles gambiae* and *Asaia*. *Microbial Ecology, 60*(3), 644–654.

Damiens, D., Benedict, M. Q., Wille, M., & Gilles, J.R.L. (2012). An inexpensive and effective larval diet for *Anopheles arabiensis* (Diptera: Culicidae): Eat like a horse, a bird, or a fish? *Journal of Medical Entomology, 49*(5), 1001–1011.

Damiens, D., Vreysen, M.J.B., & Gilles, J.R.L. (2013). *Anopheles arabiensis* sperm production after genetic manipulation, dieldrin treatment, and irradiation. *Journal of Medical Entomology, 50*(2), 314–316.

Dampf, A. (1936). El *Anopheles occidentalis* Dyar & Knab (*A. maculipennis* auct.) en la Mesa Central de Mexico. *Revista Mexicana de biología, 18*, 91–129.

Dampf, A. (1939). Los hipopigios masculinos de *Anopheles hectoris y parapunctipennis* (Insecta, Diptera, Culicidae). *Sobretiro de los anales de la Escuela Nacional de Ciencias Biológicas* (Mexico), *1*(2), 279–291.

Dampf, A. (1941). *Mochlostyrax* (*Dinoporpa*) *trifidus* Dyar nuevo miembro de la fauna culicidológica Mexicana (Insecta, Diptera, Fam. Culicidae). *Sobretiro de los anales de la Escuela Nacional de Ciencias Biológicas* (Mexico), *2*(2–3), 251–261.

Dampf, A. (1943). Distribución y ciclo anual de *Uranotaenia syntheta* Dyar & Shannon en Mexico y descripción del hipopigio masculino (Insecta, Diptera). *Revista de la Sociedad Mexicana de Historia Natural, 4*(3–4), 147–169.

Dana, A.N., Hong, Y.S., Kern, M.K., Hillenmeyer, M.E., Harker, B.W., Lobo, N.F., . . . Collins, F.H. (2005). Gene expression patterns associated with blood-feeding in the malaria mosquito *Anopheles gambiae*. *BMC Genomics, 6*(1), 5.

Dana, A.N., Hillenmeyer, M.E., Lobo, N.F., Kern, M.K., Romans, P.A., & Collins, F.H. (2006). Differential gene expression in abdomens of the malaria vector mosquito, *Anopheles gambiae*, after sugar feeding, blood feeding and *Plasmodium berghei* infection. *BMC Genomics, 7*(1), Article 119.

Danabalan, R., Ponsonby, D.J., & Linton, Y.-M. (2012). A critical assessment of available molecular identification tools for determining the status of *Culex pipiens s.l.* in the United Kingdom. *Journal of the American Mosquito Control Association, 28*(4), 68–74.

Danabalan, R., Monaghan, M.T., Ponsonby, D.J., & Linton, Y.-M. (2014). Occurrence and host preferences of *Anopheles maculipennis* group mosquitoes in England and Wales. *Medical and Veterinary Entomology, 28*, 169–178.

D'Anfreville, L. (1916). Les moustiques de Salé, Maroc. *Bulletin de la Société de pathologie exotique, 9*, 140–142.

Daniels, C.W. (1908). Notes on the mosquitoes on the river and coast districts of the eastern side of the Peninsula. *Studies of the Institute of Medical Research* (Federal Malay States), *3*(3), 262–266.

Danilov, V.N. (1974a). *Aedes* (*Ochlerotatus*) *implicatus* Vockeroth, a species new in the fauna of the USSR (Diptera, Culicidae). *Parazitologiya* (St. Petersburg), *8*(2), 133–137.

Danilov, V.N. (1974b). On the restoration of the name *Aedes* (*O.*) *mercurator* Dyar for a mosquito known in the USSR as *Aedes riparius* after Gutsevich (Diptera, Culicidae). *Parazitologiya* (St. Petersburg), *8*(4), 322–328.

Danilov, V.N. (1975). On the possible identity of the mosquitoes, *Aedes* (*Ochlerotatus*) *beklemishevi* Denisova and *A.* (*O.*) *barri* Rueger (Diptera, Culicidae). *Parazitologia* (St. Petersburg), *9*(1), 61–63.

Danilov, V.N. (1976a). *Aedes* (*Ochlerotatus*) *increpitus* Dyar, a species new in the fauna of the USSR (Diptera: Culicidae). *Newsletter of the Siberian Section Academy of Science USSR, 1*(5), 71–75.

Danilov, V.N. (1976b). Revision of some holarctic species and subspecies of the genus *Culiseta* Felt. I. *Culiseta* (*Culiseta*) *kanayamensis* Yamada as a synonym of *C.* (*C.*) *bergrothi* Edwards. *Parazitologiya* (St. Petersburg), *10*, 185–187.

Danilov, V.N. (1976c). *Aedes* (*Stegomyia*) *patriciae* Mattingly, a species new for the fauna of China, with a note on the morphology of its larva (Diptera: Culicidae). *Mosquito Systematics, 8*(4), 353–354.

Danilov, V.N. (1977). On the synonymy of species names of *Aedes* mosquitoes (subgenera *Finlaya* and *Neomelaniconion*) in the Far East fauna. *Parazitologia* (St. Petersburg), *2*(2), 181–184.

Danilov, V.N. (1978). The first finding of *Aedes* (*Finlaya*) *versicolor* Barraud (Diptera, Culicidae) in the USSR, description of its female and larva of instar IV. *Vestnik zoologii, 12*(3), 82–85.

Danilov, V.N. (1979). Nomenclature, synonomy, differential diagnosis, and distribution of *Aedes* (*Ochlerotatus*) *albineus* Seguy (Diptera). *Vestnik zoologii* (Kiev), *13*(1), 29–35.

Danilov, V.N. (1981a). On the mosquito fauna (Diptera: Culicidae) of the Ethiopian zoogeographical region 1. A subgenus of the genus *Aedes* new for the region. *Meditsinskaia parazitologiia i parizarnye bolezni, 50*(4), 86–87.

Danilov, V.N. (1981b). The identity of *Aedes* (*Ochlerotatus*) *simanini* Gutsevich with *Ae.* (*O.*) *niphadopsis* Dyar & Knab (Diptera, Culicidae). *Vestnik zoologii* (Kiev), *3*(3), 36–40.

Danilov, V.N. (1982). Mosquito fauna (Diptera, Culicidae) of the Ethiopian zoogeographical region 2. A new species of *Mansonia*

(*Coquillettidia*) and a key for adults of the genus *Mansonia*. *Meditsinskaia parazitologiia i parizarnye bolezni, 60*(4), 26–30.
Danilov, V.N. (1983a). A key for the mosquitoes of the genus *Mansonia* of the Afrotropical Region (Diptera: Culicidae). *Anais do Instituto de Hygiene e Medicina Tropical* (Lisbon), *9*, 31–36.
Danilov, V.N. (1983b). The differential diagnosis, clinal variability, larvae distribution, and synonymy of *Culiseta morsitans* (Culicidae). *Parazitologiya* (St. Petersburg), *17*(2), 156–160.
Danilov, V.N. (1984). *Culiseta* (*Culicella*) *setivalva* Maslov as a synonym of *C.* (*C.*) *fumipennis* Stephens. *Parazitologia* (St. Petersburg), *18*(4), 313–317.
Danilov, V.N. (1985). Mosquitoes of the subgenus *Edwardsaedes* of the Palearctic, with description of the larva of *Aedes* (Edw.) *bekkui*. *Parazitologia* (St. Petersburg), *19*(5), 378–381.
Danilov, V.N. (1987a). Mosquitoes of the genus *Toxorhynchites* in the fauna of the USSR and related taxa in East and South East Asia. *Parazitologiya* (St. Petersburg), *21*(2), 151–155.
Danilov, V.N. (1987b). Mosquitoes of the subgenus *Aedes* (Diptera, Culicidae) in the Fauna of the USSR. II. *Aedes dahuricus*. *Vestnik zoologii, 1987*(4), 35–41.
Danilov, V.N. (1988). Notes on the morphology of the larva of the mosquito *Culex* (*Culex*) *schwetzi* Edw. (Diptera: Culicidae) and on the geographical distribution of this species. *Garcia de Orta Série de Zoologia, 15*(2), 131–134.
Danilov, V.N. (1989). Two new Afrotropical subgenera of the mosquito genus *Culex* (Diptera, Culicidae). *Revue d'entomologie de l'URSS Entomologicseszkoje Obozrenie, 68*(4), 790–797 (Trans. pp. 112–120)
Danilov, V.N., & Filippova, V.V. (1978). A new species of mosquito, *Aedes* (*Stegomyia*) *sibiricus* sp.n. (Culicidae). *Parazitologiya* (St. Petersburg), *12*(2), 170–176.
Danilov, V.N., & Gornostaeva, R.M. (1987). A new species of mosquito, *Aedes* (*Ochlerotatus*) *intermedius* sp.n. (Diptera, Culicidae). *Parazitologiya, 21*(5), 612–619.
Danilov, V.N., & Kupriyanova, E.S. (1976). A description of the larva of *Culex* (*Neoculex*) *rubensis* Sasa et Takahashi (Diptera, Culicidae), a mosquito species new in the fauna of the USSR. *Revue d'entomologie de l'URSS Entomologicseszkoje Obozrenie, 55*, 928–930.
Danilov, V.N., & Stupin, V.I. (1982). A new mosquito species of the genus *Aedes* (Diptera, Culicidae) from southern Siberia and the Amur river basin Russian-SFSR USSR. *Revue d'entomologie de l'URSS Entomologicseszkoje Obozrenie, 61*(4), 824–829.
Danilov, V.N., & Stupin, V.I. (1986). On morphology, biology and geographical distribution of the mosquito, *Aedes* (*Ochlerotatus*) *sergievi* (Diptera, Culicidae). *Parazitologiya, 20*(3), 234–236.
Danilov, V.N., Markovich, N.Y., & Proskuryakova, A.M. (1978). A new species of mosquito, *Aedes* (*Ochlerotatus*) *sergievi* sp.n. *Parazitologiya, 12*(1), 44–49.
Danks, H.V. (2000). Measuring and reporting life-cycle duration in insects and arachnids. *European Journal of Entomology, 97*(3), 285–303.
Danks, H.V. (2004). Seasonal adaptations in arctic insects. *Integrative and Comparative Biology, 44*(2), 85–94.
Danks, H.V., & Corbet, P.S. (1973). A key to all stages of *Aedes nigripes* and *A. impiger* (Diptera: Culicidae) with a description of first-instar larvae and pupae. *The Canadian Entomologist, 105*(3), 367–376.
Dantis, F.V. (1947). The mosquitoes of Manila. *Monthly Bulletin of the Bureau of Health, 23*(3), 251–257.
Darsie, R.F. (1949). Pupae of the anopheline mosquitoes of the northeastern United States (Diptera, Culicidae). *Revista de entomología, 20*(1–3), 509–530.
Darsie, R.F., Jr. (1951). *Pupae of the culicine mosquitoes of the northeastern United States (Diptera, Culicidae, Culicini)*. Ithaca, NY: Cornell University Agricultural Experiment Station, Memoir 304.
Darsie, R.F. (1955). Notes on American mosquito pupae. I. Description of *Aedes riparius* and *Aedes pionips* (Diptera, Culicidae). *Proceedings of the Entomological Society of Washington, 57*(1), 23–29.
Darsie, R.F. (1957). Notes on American mosquito pupae. II. The *Aedes* (*Ochlerotatus*) *punctor* subgroup, with key to known Nearctic *Aedes* pupae (Diptera, Culicidae). *Annals of the Entomological Society of America, 50*, 611–620.
Darsie, R.F., Jr. (1985). Mosquitoes of Argentina. Part I. Keys for identification of adult females and fourth stage larvae in English and Spanish (Diptera, Culicidae). *Mosquito Systematics, 17*(3), 153–253.
Darsie, R.F., Jr. (1994). A revised checklist of the mosquitoes of Guatemala including a new country record, *Psorophora cyanescens*. *Journal of the American Mosquito Control Association, 10*(4), 511–514.
Darsie, R.F., Jr. (1997). Description of the pupa of *Aedes pseudotaeniatus* (Giles) (Diptera: Culicidae). *Journal of the American Mosquito Control Association, 13*(3), 286–288.
Darsie, R.F., Jr. (1998). Descriptions of the pupae of six species of *Armigeres* Theobald, subgenus *Leicesteria* Theobald (Diptera, Culicidae) from Nepal. *Proceedings of the Entomological Society of Washington, 100*(2), 234–246.
Darsie, R.F., Jr. (1999). Description of the pupa of *Aedes cretinus* Edwards, a key to the pupae of the *albopictus* subgroup, subgenus *Stegomyia* Theobald, genus *Aedes* Meigen, and characters to separate the European *Stegomyia* species (Diptera: Culicidae). *Proceedings of the Entomological Society of Washington, 101*(3), 614–618.
Darsie, R.F., Jr. (2000a). Description of the pupa of *Armigeres* (*Leicesteria*) *omissus* (Edwards) and a key to the larvae and pupae of the *Armigeres* occurring in Nepal (Diptera: Culicidae). *Proceedings of the Entomological Society of Washington, 102*(4), 964–968.
Darsie, R.F., Jr. (2000b). Description of the pupae of five species in the subgenus *Armigeres*, genus *Armigeres* Theobald, with a key to species of the known pupae of the subgenus (Diptera: Culicidae). *Proceedings of the Entomological Society of Washington, 102*(1), 108–119.
Darsie, R.F., Jr. (2000c). The pupae of three species in *Mimomyia* Theobald from Nepal and a key to the known pupae of the genus in the Oriental and Australasian regions (Diptera: Culicidae). *Proceedings of the Entomological Society of Washington, 102*(3), 625–633.
Darsie, R.F., Jr. (2002d). Redescription of the pupa of *Culex* (*Melanoconion*) *iolambdis* Dyar. *Journal of the American Mosquito Control Association, 18*(4), 277–279.
Darsie, R.F., Jr. (2003). First report of *Ochlerotatus condolescens* in the United States. *Proceedings of the Entomological Society of Washington, 105*(4), 1067–1068.
Darsie, R.F., Jr. (2005). Key to the pupae of the mosquitoes (Diptera: Culicidae) of Florida. *Proceedings of the Entomological Society of Washington, 107*(4), 892–902.
Darsie, R.F., Jr. (2006). Description of the pupae of three species of the genus *Lutzia*, a comparison of New and Old World pupae, and a key to pupae and larvae of the genus (Diptera: Culicidae). *Proceedings of the Entomological Society of Washington, 108*(1), 145–154.
Darsie, R.F., Jr. (2009). Description of the pupa of the Himalayan mosquito, *Ochlerotatus pulchriventer*. *Journal of the American Mosquito Control Association, 25*(3), 248–250.

Darsie, R.F., Jr. (2010). Description of Himalayan mosquito pupae. II. *Ochlerotatus shortti* and *Oc. elsiae*. *Journal of the American Mosquito Control Association, 26*(3), 243–248.

Darsie, R.F., Jr. (2011a). Redescription of the pupae of *Ochlerotatus dorsalis* and *Ochlerotatus squamiger*. *Journal of the American Mosquito Control Association, 27*(2), 105–110.

Darsie, R.F., Jr. (2011b). Description of Himalayan mosquito pupae. III. *Ochlerotatus gubernatoris* and *Ochlerotatus unicinctus*. *Journal of the American Mosquito Control Association, 27*(4), 353–356.

Darsie, R.F., Jr. (2011c). Descriptions of the pupae of *Ochlerotatus aloponotum* and *Ochlerotatus campestris*. *Journal of the American Mosquito Control Association, 27*(3), 201–206.

Darsie, R.F., Jr. (2011d). Description of the pupae of *Ochlerotatus flavescens* and *Ochlerotatus melanimon*. *Journal of the American Mosquito Control Association, 27*(2), 99–104.

Darsie, R.F., Jr. (2012). Description of Himalayan mosquito pupae. III. *Ochlerotatus albolateralis* and *Ochlerotatus deccanus*. *Journal of the American Mosquito Control Association, 28*(1), 9–14.

Darsie, R.F., Jr., & Cagampang-Ramos, A. (1969). *Manual of malaria entomology. A handbook of laboratory and field training exercies* (2nd ed.). Manila, Philippines: Malaria Eradication Training Center.

Darsie, R.F., Jr., & Cagampang-Ramos, A. (1971a). A subspecies of *Anopheles* new to the Philippine Islands (Diptera: Culicidae). *Proceedings of the Entomological Society of Washington, 73*(4), 399–400.

Darsie, R.F., Jr., & Cagampang-Ramos, A. (1971b). Additional species of *Anopheles* on Guam. *Mosquito Systematics Newsletter, 3*(2), 28–30.

Darsie, R.F., Jr., & Cagampang-Ramos, A. (1972). Descriptions and keys for anophelines of Guam. *Mosquito News, 32*(1), 16–22.

Darsie, R.F., Jr., & Cagampang-Ramos, A. (1977). A study of *Anopheles ludlowae* Theobald siblings, leading to new synonymy (Diptera: Culicidae). *Mosquito Systematics, 9*(2), 123–127.

Darsie, R.F., Jr., & Day, J.F. (2003). Studies of the genus *Culex* Linnaeus in Florida. I. Redescription of the pupae of *Culex nigripalpus* Theobald and *Cx. tarsalis* Coquillett, vectors of St. Louis encephalitis, and a key to pupae of *Culex* species in the eastern United States (Diptera: Culicidae). *Proceedings of the Entomological Society of Washington, 105*(1), 100–107.

Darsie, R.F., Jr., & Day, J.F. (2004). Studies of the genus *Culex* in Florida. II. Redescription of the fourth instar of *Culex nigripalpus*. *Journal of the American Mosquito Control Association, 20*(2), 110–114.

Darsie, R.F., Jr., & Day, J.F. (2005). Redescription of the pupa of *Culex (Culex) declarator* Dyar and Knab (Diptera, Culicidae), with amendments to key to the *Culex* pupae of the Eastern United States. *Journal of the American Mosquito Control Association, 21*(2), 128–130.

Darsie, R.F., Jr., & Day, J.F. (2006). Redescription of the pupa of *Culex salinarius* Coquillett and comparison with *Culex nigripalpus* Theobald. *Journal of the American Mosquito Control Association, 22*(3), 547–549.

Darsie, R.F., Jr., & Day, J.F. (2007). Redescription of the pupa of *Culex restuans* and a comparison with *Culex nigripalpus*. *Journal of the American Mosquito Control Association, 23*(2), 95–98.

Darsie, R.F., Jr., & Hobbs, J.H. (1982). First report of *Culex chrysonotum* and *Culex spissipes* in Guatemala (Diptera, Culicidae). *Mosquito Systematics, 14*(1), 73–77.

Darsie, R.F., Jr., & Pradhan, S.P. (1990). The mosquitoes of Nepal: Their identification, distribution and biology. *Mosquito Systematics, 22*(2), 69–130.

Darsie, R.F., Jr., & Pradhan, S.P. (1991). The mosquitoes of Nepal: Their identification, distribution and biology. Index and corrigendum. *Mosquito Systematics, 23*(1), 46–49.

Darsie, R.F., Jr., & Samanidou-Voyadjoglou, A. (1997). Keys for the identification of the mosquitoes of Greece. *Journal of the American Mosquito Control Association, 13*(3), 247–254.

Darsie, R.F., Jr., & Ward, R.A. (1981). Identification and geographical distribution of the mosquitoes of North America, north of Mexico. *Mosquito Systematics, 1*(Suppl.), 1–313.

Darsie, R.F., Jr., & Ward, R.A. (2000). Summary of new distribution records for mosquito species in the United States and Canada for the period 1981–99. *Journal of the American Mosquito Control Association, 16*(1), 1–4.

Darsie, R.F., Jr., & Ward, R.A. (2005). *Identification and geographical distribution of the mosquitoes of North America, north of Mexico.* Gainesville: University Press of Florida.

Darsie, R.F., Jr., Tindall, E., & Barr, A.R. (1962). Description of the pupa of *Culiseta melanura* (Diptera: Culicidae). *Proceedings of the Entomological Society of Washington, 64*(3), 167–170.

Darsie, R.F., Jr., Merino, M., & Wilton, D. (1977). Discovery of *Anopheles crucians* Subgroup on the Pacific Coast of El Salvador. *Mosquito News, 37*(4), 645–646.

Darsie, R.F., Jr., Pradhan, S.P., & Vaidya, R.G. (1991). Notes on the mosquitoes of Nepal. 1. New country records and revised *Aedes* keys (Diptera, Culicidae). *Mosquito Systematics, 23*(1), 39–45.

Darsie, R.F., Jr., Becnel, J.J., Hazard, E.I., & García, J.J. (1991). A record of mosquito species collected during a larval survey in Argentina and Uruguay (Diptera, Culicidae). *Mosquito Systematics, 23*(1), 50–52.

Darsie, R.F., Pradhan, S.P., & Vaidya, R.G. (1992). Notes on the mosquitoes of Nepal. II. New species records from 1991 collections. *Mosquito Systematics, 24*, 23–28.

Darsie, R.F., Jr., Courtney, G.W., & Pradhan, S.P. (1993). Notes on the mosquitoes of Nepal. III. Additional new records in 1992 (Diptera: Culicidae). *Mosquito Systematics, 25*(3), 186–191.

Darsie, R.F., Jr., Courtney, G.W., & Pradhan, S.P. (1996). Notes on the mosquitoes of Nepal. IV. Results of the 1994 collecting in the midwestern region, including new country records and voucher confirmation (Diptera, Culicidae). *Journal of the American Mosquito Control Association, 12*(1), 130–134.

Darsie, R.F., Jr., Vlach, J.J., & Fussell, E.M. (2002). New addition to the mosquito fauna of United States, *Anopheles grabhamii*. *Journal of Medical Entomology, 39*(4), 430–431.

Darsie, R.F., Jr., Almasi, J.R., & Day, J.F. (2006). Studies of the genus *Culex* Linnaeus in Florida. III. Redescription of the fourth-stage larva of *Culex salinarius* Coquillett and comparison with that of *Cx. nigripalpus* Theobald. *Journal of the American Mosquito Control Association, 22*(2), 179–184.

Darsie, R.F., Jr., Taylor, D.S., Prusak, Z.A., & Verna, T.N. (2010). Checklist of the mosquitoes of the Bahamas with three additions to its fauna and keys to the adult females and fourth instars. *Journal of the American Mosquito Control Association, 26*(2), 127–134.

Darwin, C. (1859). *On the origin of species by means of natural selection, or the preservation of favoured races in the struggle for life*. London, England: Murray.

Darwin, C. (1860). *Journal of the researches into the natural history and geology of the countries visited during the voyage of H.M.S. Beagle round the world, under the command of Capt. Fitz Roy, R.N.* (2nd ed.). New York, NY: Appleton.

Das, B.P. (1990). Morphological difference between sibling species of *Anopheles culicifacies* Giles in India: A preliminary report. *Mosquito-Borne Diseases Bulletin, 7*(4), 131–133.

Das, B.P., Rajagopal, R., & Akiyama, J. (1990). Pictorial key to the species of Indian anopheline mosquitoes. *Journal of Pure and Applied Zoology, 2*(3), 131–162.

Dasch, G.A., Strickman, D., Watt, G., & Eamsila, C. (1996). Measuring genetic variability in *Orientia tsutsugamushi* by PER/RFLP analysis: A new approach to questions about its epidemiology, evolution and ecology. In J. Kazar, & R. Toman (Eds.), *Rickettsiae and Rickettsial Diseases, Proceedings of the Fifth International Symposium* (pp. 79–84). Bratislava, Slovakia: Veda.

Dash, A.P., Khamre, J.S., Roy, J.R., Bendle, M.S., Das, M., & Dwived, S.R. (1983). Morphometrical variations in *Anopheles annularis*. *Journal of Communicable Diseases, 15*(1), 61–64.

Davey, T.H. (1941). The larva and pupa of *Anopheles brunnipes* Theobald. *Annals of Tropical Medicine and Parasitology, 35*(1), 5–8.

David, M.R., Lourenço-de-Oliveira, R., & Maciel-de-Freitas, R. (2009). Container productivity, daily survival rates and dispersal of *Aedes aegypti* mosquitoes in a high income dengue epidemic neighbourhood of Rio de Janeiro: Presumed influence of differential urban structure on mosquito biology. *Memórias de Instituto Oswaldo Cruz, 104*(6), 927–932.

Davidson, G., & Hunt, R.H. (1973). The crossing and chromosome characteristics of a new, sixth species in the *Anopheles gambiae* complex. *Parassitologia, 15*(1–2), 121–128.

Davies, N., & Roderick, G.K. (2005). Dipteran sex chromosomes in evolutionary developmental biology. In D.K. Yeates & B.M. Wiegman (Eds.), *The evolutionary biology of flies* (pp. 196–213). New York, NY: Columbia University Press.

Davis, N.C. (1926). Notes on the female hypopygia of anopheline mosquitoes, with special reference to some Brazilian species. *American Journal of Hygiene, 6*(March suppl.), 1–22.

Davis, N.C. (1931). A new anopheline mosquito from Para, Brazil. *American Journal of Hygiene, 13*(1), 345–348.

Davis, N.C. (1933a). Notes on some South American mosquitoes. *Annals of the Entomological Society of America, 26*(2), 277–295.

Davis, N.C. (1933b). Notes on some South American mosquitoes. II. The genus *Mansonia*, subgenus *Rhynchotaenia*. *Annals of the Entomological Society of America, 26*, 619–639.

Day, J.F., & Edman, J.D. (1983). Malaria renders mice susceptible to mosquito feeding when gametocytes are most effective. *Journal of Parasitology, 69*(1), 163–170.

Day, J.F., & Edman, J.D. (1984). Mosquito engorgement on normally defensive hosts depends on host activity patterns. *Journal of Medical Entomology, 21*(6), 732–740.

De Almeida, S.J., Ferreira, R.P.M., Eiras, Á.E., Obermayr, R.P., & Geier, M. (2010). Multi-agent modeling and simulation of an *Aedes aegypti* mosquito population. *Environmental Modelling and Software, 25*(12), 1490–1507.

de Buen, A.M. (1953a). *Orthopodomyia kummi* Edwards, 1939, mosquito nuevo para México. Descripción de la larva y de la pupa. (Diptera, Culicidae). *Anales del Instituto de Biología, Universidad Nacional Autónoma de México, 23*(for 1952)(1,2), 243–252.

de Buen, A.M. (1953b). Descripción de la pupa de *Wyeomyia* (*Wyeomyia*) *abebela* Dyar y Knab, 1908 (Diptera, Culicidae). *Anales del Instituto de Biología, Universidad Nacional Autónoma de México, 23*(for 1952)(1,2), 253–255.

De Burca, B. (1943). A new species of *Anopheles* from East Africa. *Journal of the Malaria Institute of India, 5*(2), 233–234.

de Carvalho-Pinto, C.J., & Lourenço-de-Oliveira, R. (2004). Isoenzymatic analysis of four *Anopheles* (*Kerteszia*) *cruzii* (Diptera: Culicidae) populations of Brazil. *Memórias do Instituto Oswaldo Cruz, 99*(5), 471–475.

de Charmoy, d'E. (1908). On three new species of *Culex* collected during the anti-malarial campaign in Mauritius in 1908. *Annals of Tropical Medicine and Parasitology, 2*(3), 257–264.

de Deus Palma, J., & Galvão, A.B. (1969). *Lynchiella rizzoi* sp. n., Toxorrinquitíneo bromelícola do estado de Goiás (Diptera, Culicidae). *Revista Goiana médica, 15*(1,2), 3–13.

de Freitas Barata, E.A.M., Chiaravalloti Neto, F., Dibo, M.R., Macoris, M.D.L.G., Barbosa, A.A.C., Natal, D., . . . Macoris Andriguetti, M.T. (2007). Capture of culicids in urban areas: Evaluation of the resting box method [Captura de culicideos em area urbana: avaliaçõo do metodo das caixas de repouso]. *Revista de saúde pública, 41*(3), 375–382.

de Geer, C. (1776). *Memoires pour servir a l'histoire des insectes.* Tome sixieme. Stockholm, Sweden: De l'imprimerie de Pierre Hesselberg.

de Grandpré, A.D., & de Charmoy, d'E. (1901). *Les moustiques: Anatomie et biologie. Contribution a l'étude des Culicidées et principalement des genres* Culex *et* Anopheles, *de leur rôle dans la propagation de la malaria et de le filariose et des moyens de s'en préserver* (for 1900). Port-Louis, France: Ile Maurice.

de la Rocque, S., Rioux, J.A., & Slingenbergh, J. (2008). Climate change: Effects on animal disease systems, and implications for surveillance and control. *Revue scientifique et technique* (Office International des Epizootics), *27*(2), 340–354.

de León, J.R. (1938). El anophelisino de altura en Guatemala. *Boletín Direccion General de Sanidad Pública, 9*, 411–424.

de Linero, M.Q. (1967). Nueva especie de *Culex* (*Melanoconion*). *Revista Brasileira de Biología, 27*(3), 289–291.

De Little, S.C., Bowman, D.M.J.S., Whelan, P.I., Brook, B.W., & Bradshaw, C.J.A. (2009). Quantifying the drivers of larval density patterns in two tropical mosquito species to maximize control efficiency. *Environmental Entomology, 38*(4), 1013–1021.

De Loof, A. (2008). Ecdysteroids, juvenile hormone and insect neuropeptides: Recent successes and remaining major challenges. *General and Comparative Endocrinology, 155*(1), 3–13.

de Meijere, J.C.H. (1909). Drei Myrmecophile Dipteren aus Java. *Tijdschrift voor Entomologie, 52*, 165–174.

de Meijere, J.C.H. (1910). Nepentheg-Tiere. I. Systematik. *Annales du Jardin botanique de Buitenzorg, Supplément, 3*(2), 917–940.

de Meijere, J.C.H. (1911). Zur Kenntnis niederländischer Culiciden. *Tijdschrift voor Entomologie, 54*, 137–157.

De Meillon, B. (1931a). Illustrated keys to the full-grown larvae adults of South African anopheline mosquitoes. *Publications of the South African Institute for Medical Research, 4*(28), 275–375.

De Meillon, B. (1931b). A new South African anopheline. *Journal of the Medical Association of South Africa, 5*, 482–483.

De Meillon, B. (1934). Entomological studies. Studies on insects of medical importance in South Africa. *Publications of the South African Institute for Medical Research, 6*(33), 249–307.

De Meillon, B. (1935). Entomological studies. Studies on insects of medical importance in South Africa. Part II. *Publications of the South African Institute for Medical Research, 6*(35), 323–365.

De Meillon, B. (1936a). Entomological studies. Studies on insects of medical importance in South Africa. Part III. The systematic position of *Anopheles listeri* De Meillon. *Publications of the South African Institute for Medical Research, 7*(38), 125–215.

De Meillon, B. (1936b). The eggs of some South African Anophelines. Part II. *Publications of the South African Institute for Medical Research, 7*(38), 131–132.

De Meillon, B. (1937). Studies on insects of medical importance from southern Africa and adjacent territories. Part IV. Culicidae. 2. A note on *Anopheles gambiae* and *Anopheles funestus* in northern Rhodesia. *Publications of the South African Institute for Medical Research*, *7*(40), 306–312.

De Meillon, B. (1942). New Nematocera from the Ethiopian region. *Journal of the Entomological Society of Southern Africa, 5*, 87–98.

De Meillon, B. (1943). New records, and new species of Nematocera (Diptera) from the Ethiopian region. *Journal of the Entomological Society of Southern Africa, 6*, 90–113.

De Meillon, B. (1947a). New records and species of biting insects from the Ethiopian region. II. *Journal of the Entomological Society of Southern Africa, 10*(1), 110–124.

De Meillon, B. (1947b). The Anophelini of the Ethiopian geographical region. *Publications of the South African Institute for Medical Research, 10*(49), 1–272.

De Meillon, B., & Evans, A.M. (1935). Two new anophelines from South Africa. *Annals of Tropical Medicine and Parasitology, 29*(1), 91–94.

De Meillon, B., & Hardy, F. (1953). New records and species of biting insects from the Ethiopian region. IV. *Journal of the Entomological Society of Southern Africa, 16*(1), 17–35.

De Meillon, B., & Lavoipierre, M. (1944). New records and species of biting insects from the Ethiopian region. *Journal of the Entomological Society of Southern Africa, 7*, 38–67.

De Meillon, B., & Leeson, H.S. (1940). Notes on Ethiopian Anophelini. *Bulletin of Entomological Research, 31*(1), 61–67.

De Meillon, B., & Pereira, M.D.C. (1940). Notes on some anophelines (Dipt. Culicidae) from Portuguese East Africa. *Moçambique documentario trimestral, 23*, 69–109.

De Meillon, B., & Rebêlo, A. (1941). Culicini (Diptera, Nematocera) from the colony of Moçambique. *Moçambique documentario trimestral, 27*, 69–77.

De Meillon, B., & van Eeden, G. (1976). *Anopheles* (*Cellia*) *deaconi* n. sp., from South Africa (Diptera: Culicidae). *Mosquito Systematics Newsletter, 8*(4), 335–342.

De Meillon, B., Evans, A.M., & Leeson, H.S. (1936). The characters of *Anopheles marshalli* Theobald and *A. marshalli* var. *pitchfordi* Giles in the adult, larval and pupal stages. *Annals of Tropical Medicine and Parasitology, 30*(1), 45–55.

De Meillon, B., Parent, M., & Black, L.O.C. (1945). Descriptions of new larvae and pupae of Ethiopian Culicini. *Bulletin of Entomological Research, 36*(1), 85–101.

De Meillon, B., Eeden, G.J.V., Coetzee, L., Coetzee, M., Meiswinkel, R., Toit, C.L.N.D., & Hansford, C.F. (1977). Observations on a species of the *Anopheles funestus* subgroup, a suspected exophilic vector of malaria parasites in northeastern Transvaal, South Africa. *Mosquito News, 37*(4), 657–661.

de Mello, C.F., dos Santos-Mallet, J.R., Morone, F., Guimarães, A.É., Marcondes, C.B., & Alencar, J. (2014). Ultrastructure of the egg of *Coquillettidia juxtamansonia* (Chagas, 1907) (Diptera: Culicidae). *Journal of Vector Ecology, 39*(1), 219–221.

De Mello, F. (1918). *Anopheles folquei* n. sp. (Nouvelle espècie anopheline trouvée dans notre province de Pragana). *Anais scientificos da Faculdade de Medicina do Pôrto, 4*(3–4), 363–371.

De O Gaio, A., Gusmão, D.S., Santos, A.V., Berbert-Molina, M.A., Pimenta, P.F., & Lemos, F.J. (2011). Contribution of midgut bacteria to blood digestion and egg production in *Aedes aegypti* (Diptera: Culicidae) (L.). *Parasites and Vectors, 4*(1), Article 105.

de Oliveira Castro, G.M. (1932). Estudo sobre uma especie de *Culex*, que se cria em buracos de guayamú (Dipt., Culicidae). *Revista de entomologia, 2*(1), 97–105.

de Oliveira Castro, G.M. (1935). Estudo sobre uma especie de "*Limatus*" que se cria em buracos de pau (Dipt., Culicidae). *Revista do Departamento Nacional de Produção Animal* (Rio de Janeiro), *2*, 143–155.

de Queiroz, K. (2007). Species concepts and species delimitation. *Systematic Biology, 56*(6), 879–886.

de Queiroz, K. (2012). Biological nomenclature from Linnaeus to the PhyloCode. In C.J. Bell (Ed.), *The herpetological legacy of Linnaeus: A celebration of the Linnaean Tercentary. Bibliotheca Herpetologica, 9*(1–2), 135–145.

De Réaumur, R.A.F. (1738). Histoire des cousins. *Treizième mémoire, Tome IV. Mémoires pour servir à l'histoire des insects* (pp. 573–636). Paris, France: Académie Royale des Sciences.

de Rook, H. (1929). *Malaria en Anophelinen aan de Boven-Digoel.* Mimeographed report in Dutch [English translation (1930): Malaria and *Anopheles* on the Upper Digul].

de Rook, H., & Soesilo, R. (1930). *Anopheles* (*Bironella*) *papuae*. *Geneeskundig tijdschrift voor Nederlandsch-Indië, 70*, 475–481.

de Souza, M.A., & Guedes, A.S. (1962). Sôbre *Psorophora* (*Janthinosoma*) *fiebrigi* Edwards, 1922 (Diptera-Culicidae). *Revista Brasileira de malariologia e doenças tropicais* (Rio de Janeiro), *14*, 249–254.

de Souza, M.A., & Xavier, S.H. (1964). Descrição da larva e da pupa de *Phoniomyia muhlensi* (Petrocchi,1925)—(Diptera, Culicidae). *Revista Brasileira de malariologia e doenças tropicais* (Rio de Janeiro), *16*, 459–464.

de Zulueta, J., Ramsdale, C., Cianchi, R., Bullini, L., & Coluzzi, M. (1983). Observations on the taxonomic status of *Anopheles sicaulti*. *Parassitologia* (Rome), *25*(1), 73–92.

Deane, M.P., & Causey, O.R. (1943). Viability of *Anopheles gambiae* eggs and morphology of unusual types found in Brazil. *The American Journal of Tropical Medicine and Hygiene, 23*(1), 95–103.

Deane, L.M., Deane, M.P., & Causey, O.R. (1943). Descrição do ovo, larva e pupa de *Anopheles* (*Arthuromyia*) *gilesi* (Neiva, 1908). *Papeis avulsos do Departament de Zoologia, 3*(10), 167–191.

Deane, L.M., Causey, O.R., & Deane, M.P. (1946). I. An illustrated key by adult female characteristics for the identification of thirty-five species of Anophelini, with notes on the malaria vectors (Diptera, Culicidae). In *Studies on Brazilian anophelines from the northeast and Amazon regions* (pp. 1–18). Baltimore, MD: Johns Hopkins University Press.

Deane, M.P., Causey, O.R., & Deane, L.M. (1946). III. An illustrated key by larval characteristics for the identification of thirty-two species of Anophelini, with descriptions of two larvae. In *Studies on Brazilian anophelines from the northeast and Amazon regions* (pp. 33–50). Baltimore, MD: Johns Hopkins University Press.

Deane, L.M., Deane, M.P., & Damasceno, R.G. (1949). Encontro e descrição do macho do *Anopheles* (*Lophopodomyia*) *squamifemur* Antunes, 1937 (Diptera, Culicidae). *Revista Brasileira de biologia, 9*(4), 497–504.

Debboun, M., Frances, S.P., & Strickman, D. (2007). *Insect repellents: Principles, methods, and uses.* Boca Raton, FL: CRC Press.

Decoursey, J.D., & Webster, A.P. (1952). A method of clearing the chorion of *Aedes sollicitans* (Walker) eggs and preliminary observations on their embryonic development. *Annals of the Entomological Society of America, 45*(4), 625–632.

Degallier, N., & Claustre, J. (1980). Culicidae [Diptera, Nematocera] de Guyane française: Notes faunistiques et taxonomiques. *Revue Française d'entomologie* (n.s.), *2*(3), 138–146.

Degner, E.C., & Harrington, L.C. (2016). A mosquito sperm's journey from male ejaculate to egg: Mechanisms, molecules, and methods for exploration. *Molecular Reproduction and Development, 83*(10), 897–911.

DeJong, R.J., Miller, L.M., Molina-Cruz, A., Gupta, L., Kumar, S., & Barillas-Mury, C. (2007). Reactive oxygen species detoxification by catalase is a major determinant of fecundity in the mosquito

Anopheles gambiae. *Proceedings of the National Academy of Sciences of the United States of America, 104*(7), 2121–2126.

Dekoninck, W., Hendrickx, F., Versteirt, V., Coosemans, M., De-Clercq, E.M., Hendrickx, G., . . . Grootaert, P. (2013). Changes in species richness and spatial distribution of mosquitoes (Diptera: Culicidae) inferred from museum specimen records and recent inventory: A case study from Belgium suggests recent expanded distribution of arbovirus and malaria vectors. *Journal of Medical Entomology, 50*(2), 237–243.

Del Pilar Corena, M., Van Ekeris, L., Salazar, M.I., Bowers, D., Fiedler, M.M., Silverman, D., . . . Linser, P.J. (2005). Carbonic anhydrase in the adult mosquito midgut. *Journal of Experimental Biology, 208*(17), 3263–3273.

Del Ponte, E. (1939). Observaciones sobre *Anopheles pseudopunctipennis* en la Mendieta, Provincia de Jujuy. *Revista del Instituto Bacteriológico del Departamento Nacional de Higiene, 9*(2), 149–155.

Del Ponte, E. (1940). Descripción de la termanilia macho de *Anopheles annulipalpis* L. Arr. (Dipt. Culicidae). *Revista del Instituto Bacteriológico del Departamento Nacional de Higiene, 9*(5), 602–604.

Del Ponte, E., & Castro, M. (1952). Las especies argentinas de *Psorophora* (*Psorophora*) (Dipt. Culic.). *Revista de la Sociedad Entomológica Argentina, 15*, 211–230.

Del Ponte, E., & Cerqueira, N. (1938). Alguns Sabethineos do Brasil (Diptera, Culicidae). *Revista de entomología* (Rio de Janeiro), *8*(3–4), 225–237.

Del Ponte, E., & Heredia, R.L. (1945). *Anopheles* (*Anopheles*) *holmbergi* n. sp. de Misiones (Argentina) (Diptera Culicidae). *Revista de la Sociedad Entomólogica Argentina, 12*(5), 382–386.

Del Ponte, E., Castro, M.P., & García, M. (1951). Clave para las epecies de *Psorophora* y *Aëdes* de la Argentina y comarcas vecinas. Diagnosis de *Aëdes* (*O.*) *raymondi* n. sp. (Diptera, Culicidae). *Anales de la Sociedad Cientifíca Argentina, 151*(5), 228–243.

Del Vecchio, G. (1939). Sulle varietà. de *A. claviger* (*bifurcatus*). Nota. I. *Rivista di Parassitologia, 3*(1), 27–37.

Delaunay, P., Schaffner, F., & Babinot, M. (2000). Un nouveau moustique en France *Aedes albopictus* (Skuse, 1894) (Diptera: Culicidae). On le cherchait dans les Alpes-Maritimes . . . on l'a trouvéé dans l'Orne. *Riviéra scientifique, 84*, 17–18.

Delfinado, M.D. (1966a). The culicine mosquitoes of the Philippines, tribe Culicini (Diptera, Culicidae). *Memoirs of the American Entomological Institute, 7*, 1–252.

Delfinado, M.D. (1966b). Two undescribed species of *Uranotaenia* from the Philippines (Diptera: Culicidae). *Journal of Medical Entomology, 3*(1), 36–37.

Delfinado, M.D. (1967). Contributions to the mosquito fauna of Southeast Asia. I. The genus *Aedes*, subgenus *Neomacleaya* Theobald in Thailand. *Contributions of the American Entomological Institute, 1*(8), 1–56.

Delfinado, M.D. (1968). Contributions to the mosquito fauna of Southeast Asia. III. The genus *Aedes*, subgenus *Neomacleaya* Theobald in Southeast Asia. *Contributions of the American Entomological Institute, 2*(4), 1–74.

Delfinado, M.D., & Hodges, E.R. (1968). Three new species of the genus *Tripteroides*, subgenus *Tripteroides* Giles (Diptera: Culicidae). *Proceedings of the Entomological Society of Washington, 70*(4), 361–375.

Demari-Silva, B., Vesgueiro, F.T., Sallum, M.A.M., & Marrelli, M.T. (2011). Taxonomic and phylogenetic relationships between species of the genus *Culex* (Diptera: Culicidae) from Brazil inferred from the cytochrome c oxidase I mitochondrial gene. *Journal of Medical Entomology, 48*(2), 272–279.

Demari-Silva, B., Foster, P.G., de Oliveira, T.M., Bergo, E.S., Sanabani, S.S., Pessôa, R., & Sallum, M.A. (2015). Mitochondrial genomes and comparative analyses of *Culex camposi*, *Culex coronator*, *Culex usquatus* and *Culex usquatissimus* (Diptera: Culicidae), members of the Coronator Group. *BMC Genomics, 16*, 831.

Denisova, Z.M. (1955). New type of *Aëdes* (*Ochlerotatus*). *Meditsinskaia parazitologiia i parizarnye bolezni, 24*, 58–61.

Denlinger, D.L. (1986). Dormancy in tropical insects. *Annual Review of Entomology, 31*(1), 239–264.

Denlinger, D.L., & Armbruster, P.A. (2014). Mosquito diapause. *Annual Review of Entomology, 59*, 73–93.

Derraik, J.G.B. (2004). Exotic mosquitoes in New Zealand: A review of species intercepted, their pathways and ports of entry. *Australian and New Zealand Journal of Public Health, 28*(5), 433–444.

Derua, Y.A., Alifrangis, M., Magesa, S.M., Kisinza, W.N., & Simonsen, P.E. (2015). Sibling species of the *Anopheles funestus* group, and their infection with malaria and lymphatic filarial parasites, in archived and newly collected specimens from northeastern Tanzania. *Malaria Journal, 14*, 104.

Derwesh, A.I. (1965). A preliminary list of identified insects and some arachnids of Iraq. *Ministry of Agriculture Bulletin, 112*, 1–123.

Deus, S. (2009). *Phylogeny of selected species of subgenus* Culex *mosquitoes (Diptera: Culicidae) from the U.S., Puerto Rico, and Guatemala, based on nucleotide sequences from three genes* [Doctoral dissertation, Colorado State University, Fort Collins, CO].

Dev, V., & Rai, K.S. (1985). Genetic relationships among certain species of the *Aedes* (*Stegomyia*) *scutellaris* group (Diptera: Culicidae). *Annals of Tropical Medicine and Parasitology, 79*(3), 325–331.

Dev, V., & Sharma, V.P. (2013). The dominant mosquito vectors of human malaria in India. In S. Manguin (Ed.), Anopheles *mosquitoes: New insights into malaria vectors* (pp. 239–271). Rijeka, Croatia: InTech.

Dewey, E.M., McNabb, S.L., Ewer, J., Kuo, G.R., Takanishi, C.L., Truman, J.W., & Honegger, H.W. (2004). Identification of the gene encoding bursicon, an insect neuropeptide responsible for cuticle sclerotization and wing spreading. *Current Biology, 14*(13), 1208–1213.

D'Haenens, G. (1961). Description d'un anophèle nouveau du Congo (Kwango) *A. moucheti brevoetsi* n. subsp. *Bulletin et annales de la Société Royale d'Entomologie de Belgique, 97*(7–8), 188–200.

Dhotare, B., Goswami, D., & Chattopadhyay, A. (2005). (R)-2, 3-cyc lohexylideneglyceraldehyde, a novel template for steroselective preparation of functionalized δ-lactones: Synthesis of mosquito oviposition pheromone. *Tetrahedron Letters, 46*(37), 6219–6221.

Dia, I., Guelbeogo, M.W., & Ayala, D. (2013). Advances and perspectives in the study of the malaria mosquito *Anopheles funestus*. In S. Manguin (Ed.), Anopheles *Mosquitoes: New insights into malaria vectors* (pp. 197–220). Rijeka, Croatia: InTech.

Diabaté, A. (2012). Characterization of the mating systems of *Anopheles gambiae* in malaria control perspective. *Malaria Journal, 11*(Suppl. 1), O42.

Diabaté, A., & Tripet, F. (2015). Targeting male mosquito mating behaviour for malaria control. *Parasites and Vectors, 8*(1), 347.

Diabaté, A., Dao, A., Diallo, M., Huestis, D.L., & Lehmann, T. (2011). Spatial distribution and male mating success of *Anopheles gambiae* swarms. *BMC Evolutionary Biology, 11*(1), 184.

Díaz Nájera, A. (1961). Variaciones morfológicos en larvas de dos especies de *Trichoprosopon* y datos sobre su distribución. *Revista del Instituto de Salubridad y Enfermedades Tropicales* (México), *21*, 201–219.

Díaz Nájera, A. (1965a). Mosquitos mexicanos. Clave para identificar hembras de *Mansonia* y *Psorophora*. Diptera: Culicidae. *Salud pública de México, 7*(4), 519–525.

Díaz Nájera, A. (1965b). Clave para identificar especies mexicanas de *Mansonia y Psorophora* (Diptera: Culicidae). *Revista del Instituto de Salubridad y Enfermedades Tropicales* (México), *25*(2), 127–137.

Díaz Nájera, A. (1966). *Aedes* de México. Descripción de una nueva especie del subgénero *Howardina* (Diptera: Culicidae). *Revista de investigación en salud pública, 26*(4), 331–344.

Díaz Nájera, A. (1971). Estructuras genitales de *Haemagogus* (*Longipalpifer*) *equinus* Theobald, 1903 (Diptera: Culicidae) con variaciónes morfológicas y nuevos datos sobre su ecología. *Revista de investigación en salud pública, 31*(2), 82–90.

Díaz Nájera, A. (1972). Nuevos datos morfológicos y ecológicos de *Haemagogus* (*Longipalpifer*) *equinus* Theobald, 1903 (Diptera: Culicidae). *Revista Peruana de entomología, 15*, 113–116.

Dickinson, M.H., Lehmann, F.O., & Sane, S.R. (1999). Wing rotation and the aerodynamic basis of insect flight. *Science, 284*(5422), 1954–1960.

Di Luca, M., Boccolini, D., Marinucci, M., & Romi, R. (2004). Intrapopulation polymorphism in *Anopheles messeae* (*An. maculipennis* Complex) inferred by molecular analysis. *Journal of Medical Entomology, 41*(4), 582–586.

Director General Medical Services, Land Forces Headquarters. (1943). *Notes on mosquito-borne diseases in Australasia*. Australian Military Forces, by authority: Modern Printing.

Dixit, J., Srivastava, H., Sharma, M., Das, M.K., Singh, O.P., Raghavendra, K., . . . Das, A. (2010). Phylogenetic inference of Indian malaria vectors from multilocus DNA sequences. *Infection, Genetics and Evolution, 10*, 755–763.

Djadid, N.D., Sanati, M.H., Zare, M., & Hassanzehi, A. (2003). rDNA-ITS2 identification of *Anopheles pulcherrimus* (Diptera: Culicidae): Genetic differences and phylogenetic relation with other Iranian vectors and its implications for malaria control. *Iranian Biomedical Journal, 7*(1), 1–6.

Djadid, N.D., Jazayeri, H., Gholizadeh, S., Rad, S.P., & Zakeri, S. (2009). First record of a new member of *Anopheles hyrcanus* group from Iran: Molecular identification, diagnosis, phylogeny, status of *kdr* resistance and *Plasmodium* infection. *Journal of Medical Entomology, 46*(5), 1084–1093.

Dlugosch, K.M., & Parker, I.M. (2008). Invading populations of an ornamental shrub show rapid life history evolution despite genetic bottlenecks. *Ecology Letters, 11*(7), 701–709.

Dobrotworsky, N.V. (1953). The *Culex pipiens* group in south-eastern Australia. I. *Proceedings of the Linnean Society of New South Wales, 77*(5–6), 357–360.

Dobrotworsky, N.V. (1954). The genus *Theobaldia* (Diptera, Culicidae) in Victoria. *Proceedings of the Linnean Society of New South Wales, 79*(3–4), 65–78.

Dobrotworsky, N.V. (1955a). The *Culex pipiens* group in South-Eastern Australia. III. Autogeny in *Culex pipiens* form *molestus*. *Proceedings of the Linnean Society of New South Wales, 79*(5–6), 193–195.

Dobrotworsky, N.V. (1955b). The *Culex* pipiens group in South-Eastern Australia. IV. Crossbreeding experiments within the *Culex pipiens* group. *Proceedings of the Linnean Society of New South Wales, 80*, 33–43.

Dobrotworsky, N.V. (1956). Notes on Australian mosquitoes (Diptera, Culicidae). I. Some species of the subgenus *Neoculex*. *Proceedings of the Linnean Society of New South Wales, 81*(2), 105–114.

Dobrotworsky, N.V. (1957). Notes on Australian mosquitoes (Diptera, Culicidae). II. Notes on *Anopheles stigmaticus* and description of new species of *Anopheles* from Australia and New Guinea. *Proceedings of the Linnean Society of New South Wales, 82*(2), 180–188.

Dobrotworsky, N.V. (1958a). Notes on Australian mosquitoes (Diptera, Culicidae). III. The subgenus *Lophoceraomyia* in Victoria. *Proceedings of the Linnean Society of New South Wales, 82*(for 1957)(3), 317–321.

Dobrotworsky, N.V. (1958b). Designation of the type species of the subgenus *Neotheobaldia* Dobrotworsky (genus *Theobaldia* Neveu-Lemaire, 1902) (Diptera: Culicidae). *Proceedings of the Entomological Society of Washington, 60*(4), 186.

Dobrotworsky, N.V. (1959). Notes on Australian mosquitoes (Diptera, Culicidae). IV. *Aedes alboannulatus* complex in Victoria. *Proceedings of the Linnean Society of New South Wales, 84*(1), 131–145.

Dobrotworsky, N.V. (1960a). The subgenus *Ochlerotatus* in the Australian region (Diptera: Culicidae). III. Review of the Victorian species of Perkinsi and Cunabulanus sections with descriptions of two new species. *Proceedings of the Linnean Society of New South Wales, 85*(1), 53–74.

Dobrotworsky, N.V. (1960b). The subgenus *Ochlerotatus* in the Australian region (Diptera: Culicidae). IV. Review of species of the Flavifrons Section. *Proceedings of the Linnean Society of New South Wales, 85*(2), 180–188.

Dobrotworsky, N.V. (1960c). The genus *Theobaldia* (Diptera, Culicidae) in Victoria. II. *Proceedings of the Linnean Society of New South Wales, 85*(2), 240–247.

Dobrotworsky, N.V. (1961a). Notes on Australian mosquitoes (Diptera, Culicidae). V. Subgenus *Pseudoskusea* in Victoria. *Proceedings of the Linnean Society of New South Wales, 85*(for 1960)(3), 257–267.

Dobrotworsky, N.V. (1961b). A new name for *Aedes waterhousei* Dobrotworsky (Diptera: Culicidae). *Proceedings of the Entomological Society of Washington, 62*(1960)(4), 248.

Dobrotworsky, N.V. (1962). Notes on Australian mosquitoes (Diptera, Culicidae). VI. Five new Victorian species and a description of the larva of *Aedes milsoni* (Taylor). *Proceedings of the Linnean Society of New South Wales, 87*(3), 291–302.

Dobrotworsky, N.V. (1965). *The mosquitoes of Victoria* (*Diptera, Culicidae*). New York, NY: Cambridge University Press.

Dobrotworsky, N.V. (1966). Mosquitoes of Tasmania and Bass Straits Islands. *Proceedings of the Linnean Society of New South Wales, 91*(2), 121–146.

Dobrotworsky, N.V. (1968). Notes on *Culiseta* (*Culicella*) *atra* (Lee) (Diptera: Culicidae) in Western Australia. *Journal of the Australian Entomological Society, 7*, 11–14.

Dobrotworsky, N.V. (1971). Contributions to the mosquito fauna of Southeast Asia. X. The genus *Culiseta* Felt in Southeast Asia. *Contributions of the American Entomological Institute, 7*(3), 38–61.

Dobrotworsky, N.V., & Drummond, F.H. (1953). The *Culex pipiens* group in south-eastern Australia. II. *Proceedings of the Linnean Society of New South Wales, 78*(3–4), 131–146.

Dobson, S.L., Bordenstein, S.R., & Rose, R.I. (2016). *Wolbachia* mosquito control: Regulated. *Science, 352*(6285), 526–527.

Dobzhansky, T. (1937). Genetic nature of species differences. *The American Naturalist, 71*(735), 404–420.

Dodge, H.R. (1945). Notes on the morphology of mosquito larvae. *Annals of the Entomological Society of America, 38*(2), 163–167.

Dodge, H.R. (1946). The identification of *Anopheles bradleyi* larvae. *Mosquito News, 6*(3), 125–126.

Dodge, H.R. (1947). A new species of *Wyeomyia* from the pitcher plant (Diptera, Culicidae). *Proceedings of the Entomological Society of Washington, 49*(5), 117–122.

Dodge, H.R. (1963). Studies on mosquito larvae. I. Later instars of eastern North American species. *The Canadian Entomologist, 95*(8), 796–813.

Dodge, H.R. (1964). Larval chaetotaxy and notes on the biology of *Toxorhynchites rutilus septentrionalis* (Diptera: Culicidae). *Annals of the Entomological Society of America, 57*(1), 46–53.

Dodge, H.R. (1966). Studies on mosquito larvae. II. The first-stage larvae of North American Culicidae and of world Anophelinae. *The Canadian Entomologist, 98*(4), 337–393.

Dohm, D.J., Romoser, W.S., Torell, M.J., & Linthicum, K.J. (1991). Impact of stressful conditions on the survival of *Culex pipiens* exposed to Rift Valley fever virus. *Journal of the American Mosquito Control Association, 7*(4), 621–623.

Dolbeskin, B., Gorickaya, V., & Mitrofanova, Y. (1930). Beschreibung einer neuen Art der Gattung *Aedes* (n. sp.) aus Osteuropa. *Magazin de parasitologie du Musée Zoologique de l'Academie de Sciences de l'URSS, 1*, 253–260.

Doleschall, C.L. (1856). Eerste bijdrage tot de kennis der dipterologische fauna van Nederlandsch Indië. *Natuurkundig tijdschrift voor Nederlandsch Indië, 10*, 403–414.

Doleschall, C.L. (1857). Tweede bijdrage tot de kennis der dipterologische fauna van Nederlandsche Indië. *Natuurkundig tijdschrift voor Nederlandsch Indië, 14*, 377–418.

Doleschall, C.L. (1858). Derde bijdrage tot de kennis der dipteren fauna van Nederlandsch Indië. *Natuurkunde tijdschrift voor Nederlandsch-Indië, 17*, 73–128.

Dong, L. (2000). Description of the female of *Tripteroides* (*Rachionotomyia*) *longipalpis* (Diptera: Culicidae). *Acta zootaxonomica Sinica, 25*(1), 46.

Dong, X., & Guo, X. (2008). Descriptions of female, larva and pupa of *Topomyia* (*Topomyia*) *svastii* Thurman (Diptera: Culicidae). *Entomotaxonomia, 30*(3), 204–206.

Dong, X., & Miyagi, I. (1995). A new species description of subgenus *Suaymyia* of genus *Topomyia* (Diptera: Culicidae). *Zoological Research, 16*(4), 343–347.

Dong, X., & Wang, X. (1985a). Descriptions of the larva and pupa of *Aedes* (*Finlaya*) *sintoni* Barraud, 1924. *Acta zootaxonomica Sinica, 10*(3), 334–336.

Dong, X., & Wang, X. (1985b). A new species of genus *Anopheles* (Diptera: Culicidae). *Zoological Research, 6*(2), 117–122.

Dong, X., & Wang, X. (1988). Descriptions of the female larva and pupa of *Topomyia* (*Topomyia*) *lindsayi* Thurman. *Entomotaxonomia, 10*(1–2), 159–162.

Dong, X.-S., Wang, X.-Z., Zhang, G.-C., & Wang, S.-G. (1989). Descriptions of the pupa and larva of *Anopheles* (*Anopheles*) *menglangensis* Ma. *Acta zootaxonomica Sinica, 14*(4), 500–502.

Dong, X.-S., Wang, X.-Z., & Lu, B.-L. (1990). A new species of *Topomyia* from Yunnan Province, China. *Acta zootaxonomica Sinica, 15*(1), 64–66.

Dong, X.-S., Wang, X.-Z., & Zhong, H.-N. (1993a). A new record of *Culex* from China. *Acta zootaxonomica Sinica, 18*(1), 53.

Dong, X.-S., Wang, X.-Z., & Zhou, H.-N. (1993b). Descriptions of the pupa and larva of *Armigeres* (*Leicesteria*) *longipalpis*. *Acta zootaxonomica Sinica, 18*(3), 382–384.

Dong, X., Zhou, H., Dong, L., & Wang, X. (1994). Descriptions of the pupa and larva of *Armigeres* (*Armigeres*) *theobaldi* and *Armigeres* (*Armigeres*) *seticoxitus*. *Entomotaxonomia, 16*(4), 302–305.

Dong, X., Zhou, H., & Dong, L. (1995a). A new species of the genus *Armigeres* (Diptera: Culicidae). *Entomotaxonomia, 17*(4), 281–286.

Dong, X., Zhou, H., & Dong, L. (1995b). Studies on genus *Armigeres* of Yunnan with description of a new species. (Diptera: Culicidae). *Zoological Research, 16*(2), 95–104.

Dong, X., Zhou, H., & Dong, L. (1995c). Description of the female, pupa and larva of *Culex* (*Lophoceraomyia*) *hirtipalpis*. *Acta zootaxonomica Sinica, 20*(3), 381–384.

Dong, X.-S., Wang, X.-Z., & Lu, B.-L. (1995). A new species of *Topomyia* from Yunnan Province, China (Diptera: Culicidae). *Acta zootaxonomica Sinica, 20*(4), 484–486.

Dong, X., Zhou, H., & Wang., X. (1997). A new species of the subgenus *Heizmannia*. (Diptera: Culicidae). *Zoological Research, 18*(1), 27–31.

Dong, X., Zhou, H., & Dong, L. (1997a). Study on *Topomyia* of Yunnan with one new species and one new record (Diptera: Culicidae). *Zoological Research, 18*(3), 247–251.

Dong, X.-S., Zhou, H.N., & Dong, L.M. (1997b). A new species of subgenus *Rachionotomyia* of the genus *Tripteroides* (Diptera: Culicidae). *Acta zootaxonomica Sinica, 22*(2), 197–200.

Dong, X., Zhou, H., & Dong, L. (1999). A new species and a new Chinese record of the subgenus *Bothaella* of genus *Aedes* (Diptera: Culicidae). *Entomotaxonomia, 21*(2), 133–137.

Dong, X.-S., Zhou, H.-N., & Dong, L.M. (2001). A new species of *Tripteroides* (s. str.) from Yunnan, China. *Acta zootaxonomica Sinica, 26(2)*, 221–224.

Dong, X.-S., Zhou, H.-N., & Dong, L.-M. (2002a). A new species of *Armigeres* from Yunnan, China (Diptera: Culicidae). *Jishengchong yu yixue kunchong xuebao, 9*(3), 155–159.

Dong, X.-S., Zhou, H.-N., & Dong, L.-M. (2002b). One new species and two new records of the subgenus *Kenknightia* of the genus *Aedes* in Yunnan (Diptera: Culicidae). *Acta entomologica Sinica, 45*(4), 516–521.

Dong, X.-S., Wang, X.-Z., & Zhou, H-N. (2002). A new species of genus *Heizmannia* (Diptera: Culicidae). *Entomotaxonomia, 24*(2), 131–133.

Dong, X.-S., Dong, L.-M., & Zhou, H.-N. (2003). A new species of genus *Uranotaenia* (Diptera: Culicidae). *Entomotaxonomia, 25*(3), 204–208.

Dong, X.-S., Dong, L.-Y.-M., Zhou, H.-N., Wang, X.-Z., & Lu, Y.-R. (2003). Three new species of mosquitoes from Yunnan, China (Diptera: Culidae). *Acta entomologica Sinica, 46*(4), 519–532.

Dong, X.-S., Dong, L.-M., & Wu, C. (2004). A new species of the genus *Uranotaenia* from China (Diptera, Culicidae). *Acta zootaxonomica Sinica, 29*(3), 566–568.

Dong, X.-S., Zhou, H.N., & Dong, L.M. (2004). A new species of the genus *Armigeres* (Diptera: Culicidae). *Entomotaxonomia, 26*(2), 123–125.

Dong, X., Zhou, H., & Dong, L. (2005). A new species of the genus *Heizmannia* (Diptera: Culicidae) from China. *Entomotaxonomia, 27*(3), 209–211.

Dong, X.-S., Wu, C., & Mao, X.-H. (2006). A new species of the genus *Topomyia* (Diptera: Culicidae). *Entomotaxonomia, 28*(3), 217–221.

Dong, X.-S., Zhou, H.-N., Dong, L.-M., & Mao, X.-H. (2007). A new species of genus *Anopheles* (Diptera: Culicidae). *Entomotaxonomia, 29*(1), 37–43.

Dong, X.-S., Dong, L.-M., & Wu, C. (2008). A new species of the genus *Tripteroides* (Diptera, Culicidae) from China. *Acta zootaxonomica Sinica, 33*(1), 187–190.

Dong, Y., Manfredini, F., & Dimopoulos, G. (2009). Implication of the mosquito midgut microbiota in the defense against malaria parasites. *PLOS Pathogens, 5*(5), e1000423.

Dong, Y., Simöes, M.L., Marois, E., & Dimopoulos, G. (2018). CRISPR/*Cas9*-mediated gene knockout of *Anopheles gambiae FREP1* suppresses malaria infection. *PLOS Pathogenetics, 14*(3), e1006898.

Dönitz, W. (1901). Nachrichten aus dem Berliner Entomologischen Verein. *Insekten-Börse, 18*(5), 36–38.

Dönitz, W. (1902). Beitrage zur Kenntniss der *Anopheles. Zeitschrift für Hygiene, 41*, 15–88.

Dönitz, W. (1903). Beitrage zur Kenntniss der *Anopheles*. II. Mittheilung. *Zeitschrift für Hygiene, 43*, 215–238.

Donnelly, M.J., Simard, F., & Lehmann, T. (2002). Evolutionary studies of malaria vectors. *Trends in Parasitology, 18*(2), 75–80.

Doraisamy, L. V. (1963). A subspecific variation in the larva of *Anopheles subpictus*. *Bulletin of Entomology* (Madras), *4*, 41–42.

dos Santos-Mallet, J.R., Gleiser, R.M., Alencar, J., Marques, W.A., Sarmento, J.S., Müller, G.A., & Marcondes, C.B. (2009). Scanning electron microscopy of the egg of *Ochlerotatus albifasciatus* (Diptera: Culicidae). *Journal of Medical Entomology, 46*(5), 980–985.

dos Santos-Mallet, J.R., Müller, G.A., Gleiser, R.M., Alencar, J., Marques, W.A., Sarment, J.S., & Marcondes, C.B. (2010). Scanning electron microscopy of the eggs of *Aedes scapularis* from southern South America. *Journal of the American Mosquito Control Association, 26*(2), 205–209.

dos Santos-Mallet, J., Sarmento, J.S., Alencar, J., Mueller, G.A., Oliveira, E.M., Foster, W.A., & Marcondes, C.B. (2013). Scanning electron microscopy of eggs of *Sabethes cyaneus*. *Journal of the American Mosquito Control Association, 29*(1), 64–68.

Dottorini, T., Nicolaides, L., Ranson, H., Rogers, D.W., Crisanti, A., & Catteruccia, F. (2007). A genome-wide analysis in *Anopheles gambiae* mosquitoes reveals 46 male accessory gland genes, possible modulators of female behavior. *Proceedings of the National Academy of Sciences of the United States of America, 104*(41), 16215–16220.

Doucet, J. (1949a). Étude des Culicidae du Lac Alaotra. *Mémoires de l'Institut Scientifique de Madagascar A, 3*(2), 121–145.

Doucet, J. (1949b). Recherches sur les Culicides de Madagascar. *Mémoires de l'Institut Scientifique de Madagascar A, 3*(3), 325–332.

Doucet, J. (1950). Les culicines de Madagascar (Dipt.). *Mémoires de l'Institut Scientifique de Madagascar A, 4*(1), 39–65.

Doucet, J. (1951a). Les moustiques de la région de Perinet. *Mémoires de l'Institut Scientifique de Madagascar A, 6*(1), 63–82.

Doucet, J. (1951b). *Les anophélinés de la Région Malgache*. Madagascar: Publications de l'Institut de Recherche Scientifique Tananarive-Tsimbazaza.

Doucet, J. (1951c). Étude des Culicides de la region de Vangaindrano (Diptera). *Mémoires de l'Institut Scientifique de Madagascar, 6*(1), 83–114.

Doucet, J. (1960). Moustiques forestiers de la République de Côte-d'lvoire. I. Notes biologiques et systematiques sur *Uranotaenia henrardi* Edwards, *Eretmapodites oedipodius marcellei* Hamon, *Aedes (Pseudarmigeres) kummi* Edwards et *Aedes (Pseudarmigeres) argenteoventrali*. *Bulletin de la Société de pathologie exotique, 53*, 810–814.

Doucet, J. (1962). Moustiques forestiers de la République de Côte-d'Ivoire. III. Notes biologiques et systematiques sur *Uranotaenia andreae* sp. n. et *Culex (Neoculex) vinckei* Hamon, Holstein et Rivola, 1957. *Bulletin de la Société de pathologie exotique, 54*(for 1961), 1156–1164.

Douglas, G.W. (1962). Notes on Australian mosquitoes (Diptera, Culicidae). I. The life history of *Aedomyia venustipes* (Skuse). *Proceedings of the Linnean Society of New South Wales, 86*(for 1961), 262–267.

Do-Van-Quy. (1963). Clefs pour la détermination des 25 espèces les plus communes d'*Anophèles* du Sud-Viêt-Nam (adultes femeles et larves). *Rapport annual de fonctionnement technique de l'Institute Pasteur du Viêt-Nam, 1962*, 26–34.

Dow, R.P. (1953). Notes on Iranian mosquitoes. *The American Journal of Tropical Medicine and Hygiene, 2*(4), 683–695.

Dow, R.P., Reeves, W.C., & Bellamy, R.E. (1957). Field tests of avian host preference of *Culex tarsalis* Coq. *The American Journal of Tropical Medicine and Hygiene, 6*(2), 294–303.

Downes, J.A. (1958). The feeding habits of biting flies and their significance in classification. *Annual Review of Entomology, 3*(1), 249–266.

Downes, J.A. (1966). Observations on the mating behaviour of the crab hole mosquito *Deinocerites cancer* (Diptera: Culicidae). *The Canadian Entomologist, 98*(11), 1169–1177.

Downes, J.A. (1969). The swarming and mating flight of Diptera. *Annual Review of Entomology, 14*(1), 271–298.

Downes, W.L., Jr., & Dahlem, G.A. (1987). Keys to the evolution of Diptera: Role of Homoptera. *Environmental Entomology, 16*(4), 847–854.

Downs, W. (1949). Experiments in crossing *Aedes (Stegomyia) aegypti* Linnaeus and *Aedes (Stegomyia) albopictus* Skuse. *Science, 109*(2826), 200–201.

Downs, W.G. (1950). Morphological notes on *Anopheles aquasalis* Curry and *Anopheles oswaldoi* Peryassu. *Mosquito News, 10*(1), 29–30.

Drake, E.T. (1996). *Restless genius: Robert Hooke and his earthly thoughts*. New York, NY: Oxford University Press.

Drake, L.L., Boudko, D.Y., Marinotti, O., Carpenter, V.K., Dawe, A.L., & Hansen, I.A. (2010). The aquaporin gene family of the yellow fever mosquito *Aedes aegypti*. *PLOS ONE, 5*(12), e15578.

Drake, L.L., Rodriguez, S.D., & Hansen, I.A. (2015). Functional characterization of aquaporins and aquaglyceroporins of the yellow fever mosquito, *Aedes aegypti*. *Scientific Reports, 5*, Article 7795.

Drame, P.M., Poinsignon, A., Marie, A., Noukpo, H., Dorcoure, S., Cornelie, S., & Remoue, F. (2013). New salivary biomarkers of human exposure to malaria vector bites. In S. Manguin (Ed.), Anopheles *mosquitoes: New insights into malaria vectors* (pp. 755–795). Rijeka, Croatia: InTech.

Drummond, C.L., Drobnack, J., Backenson, P.B., Ebel, G.D., & Kramer, L.D. (2006). Impact of trap elevation on estimates of abundance, parity rates, and body size of *Culex pipiens* and *Culex restuans*. *Journal of Medical Entomology, 43*(2), 177–184.

Dubitzky, A.M. (1968). A new species *Aedes montchadskyi* sp. n, from the Basin of the Ili River. *Parazitologiya* (St. Petersburg), *2*(3), 218–224.

Dubitzky, A.M. (1970a). A new species of mosquitoes *Aedes stramineus* new species (Diptera, Culicidae). *Parazitologiya* (St. Petersburg), *4*(5), 408–413.

Dubitzky, A.M. (1970b). *Bloodsucking mosquitoes (Diptera, Culicidae) of Kazakhstan*. Alma-mata.

Dubitzky, A.M. (1977). A description of the imago of a little-known species of mosquito, *Aedes (Ochlerotatus) rempeli* (Culicidae). *Parazitologiya* (St. Petersburg), *11*(1), 72–74.

Dubitzky, A.M., & Deshevykh, N.D. (1978). A new species of mosquito *Aedes (Ochlerotatus) gutzevichi* sp.n. from Central Asia (Culicidae). *Parazitologiya* (St. Petersburg),*12*(2), 177–182.

Dubitzky, A.M., & Tupitzin, J.N. (1970). Description of larvae of *Aedes (O.) rempeli* Vockeroth, a species new in the fauna of the U.S.S.R. *Parazitologiya* (St. Petersburg), *4*(2), 171–174.

Dubois, A. (1988). The genus in zoology: A contribution to the theory of evolutionary systematics. *Mémoires du Muséum National d'Historie Naturelle (Zoologie), 140*, 1–122.

DuBose, W.P., & Curtin, T.J. (1965). Identification keys to the adult and larval mosquitoes of the Mediterranean area. *Journal of Medical Entomology, 1*(4), 349–355.

Duchesne, L., Hubert, J.F., Verbavatz, J.M., Thomas, D., & Pietrantonio, P.V. (2003). Mosquito (*Aedes aegypti*) aquaporin, present in tracheolar cells, transports water, not glycerol, and forms orthogonal arrays in *Xenopus* oocyte membranes. *The FEBS Journal, 270*(3), 422–429.

Duchet, C., Allan, R., & Carnevale, P. (2013). Vector control: Some new paradigms and approaches. In S. Manguin (Ed.), Anopheles *mosquitoes: New insights into malaria vectors* (pp. 705–753). Rijeka, Croatia: InTech.

Dufourd, C., & Dumont, Y. (2012). Modeling and simulations of mosquito dispersal. The case of *Aedes albopictus*. *Biomath, 1*, 1209262, 1–7.

Dufourd, C., & Dumont, Y. (2013). Impact of environmental factors on mosquito dispersal in the prospect of sterile insect technique control. *Computers and Mathematics with Applications, 66*(9), 1695–1715.

Dukes, J.S., & Mooney, H.A. (1999). Does global change increase the success of biological invaders? *Trends in Ecology and Evolution, 14*(4), 135–139.

Dumas, E., Atyame, C.M., Milesi, P., Fonseca, D.M., Shaikevich, E.V., Unal, S., . . . Duron, O. (2013). Population structure of *Wolbachia* and cytoplasmic introgression in a complex of mosquito species. *BMC Evolutionary Biology, 13*(1), 181.

Dumas, E., Atyame, C.M., Malcolm, C.A., Le Goff, G., Unal, S., Makoundou, P., . . . Duron, O. (2016). Molecular data reveal a cryptic species within the *Culex pipiens* mosquito complex. *Insect Molecular Biology, 25*(6), 800–809.

Dumbleton, L.J. (1946). Note on *Bironella* (*Brugella*) *hollandi* Taylor. *New Zealand Journal of Science and Technology, 27*(3), 250–253.

Dumbleton, L.J. (1962). A new species and new subgenus of *Aedes* (Diptera: Culicidae) from New Zealand. *New Zealand Journal of Science, 5*(1), 17–27.

Dumbleton, L.J. (1965). Developmental stages and biology of *Culiseta tonnoiri* (Edwards) and a note on *Culex pervigilans* Bergroth (Diptera, Culicidae). *New Zealand Journal of Science, 8*(2), 137–143.

Dumbleton, L.J. (1968). A synopsis of the New Zealand mosquitoes (Diptera, Culicidae) and a key to the larvae. *Tuatara, 16*(3), 167–179.

Dunn, L.H. (1918). A new mosquito (*Aedes whitmorei*) from Colombia. *Proceedings of the Entomological Society of Washington, 20*(6), 128–130.

Dunn, R.R. (2010). Global mapping of ecosystem disservices: The unspoken reality that nature sometimes kills us. *Biotropica, 42*(5), 555–557.

Duret, J.P. (1953a). Las especies argentinas de *Culex* (*Melanoconion*) (Diptera, Culicidae). *Revista de la Sociedad Entomológica Argentina, 16*(3), 67–76.

Duret, J.P. (1953b). Notas sobre *Culex* argentinos (Diptera-Culicidae). *Revista de la sanidad militar Argentina, 52*, 272–278.

Duret, J.P. (1954a). Las especies argentinas de *Culex* (*Melanoconion*) (Diptera, Culicidae). *Revista de la Sociedad Entomológica Argentina, 16*(for 1953)(4), 99–121.,

Duret, J.P. (1954b). Nuevo mosquito argentino (Diptera, Culicidae). *Neotropica, 1*(2), 23–26.

Duret, J.P. (1967a). Dos especies nuevas de *Culex* (*Melanoconion*) (Diptera-Culicidae). *Neotropica-La Plata, 13*(41), 77–84.

Duret, J.P. (1967b). *Culex* (*Culex*) *chitae* una nueva especie de Colombia (Diptera, Culicidae). *Neotropica-La Plata, 13*(42), 113–116.

Duret, J.P. (1968a). *Culex* (*Melanoconion*) *kerri* y *C.* (*M.*) *cristovaoi* dos nuevas especies del Brasil (Diptera, Culicidae). *Neotropica-La Plata, 14*(43), 11–16.

Duret, J.P. (1968b). Cuatro especies nuevas de *Culex* (Diptera, Culicidae). *Neotropica-La Plata, 14*(44), 57–64.

Duret, J.P. (1968c). *Culex* (*Melanoconion*) *contei* una nueva especie del Brasil y Panamá (Diptera, Culicidae). *Neotropica-La Plata, 14*, 121–124.

Duret, J.P. (1968d). Dos especies nuevas de *Culex* del Brasil (Diptera, Culicidae). *Physis, 28*(76), 15–19.

Duret, J.P. (1968e). Dos especies nuevas de *Culex* (*Culex*) de la Argentina (Diptera, Culicidae). *Physis, 27*(75), 323–328.

Duret, J.P. (1968f). Cinco especies nuevas de *Culex* (*Melanoconion*) (Diptera, Culicidae). *Revista de la Sociedad Entomológica Argentina, 30*(1–4), 69–81.

Duret, J.P. (1969a). Contribución al conocimiento de los *Culex* del Paraguay (Diptera-Culicidae). *Revista de la Sociedad Entomológica Argentina, 31*(1–4), 3–13.

Duret, J.P. (1969b). *Culex* (*Melanoconion*) *bahiense* una especie nueva del Brasil (Diptera, Culicidae). *Neotropica-La Plata, 15*(4–6), 40–42.

Duret, J.P. (1969c). *Culex* (*Anoediopora*) *damascenoi* una nueva especie del Brasil (Diptera, Culicidae). *Neotropica-La Plata, 15*(4–8), 143–145.

Duret, J.P. (1969d). Estudio sobre *Culex* (*Melanoconion*) *ocellatus* y sus especies afines (Diptera; Culicidae). *Physis, 29*(78), 113–129.

Duret, J.P. (1970). *Uranotaenia riverai* una nueva especie de Colombia (Diptera, Culicidae). *Revista de la Sociedad Entomológica Argentina, 32*(1–4), 43–46.

Duret, J.P. (1971a). Cinco especies nuevas de culicidos neotropicales (Diptera, Culicidae). *Neotropica-La Plata, 17*(52), 15–28.

Duret, J.P. (1971b). Una nueva especie de *Haemagogus* de Nicaragua (Diptera, Culicidae). *Neotropica-La Plata, 17*(53), 83–86.

Duret, J.P. (1972). *Culex* (*Anoediopora*) *quasioriginator*, una nueva especie del Brasil. *Neotropica-La Plata, 18*(55), 1–5.

Duret, J.P. (1982). *Wyeomyia* (*Dendromyia*) *sirivanakarni* una especie nueva de Panamá (Diptera, Culicidae). *Revista de la Sociedad Entomológica Argentina, 41*(1–4), 167–170.

Duret, J.P., & Barreto, M.P. (1956). Notas sôbre culícidas do Estado de São Paulo, Brasil, com descrições de três novas espécies de *Culex* (Diptera, Culicidae). *Revista Brasileira de entomología, 5*, 81–99.

Duret, J.P., & Damasceno, R.G. (1955). Notas sobre *Culex* (*Tinolestes*) de Pará, Brasil (Diptera Culicidae). *Revista Ecuatoriana de entomología y parasitología Guayaquil, 2*(3–4), 393–409.

Durnez, L., & Coosemans, M. (2013). Residual transmission of malaria: An old issue for new approaches. In S. Manguin (Ed.), Anopheles *mosquitoes: New insights into malaria vectors* (pp. 671–704). Rijeka, Croatia: InTech.

Duron, O., Labbé, P., Berticat, C., Rousset, F., Guillot, S., Raymond, M., & Weill, M. (2006). High *Wolbachia* density correlates with cost of infection for insecticide resistant *Culex pipiens* mosquitoes. *Evolution, 60*(2), 303–314.

Dusfour, I., Harbach, R.E., & Manguin, S. (2004a). Bionomics and systematics of the Oriental *Anopheles sundaicus* complex in relation to malaria transmission and vector control. *The American Journal of Tropical Medicine and Hygiene, 71*(4), 518–524.

Dusfour, I., Linton, Y.-M., Cohuet, A., Harbach, R.E., Baimai, V., Trung, H.D., Chang Moh Seng, Matusop A., & Manguin, S. (2004b). Molecular evidence of speciation between island and continental populations of *Anopheles* (*Cellia*) *sundaicus* in Southeast Asia. *Journal of Medical Entomology, 41(3)*, 287–295.

Dusfour, I., Michaux, J. R., Harbach, R.E., & Manguin, S. (2007a). Speciation and phylogeography of the Southeast Asian *Anopheles sundaicus* complex. *Infection Genetics and Evolution, 7*(4), 484–493.

Dusfour, I., Londeau, J., Harbach, R.E., Vythilingham, I., Baimai, V., Trung, H. D., . . . Manguin, S. (2007b). Polymerase chain reaction identification of three members of the *Anopheles sundaicus* (Diptera: Culicidae) complex, malaria vectors in Southeast Asia. *Journal of Medical Entomology, 44*(5), 723–731.

Dusfour, I., Issaly, J., Carinci, R., Gaborit, P., & Girod, R. (2012). Incrimination of *Anopheles* (*Anopheles*) *intermedius* Peryassú, *An.* (*Nyssorhynchus*) *nuneztovari* Gabaldón, *An.* (*Nys.*) *oswaldoi* Peryassú as natural vectors of *Plasmodium falciparum* in French Guiana. *Memórias do Instituto Oswaldo Cruz, 107*(3), 429NIL–0188.

Dutta, P., Khan, S.A., Sharma, C.K., Doloi, P., Hazarika, N.C., & Mahanta, J. (1998). Distribution of potential dengue vectors in major townships along the national highways and trunk roads of northeast India. *Southeast Asian Journal of Tropical Medicine and Public Health, 29*(1), 173–176.
Duttlinger, A., Mispelon, M., & Nichols, R. (2003). The structure of the FMRFamide receptor and activity of the cardioexcitatory neuropeptide are conserved in mosquito. *Neuropeptides, 37*(2), 120–126.
Duznetsov, R.L. (1971). Recognition of *Anopheles coustani* Laveran, 1900 and *Anopheles squamosus* Theobald, 1901 in the Yemen Arab Republic. *Meditsinskaia parazitologiia i parizarnye bolezni, 40*(4), 441–443.
Dyar, H.G. (1904a). The larva of *Culex punctor* Kirby, with notes on an allied form. *Journal of the New York Entomological Society, 12*(3), 169–171.
Dyar, H.G. (1904b). Brief notes on mosquito larvae. *Journal of the New York Entomological Society, 12*(3), 172–174, 243–246.
Dyar, H.G. (1905a). Remarks on genitalic genera in the Culicidae. *Proceedings of the Entomological Society of Washington, 7*(1), 42–49.
Dyar, H.G. (1905b). A new mosquito. *Journal of the New York Entomological Society, 13*(2), 74.
Dyar, H.G. (1905c). Illustrations of the abdominal appendages of certain mosquitoes. *Journal of the New York Entomological Society, 13*(2), 53–56.
Dyar, H.G. (1907). Report on the mosquitoes of the coast region of California, with descriptions of new species. *Proceedings of the United States National Museum, 32*(1516), 121–129.
Dyar, H.G. (1916a). Mosquitoes at San Diego, California. *Insecutor Inscitiae Menstruus, 4*(4–6), 46–51.
Dyar, H.G. (1916b). New *Aedes* from the mountains of California (Diptera, Culicidae). *Insecutor Inscitiae Menstruus, 4*(7–9), 80–90.
Dyar, H.G. (1917a). The mosquitoes of the mountains of California (Diptera, Culicidae). *Insecutor Inscitiae Menstruus, 5*(1–3), 11–21.
Dyar, H.G. (1917b). The mosquitoes of the Pacific Northwest (Diptera, Culicidae). *Insecutor Inscitiae Menstruus, 5*(7–9), 97–102.
Dyar, H.G. (1917c). Notes on the *Aedes* of Montana (Diptera, Culicidae). *Insecutor Inscitiae Menstruus, 5*(7–9), 104–121.
Dyar, H.G. (1917d). A new *Aedes* from the Rocky-mountain region (Diptera, Culicdae). *Insecutor Inscitiae Menstruus, 5*(7–9), 127–128.
Dyar, H.G. (1917e). The larva of *Aedes idahoensis* (Diptera, Culicidae). *Insecutor Inscitiae Menstruus, 5*, 187–188.
Dyar, H.G. (1918a). The male genitalia of *Aedes* as indicative of natural affinities (Diptera, Culicidae). *Insecutor Inscitiae Menstruus,* 6(4–6), 71–86.
Dyar, H.G. (1918b). A revision of the American species of *Culex* on the male genitalia. *Insecutor Inscitiae Menstruus, 6*(4–6), 86–111.
Dyar, H.G. (1918c). A note on the American species of *Mansonia* (Diptera, Culicidae). *Insecutor Inscitiae Menstruus, 6*(4–6), 112–115.
Dyar, H.G. (1918d). New American mosquitoes (Diptera, Culicidae). *Insecutor Inscitiae Menstruus, 6*(7–9), 120–129.
Dyar, H.G. (1918e). A second note on the species of *Culex* of the Bahamas (Diptera, Culicidae). *Insecutor Inscitiae Menstruus, 5*(for 1917), 183–187.
Dyar, H.G. (1919a). A note on *Lesticocampa*, and a new species (Diptera, Culicidae). *Insecutor Inscitiae Menstruus, 7*, 9–11.
Dyar, H.G. (1919b). Westward extension of the Canadian mosquito fauna (Diptera, Culicidae). *Insecutor Inscitiae Menstruus, 7*(1–3), 11–39.
Dyar, H.G. (1919c). A note on Argentine mosquitoes (Diptera, Culicidae). *Insecutor Inscitiae Menstruus, 7*(4–6), 85–89.
Dyar, H.G. (1919d). A revision of the American Sabethini of the Sabethes group by the male genitalia (Diptera, Culicidae). *Insecutor Inscitiae Menstruus, 7*(7–9), 114–142.
Dyar, H.G. (1919e). A new subgenus of *Culex* Linn. (Diptera, Culicidae). *Insecutor Inscitiae Menstruus, 7*(7–9), 150.
Dyar, H.G. (1919f). The mosquitoes collected by the Canadian Arctic Expedition, 1913–18, (Diptera, Culicidae.). In *Report of the Canadian Arctic Expedition, 1913–18* (Vol. 3: Insects, Part C: Diptera) (pp. 31–33). Ottawa: J. de Labroquerie Tache.
Dyar, H.G. (1920a). The mosquitoes of British Columbia and Yukon Territory, Canada (Diptera, Culicidae). *Insecutor Inscitiae Menstruus, 8*(1–3), 1–27.
Dyar, H.G. (1920b). Notes on European mosquitoes (Diptera, Culicidae). *Insecutor Inscitiae Menstruus, 8*(4–6), 51–54.
Dyar, H.G. (1920c). The species of *Choeroporpa*, a subgenus of *Culex* (Diptera, Culicidae). *Insecutor Inscitiae Menstruus, 8*(4–6), 54–81.
Dyar, H.G. (1920d). The American *Aedes* of the *stimulans* group (Diptera, Culicidae). *Insecutor Inscitiae Menstruus, 8*, 106–120.
Dyar, H.G. (1920e). The Grabhamia group of *Psorophora* (Diptera, Culicidae). *Insecutor Inscitiae Menstruus, 8*(7–9), 140–141.
Dyar, H.G. (1920f). The *Aedes* of the mountains of California and Oregon (Diptera, Culicidae). *Insecutor Inscitiae Menstruus, 8*, 165–173.
Dyar, H.G. (1920g). A new *Culex* from Panama (Diptera, Culicidae). *Insecutor Inscitiae Menstruus, 8*(10–12), 173–174.
Dyar, H.G. (1920h). A collection of mosquitoes from the Philippine Islands (Diptera, Culicidae). *Insecutor Inscitiae Menstruus, 8*(10–12), 175–186.
Dyar, H.G. (1920i). The earliest name of the yellow fever mosquito (Diptera, Culicidae). *Insecutor Inscitiae Menstruus, 8*(10–12), 204.
Dyar, H.G. (1920j). A new mosquito from Mexico (Diptera, Culicidae). *Insecutor Inscitiae Menstruus, 8*(4–6), 81–82.
Dyar, H.G. (1920k). A second *Culex* of the subgenus *Transculicia* Dyar. Ins. *Insecutor Inscitiae Menstruus, 8*(1–3), 27–29.
Dyar, H.G. (1920l). Descriptions of hitherto unknown larvae of *Culex* (Diptera, Culicidae). *Insecutor Inscitiae Menstruus, 7*(10–12), 161–162.
Dyar, H.G. (1920m). The larva of *Aedes campestris* Dyar & Knab. *Insecutor Inscitiae Menstruus, 8*(7–9), 120.
Dyar, H.G. (1920n). The classification of American *Aedes* (Diptera, Culicidae). *Insecutor Inscitiae Menstruus, 8*(7–9), 103–106.
Dyar, H.G. (1921a). Ring-legged *Culex* in Texas (Diptera, Culicidae). *Insecutor Inscitiae Menstruus, 9*(1–3), 32–34.
Dyar, H.G. (1921b). Three new mosquitoes from Costa Rica (Diptera, Culicidae). *Insecutor Inscitiae Menstruus, 9*(1–3), 34–36.
Dyar, H.G. (1921c). Notes on the North American species of *Choeroporpa* (Diptera, Culicidae). *Insecutor Inscitiae Menstruus, 9*(1–3), 37–39.
Dyar, H.G. (1921d). A new mosquito from East Africa (Diptera, Culicidae). *Insecutor Inscitiae Menstruus, 9*(4–6), 51–52.
Dyar, H.G. (1921e). The American *Aedes* of the *punctor* group (Diptera, Culicidae). *Insecutor Inscitiae Menstruus, 9*(4–6), 69–80.
Dyar, H.G. (1921f). Two new *Culex* from Costa Rica (Diptera, Culicidae). *Insecutor Inscitiae Menstruus, 9*(4–6), 100.
Dyar, H.G. (1921g). The genus *Haemagogus* Williston (Diptera, Culicidae). *Insecutor Inscitiae Menstruus, 9*(7–9), 101–114.
Dyar, H.G. (1921h). Illustrations of certain mosquitoes (Diptera, Culicidae). *Insecutor Inscitiae Menstruus, 9*(7–9), 114–118.
Dyar, H.G. (1921i). Two new mosquitoes from China (Diptera, Culicidae). *Insecutor Inscitiae Menstruus, 9*(7–9), 147–148.

Dyar, H.G. (1921j). The species of *Finlaya* allied to *terrens* Walker (Diptera, Culicidae). *Insecutor Inscitiae Menstruus, 9*(10–12), 151–153.

Dyar, H.G. (1921k). New mosquitoes from Costa Rica (Diptera, Culicidae). *Insecutor Inscitiae Menstruus, 9*(10–12), 154–155.

Dyar, H.G. (1921l). The American *Aedes* of the *impiger* (*decticus*) group (Diptera, Culicidae). *Insecutor Inscitiae Menstruus, 10*(1–3), 3–8.

Dyar, H.G. (1921m). The male of *Psorophora coffini* Dyar & Knab. *Insecutor Inscitiae Menstruus, 9*(1–3), 31.

Dyar, H.G. (1922a). New mosquitoes from Alaska (Diptera, Culicidae). *Insecutor Inscitiae Menstruus, 10*(1–3), 1–3.

Dyar, H.G. (1922b). Note on the male genitalia of *Culex coronator* and allied forms (Diptera, Culicidae). *Insecutor Inscitiae Menstruus, 10*(1–3), 18–19.

Dyar, H.G. (1922c). The American *Aedes* of the *scapularis* group (Diptera, Culicidae). *Insecutor Inscitiae Menstruus, 10*(4–6), 51–60.

Dyar, H.G. (1922d). Mosquito notes (Diptera, Culicidae). *Insecutor Inscitiae Menstruus, 10*(4–6), 92–99.

Dyar, H.G. (1922e). The species of *Psorophora* of the *ciliata* group (Diptera, Culicidae). *Insecutor Inscitiae Menstruus, 10*(7–9), 113–117.

Dyar, H.G. (1922f). The American *Aedes* of the *serratus* group (Diptera, Culicidae). *Insecutor Inscitiae Menstruus, 10*(7–9), 157–166.

Dyar, H.G. (1922g). Notes on tropical American mosquitoes (Diptera, Culicidae). *Insecutor Inscitiae Menstruus, 10*(10–12), 188–196.

Dyar, H.G. (1922h). Illustrations of the male hypopygium of certain Sabethids (Diptera, Culicidae). *Insecutor Inscitiae Menstruus, 10*(4–6), 61–62.

Dyar, H.G. (1922i). The mosquitoes of the United States. *Proceedings of the United States National Museum, 62*(1), 1–119.

Dyar, H.G. (1923a). The mosquitoes of the Yellowstone National Park (Diptera, Culicidae). *Insecutor Inscitiae Menstruus, 11*(1–3), 36–46.

Dyar, H.G. (1923b). Mosquito notes (Diptera, Culicidae). *Insecutor Inscitiae Menstruus, 11*(4–6), 64–72.

Dyar, H.G. (1923c). Notes on American *Culex* (Diptera, Culicidae). *Insecutor Inscitiae Menstruus, 11*(7–9), 118–121.

Dyar, H.G. (1923d). A new *Culex* from Mexico (Diptera, Culicidae). *Insecutor Inscitiae Menstruus, 11*(10–12), 186–187.

Dyar, H.G. (1923e). On some of the American subgenera of *Culex* (Diptera, Culicidae). *Insecutor Inscitiae Menstruus, 11*(10–12), 187–190.

Dyar, H.G. (1924a). A new sabethid from Brazil (Diptera, Culicidae). *Insecutor Inscitiae Menstruus, 12*(7–9), 92.

Dyar, H.G. (1924b). A note on *Sabethes* Robineau-Desvoidy (Diptera, Culicidae). *Insecutor Inscitiae Menstruus, 12*(7–9), 97–100.

Dyar, H.G. (1924c). *Phoniomyia* and *Dendromyia* Theobald (Diptera, Culicidae). *Insecutor Inscitiae Menstruus, 12*, 107–113.

Dyar, H.G. (1924d). Some new mosquitoes from Colombia (Diptera, Culicidae). *Insecutor Inscitiae Menstruus, 12*, 119–124.

Dyar, H.G. (1924e). Two new mosquitoes from California (Diptera, Culicidae). *Insecutor Inscitiae Menstruus, 12*, 125–127.

Dyar, H.G. (1924f). A new mosquito from Siberia (Diptera, Culicidae). *Insecutor Inscitiae Menstruus, 12*, 127–128.

Dyar, H.G. (1924g). Mosquitoes from Chile (Diptera, Culicidae). *Insecutor Inscitiae Menstruus, 12*, 128–131.

Dyar, H.G. (1924h). The American forms of *Aedes cinereus* Meigen (Diptera, Culicidae). *Insecutor Inscitiae Menstruus, 12*, 179–180.

Dyar, H.G. (1924i). Some new mosquitoes from Colombia. II. (Diptera, Culicidae). *Insecutor Inscitiae Menstruus, 12*, 183–186.

Dyar, H.G. (1925a). A new sabethid from Panama (Diptera, Culicidae). *Insecutor Inscitiae Menstruus, 13*, 20–21.

Dyar, H.G. (1925b). Some new mosquitoes from Colombia. III. (Diptera, Culicidae). *Insecutor Inscitiae Menstruus, 13*, 21–24.

Dyar, H.G. (1925c). Note on bromelicolous *Anopheles* (Diptera, Culicidae). *Insecutor Inscitiae Menstruus, 13*, 25–27.

Dyar, H.G. (1925d). Some mosquitoes from Ecuador (Diptera, Culicidae). *Insecutor Inscitiae Menstruus, 13*, 27–31.

Dyar, H.G. (1925e). The subgenera of *Mansonia* Blanchard (Diptera, Culicidae). *Insecutor Inscitiae Menstruus, 13*(1–3), 40–43.

Dyar, H.G. (1925f). The mosquitoes of Panama (Diptera, Culicidae). *Insecutor Inscitiae Menstruus, 13*(7–9), 101–195.

Dyar, H.G. (1925g). Some mosquitoes from Venezuela (Diptera, Culicidae). *Insecutor Inscitiae Menstruus, 13*, 213–216.

Dyar, H.G. (1926). Notes on Panama mosquitoes (Diptera, Culicidae). *Insecutor Inscitiae Menstruus, 14*, 111–114.

Dyar, H.G. (1928). *The mosquitoes of the Americas*. Washington, DC: Carnegie Institution for Science.

Dyar, H.G. (1929a). A new species of mosquito from Montana with annotated list of the species known from the state. *Proceedings of the United States National Museum, 75*(2794), 1–8.

Dyar, H.G. (1929b). Remarks on the subgenus *Phalangomyia* of *Culex* Linn. (Diptera, Culicidae). *American Journal of Hygiene, 9*(2), 509–511.

Dyar, H.G. (1929c). A new mosquito from the Philippine Islands. *Proceedings of the Entomological Society of Washington, 31*(3), 61–62.

Dyar, H.G., & Barrett, H.P. (1918). Descriptions of hitherto unknown larvae of *Culex* (Diptera, Culicidae). *Insecutor Inscitiae Menstruus, 6*(7–9), 119–120.

Dyar, H.G., & Currie, R.P. (1903). The egg and young larva of *Culex perturbans* Walker. *Entomological Society of Washington, 6*, 218–220.

Dyar, H.G., & Knab, F. (1906a). Diagnoses of new species of mosquitoes. *Proceedings of the Biological Society of Washington, 19*, 133–142.

Dyar, H.G., & Knab, F. (1906b). Notes on some American mosquitoes with descriptions of new species. *Proceedings of the Biological Society of Washington, 19*, 159–172.

Dyar, H.G., & Knab, F. (1906c). The species of mosquitoes in the genus *Megarhinus*. *Smithsonian Miscellaneous Collections, 48*(3), 241–258.

Dyar, H.G., & Knab, F. (1906d). The larvae of Culicidae classified as independent organisms. *Journal of the New York Entomological Society, 14*(4), 169–230.

Dyar, H.G., & Knab, F. (1907a). Descriptions of some American mosquitoes. *Journal of the New York Entomological Society, 15*(1), 9–13.

Dyar, H.G., & Knab, F. (1907b). New American mosquitoes. *Journal of the New York Entomological Society, 15*(2), 100–101.

Dyar, H.G., & Knab, F. (1907c). Descriptions of new mosquitoes from the Panama Canal Zone. *Journal of the New York Entomological Society, 15*(4), 197–212.

Dyar, H.G., & Knab, F. (1907d). Descriptions of three new North American mosquitoes. *Journal of the New York Entomological Society, 15*(4), 213–214.

Dyar, H.G., & Knab, F. (1908). Descriptions of some new mosquitoes from tropical America. *Proceedings of the United States National Museum, 35*(1632), 53–70.

Dyar, H.G., & Knab, F. (1909a). On the identity of *Culex pipiens* Linnaeus. (Diptera, Culicidae.). *Proceedings of the Entomological Society of Washington, 11*(1), 30–39.

Dyar, H.G., & Knab, F. (1909b). Description of a new mosquito from Cuba. *Proceedings of the Entomological Society of Washington, 11*(1), 39.

Dyar, H.G., & Knab, F. (1909c). Mosquito comment. *The Canadian Entomologist, 41*(3), 101–102.

Dyar, H.G., & Knab, F. (1909d). Descriptions of some new species and a new genus of American mosquitoes. *Smithsonian Miscellaneous Collections, 52*(1822), 253–266.

Dyar, H.G., & Knab, F. (1910). Description of three new American mosquitoes. (Diptera, Culicidae). *Proceedings of the Entomological Society of Washington, 11*(for 1909)(1), 173–174.

Dyar, H.G., & Knab, F. (1913). Three new neotropical mosquitoes. *Insecutor Inscitiae Menstruus, 1*(6), 76–78.

Dyar, H.G., & Knab, F. (1914). New mosquitoes from Peru (Diptera, Culicidae). *Insecutor Inscitiae Menstruus, 2*(4), 58–62.

Dyar, H.G., & Knab, F. (1915). Notes on the species of *Culex* of the Bahamas. *Insecutor Inscitiae Menstruus, 3*(8–10), 112–115.

Dyar, H.G., & Knab, F. (1916). Eggs and oviposition in certain species of *Mansonia* (Diptera; Culicidae). *Insecutor Inscitiae Menstruus, 4*(4–6), 61–68.

Dyar, H.G., & Knab, F. (1917). Bromelicolous *Anopheles* (Diptera, Culicidae). *Insecutor Inscitiae Menstruus, 5*(1–3), 38–40.

Dyar, H.G., & Knab, F. (1918a). Bromelicolous *Anopheles*—a correction (Diptera, Culicidae). *Insecutor Inscitiae Menstruus, 6*(7–9), 140–141.

Dyar, H.G., & Knab, F. (1918b). New American mosquitoes (Diptera: Culicidae). *Insecutor Inscitiae Menstruus, 5*(10–12), 165–169.

Dyar, H.G., & Knab, F. (1918c). The genus *Culex* in the United States (Diptera, Culicidae). I. *Insecutor Inscitiae Menstruus, 5*(for 1917) (10–12), 170–183.

Dyar, H.G., & Knab, F. (1919). New species of tropical American mosquitoes (Diptera, Culicidae). *Insecutor Inscitiae Menstruus, 7*(1–3), 1–9.

Dyar, H.G., & Ludlow, C.S. (1921a). Two new American mosquitoes (Diptera, Culicidae). *Insecutor Inscitiae Menstruus, 9*(1–3), 46–50.

Dyar, H.G., & Ludlow, C.S. (1921b). A note on two Panama mosquitoes (Diptera, Culicidae). *The Military Surgeon, 48*, 677–680.

Dyar, H.G., & Ludlow, C.S. (1922). Notes on American mosquitoes (Diptera, Culicidae). *The Military Surgeon, 50*(1), 61–64.

Dyar, H.G., & Núñez Tovar, M. (1927a). Notes on biting flies from Venezuela (Diptera, Culicidae, Psychodidae). *Insecutor Inscitiae Menstruus, 14*(for 1926), 152–155, 190–191.

Dyar, H.G., & Núñez Tovar, M. (1927b). Notas sobre insectos hematófagos de Venezuela (Diptera, Culicidae, Psychodidae). *Insecutor Inscitiae Menstruus, 14*(for 1926) 3–7, 10–12.

Dyar, H.G., & Núñez Tovar, M. (1928). Description of new species of mosquitoes from Venezuela. *American Journal of Hygiene, 8*(1), 89–92

Dyar, H.G., & Shannon, R.C. (1924a). New *Culex* from Panama (Diptera, Culicidae). *Insecutor Inscitiae Menstruus, 12*(4–6), 46–48.

Dyar, H.G., & Shannon, R.C. (1924b). Another new *Culex* from Panama (Diptera, Culicidae). *Insecutor Inscitiae Menstruus, 12*, 143–144.

Dyar, H.G., & Shannon, R.C. (1924c). The American species of *Uranotaenia* (Diptera, Culicidae). *Insecutor Inscitiae Menstruus, 12*(10–12), 187–192.

Dyar, H.G., & Shannon, R.C. (1924d). The subfamilies, tribes and genera of American Culicidae. *Journal of the Washington Academy of Sciences, 14*(20), 472–486.

Dyar, H.G., & Shannon, R.C. (1925a). The types of Philippine mosquitos described by Ludlow and other notes on the fauna. (Diptera, Culicidae). *Insecutor Inscitiae Menstruus, 13*(4–6), 66–89.

Dyar, H.G., & Shannon, R.C. (1925b). New mosquitoes from Brazil (Diptera, Culicidae). *Journal of the Washington Academy of Sciences, 15*(2), 39–41.

Dyar, H.G., & Shannon, R.C. (1925c). A new sabethid mosquito from Panama (Diptera, Culicidae). *Journal of the Washington Academy of Sciences, 15*(11), 234.

Dyar, H.G., & Shannon, R.C. (1925d). The mosquitoes of Peary's North Pole expedition of 1908. (Diptera, Culicidae). *Journal of the Washington Academy of Sciences, 15*(4), 77–78.

Dye, C. (1984). Models for the population dynamics of the yellow fever mosquito, *Aedes aegypti*. *Journal of Animal Ecology, 53*(1), 247–268.

Ebi, K.L., Hartman, J., Chan, N., McConnell, J., Schlesinger, M., & Weyant, J. (2005). Climate suitability for stable malaria transmission in Zimbabwe under different climate change scenarios. *Climate Change, 73*(3), 375–393.

Eckhoff, P.A., Wenger, E.A., Godfray, H.C.J., & Burt, A. (2017). Impact of mosquito gene drive on malaria elimination in a computational model with explicit spatial and temporal dynamics. *Proceedings of the National Academy of Sciences of the United States of America, 114*(2), E255–E264.

Eckstein, F. (1918). Zur Systematik der einheimischen Stechmücken. 1. Vorläufige Mitteilung: Die Weibchen. *Zentralblatt für Bakteriologie, Abt. 1*(Originale 82), 57–68.

Edgerly, J.S., Willey, M.S., & Livdahl, T.P. (1993). The community ecology of *Aedes* egg hatching: Implications for a mosquito invasion. *Ecological Entomology, 18*(2), 123–128.

Edgerly, J.S., McFarland, M., Morgan, P., & Livdahl, T. (1998). A seasonal shift in egg-laying behaviour in response to cues of future competition in a treehole mosquito. *Journal of Animal Ecology, 67*(5), 805–818.

Edman, J.D., & Kale, H.W., II. (1971). Host behavior: Its influence on the feeding success of mosquitoes. *Annals of the Entomological Society of America, 64*(2), 513–516.

Edman, J.D., Webber, L.A., & Kale, H.W., II. (1972). Effect of mosquito density on the interrelationship of host behavior and mosquito feeding success. *The American Journal of Tropical Medicine and Hygiene, 21*(4), 487–491.

Edman, J.D., Webber, L.A., & Schmid, A.A. (1974). Effect of host defenses on the feeding pattern of *Culex nigripalpus* when offered a choice of blood sources. *Journal of Parasitology, 60*(5), 874–883.

Edman, J.D., Day, J.F., & Walker, E.D. (1984). Field confirmation of laboratory observations on the differential antimosquito behavior of herons. *The Condor, 86*(1), 91–92.

Edman, J.D., Scott, T.W., Costero, A., Morrison, A.C., Harrington, L.C., & Clark, G.G. (1998). *Aedes aegypti* (Diptera: Culicidae) movement influenced by availability of oviposition sites. *Journal of Medical Entomology, 35*(4), 578–583.

Edwards, F.W. (1911a). Some new West African species of *Anopheles* (sensu lato), with notes on nomenclature. *Bulletin of Entomological Research, 2*(2), 141–143.

Edwards, F.W. (1911b). The African species of *Culex* and allied genera. *Bulletin of Entomological Research, 2*, 241–268.

Edwards, F.W. (1911c). On the culicid genus *Eretmapodites* Theobald. *Annals and Magazine of Natural History* (Series 8), *8*(43), 67–73.

Edwards, F.W. (1912a). Notes on the British mosquitos (Culicinae). *The Entomologist, 45*(592), 260–264.

Edwards, F.W. (1912b). A key to the Australasian species of *Ochlerotatus* (Culicidae). *Annals and Magazine of Natural History* (Series 8), *9*(53), 521–527.

Edwards, F.W. (1912c). A synopsis of the species of African Culicidae, other than *Anopheles*. *Bulletin of Entomological Research, 3*(1), 1–53.

Edwards, F.W. (1912d). A key for determing the African species of *Anopheles* (sensu lato). *Bulletin of Entomological Research, 3*(3), 241–250.

Edwards, F.W. (1913a). Tipulidae and Culicidae from the lake of Tiberias and Damascus. *Journal of the Asiatic Society of Bengal, 9*, 47–51.

Edwards, F.W. (1913b). Further notes on African Culicidae. *Bulletin of Entomological Research, 4*(1), 47–59.

Edwards, F.W. (1913c). New synonymy in Oriental Culicidae. *Bulletin of Entomological Research, 4*, 221–242.

Edwards, F.W. (1914a). On the oriental culicid genus *Leicesteria*, Theobald. *Bulletin of Entomological Research, 4*(3), 255–263.

Edwards, F.W. (1914b). New species of Culicidae in the British museum, with notes on the genitalia of some African *Culex*. *Bulletin of Entomological Research* (London), *5*(1), 63–81.

Edwards, F.W. (1914c). New Culicidae from Borneo and Hong Kong. *Bulletin of Entomological Research, 5*(2), 125–128.

Edwards, F.W. (1915a). An annotated list of mosquitos occurring at Durban, Natal. *Annals of the Durban Museum, 1*(2), 160–166.

Edwards, F.W. (1915b). New and little-known East African Culicidae. *Bulletin of Entomological Research, 5*(4), 273–281.

Edwards, F.W. (1915c). Diagnoses of new Bornean Culicidae. *Bulletin of Entomological Research, 5*(4), 283–285.

Edwards, F.W. (1916). Eight new mosquitoes in the British Museum collection. *Bulletin of Entomological Research, 6*(4), 357–364.

Edwards, F.W. (1917). Notes on Culicidae, with descriptions of new species. *Bulletin of Entomological Research, 7*(3), 201–229.

Edwards, F.W. (1919). The larva and pupa of *Taeniorhynchus richiardii* Fic. (Diptera, Culicidae). *Entomologist's Monthly Magazine, 5*, 83–88.

Edwards, F.W. (1920a). Mosquito notes. *Bulletin of Entomological Research, 10*(2), 129–137.

Edwards, F.W. (1920b). Notes on the mosquitos of Madagascar, Mauritius and Reunion. *Bulletin of Entomological Research, 11*(2), 133–138.

Edwards, F.W. (1921a). H. Sauter's Pormosan collections: Culicidae. Ann. *Annals and Magazine of Natural History, 9*(8), 629–632.

Edwards, F.W. (1921b). A synonymic list of the mosquitoes hitherto recorded from Sweden, with keys for determining the genera and species. *Entomologisk tidsskrift, 42*(1), 46–52.

Edwards, F.W. (1921c). Mosquito notes. II. *Bulletin of Entomological Research, 12*(1), 69–80.

Edwards, F.W. (1921d). A revision of the mosquitos of the Palaearctic Region. *Bulletin of Entomological Research, 12*(3), 263–351.

Edwards, F.W. (1922a). A synopsis of adult Oriental culicine (including megarhinine and sabethine) mosquitoes. Part I. *Indian Journal of Medical Research* (Calcutta), *10*(1), 249–293.

Edwards, F.W. (1922b). Revised keys to the known larvae of African culicinae. Part II. *Indian Journal of Medical Research* (Calcutta), *10*(2), 430–475.

Edwards, F.W. (1922c). A revision of the genus *Harpagomyia* de Meij. (Diptera, Culicidae). *Transactions of the Royal Entomological Society of London, 1921*(3–4), 496–503.

Edwards, F.W. (1922d). Mosquito notes. III. *Bulletin of Entomological Research, 13*(1), 75–102.

Edwards, F.W. (1923a). Four new African mosquitos. *Bulletin of Entomological Research, 13*(4), 397–399.

Edwards, F.W. (1923b). Mosquito notes. IV. *Bulletin of Entomological Research, 14*(1), 1–9.

Edwards, F.W. (1923c). Oligocene mosquitoes in the British Museum; with a summary of our present knowledge concerning fossil Culicidae. *Quarterly Journal of the Geological Society, 79*(2), 139–155.

Edwards, F.W. (1923d). Mosquitoes reared by Dr. W. E. Haworth from coconut palms in east Africa. *Transactions of the Royal Society of Tropical Medicine and Hygiene, 16*(8), 498–501.

Edwards, F.W. (1924a). Some mosquitoes from Ovamboland, S. W. Africa, and from the Cape Province. *Annals of the South African Museum, 19*(1), 159–163.

Edwards, F.W. (1924b). A synopsis of the adult mosquitos of the Australasian Region. *Bulletin of Entomological Research, 14*(3), 351–401.

Edwards, F.W. (1924c). Appendix. Descriptions of two new species of mosquitoes from coconut palms in East Africa. In W.E. Haworth, *Mosquitoes and coconut palms: A mosquito survey of palm trees in East Africa and the problems resulting therefrom* (pp. 197–198). *Transactions of the Royal Society of Tropical Medicine and Hygiene, 18*(4), 162–198.

Edwards, F.W. (1925). Mosquito notes. V. *Bulletin of Entomological Research, 15*(3), 257–270.

Edwards, F.W. (1926a). Mosquito notes. VI. *Bulletin of Entomological Research, 17*(2), 101–131.

Edwards, F.W. (1926b). Diptera Nematocera from the mountains of Borneo. *Sarawak Museum Journal, 3*(8–11), 243–278.

Edwards, F.W. (1926c). Una revisione delle zanzare delle regioni paleartiche. *Rivista di malariologia, 5*(3–6), 1–152.

Edwards, F.W. (1926d). Una revisione delle zanzare delle regioni paleartiche. *Rivista di malariologia Rome, 5*, 613–652. (Translation with additions.)

Edwards, F.W. (1927a). Four new mosquitoes from the Belgian Congo. *Revue zoologique Africaine, 15*(3), 351–355.

Edwards, F.W. (1927b). New mosquitoes of the genus *Rachionotomyia* from New Guinea. *Nova Guinea, 15*(3), 352–356.

Edwards, F.W. (1928a). Mosquito notes. VII. *Bulletin of Entomological Research, 18*(3), 267–284.

Edwards, F.W. (1928b). A new mosquito from the Kivu volcanoes, Belgian Congo. *Revue zoologique Africaine (supplement botanique), 16*(3), 324–325.

Edwards, F.W. (1928c). The nematocerous Diptera of Corsica. *Encyclopedie Entomologique B, 4*(4), 157–189.

Edwards, F.W. (1929a). A new *Anopheles* from the Central Sahara. *Archives de l'Institut Pasteur d'Alǧérie, 7*(1), 82–87.

Edwards, F.W. (1929b). Mosquitoes from the Solomon Islands. 3. Descriptive notes on the material. *Bulletin of Entomological Research, 20*(3), 308–320.

Edwards, F.W. (1929c). Mosquito notes. VIII. *Bulletin of Entomological Research, 20*(3), 321–343.

Edwards, F.W. (1929d). Philippine Nematocerous Diptera II. *Notulae Entomologicae, 9*, 1–14.

Edwards, F.W. (1930a). Mosquito notes. IX. *Bulletin of Entomological Research, 21*(3), 287–306.

Edwards, F.W. (1930b). Mosquito notes. X. *Bulletin of Entomological Research, 21*(4), 541–545.

Edwards, F.W. (1932a). Diptera family Culicidae. In P. Wytsman (Ed.), *Genera insectorum* (194th Fascicule, pp. 1–258). Brussels, Belgium: Desmet-Verteneuil.

Edwards, F.W. (1932b). Mosquito notes. XI. *Bulletin of Entomological Research, 23*(4), 559–562.

Edwards, F.W. (1934). The Percy Sladen and Godman Trusts expedition to the islands in the Gulf of Guinea, October 1932–March 1933. II. Diptera, Nematocera. *Annals and Magazine of Natural History, 14*(10), 321–336.

Edwards, F.W. (1935a). Mosquito notes. XII. *Bulletin of Entomological Research, 26*(2), 127–136.

Edwards, F.W. (1935b). Results of the Oxford University expedition to Borneo, 1932. Two new species of *Megarhinus* (Diptera, Culicidae). *Annals and Magazine of Natural History* (Series 10), *15*(89), 579–581.

Edwards, F.W. (1935c). A new *Uranotaenia* from the Belgian Congo (Diptera, Culicidae). *Revue zoologique Africaine (supplement botanique), 27*(1), 96.

Edwards, F.W. (1936). New African culicine mosquitoes (Diptera, Culicidae). *Proceedings of the Royal Entomological Society of London Series B: Taxonomy, 5*(3), 49–55.

Edwards, F.W. (1938). In A.M. Evans (Ed.), *Mosquitoes of the Ethiopian Region, II: Anophelini adults and early stages*. London, England: British Museum (Natural History).

Edwards, F.W. (1939a). A new east African *Aedes* (Dipt., Culicidae). *Proceedings of the Royal Entomological Society of London, Series B: Taxonomy, 8*(2), 17.

Edwards, F.W. (1939b). A new species of *Orthopodomyia* (Diptera, Culicidae). *Proceedings of the Royal Entomological Society of London, Series B: Taxonomy, 8*(6), 121–123.

Edwards, F.W. (1940). A new plain-winged *Anopheles* from Rhodesia. *Annals of Tropical Medicine and Parasitology, 34*(2), 93–96.

Edwards, F.W. (1941). Mosquitoes of the Ethiopian Region. III. Culicine adults and pupae. *Bulletin of the British Museum (Natural History), Entomology.*

Edwards, F.W., & Gibbins, E.G. (1939). Mosquitoes. Ruwenzori Expedition 1934–5. London, England: *British Museum (Natural History), 1*(3), 29–34.

Edwards, F.W., & Given, D.H.C. (1928). The early stages of some Singapore mosquitoes. *Bulletin of Entomological Research, 18*(4), 337–357.

Egeth, M., & Kurzban, R. (2012). Artificial natural selection: Can supplemental feeding domesticate mosquitoes and control mosquito-borne diseases. *Evolutionary Psychology, 10*(3), 602–610.

Egizi, A., Fefferman, N.H., & Fonseca, D.M. (2015). Evidence that implicit assumptions of 'no evolution' of disease vectors in changing environments can be violated on a rapid timescale. *Philosophical Transactions of the Royal Society B, 370*(1665), 20140136.

Eisen, L., & Eisen, R.J. (2011). Using geographic information systems and decision support systems for the prediction, prevention, and control of vector-borne diseases. *Annual Review of Entomology, 56*, 41–61.

Eisen, L., & Lozano-Fuentes, S. (2009). Use of mapping and spatial and space-time modeling approaches in operational control of *Aedes aegypti* and dengue. *PLOS Neglected Tropical Diseases, 3*(4), e411.

Eitam, A., & Blaustein, L. (2004). Oviposition habitat selection by mosquitoes in response to predator (*Notonecta maculata*) density. *Physiological Entomology, 29*(2), 188–191.

Ekechukwu, N.E. (2015). *The genetics and ecology of male reproductive investments in the malaria mosquito* Anopheles gambiae *ss* [Doctoral dissertation, Keele University]. Keele Research Depository. Retrieved from http://eprints.keele.ac.uk/id/eprint/2914

Ekechukwu, N.E., Baeshen, R., Traorè, S.F., Coulibaly, M., Diabate, A., Catteruccia, F., & Tripet, F. (2015). Heterosis increases fertility, fecundity and survival of laboratory-produced F_1 hybrid males of the malaria mosquito *Anopheles coluzzii*. *G3: Genes, Genomes, Genetics, 5*(12), 2693–2709.

Eldridge, B.F., & Harbach, R.E. (1989). Case 2702: *Culex stigmatosoma*, Dyar 1907 and *C. thriambus*, 1921 (Insecta, Diptera): Proposed conservation of the specific names by suppression of *C. peus* Speiser, 1904. *Bulletin of Zoological Nomenclature, 46*(4), 247–249.

Eldridge, B.F., & Harbach, R.E. (1992). Conservation of the names *Culex stigmatosoma* and *Culex thriambus*. *Journal of the American Mosquito Control Association, 8*(1), 104–105.

Ellis, A.M. (2008). Linking movement and oviposition behaviour to spatial population distribution in the tree hole mosquito *Ochlerotatus triseriatus*. *Journal of Animal Ecology, 77*(1), 156–166.

Ellis, A.M., Lounibos, L.P., & Holyoak, M. (2006). Evaluating the long-term metacommunity dynamics of tree hole mosquitoes. *Ecology, 87*(10), 2582–2590.

Ellis, R.A., & Brust, R.A. (1973). Sibling species delimitation in the *Aedes communis* (De Geer) aggregate (Diptera: Culicidae). *Canadian Journal of Research, 51*(9), 915–959.

Emami, S.N., Lindberg, B.G., Hua, S., Hill, S., Mozuratis, R., Lehmann, P., . . . Faye, I. (2017). A key malaria metabolite modulates vector blood seeking, feeding, and susceptibility to infection. *Science, 355*(6329), 1076–1080.

Emerson, K.J., Merz, C.R., Catchen, P.A., Hohenlohe, P.A., Cresko, W.A., Bradshaw, W.E., & Holzapel, C.M. (2011). Resolving postglacial phylogeography using high-throughput sequencing. *Proceedings of the National Academy of Sciences of the United States of America, 107*(37), 16196–16200.

Emerson, K.J., Conn, J.E., Bergo, E.S., Randel, M.A., & Sallum, M.A.M. (2015). Brazilian *Anopheles darlingi* Root (Diptera: Culicidae) clusters by major biogeographical region. *PLOS ONE, 10*(7), e0130773.

Enderlein, G. (1920). Die culiciden-fauna Madagascars. *Wiener Entomologische Zeitung, 38*(1–3), 47–52.

Enderlein, G. (1923). Zur Kenntnis afrikanischer und orientalischer Culiciden. *Wiener Entomologische Zeitung, 40*(1–4), 25–29.

Enderlein, G. (1931). Zwei neue *Toxorhynchites* aus dem Kongogebiete. *Zoologischer Anzeiger, Leipzig, 94*(3/4), 123–125.

Endersby, N.M., White, V.L., Chan, J., Hurst, T., Rašić, G., Miller, A., & Hoffman, A.A. (2013). Evidence of cryptic genetic lineages within *Aedes notoscriptus* (Skuse). *Infection, Genetics and Evolution, 18*, 191–201.

Enfield, M.A. (1977). Additions and corrections to the records of *Aedes* mosquitoes in Alberta. *Mosquito News, 37*(1), 82–85.

Eng, K.L., Chiang, G.L., Hamidah, T., & Loong, K.P. (1990). Scanning electron microscopy of the eggs of *Mansonia uniformis*, *Mansonia indiana* and *Mansonia annulifera* (Diptera, Culicidae). *Southeast Asian Journal of Tropical Medicine and Public Health, 21*(1), 115–122.

Engel, P., & Moran, N.A. (2013). The gut microbiota of insects—diversity in structure and function. *FEMS Microbiology Reviews, 37*(5), 699–735.

Engelthaler, D.M., Fink, T.M., Levy, C.E., & Leslie, M.J. (1997). The reemergence of *Aedes aegypti* in Arizona. *Emerging Infectious Diseases, 3*(2), 241–242.

Ennos, A.R. (1989). The kinematics and aerodynamics of the free flight of some Diptera. *Journal of Experimental Biology, 142*(1), 49–85.

Enserink, M. (2002). What mosquitoes want: Secrets of host attraction. *Science, 298*(5591), 90–92.

Enserink, M. (2007). Selfish genes could help disease-free mosquitoes spread. *Science, 315*(5820), 1777–1778.

Enserink, M. (2010). Yellow fever mosquito shows up in Europe. *Science, 329*(5993), 736–736.

Entomological Society of America. (2012). Common Names of Insects Database. Retrieved from http://www.entsoc.org/common-names

Eouzan, J.-P. (1969). Description du mâle de *Culex* (*Culiciomyia*) *grenieri* sp. n. (Diptera, Culicidae). *Cahiers ORSTOM. Série entomologie médicale et parasitologie, 7*(3), 163–166.

Eouzan, J.-P., Ferrara, L., & Rickenbach, A. (1967). Description du mâle, de la larve et de la nymphe de *Ficalbia* (*Etorleptiomyia*) *xanthozona* van Someren, 1948 (Diptera, Culicidae). *Cahiers ORSTOM. Série entomologie médicale et parasitologie, 5*(4), 223–227.

Epstein, P.R., Diaz, H.F., Elias, S., Grabherr, G., Graham, N.E., Martens, W.J.M., . . . Susskind, J. (1998). Biological and physical signs of climate change: Focus on mosquito-borne diseases. *Bulletin of the American Meteorological Society, 79*(3), 409–417.

Erdelyan, C.N.G., Mahood, T.H., Bader, T.S.Y., & Whyard, S. (2012). Functional validation of the carbon dioxide receptor genes in *Aedes aegypti* mosquitoes using RNA interference. *Insect Molecular Biology, 21*(1), 119–127.

Erichson, W.F. (1842). Beitrag zur Insecten-Fauna von Vandiemensland, mit besonderer Berücksichtigung der geographischen verbreitung der Insecten, vom Herausgeber. *Archiv für Naturgeschichte, 8*(1), 83–287.

Erickson, J. (1996). *Glacial geology.* New York, NY: Facts on File.

Erickson, S.M., Xi, Z., Mayhew, G.F., Ramirez, J.L., Aliota, M.T., Christensen, B.M., & Dimopoulos, G. (2009). Mosquito infection responses to developing filarial worms. *PLOS Neglected Tropical Diseases, 3*(10), e529.

Eritja, R., & Chevillon, C. (1999). Interruption of chemical mosquito control and evaluation of insecticide resistance genes in *Culex pipiens* (Diptera: Culicidae). *Journal of Medical Entomology, 36*(1), 41–49.

Eritja, R., Aranda, C., Padrós, J., Goula, M., Lucientes, J., Escosa, R., . . . Cáceres, F. (2000). An annotated checklist and bibliography of the mosquitoes of Spain (Diptera: Culicidae). *European Mosquito Bulletin, 8*, 10–18.

Eritja, R., Escosa, R., Lucientes, J., Marques, E., Molina, R., Roiz, D., & Ruiz, S. (2005). Worldwide invasion of vector mosquitoes: Present European distribution and challenges for Spain. *Biological Invasions, 7*(1), 87–97.

Escovar, J.E., González, R., Quiñones, M.L., Wilkerson, R.C., Ruiz, F., & Harrison, B.A. (2014). Morphology of the larvae, male genitalia and DNA sequences of *Anopheles* (*Kerteszia*) *pholidotus* (Diptera: Culicidae) from Colombia. *Memórias do Instituto Oswaldo Cruz, 109*(4), 473–479.

Estrada, D.A., Quiñones, M.L., Sierra, D.M., Calle, D.A., Ruiz, F., Erazo, H.F., & Linton, Y.-M. (2003). Utilidad de la morfología de los huevos como un método indirecto para identificar *Anopheles benarrochi* Gabaldón, Cova García & López, *Anopheles oswaldoi* (Peryassú) y *Anopheles rangeli* Gabaldón, Cova García & López, (Diptera: Culicidae) en Putumayo, Colombia. *Biomédica, 23*(4), 388–395.

Estrada-Franco, J.G., Ma, M.C., Lanzaro, G.C., Gwadz, R., Galvan-Sánchez, C., Cespedes, J.L., . . . Rodriguez, R. (1992). Genetic evidence of a species complex in *Anopheles pseudopunctipennis pseudopunctipennis. Boletín de la Oficina Sanitaria Panamericana, 113*(4), 297–300.

Estrada-Franco, J.G., Lanzaro, G.C., Ma, M.C., Walker-Abbey, A., Romans, P., Galvan-Sánchez, C., . . . Gwadz, R.W. (1993a). Characterization of *Anopheles pseudopunctipennis* sensu lato from three countries of Neotropical America from variation in allozymes and ribosomal DNA. *The American Journal of Tropical Medicine and Hygiene, 49*(6), 735–745.

Estrada-Franco, J.G., Ma, M.C., Gwadz, R.W., Sakai, R., Lanzaro, G.C., Laughinghouse, A., . . . Vargassagarnaga, R. (1993b). Evidence through crossmating experiments of a species complex in *Anopheles pseudopunctipennis* sensu lato—a primary malaria vector of the American continent. *The American Journal of Tropical Medicine and Hygiene, 49*(6), 746–755.

Evans, A.M. (1922). Notes on Culicidae in Venezuela, with descriptions of new species. Part II. *Annals of Tropical Medicine and Parasitology, 16*(2), 213–222.

Evans, A.M. (1923a). A new species and A new variety of *Culex* from the Belgian Congo. *Annals of Tropical Medicine and Parasitology, 17*(1), 89–91.

Evans, A.M. (1923b). Notes on Culicidae in Venezuela, with descriptions of new species. *Annals of Tropical Medicine and Parasitology, 17*(1), 101–111.

Evans, A.M. (1923c). Note on *Aedinus amazonensis*, Lutz. *Annals of Tropical Medicine and Parasitology, 17*(3), 377–380.

Evans, A.M. (1924). Descriptions of new mosquitos from South America. *Annals of Tropical Medicine and Parasitology, 18*(3), 363–375.

Evans, A.M. (1925a). Notes on Culicidae collected in Sierra Leone, with descriptions of a new species and a new variety. *Annals of Tropical Medicine and Parasitology, 19*(1), 119–126.

Evans, A.M. (1925b). A new variety of *Anopheles marshalli* Theobald from the Belgian Congo. *Annals of Tropical Medicine and Parasitology, 19*(2), 211–213.

Evans, A.M. (1925c). A new variety of *A. marshalli* from Sierra Leone. *Annals of Tropical Medicine and Parasitology, 19*(4), 461–462.

Evans, A.M. (1926). Notes on Freetown mosquitoes, with descriptions of new and little-known species. *Annals of Tropical Medicine and Parasitology, 20*(1), 97–108.

Evans, A.M. (1927). A short illustrated guide to the anophelines of tropical and South America. *Liverpool School of Tropical Medicine Memoir, 3*, 1–79.

Evans, A.M. (1928). A new variety of *Armigeres* from Lagos, with descriptions of the larval and pupal stages. *Annals of Tropical Medicine and Parasitology, 22*(1), 39–42.

Evans, A.M. (1929a). Descriptions of the early stages of two further mosquitos collected in Southern Nigeria by Mr. L. H. Dunn. *Annals of Tropical Medicine and Parasitology, 23*(3), 407–413.

Evans, A.M. (1929b). Notes on certain varieties of *Anopheles marshalli* Theobald. *Annals of Tropical Medicine and Parasitology, 23*(3), 415–425.

Evans, A.M. (1929c). *Aedes* (*Aedimorphus*) *apicoannulatus* Edwards and yellow fever: A correction. *Annals of Tropical Medicine and Parasitology, 23*(4), 521–522.

Evans, A.M. (1931a). Notes on African anophelines. *Annals of Tropical Medicine and Parasitology, 25*(1), 129–143.

Evans, A.M. (1931b). A new subspecies of *Anopheles funestus* Giles, from Southern Rhodesia. *Annals of Tropical Medicine and Parasitology, 25*(3–4), 545–549.

Evans, A.M. (1932). Notes on African mosquitoes. *Annals of Tropical Medicine and Parasitology, 26*(1), 85–108.

Evans, A.M. (1933). *Anopheles demeilloni* sp. n., a new name for *Anopheles transvaalensis* of authors. *Annals of Tropical Medicine and Parasitology, 27*(2), 265–269.

Evans, A.M. (1934). Further notes on African anophelines, with a description of a new group of *Myzomyia*. *Annals of Tropical Medicine and Parasitology, 28*(4), 549–570.

Evans, A.M. (1935). Notes on anophelines I. Description of *Anopheles marshalli* var. *gibbinsi* from Uganda. II. The characters of *A. maculipennis* var. *messeae* in Wirral, England. *Annals of Tropical Medicine and Parasitology, 29*(4), 469–473.

Evans, A.M. (1936). Two new anopheline mosquitoes from Kenya. *Annals of Tropical Medicine and Parasitology, 30*(4), 533–540.

Evans, A.M. (1938). *Mosquitoes of the Ethiopian Region. II. Anophelini adults and early stages.* London, England: British Museum (Natural History).

Evans, A.M., & De Meillon, B. (1933). Notes on *Anopheles demeilloni* (*transvaalensis*) and *Anopheles garnhami* in South Africa with descriptions of new varieties of these species. *Annals of Tropical Medicine and Parasitology, 27*(2), 271–282.

Evans, A.M., & Leeson, H.S. (1935). The Funestus series of *Anopheles* in Southern Rhodesia, with description of a new variety. *Annals of Tropical Medicine and Parasitology, 29*(1), 33–47.

Evans, A.M., & Leeson, H.S. (1937). Notes on variation in *Anopheles rivulorum* Leeson in East Africa, with description of a new variety. *Annals of Tropical Medicine and Parasitology, 31*(3), 383–386.

Evans, A.M., & Walker, G.R. (1935). Notes on Brazilian mosquitoes: Species observed in the Amazon Valley, and record of *Aedes albifasciatus* Macq. invading a ship in harbour. *Annals of Tropical Medicine and Parasitology, 29*(4), 463–467.

Evans, T.P.O., & Bishop, S.R. (2014). A spatial model with pulsed releases to compare strategies for the sterile insect technique applied to the mosquito *Aedes aegypti*. *Mathematical Biosciences, 254*, 6–27.

Evenhuis, N.L. (1994). *Catalogue of the fossil flies of the world (Insecta: Diptera)*. Leiden, The Netherlands: Backhuys.

Evenhuis, N.L. (2008). A compendium of zoological nomenclature: A reference source. In *Bishop Museum Technical Report No. 41*. Honolulu, HI: Bishop Museum.

Evenhuis, N.L. (2013). Authors of fly names. *Bishop Museum Technical Report No. 61* (2nd ed.). Honolulu, HI: Bishop Museum.

Evenhuis, N.L., & Gon, S.M., III. (1989). Family Culicidae. In N.L. Evenhuis, *Catalog of the Diptera of the Australasian and Oceanian regions: Bishop Museum special publication 86* (pp. 191–218). Honolulu: Bishop Museum Press and E.J. Brill.

Evenhuis, N.L., & Steffan, W.A. (1986). Classification of the subgenus *Toxorhynchites* (Diptera: Culicidae). II. Revision of the *Toxorhynchites acaudatus* group. *Journal of Medical Entomology, 23*(5), 538–574.

Evenhuis, N.L., Pape, T., & Pont, A.C. (2016). Nomenclatural studies toward a world list of Diptera genus-group names. Part V. Pierre-Justin-Marie Macquart. *Zootaxa, 4172*(1), 1–211.

Eversmann, E. (1834). Diptera. Wolgam fluvium inter et montes Uralenses observata. *Bulletin de la Société Impériale des Naturalistes de Moscou, 7*, 420–432.

Eysell, A. (1903). *Aëdes cinereus* (Hoffmgg.) und *Aëdes leucopygos* n. sp. *Abhandlungen und Bericht des Vereins für Naturkunde zu Kassel, 48*, 285–306.

Eysell, A. (1905). Sind die "Culiciden" eine Familie? *Abhandlungen und Bericht des Vereins für Naturkund zu Cassel, 49*, 16–24.

Eysell, A. (1912). *Cyclophoros* (*Anopheles*) *nigripes* Staeger (nov. gen.). *Archiv für Schiffs- und Tropen-Hygiene, 16*, 421–431.

Eze, M., Labadin, J., & Lim, T. (2011). Mosquito flight model and applications in malaria control. In *International Conference on Computer Engineering and Technology, 3rd (ICCET 2011)* (pp. 59–64). ASME Press.

Fabricius, J.C. (1775). *Systema entomologiae, sistens insectorum classes, ordines, genera, species adiectis synonymis, locis, descriptionibus, observationibus*. Flensburg and Leipzig, Germany: In Officina Libraria Kortii.

Fabricius, J.C. (1787). *Mantissa insectorum: sistens eorum species nuper detectas: adjectis characteribus genericis, differentiis, specificis, emendationibus, observationibus*. Impr. Christ. Gottl. Proft (Hafniae).

Fabricius, J.C. (1794). *Entomología systematica emendata et aucta*. Impr. Christ. Gottl. Proft (Hafniae).

Fabricius, J.C. (1805). *Systema antliatorum secundum ordines, genera, species*. Brunsvigae: Apud Carolum Reichard.

Fairley, T.L., Renaud, T.M., & Conn, J.E. (2000). Effects of local geographic barriers on population structure in *Anopheles punctipennis* (Diptera: Culicidae). *Journal of Medical Entomology, 37*(5), 754–760.

Fairley, T.L., Povoa, M.M., & Conn, J.E. (2002). Evaluation of the Amazon River delta as a barrier to gene flow for the regional malaria vector, *Anopheles aquasalis* (Diptera: Culicidae) in northeastern Brazil. *Journal of Medical Entomology, 39*(6), 861–869.

Fairley, T.L., Kilpatrick, C.W., & Conn, J.E. (2005). Intragenomic heterogeneity of internal transcribed spacer rDNA in neotropical malaria vector *Anopheles aquasalis* (Diptera: Culicidae). *Journal of Medical Entomology, 42*(5), 795–800.

Falleroni, D. (1926). Fauna anofelica italiana e suo "habitat" (paludi, risaie, canali). Metodi di lotta contro la malaria. *Revista di malariologia, 5*(5–6), 553–593.

Falleroni, D. (1932). *Zooprofilassi e sua applicazione in Ardea (Argo Romano) diverse razze di* Anopheles maculipennis. Roma: Cuggiani.

Fallon, A.M., Baldridge, G.D., Carroll, E.M., & Kurtz, C.M. (2014). Depletion of host cell riboflavin reduces *Wolbachia* levels in cultured mosquito cells. *In Vitro Cellular and Developmental Biology—Animal, 50*(8), 707–713.

Fanello, C., Santolamazza, F., & Della Torre, A. (2002). Simultaneous identification of species and molecular forms of the *Anopheles gambiae* complex by PCR-RFLP. *Medical and Veterinary Entomology, 16*(4), 461–464.

Fang, J. (2010). A world without mosquitoes. *Nature, 466*(7305), 432–434.

Faraji, A., & Gaugler, R. (2015). Experimental host preference of diapause and non-diapause induced *Culex pipiens pipiens* (Diptera: Culicidae). *Parasites and Vectors, 8*(1), 389.

Farajollahi, A., Fonseca, D.M., Kramer, L.D., & Kilpatrick, A.M. (2011). "Bird biting" mosquitoes and human disease: A review of the role of *Culex pipiens* complex mosquitoes in epidemiology. *Infection, Genetics and Evolution, 11*(7), 1577–1585.

Faran, M.E. (1979). *Anopheles (Nyssorhynchus) trinkae*, a new species in the Albimanus section (Diptera: Culicidae). *Mosquito Systematics, 11*(1), 26–39.

Faran, M.E. (1980). Mosquito studies (Diptera: Culicidae). XXXIV. A revision of the Albimanus section of the subgenus *Nyssorhynchus* of *Anopheles*. *Contributions of the American Entomological Institute, 15*(7), 1–215.

Faran, M.E. (1981). Synonomy of *Anopheles (Nyssorhynchus) noroestensis* with *An.* (*Nys.*) *evansi* [*sic*], with a description of the male genitalia of the lectotype of *An.* (*Nys.*) *evansi* [*sic*] (Diptera: Culicidae). *Mosquito Systematics, 13*(1), 86–91.

Faran, M.E., & Linthicum, K.J. (1981). A handbook of the Amazonian species of *Anopheles* (*Nyssorhynchus*) (Diptera: Culicidae). *Mosquito Systematics, 13*(1), 1–81.

Farci, A., Laudani, U., & Lecis, A.R. (1973). A revision of the banding pattern of salivary chromosomes of *Anopheles atroparvus* (Diptera: Nematocera). *Zeitschrift für zoologische Systematik und Evolutionsforschung, 11*, 304–310.

Farner, D.S. (1945). A new species of *Aedes* from the Caroline Islands. (Diptera, Culicidae). *Proceedings of the Biological Society of Washington, 58*, 59–62.

Farner, D.S., & Bohart, R.M. (1944). Three new species of Australasian *Aedes* (Diptera, Culicidae). *Proceedings of the Biological Society of Washington, 57*, 117–122.

Farner, D.S., & Bohart, R.M. (1945). A preliminary revision of the *scutellaris* group of the genus *Aedes*. *Naval Medical Bulletin, 44*, 37–53.

Faulde, M.K., Rueda, L.M., & Khaireh, B.A. (2014). First record of the Asian malaria vector *Anopheles stephensi* and its possible role in the resurgence of malaria in Djibouti, Horn of Africa. *Acta Tropica, 139*, 39–43.

Fauran, P. (1961a). Notes sur les moustiques de la Guyane. *Archives de l'Institut Pasteur de la Guyane Française et du Territoire de l'Inini, 22*(464), 1–15.

Fauran, P. (1961b). Catalogue annoté des Culicidés signalés en Guyane Française. *Archives de l'Institut Pasteur de la Guyane Française et du Territoire de l'Inini, 22*(465), 1–60.

Fauran, P., & Pajot, F.X. (1974a). Complement to the catalog of the Culicidae recorded from French Guiana (South America). *Mosquito Systematics, 6*(2), 99–110.

Fauran, P., & Pajot, F.X. (1974b). *Wyeomyia ininicola* n. sp. (Diptera, Culicidae), nouveau moustique de la Guyane Française. *Cahiers ORSTOM. Série entomologie médicale et parasitologie, 12*(2), 145–148.

Favre, V.V. (1903). An attempt in the study of malaria in Russia. Kharkhov.

Fawaz, E.Y., Allan, S.A., Bernier, U.R., Obenauer, P.J., & Diclaro, J.W., II. (2014). Swarming mechanisms in the yellow fever mosquito: Aggregation pheromones are involved in the mating behavior of *Aedes aegypti. Journal of Vector Ecology, 39*(2), 347–354.

Felt, E.P. (1904). Mosquitos or Culicidae of New York State. *Bulletin of the New York State Museum, 79*, 241–400.

Felt, E.P. (1905a). *Culex brittoni* n. sp. *Entomological News, 16*(3), 79–80.

Felt, E.P. (1905b). Studies in Culicidae. *Bulletin of the New York State Museum, 97*, 442–497.

Felt, E.P., & Young, D.B. (1904). Importance of isolated rearings from culicid larvae. *Science* (n.s.), *20*(505), 312–313.

Feng, L.C. (1931). The larvae and pupae of the North China species of *Anopheles*; their structure and breeding habits. *National Medical Journal of China, 17*, 493–512.

Feng, L.C. (1938). The tree hole species of mosquitoes of Peiping, China. *Chinese Medical Journal*, Suppl. 2, 503–525.

Feng, L.C. (1941). A new mosquito, *Topomyia houghtoni* sp. nov. from Western Yunnan, China. *Bulletin of the Fan Memorial Institute of Biology, Peking (Zoology): Series 10*, 244–254.

Feng, L.C. (1958). *Chung-kuo Wen Ch'ung Miao Shu Hui Pien* [*Collection of Chinese Culicidae*]. Peking, China: Science Press.

Feng, L.P. (1964). Morphologic study of the adults of the different types of the *Anopheles sinensis* group. *Acta entomologica Sinica, 13*(4), 581–586.

Ferguson, H.M., & Read, A.F. (2002). Why is the effect of malaria parasites on mosquito survival still unresolved? *Trends in Parasitology, 18*(6), 256–261.

Ferguson, H.M., John, B., Ng'habi, K., & Knols, B.G. (2005). Redressing the sex imbalance in knowledge of vector biology. *Trends in Ecology and Evolution, 20*(4), 202–209.

Ferrara, L. (1968). Description d'une larve du troisieme stade d'*Orthopodomyia nkolbissonensis* Rickenbach et Hamon, 1965. *Cahiers ORSTOM. Série entomologie médicale et parasitology, 6*(2), 161–165.

Ferrara, L. (1970). Description complementaire d'*Eretmapodites salauni* Rickenbach Ferrara et Eouzan 1967 (Dipt. Culicidae). *Cahiers ORSTOM. Série entomologie médicale et parasitology, 8*(2), 137–139.

Ferrara, L. (1973). Une nouvelle espèce de moustique (Diptera, Culicidae) capturée au Cameroun. *Orthopodomyia aureoantennata* n. sp. *Cahiers ORSTOM. Série entomologie médicale, 11*(3), 199–203.

Ferrara, L. (1974). Description d'un nouvel *Aedes* du sous genre *Diceromyia* (Diptera, Culicidae) decouvert au Cameroun *Aedes (Diceromyia) mefouensis* n. sp. *Cahiers ORSTOM. Série entomologie médicale, 12*(1), 3–5.

Ferrara, L., & Eouzan, J.-P. (1974). Deux *Eretmapodites* nouveaux du Cameroun, appartenant au groupe *oedipodius* Graham, 1909 (Diptera, Culicidae): *Er. rickembachi*, *Er. adami*. *Cahiers ORSTOM. Série entomologie médicale et parasitologie, 12*(4), 259–262.

Ferrara, L., Germain, M., & Hervy, J.-P. (1985). *Aedes (Diceromyia) furcifer* (Edwards, 1913) et *Aedes (Diceromyia) taylori* Edwards, 1936. Le point sur la differenciation des adultes. *Cahiers ORSTOM. Série entomologie médicale et parasitologie, 22*(for 1984)(3), 179–183.

Ferreira, C.P., Yang, H.M., & Esteva, L. (2008). Assessing the suitability of sterile insect technique applied to *Aedes aegypti. Journal of Biological Systems, 16*(4), 565–577.

Ficalbi, E. (1889a). Notizie preventive sulle zanzare italiane. I. Nota preventiva. Alcune generalità. Descrizione di una specie nuova (*Culex hortensis*). *Bullettino della Società Entomologica Italiana, 21*, 20–30.

Ficalbi, E. (1889b). Notizie preventive sulle zanzare italiane. II. Nota preventiva (1). Descrizione di una specie nuova. Zanzara del Richiardi. Culex Richiardii n. sp. *Bullettino della Società Entomologica Italiana, 21*, 50–53.

Ficalbi, E. (1890a). Notizie preventive sulle zanzare italiane. IV. Nota preventiva (1). Descrizione di una specie nuova. Zanzara di colorito modesto. *Culex modestus*, n. sp. *Bullettino della Società Entomologica Italiana, 21*(for 1889), 93–94.

Ficalbi, E. (1890b). Notizie preventive sulle zanzare italiane. V. Nota preventiva (1). Descrizione di una specie nuova. Zanzara elegante. *Culex elegans n. sp. Bullettino della Società Entomologica Italiana, 21*(for 1889)(1–2), 95–100.

Ficalbi, E. (1890c). Notizie preventive sulle zanzare italiane. VI. Nota preventiva (1). Quistioni zoologiche intorno al *Culex pipiens e descrizione di una specie nuova (Culex phytophagus). Bullettino della Società Entomologica Italiana, 21*(for 1889)(1–2), 124–131.

Ficalbi, E. (1890d). Notizie preventive sulle zanzare italiane. VII. Nota preventiva (1). Descrizione di una specie nuova. Zanzara impudica. *Culex impudicus* sp. n. *Bullettino della Società Entomologica Italiana, 22*, 81–84.

Ficalbi, E. (1893). Revisione delle specie Europee. Famiglia delle zanzare. *Bullettino della Società Entomologica Italiana, 25*, 48–61, 136–144.

Ficalbi, E. (1896). Notizie sulle zanzare italiane. IX. Nota (1). Descrizione di una specie nuova. Zanzara delle saline, *Culex salinus*, sp. n. *Bullettino della Società Entomologica Italiana, 28*, 29–32.

Figueiredo, A.C. de (1904). Da existencia do *Anopheles bifurcatus* em Portugal. *A medicina contemporanea, 22*, 184–187, 223–225.

Filippova, V.V. (1978). Description of the pupa of the mosquito, *Toxorhynchites christophi* (Culicidae). *Parazitolgiia, 12*(4), 362–364.

Fischer, D., Thomas, S.M., Niemitz, F., Reineking, B., & Beierkuhnlein, C. (2011). Projection of climatic suitability for *Aedes albopictus* Skuse (Culicidae) in Europe under climate change conditions. *Global and Planetary Change, 78*(1), 54–64.

Fitch, A. (1847). Winter insects of eastern New York. *American Journal of Agricultural Science, 5*, 274–284.

Flanery, D. 1897. Immunity from mosquito bites. *Nature, 56*, 53.

Floch, H., & Abonnenc, E. (1944). *A. triannulatus* Neiva et Pinto en Guyane Française. Morphologie des oeufs et infection expérimentale par *P. falciparum* et *P. vivax*. *Publications / Institut Pasteur de la Guyane et du Territoire de l'Inini, 77*, 1–7.

Floch, H., & Abonnenc, E. (1945a). Les moustiques de la Guadeloupe. II. Genre *Anopheles*. *Publications / Institut Pasteur de la Guyane et du Territoire de l'Inini, 108*, 1–16.

Floch, H., & Abonnenc, E. (1945b). Description de *Culex* nouveaux de la Guyane Française. *Culex (Melanoconion) cauchensis* n. sp.

Culex (*Melanoconion*) *cayennensis* n. sp. *Publications / Institut Pasteur de la Guyane et du Territoire de l'Inini, 112*, 1–6.

Floch, H., & Abonnenc, E. (1945c). Description de *Culex* nouveaux de la Guyane Française. II. *Culex* (*Melanoconion*) *cavernicolus* n. sp. *Culex* (*Melanoconion*) *equinoxialis* n. sp. *Publications / Institut Pasteur de la Guyane et du Territoire de l'Inini, 114*, 1–5.

Floch, H., & Abonnenc, E. (1945d). Description d'une nouvelle espèce d'anophele: *A.* (*Stethomyia*) *canorii*. *Publications / Institut Pasteur de la Guyane et du Territoire de l'Inini, 116*, 1–3.

Floch, H., & Abonnenc, E. (1945e). Les moustiques de la Guadeloupe. II. Les genres *Megarhinus*, *Aedes*, *Culex*, *Deinocerites*, *Mansonia* et *Wyeomyia*. *Publications / Institut Pasteur de la Guyane et du Territoire de l'Inini, 110*, 1–48.

Floch, H., & Abonnenc, E. (1946a). Description de *Culex* nouveaux de la Guyane Française. III. *Culex* (*Melanoconion*) *rabanicolus* n. sp. *Culex* (*Melanoconion*) *rorotaensis* n. sp. *Publications / Institut Pasteur de la Guyane et du Territoire de l'Inini, 120*, 1–4.

Floch, H., & Abonnenc, E. (1946b). Description de *Culex* nouveaux de la Guyane Française. IV. *Culex* (*Melanoconion*) *punctiscapularis* n. sp. *Publications / Institut Pasteur de la Guyane et du Territoire de l'Inini, 122*, 1–3.

Floch, H., & Abonnenc, E. (1946c). Sur *A. nunez-tovari* et *A. pessoai* en Guyane Française. Table d'identification des *Nyssorhynchus guyanais*. *Publications / Institut Pasteur de la Guyane et du Territoire de l'Inini, 126*, 1–5.

Floch, H., & Abonnenc, E. (1946d). Presence de *Anopheles* (*Lophopodomyia*) *squamifemur* Antunes 1937 en Guyane Française. *Publications / Institut Pasteur de la Guyane et du Territoire de l'Inini, 139*, 1–4.

Floch, H., & Abonnenc, E. (1946e). Sur la validité des espèces *A. ininii* et *A. sanctielii* Senevet et Abonnenc, 1938. *Publications / Institut Pasteur de la Guyane et du Territoire de l'Inini, 124*, 1–4.

Floch, H., & Abonnenc, E. (1947). Distribution des moustiques du genre *Culex* en Guyane Française. *Publications / Institut Pasteur de la Guyane et du Territoire de l'Inini, 146*, 1–9.

Floch, H., & Abonnenc, E. (1951). *Anophèles* de la Guyane Française. *Archives de l'Institut Pasteur de la Guyane et du Territoire de l'Inini, 236*, 1–91.

Floch, H., & Fauran, P. (1954). Nouveaux *Culex* en Guyane Française. *Culex* (*Aedinus*) *amazonensis* Lutz et *Culex* (*Melanoconion*) *idottus* Dyar, *Culex* (*Melanoconion*) *vidali* n. sp. *Publications / Institut Pasteur de la Guyane et du Territoire de l'Inini, 15*(333), 1–7.

Floch, H., & Fauran, P. (1955). Nouveaux *Culex* en Guyane Française. VI. *Culex* (*Melanoconion*) *patientiae* n. sp. *Culex* (*Carrollia*) *manaensis* n. sp. *Culex* (*Microculex*) *reginae* n.sp. *Publications / Institut Pasteur de la Guyane et du Territoire de l'Inini, 16*(370), 1–7.

Floch, H., & Kramer, R. (1965). Sur *Culex* (*Melanoconion*) *vomerifer* Komp 1932, *Culex* (*Melanoconion*) *portesi* Senevet et Abonnenc 1941 et *Culex* (*Melanoconion*) *cayennensis* Floch et Abonnenc 1945, en Guyane Française. *Publications / Institut Pasteur de la Guyane et du Territoire de l'Inini, 497*, 1–6.

Floore, T.G., Harrison, B.A., & Eldridge, B.F. (1976). The *Anopheles* (*Anopheles*) *crucians* subgroup in the United States (Diptera: Culicidae). *Mosquito Systematics, 8*(1), 1–109.

Flores-Mendoza, C., & Lourenço-de-Oliveira, R. (1996). Bionomics of *Anopheles aquasalis* Curry 1932, in Guaraí, State of Rio de Janeiro, southeastern Brazil. 1. Seasonal distribution and parity rates. *Memórias do Instituto Oswaldo Cruz, 91*(3), 265–270.

Flores-Mendoza, C., Peyton, EL, Wilkerson, R.C., & Lourenço-de-Oliveira, R. (2004). *Anopheles* (*Nyssorhynchus*) *konderi* Galvão and Damasceno: neotype designation and resurrection from synonymy with *Anopheles* (*Nyssorhynchus*) *oswaldoi* (Peryassu) (Diptera: Culicidae). *Proceedings of the Entomological Society of Washington, 106*(1), 118–132.

Focks, D.A., Haile, D.G., Daniels, E., & Mount, G.A. (1993a). Dynamic life table model of a container-inhabiting mosquito, *Aedes aegypti* (L.) (Diptera: Culicidae). Part I. Analysis of the literature and model development. *Journal of Medical Entomology, 30*(6), 1003–1017.

Focks, D.A., Haile, D.G., Daniels, E., & Mount, G.A. (1993b). Dynamic life table model of *Aedes aegypti* (Diptera: Culicidae): Simulation results and validation. *Journal of Medical Entomology, 30*(6), 1018–1028.

Focks, D.A., Linda, S.B., Craig, G.B., Jr., Hawley, W.A., & Pumpuni, C.B. (1994). *Aedes albopictus* (Diptera: Culicidae): A statistical model of the role of temperature, photoperiod, and geography in the induction of egg diapause. *Journal of Medical Entomology, 31*(2), 278–286.

Focks, D.A., Brenner, R.J., Hayes, J., & Daniels, E. (2000). Transmission thresholds for dengue in terms of *Aedes aegypti* pupae per person with discussion of their utility in source reduction efforts. *The American Journal of Tropical Medicine and Hygiene, 62*(1), 11–18.

Foley, D.H., & Bryan, J.H. (1991). *Anopheles annulipes* Walker (Diptera: Culicidae) at Griffith, New South Wales. 2. Biology and behaviour of two sibling species. *Journal of the Australian Entomological Society, 30*, 109–112.

Foley, D.H., & Bryan, J.H. (1993). Electrophoretic keys to identify members of the *Anopheles punctulatus* complex of vector mosquitoes in Papua New Guinea. *Medical and Veterinary Entomology, 7*(1), 49–53.

Foley, D.H., & Torres, E.P. (2006). Population structure of an island malaria vector. *Medical and Veterinary Entomology, 20*(4), 393–401.

Foley, D.H., Paru, R., Dagoro, H., & Bryan, J.H. (1993). Allozyme analysis reveals six species within the *An. punctulatus* complex of mosquitoes in Papua New Guinea. *Medical and Veterinary Entomology, 7*(1), 37–48.

Foley, D.H., Meek, S.R., & Bryan, J.H. (1994). The *Anopheles punctulatus* group of mosquitoes in the Solomon Islands and Vanuatu surveyed by allozyme electrophoresis. *Medical and Veterinary Entomology, 8*(4), 340–350.

Foley, D.H., Cooper, R.D., & Bryan, J.H. (1995). A new species within the *Anopheles punctulatus* complex in western province, Papua New Guinea. *Journal of the American Mosquito Control Association, 11*(1), 122–127.

Foley, D.H., Bryan, J.H., Yeates, D., & Saul, A. (1998). Evolution and systematics of *Anopheles*: insights from a molecular phylogeny of Australasian mosquitoes. *Molecular Phylogenetics and Evolution, 9*(2), 262–275.

Foley, D.H., Torres, E.P., & Mueller, I. (2002). Stream-bank shade and larval distribution of the Philippine malaria vector *Anopheles flavirostris*. *Medical and Veterinary Entomology, 16*(4), 347–355.

Foley, D.H., Torres, E.P., Mueller, I., Bryan, J.H., & Bell, D. (2003). Host-dependent *Anopheles flavirostris* larval distribution reinforces the risk of malaria near water. *Transactions of the Royal Society of Tropical Medicine and Hygiene, 97*(3), 283–287.

Foley, D.H., Russell, R.C., & Bryan, J.H. (2004). Population structure of the peridomestic mosquito *Ochlerotatus notoscriptus* in Australia. *Medical and Veterinary Entomology, 18*(2), 180–190.

Foley, D.H., Bryan, J.H., & Wilkerson, R.C. (2007a). Species-richness of the *Anopheles annulipes* complex (Diptera: Culicidae) revealed by tree and model-based allozyme clustering analyses. *Biological Journal of the Linnean Society, 91*, 523–239.

Foley, D.H., Rueda, L.M., & Wilkerson, R.C. (2007b). Insight into global mosquito biogeography from country species records. *Journal of Medical Entomology, 44*(4), 554–567.

Foley, D.H., Wilkerson, R.C., Cooper, R.D., Volovsek, M.E., & Bryan, J.H. (2007c). A molecular phylogeny of *Anopheles annulipes* (Diptera: Culicidae) sensu lato: The most species-rich anopheline complex. *Molecular Phylogenetics and Evolution, 43*(1), 283–297.

Foley, D.H., Klein, T.A., Kim, H.C., Sames, W.J., Wilkerson, R.C., & Rueda, L.M. (2008a). Geographic distribution and ecology of potential malaria vectors in the Republic of Korea. *Journal of Medical Entomology, 46*(3), 680–692.

Foley, D.H., Weitzman, A.L., Miller, S.E., Faran, M.E., Rueda, L.M., & Wilkerson, R.C. (2008b). The value of georeferenced collection records for predicting patterns of mosquito species richness and endemism in the Neotropics. *Ecological Entomology, 33*(1), 12–23.

Foley, D.H., Rueda, L.M., Peterson, A.T., & Wilkerson, R.C. (2008c). Potential distribution of two species within the medically important *Anopheles minimus* complex (Diptera: Culicidae). *Journal of Medical Entomology, 45*(5), 852–860.

Foley, D.H., Klein, T.A., Kim, H.C., Brown, T., Wilkerson, R.C., & Rueda, L.M. (2010a). Validation of ecological niche models for potential malaria vectors in the Republic of Korea. *Journal of the American Mosquito Control Association, 26*(2), 210–213.

Foley, D.H., Wilkerson, R.C., Birney, I., Harrison, S., Christensen, J., & Rueda, L.M. (2010b). MosquitoMap and the Mal-area calculator: New web tools to relate mosquito species distribution with vector-borne disease. *International Journal of Health Geographics, 9*, 11.

Foley, D.H., Linton, Y.-M., Ruiz-Lopez, J.F., Conn, J.E., Sallum, M.A.M., Povoa, M.M., . . . Wilkerson, R.C. (2014). Geographic distribution, evolution, and disease importance of species within the Neotropical *Anopheles albitarsis* group (Diptera, Culicidae). *Journal of Vector Ecology, 39*(1), 168–181.

Fonseca, D.M., Lapointe, D.A., & Fleischer, R.C. (2000). Bottlenecks and multiple introductions: population genetics of the vector of avian malaria in Hawaii. *Molecular Ecology, 9*, 1803–1814.

Fonseca, D.M., Campbell, S., Crans, W.J., Mogi, M., Miyagi, I., Toma, T., . . . Wilkerson, R.C. (2001). *Aedes* (*Finlaya*) *japonicus* (Diptera: Culicidae), a newly recognized mosquito in the United States: Analysis of genetic variation in the United States and putative source populations. *Journal of Medical Entomology, 38*(2), 135–146.

Fonseca, D.M., Keyghobadi, N., Malcolm, C.A., Mehnet, C., Schaffner, F., Mogi, M., . . . Wilkerson, R.C. (2004). Emerging vectors in the *Culex pipiens* complex. *Science, 303*(5663), 1535–1538.

Fonseca, D.M., Smith, J.L., Wilkerson, R.C., & Fleischer, R.C. (2006). Pathways of expansion and multiple introductions illustrated by large genetic differentiation among worldwide populations of the southern house mosquito. *The American Journal of Tropical Medicine and Hygiene, 74*(2), 284–289.

Fonseca, D.M., Smith, J.L., Kim, H.C., & Mogi, M. (2009). Population genetics of the mosquito *Culex pipiens pallens* reveals sex-linked asymmetric introgression by *Culex quinquefasciatus*. *Infection, Genetics and Evolution, 9*(6), 1197–1203.

Fonseca, D.M., Widdel, A.K., Hutchinson, M., Spichiger, S.E., & Kramer, L.D. (2010). Fine-scale spatial and temporal population genetics of *Aedes japonicus*, a new US mosquito, reveal multiple introductions. *Molecular Ecology, 19*(8), 1559–1572.

Fonseca, D.M., Kaplan, L.R., Heiry, R.A., & Strickman, D. (2015). Density-dependent oviposition by female *Aedes albopictus* (Diptera: Culicidae) spreads eggs among containers during the summer but accumulates them in the fall. *Journal of Medical Entomology, 52*(4), 705–712.

Fontaine, M.C., Pease, J.B., Steele, A., Waterhouse, R.M., Neafsey, D.E., Sharakhov, I.V., . . . Besansky, N.J. (2015). Extensive introgression in a malaria vector species complex revealed by phylogenomics. *Science, 347*(6217), 1258524.

Fontenille, D. (1987). Description du mâle d'*Aedes* (*Diceromyia*) *coulangesi* (Rodhain et Boutonnier, 1982) moustique endemique de Madagascar. *Archives de l'Institut Pasteur de Madagascar, 53*(1), 203–208.

Fontenille, D., & Brunhes, J. (1985). Trois nouveaux culicidés de Madagascar: *Aedes* (*Aedimorphus*) *albodorsalis* n. sp., *Aedes* (*Aedimorphus*) *masoalensis* n. sp. et *Aedes* (*Aedimorphus*) *mathioti* n. sp. *Cahiers ORSTOM. Série entomologie médicale et parasitologie, 22*(for 1984)(2), 151–155.

Fontenille, D., & Jupp, P.G. (1989). Presence du moustique *Culex* (*Culex*) *neavei* à Madagascar, consequence dans la transmission des arbovirus. *Archives de l'Institut Pasteur de Madagascar, 56*(1), 287–295.

Foote, R.H. (1952). The larval morphology and chaetotaxy of the *Culex* subgenus *Melanoconion*. *Annals of the Entomological Society of America, 45*(3), 445–472.

Foote, R.H. (1953). The pupal morphology and chaetotaxy of the *Culex* subgenera *Melanoconion* and *Mochlostryax*. (Diptera, Culicidae). *Proceedings of the Entomological Society of Washington, 55*(2), 89–100.

Foote, R.H. (1954). The larvae and pupae of the mosquitoes belonging to the *Culex* subgenera *Melanoconion* and *Mochlostyrax*. *Technical Bulletin, 1091, United States Department of Agriculture.*

Foote, R.H., & Cook, D.R. (1959). Mosquitoes of medical importance. *Agricultural Handbook, U.S. Department of Agriculture, 152*, 1–158.

Forattini, O.P. (1958). Culicidae que se criam em buracos de caranguejos (Diptera). *Revista Brasileira de biologia, 18*(2), 175–179.

Forattini, O.P. (1959). Insecta Amapaensia—Diptera: Culicidae. *Studia entomologica, 2*(1–4), 285–288.

Forattini, O.P. (1961). Claves para a identificação do gênero *Anopheles* Meigen, 1818, da região Neotropical (Diptera: Culicidae). *Revista Brasileira de Entomología, 10*, 169–187.

Forattini, O.P. (1962). *Entomologia Médica*. Vol. 1. *Parte General, Diptera, Anophelini*. São Paulo, Brazil: Faculdade de Higiene e Saúde Pública, Departamento de Parasitologia.

Forattini, O.P. (1965a). *Entomologia Médica*. Vol. 2. *Culicini: Culex, Aedes e Psorophora*. São Paulo: Universidade de São Paulo.

Forattini, O.P. (1965b). *Entomologia Médica*. Vol. 3. *Culicini: Haemagogus, Mansonia, Culiseta, Sabethini, Toxorhynchitini. Arboviruses. Filariose bancroftiana*. São Paulo: Universidade de São Paulo.

Forattini, O.P. (1996). *Culicidologia médica*. Vol. 1. *Principios gerais, morfología, glossario taxonomico*. São Paulo: Editoria da Universidad de São Paulo.

Forattini, O.P. (2002). *Culicidologia Médica*. Vol. 2. *Identificação, biologia, epidemiologia*. São Paulo: Editoria da Universidad de São Paulo.

Forattini, O.P., & Coutinho, J.O. (1963). Sôbre duas espécies do subgênera *Microculex* (Dipt. Culicidae). *Studia entomologica, 6*(1–4), 481–495.

Forattini, O.P., & Lane, J. (1952). Sôbre *Toxorhynchites haemorrhoidalis separatus* Arribálzaga, 1891. *Arquivos de higiene e saúde pública* (São Paulo), *6*(1–2), 57–59.

Forattini, O.P., & Marucci, D. (1993). Scanning electron-microscopy of the eggs of two species of *Anopheles* (*Kerteszia*) (Diptera, Culicidae). *Memórias do Instituto Oswaldo Cruz, 88*(3), 349–352.

Forattini, O.P., & Rabello, E.X. (1960). Notas sôbre Culicidae (Diptera). 2. A larva, a pupa e alguns dados biológicos de *Aedes* (*Finlaya*) *fluviatilis* Lutz, 1904. *Arquivos de higiene e saúde pública da Universidade de São Paulo, 14*(1–2), 87–94.

Forattini, O.P., & Sallum, M.A.M. (1985). A new species of *Culex* (*Melanoconion*) from southern Brazil (Diptera: Culicidae). *Revista de saúde pública, 19*, 171–182.

Forattini, O.P., & Sallum, M.A.M. (1987a). Studies on some species of *Culex* (*Melanoconion*), with the description of a new one from southern Brazil (Diptera: Culicidae). *Revista de saúde pública, 21*(2), 123–156.

Forattini, O.P., & Sallum, M.A.M. (1987b). A new species of *Culex* (*Melanoconion*) from inland southern Brazil (Diptera: Culicidae). *Mosquito Systematics, 19*(2), 167–172.

Forattini, O.P., & Sallum, M.A.M. (1989a). A new species of *Culex* (*Melanoconion*) from Atlantic Tropical System of southern Brazil (Diptera: Culicidae). *Memórias do Instituto Oswaldo Cruz, 84*(1), 1–4.

Forattini, O.P., & Sallum, M.A.M. (1989b). Taxonomic study and redescription of *Culex* (*Melanoconion*) *theobaldi* (Lutz, 1904) (Diptera: Culicidae). *Memórias do Instituto Oswaldo Cruz,* 84 (Suppl. 4), 201–208.

Forattini, O.P., & Sallum, M A.M. (1989c). Redescription of *Culex* (*Melanoconion*) *delpontei* Duret, 1968 and *Cx.* (*Mel.*) *pereyrai* Duret, 1967, from southern Brazil. *Proceedings of the Entomological Society of Washington, 91*(3), 473–485.

Forattini, O.P., & Sallum, M.A.M. (1989d). Redescription of *Culex* (*Melanoconion*) *sacchettae* Sirivanakarn and Jakob 1981, with description of immature stages (Diptera: Culicidae). *Mosquito Systematics, 21*(2), 117–132.

Forattini, O.P., & Sallum, M.A.M. (1990). Redescription of *Culex* (*Melanoconion*) *lopesi* Sirivanakarn and Jakob 1979, with the description of immature stages (Diptera: Culicidae). *Mosquito Systematics, 22*(1), 57–66.

Forattini, O.P., & Sallum, M.A.M. (1992a). A new species of *Culex* (*Melanoconion*) from the Amazonian region (Diptera: Culicidae). *Memórias do Instituto Oswaldo Cruz, 87*(2), 265–274.

Forattini, O.P., & Sallum, M.A.M. (1992b). Cibarial armature as taxonomic characters for the Spissipes section of *Culex* (*Melanoconion*) (Diptera: Culicidae). *Mosquito Systematics, 24*(1), 70–84.

Forattini, O.P., & Sallum, M.A.M. (1993a). Redescription of *Culex* (*Melanoconion*) *inadmirabilis* Dyar and synonymy of *Culex* (*Melanoconion*) *mistura* Komp and Rozeboom (Diptera: Culicidae). *Mosquito Systematics, 25*(2), 77–88.

Forattini, O.P., & Sallum, M.A.M. (1993b). Taxonomic study of some species of the educator group of *Culex* (*Melanoconion*) (Diptera: Culicidae). *Mosquito Systematics, 25*(2), 89–109.

Forattini, O.P., & Sallum, M.A.M. (1995). Two new species of the Spissipes section of *Culex* (*Melanoconion*) (Diptera: Culicidae). *Mosquito Systematics, 27*(2), 125–142.

Forattini, O.P., & Toda, A. (1966). Notas sôbre Culicidae (Diptera). 11. Algumas espécies Amazônicas de *Microculex. Studia entomologica* (n.s.), *9*(1–4), 501–514.

Forattini, O.P., & Toda, A. (1967). Notas sôbre Culicidae (Diptera) 10. Nôvo encontro e alguns dados sôbre as formas imaturas de *Sabethes forattinii. Revista Brasileira de entomología* (São Paulo), *12,* 103–107.

Forattini, O.P., Rabello, E.X., & Heredia, R.L. (1956). Notas sôbre Culicidae (Diptera). *Arquivos de higiene e saúde pública, 10*(1–2), 85–88.

Forattini, O.P., Rabello, E.X., & Pattoli, D. (1960). Notas sôbre Culicidae (Diptera). 3. As larvas e as pupas de *Wyeomyia* (*W.*) *limai* Lane e Cerqueira, 1942 e de *W.* (*D.*) *personata* (Lutz, 1904). *Arquivos de higiene e saúde pública da Universidade de São Paulo, 14*(1–2), 95–109.

Forattini, P., Leon, P.M., & Rabello, E.X. (1965). Notas sôbre Culicidae (Diptera). 9. A respeito de algumas espécies de *Sabethes. Studia entomologica* (n.s.), *8,* 449–467.

Forattini, O.P., Rabello, E.X., & Souza Lopes, O. (1967). Notas sôbre Culicidae (Diptera). 12- Nova espécie do subgênero *Culex. Revista Brasileira de entomología, 12,* 13–17.

Forattini, O.P., Rabello, E.X., & Cotrim, M.D. (1970). *Catálogo das coleções entomológicas da Faculdade de Saúde Pública da Universidade de São Paulo. 1.a Série. Culicidae* (Vol. 4). São Paulo, Brazil: Universidade de São Paulo.

Forattini, O.P., Sallum, M.A.M., & Kakitani, I. (1988). Catálogo das coleções entomológicas da Faculdade de Saúde Pública da Universidade de São Paulo. 2a Série 11. Culicidae. *Revista de saúde pública* (São Paulo), *22,* 519–547.

Forattini, O.P., Sallum, M.A.M., & Flores, D.C. (1997). Description of the egg of *Anopheles* (*Anopheles*) *intermedius* (Peryassú, 1908) (Diptera: Culicidae) by scanning electron microscopy. *Revista do Instituto de Medicina Tropical de São Paulo, 39*(1), 5–9.

Forattini, O.P., Sallum, M.A.M., Marques, G.R.A.M., & Flores, D.C. (1997). Description of the eggs of *Anopheles* (*Kerteszia*) *laneanus* and *Anopheles* (*Nyssorhynchus*) *antunesi* (Diptera: Culicidae) by scanning electron microscopy. *Journal of the American Mosquito Control Association, 13*(4), 368–374.

Forattini, O.P., Sallum, M.A.M., Bergo, E.S., & Flores, D.C. (1998). Ultrastructure of eggs of *Anopheles rondoni*, *Anopheles lutzii*, and *Anopheles parvus*, three species of the subgenus *Nyssorhynchus. Journal of the American Mosquito Contol Association, 14*(3), 256–265.

Fornadel, C., Norris, L.C., Franco, V., & Norris, D.E. (2011). Unexpected anthropophily in the potential secondary malaria vectors *Anopheles coustani* s.l. and *Anopheles squamosus* in Macha, Zambia. *Vector-Borne and Zoonotic Diseases, 11*(8), 1173–1179.

Forskål, P. (1775). *Descriptiones animalium avium, amphibiorum, piscium, insectorum, vermium, quae in itinere orientali* Hauniae: Mö.lleri.

Foster, P.G., Bergo, E.S., Bourke, B.P., Oliveira, T.M.P., Nagaki, S.S., Sant'Ana, D.C., & Sallum, M.A.M. (2013). Phylogenetic analysis and DNA-based species confirmation in *Anopheles* (*Nyssorhynchus*). *PLOS ONE, 8*(2), e54063.

Foster, P.G., de Oliveira, T.M.P., Bergo, E.S., Conn, J.E., Sant'Ana, D.C., Nagaki, S.S., . . . Sallum, M.A.M. (2017). Phylogeny of Anophelinae using mitochondrial protein coding genes. *Royal Society Open Science, 4*(11), 170758.

Foster, W.A., & Takken, W. (2004). Nectar-related vs. human-related volatiles: Behavioural response and choice by female and male *Anopheles gambiae* (Diptera: Culicidae) between emergence and first feeding. *Bulletin of Entomological Research, 94*(2), 145–157.

Fouque, F., & Baumgärtner, J. (1996). Simulating development and survival of *Aedes vexans* (Diptera: Culicidae) preimaginal stages under field conditions. *Journal of Medical Entomology, 33*(1), 32–38.

Fourcroy, A. F. de (1785). *Entomologia Parisiensis; sive catalogus insectorum qu. in agro Parisiensi reperiuntur; secundam methodam Geoffroeanam in sectiones, genera & species distributus: cui addita sunt nomina trivialia & fere trecent. nov. species.* Pars prima. pp. I–VII [= 1–7], [1], 1-231. Parisiis. (Hôtel Serpente).

Fox, A.N., Pitts, R.J., Robertson, H.M., Carlson, J.R., & Zwiebel, L.J. (2001). Candidate odorant receptors from the malaria vector mosquito *Anopheles gambiae* and evidence of down-regulation in response to blood feeding. *Proceedings of the National Academy of Sciences of the United States of America, 98*(25), 14693–14697.

Fox, I. (1953). Notes on Puerto Rican mosquitoes including a new species of *Culex. Journal of Parasitology, 39*(2), 178–182.

Fox, R.M. (1958). Man-biting mosquitoes in coastal Liberia. *The American Journal of Tropical Medicine and Hygiene, 7*(2), 215–220.

Fraccaro, M., Laudani, U., Marchi, A., & Tiepolo, L. (1976). Karyotype, DNA replication and origin of sex chromosomes in *Anopheles atroparvus. Chromosoma, 55*(1), 27–36.

Frances, S.P. (2007). Efficacy and safety of repellents containing DEET. In M. Debboun, S.P. Frances, & D. Strickman (Eds.), *Insect repellents: Principles, methods and uses* (pp. 311–326). Boca Raton, FL: CRC Press.

Frank, J.H., & Curtis, G.A. (1981). On the bionomics of bromeliad-inhabiting mosquitoes. VI. A review of the bromeliad-inhabiting species. *Journal of the Florida Anti-Mosquito Association, 52*(1), 4–23.

Frank, N., & Ramus, E. (1995). *A complete guide to scientific and common names of reptiles and amphibians of the world.* Pottsville, PA: NG.

Franks, S.J., Wheeler, G.S., & Goodnight, C. (2011). Genetic variation and evolution of secondary compounds in native and introduced populations of the invasive plant *Melaleuca quinquenervia. Evolution, 66*(5), 1398–1412.

Franz, A.W., Kantor, A.M., Passarelli, A.L., & Clem, R.J. (2015). Tissue barriers to arbovirus infection in mosquitoes. *Viruses, 7*(7), 3741–3767.

Freeborn, S.B. (1926). *The mosquitoes of California* (Vol. 3). Berkeley: University of California Press, 333–460.

Freeborn, S.B. (1949). *Anophelines of the Nearctic Region.* In M.F. Boyd (Ed.), *Malariology: A comprehensive survey of all aspects of this group of diseases from a global standpoint* (Vol. 1, pp. 379–398). Philadelphia, PA: Saunders.

Freeborn, S.B., & Bohart, R.M. (1951). The mosquitoes of California. *Bulletin of the California Insect Survey, 1*(2), 25–78.

Freier, J.E., & Rosen, L. (1988). Vertical transmission of dengue virus by *Aedes mediovittatus. The American Journal of Tropical Medicine and Hygiene, 39*(2), 218–222.

Freitas, L.A., Russo, C.A.M., Voloch, C.M., Mutaquiha, O.C.F., Marques, L.P., & Schrago, C.G. (2015). Diversification of the genus *Anopheles* and a Neotropical clade from the Late Cretaceous. *PLOS ONE, 10*(8), e0134462.

Fritz, G.N., Dritz, D., Jensen, T., & Washino, R.K. (1991). Wing scale pattern variation in *Anopheles punctipennis* (Say). *Mosquito Systematics, 23*(2), 80–86.

Fritz, G.N., Conn, J., Cockburn, A., & Seawright, J. (1994). Sequence analysis of the ribosomal RNA internal transcribed spacer 2 from populations of *Anopheles nuneztovari* (Diptera: Culicidae). *Molecular Biology and Evolution, 11*(3), 406–416.

Frohne, W.C. (1953). Natural history of *Culiseta impatiens* (Wlk.), (Diptera, Culicidae), in Alaska. *Transactions of the American Miroscopical Society, 72*(2), 103–118.

Frohne, W.C. (1954). Biology of an Alaskan mosquito, *Culiseta alaskaensis* (Ludl.). *Annals of the Entomological Society of America, 47*, 9–24.

Frohne, W.C. (1955). Characteristic saddle spines of northern mosquito larvae. *Transactions of the American Microscopical Society, 74*(3), 295–302.

Frohne, W.C., & Frohne, R.G. (1954). Breeding places of *Aedes pseudodiantaeus* Smith and *diantaeus* H., D., and K. in Alaska. *Bulletin of the Brooklyn Entomological Society, 49*(4), 95–99.

Froud, M.D. (1944). *Anopheles jebudensis* sp. nov., a new anopheline mosquito from southern Nigeria. *Annals of Tropical Medicine and Parasitology, 38*(1), 73–77.

Furumizo, R.T., & Rudnick, A. (1978). Laboratory studies of *Toxorhynchites splendens* (Diptera: Culicicae): Biological observations. *Annals of the Entomology Society of America, 71*, 670–673.

Futami, K., Sonye, G., Akweywa, P., Kaneko, S., & Minakawa, N. (2008). Diving behavior in *Anopheles gambiae* (Diptera: Culicidae): Avoidance of a predacious wolf spider (Araneae: Lycosidae) in relation to life stage and water depth. *Journal of Medical Entomology, 45*(6), 1050–1056.

Gabaldón, A. (1939). On *Anopheles mattogrossensis* from Venezuela with description of the male. *The American Journal of Tropical Medicine and Hygiene, 19*(5), 457–460.

Gabaldón, A. (1940). Estudios sobre anofelinos. Serie I. Descripción de *Anopheles* (*Nyssorhynchus*) *nunez-tovari* n. sp., y consideraciones sobre una sub-división del grupo *Nyssorhynchus* (Diptera, Culicidae). *Publicaciones de la División de Malariología* (Caracas), *5*, 3–7.

Gabaldón, A. (1981). *Anopheles nuñeztovari*: importante vector y agente de malaria refractaria en Venezuela. *Boletín de al Dirección de Malariología y Saneamiento Ambiental, 31*(1), 28–38.

Gabaldón, A., & Cova García, P. (1952). Zoogeografía de los anofelinos en Venezuela. IV. Su posición en la región Neotrópica y observaciones sobre las especies de esta región. *Revista Venezolana de sanidad y asistencia social, 17*(3, 4), 171–209.

Gabaldón, A., Cova-García, P., & López, J.A. (1940a). Estudios sobre anofelinos. Serie I. 2. *Anopheles* (*Nyssorhynchus*) *rangeli*, una nueva especie de la sub-serie *Oswaldoi* (Diptera, Culicidae) de amplia distribución en Venezuela. *Publicaciones de la Division de Malariología* (Caracas), *5*, 9–23.

Gabaldón, A., Herrera, J., Pérez-Vivas, M.A., & Rausseo, J.A. (1940b). Estudio sobre anofelinos, Serie I. 6. *Chagasia bathanus* Dyar, 1928: Su hallazgo en Venezuela y nota sobre variaciones morfológicas de las pupas. *Publicaciones de la División de Malariología* (Caracas), *5*, 57–62.

Gabaldón, A., Cova-García, P., & Lopez, J.A. (1941a). Estudios sobre anofelinos. Serie II. 1. *Anopheles* (*Nyssorhynchus*) *benarrochi*, una especie de la subserie *triannulatus. Publicaciones de la División de Malariología* (Caracas), *7*, 3–24.

Gabaldón, A., Cova-García, P., & Lopez, J.A. (1941b). Estudios sobre anofelinos. Serie II. 2. *Anopheles* (*Arthuromyia*) *vargasi*, n. sp. y redescripción del subgénero *Arthuromyia* Galvão, 1941. *Publicaciones de la División de Malariología* (Caracas), *7*, 25–56.

Gad, A.M., & Salit, A. (1972). The mosquitoes of the Red Sea area, Egypt. *Journal of Medical Entomology, 9*(6), 581–582.

Gad, A.M., El Said, S., Soliman, B., Hassan, A.N., & Shoukry, A. (1987). Distribution and bionomics of Egyptian *Culex univittatus* (Theobald). *Journal of the Egyptian Society of Parasitology, 17*(1), 17–31.

Gad, A.M., Harbach, R.E., & Harrison, B.A. (2006). *Anopheles* (*Cellia*) *ainshamsi*, n. sp. (Diptera: Culicidae), a saltwater species from the Red Sea coast of Egypt. *Proceedings of the Entomological Society of Washington, 108*(2), 366–380.

Gäde, G. (2004). Regulation of intermediary metabolism and water balance of insects by neuropeptides. *Annual Review of Entomology, 49*(1), 93–113.

Gaffigan, T.V., & Ward, R.A. (1985). Index to the second supplement to "A Catalog of the Mosquitoes of the World" with corrections and additions. (Diptera: Culicidae). *Mosquito Systematics, 17*(1), 52–63.

Gage, K.L., Burkot, T.R., Eisen, R.J., & Hayes, E.B. (2008). Climate and vectorborne diseases. *American Journal of Preventive Medicine, 35*(5), 436–450.

Galindo, P. (1957a). A note on the oviposition behavior of *Sabethes* (*Sabethoides*) *chloropterus* Humboldt. *Proceedings of the Entomological Society of Washington, 59*(6), 287–288.

Galindo, P. (1957b). On the validity of *Haemagogus spegazzinii falco* Kumm et al, 1946 (Diptera, Culicidae). *Proceedings of the Entomological Society of Washington, 59*(3), 121–124.

Galindo, P. (1958). Bionomics of *Sabethes chloropterus* Humboldt, a vector of sylvan yellow fever in Middle America. *The American Journal of Tropical Medicine and Hygiene, 7*(4), 429–440.

Galindo, P. (1969). Notes on the systematics of *Culex* (*Melanoconion*) *taeniopus* Dyar and Knab and related species, gathered during arbovirus investigations in Panama. *Mosquito Systematics, 1*(4), 82–89.

Galindo, P., & Blanton, F.S. (1954). Nine new species of neotropical *Culex*, eight from Panama and one from Honduras (Diptera, Culicidae). *Annals of the Entomological Society of America, 47*(2), 231–247.

Galindo, P., & Blanton, F.S. (1955). An annotated list of the *Culex* of Panama (Diptera, Culicidae). *Proceedings of the Entomological Society of Washington, 57*(2), 68–74.

Galindo, P., & Kelley, T.F. (1943). *Culex* (*Culex*) *thriambus* Dyar a new mosquito record for California (Diptera, Culicidae). *The Pan-Pacific Entomologist, 19*(3), 87–90.

Galindo, P., & Mendez., E. (1961). Descriptions of four new species of *Culex* from Panama (Diptera: Culicidae). *Annals of the Entomological Society of America, 54*(1), 1–4.

Galindo, P., & Trapido, H. (1956). Descriptions of two new subspecies of *Haemagogus mesodentatus* Komp and Kumm 1938, from middle America (Diptera, Culicidae). *Proceedings of the Entomological Society of Washington, 58*(4), 228–231.

Galindo, P., & Trapido, H. (1967). Description of *Haemagogus aeritinctus*, a new species from British Honduras, with a note on the validity of *Haemagogus lucifer* (H.D. and K.) (Diptera: Culicidae). *Proceedings of the Entomological Society of Washington, 69*(2), 103–111.

Galindo, P., Carpenter, S.J., & Trapido, H. (1951a). Ecological observations on forest mosquitoes of an endemic yellow fever area in Panama. *The American Journal of Tropical Medicine and Hygiene, 31*(1), 98–137.

Galindo, P., Carpenter, S.J., & Trapido, H. (1951b). Descriptions of two new species of *Wyeomyia* and the male of *Sabethes tarsopus* Dyar and Knab (Diptera, Culicidae). *Proceedings of the Entomological Society of Washington, 53*(2), 86–96.

Galindo, P., Carpenter, S.J., & Trapido, H. (1953). The taxonomic status of the *Aedes leucocelaenus* complex with descriptions of two new forms (Diptera, Culicidae). *Annals of the Entomological Society of America, 45*(for 1952)(4), 529–542.

Galindo, P., Blanton, F.S., & Peyton, EL. (1954). A revision of the *Uranotaenia* of Panama with notes on other American species of the genus (Diptera, Culicidae). *Annals of the Entomological Society of America, 47*(1), 107–177.

Galindo, P., Carpenter, S.J., & Trapido, H. (1955). A contribution to the ecology and biology of tree hole breeding mosquitoes of Panama. *Annals of the Entomological Society of America, 48*(3), 158–164.

Galindo, P., Trapido, H., & Carpenter, S.J. (1956). The abundance cycles of arboreal mosquitoes during six years at a sylvan yellow fever locality in Panama. *Annals of the Entomological Society of America, 49*(6), 543–547.

Galliard, H. (1931a). *Culex brumpti*, n. sp. Moustique nouveau trouvé en Corse. *Annales de parasitologie humaine et comparée, 9*(2), 134–139.

Galliard, H. (1931b). Culicidés du Gabon. I. Culicinés, avec la description d'une espèce et de deux variétés nouvelles. *Annales de parasitologie humaine et comparée, 9*(3), 225–232.

Galliard, H. (1931c). Culicidés du Gabon. II. Culicinés (*suite*). Remarques sur la biologie des *Mansonoides* et d'*Aedes* (*Stegomyia*) *argenteus* Poiret. *Annales de parasitologie humaine et comparée, 9*(6), 514–529.

Galliard, H., & Ngu, D.V. (1940). Une espèce nouvelle d'*Anophèles* du Tonkin, *A. tonkinensis* sp. nov. *Revue médicale Française d'Extrême-Orient, 18*(9–10), 595–598.

Galliard, H., & Ngu, D.V. (1947). Culicinés du Tonkin. I. Liste des espèces et leur répartition. *Annales de parasitologie humaine et comparée, 22*(1–2), 75–80.

Galliard, H., & Ngu, D.V. (1949). Culicinés du Tonkin. II. Les genres *Ficalbia* et *Mochtogenes*. *Annales de parasitologie humaine et comparée, 24*, 495–502.

Galliard, H., & Ngu, D.V. (1950). Culicinés du Tonkin. III. Culicidés de la Baie d'Along. *Annales de parasitologie humaine et comparée, 25*(1–2), 61–68.

Galvão, A.B. (1950). Do diagnóstico diferencial entre *A.* (*N.*) *strodei* e *A.* (*N.*) *pessoai* na fase larvária. *Revista Brasileira de malariologia, 2*, 38–48.

Galvão, A.B. (1951). Um novo caráter diferencial entre *A. minor* e *A. peryassui* na fase larvária. *Revista Brasileira de malariologia e doenças tropicais* (Rio de Janeiro), *3*(2), 363–366.

Galvão, A.B. (1952). *Anopheles* (*Anopheles*) *rachoui*, uma nova espécie do grupo *Arribalzagia*. *Revista Brasileira de malariologia e doenças tropicais* (Rio de Janeiro), *4*(1), 69–74.

Galvão, A.B. (1955). *Anopheles* (*Arribalzagia*) *bustamantei*, uma nova espécie (Diptera, Culicidae). *Revista Brasileira de malariologia e doenças tropicais* (Rio de Janeiro), *7*, 255–258.

Galvão, A.L.A. (1938). Observações sobre algumas espécies do subgenera *Nyssorhynchus*, com especial referencia á morfología dos ovos. *Revista de biologia e higiene, 9*, 51–60.

Galvão, A.L.A. (1940). Contribuição ao conhecimento dos anofelinos do grupo *Nyssorhynchus* de São Paulo e regiões vizinhas (Diptera, Culicidae). *Revista do Museu Paulista, 24*, 399–484.

Galvão, A.L.A. (1941a). Notas sobre alguns anofelineos do sub-gênero *Nyssorhynchus* do norte do Brasil. *Revista de biologia e hygiene* (São Paolo), *11*(1), 92–96.

Galvão, A.L.A. (1941b). Contribuição ao conhecimento das espécies de *Myzorhynchella* (Diptera, Culicidae). *Arquivos de zoologia do estado de São Paulo, 2*(18), 505–576.

Galvão, A.L.A. (1943). Chaves para a determinação das espécies do subgênero *Nyssorhynchus* do Brasil. *Arquivos de higiene e saúde pública, 8*(19), 141–162.

Galvão, A.L.A., & Barretto, M.P. (1939). Contribuição ao conhecimento dos primeiros estádios dos anofelinos de São Paulo. *Revista de biologia e higiene, 9*(2), 110–115.

Galvão, A.L.A., & Barretto, M.P. (1941). Sobre um novo anofelino de São Paulo, *Anopheles* (*Anopheles*) *pseudotibiamaculatus* n. sp. (Diptera, Culicidae). *Revista de biologia e higiene, 11*(2), 65–70.

Galvão, A.L.A., & Damasceno, R.G. (1942a). *Anopheles* (*Nyssorhynchus*) *konderi* nova espécie de *Anopheles* do vale do Amazonas e considerações sobre as espécies do complexo *tarsimaculatus* (Diptera, Culicidae). *Folia clinica et biologica, 14*(5–6), 115–135.

Galvão, A.L.A., & Damasceno, R.G. (1942b). Sobre um novo anofelino da Ilha de Marajó, *Anopheles* (*N.*) *marajoara* N. Sp. *Anais Paulistas de medicina e cirurgia, 44*(5), 424–427.

Galvão, A.L., & Damasceno, R.G. (1942c). Sobre um novo Anofelino da Ilha Marajó, *Anopheles* (*Nyssorhynchus*) *marajoara* n.sp. (Diptera, Culicidae). *Folia clinica et biologica* (São Paulo), *14*, 60–66.

Galvão, A.L.A., & Damasceno, R.G. (1944). Observações sôbre anofelinos do complexo *albitarsis* (Diptera, Culicidae). *Annales da Faculdade de Medicina de São Paulo, 20*, 73–87.

Galvão, A.L.A., & Franco do Amaral, A.D. (1938). Sobre um novo anofelino de campos do Jordão, Estado de São Paulo, *Anopheles* (*Nyssorhynchus*) *lanei* n. sp. (Diptera, Culicidae). *Revista de biologia e higiene, 9*(1), 8–16.

Galvão, A.L.A., & Franco do Amaral, A.D. (1940). Estudios sobre os anofelinos do grupo *Myzorhynchella* com a descrição de uma espécie nova, *Anopheles* (*Nyssorhynchus*) *antunesi* n. sp. (Dipt. Culicidae). *Folia clinica et biologica* (São Paulo), *12*(5), 150–160.

Galvão, A.L.A., & Lane, J. (1937a). Notas sobre os *Nyssorhynchus* de São Paulo. II. Descrição de uma nova espécie *Anopheles* (*Nyssorhynchus*) *pessoai* (Diptera, Culicidae). *Revista de biologia e higiene, 7*(for 1936)(2), 67–79.

Galvão, A.L.A., & Lane, J. (1937b). Nota sobre os *Nyssorhynchus* de S. Paulo. VII. Estudo sobre as variedades deste grupo com a descrição de *Anopheles* (*Nyssorhynchus*) *albitarsis* Arrib., 1878 var. *limai* n. var. *Annais da Faculdade de Medicina de São Paulo, 13*, 211–238.

Galvão, A.L.A., & Lane, J. (1938). Notas sobre os *Nyssorhynchus* de São Paulo. VI. Revalidacão de *Anopheles* (*Nyssorhynchus*) *oswaldoi* Peryassú, 1922 e discussão sobre *Anopheles* (*Nyssorhynchus*) *tarsimaculatus* Goeldi, 1905. In Anonymous (Eds.). *Livro jubilar do professor Lauro Travassos*. (pp. 169–178). Rio de Janeiro: Typographia do Instituto Oswaldo Cruz.

Galvão, A.L.A., Lane, J., & Corrêa, R. (1937). Notas sobre os *Nyssorhynchus* de S. Paulo. V. Sobre os *Nyssorhynchus* de Novo Oriente. *Revista de biologia e higiene, 8*(1), 37–45.

Galvão, A.L., Lane, J., & Unti, O. (1943). Sôbre a *Anopheles noroestensis* Galvão & Lane, 1938. *Arquivos de higiene e saúde pública* (São Paulo), *19*, 35–48.

Ganeson, K., Mendki, M.J., Suryanarayana, M.V., Prakash, S., & Malhotra, R.C. (2006). Studies of *Aedes aegypti* (Diptera: Culicidae) oviposition responses to newly identified semiochemicals from conspecific eggs. *Australian Journal of Entomology, 45*(1), 75–80.

Gao, X., & Hall, D.G. (2003). 3-Boronoacrolein as an exceptional heterodyne in the highly enatio- and diasteroselective Cr (III)-catalyzed three-component [4 + 2]/allylboration. *Journal of the American Chemical Society, 125*(31), 9308–9309.

Gapuz, R.B., & Santiago, D. (1949). A supplementary guide for rapid identification of larvae of *Anopheles minimus flavirostris* (Ludlow) and *Anopheles mangyanus* (Banks). *Philippine Journal of Science, 78*, 127–133.

García, M., & Casal, O.H. (1965a). Descripción de tres pupas de mosquitos neotropicales (Diptera, Culicidae). *Revista de la Sociedad Entomológica Argentina, 27*(for 1964), 5–10.

García, M., & Casal, O.H. (1965b). Culicidae (Diptera) del Delta del Paraná. II. Apuntes sistemáticos y biológicos. *Delta del Paraná, 5*(8), 5–16.

García, M., & Casal, O.H. (1967). *Toxorhynchites* (*Lynchiella*) *cavalierii*, nueva especie bromelícola de la República Argentina (Diptera, Culicidae). *Physis, 26*(73), 435–441.

García, M., & Casal, O.H. (1968a). *Sabethes* (*Sabethinus*) *melanonymphe*; *Wyeomyia* (*W.*) *sabethea* y *W.* (*Dendromyia*) *aporonoma*, nuevas para la entomofauna Argentina (Diptera, Culicidae). *Revista de la Sociedad Entomológica Argentina, 30*(for 1967), 61–63.

García, M., & Casal, O.H. (1968b). Siete especies de Culicidae (Diptera) nuevas para la entomofauna Argentina. *Physis, 28*(76), 107–109.

García, M., & Ronderos, R.A. (1962). Mosquitos de la República Argentina. Tribu Anophelini (Diptera-Culicidae-Culicinae). *Anales de la Comisión de Investigación Científicas, Provincia de Buenos Aires, 3*, 103–212.

García, M., & Ronderos, R.A. (1963). Sobre la validez de *Aedes* (*Ochlerotatus*) *patersoni* Shannon y Del Ponte, 1927 (Diptera, Culicidae). *Revista de la Sociedad Entomológica Argentina, 25*, 29–33.

Garcia, R., Jeffery, J., & Rudnick, A. (1969). *Culiseta marchettei*, a new species of the subgenus *Climacura* Howard, Dyar, and Knab from Malaysia, with notes on its biology (Diptera: Culicidae). *Journal of Medical Entomology, 6*(3), 251–256.

García Avila, I. (1976). Insectos hematófagos de Cuba. *Poeyana, 154*, 1–15.

García-Castellanos, D., Estrada, F., Jiménez-Munt, I., Gorini, C., Fernàndez, M., Vergés, J., & De Vicente, R. (2009). Catastrophic flood of the Mediterranean after the Messinian salinity crisis. *Nature, 462*(7274), 778–782.

Garczynski, S.F., Crim, J.W., & Brown, M.R. (2005). Characterization of neuropeptide F and its receptor from the African malaria mosquito, *Anopheles gambiae*. *Peptides, 26*(1), 99–107.

Garczynski, S.F., Crim, J.W., & Brown, M.R. (2007). Characterization and expression of the short neuropeptide F receptor in the African malaria mosquito, *Anopheles gambiae*. *Peptides, 28*(1), 109–118.

Gardner, C.F., Nielsen, L.T., & Knight, K.L. (1973). Morphology of the mouthparts of larval *Aedes communis* (De Geer) (Diptera: Culicidae). *Mosquito Systematics, 5*(2), 163–182.

Garros, C., & Dujardin, J.-P. (2013). Genetic and phenetic approaches to *Anopheles* systematics. In S. Manguin (Ed.), Anopheles *mosquitoes: New insights into malaria vectors* (pp. 57–79. Rijeka, Croatia: InTech.

Garros, C., Harbach, R.E., & Manguin, S. (2005a). Morphological assessment and molecular phylogenetics of the Funestus and Minimus groups of *Anopheles* (*Cellia*). *Journal of Medical Entomology, 42*(4), 522–536.

Garros, C., Harbach, R.E., & Manguin, S. (2005b). Systematics and biogeographical implications of the phylogenetic relationships between members of the Funestus and Minimus Groups of *Anopheles* (Diptera: Culicidae). *Journal of Medical Entomology, 42*(1), 7–18.

Garros, C., Van Bortel, W., Trung, H.D., Coosemans, M., & Manguin, S. (2006). Review of the Minimus Complex of *Anopheles*, main malaria vector in Southeast Asia: From taxonomic issues to vector control strategies. *Tropical Medicine and International Health, 11*(1), 102–114.

Gary, R.E., Jr., & Foster, W.A. (2001). Effects of available sugar on the reproductive fitness and vectorial capacity of the malaria vector *Anopheles gambiae* (Diptera: Culicidae). *Journal of Medical Entomology, 38*(1), 22–28.

Gary, R.E., Jr., & Foster, W.A. (2004). *Anopheles gambiae* feeding and survival on honeydew and extra-floral nectar of peridomestic plants. *Medical and Veterinary Entomology, 18*(2), 102–107.

Gary, R.E., Jr., & Foster, W.A. (2006). Diel timing and frequency of sugar feeding in the mosquito *Anopheles gambiae*, depending on the sex, gonotrophic state and resource availability. *Medical and Veterinary Entomology, 20*(3), 308–316.

Gaschen, H. (1934). Description d'un nouvel *Aedes* du Yunnan *Finlaya yunnanensis* sp n. *Archives des Instituts Pasteur d'Indochine, 19*, 331–335.

Gaschen, H. (1955). Contribution à l'étude du *Culex* autogenicus en Suisse. *Bulletin de la Société Vaudoise des Sciences Naturelles, 66*(289), 125–140.

Gater, B.A.R. (1933). On *Anopheles baezai* n. sp., from the Malay Peninsula. *Bulletin of the Raffles Museum* (Singapore), *8*, 162–168.

Gatt, P. (1996). Mosquitoes from the Maltese Islands: New records and observations (Diptera: Culicidae). *Bollettino della Società Entomologica Italiana, 128*(1), 77–84.

Gatti, M., Santinl, G., Pimpinelli, S., & Coluzzi, M. (1977). Fluorescence banding techniques in the identification of sibling species of the *Anopheles gambiae* complex. *Heredity, 38*(1), 105–108.

Gebert, S. (1948). Notes on a new species of *Aedes*, subgenus *Mucidus* (Diptera: Culicidae) found in Mauritius. *The Entomologist, 81*, 96–99.

Geiger, J.H., Pratt, P.D., Wheeler, G.S., & Williams, D.A. (2011). Hybrid vigor for the invasive exotic Brazilian peppertree (*Schinus terebinthifolius* Raddi., Anacardiaceae) in Florida. *International Journal of Plant Sciences, 172*(5), 655–663.

Gelfand, H.M. (1954). The anopheline mosquitoes of Liberia. *West African Medical Journal, 3*, 80–88.

Gendrin, M., & Christophides, G.K. (2013). The *Anopheles* mosquito microbiota and their impact on pathogen transmission. In S. Manguin (Ed.), Anopheles *mosquitoes: New insights into malaria vectors* (pp. 525–548). Rijeka, Croatia: InTech.

Gendrin, M., Rodgers, F.H., Yerbanga, R.S., Ouédraogo, J.B., Basáñez, M.G., Cohuet, A., & Christophides, G.K. (2015). Antibiotics in ingested human blood affect the mosquito microbiota and capacity to transmit malaria. *Nature Communications, 6*, 5921.

Gentile, G., Della Torre, A., Maegga, B., Powell, J.R., & Caccone, A. (2002). Genetic differentiation in the African malaria vector, *Anopheles gambiae* s.s., and the problem of taxonomic status. *Genetics, 161*(4), 1561–1578.

Geoffroy, B. (1971a). Description du mâle de *Culex* (*Culiciomyia*) *eouzani* sp. n. (Diptera, Culicidae) nouveau moustique de la République Centrafricaine. *Cahiers ORSTOM. Série entomologie médicale et parasitologie, 9*(3), 279–283.

Geoffroy, B. (1971b). Description d'un moustique nouveau de la République Centrafricaine *Aedes* (*Aedimorphus*) *adami* sp. n. *Cahiers ORSTOM. Série entomologie médicale et parasitologie, 9*(3), 273–277.

Geoffroy, B. (1971c). Description d'un nouveau moustique de la République Centrafricaine: *Culex* (*Neoculex*) *vattieri* sp. n. *Cahiers ORSTOM. Série entomologie médicale et parasitologie, 9*(3), 265–268.

Geoffroy, B. (1974). Description du mâle de *Culex* (*Mochthogenes*) *germaini* n. sp. (Diptera, Culicidae) moustique nouveau de République Centrafricaine. *Cahiers ORSTOM. Série entomologie médicale et parasitologie, 12*(2), 93–96.

Geoffroy, B. (1987). The *Aedes* (*Aedimorphus*) *domesticus* Group (Diptera, Culicidae). I. New species, descriptions of *Aedes bambiotai* and *Aedes bancoi*. *Mosquito Systematics, 19*(1), 100–110.

Geoffroy, B., & Hervé, J.P. (1976). Description d'un nouveau moustique de République Centrafricaine, *Culex* (*Eumelanomyia*) *taufiieti* n. sp. *Cahiers ORSTOM. Série entomologie médicale et parasitologie, 14*(2), 151–154.

Gerberg, E.J. (1972). The type locality of *Aedes* (*S.*) *simpsoni* (Theobald). *Mosquito Systematics, 4*(1), 9.

Gerberg, E.J., & Arnett, R.H.J. (1976). A pictorial key to the mosquito larvae of the Seychelles. *Mosquito Systematics, 8*(4), 343–346.

Gerberg, E.J., & Hartberg, W.K. (1975). A bibliography of *Aedes simpsoni* (Theobald). *Mosquito Systematics, 7*(2), 137–147.

Gerberg, E.J., & van Someren, E.C.C. (1970). Pictorial key to the mosquitos *Aedes* (*Stegomyia*) of East Africa. Geneva, Switzerland: World Health Organization.

Gerberg, E., Barnard, D., & Ward, R.A. (1994). *Manual for mosquito rearing and experimental techniques* (Bulletin 5). Mount Laurel, NJ: Mosquito Control Association.

Germain, M., Cordellier, R., Hervé, J.-P., Geoffroy, B., Bouchite, B., Ravaonjanahary, C., & Rickenbach, A. (1975). Présence en Afrique centrale et occidentale d'*Aedes* (*Stegomyia*) *opok* Corbet et van Someren diagnose différentielle de l'espèce. *Cahiers ORSTOM. Série entomologie médicale et parasitologie, 13*(1), 41–46.

Germar, E.F. (1817). *Reise nach Dalmatien und in das Gebiet von Ragusa*. Leipzig, Germany: Brockhaus.

Gerry, B.I. (1932). Morphological studies of the female genitalia of Cuban mosquitoes. *Annals of the Entomological Society of America, 25*, 31–75.

Ghaninia, M., Larsson, M., Hansson, B.S., & Ignell, R. (2008). Natural odor ligands for olfactory receptor neurons of the female mosquito *Aedes aegypti*: Use of gas chromatography-linked single sensillum recordings. *Journal of Experimental Biology, 211*(18), 3020–3027.

Giaquinto-Mira, M. (1931). Una nueva especie de *Anopheles* en Guatemala: *Anopheles hectoris mihi*. *Boletín de la Dirección General de Salubridad, 1*(20/25), 606–615.

Giaquinto-Mira, M. (1950). Notes on the geographical distribution and biology of Anophelinae and Culicinae in Etiopia. *Rivista di malariologia* (Rome), *29*(5), 281–313.

Giebel, C.G.A. (1862). Wirbeltier und Insektenreste im Bernstein. *Zeitschrift für die Gesammten Naturwissenschaften, 20*, 311–321.

Giglioli, M.E.C. (1965). Oviposition by *Anopheles melas* and its effect on egg survival during the dry season in the Gambia, West Africa. *Annals of the Entomological Society of America, 58*(6), 885–891.

Gil Collado, J. (1936). Culicidos de la Isla de Fernando Poo recogidos por la expedición J. Gil-F. Bonet. *Revista Española de entomología, 11*(4), 311–329.

Giles, G.M. (1899). A description of the Culicidae employed by Major R. Ross, I.M.S. in his investigations on malaria. *Journal of Tropical Medicine, 2*, 62–65.

Giles, G.M. (1900a). *A handbook of the gnats or mosquitoes, giving the anatomy and life history of the Culicidae*. London, England: Bale, Sons and Danielsson.

Giles, G.M. (1900b). Addendum I. Description of two species of *Anopheles* from West Africa. *Memoirs of the Liverpool School of Tropical Medicine Memoir, 2*, 49–51.

Giles, G.M. (1901a). A plea for the collective investigation of Indian Culicidae with suggestions as to moot points for enquiry, and a prodromus of species known to the author. *Journal of the Bombay Natural History Society, 13*(4), 592–610.

Giles, G.M. (1901b). Description of four new species of *Anopheles* from India. *Entomologist's Monthly Magazine, 37*, 196–198.

Giles, G.M. (1901c). Six new species of Culicidae from India. *The Entomologist, 34*(458), 192–197.

Giles, G.M. (1902). *A handbook of the gnats or mosquitoes giving the anatomy and life history of the Culicidae, together with descriptions of all species notices up to the present date*. London, England: Bale, Sons and Danielsson.

Giles, G.M. (1904a). Notes on some collections of mosquitoes, &c., received from the Philippine Islands and Angola; with some incidental remarks upon classification. *Journal of Tropical Medicine, 7*, 365–369.

Giles, G.M. (1904b). Notes on some collections of mosquitoes received from abroad. Vol. 1, A revised tabulation of the genus *Taeniorhynchus* Arribálzaga. *Journal of Tropical Medicine, 7*, 381–384.

Giles, G.M. (1904c). *A revision of the Anophelinae, being a first supplement to the second edition of* "A Handbook of the Gnats or Mosquitoes." London, England: Bale, Sons and Danielsson.

Giles, G.M. (1906). Mosquito notes. I. Note on a small collection of mosquitoes from Bahrain in Northern Arabia (shores of Persian Gulf). *Journal of Tropical Medicine, 9*, 130–132.

Gill, F., & Donsker, D. (Eds.). (2012). *IOC world bird names* (Vol. 3). Retrieved from http://www.worldbirdnames.org

Gillett, J.D. (1945). The larva and pupa of *Taeniorhynchus* (*Coquillettidia*) *maculipennis*, Theobald. *Bulletin of Entomological Research, 35*, 395–397.

Gillett, J.D. (1946). Notes on the subgenus *Coquillettidia* Dyar (Diptera, Culicidae). *Bulletin of Entomological Research, 36*, 425–438.

Gillett, J.D. (1949). Further notes on the Ethiopian species of *Taeniorhynchus* Arribalzaga (Diptera, Culicidae). *Proceedings of the Royal Entomological Society of London, Series B: Taxonomy, 18*(5–6), 97–102.

Gillett, J.D. (1951). The larva, pupa and adult male of *Aedes* (*Stegomyia*) *ruwenzori* Haddow and van Someren (Diptera, Culicidae). *Annals of Tropical Medicine and Parasitology, 45*, 195–198.

Gillett, J.D. (1955). The male of *Anopheles* (*Myzomyia*) *distinctus* var. *ugandae* Evans (Diptera: Culicidae). *Proceedings of the Royal Entomological Society of London, Series B: Taxonomy, 24*, 36.

Gillett, J.D. (1972). *Common African mosquitoes and their medical importance*. London, England: Heinemann.

Gillett, J. D. (1975). The *Anopheles gambiae* complex; naming the sibling species. *Transactions of the Royal Society of Tropical Medicine and Hygiene, 69*(3), 366–367.

Gillies, M.T. (1955). Notes on the eggs of some east African *Anopheles*. *Annals of Tropical Medicine and Parasitology, 49*(2), 158–160.

Gillies, M.T. (1958). Notes on the biology of a new subspecies of *Anopheles wellcomei* (Diptera: Culicidae) from east Africa, and on the distribution of related forms. *Proceedings of the Royal Entomological Society of London, Series A: General Entomology, 33*(1–3), 9–14.

Gillies, M.T. (1961). Studies on the dispersion and survival of *Anopheles gambiae* Giles in East Africa, by means of marking and release experiments. *Bulletin of Entomological Research, 52*(1), 99–127.

Gillies, M.T. (1962). A new species of *Anopheles funestus* complex (Diptera: Culicidae) from east Africa. *Proceedings of the Royal Entomological Society of London, Series B: Taxonomy, 31*(7–8), 81–86.

Gillies, M.T. (1964a). Selection for host preference in *Anopheles gambiae*. *Nature, 203*(4947), 852–854.

Gillies, M.T. (1964b). A new *Anopheles* (Diptera: Culicidae) from the East African coastal lowlands. *Proceedings of the Royal Entomological Society of London, Series B: Taxonomy, 33*(3–4), 67–70.

Gillies, M.T., & Coetzee, M. (1987). A supplement to the Anophelinae of Africa south of the Sahara (Afrotropical Region). *Publications of the South African Institute for Medical Research, 55*, 1–143.

Gillies, M.T., & De Meillon, B. (1968). The Anophelinae of Africa, south of the Sahara (Ethiopian Zoogeographical Region). *Publications of the South African Institute for Medical Research, 54*, 1–343.

Gillott, C. (2003). Male accessory gland secretions: Modulators of female reproductive physiology and behavior. *Annual Review of Entomology, 48*(1), 163–184.

Gimmerthal, B.A. (1845). Erster Beitrag zu einer künftig zu bearbeitenden Dipterologie Russlands. *Bulletin de la Société Impériale des Naturalistes de Moscou , 18*(4), 287–331.

Gimnig, J.E., & Eldridge, B.F. (1999). Genetic and morphological characterization of the *Aedes* (*Ochlerotatus*) *dorsalis* (Diptera: Culicidae) group in North America. *Journal of Medical Entomology, 36*(6), 685–694.

Gimnig, J.E., Reisen, W.K., Eldridge, B.F., Nixon, K.C., & Schultz, S.J. (1999). Temporal and spatial genetic variation within and among populations of the mosquito *Culex tarsalis* (Diptera: Culicidae) from California. *Journal of Medical Entomology, 36*(1), 23–29.

Gimonneau, G., Bouyer, J., Morand, S., Besansky, N.J., Diabate, A., & Simard, F. (2010). A behavioral mechanism underlying ecological divergence in the malaria mosquito *Anopheles gambiae*. *Behavorial Ecology, 21*(5), 1087–1092.

Gingrich, J.B., & Williams, G.M. (2005). Host-feeding patterns of suspected West Nile virus mosquito vectors in Delaware, 2001–2002. *Journal of the American Mosquito Control Association, 21*(2), 194–200.

Giordano, B.V. (2019). *Transmission dynamics and epidemiology of West Nile Virus in Ontario, Canada* [Doctoral thesis, Brock University, Canada].

Girod, R. (2004). First record of *Aedes albopictus* in Mayotte Island, Comoros Archipelago. *Parasite, 11*(1), 74.

Gistel, J.N.F.X. (1831). Kerfe in Copal eingeschlossen. *Isis* (Oken's), *24*, 247–248.

Gjullin, C.M. (1937). The female genitalia of the *Aedes* mosquitoes of the Pacific coast states. *Proceedings of the Entomological Society of Washington, 39*, 252–266.

Gjullin, C.M. (1946). A key to the *Aedes* females of America north of Mexico (Diptera, Culicidae). *Proceedings of the Entomological Society of Washington, 48*(9), 215–236.

Gjullin, C.M., & Eddy, G.W. (1972). *The mosquitoes of the northwestern United States*. USDA Technical Bulletin no. 1447, Agricultural Research Service, United States Department of Agriculture, Washington, DC.

Gjullin, C.M., Yates, W.W., & Stage, H.H. (1950). Studies on *Aedes vexans* (Meig.) and *Aedes sticticus* (Meig.), flood-water mosquitoes, in the lower Columbia River Valley. *Annals of the Entomological Society of America, 43*, 262–275.

Gjullin, C.M., Sailer, R.I., Stone, A., & Travis, B.V. (1961). The mosquitoes of Alaska. *Agriculture Handbook Washington, 182*, 1–98.

Glick, J.I. (1992). Illustrated key to the female *Anopheles* of southwestern Asia and Egypt (Diptera: Culicidae). *Mosquito Systematics, 24*(2), 125–153.

Goddard, J. (2013). *Public health entomology*. Boca Raton, FL: CRC Press.

Godfray, H.C.J., & Hassell, M.P. (1987). Natural enemies may be a cause of discrete generations of tropical insects. *Nature, 327*, 144–147.

Godsey, M.S., Jr., Abdoon, A.-M.M., Savage, H.M., Al-Sharani, A.M., Al-Mazrou, Y., Al-Jeffri, M.H., . . . Miller, B.R. (2003). First record of *Aedes* (*Stegomyia*) *unilineatus* in the Kingdom of Saudi Arabia. *Journal of the American Mosquito Control Association, 19*(1), 84–86.

Goeldi, E.A. (1902). *Os mosquitos no Pará. Encarados como uma calamidade publica*. Belém: Portugal: Imprensa Official.

Goeldi, E.A. (1905). *Os mosquitos no Pará. Reunião de quatro trabalhos sobre os mosquitos indigenas, principalmente as especies que molestam o homem*. Memorias do Museu Goeldi (Museu Paraense) de Historia Natural e Ethnographia. IV. Parà, Brazil: Wiegandt.

Goff, M.L. (1980). *Distribution of mosquitoes (Diptera: Culicidae) on the east flank of Mauna Loa Volcano, Hawai'i* (Technical Report 30). Manoa, HI: Cooperative National Park Resources Studies Unit, Department of Botany, University of Hawaii at Manoa.

Goiny, H., van Someren, C.C., & Heisch, R.B. (1957). The eggs of *Aedes* (*Skusea*) *pembaensis* Theobald discovered on crabs. *East African Medical Journal, 34*(1), 1–2.

Gokberk, C. (1970). The first record of *A. pulcherrimus* Theobald 1902 in Turkey. Its morphology, ecology and distribution. *Turk Hifzissihha Tecrube Bioloji Dergisi, 30*, 20–27.

Goldberg, L.J., & Margalit, J. (1977). A bacterial spore demonstrating rapid larvicidal activity against *Anopheles sergentii, Uranotaenia unguiculata, Culex univitattus, Aedes aegypti* and *Culex pipiens*. *Mosquito News, 37*(3), 355–358.

Golding, N., Nunn, M.A., Medlock, J.M., Purse, B.V., Vaux, A.G.C., & Schäfer, S.M. (2012). West Nile virus vector *Culex modestus* established in southern England. *Parasites and Vectors, 5*, 32.

Golenda, C.F., Starkweather, W.H., & Wirtz, R.A. (1990.) The distribution of circumsporozoite protein (cs) in *Anopheles stephensi* mosquitoes infected with *Plasmodium falciparum* malaria. *The Journal of Histochemistry and Cytochemistry, 38*(4), 475–481.

Gomes, A.C., Silva, N.N., Marques, G.R.A.M., & Brito, M. (2003). Host-feeding patterns of potential human disease vectors in the Paraíba Valley Region, State of São Paulo, Brazil. *Journal of Vector Ecology, 28*(1), 74–78.

Gomes, B., Alves, J., Sousa, C.A., Santa-Ana, M., Vieira, I., Silva, T.L., . . . Pinto, J. (2012a). Hybridization and population structure of the *Culex pipiens* complex in the islands of Macaronesia. *Ecology and Evolution, 2*(8), 1889–1902.

Gomes, B., Parreira, R., Sousa, C.A., Novo, M.T., Almeida, A.P.G., Donnelly, M.F., & Pinto, J. (2012b). The *Culex pipiens* complex in continental Portugal: Distribution and genetic structure. *Journal of the American Mosquito Control Association, 28*(Suppl. 4), 75–80.

Gomez, G.F., Marquez, E.J., Gutiérrez, L.A., Conn, J.E., & Correa, M.M. (2014). Geometric morphometric analysis of Colombian *Anopheles albimanus* (Diptera: Culicidae) reveals significant effect of environmental factors on wing traits and presence of a metapopulation. *Acta Tropica, 135*, 75–85.

Gong, Z.-D. (1989a). A new species of genus *Topomyia* from Yunnan, China (Diptera: Culicidae). *Acta entomologica Sinica, 32*(1), 90–91.

Gong, Z.-D. (1989b). A new species of the genus *Topomyia* from Yunnan Province, China (Diptera: Culicidae). *Acta zootaxonomica Sinica, 14*(2), 220–223.

Gong, Z.-D. (1991). A new species of the subgenus *Topomyia* of genus *Tomomyia* [*sic*] from Yunnan, China (Diptera: Culicidae). *Acta zootaxonomica Sinica, 16*(2), 228–231.

Gong, Z.-D. (1994). A new species of the genus *Topomyia* from West Yunnan, China (Diptera: Culicidae). *Acta zootaxonomica Sinica, 19*(2), 210–213.

Gong, Z.-D. (1996). A new species of *Topomyia* from the bamboo stumps in Yunnan, China (Diptera: Culicidae). *Acta zootaxonomica Sinica, 21*(3), 362–365.

Gong, Z.-D., & Ji, S.-H (1989). A new species of genus *Tripteroides* from Yunnan. (Diptera: Culicidae). *Acta entomologica Sinica, 32*(4), 482–484.

Gong, Z.-D., & Li, Z.-H. (1999). A new species of *Heizmannia* from Yunnan, China (Diptera: Culicidae). *Acta zootaxonomica Sinica, 24*, 96–99.

Gong, Z., & Lu, B. (1985). *Culex* (*Cx.*) *hutchinsoni* Barraud, a new culicine to China. *Sichuan Journal of Zoology, 4*(2), 12.

Gong, Z., & Lu, B. (1986). A new *Aedes* of subgenus *Finlaya* from Yunnan, China. *Zoological Research, 7*(1), 1–4.

Gong, Z., & Lu, B. (1991). A new subgenus and species of the *Aedes* (Diptera: Culicidae) from Yunnan, China [in Chinese]. *Entomotaxonomia, 13*(1), 55–59.

Gong, Z., & Lu, B. (1995). Studies on the genus *Topomyia* of China (Diptera: Culicidae). *Acta entomologica Sinica, 38*(4), 475–485.

Gong, Z., Li, Z., & Li, S. (1999). A description of the male of *Topomyia* (*Suaymyia*) *dulongensis* (Diptera; Culicidae) with supplement on certain characters of the larvae. *Entomotaxonomia, 21*(3), 225–227.

González, C.R., & Sallum, M.A.M. (2009). Neotype designation and redescription of adult male and immature stages of *Anopheles* (*Nyssorhynchus*) *pictipennis* (Philippi) (Diptera: Culicidae). *Zootaxa, 2295*, 15–24.

González, C.R., & Sallum, M.A.M. (2010). *Anopheles* (*Nyssorhynchus*) *atacamensis* (Diptera: Culicidae), a new species from northern Chile. *Memórias do Instituto Oswaldo Cruz, 105*(1), 13–24.

González, C.R., Reyes, C., Rada-Chaparro, V., & Saldarriaga-Córdoba, M. (2017). A new species of *Aedes* Meigen subgenus *Ochlerotatus* Lynch Arribálzaga (Diptera: Culicidae) from the coastal wetlands of the desert in northern Chile: Morphological and molecular identification. *Zootaxa, 4273*(1), 31–49.

González, R., & Carrejo, N.S. (2009). *Introducción al estudio taxonómico de Anopheles de Colombia. Claves y notas de distribución* (2nd ed.) Colombia: Programa Editorial Universidad del Valle.

González, R., Carrejo, N., Wilkerson, R.C., Alarcon, J., Alarcon-Ormasa, J., Ruiz, F., . . . Linton, Y.-M. (2010). Confirmation of *Anopheles* (*Anopheles*) *calderoni* Wilkerson, 1991 (Diptera: Culicidae) in Colombia and Ecuador through molecular and morphological correlation with topotypic material. *Memórias do Instituto Oswaldo Cruz, 105*(8), 1001–1009.

González Broche, R. (2000). Dos nuevas especies de mosquitos de los géneros *Psorophora* (*Grabhamia*) y *Culex* (*Culex*) (Diptera, Culicidae), de Cuba. *Revista Brasileira de entomologia* (São Paulo), *44*(1–2), 27–33.

González Broche, R., & Rodríguez Rodríguez, J. (2001). Nuevo subgénero de *Culex*, descripción de la pupa y de la larva y redescripción de la hembra de *Culex nicaroensis* Duret (Díptera: Culicidae). *Avicennia, 14*, 65–74.

González-Rincones, R., & Guyon, L. (1953). *Clasificacion general de los Dipteros*. Caracas, Venezuela: Universidad Central de Venezuela, Publicaciones del Instituto de Medicina Experimental "Jose Gregorio Hernandez."

Gordeyev, M.I., Yezhov, M.N., Zvanstov, A.B., Goryacheva, I.I., Shaikevich, Y.V., Karimov, S.S., & Kadamov, D.S. (2004). Amendment to the list of *Anopheles* (Diptera: Culicidae) mosquitoes of Tadjikistan and the predominant types of vectors in the current foci of malaria in the Republic. *Meditsinskaia parazitologiia i parizarnye bolezni, 3*, 16–21.

Gordeyev, M.I., Zvantsov, A.B., Goryacheva, I.I., Shaĭkevich, Y.V., & Ezhov, M.N. (2005). Description of the new species *Anopheles artemievi* sp.n. (Diptera, Culicidae). *Meditsinskaya parazitologiya i parazitarnye bolezni*, 2005(2), 4–5.

Gordeyev, M.I., Bezzhonova, O.V., Goriacheva, I.I., Shaĭkevich, E.V., Zvantsov, A.B., Mamedov, S., Mutdalibov, N., Gasymov, E., & Ezhov, M.N. (2010). Molecular genetic analysis of malaria mosquitoes of the *Anopheles maculipennis* (Diptera, Culicidae) complex in Azerbaijan. *Meditsinskaia parazitologiia i parizarnye bolezni, 4*, 43–45.

Gordon, R.M., & Evans, A.M. (1922). Mosquitoes collected in the Manáos Region of the Amazon. *Annals of Tropical Medicine and Parasitology, 16*(3), 315–338.

Gorham, J.R., Stojanovich, C.J., & Scott, H.G. (1973). Illustrated key to the anopheline mosquitoes of western South America. *Mosquito Systematics, 5*(2), 97–156.

Gornostaeva, R.M. (2000). A revised checklist of the mosquitoes (Diptera, Culicidae) of European Russia [English trans.]. *Parazitologiya* (St. Petersburg), *34*(6), 477–485.

Gornostaeva, R.M. (2003). The status of taxa of subgenus *Aedes* (Diptera, Culicidae, *Aedes*): *esoensis* Yamada, *rossicus* Dotheskin, Gorickaja & Mitrofanova, *cinereus* Meigen, *geminus* Peus. I. Overview. *European Mosquito Bulletin, 16*, 22–26.

Gornostaeva, R.M. (2005). To the revision of the mosquitoes of the subgenus *Aedes* (Diptera, Culicidae) from Palearctic. *Parazitologiya* (St. Petersburg), *39*(6), 457–507.

Gouagna, L.C., Poueme, R.S., Dabiré, K.R., Ouédraogo, J.B., Fontenille, D., & Simard, F. (2010). Patterns of sugar feeding and host plant preferences in adult males of *An. gambiae* (Diptera: Culicidae). *Journal of Vector Ecology, 35*(2), 267–276.

Gough, L.H. (1910). On a collection of *Anopheles* made at Onderstepoort in the autumn of 1909. *Report of the Government Veterinary Bacteriologist, Union of South Africa 1908–09*, 115–121.

Gould, S.J. (1988). On replacing the idea of progress with an operational notion of directionality. In M.H. Nitecki (Ed.), *Evolutionary progress* (pp. 319–338). Chicago, IL: University of Chicago Press.

Gould, S.J. (2001). *The book of life* (p. 1). New York, NY: Norton.

Gould, S.J., & Eldredge, N. (1993). Punctuated equilibrium comes of age. *Nature, 366*(6452), 223–227.

Gould, S.J., & Vrba, E.S. (1982). Exaptation—a missing term in the science of form. *Paleobiology, 8*(1), 4–15.

Grabham, M. (1905). Notes on some Jamaican Culicidae. *The Canadian Entomologist, 37*(12), 401–411.

Grabham, M. (1906a). Notes on some mosquitoes from Newcastle, Jamaica. *The Canadian Entomologist, 38*(5), 167–173.

Grabham, M. (1906b). Four new Culicidae from Jamaica, West Indies. *The Canadian Entomologist, 38*(11), 311–320.
Grabham, M. (1907). Notes on some new mosquitoes from Jamaica, West Indies. *The Canadian Entomologist, 39*(1), 25–26.
Graham, D.H. (1929). Mosquitoes of the Auckland District. *Transactions of the Proceedings of the Royal Society of New Zealand, 60*, 205–244.
Graham, D.H., Holmes, J.L., & Black, W.C., IV. (2004). Identification of quantitative trait loci affecting sex determination in the eastern treehole mosquito (*Ochlerotatus triseriatus*). *Journal of Heredity, 95*(1), 35–45.
Graham, W.M. (1909a). Four new species of the genus *Eretmapodites* (Theobald) from Ashanti. *The Entomologist, 42*, 86–89.
Graham, W.M. (1909b). Four new species of the genus *Eretmapodites* Theobald from Ashanti. *The Entomologist, 42*, 157–169.
Graham, W.M. (1910). On new species of west African Culicidae. *Annals and Magazine of Natural History, 5*(8), 264–273.
Grassi, B. (1898). Rapporti tra la malaria e peculiari insetti (zanzaroni e zanzare palustri). *Reale Accademia dei Lincei, 7*, 163–177.
Grassi, B. (1899). Ancora sulla malaria. *Rendiconti delle seduti delle Reale Accademia dei Lincei. Classe de scienze fisiche, matematishche e naturali*, 8, 559–561.
Grassi, B. (1900). Studi di uno zoologo sulla malaria. Memoria del Socio Battista Grassi. *Reale Accademia dei Lincei. Rendiconti. Classe de scienze fisiche, matematische e naturali. Memoirs, 5*(3), 299–516.
Grassi, B., Bignami, A., & Bastianelli, G. (1899). Resoconto degli studi fatti sulla malaria durante il mese di gennaio. *Reale Accademia dei Lincei. Rendiconti. Memorie della classe de scienze fisiche, matematische e naturali. Serie quinta, 8*, 100–104.
Gray, J. (1953). Undulatory propulsion. *Journal of Cell Science, 3*(28), 551–578.
Gray, K.M., Burkett-Cadena, N.D., Eubanks, M.D., & Unnasch, T.R. (2011). Crepuscular flight activity of *Culex erraticus* (Diptera: Culicidae). *Journal of Medical Entomology, 48*(2), 167–172.
Green, C.A. (1971). The practical problem of identifying members of the *Anopheles gambiae* complex in autecological studies. *Parassitologia, 13*, 421–427.
Green, C.A. (1972). *The* Anopheles annulipes *complex of species*. Paper presented at the 14th International Congress of Entomology, Canberra, Australia.
Green, C.A. (1982). Polytene-chromosome relationships of the *Anopheles stephensi* species group from the Afrotropical and Oriental regions (Culicidae, *Anopheles* (*Cellia*), series Neocellia). In W.W.M. Steiner, W.J. Tabachnick, K.S. Rai, & S. Narang (Eds.), *Recent developments in the genetics of insect disease vectors* (pp. 49–61). Stipes.
Green, C.A. (1995). Heterogeneity in distribution of fixed chromosomal inversions in ancestor and descendant species in two groups of mosquitoes: Series Myzomyia and Neocellia of *Anopheles* (*Cellia*). In D.M. Lambert & H.G. Spencer (Eds.), *Speciation and the recognition concept: Theory and application* (pp. 225–237). Baltimore, MD: Johns Hopkins University Press.
Green, C.A., & Hunt, R.H. (1980). Interpretation of variation in ovarian polytene chromosomes of *Anopheles funestus* Giles, *A. parensis* Gillies, and *A. aruni*? *Genetica* (Dordrecht), *51*(3), 187–195.
Green, C.A., Baimai, V., Harrison, B.A., & Andre, R.G. (1985a). Cytogenetic evidence for a complex of species within the taxon *Anopheles maculatus* (Diptera, Culicidae). *Biological Journal of the Linnean Society, 24*(4), 321–328.
Green, C.A., Harrison, B.A., Klein, T., & Baimai, V. (1985b). Cladistic analysis of polytene chromosome rearrangements in anopheline mosquitoes, subgenus *Cellia*, series Neocellia. *Canadian Journal of Genetics and Cytology, 27*(2), 123–133.
Green, C.A., Cass, R.F., Munstermann, L.E., & Baimai, V. (1990). Population-genetic evidence for two species in *Anopheles minimus* in Thailand. *Medical and Veterinary Entomology, 4*(1), 25–34.
Green, C.A., Rattanarithikul, R., & Charoensub, A. (1992). Population genetic confirmation of species status of the malaria vectors *Anopheles willmori* and *Anopheles pseudowillmori* in Thailand and chromosome phylogeny of the Maculatus group of mosquitoes. *Medical and Veterinary Entomology, 6*(4), 335–341.
Green, E.E. (1905). On *Toxorhynchites immisericors* (Walker), the elephant mosquito. *Spolia Zeylanica (Ceylon Journal of Science B), 2*(8), 159–164.
Greuter, W., Garrity, G., Hawksworth, D.L., Jahn, R., Kirk, P. M., Knapp, S., . . . Tindall, B.J. (2011). Draft BioCode (2011): Principles and rules regulating the naming of organisms. *Taxon, 60*(1), 201–212.
Griffin, D.R., Webster, F.A., & Michael, C.R. (1960). The echolocation of flying insects by bats. *Animal Behavior, 8*(3–4), 141–154.
Griffith, E.E., & Pidgeon E. (1832). Supplement on the order Diptera. In G. Cuvier, E. Griffith, E. Pidgeon, & G.R. Gray (Ed.), *The animal kingdom: Arranged in conformity with its organization: Vols. 14–15. The class Insecta* (pp. 721–780). London, England: Whittaker, Treacher.
Griffith, J.S., & Turner, G.D. (1996). Culturing *Culex quinquefasciatus* mosquitoes with a blood substitute diet for females. *Medical and Veterinary Entomology, 10*(3), 265–268.
Grigoraki, L., Lagnel, J., Kioulos, I., Kampouraki, A., Morou, E., Labbé, P., . . . Vontas, J. (2015). Transcriptome profiling and genetic study reveal amplified carboxylesterase genes implicated in temephos resistance, in the Asian Tiger Mosquito *Aedes albopictus*. *PLOS Neglected Tropical Diseases, 9*(5), e0003771.
Grillet, M.E., Montañez, H., & Berti, J. (1998). Estudio biosistemático y ecológico de *Anopheles aquasalis* y sus implicaciones para el control de la malaria en el estado Sucre: 2—ecológia de sus criaderos. *Boletín de la Dirección de Malariología y Saneamiento Ambiental, 38*(1), 38–46.
Grimaldi, D.A., Engel, M.S., & Nascimbene, P.C. (2002). Fossiliferous Cretaceous amber from Myanmar (Burma): Its rediscovery, biotic diversity, and paleontological significance. In *American Museum Novitates No. 3361* (pp. 1–71). New York, NY: American Museum of Natural History.
Grimstad, P.R., Garry, C.E., & Defoliart, G.R. (1974). *Aedes hendersoni* and *Aedes triseriatus* (Diptera: Culicidae) in Wisconsin: Characterization of larvae, larval hybrids, and comparison of adult and hybrid mesoscutal patterns. *Annals of the Entomological Society of America, 67*(5), 795–804.
Grjebine, A. (1950). Un nouveau moustique forestier de la basse Côte d'Ivoire *Eretmapodites pauliani* n. sp. *Bulletin de la Société de pathologie exotique, 43*(1–2), 45–50.
Grjebine, A. (1953a). Observations sur les nematoceres vulnerants de Madagascar regions de Majunga et de la Mandraka. *Mémoires de l'Institut Scientifique de Madagascar E, 4*, 443–502.
Grjebine, A. (1953b). Une nouvelle espece d'anophele de Madagascar [Dipt. Culicidae]. *Naturaliste Malgache, 5*(2), 203–209.
Grjebine, A. (1961). Un nouvel anophèle de Madagascar. *Bulletin de la Société de pathologie exotique, 53*(for 1960)(6), 939–942.
Grjebine, A. (1964). *Anopheles* (*Neomyzomyia*) *grenieri* n. sp. (Diptera, Culicidae), nouvel anophèle de Madagascar. *Bulletin de la Société de pathologie exotique, 57*(1), 38–42.
Grjebine, A. (1966). *Faune de Madagascar* (Vol. 22; Insectes; Diptères, Culicidae, Anophelinae). Paris, France: Lahure.

Grjebine, A. (1972). Une nouvelle espèce de moustique du Congo-Brazzaville: *Eretmapodites hamoni* n. sp. (Dipt. Culicidae). *Bulletin de la Société Entomologique de France, 77*(9–10), 308–313.
Grjebine, A. (1979). Les moustiques des *Nepenthes* de Madagascar. Espèces nouvelles du genre *Uranotaenia*. (Diptera, Culicidae). *Annales de la Société Entomologique de France, 15*, 53–74.
Grjebine, A. (1986). Faune de Madagascar. 68. Insectes Diptères Culicidae Culicinae Ficalbiini. Paris, France: Muséum nationale dHistoire naturell. *Faune de Madagascar, 68*, 1–441.
Grjebine, A., & Chauvet, G. (1961). Position systématique d'*Anopheles mascarensis* de Meillon, 1947 et description de la larvae et de la nymphe. *Bulletin de la Société de pathologie exotique, 54*, 119–128.
Grjebine, A., & Lacan, A. (1953). *Anopheles (Myzomia) milloti* n. sp. In *Observations sur les Nématocères vulnérants de Madagascar Régions de Majunga et de la Mandraka*. (Vol. 4, pp. 495–501): Mémoires de l'Institut Scientifique de Madagascar Series E.
Grossbeck, J.A. (1904a). Description of a new *Culex*. *The Canadian Entomologist, 36*(11), 332.
Grossbeck, J.A. (1904b). Description of two new species of *Culex*. *Entomological News, 15*(10), 332–333.
Grossbeck, J.A. (1905). New species of Culicidae. *The Canadian Entomologist, 37*(10), 359–360.
Grossbeck, J.A. (1906). Notes on *Culex squamiger*, Coq., with description of a closely-allied species. *The Canadian Entomologist, 38*(4), 129–131.
Grothaus, R.H., Floore, T.G., Stasiak, R.S., & Miner, W.F. (1971). *A partial list of the mosquitoes of I Corps, Republic of Vietnam, with notes on bionomics* (Publication No. NMFRL-Vol-XXI/No-16). Camp Lejeune, NC: Naval Medical Field Research Lab Camp
Grünberg, K. (1902). Ein neuer *Anopheles* aus Westafrika, *Anopheles Ziemanni* nov. spec. *Zoologischer Anzeiger, 25*(677), 550–551.
Grünberg, K. (1905). Zur Kenntnis der Culicidenfauna von Kamerun und Togo. *Zoologischer Anzeiger, 29*(12), 377–390.
Grünberg, K. (1907a). Ein neuer *Toxorhynchites* aus Kamerun. (Dipt.). *Deutsche Entomologische Zeitschrift, 1907*, 405–406.
Grünberg, K. (1907b). *Die blutsaugenden Dipteren. Leitfaden zur allgemeinen Orientierung, mit besonderer Berucksichtigung der in den Deutschen Kolonien lebenden Krankheitsubertrager*. Jena, Germany: Gustav Fischer.
Grünberg, K. (1913). Zur Kenntnis der Culiciden-fFauna von Samoa. *Entomologische Rundschau, 30*(22), 130–131.
Gu, W., & Novak, R.J. (2009). Agent-based modelling of mosquito foraging behaviour for malaria control. *Transactions of the Royal Society for Tropical Medicine and Hygiene, 103*(11), 1105–1112.
Gu, W., Regens, J.L., Beier, J.C., & Novak, R.J. (2006). Source reduction of mosquito larval habitats has unexpected consequences on malaria transmission. *Proceedings of the National Academy of Sciences of the United States of America, 103*(46), 17560–17563.
Gu, W., Müller, G., Schlein, Y., Novak, R.J., & Beier, J.C. (2011). Natural plant sugar sources of *Anopheles* mosquitoes strongly impact malaria transmission potential. *PLOS ONE, 6*(1), e15996.
Gubler, D.J. (1970). Comparison of reproductive potentials of *Aedes* (*Stegomyid*) *albopictus* Skuse and *Aedes* (*Stegomyia*) *polynesiensis*. Marks. *Mosquito News, 30*(2), 201–209.
Gubler, D.J., Reiter, P., Ebi, K.L., Yap, W., Nasci, R., & Patz, J.A. (2001). Climate variability and change in the United States: Potential impacts on vector- and rodent-borne diseases. *Environmental Health Perspectives, 109*(Suppl. 2), 223–233.
Guedes, A.S., & Souza, M.A. (1964). Sôbre *Psorophora* (*Janthinosoma*) *albigenu* Lutz, 1908 e *Psorophora* (*Janthinosoma*) *albipes* (Theobald, 1907) (Diptera, Culicidae). *Revista Brasileira de malariologia e doenças tropicais* (Rio de Janeiro), *16*(4), 471–486.
Guedes, A.S., Souza, M.A., Maciel, C.S., & Xavier, S.H. (1965). Catálogo illustrado dos mosquitoes da coleção do Instituto Nacional de Endemias Rurais. I. Gênero *Psorophora* Robineau-Desvoidy, 1827. *Revista Brasileira de malariologia e doenças tropicais* (Rio de Janeiro), *17*, 3–24.
Guenther, K. (1913). Die lebenden Bewohner der Kannen der insektenfressenden Pflanze *Nepethes destillatoria* auf Ceylon. *Zeitschrift für wissenschaftliche Insektenbiologie Berlin, 9*(6–7), 259–270.
Gulia-Nuss, M., Robertson, A.E., Brown, M.R., & Strand, M.R. (2011). Insulin-like peptides and the target of rapamycin pathway coordinately regulate blood digestion and egg maturation in the mosquito *Aedes aegypti*. *PLOS ONE, 6*(5), e20401.
Gunasekaran, K. (1994). Age composition, natural survival and population growth of *Anopheles fluviatilis* James, 1902, the major malaria vector in the endemic belt of Koraput District, Orissa, India. *Southeast Asian Journal of Tropical Medicine and Public Health, 25*(1), 196–200.
Gunathilaka, N. (2017). Illustrated key to the adult female *Anopheles* (Diptera: Culicidae) mosquitoes of Sri Lanka. *Applied Entomology and Zoology, 52*(1), 69–77.
Gunathilaka, N., Fernando, T., Hapugoda, M., Abeyewickreme, W., & Wickremasinghe, R. (2014). Revised morphological identification key to the larval anopheline (Diptera: Culicidae) of Sri Lanka. *Asian Pacific Journal of Tropical Biomedicine, 4*(Suppl. 1), S222–S227.
Gunathilaka, N., Hapugoda, M., Abeyewickreme, W., & Wickremasinghe, R. (2015). Appearance of *Anopheles Jeyporiensis* James from Sri Lanka; a new species to the mosquito checklist. *Medical Entomology and Zoology, 66*(3), 1–5.
Gunay, F., Alten, B., Simsek, F., Aldemir, A., & Linton, Y.-M. (2015). Barcoding Turkish *Culex* mosquitoes to facilitate arbovirus vector, incrimination studies reveals hidden diversity and new potential vectors. *Acta Tropica, 143*, 112–120.
Gupta, S., & Chaudhry, S. (2005). SEM study on the egg surface architecture of *Anopheles* (*Cellia*) *splendidus* Koidzumi (Culicidae: Diptera). *Polskie pismo entomologiczne, 74*(1), 7–14.
Gusmão, D.S., Santos, A.V., Marini, D.C., Russo, É.D.S., Peixoto, A.M.D., Bacci Júnior, M., . . . Lemos, F.J.A. (2007). First isolation of microorganisms from the gut diverticulum of *Aedes aegypti* (Diptera: Culicidae): New perspectives for an insect-bacteria association. *Memórias do Instituto Oswaldo Cruz, 102*(8), 919–924.
Gutiérrez, L.A., Orrego, L.M., Gomez, G.F., Lopez, A., Luckhart, S., Conn, J.E., & Correa, M.M. (2010). A new mtDNA COI gene lineage closely related to *Anopheles janconnae* of the Albitarsis complex in the Caribbean region of Colombia. *Memórias do Instituto Oswaldo Cruz, 105*(8), 1019–1025.
Gutsevich, A.V. (1955). New and little known forms of mosquitoes (Diptera, Culicidae). *Trudy Zoologicheskogo Instituta Akademii Nauk, 18*, 320–324.
Gutsevich, A.V. (1962). A new species of bloodsucking mosquito of the genus *Aedes* (Diptera, Culicidae) from Kazakhstan. *Entomologicseszkoje Obozrenie, 41*(4), 886–888.
Gutsevich, A.V. (1966). A new mosquito species from Uzbekistan, *Aedes* (*Ochlerotatus*) *simanini* (Culicidae). *Zoologicheskii zhurnal, 45*(3), 457–459.
Gutsevich, A.V. (1973a). Determination of mosquito females by microscopic preparations of the head. II. A key to genera and subgenera. *Parazitologiya* (St. Petersburg), *7*, 106–111.
Gutsevich, A.V. (1973b). Determination of mosquito females (Culicidae) by microscopic preparations of the head. III. A key to species (excluding *Aedes*). *Parazitologiya* (St. Petersburg), *7*(5), 443–449.

Gutsevich, A.V. (1974). On the identification of mosquito females (Culicidae) by microscopic preparations of the head. IV. Key to species of the genus *Aedes. Parazitologiya* (St. Petersburg), *8*(4), 329–335.

Gutsevich, A.V. (1975). Structural features of the female head as systematic characters in the family Culicidae (Diptera). *Entomological Review, 54*(4), 113–114.

Gutsevich, A.V. (1976). On polytypical species of mosquitoes (Diptera, Culicidae). I. *Anopheles hyrcanus* (Pallas, 1771). *Parazitologiya* (St. Petersburg), *10*(2), 148–153.

Gutsevich, A.V., & García-Avila, I. (1969). Nuevas especies de mosquitos (Culicidae) para Cuba. *Torreia, 13*, 3–8.

Gutsevich, A.V., & Goritskaya, V.V. (1970). Occurrence of the mosquito *Aedes thibaulti* (Diptera, Culicidae) in the Soviet Union. *Parazitologiya* (St. Petersburg), *4*(1), 72–73.

Gutsevich, A.V., & Pendikova, E.G. (1972a). The bloodsucking mosquito *Aedes* (*Ochlerotatus*) *albescens* Edw. (Diptera, Culicidae) as a distinct species. *Revue d'entomologie de l'URSS Entomologicseszkoje Obozrenie, 51*, 401–402.

Gutsevich, A.V., & Pendikova, E.G. (1972b). The bloodsucking mosquitoes *Aedes* (*Ochlerotatus*) *albescens* Edw. (Diptera, Culicidae) as a species in its own right. *Revue d'entomologie de l'URSS Entomologicseszkoje Obozrenie, 51*, 241–242.

Gutsevich, A.V., Monchadsky, A.S., & Stackelberg, A.A. (1971). Mosquitoes, Family Culicidae. In *Fauna of the USSR. Diptera.* New series No. 100, 3(4). Leningrad, USSR: Zoologicheskogo Insitituta Akademii Nauk SSSR.

Gutsevich, A.V., Monchadskii, A.S., & Shtakel'berg, A.A. (1974). Mosquitoes, Family Culicidae. In *Fauna of the USSR. Diptera.* New series No. 100, 3(4). Jerusalem, Israel: Keter Publishing House. (Original work published 1971).

Guy, Y. (1959). Les *Anophèles* du Maroc. *Mémoires de la Société des Sciences Naturelles du Maroc Rabat* (n.s.), *7*, 235.

Guy, Y., & Holstein, M. (1968). Données récentes sur les *Anophèles* du Maghreb. *Archives de l'Institut Pasteur d'Algérie, 46*, 146–150.

Guy, Y., Salières, A., & Boesiger, E. (1976). Contribution a l'étude du complexe *maculipennis* (Diptera-Culicidae-Anophelinae). Mise au point en 1975. *Année biologique, 15*(5–6), 227–282.

Ha, Y.R., Lee, S.C., Seo, S.J., Ryu, J., Lee, D.K., & Lee, S.J. (2015). Comparison of the functional features of the pump organs of *Anopheles sinensis* and *Aedes togoi. Scientific Reports, 5*, 15148.

Habirov, Z., Kadamov, D., Iskandarov, F., Komilova, S., Cook, S., McAlister, E., & Harbach, R.E. (2012). Malaria and the *Anopheles* mosquitoes of Tajikistan. *Journal of Vector Ecology, 37*(2), 419–427.

Hacker, H.P. (1921). Changes in nomenclature of species. *Report of the Malaria Bureau, Federated Malay States, 2*, 1.

Hackett, L.W. (1934). The present status of our knowledge of the subspecies of *Anopheles maculipennis. Transactions of the Royal Society of Tropical Medicine and Hygiene, 28*(2), 109–128.

Hackett, L.W., & Lewis, D.J. (1935). A new variety of *Anopheles maculipennis* in southern Europe. *Rivista di malariologia, 14*(5), 377–383.

Hackett, L.W., & Missiroli, A. (1935). The varieties of *Anopheles maculipennis* and their relation to the distribution of malaria in Europe. *Rivista di malariologia, 14*(1), 45–109.

Haddow, A.D., Moulton, J.K., Gerhardt, R.R., McCuiston, L.J., & Jones, C.J. (2009). Description of the egg of *Ochlerotatus japonicus japonicus* (Diptera: Culicidae) using variable pressure scanning electron microscopy. *Journal of Medical Entomology, 46*(1), 9–14.

Haddow, A.J. (1946). The mosquitoes of Bwamba County, Uganda. IV. Studies on the genus *Eretmapodites*, Theobald. *Bulletin of Entomological Research, 37*(1), 57–82.

Haddow, A.J., & van Someren, E.C.C. (1950). A new species of *Stegomyia* Theobald from the Ruwenzori Range, Uganda. *Annals of Tropical Medicine and Parasitology, 44*(3), 281–284.

Haddow, A.J., van Someren, E.C.C., Lumsden, W.H.R., Harper, J.O., & Gillett, J.D. (1951). The mosquitoes of Bwamba County, Uganda. VIII. Records of occurrence, behaviour and habitat. *Bulletin of Entomological Research, 42*(2), 207–238.

Haddow, A.J., Casley, D.J.L., O'Sullivan, J.P., Ardoin, P.M.L., Ssenkubuge, Y., & Kitama, A. (1968). Entomological studies from a high steel tower in Zika Forest, Uganda. Part II. The biting activity of mosquitoes above the forest canopy in the hours after sunset. *Transactions of the Royal Entomological Society of London, 120*(9), 219–239.

Hadjinicolau, J. (1938). Observations on *Anopheles marteri* S.& P. (Diptera, Culicidae). *Rivista di malariologia, 17*, 44–50.

Hadjivassilis, A. (2000). Mosquitoes (Diptera: Culicidae) of Cyprus. *European Mosquito Bulletin, 7*, 38.

Haeger, J.S., & Provost, M.W. (1965). Colonization and biology of *Opifex fuscus. Transactions of the Royal Society of New Zealand, 6*(3), 21–31.

Haga, J. (1925). A new *Uranotaenia* (Culicidae, Diptera) from Alor, Lesser Sunda Islands. *Bulletin of Entomological Research, 16*(1), 43–44.

Haga, J. (1930). Tabellen voor determinatie der in Nederländsch-Oost-Indië voorkomende Anophelinen. *Geneeskundig tijdschrift voor Nederlandsch-Indië, 70*(for 1929), 363–382.

Halcrow, J.G. (1954). Catalogue of the mosquitoes of Mauritius and Rodrigues. *Bulletin of the Mauritius Institute, 3*, 234–248.

Halcrow, J.G. (1957). A new sub-species of *Anopheles gambiae* Giles from Mauritius. *East African Medical Journal, 34*(4), 133–135.

Hales, S., de Wet, N., Maindonald, J., & Woodward, A. (2002). Potential effect of population and climate on global distribution of dengue fever: An empirical model. *Lancet, 360*(9336), 830–834.

Halgoš, J., & Benková, I. (2004). First record of *Anopheles hyrcanus* (Diptera: Culicidae) from Slovakia. *Biologia Bratislava, 59*(Suppl. 15), 68.

Haliday, A.H. (1833). Catalogue of Diptera occurring about Holywood in Downshire. *Entomological Magazine, 1*, 147–180.

Hall, A.B., Basu, S., Jiang, X., Qi, Y., Timoshevskiy, V.A., Biedler, J.K., . . . Sharakhov, I.V. (2015). A male-determining factor in the mosquito *Aedes aegypti. Science, 348*(6240), 1268–1270.

Hall, C.R., Howard, T.M., & Harbach, R.E. (1999). *Sabethes* (*Peytonulus*) *luxodens*, a new species of Sabethini (Diptera: Culicidae) from Ecuador. *Memórias do Instituto Oswaldo Cruz, 94*(3), 329–338.

Hallisey, E., Brantley, D., Elmore, K., Holt, J., Lash, R., Murphy, A., & Young, R. (2012). *Cartographic guidelines for public health.* Geography and Geospatial Science Working Group, Centers for Disease Control and Prevention. Retrieved from www.cdc.gov/dhdsp/maps/GISX/resources/Cartographic_Guidelines.pdf

Hamlyn-Harris, R. (1933). Some ecological factors involved in the dispersal of mosquitos in Queensland. *Bulletin of Entomological Research, 24*(2), 229–232.

Hamon, J. (1953). Une nouvelle espècie d'*Aedes* de l'île de la Réunion. *Naturaliste Malgache, 5*, 35–41.

Hamon, J. (1954a). Contribution a l'étude des Culicidés de la région de Bobo Dioulasso (Haute-Volta). *Annales de parasitologie humaine et comparée, 29*(5–6), 573–587.

Hamon, J. (1954b). Contribution à l'étude des Culicidés de la région de Porto-Novo (Bas-Dahomey). *Annales de parasitologie humaine et comparée, 29*(5–6), 588–594.

Hamon, J. (1954c). Contribution a l'étude des Culicidés d'Afrique Occidentale. Description de *Ficalbia* (*Ficalbia*) *grenieri* sp. n. *Bulletin de la Société de pathologie exotique, 47*(6), 835–839.

Hamon, J. (1955). Contribution a l'étude des Culicidés d'Afrique Occidentale. Description de *Uranotaenia devemyi* sp. n., *Culex grahami* var. *farakoensis* var. n. et de la larve de *Harpagomyia trichorostris* Theobald. *Bulletin de la Société de pathologie exotique, 47*(for 1954)(5), 672–678.

Hamon, J. (1957). Deux moustiques nouveaux de Côte-d'Ivoire: *Aedes* (*Aedimorphus*) *falabrequesi* sp. n., *Culex* (*Culex*) *castelli* sp. n. *Bulletin de la Société de pathologie exotique, 50*(5), 685–691.

Hamon, J. (1961). Les *Eretmapodites* du groupe *Oedipodius* Graham. 1. Descriptions complémentaires et position systématique des six souse-espèces actuellement connues. *Bulletin de la Société de pathologie exotique, 54*(4), 892–906.

Hamon, J. (1962). Description de deux nouveaux *Eretmapodites* (Diptera, Culicidae) capturés en Uganda: *E. vansomereni* sp. n. et *E. corbeti* sp. n. *Bulletin de la Société de pathologie exotique, 55*(2), 253–259.

Hamon, J., & Adam, J.P. (1955). Contribution a l'étude des culicidés (Diptères) d'Afrique. Description de *Harpagomyia moucheti* sp. n. *Bulletin de la Société de pathologie exotique, 48*(6), 882–885.

Hamon, J., & Adam, J.P. (1959a). Description complémentaire d'*Eretmapodites pauliani* Grjebine, 1950 (Diptera, Culicidae). *Bulletin de la Société de pathologie exotique, 51*(for 1958)(4), 527–529.

Hamon, J., & Adam, J.P. (1959b). Description de deux nouveaux *Aedes* appartenant au groupe *tarsalis* Newstead en basse Côte-d'Ivoire (A.O.F.): *A.* (*Aedimorphus*) *reali* sp. n. et *A.* (*Aedimorphus*) *nyounae* sp. n. *Bulletin de la Société de pathologie exotique, 51*(for 1958)(1), 530–534.

Hamon, J., & Adam, J.P. (1959c). Description de deux nouveaux *Aedes* de Côte-d'Ivoire appartenant au sous-genre *Aedimorphus*: *A. ovazzai* sp. n. et *A. rickenbachi* sp. n. *Bulletin de la Société de pathologie exotique, 52*(2), 147–154.

Hamon, J., & Berre, R.L. (1961). Description de la larve et de la nymphe de *Aedes* (*Aedimorphus*) *mattinglyi* Hamon et Richenbach 1954. *Bulletin de la Société de pathologie exotique, 54*(4), 718–723.

Hamon, J., & Brengues, J. (1965). Observations sur les *Aedes* (*Aedimorphus*) d'Afrique avec description de deux nouvelles espèces: *Ae. lottei* n. sp. et *Ae. dialloi* n. sp. *Bulletin de la Société de pathologie exotique, 58*(1), 101–108.

Hamon, J., & Gandara, A. (1955). Contribution a l'étude des culícidés (Diptères) de la région Éthiopienne. Description de *Culex* (*Culiciomyia*) *cambournaci* n. sp. *Bulletin de la Société de pathologie exotique, 48*(6), 866–872.

Hamon, J., & Lambrecht, F.L. (1959). Description de *Culex* (*Culiciomyia*) *muspratti* n. sp. accompagnée de quelques notes sur les moustiques de la Forêt d'Irangi, Kivu, Congo Belge. *Bulletin de la Société de pathologie exotique, 52*(5), 582–587.

Hamon, J., & Mouchet, J. (1955). Contribution a l'étude des *Neoculex* (Diptères: Culicides) de la région éthiopienne. 2. Description de *Neoculex adami* sp. n. *Bulletin de la Société de pathologie exotique, 48*(6), 860–862.

Hamon, J., & Ovazza, M. (1954). Une nouvelle espècie de moustique (*Culex shoae* n. sp.) vivant sur les bananiers du haut plateau d'Éthiopie. *Bulletin de la Société de pathologie exotique, 47*(3), 416–421.

Hamon, J., & Ovazza, M. (1956). Contribution a l'étude des culicidés d'Afrique. Observations sur le groupe *Culex annulioris* Theobald, 1901. *Bulletin de la Société de pathologie exotique, 49*(1), 89–99.

Hamon, J., & Rickenbach, A. (1954). Contribution a l'étude des culicidés d'Afrique Occidentale. Description d'*Aedes* (*Aedimorphus*) *mattinglyi* sp. n., *Aedes* (*Banksinella*) *jamoti* sp. n. Notes complémentaires sur *Aedes* (*Aedimorphus*) *stokesi* Evans, *Aedes* (*Banksinella*) *bolensis* Edwards. *Bulletin de la Société de pathologie exotique, 47*(6), 930–941.

Hamon, J., & Rickenbach, A. (1955a). Contribution a l'étude des culicidés d'Afrique Occidentale. Description d'*Anopheles brumpti* sp. n. *Bulletin de la Société de pathologie exotique, 48*(3), 342–344.

Hamon, J., & Rickenbach, A. (1955b). Contribution a l'étude des *Neoculex* (Diptères: Culicides) de la Région éthiopienne. 1. Corrections de quelques descriptions de terminalia mâles données par Edwards, avec étude d'une nouvelle variété. *Bulletin de la Société de pathologie exotique, 48*(6), 848–859.

Hamon, J., & van Someren, E.C.C. (1961a). Description de deux nouveaux moustiques du Tanganyika: *Culex* (*Culiciomyia*) *gilliesi* sp. n. et *Eretmapodites mattinglyi* sp. n. *Bulletin de la Société de pathologie exotique, 54*(4), 724–731.

Hamon, J., & van Someren, E.C.C. (1961b). Les *Eretmapodites* du groupe Oedipodius Graham. II. Description de *E. grenieri* sp. n. et clé de détermination des espèces et sous-espècies du groupe *oedipodius*. *Bulletin de la Société de pathologie exotique, 54*(4), 907–913.

Hamon, J., Abonnenc, E., & Noel, E. (1955a). Contribution à l'étude des culicidés de l'ouest du Sénégal. *Annales de parasitologie humaine et comparée, 30*(3), 278–308.

Hamon, J., Adam, J.P., & Mouchet, J. (1955b). Contribution a l'étude des *Neoculex* (Diptères: Culicidés) de la région éthiopienne. 3. Description de *Neoculex laplantei* sp. n. *Bulletin de la Société de pathologie exotique, 48*(6), 862–865.

Hamon, J., Holstein, M., & Rivola, E. (1957). Description d'un nouveau moustique du Congo Belge: *Culex* (*Neoculex*) *vinckei* sp. n. *Bulletin de la Société de pathologie exotique, 50*(5), 681–685.

Hamon, J., Taufflieb, R., & Maillot, L. (1957). Description d'*Aedes* (*Aedimorphus*) *grjebinei* sp. n. accompagnée de quelques données nouvelles sur les moustiques d'A.E.F. *Bulletin de la Société de pathologie exotique, 50*, 692–696.

Hamon, J., Service, M.W., Adam, J.P., & Taufflieb, R. (1961). Observations sur les *Aedes* (*Aedimorphus*) du groupe *tarsalis* Newstead avec description d'une nouvelle espèce: *Aedes grenieri* n. sp. *Bulletin de la Société de pathologie exotique, 54*(2), 375–388.

Hamon, J., Taufflieb, R., & Dyemkouma, A. (1961). Observations sur la variabilité d'*Anopheles rufipes*, Gough, 1910, avec description d'une nouvelle variété. *Bulletin de la Société de pathologie exotique, 54*(1), 24–28.

Hamon, J., Maffi, M., Grenier, P., Ouédraogo, C.S., & Djimé, D. (1966). Notes sur les moustiques de la République Islamique de Mauritanie (Diptera, Culicidae). (II partie). *Annales de la Société Entomologique de France, 11*(2), 371–383.

Han, Y.S., Thompson, J., Kafatos, F.C., & Barillas-Mury, C. (2000). Molecular interactions between *Anopheles stephensi* midgut cells and *Plasmodium berghei*: The time bomb theory of ookinete invasion of mosquitoes. *The European Molecular Biology Organisation Journal, 19*(22), 6030–6040.

Hancock, P.A., Wiebe, A., Gleave, K.A., Bhatt, S., Cameron, E., Trett, A., . . . Moyes, C.L. (2018). Associated patterns of insecticide resistance in field populations of malaria vectors across Africa. *Proceedings of the National Academy of Sciences of the United States of America, 115*(23), 5938–5943.

Handlirsch, A. (1924). Systematische Übersicht. In C. Schröder, *Handbuch der Entomologie Band III* (pp. 377–1140). Jena, Germany: Gustav Fischer.

Hanney, P.W. (1959). Variations in *Anopheles flavicosta* Edwards from northern Nigeria. *Proceedings of the Royal Entomological Society of London, Series B: Taxonomy, 28*(11–12), 169–174.

Hara, J. (1957). Studies on the female terminalia of Japanese mosquitoes. *Japanese Journal of Experimental Medicine, 27*, 45–91.

Hara, J. (1958). On the newly recorded mosquito, *Aedes* (*Aedes*) *rossicus* Dolbeshkin, Goritshkaya & Mitrofanova, 1930 with the keys to the species belonging subgenus *Aedes* known from Japan (Diptera: Culicidae). Part 10. Taxonomical and ecological studies on mosquitoes of Japan. *Japanese Journal of Sanitary Zoology, 9*(1), 23–27.

Hara, J. (1959). Taxonomical notes on the female terminalia of some anopheline mosquitoes of Japan and Formosa. Part 12. Taxonomic and ecological studies on mosquitoes of Japan. *Japanese Journal of Experimental Medicine, 29*(2), 107–119.

Harant, H., Rioux, J., & Uzac, S. (1952). Un culicide nouveau pour la France: *Uranotaenia unguiculata* Edw. Étude écologique. *Annales de parasitologie humaine et comparée, 27*(4), 407–418.

Harbach, R.E. (1977). Comparative and functional morphology of the mandibles of some fourth stage mosquito larvae (Diptera: Culicidae). *Zoomorphologie, 87*, 217–236.

Harbach, R.E. (1978). Comparative structure of the labiohypopharynx of fourth stage mosquito larvae (Diptera: Culicidae), with comments on larval morphology, evolution and feeding habits. *Mosquito Systematics, 10*(3), 301–333.

Harbach, R.E. (1985a). Pictorial keys to the genera of mosquitoes, subgenera of *Culex* and the species of *Culex* (*Culex*) occurring in southwestern Asia and Egypt, with a note on the subgeneric placement of *Culex deserticola* (Diptera: Culicidae). *Mosquito Systematics, 17*(2), 83–107.

Harbach, R.E. (1985b). A new species, *Culex* (*Culex*) *litwakae* (Diptera: Culicidae), from the coastal region of Kenya. *Mosquito Systematics, 17*(3), 254–265.

Harbach, R.E. (1987). *Culex verutus*, a new species of the subgenus *Culex* (Diptera: Culicidae) from Sierra Leone. *Mosquito Systematics, 18*(for 1986)(3–4), 216–229.

Harbach, R.E. (1988). Mosquitoes of the subgenus *Culex* in southwestern Asia and Egypt (Diptera: Culicidae). *Contributions of the American Entomological Institute, 24*(1), 1–240.

Harbach, R.E. (1991a). A new subgenus of the genus *Sabethes* (Diptera: Culicidae). *Mosquito Systematics, 23*(1), 1–9.

Harbach, R.E. (1991b). Neotype designation, generic realignment and description of *Dendromyia schnusei* Martini (Diptera: Culicidae). *Mosquito Systematics, 23*(2), 175–181.

Harbach, R.E. (1991c). Ontogeny of the larval stage of *Sabethes chloropterus*, with special reference to setal development and phylogenetic implications for the family Culicidae (Diptera). *Mosquito Systematics, 23*(1), 10–24.

Harbach, R.E. (1994a). Review of the internal classification of the genus *Anopheles* (Diptera: Culicidae): The foundation for comparative systematics and phylogenetic research. *Bulletin of Entomological Research, 84*, 331–342.

Harbach, R.E. (1994b). The subgenus *Sabethinus* of *Sabethes* (Diptera: Culcidae). *Systematic Entomology, 19*(3), 207–234.

Harbach, R.E. (1995a). A new *Sabethes* of the subgenus *Peytonulus* (Diptera: Culicidae). With an unusual fourth-instar larva. *Entomologica Scandinavica, 26*(1), 87–96.

Harbach, R.E. (1995b). Two new species of the subgenus *Peytonulus* of *Sabethes* (Diptera: Culicidae) from Colombia. *Memórias do Instituto Oswaldo Cruz, 90*(5), 583–587.

Harbach, R.E. (1998). *Culex* (*Barraudius*) *pusillus*, a new occurrence record outside the Palaearctic Region. *European Mosquito Bulletin, 1*, 14.

Harbach, R.E. (1999). The identity of *Culex perexiguus* Theobald versus *Cx. univittatus* Theobald in southern Europe. *European Mosquito Bulletin, 7*, 7.

Harbach, R.E. (2004). The classification of genus *Anopheles* (Diptera: Culicidae): A working hypothesis of phylogenetic relationships. *Bulletin of Entomological Research, 94*(6), 537–553.

Harbach, R.E. (2007). The Culicidae (Diptera): A review of taxonomy, classification and phylogeny. In Z.Q. Zhang & W.A. Shear (Eds.), *Linnaeus Tercentenary: Progress in Invertebrate Taxonomy. Zootaxa 1668* (pp. 591–688).

Harbach, R.E. (2010). What's New. Mosquito Taxonomic Inventory, http://mosquito-taxonomic-inventory.info/

Harbach, R.E. (2011). Classification within the cosmopolitan genus *Culex* (Diptera: Culicidae): The foundation for molecular systematics and phylogenetic research. *Acta Tropica, 120*(1–2), 1–14.

Harbach, R.E. (2012). *Culex pipiens*: Species versus species complex—taxonomic history and perspective. *Journal of the American Mosquito Control Association, 28* (Suppl. 4), 10–23.

Harbach, R.E. (2013a). *Nyctomyia*, a replacement name for the preoccupied genus *Nyx* Harbach & Linton, 2013 (Diptera: Culicidae). *Zootaxa, 3691*(1), 199.

Harbach, R.E. (2013b). The phylogeny and classification of *Anopheles*. In S. Manguin (Ed.), Anopheles *mosquitoes: New insights into malaria vectors* (pp. 3–55). Rijeka, Croatia: InTech.

Harbach, R.E. (2016). *Anopheles* classification. Retrieved from http://mosquito-taxonomic-inventory.info/node/11358

Harbach, R.E. (2018a). An *Anopheles* by any other name . . . ? *Journal of Medical Entomology, 55*(5), 1069–1070.

Harbach, R.E. (2018b). Two new species of the subgenus *Sabethinus* of *Sabethes* (Diptera: Culicidae) from Costa Rica, first confirmation of members of the subgenus in Central America. *Zootaxa, 4429*(2), 269–280.

Harbach, R.E. (2018c). *Culicipedia: Species-group, genus-group and family-group names in Culicidae (Diptera)*. Wallingford, United Kingdom: CABI.

Harbach, R.E., & Culverwell, C.L. (2014). A new subgenus and species of *Topomyia* (Diptera: Culicidae: Sabethini) based on a remarkable male mosquito from Sabah, Malaysia. *Zootaxa, 3794*(4), 575–580.

Harbach, R.E., & Greenwalt, D. (2012a). Two Eocene species of *Culiseta* (Diptera: Culicidae) from the Kishenehn Formation in Montana. *Zootaxa, 3530*(1), 25–34.

Harbach, R.E., & Greenwalt, D. (2012b). Erratum: HARBACH, RALPH E. & GREENWALT, DALE (2012) Two Eocene species of *Culiseta* (Diptera: Culicidae) from the Kishenehn Formation in Montana. *Zootaxa*, 3530, 25–34. *Zootaxa, 3554*(1).

Harbach, R.E., & Howard, T.M. (2002). *Sabethes* (*Peytonulus*) *paradoxus*, a new species of Sabethini (Diptera: Culicidae) from Panama. *Proceedings of the Entomological Society of Washington, 104*(2), 363–372.

Harbach, R.E., & Howard, T.M. (2007). Corrections in the status and rank of names used to denote varietal forms of mosquitoes (Diptera: Culicidae). *Zootaxa, 1542*, 35–48.

Harbach, R.E., & Howard, T.M. (2009). Review of the genus *Chagasia* (Diptera: Culicidae: Anophelinae). *Zootaxa, 2210*, 1–25.

Harbach, R.E., & Kitching, I.J. (1998). Phylogeny and classification of the Culicidae (Diptera). *Systematic Entomology, 23*(4), 327–370.

Harbach, R.E., & Kitching, I.J. (2005). Reconsideration of anopheline mosquito phylogeny (Diptera: Culicidae: Anophelinae) based on morphological data. *Systematics and Biodiversity, 3*(4), 345–374.

Harbach, R.E., & Kitching, I.J. (2016). The phylogeny of Anophelinae revisited: Inferences about the origin and classification of *Anopheles* (Diptera: Culicidae). *Zoologica Scripta, 45*(1), 34–47.

Harbach, R.E., & Knight, K.L. (1978). A mosquito taxonomic glossary. XV. The egg. *Mosquito Systematics, 10*(2), 249–297.

Harbach, R.E., & Knight, K.L. (1980). *Taxonomists' glossary of mosquito anatomy*. Marlton, NJ: Plexus.

Harbach, R.E., & Knight, K.L. (1981). Corrections and additions to *Taxonomists' Glossary of Mosquito Anatomy. Mosquito Systematics, 13*(2), 201–217.

Harbach, R.E., & Mongkolpanya, K. (1988). Redescription of *Culex* (*Eumelanomyia*) *richei* (Diptera: Culicidae), with treatments of the previously unknown life stages and a record of its occurrence in Thailand. *Mosquito Systematics, 20*(3), 343–351.

Harbach, R.E., & Navarro, J.C. (1996). A new species of *Anopheles* subgenus *Kerteszia* (Diptera: Culicidae) from Venezuela. *Entomologica Scandinavica, 27*(2), 207–216.

Harbach, R.E., & Petersen, J.L. (1992). Two species previously confused under the concept of *Sabethes tarsopus* in Central America (Diptera: Culicidae). *Mosquito Systematics, 24*(2), 102–124.

Harbach, R.E., & Peyton, EL. (1990). A new subgenus in *Wyeomyia* (Diptera: Culicidae), with the reclassification and redescription of the type species, *Sabethes fernandezyepezi. Mosquito Systematics, 22*(1), 15–23.

Harbach, R.E., & Peyton, EL. (1991). Transfer of the subgenus *Davismyia* from *Wyeomyia* to *Sabethes* and description of the type species, *Miamyia petrocchiae.* (Diptera: Culicidae). *Mosquito Systematics, 22*(for 1990)(3), 149–159.

Harbach, R.E., & Peyton, EL. (1992a). A new subgenus of *Wyeomyia* (Diptera: Culicidae), with the reclassification and redescription of *Wyeomyia* (*Davismyia*) *arborea*, *Wyeomyia* (*Dendromyia*) *tarsata* and *Sabethes* (*Sabethes*) *carrilloi. Mosquito Systematics, 23*(for 1991)(2), 92–109.

Harbach, R.E., & Peyton, EL. (1992b). A new subgenus of *Culex* in the Neotropical Region (Diptera: Culicidae). *Mosquito Systematics, 24*(3), 242–252.

Harbach, R.E., & Peyton, EL. (1993). Morphology and evolution of the larval maxilla and its importance in the classification of the Sabethini (Diptera: Culicidae). *Mosquito Systematics, 25*(1), 1–16.

Harbach, R.E., & Peyton, EL. (2000). Systematics of *Onirion*, a new genus of Sabethini (Diptera: Culicidae) from the Neotropical region. *Bulletin of the Natural History Museum London* (Entomology), *69*(2), 115–159.

Harbach, R.E., & Rattanarithikul, R. (1988). A new species of the subgenus *Eumelanomyia* of *Culex* (Diptera: Culicidae) from Thailand. *Mosquito Systematics, 20*(1), 69–76.

Harbach, R.E., & Taai, K. (2014). *Nyctomyia biunguiculata*, a new cavernicolous species of tribe Aedini (Diptera: Culicidae) from southern Thailand. *Zootaxa, 3895*(3), 427–432.

Harbach, R.E., & Ward, R.A. (1994). Lectotype designation for *Anopheles* (*Cellia*) *rennellensis* Taylor and Maffi (Diptera: Culicidae). *Mosquito Systematics, 26*(2), 73–74.

Harbach, R.E., Harrison, B.A., & Gad, A.M. (1984). *Culex* (*Culex*) *molestus* Forskål (Diptera: Culicidae): neotype designation, description, variation and taxonomic status. *Proceedings of the Entomological Society of Washington, 86*(3), 521–542.

Harbach, R.E., Peyton, EL, & Harrison, B.A. (1984). A new species of *Culex* (*Melanoconion*) from southern South America (Diptera: Culicidae). *Mosquito Systematics, 16*(2), 185–200.

Harbach, R.E., Peyton, EL, & Jakob, W.L. (1984). Synonomy of *Culex* (*Culex*) *oswaldoi* with *Culex* (*Culex*) *maxi* (Diptera, Culicidae). *Mosquito Systematics, 15*(for 1983)(4), 310–317.

Harbach, R.E., Dahl, C., & White, G.B. (1985). *Culex* (*Culex*) *pipiens* Linnaeus (Diptera: Culicidae): Concepts, type designations, and descriptions. *Proceedings of the Entomological Society of Washington, 87*(1), 1–24.

Harbach, R.E., Jacob, W.L., & Peyton, EL. (1986a). Recognition of *Culex bidens* Dyar and *Culex interfor* Dyar (Diptera: Culicidae) as separate species. *Mosquito Systematics, 18*(2), 139–144.

Harbach, R.E., Rattanarithikul, R., & Peyton, EL. (1986b). Occurrence of *Uranotaenia* (*Pseudoficalbia*) *hirsutifemora* Peters (Diptera: Culicidae) in Thailand, with notes on the larval stage and species affinity. *Mosquito Systematics, 18*(3–4), 230–231.

Harbach, R.E., Harrison, B.A., Gad, A.M., Kenawy, M.A., & El-Said, S. (1988). Records and notes on mosquitoes (Diptera: Culicidae) collected in Egypt. *Mosquito Systematics, 20*(3), 317–342.

Harbach, R.E., Gaffigan, T.V., & Pecor, J.E. (1991). The J. Pedro Duret mosquito collection (Diptera: Culicidae). *Mosquito Systematics, 22*(for 1990)(3), 192–195.

Harbach, R.E., Rattanarithikul, R., & Harrison, B.A. (2005). *Baimaia*, a new subgenus for *Anopheles kyondawensis* Abraham, a unique crabhole-breeding anopheline in Southeastern Asia. *Proceedings of the Entomological Society of Washington, 107*(4), 750–761.

Harbach, R.E., Parkin, E., Chen, B., & Butlin, R.K. (2006). *Anopheles* (*Cellia*) *minimus* Theobald (Diptera: Culicidae): Neotype designation, characterization, and systematics. *Proceedings of the Entomological Society of Washington, 108*(1), 198–209.

Harbach, R.E., Garros, C., Manh, N.D., & Manguin, S. (2007). Formal taxonomy of species C of the *Anopheles minimus* sibling species complex (Diptera: Culicidae). *Zootaxa, 1654*, 41–54.

Harbach, R.E., Rattanarithikul, R., Howard, T.M., Linton, Y.-M., & Kitching, I.J. (2007). Systematics of a new genus and cavernicolous species of the mosquito tribe Aedini (Diptera: Culicidae) from Thailand. *Proceedings of the Entomological Society of Washington, 109*(2), 469–488.

Harbach, R.E., Vu, D.H., & Kitching, I.J. (2007). Systematics of *Kimia*, a new genus of Sabethini (Diptera: Culicidae) in the Oriental Region. *Proceedings of the Entomological Society of Washington, 109*(1), 102–120.

Harbach, R.E., Kitching, I.J., Culverwell, C.L., Dubois, J., & Linton, Y.-M. (2012). Phylogeny of mosquitoes of tribe Culicini (Diptera: Culicidae) based on morphological diversity. *Zoologica Scripta, 41*(5), 499–514.

Harbach, R.E., Kitching, I., Culverwell, C.L., Howard, T.M., & Linton, Y.-M. (2013). *Nyx pholeocola*, a new genus and cavernicolous species of tribe Aedini (Diptera: Culicidae) from southern Thailand based on morphological and molecular data. *Zootaxa, 3683*(2), 159–177.

Harbach, R.E., Culverwell, C.L., & Kitching, I.J. (2016). Phylogeny of the nominotypical subgenus of *Culex* (Diptera: Culicidae): Insights from analyses of anatomical data into interspecific relationships and species groups in an unresolved tree. *Systematics and Biodiversity, 15*(4), 1–11.

Harbach, R.E., Dallimore, T., Briscoe, A.G., Culverwell, C.L., Vaux, A.G.C., & Medlock, J.M. (2017). *Aedes nigrinus* (Eckstein, 1918) (Diptera, Culicidae), a new country record for England, contrasted with *Aedes sticticus* (Meigen, 1838). *Zookeys, 671*, 119–130.

Harbach, R.E., Rattanarithikul, R., & Harrison, B.A. (2017). *Anopheles prachongae*, a new species of the Gigas Complex of subgenus *Anopheles* (Diptera: Culicidae) in Thailand, contrasted with known forms of the complex. *Zootaxa, 4236*(1), 65–81.

Harbison, J.E., Metzger, M.E., Hu, R., Fritz, C.L., & Kramer, V.L. (2008). Effect of conveyance pipe dimension and orientation on mosquito oviposition in a simulated stormwater management device. *Journal of the American Mosquito Control Association, 24*(1), 98–104.

Hard, J.J., Bradshaw, W.E., & Malarkey, D.J. (1989). Resource- and density-dependent development in tree-hole mosquitoes. *Oikos, 54*(2), 137–144.

Harmston, F.C. (1949). An annotated list of mosquito records from Colorado. *Great Basin Naturalist, 9*(3–4), 65–75.

Harmston, F.C. (1969). Separation of the females of *Aedes hendersoni* Cockerell and *Aedes triseriatus* (Say) Diptera: Culicidae by the tarsal claws. *Mosquito News, 29*, 490–491.

Harrington, L.C, & Poulson, R.L. (2008). Considerations for accurate identification of adult *Culex restuans* (Diptera: Culicidae) in field studies. *Journal of Medical Entomology, 45*(1), 1–8.

Harrington, L.C., Buonaccorsi, J.P., Edman, J.D., Costero, A., Kittayapong, P., Clark, G.G., & Scott, T.W. (2001a). Analysis of survival of young and old *Aedes aegypti* (Diptera: Culicidae) from Puerto Rico and Thailand. *Journal of Medical Entomology, 38*(4), 537–547.

Harrington, L.C., Edman, J.D., & Scott, T.W. (2001b). Why do female *Aedes aegypti* (Diptera: Culicidae) feed preferentially and frequently on human blood? *Journal of Medical Entomology, 38*(3), 411–422.

Harrington, L.C., Scott, T.W., Lerdthusnee, K., Coleman, R.C., Costero, A., Clark, G.G., . . . Edman, J.D. (2005). Dispersal of the dengue vector *Aedes aegypti* within and between rural communities. *The American Journal of Tropical Medicine and Hygiene, 72*(2), 209–220.

Harris, C., Kihonda, J., Lwetoijera, D., Dongus, S., Devine, G., & Majambere, S. (2011). A simple and efficient tool for trapping gravid *Anopheles* at breeding sites. *Parasites and Vectors, 4*(1), 125.

Harris, T.W. (1835). VIII. Insects. In E. Hitchcock (Ed.), *Report; on the geology, mineralogy, botany, and zoology of Massachusetts* (2nd ed., pp. 553–602). Amherst, Massachusetts: Press of J.S. and C. Adams.

Harrison, B.A. (1972). A new interpretation of affinities within the *Anopheles hyrcanus* complex of Southeast Asia. *Mosquito Systematics, 4*(3), 73–83.

Harrison, B.A. (1973a). *Anopheles* (*An.*) *reidi*, a new species of the *Barbirostris* species complex from Sri Lanka (Diptera: Culicidae). *Proceedings of the Entomological Society of Washington, 75*(3), 365–371.

Harrison, B.A. (1973b). A lectotype designation and description for *Anopheles* (*An.*) *sinensis* Wiedemann 1828, with a discussion of the classification and vector status of this and some other oriental *Anopheles*. *Mosquito Systematics, 5*(1), 1–13.

Harrison, B.A. (1973c). Notes on some mosquito types deposited in France. *Mosquito Systematics, 5*(4), 277–279.

Harrison, B.A. (1980). Medical entomology studies. XIII. The Myzomyia Series of *Anopheles* (*Cellia*) in Thailand, with emphasis on intra-interspecific variations (Diptera: Culicidae). *Contributions of the American Entomological Institute, 17*(4), 1–195.

Harrison, B.A. (1987a). Book review. T. Ramachandra Rao (Rev. ed. 1984). The anophelines of India. *Journal of the American Mosquito Control Association, 3*(2), 338–339.

Harrison, B.A. (1987b). *Culex* subgenus *Thaiomyia* Bram, a synonym of *Culex* subgenus *Culiciomyia* Theobald (Diptera: Culicidae). *Mosquito Systematics, 19*(1), 111–116.

Harrison, B.A., & Klein, J.M. (1975). A revised list of the *Anopheles* of Cambodia. *Mosquito Systematics, 7*(1), 11–12.

Harrison, B.A., & Rattanarithikul, R. (1973). Comparative morphology of the early larval instars of *Aedes aegypti* and *A. seatoi* in Thailand. *Mosquito Systematics, 5*(4), 280–294.

Harrison, B.A., & Scanlon, J.E. (1974). *Anopheles* (*An.*) *pilinotum*, a new species name in the Aitkenii complex for *An. insulaeflorum* from the Philippines and eastern Indonesia (Diptera: Culicidae). *Mosquito Systematics, 6*(1), 32–40.

Harrison, B.A., & Scanlon, J.E. (1975). Medical entomology studies. II. The subgenus *Anopheles* in Thailand (Diptera: Culicidae). *Contributions of the American Entomological Institute, 12*(1), iv, 1–307.Harrison, B.A., Scanlon, J.E., & Reid, J.A. (1973). A new synonomy and new species name in the Southeast Asian *Anopheles hyrcanus* complex. *Mosquito Systematics, 5*(4), 263–268.

Harrison, B.A., Reinert, J.F., Sirivanakarn, S., Huang, Y.-M., Peyton, EL, & De Meillon, B. (1974). Distributional and biological notes on mosquitoes from Sri Lanka (Ceylon) (Diptera: Culicidae). *Mosquito Systematics, 6*(2), 142–162.

Harrison, B.A., Rattanarithikul, R., Peyton, EL, & Mongkolpanya, K. (1991). Taxonomic changes, revised occurrence records and notes on the Culicidae of Thailand and neighboring countries. *Mosquito Systematics, 22*(for 1990)(3), 196–227.

Harrison, B.A., Varnado, W., Whitt, P.B., & Goddard, J. (2008). New diagnostic characters for females of *Psorophora* (*Janthinosoma*) species in the United States, with notes on *Psorophora mexicana* (Bellardi) (Diptera: Culicidae). *Journal of Vector Ecology, 33*(2), 232–237.

Harrison, B.A., Ruiz-Lopez, F., Calderón Falero, G., Savage, H.M., Pecor, J.E., & Wilkerson, R.C. (2012). *Anopheles* (*Kerteszia*) *lepidotus* (Diptera: Culicidae), not the malaria vector we thought it was: Revised male and female morphology; larva, pupa, and male genitalia characters; and molecular verification. *Zootaxa, 3218*, 1–17.

Harrison, B.A., Byrd, B.D., Sither, C.B., & Whitt, P.B. (2016). *The Mosquitoes of the Mid-Atlantic Region: An Identification Guide*. Western Carolina University Mosquito and Vector-Borne Infectious Disease Laboratory.

Harrison, J.F. (2001). Insect acid-based physiology. *Annual Review of Entomology, 46*(1), 221–250.

Hartberg, W.K. (1975). Comments on reproductive isolation and phylogeny of mosquitoes of the genus *Aedes*. *Mosquito Systematics, 7*(3), 193–206.

Hartberg, W.K., & Craig, G.B., Jr. (1973). Gene-controlled morphological differences in male genitalia of *Aedes aegypti* and *Aedes mascarensis* (Diptera: Culicidae). *Mosquito News, 33*, 206–214.

Hasan, S.N., Ahmed, A., & Chari, N. (1983). Analysis of wing vibration of mosquito *Anopheles stephensi*. *Entomon, 8*, 47–51.

Hasan, S.N., Ahmed, A., & Chari, N. (1984). Aerodynamic parameters and design of flight surface of mosquito *Anopheles stephensi*. *Entomon, 9*, 247–252.

Hatori, S. (1901). Species of *Anopheles* from Formosa. *Kwampô* (*Official Reports of the Japanese Government*), *5534*, 274–275.

Haufe, W.O., & Burgess, L. (1956). Development of *Aedes* (Diptera: Culicidae) at Fort Churchill, Manitoba, and prediction of dates of emergence. *Ecology, 37*(3), 500–519.

Hausermann, W., Ray, R.W., & Hacker, C.S. (1971). Dispersal of genetically marked female *Aedes aegypti* in Mississippi. *Mosquito News, 31*(1), 37–51.

Hawaii State Department of Health. (2012). *Public asked to take action against threat of mosquito-borne illness in light of rare mosquito find at Honolulu Airport*. Retrieved from health.hawaii.gov/news/files/2013/05/12-010.pdf

Hawksworth, D.L. (2011). Introducing the *Draft Biocode* (2011). *Bionomina, 3*(1), 24–25.

Hawley, W.A. (1985). The effect of larval density on adult longevity of a mosquito *Aedes sierrensis*—epidemiological consequences. *Journal of Animal Ecology, 54*(3), 955–964.

Hawley, W.A. (1988). The biology of *Aedes albopictus*. *Journal of the American Mosquito Control Association, 4*(Supplement *1*), 1–39.

Haworth, W.E. (1924). Mosquitoes and coconut palms. A mosquito survey of palm trees in East Africa and the problems resulting

therefrom. *Transactions of the Royal Society of Tropical Medicine and Hygiene, 18*(4), 162–198.
Hay, S.I., Sinka, M.E., Okara, R.M., Kabaria, C.W., Mbithi, P.M., Tago, C.C., . . . Godfray, H.C.J. (2010). Developing global maps of the dominant *Anopheles* vectors of human malaria. *PLOS Medicine, 7*(2), e1000209.
Headlee, T.J. (1931). The biology of the important economic species of mosquitoes occurring in New Jersey. *Proceedings of the Eighteenth Annual Meeting of the New Jersey Mosquito Extermination Association, 18*, 40–69.
Heard, S.B. (1994). Wind exposure and distribution of pitcher plant mosquito (Diptera: Culicidae). *Environmental Entomology, 23*(5), 1250–1253.
Hearle, E. (1923). A new mosquito from British Columbia (Culicidae, Diptera). *The Canadian Entomologist, 55*(1), 4–5.
Hearle, E. (1927). A new Canadian mosquito (Culicidae). *The Canadian Entomologist, 59*(5), 101–103.
Hearle, E. (1929). The life history of *Aedes flavescens* Muller. *Transactions of the Royal Society of Canada, 23*, 85–102.
Hedeen, R.A. (1953). The biology of the mosquito *Aedes atropalpus* Coquillett. *Journal of the Kansas Entomological Society, 26*(1), 1–10.
Hedeen, R.A. (1954). The identification of the larva of *Psorophora (Grabhamia) signipennis* (Coquillett). *Mosquito News, 14*(3), 143–145.
Hedeen, R.A. (1959). A review of the mosquito larvae of France. II. The genus *Aedes*. *Mosquito News, 9*(3), 179–183.
Hegde, S., Rasgon, J.L., & Hughes, G.L. (2015). The microbiome modulates arbovirus transmission in mosquitoes. *Current Opinion in Virology, 15*, 97–102.
Heinemann, S.J., & Belkin, J.N. (1977a). Collection records of the project Mosquitoes of Middle America. 7. Costa Rica (CR). *Mosquito Systematics, 9*(2), 237–288.
Heinemann, S.J., & Belkin, J.N. (1977b). Collection records of the project Mosquitoes of Middle America. 8. Central America: Belize (BH), Guatemala (GUA), El Salvador (SAL), Honduras (HON), Nicaragua (Nl, NIC). *Mosquito Systematics, 9*(4), 403–454.
Heinemann, S.J., & Belkin, J.N. (1977c). Collection records of the project Mosquitoes of Middle America. 9. Mexico (MEX, MF, MT, MX). *Mosquito Systematics, 9*(4), 483–535.
Heinemann, S.J., & Belkin, J.N. (1978). Collection records of the project mosquitoes of Middle America. 2. Venezuela (VZ); Guianas: French Guiana (FG, FGC), Guyana (GUY), Surinam (SUR). *Mosquito Systematics, 10*(3), 365–459.
Heinemann, S.J., & Belkin, J.N. (1979). Collection records of the project Mosquitoes of Middle America. 13. South America: Brazil (BRA, BRAP, BRB), Ecuador (ECU), Peru (PER), Chile (CH). *Mosquito Systematics, 11*(2), 61–118.
Heinemann, S.J., Aitken, T.H.G., & Belkin, J.N. (1980). Collection records of the project Mosquitoes of Middle America. 14. Trinidad and Tobago (TR, TRM, TOB). *Mosquito Systematics, 12*(2), 179–284.
Helinski, M.E.H., & Harrington, L.C. (2011). Male mating history and body size influence female fecundity and longevity of the dengue vector *Aedes aegypti*. *Journal of Medical Entomology, 48*(2), 202–211.
Helinski, M.E.H., & Harrington, L.C. (2012). The role of male harassment on female fitness for the dengue vector mosquito *Aedes aegypti*. *Behavioral Ecology and Sociobiology, 66*(8), 1131–1140.
Helinski, M.E.H., & Knols, B.G.J. (2008). Mating competitiveness of male *Anopheles arabiensis* mosquitoes irradiated with a partially or fully sterilizing dose in small and large laboratory cages. *Journal of Medical Entomology, 45*(4), 698–705.
Helinski, M.E.H., & Knols, B.G.J. (2009). Sperm quantity and size variation in un-irradiated and irradiated males of the malaria mosquito *Anopheles arabiensis* Patton. *Acta Tropica, 109*(1), 64–69.
Helinski, M.E.H., Hood-Nowotny, R., Mayr, L., & Knols, B.G. (2007). Stable isotope-mass spectrometric determination of semen transfer in malaria mosquitoes. *Journal of Experimental Biology, 210*(7), 1266–1274.
Helinski, M.E.H., Hood, R.C., & Knols, B.G. (2008). A stable isotope dual-labelling approach to detect multiple insemination in unirradiated and irradiated *Anopheles arabiensis* mosquitoes. *Parasites and Vectors, 1*(9).
Helinski, M.E.H., Deewatthanawong, P., Sirot, L.K., Wolfner, M.F., & Harrington, L.C. (2012a). Duration and dose-dependency of female sexual receptivity responses to seminal fluid proteins in *Aedes albopictus* and *Ae. aegypti* mosquitoes. *Journal of Insect Physiology, 58*(10), 1307–1313.
Helinski, M.E.H., Valerio, L., Facchinelli, L., Scott, T.W., Ramsey, J., & Harrington, L.C. (2012b). Evidence of polyandry for *Aedes aegypti* in semifield enclosures. *The American Journal of Tropical Medicine and Hygiene, 86*(4), 635–641.
Hennig, W. (1965). Phylogenetic systematics. *Annual Review of Entomology, 10*(1), 97–116.
Hennig, W. (1966a). Einige bemerkungen ueber die typen der von Giebel 1862 angehlich aus dem Bernstein beschriebenen insektenarten. *Stuttgarter Beitraege zur Naturkunde*.
Hennig, W. (1966b). *The Diptera fauna of New Zealand as a problem in systematics and zoogeography*. Honolulu, HI: Entomology Department, Bishop Museum.
Hernandez, R., Luo, T., & Brown, D.T. (2001). Exposure to low pH is not required for penetration of mosquito cells by Sindbis virus. *Journal of Virology, 75*(4), 2010–2013.
Hernández-Martinez, S., Mayoral, J.G., Li, Y., & Noriega, F.G. (2007). Role of juvenile hormone and allatotropin on nutrient allocation, ovarian development and survivorship of mosquitoes. *Journal of Insect Physiology, 53*(3), 230–234.
Herrick, G.W. (1905). Notes on some Mississippi mosquitoes. *Entomological News, 16*(9), 281–283.
Hervé, J.-P., & Geoffroy, B. (1974). Description d'un nouveau moustique de République Centrafricaine *Culex (Neoculex) subsalisburiensis* n. sp. *Cahiers ORSTOM. Série entomologie médicale et parasitologie, 12*(3), 191–193.
Hervy, J.P., Ferrara, L., & Legros, F. (1985). *Aedes (Diceromyia) furcifer* (Edwards, 1913) et *Aedes (Diceromyia) taylori* Edwards, 1936: Diagnose des larves de quatrième stade. *Cahiers ORSTOM. Série entomologie médicale et parasitologie, 23*(1), 17–24.
Hickey, W.A., & Craig, G.B., Jr. (1966). Genetic distortion of sex ratio in a mosquito, *Aedes aegypti*. *Genetics. 53*(6), 1177–1196.
Hickner, PV., Mori, A., Zeng, E., Tan, J.C., & Severson, D.W. (2015). Whole transcriptome responses among females of the filariasis and arbovirus vector mosquito *Culex pipiens* implicate TGF-B signaling and chromatin modification as key drivers of diapause induction. *Functional and Integrative Genomics, 15*(4), 439–447.
Higa, Y., Toma, T., & Miyagi, I. (2006). Type specimens of mosquitoes (Diptera: Culicidae) of the Ryukyu Archipelago deposited in the National Science Museum and the Entomological Institute of Hokkaido University, Japan, in 2005. *Medical Entomology and Zoology, 57*(2), 135–137.
Higa, Y., Toma, T., & Miyagi, I. (2012). Description of female, pupa and larva of *Ficalbia ichiromiyagii* from Iriomote Island, Ryukyu

Archipelago, Japan. *Journal of the American Mosquito Control Association, 28*(4), 279–285.
Hii, J.L.K., Peyton, EL, & Shang, V.Y. (1988). Redescription of the adult and first description of the larva and pupa of *Anopheles (Cellia) sulawesi* Waktoedi, a species of the *Leucosphyrus* group from Sulawesi, Indonesia (Diptera, Culicidae). *Mosquito Systematics, 20*(1), 41–54.
Hill, E., & Haydon, L.G. (1907). A contribution to the study of the characteristics of larvae of species of anophelinae in South Africa. *Annals of the Natal Government Museum, 1*, 111–157.
Hill, G.F. (1925). The distribution of anopheline mosquitoes in the Australian region with notes on some culicine species. *Proceedings of the Royal Society of Victoria, 37* , 61–77.
Hill, R.B., & Hill, C.M. (1945). *Wyeomyia caracula* Dyar & Núñez Tovar, a description of the larva and pupa and a redescription of the adult (Diptera, Culicidae). *Proceedings of the Entomological Society of Washington, 47*(9), 296–299.
Hill, R.B., & Hill, C.M. (1946). *Phoniomyia hirsuta*, a new sabethine from Jamaica (Diptera, Culicidae). *Proceedings of the Entomological Society of Washington, 48*(2), 39–41.
Hillis, D.M., & Wiens, J.J. (2000). Molecules versus morphology in systematics: Conflicts, artifacts, and misconceptions. In J.J. Wiens (Ed.), *Phylogenetic analysis of morphological data* (pp. 1–19). Washington, DC: Smithsonian Institution Press.
Hinman, E.H. (1932). A description of the larva of *Anopheles atropos* D. & K., with biological notes on the species. *Proceedings of the Entomological Society of Washington, 34*(8), 138–142.
Hinton, H.E. (1968a). Observations on the biology and taxonomy of the eggs of *Anopheles* mosquitos. *Bulletin of Entomological Research, 57*(4), 495–508.
Hinton, H.E. (1968b). Structure and protective devices of the egg of the mosquito *Culex pipiens*. *Journal of Insect Physiology, 14*(2), 145–148.
Hinton, H.E. (1981). *Biology of insect eggs*. Oxford, United Kingdom: Pergamon Press.
Hinton, H.E., & Service, M.W. (1969). The surface structure of aedine eggs as seen with the scanning electron microscope. *Annals of Tropical Medicine and Parasitology, 63*(4), 409–411.
Hirao, M., Toma, T., Miyagi, I., Okazawa, T., Higa, Y., & Leh, M.U. (2010). Redescriptions of five species of the genus *Armigeres*, subgenus *Armigeres* (Diptera: Culicidae) collected from fallen coconut fruits at the coastal plains of Sarawak, East Malaysia. *Medical Entomology and Zoology, 61*(3), 281–308.
Hiwat, H., & Bretas, G. (2011). Ecology of *Anopheles darlingi* Root with respect to vector importance: A review. *Parasites and Vectors, 4*(1), 177.
Ho, C. (1931). Study of the adult culicids of Peiping. *Bulletin of the Fan Memorial Institute of Biology, 2*(8), 107–175.
Ho, C. (1938a). A new tree-hole breeding *Anopheles* from the island of Hainan. *Annals of Tropical Medicine and Parasitology, 32*(3), 279–285.
Ho, C. (1938b). On a collection of anopheline mosquitoes from the island of Hainan. *Annals of Tropical Medicine and Parasitology, 32*(3), 387–411.
Ho, K.-M. (1963). Report of a new mosquito, *Culex cheni* n. sp. *Acta entomologica Sinica, 12*, 368–371.
Hoffmann, A.A., & Sgrò, C.M. (2011). Climate change and evolutionary adaptation. *Nature, 470*(7335), 479–485.
Hoffmann, C.C. (1935). La formación de razas en los *Anopheles* mexicanos. I. *A. maculipennis* y *A. quadrimaculatus* y una raza nueva del *maculipennis*. *Anales del Instituto de Biología, Universidad nacional Autónoma de México, 6*, 3–22.
Hoffmann, C.C. (1938). La formación de razas en los *Anopheles* mexicanos. II. *Anopheles albimanus* y sus variedades en la República Mexicana. *Anales del Instituto de Biología Universidad nacional Autónoma de México, 9*, 167–180.
Hoffmann, C.C. (1939). La formación de razas de *Anopheles* mexicanos. III. *Anopheles chiriquiensis* Komp en el interior del Estado de Chiapas. *Anales del Instituto de Biología, Universidad nacional Autónoma de México, 10*, 347–352.
Hogue, C.L. (1975). A new species of bromeliad-breeding *Culex* (*Culex*) from Cocos Island. *Mosquito Systematics, 7*(4), 357–362.
Holderegger, R., & Wagner, H.H. (2008). Landscape genetics. *Bioscience, 58*(3), 199–207.
Holloway, J.E. (2003). *Dictionary of birds of the United States: Scientific and common names*. Portland, OR: Timber Press.
Holstein, M.H. (1949). Le problème d'Anophèles Gambiae. *Bulletin médical de l'Afrique-occidentale française, Numéro Spécial*, 155–160.
Holstein, M.H. (1951). Note sur l'épidémiologie du paludisme en Afrique-Occidentale française. *Bulletin of the World Health Organization, 4*, 463–473.
Holstein, M.H. (1954). *Biology of* Anopheles gambiae. *Research in French West Africa*. Geneva, Switzerland: World Health Organization.
Holt, B.G., Lessard, J.P., Borregaard, M.K., Fritz, S.A., Araújo, M.B., Dimitrov, D., . . . Rahbek, C. (2013). An update of Wallace's zoogeographic regions of the world. *Science, 339*(6115), 74–78.
Holt, R.A., Subramanian, G.M., Halpern, A., Sutton, G.G., Charlab, R., Nusskern, D.R., . . . Salzberg, S.L. (2002). The genome sequence of the malaria mosquito *Anopheles gambiae*. *Science, 298*(5591), 129–149.
Honegger, H.W., Market, D., Pierce, L.A., Dewey, E.M., Kostran, B., Wilson, M., . . . Mesce, K.A. (2002). Cellular localization of bursicon using antisera against partial peptide sequences of this insect-sclerotizing neurohormone. *Journal of Comparative Biology, 452*(2), 163–177.
Honegger, H.W., Dewey, E.M., & Ewer, J. (2008). Bursicon, the tanning hormone of insects: Recent advances following the discovery of its molecular identity. *Journal of Comparative Physiology A, 194*(12), 989–1005.
Honegger, H.W., Estévez-Lao, T.Y., & Hillyer, J.F. (2011). Bursicon-expressing neurons undergo apoptosis after adult ecdysis in the mosquito *Anopheles gambiae*. *Journal of Insect Physiology, 57*(7), 1017–1022.
Hong, H.K., & Ree, H.I. (1968). Two unreported anopheline mosquitoes in Korea. *Korean Journal of Zoology, 11*(4), 118–120.
Hongoh, V., Berrang-Ford, L., Scott, M.E., & Lindsay, L.R. (2012). Expanding geographical distribution of the mosquito, *Culex pipiens*, in Canada under climate change. *Applied Geography, 33*, 53–62.
Hood-Nowotny, R., Schwarzinger, B., Schwarzinger, C., Soliban, S., Madakacherry, O., Aigner, M., . . . Gilles, J. (2012). An analysis of diet quality, how it controls fatty acid profiles, isotope signatures and stoichiometry in the malaria mosquito *Anopheles arabiensis*. *PLOS ONE, 7*(10), e45222.
Hoogstraal, H., & Chamberlain, R.W. (1946). The larva and male of *Aedes (Skusea) amesii* (Ludlow) (Diptera, Culicidae). *Proceedings of the Entomological Society of Washington, 48*(5), 125–131.
Hoogstraal, H., & Knight, K.L. (1951). Observations on *Eretmapodites silvestris conchobius* Edwards (Culicidae) in the Anglo-Egyptian Sudan. *The American Journal of Tropical Medicine and Hygiene, 31*(5), 659–664.

Hopkins, G.H.E. (1936). *Mosquitoes of the Ethiopian Region. I. Larval bionomics of mosquitoes and taxonomy of culicine larvae*. London, England: British Museum (Natural History).
Hopkins, G.H.E. (1952). *Mosquitoes of the Ethiopian region: Larval bionomics of mosquitoes and taxonomy of culicine larvae* (2nd ed., Vol. 1). London, England: British Museum (Natural History).
Hopp, M.J., & Foley, J.A. (2001). Global-scale relationships between climate and the dengue fever vector, *Aedes aegypti*. *Climatic Change, 48*(2–3), 441–463.
Horsfall, W.R. (1942). Biology and control of mosquitoes in the rice area. *Arkansas Agricultural Experiment Station Bulletin 427*.
Horsfall, W.R. (1955). *Mosquitoes. Their bionomics and relation to disease*. New York, NY: Hafner. (Reprinted 1972)
Horsfall, W.R. (1956). A method for making a survey of floodwater mosquitoes. *Mosquito News, 16*(2), 66–71.
Horsfall, W.R. (1963). Eggs of floodwater mosquitoes (Diptera: Culicidae). IX. Local distribution. *Annals of the Entomological Society of America, 56*(4), 426–441.
Horsfall, W.R. (1972). *Mosquitoes. Their bionomics and relation to disease*. New York, NY: Hafner.
Horsfall, W.R., & Craig, G.B., Jr. (1956). Eggs of floodwater mosquitoes. IV. Species of *Aedes* common in Illinois. (Diptera: Culicidae). *Annals of the Entomological Society of America, 49*(4), 368–374.
Horsfall, W.R., & Voorhees, F.R. (1972). Eggs of floodwater mosquitoes. XIV. Northern *Aedes* (Diptera: Culicidae). *Annals of the Entomological Society of America, 65*(1), 123–126.
Horsfall, W.R., Miles, R.C., & Sokatch, J.T. (1952). Eggs of floodwater mosquitoes. I. Species of *Psorophora* (Diptera: Culicidae). *Annals of the Entomological Society of America, 45*(4), 618–624.
Horsfall, W.R., Voorhees, F.R., & Cupp, E.W. (1970). Eggs of floodwater mosquitoes. XIII. Chorionic sculpturing. *Annals of the Entomological Society of America, 63*(6), 1709–1716.
Horsfall, W.R., Fowler, H.W., Jr., Moretti, L.J., & Larsen, J.R. (1973). *Bionomics and embryology of the inland floodwater mosquito* Aedes vexans. Urbana: University of Illinois Press.
Hoshen, M.B., & Morse, A.P. (2004). A weather-driven model of malaria transmission. *Malaria Journal, 3*, 32.
Hosken, D.J., Stockley, P., Tregenza, T., & Wedell, N. (2009). Monogamy and the battle of the sexes. *Annual Review of Entomology, 54*, 361–378.
Hotchkin, P.G. (1985). The duration of larval life of *Aedes aegypti* as affected by time of hatch. *Journal of the American Mosquito Control Association, 1*(4), 489–492.
Howard, J.J., Emord, D.E., & Morris, C.D. (1983). Epizootiology of eastern equine encephalomyelitis virus in upstate New York, USA. V. Habitat preference of host-seeking mosquitoes (Diptera: Culicidae). *Journal of Medical Entomology, 20*(1), 62–69.
Howard, L.O. (1901). *Mosquitoes: How they live; how they carry disease; how they are classified; and how they may be destroyed*. New York, NY: McClure, Phillips.
Howard, L.O., Dyar, H.G., & Knab, F. (1913). *The mosquitoes of North and Central America and the West Indies* (Vol. 2)(for 1912). Washington, DC: Carnegie Institution for Science.
Howard, L.O., Dyar, H.G., & Knab, F. (1915). *The mosquitoes of North and Central America and the West Indies: Vol. 3. Systematic description (in two parts)*. Part I. Públication No. 159. Washington, DC: Carnegie Institution for Science.
Howard, L.O., Dyar, H.G., & Knab, F. (1917). *The mosquitoes of North and Central America and the West Indies. Systematic description*. Part II. Washington, DC: Carnegie Institution for Science.
Howell, P.I., & Knols, B.G. (2009). Male mating biology. *Malaria Journal, 8*(2), S8.
Hribar, L.J. (1994). Geographic variation of male genitalia of *Anopheles nuneztovari* (Diptera: Culicidae). *Mosquito Systematics, 26*(3), 132–144.
Hribar, L.J. (1995a). Costal wing spot variation within and among progeny of single female *Anopheles nuneztovari* (Diptera: Culicidae). *Mosquito Systematics, 27*(1), 1–15.
Hribar, L.J. (1995b). Larval and pupal chaetotaxy of *Anopheles nuneztovari* (Diptera: Culicidae). *Transactions of the American Entomological Society, 121*(4), 249–225.
Hribar (Leguizamón), M.N., & Hribar, L.J. (2006). Location and deposition of the type specimens of *Culex scheuberi* Carpintero & Leguizamón, *Ochlerotatus jorgi* (Carpintero & Leguizamón), and *Orthopodomyia peytoni* Liguizamón & Carpintero (Diptera: Culicidae). *Zootaxa, 1204*, 37–40.
Hsiao, C.-E. (1977). A new species of *Aedes* (Diptera, Culicidae). *Acta entomologica Sinica, 20*(4), 451–454.
Hsiao, T.-Y., & Bohart, R.M. (1946). *The mosquitoes of Japan and their medical importance. US NAVMED* (Bureau of Medicine and Surgery, Navy Department).
Hsieh, L.-K., & Liao, T.-H. (1956). A list of Amoy mosquitoes with the description of a new species and a new variety. *Acta entomologica Sinica, 6*(3), 123–127.
Hsü, C.-F. (1963). A new species of *Culex* from Hainan Island. *Acta entomologica Sinica, 12*(2), 229–232.
Hsu, K.-C. (1964). Description of a new mosquito *Tripteroides* (*Tripteroides*) *szechwanensis* sp. nov. *Acta entomologica Sinica, 13*(2), 278–282.
Huang, J., Walker, E.D., Giroux, P.Y., Vulule, J., & Miller, J.R. (2006). Ovipositional site selection by *Anopheles gambiae*: Influences of substrate moisture and texture. *Medical and Veterinary Entomology, 19*(4), 442–450.
Huang, S., Molaei, G., & Andreadis, T.G. (2011). Reexamination of *Culex pipiens* hybridization zone in the eastern United States by ribosomal DNA-based single nucleotide polymorphism markers. *American Journal of Tropical Medicine and Hygiene, 85*(3), 434–441.
Huang, Y.-M. (1968a). *Aedes* (*Verrallina*) of the Papuan subregion (Diptera: Culicidae). *Pacific Insects Monograph, 17*, 1–73.
Huang, Y.-M. (1968b). A new subgenus of *Aedes* (Diptera, Culicidae) with illustrated key to the subgenera of the Papuan subregion (Diptera, Culicidae). *Journal of Medical Entomology, 5*(2), 169–188.
Huang, Y.-M. (1968c). Neotype designation for *Aedes* (*Stegomyia*) *albopictus* (Skuse) (Diptera, Culicidae). *Proceedings of the Entomological Society of Washington, 70*(4), 297–302.
Huang, Y.-M. (1969). A new species of *Aedes* (*Stegomyia*) from Thailand (Diptera: Culicidae). *Proceedings of the Entomological Society of Washington, 71*(2), 234–239.
Huang, Y.-M. (1970a). A description of a topotypic male of *Aedes scutellaris* (Walker) (Diptera: Culicidae). *Proceedings of the Entomological Society of Washington, 71*(for 1969)(4), 471–475.
Huang, Y.-M. (1970b). A note on *Aedes aurotaeniatus* Edwards (Diptera: Culicidae). *Proceedings of the Entomological Society of Washington, 72*(3), 281–288.
Huang, Y.-M. (1970c). The subgenus *Stegomyia* in Southeast Asia. *Mosquito Systematics, 2*(1), 6–8.
Huang, Y.-M. (1971). A redescription of *Aedes* (*Stegomyia*) *scutellaris malayensis* Colless and the differentiation of the larva from that of *Aedes* (*S.*) *albopictus* (Skuse) (Diptera: Culicidae). *Proceedings of the Entomological Society of Washington, 73*(1), 1–8.
Huang, Y.-M. (1972a). Lectotype designation for *Aedes* (*Stegomyia*) *galloisi* Yamada with a note on its assignment to the Scutellaris

group of species. (Diptera: Culicidae). *Proceedings of the Entomological Society of Washington, 74*(2), 253–256.

Huang, Y.-M. (1972b). A redescription of the holotype male of *Aedes (Stegomyia) tongae* Edwards with a note on two topotypic females (Diptera: Culicidae). *Proceedings of the Entomological Society of Washington, 74*(3), 338–342.

Huang, Y.-M. (1972c). Contributions to the mosquito fauna of Southeast Asia. XIV. The subgenus *Stegomyia* of *Aedes* in Southeast Asia. I. The *scutellaris* group of species. *Contributions of the American Entomological Institute, 9*(1), 1–109.

Huang, Y.-M. (1973a). A new record of *Aedes (Stegomyia) alcasidi* Huang (Diptera: Culicidae). *Proceedings of the Entomological Society of Washington, 75*(1), 125.

Huang, Y.-M. (1973b). A new species of *Aedes (Stegomyia)* from Thailand and notes on the *mediopunctatus* subgroup (Diptera: Culicidae). *Proceedings of the Entomological Society of Washington, 75*(2), 224–232.

Huang, Y.-M. (1974a). Lectotype designation for *Aedes (Stegomyia) chemulpoensis* Yamada with a note on its assignment to the *aegypti* group of species. *Proceedings of the Entomological Society of Washington, 76*(2), 208–211.

Huang, Y.-M. (1974b). A redescription of *Aedes (Stegomyia) amaltheus* de Meillon and Lavoipierre with a note on its assignment to the *aegypti* group of species (Diptera: Culicidae). *Mosquito Systematics, 6*(1), 27–31.

Huang, Y.-M. (1974c). A new species of *Aedes (Stegomyia)* from the Andaman Islands (Diptera: Culcidae). *Mosquito Systematics, 6*(2), 137–141.

Huang, Y.-M. (1975a). A redescription of *Aedes (Stegomyia) pseudoscutellaris* (Theobald) with a note on the taxonomic status of *Aedes (Stegomyia) polynesiensis* Marks (Diptera: Culicidae). *Mosquito Systematics, 7*(1), 87–101.

Huang, Y.-M. (1975b). A new species of *Aedes (Stegomyia)* from Sri Lanka (Ceylon) (Diptera: Culicidae). *Mosquito Systematics, 7*(4), 345–356.

Huang, Y.-M. (1977a). Medical entomology studies. VII. The subgenus *Stegomyia* of *Aedes* in Southeast Asia. II. The *edwardsi* group of species. III. The *w-albus* group of species (Diptera: Culicidae). *Contributions of the American Entomological Institute, 14*(1), 1–111.

Huang, Y.-M. (1977b). Medical entomology studies. VIII. Notes on the taxonomic status of *Aedes vittatus* (Diptera: Culicidae). *Contributions of the American Entomological Institute, 14*(1), 113–132.

Huang, Y.-M. (1977c). The mosquitoes of Polynesia with a pictorial key to some species associated with filariasis and/or dengue fever. *Mosquito Systematics, 9*(3), 289–322.

Huang, Y.-M. (1977d). A new record of *Aedes (Stegomyia) patriciae* Mattingly (Diptera: Culicidae). *Proceedings of the Entomological Society of Washington, 79*(4), 625.

Huang, Y.-M. (1978a). Redescription and subgeneric position of *Aedes meronephada* (Dyar and Shannon) with notes on the subgenus *Diceromyia* (Diptera: Culicidae). *Mosquito Systematics, 10*(3), 334–350.

Huang, Y.-M. (1978b). Taxonomic status of *Aedes (Stegomyia) laffooni* Knight and Rozeboom with a redescription of *Aedes (Stegomyia) pseudalbolineatus* Brug (Diptera: Culicidae). *Mosquito Systematics, 10*(3), 351–364.

Huang, Y.-M. (1978c). The identity of two species of *Stegomyia* belonging to the *Aedes-albolineatus* group (Diptera). *Mosquito Systematics, 10*(2), 197–210.

Huang, Y.-M. (1979a). Medical entomology studies. XI. The subgenus *Stegomyia* of *Aedes* in the Oriental region with keys to the species (Diptera: Culicidae). *Contributions of the American Entomological Institute, 15*(6), 1–76.

Huang, Y.-M. (1979b). *Aedes (Stegomyia) simpsoni* complex in the Ethiopian region with lectotype designation for *simpsoni* (Theobald) (Diptera: Culicidae). *Mosquito Systematics, 11*(3), 221–234.

Huang, Y.-M. (1981). A redescription of *Aedes (Stegomyia) calceatus* Edwards and description of a new Afrotropical species, *Aedes (Stegomyia) ledgeri* (Diptera: Culicidae). *Mosquito Systematics, 13*(1), 92–113.

Huang, Y.-M. (1985). A new African species of *Aedes* (Diptera: Culicidae). *Mosquito Systematics, 17*(2), 108–120.

Huang, Y.-M. (1986a). Notes on the *Aedes (Diceromyia) furcifer* group, with a description of a new species (Diptera: Culicidae). *Proceedings of the Entomological Society of Washington, 88*(4), 634–649.

Huang, Y.-M. (1986b). *Aedes (Stegomyia) corneti*, a new species of the Africanus subgroup (Diptera: Culicidae). *Proceedings of the Entomological Society of Washington, 88*(4), 764–776.

Huang, Y.-M. (1986c). *Aedes (Stegomyia) bromeliae* (Diptera: Culicidae), the yellow fever virus vector in east Africa. *Journal of Medical Entomology, 23*(2), 196–200.

Huang, Y.-M. (1988a). *Aedes (Stegomyia) josiahae*, a new species of the *simpsoni* subgroup (Diptera: Culicidae). *Proceedings of the Entomological Society of Washington, 90*(2), 155–163.

Huang, Y.-M. (1988b). The *Aedes (Stegomyia) pseudonigeria* group with emphasis on the species from the Afrotropical Region (Diptera: Culicidae). *Mosquito Systematics, 20*(1), 1–20.

Huang, Y.-M. (1990). The subgenus *Stegomyia* of *Aedes* in the Afrotropical Region. I. The *africanus* group of species (Diptera: Culicidae). *Contributions of the American Entomological Institute, 26*(1), 1–90.

Huang, Y.-M. (1994). *Aedes (Stegomyia) mattinglyorum*, a new species of the *dendrophilus* group (Diptera: Culicidae). *Proceedings of the Entomological Society of Washington, 96*(1), 11–21.

Huang, Y.-M. (1997). The subgenus *Stegomyia* of *Aedes* in the Afrotropical Region. II. The *dendrophilus* group of species (Diptera: Culicidae). *Contributions of the American Entomological Institute, 29*(4), 1–104.

Huang, Y.-M. (2001). A pictorial key for the identification of the subfamilies of Culicidae, genera of Culicinae and subgenera of *Aedes* mosquitoes of the Afrotropical Region (Diptera: Culicidae). *Proceedings of the Entomological Society of Washington, 103*(1), 1–53.

Huang, Y.-M. (2002). A pictorial key to the mosquito genera of the world, including subgenera of *Aedes* and *Ochlerotatus* (Diptera: Culicidae). *Insecta Koreana, 19*(1), 1–130.

Huang, Y.-M. (2004). The subgenus *Stegomyia* of *Aedes* in the Afrotropical Region with keys to the species (Diptera: Culicidae). *Zootaxa, 700*, 1–120.

Huang, Y.-M. (2005). *Cornetius*, a new subgenus of *Aedes*, and a redescription of *Aedes (Cornetius) cozi* Cornet (Diptera: Culicidae). *Proceedings of the Entomological Society of Washington, 107*(3), 517–529.

Huang, Y.-M., & Hitchcock, J.C. (1980). Medical entomology studies. XII. A revision of the *Aedes scutellaris* group of Tonga (Diptera: Culicidae). *Contributions of the American Entomological Institute, 17*(3), 1–107.

Huang, Y.-M., & Rueda, L.M. (1998). Description of a paralectotype female of *Aedes (Finlaya) niveus* (Ludlow) (Diptera: Culicidae). *Proceedings of the Entomological Society of Washington, 100*(4), 824–827.

Huang, Y.-M., & Rueda, L.M. (2013). A new species of the genus *Orthopodomyia* from Ototomo, Cameroon, with notes on other *Orthopodomyia* group-species. *Journal of the American Mosquito Control Association, 29*(3), 270–274.

Huang, Y.-M., & Rueda, L.M. (2014). A pictorial key to the species of *Aedes* (*Ochlerotatus* and *Coetzeemyia*) in the Afrotropical Region (Diptera: Culicidae). *Zootaxa, 3754*(5), 592–600.

Huang, Y.-M., & Rueda, L.M. (2015). Pictorial keys to the species of *Albuginosus* and *Aedimorphus* (Grjebinei and Apicoannulatus Groups) of the genus *Aedes* Meigen in the Afrotropical Region (Diptera: Culicidae). *Zootaxa, 3925*(1), 25–36.

Huang, Y.-M., & Rueda, L.M. (2016). A pictorial key to the sections, groups, and species of the *Aedes* (*Diceromyia*) in the Afrotropical Region (Diptera: Culicidae). *Zootaxa, 4079*(2), 281–290.

Huang, Y.-M., & Rueda, L.M. (2017a). Lectotype designation for *Aedes* (*Finlaya*) *pulchrithorax* Edwards (Diptera: Culicidae), including male description. *Proceedings of the Entomological Society of Washington, 119*(2), 277–280.

Huang, Y.-M., & Rueda, L.M. (2017b). Pictorial keys to the sections, groups, and species of the *Aedes* (*Finlaya*) in the Afrotropical Region (Diptera: Culicidae). *Zootaxa, 4221*(1), 131–141.

Huang, Y.-M., & Ward, R.A. (1981). A pictorial key for the identification of the mosquitoes associated with yellow fever in Africa. *Mosquito Systematics, 13*(2), 138–149.

Huang, Y.-M., Mathis, W.N., & Wilkerson, R.C. (2010). *Coetzeemyia*, a new subgenus of *Aedes*, and a redescription of the holotype female of *Aedes* (*Coetzeemyia*) *fryeri* (Theobald) (Diptera: Culicidae). *Zootaxa, 2638*, 1–24.

Huang, Y.-M., Mathis, W.N., & Wilkerson, R.C. (2012). Subgeneric position of *Aedes dufouri* Hamon with notes on the subgenus *Levua* Stone and Bohart (Diptera: Culicidae). *Proceedings of the Entomological Society of Washington, 114*(2), 194–204.

Hubbell, S.P. (2001). *The unified neutral theory of biodiversity and biogeography*. Princeton, NJ: Princeton University Press.

Huber, B.A. (2003). Rapid evolution and species-specificity of arthropod genitalia: Fact or fiction? *Organisms Diversity and Evolution, 3*(1), 63–71.

Hudson, J.E. (1978). Canada's national mosquito? Mass-resting of *Anopheles earli* [*sic*] (Diptera: Culicidae) females in a beaver lodge in Alberta. *The Canadian Entomologist, 110*(12), 1345–1346.

Hull, D.L. (1988). Progress in ideas of progress. In M.H. Nitecki (Ed.), *Evolutionary progress* (pp. 27–48). Chicago, IL: University of Chicago Press.

Hunt, R.H., Coetzee, M., & Fettene, M. (1998). The *Anopheles gambiae* complex: A new species from Ethiopia. *Transactions of the Royal Society of Tropical Medicine and Hygiene, 92*(2), 231–235.

Hurlbut, H.S. (1938a). A study of the larval chaetotaxy of *Anopheles walkeri* Theobald. *American Journal of Hygiene, 28*(2), 149–173.

Hurlbut, H.S. (1938b). Further notes on the overwintering of the eggs of *Anopheles walkeri* Theobald with a description of the eggs. *Journal of Parasitology, 24*(6), 521–526.

Hutchings, R.S.G., & Sallum, M.A.M. (2001). Bionomics data for *Anopheles* (*Anopheles*) *forattinii* Wilkerson and Sallum, 1999. *Acta Amazonica, 31*(4), 699–701.

Hutchings, R.S.G., & Sallum, M.A.M. (2008). Two new species of *Culex* subgenus *Melanoconion* (Diptera: Culicidae) from the Amazon forest. *Zootaxa, 1920*, 41–50.

Hutton, F.W. (1902). Additions to the Diptera fauna of New Zealand. *Transactions of Zoology, 34*(for 1901), 179–196.

Hwang, U.W., Yong, T.S., & Ree, H. (2004). Molecular evidence for synonymy of *Anopheles yatsushiroensis* and *An. pullus*. *Journal of the American Mosquito Control Association, 20*(2), 99–104.

Ibáñez-Bernal, S. (1987). Nuevo registro altitudinal de *Aedes* (*Stegomyia*) *aegypti* (Linnaeus, 1762) (Diptera: Culicidae) en Mexico. *Folia entomologica Mexicana, 72*, 163–164.

Ibáñez-Bernal, S., & Martinez-Campos, C. (1994). *Aedes albopictus* in Mexico. *Journal of the American Mosquito Control Association, 10*(2, Part 1), 231–232.

Iglisch, I. (1977). Hausmücken (*Culex-pipiens*-Gruppe) als mögliche Üeberträger des Serumhepatitis-Virus. Untersuchungen zur Genitalmorphologie. *Zeitschrift für Angewandte Zoologie, 64*(3), 257–290.

Ignell, R., Dekker, T., Ghaninia, M., & Hansson, B.S. (2005). Neuronal architecture of the mosquito deutocerebrum. *Journal of Comparative Neurology, 493*(2), 207–240.

Ikeshoji, T., & Mulla, M.S. (1970). Overcrowding factors of mosquito larvae. 2. Growth-retarding and bacteriostatic effects of the overcrowding factors of mosquito larvae. *Journal of Economic Entomology, 63*(6), 1737–1743.

Ikeshoji, T., & Mulla, M.S. (1974). Overcrowding factors of mosquito larvae: Isolation and chemical identification. *Environmental Entomology, 3*(3), 482–486.

Illies, J. (Ed.) (1978). *Limnofauna Europaea: A checklist of the animals inhabiting european inland waters, with accounts of their distribution and ecology (except Protozoa)* (2nd ed.). Stuttgart. Germany: Fischer Verlag.

IMPE [Institue of Malariology, Parasitology and Entomology]. (1987). *Keys to the* Anopheles *in Vietnam (adults-pupae-larvae)* [in Vietnamese]. Hanoi, Vietnam: Department of Entomology, Institute of Malariology, Parasitology and Entomology.

Impoinvil, D.E., Kongera, J.C., Foster, W.A., Njiru, B.N., Killeen, G.F., Githure, J.I., . . . Knols, B.G.J. (2004). Feeding and survival of the malaria vector *Anopheles gambiae* on plants growing in Kenya. *Medical and Veterinary Entomology, 18*(2), 108–115.

Ingram, A., & De Meillon, B. (1927). A mosquito survey of certain parts of South Africa, with special reference to the carriers of malaria and their control. Part I. *Publications of the South African Institute for Medical Research, 4*, 1–81.

Ingram, A., & De Meillon, B. (1929). A mosquito survey of certain parts of South Africa, with special reference to the carriers of malaria and their control. Part II. *Publications of the South African Institute for Medical Research* (Johannesburg), *4*, 83–170.

Ingram, A., & Macfie, J.W.S. (1917a). Notes on some distinctive points in the pupae of West African mosquitos. *Bulletin of Entomological Research, 8*, 73–91.

Ingram, A., & Macfie, J.W.S. (1917b). The early stages of certain West African mosquitos. *Bulletin of Entomological Research, 8*, 135–154.

International Commission on Zoological Nomenclature. (1959). Opinion 548. Validation under the plenary powers of the generic name "*Toxorhynchites*" Theobald (Class Insecta, Order Diptera), as published in 1901 in the *Journal of Tropical Medicine*. *Opinions and Declarations Rendered by the International Commission on Zoological Nomenclature, 20*, 167–174.

International Commission on Zoological Nomenclature. (1959). Opinion 547. Designation under the plenary powers of a type species in harmony with accustomed usage for the nominal genus "*Anopheles*" Meigen, 1818 (Class Insecta, Order Diptera). *Opinions and Declarations Rendered by the International Commission on Zoological Nomenclature, 20*, 153–164.

International Commission on Zoological Nomenclature. (1959). Opinion 550. Suppression under the plenary powers of the generic name *Taeniorhynchus* Lynch-Arribálzaga, 1891, for the purpose of protecting the generic name *Mansonia* Blanchard, 1901 (Class Insecta, Order Diptera) and matters related to other names in the same class and in the Class Aves incidental thereto. *Opinions and Declarations Rendered by the International Commission on Zoological Nomenclature, 20*, 185–198.

International Commission on Zoological Nomenclature. (1964). Opinion 711. *Culex aegypti* Linnaeus, 1762 (Insecta, Diptera): Validated and interpreted under the plenary powers. *Bulletin of Zoological Nomenclature, 21*(4), 246–248.
International Commission on Zoological Nomenclature. (1965). Opinion 737. *Bironella gracilis* Theobald, 1905 (Insecta, Diptera): Validated under the plenary powers. *The Bulletin of Zoological Nomenclature, 22*, 164–166.
International Commission on Zoological Nomenclature. (1968). Opinion 850. *Anopheles africanus* Theobald, 1901 (Insecta, Diptera): Suppressed under the plenary powers. *Bulletin of Zoological Nomenclature, 25*, 22–23.
International Commission on Zoological Nomenclature. (1969). Opinion 874. *Cellia errabunda* Swellengrebel, 1925 (Insecta, Diptera): Suppressed under the plenary powers. *Bulletin of Zoological Nomenclature, 26*, 10–11.
International Commission on Zoological Nomenclature. (1972). Opinion 979. *Culex albirostris* Macquart, 1850 (Insecta, Diptera): Suppressed under the plenary powers. *Bulletin of Zoological Nomenclature, 29*(1), 11–12.
International Commission on Zoological Nomenclature. (1982). Opinion 1213. *Toxorhynchites brevipalpis* Theobald, 1901 (Insecta, Diptera): Conserved. *Bulletin of Zoological Nomenclature, 39*(2), 122–123.
International Commission on Zoological Nomenclature. (1991). Opinion 1644. *Culex stigmatosoma* Dyar, 1907 and *C. thriambus* Dyar, 1921 (Insecta, Diptera): Specific names conserved. *Bulletin of Zoological Nomenclature, 48*(2), 179–180.
International Commission on Zoological Nomenclature. (1994). Opinion 1775. *Banksinella luteolateralis* var. *albothorax* Theobald, 1907 (currently *Aedes* (*Neomelaniconion*) *albothorax*), *B. luteolateralis* var. *circumluteola* Theobald, 1908 (currently *A.* (*N.*) *circumluteolus*) and *A.* (*N.*) *mcintoshi* Huang, 1985 (Insecta, Diptera): specific names conserved, and *A.* (*N.*) *albothorax*: neotype designated. *Bulletin of Zoological Nomenclature, 51*(2), 179–180.
International Commission on Zoological Nomenclature. (1999). *International Code of Zoological Nomenclature* (4th ed.). London, England: International Trust for Zoological Nomenclature.
Irish, S.R., Kyalo, D., Snow, R.W., & Coetzee, M. (2020). Updated list of *Anopheles* species (Diptera: Culicidae) by country in the Afrotropical Region and associated islands. *Zootaxa, 4747*(3), 401–449.
Isely, D. (1972). The disappearance. *Taxon, 21*(1), 3–12.
Ishii, T. (1969). On the *Culex pipiens* group in Japan. Part 1. XIII. Morphological comparisons between adult females of *Culex pipiens pallens* Coquillett and *Culex vagans* Wiedemann from the northern parts of Japan (Diptera: Culicidae). *Japanese Journal of Ecology, 19*(5), 204–216.
Ishii, T. (1977). On the *Culex pipiens* group in Japan. Part III. A historical review of its research. 1. Names adopted for *Culex pipiens pallens* Coquillett (Diptera, Culicidae). *Journal of Science, University of Tokushima, 10*, 31–57.
Ishimura, K. (1952). On the early stage of *Megarhinus towadensis* Matsumura. *Japanese Journal of Sanitary Zoology, 3*(1–2), 12–19.
IUCN [International Union for Conservation of Nature]. (2011). *IUCN red list of threatened species* [Data file]. Retrieved from www.iucnredlist.org
Iversen, T.M. (1971). *Culex torrentium* Martini (Dipt. Culicidae) ny for Danmark, med en beskrivelse af larvebiotopen. *Entomologiske meddelelser, 39*, 235–239.
Iwaki, M., & Choochote, W. (1991). Scanning electron microscopy of eggs of *Mansonia uniformis*, *Ma. indiana*, *Ma. annulifera* and *Ma. annulata* (Diptera: Culicidae). *Journal of Medical Entomology, 28*(3), 334–339.
Iwaki, M., Choochote, W., & Rongsriyam, Y. (1992). Scanning electron microscopy of eggs of *Mansonia bonneae* (Diptera: Culicidae). *Journal of Medical Entomology, 29*(6), 1039–1041.
Iwaki, M., Rongsriyam, Y., Siriprasert, V., & Choochote, W. (1994). Scanning electron microscopy of the egg of *Mansonia dives* (Diptera: Culicidae). *Journal of Medical Entomology, 31*(6), 931–933.
Iyengar, M.O.T. (1922). The larva of *Anopheles annandalei*. *Indian Journal of Medical Research* (Calcutta), *10*, 526–529.
Iyengar, M.O.T. (1924). On the Indian anophelines of the *funestus* group and the description of a new species (Diptera, Culicidae). *Indian Journal of Medical Research* (Calcutta), *12*, 23–29.
Iyengar, M.O.T. (1929). Adult and larval stages of *Anopheles majidi*. *Indian Journal of Medical Research 17*(1), 1–10.
Iyengar, M.O.T. (1935). Eggs of *Ficalbia minima*, Theo., and notes on breeding habits of three species of *Ficalbia*. *Bulletin of Entomological Research, 26*(4), 423–425.
Iyengar, M.O.T. (1960). A review of the mosquito fauna of the South Pacific (Diptera, Culicidae). *South Pacific Commission Technical Paper, 130*, 1–102.
Iyengar, M.O.T. (1969). Eggs of two species of *Tripteroides* Giles from New Caledonia (Diptera: Culicidae). *Journal of the Australian Entomological Society, 8*, 214–216.
Iyengar, M.O.T., & Menon, M.A.U. (1948). Notes on *Harpagomyia genurostris* Leicester (Dipt. Culicidae). *Proceedings of the Royal Entomological Society of London, Series A: General Entomology, 23*(4–6), 39–43.
Iyengar, M.O.T, & Menon, M.A.U. (1955). Mosquitos of the Maldive Islands. *Bulletin of Entomological Research, 46*(1), 1–10.
Iyengar, M.O.T, & Menon, M.A.U. (1956). The mosquitos of South Thailand. *Bulletin of Entomological Research, 47*(4), 785–794.
Jackson, N. (1953). Observations on the feeding habits of a predaceous mosquito larva, *Culex* (*Lutzia*) *tigripes* de Grandpré and de Charmoy (Diptera). *Proceedings of the Royal Entomological Society of London, Series A: General Entomology, 28*(10–12), 153–159.
Jacob, B.G., Lampman, R.L., Ward, M.P., Muturi, E.J., Morris, J.A., Caamano, E.X., & Novak, R.J. (2009). Geospatial variability in the egg raft distribution and abundance of *Culex pipiens* and *Culex restuans* in Urbana-Champaign, Illinois. *International Journal of Remote Sensing, 30*(8), 2005–2019.
Jacob, P.G., & Bhat, H.R. (1995). Records of *Aedes* (*Finlaya*) *albocinctus* (Diptera: Culicidae) in southern India. *Mosquito Systematics, 27*(3), 177–178.
Jacquet, M., Lebon, C., Lempérière, G., & Boyer, S. (2012). Behavioural functions of grooming in male *Aedes albopictus* (Diptera: Culicidae), the Asian tiger mosquito. *Applied Entomology and Zoology, 47*(4), 359–363.
Jagge, C.L., & Pietrantonio, P.V. (2008). Diuretic hormone 44 receptor in Malpighian tubules of the mosquito *Aedes aegypti*: Evidence for transcriptional regulation paralleling urination. *Insect Molecular Biology, 17*(4), 413–426.
James, S.P. (1902). *Malaria in India: Scientific Memoirs by officers of the Medical and Sanitary Department of the government of India* (New Series No. 2). Calcutta, India: Office of the Superintendent of Government Printing.
James, S.P. (1910a). Notes on mosquitoes. *Paludism* (Simla), *1*, 33–38.
James, S.P. (1910b). A new arrangement of the India Anophelinae. *Records of the Indian Museum, 49*(5), 95–109.
James, S.P., & Liston, W.G. (1904). A monograph of the *Anopheles* mosquitoes of India. Calcutta, India: Thacker, Spink.
James, S.P., & Liston, W.G. (1911). *A monograph of the anopheline mosquitoes of India* (2nd ed.). Calcutta, India: Thacker, Spink.
Jansen, C.C., Hemmerter, S., van den Hurk, A.F., Whelan, P.I., & Beebe, N.W. (2013). Morphological versus molecular identifi-

cation of *Culex annulirostris* Skuse and *Culex palpalis* Taylor: Key members of the *Culex sitiens* (Diptera: Culicidae) subgroup in Australasia. *Australian Journal of Entomology, 52*(4), 356–362.

Jarial, M.S. (2001). Toxic effect of garlic extracts on the eggs of *Aedes aegypti* (Diptera: Culicidae): A scanning electron microscopic study. *Journal of Medical Entomology, 38*(3), 446–450.

Jayaprakash, K., & Karthikeyan, V. (2014). Does polyandrous impede mosquito control by autocidal? *International Journal of Mosquito Research, 1*(2), 10–12.

Jayewickreme, S.H. (1952). Methods of rearing the larvae of some anopheline mosquitoes of Ceylon with observations on their life history. *Ceylon Journal of Science B, 25*, 29–53.

Jeffery, J., Oothuman, P.R., & Rudnick, A. (1987). *Culex (Eumelanomyia) richardgarciai*, a new species of mosquito from Peninsular Malaysia (Diptera: Culicidae). *Mosquito Systematics, 19*(2), 139–142.

Jeffery, J., Lee, R.M.L., Tan, S.Y., Liew, C., Ng, L.C., & Lam-Phua, S.G. (2010). New mosquito species records (Diptera: Culicidae) from Singapore. *Tropical Biomedicine, 27*(1), 138–142.

Jenkins, D.W., & Carpenter, S.J. (1946). Ecology of the tree hole breeding mosquitoes of Nearctic North America. *Ecological Monographs, 16*(1), 33–47.

Jenkins, S.P., Brown, M.R., & Lea, A.O. (1992). Inactive prothoracic glands in larvae and pupae of *Aedes aegypti*: Ecdysteroid release by tissues in the thorax and abdomen. *Insect Biochemistry and Molecular Biology, 22*(6), 553–559.

Jepson, W.G., Moutia, A., & Courtois, C. (1947). The malaria problem in Mauritius: The bionomics of Mauritian anophelines. *Bulletin of Entomological Research, 38*, 177–208.

Jhumur, U.S., Dötterl, S., & Jürgens, A. (2006). Naïve and conditioned responses of *Culex pipiens pipiens* biotype *molestus* (Diptera: Culicidae) to flower odors. *Journal of Medical Entomology, 43*(6), 1164–1170.

Jhumur, U.S., Dötterl, S., & Jürgens, A. (2007). Electrophysiological and behavioural responses of mosquitoes to volatiles of *Silene otitis* (Caryophyllaceae). *Arthropod-Plant Interactions, 1*(4), 245–254.

Jhumur, U.S., Dötterl, S., & Jürgens, A. (2008). Floral odours of *Silene otites*: Their variability and attractiveness to mosquitoes. *Journal of Chemical Ecology, 34*(1), 14–25.

Jiménez-Valverde, A., Peterson, A.T., Soberón, J., Overton, J.M., Aragón, P., & Lobo, J.M. (2011). Use of niche models in invasive species risk assessments. *Biological Invasions, 13*(12), 2785–2797.

Jobling, B. (1938). *Anatomical drawings of biting flies*. London, England: British Museum (Natural History).

Jobling, B. (1987). *Anatomical drawings of biting flies*. London, England: British Museum (Natural History).

Johnson, K.N. (2015). Bacteria and antiviral immunity in insects. *Current Opinion in Insect Science, 8*, 97–103.

Johnston, C. (1855). Original communications: Auditory apparatus of the *Culex* mosquito. *Journal of Cell Science, 1*(10), 97–102.

Jones, C.E., Lounibos, L.P., Marra, P.P., & Kilpatrick, A.M. (2012). Rainfall influences survival of *Culex pipiens* (Diptera: Culicidae) in a residential neighborhood in the Mid-Atlantic United States. *Journal of Medical Entomology, 49*(3), 467–473.

Jones, P.L., Pask, G.M., Rinker, D.C., & Zwiebel, L.J. (2011a). Functional agonism of insect odorant receptor ion channels. *Proceedings of the National Academy of Sciences of the United States of America, 108*(21), 8821–8825.

Jones, V.P., Steffan, S.A., Wiman, N.G., Horton, D.R., Miliczky, E., Zhang, Q.H., & Baker, C.C. (2011b). Evaluation of herbivore-induced plant volatiles for monitoring green lacewings in Washington apple orchards. *Biological Control, 56*(1), 98–105.

Joshi, D.S. (1996). Effect of fluctuating and constant temperatures on development, adult longevity and fecundity in the mosquito *Aedes krombeini*. *Journal of Thermal Biology, 21*(3), 151–154.

Joshi, G., Pradhan, S., & Darsie, R.F., Jr. (1965). Culicine, Sabethine and Toxorhynchitine mosquitoes of Nepal including new country records (Diptera: Culicidae). *Proceedings of the Entomological Society of Washington, 67*(3), 137–146.

Joy, T.K., Arik, A.J., Corby-Harris, V., Johnson, A.A., & Riehle, M.A. (2010). The impact of larval and adult dietary restriction on lifespan, reproduction and growth in the mosquito *Aedes aegypti*. *Experimental Gerontology, 45*(9), 685–690.

Joyce, C.R., & Nakagawa, P.Y. (1963). *Aedes vexans nocturnus* (Theobald) in Hawaii. *Proceedings, Hawaiian Entomological Society, 18*(2), 273–280.

Joyce, J.D., Nogueira, J.R., Bales, A.A., Pittman, K.E., & Anderson, J.R. (2011). Interactions between La Crosse virus and bacteria isolated from the digestive tract of *Aedes albopictus* (Diptera: Culicidae). *Journal of Medical Entomology, 48*(2), 389–394.

Jozwiak, P., Rewicz, T., & Pabis, K. (2015). Taxonomic etymology—in search of inspiration. *Zookeys, 513*, 143–160.

Judd, D.D. (1996). Review of the systematics and phylogenetic relationships of the Sabethini (Diptera: Culicidae). *Systematic Entomology, 21*(2), 129–150.

Judd, D.D. (1998). Review of a bromeliad-ovipositing lineage in *Wyeomyia* and the resurrection of *Hystatomyia* (Diptera: Culicidae). *Annals of the Entomological Society of America, 91*(5), 572–589.

Judson, C.L. (1960). The physiology of hatching aedine mosquito eggs: Hatching stimulus. *Annals of the Entomological Society of America, 53*(5), 688–691.

Juliano, S.A., & Lounibos, L.P. (2005). Ecology of invasive mosquitoes: Effects on resident species and on human health. *Ecology Letters, 8*(5), 558–574.

Juliano, S.A., O'Meara, G.F., Morrill, J.R., & Cutwa, M.M. (2002). Desiccation and thermal tolerance of eggs and the coexistence of competing mosquitoes. *Oecologia, 130*(3), 458–469.

Julião, G.R., Abad-Franch, F., Lourenço-de-Oliveira, R., & Luz, S.L.B. (2010). Measuring mosquito diversity patterns in an Amazonian terra firme rain forest. *Journal of Medical Entomology, 47*(2), 121–128.

Junnila, A., Müller, G.C., & Schlein, Y. (2010). Species identification of plant tissues from the gut of *An. sergentii* by DNA analysis. *Acta Tropica, 115*(3), 227–233.

Jupp, P.G. (1970). The taxonomic status of *Culex* (*Culex*) *univittatus* Theobald (Diptera: Culicidae) in South Africa. *Mosquito Systematics, 2*(1), 9–16.

Jupp, P.G. (1971). The taxonomic status of *Culex* (*Culex*) *univittatus* Theobald (Diptera: Culicidae) in South Africa. *Journal of the Entomological Society of Southern Africa, 34*(2), 339–357.

Jupp, P.G. (1972). A morphological study of *Culex* (*Culex*) *univittatus* Theobald and *Culex* (*Culex*) *neavei* Theobald from various African countries. *Mosquito Systematics, 4*(4), 103–113.

Jupp, P.G. (1996). *Mosquitoes of southern Africa. Culicinae and Toxorhynchitinae*. Hartebeespoort, Republic of South Africa: Ekogilde

Jupp, P.G. (1998). *Aedes* (*Diceromyia*) *furcifer* (Edwards) and *Aedes* (*Diceromyia*) *cordellieri* Huang in southern Africa: Distribution and morphological differentiation. *Journal of the American Mosquito Control Association, 14*(3), 273–276.

Jupp, P.G., & Harbach, R.E. (1990). Crossmating and morphological studies of *Culex neivai* and *Culex perexiguus* (Diptera: Culicidae) to elucidate their taxonomic status. *Mosquito Systematics, 22*(1), 1–10.

Jupp, P.G., McIntosh, B.M., & Kemp, A. (1993). The *Aedes* (*Diceromyia*) *furcifer* group (Diptera: Culicidae) in southern Africa:

Identification and occurrence of *Ae. furcifer* (Edwards) and *Ae. cordellieri* Huang. *African Entomology, 1*(1), 126–128.
Kabula, B., Derua, Y.A., Tungu, P., Massue, D.J., Sambu, E., Stanley, G., . . . Kisinza, W.N. (2011). Malaria entomological profile in Tanzania from 1950 to 2010: A review of mosquito distribution, vectorial capacity and insecticide resistance. *Tanzania Journal of Health Research, 13*(5, Suppl. 1), 319–331.
Kaiser, A., Jerrentrup, H., Samanidou-Voyadoglou, A., & Becker, N. (2001). Contribution to the distribution of European mosquitoes (Diptera: Culcidae): Four new country records from northern Greece. *European Mosquito Bulletin, 10*, 9.
Kaliwal, M.B., Kumar, A.B., Shanbhag, A.B., Dash, A.P., & Javali, S.B. (2010). Spatio-temporal variations in adult density, abdominal status & indoor resting pattern of *Culex quinquefasciatus* Say in Panaji, Goa, India. *Indian Journal of Medical Research* (Calcutta), *131*(5), 711–719.
Kalpage, K.S., & Brust, R.A. (1968). Mosquitoes of Manitoba. 1. Descriptions and a key to *Aedes* eggs (Diptera: Culicidae). *Canadian Journal of Zoology, 46*(4), 699–718.
Kamimura, K. (1966). *Comments on the distribution and taxonomy of Anopheles in Japan* [in Japanese]. Paper presented at the Abstract of Speeches, 21st Annual Meeting, Western Japan Branch, Japan Society of Sanitary Zoology (p. 8).
Kamimura, K. (1968). The distribution and habit of medically important mosquitoes of Japan. *Japanese Journal of Sanitary Zoology, 19*(1), 15–34.
Kamimura, K. (1996). The mosquitoes (Diptera: Culicidae) collected from Sri Lanka. *Medical Entomology and Zoology, 47*(2), 145–153.
Kamimura, K., & Wada, Y. (1974). A new subspecies of *Culex* (*Barraudius*) *modestus* from Japan (Diptera, Culicidae). *Japanese Journal of Sanitary Zoology, 25*(1), 13–20.
Kamita, S.G., Samra, A.I., Liu, J.Y., Cornel, A.J., & Hammock, B.D. (2011). Juvenile hormone (JH) esterase of the mosquito *Culex quinquefasciatus* is not a target of the JH analog insecticide methoprene. *PLOS ONE, 6*(12), e28392.
Kanda, T., & Kamimura, K. (1967). New record of *Anopheles bengalensis* from Amami Islands, southern Japan. *Japanese Journal of Sanitary Zoology, 18*(2–3), 108–113.
Kanda, T., & Oguma, Y. (1976). Morphological variations of *Anopheles sinensis* Wiedemann, 1828 and *A. lesteri* Baisas and Hu, 1936 and frequency of clasper movements of the males of several *Anopheles* species during induced copulation. *Japanese Journal of Sanitary Zoology, 27*(4), 325–331.
Kanda, T., & Oguma, Y. (1977). Hybridization between *Anopheles sinensis* and *Anopheles sineroides*. *Mosquito News, 37*(1), 115–117.
Kanda, T., & Oguma, Y. (1978). *Anopheles engarensis*, a new species related to *sinensis* from Hokkaido Island, Japan. *Mosquito Systematics, 10*(1), 45–52.
Kang, D.S., Denlinger, D.L., & Sim, C. (2014). Suppression of allatotropin simulates reproductive diapause in the mosquito *Culex pipiens*. *Journal of Insect Physiology, 64*, 48–53.
Kang, W., Tan, J., Cao, C., Cheng, H., Yang, Y., & Huang, R. (1984). *Anopheles liangshanensis*, a new species related to *A. kweiyangensis* and *A. sinensis* from Sichuan, China (Diptera, Culicidae). *Sichuan Journal of Zoology, 3*(2), 11–14.
Kang'ethe, W., Aimanova, K.G., Pullikuth, A.K., & Gill, S.S. (2007). NHE8 mediates amiloride-sensitive Na+/H+ exchange across mosquito Malpighian tubules and catalyzes Na+ and K+ transport in reconstituted proteoliposomes. *American Journal of Physiology—Renal Physiology, 292*(5), F1501–F1512.
Kano, R., & Hayashi, S. (1949). Studies on a mosquito, *Mansonia* (*Coquillettidia*) *ochracea* (Theobald). I. Notes on the adult and egg stages. *Kontyu, 17*(3), 23–26.
Kano, R., Nitahara, M., & Awaya, J. (1954). Description of a new mosquito, *Culex* (*Culiciomyia*) *sasai* n. sp., collected in the southwestern part of Japan (Culicidae, Diptera). *Japanese Journal of Sanitary Zoology, 5*(1–2), 14–20.
Kapitskaya, M.Z., Li, C., Miura, K., Segraves, W., & Raikhel, A.S. (2000). Expression of the early-late gene encoding the nuclear receptor HR3 suggests its involvement in regulating the vitellogenic response to ecdysone in the adult mosquito. *Molecular and Celluar Endocrinology, 160*(1–2), 25–37.
Kar, I., Subbarao, S.K., Eapen, A., Ravindran, J., Satyanarayana, T.S., Raghavendra, K., . . . Sharma, V.P. (1999). Evidence for a new malaria vector species, species E, within the *Anopheles culicifacies* complex (Diptera: Culicidae). *Journal of Medical Entomology, 36*(5), 595–600.
Karas, K., Brauer, P., & Petzel, D. (2005). Actin redistribution in mosquito Malpighian tubules after a blood meal and cyclic AMP stimulation. *Journal of Insect Physiology, 51*(9), 1041–1054.
Karsch, F. (1887). Dipteriologisches von der Delagoabei. *Entomologische Nachrichten, 13*(2), 22–26.
Kato, N., Mueller, C.R., Fuchs, J.F., Wessely, V., Lan, Q., & Christensen, B.M. (2006). Regulatory mechanisms of chitin biosynthesis and roles of chitin in peritrophic matrix formation in the midgut of adult *Aedes aegypti*. *Insect Biochemistry and Molecular Biology, 36*(1), 1–9.
Kaufmann, C., & Brown, M.R. (2008). Regulation of carbohydrate metabolism and flight performance by a hypertrehalosaemic hormone in the mosquito *Anopheles gambiae*. *Journal of Insect Physiology, 54*(2), 367–377.
Kaur, R. (2003). An update on the distribution of mosquitoes of the tribe Aedini in India (Diptera: Culicidae). *Oriental Insects, 37*, 439–455.
Kaur, S. (2014). Variations in the male genitalia of *Aedes* (*Stegomyia*) *albopictus* (Skuse) from Chandigarh and its surrounding areas (Diptera: Culicidae). *International Journal of Mosquito Research, 1*(3), 55–60.
Kay, B.H. (1978). Variation in the abdominal sternal banding of *Culex annulirostris* Skuse (Diptera: Culicidae). *Journal of the Australian Entomological Society, 17*, 311–315.
Kay, B.H., & Jorgensen, W.K. (1986). Eggs of *Aedes vigilax* (Skuse) and their distribution on plants and soil in south east Queensland saltmarsh. *Journal of the Australian Entomological Society, 25*(3), 267–272.
Kearney, M., Porter, W.P., Williams, C., Ritchie, S., & Hoffmann, A.A. (2009). Integrating biophysical models and evolutionary theory to predict climatic impacts on species' ranges: The dengue mosquito *Aedes aegypti* in Australia. *Functional Ecology, 23*(3), 528–538.
Keating, W.H. (main author). 1824. *Narrative of an expedition to the source of St. Peters River, Lake Winnepeek, Lake of the Woods, performed in the year 1823, by order of the Hon. J. C. Calhoun, Secretary of War, under the command of Stephen H. Long, Major U.S. T.E.: Compiled from the notes of Major Long, Messrs. Keating, & Colhoun: Vol. II.* Philadelphia, PA: H.C. Carey & I. Lea [from Harbach 2018c].
Kebaier, C., Voza, T., & Vanderburg, J. (2010). Neither mosquito saliva nor immunity to saliva has a detectable effect on the infectivity of *Plasmodium* sporozoites injected into mice. *Infection and Immunity, 78*(1), 545–551.
Keller, S.R., & Taylor, D.R. (2010). Genomic admixture increases fitness during biological invasion. *Journal of Evolutionary Biology, 23*(8), 1720–1731.
Kelly, D.J., Richards, A.L., Temenak, J., Strickman, D., & Dasch, G.A. (2002). The past and present threat of rickettsial diseases to mili-

tary medicine and international public health. *Clinical Infectious Diseases, 34*(Suppl. 4), S145–169.

Kelly, R., Mead, D., & Harrison, B.A. (2008). Discovery of *Culex coronator* Dyar and Knab (Diptera: Culicidae) in Georgia. *Proceedings of the Entomological Society of Washington, 110*(1), 258–260.

Kengne, P., Awono-Ambene, P., Antonio-Nkondjio, C., Simard, F., & Fontenille, D. (2003). Molecular identification of the *Anopheles nili* group of African malaria vectors. *Medical and Veterinary Entomology, 17*(1), 67–74.

Kengne, P., Antonio-Nkondjio, C., Awono-Ambene, H.P., Simard, F., Awolola, T.S., & Fontenille, D. (2007). Molecular differentiation of three closely related members of the mosquito species complex, *Anopheles moucheti*, by mitochondrial and ribosomal DNA polymorphism. *Medical and Veterinary Entomology, 21*(2), 177–182.

Kennedy, J.S. (1939). The visual responses of flying mosquitoes. *Proceedings of the Zoological Society of London A, 109*(4), 221–242.

Kentzer, B. (2011). A unique inclusion of female of the mosquito in Baltic amber. Wyjatkowa inkluzja samicy komara (Diptera: Culicidae) w bursztynie baltyckim. *Dipteron Wroclaw, 31*, 40–42.

Keppler, W.J., Kitzmiller, J.B., & Rabbani, M.G. (1973). The salivary gland chromosomes of *Anopheles albimanus. Mosquito News, 33*(1), 42–49.

Kermit, H. (2002). The life of Niels Stensen. In K. Ascani, H. Kermit, & G. Skytte (Eds.), *Niccolo Stenone: Anatomista, geologo, vescovo* (Suppl. XXXI, pp. 17–22). Rome, Italy: Analecta Romana Institoti Danici, Academia di Danimarca.

Kersch, C.N., & Pietrantonio, P.V. (2011). Mosquito *Aedes aegypti* (L.) leucokinin receptor is critical for in vivo fluid excretion post blood feeding. *FEBS Letters, 585*(22), 3507–3512.

Kesavaraju, B., & Juliano, S.A. (2009). No evolutionary response to four generations of laboratory selection on antipredator behavior of *Aedes albopictus*: Potential implications for biotic resistance to invasion. *Journal of Medical Entomology, 46*(4), 772–781.

Keshishian, M.N. (1938). *Anopheles sogdianus* sp. nov. A new species of the *Anopheles* mosquito in Tadjikistan. *Meditsinskaia parazitologiia i parizarnye bolezni, 7*(6), 888–896.

Keyghobadi, N., La Pointe, D., Fleischer, R.C., & Fonseca, D.M. (2006). Fine-scale population genetic structure of a wildlife disease vector: The southern house mosquito on the island of Hawaii. *Molecular Ecology, 15*(13), 3919–3930.

Khalin, A.V. (2006). The structure of the male genitalia of *Aedes diantaeus* Howard, Dyar et Knab (Diptera, Culicidae). *Entomological Review, 86*(8), 873–888.

Khalin, A.V. (2009). Structure of the claspettes in the male genitalia of the mosquito subgenus *Ochlerotatus* Lynch Arribálzaga, genus *Aedes* Meigen (Diptera, Culicidae), and their diagnostic significance. *Revue d'entomologie de l'URSS Entomologicseszkoje Obozrenie, 88*(2), 307–313.

Khan, I., Farid, A., & Zeb, A. (2013). Development of inexpensive and globally available larval diet for rearing *Anopheles stephensi* (Diptera: Culicidae) mosquitoes. *Parasites and Vectors, 6*(1), 90.

Khatchikian, C., Sangermano, F., Kendell, D., & Livdahl, T. (2011). Evaluation of species distribution model algorithms for fine-scale container breeding mosquito risk prediction. *Medical and Veterinary Entomology, 25*(3), 268–275.

Khin-Maung-Kyi. (1970). Malaria vectors in Burma. 2. *Anopheles balabacensis balabacensis* Baisas, 1936. *Union of Burma Journal of Life Sciences, 3*, 217–225.

Khin-Maung-Kyi. (1971a). The anopheline mosquitoes of Burma. 4. Subgenus *Myzomia* Blanchard, group *Neocellia* Christophers. *Union of Burma Journal of Life Sciences, 4*(3), 485–493.

Khin-Maung-Kyi. (1971b). The anopheline mosquitoes of Burma. 1. Subgenus *Anopheles* Meigen. Series *Anopheles* and *Myzorhynchus* Edwards. *Union of Burma Journal of Life Sciences, 4*(2), 281–296.

Khin-Maung-Kyi. (1971c). The anopheline mosquitoes of Burma. 2. Subgenus *Myzomyia* Blanchard. Groups *Neomyzomyia* and *Pseudomyzomyia* Christophers. *Union of Burma Journal of Life Sciences, 4*(2), 297–305.

Khin-Maung-Kyi. (1971d). The anopheline mosquitoes of Burma. 3. Subgenus *Myzomyia* Blanchard group *Myzomyia. Union of Burma Journal of Life Sciences, 4*(3), 473–483.

Khokhar, R.R., & Tariq, Z.K. (1966). A redescription of *Aedes (Dicero-myia) periskeletus* (Giles, 1902) (Diptera: Culicidae). *Pakistan Journal of Science, 18*(3–4), 117–123.

Kidwell, M.G. (2005). Transposable elements and the evolution of dipteran genomes. In D.K. Yeates & B.M. Wiegmann (Eds.), *The evolutionary biology of flies* (pp. 145–172). New York, NY: Columbia University Press.

Kiknadze, I.I. (1967). Chromosomes of Diptera. I. Culicidae (evolutionary and practical importance of studying the karyotypes). *Genetika, 1967*(11), 145–165.

Kikuchi, K., & Mochizuki, O. (2011). Micro-PIV (micro particle image velocimetry) visualization of red blood cells (RBCs) sucked by a female mosquito. *Measurement Science and Technology, 22*(6), 064002.

Kikuchi, K., Terada, N., & Mochizuki, O. (2008). Characteristics of blood-sucking pump of a mosquito. *Transactions of the Japanese Society for Medical and Biological Engineering: BME, 46*(2), 232–237.

Killeen, G.F., & Smith, T.A. (2007). Exploring the contributions of bed nets, cattle, insecticides and excitorepellency to malaria control: A deterministic model of mosquito host-seeking behaviour and mortality. *Transactions of the Royal Society of Tropical Medicine and Hygiene, 101*(9), 867–880.

Killeen, G.F., Knols, B.G., & Gu, W. (2003). Taking malaria transmission out of the bottle: Implications of mosquito dispersal for vector-control interventions. *The Lancet Infectious Diseases, 3*(5), 297–303.

Killeen, G.F., McKenzie, F.E., Foy, B.D., Bøgh, C., & Beier, J.C. (2001). The availability of potential hosts as a determinant of feeding behaviours and malaria transmission by African mosquito populations. *Transactions of the Royal Society of Tropical Medicine and Hygiene, 95*(5), 469–476.

Kilpatrick, A.M., Kramer, L.D., Jones, M.J., Marra, P.P., & Daszak, P. (2006). West Nile virus epidemics in North America are driven by shifts in mosquito feeding behavior. *PLOS Biology, 4*(4), e82.

Kim, B.H., Kim, H.K., & Lee, S.J. (2011). Experimental analysis of the blood-sucking mechanism of female mosquitoes. *Journal of Experimental Biology, 214*(7), 1163–1169.

Kim, B.H., Ha, H., Seo, E.S., & Lee, S.J. (2013). Effect of fluid viscosity on the liquid-feeding flow phenomena of a female mosquito. *Journal of Experimental Biology, 216*(6), 952–959.

Kim, W., Koo, H., Richman, A.M., Seeley, D., Vizioli, J., Klocko, A.D., & O'Brochta, D.A. (2004). Ectopic expression of a cecropin transgene in the human malaria vector mosquito *Anopheles gambiae* (Diptera: Culicidae): effects on susceptibility to Plasmodium. *Journal of Medical Entomology, 41*(3), 447–455.

King, J.G., Vernick, K.D., & Hillyer, J.F. (2011). Members of the salivary gland surface protein (SGS) family are major immunogenic components of mosquito saliva. *Journal of Biological Chemistry, 286*(47), 40824–40834.

King, W.V. (1931). The Philippine varieties of *Anopheles gigas* and *Anopheles lindesayi. Philippine Journal of Science, 46*(4), 751–757.

King, W.V. (1932a). The Philippine *Anopheles* of the Rossi-Ludlowi group. *Philippine Journal of Science, 47*(3), 305–342.

King, W.V. (1932b). Three Philippine *Anopheles* of the *funestus-minimus* subgroup. *Philippine Journal of Science, 48*(4), 485–523.
King, W.V. (1939). Varieties of *Anopheles crucians* Wied. *The American Journal of Tropical Medicine and Hygiene, 19*(5), 461–471.
King, W.V., & Baisas, F.E. (1936). A new species and a new variety of Philippine *Anopheles* related to *Anopheles leucosphyrus* Donitz. *Proceedings of the Entomological Society of Washington, 38*(5), 79–89.
King, W.V., & Bradley, G.H. (1937). Notes on *Culex erraticus* and related species in the United States (Diptera, Culicidae). *Annals of the Entomological Society of America, 30*(2), 345–357.
King, W.V., & Bradley, G.H. (1941). General morphology of *Anopheles* and classification of the Nearctic species. *Publication of the American Association for the Advancement of Science, 15*, 63–70.
King, W.V., & Hoogstraal, H. (1946a). The New Guinea species of *Culex* (*Culiciomyia*), with descriptions of two new species. *Proceedings of the Biological Society of Washington, 59*, 143–153.
King, W.V., & Hoogstraal, H. (1946b). Two new species of *Aedes* (*Finlaya*) from New Guinea (Diptera: Culicidae). *Proceedings of the Entomological Society of Washington, 48*(2), 37–38.
King, W.V., & Hoogstraal, H. (1946c). Three new species of *Aedes* from Netherlands New Guinea. *Proceedings of the Entomological Society of Washington, 48*(4), 95–106.
King, W.V., & Hoogstraal, H. (1946d). Species of *Aedes* (*Finlaya*) of the *papuensis* group in the Australasian Region (Diptera, Culicidae). *Proceedings of the Entomological Society of Washington, 48*(6), 135–157.
King, W.V., & Hoogstraal, H. (1946e). Two new species of mosquitoes of the genus *Ficalbia* from Netherlands New Guinea. *Proceedings of the Entomological Society of Washington, 48*(7), 186–190.
King, W.V., & Hoogstraal, H. (1946f). Descriptions of three new species of mosquitoes of the genus *Aedes*, subgenus *Finlaya*, from New Guinea. *Journal of the Washington Academy of Sciences, 36*(9), 305–313.
King, W.V., & Hoogstraal, H. (1947a). New Guinea species of mosquitoes of the genus *Aedes*, subgenus *Aedes*. *Journal of the Washington Academy of Sciences, 37*(4), 113–134.
King, W.V., & Hoogstraal, H. (1947b). Two new species of *Culex* (*Neoculex*) from New Guinea (Diptera, Culicidae). *Proceedings of the Entomological Society of Washington, 49*(2), 65–69.
King, W.V., & Hoogstraal, H. (1947c). New species of New Guinea *Uranotaenia* of the Tibialis group (Diptera: Culicidae). *Annals of the Entomological Society of America, 39*(for 1946)(4), 585–596.
King, W.V., & Hoogstraal, H. (1955). Three new species of New Guinea *Culex*, subgenus *Lophoceraomyia*, with notes on other species (Diptera, Culicidae). *Proceedings of the Entomological Society of Washington, 57*(1), 1–11.
King, W.V., Bradley, G.H., Smith, C.N., & McDuffie, W.C. (1960). A handbook of the mosquitoes of the southeastern United States. *Agricultural Handbook No. 173*. Agricultural Research Service, US Department of Agriculture.
Kipyab, P.C., Khaemba, B.M., Mwangangi, J.M., & Mbogo, C.M. (2013). The bionomics of *Anopheles merus* (Diptera: Culicidae) along the Kenyan coast. *Parasites and Vectors, 6*, 37–42.
Kirby, W. (1837). Fauna Boreali-Americana. Part the fourth and last. The insects. In J. Richardson (Ed.), *Fauna Boreali-Americana; or the Zoolgy of the northern parts of British America: containing descriptions of the objects of natural collected on the late Northern Land Expeditions, under command of Captain Sir John Franklin, R.N. Josiah Fletcher, Norwich, U.K.* [from Harbach 2018c].
Kirk, H.B. (1923). Notes on the mating-habits and early life-history of the culicid *Opifex fuscus* Hutton. *Transactions of the Proceedings of the New Zealand Institute, 54*, 400–406.
Kirkpatrick, T.W. (1925a). Preliminary notes on the determination of Egyptian mosquitos, with descriptions of three species new to science. *Bulletin of the Egyptian Entomological Society, 8*(for 1924), 362–391.
Kirkpatrick, T.W. (1925b). *The mosquitoes of Egypt*. Egyptian Government Anti-Malaria Commission.
Kirkpatrick, T.W. (1926). *Culex* (*Lasiosiphon*) *adairi*, Nom. Nov. *Bulletin Société Entomologique d'Égypte, 10*, 112.
Kirti, J.S. (2014a). Description of male and female genitalia of *Anopheles* (*Cellia*) *culicifacies* Giles with the aid of scanning electron microscope. *International Journal of Pure and Applied Zoology, 2*(4), 348–350.
Kirti, J.S. (2014b). Life history studies of *Anopheles* (*Cellia*) *subpictus* Grassi (Anophelinae: Culicidae) with the aid of scanning electron microscopy (SEM) from Punjab. *Journal of Entomology and Zoology Studies, 2*(5), 345–350.
Kirti, J.S., & Kaur, J. (2004). Further studies on two Indian species of subgenus *Lutzia* Theobald of genus *Culex* Linnaeus (Diptera: Culicidae). *Entomon, 29*(1), 69–73.
Kiszewski, A.E., & Spielman, A. (1998). Spatially explicit model of transposon-based genetic drive mechanisms for displacing fluctuating populations of anopheline vector mosquitoes. *Journal of Medical Entomology, 35*(4), 584–590.
Kiszewski, A., Mellinger, A., Spielman, A., Malaney, P., Sachs, S.E., & Sachs, J. (2004). A global index representing the stability of malaria transmission. *The American Journal of Tropical Medicine and Hygiene, 70*(5), 486–498.
Kitching, I.J., Culverwell, C.L., & Harbach, R.E. (2015). The phylogenetic conundrum of *Lutzia* (Diptera: Culicidae): A cautionary account of conflict and support. *Insect Systematics and Evolution, 46*(3), 269–290.
Kittayapong, P., & Strickman, D. (1993). Distribution of container-inhabiting *Aedes* larvae (Diptera: Culicidae) at a dengue focus in Thailand. *Journal of Medical Entomology, 30*(3), 601–606.
Kittayapong, P., Baisley, K.J., Baimai, V., & O'Neill, S.L. (2000). Distribution and diversity of *Wolbachia* infections in Southeast Asian mosquitoes (Diptera: Culicidae). *Journal of Medical Entomology, 37*(3), 340–345.
Kitzmiller, J.B. (1963). Mosquito cytogenetics: A review of the literature, 1953–62. *Bulletin of the World Health Organization, 29*, 345–355.
Kitzmiller, J.B. (1967). Mosquito cytogenetics. In J.W. Wright & R. Pal (Eds.), *Genetics of insect vectors of disease* (pp. 133–150). Amsterdam, The Netherlands: Elsevier.
Kitzmiller, J.B. (1976). *Genetics, cytogenetics, and evolution of mosquitoes*. Academic Press. (Reprinted from *Advances in Genetics, 18*, 315–436)
Kitzmiller, J.B. (1977). Chromosomal differences among species of *Anopheles* mosquitoes. *Mosquito Systematics, 9*(2), 112–122.
Kitzmiller, J.B. (1982). *Anopheline names: Their derivations and histories* (Vol. VIII, Thomas Say Foundation Series). College Park, MD: Entomological Society of America.
Kitzmiller, J.B., & Baker, R.H. (1963). The salivary chromosomes of *Anopheles freeborni*. *Mosquito News, 23*(3), 254–261.
Kitzmiller, J.B., & Baker, R.H. (1965). The salivary chromosomes of *Anopheles earlei*. *Canadian Journal of Genetics and Cytology, 7*(2), 275–283.
Kitzmiller, J.B., & Chow, G.Y. (1971). The salivary gland chromosomes of *Anopheles aquasalis*. *Revista Brasileira de malariologia y doenças tropicais, 23*, 65–85.
Kitzmiller, J.B., & Laven, H. (1954). Hybridization experiments with *Culex molestus*, *Culex pipiens* and *Culex fatigans*. *Atti del IX Congresso Internazionale di Genetica, 2*, 767–771.

Kitzmiller, J.B., Baker, R.H., & Chowdaiah, B.N. (1966). The salivary gland chromosomes of *Anopheles neomaculipalpus. Caryologia, 19*(1), 1–12.

Kitzmiller, J.B., Frizzi, G., & Baker, R.H. (1967). Evolution and speciation within the *maculipennis* complex of the genus *Anopheles.* In J.W. Wright & R. Pal (Eds.), *Genetics of Insect Vectors of Disease* (pp. 151–210). Amsterdam: Elsevier.

Kitzmiller, J.B., Kreutzer, R.D., & Tallaferro, E. (1973). Chromosomal differences in populations of *Anopheles nuneztovari. Bulletin of WHO, 48*, 435–455.

Kitzmiller, J.B., Joslyn, D., & Kreutzer, R.D. (1974). The salivary gland chromosomes of *Anopheles walkeri* Theobald. *Mosquito News, 34*(1), 22–28.

Kitzmiller, J.B., Kreutzer, R.D., & Rabbani, M.G. (1976). Cytotaxonomy and salivary gland chromosomes of *Anopheles* (*Stethomyia*) *kompi. Mosquito News, 36*(4), 483–487.

Klein, J.M. (1956). Étude morphologique de l'hypopygium male de *Culex* (*Neoculex*) *rima* Theob., 1901 (Diptères: Culicides). *Bulletin de la Société de pathologie exotique, 49*, 921–935.

Klein, J.M. (1969). A new species of *Ficalbia* (*Ravenalites*) from Cambodia (Diptera: Culicidae). *Proceedings of the Entomological Society of Washington, 71*(2), 215–217.

Klein, J.M. (1970). A new species of *Culex* (*Neoculex*) from Cambodia (Diptera: Culicidae). *Proceedings of the Entomological Society of Washington, 72*(4), 504–506.

Klein, J.M. (1973). Contributions to the mosquito fauna of Southeast Asia. XVII. The Cambodian *Aedes* (*Neomacleaya*) species with some new descriptions (Diptera: Culicidae). *Contributions of the American Entomological Institute, 10*(1), 1–21.

Klein, J.M. (1977a). Deux nouvelles espècies de *Topomyia* du Cambodge (Diptera, Culicidae). *Cahiers ORSTOM. Série entomologie médicale et parasitologie, 15*(2), 123–129.

Klein, J.M. (1977b). La faune des moustiques du Cambodge. I. Anophelinae (Diptera, Culicidae). *Cahiers ORSTOM. Série entomologie médicale et parasitologie, 15*(2), 107–122.

Klein, J.M., & Marks, E.N. (1960). Australian mosquitoes described by Macquart. I. Species in the Paris Museum, *Aedes* (*Finlaya*) *alboannulatus* (Macquart), *Aedes* (*Finlaya*) *rubrithorax* (Macquart), *Aedes* (*Ochlerotatus*) *albirostris* (Macquart). New synonymy and a new species from New Zealand. *Proceedings of the Linnean Society of New South Wales, 85*(1), 107–116.

Klein, J.M., & Sirivanakarn, S. (1970). Four new species of *Culex*, subgenus *Mochthogenes* from Southeast Asia (Diptera: Culicidae). *Proceedings of the Entomological Society of Washington, 71*(for 1969)(4), 582–592.

Klein, J.M., Rivière, F., & Séchan, Y. (1984). Une nouvelle espèce de moustique (Diptera: Culicidae) des îles Marquises (Polynésie Française): *Culex* (*Culex*) *toviiensis* n.sp. Description de la larve et de la nymphe. *Cahiers ORSTOM. Série entomologie médicale et parasitologie, 21*(for 1983), 71–76.

Klein, T.A., Lima, J.B., & Toda-Tang, A.T. (1990). Colonization and maintenance of *Anopheles deaneorum* in Brazil. *Journal of the American Mosquito Control Association, 6*(3), 510–513.

Klein, T.A., Lima, J.B., & Toda-Tang, A. (1991). Hybridization evidence supporting separate species status for *Anopheles albitarsis* and *Anopheles deaneorum* (Diptera: Culicidae) in Brazil. *Journal of the American Mosquito Control Association, 7*(2), 301–303.

Kliewer, J.W. (1961). Weight and hatchability of *Aedes aegypti* eggs (Diptera: Culicidae). *Annals of the Entomological Society of America, 54*(6), 912–917.

Klowden, M.J. (2001). Sexual receptivity in *Anopheles gambiae* mosquitoes: Absence of control by male accessory gland substances. *Journal of Insect Physiology, 47*(7), 661–666.

Klowden, M.J. (2006). Switchover to the mated state by spermathecal activation in female *Anopheles gambiae* mosquitoes. *Journal of Insect Physiology, 52*(7), 679–684.

Klowden, M.J. (2007). Making generalizations about vectors: Is there a physiology of "the mosquito"? *Entomological Research, 37*(1), 1–13.

Klowden, M.J. (2009). Spermatheca. In V.H. Resh & R.T. Cardé (Eds.), *Encyclopedia of insects* (2nd ed., pp. 939–940). Academic Press.

Klowden, M.J., & Chambers, G.M. (2004). Production of polymorphic sperm by anopheline mosquitoes and their fate within the female genital tract. *Journal of Insect Physiology, 50*(12), 1163–1170.

Knab, F. (1904). The eggs of *Culex territans* Walker. *Journal of the New York Entomological Society, 12*(4), 246–248.

Knab, F. (1907a). A new species of *Megarhinus. The Canadian Entomologist, 39*(2), 50–51.

Knab, F. (1907b). A new genus and species of sabethid mosquito. *Journal of the New York Entomological Society, 15*(3), 120–121.

Knab, F. (1909). Two new Philippine Culicidae. *Entomological News, 20*(9), 386–388.

Knab, F. (1913a). A new bromelicolous *Megarhinus* (Diptera; Culicidae). *Insecutor Inscitiae Menstruus, 1*(3), 35–36.

Knab, F. (1913b). The species of *Anopheles* that transmit human malaria. *American Journal of Tropical Diseases and Preventive Medicine, 1*, 33–43.

Knab, F. (1916a). The earliest name of the yellow fever mosquito. *Insecutor Inscitiae Menstruus, 4*(4–6), 59–60.

Knab, F. (1916b). A new mosquito from the eastern United States. *Proceedings of the Biological Society of Washington, 29*, 161–163.

Knight, J.W., & Haeger, J.S. (1971). Key to adults of the *Culex* subgenera *Melanoconion* and *Mochlostyrax* of eastern North America. *Journal of Medical Entomology, 18*, 551–555.

Knight, J.W., & Nayar, J.K. (2004). Identification of four common *Culex* (*Culex*) (Diptera: Culicidae) species from Florida with isoenzyme analysis. *Florida Entomologist, 87*(1), 1–5.

Knight, K.L. (1946). The *Aedes* (*Finlaya*) *niveus* subgroup of Oriental mosquitoes. *Journal of the Washington Academy of Sciences, 36*(8), 270–280.

Knight, K.L. (1947a). The *Aedes* (*Mucidus*) mosquitoes of the Pacific (Diptera: Culicidae). *Journal of the Washington Academy of Sciences, 37*(9), 315–325.

Knight, K.L. (1947b). A new species of *Aedes* (*Christophersiomyia*) from the Philippines (Diptera: Culicidae). *Proceedings of the Biological Society of Washington, 60*, 73–76.

Knight, K.L. (1948a). The *Aedes* (*Finlaya*) *albotaeniatus* group of mosquitoes (Diptera, Culicidae). *Proceedings of the Entomological Society of Washington, 50*(1), 1–8.

Knight, K.L. (1948b). A new name for an *Aedes* (*Mucidus*) subspecies of the Solomon Islands, with two additional notes on *Aedes* (*Mucidus*) species (Diptera, Culicidae). *Proceedings of the Entomological Society of Washington, 50*(3), 68–69.

Knight, K.L. (1948c). The *Aedes* (*Finlaya*) *chrysolineatus* group of mosquitoes (Diptera: Culicidae). *Annals of the Entomological Society of America, 40*(for 1947)(4), 624–649.

Knight, K.L. (1951a). The *Aedes* (*Ochlerotatus*) *punctor* subgroup in North America (Diptera, Culicidae). *Annals of the Entomological Society of America, 4*(1), 87–99.

Knight, K.L. (1951b). A review of the *Culex pipiens* complex in the Mediterranean subregion (Diptera, Culicidae). *Transactions of the Royal Entomological Society of London, 102*, 354–364.

Knight, K.L. (1953a). The mosquitoes of the Yemen (Diptera, Culicidae). *Proceedings of the Entomological Society of Washington, 55*(5), 212–234.

Knight, K.L. (1953b). Two new species of mosquitoes from the Yemen (Diptera: Culicidae). *Journal of the Washington Academy of Sciences, 43*(10), 320–325.

Knight, K.L. (1968). Contributions to the mosquito fauna of Southeast Asia. IV. Species of the subgroup *chrysolineatus* of group D, genus *Aedes*, subgenus *Finlaya* Theobald. *Contributions of the American Entomological Institute, 2*(5), 1–45.

Knight, K.L. (1969). A new species of the genus *Aedes*, subgenus *Finlaya* Theobald, from India (Diptera: Culicidae). *Journal of the Kansas Entomological Society, 42*(4), 382–386.

Knight, K.L. (1971). Comparative anatomy of the mandible of the fourth instar mosquito larva (Diptera: Culicidae). *Journal of Medical Entomology, 8*(2), 189–205.

Knight, K.L. (1972). History of mosquito systematics. Part I. Eighteenth century. *Mosquito Systematics, 4*(1), 10–15.

Knight, K.L. (1978a). A new *Aedes* (*Finlaya*) mosquito from Thailand. *Mosquito Systematics, 10*(1), 106–116.

Knight, K.L. (1978b). *Supplement to a catalog of the mosquitoes of the world* (Diptera: Culicidae). Thomas Say Foundation, Entomological Society of America, 6.

Knight, K.L., & Abdel-Malek, A. (1951). A morphological and biological study of *Culex pipiens* in the Cairo area of Egypt (Diptera-Culicidae). *Bulletin de la Société Fouad 1er d'entomologie, 35*, 175–185.

Knight, K.L., & Chamberlain, R.W. (1948). A new nomenclature for the chaetotaxy of the mosquito pupa, based on a comparative study of the genera (Diptera: Culicidae). *Proceedings of the Helminthological Society of Washington, 15*(1), 1–18.

Knight, K.L., & Farner, D.S. (1944). A correction in anopheline nomenclature (Diptera, Culicidae). *Proceedings of the Entomological Society of Washington, 46*(5), 132–133.

Knight, K.L., & Harrison, B.A. (1988). A new *Aedes* (*Finlaya*) of the Niveus-Subgroup (Diptera: Culicidae). *Mosquito Systematics, 19*(for 1987)(3), 212–236.

Knight, K.L., & Hull, W.B. (1951a). Three new species of *Aedes* from the Philippines (Diptera, Culicidae). *Pacific Science, 5*, 197–203.

Knight, K.L., & Hull, W.B. (1951b). The *Aedes* mosquitoes of the Philippine Islands. I. Keys to species. Subgenera *Mucidus, Ochlerotatus,* and *Finlaya* (Diptera, Culicidae). *Pacific Science, 5*(2), 211–251.

Knight, K.L., & Hull, W.B. (1952). The *Aedes* mosquitoes of the Philippine Islands. II. Subgenera *Skusea, Christophersiomyia, Geoskusea, Rhinoskusea,* and *Stegomyia* (Diptera, Culicidae). *Pacific Science, 6*(2), 157–189.

Knight, K.L., & Hull, W.B. (1953). The *Aedes* mosquitoes of the Philippine Islands. III. Subgenera *Aedimorphus, Banksinella, Aedes,* and *Cancraedes* (Diptera, Culicidae). *Pacific Science, 7*(4), 453–481.

Knight, K.L., & Hurlbut, H.S. (1949). The mosquitoes of Ponape Island, eastern Carolinas. *Journal of the Washington Academy of Sciences, 39*(1), 20–34.

Knight, K.L., & Laffoon, J.L. (1946). The oriental species of the *Aedes* (*Finlaya*) *kochi* group (Diptera: Culicidae). *Transactions of the American Entomological Society, 72*(4), 203–225.

Knight, K.L., & Marks, E.N. (1952). An annotated checklist of the mosquitoes of the subgenus *Finlaya*, genus *Aedes*. *Proceedings of the United States National Museum, 101*(3288), 513–574.

Knight, K.L., & Mattingly, P.F. (1950). The *Orthopodomyia anopheloides* subgroup of mosquitoes (Diptera, Culicidae). *Proceedings of the Entomological Society of Washington, 52*(1), 1–20.

Knight, K.L., & Pugh, R.B. (1973). A bibliography of the mosquito publications of Fred V. Theobald 1868–1930. *Mosquito Systematics, 5*(3), 230–234.

Knight, K.L., & Pugh, R.B. (1974). A bibliography of mosquito writings of H.G. Dyar and Frederick Knab. *Mosquito Systematics, 6*(1), 11–26.

Knight, K.L., & Rozeboom, L.E. (1945). A new species of *Culex* from New Guinea (Diptera, Culicidae). *Proceedings of the Entomological Society of Washington, 47*, 289–295.

Knight, K.L., & Rozeboom, L.E. (1946). The *Aedes* (*Stegomyia*) *albolineatus* group (Diptera, Culicidae). *Proceedings of the Biological Society of Washington, 59*, 83–98.

Knight, K.L., & Stone, A. (1977). *A catalog of the mosquitoes of the world* (*Diptera: Culicidae*) (2nd ed.). Thomas Say Foundation, Entomological Society of America, 6.

Knight, K.L., Bohart, R.M., & Bohart, G.E. (1944). *Keys to the mosquitoes of the Australasian Region, including a synopsis of their distribution and breeding habits.* National Research Council, Division of Medical Sciences.

Knipling, E.F. (1955). Possibilities of insect control or eradication through the use of sexually sterile males. *Journal of Economic Entomology, 48*(4), 459–462.

Knols, B.G.J., & Takken, W. (2015). Alarm bells ringing: More of the same, and new and novel diseases and pests. In W. Takken & B.G.J. Knols (Eds.), *Emerging pests and vector-borne diseases in Europe* (pp. 13–19). Wageningen, The Netherlands: Wageningen Academic.

Knudsen, A.B. (1995). Global distribution and continuing spread of *Aedes albopictus*. *Parassitologia, 37*(2–3), 91–97.

Koekemoer, L.L., Misiani, E.A., Hunt, R.H., Kent, R.J., Norris, D.E., & Coetzee, M. (2009). Cryptic species within *Anopheles longipalpis* from southern Africa and phylogenetic comparison with members of the *An. funestus* group. *Bulletin of Entomological Research, 99*(1), 41–49.

Koenraadt, C.J.M., & Harrington, L.C. (2008). Flushing effect of rain on container-inhabiting mosquitoes *Aedes aegypti* and *Culex pipiens* (Diptera: Culicidae). *Journal of Medical Entomology, 45*(1), 28–35.

Koenraadt, C.J.M., Paaijmans, K.P., Githeko, A.K., Knols, B.G.J., & Takken, W. (2003). Egg hatching, larval movement and larval survival of the malaria vector *Anopheles gambiae* in desiccating habitats. *Malaria Journal, 2*(1), Article 20.

Kohn, M. (1990). A survey on indoor resting mosquito species in Phnom Penh, Kampuchea. *Folia parasitologica* (Ceske Budejovice), *37*(2), 165–174.

Koidzumi, M. (1917). On the anopheline mosquitoes from Formosa. *Dobutsugaku zasshi* (Tokyo), *29*, 131–142, 351–355.

Koidzumi, M. (1918). On the *Stegomyia* mosquitoes of Taiwan. *Transactions of the Natural History Society of Formosa, 38*, 139–145.

Koidzumi, M. (1920). Preventive medical study of mosquitoes in Formosa. Daiwan Kenkyujo Hokoku [Report of the Formosa Government] 8, 1–158.

Koidzumi, M. (1924). The anophelines of Formosa. In A.S. Hoops & J.W. Scharff (Eds.), *Far Eastern Association of Tropical Medicine—Transactions of the Fifth Biennial Congress held at Singapore, 1924* (pp. 96–101). London, England: Bale, Sons and Danielsson.

Kok, D.J. (1979). *Culex* (*Maillotia*) *pulchrithorax* Edwards: Some ecological observations and a description of the fourth-instar larva. *Journal of the Entomological Society of Southern Africa, 42*(2), 161–167.

Kokkinn, M.J., Duval, D.J., & Williams, C.R. (2009). Modelling the ecology of the coastal mosquitoes *Aedes vigilax* and *Aedes camptorhynchus* at Port Pirie, South Australia. *Medical and Veterinary Entomology, 23*(1), 85–91.

Komp, W.H.W. (1923). A guide to the identification of common mosquitoes of the Southeastern United States. *Public Health Reports, 38*(20), 1061–1080.

Komp, W.H.W. (1926). A *Culex* from Honduras. *Insecutor Inscitiae Menstruus, 14*, 44–45.

Komp, W.H.W. (1932). A new *Culex*, *Culex vomerifer*, from Panama (Dipt., Culicidae). *Psyche, 39*(3), 79–82.

Komp, W.H.W. (1935). Notes on the validity of the types of the species in the subgenera *Mochlostyrax* and *Melanoconion* in the U.S. National Museum. (Diptera, Culicidae). *Proceedings of the Entomological Society of Washington, 37*(1), 1–11.

Komp, W.H.W. (1936a). The male and larva of *Aedes dominicii* Rangel & Romero Sierra, and the male of *Aedes pseudodomincii* sp. nov., representatives of a new subgenus (*Soperia*) of the genus *Aedes*, from Colombia. *Proceedings of the Entomological Society of Washington, 38*(4), 71–75.

Komp, W.H.W. (1936b). *Anopheles* (*Anopheles*) *chiriquiensis*, a new species of *Anopheles* Panama (Diptera, Culicidae). *Proceedings of the Entomological Society of Washington, 38*(7), 156–160.

Komp, W.H.W. (1936c). Description of nine new species of *Culex*, seven from Panama and two from Venezuela (Diptera, Culicidae). *Annals of the Entomological Society of America, 29*(2), 319–334.

Komp, W.H.W. (1936d). *Anopheles* (*Nyssorhynchus*) *anomalophyllus*, a new species of *Anopheles* from Panama and Costa Rica (Dipt., Culicidae). *Proceedings of the Entomological Society of Washington, 38*(7), 160–164.

Komp, W.H.W. (1937a). *Anopheles acanthotorynus*, a new species of the subgenus *Stethomyia* from Peru (Diptera, Culicidae). *Annals of the Entomological Society of America, 30*(2), 358–360.

Komp, W.H.W. (1937b). The species of the subgenus *Kerteszia* of *Anopheles* (Diptera, Culicidae). *Annals of the Entomological Society of America, 30*(3), 492–529.

Komp, W.H.W. (1938). *Aedes leucotaeniatus*, a new species of *Aedes* allied to *A. leucocelaenus* D.&S.; and descriptions of the male and larva of *A. leucocelaenus* D. & S. (Diptera, Culicidae). *Proceedings of the Entomological Society of Washington, 40*(8), 260–266.

Komp, W.H.W. (1940). The occurrence of *Anopheles* Root in British Honduras and Guatemala. *Public Health Reports, 55*(16), 693–694.

Komp, W.H.W. (1941). The species of *Nyssorhynchus* confused under *tarsimaculatus* Goeldi, and a new name, *A. emilianus*, for one species found in Para, Brazil (Diptera, Culicidae). *Annals of the Entomological Society of America, 34*(4), 791–807.

Komp, W.H.W. (1942). The anopheline mosquitoes of the Caribbean Region. *National Institute of Health Bulletin, 179*. Washington, DC: US Government Printing Office.

Komp, W.H.W. (1943). *Anopheles clarki*, a new species of *Nyssorhynchus* of wide distribution in South America. (Diptera: Culicidae). *Proceedings of the Entomological Society of Washington, 44*(for 1942)(9), 196–201.

Komp, W.H.W. (1949). *Aedes* (*Ochlerotatus*) *nubilus* Theobald 1903 a synonym of *Aedes* (*Ochlerotatus*) *serratus* Theobald, 1901 (Diptera, Culicidae). *Proceedings of the Entomological Society of Washington, 51*(3), 105–114.

Komp, W.H.W. (1953). Notes on the larva of *Haemagogus janthinomys* Dyar (Diptera, Culicidae). *Proceedings of the Entomological Society of Washington, 57*(3), 137–138.

Komp, W.H.W. (1954a). The specific identity of two species of *Haemagogus* (Diptera, Culicidae). *Proceedings of the Entomological Society of Washington, 56*, 49–54.

Komp, W.H.W. (1954b). The male of *Haemagogus albomaculatus* Theobald (Diptera, Culicidae). *Proceedings of the Entomological Society of Washington, 56*(2), 148–153.

Komp, W.H.W. (1954c). *Haemagogus lucifer* H., D. and K., 1912, a synonym of *Haemagogus regalis* D.&K., 1906. *Proceedings of the Entomological Society of Washington, 56*(4), 193–195.

Komp, W.H.W. (1955a). The oviposition of *Haemagogus equinus* in nature (Diptera, Culicidae). *Mosquito News, 15*(3), 163–164.

Komp, W.H.W. (1955b). Notes on *Haemagogus iridicolor* Dyar (Diptera, Culicidae). *Proceedings of the Entomological Society of Washington, 57*(1), 29–31.

Komp, W.H.W. (1955c). Notes on the larva of *Haemagogus panarchys* Dyar (Diptera, Culicidae). *Proceedings of the Entomological Society of Washington, 57*(5), 237–239.

Komp, W.H.W. (1955d). The taxonomic status of *Haemagogus janthinomys* Dyar (Diptera, Culicidae). *Proceedings of the Entomological Society of Washington, 57*(6), 277–280.

Komp, W.H.W. (1956). Notes on mosquitoes from an area of endemic yellow fever in Colombia (Diptera, Culicidae). *Proceedings of the Entomological Society of Washington, 58*(1), 37–42.

Komp, W.H.W., & Brown, C.G. (1935). *Culex jubifer*, a new species of *Culex* from Panama (Diptera: Culicidae). *Annals of the Entomological Society of America, 28*(2), 254–255.

Komp, W.H.W., & Curry, D.P. (1932). A new *Culex* from Panama (Dipt., Culicidae). *Psyche, 39*(3), 82–84.

Komp, W.H.W., & Kumm, H.W. (1938). A new species of *Haemagogus*, *mesodentatus*, from Costa Rica, and a description of the larva of *Haemagogus anastasionis* Dyar (Diptera, Culicidae). *Proceedings of the Entomological Society of Washington, 40*(9), 253–259.

Komp, W.H.W., & Osorno-Mesa, E. (1936). The male and larva of *Anopheles* (*Kerteszia*) *boliviensis* Theobald. (Diptera, Culicidae). *Annals of the Entomological Society of America, 29*(3), 415–419.

Komp, W.H.W., & Rozeboom, L.E. (1951). Descriptions of eight new species of *Culex* subgenus *Melanoconion* (Diptera, Culicidae). *Proceedings of the Entomological Society of Washington, 53*(3), 121–137.

Kong, X.Q., Liu, J.L., Zhang, W.J., & Qiu, Y.D. (2015). Load-bearing ability of the mosquito tarsus on water surfaces arising from its flexibility. *AIP Advances, 5*(3), 037101.

Korvenkontio, P., Lokki, J., Saura, A., & Ulmanen, I. (1979). *Anopheles maculipenis* complex (Diptera: Culicidae) in northern Europe: Species diagnosis by egg structure and enzyme polymorphism. *Journal of Medical Entomology, 16*(2), 169–170.

Koštál, V. (2006). Eco-physiological phases of insect diapause. *Journal of Insect Physiology 52*(2), 113–127.

Kraemer, M.U., Sinka, M.E., Duda, K.A., Mylne, A.Q., Shearer, F.M., Barker, C.M., . . . Hay, S.I. (2015). The global distribution of the arbovirus vectors *Aedes aegypti* and *Ae. albopictus*. *eLife*. doi:10.7554/eLife.08347.001

Kramar, J. (1958). Komari Bodavi - Culicinae (Rad: Dvolukridi - Diptera). Fauna CSR, N. 13, 1-286 Praha.

Kreutzer, R.D. (1977). Hybridization between the Nearctic *Anopheles punctipennis* and the Paleartic *Anopheles atroparvus*. *Canadian Journal of Genetics and Cytology, 19*(2), 265–270.

Kreutzer, R.D. (1978). A mosquito with eight chromosomes: *Chagasia bathana* Dyar. *Mosquito News, 38*(4), 554–558.

Kreutzer, R.D., & Galindo, P. (1980). Isozyme studies of two *Melanoconion* mosquitoes, *Culex ocossa* and *Cx. panocossa*. *Mosquito News, 40*(4), 605–613.

Kreutzer, R.D., & Kitzmiller, J.B. (1970). Hybridization and speciation in *Anopheles crucians*. *American Society of Zoologists, 10*, 557.

Kreutzer, R.D., & Kitzmiller, J.B. (1971a). Chromosomal similarity between *Anopheles perplexens* and *Anopheles punctipennis*. *Mosquito News, 31*(3), 409–415.

Kreutzer, R.D., & Kitzmiller, J.B. (1971b). Hybridization between *Anopheles crucians* and *Anopheles bradleyi*. *Evolution, 25*(1), 195–206.

Kreutzer, R.D., Narang, S.L., & Kitzmiller, J.B. (1969). The salivary chromosomes of *Anopheles atropos*. *Mosquito News, 29*(2), 223–230.

Kreutzer, R.D., Tadano, T., Narang, S.L., & Kitzmiller, J.B. (1969). The salivary gland chromosomes of *Anopheles punctimacula. Revista Brasileira de malariologia y doencas tropicais, 21*(3), 559–570.

Kreutzer, R.D., Narang, S.L., & Kitzmiller, J.B. (1970). A comparison of the salivary gland chromosomes of *Anopheles crucians* and *Anopheles bradleyi. Cytologia, 35*, 527–551.

Kreutzer, R.D., Kitzmiller, J.B., & Ferreira, E. (1972). Inversion polymorphism in the salivary gland chromosomes of *Anopheles darlingi* Root. *Mosquito News, 32*(4), 555–565.

Kreutzer, R.D., Kitzmiller, J.B., & Rabbani, M.G. (1975). The salivary gland chromosomes of *Anopheles argyritarsis* compared with those of certain other species in the subgenus *Nyssorhynchus. Mosquito News, 35*(3), 354–365.

Kreutzer, R.D., Kitzmiller, J.B., & Rabbani, M.G. (1976). Cytogenetically distinguishable sympatric and allopatric populations of the mosquito *Anopheles albitarsis. Acta Amazonica, 6*(4), 473–481.

Krishnamoorthy, R., Munirathinam, A., Dhananjeyan, K.J., Hiriyan, J., Mariappan, T., Philip Samuel, P., & Venkatesh, A. (2013). Description of a new species, *Toxorhynchites* (*Toxorhynchites*) *tyagii* (Diptera: Culicidae), from Nilgiri hills, Western Ghats, southern India. *Zootaxa, 3701*(4), 447–459.

Krzywinski, J., & Besansky, N.J. (2003). Molecular systematics of *Anopheles*: From subgenera to subpopulations. *Annual Review of Entomology, 48*(1), 111–139.

Krzywinski, J., Wilkerson, R.C., & Besansky, N J. (2001a). Evolution of mitochondrial and ribosomal gene sequences in Anophelinae (Diptera; Culicidae): implications for phylogeny reconstruction. *Molecular Phylogenetics and Evolution, 18*(3), 479–487.

Krzywinski, J., Wilkerson, R.C., & Besansky, N.J. (2001b). Toward understanding Anophelinae (Diptera: Culicidae) phylogeny: Insights from nuclear single-copy genes and weight of evidence. *Systematic Biology, 50*(4), 540–556.

Krzywinski, J., Grushko, O.G., & Besansky, N.J. (2006). Analysis of the complete mitochondrial DNA from *Anopheles funestus*: An improved dipteran mitochondrial dimension of mosquito evolution. *Molecular Phylogenetics and Evolution, 39*(2), 417–423.

Krzywinski, J., Li, C., Morris, M., Conn, J.E., Lima, J.B., Povoa, M.M., & Wilkerson, R.C. (2011). Analysis of the evolutionary forces shaping mitochondrial genomes of a Neotropical malaria vector complex. *Molecular Phylogenetics and Evolution, 58*(3), 469–477.

Krzywinska, E., Dennison, N.J., Lycett, G.J., & Krzywinski, J. (2016). A maleness gene in the malaria mosquito *Anopheles gambiae. Science, 353*(6294), 67–69.

Kuchartshuk, L.P. (1973). New species of mosquito from the genus *Aedes* (Diptera, Culicidae). In *New and little-known species of Siberian fauna* (Part 7, pp. 98–101) [in Russian]. Nauka, Novosibirsk, USSR Academy of Sciences, Siberian Branch, Institute of Biology.

Kuhlow, F. (1962). Beobachtungen und Experimente uber den *Anopheles gambiae* Komplex, Abtrennung von *Anopheles tangensis* n. sp.? *Z Tropenmed Parasitol* (Stuttgart), *13*, 442–449.

Kulasekera, V.L., Harrison, B.A., & Amerasinghe, F.P. (1989). *Anopheles* (*Anopheles*) *peytoni* new species, the *An. insulaeflorum* auct. from Sri Lanka (Diptera: Culicidae). *Mosquito Systematics, 20*(for 1988)(3), 302–316.

Kulasekera, V., Knight, K.L., & Harbach, R.E. (1990). *Aedes* (*Finlaya*) *axitiosus*, new species of the Niveus subgroup (Diptera: Culicidae) from East Malaysia. *Mosquito Systematics, 22*(1), 26–33.

Kulkarni, M.A., Desrochers, R.E., & Kerr, J.T. (2010). High resolution niche models of malaria vectors in northern Tanzania: A new capacity to predict malaria risk? *PLOS ONE, 5*(2), e9396.

Kumar, N.P., Krishnamoorthy, N., Sahu, S.S., Rajavel, A.R., Sabesan, S., & Jambulingam, P. (2013). DNA Barcodes indicate members of the *Anopheles fluviatilis* (Diptera: Culicidae) species complex to be conspecific in India. *Molecular Ecology Resources, 13*(3), 354–361.

Kumar, R.B., Shanmugapriya, B., Thiyagesan, K., Kumar, S.R., & Xavier, S.M. (2010). A search for mosquito larvicidal compounds by blocking the sterol carrying protein, AeSCP-2, through computational screening and docking strategies. *Pharmacognosy Research, 2*(4), 247–255.

Kumar, S., Thomas, A., Samuel, T., Sahgal, A., Verma, A., & Pillai, M.K.K. (2009). Diminished reproductive fitness associated with the deltamethrin resistance in an Indian strain of dengue vector mosquito, *Aedes aegypti* L. *Tropical Biomedicine, 26*(2), 155–164.

Kumar, S., Molina-Cruz, A., Gupta L., Rodrigues, J., & Barillas-Mury, C. (2010). A peroxidase/dual oxidase system modulates midgut epithelial immunity in *Anopheles gambiae. Science, 327*(5973), 1644–1648.

Kumm, H.W. (1933). Mosquitoes breeding in bromeliads, at Bahia, Brazil. *Bulletin of Entomological Research, 24*(4), 561–573.

Kumm, H.W. (1941). The eggs of some Costa Rican anophelines. *The American Journal of Tropical Medicine and Hygiene, 21*(1), 91–102.

Kumm, H.W., & Cerqueira, N.L. (1951). The *Haemagogus* mosquitoes of Brazil. *Bulletin of Entomological Research, 42*(1), 169–181.

Kumm, H.W., & Komp, W.H.W. (1941). *Aedes* (*Howardina*) *allotecnon*, a new species of *Aedes* from Costa Rica, and a description of the larva, adult, and male terminalia of *Aedes quadrivittatus* Coq. *Proceedings of the Entomological Society of Washington, 43*(2), 17–25.

Kumm, H.W., & Novis, O. (1938). Mosquito studies on the Ilha de Marajú, Pará, Brazil. *American Journal of Hygiene, 27*(3), 498–515.

Kumm, H.W., Komp, W.H.W., & Ruiz, H. (1940). The mosquitoes of Costa Rica. *The American Journal of Tropical Medicine and Hygiene, 20*(3), 385–422.

Kumm, H.W., Osorno-Mesa, E., & Boshell-Manrique, J. (1946). Studies on mosquitoes of the genus *Haemagogus* in Columbia (Diptera, Culicidae). *American Journal of Hygiene, 43*(1), 13–28.

Kutateladze, T., Zangaladze, E., Dolidze, N., Mamatsashvili, T., Andrews, E.S., & Haddow, A.D. (2016). First record of *Aedes albopictus* in Georgia and updated checklist of reported species. *Journal of the American Mosquito Control Association, 32*(3), 230–233.

Kutz, F.W., & Darsie, R.F., Jr. (1963). Notes on American mosquito pupae. III. Description of Nearctic subgenus *Mansonia* and key to all Nearctic Species of the genus *Mansonia. Proceedings of the Entomological Society of Washington, 65*(4), 281–286.

Kwon, H., & Pietrantonio, P.V. (2013). Calcitonin receptor 1 (*Aedae*-GPCRCAL1) hindgut expression and direct role in myotropic action in females of the mosquito *Aedes aegypti* (L.). *Insect Biochemistry and Molecular Biology, 43*(7), 588–593.

Kwon, H., Lu, H-L., Longnecker, M.T., & Pietrantonio, P.V. (2012). Role in duresis of a calcitonin receptor (GPRCAL1) expressed in a distal-proximal gradient in renal organs of the mosquito *Aedes aegypti* (L.). *PLOS ONE, 7*(11), e50374.

Kyalo, D., Amratia, P., Mundia, C.W., Mbogo, C.M., Coetzee, M., & Snow, R.W. (2017). A geo-coded inventory of anophelines in the Afrotropical Region south of the Sahara: 1898–2016. *Wellcome Open Research, 2*, 57.

L'Ambert, G., Ferré, J.B., Scaffner, F., & Fontenille, D. (2012). Comparison of different trapping methods for surveillance of mosquito vectors of West Nile virus in Rhône Delta, France. *Journal of Vector Ecology, 37*(2), 269–275.

La Casse, W.J., & Yamaguti, S. (1948). *Mosquito fauna of Japan and Korea*. Office of the Surgeon, Headquarters I Corps, Japan.

La Casse, W.J., & Yamaguti, S. (1950). *Mosquito fauna of Japan and Korea*. Kyoto, Japan: Office of the Surgeon, Headquarters 8th Army, US Army.

Laarman, J.J. (1958). *Anopheles Neomyzomyia van thieli* n. sp. *Folia scientifica Africae Centralis, 4*, 18.

Laarman, J.J. (1959). A new species of *Anopheles* from a rain-forest in eastern Belgian Congo. *Tropical and Geographical Medicine, 11*(2), 147–156.

Labandeira, C.C. (1998). How old is the flower and the fly? *Science, 280*(5360), 57–59.

Labandeira, C.C. (2005). Fossil history and evolutionary ecology of Diptera and their associations with plants. In D.K. Yeates & B.M. Wiegmann (Eds.), *The evolutionary biology of flies* (pp. 217–273). New York, NY: Columbia University Press.

Labouti, M., Faraj, C., Sadak, A., Harrat, Z. Boubidi, S., Harbach, R.E., El Aouad, R., & Linton, Y.-M. (2011). DNA barcodes confirm the presence of a single member of the *Anopheles maculipennis* group in Morocco and Algeria: *An. sicaulti* is conspecific with *An. labranchiae*. *Acta Tropica, 118*(1), 6–13.

Lacan, A. (1956). Les *Anopheles* de l'Afrique Equatoriale Française et leur repartition. *Annales de parasitologie humaine et comparée, 33*(1–2), 150–170.

Lacey, L.A., & Goettel, M.S. (1995). Current developments in microbial control of insect pests and prospects for the early 21st century. *Entomophaga, 40*(1), 3–27.

Lacey, M.S., & Lake, R.W. (1972). Description of the pupa of *Aedes* (*Ochlerotatus*) *thibaulti* (Diptera: Culicidae). *Florida Entomologist, 55*(1), 89–92.

Lacroix, R., Mukabana, W.R., Gouagna, L.C., & Koella, J.C. (2005). Malaria increases attractiveness of humans to mosquitoes. *PLOS Biology, 3*(9), e298.

Lafferty, K.D. (2009). The ecology of climate change and infectious diseases. *Ecology, 90*(4), 888–890.

Laffoon, J. (1946). The Philippine mosquitoes of the genus *Aedes*, subgenus *Aedes*. *Journal of the Washington Academy of Sciences, 36*(7), 228–245.

Laffoon, J.L., & Knight, K.L. (1971). A mosquito taxonomic glossary. VI. Female genitalia. *Mosquito Systematics, 3*(2), 32–41.

Laffoon, J.L., & Knight, K.L. (1973). A mosquito taxonomic glossary. IX. The larval cranium. *Mosquito Systematics, 5*(1), 31–96.

Lahille, F. (1904). Notes sur la classification des moustiques. *Actas y Trabajos de Segundo Congreso Médico Latino-Americano* (Buenos Aires), *2*, 71–96.

Lahille, F. (1912). Sobre un *Anopheles*, una *Stegomyia* y la notación de las nervaduras de los mosquitos. *Anales del Museo Nacional de Buenos Aires, 23*, 253–263.

Lahondère, C., & Lazzari, C.R. (2012). Mosquitoes cool down during blood feeding to avoid overheating. *Current Biology, 22*(1), 40–45.

Lahondère, C., & Lazzari, C.R. (2013). Thermal stress and thermoregulation during feeding in mosquitoes. In S. Manguin (Ed.), Anopheles *mosquitoes: New insights into malaria vectors* (pp. 511–524). Rijeka, Croatia: InTech.

Laird, M. (1947). Some natural enemies of mosquitoes in the vicinity of Palmalmal, New Britain. *Transactions of the Royal Society of New Zealand, 76*(3), 453–476.

Laird, M. (1954). A mosquito survey in New Caledonia and the Belep Islands, with new locality records for two species of *Culex*. *Bulletin of Entomological Research, 45*(2), 285–293.

Laird, M. (1955). Mosquitos and malaria in the hill country of the New Hebrides and Solomon Islands. *Bulletin of Entomological Research, 46*(2), 275–289.

Laird, M. (1957). A new mosquito from Fiji, *Aedes* (*Finlaya*) *freycinetiae* n. sp. (Diptera: Culicidae). *Pacific Science, 11*(3), 342–351.

Laird, M. (1988). *The natural history of larval mosquito habitats*. London, England: Academic Press/Harcourt Brace Jovanovich.

Lake, R.W. (1963). Description of first-instar larvae of *Aedes cantator* (Coquillett) and *Aedes sollicitans* (Walker). *Entomological News, 74*(9), 253–357.

Lake, R.W. (1968). Descriptions of first -instar larvae of *Aedes aurifer* (Coquillett) and *Aedes grossbecki* Dyar & Knab. (Diptera: Culicidae). *Entomological News, 79*(2), 44–48.

Lamb, R.J., & Smith, S.M. (1980). Comparison of egg size and related life-history characteristics for two predaceous tree-hole mosquitoes (*Toxorhynchites*). *Canadian Journal of Zoology, 58*(11), 2065–2070.

Lambert, D.M. (1979). *Anopheles marshallii* (Theobald) is a complex of species. *Mosquito Systematics, 11*(3), 173–178.

Lambert, D.M. (1981). Cytogenetic evidence of a possible fourth cryptic species within the taxon *Anopheles marshallii* (Theobald) (Diptera: Culicidae) from Northern Natal. *Mosquito Systematics, 13*(2), 168–175.

Lambert, D.M., & Coetzee, M. (1982). A dual genetic and taxonomic approach to the resolution of the mosquito taxon, *Anopheles* (Cellia) *marshallii* (Culicidae). *Systematic Entomology, 7*(3), 321–331.

Lambert, D.M., & Spencer, H.G. (1995). *Speciation and the recognition concept: Theory and application*. Baltimore, MD: Johns Hopkins University Press.

Lambert, M., Pasteur, N., Rioux, J.-A., Delalbre-Belmonte, A., & Balard, Y. (1990). *Aedes caspius* (Pallas, 1771) et *A. dorsalis* (Meigen, 1830) (Diptera, Culicidae). Analyses morphologique et génétique de deux populations sympatriques. Preuves de l'isolement reproductif. *Annales de la Société Entomologique de France, 26*(3), 381–398.

Lambrecht, F.L. (1954). *Anopheles* (*Christya*) *implexus*, Theobald. Description de la variete locale de la vallee du Mosso (Urundi). *Revue zoologique Africaine (supplement botanique), 50*(3–4), 204–209.

Lambrecht, F.L., & Leleup, N. (1951). Note concernant la variablilite d'*Anopheles* (*Pyretophorus*) *christyi* Newstead et Carter. *Revue zoologique Africaine (supplement botanique), 44*(4), 328–332.

Lambrecht, F.L., & van Someren, E.C.C. (1971). Mosquitoes of the Chagos Archipelago, Indian Ocean. *Southeast Asian Journal of Tropical Medicine and Public Health, 2*(4), 483–485.

Lambrechts, L., & Scott, T.W. (2009). Mode of transmission and the evolution of arbovirus virulence in mosquito vectors. *Proceedings of the Royal Society B: Biological Sciences, 276*(1660), 1369–1378.

Lane, C.J. (1982). *Aedes* (Stegomyia) *cretinus* Edwards 1921 (Diptera: Culicidae). *Mosquito Systematics, 14*(2), 81–85.

Lane, C.J. (1992a). *Aedes* (*Ochlerotatus*) *nigrocanus* Martini, its taxonomic status (Diptera: Culicidae). *Mosquito Systematics, 23*(1), 12–15.

Lane, C.J. (1992b). On a collection of Culicinae (Diptera: Culicidae) from Brazil. *Mosquito Systematics, 24*(1), 16–22.

Lane, C.J. (1992c). *Toxorhynchites auranticauda* sp.n., a new Indonesian mosquito and potential biocontroal agent. *Medical and Veterinary Entomology, 6*(3), 301–305.

Lane, C.J. (1994). The data labels of *Aedes* (*Ochlerotatus*) *nigrocanus*. *Mosquito Systematics, 26*(1), 35.

Lane, J. (1936a). II. Notas sobre mosquitos de São Paulo. *Boletim de Inst Hygiene de São Paulo Escola de Hygiene e Saúde Pública do Estado, 60*, 3–15.

Lane, J. (1936b). Notas sobre culicideos de Matto Grosso. *Revista do Museu Paulista da Universidade de São Paulo, 20*, 173–206.

Lane, J. (1936c). Notas sobre mosquitos de São Paulo. *Revista do Museu Paulista da Universidade de São Paulo, 20*, 429–435.

Lane, J. (1939). Catálogo dos mosquitos Neotrópicos. *Boletim biológico, série monográfica, 1*, 1–218.

Lane, J. (1943). Sôbre o gênero *Uranotaenia* (Diptera, Culicidae, Culicini). *Revista de entomología, 14*(1–2), 137–161.

Lane, J. (1944). A tribo Megarhinini no Brasil Meridional. *Revista de entomología, 15*(1–2), 172–190.

Lane, J. (1945a). Os sabetineos da América. (Addenda e Corrigenda). *Revista de entomología, 16*(1–2), 132–157.

Lane, J. (1945b). Quatro espécies novas de *Culex* da Região Neotrópica. *Revista de entomología, 16*(1–2), 204–209.

Lane, J. (1947). The larva, pupa and adults of *Wyeomyia* (*Wyeomyia*) *melanopus* Dyar (Diptera, Culicidae). *Proceedings of the Entomological Society of Washington, 49*(4), 97–101.

Lane, J. (1949). Anophelines of the Neotropical Region. In M.F. Boyd (Ed.), *Malariology: A comprehensive survey of all aspects of this group of diseases from a global standpoint* (Vol. 1, pp. 399–418). Philadelphia, PA: Saunders.

Lane, J. (1951). Synonymy of Neotropical Culicidae (Diptera). *Proceedings of the Entomological Society of Washington, 53*(6), 333–336.

Lane, J. (1953). *Neotropical Culicidae* (Vols. I–II). São Paulo, Brazil: University of São Paulo.

Lane, J. (1972). The pupae of species of *Malaya* Leicester from Asia and Australia (Diptera: Culicidae). *Proceedings of the Entomological Society of Washington, 74*(4), 411–414.

Lane, J., & Antunes, P.C.A. (1937). Nota sobre o gênero *Mansonia*, subgênero *Rhynchotaenia*, com a descripção de uma nova espécie (Diptera, Culicidae). *Revista do Museu Paulista da Universidade de São Paulo, 23*, 224–232.

Lane, J., & Causey, O.R. (1955). Additional data on Sabethini (Diptera, Culicidae). *Proceedings of the Entomological Society of Washington, 57*(1), 11–17.

Lane, J., & Cerqueira, N.L. (1942). Os sabetíneos da América (Diptera, Culicidae). *Arquivos de zoologia do Estado de São Paulo, 3*, 473–849.

Lane, J., & Cerqueira, N.L. (1957). The validity and change of name of two species of *Wyeomyia* (Diptera, Culicidae). *Proceedings of the Entomological Society of Washington, 59*(5), 244.

Lane, J., & Cerqueira, N.L. (1958). The types of *Wyeomyia* (*Wyeomyia*) *dyari* Lane & Cerqueira, 1942. *Proceedings of the Entomological Society of Washington, 60*(3), 135.

Lane, J., & Coutinho, J.O. (1940). *Mansonia* subg. *Rhynchotaenia*: Descrição de duas espécies novaes dados sobre o subgênero (Dipt. Culicidae). *Revista de entomología, 11*(1–2), 589–597.

Lane, J., & Forattini, O.P. (1952). New data on Sabethini. (Diptera, Culicidae). *Proceedings of the Entomological Society of Washington, 54*(5), 254–257.

Lane, J., & Guimarães, L.R. (1937). Notas sobre culicideos de Campos do Jordao. *Anais paulistas de medicina e cirurgia, 22*, 211–217.

Lane, J., & Neghme, A. (1956). Sôbre el *Anopheles* (*Nyssorhynchus*) *pictipennis* Phillippi, 1865. *Biológica, 4*, 83–93.

Lane, J., & Ramalho, G.R. (1960). A new neotropical *Culex*. *Revista Brasileira de entomología, 9*, 173–176.

Lane, J., & Whitman, L. (1943). Novas espécies de *Culex* do Brasil (Diptera, Culicidae). *Revista de entomología, 14*(3), 389–408.

Lane, J., & Whitman, L. (1951). The subgenus *Microculex* in Brazil (Diptera, Culicidae). *Revista Brasileira de biología, 1*(3), 341–366.

Lane, J., Rabello, E.S., & Gaeta, C.H.A. (1947). Sôbre o *Anopheles pseudotibiamaculatus* Galvão e Barretto, 1941. *Arquivos de higiene e saúde pública, 1*(2), 265–269.

Lang, J.T. (1977). Contact sex pheromone in the mosquito *Culiseta inornata* (Diptera: Culicidae). *Journal of Medical Entomology, 14*(4), 448–454.

Lanzaro, G.C., & Eldridge, B.F. (1992). A classical and population genetic description of two new sibling species of *Aedes* (*Ochlerotatus*) *increpitus* Dyar. *Mosquito Systematics, 24*(2), 85–101.

Lanzaro, G.C., & Lee, Y. (2013). Speciation in *Anopheles gambiae*—the distribution of genetic polymorphism and patterns of reproductive isolation among natural populations. In S. Manguin (Ed.), Anopheles *mosquitoes: New insights into malaria vectors* (pp. 173–196). Rijeka, Croatia: InTech.

Lanzaro, G.C., Coller, T.C., & Lee, Y. (2015). Defining genetic, taxonomic, and geographic boundaries among species of the *Psorophora confinnis* (Diptera: Culicidae) complex in North and South America. *Journal of Medical Entomology, 52*(5), 907–917.

Laporta, G.Z., Ramos, D.G., Ribeiro, M.C., & Sallum, M.A.M. (2011). Habitat suitability of *Anopheles* vector species and association with human malaria in the Atlantic Forest in south-eastern Brazil. *Memórias do Instituto Oswaldo Cruz, 106*(Suppl. 1), 239–245.

Laporta, G.Z., Linton, Y.-M., Wilkerson, R.C., Bergo, E.S., Nagaki, S.S., Sant'Ana, D.C., & Sallum, M.A.M. (2015). Malaria vectors in South America: Current and future scenarios. *Parasites and Vectors, 8*, 426.

Lardeux, F., Quispe, V., Tejerina, R., Rodriguez, R., Torrez, L., Bouchité, B., & Chávez, T. (2007). Laboratory colonization of *Anopheles pseudopunctipennis* (Diptera: Culicidae) without forced mating. *Comptes rendus biologies, 330*(8), 571–575.

Larish, L.B., & Savage, H.M. (2005). Introduction and establishment of *Aedes* (*Finlaya*) *japonicus japonicus* (Theobald) on the island of Hawaii: Implications for arbovirus transmission. *Journal of the American Mosquito Control Association, 21*(3), 318–321.

Larivière, M., & Abonnenc, E. (1956). Notes biologiques et morphologie de l'oeuf, de la larve et de l'adulte de *Culex antennatus* Becker 1903. *Bulletin de l'Institut Fondamental d'Afrique Noire, 18*(4), 1191–1199.

Larrousse, F. (1925). Larve de *Culex* a branchies très développées nouvelle pour la faune Française (*Culex lavieri* n. sp.). *Annales de parasitologie humaine et comparée, 3*(1), 68–70.

Larson, S.R., DeGroote, J.P., Bartholomay, L.C., & Sugumaran, R. (2010). Ecological niche modeling of potential West Nile virus vector mosquito species in Iowa. *Journal of Insect Science, 10*(1), 110.

Lassmann, G.W. (1944). Notas sobre una nueva especie de *Psorophora*. *Psorophora pisces* n. sp. (Culicidae, Diptera). *Boletín de salubridad y asistencia, 28–29*, 3–9.

Lassmann, G.W. (1948). Distribución de los *Anopheles* en el Estado de Veracruz. *Boletín de los Servicios coordinados de salubridad y asistencia en el Estado de Veracruz, 4*(43–44), 6–11.

Lassmann, G.W. (1951). *Psorophora* (*Janthinosoma*) *totonaci*, a new species from Mexico (Diptera, Culicidae). *Proceedings of the Entomological Society of Washington, 53*(5), 284–287.

Laurence, B.R. (1960). The biology of two species of mosquito, *Mansonia africana* (Theobald) and *Mansonia uniformis* (Theobald), belonging to the subgenus *Mansonioides* Diptera, Culicidae). *Bulletin of Entomological Research, 51*, 491–517.

Laurito, M., Visintin, A.M., & Almirón, W.R. (2008). *Culex saltanensis* morphological redescription of the immature and adult stages. *Journal of the American Mosquito Control Association, 24*(2), 203–210.

Laurito, M., Almirón, W.R., & Rossi, G.C. (2009). Morphological redescription of immature and adult stages of *Culex* (*Culex*) *acharistus* Root (Diptera: Culicidae). *Zootaxa, 2263*, 21–30.

Laurito, M., Ricardo Almirón, W., & Carlos Rossi, G. (2011). Description of the immature stages of *Culex* (*Culex*) *ameliae* Casal and *Culex* (*Culex*) *articularis* Philippi (Diptera: Culicidae). *Zootaxa, 28*, 58–68.

Laurito, M., de Oliveira, T.M.P., Almirón, W.R., & Sallum, M.A.M. (2013). COI barcode versus morphological identification of *Culex* (*Culex*) (Diptera: Culicidae) species: A case study using samples from Argentina and Brazil. *Memórias do Instituto Oswaldo Cruz, 108*(Suppl. 1), 110–122.

Laurito, M., Ricardo Almirón, W., & Carlos Rossi, G. (2015). Description of the immature stages and redescription of the adults of *Culex (Culex) lahillei* Bachmann & Casal (Diptera: Culicidae). *Zootaxa, 13*, 29–38.

Laurito, M., Briscoe, A.G., Almirón, W.R., & Harbach, R.E. (2017). Systematics of the *Culex coronator* complex (Diptera: Culicidae): Morphological and molecular assessment. *Zoological Journal of the Linnean Society, 182*(4), 735–757.

Laven, H. (1951). Crossing experiments with *Culex* strains. *Evolution, 5*(4), 370–375.

Laven, H. (1959). Speciation by cytoplasmic isolation in the *Culex pipiens* complex. In *Cold Spring Harbor Symposium on Quantitative Biology* (Vol. 24, pp. 166–173). Cold Spring Harbor Laboratory Press.

Laveran, A. (1900). Sur un *Anopheles* provenant de Madagascar. *Comptes rendus hebdomadaires des séances et mémoires de la Société de Biologie , 52*, 109–110.

Laveran, A. (1901a). Au sujet de culicides recueillis à Dijbouti et à la Nouvelle-Calédonie. *Comptes rendus hebdomadaires des séances de l'Academie des Sciences D, 53*, 567–569.

Laveran, A. (1901b). Sur des culicides provenant de Hanoï (Tonkin). *Comptes rendus hebdomadaires des séances et mémoires de la Société de Biologie, 53*, 991–993.

Laveran, A. (1901c). Sur des culicides provenant du Haut-Tonkin. *Comptes rendus hebdomadaires des séances et mémoires de la Société de Biologie, 53*, 993–994.

Laveran, A. (1902a). Sur des culicides du Cambodge. *Comptes rendus hebdomadaires des séances et mémoires de la Société de Biologie, 54*, 906–908.

Laveran, A. (1902b). Sur des culicides de Nouvelles-Hebrides. *Comptes rendus hebdomadaires des séances et mémoires de la Société de Biologie, 54*, 908–910.

Laveran, A. (1902c). [Note]. In P.L. Simond (Ed.), *Description d'un moustique dont le male possede un trompe faucille* (Vol. 54, pp. 1158–1160). Serie D, sciences naturelles. Paris, France: Comptes rendus hebdomadaires des séances de l'Academie des Sciences.

Laveran, A. (1904). *Anopheles* et paludisme à Madagascar. Prophylaxie du paludisme. *Bulletin de l'Académie de Médecine* (Paris), *52*(3), 197–218.

Lavergne, S., & Molofsky, J. (2007). Increased genetic variation and evolutionary potential drive the success of an invasive grass. *Proceedings of the National Academy of Sciences, 104*(10), 3883–3888.

Lawniczak, M.K.N., Emrich, S.J., Holloway, A.K., Regier, A.P., Olson, M., White, B., . . . Besansky, N.J. (2010). Widespread divergence between incipient *Anopheles gambiae* species revealed by whole genome sequences. *Science, 330*(6003), 512–514.

Le Berre, R., & Hamon, J. (1961). Description de la larve, de la nymphe et de la femelle d'*Aedes* (*Neomelaniconion*) *jamoti* Hamon et Rickenback 1954, et revision des cles de determination concernant le sous-genre *Neomelaniconion* en Afrique au sud du Sahara. *Bulletin de la Société de pathologie exotique, 53*(for 1960), 1054–1064.

Le Goff, G., Rajaonarivelo, E., Duchemin, J.B., & Robert, V. (2002). Revue du genre *Coquillettidia* (Diptera: Culicidae) à Madagascar et description de la larve de *Cq. grandidieri* (Blanchard, 1905). *Archives de l'Institut Pasteur de Madagascar, 68*(1–2), 100–103.

Le Goff, G., Boussès, P., & Brunhes, J. (2007). Revision des *Neomelaniconion* Newstead (Diptera: Culicidae) de Madagascar: Espèces presentes et description de cinq nouvelles espèces. *Annales de la Société Entomologique de France, 43*(2), 187–200.

Le Goff, G., Boussès, P., Julienne, S., Brengues, C., Rahola, N., Rocamora, G., & Robert, V. (2012). The mosquitoes (Diptera: Culicidae) of Seychelles: Taxonomy, ecology, vectorial importance, and identification keys. *Parasites and Vectors, 5*(1), 207.

Le Goff, G., Brengues, C., & Robert, V. (2013). *Stegomyia* mosquitoes in Mayotte, taxonomic study and description of *Stegomyia pia* n. sp. *Parasite, 20*(31), 18.

Le Menach, A., McKenzie, F.E., Flahault, A., & Smith, D.L. (2005). The unexpected importance of mosquito oviposition behaviour for malaria: Non-productive larval habitats can be sources for malaria transmission. *Malaria Journal, 4*(1), 23.

Le Peletier, A.L.M., & Audinet-Serville, J.G. 1828. In P.A. Latreille (Ed.), *Histoire naturelle. Entomologie, ou histoire naturelle des crustacés, des arachnides et des insects* (pp. 345–832). In P.A. Latreille, A.L.M. Le Peletier, J.G. Audinet-Serville, & F.É. Guérin-Méneville (Eds.), *Encyclopédie methodique, ou par ordre de matieres; par une société de gens de lettres, de savans et d'artistes; Précédée d'un Vocabulaire universel, servant de Table pour tout l'Ouvrage, ornée des Portraits de M.M. Diderot & D'Alembert, premiers Éditeurs de l'Encyclopédie. Tome dixième* [from Harbach 2018c]. Paris, France: Veuve Agasse.

Leach, W.E. (1825). Description of thirteen species of *Formica* and three species of *Culex*, found in the environs of Nice. *Zoological Journal, 2*(7), 289–293.

Lee, D.J. (1937). Notes on Australian mosquitoes (Diptera, Culicidae). IV. The genus *Theobaldia*, with description of a new species. *Proceedings of the Linnean Society of New South Wales, 62*(5–6), 294–298.

Lee, D.J. (1944a). *An atlas of the mosquito larvae of the Australasian Region*. Tribes- Megarhinini and Culicini. Australian Military Forces, Headquarters.

Lee, D.J. (1944b). A new species of the genus *Anopheles* from Northern Australia (Diptera, Culicidae). *Proceedings of the Linnean Society of New South Wales, 69*(1–2), 21–25.

Lee, D.J. (1944c). Notes on Australian mosquitoes (Diptera, Culicidae). V. The genus *Armigeres* and new species of *Armigeres*, *Theobaldia* and *Culex*. *Proceedings of the Linnean Society of New South Wales, 69*(5–6), 215–225.

Lee, D.J. (1946). Notes on Australian mosquitoes (Diptera, Culicidae). Part VI. The genus *Tripteroides* in the Australasian Region. *Proceedings of the Linnean Society of New South Wales, 70*(5–6), 219–275.

Lee, D.J. (1948). Mosquitoes (Diptera, Culicidae) recorded from Tasmania. *Records of the Queen Victoria Museum, Launceston, 2*, 53–56.

Lee, D.J., & Woodhill, A.R. (1944). *The anopheline mosquitoes of the Australasian Region* (Vol. 2). Australia: University of Sydney.Lee, D.J., Dyce, A.L., & O'Gower, A.K. (1957). Blood-sucking flies (Diptera) and myxomatosis transmission in a mountain environment in New South Wales. *Australian Journal of Zoology, 5*(4), 355–401.

Lee, D.J., Hicks, M.M., Griffiths, M., Russell, R.C., & Marks, E.N. (1980). *The Culicidae of the Australasian Region* (Vol. 1). Commonwealth Department of Health, School of Public Health and Tropical Medicine Monograph Series, 2. Canberra: Australian Government.

Lee, D.J., Hicks, M.M., Griffiths, M., Russell, R.C., & Marks, E.N. (1982). *The Culicidae of the Australasian Region* (Vol. 2). Commonwealth Department of Health, School of Public Health and Tropical Medicine Monograph Series, 2. Canberra: Australian Government.

Lee, D.J., Hicks, M.M., Griffiths, M., Russell, R.C., & Marks, E.N. (1984). *The Culicidae of the Australasian Region* (Vol. 3). Commonwealth Department of Health, School of Public Health and Tropical Medicine Monograph Series, 2. Canberra: Australian Government.

Lee, D.J., Hicks, M.M., Griffiths, M., Debenham, M.L., Bryan, J.H., Russell, R.C., . . . Marks, E.N. (1987a). *The Culicidae of the Australasian Region* (Vol. 4). Commonwealth Department of Health, School of Public Health and Tropical Medicine Monograph Series, 2. Canberra: Australian Government.

Lee, D.J., Hicks, M.M., Griffiths, M., Debenham, M.L., Bryan, J.H., Russell, R.C., . . . Marks, E.N. (1987b). *The Culicidae of the Australasian Region* (Vol. 5). Commonwealth Department of Health, School of Public Health and Tropical Medicine Monograph Series, 2. Canberra: Australian Government.

Lee, D.J., Hicks, M.M., Griffiths, M., Debenham, M.L., Bryan, J.H., Russell, R.C., . . . Marks, E.N. (1988a). *The Culicidae of the Australasian Region* (Vol. 6). Commonwealth Department of Health, School of Public Health and Tropical Medicine Monograph Series, 2. Canberra: Australian Government.

Lee, D.J., Hicks, M.M., Debenham, M.L., Griffiths, M., Bryan, J.H., & Marks, E.N. (1988b). *The Culicidae of the Australasian Region* (Vol. 9). Commonwealth Department of Health, School of Public Health, and Tropical Medicine Monograph Series, 2. Canberra: Australian Government.

Lee, D.J., Hicks, M.M., Debenham, M.L., Griffiths, M., Bryan, J.H., & Marks, E.N. (1988c). *The Culicidae of the Australasian Region* (Vol. 10). Commonwealth Department of Health, School of Public Health and Tropical Medicine Monograph Series, 2. Canberra: Australian Government.

Lee, D.J., Hicks, M.M., Debenham, M.L., Griffiths, M., Marks, E.N., Bryan, J.H., & Russell, R.C. (1989a). *The Culicidae of the Australasian Region* (Vol. 7). Commonwealth Department of Health, School of Public Health and Tropical Medicine Monograph Series, 2. Canberra: Australian Government.

Lee, D.J., Hicks, M.M., Debenham, M.L., Griffiths, M., Marks, E.N., & Bryan, J.H. (1989b). *The Culicidae of the Australasian Region* (Vol. 8). Commonwealth Department of Health, School of Public Health and Tropical Medicine Monograph Series, 2. Canberra: Australian Government.

Lee, D.J., Hicks, M.M., Griffiths, M., Debenham, M.L., Marks, E.N., Bryan, J.H., & Russell, R.C. (1989c). *The Culicidae of the Australasian Region* (Vol. 11). Commonwealth Department of Health, School of Public Health and Tropical Medicine Monograph Series 2. Canberra: Australian Government.

Lee, D.K. (1998). Effect of two rice culture methods on the seasonal occurrence of mosquito larvae and other aquatic animals in rice fields of southwestern Korea. *Journal of Vector Ecology, 23*(2), 161–170.

Lee, D.W., & Pietrantonio, P.V. (2003). *In vitro* expression and pharmacology of the 5-HT-like receptor present in the mosquito *Aedes aegypti* tracheolar cells and hindgut-associated nerves. *Insect Molecular Biology, 12*(6), 561–569.

Lee, H.I., Self, L.S., Hong, H.K., & Lee, K.W. (1973). Mosquito light trap surveys in Korea 1969–1971. *Southeast Asian Journal of Tropical Medicine and Public Health, 4*(3), 382–386.

Lee, H.I., Seo, B.Y., Shin, E.H., Burkett, D.A., Lee, W.J., & Shin, Y.H. (2006). Study of flying height of culicid species in the northern part of the Republic of Korea. *Journal of the American Mosquito Control Association, 22*(2), 239–245.

Lee, K.W., & Lien., J.C. (1970). Pictorial keys to the mosquitoes of Korea. World Health Organization/Vector Biology and Control.

Lee, R.M.K.W., & Craig, D.A. (1983). Cibarial sensilla and armature in mosquito adults (Diptera: Culicidae). *Canadian Journal of Research, 61*(3), 633–646.

Lee, S.J., Kim, B.H., & Lee, J.Y. (2009). Experimental study on the fluid mechanics of blood sucking in the proboscis of a female mosquito. *Journal of Biomechanics, 42*(7), 857–864.

Lees, R.S., Knols, B., Bellini, R., Benedict, M.Q., Bheecarry, A., Bossin, H.C., . . . Egyir-Yawson, A. (2014). Improving our knowledge of male mosquito biology in relation to genetic control programmes. *Acta Tropica, 132*(Suppl.), S2–S11.

Leeson, H.S. (1930). A new variety of *Anopheles funestus* Giles, from Southern Rhodesia. *Bulletin of Entomological Research, 21*(2), 125–126.

Leeson, H.S. (1931). Anopheline mosquitoes in southern Rhodesia 1926–1928. A report on investigations made during researches on blackwater fever conducted by Dr. G.R. Ross. *Memoir Series of the London School of Hygiene and Tropical Medicine* (Vol. 4). London, England: London School of Hygiene and Tropical Medicine.

Leeson, H.S. (1935). Another anopheline of the *funestus* series from Southern Rhodesia. *Annals of Tropical Medicine and Parasitology, 29*(1), 69–71.

Leeson, H.S., & Theodor, O. (1948). Mosquitos of Socotra. *Bulletin of Entomological Research, 39*(2), 221–229.

Leguizamón, M.N., & Carpintero, D.J. (2004). Description of a new species of *Orthopodomyia* (Diptera: Culicidae) from the Neotropical region. *Studia Dipterologica 11 Heft, 1*, 203–205.

Lehmann, T., Hawley, W.A., Grebert, H. & Collins, F.H. (1998). The effective population size of *Anopheles gambiae* in Kenya: Implications for population structure. *Molecular Biology and Evolution, 15*(3), 264–276.

Lehmann, T., Hawley, W.A., Grebert, H., Danga, M., Atieli, F., & Collins, F.H. (1999). The Rift Valley complex as a barrier to gene flow for *Anopheles gambiae* in Kenya. *Journal of Heredity, 90*(6), 613–621.

Lehmann, T., Licht, M., Elissa, N., Maega, B.T.A., Chimumbwa, J.M., Watsenga, F.T., . . . Hawley, W.A. (2003). Population structure of *Anopheles gambiae* in Africa. *Journal of Heredity, 94*(2), 133–147.

Lehmann, T., Dao, A., Yaro, A.S., Adamou, A., Kassogue, Y., Diallo, M., . . . Coscarón-Arias, C. (2010). Aestivation of the African malaria mosquito, *Anopheles gambiae* in the Sahel. *The American Journal of Tropical Medicine and Hygiene, 83*(3), 601–606.

Lehr, M.A., Kilpatrick, C.W., Wilkerson, R.C., & Conn, J.E. (2005). Cryptic species in the *Anopheles* (*Nyssorhynchus*) *albitarsis* (Diptera: Culicidae) complex: Incongruence between random amplified polymorphic DNA-polymerase chain reaction identification and analysis of mitochondrial DNA COI gene sequences. *Annals of the Entomological Society of America, 98*(6), 908–917.

Lei, X. (1989a). A new species of *Aedes* (*Ochlerotatus*) (Diptera: Culicidae). *Sichuan Journal of Zoology, 8*(1), 6–8.

Lei, X. (1989b). *The mosquito fauna of Sichuan, China* [in Chinese]. Chengdu, China: Chengdu University of Science and Technology Press.

Lei, X. (1996). A new species of *Anopheles* (Diptera: Culicidae). *Sichuan Journal of Zoology, 15*(1), 1–4.

Lei, X., & Li, K. (1990). A new species of *Orthopodomyia* in Sichuan (Diptera: Culicidae). *Sichuan Journal of Zoology, 9*(1), 8–10.

Leicester, G.F. (1903). A breeding place of certain forest mosquitoes in Malaya. *Journal of Tropical Medicine, 6*, 291–292.

Leicester, G.F. (1908). The Culicidae of Malaya. *Studies from the Institute for Medical Research, Federated Malay States, 3*(3), 18–261.

Leisnham, P.T., Sala, L.M., & Juliano, S.A. (2008). Geographic variation in adult survival and reproductive tactics of the mosquito *Aedes albopictus*. *Journal of Medical Entomology, 45*(2), 210–221.

Leleup, N. (1952). Un anophèle cavernicole nouveau du Kibali-Ituri: *Anopheles feini* n. sp. *Revue de zoologie et de botanique Africaines, 46*(1–2), 151–158.

Leleup, N. (1957). *Anopheles garnhami* sbsp. *basilewskyi*, sous-espèce nouvelle de tres haute altitude (2.900 M.). *Folia Scientifica Africae Centralis, 3*(3), 69.

Leleup, N. (1960). Diptera Culicidae: gen. *Anopheles*. Mission zoologique de l'I.R.S.A.C. en Afrique orientale (P. Basilewsky et N. Leleup, 1957). *Le Musée du Congo belge à Tervueren* (Series 8), 402–404.

Leleup, N., & Lips, M. (1950). Un anophèle cavernicole nouveau du Katanga: *Anopheles rodhaini* n. sp. *Revue de zoologie et de botanique Africaines, 43*(4), 303–308.

Leleup, N., & Lips, M. (1951). Notes descriptives complementaires sur *Anopheles rodhaini*. *Revue de zoologie et de botanique Africaines, 44*(4), 169–172.

Leleup, N., & Zaghi, A. (1960). Apropos of *Anopheles* (*Myzomyia*) *faini* Leleup and of *A. faini* subsp. van Thieli Laarman. *Annales de la Société Belge de Médecine Tropicale, 40*, 925–931.

Lemen, C.A., & Freeman, P.W. (1984). The genus: A macroevolutionary problem. *Evolution, 38*(6), 1219–1237.

Leong, M.K., & Grace, J.K. (2009). Occurrence and distribution of mosquitoes (Diptera: Culicidae) of public health importance on the island of Oahu. *Proceedings of the Hawaiian Entomological Society, 41*, 57–70.

LePage, D.P., Metcalf, J.A., Bordenstein, S.R., On, J., Perlmutter, J.I., Shropshire, J.D., . . . Bordenstein, S.R. (2017). Prophage WO genes recapitulate and enhance *Wolbachia*-induced cytoplasmic incompatibility. *Nature, 543*(7644), 243–247.

Lestani, E.A., & Rossi, G.C. (2012). Description of the female, pupa, and larva of *Culex* (*Melanoconion*) *bahiensis* Duret, and redescription of the male (Diptera: Culicidae). *Zootaxa, 24*, 57–63.

Leví-Castillo, R. (1944). El complejo "*Pseudopunctipennis*" en el Ecuador (Diptera-Culicidae). Guayaquil, Ecuador: Guayaquil University, 1–7.

Leví-Castillo, R. (1945). *Los anofelinos de la República del Ecuador*. Tomo primero. Guayaquil, Ecuador: Artes Graficas Senefelder.

Leví-Castillo, R. (1949). *Atlas de los anofelinos Sudamericanos*. Guayaquil, Ecuador: Sociedad Filantropica de Guayas.

Leví-Castillo, R. (1951a). *Los mosquitos del género Haemagogus, Williston, 1896 en América del Sur*. Cuenca, Ecuador: Editorial "Don Bosco."

Leví-Castillo, R. (1951b). Nota taxonomica sobre la especie equatoriana *Aedes* (*Ochlerotatus*) *camposanus* Dyar 1918 (Dipt. Culicidae). *Revista de entomología, 22*, 383–388.

Leví-Castillo, R. (1952a). Redescription of *Aedes* (*Ochlerotatus*) *camposanus* Dyar (1918) as a valid species found in the coastal plain of Ecuador. *Pacific Science, 6*, 262–264.

Leví-Castillo, R. (1952b). Nota sobre una especie del subgenero *Phalangomyia* del genero *Culex* encontrada en la Provincia del Azuay (Ecuador): *Culex archegus* Dyar 1929 (Diptera-Culicidae). *Transactions of the Ninth International Congress of Entomology, 1*, 901–904.

Leví-Castillo, R. (1952c). *Wyeomia* [*sic*] (*Wyeomyia*) *aphobema* var. *aequetorielis* var. n., a new sabethine mosquito from Ecuador (Diptera: Culicidae). *Proceedings of the Royal Entomological Society of London, Series B: Taxonomy, 21*(9–10), 131–133.

Leví-Castillo, R. (1953a). Observations on the subgenus *Phalangomyia* of the genus *Culex* in Ecuador with description of a new species (Diptera: Culicidae). *Pacific Science, 7*(2), 187–192.

Leví-Castillo, R. (1953b). A new species of *Culex* from Ecuador (Diptera, Culicidae). *Proceedings of the Entomological Society of Washington, 55*(4), 161–163.

Leví-Castillo, R. (1953c). Dos especies nuevas de mosquitos ecuatorianos (Diptera-Culicidae). *Revista Ecuatoriana de entomología y parasitología, 1*(1), 14–18.

Leví-Castillo, R. (1953d). Dos especies nuevas de mosquitos de la sierra ecuatoriana (Diptera-Culicidae). *Revista Ecuatoriana de entomología y parasitología, 1*(4), 63–70.

Leví-Castillo, R. (1953e). Una nueva especie de *Culex* procedente de la Provincia de Los Ríos, Ecuador: *Culex* (*Carrollia*) *Babahoyensis* n. sp. *Revista Ecuatoriana de entomología y parasitología, 1*(3), 91–94.

Leví-Castillo, R. (1953f). Notas adicionales sobre la espécie *Sabethes* (*Sabethoides*) *chloropterus* Humboldt 1820, encontrado en la costa Ecuatoriana (Diptera-Culicidae). *Revista Ecuatoriana de entomología y parasitología, 1*(3), 95–100, 102 plates.

Leví-Castillo, R. (1953g). *Aedes garcíai* n. sp., un Aedino nuevo para la fauna ecuatorian—*Lanesia*, nuevo subgénero de *Aedes*. *Revista Ecuatoriana de entomología y parasitología, 1*(3), 101–106.

Leví-Castillo, R. (1953h). *Toxorhynchites aequatorianus* n. sp., especie neuva de *Toxorhynchites* hallada en Pichilingue, Los Ríos, Ecuador (Diptera-Culicidae). *Revista Ecuatoriana de entomología y parasitología, 1*(4), 71–76.

Leví-Castillo, R. (1954a). Lista provisional de los mosquitoes *Culex* del Ecuador (Diptera-Culicidae). *Revista Ecuatoriana de entomología y parasitología, 2*(1–2), 83–86.

Leví-Castillo, R. (1954b). Notas sobre *Haemagogus panarchys* Dyar, 1921 (Diptera, Culicidae). *Revista Ecuatoriana de entomología y parasitología, 2*(1–2), 247–260.

Leví-Castillo, R. (1954c). Atlas de los anofelinos Sudamericanos *Revista Ecuatoriana de entomología y parasitología, 2*(1–2), 263–267.

Leví-Castillo, R. (1955a). Nueva división sistemática del Género *Haemagogus* Williston 1896 y una nueva especie Ecuatoriana: *Haemagogus garciai* n. sp. *Revista Ecuatoriana de entomología y parasitología, 2*(3, 4), 359–364.

Leví-Castillo, R. (1955b). *Phoniomyia esmeraldasi*, a new mosquito from Ecuador. *Revista Ecuatoriana de entomología y parasitología, 2*(3–4), 389–391.

Leví-Castillo, R. (1955c). *Haemagogus soperi* n. sp. mosquito trasmisor de fiebre amarilla selvática en el Ecuador (Diptera-Culicidae). *Revista Ecuatoriana de entomología y parasitología, 2*(3–4), 479–484.

Leví-Castillo, R. (1955d). Un nuevo anofelino de altura del Ecuador: *Anopheles gomezdelatorrei* N. Sp. *Revista Ecuatoriana de entomología y parasitología, 2*(3–4), 509–514.

Leví-Castillo, R. (1956). A systematic note on *Haemagogus spegazzinii* Brèthes, 1912. (Diptera-Culicidae). *Proceedings of the Entomological Society of Washington, 58*(6), 345–347.

Levine, R.S., Peterson, A.T., & Benedict, M.Q. (2004a). Distribution of members of *Anopheles quadrimaculatus* Say s.l. (Diptera: Culicidae) and implications for their roles in malaria transmission in the United States. *Journal of Medical Entomology, 41*(4), 607–613.

Levine, R.S., Peterson, A.T., & Benedict, M.Q. (2004b). Geographic and ecologic distributions of the *Anopheles gambiae* complex predicted using a genetic algorithm. *The American Journal of Tropical Medicine and Hygiene, 70*(2), 105–109.

Lewis, D.J. (1937). A new species of *Anopheles* from the Anglo-Egyptian Sudan. *Proceedings of the Royal Entomological Society of London, Series B: Taxonomy, 6*(9), 181–183.

Lewis, D.J. (1942a). The early stages of *Aedes taylori* Edwards and *A. furcifer* Edwards (Dipt., Culicidae). *Proceedings of the Royal Entomological Society of London, Series B: Taxonomy, 11*(10), 153–154.

Lewis, D.J. (1942b). A new species of *Aëdes* (Dipt., Culicidae) from Eritrea. *Proceedings of the Royal Entomological Society of London, Series B: Taxonomy, 11*(11), 165–166.

Lewis, D.J. (1944). A new subspecies of *Aedes leesoni* Edwards (Dipt. Culicidae) from the Sudan. *Proceedings of the Royal Entomological Society of London, Series B: Taxonomy, 13*(3–4), 27–29.

Lewis, D.J. (1945). Observations on the distribution and taxonomy of Culicidae (Diptera) in the Sudan. *Transactions of the Royal Entomological Society of London, 95*(1), 1–24.

Lewis, D.J. (1949). Tracheal gills in some African culicine mosquito larvae. *Proceedings of the Royal Entomological Society of London, Series A: General Entomology, 24*(4–6), 60–66.

Lewis, D.J. (1956a). The anopheline mosquitos of the Sudan. *Bulletin of Entomological Research London, 47*(3), 475–494.

Lewis, D.J. (1956b). The *Culex* mosquitos of the Sudan. *Bulletin of Entomological Research, 47*(4), 703–721.

Li, B., Calvo, E., Marinotti, O., James, A.A., & Paskewitz, S.M. (2005). Characterization of the c-type lysosome gene family in *Anopheles gambiae. Gene, 360*(2), 131–139.

Li, C., & Wilkerson, R.C. (2005). Identification of *Anopheles* (*Nyssorhynchus*) *albitarsis* complex species (Diptera: Culicidae) using rDNA internal transcribed spacer 2-based polymerase chain reaction primes. *Memórias do Instituto Oswaldo Cruz, 100*(5), 495–500.

Li, C., & Wilkerson, R.C. (2007). Intragenomic rDNA ITS2 variation in the Neotropical *Anopheles* (*Nyssorhynchus*) *albitarsis* complex (Diptera: Culicidae). *Journal of Heredity, 98*(1), 51–59.

Li, C., Kapitskaya, M.Z., Zhu, J., Miura, K., Segraves, W., & Raikhel, A.S. (2000). Conserved molecular mechanism for the stage specificity of the mosquito vitellogenic response to ecdysone. *Developmental Biology, 224*(1), 96–110.

Li, C., Lee, J.S., Groebner, J.L., Kim, H.-C., Klein, T.A., O'Guinn, M.L., & Wilkerson, R.C. (2005). A newly recognized species in the *Anopheles hyrcanus* group and molecular identification of related species from the Republic of South Korea (Diptera: Culicidae). *Zootaxa, 939*(1), 1–8.

Li, M., Mead, E.A., & Zhu, J. (2011). Heterodimer of two bHLH-PAS proteins mediates juvenile hormone-induced gene expression. *Proceedings of the National Academy of Sciences of the United States of America, 108*(2), 638–643.

Li, S., Picimbon, J.F., Ji, S., Kan, Y., Chuanling, Q., Zhou, J.J., & Pelosi, P. (2008). Multiple functions of an odorant-binding protein in the mosquito *Aedes aegypti. Biochemical and Biophysical Research Communications, 37*(3), 464–468.

Li, Y., Hernandez-Martinez, S., & Noriega, F.G. (2004). Inhibition of juvenile hormone biosynthesis in mosquitoes: Effect of allatostatic head factors, PISCF- and YXFGL-amide-allatostatins. *Regulatory Peptides, 118*(3), 175–182.

Li, Y., Hernandez-Martinez, S., Fernandez, F., Mayoral, J.G., Topalis, P., Priestap, H., . . . Noriega, F.G. (2006). Biochemical, molecular, and functional characterization of PISCF-allostatin, a regulator of juvenile hormone biosynthesis in the mosquito *Aedes aegypti. Journal of Biological Chemistry, 281*(45), 34048–34055.

Lien, J.C. (1962). Non-anopheline mosquitoes of Taiwan: Annotated catalog and bibliography. *Pacific Insects, 4*(3), 615–649.

Lien, J.C. (1965). Mosquitoes of Taiwan: Genus *Toxorhynchites* Theobald. *Journal of Medical Entomology, 2*(1), 1–16.

Lien, J.C. (1967). New species of mosquitoes from Taiwan (Diptera: Culicidae). Part I. Two new species of *Aedes* (*Finlaya*). *Tropical Medicine* (Nagasaki), *9*(4), 177–185.

Lien, J.C. (1968a). New species of mosquitoes from Taiwan (Diptera: Culicidae). Part II. New species of *Tripteroides*, *Orthopodomyia*, *Culiseta* and *Uranotaenia*. *Tropical Medicine* (Nagasaki), *10*(1), 1–20.

Lien, J.C. (1968b). New species of mosquitoes from Taiwan (Diptera: Culicidae). Part III. Five new species of *Aedes*. *Tropical Medicine* (Japan), *10*(2), 95–115.

Lien, J C. (1968c). New species of mosquitoes from Taiwan (Diptera: Culicidae). Part IV. *Mattinglyia catesi*, n. gen., n. sp. and four new species of genus *Heizmannia* Ludlow. *Tropical Medicine* (Nagasaki), *10*(3), 127–153.

Lien, J.C. (1968d). New species of mosquitoes from Taiwan (Diptera: Culicidae) Part V. Three new subspecies of *Aedes* and seven new species of *Culex*. *Tropical Medicine* (Nagasaki), *10*(4), 217–262.

Lien, J.C., Kosman, L., Partono, F., Joesoef, A., Kosin, E., & Cross, J.H. (1975). A brief survey of mosquitoes in North Sumatra, Indonesia. *Journal of Medical Entomology, 12*(2), 233–239.

Lien, J.C., Kawengian, B.A., Partono, F., Lami, B., & Cross, J.H. (1977). A brief survey of the mosquitoes of South Sulawesi, Indonesia, with special Reference to the identity of *Anopheles barbirostris* (Diptera: Culicidae) from the Margolembo area. *Journal of Medical Entomology, 13*(6), 719–727.

Lien, J.-C., Lin, C.-C., Lin, M.-C., & Tseng, L.-F. (2008). A new species of the genus *Coquillettidia* (Diptera, Culicidae) from the Democratic Republic of São Tome and Principe. *Journal of the National Taiwan Museum, 61*(4), 55–59.

Lien, R., Lin, C., & Weng, M. (1999). *Culiseta taiwanica*, a new species of the subgenus *Climacura* Felt (Diptera: Culicidae). *Republic of China Journal of Parasitology, 12*(1–2), 1–9.

Lillie, T.H., & Nakasone, R.I. (1982). An evaluation of commercial diets for rearing *Wyeomyia smithii*. *Mosquito News, 42*(2), 225–231.

Lima, M.M. (1952). Do diagnóstico diferencial entre o *Anopheles* (*Kerteszia*) *cruzii* e o *Anopheles* (*Kerteszia*) *homunculus* na fase larvária. *Revista Brasileira de malariologia e doenças tropicais* (Rio de Janeiro), *4*, 401–411.

Limesand, K.H., Higgs, S., Pearson, L.D., & Beaty, B.J. (2000). Potentiation of vesicular stomatitis New Jersey virus infection in mice by mosquito saliva. *Parasite Immunology, 22*(9), 461–468.

Lin, C.-C., Tseng, L.-F., & Lien, J.-C. (2008). A new species of the genus *Culiseta* (Diptera, Culicidae) from the Democratic Republic of São Tome and Principe. *Journal of the National Taiwan Museum, 61*(2), 89–96.

Linam, J.H., & Nielsen, L.T. (1970). The distribution and evolution of the *Culex* mosquitoes of the subgenus *Neoculex* in the New World. *Mosquito Systematics, 2*(4), 149–157.

Lindtrop, G.T. (1924). [The white malaria mosquito *A. pulcherrimus* in Azerbaijan]. *Russian Journal of Tropical Medicine, 3*, 38–40.

Linley, J.R. (1986). Swimming behavior of the larva of *Culicoides variipennis* (Diptera: Ceratopogonidae) and its relationship to temperature and viscosity. *Journal of Medical Entomology, 23*(5), 473–483.

Linley, J.R. (1989a). Scanning electron microscopy of the egg of *Aedes* (*Protomacleaya*) *triseriatus* (Diptera: Culicidae). *Journal of Medical Entomology, 26*(5), 474–478.

Linley, J.R. (1989b). Comparative fine structure of the eggs of *Aedes albopictus*, *Ae. aegypti*, and *Ae. bahamensi* (Diptera: Culicidae). *Journal of Medical Entomology, 26*(6), 510–521.

Linley, J.R. (1989c). Egg of *Mansonia dyari* described and compared with egg of *Mansonia titillans* (Diptera: Culicidae). *Journal of Medical Entomology, 26*(1), 41–45.

Linley, J.R. (1989d). Scanning electron microscopy of the eggs of *Toxorhynchites rutilus rutilus* and *Tx. amboinensis*. (Diptera: Culicidae). *Journal of Medical Entomology, 26*(1), 1–9.

Linley, J.R. (1990). Scanning electron microscopy of the eggs of *Aedes vexans* and *Aedes infirmatus* (Diptera: Culicidae). *Proceedings of the Entomological Society of Washington, 92*(4), 685–693.

Linley, J.R. (1991). The egg of *Trichoprosopon compressum* (Diptera: Culicidae). *Mosquito Systematics, 23*(3), 161–169.

Linley, J.R. (1992). The eggs of *Anopheles atropos* and *Anopheles darlingi* (Dipteral: Culicidae). *Mosquito Systematics, 24*(1), 40–50.

Linley, J.R., & Chadee, D.D. (1990). Fine structure of the eggs of *Psorophora columbiae*, *Ps. cingulata* and *Ps. ferox* (Diptera: Culicidae). *Proceedings of the Entomological Society of Washington, 92*(3), 497–511.

Linley, J.R., & Chan, K.L. (1991). Scanning electron microscopy of the eggs of *Aedes togoi* and *Tripteroides bambusa* (Diptera: Culicidae). *Tropical Biomedicine, 8*(1), 39–47.

Linley, J.R., & Clark, G.G. (1989). Egg of *Aedes* (*Gymnometopa*) *mediovittatus*. (Diptera: Culicidae). *Journal of Medical Entomology, 26*(4), 252–255.

Linley, J.R., & Craig, G.B., Jr. (1993). The egg of *Aedes hendersoni* and a comparison of its structure with the egg of *Aedes triseriatus* (Diptera: Culicidae). *Mosquito Systematics, 25*(2), 65–76.

Linley, J.R., & Kaiser, P.E. (1994). The eggs of *Anopheles punctipennis* and *Anopheles perplexens* (Diptera: Culicidae). *Mosquito Systematics, 26*(1), 43–56.

Linley, J.R., & Lounibos, L.P. (1993). The eggs of *Anopheles* (*Nyssorhynchus*) *rangeli* and *Anopheles* (*Nyssorhynchus*) *dunhami* (Diptera: Culicidae). *Mosquito Systematics, 25*(3), 157–169.

Linley, J.R., & Lounibos, L.P. (1994). The remarkable egg of *Anopheles peryassui* (Diptera: Culicidae). *Mosquito Systematics, 26*(1), 25–34.

Linley, J.R., & Milstrey, E.G. (1995). The eggs of *Anopheles* (*Anopheles*) *mattogrossensis* and *Chagasia fajardi* (Diptera: Culicidae). *Mosquito Systematics, 27*(1), 27–39.

Linley, J.R., & Seabury, F. (1990). Egg of *Toxorhynchites splendens* described and compared with the egg of *Toxorhynchites amboinensis* (Diptera: Culicidae). *Journal of Medical Entomology, 27*(1), 9–13.

Linley, J.R., & Service, M.W. (1994). Fine structure of the eggs of *Aedes durbanensis*, *Aedes woodi*, and *Eretmapodites quinquevittatus* (Diptera: Culicidae). *Mosquito Systematics, 26*(2), 59–72.

Linley, J.R., & Service, M.W. (1995). The eggs of *Aedes polynesiensis* and *Aedes katherinensis* (Diptera: Culicidae). *Mosquito Systematics, 27*(2), 143–152.

Linley, J.R., & Turell, M.J. (1993). The eggs of *Aedes dentatus* and *Aedes fowleri* (Diptera: Culicidae). *Proceedings of the Entomological Society of Washington, 95*(1), 7–16.

Linley, J.R., & Turell, M.J. (1994). Ultrastructural descriptions of the eggs of *Aedes mcintoshi* and *Aedes circumluteolus* (Diptera: Culicidae). *Mosquito Systematics, 26*(1), 1–10.

Linley, J.R., Lounibos, L.P., & Linley, P.A. (1990). Fine structure of the egg of *Trichoprosopon digitatum* (Diptera: Culicidae) and its relationship to egg raft formation. *Journal of Medical Entomology, 27*(4), 578–585.

Linley, J.R., Geary, M.J., & Russell, R.C. (1991a). Fine structure of the eggs of *Aedes* (*Ochlerotatus*) *theobaldi*, *Ae.* (*Och.*) *sagax* and *Ae.* (*Och.*) *procax* (Diptera: Culicidae). *Mosquito Systematics, 23*(2), 144–159.

Linley, J.R., Geary, M.J., & Russell, R.C. (1991b). The eggs of *Aedes funereus*, *Aedes notoscriptus* and *Aedes alternans* (Diptera: Culicidae). *Proceedings of the Entomological Society of Washington, 93*, 592–612.

Linley, J.R., Geary, M.J., & Russell, R.C. (1991c). The eggs of *Aedes* (*Finlaya*) *alboannulatus* and *Aedes* (*Finlaya*) *rubrithorax* (Diptera: Culicidae). *Mosquito Systematics, 23*(2), 132–143.

Linley, J.R., Geary, M.J., & Russell, R.C. (1992a). The eggs of *Aedes vigilax* and *Aedes vittiger* (Diptera: Culicidae). *Proceedings of the Entomological Society of Washington, 94*(1), 48–58.

Linley, J.R., Geary, M.J., & Russell, R.C. (1992b). The eggs of *Aedes australis* and *Aedes camptorhynchus* (Diptera: Culicidae). *Mosquito Systematics, 24*(1), 29–39.

Linley, J.R., Kaiser, P.E., & Cockburn, A.F. (1993a). A description and morphometric study of the eggs of species of the *Anopheles quadrimaculatus* complex (Diptera: Culicidae). *Mosquito Systematics, 25*(2), 124–147.

Linley, J.R., Lounibos, L.P., & Conn, J. (1993b). A description and morphometric analysis of the eggs of four South American populations of *Anopheles* (*Nyssorhynchus*) *aquasalis* (Diptera: Culicidae). *Mosquito Systematics, 25*(3), 198–214.

Linley, J.R., Abo-Ghalia, A.H., Soliman, B.A., & Mukwaya, L.G. (1993c). The eggs of *Aedes caspius* and *Aedes africanus* (Diptera: Culicidae). *Mosquito Systematics, 25*(1), 25–34.

Linley, J.R., Yap, H.H., & Damar, T.B. (1995). The eggs of four species of the *Anopheles hyrcanus* group in Malaysia (Diptera: Culicidae). *Mosquito Systematics, 27*(1), 43–71.

Linley, J.R., Lounibos, L.P., Conn, J., Duzak, D., & Nishimura, N. (1996). A description and morphometric comparison of eggs from eight geographic populations of the South American malaria vector *Anopheles* (*Nyssorhynchus*) *nuneztovari* (Diptera: Culicidae). *Journal of the American Mosquito Control Association, 12*(2), 275–292.

Linnaeus, C. (1758). *Systema Naturae per Regna Tria Naturae, Secundum, Classes, Ordines, Genera, Species, cum Characteribus, Differentiis, Synonymis, Locis* (10, Rev. ed., Vol. 1). Stockholm, Sweden: Salviae; Holmiae.

Linnaeus, C. (1762). Zweyter Theil, enthalt Beschreibungen verschiedener wichtiger Naturalien. In *Reise nach Palestina in den Jahren von 1749 bis 1752*. Rostock, Germany: Koppe.

Linnaeus, C. (1767). *Systema naturae per regna tria naturae, secundum classes, ordines, genera, species, cum characteribus, differentiis, synonymis, locis* (12th ed., Vol. 1, Part 2). Holmiae.

Linnaeus, C. (1792). *Flora Lapponica . . . Editio altera, aucta et emendata sudio & cura Jacobis Edwardi Smith*. London.

Linthicum, K.J. (1988). A revision of the Argyritarsis Section of the subgenus *Nyssorhynchus* of *Anopheles* (Diptera: Culicidae). *Mosquito Systematics, 20*(2), 98–271.

Linthicum, K.J., Davies, F.G., & Kamau, J. (1985). Predation on emerging adult mosquitoes by *Brachydeutera munroi* (Diptera: Ephydridae). *Proceedings of the 72nd Annual Meeting of the New Jersey Mosquito Control Association* (pp. 178–180). Atlantic City, New Jersey, March 17–21, 1985.

Linthicum, K.J., Platt, K., Myint, K.S., Lerdthusnee, K., Innis, B.L., & Vaughn, D.W. (1996). Dengue 3 virus distribution in the mosquito *Aedes aegypti*: An immunocytochemical study. *Medical and Veterinary Entomology, 10*(1), 87–92.

Linton, Y.-M. (2004). *Systematics of the Holarctic* maculipennis *complex*. Paper presented at the 70th Annual Meeting of the American Mosquito Control Association, Savannah, Georgia.

Linton, Y.-M., Harbach, R.E., Seng, C.M., Anthony, T.G., & Matusop, A. (2001). Morphological and molecular identity of *Anopheles* (*Cellia*) *sundaicus* (Diptera: Culicidae), the nominotypical member of a malaria vector species complex in Southeast Asia. *Systematic Entomology, 26*(3), 357–366.

Linton, Y.-M., Smith, L., & Harbach, R.E. (2002a). Observations on the taxonomic status of *Anopheles subalpinus* Hackett & Lewis and *An. melanoon* Hackett. *European Mosquito Bulletin, 13*, 1–7.

Linton, Y.-M., Smith, L., & Harbach, R.E. (2002b). Molecular confirmation of sympatric populations of *Anopheles messeae* and *Anopheles atroparvus* overwintering in Kent, southeast England. *European Mosquito Bulletin, 13*, 8–16.

Linton, Y.-M., Samanidou-Voyadjoglou, A., & Harbach, R.E. (2002c). Ribosomal ITS2 sequence data for *Anopheles maculipennis* and *An. messeae* in northern Greece, with a critical assessment of previously published sequences. *Insect Molecular Biology, 11*(4), 379–383.

Linton, Y.-M., Smith, L., Koliopoulos, G., Samanidou-Voyadjoglou, A., Zounos, A.K., & Harbach, R.E. (2003). Morphological and

molecular characterization of *Anopheles* (*Anopheles*) *maculipennis* Meigen, type species of the genus and nominotypical member of the Maculipennis Complex. *Systematic Entomology, 28*(1), 39–55.
Linton, Y.-M., Lee, A.S., & Curtis, C. (2005). Discovery of a third member of the Maculipennis Group in SW England. *European Mosquito Bulletin, 19*, 5–9.
Linton, Y.-M., Dusfour, I., Howard, T.M., Ruiz L.F., Duc Manh, N., Ho Dinh, T., . . . Harbach, R.E. (2005). *Anopheles* (*Cellia*) *epiroticus* (Diptera: Culicidae), a new malaria vector species in the southeast Asian sundaicus complex. *Bulletin of Entomological Research, 95*(4), 329–339.
Linton, Y.-M., Smith, L., Koliopoulos, G., Zounos, A.K., Samanidou-Voyadjoglou, A., Patsoula, E., & Harbach, R.E. (2007). The *Anopheles* (*Anopheles*) *maculipennis* complex (Diptera: Culicidae) in Greece. *Journal of Natural History, 41*(41–44), 2683–2699.
Linton, Y.-M., Pecor, J.E., Porter, C.H., Mitchell, L.B., Garzón-Moreno, A., Foley, D H., . . . Wilkerson, R.C. (2013). Mosquitoes of eastern Amazonian Ecuador: Biodiversity, bionomics and barcodes. *Memórias do Instituto Oswaldo Cruz, 108*(Suppl. 1), 100–109.
Lioy, P. (1863). I ditteri distribuiti secondo un nuovo metodo di classificazione natural [part]. *Atti del Reale Istituto Veneto de Scienze, Lettere ed Arti* (Series 3) *9*(Part 1): 187–236.
Lips, M. (1960a). *Anophèles* du Congo Belge 2. Quelques espèces des galeries forestieres. *Rivista di malariologia, 21*(1), 39–64.
Lips, M. (1960b). *Anophèles* du Congo 3. Faune des grottes et des anfractuo sites. Références-récoltes-répartition et importance médicale actuelle. *Rivista di malariologia, 21*(4), 289–306.
Lips, M. (1962). *Anophèles* du Congo (ex Belge) 7. Quelques espèces des groupes *Cellia* et *Neocellia*. Références-récoltes-répartition et importance médicale actuelle. *Rivista di malariologia, 23*(2), 107–134.
Lips, M., & Hamon, J. (1955). Contribution a l'étude des culicidés (Diptères) de la Région Éthiopienne. Description de la larve, de la nymphe et du mâle d'*Anopheles austeni* Theobald 1905. *Bulletin de la Société de pathologie exotique, 48*(6), 872–882.
Lips, M., & Hamon, J. (1956). Contribution a l'étude des culicides d'Afrique. Description de la larve et de la nymphe de *Taeniorhynchus* (*Coquillettidia*) *flavocinctus* Edwards, 1936. *Bulletin de la Société de pathologie exotique, 49*, 86–89.
Liria, J., & Navarro, J.-C. (1999). Morfología bucal larval de *Deinocerites melanophylum* Dyar & Knab y *D. cancer* Theobald (Diptera: Culicidae). *Boletín de entomología Venezolana, 14*(1), 53–62.
Liria, J., & Navarro, J.-C. (2003a). *Psorophora* (*Psorophora*) *lineata* (Humboldt, 1819) y *Psorophora* (*Pso.*) *saeva* Dyar & Knab, 1906 (Diptera: Culicidae). Correcciones en su identificacion. *Entomotropica, 18*(2), 113–119.
Liria, J., & Navarro, J.-C. (2003b). Morphology of the larval mandible and maxilla in the subgenus *Carrollia* Lutz of *Culex* L. (Diptera: Culicidae). *Zootaxa, 186*(1), 1–8.
Liria, J., & Navarro, J.-C. (2007). Morfología del cibario en subgéneros de *Psorophora* Robineau Desvoidy (Diptera: Culicidae). *Neotropical Entomology, 36*(6), 902–909.
Liria, J., & Navarro, J.-C. (2009). Clave fotográfica para hembras de *Haemagogus* Williston 1896 (Diptera: Culicidae) de Venezuela, con nuevo registro para el país. *Boletín de malariología y salud ambiental, 49*(2), 283–292.
Liria, J., & Navarro, J.-C. (2010). Modelo de nicho ecológico en *Haemagogus* Williston (Diptera: Culicidae), vectores del virus de la fiebre amarilla. *Revista biomédica, 21*, 149–161.
Liria, J., & Navarro, J.-C. (2014). Phylogenetic relationships in the genus *Psorophora* Robineau-Desvoidy (Diptera: Culicidae), based on morphological characters. *Entomology and Applied Science Letters, 1*(4), 22–50.
Liston, W.G. (1901). A year's experience of the habits of *Anopheles* in Ellichpur. The description of the species of *Anopheles* found in Ellichpur during the year. *Indian Medical Gazette, 36*, 361–366, 441–443.
Liu, C., Pitts, R.J., Bohbot, J.D., Jones, P.L., Wang, G., & Zwiebel, L.J. (2010). Distinct olfactory signaling mechanisms in the malaria vector mosquito *Anopheles gambiae*. *PLOS Biology, 8*(8), e1000467.
Liu, K., Tsujimoto, H., Cha, S.J., Agre, P., & Rasgon, J.L. (2011). Aquaporin water channel AgAQP1 in the malaria vector mosquito *Anopheles gambiae* during blood feeding and humidity adaptation. *Proceedings of the National Academy of Sciences, 108*(15), 6062–6066.
Liu, K., Dong, Y., Huang, Y., Rasgon, J.L., & Agre, P. (2013). Impact of trehalose transporter knockdown on *Anopheles gambiae* stress adaptation and susceptibility to *Plasmodium falciparum* infection. *Proceedings of the National Academy of Sciences of the United States of America, 110*(43), 17504–17509.
Liu, S., & Lu, B.L. (1984). A new *Aedes* from China (Diptera: Culicidae). *Acta zootaxonomica Sinica, 9*(1), 76–79.
Liu, S., Lucas, K.J., Roy, S., Ha, J., & Raikhel, A.S. (2014). Mosquito-specific RNA-1174 targets *serine hydroxymethyltransferase* to control key functions in the gut. *Proceedings of the National Academy of Sciences, 111*(40), 14460–14465.
Livdahl, T.P., & Edgerly, J.S. (1987). Egg hatching inhibition: Field evidence for population regulation in the treehole mosquito. *Ecological Entomology, 12*(4), 395–399.
Loaiza, J.R., Bermingham, E., Sanjur, O.I., Scott, M.E., Bickersmith, S.A., & Conn, J.E. (2012). Review of genetic diversity in malaria vectors (Culicidae: Anophelinae). *Infection, Genetics and Evolution, 12*(1), 1–12.
Loaiza, J.R., Scott, M.E., Bermingham, E., Sanjur, O.I., Rovira, J.R., Dutari, L.C., . . . Conn, J.E. (2013). Novel genetic diversity within *Anopheles punctimacula* s.l.: Phylogenetic discrepancy between the barcode cytochrome c oxidase I (*COI*) gene and the rDNA second internal transcribed spacer (ITS2). *Acta Tropica, 128*(1), 61–69.
Loetti, M.V., Burroni, N.E., Schweigmann, N., & de Garin, A. (2007). Effect of different thermal conditions on the pre-imaginal biology of *Culex apicinus* (Philippi, 1865) (Diptera: Culicidae). *Journal of Vector Ecology, 32*(1), 106–111.
Loew, H. (1845). *Dipterologische Beiträge*. Posen: Decker.
Loew, H. (1866). Beschreibung einiger afrikanischen—Diptera, Nemocera. *Berliner Entomologische Zeitschrift, 10*, 55–62.
Loew, H. (1873). *Beschreibung europaischer Dipteren. Systematische Beschreibung der bekannten europaischen zweiflugeligen Insecten, von Johann Wilhelm Meigen* (Vol. 3). Halle, Germany: Zehnter Theil oder vierter Supplementband.
Loew, H. (1874). Diptera nova a Hug. Theod. Christopho collecta. *Zeitschrift für die gesammten Naturwissenschaft, 43*, 413–420.
Logue, K., Chan, E.R., Phipps, T., Small, S.T., Reimer, L., Henry-Halldin, C., . . . & Serre, D. (2013). Mitochondrial genome sequences reveal deep divergences among *Anopheles punctulatus* sibling species in Papua New Guinea. *Malaria Journal, 12*(1), 64.
Lomolino, M.V., Riddle, B.R., Whittaker, R.J., & Brown, J.H. (2010). *Biogeography* (4th Ed). Sunderland, MA: Sinaver.
Lopes, S.M., & Khouri, A. Estudo morfológico da estrutura genital das fêmeas de *Anopheles* (*Nyssorhynchus*) *albitarsis* Lynch Arribálzaga (1878) e *Anopheles* (*Nyssorhynchus*) *aquasalis* Curry (1932) (Diptera, Culicidae, Anophelinae). *Contribuições*

avulsas sobre a história natural do Brasil. Série zoologia, 19, 1–5.

Lorenz, C., Marques, T.C., Mureb Sallum, M.A., & Suesdek, L. (2012). Morphometrical diagnosis of the malaria vectors *Anopheles cruzii*, *An. homunculus* and *An. bellator*. *Parasites and Vectors, 5*, 257.

Lotfi, M.D. (1970). Iranian species of genus *Culex* (Culinae [*sic*]: Diptera). *Bulletin de la Société de pathologie exotique et de ses filiales, 63*(3), 399–403.

Lotfi, M.D. (1973). Iranian species of genus *Culex* (Culicinae: Diptera). II. Report of four species of larvae (including three new records) and 14 adult species. *Bulletin de la Société de Pathologie Exotique et de ses filiales, 66*(1), 204–207.

Lounibos, L.P. (1979). Mosquitoes occurring in the axils of *Pandanus rabaiensis* Rendle on the Kenya coast. *Cahiers Office de la Recherche Scientifique et Technique d'Outre-Mer, Série entomologie médicale, 17*(1), 25–29.

Lounibos, L.P. (1983). Behavioral convergence among fruit-husk mosquitoes. *Florida Entomologist, 66*(1), 32–41.

Lounibos, L.P. (2002). Invasions by insect vectors of human disease. *Annual Review of Entomology, 47*(1), 233–266.

Lounibos, L.P., & Machado-Allison, C.E. (1983). Oviposition and egg brooding by the mosquito *Trichoprosopon digitatum* in cacao husks. *Ecological Entomology, 8*(4), 475–478.

Lounibos, L.P., Duzak, D., & Linley, J.R. (1997a). Comparative egg morphology of six species of the Albimanus Section of *Anopheles* (*Nyssorhynchus*) (Diptera: Culicidae). *Journal of Medical Entomology, 34*(2), 136–155.

Lounibos, L.P., Duzak, D., Linley, J.R., & Lourenço-de-Oliveira, R. (1997b). Egg structures of *Anopheles fluminensis* and *Anopheles shannoni*. *Memórias do Instituto Oswaldo Cruz, 92*(2), 221–232.

Lounibos, L.P., Wilkerson, R.C., Conn, J.E., Hribar, L.J., Fritz, G.N., & Danoff-Burg, J.A. (1998). Morphological, molecular, and chromosomal discrimination of cryptic *Anopheles* (*Nyssorhynchus*) (Diptera: Culicidae) from South America. *Journal of Medical Entomology, 35*(5), 830–838.

Lounibos, L.P., Escher, R.L., & Lourenço-de-Oliveira, R. (2003). Asymmetric evolution of photoperiodic diapause in temperate and tropical invasive populations of *Aedes albopictus* (Diptera: Culicidae). *Annals of the Entomological Society of America, 96*(4), 512–518.

Lounibos, L.P., O'Meara, G.F., Juliano, S.A., Nishimura, N., Escher, R.L., Reiskind, M.H., . . . Greene, K. (2010). Differential survivorship of invasive mosquito species in South Florida cemeteries: Do site-specific microclimates explain patterns of coexistence and exclusion? *Annals of the Entomological Society of America, 103*(5), 757–770.

Lourenço-de-Oliveira, R. (1983). Sobre uma nova espécie de *Phoniomyia* (Diptera: Culicidae) do Rio de Janeiro, Brasil. *Memórias do Instituto Oswaldo Cruz, 78*, 501–505.

Lourenço-de-Oliveira, R., & Deane, L.M. (1984). What is *Anopheles allopha* (Lutz and Peryassú, 1921) (Diptera: Culicidae)? *Memórias do Instituto Oswaldo Cruz, 79*, 509–510.

Lourenço-de-Oliveira, R., & Silva-do-Nascimento, T.F. (1985). *Wyeomyia forcipenis*, new species of mosquito (Diptera: Culicidae) from Rio de Janeiro, Brazil. *Memórias do Instituto Oswaldo Cruz, 80*(3), 321–326.

Lourenço-de-Oliveira, R., & Silva-do-Nascimento, T.F. (1987). *Culex siphanulatus*—new species of mosquito from the coast of Rio de Janeiro State, Brazil (Diptera, Culicidae). *Memórias do Instituto Oswaldo Cruz, 82*(1), 101–110.

Lourenço-de-Oliveira, R., & Townsend, B. (1989). On *Culex spinosus* Lutz (Diptera: Culicidae), 1905. Designation of a neotype. *Memórias do Instituto Oswaldo Cruz, 84*, 431–433.

Lourenço-de-Oliveira, R., Silva-do-Nascimento, T.F., & Goncalves de Castro, M. (1991). On the immature stages of two mosquitoes (Diptera: Culicidae) originally described from Rio de Janeiro, Brazil. *Memórias do Instituto Oswaldo Cruz, 86*(2), 209–217.

Lourenço-de-Oliveira, R., Motta, M.A., & de Castro, M.G. (1992). *Wyeomyia staminifera*, a new species of mosquito from Brazil (Diptera: Cullcidae). *Memórias do Instituto Oswaldo Cruz, 87*(1), 115–121.

Lourenço-de-Oliveira, R., Harbach, R.E., Castro, M.G., Motta, M.A., & Peyton, EL. (1999). *Wyeomyia* (*Prosopolepis*) *confusa* (Lutz): subgeneric validation, species description, and recognition of *Wyeomyia flui* (Bonne-Wepster and Bonne) as the senior synonym of *Wyeomyia kerri* del Ponte and Cerqueira. *Journal of the American Mosquito Control Association, 15*(2), 200–212.

Lowe, S., Browne, M., Boudjelas, S., & DePoorter, M. (2004). *100 of the world's worst invasive alien species: A selection from the global invasive species database*. Invasive Species Specialist Group, Species Survival Commission, World Conservation Union. Retrieved from www.k-state.edu/withlab/consbiol/IUCN_invaders.pdf

Lozano-Fuentes, S., Hayden, M.H., Welsh-Rodriguez, C., Ochoa-Martinez, C., Tapia-Santos, B., Kobylinski, K.C., . . . Eisen, L. (2012a). The dengue virus mosquito vectors *Aedes aegypti* at high elevation in México. *The American Journal of Tropical Medicine and Hygiene, 87*(5), 902–909.

Lozano-Fuentes, S., Welsh-Rodriguez, C., Hayden, M.H., Tapia-Santos, B., Ochoa-Martinez, C., Kobylinski, K.C., . . . Eisen, L. (2012b). *Aedes* (*Ochlerotatus*) *epactius* along an elevation and climate gradient in Veracruz and Puebla States, México. *Journal of Medical Entomology, 49*(6), 1244–1253.

Lu, B.-L. (1981). The *Aedes niveus* group of China (Diptera: Culicidae). *Entomotaxonomia, 3*(4), 255–263.

Lu, B.-L. (1982). The study of *Aedes albopictus* subgroup of genus *Aedes* of China. I. Adults. *Zoological Research, 3*(4), 327–338.

Lu, B.-L. (1985). The study of *Aedes albopictus* subgroup of genus *Aedes* of China. II. Larvae. *Zoological Research, 6*(1), 10–18.

Lu, B.-L., & Chen, H.-B. (1983). The pupal characters and key to subfamilies and genera of pupae of Chinese mosquitoes (Diptera: Culicidae). *Entomotaxonomia, 5*(1), 83–90.

Lu, B.-L., & Gong, Z.D. (1986). Studies on the genus *Heizmannia* of China with description of two new species (Diptera: Culicidae). *Acta entomologica Sinica, 29*(2), 181–189.

Lu, B., & Ji, S. (1997). Subgenus *Finlaya* Theobald, 1903. In Lu et al. (Eds.), *Diptera: Culicidae 1. Fauna Sinica, Insecta* (Vol. 8, pp. 99–173). Beijing, China: Science Press

Lu, B.-L., & Li, B.S. (1982a). Identification of Chinese mosquitoes. In B. S. Lu (Ed.), *Identification handbook for medically important animals in China* (pp. 1–159). Beijing, China: People's Health.

Lu, B.-L., & Li, P.-S. (1982b). A note on the larvae of *Heizmannia* (*H.*) *lii* Wu, 1936. *Entomotaxonomia, 4*(1–2), 107–108.

Lu, B.-L., & Su, L. (1987). *A handbook for the identification of Chinese aedine mosquitoes*. Beijing, China: Science Press.

Lu, B.-L., & Yu, H. (2000). The description of pupa of *Aedes galloisioides* (Diptera: Culicidae). *Jishengchong yu yixue kunchong xuebao, 7*(3), 183–184.

Lu, B.-L., Dong, X.-S., & Wang, X.-W. (1986). A new species of subgenus *Topomyia* of genus *Topomyia* (Diptera: Culicidae). *Acta zoologica Sinica, 11*(4), 406–408.

Lu, B., Li, B., Ji, S., Chen, H., Meng, Q., Su, L., . . . (Eds). (1997). *Diptera: Culicidae 1. Fauna Sinica, Insecta* (Vol. 8). Beijing, China: Science Press.

Lu, H.L., Kersch, C., & Pietrantonio, P.V. (2011). The kinin receptor is expressed in the Malpighian tubule stellate cells in the mosquito

Aedes aegypti (L.): A new model needed to explain ion transport? *Insect Biochemistry and Molecular Biology, 41*(2), 135–140.
Luckhart, S., & Rosenberg, R. (1999). Gene structure and polymorphism of an invertebrate nitric oxide synthase gene. *Gene, 232*(1), 25–34.
Ludlow, C.S. (1902a). Two Philippine mosquitoes. *Journal of the New York Entomological Society, 10*(3), 127–131.
Ludlow, C.S. (1902b). Description of a new *Anopheles*. *Journal of the American Medical Association, 39*(8), 426–427.
Ludlow, C.S. (1903). Some Philippine mosquitoes. *Journal of the New York Entomological Society, 11*(3), 137–144.
Ludlow, C.S. (1904a). Concerning some Philippine mosquitoes. *The Canadian Entomologist, 36*(1), 69–72.
Ludlow, C.S. (1904b). Mosquito notes. *The Canadian Entomologist, 36*(8), 233–236.
Ludlow, C.S. (1904c). Mosquito notes. II. *The Canadian Entomologist, 36*(10), 297–301.
Ludlow, C.S. (1905a). Mosquito notes. III. *The Canadian Entomologist, 37*(3), 94–102, 129–135.
Ludlow, C.S. (1905b). A new North American *Taeniorhynchus*. *The Canadian Entomologist, 37*(3), 231–232.
Ludlow, C.S. (1905c). Mosquito notes. IV. *The Canadian Entomologist, 37*(11), 385–388.
Ludlow, C.S. (1906a). A new American mosquito. *George Washington University Bulletin, 5*, 83–84.
Ludlow, C.S. (1906b). Mosquito notes. IV. (continued). *The Canadian Entomologist, 38*(4), 132–134.
Ludlow, C.S. (1906c). An Alaskan mosquito. *The Canadian Entomologist, 38*(10), 326–328.
Ludlow, C.S. (1906d). Mosquito notes. V. *The Canadian Entomologist, 38*(11), 367–368.
Ludlow, C.S. (1907). Mosquito notes. V. (continued). *The Canadian Entomologist, 39*(4, 8, 12), 129–131, 266–268, 413–414.
Ludlow, C.S. (1908a). Mosquito notes. VI. *The Canadian Entomologist, 40*(1, 2, 9), 32–34, 50–52, 331–332.
Ludlow, C.S. (1908b). *The mosquitoes of the Philippine Islands. The distribution of certain species, and their occurrence in relation to the incidence of certain diseases* [Doctoral dissertation, George Washington University, Washington, DC].
Ludlow, C.S. (1909a). Mosquito comment. *The Canadian Entomologist*, 41(1), 21–24.
Ludlow, C.S. (1909b). New Philippine mosquitoes. *The Canadian Entomologist, 41*(3), 97.
Ludlow, C.S. (1909c). Mosquito observations. *The Canadian Entomologist, 41*(7), 233–235.
Ludlow, C.S. (1910). Mosquito observations. (continued). *The Canadian Entomologist, 42*(6), 193–196.
Ludlow, C.S. (1911a). A new Alaskan mosquito. *The Canadian Entomologist, 43*(5), 178–179.
Ludlow, C.S. (1911b). The Philippine mosquitoes. *Psyche, 18*(4), 125–133.
Ludlow, C.S. (1914a). Philippine mosquitoes. *Psyche, 21*(1), 30–32.
Ludlow, C.S. (1914b). A new anopheline. *Psyche, 21*(4), 129–130.
Ludlow, C.S. (1914c). A new aedine. *Psyche, 21*(5), 159–160.
Ludlow, C.S. (1918). *Trichoprosopon* Theobald (Diptera; Culicidae). *Psyche, 25*(3), 66–68.
Ludlow, C.S. (1919). New mosquitos from Panama. *Psyche, 26*(6), 166–169.
Ludlow, C.S. (1920a). New Siberian Culicidae (Diptera). *Insecutor Inscitiae Menstruus, 7*(for 1919)(10–12), 151–161.
Ludlow, C.S. (1920b). Siberian *Anopheles*. *Psyche, 27*(4), 74–78.
Ludlow, C.S. (1921). A new Philippine mosquito (Diptera, Culicidae). *Military Surgeon, 49*, 690–691.
Luh, P.-L., & Lee, B.S. (1975). *Aedes* (subgenus *Ochlerotatus* Lynch Arribálzaga)—new species and a description of our country's [China] *Aedes dorsalis* group. *Acta entomologica Sinica, 18*(4), 428–432 [from Harbach 2018c].
Luh, P.L., & Li, B. (1981). A new *Armigeres* from China (Diptera: Culicidae). *Acta zootaxonomica Sinica, 6*(2), 188–190.
Luh, P.L., & Shih, J.M. (1958). New *Aedes* and *Armigeres* subspecies from Guangxi [in Chinese]. *Military Medicine, 1*, 222–226.
Luh, P.-L., Chao, K., & Xu, R.-M. (1974). A new *Culiseta* from China (Diptera: Culicidae). *Acta entomologica Sinica, 17*(4), 458–462.
Lungstrom, L. (1954). Biological studies on *Culex tarsalis* (Diptera, Culicidae) in Kansas. *Transactions of the Kansas Academy of Science, 57*, 86–96.
Lungstrom, L. (1955). Additional observations on the recognition features of the *Culex tarsalis* larva and pupa (Culicidae: Diptera). *Transactions of the Kansas Academy of Science, 58*, 68–74.
Lunt, S.R. (1976). Protein band studies of fourth instar larvae in the *Aedes varipalpus* group. *Mosquito Systematics, 8*(2), 200–204.
Lunt, S.R. (1977a). Protein band studies of the subspecies in the *Aedes atropalpus* group. *Mosquito News, 37*(3), 470–473.
Lunt, S.R. (1977b). Morphological characteristics of the larvae of *Aedes triseriatus* and *Aedes hendersoni* in Nebraska. *Mosquito News, 37*(4), 654–656.
Lunt, S.R., & Nielsen, L.T. (1971a). The use of thoracic setae as a taxonomic tool and as an aid in establishing phylogenetic relationships in adult female *Aedes* mosquitoes of North America. Part I. *Mosquito Systematics Newsletter 3*(2), 69–98.
Lunt, S.R., & Nielsen, L.T. (1971b). The use of thoracic setae as a taxonomic tool and as an aid in establishing phylogenetic relationships in adult female *Aedes* mosquitoes of North America. Part II. *Mosquito Systematics Newsletter, 3*(3), 102–121.
Luntz, A.J.M. (2003). Arthropod semiochemicals: Mosquitoes, midges and sealice. *Biochemical Society Transactions, 31*(1), 128–133.
Luo, C-W., Dewey, E.M., Sudo, S., Ewer, J., Hsu, S.Y., Honegger, H-W., & Hsueh, A.J. (2004). Bursicon, the insect cuticle-hardening hormone, is a heterodimeric cystine knot protein that activates G protein-coupled receptor LGR2. *Proceedings of the National Academy of Sciences of the United States of America, 102*(8), 2420–2825.
Lutambi, A.M., Penny, M.A., Smith, T., & Chitnis, N. (2013). Mathematical modelling of mosquito dispersal in a heterogeneous environment. *Mathematical Biosciences, 241*(2), 198–216.
Lutwama, J.J., & Mukwaya, L.G. (1994). Variation in morphological characters of adults of the *Aedes* (*Stegomyia*) *simpsoni* complex from Uganda, Kenya, and South Africa (Diptera: Culicidae). *Mosquito Systematics, 26*(3), 145–157.
Lutz, A. (1904). Quadro dos gêneros da familia Culicidae [2-page folded table]. In C. Bourroul, *Mosquitos do Brasil* [Doctoral thesis, J.B. de Oliveira Costa, Salvador, Bahia].
Lutz, A. (1905). Novas espécies de mosquitos do Brasil. *Imprensa Médica* (São Paulo), *3*, 26–29, 65–69, 81–83, 101–104, 125–128, 169–172, 202–203, 269–271, 287, 290, 311–314, 347–350.
Lutz, A. (1921). Culicideos (Systematica). Chave para os Culicideos. *Folha médica, 2*, 161–163.
Lutz, A. (1928). *Estudios de zoología y parasitología Venezuelanos*. Caracas: Universidad Central de Venezuela.
Lutz, A., & Neiva, A. (1911). Notas dipterolojicas. Dipterologische Mitteilungen. *Memórias do Instituto Oswaldo Cruz, 3*(2), 295–300.
Lutz, A., & Neiva, A. (1913). Contribuições para a biolojia das megarininas com descrições de duas especies novas. Beiträge zur Biologie der Megarhinen und Beschreibung zweier neuer Arten. *Memórias do Instituto Oswaldo Cruz, 5*(2), 129–141.

Lutz, H. (1985). A fossil mosquito from the Lower Oligocene of Cereste France (Diptera: Culicidae). *Palaeontologische Zeitschrift, 59*(314), 269–276.

Lvov, D.K. (1956). Tjber die artselbestandigkeit von *Aedes esoensis* Yam. (Diptera, Culicidae). *Revue d'entomologie de l'URSS Entomologicseszkoje Obozrenie, 35*, 929–934.

Lyimo, I.N., & Ferguson, H.M. (2009). Ecological and evolutionary determinants of host species choice in mosquito vectors. *Trends in Parasitology, 25*(4), 189–196.

Lynch Arribálzaga, F. (1878). Descripción de tres nuevos Culicidae de Buenos Aires. *El naturalista Argentino, 1*(5), 149–152.

Lynch Arribálzaga, F. (1891a). Dipterologia argentine. *Revista del Museo de La Plata, 1*, 345–377.

Lynch Arribálzaga, F. (1891b). Dipterologia argentine. *Revista del Museo de La Plata, 2*, 131–174.

Ma, S. (1981a). A new species of *Anopheles* (*Anopheles*) from China. *Entomotaxonomia, 3*(4), 277–279.

Ma, S. (1981b). Studies on the *Anopheles* (*A.*) *sinensis* group of mosquitoes in China, including four new sibling species. *Sinozoologia, 1*(1), 59–70.

Ma, S. (1982). Diptera: Culicidae. In Anonymous, *Zhongguo Kexueyuan Qingzang Gaoyuan Zonghe Kexue Kaochadui* [The series of the comprehensive scientific expedition to the Qinghai-Xizang Plateau]: *Xizang Kunchong* [Insects of Xizang, Tibet] (Vol. 2). Peking, China: Science Press, 157–163.

Ma, Y., & Xu, J. (2005). The *hyrcanus* group of *Anopheles* (*Anopheles*) in China (Diptera: Culicidae): Species discrimination and phylogenetic relationships inferred by ribosomal DNA internal transcribed spacer 2 sequences. *Journal of Medical Entomology, 42*(4), 610–619.

Ma, Y., Qu, F., Cao, Y., & Yang, B. (2000a). On molecular identification and taxonomic status of *Anopheles lesteri* and *Anopheles anthropophagus* in China (Diptera: Culicidae). *Chinese Journal of Parasitology and Parasitic Diseases, 18*(6), 325–328.

Ma, Y., Qu, F., Lei, X., & Dong, X. (2000b). Comparison of rDNA-ITS2 sequences and morphological characters of *Anopheles kunmingensis* and *Anopheles liangshanensis* [*liangshanensis*] in China, with discussion on taxonomic status. *Chinese Journal of Parasitology and Parasitic Diseases, 18*(2), 65–68.

Macan, T.T. (1950). *The anopheline mosquitoes of Iraq and north Persia*. London School of Hygiene and Tropical Medicine, Research Memorandum, No. 7. London, England: H.K. & Lewis.

Macdonald, G. (1957). *The epidemiology and control of malaria*. London, England: Oxford University Press.

Macdonald, W.W. (1958). Description of a new species of *Orthopodomyia* Theobald from Malaya, with a note on the genus in that country (Diptera: Culicidae). *Proceedings of the Royal Entomological Society of London, Series B: Taxonomy, 27*(7–8), 121–126.

Macdonald, W.W. (1960). Malaysian parasites. XXXVIII. On the systematics and ecology of *Armigeres* subgenus *Leicesteria* (Diptera, Culicidae). *Studies from the Institute for Medical Research Malaya, 29*, 110–153.

Macdonald, W. W., & Mattingly, P.F. (1960). A new species of *Udaya* from Malaya and a description of the early stages of *U. argyrurus* (Edwards, 1934) (Diptera: Culicidae). *Proceedings of the Royal Entomological Society of London, Series B: Taxonomy, 29*(1–2), 22–28.

Macfie, J.W.S., & Ingram, A. (1922). On the genital armature of the female mosquito. *Annals of Tropical Medicine and Parasitology, 16*(2), 157–188.

Macfie, J.W.S., & Ingram, A. (1923). The early stages of West African mosquitoes. VI. *Bulletin of Entomological Research, 13*(4), 409–442.

Macgregor, M.E. (1919). A new mosquito of the genus *Orthopodomyia* from a beech tree-hole in England. *Journal of the Royal Army Medical Corps, 33*(6), 451–454.

Macgregor, M.E. (1924). *Aëdes* (*Stegomyia*) *mascarensis*, MacGregor: A new mosquito from Mauritius. *Bulletin of Entomological Research, 14*(4), 409–412.

MacGregor, M.E. (1932). The ova of *Aedes* (*Finlaya*) *geniculatus* Olivier. *Parasitology, 24*, 183–184.

Maciá, A., & Bradshaw, W.E. (2000). Seasonal availability of resources and habitat degradation for the western tree-hole mosquito, *Aedes sierrensis*. *Oecologia, 125*(1), 55–65.

Maciel-de-Freitas, R., Neto, R.B., Gonçalves, J.M., Codeço, C.T., & Lourenço-de-Oliveira, R. (2006). Movement of dengue vectors between the human-modified environment and an urban forest in Rio de Janeiro. *Journal of Medical Entomology, 43*(6), 1112–1120.

Mack, R., & Smith, M. (2011). Invasive plants as catalysts for the spread of human parasites. *NeoBiota, 9*, 13–29.

Mackerras, I.M. (1927). Notes on Australian mosquitoes (Diptera, Culicidae). ii. The zoogeography of the subgenus *Ochlerotatus*, with notes on the species. *Proceedings of the Linnean Society of New South Wales, 52*(3), 284–298.

Mackerras, I.M. (1937). Notes on Australian mosquitoes (Diptera, Culicidae). III. The genus *Aedomyia* Theobald. *Proceedings of the Linnean Society of New South Wales, 62*(5–6), 259–262.

Mackerras, I.M. (1970). Skeletal Anatomy. In CSIRO (Ed.), *Insects of Australia: A textbook for students and research workers* (1st ed., pp. 3–28). Carlton, Australia: Melbourne University Press.

Macquart, J. (1834). *Histoire naturelle des insectes. Diptères. Tome premier.* Paris, France: Librairie Encyclopédique de Roret.

Macquart, J. (1838a). Diptères exotiques nouveaux ou peu connus. *Mémoires de la Société Royale des Sciences, de l'Agriculture et des Arts, de Lille, 2*, 9–225.

Macquart, J. (1838b). *Diptères exotiques nouveaux ou peu connus. Tome premier: 1.re Partie.* Paris, France: Libraire Encyclopédique de Roret.

Macquart, J. (1843). Diptères exotiques nouveaux ou peu connus. *Mémoires de la Société Royale des Sciences, de l'Agriculture et des Arts, de Lille Année, 1842*, 162–460.

Macquart, J. (1846a). Diptères exotiques nouveaux ou peu connus. *Mémoires de la Société Royale des Sciences, de l'Agriculture et des Arts, de Lille Année, 1844*(Suppl.), 133–364.

Macquart, J. (1846b). *Diptères exotiques nouveaux ou peu connus. Supplément. [I]* (Extrait *des Mémoires de la Société Royale des Sciences, de l'Agriculture et des Arts, de Lille*. 1844, 133–364). Paris, France: Libraire Encyclopedique de Roret.

Macquart, J. (1847). *Diptères exotiques nouveaux ou peu connus.* 2.e Supplément (pp. 5–104). Paris, France: Libraire Encyclopedique de Roret. (Published separately.)

Macquart, J. (1850). Diptères exotiques nouveaux ou peu connus. 4.e Supplément. *Mémoires de la Société Royale des Sciences, de l'Agriculture et des Arts, de Lille Année, 1849*, 309–479.

Madakacherry, O., Lees, R.S., & Gilles, J.R.L. (2014). *Aedes albopictus* (Skuse) males in laboratory and semi-field cages: Release ratios and mating competitiveness. *Acta Tropica, 132*(Suppl.), S124–S129.

Madder, D.J., Surgeoner, G.A., & Helson, B.V. (1983). Number of generations, egg production, and developmental time of *Culex pipiens* and *Culex restuans* (Diptera: Culicidae) in southern Ontario. *Journal of Medical Entomology, 20*(3), 275–287.

Maekawa, Y., Sunahara, T., Dachlan, Y.P., Yotoranoto, S., Basuki, T., Uemura, H., Kanbara, H., & Takagi, M. 2009. First record of *Anopheles balabacensis* from western Sumbawa Island, Indonesia.

Journal of the American Mosquito Control Association, 25(2), 203–205.
Maestre-Serrano, R., Cochero, S., Bello, B., & Ferro, C. (2013). Registry and distribution update of the species of the *Haemagogus* (Diptera: Culicidae) genus in the Caribbean region of Colombia. *Biomédica, 33*, 185–189.
Maffi, M. (1962). Contributo alla conoscenza di *Anopheles* (*Myzomyia*) *lloreti* Gil Collado. *Rivista di malariologia, 41*, 63–68.
Maffi, M. (1971). On some larvae of the *Myzomyia* series collected in the Yemen. *Parassitologia, 13*(3), 449–458.
Maffi, M. (1973a). 74. Morphological observations on a population of the *punctulatus* complex of *Anopheles* (Diptera, Culicidae) from Rennell Island (Solomon Group). *The Natural History of Rennell Island, British Solomon Islands, 7*, 29–39.
Maffi, M. (1973b). 75. The mosquitoes (Diptera, Culicidae) of Rennell and Bellona. *The Natural History of Rennell Island, British Solomon Islands, 7*, 41–60.
Maffi, M., & Coluzzi, M. (1958). *Anopheles* (*Neocellia*) *salbaii* n. sp. morfología et ambiente. *Rivista di malariologia, 37*(1–3), 45–55.
Maffi, M., & Taylor, B. (1974). The mosquitoes of the Santa Cruz faunal subarea of the Southwest Pacific (Diptera: Culicidae). *Journal of Medical Entomology, 11*(2), 197–210.
Maffi, M., & Tenorio, J.A. (1977). *Culex* (*Culex*) *banksensis*, a new species of mosquito (Diptera: Culicidae) from the Banks Islands, New Hebrides. *Pacific Insects, 17*(4), 503–509.
Maffi, M., Steffan, W.A., & Supardi, P. (1979). A survey of the mosquito fauna (Diptera, Culicidae) of the Irian Jaya Province, New Guinea,Indonesia. *Pacific Insects, 21*(2–3), 203–220.
Magnarelli, L.A. (1980). Bionomics of *Psorophora ferox* (Diptera: Culicidae): Seasonal occurrence and acquisition of sugars. *Journal of Medical Entomology, 17*(4), 328–332.
Mahilum, M. (2003). *Biochemical, molecular, and microscopic studies on mosquitoes with special emphasis on* Wolbachia *infections, diagnostic assessment of Dengue viruses and biological control of Dengue vectors* [Doctoral dissertation, University of Heidelberg, Germany]. Retrieved from http://archiv.ub.uni-heidelberg.de/volltextserver/3098/
Mahon, R.J., & Miethke, P.M. (1982). *Anopheles farauti* No. 3, a hitherto unrecognized biological species of mosquito within the taxon *A. farauti* (Laveran) (Diptera: Culicidae). *Transactions of the Royal Society of Tropical Medicine and Hygiene, 76*(1), 8–12.
Mahon, R.J., Green, C.A., & Hunt, R.H. (1976). Diagnostic allozymes for routine identification of adults of the *Anopheles gambiae* complex (Diptera, Culicidae). *Bulletin of Entomological Research, 66*(1), 25–31.
Maïga, H., Niang, A., Sawadogo, S.P., Dabiré, R.K., Lees, R.S., Gilles, J.R., . . . Diabaté, A. (2014). Role of nutritional reserves and body size in *Anopheles gambiae* males mating success. *Acta Tropica, 132*(Suppl.), S102–S107.
Mains, J.W., Brelsfoard, C.L., Rose, R.I., & Dobson, S.L. (2016). Female adult *Aedes albopictus* suppression by *Wolbachia*-infected male mosquitoes. *Scientific Reports, 6*, Article 33846.
Majambere, S., Pinder, M., Fillinger, U., Ameh, D., Conway, D.J., Green, C., . . . Lindsay, S.W. (2010). Is mosquito larval source management appropriate for reducing malaria in areas of extensive flooding in The Gambia? A cross-over intervention trial. *The American Journal of Tropical Medicine and Hygiene, 82*(2), 176–184.
Mala, A.O., Irungu, L.W., Mitaki, E.K., Shililu, J.I., Mbogo, C.M., Njagi, J.K., . . . Githure, J.I. (2014). Gonotrophic cycle duration, fecundity and parity of *Anopheles gambiae* complex mosquitoes during an extended period of dry weather in a semi arid area in Baringo County, Kenya. *International Journal of Mosquito Research, 1*(2), 28–34.
Malafronte, R.S., Marrelli, M.T., & Marinotti, O. (1999). Analysis of ITS2 DNA Sequences from Brazilian *Anopheles darlingi* (Diptera: Culicidae). *Journal of Medical Entomology, 36*(5), 631–634.
Maldonado, V., Finol, H.J., & Navarro, J.C. (1997). *Anopheles aquasalis* eggs from two Venezuelan localities compared by scanning electron microscopy. *Memórias do Instituto Oswaldo Cruz, 92*(4), 487–491.
Maldonado-Capriles, J., Pippin, W.F., & Kuns, M.L. (1958). An annotated checklist of the mosquitoes of Mona Island, Puerto Rico, and the larva and male of *Aedes obturbator* D. & K. (Diptera, Culicidae). *Proceedings of the Entomological Society of Washington, 60*, 65–68.
Mallampalli, V.L. (1995). *Phylogenetic relationships among taxa in the genus* Culex, *and the phylogenetic relationships of vectors of New World alphaviruses in the subgenus* Melanoconion *(Diptera: Culicidae)* [Doctoral dissertation, University of Maryland].
Mamedniyazov O., & Yerokhin, P. (2005). *Illustrated keys of malaria mosquitoes (Culicidae, Anophelinae) in Middle Asia*. Ashgabat, Turkmenistan: Ylym.
Manalang, C. (1930). Morphology and classification of the Philippine variety of *Anopheles aconitus* Donitz, 1902, and *Anopheles minimus*, 1901. *Philippine Journal of Science, 43*(2), 247–261.
Manda, H., Gouagna, L.C., Foster, W.A., Jackson, R.R., Beier, J.C., Githure, J.I., & Hassanali, A. (2007a). Effect of discriminative plant-sugar feeding on the survival and fecundity of *Anopheles gambiae*. *Malaria Journal, 6*(1), 113.
Manda, H., Gouagna, L.C., Nyandat, E., Kabiru, E.W., Jackson, R.R., Foster, W.A., . . . Hassanali, A. (2007b). Discriminative feeding behaviour of *Anopheles gambiae* on endemic plants in western Kenya. *Medical and Veterinary Entomology, 21*(1), 103–111.
Mandoul, M., Dubos, M., Moulinier, C., & De Cournuaud, M. (1968). Une enquête sur les culicides de Charente-Maritime; présence du genre *Mansonia* Blanchard 1901, *M.* (*Coquillettidia*) *richiardii* (Ficalbi) 1896 et *M.* (*C.*) *buxtoni* (Edwards) 1923. *Bulletin de la Société de pathologie exotique, 61*(2), 282–288.
Mangkoewinoto, R.M.M. (1919). Anophelinen van West-Java. *Mededellingen van den Burgerlijken Geneeskundigen Dienst in Nederlandsch-Indië, 8*, 41–82.
Manguin, S. (Ed.) (2013). Anopheles *mosquitoes: New insights into malaria vectors*. Rijeka, Croatia: InTech.
Manguin, S., & Boëte, C. (2011). Global impact of mosquito biodiversity, human vector-borne diseases and environmental change. In J. Lapez-Pujol (Ed.), *The importance of biological interactions in the study of biodiversity* (pp. 27–50). New York, NY: InTech.
Manguin, S., Roberts, D.R., Peyton, EL, Fernandez-Salas, I., Barreto, M., Loayza, R.F., . . . Rodriguez, M.H. (1995). Biochemical systematics and population genetic structure of *Anopheles pseudopunctipennis*, vector of malaria in Central and South America. *The American Journal of Tropical Medicine and Hygiene, 53*(4), 362–377.
Manguin, S., Roberts, D.R., Andre, R.G., Rejmánková, E., & Hakre, S. (1996). Characterization of *Anopheles darlingi* (Diptera: Culicidae) larval habitats in Belize, Central America. *Journal of Medical Entomology, 33*(2), 205–211.
Manguin, S., Roberts, D.R., Peyton, EL, Rejmánková, E., & Pecor, J. (1996). Characterization of *Anopheles pseudopunctipennis* larval habitats. *Journal of the American Mosquito Control Association, 12*(4), 619–626.

Manguin, S., Garros, C., Harbach, R.E., & Coosemans, M. (2008). Bionomics, taxonomy, and distribution of the major malaria vector taxa of *Anopheles* subgenus *Cellia* in Southeast Asia: An updated review. *Infection, Genetics, and Evolution, 8*(4), 489–503.

Manguin, S., Ngo, C.T., Tainchum, K., Juntarajumnong, W., Chareonviriyaphap, T., Michon, A.-L., & Jumas-Bilak, E. (2013). Bacterial biodiversity in midguts of *Anopheles* mosquitoes, malaria vectors in Southeast Asia. In S. Manguin (Ed.), Anopheles *mosquitoes: New insights into malaria vectors* (pp. 549–576). Rijeka, Croatia: InTech.

Manh, N.D., Hinh, T.D., & Vien, N.T. (1993). *Anopheles* (*Anopheles*) *vietnamensis*: A new species in the *Anopheles hyrcanus* group from Vietnam (Diptera: Culicidae). *Mosquito Systematics, 25*(3), 215–221.

Manin, B.O., Drakeley, C.J., Chua, T.H. (2018). Mitochondrial variation in subpopulations of *Anopheles balabacensis* Baisas in Sabah, Malaysia (Diptera: Culicidae). *PLOS ONE, 13*(8), e0202905.

Mann, F.G. (1950). Dos nuevas sub-especies del *Anopheles pseudopunctipennis* Th. 1901. *Biológica, 8-11*, 33–42.

Manning, D.L., Evenhuis, N.L., & Steffan, W.A. (1982). Supplement to annotated bibliography of *Toxorhynchites* (Diptera: Culicidae). *Journal of Medical Entomology, 19*, 429–186.

Manonmani, A.M., Sadanandane, C., Sahu, S.S., Mathivanan, A., & Jambulingam, P. (2007). rDNA-ITS2-PCR assay for grouping the cryptic species of *Anopheles culicifacies* complex (Diptera: Culicidae). *Acta Tropica, 104*(1), 72–77.

Mao, Y., Xu, X., Xu, W., Ishida, Y., Leal, W.S., Ames, J.B., & Clardy, J. (2010). Crystal and solution structures of an odorant-binding protein from the southern house mosquito complexed with an oviposition pheromone. *Proceedings of the National Academy of Sciences of the United States of America, 107*(44), 19102–19107.

Marchon-Silva, V., Lourenço-de-Oliveira, R., Dolsan de Almeida, M., da Silva-Vasconcelos, A., & Costa, J. (1996). The type specimens of mosquitoes (Diptera, Culicidae) deposited in the entomological collection of the Instituto Oswald Cruz, Rio de Janeiro, Brazil. *Memórias do Instituto Oswaldo Cruz, 91*(4), 471–478.

Marcondes, C.B., & Alencar, J. (2010). Revisão de mosquitos *Haemagogus* Williston (Diptera: Culicidae) do Brasil. *Revista biomédica, 21*, 221–238.

Marcondes, C.B., Fernandes, A., Paterno, U., Müller, G.A., Pinho, L.C.D., & Struffaldi, D.V. (2003). New records of mosquitoes from the southern Brazilian States of Santa Catarina and Rio Grande do Sul, with 18 species new for the States (Diptera: Culicidae). *Zootaxa, 347*, 1–6.

Marcondes, C.B., Müller, G.A., dos Santos-Mallet, J., Alencar, J., Sarmento, J.S., dos Santos, C.B., & Rezende, H.R. (2012). Scanning electron microscopy of eggs of *Ochlerotatus taeniorhynchus* (Diptera: Culicidae, Aedini). *Zootaxa, 3509*, 77–80.

Marcondes, C.B., Contigiani, M., & Gleiser, R.M. (2017). Emergent and reemergent arboviruses in South America and the Caribbean: Why so many and why now? *Journal of Medical Entomology, 54*(3), 509–532.

Margalit, J., & Tahori, A.S. (1974). An annotated list of mosquitoes in Israel. *Israel Journal of Entomology, 9*, 77–91.

Marinotti, O., Nguyen, Q.K., Calvo, E., James, A.A., & Ribeiro, J.M.C. (2005). Microarray analysis of genes showing variable expression following a blood meal in *Anopheles gambiae*. *Insect Molecular Biology, 14*(4), 365–373.

Marinotti, O., Calvo, E., Nguyen, Q.K., Dissanayake, S., Ribeiro, J.M.C., & James, A.A. (2006). Genome-wide analysis of gene expression in adult *Anopheles gambiae*. *Insect Molecular Biology, 15*(1), 1–12.

Marks, E.N. (1947a). Studies of Queensland mosquitoes. Part I. The *Aëdes* (*Finlaya*) *kochi* group with descriptions of new species from Queensland, Bougainville and Fiji. *University of Queensland Papers, Department of Biology, 2*(5), 1–66.

Marks, E.N. (1947b). Studies of Queensland mosquitoes. Part II. New species of *Aëdes* (subgenus *Finlaya*). *University of Queensland Papers, Department of Biology, 2*(6), 1–10.

Marks, E.N. (1948). Studies of Queensland mosquitoes. Part III. The *Aëdes* (*Finlaya*) *australiensis* group. *University of Queensland Papers, Department of Biology, 2*(8), 1–42.

Marks, E.N. (1949). Studies of Queensland mosquitoes. Part IV. Some species of *Aëdes* (subgenus *Ochlerotatus*). *University of Queensland Papers, Department of Biology, 2*(11), 1–41.

Marks, E.N. (1951a). The vector of filariasis in Polynesia: A change in nomenclature. *Annals of Tropical Medicine and Parasitology, 45*(2), 137–140.

Marks, E.N. (1951b). Mosquitoes from Southeastern Polynesia. *Occasional Papers of Bernice P. Bishop Museum, 20*(9), 123–130.

Marks, E.N. (1954). A review of the *Aedes scutellaris* subgroup with a study of variation in *Aedes pseudoscutellaris* (Theobald) (Diptera: Culicidae). *Bulletin of the British Museum (Natural History) Entomology, 3*(10), 347–414.

Marks, E.N. (1955). Studies of Queensland mosquitoes. Part V. Some species of *Aedes* (subgenus *Finlaya*). *University of Queensland Papers, Department of Biology, 1*(2), 11–29.

Marks, E.N. (1956). A new species of *Anopheles* from Queensland and notes on related species (Diptera, Culicidae). *Proceedings of the Royal Society of Queensland, 67*(6), 41–52.

Marks, E.N. (1957a). The subgenus *Ochlerotatus* in the Australian region (Diptera: Culicidae). I. Notes on classification, with the description of a new species. *University of Queensland Papers, Department of Biology, 1*(5), 71–83.

Marks, E.N. (1957b). Some mosquitoes from Western Samoa, with a description of a new species of *Aedes* (*Stegomyia*). (Diptera, Culicidae). *Annals of Tropical Medicine and Parasitology, 51*(1), 50–57.

Marks, E.N. (1958a). New species and records of the *Aedes kochi* group from Eastern Australia (Diptera: Culicidae). *Proceedings of the Royal Society Queensland, 69*(6), 57–74.

Marks, E.N. (1958b). Notes on *Opifex fuscus* Hutton (Diptera: Culicidae) and the scope for further research on it. *New Zealand Entomologist, 2*(2), 20–25.

Marks, E.N. (1959a). The subgenus *Ochlerotatus* in the Australian region (Diptera: Culicidae). II. Five new species from Western Australia. *University of Queensland Papers, 1*(8) , 111–134.

Marks, E.N. (1959b). On two mosquitoes of the subgenus *Finlaya* (Diptera: Culicidae). *Proceedings of the Royal Society Queensland, 70*(4), 21–25.

Marks, E.N. (1960a). Australian mosquitoes described by Macquart. II. Species in Bigot's collection, *Aëdes* (*Ochlerotatus*) *nigrithorax* (Macquart). New synonymy. *Proceedings of the Linnean Society of New South Wales, 85*(1), 117–120.

Marks, E.N. (1960b). Taxonomy and biology of some Papuan Culicidae. I. Genus *Topomyia*. *Pacific Insects, 2*(1), 91–98.

Marks, E.N. (1963a). A revision of the subgenus *Chaetocruiomyia* Theobald (Diptera: Culicidae). *University of Queensland Papers, 1*(13), 189–211.

Marks, E.N. (1963b). The subgenus *Ochlerotatus* in the Australian region. V. The Stricklandi Section. *Journal of the Entomological Society of Queensland, 2*(1), 31–47.

Marks, E.N. (1964a). The subgenus *Ochlerotatus* in the Australian region (Diptera: Culicidae). VII. Four new species. *University of Queensland Papers, Department of Biology, 2*(3), 59–71.

Marks, E.N. (1964b). Notes on the subgenus *Chaetocruiomyia* Theobald (Diptera: Culicidae). *Proceedings of the Linnean Society of New South Wales, 89*(1), 131–147.
Marks, E.N. (1968). Northern records of the genus *Culiseta* Felt in Australia with the description of a new species (Diptera: Culicidae). *Journal of the Australian Entomological Society, 7*(1), 43–56.
Marks, E.N. (1969a). Mosquitoes (Culicidae) on Queensland's coral cays. *Queensland Naturalist, 19*(4–6), 94–98.
Marks, E.N. (1969b). The name of the Australian salt-marsh mosquito. *Mosquito Systematics, 1*(4), 81.
Marks, E.N. (1970). A list of Australian species of *Uranotaenia* Lynch Arribálzaga (Diptera: Culicidae). *Journal of the Australian Entomological Society, 9*, 165–167.
Marks, E.N. (1971a). A metapleural generic character in *Topomyia* Leicester. *Mosquito Systematics, 3*(1), 4–6.
Marks, E.N. (1971b). *Culex* (*Culex*) *barraudi* Edwards 1922 and *Culex* (*Culex*) *edwardsi* Barraud 1923. *Mosquito Systematics, 3*(4), 195–196.
Marks, E.N. (1973). *An atlas of common Queensland mosquitoes* (Rev. ed.). St. Lucia: University of Queensland.
Marks, E.N. (1977). *Aedes* (*Macleaya*) *stoneorum*, a new species from Queensland (Diptera: Culicidae). *Proceedings of the Entomological Society of Washington, 79*(1), 33–37.
Marks, E.N. (1982). Recent taxonomic studies of *Culex annulirostris* and allied species and their possible significance for arbovirus research. In T.D. St. George, & B.H. Kay (Eds.). *Arbovirus Research in Australia. Proceedings of 3rd symposium, Feb. 15–17, 1982* (pp. 146–151). Brisbane: CSIRO Division of Animal Health and Queensland Institute of Medical Research.
Marks, E.N., & Brown, P. (1970). *Culex albirostris* Macquart, 1851 (Diptera, Culicidae): Request for suppression under the plenary powers. Z. N. (S.) 1889. *Bulletin of Zoological Nomenclature, 26*, 227–229.
Marks, E.N., & Hodgkin, E.P. (1958). A new species of *Aedes* (*Finlaya*) from Northern Australia (Diptera, Culicidae). *Proceedings of the Linnean Society of New South Wales, 83*(1), 33–39.
Marks, E.N., & Nye, E.R. (1963). The subgenus *Ochlerotatus* in the Australian region (Diptera: Culicidae). VI. The New Zealand species. *Transactions of the Royal Society of New Zealand: Zoology, 4*(2), 49–60.
Marlatt, C.L. (1903). Collecting notes on mosquitoes in Oriental countries. *Proceedings of the Entomological Society of Washington, 5*(2), 111–123.
Marrelli, M.T., Li, C., Rasgon, J.L., & Jacobs-Lorena, M. (2007). Transgenic malaria-resistant mosquitoes have a fitness advantage when feeding on *Plasmodium*-infected blood. *Proceedings of the National Academy of Sciences of the United States of America, 104*(13), 5580–5583.
Marsh, F. (1933). A new species of *Anopheles* (*Myzomyia* group) from south-west Persia. *Stylops, 2*(9), 193–197.
Marshall, J.F. (1938). *The British Mosquitoes*. London, England: British Museum (Natural History).
Marshall, J.F., & Staley, J. (1933). *Theobaldia* (*Culicella*) *litorea* (Shute) n. sp. (Diptera, Culicidae). *Parasitology, 25*(1), 119–126.
Marshall, J.F., & Staley, J. (1937). Some notes regarding the morphological and biological differentiation of *Culex pipiens* Linnaeus and *Culex molestus* Forskål (Diptera, Culicidae). *Proceedings of the Royal Entomological Society of London, Series A, General Entomology 12*(1–2), 17–26.
Martin, D., Wang, S.F., & Raikhel, A.S. (2001). The vitellogenin gene of the mosquito *Aedes aegypti* is a direct target of ecdysteroid receptor. *Molecular and Cellular Endocrinology, 173*(1–2), 75–86.
Martínez, A. (1949). *Anopheles* (*Kerteszia*) *laneanus* Corrêa y Cerqueira, 1944, nueva especies para la entomofauna Boliviana. *Misión de Estudios de Patología Regional Argentina, 20*, 13–17.
Martínez, A., & Prosen, A.F. (1953a). Una especie nueva de *Uranotaenia* (Dipt. Culic.) de Argentina. *Boletín de la Sociedad Entomológica Argentina, 2*, 4.
Martínez, A., & Prosen, A.F. (1953b). Una nueva especie de mosquito de la Argentina. *Públicaciones/Misión de Estudios de Patología Regional Argentina, Jujuy, Universidad de Buenos Aires, 1*, 1–2.
Martínez, A., & Prosen, A.F. (1955). Las especies de *Aedes* (*Howardina*) de Argentina y Bolivia (Diptera, Culicidae). *Públicaciones/Misión de Estudios de Patología Regional Argentina, Jujuy, Universidad de Buenos Aires, 26*(85–86), 21–41.
Martínez, A., & Prosen, A.F. (1958). Una nueva especie de *Orthopodomyia* (Dipt. Culicidae). *Publicaciones/Misión de Estudios de Patología Regional Argentina, Jujuy, Universidad de Buenos Aires, 27*(87–88), 17–22.
Martínez, A., Prosen, A.F., & Carcavallo, R.U. (1959). Algunos culicidos interesantes de la R. Argentina (Diptera, Culicidae). *Anales del Instituto de Medicina Regional, 5*(1), 109–120.
Martínez, A., Carcavallo, R.U., & Prosen, A.F. (1960). Una nueva especie de Culicidae de Bolivia (Díptera). *Revista de la Cátedra de Microbiología y Parasitología, 28*(89), 25–29.
Martínez, A., Carcavallo, R.U., & Prosen, A.F. (1961a). El genero *Haemagogus* Williston, 1896, en la Argentina (Diptera, Culicidae). *Anales del Instituto de Medicina Regional, 5*(for 1960)(2), 63–86.
Martínez, A., Carcavallo, R.U., & Prosen, A.F. (1961b). Una nueva especie de *Culex* para la Argentina. *Anales del Instituto de Medicina Regional, 5*(3), 117–123.
Martínez, A., Prosen, A.F., & Carcavallo, R U. (1964). El alotipo macho de *Aëdes* (*Howardina*) *aurivittatus* Cerqueira (Diptera, Culicidae). *Anales del Instituto de Medicina Regional, 6*(for 1962–1963)(1–2), 53–58.
Martínez Palacios, A. (1950). Identificación de los mosquitos Mexicanos del subgénero *Culex* (Diptera: Culicidae) por la genitalia masculina. *Revista de la Sociedad Mexicana de Historia Natural, 11*(1–4), 183–187.
Martínez Palacios, A. (1952a). *Culex inflictus* Theobald y *Culex thriambus* Dyar, mosquitos nuevos para Mexico (Diptera: Culicidae). *Revista de la Sociedad Mexicana de Historia Natural, 13*(1–4), 89–95.
Martínez Palacios, A. (1952b). Nota sobre la distribución de los mosquitoes *Culex* en Mexico (Diptera, Culicidae). *Revista de la Sociedad Mexicana de Historia Natural, 13*(1–4), 75–87.
Martinez-Villegas, L., Assis-Geraldo, J., Koerich, L.B., Collier, T.C., Lee, Y., Main, B.J., . . . Pimenta, P.F.P. (2019). Characterization of the complete mitogenome of *Anopheles aquasalis*, and phylogenetic divergences among *Anopheles* from diverse geographic zones. *PLOS ONE, 14*(9), e0219523.
Martini, E. (1914). Some new American mosquitoes. *Insecutor Inscitiae Menstruus, 2*(5), 65–76.
Martini, E. (1920a). Über Stechmüken besonders deren europäische Arten und ihre Bekämpfung veröffentlicht mit Unterstützung der Hamburgischen wissenschaftlichen Stiftung. *Archiv für Schiffs- und Tropen-Hygiene Leipzig, 24*(B1), 1–267.
Martini, E. (1920b). Ueber mecklenburgische Culicinen. *Sitzungsberichte und Abhandlungen der Naturforschenden Gesellschaft zu Rostoc, 7*, 203–209.
Martini, E. (1922). Bemerkungen zu einigen neueren Mückenarbeiten, welche auch für die deutsche Fauna wichtig sind. *Entomologische Mitteilungen, 11*(4), 158–166.

Martini, E. (1924). Culiciden-Beobachtungen 1922, 1923. *Zeitschrift für angewandte Entomologie, 10*(2), 436–447.

Martini, E. (1925). Zwei bemerkenswerte Culiciden von einem eigenartigen Biotop. *Internationale Revue der gesamte Hydrobiologie und Hydrographie, 12*(5–6), 333–337.

Martini, E. (1926). Über die Stechmücken der Umgebung von Saratow. *Raboty Volzhskoi Biologicheskoi Stantsii [Arbeiten der Biologischen Wolga-Station], 8*, 189–227.

Martini, E. (1927). Über zwei neue Stechmückenarten aus Anatolien. *Archiv für Schiffs- und Tropen-Hygiene, 31*, 386–390.

Martini, E. (1929). 11. u. 12. Culicidae. In E. Lindner (Ed.), *Die Fliegen der palaerktischen Region. Band III* (pp. 1–144). Stuttgart, Germany: E. Schweizerbart'sche Verlagsbuchhandlung (Erwin Nägele).

Martini, E. (1930). 11. u. 12. Culicidae. In E. Lindner (Ed.), *Die Fliegen der palaerktischen Region. Band III* (pp. 145–320). Stuttgart, Germany: E. Schweizerbart'sche Verlagsbuchhandlung (Erwin Nägele).

Martini, E. (1931a). Ueber einige südamerikanische Culiciden. *Revista de entomologia, 1*(2), 199–219.

Martini, E. (1931b). Die Ausbeute der deutschen Chaco-Expedition 1925/26. Diptera. XXV. Culicidae. *Konowia, 10*(2), 116–120.

Martini, E. (1931c). 11. u. 12. Culicidae. In E. Lindner (Ed.), *Die Fliegen der palaerktischen Region. Band III* (pp. 321–398). Stuttgart, Germany: E. Schweizerbart'sche Verlagsbuchhandlung (Erwin Nägele).

Martini, E. (1932a). *Chagasia fajardoi* Lutz var. *stigmopteryx*, nov. var. (Dipt. Culicidae). *Revista de entomologia, 2*(3), 276–277.

Martini, E. (1932b). Dos nuevos mosquitos *Anopheles*, procedentes del Estado de Chiapas, México. *Revista Mexicana de biología, 12*, 99–102.

Martini, E. (1935). Los mosquitos de Mexico. *Departamento de Salubridad Pública, Boletines Tecnicos, Serie A: Entomología Médica y Parasitología*, No. 1.

Martins, G.F., Serrão, J.E., Ramalho-Ortigão, J.M., & Pimenta, P.F.P. (2011). Histochemical and ultrastructural studies of the mosquito *Aedes aegypti* fat body: Effects of aging and diet type. *Microscopy Research and Technique, 74*(11), 1032–1039.

Marusalin, J., Matier, B.J., Rheault, M.R., & Donini, A. (2012). Aquaporin homologs and water transport in the anal papillae of the larval mosquito, *Aedes aegypti. Journal of Comparative Physiology B, 182*(8), 1047–1056.

Masaki, S. (1980). Summer diapause. *Annual Review of Entomology, 25*(1), 1–25.

Masia, R., Aneshansley, D., Nagel, W., Nachman, R.J., & Beyenbach, K.W. (2000). Voltage clamping single cells in intact Malpighian tubules of mosquitoes. *American Journal of Physiology-Renal Physiology, 279*(4), F747–F754.

Maslov, A.V. (1963). *Mosquitoes of the group* Culiseta *(Diptera, Culicidae)* [Doctoral dissertation, Zoological Institute of the Academy of Sciences of the USSR, Leningrad].

Maslov, A.V. (1964). On the systematics of bloodsucking mosquitoes of the group *Culiseta* (Diptera, Culicidae). *Revue d'entomologie de l'URSS Entomologicseszkoje Obozrenie, 43*(1), 193–217.

Maslov, A.V. (1967). *Bloodsucking mosquitoes of the subtribe Culisetina (Diptera, Culicidae) of the world fauna*. St. Petersburg, Russia: Académie Impériale des Sciences, Musée Zoologique.

Maslov, A.V. (1989). *Bloodsucking mosquitoes of the subtribe Culisetina (Diptera, Culicidae) of the world fauna*. Washington, DC: Smithsonian Institution/National Science Foundation.

Masuoka, P., Klein, T.A., Kim, H.C., Claborn, D.M., Achee, N., Andre, R., . . . Grieco, J. (2010). Modeling the distribution of *Culex tritaeniorhynchus* to predict Japanese encephalitis distribution in the Republic of Korea. *Geospatial Health, 5*(1), 45–57.

Matheson, R. (1933). A new species of mosquito from Colorado (Diptera, Culicidae). *Proceedings of the Entomological Society of Washington, 35*(5), 69–71.

Matheson, R. (1934a). Notes on *Psorophora (Janthinosoma) horridus* Dyar and Knab. *Proceedings of the Entomological Society of Washington, 36*(2), 41–43.

Matheson, R. (1934b). Notes on mosquitoes from South America, with a description of a new species (Diptera, Culicidae). *Proceedings of the Entomological Society of Washington, 36*(5), 120–122.

Matheson, R. (1944). *Handbook of the mosquitoes of North America. Their anatomy and biology; how they can be studied and how identified; how they carry disease and how they can be controlled* (2nd ed.). Ithaca, NY: Comstock.

Mathis, M. (1938). Élévage en serie (cinq generations) d'un *Stegomyia* de Java: *Aedes annadalei* Theobald. *Bulletin de la Société de pathologie exotique, 31*, 493–497.

Matsumura, S. (1916). Additamenta. In *Thousand insects of Japan* (Vol. 2, pp. 187–474, plates 116–125). Tokyo: Keisei-sha.

Matsuo, K. (1975). Scanning electron microscopy of mosquitoes. Part IV. The egg surface structure of 3 species of *Aedes* from Japan. *Japanese Journal of Sanitary Zoology, 26*(1), 49–53.

Matsuo, K., & Iwaki, M. (1972). Morphological differences between *Culex tritaeniorhynchus* and *Culex pseudovishnui* adults from Japan. *Journal of the Kyoto Prefectural University of Medicine, 81*, 349–357.

Matsuo, K., & Ramalingam, S. (1972). Morphological characters of *Culex pseudovishnui* larvae from Malaysia, Japan and Formosa. *Southeast Asian Journal of Tropical Medicine and Public Health, 3*, 55–61.

Matsuo, K., Yoshida, Y., & Kunou, I. (1972). The scanning electron microscopy of mosquitoes (Diptera: Culicidae). I. The egg surfaces of five species of *Aedes* and *Armigeres subalbatus. Journal of the Kyoto Prefectural University of Medicine, 81*, 358–363.

Matsuo, K., Yoshida, Y., & Lien, J.-C. (1974). Scanning electron microscopy of mosquitoes. II. The egg surface structure of 13 species of *Aedes* from Taiwan. *Journal of Medical Entomology, 11*(2), 179–188.

Matsuo, K., Lien, J.-C., Kunou, I., & Ito, S. (1974). Morphological studies of the *mimeticus* group of genus *Culex* (Diptera: Culicidae). I. Pale spots on wing and wing venation. *Japanese Journal of Sanitary Zoology, 25*(1), 51–56.

Matthews, T.C., & Munstermann, L.E. (1994). Chromosomal repatterning and linkage group conservation in mosquito karyotypic evolution. *Evolution, 48*(1), 146–154.

Mattingly, P.F. (1947). Notes on the early stages of certain Ethiopian mosquitoes, with some locality records from British West Africa. *Annals of Tropical Medicine and Parasitology, 41*(2), 239–252.

Mattingly, P.F. (1949a). Anopheline pupae (Diptera, Culicidae) from West Africa. *Annals of Tropical Medicine and Parasitology, 43*(1), 23–25.

Mattingly, P.F. (1949b). *Ficalbia (Ficalbia) jacksoni* sp. n., a new species of mosquito (Diptera Culicidae) from Hong-Kong. *Proceedings of the Royal Entomological Society of London, Series B: Taxonomy, 18*(1–2), 9–11.

Mattingly, P.F. (1949c). Notes on some Oriental mosquitoes. *Proceedings of the Royal Entomological Society of London, Series B: Taxonomy, 18*(11–12), 219–228.

Mattingly, P.F. (1950). Family Culicidae. Subfamily Culicinae. In R.L. Coe, P. Freeman, & P.F. Mattingly (Eds.), *Diptera. 2. Nematocera:*

Families Tipulidae to Chironomidae (Vol. 2, pp. 102–120). London, England: Royal Entomological Society.
Mattingly, P.F. (1951a). *Culex (Culex) torrentium* Martini, a mosquito new to Great Britain. *Nature, 168*(4265), 172–173.
Mattingly, P.F. (1951b). The *Culex pipiens* complex. Introduction. *Transactions of the Royal Entomological Society of London, 102*(7), 331–342.
Mattingly, P.F. (1951c). *Culex (Culex) nakuruensis* sp. n. (Diptera, Culicidae), a remarkable new mosquito from the East African highlands. *Proceedings of the Royal Entomological Society of London, Series B: Taxonomy, 20*(3–4), 44–46.
Mattingly, P.F. (1952). The subgenus *Stegomyia* (Diptera: Culicidae) in the Ethiopian region. I. A preliminary study of the distribution of species occurring in the West African sub-region with notes on taxonomy and bionomics. *Bulletin of the British Museum (Natural History) B, 2*(5), 233–304.
Mattingly, P.F. (1953a). The subgenus *Stegomyia* (Diptera: Culicidae) in the Ethiopian region. II. Distribution of species confined to the east and south African sub-region. *Bulletin of the British Museum (Natural History). Entomology, 3*(1), 1–65.
Mattingly, P.F. (1953b). A change of name among the British mosquitoes. (Diptera, Culicidae). *Proceedings of the Royal Entomological Society of London, Series B: Taxonomy, 22*(5–6), 106–108.
Mattingly, P.F. (1953c). Notes on the Culicini of the Katanga (Diptera, Culicidae). Part I. Taxonomy. *Revue de zoologie et de botanique Africaines, 47*(3–4), 311–343.
Mattingly, P.F. (1953d). New records and a new species of the subgenus *Stegomyia* (Diptera, Culicidae) from the Ethiopian Region. *Annals of Tropical Medicine and Parasitology, 47*(3), 294–298.
Mattingly, P.F. (1954a). Notes on the subgenus *Stegomyia* (Diptera, Culicidae), with a description of a new species. *Annals of Tropical Medicine and Parasitology, 48*(3), 259–270.
Mattingly, P.F. (1954b). Notes on Ethiopian *Uranotaenia* (Diptera: Culicidae) with a description of a new species. *Proceedings of the Royal Entomological Society B, 23*(9–10), 167–171.
Mattingly, P.F. (1955a). Le sous-genre *Neoculex* (Diptera, Culicidae) dans la sous-région mediterraneenne. I.- Espèce, sous-espèce et synonymie nouvelles. *Annales de parasitologie humaine et comparée, 30*(4), 374–388.
Mattingly, P.F. (1955b). Appendix (pp. 9–10). In Iyengar, M.O.T., & Menon, M.A.U. Mosquitoes of the Maldive Islands. *Bulletin of Entomological Research, 46*, 1–10.
Mattingly, P.F. (1955c). Mosquitoes (Diptera: Culicidae) from the Tropical Institute at Hamburg. *Proceedings of the Royal Entomological Society of London, Series B: Taxonomy, 24*, 27–33.
Mattingly, P. F. (1955d). Culicidae (Diptera Nematocera). I. Mission G.F. DeWitte en collaboration avec W. Adam, A. Janssens, O. Van Meel et R. Verheyen (1946–1949). *Parck National De L'Upemba, 32*(3), 49–66.
Mattingly, P.F. (1956a). Lectotypes of mosquitoes (Diptera: Culicidae) in the British Museum. Part II. Genera *Toxorhynchites*, *Aëdes* (Subgenera *Aedimorphus, Banksinella*), *Culex* (Subgenera *Neoculex, Culiciomyia, Mocthogenes, Culex*). *Proceedings of the Royal Entomological Society of London, Series A: General Entomology, 31*(4–6), 37–44.
Mattingly, P.F. (1956b). Lectotypes of mosquitoes (Diptera: Culicidae) in the British Museum. Part I. Genera *Aedes* (subgenus *Pseudoskusea), Armigeres* and *Eretmapodites). Proceedings of the Royal Entomological Society of London, Series A: General Entomology, 31*(1–3), 25–32.
Mattingly, P.F. (1957a). *The culicine mosquitoes of the Indomalayan Area. Part I: Genus* Ficalbia *Theobald*. London, England: British Museum (Natural History).
Mattingly, P.F. (1957b). Genetical aspects of the *Aedes aegypti* problem. I. Taxonomy and bionomics. *Annals of Tropical Medicine and Parasitology, 51*(4), 392–408.
Mattingly, P.F. (1957c). Notes on the taxonomy and bionomics of certain filariasis vectors. *Bulletin of the World Health Organization, 16*, 686–696.
Mattingly, P.F. (1957d). *The culicine mosquitoes of the Indomalayan Area. Part II. Genus* Heizmannia *Ludlow*. London, England: British Museum (Natural History).
Mattingly, P.F. (1958a). *The culicine mosquitoes of the Indomalayan Area. Part III. Genus* Aedes *Meigen, subgenera* Paraedes, *Edwards,* Rhinoskusea *Edwards and* Cancraedes, *Edwards*. London, England: British Museum (Natural History).
Mattingly, P.F. (1958b). Revision of *Paraedes* Edwards and *Cancraedes* Edwards (Diptera: Culicidae). *Proceedings of the Royal Entomological Society of London, Series B: Taxonomy, 27*(5–6), 76–83.
Mattingly, P.F. (1958c). Lectotypes of mosquitoes (Diptera: Culicidae) in the British Museum. Part.III. Genera *Sabethes*, *Udaya* and *Aedes* (Subgenera *Paraedes*, *Cancraedes* and *Skusea*). *Proceedings of the Royal Entomological Society of London, Series A: General Entomology, 27*(7–8), 105–108.
Mattingly, P.F. (1959). *The culicine mosquitoes of the Indomalayan Area. Part IV. Genus* Aedes *Meigen, subgenera* Skusea *Theobald,* Diceromyia *Theobald,* Geoskusea *Edwards and* Christophersiomyia *Barraud*. London, England: British Museum (Natural History).
Mattingly, P.F. (1960). *Aëdes* subgenus *Alanstonea* subgen. nov. (Diptera: Culicidae). *Proceedings of the Royal Entomological Society of London, Series B: Taxonomy, 29*(11–12), 170–171.
Mattingly, P.F. (1961). *The culicine mosquitoes of the Indomalayan Area. Part V. Genus* Aedes *Meigen, subgenera* Mucidus *Theobald,* Ochlerotatus *Lynch Arribálzaga and* Neomelaniconion *Newstead*. London, England: British Museum (Natural History).
Mattingly, P.F. (1963). New and remarkable *Aedes* (Diptera: Culicidae) from Africa. *Proceedings of the Royal Entomological Society of London, Series B: Taxonomy, 32*(11–12), 165–170.
Mattingly, P.F. (1965a). The holotype of *Aedes (Geoskusea) tonsus* Edwards (Diptera: Culicidae). *Proceedings of the Royal Entomological Society of London, Series B: Taxonomy, 34*(1–2), 23–24.
Mattingly, P.F. (1965b). *The culicine mosquitoes of the Indomalayan Area. Part VI. Genus* Aedes *Meigen, subgenus* Stegomyia *Theobald (Groups A, B and D)*. London, England: British Museum (Natural History).
Mattingly, P.F. (1967). Taxonomy of *Aedes aegypti* and related species. *Bulletin of the World Health Organization, 36*(4), 552–554.
Mattingly, P.F. (1969a). The biology of mosquito-borne disease. *Science of Biology Series* (London), *1*, 1–184.
Mattingly, P.F. (1969b). Mosquito eggs. I. *Mosquito Systematics Newsletter, 1*(2), 13–16.
Mattingly, P.F. (1969c). Mosquito eggs. II. III. Tribe Anophelini. *Mosquito Systematics Newsletter, 1*(3), 41–50.
Mattingly, P.F. (1969d). Mosquito eggs. IV. *Mosquito Systematics Newsletter, 1*(4), 74–77.
Mattingly, P.F. (1970a). Contributions to the mosquito fauna of Southeast Asia. VI. The genus *Heizmannia* Ludlow in Southeast Asia. *Contributions of the American Entomological Institute, 5*(7), 1–104.
Mattingly, P.F. (1970b). Mosquito eggs. VI. Genera *Eretmapodites* and *Culex*. *Mosquito Systematics Newsletter, 2*(1), 17–22.
Mattingly, P.F. (1970c). Mosquito eggs. VII. Genus *Uranotaenia*. *Mosquito Systematics Newsletter, 2*(2), 61–67.
Mattingly, P.F. (1970d). Mosquito eggs. VIII. *Mosquito Systematics Newsletter, 2*(3), 87–91.
Mattingly, P.F. (1970e). Mosquito eggs. IX. Genus *Opifex* Hutton. *Mosquito Systematics Newsletter, 2*(3), 92–97.

Mattingly, P.F. (1970f). Mosquito eggs. XI. Genera *Orthopodomyia* and *Mimomyia*. *Mosquito Systematics Newsletter, 2*(4), 160–164.
Mattingly, P.F. (1970g). Mosquito larvae. II. Some undescribed first stages. *Mosquito Systematics, Newsletter, 2*(1), 34–36.
Mattingly, P.F. (1970h). Mosquito eggs. X. *Mosquito Systematics Newsletter, 2*(4), 158–159.
Mattingly, P.F. (1971a). Contributions to the mosquito fauna of Southeast Asia. XII. Illustrated keys to the genera of mosquitoes (Diptera, Culicidae). *Contributions of the American Entomological Institute, 7*(4), 1–84.
Mattingly, P.F. (1971b). Ecological aspects of mosquito evolution. *Parassitologia, 13*, 31–65.
Mattingly, P.F. (1971c). Further notes on genera *Orthopodomyia* and *Mimomyia*. *Mosquito Systematics Newsletter, 3*(2), 66–68.
Mattingly, P.F. (1971d). Mosquito eggs. XIII. Genus *Armigeres* Theobald. *Mosquito Systematics Newsletter, 3*(3), 122–129.
Mattingly, P.F. (1971e). Mosquito eggs. XIV. Genus *Armigeres* Theobald (continued) and *Aedes* subgenus *Alanstonea* Mattingly. *Mosquito Systematics Newsletter, 3*(3), 130–137.
Mattingly, P.F. (1971f). Mosquito eggs. XV. Genera *Heizmannia* Ludlow and *Haemagogus* Williston. *Mosquito Systematics Newsletter, 3*(4), 197–201.
Mattingly, P.F. (1971g). Mosquito eggs. XVI. Genus *Mansonia* (subgenus *Coquillettidia* Dyar) and genus *Ficalbia* Theobald. *Mosquito Systematics Newsletter, 3*(4), 202–210.
Mattingly, P.F. (1972a). Mosquito eggs. XIX. Genus *Mansonia* (subgenus *Mansonioides* Theobald). *Mosquito Systematics, 4*(2), 50–59.
Mattingly, P.F. (1972b). Mosquito eggs. XVIII. Genus *Mansonia* (subgenera *Rhynchotaenia* Brethes and *Mansonia* Blanchard) with a further note on genus *Ficalbia* Theobald. *Mosquito Systematics, 4*(2), 45–49.
Mattingly, P.F. (1972c). Mosquito eggs. XXI. Genus *Culiseta* Felt. *Mosquito Systematics, 4*(4), 114–127.
Mattingly, P.F. (1973a). Mosquito eggs. XXII. Eggs of two species of *Haemagogus* Williston. *Mosquito Systematics, 5*(1), 24–26.
Mattingly, P.F. (1973b). Mosquito eggs. XXIII. Eggs of *Toxorhynchites amboinensis* containing two-headed monsters. *Mosquito Systematics, 5*(2), 197–199.
Mattingly, P.F. (1973c). Mosquito eggs. XXIV. *Mosquito Systematics, 5*(3), 221–224.
Mattingly, P.F. (1973d). Contributions to the mosquito fauna of Southeast Asia. XII. Clés illustrées des genres de moustiques (J. Rageau, Trans.). *Contributions of the American Entomological Institute, 7*(4), 1–86.
Mattingly, P.F. (1974a). Mosquito eggs. XXV. Eggs of some subgenera of *Aedes* with a further note on *Haemagogus*. *Mosquito Systematics, 6*(1), 41–45.
Mattingly, P.F. (1974b). Mosquito eggs. XXVI. Further descriptions of sabethine eggs. *Mosquito Systematics, 6*(4), 231–238.
Mattingly, P.F. (1975). Mosquito eggs. XXVII. *Mosquito Systematics, 7*(1), 19–26.
Mattingly, P.F. (1976). Mosquito eggs. XXVIII. *Culex* subgenera *Melanoconion* and *Mochlostyrax*. *Mosquito Systematics, 8*(3), 223–231.
Mattingly, P.F. (1977). Names for the *Anopheles gambiae* complex. *Mosquito Systematics, 9*(3), 323–328.
Mattingly, P.F. (1980). An interim reclassification of the genus *Tripteroides* with particular reference to the Australasian subgenera. *Mosquito Systematics, 12*(2), 164–171.
Mattingly, P.F. (1981). Medical entomology studies. XIV. The subgenera *Rachionotomyia*, *Trichoteptomyia* and *Tripteroides* (Mabinii Group) of genus *Tripteroides* in the Oriental region (Diptera: Culicidae). *Contributions of the American Entomological Institute, 17*(5), 1–147.
Mattingly, P.F., & Adam, J.P. (1954). A new species of cave-dwelling anopheline from the French Cameroons. *Annals of Tropical Medicine and Parasitology, 48*(1), 55–57.
Mattingly, P.F., & Brown, E.S. (1955). The mosquitos (Diptera: Culicidae) of the Seychelles. *Bulletin of Entomological Research, 46*(1), 69–110.
Mattingly, P.F., & Bruce-Chwatt, L.J. (1954). Morphology and bionomics of *Aëdes* (*Stegomyia*) *pseudoafricanus* Chwatt (Diptera, Culicidae), with some notes on the distribution of the subgenus *Stegomyia* in Africa. *Annals of Tropical Medicine and Parasitology, 48*(2), 183–193.
Mattingly, P.F., & Grjebine, A. (1958). Revision du genre *Ficalbia* Theobald et discussion de la position systematique des *Ravenalites* Doucet (Diptera, Culicidae). *Mémoires de l'Institut Scientifique de Madagascar E, 9*, 259–290.
Mattingly, P.F., & Hamon, J. (1955). Position taxonomique et synonymie de quelques *Ficalbia* de la Region Éthiopienne (Diptera, Culicidae). *Annales de parasitologie humaine et comparée, 30*(5–6), 488–496.
Mattingly, P.F., & Knight, K.L. (1956). The mosquitoes of Arabia. *Bulletin of the British Museum* (*Natural History*), *43*(3), 91–141.
Mattingly, P.F., & Marks, E.N. (1955). Some Australian mosquitoes (Diptera, Culicidae) of the subgenera *Pseudoskusea* and *Neoculex*. *Proceedings of the Linnean Society of New South Wales, 80*(2), 163–176.
Mattingly, P.F., & Qutubuddin, M. (1952). The hitherto undescribed male of *Armigeres* (*Armigeres*) *theobaldi* Barraud (Diptera, Culicidae). *Proceedings of the Royal Entomological Society of London, Series B: Taxonomy, 21*(7–8), 92–93.
Mattingly, P.F., & Rageau, J. (1958). *Culex* (*Culex*) *iyengari* n. sp., a new species of mosquito (Diptera, Culicidae) from the South Pacific. *Pacific Science, 12*(3), 241–250.
Mattingly, P.F., Stone, A., & Knight, K.L. (1962). *Culex aegypti* Linnaeus, 1762 (Insecta, Diptera); proposed validation and interpretation under the plenary powers of the species so named. *Bulletin of Zoological Nomenclature, 19*(4), 208–219.
Mattos, S.S., & Xavier, S.H. (1991). *Culex* (*Melanoconion*) *guedesi* uma nova espécie do Brasil (Diptera, Culicidae). *Revista Brasiliera de biologia, 51*(1), 193–196.
Mattos, S.S., de Silveira Guedes, A., & Xavier, S.H. (1978). Uma espécie nova de *Culex* (*Culex*) que se cria em umbelifera (Diptera, Culicidae). *Revista Brasiliera de biologia, 38*(1), 31–35.
Mayoral, J.G., Nouzova, M., Yoshiyama, M., Shinoda, T., Hernandez-Martinez, S., Dolghih, E., . . . Noriega, F.G. (2009). Molecular and functional characterization of a juvenile hormone acid methyltransferase expressed in the corpora allata of mosquitoes. *Insect Biochemistry and Molecular Biology, 39*(1), 31–37.
Mayr, E. (1969). *Principles of systematic zoology*. New York, NY: McGraw-Hill.
Mboera, L.E.G., Takken, W., Mdira, K.Y., & Pickett, J.A. (2000). Sampling gravid *Culex quinquefasciatus* (Diptera: Culicidae) in Tanzania with traps baited with synthetic oviposition pheromone and grass infusions. *Journal of Medical Entomology, 37*(1), 172–176.
McAlpine, J.F. (1981). *Manual of Nearctic Diptera* (Vol. 1). Monograph/ Research Branch, Agriculture Canada no. 27. Canada: Agriculture and Agri-Food Canada.
McArthur, J. (1949). A new variety of *Anopheles aitkeni* from Borneo. *Bulletin of Entomological Research, 40*(1), 49–52.
McCabe, T.L., & Johnson, L.M. (1980). Catalogue of the types in the New York State Museum insect collection. *New York State Museum Bulletin, 434*, 1–38.

McCall, P.J., & Eaton, G. (2001). Olfactory memory in the mosquito *Culex quinquefasciatus*. *Medical and Veterinary Entomology, 15*(2), 197–203.

McCall, P.J., & Kelly, D.W. (2002). Learning and memory in disease vectors. *Trends in Parasitology, 18*(10), 429–433.

McCarthy, D. (2003). The trans-Pacific zipper effect: Disjunct sister taxa and matching geological outlines that link the Pacific margins. *Journal of Biogeography, 30*(10), 1545–1561.

McClelland, G.A.H. (2009). A worldwide survey of variation in scale pattern of the abdominal tergum of *Aedes aegypti* (L.). *Ecological Entomology, 126*(2), 239–259.

McClelland, G.A.H., & Green, C.A. (1970). Subtle periodicity of pupation in rapidly developing mosquitoes: With particular reference to *Aedes vittatus* and *Aedes aegypti*. *Bulletin of the World Health Organization, 42*(6), 951–955.

McClelland, G.A.H., Hartberg, W.K., & Courtois, C.M. (1973). *Aedes mascarensis* MacGregor on Mauritius. I. Ecology in relation to *Aedes albopictus* (Skuse) (Diptera: Culicidae). *Journal of Medical Entomology, 10*(6), 570–577.

McCormack, A. T. (1919). *Report of the Health Department of the Panama Canal for the calendar year 1918*. Mount Hope, Canal Zone: Panama Canal Press.

McCracken, I. (1904). *Anopheles* in California, with description of a new species. *Entomological News, 15*(1), 9–14.

McDaniel, I.N., & Horsfall, W.R. (1957). Induced copulation of aedine mosquitoes. *Science, 125*(3251), 745–745.

McDaniel, I.N., & Webb, D.R. (1974). Identification of females of the *Aedes stimulans* group in Maine including notes on larval characters and attempts at hybridization. *Annals of the Entomological Society of America, 67*(6), 915–918.

McDonald, W.A. (1957a). The adults and immature stages of *Aedes muelleri* Dyar (Diptera: Culicidae). *Annals of the Entomological Society of America, 50*(5), 505–511.

McDonald, W.A. (1957b). The adults and immature stages of *Aedes purpureipes* Aitken (Diptera: Culicidae). *Annals of the Entomological Society of America, 50*(5), 529–535.

McDowall, R.M. (1991). The "-ii's" may have it at the end—patronyms should be amended only if demonstrably incorrect. *New Zealand Natural Sciences, 18*, 25–29.

McGregor, D.D. (1963). Mouth-brush dimorphism in larvae of *Opifex fuscus* Hutton (Diptera, Culicidae). *Bulletin of Entomological Research, 54*(2), 325–327.

McGregor, D.D. (1965). Aspects of the biology of *Opifex fuscus* Hutton (Diptera: Culicidae). *Proceedings of the Royal Entomological Society of London, Series A: General Entomology, 40*(1–3), 9–14.

McIntosh, B.M. (1971). The aedine subgenus *Neomelaniconion* (Culicidae: Diptera) in southern Africa with descriptions of two new species. *Journal of the Entomological Society of Southern Africa, 34*(2), 319–333.

McIntosh, B.M. (1973). A taxonomic re-assessment of *Aedes (Ochlerotatus) caballus* (Theobald) (Diptera: Culicidae) including a description of a new species of *Ochlerotatus*. *Journal of the Entomological Society of Southern Africa, 36*(2), 261–269.

McIntosh, B.M. (1975). A taxonomic revision of certain *Aedes* species (Diptera: Culicidae) of the subgenus *Aedimorphus* in southern Africa. *Journal of the Entomological Society of Southern Africa, 38*(2), 251–287.

McIntosh, B.M., Jupp, P.G., dos Santos, I., & Barnard, B.J.H. (1980). Vector studies on Rift Valley fever virus in South Africa. *South African Medical Journal, 58*, 127–132.

McKeon, S.N., Lehr, M.A., Wilkerson, R.C., Ruiz, J.F., Sallum, M.A., Pereira, J.B.L., . . . Conn, J.E. (2010). Lineage divergence detected in the malaria vector *Anopheles marajoara* (Diptera: Culicidae) in Amazonian Brazil. *Malaria Journal, 9*, 271.

McMeniman, C.J., Hughes, G.L., & O'Neill, S.L. (2011). A *Wolbachia* symbiont in *Aedes aegypti* disrupts mosquito egg development to a greater extent when mosquitoes feed on nonhuman versus human blood. *Journal of Medical Entomology, 48*(1), 76–84.

McPhee, J. (1998). *Annals of the former world*. New York, NY: Farrar, Straus, and Giroux.

Medley, K.A. (2010). Niche shifts during the global invasion of the Asian tiger mosquito, *Aedes albopictus* Skuse (Culicidae), revealed by reciprocal distribution models. *Global Ecology and Biogeography, 19*(1), 122–133.

Medley, K.A., Jenkins, D.G., & Hoffman, E.A. (2015). Human-aided and natural dispersal drive gene flow across a range of an invasive mosquito. *Molecular Ecology, 24*(2), 284–295.

Medlock, J.M., Avenell, D., Barrass, I., & Leach, S. (2006). Analysis of the potential for survival and seasonal activity of *Aedes albopictus* (Diptera: Culicidae) in the United Kingdom. *Journal of Vector Ecology, 31*(2), 292–304.

Medlock, J.M., Hansford, K.M., Schaffner, F., Versteirt, V., Hendrickx, G., Zeller, H., & Van Bortel, W. (2012). A review of the invasive mosquitoes in Europe: Ecology, public health risks, and control options. *Vector-Borne and Zoonotic Diseases, 12*(6), 1–13.

Medschid, E. (1928). Über *Aëdes lepidonotus* Edw. und *Aëdes refiki* n. sp. *Archiv für Schiffs- und Tropen-Hygiene, 32*, 306–315.

Medschid, E. (1930). *Culex martinii* n. sp. *Archiv für Schiffs- und Tropen-Hygiene, 34*, 364–369.

Mehta, D.R. (1934). Studies on the longevity of some Indian anophelines. Part I. Survival of *Anopheles subpictus* Grassi under controlled conditions of temperature and humidity. *Records of the Malaria Survey of India, 4*(3), 261–272.

Meier, R. (2005). Role of dipterology in phylogenetic systematics: The insight of Willi Hennig. In D.K. Yeates & B.M. Wiegmann (Eds.), *The Evolutionary Biology of Flies* (pp. 45–62). New York, NY: Columbia University Press.

Meigen, J.W. (1804). *Klassifikazion und Beschreibung der europäischen Zweiflugeligen Insekten*. Braunschweig, Germany: Karl Reichard.

Meigen, J.W. (1818). *Systematische Beschreibung der bekannten europäischen zweiflügeligen Insekten*. (Vol. 1). Aachen, Germany: Bei Friedrich Forstmann, Gedrukt bei Beaufort Sohn.

Meigen, J.W. (1830a). *Abbildung aller bis jetzt bekannten europaischen zweipflügeligen Insekten*. Hamm, Germany: Schulzische Buchhandlung.

Meigen, J.W. (1830b). *Systematische Beschreibung der bekannten europäischen zweiflügeligen Insekten* (Vol. 6). Hamm, Germany: Bei Friedrich Forstmann, Gedrukt bei Beaufort Sohn.

Meigen, J.W. (1838). Zweifleugeligen, Insecten 7, besannten europaischen [100255]. Hamm, Germany: Schulzische Buchhandlung.

Mekuria, Y. (1968). A catalogue of the culicine mosquito fauna of Ethiopia and their distribution in the country. *Ethiopian Medical Journal, 6*, 73–80.

Mekuria, Y.S., Assefa, S., & Mebrahtu, Y. (1978). Culicine mosquitoes (Diptera: Culicidae) of Ethiopia: Checklist and distribution. *Ethiopian Medical Journal, 16*, 71–93.

Melero-Alcíbar, R. (2004). The pupae of Spanish *Ochlerotatus*. I. *Ochlerotatus quasirusticus* (Diptera: Culicidae). *European Mosquito Bulletin, 17*, 20–22.

Melero-Alcíbar, R. (2005a). The pupae of Spanish Culicinae: *Aedes geminus* Peus (Diptera: Culicidae). *European Mosquito Bulletin, 19*, 17–19.

Melero-Alcíbar, R. (2005b). The pupae of Spanish *Ochlerotatus*. II. *Ochlerotatus rusticus* (Rossi) (Diptera: Culicidae). *European Mosquito Bulletin, 19*, 14–16.

Melero-Alcíbar, R. (2005c). The pupae of Spanish *Ochlerotatus*. III. *Ochlerotatus punctor* (Kirby) (Diptera: Culicidae). *European Mosquito Bulletin, 20*, 10–12.

Melero-Alcíbar, R. (2005d). The pupae of Spanish *Ochlerotatus*. IV. *Ochlerotatus berlandi* (Seguy) (Diptera: Culicidae). *European Mosquito Bulletin, 20*, 6–9.

Melero-Alcíbar, R. (2006). The pupae of Spanish Culicinae. II. *Aedes vittatus* Bigot, 1861 (Diptera: Culicidae). *European Mosquito Bulletin, 21*, 19–22.

Melero-Alcíbar, R., & Molina, R. (2007). First description of the pupa of *Ochlerotatus gilcolladoi*. *Journal of the American Mosquito Control Association, 23*(3), 343–345.

Melero-Alcíbar, R., & Salom, F. (2002). Confirmation of the presence of *Ochlerotatus* (*Rusticoidus*) *quasirusticus* (Torres Canamares) in the Iberian Peninsula, and the first record from the Madrid region. *European Mosquito Bulletin, 13*, 30–32.

Melero-Alcíbar, R., & Salom-Bonet, F. (2003). Redescription of the pupa of *Ochlerotatus cantans* (Diptera: Culicidae). *European Mosquito Bulletin, 16*, 18–21.

Melero-Alcíbar, R., Ruiz, S., & Molina, R. (2009). Re-description of *Dahliana geniculata* and *Dahliana echinus* pupae (Diptera: Culicidae) and morphological key for identification of phytotelmic aedine pupae in the Palaearctic region. *European Mosquito Bulletin, 20*, 1–6.

Melero-Alcíbar, R., Aranda, C., & Molina, R. (2010). Re-description of the pupa of *Uranotaenia unguiculata* (Diptera: Culicidae) and a morphological key for identification of pupae of mosquito genera in the Palaearctic Region. *European Mosquito Bulletin, 28*, 45–50.

Mendenhall, I.H., Giraldo-Calderon, G.I., Ocampo, C.D., Wesson, D.M., & Pecor, J.E. (2011). First record of *Culex* (*Anoedioporpa*) *restrictor* from Colombia. *Journal of the American Mosquito Control Association, 27*(2), 148.

Mendenhall, I.H., Bahl, J., Blum, M.J., & Wesson, D.M. (2012). Genetic structure of *Culex erraticus* populations across the Americas. *Journal of Medical Entomology, 49*(3), 522–534.

Mendis, K.N., Ihalamulla, R.L., Peljton, E.L., & Nanayakkara, S. (1983). Biology and descriptions of the larva and pupa of *Anopheles* (*Cellia*) *elegans* James (1903). *Mosquito Systematics, 15*(4), 318–324.

Mendki, M.J., Ganesan, K., Prakash, S., Suryanarayana, M.V.S., Malhotra, R.C., & Vaidyanathaswamy, R. (2000). Heneicosane: An oviposition-attractant pheromone of larval origin in *Aedes aegypti* mosquitoes. *Journal of Current Science, 78*(11), 1295–1296.

Mendoza, J.B. (1940). New *Anopheles* of the Myzorhynchus series. *Monthly Bulletin of the Bureau of Health, 20*, (11), 353–369.

Mendoza, J.B. (1947). Two more Philippine *Anopheles* in the Myzorhynchus series. *Monthly Bulletin of the Bureau of Health, 23*(3), 171–184.

Mendoza, J.B., & Abinoja, B. (1952). Observations on supposedly unusual breeding of the Philippine malaria vector *Anopheles minimus flavirostris*. *The Philippine Journal of Science, 81*, 53–59.

Meng, C.-H. (1957). *The taxonomy, biology and control of Chinese anopheline mosquitoes*. Peking: People's Hygiene Press.

Meng, C.H. (1958). A new species of *Culex*. *Acta entomologica Sinica, 8*(4), 351–354.

Meng, C.H. (1962). A new species of *Aedes* (Diptera, Culicidae), *Aedes* (*Ochlerotatus*) *lasaensis*, sp. nov. *Acta entomologica Sinica, 11*(2), 159–163.

Meng, C.H. (1981). A proposed system of classification of *Aedes chrysolineatus* and related species. *Entomotaxonomia, 3*(4), 243–254.

Meng, C.H., & Wu, C.Y. (1962). A new species of *Theobaldia* (Diptera, Culicidae). *Acta entomologica Sinica, 11*(4), 382–387.

Meng, Q., & Chen, H. (1980). Arguments on the taxonomic positions of six species of *Culex* from China. *Entomotaxonomia, 2*(2), 161–164.

Menon, M.A.U. (1938). The egg of *Ficalbia* (*Mimomyia*) *hybrida* Leicester. *Journal of the Malaria Institute of India, 1*(2), 185–186.

Menon, M.A.U. (1940). A description and comparative study of the fourth instar larvae of *Mansonia* (*Mansonioides*) *uniformis* (Theobald), and *Mansonia* (*Mansonioides*) *annulifera* (Theobald), including a synoptic table for the identification of the larvae of the Indian species of the subgenus *Mansonoides* Theobald, 1907 (Diptera, Culicidae). *Journal of the Malaria Institute of India, 3*(1), 179–184.

Menon, M.A.U. (1944). On a new species of the subgenus *Lophoceratomyia* Theobald 1905 (Diptera, Culicidae). *Journal of the Malaria Institute of India, 5*(3), 389–393.

Menon, M.A.U. (1950). The male of a new species of the subgenus *Aedes* Meigen, 1818 (Diptera: Culicidae). *Proceedings of the Royal Entomological Society of London, Series B: Taxonomy, 19*(9–10), 139–141.

Menon, M.A.U., & Tampi, M.R.V. (1959). Notes on the feeding and egg-laying habits of *Ficalbia* (*Mimomyia*) *chamberlaini*, Ludlow 1904. (Diptera, Culicidae). *Indian Journal of Malariology, 13*, 13–18.

Merdic, E. (1992). *Aedes riparius* (Dyar & Knab, 1907), a newly recorded species of mosquitoes (Dipty., Culicidae) in Croatia. *Bioloski Vestnik, 40*, 13–17.

Metcalf, R.L., & Luckmann, W. (1975). *Introduction to insect pest management*. New York, NY: Wiley.

Meunier, F. (1902). Les Culicidae de l'ambre. *Révue scientifique du Bourbonnais, 15*, 199–200.

Meunier, F. (1915). Nouvelles récherches sur quelques insectes des platrieres d'Aix en Provence. *Verhandelingen der Koninklijke Akademie van Welenschappen le Amsterdam, 18*(5), 1–17.

Meuti, M.E., Stone, M., Ikeno, T., & Denlinger, D.L. (2015). Functional circadian clock genes are essential for the overwintering diapause of the northern house mosquito, *Culex pipiens*. *Journal of Experimental Biology, 218*(3), 412–422.

Mezenev, N.P. (1980). A possible identity of *Aedes leucomelas* Meigen, 1804 and *A. implicatus* Vockeroth, 1954 (Culicidae) . *Parassitologia, 14*, 206–209.

Mezzanotte, R., & Ferrucci, L. (1978). Recognition of the sibling species *Anopheles atroparvus* (Van Thiel) and *Anopheles labranchiae* (Falleroni) (Diptera, Culicidae) on the basis of Q and C banding. *Monitore zoologica Italiano, 12*, 211–218.

Mgeladze, V. (1991). Species composition of mosquito *Aedes* (Diptera, Culicidae) in Georgia. *Soobshcheniya Akademii Nauk Gruzii, 143*(3), 329–331.

Miao, J., Pan, J., & Jiang, W. (1988). Hybridization and chromosome observations on six species of the *Anopheles hyrcanus* group in China (Diptera: Culicidae). *Zoological Research, 9*(3), 231–237.

Miaoulis, M., Giantsis, I.A., Schaffner, F., & Chaskopoulou, A. (2018). Re-examination of the taxonomic status of *Anopheles hyrcanus* and *An. pseudopictus* using a multilocus genetic approach. *Journal of Vector Ecology, 43*(1), 179–183.

Michener, C.D. (1944). Differentiation of females of certain species of *Culex* by the cibarial armature. *Journal of the New York Entomological Society, 52*, 263–266.

Middlekauff, W.W. (1944). A new species of *Aedes* from Florida (Diptera: Culicidae). *Proceedings of the Entomological Society of Washington, 46*(2), 42–44.

Midega, J.T., Mbogo, C.M., Mwambi, H., Wilson, M.D., Ojwang, G., Mwangangi, J.M., . . . Beier, J.C. (2007). Estimating dispersal and survival of *Anopheles gambiae* and *Anopheles funestus* along the

Kenyan coast by using mark-release recapture methods. *Journal of Medical Entomology, 44*(6), 923–929.
Mihályi, V.F. (1955a). *Aedes hungaricus* n. sp. (Culicidae, Diptera). *Annales Historico-Naturales Musei Nationalis Hungarici, 6*, 343–345.
Mihályi, V.F. (1955b). Igazi Szunyogok Culicidae. *Magyarorszag Allatvilaga, 14*(5), 1–40.
Mihályi, V.F. (1961). Description of the larva of *Aedes* (*Ochlerotatus*) *hungaricus* (Mihályi) (Diptera: Culicidae). *Acta zoologica Academiae Scientiarum Hungaricae, 7*, 231–233.
Milankov, V., Petric, D., Vujic, A., & Vapa, L. (2009). Taxonomy, biology, genetic variability and medical importance of *Ochlerotatus caspius* (Pallas, 1771) and *O. dorsalis* (Meigen, 1830) (Diptera: Culicidae). *Acta entomologica Serbica, 14*(2), 195–207.
Miles, S.J., Green, C.A., & Hunt, R.H. (1983). Genetic observations on the taxon *Anopheles* (*Cellia*) *pharoensis* Theobald (Diptera: Culicidae). *The Journal of Tropical Medicine and Hygiene, 86*(4), 153–157.
Miller, B.R., Crabtree, M.B., & Savage, H.M. (1996). Phylogeny of fourteen *Culex* mosquito species, including the *Culex pipiens* complex, inferred from the internal transcribed spacers of ribosomal DNA. *Insect Molecular Biology, l5*(2), 93–107.
Miller, B.R., Crabtree, M.D., & Savage, H.M. (1997). Phylogenetic relationships of the Culicomorpha inferred from 18S and 5.8S ribosomal DNA sequences (Diptera: Nematocera). *Insect Molecular Biology, 6*(2), 105–114.
Miller, D. (1922). A remarkable mosquito, *Opifex fuscus*, Hutton. *Bulletin of Entomological Research London, 13*(1), 115–126.
Miller, D., & Phillips, W.J. (1952). *Identification of New Zealand Mosquitoes*. Nelson, New Zealand: Cawthron Institute.
Miller, J.R., Huang, J., Vulule, J., & Walker, E.D. (2007). Life on the edge: African malaria mosquito (*Anopheles gambiae* s.l.) larvae are amphibious. *Naturwissenschaften, 94*(3), 195–199.
Milne, M.A., Townsend, V.J., Jr., Smelser, P., Felgenhauer, B.E., Moore, M.K., & Smyth, F.J. (2009). Larval aquatic and terrestrial mites infesting a temperate assemblage of mosquitoes. *Experimenta Applicata Acaralogia, 47*(1), 19–33.
Minakawa, N., Githure, J.I., Beier, J.C., & Yan, G. (2001). Anopheline mosquito survival strategies during the dry period in western Kenya. *Journal of Medical Entomology, 38*(3), 388–392.
Minakawa, N., Sonye, G., Mogi, M., Githeko, A., & Yan, G. (2002). The effects of climatic factors on the distribution and abundance of malaria vectors in Kenya. *Journal of Medical Entomology, 39*(4), 833–841.
Minar, J. (1971). Mosquitoes (Diptera, Culicidae) of the North-East of Mongolia. *Folia parasitologica, 18*, 92.
Minar, J. (1976). Culiciden aus der Mongolei (Diptera). *Acta Zoologica Academiae Scientiarum Hungaricae, 22*(3–4), 335–350.
Minard, G., Mavingui, P., & Moro, C.V. (2013). Diversity and function of bacterial microbiota in the mosquito holobiont. *Parasites and Vectors, 6*(1), 146.
Mirabello, L., & Conn, J.E. (2008). Population analysis using the nuclear *white* gene detects Pliocene/Pleistocene lineage divergence within *Anopheles nuneztovari* in South America. *Medical and Veterinary Entomology, 22*(2), 109–119.
Missiroli, A. (1935). Nuova varietà de *Anopheles maculipennis*. *Annali d'igiene sperimentale, 45*(5), 333.
Mitchell, A., Sperling, F.A.H., & Hickey, D.A. (2002). Higher-level phylogeny of mosquitoes (Diptera: Culicidae): mtDNA data support a derived placement for *Toxorhynchites*. *Insect Systematics and Evolution, 33*(2), 163–174.
Mitchell, C.J., & Darsie, R.F., Jr. (1985). Mosquitoes of Argentina. Part II. Geographic distribution and bibliography (Diptera, Culicidae). *Mosquito Systematics, 17*(4), 279–360.
Mitchell, C.J., Darsie, R.F., Jr., & Monath, T.P. (1984). Occurrence of autogenous *Culex pipiens* Linnaeus, 1758, (Diptera: Culicidae) in Argentina and notes on distribution of the complex. *Mosquito Systematics, 16*(4), 308–316.
Mitchell, E.G. (1906). Mouth parts of mosquito larvae as indicative of habits. *Psyche, 13*(1), 11–21.
Mitchell, E.G. (1907a). Validity of the culicid subfamily Deinoceritinae. *Psyche 14*(1), 11–13.
Mitchell, E.G. (1907b). *Mosquito life*. New York, NY: Putnam, Knickerbocker Press.
Miyagi, I. (1971). Notes on the *Aedes* (*Finlaya*) *chrysolineatus* subgroup in Japan and Korea (Diptera: Culicidae). *Tropical Medicine* (Nagasaki), *13*, 141–151.
Miyagi, I. (1976). Description of a new species of the genus *Topomyia* Leicester from the Ryukyu Islands, Japan (Diptera: Culicidae). *Tropical Medicine* (Nagasaki), *17*(4), 201–210.
Miyagi, I. (1980). *Topomyia* (*Suaymyia*) *rausai*, a new species of mosquito from the Philippines (Diptera: Culicidae). *Journal of Medical Entomology, 17*(2), 178–181.
Miyagi, I., & Iha, S. (1970). Notes on *Culex* (*Culex*) *neovishnui* Lien, 1968 from the Ryukyus and Japan proper (Diptera: Culicidae). *Tropical Medicine* (Nagasaki), *12*, 71–78.
Miyagi, I., & Toma, T. (1979). Studies on the mosquitoes in the Yaeyama Islands, Japan. III. Description of the male, pupa and larvae of *Aedes* (*Verrallina*) *iriormotensis* (Diptera: Culicidae. *Mosquito Systematics, 11*(1), 14–25
Miyagi, I., & Toma, T. (1981). A new species of *Aedes* (*Stegomyia*) from Daito Islands, Ryukyu, Japan (Diptera: Culicidae). *Mosquito Systematics, 12*(for 1980)(4), 428–440.
Miyagi, I., & Toma, T. (1989). A new species of *Topomyia* (*Suaymyia*) *suchariti* from Thailand (Diptera: Culicidae). *Mosquito Systematics, 21*(1), 16–24.
Miyagi, I., & Toma, T. (1995). *Topomyia javaensis*: A new Indonesian species in the subgenus *Topomyia* (Diptera: Culicidae). *Mosquito Systematics, 27*(3), 179–186.
Miyagi, I., & Toma, T. (1997). Description of *Topomyia irianensis* n. sp. and new records of *To. papuensis* from Maluku and Irian Jaya, Indonesia (Diptera: Culicidae). *Journal of the American Mosquito Control Association, 13*(2), 134–139.
Miyagi, I., & Toma, T. (2001). *Tripteroides* (*Rachionotomyia*) *ponmeki* (Diptera: Culicidae): A new species from Khammouane Province, Lao PDR. *Medical and Veterinary Zoology, 52*(4), 269–277.
Miyagi, I., & Toma, T. (2005). *Topomyia roslihashimi*, a new species of the subgenus *Suaymyia* (Diptera: Culicidae) from Gomback, Peninsular Malaysia. *Medical and Veterinary Zoology, 56*(4), 275–282.
Miyagi, I., & Toma, T. (2007). A new mosquito of the genus *Topomyia* (Diptera, Culicidae) from a *Nepenthes* pitcher plant in a Bario highland of Sarawak, Malaysia. *Medical Entomology and Zoology, 58*(3), 167–174.
Miyagi, I., & Toma, T. (2008). Description of a new species *Topomyia* (*Suaymyia*) *lehcharlesi* (Diptera: Culicidae) from Sarawak, Malaysia. *Medical Entomology and Zoology, 59*(3), 163–170.
Miyagi, I., & Toma, T. (2010a). Descriptions of *Topomyia auriceps* Brug, 1939 and *Topomyia pseudoauriceps*, n. sp from Sarawak, Malaysia (Diptera: Culicidae). *Medical Entomology and Zoology, 61*(1), 27–38.
Miyagi, I., & Toma, T. (2010b). *Topomyia* (*Suaymyia*) *kelabitense* (Diptera: Culicidae), a new species from Sarawak, Malaysia. *Medical Entomology and Zoology, 61*(4), 353–361.
Miyagi, I., & Toma, T. (2013). *Uranotaenia* (*Pseudoficalbia*) *tanakai* (Diptera, Culicidae), a new species from forest swamps, lriomote Is., the Ryukyu Archipelago, Japan. *Medical Entomology and Zoology, 64*(3), 167–193.

Miyagi, I., Lha, S., & Kishimoto, T. (1969). Discovery of *Anopheles tessellatus* in Ryukyu Islands. *Tropical Medicine* (Nagasaki), *11*, 33–36.

Miyagi, I., Toma, T., & Cabrera, B.D. (1983). Synonyms of *Topomyia rausai* with *Topomyia apsarae*, with description of the larva (Diptera: Culicidae). *Mosquito Systematics, 15*(2), 111–116.

Miyagi, I., Toma, T., & Rivera., D.G. (1983). *Topomyia* (*Topomyia*) *cabrerai*, a new species from the Philippines (Diptera: Culicidae). *Mosquito Systematics, 15*(1), 1–4.

Miyagi, I., Toma, T., & Tsukamoto, M. (1983). Description of a new species, *Tripteroides* (*Tripteroides*) *riverai* from the Philippines (Diptera: Cullcidae). *Mosquito Systematics, 15*(1), 5–17.

Miyagi, I., Toma, T., Ramakrishna, K., & Ramalingam, S. (1989). Studies of the genus *Topomyia*. 3. Redescription of *spathulirostris* and transfer to the subgenus *Suaymyia*. *Mosquito Systematics, 21*(1), 40–49.

Miyagi, I., Toma, T., & Ramalingam, S. (1989). *Topomyia* (*Topomyia*) *hardini*, a new species from Sarawak, Malaysia (Diptera: Culicidae). *Tropical Biomedicine, 6*, 91–98.

Miyagi, I., Toma, T., & Ramalingam, S. (1991) *Topomyia* (*Topomyia*) *yongi*, a new species of mosquito from Peninsular Malaysia (Diptera: Culicidae). *Mosquito Systematics, 22*(for 1990)(3), 185–191.

Miyagi, I., Toma, T., Mogi, M., Martono, Yotopranoto, S., Arifin, Z., & Dachlan, Y.P. (1994). Mosquito species (Diptera: Culicidae) from Lombok Island, Indonesia. *Mosquito Systematics, 26*(1), 19–24.

Miyagi, I., Toma, T., & Higa, Y. (2004). A new species of *Mimomyia* (*Ingramia*) from Indonesia (Diptera: Culicidae). *Medical and Veterinary Zoology, 55*(1), 11–20.

Miyagi, I., Toma, T., & Lien, J.C. (2004). *Ochlerotatus* (*Geoskusea*) *timorensis* (Culicidae: Diptera), a new species from crab-holes, West Timor, Indonesia. *Medical and Veterinary Zoology, 55*(2), 107–114.

Miyagi, I., Toma, T., Okazawa, T., Mogi, M., & Hashim, R. (2005). Female *Armigeres* (*Leicesteria*) *flavus* holding an egg raft with her hind legs. *Journal of the American Mosquito Control Association, 21*(4), 466–468.

Miyagi, I., Toma, T., & Okazawa, T. (2006). Redescription of *Topomyia argenteoventralis* Leicester, 1908 (Diptera, Culicidae) from Malaysia. *Medical Entomology and Zoology, 57*(4), 347–354.

Miyagi, I., Toma, T., Okazawa, T., Leh, C., & Azirun, M.S. (2007). A redescription of *Topomyia decorabilis* Leicester, 1908 (Diptera, Culicidae) from Malaysia and Indonesia. *Medical Entomology and Zoology, 58*(4), 251–259.

Miyagi, I., Toma, T., Okazawa, T., & Leh, M.U. (2011). Description of pupa and larva of the Malaysian mosquito *Topomyia* (*Topomyia*) *rubithoracis* Leicester (Diptera, Culicidae). *Medical Entomology and Zoology, 62*(2), 93–99.

Miyagi, I., Toma, T., Tamashiro, M., Taira, K., & Endo, Y. (2012). Previously unknown adult male, pupa, larva and egg of *Heizmannia kana* (Diptera: Culicidae) from Tokunoshima, Ryukyu Archipelago, Japan. *Medical Entomology and Zoology, 63*(3), 209–216.

Miyagi, I., Toma, T., Okazawa, T., Wong, S.F., Leh, M.U., & Yong, H.S. (2012). Three new phytotelma mosquitoes of the genus *Topomyia* (Diptera: Culicidae) from Katibas, Lanjak-Entimau, Sarawak, Malaysia. *Journal of Science and Technology in the Tropics, 8*(2), 97–117.

Miyagi, I., Toma, T., Okazawa, T., Wong, S.F., Leh, M.U., & Yong, H.S. (2014). Description of a new species of *Topomyia* (*Topomyia*) *aliyusopi* (Diptera: Culicidae) from Kelabit highlands, Sarawak, Malaysia. *Journal of Science and Technology in the Tropics, 10*(1), 27–38.

Miyazaki, I. (1951). On a new anopheline mosquito *Anopheles yatsushiroensis* n. sp. found in Kyushu, with some remarks on two related species of the genus. *Kyushi Memoirs of Medical Sciences, 2*(3), 195–206.

Moberly, S.P., Lalor, C., McDonough, M., Foster, B., Estes, A., & Bentfield, D.J. (2005). Discovery of an exotic Asian mosquito *Ochlerotatus japonicus* (Diptera: Culicidae) in southern Indiana. *Proceedings of the Indiana Academy of Science, 114*(1), 62–64.

Mochizuki, D. (1913). Culicidae in the district of Fukuoka. *Fukuoka Acta Médica, 7*(1), 24–32.

Mochtar, R., & Walandouw., E.K. (1934). Een variëteit van *A. gigas*, (Uit Sumatra's Westkust). *Geneeskundig tijdschrift voor Nederlandsch-Indië, 74*, 932–935.

Moffett, A., Shackelford, N., & Sarkar, S. (2007). Malaria in Africa: Vector species' niche models and relative risk maps. *PLOS ONE, 2*(9), e824.

Moffett, A., Strutz, S., Guda, N., González, C., Ferro, M.C., Sánchez-Cordero, V., & Sarkar, S. (2009). A global public database of disease vector and reservoir distributions. *PLOS Neglected Tropical Diseases, 3*(3), e378.

Moffett, S.B., & Moffett, D.F. (2005). Comparison of immunoreactivity to serotonin, FMRFamide and SCPb in the gut and visceral nervous system of larvae, pupae and adults of the yellow fever mosquito *Aedes aegypti*. *Journal of Insect Science, 5*(1), Article 20.

Mogi, M. (1976). Notes on the northern records of *Aedes* (*Stegomyia*) *riversi* Bohart and Ingram. *Mosquito Systematics, 8*(4), 347–352.

Mogi, M. (1977). A new species of *Aedes* (Diptera: Culicidae) from Japan. *Journal of Tropical Medicine* (Nagasaki), *19*(2), 129–140.

Mogi, M. (1984). Distribution and overcrowding effects in mosquito larvae (Diptera: Culicidae) inhabiting taro axils in the Ryukyus, Japan. *Journal of Medical Entomology, 21*(1), 63–68.

Mogi, M. (1996). Overwintering strategies of mosquitoes (Diptera: Culicidae) on warmer islands may predict impact of global warming on Kyushu, Japan. *Journal of Medical Entomology, 33*(3), 438–444.

Mogi, M. (2010). Unusual life history traits of *Aedes* (*Stegomyia*) mosquitoes (Diptera: Culicidae) inhabiting *Nepenthes* pitchers. *Annals of the Entomological Society of America, 103*(4), 618–624.

Mogi, M. (2012). The forms of the *Culex pipiens* complex in East Asia, with ecological thoughts on their origin and interrelation. *Journal of the American Mosquito Control Association, 28* (Suppl. 4), 28–52.

Mogi, M., & Chan, K.L. (1996). Predatory habits of dipteran larvae inhabiting *Nepenthes* pitchers. *Raffles Bulletin of Zoology, 44*, 233–245.

Mogi, M., & Mokry, J. (1980). Distribution of *Wyeomyia smithii* (Diptera, Culicidae) eggs in pitcher plants in Newfoundland, Canada. *Tropical Medicine, 22*(1), 1–12.

Mogi, M., & Toma, T. (1999). *Culex* (*Acallyntrum*) *miyagii* (Diptera: Culicidae): New species from Seram Island, Indonesia, with keys to the species of the subgenus. *Journal of Medical Entomology, 36*(4), 473–480.

Mogi, M., Miyagi, I., & Cabrera, B.D. (1984). Development and survival of immature mosquitoes (Diptera: Culicidae) in Philippine rice fields. *Journal of Medical Entomology, 21*(3), 283–291.

Mohamed, A.H., Ali, A.M., Harbach, R.E., Reeves, R.G., Ibrahim, K.M., Ahmed Mohmed, M.A., & Azrag, R.S. (2017). *Aedes* mosquitoes in the Republic of the Sudan, with dichotomous keys for the adult and larval stages. *Journal of Natural History, 51*(9–10), 513–529.

Mohanty, A., Swain, S., Kar, S.K., & Hazra, R.K. (2009). Analysis of the phylogenetic relationship of *Anopheles* species, subgenus *Cellia* (Diptera: Culicidae) and using it to define the relationship of morphologically similar species. *Infection Genetics and Evolution, 9*(6), 1204–1224.

Mohrig, W. (1967). Die taxonomische Bedeutung der Struktur weiblicher Genitalien im Culiciden- Tribus Aedini. *Angewandte parasitologie, 8*, 67–100.

Mohrig, W. (1969). Die Culiciden Deutschlands. Untersuchungen zur Taxonomie, Biologie und Ökologie der einheimischen Stechmücken. *Parasitologische Schriftenreihe, 18*, 1–260.

Moiroux, N., Gomez, M.B., Pennetier, C., Elanga, E., Djènontin, A., Chandre, F., . . . Corbel, V. (2012). Changes in *Anopheles funestus* biting behavior following universal coverage of long-lasting insecticidal nets in Benin. *Journal of Infectious Diseases, 206*(10), 1622–1629.

Mokany, A., & Shine, R. (2003a). Competition between tadpoles and mosquito larvae. *Oecologia, 135*(4), 615–620.

Mokany, A., & Shine, R. (2003b). Oviposition site selection by mosquitoes is affected by cues from conspecific larvae and anuran tadpoles. *Austral Ecology, 28*(1), 33–37.

Molez, J.F., Desenfant, P., Pajot, F.X., Jacques, J.R., Duverseau, Y., & Saint-Jean, Y. (1987). Le paludisme en Haïti. 2. Présence d'*Anopheles* (*A.*) *pseudopunctipennis* Theobald, 1901. Première mise en évidence sur l'île d'Hispaniola. *Cahiers ORSTOM. Série entomologie médicale et parasitologie, 25*(2), 75–81.

Molina-Cruz, A., Gupta, L., Richardson, J., Bennett, K., Black, W., & Barillas-Mury, C. (2005). Effect of mosquito midgut trypsin activity on dengue-2 virus infection and dissemination in *Aedes aegypti. The American Journal of Tropical Medicine and Hygiene, 72*(5), 631–637.

Moll, R.M., Romoser, W.S., Modrakowski, M.C., Moncayo, A.C., & Lerdthusnee, K. (2001). Meconial peritrophic membranes and the fate of midgut bacteria during mosquito (Diptera: Culicidae) metamorphosis. *Journal of Medical Entomology, 38*(1), 29–32.

Molod, A., Takacs, L., Suarez, M., Bacmeister, J., Song, I.S., Eichmann, A., & Chang, Y. (2012). *The GEOS-5 atmospheric general circulation model: Mean climate and development from MERRA to Fortuna* (Vol. 28, Report No. NASA/TM-2012-104606). NASA Technical Report Series on Global Modeling and Data Assimilation. Greenbelt, Maryland, USA: National Aeronautics and Space Administration.

Monchadsky, A.S. (1936). Les larves des moustiques (Fam. Culicidae) de l'URSS et des pays limitrophes. *Tableaux analytiques de la faune de l'URSS* (*Musee Zoologique de l'Académie des Sciences*), *24*, 1–383.

Monchadskii, A.S.M. (1951). The larvae of bloodsucking mosquitoes of the USSR and adjoining countries (Subfam. Culicinae). *Tableaux analytiques de la faune de l'URSS* (*Musée Zoologique de l'Académie des Sciences*), *37*, 1–290.

Monier, H., & Treillard, M. (1935). *Anopheles* (*Myzomyia*) *funestus* var. n. de Madagascar. *Bulletin de la Société de pathologie exotique, 28*(7), 572–573.

Monnerat, A.T., Soares, M.J., Lima, J.B.P., Rosa-Freitas, M.G., & Valle, D. (1999). *Anopheles albitarsis* eggs: Ultrastructural analysis of chorion layers after permeabilization. *Journal of Insect Physiology, 45*(10), 915–922.

Montchadsky, A.S.M., & García Avila, I. (1966). Las larvas de los mosquitos (Diptera: Culicidae) de Cuba. Su biologia y determinacion. *Poeyana A, 28*, 1–92.

Montchadsky, H.S. (1926). Larva of *Aedes (Ochlerotatus) pulchritarsis*, Rond., var. *stegomyina*, Stack. & Montch., nov., from Turkestan. *Bulletin of Entomological Research, 17*(2), 151–157.

Moore, C.G., & Mitchell, C.J. (1997). *Aedes albopictus* in the United States: Ten-year presence and public health implications. *Emerging Infectious Diseases, 3*(3), 329–334.

Morales, A., & Cura, E. (1983). Descripción de la pupa de *Culex* (*Anoedioporpa*) *bamborum* Rozeboom y Komp 1948 (Diptera: Culicidae). *Biomédica, 3*(4), 114–122.

Morales-Ayala, F. (1971). A list of the mosquitoes of Peru (Diptera, Culicidae). *Mosquito Systematics, 3*(3), 138–145.

Morales-Vargas, R.E., Ya-umphan, P., Phumala-Morales, N., Komalamisra, N., & Dujardin, J.P. (2010). Climate associated size and shape changes in *Aedes aegypti* (Diptera: Culicidae) populations from Thailand. *Infection, Genetics and Evolution, 10*(4), 580–585.

Moran, N.A., & Baumann, P. (2000). Bacterial endosymbionts in animals. *Current Opinion in Microbiology, 3*(3), 270–275.

Moreira, L.A., Edwards, M.J., Adhami, F., Jasinskiene, N., James, A.A., & Jacobs-Lorena, M. (2000). Robust gut-specific gene expression in transgenic *Aedes aegypti* mosquitoes. *Proceedings of the National Academy of Sciences of the United States of America, 97*(20), 10895–10898.

Moreira, L.A., Itube-Ormaexta, I., Jeffery, J.A., Lu, G., Pyke, A.T., Hedges, L.M., . . . Hugo, L.E. (2009). A *Wolbachia* symbiont in *Aedes aegypti* limits infection with dengue, chikungunya, and *Plasmodium. Cell, 139*(7), 1268–1278.

Moreno, M., Melero-Alcíbar, R., Cano, J., Nzambo, S., Buatiche, J.N., Ondo-Esono, M., . . . Benito, A. (2006). Larvae stage description of *Anopheles* (*Cellia*) *carnevalei* from adult individuals collected in Equatorial Guinea. *Journal of the American Mosquito Control Association, 22*(2), 318–323.

Moreno, M., Marinotti, O., Krzywinski, J., Tadei, W.P., James, A.A., Achee, N.L., Conn, J.E. (2010). Complete mtDNA genomes of *Anopheles darlingi* and an approach to anopheline divergence time. *Malaria Journal, 9*, 127.

Moreno, M., Bickersmith, S.A., Harlow, W., Hildebrandt, J., McKeon, S.N., Silva-do-Nascimento, T.F., . . . Conn, J.E. (2013). Phylogeography of the neotropical *Anopheles triannulatus* complex (Diptera: Culicidae) supports deep structure and complex patterns. *Parasites and Vectors* 6, 47.

Morgan, K., O'Loughlin, S.M., Mun-Yik, F., Linton, Y.-M., Somboon, P., Min, S., . . . Walton, C. (2009). Molecular phylogenetics and biogeography of the Neocellia Series of *Anopheles* mosquitoes in the Oriental Region. *Molecular Phylogenetics and Evolution, 52*(3), 588–601.

Morgan, K., O'Loughlin, S.M., Bin, C., Linton, Y.-M., Thongwat, D., Somboon, P., . . . Walton, C. (2011). Comparative phylogeography reveals a shared impact of Pleistocene environmental change in shaping genetic diversity within nine *Anopheles* mosquito species across the Indo-Burma biodiversity hotspot. *Molecular Biology, 20*(21), 4533–4549.

Morin, C.W., & Comrie, A.C. (2010). Modeled response of the West Nile virus vector *Culex quinquefasciatus* to changing climate using the dynamic mosquito simulation model. *International Journal of Biometeorology, 54*(5), 517–529.

Morishita, K. (1946). A new variety of Formosan *Anopheles—Anopheles leucosphyrus* var. *takasagoensis* var. nov. *Journal of the Formosan Medical Association, 45*(2), 2165–2170.

Moriya, K., Yabe, T., & Harada, F. (1973). Chorionic markings of some aedine mosquitoes in Japan. 1. Preliminary observations by a scanning electron microscope and a reflected lighting microscope. *Japanese Journal of Sanitary Zoology, 24*(1), 47–55.

Moses, D.A., Howard, T.M., & Harbach, R.E. (2000). A new species of the subgenus *Sabethoides* of *Sabethes* (Diptera: Culicidae) from Venezuela and Brazil. *Proceedings of the Entomological Society of Washington, 102*(4), 991–1002.

Moss, W.W., Steffan, W.A., Evenhuis, N.L., & Manning, D.L. (1979). The genus *Toxorhynchites* (Diptera: Culicidae): Analysis of *T. splendens* and allies using techniques of numerical taxonomy. *Mosquito Systematics, 11*(4), 258–274.

Motoki, M.T., Wilkerson, R.C., & Sallum, M.A.M. (2009a). The *Anopheles albitarsis* complex with the recognition of *Anopheles oryzalimnetes* Wilkerson and Motoki, n. sp. and *Anopheles jan-*

connae Wilkerson and Sallum, n. sp. (Diptera: Culicidae). *Memórias do Instituto Oswaldo Cruz, 104*(6), 823–850.

Motoki, M.T., dos Santos, C.L., & Sallum, M.A. (2009b). Intraspecific variation on the aedeagus of *Anopheles oswaldoi* (Peryassú) (Diptera: Culicidae). *Neotropical Entomology, 38*(1), 144–148.

Motoki, M.T., Suesdek, L., Bergo, E.S., & Sallum, M.A.M. (2012). Wing geometry of *Anopheles darlingi* Root (Diptera: Culicidae) in five major Brazilian ecoregions. *Infection Genetics and Evolution, 12*(6), 1246–1252.

Motschulsky, V.I. (1859). Catalogue des insectes rapportés des environs du fleuve Amour, depuis la Schilka jusqu'a Nikolaëvsk. *Bulletin de la Société Impériale des Naturalistes de Moscou, 32*(2), 487–507.

Motta, M.A., & Lourenço-de-Oliveira, R. (1995). *Wyeomyia luteoventralis* Theobald, the type species of the subgenus *Dendromyia* Theobald (Diptera: Culicidae). *Memórias do Instituto Oswaldo Cruz, 90*, 375–385.

Motta, M.A., & Lourenço-de-Oliveira, R. (2000). The subgenus *Dendromyia* Theobald: A review with redescriptions of four species (Diptera: Culicidae). *Memórias do Instituto Oswaldo Cruz, 95*(5), 649–683.

Motta, M.A., & Lourenço-de-Oliveira, R. (2005). *Spilonympha*, a new subgenus of *Wyeomyia* (Diptera: Culicidae) and description of a new species *Wyeomyia aningae*. *Annals of the Entomological Society of America, 98*(6), 838–852.

Motta, M.A., Lourenço-de-Oliveira, R., Monteiro, F.A., & Barros, L.R. (1998). Preliminary evaluation of the genetic relatedness of three species of the subgenus *Dendromyia* Theobald and other species of the genus *Wyeomyia* Theobald (Diptera: Culicidae). *Memórias do Instituto Oswaldo Cruz, 93*(2), 189–194.

Motta, M.A., Lourenço-de-Oliveira, R., & Sallum, M.A.M. (2007). Phylogeny of genus *Wyeomyia* (Diptera: Culicidae) inferred from morphological and allozyme data. *The Canadian Entomologist, 139*(5), 591–627.

Mouchet, J., Giacomini, T., & Julvez, J. (1995). La diffusion anthropique des arthropods vecteurs de maladie dans le monde. *Cahiers sanţé, 5*(5), 293–298.

Mouchet, J., Manguin, S., Sircoulon, J., Laventure, S., Faye, O., Onapa, A.W., . . . Fontenille, D. (1998). Evolution of malaria in Africa for the past 40 years: Impact of climatic and human factors. *Journal of the American Mosquito Control Association, 14*(2), 121–130.

Mourya, D.T., Gokhale, M.D., Basu, A., Barde, P.V., Sapkal, G.N., Padbidri, V.S., & Gore, M.M. (2001a). Horizontal and vertical transmission of dengue virus type 2 in highly and lowly susceptible strains of *Aedes aegypti* mosquitoes. *Acta Virologica 45*(2), 67–71.

Mourya, D.T., Gokhale, M.D., Pidiyar, V., Barde, P.V., Patole, M., Mishra, A.C., & Shouche, Y. (2001b). Study of the effect of the midgut bacterial flora of *Culex quinquefasciatus* on the susceptibility of mosquitoes to Japanese encephalitis. *Acta Virologica, 46*(4), 257–260.

Mourya, D.T., Pidiyar, V., Patole, M., & Gokhale, M.D. (2002). Effect of midgut bacterial flora of *Aedes aegypti* on the susceptibility of mosquitoes to Dengue viruses. *Dengue Bulletin, 26*, 190–194.

Moussiegt, O. (1979). Bibliographie. In *Aedes (Aedimorphus) vexans* (Meigen, 1830) (Document E.I.D. No. 42., pp. 1–57). Entente Interdépartementale pour la démoustication du Littoral Meditérranéen Français.

Moussiegt, O. (1981). Bibliographie. In *Aedes (Ochleratatus) caspius* (Pallas, 1771) (Document E.I.D. No. 45, pp. 1–58). Entente Interdépartementale pour la démoustication du Littoral Meditérranéen Français.

Moussiegt, O. (1982). Bibliographie (Supplement). In *Aedes (Aedimorphus) vexans* (Meigen, 1830) (Document E.I.D. No. 46, pp. 1–94). Entente Interdépartementale pour la démoustication du Littoral Meditérranéen.

Moussiegt, O. (1987). Bibliographie. In *Aedes (Ochleratatus) detritus* Haliday, 1833) (Document E.I.D. No. 54, pp. 1–43). Entente Interdépartementale pour la démoustication du Littoral Meditérranéen.

Moussiegt, O. (1988a). Bibliographie, supplément. In *Aedes (Ochleratatus) caspius* (Pallas, 1771) (Document E.I.D. No. 55, pp. 1–55). Entente Interdépartementale pour la démoustication du Littoral Meditérranéen.

Moussiegt, O. (1988b). Bibliographie, 2e supplément. In *Aedes (Aedimorphus) vexans* (Meigen, 1830) (Document E.I.D. No. 56, pp. 1–57). Entente Interdépartementale pour la démoustication du Littoral Meditérranéen.

Moussiegt, O. (1990). Bibliographie. In *Aedes (Barraudius) modestus* Ficalbi, 1889 (Document E.I.D. No. 59, pp. 1–35). Entente Interdépartementale pour la démoustication du Littoral Meditérranéen.

Moussiegt, O. (1993). Bibliographie. In *Aedes* (*Ochlerotatus*) *communis* De Geer (Document E.I.D. No. 61, pp. 1–67). Entente Interdépartementale pour la démoustication du Littoral Meditérranéen.

Mudegu, J.V., & Mukiama, T.K. (1995). Electrophoretic variation in wild *Anopheles pharoensis* Theobald in Mwea Irrigation Scheme, Kenya. *Insect Science and Its Application, 16*(2), 229–231.

Muir, L.E., & Kay, B.H. (1998). *Aedes aegypti* survival and dispersal estimated by mark-release-recapture in northern Australia. *The American Journal of Tropical Medicine and Hygiene, 58*(3), 277–282.

Muirhead-Thomson, R.C. (1946). Studies on the breeding places and control of *Anopheles gambiae* and *A. gambiae var. melas* in coastal districts of Sierra Leone. *Bulletin of Entomological Research, 36*(2), 185–252.

Mukabana, W.R., Mweresa, C.K., Otieno, B., Omusula, P., Smallegange, R.C., van Loon, J.J.A., & Takken, W. (2012). A novel synthetic odorant blend for trapping of malaria and other African mosquito species. *Journal of Chemical Ecology, 38*(3), 235–244.

Mukherjee, A., Williams, D.A., Wheeler, G.S., Cuda, J.P., Pal, S., & Overholt, W.A. (2012). Brazilian peppertree (*Schinus terebinthifolius*) in Florida and South America: Evidence of a possible niche shift driven by hybridization. *Biological Invasions, 14*(7), 1415–1430.

Mukundarajan, H., Hol, F.J.H., Castillo, E.A., Newby, C., & Prakash, M. (2017). Using mobile phones as acoustic sensors for high-throughput mosquito surveillance. *eLife, 6*, e27854.

Mulcahy, D.G. (2008). Phylogeography and species boundaries of the western North American nightsnake (*Hypsiglena torquata*): Revisiting the subspecies concept. *Molecular Phylogenetics and Evolution, 46*(3), 1095–1115.

Mulla, M.S., & Chaudhury, M.F.B. (1968). Influence of some environmental factors on the viability and hatching of *Aedes aegypti* (L.) eggs. *Mosquito News, 28*(2), 217–221.

Mullen, G.R. (1975). Acarine parasites of mosquitoes. I. A critical review of all known records of mosquitoes parasitized by mites. *Journal of Medical Entomology, 12*(1), 27–36.

Mullen, G.R., & Durden, L.A. (Eds.). (2002). *Medical and Veterinary Entomology* (1st ed.). London, England: Academic Press.

Mullen, G.R., & Durden, L.A. (Eds.). (2019). *Medical and Veterinary Entomology* (3rd ed.). London, England: Academic Press.

Müller, G., & Schlein, Y. (2005). Plant tissues: The frugal diet of mosquitoes in adverse conditions. *Medical and Veterinary Entomology, 19*(4), 413–422.

Müller, G., & Schlein, Y. (2006). Sugar questing mosquitoes in arid areas gather on scarce blossoms that can be used for control. *International Journal for Parasitology, 36*(10–11), 1077–1080.
Müller, G.C., Beier, J.C., Traore, S.F., Toure, M.B., Traore, M.M., Bah, S., . . . Schlein, Y. (2010). Field experiments of *Anopheles gambiae* attraction to local fruits / seedpods and flowering plants in Mali to optimize strategies for malaria vector control in Africa using attractive toxic sugar bait methods. *Malaria Journal, 9*(1), 262.
Müller, G.C., Xue, R.D., & Schlein, Y. (2011). Differential attraction of *Aedes albopictus* in the field to flowers, fruits and honeydew. *Acta Tropica, 118*(1), 45–49.
Müller, G.C., Kravchenko, V.D., Junnila, A., & Schlein, Y. (2012). Tree-hole breeding mosquitoes in Israel. *Journal of Vector Ecology, 37*(1), 102–109.
Müller, G.C., Junnila, A., Traore, M.M., Traore, S.F., Doumbia, S., Sissoko, F., . . . Beier, J.C. (2017). The invasive shrub *Prosopis juliflora* enhances the malaria parasite transmission capacity of *Anopheles* mosquitoes: A habitat manipulation experiment. *Malaria Journal, 16*(1), 237.
Müller, O.F. (1764). *Fauna insectorum Fridrichsdalina, sive methodica descriptio insectorum agri Fridrichsdalinensis, cum characteribus genericis et specificis, nominibus trivialibus, locis natalibus, iconibus allegatis, novisque pluribus speciebus additis.* [Copenhagen and Leipzig]: Hafniae et Lipsiae.
Mulligan, H.W., & Puri, I.M. (1936). Description of *Anopheles (Anopheles) habibi* n. sp. from Ouetta, Baluchistan. *Records of the Malaria Survey of India, 6*(1), 67–71.
Mulligan, S. (2013). Aedes aegypti *(yellow fever) mosquito found in Fresno County.* County of Fresno, Department of Public Health.
Mullin-Díaz, E. (1943). Las especies de *Anopheles* del Uruguay. *Comunicaciones Zoologicas del Museo de Historia Natural de Montevideo, 1*(3), 1–8.
Munga, S., Minakawa, N., Zhou, G., Barrack, O.O.J., Githeko, A.K., & Yan, G. (2006). Effects of larval competitors and predators on oviposition site selection of *Anopheles gambiae* sensu stricto. *Journal of Medical Entomology, 43*(2), 221–224.
Munhenga, G., Brooke, B.D., Gilles, J.R., Slabbert, K., Kemp, A., Dandalo, L.C., . . . Koekemoer, L.L. (2016). Mating competitiveness of sterile genetic sexing strain males (GAMA) under laboratory and semi-field conditions: Steps towards the use of the Sterile Insect Technique to control the major malaria vector *Anopheles arabiensis* in South Africa. *Parasites and Vectors, 9*(1), 122.
Muñoz-Cabrera, L.O., Ibáñez-Bernal, S., & Corona-Vargas, M.D.C. (2006). The mosquitoes (Diptera: Culicidae) of Tlaxcala, Mexico. I. Checklist of species. Los mosquitos (Diptera: Culicidae) de Tlaxcala, México. I. Lista comentada de especies. *Folia entomológica Mexicana, 45*(3), 223–271.
Murdoch, W.W., Scott, M.A., & Ebsworth, P. (1984). Effects of the general predator, *Notonecta* (Hemiptera) upon a freshwater community. *Journal of Animal Ecology, 53*(3), 791–808.
Murillo B., C., Astaiza V., R., & Fajardo O., P. (1988). Biologia de *Anopheles (Kerteszia) neivai* H., D. & K., 1913 (Diptera: Culicidae) en la costa Pacifica de Colombia. 1. Fluctuacion de la poblacion larval y caracteristicas de sus criaderos. *Revista de saúde pública, 22*(2), 94–100.
Murillo B., C., Jaramillo S., C., Quintero C., J., & Suarez T., M. (1989). Biología de *Anopheles (Kerteszia) neivai* H., D. & K., 1913 (Diptera: Culicidae) en la costa pacifica de Colombia. 4. Estructura etarea y transmission de Malaria. *Revista de saúde pública, 23*(5), 363–367.
Murphy, M.W., Dunton, R.F., Perich, M.J., & Rowley, W.A. (2001). Attraction of *Anopheles* (Diptera: Culicidae) to volatile chemicals in Western Kenya. *Journal of Medical Entomology, 38*(2), 242–244.
Murty, U.S., Jamil, K., & Ahmed, A. (1993). Aerodynamic parameters and design of flight surface of the mosquito *Culex quinquefasciatus* (Diptera: Culicidae) Say. *Insect Science and Its Application, 14*(2), 205–209.
Muspratt, J. (1950). Notes on *Aëdes* (Diptera, Culicidae) from Natal, with a description of a new species of the subgenus *Stegomyia. Journal of the Entomological Society of Southern Africa, 13*(1), 73–79.
Muspratt, J. (1953a). Research on South African Culicini (Diptera, Culicidae). I. Descriptions of two new *Aedes (Ochlerotatus)* Lynch Arribálzaga. *Journal of the Entomological Society of Southern Africa, 16*(1), 51–58.
Muspratt, J. (1953b). Research on South African Culicini (Diptera, Culicidae). II. Taxonomy relating to eight species of *Aëdes. Journal of the Entomological Society of Southern Africa, 16*(2), 83–93.
Muspratt, J. (1955). Research on South African Culicini (Diptera, Culicidae). III. A check-list of the species and their distribution, with notes on taxonomy bionomics and identification. *Journal of the Entomological Society of Southern Africa, 18*(2), 149–207.
Muspratt, J. (1956a). The *Stegomyia* mosquitoes of South Africa and some neighbouring territories. Including chapters on the mosquito-borne virus diseases of the Ethiopian zoogeographical region of Africa. *Memoirs of the Entomological Society of Southern Africa, 4*, 1–138.
Muspratt, J. (1956b). Research on South African Culicini (Diptera, Culicidae). IV. Additional distribution records, taxonomic descriptions and miscellaneous notes. *Journal of the Entomological Society of Southern Africa, 19*(2), 37–46.
Muspratt, J. (1959). Research on South African Culicini (Diptera, Culicidae). V. A new species of and notes on the sub-genus *Mucidus*, with other observations and records. *Journal of the Entomological Society of Southern Africa, 22*(1), 64–77.
Muspratt, J. (1961). Research on South African Culicini (Diptera: Culicidae). VI. Two new sub-species and notes on bionomics and distribution. *Journal of the Entomological Society of Southern Africa, 24*(1), 92–103.
Mutero, C.M., Blank, H., Konradsen, F., & van der Hoek, W. (2000). Water management for controlling the breeding of *Anopheles* msoquitoes in rice irrigation schemes in Kenya. *Acta Tropica, 76*(3), 253–263.
Myers, C.M. (1967). Identification and description of *Aedes* eggs from California and Nevada (Diptera: Culicidae). *The Canadian Entomologist, 99*(8), 795–806.
Naccarati, C., Audsley, N., Keen, J.N., Kim, J.H., Howell, G.J., Kim, Y.J., & Isaac, R.E. (2012). The host-seeking inhibitory peptide, Aea-HP-1, is made in the male accessory gland and transferred to the female during copulation. *Peptides, 34*(1), 150–157.
Nachtigall, W. (1965). Locomotion: Swimming (hydrodynamics) of aquatic insects. *The Physiology of Insecta, 2*, 255–281.
Nagaki, S.S., de Albuquerque Motta, M., & Sallum, M.A.M. (2010). Redescription of *Anopheles (Nyssorhynchus) antunesi* Galvão & Amaral and description of a new species of the Myzorhynchella Section (Diptera: Culicidae) from Serra da Mantiqueira, Brazil. *Memórias do Instituto Oswaldo Cruz, 105*(3), 278–285.
Nagaki, S.S., da Silva, A.M., & Sallum, M.A.M. (2011). Redescription of *Anopheles (Nyssorhynchus) lutzii*, and resurrection of *Anopheles guarani* from synonymy with *An. lutzii* (Diptera: Culicidae). *Annals of the Entomological Society of America, 104*(3), 374–388.
Nagpal, B.N., & Sharma, V.P. (1995). *Indian anophelines.* New Hampshire: SciencePublishers.

Nainggolan, F.J. (1939). Over een variëteit van *Anopheles gigas* uit Oedjali Kalah (Noord Kerintji), en de variabiliteit harer vleugelteekening. *Geneeskundig tijdschrift voor Nederlandsch-Indië, 79*, 163–170.

Nakagawa, H., & Asanuma, K. (1954). Pupal chaetotaxy of mosquitoes belonging to the tribe Sabethini, with some notes of the extra setae on the abdominal segments. Studies on the pupal stage of Japanese mosquitoes. Part XI. *Miscellaneous Reports of the Research Institute for Natural Resources, 36*, 85–95.

Nakagawa, H., & Asanuma, K. (1955). On the pupa of *Aedes* (*Ochlerotatus*) *hakusanensis* Yamaguti & Tamaboko, 1954, with remarks on the pupal chaetotaxy of the subgenus *Ochlerotatus*. Studies on the pupal stage of Japanese mosquitoes. XII. *Miscellaneous Reports of the Research Institute for Natural Resources, 39*, 103–108.

Nakata, G. (1956). A new species of mosquito, *Aedes* (*Finlaya*) *kobayashii* n. sp. from Kyoto, Japan. *Japanese Journal of Sanitary Zoology, 7*(2), 135.

Nakata, G. (1959). Redescription of *Aedes* (*Finlaya*) *kobayashii* Nakata, 1956. *Japanese Journal of Sanitary Zoology, 10*(1), 16–20.

Namazov, N.D. (2014). The distribution of mosquitoes (Diptera: Culicidae) in the Republic of Azerbaijan. *Entomological Review, 92*, 280–282.

Namgay, R., Drukpa, T., Wangdi, T., Pemo, D., Harbach, R.E., & Somboon, P. (2018). A checklist of the *Anopheles* mosquito species (Diptera: Culicidae) in Bhutan. *Acta Tropica, 188*, 206–212.

Nanda, N., Joshi, H., Subbarao, S.K., Yadav, R.S., Shukla, R.P., Dua, V.K., & Sharma, V.P. (1996). *Anopheles fluviatilis* complex: Host feeding patterns of species S, T, and U. *Journal of the American Mosquito Control Association, 12*(1), 147–149.

Narang, N., Narang, S., & Kitzmiller, J.B. (1973a). The salivary chromosomes of *Anopheles subpictus*. *Parassitologia, 15*, 99–120.

Narang, N., Narang, S., Kitzmiller, J.B., Sharma, G.P., & Sharma, O.P. (1973b). Evolutionary changes in the banding patterns of salivary gland chromosomes in the genus *Anopheles*, subgenus *Cellia*. *Journal of Medical Entomology, 10*(1), 13–22.

Narang, N., Narang, S., & Kitzmiller, J.B. (1974). The salivary gland chromosomes of *Anopheles tessellatus*. *Cytologia, 39*(1), 1–10.

Narayandas, M.G. (1954). Description of the egg of *Anopheles gigas gigas* Giles, 1901. *Indian Journal of Malariology, 8*, 19–20.

Nascimento-Pereira, A.C., Talaga, S., Guimarães, A., Lourenço-de-Oliveira, R., & Motta, M.A. (2019). Taxonomic history of species without subgeneric placement in the genus *Wyeomyia* Theobald (Diptera: Culicidae) and recognition of *Wy. compta* Senevet & Abonnenc as a junior synonym of *Wy. argenteorostris* (Bonne-Wepster & Bonne). *Zootaxa, 4656*(2), 359–366.

Natarajan, R. (2019). Description of a new species of the genus *Paraedes* (Diptera: Culicidae) in Assam, India. *Zootaxa, 4691*(3), 286–292.

Natarajan, R., & Rajavel, A.R. (2009). Description of a new species of *Culex* (*Lophoceraomyia*) from Orissa, India. *Journal of the American Mosquito Control Association, 25*(4), 403–408.

Natarajan, R., Rajavel, A.R., & Jambulingam, P. (2016). Description of a new species of the genus *Hulecoeteomyia* (Diptera: Culicidae) from Meghalaya, India. *Zootaxa, 4137*(3), 330–338.

Natarajan, R., Rajavel, A.R., & Jambulingam, P. (2017). Descriptions of three new species of *Uranotaenia* (*Pseudoficalbia*) (Diptera: Culicidae) from India. *Zootaxa, 4227*(2), 251–265.

Natarajan, R., Rajavel, A.R., & Jambulingam, P. (2018). *Uranotaenia pseudostricklandi*, a new species in subgenus *Pseudoficalbia* (Diptera: Culicidae) from Kerala, India. *Zootaxa, 4429*(1), 181–188.

Natarajan, R., Eapen, A., & Jambulingham, P. (2019) Description of the adult male, larva and pupa of *Topomyia* (*Topomyia*) *aureoventer* (Diptera: Culicidae) in India. *Zootaxa, 4586*(2), 376–382.

Nation, J.L., Sr. (2008). *Insect physiology and biochemistry*. Boca Raton, FL: CRC Press.

National Institute of Allergy and Infectious Diseases. *VectorBase*. Retrieved from https://www.vectorbase.org/

Natvig, L.R. (1942). Über *Theobaldia bergrothi* Edw. *Th. borealis* Schingarew und *Th. glaphyroptera* Schiner. *Notulae Entomologicae, 22*(1–2), 1–5.

Natvig, L.R. (1948). Contributions to the knowledge of the Danish and Fennoscandian mosquitoes: Culicini. *Norsk entomologisk tidsskrift Suppl., I*, xxiii.

Natvig, L.R. (1955). The type specimens of *Aedes nigripes* (Zett.). *Norsk entomologisk tidsskrift, 9*, 86–92.

Navarro, J.C., & Liria, J. (2000). Phylogenetic relationships among eighteen neotropical Culicini species. *Journal of the American Mosquito Control Association, 16*(2), 75–85.

Navarro, J.C., & Liria, J. (2007). *Wyeomyia trujilloi* Pulido y Sutil 1981, nuevo sinonimo de *Wy. bicornis* (Root 1928) (Culicidae: Sabethini), con redescripción de la pupa y parte de la larva. *Boletín de la Dirección de Malariología y Saneamiento Ambiental, 47*(1), 89–102.

Navarro, J.C., & Machado-Allison, C. (1995). Aspectos ecológicos de *Sabethes chloropterus* (Humboldt) (Diptera: Culicidae) en un bosque humedo del Edo. Miranda, Venezuela. *Boletín de entomología Venezolana, 10*(1), 91–104.

Navarro, J.-C., & Weaver, S.C. (2004). Molecular phylogeny of the Vomerifer and Pedroi groups in the Spissipes Section of the subgenus *Culex* (*Melanoconion*). *Journal of Medical Entomology, 41*(4), 575–581.

Navarro, J.-C., Ponce, P., & Cevallos, V. (2013a). Dos nuevos registros de vectores potenciales de Fiebre Amarilla selvática y Mayaro para el Ecuador. *Boletín de malariología y salud ambiental, 53*(1), 77–81.

Navarro, J.-C., Enriquez, S., Vaca, F., & Benitez-Ortiz, W. (2013b). A new phytotelm plant, *Crinum moorei* (Asparagales: Amaryllidaceae), for the Americas and its mosquito inhabitant (Diptera: Culicidae) in Ecuador. *Florida Entomologist, 96*(3), 1224–1227.

Navarro-Silva, M.A., Marques, F.A., Duque, L., & Jonny, E. (2009). Review of semiochemicals that mediate the oviposition of mosquitoes: A possible sustainable tool for the control and monitoring of Culicidae. *Revista Brasileira de entomología, 53*(1), 1–6.

Nayar, J.K. (1982). Bionomics and physiology of *Culex nigripalpus* (Diptera: Culicidae) of Florida: An important vector of diseases. *Florida Agricultural Experiment Stations Bulletin, 827*, 1–73.

Nayar, J.K., & Sauerman, D.M. (1968). Larval aggregation formation and population density interrelations in *Aedes taeniorhynchus*, their effects on pupal ecdysis and adult characteristics at emergence. *Entomología experimentalis et applicata, 11*(4), 423–442.

Nayar, J.K., & Sauerman, D.M. (1971). The effects of diet on life-span, fecundity and flight potential of *Aedes taeniorhynchus* adults. *Journal of Medical Entomology, 8*(5), 506–513.

Ndo, C., Simard, F., Kengne, P., Awono-Ambene, P., Morlais, I., Sharakhov, I., . . . Antonio-Nkondjio, C. (2013). Cryptic genetic diversity within the *Anopheles nili* group of malaria vectors in the equatorial forest area of Cameroon (Central Africa). *PLOS ONE, 8*(3), e58862.

Neafsey, D.E., Waterhouse, R.M., Abai, M.R., Aganezov, S.S., Alekseyev, M.A., Allen, J.E., . . . Besansky, N.J. (2015). Highly evolvable malaria vectors: The genomes of 16 *Anopheles* mosquitoes. *Science, 347*(6217), 1258522.

Neiva, A. (1906). Uma nova espécie de anophelina braziliera. *Brazil-Medico, 20*, 288–289.

Neiva, A. (1908a). Contribuição ao estudo dos dipteros brazileiros. Uma nova espécie de *Sabethes*. *Brazil-Medico, 22*, 351–352.
Neiva, A. (1908b). Das anophelinas Brazileiras. *Revista medica de São Paulo, 11*(22), 455–459.
Neiva, A., & Pinto, C. (1922a). Contribuição para o conhecimento dos anophelinas do estado de Matto Grosso, com a descripção de uma nova especie. *Brazil-Medico, 36*(2), 321–322.
Neiva, A., & Pinto, C. (1922b). Considerações sobre o genero *Cellia* Theobald, com a descrição de uma nova especie. *Brazil-Medico, 36*(2), 355–357.
Neiva, A., & Pinto, C. (1922c). Commentarios sobre o genero *Uranotaenia* Arribálzaga, 1891 com descripcão de uma nova especie. *Brazil-Medico, 36*(2), 374.
Neiva, A., & Pinto, C. (1923). Sobre uma nova anophelina brazileira (*Cellia cuyabensis* nov. sp.). *Brazil-Medico, 37*(1), 235–236.
Nene, V., Wortman, J.R., Lawson, D., Haas, B., Kodira, C., Tu, Z.J., . . . Severson, D.W. (2007). Genome sequence of *Aedes aegypti*, a major arbovirus vector. *Science, 316*(5832), 1718–1723.
Neteler, M., Roiz, D., Rocchini, D., Castellani, C., & Rizzoli, A. (2011). Terra and Aqua satellites track tiger mosquito invasion: Modelling the potential distribution of *Aedes albopictus* in north-eastern Italy. *International Journal of Health Geographics, 10*(1), Article 49.
Netto, A.S. (1940). *Mosquitos do Rio Grande do Sul* [Doctoral thesis, Faculdade de Medicina de Porto Alegre, Rio Grande do Sul, Brazil].
Neveu-Lemaire, M. (1902a). Sur la classification des Culicides. *Comptes rendus hebdomadaires des séances et mémoires de la Société de Biologie , 54,* 1329–1332.
Neveu-Lemaire, M. (1902b). Description de quelques moustiques de la Guyane. *Archives de parasitologie, 6*(1), 5–25.
Neveu-Lemaire, M. (1902c). Classification de la famille des Culicidae. *Memoires de la Société Zoologique de France, 15*, 195–227.
Neveu-Lemaire, M. (1905a). Sur un nouveau moustique appartenant a la sous-famille des Anophelinae (*Nyssorhynchus Bozasi* nov. sp.). *Comptes rendus hebdomadaires des séances et mémoires de la Société de Biologie, 59*, 32–33.
Neveu-Lemaire, M. (1905b). Mission du Bourg de Bozas. Description d'une nouvelle espèce de *Stegomyia recueillie* par le Dr. Brumpt à Harar. *Bulletin de la Société Zoologique de France, 30*, 8–11.
Neveu-Lemaire, M. (1906). Étude des culicides africains (mission du Bourg de Bozas en Afrique tropical). *Archives de parasitologie, 10*(2), 238–288.
Neveu-Lemaire, M. (1923). L'évolution de la classification des *Culicidae. Annales de parasitologie humaine et comparée, 1*(1), 90–107.
Newkirk, M.R. (1947). Observations on *Megarhinus splendens* Wiedemann with reference to its value in biological control of other mosquitoes (Diptera: Culicidae). *Annals of the Entomological Society of America, 40*(3), 522–527.
Newstead, R., & Carter, H.F. (1910). Descriptions of a new genus and three new species of anopheline mosquitos. *Annals of Tropical Medicine and Parasitology, 4*(3), 377–383.
Newstead, R., & Carter, H.F. (1911a). On a new genus of Culicinae from the Amazon region. *Annals of Tropical Medicine and Parasitology, 4*(4), 553–556.
Newstead, R., & Carter, H.F. (1911b). On some new species of African mosquitos (Culicidae). *Annals of Tropical Medicine and Parasitology, 5*(2), 233–244.
Newstead, R., & Thomas, H.W. (1910). The mosquitos of the Amazon region. *Annals of Tropical Medicine and Parasitology, 4*(1), 141–150.
Newstead, R., Dutton, J.E., & Todd, J.L. (1907). Insects and other Arthropoda collected in the Congo Free State. *Annals of Tropical Medicine and Parasitology, 1*(1–5), 3–113.
Ng'habi, K.R., Huho, B.J., Nkwengulila, G., Killeen, G.F., Knols, B.G., & Ferguson, H.M. (2008). Sexual selection in mosquito swarms: May the best man lose? *Animal Behaviour, 76*(1), 105–112.
Ng'habi, K.R., Knols, B.G.J., Lee, Y., Ferguson, H.M., & Lanzaro, G.C. (2011). Population genetic structure of *Anopheles arabiensis* and *Anopheles gambiae* in a malaria endemic region of southern Tanzania. *Malaria Journal, 10*(1), Article 289.
Ngo, C.T., Harbach, R.E., Garros, C., Parzy, D., Le, H.Q., & Manguin, S. (2013). Taxonomic assessment of *Anopheles crawfordi* and *An. dangi* of the Hyrcanus Group of subgenus *Anopheles* in Vietnam. *Acta Tropica, 128*(3), 623–629.
Ngoagouni, C., Kamgang, B., Nakouné, E., Paupy, C., & Kazanji, M. (2015). Invasion of *Aedes albopictus* (Diptera: Culicidae) into central Africa: What consequences for emerging diseases? *Parasites and Vectors, 8*, 191.
Nguyen, A.T., Williams-Newkirk, A.J., Kitron, U.D., & Chaves, L.F. (2012). Seasonal weather, nutrients and conspecific presence impacts on the southern house mosquito oviposition dynamics in combined sewage overflows. *Journal of Medical Entomology, 49*(6), 1328–1338.
Nguyen, D.M., Tran, D.H., & Nguyen, T.V. (1993). *Anopheles* (*Anopheles*) *vietnamensis*: A new species in the *Anopheles hyrcanus* group from Vietnam (Diptera: Culicidae). *Mosquito Systematics, 25*(3), 215–221.
Nguyen, D.M., Tran, D.H., Harbach, R.E., Elphick, J., & Linton, Y.-M. (2000). A new species of the Hyrcanus group of *Anopheles*, subgenus *Anopheles*, a secondary vector of malaria in coastal areas of southern Vietnam. *Journal of the American Mosquito Control Association, 16*(3), 189–198.
Nguyen Thuong Hien. (1968). *The genus of* Anopheles *in Vietnam* (Military Entomology Information Service, Trans.) Saigon: Bureau of Entomology, National Malaria Program/Republic of Vietnam.
Nicolescu, G., Vladimirescu, A., & Ciolpan, O. (2003a). The distribution of mosquitoes in Romania (Diptera: Culicidae). Part II. *Culiseta*, *Coquillettidia*, *Ochlerotatus*, *Orthopodomyia* and *Uranotaenia*. *European Mosquito Bulletin, 14*, 1–15.
Nicolescu, G., Vladimirescu, A., & Ciolpan, O. (2003b). The distribution of mosquitoes in Romania (Diptera: Culicidae). Part IV. Detailed maps for *Coquillettidia*, *Culiseta*, *Ochlerotatus*, *Orthpodomyia* and *Uranotaenia*. *European Mosquito Bulletin, 15*, 16–27.
Nicolescu, G., Linton, Y.-M., Vladimirescu, A., Howard, T.M., & Harbach, R.E. (2004). Mosquitoes of the *Anopheles maculipennis* group (Diptera: Culicidae) in Romania, with the discovery and formal recognition of a new species based on molecular and morphological evidence. *Bulletin of Entomological Research, 94*, 525–535.
Nie, W.Z., Li, J.C., Li, D.X., Wang, R.J., & Gratz, N. (2004). Mosquitoes found aboard ships arriving at Qinhuangdao Port, P.R. China. *Medical Entomology and Zoology, 55*(4), 333–335.
Nielsen, E.T., & Greve, H. (1950). Studies on the swarming habits of mosquitos and other Nematocera. *Bulletin of Entomological Research, 41*, 227–258.
Nielsen, E.T., & Haeger, J.S. (1960). Swarming and mating in mosquitoes. *Miscellaneous Publications of the Entomological Society of America, 1*, 71–95.
Nielsen, E.T., & Nielsen., A.T. (1953). Field observations on the habits of *Aedes taeniorhynchus*. *Ecology, 34*(1), 141–156.
Nielsen, E.T., & Nielsen., A.T. (1958). Observations on mosquitoes in Iraq. *Entomologiske meddelelser, 28*, 282–321.
Nielsen, L.T. (1970). *Aedes cacothius* Dyar, a synonym of *Aedes ventrovittis* Dyar (Diptera: Culicidae). *Proceedings of the Entomological Society of Washington, 71*(for 1969), 530.

Nielsen, L.T. (2009). *Ochlerotatus dahlae*, a new mosquito species from Utah. *Journal of the American Mosquito Control Association, 25*(2), 123–125.
Nielsen, L.T., & Horsfall, W.R. (1973). The occurrence of *Aedes barri* Rueger in Alaska with notes on its distribution. *Mosquito News, 33*(2), 243.
Nielsen, L.T., & Rees, D.M. (1959). The mosquitoes of Utah—a revised list. *Mosquito News, 19*(2), 45–47.
Nielsen, L.T., & Rees, D.M. (1961). An identification guide to the mosquitoes of Utah. *University of Utah Biological Series, 12*(3), 1–58.
Nijhout, H.F., & Craig, G.B., Jr. (1971). Reproductive isolation in *Stegomyia* mosquitoes. III. Evidence for a sexual pheromone. *Entomología Experimentalis et Applicata, 14*(4), 399–412.
Nitecki, M.H. (1988). Introduction. In M.H. Nitecki (Ed.), *Evolutionary progress* (pp. 3–24). Chicago, IL: University of Chicago Press.
Nkya, T.E., Akhouayri, I., Kisinza, W., & David, J.P. (2013). Impact of environment on mosquito response to pyrethroid insecticides: Facts, evidences and prospects. *Insect Biochemistry and Molecular Biology, 43*(4), 407–416.
Noè, G. (1899). Contribuzione allo studio dei Culicidi. *Bollettino della Società Entomologica Italiana, 31,* 235–262.
Noeldner, E. (1953). Moustiques rares ou peu connus d'Alsace. *Encyclopedie Entomologique B, 11,* 1–45.
Norris, L.C., & Norris, D.E. (2015). Phylogeny of anopheline (Diptera: Culicidae) species in southern Africa, based on nuclear and mitochondrial genes. *Journal of Vector Ecology, 40*(1), 16–27.
Novak, R.J., Peloquin, J., & Rohrer, W. (1981). Vertical distribution of adult mosquitoes (Diptera: Culicidae) in a northern deciduous forest in Indiana. *Journal of Medical Entomology, 18*(2), 116–122.
Nurul Huda, K.M., & Harrison, B.A. (1985). Priority of the name *Anopheles pseudojamesi* for the species previously called *An. ramsayi* (Diptera: Culicidae). *Mosquito Systematics, 17*(1), 49–51.
Nyasembe, V.O., & Torto, B. (2014). Volatile phytochemicals as mosquito semiochemicals. *Phytochemistry Letters, 8,* 196–201.
Nyasembe, V.O., Teal, P.E.A., Mukabana, W.R., Tumlinson, J.H., & Torto, B. (2012). Behavioural response of the malaria vector *Anopheles gambiae* to host plant volatiles and synthetic blends. *Parasites and Vectors, 5,* 234.
Nyasembe, V.O., Teal, P.E.A., Sawa, P., Tumlinson, J.H., Borgemeister, C., & Torto, B. (2014). *Plasmodium falciparum* infection increases *Anopheles gambiae* attraction to nectar sources and sugar uptake. *Current Biology, 24*(2), 217–221.
Oaks, S.C., Ridgeway, R.L., Shirai, A., & Twartz, J.C. (1983). Scrub typhus. *Institute for Medical Research, Malaysia, Bulletin 21.* US Army Medical Research Unit.
Obenauer, P.J., Kaufman, P.E., Allan, S.A., & Kline, D.L. (2009a). Host-seeking height preferences of *Aedes albopictus* (Diptera: Culicidae) in north central Florida suburban and sylvatic locales. *Journal of Medical Entomology, 46*(4), 900–908.
Obenauer, P.J., Kaufman, P.E., Allan, S.A., & Kline, D.L. (2009b). Infusion-baited ovitraps to survey ovipositional height preferences of container-inhabiting mosquitoes in two Florida habitats. *Journal of Medical Entomology, 46*(6), 1507–1513.
Obermayr, U. (2014). Excitorepellency. In M. Debboun, S.P. Frances, & D.A. Strickman (Eds.), *Insect repellents handbook* (2nd ed., pp. 91–116). Boca Raton, FL: CRC Press.
Obsomer, V., Titeux, N., Vancustem, C., Duveiller, G., Pekel, J.-F., Connor, S., . . . Coosemans, M. (2013). From *Anopheles* to spatial surveillance: A roadmap through a multidisciplinary challenge. In S. Manguin (Ed.), Anopheles *mosquitoes: New insights into malaria vectors* (pp. 447–484). Rijeka, Croatia: InTech.
O'Connor, C.T. (1967). The distribution of anopheline mosquitoes in Ethiopia. *Mosquito News, 27*(1), 42–54.
O'Connor, K.R., & Beyenbach, K.W. (2001). Chloride channels in apical membrane patches of stellate cells of Malpighian tubules of *Aedes aegypti*. *Journal of Experimental Biology, 204*(2), 367–378.
O'Connor, L., Plichart, C., Sang, A.C., Brelsfoard, C.L., Bossin, H.C., & Dobson, S.L. (2012). Open release of male mosquitoes infected with a *Wolbachia* biopesticide: Field performance and infection containment. *PLOS Neglected Tropical Diseases, 6*(11), e1797.
Oda, T. (1968). Studies on the follicular development and overwintering of the house mosquito, *Culex pipiens pallens* in Nagasaki area. *Tropical Medicine, 10*(4), 195–216.
O'Donnell, M.J., & Spring, J.H. (2000). Modes of control of insect Malpighian tubules: Synergism, antagonism, cooperation and autonomous regulation. *Journal of Insect Physiology, 46*(2), 107–117.
Ogasawara, H. (1939a). A new mosquito from Shakujii (Tokyo). *Entomological World* (Tokyo), *7*(63), 237–239.
Ogasawara, H. (1939b). On the genus *Megarhinus* in Japan, with description of a new variety. *Entomological World* (Tokyo), *7*(63), 240–244.
Ogata, K., & Lopez Samayoa, A. (1996). Discovery of *Aedes albopictus* in Guatemala. *Journal of the American Mosquito Control Association, 12*(3), 503–506.
Ohama, S. (1947). Epidemiological study of malaria in Yaeyama. No. 2 report. On an anopheline mosquito, *Anopheles ohami* (Ishigaki Island) 1947, which is newly found in Ishigaki Island. *Records of the Public Health Department, Yaeyama Provincial Government, 4,* 1–15.
Ohmori, Y. (1957). On the male hypopygium of the *sinensis* group of anopheline mosquitoes in Japan. *Acta médica et biological, Niigata, 5,* 209–217.
Ohmori, Y. (1959). The pupae of Japanese *Anopheles*. *Japanese Journal of Sanitary Zoology, 10,* 210–225.
Okazawa, T., Hario, M., Suzuki, H., & Mogi, M. (1986a). Colonization and laboratory bionomics of *Topomyia yanbarensis* (Diptera: Culicidae). *Journal of Medical Entomology, 23*(5), 493–501.
Okazawa, T., Horio, M., Miyagi, I., Mogi, M., & Sucharit, S. (1986b). Eggs of *Tripteroides* (*Rachionotomyia*) *aranoides* (Theobald) and *Topomyia* (*Suaymia*) *yanbarensis* Miyagi. *Mosquito Systematics, 18*(3–4), 293–298.
Okazawa, T., Miyagi, I., Toma, T., Ramalingam, S., & Chang, M.S. (1991). Egg morphology and observations on the laboratory biology of *Armigeres* (*Leicesteria*) *digitatus* (Diptera: Culicidae) from Sarawak. *Journal of Medical Entomology, 28*(5), 606–610.
Okech, B.A., Gouagna, L.C., Killeen, G.F., Knols, B.G., Kabiru, E.W., Beier, J.C., . . . Githure, J.I. (2003). Influence of sugar availability and indoor microclimate on survival of *Anopheles gambiae* (Diptera: Culicidae) under semifield conditions in western Kenya. *Journal of Medical Entomology, 40*(5), 657–663.
Okuda, K., de Souza Caroci, A., Ribolla, P.E.M., De Bianchi, A.G., & Bijovsky, A.T. (2002). Functional morphology of adult female *Culex quinquefasciatus* midgut during blood digestion. *Tissue and Cell, 34*(3), 210–219.
Okuda, K., Caroci, A., Ribolla, P., Marinotti, O., de Bianchi, A.G., & Bijovsky, A.T. (2005). Morphological and enzymatic analysis of the midgut of *Anopheles darlingi* during blood digestion. *Journal of Insect Physiology, 51*(7), 769–776.
Okuda, K., de Almeida, F., Mortara, R.A., Krieger, H., Marinotti, O., & Bijovsky, A.T. (2007). Cell death and regeneration in the midgut of the mosquito, *Culex quinquefasciatus*. *Journal of Insect Physiology, 53*(12), 1307–1315.
Olagbemiro, T.O., Birkett, M.A., Mordue Luntz, A.J., & Pickett, J.A. (2004). Laboratory and field responses of the mosquito *Culex quinquefasciatus*, to plant-derived *Culex* spp. Oviposition pher-

omone and the oviposition cue skatole. *Journal of Chemical Ecology, 30*(5), 956–976.

Oliva, C.F., Benedict, M.Q., Lempérière, G., & Gilles, J. (2011). Laboratory selection for an accelerated mosquito sexual development rate. *Malaria Journal, 10*(1), 135.

Oliva, C.F., Jacquet, M., Gilles, J., Lempérière, G., Maquart, P.O., Quilici, S., . . . Boyer, S. (2012). The sterile insect techniques for controlling populations of *Aedes albopictus* (Diptera: Culicidae) on Reunion Island: Mating vigour of sterilized males. *PLOS ONE, 7*(11), e409414.

Oliva, C.F., Damiens, D., & Benedict, M.Q. (2014). Male reproductive biology of *Aedes* mosquitoes. *Acta Tropica, 132*(Suppl.), S12–S19.

Olivier, G.A. (1791). *Encyclopedique [sic] methodique. Histoire naturelle. Insectes* (Vol. 6, Part 1). Paris, France: Panckoucke.

Olsen, S.S., Cazzamali, G., Williamson, M., Grimmelikhuijzen, C.J.P., & Hauser, F. (2007). Identification of one capa and two pyrokinin receptors in the malaria mosquito *Anopheles gambiae*. *Biochemical and Biophysical Research Communications, 362*(2), 245–251.

Olson, J.K., & Meola, S.M. (1976). Variations in chorionic sculpturing of eggs of *Aedes sollicitans* (Walker). *Annals of the Entomological Society of America, 69*(1), 96–100.

Olson, S.H., Gangnon, R., Silveira, G., & Patz, J.A. (2010). Deforestation and malaria in Mâncio Lima County, Brazil. *Emerging Infectious Diseases, 16*(7), 1108–1115.

O'Meara, G.F., & Craig, G.B., Jr. (1970a). A new subspecies of *Aedes atropalpus* (Coquillett) from southwestern United States (Diptera: Culicidae). *Proceedings of the Entomological Society of Washington, 72*(4), 475–479.

O'Meara, G.F., & Craig, G.B, Jr. (1970b). Geographical variation in *Aedes atropalpus* (Diptera: Culicidae). *Annals of the Entomological Society of America, 63*(5), 1392–1400.

O'Meara, G.F., & Lounibos, L.P. (1981). Reproductive maturation in the pitcher plant mosquito, *Wyeomyia smithii*. *Physiological Entomology, 6*(4), 437–443.

O'Meara, G.F., Lounibos, L.P., & Brust, R.A. (1981). Repeated clutches without blood in the pitcher-plant mosquito. *Annals of the Entomological Society of America, 74*(1), 68–72.

Omer, S.M., & Cloudsley-Thompson, J.L. (1968). Dry season biology of *Anopheles gambiae* Giles in the Sudan. *Nature, 217*(5131), 879–880.

Omer, S.M., & Cloudsley-Thompson, J.L. (1970). Survival of female *Anopheles gambiae* Giles through a 9-month dry season in Sudan. *Bulletin of the World Health Organization, 42*(2), 319–330.

Omori, N. (1962). Morphology of undescribed male and immature mosquitoes of aedines in Japan. 1. *Aedes* (*Finlaya*) *watasei*. *Endemic Diseases Bulletin of Nagasaki University, 4*, 10–14.

Omori, N., & Fujii, S. (1953). On the feeding habits of *Aedes togoi* and some other species of mosquitoes. *Yokohama Medical Bulletin, 4*, 23–31.

Onken, H., Moffett, S.B., & Moffett, D.F. (2004). The anterior stomach of larval mosquitoes (*Aedes aegypti*): Effects of neuropeptides on transepithelial ion transport and muscular motility. *Journal of Experimental Biology, 207*(21), 3731–3739.

Oo, T.T., Storch, V., & Becker, N. (2004). Review of the *Anopheles* mosquitoes of Myanmar. *Journal of Vector Ecology, 29*(1), 21–40.

Oo, T.T., Kaiser, A., & Becker, N. (2006). Illustrated keys to the anopheline mosquitoes of Myanmar. *Journal of Vector Ecology, 31*(1), 9–16.

Országh, I., Halgoš, J., Jalili, N., & Labuda, M. (2001). Mosquitoes (Diptera, Culicidae) of Slovakia II. *European Mosquito Bulletin, 11*, 1–26.

Ortega-Morales, A.I., Elizondo-Quiroga, A., González-Villarreal, D.A., Siller-Rodriguez, Q.K., Reyes-Villanueva, F., & Fernandez-Salas, I. (2009). First record of *Culiseta melanura* in Mexico, with additional Mexican records for *Aedes sollicitans*. *Journal of the American Mosquito Control Association, 25*(1), 100–102.

Ortega-Morales, A.I., Huerta, H., Strickman, D., Sánchez Ramos, F.J., Landeros Flores, J., & Cerna Chavez, E. (2011). Registros de mosquitos en Mexico: *Culex stigmatosoma* Dyar y *Cx. thriambus* Dyar (Diptera: Culicidae) con notas taxonomicas para ambas especies. *Southwestern Entomologist, 36*(2), 177–196.

Ortega-Morales, A.I., Zavortink, T.J., Garza-Hernández, J.A., Siller-Rodríguez, Q.K., & Fernández-Salas, I. (2019). The mosquitoes (Diptera: Culicidae) of Nuevo León, Mexico, with descriptions of two new species. *PLOS ONE, 14*(8), e0217694.

Osei-Poku, J., Mbogo, C.M., Palmer, W.J., & Jiggins, F.M. (2012). Deep sequencing reveals extensive variation in the gut microbiota of wild mosquitoes from Kenya. *Molecular Ecology, 21*(20), 5138–5150.

Oshaghi, M.A., Shemshad, K., Yaghobi-Ershadi, M.R., Pedram, M., Vatandoost, H., Abaie, M.R., . . . Mohtarami, F. (2007). Genetic structure of the malaria vector *Anopheles superpictus* in Iran using mitochondrial cytochrome oxidase (*COI* and *COII*) and morphologic markers: A new species complex? *Acta Tropica, 101*(3), 241–248.

Oshaghi, M.A., Yaghobi-Ershadi, M.R., Shemshad, K., Pedram, M., & Amani, H. (2008). The *Anopheles superpictus* complex: Introduction of a new malaria vector complex in Iran. *Bulletin de la Société de pathologie exotique, 101*(5), 429–434.

Osório, H.C., Amaro, F., Zé-Zé, L., Pardal, S., Mendes, S., Ventim, R., . . . Alves, M.J. (2010). Mosquito species distribution in mainland Portugal. *European Mosquito Bulletin, 28*, 187–193.

Osorno-Mesa, E. (1944a). Organización de una colónia de *Haemagogus equinus* Theobald. *Caldasia, 3*, 39–45.

Osorno-Mesa, E. (1944b). Two new species of *Haemagogus* from Colombia, *H. andinus* and *H. boshelli* (Diptera: Culicidae). *Proceedings of the Entomological Society of Washington, 46*(7), 165–175.

Osorno-Mesa, E. (1947). Una nueva especie de *Anopheles* de Bogotá Colombia. *Caldasia, 4*, 431–446.

Osorno-Mesa, E., & Muñoz-Sarmiento, F. (1948). Una nueva variedad de *Anopheles pseudopunctipennis*. *Caldasia, 5*, 105–113.

Osta, M.A., Christophides, G.K., & Kafatos, F.C. (2004). Effects of mosquito genes on *Plasmodium* development. *Science, 303*(5666), 2030–2032.

Osten Sacken, C.R. (1868). Description of a new species of Culicidae. *Transactions of the American Entomological Society, 2*, 47–48.

Osten Sacken, C.R. (1877). Western Diptera: Descriptions of new genera and species of Diptera from the region west of the Mississippi and especially from California. *Bulletin of the United States Geological Survey, 3*(2), 189–354.

Oter, K., Gunay, F., Tuzer, E., Linton, Y.-M., Bellini, R., & Alten, B. (2013). First record of *Stegomyia albopicta* in Turkey determined by active ovitrap surveillance and DNA barcoding. *Vector-Borne and Zoonotic Diseases, 13*(10), 753–761.

Otero, M., Schweigmann, N., & Solari, H.G. (2008). A stochastic spatial dynamical model for *Aedes aegypti*. *Bulletin of Mathematical Biology, 70*(5), 1297–1325.

Otienoburu, P.E., Ebrahimi, B., Phelan, P.L., & Foster, W.A. (2012). Analysis and optimization of a synthetic milkweed floral attractant for mosquitoes. *Journal of Chemical Ecology, 38*(7), 873–881.

Otsuru, M., & Ohmori, Y. (1960). Malaria studies in Japan after World War II. Part II. The research for *Anopheles sinensis* sibling species group. *Japanese Journal of Experimental Medicine, 30*, 33–65.

Otsuru, M., Nagashima, Y., Nakamura, Y., & Kishimoto, T. (1976). Survey of eggs of the *Anopheles sinensis* sibling species group in Okinawa Is. *Japanese Journal of Sanitary Zoology, 27*(3), 301–303.

Ôuchi, Y. (1939). A new species of *Megarhinus* mosquito from Amami Oshima, the southern Japan. *Journal of the Shanghai Science Institute, 4*, 223–225.

Ovazza, M., Hamon, J., & Neri, P. (1956). Contribution a l'étude des Diptères vulnerants de l'empire d'Éthiopie I. Culicidae. *Bulletin de la Société de pathologie exotique, 49*(1), 151–182.

Overgaard, H.J., Ekbom, B., Suwonkerd, W., & Takagi, M. (2003). Effect of landscape structure on anopheline mosquito density and diversity in northern Thailand: Implications for malaria transmission and control. *Landscape Ecology, 18*(6), 605–619.

Owen, W.B. (1945). A new anopheline from the Solomon Islands with notes on its biology. *Journal of Parasitology, 31*(4), 236–240.

Owen, W.B. (1977). Morphology of the thoracic skeleton and muscles of the mosquito, *Culiseta inornata* (Williston), (Diptera: Culicidae). *Journal of Morphology, 153*(3), 427–460.

Owen, W.B. (1980). Morphology of the abdominal skeleton and muscles of the mosquito, *Culiseta inornata* (Williston) (Diptera: Culicidae). *Journal of Morphology, 166*(2), 155–178.

Pacheco, J.B., Santos-Mallet, J.R., Guimarães, A.É., Costa, J.M., & Alencar, J. (2012). Ultrastructure and morphometry of the egg of *Psorophora albigenu* Lutz, 1908 (Diptera: Culicidae). *Micron, 43*(2–3), 418–421.

Paine, R.W. (1943). An introduction to the mosquitoes of Fiji. Descriptive notes on the commoner species, their breeding places and occurrence; together with simplified keys for distinguishing the adults and larvae of Fijian mosquitoes. *Bulletin of the Department of Agriculture Fiji, 22*, 1–35.

Paine, R.W., & Edwards, F.W. (1929). Mosquitos from the Solomon Islands. *Bulletin of Entomological Research, 20*(3), 303–316.

Pajot, F.-X. (1976). Étude de l'ornémentation des imagos d'*Aedes* (*Stegomyia*) *simpsoni* (Theobald, 1905) (Diptera, Culicidae) [*sic*] en République Central africaine. *Cahiers ORSTOM. Série entomologie médicale et parasitologie, 14*, 201–207.

Pajot, F.X., & Adam, J.P. (1964). Notes morphologiques sur *Anopheles caroni* Adam, 1961 (Diptera-Culicidae). *Bulletin de la Société de pathologie exotique, 57*(3), 626–637.

Pajot, F.X., & Fauran, P. (1975). *Wyeomyia* (*Wyeomyia*) *pseudorobusta* n. sp. (Diptera. Culicidae), nouveau moustique decouvert en Guyane française. *Cahiers ORSTOM. Série entomologie médicale et parasitologie, 13*(3), 125–129.

Pajot, F.X., & Geoffroy, B. (1971). *Aedes* (*Neomelaniconion*) *bergerardi* sp. n. une nouvelle espèce de Culicidae de la République Central africaine. *Cahiers ORSTOM. Série entomologie médicale et parasitologie, 9*(3), 269–272.

Pal, R., & Ramalingam, S. (1981). *Invertebrate predators of mosquitoes*. World Health Organization WHO/VBC/81.799.

Pallas, P.S. (1771). *Reise durch verschiedene Provinzen des Russischen Reichs*. Vol. 1. St. Petersburg, Russia: Kayserliche Academie der Wissenschaften.

Palmer, A.R. (1957). *Miocene arthropods from the Mojave Desert, California*. Shorter Contribution to General Geology 1956, Geological Survey Professional Paper 294-G. Retrieved from http://pubs.er.usgs.gov/djvo/pp/pp-294.djvu

Palmer, A.R., & Bassett, A.M. (1954). Nonmarine Miocene arthropods from California. *Science, 120*(3110), 228–229.

Paluch, G., Bartholomay, L., & Coats, J. (2010). Mosquito repellents: A review of chemical structure diversity and olfaction. *Pest Management Science, 66*(9), 925–935.

Paluzzi, J.P., Vanderveken, M., & O'Donnell, M.J. (2014). The heterodimeric glycoprotein hormone, GPA/GPB5 regulates ion transport across the hindgut of the adult mosquito, *Aedes aegypti*. *PLOS ONE, 9*(1), e86386.

Panday, R.S. (1975a). Mosquito identification studies in a savanna forest in Surinam. *Mosquito News, 35*(2), 141–146.

Panday, R.S. (1975b). Mosquito identification studies in a typical coastal area in northern Surinam. *Mosquito News, 35*(3), 297–301.

Pandazis, G. (1935). La faune des Culicides de Grèce. *Acta Instituti et Musei Zoologici Universitatis Atheniensis, 1*(1–2), 1–27.

Panella, N.A., Kent Crockett, R.J., Biggerstaff, B.J., & Komar, N. (2011). The Centers for Disease Control and Prevention resting trap: A novel device for collecting resting mosquitoes. *Journal of the American Mosquito Control Association, 27*(3), 323–325.

Pankhurst, R.J. (1978). *Biological identification: The principles and practice of identification methods in biology*. London, England: Arnold.

Pantoja-Sánchez, H., Gomez, S., Velez, V., Avila, F.W., & Alfonso-Parra, C. (2019). Precopulatory acoustic interactions of the New World malaria vector *Anopheles albimanus* (Diptera: Culicidae). *Parasites and Vectors, 12*, 386.

Pao, B., & Knight, K.L. (1970a). The fourth instar larval mandible and maxilla of selected *Aedes* (*Aedimorphus*) species (Diptera, Culicidae). *Mosquito Systematics, 2*(3), 98–131.

Pao, B., & Knight, K.L. (1970b). Morphology of the fourth stage larval mouthparts of *Aedes* (*Aedimorphus*) *vexans* (Diptera: Culicidae). *Journal of the Georgia Entomological Society, 5*, 115–137.

Pape, T., & Thompson, F.C. (2013, June 13). *Systema dipterorum; The biosystematic database of world Diptera*. Retrieved from http://www.diptera.org/

Paredes-Esquival, C., Donnelly, M.J., Harbach, R.E., & Townson, H. (2009). A molecular phylogeny of mosquitoes in the *Anopheles barbirostris* subgroup reveals cryptic species: Implications for identification of disease vectors. *Molecular Phylogenetics and Evolution, 50*(1), 141–151.

Parent, M., & Vincke, I. (1945). Notes entomologiques (Diptera: Anophelinae). *Recueil des travaux des sciences et médicas Congo Belge, 4*, 23–24.

Parham, P.E., & Michael, E. (2010). Modelling climate change and malaria transmission. In E. Michael & R.C. Spear (Eds.), *Modelling parasite transmission and control* (pp. 184–199). New York, NY: Springer.

Park, L.E., & Downing, K.F. (2001). Paleoecology of an exceptionally preserved arthropod fauna from lake deposits of the Miocene Barstow Formation, southern California, U.S.A. *Palaios, 16*(2), 175–184.

Park, S.S., & Shahabuddin, M. (2000). Structural organization of posterior midgut muscles in mosquitoes, *Aedes aegypti* and *Anopheles gambiae*. *Journal of Structural Biology, 129*(1), 30–37.

Parker, B.M. (1979). Development of the mosquito *Aedes dorsalis* (Diptera: Culicidae) in relation to temperature and salinity. *Annals of the Entomological Society of America, 72*(1), 105–108.

Parker, D.J. (1977). The biology of the tree holes of Point-Pelee National Park Ontario. Part 2. *Toxorhynchites rutilus-septentrionalis* new record in Canada (Diptera, Culicidae). *The Canadian Entomologist, 109*(1), 93–94.

Parker, D.M., Zavortink, T.J., Billo, T.J., Valdez, U., & Edwards, J.S. (2012). Mosquitoes and other arthropod macro fauna associated with tank bromeliads in a peruvian cloud forest. *Journal of the American Mosquito Control Association, 28*(1), 45–46.

Parker, J.E., Angarita-Jaimes, N., Abe, M., Towers, C.E., Towers, D., & McCall, P.J. (2015). Infrared video tracking of *Anopheles gambiae* at insecticide-treated bed nets reveals rapid decisive impact after brief localised net contact. *Scientific Reports, 5*, 13392.

Parmakelis, A., Russello, M.A., Caccone, A., Marcondes, C.B., Costa, J., Forattini, O.P., . . . Powell, J.R. (2008). Historical analysis of a near disaster: *Anopheles gambiae* in Brazil. *The American Journal of Tropical Medicine and Hygiene, 78*(1), 176–178.

Parmesan, C. (2006). Ecological and evolutionary responses to recent climate change. *Annual Review of Ecology, Evolution and Systematics, 37*, 637–669.

Pasahan, S.C. (1974). The salivary chromosomes of *Anopheles barbirostris* var. *ahomi*. *Cytologia* (Japan), 39(1), 41–47.

Pascini, T.V., Ramalho-Ortigão, M., & Martins, G.F. (2012). Morphological and morphometrical assessment of spermathecae of *Aedes aegypti* females. *Memórias do Instituto Oswaldo Cruz, 107*(6), 705–712.

Pascini, T.V., Ramalho-Ortigão, J.M., & Martins, G.F. (2013). The fine structure of the spermatheca in *Anopheles aquasalis* (Diptera: Culicidae). *Annals of the Entomological Society of America, 106*(6), 857–867.

Pascoa, V., Oliveira, P.L., Dansa-Petretski, M., Silva, J.R., Alvarenga, P.H., Jacobs-Lorena, M., & Lemos, F.J. (2002). *Aedes aegypti* peritrophic matrix and its interaction with heme during blood digestion. *Insect Biochemistry and Molecular Biology, 32*(5), 517–523.

Pashley, D.P., Rai, K.S., & Pashley, D.N. (1985). Patterns of allozyme relationships compared with morphology, hybridization, and geologic history in allopatric island-dwelling mosquitoes. *Evolution, 39*(5), 985–997.

Paterson, G.C., & Shannon, R.C. (1927). Mosquitos de Embarcación (Salta) con notas sobre la zona biológica del Chaco (Chaco Life Zone). *Boletín del Instituto de Clínica Quirúrgica, 21–25*, 1–10.

Paterson, H.E., Green, C.A., & Mahon, R.J. (1976). *Aedes*. In R.C. King (Ed.), *Handbook of Genetics* (pp. 331–345). New York, NY: Springer.

Pates, H., & Curtis, C. (2005). Mosquito behavior and vector control. *Annual Review of Entomology, 50*, 53–70.

Patil, C.D., Borase, H.P., Salunke, B.K., & Patil, S.V. (2013). Alteration in *Bacillus thuringiensis* toxicity by curing gut flora: Novel approach for mosquito resistance management. *Parasitology Research, 112*(9), 3283–3288.

Patrick, M.L., Aimanova, K., Sanders, H.R., & Gill, S.S. (2006). P-type Na+/K+-ATPase and V-type H+-ATPase expression patterns in the osmoregulatory organs of larval and adult mosquito *Aedes aegypti*. *Journal of Experimental Biology, 209*(23), 4638–4651.

Patriota Nascimento, L., & Lourenço-de-Oliveira, A.C. (1984). Complexo *Anopheles* (*Nyssorhynchus*) *albitarsis*: Impossibilidade de separa—lo em duas subespécies *A. albitarsis albitarsis* e *A. albitarsis domesticus* (Diptera Culicidae). *Revista Brasileira de biologia, 44*(4), 461–465.

Patton, W.S. (1905). The culicid fauna of the Aden Hinterland, their haunts and habits. *Journal of the Bombay Natural History Society, 16*(4), 623–637.

Patton, W.S. (1933). The identity of *Culex aegypti* L. *Annals of Tropical Medicine and Parasitology, 27*(2), 182–184.

Paulian, R. (1950). Les moustiques des *Nepenthes* à Madagascar. *Naturaliste malgache, 2*, 119–121.

Paupy, C., Makanga, B., Ollomo, B., Rahola, N., Durand, P., Magnus, J., . . . Prugnolle, F. (2013). *Anopheles moucheti* and *Anopheles vinckei* are candidate vectors of ape *Plasmodium* parasites, including *Plasmodium praefalciparum* in Gabon. *PLOS ONE, 8*(2), e57294.

Pawan, J.L. (1922). The oviposition of *Joblotia digitatus* Rondani (Diptera, Culicidae). *Insecutor Inscitiae Menstruus, 10*, 63–65.

Pazos, J.H. (1908). Descripción de nuevas especies de mosquitos de Cuba. *Academia de Ciencias de la Habana, 45*, 417–432.

Pearse, A.S. (1936). Parasites from Yukatan. In A.S. Pearse, E.P. Creaser, & F.G. Hall (Eds.), *The cenotes of Yucatan: A zoological and hydrographic survey* (Publication 457, pp. 45–59). Washington, DC: Carnegie Institution for Science.

Pearson, R.G. (2010). Species' distribution modeling for conservation educators and practitioners. *Lessons in Conservation, 3*, 54–89.

Pecor, J.E., Mallampalli, V.L., Harbach, R.E., & Peyton, EL. (1992). Catalog and illustrated review of the subgenus *Melanoconion* of *Culex* (Diptera: Culicidae). *Contributions of the American Entomological Institute, 27*(2), 1–228.

Pecor, J.E., Jones, J., Turell, M.J., Fernandez, R., Carbajal, F., O'Guinn, M., . . . Klein, T.A. (2000). Annotated checklist of the mosquito species encountered during arboviral studies in Iquitos, Peru (Diptera: Culicidae). *Journal of the American Mosquito Control Association, 16*(3), 210–218.

Pecor, J.E., Harbach, R.E., Peyton, EL, Roberts, D.R., Rejmánková, E., Manguin, S., & Palanko, J. (2002). Mosquito studies in Belize, Central America: Records, taxonomic notes and a checklist of species. *Journal of the American Mosquito Control Association, 18*(4), 241–276.

Pelaez, D. (1945). Anofelinos de México. Clave para la determinación de especies y subespecies basada en los caracteres de las hembras adultas. *Ciencia, 6*, 69–77.

Pelletier, J., Guidolin, A., Syed, Z., Cornel, A.J., & Leal, W.S. (2010). Knockdown of a mosquito odorant-binding protein involved in the sensitive detection of oviposition attractants. *Journal of Chemical Ecology, 36*(3), 245–248.

Pena, C.J., & Chadee, D.D. (2004). *Mansonia dyari*, *Mansonia flaveola* and *Aedeomyia squamipennis* in the Dominican Republic, three new country records. *Journal of the American Mosquito Control Association, 20*(4), 449–450.

Pena, C.J., Gonzalvez, G., & Chadee, D. (2003). Seasonal prevalence and container preferences of *Aedes albopictus* in Santo Domingo City, Dominican Republic. *Journal of Vector Ecology, 28*(2), 208–212.

Peng, Z., & Simons, F.E.R. (2004). Mosquito allergy: Immune mechanisms and recombinant salivary allergens. *International Archives of Allergy and Immunology, 133*(2), 198–209.

Peng, Z., Beckett, A.N., Engler, R.J., Hoffman, D.R., Ott, N.L., & Simons, F.E.R. (2004a). Immune responses to mosquito saliva in 14 individuals with acute systematic allergic reactions to mosquito bites. *Journal of Allergy and Clinical Immunology, 114*(5), 1189–1194.

Peng, Z., Ho, M.K., Li, C., & Simons, F.E.R. (2004b). Evidence for natural desensitization to mosquito salivary allergens: Mosquito saliva specific IgE and IgG levels in children. *Annals of Allergy, Asthma and Immunology, 93*(6), 553–556.

Penn, G.H. (1947). The larva of *Aedes* (*Leptosomatomyia*) *aurimargo* (Diptera, Culicidae). *Proceedings of the Entomological Society of Washington, 49*(4), 103–105.

Penn, G.H. (1949a). Pupae of the Nearctic anopheline mosquitoes north of Mexico. *Journal of the National Malaria Society, 8*(1), 50–69.

Penn, G.H. (1949b). The pupae of the mosquitoes of New Guinea. *Pacific Science, 3*(1), 3–85.

Penn, G.H. (1949c). The larva and pupa of *Aedes* (*Aedimorphus*) *alboscutellatus* (Diptera, Culicidae). *Natural History Miscellanea, 40*, 1–4.

Pennetier, C., Warren, B., Dabiré, K.R., Russell, I.J., & Gibson, G. (2010). "Singing on the wing" as a mechanism for species recognition in the malarial mosquito *Anopheles gambiae*. *Current Biology, 20*(2), 131–136.

Pennington, J.E., Goldstrohm, D.A., & Wells, M.A. (2003). The role of hemolymph proline as a nitrogen sink during blood meal digestion by the mosquito *Aedes aegypti*. *Journal of Insect Physiology, 49*(2), 115–121.

Perez Vigueras, I. (1948). Notas sobre la *Psorophora johnstonii, Mansonia titillans, Anopheles atropos* y sobre la presencia en Cuba de la *Psorophora ciliata. Universidad de la Habana 73–75,* 293–302.
Perez Vigueras, I. (1956). *Los Ixodidos y culicidos de Cuba. Su historia natural y medica.* La Habana.
Pérez, A.M., & Conn, J. (1992). A polytene chromosome study of four populations of *Anopheles aquasalis* from Venezuela. *Genome, 35*(2), 327–331.
Perry, W.J. (1946). Keys to the larval and adult mosquitoes of Espiritu Santo (New Hebrides) with notes on their bionomics. *The Pan-Pacific Entomologist, 22*(1–4), 9–18.
Perry, W.J. (1949). Studies on *Mansonia xanthogaster* and its relation to filariasis in the South Pacific. *Journal of Parasitology, 35*(4), 379–382.
Perry, W.J. (1950). Biological and crossbreeding studies on *Aedes hebrideus* and *Aedes pernotatus* (Diptera, Culicidae). *Annals of the Entomological Society of America, 43*(1), 123–136.
Perty, M. (1841). *Allgemeine Naturgeschichte, als philosophische und humanitätswissenschaft für Naturforscher, Philosophen und das höher gebildete Publikum* (Vol. 3 C). Bern, Switzerland: Fischer, pp. 467–1119.
Peryassú, A.G. (1908). *Os culicideos do Brazil.* Rio de Janeiro, Brazil: Instituto de Manguinhos.
Peryassú, A.G. (1921a). Os anophelineos do Brasil. *Arquivos do Museu Nacional* (Rio de Janeiro), *23*, 5–101.
Peryassú, A.G. (1921b). Um novo anophelineo brasileiro. *Folha médica, 2*(18), 141.
Peryassú, A.G. (1922a). Uma nova especie de culicineo brasileiro. *Folha médica, 3*(15), 117–118.
Peryassú, A.G. (1922b). Duas novas especies de mosquitos do Brasil. *Folha médica, 3*, 179.
Peryassú, A.G. (1923a). Uma nova especie de mosquito do Brasil. *Folha médica, 4*, 2.
Peryassú, A.G. (1923b). Uma nova especie de anophelineo do genero *Cyclolepidopteron. A folha médica Rio de Janeiro, 4*, 68–69.
Peryassú, A.G. (1925). *Anopheles alagoanni* (n. sp.) Peryassú. *Folha médica, 6*, 258–259.
Pessoa, S.B., & Galvão, A.A. (1936). Notas sobre algumas espécies de mosquitos que se criam em bromelias (Diptera, Culicidae). *Revista de biologia e higiene, 6*(2), 79–90.
Peters, W. (1955a). The mosquitoes of Liberia (Diptera: Culicidae). *Proceedings of the Royal Entomological Society of London. Series B: Taxonomy, 24*(5–6), 81–90.
Peters, W. (1955b). New Anophelini from the southern highlands of Tanganyika and notes on some other members of the genus *Anopheles* Meigen (Diptera: Culicidae). *Proceedings of the Royal Entomological Society of London. Series B: Taxonomy, 24*(5–6), 95–103.
Peters, W. (1955c). The mosquitoes of Liberia (Diptera: Culicidae). *Proceedings of the Royal Entomological Society of London. Series B: Taxonomy, 24*(7–8), 117–128.
Peters, W. (1956). The mosquitos of Liberia (Diptera: Culicidae), a general survey. *Bulletin of Entomological Research, 47*, 525–551.
Peters, W. (1958). Mosquitoes of New Guinea (Diptera: Culicidae). I. *Proceedings of the Royal Entomological Society of London. Series B: Taxonomy, 27*(5–6), 89–92.
Peters, W. (1959). Mosquitoes of New Guinea (Diptera: Culicidae). II. Notes on *Tripteroides* Giles, 1904. *Proceedings of the Royal Entomological Society of London. Series B: Taxonomy, 28*(9–10), 135–154.
Peters, W. (1962). Mosquitoes of New Guinea (Diptera: Culicidae). III. Some undescribed pupae of Anophelini and a revised pupal key. *Proceedings of the Royal Entomological Society of London. Series B: Taxonomy, 31*, 163–169.
Peters, W. (1963a). Mosquitoes of New Guinea (Diptera: Culicidae). IV. Two new species, genera *Aedes* Meigen and *Armigeres* Theobald. *Proceedings of the Royal Entomological Society of London. Series B: Taxonomy, 32*(1–2), 1–11.
Peters, W. (1963b). Mosquitoes of New Guinea (Diptera: Culicidae). V. New species and previously undescribed pupae of the genus *Tripteroides* Giles. *Proceedings of the Royal Entomological Society B, 32*(5–6), 63–71.
Peters, W. (1963c). Mosquitoes of New Guinea (Diptera: Culicidae). VI. The genus *Tripteroides* Giles: A new species of the subgenus *Polylepidomyia* Theobald, notes on the taxonomy and distribution and a revised key to described pupae. *Proceedings of the Royal Entomological Society of London. Series B: Taxonomy, 32*(7–8), 89–100.
Peters, W. (1963d). Mosquitoes of New Guinea (Diptera: Culicidae). VII. A new species complex of *Uranotaenia* Lynch Arribálzaga, 1891. *Proceedings of the Royal Entomological Society of London. Series B: Taxonomy, 32*(9–10), 135–144.
Peters, W. (1963e). Mosquitoes of New Guinea (Diptera: Culicidae). VIII. New species of the genus *Uranotaenia* Lynch Arribálzaga, 1891. *Proceedings of the Royal Entomological Society of London. Series B: Taxonomy, 32*(11–12), 203–209.
Peters, W. (1964). Mosquitoes of New Guinea (Diptera: Culicidae). IX. Further new species of *Uranotaenia* Lynch Arribálzaga and notes on the genus. *Proceedings of the Royal Entomological Society of London. Series B: Taxonomy, 33*(1–2), 19–29.
Peters, W., & Dewar, S.C. (1956). A preliminary record of the megarhine and culicine mosquitoes of Nepal with notes on their taxonomy (Diptera: Culicidae). *Indian Journal of Malariology, 10*, 37–51.
Petersen, J.L., & Linley, J.R. (1995). Description of the egg of *Aedeomyia squamipennis* (Diptera: Culicidae). *Journal of Medical Entomology, 32*(6), 888–894.
Petersen, R.L., Faust, A., Nagawa, J., Thomas, C., Vilmenay, A. (2000). Foreign mosquito survivorship in the pitcher plant *Sarracenia purpurea*—the role of the pitcher-plant midge *Metrionemus knabi. Hydrobiologia, 439*(1–3), 13–19.
Peterson, A. (1916). The head-capsule and mouth-parts of Diptera. *Illinois Biological Monographs, 3*(2), 177–282.
Peterson, A.T. (2009). Shifting suitability for malaria vectors across Africa with warming climates. *BMC Infectious Diseases, 9*(1), 59.
Peterson, A.T., & Campbell, L.P. (2015). Global potential distribution of the mosquito *Aedes notoscriptus*, a new alien species in the United States. *Journal of Vector Ecology, 40*(1), 191–194.
Peterson, A.T., & Lieberman, B.S. (2012). Species' geographic distributions through time: Playing catch-up with changing climates. *Evolution: Education and Outreach, 5*(4), 569–581.
Peterson, A.T., Martínez-Campos, C., Nakazawa, Y., & Martínez-Meyer, E. (2005). Time-specific ecological niche modeling predicts spatial dynamics of vector insects and human dengue cases. *Transactions of the Royal Society of Tropical Medicine and Hygiene, 99*(9), 647–655.
Peterson, A.T., Soberón, J., Pearson, R.G., Anderson, R.P., Martínez-Meyer, E., Nakamura, M., & Araújo, M.B. (2011). Ecological niches and geographic distributions. In *Monographs in population biology* (Vol. 49). Princeton, NJ: Princeton University Press.
Peterson, B.V. (1971). Lectotype designation for *Aedes simici* (Diptera: Culicidae). *The Canadian Entomologist, 103*, 1199–1200.
Petrocchi, J. (1923). Estado actual de la sistemática de los Culicinae (mosquitos) en la República Argentina. Enumeración de especies. *Revista del Instituto Bacteriológico del Departamento de Higiene, 3*, 83–93.

Petrocchi [as Petrochi], J. (1925). Descripción de un nuevo *Anopheles*. *Revista del Instituto Bacteriológico del Departamento de Higiene, 4*, 69–75.

Petzel, D.H. (2000). Na^{+}/H^{+} exchange in mosquito Mapighian tubules. *American Journal of Physiology-Regulatory, Integrative and Comparative Physiology, 279*(6), R1996–R2003.

Petzel, D.H., Berg, M.M., & Beyenbach, K.W. (1987). Hormone-controlled cAMP-mediated fluid secretion in yellow-fever mosquito. *American Journal of Physiology-Regulatory, Integrative and Comparative Physiology, 253*(5), R701–R711.

Peus, F. (1930). Bemerkungen über *Theobaldia subochrea* Edw. (Dipt. Culicid.). *Mitteilungen der Deutschen Gesellschaft für Allgemeine und Angewandte Entomologie, 1*, 52–60.

Peus, F. (1933). Zur Kenntnis der *Aedes*—arten des deutschen Faunengebietes. (Dipt., Culicid.). Die weibchen der *Aedes communis*-gruppe. *Konowia, 12*, 145–159.

Peus, F. (1935). *Theobaldia* (Subg. *Culicella*) *ochroptera* sp. n., eine bisher unbekannte Stechmücke. *Markische Tierwelt: Zeitchrift für die faunistische Erforschung der Kurmark*) (Berlin), *1*, 113–121.

Peus, F. (1937). *Aedes cyprius* Ludlow (= *A. freyi* Edwards) und seine Larve (Dipt. Culicidae). *Archiv für Hydrobiologie, 31*, 242–252.

Peus, F. (1954). Über Stechmücke in Griechenland (Diptera, Culicidae). *Bonner zoologische Beiträge, 1*, 73–86.

Peus, F. (1970). Bemerkenswerte Mücken am Tegeler Fliess. *Berliner Naturschutzblätter, Special Issue*(May), 18–26.

Peus, F. (1972). Über das Subgenus *Aedes* sensu stricto in Deutschland (Diptera, Culicidae). *Zeitschrift für angewandte Entomologie, 72*(2), 177–194.

Peyton, EL. (1972). A subgeneric classification of the genus *Uranotaenia* Lynch Arribálzaga, with a historical review and notes on other categories. *Mosquito Systematics, 4*(2), 16–40.

Peyton, EL. (1974). *Uranotaenia srilankensis*. A new species of the subgenus *Pseudoficalbia* from Sri Lanka (Diptera: Culicidae). *Mosquito Systematics, 6*(3), 222–227.

Peyton, EL. (1977). Medical entomology studies. X. A revision of the subgenus *Pseudoficalbia* of the genus *Uranotaenia* in Southeast Asia (Diptera: Culicidae). *Contributions of the American Entomological Institute, 14*(3), 1–273.

Peyton, EL. (1990). A new classification for the Leucosphyrus group of *Anopheles* (*Cellia*). *Mosquito Systematics, 21*(for 1989)(3), 197–205.

Peyton, EL. (1993). *Anopheles* (*Nyssorhynchus*) *dunhami*, resurrected from synonymy with *Anopheles nuneztovari* and validated as a senior synonym of *Anopheles trinkae* (Diptera: Culicidae). *Mosquito Systematics, 25*, 151–156.

Peyton, EL, & Harbach, R.E. (1991). A new species of *Culex* (*Melanoconion*) encountered during arbovirus surveillance in Mexico (Diptera: Culcidae). *Mosquito Systematics, 23*(3), 199–204.

Peyton, EL, & Harrison, B.A. (1979). *Anopheles* (*Cellia*) *dirus*, a new species of the Leucosphyrus Group from Thailand (Diptera: Culicidae). *Mosquito Systematics, 11*(1), 40–52.

Peyton, EL, & Harrison, B.A. (1980). *Anopheles takasagoensis* an additional species in the *Anopheles balabacensis* Complex of Southeast Asia (Diptera, Culicidae). *Mosquito Systematics, 12*(3), 335–347.

Peyton, EL, & Hochman, R.H. (1968). A revised interpretation of the proctiger of male *Uranotaenia* with a related note on *Hodgesia* (Diptera: Culicidae). *Proceedings of the Entomological Society of Washington, 70*(4), 376–382.

Peyton, EL, & Klein, J.M. (1970). Five new species of *Uranotaenia* from Southeast Asia (Diptera: Culicidae). *Proceedings of the Entomological Society of Washington, 72*(2), 243–251.

Peyton, EL, & Ramalingam, S. (1988). *Anopheles* (*Cellia*) *nemophilous*, a new species of the Leucosphyrus Group from Peninsular Malaysia and Thailand (Diptera: Culicidae). *Mosquito Systematics, 20*(2), 272–299.

Peyton, EL, & Rattanarithikul, R. (1970). Five additional new species of *Uranotaenia* from Southeast Asia (Diptera: Culicidae). *Proceedings of the Entomological Society of Washington, 72*(3), 403–413.

Peyton, EL, & Scanlon, J.E. (1966). Illustrated key to the female *Anopheles* mosquitoes of Thailand. US Army Medical Component, SEATO, Bangkok, Thailand.

Peyton, EL, Jayasekera, N., & Chelliah, R.V. (1979). The biology and immature stages of *Uranotaenia* (*Pseudoficalbia*) *srilankensis* Peyton (Diptera: Culicidae). *Mosquito Systematics, 11*(3), 215–220.

Peyton, EL, Roberts, D.R., Pinheiro, F.P., Vargas, R., & Balderama, F. (1983). Mosquito collections from a remote unstudied area of southeastern Bolivia. *Mosquito Systematics, 15*(2), 61–89.

Peyton, EL, Wilkerson, R.C., & Harbach, R.E. (1992). Comparative analysis of the subgenera *Kerteszia* and *Nyssorhynchus* of *Anopheles* (Diptera: Culicidae). *Mosquito Systematics, 24*(1), 51–69.

Peyton, EL, Campbell, S.R., Candeletti, T.M., Romanowski, R.M., & Crans, W.J. (1999). *Aedes* (*Finlaya*) *japonicus japonicus* (Theobald), a new introduction into the United States. *Journal of the American Mosquito Control Association, 15*(2), 238–241.

Phan, V.T., Manh, N.D., Hinh, T.D., & Vien, N.T. (1991). *Anopheles* (*Anopheles*) *cucphuongensis*: A new species from Vietnam (Diptera: Culicidae). *Mosquito Systematics, 22*(3), 145–148.

Phanitchakun, T., Wilai, P., Saingamsook, J., Namgay, R., Drukpa, T., Tsuda, Y., . . . Somboon, P. (2017). *Culex* (*Culiciomyia*) *sasai* (Diptera: Culicidae), senior synonym of *Cx. spiculothorax* and a new country record for Bhutan. *Acta Tropica, 171*, 194–198.

Phanitchakun, T., Namgay, R., Miyagi, I., Tsuda, Y., Walton, C., Harbach, R.E., & P. Somboon. (2019). Morphological and molecular evidence for a new species of *Lutzia* (Diptera: Culicidae: Culicini) from Thailand. *Acta Tropica, 191*, 77–86.

Philip, C.B. (1931). Two new species of *Uranotaenia* (Culicidae) from Nigeria, with notes on the genus in the Ethiopian region. *Bulletin of Entomological Research, 22*(2), 183–193.

Philippi, R.A. (1865). Aufzählung der chilenischen Dipteren. *Verhandlungen der kaiserlich-königlichen zoologisch-botanischen Gesellschaft in Wien, 15*, 595–782.

Philips, K.T., Hancock, R.G., & Foster, W.A. (1996). Epigramic display and unique mating position in *Wyeomyia arthrostigma* (Diptera: Culicidae). *Journal of Insect Behavior, 9*(5), 739–753.

Pickavance, J.R., Bennett, G.F., & Phipps., J. (1970). Some mosquitoes and blackflies from Newfoundland. *Canadian Journal of Research, 48*(4), 621–624.

Pidiyar, V.J., Jangid, K., Patole, M.S., & Shouche, Y.S. (2004). Studies on cultured and uncultured microbiota of wild *Culex quinquefasciatus* mosquito midgut based on 16S ribosomal RNA gene analysis. *The American Journal of Tropical Medicine and Hygiene, 70*(6), 597–603.

Pielou, E.C. (1991). *After the Ice Age: The return of life to glaciated North America*. Chicago, IL: University of Chicago Press.

Piermarini, P.M., Hine, R.M., Schepel, M., Miyauchi, J., & Beyenbach, K.W. (2011). Role of an apical K, CI contransporter in urine formation by renal tubules of the yellow fever mosquito (*Aedes aegypti*). *American Journal of Physiology—Regulatory, Integrative and Comparative Physiology, 301*(5), R1318–R1337.

Pietrantonio, P.V., Jagge, C., Keeley, L.L., & Ross, L.S. (2000). Cloning of an aquaporin-like cDNA and *in situ* hybridization in adults of the mosquito *Aedes aegypti* (Diptera: Culicidae). *Insect Molecular Biology, 9*(4), 407–418.

Pietrantonio, P.V., Jagge, C., & McDowell, C. (2001). Cloning and expression analysis of a 5HT7-like serotonin receptor cDNA from

mosquito *Aedes aegypti* female excretory and respiratory systems. *Insect Molecular Biology, 10*(4), 357–369.
Pietrantonio, P.V., Jagge, C., Taneja-Bageshwar, S., Nachman, R.J., & Barhoumi, R. (2005). The mosquito *Aedes aegpyti* (L.) leucokinin receptor is a multiligand receptor for the three *Aedes* kinins. *Insect Molecular Biology, 14*(1), 55–67.
Pillai, J.S. (1962). A celloidin impression technique for recording egg sculpturing in *Aedes* mosquitoes. *Nature, 194*(4824), 212–213.
Pillai, J.S. (1966). *Culiseta noveezealandiae*, a new species of the subgenus *Climacura* Felt (Diptera: Culicidae: Culisetini), with notes on its ecology and development. *Transactions of the Royal Society of New Zealand: Zoology, 8*(11), 125–133.
Pillai, J.S. (1968). Notes on mosquitoes of New Zealand. II. The male terminalia of *Culiseta (Climacura) tonnoiri* and its ecology (Diptera: Culicidae: Culisetini). *Journal of Medical Entomology, 5*(3), 355–357.
Pilotte, N., Zaky, W.I., Abrams, B.P., Chadee, D.D., & Williams, S.A. (2016). A novel xenomonitoring technique using mosquito excreta/feces for the detection of filarial parasites and malaria. *PLOS Neglected Tropical Diseases, 10*(4), e004641.
Pinhão, R. (1974). Sobre a situação zoogeográfica de Timor. *Anais do Instituto de Hygiene e Medicina Tropical, 1*(1–4), 467–470.
Pinto, C. (1930). Mosquitos da região neotropica (Brasil, Estados de S. Paulo e Rio de Janeiro). II. *Psorophora genumaculata* e *P. ciliata* (Diptera: Culicidae). *Memórias do Instituto Oswaldo Cruz, 23*(4), 179–184.
Pinto, C. (1932). Alguns mosquitos do Brasil e do oriente da Bolivia (Diptera, Culicidae). *Revista médico-cirurgica do Brasil, 40*(10), 285–309.
Pinto, C. (1939). Disseminacão da malaria pela aviacão; biologia do *Anopheles gambiae* e outros anofelineos do Brasil. *Memórias do Instituto Oswaldo Cruz, 34*(3), 293–430.
Pinto, C.S., Confalonieri, U.E.C., & Mascarenhas, B.M. (2009). Ecology of *Haemagogus* sp. and *Sabethes* sp. (Diptera: Culicidae) in relation to the microclimates of the Caxiuanã National Forest, Pará, Brazil. *Memórias do Instituto Oswaldo Cruz, 104*(4), 592–598.
Piton, L. (1936). Addition à la faune entomologique des cinerites du Lac Chambon (Puy-de-Dome). *Revue scientifique du Bourbonnais, 1936*, 16–24.
Pittaluga, G. (1905). Estudios acerca de los Diptéros y de los parásitos que transmiten al hombre y á los animales domésticos. *Revista de la Real Academia de Ciencias Exactas, Físicas y Naturales de Madrid, 3*(5), 402–586.
Pittendrigh, C.S. (1950a). The ecoclimatic divergence of *Anopheles bellator* and *A. homunculus. Evolution, 4*(1), 43–63.
Pittendrigh, C.S. (1950b). The ecotopic specialization of *Anopheles homunculus*; and its relation to competition with *A. bellator. Evolution, 4*(1), 64–78.
Pitts, R.J., Liu, C., Zhou, X., Malpartida, J.C., & Zwiebel, L.J. (2014). Odorant receptor-mediated sperm activation in disease vector mosquitoes. *Proceedings of the National Academy of Sciences of the United States of America, 111*(7), 2566–2571.
Pitts, R.J., Derryberry, S.L., Zhang, Z., & Zwiebel, L.J. (2017). Variant ionotropic receptors in the malaria vector mosquito *Anopheles gambiae* tuned to amines and carboxylic acids. *Nature Scientific Reports, 7*, 40297.
Platt, K.B., Linthicum, K.J., Myint, K.S.A., & Innis, B.L. (1997). Impact of dengue virus infection on feeding behavior of *Aedes aegypti. The American Journal of Tropical Medicine and Hygiene, 57*(2), 119–125.
Podénas, S. (1999). New species of fossil mosquitoes (Diptera, Culicidae) from the Baltic Amber. *Acta Zoologica Lituanica Biodiversity, 9*(3), 111–114.
Pohl, J.E., & Kollar, V. (1832). Brasiliens vorzüglich lästige Insecten. Besonderer Abdruch aus der Reise im Innern von Brasilien von Dr. Pohl [extracted from Pohl's *Reise im Innern von Brasilien*]. Wien: Germany.
Poinar, G., Jr. (2005a). *Culex malariager* n. sp. (Diptera: Culicidae) from Dominican amber: The first fossil mosquito vector of *Plasmodium. Proceedings of the Entomological Society of Washington, 107*(3), 548–553.
Poinar, G., Jr. (2005b). *Plasmodium dominicana* n. sp. (Plasmodiidae: Haemospororida) from Tertiary Dominican amber. *Systematic Parasitology, 61*(1), 47–52.
Poinar, G.O., Jr., Zavortink, T.J., Pike, T., & Johnston, P.A. (2000). *Paleoculicis minutus* (Diptera: Culicidae) n. gen., n. sp., from Cretaceous Canadian amber, with a summary of described fossil mosquitoes. *Acta Geologica Hispanica, 35*(1–2), 119–128.
Poinar, G., Zavortink, T.J., & Brown, A. (2019). *Priscoculex burmanicus* n. gen. et sp. (Diptera: Culicidae: Anophelinae) from mid-Cretaceous Myanmar amber. *Historical Biology*, 1–6.
Poiret, J.-L.M. (1787). Mémoire sur quelque insectes de Barbarie. *Journal de physique, de chimie, d'histoire naturelle et des arts* (Paris), *30*(1), 241–245.
Pol, M., Kilama, S., Duperier, S., Soupé-Gilbert, M., Calvez, E., & Pocquet, N. (2018). Introduction of the *Anopheles bancroftii* mosquito, a malaria vector, into New Caledonia. *Emerging Infectious Diseases, 24*(3), 604–605.
Polerstock, A.R., Eigenbrode, S.D., & Klowden, M.J. (2002). Mating alters the cuticular hydrocarbons of female *Anopheles gambiae* sensu stricto and *Aedes aegypti* (Diptera: Culicidae). *Journal of Medical Entomology, 39*(3), 545–552.
Polyakova, P.E. (1970a). *Aedes fitchii* Felt and Young (Diptera, Culicidae). *Parassitologia, 4*, 389–391.
Polyakova, P.E. (1970b). Records of the bloodsucking mosquito fauna of Northern Siberia (Diptera, Culicinae). In A.I. Cherepanov (Ed.), *Fauna Sibiri* (pp. 132–137). Akademia Nauk SSR.
Polyakova, P.E., & Kukharchuk, L.P. (1970). New species in the mosquito fauna of the USSR—*Aedes rempeli* Vock. (Diptera Culicinae) in western Siberia. In A.I. Cherepanov (Ed.), *Fauna Sibiri* (pp. 130–131). Akademia Nauk SSR.
Ponlawat, A., & Harrington, L.C. (2007). Age and body size influence male sperm capacity of the dengue vector *Aedes aegypti* (Diptera: Culicidae). *Journal of Medical Entomology, 44*(3), 422–426.
Ponlawat, A., & Harrington, L.C. (2009). Factors associated with male mating success of the dengue vector mosquito *Aedes aegypti. The American Journal of Tropical Medicine and Hygiene, 80*(3), 395–400.
Ponnusamy, L., Xu, N., Nojima, S., Wesson, D.M., Schal, C., & Apperson, C.S. (2008). Identification of bacteria and bacteria-associated chemical cues that mediate oviposition site preferences by *Aedes aegypti. Proceedings of the National Academy of Sciences of the United States of America, 105*(27), 9262–9267.
Port, G.R., Wilkes, T.J., & Bryan, J.H. (1981). A note on the distribution of *Aedes* (*Diceromyia*) *furcifer*/*taylori*, a yellow fever vector, in the Gambia. *Insect Science and Its Application, 2*(3), 163–166.
Porter, C.H. (2009). *Wyeomyia* (*Hystatomyia*) *baltae*, a new species of Sabethini (Diptera: Culicidae) from Peru. *Proceedings of the Entomological Society of Washington, 111*(4), 807–825.
Porter, C.H. (2014). *Wyeomyia* (*Nunezia*) *paucartamboensis*, a new species of Sabethini (Diptera: Culicidae) from the Peruvian Andes with a diagnosis of the subgenus *Nunezia. Proceedings of the Entomological Society of Washington, 116*(3), 311–338.
Porter, C.H., & Collins, F.H. (1996). Phylogeny of Nearctic members of the *Anopheles maculipennis* species group derived from the D2

variable region of 28S ribosomal RNA. *Molecular Phylogenetics and Evolution, 6*, 178–188.

Porter, C.H., & Wolff, M.I. (2004). A new species of *Wyeomyia* (*Hystatomyia*) (Diptera: Culicidae) from Colombia and a redescription of *Wy.* (*Hystatomyia*) *intonca* Dyar and Knab. *Zootaxa, 477*, 1–31.

Porter, J.E. (1946). The larva of *Uranotaenia syntheta*. (Diptera: Culicidae). *The American Midland Naturalist, 35*(2), 535–537.

Portier, C.J., Thigpen, T.K., Carter, S.R., Dilworth, C.H., Grambsch, A.E., Gohlke, J., . . . Whung, P.Y. (2010). A human health perspective on climate change: A report outlining the research needs on the human health effects of climate change. *Environmental Health Perspectives*. Retrieved from http://www.niehs.nih.gov/health/materials/a_human_health_perspective_on_climate_change_full_report_508.pdf

Portschinsky, J. (1884). Diptera europae et asiatica nova aut minus cognita (Cum notis biologicis). *Trudy Russkago entomologicheskago obshchestva. Horae Societatis entomologicae rossicae, variis semonibus in Russia usitatis editae, 18*, 122–134.

Portschinsky, J. (1910). Overview of the distribution of the most important pest animals in Russia in 1909. *Ezhegodnik Glavnogo Upravleniya Zemleustroistva I Zemledeliya po Departmentu Zemledeliya, 3*(3), 602–614.

Portschinsky, J. (1911). Die malaria-mucke (*Anopheles claviger* F.) im Zusammenhang mit dem Sumpffieber, ihr Leben, ihre Eigenschaften und ihre Bekampfung. *Trudy Biuro po entomologii Uchenago Komiteta Glavnago Upravleniia Zemleustroistva i Zemledieliia, 5*(1), 1–137.

Postiglione, M., Smiraglia, C.B., Lavagnino, A., Gokberk, C., & Ramsdale, C.D. (1970). A preliminary note on the occurrence in Turkey of the *subalpinus* form of the *A. maculipennis* complex. *Rivista di malariologia, 31*, 155–158.

Postiglione, M., Tabanili, S., & Ramsdale, C.D. (1973). The *Anopheles* of Turkey. *Rivista di malariologia, 34*, 127–159.

Poupardin, R., Reynaud, S., Strode, C., Ranson, H., Vontas, J., & David, J.P. (2008). Cross-induction of detoxification genes by environmental xenobiotics and insecticides in the mosquito *Aedes aegypti*: Impact on larval tolerance to chemical insecticides. *Insect Biochemistry and Molecular Biology, 38*(5), 540–551.

Pozza, G.D., & Majori, G. (1992). First record of *Aedes albopictus* establishment in Italy. *Journal of the American Mosquito Control Association, 8*(3), 318–320.

Prado, A. (1931a). Contribuições ao conhecimento dos culicideos de São Paulo. I. Notas sobre *Mansonia albifera* Prado e sobre o macho *de Mansonia albicosta* (Chagas). *Memorias do Instituto Butantan, 6*, 193–199.

Prado, A. (1931b). Contribuições ao conhecimento dos culicideos de São Paulo. IV. Uma nova especie de *Uranotaenia* (Diptera, Culicidae). *Memorias do Instituto Butantan, 6*, 209–213.

Prasad, K.R., & Anbarasan, P. (2007). Stereoselective synthesis of (-)-6-acetoxyhexadecanolide: A mosquito oviposition attractant pheromone. *Tetrahedron: Asymmetry, 18*(20), 2479–2483.

Prashad, B. (1918). The description and life-history of a new species of *Anopheles* that breeds in holes in trees. *Records of the Indian Museum, 15*(3), 123–127.

Pratt, H.D. (1943). The identification of first stage larvae of Puerto Rican *Anopheles*. *Public Health Reports, 58*, 1715–1717.

Pratt, H.D. (1945). *Mansonia indubitans* Dyar and Shannon—a new mosquito addition to the United States fauna. *Journal of the Kansas Entomological Society, 18*, 121–129.

Pratt, H.D. (1947). The genus *Uranotaenia* Lynch Arribálzaga in Puerto Rico. *Annals of the Entomological Society of America, 39*(for 1946), 576–584.

Pratt, H.D. (1951). *Ficalbia minima* (Theobald) in South Indochina, with descriptions of the larva and pupa (Diptera: Culicidae). *Journal of the Washington Academy of Sciences, 41*(9), 300–302.

Pratt, H.D. (1952). Notes on *Anopheles earlei* and other American species of the *Anopheles maculipennis* complex. *The American Journal of Tropical Medicine and Hygiene, 1*(3), 484–493.

Pratt, H.D. (1953). Notes on American *Mansonia* mosquitoes (Diptera, Culicidae). *Proceedings of the Entomological Society of Washington, 55*(1), 9–19.

Pratt, H.D., & Kidwell, A.S. (1969). Eggs of mosquitoes found in *Aedes aegypti* oviposition traps. *Mosquito News, 29*(4), 545–548.

Pratt, H.D., & Seabrook, E.L. (1952). The occurrence of *Culex iolambdis* Dyar in Florida and Puerto Rico, with a description of the larva (Diptera, Culicidae). *Proceedings of the Entomological Society of Washington, 54*(1), 27–32.

Pratt, H.D., & Siren, N. (1971). *Anopheles indefinitus* and *Culex fuscanus* (Diptera: Culicidae) in Saipan. *Mosquito News, 31*(1), 114–115.

Predel, R., Neupert, S., Garczynski, S.F., Crim, J.W., Brown, M.R., Russell, W.K., . . . Nachman, R.J. (2010). Neuropeptidomics of the mosquito *Aedes aegypti*. *Journal of Proteomics Research, 9*(4), 2006–2015.

Press, F., & Siever, R. (1997). *Understanding Earth* (2nd ed.). New York, NY: Freeman.

Pressat, A. (1905). *Le paludisme et les moustiques (prophylaxie)*. Libraires de l'Académie de Médecine.

Price, R.D. (1958). A description of the larva and pupa of *Culiseta* (*Culicella*) *minnesotae* Barr (Diptera, Culicidae). *Journal of the Kansas Entomological Society, 31*, 47–53.

Price, R.D. (1960). Identification of first-instar aedine mosquito larvae of Minnesota (Diptera: Culicidae). *The Canadian Entomologist, 92*, 544–560.

Pringle, C.A.T., Azzawi, M., Riyadh, H., & Murayati, Y. (1960). The anopheline larvae of Iraq with a Report on *A. culicifacies*, *A. multicolor* and *A. apoci* breeding in Syrian desert spring. *Bulletin of Endemic Diseases* (Baghdad), *3*, 36–56.

Pringle, G. (1952). The identification of the larvae of *Anopheles stephensi* Liston and *Anopheles superpictus* Grassi in Iraq. *Bulletin of Entomological Research, 42*(4), 779–783.

Pringle, G. (1954). The identification of the adult anopheline mosquitoes of Iraq and neighbouring territories. *Bulletin of Endemic Diseases, 1*(1), 53–76.

Pringle, G., & Garrett-Jones, C. (1957). Additions to the *Anopheles* fauna of Iraq: *A. sergenti* Theobald and *A. apoci* Marsh. *Bulletin of Endemic Diseases, 2*, 88–91.

Prosen, A., Carcavallo., R.U., & Martínez, A. (1963). Culicidae de Bolivia (Diptera). *Anales del Instituto de Medicina Regional, 6*(1–2), 59–124.

Pruess, K.P. (1983). Day-degree methods for pest management. *Environmental Entomology, 12*(3), 613–619.

Pulido, F., & Sutil Oramas, E. (1981). *Wyeomyia* (*Wyeomyia*) *trujilloi* (Diptera, Culicidae) nueva especie de Venezuela. *Boletín de la Dirección de Malariología y Saneamiento Ambiental, 21*(3–4), 219–226.

Puri, I. M. (1928). On two species of Indian anopheline mosquitoes, *A. jeyporiensis* James and *A. moghulensis* Christophers. *Indian Journal of Medical Research, 16*(2), 513–518.

Puri, I.M. (1929a). Description of the male, female, egg and larva of *Anopheles annandalei* var. *interruptus* nov. var. with corrections for the previous descriptions of the type species. *Indian Journal of Medical Research* (Calcutta), *17*(2), 385–395.

Puri, I.M. (1929b). A new tree-hole breeding *Anopheles* from South India—*Anopheles sintoni* sp. nov. and a revised description of the larva of *A. culiciformis* Cogill. *Indian Journal of Medical Research* (Calcutta), *17*(2), 397–404.

Puri, I.M. (1930). A note on two species of Indian anopheline mosquitoes—*A. insulaeflorum* Swellengrebel and *A. aitkenii* James, with its variety *bengalensis* nov. var. *Indian Journal of Medical Research* (Calcutta), *17*(3), 953–956.

Puri, I.M. (1949). Anophelines of the Oriental Region. In M.F. Boyd (Ed.), *Malariology: A comprehensive survey of all aspects of this group of diseases from a global standpoint, by 65 contributors*. Philadelphia, PA: Saunders.

Puslednik, L., Russell, R.C., & Ballard, J.W.O. (2012). Phylogeography of the medically important mosquito *Aedes* (*Ochlerotatus*) *vigilax* (Diptera: Culicidae) in Australasia. *Journal of Biogeography, 39*(7), 1333–1346.

Qu, F., & Qian, G. (1993). The morphology of female cibarial armature of ten species of Uranotaenine mosquitoes from China (Diptera, Culicidae). *Acta parasitology et médica entomologica Sinica, 1*, 32–38.

Qu, F., & Zhu, H. (2008). On a checklist of the anopheline mosquitoes in China with rectification for some specific names. *Chinese Journal of Parasitology and Parasitic Diseases, 26*(3), 210–216.

Qualls, W.A., Day, J.F., Xue, R.D., & Bowers, D.F. (2011). Altered response to DEET repellent after infection of *Aedes aegypti* (Diptera: Culicidae) with Sindbis virus. *Journal of Medical Entomology, 48*(6), 1226–1230.

Quiñones, M.L., Harbach, R.E., Calle, D.A., Ruiz, F., Erazo, H.F., & Linton, Y.-M. (2001). Variante morfológica de adultos hembras de *Anopheles benarrochi* (Diptera: Culicidae) en Putumayo, Colombia. *Biomédica, 21*, 351–359.

Qutubuddin, M. (1945). A description of the hitherto unknown female of *Aedes* (*Diceromyia*) *periskeletus* (Giles) 1902. *Proceedings of the National Academy of Sciences* (India), *15*(2), 33–35.

Qutubuddin, M. (1947). A new species of *Uranotaenia*, Culicidae, Diptera from Hyderabad (Deccan). *Indian Journal of Entomology, 8*(for 1946)(1), 117–118.

Qutubuddin, M. (1951). *Uranotaenia mattinglyi* sp. n., a new species of mosquito (Diptera, Culicidae) from Hyderabad (Deccan), India. *Proceedings of the Royal Entomological Society of London. Series B: Taxonomy, 20*(9–10), 107–109.

Qutubuddin, M. (1952). A description of the hitherto unknown larva of *Theobaldia indica* Edwards (Diptera, Culicidae), with a key to larvae of the Pakistan species of *Theobaldia* Neveu-Lemaire, 1902. *Proceedings of the Royal Entomological Society of London. Series B: Taxonomy, 21*(3–4), 37–40.

Qutubuddin, M. (1954). The early stages and male terminalia of *Aedes* (*Finlaya*) *lophoventralis* (Theobald) (Diptera, Culicidae). *Annals of Tropical Medicine and Parasitology, 48*, 135–141.

Qutubuddin, M. (1956). *Culex* (*Culex*) *afridii* sp. n. A new species of mosquito (Diptera, Culicidae) from Dacca, East Pakistan. *Mosquito News, 16*(2), 140–142.

Qutubuddin, M. (1959). A new species of *Aedimorphus* (Diptera: Culicidae) from Sudan Republic. *Annals and Magazine of Natural History, 2*(13), 21–23.

Qutubuddin, M. (1972). A new species of *Finlaya* (Diptera: Culicidae) from Papua. *Pakistan Journal of Health, 22*(1–4), 119–120.

Rachou, R.G., & Sousa, J.D. (1950). Do encontro de um exemplar de *A.* (*M.*) *nigritarsis* (Chagas, 1907) no estado do Paraná. *Revista Brasileira de malariologia e doenças tropicais* (Rio de Janeiro), *2*, 234–238.

Raffaele, G. (1928). Una nuova specie di *Anopheles*. *Rivista di malariologia, 7*, 11–17.

Rageau, J., & Hamon, J. (1957). *Aedes* (Diptera, Culicidae) appartenant au sous-genre *Mucidus* en Nouvelle-Caledonie. *Bulletin de la Société de pathologie exotique, 50*, 372–378.

Rageau, J., Mouchet, J., & Abonnenc, E. (1972). Repartition géographique des moustiques (Diptera: Culicidae) en France. *Proceedings of the 13th International Congress on Entomology* (Moscow), *1968*(3), 230–232.

Rahman, S.J., Choudhury, D.S., & Kalra, N.L. (1969). Description of *Culex* (*Lophoceraomyia*) *raghavanii* sp. n. (Diptera: Culicidae) from Nilgiri Hills, South India. *Bulletin of the Indian Society for Malaria and Other Communicable Diseases, 5*(for 1968)(1–2), 28–34.

Rahman, W.A., Che'Rus, A., & Ahmad, A.H. (1997) Review: Malaria and *Anopheles* mosquitoes in Malaysia. *Southeast Asian Journal of Tropical Medicine and Public Health, 28*(3), 599–605.

Rahola, N., Makanga, B., Yangari, P., Jiolle, D., Fontenille, D., Renaud, F., . . . Paupy, C. (2014). Description of *Anopheles gabonensis*, a new species potentially involved in rodent malaria transmission in Gabon, Central Africa. *Infection, Genetics and Evolution, 28*, 628–634.

Raikhel, A.S., Kokoza, V.A., Zhu, J., Martin, D., Wang, S.F., Li, C., . . . Attardo, G. (2002). Molecular biology of mosquito vitellogenesis: From basic studies to genetic engineering of antipathogen immunity. *Insect Biochemistry and Molecular Biology, 32*(10), 1275–1286.

Rainey, S.M., Shah, P., Kohl, A., & Dietrich, I. (2014). Understanding the *Wolbachia*-mediated inhibition of arboviruses in mosquitoes: Progress and challenges. *Journal of General Virology, 95*(3), 517–530.

Rajagopalan, P.K. (1956). Larva and pupa of *Aedes* (*Stegomyia*) *w-albus* Theobald, 1905 (Diptera-Culicidae). *Indian Journal of Medical Research* (Calcutta), *44*(3), 481–489.

Rajapaksa, N. (1971). Field and laboratory observations in Sabah, East Malaysia on the proportion of *Anopheles balabacensis balabacensis* eggs hatching after holding in a humid atmosphere. *Bulletin of the World Health Organization, 45*(2), 263–265.

Rajavel, A.R., Natarajan, R., & Munirathinam, A. (1998). Occurence of *Culex* (*Culiciomyia*) *spathifurca* (Edwards) (Diptera: Culicidae) in India. New country record. *Journal of the American Mosquito Control Association, 14*(3), 346–347.

Rajavel, A.R., Natarajan, R., & Vaidyanathan, K. (2004). A checklist of mosquitoes (Diptera: Culicidae) of Pondicherry, India with notes on new area records. *Journal of the American Mosquito Control Association, 20*(3), 228–232.

Ramalingam, S. (1969). New record of *Aedes* (*Finlaya*) *togoi* (Theobald) in West Malaysia. *Medical Journal of Malaya, 23*, 288–292.

Ramalingam, S. (1972). A new species of *Armigeres* from Sabah, Borneo (Diptera: Culicidae). *Proceedings of the Entomological Society of Washington, 74*(4), 459–467.

Ramalingam, S. (1974). Some new records of *Anopheles* from Sabah, Malaysia. *Southeast Asian Journal of Tropical Medicine and Public Health, 5*(1), 147–148.

Ramalingam, S. (1975). A new species of *Topomyia* from Peninsular Malaysia (Diptera: Culicidae. *Mosquito Systematics, 7*(3), 185–192.

Ramalingam, S. (1976). An annotated checklist and keys to the mosquitoes of Samoa and Tonga. *Mosquito Systematics, 8*(3), 298–318.

Ramalingam, S. (1983). *Topomyia houghtoni* Feng, a new record in Malysia and a redescription of the adult and immature stages. *Mosquito Systematics, 15*(1), 33–49.

Ramalingam, S. (1987). On the restriction of *Armigeres durhami* Edwards and the description of *Armigeres besseli* n. sp. (Diptera: Culicidae). *Tropical Biomedicine, 4*(1), 55–65.

Ramalingam, S. (1988). Description of a new species of *Topomyia* from Peninsular Malaysia (Diptera: Culicidae). *Mosquito Systematics, 19*(for 1987)(3), 245–250.

Ramalingam, S., & Banu, Q. (1987). Studies on the genus *Topomyia*: 2. Description of a new species from Sabah, Malaysia (Diptera: Culicidae). *Tropical Biomedicine, 4*(1), 119–124.

Ramalingam, S., & Belkin, J.N. (1964). Vectors of sub-periodic bancroftian filariasis in the Samoa-Tonga area. *Nature, 201*, 105–106.

Ramalingam, S., & Belkin, J.N. (1965). Mosquito studies (Diptera: Culicidae). III. Two new species of *Aedes* from Tonga and Samoa. *Contributions of the American Entomological Institute, 1*(4), 1–10.

Ramalingam, S., & Belkin, J.N. (1976). The immature stages of *Aedes* (*F.*) *samoanus* and the status of *Toxorhynchites* in American Samoa. *Mosquito Systematics, 8*(2), 194–199.

Ramalingam, S., & Pillai, A.G. (1972). A new species of Malaya from West Malaysia (Diptera: Culicidae). *Proceedings of the Entomological Society of Washington, 74*(4), 451–459.

Ramalingam, S., & Pillai, A.G. (1973). Ten new records of mosquitoes occurring in West Malaysia. *Southeast Asian Journal of Tropical Medicine and Public Health, 4*(2), 271–272.

Ramalingam, S., & Ramakrishna, K. (1971). Redescription of *Aedes* (*Alanstonea*) *brevitibia* (Edwards) from Brunei, Borneo (Diptera: Culicidae). *Proceedings of the Entomological Society of Washington, 73*, 231–238.

Ramalingam, S., & Ramakrishna, K. (1988). Studies on the genus *Topoymia*. 1. A new species from Sabah, Malaysia. *Mosquito Systematics, 20*(1), 33–40.

Ramirez, C.C., & Dessen, E.M. (2000a). Chromosomal evidence for sibling species of the malaria vector *Anopheles cruzii*. *Genome, 43*(1), 143–151.

Ramirez, C.C.L., & Dessen, E.M.B. (2000b). Chromosome differentiated populations of *Anopheles cruzii*: Evidence for a third sibling species. *Genetica* (Dordrecht), *108*(1), 73–80.

Ramirez, J.L., Souza-Neto, J., Cosme, R.T., Rovira, J., Ortiz, A., Pascale, J.M., & Dimopoulos, G. (2012). Reciprocal tripartite interactions between the *Aedes aegypti* midgut microbiota, innate immune system, and dengue virus influences vector competence. *PLOS Neglected Tropical Diseases, 6*(3), e1561.

Ramirez, J.L., de Almeida Oliveira, G., Calvo, E., Dolli, J., Colas, R.A., Serhan, C.N., . . . Barillas-Mury, C. (2015). A mosquito lipoxin/lipocalin complex mediates innate immune pruning in *Anopheles gambiae*. *Nature Communications, 6*, 7403.

Ramos, A.S. (1937). Descripção do macho de *Nyssorhynchus* (*Myzorhynchella*) *lutzi* (Cruz, 1901). *Archivos de higiene e saúde pública* (São Paulo), *2*(3), 215–218.

Ramos, A.S. (1942). Sobre uma variedade nova de *Anopheles* (*Nyssorhynchus*) *oswaldoi* Peryassú, 1922 *(Diptera-Culicidae). Arquivos de higiene e saúde pública* (São Paulo), *7*(15), 61–63.

Ramsdale, C.D., & Snow, K. (2000). Distribution of the genus *Anopheles* in Europe. *European Mosquito Bulletin, 7*, 1–26.

Ramsdale, C.D., & Snow, K. (2001). Distribution of the genera *Coquillettidia*, *Orthopodomyia* and *Uranotaenia* in Europe. *European Mosquito Bulletin, 10*, 25.

Ramsdale, C.D., Alten, B., Caglar, S.S., & Ozer, N. (2001). A revised, annotated checklist of the mosquitoes (Diptera, Culicidae) of Turkey. *European Mosquito Bulletin, 9*, 18–28.

Randolph, N.M., & O'Neill, K. (1944). *The mosquitoes of Texas*. Texas State Health Department, pp. 1–100.

Rangel, R., & Romero Sierra, J.M. (1907). Mosquito número 9. In J.M. Romero Sierra (Ed.), *Contribucion al estudio de los mosquitos de Caracas. Tesis de Doctorado No. 67* (pp. 45–48). Caracas: Facultad de Ciencias Médicas, Universidad Central de Venezuela.

Rani, A., Sharma, A., Rajagopal, R., Adak, T., & Bhatnagar, R.K. (2009). Bacterial diversity analysis of larvae and adult midgut microflora using culture-dependent and culture-independent methods in lab-reared and field-collected *Anopheles stephensi*—an Asian malarial vector. *BMC Microbiology, 9*(96), 1–22.

Rao, B.A., Sweet, W.C., & Rao, A.M.S. (1938). Ova measurements of *A. stephensi* type and *A. stephensi* var. *mysorensis*. *Journal of the Malaria Institute of India, 1*, 261–266.

Rao, T.R. (1984). *The anophelines of India* (Rev.). New Delhi, India: Malaria Research Centre, Indian Council of Medical Research.

Rao, T.R., Dhanda, V., Bhat, H.R., & Kulkarni, S.M. (1973). A survey of haematophagous arthropods in western Himalayas, Sikkim and hill districts of West Bengal, a general account. *Indian Journal of Medical Research* (Calcutta), *61*, 1421–1458.

Rao, V.V. (1947). On gonotrophic discordance among certain Indian *Anopheles*. *Indian Journal of Malariology, 1*(1), 43–50.

Rao, V.V., & Ramakrishna, V. (1940). A note on the larva of *A. varuna* (Iyengar). *Journal of the Malaria Institute of India, 3*, 509–512.

Raphemot, R., Rouhier, M.F., Hopkins, C.R., Gogliotti, R.D., Lovell, K.M., Hine, R.M., . . . Piermarini, P.M. (2013). Eliciting renal failure in mosquitoes with a small-molecule inhibitor of inward-rectifying potassium channels. *PLOS ONE, 8*(5), e64905.

Rascón, A.A., Jr. (2010). Enzyme kinetics and biochemical properties of serine proteases from the midgut of the Dengue mosquito, *Aedes aegypti* [Doctoral dissertation, University of Arizona]. Retrieved from http://hdl.handle.net/10150/194426

Rasgon, J.L., & Scott, T.W. (2003). *Wolbachia* and cytoplasmic incompatibility in the California *Culex pipiens* mosquito species complex: Parameter estimates and infection dynamics in natural populations. *Genetics, 165*(4), 2029–2038.

Rasgon, J.L., Cornel, A.J., & Scott, T.W. (2006). Evolutionary history of a mosquito endosymbiont revealed through mitochondrial hitchhiking. *Proceedings of the Royal Society of London B: Biological Sciences, 273*(1594), 1603–1611.

Rasnitsyn, S.P., & Lebedeva, N.N. (1998). Frequency of horizontal movements in mosquito larvae (Diptera: Culicidae). *Entomological Review, 78*(4), 483–487.

Rasnitsyn, S.P., & Lebedeva, N.N. (2000). Direction of horizontal movements of mosquito larvae (Diptera: Culicidae). *Entomological Review, 80*(9), 1027–1031.

Rasnitsyn, S.P., & Lebedeva, N.N. (2002). Speed of horizontal movements of mosquito larvae (Diptera: Culicidae). *Entomological Review, 82*(8), 933–937.

Rattanarithikul, R., & Green, C.A. (1987). Formal recognition of the species of the *Anopheles maculatus* group (Diptera: Culicidae) occurring in Thailand, including the descriptions of two new species and a preliminary key to females. *Mosquito Systematics, 18*(for 1986)(3–4), 246–278.

Rattanarithikul, R., & Harbach, R.E. (1991). *Anopheles maculatus* (Diptera: Culicidae) from the type locality of Hong Kong and two new species of the *maculatus* complex from the Philippines. *Mosquito Systematics, 22*(for 1990)(3), 160–183.

Rattanarithikul, R., & Harrison, B.A. (1973). An illustrated key to the *Anopheles* larvae of Thailand. US Army Medical Component, SEATO, Bangkok, Thailand.

Rattanarithikul, R., & Harrison, B.A. (1988). *Aedes* (*Finlaya*) *reinerti*, a new species from northern Thailand related to *Aedes* (*Finlaya*) *formosensis* Yamada (Diptera: Culicidae). *Mosquito Systematics, 20*(1), 77–96.

Rattanarithikul, R., & Panthusiri, P. (1994). Illustrated keys to the medically important mosquitoes of Thailand. *Southeast Asian Journal of Tropical Medicine and Public Health, 66*(Suppl. 1), 1.

Rattanarithikul, R., Harrison, B.A., Panthusiri, P., & Coleman, R.E. (2005a). Illustrated keys to the mosquitoes of Thailand. I. Background; geographic distribution; lists of genera, subgenera, and species; and a key to the genera. *Southeast Asian Journal of Tropical Medicine and Public Health, 36*(1), 1–80.

Rattanarithikul, R., Harbach, R.E., Harrison, B.A., Panthusiri, P., Jones, J.W., & Coleman, R.E. (2005b). Illustrated keys to the mosquitoes of Thailand. II. Genera *Culex* and *Lutzia. Southeast Asian Journal of Tropical Medicine and Public Health, 36*(Suppl. 2), 1–97.

Rattanarithikul, R., Harrison, B.A., Panthusiri, P., Peyton, EL, & Coleman, R.E. (2006a). Illustrated keys to the mosquitoes of Thailand. III. Genera *Aedeomyia, Ficalbia, Mimomyia, Hodgesia, Coquillettidia, Mansonia,* and *Uranotaenia. Southeast Asian Journal of Tropical Medicine and Public Health, 85*(Suppl. 1), 1.

Rattanarithikul, R., Harrison, B.A., Harbach, R.E., Panthusiri, P., & Coleman, R.E. (2006b). Illustrated keys to the mosquitoes of Thailand. IV. *Anopheles. Southeast Asian Journal of Tropical Medicine and Public Health, 128*(Suppl. 2), 2.

Rattanarithikul, R., Harbach, R.E., Harrison, B.A., Panthusiri, P., & Coleman, R.E. (2007). Illustrated keys to the mosquitoes of Thailand. V. *Genera Orthopodomyia, Kimia, Malay, Topomyia Tripteroides,* and *Toxorhynchites. Southeast Asian Journal of Tropical Medicine and Public Health, 38*(Suppl. 2), 1–65.

Rattanarithikul, R., Harbach, R.E., Harrison, B.A., Panthusiri, P., Coleman, R.E., & Richardson, J.H. (2010). Illustrated keys to the mosquitoes of Thailand. VI. Tribe Aedini. *Southeast Asian Journal of Tropical Medicine and Public Health, 41*(1), 1–225.

Raup, D.M. (1988). Testing the fossil record for evolutionary progress. In M.H. Nitecki (Ed.), *Evolutionary progress* (pp. 293–317). Chicago, IL: University of Chicago Press.

Ravaonjanahary, C. (1978). Les *Aedes* de Madagascar (Diptera - Culicidae). 1- Étude monographique du genre. 2- Biologie d'*Aedes (Diceromyia) tiptoni. Travaux et Documents de l'ORSTOM, 87,* 1–211.

Ravaonjanahary, C. (1979). Presence à Madagascar de *Culex* (*Culex*) *watti* Edwards, 1920. Description de la larve et de la nymphe de l'espèce. *Cahiers ORSTOM. Série entomologie médicale et parasitologie, 17*(3), 193–194.

Ravaonjanahary, C., & Bruhnes [*sic*], J. (1977). Un nouvel *Aedes* du genre *Skusea* découvert à Madagascar: *Aedes* (*Skusea*) *moucheti* sp.n. *Cahiers ORSTOM. Série entomologie médicale et parasitologie, 15*(3), 213–215.

Rawat, S.S., Viard, M., Gallo, S.A., Rein, A., Blumenthal, R., & Puri, A. (2003). Modulation of entry of enveloped viruses by cholesterol and sphingolipids (Review). *Molecular Membrane Biology, 20*(3), 243–254.

Ree, H.-I. (2003). Taxonomic review and revised keys of the Korean mosquitoes (Diptera: Culicidae). *Korean Journal of Entomology, 33*(1), 39–52.

Reed, W., & Carroll, J. (1901). The prevention of yellow fever. *Public Health Papers and Reports, 27,* 113–129.

Rees, D.M., & Collett, G.C. (1954). The biology of *Aedes niphadopsis* Dyar and Knab (Diptera, Culicidae). *Proceedings of the Entomological Society of Washington, 56*(4), 207–214.

Rees, D.M., & Onishi, K. (1951). Morphology of the terminalia and internal reproductive organs, and copulation in the mosquito, *Culiseta inornata* (Williston) (Diptera, Culicidae). *Proceedings of the Entomological Society of Washington, 53*(5), 233–246.

Reeves, L.E., Holderman, C.J., Blosser, E.M., Gillett-Kaufman, J.L., Kawahara, A.Y., Kaufman, P.E., & Burkett-Cadena, N.D. (2018). Identification of *Uranotaenia sapphirina* as a specialist of annelids broadens known mosquito host use patterns. *Communications Biology, 1*(1), Article 92.

Reeves, W.K., Breidenbaugh, M.S., Thomas, E.E., & Glowacki, M.N. (2013). Mosquitoes of Thule Air Base, Greenland. *Journal of the American Mosquito Control Association, 29*(4), 383–384.

Reid, J.A. (1947). Type specimens of Culicidae described by Laveran (Diptera: Culicidae). *Proceedings of the Royal Entomological Society of London. Series B: Taxonomy, 16*(7–8), 86–91.

Reid, J.A. (1949). A preliminary account of the forms of *Anopheles leucosphyrus* Doenitz (Diptera: Culicidae). *Proceedings of the Royal Entomological Society of London. Series B: Taxonomy, 18*(3–4), 42–53.

Reid, J.A. (1950). The *Anopheles umbrosus* group (Diptera: Culicidae). I: Systematics, with descriptions of two new species. *Proceedings of the Royal Entomological Society of London. Series B: Taxonomy, 101*(9), 281–318.

Reid, J.A. (1953). The *Anopheles hyrcanus* group in south-east Asia. (Diptera: Culicidae). *Bulletin of Entomological Research, 44*(1), 5–76.

Reid, J.A. (1959). A note on the larvae of *Anopheles subpictus* and *sundaicus. Mosquito News, 19,* 101–102.

Reid, J.A. (1962). The *Anopheles barbirostris* group (Diptera, Culicidae). *Bulletin of Entomological Research, 53*(1), 1–57.

Reid, J.A. (1963a). Notes on anopheline mosquitoes from Malaya, with descriptions of three new species. *Annals of Tropical Medicine and Parasitology, 57*(1), 97–116.

Reid, J.A. (1963b). A note on the structure of the pupal trumpet in some mosquitoes. *Proceedings of the Royal Entomological Society of London. Series A: General Entomology, 38,* 32–38.

Reid, J.A. (1965). A revision of the *Anopheles aitkenii* group in Malaya and Borneo. *Annals of Tropical Medicine and Parasitology, 59*(1), 106–125.

Reid, J.A. (1966). A note on *Anopheles subpictus* Grassi and *A. indefinitus* Ludlow (Diptera: Culicidae). *Journal of Medical Entomology, 3*(3–4), 327–331.

Reid, J.A. (1967). Two forms of *Anopheles philippinensis* in Malaya. *Journal of Medical Entomology, 4*(2), 175–179.

Reid, J.A. (1968). Anopheline mosquitoes of Malaya and Borneo. *Studies from the Institute for Medical Research Malaysia, 31,* 1–520.

Reid, J.A. (1973). Larval differences between sympatric populations from Kaduna, West Africa, of species A and B of the *Anopheles gambiae* group. *Parassitologia, 15*(1–2), 87–98.

Reid, J.A. (1975a). Pupal differences between species A and B of the *Anopheles gambiae* group from Kisumu, East Africa. *Mosquito Systematics, 7*(1), 1–7.

Reid, J.A. (1975b). Pupal differences between species A and B of the *Anopheles gambiae* group from Kaduna, West Africa. *Mosquito Systematics, 7*(4), 299–302.

Reid, J.A. (1976). A lectotype for *Anopheles* (*Cellia*) *pampanai* Buettiker and Beales, 1959. *Mosquito Systematics, 8*(1), 11–113.

Reid, J.A. (1980). *Anopheles donaldi*: Some recollections and speculations. *Medical Journal of Malaysia, 34*(4), 399–402.

Reid, J.A., & Knight, K.L. (1961). Classification within the subgenus *Anopheles* (Diptera, Culicidae). *Annals of Tropical Medicine and Parasitology, 55*(4), 474–488.

Reidenbach, K.R., Cook, S., Bertone, M.A., Harbach, R.E., Wiegmann, B.M., & Besansky, N.J. (2009). Phylogenetic analysis and temporal diversification of mosquitoes (Diptera: Culicidae) based on nuclear genes and morphology. *BMC Evolutionary Biology, 9*(1), Article 298.

Reinert, J.F. (1970a). Contributions to the mosquito fauna of South-East Asia. 5. Genus *Aedes,* subgenus *Diceromyia,* Theobald in South-East Asia. *Contributions of the America Institute of Entomology, 5*(4), 1–43.

Reinert, J.F. (1970b). Description of the pupa of *Aedes* (*Ochlerotatus*) *dupreei* (Diptera: Culicidae). *Florida Entomologist, 53*(4), 243–247.

Reinert, J.F. (1970c). Description of the pupa of *Aedes* (*Ochlerotatus*) *fulvus pallens* (Diptera: Culicidae). *Florida Entomologist, 53*(1), 47–50.

Reinert, J.F. (1970d). Description of the pupa of *Aedes* (*Ochlerotatus*) *infirmatus* (Diptera: Culicidae). *Florida Entomologist, 53*(3), 147–151.
Reinert, J.F. (1970e). Description of the pupa of *Aedes* (*Ochlerotatus*) *mitchellae* (Diptera: Culicidae). *Florida Entomologist, 53*(2), 73–77.
Reinert, J.F. (1970f). Description of the pupa of *Deinocerites pseudes* (Diptera: Culicidae). *Florida Entomologist, 53*(1), 17–21.
Reinert, J.F. (1970g). Description of the pupa of *Toxorhynchites* (*Lynchiella*) *rutilus rutilus* (Diptera: Culicidae). *Florida Entomologist, 53*(4), 209–213.
Reinert, J.F. (1970h). Description of the pupae of *Wyeomyia* (*Wyeomyia*) *vanduzeei* and *W.* (*W.*) *mitchellii* (Diptera, Culicidae). *Florida Entomologist, 53*(3), 163–169.
Reinert, J.F. (1970i). The zoogeography of *Aedes* (*Diceromyia*) Theobald (Diptera: Culicidae). *Journal of the Entomological Society of Southern Africa, 33*(1), 129–141.
Reinert, J.F. (1971). Descriptions of the pupae of *Aedes* (*Ochlerotatus*) *atlanticus* and *Aedes* (*O.*) *tormentor* (Diptera: Culicidae). *Florida Entomologist, 54*(2), 171–178.
Reinert, J.F. (1972a). Description of the egg of *Aedes* (*Aedimorphus*) *domesticus* (Theobald) (Diptera: Culicidae). *Mosquito Systematics, 4*(2), 60–62.
Reinert, J.F. (1972b). Description of the egg of *Aedes* (*Diceromyia*) *furcifer* (Edwards) (Diptera: Culicidae). *Mosquito Systematics, 4*(3), 87–89.
Reinert, J.F. (1972c). Contributions to the mosquito fauna of Southeast Asia. XV. Genus *Aedes* Meigen, subgenus *Ayurakitia* Thurman. *Contributions of the American Entomological Institute, 9*(2), 1–42.
Reinert, J.F. (1972d). Description of the egg of *Aedes* (*Levua*) *suvae* Stone and Bohart (Diptera: Culicidae). *Mosquito Systematics, 4*(4), 128–130.
Reinert, J.F. (1972e). *Aedes gouldi*, a new species of the subgenus *Aedimorphus* Theobald from West Pakistan (Diptera: Culicidae). *Proceedings of the Entomological Society of Washington, 74*(4), 357–362.
Reinert, J.F. (1973a). *Aedes wainwrighti* Baisas, a synonym of *Aedes* (*Stegomyia*) *meronephada* (Dyar and Shannon), with notes on the subgenus *Stegomyia* Theobald (Diptera: Culicidae). *Mosquito Systematics, 5*(1), 27–30.
Reinert, J.F. (1973b). Contributions to the mosquito fauna of Southeast Asia. XVI. Genus *Aedes* Meigen, subgenus *Aedimorphus* Theobald in Southeast Asia. *Contributions of the American Entomological Institute, 9*(5), 1–218.
Reinert, J.F. (1973c). *Aedes consonensis*, a new species of the subgenus *Neomacleaya* from South Vietnam (Diptera: Culicidae). *Mosquito Systematics, 5*(3), 252–262.
Reinert, J.F. (1973d). Contributions to the mosquito fauna of Southeast Asia. XVIII. A reconsideration of *Diceromyia* Theobald with the inclusion of *Aedes nummatus* Edwards and *Aedes pseudonummatus* new species (Diptera: Culicidae). *Contributions of the American Entomological Institute, 10*(1), 22–40.
Reinert, J.F. (1973e). Contributions to the mosquito fauna of Southeast Asia. XIX. *Bothaella*, A new subgenus of *Aedes* Meigen. *Contributions of the American Entomological Institute, 10*(3), 1–51.
Reinert, J.F. (1973f). Notes on the genus *Heizmannia* and the inclusion in subgenus *Mattinglyia* of two species previously placed in *Aedes* subgenus uncertain (Diptera: Culicidae). *Mosquito Systematics, 5*(1), 14–23.
Reinert, J.F. (1974a). A new intepretation of the subgenus *Verrallina* of the genus *Aedes* (Diptera: Culicidae). *Contributions of the American Entomological Institute, 11*(1), 1–249.
Reinert, J.F. (1974b). Terminology and preparation of the female genitalia of aedine mosquitoes (Diptera: Culicidae). *Mosquito Systematics, 6*(1), 46–56.
Reinert, J.F. (1976a). A ventromedian cervical sclerite of mosquito larvae (Diptera: Culicidae). *Mosquito Systematics, 8*(2), 205–208.
Reinert, J.F. (1976b). A new man-biting species of *Aedes* (*Paraedes*) from southeast Asia (Diptera, Culicidae). *Mosquito Systematics, 8*(3), 319–331.
Reinert, J.F. (1976c). Medical entomology studies. IV. The subgenera *Indusius* and *Edwardsaedes* of the genus *Aedes* (Diptera: Culicidae). *Contributions of the American Entomological Institute, 13*(1), 1–45.
Reinert, J.F. (1976d). Medical entomology studies .V. The subgenus *Rhinoskusea* of the genus *Aedes* (Diptera: Culicidae). *Contributions of the American Entomological Institute, 13*(2), 1–60.
Reinert, J.F. (1977). Description of a new species of *Aedes* (*Verrallina*) from Sri Lanka (Diptera: Culicidae). *Mosquito Systematics, 9*(3), 366–371.
Reinert, J.F. (1978). First description of the immature stages of *Aedes* (*Verrallina*) *pseudomediofasciatus* and *Ae. yerburyi* (Diptera: Culicidae). *Mosquito Systematics, 10*(2), 225–238.
Reinert, J.F. (1979). A description of *Isoaedes*, A new subgenus of *Aedes* Meigen, and its type-species, *Ae.* (*Isa.*) *cavaticus* new species (Diptera: Culicidae). *Mosquito Systematics, 11*(2), 144–162.
Reinert, J.F. (1981). Medical entomology studies. XV. A revision of the subgenus *Paraedes* of the genus *Aedes* (Diptera: Culicidae). *Contributions of the American Entomological Institute, 18*(4), 1–91.
Reinert, J.F. (1982). *Belkinius*, a new subgenus of *Aedes* and a description of *Ae.* (*Blk.*) *aurotaeniatus* Edwards (Diptera: Culicidae). *Mosquito Systematics, 14*(2), 101–121.
Reinert, J.F. (1984). Medical entomology studies. XVI. A review of the species of subgenus *Verrallina*, genus *Aedes*, from Sri Lanka and a revised description of the subgenus (Diptera: Culicidae). *Mosquito Systematics, 16*(1), 1–130.
Reinert, J.F. (1985). A description of *Scutomyia*, a subgenus resurrected from the Albolineatus group of the genus *Aedes* (Diptera: Culicidae). *Mosquito Systematics, 17*(2), 121–131.
Reinert, J.F. (1987). *Albuginosus*, a new subgenus of *Aedes* Meigen (Diptera: Culicidae) described from the Afrotropical Region. *Mosquito Systematics, 18*(for 1986)(3–4), 307–326.
Reinert, J.F. (1988). Descriptions of the holotypes of *Aedes* (*Finlaya*) *purpureus* and *Ae.* (*Fin.*) *pecuniosus* with a revalidation of the latter species (Diptera: Culicidae). *Mosquito Systematics, 20*(1), 55–68.
Reinert, J.F. (1989). *Udaya subsimilis*, a new generic combination, a redescription of the holotype, and notes on the genus *Udaya* (Diptera: Culicidae). *Mosquito Systematics, 21*(3), 206–215.
Reinert, J.F. (1990). Medical entomology studies. XVII. Biosystematics of *Kenknightia*, a new subgenus of the mosquito genus *Aedes* Meigen from the Oriental region (Diptera: Culicidae). *Contributions of the American Entomological Institute, 26*(2), 1–119.
Reinert, J.F. (1993). Redescription of *Molpemyia*, and its revalidation as a subgenus of *Aedes* (Diptera: Culicidae). *Mosquito Systematics, 25*(1), 41–63.
Reinert, J.F. (1997). Bibliography of *Anopheles quadrimaculatus* Say sensu lato. *Journal of the American Mosquito Control Association, 13*(Suppl.), 112–161.
Reinert, J.F. (1999a). Descriptions of *Zavortinkius*, a new subgenus of *Aedes*, and the eleven included species from the Afrotropical region (Diptera: Culicidae). *Contributions of the American Entomological Institute, 31*(2), 1–105.
Reinert, J.F. (1999b). Lectotype selection for *Aedes quasirusticus* (Diptera: Culicidae). *European Mosquito Bulletin, 5*, 35–36.

Reinert, J.F. (1999c). Restoration of *Verrallina* to generic rank in tribe Aedini (Diptera: Culicidae) and descriptions of the genus and three included subgenera. *Contributions of the American Entomological Institute, 31*(3), 1–83.

Reinert, J.F. (1999d). The subgenus *Rusticoidus* of genus *Aedes* (Diptera: Culicidae) in Europe and Asia. *European Mosquito Bulletin, 4*, 1–7.

Reinert, J.F. (2000a). Comparative anatomy of the female genitalia of genera and subgenera in tribe Aedini (Diptera: Culicidae). Part I. Introduction, preparation techniques and anatomical terminology. *Contributions of the American Entomological Institute, 32*(2), 1–18.

Reinert, J.F. (2000b). Comparative anatomy of the female genitalia of genera and subgenera in tribe Aedini (Diptera: Culicidae). Part II. Genus *Psorophora* Robineau-Desvoidy. *Contributions of the American Entomological Institute, 32*(2), 19–35.

Reinert, J.F. (2000c). Comparative anatomy of the female genitalia of genera and subgenera in tribe Aedini (Diptera: Culicidae). Part III. Genus *Udaya* Thurman. *Contributions of the American Entomological Institute, 32*(2), 37–44.

Reinert, J.F. (2000d). Comparative anatomy of the female genitalia of genera and subgenera in tribe Aedini (Diptera: Culicidae). Part IV. Genus *Zeugnomyia* Leicester. *Contributions of the American Entomological Institute, 32*(2), 45–52.

Reinert, J.F. (2000e). Comparative anatomy of the female genitalia of genera and subgenera in tribe Aedini (Diptera: Culicidae). Part V. Genus *Aedes* Meigen. *Contributions of the American Entomological Institute, 32*(3), 1–102.

Reinert, J.F. (2000f). Assignment of two North American species of *Aedes* to subgenus *Rusticoidus*. *Journal of the American Mosquito Control Association, 16*(1), 42–43.

Reinert, J.F. (2000g). Descriptions of *Fredwardsius*, a new subgenus of *Aedes* (Diptera: Culicidae). *European Mosquito Bulletin, 6*, 1–7.

Reinert, J.F. (2000h). New classification for the composite genus *Aedes* (Diptera: Culicidae: Aedini), elevation of subgenus *Ochlerotatus* to generic rank, reclassification of the other subgenera and notes on certain subgenera and species. *Journal of the American Mosquito Control Association, 16*(3), 175–188.

Reinert, J.F. (2000i). Restoration of *Ayurakitia* to generic rank in tribe Aedini and a revised definition of the genus. *Journal of the American Mosquito Control Association, 16*(2), 57–65.

Reinert, J.F. (2000j). Selection of a lectotype for *Aedes refiki* Medschid (Diptera: Culicidae) and redescription of the male, females and fourth-instar larvae of the type series. *European Mosquito Bulletin, 8*,1–6.

Reinert, J.F. (2000k). Synonymy of subgenus *Sinoaedes* of genus *Aedes* with subgenus *Mattinglyia* of genus *Heizmannia*. *Journal of the American Mosquito Control Association, 16*(1), 38–39.

Reinert, J.F. (2001a). Comparative anatomy of the female genitalia of genera and subgenera in tribe Aedini (Diptera: Culicidae). Part VI. Genus *Ayurakitia* Thurman. *Contributions of the American Entomological Institute, 32*(7), 1–7.

Reinert, J.F. (2001b). Comparative anatomy of the female genitalia of genera and subgenera in tribe Aedini (Diptera: Culicidae). Part VII. Genus *Opifex* Hutton. *Contributions of the American Entomological Institute, 32*(4), 9–15.

Reinert, J.F. (2001c). Comparative anatomy of the female genitalia of genera and subgenera in tribe Aedini (Diptera: Culicidae). Part VIII. Genus *Verrallina* Theobald. *Contributions of the American Entomological Institute, 32*(4), 17–39.

Reinert, J.F. (2001d). Comparative anatomy of the female genitalia of genera and subgenera in tribe Aedini (Diptera: Culicidae). Part IX. Genus *Eretmapodites* Theobald. *Contributions of the American Entomological Institute, 32*(4), 41–49.

Reinert, J.F. (2001e). Revised list of abbreviations for genera and subgenera of Culicidae (Diptera) and notes on generic and subgeneric changes. *Journal of the American Mosquito Control Association, 17*(1), 51–55.

Reinert, J.F. (2002a). A comparative analysis of the ventral brush and its attachment to segment X of fourth-stage larvae in tribe Aedini (Diptera: Culicidae). *Journal of the American Mosquito Control Association, 18*(1), 16–25.

Reinert, J.F. (2002b). Comparative anatomy of the female genitalia of genera and subgenera in tribe Aedini (Diptera: Culicidae). Part X. Genus *Heizmannia* Ludlow. *Contributions of the American Entomological Institute, 32*(5), 1–15.

Reinert, J.F. (2002c). Comparative anatomy of the female genitalia of genera and subgenera in tribe Aedini (Diptera: Culicidae). Part XI. Genus *Haemagogus* Williston. *Contributions of the American Entomological Institute, 32*(5), 17–30.

Reinert, J.F. (2002d). Comparative anatomy of the female genitalia of genera and subgenera in tribe Aedini (Diptera: Culicidae). Part XII. Genus *Armigeres* Theobald. *Contributions of the American Entomological Institute, 32*(5), 31–44.

Reinert, J.F. (2002e). Comparative anatomy of the female genitalia of genera and subgenera in tribe Aedini (Diptera: Culicidae). Part XIII. Genus *Ochlerotatus* Lynch Arribálzaga. *Contributions of the American Entomological Institute, 33*(1), 1–111.

Reinert, J.F. (2002f). Comparative anatomy of the female genitalia of genera and subgenera in tribe Aedini (Diptera: Culicidae). Part XIV. Key to genera. *Contributions of the American Entomological Institute, 33*(1), 113–117.

Reinert, J.F. (2003a). Description of *Bruceharrisonius*, a new subgenus of *Ochlerotatus*, and a redescription of its type species *Oc. (Brh.) greenii*. *Journal of the American Mosquito Control Association, 19*(4), 309–322.

Reinert, J.F. (2003b). Transfer of *Heizmannia thelmae* from subgenus *Mattinglyia* to subgenus *Heizmannia* of genus *Heizmannia* and new characters for separating the two subgenera. *Journal of the American Mosquito Control Association, 19*(4), 433–434.

Reinert, J.F. (2005a). Redescription of the holotype of *Ochlerotatus eatoni* Edwards, 1916 (Diptera: Culicidae: Aedini). *European Mosquito Bulletin, 20*, 1–2.

Reinert, J.F. (2005b). List of species described in the egg stage of tribe Aedini (Diptera: Culicidae) with their literature citations. *Journal of the American Mosquito Control Association, 21*(3), 252–262.

Reinert, J.F. (2006a). *Tewarius* Reinert, a new genus of Aedini (Diptera: Culicidae). *Proceedings of the Entomological Society of Washington, 108*(3), 639–645.

Reinert, J.F. (2006b). Description of subgenus *Pardomyia* Theobald and comparison with the nominotypical subgenus of *Mucidus* Theobald (Diptera: Culicidae: Aedini). *Journal of the American Mosquito Control Association, 22*(4), 586–600.

Reinert, J.F. (2008a). Comparative anatomy of the female genitalia of generic-level taxa in tribe Aedini (Diptera: Culicidae). Part XV. Genus *Georgecraigius* Reinert, Harbach and Kitching. *Contributions of the American Entomological Institute, 35*(2), 1–10.

Reinert, J.F. (2008b). Comparative anatomy of the female genitalia of generic-level taxa in tribe Aedini (Diptera: Culicidae). Part XVI. Genus *Phagomyia* Theobald. *Contributions of the American Entomological Institute, 35*(2), 11–19.

Reinert, J.F. (2008c). Comparative anatomy of the female genitalia of generic-level taxa in tribe Aedini (Diptera: Culicidae). Part XVII. Genus *Dahliana* Reinert, Harbach and Kitching. *Contributions of the American Entomological Institute, 35*(2), 21–29.

Reinert, J.F. (2008d). Comparative anatomy of the female genitalia of generic-level taxa in tribe Aedini (Diptera: Culicidae). Part XVIII. Genus *Hulecoeteomyia* Theobald. *Contributions of the American Entomological Institute, 35*(2), 31–39.

Reinert, J.F. (2008e). Comparative anatomy of the female genitalia of generic-level taxa in tribe Aedini (Diptera: Culicidae). Part XIX. Genus *Danielsia* Theobald. *Contributions of the American Entomological Institute, 35*(3), 1–9.

Reinert, J.F. (2008f). Comparative anatomy of the female genitalia of generic-level taxa in tribe Aedini (Diptera: Culicidae). Part XX. Genus *Rampamyia* Reinert, Harbach and Kitching. *Contributions of the American Entomological Institute, 35*(3), 11–18.

Reinert, J.F. (2008g). Comparative anatomy of the female genitalia of generic-level taxa in tribe Aedini (Diptera: Culicidae). Part XXI. Genus *Patmarksia* Reinert, Harbach and Kitching. *Contributions of the American Entomological Institute, 35*(3), 19–27.

Reinert, J.F. (2008h). Comparative anatomy of the female genitalia of generic-level taxa in tribe Aedini (Diptera: Culicidae). Part XXII. Genus *Downsiomyia* Vargas. *Contributions of the American Entomological Institute, 35*(3), 29–37.

Reinert, J.F. (2008i). Comparative anatomy of the female genitalia of generic-level taxa in tibe Aedini (Diptera: Culicidae). Part XXIII. Genus *Tanakaius* Reinert, Harbach and Kitching. *Contributions of the American Entomological Institute, 35*(4), 1–9.

Reinert, J.F. (2008j). Comparative anatomy of the female genitalia of generic-level taxa in tribe Aedini (Diptera: Culicidae). Part XXIV. Genus *Vansomerensis* Reinert, Harbach and Kitching. *Contributions of the American Entomological Institute, 35*(4), 11–19.

Reinert, J.F. (2008k). Comparative anatomy of the female genitalia of generic-level taxa in tribe Aedini (Diptera: Culicidae). Part XXV. Genus *Dobrotworskyius* Reinert, Harbach and Kitching. *Contributions of the American Entomological Institute, 35*(4), 21–30.

Reinert, J.F. (2008l). Comparative anatomy of the female genitalia of generic-level taxa in tribe Aedini (Diptera: Culicidae). Part XXVI. Genus *Collessius* Reinert, Harbach and Kitching. *Contributions of the American Entomological Institute, 35*(4), 31–41.

Reinert, J.F. (2008m). Redescription of the holotype of *Finlaya lepchana* Barraud, 1923 (Diptera: Culicidae: Aedini). *Zootaxa, 1767*, 64–66.

Reinert, J.F. (2009a). Comparative anatomy of the female genitalia of generic-level taxa in tribe Aedini (Diptera: Culicidae). Part XXVII. Genus *Hopkinsius* Reinert, Harbach and Kitching. *Contributions of the American Entomological Institute, 36*(1), 1–11.

Reinert, J.F. (2009b). Comparative anatomy of the female genitalia of generic-level taxa in tribe Aedini (Diptera: Culicidae). Part XXVIII. Genus *Luius* Reinert, Harbach and Kitching. *Contributions of the American Entomological Institute, 36*(1), 13–21.

Reinert, J.F. (2009c). Comparative anatomy of the female genitalia of generic-level taxa in tribe Aedini (Diptera: Culicidae). Part XXIX. Genus *Jihlienius* Reinert, Harbach and Kitching. *Contributions of the American Entomological Institute, 36*(1), 23–32.

Reinert, J.F. (2009d). Comparative anatomy of the female genitalia of generic-level taxa in tribe Aedini (Diptera: Culicidae). Part XXX. Genus *Gilesius* Reinert, Harbach and Kitching. *Contributions of the American Entomological Institute, 36*(1), 33–42.

Reinert, J.F. (2009e). Descriptions of *Danielsia* Theobald, a genus recently resurrected from synonymy (Diptera: Culicidae: Aedini), and its type species *Dn. albotaeniata* Leicester. *Zootaxa, 1976*, 25–38.

Reinert, J.F. (2009f). List of abbreviations for currently valid generic-level taxa in family Culicidae (Diptera). *European Mosquito Bulletin, 27*, 68–76.

Reinert, J.F. (2010a). Comparative anatomy of the female genitalia of generic-level taxa in tribe Aedini (Diptera: Culicidae). Part XXXI. Genus *Sallumia* Reinert, Harbach and Kitching. *Contributions of the American Entomological Institute, 36*(2), 1–10.

Reinert, J.F. (2010b). Comparative anatomy of the female genitalia of generic-level taxa in tribe Aedini (Diptera: Culicidae). Part XXXII. Genus *Jarnellius* Reinert, Harbach and Kitching. *Contributions of the American Entomological Institute, 36*(2), 11–20.

Reinert, J.F. (2010c). Comparative anatomy of the female genitalia of generic-level taxa in tribe Aedini (Diptera: Culicidae). Part XXXIII. Genus *Lewnielsenius* Reinert, Harbach and Kitching. *Contributions of the American Entomological Institute, 36*(2), 21–30.

Reinert, J.F. (2010d). Comparative anatomy of the female genitalia of generic-level taxa in tribe Aedini (Diptera: Culicidae). Part XXXIV. Genus *Catageiomyia* Theobald. *Contributions of the American Entomological Institute, 36*(2), 31–42.

Reinert, J.F. (2010e). Comparative anatomy of the female genitalia of generic-level taxa in tribe Aedini (Diptera: Culicidae). Part XXXV. Genus *Elpeytonius* Reinert, Harbach and Kitching. *Contributions of the American Entomological Institute, 36*(3), 1–11.

Reinert, J.F. (2010f). Comparative anatomy of the female genitalia of generic-level taxa in tribe Aedini (Diptera: Culicidae). Part XXXVI. Genus *Polyleptiomyia* Theobald. *Contributions of the American Entomological Institute, 36*(3), 13–22.

Reinert, J.F. (2010g). Comparative anatomy of the female genitalia of generic-level taxa in tribe Aedini (Diptera: Culicidae). Part XXXVII. Genus *Bifidistylus* Reinert, Harbach and Kitching. *Contributions of the American Entomological Institute, 36*(3), 23–34.

Reinert, J.F. (2010h). Comparative anatomy of the female genitalia of generic-level taxa in tribe Aedini (Diptera: Culicidae). Part XXXVIII. Genus *Petermattinglyius* Reinert, Harbach and Kitching. *Contributions of the American Entomological Institute, 36*(3), 13–22.

Reinert, J.F. (2010i). Comparative anatomy of the female genitalia of generic-level taxa in tribe Aedini (Diptera: Culicidae). Part XXXIX. List of descriptions of generic-level taxa. *Contributions of the American Entomological Institute, 36*(3), 49–67.

Reinert, J.F. (2010j). List of aedine species with published illustrations and/or descriptions of female genitalia (Diptera: Culicidae: Aedini). *European Mosquito Bulletin, 28*, 1–31.

Reinert, J.F. (2010k). Species of Culicidae (Diptera) with published illustrations and/or descriptions of female genitalia—summary. *European Mosquito Bulletin, 28*, 98–100.

Reinert, J.F. (2010l). Species of mosquitoes (Diptera: Culicidae) with published illustrations and/or descriptions of eggs—summary. *European Mosquito Bulletin, 28*, 182–186.

Reinert, J.F., & Harbach, R.E. (1992). Description of the larva of *Aedes* (*Verrallina*) *lugubris* (Diptera: Culicidae). *Mosquito Systematics, 24*(3), 253–256.

Reinert, J.F., & Harbach, R.E. (2005a). Corrections to the spelling of species in genera *Psorophora* and *Verrallina*, with a note on the correct gender of *Udaya*. *Journal of the American Mosquito Control Association, 21*(2), 227–228.

Reinert, J.F., & Harbach, R.E. (2005b). Description of *Finlaya* Theobald, a genus in tribe Aedini (Diptera: Culicidae) and its type species *Fl. kochi* (Doenitz). *Journal of the American Mosquito Control Association, 21*(2), 154–170.

Reinert, J.F., & Harbach, R.E. (2006). Description of genus *Downsiomyia* Vargas (Diptera: Culicidae: Aedini) and its type species *Do. nivea* (Ludlow). *Zootaxa, 1196*, 33–61.

Reinert, J.F., & Ramalingam, S. (1983). *Aedes* (*Stegomyia*) *platylepidus*, first description of the pupa and larva (Diptera: Culicidae). *Mosquito Systematics, 15*(4), 337–344.

Reinert, J.F., Kaiser, P.E., & Seawright, J.A. (1997). Analysis of the *Anopheles* (*Anopheles*) *quadrimaculatus* complex of sibling species (Diptera: Culicidae) using morphological, cytological, molecular, genetic, biochemical, and ecological techniques in an integrated approach. *Journal of the American Mosquito Control Association, 13*(Suppl.), 1–102.

Reinert, J.F., Harbach, R.E., & Kitching, I.J. (2004). Phylogeny and classification of Aedini (Diptera: Culicidae), based on morphological characters of all life stages. *Zoological Journal of the Linnean Society, 142*, 289–368.

Reinert, J.F., Harbach, R.E., & Huong, V.D. (2006). *Aedes* (*Finlaya*) *alongi* Galliard and Ngu (Diptera: Culicidae: Aedini): Description of the female, redescription of the fourth-instar larva and reassignment to genus *Bothaella* Reinert. *Zootaxa, 1135*, 49–55.

Reinert, J.F., Harbach, R.E., & Kitching, I.J. (2006). Phylogeny and classification of *Finlaya* and allied taxa (Diptera: Culicidae: Aedini) based on morphological data from all life stages. *Zoological Journal of the Linnaean Society, 148*(1), 1–101.

Reinert, J.F., Harbach, R.E., & Kitching, I.J. (2008). Phylogeny and classification of *Ochlerotatus* and allied taxa (Diptera: Culicidae: Aedini) based on morphological data from all life stages. *Zoological Journal of the Linnean Society, 153*, 29–114.

Reinert, J.F., Harbach, R.E., & Kitching, I.J. (2009). Phylogeny and classification of tribe Aedini (Diptera: Culicidae). *Zoological Journal of the Linnean Society, 157*(4), 700–794.

Reisen, W.K. (2012). The contrasting bionomics of *Culex* mosquitoes in western North America. *Journal of the American Mosquito Control Association, 28*(Suppl. 4), 82–91.

Reisen, W.K. (2013). Ecology of West Nile virus in North America. *Viruses, 5*(9), 2079–2105.

Reisen, W.K., & Lothrop, H.D. (1995). Population ecology and dispersal of *Culex tarsalis* (Diptera: Culicidae) in the Coachella Valley of California. *Journal of Medical Entomology, 32*(4), 490–502.

Reisen, W.K., Burns, J.P., & Basio, R.G. (1972). A mosquito survey of Guam, Marianas Islands with notes on the vector borne disease potential. *Journal of Medical Entomology, 9*(4), 319–324.

Reisen, W.K., Milby, M.M., & Meyer, R.P. (1992). Population dynamics of adult *Culex* mosquitoes (Diptera: Culicidae) along the Kern River, Kern County, California, in 1990. *Journal of Medical Entomology, 29*(3), 531–543.

Reisen, W.K., Thiemann, T., Barker, C.M., Lu, H., Carroll, B., Fang, Y., & Lathrop, H.D. (2010). Effects of warm winter temperature on the abundance and gonotrophic activity of *Culex* (Diptera: Culicidae) in California. *Journal of Medical Entomology, 47*(2), 230–237.

Reiskind, M.H., Walton, E.T., & Wilson, M.L. (2004). Nutrient-dependent reduced growth and survival of larval *Culex restuans* (Diptera: Culicidae): Laboratory and field experiments in Michigan. *Journal of Medical Entomology, 41*(4), 650–656.

Reiter, M.E., & LaPointe, D.A. (2007). Landscape factors influencing the spatial distribution and abundance of mosquito vector *Culex quinquefasciatus* (Diptera: Culicidae) in a mixed residential-agricultural community in Hawaii. *Journal of Medical Entomology, 44*(5), 861–868.

Reiter, P. (2007). Oviposition, dispersal, and survival in *Aedes aegypti*: Implications for the efficacy of control strategies. *Vector-Borne and Zoonotic Diseases, 7*(2), 261–273.

Reiter, P., Amador, M.A., Anderson, R.A., & Clark, G.G. (1995). Short report: Dispersal of *Aedes aegypti* in an urban area after blood feeding as demonstrated by rubidium-marked eggs. *The American Journal of Tropical Medicine and Hygiene, 52*(2), 177–179.

Rejmánková, E., Savage, H.M., Rejmanek, M., Arredondo-Jimenez, J.I., & Roberts, D.R. (1991). Multivariate analysis of relationships between habitats, environmental factors and occurrence of anopheline mosquito larvae *Anopheles albimanus* and *A. pseudopunctipennis* in southern Chiapas, Mexico. *Journal of Applied Ecology, 28*(3), 827–841.

Rejmánková, E., Pope, K.O., Roberts, D.R., Lege, M.G., Andre, R., Greico, J., & Alonzo, Y. (1998). Characterization and detection of *Anopheles vestitipennis* and *Anopheles punctimacula* (Diptera: Culicidae) larval habitats in Belize with field survey and SPOT satellite imagery. *Journal of Vector Ecology, 23*(1), 74–88.

Rejmánková, E., Grieco, J., Achee, N., & Roberts, D.R. (2013). Ecology of larval habitats. In S. Manguin (Ed.), Anopheles *mosquitoes: New insights into malaria vectors* (pp. 397–446). Rijeka, Croatia: InTech.

Rempel, J.G. (1950). A guide to the mosquito larvae of western Canada. *Canadian Journal of Research, 28*, 207–248.

Rempel, J.G. (1953). The mosquitoes of Saskatchewan. *Canadian Journal of Research, 31*, 433–509.

Reuben, R. (1967). *Aedes* (*Diceromyia*) *ramachandrai* n. sp. (Diptera: Culicidae) from southern India. *Journal of Medical Entomology, 4*(2), 234–236.

Reuben, R. (1969). A redescription of *Culex vishnui* Theo., with notes on *C. pseudovishnui* Colless and C. *tritaeniorhynchus* Giles, from southern India. *Bulletin of Entomological Research, 58*(3), 643–652.

Reuben, R., & Suguna, S.G. (1983). Morphological differences between sibling species of the taxon *Anopheles subpictus* Grassi in India, with notes on relationships with known forms. *Mosquito Systematics, 15*(2), 117–126.

Reuben, R., Tewari, S.C., Hiriyan, J., & Akiyama, J. (1994). Illustrated keys to species of *Culex* (*Culex*) associated with japanese encephalitis in Southeast Asia (Diptera: Culicidae). *Mosquito Systematics, 26*(2), 75–96.

Reynolds, D.R., Smith, A.D., Mukhopadhyay, S., Chowdhury, A.K., De, B.K., Nath, P.S., . . . Mukhopadhyay, S. (1996). Atmospheric transport of mosquitoes in northeast India. *Medical and Veterinary Entomology, 10*(2), 185–186.

Riaz, M.A., Poupardin, R., Reynaud, S., Strode, C., Ranson, H., & David, J.P. (2009). Impact of glyphosate and benzo[a]pyrene on the tolerance of mosquito larvae to chemical insecticides. Role of detoxification genes in response to xenobiotics. *Aquatic Toxicology, 93*(1), 61–69.

Ribeiro, H. (1966). Research on the mosquitoes of Angola (Diptera: Culicidae). II. Some new culicine records. *Anais do Instituto de Hygiene e Medicina Tropical, 23*, 163–166.

Ribeiro, H. (1969). Research on the mosquitoes of Angola. IV. Description of *Anopheles* (*Cellia*) *azevedoi* sp. nov. (Diptera, Culicidae). *Anais da Escola Nacional de Saude Pública e de Medicina Tropical, 3*(1–4), 113–123.

Ribeiro, H. (1991). Research on the mosquito subfamily Toxorhynchitinae (Diptera: Culicidae). I. The Afrotropical group *brevipalpis* (adults) (1). *Arquivos do Museu Bocage* [Universidade de Lisboa; Museu e Laboratório Zoológico e Antropológico], *2*(3), 31–62.

Ribeiro, H. (1992a). On the terrestrial and riverine fauna of the Democratic Republic of São Tome and Principe. 7. Description of *Toxorhynchites capelai* sp. nov. (Diptera, Culicidae). *García de Orta: Serie de Zoologia, 19*(1–2), 135–138.

Ribeiro, H. (1992b). Research on the mosquito subfamily *Toxorhynchitinae* (Diptera: Culicidae) II. Description of the Afrotropical subgenus *Afrorhynchus* subgen. nov. *Mosquito Systematics, 23*(for 1991)(3), 195–198.

Ribeiro, H. (1992c). Research on the mosquito subfamily Toxorhynchitinae (Diptera: Culicidae). 2. Description of two new species from Angola. *Actas do Congresso Iberico de Entomología, 5*(2), 187–196.

Ribeiro, H. (1997). New species of *Toxorhynchites* (Diptera: Culicidae) from Macau (China). *Journal of the American Mosquito Control Association, 13*(3), 213–217.

Ribeiro, H. (2004). The *Toxorhynchites* Theobald of Madagascar (Diptera: Culicidae). *Annales de la Société Entomologique de France, 40*(3–4), 243–257.

Ribeiro, H. (2005a). Three new species and one new subspecies of *Toxorhynchites* (Diptera: Culicidae) of the Afrotropical region. *Journal of Medical Entomology, 42*(3), 224–232.

Ribeiro, H. (2005b). Systematics and identification of Afrotropical Toxorhynchitinae (Diptera: Culicidae). *Journal of Medical Entomology, 42*(6), 953–958.

Ribeiro, H., & da Cunha Ramos, H. (1973). Research on the mosquitoes of Angola. 8. The genus *Aedes* Meigen, 1818 (Diptera: Culicidae). Check-list with new records, keys to females and larvae, distribution and taxonomic and bioecological notes. *Anais do Instituto de Higiene e Medicina Tropical, 1*(1–4), 107–138.

Ribeiro, H., & da Cunha Ramos, H. (1974). Research on the mosquitoes of Angola (Diptera, Culicidae) VII. Redescription of the larva of *Aedes durbanensis durbanensis* (Theo., 1903) and description of *Aedes durbanensis angolae* subsp. nov. *Mosquito Systematics, 6*(2), 93–98.

Ribeiro, H., & da Cunha Ramos, H. (1975). Research on the mosquitoes of Angola. VI. The genus *Anopheles* Meigen 1818 (Diptera. Culicidae). Check-list with new records, keys to the females and larvae, distribution and bioecological notes. *García de Orta: Serie de zoologica, 4*(1), 1–40.

Ribeiro, H., & da Cunha Ramos, H. (1980). Research on the mosquitoes of Angola (Diptera: Culicidae). X. The genus *Culex* L., 1758. Check-list with new records, keys to females and larvae, distribution, and taxonomic and bioecological notes. *Junta de Investigaciones Ultlramar, Estudos, Ensaios et Documentos, 134,*13–175.

Ribeiro, H., & Mexia, J.T. (1966). Research on the mosquitoes of Angola (Diptera: Culicidae). III. A culicine survey in the Lobito-Catumbela region. *Anais do Instituto de Hygiene e Medicina Tropical, 23*(1–2), 167–182.

Ribeiro, H., da Cunha Ramos, H., Antunes Capela, R., & Alves Pires, C. (1977). Research on the mosquitoes of Portugal (Diptera, Culicidae). III. Further five new mosquito records. *Garcia de Orta: Série de Zoologia, 6*(1–2), 51–60.

Ribeiro, H., da Cunha Ramos, H., Pires, C.A., & Capela, R.A. (1979). Description and biometric study of *Anopheles* (*Cellia*) *quadriannulatus davidsoni* ssp. n., a seventh member of the *Anopheles gambiae* Giles complex (Diptera, Culicidae) endemic to the Cape Verde archipelago. *Garcia de Orta: Série de Zoologia, 8*(1–2), 75–88.

Ribeiro, H., da Cunha Ramos, H., Capela, R.A., & Pires, C.A. (1980a). *Os mosquitos de Cabo Verde* (*Diptera: Culicidae*). *Sistematica, distribuição, bioecologia e importancia médica*. Ensaios e Documentos. Lisboa: Junta de Investigações Científicas do Ultramar.

Ribeiro, H., da Cunha Ramos, H., Pires, C.A., & Capela, R.A. (1980b). Research on the mosquitos of Portugal (Diptera, Culicidae). 4. Two new anopheline records. *García de Orta: Série de zoologia, 9*(1–2), 129–137.

Ribeiro, H., Danilov, V.N., & da Cunha Ramos, H. (1982). Description of the fourth instar larva of *Culex* (*Culex*) *trifilatus trifilatus* Edwards, 1914 (Diptera: Culicidae). *Mosquito Systematics, 14*(2), 94–101.

Ribeiro, H., da Cunha Ramos, H., Pires, C.A., & Capela, R.A. (1987). Research of the mosquitoes of Portugal (Diptera, Culicidae). IX. A new anopheline record. *García De Orta: Serie de Zoologia, 12*(for 1985)(1–2), 105–112.

Ribeiro, H., da Cunha Ramos, H., Pires, C.A., & Capela, R.A. (1988). An annotated checklist of the mosquitoes of continental Portugal (Diptera, Culicidae). *Actas Congreso Ibérico de Entomología, Granada, 3,* 233–254.

Ribeiro, H., da Cunha Ramos, H., Capela, R.A., & Pires, C.A. (1998). Os mosquitos (Diptera: Culicidae) da Îlha de São Tome. *García de Orta: Série de zoologia, 22*(1–2), 1–20.

Ribeiro, J.M.C. (2000). Blood-feeding in mosquitoes: Probing time and salivary gland anti-haemostatic activities in representatives of three genera (*Aedes, Anopheles, Culex*). *Medical and Veterinary Entomology, 14*(2), 142–148.

Ribeiro, J.M.C., & Francischetti, I.M. (2003). Role of arthropod saliva in blood feeding: *Sialome* and post-*sialome* perspectives. *Annual Review of Entomology, 48*(1), 73–88.

Ribeiro, J.M.C., Charlab, R., Pham, V.M., Garfield, M.K., & Valenzuela, J.G. (2004). An insight into the salivary transcriptome and proteome of the adult female mosquito *Culex pipiens quinquefasciatus*. *Insect Biochemistry and Molecular Biology, 34*(6), 543–563.

Ribeiro, J.M.C., Arcà, B., Lombardo, F., Calvo, E., Phan, V.M., Chandra, P.K., & Wikel, S.K. (2007). An annotated catalogue of salivary gland transcripts in the adult female mosquito, *Aedes aegypti*. *BMC Genomics, 8*(6), Article 1.

Ricci, I., Damiani, C., Scuppa, P., Mosca, M., Crotti, E., Rossi, P., . . . Chouaia, B. (2011a). The yeast *Wickerhamomyces anomalus* (*Pichia anomala*) inhabits the midgut and reproductive system of the Asian malaria vector *Anopheles stephensi*. *Environmental Microbiology, 13*(4), 911–921.

Ricci, I., Mosca, M., Valzano, M., Damiani, C., Scuppa, P., Rossi, P., . . . Esposito, F. (2011b). Different mosquito species host *Wickerhamomyces anomalus* (*Pichia anomala*): Perspectives on vector-borne diseases symbiotic control. *Antonie Van Leeuwenhoek, 99*(1), 43–50.

Ricci, I., Damiani, C., Capone, A., DeFreece, C., Rossi, P., & Favia, G. (2012). Mosquito/microbiota interactions: From complex relationships to biotechnological perspectives. *Current Opinion in Microbiology, 15*(3), 278–284.

Richards, C.S. (1956). *Aedes melanimon* Dyar and related species. *The Canadian Entomologist, 88*, 261–269.

Rickenbach, A. (1967). Description d'un nouvel *Eretmapodites* (Diptera, Culicidae) capturé au Cameroun. *Bulletin de la Société de pathologie exotique, 59*(for 1966)(3), 374–376.

Rickenbach, A., & Eouzan, J.-P. (1970). Description de quatre *Eretmapodites* nouveaux du groupe *plioleucus* Edwards 1941, capturés au Cameroun (Diptera, Culicidae). *Cahiers ORSTOM. Série entomologie médicale et parasitologie, 8*(2), 131–135.

Rickenbach, A., & Ferrara, L. (1964). Description de la larve et de la nymphe de *Aedes* (*Aedimorphus*) *rickenbachi* Hamon et Adam 1959 (Diptera: Culicidae). *Bulletin de la Société de pathologie exotique, 57,* 972–978.

Rickenbach, A., & Ferrara, L. (1965). Description de deux nouveaux *Aedes* du Cameroun appartenant au sous-genre *Aedimorphus* (Diptera, Culicidae). *Bulletin de la Société de pathologie exotique, 58*(1), 24–29.

Rickenbach, A., & Hamon, J. (1966). Description d'une nouvelle espèce *d'Orthopodomyia* (Diptera, Culicidae) capturée au Cameroun: *O. nkolbissonensis* n. sp. *Bulletin de la Société de pathologie exotique, 58*(for 1965)(6), 1112–1117.

Rickenbach, A., & Lombrici, G. (1974). Un nouvel *Eretmapodites* Theobald capturé au Cameroun: *E. eouzani* n. sp. (Dipt. Culicidae). *Bulletin de la Société Entomologique de France, 79*(7–8), 218–222.

Rickenbach, A., & Lombrici, G. (1975). Les *Eretmapodites* du groupe *leucopus* Graham, 1909. Position systematique des deux sous-espèces actuellement connues et description de deux espèces nouvelles. *Bulletin de la Société Entomologique de France, 80*(1–2), 39–42.

Rickenbach, A., Ferrara, L., & Eouzan, J.P. (1968). Deux *Eretmapodites* nouveaux du Cameroun, appartenant au groupe *Oedipodius* Graham, 1909 (Dipt. Culicidae). *Bulletin de la Société Entomologique de France, 72*(for 1967)(10), 297–300.

Rickenbach, A., Eouzan, J.-P., Ferrara, L., & Bailly-Choumara, H. (1976). Données nouvelles sur la presence, la frequence et la repartition des Toxorhynchitinae et Culicinae (Diptera, Culicidae) au Cameroun. 2. Genres *Eretmapodites* et *Culex*. *Cahiers ORSTOM. Série entomologie médicale et parasitologie, 14*(2), 93–100.

Riddiford, L.M. (2013). Microarrays reveal discrete phases in juvenile hormone regulation of mosquito reproduction. *Proceedings of the National Academy of Sciences of the United States of America, 110*(24), 9623–9624.

Riddiford, L.M., Cherbas, P., & Truman, J.W. (2000). Ecdysone receptors and their biological actions. *Vitamins and Hormones, 60*, 1–73.

Riehle, M.A., Garczynski, S.F., Crim, J.W., Hill, C.A., & Brown, M.R. (2002). Neuropeptides and peptide hormones in *Anopheles gambiae*. *Science, 298*(5591), 172–175.

Rings, R.W., & Hill, S.O. (1946). The larva of *Aedes* (*Ochlerotatus*) *mathesoni* Middlekauff (Diptera, Culicidae). *Proceedings of the Entomological Society of Washington, 48*, 237–240.

Rings, R.W., & Hill, S.O. (1948). The taxonomic status of *Aedes mathesoni* (Diptera, Culicidae). *Proceedings of the Entomological Society of Washington, 50*(2), 41–49.

Rioux, J.A. (1958). Les culicides du Midi Mediterraneen étude systematique et écologique. *Encyclopedie entomologique A, 35*, 1–303.

Rioux, J.A. (1959). Liste sommaire des Culicides du Nord-Tchad. *Annales de parasitologie humaine et comparée,34*(5–6), 733.

Rioux, J.A. (1961). Contribution a l'étude des culicides (Diptera-Culicidae) du Nord-Tchad. In J.A. Rioux (Ed.), *Mission epidemiologique au Nord Tchad*. Paris, France: Arts et Métiers Graphiques

Rioux, J.A., & Arnold, M. (1955). Les culicides de Camargue (étude systématique et écologique). *La Terre et la Vie. Revue d'Histoire Naturelle, 4*, 244–286.

Rioux, J.A., & Pech, J.M. (1959). Le biotype autogène de *Culex pipiens* L. ne doit pas être nomme *Culex molestus* Forskål (Diptera, Culicidae). *Cahiers des naturalites: Bulletin de naturalistes parisiens 15*, 115–117.

Rioux, J.A., & Quezel, P. (1957). Presence de *Culex* (*Lasiosiphon*) *adairi* Kirkpatrick 1924 au Tibesti. *Bulletin de la Société d'Histoire Naturelle de l'Afrique du Nord, 18*, 461–470.

Rioux, J.A., Guilvard, E., & Pasteur, N. (1998). Description of *Aedes* (*Ochlerotatus*) *coluzzii* n. sp. (Diptera, Culicidae) sibling species A from the *detritus* complex. *Parassitologia, 40*(3), 353–360.

Ritchie, S.A., & Montague, C.L. (1995). Simulated populations of the black salt marsh mosquito (*Aedes taeniorhynchus*) in a Florida mangrove forest. *Ecological Modelling, 77*(2), 123–141.

Rivola, E., & Holstein, M.H. (1957). Note sur une variété d'*Anopheles nili* Theo. *Bulletin de la Société de pathologie exotique, 50*(3), 382–387.

Rjazantzeva, A.E. (1970). The structure of female genitalia of bloodsucking mosquitoes of the genus *Aedes* (Diptera, Culicidae). *Parassitologia, 4*(5), 401–407.

Rjazantzeva, A.E. (1972). The structure of female genitalia in bloodsucking mosquitoes of the subgenus *Ochlerotatus* (Diptera, Culicidae). *Parassitologia, 6*(1), 35–47.

Rjazantzeva, A.E. (1985). Morphological features of the female genitalia of bloodsucking Diptera (Culicidae) and their use in systematics. In O.A. Skarlato (Ed.), *Systematics of Diptera (Insecta) ecological and morphological principles* (pp. 129–131). Academy of Science of the USSR, Institute of Zoology. (Translated from Russian, Amerind, New Delhi, India)

Robert, L.L., Schneider, I., & Wirtz, R.A. (1991). Deet and permethrin as protectants against malaria-infected and uninfected *Anopheles stephensi* mosquitoes. *Journal of the American Mosquito Control Association, 7*(2), 304–306.

Robert, V., Günay, F., Le Goff, G., Boussès, P., Sulesco, T., Khalin, A., Medlock, J.M., Kampen, H., Petrić, D., & Schaffner, F. (2019). Distribution chart for Euro-Mediterranean mosquitoes (western Palaearctic region). *Journal of the European Mosquito Control Association, 37*, 1–28.

Roberts, D.R., & Scanlon, J.E. (1975). The ecology and behavior of *Aedes Atlanticus* D. & K. and other species with reference to Keystone virus in the Houston area Texas USA. *Journal of Medical Entomology, 12*(5), 537–546.

Roberts, D.R., & Scanlon, J.E. (1979). An evaluation of morphological characters for separating females of *Aedes (Ochlerotatus) atlanticus* Dyar and Knab, and *Aedes (Ochlerotatus) tormentor* Dyar and Knab (Diptera: Culicidae). *Mosquito Systematics, 11*(3), 203–208.

Roberts, D.R., Tren, R., & Hess, K. (2013). Perspectives on barriers to control of *Anopheles* mosquitoes and malaria. In S. Manguin (Ed.), Anopheles *mosquitoes: New insights into malaria vectors* (pp. 635–669). Rijeka, Croatia: InTech.

Robertson, H.M., Navik, J.A., Waldens, K.K.O., & Honegger, H.W. (2007). The bursicon gene in mosquitoes: An unusual example of mRNA trans-splicing. *Genetics, 176*(2), 1351–1353.

Robineau-Desvoidy, A.J.B. (1827). Essai sur la tribu des culicides. *Société d'histoire naturelle, 3*, 390–413.

Robinson, G.G. (1948). Mosquitoes caught in Northern Rhodesia at Balovale and Livingstone. *Journal of the Entomological Society of Southern Africa, 11*, 63–67.

Robinson, G.G. (1950). A new species of *Aedes* (*Finlaya*) from Northern Rhodesia. *Journal of the Entomological Society of Southern Africa, 13*, 80–82.

Rocha, G.P., Lourenço-de-Oliveira, R., & Motta, M.A. (2012). *Wyeomyia exallos*, a new species of sylvatic mosquito (Diptera: Culicidae) from Brazil. *Memórias do Instituto Oswaldo Cruz, 107*(7), 928–934.

Rochlin, I., Gaugler, R., Williges, E., & Farajollahi, A. (2012). The rise of the invasives and decline of the natives: Insights revealed from adult populations of container-inhabiting *Aedes* mosquitoes (Diptera: Culicidae) in temperate North America. *Biological Invasions, 15*(5), 991–1003.

Rochlin, I., Ninivaggi, D.V., Hutchinson, M.L., & Farajollahi, A. (2013). Climate change and range expansion of the Asian tiger mosquito (*Aedes albopictus*) in northeastern USA: Implications for public health practitioners. *PLOS ONE, 8*(4), e60874.

Rodenwaldt, E. (1925). Entomologische notities. III. *Geneeskundig tijdschrift voor Nederlandsch-Indië, 65*, 173–201.

Rodenwaldt, E. (1926). Entomologische notitides. IV. *Geneeskundig tijdschrift voor Nederlandsch-Indië, 66*, 787–799.

Rodhain, F., & Boutonnier, A. (1983a). Description complementaire d'*Aedes madagascarensis* van Someren 1949. *Archives de l'Institut Pasteur de Madagascar, 49*(1982), 197–202.

Rodhain, F., & Boutonnier, A. (1983b). Description d'un nouvel *Aedes* endemique de Madagascar: *Ae.* (*Diceromyia*) *coulangesi* nova species. *Archives de l'Institut Pasteur de Madagascar, 49*(for 1982), 193–196.

Rodhain, F., & Boutonnier, A. (1983c). Description d'un nouvel *Aedes* du sous genre *Ochlerotatus* (Diptera: Culicidae): *Aedes ambreensis* nova species et considerations generales sur les femelles de sous-genre *Ochlerotatus* dans la région afro-tropicale. *Bulletin de la Société de pathologie exotique, 76*, 825–833.
Rodhain, F., & Boutonnier, A. (1983d). *Aedes*, subgenus *Diceromyia* (Diptera: Culicidae) in Madagascar. *Mosquito Systematics, 15*(4), 330–336.
Rodhain, F., & Boutonnier, A. (1984a). Checklist of the type specimens of mosquitoes (Diptera: Culicidae) in the medical entomology collections of the Pasteur Institute in Paris. *Mosquito Systematics, 16*(4), 271–281.
Rodhain, F., & Boutonnier, A. (1984b). Le genre *Orthopodomyia* (Diptera: Culicidae) à Madagascar. *Archives de l'Institut Pasteur de Madagascar, 51*(for 1983)(1), 203–248.
Rodhain, F., & Boutonnier, A. (1985). Checklist of the type specimens of mosquitoes (Diptera: Culicidae) in the medical entomology collections of the Pasteur Institute in Paris. *Mosquito Systematics, 16*(for 1984)(4), 271–281.
Rodhain, F., & Gaxotte, P. (1977). An entomological survey on the mosquitoes of Wuvulu Island, Papua-New Guinea. *Southeast Asian Journal of Tropical Medicine and Public Health, 8*(1), 77–79.
Rodhain, F., Boutonnier, A., & Coulanges, P. (1977). Bibliographie des Culicides de Madagascar. *Archives de l'Institut Pasteur de Madagascar, 46*, 485–495.
Rodriguez, M.H., Chavez, B., Orozco, A., Loyola, E.G., & Martinez-Palomo, A. (1992). Scanning electron microscopic observations of *Anopheles albimanus* (Diptera: Culicidae) eggs. *Journal of Medical Entomology, 29*(3), 400–406.
Rodriguez, M.H., Chávez, B., Ulloa, A., & Arredondo-Jiménez., J.I. (2002). Fine structure of the eggs of *Anopheles* (*Anopheles*) *punctimacula*. *Journal of the American Mosquito Control Association, 18*(1), 1–9.
Rogers, D.J., & Randolph, S.E. (2000). The global spread of malaria in a future, warmer world. *Science, 289*(5485), 1763–1766.
Rogers, D.W., Baldini, F., Battaglia, F., Panico, M., Dell, A., Morris, H.R., & Catteruccia, F. (2009). Transglutaminase-mediated semen coagulation controls sperm storage in the malaria mosquito. *PLOS Biology, 7*(2), e1000272.
Roitberg, B.D., Mondor, E.B., & Tyerman, J.G. (2003). Pouncing spider, flying mosquito: Blood acquisition increases predation risk in mosquitoes. *Behavioral Ecology, 14*(5), 736–740.
Romeo Viamonte, J.M. (1950). Los anofelinos de España y de la zona Española del Protectorado de Marruecos. Su relacioncon la difusion del paludismo. *Revista de sanidad e higiene pública* (Madrid), *24*, 213–295.
Romeo Viamonte, J.M., & Castro, M. (1951). Estudio de la morfología de la armadura faringea de algunos anofilinos. (Dipt. Culic.). *Revista de sanidad e higiene pública, 25*, 313–330.
Romero Sierra, J.M. (1907). *Contribucion al estudio de los mosquitos de Caracas* [Doctoral dissertation, Universidad Central de Venezuela, Caracas].
Romi, R., Pontuale, G., & Sabatinelli, G. (1997). Identification of the immature stages of the Italian mosquitoes (Diptera, Culicidae). Le zanzare italiane: generalità e identificazione degli stadi preimaginali (Diptera, Culicidae). *Fragmenta entomologica, 29*(Suppl.), 1–141.
Romoser, W.S., & Oviedo, M.V.N. (2011). Function of the dendritic setae in *Aedes aegypti* mosquito pupae: Float hairs don't float. *Open Access Insect Physiology, 2011, 3*, 7–11.
Romoser, W.S., Wasieloski, L.P., Pushko, P., Kondig, J.P., Lerdthusnee, K., Neira, M., & Ludwig, G.V. (2004). Evidence for arbovirus dissemination conduits from the mosquito (Diptera: Culicidae) midgut. *Journal of Medical Entomology, 41*(3), 467–475.
Romoser, W.S., Turell, M.J., Lerdthusnee, K., Neira, M., Dohm, D., Ludwig, G., & Wasieloski, L. (2005). Pathogenesis of Rift Valley fever virus in mosquitoes—tracheal conduits & the basal lamina as an extra-cellular barrier. In C.J. Peters & C.H. Calisher (Eds.), *Infectious diseases from nature: Mechanisms of viral emergence and persistence. Archives of virology.* (Suppl. 19, pp. 89–100). Vienna, Austria: Springer Verlag.
Rona, L.D.P., Carvalho-Pinto, C.J., Gentile, C., Grisard, E.C., & Peixoto, A.A. (2009). Assessing the molecular divergence between *Anopheles* (*Kerteszia*) *cruzii* populations from Brazil using the timeless gene: Further evidence of a species complex. *Malaria Journal, 8*, 60.
Rona, L.D.P., Carvalho-Pinto, C.J., & Peixoto, A.A. (2010a). Molecular evidence for the occurrence of a new sibling species within the *Anopheles* (*Kerteszia*) *cruzii* complex in southeast Brazil. *Malaria Journal, 9*, 33.
Rona, L.D.P., Carvalho-Pinto, C.J., Mazzoni, C.J., & Peixoto, A.A. (2010b). Estimation of divergence time between two sibling species of the *Anopheles* (*Kerteszia*) *cruzii* complex using a multilocus approach. *BMC Evolutionary Biology, 10*, 91.
Rona, L.D.P., Carvalho-Pinto, C.J., & Peixoto, A.A. (2013). Evidence for the occurrence of two sympatric sibling species within the *Anopheles* (*Kerteszia*) *cruzii* complex in southeast Brazil and the detection of asymmetric introgression between them using a multilocus analysis. *BMC Evolutionary Biology, 13*, 207.
Rondani, C. (1848). Esme di varie specie d'*Äôinsetti ditteri brasiliani. Studi entomologici* (Turino), *1*, 63–112.
Rondani, C. (1872). Sulle specie Italiane del genere *Culex* Lin. *Bolotino della Societa Entomologica Italiana, 4*, 29–31.
Ronderos, R.A. (1965). Notas sobre Culicidae Venezolanos (Diptera). *Physis, 25*(70), 249–253.
Ronderos, R.A., & Bachmann, A.O. (1963a). A proposito del complejo *Mansonia* (Diptera-Culicidae). *Revista de la Sociedad Entomológica Argentina, 25*(for 1962), 43–51.
Ronderos, R.A., & Bachmann, A.O. (1963b). Mansoniini neotropicales. I. (Diptera, Culicidae). *Revista de la Sociedad Entomológica Argentina, 26*, 57–65.
Ronderos, R.A., & Bachmann, A O. (1964). Mansoniini neotropicales. II. (Diptera, Culicidae). *Provincia de Buenos Aires gobernacion. Comision de Investigacion Cientifica, Notas, 2*(6), 1–8.
Ronderos, R.A., & García, M. (1963a). Description de pupas de *Aedes* argentinos (Diptera-Culicidae). *Revista de la Sociedad Entomolögica Argentina, 25*(for 1962), 35–37.
Ronderos, R.A., & García, M. (1963b). Nota sinonimica en Culicidae (Diptera). *Revista de la Sociedad Entomologica Argentina, 25*(1962), 38.
Rongnoparut, P., Sirichotpakorn, N., Rattanarithikul, R., & Linthicum, K.J. (1998). Sequence heterogeneity in *copia*-like retrotransposons in *Anopheles* (Diptera: Culicidae) in Thailand. *Journal of Medical Entomology, 35*(5), 771–777.
Root, F.M. (1922a). The classification of American *Anopheles* mosquitoes. *American Journal of Hygiene, 2*, 321–322.
Root, F.M. (1922b). Notes on mosquitoes and other blood-sucking flies from Porto Rico. *American Journal of Hygiene, 2*(4), 394–405.
Root, F.M. (1923). The male genitalia of some American *Anopheles. American Journal of Hygiene, 3*(3), 264–279.
Root, F.M. (1926). Studies on Brazilian mosquitoes. I. The anophelines of the *Nyssorhynchus* group. *American Journal of Hygiene, 6*(5), 684–717.

Root, F.M. (1927a). Note on the mosquito fauna of the Republic of Haiti. *American Journal of Hygiene, 7*(4), 463–469.

Root, F.M. (1927b). Studies on Brazilian mosquitoes. II. *Chagasia fajardoi. American Journal of Hygiene, 7*(4), 470–480.

Root, F.M. (1927c). Studies on Brazilian mosquitoes. III. The genus *Culex. American Journal of Hygiene, 7*(5), 574–598.

Root, F.M. (1927d). Studies on Brazilian mosquitoes. IV. Notes on some Brazilian species of *Anopheles. American Journal of Hygiene, 7*(5), 599–605.

Root, F.M. (1937). A new species of *Uranotaenia* (Diptera, Culicidae) from Haiti. *Journal of Parasitology, 23*(1), 98–100.

Roper, R. (1914). An account of some anopheline mosquitos found in British North Borneo, with description of a new species. *Bulletin of Entomological Research, 5*(2), 137–147.

Roque, A.B. (1903). Contribuiço para o estudio da malaria e dos mosquitos de Angola. *Medicina Contemporanea, 6*, 110–114.

Rosa-Freitas, M.G. (1989). *Anopheles (Nyssorhynchus) deaneorum*: A new species in the Albitarsis complex (Diptera: Culicidae). *Memórias do Instituto Oswaldo Cruz, 84*(4), 535–543.

Rosa-Freitas, M.G., & Deane, L.M. (1989). The neotype of *Anopheles albitarsis* (Diptera: Culicidae). *Memórias do Instituto Oswaldo Cruz, 84*(3), 289–302.

Rosa-Freitas, M.G., Deane, L.M., & Momen, H. (1990). A morphological, isoenzymatic and behavioural study of ten populations of *Anopheles (Nyssorhynchus) albitarsis* Lynch-Arribálzaga, 1878 (Diptera: Culicidae) including from the type-locality—Baradero, Argentina. *Memórias do Instituto Oswaldo Cruz, 85*(3), 275–289.

Rosa-Freitas, M.G., Tsouris, P., Peterson, A.T., Honoria, N.A., Montelrode de Barros, F.S., Barros de Aguiar, D., . . . Luitgards-Moura, J.F. (2007). An ecoregional classification for the state of Roraima, Brazil. The importance of landscape in malaria biology. *Memórias do Instituto Oswaldo Cruz, 102*(3), 349–357.

Rosen, L., & Rozeboom, L.E. (1954). Morphologic variations of larvae of the *scutellaris* group of *Aedes* (Diptera, Culicidae) in Polynesia. *American Journal of Tropical Medicine and Hygiene, 3*(3), 529–538.

Rosenberg, R. (1982). Forest malaria in Bangladesh. III. Breeding habits of *Anopheles dirus. American Journal of Tropical Medicine and Hygiene, 31*(2), 192–201.

Rosenberg, R., & Evenhuis, N.L. (1985). A new species of *Toxorhynchites* from Bangladesh (Diptera: Culicidae). *Mosquito Systematics, 17*(1), 36–43.

Rosenstiel, R.G. (1947). Dispersion and feeding habits of *Anopheles freeborni. Journal of Economic Entomology, 40*(6), 795–800.

Rosero, D.A., Jaramillo, L.M., Gutiérrez, L.A., Conn, J.E., & Correa, M.M. (2012). Genetic diversity of *Anopheles triannulatus* s.l. (Diptera: Culicidae) from Northwestern and Southeastern Colombia. *American Journal of Tropical Medicine and Hygiene, 87*(5), 910–920.

Ross, E.S. (1943). The identity of *Aedes bimaculatus* (Coquillett) and a new subspecies of *Aedes fulvus* (Wiedemann) from the United States (Diptera, Culicidae). *Proceedings of the Entomological Society of Washington, 45*(6), 143–151.

Ross, E.S. (1950). Additional illustrations and notes on *Aedes bambusicolus* Knight and Rozeboom. *Journal of the Washington Academy of Sciences, 40*, 79–82.

Ross, E.S., & Roberts, H.R. (1943a). *Mosquito atlas. Part I. The Nearctic* Anopheles, *important malaria vectors of the Americas, and* Aëdes aegypti, Culex quinquefasciatus. Philadelphia: The American Entomological Society, the Academy of Natural Sciences.

Ross, E.S., & Roberts, H.R. (1943b). *Mosquito atlas. Part II. Eighteen old world anophelines important to malaria.* Philadelphia: The American Entomological Society, the Academy of Natural Sciences.

Ross, H.H. (1947). The mosquitoes of Illinois (Diptera, Culicidae). *Bulletin of the Illinois Natural History Survey, 24*(1), 1–96.

Ross, H.H. (1951). Conflict with *Culex. Mosquito News, 11*, 128–132.

Ross, H.H. (1964). The colonization of temperate North America by mosquitoes and man. *Mosquito News, 24*(2), 103–118.

Ross, H.H., & Horsfall, W.R. (1965). A synopsis of the mosquitoes of Illinois (Diptera, Culicidae). *Illinois Natural History Survey Biological Notes, 52*, 1–50.

Ross, R. (1898). Report on a preliminary investigation into malaria in the Sigur Ghat, Ootacamund. *Indian Medical Gazette, 33*(4–5), 133–136, 170–175.

Ross, R. (1905). The logical basis of the sanitary policy of mosquito reduction. *Science, 22*(570), 689–699.

Rossi, G.C. (2006). Redescription of *Culex (Culex) brethesi* Dyar, 1919 (Diptera: Culicidae). *Zootaxa, 1312*, 25–35.

Rossi, G.C. (2017). Redescription of larva, pupa and adult of *Anopheles (Anopheles) annulipalpis* (Diptera: Culicidae) and the removal of the species of the Cycloleppteron Series. *Iheringia, Serie Zoologia, 107*, e2017025.

Rossi, G.C., & Harbach, R.E. (2008). *Phytotelmatomyia*, a new neotropical subgenus of *Culex* (Diptera: Culicidae). *Zootaxa, 1879*, 1–17.

Rossi, G.C., & Martínez, M. (2003). Mosquitoes (Diptera: Culicidae) from Uruguay. *Entomon Vect, 10*(4), 469–478.

Rossi, G.C., & Vezzani, D. (2011). An update of mosquitoes of Argentine Patagonia with new distribution records. *Journal of the American Mosquito Control Association, 27*(2), 93–98.

Rossi, G.C., Scheibler, E., & Dominguez, M.C. (2006). Morphological descriptions of the female, pupa, and larva of *Culex (Culex) cuyanus* Duret (Diptera: Culicidae). *Zootaxa, 1165*, 47–55.

Rossi, G.C., Laurito, M., & Almirón, W.R. (2008). Morphological description of the pupa and redescription of the adults and larva of *Culex (Culex) apicinus* Philippi (Diptera: Culicidae). *Zootaxa, 1941*, 31–42.

Rossi, G.C., Stein, M., & Almirón, W.R. (2008). *Psorophora (Grabhamia) varinervis* (Diptera: Culicidae) morphological description including pupa and fourth-stage larva previously unknown. *Journal of Medical Entomology, 45*(3), 342–346.

Rossi, P. (1790). *Fauna Etrusca: Sistens Insecta quce in provinciis Florentina et Pisana praesertim collegit* (Vol. 2). Liburni: Masi and Sociorum.

Rossignol, P.A., & McIver, S.B. (1977). Fine structure and role in behavior of sensilla on the terminalia of *Aedes aegypti* (L.) (Diptera: Culicidae). *Journal of Morphology, 151*(3), 419–437.

Rossignol, P.A., & Shieh, J.N. (1993). Feeding success of vectors on infected hosts. *Parasitology Today, 9*(12), 442–443.

Roth, L.M. (1944). A key to the *Anopheles* of the southeastern United States, by male genitalia (Diptera, Culicidae). *The American Midland Naturalist, 31*, 96–110.

Roth, L.M. (1945). The male and larva of *Psorophora (Janthinosoma) horrida* (Dyar and Knab) and a new species of *Psorophora* from the United States (Diptera: Culicidae). *Proceedings of the Entomological Society of Washington, 47*(1), 1–23.

Roth, L.M. (1946a). The female genitalia of the *Wyeomyia* of North America (Diptera: Culicidae). *Annals of the Entomological Society of America, 39*, 292–297.

Roth, L.M. (1946b). The larva and pupa of *Uranotaenia bimaculata* Leicester on Okinawa Shima, Ryukyu Retto (Diptera: Culicidae). *Proceedings of the Entomological Society of Washington, 48*(3), 67–75.

Rothwell, S. (1907). A new mosquito from India. *Entomologist, 40*(525), 34–36.

Roubaud, M.E. (1923). Les disharmonies de la function rénale et leurs consequences biologiques chez les moustiques. *Annales de l'Institut Pasteur, 37*(7), 43–679.

Roubaud, M.E. (1934). Un type racial nouveau de l'*Anopheles maculipennis. Bulletin de la Société de pathologie exotique, 27,* 737–740.

Roubaud, M.E. (1935a). Variété nouvelle de l'*Anopheles maculipennis* au Maroc, *A. maculipennis sicaulti* (n. var.). *Bulletin de la Société de pathologie exotique, 28*(2), 107–111.

Roubaud, M.E. (1935b). La microstructure du flotteur de l'oeuf dans les races biologiques de *Culex pipiens* L. *Bulletin de la Société de pathologie exotique, 28,* 443–445.

Roubaud, M.E. (1945). Le problème de l'espèce chez le moustique commun *Culex* pipiens L. *Bulletin de la Société de pathologie exotique, 38*(1–2), 47–60.

Roubaud, M.É. (1957). Biologie des moustiques—sur l'existence de deux entités biologiques distinctes dans la définition du moustique commun rural, *Culex pipiens pipiens. Comptes rendus des seances de l'Academie des Sciences* (Paris), *244,* 3115–3116.

Roubaud, M.E., & Ghelelovitch, S. (1956). Observations sur les moustiques anthropophile meditérranéen du groupe *pipiens, Culex berbericus* Roub. *Comptes rendus d'Academie des Sciences* (Paris), *242,* 2900–2903.

Roubaud, M.E., & Ghelelovitch, S. (1957). Sur la différenciation des elements du complexe par la morphologie du phallosome, chez le moustique commun palearctique *Culex pipiens. Comptes rendus des seances de l'Academie des Sciences* (Paris), *245,* 7–9.

Roubaud, M.E., & Treillard, M. (1936). Sur une variété portugaise do l'*Anopheles maculipennis* (groupe *atroparvus*). *Bulletin de la Société de pathologie exotique, 29,* 726–730.

Roubaud, M.E., & Treillard, M. (1940). Diversités biologiques et différenciations raciales chez l'*Anopheles claviger* Mg (*bifurcatus*). *Bulletin de la Société de pathologie exotique, 33,* 21–25.

Rouhier, M.F., & Piermarini, P.M. (2014). Identification of life-stage and tissue-specific splice variants of an inward rectifying potassium (Kir) channel in the yellow fever mosquito *Aedes aegypti. Insect Biochemistry and Molecular Biology, 48,* 91–99.

Rouhier, M.F., Raphemot, R., Denton, J.S., & Piermarini, P.M. (2014). Pharmacological validation of an inward-rectifier potassium (Kir) channel as an insecticide target in the yellow fever mosquito *Aedes aegypti. PLOS ONE, 9*(6), e100700.

Roy, D.N., & Siddons, L.B. (1939). Egg of *A. philippinensis* Ludl. *Journal of the Malaria Institute of India, 2,* 159–164.

Roy, S., Saha, T.T., Johnson, L. Zhao, B., Ha, J., White, K.P., . . . Raikhel, A.S. (2015). Regulation of gene expression patterns in mosquito reproduction. *PLOS Genetics, 11*(8), e100545.

Roy, S.G., & Raikhel, A.S. (2012). Nutritional and hormonal regulation of the TOR effector 4E-binding protein (4E-BP) in the mosquito *Aedes aegypti. The FASEB Journal, 26*(3), 1334–1342.

Roy, S.G., Hansen, I.A., & Raikhel, A.S. (2007). Effect of insulin and 20-hydroxyecdysone in the fat body of the yellow fever mosquito, *Aedes aegypti. Insect Biochemistry and Molecular Biology, 37*(12), 1317–1326.

Rozeboom, L.E. (1935). *Culex rooti,* a new *Culex* from Panama (Diptera: Culicidae). *Annals of the Entomological Society of America, 28*(2), 251–253.

Rozeboom, L.E. (1936). The larva and adult of *Culex rooti* Rozeboom. *Annals of the Entomological Society of America, 29*(2), 266–267.

Rozeboom, L.E. (1937). The egg of *Anopheles pseudopunctipennis* in Panama. *Journal of Parasitology, 23*(5), 538–539.

Rozeboom, L.E. (1938). The eggs of the *Nyssorhynchus* group of *Anopheles* (Culicidae) in Panama. *American Journal of Hygiene, 27,* 95–107.

Rozeboom, L.E. (1939). The larva of *Psorophora* (*Janthinosoma*) *horrida* Dyar & Knab (Diptera: Culicidae). *Journal of Parasitology, 25*(2), 145–147.

Rozeboom, L.E. (1941). Distribution and ecology of the *Anopheles* mosquitoes of the Caribbean Region. *Publication of the American Association for the Advancement of Science, 15,* 98–107.

Rozeboom, L.E. (1942a). Subspecific variations among Neotropical *Anopheles* mosquitoes, and their importance in the transmission of malaria. *The American Journal of Tropical Medicine and Hygiene, 22*(3), 235–255.

Rozeboom, L.E. (1942b). The mosquitoes of Oklahoma. *Technical Bulletin of the Oklahoma Agricultural and Mechanical College Agricultural Experiment Station, T-16.*

Rozeboom, L.E. (1946). Three new *Finlayas* of the *dissimilis* group (Diptera: Culicidae) from the Philippine Islands. *Journal of Parasitology, 32*(6), 587–595.

Rozeboom, L.E. (1951a). A new species of *Anopheles* from the Philippine Islands (Diptera: Culicidae). *Journal of Parasitology, 37*(5), 502–506.

Rozeboom, L.E. (1951b). The *Culex pipiens* complex in North America. *Transactions of the Royal Entomological Society of London, 102*(7), 343–353.

Rozeboom, L.E. (1952). *Anopheles* (*A.*) *earlei* Vargas, 1943, in Montana: Identity and adaptation to laboratory conditions (Diptera: Culicidae). *The American Journal of Tropical Medicine and Hygiene, 1*(3), 477–483.

Rozeboom, L.E. (1953). Sexual isolation in some North American *Anopheles* mosquitoes. *American Journal of Tropical Medicine and Hygiene, 2*(4), 677–682.

Rozeboom, L.E., & Gabaldon, A. (1941). A summary of the "*tarsimaculatus*" complex of *Anopheles* (Diptera: Culicidae). *American Journal of Hygiene, 33*(3)(Sec. C), 88–100.

Rozeboom, L.E., & Knight, K.L. (1946). The *punctulatus* complex of *Anopheles* (Diptera: Culicidae). *Journal of Parasitology, 32*(2), 95–131.

Rozeboom, L.E., & Komp, W.H.W. (1948). Three new species of *Culex* (Diptera: Culicidae) from Colombia. *Journal of Parasitology, 34*(5), 396–406.

Rozeboom, L.E., & Komp, W.H.W. (1950a). A review of the species of *Culex* of the subgenus *Melanoconion* (Diptera, Culicidae). *Annals of the Entomological Society of America, 43*(1), 75–114.

Rozeboom, L.E., & Komp, W.H.W. (1950b). A new *Microculex elongatus,* from Colombia, with notes on the subgenus (Diptera, Culicidae). *Proceedings of the Entomological Society of Washington, 52*(3), 147–157.

Rubio-Palis, Y. (2000). *Anopheles* (*Nyssorhynchus*) *de Venezuela taxonomía, bionomía, ecología e importancia médica.* Maracay, Venezuela: Escuela de Malariología y Saneamiento Ambiental "Dr. Arnoldo Gabaldon" y el Proyecto Control de Enfermedades Endémicas.

Rubio-Palis, Y., & Moreno, J. (2003). Primer registro de *Anopheles* (*Anopheles*) *calderoni* (Diptera: Culicidae) en Venezuela. *Entomotropica, 18*(2), 159–161.

Rubio-Palis, Y., & Zimmerman, R.H. (1997). Ecoregional classification of malaria vectors in the Neotropics. *Journal of Medical Entomology, 34*(5), 499–510.

Rubio-Palis, Y., Wilkerson, R.C., & Guzmán, H. (2003). Morphological characters of adult *Anopheles* (*Nyssorhynchus*) *marajoara* in Venezuela. *Journal of the American Mosquito Control Association, 19*(2), 107–114.

Rueanghiran, C., Apiwathnasorn, C., Sangthong, P., Samung, Y., & Ruangsittichai, J. (2011). Utility of a set of conserved mitochondrial cytochrome oxidase subunit I gene primers for *Mansonia annulata* identification. *Southeast Asian Journal of Tropical Medicine and Public Health, 42*(6), 1381–1387.

Rueda, L.M. (2004). Pictorial keys for the identification of mosquitoes (Diptera: Culicidae) associated with dengue virus transmission. *Zootaxa, 589*, 1–60.

Rueda, L.M. (2005). Two new species of *Anopheles (Anopheles) hyrcanus* group (Diptera: Culicidae) from the Republic of South Korea. *Zootaxa, 941*, 1–26.

Rueda, L.M. (2008). Global diversity of mosquitoes (Insecta: Diptera: Culicidae) in freshwater. *Hydrobiologia, 595*(1), 477–487.

Rueda, L.M., Patel, K.J., Axtell, R.C., & Stinner, R.E. (1990). Temperature-dependent development and survival rates of *Culex quinquefasciatus* and *Aedes aegypti* (Diptera: Culicidae). *Journal of Medical Entomology, 27*(5), 892–898.

Rueda, L.M., Peyton, EL, & Manguin, S. (2004). *Anopheles (Anopheles) pseudopunctipennis* Theobald (Diptera: Culicidae): Neotype designation and description. *Journal of Medical Entomology, 41*(1), 12–22.

Rueda, L.M., Pecor, J.E., Yuen, M.-C., & Lee, M.-W. (2004). New record, habitats and updated checklist of the mosquitoes of Hong Kong. *Journal of the American Mosquito Control Association, 20*(2), 204–207.

Rueda, L.M., Wilkerson, R.C., & Li, C. (2005). *Anopheles (Anopheles) lesteri* Baisas and Hu (Diptera: Culicidae): Neotype designation and description. *Proceedings of the Entomological Society of Washington, 107*(3), 604–622.

Rueda, L. M., Ma, Y.-J., Song, G.-H., & Gao, Q. (2005). Notes on the distribution of *Anopheles (Anopheles) sinensis* Wiedemann (Diptera: Culicidae) in China and the Status of Some *Anopheles hyrcanus* group type specimens from China. *Proceedings of the Entomological Society of Washington, 107*(1), 235–238.

Rueda, L.M., Pecor, J.E., Lowen., R. G., & Carder, M. (2008). New record and updated checklist of the mosquitoes of Afghanistan and Iraq. *Journal of Vector Ecology, 3*(2), 397–402.

Rueda, L.M., Rodriguez, J.A., Bertugio, M.C., Pecor, J.E., Li, C., & Wilkerson, R.C. (2008). *Anopheles (Nyssorhynchus) pictipennis*: A new mosquito record from the Atacama region of northern Chile. *Journal of the American Mosquito Control Association, 24*(3), 448–449.

Rueda, L.M., Brown, T.L., Kim, H.C., Klein, T.A., Thongkukiatkul, A., & Sherwood, V. (2009). Description and comparison of morphological structures of the eggs of *Anopheles hyrcanus* group and related species (Diptera: Culicidae) from the Republic of Korea. *Zootaxa, 2268*, 23–40.

Rueger, M.E. (1958). *Aedes (Ochlerotatus) barri*, a new species of mosquito from Minnesota (Diptera, Culicidae). *Journal of the Kansas Entomological Society, 31*(1), 34–46.

Ruiz, F., Quiñones, M.L., Erazo, H.F., Calle, D.A., Alzate, J.F., & Linton, Y.-M. (2005). Molecular differentiation of *Anopheles (Nyssorhynchus) benarrochi* and *An. (N.) oswaldoi* from Southern Colombia. *Memórias do Instituto Oswaldo Cruz, 100*(2), 155–160.

Ruiz, F., Linton, Y.-M., Ponsonby, D.J., Conn, J.E., Herrera, M., Quinones, M.L., . . . Wilkerson, R.C. (2010). Molecular comparison of topotypic specimens confirms *Anopheles (Nyssorhynchus) dunhami* Causey (Diptera: Culicidae) in the Colombian Amazon. *Memórias do Instituto Oswaldo Cruz, 105*(7), 899–903.

Ruiz-Lopez, F., Wilkerson, R.C., Conn, J.E., McKeon, S.N., Levin, D.M., Quinones, M.L., . . . Linton, Y.-M. (2012). DNA barcoding reveals both known and novel taxa in the Albitarsis Group (*Anopheles: Nyssorhynchus*) of Neotropical malaria vectors. *Parasites and Vectors, 5*, 44.

Ruiz-Lopez, F., Wilkerson, R.C., Ponsonby, D.J., Herrera, M., Sallum, M.A.M., Velez, I.D., . . . Linton, Y.-M. (2013). Systematics of the Oswaldoi Complex (*Anopheles, Nyssorhynchus*) in South America. *Parasites and Vectors, 6*, 13.

Russell, C., & Hunter, F. (2010). Influence of elevation and avian or mammalian hosts on attraction of *Culex pipiens* (Diptera: Culicidae) in southern Ontario. *The Canadian Entomologist, 142*(3), 250–255.

Russell, P.F., Rozeboom, L.E., & Stone, A. (1943). *Keys to the anopheline mosquitoes of the world with notes on their identification, distribution, biology, and relation to malaria.* Philadelphia, PA: American Entomological Society, Academy of Natural Sciences.

Russell, P.F., West, L.S., Manwell, R.D., & Macdonald, G. (1963). *Practical Malariology* (2nd ed.). London, England: Oxford University.

Russell, R.C. (1986a). *Culex annulirostris* Skuse (Diptera: Culicidae) at Appin, N.S.W.—bionomics and behaviour. *Journal of the Australian Entomological Society, 25*(2), 103–109.

Russell, R.C. (1986b). Larval competition between the introduced vector of dengue fever in Australia, *Aedes aegypti* (L.), and a native container-breeding mosquito, *Aedes notoscriptus* (Skuse) (Diptera: Culicidae). *Australian Journal of Zoology, 34*(4), 527–534.

Russell, R.C. (1987). Age composition and overwintering of *Culex annulirostris* Skuse (Diptera: Culicidae) near Deniliquin, in the Murray Valley of New South Wales. *Journal of the Australian Entomological Society, 26*(1), 93–96.

Russell, R.C. (2012). A review of the status and significance of the species within the *Culex pipiens* group in Australia. *Journal of the American Mosquito Control Association, 48*(Suppl. 4), 24–27.

Russell, R.C., Webb, C.E., Williams, C.R., & Ritchie, S.A. (2005). Mark–release–recapture study to measure dispersal of the mosquito *Aedes aegypti* in Cairns, Queensland, Australia. *Medical and Veterinary Entomology, 19*(4), 451–457.

Rutledge, L.C., Ward, R.A., & Bickley, W.E. (1970). Experimental hybridization of geographic strains of *Anopheles stephensi* (Diptera: Culicidae). *Annals of the Entomological Society of America, 63*(4), 1024–1030.

Ryba, J., Hajkova, Z., & Kaftan, M. (1974). Occurrence of *Uranotaenia unguiculata* Edwards, 1913 (Diptera, Culicidae) in Czechoslovakia. *Folia parasitologica, 21*, 142.

Rydzanicz, K., Czułowska, A., Manz, C., & Jawień, P. (2017). First record of *Anopheles daciae* (Linton, Nicolescu & Harbach, 2004) in Poland. *Journal of Vector Ecology, 41*(1), 196–199.

Sabrosky, C.W. (1999). *Family-group names in Diptera: An annotated catalog* (Myia, Vol. 10). Leiden, The Netherlands: Backhuys.

Saeung, A., Baimai, V., Thongsahuan, S., Otsuka, Y., Srisuka, W., Taai, K., . . . Choochote, W. (2014). Cytogenetic, cross-mating and molecular evidence of four cytological races of *Anopheles crawfordi* (Diptera: Culicidae) in Thailand and Cambodia. *Comptes rendus biologies, 337*(11), 625–634.

Sakakibara, M. (1959). On *Anopheles (Anopheles) omorii* n. sp. *Endemic Diseases Bulletin of Nagasaki University, 1*, 288–295.

Sakakibara, M., & Omori, N. (1962). Morphology of undescribed male and immature mosquitoes of aedines in Japan. 2. *Aedes (Finlaya) koreicoides. Endemic Diseases Bulletin of Nagasaki University, 4*(1), 15–21.

Salem, H.H. (1938). The mosquito fauna of Sinai Peninsula (Egypt) with a description of two new species. *Publication of the Egyptian University, Faculty of Medicine, 16*, 1–31.

Saliternik, Z. (1955). The specific biological characteristics of *Anopheles (Myzomyia) sergentii (Theo.)* and their correlation with malaria control in Israel. *Bulletin of Entomological Research, 46*(2), 445–462.

Saliternik, Z., & Theodor, O. (1942). On a new variety of *Anopheles turkhudi* from Palestine. *Journal of the Malaria Institute of India, 4*, 429–434.
Sallum, M.A.M., & Flores, D.C. (2004). Ultrastructure of the eggs of two species of *Anopheles* (*Anopheles*) Meigen (Diptera: Culicidae). *Revista Brasileira de entomología, 48*(2), 185–192.
Sallum, M.A.M., & Forattini, O.P. (1996). Revision of the Spissipes section of *Culex* (*Melanoconion*) (Diptera: Culicidae). *Journal of the American Mosquito Control Association, 12*(3), 517–600.
Sallum, M.A.M., & Hutchings, R. (2003). Taxonomic studies on *Culex* (*Melanoconion*) *coppenamensis* Bonne-Wepster & Bonne (Diptera: Culicidae), and description of two new species from Brazil. *Memórias do Instituto Oswaldo Cruz, 98*(5), 615–622.
Sallum, M.A.M., & Wilkerson, R.C. (1997). Description of the immature stages of *Anopheles* (*Nyssorhynchus*) *rondoni* (Neiva and Pinto) (Diptera: Culicidae). *Memórias do Instituto Oswaldo Cruz, 92*(3), 365–372.
Sallum, M.A.M., Uramoto, K., & Forattini, O.P. (1988). Redescription, and resurrection from synonymy, of *Aedes* (*Ochlerotatus*) *rhyacophilus* Costa Lima, 1933. *Memórias do Instituto Oswaldo Cruz, 83*(1), 67–77.
Sallum, M.A.M., Almirón, W.R., & Forattini, O.P. (1996). *Culex* (*Culex*) *interfor* Dyar (Diptera: Culicidae), morphological description including previously unknown life stages. *Memórias do Instituto Oswaldo Cruz, 9*(5), 563–570.
Sallum, M.A.M., Hutchings, R.S.G., & Ferreira, R.L.M. (1997). *Culex gnomatos*, a new species of the Spissipes section of *Culex* (*Melanoconion*) (Diptera: Culicidae) from the Amazon region. *Memórias do Instituto Oswaldo Cruz, 92*(2), 215–219.
Sallum, M.A.M., Wilkerson, R.C., & Forattini, O.P. (1999). Taxonomic study of species formerly identified as *Anopheles mediopunctatus* and resurrection of *An. costai* (Diptera: Culicidae). *Journal of Medical Entomology, 36*(3), 282–300.
Sallum, M.A.M., Forattini, O.P., & Wilkerson, R.C. (2000a). Redescription of the adult and larva and first description of the pupa of *Anopheles* (*Kerteszia*) *laneanus*. *Journal of the American Mosquito Control Association, 16*(2), 86–92.
Sallum, M.A.M., Schultz, T., & Wilkerson, R.C. (2000b). Phylogeny of Anophelinae (Diptera, Culicidae) based on morphological characters. *Annals of the Entomological Society of America, 93*(4), 745–775.
Sallum, M.A.M., Kobayashi, K.M., & Forattini, O.P. (2001). Description of immature stages of *Culex ocossa* Dyar & Knab, *Culex delpontei* Duret and *Culex pereyrai* Duret of the *Melanoconion* subgenus (Diptera: Culicidae). *Memórias do Instituto Oswaldo Cruz, 96*(7), 927–943.
Sallum, M.A.M., Bergo, E.S., Flores, D.C., & Forattini, O.P. (2002a). Systematic studies on *Anopheles galvaoi* Causey, Deane & Deane from the subgenus *Nyssorhynchus* Blanchard (Diptera: Culicidae). *Memórias do Instituto Oswaldo Cruz, 97*(8), 1177–1189.
Sallum, M.A.M., Bergo, E.S., Forattini, O.P., & Flores, D.C. (2002b). The eggs of *Anopheles galvaoi* and *Anopheles evansae*, two species of the subgenus *Nyssorhynchus*. *Journal of the American Mosquito Control Association, 18*(1), 10–15.
Sallum, M.A.M., Schultz, T.R., Foster, P.G., Aronstein, K., Wirtz, R.A., & Wilkerson, R.C. (2002c). Phylogeny of Anophelinae (Diptera: Culicidae) based on nuclear ribosomal and mitochondrial DNA sequences. *Systematic Entomology, 27*(3), 361–382.
Sallum, M.A.M., Peyton, EL, & Wilkerson, R.C. (2005a). Six new species of the *Anopheles leucosphyrus* group, reinterpretation of *An. elegans* and vector implications. *Medical and Veterinary Entomology, 19*, 158–199.
Sallum, M.A.M., Peyton, EL, Harrison, B.A., & Wilkerson, R.C. (2005b). Revision of the Leucosphyrus Group of *Anopheles* (*Cellia*) (Diptera: Culicidae). *Revista Brasileira de entomología, 49*(Suppl. 1), 1–152.
Sallum, M.A.M., Foster, P.G., Li, C., Sithiprasasna, R., & Wilkerson, R.C. (2007). Phylogeny of the Leucosphyrus Group of *Anopheles* (*Cellia*) (Diptera: Culicidae) based on mitochondrial gene sequences. *Annals of the Entomological Society of America, 100*(1), 27–35.
Sallum, M.A.M., Marrelli, M.T., Nagaki, S.S., Laporta, G.Z., & dos Santos, C.L.S. (2008). Insight into *Anopheles* (*Nyssorhynchus*) (Diptera: Culicidae) species from Brazil. *Journal of Medical Entomology, 45*(6), 970–981.
Sallum, M. A. M., Santos, C. L. S. D., & Wilkerson, R. C. (2009). Studies on *Anopheles* (*Kerteszia*) *homunculus* Komp (Diptera: Culicidae). *Zootaxa*, 2299, 1–18.
Sallum, M.A.M., Foster, P.G., dos Santos, C.L.S, Flores, D.C., Motoki, M.T., & Bergo, E.S. (2010). Resurrection of two species from synonymy of *Anopheles* (*Nyssorhynchus*) *strodei* Root, and characterization of a distinct morphological form from the Strodei Complex (Diptera: Culicidae). *Journal of Medical Entomology, 47*(4), 504–526.
Samanidou, A., & Harbach, R.E. (2003). *Culex* (*Culex*) *tritaeniorhynchus* Giles, a newly discovered potential vector of arboviruses in Greece. *European Mosquito Bulletin, 16*, 15–17.
Samanidou-Voyadjoglou, A., & Darsie, R.F., Jr. (1993). An annotated checklist and bibliography of the mosquitoes of Greece (Diptera: Culicidae). *Mosquito Systematics, 25*(3), 177–185.
Samanidou-Voyadjoglou, A., & Harbach, R.E. (2001). Keys to the adult female mosquitoes (Culicidae) of Greece. *European Mosquito Bulletin, 10*, 13–20.
Sames, W.J., Herman, W.E., Florin, D.A., & Maloney, F.A. (2004). Distribution of *Ochlerotatus togoi* along the Pacific Coast of Washington. *Journal of the American Mosquito Control Association, 20*(2), 105–109.
Sames, W.J., Kim, H.C., Chong, S.T., Harrison, B.A., Lee, W.J., Rueda, L.M., & Klein, T.A. (2008). *Anopheles lindesayi japonicus* Yamada (Diptera, Culicidae) in Korea: Comprehensive review, new collection records and description of larval habitats. *Journal of Vector Ecology, 33*(1), 99–106.
Sánchez, J. (1885). Nota los moscos zancudos que han invadido la capital en el presente año. *La Naturaleza* (Mexico), *7*, 203–213.
Sánchez, J. (1886). Revista de historia natural. 4. Los mosquitos zancudos. *La Naturaleza,* (Mexico), *7*, 323–330.
Sánchez-Covisa Villa, A., Rodriguez Rodriguez, J.A., & Guillen Llera, J.L. (1985). Estudio de la larva de cuarto estado de las especies del subgenero *Finlaya* (Diptera: Culicidae) de España peninsular. *Actas do Congresso Iberico de Entomología, 2*(1), 439–448.
Sánchez-Trinidad, A., Ordenez-Sánchez, F., Valdes-Pérezgasga, M.T., Sánchez-Ramos, F.J., Zavortink, T.J., Cortes-Guzmán, A.J., & Ortega-Morales, A.I. (2014). Geographical distribution of the Aedes Triseriatus Group (Diptera: Culicidae) in Mexico. *Journal of Vector Ecology, 39*(1), 134–137.
Sanders, H.R., Evans, A.M., Ross, L.S., & Gill, S.S. (2003). Blood meal induces global changes in midgut gene expression in the disease vector, *Aedes aegypti*. *Insect Biochemistry and Molecular Biology, 33*(11), 1105–1122.
Sandosham, A.A. (1944). *Malaria in Malai. A handbook for antimalaria students*. Singapore: Syonan Tokubetu-si, pp. 129–132.
Sandosham, A.A. (1959). *Malariology with special reference to Malaya*. Singapore: University of Malaya.
Sanford, J.L., Shields, V.D.C., & Dickens, J.C. (2013). Gustatory receptor neuron responds to DEET and other insect repellents in the yellow-fever mosquito, *Aedes aegypti*. *Naturwissenschaften, 100*(3), 269–273.

Sanogo, Y.O., Dobson, S.L., Bordenstein, S.R., & Novak, R.J. (2007). Disruption of the *Wolbachia* surface protein gene wspB by a transposable element in mosquitoes of the *Culex pipiens* complex (Diptera, Culicidae). *Insect Molecular Biology, 16*(2), 143–154.

Santa-Ana, M., Capela, R.A., & Christensen, B.M. (2007). Reproductive costs of the immune response of the autogenous mosquito *Culex pipiens molestus* against inoculated *Dirofilaria immitis*. *Vector-Borne and Zoonotic Diseases, 7*(4), 541–546.

Sant'Ana, D.C., & Sallum, M.A.M. (2017). *Anopheles* (*Nyssorhynchus*) *striatus*, a new species of the Strodei Subgroup (Diptera, Culicidae). *Revista Brasileira de entomología, 61*(2016)(2), 136–145.

Sant'Ana, D.C., Bergo, E.S., & Sallum, M.A.M. (2015). *Anopheles goeldii* Rozeboom & Gabaldón (Diptera, Culicidae): A species of the Nuneztovari Complex of *Anopheles* Meigen. *Revista Brasileira de entomología, 59*, 68–76.

Santiago, A.T., & Claveria, F.G. (2008). Larval mosquito fauna (family Culicidae) of Sta. Ignacia, Tarlac, Province, Philippines. *Journal of Protozoology Research, 18*(1), 17–25.

Santiago, A.T., & Claveria, F. (2012). Medically important mosquitoes (Diptera: Culicidae) identified in rural Barangay Binubusan, Lian, Batangas Province, Philippines. *Philippine Journal of Science, 141*(1), 103–109.

Santos, J.N., Lanfredi, R.M., & Pimenta, P.F.P. (2006). The invasion of the midgut of the mosquito *Culex* (*Culex*) *quinquefasciatus* Say, 1823 by the helminth *Litomosoides chagasfilhoi*. *Journal of Invertebrate Pathology, 93*(1), 1–10.

Sarala, K.S., Nutan, N., Vasantha, K., Dua, V.K., Malhotra, M.S., Yadav, R.S., & Sharma, V.P. (1994). Cytogenetic evidence for three sibling species in *Anopheles fluviatilis* (Diptera: Culicidae). *Annals of the Entomological Society of America, 87*, 116–121.

Saridaki, A., & Bourtzis, K. (2010). *Wolbachia*: More than just a bug in insects['] genitals. *Current Opinion in Microbiology, 13*(1), 67–72.

Sasa, M. (1949). Redescription of *Aedes nobukonis* Yamada, 1932, with designation of its subgenus. *Japanese Journal of Experimental Medicine, 20*, 409–411.

Sasa, M., & Ishimura, K. (1951). A new species of mosquito, *Aedes* (*Finlaya*) *bunanoki*, n. sp: from Aomori, Japan. *Japanese Journal of Experimental Medicine, 21*, 103–108.

Sasa, M., & Kamimura, K. (1971). Index and consideration on taxonomy of the Japanese mosquitoes . In M. Sasa (Ed.), *Progress in medical zoology* (Vol. I, pp. 1–47). Tokyo: Gakujutsusho Shuppankai.

Sasa, M., & Kano, R. (1951). Description and comparative studies on the larva of *Aedes* (*Stegomyia*) *galloisi* Yamada. (Diptera, Culicidae). *Japanese Journal of Experimental Medicine, 21*, 109–113.

Sasa, M., & Nakahashi, Y. (1952). A new species of the *Aedes* (*Finlaya*) *niveus* subgroup of mosquito from Japan. *Japanese Journal of Experimental Medicine, 22*(3), 257–265.

Sasa, M., & Takahasi, H. (1948). Descriptions of *Culex rubensis* n. sp. and the male form of adult of *Theobaldia kayamensis* Yamada. *Japanese Journal of Bacteriology, 3*(2), 51–52.

Sasa, M., & Takahasi, H. (1949). Some new species and new records of mosquitoes of Japan. *Japanese Medical Journal, 2*(1), 50–54.

Sasa, M., Kano, R., & Hayashi, S. (1950). A new species of mosquito, *Aedes koreicoides* n. sp., from Hokkaido. *Japanese Journal of Experimental Medicine, 20*, 627–629.

Sasa, M., Kano, R., & Takahasi, H. (1950). A revision of the adult Japanese mosquitoes of the genus *Aedes*, subgenus *Aedes* with description of two new species. *Japanese Journal of Experimental Medicine, 20*, 631–640.

Sathe, T.V., & Girhe, B.E. (2002). Biodiversity of mosquitoes in Kolhapur District, Maharashtra. *Revista di Parassitologia, 18*(3), 193–198.

Sathe, T.V., & Jagtap, M. (2012). Taxonomy and molecular phylogeny of new species of the genus *Anopheles* (*Cellia*) Theobald (Culicidae: Diptera) from India. *Journal of Entomological Research* (New Delhi), *36*(1), 71–75.

Satoto, T.B.T. (2001). *Cryptic species within* Anopheles barbirostris *van der Wulp, 1884, inferred from nuclear and mitochondrial gene sequence variation* [Doctoral dissertation, University of Liverpool, England].

Saugstad, E.S. (1973). Two mosquito species new to the Ryukyu Islands. *Japanese Journal of Sanitary Zoology, 23*(4), 284.

Saul, S.H., Sinsko, M.J., Grimstad, P.R., & Craig, G.B., Jr. (1977). Identification of sibling species, *Aedes triseriatus* and *Ae. hendersoni*, by electrophoresis. *Journal of Medical Entomology* (Honolulu), *13*(6), 705–708.

Savage, H.M. (1986). Identification and location of the holotype and paratypes of *Anopheles* (*Nyssorhynchus*) *nuneztovari* Gabaldon (Diptera: Culicidae). *Mosquito Systematics, 18*(3–4), 279–284.

Savage, H.M. (2005). Classification of mosquitoes in tribe Aedini (Diptera: Culicidae): Paraphylyphobia, and classification versus cladistic analysis. *Journal of Medical Entomology, 42*(6), 923–927.

Savage, H.M., & Kothera, L. (2012). The *Culex pipiens* complex in the Mississippi River Basin: Identification, distribution, and bloodmeal hosts. *Journal of the American Mosquito Control Association, 28*(Suppl. 4), 93–99.

Savage, H.M., & Strickman, D. (2004). The genus and subgenus categories within Culicidae and placement of *Ochlerotatus* as a subgenus of *Aedes*. *Journal of the American Mosquito Control Association, 20*(2), 208–214.

Sawadogo, S.P., Diabaté, A., Toé, H.K., Sanon, A., Lefevre, T., Baldet, T., . . . Dabiré, R.K. (2013). Effects of age and size on *Anopheles gambiae* ss male mosquito mating success. *Journal of Medical Entomology, 50*(2), 285–293.

Say, T. (1823). Descriptions of dipterous insects of the United States. *Journal of the Academy of Natural Sciences* (Philadelphia), *3*, 9–54.

Say, T. (1824). Appendix. Part I. Natural history. 1. Zoology. E. Class Insecta. In W.H. Keating, *Narrative of an expedition to the source of St. Peter's River* (Vol. 2, pp. 268–378). Philadelphia, PA: Carey and Lea.

Say, T. (1829). Descriptions of North American dipterous insects. *Journal of the Academy of Natural Sciences of Philadelphia, 6*,(1), (149–178).

Sazanova, O.N. (1955). Description of a new species of blood-sucking mosquito, *Aedes* (*Ochlerotatus*) *grandilarva* n. sp. *Bulletin de la Société Impériale des Naturalistes de Moscou: Section Biologique, 60*(4), 99–102.

Sazanova, O.N. (1956). A new species of blood sucking mosquito of the genus *Aedes* Arr. (Diptera, Culicidae) from the European part of the USSR. *Magazin de parasitologie du Musée Zoologique de l'Académie de Sciences de l'URSS, 16*, 145–151.

Sazanova, O.N. (1958). The key for the identification of females of the genus *Aedes* (Diptera, Culicidae) of the forest zone of the USSR. *Revue d'entomologie de l'URSS, 37*, 741–752.

Scanlon, J.E., & Peyton, EL. (1967). *Anopheles* (*Anopheles*) *tigertti*, a new species of the Aitkenii group from Thailand (Diptera: Culicidae). *Proceedings of the Entomological Society of Washington, 69*(1), 18–23.

Scanlon, J.E., & Yates, M.M. (1970). Preliminary observations on *Aedes* (*Ochlerotatus*) *atlanticus* Dyar and Knab 1906, and other members of the Scapularis species group in Texas. *Proceedings of the Texas Mosquito Control Association, December* 9–10, *1970*, 1.

Scanlon, J.E., Peyton, EL, & Gould, D.J. (1968). An annotated checklist of the *Anopheles* of Thailand (Diptera: Culicidae). *Thai Na-*

tional Scientific Papers Fauna Series No. 2. Bangkok: Applied Scientific Research Corporation of Thailand.

Scarpassa, V.M., & Conn, J.E. (2007). Population genetic structure of the major malaria vector *Anopheles darlingi* (Diptera: Culicidae) from the Brazilian Amazon, using microsatellite markers. *Memórias do Instituto Oswaldo Cruz, 102*(3), 319–327.

Scarpassa, V.M., & Conn, J.E. (2011). Mitochondrial DNA detects a complex evolutionary history with Pleistocene Epoch divergence for the neotropical malaria vector *Anopheles* nuneztovari Sensu Lato. *The American Journal of Tropical Medicine and Hygiene, 85*(5), 857–867.

Scarpassa, V.M., Cunha-Machado, A.S., & Saraiva, J.F. (2016). Evidence of new species for malaria vector *Anopheles nuneztovari* sensu lato in the Brazilian Amazon region. *Malaria Journal, 15*, 205.

Schaeffer, B., Mondet, B., & Touzeau, S. (2008). Using a climate-dependent model to predict mosquito abundance: Application to *Aedes* (*Stegomyia*) *africanus* and *Aedes* (*Diceromyia*) *furcifer* (Diptera: Culicidae). *Infection, Genetics and Evolution, 8*(4), 422–432.

Schäfer, M., & Lundström, J.O. (2006). Different responses of two floodwater mosquito species, *Aedes vexans* and *Ochlerotatus sticticus* (Diptera: Culicidae), to larval habitat drying. *Journal of Vector Ecology, 31*(1), 123–128.

Schaffner, F. (2003). Mosquitoes in used tyres in Europe: Species list and larval key. *European Mosquito Bulletin, 16*, 7–12.

Schaffner, F., & van Bortel, W. (2010). Current status of invasive mosquitoes in Europe. *European Centre for Disease Prevention and Control (ECDC), VBORNET Newsletter, 2*, 6–8.

Schaffner, F., Chouin, S., & Guilloteau, J. (2003). First record of *Ochlerotatus japonicus japonicus* in metropolitan France. *Journal of the American Mosquito Control Association, 19*(1), 1–5.

Schaffner, F., Kaufmann, C., Hegglin, D., & Mathis, A. (2009). The invasive species *Aedes japonicus* in Central Europe. *Medical and Veterinary Entomology, 23*(4), 448–451.

Scheffer, S.J. (2005). Invasive Diptera: Using molecular markers to investigate cryptic species and the global spread of introduced flies. In D.K. Yeates & B.M. Wiegmann (Eds.), *The evolutionary biology of flies* (pp. 371–387). New York, NY: Columbia University Press.

Schick, R.X. (1970a). Mosquito studies (Diptera, Culicidae). XX. The Terrens group of *Aedes* (*Finlaya*). *Contributions of the American Entomological Institute, 5*(3), 1–158.

Schick, R.X. (1970b). Mosquito studies (Diptera, Culicidae). XXIII. Additions and corrections to the revision of the *Aedes terrens* group. *Contributions of the American Entomological Institute, 7*(1), 13–40.

Schierenbeck, K.A., & Ellstrand, N.C. (2009). Hybridization and the evolution of invasiveness in plants and other organisms. *Biological Invasions, 11*(5), 1093–1105.

Schiner, J.R. (1864). *Fauna Austriaca. Die Fliegen* (*Diptera*) (Vol. 2). Wien, Germany: Carl Jerold's Sohn.

Schiner, J.R. (1868). Diptera. In V. Wullerstorf-Urbair (Ed.), *Reise der Osterreichischen Fregatte Novara um die Erde in den Jahren 1857, 1858, 1859* (Vol. 2, 1 abt., B. 1). Wien, Germany: Zoologischer Theil.

Schlein, Y., & Müller, G.C. (1995). Assessment of plant tissue feeding by sand flies (Diptera: Psychodidae) and mosquitoes (Diptera: Culicidae). *Journal of Medical Entomology, 32*(6), 882–887.

Schlein, Y., & Müller, G.C. (2008). An approach to mosquito control: Using the dominant attraction of flowering *Tamarix jordanis* trees against *Culex pipiens*. *Journal of Medical Entomology, 45*(3), 384–390.

Schlein, Y., & Müller, G.C. (2012). Diurnal resting behavior of adult *Culex pipiens* in an arid habitat in Israel and possible control measurements with toxic sugar baits. *Acta Tropica, 124*(1), 48–53.

Schmidt, E.R., Foley, D.H., Hartel, G.F., Williams, G.M., & Bryan, J.H. (2001). Descriptions of the *Anopheles* (*Cellia*) *farauti* complex of sibling species (Diptera: Culicidae) in Australia. *Bulletin of Entomological Research, 91*(5), 389–410.

Schmidt, E.R., Foley, D.H., Bugoro, H., & Bryan, J.H. (2003). A morphological study of the *Anopheles punctulatus* group in the Solomon Islands, with a description of *Anopheles irenicus* Schmidt, sp. n. *Bulletin of Entomological Research, 93*(6), 515–526.

Schmidt, J.M., & Friend, W.G. (1991). Ingestion and diet destination in the mosquito *Culiseta inornata*: Effects of carbohydrate configuration. *Journal of Insect Physiology, 37*(11), 817–828.

Schmidt, J.P., Springborn, M., & Drake, J.M. (2012). Bioeconomic forecasting of invasive species by ecological syndrome. *Ecosphere, 3*(5), 1–19.

Schneider, B.S., & Higgs, S. (2008). The enhancement of arbovirus transmission and disease by mosquito saliva is associated with modulation of the host immune response. *Transactions of the Royal Society of Tropical Medicine and Hygiene, 102*(5), 400–408.

Schneider, P. (1913). Beitrag zur Kenntnis der Culiciden in der Umgebung von Bonn. *Verhandlungen des Naturhistorischen Vereins der preufsischen Rheinlande und Westfalens, 70*, 1–54.

Schöller, S., Belmont, M., Cazzamali, G., Hauser, F., Williamson, M., & Grimmelikhuijzen, C.J.P. (2005). Molecular identification of the myosuppresin receptor from the malaria mosquito *Anopheles gambiae*. *Biochemical and Biophysical Research Communications, 327*(1), 29–34.

Scholte, E.J., Knols, B.G., & Takken, W. (2006). Infection of the malaria mosquito *Anopheles gambiae* with the entomopathogenic fungus *Metarhizium anisopliae* reduces blood feeding and fecundity. *Journal of Invertebrate Pathology, 91*(1), 43–49.

Scholte, E.J., Jacobs, F., Linton, Y.-M., Dijkstra, E., Fransen, J., & Takken, W. (2007). First record of *Aedes* (*Stegomyia*) *albopictus* in the Netherlands. *European Mosquito Bulletin, 22*, 5–9.

Scholte, E.J., Den Hartog, W., Braks, M., Reusken, C., Dik, M., & Hessels, A. (2009). First report of a North American invasive mosquito species *Ochlerotatus atropalpus* (Coquillett) in the Netherlands, 2009. *Euro Surveillance, 14*(45), 429–433.

Scholte, E.J., Braks, M., & Schaffner, F. (2010a). Aircraft-mediated transport of *Culex quinquefasciatus*: A case report. *European Mosquito Bulletin, 28*, 208–212.

Scholte, E.J., Den Hartog, W., Dik, M., Schoeltz, B., Brooks, M., Schaffner, F., . . . Beeuwkes, J. (2010b). Introduction and control of three invasive mosquito species in the Netherlands, July-October 2010. *Euro Surveillance, 15*(45), 1817–1824.

Scholte, E.J., Dik, M., Ibanez Justicia, A., Den Hartog, W., Schoelitsz, B., Brooks, M., . . . Steeghs, M. (2012). Findings and control of two invasive exotic mosquito species, *Aedes albopictus* and *Ae. atropalpus* (Diptera: Culicidae) in the Netherlands, 2011. *European Mosquito Bulletin, 30*, 1–14.

Scholte, E.J., Ibáñez-Justicia, A., Stroo, A., De Zeeuw, J., den Hartog, W., & Reusken, C.B.E.M. (2014). Mosquito collections on incoming intercontinental flights at Schiphol International Airport, the Netherlands, 2010–2011. *Journal of the European Mosquito Control Association, 32*, 17–21.

Schouteden, H. (1912). Insectes recueillis au Congo au cours du voyage de S. A. R. le Prince Albert de Belgique. *Revue zoologique Africaine, 2*, 63–90.

Schrank, Francisci de Pavla. (1781). *Enumeratio Insectorum Austriae indigenorum*. Vindelicorum [Augsburg]: Klett.

Schrank, Franz von Paula. (1776). *Beytrage zur Naturgeschichte*. Leipzig, Germany: Fritsch.

Schreiber, E.T., Meek, C.L., & Yates, M.M. (1988). Vertical distribution and species coexistence of tree hole mosquitoes in Louisiana. *Journal of the American Mosquito Control Association, 4*(1), 9–14.

Schwardt, H.H. (1939). Biologies of Arkansas rice field mosquitos. *University of Arkansas Agricultural Experiment Station Bulletin No. 377,* 1–22.

Scott, B.N., Yu, M.J., Lee, L.W., & Beyenbach, K.W. (2004). Mechanisms of K^+ transport across basolateral membranes of principal cells in Malpighian tubules of the yellow fever mosquito, *Aedes aegypti*. *Journal of Experimental Biology, 207*(10), 1655–1663.

Scott, T.W., & Takken, W. (2012). Feeding strategies of anthropophilic mosquitoes result in increased risk of pathogen transmission. *Trends in Parasitology, 28*(13), 114–121.

Scott, T.W., Amerasinghe, P.H., Morrison, A.C., Lorenz, L.H., Clark, G.G., Strickman, D., . . . Edman, J.D. (2000). Longitudinal studies of *Aedes aegypti* (Diptera: Culicidae) in Thailand and Puerto Rico: Blood feeding frequency. *Journal of Medical Entomology, 37*(1), 89–101.

Scott, T.W., Takken, W., Knols, B.G., & Boëte, C. (2002). The ecology of genetically modified mosquitoes. *Science, 298*(5591), 117–119.

Scott, T.W., Harrington, L.C., Knols, B.G., & Takken, W. (2008). Applications of mosquito ecology for successful insect transgenesis-based disease prevention programs. In S. Aksoy (Ed.), *Transgenesis and the management of vector-borne disease* (pp. 151–168). New York, NY: Springer.

Scotton, G.L., & Axtell, R.C. (1979). *Aedes taeniorhynchus* and *Aedes sollicitans* (Diptera, Culicidae) oviposition on coastal dredge spoil. *Mosquito News, 39*(1), 97–110.

Scudder, S.H. (1877). The first discovered traces of fossil insects in the American tertiaries. *Bulletin of the United States Geological Survey, 4*(3), 741–762.

Scudder, S.H. (1882). Nomenclature zoologicus. An alphabetical list of all generic names that have been employed by naturalists for recent and fossil animals from the earliest times to the close of the year 1879. I. Supplemental list. *Bulletin of the United States National Museum, 21*(19), 1–376.

Scudder, S.H. (1890). *The tertiary insects of North America* (Vol. 13). Washington, DC: Government Printing Office.

Sedaghat, M.M., Linton, Y.-M., Nicolescu, G., Smith, L., Koliopoulos, G., Zounos, A., . . . Harbach, R.E. (2003a). Morphological and molecular characterization of *Anopheles* (*Anopheles*) *sacharovi* Favre, a primary vector of malaria in the Middle East. *Systematic Entomology, 28*, 241–256.

Sedaghat, M.M., Linton, Y.-M., Oshaghu, M.A., Vatandoost, H., & Harbach, R.E. (2003b). The *Anopheles maculipennis* complex in Iran: Molecular characterization and recognition of a new species. *Bulletin of Entomological Research, 93*(6), 527–535.

Sedaghat, M.M., Howard, T., & Harbach, R.E. (2009). Morphological study and description of *Anopheles* (*Anopheles*) *persiensis*, a member of the Maculipennis Group (Diptera: Culicidae: Anophelinae) in Iran. *Journal of Entomological Society of Iran, 28*(2), 25–35.

Seetharam, P.L., & Chowdaiah, B.N. (1971). Chromosome studies of oriental anophelines—II. The salivary gland chromosomes of *Anopheles nigerrimus*. *Parassitologia, 13*, 429–434.

Segata, N., Baldini, F., Pompon, J., Garrett, W.S., Truong, D.T., Dabiré, R.K., . . . Catteruccia, F. (2016). The reproductive tracts of two malaria vectors are populated by a core microbiome and by gender- and swarm-enriched microbial biomarkers. *Scientific Reports, 6*, Article 24207.

Segerman, J. (1990). A list of the type specimens of Culicidae housed in the Department of Medical Entomology at the South African Institute for Medical Research, Johannesburg. *Mosquito Systematics, 22*(1), 53–56.

Séguy, E. (1920). Les moustiques de France. *Bulletin Museum d'Histoire Naturelle, 26*, 51–58, 141–147, 223–230, 322–329, 407–414, 512–519.

Séguy, E. (1921a). Description d'un nouveau moustique français du groupe de l'*Aedes mariae* et synopsis des espèces de ce groupe. *Bulletin de la Société Entomologique de France* (Paris), *1921*(13), 192–195.

Séguy, E. (1921b). Note sur l'armure genitale du *Culex jugorum* Vilien. (Dipt. Culicidae). *Bulletin de la Société Entomologique de France, 26*(3), 39–40.

Séguy, E. (1923). Remarques sur quelques moustiques. *Annales de la Société Entomologique de France, 92*, 205–208.

Séguy, E. (1924). Les moustiques de l'Afrique Mineure, de l'Égypte et de la Syrie. *Encyclopédie Entomologique A, 1*, 1–257.

Séguy, E. (1925). Notes sur les moustiques de l'Afrique Mineure, de l'Égypte et de la Syrie. I. *Encyclopédie Entomologique B : Mémoires et Notes 4*(1), 13–21.

Séguy, E. (1927). Notes sur les moustiques de l'Afrique Mineure, de l'Égypte et de la Syrie. II. *Encyclopédie Entomologique B 4*(1), 27–28.

Séguy, E. (1950). La biologie des Diptères. *Encyclopédie Entomologique A, 26*, 1–609.

Seidahmed, O.M.E. (2015). *Mating biology of male* Anopheles arabiensis *Patton (Diptera: Culicidae): Implications in the use of sterile insect technique in vector control* [Doctoral dissertation, University of Khartoum, Sudan]. Retrieved from http://khartoumspace.uofk.edu/

Sen, P. (1938). The early stages of *Aedes lophoventralis* (Theobald). *Records of the Indian Museum, 40*, 359–361.

Senevet, G. (1930). Contribution à l'étude des nymphes de culicides. Description de celles de certain anophelines et plus spécialement des espèces européennes et meditérranéennes. *Archives de l'Institut Pasteur d'Algérie, 8*(3–4), 297–382.

Senevet, G. (1931). Contribution à l'étude des nymphes d'anophelines (2e Memoire). *Archives de l'Institut Pasteur d'Algérie, 9*(1), 17–112.

Senevet, G. (1932a). Contribution à l'étude des nymphes d'anophelines (3e Memoire). *Archives de l'Institut Pasteur d'Algérie, 10*(2), 204–254.

Senevet, G. (1932b). À propos d'*Anopheles coustani* Laveran. *Archives de l'Institut Pasteur d'Algérie, 10*(4), 478–484.

Senevet, G. (1933). Notes sur les moustiques. II. *Archives de l'Institut Pasteur d'Algérie, 11*(2), 198–207.

Senevet, G. (1934). Contribution a l'étude des nymphes d'anophelines (4e Memoire). *Archives de l'Institut Pasteur d'Algérie, 12*(1), 29–76.

Senevet, G. (1936a). Les moustiques de la Martinique. *Archives de l'Institut Pasteur d'Algérie, 14*(2), 123–134.

Senevet, G. (1936b). Notes sur les moustiques. IV. Quelques culicides de la région de l'Aures (Algérie). *Archives de l'Institut Pasteur d'Algérie, 14*(4), 432–448.

Senevet, G. (1937). Les moustiques de la Guyane Française (mission 1934). *Archives de l'Institut Pasteur d'Algérie, 15*(3), 352–382.

Senevet, G. (1938). Les moustiques de la Guadeloupe (mission 1936). *Archives de l'Institut Pasteur d'Algérie, 16*(2), 176–190.

Senevet, G. (1946). Contribution à l'étude des larves de culicides. La structure du prementum. *Archives de l'Institut Pasteur d'Algérie, 24*(3–4), 315–329.

Senevet, G. (1947a). Le genre *Culex* en Afrique du nord. I. Les larves. *Archives de l'Institut Pasteur d'Algérie, 25*(2), 109–136.

Senevet, G. (1947b). Le genre *Culex* en Afrique du Nord. 1. Clef de determination des larves. *Archives de Institut Pasteur de Algérie, 25*, 212–213.

Senevet, G. (1947c). *Anopheles tonkinensis* synonyme de *A. alongensis. Archives de l'Institut Pasteur d'Algérie, 25*(3–4), 214–217.
Senevet, G. (1948a). A propos d'*Anopheles* (*Nyssorhynchus*) *ininii* Senevet et Abonnenc, 1938. *Archives de l'Institut Pasteur d'Algérie, 26*(3), 277–279.
Senevet, G. (1948b). *Anopheles pessoai* en Guyane Française. Description de la nymphe. *Archives de l'Institut Pasteur d'Algérie, 26*(3), 280–287.
Senevet, G. (1948c). À propos de la nymphe d'*Anopheles ininii. Archives de l'Institut Pasteur d'Algérie, 26*(4), 431–432.
Senevet, G. (1948d). Au sujet de quelques nymphes de *Nyssoryhnchus. Archives de l'Institut Pasteur d'Algérie, 26*(4), 433–440.
Senevet, G. (1949). Le genre *Culex* en Afrique du Nord. 2. Les armures genitales mâles. *Archives de l'Institut Pasteur d'Algérie, 27*(1), 48–65.
Senevet, G. (1958). Les *Anopheles* du globe. Révision generale. *Encyclopedie Entomologique A, 36*, 1–215.
Senevet, G., & Abonnenc, E. (1938). Quelques anophelines de la Guyane Française. *Archives de l'Institut Pasteur d'Algérie, 16*(4), 486–512.
Senevet, G., & Abonnenc, E. (1939a). Les moustiques de la Guyane Française. II. Le genre *Culex. Archives de l'Institut Pasteur d'Algérie, 17*(1), 62–134.
Senevet, G., & Abonnenc, E. (1939b). Les moustiques de la Guyane Française. III. Les sabethines. *Archives de l'Institut Pasteur d'Algérie, 17*, 247–281.
Senevet, G., & Abonnenc, E. (1939c). Les moustiques de la Guyane. IV. Le genre *Aedes. Archives de l'Institut Pasteur d'Algérie, 17*(2), 467–480.
Senevet, G., & Abonnenc, E. (1939d). Les moustiques de la Guyane. V. Les genres *Mansonia*, *Orthopodomyia*, *Aedeomyia*, *Psorophora*, *Uranotaenia. Archives de l'Institut Pasteur d'Algérie, 17*(4), 585–597.
Senevet, G., & Abonnenc, E. (1940). Les moustiques de la Guyane Française. VI. Les anophelines. 2. Le sous-genre *Stethomyia. Archives de l'Institut Pasteur d'Algérie, 18*(4), 441–442.
Senevet, G., & Abonnenc, E. (1941). Les moustiques de la Guyane Française. Le genre *Culex*. 2. Nouvelle espèce du sous-genre *Melanoconion. Archives de l'Institut Pasteur d'Algérie, 19*(1), 41–44.
Senevet, G., & Abonnenc, E. (1946). Les moustiques de la Guyane Française. X. Le genre *Culex*. 3. Nouvelle espèce du sous-genre *Culex. Archives de l'Institut Pasteur d'Algérie, 24*(2), 135–140.
Senevet, G., & Abonnenc, E. (1958a). Quelques nymphes du sous-genre *Carollia*, g. *Culex*- Culicidae. *Archives de l'Institut Pasteur d'Algérie, 36*(2), 287–293.
Senevet, G., & Abonnenc, E. (1958b). A propos de quelques *Culex* sud-américains. *Archives de l'Institut Pasteur d'Algérie, 36*(3), 343–350.
Senevet, G., & Andarelli, L. (1954a). Le genre *Culex* Afrique du Nord. III. Les adultes. *Archives de l'Institut Pasteur d'Algérie, 32*(1), 36–70.
Senevet, G., & Andarelli, L. (1954b). Le genre *Aedes* en Afrique du Nord. I. Les larves. *Archives de l'Institut Pasteur d'Algérie, 32*(4), 310–351.
Senevet, G., & Andarelli, L. (1955a). *Les* Anopheles *de l'Afrique du Nord et du Bassin Meditérranéen*. P. Lechevallier (Ed.), Encyclopédie Entomologique, Paris, *33*, 1–280.
Senevet, G., & Andarelli, L. (1955b). Races et variétés de l'*Anopheles claviger* Meigen, 1804. *Archives de l'Institut Pasteur d'Algérie, 33*(2), 128–137.
Senevet, G., & Andarelli, L. (1955c). Les soies antepalmees chez les larves d'*Anopheles*, leur utilisation comme caracteres de groupe (IIe note). *Archives de l'Institut Pasteur d'Algérie, 33*(4), 322–343.
Senevet, G., & Andarelli, L. (1956a). Presence en Algérie de *Theobaldia subochrea* Edwards, 1921. *Archives de l'Institut Pasteur d'Algérie, 34*(2), 223–226.
Senevet, G., & Andarelli, L. (1956b). Presence en Algérie de *Theobaldia litores* (Shute). *Archives de l'Institut Pasteur d'Algérie, 34*(3), 400–403.
Senevet, G., & Andarelli, L. (1956c). *Les Anopheles de l'Afrique du Nord et du bassin mediterraneen*. Paris, France: Lechevalier.
Senevet, G., & Andarelli, L. (1958). Le genre *Aedes* en Afrique du Nord. II. Les nymphes. *Archives de l'Institut Pasteur d'Algérie, 36*(2), 266–286.
Senevet, G., & Andarelli, L. (1959). *Les moustiques de l'Afrique du Nord et du Bassin Meditérranéen. Les genres* Culex, Uranotaenia, Theobaldia, Orthopodomyia *et* Mansonia. P. Lechevallier (Ed.), *Encyclopédie Entomologique* (Paris), *35*, 1–383.
Senevet, G., & Prunnelle, M. (1927). Une nouvelle espèce d'*Anopheles* en Algérie, *Anopheles marteri* n. sp. *Archives de l'Institut Pasteur d'Algérie, 5*(4), 529–533.
Senevet, G., & Quievreux, L. (1941). Les moustiques de la Martinique. (2e Memoire). *Archives de l'Institut Pasteur d'Algérie, 19*(2), 248–264.
Senevet, G., & Rioux, J. (1960). *Anopheles* (*Myzomyia*) *hispaniola* Theobald 1903 simple sous-espèce de *Anopheles* (*Myzomyia*) *cinereus* Theobald, 1901. *Archives de l'Institut Pasteur d'Algérie, 38*(4), 530–535.
Senevet, G., Chabelard, R., & Abonnenc, E. (1942). Les moustiques de la Guyane. III. Les Sabethines 2. *Archives de l'Institut Pasteur d'Algérie, 20*(4), 336–348.
Senevet, G., Gaud, J., & Millet, A. (1949). Validite de l'espèce *Culex mauritanicus* Callot, 1940. *Archives de l'Institut Pasteur d'Algérie, 27*(1), 42–47.
Senevet, G., Mandoul, R., & Jacquemin, P. (1949). Sur une forme aberrante de l'*Aedes caspius* Pallas. *Archives de l'Institut Pasteur d'Algérie, 27*(3), 277–280.
Senevet, G., Andarelli, L., & Duzer, A. (1954). Sur la presence en Algérie de *Aedes longitubus* Cambournac et sur quelques espèces de moustiques peu communes en Afrique du Nord. *Archives de l'Institut Pasteur d'Algérie, 32*(3), 266–275.
Senevet, G., Andarelli, L., Buffart, G., & Duzer, A. (1955). Quelle est la limite septentrionale de *Culex deserticola* Kirk, en Afrique du Nord? *Archives de l'Institut Pasteur d'Algérie, 33*(1), 51–53.
Senevet, G., Andarelli, L., & Duzer, A. (1956). Études sur les stations a *Anopheles multicolor* des environs de Tenes (Algérie). *Archives de l'Institut Pasteur d'Algérie, 34*(1), 119–128.
Senevet, G., Andarelli, L., & Graells, R. (1957). Les epines preclypeales des larves de moustiques; leur utilisation dans la diagnose. Les soies dorsales prothoraciques chez les larves de *Culex. Archives de l'Institut Pasteur d'Algérie, 35*(2), 83–89, 90–98.
Senevet, G., Andarelli, L., & Lieutaud, A. (1957). A propos de *Culex univittatus* Theob. *Archives de l'Institut Pasteur d'Algérie, 35*(1), 52–53.
Senevet, G., Clastrier, J., & Andarelli, L. (1959). Les moustiques du Tassili des Ajjer. V. *Archives de Institut Pasteur d'Algérie, 37*(4), 598–602.
Senior-White, R. (1927). Notes on Ceylon mosquitoes. II. The larvae of the commoner non-anopheline mosquitoes. *Spolia Zeylanica* (*Ceylon Journal of Science B*), *14*, 61–76.
Senise, L.V., & Sallum, M.A.M. (2007). Systematics of *Anopheles lanei* Galvao and Amaral, a species of the subgenus *Nyssorhynchus* Blanchard (Diptera: Culicidae). *Memórias do Instituto Oswaldo Cruz, 102*(8), 959–967.

Senise, L.V., & Sallum, M.A.M. (2008). Redescription of *Culex (Culex) dolosus* (Lynch-Arribálzaga) (Diptera: Culicidae) based on specimens from Pico do Itapeva, Serra da Mantiqueira, São Paulo, Brasil. *Zootaxa, 1683*, 51–62.

Senise, L., Motoki, M.T., & Sallum, M.A.M. (2006). Notes on the holotype of *Anopheles marajoara* Galvão & Damasceno (Diptera, Culicidae). *Revista Brasileira de entomología, 50*(4), 453–457.

Sepkoski, J.J., Jr. (2001). Foundations of life in the oceans. In S.J. Gould (Ed.), *The book of life* (pp. 36–63). New York, NY: Norton.

Serafim, J., Jr., & Davis, N.C. (1933). Distribution of *Aedes (Taeniorhynchus) taeniorhynchus* (Wiedemann). *Aedes (Taeniorhynchus) jacobinae*, new species. *Annals of the Entomological Society of America, 26*(1), 13–19.

Sergent, E. (1909). Liste des moustiques de l'Afrique du Nord (lre note). *Annales de la Société Entomologique de France, 78*(3), 440–448.

Sergent, E. (1937). Les oeufs d'*Anopheles hispaniola* Theo. *Archives de l'Institut Pasteur d'Algérie, 15*(1), 102–103.

Sergent, E., & Sergent, E. (1903). Observations sur les moustiques des environs d'Alger. *Annales de l'Institut Pasteur d'Algérie, 17*(1), 60–67.

Sergent, E., & Sergent, E. (1905). Sur un culicide nouveau, très commun a Biskra (*Grabhamia subtilis*). *Comptes rendus Société de Biologie, 58*, 673–674.

Service, M.W. (1958). A new *Anopheles* from the Southern Cameroons, *Anopheles buxtoni* sp. nov. *Annals of Tropical Medicine and Parasitology, 52*(2), 176–180.

Service, M.W. (1959a). The morphology of some new culicine larvae from the Southern Cameroons, with notes on *Culex (Culex) ingrami* Edwards and *Culex (Culex) toroensis* Edwards. *Proceedings of the Royal Entomological Society of London. Series B: Taxonomy, 28*, 1–6.

Service, M.W. (1959b). A new species of the genus *Aedes* from southern Nigeria. *Proceedings of the Royal Entomological Society of London. Series B: Taxonomy, 28*(5–6), 73–76.

Service, M.W. (1959c). A new culicine mosquito, *Aedes (Aedimorphus) lokojensis* sp. nov., from northern Nigeria. *Revue zoologique Africaine (supplement botanique), 59*(3–4), 243–248.

Service, M.W. (1960a). Taxonomic notes on some larvae of the genus *Aedes* (Diptera: Culicidae) from Lokoja in northern Nigeria. *Entomological Monthly Magazine, 96*, 229–230.

Service, M.W. (1960b). A taxonomic study of *Anopheles funestus funestus* Giles (Diptera: Culicidae) from southern and northern Nigeria, with notes on its varieties and synonyms. *Proceedings of the Royal Entomological Society of London. Series B: Taxonomy, 29*(5–6), 77–84.

Service, M.W. (1960c). A new species and variety of *Anopheles* from Nigeria, with notes on *Anopheles flavicosta* Edwards and *Aedes (Ae.) wendyae* Service. *Proceedings of the Royal Entomological Society of London, Series B: Taxonomy, 29*(7–8), 85–90.

Service, M.W. (1962). Keys to the West African Anophelini. *Acta Tropica, 19*(2), 120–158.

Service, M.W. (1968). The taxonomy and biology of two sympatric sibling species of *Culex*, *C. pipiens* and *C. torrentium* (Diptera, Culicidae). *Journal of Zoology, 156*, 313–323.

Service, M.W. (1970a). A new *Anopheles* (Dipt., Culicidae) from Nigeria and a note on *A. pretoriensis* (Theobald). *Bulletin of Entomological Research, 59*, 647–649.

Service, M.W. (1970b). Identification of the *Anopheles gambiae* complex in Nigeria by larval and adult chromosomes. *Annals of Tropical Medicine and Parasitology, 64*(2), 131–136.

Service, M.W. (1970c). Studies on the biology and taxonomy of *Aedes (Stegomyia) vittatus* (Bigot) (Diptera: Culicidae) in northern Nigeria. *Transactions of the Royal Entomological Society of London, 122*, 101–143.

Service, M.W. (1970d). The taxonomy and distribution of two sibling mosquitoes, *Culiseta morsitans* (Theobald) and *C. litorea* (Shute). *Journal of Natural History, 4*, 481–491.

Service, M.W. (1973). The biology of *Anopheles claviger* (Mg.) (Dipt., Culicidae) in southern England. *Bulletin of Entomological Research, 63*(2), 347–359.

Service, M.W. (1976a). Contribution to the knowledge of the mosquitoes (Diptera, Culicidae) of Gabon. *Cahiers ORSTOM. Série entomologie médicale et parasitologie, 14*(3), 259–263.

Service, M.W. (1976b). *Mosquito ecology. Field sampling methods.* New York, NY: Wiley.

Service, M.W. (1977). A new *Anopheles* (Dipt., Culicidae) from Nigeria together with a note on *An. brohieri* Edwards. *Annals of Tropical Medicine and Parasitology, 71*(2), 233–236.

Service, M.W. (1980). Effects of wind on the behavior and distribution of mosquitoes and blackflies. *International Journal of Biometeorology, 24*(4), 347–353.

Service, M.W. (1990). *Handbook to the Afrotropical toxorhynchitine and culicine mosquitoes, excepting* Aedes *and* Culex. London, England: British Museum (Natural History).

Service, M.W. (1993). Mosquitoes (Culicidae). In R.P. Lane & R.W. Crosskey (Eds.), *Medical insects and arachnids* (pp. 120–240). New York, NY: Chapman and Hall.

Service, M.W. (1997). Mosquito (Diptera: Culicidae) dispersal—the long and short of it. *Journal of Medical Entomology, 34*(6), 579–588.

Service, M.W. (2010). Sequel to Kitzmiller's *Anopheline Names: Their Derivations and Histories. Journal of Vector Ecology, 35*(2), 213–266.

Shaalan, E.A.S., & Canyon, D.V. (2009). Aquatic insect predators and mosquito control. *Tropical Biomedicine, 26*(3), 223–261.

Shahgudian, E.R. (1956). Notes on *Anopheles marteri* Senevet and Prunelle, 1927. *Proceedings of the Royal Entomological Society of London. Series A: General Entomology, 31*(4–6), 71–75.

Shalaby, A.M. (1972). Survey of the mosquito fauna of Fezzan southwestern Libya (Diptera Culicidae). *Bulletin de la Société Entomologique d'Egypte, 56*, 301–312.

Shaman, J., Stieglitz, M., Stark, C., Le Blanq, S., & Cane, M. (2002). Using a dynamic hydrology model to predict mosquito abundance in flood and swamp water. *Emerging Infectious Diseases, 8*(1), 6–13.

Shaman, J., Spiegelman, M., Cane, M., & Stieglitz, M. (2006). A hydrologically driven model of swamp water mosquito population dynamics. *Ecological Modelling, 194*(4), 395–404.

Shamrai, A.F., & Gutsevich, A.V. (1974). Discovery of the mosquito *Toxorhynchites christophi* (Portschinsky, 1844) in the Khabarovsk territory (with a description of the male). *Revue d'entomologie de l'URSS Entomologicseszkoje Obozrenie, 53*(2), 427–428.

Shannon, R.C. (1927). Contribucion a los estudios de las zonas biologicas de la Republica Argentina. *Revista de la Sociedad Entomologica Argentina, 1*(4), 1–14.

Shannon, R.C. (1931). On the classification of Brazilian Culicidae with special reference to those capable of harboring the yellow fever virus. *Proceedings of the Entomological Society of Washington, 33*(6), 125–164.

Shannon, R.C. (1932). *Anopheles gambiae* in Brazil. *American Journal of Hygiene, 15*(3), 633–663.

Shannon, R. C. (1933a). Correction. *Proceedings of the Entomological Society of Washington, 35*(4), 58.

Shannon, R.C. (1933b). Anophelines of the Amazon Valley. *Proceedings of the Entomological Society of Washington, 35*(7), 117–143.

Shannon, R.C. (1934). The genus *Mansonia* (Culicidae) in the Amazon Valley. *Proceedings of the Entomological Society of Washington, 36*(5), 99–110.
Shannon, R.C., & Cerqueira, N.L. (1943). *Psorophora lanei*, um novo mosquito do Brasil e da Bolivia (Diptera, Culicidae). *Revista de entomología* (Rio de Janeiro), *14,* 135–137.
Shannon, R.C., & Davis, N.C. (1930). Observations on the Anophelini (Culicidae) of Bahia, Brazil. *Annals of the Entomological Society of America, 23*(3), 467–505.
Shannon, R.C., & Del Ponte, E. (1928). Cuatro notas sobre especies nuevas de Dípteros Nematóceros, Hematófagos o no, de la República Argentina. *Revista del Instituto Bacteriologico Buenos Aires, 4*(for 1927)(7), 724–736.
Shao, L., Devenport, M., & Jacobs-Lorena, M. (2001). The peritrophic matrix of hematophagous insects. *Archives of Insect Biochemistry and Physiology, 47*(2), 119–125.
Sharakhov, I.V., Serazin, A.C., Grushko, O.G., Dana, A., Lobo, N., Hillenmeyer, M.E., . . . Besansky, N.J. (2002). Inversions and gene order shuffling in *Anopheles gambiae* and *A. funestus. Science, 298*(5591), 182–185.
Sharakhova, M.V., & Sharakhov, I.V. (2013). Progress in mapping the yellow fever mosquito genome. *Cytology* (Russia), *55*(4), 241–243.
Sharma, A.K., Tyagi, V., Singh, S., Veer, V., Agrawal, O.P., & Sukumaran, D. (2014). Distribution of *Anopheles culicifacies* and detection of its sibling species E from Madhya Pradesh: Central India. *Journal of Arthropod-Borne Diseases, 8*(2), 186–196.
Sharma, G.P., & Chaudhry, S. (1974). A strain of *Aedes* (*Stegomyia*) *vittatus* Bigot with an unusual pattern on the tarsi (Diptera: Culicidae). *Mosquito Systematics, 6*(2), 163–166.
Sharma, K.R., Seenivasagan, T., Rao, A.N., Ganeson, K., Agrawal, O.P., & Prakash, S. (2009). Mediation of oviposition responses in the malaria mosquito *Anopheles stephensi* Liston by certain fatty acid esters. *Parasitology Research, 104*(2), 281–286.
Sharma, P., Sharma, S., De Maurya, R.K., Das de T., Thomas, T., Lata, S., . . . Dixit, R. (2014). Salivary glands harbor more diverse microbial communities than gut in *Anopheles culicifacies. Parasites and Vectors, 7*(1), 1.
Shaw, W.R., Teodori, E., Mitchell, S.N., Baldini, F., Gabrieli, P., Rogers, D.W., & Catteruccia, F. (2014). Mating activates the heme peroxidase HPX15 in the sperm storage organ to ensure fertility in *Anopheles gambiae. Proceedings of the National Academy of Sciences of the United States of America, 111*(16), 5854–5859.
Shaw, W.R., Attardo, G.M., Aksoy, S., & Catteruccia, F. (2015). A comparative analysis of reproductive biology of insect vectors of human disease. *Current Opinion in Insect Science, 10,* 142–148.
Shaw, W.R., Marcenac, P., Childs, L.M., Buckee, C.O., Baldini, F., Sawadogo, S.P., . . . Catteruccia, F. (2016). *Wolbachia* infections in natural *Anopheles* populations affect egg laying and negatively correlate with *Plasmodium* development. *Nature Communications, 7,* 11772.
Shepard, J.J., Andreadis, T.G., & Vossbrinck, C.R. (2006). Molecular phylogeny and evolutionary relationships among mosquitoes (Diptera: Culicidae) from the northeastern United States based on small subunit ribosomal DNA (18S rDNA) sequences. *Journal of Medical Entomology, 43*(3), 443–454.
Sheppard, P.M., Macdonald, W.W., Tonn, R.J., & Grab, B. (1969). The dynamics of an adult population of *Aedes aegypti* in relation to dengue haemorrhagic fever in Bangkok. *Journal of Animal Ecology, 38*(3), 661–702.
Sheremet, V.P. (1975). Detection of *Aedes vexans nipponi* New-Record (Diptera, Culicidae) in the Ukrainian-SSR (USSR). *Meditsinskaya parazitologiya i parazitarnye bolezni, 44*(1), 100.
Shestakov, V.I. (1961). Contribution to the biology of the mosquito *Aedes togoi* Theobald. *Zoologicheskii zhurnal, 40*(2), 284–285.
Shiao, S.H., Hansen, I.A., Zhu, J., Sieglaff, D.H., & Raikhel, A.S. (2008). Juvenile hormone connects larval nutrition with target of rapamycin signaling in the mosquito *Aedes aegypti. Journal of Insect Physiology, 54*(1), 231–239.
Shidrawi, G.R., & Gillies, M.T. (1988). *Anopheles paltrinierii*, n. sp., (Culicidae: Diptera) from the Sultanate of Oman. *Mosquito Systematics, 19*(for 1987)(3), 201–211.
Shililu, J., Ghebremeskel, T., Mengistu, S., Fekadu, H., Zerom, M., Mbogo, C., . . . Beier, J.C. (2003). Distribution of anopheline mosquitoes in Eritrea. *The American Journal of Tropical Medicine and Hygiene, 69*(3), 295–302.
Shin, E.-H., & Hong, H.K. (2001). A new synonym of *Anopheles* (*Anopheles*) *pullus* Yamada, 1937—*A.* (*A.*) *yatsushiroensis* Miyazaki, 1951. *Korean Journal of Entomology, 31*(1), 1–5.
Shin, S.W., Zou, Z., Saha, T.T., & Raikhel, A.S. (2012). bHLH-PAS heterodimer of methoprene-tolerant and cycle mediates circadian expression of juvenile hormone-induced mosquito genes. *Proceedings of the National Academy of Sciences of the United States of America, 109*(41), 16576–16581.
Shingarev, N.I. (1926). New information on Culicidae of U.S.S.R. *Tropichenskai meditsina I: Veterinariia* (Moscow), *2*, 47–48.
Shingarev, N.I. (1927). Notes on Culicidae. II. *Tropichenskai meditsina I: Veterinariia, 5*, 545–550.
Shingarev, N.I. (1928). Notes on Culicidae. III. *Tropichenskai meditsina I: Veterinariia* (Moscow), *6*, 47–53.
Shishika, D., Manoukis, N.C., Butail, S., & Paley, D.A. (2014). Male motion coordination in anopheline mating swarms. *Scientific Reports, 4*, Article 6318.
Shone, S.M., Curriero, F.C., Lesser, C.R., & Glass, G.E. (2006). Characterizing population dynamics of *Aedes sollicitans* (Diptera: Culicidae) using meteorological data. *Journal of Medical Entomology, 43*(2), 393–402.
Shousha, A.T. (1948). Species-eradication: The eradication of *Anopheles gambiae* from Upper Egypt, 1942–1945. *Bulletin of the World Health Organization, 1*(2), 309–352.
Shroyer, D.A. (1981). Establishment of *Wyeomyia mitchellii* on the island of Oahu, Hawaii. *Mosquito News, 41*(4), 805–806.
Shroyer, D.A., Harrison, B.A., Bintz, B.J., Wilson, M.R., Sither, C.B., & Byrd, B.D. (2015). *Aedes pertinax*, a newly recognized mosquito species in the United States. *Journal of the American Mosquito Control Association, 31*(1), 97–100.
Shu-Po, L., & Chen, H.-B. (1986). A new record of *Culex* from China. *Acta zootaxonomica Sinica, 11*(2).
Shute, P.G. (1928). A new variety of *Culicella morsitans* (Theobald) (Diptera). *Entomologist, 61*, 186.
Shutt, B., Stables, L., Aboagye-Antwi, F., Moran, J., & Tripet, F. (2010). Male accessory gland proteins induce female monogamy in anopheline mosquitoes. *Medical and Veterinary Entomology, 24*(1), 91–94.
Sicart, M. (1940). Note sur la presence de *Culex* (*Culiciomyia*) *impudicus* Ficalbi en Tunisie. *Archives de l'Institut Pasteur de Tunis, 29*, 471–474.
Sicart, M. (1952a). Description et étude de *Aedes pulchritarsis* (Rondani, 1872) recolte en Tunisie. *Bulletin de la Société des Sciences Naturelles de Tunisie, 5*, 95–101.
Sicart, M. (1952b). Note sur la differenciation de *C. hortensis* (Fic. 1889) et *C. deserticola* (Kirkp. 1924) au stade IV. *Bulletin de la Société des Sciences Naturelles de Tunisie, 5*, 149–150.
Sicart, M. (1954). Contribution a l'étude des nymphes du genre *Culex* en Tunisie. *Bulletin de la Société des Sciences Naturelles de Tunisie*, 27–29.

Sicart, M., & Ruffie, J. (1960). Étude sur les nymphes des culicines du sous-bassin Aquitain. *Annales de parasitologie humaine et comparée, 35*, 631–647.

Sichinava, S.G., Stegni, V.N., & Sipovich, N.G. (1983). The prevalence and biology of *Anopheles maculipennis* Meig., 1818 in Western Georgia. *Bulletin of the Academy of Sciences of the Georgian SSR, 112*(2), 417–420.

Sierra, D.M., Velez, I.D., & Linton, Y.-M. (2004). Malaria vector *Anopheles (Nyssorhynchus) nuneztovari* comprises one genetic species in Colombia based on homogeneity of nuclear ITS2 rDNA. *Journal of Medical Entomology, 41*(3), 302–307.

Siju, K.P., Hill, S.R., Hansson, B.S., & Ignell, R. (2010). Influence of blood meal on the responsiveness of olfactory receptor neurons in antennal sensilla trichodea of the yellow fever mosquito, *Aedes aegypti*. *Journal of Insect Physiology, 56*(6), 659–665.

Siju, K.P., Reifenrath, A., Scheiblich, H., Neupert, S., Predel, R., Hansson, B.S., . . . Ignell, R. (2014). Neuropeptides in the antennal lobe of the yellow fever mosquito, *Aedes aegypti*. *Journal of Comparative Neurology, 522*(3), 592–608.

Silberbush, A., & Blaustein, L. (2008). Oviposition habitat selection by a mosquito in response to a predator: Are predator-released kairomones air-borne cues? *Journal of Vector Ecology, 33*(1), 208–211.

Silberbush, A., & Blaustein, L. (2011). Mosquito females quantify risk of predation to their progeny when selecting an oviposition site. *Functional Ecology, 25*(5), 1091–1095.

Silberbush, A., Markman, S., Lewinsohn, E., Bar, E., Cohen, J.E., & Blaustein, L. (2010). Predator-released hydrocarbons repel oviposition by a mosquito. *Ecology Letters, 13*(9), 1129–1138.

Silva-do-Nascimento, T.F., & Lourenço-de-Oliveira, R. (2002). *Anopheles halophylus*, a new species of the subgenus *Nyssorhynchus* (Diptera: Culicidae) from Brazil. *Memórias do Instituto Oswaldo Cruz, 97*(6), 801–811.

Silva-do-Nascimento, T.F., & Lourenço-de-Oliveira, R. (2007). Diverse population dynamics of three *Anopheles* species belonging to the Triannulatus Complex (Diptera: Culicidae). *Memórias do Instituto Oswaldo Cruz, 102*, 975–982.

Silva-do-Nascimento, T.F., Wilkerson, R.C., Lourenço-de-Oliveira, R., & Monteiro, F.A. (2006). Molecular confirmation of the specific status of *Anopheles halophylus* (Diptera: Culicidae) and evidence of a new cryptic species within *An. triannulatus* in Central Brazil. *Journal of Medical Entomology, 43*(3), 455–459.

Silva-do-Nascimento, T.F., Rona Pitaluga, L.D., Peixoto, A.A., & Lourenço-de-Oliveira, R. (2011). Molecular divergence in the *timeless* and *cpr* genes among three sympatric cryptic species of the *Anopheles triannulatus* complex. *Memórias do Instituto Oswaldo Cruz, 222*(Suppl. 1), 1.

Sim, C., & Denlinger, D.L. (2009). A shut-down in expression of an insulin-like peptide, ILP-1, halts ovarian maturation during the overwintering diapause of the mosquito *Culex pipiens*. *Insect Molecular Biology, 18*(3), 325–332.

Sim, C., & Denlinger, D.L. (2013). Juvenile hormone III suppresses forkhead of transcription factor in the fat body and reduces fat accumulation in the diapausing mosquito, *Culex pipiens*. *Insect Molecular Biology, 22*(1), 1–11.

Sim, C., Kang, D.S., Kim, S., Bai, X., & Denlinger, D.L. (2015). Identification of FOXO targets that generate diverse features of the diapause phenotype in the mosquito *Culex pipiens*. *Proceedings of the National Academy of Science, 112*(2), 3811–3816.

Simard, F., Ayala, D., Kamden, G.C., Pombi, M., Etouna, J., Ose, K., . . . Constantini, C. (2009). Ecological niche partitioning between *Anopheles gambiae* molecular forms in Cameroon: The ecological side of speciation. *BMC Ecology, 9*(1), 17.

Simmons, J.S. (1937). Observations on the importance of *Anopheles punctimacula* as a malaria vector in Panama, and report of experimental infections in *A. neomaculipalpus*, *A. apicimacula*, and *A. eiseni*. *The American Journal of Tropical Medicine and Hygiene, 17*, 191–212.

Simmons, J.S., & Aitken, T.H.G. (1942). *The anopheline mosquitoes of the northern half of the Western Hemisphere and of the Philippine Islands. (Distribution, habits, identification, importance as vectors, and control)* (1st ed). Medical Bulletin 59. Department of the Army, Medical Field Service School.

Simon, C. (1983). Application of numerical techniques to the systematics of *Toxorhynchites* (Diptera: Culicidae). In *Numerical Taxonomy* (Vol. 1, pp. 527–531). NATO ASI Series. Retrieved from: https://rd.springer.com/chapter/10.1007/978-3-642-69024-2_58

Simons, F.E.R., & Peng, Z. (2001). Mosquito allergy: Recombinant mosquito salivary antigens for new diagnostic tests. *International Archives of Allergy and Immunology, 124*(1–3), 403–405.

Sinclair, B.J., Borkent, A., & Wood, D.M. (2007). The male genital tract and aedeagal components of the Diptera with a discussion of their phylogenetic significance. *Zoological Journal of the Linnaean Society, 150*(4), 711–742.

Singer, C. (1950). *A history of biology* (Rev. ed.). New York, NY: Schuman.

Singh, G., & Prakash, S. (2008). A new species of *Lutzia* (Diptera: Culicidae) comparable with *halifaxii* Theobald 1903 in a semi-arid zone of India. In B.K. Tyagi (Ed.), *Vector-borne diseases: Epidemiology and control* (Ch. 24, pp. 195–204). Jodhpur, India: Scientific.

Singh, S.P., Raghavendra, K., Kumar, R., Nanda, N., & Subbarao, S.K. (2004). Morphotaxonomic studies to identify the members of the *Anopheles subpictus* Grassi (Diptera: Culicidae) species complex in riverine villages of District Sonepat, Haryana State, India. *Journal of Communicable Diseases, 36*(1), 35–40.

Singh, S., Prakash, A., Yadav, R.N.S., Mohapatra, P.K., Sarma, N.P., Sarma, D.K., . . . Bhattacharyya, D.R. (2012). *Anopheles* (*Cellia*) *maculatus* group: Its spatial distribution and molecular characterization of member species in north-east India. *Acta Tropica, 124*(1), 62–70.

Sinka, M.E. (2013). Global distribution of the dominant vector species of malaria. In S. Manguin (Ed.), Anopheles *mosquitoes: New insights into malaria vectors* (pp. 109–143). Rijeka, Croatia: InTech.

Sinka, M.E., Bangs, M.J., Manguin, S., Coetzee, M., Mbogo, C.M., Hemingway, J., . . . Hay, S.I. (2010a). The dominant *Anopheles* vectors of human malaria in Africa, Europe and the Middle East: Occurrence, data, distribution maps and bionomic précis. *Parasites and Vectors, 3*(117), 1–34.

Sinka, M.E., Rubio-Palis, Y., Manguin, S., Patil, A.P., Temperley, W.H., Gething, P.W., . . . Hay, S.I. (2010b). The dominant *Anopheles* vectors of human malaria in the Americas: Occurrence data, distribution maps and bionomic précis. *Parasites and Vectors, 3*(1), 72.

Sinka, M.E., Bangs, M.J., Manguin, S., Chareonviriyaphap, T., Patil, A.P., Temperley, W.H., . . . Hay, S.I. (2011). The dominant *Anopheles* vectors of human malaria in the Asia-Pacific region: Occurrence data, distribution maps and bionomic précis. *Parasites and Vectors, 4*(1), 89.

Sinkins, S.P. (2016). Yob makes mosquitoes male. *Science, 353*(6294), 33–34.

Sinkins, S.P., Walker, T., Lynd, A.R., Steven, A.R., Makepeace, B.L., Godfray, H.C.J., & Parkhill, J. (2005). *Wolbachia* variability and host effects on crossing type in *Culex* mosquitoes. *Nature, 436*(7048), 257.

Sirivanakarn, S. (1968). The *Culex* subgenus *Lophoceraomyia* in New Guinea and Bismarck Archipelago (Diptera, Culicidae). *Pacific Insects Monograph, 17*, 75–186.

Sirivanakarn, S. (1969). Note on the change of a specific name of New Guinea *Culex* (*Lophoceraomyia*) (Diptera: Culicidae). *Journal of Medical Entomology, 6*(2), 119.

Sirivanakarn, S. (1971a). Contributions to the mosquito fauna of Southeast Asia. XI. A proposed reclassification of *Neoculex* Dyar based primarily on the male terminalia. *Contributions of the American Entomological Institute, 7*(3), 62–85.

Sirivanakarn, S. (1971b). *Culex* (*Neoculex*) *nematoides* (Dyar and Shannon 1925) an erroneous record from the Philippines (Diptera: Culicidae). *Mosquito Systematics, 3*(1), 1–3.

Sirivanakarn, S. (1972). The genus *Culex*, subgenus *Eumelanomyia* Theobald in southeast Asia and adjacent areas. *Contributions of the American Entomological Institute, 8*(6), 1–86.

Sirivanakarn, S. (1973a). Descriptions of two new species of *Culex* (*Lophoceraomyia*) with notes on three other species from the Papuan subregion (Diptera: Culicidae). *Journal of Medical Entomology, 10*(2), 212–216.

Sirivanakarn, S. (1973b). Three new species of *Culex* subgenus *Culiciomyia* Theobald from Southeast Asia and a redescription of the type of *C. tricuspis* Edwards from Alor, Lesser Sunda Islands, Indonesia (Diptera: Culicidae). *Proceedings of the Entomological Society of Washington, 75*(1), 112–124.

Sirivanakarn, S. (1973c). The forms of *Culex* (*Culex*) *bitaeniorhynchus* Giles in Southeast Asia. *Mosquito Systematics, 5*(3), 235–251.

Sirivanakarn, S. (1974). Redescription of *Culex* (*Culex*) *bihamatus* Edwards with a discussion of its affinity (Diptera: Culicidae). *Mosquito Systematics, 6*(4), 259–262.

Sirivanakarn, S. (1975a). The systematics of *Culex vishnui* complex in Southeast Asia with the diagnosis of three common species (Diptera: Culicidae). *Mosquito Systematics, 7*(1), 69–85.

Sirivanakarn, S. (1975b). A new species of *Culex* (*Eumelanomyia*) Theobald from Manus Island, Papua-New Guinea (Diptera: Culicidae). *Mosquito Systematics, 7*(4), 394–400.

Sirivanakarn, S. (1976). Medical entomology studies. III. A revision of the subgenus *Culex* in the Oriental region (Diptera: Culicidae). *Contributions of the American Entomological Institute, 12*(2), 1–272.

Sirivanakarn, S. (1977a). Medical entomology studies. VI. A revision of the subgenus *Lophoceraomyia* of the genus *Culex* in the Oriental region (Diptera: Culicidae). *Contributions of the American Entomological Institute, 13*(4), 1–245.

Sirivanakarn, S. (1977b). Additional descriptions of three species of *Culex* (*Eumelanomyia*) with the description of a new species from Peninsular Malaysia (Diptera: Culicidae). *Mosquito Systematics, 9*(1), 73–87.

Sirivanakarn, S. (1977c). Redescription of four Oriental species of *Culex* (*Culiciomyia*) and the description of a new species from Thailand (Diptera: Culicidae). *Mosquito Systematics, 9*(2), 93–111.

Sirivanakarn, S. (1977d). A new species of *Culex* (*Eumelanomyia*) from India with descriptions of pupae and larvae of *Cx. pluvialis* Barraud and *Cx. iphis* Barraud (Diptera: Culicidae). *Mosquito Systematics, 9*(4), 537–547.

Sirivanakarn, S. (1978a). Revalidation of *Culex* (*Melanoconion*) *invocator* Pazos with a redescription of adults and illustration of male genitalia (Diptera: Culicidae). *Mosquito Systematics, 10*(2), 239–245.

Sirivanakarn, S. (1978b). The female cibarial armature of new world *Culex*, subgenus *Melanoconion* and related subgenera with notes on this character in subgenera *Culex*, *Lutzia* and *Neoculex* and genera *Galindomyia* and *Deinocerites* (Diptera: Culicidae). *Mosquito Systematics, 10*(4), 474–492.

Sirivanakarn, S. (1979). A new species of *Culex* (*Melanoconion*) from Bolivia and Ecuador (Diptera: Culicidae). *Mosquito Systematics, 11*(2), 135–138.

Sirivanakarn, S. (1983). A review of the systematics and a proposed scheme of internal classification of the New World subgenus *Melanoconion* of *Culex* (Diptera: Culicidae). *Mosquito Systematics, 14*(for 1982)(4), 265–333.

Sirivanakarn, S., & Belkin, J.N. (1980). The identity of *Culex* (*Melanoconion*) *taeniopus* Dyar and Knab and related species with notes on the synonymy and description of a new species (Diptera: Culicidae). *Mosquito Systematics, 12*(1), 7–24.

Sirivanakarn, S., & Degallier, N. (1982). Redescription of *Culex* (*Melanoconion*) *portesi* Senevet & Abonnenc, 1941, with notes on synonymy (Diptera: Culicidae). *Mosquito Systematics, 13*(for 1981)(2), 153–167.

Sirivanakarn, S., & Galindo, P. (1980). *Culex* (*Melanoconion*) *adamesi*, a new species from Panama (Diptera, Culicidae). *Mosquito Systematics, 12*(1), 25–34.

Sirivanakarn, S., & Heinemann, S. (1980). Description of the hitherto unknown adult and pupa of *Culex* (*Melanoconion*) *simulator* Dyar and Knab, and redescription of its larva (Diptera: Culicidae). *Mosquito Systematics, 12*(1), 41–49.

Sirivanakarn, S., & Jakob, W.L. (1979). A new species of *Culex* (*Melanoconion*) from southern Brazil (Diptera: Culicidae). *Mosquito Systematics, 11*(2), 139–143.

Sirivanakarn, S., & Jakob, W.L. (1982). *Culex* (*Melanoconion*) *sacchettae*, a new species from the state of São Paulo, Brazil (Diptera: Culicidae). *Mosquito Systematics, 13*(for 1981)(2), 191–194.

Sirivanakarn, S., & Kurihara, T. (1973). A new species of *Culex*, subgenus *Culiciomyia* Theobald from Ceram, Indonesia (Diptera: Culicidae). *Proceedings of the Entomological Society of Washington, 75*(2), 220–224.

Sirivanakarn, S., & Ramalingam, S. (1976). A new species of *Culex* (*Eumelanomyia*) Theobald with notes on three other species from Malaysia (Diptera: Culicidae). *Mosquito Systematics, 8*(2), 209–216.

Sirivanakarn, S., & White, G.B. (1978). Neotype designation of *Culex quinquefasciatus* Say (Diptera: Culicidae). *Proceedings of the Entomological Society of Washington, 80*(3), 360–372.

Sirot, L.K., Poulson, R.L., McKenna, M.C., Girnary, H., Wolfner, M.F., & Harrington, L.C. (2008). Identity and transfer of male reproductive gland proteins of the dengue vector mosquito, *Aedes aegypti*: Potential tools for control of female feeding and reproduction. *Insect Biochemistry and Molecular Biology, 38*(2), 176–189.

Sirot, L.K., Hardstone, M.C., Helinski, M.E., Ribeiro, J.M., Kimura, M., Deewatthanawong, P., . . . Harrington, L.C. (2011). Towards a semen proteome of the dengue vector mosquito: Protein identification and potential functions. *PLOS Neglected Tropical Diseases, 5*(3), e989.

Skierska, B. (1970). *Aedes* (*Aedes*) *rossicus* D. G. M. und *Aedes* (*Ochlerotatus*) *beklemishevi* Den. Zwei neue fuer Polen Stechmueckenarten (Deptera, Culicinae). *Polskie Pismo entomologiczne, 40*, 611–617.

Skierska, B. (1971). Klucze Do Oznacaznia Owadow Polski. Czesc XXVIII. (Diptera: Culicidae). *Polskie Towarzystwo Entomologiczne, 69*, 3–138.

Skierska, B. (1977). [Culicidae]. *Klucze do oznaczania owadow Polski, 28*(96), 1–119.

Skuse, F.A.A. (1889). Diptera of Australia. Part V. The Culicidae. *Proceedings of the Linnean Society of New South Wales, 3*, 1717–1764.

Skuse, F.A.A. (1895). The banded mosquito of Bengal. *Indian Museum Notes, 5*(for 1984)(3), 20.

Slater, J.D., & Pritchard, G. (1979). A stepwise computer program for estimating development time and survival of *Aedes vexans* (Diptera: Culicidae) larvae and pupae in field populations in southern Alberta. *The Canadian Entomologist, 111*(11), 1241–1253.

Slooff, R. (1961). *Tripteroides (Rachisoura) tityae* (Diptera; Culicidae), a new species from Netherlands New Guinea. *Entomologische Berichten* (Amsterdam), *21*(1), 22–23.

Slooff, R. (1963a). Some notes on *Bironella (Bironella) confusa* Bonne-Wepster (Diptera; Culicidae; Anophelinae). *Entomologische Berichten* (Amsterdam), *23*(1), 137–142.

Slooff, R. (1963b). *Uranotaenia gerdae*, a new species of the Tibialis group from West New Guinea (Diptera: Culicidae). *Annals of the Entomological Society of America, 56*(1), 82–86.

Slooff, R. (1972). Mosquitoes collected in the British Solomon Islands Protectorate between March 1964 and October 1968 (Diptera: Culicidae). *Entomologische Berichteno* (Amsterdam), *32*, 171–181.

Slooten, E., & Lambert, D.M. (1983). Evolutionary studies of the New Zealand coastal mosquito *Opifex fuscus* (Hutton). I. Mating behaviour. *Behaviour, 84*(1), 157–171.

Slooten, E., & Lambert, D.M. (1984). Evolutionary studies of the New Zealand coastal mosquito *Opifex fuscus* (Hutton). II. Competition for mates. *Behaviour, 88*(1–2), 1–12.

Smallegange, R.C., Qiu, Y.T., van Loon, J.J.A., & Takken, W. (2005). Synergism between ammonia, lactic acid and carboxylic acids as kairomones in the host-seeking behavior of the malaria mosquito *Anopheles gambiae sensu stricto* (Diptera: Culicidae). *Chemical Senses, 30*(2), 145–152.

Smallegange, R.C. Bulkovinskiné-Kiss, G., Otieno, B., Mbadi, P.A., Takken, W., . . . van Loon, J.J.A. (2012). Identification of candidate volatiles that affect the behavioural response of the malaria mosquito *Anopheles gambiae* sensu stricto to an active kairomone blend: Laboratory and semi-field assays. *Physiological Entomology, 37*(1), 60–71.

Smith, D.R., Arrigo, N.C., Leal, G., Muehlberger, L.E., & Weaver, S.C. (2007). Infection and dissemination of Venezuelan equine encephalitis virus in the epidemic mosquito vector, *Aedes taeniorhynchus. The American Journal of Tropical Medicine and Hygiene, 77*(1), 176–187.

Smith, D.R., Adams, A.P., Kenney, J.L., Wang, E., & Weaver, S.C. (2008). Venezuelan equine encephalitis virus in the mosquito vector *Aedes taeniorhynchus*: Infection initiated by a small number of susceptible epithelid cells and a population bottleneck. *Virology, 372*(1), 176–186.

Smith, M.E. (1952). A new northern *Aedes* mosquito, with notes on its close ally, *Aedes diantaeus* H., D. & K. (Diptera, Culicidae). *Bulletin of the Brooklyn Entomological Society, 47*(1–2), 19–40.

Smith, M.E. (1958). The *Aedes* mosquitoes of New England (Diptera, Culicidae). I. Key to adult females. *Bulletin of the Brooklyn Entomological Society, 53*(2), 39–47.

Smith, M.E. (1969a). The *Aedes* mosquitoes of New England (Diptera, Culicidae). II. Larvae: Keys to instars and to species exclusive of first instar. *The Canadian Entomologist, 101*(1), 41–51.

Smith, M.E. (1969b). The *Aedes* mosquitoes of New England (Diptera, Culicidae). III. Saddle hair position in 2nd and 3rd instar larvae, with particular reference to instar recognition and species relationships. *Mosquito Systematics, 1*(3), 57–62.

Smith, S.M., & Brust, R.A. (1970). Autogeny and stenogamy of *Aedes rempeli* (Diptera: Culicidae) in arctic Canada. *The Canadian Entomologist, 102*(2), 253–256.

Smith, S.M., & Trimble, R.M. (1973). The biology of tree-holes of Point Pelee National Park, Ontario. I. New mosquito records for Canada (Diptera: Culicidae). *The Canadian Entomologist, 105*, 1585–1586.

Smithson, T.W. (1972). Species rank for *Anopheles franciscanus* based on failure of hybridization with *Anopheles pseudopunctipennis pseudopunctipennis. Journal of Medical Entomology, 9*(6), 501–505.

Smithson, T.W., & McClelland., G.A.H. (1972). Salivary gland chromosome map of *Anopheles pseudopunctipennis franciscanus. Mosquito News, 32*(1), 80–87.

Smith-White, S., & Woodhill, A.R. (1954). The nature and significance of nonreciprocal fertility in *Aedes scutellaris* and other mosquitoes. *Proceedings of the Linnean Society of New South Wales, 79*, 163–176.

Sneath, P.H.A. (1995). Thirty years of numerical taxonomy. *Systematic Biology, 44*(3), 281–298.

Snell, A.E. (2005). Identification keys to larval and adult female mosquitoes (Diptera: Culicidae) of New Zealand. *New Zealand Journal of Zoology, 32*(2), 99–110.

Snell, A.E., Knox, R.L., & Cane, R.P. (2010). Aspects of nutrition and oviposition in the endemic rockpool mosquito *Opifex fuscus* Hutton (Diptera: Culicidae). *New Zealand Entomologist, 33*(1), 79–83.

Snodgrass, R.E. (1959). The anatomical life of the mosquito. *Smithsonian Miscellaneous Collections, 139*(8), 1–87.

Snow, K.R. (1986). A note on the spelling of the name *Orthopodomyia pulcripalpis* (Rondani, 1872). *Mosquito Systematics, 17*(4), 361–362.

Snow, K.R. (2003). A summary of the names of European mosquitoes. *European Mosquito Bulletin, 14*, 16–20.

Snow, K.R. (2010). Names of European mosquitoes—an update. *European Mosquito Bulletin, 28*, 101–102.

Snow, K.R., & Medlock, J.M. (2008). The mosquitoes of Epping Forest, Essex, UK. *European Mosquito Bulletin, 26*, 9–17.

Snow, K.R., & Ramsdale, C. (1999). Distribution chart for European mosquitoes. *European Mosquito Bulletin, 3*, 14–31.

Snow, K.R., Rees, A.T., & Brooks, J.A. (1997). A revised bibliography of the mosquitoes of the British Isles. London, England: Department of Environmental Science and Mathematics, University of East London.

Snow, W.F., & Wilkes, T.J. (1977). Age composition and vertical distribution of mosquito populations in The Gambia, West Africa. *Journal of Medical Entomology, 13*(4–5), 507–513.

Sobti, S.K. (1968). A new species of the *Anopheles funestus* complex (Diptera: Culicidae) from Zanzibar, United Republic of Tanzania. *Bulletin of the World Health Organization, 38*, 481–483.

Soesilo, R. (1932). *Anopheles (Bironella) walchi* n. sp. *Geneeskundig tijdschrift voor Nederlandsch-Indië, 72*, 1035–1036.

Soesilo, R., & van Slooten, J. (1931a). Verspreide aanteekeningen over Anophelinen in de Oost-Indischen Archipel. *Geneeskundig tijdschrift voor Nederlandsch-Indië, 71*(7–18), 1254–1259.

Soesilo, R., & van Slooten, J. (1931b). Miscellaneous notes on anopheline mosquitoes in the Dutch East Indies. *Mededeelingen van den Dienst der Volksgezondbeid in Nederlandsch-Indië, 20*, 124–128.

Soghigian, J., Andreadis, T.G., & Livdahl, T.P. (2017). From ground pools to treeholes: Convergent evolution of habitat and phenotype in *Aedes* mosquitoes. *BMC Evolutionary Biology, 17*, 262.

Solarte, Y., Hurtado, C., González, R., & Alexander, B. (1996). Man-biting activity of *Anopheles (Nyssorhynchus) albimanus* and *An. (Kerteszia) neivai* (Diptera: Culicidae) in the Pacific lowlands of Colombia. *Memórias do Instituto Oswaldo Cruz, 91*(2), 141–146.

Somboon, P., Tuno, N., Tsuda, Y., & Takagi, M. (2000). Evidence of the specific status of *Anopheles flavirostris* (Diptera: Culicidae). *Journal of Medical Entomology, 37*(3), 476–479.

Somboon, P., Thongwat, D., Choochote, W., Walton, C., & Takagi, M. (2005). Crossing experiments of *Anopheles minimus* species C and putative species E. *Journal of the American Mosquito Control Association, 21*(1), 5–9.

Somboon, P., Rory, A., Tsuda, Y., Takagi, M., & Harbach, R.E. (2010). Systematics of *Anopheles (Cellia) yaeyamaensis* sp. n., alias spe-

cies E of the *An. minimus* complex in southeastern Asia (Diptera: Culicidae). *Zootaxa, 2651*, 43–51.
Somboon, P., Thongwat, D., & Harbach, R.E. (2011). *Anopheles (Cellia) rampae* n. sp., alias chromosomal form K of the Oriental Maculatus Group (Diptera: Culicidae) in Southeast Asia. *Zootaxa, 2810*, 47–55.
Somboon, P., Phanitchakun, T., Namgay, R., & Harbach, R.E. (2020). Description of *Aedes (Hulecoeteomyia) bhutanensis* n. sp. (Diptera: Culicidae) from Bhutan. *Acta Tropica, 203*, 105280.
Songsawatkiat, S., Baimai, V., Saeung, A., Thongsahuan, S., Otsuka, Y., Srisuka, W., & Choochote, W. (2013). Cytogenetic, hybridization and molecular evidence of four cytological forms of *Anopheles nigerrimus* (Hyrcanus Group) in Thailand and Cambodia. *Journal of Vector Ecology, 38*(2), 266–276.
Soper, F.L. (1963). The elimination of urban yellow fever in the Americas through the eradication of *Aedes aegypti. American Journal of Public Health and the Nation's Health, 53*(1), 7–16.
Soper, F.L. (1966). Paris green in the eradication of *Anopheles gambiae*: Brazil, 1940; Egypt, 1945. *Mosquito News, 26*(4), 470–476.
Soper, F.L., & Wilson, D.B. (1943). Anopheles gambiae *in Brazil, 1930–1940.* New York, NY: The Rockefeller Foundation.
Soper, F.L., Wilson, D.B., Lima, S., & Antunes, W.S. (1943). *The organization of permanent nation-wide anti-*Aedes aegypti *measures in Brazil.* New York, NY: The Rockefeller Foundation.
Sota, T., & Mogi, M. (1992). Interspecific variation in desiccation survival time of *Aedes* (*Stegomyia*) mosquito eggs is correlated with habitat and egg size. *Oecologia, 90*(3), 353–358.
South, A., & Catteruccia, F. (2016). Sexual selection and the evolution of mating systems in mosquitoes. In A.S. Raikhel (Ed.), *Advances in insect physiology,* (pp. 67–92). San Diego, CA: Academic Press.
South, S.H., & Arnqvist, G. (2008). Evidence of monandry in a mosquito (*Sabethes cyaneus*) with elaborate ornaments in both sexes. *Journal of Insect Behavior, 21*(6), 451–459.
South, S.H., Steiner, D., & Arnqvist, G. (2009). Male mating costs in a polygynous mosquito with ornaments expressed in both sexes. *Proceedings of the Royal Society of London B: Biological Sciences, 276*(1673), 3671–3678.
Southwood, T.R.E. (1976). Bionomic strategies and population parameters. In R.M. May (Ed.), *Theoretical ecology* (pp. 26–48). London, England: Blackwell.
Souza-Neto, J.A., Machado, F.P., Lima, J.B., Valle, D., & Ribolla, P.E. (2007). Sugar digestion in mosquitoes: Identification and characterization of three midgut α-glucosidases of the neo-tropical malaria vector *Anopheles aquasalis* (Diptera: Culicidae). *Comparative Biochemistry and Physiology Part A: Molecular and Integrative Physiology, 147*(4), 993–1000.
Sparks, J.T., Vinyard, B.T., & Dickens, J.C. (2013). Gustatory receptor expression in the labella and tarsi of *Aedes aegypti. Insect Biochemistry and Molecular Biology, 43*(12), 1161–1171.
Speiser, P. (1904). Zur Nomenclatur blutsaugender Dipteren Amerikas. *Insekten-Börse 21*(19), 148.
Speiser, P. (1909). 10. Diptera 4. Orthorhapha Nematocerca. In Y. Sjostedt (Ed.), *Wissenschaftliche Ergebnisse der Schwedischen Zoologischen Expedition nach Demkilimandjaro, dem Memu und den umgebenden Massaisteppen Deutsch-Ostafrikas 1905–1906* (pp. 31–65). Stockholm, Sweden: Tryckt hos P. Palmquists aktiebolag.
Spencer, M., Bloustein, L., & Cohen, J.E. (2002). Oviposition habitat selection by mosquitoes (*Culiseta longireolata*) and consequences for population size. *Ecology, 83*(3), 669–679.
Spielman, A. (1964). The mechanics of copulation in *Aedes aegypti. The Biological Bulletin, 127*(2), 324–344.
Spielman, A. (1971). Bionomics of autogenous mosquitoes. *Annual Review of Entomology, 16*(1), 231–248.
Spielman, A. (1974). Effect of synthetic juvenile hormone on ovarian diapause of *Culex pipiens* mosquitoes. *Journal of Medical Entomology, 11*(2), 223–225.
Spielman, A. (2001). *Mosquito: The story of man's deadliest foe.* Hachette Books.
Spielman, A., & Wong, J. (1973). Studies on autogeny in natural populations of *Culex pipiens.* III. Midsummer preparation for hibernation in anautogenous populations. *Journal of Medical Entomology, 10*(4), 319–324.
Spillings, B.L., Brooke, B.D., Kockemoer, L.L., Chiphwanya, J., Coetzee, M., & Hunt, R.H. (2009). A new species concealed by *Anopheles funestus* Giles, a major malaria vector in Africa. *The American Journal of Tropical Medicine and Hygiene, 81*(3), 510–515.
Spring, J.H., Robichaux, S.R., & Hamlin, J.A. (2009). The role of aquaporins in excretion in insects. *Journal of Experimental Biology, 212*(3), 358–362.
Spungis, V. (2000). A checklist of Latvian mosquitoes (Diptera: Culicidae). *European Mosquito Bulletin, 6*, 8–11.
St John, O. (2007). *Phylogeny of the genus* Culex *(Diptera: Culicidae)* [Master's thesis, Imperial College, London].
St. Laurent, B., Oy, K., Miller, B., Gasteiger, E.B., Lee, E., Sovannarothi, S., . . . Fairhurst, R.M. (2016). Cow-baited tents are highly effective in sampling diverse *Anopheles* malaria vectors in Cambodia. *Malaria Journal, 15*(1), 440.
Stackelberg, A. (1927). Les culicides de l'URSS et des pays limitrophes. *Tableaux analytiques de la Faune de l'URSS (Musée Zoologique de l'Academie des Sciences), 1*, 1–169.
Stackelberg, A.A. (1937). Faune de l'URSS. Insectes, Diptères. Vol. III no. 4. Fam. Culicidae (Subfam. Culicinae). Institut zoologique de l'Academie des Sciences de l'URSS, Nouvelle Série No. 11, Moscow, 1–258.
Stackelberg, A.A. (1943). A new species of *Finlaya* Theo. (Diptera, Culicidae) from Ussuri Land. *Bulletin of Entomological Research, 34*(4), 311.
Staeger, R.C. (1839). Systematik fortegnelse over de hidtil i Denmark fundne Diptera. *Naturhistorisk tidsskrift, 1*(2), 549–600.
Stanek, D.M., Pohl, J., Crim, J.W., & Brown, M.R. (2002). Neuropeptide F and its expression in the yellow fever mosquito, *Aedes aegypti. Peptides, 23*(8), 1367–1378.
Stanton, A.T. (1912). The *Anopheles* mosquitoes of Malaya and their larvae, with some notes on malaria-carrying species. *Journal of the London School of Tropical Medicine, 2*(1), 3–11.
Stanton, A.T. (1915). A new anopheline mosquito from Sumatra. *Bulletin of Entomological Research, 5*(4), 373–375.
Stanton, A.T., & Hacker, H.P. (1917). The *Anopheles* of Malaya. III. A new variety of *A. albotaeniatus*, Theo. *Bulletin of Entomological Research, 7*(3), 273–275.
Starkey, G.S., & J.E. Webb. (1946). A new species of *Aedes* of the *Finlaya* group from Angaur Island. *Proceedings of the Entomological Society of Washington, 48*(7), 179–182.
Statz, G. (1941). Plagegister aus dem Reinischen Braunkohlewnwalde. *Rheinischer Naturfreund, 5*(1), 1–16.
Statz, G. (1944). Neue Dipteren (Nematocera) aus dem Oberoligozän von Rott. II. Teil. V. Familie Culicidae (Stechmücken). *Palaeontographica, 95*(A), 108–120.
Steffan, W.A. (1966). A checklist and review of the mosquitoes of the Papuan subregion (Diptera: Culicidae). *Journal of Medical Entomology, 3*(2), 179–237.
Steffan, W.A. (1968a). Hawaiian *Toxorhynchites* (Diptera: Culicidae). *Proceedings of the Hawaiian Entomological Society, 20*(1), 141–155.

Steffan, W.A. (1968b). *Armigeres* of the Papuan subregion (Diptera: Culicidae). *Journal of Medical Entomology, 5*(2), 135–159.
Steffan, W.A. (1968c). Mosquitoes of the Papuan subregion (Diptera: Culicidae). *Toxorhynchites* (*Toxorhynchites*) *amboinensis* (Doleschall). *Journal of Medical Entomology, 5*(3), 310–313.
Steffan, W.A. (1979). New *Culex* (*Acallyntrum*) from the Papuan Subregion (Diptera, Culicidae). *Journal of Medical Entomology, 16*(2), 133–139.
Steffan, W.A. (1981). The type-specimens of *Toxorhynchites* (Diptera: Culcidae) at the National Museum of Natural History. *Mosquito Systematics, 12*(for 1980)(4), 379–385.
Steffan, W.A., & Evenhuis, N.L. (1980). Description of the pupa of *Toxorhynchites* (*Lynchiella*) *rutilus septentrionalis* (Diptera: Culicidae). *Mosquito Systematics, 12*(2), 175–178.
Steffan, W.A., & Evenhuis, N.L. (1981). Biology of *Toxorhynchites. Annual Review of Entomology, 26*, 159–181.
Steffan, W.A., & Evenhuis, N.L. (1982). A new species of *Toxorhynchites* from Papua New Guinea (Diptera: Culicidae). *Mosquito Systematics, 14*(1), 1–13.
Steffan, W.A., & Evenhuis, N.L. (1985). Classification of the subgenus *Toxorhynchites* (Diptera: Culicidae). 1. Australasian, eastern Palaearctic, and Oriental species groups. *Journal of Medical Entomology, 22*(4), 421–446.
Steffan, W.A., & White, G.B. (1981). The type specimens of *Toxorhynchites* (Diptera: Culicidae) at the British Museum (Natural History) and Oxford University. *Mosquito Systematics, 13*(1), 114–123.
Steffan, W.A, Evenhuis, N.L., & Manning, D.L. (1980). Annotated bibliography of *Toxorhynchites* (Diptera: Culicidae). *Journal of Medical Entomology, 17*(Suppl. 3), 1–140.
Steffan, W.A., Kodani, A.H., & Evenhuis, N.L. (1982). Male and female genitalia of *Toxorhynchites amboinensis* (Diptera: Culicidae). *Mosquito Systematics, 14*(1), 14–33.
Stegnii, V.N., & Kabanova, V.M. (1976). Cytoecological Study of indigenous populations of the malaria mosquito in the territory of the U.S.S.R. I. Identification of a new species of *Anopheles* in the Maculipennis Complex by the cytodiagnostic method. *Meditsinskaia parazitologiia i parizarnye bolezni, 45*(2), 192–198.
Stegnii, V.N., Kabanova, V.M., Novikov Yu, M., & Pleshkova, G.N. (1976). Inversional polymorphism in the malaria mosquito *Anopheles messeae.* Part 2. Distribution of inversions over the range of the species. *Genetika, 12*(4), 47–55.
Stein, M., Laurito, M., Rossi, G.C., & Almirón, W.R. (2009). Morphological description of the pupa and fourth-instar larva and redescription of the adults of *Psorophora* (*Psorophora*) *pallescens* Edwards (Diptera: Culicidae). *Zootaxa, 2306*, 51–58.
Stein, M., Rossi, G.C., & Almirón, W.R. (2013). Morphological description of male pupa and larva of *Psorophora* (*Grabhamia*) *paulli* and redescription of female (Diptera: Culicidae). *Zootaxa, 3686*(3), 389–395.
Steiner, W.W.M., Kitzmiller, J.B., & Osterbur, D.L. (1980). Gene differentiation in chromosome races of *Anopheles nuneztovari* (Gabaldon). *Mosquito Systematics, 12*(3), 306–319.
Steiner, W.W.M., Narang, S., Kitzmiller, J.B., & Swofford, D.L. (1982). Genetic divergence and evolution in Neotropical *Anopheles* (subgenus *Nyssorhynchus*). In W.W.M. Steiner, W.J. Tabachnick, K.S. Rai, & S. Narang (Eds.), *Recent developments in the genetics of insect disease vectors* (pp. 523–550). Illinois: Stipes.
Stephens, J.F. (1825). Some observations on the British Tipulidae, together with descriptions of the species of *Culex* and *Anopheles* found in Britain. *Zoological Journal, 1*, 448–457.
Stephens, J.F. (1828). Art. LIV. Note on the foregoing paper, with a description of a new species of *Anopheles. Zoological Journal, 3*, 502–504.
Stephens, J.F. (1846). *Illustrations of British entomology; or, a synopsis of indigenos insects: containing their generic and specfic distinctions; with an account of their metamorphoses, times of appearance, localities, food, and economy, as far as practicable. Supplement.* London, England: Henry G. Bohn.
Stephens, J.W.W., & Christophers, S.R. (1908). *The practical study of malaria and other blood parasites* (3rd ed.). London, England: Willams and Norgate.
Steward, C.C., & McWade, J.W. (1961). The mosquitoes of Ontario (Diptera, Culicidae) with keys to the species and notes on distribution. *Proceedings of the Entomological Society of Ontario, 91*(for 1960), 121–188.
Stewart, W.W.A. (1974). The rate of larval development of *Aedes implicatus* Vockeroth in field and laboratory. *Mosquito News, 34*(3), 283–285.
Stocker, T.F., Plattner, G.K., Tignor, M., Allen, K., Boschung, J., Navels, A., . . . Midgley, P.M. (Eds.). (2013). Summary for policymakers. In *Climate change 2013: The physical science basis.* Contribution of Working Group I to the Fifth Assessment Report of the Intergovernmental Panel on Climate Change. Cambridge, United Kingdom: Cambridge University Press.
Stoffolano, J.G., Jr., & Haselton, A.T. (2013). The adult Dipteran crop: A unique and overlooked organ. *Annual Review of Entomology, 58*, 205–225.
Stojanovich, C.J. (1960). *Illustrated key to common mosquitoes of southeastern United States.* Atlanta, GA: Cullom & Ghertner.
Stojanovich, C.J. (1961). *Illustrated key to common mosquitoes of northeastern United States.* [Self-published].
Stoker, W.J., & Waktoedi Koesoemawinangoen, R. (1949). *Illustrated map of the anopheline imagines of Indonesia* [English trans.]. Djakarta, Indonesia: Ministry of Health.
Stone, A. (1939). Two new *Aedes* from Guam (Diptera, Culicidae). *Proceedings of the Entomological Society of Washington, 41*(5), 162–165.
Stone, A. (1944). Notes on the genus *Trichoprosopon* (Diptera, Culicidae). *Revista de entomología, 15*(3), 335–341.
Stone, A. (1945a). A mosquito synonym (Diptera: Culicidae). *Proceedings of the Entomological Society of Washington, 47*(2), 38–39.
Stone, A. (1945b). A new species of *Aedes* from Saipan and the larva of *Aedes pandani* (Diptera: Culicidae). *Proceedings of the Entomological Society of Washington, 47*(3), 65–69.
Stone, A. (1947). A topotypic male of *Aedes scutellaris* (Walker) (Diptera, Culicidae). *Proceedings of the Entomological Society of Washington, 49*(3), 85.
Stone, A. (1948). A change of name in mosquitoes. *Proceedings of the Entomological Society of Washington, 50*(6), 161.
Stone, A. (1950). The larva of *Culex foliaceus* Lane (Diptera, Culicidae). *Revista de entomología, 21*(1–2), 237–239.
Stone, A. (1957a). Corrections in the taxonomy and nomenclature of mosquitoes (Diptera, Culicidae). *Proceedings of the Entomological Society of Washington, 58*(for 1956), 333–344.
Stone, A. (1957b). Notes on types of mosquitoes in the Hungarian National Museum (Diptera, Culicidae). *Annals of the Entomological Society of America, 50*(2), 171–174.
Stone, A. (1958a). The identity of *Culex aestuans* Wiedemann (Diptera, Culicidae). *Proceedings of the Entomological Society of Washington, 60*(4), 186.
Stone, A. (1958b). Types of mosquitoes described by C. F. Adams in 1903 (Diptera, Culicidae). *Journal of the Kansas Entomological Society, 31*(3), 235–237.
Stone, A. (1961a). A new subspecies of *Uranotaenia unguiculata* Edwards from Arabia (Diptera: Culicidae). *Proceedings of the Entomological Society of Washington, 62*(for 1960), 249.

Stone, A. (1961b). A correction in mosquito nomenclature. *Proceedings of the Entomological Society of Washington, 63*(4), 246.

Stone, A. (1961c). A synoptic catalog of the mosquitoes of the world, supplement I (Diptera: Culicidae). *Proceedings of the Entomological Society of Washington, 63*(1), 29–52.

Stone, A. (1963). A synoptic catalog of the mosquitoes of the world, supplement II (Diptera: Culicidae). *Proceedings of the Entomological Society of Washington, 65*(2), 117–140.

Stone, A. (1966). A new species of *Coquillettidia* from Samoa (Diptera: Culicidae). *Proceedings of the Entomological Society of Washington, 68*(4), 331–333.

Stone, A. (1967). A synoptic catalog of the mosquitoes of the world. Supplement III (Diptera: Culicidae). *Proceedings of the Entomological Society of Washington, 69*(3), 197–224.

Stone, A. (1969). Bredin-Archbold-Smithsonian Survey of Dominica: The mosquitoes of Dominica (Diptera: Culicidae). *Smithsonian Contributions to Zoology, 16*, 1–8.

Stone, A. (1970). A synoptic catalog of the mosquitoes of the world, suppl. IV (Diptera: Culicidae). *Proceedings of the Entomological Society of Washington, 72*(2), 137–171.

Stone, A., & Barreto, P. (1969). A new genus and species of mosquito from Colombia, *Galindomyia leei* (Diptera, Culicidae, Culicini). *Journal of Medical Entomology, 6*(2), 143–146.

Stone, A., & Bohart, R.M. (1944). Studies on mosquitoes from the Philippine Islands and Australasia (Diptera: Culicidae). *Proceedings of the Entomological Society of Washington, 46*(8), 205–225.

Stone, A., & Delfinado, M.D. (1973). Family Culicidae. In M.D. Delfinado & D.E. Hardy (Eds.), *A catalog of the Diptera of the Oriental Region: Vol. I. Suborder Nematocera* (pp. 266–343). Honolulu: University Press of Hawaii.

Stone, A., & Farner, D.S. (1945). Further notes on the *Aedes scutellaris* group (Diptera, Culicidae). *Proceedings of the Biological Society of Washington, 58*, 155–162.

Stone, A., & Farner, D.S. (1946). Further notes on the *Aedes scutellaris* group (Diptera, Culicidae). *Proceedings of the Biological Society of Washington, 58*, 166–162.

Stone, A., & Hair, J.A. (1968). A new *Culex* (*Melanoconion*) from Florida (Diptera, Culicidae). *Mosquito News, 28*(1), 39–41.

Stone, A., & Knight, K.L. (1955). Type specimens of mosquitos in the United States National Museum. I. The genera *Armigeres*, *Psorophora* and *Haemagogus* (Diptera, Culicidae). *Journal of the Washington Academy of Sciences, 45*(9), 282–289.

Stone, A., & Knight, K.L. (1956a). Type specimens of mosquitoes in the United States National Museum. II. The genus *Aedes* (Diptera, Culicidae). *Journal of the Washington Academy of Sciences, 46*(7), 213–228.

Stone, A., & Knight, K.L. (1956b). Type specimens of mosquitoes in the United States National Museum. III. The genera *Anopheles* and *Chagasia* (Diptera, Culicidae). *Journal of the Washington Academy of Sciences, 46*(9), 276–280.

Stone, A., & Knight, K. L. (1956c). A new name in Culicidae (Diptera). *Proceedings of the Entomological Society of Washington, 58*(1), 34.

Stone, A., & Knight, K.L. (1957a). Type specimens of mosquitoes in the United States National Museum. IV. The genus *Culex* (Diptera, Culicidae). *Journal of the Washington Academy of Sciences, 47*(2), 42–59.

Stone, A., & Knight, K.L. (1957b). Type specimens of mosquitoes in the United States National Museum. V. The Sabethini (Diptera, Culicidae). *Journal of the Washington Academy of Sciences, 47*(4), 117–126.

Stone, A., & Knight, K.L. (1957c). Type specimens of mosquitoes in the United States National Museum. VI. Miscellaneous genera, addenda and summary. (Diptera, Culicidae). *Journal of the Washington Academy of Sciences, 47*(6), 196–202.

Stone, A., & Knight, K.L. (1957d). Proposed use of the plenary powers to validate the generic name *Toxorhynchites* Theobald as published in 1901 in the *J. Tropical Medicine* (Class Insecta, Order Diptera). *Bulletin of Zoological Nomenclature, 13*(4), 128–131.

Stone, A., & Knight, K.L. (1958). Two new names in mosquitoes (Diptera, Culicidae). *Proceedings of the Entomological Society of Washington, 60*(2), 69.

Stone, A., & Penn, G.H. (1947). A new species of *Culex* from New Guinea (Diptera, Culicidae). *Journal of the Washington Academy of Sciences, 37*(3), 89–91.

Stone, A., & Penn, G.H. (1948). A new subgenus and two new species of the genus *Culex* L. (Diptera, Culicidae). *Proceedings of the Entomological Society of Washington, 50*(5), 109–120.

Stone, A., & Rosen, L. (1952). A new species of *Aedes* from Tahiti (Diptera: Culicidae). *Proceedings of the Hawaiian Entomological Society, 14*(3), 425–428.

Stone, A., & Rosen, L. (1953). A new species of *Culex* from the Marquesas Islands and the larva of *Culex atriceps* Edwards (Diptera: Culicidae). *Journal of the Washington Academy of Sciences, 43*(11), 354–358.

Stone, A., & Thurman, E.B. (1958). *Armigeres* (*Armigeres*) *baisasi*, a new mosquito from the Philippine Islands (Diptera: Culicidae). *Journal of the Washington Academy of Sciences, 48*(7), 240–243.

Stone, A., Knight, K.L., & Starcke, H. (1959). *A synoptic catalog of the mosquitoes of the world* (*Diptera, Culicidae*) (Vol. 6). Washington, DC: Thomas Say Foundation, Entomological Society of America.

Stone, C.M., & Foster, W.A. (2013). Plant-sugar feeding and vectorial capacity. In W. Takken & C.J.M. Koenraadt (Eds.), *Ecology of parasite-vector interactions* (pp. 35–79). Wageningen, The Netherlands: Wageningen Academic.

Stone, C.M., Jackson, B.T., & Foster, W.A. (2012a). Effects of plant-community composition on the vectorial capcity and fitness of the malaria mosquito *Anopheles gambiae*. *The American Journal of Tropical Medicine and Hygiene, 87*(4), 727–736.

Stone, C.M., Jackson, B.T., & Foster, W.A. (2012b). Effects of bed net use, female size, and plant abundance on the first meal choice (blood vs. sugar) of the malaria mosquito *Anopheles gambiae*. *Malaria Journal, 11*(1), 3.

Stone, C.M., Witt, A.B.R., Walsh, G.C., Foster, W.A., & Murphy, S.T. (2018). Would the control of invasive alien plants reduce malaria transmission? A review. *Parasites and Vectors, 11*(1), 76.

Stone, W.S., & Reynolds, F.H.K. (1939). Hibernation of anopheline eggs in the tropics. *Science, 90*(2338), 371–372.

Stracker, T.H., Thompson, S., Grossman, G.L., Riehle, M.A., & Brown, M.R. (2002). Characterization of the AeaHP gene and its expression in the mosquito *Aedes aegypti* (Diptera: Culicidae). *Journal of Medical Entomology, 39*(2), 331–342.

Strand, E. (1928). Miscellanea nomenclatorica zoologica et palaeontologica. *Archiv für Naaturgeschichte. Abteilung A, 92*(for 1926)(8), 30–75.

Strickland, C. (1914). Short description of the larva of *Lophoscelomyia asiatica* Leicester, 1905, and notes on the species. *Parasitology, 7*(1), 12–16.

Strickland, C. (1916a). A new species of protanopheline from Malaya, *Myzorhynchus hunteri*. *Indian Journal of Medical Research* (Calcutta), *4*(2), 263–270.

Strickland, C. (1916b). An *umbrosus*-like anopheline from Malaya—*Myzorhynchus novumbrosus* [*sic*]. *Indian Journal of Medical Research* (Calcutta), *4*(2), 271–273.

Strickland, C. (1917). A new species of anopheline, *Myzorhynchus similis* from the Malay Peninsula. *Indian Journal of Medical Research* (Calcutta), *4*(3), 611.
Strickland, C., & Chowdhury, K.L. (1927a). *An illustrated key to the identification of the anopheline larvae of India, Ceylon, and Malaya, west of Wallace's Line with practical notes on their collection.* Calcutta, India: Thacker, Spink.
Strickland, C., & Chowdhury, K.L. (1927b). A new species of anopheline *A. pseudojamesi* common in Bengal. *Indian Medical Gazette, 62*(5), 240–243.
Strickland, C., & Chowdhury, K.L. (1931). The anopheline larvae of the countries from India and the Orient to the Antipodes. Calcutta, India: Thacker, Spink.
Strickland, E.H. (1911a). Some new Culicidae from Western Australia, South Queensland, and Tasmania. *The Entomologist, 44*(576, 577, 578), 179–182, 201–204, 249–251.
Strickland, E.H. (1911b). A new mosquito from Paraguay. *The Entomologist, 44*(579), 268–269.
Strickland, E.H. (1915). Note on *Anopheles brevipalpis* Roper, and description of its egg and larva. *Indian Journal of Medical Research* (Calcutta), *3*(1), 201–204.
Strickman, D. (1977). Observation of penetration of a feeding mosquito by the mouthparts of another mosquito. *Mosquito News, 37*(2), 302–302.
Strickman, D. (1980a). Observations on adults and eggs of *Chaoborus flavicans* (Diptera: Culicidae). *Hydrobiologia, 74*(3), 195–197.
Strickman, D. (1980b). Stimuli affecting selection of oviposition sites by *Aedes vexans*: Conditioning of the soil. *Mosquito News, 40*(3), 413–417.
Strickman, D. (1980c). Stimuli affecting selection of oviposition sites by *Aedes vexans* (Diptera: Culicidae): Light. *Journal of Medical Entomology, 19*(2), 181–184.
Strickman, D. (1980d). Stimuli affecting selection of oviposition sites by *Aedes vexans* (Diptera: Culicidae): Moisture. *Mosquito News, 40*(2), 236–245.
Strickman, D. (1988a). Redescription of the holotype of *Culex* (*Culex*) *peus* Speiser and taxonomy of *Culex* (*Culex*) *stigmatosoma* Dyar and *thriambus* Dyar (Diptera: Culicidae). *Proceedings of the Entomological Society of Washington, 90*(4), 484–494.
Strickman, D. (1988b). Rate of oviposition by *Culex quinquefasciatus* in San Antonio, Texas, during three years. *Journal of the American Mosquito Control Association, 4*(3), 339–344.
Strickman, D. (1989). Biosystematics of larval movement of Central American mosquitoes and its use for field identification. *Journal of the American Mosquito Control Association, 5*(2), 208–218.
Strickman, D. (1990). *Culex pseudostigmatosoma, Cx. yojoae*, and *Cx. aquarius*: New Central American species in the subgenus *Culex* (Diptera: Culicidae). *Mosquito Systematics, 21*(for 1989)(3), 143–177.
Strickman, D. (2006). Longevity of *Aedes aegypti* (Diptera: Culicidae) compared in cages and field under ambient conditions in rural Thailand. *Southeast Asian Journal of Tropical Medicine, 37*(3), 456–462.
Strickman, D. (2019). Perspectives: Olfactory targeting of integrated vector management. In J.-F. Picimbon (Ed.), *Olfactory concepts of insect control—Alternative to insecticides, Volume 1.* Zug, Switzerland: Springer Nature Switzerland AG.
Strickman, D., & Darsie, R.F., Jr. (1988). The previously undetected presence of *Culex restuans* (Diptera: Culicidae) in Central America, with notes on identification. *Mosquito Systematics, 20*(1), 21–27.
Strickman, D., & Fonseca, D.M. (2012). Autogeny in *Culex pipiens* complex mosquitoes from the San Francisco Bay area. *The American Journal of Tropical Medicine and Hygiene, 87*(4), 719–726.
Strickman, D., & Kittayapong, P. (1993). Laboratory demonstration of oviposition by *Aedes aegypti* (Diptera: Culicidae) in covered water jars. *Journal of Medical Entomology, 30*(5), 947–949.
Strickman, D., & Kittayapong, P. (2003). Dengue and its vectors in Thailand: Calculated transmission risk from total pupal counts of *Aedes aegypti* and association of wing-length measurements with aspects of the larval habitat. *The American Journal of Tropical Medicine and Hygiene, 68*(2), 209–217.
Strickman, D., & Lang, J.T. (1986). Activity of *Culex quinquefasciatus* in an underground storm drain in San Antonio, Texas. *Journal of the American Mosquito Control Association, 2*(3), 379–381.
Strickman, D., & Pratt, J. (1989). Redescription of *Cx. corniger* Theobald and elevation of *Culex* (*Culex*) *lactator* Dyar and Knab from synonymy based on specimens from Central America (Diptera: Culicidae). *Proceedings of the Entomological Society of Washington, 91*(4), 551–574.
Strickman, D., Sithiprasasna, R., & Southard, D. (1997). Bionomics of the spider, *Crossopriza lyoni* (Araneae: Pholcidae), a predator of dengue vectors in Thailand. *The Journal of Arachnology, 25*(2), 194–201.
Strickman, D., Miller, M.E., Kim, H.C., & Lee, K.W. (2000a). Mosquito surveillance in the Demilitarized Zone, Republic of Korea, during an outbreak of *Plasmodium vivax* malaria in 1996 and 1997. *Journal of the American Mosquito Control Association, 16*(2), 100–113.
Strickman, D., Timberlake, B., Estrada-Franco, J., Weissman, M., Fenimore, P.W., & Novak, R.J. (2000b). Effects of magnetic fields on mosquitoes. *Journal of the American Mosquito Control Association, 16*(2), 131–137.
Strickman, D., Frances, S.P., & Debboun, M. (2009). *Prevention of bug bites, stings, and diseases.* New York, NY: Oxford University Press.
Strode, C., Steen, K., Ortelli, F., & Ranson, H. (2006). Differential expression of the detoxification genes in the different life stages of the malaria vector *Anopheles gambiae. Insect Molecular Biology, 15*(4), 523–530.
Strode, C., Wondji, C.S., David, J.P., Hawks, N.J., Lumjuan, N., Nelson, D.R., . . . Ranson, H. (2008). Genomic analysis of detoxification genes in the mosquito *Aedes aegypti. Insect Biochemistry and Molecular Biology, 38*(1), 113–123.
Stschelkanovzev, J. (1926). Zur Kenntnis der Steckmucken Fauna von Sud-Ost RSFSR. *Bulletin de la Société Naturelle de Voroneje, 1*, 123–134.
Styer, L.M., Carey, J.R., Wang, J.L., & Scott, T.W. (2007). Mosquitoes do senesce: Departure from the paradigm of constant mortality. *The American Journal of Tropical Medicine and Hygiene, 76*(1), 111–117.
Styer, L.M., Lim, P.Y., Louie, K.L., Albright, R.G., Kramer, L.D., & Bernard, K.A. (2010). Mosquito saliva causes enhancement of West Nile virus infection in mice. *Journal of Virology, 85*(4), 1517–1527.
Su, L., & Wang, C. (1981). A new species of the genus *Toxorhynchites* Theobald (Diptera: Culicidae). *Acta entomologica Sinica, 24*(3), 327–331.
Su, L., & Zhang, Y.H. (1987). A new species of the subgenus *Ochlerotatus* Lynch Arribálzaga (Diptera: Culicidae). *Acta entomologica Sinica, 30*(4), 450–452.
Su, L., & Zhang, Y. (1988). A new species of *Aedes* (Diptera: Culicidae). *Entomotaxonomia, 10*(1–2), 11–13.
Su, L., Wang, C., & Li, F. (1978). A new species of *Aedes, Aedes* (*Edwardsaedes*) *antuensis* sp. nov. *Acta entomologica Sinica, 21*(4), 439–444.

Suárez, M F., Quiñones, M., Fleming, G.A., & Robayo, M. (1988). *Guía introductoria a la morfología de* Anopheles *y clave para determinación de las principales especies de Colombia, D.C.* Bogotá, Colombia: Ministerio de Salud.

Suarez, O.M., & García, P.C. (1961). Estudio morfologico de *Aedes (H.) ioliota* Dyar & Knab, 1913. *Revista Brasileira de entomología, 10*, 17–24.

Subbarao, S.K., Kumar, K.V., Nanda, N., Nagpal, B.N., Dev, V.,& Sharma, V.P. (2000). Cytotaxonomic evidence for the presence of *Anopheles nivipes* in India. *Journal of the American Mosquito Control Association, 16*(2), 71–74.

Suguna, S.G. (1982). Cytological and morphological evidences for sibling species in *Anopheles subpictus* Grassi. *Journal of Communicable Diseases, 14*, 1–8.

Suguna, S.G., Rathinam, K.G., Rajavel, A.R., & Dhanda, V. (1994). Morphological and chromosomal descriptions of new species in the *Anopheles subpictus* complex. *Medical and Veterinary Entomology, 8*(1), 88–94.

Sukowati, S., & Baimai, V. (1996). A standard cytogenetic map for *Anopheles sundaicus* (Diptera: Culicidae) and evidence for chromosomal differentiation in populations from Thailand and Indonesia. *Genome, 39*(1), 165–173.

Sukowati, S., Baimai, V., Harun, S., Dasuki, Y., Andris, H., & Efriwati, M. (1999). Isozyme evidence for three sibling species in the *Anopheles sundaicus* complex from Indonesia. *Medical and Veterinary Entomology, 13*, 408–414.

Suleman, M., Khan, K., & Khan, S. (1993). Ecology of mosquitoes in Peshawar Valley and adjoining areas: Species composition and relative abundance. *Pakistan Journal of Zoology, 25*(4), 321–328.

Suman, D.S., Shrivastava, A.R., Parashar, B.D., Pant, S.C., Agrawal, O.P., & Prakash, S. (2008). Scanning electron microscopic studies on egg surface morphology and morphometrics of *Culex tritaeniorhynchus* and *Culex quinquefasciatus* (Diptera: Culicidae). *Parasitology Research, 104*, 173–176.

Suman, D.S., Tikar, S.N., Mendki, M.J., Sukumaran, D., Agrawal, O.P., Parashar, B.D., & Prakash, S. (2011). Variations in life tables of geographically isolated strains of the mosquito *Culex quinquefasciatus*. *Medical and Veterinary Entomology, 25*(3), 276–288.

Suman, D.S., Wang, Y., & Gaugler, R. (2015). The insect growth regulator pyriproxyfen terminates egg diapause in the Asian tiger mosquito, *Aedes albopictus*. *PLOS ONE, 10*(6), e0130499.

Sumanochitrapon, W., Strickman, D., Sithiprasasna, R., Kittayapong, P., & Innis, B.L. (1998). Effect of size and geographic origin of *Aedes aegypti* on oral infection with dengue-2 virus. *The American Journal of Tropical Medicine and Hygiene, 58*(3), 283–286.

Sumba, L.A., Okoth, K., Deng, A.L., Githure, J., Knols, B.G., Beier, J.C., & Hassanali, A. (2004). Daily oviposition patterns of the African malaria mosquito *Anopheles gambiae* Giles (Diptera: Culicidae) on different types of aqueous substrates. *Journal of Circadian Rhythms, 2*(1), Article 6.

Summers Connal, S.L.M. (1931). The larva and pupa of *Uranotaenia ornata*, Theo. *Bulletin of Entomological Research, 22*(4), 459–460.

Sun, G., Zhu, J., Li, C., Tu, Z., & Raikhel, A.S. (2002). Two isoforms of the early E74 gene, an Ets transcription factor homologue, are implicated in the ecdysteroid hierarchy governing viellogenesis of the mosquito *Aedes aegypti*. *Molecular and Cellular Endocrinology, 190*(1), 147–157.

Sunahara, T., & Mogi, M. (1998). Distribution and turnover of a mosquito (*Tripteroides bambusa*) metapopulation among bamboo groves. *Ecological Research, 13*(3), 291–299.

Surcouf, J. (1912). Note sur les culicides. *Bulletin du Muséum National d'Histoire Naturelle, 18*(2), 59–61.

Surcouf, J.M.R., & González-Rincones, R. (1911). *Essai sur les diptères vulnerants du Venezuela. Matériaux pour servir à l'étude des diptères piqueurs et suceurs de sang de l'Amérique intertropicale.* Paris, France: A. Malone.

Surcouf, J., & González-Rincones, R. (1912). Diptères piqueurs et suceurs de sang actuellement connus de la République de Venezuela. *Archives de parasitologie Paris, 15*(2),248–314.

Surendran, S.N., Sarma, D.K., Jude, P.J., Kemppainen, P., Kanthakumaran, N., Gajapathy, K., . . . Walton, C. (2013). Molecular characterization and identification of members of the *Anopheles subpictus* complex in Sri Lanka. *Malaria Journal, 12*, 304.

Surtees, G. (1958). A description of the larva of *Toxorhynchites erythrurus* Edwards, with a note on the larva of *Ficalbia (Mimomyia) lacustris* Edwards from southern Nigeria (Diptera: Culicidae). *Proceedings of the Royal Entomological Society of London, Series B, 27*, 63–64.

Sutcliffe, J.F., & Yin, S. (2014). Behavioural responses of females of two anopheline mosquito species to human-occupied, insecticide-treated and untreated nets. *Malaria Journal, 13*(1), 294.

Sutil Oramas, E. (1976). Redescripción de la especie *Anopheles (Nyssorhynchus) nunez-tovari* Gabaldon, 1940, y su distribución geográfica en Venezuela. *Boletín de la Dirección de Malariología y Saneamiento Ambiental, 16*(1), 33–45.

Sutil Oramas, E. (1980). *Enumeración histórica y geográfica de las especies de Culicida de Venezuela ordenadas según su taxonomía.* Publicación de la División de Endemias Rurale, Dirección de Malariología y Saneamiento Ambiental, Ministerio de Sanidad y Asistencia Social, Maracay, Venezuela.

Sutil Oramas, E., & Pulido, F.J. (1974). *Wyeomyia cova-garciai* n. sp. *Boletín de la Dirección de Malariología y Saneamiento Ambiental, 14*(1–2), 24–26.

Sutil Oramas, E., & Pulido, F.J. (1978). *Sabethes (Sabethes) carrilloi* sp. n. de Venezuela. *Boletín de la Dirección de Malariología y Saneamiento Ambiental, 28*(3), 199–204.

Sutil Oramas, E., Pulido, F.J., & Amarista, M.J.R. (1987). Dos especies nuevas de *Culex* de Venezuela (Diptera: Culicidae). *Boletín de la Dirección de Malariología y Saneamiento Ambiental, 27*(1–4), 81–85.

Suwannachote, N., Grieco, J.P., Achee, N.L., Suwonkerd, W., Wongtong, S., & Chareonviriyaphap, T. (2009). Effects of environmental conditions on the movement patterns of *Aedes aegypti* (Diptera: Culicidae) into and out of experimental huts in Thailand. *Journal of Vector Ecology, 34*(2), 267–275.

Suwannamit, S., Baima, V., Otsuka, Y., Saeung, A., Thongsahuan, S., Tuetun, B., Apiwathnasorn, C., Jariyapan, N., Somboon, P., Takaoka, H., & Choochote, W. (2009). Cytogenetic and molecular evidence for an additional new species within the taxon *Anopheles barbirostris* (Diptera: Culicidae) in Thailand. *Parasitology Research, 104*, 905–918.

Suwonkerd, W., Mongkalangoon, P., Parbaripai, A., Grieco, J.P., Achee, N., Roberts, D., & Chareonviriyaphap, T. (2006). The effect of host type on movement patterns of *Aedes aegypti* (Diptera: Culicidae) into and out of experimental huts in Thailand. *Journal of Vector Ecology, 31*(2), 311–318.

Suwonkerd, W., Ritthison, W., Ngo, C.T., Tainchum, K., Bangs, M.J., & Chareonviriyaphap, T. (2013). Vector biology and malaria transmission in Southeast Asia. Anopheles *mosquitoes: New insights into malaria vectors* (pp. 273–325). Rijeka, Croatia: InTech.

Suzuki, K. (1959). Notes on two species of the subgenus *Ochlerotatus* (gen. *Aedes*) from Hokkaido, Japan (Culicidae, Diptera). *Zoological Magazine of the Zoological Society of Japan, 68*, 291–296.

Suzuki, Y., Tanimura, S., Miyagawa, M., & Murata, S. (1953). Species of mosquitoes collected in Tokushima City. III [in Japa. *Igaku to seibutsugaku* [Medicine and biology], *28*, 8–10.

Sweeney, A.W. (1987). Larval salinity tolerances of the sibling species of *Anopheles farauti*. *Journal of the American Mosquito Control Association, 3*(4), 589–592.
Sweeney, A.W., Beebe, N.W., Cooper, R.D., Bauer, J.T., & Peterson, A.T. (2006). Environmental factors associated with distribution and range limits of malaria vector *Anopheles farauti* in Australia. *Journal of Medical Entomology, 43*(5), 1068–1075.
Sweeney, A.W., Beebe, N.W., & Cooper, R.D. (2007). Analysis of environmental factors influencing the range of anopheline mosquitoes in northern Australia using a genetic algorithm and data mining methods. *Ecological Modelling, 203*(3), 375–386.
Sweet, W.C., & Rao, B.A. (1937). Races of *A. stephensi* Liston 1901. *Indian Medical Gazette, 72*(11) 665–674.
Swellengrebel, N.H. (1914). Een nieuwe Anopheline voor Deli. *Myzorhynchus argyropus*. n. sp. [eine neue Anopheline in Deli.]. *Geneeskundig tijdschrift voor Nederlandsch-Indië, 54,* 334–338.
Swellengrebel, N.H. (1917). *Myzomyia flava* n.sp., een nieuwe Anopheline voor Ned.-Indië. *Geneeskundig tijdschrift voor Nederlandsch-Indië, 57,* 807–809.
Swellengrebel, N.H. (1919). Eenige voor Nederl.-Indie nieuwe anophelinen. *Geneeskundig tijdschrift voor Nederlandsch Indië, 59,* 1–12.
Swellengrebel, N.H. (1921). Die anophelien van Nederlandsch Oost-Indie. *Koloniaal Instituut Amsterdam Mededeeling No. XV. Afdeeling Tropische Hygiene, 10* (15), 1–155.
Swellengrebel, N.H., & Rodenwaldt, E. (1932). *Die anophelen von Niedeländisch-Ostindien*. 3te auflage. Gustav Fischer, Jena.
Swellengrebel, N.H., & Swellengrebel de Graaf, J.M.H. (1920). Addenda to bescription [*sic*] of larvae of Netherlands Indian anophelines. *Burgerlijken Geneeskundigen Dienst in Nederlandsch-Indië* (Mededeeligen), *9*(for 1919), 1–3.
Syed, Z., & Leal, W.S. (2008). Mosquitoes smell and avoid the insect repellent DEET. *Proceedings of the National Academy of Sciences of the United States of America, 105*(36), 13598–13603.
Symes, C.B. (1955). Filarial infections in mosquitoes in Fiji. *Transactions of the Royal Society of Tropical Medicine and Hygiene, 49,* 280–282; with comments by Sir Philip Manson-Bahr, 282–284.
Szadziewski, R. (1998). New mosquitoes from Baltic amber (Diptera: Culicidae). *Polish Journal of Entomology, 67*(3–4), 233–244.
Szadziewski, R., & Gilka, W. (2011). A new fossil mosquito, with notes on the morphology and taxonomy of other species reported from Eocene Baltic amber (Diptera: Culicidae). *Polish Journal of Entomology, 80*(4), 765–777.
Szadziewski, R., & Szadziewska, M.M. (1985). *Culex erikae* sp. n. (Diptera, Culicidae) from the Baltic amber. *Polish Journal of Entomology, 55,* 513–518.
Szadziewski, R., Sontag, E., & Szwedo, J. (2019). Mosquitoes of the extant avian malaria vector *Coquillettidia* Dyar, 1905 from Eocene Baltic amber (Diptera: Culicidae). *Zootaxa, 2*(6), 650–656.
Taai, K., & Harbach, R.E. (2015). Systematics of the *Anopheles barbirostris* species complex (Diptera: Culicidae: Anophelinae) in Thailand. *Zoological Journal of the Linnean Society, 174*(2), 244–264.
Taai, K., Baimai, V., Saeung, A., Thongsahuan, S., Min, G.-S., Otsuka, Y., . . . Choochote, W. (2013). Genetic compatibility between *Anopheles lesteri* from Korea and *Anopheles paraliae* from Thailand. *Memórias do Instituto Oswaldo Cruz, 108*(3), 312–320.
Tabachnick, W.J. (1991). Evolutionary genetics and arthropod-borne disease: The yellow fever mosquito. *American Entomologist, 37*(1), 14–26.
Tabachnick, W.J. (2010). Challenges in predicting climate and environmental effects on vector-borne disease episystems in a changing world. *Journal of Experimental Biology, 213*(6), 946–954.
Tajima, K. (1952). Some notes on the eggs of *Anopheles lindesayi* var. *japonicus* Yamada. *Japanese Journal of Sanitary Zoology, 3,* 90–94.
Takahashi, M. (1968). Taxonomic and ecological notes on *Culex* (*Melanoconion*) *spissipes* (Theobald). *Journal of Medical Entomology, 5*(3), 329–331.
Takano, K.T., Nguyen, N.T.H., Nguyen, B.T.H., Sunahara, T., Yasunami, M., Nguyen, M.D., & Takagi, M. (2010). Partial mitochondrial DNA sequences suggest the existence of a cryptic species within the Leucosphyrus group of the genus *Anopheles* (Diptera: Culicidae), forest malaria vectors, in northern Vietnam. *Parasites & Vectors, 3,* 41.
Takei, H. (1941). Kinds and distribution of *Anopheles* in Hainan Island. *Bulletin of the Naval Medical Association of Japan, 3,* 90–94.
Takken, W., & Verhulst, N.O. (2013). Host preferences of blood-feeding mosquitoes. *Annual Review of Entomology, 58,* 443–453.
Takken, W., van Loon, J.J.A., & Adam, W. (2001). Inhibition of host-seeking response and olfactory responsiveness in *Anopheles gambiae* following blood feeding. *Journal of Insect Physiology, 47*(3), 303–310.
Takken, W., Greene, R., Adam, W., Jetten, T.H., & van der Velden, J.A. (2002). Distribution and dynamics of larval populations of *Anopheles messeae* and *A. atroparvus* in the delta of the Rivers Rhine and Meuse, The Netherlands. *AMBIO: A Journal of the Human Environment, 31*(3), 212–218.
Takken, W., Costantini, C., Dolo, G., Hassanali, A., Sagnon, N.F., & Osir, E. (2006). Mosquito mating behaviour. *Bridging Laboratory and Field Research for Genetic Control of Disease Vectors, 11,* 183–188.
Takumi, K., Scholte, E.J., Braks, M., Reusken, C., Avenell, D., & Medlock, J.M. (2009). Introduction, scenarios for establishment and seasonal activity of *Aedes albopictus* in the Netherlands. *Vector-Borne and Zoonotic Diseases, 9*(2), 191–196.
Talyuli, O.A., Bottino-Rojas, V., Taracena, M.L., Soares, A.L.M., Oliveira, J.H.M., & Oliveira, P.L. (2015). The use of a chemically defined artificial diet as a tool to study *Aedes aegypti* physiology. *Journal of Insect Physiology, 83,* 1–7.
Tamayo, M.O., & García, C.A. (1907). Las aguas de Huacachina informe presentado a la Socieded Geográfica de Lima. *Memórias de la Municipalidad de Lima, 1906,* 1–63 (Reproduced in part September 30, 1907, as "Las Aguas do Huacachina" in *Boletín de la Sociedad Geográfica de Lima, 21,* 241–251, including the description of *Anopheles peruvianus* Tamayo [from Harbach 2018c].
T'an, C.-C. (1945). On the larval structures of *Armigeres obturbans* Walker and *Lutzia fuscanus* Wiedemann. *Sinensia, 16,* 45–49.
Tan, C.H., Vythilingam, I., Matusop, A., Chan, S.T., & Singh, B. (2008). Bionomics of *Anopheles latens* in Kapit, Sarawak, Malaysian Borneo in relation to the transmission of zoonotic simian malaria parasite *Plasmodium knowlesi*. *Malaria Journal, 7,* 52.
Tanaka, K. (1971). Mosquitoes of the Ryukyu Islands. *Abstracts 23rd Annual Meeting of the Japanese Society of Sanitary Zoology,* 4.
Tanaka, K. (1972). An aedine mosquito new to the Japanese fauna. *Japanese Journal of Sanitary Zoology, 22*(4), 229.
Tanaka, K. (1998). A new species of mosquitoes of *Aedes* (*Ochlerotatus*) (Diptera: Culicidae) from Japan. *Japanese Journal of Systematic Entomology, 4*(2), 215–225.
Tanaka, K. (1999). Studies on the pupal mosquitoes (Diptera: Culicidae) of Japan. 1. *Aedes* (*Ochlerotatus*). *Japanese Journal of Systematic Entomology, 5*(1), 105–124.

Tanaka, K. (2000a). Studies on the pupal mosquitoes (Diptera: Culicidae) of Japan. 2. *Aedes* (*Aedes*). *Japanese Journal of Systematic Entomology, 6*(1), 1–9.
Tanaka, K. (2000b). Studies on the pupal mosquitoes (Diptera, Culicidae) of Japan. 3. *Aedes* (*Stegomyia*). *Japanese Journal of Systematic Entomology, 6*(2), 225–247.
Tanaka, K. (2001). On the pupa of *Aedes* (*Stegomyia*) *chemulpoensis* Yamada from Korea (Diptera: Culididae). *Japanese Journal of Systematic Entomology, 7*(2), 189–193.
Tanaka, K. (2002a). Studies on the pupal mosquitoes of Japan. 5. Four subspecies of *Aedes* (*Finlaya*) *japonicus* Theobald, including subsp. *shintienensis* from Taiwan (Diptera, Culicidae). *Japanese Journal of Systematic Entomology, 8*(1), 63–77.
Tanaka, K. (2002b). Studies on the pupal mosquitoes of Japan. 6. *Aedes* (*Finlaya*) (Diptera, Culicidae). *Japanese Journal of Systematic Entomology, 8*(2), 137–177.
Tanaka, K. (2003a). Studies of the pupal mosquitoes of Japan. 7. Subgenera *Aedimorphus*, *Geoskusea*, *Neomelaniconion* and *Neomacleaya* of genus *Aedes*. (Diptera: Culicidae). *Japanese Journal of Systematic Entomology, 9*(1), 11–28.
Tanaka, K. (2003b). Studies on the pupal mosquitoes of Japan. 8. A key to species of the genus *Aedes*. *Medical and Veterinary Zoology, 54*(1), 105–111.
Tanaka, K. (2003c). Studies on the pupal mosquitoes of Japan. 9. Genus *Lutzia*, with establishment of two new subgenera, *Metalutzia* and *Insulalutzia* (Diptera, Culicidae). *Japanese Journal of Systematic Entomology, 9*(2), 159–169.
Tanaka, K. (2004a). Studies on the pupal mosquitoes of Japan. 10. *Culex* (*Culex*) (Diptera, Culicidae). *Japanese Journal of Systematic Entomology, 10*(1), 9–42.
Tanaka, K. (2004b). Studies on the pupal mosquitoes of Japan. 11. Subgenera *Oculeomyia* (stat. nov.) and *Sirivanakarnius* (nov.) of the genus *Culex*, with a key of pupal mosquitoes from Ogasawara-Gunto (Diptera: Culicidae). *Medical and Veterinary Zoology, 55*(3), 217–231.
Tanaka, K. (2014). Family Culicidae. In T. Nakamura, T. Saigusa, & M. Suwa (Eds.), *Catalogue of the insects of Japan: Vol. 8, Diptera: Part 1, Nematocera—Brchycera Aschiza* (pp. 181–201). Tokyo, Japan: Entomological Society of Japan.
Tanaka, K. (2018). A taxonomic catalogue of the mosquitoes from Japan, with a key to Japanese genera and subgenera of the subtribe Aedina (Diptera: Culicidae). *MAKUNAGI/Acta Dipterologica, 11*(6), Supplement No. 3, 1–157.
Tanaka, K., & Mizusawa, K. (1973). Two new species of the genus *Aedes* (*Neomacleaya*) from the Ryukyu Islands (Diptera, Culicidae). *Bulletin of the National Science Museum, Tokyo, 16*, 625–638.
Tanaka, K., Saugstad, E.S., & Mizusawa, K. (1974). Crab-hole mosquitoes of the Ryukyus. [abstract.]. *Japanese Journal of Sanitary Zoology, 24*(4), 308.
Tanaka, K., Mizusawa, K., & Saugstad, E.S. (1975a). Two new crab-hole mosquitoes of the genus *Uranotaenia* (*Pseudoficalbia*) from the Ryukyu Islands (Diptera, Culicidae). *Mosquito Systematics, 7*(1), 27–40.
Tanaka, K., Mizusawa, K., & Saugstad, E.S. (1975b). A new species of the genus *Aedes* (*Aedes*) from Japan, with synonymical notes on Japanese species of the subgenus *Aedes* (Diptera, Culicidae). *Mosquito Systematics, 7*(1), 41–58.
Tanaka, K., Saugstad, E.S., & Mizusawa, K. (1975c). Mosquitoes of the Ryukyu Archipelago (Diptera: Culicidae). *Mosquito Systematics, 7*(3), 207–233.
Tanaka, K., Mizusawa, K., & Saugstad, E.S. (1975d). Addendum to: A new species of the genus *Aedes* (*Aedes*) from Japan, with synonymical notes on Japanese species of the subgenus *Aedes* (Diptera, Culicidae). *Mosquito Systematics, 7*(l), 41–58; *Mosquito Systematics, 7*(2), 174–177.
Tanaka, K., Mizusawa, K., & Saugstad, E.S. (1979). A revision of the adult and larval mosquitoes of Japan (including the Ryukyu Archipelago and Ogasawara Islands) and Korea (Diptera: Culicidae). *Contributions of the American Entomological Institute, 16*, 1–987.
Tanaka, S. (2000). The role of moisture in the control of diapause, mating and aggregation in a tropical insect. *Entomological Science, 3*(1), 147–155.
Tanimura, S. (1952). On the variation of the number of white bands of legs of *Aedes japonicus* Theobald collected in Tokushima City. *Likken Seibutsu Gakuho, 2*(1), 173–178.
Tate, P. (1932). The larval instars of *Orthopodomyia pulchripalpis* Rond. (Diptera, Nematocera). *Parasitology, 24*(1), 111–120.
Tatem, A.J., Hay, S.I., & Rogers, D.J. (2006). Global traffic and disease vector dispersal. *Proceedings of the National Academy of Sciences of the United States of America, 103*(16), 6242–6247.
Tauthong, P., & Brust, T.A. (1977). The effect of temperature on the development and survival of two populations of *Aedes campestris* Dyar and Knab (Diptera: Culicidae). *Canadian Journal of Zoology, 55*(1), 135–137.
Taylor, B. (1972). A new species of *Aedes* from San Cristobal, British Solomon lslands Protectorate. *Journal of Medical Entomology* (Honolulu), *9*(4), 317–318.
Taylor, B. (1973). The mosquitoes (Diptera, Culicidae) of Rennell and Bellona, a further contribution. *Natural History of Rennell Island, British Solomon Islands, 7*, 61–71.
Taylor, B., & Maffi, M. (1971). Contribution to the knowledge of *Aedes* (*Stegomyia*) *hoguei* Belkin 1962 (Diptera: Culicidae). *Pacific Insects, 13*(1), 119–121.
Taylor, B., & Maffi, M. (1978). A review of the mosquito fauna of the Solomon Islands. *Pacific Insects, 19*(3–4), 165–248.
Taylor, B., & Maffi, M. (1991). *Anopheles* (*Cellia*) *rennellensis*, a new species within the *punctulatus* complex of *Anopheles* (Diptera, Culicidae) from Rennell Island. *Natural History of Rennell Island, British Solomon Islands, 8*, 195–198.
Taylor, B., & Tenorio, J.A. (1974). *Aedes* (*Finlaya*) *maffii*, a new species of mosquito from the British Solomon Islands (Diptera: Culicidae). *Journal of Medical Entomology, 11*(5), 577–581.
Taylor, C.E., Touré, Y.T., Coluzzi, M., & Petrarca, V. (1993). Effective population size and persistence of *Anopheles arabiensis* during the dry season in west Africa. *Medical and Veterinary Entomology, 7*(4), 351–357.
Taylor, D.B. (1990). Genetics of interspecific hybridization in the Triseriatus and Zoosophus groups of *Aedes* (*Protomacleaya*) (Diptera: Culicidae). *Annals of the Entomological Society of America, 83*(6), 1181–1191.
Taylor, D.S., & Turner, R.L. (1998). Notes on mosquito collections from Utila, Bay Islands, Honduras. *Journal of the American Mosquito Control Association, 14*(2), 214–215.
Taylor, F.H. (1912a). Description of mosquitoes collected in the Northern Territory during the expedition, 1911. *Bulletin of the Northern Territory of Australia, 1a*, 25–34.
Taylor, F.H. (1912b). Contribution of a knowledge of the Culicidae and Ixodidea of North Queensland and the Torres Straits. *Annual Report of Community Public Health Queensland 1912*, 27–29.
Taylor, F.H. (1913). Report of the entomologist. *Report of the Australian Institute of Tropical Medicine, 1911*, 49–74.
Taylor, F.H. (1914a). A revision of the Culicidae in the Macleay Museum, Sydney. *Proceedings of the Linnean Society of New South Wales, 38*(for 1913)(4), 747–760.

Taylor, F.H. (1914b). Contributions to a knowledge of Australian Culicidae. No. I. *Proceedings of the Linnean Society of New South Wales, 39*(3), 454–468.

Taylor, F.H. (1914c). The Culicidae of Australia. I. *Transactions of the Entomological Society of London, 1913,* 683–708.

Taylor, F.H. (1914d). Culicidae from Papua. *Transactions of the Royal Entomological Society of London, 1914,* 185–205.

Taylor, F.H. (1915). Contributions to a knowledge of Australian Culicidae. No. II. *Proceedings of the Linnean Society of New South Wales, 40*(3), 176–184.

Taylor, F.H. (1916). Contributions to a knowledge of Australian Culicidae. No. III. *Proceedings of the Linnean Society of New South Wales, 41*(3), 564–574.

Taylor, F.H. (1919). Contributions to a knowledge of Australian Culicidae. No. IV. *Proceedings of the Linnean Society of New South Wales, 43*(for 1918)(4), 826–843.

Taylor, F.H. (1920). Australian phlebotomic Diptera. New Culicidae, Tabanidae and synonymy. *Proceedings of the Royal Society of Victoria, 32*(2), 164–167.

Taylor, F.H. (1927). Two new species of Australian Culicidae (Diptera). *Bulletin of Entomological Research, 18*(1), 67–69.

Taylor, F.H. (1929). Notes on Australian Culicidae (Dipt.). *Bulletin of Entomological Research, 20*(3), 271–277.

Taylor, F.H. (1933). A new species of *Finlaya* (order Diptera: family Culicidae) from Papua. *Australian Journal of Zoology, 7*(5), 370.

Taylor, F.H. (1934). The Diptera of the territory of New Guinea. I. Family Culicidae. *Proceedings of the Linnean Society of New South Wales, 59*(3–4), 229–236.

Taylor, F.H. (1943a). Contributions to a knowledge of Australian Culicidae. No. VI. *Proceedings of the Linnean Society of New South Wales, 68*(3–4), 153–157.

Taylor, F.H. (1943b). Mosquito intermediary hosts of disease in Australia and New Guinea. *Commonwealth of Australia, Department of Health. Service Publication (School of Public Health and Tropical Medicine), 4.*

Taylor, F.H. (1944). Contributions to a knowledge of Australian Culicidae. No. VII. *Proceedings of the Linnean Society of New South Wales, 69*(3–4), 120–128.

Taylor, F.H. (1946). Contributions to a knowledge of Australian Culicidae. No. VIII. *Proceedings of the Linnean Society of New South Wales, 70*(for 1945) (5–6), 211–212.

Taylor, K.E., Stouffer, R.J., & Meehl, G.A. (2012). An overview of CMIP5 and the experimental design. *Bulletin of the American Meteorological Society, 93*(4), 485–498.

Tchioffo, M.T., Boissière, A., Churcher, T.S., Abate, L., Gimonneau, G., Nsango, S.E., . . . Morlais, I. (2013). Modulation of malaria infection in *Anopheles gambiae* mosquitoes exposed to natural midgut bacteria. *PLOS ONE, 8*(12), e81663.

Tchouassi, D.P., Sang, R., Sole, C.L., Bastos, A.D.S., Teal, P.E.A., Borgemeister, C., & Torto, B. (2013). Common host-derived chemicals increase catches of disease-transmitting mosquitoes and can improve early warning systems for Rift Valley fever virus. *PLOS Neglected Tropical Diseases, 7*(1), e2007.

Teesdale, C. (1955). Studies on the bionomics of *Aedes aegypti* (L.) in its natural habitats in a coastal region of Kenya. *Bulletin of Entomological Research, 46,* 711–742.

Telang, A., & Wells, M.A. (2004). The effect of larval and adult nutrition on successful autogenous egg production by a mosquito. *Journal of Insect Physiology, 50*(7), 677–685.

Teleman, A.A., Chen, Y.W., & Cohen, S.M. (2005). 4E-BP functions as a metabolic brake used under stress condition but not during normal growth. *Genes and Development, 19*(6), 1844–1848.

Templeton, T.J., Kaslow, D.C., & Fidock, D.A. (2000). Developmental arrest of the human malaria parasite *Plasmodium falciparum* within the mosquito midgut via CTRP gene disruption. *Molecular Microbiology, 36*(1), 1–9.

Temu, E.A., Minjas, J.N., Tuno, N., Kawada, H., & Takagi, M. (2007). Identification of four members of the *Anopheles funestus* (Diptera: Culicidae) group and their role in *Plasmodium falciparum* transmission in Bagamoyo, coastal Tanzania. *Acta Tropica, 102*(2), 119–125.

Tennent, J.E. (1859). *Ceylon: an account of the island; physical, historical, and topgraphical; with notices of its natural history, antiquities and productions* (2nd ed., vol. 1). Longman, Green, Longman, and Roberts, London.

Tennent, J.E. (1861). *Sketches of the natural history of Ceylon with narratives and anecdotes illustrative of the habits and instincts of the mammalia, birds, reptiles, fishes, insects, etc. including a monograph of the elephant and a description of the modes of capturing and training it.* London: Longman, Green, Longman, & Roberts.

Tenorio, J.A. (1975). *Bironella (Brugella) obscura*, a new species of mosquito from New Guinea (Diptera: Culicidae). *Journal of Medical Entomology, 11*(for 1974)(6), 734–738.

Tenorio, J.A. (1977). Revision of the genus *Bironella* (Diptera: Culicidae). *Journal of Medical Entomology, 14*(3), 317–361.

Terenius, O., Lindh, J.M., Eriksson-Gonzales, K., Bussière, L., Laugen, A.T., Bergquist, H., . . . Faye, I. (2012). Midgut bacterial dynamics in *Aedes aegypti*. *FEMS Microbiology Ecology, 80*(3), 556–565.

Tewari, S.C., & Hiriyan, J. (1988). Description of the male of *Aedes (Verrallina) agrestis* (Diptera: Culicidae) with notes on the female. *Mosquito Systematics, 20*(1), 28–32.

Tewari, S.C., & Hiriyan, J. (1991). Description of the pupa of *Anopheles (Anopheles) sintoni* (Diptera: Culicidae). *Mosquito Systematics, 23*(2), 170–174.

Tewari, S.C., & Hiriyan, J. (1992a). Description of a new species of *Aedes (Rhinoskusea)* from South India. *Mosquito Systematics, 23*(for 1991)(2), 123–131.

Tewari, S.C., & Hiriyan, J. (1992b). Description of two new species of *Aedes (Diceromyia)* from South India (Diptera: Culicidae). *Mosquito Systematics, 24*(2), 154–176.

Tewari, S.C., & Hiriyan, J. (1994a). Description of the larva and pupa of *Aedes (Verrallina) agrestis* (Diptera: Culicidae). *Mosquito Systematics, 26*(1), 36–42.

Tewari, S.C., & Hiriyan, J. (1994b). Description of the female, pupa, and larva of *Aedes (Paraedes) barraudi* and the pupa and larva of *Aedes (Paraedes) menoni* (Diptera: Culicidae). *Mosquito Systematics, 26*(3), 105–115.

Tewari, S.C., & Hiriyan, J. (1995). Description of *Aedes (Finlaya) niveus* (Diptera: Culicidae) from Andaman and Nicobar, India. *Mosquito Systematics, 27*(3), 167–176.

Tewari, S.C., & Hiriyan, J. (1996). Revalidation and description of *Aedes (Finlaya) pallirostris* from northeastern India (Diptera: Culicidae). *Journal of the American Mosquito Control Association, 12*(4), 713–720.

Tewari, S.C., Hiriyan, J., & Reuben, R. (1987). Description of male, larva and pupa of *Heizmannia Discrepans* [*sic*] (Diptera: Culicidae) with notes on variations in female morphology. *Mosquito Systematics, 19*(2), 127–134.

Tewari, S.C., Hiriyan, J., & Reuben, R. (1990). Description of the male, larva and pupa of *Aedes (Diceromyia) kanarensis* (Diptera: Culicidae). *Mosquito Systematics, 22*(1), 41–46.

Tewari, S.C., Munirathinam, A., Venkatesh, A., Hiriyan, J., & Reuben, R. (2008). Description of the fourth-stage larva and pupa of

Stegomyia edwardsi Barraud (Diptera: Culicidae). *Zootaxa, 1896*, 51–57.

Tewari, S.C., Krishnamoorthy, R., Ayanar, K., Munirathinam, A., Venkatesh, A., & Tyagi, B.K. (2012). Redescription of *Heizmannia (Heizmannia) chandi* Edwards (Diptera: Culicidae). *Zootaxa 3255*, 62–64.

Tewfik, H.P., & Barr, A.R. (1974). The salivary gland chromosomes of *Culex pipiens* L. *Mosquito News, 34*(1), 47–54.

Thailayil, J., Magnusson, K., Godfray, H.C.J., Crisanti, A., & Catteruccia, F. (2011). Spermless males elicit large-scale female responses to mating in the malaria mosquito *Anopheles gambiae*. *Proceedings of the National Academy of Sciences of the United States of America, 108*(33), 13677–13681.

Thaler, C.D., Miyata, H., Haimo, L.T., & Cardullo, R.A. (2013). Waveform generation is controlled by phosphorylation and swimming direction is controlled by Ca2+ in sperm from the mosquito *Culex quinquefasciatus*. *Biology of Reproduction, 89*(6), 135, 1–11.

Thavara, U., Tawatsin, A., & Chompoosri, J. (2004). Evaluation of attractants and egg-laying substrate preference for oviposition by *Aedes albopictus* (Diptera: Culicidae). *Journal of Vector Ecology, 29*(1), 66–72.

Theobald, F.V. (1900a). *Report on the collections of mosquitoes (Culicidae) received at the British Museum (Natural History) from various parts of the world in connection with the investigations into the causes of malaria, conducted by the Colonial Office and the Royal Society*. London, England: British Museum (Natural History).

Theobald, F.V. (1900b). A new *Anopheles* (*A. paludis*) from Sierra Leone. *Reports to the Malaria Committee of the Royal Society, 1*, 75–76.

Theobald, F.V. (1901a). *A monograph of the Culicidae or mosquitoes* (Vol. 1). London, England: British Museum (Natural History).

Theobald, F.V. (1901b). The classification of mosquitoes. *Journal of Tropical Medicine, 4*, 229–235.

Theobald, F.V. (1901c). *A monograph of the Culicidae or mosquitoes* (Vol. 2). London, England: British Museum (Natural History).

Theobald, F.V. (1901d). Notes on a collection of mosquitoes from West Africa, and descriptions of new species. Appendix. *Memoirs of the Liverpool School of Tropical Medicine 4*, i–xiv.

Theobald, F.V. (1902a). A short description of the Culicidae of India, with descriptions of new species of *Anopheles*. *Proceedings of the Royal Entomological Society of London, Series B: Taxonomy, 69*, 367–394.

Theobald, F.V. (1902b). The classification of the Anophelinae. *Journal of Tropical Medicine, 5*, 181–183.

Theobald, F.V. (1903a). *A monograph of the Culicidae of the World* (Vol. 3). London, England: British Museum (Natural History).

Theobald, F.V. (1903b). Report on a collection of mosquitoes and other flies from equatorial East Africa and the Nile provinces of Uganda. *Reports of the Sleeping Sickness Commission of the Royal Society, 7*(3), 33–42.

Theobald, F.V. (1903c). Report on a collection of mosquitoes or Culicidae, etc., from Gambia, and descriptions of new species. Appendix. *Memoirs of the Liverpool School of Tropical Medicine, 10*, i–xi.

Theobald, F.V. (1903d). A new culicid from Senegal and notes on the species of mosquitoes, etc. Appendix. *Memoirs of the Liverpool School of Tropical Medicine, 11* (Part 2), i–iii.

Theobald, F.V. (1903e). Two new Australian culicids. *The Entomologist, 36*(481), 154–157.

Theobald, F.V. (1903f). New Culicidae from the Federated Malay States. *The Entomologist, 36*(485), 256–259.

Theobald, F.V. (1903g). Two new Jamaican Culicidae. *The Entomologist, 36*(486), 281–283.

Theobald, F.V. (1903h). Notes on Culicidae and their larvae from Pecos, New Mexico, and description of a new *Grabhamia*. *The Canadian Entomologist, 35*(11), 311–316.

Theobald, F.V. (1903i). Description of a new North American *Culex*. *The Canadian Entomologist, 35*(8), 211–213.

Theobald, F.V. (1904a). A new culicid genus from Uganda. *Journal of Tropical Medicine, 7*, 17–18.

Theobald, F.V. (1904b). The mosquitoes of Egypt, the Sudan, and Abyssinia. In A. Balfour (director), *First Report of the Wellcome Research Laboratories at the Gordon Memorial College Khartoum* (pp. 62–83). Department of Education, Sudan Government, Khartoum.

Theobald, F.V. (1904c). New Culicidae from the Federated Malay States. *The Entomologist, 37*(488, 489, 490, 491, 492, 493, 494, 495, 496), 12–15, 36–39, 77–78, 111–113, 163–165, 211–213, 236–239.

Theobald, F.V. (1905a). A new genus of Culicidae. *The Entomologist, 38*(501), 52–56.

Theobald, F.V. (1905b). New Culicidae from the west coast of Africa. *The Entomologist, 38*(503), 101–104, 154–158.

Theobald, F.V. (1905c). A new *Stegomyia* from the Transvaal. *The Entomologist, 38*(508), 224–225.

Theobald, F.V. (1905d). The mosquitoes or Culicidae of Jamaica. Institute of Jamaica, Kingston.

Theobald, F.V. (1905e).*Genera Insectorum. Diptera, Fam. Culicidae.* Fascicle 26. Brussels, Belgium: Verteneuil & L. Desmet.

Theobald, F.V. (1905f). A catalogue of the Culicidae in the Hungarian National Museum with description of new genera and species. *Annales Musei Nationalis Hungarici, 3*(1), 61–119.

Theobald, F.V. (1905g). New Culicidae from India, Africa, British Guiana, and Australia. *Journal of Economic Entomology, 1*, 17–36.

Theobald, F.V. (1905h). Some new mosquitoes from Ceylon. *Journal of the Bombay Society, 16*(2), 237–250.

Theobald, F.V. (1905i). A new *Ficalbia* from West Africa (*Ficalbia nigripes*). *Annals and Magazine of Natural History* (Series 7), *15*(86), 199–200.

Theobald, F.V. (1906a). A new *Megarhinus*. *The Entomologist, 39*(522), 241.

Theobald, F.V. (1906b). Report on economic entomology. Part I. Second report on the mosquitoes or Culicidae of the Sudan. *Report Wellcome Laboratories, Gordon College, Khartoum, 2*, 67–83.

Theobald, F.V. (1907). *A monograph of the Culicidae of the World* (Vol. 5). London, England: British Museum (Natural History).

Theobald, F.V. (1908a). Notes on some Transvaal mosquitoes, including two new species and a new variety. *The Entomologist, 41*(540), 106–109.

Theobald, F.V. (1908b). First report on the collection of Culicidae and Corethridae in the Indian Museum, Calcutta, with descriptions of new genera and species. *Records of the Indian Museum, 2*(3), 287–302.

Theobald, F.V. (1908c). New mosquitoes from the Sudan with list and synoptic table of all the known Sudanese species. In A. Balfour (director), *Third Report of the Wellcome Research Laboratories at the Gordon Memorial College, Khartoum* (pp. 249–267). London, England: Baillière, Tindall & Cox.

Theobald, F.V. (1909a). A new culicid genus. *Annals of Tropical Medicine and Parasitology, 2*(for 1908), 297.

Theobald, F.V. (1909b). *Descriptions of the new mosquitoes collected by Dr. Graham in Ashanti*. Gold Coast Miscellaneous Publication No. 237. London, England: Colonial Office.

Theobald, F.V. (1910a). Second report on the collection of Culicidae in the Indian Museum, Calcutta, with descriptions of new genera and species. *Records of the Indian Museum, 4*(1), 1–33.

Theobald, F.V. (1910b). The Culicidae of Fiji, including two new species. *The Entomologist, 43*(565), 155–159.

Theobald, F.V. (1910c). *A monograph of the Culicidae of the World*. (Vol. 5). London, England:. British Museum (Natural History).

Theobald, F.V. (1910d). Five new Culicidae from Ashanti. *Annals and Magazine of Natural History, 5*, 373–378.

Theobald, F.V. (1911a). Uganda Culicidae. In F.V. Theobald, *Novae Culicidae* (Part 1). Self-published, Wye, Kent, United Kingdom.

Theobald, F.V. (1911b). The Culicidae or mosquitoes of the Transvaal. *Report of the Director of Veterinary Research, 1*, 232–272.

Theobald, F.V. (1911c). A new genus and two new species of Culicidae from the Sudan. In A. Balfour (Director), *Fourth Report of the Wellcome Tropical Research Laboratories at the Gordon Memorial College, Khartoum. Volume B, General Science* (pp. 151–156). London, England: Baillière, Tindall and Cox.

Theobald, F.V. (1911d). Culicidae of the R. Zool. Soc. Natura Artis Magistra, Amsterdam and description of three new species. *Tijdschrift voor Entomologie, 54*, 233–240.

Theobald, F.V. (1912a). Three new Culicidae from the Transvaal. *The Entomologist, 45*(586), 92–93.

Theobald, F.V. (1912b). No. V. Diptera, Culicidae. *Transactions of the Linnean Society of London: Zoology, 15*(1), 81–94.

Theobald, F.V. (1913a). Second report on the mosquitoes of the Transvaal. In A. Theiler (Ed.), *Second Report of the Director of Veterinary Research* (pp. 315–341). Cape Town: Cape Times.

Theobald, F.V. (1913b). Culicidae from New Caldeonia and the Loyalty Islands. In F. R. Sarasin andJ. Roux (Eds.), *Nova Caledonia. Forschungen in Neu-Caledonien und auf den Loyalty-Inseln. Recherches scientifiques en Nouvelle-Calédonie et aux iles Loyalty. A. Zoologie* (Vol. 1, L. III, pp. 163–164). Wiesbaden, Germany: C.W. Kreidels Verlag.

Theobald, F.V. (1913c). A new mosquito from northern China. *The Entomologist, 46*(601), 179–180.

Theobald, F.V. (1913d). New Culicidae from the Sudan. *Annals of Tropical Medicine and Parasitology* (Liverpool), *7*(4), 591–602.

Theobald, F.V. (1914). A new mosquito from Samoa. *The Entomologist, 47*(608), 36–37.

Theodor, O. (1925). Observations on Palestinian *Anopheles*. *Bulletin of Entomological Research, 15*(4), 377–382.

Thielman, A., & Hunter, F.F. (2006). Establishment of *Ochlerotatus japonicus* (Diptera: Culicidae) in Ontario, Canada. *Journal of Medical Entomology, 43*(2), 138–142.

Thien, L.B. (1969). Mosquito pollination of *Habenaria obtusata* (Orchidaceae). *American Journal of Botany, 56*(2), 232–237.

Thompson, G.A. (1956). *Aedes stenei*, a new species of mosquito from Jamaica, BWI, with redescription of a closely related species (Diptera, Culicidae). *Mosquito News, 16*(3), 196–200.

Thomson, C.G. (1869). Diptera. Species nova descripsit, In *K. Svenska VetenskapsAkademien, Kongliga Svenska Fegatten Eugenies resa omkring Jorden. Vetenskapliga Iakitagelser. II. Zoologi. 1. Insecta* (pp. 443–614). Stockholm, Sweden: P.A. Norstedt & Soner, Stockholm.

Thomson, F.W. (1903). Notes on the Culicidae of Dehra Dun. With a description of a new *Mansonia*, which mimics *Anopheles*. *Journal of Tropical Medicine, 6*, 314–315.

Thomson, M.C., Deblas-Reyes, F.J., Mason, S.J., Hagedorn, R., Connor, S.J., Pindela, T., . . . Palmer, T.N. (2006). Malaria early warnings based on seasonal climate forecasts from multi-model ensembles. *Nature, 439*(7076), 576–579.

Thongsahuan, S., Baimai, V., Otsuka, Y., Saeung, A., Tuetun, B., Jaryapan, N., . . . Choochote, W. (2009). Karyotypic variation and geographic distribution of *Anopheles campestris*-like (Diptera: Culicidae) in Thailand. *Memórias do Instituto Oswaldo Cruz, 104*(4), 558–566.

Thurman, D.C. (1954a). *Ayurakitia*, a new genus of mosquito from northern Thailand (Diptera: Culicidae). *Journal of the Washington Academy of Sciences, 44*(6), 197–200.

Thurman, D.C. (1954b). The discovery of *Paraedes (Udaya) argyrurus* Edwards, 1934, in Thailand (Diptera: Culicidae). *Mosquito News, 14*(2), 83–86.

Thurman, D.C. (1955). *Culex (Culiciomyia) termi*, an unusual new mosquito from Thailand. *Proceedings of the Entomological Society of Washington, 57*(1), 17–23.

Thurman, D.C., & Thurman, E.B. (1958). Two new species in the genus *Armigeres* Theobald, 1901, from northern Thailand (Diptera: Culicidae). *Journal of the Washington Academy of Sciences, 48*(6), 186–187.

Thurman, E.B. (1958). *Malaya jacobsoni* (Edwards, 1930), a new occurrence record for northern Thailand (Diptera: Culicidae). *Proceedings of the Entomological Society of Washington, 60*(1), 15.

Thurman, E.B. (1959a). Revalidation of three species of *Armigeres* Theobald, 1901 (Diptera: Culicidae). *Journal of the Washington Academy of Sciences, 48*(for *1958*), 389–393.

Thurman, E.B. (1959b). A contribution to a revision of the Culicidae of northern Thailand. *Bulletin of the University of Maryland Agriculture Experimental Station A, 100*, 1–177.

Thurman, E.B., & Winkler, E.C. (1950). A new species of mosquito in California, *Aedes (Ochlerotatus) bicristatus* (Diptera, Culicidae). *Proceedings of the Entomological Society of Washington, 52*(5), 237–250.

Thurman, E.B., Haeger, J.S., & Mulrennan, J.A. (1949). The occurrence of *Aedes (Ochlerotatus) thelcter* Dyar in the Florida Keys. *Mosquito News, 9*(4), 171–172.

Thurman, E.B., Haeger, J.S., & Mulrennan, J.A. (1951). The taxonomy and biology of *Psorophora (Janthinosoma) johntonii* (Grabham, 1905) (Diptera: Culicidae). *Annals of the Entomological Society of America, 44*, 144–157.

Toma, T., & Higa, Y. (2004). A new species of *Ficalbia* (Diptera: Culicidae) from Iriomote Island, Okinawa, Ryukyu Archipelago, Japan. *Medical and Veterinary Zoology, 55*(3), 195–199.

Toma, T., & Miyagi, I. (1986). The mosquito fauna of the Ryukyu archipelago with identification keys, pupal descriptions and notes on biology, medical importance and distribution. *Mosquito Systematics, 18*(1), 1–109.

Toma, T., & Miyagi, I. (2002). *Armigeres (Armigeres) sembeli* (Diptera: Culicidae), a new species from Sulawesi, Indonesia and redescription of *Armigeres (Armigeres) giveni* Edwards. *Medical and Veterinary Zoology, 53*(4), 257–265.

Toma, T., & Miyagi, I. (2003). *Armigeres (Armigeres) laoensis* sp. nov. (Culicidae) from Khammouane Province, Lao PDR. *Medical and Veterinary Zoology, 54*(2), 169–172.

Toma, T., & Miyagi, I. (2005a). Notes on mosquitoes in Chichi-Jima, Ogasawara Archipelago, Japan and biology of *Culex (Sirivanakarnius) boninensis* (Diptera: Culicidae). *Medical Entomology and Zoology, 56*(3), 237–241.

Toma, T., & Miyagi, I. (2005b). Redescription of *Armigeres (Armigeres) conjungens* Edwards (Diptera: Culicidae) collected from the Peninsular Malaysia. *Medical and Veterinary Zoology, 56*(1), 1–9.

Toma, T., & Mogi, M. (2003). *Topomyia (Suaymyia) miyagii* (Diptera: Culicidae): A new species from Flores Is., Indonesia. *Medical and Veterinary Zoology, 54*(1), 25–30.

Toma, T., Miyagi, I., & Cabrera, B.D. (1984). *Culex (Culiciomyia) azurini*, a new crab hole mosquito species from the Philippines. *Mosquito Systematics, 16*(2), 172–182.

Toma, T., Miyagi, I., & Tanaka, K. (1990). A new species of *Toxorhynchites* mosquito (Diptera: Culicidae) from the Ryukyu Archipelago, Japan. *Journal of Medical Entomology, 27*, 344–355.

Toma, T., Miyagi, I., & Benjaphong, N. (1994). Redescription of *Armigeres (Armigeres) theobaldi* (Diptera: Culicidae). *Mosquito Systematics, 26*(1), 11–18.

Toma, T., Miyagi, I., & Syafruddin. (1995). Description of *Armigeres (Armigeres) alkatirii*, a new species from Sulawesi, Indonesia (Diptera: Culicidae). *Mosquito Systematics, 27*(2), 110–117.

Toma, T., Miyagi, I., Murakami, H., Nerome, H., Yonamine, M., Higa, Y., & Tokuyama, Y. (2003). Distribution and seasonal prevalence of *Anopheles minimus* Theobald (Diptera: Culicidae) in the Yaeyama Island group (except Ishigaki Island), Ryukyu Archipelago, Japan, 1999–2000. *Medical and Veterinary Zoology, 54*(3), 267–274.

Toma, T., Miyagi, I., Okazawa, T., & Leh, C. (2007). Redescription of *Armigeres (Armigeres) setifer* Delfinado, 1966 (Diptera: Culicidae) collected from Kuching, Sarawak, Malaysia. *Medical Entomology and Zoology, 58*(4), 283–290.

Toma, T., Miyagi, I., Okazawa, T., Leh, M.U., & Wong, S.F. (2011). Redescription of *Armigeres (Armigeres) hybridus* (Diptera: Culicidae) based on specimens collected in Lanjak, Sarawak, Malaysia. *Medical Entomology and Zoology, 62*(4), 225–233.

Toma, T., Miyagi, I., Okazawa, T., Higa, Y., Wong, S.F., Leh, M.U., & Yong, H.S. (2012). Redescriptions of *Armigeres annulipalpis* and *Armigeres flavus* (Diptera: Culicidae) from Sarawak, Malaysia. *Journal of Science and Technology in the Tropics, 8*(1), 5–19.

Tomberlin, J.K., Rains, G.C., Allan, S.A., Sanford, M.R., & Lewis, W.J. (2006). Associative learning of odor with food- or blood-meal by *Culex quinquefasciatus* Say (Diptera: Culicidae). *Naturwissenschaften*, *93*(11), 551.

Tomé, H.W., Pascini, T.V., Dâgelo, R.A.C., Guedes, R.N.C., & Martins, G.F. (2014). Survival and swimming behavior of insecticide-exposed larvae and pupae of the yellow fever mosquito *Aedes aegypti*. *Parasites and Vectors, 7*(1), Article 195.

Tomita, T., Kono, Y., & Shimada, T. (1996). Chromosomal localization of amplified esterase genes in insecticide resistant *Culex* mosquitoes. *Insect Biochemistry and Molecular Biology, 26*(8–9), 853–857.

Topp, W. (1990). Selection for optimal monovoltine life cycle in an unpredictable environment. Studies on the beetle *Catops nigricans* Spence (Col., Catopidae). *Oecologia, 84*(1), 134–141.

Torres, E.P., Foley, D.H., & Bryan, J.H. (2006). Molecular systematics of the Philippine malaria vector *Anopheles flavirostris*. *Medical and Veterinary Entomology, 20*(1), 44–52.

Torres Cañamares, F. (1945). Contribucion al conocimiento del *Anopheles claviger* Mg. de España (Dip. Cul.). *EOS Revista Española de Entomología, 20*, 233–245.

Torres Cañamares, F. (1951). Una nueva especie de *Aedes* (Dipt. Cul.). *EOS Revista Española de Entomología, 27*, 79–92.

Torres-Estrada, J.L., Rodríguez, M.H., Cruz-López, L., & Arredondo-Jimenez, J.I. (2001). Selective oviposition by *Aedes aegypti* (Diptera: Culicidae) in response to *Mesocyclops longisetus* (Copepoda: Cyclopoidea) under laboratory and field conditions. *Journal of Medical Entomology, 38*(2), 188–192.

Torres-Gutierrez, C., & Sallum, M.A.M. (2015). Catalog of the subgenus *Melanoconion* of *Culex* (Diptera: Culicidae) for South America. *Zootaxa, 4028*(1), 1–50.

Toto, J.C., Abaga, S., Carnevale, P., & Simard, F. (2003). First report of the oriental mosquito *Aedes albopictus* on the west African island of Bioko, Equatorial Guinea. *Medical and Veterinary Entomology, 17*(3), 343–346.

Toumanoff, C. (1931). Sur une variété nouvelle d'*Anopheles aconitus* Dönitz observée au Tonkin. *Comptes rendus des seances de la Société de Biologie et de ses filiales, 107*, 575–576.

Townsend, B.C. (1990). Culicidae. In B.C. Townsend, J. E. Chainey, R.W. Crosskey, A.C. Pont, R.P. Lane, J.P.T. Boorman, & C.A. Crouch (Eds.), *A catalogue of the types of bloodsucking flies in the British Museum (Natural History)* (pp. 35–152). London, England: British Museum (Natural History).

Townsend, C.H.T. (1934). Mosquitoes of the Rio Tapajós. *Revista de entomología, 4*(4), 486–499.

Townson, H., Dyer, E., McAlister, E., Sasoto, T.B.T., Bangs, M.J., & Harbach, R.E. (2013). Systematics of *Anopheles barbirostris* van der Wulp and a sibling species of the Barbirostris Complex (Diptera: Culicidae) in eastern Java, Indonesia. *Systematic Entomology, 38*(1), 180–191.

Trari, B., Dakki, M., Himmi, O., & El Agbani, M.A. (2002). Le moustiques (Diptera: Culicidae) du Maroc: Revue bibliographique (1916–2001) et inventaire des espèces. *Bulletin de la Société de pathologie exotique, 96*(4), 329–334.

Trawinski, P.R., & MacKay, D.S. (2008). Meteorologically conditioned time-series predictions of West Nile virus vector mosquitoes. *Vector-Borne and Zoonotic Diseases, 8*(4), 505–522.

Trebatoski, A.M., & Haynes, J.F. (1969). Comparison of enzymes of twelve species of mosquitoes. *Annals of the Entomological Society of America, 62*(2), 327–335.

Trembley, H.L. (1947). Biological characteristics of laboratory-reared *Aëdes atropalpus*. *Journal of Economic Entomology, 40*(2), 244–250.

Trimble, R.M., & Smith, S.M. (1975). A bibliography of *Toxorhynchites rutilus* (Coquillett) (Diptera: Culicidae). *Mosquito Systematics, 7*(2), 115–126.

Trimble, R.M., & Smith, S.M. (1978). Geographic variation in development time and predation in the three-hole mosquito, *Toxorhynchites rutilus septentrionalis* (Diptera: Culicidae). *Canadian Journal of Zoology, 56*(10), 2156–2165.

Tripet, F., Touré, Y.T., Taylor, C.E., Norris, D.E., Dolo, G., & Lanzaro, G.C. (2001). DNA analysis of transferred sperm reveals significant levels of gene flow between molecular forms of *Anopheles gambiae*. *Molecular Ecology, 10*(7), 1725–1732.

Tripet, F., Touré, Y.T., Dolo, G., & Lanzaro, G.C. (2003). Frequency of multiple inseminations in field-collected *Anopheles gambiae* females revealed by DNA analysis of transferred sperm. *The American Journal of Tropical Medicine and Hygiene, 68*(1), 1–5.

Tripet, F., Lounibos, L.P., Robbins, D., Moran, J., Nishimura, N., & Blosser, E.M. (2011). Competitive reduction by satyrization? Evidence for interspecific mating in nature and asymmetric reproductive competition between invasive mosquito vectors. *The American Journal of Tropical Medicine and Hygiene, 85*(2), 265–270.

Trpis, M., & Hausermann, W. (1986). Dispersal and other population parameters of *Aedes aegypti* in an African village and their possible significance in epidemiology of vector-borne diseases. *The American Journal of Tropical Medicine and Hygiene, 35*(6), 1263–1279.

Trpis, M., & Shemanchuk, J.A. (1969). The effect of temperature on pre-adult development of *Aedes flavescens* (Diptera: Culicidae). *The Canadian Entomologist, 101*(2), 128–132.

Tsai, C., & Lien, J.C. (1950). A new species of *Aedes (Finlaya)* found in Taiwan. *Journal of the Formosan Medical Association, 49*(5), 177–183.

Tsuchimoto, S. (1948). On a variety of *Anopheles koreicus*. *Medicine and Biology* (Japan), 6(for 1944)(4–6), 364–366.

Tsuda, Y., Takagi, M., Wang, S., Wang, Z., & Tang, L. (2001). Movement of *Aedes aegypti* (Diptera: Culicidae) released in a small isolated village in Hainan Island, China. *Journal of Medical Entomology, 38*(1), 93–98.

Tsujimoto, H., Liu, K., Linser, P.J., Agre, P., & Rasgon, J.L. (2013). Organ-specific splice variants of aquaporin water channel AgAQP1 in the malaria vector *Anopheles gambiae. PLOS ONE, 8*(9), e75888.

Tsukamoto, M. (1989). Two new mosquito species from a pitcher plant of Mt. Kinabalu, Sabah, Malaysia: *Culex rajah* and *Toxorhynchites rajah* (Diptera: Culicidae). *Japanese Journal of Tropical Medicine and Hygiene, 17*(3), 215–228.

Tsukamoto, M., Toma, T., & Miyagi, I. (1989). *Culex* (*Culiciomyia*) *javanensis* (Diptera: Culicidae): New distribution records and redescription of the larva, with a larval key to species of the subgenus *Culiciomyia* in Asia and Oceania. *Japanese Journal of Sanitary Zoology, 40*(4), 295–301.

Tsuzuki, J. (1902a). A report of an investigation on malaria in Taiwan. *Saikingaku zasshi, 75*, 25–51.

Tsuzuki, J. (1902b). Malaria und ihre Vermittler in Japan. *Archiv für Schiffs- und Tropen-Hygiene, 6*, 285–295.

Tsuzuki, J. (1902c). Ueber die Ergebnisse meiner Malariaforschung in Hokkaido (Japan). *Zentralblatt für Bakteriologie, Parasitenkunde und Infektionskrankheiten, 1*, 763–768.

Tsuzuki, J. (1907). Über die *Anopheles*-arten in Japan und einige Beiträge zur Kenntnis des Entwicklungsganges der *Anopheles*-larven. *Zoologische Jahrbücher Abteilung, 1, 25*(5–6), 525–556.

Tung, L.H. (1955). On three new species of mosquitoes from Kiangsi Province, China. *Acta entomologica Sinica, 5*(9), 309–315.

Tun-Lin, W., Burkot, T.R., & Kay, B.H. (2000). Effects of temperature and larval diet on development rates and survival of the dengue vector *Aedes aegypti* in north Queensland, Australia. *Medical and Veterinary Entomology, 14*(1), 31–37.

Tupitsyn, Y.N., & Dubitskii, A.M. (1972). Detection in Kazakhstans fauna of a new subgenus and species of blood sucking mosquito *Aedes* (*Stegomyia*) *galloisi. Meditsinskaya parazitologiya i parazitarnye bolezni, 41*(1), 106–107.

Turley, A.P., Moreira, L.A., O'Neill, S.L., & McGraw, E.A. (2009). *Wolbachia* infection reduces blood-feeding success in the dengue fever mosquito, *Aedes aegypti. PLOS Neglected Tropical Diseases, 3*(9), e516.

Turner, R.L. (1924). A new mosquito from Texas (Diptera, Culicidae). *Insecutor Inscitiae Menstruus, 12*, 84.

Turner, S.L., Li, N., Guda, T., Githure, J., Carde, R.T., & Ray, A. (2011). Ultra-prolonged activation of CO_2-sensing neurons disorients mosquitoes. *Nature, 474*(7349), 87–91.

Tuten, H.C., Stone, C.M., & Dobson, S.L. (2013). Swarming behavior of *Aedes polynesiensis* (Diptera: Culicidae) and characterization of swarm markers in American Samoa. *Journal of Medical Entomology, 50*(4), 740–747.

Ty, Dao Van. (1945). Sur la biologie d'*Armigeres obturbans* (Walker). *Bulletin de la Société de pathologie exotique, 38*, 304–309.

Tyagi, B.K., Hiriyan, J., Tewari, S.C., Ayanar, K., Samuel, P.P., Arunachalam, N., . . . Rajendran, R. (2009). Description of a new species, *Anopheles pseudosundaicus* (Diptera: Culicidae) from Kerala, India. *Zootaxa, 2219*, 49–60.

Tyagi, B.K., Munirathinam, A., Krishnamoorthy, R., Baskaran, G., Govindarajan, R., Krishnamoorthi, R., . . . Venkatesh, A. (2015). A revision of genus *Toxorhynchites* Theobald, 1901, in the South-East Asian countries, with description of a new species *Toxorhynchites* (*Toxorhynchites*) *darjeelingensis* from West Bengal, India (Diptera, Culicidae). *Halteres, 6*, 13–32.

Tyson, W.H. (1970). Notes on African *Aedes*, subgenus *Mucidus* (Diptera: Culicidae). *Journal of the Entomological Society of Southern Africa, 33*(1), 81–88.

Tyson, W.H. (1970a). Contributions to the mosquito fauna of Southeast Asia. VII. Genus *Aedeomyia* Theobald in Southeast Asia. *Contributions of the American Entomological Institute, 6*(2), 1–27.

Tyson, W.H. (1970b). Contributions to the mosquito fauna of Southeast Asia. VIII. Genus *Aedes*, subgenus *Mucidus* Theobald in Southeast Asia. *Contributions of the American Entomological Institute, 6*(2), 28–80.

Tyson, W.H. (1970c). Notes on African *Aedes*, subgenus *Mucidus* (Diptera: Culicidae). *Journal of the Entomological Society of Southern Africa, 33*(1), 81–88.

Uchida, K., Moribayashi, A., Matsuoka, H., & Oda, T. (2003). Effects of mating on oogenesis induced by amino acid infusion, amino acid feeding, blood feeding in the mosquito *Anopheles stephensi* (Diptera: Culicidae). *Journal of Medical Entomology, 40*(4), 441–446.

Ulloa, A., Rodriguez, M.H., Arredondo-Jimenez, J.I., & Fernandez-Salas, I. (2005). Biological variation in two *Anopheles vestitipennis* populations with different feeding preferences in southern Mexico. *Journal of the American Mosquito Control Association, 21*(4), 350–354.

Ulyshen, M.D. (2011). Arthropod vertical stratification in temperate deciduous forests: Implications for conservation-oriented management. *Forest Ecology and Management, 261*(9), 1479–1489.

Ungureanu, E. (1956). Noi contributii la studiul geului *Theobaldia* in R.P.R. *Acaedmiă Republicii Poulare Romine Filialăl Lăşi, Studii şi Cercetări Ştiinte Agricole 7*(2), 1–9.

Unlu, I., Farajollahi, A., Strickman, D., & F onseca, D.M. (2013). Crouching tiger, hidden trouble: Urban sources of *Aedes albopictus* (Diptera: Culicidae) refractory to source-reduction. *PLOS ONE, 8*(10), e77999.

Unti, O. (1940a). Anofelinos do vale do Paraiba. Nota III. Biologia do *Anofeles* [*sic*] (*Nyssorhynchus*) *strodei* Root, 1926 com a descrição d'uma variedade nova. *Anofeles* [*sic*] (*Nyssorhynchus*) *strodei ramosi* var. n. *Anais paulistas de medicina e cirurgia, 40*(12), 489–505.

Unti, O. (1940b). Anofelinos do vale do Paraiba. Nova variedade e ciclo evolutivo do *lb, n* [*sic*] (*Nyssorhynchus*) *oswaldoi* var. *ayrozai* n. var. *Anais paulistas de medicina e cirurgia, 40*(11), 377–392.

Unti, O. (1941). Anofelinos do vale do Rio Paraiba, *Anopheles* (*Nyssorhynchus*) *strodei* Root 1926, com a descrição de tres variedades novas. *Serviço público de profilaxia en Malaria, 33*, 3–18.

Urbanelli, S., Bellini, R., Carrieri, M., Sallicandro, P., & Celli, G. (2000). Population structure of *Aedes albopictus* (Skuse): The mosquito which is colonizing Mediterranean countries. *Heredity, 84*(3), 331–337.

Urbanski, J.M., Benoit, J.B., Michaud, M.R., Denlinger, D.L., & Armbruster, P. (2010). The molecular physiology of increased egg desiccation resistance during diapause in the invasive mosquito, *Aedes albopictus. Proceedings of the Royal Society of London B: Biological Sciences, 277*(1694), 2683–2692.

Urbino, C.M. (1936). The eggs of some Philippine *Anopheles. Monthly Bulletin of the Bureau of Health, 16*, 261–275.

Utrio, P. (1975). *Anopheles claviger* (Meig.), *Aedes pionips* Dyar and *A. beklemishevi* Den. (Diptera, Culicidae) found in Finland. *Notulae entomologicae, 55*, 63–64.

Utrio, P. (1976). Identification key to Finnish mosquito larvae (Diptera, Culicidae). *Annales agriculturae Fenniae, 15*, 128–136.

Valencia, J.D. (1973). Mosquito studies (Diptera, Culicidae). XXXI. A revision of the subgenus *Carollia* of *Culex. Contributions of the American Entomological Institute, 9*(4), 1–134.

Valenzuela, J.G., Pham, V.M., Garfield, M.K., Francischetti, I.M.B., & Ribeiro, J.M.C. (2002). Toward a description of the sialome of the adult female mosquito *Aedes aegypti. Insect Biochemistry and Molecular Biology, 32*(9), 1101–1122.

Van Aken, W. (1961). Systematisch onderzbek van enkele Culicidae—exuvien van Belgie. *Biologisch Jaarboek, 29*, 246–283.

Van Bortel, W., Harbach, R.E., Trung, H.-D., Roelants, P., Backeljau, T., & Coosemans, M. (2001). Confirmation of *Anopheles varuna* in Vietnam, previously misidentified and mistargeted as the malaria vector *Anopheles minimus. The American Journal of Tropical Medicine and Hygiene, 65*(6), 729–732.

Van Dam, A.R., & Walton, W.E. (2008). The effect of predatory fish exudates on the ovipositional behaviour of three mosquito species: *Culex quinquefasciatus, Aedes aegypti* and *Culex tarsalis. Medical and Veterinary Entomology, 22*(4), 399–404.

van den Assem, J. (1958a). Some notes on *Ficalbia modesta* King & Hoogstraal (Culicidae). *Entomologische Berichten* (Amsterdam), *18*, 131–132.

van den Assem, J. (1958b). *Mansonia (Mansonioides) bonnewepsterae*, spec. nov. (Culicidae) with notes on habits and breeding place. *Tropical and Geographical Medicine, 10*(3), 205–212.

van den Assem, J. (1958c). The larva and female of *Culex (Lophoceraomyia) marksae* King Hoogstraal (Diptera, Culicidae). *Entomologische Berichten* (Amsterdam), *18*, 218–219.

van den Assem, J. (1959). Notes on new Guinean species of *Tripteroides*, subgenus *Rachisoura* (Diptera, Culicidae), with descriptions of two new species. *Tijdschrift voor Entomologie, 102*(1), 35–55.

van den Assem, J. (1960). A description of male and larva of *Tripteroides (Mimeteomyia) microlepis* (Edwards), (Diptera, Culicidae). *Entomologische Berichten* (Amsterdam), *20*(1), 13–14.

van den Assem, J., & Bonne-Wepster, J. (1964). New Guinea Culicidae, a synopsis of vectors, pests, and common species. *Zoologische Bijdragen No. 6*, 1–136.

van der Kuyp, E. (1954). Mosquitoes of the Netherlands Antilles and their hygienic importance. *Studies on the Fauna of Curaçao and Other Caribbean Islands, 5*, 37–114.

van der Wulp, F.M. (1867). Eenige Noord-Americaansch Diptera. *Tijdschrift voor Entomologie, 10*(2), 125–164.

van der Wulp, F.M. (1881). Diptera. In P.J. Veth (Ed.), *Midden-Sumatra, Reizenen onderzockingen der Sumatra Expeditie* (Vol. IV, 2de Afd., pp. 1–60). Leiden, The Netherland: E.J. Brill

van der Wulp, F.M. (1884). Note XXXVIII. On exotic Diptera. Part 1. *Notes from the Leyden Museum, 6*(4), 248–256.

Van Handel, E. (1969). The equilibrium reaction sorbitol⇆fructose in the intact mosquito. *Comparative Biochemistry and Physiology, 29*(3), 1023–1030.

Van Handel, E. (1984). Metabolism of nutrients in the adult mosquito. *Mosquito News, 44*(4), 573–579.

Van Loy, T., Van Hiel, M.B., Vandersmissen, H.P., Poels, J., Mendive, F., Vassart, G., & Broeck, J.V. (2007). Evolutionary conservation of bursicon in the animal kingdom. *General and Comparative Endocrinology, 153*(1–3), 59–63.

van Someren, E.C.C. (1945). Ethiopian Culicidae: Descriptions of two East African culicine larvae, not hitherto described, and a new subspecies from Abyssinia (Diptera). *Proceedings of the Royal Entomological Society B: Taxonomy, 14*(5–6), 77–80.

van Someren, E.C.C. (1946a). Ethiopian Culicidae: Notes and descriptions of some new species and hitherto unknown larvae and pupae (Diptera). *Transactions of the Royal Entomological Society of London B: Taxonomy, 96*(7), 109–124, 112 figs.

van Someren, E.C.C. (1946b). Ethiopian Culicidae. Tribe Megarhinini—notes and descriptions. *Transactions of the Royal Entomological Society of London Series B: Taxonomy, 97*(7), 177–186.

van Someren, E.C.C. (1946c). Ethiopian Culicidae: Descriptions of the adults and pupae of two new species and two subspecies of the genus *Aedes* from East Africa (Diptera). *Proceeding of the Royal Entomological Society of London, Series B: Taxonomy, 15*(1–2), 1–6.

van Someren, E.C.C. (1946d). Ethiopian Culicidae: Notes and descriptions. *Proceedings of the Royal Entomological Society of London, Series B: Taxonomy, 16*, 129–132.

van Someren, E.C.C. (1947a). The description of a new mosquito from the Seychelles. *East African Medical Journal, 24*(1), 29–35.

van Someren, E.C.C. (1947b). The description of a new *Anopheles* of the Myzorhynchus series from Madagascar with notes on its systematic position in relation to the Ethiopian species of this group. *East African Medical Journal, 24*(1), 42–46.

van Someren, E.C.C. (1947c). Ethiopian Culicidae: the description of a new *Eretmapodites* Theobald. *Proceedings of the Royal Entomological Society of London, Series B: Taxonomy, 16*(1–2), 13–15.

van Someren, E.C.C. (1947d). Ethiopian Culicidae: Notes and descriptions. *Proceedings of the Royal Entomological Society of London, Series B: Taxonomy, 16*, 129–132.

van Someren, E.C.C. (1948). Ethiopian Culicidae: Some new mosquitoes from Uganda. *Proceedings of the Royal Entomological Society of London, Series B: Taxonomy, 17*(9–10), 128–132.

van Someren, E.C.C. (1949a). Ethiopian Culicidae: Descriptions of four new mosquitoes from Madagascar. *Proceedings of the Royal Entomological Society of London, Series B: Taxonomy, 18*(1–2), 3–8.

van Someren, E.C.C. (1949b). Ethiopian Culicidae—*Eretmapodites* Theobald: Description of four new species of the *chrysogaster* group with notes on the five known species of this group. *Proceedings of the Royal Entomological Society of London, Series B: Taxonomy, 18*(7–8), 119–129.

van Someren, E.C.C. (1950a). Ethiopian Culicidae: Description of a new *Aëdes* from East Africa and notes on two known *Aëdes* with which it might be confused. *Proceedings of the Royal Entomological Society of London, Series B: Taxonomy, 19*(1–2), 7–10.

van Someren, E.C.C. (1950b). Ethiopian Culicidae: A description of two new *Aëdes* of the subgenus *Aëdimorphus* Theobald, from Uganda. *Proceedings of the Royal Entomological Society of London, Series B: Taxonomy, 19*(3–4), 39–41.

van Someren, E.C.C. (1950c). Ethiopian Culicidae: A description of a new sub-species of *Aëdes (Aëdimorphus) tricholabis* Edwards. *Proceedings of the Royal Entomological Society of London, Series B: Taxonomy, 19*(5–6), 67–69.

van Someren, E.C.C. (1950d). Ethiopian Culicidae: A description of the larva and pupa of *Aëdes (Aëdimorphus) mutilus* Edwards. *Proceedings of the Royal Entomological Society of London, Series B: Taxonomy, 19*(5–6), 65–66.

van Someren, E.C.C. (1951). New Culicini from Kenya and Uganda. *Proceedings of the Royal Entomological Society of London, Series B: Taxonomy, 20*(1–2), 1–9.

van Someren, E.C.C. (1954a). New Culicini from the Kenya Coast. *Proceedings of the Royal Entomological Society of London, Series B: Taxonomy, 23*(5–6), 103–109.

van Someren, E.C.C. (1954b). Ethiopian Culicidae: Descriptions of a new *Culex*, the female of *Eretmapodites tonsus* Edwards and the early stages of two *Aëdes* of the subgenus *Banksinella* Theobald. *Proceedings of the Royal Entomological Society of London, Series B: Taxonomy, 23*(7–8), 119–126.

van Someren, E.C.C. (1956). Undescribed culicine larvae and pupae from Uganda. *Proceedings of the Royal Entomological Society of London, Series B: Taxonomy, 25*(1–2), 3–12.

van Someren, E.C.C. (1958). Ethiopian Culicidae: A description of the early stages of *Culex (Culex) terzii* Edwards. *Proceedings of the*

Royal Entomological Society of London, Series B: Taxonomy, 27(5–6), 84–88.
van Someren, E.C.C. (1962). Ethiopian Culicidae: Three new *Aedes* from Tanganyika, with a description of the male of *Aedes usambara* Mattingly and the female of *Uranotaenia henrardi* Edwards. *Proceedings of the Royal Entomological Society of London, Series B: Taxonomy, 31*(1–2), 19–26.
van Someren, E.C.C. (1967). The female and early stages of *Culex* (*Culex*) *nakuruensis* Mattingly, with a description of a new subspecies of *Culex* (*Culex*) *shoae* Hamon & Ovazza. *Proceedings of the Royal Entomological Society of London, Series B: Taxonomy, 36*(1–2), 11–16.
van Someren, E.C.C. (1971). The description of a new *Aedes* mosquito of the subgenus *Skusea* from the Seychelles. *Journal of Entomology B, 40*(1), 21–25.
van Someren, E.C.C. (1972). The male of *Aedes* (*Stegomyia*) *masseyi* Edwards. *Mosquito Systematics, 4*(3), 91–92.
van Someren, E.C.C., & Hamon, J. (1964). Ethiopian Culicidae: A new species of *Culex* from Tanganyika, the descriptions of the pupa of *Aedes usumbara* Mattingly and the early stages of *Eretmapodites tonsus* Edwards. *Journal of the Entomological Society of Southern Africa, 27*(1), 78–85.
van Someren, E.C.C., & Harper, J.O. (1955). Ethiopian Culicidae: A description of the larva and pupa of *Aëdes* (*Stegomyia*) *woodi* Edwards. *Proceedings of the Royal Entomological Society of London B: Taxonomy, 24*(5–6), 77–80.
van Someren, E.C.C., Teesdale, C., & Furlong, M. (1955). The mosquitos of the Kenya Coast; records of occurrence, behaviour and habitat. *Bulletin of Entomological Research, 46*(3), 463–493.
van Thiel, P.H. (1927). Sur l'origine des variations de taille de l'*Anopheles maculipennis* dans les Pays-Bas. *Bulletin de la Société de pathologie exotique, 20,* 366–390.
Vargas, D.L. (1979a). *Anopheles* (*Anopheles*) *veruslanei* Vargas, n.sp. (Diptera: Culicidae). *Gaceta medica de Mexico, 115*(2), 87–88.
Vargas, D.L. (1979b). Caracteres adicionales para la identificación de *Anopheles* (*Anopheles*) *veruslanei* Vargas, 1979. *Prensa medical Mexicana, 44*(9–10), 194–195.
Vargas, L. (1939). Datos acerca del *A. pseudopunctipennis* y de un *Anopheles* nuevo de California. *Medicina*: *Revista Mexicana Suplemento, 19,* 356–362.
Vargas, L. (1940a). Clave para identificar las larvas de *Anopheles* mexicanos. *Ciencia, Mexico, 1,* 66–68.
Vargas, L. (1940b). Clave para identificar algunos generos de mosquitos amerianos, utilizando los caracteres de los adultos. *Revista del Instituto de Salubridad y Enfermedades Tropicales* (México), *1,* 189–198.
Vargas, L. (1941a). El problema de las variedades de *A. punctipennis*. Descripción de *A. punctipennis stonei* n. var. (Diptera, Culicidae). *Revista de la Sociedad Mexicana de Historia Natural, 2*(2–3), 175–180.
Vargas, L. (1941b). *Aedes* (*Ochlerotatus*) *rozeboomi* nueva especie (Dipt. Culicidae). *Gaceta médica de México, 71*(3), 393–395.
Vargas, L. (1941c). *Anopheles* (*Anopheles*) *gabaldoni* n. sp. del Estado de Tabasco, México (Dipt. Culicidae). *Medicina: Revista Mexicana, 21*(395), 385–394.
Vargas, L. (1941d). Nota sobre los huevecillos de *Anopheles* mexicanos. *Gaceta medica de Mexico, 71,* 107–123.
Vargas, L. (1941e). New variety of *Anopheles pseudopunctipennis* (Diptera, Culicidae). *Bulletin of the Brooklyn Entomological Society, 36*(2), 73–74.
Vargas, L. (1941f). *Anopheles pseudopunctipennis willardi* n. var. (Dipt, Culicidae). *Revista de la Sociedad Mexicana de Historia Natural, 2*(1), 47–49.
Vargas, L. (1942a). Nota sobre la validez del *Psorophora* (*Psorophora*) *virescens* Dyar y Knab, 1906. *Revista medicina de México, 22*(407), 81–84.
Vargas, L. (1942b). Las hembras americanas del subgenero *Anopheles* (Dipt, Culicidae, *Anopheles*). *Revista del Instituto de Salubridad y Enfermedades Tropicales* (México), *3*(1), 67–74.
Vargas, L. (1942c). *Anopheles xelajuensis* Romeo de Leon, 1938 en Mexico. *Revista del Instituto de Salubridad y Enfermedades Tropicales* (México), *3*(2), 169–175.
Vargas, L. (1942d). El huevo de *Anopheles barberi* Coquillet, 1903. *Revista del Instituto de Salubridad y Enfermedades Tropicales* (México), *3,* 329–331.
Vargas, L. (1942e). El huevecillo de *Anopheles* (*Anopheles*) *eiseni* Coquillet, 1902. *Revista del Instituto de Salubridad y Enfermedades Tropicales* (México), *3*(2), 185–187.
Vargas, L. (1943a). Los subgeneros americanos de *Anopheles* (Diptera, Culicidae). *Anopheles* (*Russellia*) *xelajuensis* Leon [*sic*], 1938 n. subgn. y *Anopheles* (*Coelodiazesis*) *fausti* n. sp. *Revista del Instituto de Salubridad y Enfermedades Tropicales* (México), *4*(1), 57–77.
Vargas, L. (1943b). *Anopheles earlei* Vargas, 1942, n. sp. norteamericana del grupo *maculipennis*. *Boletín de la Oficina Sanitaria Panamericana, 22*(1), 8–12.
Vargas, L. (1943c). El "grupo *maculipennis*" del Nuevo Mundo y el *Anopheles earlei*. *Revista del Instituto de Salubridad y Enfermedades Tropicales* (México), *4*(3), 279–284.
Vargas, L. (1948). Nota sobre la antigua presencia de *Anopheles* (*Nyssorhynchus*) *albimanus* Wiedemann, 1821 en Brasil. *Revista del Instituto de Salubridad y Enfermedades Tropicales* (México), *9*(2), 153–163.
Vargas, L. (1949a). Caracteres morfológicos diferenciales de algunas pupas de anofelinos neotropicales (Diptera: Culicidae). *Rivista di malariologia, 10*(4), 231–235.
Vargas, L. (1949b). Nueva larva que corresponde probablemente a *Aedes purpureipes* Aitken, 1941. *Revista del Instituto de Salubridad y Enfermedades Tropicales* (México), *10*(3), 261–265.
Vargas, L. (1950a). Nota sobre la distribución holártica del grupo *maculipennis* y del subgénero *Coelodiazesis*. *Missirolimyie*, nov. subg. de *Anopheles*. (Diptera: Culicidae). *Rivista di malariologia, 11*(2), 73–84.
Vargas, L. (1950b). Nota sobre *Aedes*. Revisión de algunos *Ochlerotatus* americanos. *Kummyia* n. subgen. (Diptera, Culicidae). *Revista de la Sociedad Mexicana de Historia Natural, 11*(1–4), 149–159.
Vargas, L. (1950c). Los subgéneros de *Aëdes*. *Downsiomyia* n. subgen. (Diptera: Culicidae). *Revista del Instituto de Salubridad y Enfermedades Tropicales* (México), *11*(1), 61–69.
Vargas, L. (1951). Las especies de *Psorophora* (Diptera: Culicidae). *Psorophora* (*Janthinosoma*) *fiebrigi* Edwards, 1922. *Revista del Instituto de Salubridad y Enfermedades Tropicales* (México), *12*(1–4), 53–58.
Vargas, L. (1953a). *Megarhinus* de Norteamerica (Diptera: Culicidae). *Ciencias biológicas: Zoologiea* (*Memorias Congresso Ciencias Mexico*), 7, 308–315.
Vargas, L. (1953b). *Megarhinus* de Norteamerica (Diptera: Culicidae). *Revista del Instituto de Salubridad y Enfermedades Tropicales* (México), *13,* 27–32.
Vargas, L. (1956a). *Psorophora* (*Psorophora*) *stonei* n. sp. *Revista del Instituto de Salubridad y Enfermedades Tropicales* (México), *16*(2), 15–16.
Vargas, L. (1956b). Especies y distribución de mosquitos mexicanos no anofelinos (Insecta, Diptera). *Revista del Instituto de Salubridad y Enfermedades Tropicales* (México), *16*(1), 19–36.

Vargas, L. (1959). Lista de *Anopheles* de las Americas y su identificacion por caracteres masculinos. (Diptera: Culicidae). *Revista del Instituto de Salubridad y Enfermedades Tropicales* (México), *19*(4), 367–386.

Vargas, L. (1984). The *Anopheles pseudopunctipennis* complex (summary). *Proceedings and Papers of the Annual Meeting of the Utah Mosquito Abatement Association,* 35(for 1983).

Vargas, L., & Diaz Nájera, A. (1959). Descripción del macho de *Sabethes (Sabethes) belisarioi* Neiva, 1908. Nueva especie para México (Insecta: Diptera). *Revista del Instituto de Salubridad y Enfermedades Tropicales* (México), *19*(3), 299–308.

Vargas, L., & Diaz Nájera, A. (1961). Nueva especie de *Sabethes* de Mexico (Diptera, Culicidae). *Revista del Instituto de Salubridad y Enfermedades Tropicales* (México), *21*(1–2), 73–77.

Vargas, L., & Downs, W.G. (1948). Dos especies de *Wyeomyia* nuevas para Mexico. Descripción de la pupa de *Wyeomyia (Dendromyia) pseudopecten. Revista del Instituto de Salubridad y Enfermedades Tropicales* (México), *9*(2), 95–97.

Vargas, L., & Downs, W.G. (1950). Tres especies nuevas de *Aedes* (Diptera, Culicidae). *Revista de la Sociedad Mexicana de Historia Natural, 11*(1–4), 161–172.

Vargas, L., & Martínez Palacios, A. (1942). *Anopheles hectoris* Mira, 1931. *Revista del Instituto de Salubridad y Enfermedades Tropicales* (México), *3*(2), 177–184.

Vargas, L., & Martínez Palacios, A. (1946). Descripción del huevo, larva y pupa de *Anopheles gabaldoni* Vargas, 1941. *Revista del Instituto de Salubridad y Enfermedades Tropicales* (México), *7*(1), 19–27.

Vargas, L., & Martínez Palacios, A. (1953). Descripción de *Wyeomyia (Wyeomyia) stonei,* n. sp., y notas sobre otros Sabethini de Mexico. *Revista del Instituto de Salubridad y Enfermedades Tropicales* (México), *13*(4), 293–307.

Vargas, L., & Martínez Palacios, A.M. (1954). *Culex (Culex) coronator mooseri* n. subsp. de México. *Revista del Instituto de Salubridad y Enfermedades Tropicales* (México), *14*(1), 33–37.

Vargas, L., & Martínez Palacios, A.M. (1955). Distribución de los Anofelinos de Mexico. *Revista del Instituto de Salubridad y Enfermedades Tropicales* (México), *15*(2), 81–123.

Vargas, L., & Martínez Palacios, A. (1956). *Anofelinos mexicanos. Taxonomía y distribución.* Distrito Federal, México: Secretaría de Salubridad y Asistencia, Comisión Nacional para la Erradicación del Paludismo (CNEP).

Vargas, V.M. (1975). Clave para anofelinos adultos (hembras) de Costa Rica (Diptera, Culicidae). *Brenesia, 6,* 77–80.

Vattier, G., & Adam, J.P. (1963). Description de la nymphe d'*Uranotaenia cavernicola* Mattingly, 1954. Notes morphologiques et écologique sur la larve et l'adulte. *Bulletin de la Société de pathologie exotique, 55*(for 1962), 911–918.

Vattier, G., & Hamon, J. (1962). Description de la larve et de la nymphe de *Culex (Culiciomyia) gilliesi* Hamon et van Someren, 1961, clef des larves du sous-genre *Culiciomyia* connues en Afrique au sud du Sahara. *Bulletin de la Société de pathologie exotique, 55,* 246–252.

Vences, M., Guayasamin, J.M., Miralles, A., & de la Riva, I. (2013). To name or not to name: Criteria to promote economy of change in Linnaean classification schemes. *Zootaxa, 3636*(2), 201–244.

Venhuis, W.G. (1932). Een nieuwe anopheline, *Anopheles meraukensis. Geneeskundig Tijdschrift voor Nederlandsch-Indië Batavia, 72,* 1040–1042.

Venhuis, W.G. (1933). Een nieuwe varieteit van de superspecies *An. punctulatus. Geneeskundig Tijdschrift voor Nederlandsch-Indië Batavia, 73,* 203–206.

Venhuis, W.G. (1940). Een nieuwe *Anopheles* larveuit Indo-China, *Anopheles alongensis* nov. sp. *Geneeskundig Tijdschrift voor Nederlandsch-Indië Buitenzorg, 80,* 173–178.

Venkatesan, M., & Rasgon, J.L. (2010). Population genetic data suggest a role for mosquito-mediated dispersal of West Nile virus across the western United States. *Molecular Ecology, 19*(8), 1573–1584.

Ventrillon, E. (1904). Description de culicides de Madagascar. *Bulletin du Muséum National d'Histoire Naturelle* (Paris), *10*(1), 550–555.

Ventrillon, E. (1905a). Culicides nouveaux de Madagascar. *Archives de parasitology* (Paris), *9*(4), 441–450.

Ventrillon, E. (1905b). Culicides nouveaux de Madagascar. *Bulletin du Muséum National d'Histoire Naturelle* (Paris), *11*(6), 427–431.

Ventrillon, E. (1906a). *Culex* nouveaux de Madagascar. *Bulletin Museum d'Histoire Naturelle, 12*(2), 100–106.

Ventrillon, E. (1906b). *Stegomyia cartroni*, culicide nouveau de Madagascar. *Bulletin du Muséum National d'Histoire Naturelle* (Paris), *12*(3), 143–145.

Ventrillon, E. (1906c). *Cellia tananariviensis*, Culicide nouveau de Madagascar. 9. Genre de la sous-famille des Anophelinae. *Bulletin du Muséum National d'Histoire Naturelle* (Paris), *12*(4), 198–202.

Verdonschot, P.F.M., & Besse-Lototskay, A.A. (2014). Flight distance of mosquitoes (Culicidae): A metadata analysis to support the management of barrier zones around rewetted and newly constructed wetlands. *Limnologica—Ecology and Management of Inland Waters, 45,* 69–79.

Versteirt, V., DeClerq, E.M., Fonseca, D.M., Pecor, J., Schaffner, F., Coosemans, M., & Van Bortel, W. (2012). Bionomics of the established exotic mosquito species *Aedes koreicus* in Belgium, Europe. *Journal of Medical Entomology, 49*(6), 1226–1232.

Versteirt, V., Pecor, J.E., Fonseca, D.M., Coosemans, M., & Van Bortel, W. (2012). Confirmation of *Aedes koreicus* (Diptera: Culicidae) in Belgium and description of morphological differences between Korean and Belgian specimens validated by molecular identification. *Zootaxa, 3191,* 21–32.

Vesgueiro, F.T., Demari-Silva, B., dos Santos Malafronte, R., Sallum, M.A.M., & Marrelli, M.T. (2011). Intragenomic variation in the second internal transcribed spacer of the ribosomal DNA of species of the genera *Culex* and *Lutzia* (Diptera: Culicidae). *Memórias do Instituto Oswaldo Cruz, 106*(1), 1–8.

Vezenegho, S.B., Chiphwanya, J., Hunt, R.H., Coetzee, M., Bass, C., & Koekemoer, L.L. (2013). Characterization of the *Anopheles funestus* group, including *Anopheles funestus*-like, from northern Malawi. *Transactions of the Royal Society of Tropical Medicine and Hygiene, 107*(12), 753–762.

Vijayavenkataraman, S., Iniyan, S., & Goic, R. (2012). A review of climate change, mitigation and adaptation. *Renewable and Sustainable Energy Reviews, 16*(1), 878–897.

Villalon, J.M., Ghosh, A., & Jacobs-Lorena, M. (2003). The peritrophic matrix limits the rate of digestion in adult *Anopheles stephensi* and *Aedes aegypti* mosquitoes. *Journal of Insect Physiology, 49*(10), 891–895.

Villanueva Rodriguez, C. (1961). Distribución geográfica de *Chagasia bathanus* (Dyar, 1928) y algunos aspectos bionómicos de la larva. *Revista Brasileira de entomologia, 10,* 217–225.

Villeneuve, J. (1919). Deuxieme note sur les Nematoceres vulnerants (espèces françaises). *Bulletin de la Société Entomologique de France, 1919*(2), 54–60.

Villiard, A., & Gaugler, R. (2015). Long-term effects of carbohydrate availability on mating success of newly eclosed *Aedes albopictus* (Diptera: Culicidae) males. *Journal of Medical Entomology, 52*(3), 308–314.

Vincke, I., & Leleup, N. (1949a). Description de la pupa, de la larve et de l'oeuf de *Anopheles concolor* Edwards. *Revue de zoologie et de botanique Africaines* (Brussels), *42*(3–4), 245–247.

Vincke, I., & Leleup, N. (1949b). Notes biologiques et ethologiques sur les *Anophèles* dendrophiles du Katanga. *Revue de zoologie et de botanique Africaines, 42*(3–4), 250–258.

Vincke, I., & Leleup, N. (1949c). Un nouvel Anophèle dendrophil de Katanga: *Anopheles berghei* sp. nov. *Revue de zoologie et de botanique Africaines, 42*(3–4), 248–249.

Vincke, I., & Leleup, N. (1950). Notes ethologiques et biologiques sur *Anopheles argenteolobatus* Gough. *Annales de la Société Belge de Médécine Tropicale, 30*(6), 1601–1603.

Vinogradova, E.B. (2000). Culex pipiens pipiens *mosquitoes: Taxonomy, distribution, ecology, physiology, genetics, applied importance and control* (No. 2). Pensoft.

Viswanathan, D.K., Rao, T.R., & Halgeri, A.V. (1955). Observations on some aspects of the nocturnal behavior of *Anopheles culicifacies*. *Indian Journal of Malariology, 9*(4), 371–384.

Vizioli, J., Catteruccia, F., della Torre, A., Reckmann, I., & Müller, H.M. (2001). Blood digestion in the malaria mosquito *Anopheles gambiae*. *European Journal of Biochemistry, 268*(14), 4027–4035.

Vo, M., Linser, P.J., & Bowers, D.F. (2010). Organ-associated muscles in *Aedes albopictus* (Diptera: Culicidae) respond differentially to Sindbis virus. *Journal of Medical Entomology, 47*(2), 215–225.

Vockeroth, J.R. (1950). Specific characters in tarsal claws of some species of *Aedes* (Diptera: Culicidae). *The Canadian Entomologist, 82*(7), 160–162.

Vockeroth, J.R. (1952). The specific status of *Aedes pionips* Dyar (Diptera: Culicidae). *The Canadian Entomologist, 84*(8), 243–247.

Vockeroth, J.R. (1954a). Notes on northern species of *Aedes*, with descriptions of two new species (Diptera: Culicidae). *The Canadian Entomologist, 86*(3), 109–116.

Vockeroth, J.R. (1954b). Notes on the identities and distribution of *Aedes* species of northern Canada, with a key to the females (Diptera: Culicidae). *The Canadian Entomologist, 86*(6), 241–255.

von Bergroth, E. (1889). Zwei neue Diptera. *Wiener Entomologische Zeitung, 8*, 295–298.

von Dungern, P., & Briegel, H. (2001). Enzymatic analysis of uricotelic protein catabolism in the mosquito *Aedes aegypti*. *Journal of Insect Physiology, 47*(1), 73–82.

Von Eichwald, C.E. (1837). *Reise auf dem Caspischen Meere, und in den Kaukasus: Unternommen in den jahren 1825–1826*. Erster band. Zweite Abtheilung, den historishen Bericht der Reise in den Kaudasus enthaltend. Stuttgart, Germany: J.G. Cotta'schen Buchhandlung.

Vonesh, J.R., & Blaustein, L. (2010). Predator-induced shifts in mosquito oviposition site selection: A meta-analysis and implications for vector control. *Israel Journal of Ecology and Evolution, 56*(3–4), 263–279.

von Frauenfeld, G. (1867). *Zoologische Miscellen* (Vol. 9). *Verhandlungen der kaiserlich-königlichen zoologisch-botanischen Gesellschaft in Wien 17*(2–3), 425–502.

von Heyden, L. (1870). Fossile Dipteren aus der Braunkohle van Rott im Siebengebirge. *Palaeontographica, 17*(6), 237–266, 232.

von Humboldt, F.H.A. (1819). Voyage aux régions équinoxiales du Nouveau Continent, fait en 1799, 1800, 1801, 1802, 1803 et 1804 par Al. de Humboldt et A. Bonpland. In F.H.A. Humboldt (Ed.), *Voyage de Humboldt et Bonpland* (2nd ed.). Premiére Partie. Paris, France: Relation Historique.

Von Röder, V. (1885). Dipteren von der Insel Portorico. *Stettiner Entomologische Zeitung, 46*(7–9), 337–349.

Voyadjoglou, A.S., & Harbach, R.E. (2001). Keys to the adult female mosquitoes (Culicidae) of Greece. *European Mosquito Bulletin, 10*, 13.

Vrzal, E.M., Allan, S.A., & Hahn, D.A. (2010). Amino acids in nectar enhance longevity of female *Culex quinquefasciatus* mosquitoes. *Journal of Insect Physiology, 56*(11), 1659–1664.

Vu, T.P., Nguyen, D.M., Tran, D.H., & Nguyen, T.V. (1991). *Anopheles (Anopheles) cucphuongensis*: A new species from Vietnam (Diptera: Culicidae). *Mosquito Systematics, 22*(for 1990), 145–148.

Vythilingam, I., & Hii, J. (2013). Simian malaria parasites: Special emphasis on *Plasmodium knowlesi* and their *Anopheles* vectors in Southeast Asia. In S. Manguin (Ed.), Anopheles *mosquitoes: New insights into malaria vectors* (pp. 487–510). Rijeka, Croatia: InTech.

Vythilingam, I., Phetsouvanh, R., Keokenchanh, K., Yengmala, V., Vanisaveth, V., Phompida, S., & Hakim, S.L. (2003). The prevalence of *Anopheles* (Diptera: Culicidae) mosquitoes in Sekong Province, Lao PDR in relation to malaria transmission. *Tropical Medicine and International Health, 8*(6), 535–535.

Vythilingam, I., Tan, C.H., Asmad, M., Chan, S.T., Lee, K.S., & Singh, B. (2006). Natural transmission of *Plasmodium knowlesi* to humans by *Anopheles latens* in Sarawak, Malaysia. *Transactions of the Royal Society of Tropical Medicine and Hygiene, 100*(1), 1087–1088.

Vythilingam, I., Jeffery, J., Harbach, R.E., Chan, S.T., Tan, C.H., & Asmad, M. (2007). *Anopheles (Cellia) kokhani* n. sp. (Diptera: Culicidae) from Kapit, Sarawak, East Malaysia. *Proceedings of the Entomological Society of Washington, 109*(2), 463–468.

Wada, Y., Takahashi, S., & Hori, E. (1973). The new description of male, pupa and larva of *Aedes (Finlaya) savoryi* Bohart, 1956, with a note on its bionomics. *Japanese Journal of Sanitary Zoology, 24*(2), 129–134.

Waktoedi Koesoemawinangoen, R. (1954). *Anophelini di Indonesia*. Kjilid I. Djakarta, Indonesia: Djakarta Perpustakaan Perguruan Kementerian.

Waku, Y. (1950). A redescription of *Aedes (Banksinella) imprimens* (Walker) from specimens taken in Japan. *Japanese Journal of Sanitary Entomology, 1*(3), 71–72.

Walker, E.D., & Archer, W.E. (1988). Sequential organization of grooming behavior in the mosquito, *Aedes triseriatus*. *Journal of Insect Behavior, 1*(1), 97–109.

Walker, E.D., & Merritt, R.W. (1991). Behavior of larval *Aedes triseriatus* (Diptera: Culicidae). *Journal of Medical Entomology, 28*(5), 581–589.

Walker, E.D., Olds, E.J., & Merritt, R.W. (1988). Gut content analysis of mosquito larvae (Diptera: Culicidae) using DAPI stain and epifluorescence microscopy. *Journal of Medical Entomology, 25*(6), 551–554.

Walker, F. (1848). *List of the specimens of dipterous insects in the collection of the British Museum* (Part I). London, England: British Museum.

Walker, F. (1856a). *Insecta Saundersiana: On characters of undescribed insects in the collection of William Wilson Saunders, Esq., F.R.S., F.L.S., &c.* Vol. I: *Diptera*. London, England: van Voorst.

Walker, F. (1856b). Catalogue of the dipterous insects collected at Singapore and Malacca by Mr. A. R. Wallace, with descriptions of new species. *Journal of the Proceedings of the Linnean Society of London: Zoology, 1*, 4–39.

Walker, F. (1858). Catalogue of the dipterous insects collected in the Aru Islands by Mr. A. R. Wallace, with descriptions of new species. *Journal of the Proceedings of the Linnean Society of London: Zoology, 3*, 77–131.

Walker, F. (1859). Catalogue of the dipterous insects collected at Makessar, in Celebes, by Mr. A. R. Wallace, with descriptions of new species. *Journal of the Proceedings of the Linnean Society of London: Zoology, 4*(15), 90–92.

Walker, F. (1860a). Catalogue of the dipterous insects collected at Makessar, in Celebes, by Mr. A. R. Wallace, with descriptions of new species (continued). *Journal of the Proceedings of the Linnean Society of London: Zoology, 4*(16), 90–172.

Walker, F. (1860b). Catalogue of the dipterous insects collected in Amboyna by Mr. A. R. Wallace, with descriptions of new species. *Journal of the Proceedings of the Linnean Society of London: Zoology, 5*, 144–168.

Walker, F. (1861a). Catalogue of the dipterous insects collected at Dorey, New Guinea, by Mr. A. R. Wallace, with descriptions of new species. *Journal of the Proceedings of the Linnean Society of London: Zoology, 5*, 229–254.

Walker, F. (1861b). Insects. In J.E. Tennent (Ed.), *Sketches of the natural history of Ceylon with narratives and anecdotes illustrative of the habits and instincts of the . . . insects . . .* (pp. 403–479). London: Longman, Green, Longman, & Roberts.

Walker, F. (1865). Descriptions of new species of the dipterous insects of New Guinea. *Journal of the Proceedings of the Linnean Society of London: Zoology, 89*(30–32), 102–130.

Walker, F. (1866). List of Diptera. In J.K. Lord (Ed.), *The naturalist in Vancouver Island and British Colombia* (Vol. 2, pp. 337–340). London, England: Bentley.

Walker, F. (1874). *Notes on Diptera and lists of species*. First Series. London, England: Janson.

Walker, J.D., Geissman, J.W., Bowring, S.A., & Babcock, L.E. (2013). The Geological Society of America geologic time scale. *Bulletin of the Geological Society of America, 125*(3–4), 259–272.

Walker, K.R., Joy, T.K., Ellers-Kirk, C., & Ramberg, F.B. (2011). Human and environmental factors affecting *Aedes aegypti* distribution in an arid urban environment. *Journal of the American Mosquito Control Association, 27*(2), 135–141.

Wallace, A.R. (1823–1913). *The works of Alfred Russell Wallace* [Kindle DX version]. Retrieved from https://www.amazon.com/Works-Alfred-Russel-Wallace-ebook/dp/B00CIY2IAG

Wallis, R.C., & Whitman, L. (1971). New collection records for *Psorophora ciliata* (Fabricius), *Psorophora ferox* (Humboldt) and *Anopheles earlei* Vargas in Connecticut (Diptera: Culicidae). *Journal of Medical Entomology, 8*(3), 336–337.

Walsh, A.S., Glass, G.E., Lesser, C.R., & Curriero, F.C. (2008). Predicting seasonal abundance of mosquitoes based on off-season meteorological conditions. *Environmental and Ecological Statistics, 15*(3), 279–291.

Waltl, J. (1835). *Reise durch Tyrol, Oberitalien und Piemont nach dem sudlichen Spainen . . . nebst einem Anhang zoologischen Inhalts* (Vol. 2). Passau, Germany: Pustet.

Wang, H.-H., & Feng, C.-Y. (1964). Description of three new species of *Culex* (*Lophoceratomyia*). *Acta zootaxonomica Sinica, 1*(1), 37–41.

Wang, H.-H., & Feng, C.-Y. (1965). A new mosquito, *Uranotaenia lishanensis* sp. nov., with notes on the adult male and larva of *U. hebes* Barraud. *Acta zootaxonomica Sinica, 2*(3), 206–210.

Wang, S.F., Li, C., Zhu, J., Miura, K., Miksicek, R.J., & Raikhel, A.S. (2000). Differential expression and regulation by 20-hydroxyecdysone of mosquito ultraspiracle isoforms. *Developmental Biology, 218*(1), 99–113.

Wang, S.F., Li, C., Sun, G., Zhu, J., & Raikhel, A.S. (2002). Differential expression and regulation by 20-hydroxyecdysone of mosquito ecdysteroid receptor isoforms A and B. *Molecular Cellular Endocrinology, 196*(1), 29–42.

Wang, X.-X. (1984). Description of a new *Culex* species (Diptera: Culicidae). *Acta entomologica Sinica, 27*(2), 217–220.

Wang, Y., Gilbreath, T.M., III, Kukutla, P., Yan, G., & Xu, J. (2011). Dynamic gut microbiome across life history of the malaria mosquito *Anopheles gambiae* in Kenya. *PLOS ONE, 6*(9), e24767.

Wanson, M., & Berteaux, M. (1954). La larve de l'*Anopheles mortiauxi* Edwards. *Revue de zoologie et de botanique Africaines, 49*, 33–38.

Wanson, M., & Lebied, B. (1945). Un nouvel Anophèle cavernicole du Congo belge: *Anopheles* (*Myzomyia*) *vanhoofi* spec. nov. *Revue de zoologie et de botanique Africaines* (Brussels), *39*(1), 118–129.

Wanson, M., Wolfs, J., & Lebied, B. (1947). Comportement de l'*Anopheles* (*Myzomyia*) *moucheti* Evans. *Recueil de travaux de sciences médicales au Congo Belge, 6*, 39–62.

Ward, R.A. (1972). Mosquitoes of Afghanistan: An annotated checklist. *Mosquito Systematics, 4*(3), 93–97.

Ward, R.A. (1979). Corrections in the nomenclature of three anophelines (Diptera: Culicidae). *Mosquito Systematics, 11*(2), 130–134.

Ward, R.A. (1984a). Mosquito fauna of Guam: Case history of an introduced fauna. In M. Laird (Ed.), *Commerce and the spread of pests and disease vectors* (pp. 143–162). New York, NY: Praeger.

Ward, R.A. (1984b). Second supplement to a catalog of the mosquitoes of the World (Diptera: Culicidae). *Mosquito Systematics, 16*(3), 227–270.

Ward, R.A. (1987). Nomenclatural status and some additions to the species listed in the publication, new species of mosquitoes in the fauna of the USSR by A.V. Gutsevich and A.M. Dubitskiy (1981). *Mosquito Systematics, 19*(1), 93–99.

Ward, R.A. (1992). Third supplement to "A Catalog of the Mosquitoes of the World (Diptera: Culicidae)." *Mosquito Systematics, 24*(3), 177–231.

Ward, R. A., & Darsie, R.F., Jr. (1982). Corrections and additions to the publication "Identification and Geographical Distribution of the Mosquitoes of North America, North of Mexico." *Mosquito Systematics, 14*(3), 209–219.

Ward, R.A., Jordan, B., Gillogly, A.R., & Harrison, F.J. (1976). *Anopheles litoralis* King and *A. barbirostris* group on the island of Guam. *Mosquito News, 36*(1), 99–100.

Washino, R.K., & Bailey, S.F. (1970). Overwintering of *Anopheles punctipennis* (Diptera: Culicidae) in California. *Journal of Medical Entomology, 7*(1), 95–98.

Waterhouse, C.O. (1905). Notes on some British Culicidae. *Annals and Magazine of Natural History* (Series 7), *16*, 674–676.

Waterston, J. (1918). On the mosquitos of Macedonia. *Bulletin of Entomological Research, 9*(1), 1–12.

Watson, M. (1910). A new anopheline from the Federated Malay States. *Annals of Tropical Medicine and Parasitology, 4*(2), 251–253.

Watt, J.C. (1979). Abbreviations for entomological collections. *New Zealand Journal of Zoology, 6*(3), 519–520.

Wattal, B.L., & Kalra, N.L. (1965). An entomological survey of Dehra Dun Valley (Uttar Pradesh). Part II. Description of *Culex* (*Culiciomyia*) *ramakrishnii* sp. n., and allotype (M) of *Uranotaenia maculipleura* Leicester, 1908. *Bulletin of the Indian Society for Malaria and Other Communicable Diseases, 2*(4), 309–313.

Wattal, B.L., Bhatia, M.L., & Kalra, N.L. (1958). Some new records of culicines of Dehra Dun (Uttar Pradesh) with a description of a new variety. *Indian Journal of Malariology* (Delhi), *12*(3), 217–230.

Wattal, B.L., Kalra, N.L., & Gopal, R.R. (1962). A study of *Anopheles barbirostris* van der Wulp in the collection of the Malaria Institute of India with first record of *A. barbumbrosus* Strickland and Chowdhury and *A. vanus* Walker in India. *Indian Journal of Malariology, 16*(1), 63–74.

Wattal, B.L., Kalra, N.L., & Krishnan, K.S. (1966). Description of *Culex* (*Mochthogenes*) *shivastavii* sp. n. from Andaman Islands, India. *Bulletin of the Indian Society for Malaria and Other Communicable Diseases, 3*(2), 159–161.

Weaver, S.C., Scherer, W.F., Taylor, C.D., Castello, D.A., & Cupp, E.W. (1986). Laboratory vector competence of *Culex* (*Melanoconion*) *cedecei* for sympatric and allopatric Venezuelan equine encephalomyelitis viruses. *The American Journal of Tropical Medicine and Hygiene, 35*(3), 619–623.

Weiss, B., & Aksoy, S. (2011). Microbiome influences on insect host vector competence. *Trends in Parasitology, 27*(11), 514–522.

Weldon, P.J., & Carroll, J.F. (2007). Vertebrate chemical defense: Secreted and topically acquired deterrents of arthropods. In M. Debboun, S.P. Frances, & D. Strickman (Eds.), *Insect repellents: Principles, methods and uses* (pp. 47–76). Boca Raton, FL: CRC Press.

Weng, X.H., Huss, M., Wieczorek, H., & Beyenbach, K.W. (2003). The V-type H^+-ATPase in malpighian tubules of *Aedes aegypti*: Localization and activity. *Journal of Experimental Biology, 206*(13), 2211–2219.

Werren, J.H. (1997). Biology of *Wolbachia*. *Annual Review of Entomology, 42*(1), 587–609.

Wesenberg-Lund, C. (1921). Contributions to the biology of the Danish Culicidae. *Det Kongelige Danske Videnskabernes Selskabs Skrifter. Naturvidenskabelig og mathematisk Afdeling* (Series 8), *7*(1), 1–120.

Westwood, J.L.S. (1835). Insectorum nonnullorum novum. (Ex ordine Dipterorum) descriptiones. *Annales de la Société Entomologique de Belgique, 4*, 681.

Weyenbergh, S.H. (1882). Los habitantes del Río Primero. *Anales del Ateneo* (Montevideo), *1882*, 1–27.

Weyer, F. (1954). Bestimmungsschlüssel für die *Anopheles*-Weibchen und-Larven in Europa, Nordafrika und Westasien. (2nd Suppl.). *Merkblätter de Bernhard-Nocht-Instuts für Schiffs-und Tropenkrankheiten* (Hamburg), *12*, 3–38.

Wharton, R.H. (1947). Notes on Australian mosquitoes (Diptera, Culicidae). VII. The genus *Harpagomyia*. *Proceedings of the Linnean Society of New South Wales, 72*(1–2), 58–68.

Wharton, R.H. (1962). The biology of *Mansonia* mosquitoes in relation to the transmission of filariasis in Malaya. *Bulletin of the Institute for Medical Research, Kuala Lumpur, Federation of Malaya, 11*, 1–114.

Wheelock, G.D., Petzel, D.H., Gillett, J.D., Beyenbach, K.W., & Hagedom, H.H. (1988). Evidence for hormonal control of diuresis after a blood meal in the mosquito *Aedes aegypti*. *Archives of Insect Biochemistry and Physiology, 7*(2), 75–89.

Whelan, P., & Hapgood, G. (2000). A mosquito survey of Dili, East Timor, and implications for disease control. *Arbovirus Research in Australia, 8*, 405–416.

Whisenton, L.R., & Bollenbacher, W.E. (1986). Presence of gonadotropic and prothoracicotropic factors in pupal and adult heads of mosquitoes. In A.R. Borkovec, & D.B. Gelman (Eds.), *Insect Neurochemistry and Neurophysiology* (pp. 339–342). Clifton, NJ: Humana Press.

White, B.J., Collins, F.H., & Besansky, N.J. (2011a). Evolution of *Anopheles gambiae* in relation to humans and malaria. *Annual Review of Ecology, Evolution, and Systematics, 42*, 111–132.

White, B.J., Lawniczak, M.K.N., Cheng, C., Coulibaly, M.B., Wilson, M.D., Sagnon, N., . . . Besansky, N.J. (2011b). Adaptive divergence between incipient species of *Anopheles gambiae* increases resistance to *Plasmodium*. *Proceedings of the National Academy of Sciences of the United States of America, 108*(1), 234–249.

White, G. (2007). Terminology of insect repellents. In M. Debboun, S.P. Frances, & D. Strickman (Eds.), *Insect repellents: Principles, methods and uses* (pp. 31–46). Boca Raton, FL: CRC Press.

White, G.B. (1973). Cytogenetics, systematics and importance of *Anopheles gambiae* species. D. *9th International Congress of Tropical Medicine and Malaria* (Athens), *9*(1), 249–250.

White, G.B. (1974). Priority of *Ingramia* Edwards, 1912, over *Ravenalites* Doucet 1957 as the name for a subgenus of *Mimomyia* Theobald, 1903 (Diptera: Culicidae). *Mosquito Systematics, 6*(4), 239–242.

White, G.B. (1975). Notes on a catalogue of the Ethiopian Region. *Mosquito Systematics, 7*(4), 303–344.

White, G.B. (1977). The place of morphological studies in the investigation of *Anopheles* species complexes. *Mosquito Systematics, 9*(1), 1–24.

White, G.B. (1978). Systematic reappraisal of the *Anopheles maculipennis* complex. *Mosquito Systematics, 10*(1), 13–44.

White, G.B. (1979). Identification of mosquitoes as vectors of malaria and filariasis. In A.E.R. Tayor and R. Muller (Eds.), *Symposia of the British Society for Parasitology (Problems in the identification of parasites and their vectors), 17*, 103–143.

White, G.B. (1980). Chapter 8. Culicidae. In R.W. Crosskey (Ed.), *Catalogue of the Diptera of the Afrotropical Region* (pp. 114–148). London, England: British Museum (Natural History).

White, G.B. (1985). *Anopheles bwambae* sp. n., a malaria vector in the Semliki Valley, Uganda, and its relationships with other sibling species of the *An. gambiae* complex (Diptera: Culicidae). *Systematic Entomology, 10*(4), 501–522.

White, G.B., & Muniss, J.N. (1972). Taxonomic value of spermatheca size for distinguishing four members of the *Anopheles gambiae* complex in East Africa. *WHO Bulletin, 46*, 793–799.

Whiting, M.F. (2005). Phylogenetic position of Diptera: Review of the evidence. In D.K. Yeates & B.M. Wiegmann (Eds.), *The evolutionary biology of flies* (pp. 3–13). New York, NY: Columbia University Press.

Whitten, M.M.A., Shiao, S.H., & Levashina, E.A. (2006). Mosquito midguts and malaria: Cell biology, compartmentalization and immunology. *Parasite Immunology, 28*(4), 121–130.

Wichard, W., Gröhn, C., & Seredszus, F. (2009). Aquatic insects in Baltic amber. Wasserinsekten im Baltischen Bernstein. Remagen, Germany: Verlag Kessel.

Wiebe, A., Longbottom, J., Gleave, K., Shearer, F.M., Sinka, M.E., Massey, N.C., . . . Smith, D. L. (2017). Geographical distributions of African malaria vector sibling species and evidence for insecticide resistance. *Malaria Journal, 16*(1), 85.

Wiedemann, C.R.G. (1819). Beschreibung neuer Zweiflüger aus Ostindien und Afrika. *Zoologisches magazin, 1*(3), 1–39.

Wiedemann, C.R.G. (1820). *Diptera exotica. Section I. antennis multiarticulatus*. Kiel, Germany: Kiliyae.

Wiedemann, C.R.G. (1821). *Diptera exotica, Part I*. Kiel, Germany: Kiliyae.

Wiedemann, C.R.G. (1828). *Aussereuropäische zweiflägelige insekten*. (Vol. 1). Germany: Hamm.

Wiegmann, B.M., Trautwein, M.D., Winkler, I.S., Barr, N.B., Kim, J.W., Lambkin, C., . . . Yeates, D.K. (2011). Episodic radiations in the fly tree of life. *Proceedings of the National Academy of Sciences of the United States of America, 108*(14), 5690–5695.

Weinbren, M.P., & Weinbren, B.M. (1970). Observations on the mosquito population in the irradiated forest at El Verde. In H.T. Odum & K.F. Pigeon (Eds.), *A tropical rain forest: A study of irradiation and ecology at El Verde, Puerto Rico* (pp. E-159–167). Springfield, Virginia: National Technical Information Service.

Wiens, J.J. (Ed.) (2000). *Phylogenetic analysis of morphological data*. Washington, DC: Smithsonian Institution Press.

Wigglesworth, V.B. (1930). Some notes on the physiology of insects related to human disease. *Proceedings of the Royal Society of Tropical Medicine and Hygiene, 23*(6), 553–577.

Wigglesworth, V.B. (1933). The function of the anal gills of the mosquito larva. *Journal of Experimental Biology, 10*(1), 16–26.

Wijesundara, D.P. (1951). Notes on species of the subgenus *Aädes* Meigen (*sens. str.*) in Ceylon, with descriptions of three new species. *Ceylon Journal of Science B: Zoology, 24*(3), 173–179.

Wijeyaratne, P.M., Seawright, J.A., & Weidhaas, D.E. (1974). Development and survival of a natural population of *Aedes aegypti. Mosquito News, 34*(1), 36–42.

Wikström, N., Savolainen, V., & Chase, M.W. (2001). Evolution of the angiosperms: Calibrating the family tree. *Proceedings of the Royal Society B: Biological Sciences, 268*(1482), 2211–2220.

Wiley, E.O. (1979). An annotated Linnaean hierarchy, with comments on natural taxa and competing systems. *Systematic Zoology, 28*(3), 308–337.

Wiley, E.O. (1981). *Phylogenetics: The theory of phylogenetic systematics.* New York, NY: Wiley-Interscience.

Wiley, E.O., & Liebermann, B.S. (2011). *Phylogenetic systematics: Theory and practice of phylogenetic systematics* (2nd ed.). Hoboken, NJ: Wiley-Blackwell.

Wilkerson, R.C. (1988). Notes and redescriptions of some *Anopheles* (Series Arribalzagia) holotypes (Diptera, Culicidae) in the British Museum Natural History, UK. *Proceedings of the Entomological Society of Washington, 90*(4), 411–421.

Wilkerson, R.C. (1990a). Elevation of *Anopheles chiriquiensis* from synonymy with *Anopheles parapunctipennis* and designation of name bearing types for *Anopheles parapunctipennis* and *Anopheles parapunctipennis parapunctipennis* (Diptera: Culicidae). *Mosquito Systematics, 21*(for 1989)(3), 178–187.

Wilkerson, R.C. (1990b). Redescriptions of *Anopheles punctimacula* and *An. malefactor* (Diptera: Culicidae). *Journal of Medical Entomology, 27*(2), 225–247.

Wilkerson, R.C. (1991). *Anopheles* (*Anopheles*) *calderoni* n. sp., a malaria vector of the Arribalzagia Series from Peru (Diptera: Culicidae). *Mosquito Systematics, 23*(1), 25–38.

Wilkerson, R.C., & Linton, Y.-M. (2015). Elevation of *Pseudoskusea*, *Rusticoidus* and *Protomacleaya* to valid subgenera in the mosquito genus *Aedes* based on taxon naming criteria recently applied to other members of the Tribe Aedini (Diptera: Culicidae). *Parasites & Vectors, 8*, 668.

Wilkerson, R.C., & Peyton, EL. (1990). Standardized nomenclature for the costal wing spots of the genus *Anopheles* and other spotted-winged mosquitoes (Diptera: Culicidae). *Journal of Medical Entomology, 27*(2), 207–224.

Wilkerson, R.C., & Peyton, EL. (1991). The Brazilian malaria vector *Anopheles* (*Kerteszia*) *cruzii*: Life stages and biology (Diptera: Culicidae). *Mosquito Systematics, 23*(2), 110–122.

Wilkerson, R.C., & Sallum, M.A.M. (1999). *Anopheles* (*Anopheles*) *forattinii*: A new species in series Arribalzagia (Diptera: Culicidae). *Journal of Medical Entomology, 36*(3), 345–354.

Wilkerson, R.C., & Strickman, D. (1990). Illustrated key to the female anopheline mosquitoes of Central America and Mexico. *Journal of the American Mosquito Control Association, 6*(1), 7–34.

Wilkerson, R.C., Strickman, D., Fernández-Salas, I., & Ibáñez-Bernal, S. (1993). *Clave ilustrada para la identificacion de las hembras de mosquitos de Mexico y Centroamerica.* Mexico: Secretaria de Salud, Subsecretaria de Coordinacion y Desarrollo.

Wilkerson, R.C., Gaffigan, T.V., & Lima, J.B. (1995a). Identification of species related to *Anopheles* (*Nyssorhynchus*) *albitarsis* by random amplified polymorphic DNA-polymerase chain reaction (Diptera: Culicidae). *Memórias do Instituto Oswaldo Cruz, 90*(6), 721–732.

Wilkerson, R.C., Hribar, L.J., Milstrey, E.G., & Falero, G.C. (1995b). The identification of *Anopheles* (*Nyssorhynchus*) *rondoni* (Diptera: Culicidae) in Mato Grosso State, Brazil: An analysis of key character variability. *Memórias do Instituto Oswaldo Cruz, 90*(5), 575–582.

Wilkerson, R.C., Parsons, T.J., Klein, T.A., Gaffigan, T.V., Bergo, E., & Consolim, J. (1995c). Diagnosis by random amplified polymorphic DNA polymerase chain reaction of four cryptic species related to *Anopheles* (*Nyssorhynchus*) *albitarsis* (Diptera: Culicidae) from Paraguay, Argentina, and Brazil. *Journal of Medical Entomology, 32*(5), 697–704.

Wilkerson, R.C., Sallum, M.A.M., & Forattini, O.P. (1997). Redescription of *Anopheles* (*Anopheles*) *shannoni* Davis; a member of the Arribalzagia series from the Amazon Basin (Diptera: Culicidae). *Proceedings of the Entomological Society of Washington, 99*(3), 461–471.

Wilkerson, R.C., Li, C., Rueda, L.M., Kim, H., Klein, T.A., Song, G.-H., & Strickman, D. (2003). Molecular confirmation of *Anopheles* (*Anopheles*) *lesteri* from the Republic of South Korea and its genetic identity with *An.* (*Ano.*) *anthropophagus* from China (Diptera: Culicidae). *Zootaxa, 378*, 1–14.

Wilkerson, R.C., Reinert, J.F., & Li, C. (2004). Ribosomal DNA ITS2 sequences differentiate six species in the *Anopheles crucians* complex (Diptera: Culicidae). *Journal of Medical Entomology, 41*(3), 392–401.

Wilkerson, R.C., Foster, P.G., Li, C., & Sallum, M.A.M. (2005). Molecular phylogeny of neotropical *Anopheles* (*Nyssorhynchus*) *albitarsis* complex (Diptera; Culicidae). *Annals of the Entomological Society of America, 98*(6), 918–925.

Wilkerson, R.C., Linton, Y.-M., Fonseca, D., Schultz, T., Price, D., & Strickman, D. (2015). Making mosquito taxonomy useful: A stable classification of tribe Aedini that balances utility with current knowledge of evolutionary relationships. *PLOS ONE, 10*(7), e0133602.

Wilkerson, R.C., Linton, Y.-M., Strickman, D. (2020). *Mosquitoes of the World.* In 2 Volumes. Baltimore, MD: Johns Hopkins University Press.

Wilkins, O.P., & Breland, O.P. (1951). The larval stages and the biology of the mosquito, *Orthopodomyia alba* Baker (Diptera: Culicidae). *Journal of the New York Entomological Society, 59*(4), 225–240.

Willcocks, F.C. (1910). A preliminary note on the prevalence of mosquitoes in Cairo and its environs. *Annals of Tropical Medicine and Parasitology, 3*(5), 583–589.

Williams, C.R., & Kokkinn, M.J. (2005). Daily patterns of locomotor and sugar-feeding activity of the mosquito *Culex annulirostris* from geographically isolated populations. *Physiological Entomology, 30*(4), 309–316.

Williams, C.R., Long, S.A., Russell, R.C., & Ritchie, S.A. (2006). Optimizing ovitrap use for *Aedes aegypti* in Cairns, Queensland, Australia: Effects of some abiotic factors on field efficacy. *Journal of the American Mosquito Control Association, 22*(4), 635–640.

Williams, C.R., Leach, K.J., Wilson, N.J., & Swart, V.R. (2008). The Allee effect in site choice behaviour of egg-laying dengue vector mosquitoes. *Tropical Biomedicine, 25*(2), 140–144.

Williams, F.X. (1943). Mosquitoes and some other noxious flies that occur in New Caledonia. *Hawaiian Planters' Record, 47*(4), 205–222.

Williams, T. (2015). *A dictionary for entomology.* Lulu.com Publisher, ISBN 9781447561095, http://www.lulu.com/shop/tim-williams/a-dictionary-for-entomology/paperback/product-5426918.html.

Williston, S.W. (1893). List of Diptera of the Death Valley Expedition. In C.V. Riley (Ed.), *Report on a small collection of insects made during the Death Valley Expedition.* North American Fauna No. 7, (pp. 235–268). The Death Valley expedition. A biological survey of parts of California, Nevada, Arizona, and Utah. Part 2. Washington, DC: U.S. Government Printing Office.

Williston, S.W. (1896). On the Diptera of St. Vincent (West Indies). *Transactions of the Royal Entomological Society,London, 44*, 253–446.

Williston, S.W. (1900). Culicidae. In C.R. Osten Sacken (Ed.), *Biologia Centrali-Americana. Insecta. Diptera* (Vol. I, pp. 224). The Supplement by S.W. Williston, J.M. Aldrich, W.M. Wheeler and A.L. Melander. R.H. Porter, London, England: R.H. Porter.

Williston, S.W. (1906). The classification of the Culicidae. *The Canadian Entomologist, 38*(12), 384–388.

Willmann, R. (1989). Rediscovered: *Permotipula patricia*, the oldest fly. *Naturwissenschaften, 76*(8), 375–377.

Wimberly, M.C., Hildreth, M.B., Boyte, S.P., Lindquist, E., & Kightlinger, L. (2008). Ecological niche of the 2003 West Nile virus epidemic in the northern Great Plains of the United States. *PLOS ONE, 3*(12), e3744.

Wimsatt, W.C., & Schank, J.C. (1988). Two constraints on the evolution of complex adaptations and the means of their avoidance. In M.H. Nitecki (Ed.), *Evolutionary progress* (pp. 231–273). Chicago, IL: University of Chicago Press.

Winston, J.E. (1999). *Describing species: Practical taxonomic procedure for biologists.* New York, NY: Columbia University Press.

Wirth, W.W. (1948). Diptera. In R.L. Usinger, I. La Rivers, H.P. Chandler, & W.W. Wirth (Eds.), *Entomology 133: Biology of aquatic and littoral insects.* (pp. 86–136, keys 223–239). Berkeley: Syllabus Series, University of California Press.

Wolff, T.A., & Nielsen, L.T. (1976). The distribution of snowpool *Aedes* mosquitoes in the southwestern states of Arizona and New Mexico with notes on biology and past dispersal patterns. *Mosquito Systematics, 8*(4), 413–439.

Wolff, T.A., & Nielsen, L.T. (1977). A chaetotaxic study of snowpool *Aedes* larvae and pupae with an analysis of variance of the larvae of eight species. *Mosquito Systematics, 9*(2), 176–236.

Wolfinsohn, M., & Galun, E. (1953). A method for determining the flight range of *Aedes aegypti* (Linn.). *Bulletin of the Research Council of Israel, 2*(4), 433–436.

Wolfs, J. (1945a). *Aedes* (*Mucidus*) *grahami* Theobald. Description de la larve et de la nymphe. *Recueil des travaux des sciences et Médicas Congo Belge, 4*, 64–66.

Wolfs, J. (1945b). *Anopheles* (*Anopheles*) *implexus* var. *henrardi* var. nov. *Recueil des travaux des sciences et Médicas Congo Belge, 3*, 174–176.

Wolfs, J. (1945c). *Culex* (*Neoculex*) *wansoni* sp. n. *Annales de la Société Belge de Médicine Tropicale, 25*, 221–224.

Wolfs, J. (1947a). Un culicide nouveau du Katanga, *Aedes* (*Banksinella*) *bequaerti*, sp. n. *Revue zoologique Africaine (supplement botanique), 40*(1), 40–41.

Wolfs, J. (1947b). *Culex* (*Culex*) *bukavuensis* sp. n. *Revue zoologique Africaine (supplement botanique), 40*(2–3), 159–164.

Wolfs, J. (1947c). *Culex* (*Culex*) *seldeslachtsi*, sp. nov. *Revue zoologique Africaine (supplement botanique), 40*(2–3), 224–228.

Wolfs, J. (1947d). Description des larves et des nymphes de quelques megarhines. *M. phytophagus* Theobald, *M. viridibasis* Edwards, et *M. aeneus* Theobald [*sic*, Evans]. *Recueil des travaux des sciences et Médicas Congo Belge, 6*, 63–72.

Wolfs, J. (1948a). *Taeniorhynchus* (*Coquillettidia*) *schoutedeni* n. sp. (Culic.). *Revue zoologique Africaine (supplement botanique), 40*(4), 244–247.

Wolfs, J. (1948b). *Taeniorhynchus* (*Coquillettidia*) *vanoyei* sp. n. (Dipt. Culicidae). *Revue zoologique Africaine (supplement botanique), 41*(1), 83–86.

Wolfs, J. (1948c). *Culex* (*Culex*) *zombaensis* Theobald. Description de la larve et nymphe. *Revue zoologique Africaine (supplement botanique), 41*(1), 87–90.

Wolfs, J. (1949). *Aedes* (*Stegomyia*) *apicoargenteus* var. *denderensis* var. n. (Culicidae). *Revue zoologique Africaine (supplement botanique), 42*(2–3), 190–192.

Wolfs, J. (1951). Un culicide nouveau de la région Éthiopienne: *Taeniorhynchus* (*Coquillettidia*) *karandalaensis* sp. n. *Revue zoologique Africaine (supplement botanique), 45*(1–2), 12–14.

Wolfs, J. (1952). Note sur la faune culicidienne de Costermansville. *Annales de la Société Belge de Medicine Tropicale, 32*(25), 531–537.

Wolfs, J. (1958). Deux Culicidae nouveaux du Congo Belge (Diptera Nematocera). *Revue de zoologie et de botanique Africaines, 58*(3–4), 297–298.

Wolfs, J. (1964). *Uranotaenia benoiti* Mwendua River, Kasongo, Congo. In P.L.G. Benoit (1964) Diptera - Stomoxydae, Calliphoridae d'interet medical. Streblidae et Hippoboscidae. Tabanidae. Culicidae (pp. 309–310). Mission de Zoologie medicale au Maniema (Congo, Leopoldville) (P.L.G. Benoit, 1959): Resultats scientifiques. Musée Royal de l'Afrique centrale - Tervuren, Belgique. Annales - Serie in-8 - Sciences Zoologiques - no. 132.

Wong, J., Stoddard, S.T., Astete, H., Morrison, A.C., & Scott, T.W. (2011). Oviposition site selection by the dengue vector *Aedes aegypti* and its implications for dengue control. *PLOS Neglected Tropical Diseases, 5*(4), e1015.

Wood, D.M. (1977). Notes on the identities of some common Nearctic *Aedes* mosquitoes. *Mosquito News, 37*(1), 71–81.

Wood, D.M. (1990). *Homology and phylogenetic implications of male genitalia in Diptera. The ground plan.* Paper presented at the Second International Congress of Dipterology. Bratislava, Czechoslovakia.

Wood, D.M. (1991). Homology and phylogenetic implications of male genitalia in Diptera. The ground plan. In L. Weissman, I. Orszagh, & A.C. Pont (Eds.), *Proceedings of the Second International Congress of Dipterology* (p. 255). Bratislava, Czechoslovakia: SPB Academic.

Wood, D.M., & Borkent, A. (1989). Phylogeny and classification of the Nematocera. Research Branch Agriculture Canada Monograph No. 32. In J.F. McAlpine & D.M. Wood (Eds.), *Manual of Nearctic Diptera* (Vol. 3, pp. 1333–1370). Hull, Canada: Canadian Government.

Wood, D.M., Dang, P.T., & Ellis, R.A. (1979). *The Insects and Arachnids of Canada. Part 6. The Mosquitoes of Canada (Diptera: Culicidae).* Hull: Canadian Government.

Woodhill, A.R. (1946). Observations on the morphology and biology of the subspecies of *Anopheles punctulatus* Doenitz. *Proceedings of the Linnean Society of New South Wales, 70*(5–6), 276–287.

Woodhill, A.R. (1949). A new subspecies of *Aedes* (*Stegomyia*) *scutellaris* Walker (Diptera, Culicidae) from northern Australia. *Proceedings of the Linnean Society of New South Wales, 74*(3–4), 140–144.

Woodhill, A.R. (1950). Further notes on experimental crossing within the *Aedes scutellaris* group of species (Diptera: Culicidae). *Proceedings of the Linnaean Society of New South Wales, 75*(5–6), 251–253.

Woodhill, A.R. (1954). Experimental crossing of *Aedes* (*Stegomyia*) *pseudoscutellaris* Theobald and *Aedes* (*Stegomyia*) *polynesiensis* Marks (Diptera, Culicidae). *Proceedings of the Linnean Society of New South Wales, 79*, 19–20.

Woodhill, A.R., & Lee, D.J. (1944). The subspecies of *Anopheles amictus* Edwards (Diptera, Culicidae). *Proceedings of the Linnean Society of New South Wales, 69*(1–2), 62–66.

Woodhill, A.R., & Pasfield, G. (1941). An illustrated key to some common Australian culicine mosquito larvae, with notes on the morphology and breeding places. *Proceedings of the Linnean Society of New South Wales, 66*, 201–214.

Woodruff, R.E., Guest, C.S., Garner, M.G., Becker, N., Lindesay, J., Carvan, T., & Ebi, K. (2002). Predicting Ross River virus epidemics from regional weather data. *Epidemiology, 13*(4), 384–393.

Workman, P.D., & Walton, W.E. (2003). Larval behavior of four *Culex* (Diptera: Culicidae) associated with treatment wetlands in the southwestern United States. *Journal of Vector Ecology, 28*(2), 213–228.

World Health Organization. (1972). A system of world-wide surveillance for vectors. *Weekly Epidemiological Record, 47*(2), 73–84.

Worth, C.B. (1960). Description of a new species of the aedine subgenus *Neomelaniconion* from Tongaland, South Africa (Diptera: Culicidae). *Journal of the Entomological Society of Southern Africa, 23*(2), 312–313.

Worth, C.B., & De Meillon, B. (1960). Culicine mosquitoes (Diptera: Culicidae) recorded from the province of Moçambique (Portuguese East Africa) and their relationship to arthropod-borne viruses. *Anais do Instituto de Hygiene e Medicina Tropical, 17*(1–2), 231–256.

Worth, C.B., de Sousa, J.D., & Wienbren, M.P. (1961). Studies on the life history of *Aedes* (*Skusea*) *pembaensis* (Theobald) (Diptera, Culicidae). *Bulletin of Entomological Research, 52*(2), 257–261.

Wright, G.A., & Schiestl, F.P. (2009). The evolution of floral scent: The influence of olfactory learning by insect pollinators on the honest signaling of floral rewards. *Functional Ecology, 23*(5), 841–851.

Wu, C.W., Kong, X.Q., & Wu, D. (2007). Micronanostructures of the scales on a mosquito's legs and their role in weight support. *Physical Review E, 75*(1), 017301.

Wu, S.C. (1936). Further notes on the mosquitoes of Hangchow, Chekiang, with description of one new species. *Yearbook of the Bureau of Entomology Hangchow, 5*(for 1935), 46–53.

Wulf, A. (2009). *The brother gardeners*. New York, NY: Knopf.

Xavier, S.H. (1964). Descrição de larva e pupas e *Psorophora* Robineau Desvoidy, 1827 (Diptera, Culicidae). *Revista Brasileira de biologia, 24*(2), 235–238.

Xavier, S.H. (1967). Descrição da larva e da pupa de *Aedes* (*Ochlerotatus*) *oligopistus* Dyar, 1918 (Diptera, Culicidae). *Revista Brasileira de malariologia e doenças tropicais* (Rio de Janeiro), *19*(2), 227–230.

Xavier, S.H. (1973). Mosquito types of Brazil. *Mosquito Systematics, 5*(2), 159–160.

Xavier, S.H., & Mattos, S.S. (1967). Descrição da pupa *Aedes* (*Ochlerotatus*) *serratus* (Theobald, 1901) e da larva e pupa de *Trichoprosopon* (*Runchomyia*) *similis* Lane & Cerqueira, 1942 (Diptera, Culicidae). *Revista Brasileira de malariologia e doenças tropicais* (Rio de Janeiro), *19*(4), 617–624.

Xavier, S.H., & Mattos, S.S. (1972). *Culex* (*Melanoconion*) *milwardi* sp. n. (Diptera, Culicidae). *Journal of Cell Biology, 32*(4), 569–571.

Xavier, S.H., & Mattos, S.S. (1989). Mosquito types of Brazil—supplement. *Mosquito Systematics, 21*(2), 50.

Xavier, S.H., Mattos, S.S., & da Silva, J.E. (1970). *Culex* (*Melanoconion*) *olimpioi* sp. n. (Diptera, Culicidae). *Revista Brasileira de malariologia e doenças tropicais* (Rio de Janeiro), *22*(2–4), 183–188.

Xavier, S.H., Mattos, S.S., & da Silva, J.E. (1970). *Uranotaenia* (*Uranotaenia*) *rachoui*, uma nova espécie do Brasil (Diptera, Culicidae). *Revista Brasileira de malariologia e doenças tropicais* (Rio de Janeiro), *22*(2–4), 411–416.

Xavier, S.H., Mattos, S.S., & da Silveira Guedes, A. (1980). Descrição da femea de *Culex* (*Culex*) *machadoi* Mattos, Guedes e Xavier, 1978 (Diptera, Culicidae). *Ciencia e cultura—Sociedade Brasileira para o Progresso da Ciencia* (São Paulo), *32*(8), 1079–1080.

Xi, Z., Ramirez, J.L., & Dimopoulos, G. (2008). The *Aedes aegypti* toll pathway controls dengue virus infection. *PLOS Pathogens, 4*(7), e1000098.

Xu, C.Y., Julien, M.H., Faterni, M., Girod, C., Van Klinken, R.D., Gross, C.L., & Novak, S.J. (2010). Phenotypic divergence during the invasion of *Phyla canescens* in Australia and France: Evidence for selection-driven evolution. *Ecology Letters, 13*(1), 32–44.

Xu, J.-J., & Feng, L.-C. (1975). Studies on the *Anopheles hyrcanus* group of mosquitoes in China. *Acta entomologica Sinica, 18*(1), 77–98, 6pls.

Xu, J.-J., & Luo, X.-F. (1998). A new species of *Anopheles hyrcanus* group from Nei Mongol Autonomous Region (Diptera: Culicidae). *Chinese Journal of Parasitology and Parasitic Diseases, 16*(1), 45–52.

Yadav, K., Sarkar, P.K., & Veer, V. (2014). Utility of male mosquito hypopgygium in species identification. *International Journal of Mosquito Research, 1*(4), 50–54.

Yamada, M. (1937). A new species of *Anopheles* in Chosen (Korea). *Keijo Journal of Medicine, 8*, 237–255, 5pls.

Yamada, S. (1917). On two new species of mosquitoes from Japan. *Dubutsugaku zasshi* (*Zoological Magazine*] (Tokyo), *29*(341), 61–72.

Yamada, S. (1918). A new species of *Anopheles* collected from Hokkaido. *Eiseigaku Densenbryogaku zasshi (Journal of Hygiene and Infectious Diseases), 13*, 689–691.

Yamada, S. (1921). Description of ten new species of *Aedes* found in Japan, with notes on the relation between some of these mosquitoes and the larva of *Filaria bancrofti* Cobbold. *Annotationes zoologicae Japonenses, 10*(6), 45–81.

Yamada, S. (1924). A revision of the adult anopheline mosquitoes of Japan: Systematic descriptions, their habits and their relations to human diseases, together with an account of three new species. *Scientific Reports from the Government Institute for Infectious Diseases, Tokyo Imperial University, 3*, 215–241.

Yamada, S. (1925). A revision of the adult anopheline mosquitoes of Japan: Systematic descriptions, their habits and their relations to human diseases. Part II. *Scientific Reports from the Government Institute for Infectious Diseases, Tokyo Imperial University, 4*, 447–493.

Yamada, S. (1927). An experimental study on twenty-four species of Japanese mosquitoes regarding their suitability as intermediate hosts for *Filaria bancrofti* Cobbold. *Scientific Reports of the Government Institute for Infectious Diseases, Tokyo Imperial University, 6*, 559–622, 3 pls.

Yamada, S. (1932). Culicidae. In T. Esaki (Ed.), *Nippon konchu zukan* (1st ed.). Tokyo, Japan: Hokuryukan.

Yamada, S., & Watanabe, O. (1918). A new species of Korean *Anopheles*. *Japanese Journal of Experimental Medicine, 2*, 206–209.

Yamaguti, S., & LaCasse, W.J. (1950a). *Mosquito fauna of Guam*. Office of the Surgeon, Headquarters, 8th Army, APO 343. Office of the Surgeon-General. 207th Malaria Survey Detachment.

Yamaguti, S., & LaCasse, W.J. (1950b). *Mosquito fauna of North America*. Part I. *Genus* Anopheles. Office of the Surgeon, Headquarters, 8th Army, US Army, United States.

Yamaguti, S., & LaCasse, W.J. (1951a). *Mosquito fauna of North America*. Part II. *Genera* Megarhinus, Wyeomyia, Uranotaenia *and* Culiseta. Office of the Surgeon, Headquarters, 8th Army, APO 343. Office of the Surgeon-General. 207th Malaria Survey Detachment.

Yamaguti, S., & LaCasse, W.J. (1951b). *Mosquito fauna of North America*. Part III. *Genera* Orthopodomyia, Mansonia *and* Psorophora. Office of the Surgeon, Headquarters, 8th Army, APO 343. Office of the Surgeon-General. 207th Malaria Survey Detachment.

Yamaguti, S., & LaCasse, W.J. (1951c). *Mosquito fauna of North America*. Part IV. *Genera* Culex *and* Deinocerites. Office of the Surgeon, Headquarters, 8th Army, APO 343. Office of the Surgeon-General. 207th Malaria Survey Detachment.

Yamaguti, S., & LaCasse, W.J. (1951d). *Mosquito fauna of North America*. Part V. *Genus* Aedes. Office of the Surgeon, Headquarters, 8th Army, APO 343, US Army. Office of the Surgeon-General, 207th Malaria Survey Detachment.

Yamaguti, S., & LaCasse, W.J. (1952). *Culex* (*Culiciomyia*) *kyotoensis* n. sp. (Culicidae, Diptera) from Japan. Self-published, Kyoto, Japan.

Yamaguti, S., & Tamaboko, R. (1954). Two new species of *Aedes* (Culicidae, Diptera) from Japan. *Acta medicinae Okayama, 8*(for 1952)(4), 414–422.

Yan, G., Severson, D.W., & Christensen, B.M. (1997). Costs and benefits of mosquito refractoriness to malaria parasites: Implications for genetic variability of mosquitoes and genetic control of malaria. *Evolution, 51*(2), 441–450.

Yan, J., Avadhanula, S.A., Birch, J., Dickinson, M.H., Sitti, M., Su, T., & Fearing, R.S. (2001). Wing transmission for a micromechanical flying insect. *Journal of Micromechanics, 1*(3), 221–237.

Yan, Z.C., & Zhong, H. (2005). Comparison of adult mosquito community structure on various habitats. *Insect Science, 12*(3), 193–197.

Yang, G.J., Bradshaw, C.J.A., Whelan, P.I., & Brook, B.W. (2008a). Importance of endogenous feedback controlling the long-term abundance of tropical mosquito species. *Population Ecology, 50*(3), 293–305.

Yang, G.J., Brook, B.W., Whelan, P.I., Cleland, S., & Bradshaw, C.J.A. (2008b). Endogenous and exogenous factors controlling temporal abundance patterns of tropical mosquitoes. *Ecological Applications, 18*(8), 2028–2040.

Yang, H.M., Macoris, M.L.G., Galvani, K.C., Andrighetti, M.T.M., & Wanderley, D.M.V. (2009). Assessing the effects of temperature on the population of *Aedes aegpyti*, the vector of dengue. *Epidemiology and Infection, 137*(8), 1188–1202.

Yang, J.-L., Li, F.-H., & Chen., H.-B. (1993). A new species of *Culex* from Yunan (Diptera, Culicidae). *Acta zootaxonomica Sinica, 18*(2), 225–228.

Yang, P., Furumizo, R., Tangalin, L., Takekuma, C., & Hall, K. (2003). Mosquito species breeding in bromeliad axils on the island of Kauai, Hawaii. *Proceedings of the Hawaiian Entomological Society, 36*, 95–101.

Yang, Y.Y., Liu, Y., Teng, H.J., Sauman, I., Sehnal, F., & Lee, H.J. (2010). Circadian control of permethrin-resistance in the mosquito *Aedes aegypti*. *Journal of Insect Physiology, 56*(9), 1219–1223.

Yao, Y.-C., & Wu, C.-C. (1944). A new anopheline mosquito, *Anopheles kweiyangensis*, sp. n. from Kweiyang, Kweicho. *National Medical Journal of China* (Chungking), *30*(1), 1–10.

Yasuno, M., & Tonn, R.J. (1970). Bionomics of *Toxorhynchites splendens* in the larval habitats of *Aedes aegypti* in Bangkok, Thailand. *Bulletin of the World Health Organization, 43*(5), 762–766.

Ye, Y.H., Carrasco, A.M., Frentia, F.D., Chenometh, S.F., Beebe, N.W., van den Hurk, A.F., . . . McGow, E.A. (2015). *Wolbachia* reduces the transmission potential of dengue-infected *Aedes aegypti*. *PLOS Neglected Tropical Diseases, 9*(6), e0003894.

Yeates, D.K., & Wiegmann, B.M. (1999). Congruence and controversy: Toward a higher-level phylogeny of Diptera. *Annual Review of Entomology, 44*(1), 397–428.

Yeates, D.K., & Wiegmann, B.M. (2005). Phylogeny and evolution of Diptera: Recent insights and new perspectives. In D.K. Yeates & B.M. Wiegmann (Eds.), *The evolutionary biology of flies* (pp. 14–44). New York, NY: Columbia University Press.

Yébakima, A., Marquine, M., Rosine, J., Yp-Tcha, M.M., & Pasteur, N. (2004). Evolution of resistance under insecticide selection pressure in *Culex pipiens quinquefasciatus* (Diptera: Culicidae) from Martinique. *Journal of Medical Entomology, 41*(4), 718–725.

Yen, J.H., & Barr, A.R. (1973). The etiological agent of cytoplasmic incompatibility in *Culex pipiens*. *Journal of Invertebrate Pathology, 22*(2), 242–250.

Yin Yin May. (1972). Some of the culicine mosquitoes of Rangoon area. *Union of Burma Journal of Life Sciences, 5*(1), 1–80.

Yong, H.S., Cheong, W.H., Chiang, G.L., Dhaliwal, S.S., Loong, K.P., & Sarjan, R. (1983). Biochemical genetic relationship of three taxa of the malarial mosquito *Anopheles balabacensis* complex. *Comparative Biochemistry and Physiology, 76B*, 611–613.

Yoshimeki, M. (1955). Morphological studies on the tracheal system of two Culicini larvae, *Culex pipiens* L. var. *pallens* Coquillett and *Aedes vexans nipponii* Theobald. *Ecological Review* (Japan), *14*(1), 81–89.

Yoshioka, M., Couret, J., Kim, F., McMillan, J., Burkot, T.R., Dotson, E.M., . . . Vazquez-Prokopec, G.M. (2012). Diet and density dependent competition affect larval performance and oviposition site selection in the mosquito species *Aedes albopictus* (Diptera: Culicidae). *Parasites & Vectors, 5*(1), 225.

Young, T.C.M., & Majid, S.A. (1928). A variety of Anopheles [sic] *karwari* collected in Coorg, S. India. *Indian Journal of Medical Research 16*(2): 469–471.

Yu, Y., & Zhang, B. (1981). Supplemental descriptions of *Anopheles (Cellia) pattoni* Christophers, 1926. *Entomotaxonomia, 3*(2), 132–136.

Yuval, B. (2006). Mating systems of blood-feeding flies. *Annual Review of Entomology, 51*, 413–440.

Zahiri, N., Dunphy, G.B., & Rau, M.E. (1998). Serum composition of *Aedes aegypti* (Diptera: Culicidae) larvae and the production of an oviposition repellent are influenced by infection with the entomopathogenic digenean *Plagiorchis elegans* (Trematoda: Plagiorchidae), starvation, and crowding. *Journal of Medical Entomology, 35*(2), 162–168.

Zaim, M. (1987). The distribution and larval habitat characteristics of Iranian Culicinae. *Journal of the American Mosquito Control Association, 3*(4), 568–573.

Zaim, M., Lewandowski, H.B., Newson, H.D., & Hooper, G.R. (1977). Differentiation of *Aedes triseriatus* and *Ae. hendersoni* (Diptera: Culicidae) based on the surface structure of the egg shell. *Journal of Medical Entomology, 14*(4), 489–490.

Zaim, M., Subbarao, S.K., Manouchehri, A.V., & Cochrane, A.H. (1993) Role of *Anopheles culicifacies* s.l. and *An. pulcherrimus* in malaria transmission in Ghassreghand (Baluchistan), Iran. *Journal of the American Mosquito Control Association, 9*(1), 23–26.

Zalunin, I.A., Chaĭka, S., Dronina, M.A., & Revina, L.P. (2001). Cytopathological effect of *Bacillus thuringiensis israelensis* endotoxins on the intestines of *Aedes aegypti* mosquito larvae. *Parazitologiia, 36*(5), 337–344.

Zarowiecki, M., Walton, C., Torres, E., McAlister, E., Htun, P.T., Sumrandee, C., . . . Linton, Y.-M. (2011). Pleistocene genetic connectivity in a widespread, open-habitat-adapted mosquito in the Indo-Oriental region. *Journal of Biogeography, 38*(7), 1422–1432.

Zavortink, T.J. (1968). Mosquito studies (Diptera, Culicidae). VIII. A prodrome of the genus *Orthopodomyia*. *Contributions of the American Entomological Institute, 3*(2), 1–221.

Zavortink, T.J. (1969). Mosquito studies (Diptera, Culicidae). XV. A new species of treehole breeding *Anopheles* from the southwestern United States. *Contributions of the American Entomological Institute, 4*(4), 27–38.

Zavortink, T.J. (1970a). Mosquito studies (Diptera, Culicidae). XVI. A new species of treehole breeding *Aedes* (*Ochlerotatus*) from southern California. *Contributions of the American Entomological Institute, 5*(for 1969)(1), 1–7.

Zavortink, T.J. (1970b). Mosquito studies (Diptera, Culicidae). XIX. The treehole *Anopheles* of the New World. *Contributions of the American Entomological Institute, 5*(2), 1–35.

Zavortink, T.J. (1970c). Mosquito studies (Diptera, Culicidae). XXII. A new subgenus and species of *Aedes* from Arizona. *Contributions of the American Entomological Institute, 7*(1), 1–11.

Zavortink, T.J. (1971). Contributions to the mosquito fauna of Southeast Asia. IX. The genus *Orthopodomyia* Theobald in Southeast Asia. *Contributions of the American Entomological Institute, 7*(3), 1–37.

Zavortink, T.J. (1972). Mosquito studies (Diptera, Culicidae). XXVIII. The new world species formerly placed in *Aedes* (*Finlaya*). *Contributions of the American Entomological Institute, 8*(3), 1–206.

Zavortink, T.J. (1973). Mosquito studies (Diptera, Culicidae). XXIX. A review of the subgenus *Kerteszia* of *Anopheles*. *Contributions of the American Entomological Institute, 9*(3), 1–54.

Zavortink, T.J. (1979a). A reclassification of the sabethine genus *Trichoprosopon*. *Mosquito Systematics, 11*(4), 255–257.

Zavortink, T.J. (1979b). The new sabethine genus *Johnbelkinia* and a preliminary reclassification of the composite genus *Trichoprosopon*. *Contributions of the American Entomological Institute, 17*(1), 1–61.

Zavortink, T.J. (1981). Species complexes in the genus *Trichoprosopon*. *Mosquito Systematics, 13*(1), 82–85.

Zavortink, T.J. (1986). *Zinzala*, A new subgenus of *Wyeomyia* with two new species from pitcher-plants in Venezuela (Diptera, Culicidae. Sabethini). *Wassman Journal of Biology, 43*(for 1985)(1–2), 46–59.

Zavortink, T.J. (1987). *Biosystematics of* Aedes (Neomelaniconion) *for the period 5 May 1986–4 May 1987* (Publication No. DAMD17-86-C-6134). Retrieved from https://archive.org/details/DTIC_ADA183519/page/n2

Zavortink, T.J. (2004). Conservation of the names *Aedes albothorax, Ae. circumluteolus*, and *Ae. mcintoshi*. *Journal of the American Mosquito Control Association, 20*(2), 197–198.

Zavortink, T.J., & Chaverri, L.G. (2009). Resurrection of the names *Toxorhynchites moctezuma* (Dyar & Knab) and *Toxorhynchites hypoptes* (Knab) from synonymy with *Toxorhynchites theobaldi* (Dyar & Knab) (Diptera: Culicidae). *Proceedings of the Entomological Society of Washington, 111*(4), 890–897.

Zavortink, T.J., & O'Meara, G.F. (1999). *Culex* (*Micraedes*) *biscaynensis* n. p. from Florida (Diptera: Culicidae). *Journal of the American Mosquito Control Association, 15*(3), 263–270.

Zavortink, T.J., & Poinar, G.O., Jr. (2000). *Anopheles* (*Nyssorhynchus*) *dominicanus* sp. n. (Diptera: Culicidae) from Dominican Amber. *Annals of the Entomological Society of America, 93*(6), 1230–1235.

Zavortink, T.J., & Poinar, G.O., Jr. (2008). *Toxorhynchites* (*Toxorhynchites*) *mexicanus* n. sp. (Diptera: Culicidae) from Mexican amber: A New World species with Old World affinities. *Proceedings of the Entomological Society of Washington, 110*(1), 116–125.

Zavortink, T.J., Roberts, D.R., & Hoch, A.L. (1983). *Trichoprosopon digitatum*—morphology, biology, and potential medical importance. *Mosquito Systematics, 15*(2), 141–149.

Zayed, A.B.B. (2004). Impact of larval crowding on some biological and physiological attributes of female *Culex pipiens* (L.) (Diptera: Culicidae) in Egypt. *International Journal of Dipterology Research, 15*(1), 99–106.

Zayed, M.E., & Bream, A.S. (2003). Biodiversity of the microbial flora associated with two strains of *Culex pipiens* (Diptera: Culicidae). *Communications in Agricultural and Applied Biological Sciences, 69*(3), 229–234.

Zelditch, M.L., Swiderski, D.L., & Fink, W.L. (2000). Discovery of phylogenetic characters in morphometric data. In J.J. Wiens (Ed.), *Phylogenetic analysis of morphological data* (pp. 37–83). Washington, DC: Smithsonian Institution Press.

Zetterstedt, J.W. (1838). *Insecta Lapponica descripta*. Voss, Norway: Lipsiae.

Zetterstedt, J.W. (1850). Diptera Scandinaviae, disposita et descripta. *Officina Lundbergiana, Lund, Sweden, 9*, 3367–4090.

Zhang, J., Zhang, X., Arakane, Y., Muthukrishnan, S., Kramer, K.J., Ma, E., & Zhu, K.Y. (2011). Comparative genomic analysis of chitinase and chitinase-like genes in the African malaria mosquito (*Anopheles gambiae*). *PLOS ONE, 6*(5), e19899.

Zhang, P.X., & Zhang, S.M. (1982). A new record of *Culex* from China. *Acta zootaxonomica Sinica, 7*(2), 202.

Zhang, X., Zhang, J., Park, Y., & Zhu, K.Y. (2012). Identification and characterization of two chitin synthase genes in African malaria mosquito, *Anopheles gambiae*. *Insect Biochemistry and Molecular Biology, 42*(9), 674–682.

Zhelochovtsev, A.N. (1937). Notes on the synonymy of the Palaearctic *Anopheles*. *Meditsinskaia Parazitologiia i Parazitarnye Bolezni 6*(5) , 707–709.

Zhong, D., Lo, E., Hu, R., Metzger, M.E., Cummings, R., Bonizzoni, M., . . . Yan, G. (2013). Genetic analysis of invasive *Aedes albopictus* populations in Los Angeles County, California and its potential health impact. *PLOS ONE, 8*(7), e68586.

Zhou, G., & Miesfeld, R.L. (2009). Energy metabolism during diapause in *Culex pipiens* mosquitoes. *Journal of Insect Physiology, 55*(1), 40–46.

Zhou, H., Zhu, G., & Lu, Y. (1999). One new species of the genus *Topomyia* (Diptera: Culicidae). *Acta parasitologica et medica entomologica Sinica, 6*(2), 107–111.

Zhu, F., Xu, P., Barbosa, R.M., Choo, Y.M., & Leal, W.S. (2013). RNAi-based demonstration of direct link between specific odorant receptors and mosquito oviposition behavior. *Insect Biochemistry and Molecular Biology, 43*(10), 916–923.

Zhu, J., Miura, K., Chen, L., & Raikhel, A.S. (2000). AHR38, a homolog of NGFI-B, inhibits formation of the functional ecdysteroid receptor in the mosquito *Aedes aegypti*. *Journal of the European Molecular Biology Organisation, 19*(2), 253–262.

Zhu, J., Chen, L., & Raikhel, A.S. (2003). Posttranscriptional control of the competence factor ßFTZ-F1 by juvenile hormone in the mosquito *Aedes aegypti*. *Proceedings of the National Academy of Sciences of the United States of America, 100*(23), 13338–13343.

Zhu, J., Busche, J.M., & Zhang, X. (2010). Identification of juvenile hormone target genes in the adult female mosquito. *Insect Biochemistry and Molecular Biology, 40*(1), 23–29.

Ziegler, R., & Ibrahim, M.M. (2001). Formation of lipid reserves in fat body and eggs of the yellow fever mosquito, *Aedes aegypti*. *Journal of Insect Physiology, 47*(6), 623–627.

Zieler, H., Garon, C.F., Fischer, E.R., & Shahabuddin, M. (2000). A tubular network associated with the brush-border surface of the *Aedes aegypti* midgut: Implications for pathogen transmission by mosquitoes. *Journal of Experimental Biology, 203*(10), 1599–1611.

Zimmerman, R.H., & Turner, E.C., Jr. (1982). Mosquito distribution and abundance in an inland salt marsh, Saltville, Virginia. *Mosquito News, 42*(1), 105–109.

Zimmerman, R.H., Galardo, A.K., Lounibos, L.P., Arruda, M., & Wirtz, R. (2006). Bloodmeal hosts of *Anopheles* species (Diptera: Culicidae) in a malaria-endemic area of the Brazilian Amazon. *Journal of Medical Entomology, 43*(5), 947–956.

Zou, Z., Saha, T.T., Roy, S., Shin, S.W., Backman, T.W., Girke, T., . . . Raikhel, A.S. (2013). Juvenile hormone and its receptor, methoprene-tolerant, control the dynamics of mosquito gene expression. *Proceedings of the National Academy of Sciences of the United States of America, 110*(24), e2173–2181.

Zouache, K., Voronin, D., Tran-Van, V., Mousson, L., Failloux, A.B., & Mavingui, P. (2009). Persistent *Wolbachia* and cultivable bacteria infection in the reproductive and somatic tissues of the mosquito vector *Aedes albopictus*. *PLOS ONE, 4*(7), e6388.

Zouache, K., Raharimalala, F.N., Raquin, V., Tran-Van, V., Raveloson, L.H.R., Ravelonandro, P., & Mavingui, P. (2011). Bacterial diversity of field-caught mosquitoes, *Aedes albopictus* and *Aedes aegypti*, from different geographic regions of Madagascar. *FEMS Microbiology Ecology, 75*(3), 377–389.

Zsemlye, J.L., Hancock, R.G., & Foster, W.A. (2005). Analysis of a complex vertical copulatory-courtship display in the yellow fever vector *Sabethes chloropterus*. *Medical and Veterinary Entomology, 19*(3), 276–285.

Zwiebel, L.J., & Takken, W. (2004). Olfactory regulation of mosquito–host interactions. *Insect Biochemistry and Molecular Biology, 34*(7), 645–652.

Index